Y0-BEE-991

American Men & Women of Science

1992-93 • 18th Edition

The 18th edition of *AMERICAN MEN & WOMEN OF SCIENCE* was prepared by the R.R. Bowker Database Publishing Group.

Stephen L. Torpie, Managing Editor
Judy Redel, Managing Editor, Research
Richard D. Lanam, Senior Editor
Tanya Hurst, Research Manager
Karen Hallard, Beth Tanis, Associate Editors

Peter Simon, Vice President, Database Publishing Group
Dean Hollister, Director, Database Planning
Edgar Adcock, Jr., Editorial Director, Directories

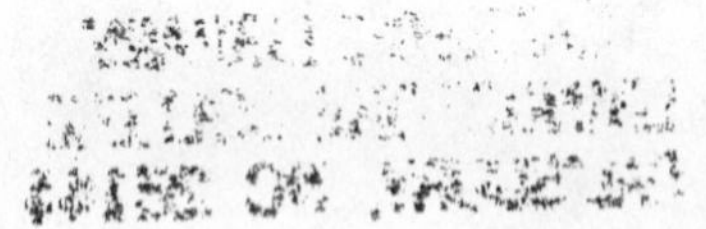

American Men & Women of Science

1992-93 • 18th Edition

A Biographical Directory of Today's Leaders in Physical, Biological and Related Sciences.

Volume 5 • M-P

R. R. BOWKER
New Providence, New Jersey

Published by R.R. Bowker, a division of Reed Publishing, (USA) Inc.

International Standard Book Number

Set:	0-8352-3074-0
Volume I:	0-8352-3075-9
Volume II:	0-8352-3076-7
Volume III:	0-8352-3077-5
Volume IV:	0-8352-3078-3
Volume V:	0-8352-3079-1
Volume VI:	0-8352-3080-5
Volume VII:	0-8352-3081-3
Volume VIII:	0-8352-3082-1

International Standard Serial Number: 0192-8570
Library of Congress Catalog Card Number: 6-7326
Printed and bound in the United States of America.

8 Volume Set

ISBN 0-8352-3074-0

9 780835 230742

Contents

Advisory Committee

Dr. Robert F. Barnes
Executive Vice President
American Society of Agronomy

Dr. John Kistler Crum
Executive Director
American Chemical Society

Dr. Charles Henderson Dickens
Section Head, Survey & Analysis Section
Division of Science Resource Studies
National Science Foundation

Mr. Alan Edward Fechter
Executive Director
Office of Scientific & Engineering Personnel
National Academy of Science

Dr. Oscar Nicolas Garcia
Prof Electrical Engineering
Electrical Engineering & Computer Science Department
George Washington University

Dr. Charles George Groat
Executive Director
American Geological Institute

Dr. Richard E. Hallgren
Executive Director
American Meteorological Society

Dr. Michael J. Jackson
Executive Director
Federation of American Societies for Experimental Biology

Dr. William Howard Jaco
Executive Director
American Mathematical Society

Dr. Shirley Mahaley Malcom
Head, Directorate for Education and Human Resources Programs
American Association for the Advancement of Science

Mr. Daniel Melnick
Sr Advisor Research Methodologies
Sciences Resources Directorate
National Science Foundation

Ms. Beverly Fearn Porter
Division Manager
Education & Employment Statistics Division
American Institute of Physics

Dr. Terrence R. Russell
Manager
Office of Professional Services
American Chemical Society

Dr. Irwin Walter Sandberg
Holder, Cockrell Family Regent Chair
Department of Electrical & Computer Engineering
University of Texas

Dr. William Eldon Splinter
Interim Vice Chancellor for Research,
Dean, Graduate Studies
University of Nebraska

Ms. Betty M. Vetter
Executive Director, Science Manpower Comission
Commission on Professionals in Science & Technology

Dr. Dael Lee Wolfe
Professor Emeritus
Graduate School of Public Affairs
University of Washington

Preface

American Men and Women Of Science remains without peer as a chronicle of North American scientific endeavor and achievement. The present work is the eighteenth edition since it was first compiled as *American Men of Science* by J. Mckeen Cattell in 1906. In its eighty-six year history *American Men & Women of Science* has profiled the careers of over 300,000 scientists and engineers. Since the first edition, the number of American scientists and the fields they pursue have grown immensely. This edition alone lists full biographies for 122,817 engineers and scientists, 7021 of which are listed for the first time. Although the book has grown, our stated purpose is the same as when Dr. Cattell first undertook the task of producing a biographical directory of active American scientists. It was his intention to record educational, personal and career data which would make "a contribution to the organization of science in America" and "make men [and women] of science acquainted with one another and with one another's work." It is our hope that this edition will fulfill these goals.

The biographies of engineers and scientists constitute seven of the eight volumes and provide birthdates, birthplaces, field of specialty, education, honorary degrees, professional and concurrent experience, awards, memberships, research information and adresses for each entrant when applicable. The eighth volume, the discipline index, organizes biographees by field of activity. This index, adapted from the National Science Foundation's Taxonomy of Degree and Employment Specialties, classifies entrants by 171 subject specialties listed in the table of contents of Volume 8. For the first time, the index classifies scientists and engineers by state within each subject specialty, allowing the user to more easily locate a scientist in a given area. Also new to this edition is the inclusion of statistical information and recipients of theNobel Prizes, the Craaford Prize, the Charles Stark Draper Prize, and the National Medals of Science and Technology received since the last edition.

While the scientific fields covered by *American Men and Women Of Science* are comprehensive, no attempt has been made to include all American scientists. Entrants are meant to be limited to those who have made significant contributions in their field. The names of new entrants were submitted for consideration at the editors' request by current entrants and by leaders of academic, government and private research programs and associations. Those included met the following criteria:

1. Distinguished achievement, by reason of experience, training or accomplishment, including contributions to the literature, coupled with continuing activity in scientific work;

   or

2. Research activity of high quality in science as evidenced by publication in reputable scientific journals; or for those whose work cannot be published due to governmental or industrial security, research activity of high quality in science as evidenced by the judgement of the individual's peers;

   or

3. Attainment of a position of substantial responsibility requiring scientific training and experience.

This edition profiles living scientists in the physical and biological fields, as well as public health scientists, engineers, mathematicians, statisticians, and computer scientists. The information is collected by means of direct communication whenever possible. All entrants receive forms for corroboration and updating. New entrants receive questionaires and verification proofs before publication. The information submitted by entrants is included as completely as possible within the boundaries of editorial and space restrictions. If an entrant does not return the form and his or her current location can be verified in secondary sources, the full entry is repeated. References to the previous edition are given for those who do not return forms and cannot be located, but who are presumed to be still active in science or engineering. Entrants known to be deceased are noted as such and a reference to the previous edition is given. Scientists and engineers who are not citizens of the United States or Canada are included if a significant portion of their work was performed in North America.

The information in AMWS is also available on CD-ROM as part of *SciTech Reference Plus*. In adition to the convenience of searching scientists and engineers, *SciTech Reference Plus* also includes *The Directory of American Research & Technology*, *Corporate Technology Directory*, sci-tech and medical books and serials from *Books in Print* and *Bowker International Series*. *American Men and Women Of Science* is available for online searching through the subscription services of DIALOG Information Services, Inc. (3460 Hillview Ave, Palo Alto, CA 94304) and ORBIT Search Service (800 Westpark Dr, McLean, VA 22102). Both CD-Rom and the on-line subscription services allow all elements of an entry, including field of interest, experience, and location, to be accessed by key word. Tapes and mailing lists are also available through the Cahners Direct Mail (John Panza, List Manager, Bowker Files 245 W 17th St, New York, NY, 10011, Tel: 800-537-7930).

A project as large as publishing *American Men and Women Of Science* involves the efforts of a great many people. The editors take this opportunity to thank the eighteenth edition advisory committee for their guidance, encouragement and support. Appreciation is also expressed to the many scientific societies who provided their membership lists for the purpose of locating former entrants whose addresses had changed, and to the tens of thousands of scientists across the country who took time to provide us with biographical information. We also wish to thank Bruce Glaunert, Bonnie Walton, Val Lowman, Debbie Wilson, Mervaine Ricks and all those whose care and devotion to accurate research and editing assured successful production of this edition.

Comments, suggestions and nominations for the nineteenth edition are encouraged and should be directed to The Editors, *American Men and Women Of Science*, R.R. Bowker, 121 Chanlon Road, New Providence, New Jersey, 07974.

Edgar H. Adcock, Jr.
Editorial Director

Major Honors & Awards

Nobel Prizes

Nobel Foundation

The Nobel Prizes were established in 1900 (and first awarded in 1901) to recognize those people who "have conferred the greatest benefit on mankind."

1990 Recipients

Chemistry:

Elias James Corey

Awarded for his work in retrosynthetic analysis, the synthesizing of complex substances patterned after the molecular structures of natural compounds.

Physics:

Jerome Isaac Friedman

Henry Way Kendall

Richard Edward Taylor

Awarded for their breakthroughs in the understanding of matter.

Physiology or Medicine:

Joseph E. Murray

Edward Donnall Thomas

Awarded to Murray for his kidney transplantation achievements and to Thomas for bone marrow transplantation advances.

1991 Recipients

Chemistry:

Richard R. Ernst

Awarded for refinements in nuclear magnetic resonance spectroscopy.

Physics:

Pierre-Gilles de Gennes*

Awarded for his research on liquid crystals.

Physiology or Medicine:

Erwin Neher

Bert Sakmann*

Awarded for their discoveries in basic cell function and particularly for the development of the patch clamp technique.

Crafoord Prize

Royal Swedish Academy of Sciences
(Kungl. Vetenskapsakademien)

The Crafoord Prize was introduced in 1982 to award scientists in disciplines not covered by the Nobel Prize, namely mathematics, astronomy, geosciences and biosciences.

1990 Recipients

Paul Ralph Ehrlich

Edward Osborne Wilson

Awarded for their fundamental contributions to population biology and the conservation of biological diversity.

1991 Recipient

Allan Rex Sandage

Awarded for his fundamental contributions to extragalactic astronomy, including observational cosmology.

Charles Stark Draper Prize

National Academy of Engineering

The Draper Prize was introduced in 1989 to recognize engineering achievement. It is awarded biennially.

1991 Recipients

Hans Joachim Von Ohain

Frank Whittle

Awarded for their invention and development of the jet aircraft engine.

National Medal of Science

National Science Foundation

The National Medals of Science have been awarded by the President of the United States since 1962 to leading scientists in all fields.

1990 Recipients:

Baruj Benacerraf
Elkan Rogers Blout
Herbert Wayne Boyer
George Francis Carrier
Allan MacLeod Cormack
Mildred S. Dresselhaus
Karl August Folkers
Nick Holonyak Jr.
Leonid Hurwicz
Stephen Cole Kleene
Daniel Edward Koshland Jr.
Edward B. Lewis
John McCarthy
Edwin Mattison McMillan**
David G. Nathan
Robert Vivian Pound
Roger Randall Dougan Revelle**
John D. Roberts
Patrick Suppes
Edward Donnall Thomas

1991 Recipients

Mary Ellen Avery
Ronald Breslow
Alberto Pedro Calderon
Gertrude Belle Elion
George Harry Heilmeier
Dudley Robert Herschbach
George Evelyn Hutchinson**
Elvin Abraham Kabat
Robert Kates
Luna Bergere Leopold
Salvador Edward Luria**
Paul A. Marks
George Armitage Miller
Arthur Leonard Schawlow
Glenn Theodore Seaborg
Folke Skoog
H. Guyford Stever
Edward Carroll Stone Jr
Steven Weinberg
Paul Charles Zamecnik

National Medal of Technology

U.S. Department of Commerce,
Technology Administration

The National Medals of Technology, first awarded in 1985, are bestowed by the President of the United States to recognize individuals and companies for their development or commercialization of technology or for their contributions to the establishment of a technologically-trained workforce.

1990 Recipients

John Vincent Atanasoff
Marvin Camras
The du Pont Company
Donald Nelson Frey
Frederick W. Garry
Wilson Greatbatch
Jack St. Clair Kilby
John S. Mayo
Gordon Earle Moore
David B. Pall
Chauncey Starr

1991 Recipients

Stephen D. Bechtel Jr
C. Gordon Bell
Geoffrey Boothroyd
John Cocke
Peter Dewhurst
Carl Djerassi
James Duderstadt
Antonio L. Elias
Robert W. Galvin
David S. Hollingsworth
Grace Murray Hopper
F. Kenneth Iverson
Frederick M. Jones**
Robert Roland Lovell
Joseph A. Numero**
Charles Eli Reed
John Paul Stapp
David Walker Thompson

*These scientists' biographies do not appear in *American Men & Women of Science* because their work has been conducted exclusively outside the US and Canada.

**Deceased [Note that Frederick Jones died in 1961 and Joseph Numero in May 1991. Neither was ever listed in *American Men and Women of Science*.]

Statistics

Statistical distribution of entrants in *American Men & Women of Science* is illustrated on the following five pages. The regional scheme for geographical analysis is diagrammed in the map below. A table enumerating the geographic distribution can be found on page xvi, following the charts. The statistics are compiled by tallying all occurrences of a major index subject. Each scientist may choose to be indexed under as many as four categories; thus, the total number of subject references is greater than the number of entrants in *AMWS*.

All Disciplines

	Number	Percent
Northeast	58,325	34.99
Southeast	39,769	23.86
North Central	19,846	11.91
South Central	12,156	7.29
Mountain	11,029	6.62
Pacific	25,550	15.33
TOTAL	**166,675**	**100.00**

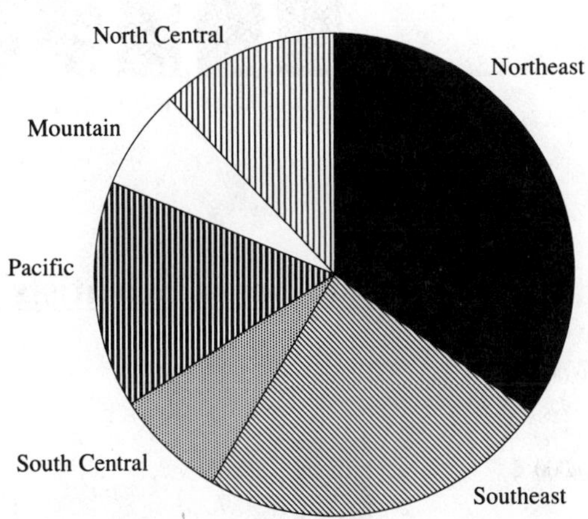

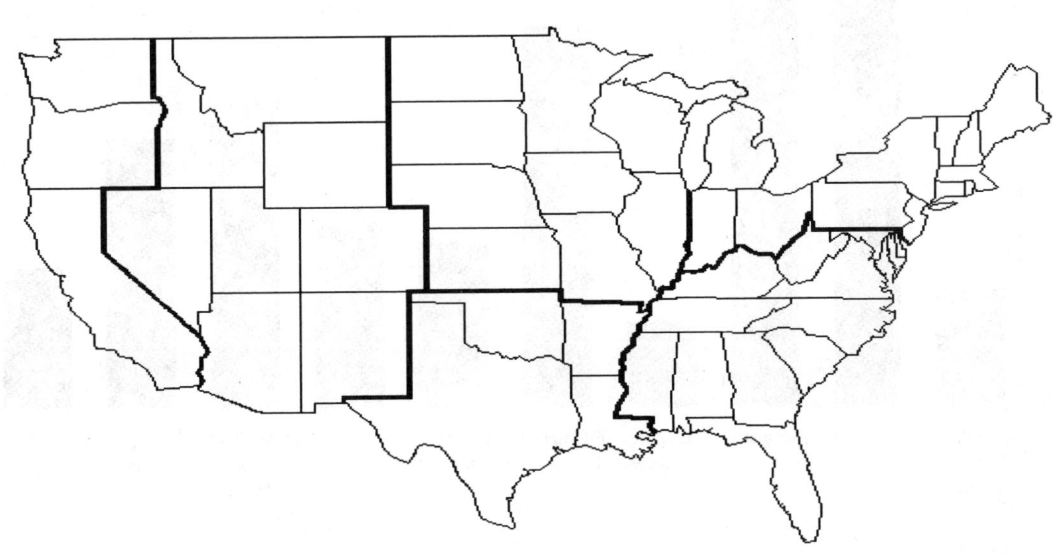

Age Distribution of American Men & Women of Science

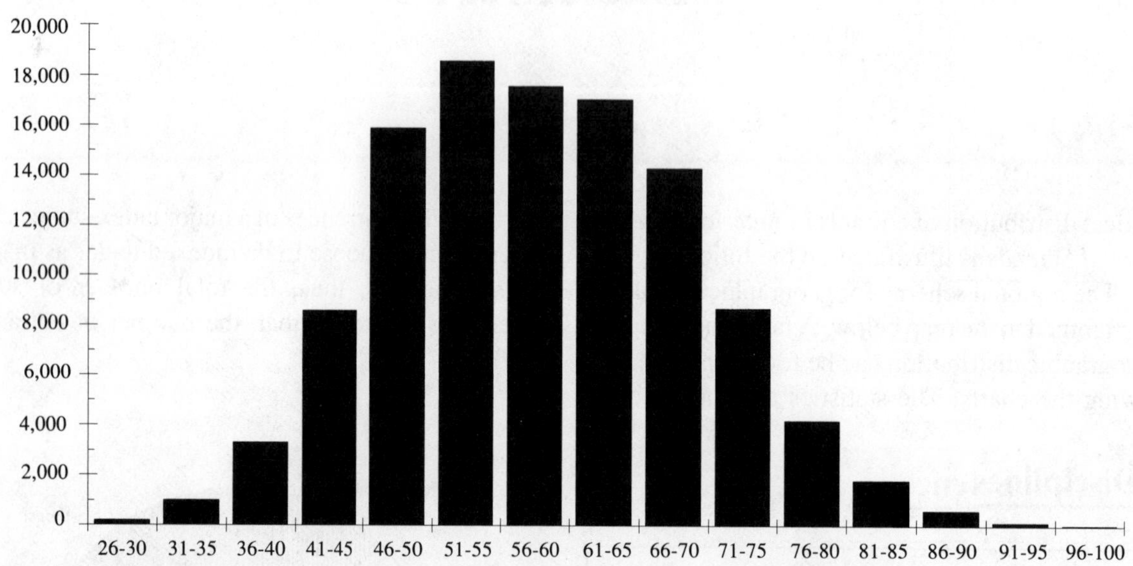

Number of Scientists in Each Discipline of Study

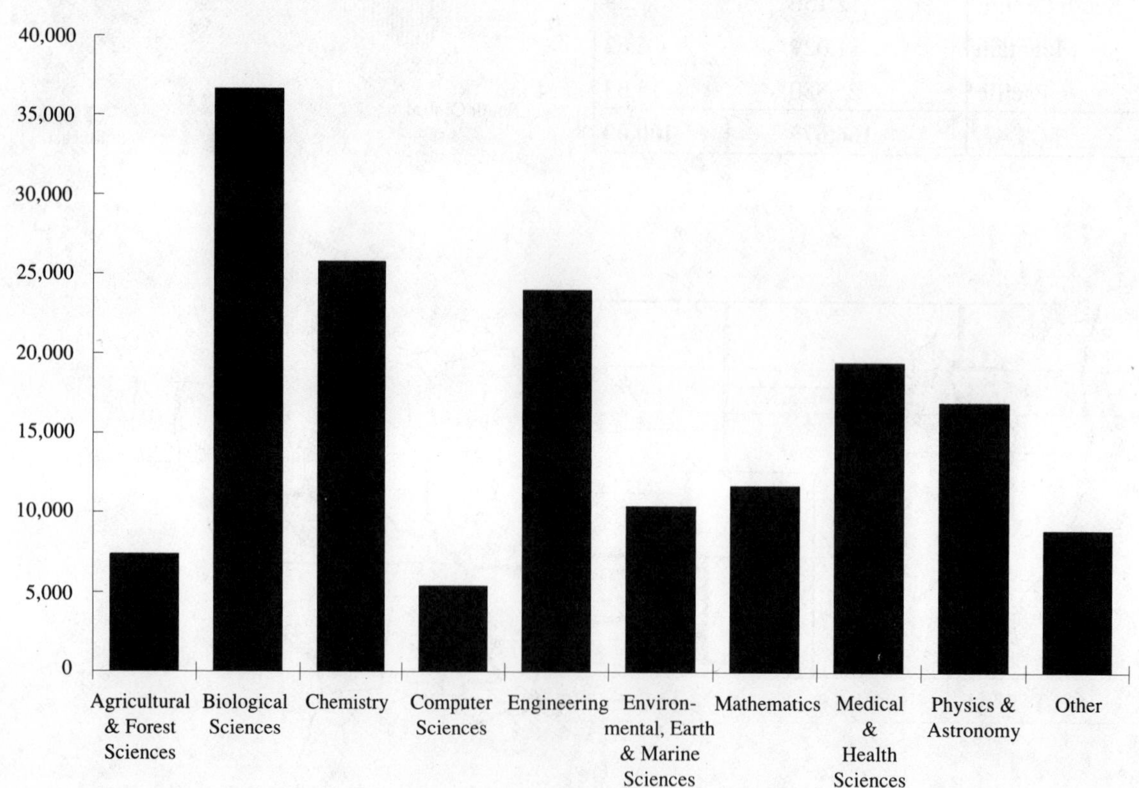

Agricultural & Forest Sciences

	Number	Percent
Northeast	1,574	21.39
Southeast	1,991	27.05
North Central	1,170	15.90
South Central	609	8.27
Mountain	719	9.77
Pacific	1,297	17.62
TOTAL	**7,360**	**100.00**

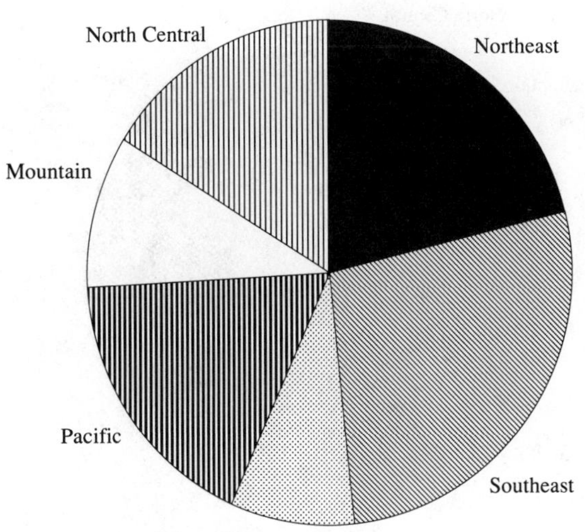

Biological Sciences

	Number	Percent
Northeast	12,162	33.23
Southeast	9,054	24.74
North Central	5,095	13.92
South Central	2,806	7.67
Mountain	2,038	5.57
Pacific	5,449	14.89
TOTAL	**36,604**	**100.00**

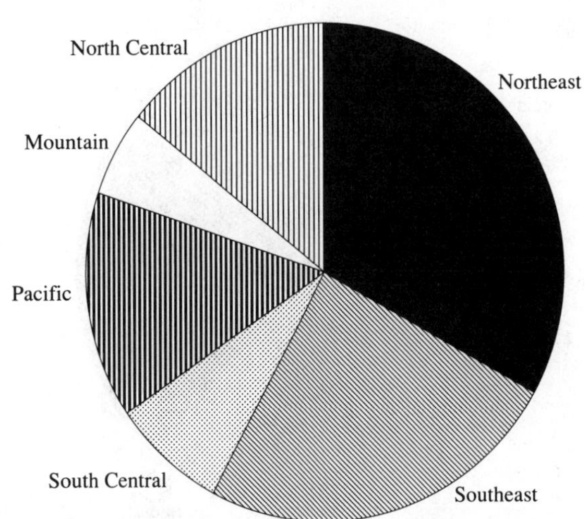

Chemistry

	Number	Percent
Northeast	10,343	40.15
Southeast	6,124	23.77
North Central	3,022	11.73
South Central	1,738	6.75
Mountain	1,300	5.05
Pacific	3,233	12.55
TOTAL	**25,760**	**100.00**

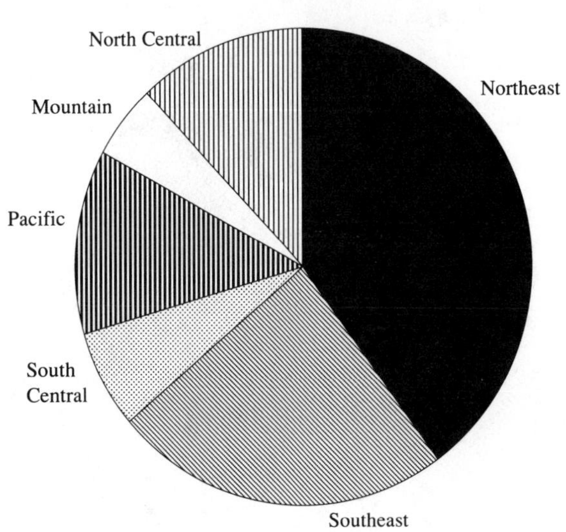

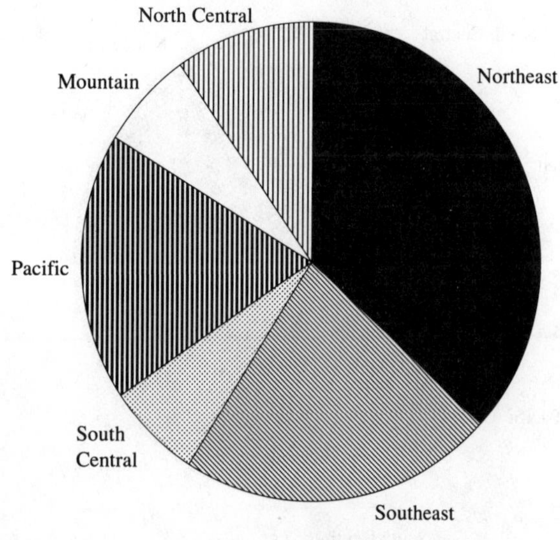

Computer Sciences

	Number	Percent
Northeast	1,987	36.76
Southeast	1,200	22.20
North Central	511	9.45
South Central	360	6.66
Mountain	372	6.88
Pacific	976	18.05
TOTAL	**5,406**	**100.00**

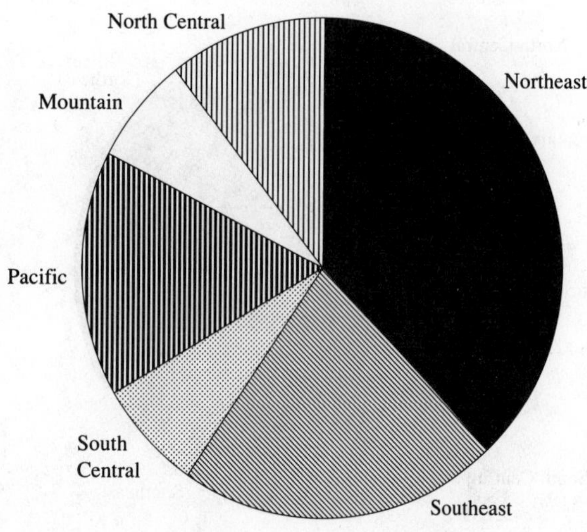

Engineering

	Number	Percent
Northeast	9,122	38.01
Southeast	5,202	21.68
North Central	2,510	10.46
South Central	1,710	7.13
Mountain	1,646	6.86
Pacific	3,807	15.86
TOTAL	**23,997**	**100.00**

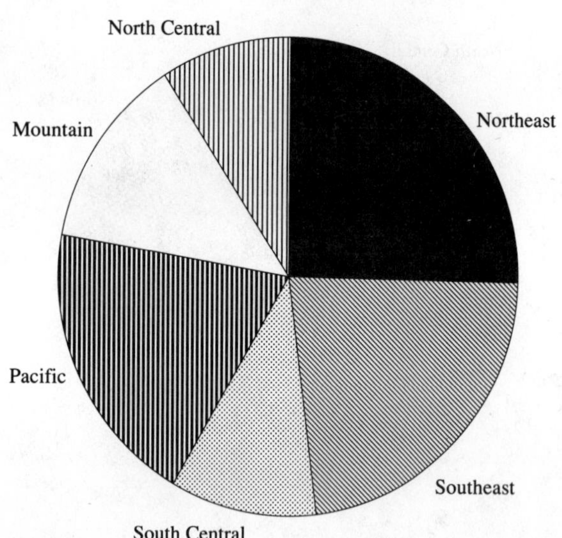

Environmental, Earth & Marine Sciences

	Number	Percent
Northeast	2,657	25.48
Southeast	2,361	22.64
North Central	953	9.14
South Central	1,075	10.31
Mountain	1,359	13.03
Pacific	2,022	19.39
TOTAL	**10,427**	**100.00**

Mathematics

	Number	Percent
Northeast	4,211	35.92
Southeast	2,609	22.26
North Central	1,511	12.89
South Central	884	7.54
Mountain	718	6.13
Pacific	1,789	15.26
TOTAL	**11,722**	**100.00**

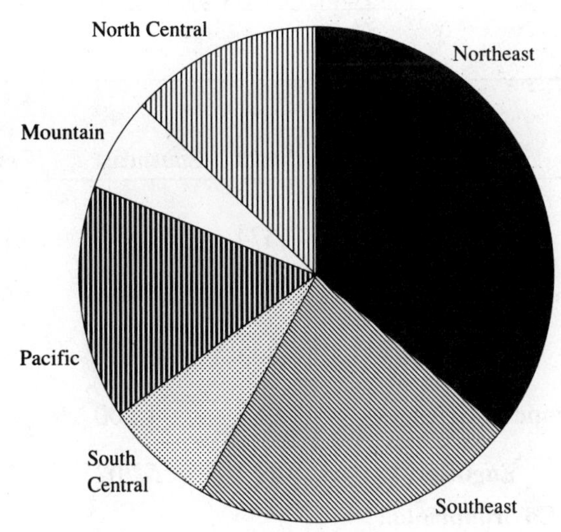

Medical & Health Sciences

	Number	Percent
Northeast	7,115	36.53
Southeast	5,004	25.69
North Central	2,577	13.23
South Central	1,516	7.78
Mountain	755	3.88
Pacific	2,509	12.88
TOTAL	**19,476**	**100.00**

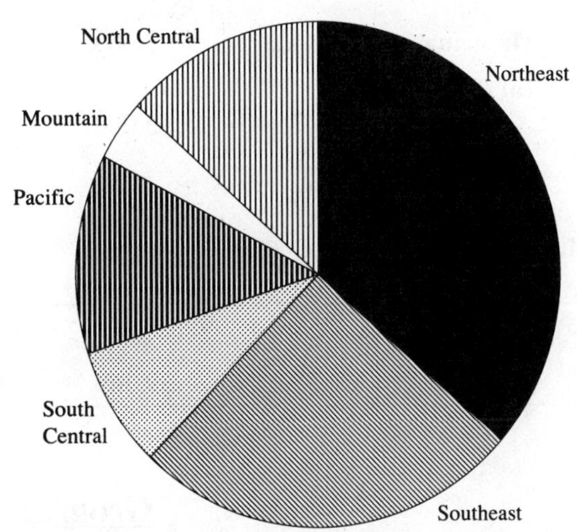

Physics & Astronomy

	Number	Percent
Northeast	5,961	35.12
Southeast	3,670	21.62
North Central	1,579	9.30
South Central	918	5.41
Mountain	1,607	9.47
Pacific	3,238	19.08
TOTAL	**16,973**	**100.00**

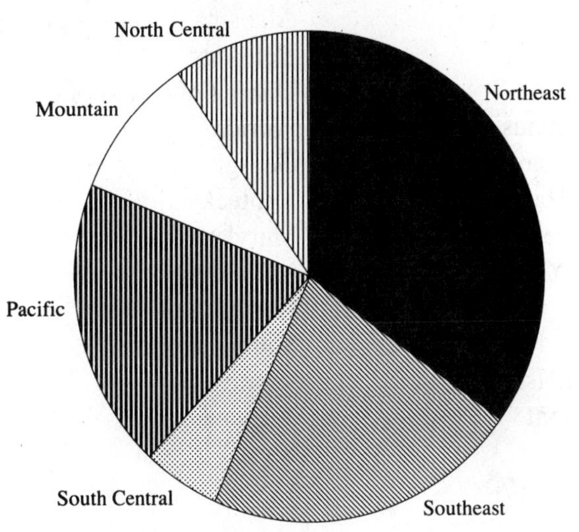

Geographic Distribution of Scientists by Discipline

	Northeast	Southeast	North Central	South Central	Mountain	Pacific	TOTAL
Agricultural & Forest Sciences	1,574	1,991	1,170	609	719	1,297	**7,360**
Biological Sciences	12,162	9,054	5,095	2,806	2,038	5,449	**36,604**
Chemistry	10,343	6,124	3,022	1,738	1,300	3,233	**25,760**
Computer Sciences	1,987	1,200	511	360	372	976	**5,406**
Engineering	9,122	5,202	2,510	1,710	1,646	3,807	**23,997**
Environmental, Earth & Marine Sciences	2,657	2,361	953	1,075	1,359	2,022	**10,427**
Mathematics	4,211	2,609	1,511	884	718	1,789	**11,722**
Medical & Health Sciences	7,115	5,004	2,577	1,516	755	2,509	**19,476**
Physics & Astronomy	5,961	3,670	1,579	918	1,607	3,238	**16,973**
Other Professional Fields	3,193	2,554	918	540	515	1,230	**8,950**
TOTAL	**58,325**	**39,769**	**19,846**	**12,156**	**11,029**	**25,550**	**166,675**

Geographic Definitions

Northeast
Connecticut
Indiana
Maine
Massachusetts
Michigan
New Hampshire
New Jersey
New York
Ohio
Pennsylvania
Rhode Island
Vermont

Southeast
Alabama
Delaware
District of Columbia
Florida
Georgia
Kentucky
Maryland
Mississippi
North Carolina
South Carolina
Tennessee
Virginia
West Virginia

North Central
Illinois
Iowa
Kansas
Minnesota
Missouri
Nebraska
North Dakota
South Dakota
Wisconsin

South Central
Arkansas
Louisiana
Texas
Oklahoma

Mountain
Arizona
Colorado
Idaho
Montana
Nevada
New Mexico
Utah
Wyoming

Pacific
Alaska
California
Hawaii
Oregon
Washington

Sample Entry

American Men & Women of Science (AMWS) is an extremely useful reference tool. The book is most often used in one of two ways: to find more information about a particular scientist or to locate a scientist in a specific field.

To locate information about an individual, the biographical section is most helpful. It encompasses the first seven volumes and lists scientists and engineers alphabetically by last name. The fictitious biographical listing shown below illustrates every type of information an entry may include.

The Discipline Index, volume 8, can be used to easily find a scientist in a specific subject specialty. This index is first classified by area of study, and within each specialty entrants are divided further by state of residence.

Name — Date(s) of marriage — Degrees Earned — Professional Experience — Current Position

CARLETON, PHYLLIS B(ARBARA), b Glenham, SDak, April 1, 30. m 53, 69; c 2. ORGANIC CHEMISTRY. *Educ:* Univ Notre Dame, BSc, 52, MSc, 54, Vanderbilt Univ, PhD(chem), 57. *Hon Degrees:* DSc, Howard Univ, 79. *Prof Exp:* Res chemist, Acme Chem Corp, 54-59, sr res chemist, 59-60; from asst to assoc prof chem 60-63, prof chem, Kansas State Univ, 63-72; prof chem, Yale Univ, 73-89; CONSULT, CARLETON & ASSOCS, 89-. *Concurrent Pos:* Adj prof, Kansas State Univ 58-60; vis lect, Oxford Univ, 77, consult, Union Carbide, 74-80. *Honors & Awards:* Gold Medal, Am Chem Society, 81; *Mem:* AAAS, fel Am Chem Soc, Sigma Chi. *Res:* Organic synthesis, chemistry of natural products, water treatment and analysis. *Mailing Address:* Carleton & Assocs 21 E 34th St Boston MA 02108

Birthplace & Date — Number of Children — Field of Specialty — Honorary Degrees — Concurrent Positions — Membership — Areas of research — Address

Abbreviations

AAAS—American Association for the Advancement of Science
abnorm—abnormal
abstr—abstract
acad—academic, academy
acct—Account, accountant, accounting
acoust—acoustic(s), acoustical
ACTH—adrenocorticotrophic hormone
actg—acting
activ—activities, activity
addn—addition(s), additional
Add—Address
adj—adjunct, adjutant
adjust—adjustment
Adm—Admiral
admin—administration, administrative
adminr—administrator(s)
admis—admission(s)
adv—adviser(s), advisory
advan—advance(d), advancement
advert—advertisement, advertising
AEC—Atomic Energy Commission
aerodyn—aerodynamic
aeronaut—aeronautic(s), aeronautical
aerophys—aerophsical, aerophysics
aesthet—aesthetic
AFB—Air Force Base
affil—affiliate(s), affiliation
agr—agricultural, agriculture
agron—agronomic, agronomical, agronomy
agrost—agrostologic, agrostological, agrostology
agt—agent
AID—Agency for International Development
Ala—Alabama
allergol—allergological, allergology
alt—alternate
Alta—Alberta
Am—America, American
AMA—American Medical Association
anal—analysis, analytic, analytical
analog—analogue
anat—anatomic, anatomical, anatomy
anesthesiol—anesthesiology
angiol—angiology
Ann—Annal(s)
ann—annual
anthrop—anthropological, anthropology
anthropom—anthropometric, anthropometrical, anthropometry
antiq—antiquary, antiquities, antiquity
antiqn—antiquarian
apicult—apicultural, apiculture
APO—Army Post Office
app—appoint, appointed
appl—applied
appln—application
approx—approximate(ly)
Apr—April
apt—apartment(s)
aquacult—aquaculture
arbit—arbitration
arch—archives
archaeol—archaeological, archaeology
archit—architectural, architecture
Arg—Argentina, Argentine
Ariz—Arizona
Ark—Arkansas
artil—artillery
asn—association
assoc(s)—associate(s), associated
asst(s)—assistant(s), assistantship(s)
assyriol—Assyriology
astrodyn—astrodynamics
astron—astronomical, astronomy
astronaut—astonautical, astronautics
astronr—astronomer
astrophys—astrophysical, astrophysics
attend—attendant, attending
atty—attorney
audiol—audiology
Aug—August
auth—author
AV—audiovisual
Ave—Avenue
avicult—avicultural, aviculture

b—born
bact—bacterial, bacteriologic, bacteriological, bacteriology
BC—British Colombia
bd—board
behav—behavior(al)
Belg—Belgian, Belgium
Bibl—biblical
bibliog—bibliographic, bibliographical, bibliography
bibliogr—bibliographer
biochem—biochemical, biochemistry
biog—biographical, biography
biol—biological, biology
biomed—biomedical, biomedicine
biomet—biometric(s), biometrical, biometry
biophys—biophysical, biophysics
bk(s)—book(s)
bldg-building
Blvd—Boulevard
Bor—Borough
bot—botanical, botany
br—branch(es)
Brig—Brigadier
Brit—Britain, British
Bro(s)—Brother(s)
byrol—byrology
bull—Bulletin
bur—bureau
bus—business
BWI—British West Indies

c—children
Calif—California
Can—Canada, Canadian
cand—candidate
Capt—Captain
cardiol-cardiology
cardiovasc—cardiovascular
cartog—cartographic, cartographical, cartography
cartogr—cartographer
Cath—Catholic
CEngr—Corp of Engineers
cent—central
Cent Am—Central American
cert—certificate(s), certification, certified
chap—chapter
chem—chemical(s), chemistry
chemother—chemotherapy
chg—change
chmn—chairman
citricult—citriculture
class—classical
climat—climatological, climatology
clin(s)—clinic(s), clinical
cmndg—commanding
Co—County
co—Companies, Company
co-auth—coauthor
co-dir—co-director
co-ed—co-editor
co-educ—coeducation, coeducational
col(s)—college(s), collegiate, colonel
collab—collaboration, collaborative
collabr—collaborator
Colo—Colorado
com—commerce, commercial
Comdr—Commander

commun—communicable, communication(s)
comn(s)—commission(s), commissioned
comndg—commanding
comnr—commissioner
comp—comparitive
compos—composition
comput—computation, computer(s), computing
comt(s)—committee(s)
conchol—conchology
conf—conference
cong—congress, congressional
Conn—Connecticut
conserv—conservation, conservatory
consol—consolidated, consolidation
const—constitution, constitutional
construct—construction, constructive
consult(s)—consult, consultant(s), consultantship(s), consultation, consulting
contemp—contemporary
contrib—contribute, contributing, contribution(s)
contribr—contributor
conv—convention
coop—cooperating, cooperation, cooperative
coord—coordinate(d), coordinating, coordination
coordr—coordinator
corp—corporate, corporation(s)
corresp—correspondence, correspondent, corresponding
coun—council, counsel, counseling
counr—councilor, counselor
criminol—criminological, criminology
cryog—cryogenic(s)
crystallog—crystallographic, crystallographical, crystallography
crystallogr—crystallographer
Ct—Court
Ctr—Center
cult—cultural, culture
cur—curator
curric—curriculum
cybernet—cybernetic(s)
cytol—cytological, cytology
Czech—Czechoslovakia

DC—District of Columbia
Dec—December
Del—Delaware
deleg—delegate, delegation
delinq—delinquency, delinquent
dem—democrat(s), democratic
demog—demographic, demography
demogr—demographer
demonstr—demontrator
dendrol—dendrologic, dendrological, dendrology
dent—dental, dentistry
dep—deputy
dept—department
dermat—dermatologic, dermatological, dermatology
develop—developed, developing, development, developmental
diag—diagnosis, diagnostic
dialectol-dialectological, dialectology
dict—dictionaries, dictionary
Dig—Digest

dipl—diploma, diplomate
dir(s)—director(s), directories, directory
dis—disease(s), disorders
Diss Abst—Dissertation Abstracts
dist—district
distrib—distributed, distribution, distributive
distribr—distributor(s)
div—division, divisional, divorced
DNA—deoxyribonucleic acid
doc—document(s), documentary, documentation
Dom—Dominion
Dr—Drive
E—east
ecol—ecological, ecology
econ(s)—economic(s), economical, economy
economet—econometric(s)
ECT—electroconvulsive or electroshock therapy
ed—edition(s), editor(s), editorial
ed bd—editorial board
educ—education, educational
educr—educator(s)
EEG—electroencephalogram, electroencephalographic, electroencephalography
Egyptol—Egyptology
EKG—electrocardiogram
elec—elecvtric, electrical, electricity
electrochem-electrochemical, electrochemistry
electroph—electrophysical, electrophysics
elem—elementary
embryol—embryologic, embryological, embryology
emer—emeriti, emeritus
employ—employment
encour—encouragement
encycl—encyclopedia
endocrinol—endocrinologic, endocrinology
eng—engineering
Eng—England, English
engr(s)—engineer(s)
enol—enology
Ens—Ensign
entom—entomological, entomology
environ-environment(s), environmental
enzym—enzymology
epidemiol—epideiologic, epidemiological, epidemiology
equip—equipment
ERDA—Energy Research & Development Administration
ESEA—Elementary & Secondary Education Act
espec—especially
estab—established, establishment(s)
ethnog—ethnographic, ethnographical, ethnography
ethnogr—ethnographer
ethnol—ethnologic, ethnological, ethnology
Europ—European
eval—evaluation
Evangel—evangelical
eve—evening
exam—examination(s), examining
examr—examiner
except—exceptional
exec(s)—executive(s)

exeg—exegeses, exegesis, exegetic, exegetical
exhib(s)—exhibition(s), exhibit(s)
exp—experiment, experimental
exped(s)—expedition(s)
explor—exploration(s), exploratory
expos—exposition
exten—extension

fac—faculty
facil—facilities, facility
Feb—February
fed—federal
fedn—federation
fel(s)—fellow(s), fellowship(s)
fermentol—fermentology
fertil—fertility, fertilization
Fla—Florida
floricult—floricultural, floriculture
found—foundation
FPO—Fleet Post Office
Fr—French
Ft—Fort

Ga—Georgia
gastroenterol—gastroenterological, gastroenterology
gen—general
geneal—genealogical, genealogy
geod—geodesy, geodetic
geog—geographic, geographical, geography
geogr—geographer
geol—geologic, geological, geology
geom—geometric, geometrical, geometry
geomorphol—geomorphologic, geomorphology
geophys—geophysical, geophysics
Ger—German, Germanic, Germany
geriat—geriatric
geront—gerontological, gerontology
GES—Gesellschaft
glaciol—glaciology
gov—governing, governor(s)
govt—government, governmental
grad—graduate(d)
Gt Brit—Great Britain
guid—guidance
gym—gymnasium
gynec—gynecologic, gynecological, gynecology

handbk(s)—handbook(s)
helminth—helminthology
hemat—hematologic, hematological, hematology
herpet—herpetologic, herpetological, herpetology
HEW—Department of Health, Education & Welfare
Hisp—Hispanic, Hispania
hist—historic, historical, history
histol—histological, histology
HM—Her Majesty
hochsch—hochschule
homeop—homeopathic, homeopathy
hon(s)—honor(s), honorable, honorary
hort—horticultural, horticulture
hosp(s)—hospital(s), hospitalization
hq—headquarters

ABBREVIATIONS

HumRRO—Human Resources Research Office
husb—husbandry
Hwy—Highway
hydraul—hydraulic(s)
hydrodyn—hydrodynamic(s)
hydrol—hydrologic, hydrological, hydrologics
hyg—hygiene, hygienic(s)
hypn—hypnosis

ichthyol—ichthyological, ichthyology
Ill—Illinois
illum—illuminating, illumination
illus—illustrate, illustrated, illustration
illusr—illustrator
immunol—immunologic, immunological, immunology
Imp—Imperial
improv—improvement
Inc—Incorporated
in-chg—in charge
incl—include(s), including
Ind—Indiana
indust(s)—industrial, industries, industry
Inf—infantry
info—information
inorg—inorganic
ins—insurance
inst(s)—institute(s), institution(s)
instnl—institutional(ized)
instr(s)—instruct, instruction, instructor(s)
instrnl—instructional
int—international
intel—intellligence
introd—introduction
invert—invertebrate
invest(s)—investigation(s)
investr—investigator
irrig—irrigation
Ital—Italian

J—Journal
Jan—January
Jct—Junction
jour—journal, journalism
jr—junior
jurisp—jurisprudence
juv—juvenile

Kans—Kansas
Ky—Kentucky

La—Louisiana
lab(s)—laboratories, laboratory
lang—language(s)
laryngol—larygological, laryngology
lect—lecture(s)
lectr—lecturer(s)
legis—legislation, legislative, legislature
lett—letter(s)
lib—liberal
libr—libraries, library
librn—librarian
lic—license(d)
limnol—limnological, limnology
ling—linguistic(s), linguistical
lit—literary, literature
lithol—lithologic, lithological, lithology

Lt—Lieutenant
Ltd—Limited

m—married
mach—machine(s), machinery
mag—magazine(s)
maj—major
malacol—malacology
mammal—mammalogy
Man—Manitoba
Mar—March
Mariol—Mariology
Mass—Massechusetts
mat—material(s)
mat med—materia medica
math—mathematic(s), mathematical
Md—Maryland
mech—mechanic(s), mechanical
med—medical, medicinal, medicine
Mediter—Mediterranean
Mem—Memorial
mem—member(s), membership(s)
ment—mental(ly)
metab—metabolic, metabolism
metall—metallurgic, metallurgical, metallurgy
metallog—metallographic, metallography
metallogr—metallographer
metaphys—metaphysical, metaphysics
meteorol—meteorological, meteorology
metrol—metrological, metrology
metrop—metropolitan
Mex—Mexican, Mexico
mfg—manufacturing
mfr—manufacturer
mgr—manager
mgt—management
Mich—Michigan
microbiol—microbiological, microbiology
micros—microscopic, microscopical, microscopy
mid—middle
mil—military
mineral—mineralogical, mineralogy
Minn—Minnesota
Miss—Mississippi
mkt—market, marketing
Mo—Missouri
mod—modern
monogr—monograph
Mont—Montana
morphol—morphological, morphology
Mt—Mount
mult—multiple
munic—municipal, municipalities
mus—museum(s)
musicol—musicological, musicology
mycol—mycologic, mycology

N—north
NASA—National Aeronautics & Space Administration
nat—national, naturalized
NATO—North Atlantic Treaty Organization
navig—navigation(al)
NB—New Brunswick
NC—North Carolina
NDak—North Dakota
NDEA—National Defense Education Act
Nebr—Nebraska

nematol—nematological, nematology
nerv—nervous
Neth—Netherlands
neurol—neurological, neurology
neuropath—neuropathological, neuropathology
neuropsychiat—neuropsychiatric, neuropsychiatry
neurosurg—neurosurgical, neurosurgery
Nev—Nevada
New Eng—New England
New York—New York City
Nfld—Newfoundland
NH—New Hampshire
NIH—National Institute of Health
NIMH—National Institute of Mental Health
NJ—New Jersey
NMex—New Mexico
No—Number
nonres—nonresident
norm—normal
Norweg—Norwegian
Nov—November
NS—Nova Scotia
NSF—National Science Foundation
NSW—New South Wales
numis—numismatic(s)
nutrit—nutrition, nutritional
NY—New York State
NZ—New Zealand

observ—observatories, observatory
obstet—obstetric(s), obstetrical
occas—occasional(ly)
occup—occupation, occupational
oceanog—oceanographic, oceanographical, oceanography
oceanogr—oceanographer
Oct—October
odontol—odontology
OEEC—Organization for European Economic Cooperation
off—office, official
Okla—Oklahoma
olericult—olericulture
oncol—oncologic, oncology
Ont—Ontario
oper(s)—operation(s), operational, operative
ophthal—ophthalmologic, ophthalmological, ophthalmology
optom—optometric, optometrical, optometry
ord—ordnance
Ore—Oregon
org—organic
orgn—organization(s), organizational
orient—oriental
ornith—ornithological, ornithology
orthod—orthodontia, orthodontic(s)
orthop—orthopedic(s)
osteop—osteopathic, osteopathy
otol—otological, otology
otolaryngol—otolaryngological, otolaryngology
otorhinol—otorhinologic, otorhinology

Pa—Pennsylvania
Pac—Pacific
paleobot—paleobotanical, paleontology
paleont—paleontology

Pan-Am—Pan-American
parisitol—parasitology
partic—participant, participating
path—pathologic, pathological, pathology
pedag—pedagogic(s), pedagogical, pedagogy
pediat—pediatric(s)
PEI—Prince Edward Islands
penol—penological, penology
periodont—periodontal, periodontic(s)
petrog—petrographic, petrographical, petrography
petrogr—petrographer
petrol—petroleum, petrologic, petrological, petrology
pharm—pharmacy
pharmaceut—pharmaceutic(s), pharmaceutical(s)
pharmacog—pharmacognosy
pharamacol—pharmacologic, pharmacological, pharmacology
phenomenol—phenomenologic(al), phenomenology
philol—philological, philology
philos—philosophic, philosophical, philosophy
photog—photographic, photography
photogeog—photogeographic, photogeography
photogr—photographer(s)
photogram—photogrammetric, photogrammetry
photom—photometric, photometrical, photometry
phycol—phycology
phys—physical
physiog—physiographic, physiographical, physiography
physiol—physiological, phsysiology
Pkwy—Parkway
Pl—Place
polit—political, politics
polytech—polytechnic(s)
pomol—pomological, pomology
pontif—pontifical
pop—population
Port—Portugal, Portuguese
Pos:—Position
postgrad—postgraduate
PQ—Province of Quebec
PR—Puerto Rico
pract—practice
practr—practitioner
prehist—prehistoric, prehistory
prep—preparation, preparative, preparatory
pres—president
Presby—Presbyterian
preserv—preservation
prev—prevention, preventive
prin—principal
prob(s)—problem(s)
proc—proceedings
proctol—proctologic, proctological, proctology
prod—product(s), production, productive
prof—professional, professor, professorial
Prof Exp—Professional Experience
prog(s)—program(s), programmed, programming
proj—project(s), projection(al), projective
prom—promotion
protozool—protozoology
Prov—Province, Provincial
psychiat—psychiatric, psychiatry
psychoanal—psychoanalysis, psychoanalytic, psychoanalytical
psychol—psychological, psychology
psychomet—psychometric(s)
psychopath—psychopathologic, psycho pathology
psychophys—psychophysical, psychophysics
psychophysiol—psychophysiological, psychophysiology
psychosom—psychosomtic(s)
psychother—psychoterapeutic(s), psychotherapy
Pt—Point
pub—public
publ—publication(s), publish(ed), publisher, publishing
pvt—private

Qm—Quartermaster
Qm Gen—Quartermaster General
qual—qualitative, quality
quant—quantitative
quart—quarterly
Que—Quebec

radiol—radiological, radiology
RAF—Royal Air Force
RAFVR—Royal Air Force Volunteer Reserve
RAMC—Royal Army Medical Corps
RAMCR—Royal Army Medical Corps Reserve
RAOC—Royal Army Ornance Corps
RASC—Royal Army Service Corps
RASCR—Royal Army Service Corps Reserve
RCAF—Royal Canadian Air Force
RCAFR—Royal Canadian Air Force Reserve
RCAFVR—Royal Canadian Air Force Volunteer Reserve
RCAMC—Royal Canadian Army Medical Corps
RCAMCR—Royal Canadian Army Medical Corps Reserve
RCASC—Royal Canadian Army Service Corps
RCASCR—Royal Canadian Army Service Corps Reserve
RCEME—Royal Canadian Electrical & Mechanical Engineers
RCN—Royal Canadian Navy
RCNR—Royal Canadian Naval Reserve
RCNVR—Royal Canadian Naval Volunteer Reserve
Rd—Road
RD—Rural Delivery
rec—record(s), recording
redevelop—redevelopment
ref—reference(s)
refrig—refrigeration
regist—register(ed), registration
registr—registrar
regt—regiment(al)
rehab—rehabilitation
rel(s)—relation(s), relative
relig—religion, religious
REME—Royal Electrical & Mechanical Engineers
rep—represent, representative
Repub—Republic
req—requirements
res—research, reserve
rev—review, revised, revision
RFD—Rural Free Delivery
rhet-rhetoric, rhetorical
RI—Rhode Island
Rm—Room
RM—Royal Marines
RN—Royal Navy
RNA—ribonucleic acid
RNR—Royal Naval Reserve
RNVR—Royal Naval Volunteer Reserve
roentgenol—roentgenologic, roentgenological, roentgenology
RR—Railroad, Rural Route
Rte—Route
Russ—Russian
rwy—railway

S—south
SAfrica—South Africa
SAm—South America, South American
sanit—sanitary, sanitation
Sask—Saskatchewan
SC—South Carolina
Scand—Scandinavia(n)
sch(s)—school(s)
scholar—scholarship
sci—science(s), scientific
SDak—South Dakota
SEATO—Southeast Asia Treaty Organization
sec—secondary
sect—section
secy—secretary
seismog—seismograph, seismographic, seismography
seismogr—seismographer
seismol—seismological, seismology
sem—seminar, seminary
Sen—Senator, Senatorial
Sept—September
ser—serial, series
serol—serologic, serological, serology
serv—service(s), serving
silvicult—silvicultural, silviculture
soc(s)—societies, society
soc sci—social science
sociol—sociologic, sociological, sociology
Span—Spanish
spec—special
specif—specification(s)
spectrog—spectrograph, spectrographic, spectrography
spectrogr—spectrographer
spectrophotom—spectrophotometer, spectrophotometric, spectrophotometry
spectros—spectroscopic, spectroscopy
speleol—speleological, speleology
Sq—Square
sr—senior
St—Saint, Street(s)
sta(s)—station(s)
stand—standard(s), standardization
statist—statistical, statistics
Ste—Sainte
steril—sterility

stomatol—stomatology
stratig—stratigraphic, stratigraphy
stratigr—stratigrapher
struct—structural, structure(s)
stud—student(ship)
subcomt—subcommittee
subj—subject
subsid—subsidiary
substa—substation
super—superior
suppl—supplement(s), supplemental, supplementary
supt—superintendent
supv—supervising, supervision
supvr—supervisor
supvry—supervisory
surg—surgery, surgical
surv—survey, surveying
survr—surveyor
Swed—Swedish
Switz—Switzerland
symp—symposia, symposium(s)
syphil—syphilology
syst(s)—system(s), systematic(s), systematical

taxon—taxonomic, taxonomy
tech—technical, technique(s)
technol—technologic(al), technology
tel—telegraph(y), telephone
temp—temporary
Tenn—Tennessee
Terr—Terrace
Tex—Texas
textbk(s)—textbook(s)
text ed—text edition
theol—theological, theology
theoret—theoretic(al)
ther—therapy
therapeut—therapeutic(s)
thermodyn—thermodynamic(s)
topog—topographic, topographical, topography
topogr—topographer
toxicol—toxicologic, toxicological, toxicology
trans—transactions
transl—translated, translation(s)
translr—translator(s)
transp—transport, transportation
treas—treasurer, treasury
treat—treatment
trop—tropical
tuberc—tuberculosis
TV—television
Twp—Township

UAR—United Arab Republic
UK—United Kingdom
UN—United Nations
undergrad—undergraduate
unemploy—unemployment
UNESCO—United Nations Educational Scientific & Cultural Organization
UNICEF—United Nations International Childrens Fund
univ(s)—universities, university
UNRRA—United Nations Relief & Rehabilitation Administration
UNRWA—United Nations Relief & Works Agency
urol—urologic, urological, urology
US—United States
USAAF—US Army Air Force
USAAFR—US Army Air Force Reserve
USAF—US Air Force
USAFR—US Air Force Reserve
USAID—US Agency for International Development
USAR—US Army Reserve
USCG—US Coast Guard
USCGR—US Coast Guard Reserve
USDA—US Department of Agriculture
USMC—US Marine Corps
USMCR—US Marine Corps Reserve
USN—US Navy
USNAF—US Naval Air Force
USNAFR—US Naval Air Force Reserve
USNR—US Naval Reserve
USPHS—US Public Health Service
USPHSR—US Public Health Service Reserve
USSR—Union of Soviet Socialist Republics

Va—Virginia
var—various
veg—vegetable(s), vegetation
vent—ventilating, ventilation
vert—vertebrate
Vet—Veteran(s)
vet—veterinarian, veterinary
VI—Virgin Islands
vinicult—viniculture
virol—virological, virology
vis—visiting
voc—vocational
vocab—vocabulary
vol(s)—voluntary, volunteer(s), volume(s)
vpres—vice president
vs—versus
Vt—Vermont

W—west
Wash—Washington
WHO—World Health Organization
WI—West Indies
wid—widow, widowed, widower
Wis—Wisconsin
WVa—West Virginia
Wyo—Wyoming

Yearbk(s)—Yearbook(s)
YMCA—Young Men's Christian Association
YMHA—Young Men's Hebrew Association
Yr(s)—Year(s)
YT—Yukon Territory
YWCA—Young Women's Christian Association
YWHA—Young Women's Hebrew Association

zool—zoological, zoology

American Men & Women of Science

M

MA, BENJAMIN MINGLI, b Nanking, China, Feb 2, 24; nat US; m 58; c 2. NUCLEAR ENGINEERING, AERONAUTICAL ENGINEERING. *Educ:* Nat Cent Univ, China, BS, 45; Stanford Univ, MS, 47, EngD(aeronaut & astronaut eng), 50; Iowa State Univ, PhD(nuclear sci & eng), 62. *Prof Exp:* Asst, Stanford Univ, 47-49, res assoc, 49-51; asst prof mech eng, Ohio Northern Univ, 55-56; assoc prof, SDak State Univ, 56-60; assoc prof, Univ Mich, 60-61; from asst prof to assoc prof, 61-80, FULBRIGHT PROF NUCLEAR ENG, IOWA STATE UNIV, 80- *Concurrent Pos:* Consult, Argonne Nat Lab, Hanford Eng Develop Lab; Fulbright prof & invited lectr, Univ Istanbul, Israel Inst Technol, Univ Stockholm, Univ Copenhagen, Athens Tech Univ & Cambridge Univ. *Mem:* Am Nuclear Soc; Am Inst Aeronaut & Astronaut; Am Soc Mech Eng; Am Soc Eng Educ; Am Phys Soc; Sigma Xi. *Res:* Materials and strength at elevated temperature; heat transfer; radiation effects; reactor fuel elements; nuclear energy conversion systems; nuclear fusion theory and technology. *Mailing Add:* 701 Ash Ames IA 50011

MA, CYNTHIA SANMAN, b Hong Kong, May 16, 40; US citizen; m 66; c 2. STATISTICS, COMPUTER SCIENCE. *Educ:* Siena Col, BS, 62; Fla State Univ, MS, 66, PhD(systs anal), 69. *Prof Exp:* Statistician, Community Studies Inst, Kansas City, Mo, 62-63; res asst statist, Fla State Univ, 63-66, comput programmer & analyst, Dept Physics, 66-67, systs design analyst, Systs Planning & Develop Ctr, 67-69; ASSOC PROF DATA PROCESSING & STATIST, BALL STATE UNIV, 69- *Concurrent Pos:* Consult, 70- *Mem:* Am Statist Asn; Asn Systs Mgt; Asn Comput Mach. *Res:* Design of computer-based systems for finance analysis and teaching; design and alocation resource model for planning; use of graphics for information systems. *Mailing Add:* Dept Mgt & Sci Ball State Univ Muncie IN 47306

MA, ER-CHIEH, b Jukao, China, Dec 7, 22; m 46; c 3. ENGINEERING MECHANICS, APPLIED MATHEMATICS. *Educ:* Chiao-Tung Univ, BS, 46; Kans State Univ, MS, 59, PhD(appl mech), 62. *Prof Exp:* Res assoc design, Ord Res Inst, 47-55; sr expert sci educ, Ministry Educ, 55-58; res assoc appl mech, Kans State Univ, 60-62; assoc prof, Wichita State Univ, 62-64; from assoc prof to prof eng, 64-81, PROF MATH, UNIV ARK, LITTLE ROCK, 81- *Concurrent Pos:* Chinese del, Study Conf Sci Teaching, UNESCO, 56. *Mem:* Am Soc Eng Educ; Am Soc Mech Engrs. *Res:* Elasticity; viscoelasticity; vibrations; celestial mechanics; applied analysis. *Mailing Add:* Box 3017 Little Rock AR 72203

MA, FAI, b Canton, People's Repub China, Aug 6, 54; US citizen. RANDOM VIBRATIONS, STATISTICAL SIMULATION. *Educ:* Univ Hong Kong, BS, 77; Calif Inst Technol, MS & PhD(appl math & eng sci), 81. *Prof Exp:* Sr res engr, Weidlinger Assocs, 81-82; res fel, Int Bus Mach, 82-83; sr engr, Standard Oil Co, 83-86; ASSOC PROF, DEPT MECH ENG, UNIV CALIF, BERKELEY, 86- *Concurrent Pos:* Consult, var co, 86-; NSF presidential young investr award, 87. *Mem:* Am Soc Mech Engrs; fel Inst Diag Engrs UK. *Res:* Random vibrations; statistical simulation; system analysis and control. *Mailing Add:* Dept Mech Eng Univ Calif Berkeley CA 94720

MA, JOSEPH T, b Fuzhou, China, Aug 15, 25; US citizen; m 49; c 1. ADVANCED TECHNOLOGIES DISK DRIVES STORAGE & PRINTERS. *Educ:* Nat Sun-Yet-Sen Univ, BS, 48; Tex Tech Col, BS, 49; Purdue Univ, MS, 51; Iowa State Univ, PhD(mech eng), 59. *Prof Exp:* Sr engr prod develop, Advan Systs Develop Div, Los Gatos Lab, IBM, 63-68, engr mgr advan develop, 70-73, syst mgr L-5 store, Boulder Lab, IBM Corp, 74-75, sr tech staff, Gen Prod Div, 76-79, mgr advan tech, 80-84; adj prof mech eng, 85-89, DIR EXTEN PROG CONTINUING EDUC, SCH ENG, SANTA CLARA UNIV, 89- *Concurrent Pos:* Asst prof, Col Eng, Iowa State Univ, 55-59; res staff mem, Res Div, IBM Corp, 59-62; vis prof, San Jose State Univ, Calif, 60-; fel systs eng, Advan Eng Study, Mass Inst Technol, 69-70; mem, Bd Link Technologies, San Jose, Calif, 85-87; consult, Digital Equip Corp, Cupertino, Calif, 87- & Hambrech & Quist, San Francisco, Calif, 90-; lectr, Shanghai Jiao Tong Univ, 88- *Mem:* Fel Am Soc Mech Engrs; sr mem Inst Elec & Electronics Engrs. *Res:* Electro-mechanical servo and control; tribology; fluid mechanics and air bearings research. *Mailing Add:* 15541 Toyon Dr Los Gatos CA 95030

MA, MARK T, b Kiangsu, China, Mar 21, 33; m 58; c 2. ELECTROMAGNETISM, TELECOMMUNICATION. *Educ:* Nat Taiwan Univ, BS, 55; Univ Ill, MS, 57; Syracuse Univ, PhD(elec eng), 61. *Prof Exp:* Lectr elec eng, Nat Taiwan Univ, 55-56; engr electronics, Int Bus Mach Corp, 57-58 & Gen Elec Co, 61-64; RES SCIENTIST & GROUP LEADER, NAT INST STANDARDS & TECHNOL, US DEPT COMMERCE, 64- *Concurrent Pos:* Adj prof, Univ Colo, 64-; assoc ed, Radio Sci, 66-67. *Honors & Awards:* US Dept Com Bronze Medal, 90. *Mem:* Fel Inst Elec & Electronics Engrs. *Res:* Electromagnetic theory; electromagnetic interference; antenna theory; telecommunications. *Mailing Add:* Nat Inst Standards & Technol US Dept Com Boulder CO 80303

MA, MAW-SUEN, b Peking, China, July 18, 47, US citizen; m 71; c 2. RADIO ANALYTICAL CHEMISTRY, GEOCHEMISTRY. *Educ:* Cheng Kung Univ, Taiwan, BS, 68; Univ Ky, PhD(radio anal chem), 75. *Prof Exp:* Teaching asst chem, Cent Univ, Taiwan, 70-71; teaching asst, Univ Ky, 71-72, res fel, 72-75; co-investr NASA lunar prog geochem, Ore State Univ, 75-81, res asst prof, 81; res scientist, US Testing Co, Inc, 81-82; chem supvr, Consumers Power Co, 82-84; radiol serv mgr, Impell Corp, 84-87; radiochem sect head, Lilco, 87-89, rad control div mgr, 89-90; GEN MGR, SUN-HSING ENTERPRISES, 90- *Mem:* Am Nuclear Soc; Health Phys Soc. *Res:* Applications of analytical chemistry; radio chemical methods of analysis; neutron activation analysis to samples of geological, environmental and industrial interests; trace element geochemistry; nuclear power plant chemistry. *Mailing Add:* 18 Mink Lane East Setauket NY 11733

MA, NANCY SHUI-FONG, US citizen. CYTOGENETICS. *Educ:* Mt St Vincent Col, Can, BSc, 64; Northeastern Univ, MS, 68; Boston Col, PhD(biol), 73. *Prof Exp:* ASSOC PATH, NEW ENGLAND REGIONAL PRIMATE RES CTR, HARVARD UNIV, 73- *Concurrent Pos:* Cytogenetics consult, Pathobiol Inc, 75- *Res:* Cytogenetic studies of the New World monkeys. *Mailing Add:* New Eng Regional Primate Res Ctr Med Sch Harvard Univ Southborough MA 01772

MA, TE HSIU, b Hopei, China, Aug 24, 24; nat US; m 60; c 2. CYTOGENETICS. *Educ:* Cath Univ Peking, BS, 48; Nat Taiwan Univ, MS, 50; Univ Va, PhD(biol), 59. *Prof Exp:* Asst cytologist, Sugarcane Cytol, Taiwan Sugar Exp Sta, 50-55; from asst prof to assoc prof biol, Emory & Henry Col, 59-64; asst prof, Western Ill Univ, 64-66; geneticist, Radiation Biol Lab, Smithsonian Inst, Wash, DC, 66-69; assoc prof, 69-73, PROF BIOL SCI, WESTERN ILL UNIV, 73- *Concurrent Pos:* Res partic, Oak Ridge Nat Lab, 63-64; res consult, Oak Ridge Inst Nuclear Studies, 63-; Atomic Energy Comn res grant, 65; Environ Protection Agency res grant, 71-73; radiation safety officer, Western Ill Univ, 71- *Mem:* AAAS; Am Genetics Soc; Am Soc Cell Biol; NY Acad Sci. *Res:* Cytogenetics of sugarcane, corn and Tradescantia; radiation effects on chromosomes of Tradescantia and Vicia; air pollutant effects on chromosomes of Tradescantia; radioactive pollutant effect on fish. *Mailing Add:* Dept of Biol Sci Western Ill Univ Macomb IL 61455

MA, TERENCE P, b Hong Kong, June 27, 58. NEUROBIOLOGY. *Educ:* Johns Hopkins Univ, BS, 81; Wayne State Univ, PhD(anat), 88. *Prof Exp:* Nat res serv award fel, Nat Eye Inst, NIH, 87-89, intramural res training award fel, 89-90; ASST PROF, DEPT ANAT, UNIV MISS MED CTR, 90- *Mem:* AAAS; Asn Res Vision & Ophthal; Am Asn Anatomists; Int Basil Ganglia Soc; Int Brain Res Orgn; Sigma Xi; Soc Neurosci. *Mailing Add:* Dept Anat Univ Miss Med Ctr 2500 N State St Jackson MS 39216

MA, TSO-PING, b Lan-Chov, China, Nov 13, 45. SOLID STATE PHYSICS, MICRO ELECTRONICS. *Educ:* Nat Taiwan Univ, BS, 68; Yale Univ, MS, 70, PhD(appl physics), 74. *Prof Exp:* Sr asst, Inst Bus Mach, 74-75, staff eng, 75-77; res asst physics, 70-74, asst prof physics, 77-80, PROF MICRO-ELEC & SOLID STATE PHYSICS, YALE UNIV, 80- *Honors & Awards:* G E Whitney Award, 85. *Mem:* Mat Res Soc; fel Am Phys Soc; Inst Elec & Electronic Engrs. *Res:* Conductor physics; mos-interface newgate dielectric radiation; hot electron plasma interface; annealing of auto electric devices. *Mailing Add:* Dept Eas-Becton Ctr Yale Univ 15 Prospect St PO Box 2157 Yale Sta New Haven CT 06520

MA, TSU SHENG, b Canton, China, Oct 15, 11; nat US; m 42; c 2. MICROCHEMISTRY, ORGANIC CHEMISTRY. *Educ:* Tsing Hua Univ, China, BS, 31; Univ Chicago, PhD(chem), 38. *Prof Exp:* Instr in charge microchem lab, Univ Chicago, 38-46; prof, Peking Univ, 46-49; sr lectr microchem, Univ Otago, NZ, 49-51; asst prof chem, NY Univ, 51-54; assoc prof, 54-58, PROF CHEM, CITY UNIV NEW YORK, 58- *Concurrent Pos:* Vis prof, Tsinghua Univ, Peking, 47, Lingnan Univ, 49, NY Univ, 54-60, Univ Singapore, 75-76; Fulbright-Hays lectr, China, Japan, Hong Kong, India, Malaya, Australia & NZ, 61-62, Korea, Thailand, Hong Kong & Iran, 68-69; Am specialist, Bur Educ & Cult Affairs, US State Dept, Ceylon, Burma, Thailand, Hong Kong & Philippines, 64; mem comn reagents & reactions, Int Union Pure & Appl Chem, 64-69; ed, Mikrochimica Acta, 65- 89. *Honors & Awards:* Benedetti-Pichler Award in Microchem, 76. *Mem:* AAAS; Am Inst Chem; Am Chem Soc; Royal Soc Chem; NY Acad Sci; Soc Appl Spectros. *Res:* Synthetic drugs; medicinal plants; microchemical analysis; microtechniques in organic synthesis; small scale experiments for teaching general and organic chemistry. *Mailing Add:* Seven Banbury Lane Chapel Hill NC 27514

MA, WILLIAM HSIOH-LIEN, b China, Jan 31, 47; US citizen; m 70; c 1. ELECTRICAL ENGINEERING, PHYSICAL SCIENCE. *Educ:* Univ Ark, BSEE, 70; Ga Inst Technol, MSEE, 70; Purdue Univ, PhD(elec eng), 74. *Prof Exp:* ENGR RES & DEVELOP MGT, IBM CORP, 74- *Mem:* Inst Elec & Electronics Engrs; Am Phys Soc; Am Vacuum Soc. *Res:* Research and development in semiconductor device fabrication process; dry etching, plasma etching, reactive ion etching and device processing development. *Mailing Add:* IBM Ltd Zip 77n IBM E Fishkill Hopewell Junction NY 12533

MA, YI HUA, b Nanking, China, Nov 7, 36; m 63; c 3. CHEMICAL ENGINEERING. *Educ:* Nat Taiwan Univ, BS, 59; Univ Notre Dame, MS, 63; Mass Inst Technol, ScD(chem eng), 67. *Prof Exp:* From asst prof to assoc prof, 67-76, dept head, 79-89, PROF CHEM ENG, WORCESTER POLYTECH INST, 76- *Concurrent Pos:* Consultant. *Mem:* Am Chem Soc; Am Inst Chem Engrs; Am Asn Univ Professors. *Res:* Inorganic membranes; transport phenomena in porous catalyst; applied mathematics in chemical engineering; drying; adsorption and diffusion in zeolites. *Mailing Add:* Dept Chem Eng Worcester Polytech Inst Worcester MA 01609

MA, Z(EE) MING, b Shanghai, China, Feb 4, 42; m 64; c 2. HIGH ENERGY PHYSICS. *Educ:* Southwestern Univ, Tex, BS, 62; Duke Univ, PhD(physics), 67. *Prof Exp:* From instr to asst prof physics, Mich State Univ, 67-74, assoc prof, 74-80; MEM STAFF, BELL LABS, 80- *Concurrent Pos:* Prin investr, US Dept Energy. *Mem:* Am Phys Soc. *Res:* Experimental high energy physics. *Mailing Add:* AT&T Bell Labs 2A249 One Whippany Rd Whippany NJ 07981

MAA, JER-SHEN, b Shanghai, July 6, 44; m. THIN FILM SCIENCE, MATERIALS CHARACTERIZATION. *Educ:* Nat Cheng Kung Univ, Taiwan, BS, 68; Univ Minn, PhD(mat sci), 74. *Prof Exp:* Res asst, Mat Sci Dept, Univ Minn, 69-74; res assoc, Phys Div, Argonne Nat Lab, 74-75; assoc prof, Mat Sci Dept, Nat Tsing Hua Univ, Taiwan, 75-79; mem tech staff, Solid State Technol Ctr, Solid State Div, 79-81; mem tech staff, RCA Labs, Princeton, 81-87; MEM TECH STAFF, DAVID SARNOFF RES CTR, PRINCETON, 87- *Concurrent Pos:* Mem staff, Thin Film Technol Group, RCA Labs, 79. *Mem:* Am Vacuum Soc; Electrochem Soc Am; Inst Elec & Electronic Engrs. *Res:* Thin film processes, material characterization, dry etching technology; metallgation technology, multilevel metalligation processes; refractory metal silicide technology, bias sputtering; chemical vapor deposition of tungsten and silicide; dry process for large area electroluminescence display devices. *Mailing Add:* Seven Linden Lane Plainsboro NJ 08536

MAACK, CHRISTOPHER A, b Chicago, Ill, Aug 23, 49; m; c 1. PROTEIN CHEMICALS. *Educ:* Univ Calif, Berkeley, BA, 71, PhD(biochem), 78. *Prof Exp:* Postdoctoral fel, Dept Epidemiol, Univ Calif Med Ctr, San Francisco, 77-79; postdoctoral fel & med res chem staff scientist, Lab Develop Biochem, Nat Inst Med Res, London, UK, 80-82; res scientist, Dept Biol Sci, Stauffer Chem Co, Richmond, Calif, 82-85, sr res biologist, 85-87; SR SCIENTIST PROTEIN CHEM, BIOGROWTH INC, RICHMOND, CALIF, 87- *Concurrent Pos:* Postdoctoral fel, NIH, 80-82. *Mem:* AAAS; Am Soc Biochem & Molecular Biol. *Res:* Protein chemistry including purification and characterization; testing the potential of human therapeutic drugs in vivo. *Mailing Add:* Dept Protein Chem BioGrowth Inc 3065 Richmond Pkwy Suite 117 Richmond CA 94806

MAAG, URS RICHARD, b Winterthur, Switz, Jan 20, 38; m 65; c 3. STATISTICS. *Educ:* Swiss Fed Inst Technol, Dipl Math, 61; Univ Toronto, MA, 62, PhD, 65. *Prof Exp:* From lectr to assoc prof math, 64-73, assoc prof, 73-78, PROF COMPUT SCI & OPERS RES, UNIV MONTREAL, 78- *Concurrent Pos:* Statist consult. *Mem:* Am Statist Asn; Inst Math Statist; Statist Soc Can (pres, 80); Int Statist Inst. *Res:* Nonparametric statistics, robust methods in multivariate analysis; applications in the biomedical sciences; data analysis. *Mailing Add:* Dept Comput Sci & Opers Res Univ Montreal Box 6128 Montreal PQ H3C 3J7 Can

MAAHS, HOWARD GORDON, b Los Angeles, Calif, May 16, 39; m 62; c 4. HIGH TEMPERATURE CHEMISTRY, KINETICS. *Educ:* Stanford Univ, BS, 59; Univ Wash, PhD(chem eng), 64. *Prof Exp:* Chem engr, Univ Wash, 60-61; AEROSPACE TECHNOLOGIST, LANGLEY RES CTR, NASA, 64- *Mem:* Am Chem Soc; Am Carbon Soc. *Res:* Advanced composite materials and high temperature materials and coatings for aircraft and spacecraft structural applications; chemical kinetics and catalysis; aqueous chemistry of tropospheric cloud droplets; combustion kinetics and pollution formation; heat transfer; graphite ablation and sublimation. *Mailing Add:* 22 Ridgewood Pkwy Newport News VA 23602

MAAR, JAMES RICHARD, b Wellsville, NY, Oct 7, 43. MATHEMATICAL STATISTICS, STATISTICAL ANALYSIS. *Educ:* Eckerd Col, BS, 65; Brown Univ, ScM, 67; George Washington Univ, PhD(math statist), 73. *Prof Exp:* MATHEMATICIAN, US DEPT DEFENSE, FT GEORGE G MEADE, 67- *Concurrent Pos:* Assoc prof lectr mgt sci, Col Gen Studies, George Washington Univ, 73-78; lectr math, Univ Md, 79-88. *Mem:* Sigma Xi; Inst Math Statist; Am Statist Asn. *Res:* Multivariate data analysis, including cluster analysis, discriminant analysis, statistical computing and graphical data analysis; new counterexamples to plausible but false statements in probability theory and mathematical statistics. *Mailing Add:* 9608 Basket Ring Columbia MD 21045

MAAS, EUGENE VERNON, b Jamestown, NDak, Dec 18, 36; m 61; c 3. PLANT STRESS PHYSIOLOGY, CROP SALT TOLERANCE. *Educ:* Jamestown Col, BS, 58; Univ Ariz, MS, 61; Ore State Univ, PhD(soils), 66. *Prof Exp:* Res assoc soil physics, Univ Ariz, 61; asst soils, Ore State Univ, 61-66; plant physiologist, Mineral Nutrit Lab, 66-68, res leader, 75-89, PLANT PHYSIOLOGIST, AGR RES SERV, US SALINITY LAB, USDA, 68- *Concurrent Pos:* Nat Acad Sci-Nat Res Coun resident res associateship, 66-68; adj prof, Dept Soil & Environ Sci, Univ Calif, Riverside, 80-; vis scientist, Agr Res Serv, Systs Res Lab, USDA, Beltsville, MD, 88-89. *Mem:* Am Soc Plant Physiol; Am Soc Agron; Sigma Xi. *Res:* Ion absorption and transport in plants; environmental physiology of plants; tolerance to salts and air pollution; crop growth simulation modeling. *Mailing Add:* US Salinity Lab 4500 Glenwood Dr Riverside CA 92501

MAAS, JAMES WELDON, b St Louis, Mo, Oct 26, 29; m 53; c 3. PHARMACOLOGY. *Educ:* Wash Univ, BA, 50, MD, 54. *Hon Degrees:* MA, Yale Univ, 72. *Prof Exp:* Intern med, Grady Mem Hosp, Atlanta, Ga, 54-55; resident psychiat, Cincinnati Gen Hosp, Ohio, 55-56 & 58-60; chief sect psychosom med, NIMH, 60-66; prof psychiat, Univ Ill Col Med, 66-72, & Sch Med, Yale Univ, 72-82; PROF, DEPT PSYCHIAT, UNIV TEX HEALTH SCI CTR, 82- *Concurrent Pos:* Dir res, Ill State Psychiat Inst, 66-72. *Honors & Awards:* Anna Maika Award, 73. *Mem:* AAAS; Am Psychiat Asn; Am Psychosom Soc; fel Am Col Neuropsychopharmacology; Am Soc Pharmacol & Exp Therapeut. *Res:* Relationship between biology and behavior; brain chemistry and behavior; biochemistry of synapse; biogenic amines in brain; neurochemistry. *Mailing Add:* Univ Texas Health Sci Ctr 7703 Floyd Curl Dr San Antonio TX 78284

MAAS, JOHN LEWIS, b Detroit, Mich, Aug 13, 40; m 62; c 3. PLANT PATHOLOGY, PLANT PHYSIOLOGY. *Educ:* Mich State Univ, BS, 62; Univ Wash, MS, 64; Ore State Univ, PhD(plant path), 68. *Prof Exp:* RES PLANT PATHOLOGIST, FRUIT LAB, HORT SCI INST, AGR RES CTR, NORTHEAST REGION, USDA, 68- *Concurrent Pos:* ed, Compendium Strawberry Dis, Am Phytopath Soc, 84, Advances in Strawberry Prod, 86-; Consult, strawberry prod progs, China, Lectr Prog anal, 86. *Honors & Awards:* Award of Achievement, NAm Strawberry Growers Asn, 84, 88. *Mem:* Am Phytopath Soc; Am Soc Horticult Sci; North Am Strawberry Growers Asn. *Res:* Fungus diseases of small fruit crops; etiology and control of Phytophthora fragariae root rot and Botrytis cinerea fruit rot of strawberry; genetic sources of disease resistance host physiological factors contributing to disease resistance. *Mailing Add:* Hort Sci Inst Agr Res Ctr NE Region US Dept of Agr Beltsville MD 20705

MAAS, KEITH ALLAN, b Burlington, Wis, Apr 7, 36; div; c 2. PHOTOGRAPHIC CHEMISTRY. *Educ:* Mass Inst Technol, SB, 58; Univ Vt, MS, 60; Univ Calif, Davis, PhD(chem), 63. *Prof Exp:* Res chemist, tech rep & col rels rep, E I du Pont de Nemours & Co, Inc, 63-71; SR STAFF SCIENTIST, PHOTOGRAPHIC SYSTS, TECHNICOLOR GRAPHIC SERVS, INC, 72- *Mem:* Am Chem Soc; Soc Photog Scientists & Engrs. *Res:* Photographic materials and processes; techniques of image presentation. *Mailing Add:* 1526 S Center St Redlands CA 92373

MAAS, PETER, b Evanston, Ill, Apr 9, 39; m 64; c 2. SOLID STATE PHYSICS, BIOPHYSICS. *Educ:* Mass Inst Tech, BS, 62; Stanford Univ, MS, 64; Univ Colo, PhD(physics), 69. *Prof Exp:* Asst engr, Lockheed Missile & Space Co, 62-63, grad study scientist, 63-65, scientist, 64-65; asst physics, Univ Colo, Boulder, 65-67, solid state physics, 67-69, res assoc & instr biophys, Med Ctr, 69-70; lectr appl physics, 70-76, SR LECTR APPL PHYSICS, UNIV STRATHCLYDE, 76- *Concurrent Pos:* Consult to various co. *Mem:* Asn Comput Mach; fel Inst Physics; IEEE Comput Soc. *Res:* Physics applications in biology; magneto encephalography & biomagnetism; computer uses in science. *Mailing Add:* Dept of Physics & Appl Physics Univ of Strathclyde 107 Rottenrow Glasgow G40NG Scotland

MAAS, STEPHEN JOSEPH, b Temple, Tex, Nov 17, 50. AGRICULTURAL REMOTE SENSING. *Educ:* Tex A&M Univ, BS, 73, MS, 75, PhD(agron), 85. *Prof Exp:* Grad asst, Tex A&M Univ, 73-75, grad asst meteorol & soil sci, 82-84; sr meteorologist, Meteorol Res Inc, 75-77; res scientist, Tex Agr Exp Sta, 77-82; PLANT PHYSIOLOGIST REMOTE SENSING RES, AGR RES SERV, USDA, 84- *Mem:* Am Soc Agron; Am Meteorol Soc; Sigma Xi. *Res:* Quantifying the effects of the climatic and edaphic environment on plant growth and development; crop simulation modeling. *Mailing Add:* Blackland Agr Exp Ctr Box 748 Temple TX 76501

MAAS, WERNER KARL, b Kaiserslautern, Ger, Apr 27, 21; nat US; m 60; c 3. MOLECULAR GENETICS. *Educ:* Harvard Univ, BA, 43; Columbia Univ, PhD(zool), 48. *Prof Exp:* Asst zool, Columbia Univ, 43-45; mem staff, Med Col, Cornell Univ, 49-54; from asst prof pharmacol to assoc prof microbiol, 54-63, adv, Grad Dept, 64-70, chmn, Dept Basic Med Sci, 75-80, PROF MICROBIOL, SCH MED, NY UNIV, 63- *Concurrent Pos:* Vis investr, Mass Gen Hosp, 52-53; dir honors prog, Sch Med, NY Univ, 58-61, co-dir, USPHS Genetics Training Grant, 61-68, dir, 68-75, dir, Microbiol Training Grant, 64-69; mem, NIH Genetics Training Grants Comt, 61-65, mem study sect microbial chem, 68-72, chmn, 70-72; mem staff, Univ Brussels, 63; mem test comt microbiol, Nat Bd Med Exam, 71-75. *Mem:* Am Soc Microbiol; Am Soc Biol Chem; Genetics Soc Am. *Res:* Microbial genetics and physiology, with emphasis on regulatory mechanisms, especially of protein synthesis; amino acid permeases; polyamine metabolism; genetics of extrachromosomal elements. *Mailing Add:* Dept of Microbiol NY Univ Sch of Med 550 1st Ave New York NY 10016

MAASBERG, ALBERT THOMAS, b Bronx, NY, Feb 10, 15; m 41; c 2. CHEMISTRY, CHEMICAL ENGINEERING. *Educ:* State Univ NY, BS, 36. *Prof Exp:* Res dir cellulose prod dept, Dow Chem USA, 46-50, asst prod mgr, 50-52, mgr, 52-54, tech dir plastics prod dept, 54-56, dir res & develop, Midland Div, 56-63, dir contract res, develop & eng, 63-79. *Mem:* Am Chem Soc; Sigma Xi; Tech Asn Pulp & Paper Indust; fel Am Inst Chem Eng. *Res:* Pulp and paper manufacturing; cellulose ethers; water soluble polymers; chemicals and plastics process and product; government and industrial contract research and development administration. *Mailing Add:* 3220 Noeske Midland MI 48640

MAASS, ALFRED ROLAND, b Plymouth, Wis, Apr 14, 18; m 47; c 3. BIOCHEMISTRY. *Educ:* Antioch Col, BS, 42; Univ Wis, MS, 47, PhD(biochem), 50. *Prof Exp:* Consult, Argonne Nat Labs, 51; sr res biochemist, 51-56, group leader, Biochem Sect, 56-57, asst sect head, 57-62, sect head, 62-67, assoc dir biochem, 67-74, dir develop proj, 74-78, dir anal biochem res, 78-81, dir sci admin pre-clin develop, Smith, Kline & French Labs, 81-82; CONSULT, MAASS ASSOC, 82- *Concurrent Pos:* Chmn civilian adv comt radiation safety, City Philadelphia, 64-73. *Mem:* AAAS; Health Physics Soc; Am Chem Soc; Am Acad Neurol; NY Acad Sci. *Res:* Isotope tracers; drug metabolism; enzymes; neurobiochemistry; gastric and renal physiology; anti-hypertensive, diuretic, uricosuric and anti-lipemic agents. *Mailing Add:* PO Box 50 Rd 2 New Milford PA 18834

MAASS, WOLFGANG SIEGFRIED GUNTHER, b Helsinki, Finland, Oct 23, 29; m 60; c 2. PLANT PHYSIOLOGY, PLANT BIOCHEMISTRY. *Educ:* Univ Tübingen, Dr rer nat, 57. *Prof Exp:* Asst bot, Univ Tübingen, 57; sci collabr, Max-Planck Inst Protein & Leather Res, 58-60; fel biol, Dalhousie Univ, 60-62; asst res officer, Atlantic Regional Lab, Nat Res Coun Can, 62-65, assoc res officer, 66-; AT DEPT MATH, STAT & COMPUT SCI, UNIV ILL, CHICAGO. *Concurrent Pos:* Mem, Plant Phenolics Group NAm. *Mem:* Can Soc Plant Physiol; Ger Bot Soc. *Res:* Taxonomy and distribution of Sphagnum; chemical taxonomy; biochemical aspects of cellular development; biosynthesis of pulvinic acid derivatives and phenolic compounds in general. *Mailing Add:* Dept Math Univ Ill Chicago IL 60680

MAASSAB, HUNEIN FADLO, b Damascus, Syria, June 11, 28; nat US; m 59; c 2. EPIDEMIOLOGY, VIROLOGY. *Educ:* Univ Mo, BA, 50, MA, 52; Univ Mich, MPH, 55, PhD(epidemiol sci), 56; Am Bd Med Microbiol, dipl. *Prof Exp:* Res assoc, 57-60, from asst prof to assoc prof, 60-72, PROF EPIDEMIOL, UNIV MICH, ANN ARBOR, 72- *Mem:* AAAS; Am Asn Immunol; Brit Soc Gen Microbiol; Tissue Cult Asn; Soc Exp Biol & Med. *Res:* Metabolism of infection; host-virus interaction; immunology; tissue culture; tumors; biology of myxoviruses. *Mailing Add:* Dept Epidemiol Univ Mich Sch Pub Health Ann Arbor MI 48109

MAASS-MORENO, ROBERTO, b Philadelphia, Pa, Aug 11, 52. PHYSIOLOGY, BIOPHYSICS. *Educ:* Univ Mex, BS, 79; Case Western Reserve Univ, MS, 83, PhD(biomed eng), 88. *Prof Exp:* Postdoctoral fel, 87-91, RES ASSOC, DEPT PHYSIOL & BIOPHYS, SCH MED, IND UNIV, 91- *Mem:* Am Phys Soc; Am Eng Soc; Inst Elec & Electronics Engrs. *Mailing Add:* Physiol & Biophys Dept Sch Med Ind Univ 635 Barnhill Dr Rm 330 Indianapolis IN 46202-5120

MAATMAN, RUSSELL WAYNE, b Chicago, Ill, Nov 7, 23; m 48; c 5. PHYSICAL CHEMISTRY. *Educ:* Calvin Col, AB, 46; Mich State Col, PhD(chem), 50. *Prof Exp:* Asst prof chem, DePauw Univ, 49-51; sr technologist, Socony-Mobil Oil Co, 51-58; assoc prof chem, Univ Miss, 58-63; prof chem, Dordt Col, 63-90, emer prof, 90-; RETIRED. *Mem:* Am Chem Soc; Am Sci Affil. *Res:* Catalysis; solution-solid reactions; ion solvation. *Mailing Add:* Dordt Col Sioux Center IA 51250

MABEY, WILLIAM RAY, b Los Angeles, Calif, Oct 16, 41; m 81; c 1. PHYSICAL ORGANIC CHEMISTRY. *Educ:* Univ Calif, Riverside, BA, 65; San Diego State Univ, MS, 68; Univ Ore, PhD(chem), 72. *Prof Exp:* Phys org chemist, SRI Int, 72-84; sr environ chemist, Kennedy Jenks & Chilton Inc Consult Engrs, 84-90; PRIN CHEMIST, JAMES M MONTGOMERY CONSULT ENGRS, 90- *Mem:* Am Chem Soc; AAAS; Soc Environ Toxicol Chem. *Res:* Environmental chemistry; kinetics and mechanisms of processes in environment; persistence and fate of chemicals in environment; hazardous chemicals investigation and site cleanup. *Mailing Add:* 240 Aptos Place Danville CA 94526-5430

MABIE, CURTIS PARSONS, b Memphis, Tenn, Feb 26, 32; m 59; c 1. DENTAL MATERIALS, MICROSCOPY. *Educ:* Western Reserve Univ, BA, 54; Univ Mich, MS, 58. *Prof Exp:* Res geologist, US Bur Mines, 58-61; anthracologist, Appl Res Ctr, US Steel Corp, 61-63; res microscopist, IRC, 63-65; res engr, Norton Co, 65-66; ceramic technologist, NL Indust Inc, 66-67; chief scientist, Ceramics Div, Am Dent Asn, Nat Bur Standards, 68-80; MGR CERAMIC RES, SEMIX INC, 80- *Mem:* Am Ceramic Soc; Am Soc Testing & Mat; Sigma Xi. *Res:* Dental porcelains, investments, fillers, and cements; microscopy of dental materials and biological calcifications. *Mailing Add:* 20 Franklin St Frederick MD 21701

MABIE, HAMILTON HORTH, b Rochester, NY, Oct 21, 14; m 41; c 3. MECHANICAL ENGINEERING. *Educ:* Univ Rochester, BS, 40; Cornell Univ, MS, 43; Pa State Univ, PhD(mech eng), 54. *Prof Exp:* From instr to assoc prof mech eng, Cornell Univ, 41-60; res & develop engr, Sandia Corp, NMex, 60-64; PROF MECH ENG, VA POLYTECH INST & STATE UNIV, 64- *Concurrent Pos:* Consult, Sandia Corp, NMex, 58-59 & 64-67 & Westinghouse Elec Corp, Va, 69-70. *Mem:* Fel Am Soc Mech Engrs; Am Soc Lubrication Engrs; Am Soc Eng Educ; Soc Exp Stress Anal. *Res:* Mechanisms and kinematics; fatigue of mechanical components; instrument bearings and lubrication. *Mailing Add:* 840 Hutcheson Dr Blacksburg VA 24060

MABROUK, AHMED FAHMY, b Cairo, UAR, Sept 30, 23; m 54; c 3. AGRICULTURAL CHEMISTRY. *Educ:* Univ Cairo, BSc, 45, MSc, 50; Ohio State Univ, PhD(lipid chem), 54. *Prof Exp:* Chemist, Ministry Agr, Egypt, 45-46; instr chem, Fac Agr, Univ Cairo, 46-48, instr agr indust, 48-51; fel agr chem, Ohio State Univ, 54-55; lectr & asst prof chem, Univ Cairo, 55-58, lectr, Grad Sch, 56-58; res org chemist, Am Meat Inst Found, Univ Chicago, 58-61; prin org chemist, Northern Utilization Res & Develop Div, USDA, 61-65; sr res chemist, Food Sci Lab, 65-80, SR RES CHEMIST, SCI & ADVAN TECHNOL LAB, US ARMY NATICK RES & DEVELOP LABS, 80- *Concurrent Pos:* Fel physiol chem, Ohio State Univ, 56; lectr to postgrads, Agr Schs, Egypt; consult, Tahreer Prov Authority, Egypt, 55-57; abstractor, Chem Abstr, 55-65. *Honors & Awards:* Sci Dir Silver Key Res Award, 68. *Mem:* Am Oil Chemists Soc; Am Chem Soc; Inst Food Technologists; Royal Soc Chem; Brit Soc Chem Indust. *Res:* Chromatography of organic compounds; gel permeation chromatography; gas chromatography; ultrafiltration; fat and flavor chemistry; heterogeneous and homogenous hydrogenation of fats and organic compounds; chemical kinetics of oxidative rancidity; antioxidants; isolation and synthesis of naturally occurring compounds; polymer chemistry. *Mailing Add:* PO Box 2494 Framingham MA 01701-0405

MABRY, JOHN WILLIAM, b Batesville, Ark, Oct 17, 50. ANIMAL BREEDING, SWINE MANAGEMENT. *Educ:* Okla State Univ, BS, 72; Iowa State Univ, MS, 74, PhD(animal breeding), 76. *Prof Exp:* Grad asst animal breeding, Iowa State Univ, 72-76; asst prof, Univ Wis, River Falls, 76-79; ASST PROF ANIMAL BREEDING, UNIV GA, 79- *Mem:* Am Soc Animal Sci; Nat Asn Col Teachers Agr. *Res:* Effects of photoperiod, breeding herd management and the relationships between compositional and sexual maturity. *Mailing Add:* 208 Livestock-Poultry Bldg Univ Ga Athens GA 30602

MABRY, PAUL DAVIS, b Meridian, Miss, Sept 28, 43. NEUROPSYCHOPHARMACOLOGY, DEVELOPMENTAL PSYCHOBIOLOGY. *Educ:* Millsaps Col, BS, 65; Univ Miss, MS, 67, PhD(psychobiol), 70. *Prof Exp:* Res fel, Nat Inst Neurol Dis & Stroke, NIH, dept neurosurg, Univ Miss Med Ctr, 69-70; res assoc, Neurosci & Behav Prog, Princeton Univ, 70-76; assoc prof psychol, chmn dept & head, Div Behav Sci, Sacred Heart Col, 76-86; ASSOC PROF PSYCHOL & CHMN DEPT, UNIV ST THOMAS, 86- *Mem:* Soc Neurosci; AAAS; Am Psychol Asn; Sigma Xi. *Res:* The functional development of neurotransmitter systems in the brain; the role of monoamines in behavioral arousal motivation and emotion. *Mailing Add:* Dept Psychol Col St Thomas 2115 Summit Ave St Paul MN 55105

MABRY, TOM JOE, b Commerce, Tex, June 6, 32; m 54, 71; c 1. ORGANIC CHEMISTRY, PHYTOCHEMISTRY. *Educ:* ETex State Univ, BS & MS, 53; Rice Univ, PhD(chem), 60. *Prof Exp:* NIH fel chem, Org Chem Inst, Univ Zurich, 60-61; res scientist, 62, from asst prof to prof phytochem, 63-73, PROF BOT, UNIV TEX, AUSTIN, 73- *Concurrent Pos:* Guggenheim fel, Univ Freiburg, 71. *Mem:* Am Chem Soc; Royal Soc Chem; Bot Soc Am; Phytochem Soc NAm (vpres, 65, pres, 66-67). *Res:* Natural products chemistry; biochemical systematics; molecular evolution. *Mailing Add:* 3010 Oakhurst Ave Austin TX 78703-1428

MABUCHI, KATSUHIDE, b Japan, Feb 25, 42. MICROSCOPY. *Educ:* Hungary Univ, PhD(biochem), 68. *Prof Exp:* RES ASSOC, BOSTON BIOMED RES INST, 71- *Mem:* Am Biophys Soc. *Mailing Add:* Boston Biomed Res Inst 20 Staniford St Boston MA 02114

MAC, MICHAEL JOHN, b Dearborn, Mich, Jan 15, 51; m 75. AQUATIC TOXICOLOGY. *Educ:* Wayne State Univ, BS, 73; Univ Mich, MS, 80; Univ Wyo, PhD(phys & zool), 86. *Prof Exp:* PROJ LEADER & RES BIOLOGIST, US FISH & WILDLIFE SERV, NAT FISHERIES RES CTR, GREAT LAKES, 73- *Concurrent Pos:* Adj prof, Mich State Univ, 90-91. *Mem:* Int Asn Great Lakes Res (secy, 89-92); Soc Environ Toxicol & Chem; Am Fisheries Soc. *Res:* Research on responses of freshwater fish to environmental contaminants; focus on organic contaminant accumulation and effects on reproduction. *Mailing Add:* US Fish & Wildlife Serv 1451 Green Rd Ann Arbor MI 48105

MACADAM, DAVID LEWIS, b Philadelphia, Pa, July 1, 10; m 38; c 4. COLOR MEASUREMENT. *Educ:* Lehigh Univ, BS, 32; Mass Inst Technol, PhD, 36. *Prof Exp:* Sr res assoc, Res Labs, Eastman Kodak Co, 36-75; PROF OPTICS, UNIV ROCHESTER, 76- *Concurrent Pos:* Mattiello mem lectr, Fedn Socs Paint Technol, 65; Hurter & Driffield mem lectr, Royal Photog Soc, 66; hon mem, Inter-Soc Color Coun; ed jour, Optical Soc Am, 64-75; chmn, Color Measurement Comt, Mil Personnel Supplies Adv Bd, Nat Res Coun, 78-82. *Honors & Awards:* Godlove Award, Inter-Soc Color Coun, 63; Lomb Medal, Optical Soc Am, 40, Ives Medal, 74; Judd Medal Int Asn Color, 83; Sir Isaac Newton Medal, Gt Brit Color Group, 85. *Mem:* Fel Optical Soc Am (pres, 62); Int Comn Illum. *Res:* Optics; color photography; influence of color contrast on visual acuity; spectroradiometry; color television; photographic image structure; visual sensitivity to small color differences. *Mailing Add:* Village at Park Ridge 1471 Long Pond Rd Apt 305 Rochester NY 14626

MACADAM, KEITH BRADFORD, b Rochester, NY, Feb 27, 44; m 69; c 2. HIGHLY EXCITED ATOMS, ATOMIC COLLISIONS. *Educ:* Swarthmore Col, BA, 65; Harvard Univ, MA, 67, PhD(physics), 71. *Prof Exp:* Res assoc, Univ Stirling, Scotland, 71-73, Yale Univ, 73-74 & Univ Ariz, 74-77; from asst prof to assoc prof, 77-86, PROF, DEPT PHYSICS & ASTRON, UNIV KY, 86- *Concurrent Pos:* Mem, gen comt, Int Conf Physics Electronic & Atomic Collisions, 81-85 & prog comt, Div Atomic Molecular & Optical Physics, Am Phys Soc, 90-93; vis scientist, Inst Physics, Univ Aarhus, Denmark, 83; res prof, Univ Ky, 90-91; vis fel, Joint Inst Lab Astrophys, Univ Colo, 91-92. *Mem:* Fel Am Phys Soc; Inst Physics Eng. *Res:* Experimental atomic physics research; orientation and alignment in electron-atom scattering; spectroscopy and collisions of highly excited atoms; electron capture by highly charged ions; analysis of final states in atomic collisions. *Mailing Add:* Dept Physics & Astron Univ Ky Lexington KY 40506-0055

MCADAM, WILL, b Wheeling, WVa, Oct 22, 21; m 45; c 2. CONTROL SYSTEMS ENGINEERING. *Educ:* Case Inst Technol, BS, 42; Univ Pa, MS, 59. *Prof Exp:* Asst, 45-49, res technologist, 49-51, sr res technologist, 51-57, head elec sect, Res Div, 57-68, assoc dir res opers, Tech Ctr, 68-80, mgr chem inst res & develop, Leeds & Northrup Co, 80-; CONSULT. *Mem:* Fel Inst Elec & Electronics Engrs. *Res:* Electrical instruments for measurement industrial recording and process control. *Mailing Add:* PO Box 470 Worcester PA 19490

MCADAMS, ARTHUR JAMES, b Santa Barbara, Calif, July 16, 23; m 46; c 4. PEDIATRICS, PATHOLOGY. *Educ:* Johns Hopkins Univ, MD, 48. *Prof Exp:* Instr pediat path, Col Med, Univ Cincinnati, 56-58; dir labs, Children's Hosp East Bay, Oakland, Calif, 58-63; assoc prof, 63-69, PROF PEDIAT & PATH, COL MED, UNIV CINCINNATI, 69-; DIRECTING PATHOLOGIST, CHILDREN'S HOSP & RES FOUND, 63- *Concurrent Pos:* Asst pathologist, Cincinnati Children's Hosp, 56-58; clin asst prof, Sch Med, Univ Calif, 62-63. *Mem:* AAAS; Am Asn Path & Bact; Soc Pediat Res; Int Acad Path; Am Pediat Soc. *Res:* Role of complement in nephritis; pulmonary disease in the newborn; anoxic brain damage. *Mailing Add:* Dept of Path Children's Hosp Med Ctr Elland & Bethesda Ave Cincinnati OH 45229

MCADAMS, ROBERT ELI, b Hudson, Colo, Jan 2, 29; m 50. NUCLEAR PHYSICS. *Educ:* Colo State Univ, BS, 57; Iowa State Univ, PhD(physics), 64. *Prof Exp:* Fel physics, Iowa State Univ, 64-65; asst prof, 65-73, ASSOC PROF PHYSICS, UTAH STATE UNIV, 73- *Mem:* Am Phys Soc. *Res:* Intermediate-energy physics. *Mailing Add:* 820 N 600 E Logan UT 84321

MCADIE, HENRY GEORGE, b Montreal, Que, May 12, 30; m 56; c 4. ENVIRONMENTAL CHEMISTRY, ANALYTICAL CHEMISTRY. *Educ:* McGill Univ, BSc, 51; Queen's Univ, Ont, MA, 53, PhD(phys chem), 56. *Prof Exp:* Res fel, Ont Res Found, 56-63, sr res scientist, 64-67, prin res scientist, 67-70, from asst dir to actg dir, dept phys chem, 70-71, dir, dept environ chem, 72-83, assoc dir, Div Environ & Chem Eng, 83-84; FOUNDER & CONSULT, H G MCADIE ASSOCS, 84- *Concurrent Pos:* Chmn, Can Adv Comat, Int Stand Orgn/Tech Comt-146 Air Qual, 74-77, 83-90; chmn bd dirs, Chem Inst Can, 77-79; designated chartered chemist, Ont, 84; fel NAm Thermal Anal Soc, 83. *Honors & Awards:* J Charles Honey Award, 78; N S Kurnakov Medal, 85; Outstanding Serv Award, NAm Thermal Anal Soc, 86. *Mem:* Fel Chem Inst Can; Int Conf Thermal Anal (vpres, 74-77, pres, 77-80); Am Chem Soc; Air & Waste Mgt Asn; fel Royal Soc Chem; Am Soc Testing & Mat. *Res:* Thermoanalytical methods and applications; heterogeneous processes; air pollution instrumentation; ambient and work-room monitoring; emissions testing; emission control processes; long-range transport. *Mailing Add:* H G McAdie Assocs 104 Golfdale Rd Toronto ON M4N 2B7 Can

MCADOO, DAVID J, b Washington, Pa, Aug 11, 41; m 67; c 2. NEUROSCIENCES. *Educ:* Lafayette Col, BA, 63; Cornell Univ, PhD(chem), 70. *Prof Exp:* Res asst, Union Carbide Res Inst, 66-67; postdoctoral, Chem Dept, Johns Hopkins, 70-71; sr scientist, Jet Propulsion Lab, 71-73; from asst prof to assoc prof, 73-81, PROF, MED BR, UNIV TEX, 81- *Mem:* Am Chem Soc; Am Soc Mass Spectrometry; Am Asn Sci; Sigma Xi; Soc Neurosci; Neurotrauma Soc. *Res:* Mechanisms and dynamics of reactions of ions in the gas phase by mass spectrometry; neurochemistry and neurophysiology of communication between neurons; biochemistry of spinal cord injury. *Mailing Add:* Univ Tex Med Br 200 University Blvd Galveston TX 77550

MCADOO, JOHN HART, b Baltimore, Md, Mar 16, 45; m 76. CHARGED PARTICLE BEAMS. *Educ:* Princeton Univ, BSE, 67; Univ Rochester, MS, 76, PhD(plasma physics), 81. *Prof Exp:* Res physicist optics, Sachs/Freeman Assoc, Inc, 81-83; res physicist optics, Naval Res Lab, 83; RES PHYSICIST OPTICS, UNIV MD, 83- *Mem:* Am Phys Soc; AAAS; Astron Soc Pac; Inst Elec & Electronics Engrs. *Res:* Propagation of high current density electron beams through periodic focusing channels; development of a 30 Megawatt, 10 Gigahertz scource of microwave radiation for driving electron accelerators. *Mailing Add:* Mission Res Corp 8560 Cinderbed Rd Suite 700 Newington VA 22122

MCAFEE, DONALD A, b Macon, Ga, June 8, 41; m. RESEARCH. *Educ:* Portland State Univ, BS, 63; Univ Ore, Portland, PhD(physiol), 69. *Prof Exp:* Res trainee, Dept Physiol, Med Sch, Univ Ore, Portland, 63-69, instr, 68-69; postdoctoral fel, Dept Pharmacol Sch Med, Yale Univ, New Haven, Conn, 69-72, asst prof, 72-73; asst prof, Dept Physiol & Biophys, Sch Med, Univ Miami, Fla, 73-77; assoc res scientist & head, Sect Receptor Physiol, Div Neurosci, City of Hope, Duarte, Calif, 77-83; res scientist, Div Neurosci, Beckman Res Inst, City of Hope, Duarte, Calif, 83-89; VPRES RES, WHITBY RES, INC, 86- *Concurrent Pos:* Spec fel, Nat Inst Nervous Dis & Stroke, 70-71; fel, Conn Heart Asn, 71-72; NIH res grant, 74-91; grant, Fla Heart Asn, 74-76, NSF, 79-85; coordr, Life Sci Interdisciplinary Grad Prog, Sch Med, Univ Miami & mem, Grad Student Affairs Comt, 74-76; chmn, Prog Comt, Res Staff Orgn, Beckman Res Inst & chmn, Res Serv Comt, 82-84, mem, Data Processing Comt & Libr Support Comt; adj prof biol, Prog Neutral, Informational & Behav Sci, Univ Southern Calif, Los Angeles, 82-86; chmn, Div Neurosci, Beckman Res Inst, City of Hope, Duarte, Calif, 83-86 & adj head, Sect Receptor Physiol, 86-; adj prof pharmacol, Univ Calif, Irvine, 86- & Dept Pharmacol & Toxicol, Med Col Va, Va Commonwealth Univ, Richmond, 90- *Mem:* Am Physiol Soc; Soc Neurosci; Sigma Xi; Soc Gen Physiologists; Am Asn Pharmaceut Scientists. *Res:* Author of various publications. *Mailing Add:* Whitby Res Inc 2801 Reserve St PO Box 27426 Richmond VA 23261-7426

MCAFEE, JERRY, b Port Arthur, Tex, Nov 3, 16; m 40; c 4. CHEMICAL ENGINEERING. *Educ:* Univ Tex, BS, 37; Mass Inst Technol, ScD(chem eng), 40. *Hon Degrees:* LLD, Waynesburg Col, 81. *Prof Exp:* Res chem engr, Universal Oil Prod Co, Ill, 40-45; tech specialist, Gulf Oil Corp, 45-50, dir chem div, Gulf Res & Develop Co, 50-51, asst dir res, 51-54, vpres & assoc dir res, 54-55; vpres eng, Gulf Oil Corp, 55-60, vpres & exec tech adv, 60-62, vpres & dir planning & econs, 62-64, sr vpres, 64-67; exec vpres, Brit Am Oil Co Ltd, 67-69; pres, Gulf Oil Can Ltd, 69-76, chmn bd, Gulf Oil Corp, 76-81; RETIRED. *Honors & Awards:* Gold Medal, Am Petrol Inst. *Mem:* Nat Acad Eng; Am Chem Soc; Am Inst Chem Engrs (vpres, 59, pres, 60). *Res:* Products; crude oil, refined oil products, and natural gas. *Mailing Add:* 1150 Beach Rd Apt 3L Vero Beach FL 32963

MCAFEE, JOHN GILMOUR, b Toronto, Ont, June 11, 26; nat US; wid; c 3. RADIOLOGY, NUCLEAR MEDICINE. *Educ:* Univ Toronto, MD, 48. *Prof Exp:* Jr intern, Victoria Hosp, London, Ont, 48-49, asst resident radiol, 50-51; sr intern, Westminster Hosp, 49-50; resident radiol, Johns Hopkins Hosp, 51-52; from instr to assoc prof, Johns Hopkins Univ, 53-65; prof radiol, State Univ NY Health Sci Ctr, 65-90, fac scholar, 65-90; PROF RADIOL, GEORGE WASH UNIV MED CTR, 90- *Concurrent Pos:* Fel radiol, Johns Hopkins Univ, 52-53. *Honors & Awards:* John G McAfee Lectr, 78; Paul C Aebersold Award, Soc Nuclear Med, 79, Charles der Hevesy Nuclear Med Pioneer Award, 89; Diamond Jubilee Lectr Nuclear Med, Radiol Soc NAm, 89. *Mem:* Soc Nuclear Med; Radiol Soc NAm; Am Roentgen Ray Soc; Royal Col Physicians & Surgeons Can; Asn Univ Radiologists. *Res:* Use of radioactive tracers in clinical diagnosis; nuclear instrumentation and radiochemistry; magnetic resonance imaging. *Mailing Add:* Nuclear Med George Wash Univ Med Ctr 901 23rd St NW Washington DC 20037-0001

MCAFEE, KENNETH BAILEY, JR, b Chicago, Ill, June 22, 24; m 59; c 2. CHEMICAL PHYSICS, ATMOSPHERIC CHEMISTRY. *Educ:* Harvard Univ, BS, 46, MA, 47, PhD(chem physics), 50. *Prof Exp:* Mem tech staff, AT&T Bell Labs, 50-66, head atmospheric res dept, 66-87; RETIRED. *Concurrent Pos:* Mem, Defense Sci Bd, Dept Defense, 65-66; mem, sci adv comt, Advan Res Projs Agency, 67-68; mem, Health Res Coun, New York, 71-74; mem, Delphi panel on sulfur oxides forecasting, US Environ Protection Agency, 72-74; chmn comt sulfur oxides control technol, Nat Acad Sci-Nat Acad Eng, sociotech systs, Nat Res Coun, 75-78 & gov's sci adv comt, State of NJ, 79-83; consult, 87- *Honors & Awards:* Distinguished Scientist Award, AT&T Bell Labs, 82. *Mem:* Fel Am Phys Soc; Am Chem Soc; Sigma Xi. *Res:* Semiconductors; gaseous diffusion and separation; atomic collision processes; reentry physics; upper atmosphere; atmospheric reactions and dispersion of contaminants; environmental physics and chemistry; halogen photochemistry and excited state reactions; atomic and molecular physics. *Mailing Add:* 41 Ellis Dr Basking Ridge NJ 07920

MCAFEE, ROBERT DIXON, b Zamboanga City, Philippines, Sept 9, 25; US citizen; m 53; c 3. PHYSIOLOGY, BIOPHYSICS. *Educ:* Cent Col, AB, 48; Univ Tenn, MS, 51; Tulane Univ, PhD(physiol), 53. *Prof Exp:* Sr scientist, Vet Admin Hosp, New Orleans, 59-85; res physiologist, Sch Med, Tulane Univ 54-59, assoc physiol, 59-85; CONSULT PROF, COL ENG, UNIV NEW ORLEANS, 72- *Concurrent Pos:* adj prof, Delta Regional Primate Res Ctr, Covington La, 75-, adj prof, Dept Ophthal, Sch Med, Tulane Univ, 76- *Mem:* Am Physiol Soc; Biophysics Soc; Bioelectromagnetics Soc. *Res:* Physiological and behavioral effects of microwave irradiation and physiological correlates of operant blood pressure conditioning; biomedical engineering; radiation safety. *Mailing Add:* Dept Elec Eng Col Eng Univ New Orleans Lakefront New Orleans LA 70148

MCAFEE, WALTER SAMUEL, b Ore City, Tex, Sept 2, 14; m 41; c 2. THEORETICAL PHYSICS. *Educ:* Wiley Col, BS, 34; Ohio State Univ, MS, 37; Cornell Univ, PhD(physics), 49. *Hon Degrees:* DSc, Monmouh Col, 85. *Prof Exp:* Teacher jr high sch, 37-42; physicist theoret studies unit, Eng Labs, US Army Electronics Res Command, 42-45, physicist & supvr, 45-46, physicist radiation physics, 48-53, chief sect electro-magnetic wave propagation, 53-57, consult physicist, Appl Physics Div, 58-65, tech dir, Passive Sensing Tech Area, 65-71, sci adv to dir res, Develop & Eng, Eng Labs, 71-78, sci adv, 78-; RETIRED. *Concurrent Pos:* Secy of Army fel, Harvard Univ, 57-58; lectr, West Long Br, Monmouth Col, NJ, 58-75. *Honors & Awards:* Stevens Award, Stevens Inst Tech, 85. *Mem:* AAAS; Am Astron Soc; Am Phys Soc; Am Asn Physics Teachers; sr mem Inst Elec & Electronics Engrs. *Res:* Theoretical nuclear physics; electromagnetic theory. *Mailing Add:* 723 17th Ave South Belmar NJ 07719

MACAIONE, DOMENIC PAUL, b Cambridge, Mass, Feb 12, 37; m 70; c 1. FIRE RETARDANT ADDITIVES FOR ORGANIC POLYMERS, THERMALLY STABLE MONOMERS & POLYMERS. *Educ:* Boston Col, BS, 59, MS, 61. *Prof Exp:* Chemist, Info Technol Labs, Lexington, Mass, 61-62; res chemist, Natick Lab, US Army, Natick Mass, 62-68; RES CHEMIST, ARMY MAT TECHNOL LAB, WATERTOWN, MASS, 68- *Mem:* Am Chem Soc; fel Am Inst Chemists. *Res:* Synthesis of high temperature monomers/polymers; thermal resistance; fire resistant organic polymers and reinforced composites. *Mailing Add:* 46 S Walnut St Mansfield MA 02048

MCALACK, ROBERT FRANCIS, b Camden, NJ, June 1, 40. MICROBIOLOGY, IMMUNOLOGY. *Educ:* Drexel Univ, BS, 64; Thomas Jefferson Univ, MS, 66, PhD(microbiol), 68. *Prof Exp:* Assoc immunol, 68-71, TRANSPLANT IMMUNOLOGIST, DEPT SURG, ALBERT EINSTEIN MED CTR, PHILADELPHIA, 71- *Mem:* AAAS; Am Soc Microbiol; Reticuloendothelial Soc. *Res:* Transplant immunology; autoimmune diseases; immune cell interactions; leukemia immune expressions. *Mailing Add:* Off Dir Transplant Immunol Lab Albert Einstein Med Ctr 5501 Old York Rd Philadelphia PA 19141

MACALADY, DONALD LEE, b Shamokin, Pa, Apr 19, 41; m 64; c 5. PHYSICAL INORGANIC CHEMISTRY, ENVIRONMENTAL CHEMISTRY. *Educ:* Pa State Univ, BS, 63; Univ Wis-Madison, PhD(chem), 69. *Prof Exp:* Jr engr, H R B Singer, Inc, 63; asst prof chem, Grinnell Col, 68-70; asst prof chem, Northern Mich Univ, 70-74, assoc prof, 74-; AT DEPT CHEM & GEOCHEM, COLO SCH MINES. *Concurrent Pos:* NSF Fac fel sci, Rosenstiel Sch Marine & Atmospheric Sci, Univ Miami, 75-76. *Mem:* AAAS; Am Chem Soc; Water Pollution Control Fedn. *Res:* Aquatic chemistry. *Mailing Add:* Dept Chem & Geochem Colo Sch Mines Golden CO 80401-1888

MCALDUFF, EDWARD J, b Alberton, PEI, Dec 3, 39; m; c 3. ELECTRONIC STRUCTURE. *Educ:* St Francis Xavier Univ, BSc, 61; Univ Toronto, PhD(phys chem), 65. *Prof Exp:* Demonstr, Univ Toronto, 61-65; res fel, Univ Wash, 66-67; from asst prof to assoc prof, 67-82, assoc dean, 80-81, dean sci, 81-87, PROF CHEM, ST FRANCIS XAVIER UNIV, 82- *Concurrent Pos:* Nat Res Coun grant, 67-70; res assoc, La State Univ, 75-76; Nat Sci & Eng Res Coun Can grants, 76-83. *Mem:* Am Chem Soc; Chem Inst Can; Can Soc Study Higher Educ. *Res:* Photoelectron spectroscopy of organic molecules and correlation with chemical reactivity. *Mailing Add:* Dept of Chem St Francis Xavier Univ Antigonish NS B2G 1C0 Can

MCALEECE, DONALD JOHN, b Detroit, Mich; m 54; c 2. ENGINEERING MATERIALS & THEIR APPLICATIONS, STUDENT DESIGN ACTIVITIES. *Educ:* Purdue Univ, BS, 52; Ball State Univ, MA, 68. *Prof Exp:* Mech engr, Gen Elec Co, Ft Wayne, 52-53, res prof engr, Linton, 53-61, mech engr, Ft Wayne, 61-66; prof, 66-88, EMER PROF MECH ENG TECHNOL, PURDUE UNIV, 88- *Concurrent Pos:* Vis prof mech eng technol, Univ Ark, Little Rock, 89; instr mech eng technol, Trident Tech Col, 89-91. *Honors & Awards:* Ralph R Teetor Award, Soc Automotive Engrs, 72. *Mem:* Am Soc Mech Engrs; Am Soc Heating, Refrig & Air Conditioning Engrs; Soc Automotive Engrs; Soc Mfg Engrs & Robotics Int. *Res:* Engineering materials and their applications. *Mailing Add:* 4426 Dicke Rd Ft Wayne IN 46804

MCALEER, WILLIAM JOSEPH, organic chemistry, biochemistry, for more information see previous edition

MCALESTER, ARCIE LEE, JR, b Dallas, Tex, Feb 3, 33; m 77; c 2. PALEOBIOLOGY. *Educ:* Southern Methodist Univ, BA, 54, BBA, 54; Yale Univ, MS, 57, PhD(geol), 60. *Prof Exp:* From instr to prof geol, Yale Univ, 59-74, from asst cur to cur, Peabody Mus, 59-74; dean, Sch Humanities & Sci, 74-77, PROF GEOL SCI, SOUTHERN METHODIST UNIV, 74- *Concurrent Pos:* Guggenheim fel, Glasgow Univ, 64-65. *Mem:* AAAS; Geol Soc Am; Paleont Soc; Soc Syst Zool; Sigma Xi. *Res:* Paleozoic geology; marine ecology and paleocology; petroleum geology. *Mailing Add:* Dept Geol Sci Southern Methodist Univ Dallas TX 75275

MCALICE, BERNARD JOHN, b Providence, RI, Apr 20, 30; m 55; c 3. OCEANOGRAPHY. *Educ:* Univ RI, BS, 62, PhD(biol oceanog), 69. *Prof Exp:* Asst prof, Univ Maine, 67-75, actg dir, Ira C Darling Ctr, 76-77, assoc prof oceanog & zool, 75-81, assoc prof bot & oceanog, 81-90, ASSOC PROF OCEANOG, IRA C DARLING CTR, UNIV MAINE, 90- *Concurrent Pos:* Assoc ed, Estuaries, 78-81; mem, gov bd, Estuarine Res Fedn, 78-80. *Mem:* Am Soc Limnol & Oceanog; Estuarine Res Fedn; Crustacean Soc; New Eng Estuarine Res Soc (secy-treas, 76-78, pres, 78-80). *Res:* Ecology, distribution and succession of estuarine plankton. *Mailing Add:* Ira C Darling Ctr Univ Maine Walpole ME 04573

MCALISTER, ARCHIE JOSEPH, solid state physics, for more information see previous edition

MCALISTER, DEAN FERDINAND, b Logan, Utah, Sept 8, 10; m 32; c 3. AGRONOMY. *Educ:* Utah State Col, BS, 31, MS, 32; Univ Wis, PhD(plant physiol), 36. *Prof Exp:* Field asst agron, Exp Sta, Utah State Col, 30-32; asst plant physiol, Univ Wis, 32-36; asst physiologist, Bur Plant Indust, 36-42, from assoc plant physiologist to physiologist, Regional Soybean Lab, 46-52, agronomist, Agr Exp Sta, USDA, 52-77, asst dir sta, 58-77, EMER PROF AGRON, UNIV ARIZ, 77- *Concurrent Pos:* Prof agron & head dept, Univ Ariz, 52-66, chief of party, Univ Ariz-Univ Ceara, Brazil Proj, US AID, 66-68. *Mem:* Fel AAAS; Am Soc Plant Physiol; fel Am Soc Agron; Sigma Xi; fel, Crop Sci Soc Am. *Res:* Crop physiology and production; agricultural administration. *Mailing Add:* 3428 E 4th Tucson AZ 85716

MCALISTER, HAROLD ALISTER, b Chattanooga, Tenn, July 1, 49; m 72; c 1. ASTRONOMY. *Educ:* Univ Tenn, Chattanooga, BA, 71; Univ Va, MA, 74, PhD(astron), 75. *Prof Exp:* Res asst, Dept Astron, Univ Va, 71-75; res assoc astron, Kitt Peak Nat Osberv, 75-77; PROF, DEPT PHYSICS & ASTRON, GA STATE UNIV, 77- *Concurrent Pos:* Dir, Ctr for High Angular Resolution Astron, 85- *Mem:* Astron Soc Pac; Int Astron Union; Am Astron Soc. *Res:* Astrometry; speckle interferometry of binary stars; astronomical instrumentation; extra-solar planets. *Mailing Add:* Dept Physics & Astron University Plaza Atlanta GA 30303

MCALISTER, SEAN PATRICK, b Durban, SAfrica, Apr 1, 45; Can citizen; div; c 1. SOLID STATE PHYSICS. *Educ:* Univ Natal, BSc, 66, MSc, 68; Univ Cambridge, PhD(physics), 71. *Prof Exp:* Fel exp physics, Simon Fraser Univ, 71-75; res assoc, 75-77, asst res officer, 77-85, SR RES OFFICER, DIV CHEM, NAT RES COUN CAN, 86- *Mem:* Am Phys Soc. *Res:* Electronic properties of GaAs devices; device simulations; magnetism. *Mailing Add:* Inst Microstruct Sci Nat Res Coun Can Ottawa ON K1A 0R9 Can

MCALISTER, WILLIAM BRUCE, b Seattle, Wash, Aug 11, 29. MARINE ECOLOGY. *Educ:* Univ Wash, BS, 49, MS, 58, JD, 80; Ore State Univ, PhD(oceanog), 62. *Prof Exp:* From instr to asst prof oceanog, Ore State Univ, 58-64; proj dir, Bur Com Fisheries, Wash, 64-70; prog dir phys oceanog, NSF, 70-72; dep dir marine fish, Northwest Fisheries Ctr, Nat Marine Fisheries Serv, 72-74, dep dir marine mammal div, 74-80; PARTNER, AQUASCI MARINE RESOURCE CONSULTS, 81- *Concurrent Pos:* Fulbright res fel, Water Res Inst, Oslo, Norway, 63-64; asst prof, Univ Wash, 65-70; attorney, 80- *Mem:* AAAS; Am Inst Fish Res Biologists; Sigma Xi. *Res:* Resource management, including legal problems; ecosystem dynamics, marine mammals and fisheries. *Mailing Add:* 14014 81st Pl NE Bothell WA 98011

MCALLISTER, ALAN JACKSON, b Shelbyville, Ky, Aug 19, 45; m 65; c 2. ANIMAL BREEDING. *Educ:* Univ Ky, BS, 67; Ohio State Univ, MS, 70, PhD(animal breeding), 75. *Prof Exp:* Biomet geneticist poultry, DeKalb Agr Res Inc, 72-75; fel poultry genetics, 75-76, RES SCIENTIST & CHMN, DAIRY CATTLE BREEDING PROG, ANIMAL RES CTR, AGR CAN, 76- *Mem:* Am Soc Animal Sci; Am Dairy Sci Asn; Int Biomet Soc; Can Soc Animal Sci. *Res:* Genetic improvement of animal populations through application of optimal genetic evaluation procedures; where performance includes both directly observable economic traits and basic physiological measurements. *Mailing Add:* Res Br Agr Can Cent Exp Firm Bldg 60 Birch Dr Ottawa ON K1A 0C6 Can

MCALLISTER, ARNOLD LLOYD, b Petitcodiac, NB, Dec 24, 21. GEOLOGY. *Educ:* Univ NB, BSc, 43; McGill Univ, MSc, 48, PhD, 50. *Prof Exp:* Res geologist, Int Nickel Co, 50-52; assoc prof, 52-62, head dept, 62-74, PROF GEOL, UNIV NB, 62- *Mem:* Royal Soc Can; Geol Asn Can; Can Inst Mining & Metall. *Res:* Economic and structural geology. *Mailing Add:* Dept Geol Univ NB Col Hill Box 4400 Fredericton MB E3B 5A3 Can

MCALLISTER, BYRON LEON, b Midvale, Utah, Apr 29, 29; m 57; c 3. GENERAL TOPOLOGY, HISTORY MATHEMATICS. *Educ:* Univ Utah, BA, 51, MA, 55; Univ Wis, PhD(math), 66. *Prof Exp:* From asst prof to assoc prof math, SDak Sch Mines & Technol, 58-67; from assoc prof to prof, 67-91, EMER PROF MATH, MONT STATE UNIV, 91- *Concurrent Pos:* Instr, Fox Valley Ctr, Univ Wis, 61-62; consult math, Headlands Indian Health Careers, Univ Okla Health Sci Center, 83- *Mem:* Am Math Soc; Math Asn Am; Asn Women Math. *Res:* General topology, particularly Whyburn cyclic element theory and extensions, multifunctions and hyperspaces. *Mailing Add:* 606 S Fifth Ave Bozeman MT 59715

MCALLISTER, DAVID FRANKLIN, b Richmond, Va, July 2, 41; c 1. GRAPHICS, SOFTWARE RELIABILITY. *Educ:* Univ NC, Chapel Hill, BS, 63, PhD(comput sci), 72; Purdue Univ, MS, 67. *Prof Exp:* Asst math, Purdue Univ, 65-67; instr, Univ NC, Greensboro, 67-72; from asst prof to assoc prof, 72-82, PROF COMPUT SCI, NC STATE UNIV, 82- *Concurrent Pos:* Consult, Environ Protection Agency, 74-81, Monsanto Corp, 80-81, Boeing Computer Serv, 82- & Potters Indust, 81-82; prin investr, NASA, 78-, NSF, 79-, Regiment Air Defense Ctr & ARO. *Mem:* Asn Comput Mach; Inst Elec & Electronic Engrs; Am Asn Univ Prof; Soc Photo-Optical Instrumentation Engrs. *Res:* Software reliability modeling; computer graphics. *Mailing Add:* Dept Comput Sci NC State Univ Raleigh NC 27695

MCALLISTER, DONALD EVAN, b Victoria, BC, Aug 23, 34; c 5. SYSTEMATIC ICHTHYOLOGY. *Educ:* Univ BC, BA, 55, MA, 57, PhD(ichthyol), 64. *Prof Exp:* RES CUR FISHES, CAN MUS NATURE, 58- *Concurrent Pos:* Adj prof, Univ Ottawa, 64- & Carleton Univ, 81-; pres & ed, Int Marinelife Alliance Can. *Mem:* Am Soc Ichthyol & Herpet; Can Soc Zool; Can Soc Wildlife & Fishery Biol; Japanese Soc Ichthyol; Soc Syst Zool. *Res:* Fish systematics, evolution, arctic, Canada and world; biogeography; Osmeridae; Cottidae; endangered ichthyofauna; threatened tropical marine environments in developing countries; list of fishes of the world; coral reef fishes; biodiversity. *Mailing Add:* Ichthyol Sect PO Box 3443 Sta D Can Mus Nature Ottawa ON K1P 6P4 Can

MCALLISTER, GREGORY THOMAS, JR, b Boston, Mass, July 6, 34; m 61; c 3. MATHEMATICS. *Educ:* St Peters Col, NJ, BS, 56; Univ Calif, Berkeley, PhD(math), 62. *Prof Exp:* Res mathematician, US Army Ballistic Res Labs, 63-65; PROF MATH, CTR APPLN MATH, LEHIGH UNIV, 65- *Concurrent Pos:* Consult, Gen Motors Tech Ctr. *Mem:* Am Math Soc; Soc Indust & Appl Math; Sigma Xi. *Res:* Calculus of variations; partial differential equations; numerical methods. *Mailing Add:* Dept Math Lehigh Univ Bethlehem PA 18015-9988

MCALLISTER, HOWARD CONLEE, b Cheyenne, Wyo, Mar 14, 24; m 44; c 4. PHYSICS. *Educ:* Univ Wyo, BS, 48, MS, 50; Univ Colo, PhD(physics), 59. *Prof Exp:* From asst prof to assoc prof, 59-70, PROF PHYSICS, UNIV HAWAII, 70-; PRIN INVESTR, INST ASTRON, 69- *Concurrent Pos:* Consult, Univ Colo, 61 & Lincoln Lab, Mass Inst Technol, 63-64; Nat Acad Sci-Nat Res Coun sr resident res assoc, Goddard Space Flight Ctr, 65-66. *Mem:* Am Phys Soc; Optical Soc Am. *Res:* Solar ultraviolet spectroscopy and upper atmospheric physics; atomic and molecular spectroscopy; spectroscopic instrumentation for space research. *Mailing Add:* 6439 Hawaii Kai Dr Honolulu HI 96825

MCALLISTER, JEROME WATT, b Bay City, Tex, Mar 19, 44; m 67. PHYSICAL CHEMISTRY. *Educ:* Tex Christian Univ, BS, 67; Univ Tex, Austin, PhD(chem physics), 71. *Prof Exp:* Res assoc phys chem, Iowa State Univ, 71-72; sr res chemist, 72-75, develop specialist, 75-77, supvr, 77-79, MGR RES & DEVELOP, 3M CO, 79- *Mem:* Am Chem Soc; Am Indust Hyg Asn. *Res:* Development of adsorbents and indicating chemistries for use in respirators and personal monitors for gases and vapors. *Mailing Add:* Krattley Lane 412 Krattley Lane Hudson WI 54016-7102

MCALLISTER, MARIALUISA N, b Milan, Italy, Aug 22, 39; US citizen; m 61; c 3. MATHEMATICS, COMPUTER SCIENCES. *Educ:* Univ Rome, PhD(math). *Prof Exp:* From asst prof to assoc prof math, 66-81, chmn dept math, 88-89, PROF MATH, MORAVIAN COL, 81- *Concurrent Pos:* Consult, Opers Res Computational Methods; NSF grant, 85-86, Fulbright grant, 88. *Mem:* Math Asn Am; Int Fuzzy Systs Asn; Int Asn Math Modeling. *Res:* Numerical analysis; operations research; graph theory, fuzzy sets and complexity of numerical computations in expert systems. *Mailing Add:* Dept Math Moravian Col Bethlehem PA 18018

MCALLISTER, RAYMOND FRANCIS, b Ithaca, NY, June 26, 23; div; c 3. OCEANOGRAPHY, OCEAN ENGINEERING. *Educ:* Cornell Univ, BS, 50; Univ Ill, MS, 51; Agr & Mech Col, Tex, PhD(geol oceanog), 58. *Prof Exp:* Instr geol, Univ Ill, 50-51; res oceanogr, Scripps Inst, Univ Calif, 51-54; marine res geologist & instr geol, Agr & Mech Col, Tex, 54-58; sr oceanogr, Geophys Field Sta, Columbia Univ, 58-63; asst dir marine technol group, NAm Aviation Inc, Miami, 64-65; PROF OCEAN ENG, FLA ATLANTIC UNIV, 65-; PRES, MCALLISTER MARINE, LIGHTHOUSE PT, FLA, 75- *Concurrent Pos:* Mem, Man in the Sea Panel, Nat Acad Eng, 71-; comnr, Hillsboro Inlet Improv & Maintenance Dist, 87- *Mem:* Am Soc Eng Educ.

Res: Ocean engineering and marine geology; beach erosion mitigation; remote sensing in underwater archaeology; environmental surveys; ocean outfall studies. *Mailing Add:* Dept Ocean Eng Fla Atlantic Univ Boca Raton FL 33431

MCALLISTER, ROBERT MILTON, b Philadelphia, Pa, June 10, 22; m 49; c 6. PEDIATRICS. *Educ:* Ursinus Col, BS, 42; Univ Pa, MD, 45, MS, 55; Am Bd Pediat, dipl, 53. *Prof Exp:* From instr to asst prof pediat, Sch Med, Univ Pa, 51-59; assoc prof, 59-64, PROF PEDIAT, CHILDREN'S HOSP LOS ANGELES, SCH MED, UNIV SOUTHERN CALIF, 64-, DIR RES, 75- *Concurrent Pos:* Vchmn Acad Affairs, Children's Hosp Los Angeles, Sch Med, Univ Southern Calif. *Mem:* Am Pediat Soc; Soc Pediat Res; Am Asn Cancer Res; fel Am Acad Pediat; AMA. *Res:* Microbiology and oncology in the pediatric age group; tissue culture. *Mailing Add:* 4650 Sunset Blvd Box 93 Los Angeles CA 90027

MCALLISTER, ROBERT WALLACE, b Hermosa Beach, Calif, Feb 16, 29; div; c 4. PHYSICS, INSTRUMENTATION. *Educ:* Occidental Col, BA, 51; Stanford Univ, PhD(physics), 60. *Prof Exp:* Asst physics, Univ Zurich, 56-60; instr, Cornell Univ, 60-64; asst prof, 64-66, ASSOC PROF PHYSICS, COLO SCH MINES, 66- *Mem:* Sigma Xi. *Res:* Application and interfacing of microcomputers to control experiments and record data. *Mailing Add:* Dept Physics Colo Sch Mines Golden CO 80401

MCALLISTER, WARREN ALEXANDER, b Augusta, Ga, Mar 12, 41; m 62. INORGANIC CHEMISTRY. *Educ:* Mercer Univ, BA, 63; Univ SC, PhD(inorg chem), 67. *Prof Exp:* Res assoc, Vanderbilt Univ, 66-67; from asst prof to prof chem, E Carolina Univ, 67-80; AT BURROUGHS WELLCOME CO, GREENVILLE. *Concurrent Pos:* Grants, USPHS, Environ Protection Agency & NC Bd Sci & Technol; res assoc, Brown Univ, 73; adj prof, East Carolina Univ, 80- *Mem:* Am Chem Soc; Sigma Xi. *Res:* Infrared and Raman studies of inorganic compounds. *Mailing Add:* 408 Queen Anne's Rd Greenville NC 27858

MCALLISTER, WILLIAM ALBERT, physical chemistry, for more information see previous edition

MCALLISTER, WILLIAM T, b Apr 25, 44; m; c 3. MICROBIOLOGY. *Educ:* Lehigh Univ, Bethlehem, Pa, BA, 66; Univ NH, Durham, PhD(biochem), 70. *Prof Exp:* Teaching asst, Dept Microbial & Molecular Biol, Univ Pittsburgh, 66-67; res asst, Dept Biochem, Univ NH, 67-70; NIH postdoctoral fel, Inst Molecular Genetics, Univ Heidelberg, Ger, 70-72, res assoc, 72-73; from asst prof to prof, Dept Microbiol, Univ Med & Dent NJ, Rutgers Med Sch, Piscataway, NJ, 73-86; PROF & CHMN, DEPT MICROBIOL & IMMUNOL, HEALTH SCI CTR, STATE UNIV NY, BROOKLYN, 86- *Concurrent Pos:* Co-investr, NIH grant, 87-92, prin investr, 74-93; prin investr, NJ Comn Cancer Res, 85-87, Pharmacia P-L Biochem, Inc, 87-90 & NSF, 88-89; adj prof, Dept Molecular Genetics & Microbiol, Robert Wood Johnson Med Sch, Piscataway, NJ, 86-90; Nat Lectr, Am Soc Microbiol, 86-87, chair-elect, Div M, 91; couns, Harvey Soc, 87-; reviewer, Div Cellular & Molecular Biol, NSF & Spec Study Sect Small Bus & Innovative Res, NIH, 89. *Mem:* AAAS; Am Soc Microbiol; Am Soc Biochem & Molecular Biol; Am Chem Soc; Sigma Xi; NY Acad Sci. *Res:* Structure and function of DNA-dependent RNA polymerases and the regulation of transcription; genetic and biochemical analyses of a human folate binding protein; homologous recombination in mammalian cells; author of various publications. *Mailing Add:* Dept Microbiol & Immunol Health Sci Ctr State Univ NY 450 Clarkson Ave Box 44 Brooklyn NY 11203

MCALPIN, JOHN HARRIS, b Natchez, Miss, Dec 7, 33; m 72; c 4. SOFTWARE ENGINEERING, SIMULATION. *Educ:* Columbia, AB, 58, PhD(math), 65; NY Univ, MS, 60. *Prof Exp:* Instr math, Brown Univ, 64-66; asst prof math, Univ Colo, 66-69; opers res analyst, US Navy, 78-82; prof math, SC State Col, 69-78; PROF COMPUTER SCI, BRADLEY UNIV, 82- *Concurrent Pos:* Instr, Columbia Col Pharm, 61-62; vis asst prof, Miss State Univ, 66 & Columbia, 69; vis prof, Columbia Univ, 70; mathematician, Lawrence Livermore Lab, 71; vis scientist, Lincoln Lab, Mass Inst Technol, 74; statistician, Social Security Admin, 75; mgt analyst, Dept Treasury, 77. *Mem:* Am Asn Artificial Intel; Inst Elec & Electronics Engrs Computer Soc. *Res:* Fault-tolerant signalling in distributed systems; coding theory; artificial intelligence; inventory control; logistics; discriminant analysis; differential geometry; topology. *Mailing Add:* Dept Computer Sci Bradley Univ Peoria IL 61625

MCALPINE, GEORGE ALBERT, b Tampa, Fla, Dec 11, 33; m 52; c 4. ELECTRICAL ENGINEERING, RESEARCH ADMINISTRATION. *Educ:* Univ Va, BEE, 57; Stanford Univ, MS, 61; Univ Va, DSc, 67. *Prof Exp:* Sr engr, Martin Co, Fla, 57-59, Sylvania Electronic Defense Labs, Calif, 59-61 & Martin Co, Fla, 61-62; proj engr, Sperry, Piedmont, Va, 62-64; exec dir indust res & develop ctr, Univ Va, 64-, dir res labs eng sci, 67-, assoc prof elec eng, 68-; mgr corp develop, Lord Corp, 87-88; vpres, Quorum Int Inc, 87-88; PRES, DANBY NAM INC, 88- *Concurrent Pos:* Consult, Sperry Marine Systs Div, Sperry Rand Corp, 64, Radio Corp Am, 65-66, Bio-Space Technol Training Prog, 65-70, Deco Electronics, Inc, Va, 65- & Electronic Commun, Inc, 68; actg assoc provost res, Univ Va, 76-77. *Mem:* Inst Elec & Electronics Engrs; Sigma Xi; Am Soc Eng Educ. *Res:* Communications and systems theory; digital filters; stochastic processes. *Mailing Add:* Danby NAm, Inc 1135 Kildaire Farm Rd Suite 311-80 Cary NC 27511

MACALPINE, GORDON MADEIRA, b Bozeman, Mont, Feb 23, 45; m 67; c 2. QUASARS, SUPERNOVAE. *Educ:* Earlham Col, AB, 67; Univ Wis, PhD(astron), 71. *Prof Exp:* Mem, Inst Advan Study, 71-72; from asst prof to assoc prof, 72-84, PROF ASTRON, UNIV MICH, ANN ARBOR, 84- *Mem:* Am Astron Soc; Int Astron Union. *Res:* Detailed theoretical and observational investigation of the emission-line regions in quasi-stellar objects, Seyfert galaxies, and supernova remnants. *Mailing Add:* Dept of Astron Univ of Mich Ann Arbor MI 48109

MCALPINE, JAMES BRUCE, b Ingham, Queensland, Australia, Dec 26, 39; m 68; c 3. ORGANIC CHEMISTRY. *Educ:* Univ New Eng, Australia, BSc, 62, MSc, 64, PhD(org chem), 69; Lake Forest Sch Mgt, Ill, MBA, 83. *Prof Exp:* Fel biochem, Med Sch, Northwestern Univ, 69-71, asst prof biochem, 71-72; sr res chemist, 72-75, proj leader antibiotics modification, 75-82, SR PROJ LEADER MICROBIAL SCREENING, ABBOTT LABS, 83- *Mem:* Royal Soc Chem; Am Chem Soc; Sigma Xi; Am Soc Microbiol. *Res:* The chemistry, mode of action and toxicity of antimicrobial agents; discovery and structural elucidation of biologically active natural products. *Mailing Add:* 211 W Rockland Rd Libertyville IL 60048

MCALPINE, JAMES FRANCIS, b Maynooth, Ont, Sept 25, 22; m 50; c 6. PALEOENTOMOLOGY. *Educ:* Univ Toronto, BSA, 50; Univ Ill, MSc, 54, PhD(entom), 62. *Prof Exp:* Tech officer, 50-53, res officer, 53-85, RES SCIENTIST, BIOSYSTS RES CTR, AGR CAN, 50-, RES ASSOC, 85- *Honors & Awards:* Thomas Say Award, Entomol Soc Am, 89. *Mem:* Entom Soc Can. *Res:* Systematics of two-winged flies; author, editor and coordinator of publications. *Mailing Add:* 524 Evered Ottawa ON K1Z 5K8 Can

MCALPINE, KENNETH DONALD, b Morenci, Ariz, Oct 31, 53; m 82; c 1. ROCKET PROPULSION, SYSTEMS ENGINEERING. *Educ:* Univ Ariz, BSEE, 75, MSEE, 77. *Prof Exp:* Res asst solid state physics, Eng Sta, Univ Ariz, 75-77; radar-laser engr, Goodyear Aerospace, 77-79; radar-laser metallurgist, Boeing Mil Co, 79-81; mem tech staff, Bell Tech Opers, 81-85; consult, Pine Cap Assocs, 85-91; sr staff engr, Martin Marietta Missiles & Electronics, 86-90; SR SYSTS ANALYST, CTR SPACE RES, PATRICK AFB, FLA, 90- *Concurrent Pos:* Sr laser engr, USAF & USN, 81-85. *Mem:* Soc Photo-Optical Instrumentation Engrs; Soc Advan Mat & Process Eng; Nat Asn Rocketry. *Res:* Ultraviolet calibration for use on Voyager 2; high efficiency concentration photo voltaics for use around Venus; prototype of ultraviolet, four color solar cell and high voltage solar cells; aluminum alloys, titanium and tungsten alloys that are forgeable, castable, and hemp; cryogenics. *Mailing Add:* PO Box 5279 Winter Park FL 32793

MCALPINE, PHYLLIS JEAN, b Petrolia, Ont, Aug 29, 41. HUMAN GENETICS, MEDICAL GENETICS. *Educ:* Univ Western Ont, BSc, 63; Univ Toronto, MA, 66; Univ London, PhD(human genetics), 70. *Prof Exp:* Med Res Coun Can fel, Queen's Univ, 70-72; res assoc human genetics, Health Sci Children's Ctr, 72-74; asst prof 74-81, ASSOC PROF PEDIAT, UNIV MAN, 81-, PROF HUMAN GENETICS, 84-; MEM SCI STAFF, HEALTH SCI CHILDREN'S CTR, 75- *Concurrent Pos:* Consult, Howard Hughes Med Inst Human Gene Mapping Libr, Yale Univ, 87-89. *Mem:* Genetics Soc Can; Am Soc Human Genetics; AAAS. *Res:* Mapping the human genome by use of somatic cell hybrids and family studies; expression of genes in human tissues. *Mailing Add:* Dept Human Genetics 700 Bannatyne Ave Winnipeg MB R3E 0W3 Can

MACALUSO, ANTHONY, SR, b New Orleans, La, Oct 4, 39; m 61; c 3. CHEMISTRY. *Educ:* Loyola Univ, BS, 61; Tulane Univ, MS, 63, PhD(chem), 65. *Prof Exp:* SR CHEMIST, RES & TECH DEPT, TEXACO INC, 67- *Mem:* Am Chem Soc. *Res:* Exploratory research in organic and petroleum chemistry. *Mailing Add:* 4135 42nd St Pt Arthur TX 77642

MACALUSO, PAT, b New York, NY, Aug 29, 16; m 41; c 2. PHYSICAL CHEMISTRY, COMPUTER SCIENCE. *Educ:* City Col NY, BS, 39; Polytech Inst Brooklyn, MS, 46, 78. *Prof Exp:* Chemist indust & consumer specialties, Foster D Snell, Inc, 40-42, group leader, 42-45, acct exec, 45-47; group leader agr chem formulations, 47-54, head prod develop & res sect, 54-65, head polymer prod sect, 65-67, mgr info serv, 67-80, MGR COMPUT SERV, EASTERN RES CTR, STAUFFER CHEM CO, 80- *Mem:* Am Chem Soc; Am Soc Testing & Mat; NY Acad Sci; Asn Comput Mach. *Res:* Polymers, elastomers; polyvinyl chloride; industrial chemical applications; chemical specialties; chemistry of sulfur; physical chemistry of proteins; computer applications in chemical research. *Mailing Add:* Nine Church Ct White Plains NY 10603

MACALUSO, SISTER MARY CHRISTELLE, b Lincoln, Nebr, July 9, 31. HEALTH PROMOTION. *Educ:* Col St Mary, Nebr, BS, 56; Univ Notre Dame, MS, 61; Univ Nebr, PhD(anat), 66. *Prof Exp:* From instr to assoc prof, 61-73, chmn dept, 67-73, prof biol, 73-80, PROF SPEAKER, COL ST MARY, 80- *Mem:* Am Asn Therapeut Humor. *Res:* Sex education; ultrastructure of the corpus luteum of pregnancy and the corpus luteum of lactation in Swiss mice during the first nineteen days postpartum; human development. *Mailing Add:* Col of St Mary Omaha NE 68124

MCANALLY, JOHN SACKETT, b Indianapolis, Ind, Apr 15, 18; m 43; c 2. ANALYTICAL BIOCHEMISTRY. *Educ:* Ind Univ, BS, 38, AM, 40, PhD(chem), 50. *Prof Exp:* Res asst prof biochem, Med Res Unit, Univ Miami, 50-52, asst prof, Med Sch, 52-57; from asst prof to prof biochem, Occidental Col, 57-82, dean students, 65-68; RETIRED. *Mem:* AAAS; Am Chem Soc; NY Acad Sci. *Res:* Vitamin A; urinary estrogens and androgens; trace element analysis; seawater and marine organisms. *Mailing Add:* 1530 Holcomb St Port Townsend WA 93868-8408

MCANDREW, DAVID WAYNE, b Brandon, Man, Feb 8, 52; m 87. SOIL FERTILITY, SOIL CHEMISTRY. *Educ:* Univ Man, BSc, 77, MSc, 80; Ore State Univ, PhD(soil fertil), 83. *Prof Exp:* Res scientist soil fertil, Swift Current Res Sta, Agr Can, 83-84; OFFICER-IN-CHARGE & SOIL AGRONOMIST SOIL MGMT & CONSERV, SOILS & CROPS SUBSTA, AGR CAN, 84- *Mem:* Agr Inst Can; Can Soc Soil Sci; Am Soc Soil Sci; Am Soc Agron; Sigma Xi. *Res:* Ameliorative methods for solonetzic (sodium affected natural soils); zero-tillage for northern temperate climate zone. *Mailing Add:* PO Box 1408 Agr Can Res Br Vegreville AB T0B 4L0 Can

MCANDREWS, JOHN HENRY, b Minneapolis, Minn, Jan 16, 33; Can citizen; m 58, 77; c 5. BOTANY, PLANT ECOLOGY. *Educ:* Col St Thomas, BS, 57; Univ Minn, MS, 59, PhD(bot), 64. *Prof Exp:* Res assoc paleoecol, Inst Bio-Archeol, Groningen, Neth, 63-64; asst prof biol, Jamestown Col, 64-66

& Cornell Col, 66-67; cur geol, 67-78, CUR BOT ROYAL ONT MUS, 78-; PROF BOT & GEOL, UNIV TORONTO, 82- *Concurrent Pos:* Assoc prof bot, Univ Toronto, 68-82, assoc prof geol, 80-82; vis prof ecol, Univ Minn, 74; vis lectr anthrop, Univ Man, 75. *Res:* Vegetation history; climatic change; pollen analysis; pollen morphology. *Mailing Add:* Bot Royal Ont Mus 100 Queens Park Toronto ON M5S 2C6 Can

MCANELLY, JOHN KITCHEL, b Logansport, Ind, June 22, 31; m 53; c 2. FOOD MICROBIOLOGY, QUALITY ASSURANCE. *Educ:* Iowa State Univ, BS, 53; NC State Univ, MS, 56; Univ Wis, PhD(bact), 60. *Prof Exp:* Res biochemist, Res & Develop Ctr, Swift & Co, Ill, 59-61, head biochem div, 61-66; dir res & develop, Rival Pet Foods, 66-68, vpres tech develop, 68-75; dir corp qual assurance, Nabisco Brands Inc, 75-78, dir food proctection, 78-82, group dir corp qual assurance, 82-89; PRES, MCANELLY ASSOCS INC, 89- *Mem:* Am Chem Soc; Am Soc Microbiol; Inst Food Technol; Am Soc Qual Control; Sigma Xi. *Res:* Biochemistry and microbiology of food products and processes; thermal processing of foods; utilization of proteins. *Mailing Add:* 15 Dixon Pl Wayne NJ 07470

MCANENY, LAURENCE RAYMOND, b Seattle, Wash, Apr 12, 26; m 46; c 3. PHYSICS. *Educ:* Univ Kans, BS, 46, PhD(physics), 57; Univ Calif, MA, 48. *Prof Exp:* Asst prof physics, Park Col, 51-57; from asst prof to assoc prof physics, 57-67, asst dean acad affairs, 63-67, dean div sci & technol, 67-73, prof, 67-88, EMER PROF PHYSICS, SOUTHERN ILL UNIV, EDWARDSVILLE, 88- *Concurrent Pos:* Pres bd dirs, Cent States Univs, Inc, 75 & 85. *Mem:* Am Phys Soc; Am Asn Physics Teachers. *Res:* Theoretical physics; statistical mechanics. *Mailing Add:* 3000 Mockingbird Ln Granite City IL 62040

MCANINCH, LLOYD NEALSON, b Guelph, Ont, June 6, 20; m 44; c 3. UROLOGY. *Educ:* Univ Western Ont, MD, 45; FRCPS(C), 53. *Prof Exp:* Intern, Victoria Hosp, London, Ont, 45-46; asst to Dr E D Busby, 46-49; instr anat, Fac Med, Univ Western Ont, 49-50; asst resident surg, Westminster Vet Hosp, Ont, 50-51; resident urol, Toronto Gen Hosp, 51-52; resident, Sunnybrook Hosp, 52-53; asst resident path, Westminster Hosp, 53; from instr to asst prof surg, Univ Western Ont, 53-63, from clin assoc prof to clin prof urol, 63-70, chief urol, 70-73, prof surg, Fac Med, 70-75, chief urol, Univ Hosp, 73-75; RETIRED. *Concurrent Pos:* Consult, Westminster Vet Hosp, 54-; chief urol, Victoria Hosp, Ont, 56-73, consult; attend urologist & consult, St Joseph's Hosp, Ont. *Mem:* Fel Am Col Surg; Can Med Asn; Can Urol Asn (pres, 74-75); fel Can Asn Clin Surg; Can Acad Urol Surg (treas, 60-65, pres, 66). *Res:* Vesico-ureteral reflux; renal trauma; retroperitoneal tumors; urological emergencies; chemotherapy in urology; ureteral substitutions; external meatotomy. *Mailing Add:* 1 Grosvenor St Apt 1423 London ON N6A 1Y2 Can

MCARDLE, EUGENE W, b Chicago, Ill, Oct 26, 31; m 61; c 6. ZOOLOGY. *Educ:* St Mary's Col, Minn, BS, 53; Marquette Univ, MS, 56; Univ Ill, PhD(zool), 60. *Prof Exp:* Asst prof biol, St Benedict's Col, Kans, 60-61; assoc prof, St Mary's Col, Minn, 61-69; PROF BIOL, NORTHEASTERN ILL UNIV, 69- *Concurrent Pos:* Lectr, NSF In-Serv-Insts, Winona State Col, 65; visitor, Argonne Nat Lab, 68- *Mem:* Soc Protozoologists. *Res:* Cytology and behavior of ciliate protozoan Tetrahymena rostrata. *Mailing Add:* Dept of Biol Northeastern Ill Univ Chicago IL 60625

MCARDLE, JOSEPH JOHN, b Wilmington, Del, July 21, 45. NEUROPHYSIOLOGY, NEUROPHARMACOLOGY. *Educ:* Univ Del, BA, 67; State Univ NY Buffalo, PhD(pharmacol), 72. *Prof Exp:* Vis asst prof pharmacol, Sch Med & Dent, State Univ NY Buffalo, 71-72; from asst prof to assoc prof, 72-82, PROF PHARMACOL, COL MED & DENT, NJ, 82- *Concurrent Pos:* Nat Inst Neurol Comn Dis & Stroke grant, Col Med & Dent NJ, 73-82; vis scientist, Lab de Neurbiol, Gif-sur-Yvette, France, 79-80; fel, La Found de L'Indust Pharmaceut Pour La Recherche, France; res gr, NJ State Comn Cancer, 85-87; vis fel, Dept Physiol, John Curtin Sch Med Res, Australia, 87; vis scientist, lab develop, neurobiol, Nat Inst Neurol Commun Dis & Stroke, Bethesda, MD, 87; Nat Inst on Alcohol Abuse & Alcoholic Grant, 89-92, Am Heart Asn, 90-92; mem, Cellular Neurosci Panel, 90- *Honors & Awards:* NSF, 88. *Mem:* Am Soc Pharmacol & Exp Therapeut; Soc Neurosci; Biophys Soc; Int Soc Myochem. *Res:* Trophic influences of nerve upon muscle and the effects of drugs upon these phenomena; neuro-muscular regeneration; neuro-muscular development; excitation-secretion coupling; uses of computers in the study of neurobiological problems; cardiac pharmacology; direct recording of single channel conductance from cultured and freshly dissected cells; drug abuse and addiction; calcium channels; biophysics. *Mailing Add:* Dept of Pharmacol NJ Med Sch Med & Dent Univ of NJ Newark NJ 07103

MACARIO, ALBERTO J L, b Naschel, San Luis, Arg, Dec 1, 35; US citizen; m 63; c 2. MOLECULAR GENETICS, IMMUNOCHEMISTRY. *Educ:* Nat Univ Buenos Aires, MD, 61. *Prof Exp:* Eleanor Roosevelt fel immunol, Karolinska Inst, Stockholm, Sweden, 69-71; mem staff immunobiol, Lab Cell Biol, Rome, Italy, 71-73; head immunol, tumor immunol, Int Agency Res Cancer, WHO, France, 73-74; vis scientist, Brown Univ, RI, 74-76; res scientist hemat, NY State Dept Health, 76-81, chief hemat, Div Labs & Res, 79-81, PROF SCH PUB HEALTH SCI, STATE UNIV NY-NY STATE DEPT HEALTH, 85-, RES PHYSICIAN, WADSWORTH CTR LABS & RES, 81- *Honors & Awards:* B Rivadavia Prize, Nat Acad Med, Buenos Aires, 67. *Mem:* Scand Soc Immunol; Ital Asn Immunologists; France Soc Immunologists; Am Asn Immunologists; Am Soc Microbiol; Am Asn Pathologists. *Res:* Molecular genetics of differentiation and morphogenesis; lymphocyte engineering; monoclonal antibodies; molecular immunology of bacterial envelopes; molecular pathology of infectious diseases; microbiol consortia and biofilms; author of over 150 articles published in scientific referred journals; editor of multivolume treatise and a book. *Mailing Add:* Wadsworth Ctr Labs & Res NY State Dept Health ESP PO Box 509 Albany NY 12201-0509

MCARTHUR, CHARLES STEWART, biochemistry; deceased, see previous edition for last biography

MCARTHUR, CHARLES WILSON, b New Orleans, La, Nov 4, 21; m 43; c 6. MATHEMATICS. *Educ:* La State Univ, BS, 47; Brown Univ, MS, 50; Tulane Univ, PhD(math), 54. *Prof Exp:* Instr math, Univ Md, 52-53; asst prof, Ala Polytech Inst, 53-56; assoc prof, 56-64, chmn dept, 74-80, PROF MATH, FLA STATE UNIV, 64- *Mem:* AAAS; Am Math Soc; Math Asn Am. *Res:* Functional analysis, particularly biorthogonal systems and Schauder bases; ordered topological vector spaces. *Mailing Add:* Dept Math Fla State Univ Tallahassee FL 32306

MCARTHUR, COLIN RICHARD, b Beamsville, Ont, July 18, 35; m 59; c 2. ORGANIC SYNTHESIS METHODOLOGY. *Educ:* Univ Western Ont, BSc, 57, MSc, 58; Univ Ill, PhD(org chem), 61. *Prof Exp:* Sr res chemist, Allied Chem Corp, NY, 61-67; asst prof natural sci, 67-70 & chem, 70-71, dir, Liberal Sci Prog, 77-82, chmn, Dept Chem, 82-85, ASSOC PROF CHEM, YORK UNIV, 71- *Mem:* fel Chem Inst Can; Can Soc Chem; Int Union Pure & Appl Chem; Soc Chem Indust. *Res:* Organic halogen compounds; organometallic reagents; asymmetric syntheses; syntheses on polymer supports; carbenes; azomethines; heterocycles. *Mailing Add:* Dept Chem York Univ 4700 Keele St Toronto ON M3J 1P3 Can

MCARTHUR, DAVID SAMUEL, b Nelson, NZ, May 3, 41. GEOMORPHOLOGY. *Educ:* Univ NZ, BSc, 62; Univ Canterbury, MSc, 64; Christchurch Teachers' Col, dipl, 64; La State Univ, PhD(geog), 69. *Prof Exp:* Res instr coastal geomorphol, Coastal Studies Inst, La State Univ, 68; vis asst prof geol, Mich State Univ, 69; vis asst prof geog, La State Univ, 69-70; asst prof, Univ Calif, Davis, 70-73; PROF GEOG, SAN DIEGO STATE UNIV, 73- *Mem:* Asn Am Geog. *Res:* Coastal geomorphology, especially beach sedimentation and morphology. *Mailing Add:* Dept Geog San Diego State Univ San Diego CA 92182

MACARTHUR, DONALD M, environmental management, health sciences; deceased, see previous edition for last biography

MACARTHUR, DUNCAN W, b Ft Huachuca, Ariz, Feb 23, 56; m 77; c 1. PARTICLE DETECTOR DESIGN. *Educ:* Carleton Col, Northfield, Minn, BA, 77; Princeton Univ, MA, 79, PhD(physics), 82. *Prof Exp:* Postdoctoral fel, Group MP-4, 82-85, STAFF MEM, GROUP N-2, LOS ALAMOS NAT LAB, 85- *Mem:* Am Phys Soc. *Res:* Alpha and neutron detector design and development; relativity tests and high-velocity atomic beam experiments; parity violation in atomic systems and pion double charge exchange; electric dipole moment searches and low-velocity atomic beams. *Mailing Add:* Los Alamos Nat Lab MS J562 Los Alamos NM 87545

MCARTHUR, ELDON DURANT, b Hurricane, Utah, Mar 12, 41; m 63; c 4. PLANT GENETICS. *Educ:* Univ Utah, BS, 65, MS, 67, PhD(plant genetics), 70. *Prof Exp:* Teaching asst biol, Univ Utah, 66-70; Agr Res Coun Gt Brit fel, Sigma Xi grant & demonstr, Univ Leeds, 70-71; teaching fel biol, Univ Utah, 71; res geneticist, Great Basin Exp Area, 72-75, RES GENETICIST, SHRUB SCI LAB, INTERMOUNTAIN RES STA, FOREST SERV, USDA, 75-, PROJ LEADER, 83- *Concurrent Pos:* Adj fac mem bot & range sci, Brigham Young Univ, 75-; adj fac mem biol sci, Wayne State Univ, 85-87. *Mem:* Soc Range Mgt; Bot Soc Am; Soc Study Evolution; Am Genetic Asn. *Res:* Genetics and cytology of Mimulus; cytology of Brassiceae; genetics, cytology, breeding and selection of intermountain shrubs; rangeland rehabilitation. *Mailing Add:* Shrub Sci Lab USDA 735 North 500 East Provo UT 84606

MCARTHUR, JANET W, b Bellingham, Wash, June 25, 14. RESEARCH ADMINISTRATION. *Educ:* Univ Wash, AB, 35, MS, 37; Northwestern Univ, MD, 42; Am Bd Internal Med, dipl, 48. *Hon Degrees:* DSc, Mt Holyoke Col, 62. *Prof Exp:* Res fel pediat, 48-50, instr, 50-51, instr gynec, 51-57, clin assoc med, 57-60, from asst clin prof to assoc clin prof, 60-71, assoc prof, 71-73, assoc prof obstet, 73-77, prof obstet & gynec, 73-84, EMER PROF OBSTET & GYNEC, HARVARD MED SCH, 84-; ADJ PROF HEALTH SCI, SARGENT COL, BOSTON UNIV, 82- *Concurrent Pos:* Asst, Children's Med Serv, Mass Gen Hosp, 48-50, asst med, 50-51, asst physician, 51-60, assoc physician, 60-84; sr gynecologist, Mass Gen Hosp, 80- *Mem:* AAAS; Am Fertil Soc; Endocrine Soc; AMA; Am Col Physicians. *Res:* Bioassay of pituitary hormones; identification of pituitary hormones in human plasma; reproductive endocrinology; effects of exercise on female reproductive system. *Mailing Add:* Boston Univ 635 Commonwealth Ave Boston MA 02215

MACARTHUR, JOHN DUNCAN, b Toronto, Ont, Apr 13, 36; m 59; c 2. NUCLEAR PHYSICS. *Educ:* Univ Western Ont, BSc, 58; McMaster Univ, PhD(nuclear physics), 62. *Prof Exp:* Asst prof, 62-67, ASSOC PROF PHYSICS, QUEEN'S UNIV, ONT, 67- *Mem:* Can Asn Physicists; Am Asn Physics Teachers. *Res:* Low energy nuclear physics. *Mailing Add:* Dept of Physics Queen's Univ Kingston ON K7L 3N6 Can

MACARTHUR, JOHN WOOD, b Chicago, Ill, Sept 1, 22; m 47; c 5. PHYSICS. *Educ:* Univ Toronto, BA, 45; Rensselaer Polytech Inst, PhD(physics), 53. *Prof Exp:* MEM FAC PHYSICS, MARLBORO COL, 48- *Concurrent Pos:* Res div, Mass Inst Technol, 57-58; lunar & planetary lab, Univ Ariz, 66-67. *Mem:* Am Phys Soc; Am Asn Physics Teachers; Am Geophys Union; Sigma Xi. *Res:* Astronomy; geophysics; population biology. *Mailing Add:* PO Box 15 Marlboro VT 05344

MCARTHUR, RICHARD EDWARD, b Bradford, Pa, Dec 29, 15; m 42; c 3. ORGANIC CHEMISTRY. *Educ:* Temple Univ, AB, 37; Pa State Col, MS, 39, PhD(org chem), 41. *Prof Exp:* Instr chem, Exten Sch, Temple Univ, 36-37; chem operator, Pa State Univ, 39; res chemist, Niagara Alkali Co, NY, 40; res chemist, Sherwood Refining Co, Pa, 41-45, chief chemist & operating supt, 45-46; res chemist, Olin Corp, Conn, 46-52, sect head chem res, 52-56, mgr, 56-60, assoc dir, 60-64, planning scientist, 65-69, sect mgr customer serv-urethanes, 69-74; CHEM CONSULT, 74- *Mem:* Am Chem Soc; Sigma Xi. *Res:* Organic fluorine compounds; laboratory process development; high pressure reactions; economic analyses and chemical feasibility studies; flexible and rigid urethane foams; polyether and isocyanate synthesis and process development. *Mailing Add:* 104 Walter Lane Hamden CT 06514

MCARTHUR, WILLIAM GEORGE, b Kearney, NJ, July 1, 40; m 66; c 3. COMPUTER SCIENCE. *Educ:* Villanova Univ, BS, 66; Pa State Univ, PhD(math), 69. *Prof Exp:* From asst prof to assoc prof math, 69-74, PROF MATH & COMPUT SCI, SHIPPENSBURG UNIV, 74- *Mem:* Asn Comput Mach; Math Asn Am; Sigma Xi; Inst Elec & Electronics Engrs Comput Soc; Am Asn Artificial Intel. *Res:* Use of microcomputers in education. *Mailing Add:* 515 Brenton St Shippensburg PA 17257

MCARTHUR, WILLIAM P, b Mason City, Iowa, Mar 31, 43. IMMUNOLOGY, MICROBIOLOGY. *Educ:* Univ Kans, BA, 65; Purdue Univ, PhD(immunol), 69. *Prof Exp:* From asst prof to assoc prof, Dept Dent, Col Dent, Univ Pa, 72-81; assoc prof, Dept Dent, 81-83, PROF, DEPT ORAL BIOL, COL DENT, UNIV FLA, 83- *Concurrent Pos:* Res career develop award, Dent Res, NIH, 78-83. *Mem:* Am Asn Immunol; Am Asn Dent Res; Int Asn Dent Res; Am Soc Microbiol. *Mailing Add:* Oral Biol Dept Col Dent Univ Fla JHMHSC Box J-405 Gainesville FL 32610

MACARTNEY, LAWSON, b Irvine, Scotland, Oct 4, 57. PATHOLOGY. *Educ:* Glasgow Univ, DVM, 79, PhD(molecular path), 83. *Prof Exp:* Assoc dir path, Smith, Kline French Labs, Eng, 88-89, PATHOLOGIST, DEPT EXP PATH, SMITH KLINE BEECHAM PHARMACEUTICALS, 89- *Honors & Awards:* Lawson Walley & Williams Prize, Royal Col Vet Surgeons, 79. *Mem:* Royal Col Vet Surgeons; Am Asn Path; Royal Col Pathologists; Soc Toxicol Path. *Mailing Add:* Dept Exp Path Smith Kline Beecham Pharmaceut PO Box 1539 King of Prussia PA 19406

MCASSEY, EDWARD V, JR, b New York, NY, Sept 17, 35; m 58; c 4. MECHANICAL ENGINEERING. *Educ:* Polytech Inst Brooklyn, BME, 56, MME, 59; Univ Pa, PhD(mech eng), 68. *Prof Exp:* Engr, Grumman Aircraft Corp, 59-60 & RCA-Astro Electronic Div, 60-65; from asst prof to assoc prof, 67-75, PROF MECH ENG, VILLANOVA UNIV, 75-, chmn, 82-88. *Concurrent Pos:* Consult, Perkin Elmer Co, 68-; NSF fac fel, 75-76. *Mem:* Am Inst Aeronaut & Astronaut; Am Soc Mech Engrs; Am Soc Eng Educ. *Res:* Fluid mechanics, thermodynamics and heat transfer; nuclear heat transfer and two phase flow; numerical heat transfer. *Mailing Add:* 1211 Muhlenberg Dr Wayne PA 19087

MCATEE, JAMES LEE, JR, b Waco, Tex, Aug 29, 24; m 47; c 4. COLLOID CHEMISTRY. *Educ:* Tex A&M Univ, BS, 47; Rice Univ, MS, 49, PhD, 51. *Prof Exp:* Supvr tech serv labs, Baroid Div, Nat Lead Co, 51-59; from asst prof to assoc prof, 59-71, from actg chmn to chmn, 81-84, assoc dean col Arts & Sci, 84-86, PROF CHEM, BAYLOR UNIV, 71- *Concurrent Pos:* Consult, NL Indust, Southern Clay Prods, Mobil, Slurry Seal. *Mem:* Am Crystallog Asn; Am Chem Soc; Mineral Soc Am; NAm Thermal Anal Soc; Clay Minerals Soc. *Res:* Clay minerals, especially montmorillonite; crystal structure of montmorillonite and organic and metal-ligand-montmorillonite complexes by means of x-ray diffraction, differential thermal techniques and electron microscopy; use of thermoluminescence for dating pottery and ceramic archaeological artifacts. *Mailing Add:* Dept Chem Baylor Univ Waco TX 76703

MCATEE, LLOYD THOMAS, b Lexington, Ky, July 4, 39; m 66; c 2. CELL BIOLOGY, ZOOLOGY. *Educ:* Hanover Col, BA, 61; Drake Univ, MA, 63; Univ Md, PhD, 69. *Prof Exp:* Assoc prof, 68-75, chmn dept biol sci, 74-81, PROF MICROBIOL, PRINCE GEORGE'S COMMUNITY COL, 75-, ASSOC DEAN SCI/MATH/ENG, 82- *Mem:* AAAS; Am Inst Biol Sci; Am Soc Microbiol. *Res:* Cytogenetic and kinetic effects of sublethal heat shocks on cell suspension cultures. *Mailing Add:* Div Sci & Math/Eng Prince George's Community Col Upper Marlboro MD 20772

MACAULAY, ANDREW JAMES, b Cazenovia, NY, Oct 31, 37; m 60; c 3. NUCLEAR ENGINEERING, WATER CHEMISTRY. *Educ:* Univ Mo, BS, 60, Xavier Univ, MBA, 88. *Prof Exp:* Design engr, Bettis Lab, 63-67, engr, Plant Opers & Training, Naval Reactor Facil, 67-69, supvr & mgr, 69-74, mgr testing, Bettis Lab, 74-77, dir sr officer training, Naval Reactor Facil, Westinghouse Elec Corp, 77-85, CAPITAL PROJ, WESTINGHOUSE MAT COOP, OHIO, 86- *Mailing Add:* 129 Laurel Oak Dr Akin SC 29803

MCAULAY, ROBERT J, b Toronto, Ont, Oct 23, 39; m 62. ELECTRICAL ENGINEERING. *Educ:* Univ Toronto, BASc, 62; Univ Ill, Urbana, MSc, 63; Univ Calif, Berkeley, PhD, 67. *Prof Exp:* RES ENGR, LINCOLN LAB, MASS INST TECHNOL, 67- *Concurrent Pos:* Vis assoc prof, McGill Univ, Montreal, 76. *Honors & Awards:* M Barry Carlton Award, 78; Sr Award, Signal Processing Soc, 90. *Mem:* Inst Elec & Electronic Engrs. *Res:* Statistical communication theory; optimization techniques; optimal control; radar theory; signal design; monopulse analysis; interferometer design; robust speech processing in noise; low-rate vocoder design. *Mailing Add:* Lincoln Labs Mass Inst Technol PO Box 73 Lexington MA 02173

MCAULEY, ALEXANDER, b Glasgow, Scotland; Can citizen. COORDINATION CHEMISTRY. *Educ:* Univ Glasgow, BSc, 58, PhD(chem), 62, DSc, 77. *Prof Exp:* Asst lectr chem, Univ Glasgow, 61-62, sr lectr, 70-75; res assoc, Mich State Univ, 62-63; ICI fel, Univ Strathclyde, Scotland, 63-64, lectr, 64-69; prof chem, 75-86, DEAN, FAC GRAD STUDIES, UNIV VICTORIA, 86- *Concurrent Pos:* Chmn, Chem Grants Selection Comt, Natural Sci & Eng Res Coun, 83-86; chmn, Div Inorg Chem, Chem Inst Can, 84-86; group chmn, scholars & fels comt, Nat Sci & Eng Coun, Can; mem, scholars & fels comt, Sci Coun, BC. *Mem:* Can Asn Coop Educ (pres, 78-80); fel Chem Inst Can; Royal Soc Chem. *Res:* Kinetics and mechanisms of inorganic reactions; electron transfer process; use of macrocyclic ligands in the stabilization of unusual oxidation states. *Mailing Add:* Dept Chem Univ Victoria PO Box 1700 Victoria BC V8W 2Y2 Can

MCAULEY, LOUIS FLOYD, b Travelers Rest, SC, Aug 21, 24; m 65; c 3. MATHEMATICS. *Educ:* Okla State Univ, BS, 49, MS, 50; Univ NC, PhD(math), 54. *Prof Exp:* Asst math, Okla State Univ, 49-50, instr, 50; instr Univ NC, 51-54 & Univ Md, 54-56; from instr to assoc prof, Univ Wis, 56-63; prof, Rutgers Univ, New Brunswick, 63-69; chmn dept, 69-77, LEADING PROF MATH, STATE UNIV NY BINGHAMTON, 69- *Concurrent Pos:* Vis assoc prof, La State Univ, 59-60; Off Naval Res fel, Univ Va, 62-63; mem, Inst Advan Study, 66-67; vis, Inst Advan Study, Princeton Univ, 78-79. *Mem:* Am Math Soc; Math Asn Am; Sigma Xi. *Res:* Topology; point sets; structure of continua, upper semicontinuous collections; abstract spaces; fiber spaces; mappings equivalent to orbit map of group actions; light open mappings; manifolds; regular mappings and generalizations. *Mailing Add:* Dept of Math Sci State Univ of NY Binghamton NY 13901

MCAULEY, PATRICIA TULLEY, b Middlebury, Vt, May 23, 35; m 65; c 3. TOPOLOGY. *Educ:* Vassar Col, AB, 55; Univ Wis-Madison, MS, 58, PhD(math), 62. *Prof Exp:* Asst prof math, Univ Md, 62-65; asst prof, Rutgers Univ, 65-68, assoc prof & chmn dept, Douglass Col, 68-69; ASSOC PROF MATH, STATE UNIV NY BINGHAMTON, 69- *Concurrent Pos:* Grant-dir undergrad res partic proj, NSF, 73. *Mem:* Am Math Soc. *Res:* Fiber spaces; shape theory; open maps; fixed point problems. *Mailing Add:* Dept Math State Univ NY Binghamton NY 13903

MCAULEY, VAN ALFON, b Travelers Rest, SC, Aug 28, 26. CONTINUOUS CONTROL THEORY, NUMERICAL MATHEMATICAL ANALYSIS. *Educ:* Univ NC, BA, 51. *Prof Exp:* Mathematician, Army Ballistic Missile Agency, Huntsville, Ala, 56-59; physicist, guid & control div, Marshall Space Flight Ctr, NASA, Huntsville, Ala, 60-61, res mathematician, Astrionics Lab, 62-70, Comput Lab, 70-81; RETIRED. *Honors & Awards:* Cert Recognition Award, NASA Inventions & Contbr Bd, 77. *Mem:* NY Acad Sci; Am Math Soc; Math Asn Am; Soc Indust & Appl Math; AAAS. *Res:* Control theory and new electrical control network inventions; new numerical methods in stability analysis and partial differential equations. *Mailing Add:* 3529 Rosedale Dr Huntsville AL 35810-2573

MCAULIFFE, CLAYTON DOYLE, b Chappell, Nebr, Aug 18, 18; m 43; c 4. FATES & EFFECTS OF SPILLED OIL & METHODS OF CONTROL, PETROLEUM MIGRATION. *Educ:* Nebr Wesleyan Univ, AB, 41; Univ Minn, MS, 42; Cornell Univ, PhD(soil sci), 48. *Prof Exp:* Res chemist, Div War Res, Columbia Univ, 43-44 & Carbide & Carbon Chem Corp, 44-46; asst soil scientist, Bur Plant Indust, USDA, 47-48; res assoc agron, Cornell Univ, 48-50; res assoc prof, Stable Isotopes Lab, NC State Univ, 50-56; sr res chemist, Chevron Oil Field Res Co, 56-67, sr res assoc, 67-86; PRES, CLAYTON MCAULIFFE & ASSOC, INC, 86- *Concurrent Pos:* chmn, Orange County Sect, Am Chem Soc, 63-64; vis scientist lectr, Soil Sci Soc Am, 64-67; mem, steering comt petrol marine environ, Nat Acad Sci, 72-75, comt energy & environ, 75-77, ocean Sci bd, 75-78; mem, comt Effectiveness Oil Spill Dispersants, Nat Res Coun, 85-89; mem & chmn, several comts concerned with fates & effects of oil & spill control, Am Petrol Inst. *Honors & Awards:* Meritorious Serv Citations, Am Petrol Inst, 82. *Mem:* Fel AAAS; Am Chem Soc; Soil Sci Soc Am; Am Soc Agron; Soc Petrol Eng; Am Asn Petrol Geologists. *Res:* Environmental studies; analysis methods (trace organics); solubility of hydrocarbons in water; multiphase fluid flow; geochemistry in petroleum exploration; soil chemistry; radio isotopes and stable isotopes in soil-plant investigations; stable isotope in surface area measurements; isotopic analysis of uranium. *Mailing Add:* 1220 Frances Ave Fullerton CA 92631-1807

MCAULIFFE, WILLIAM GEOFFREY, b Santa Monica, Calif, Aug 9, 48. HISTOCHEMISTRY, ELECTRON MICROSCOPY. *Educ:* Calif State Univ, Long Beach, BS, 71, MA, 75; Univ Cincinnati, PhD(anat), 78. *Prof Exp:* Instr, Sch Med & Dent, Univ Louisville, 78-80, asst prof anat, 80-85; AT ROBERT W JOHNSON MED SCH, UNIV MED & DENT NJ, 86- *Mem:* AAAS; Sigma Xi. *Res:* Histochemistry and electron microscopy of neurons and glia; cell and tissue biology. *Mailing Add:* Neurosci & Cell Biol Dept Robert W Johnson Med Sch Univ Med & Dent NJ Piscataway NJ 08854

MCAVOY, BRUCE RONALD, b Jamestown, NY, Jan 30, 33. PHYSICS. *Educ:* Univ Rochester, BS, 56. *Prof Exp:* Jr engr, Air Arm Div, Westinghouse Elec Corp, 56, assoc engr, 57, res engr, Res Lab, 57-66, sr res engr, 67-78, fel engr, 78-81, group leader microwave acoust, Res Lac, 81-85, ADV SCIENTIST, WESTINGHOUSE ELEC CORP, 85- *Concurrent Pos:* Lectr, Dept Elec Eng, Carnegie-Mellon Univ, 68-; mem, Nat Patent Coun. *Honors & Awards:* Centennial Medal, Inst Elec & Electronic Engrs. *Mem:* Fel Inst Elec & Electronic Engrs. *Res:* Optical physics; masers and lasers; microwave bulk effects in solids; microwave acoustics; microwave superconductors. *Mailing Add:* Res Lab Westinghouse Elec Corp Pittsburgh PA 15235

MACAVOY, THOMAS COLEMAN, b Jamaica, NY, Apr 24, 28; m 52; c 4. ORGANIC CHEMISTRY. *Educ:* Queens Col, NY, BS, 50; St Johns Univ, MS, 52; Univ Cincinnati, PhD(chem), 57. *Prof Exp:* Anal chemist, Chas Pfizer & Co, Inc, 53-54; sr chemist, Corning Inc, 57-61, mgr electronic res, 61-64, dir phys res, 64-66, gen mgr & vpres, Electronic Prod Div, 66-69, gen mgr tech prod group, 69-71, PRES & DIR, CORNING INC, 71- *Concurrent Pos:* Med bd dirs, Quaker Oats Co, Chicago, Dow Corning Corp, Midland, Mich & Chubb Corp, New York. *Mem:* Sigma Xi. *Res:* Composition and properties of glass; complex phosphates and phosphate glasses; ion exchange equilibria; high temperature inorganic chemistry; growth of single crystals of refractory compounds. *Mailing Add:* Corning Inc MPQX-3-1 Corning NY 14831

MCAVOY, THOMAS JOHN, b New York, NY, Apr 25, 40; m 62; c 3. CHEMICAL ENGINEERING. *Educ:* Polytech Inst Brooklyn, BSChE, 61; Princeton Univ, MA, 63, PhD(chem eng), 64. *Prof Exp:* From asst prof to prof chem eng, Univ Mass, Amherst, 64-80; PROF CHEM ENG, UNIV MD, 80- *Concurrent Pos:* NSF res grants, 67-88; res assoc, dept appl physics, Tech Univ Delft, 71-72. *Honors & Awards:* Donald Eckman Award, Inst Soc Am, 87. *Mem:* Am Inst Chem Engrs; Am Chem Soc; Inst Soc Am. *Res:* Process dynamics and control; artificial intelligence; neural networks. *Mailing Add:* Dept Chem Eng Univ Md College Park MD 20742

MCBAY, ARTHUR JOHN, b Medford, Mass, Jan 6, 19; m 46; c 2. FORENSIC TOXICOLOGY. *Educ:* Mass Col Pharm, BS, 40, MS, 42; Purdue Univ, PhD(chem), 48. *Prof Exp:* From asst prof to assoc prof chem, Mass Col Pharm, 48-55; supvr chem lab, Mass State Police, 55-63; supvr lab, Mass Dept Pub Safety, 63-69; chief toxicologist, Off Chief Med Examr, NC, 69-89; assoc prof path, Med Sch & assoc prof toxicol, Sch Pharm, 69-74, prof path & pharm, 74-89, EMER PROF, SCH PHARM & ADJ PROF DEPT PATH, UNIV NC, CHAPEL HILL, 89-; CONSULT FORENSIC TOXICOL, 89- *Concurrent Pos:* Asst, Harvard Med Sch, 52-63; consult, Mass Col Pharm, 55-69; assoc prof, Law-Med Inst & Med Sch, Boston Univ, 63. *Mem:* Am Pharmaceut Asn; Am Acad Forensic Sci. *Res:* Organic pharmaceutical chemistry; spectrophotometric assays; toxicology-barbiturates and carbon monoxide; analytical chemistry; marijuana, other drugs and driving. *Mailing Add:* 102 Kings Mountain Ct Chapel Hill NC 27516

MCBAY, HENRY CECIL, b Mexia, Tex, May 29, 14; m 54; c 2. CHEMISTRY. *Educ:* Wiley Col, BS, 34; Atlanta Univ, MS, 36; Univ Chicago, PhD(chem), 45. *Hon Degrees:* DSc, Atlanta Univ, 87. *Prof Exp:* Instr chem, Wiley Col, 36-38 & Western Univ Kansas City, 38-39; from instr to assoc prof, 45-48, PROF CHEM, MOREHOUSE COL, 48-, CHMN DEPT, 56- *Concurrent Pos:* Tech expert, UNESCO, 51. *Mem:* Am Chem Soc. *Res:* Organic and inorganic chemistry; free radicals. *Mailing Add:* 1719 Detroit Ct NW Atlanta GA 30314

MCBEAN, EDWARD A, b Vancouver, BC, Nov 25, 45; m 68; c 3. ACID RAIN, HAZARDOUS WASTE DISPOSAL. *Educ:* Univ Brit Columbia, BASc, 68; MIT, SM, 70, CE, 72 & PhD(civil eng), 73. *Prof Exp:* Eng, Acres Consult Serv Ltd, 70-71; proj eng, Meta Systs Inc, 72-76; res assoc water res, Cornell Univ, 73-74,; from asst prof to assoc prof, 74-81, PROF, UNIV WATERLOO, 81-; SR TECH ADV, CONE STOGA-ROVERS & ASSOC, 76- *Concurrent Pos:* Assoc ed, Am Geophys Union, 79-87; vis prof, Univ Roorkee, India, 83; Canadian rep, UN Environ Prog, 85; bilateral exchange partic, Japan Soc Prom Sci, 86; co-prin investr, Soil Proj, 87-; mem, Grant Select Comt, Nat Sci & Eng Res Coun, 88- *Honors & Awards:* Gold Medal Award, India Inst Engrs, 86-87. *Mem:* Am Soc Eng; Am Geophys Union; Nat Water Well Assoc. *Res:* Environmental engineering involving utilization of system analysis and statistical interpretation of engineering data, these concerns include attention to economic assessment for remediation and alternative operating policies. *Mailing Add:* Dept Civil Eng Univ Waterloo Waterloo ON N2L 3G1 Can

MCBEAN, LOIS D, CALCIUM RESEARCH. *Educ:* Cornell Univ, MA, 68. *Prof Exp:* NUTRIT INFO SPECIALIST, MICH DAIRY COUN, 79- *Mem:* Am Inst Nutrit. *Mailing Add:* Mich Dairy Coun 1654 Morehead Dr Ann Arbor MI 48103

MCBEAN, ROBERT PARKER, b Chilliwack, BC, May 6, 39; m 62; c 3. STRUCTURAL ENGINEERING. *Educ:* Univ BC, BASc, 62, MASc, 65; Stanford Univ, PhD(civil eng), 68. *Prof Exp:* Structural engr, H A Simons, Ltd, Int, Vancouver, Can, 62-63, Phillips, Barratt & Partners, 64-65 & Hooley Eng, 65; res engr, John A Blume & Assocs, Calif, 66; instr civil eng, Stanford Univ, 67; asst prof, Univ Missouri-Columbia, 68-71, assoc prof, 71-74; PROJ ENGR, BLACK & VEATCH, CONSULT ENGRS, 74- *Mem:* Am Soc Civil Engrs; Nat Soc Prof Engrs; Sigma Xi. *Res:* Applied mechanics; computer applications, static and dynamic analysis of complex structures. *Mailing Add:* 8221 W 101 Overland Park KS 66212

MCBEATH, ELENA R, IMMUNOCYTOCHEMISTRY. *Prof Exp:* FEL, DEPT ANAT & CELL BIOL, SCH MED, HARVARD UNIV, 84- *Mailing Add:* Dept Anat & Cell Biol Nat Cardiovasc Ctr Res Inst 5-125 Fu-jishiro-dai Suita Osaka 565 Japan

MCBEE, FRANK W(ILKINS), JR, b Ridley Park, Jan 22, 20; m 43; c 2. MECHANICAL ENGINEERING. *Educ:* Univ Tex, BS, 47, MS, 50. *Hon Degrees:* PhD(humane lett), St Edward's Univ, 86. *Prof Exp:* From instr to asst prof mech eng, Univ Tex, 47-53, supvr, Defense Res Lab, 50-59; treas & sr vpres, Tracor, Inc, 55-67, exec vpres, 67-70, pres, 70-86, chief exec officer, 70-87, chmn bd, 72-87, EXEC CONSULT, TRACOR, INC, 88- *Concurrent Pos:* Sr chmn, MBank, Austin; mem bd dirs, MCorp, Dallas, KMW Systs Corp, Austin, Med Systs Support, Inc, Dallas, Radian Corp, Austin & Pritronix, Dallas; chmn comput sci develop comt & mem chancellor's coun, Univ Tex; mem corp cabinet, Am Heart Asn & Tex Affil, Inc; mem finance coun, Seton Med Ctr; emer trustee, Southwest Tex Pub Broadcasting Coun; vchmn, Headliners Found. *Honors & Awards:* Clara Driscoll Award, Laguna Gloria Art Mus, 87. *Mem:* Nat Acad Eng; Sigma Xi; Nat Soc Prof Engrs; Am Electronics Asn. *Res:* Resistance welding; rare metals and mechanical design of underwater devices; industrial engineering and relations; management of scientific companies. *Mailing Add:* 705 San Antonio Austin TX 78701

MCBEE, GEORGE GILBERT, b Eastland, Tex, Aug 15, 29; m 54; c 2. AGRONOMY. *Educ:* Tex A&M Univ, BS, 51, MS, 56, PhD(plant physiol), 65. *Prof Exp:* Asst county agr agent, Agr Exten Serv, Tex A&M Univ, 53-54, res asst soil chem, 54-56, area agron specialist, Agr Exten Serv, 56-60, state agron specialist, 60-62, res asst plant physiol & biochem, 62-64, asst prof turf physiol, 64-69, resident dir res, Tex Agr Exp Sta, 69-75, PROF SOIL & CROP SCI, TEX A&M UNIV, 75- *Mem:* Am Soc Plant Physiol; Am Soc Agron. *Res:* Carbohydrate metabolism in sorghum; methanogenesis studies for conversion of sorghum biomass; related crop physiology studies on sorghum. *Mailing Add:* 3704 Oak Ridge Dr Bryan TX 77801

MCBEE, W(ARREN) D(ALE), b Toledo, Ohio, Mar 18, 25; m 50; c 2. ELECTRICAL ENGINEERING. *Educ:* Marquette Univ, BEE, 45; Univ Mich, MS, 48, PhD(elec eng), 51. *Prof Exp:* Proj engr, Sperry Gyroscope Co, 50-53, sr proj engr, 53-54, sect head, 54-56, eng dept head, 56-61, res staff mem, Sperry Rand Res Ctr, 61-64, head radiation sci dept, 64-67, mgr radiation & info sci lab, 67-70, DIR SYSTS LAB, SPERRY RES CTR, SUDBURY, 70- *Mem:* Fel Inst Elec & Electronics Engrs; Am Phys Soc; Sigma Xi. *Res:* Research management; electromagnetics; plasma physics. *Mailing Add:* 23 Lakin St PO Box 347 Pepperell MA 01463

MACBETH, ROBERT ALEXANDER, b Edmonton, Alta, Aug 26, 20: Can citizen; m 49; c 4. SURGERY, MEDICAL HISTORY. *Educ:* Univ Alta, BA, 42, MD, 44; McGill Univ, MSc, 47, dipl surg, 52; FRCS(C), 52; FACS, 58. *Hon Degrees:* DSc, Univ Alta, 88. *Prof Exp:* Assoc prof, Univ Alta, 57-60, prof surg & head dept, 60-75, dir surg serv, Univ Hosp, 60-75; prof surg & assoc dean, Dalhousie Univ, 75-77; exec vpres, Nat Cancer Inst Can, 77-85; exec vpres, Can Cancer Soc, 77-85; EXEC DIR, HANNAH INST HIST MED, 87- *Concurrent Pos:* Res fel endocrinol, McGill Univ, 46-47, teaching fel anat, 47-48; Nuffield Found traveling fel surg, Brit Postgrad Med Sch, 50-51; consult, Dept Vet Affairs, Col Mewburn Pavillion, Edmonton, 57-75; dir med educ, Prov of New Brunswick, 75-77; mem coun, Int Union Against Cancer, 78-86; mem bd dirs, Assoc Med Serv, 84-87. *Honors & Awards:* Award Merit, Int Union Against Cancer, 86. *Mem:* Am Surg Asn; fel Am Col Surg; Can Soc Clin Invest (pres, 70-71); Int Surg Group (tres, 82-88); James IV Asn Surgeons (dir, 82-87, vpres, 84-87). *Res:* Metabolic adaption to cold; magnesium metabolism; serum and tissue glycoproteins in relation to malignant and inflammatory disease; Canadian medical history. *Mailing Add:* Seven Gordon Rd Willowdale ON M2P 1E2 Can

MCBIRNEY, ALEXANDER ROBERT, b Sacramento, Calif, July 18, 24; m 47; c 4. IGNEOUS GEOLOGY, PETROLOGY. *Educ:* US Mil Acad, BS, 46; Univ Calif, Berkeley, PhD(geol), 61. *Prof Exp:* Asst prof geol, Univ Calif, San Diego & staff mem, Scripps Inst Oceanog, 62-65; dir, Ctr Volcanology, Univ Ore, 65-68, assoc prof, 65-70, chmn dept, 68-71, PROF GEOL, UNIV ORE, 70- *Honors & Awards:* Bowen Medal, Am Geophys Union, 90. *Mem:* AAAS; Geol Soc Am; Am Geophys Union. *Res:* Geology and igneous petrology of Circum-Pacific orogenic regions; layered intrusions; volcanology, Galapagos Islands. *Mailing Add:* Dept Geol Univ Ore Eugene OR 97403

MCBLAIR, WILLIAM, b San Diego, Calif, Apr 19, 17; m 57; c 3. PHYSIOLOGY, OCEANOGRAPHY. *Educ:* San Diego State Univ, BA, 47; Univ Calif, Los Angeles, PhD(zool), 56. *Prof Exp:* Asst instr chem, San Diego State Univ, 47; asst zool, Univ Calif, Berkeley, 47-48; instr biol, San Diego State Univ, 48-51, asst prof zool, 51-62, from assoc prof to prof biol, 62-82, EMER PROF BIOL, SAN DIEGO STATE UNIV, 82- *Concurrent Pos:* Asst, Scripps Inst Oceanog, Univ Calif, San Diego, 48-52. *Mem:* AAAS; Am Soc Zool; NY Acad Sci; fel Int Oceanog Found; Nat Audubon Soc; Am Asn Univ Prof. *Res:* Active uptake; marine and environmental physiology; biology of spiders, especially prey-predator relationships. *Mailing Add:* Box 877 La Mesa CA 91944-0877

MCBRADY, JOHN J, b St Paul, Minn, Feb 1, 16; m 44. PHYSICAL CHEMISTRY. *Educ:* Univ Minn, Minneapolis, BChem, 38, PhD(phys chem), 44. *Prof Exp:* Res chemist, Donnelley & Sons Co, 46 & Celanese Corp, 47-52; sr chemist, Minn Mining & Mfg Co, 52-65, res specialist, 65-70, sr res specialist, 70-81; RETIRED. *Mem:* Am Chem Soc. *Res:* Molecular spectroscopy; infrared and nuclear magnetic resonance; photochemistry. *Mailing Add:* 1555 Burns Ave St Paul MN 55106

MCBRAYER, JAMES FRANKLIN, b Rowan Co, Ky, July 12, 41; m 67; c 2. ECOLOGY, SCIENCE POLICY. *Educ:* Miami Univ, BS, 63; Purdue Univ, West Lafayette, MS, 70; Univ Tenn, Knoxville, PhD(ecol), 73. *Prof Exp:* Teacher biol pub schs, Ohio, 63-67; asst res biologist, Lab Nuclear Med, Univ Calif, 73-74; asst prof ecol, Univ Minn, St Paul, 74-76; ecologist, 76-84, SR STAFF, HAZARDOUS WASTE REMEDIAL ACTIONS PROG, OAK RIDGE NAT LAB, 84- *Concurrent Pos:* Consult, Desert Biome, US Int Biol Prog, 74-75; mem, Interbiome Specialist Comt Elemental Cycling, 74-75. *Mem:* AAAS; Ecol Soc Am; Sigma Xi. *Res:* Ecosystem analysis; decompostition and elemental cycling in terrestrial ecosystems; synergistic relationships between microflora and decomposer invertebrates; role of fossorial animals in community development; environmental impact assessment; technical management. *Mailing Add:* 110 Normandy Rd Oak Ridge TN 37830

MCBREEN, JAMES, b Cavan, Ireland, Sept 5, 38; US citizen; m 66; c 2. ELECTROCHEMISTRY. *Educ:* Nat Univ Ireland, BSc, 61; Univ Pa, PhD(phys chem), 65. *Prof Exp:* Res chemist, Yardney Elec Corp, 65-68; sr res chemist, Res Labs, Gen Motors Corp, 68-77; assoc chemist, 77-80, CHEMIST, BROOKHAVEN NAT LAB, 80- *Concurrent Pos:* Chmn, Educ Comt, Electrochem Soc, 80-82, Energy Technol Group, 81-83, Battery Div, 90- *Honors & Awards:* Battery Div Res Award, Electrochem Soc, 74. *Mem:* AAAS; Electrochem Soc; NY Acad Sci; Am Chem Soc. *Res:* Ambient-temperature aqueous batteries, including work on the zinc, nickel oxide and manganese dioxide electrodes; sealed zinc-nickel oxide batteries; hydrogen-halogen batteries and zinc-bromine and zinc-chlorine batteries; fuel cells; metal deposition; industrial water electrolysis; organic electrochemistry; electrocatalysis; conductive polymers; membranes; x-ray absorption studies of battery and fuel cell materials; hydrogen in metals. *Mailing Add:* Dept Appl Sci Brookhaven Nat Lab Upton NY 11973

MCBRIDE, ANGELA BARRON, b Baltimore, Md, Jan 16, 41; m 65; c 2. NURSING. *Educ:* Georgetown Univ, BS, 62; Yale Univ, MS, 64; Purdue Univ, PhD(psychol), 78. *Hon Degrees:* DPS, Univ Cincinnati, 83, Eastern Ky Univ, 91. *Prof Exp:* Instr, Yale Univ Sch Nursing, 64-68, res asst, 68-71, asst prof, 71-73; assoc dean res, 85-90, PROF NURSING, IND UNIV SCH NURSING, 78-, ADJ ASSOC PROF, 80-, EXEC ASSOC DEAN RES, 90-, INTERIM DEAN, 91- *Concurrent Pos:* Adj prof, Purdue Univ Sch Sci, Indianapolis, 81, Indianapolis Univ Sch Med, 81- *Mem:* Nat Acad Practice fel; Nat Adv Mental Health Coun; Am Nursing Assoc; Am Psychol Assoc; Soc Res Child Develop; AAAS. *Res:* Experience of parenthood, womens health, adult development, especially combining work and family, and assessment of the chronically mentally ill. *Mailing Add:* Ind Univ Sch Nursing 1111 Middle Dr Indianapolis IN 46202-5107

MCBRIDE, BARRY CLARKE, b Victoria, BC, June 22, 40; m 63; c 2. MICROBIOLOGY. *Educ:* Univ BC, BSc, 63, MSc, 65; Univ Ill, Urbana, PhD(microbiol), 70. *Prof Exp:* From asst prof to assoc prof, 70-81, head microbiol, 86-90, PROF MICROBIOL, UNIV BC, 81-, DEAN SCI, 90-

Concurrent Pos: Coun mem, Med Res Coun, 73-79; assoc ed, Oral Microbiol & Immunol, 88- *Mem:* Int Asn Dent Res; Am Soc Microbiol; Can Asn Dent Res (pres, 90-92). *Res:* Adhesive properties of bacteria, identification, characterization and biological function of adhesions; characterization and identification of bacterial revelence determinants. *Mailing Add:* Dept Microbiol Univ BC Vancouver BC V6T 1W5 Can

MCBRIDE, CLIFFORD HOYT, b Massena, Iowa, June 14, 26; m 50; c 6. CHEMISTRY. *Educ:* Iowa State Univ, BS, 48; St Louis Univ, MS, 57. *Prof Exp:* Chemist, Mallinckrodt Chem Works, 49-55, supvr, 55-62; sr res chemist, Armour Agr Chem Co, 62-63, anal res mgr, 63-68; sect head, 68-72, MGR ANAL SERV, AGRI-CHEM DIV, US STEEL CORP, 72- *Mem:* Am Chem Soc; Asn Off Anal Chemists. *Res:* Analytical chemistry; chromatography; spectrometry. *Mailing Add:* 4417 Cain Circle Tucker GA 30084

MCBRIDE, DUNCAN ELDRIDGE, b Chicago, Ill, Oct 26, 45; m 68; c 2. CONDENSED MATTER PHYSICS, SURFACE PHYSICS. *Educ:* Carleton Col, BA, 67; Univ Calif, Berkeley, MA, 69, PhD(physics), 73. *Prof Exp:* US-France exchange fel physics, Lab Solid State Physics, Univ Paris VII, 73-74; asst prof physics, Swarthmore Col, 74-79; from asst prof to assoc prof physics & chmn dept, Kenyon Col, 79-87; PROG DIR, NAT SCI FOUND, 85- *Concurrent Pos:* Vis res physicist, Univ Calif, Santa Barbara, 77-78; vis scientist, IBM Res Lab, San Jose, Calif, 79, 80 & 81; assoc prog dir, NSF, 85-86. *Mem:* AAAS; Am Phys Soc; Sigma Xi. *Res:* Surface science; electron tunneling; inelastic electron tunneling spectroscopy; catalysis. *Mailing Add:* Nat Sci Found 1800 G St NW Washington DC 20550

MCBRIDE, EARLE FRANCIS, b Moline, Ill, May 25, 32; m 56; c 2. SEDIMENTARY PETROGRAPHY. *Educ:* Augustana Col, AB, 54; Univ Mo, MA, 56; Johns Hopkins Univ, PhD(geol), 60. *Prof Exp:* From instr to prof geol, Univ Tex, Austin, 59-82, chmn dept, 80-84, Wilton E Scott Centennial prof, 82-90, J NALLE GREGORY CHAIR, UNIV TEX, AUSTIN, 90-; PRES, SANDSTONES, INC, 79- *Concurrent Pos:* NATO vis prof, Univ Perugia, Italy, 77; Merril W Hass Distinguished Prof, Univ Kans, 75. *Mem:* Geol Soc Am; Am Asn Petrol Geol; hon mem Soc Econ Paleont & Mineral (pres, 79-80); Int Asn Sedimentol; Nat Asn Geol Teachers. *Res:* Sedimentary petrology; Cretaceous rocks of northern Mexico; Paleozoic rocks of Marathon Region, Texas; origin of sedimentary rocks, chiefly clastic rocks and chert; origin of bedding; sandstone diagenesis; evolution of porosity. *Mailing Add:* Dept Geol Sci Univ Tex Austin TX 78713-7909

MCBRIDE, GORDON WILLIAMS, b Washington, DC, Nov 24, 10; m 38; c 2. CHEMICAL ENGINEERING, ASSOCIATION MAMAGEMENT. *Educ:* Yale Univ, BS, 31, ChE, 34. *Prof Exp:* Chem engr, Procter & Gamble Co, Cincinnati, Ohio, 34-38; consult engr, Washington, DC, 39-42 & 45-56; prin chem engr, USDA, 42-45; tech coordr, Union Carbide Corp, 57-65; exec dir, Indust Res Inst, New York, 65-75; RETIRED. *Concurrent Pos:* Secy, Alcohol Res Adv Group, War Prod Bd, USDA, 43-45. *Mem:* Am Chem Soc; AAAS; Am Defense Preparedness Asn. *Res:* Processing and utilization of fats and oils; production and utilization of agricultural chemicals, including alcohol; technical and management information exchange. *Mailing Add:* 3323 Stuyvesant Pl NW Washington DC 20015

MCBRIDE, J(AMES) W(ALLACE), b Winnipeg, Man, Feb 26, 15; nat US; m 46; c 1. AIRBREATHING PROPULSION SYSTEMS, COMPONENT AERODYNAMICS. *Educ:* Univ Man, BSc, 38; Mass Inst Technol, SM, 40. *Prof Exp:* Res assoc aerodyn, Mass Inst Technol, 38-43; chief exp engr, TurboRes, Ltd, 43-46, prof aeronaut, Purdue Univ, 47-48; chief res engr, govt & spec prod develop, Carrier Corp, 48-54; pres & gen mgr, J W McBride & Assocs, Tech Consult, 54-56; from mgr component develop to chmn div develop bd & mgr, TENEC prog, 56-65, consult engr & mgr advan propulsion systs & progs, 65-80, consult engr & mgr advan engine progs, Group Eng Div, Gen Elec Co, 80-85; RETIRED. *Concurrent Pos:* Mem aerodyn subcomt, Aeronaut Res Comt, Nat Res Coun Can, 44-47; lectr, Columbia Univ, 52-55 & 71-; vis prof, Univ Tenn Space Inst, Tullahoma; consult, prop systs & assoc components develop, J W McBride & Assoc, 85- *Mem:* Am Inst Aeronaut & Astronaut. *Res:* Aerodynamics of turbomachinery, especially unusual gases and vapors; methodology of research and development for effective product development; gas dynamics. *Mailing Add:* 6750 Camaridge Lane Cincinnati OH 45243-3802

MCBRIDE, JAMES MICHAEL, b Lima, Ohio, Feb 25, 40; m 64; c 2. PHYSICAL ORGANIC CHEMISTRY, ORGANIC SOLID STATE CHEMISTRY. *Educ:* Harvard Univ, BA, 62, PhD(chem), 67. *Prof Exp:* From asst prof to assoc prof, 66-80, PROF CHEM, YALE UNIV, 80- *Mem:* Am Chem Soc. *Res:* Free radical reactions; influence of viscous and rigid media on the course of organic reactions; solid state chemistry; electron paramagnetic resonance; isotope effects; structural understanding of bulk properties of solids; crystal growth. *Mailing Add:* Dept Chem Yale Univ Box 6666 New Haven CT 06511-8118

MCBRIDE, JOHN BARTON, b Philadelphia, Pa, July 6, 43. PHYSICS. *Educ:* St Joseph's Col, Pa, BS, 65; Dartmouth Col, MA, 67, PhD(physics), 69. *Prof Exp:* Res instr physics, Dartmouth Col, 69; res assoc, Princeton Univ, 69-70; mem staff, US Naval Res Lab, 70-74; MEM STAFF, LAB APPL PLASMA SCI, SCI APPLN INC, 74- *Mem:* Am Phys Soc. *Res:* Theoretical plasma physics. *Mailing Add:* Lab Appl Plasma Sci Sci Appl Inc 10260 Camppoint Dr San Diego CA 92121

MCBRIDE, JOSEPH JAMES, JR, b Philadelphia, Pa, Dec 10, 22; m 48; c 7. ORGANIC CHEMISTRY. *Educ:* St Joseph's Col, BS, 43; Univ Del, MS, 47, PhD(chem), 50. *Prof Exp:* Res chemist, Tidewater Assoc Oil Co, 49-54 & citrus exp sta, Univ Fla, 54-59; sect head org res, Armour Indust Chem Co, 59-61; dir develop, 61-88, CONSULT, ARIZ CHEM CO, 88- *Mem:* Am Chem Soc; Am Oil Chemist Soc; Pulp Chem Asn. *Res:* Organosilicon compounds; organic chemistry of nitrogen, fatty acids; rosin; terpenes. *Mailing Add:* 807 Wood Ave Panama City FL 32401-1799

MCBRIDE, LYLE E(RWIN), JR, b Omaha, Nebr, Oct 18, 29; m 52; c 6. AUTOMATIC CONTROL SYSTEMS, ELECTOMAGNETIC SHIELDING. *Educ:* Cornell Univ, BEE, 52; Harvard Univ, AM, 61, PhD(appl physics), 64. *Prof Exp:* Engr, Overseas Div, Procter & Gamble Co, Ohio, 52-59; asst prof elec eng, Princeton Univ, 64-67; engr supvr electronic develop & eng, Control Prod Div, Texas Instruments, Inc, 67-71, mgr systs develop, 71-73, elec appln engr, Metall Mat Div, 73-79, mgr appln eng, Metal Systs Div, sr mem tech staff, 80-85; PROF & CHMN EE ENGR, CALIF STATE UNIV, CHICO, 85- *Concurrent Pos:* Tech translator, 60-64; energy consult, 70- *Mem:* Inst Elec & Electronic Engrs. *Res:* Automatic control theory; multiparameter self-optimizing systems; system identification from operating records; nonlinear system modeling; composite metal shield theory. *Mailing Add:* Elec Eng Calif State Univ Chico Chico CA 95929-0930

MCBRIDE, MOLLIE ELIZABETH, b Montreal, Que, May 7, 29; Can; m 51; c 5. MEDICAL MICROBIOLOGY. *Educ:* Dalhousie Univ, BSc, 55; Bryn Mawr Col, MA, 57; McGill Univ, PhD(bact, immunol), 59. *Prof Exp:* Res fel, McGill Univ, 59-60; bacteriologist, Halifax Children's Hosp, 60-63; asst prof, 64-75, ASSOC PROF MICROBIOL, BAYLOR COL MED, 75- *Mem:* AAAS; Brit Soc Gen Microbiol; Can Soc Microbiol; Am Soc Microbiol; Soc Invest Dermat. *Res:* Microbial ecology of the skin; comparison in health and disease; gram negative colonization skin disinfection; microbial ecology of cervix and bacterial interference. *Mailing Add:* Dept Dermat & Microbiol Baylor Col Med One Baylor Plaza Houston TX 77030

MCBRIDE, ORLANDO W, b Cincinnati, Ohio, Sept 18, 32. HUMAN GENETICS. *Educ:* Muskingum Col, BS, 54; Johns Hopkins Univ, MD, 58. *Prof Exp:* Clin assoc biochem, Nat Heart Inst, NIH, 60-63; fel, Dept Biol, Johns Hopkins Univ, 63-65; SR INVESTR, NAT CANCER INST, NIH, 66- *Mem:* Am Soc Biochem & Molecular Biol; Am Soc Human Genetics. *Mailing Add:* Lab Biochem Nat Cancer Inst NIH Bldg 37 Rm 4D06 Bethesda MD 20892

MCBRIDE, RALPH BOOK, b Slippery Rock, Pa, Feb 1, 28; m 54; c 3. MATHEMATICS EDUCATION. *Educ:* Slippery Rock State Col, BS, 51; Indiana Univ Pa, ME, 65; Univ Mich, PhD(math educ), 70. *Prof Exp:* Teacher math, North Butler County Schs, 51-52, Apollo Area Schs, 54-56 & Kiski Area Schs, 56-65; assoc prof, 65-76, chmn dept, 71-80, PROF MATH, MANCHESTER COL, 76-, ISAAC & ETTA H OPPENHEIM PROF, 71- *Res:* General Mathematics. *Mailing Add:* Dept Math Sci Manchester Col North Manchester IN 46962

MCBRIDE, RAYMOND ANDREW, b Houston, Tex, Dec 27, 27; div; c 4. IMMUNOGENETICS, PATHOLOGY. *Educ:* Tulane Univ, BS, 52, MD, 56; Am Bd Path, dipl, 61. *Prof Exp:* Intern surg, Baylor Col Med, 56-57; asst resident pathologist, Peter Bent Brigham Hosp, 57-58, asst in path, 58-60, sr resident pathologist, 60-61; asst prof path, Col Physicians & Surgeons, Columbia Univ, 63-65; assoc prof surg & immunogenetics, Mt Sinai Sch Med, 65-68; prof path, New York Med Col, Flower & Fifth Ave Hosps, 68-78; exec dean, NY Med Col, Valhalla, 73-75, prof path, 68-78; PROF PATH, BAYLOR COL MED, TEX MED CTR, 78- *Concurrent Pos:* Teaching fel path, Harvard Med Sch, 58-61; Nat Cancer Inst spec fel, McIndoe Mem Res Unit, East Grinstead, Eng, 61-63; resident pathologist, Free Hosp for Women, 59; asst resident pathologist, Children's Hosp Med Ctr, 60; asst attend pathologist, Presby Hosp, New York, 63-65; career scientist, NY Health Res Coun, 67-73; attend pathologist, New York Med Col, Flower & Fifth Ave Hosps, 68-78 & Metrop Hosp, 68-78; exec dir & chief operating officer, Westchester Med Ctr Develop Bd, Inc, 74-76; mem active staff, Methodist Hosp, Houston, 78-; attending pathologist, Harris County Hosp Dist, 78-; res assoc, Mt Sinai Sch Med, NY; vis grad fac, Tex A&M Univ, Col Sta, 79-; clin prof path, Grad Sch Biomed Sci, Univ Tex & dept path, Med Br, 82- *Mem:* AAAS; Am Asn Pathologists; Transplantation Soc; Am Asn Immunol; Am Asn Clin Pathologists. *Res:* Antigen and antibody interactions in isoimmune systems; population dynamics of antibody forming cells; factors involved in the recognition of immunogenicity; immunogenetics; DNA and RNA tumor viruses; tumor immunology; molecular genetics; immunogenetics. *Mailing Add:* Dept Path Baylor Col Med Houston TX 77030

MCBRIDE, WILLIAM H, b Worcester, Eng, Apr 14, 44. RADIATION ONCOLOGY. *Educ:* Univ Edinburgh, Scotland, BSc, 66, PhD, 71, DSc, 87. *Prof Exp:* Lectr, Dept Bact, Med Sch, Univ Edinburgh, Scotland, 71-82, sr lectr, 82-84; adj prof, 84-87, div chief, 89-91, PROF, DEPT RADIATION ONCOL, UNIV CALIF, LOS ANGELES, 87- *Concurrent Pos:* Proj investr, Depts Radiother & Exp Radiother, Univ Tex Cancer Ctr, M D Anderson Hosp & Tumor Inst, Houston, 75, 79 & 82. *Mem:* Am Asn Immunologists; Am Asn Cancer Res; Am Soc Therapeut Radiol & Oncol; Brit Asn Cancer Res; Radiation Res Soc; fel Royal Col Pathologists; Reticuloendothelial Soc. *Mailing Add:* Radiation Oncol Dept Med Ctr Univ Calif 10833 LeConte Ave Los Angeles CA 90024-1714

MCBRIDE, WILLIAM JOSEPH, b Philadelphia, Pa, Dec 24, 38; m 62; c 4. NEUROCHEMISTRY, NEUROBIOLOGY. *Educ:* Rutgers Univ, BA, 64; State Univ NY Buffalo, PhD(biochem), 68. *Prof Exp:* PROF NEUROBIOL & BIOCHEM, SCH MED, IND UNIV, INDIANAPOLIS, 71- *Concurrent Pos:* NIH fel neurobiol, Ind Univ, Bloomington, 68-71; res scientist develop award, NIMH, 79-83. *Mem:* Int Soc Neurochem; Soc Neurosci; Am Soc Neurochem; Res Soc Alcoholism. *Res:* Study of the mechanisms of how nerve cells communicate with one another and how alterations in this process may have an effect on the behavior of animals and man; neurobiological basis of alcoholism. *Mailing Add:* Inst Psychiat Res Ind Univ Med Ctr Indianapolis IN 46202-4887

MCBRIDE, WILLIAM ROBERT, b Topeka, Kans, May 9, 28; m 52; c 3. INORGANIC CHEMISTRY. *Educ:* Univ Calif, Los Angeles, BS, 50; Univ Tex, PhD(inorg chem), 55. *Prof Exp:* Res chemist, US Naval Ord Test Sta, 50-51 & 55-59, head inorg chem br, 59-80, SR RES SCIENTIST, NAVAL WEAPONS CTR, 80- *Mem:* Am Chem Soc; Sigma Xi. *Res:* Chemistry of hydronitrogens and derivatives; nonaqueous solutions and solvents; propellants; absorption spectroscopy of optical materials and filters; crystal growth of inorganic compounds; reaction kinetics. *Mailing Add:* 933 Jessica St Ridgecrest CA 93555

MCBRIDE, WOODROW H, b Milton, NDak, May 23, 18; m 44; c 1. MATHEMATICS. *Educ:* Jamestown Col, BA, 40; Univ NDak, MS, 47. *Prof Exp:* Teacher & prin, High Schs, NDak, 40-44; prin, High Schs, NDak & Minn, 44-46; from instr to assoc prof, 47-69, ASSOC PROF MATH, UNIV NDAK, 69- *Mem:* Math Asn Am. *Mailing Add:* 126 Columbia Ct Grand Forks ND 58201

MCBRIEN, VINCENT OWEN, b Attleboro, Mass, Apr 21, 16; m 48; c 4. MATHEMATICS. *Educ:* Providence Col, BS, 37; Cath Univ Am, MS, 40, PhD(math), 42. *Prof Exp:* Physicist, David Taylor Model Basin, US Dept Navy, 42 & Off Sci Res & Develop, 42-43; asst prof math, Hamilton Col, 43-44; from asst prof to prof math, Col Holy Cross, 43-78, chmn dept, 60-70. *Concurrent Pos:* Ford Found fel, Harvard Univ, 52-53; NSF fac fel, Univ Calif, Berkeley, 60-61; vis prof, Trinity Col, Dublin, 71-72; res assoc, Univ Calif, Berkeley, 76-77 & Harvard Univ, 78-79 & 79-91; res, algebraic geom; informal vis, Harvard Univ, 79- *Mem:* Am Math Soc; Math Asn Am. *Res:* Algebraic geometry. *Mailing Add:* 14 Saratoga Rd Auburn MA 01501

MCBROOM, MARVIN JACK, b Cherokee, Okla, Apr 13, 41; m 81; c 1. PHYSIOLOGY. *Educ:* Northwestern State Col, BS, 63; Okla State Univ, MS, 64; Univ Okla, PhD(physiol), 68. *Prof Exp:* From asst prof to assoc prof physiol, Sch Med, Univ SDak, Vermillion, 68-76; PROF PHYSIOL DEPT, FAC MED, KUWAIT UNIV, 76- *Mem:* AAAS; fel Geront Soc; Am Physiol Soc; Sigma Xi; Soc Exp Biol & Med. *Res:* Physiology of aging of soft tissues of mammals as related to fluid and electrolyte metabolism and taurine. *Mailing Add:* Dept Physiol Fac Med Kuwait Univ PO Box 24923 SAFAT Kuwait Kuwait

MCBRYDE, F(ELIX) WEBSTER, b Lynchburg, Va, April 23, 08; m 34; c 3. CARTOGRAPHIC DESIGN, DEVELOPMENT OF MAP PROJECTIONS. *Educ:* Tulane Univ, New Orleans, BA, 30; Univ Calif, Berkeley, PhD(geog), 40. *Hon Degrees:* LLD, Tulane Univ, 67. *Prof Exp:* Teaching asst, Univ Calif, Berkeley, 33-35 & 37; inst geog, Ohio State Univ, Columbus, 37-42; expert consult, 42; sr geographer, Mil Intel, Wash Dept, 42-45; dir, Peruvian off, Inst Social Anthrop, Smithsonian Inst, 45-47; geographer, off Coordr Int Statist, US Bur Census, 48-56; dir, regional demog planning, Gordon A Friesen Assocs Hosp Consults, Wash & Costa Rica, 56-58; pres, F W McBryde Assocs, Inc, Wash & Guatemala, 58-64; chief, Phys & Cult Geog Br, Natural Resources Div, Inter-Am Geod Surv, US Army, Fort Clayton, 64-65; field dir, Bioenviron Prog & Andean Ecol Proj, Battelle Mem Inst, Columbus, Ohio, 65-70; DIR, MCBRYDE CTR HUMAN ECOL, POTOMAC, MD, 69- *Concurrent Pos:* Fel, various insts & univs, 30-41; photogr-geogr exped, Smithsonian Inst, Univ Utah, 31; lectr, Western Reserve Univ, Cleveland, 44 & Foreign Serv Inst, US Dept State, 49-53; consult, 45-; coordr, Viru Valley Exped, Northern Peru, 46-47; spec rep, Inst Andean Res, Lima, Peru, 47-48; instr to prof, Univ Md, 48-63; chief, US census mission & tech adv, First Nat Pop census Ecuador, 49-51 & Geog Div, Transemantic, Inc, Wash, 75-; dir, Hosp Surv Guatemala, 58-59, Proj Develop Prog, Battelle Mem Inst, Cent Am & Mex, 68-69; founder-pres, Inter-Am Inst Mod Lang, 62-66; ecologist, Bayano River Hydroelect Proj, World Bank, 73 & US Army, Wash, 74; mem, Am Cong Surv & Mapping. *Mem:* AAAS; Asn Am Geogrs; Am Geophys Union; Am Inst Biol Sci; fel Explorers Club; Am Cong Surv & Mapping. *Res:* Thematic cartography; equal-area world map plottings and oceans atlas; geopolitics of Central America; interoceanic canal studies in Panama; ecology of Latin American native cultures. *Mailing Add:* 10100 Falls Rd Potomac MD 20854

MCBRYDE, VERNON E(UGENE), b Palmyra, Ark, Feb 3, 33; m 54; c 2. INDUSTRIAL ENGINEERING, APPLIED STATISTICS. *Educ:* Univ Ark, BSIE, 58, MSIE, 60; Ga Inst Technol, PhD(indust eng), 64. *Prof Exp:* Indust engr, Clary Corp, Ark, 57-58; from instr to assoc prof indust eng, Univ Ark, 58-67; prof indust & mgt eng & head dept, Mont State Univ, 67-70; dean sch systs sci, Ark Polytech Col, 70-71, vpres acad affairs, 71-73; ASSOC DEAN, COL ENG, UNIV ARK, 77- *Concurrent Pos:* Pres, Productivity Int, Mgt Consult, 63-; NIH res assoc drug systs, Med Ctr, Univ Ark, 63-65; pres, Ozark Mfg Co, 83- *Mem:* Am Inst Indust Engrs; Am Soc Eng Educ; Am Soc Qual Control; Nat Soc Prof Engrs; fel Am Prod & Inventory Control Soc. *Res:* Manufacturing management; quality control; production control; cost control. *Mailing Add:* 5790 W Wheeler Rd Fayetteville AR 72703

MCBRYDE, WILLIAM ARTHUR EVELYN, b Ottawa, Ont, Oct 20, 17; m 49; c 2. CHEMISTRY. *Educ:* Univ Toronto, BA, 39, MA, 40; Univ Va, PhD(chem), 47. *Prof Exp:* Asst chem, Univ Toronto, 39-41; chemist, Welland Chem Works, 42-44; asst prof chem, Univ Va, 47-48; from asst prof to assoc prof, Univ Toronto, 48-60; prof & chmn dept, Univ Waterloo, 60-64, dean fac sci, 61-69; vis fel, Australian Nat Univ, 69-70; chmn dept, 71-77, prof, 71-86, EMER PROF CHEM, UNIV WATERLOO, 86- *Honors & Awards:* Chem Educ Award, Chem Inst Can, 67; Fisher Lect Award, Anal Chem, 73. *Mem:* Chem Inst Can. *Res:* Chemistry of precious metals; colorimetric analysis; coordination chemistry; history of chemistry. *Mailing Add:* Dept of Chem Univ of Waterloo Waterloo ON N2L 3G1 Can

MCCAA, CONNIE SMITH, b Lexington, Miss, Dec 6, 37; m 57; c 4. BIOCHEMISTRY. *Educ:* Miss Col, BS, 58; Univ Miss, PhD(biochem), 63, MD, 77. *Prof Exp:* Asst, 59-63. from instr to assoc prof, 63-73, PROF BIOCHEM, MED CTR, UNIV MISS, 73-, ASSOC PROF PHYSIOL & BIOPHYS, 70-, ASSOC PROF OPHTHAL, 81- *Concurrent Pos:* Nat Heart & Lung Inst res grant, 63-74, spec res fel, 67-70; Miss Heart Asn grant, 63-66; Miss Cancer Soc grant, 65-66; mem cardiovasc & renal study sect, NIH, 74-78; corneal fel, LSU Eye Ctr, 84-85, res fel, 84-85. *Mem:* Am Chem Soc; Am Physiol Soc; Am Heart Asn; Endocrine Soc; Am Soc Nephrology; Am Acad Ophthal. *Res:* Epidermal growth factor and corneal endothelial repair. *Mailing Add:* Dept Biochem Univ Miss Med Ctr Jackson MS 39216

MCCABE, BRIAN FRANCIS, b Detroit, Mich, June 16, 26; m 51; c 2. OTOLARYNGOLOGY. *Educ:* Univ Detroit, BS, 50; Univ Mich, MD, 54. *Prof Exp:* From instr to assoc prof otolaryngol, Med Sch, Univ Mich, 59-64; PROF OTOLARYNGOL & CHMN DEPT, COL MED, UNIV IOWA, 64- *Concurrent Pos:* Consult to Surgeon Gen, USPHS, 66; mem bd dirs, Am Bd Otolaryngol, 67. *Honors & Awards:* Mosher Award, Am Laryngol, Rhinol & Otol Soc, 65. *Mem:* Am Laryngol Asn; Am Otol Soc; Am Acad Ophthal & Otolaryngol; Am Laryngol, Rhinol & Otol Soc; Col Oto-Rhino-Laryngol Amicitiae Sacrum. *Res:* Maxillofacial surgery; neurophysiology of the vestibular apparatus; mechanism of the quick component of nystagmus; surgery of the major salivary glands, particularly cancer. *Mailing Add:* Dept Otolaryngol Univ Iowa Hosp & Clin Iowa City IA 52242

MCCABE, DONALD LEE, general practice & psychiatry, for more information see previous edition

MCCABE, GEORGE PAUL, JR, b Brooklyn, NY, Apr 2, 45; m 67; c 4. STATISTICS. *Educ:* Providence Col, BS, 66; Columbia Univ, PhD(math statist), 70. *Prof Exp:* From asst prof to assoc prof, 70-83, HEAD STATIST CONSULT, PURDUE UNIV, WEST LAFAYETTE, 70-, PROF STATIST, 83- *Concurrent Pos:* Vis assoc prof statist, Princeton Univ, 76-77. *Mem:* Sigma Xi; Inst Math Statist; Am Statist Asn. *Res:* Mathematical statistics; applied statistics; statistical computing. *Mailing Add:* Dept Statist Purdue Univ West Lafayette IN 47907-1399

MCCABE, GREGORY JAMES, JR, b Wilmington, Del, Aug 13, 56; m 81; c 4. SYNOPTIC CLIMATOLOGY, WATER BALANCE CLIMATOLOGY. *Educ:* Univ Del, BA, 80, MS, 84; La State Univ, PhD(phys geog), 86. *Prof Exp:* Asst prof phys geog, Memphis State Univ, 86-88; PHYS SCIENTIST, US GEOL SURV, 88- *Concurrent Pos:* Adj prof, Rider Col, 89, Delaware Valley Col, 90-91. *Mem:* Sigma Xi; Asn Am Geographers; Am Meteorol Soc. *Res:* Water balance modeling; weather pattern classification; climate change; sensitivity of water resources to climate change. *Mailing Add:* US Geol Survey 810 Bear Tavern Rd Suite 206 West Trenton NJ 08628

MACCABE, JEFFREY ALLAN, b Oakland, Calif, Jan 30, 43; m 71; c 2. DEVELOPMENTAL BIOLOGY. *Educ:* Univ Calif, Davis, BS, 64, PhD(genetics), 69. *Prof Exp:* Res assoc, State Univ NY Albany, 69-71; from asst prof to assoc prof, 72-84, PROF ZOOL, UNIV TENN, KNOXVILLE, 84- *Concurrent Pos:* Res career develop award, NIH, 78-83. *Mem:* AAAS; Am Genetic Asn; Am Soc Zool; Soc Develop Biol; NY Acad Sci; Soc Cell Biol. *Res:* Morphogenesis of the vertebrate limb. *Mailing Add:* Dept Zool Univ Tenn Knoxville TN 37916-0810

MCCABE, JOHN PATRICK, b New York, NY, Aug 17, 35. MATHEMATICS. *Educ:* Manhattan Col, BS, 57; Harvard Univ, AM, 58, PhD(math), 68. *Prof Exp:* Systs analyst, Gen Elec Co, 62-66; from instr to asst prof, 66-74, chmn dept, 74-77, ASSOC PROF MATH, MANHATTAN COL, 74- *Mem:* Am Math Soc; Soc Indust & Appl Math. *Res:* Abelian varieties over local fields. *Mailing Add:* 415 Plaza Rd Fair Lawn NJ 07410

MCCABE, R TYLER, b Brigham City, Utah, Aug 20, 59; m 88. MOLECULAR PHARMACOLOGY, RECEPTOR PHARMACOLOGY. *Educ:* Univ Utah, BS, 82, MS, 86, PhD(neuropharmacol), 87. *Prof Exp:* Postdoctoral fel, Dept Psychiat, Sch Med, Univ Utah, 87-88; pharmacol res assoc training fel, Lab Neurosci, Nat Inst Diabetes, Digestive & Kidney Dis, NIH, Bethesda, Md, 88-90, staff fel, 90-91; VPRES DRUG DISCOVERY SCI, PHARMACEUT DEVELOP ASSOCS, INC, ELMSFORD, NY, 91- *Concurrent Pos:* Vis scientist, Dept Pharmacol, Sch Med, Univ Calif, Los Angeles, 86; supvr, Dept Psychiat, Univ Utah, 86-88, instr, Dept Pharmacol, Sch Med, 87, Gen Pharmacol, Col Pharm, 87, Cath Univ Am, 89-91; consult, Clin Technol Assocs, Inc, 89, Weinberg Consult Group, 89, John R Vogel Assocs, Inc, 90-; supvr, Lab Neurosci, Nat Inst Diabetes, Digestive & Kidney Dis, 89-90. *Mem:* Soc Neurosci; Am Soc Pharmacol & Exp Therapeut. *Res:* Neurotransmitter receptors and their interactions with centrally active drugs; analysis of drug-receptor interactions using state-of-the-art techniques including: fluorescent ligand binding, radioreceptor binding, autoradiography and molecular pharmacology; neurological disorders and therapeutic means for treating diseases. *Mailing Add:* Pharmaceut Develop Assocs Inc Seven Westchester Plaza Elmsford NY 10523

MCCABE, ROBERT ALBERT, b Milwaukee, Wis, Jan 11, 14; m 41; c 4. WILDLIFE MANAGEMENT. *Educ:* Carroll Col, BA, 39; Univ Wis, MS, 43, PhD(wildlife mgt, zool), 49. *Hon Degrees:* LLD, Nat Univ Ireland, 88; DHL, Carroll Col, 89. *Prof Exp:* Biologist, Arboretum, 43-46, from instr to prof, 46-84, chmn dept, 52-79, EMER PROF WILDLIFE MGT, UNIV WIS-MADISON, 84- *Concurrent Pos:* Mem res adv comn, Wis Dept Natural Resources, 54-; secy adv comt, Wis Dept Resource Develop, 60-64; mem bd dirs, Wis Expos Dept, 60-68; chmn subcomt vertebrates as pests, Nat Acad Sci, 64-; Fulbright prof, Univ Col, Univ Dublin, 69-70; adv to Irish Nat Parks, 70-81. *Honors & Awards:* Aldo Leopold Medal, Wildlife Soc. *Mem:* Wildlife Soc (pres, 76-77); Am Soc Mammal; Wilson Ornith Soc; Cooper Ornith Soc; fel Am Ornith Union; Brit Ornith Union; Brit Ecol Soc. *Res:* Wildlife ecology; techniques in wildlife management; land use in wildlife relationship; ornithology. *Mailing Add:* Dept Wildlife Ecol Col Agr & Life Sci Univ Wis Madison WI 53706

MCCABE, ROBERT LYDEN, b Tarrytown, NY, Mar 5, 36; m 59; c 3. MATHEMATICS. *Educ:* Union Col, NY, BS, 57; San Diego State Col, MA, 60; Boston Univ, PhD(math), 71. *Prof Exp:* Asst prof, 64-68, ASSOC PROF MATH, SOUTHEASTERN MASS UNIV, 68- *Concurrent Pos:* Instr, Upperward Bound, 68. *Res:* Ergodic and dilation theories; Markov processes. *Mailing Add:* Southeastern Mass Univ N Dartmouth MA 02747-3409

MCCABE, STEVEN LEE, b Denver, Colo, July 11, 50; m 74; c 1. STRUCTURAL ENGINEERING, EARTHQUAKE ENGINEERING. *Educ:* Colo State Univ, BS, 72, MS, 74; Univ Ill, Urbana, PhD(civil eng), 87. *Prof Exp:* Engr, Pub Serv Co Colo, 74-77; sr engr, R W Beck & Assocs, 77-78; engr & proj engr, Black & Veatch Consult Engrs, 78-81; asst prof struct, 85-91, ASSOC PROF ENG TOPICS, CIVIL ENG DEPT, UNIV KANS, 91- *Concurrent Pos:* Consult engr, S L McCabe Eng Consults, 85-; state bd dirs & vpres, Kans Sect, Am Concrete Inst; mem var tech comts, Am Soc Mech

Engrs, Am Soc Civil Engrs, Am Concrete Inst & Earthquake Eng Res Inst. *Honors & Awards:* Press Vessels & Piping Serv Award, Am Soc Mech Engrs, 89. *Mem:* Am Concrete Inst; Am Soc Mech Engrs; Am Soc Civil Engrs; Earthquake Eng Res Inst. *Res:* Structural dynamics; earthquake engineering; computational mechanics. *Mailing Add:* 2015 Learned Hall Dept Civil Eng Univ Kans Lawrence KS 66045-2225

MCCABE, WILLIAM R, b Hugo, Okla, Sept 13, 28; m 51; c 3. INFECTIOUS DISEASES, MICROBIOLOGY. *Educ:* Univ Okla, BS, 49, MD, 53. *Prof Exp:* Intern med, Univ Okla Hosps, 53-54, resident, 56-58; res asst & chief med resident infectious dis, Col Med, Univ Ill, 58-60, asst prof med, 62-63; from asst prof to assoc prof, Sch Med, 63-71, asst prof microbiol, 63-71, PROF MED & MICROBIOL, MED CTR, BOSTON UNIV, 71-, DIR, DIV INFECTIOUS DIS, 63- *Concurrent Pos:* NIH trainee cardiovasc dis, Univ Okla Hosps, 57-58; res fel infectious dis, Univ Ill Res & Educ Hosp, 58-60; clin investr, West Side Vet Admin Hosp, Chicago, 60-63; vis physician, Boston City Hosp, 68-; consult physician, Vet Admin Hosp, Boston, 68-; mem drug eval panel, Nat Acad Sci-Nat Res Coun, 69-71; mem & chmn bact & mycol study sect, Nat Inst Allergy & Infectious Dis, 69-73; assoc ed, J Infectious Dis, 69-; dir, Maxwell Finland Lab for Infectious Dis, Boston City Hosp, 77-; head med bact & vis physician, Univ Hosp, 63-77. *Mem:* AAAS; Am Soc Clin Invest; Infectious Dis Soc Am; Am Soc Microbiol; Asn Am Physicians. *Res:* Host defense mechanisms and bacterial virulence factors in clinical infections. *Mailing Add:* Lab for Infectious Dis 774 Albany St Boston MA 02118

MACCABEE, BRUCE SARGENT, b Rutland, Vt, May 6, 42. ELECTRONICS, ELECTROOPTICS. *Educ:* Worcester Polytech Inst, BS, 64; Am Univ, MS, 67, PhD(physics), 70. *Prof Exp:* Res assoc physics, Am Univ, 67-72; RES PHYSICIST, WHITE OAK LAB, NAVAL SURFACE WEAPONS CTR, 72- *Concurrent Pos:* Consult, Nat Invests Comt Aerial Phenomena, 66-, Tracor, Inc, 70-71, Compackager Corp, 70-73, Sci Appln, Inc, 73-74 & Radix II, Inc, 78; mem sci bd, Ctr UFO Studies; chmn, Fund for Unidentified Flying Objects, 79-; prof lectr, Am Univ, 85- *Mem:* Am Phys Soc; AAAS; Am Optical Soc. *Res:* High energy lasers; optics; underwater sound; high power laser phenomenology; UFO phenomena and history. *Mailing Add:* 6962 Eyler Valley Flint Rd Sabillasville MD 21780

MCCAFFERTY, EDWARD, b Swoyerville, Pa, Nov 28, 37; m 66; c 2. CHEMISTRY. *Educ:* Wilkes Col, BS, 59; Lehigh Univ, MS, 64, PhD(chem), 68. *Prof Exp:* Res engr, Bethlehem Steel Corp, Pa, 59-64; Robert A Welch fel chem, Univ Tex, Austin, 68-70; sci off, Off Naval Res, 81-82, RES CHEMIST MAT SCI & TECH, NAVAL RES LAB, 70- *Honors & Awards:* Victor K LaMer Award, Am Chem Soc, 71. *Mem:* Electrochem Soc; Am Chem Soc. *Res:* Surface chemistry; corrosion science; electrochemistry of corrosion processes, kinetics and inhibition; adsorption on metals and oxides. *Mailing Add:* Mat Div Naval Res Lab 4555 Overlook Ave Washington DC 20375

MCCAFFERTY, WILLIAM PATRICK, b Murray, Utah, Oct 11, 45; m 64; c 3. AQUATIC ENTOMOLOGY, SYSTEMATICS. *Educ:* Univ Utah, BS, 67, MA, 69; Univ Ga, PhD(entom), 71. *Prof Exp:* Asst prof biol, Dixie Col, 70-71; PROF ENTOM, PURDUE UNIV, 71- *Concurrent Pos:* Prin investr grants, US Environ Protection Agency, NSF & Off Water Resources, 71-; consult, State Minn Dept Natural Resources, 76-78 & Inst Paper Chem, 79-81; vis prof, Univ Utah, 77 & Fla A&M Univ, 78; instr res & develop, US Environ Protection Agency, 78; ed, Int Quart Entom, J NAm Benthological Soc; res fel, Albany Mus, SAfrica, 90. *Mem:* Entom Soc Am; NAm Benthological Soc. *Res:* The biology of aquatic insects with special attention to systematics, ecology and behavior of benthic forms; classification and functional morphology of world ephemeroptera; freshwater resource protection; fly-fishing entomology. *Mailing Add:* Dept Entom Purdue Univ West Lafayette IN 47907

MCCAFFREY, DAVID SAXER, JR, b Mt Kisco, NY, Oct 23, 42; m 67; c 4. CHEMICAL ENGINEERING. *Educ:* Univ Notre Dame, BS, 64, MS, 66, PhD(chem eng), 67. *Prof Exp:* Eng assoc, 69-80, HEAD SECT, EXXON RES & ENG CO, 80- *Mem:* Am Inst Chem Engrs; Am Chem Soc; Sigma Xi. *Res:* Fluidization; air pollution control technology; atmospheric dispersion modeling; engineering thermodynamics; synthetic fuels. *Mailing Add:* 20 Arlington Dr Denville NJ 07834

MCCAFFREY, JOSEPH PETER, b Providence, RI, Apr 7, 51; m 84. BIOLOGICAL WEED CONTROL. *Educ:* Univ RI, BA, 74; Va Polytech Inst & State Univ, MS, 78, PhD(entom), 81. *Prof Exp:* ASST PROF, DEPT PLANT SOIL & ENTOM SCI, UNIV IDAHO, 81-, ASSOC PROF, 87- *Mem:* Entom Soc Am; Int Orgn Biol Control; Sigma Xi. *Res:* Introduction, establishment, and evaluation of biological central agents for noxious weed control; development of integrated pest management programs for weeds; integrated pest management of winter rape seed. *Mailing Add:* Dept Entom Univ Idaho Moscow ID 83843

MCCAFFREY, THOMAS VINCENT, b Chicago, Ill, Nov 29, 48; m. HEAD & NECK SURGERY, UPPER AIRWAY PHYSIOLOGY. *Educ:* Loyola Univ, Chicago, BS, 70, MD, 74, PhD(physiol), 81; Univ Minn, PhD(otolaryngol), 81. *Prof Exp:* Res assoc physiol, Stritch Sch Med, Loyola Univ, 74-75; intern, Mayo Grad Sch Med, 75-76, resident otolaryngol, 76-80, instr otolaryngol, 80-84, asst prof, 84-87, ASSOC PROF OTOLARYNGOL, MAYO CLIN, 87- *Honors & Awards:* Maurice Cottle Award , Am Rhinologic Soc, 79; Mosher Award, Triological Soc, 88. *Mem:* Am Physiol Soc; Asn Res Otolaryngol; Am Acad Otolaryngol; Am Bronchoesophagological Asn; Soc Head & Neck Surgeons; Am Soc Head & Neck Surg; Am Laryngol, Rhinologic & Otol Soc; Am Laryngol Asn. *Res:* Physiology of upper airway and relation between upper and lower airways; respiratory reflexes and their relation to pulmonary function. *Mailing Add:* Dept Otolaryngol Mayo Clin Rochester MN 55905

MCCAIN, ARTHUR HAMILTON, b San Francisco, Calif, Aug 31, 25; m 59; c 1. PLANT PATHOLOGY. *Educ:* Univ Calif, BS, 49, PhD, 59. *Prof Exp:* LECTR PLANT PATH, UNIV CALIF, BERKELEY & EXTEN PLANT PATHOLOGIST, AGR EXP STA, 59- *Concurrent Pos:* Consult, US Forest Serv. *Mem:* Am Phytopath Soc; Sigma Xi. *Res:* Forest tree diseases; control of plant dieases; ornamental diseases; biological control. *Mailing Add:* Dept Plant Path Univ Calif Berkeley CA 94720

MCCAIN, DOUGLAS C, PHYSICAL CHEMISTRY. *Educ:* Univ Calif, Berkeley, PhD(chem), 66. *Prof Exp:* Assoc prof, 76-84, PROF CHEM, UNIV SOUTHERN MISS, 84- *Mailing Add:* Dept Chem SS Box 9281 Univ Southern Miss Box 9281 Southern Sta Hattiesburg MS 39401

MCCAIN, FRANCIS SAXON, b Ashland, Ala, Aug 13, 21; m 50; c 1. PLANT BREEDING. *Educ:* Auburn Univ, BS, 42, MS, 48; Purdue Univ, PhD, 50. *Prof Exp:* Assoc prof agron & soils & assoc plant breeder, Auburn Univ, 50-59, prof agron & soils & plant breeder, 59-66; PROF AGR & CHMN DIV AGR, HOME ECON & FORESTRY, ABRAHAM BALDWIN AGR COL, 66-, ASST DIR RURAL DEVELOP CTR, 69- *Mem:* Am Soc Agron; Sigma Xi. *Res:* Corn breeding. *Mailing Add:* 2205 Murray Ave Tifton GA 31794

MCCAIN, GEORGE HOWARD, b Flora, Ind, Nov 15, 24; m 48; c 1. ORGANIC CHEMISTRY, POLYMER SYNTHESIS. *Educ:* Franklin Col, AB, 49; Univ Ill, MS, 50, PhD(chem), 53. *Prof Exp:* Res chemist, Electrochem Dept, E I du Pont de Nemours & Co, Inc, 52-55; sr chemist, Diamond Shamrock Corp, 55-59, res assoc, 59-83; consult, Biospecific Technol Inc, 84-88; CONSULT, 88- *Mem:* Am Chem Soc; Sigma Xi. *Res:* Vinyl polymerization; ionogenic polymers; membrane chemistry; biologically active polymers, fluorinated polymers. *Mailing Add:* 250 Orton Rd Painesville OH 44077

MCCAIN, JAMES HERNDON, b Little Rock, Ark, Sept 18, 41; m 63; c 2. ORGANIC CHEMISTRY. *Educ:* Southwestern at Memphis, BS, 63; Northwestern Univ, PhD(org chem), 67. *Prof Exp:* CHEMIST, UNION CARBIDE CORP, 67- *Mem:* Am Chem Soc. *Res:* Applied research; development; plant problems. *Mailing Add:* 1987 Parkwood Rd Charleston WV 25314-2241

MCCAIN, JOHN CHARLES, b Ft Worth, Tex, Aug 11, 39; m 56, 77; c 2. MARINE BIOLOGY. *Educ:* Tex Christian Univ, BA, 62; Col William & Mary, MA, 64; George Washington Univ, PhD(zool), 67; Univ Hawaii, MPH, 75. *Prof Exp:* Res asst, Va Inst Marine Sci, 62-64; mus technologist syst zool, Smithsonian Inst, 64-65, mus specialist, 65, asst cur, 65-67; res assoc oceanog, NSF Antarctic Prog res grant to Dr Joel W Hedgpeth, Ore State Univ, 67-69; in-chg benthic invert div, Oceanog Sorting Ctr, Smithsonian Inst, Washington, DC, 69-70; aquatic ecologist, TRW/Hazleton Labs, Inc, 70-71; sr marine biologist, Hawaiian Elec Co, 71-77, mgr, Environ Dept, 77-80; prin biologist, Tetra Tech Inc, 80-83; SR RES SCI PROF, KING FAHD UNIV OF PETROL & MINERALS RES INST, 83- *Concurrent Pos:* Res assoc, Bernice P Bishop Mus, 73-; res fel, Smithsonian Inst, 66; vpres, Aecos Inc, 86- *Mem:* Soc Syst Zool; Am Fisheries Soc; Western Soc Naturalists; Biomet Soc; Nat Asn Underwater Instr. *Res:* Taxonomy and ecology of invertebrates, particularly Amphipoda; marine ecology; pollution biology; biostatistics. *Mailing Add:* Res Inst King Fand Univ Petrol & Minerals Dhahran 31261 Saudi Arabia

MCCALDEN, THOMAS A, PHARMACOLOGY. *Prof Exp:* ASST DEAN RES, MED SCH, UNIV NEV, 89- *Mailing Add:* Pharmacol Dept Med Sch Univ Nev Reno NV 89557

MCCALDIN, J(AMES) O(ELAND), b Shreveport, La, Apr 3, 22. SOLID STATE PHYSICS, ELECTRONIC MATERIALS. *Educ:* Univ Tex, BA, 44; Calif Inst Technol, PhD(mech eng), 54. *Prof Exp:* Instr math & physics, San Antonio Col, 46-48; engr telemetering, Arabian-Am Oil Co, 52; sr engr physics & metall, Res Labs, Gen Motors Corp, 54-55; mem tech staff, Res Lab, Hughes Aircraft Co, 56-60; group leader semiconductors, NAm Aviation Sci Ctr, 61-68; assoc prof eng, 68-73, prof, 73-83, EMER PROF APPL PHYSICS, CALIF INST TECHNOL, 83- *Concurrent Pos:* Lectr, Eng Exten, Univ Calif, Los Angeles, 57-59. *Mem:* Am Phys Soc; Inst Elec & Electronics Engrs; Electrochem Soc. *Res:* Semiconductor interfaces and thin films; physical chemistry of semiconductors. *Mailing Add:* Dept Appl Physics Calif Inst Technol Pasadena CA 91125

MCCALEB, KIRTLAND EDWARD, b Brighton, Mass, Sept 10, 26; m 49; c 2. ORGANIC CHEMISTRY. *Educ:* Dartmouth Col, AB, 46; Univ Wis, PhD(org chem), 49. *Prof Exp:* Jr chemist, Eastman Kodak Co, 46; fel, Univ Wis, 49-50; sr org chemist, Res Labs, Gen Mills, Inc, Minn, 50-53, leader nitrogen prod sect, 53-60; mgr tech sales serv, Foremost Chem Prod Co, Calif, 60-65; indust economist, Chem Econ Handbook, 65-68, ed, 68-72, DIR CHEM-ENVIRON PROG, SRI INT, 72- *Mem:* Am Chem Soc; Am Oil Chem Soc. *Res:* Surface-active agents; chemical market research; environmental chemicals. *Mailing Add:* Chem Environ Prog SRI Int Menlo Park CA 94025

MCCALEB, STANLEY B, b Santa Barbara, Calif, Oct 29, 19; m 52; c 3. CLAY MINERALOGY. *Educ:* Univ Calif, BS, 42; Cornell Univ, MS, 48, PhD(soils), 50. *Prof Exp:* From asst prof to assoc prof soil genesis, NC State Col, 50-56; sr soil correlator, Soil Conserv Serv, USDA, Calif, 56-58; sr res geologist & head clay mineral res sect, 58-70, supvr chem eng, Prod Serv Lab, 70-74, mgr prod serv lab, Sun Co, Inc, 74-85; CONSULT, 85- *Mem:* Clay Minerals Soc (pres, 75-76); Soil Sci Soc Am; Am Soc Agron; Int Soc Soil Sci; Asn Inst Mining, Metall & Petrol Engrs. *Res:* Soil morphology, genesis and classification; mineral weathering; land use and management; analysis labs; pollution evaluation; petroleum exploration and production. *Mailing Add:* 200 Westshore Dr Richardson TX 75080

MCCALL, CHARLES B, b Memphis, Tenn, Nov 2, 28; m 51; c 4. MEDICINE. *Educ:* Vanderbilt Univ, BA, 50, MD, 53. *Prof Exp:* Nat Acad Sci-Nat Res Coun pulmonary fel, 57-58; instr med, Med Col Ala, 58-59; from asst prof to assoc prof, Univ Tenn, 59-69; prof, Univ Tex Med Br, Galveston, 69-72, asst vchancellor & coordr regional med prog, Univ Tex Syst, 69-72; assoc dean clin affairs, Univ Tex Southwestern Med Sch, 72-75; dean col med, Ctr Health Sci, Univ Tenn, Memphis, 75-77; dean, Oral Roberts Univ Sch Med, 77-78; consult in residence, Univ Okla, Tulsa Med Col, 79-; AT COL MED, UNIV OKLA HEALTH SCI CTR. *Mem:* Am Fedn Clin Res; fel Am Col Physicians; fel Am Col Chest Physicians; Am Thoracic Soc. *Res:* Pulmonary physiology and mechanics. *Mailing Add:* Univ Tex M D Anderson Cancer Ctr 1515 Holcombe Blvd Houston TX 77030

MCCALL, CHARLES EMORY, b Lenoir, NC, Jan 30, 35; m 57; c 3. INFECTIOUS DISEASES. *Educ:* Wake Forest Col, BS, 57; Bowman Gray Sch Med, MD, 61. *Prof Exp:* From intern to resident, Harvard Med Sch, 61-66, teaching asst, 66-68; from asst prof to assoc prof, 68-75, PROF MED, BOWMAN GRAY SCH MED, 75-, DIR INFECTIOUS DIS, 71-, ASSOC PHARM, 80- *Concurrent Pos:* Bowman Gray Sch Med fac award, 61; NIH fel, Thorndike Res Lab, Harvard Med Sch, 66-68; NIH res career develop award, 74. *Mem:* Fel Royal Soc Med; Infectious Dis Soc Am; Am Fedn Clin Res; Am Asn Immunologists; Am Soc Exp Biol; Sigma Xi. *Res:* Neutrophil biology and host defense. *Mailing Add:* Dept of Med Bowman Gray Sch Med Winston-Salem NC 27103

MCCALL, CHESTER HAYDEN, JR, b Vandergrift, Pa, Aug 6, 27; m 52, 72; c 3. APPLIED STATISTICS, RESEARCH METHODS. *Educ:* George Wash Univ, AB, 50, AM, 52, PhD, 57. *Prof Exp:* Asst prof statist, George Wash Univ, 52-56; res dir, Booz-Allen Appl Res, Inc, Md, 59-63, vpres & dir western opers, 63-69, managing vpres, 69-71; pres, Int Careers Inst, Inc, 71-74; dept mgr, Systs Eval Dept, Caci, Inc, 74-78; SR ASSOC, MM ASSOCS, 78-; PROF RES METHODS, PEPPERDINE UNIV, 82- *Concurrent Pos:* Consult, Hawaii Dept Health, 79-,; Santa Clara County, Dept of Health, 80- *Mem:* Am Statist Asn; Opers Res Soc Am; fel Am Soc Qual Control; Sigma Xi; Nat Educ Asn; Am Educ Res Asn. *Res:* Survey sampling; experimental design; micro-computer systems; operations research; audio-visual and audio-manual education, training, and information transfer; application of basic statistical methods to nationwide survey polling and the evaluation of transportation and health care systems. *Mailing Add:* Grad Sch Educ & Psychol Pepperdine Univ Plaza 400 Corp Pt Culver City CA 90230

MCCALL, DAVID WARREN, b Omaha, Nebr, Dec 1, 28; m 55; c 2. PHYSICAL CHEMISTRY. *Educ:* Univ Wichita, BS, 50; Univ Ill, MS, 51, PhD(chem), 53. *Prof Exp:* Head phys chem, 61-69, asst chem dir, 69-73, MEM TECH STAFF, BELL TEL LABS, INC, 53-, CHEM DIR, 73- *Concurrent Pos:* Chmn, Nat Comn Superconductivity. *Mem:* Nat Acad Eng; Am Chem Soc; fel Am Phys Soc; fel Royal Soc Chem; Am Inst Chem Engrs; fel AAAS. *Res:* Nuclear magnetic resonance; diffusion in liquids; polymer relaxation; dielectric properties; materials for communications systems. *Mailing Add:* Bell Tel Labs Inc 600 Mountain Ave Murray Hill NJ 07974

MCCALL, JAMES LODGE, b Morristown, NJ, June 3, 35; m 56; c 2. METALLURGICAL ENGINEERING. *Educ:* Ohio State Univ, BMetE, 58, MS, 61. *Prof Exp:* Assoc div chief, Metallog Div, 65-67, chief, Struct of Metals Div, 67-76, mgr mat resources & process metall, Metal Div, 76-84, SR MGR, MAT RES, DEPT METALL, 84- *Honors & Awards:* Pres Award, Int Metallog Soc. *Mem:* Fel Am Soc Metals; Am Inst Mining, Metall & Petrol Engrs; Electron Micros Soc Am; Inst Microstruct Anal Soc. *Res:* Metallographic research on metals and ceramics; process metallurgical research on minerals processing; primary ferrous and non-ferrous metals processing; recycling, energy and environmental concerns of metallurgical operations. *Mailing Add:* Metall Dept Columbus Labs Battelle Mem Inst 505 King Ave Columbus OH 43201

MCCALL, JERRY C, b Oxford, Miss, June 30, 27; m 52; c 3. ELECTRONICS ENGINEERING. *Educ:* Univ Miss, Oxford, BA & MA, 51; Univ Ill, Urbana-Champaign, MS, 56, PhD(math), 59. *Prof Exp:* Grad asst math, Univ Miss, 50-63; res assoc, Control Systs Lab, Univ Ill, Urbana, 53-57; appl sci rep, IBM Corp, 57-58; exec vpres, Midwest Comput Serv, Inc, 58-59; dep dir res & develop opers, Nat Aeronaut & Space Admin, 59-66; dir develop, Space Transp Systs & tech adv, IBM Electronics Systs Ctr, 66-72; dir, Info Br, NASA, 72-73; exec vchancellor & prof math, Univ Miss, 73-76; pres, First State Bank & Trust Co, Gulfport, Miss, 76-77; DIR, NAT DATA BUOY CTR, STENNIS SPACE CTR, 77- *Concurrent Pos:* Consult, Gen Elec Co, Bay St Louis, 73-74. *Honors & Awards:* Apollo Achievement Award, NASA. *Mem:* Am Math Soc; Am Rocket Soc; Asn Comput Mach; Am Inst Aeronaut & Astronaut; Sigma Xi; Marine Technol Soc. *Mailing Add:* Nat Data Buoy Ctr Stennis Space Ctr MS 39529

MCCALL, JOHN MICHAEL, b Dallas, Tex, Jan 11, 44; m 66; c 2. ORGANIC CHEMISTRY, MEDICINAL CHEMISTRY. *Educ:* DePauw Univ, BA, 66; Univ Wis, PhD(chem), 70. *Prof Exp:* Res chemist org chem, Monsanto Co, 65-66; NIH fel, Univ Wis, 67-70; fel chem, Harvard Univ, 70-72; RES CHEMIST, DIR CENTRAL NERVOUS SYST RES, UPJOHN CO, 72- *Mem:* Am Chem Soc. *Res:* Medicinal chemistry in the CNS cardiovascular area; heterocyclic chemistry, particularly of steroids, pyrimidines, quinolines, pyridines and pyrans. *Mailing Add:* 3822 Edenburgh Kalamazoo MI 49007-5445

MCCALL, JOHN TEMPLE, b Davenport, Iowa, May 1, 21; m 50; c 3. BIOCHEMISTRY. *Educ:* Rollins Col, BS, 48; Univ Fla, MS, 56, PhD(animal nutrit), 58. *Prof Exp:* Lab asst animal husb & nutrit, Univ Fla, 50-51, asst, 51-58, asst chemist, Agr Exp Sta, 58-60; asst prof, Iowa State Univ, 60-64; assoc prof biochem & prof lab med, Mayo Grad Sch Med, Univ Minn, 65-85, biochemist, Mayo Clin, 65-84; RETIRED. *Mem:* Am Chem Soc; Am Asn Clin Chemists; Am Inst Nutrit. *Res:* Trace mineral metabolism, especially interactions among trace mineral elements; effects of protein-mineral chelates on mineral metabolism and mechanism of mineral transport. *Mailing Add:* Mayo Clin 200 Second St Rochester MN 55905

MCCALL, KEITH BRADLEY, b Oak Park, Ill, Dec 29, 19; m 42; c 5. BIOCHEMISTRY. *Educ:* Univ Nebr, BSc, 41, MA, 43; Univ Wis, PhD(biochem), 47. *Prof Exp:* Asst biochem, Univ Wis, 43-46; instr chem, Mich State Univ, 47-48, asst prof, 48-51; chief blood derivatives unit, Div Labs, State Dept Health, Mich, 51-62; head biol prods develop, Squibb Inst Med Res, NJ, 62-65; asst chief biol prods div, Bur Labs, 65-76, CHIEF, BIOL PROD DIV, MICH DEPT PUB HEALTH, 76- *Mem:* AAAS; Am Chem Soc. *Res:* Blood plasma protein fractionation; biologic products development. *Mailing Add:* 232 Lexington Ave East Lansing MI 48823

MCCALL, MARVIN ANTHONY, b Pitts, Ga, Feb 7, 18; m 45; c 2. ORGANIC CHEMISTRY. *Educ:* Univ Ga, BS, 42, MS, 44; Univ Rochester, PhD(org chem), 51. *Prof Exp:* Instr chem, Univ Ga, 43-44; res chemist, Eastman Kodak Co, 44-51, sr res chemist, Tenn Eastman Co, 51-69, res assoc, 69-81; RETIRED. *Mem:* Am Chem Soc. *Res:* Synthetic organic, polymer and organophosphorus chemistry; organometallic compounds; synthesis of organic polymer intermediates, additives and flame retardants. *Mailing Add:* 1720 Savona Pkwy Cape Coral FL 33904

MCCALL, PETER LAW, b Evanston, Ill, Dec 26, 48; m 78. BENTHIC ECOLOGY, PALEOBIOLOGY. *Educ:* Washington Univ, AB, 70; Yale Univ, MPhil, 72, PhD(geol), 75. *Prof Exp:* ASST PROF EARTH SCI, CASE WESTERN RESERVE UNIV, 75- *Mem:* Paleontol Soc; Int Asn Great Lakes Res; Sigma Xi. *Res:* Marine ecology; freshwater benthic ecology; paleontology; invertebrate ecology and evolution. *Mailing Add:* Dept Geol Case Western Reserve Univ Cleveland OH 44106

MCCALL, RICHARD C, b Inavale, Nebr, Sept 13, 29; m 56; c 3. RADIOLOGICAL PHYSICS. *Educ:* Mass Inst Technol, BS, 52, PhD, 57. *Prof Exp:* Physicist, Hanford Atomic Prod Oper, Gen Elec Co, 56-59; NIH fel, Univ Lund, 59-61; sr physicist, Controls Radiation, Inc, Mass, 61-64; head health physics group, 64-76, RADIATION PHYSICS GROUP LEADER & RADIATION SAFETY OFFICER, STANFORD LINEAR ACCELERATOR CTR, 76- *Concurrent Pos:* Consult, Varian Assocs and others. *Mem:* Am Asn Physicists in Med; fel, Health Physics Soc. *Res:* Dosimetry; radiation shielding; electron accelerators; health physics and medical physics instrumentation. *Mailing Add:* Stanford Linear Accelerator Ctr Stanford Univ PO Box 4349 Stanford CA 94309

MCCALL, ROBERT B, ENTOMOLOGY. *Educ:* Mich State Univ, PhD, 77. *Prof Exp:* RES SCIENTIST, UPJOHN CO, 79- *Mailing Add:* Upjohn Co Kalamazoo MI 49001

MCCALL, ROBERT G, b Follansbee, WVa, Dec 20, 13; m 40; c 2. ENVIRONMENTAL HEALTH ENGINEERING. *Educ:* Univ WVa, BSCE, 35; Univ NC, MSSE, 50; Am Acad Environ Engrs Intersoc, dipl. *Prof Exp:* Sanit engr, Brooke County Health Dept, WVa, 35-38; sanit engr, State Health Dept, WVa, 38-43 & 46-48, actg dir, Sanit Eng Div, 48-50; actg dir, Sanit Eng Div, US Army, 43-46 & 50-69, sanit engr, Hq, 5th Army, Ill, 50-52, chief, Sanit Eng Div, Environ Health Lab, Army Chem Ctr, Md, 52-55, consult pub health eng, Hq US Army Europe, Ger, 55-58, chief sanit eng, Off Surg Gen, 58-62, consult, 58-68, chief, Sanit Eng Br, Med Field Serv Sch, Ft Sam Houston, Tex, 68-69; state sanit engr & dir, Environ Health Serv, WVa State Health Dept, 69-80; SPEC CONSULT, KELLY, GIDLEY, BLAIR & WOLFE, CONSULT ENGRS, 81- *Concurrent Pos:* Mem comt food serv equip, Nat Sanit Found, 58-66, swimming pool equip, 63-, plastics, 63-; mem study sect, sanit eng & occup health, NIH, 58-62; secy, Conf Fed Sanit Eng, 58-59, pres, 60; liaison rep, Comt Sanit Eng & Environ, Nat Res Coun, 58-62; alt comnr, Interstate Comn Potomac River Basin, 69-74, mem exec bd, 71-; secy-treas, Conf State Sanit Eng, 71-73, chmn, 74-; mem, effluent standards & water qual adv comt, Environ Protection Agency, 72-76, Nat Drinking Water Adv Coun, 77-80; mem, Coun Pub Health Consult, NSF, 75-; dir, Am Water Works Asn, 77-79; trustee, Am Acad Environ Engrs, 77-79, trustee, Water Qual Div, Am Water Works Asn, 80-86. *Honors & Awards:* Bedell Award, Fedn Sewage Works Asn, 50; Fuller Award, Am Water Works, 79. *Mem:* Fel Am Pub Health Asn; Nat Soc Prof Engrs; hon mem Am Water Works Asn; Fed Water Poll Control Asn. *Res:* Industrial wastes; stream pollution control; food service equipment; small water systems. *Mailing Add:* Kelly Gidley Blair & Wolfe Consult Engrs 550 Eagan St Charleston WV 25305

MCCALLA, DENNIS ROBERT, b Edmonton, Alta, July 30, 34; m 57; c 2. BIOCHEMISTRY, MUTAGENESIS. *Educ:* Univ Alta, BSc, 57; Univ Sask, MSc, 58; Calif Inst Technol, PhD(biochem, genetics), 61. *Prof Exp:* From asst prof to prof biochem, McMaster Univ, 61-89, dean fac sci, 67-75, vpres health sci, 82-89, EMER PROF BIOCHEM, MCMASTER UNIV, 89- *Concurrent Pos:* Chmn, Ont Coun Occup Health & Safety, 82-89. *Mem:* Can Soc Cell Biol; Can Asn Res Toxicol; Physiol; fel Chem Inst Can; Am Soc Biol Chem; Sigma Xi. *Res:* Biochemical genetics; mechanism of action of mutagens and carcinogens. *Mailing Add:* RR No 1 Clarksburg ON N0H 1S0 Can

MCCALLEY, ROBERT B(RUCE), JR, b Miami, Fla, Dec 17, 22; m 51. MECHANICAL ENGINEERING, MECHANICS. *Educ:* Univ SC, BS, 47; Cornell Univ, MCE, 49, PhD(struct, civil eng), 52. *Prof Exp:* Res assoc civil eng, Cornell Univ, 48-51; engr, Sr Staff, Appl Physics Lab, Johns Hopkins Univ, 51-55; struct engr, Knolls Atomic Power Lab, Gen Elec Co, 55-58; reactor containment engr, USAEC, 58-59; consult engr, Knolls Atomic Power Lab, Gen Elec Co, 60-62, mgr stress anal, 63-68, mgr struct mech suboper, Aircraft Engine Group, 68-71, mgr tech opers mach apparatus, 71-76, mgr eng support, Mach Apparatus Oper, 76-87; RETIRED. *Mem:* Am Soc Civil Engrs; fel Am Soc Mech Engrs. *Res:* Elasticity; applied mechanics; numerical analysis. *Mailing Add:* 826 Karenwald Lane Schenectady NY 12309-6414

MCCALLEY, RODERICK CANFIELD, b Portland, Ore, Aug 2, 43; m 76; c 2. MAGNETIC RESONANCE. *Educ:* Calif Inst Technol, BS, 64; Harvard Univ, PhD(phys chem), 71. *Prof Exp:* NSF fel phys chem, Stanford Univ, 71-72; asst prof chem, Dartmouth Col, 72-79; MEM STAFF, LOCKHEED PALO ALTO RES LAB, 79- *Mem:* Sigma Xi; Am Chem Soc. *Res:* Electron spin resonance; magnetic relaxation; electron structure of free radicals; leak-test interpretation and development. *Mailing Add:* Dept 93-50 Bldg 204 Lockheed Res & Develop 3251 Hanover St Palo Alto CA 94304-1187

MCCALLION, WILLIAM JAMES, b Toronto, Ont, Aug 20, 18; m 44; c 4. MATHEMATICS, ASTRONOMY. *Educ:* McMaster Univ, BA, 43, MA, 46. *Prof Exp:* Lectr math, 43-52, from asst prof to assoc prof, 53-70, asst dir exten, 52-53, dir, 55-70, dir educ servs, 62-70, dean, Sch Adult Educ, 70-78, PROF MATH, McMASTER UNIV, 70-, DEAN, ADULT EDUC. *Mem:* Am Math Soc; Math Asn Am; Royal Astron Soc Can; Can Math Cong. *Res:* Modern algebra; mathematics of business. *Mailing Add:* 84 Dalewood Crescent Hamilton ON L8S 4B7 Can

MCCALLISTER, LAWRENCE P, b Chicago, Ill, Mar 27, 43; m 73. MICROSCOPIC ANATOMY. *Educ:* Univ Ill, Champaign, BS, 66; Loyola Univ Chicago, PhD(anat), 71. *Prof Exp:* USPHS fel med & physiol, Pritzker Sch Med, Univ Chicago, 71-72; ASST PROF ANAT, MILTON S HERSHEY MED SCH, PA STATE UNIV, 72- *Concurrent Pos:* Nat Heart & Lung Inst grant, Milton S Hershey Med Sch, Pa State Univ, 72-75 & 76-79, Nat Heart & Lung Inst res career develop award, 76- *Mem:* Am Asn Anatomists; Sigma Xi. *Res:* Ultrastructure and function of muscle and nerve. *Mailing Add:* 1304 Alton Ave Pittsburgh PA 15216

MCCALLUM, CHARLES ALEXANDER, JR, b North Adams, Mass, Nov 1, 25; m 55; c 4. ORAL SURGERY. *Educ:* Tufts Univ, DMD, 51; Univ Ala, MD, 57; Am Bd Oral & Maxillofacial Surg, dipl, 57. *Hon Degrees:* DSc, Univ Ala, 75, Georgetown Univ, 82, Tufts Univ Sch Dent Med, 88. *Prof Exp:* Intern-resident oral surg, Univ Hosp & Hillman Clin, 51-54, intern med, 57-58, from instr to assoc prof dent, dept oral surg, Sch Dent, 56-59, chmn dept, 58-65 & 68-69, asst dean, 61-62, dean, 62-77, chief, dept oral surg, Sch Med, 65-69, dep vpres health affairs-dent, Med Ctr, 73-77 & vpres health affairs, 77-82, dir med ctr, 77-87, asst to chancellor med progs, 79-87, sr vpres, 82-87, PROF DENT, SCH DENT, UNIV ALA, 59-, PROF SURG, SCH MED, 65-, PRES, 87- *Concurrent Pos:* Consult, Vet Admin Hosps, Birmingham, Tuscaloosa & Tuskegee, 58-, Coun Dent Educ, Am Dent Asn, 66-70 & Viet Nam Proj, 73-74, Surgeon Gen of Army, 66-72, Nat Cheng Kung Univ, Tainan, Taiwan, 82-; vis lectr, Med Univ, SC Sch Dent, 73; vis prof, Sch Dent, Univ Baghdad, Iraq, 75-77; Percy T Phillips vis prof, Sch Dent & Oral Surg, Columbia Univ, 81. *Honors & Awards:* William J Gies Award Oral Surg, William J Gies Found Advan Dent & Am Asn Oral & Maxillofacial Surgeons, 80; Pierre Fauchard Gold Medal, Pierre Fauchard Acad, 82. *Mem:* Inst Med-Nat Acad Sci; Am Dent Asn; AMA; Sigma Xi; fel Am Col Dentists; Int Asn Oral Surgeons; Am Asn Oral & Maxillofacial Surgeons (vpres, 73-74, pres-elect, 74-75, pres, 75-76); fel Am Dent Soc Anesthesiol; AAAS; Am Asn Dent Res. *Res:* Dental infection and effects of various systemic drugs on control of hemorrhage. *Mailing Add:* 4101 Altamont Rd Birmingham AL 35213

MCCALLUM, CHARLES JOHN, JR, b Tacoma, Wash, Apr 26, 43; m 67; c 2. OPERATIONS RESEARCH. *Educ:* Mass Inst Technol, SB, 65; Stanford Univ, MS, 67, PhD(opers res), 70. *Prof Exp:* Mem tech staff opers res, 70-75, supvr opers res appl/methods groups, 75-81, HEAD OPERS RES DEPT, AT&T BELL LABS, 81- *Concurrent Pos:* Coadjutant, opers res, Rutgers Univ, 75-76. *Honors & Awards:* Full mem, Sigma Xi. *Mem:* Opers Res Soc Am (treas, 86-89, vpres, 90-91, pres, 91-92); Inst Mgt Sci, Mgt Sci Roundtable; Math Prog Soc. *Res:* Mathematical programming; network flows; the linear complementarity problem; facility location models; manufacturing. *Mailing Add:* 54 Stonehenge Dr Lincroft NJ 07738-1539

MACCALLUM, CRAWFORD JOHN, b New York, NY, May 28, 29; m 51; c 6. GAMMA RAY ASTRONOMY. *Educ:* Princeton Univ, BA, 51; Univ NMex, PhD(physics), 62. *Prof Exp:* STAFF MEM, SANDIA CORP, 57- *Concurrent Pos:* Fulbright lectr, Univ Cairo, 64-65. *Mem:* Am Phys Soc; Am Astron Soc. *Res:* Radiation physics; analytical methods electron penetration in solids; gamma ray astronomy. *Mailing Add:* Sandia Lab 9231 Albuquerque NM 87185

MACCALLUM, DONALD KENNETH, b Los Angeles, Calif, Apr 13, 39; m 62; c 2. ANATOMY, HISTOLOGY. *Educ:* Pomona Col, BA, 61; Univ Southern Calif, MS, 64, PhD(anat), 66. *Prof Exp:* Staff mem exp path, Walter Reed Army Inst Res, 66-68; from asst prof to assoc prof anat, Med Sch, 69-81, from asst prof to assoc prof oral biol, Dent Sch, 69-85, res scientist, Dent Res Inst, 69-88, PROF ANAT & CELL BIOL, MED SCH, UNIV MICH, ANN ARBOR, 81- *Concurrent Pos:* Nat Inst Dent Res fel, Case Western Reserve Univ, 68-69; asst prof lectr, Sch Med, George Washington Univ, 67-68; res scientist, Proteoglycan Sect, Lab Biochem, Nat Inst Dent Res & NIH, Bethesda, MD,. *Mem:* Am Asn Anat; Am Soc Cell Biol. *Res:* Oral mucosal ultrastructure; epithelial-connective tissue interaction; experimental cytology; corneal endothelial structure and function. *Mailing Add:* Dept of Anat & Cell Biol Univ of Mich Med Sch Ann Arbor MI 48109-0616

MCCALLUM, KEITH STUART, b Fond du Lac, Wis, June 21, 19; m 42; c 2. ANALYTICAL CHEMISTRY. *Educ:* Univ Wis, BS, 42, PhD(chem), 50. *Prof Exp:* Group leader anal chem, Redstone Arsenal Res Div, 50-59, proj leader, Rhom and Haas Co, 59-84; RETIRED. *Mem:* Am Chem Soc. *Res:* Trace analysis of air and water pollutants, and impurities from chemical and biodegradation. *Mailing Add:* 659 Pine Tree Rd Jenkintown PA 19046

MCCALLUM, KENNETH JAMES, b Scott, Sask, Apr 25, 18; m 50; c 2. PHYSICAL CHEMISTRY. *Educ:* Univ Sask, BSc, 36, MSc, 39; Columbia Univ, PhD(chem), 42. *Prof Exp:* From asst prof to prof chem, 43-85, head, dept chem, 59-70, dean, Col Grad Studies & Res, 70-85, EMER DEAN GRAD STUDIES & RES, UNIV SASK, 85- *Concurrent Pos:* Nuffield traveling fel natural sci, Cambridge Univ, 50-51. *Mem:* Fel AAAS; fel Royal Soc Can; Chem Inst Can. *Res:* Chemical effects of nuclear reactions; radiation chemistry; isotopic exchange reactions. *Mailing Add:* 1622 Park Ave Saskatoon SK S7H 2P3 Can

MCCALLUM, MALCOLM E, b Springfield, Mass, July 28, 34; m 63; c 2. GEOLOGY. *Educ:* Middlebury Col, AB, 56; Univ Tenn, MS, 58; Univ Wyo, PhD(geol), 64. *Prof Exp:* From asst prof to assoc prof geol, 62-71, PROF GEOL, COLO STATE UNIV, 71-; GEOLOGIST, US GEOL SURV, 56- *Concurrent Pos:* Res grants, Wyo Geol Surv, 59-61, Geol Soc Am, 60 & 70 & NSF, 74-76, 77-79, 79-81, 89-91, Winston Found, 80-83, SAfrican Coun Sci & Indust Res, 85; contract geologist, Can Geol Surv & Minerals Explor Cos, 87- 90. *Mem:* AAAS; fel Geol Soc Am; fel Mineral Soc Am; Geochem Soc; Soc Explor Geochem; Soc Econ Geologists; Mineral Asn Can. *Res:* Petrology, structure, and mineral resources of Precambrian crystalline rocks, northern Colorado and southern Wyoming; Precambrian-Tertiary structure and stratigraphy of disturbed belt in northern Big Belt Mountains, Montana; petrology and chemistry of kimberlite and upper mantle-lower crustal rocks and diamonds included in diatremes of northern Colorado and southern Wyoming; lamprophyric diatremes and gold deposits in British Columbia. *Mailing Add:* Dept Earth Resources Colo State Univ Ft Collins CO 80523

MCCALLUM, RODERICK EUGENE, b Denver, Colo, Aug 14, 44; m 67; c 2. MICROBIOLOGY, BACTERIOLOGY. *Educ:* Univ Kans, BA, 67, PhD(microbiol), 70. *Prof Exp:* Instr microbiol, Univ Kans, 70; res assoc, Univ Tex, 70-72, instr, 71; from asst prof to assoc prof, 72-84, PROF MICROBIOL, UNIV OKLA, 84-, ADJ ASSOC PROF ORAL PATH, 77- *Concurrent Pos:* USPHS trainee, Univ Tex, Austin, 70-72; guest prof, Univ Heidelberg, Ger, 80 & 85. *Honors & Awards:* Masua Hon Lectr, 88. *Mem:* Shock Soc; Sigma Xi; Am Soc Microbiol; Reticuloendothelial Soc; Int Endotoxin Soc; AAAS. *Res:* Metabolism in disease; endotoxin; mixed anaerobic infections; role of cytokines in disease. *Mailing Add:* Dept of Microbiol & Immunol Univ of Okla Health Sci Ctr Oklahoma City OK 73190

MCCALLY, RUSSELL LEE, b Marion, Ohio, Sept 27, 40; m 61; c 2. PHYSICS, BIOPHYSICS. *Educ:* Ohio State Univ, BSc, 64; Johns Hopkins Univ, MSc, 73, MA, 83. *Prof Exp:* Assoc physicist, 65-76, SR PHYSICIST, JOHNS HOPKINS UNIV APPL PHYSICS LAB, 76- *Concurrent Pos:* Co-prin investr, Nat Heart, Lung & Blood Inst Grant, NIH, 76-79 & Nat Eye Inst grant; William S Parsons fel, Johns Hopkins Univ, 79-80. *Mem:* Am Phys Soc; Asn Res Vision & Ophthal. *Res:* Application of light scattering methods to biological structural investigations especially corneal structure; laser-tissue interactions-especially cornea; magnetism in amorphous alloys. *Mailing Add:* Appl Physics Lab Johns Hopkins Univ Johns Hopkins Rd Laurel MD 20707

MCCALMON, ROBERT T, JR, b Bremerton, Wash, May 5, 43. IMMUNOLOGY. *Educ:* Univ Colo, BA, 67; Drake Univ, MS, 69; Univ Ariz, PhD(immunol), 73. *Prof Exp:* Asst prof, Dept Surg, Univ Colo, 75-79; PRES, IMMUNOL ASSOCS, DENVER, 79- *Concurrent Pos:* Dir surg-immunol, Denver Vet Hosp, 75-79. *Mem:* Am Asn Immunologists; Am Soc Histocompatibility; Am Col Rheumatology; Am Soc Transplant Physicians. *Mailing Add:* Immunol Assocs Denver 101 University Blvd PO Box 61159 Denver CO 80206

MCCALPIN, JAMES P, b Beaumont, TX, Oct 13, 50; m 82. NEOTECTONICS, GEOLOGIC HAZARDS. *Educ:* Univ Tex, Austin, BA, 72; Univ Colo, MS, 75; Colo Sch Mines, PhD(geol), 81. *Prof Exp:* Asst prof, 82-87, ASSOC PROF GEOL, UTAH STATE UNIV, 87- *Concurrent Pos:* Fulbright fel, Victoria Univ Wellington, NZ 87-88. *Mem:* Geol Soc Am; Asn Eng Geologists; Am Quaternary Assoc; Nat Asn Geol Teachers. *Res:* Tectonic geomorphology of active fault zones and dtermination of paleoseismic magnitude recurrence and surface-faulting style via trenching; similar studies of landslides; application of geologic hazard data to land-use planning. *Mailing Add:* Dept Geol Utah State Univ Logan UT 84322-4505

MCCAMAN, MARILYN WALES, b Oak Park, Ill, Oct 10, 28; div; c 5. NEUROSCIENCES. *Educ:* Grinnell Col, AB, 50; Wash Univ, PhD(pharmacol), 57. *Prof Exp:* Asst pharmacol, Wash Univ, 56-57; from res assoc to asst prof neurol, Sch Med & Inst Psychiat Res, Ind Univ, Indianapolis, 59-70; ASSOC RES SCIENTIST, CITY OF HOPE MED CTR, 70- *Mem:* Am Soc Mass Spectrometry. *Res:* Biochemistry correlates of long term potentiation in mammalian nerve cells; metabolism of neurotransmitters in the invertebrate central nervous system. *Mailing Add:* Dept Neurosci City of Hope Med Ctr Duarte CA 91010-0269

MCCAMAN, RICHARD EUGENE, b Barberton, Ohio, Mar 28, 30; m 53; c 5. BIOCHEMISTRY, MOLECULAR BIOLOGY. *Educ:* Wabash Col, AB, 52; Wash Univ, PhD(pharmacol), 57. *Prof Exp:* Res assoc & instr biochem & pharmacol, Inst Psychiat Res, Sch Med, Ind Univ, Indianapolis, 57-62, from asst prof to assoc prof pharmacol, 62-68, prin investr neurochem, 62-68; CHIEF NEUROPHARMACOL, CITY OF HOPE MED CTR, 68- *Mem:* Soc Neurosci; Am Soc Neurochem; Int Soc Neurochem; Am Soc Biol Chem. *Res:* Biochemistry, physiology and chemical pathology of nervous system; chemistry, pharmacology and physiology of neurotransmitters; invertebrate neurobiology; intracellular electrophysiology. *Mailing Add:* Beckman Res Inst City of Hope Div Neurosci 1450 E Duarte Rd Duarte CA 91010

MCCAMMON, DAN, b Los Angeles, Calif, Aug 6, 44; m 76; c 1. ASTRONOMY. *Educ:* Calif Inst Technol, BS, 66; Univ Wis, PhD(physics), 71. *Prof Exp:* Foreign specialist physics, Prince Songkla Univ, 74-76; assoc scientist astron, 76-80, asst prof, 80-86, ASSOC PROF PHYSICS, UNIV WIS-MADISON, 86- *Mem:* Am Astron Soc; Int Astron Union. *Res:* X-ray astronomy. *Mailing Add:* Dept Physics Univ Wis 1150 University Ave Madison WI 53706

MCCAMMON, HELEN MARY, b Winnipeg, Man, Aug 16, 33; US citizen; m 34; c 2. GENERAL ENVIRONMENTAL SCIENCES, GENERAL EARTH SCIENCES. *Educ:* Univ Man, BSc, 55; Univ Mich, MS, 56; Ind Univ, PhD(geol), 59. *Prof Exp:* Res technician stratig, Man Dept Mines & Natural Resources, 52-59; lectr geog, Univ NDak, Grand Fords, 61; from asst prof to assoc prof geol, Dept Earth Sci, Univ Pittsburgh, 63-68; vis assoc prof geol, Dept Geol, Univ Ill, Chicago, 68-70; res assoc geol, Field Mus Natural Hist, Chicago, 69-72; dir environ sci, Off Res Prog, Environ Protection Agency, Boston, 73-76; sr oceanog marine sci, Environ Res Div, 77-79, DIR ENVIRON SCI, ECOL RES DIV, DEPT ENERGY, WASHINGTON, DC, 79- *Mem:* Fel AAAS; Am Geol Inst; Ecol Soc Am; Am Soc Zoologists; Oceanog Soc; Am Soc Limnol & Oceanog. *Res:* Identify, anticipate and ameliorate impacts from energy activities in coastal oceans and terrestrial environments ranging from arctic tundra to temperate forests and desert regions; marine physiology and ecology. *Mailing Add:* 8430 Bradley Blvd Bethesda MD 20817

MCCAMMON, JAMES ANDREW, b Lafayette, Ind, Feb 8, 47; m 69. STATISTICAL MECHANICS, BIOPOLYMER THEORY. *Educ:* Pomona Col, AB, 69; Harvard Univ, AM, 70, PhD(chem physics), 76. *Prof Exp:* NSF res fel chem, Harvard Univ, 76-77, NIH res fel, 77-78; asst prof, 78-81, chmn, phys chem div, 79-84, M D ANDERSON PROF CHEM, 81-, DIR, INST MOLECULAR DESIGN, UNIV HOUSTON, 87- *Concurrent Pos:* Alfred P Sloan Found fel, 80; Res Career Develop Award, NIH, 80; auth; chmn, Theor Chem Div, Univ Houston, 84-; adj prof physiol & molecular biophys, Baylor Col Med, 86-; mem, Study Sect on Molecular & Cellular Biophys, NIH, 87- *Honors & Awards:* George H Hitchings Award, Innovative Methods in Drug Design, Burroughs Wellcome Fund, 87. *Mem:* Fel Am Phys Soc; Am Chem Soc; Biophys Soc; AAAS. *Res:* Statistical mechanics of macromolecules and liquids; theory of protein structure, dynamics and function; development and application of computer models and simulation methods for molecular systems. *Mailing Add:* Dept Chem Univ Houston Houston TX 77204-5641

MCCAMMON, LEWIS B(ROWN), JR, b Wheeling, WVa, Mar 28, 20; m 42; c 3. STRUCTURAL ENGINEERING. *Educ:* Purdue Univ, BS, 41, MS, 48, PhD(struct eng), 51. *Prof Exp:* Stud engr, Long Lines Dept, Am Telephone & Telegraph Co, 41-42; instr civil eng, Purdue Univ, 46-48, instr struct eng, 50-51; struct designer, CF Braun & Co, 51-53, sr designer, 55-59; assoc prof civil eng, Northwestern Technol Inst, 54-55; mem sr staff, Nat Eng Sci Co, 59-64, chief adminr, 64-66, vpres eng, 66-68; mgr ocean technol, W Offshore Drilling & Explor Co, 68-74; head, struct eng sect, C F Braun & Co, 74-82, eng adv, 82-84; RETIRED. *Concurrent Pos:* Reserve instr, US Mil Acad, 52-53; lectr, Univ Calif, Los Angeles, 53 & Univ Southern Calif, 57-58. *Mem:* Am Soc Civil Engrs; Am Soc Eng Educ. *Res:* Buckling of column elements; steel; stress distribution in actual bridge members; bolted structural joints. *Mailing Add:* 105 N Cordova St Alhambra CA 91801

MCCAMMON, MARY, b Eng, Aug 23, 27; m 60. MATHEMATICS. *Educ:* Univ London, BSc, 49, MSc, 50, PhD(math), 53. *Prof Exp:* Res assoc math, Mass Inst Technol, 53-54; asst prof, 54-60, ASSOC PROF MATH & COMPUT SCI, PA STATE UNIV, UNIVERSITY PARK, 60- *Mem:* Am Math Soc; Asn Comput Mach. *Res:* Numerical analysis; uses of digital computers. *Mailing Add:* Dept Math Pa State Univ 328 McAllister Bldg University Park PA 16802

MCCAMMON, RICHARD B, b Indianapolis, Ind, Dec 3, 32; m 56; c 2. GEOLOGY. *Educ:* Mass Inst Technol, BSc, 55; Univ Mich, MSc, 56; Ind Univ, PhD(geol), 59. *Prof Exp:* Rockefeller fel statist, Univ Chicago, 59-60; asst prof geol, Univ NDak, 60-61; res geologist, Gulf Res & Develop Co, 61-68; assoc prof, Univ Ill, Chicago Circle, 68-70, prof geol, 70-76; GEOLOGIST, OFF RESOURCE ANAL, US GEOL SURV, 76- *Mem:* Geol Soc Am; Am Statist Asn; Soc Econ Paleont & Mineral. *Res:* Statistical application to problems in petroleum exploration; multivariate methods in biostratigraphy and paleoecology; computer applications in geology. *Mailing Add:* GEO OMR BORA MS920 12201 Sunrise Valley Dr Reston VA 22092

MCCAMMON, ROBERT DESMOND, b Northern Ireland, July 23, 32; m 60, 78. SOLID STATE PHYSICS. *Educ:* Queen's Univ, Ireland, BSc, 53; Oxford Univ, DPhil(physics), 57. *Prof Exp:* Vis res assoc, 57-59, asst prof, 59-69, ASSOC PROF PHYSICS, PA STATE UNIV, 69- *Concurrent Pos:* Vis fel nat standards lab, Commonwealth Sci & Indust Res Orgn, Australia, 62-63. *Res:* Plastic deformation of metals at low temperatures; internal friction and dielectric behavior of high polymers at low temperatures; thermal expansion of solids; liquid helium-3 and helium-4 solutions; thermal properties of disordered solids at low temperatures. *Mailing Add:* Box 25 Aaronsburg PA 16820

MCCAMMOND, DAVID, b Belfast, UK, Aug 23, 41; m 67; c 2. MECHANICAL ENGINEERING, BIOENGINEERING. *Educ:* Queen's Univ Belfast, BSc, 65, PhD(mech eng), 68. *Prof Exp:* Design draftsman, Harland & Wolff Ltd, 56-61; asst lectr mech eng, Queen's Univ Belfast, 67-68, lectr, 68-69; from asst prof to assoc prof, 69-81, assoc dean appl sci & eng, 82-86, PROF MECH ENGR, UNIV TORONTO, 81-, ASSOC DEAN GRAD STUDIES, 87- *Res:* Fracture & fatigue of composite materials and high strength metal alloys. *Mailing Add:* Dept of Mech Eng Univ of Toronto Toronto ON M5S 1A4 Can

MCCAMY, CALVIN S, b St Joseph, Mo, Sept 22, 24; m 45; c 3. SCIENCE OF COLOR & APPEARANCE, PHOTOGRAPHIC SCIENCE. *Educ:* Univ Minn, BChE, 45, MS, 50. *Prof Exp:* Instr math, Univ Minn, 47-50; instr physics, Clemson Univ, 50-52; physicist, Nat Bur Standards, 52-57, chief image optics & photog, 57-70; VPRES RES COLOR SCI, MACBETH DIV, KOLLMORGEN CORP, 70- *Concurrent Pos:* Mem, Nat Res Coun, 67-73; chmn, Image Technol Standards Mgt Bd, Am Nat Standards Inst, 70-76; chmn color group, Optical Soc Am, 75-77; chmn image technol comt, Inter-Soc Color Coun, 78-82; adj prof, Rensselaer Polytech Inst, 79-83; pres, Kollmorgen Found, 78-; mem color comt, Illuminating Eng Soc, 82-; mem color comt, Illuminating Eng Soc, 82-; consult, Graphic Arts Tech Found, 87- *Mem:* Fel Optical Soc Am; fel Soc Photog Scientists & Engrs; fel Royal Photog Soc Gt Brit; fel Soc Motion Picture & TV Engrs; Inter-Soc Color Coun. *Res:* Methods of densitometry, spectrophotometry and goniophotometry for industrial management of the color and other attributes of appearance of materials and objects. *Mailing Add:* Macbeth Div Kollmorgen Instruments Corp Box 230 Newburgh NY 12550

MACCAMY, RICHARD C, b Spokane, Wash, Sept 26, 25; m 49; c 3. MATHEMATICS. *Educ:* Reed Col, AB, 49; Univ Calif, Berkeley, PhD(math), 55. *Prof Exp:* Res engr, Univ Calif, Berkeley, 50-54, res mathematician, 54-56; from asst prof to assoc prof, 56-64, PROF MATH, CARNEGIE-MELLON UNIV, 64-, ASSOC CHMN DEPT, 74- *Concurrent Pos:* Air Force Off Sci Res fel, 56-66; Soc Naval Archit & Marine Engrs fel, 57-63; Boeing Sci Res Lab grant, 63-64. *Mem:* Am Math Soc. *Res:* Partial differential and integral equations; fluid dynamics; elasticity; electromagnetic theory. *Mailing Add:* Dept of Math Carnegie-Mellon Univ 5000 Forbes Ave Pittsburgh PA 15213

MCCANDLESS, DAVID WAYNE, b Dayton, Ohio, Dec 16, 41. BIOLOGY CHEMISTRY. *Educ:* Univ Cincinnati, BS, 67; George Washington Univ, MS, 69, MPhil, 72, PhD, 73. *Prof Exp:* Teaching fel anat, Med Sch, George Washington Univ, 68-73; asst prof anat, Med Sch, Univ Vt, Burlington, 73-76; staff fel, LNC, Nat Inst Neurol & Commun Dis & Stroke, NIH, Bethesda, Md, 76-78; from asst prof to assoc prof, Dept Neurobiol & Anat, Med Sch, Univ Tex, Houston, 78-86; JOHN J SHEININ PROF ANAT, DEPT CELL BIOL & ANAT, CHICAGO MED SCH, 86-, INTERIM CHMN, 88- *Concurrent Pos:* Ed-in-chief, Metab Brain Dis, 85- *Mem:* Am Fedn Clin Res; Am Asn Anatomists; Soc Exp Biol & Med; Am Inst Nutrit; Fedn Am Socs Exp Biol; Int Soc Cerebral Blood Flow & Metab; Am Soc Neurochem. *Mailing Add:* Cell Biol & Anat Dept Univ Health Sci Chicago Med Sch 3333 Green Bay Rd North Chicago IL 60064

MCCANDLESS, FRANK PHILIP, b Florence, Colo, Apr 27, 32; m 55; c 2. CHEMICAL ENGINEERING. *Educ:* Univ Colo, BS, 59, MS, 63; Mont State Univ, PhD(chem eng), 66. *Prof Exp:* Process engr, Chevron Res Corp, 59-61 & Marathon Oil Co, 61-62; res engr, 66-68, from asst prof to assoc prof, 68-75, PROF CHEM ENG, MONT STATE UNIV, 75- *Mem:* Am Inst Chem Engrs; Am Chem Soc. *Res:* Kinetics; catalysis; extractive crystallization; membrane separations. *Mailing Add:* Dept Chem Eng Mont State Univ Bozeman MT 59717

MCCANDLISS, RUSSELL JOHN, b Pittsburgh, Pa, Jan 18, 48; m 73; c 3. ENZYMOLOGY, PARASITOLOGY. *Educ:* Univ Tenn, Knoxville, BS, 69; Purdue Univ, MS, 72, PhD(biochem), 78. *Prof Exp:* Fel, Roche Inst Molecular Biol, 78-80; sr res scientist, Genex Corp, 81-82; prin scientist, Genetic Res Corp, 82-83; prin res scientist, Genex Corp, 83-86; RES DIR, CHEMGEN CORP, 86- *Mem:* Am Chem Soc; Am Soc Microbiol. *Res:* Production of chemicals and proteins by microorganisms, using biochemical genetics and recombinant DNA technology; vaccine development. *Mailing Add:* 939 Pointer Ridge Dr Gaithersburg MD 20878-1132

MCCANN, DAISY S, b Hamburg, Germany, Mar 8, 27; Can citizen; m 50; c 1. CLINICAL CHEMISTRY. *Educ:* Univ Toronto, BA, 50; Wayne State Univ, MS, 56, PhD(biochem), 58. *Prof Exp:* Res assoc chem, Wayne State Univ, 60-67, adj assoc prof, 67-70; dir, res labs, Wayne County Gen Hosp, 70-81; asst prof, 70-77, ASSOC PROF BIOCHEM, UNIV MICH, 77-; DIR, RES LABS, UNIV MED AFFIL, PC, 81- *Concurrent Pos:* A J Boyle fel, Wayne State Univ, 59-60; consult, dept rheumatology, Wayne State Univ, Detroit, 60-65; consult, Jordan Clin Lab, Detroit, 66-68, assoc dir, 68-72; dir, Newman Clin Labs, Inc, 84- *Mem:* Am Chem Soc; Soc Advan Sci; Am Asn Clin Chem; Endocrine Soc; Clin Ligand Assay Soc; NY Acad Sci; Sigma Xi. *Res:* Development of clinical laboratory techniques, including radioimmunoassay and high performance liquid chromatography; clinical and physiological prostaglandin and catecholamine studies; hepatocyte cultures; catecholamine metabolism. *Mailing Add:* McCann Assoc Inc 3139 S Wayne Rd Wayne MI 48184

MCCANN, FRANCES VERONICA, b Manchester, Conn, Jan 15, 27; m 62, 70. PHYSIOLOGY, ELECTROPHYSIOLOGY. *Educ:* Univ Conn, AB, 52, PhD(physiol), 59; Univ Ill, MS, 54. *Hon Degrees:* MA, Dartmouth Col, 73. *Prof Exp:* Res assoc, Marine Biol Lab, Woods Hole, 52-58; from instr to assoc prof, 59-73, PROF PHYSIOL, DARTMOUTH MED SCH, 73-, CONSULT, 80- *Concurrent Pos:* Estab investr, Am Heart Asn, 65-70, mem coun basic sci, 72- & New Eng res rev comt, 74-80; mem physiol study sect, NIH, 73-75 & Gen Res Suppl Rev Comn, 78-; consult, Hitchcock Hosp, 75; sr staff, Norris Colton Cancer Ctr, 80- *Mem:* Soc Gen Physiol; Am Physiol Soc; Soc Neurosci; Am Heart Asn. *Res:* Electrophysiology of neuro-muscular systems; cardiac electrophysiology, insect flight mechanisms; electrophysiology of cancer cells, B lymphocytes and other cells of immune system. *Mailing Add:* Dept Physiol Dartmouth Med Sch Hanover NH 03755

MCCANN, GILBERT DONALD, JR, b Glendale, Calif, Jan 12, 12; m 36; c 2. INFORMATION SCIENCE. *Educ:* Calif Inst Technol, BS, 34, MS, 35, PhD(elec eng), 39. *Prof Exp:* Engr cent sta, Westinghouse Elec Corp, Pa, 38-41, consult transmission engr, 41-46; prof elec eng, 47-65, prof appl sci, 65-80, dir, Willis H Booth Comput Ctr, 61-71, EMER PROF APPL SCI, CALIF INST TECHNOL, 80- *Concurrent Pos:* Westinghouse lectr, Univ Pittsburgh, 40-46; mem adv comt lightning protection, US Army Ord; mem, Int Cong High Tension Transmission, 47. *Mem:* Am Soc Mech Eng; Am Soc Eng Educ; fel Inst Elec & Electronics Eng. *Res:* Electrical transmission and lightning research; large-scale computing devices; fulchronograph; Westinghouse transients analyzer; electric analog computer; bioengineering; biomedical engineering. *Mailing Add:* Engr & Applied Sci 286-80 Calif Inst Technol Pasadena CA 91125

MCCANN, JAMES ALWYN, b Boston, Mass, May 20, 34; m 57; c 3. FISHERIES ECOLOGY, BIOMETRICS. *Educ:* Univ Mass, BS, 56; Iowa State Univ, MS, 58, PhD(fishery mgt, biomet), 60. *Prof Exp:* Asst, Iowa State Univ, 56-60; marine biologist, US Bur Commercial Fisheries, Mass, 60-63, leader fisheries res, Mass Coop Fishery Unit, US Bur Sport Fisheries & Wildlife, 63-72; chief, Br Fish Ecosyst Res, US Fish & Wildlife Serv, Gainesville, Fla, 72-73, chief, Div Prop Ecol, 73-76, Div Fishery Res, 76-77, dir, Nat Fisheries Ctr-Leetown, Kearneysville, WVa, 82-85, DIR, NAT FISHERY RES LAB, US FISH & WILDLIFE SERV, GAINESVILLE, FLA, 77-82 & 87- *Concurrent Pos:* Assoc prof, Univ Mass, 63-72. *Mem:* Am Soc Ichthyologists & Herpetologists; Am Fisheries Soc; Am Inst Fishery Res Biol. *Res:* Fishery research and biometrics; aquatic ecology; exotic fishes; fishery ecosystems. *Mailing Add:* Nat Fisheries Res Ctr 7920 NW 71st St Gainesville FL 32606

MCCANN, LESTER J, b Minneapolis, Minn, Sept 16, 15; m 44; c 7. BIOLOGY, ANATOMY. *Educ:* Univ Minn, BS, 37, MS, 38; Univ Utah, PhD(vert zool), 55. *Prof Exp:* Instr biol, Boise Col, 49-55; PROF BIOL, COL ST THOMAS, 55- *Concurrent Pos:* NSF res grant, 60-61; head dept natural sci, St Mary's Jr Col, 61-69, assoc dean tech educ, 68-69. *Mem:* Creation Res Soc; Sigma Xi. *Res:* Scientific flaws of Darwinist doctrine and their societal effects; present status of our wildlife resources and how prevailing attitudes towards the predator species adversely affects many prey populations; role of predation in wildlife ecology; embryology. *Mailing Add:* 7555 Co Rd #10 Waconia MN 55387

MCCANN, MICHAEL FRANCIS, b Toronto, Ont, Jan 19, 43; m 84. INDUSTRIAL HYGIENE. *Educ:* Univ Calgary, BSc, 64; Columbia Univ, NY, PhD(chem), 72; Am Bd Indust Hygiene, certif, 79. *Prof Exp:* Sr tech writer & prod safety coordr, GAF Corp, 72-75; dir Art Hazards Resource Ctr, Found for Community Artists, 75-77; EXEC DIR, CTR SAFETY THE ARTS, 77- *Concurrent Pos:* Ed, Art Hazards News, 78-; adj fac, NY State Sch Indust & Labor Rels, Cornell Univ, 78-79; lectr, Columbia Univ Sch Public Health, 81-; instr, Univ Man, 82-83; mem, Health & Welfare, Can Ad Hoc Comt Health Hazards Arts & Crafts Mats, 81- *Mem:* Am Indust Hyg Asn; Am Bd Indust Hygiene; Am Public Health Asn; Am Chem Soc; Int Arts Med Asn. *Res:* Writing, lecturing, and consulting on occupational hazards in the arts. *Mailing Add:* Ctr Safety the Arts Five Beekman St Suite 1030 New York NY 10038

MCCANN, PETER PAUL, b Alexandria, La, Nov 21, 43; m 71. CELL BIOLOGY, CELL PHYSIOLOGY. *Educ:* Columbia Col, Columbia Univ, AB, 65; Syracuse Univ, PhD(molecular biol), 70. *Prof Exp:* Fel molecular biol, Lab Molecular Biol, Nat Inst Arthritis, Metabolism & Digestive Dis, NIH, 70-73; sr scientist biochem, Ctr Rech Merrell Int, Strasbourg, France, 73-79; dir sci admin, Merrell Dow Res Inst, Cincinnati, Ohio, 79-90, dir sci & acad liasion, 79-90, SR DIR, MARION MERRELL DOW INC, INDIANAPOLIS, 90- *Concurrent Pos:* Prof, Dept Anat & Cell Biol, Col Med, Univ Cincinnati, 81- *Mem:* Am Soc Cell Biol; Am Soc Tropical Med & Hygiene; Am Soc Biochem & Molecular Biol; Biochem Soc; fel Royal Soc Trop Med & Hyg. *Res:* Role of the induction of ornithine decarboxylase and putrescine biosynthesis in cell replication and cell differentiation. *Mailing Add:* Marion Merrell Dow Inc PO Box 68470 Indianapolis IN 46268-0470

MCCANN, ROGER C, b Lakewood, Ohio, Aug 30, 42. MATHEMATICS. *Educ:* Pomona Col, BA, 64; Case Inst Technol, MS, 66; Case Western Reserve Univ, PhD(math), 68. *Prof Exp:* Asst prof math, Calif State Col, Los Angeles, 68-69; vis asst prof, 69-70, asst prof math, Case Western Reserve Univ, 70-77; from asst prof to prof math, Miss State Univ, 77-85; SR RES MATHAMATICIAN, MOBIL RES & DEVELOP, 85- *Mem:* Am Math Soc. *Res:* Applied math for petroleum industry. *Mailing Add:* Mobil Res & Develop 13977 Midway Rd Dallas TX 75244

MCCANN, SAMUEL MCDONALD, b Houston, Tex, Sept 8, 25; m 50; c 3. NEUROENDOCRINOLOGY, PHYSIOLOGY. *Educ:* Rice Inst, 42-44; Univ Pa, MD, 48. *Prof Exp:* Intern internal med, Mass Gen Hosp, 48-49, asst resident, 49-50; instr physiol, Sch Med, Univ Pa, 48 & 52-53, assoc, 53-54, from asst prof to prof, 54-64; prof & chmn, Dept Physiol, 65-85, PROF PHYSIOL & DIR, NEUROPEPTIDE DIV, UNIV TEX SOUTHWESTERN MED CTR, DALLAS, 85- *Concurrent Pos:* Rockefeller Found traveling grant, Royal Vet Col, Sweden, 54-55; sr asst, Royal Vet Col, Sweden, 54-55; consult, Schering Corp, NJ, 58; mem gen med B study sect, NIH, 65-67, mem endocrinol study sect, 67-69, mem corpus luteum panel, 71-74; mem, Pop Res Comt, Nat Inst Child Health & Human Develop, 74-76, Reproductive Biol Study Sect, 77-82, chmn, 80-82; mem sci adv bd, Wis Regional Primate Ctr, 74- *Honors & Awards:* Oppenheimer Award, Endocrine Soc, 66; Fred Conrad Koch Medal, Endocrine Soc, 79; Harris Mem lectr, Melbourne, 80; First James Law Honor Lectr, Cornell Univ, 82; Hartman Award, Soc Study Reprod, 86. *Mem:* Nat Acad Sci; NY Acad Sci; Am Physiol Soc; Am Soc Clin Invest; Int Brain Res Orgn; Int Soc Res Reproduction; Int Soc Neuroendocrinol (vpres, 80-84, pres, 84-88); fel AAAS. *Res:* Experimental neurogenic hypertension; hypothalamic regulation of thirst and pituitary function. *Mailing Add:* Dept Physiol Univ Tex Southwestern Med Ctr Dallas TX 75235

MCCANN, WILLIAM PETER, b Baltimore, Md, Dec 12, 24; m 54; c 2. PHARMACOLOGY, INTERNAL MEDICINE. *Educ:* Princeton Univ, AB, 49; Cornell Univ, MD, 49. *Prof Exp:* Intern internal med, Barnes Hosp, St Louis, Mo, 49-50; instr clin pharmacol & med, Johns Hopkins Univ, 57-58; asst prof pharmacol, Sch Med, Univ Colo, 58-63, asst prof med, 58-63; assoc prof pharmacol, 63-67, asst prof med, 63-71, PROF PHARMACOL, UNIV ALA, BIRMINGHAM, 67-, ASSOC PROF MED, 71- *Concurrent Pos:* Res fel pharmacol, Sch Med, Wash Univ, 52-54; res fel physiol chem, Sch Med, Johns Hopkins Univ, 54-55, res fel clin pharmacol & med, 55-57. *Mem:* AAAS; Am Soc Pharmacol & Exp Therapeut; Am Soc Nephrology; Am Soc Clin Pharmacol & Therapeut. *Res:* Human pharmacology; pharmacokinetics. *Mailing Add:* 3875 S Cove Dr Mountain Brook AL 35213

MCCANN-COLLIER, MARJORIE DOLORES, b Vick, La, Aug 27, 37; m; c 1. ELECTRON MICROSCOPY. *Educ:* La State Univ, BS, 58; City Univ, NY, PhD(biol), 75. *Prof Exp:* Assoc prof, 76-88, PROF BIOL, ST PETERS COL, 76- *Concurrent Pos:* Corp mem, Marine Biol Lab. *Mem:* Sigma Xi; Soc Develop Biol; Am Soc Zoologists; Am Soc Cell Biol. *Res:* Oogenesis and early development of invertebrates through cytoplasmic determinants in the embryos of Ilyanassa obsoleta the marine mud snail. *Mailing Add:* 778 E 34th St Brooklyn NY 11210

MACCANNELL, KEITH LEONARD, b Transcona, Man, Jan 8, 34; m 72; c 3. CLINICAL PHARMACOLOGY, GASTROENTEROLOGY. *Educ:* Univ Man, BSc & MD, 58, PhD(pharmacol), 63; FRCP(C), 65; FACP, 78. *Prof Exp:* Vis asst prof pharmacol & internal med, Sch Med, Emory Univ, 64-66; assoc prof pharmacol & asst prof internal med, Univ BC, 67-69; chmn dept pharmacol & therapeut, 69-74, assoc dean res, 72-75, PROF PHARMACOL, FAC MED, UNIV CALGARY, 69-, PROF INTERNAL MED, 74- *Concurrent Pos:* Can Found for Advan Therapeut fel, Emory Univ, 64-66; Markle scholar & Med Res Coun Can scholar, 66-71. *Mem:* Pharmacol Soc Can; Am Soc Pharmacol & Exp Therapeut; Can Med Asn. *Res:* Cardiovascular and gastrointestinal physiology-pharmacology. *Mailing Add:* Dept Pharmacol & Internal Med Univ Calgary Fac Med 3330 Hosp Dr NW Calgary AB T2N 4N1 Can

MACCANON, DONALD MOORE, b Norwood, Iowa, June 17, 24; m 46; c 2. CARDIOVASCULAR PHYSIOLOGY, PHARMACOLOGY. *Educ:* Drake Univ, BA, 48; Univ Iowa, MS, 51, PhD(physiol), 53. *Prof Exp:* Asst, Drake Univ, 42-43; asst physiol, Univ Iowa, 50-51, res assoc, 53-54; asst prof physiol & pharmacol, Sch Med Sci, Univ SDak, 54-60; asst prof physiol & pharmacol & chief exp cardiol, Chicago Med Sch, 60-61, from assoc prof to prof cardiovasc res, 61-71, assoc dir, 62-71, from assoc prof to prof physiol, 64-71; health scientist adminr, Training Grants & Awards Br, Nat Heart & Lung Inst, 71-72, actg head postdoctoral sect, Fel Br, Nat Inst Gen Med Sci, 72-73; health scientist adminr, Cardiac Functions Br, Nat Heart, Lung & Blood Inst, 73-74, chief, Manpower Br, 74-80, chief, Res Training & Develop Br, Div Heart & Vascular Dis, 80-89; RETIRED. *Concurrent Pos:* Life ins res fel physiol, Univ Iowa, 51-54; NIH career prog awardee, 61-71; consult, Physiol Fels Rev Panel, USPHS, 65-68, Anesthesiol Training Comt, NIH, 68-70, Chicago Heart Res Comt, 69-71 & Career Opportunities Physiol Comt, 79-82, 85-88. *Honors & Awards:* Morris L Parker Award Meritorious Res, 64; Res Award, Interstate Postgrad Med Asn, 67. *Mem:* Fel AAAS; fel Am Col Cardiol; Am Physiol Soc; Am Heart Asn. *Res:* Cardiac function; pulmonary circulation; hemodynamics; cardiac vibrations. *Mailing Add:* 401 Nina Pl Rockville MD 20852-4110

MCCANTS, CHARLES BERNARD, b Andrews, SC, Sept 14, 24; m 47; c 2. SOIL SCIENCE. *Educ:* NC State Col, BS, 49, MS, 50; Iowa State Univ, PhD(soils), 55. *Prof Exp:* Instr soils, NC State Col, 50-52; assoc agronomist, Clemson Col, 55-56; from asst prof to assoc prof, 56-63, prof, 63-71, PROF & HEAD DEPT SOIL SCI, NC STATE UNIV, 71- *Mem:* Am Soc Agron; Soil Sci Soc Am. *Res:* Soil fertility; ion absorption; plant nutrition; tobacco production. *Mailing Add:* Dept Soil Sci NC State Univ Raleigh NC 27650

MCCARDELL, W(ILLIAM) M(ARKHAM), b Houston, Tex, Nov 1, 23; m 49; c 2. CHEMICAL ENGINEERING. *Educ:* Rice Univ, BS, 48; Calif Inst Technol, MS, 49. *Prof Exp:* Res engr, Humble Oil & Refining Co, 49-59, chief res engr, Prod Res Div, 59-61, mgr econ & planning, 61-63, regional mkt mgr, 63-68, mkt vpres, 68-70; mkt vpres, Standard Oil Co NJ, 70-77, vpres mining & synthetic fuels, Exxon Corp, 77-80, PRES, EXXON MINERALS CO, 80- *Concurrent Pos:* Mem adv mgt prog, Harvard Univ, 66; exec vpres, Esso Eastern Inc, 75-77. *Mem:* Am Meteorol Soc; Am Inst Mining, Metall & Petrol Engrs; Am Inst Chem Engrs. *Res:* Management; petroleum and chemical engineering; physical chemistry. *Mailing Add:* 246 Dolphin CV Quay Stanford CT 06902

MCCARL, HENRY NEWTON, b Baltimore, Md, Jan 24, 41; m 63, 87; c 4. MINERAL ECONOMICS & RESOURCE MANAGEMENT, ENERGY CONSERVATION. *Educ:* Mass Inst Technol, SB, 62; Pa State Univ, MS, 64, PhD(mineral econ), 69. *Prof Exp:* Mineral specialist, US Bur Mines, 65-66; mkt res analyst, Vulcan Mat Co, 66-68, sr analyst, 68-69; from asst prof to assoc prof econ & geol, 69-90, DIR, CTR ECON EDUC, UNIV ALA, BIRMINGHAM, 88-, PROF ECON & GEOL, 91- *Concurrent Pos:* Fulbright-Hays sr lectr energy, environ & resource econ, Acad Econ Studies, Bucharest, Romania, 77-78; mem bd dirs, Soc Mining Engrs, 79-81; vis fel, Grad Sch Arts & Sci, Dept Econ, Harvard Univ, 86-87. *Mem:* Soc Mining Engrs; Am Inst Prof Geologists; Nat Asn Econ Educators; Int Asn Commodity Sci & Technol. *Res:* Mineral deposit valuation; mineral economics; mineral resource management; energy resources, conservation and environmental economic analysis; author of numerous publications. *Mailing Add:* 1828 Misson Rd Birmingham AL 35216-2229

MCCARL, RICHARD LAWRENCE, b Grove City, Pa, July 6, 27; m 51; c 2. BIOCHEMISTRY. *Educ:* Grove City Col, BS, 50; Pa State Univ, MS, 58, PhD(biochem), 61; UCLA, Post Doc, 63. *Prof Exp:* Teacher high sch, Pa, 50-55; asst biochem, 55-57, instr, 57-61, from asst prof to assoc prof, 61-74, PROF BIOCHEM, PA STATE UNIV, 74-, ASSOC DEAN, GRAD SCH, 82- & DIR INTERCOL RES PROGS, 85- *Concurrent Pos:* NIH grants, 66-78. 70-73. *Mem:* Am Chem Soc; Am Soc Biol Chem; Tissue Cult Asn; AAAS. *Res:* Lipid biosynthesis; animal tissue culture; physiology of cultured heart cells; lipid, RNA, DNA and protein patterns during differentiation and aging. *Mailing Add:* 1110 Westerly Pkwy State College PA 16801

MCCARLEY, ROBERT EUGENE, b Denison, Tex, Aug 17, 31; m 52; c 4. INORGANIC CHEMISTRY. *Educ:* Univ Tex, BS, 53, PhD(chem), 56. *Prof Exp:* From instr to assoc prof chem, Univ, 56-70, res assoc, Ames Lab, 56-57, from assoc chemist to chemist, 57-74, chmn dept, 73-77, PROF CHEM, IOWA STATE UNIV, 70-, SR CHEMIST, AMES LAB, 70- *Mem:* Am Chem Soc. *Res:* Chemistry of transition elements; metal halide, oxide, sulfide and nitride compounds; metal cluster compounds; metal-metal bonding; synthetic solid state chemistry. *Mailing Add:* Dept Chem Iowa State Univ Ames IA 50011

MCCARLEY, ROBERT WILLIAM, b Mayfield, Ky, Aug 17, 37; m 68; c 2. NEUROPHYSIOLOGY. *Educ:* Harvard Col, AB, 59; Harvard Med Sch, MD, 64. *Prof Exp:* Intern, Peter Bent Brigham Hosp, Boston, 64-65; resident psychiat, Mass Ment Health Ctr, 65-68, from instr to assoc prof, 68-84, PROF PSYCHIAT, HARVARD MED SCH, 84-, DIR LAB NEUROSCI, 85-, CO-DIR CLIN RES TRAINING PROG, 80- *Concurrent Pos:* NIH spec fel, 68-71; Nat Ctr Sci Res fel, Univ Paris, 75; investr res grants, Nat Inst Ment Health, 72-, Milton Fund, Harvard Univ, 75-80, NSF, 79 & NIMH res scientist develop award, 80-85, Vet Admin, 85-; assoc chief psychiat, Brockton Vet Admin Med Ctr, 85- *Mem:* Sleep Res Soc (exec secy); AAAS; Neurosci Soc; Am Psychiat Asn; Asn Prof Sleep Soc (vchmn). *Res:* Physiology of sleep; mathematical modeling and computer processing of physiological data; electroencephalogram and evoked potential studies in schizophrenia. *Mailing Add:* Harvard Med Sch Brockton Vet Admin Med Ctr Brockton MA 02401

MCCARLEY, WARDLOW HOWARD, b Pauls Valley, Okla, May 2, 26; m 50; c 3. VERTEBRATE ZOOLOGY. *Educ:* Austin Col, AB, 48; Univ Tex, MA, 50, PhD(ecol), 53. *Prof Exp:* Instr zool, Stephen F Austin State Col, 50-52, from asst prof to assoc prof, 53-59; prof biol & head dept, Southeastern State Col, 59-61; prof biol, 61-91, chmn dept, 63-71, EMER PROF BIOL, AUSTIN COL, 91- *Concurrent Pos:* AEC res grant, 54-58; vis prof biol sta, Univ Okla, 58-; NSF res grants, 59-62 & 64-71. *Mem:* AAAS; Am Soc Mammal; Ecol Soc Am. *Res:* Speciation; behavioral ecology. *Mailing Add:* Dept Biol Austin Col Sherman TX 75090

MCCARN, DAVIS BARTON, b Chicago, Ill, Oct 15, 28; m 73; c 5. INFORMATION RETRIEVAL, LIBRARY COMPUTER SYSTEMS. *Educ:* Haverford Col, Pa, BA, 51. *Prof Exp:* Mathematician, US Air Force Intel Ctr, 51-62; chief, Plans & Mgt Div, Automatic Data Processing Systs Ctr, Defense Intel Agency, 62-67; dep dir, Lister Hill Ctr, Nat Libr Med, 67-72, assoc dir, comput & commun systs, 72-77 & sci commun planning, 77-78; pres, Online Info Int, Inc, 78-80; dir, computerized biblio serv, H W Wilson Co, 80-82; PRES, ONLINE INFO INT, INC, 82- *Concurrent Pos:* Adj prof, Col Libr & Info Sci, Univ Md, 75-77. *Mem:* Fel AAAS; Med Libr Asn; Am Soc Info Sci. *Res:* Development of scientific information systems; conception and development of MEDLINE, the first nationwide online service; direction of the computerization of Readers' Guide and related indexes; development of IBM PC front end assistance systems for WILSONLINE and MEDLINE. *Mailing Add:* 6455 Windermere Circle Rockville MD 20852

MCCARROLL, BRUCE, b 1933. PHYSICAL CHEMISTRY. *Educ:* Univ Calif, Los Angeles, BS, 55; Univ Chicago, PhD(phys chem), 58. *Prof Exp:* Asst, AEC, Univ Chicago, 55-56, res assoc, Dept Chem & Enrico Fermi Inst Nuclear Studies, 58-60; phys chemist, Metall & Ceramics Lab, Gen Elec Res & Develop Ctr, NY, 60-71, sr develop engr, 71-74, MGR MAT ANALYSIS & TEST, GEN ELEC CO, 74-, MGR, COMPUT AIDED DESIGN & COMPUT AIDED MANUFACTURING PROJ, 80- *Mem:* Am Phys Soc; Am Chem Soc. *Res:* Advanced development; electro-mechanical systems; food preparation systems; heat transfer; technological forecasting; materials testing; quality control; computer aided manufacturing. *Mailing Add:* 44 Flower House Dr Fairfield CT 06430

MCCARROLL, WILLIAM HENRY, b Brooklyn, NY, Mar 19, 30; m 58; c 4. INORGANIC CHEMISTRY. *Educ:* Univ Conn, BA, 53, MS, 55, PhD(chem), 57. *Prof Exp:* Asst chem, Univ Conn, 54-56; mem tech staff, RCA Labs, 56-67; from asst prof to assoc prof, 67-78, chmn dept, 72-80, PROF CHEM, RIDER COL, 79- *Concurrent Pos:* Consult, RCA Labs, NJ, 56-67. *Mem:* Am Chem Soc; Sigma Xi. *Res:* Inorganic solid state; crystal structures; photoelectronic materials; transition metal oxides in low valence states; oxides with metal atom clusters; research and development of photoelectronic materials. *Mailing Add:* Dept Chem Rider Col Lawrenceville NJ 08648

MCCARRON, MARGARET MARY, b Chicago, Ill. INTERNAL MEDICINE, PHARMACY. *Educ:* St Xavier Col, Ill, BA, 50; Loyola Univ Chicago, MD, 54. *Prof Exp:* Asst prof, 60-68, assoc prof pharm, 68-76, ASSOC PROF MED, UNIV SOUTHERN CALIF, 60-; ASST MED DIR, LOS ANGELES COUNTY-UNIV SOUTHERN CALIF MED CTR, 62- *Concurrent Pos:* NIH fel diabetes, Univ Southern Calif, 58-59; consult, Vet Admin & Indian Health Serv; consult task force on prescription drugs, Dept Health, Educ & Welfare, consult rev comt task force on drugs; consult comt rev, US Pharmacopeia; consult task force on role of pharmacist, Nat Ctr Health Serv Res & Develop. *Honors & Awards:* Dart Award, Univ Southern Calif, 70. *Mem:* AMA; Am Col Physicians; Am Soc Clin Pharmacol & Chemother; Am Asn Cols Pharm; Am Pharmaceut Asn. *Mailing Add:* 1200 N State St Los Angeles CA 90033

MCCARRON, RICHARD M, b Ludlow, Mass. NEUROIMMUNOLOGY, NEUROPHARMACOLOGY. *Educ:* Univ Mass, BS, 72; Univ Md, MS, 75, PhD(immunol & microbiol), 79. *Prof Exp:* Sr staff fel immunol, Neuroimmunol Br, 83-87, SPEC EXPERT, STROKE BR, NAT INST NEUROL DIS & STROKE, NIH, 87- *Mem:* Am Asn Immunologists; AAAS; Am Soc Microbiol. *Res:* Cerebral vascular endothelial cells in normal and pathological conditions and role in blood-brain barrier function. *Mailing Add:* Stroke Br Nat Inst Neurol Dis & Stroke NIH 36 4D-04 9000 Rockville Pike Bethesda MD 20892

MCCART, BRUCE RONALD, b Omaha, Nebr, Dec 21, 38; m 61; c 4. PHYSICS. *Educ:* Carleton Col, BA, 60; Iowa State Univ, PhD(physics), 65. *Prof Exp:* From asst prof to assoc prof, 65-79, assoc dean & dir instnl res, 70-76, PROF PHYSICS, AUGUSTANA COL, ILL, 79- *Mem:* Am Phys Soc; Sigma Xi. *Res:* Nuclear magnetic resonance in solids; x-ray studies of liquid crystals; optimal structural design. *Mailing Add:* Dept of Physics Augustana Col Rock Island IL 61201

MCCARTAN, LUCY, b Miami, Fla, Oct 4, 42. INTERPRETATION OF WEATHERING PROFILES. *Educ:* Occidental Col, BA, 65; Lehigh Univ, MS, 67, PhD(geol), 72. *Prof Exp:* GEOLOGIST STRATIG, US GEOL SURV, 73. *Mem:* Soc Econ Mineral & Paleontologists; Geol Soc Am; Am Inst Prof Geologists; Asn Women Geol Scientists. *Res:* Petrology mapping and environmental interpretation of southeastern United States coastal plain sediments (Cretaceous to Holocene); mapping and structural interpretation of Triassic and Permian sedimentary rocks in New Zealand and Devonian sedimentary rocks in Antarctica. *Mailing Add:* US Geol Surv 926 12201 Sunrise Valley Dr Reston VA 22092

MCCARTER, JAMES THOMAS, b Greenville, SC, Sept 24, 32; m 59; c 3. AIR CONDITIONING SYSTEM DESIGN, PROCESS PIPING. *Educ:* Clemson Univ, BS, 54. *Prof Exp:* Mech engr, Western Elec, 54-55 & J E Sirrine Co, 57-58; resident mech engr, US Army, 58-59; proj mgr mech eng, Davis Mech Contractors, 59-64; exec vpres, Piedmont Engrs Archit & Planners, 64-77, Universal Ser SC, 77-83; vpres, Ballenger Group, 83-86; PRES MECH ENG, BALMONT CORP, 86- *Concurrent Pos:* Nat dir, Nat Soc Prof Engrs, 76- *Mem:* Nat Soc Prof Engrs; Am Soc Mech Engrs; Am Soc Heating, Refrig & Air Conditioning Engrs. *Res:* Air conditioning equipment that resulted in issue of two patents; packaging of plastic film that resulted in the issue of a patent. *Mailing Add:* 228 McSwain Dr Greenville SC 29615

MCCARTER, JOHN ALEXANDER, b Eng, Jan 25, 18; nat Can; m 41; c 4. VIROLOGY. *Educ:* Univ BC, BA, 39, MA, 41; Univ Toronto, PhD(biochem), 45. *Prof Exp:* Asst res officer, Nat Res Coun Can, 45-48; from assoc prof to prof biochem, Dalhousie Univ, 48-65; prof biochem & dir cancer res, Univ Western Ont, 65-80; prof biochem & microbiol, Univ Victoria, 80-90; mem fac, Dept Biochem, Med Sch, Univ Western Ont; RETIRED. *Concurrent Pos:* Brit Empire Cancer Campaign fel, 59-60. *Mem:* Can Biochem Soc; Royal Soc Can. *Res:* Carcinogenesis. *Mailing Add:* 3171 Henderson Rd Victoria BC V8P 5A3 Can

MCCARTER, ROGER JOHN MOORE, b Bloemfontein, SAfrica, Feb 18, 41; US citizen; m 70; c 1. PHYSIOLOGY, BIOPHYSICS. *Educ:* Univ Witwatersrand, BSc Hons, 64, MSc, 65; Med Col Va, PhD(physiol), 68. *Prof Exp:* Lectr physiol & physics, Univ Witwatersrand, 70-72; asst prof, 72-76, ASSOC PROF PHYSIOL, UNIV TEX HEALTH SCI CTR, 77- *Concurrent Pos:* Can Med Res fel, Univ Western Ont, 68-69; co-investr NIH grants aging & nutrit, Univ Tex Med Sch, San Antonio, 74-78, prin investr aging of skeletal muscle, 79- *Mem:* Geront Soc; Am Physiol Soc; SAfrican Inst Physics. *Res:* Aging process in skeletal muscle; biophysics of muscular contraction. *Mailing Add:* Dept Physiol Health Sci Ctr Univ Tex 7703 Floyd Curl Dr San Antonio TX 78284

MCCARTER, STATES MARION, b Clover, SC, Sept 30, 37; m 63; c 2. PLANT PATHOLOGY, PHYTOBACTERIOLOGY. *Educ:* Clemson Univ, BS, 59, MS, 61, PhD(plant path), 65. *Prof Exp:* Exten plant pathologist & nematologist, Auburn Univ, 65-66; res plant pathologist, Ga Coastal Plain Exp Sta, Agr Res Serv, USDA, 66-68; from asst prof to assoc prof, 68-78, PROF PLANT PATH, UNIV GA, 78- *Mem:* Am Phytopath Soc. *Res:* Bacterial diseases of plants. *Mailing Add:* Dept Plant Path Col Agr Univ Ga Athens GA 30602

MCCARTHY, CHARLES ALAN, b Rochester, NY, Aug 7, 36; div; c 3. MATHEMATICAL ANALYSIS. *Educ:* Univ Rochester, BA, 56; Yale Univ, PhD(math), 59. *Prof Exp:* CLE Moore instr math, Mass Inst Technol, 59-61; from asst prof to assoc prof, 61-67, PROF MATH, UNIV MINN, MINNEAPOLIS, 67- *Concurrent Pos:* Alfred P Sloan Res Found fel, 62-64. *Mem:* Am Math Soc; Math Asn Am. *Res:* Functional analysis, especially the theory of operators. *Mailing Add:* Sch Math Univ Minn Minneapolis MN 55455

MCCARTHY, CHARLES R, b St Paul, Minn, May 26, 26; m 71. MEDICAL RESEARCH. *Educ:* Col St Thomas, BA, 47; Univ Toronto, MA, 57, PhD, 61. *Prof Exp:* Assoc prof philos & polit sci, St Paul Col, 61-71; adj assoc prof, Dept Polit, Cath Univ Am, 65-71; prog analyst, Div Legis Analysis, 71-75, chief, Legis Develop Br, 75-78, DIR, OFF PROTECTION RES RISKS, NIH, 78- *Concurrent Pos:* Mem, Privacy Act Coord Comt, NIH, 75, Pub Health Serv Task Force AIDS, 85-; chmn, Pub Health Steering Comt, Protection Human Subj, 80-84, Fedn Coord Coun Sci, Eng & Technol, 82, Interagency Coord Comt, 86-; ed, J Clin Res & Drug Develop, 88- *Mem:* Acad Polit Sci; Am Asn Lab Animal Sci; Am Polit Sci Asn; Am Cath Philos Asn; Found Advan Educ Sci; Soc Health & Human Values. *Res:* Genetics and human reproduction. *Mailing Add:* Off Dir-Off Protection Res Risks Bldg 31 Rm SB59 NIH 9000 Rockville Pike Bethesda MD 20892

MCCARTHY, CHARLOTTE MARIE, b Watford City, NDak, Sept 7, 37. MEDICAL MICROBIOLOGY, MYCOBACTERIOLOGY. *Educ:* Idaho State Univ, BS, 58; Ore State Univ, MS, 61; Univ Wash, PhD(microbiol), 67. *Prof Exp:* Res microbiologist, Nat Jewish Hosp & Res Ctr, Denver, 68-71; from asst prof to assoc prof, 72-83, PROF BIOL, NMEX STATE UNIV, 83- *Concurrent Pos:* NIH res grants, 70-76, 78-84 & 84-87. *Mem:* Am Soc Microbiol; AAAS; Sigma Xi. *Res:* Physiology of pathogenic bacteria, particularly mycobacteria which are drug-resistant. *Mailing Add:* Dept Biol Dept 3AF NMex State Univ Box 30001 Las Cruces NM 88003-0001

MCCARTHY, DANNY W, b New Orleans, La, Sept 6, 50; m 73; c 2. ON-LINE COMPUTER CONTROL. *Educ:* Tulane Univ, BS, 72, ME, 75, PhD(chem eng), 76. *Prof Exp:* Prof chem eng, Tulane Univ, 75-84; SPECIALIST, ENTERGY CORP, 84- *Mem:* Am Inst Chem Engrs; Inst Elec & Electronics Engrs; Soc Petrol Engrs. *Res:* On-line computer control; advanced optimization. *Mailing Add:* 87 McKinley Dr Kenner LA 70065

MCCARTHY, DENNIS DEAN, b Oil City, Pa, Sept 22, 42; m 66; c 2. ASTRONOMY. *Educ:* Case Inst Technol, BS, 64; Univ Va, MA, 70, PhD(astron), 72. *Prof Exp:* ASTRONR, US NAVAL OBSERV, WASHINGTON, DC, 65- *Concurrent Pos:* Consult, Cagigal Observ, 69; prin investr, Crustal Dynamics Prog, NASA, 81-; vpres, Comm 31, Int Astron Union, 82-85, pres, 85-88; secy, sect 5, Int Asn Geodesy; mem proj merit working group, Int Astron Union/Int Asn Geodesy. *Honors & Awards:* Simon Newcomb Res Award. *Mem:* Am Astron Soc; Am Geophys Union; Int Astron Union; Royal Astron Soc. *Res:* Astronomical research into rotational speed of the Earth; variation of astronomical latitude; precise star positions and proper motions; fundamental astronomical constants. *Mailing Add:* US Naval Observ 34th & Massachusetts Aves NW Washington DC 20392

MCCARTHY, DENNIS JOSEPH, b Syracuse, NY, Oct 10, 53; m 80. BIOLOGY. *Educ:* State Univ NY at Oswego, BA, 75, Syracuse Univ, PhD(biol), 81. *Prof Exp:* RES BIOCHEMIST, SOUTHERN RES INST, 81- *Mem:* Am Chem Soc; Am Asn Cancer Res. *Res:* Isolation and identification of drug metabolites including synthesis of proposed metabolites; identification of organic compounds using mass spectroscopy, infrared spectroscopy, and nuclear magnetic resonance spectroscopy; high-pressure liquid chromatography and gas chromatography; investigation into the mechanism of retinoid toxicity. *Mailing Add:* 2308 Savoy St Birmingham AL 35226-1528

MCCARTHY, DONALD JOHN, b New York, NY, Dec 22, 38; m 63; c 4. ALGEBRA. *Educ:* Manhattan Col, BS, 61; NY Univ, MS, 63, PhD(math), 65. *Prof Exp:* Asst prof math, Fordham Univ, 65-70; asst prof, 70-72, ASSOC PROF MATH, ST JOHN'S UNIV, NY, 72- *Mem:* Am Math Soc; Soc Indust & Appl Math; NY Acad Sci; Parapsychol Asn; Math Asn Am. *Res:* Group theory; general algebra; parapsychology; combinatorial mathematics. *Mailing Add:* Dept Math & Comput Sci St John's Univ Grand Cent & Utopia Pkwys Jamaica NY 11439

MCCARTHY, DOUGLAS ROBERT, b New York, NY, July 10, 43; m 64; c 1. APPLIED & PHYSICAL MATHEMATICS, COMPUTER SCIENCES. *Educ:* Rensselaer Polytech Inst, BS, 64; Rutgers Univ, MS, 66, PhD(math), 70. *Prof Exp:* Asst prof math, Purdue Univ, Ft Wayne, 69-75, assoc prof math, 75-80, vis assoc prof, West Lafayette, 80-83, head, dept math sci, Calumet, 83-84; PRIN ENGR, BOEING COM AIRPLANE GROUP, 84- *Concurrent Pos:* Res analyst, Boeing Com Airplane Co, 78-79. *Mem:* Math Asn Am; Soc Indust & Appl Math; Am Inst Aeronaut & Astronaut. *Res:* Computational fluid dynamics; numerical optimization; numerical solution of partial differential equations. *Mailing Add:* Boeing Com Airplane Group PO Box 3707 MS 6M-98 Seattle WA 98124-2207

MCCARTHY, EUGENE GREGORY, b Boston, Mass, Nov 8, 34; m 58; c 3. PUBLIC HEALTH. *Educ:* Boston Col, AB, 56; Yale Univ, MD, 60; Johns Hopkins Univ, MPH, 62. *Hon Degrees:* Dr, Univ Asuncion, Paraguay, 64. *Prof Exp:* Asst prof pub health, Med Sch, Columbia Univ, 65-69; assoc prof, 70-78, CLIN PROF PUB HEALTH, MED SCH, CORNELL UNIV, 78- *Concurrent Pos:* Consult health, AID, Dept State, 64-67, US Social Security Admin, 65-68, Roman Catholic Diocese, Brooklyn, 65-69, Catholic Hosp Asn US & Can, 66-69, Roman Catholic Archdiocese New York, 66-70, Mercy Catholic Med Ctr, 66-78, Tufts Med Sch, 69-70, City New York, 75-78, Gen Motors, Ford, Chrysler and United Auto Workers, 76- & Off Health Serv Res, HEW, 79-; vis prof, Med Sch, Johns Hopkins Univ, 79. *Honors & Awards:* Eisenberg Lectr, Peter Bent Brigham Hosp, Harvard Med Sch, 76. *Mem:* Am Fedn Clin Investrs; NY Acad Med; Am Pub Health Asn. *Res:* Delivery of health services; surgical health care; established second opinion elective surgical consultations as a medical care instrument for improvement in quality of care and reduction of in-hospital elective surgery. *Mailing Add:* Dept Pub Health Cornell Univ Med Col 1300 New York Ave New York NY 10021

MCCARTHY, F D, b Buffton, Ohio, July 26, 53. GROSS BIOLOGY, NUTRITION. *Educ:* Mich State Univ, PhD, 81. *Prof Exp:* ASST PROF ANIMAL SCI, VA POLYTECH INST & STATE UNIV, 81- *Mailing Add:* Vigortone Agr Prod 1632 Cascade Dr Marion OH 43302

MCCARTHY, FRANCIS DAVEY, b Sioux City, Iowa, Apr 30, 43; m 68; c 1. BIOLOGICAL OCEANOGRAPHY. *Educ:* Marquette Univ, BS, 68; Tex A&M Univ, PhD(biol oceanog), 73. *Prof Exp:* Asst prof, 73-78, ASSOC PROF BIOL SCI, CALIF STATE UNIV, DOMINGUEZ HILLS, 78- *Mailing Add:* Dept Biol Sci Calif State Univ Dominguez Hills 1000 E Victoria St Carson CA 90747

MCCARTHY, FRANCIS DESMOND, b Cork, Ireland, Dec 2, 36. ELECTRICAL ENGINEERING. *Educ:* Univ Col Cork, BE, 58; NY Univ, MEE, 63, PhD(elec eng), 66. *Prof Exp:* Grad appt, Gen Elec Co Ltd, 58-60, design standards engr, 60-61; teaching fel & res assoc elec eng, NY Univ, 62-66; staff scientist, Draper Lab, Mass Inst Technol, 66-69; asst prof, 69-76, ASSOC PROF ELEC ENG, NORTHEASTERN UNIV, 76- *Concurrent Pos:* Adj lectr, Northeastern Univ, 68-69; staff scientist, Draper Lab, Mass Inst Technol, 69- *Mem:* Brit Inst Elec Engrs. *Res:* Statistical analysis; data reduction; control systems; gyrators; mathematical modeling of physical processes. *Mailing Add:* 6911 Via Carona Dr Huntington Beach CA 92647

MCCARTHY, FRANK JOHN, b Mt Carmel, Pa, Mar 7, 28; m 51; c 3. MICROBIOLOGY. *Educ:* Mt St Mary's Col, Md, BS, 49; Fordham Univ, MS, 51; Lehigh Univ, PhD(biol), 60. *Prof Exp:* Asst biol, Fordham Univ, 49-51; res assoc, Merck Inst Therapeut Res, 53-63; mgr virus prods, 63-75, ASSOC DIR BIOL PRODS, WYETH LABS, INC, 75- *Mem:* Am Soc Microbiol; Tissue Cult Asn; NY Acad Sci; Sigma Xi. *Res:* Viral and bacterial vaccines; tissue culture; cytogenetics; anti-viral and anti-tumor agents; purification of virus and bacterial products. *Mailing Add:* 241 N Charlotte St Lancaster PA 17603

MCCARTHY, FRANK MARTIN, dentistry, anesthesiology, for more information see previous edition

MCCARTHY, GERALD T(IMOTHY), civil engineering; deceased, see previous edition for last biography

MCCARTHY, GREGORY JOSEPH, b Waltham, Mass, Apr 6, 43; m 67; c 2. SOLID STATE CHEMISTRY, X-RAY CRYSTALLOGRAPHY. *Educ:* Boston Col, BS, 64; Pa State Univ, PhD(solid state sci), 69. *Prof Exp:* Asst prof solid state sci, Pa State Univ, 69-74, sr res assoc solid state chem, 74-78, assoc prof mat res, 78-79; PROF CHEM & GEOL, NDAK STATE UNIV, 79- *Concurrent Pos:* Dir & chmn tech comt, Joint Comt Powder Diffraction Stand, Int Ctr Diffraction Data, 73-82, chmn bd dir, 82-86, past chmn, 86- *Mem:* Am Chem Soc; Am Ceramic Soc; AAAS; Mineral Soc Am; Mat Res Soc. *Res:* Crystal chemistry, phase equilibria, geochemistry of rare earths and transition metal oxide systems; nuclear waste solidification and geologic isolation; coal by-products utilization and disposal; x-ray diffraction analysis. *Mailing Add:* Chem Dept NDak State Univ Fargo ND 58105

MCCARTHY, JAMES BENJAMIN, PATHOLOGY, CELL ADHESION. *Educ:* Catholic Univ, PhD(microbiol), 81. *Prof Exp:* Fel, 81-85, ASST PROF PATH, MED SCH, UNIV MINN, 85- *Mem:* Am Soc Cell Biol. *Res:* Extracellular matrix. *Mailing Add:* Dept Lab Med & Pathol Rm 6-159 Jackson Hall Univ Minn 321 Church St SE Minneapolis MN 55455

MCCARTHY, JAMES FRANCIS, organic chemistry; deceased, see previous edition for last biography

MCCARTHY, JAMES FRANCIS, b Waterbury, Conn, Sept 28, 49. DEMOGRAPHY, FAMILY SOCIOLOGY. *Educ:* Col Holy Cross, BA, 71; Ind Univ, MA, 72; Princeton Univ, MA, 76, PhD(sociol), 77. *Prof Exp:* Res assoc demog, Off Pop Res, Princeton Univ, 77-79; asst prof, 79-82, ASSOC PROF DEMOG, DEPT POP DYNAMICS, JOHNS HOPKINS UNIV, 82-, DIR, HOPKINS POP CTR, 84- *Mem:* Pop Asn Am; Int Union Sci Study Pop; Am Pub Health Asn. *Res:* Determinants and consequences of demographic behavior, especially fertility related behavior in the United States and in less developed countries. *Mailing Add:* Columbia Univ Ctr Pop & Family Health 60 Haven Ave New York NY 10032

MCCARTHY, JAMES JOSEPH, b Ashland, Ore, Jan 25, 44; m 69; c 2. BIOLOGICAL OCEANOGRAPHY. *Educ:* Gonzaga Univ, BS, 66; Scripps Inst Oceanog, Univ Calif, San Diego, PhD(biol oceanog), 71. *Prof Exp:* Res assoc biol oceanog, Chesapeake Bay Inst, Johns Hopkins Univ, 71-72, assoc res scientist, 72-74; asst prof, 74-77, assoc prof, 77-80, PROF BIOL OCEANOG, HARVARD UNIV, 80-, DIR, MUS COMP ZOOL, 82- *Mem:* AAAS; Phycol Soc Am; Sigma Xi; Am Soc Limnol & Oceanog; Am Geophys Union. *Res:* Biogeochemical cycling of nitrogen and phosphorous employing isotopic techniques to quantitate the rate of flux of elemental material via various processes within planktonic ecosystems. *Mailing Add:* Dept Rehab Psychol Univ Wis Madison Ed/432N Murray Madison WI 53706

MACCARTHY, JEAN JULIET, b Rathcoole, Ireland. CELL BIOLOGY. *Educ:* Nat Univ, Ireland, BSc, 63, MSc, 64; Univ Md, PhD(plant physiol), 72. *Prof Exp:* Fel plant tissue, Nat Res Coun Can, 73-75; res assoc, Bot Lab, Univ Leicester, Eng, 75-77; res biochemist, Dept Biochem & Biophys, Univ Calif, Davis, 77-80; RES SPECIALIST PLANT TISSUE CULTURES, COL PHARM HEALTH SCI, UNIV MINN, 80- *Mem:* Int Asn Plant Tissue Culture; Sigma Xi. *Res:* Plant cell and tissue culture; lipid biosynthesis and secondary product metabolism; steroidal glycoside production and biosynthesis; algal physiology. *Mailing Add:* Rathcoole House Rathcoole Cork-Mallow County Ireland

MCCARTHY, JOHN, b Boston, Mass, Sept 4, 27; m 54; c 3. COMPUTER SCIENCE. *Educ:* Calif Inst Technol, BS, 48; Princeton Univ, PhD(math), 51. *Prof Exp:* Res instr math, Princeton Univ, 51-53; actg asst prof, Stanford Univ, 53-55; asst prof, Dartmouth Col, 55-58; from asst prof to assoc prof commun sci, Mass Inst Technol, 58-61; PROF COMPUT SCI, STANFORD UNIV, 62- *Honors & Awards:* A M Turing Award, Asn Comput Mach, 71; Kyoto Prize, 88; Nat Medal Sci, 90. *Mem:* Nat Acad Sci; Asn Comput Mach; Am Acad Arts & Sci; Nat Acad Eng; Am Math Soc. *Res:* Programming languages; time-sharing; mathematical theory of computation; artificial intelligence. *Mailing Add:* 885 Allardice Way Stanford CA 94305

MCCARTHY, JOHN, b New Orleans, La, July 3, 42. METEOROLOGY. *Educ:* Grinnell Col, BA, 64; Univ Okla, MS, 67; Univ Chicago, PhD(geophys sci), 73. *Prof Exp:* Res meteorologist, Weather Sci, Inc, 66-67 & Cloud Physics Lab, Univ Chicago, 67-68; asst prof meteorol, Univ Okla, 73-78; assoc prof, 78-80, staff scientist, 80-89, DIR APPL PROG, NAT CTR ATMOSPHERIC RES, 89- *Concurrent Pos:* Mem, Univ Rels Comt, Univ Corp Atmospheric Res, 74- *Mem:* Am Meteorol Soc; Sigma Xi. *Res:* Radar and aircraft studies of severe thunderstorms and tornadoes; source of thunderstorm rotation; applied aircraft hazards near thunderstorms; weather modification. *Mailing Add:* 400 Fairfield Lane Louisville CO 80027

MCCARTHY, JOHN F(RANCIS), b St Louis, Mo, May 5, 20; m 56; c 4. CIVIL ENGINEERING. *Educ:* Univ Mo-Rolla, BS, 48, MS, 50. *Prof Exp:* Instr civil eng, Mo Sch Mines, 48-51; chief struct engr, Fruin-Colnon Contracting Co, 51-53 & W R Bendy, Cement Engrs, 53-55; from asst prof to prof civil eng, St Louis Univ, 55-69, dir dept, 56-69; supt, Bissell Point, 69-76, TECH COORDR, METROP ST LOUIS SEWER DIST, 76- *Concurrent Pos:* Chmn bldg code rev comt, St Louis County, 62- *Mem:* Am Soc Civil Engrs; Nat Soc Prof Engrs; Am Soc Eng Educ; Water Pollution Control Fedn. *Res:* Structural engineering. *Mailing Add:* MSD St Louis Sewer Dist St Louis MO 63147

MCCARTHY, JOHN F, b New York, NY, June 16, 47; m 69, 87; c 1. ENVIRONMENTAL SCIENCES. *Educ:* Fordham Univ, NY, BS, 69; Univ RI, PhD(oceanog), 74. *Prof Exp:* Fel, Biol Div, Oak Ridge Nat Lab, 75-78, res assoc, 78-85, res staff mem, 85-90, group leader, Environ Sci Div, 86-90, SR SCIENTIST, ENVIRON SCI DIV, OAK RIDGE NAT LAB, 90- *Concurrent Pos:* vis prof, Bhzba Atomic Res Ctr, Bombay, India, 73, New York Univ, 72. *Mem:* Sigma Xi; Am Chem Soc; AAAS; Soc Environ Toxicol & Chem. *Res:* Fate and transport of organic contaminants in surface water and groundwater; Aquatic ecology and toxicology; uptake and metabolism of organic pollutants by aquatic organisms. *Mailing Add:* Environ Sci Div Oak Ridge Nat Lab PO Box X Oak Ridge TN 37831

MCCARTHY, JOHN JOSEPH, b Troy, NY, June 24, 23; m 48; c 1. PHYSICAL METALLURGY. *Educ:* Rensselaer Polytech Inst, BMetE, 48, MMetE, 51, PhD(metall), 58. *Prof Exp:* From instr to asst prof, 48-60, ASSOC PROF METALL ENG, RENSSELAER POLYTECH INST, 60-, EXEC OFFICER, DEPT MAT ENG, 76- *Mem:* Am Soc Metals; Am Welding Soc. *Res:* Welding, brazing and mechanical property determinations. *Mailing Add:* Dept Mat Eng Rensselaer Polytech Inst Troy NY 12180-3590

MCCARTHY, JOHN LAWRENCE, JR, b New Haven, Conn, Aug 13, 29; m 55; c 2. PHYSIOLOGICAL CHEMISTRY. *Educ:* Univ Miami, Fla, BS, 51; Purdue Univ, MA, 53, PhD(endocrinol), 58. *Prof Exp:* From asst prof to assoc prof biol, 58-69, PROF BIOL, SOUTHERN METHODIST UNIV, 69- *Concurrent Pos:* NIH fel, 64-65; vis prof biochem, Univ Edinburgh, 72; bd dir, Oak Ridge Assoc Univ, 81-84. *Mem:* AAAS; Am Soc Zool; Soc Exp Biol Med; Am Physiol Soc; Endocrine Soc. *Res:* Physiological and chemical relationship between the thyroid and adrenal glands; adrenal steroid biosynthesis; steroid metabolism. *Mailing Add:* 3424 Purdue St Dallas TX 75225

MCCARTHY, JOHN LOCKHART, b San Francisco, Calif, Jan 24, 43; m 77; c 2. INFORMATION MANAGEMENT. *Educ:* Stanford Univ, BA, 64; Yale Univ, PhD(hist), 70. *Prof Exp:* Actg instr hist, Yale Univ, 68-69, lectr, 69-70, asst prof, 70-75; dir data anal, Surv Res Ctr, Univ Calif, Berkeley, 75-80; STAFF SCIENTIST, LAWRENCE BERKELEY LAB, 80- *Concurrent Pos:* Morse fel, Yale Univ, 72-73; Am Coun Learned Soc grant comput oriented res in humanities, 73-74; mem assembly behav & soc sci, Nat Res Coun, Nat Acad Sci, 73-; lectr polit sci, Univ Calif, Berkeley, 74-77; prin investr, Mat Info Sci & Technol proj, 84-89. *Mem:* Asn Comput Mach; Inst Elec & Electronics Engrs Comput Soc. *Res:* Computer science-database management; quantitative data analysis; social science computer applications; information systems and knowledge base systems; material property databases and data exchange formats. *Mailing Add:* Bldg 50B Rm 3238 Lawrence Berkeley Lab Univ Calif Berkeley CA 94720

MCCARTHY, JOHN MICHAEL, b Pittsburgh, Pa, June 24, 52; m 76; c 4. MACHINE DESIGN & ROBOTICS COMPUTER AIDED DESIGN. *Educ:* Loyola Marymount Univ, BS, 74; Stanford Univ, MS, 75, PhD(mech eng), 79. *Prof Exp:* Sr engr & consult, Cent Eng Lab, F M C Corp, 79-80; asst prof teaching, Loyola Marymount Univ, 80-83; asst prof teaching & res, Univ Pa, 83-86; ASST & ASSOC PROF TEACHING & RES, UNIV CALIF, IRVINE, 86- *Concurrent Pos:* Legal consult, Noise & Vibration, Mach Design, 80-; prin investr, NSF grants, 83-91; ed, Kinematics of Robot Manipulators, Mass Inst Technol Press, 84-85; assoc ed, J Mech Design, Am Soc Mech Engrs, 88-91; mem, Organizing Comt, Robotics & Automation Conf, Inst Elec & Electronic Engrs, 89; assoc dean undergrad studies, Sch Eng, Univ Calif, Irvine. *Mem:* Am Soc Mech Engrs; Inst Elec & Electronic Engrs. *Res:* Design and control of spatial mechanisms, with applications to two armed robots, mechanical hands and walking machines; development of computer methods for design, path planning and optimal control for maximum performance; editor-author of two books. *Mailing Add:* Dept Mech Eng Univ Calif Irvine CA 92717

MCCARTHY, JOHN RANDOLPH, b Huntington, WVa, Mar 1, 15; m 49. ORGANIC CHEMISTRY. *Educ:* Columbia Univ, PhD(chem), 49. *Prof Exp:* Anal chemist, Stand Ultramarine Co, 39-41; res chemist, Del, 49-74, SUPVR, MED LAB, CHAMBERS WORKS, E I DU PONT DE NEMOURS & CO, DEEPWATER, NJ, 74- *Mem:* Am Chem Soc. *Res:* Process control; elastomers; petroleum; dyes and intermediates; urine and blood analyses for biological monitoring of workplace exposure. *Mailing Add:* 230 Meredith St Kennett Square PA 19348

MCCARTHY, JOSEPH L(EPAGE), b Spokane, Wash, Oct 19, 13; m 45; c 2. CHEMICAL ENGINEERING, WOOD CHEMISTRY. *Educ:* Univ Wash, Seattle, BS, 34; Univ Idaho, MS, 36; McGill Univ, PhD(chem), 38. *Prof Exp:* Sessional lectr & res fel, McGill Univ, 38-40; dir res, Fraser Cos, Ltd, 40-41; res assoc chem, 41-42, from instr to assoc prof, 42-52, prin investr, Pulp Mills Res, 44-74, dean, Grad Sch, 59-74, prof chem eng, 52-84, prof forest resources, 78-84, EMER PROF CHEM ENG, UNIV WASH, 84, EMER PROF FOREST RESOURCES, 84, EMER DEAN, GRAD SCH, 74- *Concurrent Pos:* Prof, Nanjing Forestry Univ, 85- *Honors & Awards:* Johan Richter Prize, 78. *Mem:* Am Soc Eng Educ; Tech Asn Pulp & Paper Indust: Am Inst Chem Engrs; Can Pulp & Paper Asn; fel Int Acad Wood Sci. *Res:* Wood chemistry, lignin, cellulose, pulp and paper; chemical engineering thermodynamics; lignin and cellulose chemistry; thermodynamics; biochemical engineering. *Mailing Add:* Dept Chem Eng BF-10 Univ Wash Seattle WA 98195

MCCARTHY, KATHRYN AGNES, b Lawrence, Mass, Aug 7, 24. SOLID STATE PHYSICS. *Educ:* Tufts Univ, AB, 45, MS, 46; Radcliffe Col, PhD(appl physics), 57. *Hon Degrees:* DSc, Col Holy Cross, 78; DHL, Merrimack Col, 79. *Prof Exp:* Asst physics, 45-46, instr, 46-53, from asst prof to assoc prof, 57-62, actg chmn dept, 61-62, dean grad sch arts & sci, 69-73; provost/sr vpres, 73-79, PROF PHYSICS, TUFTS UNIV, 62- *Concurrent Pos:* Res physicist, Baird Assocs, Inc, 47-49 & 51; res fel phys metall, Harvard Univ, 57-59, vis scholar, 79-80; assoc res engr & res consult, Univ Mich, 57-59; dir, Doble Eng Co, 67-70; dir, Mass Elec Co, 73- & State Mutual Assurance, 76- mem exec bd, New Eng Conf Grad Educ, 70-74; mem comn inst higher educ, New Eng Asn Cols & Sec Schs, Inc, 72-77, vchmn, 73-74, chmn, 74-77, mem comt disadvantaged students, Coun Grad Schs, 72-74, mem exec comt, 73-75; mem comn inst affairs, Asn Am Cols, 73-; mem grad adv acad comt, Bd High Educ Mass, 72-74; mem comn admin affairs & educ statist, Am Coun Educ, 73-; trustee, Southern Mass Univ, 72-74, Merrimack Col, 74-83 & Col Holy Cross, 80-; mem bd dir, Lawrence Mem Hosp, Medford, 77-, vchmn, 84-91, chmn, 91- *Mem:* Fel Am Phys Soc; fel Optical Soc Am; sr mem Soc Women Engrs (dir-at-large, 82-84). *Res:* Low-temperature properties; infrared optical materials, both single crystal and amorphous. *Mailing Add:* Dept Physics Tufts Univ Medford MA 02155

MCCARTHY, LAWRENCE E, PHARMACOLOGY. *Educ:* Dartmouth Col, PhD, 79. *Prof Exp:* ASSOC PROF PHARMACOL & TOXICOL, SCH MED, DARTMOUTH COL, 79- *Mailing Add:* Dept Pharmacol Dartmouth Med Sch Hanover NH 03755

MCCARTHY, MARTIN, b Lowell, Mass, July 10, 23. ASTRONOMY. *Educ:* Boston Col, AB, 46, MA, 47; Georgetown Univ, PhD(astron), 51; Weston Col, STL, 55. *Prof Exp:* MEM STAFF ASTRON, VATICAN OBSERV, 56- *Concurrent Pos:* Vis prof, Georgetown Univ, 62-63; guest investr, Carnegie Inst, 65-66, 71, 73 & 74; observ consult, Cerro Tololo Interam Observ, Asn Univs Res in Astron, 72 & consult adv, 74; observer for Vatican City State at World Meteorol Orgn, 80. *Mem:* AAAS; Royal Astron Soc; Sigma Xi; Int Astron Union. *Res:* Spectral classification of stars; galactic structure; photoelectric and photographic photometry. *Mailing Add:* Vatican Observ Vatican City I-00120 Vatican City State Italy

MCCARTHY, MARY ANNE, b Tarentum, Pa, May 26, 39; div. NUMERICAL ANALYSIS, SYSTEM ANALYSIS. *Educ:* Wellesley Col, BA, 61; Rice Univ, MA, 66, PhD(math sci), 73. *Prof Exp:* Comput programmer, Bell Tel Labs Inc, 61-63; physicist, Shell Develop Co, Shell Oil Co, 67-69; assoc syst analyst, Lulejian & Assoc Inc, 73-75; researcher, R & D Assoc Inc, 75-81; vpres, 81-84, PRES, PARAGON SOFTWARE INC, 84- *Mem:* Am Math Soc; Asn for Women in Math; Soc Indust & Appl Math; Sigma Xi. *Res:* Development of a mathematical method which yields closed form solutions to various types of probabilistic problems which previously have required Monte Carlo simulation. *Mailing Add:* 2622 Laurel Ave Manhattan Beach CA 90266

MCCARTHY, MILES DUFFIELD, b Camden, NJ, Oct 12, 14; m 59. REPRODUCTIVE BIOLOGY. *Educ:* Pa State Teachers Col, BS, 36; Univ Pa, PhD(zool), 43. *Prof Exp:* Asst, Inst Med Res, Pa, 37-39; instr zool, Univ Pa, 39-42, tech assoc, Med Sch, 42, instr vert anat, Univ, 44-45, genetics, Vet Sch, 45-46; asst prof zool, Pomona Col, 46-49, from assoc prof to prof, 49-59; vpres acad affairs, 70-74, chmn div sci & math, 59-64, dean sch letters, arts & sci, 64-70, actg pres, 81, PROF BIOL, CALIF STATE UNIV, FULLERTON, 59- *Concurrent Pos:* Harrison res fel, Med Sch, Univ Pa, 44-46, vis asst prof, 51-52; responsible investr, US Dept Army contract, Pomona Col, 54-56; NIH fel, 57-61. *Mem:* AAAS; Soc Exp Biol & Med; Am Physiol Soc. *Res:* Growth and development; burn shock and infusion fluids; hemolytic and hematopoietic changes in relation to survival following severe thermal injury; reproduction and human sexuality. *Mailing Add:* Dept of Biol Calif State Univ Fullerton CA 92634

MCCARTHY, NEIL JUSTIN, JR, b Chicago, Ill, Oct 25, 39; m 63; c 2. ORGANIC POLYMER CHEMISTRY, ADVANCED COMPOSITES. *Educ:* Georgetown Univ, BS, 62; Cornell Univ, PhD(org chem), 69. *Prof Exp:* Res chemist, Res Lab & Develop Lab, Union Carbide Corp, Bound Brook, 66-76, proj scientist org polymer chem, 71-76, financial analyst specialty chem, 76-80, market mgr, Epoxies & Plasticizers, 80-85; MKT MGR, ADVAN COMPOSITES, AMOCO PERFORMANCE PROD, 85- *Mem:* Am Chem Soc; Soc Plastics Engrs; Soc Advan Mat & Process Eng. *Res:* Physical-organic and physical polymer chemistry and composite science; specialty chemicals. *Mailing Add:* Amoco Performance Prod 375 N Ridge Rd Atlanta GA 30350

MACCARTHY, PATRICK, b Galway, Ireland, Jan 13, 46; m 72; c 5. ANALYTICAL CHEMISTRY, SOIL CHEMISTRY. *Educ:* Univ Col Galway, Ireland, BSc, 68, MSc, 71; Northwestern Univ, MS, 71; Univ Cincinnati, PhD(anal chem), 75. *Prof Exp:* Vis lectr anal chem, Univ Ga, 75-76; asst prof chem, 76-79, assoc prof, 79-85, PROF CHEM & GEOCHEM, COLO SCH MINES, 85- *Concurrent Pos:* Chmn, Humic Acid Working Group, Int Soc Soil Sci; US Geol Surv, Water Resources Div, Denver, Colo, 85-86. *Mem:* Int Humic Substances Soc (secy-treas); Am Chem Soc; Soil Sci Soc Am; Int Soc Soil Sci. *Res:* Metal complexation by humic substances; chemistry of humic substances; use of chemically modified peat for removal of pollutants; mathematical modeling and computer simulation of multimetal-multiligand equilibrium; analytical chemistry. *Mailing Add:* Dept of Chem & Geochem Colo Sch of Mines Golden CO 80401

MCCARTHY, PATRICK CHARLES, b Cadillac, Mich, Mar 27, 40; m 66; c 3. GENETICS, MOLECULAR BIOLOGY. *Educ:* Wayne State Univ, BS, 65, MS, 66, PhD(biol), 70. *Prof Exp:* Instr, Wayne State Univ, 68-70; from asst prof to assoc prof biol, 70-85, chmn biol, 85-89, PROF BIOL, WESTMINSTER COL, 89- *Concurrent Pos:* J S Mack Found researcher, Westminster Col, 71-72, 84-85, 90-91. *Mem:* Nat Asn Biol Teachers; Genetics Soc Am; Electron Microscopy Soc Am; Sigma Xi; AAAS. *Res:* Physiology and genetics of tissue proteins in the Mongolian gerbil, Meriones unguiculatus; immunology of gerbils, particularly tissue transplantation phenomena and cytology of gerbil immune responses; molecular biology of alcohol dehydrogenase in Drosophila Melanogaster; aging studies in Meriones Unguiculatus and Drosophila Melanogaster, focusing on lipofuscin and free radical phenomena. *Mailing Add:* Dept Biol Westminster Col New Wilmington PA 16172-0136

MCCARTHY, PAUL JAMES, b Rochester, NY, June 15, 24. PHYSICAL INORGANIC CHEMISTRY. *Educ:* Spring Hill Col, AB, 50; Col Holy Cross, MS, 52; Clark Univ, PhD(chem), 55. *Prof Exp:* Instr high sch, NY, 50-51; instr chem, St Peters Col, 59-60; instr, 60-62, from asst prof to assoc prof, 62-72, PROF CHEM, CANISIUS COL, 72- *Concurrent Pos:* NATO fel, Copenhagen Univ, 66-67; vis prof, Univ Bern, Switz, 82-83. *Mem:* Am Chem Soc. *Res:* Coordination complexes, especially their preparation, structure and absorption spectra. *Mailing Add:* Dept Chem Canisius Col 2001 Main St Buffalo NY 14208

MCCARTHY, PAUL JOSEPH, b Chicago, Ill, Oct 23, 28; m 52; c 2. COMBINATORICS & FINITE MATHEMATICS. *Educ:* Univ Notre Dame, BS, 50, MS, 52, PhD(math), 55. *Prof Exp:* Instr math, Univ Notre Dame, 54-55; instr, Col Holy Cross, 55-56; from asst prof to assoc prof, Fla State Univ, 56-61; assoc prof, 61-65, PROF MATH, UNIV KANS, 65- *Mem:* Am Math Soc; Math Asn Am. *Res:* Rings; graph theory; number theory. *Mailing Add:* Dept Math Univ Kans Lawrence KS 66045

MCCARTHY, PHILIP JOHN, b Friendship, NY, Feb 9, 18; m 42; c 4. STATISTICS. *Educ:* Cornell Univ, AB, 39; Princeton Univ, MA, 41, PhD(math), 47. *Prof Exp:* Res assoc, Columbia Univ, 41-42; opers res, US Dept Navy, 42-43; Princeton Univ, 43-46, Soc Sci Res Coun fel, 46-48; prof, 47-90, EMER PROF APPL STATIST, STATE UNIV NY SCH INDUST & LABOR RELS, CORNELL UNIV, 90- *Concurrent Pos:* Consult, Nat Ctr Health Statist, 64- & Comt Nat Statist, 74- *Mem:* Fel Am Statist Asn; Inst Math Statist; Int Asn Surv Statisticians; Sigma Xi. *Res:* Sampling methods; application of statistics in the social sciences. *Mailing Add:* 358 Ives Hall Cornell Univ Ithaca NY 14850

MCCARTHY, RAYMOND LAWRENCE, b Jersey City, NJ, Sept 6, 20; m 43; c 4. FREE RADICAL CHEMISTRY. *Educ:* Fordham Univ, AB, 41, MA, 42; Yale Univ, MS, 47, PhD(physics), 48. *Prof Exp:* Physicist, (USNR) Naval Ord Lab, 42-44; asst, Yale Univ, 46-48; res physicist, 48-55, res mgr, Radiation Physics Lab, 55-59, asst dir, 59-61, process supt, Org Chem Dept, 61-64, dir Freon Prod Lab, E I DU Pont De Nemours & Co, Inc, 64-84; RETIRED. *Concurrent Pos:* Adj prof physics, Univ Del, 68-72; ADHOC Faculty Comm, Harvard Univ, 77; Del UNEP, 78-82. *Mem:* Sigma Xi; AAAS; Am Phys Soc; Am Chem Soc. *Res:* Fluorocarbon compounds and their environmental effects; atmospheric chemistry and physics; chemical and radiation physics; process control. *Mailing Add:* LENAPE Westtown PA 19395

MCCARTHY, ROBERT ELMER, b Washington, DC, May 19, 26; m 49; c 1. IMMUNOLOGY, MICROBIOLOGY. *Educ:* Univ Md, BS, 51, MS, 53; Brown Univ, PhD(biol), 56. *Prof Exp:* Asst poultry, Univ Md, 51-53; asst biol, Brown Univ, 53-55; asst path, Harvard Med Sch, 57-62, res assoc, 62-69, assoc, 69-70; ASSOC PROF MED MICROBIOL & IMMUNOL, UNIV NEBR MED CTR, OMAHA, 70- *Concurrent Pos:* USPHS fel, Inst Cell Res & Genetics, Karolinska Inst, Sweden, 62-63; asst path, Children's Hosp, Boston, 56-59, res assoc, 59-70; asst path, Children's Cancer Res Found, 56-59, res assoc, 59-70, chief lab of immunol, 68-69; assoc oral path, Sch Dent Med, Harvard Univ, 69-70. *Mem:* Am Soc Exp Path; NY Acad Sci; Am Asn Immunol; Am Soc Cancer Res. *Res:* Radiation biology, antigen structure, cytochemistry and cytogenetics of mammalian cells in culture; study of inhibition of allograft rejection by ascites tumors; immunochemistry of synthetic polypeptides; immunoregulators; effect of pesticides on immune response. *Mailing Add:* 353 Garnsey Bakersfield CA 93309

MCCARTHY, SHAUN LEAF, b Berkeley, Calif, Oct 20, 39; m 63; c 2. EXPERIMENTAL SOLID STATE PHYSICS. *Educ:* Univ Calif, Berkeley, BA, 62; Univ Chicago, MS, 63; Univ Calif, San Diego, PhD(physics), 70. *Prof Exp:* Res physicist, Univ Calif, San Diego, 70-71; res physicist, State Univ NY, Stony Brook, 71-73; RES SCIENTIST, FORD MOTOR CO, 73- *Mem:* Am Phys Soc. *Res:* High temperature superconductivity; thin-film Josephson phenomena; optical properties of thin-metal films; electroluminescence anisotropic etching of silicon. *Mailing Add:* 2600 Gladstone Ave Ann Arbor MI 48104

MCCARTHY, WALTER CHARLES, b Brooklyn, NY, Oct 25, 22; m 45; c 3. MEDICINAL CHEMISTRY. *Educ:* Mass Inst Technol, BS, 43; Univ Ind, PhD(org chem), 49. *Prof Exp:* From asst prof to assoc prof, 49-65, PROF PHARMACEUT CHEM, UNIV WASH, 65- *Mem:* Am Chem Soc; Am Pharmaceut Asn; The Chem Soc; Swiss Chem Soc. *Res:* Sympathomimetic amines; organic sulfur compounds; thiophene and furan derivatives. *Mailing Add:* 16202 Beach Dr NE Seattle WA 98155

MCCARTHY, WALTER J, JR, b Apr 20, 25; m 88; c 8. NUCLEAR ENGINEERING & MANAGEMENT. *Educ:* Cornell Univ, BME, 49; Oakridge Sch Reactor Tech, MA, 52. *Hon Degrees:* LHD, Wayne State Univ, Eastern Mich Univ, Alma Col; DEng, Lawrence Inst Technol. *Prof Exp:* Chmn bd & chief exec officer, Detroit Edison Co, 85-90; RETIRED. *Concurrent Pos:* Mem bd dir, Comerica, Inc, Fed-Mogul, Perry Drugs, Inc & Detroit Edison, Co. *Mem:* Nat Acad Eng; fel Am Nuclear Soc. *Res:* Nuclear reactor safety. *Mailing Add:* 776-4 St Josephs Sutton's Bay MI 49682

MCCARTHY, WILLIAM CHASE, chemical engineering, for more information see previous edition

MCCARTHY, WILLIAM JOHN, b Mamaroneck, NY, June 27, 41; m 77; c 2. VIROLOGY, BIOCHEMISTRY. *Educ:* Univ Del, BA, 63; NY Univ, MS, 68, PhD(biol), 70. *Prof Exp:* Fel virol, Boyce Thompson Inst Plant Res, 70-73; assoc res scientist, Pesticide Res Lab, 74-78, ASST PROF PLANT PATH, PA STATE UNIV, 78- *Mem:* Am Soc Microbiol; Soc Invert Path; Tissue Cult Asn. *Res:* Cellular and viral protein and nucleic acid synthesis in Lepidoptera and insect tissue cultures; isolation and characterization of insect viruses; proteins and nucleic acids; characterization of nucleic acid and proteins of baculoviruses, entomopoxviruses, and cytoplasmic polyhedrosis viruses; genetics of bawloviruses. *Mailing Add:* Pesticide Res Lab Pa State Univ University Park PA 16802

MCCARTIN, BRIAN JAMES, b Providence, RI, Aug 26, 51; m 71; c 2. NUMERICAL ANALYSIS, COMPUTATIONAL ELECTROMAGNETICS. *Educ:* Univ RI, BS, 76, MS, 77; New York Univ, PhD(math), 81. *Prof Exp:* Sci analyst, Pratt & Whitney Aircraft, 77-81; appl mathematician, Gerber Systs Technol, 81; appl mathematician, United Technol Res Ctr, 81-89; CHMN & PROF, DEPT COMPUTER SCI, HARTFORD GRAD CTR, 89- *Concurrent Pos:* Instr math, Manchester Community Col, 80-81. *Mem:* Asn Comp Machinery; Inst Elec & Electronic Engrs Computer Soc; Math Asn Am; Soc Indust & Appl Math. *Res:* Numerical solution of partial differential equations, including fluid dynamics, semiconductor physics, and electromagnetics; computational algorithms in approximation theory, particularly splines. *Mailing Add:* 110 Margaret Dr South Windsor CT 06074

MCCARTNEY, MORLEY GORDON, b Picton, Ont, Mar 4, 17; nat US; m 46; c 5. GENETICS. *Educ:* Ont Agr Col, BSA, 40; Univ Md, MS, 47, PhD, 49. *Prof Exp:* Poultry specialist, Ont Agr Col, 40-41; asst poultry husb, Univ Md, 46-49; poultry physiologist, USDA, Md, 49; from asst prof to assoc prof poultry, Pa State Univ, 49-55; prof & assoc chmn dept, Agr Exp Sta, Ohio State Univ, 55-64; PROF POULTRY SCI, HEAD DEPT & CHMN DIV POULTRY, UNIV GA, 64- *Honors & Awards:* Award, Nat Turkey Fedn, 59. *Mem:* World's Poultry Sci Asn; Poultry Sci Asn. *Res:* Population genetics and physiology of reproduction of poultry. *Mailing Add:* 370 Ashton Dr Athens GA 30606

MCCARTNEY-FRANCIS, NANCY L, b Bridgeport, Nebr, June 5, 50. IMMUNOLOGY. *Educ:* Univ Tex, Austin, MA, 77, PhD(microbiol), 80. *Prof Exp:* Postdoctoral fel, Nat Inst Allergy & Infectious Dis, 81-85, SR STAFF FEL, LAB IMMUNOL, NAT INST DENT RES, NIH, 85- *Concurrent Pos:* Postdoctoral fel, Nat Cancer Soc, 81-82. *Mem:* Am Asn Immunologists; AAAS. *Mailing Add:* Lab Immunol Nat Inst Dent Res NIH Bldg 30 Rm 327 Bethesda MD 20892

MCCARTY, BILLY DEAN, b Wymore, Nebr, Oct 31, 23; m 46; c 2. ANALYTICAL CHEMISTRY. *Educ:* Nebr Wesleyan Univ, AB, 48. *Prof Exp:* Anal chemist oil shale demonstration plant, US Bur Mines, Colo, 50-53; spectrographer, Los Alamos Sci Lab, 53-58 & Union Carbide Nuclear Co, Colo, 58-62; spectrographer, Marathon Oil Co Res Ctr, 62-66, analysis unit supvr, 66-70, advan scientist, 70-78, sr scientist, 78-81, instrumental analysis sect supvr, 81-87; RETIRED. *Honors & Awards:* Outstanding Serv Award, Rocky Mt Sect, Soc Appl Spectros, 72. *Mem:* Soc Appl Spectros; Sigma Xi. *Res:* Optical emission spectroscopy; x-ray diffraction and spectroscopy; scanning electron microscopy; isotopic ratio mass spectroscopy. *Mailing Add:* 2001 Plum Tree Ln Hendersonville NC 28739

MCCARTY, CLARK WILLIAM, b Kansas City, Mo, Feb 27, 16; m 40; c 5. TEACHING, FORECASTING. *Educ:* Univ Kans City, AB, 37; Cent Mo State Col, BSE, 40; Univ Nebr, MS, 39; Univ Mo, AM, 47, PhD(chem), 53. *Prof Exp:* Asst chem, Univ Nebr, 37-39; chemist & supvr bact, Cerophyl Labs, Inc, 40-42; instr chem & math, Kemper Mil Sch, 42-43; instr, Jefferson City Jr Col, 46; instr chem, Univ Mo, 46-47,instr math, 48-49; instr chem, Southwest Mo State Col, 47-48; instr math, Univ Mo, 48-49; assoc prof chem & math, 50 & 52-58, prof & chmn, Dept Physics, 58-81, EMER PROF PHYSICS, OUACHITA BAPTIST UNIV, 81- *Mem:* Am Asn Physics Teachers; Nat Weather Asn; Optical Soc Am; Am Meteorol Soc. *Res:* Partial vapor pressures of ternary systems. *Mailing Add:* 962 N Eighth St Arkadelphia AR 71923

MACCARTY, COLLIN STEWART, b Rochester, Minn, Sept 20, 15; m 40; c 3. NEUROSURGERY. *Educ:* Dartmouth Col, AB, 37; Johns Hopkins Univ, MD, 40; Univ Minn, MS, 44; Am Bd Neurol Surg, dipl, 47. *Prof Exp:* From instr to prof neurosurt, Mayo Grad Sch Med, Univ Minn, 61-73, prof, 73-80, mem dept neurosurg, Mayo Clin, 46-80, chmn, 63-80; RETIRED. *Concurrent Pos:* Pres-elect med staff, Mayo Clin, 65, pres, 66; secy cong affairs, Liaison & Admin Coun & chmn prog comt, World Fedn Neurosurg Socs, 65-69; nat consult neurol surg, US Air Force, 72- *Mem:* Am Col Surg; Soc Neurol Surg; Neurosurg Soc Am (vpres, 54, pres, 59); Am Asn Neurol Surg (vpres, 6S-66, pres, 70-71); Sigma Xi. *Mailing Add:* Northome Rte 2 Box 71A Cable WI 54821

MCCARTY, DANIEL J, JR, b Philadelphia, Pa, Oct 31, 28; m 54; c 5. INTERNAL MEDICINE. *Educ:* Villanova Univ, BS, 50; Univ Pa, MD, 54. *Prof Exp:* Intern med, Fitzgerald-Mercy Hosp, Darby, Pa, 54-55; resident internal med, Hosp, Univ Pa, 57-58; resident, Philadelphia Vet Admin Hosp, 58-59; from asst prof to assoc prof internal med, Hahnemann Med Col, 60-67, head sect rheumatol, 60-67; prof internal med, Univ Chicago, 67-74, head sect arthritis & metab dis, 67-74; chief med, Milwaukee County Gen Hosp, 74-89; Will & Cara Ross prof med & chmn dept, 74-89, DIR, ARTHRITIS INST, MED COL WIS, 89- *Concurrent Pos:* Nat Inst Arthritis & Metab Dis fel rheumatol, Hosp, Univ Pa, 59-60; Markle scholar, 62-67; attend physician & chief arthritis clin, Div B, Philadelphia Gen Hosp, 60-67; consult, Vet Admin Hosp, Philadelphia, 60-67; consult arthritis training progs, Surgeon Gen, USPHS, 67-71; mem fel subcomt, Arthritis Found, 67-71, res comt chmn nat reference ctr rheumatol, 69; mem subcomt rheumatol, Am Bd Internal Med, 70-76; chief ed, Arthritis & Rheumatism, Am Rheumatism Asn, 65-70; vchmn bd trustees, Arthritis Found, 79-80. *Honors & Awards:* Hektoen Silver Medal, AMA, 63; Russell S Cecil Award, Arthritis Found & Gairdner Found Award, 65; Ciba-Geigy Int Rheumatism Award, 81; Heberdenn Oration & Gold Medal, 82; Highton Rheumatism Prize, 82; Van Brennen Award, 82; Bunim lectr & Gold Medal, 83. *Mem:* AAAS; Am Fedn Clin Res; Am Rheumatism Asn (vpres, 78-79, pres, 79-80); Am Soc Clin Pharmacol & Therapeut; NY Acad Sci; Am Col Physicians; Am Soc Clin Invest; Asn Am Physicians; Asn Am Prof Med. *Res:* Rheumatic diseases; crystallography, mechanisms of inflammation; measurement of inflammation, clinical, physiological and biochemical parameters; inorganic pyrophosphate metabolism. *Mailing Add:* Dept Med Med Col Wis 8701 Watertown Plank Rd Milwaukee WI 53226

MCCARTY, FREDERICK JOSEPH, b Chillicothe, Ohio, May 2, 27; m 61; c 4. ORGANIC CHEMISTRY, MEDICINAL CHEMISTRY. *Educ:* Univ Ohio, BS, 51; Mich State Univ, MS, 53; Univ Mich, PhD(med chem), 59. *Prof Exp:* Asst res chemist, Wm S Merrell Co Div, Richardson-Merrell, Inc, 53-56, proj leader med chem res, 59-64; sect head org res, Aldrich Chem Co, 64-65; group leader med chem res, Nat Drug Co Div, 65-70, sect head chem develop, Richard-Merrell, Inc, 70-81, SECT HEAD CHEM DEVELOP, MERRELL-DOW RES INST, DOW CHEM CO, 81- *Mem:* Am Chem Soc. *Res:* Mannich reaction. *Mailing Add:* 7141 Juniperview Lane Cincinnati OH 45243-2556

MCCARTY, GALE A, RHEUMATOLOGY, IMMUNOLOGY. *Prof Exp:* ASST PROF RHEUMATOLOGY & IMMUNOL, GEORGETOWN UNIV MED CTR, 83- *Mailing Add:* Dept Med Georgetown Univ Med Ctr 3900 Reservoir Rd NW Washington DC 20007

MCCARTY, JOHN EDWARD, b Iowa City, Iowa, Aug 27, 28; m 55; c 2. ORGANIC CHEMISTRY. *Educ:* Univ Iowa, BS, 50; Univ Calif, Los Angeles, PhD, 57. *Prof Exp:* Res fel, Univ Kans, 56-59; from asst prof to prof chem, Mankato State Univ, 59-67, prof, Environ Inst, 72-75, emer prof, 87-; RETIRED. *Mem:* Am Chem Soc. *Res:* Mechanism of organic reactions; heterocyclic compounds; environmental chemistry. *Mailing Add:* 1708 B Friendship Rd Waldoboro ME 04572-9410

MCCARTY, JON GILBERT, b Keokuk, Iowa, Nov 26, 46; m 69; c 1. CATALYSIS, REACTION KINETICS. *Educ:* Iowa State Univ, BS, 69; Stanford Univ, MS, 71, PhD(chem eng), 74. *Prof Exp:* Chem engr process develop, E I du Pont de Nemours & Co, 74-75; assoc prof chem eng, San Jose State Univ, 82-84; CHEM ENGR & CHEMIST, SRI INT, 75- *Mem:* Am Chem Soc; NAm Catalysis Soc; Am Inst Chem Engrs; Am Vacuum Soc. *Res:* Heterogeneous catalysis; chemisorption; surface analysis. *Mailing Add:* 15 Berenda Way Portola Valley CA 94025-7528

MCCARTY, KENNETH SCOTT, b Dallas, Tex, June 20, 22; m 44; c 2. BIOCHEMISTRY. *Educ:* Georgetown Univ, BS, 44; Columbia Univ, PhD, 58. *Prof Exp:* Res biochemist, US Food & Drug Admin, 44-45; asst to dir cancer res, Hyde Found, Interchem Corp, NY, 45-52; mem staff tissue cult course, NY & Colo, 54-58; instr biochem, Col Physicians & Surgeons, Columbia Univ, 57-59; assoc prof, 59-68, PROF BIOCHEM, SCH MED, DUKE UNIV, 68- *Concurrent Pos:* Electron micros consult, Vet Admin Hosp, NY, 49-54; consult, Organon Corp, NJ, 56-60 & Parke, Davis & Co, Mich, 60-61. *Mem:* Am Chem Soc; Am Soc Biol Chemists; Electron Micros Soc Am; Harvey Soc; Am Asn Cancer Res. *Res:* Molecular mechanisms in the control of RNA transcription, RNA transport and RNA translation; development of procedures to maintain homeostasis in tissue cultures to study malignant transformation, nucleic acid synthesis, amino acid metabolism and hormone induction in tissue culture. *Mailing Add:* Dept Biochem Duke Med Ctr Box 3711 2511 Perkins Rd Durham NC 27710

MCCARTY, LESLIE PAUL, b Detroit, Mich, May 30, 25; m 48; c 4. PHARMACOLOGY. *Educ:* Salem Col, BS, 47; Ohio State Univ, MSc, 49; Univ Wis, PhD(pharmacol), 60; Am Bd Toxicol, dipl. *Prof Exp:* Res chemist, Upjohn Co, 49-55; res pharmacologist, 60-64, sr res pharmacologist, 64-80, RES LEADER, DOW CHEM CO, 80- *Mem:* Am Soc Pharmacol & Exp Therapeut; Sigma Xi. *Res:* Cardiovascular, autonomic and neuromuscular pharmacology. *Mailing Add:* Occup Health Bldg 1803 Dow Chem Co Midland MI 48674

MCCARTY, LEWIS VERNON, b Buffalo, NY, June 7, 19; m 46; c 1. CHEMISTRY. *Educ:* Oberlin Col, AB, 41; Univ Rochester, PhD(phys chem), 45. *Prof Exp:* Res chemist, 45-64, mgr feasibility invests, 64-66, res adv inorg chem, Gen Elec Co, 66-81; ADJ CUR MAT CONSERV, CLEVELAND MUSEUM NATURAL HIST, 83- *Mem:* Am Chem Soc; Electrochem Soc. *Res:* Reaction kinetics and mechanisms; inorganic syntheses; high temperature chemistry; corrosion chemistry. *Mailing Add:* 3354 Rumson Rd Cleveland Heights OH 44118

MCCARTY, MACLYN, b South Bend, Ind, June 9, 11; m 34, 66; c 4. MEDICAL BACTERIOLOGY. *Educ:* Stanford Univ, AB, 33; Johns Hopkins Univ, MD, 37. *Hon Degrees:* DSc, Columbia Univ, 76, Univ Fla, 77, Rockefeller Univ, 82, Med Col Ohio, 85, Emory Univ, 87, Univ Köln, 88 & Wittenberg Univ, 89. *Prof Exp:* Asst pediat, Sch Med, Johns Hopkins Univ, 39-40; fel sulfonamide drugs, NY Univ, 40-41; Nat Res Coun fel, Rockefeller Univ, 41-42, assoc med, 46-48, assoc mem, 48-50, mem & physician, 50-60, physician-in-chief, Hosp, 60-74, prof, 60-81, vpres, 65-78, EMER PROF BACT & IMMUNOL, ROCKEFELLER UNIV, 81- *Concurrent Pos:* Vpres & chmn sci adv comt, Helen Hay Whitney Found; chmn, Health Res Coun New York, 72-75; chmn bd trustees, Pub Health Res Inst New York. *Honors & Awards:* Eli Lilly Award, 46; Waterford Biomed Sci Award, 77; Robert Koch Gold Medal, 81; Koralenko Medal, Nat Acad Sci, 88; Wolf Prize Med, Israel, 90. *Mem:* Nat Acad Sci; Nat Inst Med; Am Soc Microbiol; Am Soc Clin Invest; Soc Exp Biol & Med (pres, 74-75); Am Phil Soc. *Res:* Transformation of pneumococcal types; biology of hemolytic streptococci; rheumatic fever. *Mailing Add:* Rockefeller Univ York Ave & E 66th St New York NY 10021-6399

MCCARTY, PERRY L(EE), b Grosse Pointe, Mich, Oct 29, 31; m 53; c 4. ENVIRONMENTAL ENGINEERING. *Educ:* Wayne State Univ, BS, 53; Mass Inst Technol, SM, 57, ScD(sanit eng), 59. *Prof Exp:* Instr civil eng, Wayne State Univ, 53-54; from instr to asst prof sanit eng, Mass Inst Technol, 58-62; assoc prof, 62-67, prof, 67-75, chmn dept, 80-85, SILAS H PALMER PROF CIVIL ENG, STANFORD UNIV, 75-; DIR, WESTERN REGION HAZARDOUS SUBSTANCES RES CTR, 89- *Concurrent Pos:* Vchmn, Environ Studies Bd, Nat Res Coun, 77-80, mem, Comn on Natural Resources, 77-80, chmn, Comt to Rev Potomac Estuarf Exp Water Treatment Plant, 77-80; consult, Interagency Agr Waste Water Treatment Study, Fed Water Pollution Control Admin, US Bur Reclamation & Calif Dept Water Resources, 66-71; NSF sci fac fel, Harvard Univ, 68-69; fac mem, postgrad course hydrol eng, Ministry Pub Works, Venezuela, 69-; consult, training grants div, Environ Protection Agency, 70-; chmn, Gordon Res Conf on Environ Sci-Water, 72; mem, Nat Res Coun-Comt on Phys Sci, Math, Resources, 85-88; chmn, Comt Remedial Actions Priorities for Hazardous Waste Sites, Nat Res Coun, 91. *Honors & Awards:* Eddy Medal, Water Pollution Control Fedn, 62 & 67, Thomas Camp Award, 75; Huber Res Prize, Am Soc Civil Engrs, 69, Simons W Freese Environ Eng Lectr Award, 79. *Mem:* Nat Acad Eng; Am Soc Civil Engrs; hon mem Water Pollution Control Fedn; fel AAAS; hon mem Am Water Works Asn; Int Asn Water Pollution. *Res:* Biological waste treatment processes, treatment and fate of trace contaminants. *Mailing Add:* Dept Civil Eng Stanford Univ Stanford CA 94305

MCCARTY, RICHARD CHARLES, b Portsmouth, Va, July 12, 47; m 65; c 4. EXPERIMENTAL HYPERTENSION, PHYSIOLOGY OF STRESS. *Educ:* Old Dominion Univ, BS, 70, MS, 72; Johns Hopkins Univ, PhD(pathobiol), 76. *Prof Exp:* Res assoc, Nat Inst Ment Health, 76-78; from asst prof to assoc prof, 78-88, PROF PSYCHOL, UNIV VA, 88-, CHAIR PSYCHOL, 90- *Concurrent Pos:* Sr fel, Nat Heart, Lung & Blood Inst, 84-85; Prin Investr, Nat Heart, Lung & Blood Inst, 84-87, Nat Inst Mental Health grant, 85-88, res scientist develop award, 85-90 & NSF grant, 88-92. *Mem:* Fel Soc Behav Med; fel Acad Behav Med Res; Am Psychol Soc. *Res:* The role of early environmental factors in the development of experimental hypertension; effects of acute and chronic stress on regulation of the sympathetic nervous system. *Mailing Add:* Dept Psychol Univ Va Gilmer Hall Charlottesville VA 22903-2477

MCCARTY, RICHARD EARL, b Baltimore, Md, May 3, 38; m 61; c 3. BIOCHEMISTRY. *Educ:* Johns Hopkins Univ, BA, 60, PhD(biochem), 64. *Prof Exp:* NIH fel biochem, Pub Health Res Inst, New York, 64-66; from asst prof to assoc prof, 66-77, chmn, Sect Biochem, Molecular & Cell Biol, 80-85, dir, Biotechnol Prog, 88-90, PROF BIOCHEM, CORNELL UNIV, 77-; PROF & CHMN, DEPT BIOL, JOHNS HOPKINS UNIV, 90- *Concurrent Pos:* NIH career develop award, 69. *Mem:* Am Soc Biol Chem; Am Soc Plant Physiol; AAAS. *Res:* Photophosphorylation and electron flow in chloroplasts. *Mailing Add:* Dept Biol Johns Hopkins Univ Baltimore MD 21218

MCCARVILLE, MICHAEL EDWARD, b Moorland, Iowa, Aug 27, 36; m 62; c 3. ENZYMOLOGY. *Educ:* Loras Col, BS, 58; Iowa State Univ, PhD(biophys), 67. *Prof Exp:* Fel chem, La State Univ, Baton Rouge, 67-68; from asst prof to assoc prof, 68-84, PROF CHEM, WESTERN MICH UNIV, 84-, DEPT CHMN, 86- *Concurrent Pos:* Sabbatical leave, Upjohn Co, 75-76. *Mem:* AAAS; Am Chem Soc. *Res:* Purification and characterization of microbial enzymes involved in waste treatment; remediation of groundwater contamination. *Mailing Add:* Dept Chem Western Mich Univ Kalamazoo MI 49008

MCCARY, RICHARD O, b Sherman, Tex, Mar 21, 26; m 47; c 4. EDDY CURRENT TESTING, MAGNETIC DEVICES & PROCESSES. *Educ:* Univ Louisville, BEE, 46; Syracuse Univ, MEE, 57. *Prof Exp:* Test engr, Gen Elec Co, 46-48, develop engr, Electronics Dept, 48-49; instr elec eng, Syracuse Univ, 49-51; develop engr, Adv Electronics Ctr, 53-56, prin engr, Comput Dept, Univ Ariz, 56-61, consult engr, Light Mil Electronics Dept, NY, 61-64, STAFF MEM, RES & DEVELOP CTR, GEN ELEC CO, NY, 64- *Concurrent Pos:* Spec lectr, Ariz State Univ, 57-59. *Honors & Awards:* Centenial Award, Inst Elec & Electronic Engrs. *Mem:* Inst Elec & Electronics Engrs; Magnetic Soc; Comput Soc; Sigma Xi. *Res:* Nonlinear magnetic devices and circuits; computer memories; magnetic recording; eddy current testing. *Mailing Add:* 1355 Valencia Rd Schenectady NY 12309

MCCASKEY, A(MBROSE) E(VERETT), civil engineering; deceased, see previous edition for last biography

MCCASKEY, THOMAS ANDREW, b Jacksonville, Ohio, Mar 23, 28; m 60; c 1. MICROBIOLOGY. *Educ:* Ohio Univ, BS, 60; Purdue Univ, MS, 63, PhD(dairy microbiol), 66. *Prof Exp:* Teaching asst microbiol, Purdue Univ, 60-62, teaching assoc, 62-63, res asst dairy microbiol, 63-66, res assoc pesticide anal, 66-67; asst prof, 67-74, ASSOC PROF MICROBIOL, AUBURN UNIV, 74- *Mem:* Inst Food Technologists; Am Dairy Sci Asn. *Res:* Pesticide residues in poultry and processed milk; microbial spoilage of milk; prevalence of salmonellae in food; water and soil pollution by dairy farm wastes. *Mailing Add:* 1030 Crestwood Dr Auburn AL 36849

MCCASLIN, DARRELL, b Cushing, Okla, Nov 19, 51. PHYSICAL BIOCHEMISTRY, PROTEIN. *Educ:* Okla State Univ, BS, 73; Duke Univ, PhD (phys biochem), 79. *Prof Exp:* Med res asst prof anat, Duke Univ Med Ctr, 84-87; ASST PROF CHEM, RUTGERS UNIV, 87- *Mem:* Am Chem Soc; Biophys Soc; Protein Soc; Am Soc Biochem & Molecular Biol; Asn Comput Mach. *Res:* Structural function relationships in protein with primary emphasis on membrane proteins. *Mailing Add:* Dept Chem Rutgers Univ 73 Warren St Newark NJ 07102

MCCAUGHAN, DONALD, b Dublin, Ireland, Mar 10, 29; US citizen; c 3. MEDICINE, CARDIOLOGY. *Educ:* Dublin Univ, BA, 52, MB, BCh & BAO, 53, MD & MA, 61; Am Bd Int Med, cert, 73. *Prof Exp:* Asst chief cardiol, Vet Admin Hosp, West Roxbury, Mass, 63-76; vet admin, comput data processing study, Worcester, Vet admin, hosp, 63-76, vet admin anti hypertensive study, 63-76; CHIEF MED OFFICER, WORCESTER VET ADMIN CLIN, 76- *Concurrent Pos:* Prin investr coronary drug proj, Vet Admin Hosp, West Roxbury, Mass, 67-75, prin investr persantine aspirin reinfarction study, 75-80; asst prof med, Harvard Med Sch, 73-84. *Mem:* AAAS. *Res:* Coronary heart disease; hyperlipidemia; hypertension; electrocardiography; lipidology. *Mailing Add:* Vet Admin Clinic 595 Main St Worester MA 01601

MCCAUGHEY, JOSEPH M, b Philadelphia, Pa, Nov 23, 21; m 50; c 3. METALLURGY, PHYSICS. *Educ:* Drexel Univ, BSChE, 48. *Prof Exp:* Physicist, Frankford Arsenal, Philadelphia, 64-55, res adv, 55-69, assoc tech dir, 69-73, civilian exec, 73-77; CONSULT, 77- *Honors & Awards:* Eisenman Award, Am Soc Metals, 74. *Mem:* Fel Am Soc Metals. *Res:* Administration of programs in basic and applied research, materials engineering and exploratory development. *Mailing Add:* 719 Oak Terrace Dr Ambler PA 19002

MCCAUGHEY, WILLIAM FRANK, b Chicago, Ill, Oct 1, 21; m 43; c 1. NUTRITIONAL BIOCHEMISTRY. *Educ:* Purdue Univ, BS, 42; Northwestern Univ, MS, 48; Univ Ariz, PhD(agr chem), 51. *Prof Exp:* Asst biochem & nutrit, 51-55, asst agr biochemist, 55-57, from asst prof to assoc prof biochem & nutrit, 57-64, PROF BIOCHEM & NUTRITION, UNIV ARIZ, 64- *Mem:* Am Assoc Univ Prof; AAAS; Sigma Xi. *Res:* Honey bee nutrition; intermediary metabolism of amino acids. *Mailing Add:* Dept Nutrit & Food Sci Univ Ariz Tucson AZ 85721

MCCAUGHRAN, DONALD ALISTAIR, b Vancouver, BC, July 9, 32; m 52; c 3. BIOMETRICS, STATISTICS. *Educ:* Univ BC, BSc, 59, MSc, 62; Cornell Univ, PhD(biomet, statist), 70. *Prof Exp:* Regional wildlife biologist, BC Fish & Game Br, 61-65; asst prof statist & biomet, Univ Wash, 69-77; statist consult, 77-78; DIR INT PAC HALIBUT COMN, 78- *Mem:* Am Statist Asn; Biomet Soc; Am Fisheries Soc. *Res:* Experimental design; mathematical models in biology; estimation problems in fisheries and wildlife. *Mailing Add:* 6148 133 NE Kirkland WA 98033

MCCAULEY, DAVID EVAN, b Baltimore, Md, Aug 4, 50; m 80; c 1. EVOLUTIONARY BIOLOGY, POPULATION GENETICS. *Educ:* Univ Md, BS, 72; State Univ NY Stony Brook, PhD(ecol, evolution), 76. *Prof Exp:* Teach asst ecol, State Univ NY Stony Brook, 72-74, res asst population biol, 74-76; res assoc population biol, Univ Chicago, 76-80; DEPT BIOL, VANDERBILT UNIV, 80- *Mem:* Soc Study Evolution; Ecol Soc Am; Entom Soc Am; Am Soc Naturalists. *Res:* Population ecology and genetics of insect populations, including laboratory and field studies of population structure, dispersal, and mating behavior in flour, milkweed and soldier beetles. *Mailing Add:* Dept Biol Vanderbilt Univ Nashville TN 37235

MCCAULEY, GERALD BRADY, b Missoula, Mont, Apr 18, 35; m 58; c 1. ANALYTICAL CHEMISTRY. *Educ:* Univ Mont, BA, 57; Univ of the Pac, PhD(inorg chem), 67. *Prof Exp:* Physicist, US Naval Ord Lab, Calif, 57-63; res scientist anal chem, Lockheed Aircraft Corp, 67-70, RES SCIENTIST ANAL CHEM, LOCKHEED RES LAB, LOCKHEED MISSILES & SPACE CO, 70- *Mem:* Am Chem Soc. *Res:* Macrocyclic complexes of group eight metals; trace contamination analysis; vacuum degassing analysis using thermogravimetric techniques. *Mailing Add:* 163 Caymus Ct Sunnyvale CA 94086

MCCAULEY, H(ENRY) BERTON, b Duluth, Minn, Dec 20, 13; m 37. DENTISTRY. *Educ:* Univ Md, DDS, 36. *Prof Exp:* Instr oral roentgenology, Univ Md, 36-40; Carnegie fel dent & radiol, Univ Rochester, 40-43, asst prof dent, 43-45; res officer, USPHS, NIH, 45-49; dir dent care, Baltimore City Health Dept, 45-77, gen adminr, 77-80; health adv, Mayor & City Coun, Baltimore, 75-77; RETIRED. *Concurrent Pos:* Consult, War Dept, Manhattan Proj & Eastman Kodak, 43-45; pres, Md Soc Dent for Chldrn, 54-55; historian Md St Dent Asn, 58-87; lectr, Dent Sch, Univ Md, 63-78; pres, Md Pub Hlth Asn, 67-68; pres, Balto City Dent Soc, 73; Svc Award, Am Soc Dent for Chldrn, 78 & Md State Dent Asn, 86; pres, N Balto (mental hlth) Ctr, 80. *Honors & Awards:* Hayden-Harris Award, Am Acad Hist Dent, 88. *Mem:* Am Dent Asn; fel Am Pub Health Asn; fel AAAS; fel Am Col Dentists; Sigma Xi; Am Asn Dent Res. *Res:* Radioactive tracer studies of dental tissue metabolism; radiographic anatomy of teeth and jaws; epidemiology of fluoride and its effects on teeth and bones; history of dentistry and public health; observations of poliomyelitis as a function of dental caries experience. *Mailing Add:* Pres Am Acad Hist Dent 3804 Hadley Sq E Baltimore MD 21218-1807

MCCAULEY, HOWARD W, b Superior, Wis, Nov 27, 19; m 42; c 3. STRUCTURAL & CIVIL ENGINEERING. *Educ:* NDak State Univ, BS, 48; Univ Minn, MS, 49. *Prof Exp:* Assoc prof civil eng, NDak State Univ, 50-56; prof civil eng, 56-77, PROF MAT ENG, UNIV IOWA, 77- *Mem:* Am Soc Civil Engrs; Am Soc Eng Educ. *Res:* Response to reinforced concrete beams to torsional and bending loads. *Mailing Add:* 3344 Hanover Ct Iowa City IA 52245

MCCAULEY, JAMES A, b New York, NY, Jan 15, 41; m 68; c 2. PHYSICAL CHEMISTRY. *Educ:* St Vincent Col, BS, 62; Fordham Univ, PhD(chem), 68. *Prof Exp:* Sr res chemist, Merck Sharp & Dohme Res Labs, 67-78, res fel, 78-84, sr res fel, 84-90, SR INVESTR, MERCK SHARP & DOHME RES LABS, 90- *Mem:* Am Chem Soc. *Res:* Analytical and physical chemistry of immunological polysaccharides; the application of thermodynamics to the evaluation of the purity of biologically active compounds; isolation and identification of impurities in these compounds; polymorphism. *Mailing Add:* 54 Riverview Terr RD 1 Belle Mead NJ 08502

MCCAULEY, JAMES WEYMANN, b Philadelphia, Pa, Mar 21, 40; m 64; c 3. CHARACTERIZATION, PROCESSING. *Educ:* St Joseph's Col, Ind, BS, 61; Pa State Univ, MS, 65, PhD(solid state sci), 68. *Prof Exp:* Res asst solid state sci, Pa State Univ, 66-68; res chemist, 68-76, supvry mat res engr, mat & mech res ctr, 76-80, Chief Army Mats Characterization Div, 80-87, CHIEF, MAT SCI BR, 87- *Honors & Awards:* FH Norton Award; James I Mueller Mem Lectr. *Mem:* Am Crystallog Asn; fel Am Ceramic Soc; Mineral Soc Am; Mats Res Soc; Am Ceramic Soc (vpres, 87-88, trustee, 88-91). *Res:* Microstructural control of material properties; powder characterization; materials characterization; oxides; oxynitrides; micas; carbonates. *Mailing Add:* Mat Sci Br Subordinate Lab Command Mat Technol-Emerging Mat Div US Army Mat Technol Lab Watertown MA 02172

MCCAULEY, JOHN CORRAN, JR, orthopedic surgery, for more information see previous edition

MCCAULEY, JOHN FRANCIS, b New York, NY, Apr 2, 32; m 56; c 4. ASTROGEOLOGY. *Educ:* Fordham Univ, BS, 53; Columbia Univ, MA, 57, PhD(geol), 59. *Prof Exp:* Coop geologist, Pa Geol Surv, 56-58; from asst prof to assoc prof, Univ Sc, 58-63; geologist, 63-70, br chief, Br Astrogeol Studies, 70-75, geologist, 75-86, EMER GEOLOGIST, US GEOL SURV, 86- *Concurrent Pos:* Lectr, Columbia Univ, 57-58; proj geologist, Div Geol, SC Develop Bd, 58-63; consult geologist, 58-63; geol team leader, Mars Mariner TV Team, 71, prin investr, NASA Lunar & Planetary, 70-85, Viking & Voyager Progs, 76-82, Shuttle Imaging Radar, NASA, 82-91; chief geologist, Mineral Petrol Ground Water Explor Prog, USAID, Egypt, 82; adj prof, N Ariz Univ, 73- *Honors & Awards:* Autometric Award, Am Soc Photogram, 82. *Mem:* Fel Geol Soc Am. *Res:* Uranium occurrences in Pennsylvania; metamorphism, structure and mineral deposits of the Southern Appalachian Piedmont; lunar structure and stratigraphy; use of television and radar systems for planetary exploration; geology of Mars; geology of desert regions and eolian processes; geology and resources of Eastern Sahara. *Mailing Add:* 189 Wilson Canyon Rd Sedona AZ 86336

MCCAULEY, JOSEPH LEE, b Lexington, Ky, Feb 22, 43; m 62; c 3. MATH PHYSICS, PHYSICAL MATHEMATICS. *Educ:* Univ Ky, BSS, 65; Yale Univ, MS, 66, MPh, 68, PhD(physics), 72. *Prof Exp:* Asst 74-80, assoc prof, 74-89, PROF PHYSICS, UNIV HOUSTON, 89- *Concurrent Pos:* Consult, Inst Energit Chemikk, Kjeller, Norway; Am-Scand Found fel, 85. *Mem:* Am Phys Soc. *Res:* Nonlinear dynamics & tractals chaos theory; statistical mechanics and fluids theory of computability of chaos vortices in superfluids; vortex dynamics; nonlinear dynamics and chaos theory. *Mailing Add:* Physics Dept Univ Houston Houston TX 77204-5504

MCCAULEY, ROBERT F(ORRESTELLE), b Reno, Nev, Aug 31, 13; m 45; c 2. CIVIL & SANITARY ENGINEERING. *Educ:* NMex State Col, BS, 39; Mich State Univ, MS, 49; Mass Inst Technol, DSc, 52. *Prof Exp:* Assoc prof civil & sanit eng, 47-74, EMER ASSOC PROF, MICH STATE UNIV, 74-; pres, Wolverine Engrs & Survr, 69-87; RETIRED. *Concurrent Pos:* Vis prof, Sch Pub Health, Hawaii Univ, 66-67; recipient res grants & prin investr, NIH; eng consult, Mich State Univ. *Mem:* Am Water Works Asn; Water Pollution Control Fedn. *Res:* Control of corrosion in water distribution systems; removal of radioactivity and pollution from water. *Mailing Add:* 2803 LaSalle Gardens Lansing MI 48912

MCCAULEY, ROBERT WILLIAM, b Toronto, Ont, July 8, 26; m 56; c 3. FISH BIOLOGY. *Educ:* Univ Toronto, BA, 50, MA, 57; Univ Western Ont, PhD(zool), 62. *Prof Exp:* Res scientist, Fisheries Res Bd, Can, 55-62; biologist in charge fish cultural prog, Ont Dept Lands & Forests, 62-63, res scientist, 63-65; assoc prof, 65-80, PROF BIOL, WILFRID LAURIER UNIV, 80- *Concurrent Pos:* Consult, Ont Hydro-Prov of Ont, 74-76. *Mem:* Am Fisheries Soc; Am Inst Biol Sci; Can Soc Zool. *Res:* Fish physiology; thermal ecology; effects of warm effluents on fish. *Mailing Add:* 118 Forest Hill Dr Kitchener ON N2M 4G3 Can

MCCAULEY, ROY B, JR, b Chicago, Ill, Feb 9, 19; m 41; c 4. ENGINEERING. *Educ:* Cornell Col, BA, 40; Ill Inst Technol, MS, 43. *Prof Exp:* Res metallurgist, 38-39; asst metall, Ill Inst Technol, 40-43, from instr to asst prof, 43-50, actg chmn dept metall eng, 44-46; instr welding eng, 50-54, assoc prof, 54-60, res supvr, Eng Exp Sta, 56-62, chmn dept, 56-79, PROF WELDING ENG, OHIO STATE UNIV, 56-, PROF METALL ENG, 71-, DIR CTR WELDING RES, 79- *Concurrent Pos:* Robert Mehl hon lectr, Soc Nondestructive Testing, 65, lectr, Assembly, 67; vis scientist, Yugoslavia Welding Soc, 65; Nat Acad Sci vis scientist, Romania, 69; consult, govt & various indust concerns; mem comn pres educ, Int Inst Welding, 64-81, sub-comt pres testing, 77- *Honors & Awards:* Award, Am Welding Soc, 59. *Mem:* Am Soc Eng Educ; Am Soc Metals; Soc Nondestructive Testing; Am Welding Soc (vpres, 63-66, pres, 66-67). *Res:* Welding engineering education and metallurgy; development of phosfide alloys; casting metallurgy; industrial radiography with isotopes and x-rays; state of stress; bond mechanisms; discontinuity evaluations of manufactured metals and alloys. *Mailing Add:* 845 Linworth Rd E Worthington OH 43235

MCCAULEY, ROY BARNARD, b Chicago, Ill, Sept 20, 43; m 86; c 2. BIOCHEMICAL PHARMACOLOGY. *Educ:* Ohio State Univ, BS, 65, PhD(pharmacol), 70. *Prof Exp:* Res assoc biochem, Cornell Univ, 70-72; asst prof pharmacol, State Univ NY Downstate Med Ctr, 72-76; ASSOC PROF PHARMACOL, WAYNE STATE UNIV, 76- *Mem:* Am Soc Pharmacol & Exp Therapeut. *Res:* Biochemistry of drug action and the enzymatic organization of mammalian cells; enzymes involved in detoxification like the monoamine oxidases and phosphorus 450s. *Mailing Add:* Dept Pharmacol 6374 Scott Hall Wayne State Univ 540 E Canfield Detroit MI 48201

MCCAULEY, WILLIAM JOHN, b Ray, Ariz, Aug 11, 20; m 47; c 2. PHYSIOLOGY, ANATOMY. *Educ:* Univ Ariz, BS, 47; Univ Southern Calif, PhD(zool), 55. *Prof Exp:* Res assoc pharmacol, Lederle Labs, Am Cyanamid Co, NY, 47-49; lab assoc zool, Univ Southern Calif, 49-51, instr anat, Sch Dent, 52-54, Sch Med, 54-55; instr zool, 55-56, from asst prof to prof biol, 56-78, assoc dean, grad col, 74-78, prof, 78-82, EMER PROF GEN BIOL, UNIV ARIZ, 82- *Res:* Vertebrate physiology; biological education. *Mailing Add:* 1742 N Louis Lane Tucson AZ 85712

MCCAULEY CHAPMAN, KAREN, DIETETICS. *Educ:* Univ Ill, Champaign-Urbana, BS, 79, PhD(nutrit sci), 87; Eastern Ill Univ, Charleston, MS, 83. *Prof Exp:* Clin dietitian, 80-81, computer appl coordr, Dietetic Serv, 88-89, STAFF DEVELOP COORDR, DIETETIC SERV, VET ADMIN MED CTR, DANVILLE, 90-; ASST PROF & NUTRIT SPECIALIST FOODS & NUTRIT, COOP EXTENSION SERV, UNIV ILL, CHAMPAIGN-URBANA, 91- *Concurrent Pos:* Nutrit Support Team dietitian, Vet Admin Med Ctr, Danville, 81-, Equal Employment Officer counr, 90-; mem, Human Studies Subcomt, Danville Vet Admin Med Ctr, 86-89, nutrit comt, 89, res & develop comt, 90-, chairperson, 90-; consult, Crosspoint Human Serv, Danville, 88-; clin instr, Col Med & Div Nutrit Sci, Univ Ill, 89- *Mem:* Am Dietetic Asn; Am Soc Parenteral & Enteral Nutrit; Am Inst Nutrit; Am Soc Clin Nutrit. *Res:* Nutritional aspects of cancer; nutrition in critical care nursing. *Mailing Add:* Dept Dietetics Vet Admin Med Ctr 1900 E Main St Danville IL 61832

MCCAULLY, RONALD JAMES, b West Reading, Pa, Dec 25, 36; m 87; c 6. ORGANIC CHEMISTRY, MEDICINAL CHEMISTRY. *Educ:* Mass Inst Technol, SB, 58; Harvard Univ, PhD(org chem), 65. *Prof Exp:* Chemist, Arthur D Little, Inc, 60-61; sr res chemist, Wyeth Labs, Am Home Prod Corp, 63-68, group leader, 68-77, sect mgr, 77-84, assoc dir, 84-87, assoc dir, 87-89, DIR, WYETH-AYERST RES, AM HOME PROD CORP, 89- *Mem:* Am Chem Soc; AAAS. *Res:* Synthesis of new organic compounds as potential medicinal products; investigation of novel chemical reactions; syntheses of cardiovascular agents, central nervous system agents, and respiratory drugs. *Mailing Add:* Wyeth-Ayerst Res Cn 8000 Princeton NJ 08543-8000

MCCAUSLAND, IAN, b Lisbellaw, Northern Ireland, Apr 10, 29; m 65; c 2. ELECTRICAL ENGINEERING. *Educ:* Queen's Univ Belfast, BSc, 49, MSc, 50; Univ Toronto, PhD(elec eng), 58; Univ Cambridge, PhD(control eng), 64. *Prof Exp:* Design engr, Ferranti Ltd, Eng, 50-52; distribution engr, Elec Supply Bd, Ireland, 52-53; develop engr, Can Gen Elec, Toronto, 53-54 & Guelph, 54-55; demonstr elec eng, Univ Toronto, 55-56, lectr, 58-60, asst prof, 60-61; NATO sci fel, Nat Res Coun Can, Churchill Col, Univ Cambridge, 61-63; from asst prof to assoc prof, 64-71, assoc chmn dept, 72-76, PROF ELEC ENG, UNIV TORONTO, 71-, ASSOC CHMN GRAD STUDIES, 85-90, 91- *Concurrent Pos:* Vis scientist, Univ Cambridge, 71-72; ed, Can Elec Eng J, 79-80. *Mem:* Inst Elec & Electronic Engrs. *Res:* Electric circuits; control engineering; special relativity. *Mailing Add:* Dept Elec Eng Univ Toronto Toronto ON M5S 1A4 Can

MCCAWLEY, ELTON LEEMAN, b Long Beach, Calif, Nov 1, 15; m 40; c 2. PHARMACOLOGY. *Educ:* Univ Calif, AB, 38, MS, 39, PhD(pharmacol), 42. *Prof Exp:* Asst pharmacol, Sch Med, Univ Calif, 39-42, Int Cancer Res Found fel, lectr & res assoc, 42-43; from instr to asst prof pharmacol, Sch Med, Yale Univ, 43-49; assoc prof, 49-60, PROF PHARMACOL, MED SCH, UNIV ORE, 60- *Concurrent Pos:* Dir, Ore Poison Control Ctr, 57-64; consult, Providence Hosp; trustee, Ore Mus Sci & Indust, 60-66; mem bd dirs, Portland Ctr Speech & Hearing, 63-66; chmn, Gov Adv Comt Methadone Treatment Heroin Addicts, 69-72; inst rev comt drug studies in humans, Ment Health Div, 71-72. *Mem:* Am Soc Pharmacol & Exp Therapeut; AMA; Am Fedn Clin Res. *Res:* Cardiovascular pharmacology; morphine derivatives; drugs in cardiac arrhythmias; anesthetics. *Mailing Add:* Dept Pharmacol 7181 SW Jackson Park Rd Portland OR 97201

MCCAWLEY, FRANK X(AVIER), b Scranton, Pa, May 18, 24; m 54; c 5. METALLURGY, INORGANIC CHEMISTRY. *Educ:* Scranton Univ, BS, 49. *Prof Exp:* Chemist, Chicago Develop Corp, Md, 49-51, lab supvr, 51-59; metallurgist, 59-61, proj leader electroplating & electrometall, 61-73, proj leader, Corrosion & Electrometall Res Group, College Park, 73-79, SR COORDR, MINERAL LAND ASSESSMENT & FIELD OPER, US BUR MINES, WASHINGTON, DC, 79- *Mem:* Am Soc Metals; Am Inst Mining, Metall & Petrol Engrs; Electrochem Soc; fel Am Inst Chemists; AAAS. *Res:* Electrorefining and electroplating of refractory type metals from fused salts; electroplating of platinum metals from fused salts; electrorefining of copper from sulphate solutions; corrosion studies of construction materials in hypersaline geothermal brine; mineral land assessment management; resource project management. *Mailing Add:* 2309 Cheverly Ave Cheverly MD 20785

MCCAY, PAUL BAKER, b Tulsa, Okla, June 5, 24. BIOCHEMISTRY. *Educ:* Univ Okla, BS, 48, MS, 50, PhD(physiol, biochem), 55. *Prof Exp:* USPHS fel biochem, Okla Med Res Inst, 56-58, mem biochem sect, 58-71; head Biomembrane Res Lab, Okla Med Res Found, 72-87; from asst prof to assoc prof, 59-68, PROF BIOCHEM, SCH MED, UNIV OKLA, 68- *Concurrent Pos:* NIH grants, 56-; mem nutrit study sect, NIH, 74-78; fel Nat Sci Found, 52-54. *Mem:* Am Inst Nutrit; Am Soc Biol Chemists; Brit Biochem Soc; Am Oil Chemists Soc. *Res:* Role of membrane-bound electron transport systems in promoting lipid peroxidation in biological membranes; alterations of membrane function caused by free radicals generated by the activity of oxidoreductases; role of dietary antioxidants and fat in chemical carcinogenesis; influence of dietary states on carcinogen metabolism; free radical processes in biology and medicine, free radical detection in vivo and in vitro using spin trapping and electron spin resonance spectroscopy; tissue injury processes associated with ischemia/reperfusion, septic shock, alcohol consumption, herbicide poisoning and cocaine toxicity, role of tissue antioxidant and anti-free radical systems in controlling tissue injury by the conditions listed above. *Mailing Add:* 825 NE 13th Okla Med Res Found Oklahoma City OK 73104

MCCHESNEY, JAMES DEWEY, b Hatfield, Mo, Aug 27, 39; m 59; c 7. PLANT SCIENCE, BIO-ORGANIC CHEMISTRY. *Educ:* Iowa State Univ, BS, 61; Ind Univ, MA, 64, PhD(org chem), 65. *Prof Exp:* Chemist, Battelle Mem Inst, summer, 61; from asst prof to prof bot & med chem, Univ Kans, 65-78; prof & chmn, dept pharmacognosy, 78-86, DIR, RES INST PHARMACEUT SCI, UNIV MISS, 87- *Mem:* AAAS; Am Chem Soc; Am Soc Plant Physiol; NY Acad Sci; Am Asn Cols Pharm; Am Soc Pharmacog; Int Soc Chem Ecol; fel Am Inst Chem. *Res:* Chemistry and biochemistry of biologically active secondary plant substances; chemotherapy of tropical diseases. *Mailing Add:* RR #1 Box 340 Etta MS 38627

MACCHESNEY, JOHN BURNETTE, b Glen Ridge, NJ, July 8, 29; m 53; c 1. MATERIAL SCIENCE, CERAMICS ENGINEERING. *Educ:* Bowdoin Col, BA, 51; Pa State Univ, PhD(geochem), 59. *Prof Exp:* mem tech staff, 59-81, LAB FEL, AT&T BELL LABS, 82- *Honors & Awards:* George M Morey Award, Am Ceramic Soc; Morris N Liebman Award, Inst Elec & Electronic Engrs; Technol Achievement Award, Soc Photo-Optical Instrumentation Engrs; Commonwealth Award Sci & Invention, Sigma Xi; Int Prize New Mat, Am Phys Soc; Thomas A Edison Patent Award Res & Develop, Coun NJ. *Mem:* Nat Acad Eng; Am Ceramics Soc; Inst Elec & Electronic Engrs; Sigma Xi. *Res:* Investigation and properties of glass and other materials for use in optical communications. *Mailing Add:* AT&T Bell Labs 600 Mountain Ave Murray Hill NJ 07974

MACCHI, I ALDEN, b Bologna, Italy, Feb 21, 22; nat US; m 53; c 1. ENDOCRINOLOGY. *Educ:* Clark Univ, BA, 47, MA, 50; Boston Univ, PhD(endocrinol), 54. *Prof Exp:* Mem res staff, Worcester Found Exp Biol, 50-54; asst prof physiol, Clark Univ, 54-56; from asst prof to assoc prof, 56-64, exec asst res, Biol Sci Ctr, 56-67, prof biol, 64-83, chmn ad interim dept biol, 74-76, chmn ad assoc chmn dept, 76-77, EMER PROF BIOL, COL LIB ARTS, BOSTON UNIV, 83- *Concurrent Pos:* Lalor Found fel, 55; vis lectr & Dept Sci & Indust Res Eng sr res fel, Univ Sheffield, 62-63. *Mem:* Am Physiol Soc; Am Soc Zool; Endocrine Soc; Soc Exp Biol & Med; fel AAAS. *Res:* Comparative aspects of corticosteroid biogenesis; regulation of adrenocortical and pancreatic endocrine secretion; transplantation of adrenal and endocrine pancreas; pancreatic tissue culture. *Mailing Add:* 52 Roundwood Rd Newton MA 02164

MACCHIA, DONALD DEAN, b Gary, Ind, May 17, 48; m 69; c 2. CARDIAC ELECTROPHYSIOLOGY, TRANSPORT PHYSIOLOGY. *Educ:* Ind Univ, AB, 71; Ball State Univ, MA, 72; Univ Ill, Urbana, MS, 74, PhD(physiol & biophys), 77. *Prof Exp:* Teaching fel anat, Univ Ill, 72-74; instr physiol, Columbia Univ, 74-75; USPHS trainee, Univ Ill, 75-76; NIH fel, Univ Chicago, 76-77, asst prof physiol & med, 79-80; asst prof physiol, 80-85, ASSOC PROF PHARMACOL & PHYSIOL, IND UNIV, 85- *Mem:* Am Physiol Soc; Biophys Soc; Int Soc Heart Res; Am Heart Asn; Soc Gen Physiologists; Sigma Xi. *Res:* Cardiac electrophysiology; membrane physiology and biophysics; transport physiology. *Mailing Add:* Dept Pharmacol Sch Med NWCME Ind Univ 3400 Broadway Gary IN 46408

MACCINI, JOHN ANDREW, b Boston, Mass, July 9, 28; m 61; c 4. GEOLOGY. *Educ:* Boston Univ, BA, 52, MA, 54; Ohio State Univ, PhD(earth sci educ), 69. *Prof Exp:* Sr tech officer, Nfld Geol Surv, 53; construct & soils engr, Thompson & Lichtner Co, Brookline, Mass, 54-58; teacher earth sci, Lincoln-Sudbury Regional High Sch, Mass, 58-66; teaching assoc geol, Ohio State Univ, 66-69; assoc prof geol & sci educ, Univ Md, College Park, 69-71; prog mgr sci educ, 71-82, PROG MGR GEOL, NSF, 82- *Concurrent Pos:* Regional dir skylab student proj, NASA, 70-71. *Mem:* Nat Sci Teachers Asn; Nat Asn Geol Teachers; Nat Asn Res Sci Teaching. *Res:* Improvement of undergraduate science instruction with specific focus on audio-visual tutorial laboratory development for undergraduate geology courses. *Mailing Add:* Div Earth Sci Rm 602 GEO 1800 G St NW Washington DC 20550

MCCLAIN, EUGENE FREDRICK, b Salmon, Idaho, Nov 24, 26; m 48; c 2. FORAGE CROP BREEDING, FORAGE CROP MANAGEMENT. *Educ:* Univ Idaho, BS, 54; Univ Calif, MS, 56; Univ Ga, PhD(agron), 73. *Prof Exp:* Asst prof, 56-75, ASSOC PROF AGRON, CLEMSON UNIV, 75- *Mem:* Sigma Xi. *Res:* Breeding, genetic studies and management of forage grasses, legumes and sunflowers; adaptation of new crops; performance testing of forage, field corn and grain sorghum cultivers. *Mailing Add:* DRTE 3 Box 75 Pendleton SC 29670

MCCLAIN, GERALD RAY, b Chickasha, Okla, Dec 27, 41; m 72. INVENTION, CREATIVITY. *Educ:* Okla State Univ, BS, 64, MS, 74. *Prof Exp:* Designer, Hughes Tool Co, 61 & Gen Elec, 64-65; tech teacher design & drafting, Okla City Pub Sch, 65-68; HEAD & ASSOC PROF MECH DESIGN & MFG, ENG, OKLA STATE UNIV, 68- *Concurrent Pos:* Computer aided design training, autocad, cadkey, Screen Prod, 74-89, York Int Inc, 90-91, Mercruiser Inc, 91. *Mem:* Soc Mfg Engrs; Am Soc Eng Educators; Am Soc Mech Engrs. *Res:* Manufacturing education and computer aided design. *Mailing Add:* 25 Liberty Circle Stillwater OK 74075

MCCLAIN, JOHN WILLIAM, b Dayton, Ohio, Nov 3, 28. PHYSICS. *Educ:* Antioch Col, BS, 52; Princeton Univ, MA, 53, PhD(physics), 57. *Prof Exp:* Instr physics, Princeton Univ, 55-56; asst lectr, Univ Manchester, 56-57; from asst prof to assoc prof, Am Univ Beirut, 57-76; assoc prof, Yarmouk Univ, 78-85; vis assoc prof physics, Birzeit Univ, 85-88; VIS PROF PHYSICS, WESTERN KY UNIV, 88- *Concurrent Pos:* Res assoc, Harvard Univ, 67. *Mem:* Am Phys Soc; Sigma Xi; Am Asn Physics Teachers; NY Acad Sci. *Res:* Science education; history of science; participation in project to improve science teaching in West Bank schools: curriculum revision, construction of indigenous equipment, in-service teacher training. *Mailing Add:* Dept Physics & Astron Western KY Univ Bowling Green KY 42101

MCCLAMROCH, N HARRIS, b Houston, Tex, Oct 7, 42; m 63. APPLIED MATHEMATICS. *Educ:* Univ Tex, Austin, BS, 63, MS, 65, PhD(eng mech), 67. *Prof Exp:* PROF COMPUT INFO & CONTROL ENG, UNIV MICH, ANN ARBOR, 67- *Mem:* Math Asn Am; fel Inst Elec & Electronics Eng; Soc Indust & Appl Math; AAAS; Am Inst Aeronaut & Astronaut. *Res:* Optimal control theory; theory of differential equations; stability theory; applications of control to mechanical systems robotics and advanced automation. *Mailing Add:* Dept of Aerospace Univ of Mich Ann Arbor MI 48109

MCCLARD, RONALD WAYNE, b Brawley, Calif, Aug 12, 51; m. METABOLIC DISEASE. *Educ:* Cent Col, Iowa, BA, 72; Univ Calif, Los Angeles, PhD(biochem), 78. *Prof Exp:* NIH fel, Dept Biochem, Univ NC, 79-82; asst prof, dept chem, Boston Col, 82-84; ASSOC PROF, DEPT CHEM, REED COL, 87- *Concurrent Pos:* NSF Biochem Panel, 90. *Honors & Awards:* Dreyfus Teaching/Res Award, 87. *Mem:* Am Chem Soc. *Res:* Design of ribosyl-(C1)-phosphonate analog/inhibitors and metabolism of ribosyl phosphates; multifunctional proteins; inhibitors of nucleic acid precursor biosynthesis; regulation of metabolism; phosphorus synthetic methodology. *Mailing Add:* Dept Chem Reed Col 3203 SE Woodstock Blvd Portland OR 97202

MACCLAREN, ROBERT H, b Scranton, Pa, Aug 24, 13; m 35, 48, 76; c 6. CHEMICAL ENGINEERING. *Educ:* Wayne State Univ, BS, 34; Univ Mich, MS, 35. *Prof Exp:* Admin supvr wood cellulose, Eastman Kodak Co, 35-65; mgr paper technol, Xerox Corp, Webster, 65-76; CHIEF CHEMIST, NAT ARCHIVES & RECORDS SERV, 76- *Mem:* AAAS; Am Inst Chem Engrs; Am Chem Soc; Tech Asn Pulp & Paper Indust; Can Pulp & Paper Asn. *Res:* Cellulose morphology, water relationships and wood analysis; paper, film and image permanence; paper and parchment restoration and conservation. *Mailing Add:* 8380 Greensboro Dr No 511 McLean VA 22102

MCCLARY, ANDREW, b Chicago, Ill, Apr 15, 27; m 54; c 2. SCIENCE EDUCATION, SCIENCE IN MEDIA. *Educ:* Dartmouth Col, AB, 50; Univ Mich, MA, 54, PhD(zool), 60. *Prof Exp:* Res asst city planning, Chicago Plan Comn, 50-51; shrimp fishery, 51-52; from asst prof to assoc prof zool, Univ Wis-Milwaukee, 59-64; from asst prof to assoc prof natural sci, 64-69, PROF NATURAL SCI, MICH STATE UNIV, 69- *Mem:* AAAS; Am Inst Biol Sci. *Res:* History of hygiene, popular and consumer health and sanitation; science, especially health science education; health sciences and the media. *Mailing Add:* Dept Natural Sci 215 N Kedzie Lab Mich State Univ East Lansing MI 48824

MCCLARY, CECIL FAY, b Grayson, La, June 22, 13; m 40; c 2. GENETICS. *Educ:* La State Univ, BS, 36, MS, 38; Univ Calif, PhD(genetics), 50. *Prof Exp:* Asst poultryman, Western Wash Exp Sta, Wash State Univ, 38-52, assoc poultry scientist, 52-56; geneticist, Heisdorf & Nelson Farms, Inc, 56-71; dir genetics res, H&N Inc, 72-77; RETIRED. *Concurrent Pos:* Asst prof, Wash State Univ, 46; asst, Univ Calif, 46-49; consult, poultry breeding and prod mgt. *Mem:* Poultry Sci Asn. *Res:* Disease resistance and heredity; genetics of egg quality. *Mailing Add:* 8810-172nd Ave NE Redmond WA 98052

MCCLATCHEY, KENNETH D, Detroit, Mich, Apr 28, 42. HEAD & NECK PATHOLOGY. *Educ:* Univ Mich, MD, 75, DDS, 68. *Prof Exp:* ASSOC PROF PATH, UNIV MICH, 82- & ASSOC CHMN DEPT, 83- *Mem:* Col Am Path; Am Asn Pathologists; Am Soc Clin Pathologists. *Mailing Add:* Dept Path Rm G332 Box 0054 Univ Mich 1500 E Med Ctr Dr Ann Arbor MI 48109-0054

MCCLATCHEY, ROBERT ALAN, b Rockville, Conn, July 26, 38; m 61; c 2. ATMOSPHERIC PHYSICS. *Educ:* Mass Inst Technol, BS, 60, MS, 61; Univ Calif, Los Angeles, PhD(meteorol), 66. *Prof Exp:* Sr scientist, Jet Propulsion Lab, Calif Inst Technol, 63-67; sr scientist, AVCO Space Systs Div, 67-68; res scientist, 68-75, br chief, 75-80, DIR, ATMOSPHERIC SCI DIV, GEOPHYS DIRECTORATE, PHILLIPS LAB, HANSCOM AFB, 80- *Honors & Awards:* Losey Award, Am Inst Aeronaut & Astronaut, 78. *Mem:* AAAS; fel Am Meteorol Soc. *Res:* Atmospheric radiation, the transmission and emission of radiation in the atmosphere; molecular spectroscopy involved in the problem of atmospheric transmission; radiative transfer in the atmosphere and its relationship to meteorology; satellite-based atmospheric remote sensing. *Mailing Add:* Geophys Directorate (LY) Phillips Lab Hanscom AFB MA 01731

MCCLATCHY, JOSEPH KENNETH, b Brownwood, Tex, July 5, 39; m 60; c 4. MEDICAL MICROBIOLOGY. *Educ:* Tex Tech Col, BS, 61; Univ Tex, MA, 63, PhD(microbiol), 66. *Prof Exp:* Fel, Ind Univ, 65-66; fel, Nat Jewish Hosp & Res Ctr, 66-67, chief clin labs, 67-77, asst dir prof serv, 77-81, res scientist, 81-88; lab dir, Colo Clin Lab, 81-88; TECH DIR, NAT HEALTH LABS, 88- *Concurrent Pos:* Mem adv comt, Trudeau Mycobacterial Cult Collection, 77-80; consult microbiol, US Army, 77-81; vpres, Arvada Mgt Co, 81-82. *Mem:* AAAS; Am Soc Microbiol; Conf Pub Health Lab Dirs. *Res:* Modes of action of antituberculosis drugs; diagnostic microbiology; resistance to microbial infections; immunotherapy of cancer; mechanisms of drug resistance; gerontology. *Mailing Add:* Nat Health Labs 6665 S Kenton Suite 210 Englewood CO 80111

MCCLEARY, HAROLD RUSSELL, b Huntsville, Ohio, Oct 11, 13; m 41; c 3. PHYSICAL CHEMISTRY. *Educ:* Monmouth Col, BS, 37; Columbia Univ, PhD(chem), 41. *Prof Exp:* Asst, Columbia Univ, 37-40; res chemist, Am Cyanamid Co, 41-45, asst chief chemist, 45-52, asst mgr pigments res, 52-54, mgr, 54-57, dir res serv, 57-64, mgr dyes res & develop, 64-71, mgr dyes & textiles chem res & develop, 71-74, asst dir, Bound Brook Res Lab, 75-78; RETIRED. *Mem:* Am Chem Soc. *Res:* Kinetics and mechanism of organic reactions; spectrophotometric methods; dyeing and textile chemistry; analytical methods; pigments. *Mailing Add:* Box 76 Rosemont NJ 08556

MCCLEARY, JAMES A, b Bridgeport, Ohio, Mar 26, 17; m 38; c 4. BOTANY. *Educ:* Asbury Col, AB, 38; Univ Ohio, MS, 46; Univ Mich, PhD, 52. *Prof Exp:* Asst bot, Univ Ohio, 39-40; asst prof, Ariz State Univ, 47-51, from assoc prof to prof, 52-60; prof, Calif State Univ, Fullerton, 60-69; chmn dept biol sci, Northern Ill Univ, 69-78, prof bot, 69-82; RETIRED. *Concurrent Pos:* Instr, Univ Mich, 51; NSF fac fel, 59-60; assoc prog dir, NSF Undergrad Res Participation Prog, Washington, DC. *Mem:* AAAS; Am Bryol & Lichenological Soc; Ecol Soc Am; Int Soc Plant Taxon; Bot Soc Am. *Res:* Bryophytes; desert plants. *Mailing Add:* 118 Augusta Apt 118 DeKalb IL 60115

MCCLEARY, JEFFERSON RAND, b Reno, Nev, Aug 31, 48; m 68. GEOLOGY, GEOLOGICAL ENGINEERING. *Educ:* Univ Nev, BS, 70, MS, 75. *Prof Exp:* Geologist remote sensing, NASA Earth Res Prog, Univ Nev, 68; geologist eng geol, Nev State Hwy Dept, 69; teaching fel geol, Univ Nev, 71, teaching asst, 71-73; GEOLOGIST, WOODWARD-CLYDE CONSULTS, 73- *Mem:* Geol Soc Am; Sigma Xi. *Res:* Manifestations of active tectonic processes in areas of low and moderate seismscicity as determined from detailed studies of structure and stratigraphy. *Mailing Add:* 367 E Center St Moab UT 84532-2432

MCCLEARY, STEPHEN HILL, b Houston, Tex, Feb 6, 41; div. ALGEBRA. *Educ:* Rice Univ, BA, 63; Univ Wis, MS, 64, PhD(math), 67. *Prof Exp:* Instr math, Univ Wis, 67; vis asst prof, Tulane Univ, 67-68; ASSOC PROF MATH, UNIV GA, 68- *Mem:* Am Math Soc. *Res:* Ordered groups. *Mailing Add:* Bowling Green State Univ Bowling Green OH 43403

MCCLEAVE, JAMES DAVID, b Atchison, Kans, Dec 17, 39; m 62; c 1. ZOOLOGY. *Educ:* Carleton Col, AB, 61; Mont State Univ, MS, 63, PhD(zool), 67. *Prof Exp:* Instr zool, Mont State Univ, 63-64; asst prof, Western Ill Univ, 67-68; assoc prof, 68-77, PROF ZOOL, UNIV MAINE, ORONO, 77- *Concurrent Pos:* NSF grants, Mont State Univ, 68-70 & Univ Maine, 70-71; Off Naval Res grant, Univ Maine, 71-72. *Mem:* AAAS; Am Soc Zool; Am Fisheries Soc; Animal Behav Soc; Sigma Xi. *Res:* Orientation of migratory fishes; effects of thermal pollution on ecology of fishes. *Mailing Add:* Dept of Zool Univ of Maine Orono ME 04469

MCCLELLAN, AUBREY LESTER, b Oklahoma City, Okla, Feb 5, 23; m 46; c 2. PHYSICAL CHEMISTRY. *Educ:* Centenary Col, BS, 43; Univ Tex, MS, 46, PhD(phys chem), 49. *Prof Exp:* Res fel, Univ Calif, 48-50; res fel, Mass Inst Technol, 50-51; res chemist, Chevron Res Co, Standard Oil Co Calif, 51-54, staff asst to vpres res, 54-56, mem president's staff, 56-57, sr res chemist, 57-65, sr res assoc, 65-72, group leader, molecular identification, 72-85; RETIRED. *Concurrent Pos:* Chem study, Curric Revision Team, 63-65. *Mem:* Am Chem Soc; AAAS. *Res:* Hydrogen bonding; chemical education in secondary schools; dipole moments. *Mailing Add:* 8636 Don Carol Dr El Cerrito CA 94530

MCCLELLAN, BETTY JANE, b Little Rock, Ark, Oct 26, 32. PATHOLOGY. *Educ:* Southern Methodist Univ, BS, 55; Univ Ark, MD, 57; Am Bd Path, dipl, 67. *Prof Exp:* From intern to asst resident path, Mass Gen Hosp, 57-60; from instr to asst prof, Johns Hopkins Univ, 61-64; dir surg path, Health Sci Ctr, 64-75, assoc prof, 64-69, PROF PATH, SCH MED, UNIV OKLA, 69-, PROF CLIN LAB SCI, COL HEALTH 70-, ASSOC DIR SURG PATH, UNIV HOSP & CLINS, 75-, MED DIR, UNIV HOSP & CLINS SCH CYTOTECHNOL, 77-, ACTG DIR SURG PATH & DIR PROG CYTOTECHNOL, COL HEALTH, 78- *Concurrent Pos:* Pathologist, Johns Hopkins Hosp, 62-64, assoc dir, Sch Cytotechnol, 63-64; consult & mem sci adv bd consults, Armed Forces Inst Path, Washington, DC, 78-80. *Mem:* Am Soc Clin Pathologists; Col Am Pathologists; Am Soc Cytol; Int Acad Path. *Res:* Anatomic, surgical pathology; cytopathology; cancer detection. *Mailing Add:* Dept Cytopath Okla Mem Hosp PO Box 26307 Oklahoma City OK 73126

MCCLELLAN, BOBBY EWING, b Cayce, Ky, July 5, 37; m 59. ANALYTICAL CHEMISTRY. *Educ:* Murray State Univ, BA, 59; Univ Miss, PhD(anal chem), 63. *Prof Exp:* Res fel, Univ Ariz, 63-64; group leader analysis chem, PPG Industs, 64-65; Comt Instnl Studies & Res Grant, 66-67, PROF CHEM, MURRAY STATE UNIV, 65- *Mem:* Am Chem Soc. *Res:* Solvent extraction separations; extraction kinetics; atomic absorption spectrometry; nuclear chemistry as applied to analytical chemistry. *Mailing Add:* Dept Chem Murray State Univ Box 3226 University Sta Murray KY 42071

MCCLELLAN, GUERRY HAMRICK, b Gainesville, Fla, Nov 1, 39; m 62; c 2. MINERALOGY, GEOLOGY. *Educ:* Univ Fla, BS, 61, MS, 62; Univ Ill, PhD(geol, clay mineral, chem), 64. *Prof Exp:* Res chemist, Nat Fertilizer Res Ctr, Int Fertilizer Develop Ctr, 65-66, proj leader, Nat Fertilizer Develop Ctr, 66-73, sr proj leader, 73-76, res coordr, 76-87. *Concurrent Pos:* Fel, Nat Petrol Co of Aquitaine & Univ Bordeaux, France, 64-65. *Mem:* Mineral Soc Am; Clay Mineral Soc; Mineral Soc Gt Brit & Ireland. *Res:* Evaluation of phosphate rocks for industrial and agricultural uses; modification of elemental sulfur melts; sludge from stack gas removal processes; crystallography; geochemistry, inorganic and organic; phosphate rock beneficiation. *Mailing Add:* 5400 NW 159th St 121 Hialeah FL 33014

MCCLELLAN, JAMES HAROLD, b Guam, Oct 5, 47; US citizen; m 69; c 2. DIGITAL SIGNAL PROCESSING, COMPUTER ENGINEERING. *Educ:* La State Univ, BS, 69; Rice Univ, MS, 72, PhD(elec eng), 73. *Prof Exp:* Tech staff, Lincoln Lab, Mass Inst Technol, 73-75, assoc prof elec eng, 75-82; tech consult, Schlumberger Well Servs, 82-87; PROF ELEC ENG SCH ELEC ENG, GA TECH, 87- *Concurrent Pos:* Consult, Lincoln Lab, Mass Inst Technol, 75-82; res eng, Lawrence Livermore Nat Lab, 79. *Honors & Awards:* Tech Achievement Award, Inst Elec & Electronic Engrs Accoust, Speech & Signal Processing Soc, 87. *Mem:* Fel Inst Elec & Electronic Engrs. *Res:* Theory & applications of digital signal processing techniques for beam forming, spectrum analysis and multi-dimensional systems; high-speed algorithms and implementations on special purpose computer architectures. *Mailing Add:* Sch Elec Eng Ga Tech Atlanta GA 30332

MCCLELLAN, JOHN FORBES, b Pembrook, Ont, Aug 30, 17; nat US; m 50; c 3. PROTOZOOLOGY, SOIL PROTOZOA. *Educ:* Univ Ill, Urbana, BS, 47, MS, 48, PhD(zool), 54. *Prof Exp:* Asst instr zool, Univ Ill, Urbana, 47, asst instr protozool, 47-49; from instr to assoc prof, Univ Detroit, 51-63; ASSOC PROF ZOOL, COLO STATE UNIV, 63- *Concurrent Pos:* Grants, Nat Cancer Inst, Univ Detroit, 57-60, AEC, 62, NSF co-investr, Belowground Ecosyst Processes, 75-83; consult, Wayne County Rd Comnr, Mich, 62; grant consult, NSF Undergrad Sci Equip Prog, 64-65, dir, NSF Summer Insts in Field Biol for Col Teachers, Colo State Univ, Mountain Campus, 68-72, proposal consult, NSF Col Teachers Progs, 69-71; consult, Colo Utility Power Co, 70-71. *Mem:* Fel AAAS; Soc Protozool; Sigma Xi; Am Soc Zool. *Res:* Role of Protozoa in soil; interaction of protozoa with other microbiol soil organisms including the increase in soil mineralization due to associations of microbial organisms and plant species; cytology. *Mailing Add:* Dept of Zool & Entom Colo State Univ Ft Collins CO 80523

MCCLELLAN, ROGER ORVILLE, b Tracy, Minn, Jan 5, 37; m 62; c 3. INHALATION TOXICOLOGY, RADIATION TOXICOLOGY. *Educ:* Wash State Univ, DVM, 60; Am Bd Vet Toxicol, dipl, 67; Am Bd Toxicol, cert, 80; Univ NMex, MMS, 80. *Prof Exp:* Biol scientist, Biol Labs, Gen Elec Co, 59-62, sr scientist, Hanford Labs, 63-64; sr scientist, Pac Northwest Labs, Battelle Mem Inst, 65; scientist div biol & med, US AEC, 65-66; dir fission prod inhalation prog & asst dir res, 66-73, vpres & dir res admin, 73-76, pres/dir Inhalation Toxicol Res Inst, Lovelace Biomed & Environ Res Inst, 76-88; PRES, CHEM INDUST INST TOXICOL, 88- *Concurrent Pos:* Mem div biol & med, Adv Comt Space Nuclear Systs Radiol Safety, 67-73; adv, Lab Animal Biol & Med Training Prog, Univ Calif, Davis, 68-70; consult, Nat Inst Environ Health Sci, NIH, 68-71, mem toxicol study sect, NIH, 69-73; mem, Nat Coun Radiation Protection & Measurements, 69-, chmn sci comt, 69-79, mem sci comt 57, 77-; adj prof, Dept Radiation Sci, Univ Ark, 70-, adj prof toxicol, Duke Univ, 88-, Univ NC, Chapel Hill, 89-, clin prof, Col Pharm, Univ NMex, 86-; pres, Am Bd Vet Toxicol, 70-73; clin assoc sch med, Univ NMex, 71-, adj prof biol, 73-83; mem, Transuranium Tech Group, Adv to US AEC Div Biomed & Environ Res, 72-77; mem environ radiation exposure adv comt, Environ Protection Agency, 72-77, chmn, 74-77, mem exec comt, sci adv bd, 74-80 & 81-; mem adv bd vet specialities, 73-76; mem NIH animal res adv comt, 74-78; mem, ad hoc comt on biol effects of ionizing radiations, Nat Res Coun-Nat Acad Sci, 74-76, mem comt animal models for res on aging & chmn, subcomt on carnivores, Inst Lab Animal Res, 77-80, comt on Animal Models for Res on Aging, Inst Lab Animal Resources, 77-80, ad hoc mem, Bd Toxicol & Environ Health Hazards, 80-87; mem, biomed adv comt on health effects for reactor safety study to Nuclear Regulatory Comt, 75-76; chmn, Ad Hoc Rev Comt, US Environ Protection Agency, 78-80, co-chmn, Health Effects Res Prog, 78-79, mem, Health Effects Comt, 80-83, Ad Hoc Clear Air Sci Adv Comt, 80-83; chmn, ad hoc review comt on sci criteria for environ lead, Environ Protection Agency, 77-78; mem, Enewetak Atoll cleanup adv group, 77-80; chmn, health effects panel, Environ Impact of Oil Shale Technol Workshop, 77-78; mem, comt legislative assistance; adj prof, Dept Vet & Comp Anat Pharmacol & Physiol & mem, Grad Fac Prog Vet Sci, Col Vet Med, Wash State Univ, Pullman, 80-; mem, Nat Coun Radiation Protection & Measurements, 71-, Prog Comt, 77-81, chmn, 81, Sci Comt #57, 77-, Comt #1-2, 88-; mem health res comt, Health Effects Inst, 81-;

mem, Dose Assessment Adv Group, Dept Energy, 80-87; mem, Adv Coun, Ctr Risk Mgt, Resources for the Future, 87-, Nat Adv Environ Health Sci Coun, NIH, 87-, bd dirs, Lovelace- Anderson Endowment Found, 88-, Nat Acad Sci/Nat Res Coun Comt Toxicol, 79-89, chmn, 80-87; assoc ed, Lab Animal Sci, 76-80, J Toxicol & Environ Health, 82-, J Fundamental & Appl Toxicol, 87-88, Inhalation Toxicol J, 87-89, ed, CRC Critical Rev Toxicol, 87-, ed adv, Environ Bus J, 89- *Honors & Awards:* Elda Anderson Award, Health Physics Soc, 74; Herbert E Stokinger Award, Am Conf Govt Indust Hygienists, 85; Frank R Blood Award, Soc Toxicol, 89. *Mem:* Inst Med-Nat Acad Sci; fel AAAS; Am Vet Med Asn; Health Physics Soc; Soc Risk Anal; Soc Toxicol; Radiation Res Soc; Am Asn Aerosol Res; Am Thoracic Soc; Soc Exp Biol & Med. *Res:* Inhalation toxicology; occupational and environmental exposure standards; comparative medicine; health effects of automotive emissions, effects of inhaled radioactivity; radiation toxicology. *Mailing Add:* Chem Indust Inst Toxicol PO Box 12137 Research Triangle Park NC 27709

MCCLELLAN, WILLIAM ALAN, petroleum geology, for more information see previous edition

MCCLELLAND, ALAN LINDSEY, b Galesburg, Ill, Sept 19, 25; m 47; c 4. INORGANIC CHEMISTRY, RESEARCH ADMINISTRATION. *Educ:* Northwestern Univ, BS, 45; Univ Ill, PhD(chem), 50. *Prof Exp:* Asst, Univ Ill, 47-49; Nat Res Coun fel, Univ Birmingham, 50-51; instr chem, Univ Conn, 51-54; res chemist, Cent Res Dept, E I du Pont de Nemours & Co, Inc, 54-60, col rels rep, Employee Rels Dept, 60-64; vpres eng, Cherry-Burrell Corp, Iowa, 64-67; staff asst, Personnel Div, Employee Rels Dept, E I Du Pont de Nemours & Co, Inc, 67-69, personnel adminr, Cent Res Dept, 69-87; NAT SCI FOUND, 87- *Mem:* AAAS; Am Chem Soc; Sigma Xi. *Res:* Technical personnel administration. *Mailing Add:* 22 Guyencourt Rd Wilmington DE 19807

MCCLELLAND, BERNARD RILEY, b Denver, Colo, Jan 23, 35. AVIAN ECOLOGY. *Educ:* Colo A&M Col, BS, 56; Colo State Univ, MS, 68; Univ Mont, PhD(habitat mgt), 77. *Prof Exp:* Ranger & naturalist protection interpretation, Nat Park Serv, US Dept Interior, 56-73; instr resource mgt, 73-77, ASST PROF RESOURCE MGT, UNIV MONT, 77- *Mem:* Wildlife Soc; Am Inst Biol Scientists; Sigma Xi. *Res:* Ecology of hole nesting birds; long range movements of bald eagles; fire ecology in the northern Rocky Mountains. *Mailing Add:* Box 366 West Glacier MT 59936

MCCLELLAND, BRAMLETTE, b Newnan, Ga, Dec 16, 20; c 5. GEOTECHNICAL ENGINEERING. *Educ:* Univ Ark, BS, 40; Purdue Univ, MS, 42. *Hon Degrees:* DEng, Purdue Univ, 84. *Prof Exp:* Design engr, San Jacinto Water Dist, Houston, Tex, 43-46; partner, Greer & McClelland, 46-55; pres & chmn, McClelland Engrs Inc, 55-87; chmn, Fugro-McClelland BV, 87-90; RETIRED. *Concurrent Pos:* Consult & mem supvry bd, Fugro-McClelland BV, 90- *Honors & Awards:* James Laurie Prize, Am Soc Civil Engrs, 59. *Mem:* Nat Acad Eng; Am Soc Civil Engrs. *Res:* Geotechnical and geoenvironmental engineering; author of over 30 technical papers on engineering geology, pile foundations, foundations for hot-process structures, mat foundations for high-rise buildings and in situ testing devices. *Mailing Add:* Fugro-McClelland Inc 6100 Hillcroft Houston TX 77081

MCCLELLAND, GEORGE ANDERSON HUGH, b Bushey, Eng, May 12, 31; m 58, 75; c 4. MEDICAL ENTOMOLOGY, ECOLOGY. *Educ:* Univ Cambridge, BA, 55; Univ London, PhD(zool, entom), 62. *Prof Exp:* Sci officer, EAfrican Virus Res Inst, Entebbe, Uganda, 55-59; res assoc entom, London Sch Hyg & Trop Med, 59-62; res assoc, Univ Notre Dame, 62-63; from asst prof to assoc prof, 63-75, PROF ENTOM, UNIV CALIF, DAVIS, 75- *Concurrent Pos:* Proj leader, EAfrican Aedes Res Univ, WHO, Tanzania, 69-70. *Mem:* Am Soc Naturalists; Am Soc Trop Med & Hyg; Royal Entom Soc London; Zool Soc London. *Res:* Biology and ecology of mosquitoes; yellow fever mosquito Aedes aegypti. *Mailing Add:* Dept Entom Univ Calif Davis CA 95616

MCCLELLAND, JOHN FREDERICK, b Elmira, NY, July 22, 41; m; c 2. PHOTOACOUSTIC & INFRARED SPECTROSCOPY, PHOTOACOUSTIC MICROSCOPY. *Educ:* Dickinson Col, BS, 65; Iowa State Univ, PhD(physics), 76. *Prof Exp:* Physicist, Nat Bur Standards, 62-74; res physics, Iowa State Univ, 67-76; res scientist electro-optics, Honeywell Electro-Optics Ctr, 76-77; PHYSICIST, AMES LAB, US DEPT ENERGY, 77-, DIR MACH TECHNOL PHOTOACOUST, 84-; PRES, MTEC PHOTOACOUST INC, 84- *Honors & Awards:* IR-100 Award, 85. *Mem:* Am Phys Soc; Optical Soc Am. *Res:* Optical properties of condensed matter; infrared spectroscopy; photoacoustic and modulation spectroscopy. *Mailing Add:* Ames Lab US Dept Energy Iowa State Univ Ames IA 50011

MCCLELLAND, MICHAEL, b Ashton-Under-Lyne, Eng, Aug 20, 56; m 90. RESTRICTION ENDONUCLEASES & METHYLASES, GENETIC FINGERPRINTING. *Educ:* Bristol Univ, BSc, 78; Univ Ga, Athens, PhD(molecular & pop genetics), 83. *Prof Exp:* Postdoctoral fel biochem, Univ Calif, Berkeley, Ca, 83-84; Markey scholar genetics, Columbia Univ, NY, 84-86; asst prof biochem & molecular biol, Univ Chicago, Ill, 86-89; RES PROG DIR MOLECULAR BIOL, CALIF INST BIOL RES, 89- *Concurrent Pos:* Nato postdoctoral fel, SERC, UK, 84; Markey biomed scholar, Markey Found, Fla. *Mem:* Am Soc Biochem & Molecular Biol; Am Soc Microbiol; AAAS. *Res:* DNA cleavage enzymology for pulsed field gel electrophoresis; physical mapping in bacterial genomes; genome fingerprinting by arbitrarily primer polymerase chain reaction for phylogeny and for genetic mapping. *Mailing Add:* Calif Inst Biol Res 11099 N Torrey Pines Rd La Jolla CA 92037

MCCLELLAND, NINA IRENE, b Columbus, Ohio, Aug 21, 29. ENVIRONMENTAL CHEMISTRY. *Educ:* Univ Toledo, BS, 51, MS, 63; Univ Mich, MPH, 64, PhD(environ chem), 68. *Prof Exp:* Chemist-bacteriologist, City of Toledo, 51-56, from chemist to chief chemist, 56-63; prof dir water, 68-74, VPRES TECH SERV, NAT SANIT FOUND, 74- *Concurrent Pos:* Consult, Ann Arbor Sci Publ, 70-; resident lectr, Univ Mich, 70- *Mem:* Am Chem Soc; Am Pub Health Asn; Am Water Works Asn; Nat Environ Health Asn; Water Pollution Control Fedn. *Res:* Development of electrochemical instrumentation for potable water quality characterization, continuous monitoring in distribution systems and in-plant treatment process control; chemical leaching from plastics pipe and piping systems for potable water applications. *Mailing Add:* 5 Ridgemor Dr Ann Arbor MI 48103

MCCLELLAND, ROBERT NELSON, b Gilmer, Tex, Nov 20, 29; m 58; c 3. SURGERY. *Educ:* Univ Tex, BA, 52; Univ Tex Med Br Galveston, MD, 54. *Prof Exp:* Intern, Med Ctr, Univ Kans, 54-55; resident gen surg, Parkland Mem Hosp, Dallas, 57-59; gen practitioner, Neblett Hosp, Canyon, Tex, 59-60; resident gen surg, Parkland Mem Hosp, Dallas, 60-62; from instr to assoc prof, 62-71, PROF GEN SURG, UNIV TEX HEALTH SCI CTR, DALLAS, 71- *Concurrent Pos:* Nat Inst Gen Med Sci res grant, Univ Tex Health Sci Ctr Dallas, 65-67, NIH res grant, 67-; surg consult, 4th Army, US Darnall Gen Hosp, 68-; ed, Audio-Jour Rev-Gen Surg, 71-; chmn ad hoc comt study liver injuries in Viet Nam, Vet Admin, 71- *Mem:* Soc Surg Alimentary Tract; Am Gastroenterol Asn; Am Surg Asn; fel Am Col Surg. *Res:* Gastroenterology, especially splanchnic blood flow and gastroduodenal stress ulceration. *Mailing Add:* 5323 Harry Hines Blvd Dallas TX 75235

MCCLEMENT, JOHN HENRY, pulmonary diseases; deceased, see previous edition for last biography

MCCLENACHAN, ELLSWORTH C, b Chicago, Ill, Mar 13, 34; m 58; c 1. ORGANIC CHEMISTRY. *Educ:* Univ Chicago, BA, 54, SM, 55; Univ Mich, PhD(org chem), 59. *Prof Exp:* Res chemist, Am Cyanamid Co, 59-61; res dir plastics, Miller-Stephenson Chem Co, 61-64; vpres, 64-67, PRES, R H CARLSON CO, 67- *Mem:* Am Chem Soc; Elec-Electronics Mats Distribrs Asn. *Res:* Organic reaction mechanisms; thermosetting resins to use in electronic and aerospace applications; epoxy resins, polyesters and silicones. *Mailing Add:* 55 North St Greenwich CT 06830

MCCLENAGHAN, LEROY RITTER, JR, b Kansas City, Mo, Nov 10, 48; m 69; c 2. POPULATION ECOLOGY, ECOLOGICAL GENETICS. *Educ:* Colo State Univ, BS, 70; Univ Kans, PhD(biol), 77. *Prof Exp:* PROF BIOL, SAN DIEGO STATE UNIV, 77- *Mem:* Am Soc Mammalogists; Ecol Soc Am; Soc Study Evolution. *Res:* Evolution and ecology of small mammals. *Mailing Add:* Dept of Biol San Diego State Univ San Diego CA 92182-0057

MCCLENAHAN, WILLIAM ST CLAIR, b Brainerd, Minn, Sept 24, 12; m 41; c 4. ORGANIC CHEMISTRY. *Educ:* Carleton Col, BA, 33; Mass Inst Technol, PhD(org chem), 38. *Prof Exp:* Chemist, Northwest Paper Co, Minn, 33-35; res assoc div chem, NIH, 38-40; indust fel, Mellon Inst, 40-43, sr fel, 43-55; coordr chem & phys res div, Standard Oil Co (Ohio), 55-60, chief spec projs mkt res & prod develop div, Chem Dept, 60-63, sr mkt res analyst, Corp Planning & Develop Dept, 63; assoc, Skeist Labs, 64-65; chief chem resources group, Inst Paper Chem, 65-69, dir, Div Info Serv, 69-81; RETIRED. *Concurrent Pos:* Corn Industs Res Found fel, NIH, 38-40; lectr, Univ Pittsburgh, 47-48. *Mem:* Tech Asn Pulp & Paper Indust; Am Chem Soc; Chem Mkt Res Asn. *Res:* Oxidation of glycosides with lead tetraacetate; chemistry of starch and dextrins; patents; market research; pulp and paper; information storage and retrieval. *Mailing Add:* 2600 Heritage Woods Dr Appleton WI 54915

MCCLENDON, JAMES FRED, b Alexandria, La, Jan 1, 38; div; c 3. MATHEMATICS. *Educ:* Tulane Univ, BA, 58; Univ Calif, Berkeley, MS, 63, PhD(math), 66. *Prof Exp:* Instr math, Yale Univ, 66-68; from asst prof to assoc prof, 68-87, PROF MATH, UNIV KANSAS, 87- *Mem:* Am Math Soc. *Res:* Algebriac topology. *Mailing Add:* Dept Math Univ Kans Lawrence KS 66045

MCCLENDON, JOHN HADDAWAY, b Minneapolis, Minn, Jan 17, 21; m 47; c 3. PLANT PHYSIOLOGY. *Educ:* Univ Minn, BA, 42; Univ Pa, PhD(bot), 51. *Prof Exp:* Res assoc, Hopkins Marine Sta, Stanford Univ, 51-52; res assoc bot, Univ Minn, 52-53; asst res prof agr biochem & food tech, Univ Del, 53-64, actg chmn dept, 59-64; from assoc prof to prof bot, Univ Nebr, Lincoln, 65-89; RETIRED. *Concurrent Pos:* Civilian with sci & tech div, Supreme Comdr Allied Powers, Tokyo, 46-47. *Mem:* AAAS; Bot Soc Am; Am Soc Plant Physiol; Sigma Xi. *Res:* Photosynthesis, leaf structure and optics; origin of life. *Mailing Add:* 105 Bush St Ashland OK 97520

MCCLENON, JOHN R, b Grinnell, Iowa, May 1, 37; m 59; c 3. ORGANIC CHEMISTRY. *Educ:* Grinnell Col, BA, 59; Univ Calif, Los Angeles, PhD(org chem), 64. *Prof Exp:* Asst prof chem, Milton Col, 63-65; asst prof, 65-71, assoc prof, 71-76 PROF CHEM, SWEET BRIAR COL, 76 - *Concurrent Pos:* Pres, FBN Microcomputing. *Mem:* Am Asn Univ Prof. *Res:* Chemistry of allenes; use of differential thermal analysis for analysis of organic compounds; construction of inexpensive instruments for teaching and research; use of computers in teaching and research. *Mailing Add:* Dept Chem Sweet Briar Col Sweet Briar VA 24595

MCCLINTIC, JOSEPH ROBERT, b Fayette, Mo, July 13, 28. PHYSIOLOGY. *Educ:* San Diego State Col, BA, 49; Univ Calif, PhD(physiol), 55. *Prof Exp:* Asst instr biol, San Diego State Col, 49-50; res physiologist, Univ Calif, 53-54; from instr to assoc prof biol, 54-70, PROF BIOL, CALIF STATE UNIV, FRESNO, 70- *Res:* Sodium metabolism; endocrinology. *Mailing Add:* Dept Biol Calif State Univ 6241 N Maple Ave Fresno CA 93740

MCCLINTOCK, BARBARA, b Hartford, Conn, June 16, 02. GENETICS. *Educ:* Cornell Univ, BS, 23, MA, 25, PhD, 27. *Hon Degrees:* Numerous from US cols and univs, 47-83. *Prof Exp:* Instr bot, Cornell Univ, 27-31, res assoc, 34-36; fel, Nat Res Coun, 31-33 & Guggenheim Found, 33-34; asst prof bot, Univ Mo, Columbia, 36-41; staff mem, 42-67, DISTINGUISHED SERV MEM, CARNEGIE INST WASH, COLD SPRING HARBOR, NY, 67- *Honors & Awards:* Nobel Prize Philos Med, Carnegie Inst Wash, 83; Charles Leopold Mayer Prize, Acad Sci, Inst France, 82; Louisa Gross Horwitz Prize Biol & Biochem, Columbia Univ, 82. *Mem:* Nat Acad Sci; AAAS; Am Acad

Arts & Sci; Am Philos Soc; Am Soc Naturalists; Genetics Soc Am (vpres, 39, pres, 45); foreign mem Royal Soc Eng. *Res:* Mutation in kernels of maize (corn); transposable genetic elements; molecular and microbial genetics. *Mailing Add:* Cold Spring Harbor Lab PO Box 100 Cold Spring Harbor NY 11724

MACCLINTOCK, COPELAND, b Princeton, NJ, Dec 3, 30; m 56; c 2. INVERTEBRATE PALEONTOLOGY. *Educ:* Franklin & Marshall Col, BS, 54; Univ Wyo, MA, 57; Univ Calif, Berkeley, PhD(paleont), 64. *Prof Exp:* Field asst, US Geol Surv, 54-55; res asst, 63-65, RES ASSOC, PEABODY MUS, YALE UNIV, 65-, ASST TO DIR, 68- *Concurrent Pos:* NASA res grant, 65-68. *Mem:* Geol Soc Am; Paleont Soc; Sigma Xi. *Res:* Microstructure and growth of fossil and recent mollusk shells; relationship between shell structures and classification, phylogeny and ecology of mollusks. *Mailing Add:* 33 Rogers Rd Hampden CT 06520

MCCLINTOCK, DAVID K, b Springfield, Ohio, May 1, 38; m 61; c 2. BIOCHEMISTRY, IMMUNOLOGY. *Educ:* State Univ NY Buffalo, BA, 64, PhD(biochem), 69. *Prof Exp:* Biochemist fibrinolysis, 70-74, group leader atherosclerosis, 74, head dept cardiovasc-renal pharmacol, 74-75, sect dir med prod, 75-78, SECT DIR METAB DIS RES, LEDERLE LABS DIV, AM CYANAMID CO, 78- *Mem:* AAAS. *Res:* New drug discovery; Immunologic diseases research, arthritis, cancer, asthma, allergy, diabetes. *Mailing Add:* Lederle Labs 401 N Middletown Rd Pearl River NY 10965

MCCLINTOCK, ELIZABETH, b Los Angeles, Calif, July 7, 12. SYSTEMATIC BOTANY. *Educ:* Univ Calif, Los Angeles, BA, 37, MA, 39; Univ Mich, PhD, 56. *Prof Exp:* Herbarium botanist, Univ Calif, Los Angeles, 41-47; assoc cur, 48-69, cur bot, Calif Acad Sci, 69-77; RETIRED. *Concurrent Pos:* Res assoc, Herbarium, Univ Calif, Berkeley, 78- *Mem:* Am Inst Biol Sci; Am Soc Plant Taxon; Sigma Xi. *Res:* Monographic studies in Labiatae and Hydrangeaceae; taxonomy of woody ornamentals. *Mailing Add:* Herbarium Univ Calif Berkeley CA 94720

MCCLINTOCK, FRANK A(MBROSE), b St Paul, Minn, Jan 2, 21; m 44; c 4. MECHANICAL ENGINEERING. *Educ:* Mass Inst Technol, BS & MS, 43; Calif Inst Technol, PhD(mech eng), 50. *Hon Degrees:* LLD, Univ Glasgow, 81. *Prof Exp:* Asst proj engr, United Aircraft Corp, 43-46; instr mech eng, Calif Inst Technol, 48-49; from asst prof to prof, 49-91, EMER PROF MECH ENG, MASS INST TECHNOL, 91- *Honors & Awards:* Nadai Award, Am Soc Mech Engrs, 78; Howe Medal, Am Soc Metals, 91. *Mem:* Nat Acad Eng; fel Am Acad Arts & Sci. *Res:* Plastic flow and fracture. *Mailing Add:* Rm 1-304C Mass Inst Technol Cambridge MA 02139

MCCLINTOCK, PATRICK RALPH, VIROLOGY, HYBRIDOMAS. *Educ:* Univ Tenn, Oakridge, PhD(biomed sci), 75. *Prof Exp:* RES IMMUNOLOGIST, AM TYPE CULT COLLECTION, 85- *Res:* Ticornaviruses. *Mailing Add:* Am Type Cult Collection 12301 Parklawn Dr Rockville MD 20852

MCCLOSKEY, ALLEN LYLE, b Granville, NDak, Aug 25, 22; m 47; c 3. INDUSTRIAL CHEMISTRY. *Educ:* Whittier Col, AB, 46; Univ Wis, PhD(org chem), 51. *Prof Exp:* Instr chem, Univ Calif, Los Angeles, 51-52; instr, Univ Pa, 52-54; res chemist, US Borax & Chem Corp, 55-57, group leader, US Borax Res Corp, 57-59, from assoc dir chem res to dir chem res, 59-69, VPRES & DIR RES, US BORAX RES CORP, 69- *Mem:* Am Chem Soc. *Res:* Steroid synthesis; glyicodnitriles; acyloin reaction in ammonia; mechanism of Stobbe reaction; addition of carboxylic acids to olefins; boron and organoboron chemistry. *Mailing Add:* 18422 Taft Ave Villa Park CA 92667

MCCLOSKEY, CHESTER MARTIN, b Fresno, Calif, July 21, 18; m 44; c 2. ORGANIC CHEMISTRY. *Educ:* Whittier Col, BA, 40; Univ Iowa, MS, 42, PhD(org chem), 44. *Prof Exp:* With Nat Defense Res Comt Proj, Iowa, 44-45 & comt med res proj, Calif Inst Technol, 45-46; chief chemist, Alexander H Kerr & Co, Inc, 46-48; phys scientist, Off Naval Res, 48-54, chief scientist, Calif, 55-57; exec dir & sr res fel, Off Indust Assocs, Calif Inst Technol, 57-62; PRES & TECH DIR, NORAC CO, INC, 62- *Honors & Awards:* Bartow Award, Am Chem Soc. *Mem:* Am Chem Soc; Soc Plastics Indust; Nat Fire Protection Asn. *Res:* Organic peroxides; vinyl polymerization; photopolymerization; carbohydrates. *Mailing Add:* Norac Co Inc 405 S Motor Ave Azusa CA 91702

MCCLOSKEY, JAMES AUGUSTUS, JR, b San Antonio, Tex, June 25, 36; m 60; c 4. CHEMISTRY. *Educ:* Trinity Univ, Tex, BS, 57; Mass Inst Technol, PhD(chem), 63. *Prof Exp:* Asst prof chem, Baylor Col Med, 64-67, from assoc prof to prof, 67-74; PROF BIOMED CHEM, UNIV UTAH, 74-, PROF BIOCHEM, 75- *Concurrent Pos:* NIH fel, Nat Ctr Sci Res, Ministry Educ, France, 63-64; sabbatical, Nat Cancer Ctr Res Inst, Tokyo, 71 & vis prof, Univ Utah, 72. *Mem:* Am Chem Soc; Am Soc Biol Chem; Am Soc Mass Spectrometry (pres, 78-80). *Res:* Mass spectrometry and its applications to structural problems in organic chemistry and biochemistry. *Mailing Add:* Dept Med Chem Univ of Utah Salt Lake City UT 84112

MCCLOSKEY, JOHN W, b Dayton, Ohio, Mar 2, 38; m 60; c 3. MATHEMATICAL STATISTICS. *Educ:* Univ Dayton, BS, 60; Mich State Univ, MS, 62, PhD(statist), 65. *Prof Exp:* Asst prof math, 65-69, ASSOC PROF MATH, UNIV DAYTON, 69-, CHMN DEPT, 76- *Mem:* Am Statist Asn; Sigma Xi. *Res:* Computer simulation; error analysis; mathematical modeling; regression techniques. *Mailing Add:* Dept of Math Univ of Dayton Dayton OH 45409

MCCLOSKEY, LAWRENCE RICHARD, b Philadelphia, Pa, May 5, 39. ZOOLOGY. *Educ:* Atlantic Union Col, AB, 61; Duke Univ, MA, 65, PhD(zool), 68. *Prof Exp:* Res assoc, Friday Harbor Labs, Univ Wash, 67-68; res fel, Systs-Ecol Prog, Marine Biol Lab, Woods Hole, 69-71; asst prof, 70-76, assoc prof, 76-79, PROF BIOL, WALLA WALLA COL, 79- *Mem:* AAAS; Am Soc Limnol & Oceanog; Ecol Soc Am; Am Inst Biol Sci; Sigma Xi. *Res:* Marine and community ecology; coral biology; effects of various pollutants on marine animals; coral respiration. *Mailing Add:* Dept Biol Walla Walla Col College Place WA 99324

MCCLOSKEY, RICHARD VENSEL, b Wilkinsburg, Pa, Dec 27, 33; m 56; c 3. INFECTIOUS DISEASES. *Educ:* Washington & Jefferson Col, AB, 55; Univ Rochester, MS & MD, 60; Am Bd Internal Med, dipl, 67; Am Bd Infectious Dis, dipl, 72. *Prof Exp:* Intern med, Med Ctr, Duke Univ, 60-61; res assoc, Lab Trop Virol, Nat Inst Allergy & Infectious Dis, 61-63; jr & sr asst resident med, Med Ctr, Duke Univ, 63-65, assoc med & immunol, 65-66, asst prof med & assoc immunol, 66-67; from assoc prof to prof physiol & med, Med Sch, Univ Tex San Antonio, 67-72, head sect infectious dis, 67-72; chief Infectious Dis, Presby-Univ Pa Med Ctr, 86-90; CLIN PROF MED, UNIV PA SCH MED, 87-; VPRES MED AFFAIRS, CENTOCOR, 90- *Concurrent Pos:* Am Col Physicians-Mead Johnson scholar, 64-65; head sect infectious dis & immunol, Durham Vet Admin Hosp, NC, 66-67. *Mem:* Am Soc Microbiol; Am Fedn Clin Res; Infectious Dis Soc Am. *Res:* Monoclonal antibodies; immunology of infectious diseases as one of the determinants of infection. *Mailing Add:* Centocor 244 Great Valley Pkwy Malvern PA 19355

M'CLOSKEY, ROBERT THOMAS, b Los Angeles, Calif, Nov 10, 40; m 64; c 2. ECOLOGY. *Educ:* Univ Calif, Los Angeles, BA, 64; Calif State Col, Los Angeles, MA, 66; Univ Calif, Irvine, PhD(pop biol), 70. *Prof Exp:* Asst prof ecol, 70-75, ASSOC PROF BIOL, UNIV WINDSOR, 75- *Concurrent Pos:* Nat Res Coun Can fel, Univ Windsor, 70-; Dept Indian Affairs Nat Parks Br fel, 71-72. *Mem:* AAAS; Am Inst Biol Sci; Ecol Soc Am; Brit Ecol Soc; Am Soc Mammal. *Res:* Species diversity and coexistence; habitat selection. *Mailing Add:* Dept Biol Sci Univ Windsor Windsor ON N9B 3P4 Can

MCCLOSKEY, TERESEMARIE, b Toledo, Ohio, July 1, 26. MATHEMATICAL LOGIC. *Educ:* Notre Dame Col, Ohio, BS, 49; John Carroll Univ, MS, 53; Case Western Reserve Univ, PhD(math), 72. *Prof Exp:* Teacher math & sci, St Mary High Sch, Warren, Ohio, 49-50, Notre Dame Acad, Cleveland, 50-53 & Elyria Dist Cath High Sch, 53-58, 60-63; instr math, Notre Dame Col, 63-65; teacher math & chmn dept, Elyria Dist Cath High Sch, 67-68; from instr to assoc prof, 68-78, chmn dept, 74-78, PROF MATH, NOTRE DAME COL, 78- *Concurrent Pos:* supvr, Student Computer Opers, Notre Dame Col, 81- *Mem:* Math Asn Am; Asn Comput Mach. *Res:* Abstract families of languages, leading to more complete treatment of context sensitive languages; computer simulation techniques for small machines; cognitive science based on Newman's Grammar of Assent. *Mailing Add:* Notre Dame Col 4545 College Rd Cleveland OH 44121

MCCLOUD, DARELL EDISON, b Cass Co, Ind, Mar 7, 20; m 40; c 3. AGRONOMY, AGRICULTURE. *Educ:* Purdue Univ, BSA, 45, MS, 47, PhD(agron), 49. *Prof Exp:* Asst prof agron, Univ Fla, 48-54, assoc agronomist, 54-57; head humid pasture & range res sect, Agr Res Serv, USDA, 57-65; chmn dept, 6574, Univ Fla, prof agron, 74-80; chief party, Malawi Proj, USAID, 80-83; Malawi proj prof, 83-84, prof, 85-89, EMER PROF AGRON, INST FOOD & AGR SCI, UNIV FLA, 89- *Concurrent Pos:* Consult, Lab Climat, Johns Hopkins Univ, 53; consult bioclimat sect, US Weather Bur, 55. *Honors & Awards:* Medallion Award, Am Forage & Grassland Coun, 74. *Mem:* Crop Sci Soc Am (pres, 70-71); fel Am Soc Agron (pres, 73-74). *Res:* Agricultural administration; agroclimatology; crop ecology; field and forage crop production and management research. *Mailing Add:* 304 Newell Hall Dept of Agron Univ Fla Gainesville FL 32611

MCCLOUD, HAL EMERSON, JR, b Kansas City, Mo, Feb 15, 38; m 66. SOLID STATE PHYSICS, ELECTRICAL ENGINEERING. *Educ:* Univ Kans, BS, 61; Univ Mo, Rolla, MS, 64, PhD(eng physics), 67. *Prof Exp:* Asst prof, 66-71, ASSOC PROF PHYSICS, ARK STATE UNIV, 71- *Mem:* Am Phys Soc. *Res:* Electromagnetic theory. *Mailing Add:* Dept of Physics Box 70 Ark State Univ PO Box 1990 State University AR 72467

MCCLOY, JAMES MURL, b Hollywood, Calif, Mar 28, 34; m 64. EPIDEMIOLOGY, MARINE SCIENCES. *Educ:* State Col, Los Angeles, BA, 61; La State Univ, PhD(geog), 69. *Prof Exp:* Res asst, Coastal Studies Inst, La State Univ, 64-68; asst prof geog, Univ Mt, 68-71; vis asst prof geog, 71-72, asst dean acad affairs, 72-75, head, dept gen acad, 75-77, dir, Coastal Zone Lab, 78-88, VPRES ACAD AFFAIRS, TEX A&M UNIV, GALVESTON, 88- *Concurrent Pos:* Vis asst prof, dept environ sci, Univ Va, 69 & 71; asst prog dir, Nat Oceanic & Atmospheric Admin, 77-78; mem water safety comn, Nat Safety Coun, 80-; consult bd dirs, Coun Nat Coop Aquatics, 81-90; mem bd dirs, Coastal Soc, 82-84; chair, coastal & marine group, Asn Am Geographers, 83-85; exec bd & int liason officer, US Lifesaving Asn, 87- *Mem:* Asn Am Geographers; Coastal Soc; Nat Safety Coun; Marine Technol Soc; US Lifesaving Asn; World Lifesaving. *Res:* Epidemiology and etiology of water-related accidents and fatalities and the risk management of public open-water recreational beaches. *Mailing Add:* Tex A&M Univ PO Box 1675 Galveston TX 77553

MACCLUER, JEAN WALTERS, b Columbus, Ohio, Mar 30, 37. HUMAN GENETICS, POPULATION GENETICS. *Educ:* Ohio State Univ, BSc, 59; Univ Mich, MSc, 63, PhD(human genetics), 68. *Prof Exp:* Res asst mech eng, Battelle Mem Inst, 59-60; elec eng, Antenna Lab, Ohio State Univ, 60-62; res assoc human genetics, Univ Mich, 68-71; res assoc anthrop, Univ Pa, University Park, 71-72, asst prof, 72-74, assoc prof biol, 74-; assoc scientist, 81-85, SCIENTIST, SOUTHWEST FOUND BIOMED RES, 85-; PROF CELLULAR & STRUCT BIOL, UNIV TEX HEALTH SCI CTR, SAN ANTONIO, 85- *Concurrent Pos:* Assoc prof cellular & struct biol, Univ Tex Health Sci Ctr, San Antonio, 81-85. *Mem:* Am Soc Human Genetics; Am Heart Asn; Int Soc for Animal Blood Group Res; Soc Study Evolution; Soc Study Social Biol; Sigma Xi; fel Ctr Adv Study, Behav Sci, 77-78. *Res:* Population genetics; genetic epidemiology; computer simulation. *Mailing Add:* Southwest Found Biomed Res PO Box 28147 San Antonio TX 78228-0147

MCCLUER, ROBERT HAMPTON, b San Angelo, Tex, Apr 13, 28; m 49; c 4. BIOCHEMISTRY. *Educ:* Rice Inst, BA, 49; Vanderbilt Univ, PhD(biochem), 55. *Prof Exp:* Res fel biochem, Univ Ill, 55-57; from instr to assoc prof physiol chem & psychiat, Ohio State Univ, 57-68, prof physiol chem, 68-69; asst dir biochem res, 70-75, DIR BIOCHEM RES, EUNICE

KENNEDY SHRIVER CTR MENT RETARDATION, 75-; PROF BIOCHEM, BOSTON UNIV, 71- *Mem:* Am Chem Soc; NY Acad Sci; Am Soc Neurochem; Int Soc Neurochem; Soc Neurosci; Sigma Xi. *Res:* Chemistry and metabolism of gangliosides; chemistry and metabolism of neutral glycolipids; neurochemistry; high performance liquid chromatography of lipids; kidney glycolipids. *Mailing Add:* Eunice Kennedy Shriver Ctr for Ment Retardation 200 Trapelo Rd Waltham MA 02178

MCCLUNG, ANDREW COLIN, b Morgantown, WVa, Oct 15, 23; m 47; c 2. SOIL FERTILITY. *Educ:* Univ WVa, BS, 47; Cornell Univ, MS, 49, PhD(soils), 50. *Prof Exp:* Res assoc prof agron, NC State Col, 50-56; agronomist, IBEC Res Inst, Sao Paulo, Brazil, 56-60; soil scientist, Rockefeller Found, 60-64; assoc dir, Int Rice Res Inst, Philippines, 64-71; dep dir gen, Int Ctr Trop Agr, 72-73; dep dir agr sci, Rockefeller Found, 73-76; exec officer, Int Agr Develop Serv, 76-80, pres, 80-85; regional rep, Asia & Winrock Int, Inst Agri Develop, 85-89; INDIA COORDR, WINROCK INT INST AGR DEVELOP, 89- *Mem:* Soil Sci Soc Am; Am Soc Agron. *Res:* Mineral nutrition of tree crops; methods of assessing the fertility status of soils; tropical agriculture; rural development; farming systems. *Mailing Add:* PO Box 12706 Arlington VA 22209

MCCLUNG, J KEITH, b Parkersburg, WVa, July 13, 54; c 1. CELLULAR AGING. *Educ:* Univ WVa, PhD(biochem), 83. *Prof Exp:* Fel, Baylor Col Med, 83-86; ASST SCIENTIST CELL BIOL, SAMUEL ROBERTS NOBLE FOUND, 86- *Mem:* Am Soc Cell Biol; Tissue Cult Asn. *Res:* Inhibitor of DNA synthesis and its relationship to aging. *Mailing Add:* Samuel Roberts Noble Found PO Box 2180 Ardmore OK 73402

MCCLUNG, LELAND SWINT, b Atlanta, Tex, Aug 4, 10; m 44. MICROBIOLOGY. *Educ:* Univ Tex, AB, 31, AM, 32; Univ Wis, PhD(bact), 34. *Prof Exp:* Res bacteriologist, Res Div, Am Can Co, 34-36; instr fruit prod & jr bacteriologist, Col Agr, Univ Calif, 36-37, instr res med, Hooper Found, Med Sch, 37-39; Guggenheim fel, Harvard Med Sch, 39-40; asst prof in chg dept, Ind Univ, 40-44, assoc prof bact, 44-48, chmn dept, 47-66, asst dir div biol sci, 65-68, prof, 48-81, EMER PROF MICROBIOL, IND UNIV, BLOOMINGTON, 81- *Concurrent Pos:* Secy, Pub Health & Nutrit Sect, Pac Sci Cong, 39; mem comt educ, Am Inst Biol Sci, 59-65, bd gov, 67-74; archivist, Am Soc Microbiol, 53-81; prog comt, Am Acad Microbiol, 41-58, chmn, 43-48, comt bacteriol tech, 42-54, comt educ, 58-64, chmn, 58-62, bd gov, 61, 67, 73-74, homor mem, 81- *Honors & Awards:* Cohen Award, Am Soc Microbiol, 84. *Mem:* AAAS; Soc Indust Microbiol (vpres, 58); Soc Exp Biol & Med; Nat Asn Biol Teachers (pres, 65); fel Am Acad Microbiol. *Res:* General and applied microbiology; history of microbiology; bacteriophage; Clostridium; science education; microbial food poisoning; studies on anaerobic spore-forming bacteria (genus Clostridium) including chromogenic pectin-fermenting types, bacterial viruses, and Clostridium perfringens as an agent of human food poisoning and ecological and taxonomic problems. *Mailing Add:* Dept Biol Jordan Hall 138 Ind Univ Bloomington IN 47405

MCCLUNG, MARVIN RICHARD, b Bamboo, WVa, Apr 3, 17; m 43; c 2. ANIMAL BREEDING. *Educ:* Univ WVa, BS, 41; Univ Md, MS, 42; Iowa State Univ, PhD(animal breeding, statist), 53. *Prof Exp:* County agent, Exten Serv, WVa, 42-45; agr counr, Monongahela Power Co, 45-47; asst prof animal indust & vet sci, Univ Ark, 47-52; geneticist & dir poultry div, Southern Farms Asn, 52-57; chmn, Dept Poultry Sci, Univ RI, 57-64; chmn, Dept Animal Indust & Vet Sci, 64-70, prof animal sci & animal scientist, 70-79, EMER PROF ANIMAL SCI, WVA UNIV, 79- *Concurrent Pos:* Dir, Dist E, Southern States Coop, Inc, Richmond, Va, 80-; secy, WVa Beef Bd, 84-; chmn, corp bd dir, Southern States Coop, Inc, Richmond, 90- *Mem:* Fel AAAS; Poultry Sci Asn; Am Soc Animal Sci; Genetics Soc Am; Am Genetic Asn; Sigma Xi. *Res:* Population genetics; selection experiments in poultry and beef cattle and correlated responses. *Mailing Add:* 1150 Charles Ave Morgantown WV 26505

MCCLUNG, NORVEL MALCOLM, b Bingham, WVa, June 9, 16; m 45, 66; c 4. MICROBIOLOGY. *Educ:* Glenville State Col, AB, 36; Univ Mich, MS, 40, PhD(bot), 49. *Prof Exp:* Teacher high sch, WVa, 36-41; asst prof bot, Univ Kans, 48-57; assoc prof bact, Univ Ga, 57-66; PROF BIOL, UNIV SOUTH FLA, 66- *Concurrent Pos:* Fulbright res scholar, Japan, 62-63. *Honors & Awards:* Fulbright Res Award, Japan, 62. *Mem:* AAAS; fel Am Acad Microbiol; Am Soc Microbiol; Mycol Soc Am; Bot Soc Am. *Res:* Biology of Actinomycetes, especially the genus Nocardia; ultrastructure of microorganisms; medical mycology. *Mailing Add:* Dept Biol Univ S Fla 4202 Fowler Ave Tampa FL 33620

MCCLUNG, ROBERT W, b Memphis, Tenn, Oct 6, 20. NONDESTRUCTIVE TESTING, QUALITY CONTROL. *Educ:* Univ Tenn, BS, 50. *Prof Exp:* Develop & appl high voltage radiography, Y-12 Plant, 50-52, develop engr & group leader, Nondestructive Testing Eval, 55-88, CONSULT, NONDESTRUCTIVE TESTING TECH, OAK RIDGE NAT LAB, 88- *Concurrent Pos:* Chmn comt E7, Am Soc Testing & Mat, 72-76. *Honors & Awards:* Coolidge Award, Am Soc Nondestructive Testing, Lester Honor Lectr, Gold Medal; Award of Merit, Am Soc Testing & Mat. *Mem:* Hon mem Am Soc Nondestructive Testing (pres, 69-70); fel Am Soc Testing & Mat; fel Am Soc Metals Int. *Res:* Extensive ultra and sonics of phases and nondestructive testing. *Mailing Add:* Oak Ridge Nat Lab Bldg 4500-S Mail Stop 151 PO Box X Oak Ridge TN 37831-6151

MCCLUNG, RONALD EDWIN DAWSON, b Victoria, BC, Oct 27, 41; m 64; c 2. CHEMISTRY, PHYSICS. *Educ:* Univ Alta, BSc, 63; Univ Calif, Los Angeles, PhD(chem), 67. *Prof Exp:* Nat Res Coun Can fel, Univ Leeds, 68-69; from asst prof to assoc prof, 69-81, PROF CHEM, UNIV ALTA, 81- *Mem:* Chem Inst Can. *Res:* Molecular motion in fluids; spectroscopy; nuclear magnetic resonance; transition metal chemistry. *Mailing Add:* Dept of Chem Univ of Alta Edmonton AB T6G 2G2 Can

MCCLURE, BENJAMIN THOMPSON, b Rochester, Minn, Oct 14, 25; m 51; c 3. SOLID STATE PHYSICS. *Educ:* Univ Minn, BA, 45, MA, 47; Harvard Univ, PhD(physics), 51. *Prof Exp:* Physicist, Tracerlab Inc, 48-49; mem tech staff, Bell Tel Labs, Inc, 51-59; prin res scientist, Honeywell, Inc, 60-83; SCIENTIST, US BUR MINES, 84- *Concurrent Pos:* Instr, Augsburg Col, 62-63 & Macalester Col, 64-65. *Mem:* Am Phys Soc; Am Meteorol Soc; Sigma Xi; Soc Automotive Engr. *Res:* Diesel engine testing (emissions); charge transport in dielectrics; atmospheric impurity detection. *Mailing Add:* 140 Meadowbrook Rd Hopkins MN 55343-8535

MCCLURE, CLAIR WYLIE, b Greenville, Pa, Oct 17, 27; m 48; c 3. MATHEMATICS. *Educ:* Thiel Col, BS, 50; Ohio State Univ, MA, 62, PhD(math educ), 71. *Prof Exp:* Teacher math, Mercer Area Sch Dist, Pa, 51-63; from assoc prof to prof math, 63-85, chmn dept, 72-75; RETIRED. *Mem:* Math Asn Am. *Res:* Mathematics laboratories for junior high school students; non-Euclidean geometries for high school students. *Mailing Add:* 3212 Crabwood Ct Port St Lucie FL 34952

MCCLURE, DAVID WARREN, b Yakima, Wash, Sept 12, 36; m 81; c 2. PHYSICAL CHEMISTRY, NUMERICAL METHODS. *Educ:* Wash State Univ, BS, 58; Univ Wash, PhD(phys chem), 63. *Prof Exp:* Gatty fel, Cambridge Univ, 63-64, Oppenheimer res grant, 64-65; chemist, Shell Develop Co, 65-66; assoc prof, 66-80, head dept, 77-86, PROF CHEM, PORTLAND STATE UNIV, 80- *Concurrent Pos:* Asst prof, Med Sch, Univ Ore. *Mem:* Am Phys Soc. *Res:* Thermodynamics and statistical mechanics of non-ideal gases and solutions; computational methods. *Mailing Add:* Dept Chem Portland State Univ Box 751 Portland OR 97207

MCCLURE, DONALD ERNEST, b Portland, Ore, Oct 22, 44; m 71; c 2. APPLIED MATHEMATICS. *Educ:* Univ Calif, Berkeley, AB, 66; Brown Univ, PhD(appl math), 70. *Prof Exp:* From instr to assoc prof appl math, 69-82, chmn dept, 85-88, PROF APPL MATH, BROWN UNIV, 82-, ASSOC DIR, CTR INTEL CONTROL SYSTS, 86- *Concurrent Pos:* Consult, Math Technol Inc & US Army Night Vision & Electro-Optics Lab. *Mem:* Am Math Soc; Inst Math Statist; Soc Indust & Appl Math; Math Asn Am; Am Statist Asn. *Res:* Pattern analysis; approximation theory; mathematical statistics; image processing; computer vision; applications of computer vision in semiconductor manufacturing. *Mailing Add:* Div Appl Math Brown Univ PO Box F Providence RI 02912

MCCLURE, DONALD STUART, b Yonkers, NY, Aug 27, 20; m 49; c 3. ELECTRONIC SPECTROSCOPY. *Educ:* Univ Minn, BCh, 42; Univ Calif, Berkeley, PhD(phys chem), 48. *Prof Exp:* Res scientist, War Res Div, Columbia Univ, 42-44, SAM Labs & Carbide & Carbon Chem Corp, 44-46; asst chem, Univ Calif, Berkeley, 46-47, from instr to asst prof, 48-55; mem tech staff, RCA Labs, Inc, 55-62; prof chem, Univ Chicago, 62-67; PROF CHEM, PRINCETON UNIV, 67- *Concurrent Pos:* Guggenheim fel, 72 & Humboldt fel, 81. *Honors & Awards:* Langmuir Award chem physics, 79. *Mem:* Nat Acad Sci; Am Chem Soc; fel Am Phys Soc; Am Acad Arts & Sci. *Res:* Ultraviolet spectra and triplet states of organic molecules; spectra and electronic processes in molecular crystals; crystal field theory and spectra of inorganic ions and crystals; photochemistry. *Mailing Add:* Dept Chem Princeton Univ Princeton NJ 08544

MCCLURE, ELDON RAY, b Carson, NDak, Dec 31, 33; m 56; c 2. MECHANICAL ENGINEERING. *Educ:* Wash State Univ, BS, 55; Ohio State Univ, MS, 59; Univ Calif, Berkeley, DEng(mech eng), 69. *Prof Exp:* Flight test engr, Boeing Co, Wash, 55-57; res engr, Battelle Mem Inst, Ohio, 57-58; from instr to asst prof mech eng, Ore State Univ, 58-63; mech engr, 65-71, dep head dept mech eng, 71-72, div leader, Energy Systs Eng Div, 72-78, PROG LEADER, PRECISION ENG PROG, LAWRENCE LIVERMORE LAB, 78- *Concurrent Pos:* Consult, Gerlinger Carrier Co, Ore, 61-62 & US Bur Mines, 62-63. *Mem:* Am Soc Mech Engrs; Soc Mfg Engrs; AAAS; Japan Soc Precision Eng; Am Soc Precision Eng (pres, 88). *Res:* Machine tool metrology; manufacturing engineering. *Mailing Add:* 5721 Crestmont Ave Livermore CA 94550

MCCLURE, GORDON WALLACE, b Mt Pleasant, Iowa, Mar 22, 23; m 44; c 3. PHYSICS. *Educ:* Univ Ill, BS, 44; Univ Chicago, PhD(physics), 50. *Prof Exp:* Mem staff, Radiation Lab, Mass Inst Technol, 44-45; physicist, Bartol Res Found, Franklin Inst, 49-55; PHYSICIST, SANDIA LABS, 55- *Mem:* Am Phys Soc. *Res:* Gaseous electronics; atomic collisions; physical electronics. *Mailing Add:* 3901 Indian Sch NE Apt A408 Albuquerque NM 87110

MCCLURE, HAROLD MONROE, b Hayesville, NC, Oct 2, 37; m 58; c 3. VETERINARY PATHOLOGY, COMPARATIVE PATHOLOGY. *Educ:* NC State Col, BS, 63; Univ Ga, DVM, 63. *Prof Exp:* NIH fel exp path, Univ Wis-Madison, 63-66; clin instr & pathologist, 66-80, ASST PROF PATH, RES PROF & CHIEF, DIV PATH & IMMUNOL, EMORY UNIV, 80-, ASSOC DIR SCI PROG, YERKES PRIMATE CTR, 82- *Concurrent Pos:* Grants, USDA, Emory Univ, 69-72 & Nat Cancer Inst, 71-82, Nat Inst Dent Res, 75-80, Cystic Fibrosis Found, 79-85, NASA, 79-84, NIH, 82-88, CDC, 83-88; assoc ed, Lab Animal Sci & Am J Primatology. *Mem:* Am Soc Primatologists; Am Vet Med Asn; Am Asn Lab Animal Sci; Int Acad Path; Am Soc Vet Clin Pathologists; Int Primatological Soc; Am Soc for Microbiol. *Res:* Comparative and clinical pathology; electron microscopy; cancer research; animal models for human diseases; AIDS animal models. *Mailing Add:* Yerkes Primate Ctr Emory Univ Atlanta GA 30322

MCCLURE, J DOYLE, b Clayton, NMex, June 23, 35; m 57; c 2. FLUID MECHANICS. *Educ:* Univ Wash, Seattle, BS, 56, MS, 58; Mass Inst Technol, PhD(fluid mech), 62. *Prof Exp:* Res engr, Boeing Airplane Co, 57-58, res specialist, Boeing Sci Res Labs, 62-72, sr res engr laser technol, 72-81, mgr pulsed laser res & develop, 81-86, proj mgr, Infrared & Visable Sensors, Aerospace & Electronics High Tech Ctr, 86-90; RETIRED. *Mem:* Am Phys Soc. *Res:* Theoretical and experimental investigations of the interaction kinetics of gas-solid scattering processes; chemical and physical processes in pulsed gas lasers; high power pulsed laser research. *Mailing Add:* 21831 Oak Way Lynnwood WA 98036-8100

MCCLURE, JERRY WELDON, b Floydada, Tex, May 3, 33; m 54; c 2. PLANT BIOCHEMISTRY. *Educ:* Tex Tech Col, BS, 54, MS, 61; Univ Tex, PhD(bot), 64. *Prof Exp:* From asst prof to assoc prof, 64-73, PROF BOT, MIAMI UNIV, 73- *Concurrent Pos:* NSF grants, 65, 70, 75, 83 & 85; Fulbright hon res fel, WGer, 74-75; Humboldt sr scientist award, WGer Govt, 74-75; mem adv screening comt Life Sci, Coun Int Exchange Scholars, 75-78. *Mem:* AAAS; Bot Soc Am; Phytochem Soc NAm (secy, 71-74, pres elect, 75-76, pres, 76-77); Am Soc Plant Physiol. *Res:* Phytochemistry, especially biochemistry of plant phenolics; physiology and functions of flavonoids. *Mailing Add:* Dept of Bot Miami Univ Oxford OH 45056

MCCLURE, JOEL WILLIAM, JR, b Irvine, Ky, Aug 8, 27; m 53; c 2. THEORETICAL PHYSICS, SOLID STATE PHYSICS. *Educ:* Northwestern Univ, BS, 49, MS, 51; Univ Chicago, PhD, 54. *Prof Exp:* Asst prof physics, Univ Ore, 54-56; res physicist, Nat Carbon Co Div, Union Carbide Corp, 56-61; from assoc prof to prof, 61-84, EMER PROF PHYSICS, UNIV ORE, 84- *Honors & Awards:* Charles Pettinos Award, Am Carbon Soc, 85. *Mem:* Am Phys Soc. *Res:* Transport and magnetic properties; carbons and graphite. *Mailing Add:* Dept Physics Univ Ore Eugene OR 97403

MCCLURE, JOHN ARTHUR, b Belle Center, Ohio, Jan 19, 34; m 60; c 3. MATHEMATICAL ANALYSIS. *Educ:* Geneva Col, BS, 56; Univ Rochester, MS, 57; Va Polytechnic Inst, PhD(physics), 62. *Prof Exp:* Scientist reactor physics, Phillips Petrol Co, 62-69; assoc scientist comput sci, Aerojet Nuclear Corp, 69-73; sr scientist physics, 73-75, PRIN ANALYST PHYSICS, ENERGY INC, 75- *Mem:* Am Phys Soc; Soc Indust & Appl Math; Sigma Xi. *Res:* Application of numerical methods to problems in science and engineering. *Mailing Add:* 1840 Avalon St Idaho Falls ID 83402

MCCLURE, JOSEPH A, b Ga, May 8, 51. EXPERIMENTAL BIOLOGY. *Educ:* Univ Fla, BS, 73, MD, 77, DPhil, 78; Am Bd Internal Med, dipl. *Prof Exp:* Intern Med, Univ Utah Affil Hosps, Salt Lake City, 78-79, resident internal med, 79-81; med staff fel, Nat Heart, Lung & Blood Inst, NIH, Bethesda, Md, 81-83; fel hemat & oncol, Dept Med, Univ Fla, Gainesville, 83-84; HEMATOLOGIST & ONCOLOGIST, MELBOURNE INTERNAL MED CTR, 86- *Mem:* Fel Am Col Physicians; Am Soc Cell Biol; Am Col Physicians Health Execs; Am Soc Clin Oncol; Am Soc Hemat; Acad Hospice Physicians; Am Cancer Soc. *Res:* Author of various publications. *Mailing Add:* Melbourne Internal Med Ctr 200 E Sheridan Rd Melbourne FL 32901

MCCLURE, JOSEPH ANDREW, JR, b Sumter, SC, Jan 22, 34; m 57; c 2. THEORETICAL PHYSICS. *Educ:* Univ NC, BS, 56; Vanderbilt Univ, MA, 60, PhD(theoret physics), 63. *Prof Exp:* Res assoc & NSF fel theoret physics, Stanford Univ, 63-64; res assoc, Tufts Univ, 64-67; asst prof, 67-72, ASSOC PROF THEORET PHYSICS, GEORGETOWN UNIV, 72- *Mem:* Am Phys Soc; Sigma Xi. *Res:* High energy theoretical physics. *Mailing Add:* Dept Physics 548 Reiss Georgetown Univ 37th & O Sts NW Washington DC 20057

MCCLURE, JUDSON P, b Longmont, Colo, Feb 7, 34; m 61; c 3. ORGANIC CHEMISTRY. *Educ:* Bob Jones Univ, BS, 55; Univ Colo, PhD(chem), 61. *Prof Exp:* Res chemist, Esso Res & Eng Co, NJ, 61-63; from asst prof to assoc prof chem, Ohio Northern Univ, 63-68; fel, Case Western Reserve Univ, 68-69; assoc prof, Ohio Northern Univ, 69-70; from asst prof to assoc prof, 70-75, chmn, Dept Nat Sci, 77-83 & 90-91, PROF CHEM, MERCY COL NY, 75- *Concurrent Pos:* NSF-Acad Year Exten fel, 64-66. *Mem:* Am Chem Soc. *Res:* Application of HPLC to undergraduate organic chemistry laboratory. *Mailing Add:* Dept Natural Sci Mercy Col 555 Broadway Dobbs Ferry NY 10522

MCCLURE, MARK STEPHEN, b Northbridge, Mass, Oct 27, 48; m 71; c 2. HOMOPTERA. *Educ:* Univ Mass, Boston, BA, 70; Univ Ill, Urbana, MS, 73, PhD(entom), 75. *Prof Exp:* Med technician, Leonard Morse Hosp, Mass, 70-71; USPHS trainee entom, Univ Ill, Urbana, 71-75, teaching asst insect ecol, 72-73; from asst agr scientist to assoc agr scientist, 75-81, agr scientist, 81-87, CHIEF AGR SCIENTIST, CONN AGR EXP STA, 87- *Concurrent Pos:* Mem integrated pest mgt deleg, USDA, People's Repub of China, 82. *Mem:* AAAS; Ecol Soc Am; Entom Soc Am; Nat Arborist Assn; Org Trop Studies. *Res:* The biology, ecology and control of Homoptera, expecially Adelgidae, Coccoidea & Cicadellidae; ecology and control of Matsucoccus resinosae and Pineus boerneri on red pine, Fiorinia externa, Nuculaspis tsugae and Adelges tsugae on hemlock, and leafhopper vectors of peach X-disease. *Mailing Add:* Conn Agr Exp Sta Valley Lab PO Box 248 Windsor CT 06095

MCCLURE, MICHAEL ALLEN, b San Diego, Calif, Jan 19, 38; m 56; c 3. NEMATOLOGY. *Educ:* Univ Calif, Davis, BS, 59, PhD(nematol), 64. *Prof Exp:* Nematologist, Univ Calif, Davis, 64-65; asst prof nematol, Rutgers Univ, 65-68; assoc prof plant path & assoc plant pathologist, 68-74, PROF PLANT PATH, UNIV ARIZ, 74-, RES SCIENTIST PLANT PATH, AGR EXP STA, 76- *Mem:* Soc Nematol. *Res:* Nematode biology and physiology; culture and nutrition of plant parasitic nematodes; physiology of host-parasite relationships. *Mailing Add:* Dept of Plant Path Univ of Ariz Tucson AZ 85721

MCCLURE, MICHAEL EDWARD, b Goshen, Ind, Aug 15, 41; m 62. CELL BIOLOGY, BIOCHEMISTRY. *Educ:* Purdue Univ, BS, 63; Univ Tex Grad Sch Biomed Sci, Houston, MS, 66, PhD(cell biochem), 70. *Prof Exp:* Univ Tex M D Anderson Hosp & Tumor Inst Houston fel, 69-71, asst biochemist, 71-72, asst prof biochem, 72-73; asst prof cell biol, Baylor Col Med, 73-76; asst prof develop therapeut, Syst Cancer Ctr, Univ Tex, 76-79; rep sci br, 79-88, CHIEF, NICHD, NIH, HSA, 88- *Concurrent Pos:* Peer Rev Consult, NIH, USDA, AID, WHO; fel, PHS, 63-68, R B Hite, 69-70. *Mem:* Biochem Soc London; NY Acad Sci; Am Soc Cell Biologists; Am Soc Immunol Reproduction. *Res:* Biochemistry of the cell cycle, chromatin, genetic control mechanisms and development; Hormonal regulatory mechanisms of cell growth and reproduction. *Mailing Add:* 303 Epping Way Annapolis MD 21401-6615

MCCLURE, POLLEY ANN, b Austin, Tex, Apr 5, 43; m. REPRODUCTION, GROWTH. *Educ:* Univ Tex, Austin, BA, 65, PhD(ecol), 70; Univ Mont, MA, 67. *Prof Exp:* Asst prof zool, 70-76, assoc prof, 77-83, PROF BIOL, IND UNIV, 83-, assoc dean res, 83-87, DEAN ACAD COMPUT, 87- *Mem:* Ecol Soc Am; AAAS; Am Soc Zoologists; Asn Study Animal Behav; Asn Women Sci. *Res:* Physiological mechanisms and functional significance of life history features; empirical analysis of reproductive costs and reproductive strategies. *Mailing Add:* Ind Univ Bloomington IN 47401

MCCLURE, RICHARD MARK, b Rutland, Pa, May 4, 34; m 56; c 1. CIVIL ENGINEERING. *Educ:* Pa State Univ, BSCE, 58, MSCE, 66, PhD(civil eng), 69. *Prof Exp:* Civil engr, US Forest Serv, 58-63; from instr to asst prof struct, 63-77, ASSOC PROF CIVIL ENG, PA STATE UNIV, 77- *Mem:* Am Soc Civil Engrs; Am Soc Eng Educ. *Res:* Combined bending and torsion in prestressed concrete box beams; computer applications in structural analysis and design. *Mailing Add:* Dept Civil Eng Pa State Univ 212 Sackett Bldg University Park PA 16802

MCCLURE, ROBERT CHARLES, b Grinnell, Iowa, Feb 15, 32; m 54; c 2. VETERINARY ANATOMY. *Educ:* Iowa State Univ, DVM, 55; Cornell Univ, PhD(vet anat surg physiol), 64. *Prof Exp:* Instr vet anat, Iowa State Univ, 55-56; instr vet anat, Cornell Univ, 56-60, asst comp neurol, 60; chmn dept, 60-69 & 81-83, PROF VET ANAT, UNIV MO-COLUMBIA, 60- *Concurrent Pos:* Mem, Int Comt Vet Anat Nomenclature, 64-; Fulbright lectr, Inst Vet Med, Austria, 72-73; chmn, Nomina Embryologia Vet Comt, World Asn Vet Anatomists, 75-80; vis prof anat, Tufts Univ, 79-80, 86. *Mem:* Am Asn Vet Anat (secy-treas, 65-66, pres elect, 66-67, pres, 67-68); Am Asn Anat; World Asn Vet Anat; Am Vet Med Asn; Am Asn Lab Animal Sci. *Res:* Comparative gross veterinary and developmental anatomy; laboratory animal anatomy; tooth development; comparative neuroanatomy; comparative anatomical nomenclature; veterinary medical and anatomical education and history. *Mailing Add:* Dept Vet Biomed Sci Col Vet Med Univ Mo Columbia MO 65211

MCCLURE, THEODORE DEAN, b New Virginia, Iowa, Oct 10, 36; m 59; c 3. NEUROANATOMY. *Educ:* Simpson Col, BA, 62; Univ Okla, MS, 65, PhD(med sci), 70. *Prof Exp:* Instr anat, Med Ctr, Univ Okla, 66-70, asst prof anat sci, Health Sci Ctr, 70-74, prof anat, 74-75; PROF ANAT, MED SCH, UNIV MO-KANS CITY, 75- *Mem:* Am Asn Anat; Sigma Xi. *Res:* Neuroanatomy; histology; embryology; teratology; behavioral studies. *Mailing Add:* Sch Med Univ Mo 2411 Holmes Kansas City MO 64108

MCCLURE, WILLIAM OWEN, b Yakima, Wash, Sept 29, 37; m 80; c 2. NEUROCHEMISTRY. *Educ:* Calif Inst Technol, BSc, 59; Univ Wash, PhD(biochem), 64. *Prof Exp:* Guest investr biochem, Rockefeller Univ, 64-65, res assoc, 65-66, asst prof, 66-68; asst prof biochem, Univ Ill, Urbana, 68-75; vpres, Nelson Res, Irvine, Calif, 81-82; assoc prof, 75-78, dir cellular biol, 77-81, PROF BIOL SCI, UNIV SOUTHERN CALIF, 78-, PROF NEUROL, 79- *Concurrent Pos:* USPHS fel, 64-66; fel neurosci, Alfred P Sloan Found, 72-76; mem res coun, Nelson Res & Develop, 72-; fel, Intersci Res Inst, 89-91. *Mem:* Am Chem Soc; Am Soc Neurochem; Soc Neurosci; NY Acad Sci; Am Soc Biol Chemists; Sigma Xi. *Res:* Schizophrenia; mechanism of neurotransmitter release; mechanism of action of psychoactive drugs; mechanism of action of neurotoxins. *Mailing Add:* Dept Biol Sci Univ Southern Calif Los Angeles CA 90089-2520

MCCLURE, WILLIAM ROBERT, b Detroit, Mich, Dec 30, 41; m 66; c 2. ENZYMOLOGY. *Educ:* Univ Alaska, BS, 66; Univ Wis, PhD(biochem), 70. *Prof Exp:* Fel, Max-Planck Inst, Göttingen, 70-73; from asst prof to assoc prof, Harvard Univ, 73-81; PROF BIOCHEM, CARNEGIE MELLON UNIV, 81- *Concurrent Pos:* Consult, Study Sect, NIH, 86-90; ed, Nucleic Acids Res. *Mem:* Am Soc Biol Chemists; Am Chem Soc; Am Soc Microbiol; AAAS; Biophys Soc. *Res:* Polynucleotide enzymology: The mechanism and regulation of E coli RNA polymerase; the effect of DNA sequence and structure on promoter function; interaction of protein activators and repressors with RNA polymerase during initiation of RNA synthesis. *Mailing Add:* Dept Biol Sci Carnegie Mellon Univ Pittsburgh PA 15213-3890

MCCLURG, CHARLES ALAN, b Ames, Iowa, Aug 14, 44; m 79; c 2. HORTICULTURE. *Educ:* Iowa State Univ, BS, 66; Pa State Univ, MS, 68, PhD(hort), 70. *Prof Exp:* Asst prof, 71-78, ASSOC PROF HORT, UNIV MD, 78- *Mem:* Am Soc Hort Sci. *Res:* Vegetable growth, development and processing. *Mailing Add:* Dept Hort Univ Md Col Park MD 20742

MCCLURG, JAMES EDWARD, b Bassett, Nebr, Mar 23, 45; m 89. HUMAN PHARMACOLOGY, CLINICAL RESEARCH. *Educ:* Nebr Wesleyan Univ, BS, 67; Univ Nebr, PhD(biochem), 73. *Prof Exp:* Instr biochem & res instr obstet & gynec, Univ Nebr Med Ctr, Omaha, 73-76; tech dir, 76-78, vpres, 78-84, PRES & CEO, HARRIS LABS, INC. *Mem:* AAAS; Am Chem Soc (secy, Div Small Chem Bus); Am Soc Microbiol; Sigma Xi; Am Asn Clin Chemists. *Res:* Local immune response; biopharmaceutics; normal and malignant cell growth. *Mailing Add:* 2030 Surfside Dr Lincoln NE 68528

MCCLURKIN, ARLAN WILBUR, b Clay Center, Kans, July 8, 17; m 54; c 4. VETERINARY MEDICINE, VIROLOGY. *Educ:* Kans State Univ, DVM, 43; Univ Wis, PhD(virol), 56. *Prof Exp:* Asst vet, Pvt Vet Hosp, 43; vet & prof animal husb, Allahabad Agr Inst, India, 47-51; res asst & instr, Univ Wis, 54-56; VET VIROLOGIST, NAT ANIMAL DIS LAB, SCI & EDUC ADMIN-AGR RES, USDA, 56- *Mem:* Am Vet Med Asn; Conf Res Workers Animal Dis; NY Acad Sci; Sigma Xi. *Res:* Virus diseases of animals; characterization of the virus, epizootiology, pathogenesis and pathology produced in the animal. *Mailing Add:* 1612 Duff Ave Ames IA 50010

MCCLURKIN, IOLA TAYLOR, b Kinston, NC, May 22, 30; m 58; c 3. BIOLOGY, HISTOLOGY. *Educ:* Duke Univ, BA, 52; E Carolina Col, MA, 57; Univ Miss, PhD(physiol), 65. *Prof Exp:* Asst zool, Duke Univ, 52-53; med technologist, Kafer Mem Hosp, New Bern, NC, 53-54; teacher high sch, NC, 54-57; asst zool, Duke Univ, 57-58; from instr to PROF BIOL, 58-, CHMN, BIOL DEPT, UNIV MISS, 87- *Concurrent Pos:* Fac res grants, Univ Miss, 64-67, 78. *Mem:* AAAS; Histochem Soc; Electron Micros Soc Am; Sigma Xi. *Res:* Cytochemical studies involving transport enzyme locations in cellular membrane systems; histology of avian kidney; histology; histochemistry. *Mailing Add:* Dept Biol Univ Miss University MS 38677

MCCLUSKEY, EDWARD JOSEPH, b New York, NY, Oct 16, 29; m 81; c 6. COMPUTER SCIENCE, ELECTRICAL ENGINEERING. *Educ:* Bowdoin Col, AB, 53; Mass Inst Technol, BS & MS, 53, ScD(elec eng), 56. *Prof Exp:* Instr, Mass Inst Technol, 55; mem tech staff, Bell Tel Labs, Inc, 55-59; from assoc prof to prof elec eng, Princeton Univ, 59-67; PROF COMPUT SCI & ELEC ENG, STANFORD UNIV, 67- *Concurrent Pos:* Dir, Comput Ctr, Princeton Univ, 61-66; ed, J Asn Comput Mach, 63-69; dir, Digital Systs Lab, Stanford Univ, 69-78; ed, Comput Design & Archit Series, 73-; ed, Annals of Hist Comput, 78-; res fel, Japan Soc Prom Sci, 78; Signetics consult, Digital Equip Corp, Xerox Corp, Gen Precision Co, IBM & Honeywell. *Mem:* Fel AAAS; Inst Elec & Electronics Engrs; Asn Comput Mach. *Res:* Computer reliability; digital design and testing; multiprocessors; computer design automation. *Mailing Add:* Dept Comput Sci Elec Eng Erl 460 Stanford Univ Ctr Reliable Computing Stanford CA 94305

MCCLUSKEY, ELWOOD STURGES, b Loma Linda, Calif, June 1, 25; m 48; c 4. COMPARATIVE PHYSIOLOGY, CIRCADIAN RHYTHMS. *Educ:* Walla Walla Col, BA, 50, MA, 52; Stanford Univ, PhD(biol), 59. *Prof Exp:* Asst bot, Walla Walla Col, 50-51; asst embryol, Stanford Univ, 51; asst prof biol, Atlantic Union Col, 52-54; asst instr physiol, 54-59, from instr to asst prof, 59-75, ASSOC PROF PHYSIOL & BIOL, LOMA LINDA UNIV, 75- *Concurrent Pos:* NSF fel, Harvard Univ, 59-60. *Mem:* AAAS; Am Soc Zoologists; Ecol Soc Am; Entom Soc Am. *Res:* Comparative physiology, periodicity and ant biology; entomology. *Mailing Add:* Physiol Dept Loma Linda Univ Loma Linda CA 92350

MCCLUSKEY, GEORGE E, JR, b Hammonton, NJ, Aug 28, 38; m 79. ASTROPHYSICS. *Educ:* Univ Pa, AB, 60, MS, 63, PhD(astron), 65. *Prof Exp:* Asst prof, 65-68, assoc prof, 68-76, PROF ASTRON, LEHIGH UNIV, 76- *Concurrent Pos:* NASA grant, Lehigh Univ, 71-75; guest investr, Copernicus Satellite. *Mem:* AAAS; Am Astron Soc; Int Astron Union. *Res:* Eclipsing binary stars; double stars; x-ray and gamma-ray generation in close binaries; general relativistic effects in close binary systems containing neutron stars. *Mailing Add:* Dept of Math Bldg #14 Lehigh Univ Bethlehem PA 18015

MCCLUSKEY, ROBERT TIMMONS, b New Haven, Conn, Jan 16, 23; m 57; c 2. PATHOLOGY. *Educ:* Yale Univ, AB, 44; NY Univ, MD, 47. *Prof Exp:* Intern, Kings County Hosp, Brooklyn, NY, 47-49, asst resident path, 49-50; asst, Sch Med, NY Univ, 50-52, from instr to prof & dir univ hosp labs, 52-68; prof path & chmn dept, Sch Med & Dent, State Univ NY Buffalo, 68-71; S Burt Wolbach prof, 71-74, Mallinckrodt prof path, 75-82, BENJAMIN CASTLEMAN PROF PATH, HARVARD MED SCH, 82-; CHIEF PATH, MASS GEN HOSP, 74- *Concurrent Pos:* Resident, Bellevue Hosp, NY, 50-52, asst pathologist, 52-53, 55-60, assoc pathologist, 60; consult, Manhattan Vet Admin Hosp, NY, 59-68; pathologist-in-chief, Children's Hosp Med Ctr, 71-74; co-ed, Clin Immunol & Immunopath. *Mem:* Am Soc Exp Path; AMA; Asn Path & Bact; Int Acad Path; Am Soc Nephrology; Am Asn Immunologists. *Res:* Pathogenesis of glomerulonephritis; pathogenesis of cell mediated reactions; identification of lymphocyte subsets in tissue sections with monoclonal antibodies. *Mailing Add:* Dept Path Mass Gen Hosp Fruit St Boston MA 02114

MCCLYMER, JAMES P, b Wilmington, Del, Sept 18, 58; m 78; c 3. LIQUID CRYSTALS, DYNAMIC LIGHT SCATTERING. *Educ:* Univ Del, BS, 80, PhD(physics), 86. *Prof Exp:* Bartol grad fel, Bartol Res Found, Franklin Inst, 80-85; postdoctoral, Dept Chem, Temple Univ, 85-87; ASST PROF PHYSICS, DEPT PHYSICS & ASTROPHYS, UNIV MAINE, 87- *Mem:* Am Phys Soc. *Res:* Equilibrium and non-equilibrium phase transitions in liquid crystals; dynamic light scattering; computer aided microscopy; lyotropic liquid crystals. *Mailing Add:* Dept Physics & Astron Bennett Hall Univ Maine Orono ME 04469

MCCLYMONT, JOHN WILBUR, b Geneseo, NY, Feb 13, 25; m 52; c 1. BOTANY. *Educ:* Syracuse Univ, AB, 49, MS, 50; Univ Mich, PhD(bot), 55. *Prof Exp:* Asst prof biol, Milwaukee-Downer Col, 54-59; assoc prof, Morris Harvey Col, 59-61; prof, Findlay Col, 61-65; PROF BIOL, SOUTHERN CONN STATE COL, 65- *Mem:* Am Bryol & Lichenological Soc; Bot Soc Am. *Res:* Spores of the Bryophyta. *Mailing Add:* Dept Biol Southern Conn State Col 501 Crescent St New Haven CT 06515

MCCLYMONT, KENNETH, b Toronto, Ont, Nov 13, 24. CONTROL SYSTEMS. *Educ:* Univ Toronto, BA, 47. *Prof Exp:* Res engr, Ont Hydro, 49-59, sr syst planning engr, 59-72, mgr, Generation & Transmission Planning, 72-80, group mgr, Bulk Elec Syst, Ont Hydro, 80-83; CONSULT, 83- *Concurrent Pos:* Adv, Nat Energy Bd, 86. *Honors & Awards:* Centennial Medal, Inst Elec & Electronic Engrs, 84. *Mem:* Fel Inst Elec & Electronic Engrs. *Res:* Interconnection Study for 5 countries in the Middle East. *Mailing Add:* 2045 Lakeshore Blvd W Suite 203 Toronto ON M8V 2Z6 Can

MCCOLL, JAMES RENFREW, b Albany, NY, Oct 30, 40; m 62; c 3. EXPERIMENTAL SOLID STATE PHYSICS. *Educ:* Univ Rochester, BS, 62; Univ Calif, Berkeley, PhD(physics), 67. *Prof Exp:* NATO fel, Clarendon Lab, Oxford, Eng, 67-68; asst prof physics, Yale Univ, 68-73, assoc prof, 73-76; MEM TECH STAFF, GTE LABS, WALTHAM, MASS, 76- *Res:* Liquid crystals; amorphous solids; phosphors. *Mailing Add:* 35 Wolf Pine Way Concord MA 01742

MCCOLL, JOHN DUNCAN, b London, Ont, Nov 11, 25; m 54; c 3. PHARMACOLOGY. *Educ:* Univ Western Ont, BA, 46, MSc, 50; Univ Toronto, PhD(pharmacol), 53. *Prof Exp:* Asst res chemist, Parke, Davis & Co, Mich, 50-51; dir pharmacol res, Frank W Horner, Ltd, Can, 53-62, asst dir res, 62-69; vpres biol sci, Mead Johnson Res Ctr, 70-75; vpres & dir res & develop, Chattem Inc, 75-83; PRES, MCCOLL & ASSOCS INC, 83- *Concurrent Pos:* Mem toxicol panel, Defence Res Bd Can, 69-72. *Mem:* Am Soc Pharmacol & Exp Therapeut; Am Soc Clin Pharmacol & Therapeut; Am Chem Soc; Pharmacol Soc Can (secy, 61-63); Can Biochem Soc; Sigma Xi. *Res:* Toxicology; neuropharmacology; pharmacology; biochemistry. *Mailing Add:* 901 Brynwood Dr Chattanooga TN 37415

MCCOLL, MALCOLM, b Detroit, Mich, Oct 6, 33; m 54; c 2. ELECTRONICS ENGINEERING. *Educ:* Wayne State Univ, BS, 57; Calif Inst Technol, MS, 58, PhD(elec eng), 64. *Prof Exp:* RES SCIENTIST, IVAN A GETTING LABS, AEROSPACE CORP, 62- *Concurrent Pos:* Res fel, Calif Inst Technol, 64-66. *Mem:* Inst Elec & Electronic Engrs. *Res:* Transport mechanisms in electronic devices; development of millimeter and submillimeter-wave detectors; solid-state device fabrication and diagnostic techniques. *Mailing Add:* Aerospace Corp 2350 E El Segundo Blvd Mail Sta M1-111 El Segundo CA 90245

MACCOLL, ROBERT, b Brooklyn, NY, Mar 27, 42; c 3. PROTEINS, PHOTOSYNTHESIS. *Educ:* Queens Col, NY, BA, 63; Univ Miss, MS, 67; Adelphi Univ, PhD(phys chem), 69. *Prof Exp:* Fel, 69-70, RES SCIENTIST, WADSWORTH CTR, NY STATE DEPT HEALTH, 70-, CHIEF, LAB BIOPHYS, 86- *Concurrent Pos:* Prof, Dept Biomed Sci, State Univ NY, Albany, 89- *Mem:* Am Chem Soc; Am Soc Photobiol. *Res:* Biochemical and spectrocopic approaches in the study of proteins; study of biliproteins and their roles in light harvesting and exciton migration in photosynthesis. *Mailing Add:* Wadsworth Ctr NY State Dept Health PO Box 509 Albany NY 12201-0509

MCCOLLESTER, DUNCAN L, b Tyringham, Mass, July 2, 25; m 57; c 1. CANCER. *Educ:* Harvard Univ, BA, 48; Tufts Univ, MD, 53; Univ Cambridge, PhD(biochem), 64. *Prof Exp:* From intern to resident med, Bellevue Hosp, New York, 53-56; clin assoc, Nat Heart Inst, 56-58; RES ASSOC COLUMBIA-PRESBY MED CTR, 65- *Mem:* AAAS; Brit Biochem Soc; Biophys Soc; Am Chem Soc. *Res:* Separation and properties of cell surface membranes; cancer immunology; cancer specific immunotherapy. *Mailing Add:* 10 N Broadway Irvington NY 10533-1803

MCCOLLISTER, ROBERT JOHN, b Iowa City, Iowa, July 27, 28; m 58; c 2. INTERNAL MEDICINE. *Educ:* State Univ Iowa, BA, 49, MD, 52. *Prof Exp:* Instr med, Sch Med, Univ Minn, Minneapolis, 59-61; instr, Duke Univ, 61-62; from instr to asst prof, 62-69, asst dean sch, 64-76, ASSOC PROF MED, SCH MED, UNIV MINN, MINNEAPOLIS, 69-, ASSOC DEAN SCH, 76- *Concurrent Pos:* Asst secy, Liason Comt Med Educ, 78- *Mem:* Am Fedn Clin Res; Am Soc Hemat. *Res:* Hematology; medical education; medical accreditation. *Mailing Add:* 412 Union St SE Minneapolis MN 55455

MCCOLLOCH, ROBERT JAMES, b Manhattan, Kans, July 26, 20; m 48; c 2. PHYSICAL CHEMISTRY, BIOCHEMISTRY. *Educ:* Kans State Col, BS, 41, PhD(phys biochem), 48. *Prof Exp:* Asst & asst chemist agr biochem, Purdue Univ, 42-44; from investr to res assoc, Plant Biochem, Cornell Univ, 44-48; chemist, Fruit & Veg Chem Lab, USDA, 48-52, unit supvr agr res serv, 52-56; assoc prof agr res chem, 56-58, PROF BIOCHEM & HEAD DIV, UNIV WYO, 58-, DEAN GRAD SCH, 72- *Concurrent Pos:* Instr Kans State Col, 46-47. *Mem:* AAAS; Am Chem Soc; Am Soc Plant Physiol; Am Soc Animal Sci; NY Acad Sci; Sigma Xi. *Res:* Plant biochemistry; citrus products biochemistry and technology; pectic substances and pectic enzymes; instrumentation. *Mailing Add:* 1511 Steele Laramie WY 82070

MACCOLLOM, GEORGE BUTTERICK, b Boston, Mass, June 10, 25; m 53; c 4. ENTOMOLOGY. *Educ:* Univ Mass, BS, 50; Cornell Univ, PhD, 54. *Prof Exp:* Entomologist, Exp Sta, 54-66, PROF ENTOM, UNIV VT, 66- *Mem:* Entom Soc Am. *Res:* Biology and control of apple insects; detection and measurement of environmental contamination by insecticides. *Mailing Add:* Dept Plant & Soil Sci Univ Vt Agr Col 85 S Prospect St Burlington VT 05405

MCCOLLOM, KENNETH A(LLEN), b Sentinel, Okla, June 17, 22; m 44; c 2. ELECTRICAL ENGINEERING, NUCLEAR ENGINEERING. *Educ:* Okla State Univ, BS, 48; Univ Ill, MS, 49; Iowa State Univ, PhD(elec eng), 64. *Prof Exp:* Jr engr, Res Div, Phillips Petrol Co, 49-51, group leader nuclear instrumentation design, Atomic Energy Div, 51-54, engr, Res Div, 54-57, br mgr, Instrument Develop Br, Atomic Energy Div, 57-62; assoc nuclear reactor control, Ames Lab, Iowa State Univ, 62-64; assoc prof, 64-67, from asst dean to assoc dean eng, 68-77, dean eng, 77-86, PROF ELEC ENG, OKLA STATE UNIV, 67- *Concurrent Pos:* Consult, Ames Lab, Iowa State Univ, 64-65, Babcock & Wilcox Corp, 64-65 & Atlantic Refining Co, 65-68; admin judge, Atomic Safety & Licensing Bd Panel, US Nuclear Regulatory Comn, 72-; Okla Bd Regist For Engrs & Land Surveyors, 86-; coun mem, Nat Coun Examr Eng & Surv, 86- *Honors & Awards:* Chester F. Carlson Award, Am Soc Eng Educ, 73. *Mem:* Sr mem Inst Elec & Electronic Engrs; Am Nuclear Soc; Nat Soc Prof Engrs; fel Am Soc Eng Educ; Nat Coun Eng Examiners. *Res:* Nuclear reactor instrumentation and control design; design of physical and chemical measuring instrumentation; use of computers in control system design; implementation and design of innovative educational methods. *Mailing Add:* 1107 W Knapp St Stillwater OK 74075

MCCOLLOUGH, FRED, JR, b Crawfordsville, Ind, July 19, 28; m 52; c 2. INORGANIC CHEMISTRY. *Educ:* Wabash Col, AB, 50; Univ Ill, MS, 52, PhD(chem), 55. *Prof Exp:* Sr res chemist, Victor Chem Works, 55-57, supvr inorg res, 57-64; from assoc prof to prof, 64-90, chmn dept, 69-90, EMER PROF CHEM, MACMURRAY COL, 90- *Res:* Coordination and phosphorus chemistry; computer application to undergraduate teaching. *Mailing Add:* 1625 Engman Lake Rd Skandia MI 49885

MCCOLLUM, ANTHONY WAYNE, b Macomb, Ill, Sept 3, 44; c 3. ORGANIC CHEMISTRY. *Educ:* Western Ill Univ, BS, 65; Univ Ark, Fayetteville, PhD(org chem), 69. *Prof Exp:* Chemist, 69-74, group leader appl chem res & sr chemist, 74-79, head res & develop planning & serv, 77-79, res assoc, 79, HEAD, CHEM RES DEPT, TEX EASTMAN CO, DIV EASTMAN KODAK CO, 79- *Mem:* Am Chem Soc. *Res:* Organic synthesis; new monomer synthesis; natural products synthesis; aldol chemistry; high solids and water-borne coatings. *Mailing Add:* 15 Starwood Dr Longview TX 75601-8811

MCCOLLUM, CLIFFORD GLENN, b Macon Co, Mo, May 12, 19; m 40; c 2. ZOOLOGY, HISTORY OF SCIENCE. *Educ:* Univ Mo, BS, 39, AM, 47, EdD(zool), 49. *Prof Exp:* Instr high schs, Mo, 38-43; asst prof, State Col Iowa, 49-55; prof, State Univ NY Col Oneonta, 55-56; assoc prof sci, 56-59, head dept, 57-68, prof biol, 59-84, dean, 68-84, EMER PROF BIOL & EMER DEAN, COL NATURAL SCI, UNIV NORTHERN IOWA, 84- *Concurrent Pos:* Consult, schs & cols sci curricula. *Mem:* AAAS; Nat Sci Teachers Asn; Nat Asn Res Sci Teaching; Nat Asn Biol Teachers; Hist Sci Soc; Sigma Xi. *Res:* Small mammal populations; science curricula in elementary and secondary schools. *Mailing Add:* 2002 Chapel Hill Rd Columbia MO 65203-1916

MCCOLLUM, DONALD CARRUTH, JR, b Baltimore, Md, Nov 6, 30; m 52; c 1. PHYSICS. *Educ:* Univ Calif, BA, 53, MA, 54, PhD(physics), 60. *Prof Exp:* From instr to assoc prof, 58-72, PROF PHYSICS, UNIV CALIF, RIVERSIDE, 72- *Concurrent Pos:* NSF fac fel, 65. *Mem:* Am Phys Soc; Am Asn Physics Teachers. *Res:* Low temperature physics. *Mailing Add:* Dept Physics Univ Calif 900 Univ Ave Riverside CA 92502

MCCOLLUM, DONALD E, b Winston Salem, NC, Dec 7, 27; m 53; c 1. ORTHOPEDIC SURGERY. *Educ:* Wake Forest Col, BS, 49; Bowman Gray Sch Med, MD, 53. *Prof Exp:* Resident orthop surg, 58-62, from asst prof to assoc prof, 62-72, PROF ORTHOP SURG, MED CTR, DUKE UNIV, 72- *Concurrent Pos:* NIH trainee rheumatology, Med Ctr, Duke Univ, 57-58; consult, Vet Admin Hosp, 62-; orthop consult, US Air Force & Voc Rehab, 64- *Mem:* Am Acad Orthop Surg; Am Rheumatism Asn; Am Orthop Asn; Asn Bone & Joint Surgeons. *Res:* Rheumatology. *Mailing Add:* Dept Orthop Surg Med Ctr Duke Univ Durham NC 27710

MCCOLLUM, GILBERT DEWEY, JR, b Bellingham, Wash, Aug 25, 29. PLANT CYTOGENETICS. *Educ:* Wash State Univ, BS, 51, MS, 53; Univ Calif, PhD(genetics), 58. *Prof Exp:* RES GENETICIST, AGR RES SERV, USDA, 58- *Mem:* Bot Soc Am. *Res:* Botany; cytology; genetics; species relationships. *Mailing Add:* 7052 Goodwin Rd Everson WA 98247

MCCOLLUM, GIN, b Washington, DC. NEUROBIOLOGY. *Educ:* Brandeis Univ, BA, 69; Yeshiva Univ, PhD(physics), 75. *Prof Exp:* MEM STAFF, NEUROL SCI INST, GOOD SAMARITAN HOSP & MED CTR, 81- *Mem:* Soc Neurosci. *Mailing Add:* R S Dow Neurol Sci Inst 1120 NW 20th Ave Portland OR 97209-1595

MCCOLLUM, JOHN DAVID, b Evanston, Ill, Jan 8, 29; m 50; c 2. CHEMISTRY. *Educ:* Univ Ill, Urbana, BSc, 49; Harvard Univ, AM, 51, PhD(chem), 57. *Prof Exp:* Sr res scientist, Res & Develop Dept, Am Oil Co, 53-72, RES ASSOC, AMOCO OIL CO, 72- *Mem:* Am Chem Soc; Am Inst Chemists; AAAS. *Res:* Homogeneous and heterogeneous catalysis of hydrocarbon reactions; organometallic chemistry; photo and radiation chemistry of hydrocarbons; high pressure hydrothermal reactions; fuel technology. *Mailing Add:* 1113 N Brainard St Naperville IL 60563

MCCOLLUM, ROBERT EDMUND, b Reidsville, NC, Jan 3, 22; m 46; c 6. SOILS, PLANT BIOCHEMISTRY. *Educ:* NC State Col, BS, 52, MS, 53; Univ Ill, PhD, 57. *Prof Exp:* Asst, NC State Col, 52-53 & Univ Ill, 53-57; assoc prof, 57-87, EMER PROF SOIL FERTIL, NC STATE UNIV, 87- *Concurrent Pos:* NC State Univ-USAID Proj, Peru, 63-66 & 82-85. *Mem:* Am Soc Agron. *Res:* Soil chemistry and fertility; tropical agriculture. *Mailing Add:* Dept Soil Sci NC State Univ Raleigh NC 27695-7619

MCCOLLUM, ROBERT WAYNE, b Waco, Tex, Jan 29, 25; m 54; c 2. EPIDEMIOLOGY. *Educ:* Baylor Univ, AB, 45; Johns Hopkins Univ, MD, 48; Univ London, DPH, 58. *Prof Exp:* Asst prev med, Sch Med, Yale Univ, 51-52, from asst prof to assoc prof epidemiol & prev med, 54-65, prof, epidemiol, 65-81, chmn dept epidemiol & pub health, 69-81; dean, 82-90, PROF EPIDEMIOL, DARTMOUTH MED SCH, 82- *Concurrent Pos:* Assoc mem comm viral infections, Armed Forces Epidemiol Bd, 59-61, mem, 61-72; consult, Surgeon Gen, 61-81 & WHO, 62-63, 70, 72, 74 & 79. *Mem:* Soc Epidemiol Res; Int Epidemiol Asn; Am Epidemiol Soc; Infectious Dis Soc Am; fel Am Col Epidemiol. *Res:* Infectious diseases; viral hepatitis. *Mailing Add:* Dartmouth Med Sch Hanover NH 03756

MCCOLLUM, WILLIAM HOWARD, b Traveller's Rest, Ky, Aug 17, 23. VIROLOGY. *Educ:* Univ Ky, BS, 47, MS, 49; Univ Wis, PhD(bact, biochem), 54. *Prof Exp:* Bacteriologist virol, 54-60, prof animal path, 60, PROF VET SCI, UNIV KY, 63- *Mem:* Am Soc Microbiol; Poultry Sci Asn; Tissue Cult Asn. *Res:* Virus diseases of animals, especially the horse; immunology. *Mailing Add:* Dept of Vet Sci Univ of KY Lexington KY 40506

MCCOLLY, HOWARD F(RANKLIN), b Ames, Iowa, Apr 8, 02; m 26; c 3. AGRICULTURAL ENGINEERING. *Educ:* Iowa State Univ, BS, 25, MS, 26. *Prof Exp:* Mem staff, Deere & Co, Ill, 25-28; asst agr engr, NDak Agr Col, 29-31, head, Agr Eng Dept, 31-39; secy & chief engr, State Water Conserv Comn, NDak, 39-41; construct engr, USDA, Colo, 41-42, chief water facilities eng, 42-46, res agr engr, Comt Agr Eng to China, 46-49; prof, 49-70, EMER PROF AGR ENG, MICH STATE UNIV, 70- *Concurrent Pos:* Foreign assignments, Taiwan, 60-62, Saudi Arabia, 66, Asian studies, 67 & Peru, 69; Div ed, Am Soc Agr Engrs Transactions, 72-80. *Honors & Awards:* McCormick Gold Medal, Am Soc Agr Engrs, 68. *Mem:* Fel Am Soc Agr Engrs. *Mailing Add:* 225 Kensington Rd East Lansing MI 48823

MCCOLM, DOUGLAS WOODRUFF, b Kisaran, Sumatra, Sept 26, 33; US citizen; m 55; c 3. PHYSICS. *Educ:* Oberlin Col, BA, 55; Yale Univ, PhD(physics), 61. *Prof Exp:* Staff physicist, Lawrence Radiation Lab, Univ Calif, Berkeley, 61-66; ASSOC PROF PHYSICS, UNIV CALIF, DAVIS, 66- *Mem:* Am Phys Soc. *Res:* Atomic physics. *Mailing Add:* Dept of Physics Univ Calif 1000 Plum Lane Davis CA 95616

MCCOMAS, DAVID JOHN, b Milwaukee, Wis, May 22, 58; m 81; c 1. SPACE PHYSICS, MAGNETOSPHERIC PHYSICS. *Educ:* Mass Instit Technol, BS, 80; Univ Calif, Los Angeles, MS, 85, PhD(space physics), 86. *Prof Exp:* STAFF MEM, LOS ALAMOS NAT LAB, 80- *Concurrent Pos:* Co-investr, Cassini Ulysses, Cluster & Polar missions, NASA, 87- *Mem:* Am Geophys Union. *Res:* Design and develop instrumentation for magnetospheric, planetary, and solar wind space missions, analyzed space data from magnetosperic, commetary, planetary, and solar wind spacecraft particularly in the areas of solar wind and planetary magnetospheric physics. *Mailing Add:* D438 Los Alamos Nat Lab Los Alamos NM 87545

MCCOMAS, STUART T, b Shelbyville, Ind, Apr 23, 32; m 56; c 2. HEAT TRANSFER, FLUID MECHANICS. *Educ:* Marquette Univ, BSME, 56; Univ Minn, MS, 60, PhD, 64. *Prof Exp:* Instr mech eng, Marquette Univ, 56-58 & Univ Minn, 60-63; from asst prof to assoc prof, 63-71, asst dean, 70-76, PROF MECH ENG, UNIV NOTRE DAME, 71- *Concurrent Pos:* Consult, Inc, 66-67 & Miles Lab Inc, 69; vis engr, US Environ Protection Agency, 74-75. *Mem:* Am Soc Mech Engrs; Sigma Xi. *Res:* Thermodynamics and heat transfer; solar energy; energy conservation; energy conversion. *Mailing Add:* Col Eng Univ of Notre Dame Notre Dame IN 46556

MCCOMBE, BRUCE DOUGLAS, b Sanford, Maine, Mar 2, 38; m 63; c 1. SOLID STATE & SEMICONDUCTOR PHYSICS. *Educ:* Bowdoin Col, AB, 60; Brown Univ, PhD(physics), 66. *Prof Exp:* Nat Res Coun/Naval Res Lab res assoc semiconductor physics, Naval Res Lab, 65-67, res physicist, 67-70, actg head surface & transp sect, 70, head semiconductor & appln sct, 70-73, head optical interactions sect, 73-74, head, Semiconductors Br, 74-80 & supt, Electronics Technol Div, 80-82; PROF PHYS & ASTRON, STATE UNIV NY, BUFFALO, 82-, CHMN DEPT, 87- *Concurrent Pos:* Mem ad hoc comt elec properties, Nat Res Coun/Nat Acad Sci, 70, mem ad hoc panel high magnetic field res & facil, 77-78; Naval Res Lab sabbatical study prog, Max Planck Inst Solid State Res, Stuttgart, 72-73; Navy mem tech rev comt, Joint Serv Electronics Prog, Dept Defense, 74-82; mem var prog comts int conf, 74-; mem Mat Res adv comt, NSF, 78-82; mem, Nat Acad Sci-Nat Res Coun Panel on Artificially Structured Mat, 85. *Honors & Awards:* Pure Sci Award, Sci Res Soc Am, 72. *Mem:* Fel Am Phys Soc; Sigma Xi; Int Soc Optical Eng. *Res:* Semiconductors; infrared and far infrared properties of solids; magneto-optical studies of semiconductor hetero structures, quantum wells and superlattices; metal-insulator-semiconductor structures; photoluminescence in semiconductors; electronic materials characterization; properties of electronic systems of reduced dimensionality. *Mailing Add:* Dept Phys & Astron Fronczak Hall SUNY Buffalo NY 14260

MCCOMBS, CANDACE CRAGEN, b Springfield, Ill, 1947; m 85; c 2. IMMUNOLOGY, GENETICS. *Educ:* Stanford Univ, BS, 69, AB, 69; Univ Calif, Berkeley, PhD(genetics), 74. *Prof Exp:* Postdoctoral fel genetics, Univ Calif, San Francisco, 74-78; adj asst prof res immunol, Univ Calif, Long Beach, 78-82; from asst prof to prof med, Med Ctr, La State Univ, New Orleans, 82-90; PROF MED, UNIV SALA, MOBILE, 90- *Concurrent Pos:* Vis prof med, Univ Col, Galway, Ireland, 88-89; mem basic res coun, Am Heart Asn. *Mem:* Am Heart Asn; Am Asn Immunologists; AAAS; Am Soc Human Genetics; Genetics Soc Am; Fedn Am Socs Exp Biol. *Res:* Genetic factors predisposing to disease, especially HLA and disease associations, immune regulation, autoimmunity, cytokines and protease inhibitors. *Mailing Add:* Univ SAla CC CB Rm 478 Mobile AL 36688

MCCOMBS, CHARLES ALLAN, b Oklahoma City, Okla, Aug 28, 48; m 71. ORGANIC CHEMISTRY. *Educ:* Calif State Univ, Long Beach, BS, 73; Univ Calif, Los Angeles, PhD(org chem), 78. *Prof Exp:* RES CHEMIST ORG CHEM, RES LABS, TENN EASTMAN CO, 78- *Mem:* Am Chem Soc. *Res:* Development of new synthetic reagents; new methodology for total synthesis; transition metal catalysis for organic synthesis. *Mailing Add:* 434 Coralwood Dr Kingsport TN 37663

MCCOMBS, FREDA SILER, b Asheville, NC, Feb 26, 33; m 62; c 3. SCIENCE EDUCATION. *Educ:* Salem Col, BS, 55; Univ NC, Chapel Hill, MEd, 59, EdD(sci educ), 63. *Prof Exp:* Teacher gen sci & biol city schs, Va, 55-56; teacher biol & chem pub schs, NC, 57-58; instr biol & phys sci, Longwood Col, 61-63; assoc prof biol, Cent Va Community Col, 68-70; asst prof, 70-71, ASSOC PROF SCI EDUC, LONGWOOD COL, 71- *Res:* Curriculum materials; teaching aids and student resource materials for elementary science, particularly kindergarten and primary grades. *Mailing Add:* Dept Biol Longwood Col Farmville VA 23901

MCCOMBS, H LOUIS, b Pittsburgh, Pa, Oct 22, 32. ELECTRON MICROSCOPY. *Educ:* Hahnemann Univ, MD, 58. *Prof Exp:* PRES, CHARLES RIVER PATH ASSOCS, INC, 70- *Concurrent Pos:* Mem, Coun Atherosclerosis, Am Heart Asn. *Mem:* Col Am Pathologists; Am Asn Pathologists; Am Soc Clin Pathologists; Am Heart Asn; AMA. *Mailing Add:* 230 North St 1678 Farm Medfield MA 02052

MCCOMBS, ROBERT MATTHEW, HEPATITIS, HERPES. *Educ:* Syracuse Univ, PhD(microbiol), 64. *Prof Exp:* PROF MICROBIOL, EASTERN VA MED SCH, 73- *Mailing Add:* Dept Microbiol Eastern Vet Admin Med Sch PO Box 1980 Norfolk VA 23501

MCCOMIS, WILLIAM T(HOMAS), b Chicago, Ill, Aug 20, 38; m 62; c 1. FOOD SCIENCE, BIOMEDICAL ENGINEERING. *Educ:* Ill Inst Technol, BS, 60; Mass Inst Technol, PhD(food sci), 64. *Prof Exp:* Res food technologist, Div Biochem & Microbiol, Columbus Div, Battelle Mem Inst, 64-67, div chief, Div Microbiol & Environ Biol, 67-69, div chief, 69-71, assoc chief, Div Food & Biol Sci, 71-73, sect mgr, 73-74, assoc mgr, Bioeng-Health Sci Sect, 74-79, prog mgr, Bioprocess & Med Technol Sect, 79-88, PROG MGR, ORG & POLYMER CHEM DEPT, COLUMBUS DIV, BATTELLE MEM INST, 88- *Concurrent Pos:* Jr technician, Post Cereal Div, Gen Foods Corp, 60; adj prof, Ohio State Univ. *Mem:* Controlled Release Soc; Inst Food Technol; Soc Rheol. *Res:* Food product and process development; biomedical engineering. *Mailing Add:* 6562 Hawthrone St Worthington OH 43085

MCCONAGHY, JOHN STEAD, JR, b Philadelphia, Pa, Feb 3, 42; m 65; c 2. PHYSICAL ORGANIC CHEMISTRY, CATALYSIS. *Educ:* Haverford Col, BA, 63; Yale Univ, MS, 64, PhD(org chem), 66. *Prof Exp:* Sr res chemist, Cent Res Dept, 66-69 & Petrochem Res Dept, 69-75, res group leader, 76-80, sr res group leader, 80-87, SCIENCE FEL, SPECIALTY CHEM TECHNOL, MONSANTO CHEM CO, 88- *Mem:* Am Chem Soc; Sigma Xi. *Res:* Mechanism; catalysis; vapor phase reactions; organometallic chemistry; process development; functional fluids. *Mailing Add:* Monsanto Co 800 N Lindbergh Blvd St Louis MO 63167

MCCONAHEY, WILLIAM MCCONNELL, JR, b Pittsburgh, Pa, May 7, 16; m 40; c 3. INTERNAL MEDICINE, ENDOCRINOLOGY. *Educ:* Washington & Jefferson Col, AB, 38; Harvard Med Sch, MD, 42; Univ Minn, MS, 48. *Prof Exp:* Intern, Philadelphia Gen Hosp, 42-43; fel internal med, Mayo Grad Sch Med, 46-48; from instr to prof, Mayo Grad Sch Med, 50-73, chmn dept endocrinol, 67-74, PROF MED, MAYO MED SCH, 73- *Concurrent Pos:* Consult, Mayo Clin & Hosps, 49-85; emer consult, 86. *Honors & Awards:* Distinguished Serv Award, Am Thyroid Asn, 73; Laureate Award, Minn Chap, Am Col Physicians, 88. *Mem:* Endocrine Soc; Am Diabetes Asn; Am Thyroid Asn (treas, 61-65, secy, 66-70, pres, 75-76); Am Fedn Clin Res; fel Am Col Physicians; fel Am Med Asn; Cent Soc Clin Res. *Res:* Diseases of the thyroid gland; author of over 100 medical articles or book chapters published in medical journals or textbooks on endocrine subjects, primarily thyroid related. *Mailing Add:* Emer Off Mayo Clin Rochester MN 55905

MCCONATHY, WALTER JAMES, b McAlester, Okla, Nov 3, 41; m 66; c 1. LIPID CHEMISTRY, IMMUNOLOGY. *Educ:* Univ Okla, Norman, BA, 64, BS, 66; Univ Okla, Oklahoma City, PhD(biochem), 71. *Prof Exp:* Fel, 71-74, staff scientist, 74-75, asst mem, Lab Lipid & Lipoprotein Studies, 75-81, ASSOC MEM, LIPOPROTEIN & ATHEROSCLEROSIS RES PROG, OKLA MED RES FOUND, 81-, DIR CORE LAB, 78- *Concurrent Pos:* Adj assoc prof, Dept Biochem & Molecular Biol, Sch Med, Univ Okla, 85- *Honors & Awards:* Merrick Award, Okla Med Res Found, 82. *Mem:* Am Heart Asn; Sigma Xi; AAAS; Am Oil Chemists; Am Chem Soc Soc; Am Soc Biol Chemists. *Res:* Investigating the structure, functions and interrelations of human plasma lipoproteins and clotting factors in both normal and pathological conditions for potential application as diagnostic or prognostic tools in human disease states. *Mailing Add:* 2657 Lakeside Dr Oklahoma City OK 73120

MCCONKEY, JOHN WILLIAM, b Portadown, North Ireland. PHYSICS. *Educ:* Queen's Univ Belfast, BSc, 58, PhD(physics), 62. *Prof Exp:* Lectr physics, Queen's Univ Belfast, 63-70; PROF PHYSICS, UNIV WINDSOR, 70- *Concurrent Pos:* Commonwealth fel, Univ Manchester, 76-77; NASA sr fel, Jet Propulsion Lab, 83-84 & 90-91; Can Coun Killam Fel, 86-88. *Mem:* Fel Brit Inst Physics; fel Am Phys Soc; Can Asn Physicists. *Res:* Electron collisions; atmospheric processes; spectroscopy. *Mailing Add:* Dept Physics Univ Windsor Windsor ON N9B 3P4 Can

MCCONNACHIE, PETER ROSS, b Montreal, Que, Mar 31, 40; m 64; c 2. IMMUNOBIOLOGY. *Educ:* Univ BC, BSc, 62, MSc, 65; Univ Alta, PhD(immunol), 70. *Prof Exp:* Sr scientist tissue typing, Dept Clin Path, Univ Hosp, Edmonton, Alta, 70-74, sr scientist, Transplant Lab, 74-75; DIR TRANSPLANT LAB, MEM MED CTR, 75- *Concurrent Pos:* Immunologist II, Dept Surg, Univ Calif, Los Angeles, 72-73; sessional lectr path, Med Lab Sci, Univ Alta, 73-, prof asst, Med Res Coun Can Transplant Immunol Unit, 74-75; assoc dir, Immunotransplant Lab, Vancouver Gen Hosp, clin assoc prof, Dept of Pathology, Univ of BC, 84-85; coun, Am Soc Histo repro immunogenetics, 85-88. *Mem:* AAAS; Can Soc Clin Invest; Sec Treas Am Soc for Repro Immunol (86-89); Am Soc for Histocompatibility and Immunogenesis; Am Soc for Reproductive Immunol. *Res:* Transplantation immunology; histo compatibility; reproductive immunology; immuno toxicology. *Mailing Add:* Dept Med Microbiol & Immunol S Ill Univ Med Sch 800 N Rutledge Springfield IL 62781

MCCONNAUGHEY, BAYARD HARLOW, b Pittsburgh, Pa, Apr 21, 16; m 49; c 5. MARINE BIOLOGY. *Educ:* Pomona Col, BA, 38; Univ Hawaii, MA, 41; Univ Calif, PhD(zool), 48. *Prof Exp:* Asst zool, Univ Calif, 38-40 & Univ Hawaii, 40-41; asst, Scripps Inst, Univ Calif, 47-48, res assoc, 48; from asst prof to assoc prof, 48-72, PROF BIOL, UNIV ORE, 72- *Concurrent Pos:* Ky Contract Team assoc prof, Inst Agr Sci, Fac Fisheries, Univ Indonesia, 63-66; guest scientist, Coun Sci, Indonesia, 66; mem fac, World Campus Afloat, 75. *Mem:* Fel AAAS; Soc Syst Zool; Am Soc Parasitol; Am Soc Limnol & Oceanog; Sigma Xi. *Res:* Taxonomy and life history of the Mesozoa; microbiology; association analysis; plankton communities; delimiting and characterizing marine biotic communities. *Mailing Add:* 1653 Fairmount Blvd Eugene OR 97403

MCCONNAUGHEY, MONA M, b Larson AFB, Wash, Oct 9, 51. HYPERTENSION. *Educ:* Indiana Univ, PhD(pharmacol), 79. *Prof Exp:* INSTR PHARMACOL, E CAROLINA SCH MED, 81- *Mem:* Sigma Xi; Am Soc Pharmacol & Exp Therapeut. *Mailing Add:* Dept Pharmacol E Carolina Univ Sch Med Greenville NC 27834

MCCONNEL, ROBERT MERRIMAN, b Rochester, Pa, Mar 19, 36; m 62. MATHEMATICS. *Educ:* Washington & Jefferson Col, BA, 58; Duke Univ, PhD(math), 62. *Prof Exp:* Asst prof math, Univ Ariz, 62-64; asst prof, 64-66, assoc prof, 66-76, PROF MATH, UNIV TENN, KNOXVILLE, 76- *Mem:* Am Math Soc; Math Asn Am. *Res:* Number theory; finite field theory; polynomials over finite fields. *Mailing Add:* Univ Tenn Knoxville TN 37996

MCCONNELL, BRUCE, b Pittsburgh, Pa, Sept 12, 32; m 64; c 1. BIOCHEMISTRY. *Educ:* Grove City Col, BS, 54; Univ Vt, PhD(biochem), 66. *Prof Exp:* Tech rep, Atlas Powder Co, Del, 54-56; pilot plant supvr, Koppers Co, Inc, Pa, 56-59; NIH fel molecular biol, Dartmouth Med Sch, 66 & Univ Ore, 66-69; asst prof, 69-74, ASSOC PROF BIOPHYS, UNIV HAWAII, HONOLULU, 74- *Concurrent Pos:* NSF res grant. *Mem:* Am Chem Soc. *Res:* Physical chemistry and biosynthesis of macromolecules; protein chemistry and isolation; molecular configurations of DNA by hydrogen exchange and physical-chemical; nuclear magnetic resonance-proton exchange. *Mailing Add:* 1224 Aulepe St Kailua HI 96734-4101

MCCONNELL, C(HARLES) W(ILLIAM), b Salem, Ohio, July 24, 12; m 37; c 2. CHEMICAL ENGINEERING. *Educ:* Carnegie Inst Technol, BS, 34, ChE, 38. *Prof Exp:* Res chem engr, Conewango Refining Co, 34-36, chief chemist, 36-39; chem engr, Linde Air Prod Co, 39-44, patent engr, 44-53, engr, Patent Dept, Union Carbide & Carbon Corp, 53-58, asst div attorney, Union Carbide Corp, 58-61, Prom & Advert Dept, Chem Div, 61-63, ed, Chem Prog, 63-67, publ mgr, Chem & Plastics Div, 67-77; RETIRED. *Res:* Lube oil refining; antifreeze; synthetic lubricants. *Mailing Add:* 17 Ann Dr Tappan NY 10983

MCCONNELL, DAVID GRAHAM, b Bronxville, NY, Dec 3, 26; m 48; c 6. BIOPHYSICS, BIOCHEMISTRY. *Educ:* Columbia Univ, AB & AM, 49; Ind Univ, PhD(exp psychol), 57. *Prof Exp:* Res assoc psychol, Ohio State Univ, 56-61, res assoc chem, 60-61; proj dir med prog, Britannica Ctr, Calif, 61-62; assoc prof physiol optics, Ohio State Univ, 62-65, assoc prof biophys, 65-71, assoc prof biochem, 67-71, prof biochem & biophys, 71-72; PROF BIOCHEM, MICH STATE UNIV, 73- *Concurrent Pos:* Vis assoc prof, Enzyme Inst, Univ Wis, 64-66. *Mem:* Biophys Soc; Am Soc Cell Biol; Am Soc Biol Chemists; Am Soc Neurochem. *Res:* Biochemistry of the retina; neurochemistry; RAS gene expression. *Mailing Add:* Biochem Bldg Mich State Univ East Lansing MI 48824

MCCONNELL, DENNIS BROOKS, b Waupun, Wis, Aug 18, 38; m 64; c 2. ORNAMENTAL HORTICULTURE. *Educ:* Wis State Univ, River Falls, 66; Univ Wis-Madison, MS, 67, PhD(bot, hort), 70. *Prof Exp:* Asst prof ornamental hort, Agr Res Ctr, 70-73, assoc prof, 73-86, PROF ORNAMENTAL HORT, UNIV FLA, 86- *Mem:* Am Soc Hort Sci; Am Hort Soc; Bot Soc Am. *Res:* Morphology and anatomy of normal and experimentally modified subtropical and tropical plants. *Mailing Add:* Dept Ornamental Hort Univ Fla Gainesville FL 32611

MCCONNELL, DUNCAN, b Chicago, Ill, Jan 30, 09; m 34; c 3. DENTAL RESEARCH, MATERIALS SCIENCE. *Educ:* Washington & Lee Univ, BS, 31; Cornell Univ, MS, 32; Univ Minn, PhD(mineral), 37. *Prof Exp:* Teaching fel mineral, Stanford Univ, 34-35; instr mineral, Univ Tex, Austin, 37-41; chemist-petrographer, US Bur Reclamation, Denver, 41-47; actg asst div chief geol, Gulf Res & Develop Co, 47-50; prof mineral, 50-56 & 64-76, prof dent res, 57-76, EMER PROF DENT RES & GEOL MINERAL, OHIO STATE UNIV, 76- *Concurrent Pos:* USPHS spec res fel, Ohio State Univ, 57-61, USPHS grants, 57-68, US Army Med Res & Develop Command grant, 66-67; consult & site visitor, Nat Inst Dent Res, 73; examr, Dept Orthop Surg, Yale Univ, 73; consult, NSF, 73 & 90, Nat Inst Dent Res, 74 & Prev Dent Res Inst, Ind Univ, Ft Wayne, 74; invited lectr, Inst Inorganic Chem, Univ Munich & Nat Univ Mexico, 73. *Mem:* Fel AAAS; Electron Micros Soc Am; fel Mineral Soc Am; Brit Mineral Soc; fel Soc Econ Geologists; fel Royal Soc Arts. *Res:* Crystal chemistry of inorganic component of teeth and bones, restorative materials, mineralization process and biogeochemistry of phosphate minerals. *Mailing Add:* 4312 S 31st St Apt 129 Temple TX 76502-3360

MCCONNELL, ERNEST EUGENE, b Orrville, Ohio, Nov 14, 37; m 58; c 3. PATHOLOGY, TOXICOLOGY. *Educ:* Ohio State Univ, DVM, 61; Mich State Univ, MS, 66; Am Col Vet Pathologists, dipl, 68; Am Bd Toxicol, dipl, 80. *Prof Exp:* Base vet, Hill AFB, Utah, 61-64; resident path, Armed Forces Inst Path, 65-67 & Aerospace Path Div, 67-69; researcher, Onderstepoort Vet Inst, S Africa, 69-72, path br chief, Inhalation Toxicol Lab, Wright Patterson AFB, 72-74; researcher, NIH, 74-80, chief, Chem Path Br, 80-83, dir, Toxicol Res Testing Prog & Nat Inst Environ Health Sci, 83-88; CONSULT TOXICOL & PATH, 88- *Concurrent Pos:* Consult, Zool Park, NC, 67-71; mem biohazards safety comt, NIH, 76-79; mem exp design subgroup, Nat Cancer Inst, 77-80; adj asst prof vet sci, NC State Univ, 77-; mem, Agent Orange Working Group, Vet Admin, 82-; chmn, Natural Exposure Study Animals, Nat Res Coun, Nat Acad Sci, 85; Mem bd dirs, Am Bd Toxicol, 85; pres, Am Bd Toxicol, 88; mem, Animals as Sentinals Chem Exposure, Nat Acad Sci, 88-91, Comt Toxicol, 89- & Sci Adv Bd, Environ Protection Agency, 89- *Honors & Awards:* Commendation Medal, USPHS, 78; Outstanding Serv Medal, NIH, 85. *Mem:* Am Col Vet Pathologists; Am Vet Med Asn; Soc Toxicol Path; Soc Toxicol. *Res:* Pathology of toxic chemicals of environmental interest, especially halogenated hydrocarbons; spontaneous diseases of primates; asbestos; veterinary medicine. *Mailing Add:* 3028 Ethan Lane Raleigh NC 27613

MCCONNELL, FREEMAN ERTON, b West Point, Ind, Apr 20, 14; m 41; c 2. AUDIOLOGY. *Educ:* Univ Ill, BS, 39, MA, 46; Northwestern Univ, PhD(audiol), 50. *Prof Exp:* Instr, Ill High Sch, 39-43; high sch & jr col instr, 46-48; lectr speech correction & audiol, Northwestern Univ, 48-50; asst prof audiol, Inst Logopedics Wichita & Wichita Univ, 50-51; prof & head dept, Univ Tenn, 60-63; from asst prof to prof & head dept, 51-60, head dept, 63-76, prof audiol & dir, Bill Wilkerson Hearing & Speech Ctr, 63-79, EMER PROF AUDIOL, VANDERBILT UNIV, 79- *Concurrent Pos:* Consult, Arnold Eng Develop Ctr, US Air Force, 53-71 & Bur Res, US Off Educ, 65-73; chmn bd dirs, Educ & Auditory Res Found, Nashville, 78-79; mem, US-Can Adv Comt, Ann Surv Hearing Impaired Children & Youth, 72-78. *Mem:* Coun Except Children; Acoust Soc Am; fel Am Speech-Lang-Hearing Asn; Acad Rehab Audiol (pres, 71). *Res:* Clinical audiology; language impaired children; hereditary deafness. *Mailing Add:* Dept Hearing & Speech Sci Vanderbilt Univ Med Ctr Nashville TN 37232

MCCONNELL, H(OWARD) M(ARION), b Dallas Co, Iowa, July 18, 24; c 4. ELECTRICAL ENGINEERING, ENGINEERING MANAGEMENT. *Educ:* Marquette Univ, BEE, 47; Carnegie Inst Technol, MS, 48, DSc(elec eng), 50. *Prof Exp:* Instr elec eng, Carnegie Inst Technol, 49-50, asst prof, 50-55; mgr res & develop, Jack & Heintz, Inc, 55-58, chief engr, 58-62; dir eng, Power Equip Div, Lear-Siegler, Inc, 62-63; prod mgr, Harris-Intertype Corp, 64-67, chief printing systs res, 67-68, chief elec engr, Harris-Seybold Co, 68-74; consult, 74-78; consult programmer-analyst, 78-80; staff engr, Avtron Mfg Inc, 80-84, chief engr aircraft equip, 84-87,; CONSULT, 87- *Concurrent Pos:* Lectr, Exten Div, Univ Pittsburgh, 50-51; design engr, Westinghouse Elec Corp, 50-51; lectr, Cleveland State Univ, 81-91. *Mem:* Sigma Xi; sr mem Inst Elec & Electronic Engrs. *Res:* Electrical machinery; magnetic amplifiers; magnetic materials; aircraft accessories; graphic arts equipment; food processing machinery; computer programming; aerospace test equipment. *Mailing Add:* 1082 Lakeland Ave Lakewood OH 44107-1228

MCCONNELL, HARDEN MARSDEN, b Richmond, Va, July 18, 27; m 56; c 3. BIOPHYSICAL CHEMISTRY. *Educ:* George Washington Univ, BS, 47; Calif Inst Technol, PhD(chem), 51. *Prof Exp:* Nat Res Coun fel physics, Univ Chicago, 50-52; res chemist, Shell Develop Co, 52-56; from asst prof to prof chem, Calif Inst Technol, 56-64; prof, 64-79, ROBERT ECKLES SWAIN PROF CHEM, STANFORD UNIV, 79-, CHMN DEPT, 89- *Concurrent Pos:* Consult, PAST, Varian Assocs, Syva Inc, Becton Dickinson & Co, NJ, Exxon Corp, DNAX Res Inst, Calif & Liposome Technol, Inc; mem bd dirs, Biospan Corp, Menlo Park, Calif, 81-; Founder & dir, Molecular Devices Corp, Palo Alto. 83-, consult & mem, bd dirs; pres, Found Basic Res Chem, 90-; mem & hon assoc, Neurosci Res Prog, Mass Inst Technol. *Honors & Awards:* Pure Chem Award, Am Chem Soc, 62, Harrison Howe Award, 68, Irving Langmuir Award, 71 & Remsen Mem lectr, 82; Harkins lectr, Univ Chicago, 67; Falk-Plaut lectr, Columbia Univ, 67; Debeye lectr, Cornell Univ, 71; Harvey lectr, Rockefeller Univ, 77; A L Patterson lectr, Inst Cancer Res, Univ Pa, 78; Pauling lectr, Stanford Univ, 81; Wolf Prize, 84; Nat Medal Sci, 89. *Mem:* Nat Acad Sci; fel Am Phys Soc; Am Soc Biol Chemists; Biophys Soc; Am Chem Soc; hon mem Swed Biophys Soc; fel Am Acad Arts & Sci; AAAS. *Res:* Spin distributions and hyperfine interactions in organic radicals, electron and nuclear magnetic resonance spectroscopy; excitons in molecular crystals; spin labels; biological membranes; physical chemistry of immune recognition. *Mailing Add:* Dept Chem Stanford Univ Stanford CA 94305

MCCONNELL, JACK BAYLOR, b Crumpler, WVa, Feb 1, 25; m 58; c 3. MEDICINE. *Educ:* Univ Tenn, MD, 49. *Hon Degrees:* DSc, Emory & Henry Col, 82. *Prof Exp:* Resident, Baylor Univ, 49-52; assoc dir prof serv, Lederle Labs, Am Cyanamid Co, NY, 53-57, dir clin invest, 57-59, dir clin develop, 59-61; exec dir new prod div, McNeil Labs, Inc, 61-67, vpres com develop, 67-69, CORP DIR ADVAN TECHNOL, JOHNSON & JOHNSON, 69- *Mem:* AAAS; Am Acad Pediat; Am Thoracic Soc; AMA; Am Med Writers' Asn; Am Acad Med Dirs; Am Inst Ultrasound Med; Am Soc Neuroimaging; Asn Advan Med Instrumentation. *Res:* Biomedical instruments and materials in the field of respiratory care, laboratory instruments and patient monitoring. *Mailing Add:* 29 Oldford Dr Hilton Head Island SC 29926

MCCONNELL, JAMES FRANCIS, b Syracuse, NY, Dec 15, 36; m 59; c 4. ORGANIC CHEMISTRY. *Educ:* Le Moyne Col, NY, BS, 58; Syracuse Univ, PhD(org chem), 59. *Prof Exp:* Asst prof org chem, Mansfield State Col, 62-64; asst prof, 64-74, ASSOC PROF ORG CHEM, STATE UNIV NY COL CORTLAND, 74- *Mem:* Am Chem Soc. *Res:* Stereochemical approach to the mechanism of the Mannich reaction; rate of loss of optical activity of nitroalkanes compared to the rate in which they form Mannich base products in Mannich reaction. *Mailing Add:* Dept Chem State Univ NY PO Box 2000 Cortland NY 13045-0900

MCCONNELL, JOHN CHARLES, b Belfast, Northern Ireland, Sept 11, 45; m; c 3. AERONOMY, SPACE SCIENCE. *Educ:* Queens Univ Belfast, BSc, 66, PhD(quantum mech), 69. *Prof Exp:* Res asst planetary sci, Kitt Peak Nat Observ, 69-70; res asst atmospheric physics, Ctr Earth Planetary Physics, Harvard Univ, 70-72; from asst prof to prof physics, 72-82, chmn dept earth & atmospheric sci, 82-86, PROF ATMOSPHERIC PHYSICS, YORK UNIV, 82- *Mem:* Am Astron Soc; Can Meterol & Oceanog Soc; Am Geophys Union; Can Asn Physicists. *Res:* Tropospheric and stratospheric chemistry; auroral physics; planetary atmospheres; ionospheric physics; radiative transfer; atmospheric modelling. *Mailing Add:* Dept of Earth & Atmospheric Sci York Univ 4700 Keele St North York ON M3J 1P3 Can

MACCONNELL, JOHN GRIFFITH, b Chicago, Ill, Oct 14, 42; m 75. AGRICULTURAL PESTICIDE RESEARCH. *Educ:* Univ Ill, Urbana, BS, 64; Univ Mich, Ann Arbor, MS, 68, PhD(org chem), 69. *Prof Exp:* Res assoc org chem, Univ Ga, 69-70, NY State Col Forestry, Syracuse Univ, 70-72 & Univ Ga, 72-73; asst prof chem, Dalton Jr Col, 73-74; instr psychiat & Chem, Emory Univ, 74-75; SR RES CHEMIST, MERCK SHARP & DOHME RES LABS, 75- *Mem:* AAAS; Am Chem Soc. *Res:* Planning and execution of agricultural pesticide research and development programs. *Mailing Add:* Merck & Co 80Y-2A 62 Rahway NJ 07065

MCCONNELL, KENNETH G, b East Ellsworth, Wis, May 11, 34; m 59; c 4. ENGINEERING MECHANICS. *Educ:* Col St Thomas, BA, 57; Univ Notre Dame, BS, 57; Iowa State Univ, MS, 60, PhD(eng mech), 63. *Prof Exp:* From instr to assoc prof eng mech, 57-74, PROF AEROSPACE ENG & ENG MECH, IOWA STATE UNIV, 74- *Concurrent Pos:* Consult, sound & vibration to local and nat industs, Korean Adv Inst Sci & Technol, 84, Korean Inst Mach, 84, Univ Uberlandis, Brazil, 86, Seoul Nat Univ, 88, Sandia Nat Labs, 88 & 89, Cummins Engine Co, 90, Imperial Col London, 90-91; assoc ed, Int J Analysis & Exp Modal Analysis, 85-90. *Honors & Awards:* M M Frocht Award, Soc Exp Mech, 85; fel Soc Exp Mech. *Mem:* Am Soc Eng Educ; Soc Exp Stress Analysis. *Res:* Vibrations; acoustics; dynamic instrumentation; fluid structure interaction; vibration testing and measurements; damping measurements; modal analysis. *Mailing Add:* RR 5 2590 Meadow Glen Ames IA 50010

MCCONNELL, KENNETH PAUL, b Rochester, NY, June 19, 11; m 50. BIOCHEMISTRY. *Educ:* Univ Rochester, AB, 35, MS, 37, PhD(biochem, nutrit), 42. *Prof Exp:* Asst, Sch Med & Dent, Univ Rochester, 35-37, instr, Dept Radiation Biol & assoc, Atomic Energy Proj, 48-49; asst, Med Sch, Ohio State Univ, 37-38; asst, Univ Iowa, 38-39; biochemist, USPHS, 46; biochemist, Med Dept, Field Res Lab, Ft Knox, Ky, 46-48; asst prof biochem & nutrit & dir med physics lab, Univ Tex Med Br, 50-52; res assoc, 52-60, from assoc prof to prof, 60-78, EMER PROF BIOCHEM, SCH MED, UNIV LOUISVILLE, 78- *Concurrent Pos:* Asst dir radioisotope serv, Vet Admin Hosp, 52-60, biochemist & prin scientist, 52-78; consult, Oak Ridge Inst Nuclear Studies, 52-59. *Mem:* Fel AAAS; Am Chem Soc; Am Inst Nutrit; Am Soc Biol Chemists; Soc Exp Biol & Med. *Res:* Metabolism of trace elements in the mammalian organism in health and disease; use of isotopes in research. *Mailing Add:* 705 Victoria Pl Louisville KY 40207

MCCONNELL, RICHARD LEON, b Gate City, Va, Mar 23, 26; m 48; c 3. HOT-MELT ADHESIVES, POLYMER STABILIZATION. *Educ:* Univ Ky, BS, 48; Univ Va, MS, 50, PhD(org chem), 52. *Prof Exp:* Res chemist, res labs, Tenn Eastman Co, 51-57, sr res chemist, 57-70, res assoc, 70-80; sr res assoc, Res Labs, Eastman Chem Div, Eastman Kodak Co, 80-87; RETIRED. *Concurrent Pos:* Tour speaker, Am Chem Soc, 67-68. *Mem:* Am Chem Soc; Sigma Xi. *Res:* Chemically modified polymers; polyester copolymers; polymer blends; vinyl polymers; stabilization of polymers; hot-melt adhesives; synthetic waxes; nonwoven fabrics; colored polymers; polyolefin homo- and copolymers; organophosphorus chemistry. *Mailing Add:* 421 Manderley Rd Kingsport TN 37660

MCCONNELL, ROBERT A, b McKeesport, Pa, Apr, 14. BIOPHYSICS, PARAPSYCHOLOGY. *Educ:* Carnegie Inst Technol, BS, 35; Univ Pittsburgh, PhD(physics), 47. *Prof Exp:* Asst geophysicist, Gulf Res Develop Co, 37-39; asst physicist, US Naval Aircraft Factory, 39-41; mem staff, Radiation Lab, Mass Inst Technol, 41-44, group leader, 44-46; asst prof physics, Univ Pittsburgh, 47-53, from asst res prof to assoc res prof, 53-63, res prof biophys, 64-84, EMER RES PROF BIOL SCI, UNIV PITTSBURGH, 84- *Mem:* Inst Elec & Electronics Engrs; Parapsychol Asn (pres, 57-58); Am Psycho Soc. *Res:* Radar moving target indication; theory of the iconoscope; ultrasonic microwaves; extrasensory perception; psychokinesis. *Mailing Add:* Dept Biol Sci Univ Pittsburgh Pittsburgh PA 15260

MCCONNELL, ROBERT KENDALL, b Toronto, Ont, Mar 1, 37; m 58; c 2. INFORMATION THEORY, GEOPHYSICS. *Educ:* Princeton Univ, BScE, 58; Univ Toronto, MASc, 60, PhD(physics), 63. *Prof Exp:* Geologist, Invex Corp, Ltd, 57-58; geophysicist, Hunting Surv Corp, 59-60; geophysicist, Arthur D Little, Inc, 63-69; pres, Earth Sci Res, Inc, 70-75; res geophysicist, Boston Col, 75-78; CONSULT, 77-; CHMN, WAYLAND RES, INC, 82- *Concurrent Pos:* Vis prof, Univ RI, 70. *Mem:* Am Geophys Union; AAAS; Seismol Soc Am. *Res:* Computer pattern recognition for industrial and scientific applications; planetary evolution; geological and geophysical interpretation; tectonophysics. *Mailing Add:* 14 Stevens Terr Arlington MA 02174

MCCONNELL, STEWART, b El Paso, Tex, Apr 25, 23; m 49; c 3. VIROLOGY, IMMUNOLOGY. *Educ:* Ohio State Univ, MS, 60; Tex A&M Univ, DVM, 50. *Prof Exp:* Vet virol & res, Walter Reed Army Inst Res, Vet Corps, US Army, 60-63, adv, US Agency Int Develop, 63-65, vet malaria res, Walter Reed Army Inst Res, 65-66, vet infectious dis res, US Army Med Res Inst Infectious Dis, 66-68; from assoc prof to prof vet microbiol, Tex A&M Univ, 68-89; RETIRED. *Concurrent Pos:* Mem subcomt fish standards, comt standards, Inst Lab Animal Resources, Nat Res Coun-Nat Acad Sci, foreign animal dis comt, US Animal Health Asn, Western Hemisphere Comt Animal Virus Characterization & bd comp virol, WHO-Food & Agr Orgn. *Honors & Awards:* Medal, Bolivian Ministry Agr, 64; Spec Commendation, Bolivian Hemorrhagic Fever Comn, 65. *Mem:* Am Asn Lab Animal Sci; Am Soc Lab Animal Practitioners; Am Soc Microbiol; fel NY Acad Sci; Sigma Xi. *Res:* Virus that play a role in respiratory diseases of feedlot cattle; poxvirus of animals; virus diseases of marine fish and shellfish. *Mailing Add:* 1610 Dominik College Station TX 77840

MACCONNELL, WILLIAM PRESTON, b NB, Can, June 15, 18; nat US; m 43; c 3. FORESTRY. *Educ:* Univ Mass, BS, 43; Yale Univ, MF, 48. *Prof Exp:* Assoc prof, 48-63, PROF FORESTRY, UNIV MASS, AMHERST, 63- *Mem:* Soc Am Foresters; Am Soc Photogram. *Res:* Forest management; aerial photogrammetry. *Mailing Add:* Dept Forestry & Wildlife Mgt Hadsworth Hall Univ Mass Amherst MA 01003

MCCONNELL, WILLIAM RAY, b Wise, Va, Oct 15, 43; m 67; c 1. TOXICOLOGY, PHARMACOLOGY. *Educ:* Va Polytech Inst & State Univ, BS, 65; Med Col Va, MS, 73, PhD(pharmacol), 76. *Prof Exp:* Res pharmacologist, Southern Res Inst, 76-79; sr res toxicologist, A H Robins Pharmaceut Co, Inc, 79-90; MGR, TOXICOL & SAFETY ASSESSMENT, WHITBY RES, INC, 90- *Concurrent Pos:* Adj asst prof pharmacol, Univ Ala, Birmingham, 77-79. *Mem:* Am Soc Pharm & Exp Therapeut; Am Soc Toxicol. *Res:* Biochemical pharmacology as related to pharmacokinetics and metabolism of anti-cancer and anti-malarial drugs; analytical and biochemical methodology for measuring drugs; toxicology of pharmaceutical drugs. *Mailing Add:* 1461 King WM Woods Rd Midlothian VA 23113-9118

MCCONOUGHEY, SAMUEL R, b Iowa, Feb 14, 24. TELECOMMUNICATIONS. *Educ:* Iowa State Col, BS, 50. *Prof Exp:* Elec engr, Fed Commun Comn, 65-82; CONSULT, MOBILE COMN, 82- *Mem:* Fel Inst Elec & Electronic Engrs. *Mailing Add:* 13017 Chestnut Oak Dr Rm 200B Gaithersburg MD 20878

MCCONVILLE, BRIAN JOHN, b Queenstown, NZ, May 11, 33; m 59; c 3. PSYCHIATRY. *Educ:* Univ Otago, NZ, MB, ChB, 57; Royal Col Physicians Can, cert psychiat, 66, fel psychiat, 72. *Prof Exp:* Lectr psychiat, 65-67, lectr pediat, 65-69, from asst prof to assoc prof psychiat, 67-77, ASST PROF

PEDIAT, QUEEN'S UNIV, ONT, 69-, DIR CHILD PSYCHIAT TRAINING, 70-, PROF PSYCHIAT, 77- *Concurrent Pos:* Dir, Regional Children's Centre, Kingston Psychiat Hosp, 67-; liaison off, Can Psychiat Asn-Can Pediat Soc, 70-; chmn child-adolescent comt, Child, Adolescent & Retardate Sect, Can Psychiat Asn, 70-, mem sci coun, 72- *Mem:* Can Psychiat Asn; Can Med Asn; Soc Psychophysiol Res; Am Psychiat Asn; Am Acad Child Psychiat. *Res:* Child psychiatry; measurement of behavioral change in children's inpatient units; phenomena of childhood depression-aggression; family therapy; illness after depression. *Mailing Add:* Psychiat Dept ML-559 Univ Cincinnati 231 Bethesda Ave Cincinnati OH 45236

MCCONVILLE, DAVID RAYMOND, b Benson, Minn, Feb 9, 46; m; c 3. LIMNOLOGY, FISHERIES. *Educ:* St Cloud State Univ, BS, 68, MA, 69; Univ Minn, St Paul, PhD(fisheries), 72. *Prof Exp:* Asst, St Cloud State Univ, 68-70, res assoc, 70-71; instr biol, Univ Minn Tech Col-Waseca, 71-73; assoc prof, 73-85, DIR AQUATIC STUDIES, ST MARY'S COL, MINN, 73-, PROF BIOL, 85- *Concurrent Pos:* Consult ecol studies, Northern States Power Co, Minneapolis, 73-75; prin investr, Off Water Resources res grant fisheries studies, St Mary's Col, 73-75, dir, NSF environ equip grant, 75-77, dir, US Fish & Wildlife Serv biol res contract, 75-79, dir, US Army Corps Engrs biol res contract, 75-76; dir, US Environ Protection Agency Res Contract, 76-78; river biologist, US Army Corps, 80-81; toxicol intern, US Fish & Wildlife Serv, 83; dir, US Army Corps Engrs Biol Res Contract, 83-84; consult, ecol studies Energy Res Mgt NCent, Minneapolis, 85-; prin investr, US Fish & Wildlife Serv biol res contract, 88, sabbatical appointment, geog info systs river anal, La Crosse, Wis, 91. *Mem:* Am Fisheries Soc; NAm Benthological Soc; Sigma Xi. *Res:* Current flows and fisheries; use of artificial substrates; emergence studies of insects; relative abundance, age and growth, food habits and selected movement studies of Mississippi River fish; riverine habitat improvement; benthic invertebrate ecology; geographic information systems analysis of aquatic habitats using arc-info, pc arc-info and eppl7. *Mailing Add:* Dept Aquatic Studies Biol St Mary's Col Terrace Heights Winona MN 55987

MCCONVILLE, GEORGE T, b St Paul, Minn, Sept 23, 34; m 59; c 1. PHYSICS. *Educ:* Carleton Col, AB, 56; Purdue Univ, MS, 59; Rutgers Univ, PhD(physics), 64. *Prof Exp:* Tech staff assoc low temperature physics, RCA Labs, 59-61; sr res physics, Monsanto Res Corp, 64-69, from res specialist to sr res specialist, 69-86, Monsanto fel, 86-88; EG&G SCI FEL, 88- *Concurrent Pos:* Vis scientist, Kanierlingh Ounes Lab, Univ Leiden, Neth, 85; G Colonnetti Inst Meterol, Turin, Italy, 86, vis prof, 90. *Mem:* Am Phys Soc; AAAS. *Res:* Low temperature physics; superconductivity; calorimetry; gas transport. *Mailing Add:* EG&G Mound Appl Technol PO Box 3000 Miamisburg OH 45343

MCCONVILLE, JOHN THEODORE, b Centerville, Iowa, Nov 15, 27; m 54; c 3. APPLIED PHYSICAL ANTHROPOLOGY. *Educ:* Univ NMex, BA, 53; Univ Ariz, MA, 59; Univ Minn, PhD, 69. *Prof Exp:* Res analyst, US Army Electronic Proving Grounds, 56-59; assoc dir anthrop res proj, Antioch Col, 59-69 & Webb Assocs, Inc, 70-77; PRES, ANTHROP RES PROJ, INC, 78- *Mem:* Am Asn Phys Anthrop. *Res:* Human biology, particularly man as a component of complex systems; research in body size variability and human physical characteristics such as body strength and composition. *Mailing Add:* Anthrop Res Proj 503 Xenia Ave PO Box 307 Yellow Springs OH 45387

MCCOOK, GEORGE PATRICK, b July 23, 37; US citizen; m 61; c 3. ASTRONOMY. *Educ:* Villanova Univ, BS & MA, 61; Univ Pa, MS, 65, PhD(astron), 68. *Prof Exp:* From asst prof to assoc prof, 61-81, PROF ASTRON, VILLANOVA UNIV, 81-, CHMN, DEPT ASTRON, 74- *Mem:* Am Astron Soc. *Res:* Microcomputer based instrumentation and photometric studies of variable stars. *Mailing Add:* Dept of Astron & Astrophysics Villanova Univ Mendel Hall Villanova PA 19085

MCCOOK, ROBERT DEVON, b Columbus, Ga, Jan 23, 29; m 57. CLINICAL ENGINEERING. *Educ:* La State Univ, BS, 50, MS, 52; Loyola Univ, Ill, PhD(physiol), 64. *Prof Exp:* Instr chem, Concord Col, 53-54; asst biochem, Stritch Sch Med, Loyola Univ Chicago, 54-57, asst physiol, 57-62, from instr to asst prof physiol, 62-74; CHIEF BIOMED ENG, VET ADMIN HOSP, LEXINGTON, 74- *Concurrent Pos:* Res physiologist, Vet Admin Hosp, Hines, Ill, 71-74. *Mem:* AAAS; Am Physiol Soc; Inst Elec & Electronics Engrs; Sigma Xi. *Res:* Central control of temperature regulation; physiological instrumentation; computer applications to medicine and biology. *Mailing Add:* Lexington Vet Admin Hosp 3136 Hyde Park Dr Lexington KY 40503

MCCOOL, D(ONALD) K, b St Joseph, Mo, May 22, 37; m 72; c 3. SOIL & WATER. *Educ:* Univ Mo, BS(agr) & BS(agr eng), 60, MS, 61; Okla State Univ, PhD(agr eng), 65. *Prof Exp:* Agr engr, Water Conserv Struct Lab, Southern Plains Br, Soil & Water Conserv Res Div, 61-71, AGR ENGR, PALOUSE CONSERV FIELD STA, PAC WEST AREA, AGR RES SERV, USDA, 71-; ASSOC AGR ENGR, DEPT AGR ENG, WASH STATE UNIV, 71- *Concurrent Pos:* Asst prof, agr eng dept, Okla State Univ, 66-71. *Mem:* Am Soc Agr Engrs; Am Soc Civil Engrs; Soil & Water Conserv Soc; Sigma Xi. *Res:* Hydrologic conditions leading to runoff and erosion-sedimentation events; effective hydraulic roughness of soil-crop surfaces and their resistance to erosion; prediction model for estimating erosion, sedimentation and downstream water quality for different crop managements under Pacific Northwest conditions; controlling snow deposition and soil freezing. *Mailing Add:* Dept Agr Eng Smith Eng Bldg Wash State Univ Pullman WA 99164-6120

MCCOOL, JOHN MACALPINE, b San Mateo, Calif, Apr 24, 30; m 52. ELECTRICAL ENGINEERING, ACOUSTICS. *Educ:* Calif Inst Technol, BS, 52. *Prof Exp:* Physicist, US Naval Ord Test Sta, 52-56, head electronics br, 56-71, head electronics div, US Naval Undersea Ctr, 71-74, consult, Naval Ocean Systs Ctr, 74-81; CONSULT, LINEAR MEASUREMENTS CORP, 81- *Mem:* Inst Elec & Electronics Engrs. *Res:* Extension of statistical detection theory to a back scattering medium; research in adaptive systems. *Mailing Add:* 11624 Fuerte Dr El Cajon CA 92020

MCCORD, COLIN WALLACE, b Chicago, Ill, May 15, 28; m 54; c 3. PUBLIC HEALTH, SURGERY. *Educ:* Williams Col, BA, 49; Columbia Univ, MD, 53. *Prof Exp:* Instr & asst prof surg, Sch Med, Univ Ore, 61-65; asst prof & assoc prof, Columbia Univ, 65-71; ASSOC PROF INT HEALTH, SCH HYG, JOHNS HOPKINS UNIV, 71- *Concurrent Pos:* Attend clin surgeon, St Lukes Hosp, NY, 65-71; resident dir, Rural Health Res Ctr, Narangwal, Punjab, India, 71-72; dir, Companiganj Health Proj, Noakhali, Bangladesh, 72-75. *Mem:* Am Col Surgeons; Am Asn Thoracic Surgeons. *Res:* Cost and effectiveness of health programs in developing countries; cost and effectiveness of nutrition interventions. *Mailing Add:* Harlem Hosp Ctr 136th & Lennox Ave New York NY 10037

MCCORD, JOE M, b Memphis, Tenn, Mar 3, 45. MEDICINE, BIOCHEMISTRY. *Educ:* Rhodes Col, BS, 66; Duke Univ, PhD(biochem), 70. *Prof Exp:* Asst res prof, Dept Med, Duke Univ, 72-76; from assoc prof to prof biochem, Univ Ala, 76-90, chmn, 81-90; HEAD, DEPT BIOCHEM & MOLECULAR CHEM, WEBB-WARING LUNG INST, UNIV COLO, 90- *Mem:* Am Soc Biochem & Molecular Biol; Am Physiol Soc; Oxygen Soc. *Mailing Add:* Webb-Waring Lung Inst Univ Colo 4200 E Ninth Ave Box C-321 Denver CO 80262

MCCORD, JOE MILTON, b Memphis, Tenn, Mar 3, 45. FREE RADICAL PATHOLOGY. *Educ:* Southwestern Univ Memphis, BS, 66; Duke Univ, PhD (biochem), 70. *Prof Exp:* Fel biochem, Duke Univ Med Ctr, 70-71, assoc med & biochem, 72-76, res asst prof, 76; assoc prof, 76-80, co-chmn, 80-81, PROF BIOCHEM, UNIV SOUTH ALA, 80-, CHMN DEPT, 81 - *Concurrent Pos:* Consult, Diag Data Inc, 71-77, Pharmacia AB, 80-81, Ciba Geigy, 84- & Biotechnol Gen, 85 - *Mem:* Am Soc Biol Chemists. *Res:* Biology of superoxide free radical and superoxide dismutase; pathophysiological roles of oxygen-derived metabolites, especially in inflammatory and ischemic disease states. *Mailing Add:* Dept Biochem Univ S Ala Mobile AL 36688

MCCORD, MICHAEL CAMPBELL, b Knoxville, Tenn, Apr 28, 36; m 58; c 3. COMPUTER SCIENCE. *Educ:* Univ Tenn, BA, 58; Univ Tenn, MA, 60; Yale Univ, PhD(math), 63. *Prof Exp:* Actg instr math, Yale Univ, 62-63; asst prof, Univ Wis, 63-64; from asst prof to assoc prof, Univ Ga, 64-69; assoc prof math, 69-75, ASSOC PROF COMPUT SCI, UNIV KY, 75- *Concurrent Pos:* Mem, Inst Advan Study, NJ, 67-68. *Mem:* Asn Comput Mach; Asn Computational Ling. *Res:* Artificial intelligence and logic programming; computational linguistics; syntax and semantics. *Mailing Add:* 102 W Lease Way Dr Lexington KY 40506

MCCORD, THOMAS BARD, b Elverson, Pa, Jan 18, 39; m 62. ASTRONOMY. *Educ:* Pa State Univ, BS, 64; Calif Inst Technol, MS, 66, PhD(planetary sci & astron), 68. *Prof Exp:* Res fel planetary sci, Calif Inst Technol, 68; asst prof, Mass Inst Technol, 68-71, assoc prof planetary physics & dir, George R Wallace Astrophys Observ, 71-77; asst dir Inst Artron, 76-79, head planetary geosci div, Hawaii Inst Geophysics, 79-90, PROF PLANETARY SCI, DEPT GEO & GEOPHYS, UNIV HAWAII, 79-; AT DEPT ASTRON GEOL, UNIV HAWAII AT MANOA. *Concurrent Pos:* Vis assoc, Calif Inst Technol, 69-72; sr scientist, Ctr Space Res Mass Inst Technol, 77-88; team leader, VIMS invest, NASA space mission comet; team mem, NIMS invest, NASA space mission Jupiter (Galileo); chmn & co-found, SETS Tech, Inc. *Mem:* Fel AAAS; fel Am Geophys Union (pres, planetology sec, 88-89); Am Astron Soc (pres, div planetary sci, 82-84); Am Astronomical Soc. *Res:* Composition and structure of surfaces of solar system objects using remote sensing techniques; development of instrumentation with which to make such studies. *Mailing Add:* Planetary Geosci Div Sch Ocean & Earth Sci 2525 Correa Rd Honolulu HI 96822

MCCORD, TOMMY JOE, b Lubbock, Tex, Feb 19, 32; m 51; c 4. BIOCHEMISTRY. *Educ:* Abilene Christian Col, BS, 54; Univ Tex, MA, 58, PhD(biochem, org chem), 59. *Prof Exp:* Asst, Abilene Christian Col, 53-54; res scientist, Biochem Inst, Univ Tex, 56-58; from asst prof to assoc prof chem, 58-64, asst head dept, 65-67, PROF CHEM, ABILENE CHRISTIAN UNIV, 64-, HEAD DEPT, 67- *Concurrent Pos:* Res grants, USPHS, NSF & Welch Found. *Mem:* AAAS; Am Chem Soc; Sigma Xi. *Res:* Preparation and biological testing of structural analogs of important biological compounds as potential metabolic antagonists. *Mailing Add:* Sta ACC PO Box 8132 Abilene TX 79601

MCCORD, WILLIAM MELLEN, b Durban, SAfrica, Jan 24, 07; m 30; c 2. BIOCHEMISTRY. *Educ:* Oberlin Col, AB, 28; Yale Univ, PhD(org chem), 31; La State Univ, MB, 39, MD, 40. *Prof Exp:* From instr to assoc prof biochem, Sch Med, La State Univ, 31-45; prof, 45-77, EMER PROF BIOCHEM, MED UNIV SC, 77-, pres, 64-75. *Concurrent Pos:* Intern, Charity Hosp, New Orleans, 39-40, vis biochemist, 40-42. *Mem:* AAAS; Am Chem Soc; Am Fedn Clin Res. *Res:* Aldehyde condensations; blood chemistry; pathological chemistry; toxicology; clinical biochemistry; porphyrin determinations; sickle cell anemia. *Mailing Add:* PO Box 40 Lodge SC 29082-0040

MCCORKLE, GEORGE MASTON, b Wise, Va, Aug 1, 21. CATALYTIC RNA, PARALLEL COMPUTING APPLICATIONS. *Educ:* Yale Univ, BA, 42, MPhil, 73, PhD(biol), 75. *Prof Exp:* Vpres & dir, Charles Scribner's Sons, 55-60; vpres finance & treas, New Am Libr, 60-67; pres, R R Bowker Co, 68-70; res assoc genetics, Dept Biol Sci, Purdue Univ, 75-79; ASSOC RES SCIENTIST, BIOL DEPT, YALE UNIV, 79-, ASSOC RES SCIENTIST, MED INFO PROG, YALE MED SCH, 89- *Concurrent Pos:* Trustee, New Eng Inst Med Res, 63-73. *Mem:* Soc Develop Biol; AAAS; Sigma Xi. *Res:* Control of bacterial and eukaryotic gene expression; regulatory and recognition sequences in DNA; comparative analysis of sequence and structural models of the RNA subunit of ribonuclease P and other catalytic RNAs. *Mailing Add:* 45 Mill Rock Rd Hamden CT 06517-4021

MCCORKLE, RICHARD ANTHONY, b Gastonia, NC, Aug 6, 40; m 64; c 1. PLASMA PHYSICS. *Educ:* NC State Univ, BS, 62, PhD(physics), 70. *Prof Exp:* Asst prof physics, ECarolina Univ, 68-73; RES SCIENTIST, IBM, THOMAS J WATSON RES CTR, 73- *Concurrent Pos:* Res dir, NSF grant, 69-70. *Honors & Awards:* Bisplinghoff Award, 72. *Mem:* Am Phys Soc. *Res:* Ion sources; short wavelength sources, x-ray; electron streams. *Mailing Add:* RR 1 Box 210 South Salem NY 10590

MCCORKLE, RUTH, NURSING. *Educ:* Univ Md, BS, 68; Univ Iowa, MA, 72, PhD(mass commun), 75. *Prof Exp:* Staff nurse, Coronary Care Unit, Vancouver Mem Hosp, Wash, 68-69; oncol clin nurse specialist, Univ Iowa Hosps & Clins, Iowa City, 71-73, instr, psychiat nursing & oncol nursing, Sch Nursing, Univ Iowa, 74-75; from asst prof to prof, Dept Community Health Care Systs, Sch Nursing, Univ Wash, Seattle, 75-85; chairperson, 88-89, PROF, ADULT HEALTH & ILLNESS DIV, SCH NURSING, UNIV PA, 86-, DIR, CTR ADVAN CARE SERIOUS ILLNESS, 89- *Concurrent Pos:* Mem, Inst Cancer Res Comt, Am Cancer Soc, Univ Wash, 79-85; mem, res comt, Oncol Nursing Soc, 81-82, dir-at-large, 83-85; dir-at-large, Int Soc Nurses Cancer Care, 83-; mem, Nat Nursing Adv Comt, Am Cancer Soc, 84-, Inst Cancer Res Comt, Univ Pa, 86-; mem, Nursing Sci Rev Comt, Nat Ctr Nursing Res, 88-; dir, Cancer Control, Cancer Comprehensive Ctr, Univ Pa, 90- *Mem:* Inst Med-Nat Acad Sci; fel Am Acad Nursing; Am Nurses' Asn; Int Soc Nurses Cancer Care; Am Asn Cancer Educ; Oncol Nursing Soc. *Res:* Author or co-author of over 30 publications. *Mailing Add:* Sch Nursing Univ Pa 420 Guardian Dr Phildelphia PA 19104-6096

MCCORMAC, BILLY MURRAY, b Zanesville, Ohio, Sept 8, 20; m 48, 69; c 5. ASTROPHYSICS. *Educ:* Ohio State Univ, BS, 43; Univ Va, MS, 56, PhD(nuclear physics), 57. *Prof Exp:* Physicist, Off Spec Weapons Develop, US Army, 57-60, scientist, Off Chief Staff, 60-61, physicist, Defense Atomic Support Agency, 61-62, chief electromagnetic br, 62-63; sci adv, IIT Res Inst, 63, dir geophys div, 63-68; sr consult scientist, 68-69, mgr, Radiation Physics Lab, 69-74, mgr, Electro-Optics Lab, 74-76, mgr solar physics, 76-80, MGR SOLAR & OPTICAL PHYSICS LAB, LOCKHEED PALO ALTO RES LAB, 80- *Concurrent Pos:* Ed, J Water, Air & Soil Pollution & Geophys & Astrophys Monographs; chmn adv study inst radiation trapped in earth's magnetic field, Norway, 65, aurora & airglow, Eng, 66, Norway, 68, Can, 70, France, 72 & Belg, 74, earth's particles & fields, Ger, 67, Calif, 69, Italy, 71, Eng, 73, Austria, 75, solar terrestrial influences on weather & climate, Ohio, 78 & Colo, 82, shuttle environ & opers, Washington, DC, 83 & Houston 85, space sta twenty first century, Reno, space sta utilization, Arlington, Va, 88. *Mem:* Am Astronomical Soc; Am Geophys Union; Marine Technol Soc; assoc fel Am Inst Aeronaut & Astronaut; Am Phys Soc. *Res:* Nuclear weapons effects, especially high altitude effect; multidisciplinary research in aeronomy and the earth's magnetosphere; solar physics; electro-optical systems. *Mailing Add:* 12861 Alta Tierra Rd Los Altos Hills CA 94022

MCCORMAC, JACK CLARK, b Columbia, SC, July 6, 27; m 57; c 2. CIVIL ENGINEERING. *Educ:* The Citadel, BS, 48; Mass Inst Technol, SM, 49. *Prof Exp:* Instr civil eng, Clemson Col, 49-51; civil engr, E I du Pont de Nemours & Co, 51-54 & Finfrock Industs, 54-55; from asst prof to prof, 55-89, EMER PROF CIVIL ENG, CLEMSON UNIV, 89- *Mem:* Am Soc Civil Engrs. *Res:* Bridge vibrations; prestressed concrete and steel structures. *Mailing Add:* Dept Civil Eng Clemson Univ Clemson SC 29634-0911

MCCORMACK, CHARLES ELWIN, b Gurnee, Ill, Jan 3, 38; m 61; c 2. PHYSIOLOGY. *Educ:* Carroll Col, Wis, BS, 59; Univ Wis-Madison, PhD(endocrinol), 63. *Prof Exp:* Fel zool, Univ Wis, 63-64; from instr to asst prof physiol, 64-70, assoc prof, 70-76, PROF PHYSIOL, CHICAGO MED SCH, 76- *Mem:* Am Soc Zoologists; Endocrine Soc; Am Physiol Soc; Int Soc Neuroendocrinol; Soc Study Reproduction. *Res:* Influence of the brain and gonadal steroids on ovulation. *Mailing Add:* Dept Physiol Chicago Med Sch 3333 Green Bay Rd North Chicago IL 60064

MCCORMACK, DONALD EUGENE, soil science; deceased, see previous edition for last biography

MCCORMACK, ELIZABETH FRANCES, b Boston, Mass, Dec 31, 60. LASER SPECTROSCOPY. *Educ:* Wellesley Col, BA 83; Yale Univ, PhD(physics), 89. *Prof Exp:* Postdoctoral appointee, 89-90, ALEXANDER HOLLAENDER POSTDOCTORAL FEL, DEPT ENERGY, ARGONNE NAT LAB, 90- *Mem:* Am Phys Soc. *Res:* Atomic and molecular physics; Ryberg and excited stat dynbamics; near-threshold photo-dissociation, photoionization, multiphoton laser spectroscopy; narrow-band and tunable ultraviolet light generation. *Mailing Add:* Argonne Nat Lab 9700 S Cass Bldg 203 Argonne IL 60439

MCCORMACK, FRANCIS JOSEPH, b Mobile, Ala, Dec 6, 38; m 67. PLASMA PHYSICS. *Educ:* Spring Hill Col, BS, 60; Fla State Univ, PhD(physics), 64. *Prof Exp:* From asst prof to assoc prof, 67-77, PROF PHYSICS, UNIV NC, GREENSBORO, 77- *Mem:* Am Phys Soc; Am Asn Physics Teachers. *Res:* Kinetic theory of ordinary and ionized gases. *Mailing Add:* Dept Physics Univ NC 1000 Spring Garden St Greensboro NC 27412

MCCORMACK, GRACE, b Rochester, NY, Feb 16, 08. MICROBIOLOGY. *Educ:* Univ Rochester, AB, 41; Univ Md, MS, 51; Nat Registry Microbiologists, regist, 75. *Prof Exp:* Technician endocrinol, Sch Med & Dent, Univ Rochester, 42-48; bacteriologist fish & wildlife serv, US Dept Interior, 48-53; assoc bacteriologist, Md State Dept Health, 53-55; bacteriologist-technologist, Vet Admin Hosp, Canandaigua, NY, 55-66; from asst prof to prof, Monroe Community Col, 66-77, emer prof biol sci, 77-85; RETIRED. *Concurrent Pos:* Lectr, Univ Md, 50; teacher microbiol, Community Col Finger Lakes, 81- *Honors & Awards:* Sustained Superior Serv Award, Vet Admin Hosp, 63. *Mem:* AAAS; fel Am Inst Chem; Am Soc Microbiol; Am Chem Soc; Am Pub Health Asn. *Res:* Assays of gonadotropic hormone; enterococci; antibiotic study of preservation of crabmeat; pollution studies of clam and oyster beds; comparative study of various media; cause of discoloration of oysters; collaborating on research problem; testing of viability of brewery yeast. *Mailing Add:* 162 Raleigh Rochester NY 14606

MCCORMACK, HAROLD ROBERT, b St Louis, Mo, Nov 25, 22; m 46; c 4. GEOPHYSICS. *Educ:* St Louis Univ, BS, 46. *Prof Exp:* Geophysicist, Sun Oil Co, 43-53, dir geophys, 54-67, mgr geophys res, 68-75, mgr geophys data processing, 75-77; MEM STAFF, SUNMARK EXPLOR CO, 77- *Mem:* Europeans Asn Explor Geophysicists; Soc Explor Geophys (secy-treas, 75-76); Seismol Soc Am. *Res:* Geophysical research applied to petroleum exploration. *Mailing Add:* 4456 Mill Run Rd Dallas TX 75244

MCCORMACK, JOHN JOSEPH, JR, b Boston, Mass, Jan 20, 38; m 62; c 4. MEDICINAL CHEMISTRY, PHARMACOLOGY. *Educ:* Boston Col, BS, 59; Yale Univ, PhD(pharmacol), 64. *Prof Exp:* Nat Cancer Inst res fel med chem, Australian Nat Univ, 64-66; from asst prof to assoc prof, 66-77, PROF PHARMACOL, UNIV VT, 77- *Concurrent Pos:* Res pharmacologist, Nat Cancer Inst, 78-79, 87-88, also mem Develop Ther Comt; Grant Rev Comt, Am Cancer Soc. *Mem:* AAAS; Am Chem Soc; Am Asn Cancer Res; Am Soc Pharmacol & Exp Therapeut; Am Soc Clin Oncol; Asn Off Anal Chemists. *Res:* Heterocyclic and medicinal chemistry; chemotherapy of neoplastic and protozoal diseases. *Mailing Add:* Dept Pharmacol Given Med Bldg Univ Vt Burlington VT 05405

MCCORMACK, MIKE, b Basil, Ohio, Dec 14, 21; m 47; c 3. INORGANIC CHEMISTRY. *Educ:* Wash State Univ, BS, 48, MS, 49, DEng, Stevens Inst Technol. *Hon Degrees:* JD, Salisburg State Col, 81. *Prof Exp:* Res scientist, Hanford AEC Facility, 50-70; state legislator, Washington State House Reps, 57-70, congressman, US Congress, 71-81, PRES & CONSULT, SCI, ENERGY & GOVT, MCCORMACK ASSOC INC, 82-, ENVIRON HEALTH & HUMAN SAFETY, CHELAN ASSOC INC, 85-; PRES & CONSULT, MC CORMACK ASSOC INC, 82-, CHELAN ASSOC INC, 85- *Concurrent Pos:* Bd dirs, Universal Voltronics Corp & consult, Am Chem Soc, 81-; adv, Coun Sci Soc Pres, 87-; appointee, Space Telescope Inst Coun, Asn Univ Res in Astron, 88; consult, Am Chem Soc, 81- *Mem:* Fel, Am Asn Advan Sci, 80; fel, Am Nuclear Soc; fel, Am Inst Chemists; Am Chem Soc. *Res:* Numerous public papers and statements on energy technologies, energy conservation, energy policy and environmental policy. *Mailing Add:* 508 A Street SE Washington DC 20003

MCCORMACK, WILLIAM BREWSTER, b Mirror, Alta, Nov 20, 23; US citizen; m 45; c 2. ORGANIC CHEMISTRY. *Educ:* Univ Alta, BSc, 44; Univ Wis, PhD(chem), 48. *Prof Exp:* Res chemist, 48-53, res supvr, 53-55, RES ASSOC, JACKSON LAB, E I DU PONT DE NEMOURS & CO, INC, 55-, INTERNAL TECH CONSULT, 70- *Mem:* AAAS; Am Chem Soc; Combustion Inst; Royal Soc Chem; Sigma Xi. *Res:* Phosphorus; fluorine; organometallics; dyes and textile chemicals; nucleophilic and free radical processes; aromatic intermediates; scientific philosophy; combustion; permeation separation. *Mailing Add:* 3 Wordsworth Dr Hyde Park Wilmington DE 19808

MCCORMICK, ANNA M, b Pittsburgh, Pa, Apr 14, 49. RETINOID ACTION. *Educ:* WVa Univ, PhD(biochem), 75. *Prof Exp:* Asst prof biochem, Health Sci Ctr, Univ Tex, 81-88; GENETICS PROG DIR, BIOL OF AGING PROG, NAT INST, AGING, NIH, 88- *Mem:* Sigma Xi; Am Inst Nutrit; Am Soc Biol Chemists. *Mailing Add:* Genetics Prog Molecular & Cell Biol Br NIA NIH Bldg 31 Rm 5C-21 Bethesda MD 20892

MCCORMICK, BAILIE JACK, b Amarillo, Tex, Aug 20, 37; m 58; c 2. INORGANIC CHEMISTRY. *Educ:* West Tex State Univ, BS, 59; Okla State Univ, PhD(chem), 62. *Prof Exp:* Robert A Welch Found fel, Univ Tex, 62-64; from asst prof to assoc prof chem, WVa Univ, 64-73, prof, 73-79; PROF CHEM & CHMN DEPT, WICHITA STATE UNIV, 79- *Mem:* Am Chem Soc. *Res:* Structure, spectra and reactions of coordination compounds; inorganic geochemistry; sulfide minerals; donor properties of sulfur; biological aspects of metal binding. *Mailing Add:* Chem Dept Wichita State Univ Wichita KS 67208

MCCORMICK, BARNES W(ARNOCK), JR, b Waycross, Ga, July 15, 26; m 46; c 1. AERONAUTICAL ENGINEERING. *Educ:* Pa State Univ, BS, 48, MS, 49, PhD(aeronaut eng), 54. *Prof Exp:* Assoc prof eng res, Pa State Univ, 54-56; chief aerodyn, Vertol Aircraft Corp, 56-58; assoc prof aeronaut eng & head dept, Univ Wichita, 58-59; assoc prof aeronaut eng, 59-65, head dept, 69-85, PROF AEROSPACE ENG, PA STATE UNIV, 65-, BOEING PROF AEROSPACE ENG, 85- *Concurrent Pos:* Consult, Boeing Helicopters, Boeing Mil, Cessna, NAm Aviation, TRW Martin-Marietta, Vitro Labs, Outboard Marine Corp, HRB-Singer, US Army Aviation Systs, US Naval Surface Weapons Ctr; researcher, Ord Res Lab, Pa State Univ, 59-; pres, Aero Eng Assocs, Inc, 66-; ed, J Am Helicopter Soc, 71-72; tech dir & ed, J, Am Helicopter Soc; vpres educ & assoc ed, Am Inst Aeronaut & Astronaut. *Honors & Awards:* Educ Achievement Award, Am Soc Eng Educ & Am Inst Aeronaut & Astronaut, 76. *Mem:* Am Inst Aeronaut & Astronaut; Am Helicopter Soc; Am Soc Eng Educ; Soc Automotive Engrs. *Res:* Aerodynamics; hydrodynamics; helicopters; standard and vertical take-off and landing aircraft; hydroballistics. *Mailing Add:* Dept of Aerospace Eng Pa State Univ 233 Hammond Bldg University Park PA 16802

MCCORMICK, CLYDE TRUMAN, b Oblong, Ill, Mar 6, 08; m 37; c 3. APPLIED MATHEMATICS. *Educ:* Univ Ill, AB, 30, AM, 32; Ind Univ, PhD(math physics), 37. *Prof Exp:* Prof math & physics, Greenville Col, 37-38; from asst prof to prof math, Ft Hays State Univ, 38-44, head dept, 41-44; assoc prof, Ill State Univ, 44-48, prof math, 48-54, head dept, 54-71; from vis prof to distinguished vis prof, Palm Beach Atlantic Col, 71-86; RETIRED. *Mem:* Am Math Soc; Math Asn Am. *Res:* Applied mathematics in mechanics and sound. *Mailing Add:* 1701 Flagler Dr Apt 405 West Palm Beach FL 33405

MCCORMICK, DAVID LOYD, b Abington, Pa, Jan 19, 53; m 78; c 1. CANCER CHEMOPREVENTION, PRECLINICAL TOXICOLOGY. *Educ:* Middlebury Col, AB, 74; NY Univ, MS, 76, PhD(environ med), 79. *Prof Exp:* Asst res scientist cancer res, Med Ctr, NY Univ, 77-79; assoc physiologist, IIT Res Inst, 79-80, res physiologist, 80-82, sr physiologist-toxicologist, 82-90, HEAD, TOXICOL & ENVIRON HEALTH, IIT RES INST, 90- *Concurrent Pos:* New investr res award, US Nat Cancer Inst, 81; adj assoc prof, Ill Inst Technol, 82- *Mem:* Am Asn Cancer Res; Soc Toxicol; Int Asn Breast Cancer Res; Soc Exp Biol & Med. *Res:* Development of drugs for cancer prevention, especially breast cancer, including studies of efficacy, preclinical toxicology and mechanisms of action. *Mailing Add:* IIT Res Inst Ten W 35th St Chicago IL 60616

MCCORMICK, DONALD BRUCE, b Front Royal, Va, July 15, 32; m 55; c 3. NUTRITION. *Educ:* Vanderbilt Univ, BA, 53, PhD(biochem), 58. *Prof Exp:* Asst biochem, Vanderbilt Univ, 53-58; consult biochemist, Interdepartment Comt Nutrit Nat Defense, Madrid, 58; USPHS res fel, Univ Calif, 58-60; asst prof nutrit, Cornell Univ, 60-65, from assoc prof to prof biochem & biol sci, 65-78, L H Bailey prof nutrit, biochem & molecular Biol, 78-79; exec assoc dean sci, Med Sch, 85-89, PROF & CHMN BIOCHEM, EMORY UNIV, 79- *Concurrent Pos:* Guggenheim Mem Found fel, 66-67; vis prof, Wellcome, 85-86 & 88-89. *Honors & Awards:* Mead Johnson Award, Am Inst Nutrit, 70; Osborne & Mendel Award, Am Inst Nutrit, 78. *Mem:* AAAS; Am Soc Biol Chemists; Am Inst Nutrit; Am Chem Soc; Am Soc Microbiol; Biophys Soc; Soc Exp Biol Med; Food & Nutrit Bd, Nat Res Coun/Nat Acad Sci. *Res:* Chemistry and biochemistry of cofactors; enzymology; chemistry, metabolism and enzymology of vitamins, coenzymes and metal ions. *Mailing Add:* Dept Biochem Emory Univ Atlanta GA 30322

MCCORMICK, FRANCIS B, agricultural economics, for more information see previous edition

MCCORMICK, FRED C(AMPBELL), b Waynesboro, Va, Jan 18, 26; m 54; c 1. STRUCTURAL ENGINEERING, FORENSIC ENGINEERING. *Educ:* Univ Va, BCE, 51; Univ Mich, MSE, 52, PhD, 64. *Prof Exp:* Construct engr, E I du Pont de Nemours & Co, SC, 52-54; design engr, Polglaze & Basenberg Engrs, Ala, 54-55; from asst prof to assoc prof, 55-72, PROF CIVIL ENG, UNIV VA, 72- *Concurrent Pos:* Hwy res engr & indust consult, 72- *Mem:* Am Soc Civil Engrs; Am Soc Chem Engrs; Struct Plastics Res Coun. *Res:* Experimental stress analysis; materials of construction and composite materials. *Mailing Add:* Dept Civil Eng Thornton Hall D206 Univ Va Charlottesville VA 22903

MCCORMICK, GEORGE R, b Columbus, Ohio, Apr 12, 36; m 62; c 2. MINERALOGY, PETROLOGY. *Educ:* Ohio Wesleyan Univ, BA, 58; Ohio State Univ, MSc, 60, PhD(mineral), 64. *Prof Exp:* Geologist, Ohio Div, Geol Surv, 61-62, Norris Lab, US Bur Mines, 62-65; asst prof geol, Boston Univ, 65-68; PROF GEOL, UNIV IOWA, 68- *Concurrent Pos:* Consult, Kennecott Copper Co, 67-69, Humble Minerals, 71-72; geologist, US Geol Surv, 72-78; Fulbright Scholar, 83-84. *Mem:* Clay Minerals Soc; Mineral Soc Am; Mineral Soc Gt Brit & Ireland; Geol Soc Am; AAAS. *Res:* Microchemical study of igneous and metamorphic complexes; chemical and mineralogic study of clays; geochemical studies of copper, lead, zinc and silver deposits of Utah, Nevada and Arizona; carbonatite studies in central Arkansas; petrology of volcanic and ultrabasic rocks of South Asia. *Mailing Add:* 230 E Fairchild Iowa City IA 52245

MCCORMICK, J FRANK, b Indianapolis, Ind, Oct 25, 35; m 59; c 3. ECOLOGY. *Educ:* Butler Univ, BS, 58; Emory Univ, MS, 60, PhD(biol), 61. *Prof Exp:* Asst prof biol, Vanderbilt Univ, 61-63; asst prof zool, Univ Ga, 63-64; asst prof bot, Univ NC, Chapel Hill, 64-71, prof bot & ecol, 71-74; PROF & DIR GRAD PROG ECOL, UNIV TENN, 74- *Concurrent Pos:* Grants, NSF, Natural Sci Comt, Vanderbilt Univ, 62; AEC, 64-, radiation ecol training, 66-; Univ NC fac res coun, 65 & 66; USPHS ecol training prog, 66-; consult, Oak Ridge Nat Lab, 62-; PR Nuclear Ctr, 64- *Honors & Awards:* Sigma Xi Res Award, Emory Univ, 61. *Mem:* AAAS; Ecol Soc Am. *Res:* Environmental and radiation biology; evolution and natural selection; environmental assessment; population; pollution studies; natural resource management. *Mailing Add:* 108A Hoskins Libr Knoxville TN 37996

MCCORMICK, J JUSTIN, b Detroit, Mich, Sept 26, 33. CANCER, MOLECULAR BIOLOGY. *Educ:* St Paul's Col, Washington, DC, BA, 57, MA, 59; Cath Univ Am, MS, 61, PhD(biol), 67. *Prof Exp:* Fel, McArdle Lab Cancer Res, Univ Wis, 67-70, res assoc, 70-71; res scientist, Mich Cancer Found, 71-73, chief, Molecular Biol Lab, 73-76; assoc prof, Carcinogenesis Lab, 76-80, assoc prof microbiol & dept biochem, 76-80, PROF MICROBIOL, PUB HEALTH & DEPT BIOCHEM, COL OSTEOP MED, MICH STATE UNIV, 80-, CO-DIR, CARCINOGENESIS LAB, 76-, ASSOC DEAN RES, 87- *Concurrent Pos:* Assoc dir, Cancer Etiology Prog, Comprehensive Breast Cancer Ctr, Mich State Univ, 90- *Mem:* Am Asn Cancer Res; Am Soc Cell Biol; Environ Mutagen Soc; Am Soc Photobiol; Sigma Xi. *Res:* Mutagenic and transforming effect of chemical and physical carcinogens on human cells in culture; DNA repair; application of biophysical techniques to the problems of carcinogenesis. *Mailing Add:* Carcinogenesis Lab Fee Hall Mich State Univ East Lansing MI 48824

MCCORMICK, J ROBERT D, b St Albans, WVa, Feb 24, 21; m 53; c 1. FERMENTATION, BIOCHEMISTRY. *Educ:* Rensselaer Polytech Inst, BS, 46; Univ Calif, Los Angeles, PhD(chem), 54. *Prof Exp:* Res chemist, Winthrop Chem Co, 43-46; develop chemist, Lederle Labs, Am Cyanamid Co, 49-54, biochem group leader, 54-58, head biochem dept, 58-60, res assoc fermentation biochem, 60-65, res fel fermentation biochem, 65-84; CONSULT FERMENTATION TECHNOL, 84- *Concurrent Pos:* Am Cyanamid award advan study sr res fel, Univ Leicester, 70-71. *Mem:* Fel AAAS; fel Am Inst Chem; Am Chem Soc; NY Acad Sci; Tissue Cult Asn. *Res:* Natural products isolation and structure; fermentation methodology. *Mailing Add:* McNamara Rd Spring Valley NY 10977

MCCORMICK, JOHN E, b Manitowoc, Wis, Aug 18, 23; m 58; c 4. CHEMICAL ENGINEERING, MATHEMATICS. *Educ:* Iowa State Univ, BSc, 48; Univ Cincinnati, PhD(chem eng), 57. *Prof Exp:* Jr engr, Linde Air Prod Div, Union Carbide Corp, NY, 48-53; asst, Univ Cincinnati, 53-55, teaching fel quant chem, 55-56; engr, Esso Res & Eng Co, NJ, 57-62; from asst prof to assoc prof chem eng, NJ Inst Technol, 62-88, assoc chmn dept, 79-88; RETIRED. *Concurrent Pos:* Consult, Am Cyanamid Co, NJ, 63-65, Nopco Chem Div, Diamond Shamrock Chem Co, 66-67 & 70- & Merck & Co Inc, 68-70 & 73-78. *Mem:* Am Inst Chem Engrs. *Res:* Molecular sieve preparation; acidic alkylation; mathematics of engineering processes; applications of computers. *Mailing Add:* 344 Parkview Dr Scotch Plains NJ 07076

MCCORMICK, JOHN PAULING, b Lansing, Mich, May 26, 43; m 68; c 2. BIO-ORGANIC CHEMISTRY. *Educ:* DePauw Univ, BA, 65; Stanford Univ, PhD(chem), 70. *Prof Exp:* Res assoc, Swiss Fed Inst Technol, 70-71; vis prof org chem, State Univ Groningen, 71-72; from asst prof to assoc prof, 72-83, PROF ORG CHEM, UNIV MO-COLUMBIA, 83-, ASSOC VPROVOST RES, 87- *Concurrent Pos:* Assoc grad dean res, Univ Mo-Columbia, 84-87. *Mem:* Am Chem Soc; AAAS. *Res:* Terpene synthesis and biosynthesis; mechanism of biochemical reactions; structure of physiologically active natural products; invertebrate chemical ecology; insect hormones and defensive substances. *Mailing Add:* 1424 Bradford Dr Columbia MO 65203

MCCORMICK, JON MICHAEL, b Portland, Ore, Feb 8, 41; m 71. MARINE ECOLOGY. *Educ:* Portland State Univ, BS, 63; Ore State Univ, MS, 65, PhD(biol sci), 69. *Prof Exp:* Asst prof biol, Millersville State Col, 68-70; from asst prof to assoc prof, 70-85, PROF BIOL, MONTCLAIR STATE COL, 85- *Concurrent Pos:* NSF fel, Marine Biol Lab, Woods Hole, 69, partic, NSF instr Sci Equip Prog, Millersville State Col, 69-70; res ecol New York Harbor, Nat Oceanic & Atmospheric Admin, 72-; consult estravine ecol, 72-; instr, NJ Marine Sci Consortium, 71-75. *Mem:* AAAS; Ecol Soc Am; Am Soc Limnol & Oceanog; Atlantic Estuarine Res Soc. *Res:* Estuarine ecology; marine zooplankton; marine benthos, hydrozoa, population dynamics, trophic ecology and environmental toxicology. *Mailing Add:* Dept of Biol Montclair State Col Upper Montclair NJ 07043

MCCORMICK, KATHLEEN ANN, b Manchester, NH, June 27, 47; m 71. PHYSIOLOGY, NURSING. *Educ:* Univ Wis, PhD(physiol), 78. *Prof Exp:* Asst for res to the chief, nursing dept, 78-, AT GERONTOL CTR, NAT INST AGING, NIH. *Mem:* Am Thoracic Soc; Am Nurses' Asn. *Res:* Lung pathophysiology; adult respiratory distress syndrome; nursing res; exercise physiology; blood coagulation pathophysiology. *Mailing Add:* NIA-LBS GRC 4940 Eastern Ave Baltimore MD 21224

MCCORMICK, KENNETH JAMES, b Toledo, Ohio, Sept 11, 37; m 63; c 3. CANCER. *Educ:* Univ Toledo, BS, 59; Univ Mich, Ann Arbor, MS, 62, PhD(microbiol), 65. *Prof Exp:* From instr to assoc prof exp biol, Dept Surg, Baylor Col Med, 65-75; head sect tumor immunol, Lab Cancer Res, St Joseph Hosp, Houston, 75-76, asspc lab dir, Stehlin Found Cancer Res, 76-; RES SCIENTITST OTOLARYNGOL, HEAD & NECK SURG, HOSP & CLINS, IOWA CITY. *Concurrent Pos:* Am Cancer Soc instnl grants, Baylor Col Med, 67-68 & 70-71, adj assoc prof exp biol, 75- *Mem:* AAAS; Am Asn Cancer Res; Am Asn Immunologists; Tissue Cult Asn; Am Soc Microbiol. *Res:* Tumor immunology and virology. *Mailing Add:* 5555 S Everet Ave Chicago IL 60637

MCCORMICK, MICHAEL EDWARD, b Washington, DC, Sept 11, 36; m; c 4. FLUID MECHANICS, APPLIED MECHANICS. *Educ:* Am Univ, BA, 59; Cath Univ Am, MS, 61, PhD(fluid mech), 66; Trinity Col, Dublin, PhD (hydraul struct), 85. *Prof Exp:* Physicist, David Taylor Model Basin, 58-61; instr mech eng, Swarthmore Col, 61-62; instr, Cath Univ Am, 62-63, asst prof, 63-64, res engr, 62-64; res mech engr, David Taylor Model Basin, 64-65; asst prof mech, Trinity Col, Conn, 65-68; assoc prof ocean eng, 68-74, chmn, Naval Syst Eng Dept, 70-77, DIR OCEAN ENG, US NAVAL ACAD, 68-70 & 78-, PROF OCEAN ENG, 74- & NAVAL FAC ENG COMMAND PROF, 84-; RES PROF CIVIL ENG, JOHNS HOPKINS UNIV, 87- *Concurrent Pos:* Vis scholar, Swarthmore Col, 76-77. *Mem:* Am Soc Mech Engrs; fel Marine Technol Soc; fel Am Soc Civil Engrs; Soc Underwater Technol. *Res:* Wave energy; wave mechanics; hydromechanics; drag reduction; coastal engineering. *Mailing Add:* Dept Naval Systs Eng US Naval Acad Annapolis MD 21402

MCCORMICK, MICHAEL PATRICK, b Canonsburg, Pa, Nov 23, 40; m 62; c 2. ATMOSPHERIC PHYSICS. *Educ:* Washington & Jefferson Col, BA, 62; Col William & Mary, MA, 64, PhD(physics), 67. *Hon Degrees:* DSc, Washington & Jefferson Col, 81. *Prof Exp:* Head photoelectronic instrumentation sect, 67-75, CHIEF, AEROSOL MEASUREMENTS RES BR, LANGLEY RES CTR, NASA, 75- *Honors & Awards:* Arthur S Fleming Award, 80. *Mem:* Am Geophys Union. *Res:* Design, development and application of advanced sensors to atmospheric research; light scattering and atmospheric optics; electrooptics; laser radars. *Mailing Add:* Langley Res Ctr M/S 475 Bldg 1250A NASA Hampton VA 23665

MCCORMICK, NEIL GLENN, b Everett, Wash, Aug 17, 27; m 57; c 3. MICROBIOLOGY, FOOD SCIENCE & TECHNOLOGY. *Educ:* Univ Wash, BS, 51, MS, 57, PhD(microbiol), 60. *Prof Exp:* Anal chemist, Wash State Dept Agr, 51-53; res assoc microbiol, Univ Wash, 60-61; USPHS fel, Univ Wis, 61-63; asst prof, Univ Va, 63-68; RES MICROBIOLOGIST, SCI ADV TECH LAB, US ARMY NATICK RES & DEVELOP CTR, 68- *Concurrent Pos:* Nat Inst Arthritis & Infectious Dis grants, 64-66. *Mem:* AAAS; Am Soc Microbiol; Soc Indust Microbiol. *Res:* Heat resistance of thermophilic sporeformers; aseptic processing of particulates. *Mailing Add:* 11 Duggan Dr Framingham MA 01701

MCCORMICK, NORMAN JOSEPH, b Hays, Kans, Dec 9, 38; m 61; c 2. RADIATIVE TRANSFER. *Educ:* Univ Ill, BS, 60, MS, 61; Univ Mich, PhD(nuclear eng), 65. *Prof Exp:* NSF fel, Univ Ljubljana, 65-66, Nat Acad Sci fel, 71; from vis asst prof to assoc prof, 66-75, prof nuclear eng, 75-89, PROF MECH ENG & NUCLEAR ENG, UNIV WASH, 89- *Concurrent Pos:* Res scientist, Sci Applns, Palo Alto, 74-75; consult, Westinghouse Hanford Co, 74-81 & Battelle Pac Northwest Labs, 84-86; exec ed, Progress in Nuclear Energy, 80-85. *Mem:* Fel Am Nuclear Soc; Optical Soc Am; Soc Risk Analysis. *Res:* Reliability and risk analysis, methods and nuclear and non-nuclear power applications; neutron and photon transport, both direct and inverse problems. *Mailing Add:* Dept Mech Eng FU-10 Univ Wash Seattle WA 98195

MCCORMICK, PAUL R(ICHARD), b Johnstown, Pa, Aug 5, 25; m 47; c 3. ELECTRICAL ENGINEERING. *Educ:* Cornell Univ, BS, 47; Univ Pittsburgh, MS, 53, PhD, 62. *Prof Exp:* From instr to asst prof, 47-62, assoc chmn dept, 64-66, ASSOC PROF ELEC ENG, UNIV PITTSBURGH, 62-, DIR LOWER DIV PROGS, 69-, ASSOC CHMN DEPT, 81- *Concurrent Pos:* Grants, Danforth Found, 48 & NSF, 59; instr, Duquesne Light Co, 55-58; USAID chief of party & consult, Valparaiso Tech Univ, Chile, 62-64. *Mem:* Inst Elec & Electronics Engrs; Am Soc Eng Educ. *Res:* Power system development and planning; machinery; control; direct energy conversion. *Mailing Add:* 510 Horizonview Dr Pittsburgh PA 15235

MCCORMICK, PAULETTE JEAN, b Buffalo, NY, Jan 26, 51. DEVELOPMENTAL GENETICS, ONCOLOGY. *Educ:* Barnard Col, AB, 72; State Univ NY-Albany, PhD(biol), 79. *Prof Exp:* Fel, Sloan Kettering Inst, 79-82, res assoc & instr develop genetics, 82-85; ASST PROF DEVELOP BIOL, STATE UNIV NY, ALBANY, 85- *Mem:* AAAS; Sigma Xi; Am Women Sci. *Res:* Examination of the genetic regulation of both mammalian embryo genesis and oncogenesis using terato carcinoma stem cells as the model system and retroviral insertion to create mutants in either of the above processes. *Mailing Add:* Ctr Cellular Differentiation Dept Biol Sci SUNY Albany 1400 Washington Ave Albany NY 12222

MCCORMICK, PHILIP THOMAS, b Oak Park, Ill, Nov 28, 26; m 57; c 5. THEORETICAL PHYSICS, PLANETARY SCIENCE. *Educ:* Univ Notre Dame, BS, 48, PhD(physics), 54. *Prof Exp:* Instr physics, DePaul Univ, 52-54; head sidewinder simulation sect, Naval Ord Test Sta, 54-57, physicist weapons planning group, 57-58; from asst prof to assoc prof, 58-69, PROF PHYSICS, UNIV SANTA CLARA, 70- *Concurrent Pos:* Sr eng specialist advan progs sect, Philco Corp, 59-61; Fulbright lectr, Univ Sind, Pakistan, 64-65; consult weapons planning group, Naval Weapons Ctr, 67-68. *Mem:* AAAS; Am Geophys Union; Am Phys Soc; Am Asn Physics Teachers; Sigma Xi. *Res:* Space science; electromagnetic interactions with nuclei; weapons and communications systems analysis; planetary environments; radio wave investigations of planetary atmospheres and ionospheres; aeronomy; mathematical modeling of planetary ionospheres. *Mailing Add:* Dept of Physics Univ of Santa Clara Santa Clara CA 95053

MCCORMICK, ROBERT BECKER, geology, mining engineering, for more information see previous edition

MCCORMICK, ROBERT H(ENRY), b Centre Co, Pa, Apr 28, 14; m 39; c 1. CHEMICAL ENGINEERING, TECHNOLOGY TRANSFER. *Educ:* Pa State Univ, BS, 35, MS, 42. *Prof Exp:* Asst petrol ref, 35-44, from instr to asst prof, 44-49, from asst res prof to assoc res prof, 49-64, mem grad fac, 53-79, prof chem eng, 64-79, EMER PROF CHEM ENG, PA STATE UNIV, 79- *Concurrent Pos:* Tech field specialist, Pa Tech Assistance Prog, 71-79; consult to indust & govt agencies. *Mem:* Fel Am Inst Chem Engrs; Am Chem Soc; fel Am Inst Chem; Am Soc Eng Educ; Am Asn Univ Prof; Am Asn Engr Educr. *Res:* Vapor-liquid and liquid-liquid equilibrium data on multi-component systems; equipment, methods and processes for making separations on multicomponent mixtures; air and water pollution; solid waste disposal. *Mailing Add:* Dept of Chem Eng Pa State Univ 112 Chem Eng Bldg University Park PA 16802

MCCORMICK, ROBERT T, b Brooklyn, NY, Jan 20, 51. IMMUNOLOGY. *Educ:* Univ Minn, MS, 78, PhD(path biol), 86. *Prof Exp:* Med technologist genetics, Univ Minn, 74-75, res scientist, Dept Surg, 75-78; pharmaceut sales, Pfizer Pharmaceuticals, 78-80; res scientist, 3M Corp, 80-83; SR SCIENTIST IMMUNOL, HYBRITECH INC, 86- *Mem:* Am Asn Immunol. *Mailing Add:* Hybritech Inc 11085 Torreyana Rd San Diego CA 92121

MCCORMICK, ROY L, b Hillsboro, Ind, Feb 8, 29; m 49; c 2. MATHEMATICS. *Educ:* Wabash Col, AB, 50; Wash Univ, MA, 60; Purdue Univ, PhD(math ed), 65. *Prof Exp:* Teacher high schs, Ind, 50-51 & 56-60; instr math ed, Purdue Univ, 60-62; from asst prof to assoc prof, 62-71, PROF MATH, BALL STATE UNIV, 71- *Concurrent Pos:* Dir, NSF Inst, 67-72; chmn state adv coun, Elem & Sec Educ Act Title III, 70-72. *Mem:* Math Asn Am. *Res:* Training of secondary and elementary teachers of mathematics. *Mailing Add:* Dept Comp Sci/cp 204F Ball State Univ Muncie IN 47306

MCCORMICK, STEPHEN DANIEL, b Schenectady, NY, June 11, 55; m 83; c 2. ANIMAL PHYSIOLOGY. *Educ:* Bates Col, BS, 77, Mass Inst Technol, PhD(biol oceonog), 83. *Prof Exp:* Fel, Natural Sci & Eng Res Coun, Dept Fisheries & Oceans Biol Sta, St Andrews, NB, Can, 83-86; NIH fel, Univ Calif, Berkeley, 86-89; postdoctoral res, Dept Integrative Biol, Univ Calif, Berkeley, 89-90; SECT LEADER PHYSIOL, NORTHEAST ANADROMOUS FISH RES LAB, US FISH & WILDLIFE SERV, TURNERS FALLS, MA, 90- *Concurrent Pos:* Res fel, Dept Zoophysiol, Univ Goteborg, 88, Ocean Res Inst, Univ Tokyo, 89-90. *Mem:* Am Soc Zoologists; Am Fisheries Soc. *Res:* Environmental physiology and endocrinology of growth metabolism and osmoregulation in teleosts especially anadromous fishes. *Mailing Add:* Northeast Anadromous Fish Res Lab US Fish & Wildlife Serv PO Box 796 Turners Falls MA 01376

MCCORMICK, STEPHEN FAHRNEY, b Glendale, Calif, Sept 16, 44; m 68. NUMERICAL ANALYSIS. *Educ:* San Diego State Col, BA, 66; Univ Southern Calif, PhD(math), 71. *Prof Exp:* Mem tech staff data analysis, Hughes Aircraft Co, 66-68; asst prof math, Pomona Col, 70-72; asst dir, Inst Educ Comput, Claremont Cols, 70-73; asst prof numerical analysis, 73-77, ASSOC PROF MATH, COLO STATE UNIV, 77- *Concurrent Pos:* Partner & dir comput, Solar Environ Eng Co, 74- *Mem:* Soc Indust & Appl Math; Am Math Soc; Asn Comput Mach. *Res:* Matrix linear algebra, optimization, approximation theory, and algebraic iterative methods in general. *Mailing Add:* Prof Math Univ Colo 1200 Larimer St Denver CO 80204

MCCORMICK, WILLIAM DEVLIN, b Tacoma, Wash, May 9, 31; m 70. PHYSICS. *Educ:* Calif Inst Technol, BS, 53; Duke Univ, PhD(physics), 59. *Prof Exp:* Fulbright travel grant, Univ Padua, 59-60; asst prof physics, Univ Wash, 63-68; ASSOC PROF PHYSICS, UNIV TEX, AUSTIN, 68- *Concurrent Pos:* Sloan fel, Univ Wash, 63-65; Richland fac fel, Battelle-Northwest, 67-68. *Mem:* Am Inst Physics; Sigma Xi. *Res:* Low temperature and solid state physics; liquid helium; phase transitions; nuclear magnetic resonance and calorimetry of molecular solids; surface physics. *Mailing Add:* Dept of Physics Univ of Tex Austin TX 78712

MCCORMICK, WILLIAM F, b Riverton, Va, Sept 9, 33; m 54; c 2. PATHOLOGY, NEUROPATHOLOGY. *Educ:* Univ Chattanooga, BS, 53; Univ Tenn, MD, 55, MS, 57; Am Bd Path, dipl, 60, 66, 79. *Prof Exp:* Asst path, Univ Tenn, 57-60, from instr to asst prof, 60-64; from assoc prof neuropath to prof path, Univ Iowa, 64-71, prof neurol, 71-73; prof neurol & path, Univ Tex Med Br, Galveston, 73-84; prof path, 85-89, PROF & HEAD DEPT FORENSIC PATH, E TENN STATE UNIV, 89-; DEP CHIEF MED EXAMR, STATE OF TENN. *Concurrent Pos:* NIH spec fel & instr neuropath, Col Physicians & Surgeons, Columbia Univ, 60-61; scientist, Armed Forces Inst Path, 65-66; consult, Vet Admin Hosp, Iowa City, 64-73; dep chief med examiner, Galveston Co, 78-84. *Honors & Awards:* Third Annual Milton Helpern Mem Lectr, Nat Asn Med Examrs, 85; First George Peret Mem Lectr, Univ Iowa, 90. *Mem:* Am Asn Path & Bact; Am Soc Exp Pathologists; Am Soc Human Genetics; Am Asn Neuropath; fel Am Col Pathologists; Am Asn Phys Anthropol; Nat Asn Med Examrs; Acad Forensic Sic. *Res:* Aneurysms, angiomas and infections of the central nervous system; diseases of muscle, brain death; forensic anthropology; forensic pathology. *Mailing Add:* Dept Forensic Path E Tenn State Univ Box 19250A Johnson City TN 37614

MCCORMICK-RAY, M GERALDINE, b San Diego, Calif, Nov 8, 43; m 78. MARINE INVERTEBRATE PATHOLOGY. *Educ:* Wake Forest Univ, BA, 72; Univ Alaska, Fairbanks, MS, 83. *Prof Exp:* Asst pathobiol, Johns Hopkins Univ, 76-78, assoc, 78-80; res scientist, Marine Ecol, 80-86, RES SCIENTIST, DEPT ENVIRON SCI, UNIV VA, 86- *Concurrent Pos:* Coun Environ Qual, Nature Conservancy, 83-87; res coordr, Nat Aquarium, Baltimore, 89- *Mem:* Soc Invert Path; Int Asn Aquatic Animal Med; Nat Estuarine Fed; Nat Shellfisheries Asn. *Res:* Visual acuity and color discrimination testing in seals; effects of crude oil on mussel blood cells; marine conservation; oyster hemocytes; classification and motility; proposed Estuarine Salinity classification. *Mailing Add:* Dept Environ Sci Clark Hall Univ Va Charlottesville VA 22903

MCCORQUODALE, DONALD JAMES, b Winnipeg, Man, Aug 27, 27; nat US; m 51; c 4. BIOCHEMISTRY. *Educ:* Univ BC, BA, 50; Univ Wis, PhD(cytol, biochem), 55. *Prof Exp:* Am Cancer Soc fels, Dept Oncol, McArdle Mem Inst, Univ Wis, 55-57 & Max Planck Inst Biochem, Munich, 57-58; from asst prof to assoc prof biochem, Emory Univ, 58-67; assoc prof, Southwest Ctr Advan Studies, 68-69, from assoc prof to prof, Univ Tex, Dallas, 69-75; PROF BIOCHEM, MED COL OHIO, 75- *Concurrent Pos:* USPHS res career develop award, 60-67; vis assoc prof, Mass Inst Technol, 65-66. *Mem:* AAAS; Am Chem Soc; Am Soc Biol Chemists; Am Soc Microbiol. *Res:* Biochemical virology; macromolecular biosynthesis; cellular control mechanisms. *Mailing Add:* Dept of Biochem Med Col of Ohio C S 10008 Toledo OH 43699

MCCORQUODALE, JOHN ALEXANDER, b Woodstock, Ont, June 21, 38; m 65; c 4. CIVIL ENGINEERING. *Educ:* Univ Western Ont, BES, 62; Glasgow Univ, MS, 64; Univ Windsor, PhD(civil eng), 70. *Prof Exp:* Hydraul engr, H G Acres & Co Ltd, 64-66; lectr, 66-68, from asst prof to assoc prof, 68-76, PROF HYDROL & HYDRAUL, UNIV WINDSOR, 76- *Concurrent Pos:* Consult hydraulics. *Honors & Awards:* Galbraith Prize, Eng Inst Can. *Mem:* Eng Inst Can; Am Soc Civil Engrs; Asn Professional Engrs Ont. *Res:* Hydraulics; hydrology; numerical analysis. *Mailing Add:* Dept Civil Eng Univ Windsor Windsor ON N9B 3P4 Can

MCCORRISTON, JAMES ROLAND, b Sask, July 27, 19; wid; c 3. SURGERY. *Educ:* Univ Sask, BA, 39; Queen's Univ, Ont, MD & CM, 43; McGill Univ, MSc, 48; FRCPS(C), 50; Am Bd Surg, dipl, 52. *Prof Exp:* Demonstr, McGill Univ, 51-53, lectr, 53-56, Markle scholar med sci, 53-58, from asst prof to assoc prof surg, 56-63; head dept, 63-73, PROF SURG, QUEEN'S UNIV, ONT, 63-; surgeon, Kingston Gen Hosp, 73-85; EMER PROF, 85- *Concurrent Pos:* Clin asst surg, Royal Victoria Hosp, Montreal, 51-53, asst surgeon, 53-59, assoc surgeon, 59-63, hon consult, 63-; surgeon-in-chief, Kingston Gen Hosp, 63-73; consult, Hotel Dieu Hosp, Kingston. *Mem:* Fel Am Col Surgeons; Whipple Surg Soc; Can Med Asn; Can Asn Clin Surg; Can Asn Gastroenterol. *Res:* Clinical, general and experimental surgery; peptic ulcer; blood volume; electrolyte balance; homotransplantation; esophageal physiology. *Mailing Add:* PO Box 370 Sharbot Lake ON K0H 2P0 Can

MCCOSKER, JOHN E, b Los Angeles, Calif, Nov 17, 45. AQUATIC BIOLOGY, ICHTHYOLOGY. *Educ:* Occidental Col, BA, 67; Scripps Inst Oceanog, PhD(marine biol), 73. *Hon Degrees:* DSc, Occidental Col, 87. *Prof Exp:* Res fel ichthyol, Smithsonian Trop Res Inst, 70-71; lectr marine biol, Univ Calif, San Diego, 73; DIR, STEINHART AQUARIUM, CALIF ACAD SCI, 73- *Concurrent Pos:* Adj prof, San Francisco State Univ, 75- *Mem:* AAAS; Am Soc Ichthyologists & Herpetologists; Soc Protection Old Fishes; Am Asn Zool Parks & Aquariums. *Res:* Systematics of marine fishes; aquarium maintenance of mesopelagic animals; behavior of sea snakes; solar energy applications to aquarium design; biology of primitive fishes, particularly the coelacanth; biology of the white shark. *Mailing Add:* Steinhart Aquarium Cal Acad Sci Golden Gate Park San Francisco CA 94118

MACCOSS, MALCOLM, b Cleator, Eng, June 2, 47; m 71; c 2. MEDICINAL CHEMISTRY. *Educ:* Univ Birmingham, UK, BSc, 68, PhD(chem), 71. *Prof Exp:* Fel chem, Univ Alta, 72-74, res assoc, 74-76; asst scientist, 76-80, scientist bio-org chem, Argonne Nat Lab, 80-82; res fel, 82-87, ASST DIR,

MERCK & CO, 87- *Concurrent Pos:* Adj assoc prof med chem, Dept Med Chem, Col Pharm, Univ Ill Med Ctr, 81-82. *Mem:* Royal Soc Chem; Am Chem Soc; NY Acad Sci. *Res:* Design of protease inhibitors; design of oral hypoglycemic agents; design of angiotensin II receptor antagonists; design of antiviral agents; nucleoside and nucleotide chemistry; heterocyclic chemistry; peptide chemistry; carbohydrate chemistry. *Mailing Add:* Merck Sharpe Dohme Res Labs PO Box 2000 Rahway NJ 07065-0900

MCCOUBREY, ARTHUR ORLANDO, b Regina, Sask, Mar 11, 20; nat US; m 42; c 3. PHYSICS, SCIENCE ADMINISTRATION. *Educ:* Calif Inst Technol, BS, 43; Univ Pittsburgh, PhD(physics), 53. *Prof Exp:* Res engr, Res Labs, Westinghouse Elec Corp, 43-47, res physicist, 47-52, adv physicist, 52-57; mgr physics dept, Nat Co, Inc, 57-60; mgr res & adv develop, Bomac Labs, Inc, 60-63; mgr atomic frequency stand, Varian Assocs, Mass, 64-67, mgr res & develop, Quantum Electronics Div, Calif, 67-68, dir cent res labs, 68-72; vpres, Frequency & Time Systs, Inc, Danvers, Mass, 72-74; dir, Inst Basic Stand, Nat Bur Standards, 74-78, assoc dir measurement serv, 78-84; MGR, TECH APPLICATIONS MEASUREMENT STANDARDS, NAT INST STANDARDS & TECHNOL, 84- *Mem:* Am Phys Soc; fel Inst Elec & Electronics Eng; fel Instrument Soc Am; Am Soc Qual Control. *Res:* Microwaves; resonance physics; atomic frequency standards; quantum devices; basic measurement standards. *Mailing Add:* 8113 Whirlwind Ct Gaithersburg MD 20882-4469

MCCOURT, A(NDREW) W(AHLERT), b St Louis, Mo, Jan 13, 24; m 49; c 4. AEROSPACE & DEFENSE SYSTEMS. *Educ:* Washington Univ, St Louis, BS, 43; Calif Inst Technol, MS, 47; Univ Pittsburgh, PhD(elec eng), 56. *Prof Exp:* Engr, Spec Prod Div, Westinghouse Elec Corp, 47-51 & Air Arm Div, 52-63, eng sect mgr, Defense & Space Ctr, 63-67, dir syst analysis, Ctr Advan Studies & Analysis, 67-74, adv engr, Defense & Electronic Syst Ctr, 75-82. *Concurrent Pos:* Adj assoc prof, Univ Pittsburgh, 57-65. *Mem:* Am Inst Aeronaut & Astronaut; Inst Elec & Electronics Engrs. *Res:* Weapon system analysis; aircraft dynamics and control; computers. *Mailing Add:* 468 Arundel Beach Rd Severna Park MD 21146

MCCOURT, FREDERICK RICHARD WAYNE, b New Westminster, BC, Jan 16, 40; m 71. MOLECULAR PHYSICS, CHEMICAL PHYSICS. *Educ:* Univ BC, BSc, 63, PhD(chem), 66. *Prof Exp:* Nat Res Coun Can fel, Swiss Fed Inst Technol, 66-67; Nat Res Coun Can fel, Kamerlingh Onnes Lab, Leiden, 67-68, Found Fundamental Res Matter sci co-worker, 68-69; vis asst prof, Inst Physics, Univ Genoa, 69; from asst prof to assoc prof, 70-78, assoc prof appl math, 73-78, PROF CHEM, UNIV WATERLOO, 78-, PROF APPL MATH, 78- *Concurrent Pos:* Alfred P Sloan Found fel, 73-75; SERC sr res fel, UK, 83-84. *Mem:* Am Phys Soc; Can Soc Chem. *Res:* Nuclear magnetic relaxation; light scattering; spectral line broadening; kinetic theory of polyatomic gases; molecular collision theory; energy miration in polymer films. *Mailing Add:* Dept Chem Univ Waterloo Waterloo ON N2L 3G1 Can

MCCOURT, ROBERT PERRY, b Highland Park, Mich, Apr 27, 29; m 50; c 7. VERTEBRATE ZOOLOGY, RADIATION BIOLOGY. *Educ:* Univ Mich, AB, 50; Ohio State Univ, MS, 52, PhD, 54. *Prof Exp:* Asst instr zool, Ohio State Univ, 52-54; from instr to prof, 54-91, PROF, HOFSTRA UNIV, 91- *Concurrent Pos:* Dir, Artemia Bibliographic Ctr, 80- *Mem:* AAAS; Sigma Xi. *Res:* Radiation biology of evolution of artemia; computers in biology. *Mailing Add:* Dept Biol Hofstra Univ 1000 Fulton Ave Hempstead NY 11550

MCCOWAN, JAMES ROBERT, b Nampa, Idaho, Feb 24, 23; m 46; c 3. PHARMACY. *Educ:* Univ Colo, BS, 48, MS, 49; Univ Fla, PhD(pharm), 54. *Prof Exp:* From instr to asst prof pharm, Loyola Univ La, 49-55; from asst prof to prof, St Louis Col Pharm, 55-68; asst dean, Col Pharm, Univ Ark Med Sci, 70-77, prof pharmaceut, 68-88, assoc dean acad affairs, 77-88; RETIRED. *Mem:* Am Pharmaceut Asn; Sigma Xi. *Res:* Ophthalmic solutions; aerosols. *Mailing Add:* 7207 Kingwood Rd Little Rock AR 72207

MCCOWAN, OTIS BLAKELY, b Monterey, Tenn, June 17, 34. MATHEMATICS. *Educ:* Tenn Technol Univ, BS, 59; La State Univ, MA, 66; George Peabody Col, PhD, 75. *Prof Exp:* Mathematician, Air Force Missile Develop Ctr, Holloman AFB, NMex, 62-63; math teacher, Rhea Cent High Sch, Dayton, Tenn, 63-65; instr math, Kilgore Col, Tex, 66-67; from asst prof to assoc prof math, 67-75, PROF MATH, BELMONT COL, NASHVILLE, 75- *Mem:* Math Asn Am; Am Math Soc; Nat Coun Teachers Math. *Res:* Mathematics in education; mathematics teacher education. *Mailing Add:* Dept Math Belmont Col Nashville TN 37212

MCCOWEN, MAX CREAGER, b Sullivan, Ind, July 4, 15; m 85. PARASITOLOGY. *Educ:* Ind State Univ, BS, 37, MS, 38. *Prof Exp:* Instr high sch, Ind, 38-42; res assoc, Res Labs, Eli Lilly & Co, 46-47, asst chief parasitol res, 47-48, head dept, 48-58, sr parasitologist, 58-65, chief parasitologist, Lilly Res Ctr, Surrey, Eng, 65-68, head parasitol res, 68-71, res scientist aquatic biol, 71-80, RES SCIENTIST PLANT MGT RES, GREENFIELD LABS, ELI LILLY & CO, 80- *Concurrent Pos:* Lectr, Sch Med, Ind Univ, 60. *Mem:* AAAS; Am Soc Parasitol; Aquatic Plant Mgt Soc; Am Soc Trop Med & Hyg; Sigma Xi. *Res:* Chemotherapy of parasitic diseases; immunology of parasitic infections; diagnosis of protozoan and helminthic diseases; aquatic research; aquatic herbicides, algicides, molluscicides and aquatic larvacides; environmental research. *Mailing Add:* 580 Conner Creek Dr Noblesville IN 46060

MCCOWEN, SARA MOSS, b Washington, NC, Jan 7, 44; m 69. MICROBIAL PHYSIOLOGY. *Educ:* Duke Univ, BA, 66; Univ NC, MAT, 68; Va Commonwealth Univ, MS, 73, PhD(microbiol), 75. *Prof Exp:* ASST PROF BIOL, VA COMMONWEALTH UNIV, 75- *Mem:* Am Soc Microbiol. *Res:* Carbohydrate metabolism in Pseudomonas species. *Mailing Add:* Dept Biol Va Commonwealth Univ 816 Park Ave Richmond VA 23284

MCCOWN, BRENT HOWARD, b Chicago, Ill, Feb 21, 43; m 68; c 1. PHYSIOLOGICAL ECOLOGY, HORTICULTURE. *Educ:* Univ Wis-Madison, BS, 65, MS, 67, PhD(bot, hort), 69. *Prof Exp:* Res & develop coordr, US Army Cold Regions, Res & Eng Lab, Hanover, NH, 70-72; asst prof, 72-77, ASSOC PROF HORT, INST ENVIRON STUDIES, UNIV WIS-MADISON, 78- *Concurrent Pos:* Adj asst prof, Inst Arctic Biol, Univ Alaska, 70-72. *Mem:* AAAS; Am Soc Plant Physiol; Am Soc Hort Sci; Nat Parks & Conserv Org; Int Plant Propagations Soc; Sigma Xi. *Res:* Adaption of plants to environment; plant juvenility; resource analysis and carrying capacity; influence of man's activities on plant growth and survival; petroleum toxicity to terrestrial ecosystems; plant growth regulation; microculture of plant cells and organs. *Mailing Add:* Dept of Hort Univ of Wis 1575 Linden Madison WI 53706

MCCOWN, JOHN JOSEPH, b Cleveland, Ohio, Mar 14, 29; m; c 5. SODIUM TECHNOLOGY, RADIO CHEMISTRY. *Educ:* Drury Col, BS, 51; Univ Tenn, MS, 55. *Prof Exp:* Jr chemist, USDA, 51-52; from jr chemist to assoc chemist anal chem div, Oak Ridge Nat Lab, 52-56; from asst chemist to assoc chemist chem eng div, Argonne Nat Lab, 56-60, assoc chemist, Idaho Div, 60-63; sr scientist, Nev Test Opers, Westinghouse Astronuclear Lab, 63-65, fel scientist, 65-68; sr res scientist, Pac Northwest Labs, Battelle Mem Inst, 68-70; fel scientist, Hanford Eng Develop Lab, Westinghouse Hanford Co, 70-74, mgr sodium systs anal, 74-76, mgr sodium systs tech, 76-81, mgr chem & anal, 81-88; prog mgr, Chem Sci Dept, Battelle Pac Northwest Lab, 88-89; ADV SCIENTIST, WESTINGHOUSE HANFORD CO, 90- *Mem:* Am Chem Soc; Am Soc Testing & Mat. *Res:* Remote analyses of highly radioactive materials; remote analytical techniques and equipment; analytical radiochemistry; reactor core analysis by gamma spectrometry methods; sodium systems eng; sodium technology; reactor coolant and cover gas characterization and fuel failure monitoring; energy dispersive x-ray fluorescent spectroscopy; environmental mixed waste analysis. *Mailing Add:* US Dept Energy EM532 Trevion II Washington DC 20585-0002

MCCOWN, JOSEPH DANA, b Moscow, Idaho, Aug 31, 40; m 69; c 2. INDUSTRIAL ORGANIC CHEMISTRY, RESEARCH ADMINISTRATION. *Educ:* Univ Idaho, BS, 62, MS, 64; Univ Iowa, PhD(org chem), 68. *Prof Exp:* Sr res chemist, 68-71, info scientist, 71-73, res supvr, Minn Mining & Mfg Co, 73-78; mgr explor imaging & appln, Minn 3M Res Ltd, Harlow, England, 78-81, lab mgr, Cent Res Labs, St Paul, Minn, 81-85, TECH DIR, STATIC & ELECTROMAGNETIC CONTROL DIV, AUSTIN, TEX, 3M CO, 85- *Mem:* Am Chem Soc; Sigma Xi. *Res:* Aryl sulfonyl isocyanates; hydroxamic acid esters; beta-diketones; N-nitrosoenamines; thermal rearrangement reactions; fluorocarbon chemistry. *Mailing Add:* 10521 Glass Mountain Trail Austin TX 78750-2510

MCCOWN, MALCOLM G, b Prairie Lea, Tex, Aug 14, 19; m 51; c 1. ALGEBRA. *Educ:* Trinity Univ, BS, 41, MEd, 51. *Prof Exp:* Instr math, Trinity Univ, 55-60, asst prof, 60-69; instr math, Jourdanton Tex High Sch, 79-87; INSTR MATH, BEE CO JR COL, 80- *Mem:* Am Math Soc; Nat Coun Teachers Math. *Res:* Geometry; theory of equations. *Mailing Add:* 218 Brian Dr Pleasanton TX 78064

MCCOWN, ROBERT BRUCE, b Portland, Ore, Nov 7, 39; m 73. PHYSICS. *Educ:* Hastings Col, BA, 64, MS, 66, PhD(physics), 75. *Prof Exp:* Asst prof physics, Univ South Ala, 76-80; mem staff, Intermagnetics Gen Corp, 80-; PRIN RES SCIENTIST, BATTELLE COLUMBUS LABS. *Mem:* AAAS; Inst Elec & Electronics Engrs. *Res:* Use of microcomputers in scientific teaching and in devices for the disabled. *Mailing Add:* Battelle Columbus Labs 505 King Ave Columbus OH 43201

MCCOY, BARRY, b Trenton, NJ, Dec 14, 40; m 70. THEORETICAL PHYSICS. *Educ:* Calif Inst Technol, BS, 63; Harvard Univ, PhD(physics), 67. *Prof Exp:* Res assoc, 67-69, from asst prof to assoc prof, 69-80, PROF PHYSICS, STATE UNIV NY STONY BROOK, 80- *Mem:* Inst Theoret Physics. *Res:* Statistical mechanics. *Mailing Add:* Dept Physics State Univ NY Stony Brook NY 11794

MCCOY, BENJAMIN J(OE), b Fairview, WVa, May 9, 41; m 70; c 1. CHEMICAL ENGINEERING. *Educ:* Ill Inst Technol, BS, 63; Univ Minn, Minneapolis, MS, 64, PhD(chem eng), 67. *Prof Exp:* From assoc prof to prof chem eng, 67-88, chmn, 80-88, ASSOC DEAN RES, COL ENG, UNIV CALIF, DAVIS, 88- *Mem:* Am Inst Chem Engrs; Am Phys Soc; Am Soc Eng Educ; Am Chem Soc. *Res:* Separation processes; transport phenomena. *Mailing Add:* Dept Chem Eng Univ Calif Davis CA 95616

MCCOY, CHARLES RALPH, b Freeport, Ill, June 14, 27; m 55; c 1. PHYSICAL CHEMISTRY. *Educ:* Roosevelt Univ, BS, 56; Northwestern Univ, PhD(chem), 60. *Prof Exp:* From instr to asst prof chem, Loyola Univ Chicago, 59-63; prof chem, 63-76, MEM FAC CHEM, UNIV WIS-WHITEWATER, 76- *Mem:* Am Chem Soc; Am Phys Soc. *Res:* Kinetics. *Mailing Add:* Box 254 Palmyra WI 53156

MCCOY, CLARENCE JOHN, JR, b Lubbock, Tex, July 25, 35; m 57; c 2. HERPETOLOGY. *Educ:* Okla State Univ, BS, 57, MS, 60; Univ Colo, PhD(zool), 65. *Prof Exp:* From asst cur to assoc cur, 64-72, CUR AMPHIBIANS & REPTILES, CARNEGIE MUS NATURAL HIST, 72-, ED, SCI PUBL, 90- *Concurrent Pos:* Vis prof, Ariz State Univ, 69-70; adj assoc prof, Univ Pittsburgh, 72-; chmn comt syst resources herpet, 75-77; chmn, Pa Fish Comn Adv Comt, 74-, Pa Biol Survey Steering Comt, 88-; dir, Western Pa Conservancy, 90- *Mem:* Am Soc Ichthyologists & Herpetologists; Am Soc Mammal; Soc Syst Zool; Soc Study Amphibians & Reptiles(pres, 72); Herpetologists League (vpres, 84-85, pres, 86-87). *Res:* Natural history and systematics of reptiles and amphibians; reptilian reproductive cycles; biogeography of Mexico & Cen Am; general vertebrate zoology. *Mailing Add:* Carnegie Mus Natural Hist 4400 Forbes Ave Pittsburgh PA 15213

MCCOY, CLAYTON WILLIAM, b Rochester, Minn, June 22, 38; m 63; c 1. ENTOMOLOGY. *Educ:* Gustavus Adolphus Col, BS, 60; Univ Nebr, MSc, 63; Univ Calif, PhD(entom), 67. *Prof Exp:* Res entomologist, USDA, 67-72; ASSOC ENTOMOLOGIST, AGR RES & EDUC CTR, UNIV FLA, 72- *Mem:* Entom Soc Am; Int Orgn Biol Control; Int Soc Invert Path; Sigma Xi. *Res:* Biological control of economic pests of citrus. *Mailing Add:* Ifas PO Box 1088 Lake Alfred FL 33850

MACCOY, CLINTON VILES, b Brookline, Mass, Mar 27, 05; m 36; c 2. ECOLOGY. *Educ:* Harvard Univ, AB, 28, AM, 29, PhD(biol), 34. *Prof Exp:* Cur fishes & mammals, Secy & ed, Boston Soc Natural Hist, 29-39; asst prof zool, Mass State Col, 39-44; from assoc prof to prof, 44-70, EMER PROF BIOL, WHEATON COL, MASS, 70-; DIR NORWELL LABS, 70- *Concurrent Pos:* Asst biologist, State Fish & Game Dept, NH, 39; res assoc, Woods Hole Oceanog Inst, 55-64. *Mem:* Am Soc Limnol & Oceanog; Int Soc Limnol; AAAS. *Res:* Limnology; oceanography, especially estuarine areas; intertidal in-fauna, especially meiofauna. *Mailing Add:* Norwell Labs 77 Winter St Norwell MA 02061

MCCOY, DAVID ROSS, b Culver City, Calif, July 21, 42; m 66; c 3. ORGANIC CHEMISTRY. *Educ:* Univ Calif, Los Angeles, BS, 63; Univ Ore, PhD(org chem), 67. *Prof Exp:* From chemist to sr chemist, Texaco, Inc, NY, 67-72, res chemist, 72-76; sr proj chemist, 76-82, supvr appl res, 82-85, supvr, Urethane Res, 85-87, MGR, RES DEPT, TEXACO CHEM CO, 87- *Mem:* Am Chem Soc. *Res:* Chemistry of heterocyclic compounds; organophosphorus chemistry; mechanisms of heterogeneous catalysis; petrochemical synthesis; enhanced oil recovery; demulsifiers; polyurethanes. *Mailing Add:* 9202 Parkfield Dr Austin TX 78758

MCCOY, DOROTHY, b Waukomis, Okla, Aug 9, 03. MATHEMATICS. *Educ:* Baylor Univ, BA, 25; Univ Iowa, MS, 27, PhD(math), 29. *Prof Exp:* Teacher high sch, Iowa, 25-26; asst, Univ Iowa, 28-29; prof math & head dept, Belhaven Col, 29-40; chmn div phys & biol sci, 49-72, prof math & head dept, 49-75, DISTINGUISHED PROF & EMER MATHEMETICIAN, WAYLAND BAPTIST COL, 75- *Concurrent Pos:* Fulbright prof, Col Sci, Univ Baghdad, 53-54. *Mem:* Am Math Soc; Math Asn Am. *Res:* Geometry; complete existential theory of eight fundamental properties of topological spaces. *Mailing Add:* 2102 W 8th St Plainview TX 79072

MCCOY, EARL DONALD, b Hamilton, Ohio, May 14, 48; m 71; c 2. BIOGEOGRAPHY RESEARCH, COLLEGE LECTURING. *Educ:* Fla State Univ, BS, 70, PhD(biol), 77; Univ Miami, MS, 73. *Prof Exp:* Res ecologist, Fla Med Entomol Lab, 77-78; from asst prof to assoc prof, 78-89, PROF BIOL, UNIV SFLA, 89- *Concurrent Pos:* From vis asst prof to vis assoc prof, Mountain Lake Biol Sta, Univ Va, 81-87. *Mem:* Fel Am Soc Naturalists; Asn Trop Biol; Ecol Soc Am; Sigma Xi. *Res:* Population and community ecology, especially of insects; biogeography; statistical ecology; conservation ecology; designing schemes for conserving the biologically-rich but extremely fragile upland habitats of Florida. *Mailing Add:* Dept Biol Univ SFla Tampa FL 33620-5150

MCCOY, ERNEST E, b Victoria, BC, Sept 3, 23; m 50; c 4. MEDICINE. *Educ:* Univ Alta, BSc, 47, MD, 49. *Prof Exp:* Jr intern med, Royal Alexandria Hosp, Edmonton, Alta, 49-50; sr intern, Col Belcher Hosp, Calgary, 50-51; jr resident pediat, St Louis Children's Hosp, Mo, 51-52, asst resident, 52-53; clin instr, Univ BC, 54-57; res assoc microbiol, Vanderbilt Univ, 57-59, asst prof pediat, 59-61; assoc prof, Univ Mo-Columbia, 61-66; prof, Univ Va, 66-69; prof & chmn dept, 70-84, PROF PEDIAT, UNIV ALTA, 84- *Concurrent Pos:* Res fel pediat, Wash Univ, 53; Markle scholar, 59. *Mem:* Am Pediat Soc; Soc Pediat Res. *Res:* Metabolic studies of down syndrome. *Mailing Add:* Dept Pediat Univ Alta Sch Med Edmonton AB T6G 2G3 Can

MCCOY, FLOYD W, JR, b Seattle, Wash, Aug 21, 36; m 68; c 4. VOLCANICLASTIC SEDIMENTATION, PALEOSEDIMENTATION PATTERNS. *Educ:* Univ Hawaii, BS, 62, MS, 65; Harvard Univ, PhD(Geol), 74. *Prof Exp:* Geophysicist, Hawaii Inst Geophysics, Univ Hawaii, 65; researcher, Woods Hole Oceanog Inst, 66-69; vis invest, Woods Lake Oceanog Inst, 70-73; res sedimentologist, deep sea drilling, Scripps Inst Oceanog, Univ Calif, 74; curator & res assoc, 75-84, ADJ RES ASSOC, LAMONT-DOHERTY GEOL OBSERV, COLUMBIA UNIV, 84-; SR SCIENTIST, ASSOC SCIENTISTS WOODS HOLE, 84- *Concurrent Pos:* mem, Govs Comt Hudson River Estuarine Sanctuaries, 82-86; lectr geol & oceanog, Windward Col, Univ Hawaii, 90-; dir, Woods Hole Hist Collection & Woods Hole Community Asn. *Mem:* fel Geol Soc Am; Am Geophysical Union; Int Asn Sedimentologists; Soc Econ Paleontologists & Minerologists; Sigma Xi; Int Comt for Scientific Exploration of Mediteranean Sea. *Res:* Marine geology, marine sedimentology and abyssal stratigraphy, paleoceanography, volcaniclastic sedimentology, tephrochronology, geoarchaeology. *Mailing Add:* PO Box 721 Woods Hole MA 02543

MCCOY, JAMES ERNEST, b Glendale, Calif, May 4, 41; m 62; c 2. MAGNETOSPHERIC PHYSICS. *Educ:* Calif Inst Technol, BS, 63; Rice Univ, PhD(space sci), 69. *Prof Exp:* PHYSICIST, MANNED SPACECRAFT CTR, NASA, 63- *Concurrent Pos:* Consult, Appl Sci Inc. *Mem:* Am Geophys Union; Inst Elec & Electronics Engrs; Am Inst Aeronaut & Astronaut. *Res:* Electrodynamic tethers; high voltage interactions with ionospheric plasmas; geophysical instrumentation; hollow cathode devices; digital communications. *Mailing Add:* SN3 NASA L B Johnson Space Ctr Houston TX 77058

MCCOY, JIMMY JEWELL, b Corsicana, Tex, Mar 18,43; m 65. MATHEMATICAL PHYSICS, SOLID STATE PHYSICS. *Educ:* Baylor Univ, BS, 65, PhD(physics), 69. *Prof Exp:* AEC grant, Baylor Univ, 69-70; ASST PROF PHYSICS, TARLETON STATE COL, 70- *Res:* Particle size effects in x-ray scattering. *Mailing Add:* Dept Physics Tarleton State State Col Tarleton Station Stephenville TX 76402

MCCOY, JOHN HAROLD, b Conroe, Tex, May 16, 35; m 61; c 3. COMPUTER SCIENCE. *Educ:* Sam Houston State Univ, BS, 63, MA, 64; Ohio State Univ, PhD(elec eng), 68. *Prof Exp:* Exp physicist, United Aircraft Res Labs, 68-70; assoc prof math, 70-72, PROF COMPUT SCI, SAM HOUSTON STATE UNIV, 72-, DIR COMPUT FACIL, 71- *Mem:* Asn Comput Mach; Inst Elec & Electronics Engrs. *Mailing Add:* Computer Sci Dept Sam Houston State Univ Box 2206 Huntsville TX 77341

MCCOY, JOHN J, b New York, NY, Sept 18, 36; m 62; c 3. ENGINEERING MECHANICS. *Educ:* Cooper Union, BCE, 58; Columbia Univ, MS, 59, ScD(eng mech), 61. *Prof Exp:* Staff engr, Int Bus Mach Corp, 61-63; asst prof appl mech, Univ Pa, 63-69; assoc prof, 69-74, PROF CIVIL ENG, CATH UNIV AM, 74- *Concurrent Pos:* Consult, Statcon, Inc, Silver Spring, Md, 85-; assoc ed, Jour Appl Mech, 76-82. *Mem:* Fel Am Soc Mech Engrs; Acoust Soc Am; Am Acad Mech; Am Soc Civil Engrs; Am Geophys Acad Union. *Res:* Stochastic processes; propagation through random media; theory of heterogeneous media; waves in fluids and solids; applied mathematics; geophysics. *Mailing Add:* 705 Pine Ridge Rd Media PA 19063

MCCOY, JOHN J, ornithology, for more information see previous edition

MCCOY, JOHN PHILIP, JR, FLOW CYTOMETRY, EXTRA CELLULAR MATRIX. *Educ:* Univ Miami, PhD(microbiol), 79. *Prof Exp:* Asst prof, Sch Med, Univ Mich, 81-87; PITTSBURGH CANCER CTR, 87- *Mailing Add:* Pittsburgh Cancer Ctr 230 Lathrop St Pittsburgh PA 15213-2592

MCCOY, JOHN ROGER, b Trenton, NJ, June 11, 16; m 45; c 1. CHEMOTHERAPY, COMPARATIVE PATHOLOGY. *Educ:* Univ Pa, VMD, 40. *Prof Exp:* Res specialist, Bur Biol Res, Rutgers Univ, 50-69, adj res prof, 70-83; prof comp path & dir vivarium, 70-83, EMER PROF PATH, RUTGERS MED SCH, UNIV MED & DENT NJ, 83- *Concurrent Pos:* Consult to govt & var pharmaceut concerns; mem grant adv coun, Seeing Eye Found, 74-77; mem vet adv coun, Vet Cancer Unit, Sloan-Kettering Inst, 74-76; mem subcomt org contaminants, Safe Drinking Water Comt, Nat Res Coun-Nat Acad Sci; mem comt substances generally recognized as safe, Fedn Am Soc Exp Biol, 72-83; expert pan eval evidence of carcinogenicity & genotoxicity drugs & cosmetics ingreds, Fedn Am Soc Exp Biol, 83-85. *Honors & Awards:* Am Vet Med Asn Award, 81. *Mem:* AAAS; Am Vet Med Asn (pres, 71-72); Soc Toxicol; fel NY Acad Sci; Soc Toxicol Pathologists (pres, 77-78). *Res:* Spontaneous canine cancer; pathology of leishmaniasis; food additives; steroids; cancer chemotherapy; drug safety; periodontal disease; cardiovascular disease. *Mailing Add:* 1007 River Rd Piscataway NJ 08854

MCCOY, JOSEPH HAMILTON, b Johnson City, Tenn, May 7, 34; m; c 2. ANALYTICAL CHEMISTRY. *Educ:* E Tenn State Univ, BS, 56; Mich State Univ, PhD(chem), 63. *Prof Exp:* Res chemist, E I du Pont de Nemours & Co, 62-64; asst prof, 64-67, ASSOC PROF CHEM, EMORY & HENRY COL, 67- *Concurrent Pos:* Vis prof, NC State Univ, 83-84. *Mem:* AAAS; Am Chem Soc. *Res:* Polymer solution studies; gas chromatography. *Mailing Add:* PO Box EE Emory VA 24327

MCCOY, JOSEPH WESLEY, organic chemistry, for more information see previous edition

MCCOY, LAYTON LESLIE, b Seattle, Wash, Mar 11, 27. ORGANIC CHEMISTRY. *Educ:* Univ Wash, BS, 47, PhD(chem), 51. *Prof Exp:* Instr chem, Columbia Univ, 51-53; NSF fel, Univ Wash, 57-58; from instr to asst prof chem, Columbia Univ, 58-62; from asst prof to assoc prof, 62-67, asst chmn dept, 73-75, PROF CHEM, UNIV MO-KANSAS CITY, 67- *Mem:* AAAS; Am Chem Soc. *Res:* Synthesis and reactions of cyclopropanes; stereospecific and stereoselective syntheses and reactions; correlation of structure and relative acidity of acids. *Mailing Add:* Dept Chem Univ Mo Kansas City MO 64110

MCCOY, LOWELL EUGENE, b Hillsboro, Ohio, June 1, 37; m 59; c 4. PHYSIOLOGY, BIOCHEMISTRY. *Educ:* Miami Univ, BA, 60, MA, 61; Wayne State Univ, PhD(physiol), 66. *Prof Exp:* Res asst cancer chemother & toxicol, Christ Hosp Inst Med Res, 61-63; res assoc blood coagulation biochem, 66-67, asst prof, 68-73, PROF PHYSIOL, SCH MED, WAYNE STATE UNIV, 78- *Concurrent Pos:* Mem coun thrombosis, Am Heart Asn. *Mem:* Am Inst Chem; Am Chem Soc; Int Soc Thrombosis & Haemostasis; NY Acad Sci. *Res:* Blood coagulation protein biochemistry, including physical and enzymologic properties; hematology; erythropoiesis; peptide synthesis; amino acid sequence. *Mailing Add:* 22175 Thorofare Grosse Ile MI 48138-1449

MCCOY, RALPH HINES, b Crowell, Tex, Nov 22, 40; m 74; c 1. WILDLIFE DISEASES, MEDICAL BACTERIOLOGY. *Educ:* Baylor Univ, BS, 63, MS, 65; Ore State Univ, PhD(microbiol), 73. *Prof Exp:* asst prof biol, Tex A&I Univ, 74-76; asst prof, 77-80, ASSOC PROF BIOL, AUSTIN PEAY STATE UNIV, 80- *Concurrent Pos:* Consult, Garcia & Rangel, Attys, 75-76 & Mem Hosp, Clarksville, Tenn, 77- *Mem:* Wildlife Dis Asn; Am Soc Microbiol; Am Soc Mammalogists; Wildlife Soc. *Res:* Zoonoses; infectious diseases of wild and domestic animals. *Mailing Add:* Dept Biol Austin Peay State Univ Clarksville TN 37044

MCCOY, RAWLEY D, b Nutley, NJ, Dec 21, 14; m 47; c 3. COMPUTER & SPACE SCIENCE. *Educ:* Stevens Inst Technol, ME, 37. *Prof Exp:* Instr elec eng, Stevens Inst Technol, 37-38; radio engr, Lustron Lights, Inc, 38-40; proj engr, Sperry Gyroscope Co, 41-46; chief engr, Reeves Instrument Corp, 46-58, vpres eng, 58-66, sr vpres, 66-67; vpres, RC-95, Inc, 67-70; consult, 71-85; RETIRED. *Mem:* Fel Inst Elec & Electronics Engrs. *Res:* Servomechanisms; electronic analogue computers. *Mailing Add:* 16 Strathmore Ln Madison CT 06443-3306

MCCOY, RAYMOND DUNCAN, physical chemistry, instrumentation, for more information see previous edition

MCCOY, ROBERT A, b Springfield, Ill, Oct 1, 42; m 64; c 2. TOPOLOGY. *Educ:* Southern Ill Univ, BA, 64; Iowa State Univ, PhD(math), 68. *Prof Exp:* Asst prof, 68-73, assoc prof, 73-79, PROF MATH, VA POLYTECH INST & STATE UNIV, 79- *Mem:* Am Math Soc; Math Asn Am. *Res:* General topology. *Mailing Add:* Dept Math Va Polytech Inst & State Univ Blacksburg VA 24061

MCCOY, ROBERT ALLYN, b Columbus, Ohio, Oct 12, 39; m 67; c 4. FAILURE ANALYSIS, MECHANICAL TESTING OF MATERIALS. *Educ:* Ohio State Univ, BS, 62, MS, 63; Univ Calif, Berkeley, DEngr, 71. *Prof Exp:* Metallurgist, Crucible Res Lab, 66-68 & Lawrence Berkeley Lab, 68-72; mat engr, Ocean Eng Proj Off, US Navy, 72-74 & Am Sterilizer Res & Develop, 76-80; asst prof mat sci, US Naval Acad, 74-76; PROF MAT SCI, YOUNGSTOWN STATE UNIV, 80- *Concurrent Pos:* Failure anal consult, 80-; acad adv, Youngstown State Univ Student Chap, Am Soc Metals, 81- & Soc Advan Mat & Process Eng, 90- *Mem:* Am Soc Metals; Soc Advan Mat & Process Eng; Am Inst Mech Engrs. *Res:* Failure analysis and mechanical testing of materials using a tensile tester, an instrumented Charpy impact tester, and a scanning electron microscopy with electrodiagnosis capabilities. *Mailing Add:* Mat Eng Dept Youngstown State Univ Youngstown OH 44555

MCCOY, ROGER MICHAEL, b Dewey, Okla, Feb 3, 33; m 55; c 2. PHYSICAL GEOGRAPHY. *Educ:* Univ Okla, BS, 57; Univ Colo, MA, 64; Univ Kans, PhD(geog), 67. *Prof Exp:* Asst prof geog, Univ Ill, Chicago, 67-69; asst prof, Univ Ky, 69-72; assoc prof, 72-80, PROF GEOG, UNIV UTAH, 80- *Concurrent Pos:* Consult remote sensing. *Honors & Awards:* Ford-Bartlett Award, Am Soc Photogram. *Mem:* Am Soc Photogram; Asn Am Geog. *Res:* Fluvial geomorphology; remote sensing. *Mailing Add:* Dept Geog Univ Utah Salt Lake City UT 84112

MCCOY, SCOTT, JR, b Indianapolis, Ind, Oct 8, 39; div; c 1. BIO-STRATIGRAPHY. *Educ:* Wheaton Col, BS, 61; Univ Ariz, MS, 64. *Prof Exp:* Field geologist, Phillips Petrol 64-77; sr staff geologist, Amoco Prod, 77-85; OWNER, WAVE VIDEO, 85- *Mem:* Fel Geol Soc Am; Am Asn Petrol Geologists. *Res:* Southern Alaska biostratigraphy. *Mailing Add:* 71-1 Puapake Place Lahaina HI 96761

MCCOY, SUE, b Charlottesville, Va, Nov 14, 35. OXYGEN TRANSPORT TO TISSUE. *Educ:* Radcliffe Col, AB, 57; Johns Hopkins Univ, Baltimore, PhD(physiol chem), 64; Univ Va, Charlottesville, MD, 80. *Prof Exp:* Postdoctoral fel physiol chem, Johns Hopkins Univ, 64-67; asst prof chem, Univ SFla, 67-69; asst prof orthop, Univ Va Sch Med, 69-73, asst prof surg, 73-78; resident surg, Hosp Univ Pa, 80-83 & Cooper Hosp, Rutgers Med Sch, Camden, 83-85, asst prof surg, 85-86; ASST PROF SURG, E TENN STATE UNIV COL MED, 86- *Concurrent Pos:* Staff surgeon, Mountain Home Vet Affairs Med Ctr, Johnson City, 86- *Mem:* Am Chem Soc; Royal Soc Chem; Asn Acad Surg; Am Fedn Clin Res. *Res:* Affect of aging on the metabolic responses to hemorrhagic shock; oxygen transport to tissue; elderly surgical patient. *Mailing Add:* Box 265 Mountain Home TN 37684

MCCOY, THOMAS JOSEPH, b Janesville, Wis, Oct 18, 50; m 83; c 1. PLANT CELL & TISSUE CULTURE, CYTOGENETICS & MOLECULAR GENETICS. *Educ:* Univ Wis-Madison, BS, 73, MS, 76; Univ Minn, PhD(plant breeding), 80. *Prof Exp:* Res geneticist, Res Unit, Univ Nev-Reno, 80-84; res leader, alfalfa prod, Agr Res Serv, USDA, 84-86; assoc prof plant cytogenetics & somatic cell genetics, Univ Ariz, Tucson, 86-89; DEPT HEAD, MONT STATE UNIV, 89- *Mem:* AAAS; Am Soc Agron; Int Soc Plant Molecular Biol; Crop Sci Soc Am; Genetic Soc Can. *Res:* Genetics and cytogenetics of plants, specifically Medicago species; genetic engineering of alfalfa for transfer of useful genes from other genera or kingdoms; interspecific hybridization of Medicago species using in vitro approaches, including embryo culture and somatic cell hybridization. *Mailing Add:* Dept Plant & Soil Sci Mont State Univ Bozeman MT 59717-0312

MCCOY, THOMAS LARUE, b Seville, Ohio, Jan 16, 33; m 88; c 4. MATHEMATICS. *Educ:* Oberlin Col, BA, 54; Univ Wis, MS, 56, PhD(math), 61. *Prof Exp:* Instr math, Ill Inst Technol, 60, asst prof, 64; from asst prof to assoc prof, 64-77, PROF MATH, MICH STATE UNIV, 77- *Mem:* Am Math Soc; Math Asn Am; Soc Indust & Appl Math. *Res:* Functions of a complex variable; differential equation; calculus of variation. *Mailing Add:* Dept Math A303 Wells Hall Mich State Univ East Lansing MI 48824

MCCOY, V EUGENE, JR, b Chicago, Ill, July 7, 33; m 57; c 3. PHYSICAL ORGANIC CHEMISTRY. *Educ:* Princeton Univ, AB, 55; Harvard Univ, PhD(chem), 65. *Prof Exp:* Res chemist, Pioneering Res Lab, Textile Fibers Dept, 64-67, sr res chemist, 67-68 & 69-73, SR RES CHEMIST, TEXTILE RES LAB, E I DU PONT DE NEMOURS & CO, INC, 73- *Mem:* Am Chem Soc; Sigma Xi; Textured Yarn Asn Am. *Res:* Electron exchange reactions; chemistry of organophosphorus compounds; polymers and chemical reactions for textile fibers; feed yarn and textured yarn development. *Mailing Add:* 2641 Majestic Dr Wilmington DE 19810

MCCRACKEN, ALEXANDER WALKER, b Motherwell, Scotland, Nov 24, 31; m 60; c 2. MEDICAL MICROBIOLOGY. *Educ:* Univ Glasgow, MB, ChB, 55; Univ London, DCP, 62; Univ Liverpool, DTM, 66. *Prof Exp:* Pathologist, Royal Air Force Hosp, Akrotiri, Cyprus, 58-61; chief clin microbiol, Royal Air Force Inst Path & Trop Med, Halton, Eng, 62-68; assoc dir labs, Bexar County Teaching Hosp, 71-72; dir labs, Herman Hosp, Houston, 72-73; chief microbial path, Univ Tex Med Sch San Antonio, 68-71, assoc prof path & microbiol, 68-73, actg chmn dept, 71-73, prof path, 72-73, adj prof microbiol, 73-80, prof allied health sci, 78-80; prof & chmn dept path, assoc chief path serv & dir clin microbiol & virol, Baylor Univ Med Ctr, 73-81, prof path, 73-81; CHIEF PATH SERV & DIR LABS, METHODIST HOSPS, DALLAS, 81-; CLIN PROF PATH, UNIV TEX, SOUTHWESTERN MED SCH, DALLAS. *Concurrent Pos:* Adj prof path, Baylor Col Dent, Dallas, 81- *Mem:* Brit Asn Clin Path; Brit Soc Gen Microbiol; Col Path Eng; Int Acad Path; Am Col Pathologists; Am Med Asn. *Res:* Rapid diagnostic techniques in clinical microbiology; fluorescent antibody methodology; respiratory virus isolation methods and epidemiology; automation in microbiology; pathogenesis of viral infections. *Mailing Add:* Dept of Path Methodist Med Ctr 301 W Colorado Dallas TX 75208

MCCRACKEN, DEREK ALBERT, b Eng, Feb 11, 43; m 63; c 3. PLANT PHYSIOLOGY. *Educ:* McMaster Univ, BA, 63; Univ Toronto, MA, 66, PhD(bot), 69. *Prof Exp:* asst prof, 69-77, ASSOC PROF BIOL SCI, ILL STATE UNIV, 77- *Mem:* AAAS; Am Soc Plant Physiol; Can Soc Plant Physiol; Scand Soc Plant Physiol. *Res:* Starch biosynthesis; interaction between nuclear and cytoplasmic genes. *Mailing Add:* Dept Biol Sci FSA 153A Ill State Univ Normal IL 61761

MCCRACKEN, FRANCIS IRVIN, b Bicknell, Ind, Aug 19, 35; m 59; c 2. FOREST PATHOLOGY. *Educ:* Ariz State Univ, BS, 57; Okla State Univ, MS, 59; Wash State Univ, PhD(plant path), 72. *Prof Exp:* Instr bot, Univ Idaho, 62-67; RES PLANT PATH, US FOREST SERV, USDA, 67- *Mem:* Am Phytopath Soc; Sigma Xi. *Res:* Etiology and control of diseases effecting southern bottomland hardwood tree species. *Mailing Add:* US Forest Serv PO Box 227 Stoneville MS 38776

MCCRACKEN, JOHN AITKEN, b Glasgow, Scotland, Aug 16, 34. ENDOCRINOLOGY. *Educ:* Glasgow Univ, BVMS, 58, PhD(endocrinol), 63. *Prof Exp:* House surgeon, Glasgow Univ, 58-59; vis res fel, Cambridge Univ, 60; Agr Res Coun res fel endocrine physiol, Univ Birmingham, 63-64; staff scientist, 64-70, SR SCIENTIST, WORCESTER FOUND EXP BIOL, 70- *Concurrent Pos:* Vis prof, Cornell Univ, 69-70; adj prof physiol, Med Sch, Univ Mass, 78- *Mem:* AAAS; Soc Study Reproduction; NY Acad Sci; Royal Col Vet Surg; Endocrine Soc. *Res:* Steroid biochemistry; physiology and biochemistry of prostaglandins; transplantations of ovary, uterus and adrenal glands in the sheep and their various interrelationships; interrelationships between steroids, prostaglandins and polypeptide hormones; utero-ovarian-pituitary axis. *Mailing Add:* Worchester Found Exp Biol 222 Maple Ave Shrewsbury MA 01545

MCCRACKEN, JOHN DAVID, b Fairfield, Iowa, Sept 17, 39; m 61; c 2. AGRICULTURAL EDUCATION, RESEARCH DESIGN. *Educ:* Iowa State Univ, BS, 61, MS, 62; Ohio State Univ, PhD(agr educ), 70. *Prof Exp:* Teacher voc agr, Charles City Public Schs, 64-68; res assoc voc educ, 68-70, res specialist, 70-73, from asst prof to assoc prof, 73-79, PROF AGR EDUC, OHIO STATE UNIV, 79- *Concurrent Pos:* ed, Summaries res & develop activities Agr Educ, 74-76, & J Voc Educ Res, 79; policy comn, Agr Educ Div, Am Voc Asn, 78-80; vis prof, Kans State Univ, 81 & 90; sect ed, Annual Rev Res Voc Educ, Univ Ill, 81; distinguished lectr, Am Asn Teacher Educr Agr, 82; consult, agr & educ trends, Iowa State Univ, 83, Nat Agr Educ Workshop, 84; Fulbright scholar, Malaysia, 85-86; consult, Malaysia & Thailand, 88; US & Japan Found Work, Japan, 90. *Honors & Awards:* Distinguished Serv Award, Am Asn Agr Educ, 90. *Mem:* Am Asn Teacher Educr Agr (vpres, 76-77, pres, 79, past pres, 80); Am Voc Asn; Am Voc Educ Res Asn (secy, 84, vpres, 89, pres, 90); Am Educ Res Asn. *Res:* Rural schools, agricultural education; internationalizing the agricultural education secondary curriculum. *Mailing Add:* 6641 Millbrae Rd Columbus OH 43235

MCCRACKEN, LESLIE G(UY), JR, b Philadelphia, Pa, June 8, 25; m 47; c 1. ELECTRICAL ENGINEERING. *Educ:* Mass Inst Technol, SB, 45; Lehigh Univ, MS, 47; Pa State Univ, PhD(elec eng), 52. *Prof Exp:* Asst elec eng, Lehigh Univ, 46-47; asst ord res lab, Pa State Univ, 47-48, instr elec eng, 49-51; electronic scientist, Naval Res Lab, Wash, 51-56; ASSOC PROF ELEC ENG, LEHIGH UNIV, 56- *Concurrent Pos:* Staff mem, Gen Atronics Corp, Pa, 58-59 & Bell Aerospace Corp, 63; Off Naval Res grant, 60; Int Bus Mach Corp fel, 61; NASA-Am Soc Eng Educ fel, Univ Md, 65. *Mem:* Am Soc Eng Educ; Sigma Xi; sr mem Inst Elec & Electronic Engrs. *Res:* Electromagnetic boundary value problems with cylindrical symmetry; reflection and scattering of impulse waves by random boundaries; nonlinear control system analysis and synthesis; dynamic and statistical design of multivariable control systems; multivariable stochastic processes; synthesis of sequential machines. *Mailing Add:* 1782 W Union Blvd Bethlehem PA 18018

MACCRACKEN, MICHAEL CALVIN, b Schenectady, NY, May 20, 42; m 67; c 2. CLIMATE CHANGE, AIR POLLUTION. *Educ:* Princeton Univ, BSE, 64; Univ Calif, Davis, MS, 66, PhD(appl sci), 68. *Prof Exp:* PHYSICIST, LAWRENCE LIVERMORE LAB, UNIV CALIF, 68-, DIV LEADER, ATMOSPHERIC & GEOPHYS SCI, 87- *Concurrent Pos:* Dep div leader, 74-87, proj dir, Multistate Atmospheric Power Prod Pollution Study, US Dept Energy, 76-79; sci dir, Climate & Carbon Dioxide Res Prog, US Dept Energy, 79-; chmn, comt climate var, Am Meteorol Soc, 83-85, Int Comn Climate, 87-; assoc ed, J Climate, 88-; mem-at-large, Sect Atmospheric & Hydrospheric Sci, AAAS, 91-95. *Mem:* Am Meteorol Soc; Am Geophys Union; Am Quaternary Asn; AAAS; Oceanog Soc. *Res:* Numerical simulation of processes governing the global climate and the factors causing climatic change (including CO2, nuclear war, etc) and regional air quality for land use planning and control strategy assessment. *Mailing Add:* Lawrence Livermore Nat Lab PO Box 808 (L-262) Livermore CA 94550

MCCRACKEN, MICHAEL DWAYNE, b Terrell, Tex, Feb 15, 41. BOTANY. *Educ:* Tex Tech Univ, BS, 63, MS, 65; Ind Univ, Bloomington, PhD(bot), 69. *Prof Exp:* Asst prof bot, Univ Wis-Madison, 69-71; asst prof, 71-75, chmn, 75-81, ASSOC PROF BIOL, TEX CHRISTIAN UNIV, 75-, DEAN ARTS & SCI, 81- *Concurrent Pos:* NSF fel, Univ Wis-Madison, 70-72. *Mem:* Phycol Soc Am; Int Phycol Soc; Am Soc Limnol & Oceanog; Int Soc Theoret & Appl Limnol; Sigma Xi. *Res:* Algology, differentiation and development in Volvox; algal primary productivity. *Mailing Add:* Dept of Biol Tex Christian Univ Ft Worth TX 76129

MCCRACKEN, PHILIP GLEN, b Santa Cruz, NMex, Feb 11, 28; m 49; c 4. CHEMICAL ENGINEERING. *Educ:* NMex State Univ, BS, 51; Purdue Univ, PhD(chem eng), 56. *Prof Exp:* Mem staff, Com Solvents Corp, 51-52 & Dow Chem Co, 56-61; head develop, Ciba-Geigy Chem Corp, McIntosh, 61-68, tech mgr, St Gabriel Plant, 68-71, tech dir, Agr Chem Div, 71-86; RETIRED. *Mem:* Sigma Xi. *Res:* Research and development of chemical processes; chronological history; process engineering; project engineering. *Mailing Add:* 4016 Friendly Ave Greensboro NC 27410

MCCRACKEN, RALPH JOSEPH, b Guantanamo, Cuba, July 3, 21; nat US; m 49; c 2. SOIL SCIENCE. *Educ:* Earlham Col, AB, 42; Cornell Univ, MS, 51; Iowa State Univ, PhD(soils), 56. *Prof Exp:* Soil scientist, USDA, 47-49, 51-55; assoc agronomist, Univ Tenn, 55-56; assoc prof, Soil Sci, N C State Univ, 56-62, prof, 62-73, asst dir res, Sch Agr & Life Sci, 70-73; assoc admin, USDA Agr Res Serv, 73-78; actg assoc dir, Sci & Educ Admin, 78-80, dep chief, Soil Conservation Serv, USDA, 81-85; ADJ PROF SOIL SCI, NC STATE UNIV & NC AGR & TECH STATE UNIV, 86- *Honors & Awards:* Distinguished Serv Award, Soil Sci Soc Am, 87. *Mem:* Fel AAAS; fel Soil Sci Soc Am; fel Am Soc Agron; Soil Conserv Soc Am; Int Soc Soil Sci. *Res:* Genesis, classification and mineralogy of soils. *Mailing Add:* Dept Plant Sci NC Agri Tech State Univ 1601 E Market St Greenboro NC 27411

MCCRACKEN, RICHARD OWEN, b Racine, Wis, Mar 26, 39; m 65; c 2. PARASITOLOGY. *Educ:* Univ Wis-Whitewater, BS, 65, MST, 67; Iowa State Univ, PhD(zool), 72. *Prof Exp:* Instr natural sci, Milton Col, 67-68; sr res parasitologist, Merck Inst Therapeut Res, Merck & Co, Inc, 74-75; lectr, Tex A&M Univ, 75-77; asst prof, 77-81, ASSOC PROF BIOL, IND UNIV-PURDUE UNIV, INDIANAPOLIS, 81- *Concurrent Pos:* NIH fel, Tulane Univ, 72-74. *Mem:* Am Soc Parasitologists; Am Inst Biol Sci; AAAS; Sigma Xi. *Res:* Biochemistry; physiology and chemotherapy of parasitic helminths; structure and function of parasite surface membranes; membrane transport. *Mailing Add:* Biology Dept Kb 313 Purdue Univ Sch Sci 1125 E 38th St Indianapolis IN 46205

MCCRACKEN, WALTER JOHN, b Bucyrus, Ohio, Sept 24, 53. MATERIALS SCIENCE ENGINEERING. *Educ:* Gen Motors Inst, Flint, Mich, BA, 76; Univ Fla, MA, 78, PhD(mat sci), 81. *Prof Exp:* Assoc process develop engr, Frigidaire Div, Gen Motors Corp, 76-77; grad res assoc, dept mat sci & eng, Univ Fla, 77-81; ENGR CERAMICS, NEUTRON DEVICES DEPT, GEN ELEC CO, 81- *Concurrent Pos:* Fel, Dept Mat Sci & Eng, Univ Fla, 81; pres, Mc-R Consults Inc, Gainesville, Fla, 81- *Mem:* Am Ceramic Soc; Nat Asn Corrosion Engrs; Electron Micros Soc Am. *Res:* Development and characterization of glasses and glass-ceramics for sealing to metals; sol-gel processing of glass films and monoliths; electron beam channeling of single crystal ceramics. *Mailing Add:* Gen Elec Co 7887 Bryan Dairy Rd Largo FL 34643

MCCRADY, EDWARD, III, b Trenton, NJ, Sept 24, 33; m 55; c 3. DEVELOPMENTAL BIOLOGY. *Educ:* Univ of the South, BS, 55; Univ Va, MA, 61, PhD(biol), 64. *Prof Exp:* Asst biol, Univ Va, 60-63; asst prof, 64-70, ASSOC PROF BIOL, UNIV NC, GREENSBORO, 70- *Mem:* AAAS; Am Soc Zool; Develop Biol Soc. *Res:* Insect tissue culture; development of imaginal discs. *Mailing Add:* Dept Biol 312-C Life Sci Bldg Univ NC 1000 Spring Garden St Greensboro NC 27412

MCCRADY, JAMES DAVID, b Beaumont, Tex, June 26, 30; m 51; c 3. VETERINARY PHYSIOLOGY. *Educ:* Tex A&M Univ, BS, 52, DVM, 58; Baylor Univ, PhD(physiol), 65. *Prof Exp:* From instr to asst prof physiol, Tex A&M Univ, 58-62; fel, instr & dir animal resources, Col Med, Baylor Univ, 62-64; assoc prof physiol, 64-65, prof vet physiol & pharmacol & head dept, 65-89, DIR, SPECIAL PROGS, COL VET MED, TEX A&M UNIV, 89- *Mem:* Am Soc Vet Physiol & Pharmacol; Am Physiol Soc; Am Col Clin Pharmacol. *Res:* Pulmonary hypertension; atrial fibrillation; electrocardiography. *Mailing Add:* Dept of Vet Physiol & Pharmacol Tex A&M Univ College Station TX 77843

MCCRADY, WILLIAM B, b Forreston, Tex, July 25, 33; m 54; c 3. GENETICS. *Educ:* ETex State Col, BS, 54, MS, 58; Univ Nebr, PhD(zool), 61. *Prof Exp:* Asst prof biol, Ark State Col, 61-62; from asst prof to assoc prof, 62-72, PROF BIOL, UNIV TEX, ARLINGTON, 72-, DIR, SCI LEARNING CTR, 78- *Mem:* Genetics Soc Am; AAAS. *Res:* Carbon dioxide sensitivity in Drosophila. *Mailing Add:* Dept Biol Box 19498 Univ Tex Arlington TX 76019

MCCRAE, JOHN LEONIDAS, b Lexington, Miss, Sept 16, 17; m 40; c 2. BITUMINOUS PAVEMENTS, SOIL MECHANICS. *Educ:* Northwestern Univ, BS, 48. *Prof Exp:* Chief, bituminous & chem lab, US Army Eng Waterways Exp Sta, Vicksburg, Miss, 50-61, res engr, Mobility & Environ Div, 61-72; PRES, EDCO INC, 60- *Mem:* Nat Soc Prof Engrs; Am Soc Testing Mat; Asn Asphalt Paving Technologists; fel Am Soc Civil Engrs. *Res:* Geotechnical research and development testing in soils and bituminous pavements; testing machine development; awarded five patents; numerous technical publications. *Mailing Add:* 416 Groome Dr Vicksburg MS 39180

MCCRANIE, ERASMUS JAMES, b Milan, Ga, July 24, 15; m 45; c 3. PSYCHIATRY. *Educ:* Emory Univ, AB, 37, MS, 38; Univ Mich, PhD(bot), 42, MD, 45. *Prof Exp:* Instr biol, SGa Col, 41-42; from instr to assoc prof psychiat, Univ Tex Southwestern Med Sch, 51-56; chmn dept psychiat, Med Col Ga, 57-79, prof, 56-84; RETIRED. *Concurrent Pos:* Consult, Vet Admin Hosp, Augusta, 56-84 & Ga Regional Hosp, 70-84. *Mem:* AMA; Am Psychiat Asn. *Res:* Psychosomatic medicine; medical applications of hypnosis; psychodynamics. *Mailing Add:* 742 Lancaster Rd Augusta GA 30901

MCCRARY, ANNE BOWDEN, b Wilmington, NC, Oct 25, 26; m 44; c 2. INVERTEBRATE ZOOLOGY, MARINE BIOLOGY. *Educ:* Univ NC, Chapel Hill, AB, 61, MA, 65, PhD(zool), 69. *Prof Exp:* From asst prof to assoc prof biol, Univ NC, Wilmington, 70-85, prof, 85-90; RETIRED. *Mem:* Am Soc Zool; Am Inst Biol Sci; Estuarine Res Fedn; Am Malacological Union; Crustacean Soc. *Res:* Marine Invertebrate Larvae; zooplankton. *Mailing Add:* 411 Summer Rest Rd Wilmington NC 28405

MCCRAY, RICHARD A, b Los Angeles, Calif, Nov 24, 37. ASTRONOMY. *Educ:* Stanford Univ, BS, 59; Univ Calif, Los Angeles, MA, 62, PhD(theoret physics), 67. *Prof Exp:* High sch teacher, Harvard Sch, North Hollywood, Calif, 60-62; res fel physics, Calif Inst Technol, 67-68; asst prof astron, Harvard Col Observ, 68-71; assoc prof, Dept Physics & Astrophys, 71-75, chmn, Ctr Astrophys & Space Astron, 84-86, FEL, JOINT INST LAB ASTROPHYS, UNIV COLO, BOULDER, 71-, PROF, DEPT ASTROPHYS PLANETARY & ATMOSPHERIC SCI, 75- *Concurrent Pos:* John Simon Guggenheim Found fel, 75-76; chmn, Astron Surv Comt, Panel Theoret & Lab Astrophys, Nat Acad Sci, 79-82, Astron & Astrophys Surv Comt, Panel Policy Opportunities, 90, mem, Comt Space Astron & Astrophys, 81-84; counr, Am Astron Soc, 80-83, chmn, High Energy Astrophys Div, 86-87, mem, Comt Pub Policy, 89-; mem, Astron Adv Comt, NSF, 81-83; chmn, Joint Inst Lab Astrophys, Univ Colo & Nat Inst Standards & Technol, 81-82; vis sr res assoc, NASA-Goddard Space Flight Ctr, 83-84; vis prof, Dept Astron, Columbia Univ, 89-90. *Honors & Awards:* Dannie S Heineman Prize for Astrophys, Am Phys Soc, 90. *Mem:* Nat Acad Sci; Am Astron Soc; Int Astron Union. *Res:* Theoretical astrophysics: gas dynamics, kinetic theory, radiative transfer, cosmic x-ray sources, interstellar matter, supernovae; infrared, ultraviolet and x-ray space astronomy; science policy. *Mailing Add:* Joint Inst Lab Astrophys Univ Colo Boulder CO 80309-0440

MCCREA, KENNETH DUNCAN, b New Haven, Conn, Dec 9, 53. PLANT ECOLOGY, COEVOLUTION. *Educ:* Bucknell Univ, BS, 75; Purdue Univ, PhD(ecol), 81. *Prof Exp:* Assoc, Purdue Univ, 81-82; ASSOC, BUCKNELL UNIV, 82- *Res:* Life history strategies of clonal plants; plant-animal interactions; character displacement in insect pollinated plants; coevolution. *Mailing Add:* Dept Biol Bucknell Univ Lewisburg PA 17837

MCCREA, PETER FREDERICK, instrumentation, process control, for more information see previous edition

MACCREADY, PAUL BEATTIE, JR, b New Haven, Conn, Sept 29, 25; m 57; c 3. AIR POLLUTION. *Educ:* Yale Univ, BS, 47; Calif Inst Technol, MS, 48, PhD(aeronaut), 52. *Prof Exp:* Meteorol consult, Salt River Valley Water Users' Asn, 50-51; pres, Meteorol Res, Inc, 51-70 & Atmospheric Res Group, 58-70; pres, 71-90, CHMN, AERO VIRONMENT INC, 71- *Concurrent Pos:* Consult, various orgns, 73- *Honors & Awards:* Reed Aeronautical Award, Am Inst Aeronaut & Astronaut, 79; Edward Longstreth Medal, Franklin Inst, 79; Kremer Prize, 77 & 79; Guggenheim Medal, Am Inst Aeronaut & Astron, Am Soc Mech Eng, SAE, 88. *Mem:* Nat Acad Eng; AAAS; fel Am Inst Aeronaut & Astronaut; fel Am Meteorol Soc; Am Acad Arts & Sci. *Res:* Instrumentation development in aeronautics and atmospheric science; basic and applied studies in turbulence and diffusion; cloud physics, cloud electrification and weather modification; developer of human-powered, electric powered and solar powered, aircraft and surface vehicles. *Mailing Add:* Aero Vironment Inc 222 E Huntington Dr Monrovia CA 91016

MCCREADY, RONALD GLEN LANG, b Calgary, Alta, Jan 19, 39; m 72; c 1. MICROBIOLOGY. *Educ:* Univ Alta, BSc, 61, MSc, 63; Univ Calgary, PhD(microbiol), 73. *Prof Exp:* Res assoc microbiol, Univ Calgary Interdisciplinary Sulphur Res, 73-75; staff microbiologist, CANMET, 75-78; microbiologist, Lethbridge Res Sta Agr Can, 78-82; HEAD BIOTECH SECT, CANMET, ENERGY, MINES & RESOURCES, 85- *Concurrent Pos:* Microbiologist indust biol, Dalhousie Univ, 82-85. *Mem:* Can Soc Microbiol; Am Soc Microbiol; Soc Indust Microbiol; Can Pharmaceut Asn; Can Col Microbiologists. *Res:* Stable isotope metabolism by microorganisms; microbial sulphur cycle; microbial leaching of metals, microbial treatment of mine effluents. *Mailing Add:* 66 Gowrie Dr Kanata ON K2L 2S5 Can

MCCREADY, THOMAS ARTHUR, b Pueblo, Colo, Sept 1, 40; m 65; c 2. MATHEMATICS. *Educ:* Univ Calif, Berkeley, AB, 62; Stanford Univ, PhD(math), 68. *Prof Exp:* Sr assoc programmer, Eng & Sci Comput Lab, Int Bus Mach Corp, Calif, 66-68; assoc prof, 68-73, PROF MATH, CALIF STATE UNIV, CHICO, 85-, CHMN DEPT MATH, 85- *Res:* Fluid dynamics; numerical analysis; elliptic partial differential equations. *Mailing Add:* Dept Math Calif State Univ Chico CA 95929-0525

MCCREDIE, JOHN A, b Anahilt, Northern Ireland, Sept 8, 23; m 54; c 5. CANCER, SURGERY. *Educ:* Queen's Univ Belfast, MB, 46, MCh, 57; FRCS(E) & FRCS, 51; FRCS(C), 60. *Prof Exp:* Registr surg, Royal Victoria Hosp, Belfast, 49-56; instr, Univ Ill, Chicago, 57-58; lectr, 59-68, from assoc prof to prof, 68-86, EMER PROF SURG & RADIATION ONCOL, UNIV WESTERN ONT, 86-; RES ASSOC, ONT CANCER FOUND, 59- *Honors & Awards:* Jacksonian Prize, Royal Col Surgeons, Eng, 56. *Mem:* fel Am Col Surgeons. *Res:* Immunology of tumors. *Mailing Add:* 970 Wellington St N London ON N6A 3T2 Can

MCCREDIE, KENNETH BLAIR, hematology & oncology, leukemia research; deceased, see previous edition for last biography

MCCREERY, RICHARD LOUIS, b Los Angeles, Calif, Oct 8, 48; m 74; c 3. ANALYTICAL CHEMISTRY. *Educ:* Univ Calif, Riverside, BS, 70; Univ Kans, PhD(chem), 74. *Prof Exp:* From asst prof to assoc prof, 74-83, PROF CHEM, OHIO STATE UNIV, 83 - *Mem:* Am Chem Soc; AAAS; Am Phys Soc. *Res:* Electrochemistry; spectroscopy; reactions of electrogenerated species. *Mailing Add:* Chem Dept Ohio State Univ 120 W 18th St Columbus OH 43210

MCCREIGHT, CHARLES EDWARD, b Camden, SC, Mar 17, 13. ANATOMY. *Educ:* George Washington Univ, BS, 48, MS, 50, PhD(anat), 54. *Hon Degrees:* DSc, Lenoir Rhynea Col, 88. *Prof Exp:* Lab instr zool, George Washington Univ, 48-50, asst anat, 50-52; assoc, 52-53, from instr to assoc prof, 54-83, EMER PROF ANAT, BOWMAN GRAY SCH MED, 83- *Mem:* Fel Geront Soc; Am Asn Anatomists; Am Asn Univ Prof; Sigma Xi. *Res:* Experimental studies in growth, development and regeneration of the kidney; epidermal cytology; cytology of lung. *Mailing Add:* Dept of Anat Bowman Gray Sch of Med Winston-Salem NC 27103

MCCREIGHT, EDWARD M, US citizen. COMPUTER SCIENCES. *Educ:* Col Wooster, AB, 66; Carnegie-Mellon Univ, PhD(comput sci), 70. *Prof Exp:* Res scientist comput sci, Boeing Sci Res Labs, Wash, 69-71; MEM RES STAFF COMPUT SCI, PALO ALTO RES CTR, XEROX CORP, 71-, SR RES FEL. *Mem:* Asn Comput Mach; Inst Elec & Electronics Engrs. *Res:* Data structures; analysis of algorithms; theory of computing; computer architecture. *Mailing Add:* Adobe Syst Inc 1585 Charleston Rd Mountain View CA 94039

MCCREIGHT, LOUIS R(ALPH), b Zion, Ill, Nov 26, 22; m 49; c 2. CERAMIC ENGINEERING. *Educ:* Univ Ill, BS, 46, MS, 49. *Prof Exp:* Res asst, Manhattan Proj, 44-45 & Air Force Proj, Univ Ill, 45-49; ceramist, Knolls Atomic Power Lab, 49-55, mgr mat sci, Space Sci Lab, Missiles & Space Div, Gen Elec Co, 55-81; sr eng, Aerospace Corp, Los Angeles, 81-91; CHMN ORGANIZING BD, NAT ADVAN CERAMICS RES ASN, 88- *Concurrent Pos:* Mem panels mat adv bd, Nat Acad Sci-Nat Res Coun, 59-; mem, Mat Res Adv Comt, NASA, 59-69; mem, Gov Sci Adv Comt & Mat Panel, Pa, 60-70. *Mem:* Fel Am Ceramic Soc; Mat Res Soc; Am Inst Aeronaut & Astronaut; fel Inst Adv Eng. *Res:* Refractory materials; nuclear fuel elements; space vehicle materials; space processing; advanced manufacturing methods. *Mailing Add:* 2763 San Ramon Dr Rancho Palos Verdes CA 90274

MCCRELESS, THOMAS GRISWOLD, b San Antonio, Tex, Aug 20, 27; m 51; c 2. NUCLEAR ENGINEERING. *Educ:* US Naval Acad, BS, 51; Univ Md, MS, 65, PhD(nuclear eng), 77. *Prof Exp:* Staff engr, US AEC, 60-75; asst exec dir, US Nuclear Regulatory Comm, 80-88, sr staff engr & br chief adv comt Reactor Safeguards, 75-88; RETIRED. *Concurrent Pos:* Pres, Bd Dirs, Cyprus Cove Community Asn. *Honors & Awards:* Meritorious Serv Award, US Nuclear Regulatory Comn, 88. *Mem:* Sigma Xi; Am Nuclear Soc. *Res:* Applications of invariant imbedding techniques to classical engineering problems. *Mailing Add:* 3817 Calle de las Focas San Clemente CA 92672

MCCRIMMON, DONALD ALAN, JR, b Tampa, Fla, Feb 2, 44; m 69; c 3. RESOURCE MANAGEMENT. *Educ:* Univ SFla, BA, 64; Vanderbilt Univ, Nashville, MA, 67; NC State Univ, Raleigh, PhD(zool), 75. *Prof Exp:* Instr psychol, Univ NC, Asheville, 66-70; biologist, 75-79, sr staff scientist, Nat Audubon Soc, 79-84; exec dir, Pt Reyes Bird Observ, 84-86; assoc dir, Mt Desert Island Biol Lab, 87-91; DIR, OFF RES, OAKLAND UNIV, ROCHESTER, MICH, 91- *Concurrent Pos:* Asst prof, Cornell Univ, 79-84; fac assoc, Col Atlantic, 88-91; coop prof, Univ Maine, 90-91. *Mem:* Am Ornithologists Union. *Res:* Ecology and reproductive biology of avian species, especially colonially nesting birds; design and management of nature preserves; biostatistics. *Mailing Add:* Off Res Oakland Univ Rochester MI 48309-4401

MACCRIMMON, HUGH ROSS, b Hamilton, Ont, Oct 4, 23; m 49; c 3. FISHERIES, WILDLIFE & ENVIRONMENTAL SCIENCES. *Educ:* McMaster Univ, BA, 46; Univ Toronto, PhD(zool), 49. *Prof Exp:* Lectr zool, Univ Toronto, 46-48; fish & wildlife supvr, Ont Dept Lands & Forests, 49-57; prof life sci, 57-89, ASSOC CTR PROGS, UNIV GUELPH, 86-, EMER PROF LIFE SCI, 89- *Concurrent Pos:* Pres, Faunaquatics Can Ltd, 79-88; vis scientist, Fisheries Inst, Alvkarleo, Sweden, 83; vis prof, Univ Umea, Sweden, 83 & Univ Stirling, Scotland, 84; consult, 57-88. *Mem:* Fel Int Acad Fisheries Scientist; fel Am Inst Fisheries Res Biologists (pres, 84-86); fel Soc Antiquaries Scotland; fel Inst Fisheries Mgt Gt Brit; Can Soc Zoologists; Am Fisheries Soc; Chem Inst Can. *Res:* Fundamental limnology, fisheries science & environmental quality; appliction of these & other findings to fisheries management, aquaculture, & environmental assessment; integrated natural resource developoment of aquatic & terrestrial ecosystems. *Mailing Add:* Col Biol Sci Univ Guelph Guelph ON N1G 2W1 Can

MCCRONE, ALISTAIR WILLIAM, b Regina, Sask, Oct 7, 31; US citizen; m 58; c 3. GEOLOGY, PALEOECOLOGY. *Educ:* Univ Sask, BA, 53; Univ Nebr, MSc, 55; Univ Kans, PhD(geol), 61. *Prof Exp:* From instr to prof geol, NY Univ, 59-70, chmn dept, 66-69, assoc dean, Grad Sch Arts & Sci, 69-70; prof, Univ of the Pac, 70-74, acad vpres, 70-71, 71-74; PROF GEOL & PRES, HUMBOLDT STATE UNIV, 74- *Concurrent Pos:* Wellsite geologist, Brit Am Oil Co, Sask, 53; field geologist, Shell Oil Co, Can, 54-55, field party chief, 56-58; mem fac comt educ policy, NY Univ, 65-69, fac comt discipline, 63-64; mem, Chancellor's Comt Innovative Progs, Calif State Univ Syst, 74-76, Trustees' Task Force Off-Campus Instruct, 75-76, Presidential Search Comts, 76-80, Comn Educ Telecommun, 83-87 & Systemwide Comt Earthquake & Emergency Preparedness, chair, 85-; mem bd dirs, Redwood Empire Asn, 83-86, Am Asn Univ Administrators, 85-88, Asn Am Cols, 88-, chair, 91; chair prog comt, Western Col Asn, 83-84; mem, site vis teams, USPHS res contracts; foreign fac Fulbright coordr, NY City, 69-70; mem, bd trustees, Presbyterian Hos, Pac Med Ctr, San Francisco, 71-74; mem, Comt Environ & Energy, Am Asn State Cols & Univs, 74-81, vchair, 78-79, chair, 80-81; mem, Calif Coun Humanities, 77-82, prog comt chair, 81-82. *Mem:* Fel AAAS; fel Geol Soc Am; Am Asn Petrol Geologists; Sigma Xi; Soc Econ Paleontologists & Mineralogists. *Res:* Paleozoic and mesozoic stratigraphy; marine ecology and paleoecology; marine and estuarine sedimentation; estuarine and aquatic geochemistry; sedimentary facies analysis; author of numerous publications. *Mailing Add:* Humboldt State Univ Arcata CA 95521-4957

MCCRONE, JOHN DAVID, b Somerville, Mass, Nov 9, 34; m 57; c 2. ZOOLOGY, ADMINISTRATIVE SCIENCES. *Educ:* Univ Fla, BS, 56, PhD(biol), 61. *Prof Exp:* Asst prof biol, Fairleigh Dickinson Univ, 61-62; from asst prof to assoc prof, Fla Presby Col, 62-66; assoc prof zool, Univ Fla, 66-69; assoc dean grad sch, Univ of the Pac, 69-71; dir res, Grad Sch, 71-73, assoc vpres educ develop & res, Univ Iowa, 73-75; DEAN, SCH ARTS & SCI, WESTERN CAROLINA UNIV, 75- *Mem:* Fel AAAS; Soc Syst Zool; Soc Study Evolution; Int Soc Toxinol. *Res:* Arachnology; systematics; ecology; toxinology. *Mailing Add:* Grad Sch Dean's Off Pittsburgh State Univ Pittsburgh KS 66762

MACCRONE, ROBERT K, b Johannesburg, SAfrica; m; c 3. PHYSICS, CERAMICS. *Educ:* Univ Witwatersrand, BSc, 54, Hons, 55, MSc, 56; Oxford Univ, DPhil(physics), 59. *Prof Exp:* Fel metall eng, Univ Pa, 59-62, asst prof, 62-67; PROF MAT SCI, RENSSELAER POLYTECH INST, 67- *Mem:* Am Ceramic Soc; Am Physical Soc; Bot Soc SAfrica. *Res:* Electric and magnetic properties of oxides, glasses and polymers; paramagnetic resonance; protective coatings; ion implantation; super conductivity. *Mailing Add:* Rensselaer Polytech Inst 112 MRC Troy NY 12181

MCCRONE, WALTER C, b Wilmington, Del, June 9, 16; m 57. CHEMICAL MICROSCOPY. *Educ:* Cornell Univ, BChem, 38, PhD(chem micros), 42. *Prof Exp:* Chem microscopist, Off Sci Res & Develop Proj, Cornell Univ, 42-44; chem microscopist, Armour Res Found, Ill Inst Technol, 44- 45, supvr anal chem, 45-46, asst chmn chem & chem eng, 46-52, sr scientist, 52-56; founding mem, McCrone Assocs, Inc, 56-79, chmn bd, 67-81, PRES, MCCRONE RES INST, 61- *Concurrent Pos:* Chmn bd, ADA S McKinley community serv & Vandercook Col; adj prof, NY Univ, Cornell Univ, Ill Inst Technol & Univ Ill. *Honors & Awards:* Benedetti-Pichler Award, 71; Ernst Abbe Award, 77. 71; Anachem Award, 81; Franklin Inst Award, 82; Selikoff Award, 90. *Mem:* Am Chem Soc; Am Phys Soc; Am Soc Test & Mat; Am Acad Forensic Sci; Royal Micros Soc; Can Micros Soc; Australian Micros Soc; Sigma Xi. *Res:* Crystallography; chemical microscopy; polymorphism; crystal growth; correlation of solid state properties and performance; ultramicroanalysis; authentication of paintings. *Mailing Add:* McCrone Res Inst Inc 2820 S Michigan Ave Chicago IL 60616-3292

MCCROREY, HENRY LAWRENCE, b Philadelphia, Pa, Mar 13, 27; m 57; c 4. PHYSIOLOGY. *Educ:* Univ Mich, BS, 49, MS, 50; Univ Ill, MS, 58, PhD(physiol), 63. *Prof Exp:* Res assoc pharmacol, Sharp & Dohme, Pa, 51-55; asst physiol, Col Med, Univ Ill, 56-61, from instr to asst prof, 61-66; from asst prof to assoc prof, 66-73, PROF PHYSIOL & BIOPHYS, COL MED, UNIV VT, 73- *Mem:* Am Physiol Soc. *Res:* Muscle contraction; experimental hypertension; renal physiology. *Mailing Add:* Dept Physiol Sch Allied Health Univ Vt 85 S Prospect St Burlington VT 05405

MCCRORY, ROBERT LEE, JR, b Lawton, Okla, Apr 30, 46; m 69; c 3. PLASMA PHYSICS, HYDRODYNAMICS. *Educ:* Mass Inst Technol, ScB, 68, PhD(plasma physics), 73. *Prof Exp:* Scientist geophys, Pan Am Petrol Res, Standard Oil Ind, 68-73; staff scientist theoret physics, theoret div, Los Alamos Sci Lab, Univ Calif, 73-76; group leader theory & comput group, Lab Laser Energetics, 76-78, assoc prof phys & mech eng, 80-84, DIR THEORET DIV, UNIV ROCHESTER, 78-, LAB DIR, 83-, PROF MECH ENG, UNIV ROCHESTER, 84- *Concurrent Pos:* Sr sci programmer, Res Lab Electronics, Mass Inst Technol, 68-73, res assoc, Dept Nuclear Eng/Dept Aerodyn & Astrophys, 72-73; US deleg, Conf Plasma Physics & Controlled Fusion Res, Int Atomic Energy Agency, Vienna, Austria, 74, 84 & 86; vis staff mem, theoret div, Los Alamos Sci Lab, Univ Calif, 76-; consult, Cambridge Hydrodynamics, Inc, Boston; exec comt & trustee, Consortium Sci Comput, Princeton, NJ. *Mem:* Fel Am Phys Soc; Sigma Xi. *Res:* Symmetry-stability studies of laser induced implosions; plasma physics/hydrodynamics of laser-initiated fusion; microinstabilities of magnetically confined plasma; weapons physics; laser-plasma interaction studies; radiation hydrodynamics. *Mailing Add:* 93 Kirklees Rd Pittsford NY 14534

MCCRORY, WALLACE WILLARD, b Racine, Wis, Jan 19, 20; m 43; c 3. PEDIATRICS. *Educ:* Univ Wis, BS, 41, MD, 44. *Prof Exp:* Instr path, Med Sch, Univ Wis, 42-43; from asst prof to assoc prof, Sch Med, Univ Pa, 53-58; prof & head dept, Col Med, Univ Iowa, 58-61; PROF PEDIAT & CHMN DEPT, MED CTR, CORNELL UNIV, 61-; PEDIATRICIAN-IN-CHIEF, NEW YORK HOSP, 61- *Concurrent Pos:* Ledyard fel pediat, Cornell Univ, 49-50; dir, Clin Chem Lab, Children's Hosp, Philadelphia, 53-58. *Mem:* AAAS; Am Pediat Soc; NY Acad Sci; fel Royal Soc Med. *Res:* Renal disease and clinical biochemistry. *Mailing Add:* 525 E 68th St New York NY 10021

MCCROSKEY, ROBERT LEE, b Richwood, WVa, Feb 22, 24; m 50; c 4. AUDIOLOGY, SPEECH PATHOLOGY. *Educ:* Ohio State Univ, BS, 48, MA, 52, PhD(speech sci), 56. *Prof Exp:* Res assoc speech sci, Ohio State Res Found, Pensacola, Fla, 55-56; prof speech path, Miss Southern Col, 56-59; prof speech & hearing, Emory Univ, 59-67; prof commun disorders & sci, Wichita State Univ, 67-87, dean, Col Educ, 87-88; RETIRED. *Concurrent Pos:* Dir, Atlanta Speech Sch, Ga, 59-67; Am Speech & Hearing Asn int travel grant, 63 & 66; consult, Am Hearing Soc, 64-67, Div Voc Rehab, Ga, 66, Bur Educ of Handicapped, US Off Educ, 71- & Glenrose Hosp, Edmonton, Ont, 73-74; dir prof servs, Inst Logopedics, 68-70; reviewer div int activities, Dept Health, Educ & Welfare, 71-75, consult, Indust Noise, 75- *Mem:* Speech Commun Asn; fel Am Speech & Hearing Asn; Am Auditory Soc. *Res:* Early diagnosis of hearing impairments; home training programs for deaf; auditory temporal processing for children with learning disabilities; effects of speech expansion upon comprehension; auditory localization. *Mailing Add:* 7413 Patchent Lane Wichita KS 67206

MCCROSKY, RICHARD EUGENE, b Akron, Ohio, Apr 28, 24; m 52; c 4. ASTRONOMY. *Educ:* Harvard Univ, BS, 52, PhD(astron), 56. *Prof Exp:* Sr res asst, 56, LECTR ASTRON, HARVARD UNIV, 56-, RES ASSOC, SMITHSONIAN ASTRON OBSERV, 57- *Concurrent Pos:* Astronr, Smithsonian Astrophys Observ, 57-58 & 60-, consult, 58-59; scientist chg, Meteorite Photog & Recovery Proj, 62-; mem, Int Astron Union. *Mem:* Fel AAAS; Meteoritical Soc; Am Astron Soc; Am Geophys Union; Sigma Xi. *Res:* Physics of meteors; meteor statistics and orbits; optical instrumentation; meteor photography. *Mailing Add:* Harvard Observ 60 Garden St Cambridge MA 02138

MCCROSSON, F JOSEPH, b Brooklyn, NY, June 27, 40; m 65; c 3. REACTOR PHYSICS. *Educ:* Fairfield Univ, BS, 62; Va Poly Inst & State Univ, MS, 65, PhD(physics), 68. *Prof Exp:* Proj mgr energy technol, SC Energy Res Inst, 78-82; physicist, Savannah River Lab, E4 DuPont de Nemours & Co, 65-67, res physicist, 68-72, staff physicist, 72-74, res supvr reactor physics, 74-78 & 82-88; MGR, NEW PROD REACTOR DEVELOP, WESTINGHOUSE SAVANNAH RIVER CO, 88- *Mem:* Sigma Xi; Am Nuclear Soc; AAAS. *Res:* Neutron cross-sections; shielding; charge design; accident analysis; reactor materials research; thermal hydraulics. *Mailing Add:* 159 Gatewood Dr Aiken SC 29801

MCCRUMM, J(OHN) D(OENCH), b Colorado Springs, Colo, Apr 17, 12; m 38, 69; c 2. ELECTRICAL ENGINEERING. *Educ:* Univ Colo, BS, 33, MS, 34. *Prof Exp:* Test engr, Gen Elec Co, NY, 34-35, res engr, Pa, 38; from instr to asst prof elec eng, Swarthmore Col, 35-44; res engr, Douglas Aircraft Co,

Calif, 44-46; from assoc prof elec eng to prof & chmn dept eng, 46-59, Eavenson prof, 59-78, chmn dept elec eng, 70-78, EMER PROF ENG, SWARTHMORE COL, 78- *Concurrent Pos:* Consult to various indust concerns, 46-; asst to vpres, Burroughs Corp, 57-58; NSF fel, Athens, Greece, 65-66. *Mem:* Sigma Xi; Inst Elec & Electronics Engrs. *Res:* Control theory; systems engineering; research management. *Mailing Add:* 606 Ogden Ave Swarthmore PA 19081

MCCUBBIN, DONALD GENE, b Glencoe, Okla, Oct 24, 30; m 62; c 2. PETROLEUM GEOLOGY, SEDIMENTOLOGY. *Educ:* Okla State Univ, BS, 52; Harvard Univ, MA, 55, PhD(geol), 61. *Prof Exp:* Res geologist, 57-65, adv res geologist, 65-70, SR RES GEOLOGIST & ADVAN SR GEOLOGIST, DENVER RES CTR, MARATHON OIL CO, 70- *Mem:* Geol Soc Am; Am Asn Petrol Geol; Soc Econ Paleont & Mineral. *Res:* Sandstone petrology and facies; sedimentary structures; nearshore sediments and processes; sandstone petroleum reservoirs; diagenesis of clay sediments; hydrocarbon generation and migration; thermal history of sedimentary basins. *Mailing Add:* Petrol Technol Ctr Marathon Oil Co Box 269 Littleton CO 80160

MCCUBBIN, THOMAS KING, JR, b Baltimore, Md, June 1, 25; m 51; c 3. PHYSICS. *Educ:* Univ Louisville, BEE, 46; Johns Hopkins Univ, PhD(physics), 51. *Prof Exp:* Fel physics, Johns Hopkins Univ, 51-53; mem res staff phys chem, Mass Inst Technol, 53-57; from asst prof to assoc prof, 57-64, PROF PHYSICS, PA STATE UNIV, 64- *Mem:* Fel Optical Soc Am; fel Am Phys Soc. *Res:* Optics and spectroscopy; molecular spectra. *Mailing Add:* Dept Physics Davey Lab Pa State Univ University Park PA 16802

MCCUE, CAROLYN M, b Richmond, Va, June 26, 16; m 41; c 2. PEDIATRICS. *Educ:* Stanford Univ, BA, 37; Med Col Va, MD, 41; Am Bd Pediat, dipl, 48, cert pediat cardiol, 59. *Prof Exp:* Instr, Med Col Va, 46-49, assoc, 49-52, from asst prof to prof pediat, 52-90, interim chmn dept, 58-61; RETIRED. *Concurrent Pos:* Clin dir, Richmond Rheumatic Fever & Congenital Heart Clins, 47-; mem staff, Dept Pediat, VA Commonwealth Univ. *Mem:* Am Heart Asn; fel Am Col Physicians; fel Am Acad Pediat; fel Am Col Cardiol. *Res:* Pediatric cardiology. *Mailing Add:* 12 Huntly Rd Richmond VA 23296

MCCUE, EDMUND BRADLEY, b Worcester, Mass, Mar 8, 29. MATHEMATICS. *Educ:* Union Col, AB, 50; Univ Mich, MS, 51; Carnegie-Mellon Univ, PhD(math), 60. *Prof Exp:* Mathematician, US Dept Defense, 51-54; asst prof math, Ohio Univ, 58-63; prof statist, Inter-Am Statist Inst, 63-64; ASSOC PROF MATH & STATIST, AM UNIV, 64- *Mem:* Inst Math Statist; Am Statist Asn; Math Asn Am. *Res:* Mathematical statistics. *Mailing Add:* c/o Henriques-Carlsen Bregnevej 19 Gentofte 2840 Denmark

MCCUE, JOHN FRANCIS, b Milford, NH, Nov 22, 33; m 62; c 4. PARASITOLOGY. *Educ:* St John's Univ, Minn, BS, 60; Univ Notre Dame, MS, 62, PhD(biol), 64. *Prof Exp:* NIH fel, Sch Med, Nat Univ Mex, 64-65; res assoc parasitol, Sch Vet Med, Auburn Univ, 65-67; from asst prof to assoc prof, 67-77, PROF BIOL SCI, ST CLOUD STATE UNIV, 77- *Mem:* Am Soc Parasitologists; Sigma Xi. *Res:* Parasitic immunology; parasite life cycles; host-parasite relationships. *Mailing Add:* 344 Riverside Dr NE St Cloud St Univ St Cloud MN 56301

MCCUE, ROBERT OWEN, b Kearney, Nebr, May 20, 47; m 70; c 3. BIOLOGY. *Prof Exp:* Asst prof, Univ Nev, 77-78; PROF BIOL, WAYNE STATE COL, 78- *Mem:* Nat Asn Biol Teachers. *Res:* Embryonic development and differentiation and regeneration; viral carcinogenesis. *Mailing Add:* Math/Sci Div Wayne State Col Wayne NE 68787

MCCUEN, ROBERT WILLIAM, b Darby, Pa, Apr 14, 40; m 65; c 2. MICROBIOLOGY, GENETIC TOXICOLOGY. *Educ:* Drexel Univ, BS, 63; Temple Univ, MS, 66; Univ Mass, PhD(microbiol), 71. *Prof Exp:* Res fel, Sloan-Kettering Inst Cancer Res, 71-73; res scientist, 73-80, ASSOC SR SCIENTIST BIOL EFFECTS OF SMOKE, PHILIP MORRIS RES CTR, 80- *Mem:* Am Asn Cancer Res; Am Soc Microbiol; Environ Mutagen Soc; Genetic Toxicol Asn. *Res:* In vitro short term bioassays as predictors of chemical carcinogens. *Mailing Add:* PO Box 26583 Richmond VA 23261

MCCUISTION, WILLIS LLOYD, b Fowler, Colo, May 28, 37; m 59; c 2. GENETICS, PLANT BREEDING. *Educ:* Colo State Univ, BS, 59. *Prof Exp:* asst plant breeding & genetics, Okla State Univ, 62-67; plant breeder, Rockefeller Found, India, 67-68; dir & plant breeder, Int Maize & Wheat Improv Ctr, Tunisia, 68-71 & Algeria, 71-75; ASSOC PROF, CROP SCI DEPT, ORE STATE UNIV, 75- *Concurrent Pos:* Rockefeller Found fel, All-India Mex Wheat Prog & Tunisia Wheat Prog, 67-69. *Mem:* Am Soc Agron; Crop Sci Soc Am; Am Inst Biol Sci; Sigma Xi. *Res:* Increasing agricultural production in general, especially in developing countries; training local scientists in applied scientific methods. *Mailing Add:* NARP Box 3567 Las Cruces NM 88003-3567

MCCULLA, WILLIAM HARVEY, b Birmingham, Ala, Sept 19, 41; m 63; c 2. CHEMISTRY. *Educ:* Auburn Univ, BS, 65, MS, 68; Univ Tenn, PhD(chem, physics chem), 81. *Prof Exp:* Res assoc process chem group, Union Carbide Corp, 67-70, res chemist, 70-74, res chemist advan isotope separation group, 74-76, prin investr Laser Isotope Separation Exp Advan Isotope Chem Group, 76-80, group leader, Laser Spectros Group, Nuclear Div, 80-85; proj mgr, 85-88, DEP GROUP LEADER, PROCESS DEVELOP GROUP, LOS ALAMOS NAT LAB, 88- *Mem:* AAAS; Optical Soc Am. *Res:* Laser isotope separation including uranium enrichment, laser raman spectroscopy and laser induced chemistry. *Mailing Add:* Los Alamos Nat Lab CLS3 MSJ565 Los Alamos NM 87545

MCCULLEN, JOHN DOUGLAS, b Sioux Falls, SDak, Oct 4, 32; m 54; c 3. NUCLEAR PHYSICS. *Educ:* Univ Colo, BA, 54, MS, 58, PhD(nuclear physics & spectros), 60. *Prof Exp:* Asst physics, Univ Colo, 56-60; res assoc, Princeton Univ, 60-62, asst prof, 62-65; assoc prof, 65-67, PROF PHYSICS, UNIV ARIZ, 67- *Concurrent Pos:* Sloane Found fel, 66-68. *Mem:* Am Phys Soc. *Res:* Nuclear spectroscopy and structure theory; atomic and polarized beams. *Mailing Add:* Dept of Physics Univ of Ariz Tucson AZ 85721

MCCULLOCH, ARCHIBALD WILSON, b Troon, Scotland, Oct 31, 40; Can citizen; m 71; c 2. ORGANIC CHEMISTRY. *Educ:* Univ Glasgow, BSc, 62, PhD(chem), 65. *Prof Exp:* Fel chem, Inst Marine Biosci, Nat Res Coun Can, 65-66, asst res officer, 66-70, assoc res officer, 71-79, SR RES OFFICER CHEM, INST MARINE BIOSCI, NAT RES COUN CAN, HALIFAX, 80-, HEAD, BIOL CHEM SECT, 87- *Mem:* Fel Chem Inst Can; Royal Soc Chem. *Res:* Biological chemistry; natural products chemistry; nuclear magnetic resonance; marine chemistry. *Mailing Add:* Inst Marine Biosci Nat Res Coun 1411 Oxford St Halifax NS B3H 3Z1 Can

MCCULLOCH, ERNEST ARMSTRONG, b Toronto, Ont, Apr 27, 26; m 53; c 5. MEDICINE. *Educ:* Univ Toronto, MD, 48; FRCP(C), 54. *Prof Exp:* From jr intern to sr intern, Toronto Gen Hosp, 49-52; asst resident, Sunnybrook Hosp, 52-53; clin teacher med, Univ Toronto, 54-60, from asst prof to prof med biophys, 59-88, assoc med, 60-67, from asst prof to prof, Inst Med Sci, 68-88, univ prof, 88-91, EMER UNIV PROF, UNIV TORONTO, 91- *Concurrent Pos:* Ellen Mickle fel, Lister Inst, Eng, 48-49; Nat Res Coun Can res fel path, Univ Toronto, 50-51; Nat Cancer Inst Can fel, 54-57; asst physician, Toronto Gen Hosp, 54-60, physician, 60-67; head subdiv hemat, Div Biol Res, Int Cancer Inst, 57-67; mem grants comt microbiol & path, Med Res Coun Can, 66-67, immunol & transplantation, 67-69; mem panel B, Grants Comt, Nat Cancer Inst Can, 70-74; dir Inst Med Sci, 75-79; asst dean, Sch Grad Studies, Div Biol Res, Ont Cancer Inst, 79-82, head div, 82-89, lead, Div Cellular & Molecular Biol, 89-91, emer sr scientist, 91-; ed, J Cell Physiol, 69-; sr physician, Princess Margaret Hosp, 70- *Honors & Awards:* Starr Medal, Univ Toronto, 57, Goldie Prize in Med, 64; Annael Gairdner Award, 69; Silver Jubillee Medal, 77; Stratton Lectr, Am Soc Hemat, 82; Officer, Order of Can, 88; Eadie Medal, Royal Soc Can, 91. *Mem:* Am Asn Cancer Res; Am Soc Exp Path; Soc Exp Biol & Med; Can Soc Cell Biol; Royal Soc Can. *Res:* Hematology; stem cell functions, especially physiologic and genetic control mechanisms; leukemia research. *Mailing Add:* Depts Med & Med Biophys Univ Toronto Ont Cancer Inst 500 Sherbourne St Toronto ON M5S 1A8 Can

MCCULLOCH, JOSEPH HOWARD, b Durham, NC, Feb 17, 46; m 70; c 2. BIOLOGY, BOTANY. *Educ:* Eastern Mich Univ, BS, 68, MS, 71; Mich State Univ, PhD(bot), 77. *Prof Exp:* Instr, Eastern Mich Univ, 70-73 & Mich State Univ, 73-74; PROF BIOL, NORMANDALE COMMUNITY COL, 74- *Mem:* Am Bot Soc; Am Fern Soc; Am Mus Natural Hist. *Res:* Cheilanthoid fern rhizome anatomy; ontogeny of stelar anatomy in the pteridophyta; educational biology. *Mailing Add:* Dept Bot Normandale Community Col Bloomington MN 55431

MCCULLOH, THANE H, b Glendale, Calif, July 25, 26; m 49, 63; c 5. PETROLEUM GEOLOGY. *Educ:* Pomona Col, AB, 49; Univ Calif, Los Angeles, PhD(geol), 52. *Prof Exp:* Res assoc geol, Univ Calif, Los Angeles, 52-53; asst prof, Calif Inst Technol, 53-55; from assoc prof to prof, Univ Calif, Riverside, 55-64; geologist, DC, 64-72, chief, Off Energy Res, 72-73, res geologist, US Geol Surv, 73-82; sr res geologist, 82-83, mgr geochem explor res, Mobil Res & Develop, 85-87; MGR STRATIG SERV, MOBIL EXPL & PROD SERV INC, 87- *Concurrent Pos:* NSF fel, 52-53; Guggenheim fel, 60; affil prof oceanog, Univ Wash, 73-82. *Honors & Awards:* PL 313 Position Award, US Geol Surv, 74. *Mem:* Geol Soc Am; Soc Explor Geophys; Am Asn Petrol Geol; Norweg Geol Asn; Soc Econ Petrologists & Mineralogists. *Res:* Areal geology of the central Mojave Desert; igneous and metamorphic petrology; gravimetry; structural geology of the Los Angeles Basin; petrophysics; petroleum exploration research and sedimentary basin resource studies; thermal history of sedimentary rocks; geochemistry. *Mailing Add:* 7136 Aberdeen Ave Dallas TX 75230

MCCULLOUGH, BENJAMIN FRANKLIN, b Austin, Tex, Mar 25, 34; m 56; c 5. CIVIL ENGINEERING. *Educ:* Univ Tex, BS, 57, MS, 62; Univ Calif, PhD(transp eng), 69. *Prof Exp:* Engr in training, US Bur Reclamation, 56; testing engr, Convair Aircraft Co, 57; supv design & res engr, Tex Hwy Dept, 57-66; sr engr, Mat Res & Develop, Inc, Calif, 66-68; asst prof civil eng, Ctr Hwy Res, Univ Tex, Austin, 68-71; design engr, Dallas-Ft Worth Regional Airport, 71-73; PROF CIVIL ENG, UNIV TEX, AUSTIN, 76-, DIR, CTR TRANSP RES, 80- *Concurrent Pos:* Chmn comt A2-B01, Transp Res Bd, Nat Acad Sci-Nat Res Coun, 66-77; proj engr, US Air Force, Palmdale AFB, Calif, 69-70; consult, Bd Consult, Dept Transp, 76-, Fed Aviation Admin, 77- & Area 5-E Zero Maintenance, Transp Res Bd, Fed Hwy Admin, Washington, DC, 77- *Mem:* Am Soc Civil Engrs; Am Concrete Inst. *Res:* Highway and airport pavement design; pavement rehabilitation; design of military airport in Saudi Arabia. *Mailing Add:* Univ Tex 913 Castlenridge Rd Austin TX 78746

MCCULLOUGH, C BRUCE, b Kenton, Ohio, May 3, 44. VETERINARY PATHOLOGY. *Educ:* Ohio State Univ, DVM, 69. *Prof Exp:* MGR TOXICOL SECT, SHELL DEVELOP CO, 79- *Mem:* Int Acad Path; Am Col Vet Pathologists; Am Asn Anatomists. *Mailing Add:* Drug Safety & Metab Dept Schereng Plough Res PO Box 32 Lafayette NJ 07848

MCCULLOUGH, DALE RICHARD, b Sioux Falls, SDak, Dec 5, 33; m 58, 74; c 5. WILDLIFE MANAGEMENT, ECOLOGY. *Educ:* SDak State Univ, BS, 57; Ore State Univ, MS, 60; Univ Calif, Berkeley, PhD(zool), 66. *Prof Exp:* Instr zool, Univ Calif, Berkeley, 65-66; asst prof wildlife mgt, Univ Mich, Ann Arbor, 66-69, assoc prof wildlife & fisheries, 69-74, chmn resource ecol prog, 71-74, prof natural resources, 74-80; MEM FAC FORESTRY & RES MGT, UNIV CALIF, BERKELEY, 80- *Mem:* Wildlife Soc; Ecol Soc Am; Am Soc Mammalogists. *Res:* Ecology and population dynamics of large herbivores; ecosystem processes. *Mailing Add:* Dept Forestry & Res Mgt Univ Calif Eight Mulford Hall Berkeley CA 94720

MCCULLOUGH, DARRYL J, b Columbus, Ohio, Dec 20, 51; m 76. HOMEOMORPHISMS OF THREE-MANIFOLDS. *Educ:* Ohio State Univ, BA, 72; Univ Mich, MA, 74, PhD(math), 78. *Prof Exp:* From asst prof to assoc prof, 78-88, PROF MATH, UNIV OKLA, 88- *Concurrent Pos:* Mem, Math Sci Res Inst, Berkeley, Calif, 84. *Mem:* Am Math Soc. *Res:* Low-dimensional geometric topology; automorphisms of manifolds and cell complexes; connections between topology and group theory; Fuchsian and Kleinian groups. *Mailing Add:* Dept Math Univ Okla Norman OK 73019

MCCULLOUGH, EDGAR JOSEPH, JR, b Charleston, WVa, Nov 29, 31; c 2. STRUCTURAL GEOLOGY. *Educ:* Univ WVa, AB, 53, MS, 55; Univ Ariz, PhD(geol), 63. *Prof Exp:* From instr to assoc prof geol, 58-69, prof geosci & head dept, 70-84, DEAN FAC SCI, UNIV ARIZ, 84- *Concurrent Pos:* Acting asst dir geol surv bur & acting state geologist, Ariz Bur Geol & Mineral Technol, 77-79, dir, 87. *Mem:* AAAS; Nat Asn Geol Teachers; Geol Soc Am; Sigma Xi. *Res:* Geologic hazards; science education. *Mailing Add:* Univ Ariz 336 Calle De Madrid Tucson AZ 85711

MCCULLOUGH, EDWIN CHARLES, b New York, NY, June 2, 42; m 70; c 2. MEDICAL PHYSICS. *Educ:* State Univ NY Stony Brook, BS, 64; Univ Md, MS, 67; Univ Wis-Madison, PhD(radiol physics), 71. *Prof Exp:* Scientist neutron physics, Med Res Coun Cyclotron Unit, Hammersmith Hosp, 71; res scientist radiation physics, Univ Wis-Madison, 71-73; STAFF PHYSICIST, MAYO CLIN, 73- *Mem:* Am Asn Physicists Med; Radiol Soc NAm; Sigma Xi. *Res:* Radiation therapy dosimetry and treatment planning; diagnostic radiology imaging; computer assisted tomography performance evaluation and quality assurance. *Mailing Add:* Mayo Clin 200 First St SW Rochester MN 55901

MCCULLOUGH, ELIZABETH ANN, b Spartanburg, SC, Feb 14, 52; div. TEXTILE SCIENCE. *Educ:* Ohio State Univ, BS, 74; Univ Tenn, Knoxville, MS, 75, PhD(textiles), 78. *Prof Exp:* Grad asst clothing & textiles, Univ Tenn, Knoxville, 74-78; from asst prof to assoc prof, 78-87, PROF TEXTILE SCI, KANS STATE UNIV, 87- *Concurrent Pos:* Assoc dir, Inst Environ Res, Kans State Univ, 85- *Honors & Awards:* Ralph G Nevins Res Award, Am Soc Heating, Refrigerating & Air-Conditioning Engrs, 85; Manmade Fiber Res Award, 90. *Mem:* Am Soc Testing & Mat; Asn Col Prof Textiles & Clothing; Am Asn Textile Chemists & Clorists; Indust Fabrics Asn Int; Asn Nonwoven Fabrics Indust. *Res:* Measuring the heat transfer characteristics of fabrics using a guarded hot plate and of clothing systems using an electrically heated mannequin; factors which affect the heat exchange permitted by clothing between a person and the environment. *Mailing Add:* Kans State Univ 219 Justin Hall Manhattan KS 66506

MCCULLOUGH, HERBERT ALFRED, biology; deceased, see previous edition for last biography

MCCULLOUGH, JACK DENNIS, b San Antonio, Tex, Aug 8, 31; m 53; c 2. LIMNOLOGY. *Educ:* Univ Tex, BS, 55; Stephen F Austin State Univ, MA, 62; Tex A&M Univ, PhD(biol), 70. *Prof Exp:* Teacher pub schs, Tex, 55-63; instr biol, Tex A&M Univ, 63-64; assoc prof, 64-77, PROF BIOL, STEPHEN F AUSTIN STATE UNIV, 77- *Concurrent Pos:* Researcher, US Army Corps Engrs, 71-73 & 75; mem, Tex Gov 2000 planning comt; mem eval panel, NSF; pres, Tex Acad Sci. *Mem:* Sigma Xi; Am Soc Limnol & Oceanog; Int Soc Limnol. *Res:* Primary productivity in aquatic environments. *Mailing Add:* Dept Biol Stephen F Austin State Univ Nacogdoches TX 75961

MCCULLOUGH, JAMES DOUGLAS, JR, b Long Beach, Calif, Nov 28, 38; m 75; c 1. ORGANIC CHEMISTRY, PHYSICAL CHEMISTRY. *Educ:* Univ Calif, Los Angeles, BS, 63; San Diego State Col, MS, 65; Univ Ill, Urbana, PhD(chem), 70. *Prof Exp:* STAFF RES CHEMIST, WESTHOLLOW RES CTR, SHELL DEVELOP CO, HOUSTON, 76- *Res:* Plastics technology; polymer science; petrochemical additives. *Mailing Add:* Shell Develop Co Westhollow Res Ctr PO Box 1380 Houston TX 77251-1380

MCCULLOUGH, JOHN JAMES, b Belfast, Northern Ireland, Sept 27, 37; m 63. ORGANIC CHEMISTRY. *Educ:* Queens Univ, Belfast, BSc, 59, PhD(chem), 62. *Prof Exp:* Dept Sci & Indust Res fel, 62-63; res assoc, Univ Wis, 63-65; from asst prof to assoc prof, 65-75, PROF CHEM, MCMASTER UNIV, 75- *Mem:* Am Chem Soc; Chem Inst Can; Royal Soc Chem; Sigma Xi. *Res:* Stereochemistry; photochemistry; physical organic chemistry. *Mailing Add:* 59 Wellesley Ave Belfast BT9-6DG Northern Ireland

MCCULLOUGH, JOHN JEFFREY, b Boston Mass, 38. BLOOD BANKING. *Educ:* Northwestern Univ, BA, 59; Ohio State Univ, MD, 63. *Prof Exp:* DIR BLOOD BANK & PROF IMMUNOL HEMAT, DEPT LAB MED & PATH, UNIV MINN, 68- *Res:* Blood transfusion and transplantation. *Mailing Add:* Dept Lab Med Univ Minn Box 198 Mayo Mem Bldg 420 Delaware St SE Minneapolis MN 55455

MCCULLOUGH, JOHN MARTIN, b Chicago, Ill, Mar 9, 40; m 71; c 3. BEHAVIOR ETHOLOGY, GENETICS. *Educ:* Pa State Univ, BA, 62; Univ Ill, PhD(anthrop), 72. *Prof Exp:* Asst prof, 69-75, chmn dept, 78-84, ASSOC PROF ANTHROP, UNIV UTAH, 75- *Concurrent Pos:* Forensic anthropologist, Off Med Examr, Utah State Bd Health, 69-; vis fel, human genetics, Univ Newcastle-upon-Tyne, UK, 84-85. *Mem:* AAAS; Am Asn Phys Anthrop; Am Anthrop Asn; Brit Soc Study Human Biol; Int Asn Human Biol. *Res:* Human biology; adaptability; ecological genetics; ecological demography; Latin America, especially Mexico, Yucatan and Europe. *Mailing Add:* Dept Anthrop Univ Utah Salt Lake City UT 84112

MCCULLOUGH, JOHN PRICE, b Dallas, Tex, May 10, 25; m 46; c 3. PHYSICAL CHEMISTRY. *Educ:* Univ Okla, BS, 45; Ore State Col, MS, 48, PhD(chem), 49. *Prof Exp:* Phys chemist, Thermodyn Lab, Bartlesville Petrol Res Ctr, US Bur Mines, Okla, 49-56, chief, 56-63; mgr cent res div, Mobil Oil Corp, 63-69, appl res & develop, Mobil Res & Develop Corp, 69-71, gen mgr, res & develop, Mobil Chem Co, 71-78; VPRES, ENVIRON AFFAIRS & TOXICOL, MOBIL RES & DEVELOP CORP, 78- *Concurrent Pos:* Adj Prof, Okla State Univ, 61-63; dir, Mobil Solar Energy Co, 74-86 & Chem Indust Inst Toxicol, 79-, Int Petrol Indust Environ Conserv Asn, 81-, HEICO Corp, 87-; chmn, 85-88, World Environ Ctr, 86-; mem adv bd, Georgetown Univ, Inst Health Policy Anal, 86- *Honors & Awards:* Distinguished Serv Award, US Dept Interior, 62; Award in Petrol Chem, Am Chem Soc, 63, Leo Friend Award, 77. *Mem:* AAAS; Am Chem Soc. *Res:* Thermodynamics and molecular structure of hydrocarbons and related substances; research administration. *Mailing Add:* 30 Boudenot St PO Box 1031 Princeton NJ 08540

MCCULLOUGH, MARSHALL EDWARD, b Wick, WVa, July 24, 24. DAIRY NUTRITION. *Educ:* Berea Col, BS, 49; Univ Ky, MS, 51. *Prof Exp:* Asst dairy, Univ Ky, 50-51; asst dairy nutritionist, Agr Exp Sta, Univ Ga, 51-59, assoc nutritionist, 59-84, prof animal nutrit, 68-84, head dept animal sci, 71-84; RETIRED. *Concurrent Pos:* Consult nutritionist, 84- *Honors & Awards:* Zur Craine Medal; George Raithby Lectr, Univ Guelph, 87. *Mem:* Fel AAAS; Soc Range Mgt; Am Soc Animal Sci; Am Dairy Sci Asn; Am Inst Nutrit. *Res:* Plant and animal complex in grassland utilization; calf nutrition; systems for forage evaluation; silage production and evaluation optimum ruminant rations. *Mailing Add:* PO Box 672 Experiment GA 30212

MCCULLOUGH, RICHARD DONALD, b Los Angeles, Calif, Mar 8, 41; m 64; c 2. PSYCHOPHYSIOLOGY. *Educ:* Loyola Univ, BS, 63; Purdue Univ, MS, 69; US Int Univ, PhD(human behav & psychophysiol), 75. *Prof Exp:* PROF HUMAN PHYSIOL & CELLULAR & MOLECULAR BIOL, SADDLEBACK COL, 71-, CHMN DEPTS BIOL & MARINE SCI, 71- *Mem:* AAAS; Int Orgn Psychophysiol. *Res:* Use of both TEM and SEM in the study of animal tissue. *Mailing Add:* Dept Biol Sci Saddelback Col 28000 Marguerite Pkwy Mission Viejo CA 92692

MCCULLOUGH, ROBERT WILLIAM, b Washington, DC, Oct 16, 47; m 67; c 1. MECHANICAL ENGINEERING, PHYSICS. *Educ:* Occidental Col, BA, 70; Columbia Univ, BS, 70; Stanford Univ, MS, 71, PhD(mech eng), 75. *Prof Exp:* Res fel high temperature chem kinetics, High Temperature Gas Dynamics Lab, 70-75; proj scientist solar energy, Linde Res, Union Carbide Corp, 75-78; assoc consult radiation transp, 78-81, VPRES AEROPHYSICS, AERONAUT RES ASSOC PRINCETON, 81- *Mem:* Am Soc Mech Engrs; Am Inst Aeronaut & Astronaut; Int Solar Energy Soc; Sigma Xi; Optical Soc Am. *Res:* Thermosciences; heat transfer; thermodynamics; fluid mechanics; molecular physics; chemical kinetics; radiation transport. *Mailing Add:* 22 Berkshire Dr Princeton NJ 08550

MCCULLOUGH, ROY LYNN, b Hillsboro, Tex, Mar 20, 34; m 58; c 3. COMPOSITE MATERIALS. *Educ:* Baylor Univ, BS, 55; Univ NMex, PhD(chem), 60. *Prof Exp:* Mem staff, Los Alamos Sci Lab, 58-59, Chemstrand Res Ctr, NC, 60-69 & Polymer Sci Lab, Boeing Sci Res Lab, Wash, 69-71; assoc dir, Ctr Composite Mat, 77-90, PROF CHEM ENG, UNIV DEL, 71-, DIR, CTR COMPOSITE MAT, 90- *Mem:* Am Chem Soc; Am Phys Soc; Am Inst chem Engrs. *Res:* Structure-property relationships; composite materials. *Mailing Add:* Dept of Chem Eng Univ of Del Newark DE 19711

MCCULLOUGH, THOMAS F, b Los Angeles, Calif, Nov 12, 22. ORGANIC CHEMISTRY. *Educ:* Univ Notre Dame, BS, 48, MS, 49; Univ Utah, PhD(chem), 55. *Prof Exp:* Analytical chemist & chem engr, Union Oil Co, 50-51; instr chem, Long Beach City Col, 55-56; INSTR CHEM, ST EDWARD'S UNIV, 57- *Mem:* Am Chem Soc; Entom Soc Am. *Res:* Analytical organic chemistry in natural products. *Mailing Add:* Dept Chem St Edward's Univ 3000 S Congress Ave Austin TX 78704

MCCULLOUGH, WILLARD GEORGE, b Brighton, Mich, Nov 13, 14; m 44; c 2. GROWTH FACTORS, IRON CHELATES. *Educ:* Mich State Col, BS, 41, MS, 42; Univ Wis, PhD(biochem), 49. *Prof Exp:* Res asst bacteriol, Mich State Col, 41-42; bacteriologist, Mich State Exp Sta, 42-43; res asst biochem, Univ Wis, 46-49; instr bacteriol, Univ Mich, 49-50; from asst prof to assoc prof, Col Med, Wayne State Univ, 50-60; res chemist, Nat Animal Dis Ctr, USDA, 60-78; RETIRED. *Mem:* Fel AAAS; Am Chem Soc; Am Soc Microbiol. *Res:* Biochemistry; nutrition and metabolism of microorganisms. *Mailing Add:* 11700 S 50th Ave Sears MI 49679-9728

MCCULLY, JOSEPH C, b Kalamazoo, Mich, Feb 6, 24; m 47; c 4. MATHEMATICS. *Educ:* Western Mich Univ, BA, 47; Univ Mich, MA, 49, PhD(math), 57. *Prof Exp:* Asst prof math, Univ SDak, 53-55; asst prof, Univ RI, 55-56; assoc prof, 56-66, PROF MATH, WESTERN MICH UNIV, 66- *Mem:* Am Math Soc; Math Asn Am. *Res:* Integral transforms; operational calculus. *Mailing Add:* 1942 Hazel Ave Kalamazoo MI 49001

MCCULLY, KILMER SERJUS, b Daykin, Nebr, Dec 23, 33; m 55; c 2. EXPERIMENTAL PATHOLOGY, BIOCHEMISTRY. *Educ:* Harvard Univ, AB, 55; Harvard Med Sch, MD, 59; Brown Univ, MA, 83. *Prof Exp:* Intern, Mass Gen Hosp, 59-60; biochemist, NIH, 60-62; USPHS res fel med, Harvard Med Sch, 62-63, res fel biol, Harvard Univ, 63-65, Am Cancer Soc fac res assoc, 63-68, asst path, Harvard Med Sch, 65-68, instr, 68-70; asst pathologist, 68-73, assoc pathologist, Mass Gen Hosp, 74-; asst prof path, Harvard Med Sch, 70-79; assoc prof, Brown Univ, 81-89; PATHOLOGIST, LAB SERV, VET ADMIN MED CTR, PROVIDENCE, RI, 81- *Concurrent Pos:* Res assoc, Glasgow Univ, 63-64; fac res assoc award, Am Cancer Soc, 63-68; clin & res fel path, Mass Gen Hosp, 65-68; vis prof, Lab Med, Univ Conn, 80-81. *Honors & Awards:* Career Develop Award, NIH, 71-76. *Mem:* AAAS; Am Asn of Pathologists. *Res:* Nucleic acid structure; protein biosynthesis; microbial genetics; amino acid metabolism; arteriosclerosis; growth hormone; ascorbate; pyridoxine; somatomedin; cancer; synthetic derivatives and discovery of abnormalities of homocysteine thiolactone metabolism in arteriosclerosis and cancer and their chemopreventive activities; homocysteine theory of arterio sclerosis. *Mailing Add:* Lab Serv Vet Admin Ctr Davis Park Providence RI 02908

MCCULLY, MARGARET E, b St Marys, Ont, July 25, 34. BIOLOGY. *Educ:* Univ Toronto, BSA, 56, MSA, 60; Harvard Univ, PhD(biol), 66. *Prof Exp:* Res assoc microbiol, Parke Davis Co, 56-57; teacher high sch, Ont, 57-58 & St Felix Sch, Eng, 60-62; from asst prof to assoc prof, 66-78, PROF BIOL, CARLETON UNIV, 78- *Concurrent Pos:* External prof, Univ Ottawa, 77- *Mem:* Soc Develop Biol; Histochem Soc; Soc Exp Biol; Bot Soc Am; Royal Micros Soc; Sigma Xi. *Res:* Cytology, histology and histochemistry in relation to plant development. *Mailing Add:* Dept of Biol Carleton Univ Ottawa ON K1S 5B6 Can

MCCULLY, WAYNE GUNTER, b New Cambria, Mo, Jan 3, 22; m 44. RANGE SCIENCE. *Educ:* Colo State Univ, BS, 47; Tex A&M Univ, MS, 50, PhD, 58. *Prof Exp:* From asst prof to prof range mgt, Tex A&M Univ, 47-73, acting head dept, 71-72, prof & resident dir res, 73-79, RANGE SCIENTIST & HEAD VEG MGT, TEX A&M UNIV, 62- *Concurrent Pos:* Mem roadside maintenance comt, Transp Res Bd, Nat Res Coun, Nat Acad Sci, chmn, 89- *Mem:* Soc Range Mgt; Am Soc Plant Physiol; Weed Sci Soc Am. *Res:* Noxious plant control; range ecology; grassland management; roadside vegetation management. *Mailing Add:* Tex A&M Univ Tex Transp Inst College Station TX 77843-3135

MCCUMBER, DEAN EVERETT, b Rochester, NY, Nov 25, 30; m 57; c 2. TECHNOLOGY ASSESSMENT & APPLICATION, PROGRAM & QUALITY MANAGEMENT. *Educ:* Yale Univ, BEEE, 52, MEEE, 55; Harvard Univ, AM, 56, PhD(physics), 60. *Prof Exp:* NSF postdoctoral fel, Paris/Copenhagen, 59-61; physicist, AT&T Bell Labs, 61-89, head, Crystal Electronics Res Dept, 65-69, dir, Mil Phys Res Lab, 69-71, Interconnection Technol Lab, 71-77, Network Performance Planning Lab, 77-80 & Energy Systs & Power Technol Lab, 83-89; SR MGT CONSULT, COMPETITIVE TECHNOL STRATEGIES, 90- *Concurrent Pos:* Eng officer, US Navy, 52-54; dir oper planning, AT&T, 80-83; mem rev comt, Nat Acad Sci/Nat Res Coun, 89-91 & NSF grad fel rev comt, Nat Res Coun, 91- *Honors & Awards:* Baker Prize, Inst Elec Electronic Eng, 69. *Mem:* Sr mem Inst Elec & Electronics Eng; fel Am Phys Soc. *Res:* Lasers; vibrational structure in optical spectra of solids; electrical instabilities in solid-state materials; Gunn effect; electronic interconnection systems; communications systems performance; electronic power systems. *Mailing Add:* Competitive Technol Strategies 58 Dale Dr Summit NJ 07901-3129

MCCUNE, AMY REED, b Cincinnati, Ohio, May 31, 54; m 86. EVOLUTIONARY BIOLOGY, ICHTHYOLOGY. *Educ:* Brown Univ, AB, 76; Yale Univ, MPhil, 78, PhD(biol), 82. *Prof Exp:* Res fel, Miller Inst for Basic Res in Sci, Univ Calif, Berkeley, 82-83; asst prof, 83-89, ASSOC PROF, ECOL & SYSTS SECT & CUR ICHTHYOL, CORNELL UNIV, 89- *Concurrent Pos:* Bd trustees, Paleontol Res Inst, 85-; bd gov, Am Soc Ichthyol & Herpetologists, 89-; coun, Soc Syst Zool, 89- *Mem:* Am Soc Ichthyologists & Herpetologists; AAAS; Am Soc Zoologists; Soc Study Evolution; Soc Syst Zool; Soc Vert Paleont. *Res:* Systematics and evolution of semionotid fishes from Mesozoic lakes; Newark Supergroup deposits in eastern North America; development and evolution of fishes; evolutionary theory; systematics; paleontology. *Mailing Add:* Ecol & Systs Sect Cornell Univ Corson Hall Ithaca NY 14853

MCCUNE, CONWELL CLAYTON, b Salt Lake City, Sept 8, 32; m 57; c 6. CHEMICAL ENGINEERING, PHYSICAL CHEMISTRY. *Educ:* Univ Utah, BS, 53, MS, 58, PhD(chem eng), 61. *Prof Exp:* Assoc technologist, Appl Res Lab, US Steel Corp, 60-62; res engr, Gulf Res & Develop Co, 62-66; sr chem engr, Houdry Process & Chem Co, Pa, 66-67; res adv engr, Westinghouse Elec Corp, 67-68; res engr, 68-70, sr res engr, 70-77, SR ENG ASSOC, CHEVRON OILFIELD RES CO, 77- *Concurrent Pos:* Eve instr continuing educ prog, Pa State Univ, 64-66; eve instr petrol eng, Univ Southern Calif, 77-83. *Mem:* AAAS; Am Inst Chem Engrs; Soc Petrol Engrs; Am Asn Physics Teachers; Am Soc Testing & Mat. *Res:* High pressure reaction kinetics; shock tube and propellant ignition; refinery process; hydrocarbon oxidation; oil well stimulation processes; oil field water treating. *Mailing Add:* Chevron Res Co PO Box 446 Lattabre CA 90633

MCCUNE, DELBERT CHARLES, b Los Angeles, Calif, Aug 21, 34. PLANT PHYSIOLOGY, AIR POLLUTION. *Educ:* Calif Inst Technol, BS, 56; Yale Univ, MS, 57, PhD(bot), 60. *Prof Exp:* From asst plant physiologist to assoc plant physiologist, 60-68, PLANT PHYSIOLOGIST, BOYCE THOMPSON INST PLANT RES, 68- *Mem:* AAAS; Am Soc Plant Physiol; NY Acad Sci. *Res:* Plant growth regulators; air pollution; effects on vegetation. *Mailing Add:* Boyce Thompson Inst for Plant Res Cornell Univ Tower Rd Ithaca NY 14853-1801

MCCUNE, DUNCAN CHALMERS, b Chicago, Ill, Mar 16, 25; m 47; c 2. QUALITY SYSTEMS, QUALITY CONTROL. *Educ:* Col of Wooster, AB, 48; Purdue Univ, MS, 50. *Prof Exp:* Res assoc, Purdue Univ, 50-52; statistician, Tubular Prods Div, Babcock & Wilcox Co, 52-55; sr statistician, Jones & Laughlin Steel Corp, 55-64, supvr appl math, 64-68, asst to dir tech serv, 69-71, mgr qual assurance, 71-80; mgr statist process control, LTV Steel Co, 80-85; CONSULT, QUAL/STATIST, 86- *Concurrent Pos:* Sr lectr, Carnegie Mellon Inst, 60-64; instr & adj prof, Penn State Univ, Beaver Campus, 64-70. *Mem:* Fel Am Soc Qual Control; fel Am Soc Test & Mat; fel Royal Statist Soc. *Mailing Add:* 1236 Second St Beaver PA 15009

MCCUNE, EMMETT L, b Cuba, Mo, Jan 2, 27; m 54; c 3. VETERINARY MICROBIOLOGY, AVIAN PATHOLOGY. *Educ:* Univ Mo, BS & DVM, 56, MS, 61; Univ Minn, PhD(microbiol), 67. *Prof Exp:* Instr vet bact, 56-61, from asst prof to assoc prof, 61-77, PROF VET MICROBIOL, COL VET MED, UNIV MO-COLUMBIA, 77- *Concurrent Pos:* Consult on com avian health mgt. *Mem:* Am Col Vet Microbiol; Am Vet Med Asn; Am Asn Avian Path; Am Soc Microbiol. *Res:* Microbiology, expecially mycoplasma, Pasteurella and Escherichia coli as related to avian pathology and metabolism; resistance factor in Salmonella and Escherichia coli. *Mailing Add:* Vet Med Diag Lab Univ Mo Columbia MO 65211

MCCUNE, FRANCIS K(IMBER), b Santa Barbara, Calif, Apr 10, 06; m 69; c 1. ENGINEERING. *Educ:* Univ Calif, BS, 28. *Prof Exp:* Engr, Gen Elec Co, 28-45, asst works engr, Mass, 45-46, asst to mgr eng, NY, 46-48, asst gen mgr nucleonics, Wash, 48-51, mgr eng, Apparatus Div, NY, 51, vpres, Atomics Prod Div, 53-54, vpres mkt serv, 54-60, vpres eng, 60-65, vpres bus studies serv, 65-67; pres, Am Nat Standards Inst, 66-68, McCune Realty & Oil Co, 68-75; RETIRED. *Concurrent Pos:* Past pres & hon dir, Atomic Indust Forum; past mem adv bd for residencies in eng, Ford Found; mem, Nat Res Coun, 69-72; hon mem, Woods Hole Oceonog Inst. *Honors & Awards:* Howard Coonley Medal, Am Standards Inst, 69. *Mem:* Nat Acad Eng; fel Inst Elec & Electronic Engrs; Am Soc Mech Engrs; Am Nuclear Soc. *Res:* Nuclear, electrical and mechanical engineering. *Mailing Add:* 470 Magellan Dr Sarasota FL 34243

MCCUNE, HOMER WALLACE, b Grove City, Pa, Sept 5, 23; m 49; c 3. CHEMISTRY. *Educ:* Grove City Col, BS, 44; Cornell Univ, PhD(inorg chem), 49. *Prof Exp:* Mem staff, Coal Res Lab, Carnegie Inst Technol, 44-45; res chemist, Chem Div, Procter & Gamble Co, 49-54, sect head, 54-56, dept head, 56-58, patent div, 58-59, assoc dir, Res Div, 59-71, assoc dir paper & toilet goods technol div, 71-74, assoc dir soap & toilet goods technol div, 74, assoc dir, Food, Coffee & Toilet Goods Technol Div, Res & Develop Dept, 81-84; RETIRED. *Mem:* AAAS; Am Chem Soc. *Res:* Phosphates; corrosion; phosphorus compounds; surface chemistry; detergency; administration. *Mailing Add:* 580 Larchmont Dr Cincinnati OH 45215

MCCUNE, JAMES E, b Tulsa, Okla, Apr 28, 31; c 4. PHYSICS. *Educ:* Carnegie Inst Technol, BS, 53; Cornell Univ, PhD(philos, aeronaut eng), 59. *Prof Exp:* PROF AERONAUT & ASTRONAUT, MASS INST TECHNOL, 63- *Concurrent Pos:* Mem, Max Planck Soc, W Ger, 68. *Honors & Awards:* Bausch & Lomb Sci Award, 49; Laurence Sperry Award, Am Inst Aeronaut & Astronaut, 60. *Mem:* Fel Nat Sci Found; fel Am Phys Soc; fel Am Inst Aeronaut & Astronaut. *Res:* Author of 77 articles. *Mailing Add:* Ten Eustis Ave Wakefield MA 01880

MCCUNE, LEROY K(ILEY), b Chicago, Ill, June 30, 17; m 42, 68; c 8. CHEMICAL ENGINEERING. *Educ:* Princeton Univ, BS, 40, PhD(chem eng), 50. *Prof Exp:* Res engr, Textile Fibers Dept, E I DuPont de Nemours & Co, Inc, 48-50, res supvr, 50-52, res mgr, 52-53, tech supt, Film Dept, 53-55, lab dir, Textile Fiber Dept, 56-59, plant tech mgr, 59-63, tech dir, 63-65, mfg dir, 65-67, dir nylon mfg div, 67-71, dir, Prod Div & gen dir mfg, 71-79, dir dept plans, Textile Fibers Dept, 79-81; mgt consult, 81-88; RETIRED. *Mem:* Am Inst Chem Engrs. *Res:* Polymer and fiber manufacture. *Mailing Add:* Three Barley Mill Dr Greenville DE 19807

MCCUNE, MARY JOAN HUXLEY, b Lewistown, Mont, Jan 14, 32; m 65; c 2. MICROBIOLOGY. *Educ:* Mont State Univ, BS, 53; Wash State Univ, MS, 55; Purdue Univ, PhD(microbiol), 65. *Prof Exp:* Res technician, Vet Admin Hosp, Oakland, Calif, 55-59; bacteriologist, US Naval Radiol Defense Lab, Calif, 59-61; teaching assoc microbiol, Purdue Univ, 61-65, vis asst prof biol, 65-66; asst prof microbiol, Occidental Col, 66-69; fel, Univ Calif, Los Angeles, 69-70; affil asst prof, 70-80, ASST PROF MICROBIOL, IDAHO STATE UNIV, 80- *Mem:* AAAS; Am Soc Microbiol; NY Acad Sci; Sigma Xi. *Res:* Microbial physiology; genetics and taxonomy; microbial regulation of protein synthesis; ethanol production from renewable resources. *Mailing Add:* Dept Microbiol & Biochem Idaho State Univ Box 8007 Pocatello ID 83209

MCCUNE, RONALD WILLIAM, b Glade, Kans, Sept 23, 38; m 65; c 2. BIOCHEMISTRY. *Educ:* Kans State Univ, BS, 61; Purdue Univ, MS, 64, PhD(biochem), 66. *Prof Exp:* Trainee, Univ Calif, Los Angeles, 66-67, res biol chemist, 67-70; asst prof, 70-73, chmn dept microbiol & biochem, 71-77, assoc prof, 73-79, PROF BIOCHEM, IDAHO STATE UNIV, 79- *Concurrent Pos:* Vis assoc res biochemist, Univ Calif, San Diego, 77-78. *Mem:* Am Soc Microbiol; Sigma Xi; AAAS; Am Chem Soc; NY Acad Sci; Nat Asn Adv Health Professions. *Res:* Effect of antibiotics on short-chain fatty acid metabolism by rumen microorganisms; urea-N metabolism in the ruminant animal; metabolic control of adrenal steroid biosynthesis; mechanisms of hormonal action; cAMP and cGMP-dependent protein kinases; mitochondrial metabolism. *Mailing Add:* Dept Microbiol & Biochem Idaho State Univ 8007 Pocatello ID 83209-0009

MCCUNE, SUSAN K, b Takoma Park, Md, Jan 18, 58; m 83; c 1. PEDIATRICS. *Educ:* Harvard Univ, BA, 79; George Wash Univ Med Sch, MD, 83. *Prof Exp:* Nat Res Ser Award fel molecular neurosci, Lab Develop Neurosci, Nat Inst Child Health & Human Develop, NIH, 87-90; ASST PROF PEDIAT MED, CHILDREN'S NAT MED CTR, 90- *Concurrent Pos:* Guest researcher, Lab Develop Neurosci, Nat Inst Child Health & Human Develop, NIH, 90- *Mem:* Soc Neurosci; Soc Cell Biol; Am Acad Pediat. *Res:* Investigating the regional brain distribution and peripheral tissue distribution of multiple subtypes of alpha adrenergic receptor mRNAs during development. *Mailing Add:* Lab Develop Neurobiol NICHD NIH Bldg 36 Rm 2A21 Bethesda MD 20892

MCCUNE, SYLVIA ANN, b June 29, 44; m; c 2. LIPID SYNTHESIS, OBESITY. *Educ:* Ind Univ, PhD(med genetics), 74. *Prof Exp:* asst prof biochem, 81-87, ASSOC PROF FOOD SCI & TECHNOL, 87- *Mem:* Am Inst Nutrit; Am Soc Biochem & Molecular Biol; Am Diabetes Asn; AAAS. *Res:* Genetic and biochemical studies in obese, diabetic, hypertensive rat model; studies on sex differences in disease. *Mailing Add:* 7757 Tenbury Dr Dublin OH 43017

MCCUNE, WILLIAM JAMES, JR, b Glen Falls, NY, June 2, 15. ENGINEERING. *Educ:* Mass Inst Technol, BS, 37. *Prof Exp:* Staff engr, overseas opers, Gen Motors Corp, 37-39; staff engr, Polaroid Corp, 39-50, gen mgr, 50-54, vpres eng, 54-69, asst gen mgr, 63-69, exec vpres, 69-75, dir, 75-82, pres, 75-83, chief exec officer, 75-86, chmn, 82-91; RETIRED. *Honors & Awards:* Lifetime Achievement Award, Soc Imaging Scientists & Engrs. *Mem:* Nat Acad Eng; fel Am Acad Arts & Sci. *Mailing Add:* Polaroid Corp 549 Technology Sq 2nd Floor Cambridge MA 02139

MCCURDY, ALAN HUGH, b Princeton, NJ, Feb 15, 59. COHERENT RADIATION GENERATION, FREE ELECTRON DEVICES. *Educ:* Carnegie-Mellon Univ, BS, 81, BS, 82; Yale Univ, MA, 83, MPhil, 86, PhD(appl physics), 87. *Prof Exp:* Res scientist, Omega-P, Inc, 85-88; ASST PROF ELEC ENG & ELECTRO-PHYSICS, UNIV SOUTHERN CALIF, 89- *Honors & Awards:* Simon Ramo Award, Am Phys Soc, 88. *Mem:* Am Phys Soc. *Res:* Theory and operation of free electron microwave devices; phase locking, priming and mode control of high frequency sources of coherent radiation; applied nonlinear oscillator dynamics. *Mailing Add:* Dept Elec Eng Electro Physics MC 0271 Univ Southern Calif Los Angeles CA 90089

MCCURDY, DAVID HAROLD, b East Orange, NJ, Apr 28, 30; m 56; c 1. PHARMACOLOGY. *Educ:* Dalhousie Univ, BSc, 53, MSc, 55; Univ Toronto, PhD(pharmacol), 58. *Prof Exp:* Asst pharmacol, Dalhousie Univ, 53-55; asst, Univ Toronto, 55-58; pharmacologist, Chem Res & Develop Labs, US Army Chem Ctr, Md, 58-62; sr res pharmacologist, 62-65, mgr pharmacol sect, 65-75, asst dir drug design & eval, 73-76, DIR RES ADMIN, ICI AMERICAS INC, 76- *Mem:* AAAS; Sigma Xi; NY Acad Sci; Pharmacol Soc Can; Am Soc Pharmacol & Exp Therapeut. *Res:* Basic pharmacology, including the molecular basis of drug action; drug screening and development; central nervous system pharmacology research and management. *Mailing Add:* 139 E Third St New Castle DE 19720

MCCURDY, HOWARD DOUGLAS, JR, b London, Ont, Can, Dec 10, 32; m 56, 78; c 4. MICROBIOLOGY, BIOCHEMISTRY. *Educ:* Western Ontario Univ, BA, 53; Assumption Univ, BSc, 54; Mich State Univ, MSc, 55, PhD(microbiol), 59. *Prof Exp:* Asst, Mich State Univ, 55-59; from assoc to prof biol, 59-71, dept head, 74-79. *Concurrent Pos:* Elected to House of Commons, 84. *Honors & Awards:* Centennial Medal, 87. *Mem:* AAAS; Am Soc Microbiol; Can Soc Microbiol; Can Asn Univ Teachers (pres, 68-69); Can Col Microbiologists, 76-78. *Res:* Ecology of prokaryotes; biology of myxobacteria. *Mailing Add:* House of Commons Ottawa ON K1A 0A6 Can

MCCURDY, JOHN DENNIS, b Irvington, NJ, Aug 17, 42; c 2. BIOCHEMISTRY. *Educ:* Fairleigh Dickinson Univ, BSc, 66; Am Univ, MSc, 72, PhD(chem), 74; Am Bd Toxicol, dipl. *Prof Exp:* Lab officer biochem, US Army, 66-69; res asst, Am Univ, 69-74; new drug chemist analysis & biochem, Food & Drug Admin, 74-80, biochemist, Animal Feed Safety, Bur Vet Med, 80-83, sr rev scientist, petitions rev staff, Ctr Vet Med, 84-85, actg supvr, 85-86, sr rev scientist, regulatory chem/toxicol petitions rev & med feeds br, 86-90, REGULATORY CHEMIST/TOXICOLOGIST, CTR VET MED, FOOD & DRUG ADMIN, 90- *Concurrent Pos:* Res scientist chem, Am Univ, 75-; vis scientist, Dept Pharmacol, Col Med, Howard Univ, Washington, DC, 83-84. *Mem:* Am Chem Soc; NY Acad Sci; AAAS; Asn Off Analytical Chemists; fel Am Inst Chemists. *Res:* Biochemistry of pharmacologically active compounds; thymic hormone research. *Mailing Add:* 3949 Sugarloaf Dr Monrovia MD 21770

MCCURDY, KEITH G, b Lampman, Sask, Can, Dec 4, 37; m 60; c 3. PHYSICAL CHEMISTRY. *Educ:* Univ Sask, BA, 59, MA, 61; Univ Ottawa, PhD(chem), 64. *Prof Exp:* Asst prof chem, Univ Guelph, 63-67; from asst prof to assoc prof, 67-77, PROF CHEM, UNIV LETHBRIDGE, 77- *Res:* Chemical kinetics. *Mailing Add:* Dept of Chem Univ Lethbridge 4401 University Dr Lethbridge AB T1K 3M4 Can

MCCURDY, LAYTON, b Florence, SC, Aug 20, 35; m 58; c 2. PSYCHIATRY. *Educ:* Med Univ SC, MD, 60. *Prof Exp:* Rotating intern, Med Col SC, 60-61; resident psychiat, NC Mem Hosp, 61-64; asst prof psychiat, Sch Med, Emory Univ, 66-68; prof psychiat & chmn dept, Med Univ SC, 68-; AT INST PSYCHOANAL PSYCHOTHER. *Concurrent Pos:* Consult, NIMH, 66-67, Clayton County Ment Health Clin, Ga, 66-68, Volunteers in Serv to Am, Washington, DC, 67-68, Charleston Vet Admin Hosp, Va & SC Dept Ment Health, 68- & Va Med Res Rev Bd Behav Sci, NIMH, 74-77; psychoanal trainee, Columbia Univ Psychoanal Clin, Atlanta, Ga, 67-68; mem res rev comt, Appl Res Br, NIMH, 70-74; NIMH res fel, Maudsley Hosp, London, Eng, 74-75; mem, Nat Adv Ment Health Coun, 80-83. *Mem:* Fel Am Col Psychiat; fel Am Psychiat Asn; fel Am Psychosom Soc; Asn Am Med Cols; fel Asn Acad Psychiat (pres, 70-72). *Res:* Behavioral sciences; medical education. *Mailing Add:* 171 Ashley Ave Charleston SC 29425-2201

MCCURDY, PAUL RANNEY, b Middletown, Conn, Sept 26, 25; m 52; c 5. INTERNAL MEDICINE, HEMATOLOGY. *Educ:* Wesleyan Univ, AB, 46; Harvard Med Sch, MD, 49. *Prof Exp:* Fel med, Sch Med, Georgetown Univ, 54-55; instr, 55-60, assoc prof, 60-72, PROF MED, SCH MED, GEORGETOWN UNIV, 72- *Concurrent Pos:* Consult; dir, Am Red Cross Blood Serv, Washington, DC, Region, 75- *Mem:* Am Soc Hemat; Int Soc Hemat; Am Col Physicians; sr mem Am Fedn Clin Res; Am Asn Blood Banks. *Res:* Red blood cell, including enzymes G-6-PD and abnormal hemoglobins. *Mailing Add:* 6452 Elmdale Rd Alexandria VA 22312

MCCURDY, WALLACE HUTCHINSON, JR, b Pittsburgh, Pa, Nov 16, 26; m 54; c 3. ANALYTICAL CHEMISTRY. *Educ:* Pa State Univ, BS, 47; Univ Ill, MS, 47, PhD(chem), 51. *Prof Exp:* From instr to asst prof chem, Princeton Univ, 51-59; asst prof, 59-64, ASSOC PROF CHEM, UNIV DEL, 64- *Concurrent Pos:* Res chemist, Bell Tel Labs, Inc, NJ, 53; Textile Res Inst, 56; res fel, Nat Bur Standards, 70-71. *Mem:* Am Chem Soc. *Res:* Coordination complexes; nonaqueous solvent systems; environmental analysis; high performance liquid chromatography. *Mailing Add:* Dept of Chem Univ of Del Newark DE 19711

MCCURRY, PATRICK MATTHEW, JR, b Homestead, Pa, Nov 2, 44; m 67; c 2. CHEMISTRY. *Educ:* Univ Pittsburgh, BS, 65; Columbia Univ, PhD, 70. *Prof Exp:* Res chemist, US Bur Mines, 65 & 66; NIH fel, Stanford Univ, 70-71; asst prof chem, Mellon Inst Sci, Carnegie-Mellon Univ, 71-76, assoc prof, 76-81; sr res chemist, A E Staley Corp, Decatur, Ill, 81-88; RES FEL, HENKEL CORP, 88- *Concurrent Pos:* Res fel, Alfred P Sloan Found, 75-78. *Mem:* Am Chem Soc; Royal Soc Chem. *Res:* Natural product synthesis; novel synthetic methods. *Mailing Add:* Henkel Corp 300 Brookside Ave Bldg 23 Ambler PA 19002

MCCUSKER, JANE, b London, Eng, Aug 27, 43; m 67; c 3. EPIDEMIOLOGY. *Educ:* McGill Univ, MD, CM, 67; Columbia Univ, MPH, 69, DrPH(epidemiol), 74. *Prof Exp:* Lectr epidemiol statist, Fac Med, Univ Dar es Salaam, 73-74; asst prof prev med, Sch Med & Dent, Univ Rochester, 75-81; ASSOC PROF, DIV PUB HEALTH, SCH HEALTH SCI, UNIV MASS, AMHERST, 81- *Mem:* Am Pub Health Asn; Soc Epidemiol Res; Int Epidemiol Asn. *Res:* Epidemiology in health services research; evaluation of health care of chronically and terminally ill; general epidemiologic research; acquired immune deficiency syndrome epidemiology. *Mailing Add:* Div Pub Health Sch Health Sci Univ Mass Amherst MA 01003

MCCUSKEY, ROBERT SCOTT, b Cleveland, Ohio, Sept 8, 38; div; c 3. MICROCIRCULATION, HEPATOLOGY. *Educ:* Western Reserve Univ, AB, 60, PhD(anat), 65. *Prof Exp:* From instr to prof anat, Col Med, Univ Cincinnati, 65-78; prof anat & chmn dept, Sch Med, WVA Univ, 78-86; PROF ANAT & HEAD DEPT COL MED, UNIV ARIZ, 86-, PROF PHYSIOL, 87- *Concurrent Pos:* Consult, Procter & Gamble Co, 66-68, 71-75 & 85-86 & Hoffmann-La Roche, 72-75; NIH res grants, Univ Cincinnati, 66-78, Southwest Ohio Heart Asn res grant, 68-69 & 73-77, Akron Heart Asn res grant, 70-71 & Am Heart Asn res grant, 77-86, WVa Univ, 78-86; NIH res career develop award, 69-74; assoc ed, Microvascular Res, 74-85, Microcirculation, Endothelium & Lymphatics, 88-; vis prof, Univ Heidelberg, 81-83, 87 & 88. *Honors & Awards:* Alexander von Humboldt US Sr Scientist Prize, W Ger, 82; Nishimaru Award, Japan, 87. *Mem:* Microcirc Soc; Am Asn Anatomists; Am Soc Hemat; Int Soc Exp Hemat; Am Asn Study Liver Dis. *Res:* In vivo microscopic anatomy of living organs; microcirculation; liver structure and function; hematology; endotoxins and host defense mechanisms. *Mailing Add:* Dept Anat Col Med Univ Ariz Tucson AZ 85724

MCCUTCHAN, ROY T(HOMAS), b Dallas, Tex, Sept 19, 18; m 41. CHEMICAL ENGINEERING, PHYSICAL CHEMISTRY. *Educ:* Univ Tex, BS, 40, MS, 41, PhD(chem eng), 49. *Prof Exp:* Engr, Dow Chem Co, 42-44; mem staff, Los Alamos Sci Lab, 49-52; chem engr, Southwest Res Inst, 52-56, mgr inorg process develop, 56-59; asst prof chem, St Mary's Univ, Tex, 59-62; aerospace technologist, Manned Spacecraft Ctr, NASA, 62-64; supvr process develop, Thiokol Chem Corp, 64-66; assoc prof, 66-69, PROF ENG, ST MARY'S UNIV, TEX, 69- *Mem:* Am Inst Chem Engrs. *Res:* Solid propellant rocket process development; military pyrotechnics; high energy boron fuel; aromatic mercury compounds; thermodynamics; water quality surveys. *Mailing Add:* Div Eng St Mary's Univ San Antonio TX 78284

MCCUTCHEN, CHARLES WALTER, b Princeton, NJ, Mar 9, 29. PHYSICS, BIOLOGY. *Educ:* Princeton Univ, BA, 50; Brown Univ, MSc, 52; Cambridge Univ, PhD, 57. *Prof Exp:* Res physicist, Cambridge Univ, 57-62; RES PHYSICIST, LAB EXP PATH, NAT INST DIABETES & DIGESTIVE DIS. *Mem:* Am Phys Soc; Optical Soc Am. *Res:* Lubrication of animal joints; optical diffraction theory; applied optics; hydrodynamics. *Mailing Add:* 5213 Acacia Ave Bethesda MD 20814

MCCUTCHEN, SAMUEL P(ROCTOR), b Corinth, Miss, May 2, 28; m 53, 72; c 6. ELECTRICAL ENGINEERING. *Educ:* Vanderbilt Univ, BE, 50; LaSalle Exten Univ, LLB, 66. *Prof Exp:* Prod engr magnetic head design, Brush Electronics Co, 53-55; proj engr infrared syst, Aerojet-Gen Corp, 55-58; design supvr Polaris syst, Lockheed Aircraft Corp, 58-59; supvr electronic systs studies, Sylvania Elec Prod Co, 59-60; mem staff, Inst Defense Anal, 60-63; mgr systs eng dept, NAm Air Defense Combat Opers Ctr Proj, Burroughs Corp, 63-66; sr staff mem, ASW Systs Studies, TRW Systs, Washington, 66-69; dir, mgt info systs, US Army Mobility Equip Res & Develop Command, 69-85; sr staff engr, TRW, 85-88; COMPUTER SCIENTIST, FED BUR INVEST, 88- *Res:* Systems management; computer center management; patent-magnetic head design-1955. *Mailing Add:* 6022 Florence Lane Alexandria VA 22310

MCCUTCHEON, ERNEST P, b Durham, NC, May 7, 33; m 56; c 3. MEDICAL PHYSIOLOGY, PUBLIC HEALTH & EPIDEMIOLOGY. *Educ:* Davidson Col, BS, 55; Duke Univ, MD, 59. *Prof Exp:* Intern med, Grady Mem Hosp, Atlanta, Ga, 59-60; res med, Duke Univ, 60-61; head bioinstrumentation & physiol data sect, Launch Site Med Opers, NASA, 61-63; NIH sr res fel physiol & biophys, Univ Wash, 63-66; from asst prof to assoc prof, Univ Ky, 66-74; sr res assoc, Nat Acad Sci-Nat Res Coun, NASA, 73-74, res med officer, Biomed Res Div, Ames Res Ctr, 74-77; ASSOC PROF PHYSIOL, SCH MED, UNIV SC, 77- *Concurrent Pos:* Vis assoc prof, Stanford Univ Med Ctr, 73-74. *Mem:* Am Physiol Soc; AAAS; Inst Elec & Electronics Engrs; Am Heart Asn; Biomed Eng Soc. *Res:* Cardiovascular dynamics; instrumentation; indirect blood pressure determination; ultrasonics. *Mailing Add:* Dept Physiol Sch Med Univ SC 217 Northlake Rd Columbia SC 29223

MCCUTCHEON, MARTIN J, b Little Rock, Ark, Dec 23, 41; m 63; c 2. INSTRUMENTATION, MEASUREMENT. *Educ:* Univ Ark, BSEE, 64, MSEE, 65, PhD(elec eng), 67. *Prof Exp:* Asst prof, 67-71, assoc prof, 71-76, PROF ENG, UNIV ALA, BIRMINGHAM, 77-, ASSOC PROF BIOCOMMUN, 75- *Concurrent Pos:* Consult, NIH, 80 & Zurn Air Systs, 81. *Mem:* Inst Elec & Electronics Engrs; Eng Med & Biol Soc; Acoust Soc Am. *Res:* Development of instrumentation for speech research and therapy; applications of computers and microprocessors in engineering in medicine and biology. *Mailing Add:* Dept Biomed Eng Univ Ala 3035 Taralane Dr Birmingham AL 35216

MCCUTCHEON, ROB STEWART, b Idaho Falls, Idaho, May 10, 08; m 29; c 3. PHARMACOLOGY. *Educ:* Univ Idaho, BS, 33; Univ Wash, MS, 46, PhD(pharmaceut chem), 48. *Prof Exp:* Prof pharmacol, Ore State Univ, 48-64; scientist admin pharmacol, NIH, 64-65, exec secy, toxicol study sect, div res grants, 65-78; RETIRED. *Concurrent Pos:* Res fel cardiovasc pharmacol, Med Col Ga, 55-56; NIH spec fel, Sch Med, WVa Univ, 62-63; ed, J Toxicol & Environ Health, 78-83; health & hearing impaired consult, 78- *Mem:* Fel AAAS; Am Pharmaceut Asn; Am Soc Pharmacol & Exp Therapeut; Soc Toxicol. *Res:* Cardiovascular pharmacology; toxicology. *Mailing Add:* 703 Morgan Creek Rd Chapel Hill NC 27514

MCCUTCHEON, WILLIAM HENRY, b Toronto, Ont, Aug 26, 40; m 65; c 2. RADIO ASTRONOMY. *Educ:* Queen's Univ, Ont, BSc, 62, MSc, 64; Univ Manchester, PhD(radio astron), 69. *Prof Exp:* Res asst radio astron, Nuffield Radio Astron Labs, Jodrell Bank, Eng, 67-69; fel, Univ BC, 69-71, res assoc, 71-72, vis asst prof, 72-73, from asst prof to assoc prof, 73-90, PROF DEPT PHYSICS, UNIV BC, 90- *Concurrent Pos:* Vis scientist, Commonwealth Sci & Indust Res Orgn, Div Radiophysics, Sydney, Australia, 80-81 & 87-88. *Mem:* Royal Astron Soc; Can Astron Soc (treas, 78-83); Am Astron Soc; Int Astron Union. *Res:* Spectral line studies in radio astronomy; molecular clouds, star forming regions, galactic structure. *Mailing Add:* Dept Physics Univ BC 2075 Wesbook Mall Vancouver BC V6T 1Z2 Can

MCDADE, JOSEPH EDWARD, b Cumberland, Md, Feb 4, 40; m 64; c 2. MICROBIOLOGY, RICKETTSIOLOGY. *Educ:* Western Md Col, BA, 62; Univ Del, MA, 65, PhD(microbiol), 67. *Hon Degrees:* SDc, Western Md Col, 77. *Prof Exp:* Captain, US Army Biol Labs, 67-69; dir cell prod dept, Microbiol Assocs, Inc, 69-71; res assoc, Sch Med, Univ Md, 71-75; res microbiologist, 75-84, chief viral & rickettsial br, Ctr Dis Control, 84-89; ASSOC DIR, CTR INFECTIOUS DIS, 89- *Honors & Awards:* Richard & Hinda Rosenthal Award, Am Col Physicians, 79. *Mem:* Am Soc Microbiol; AAAS; Am Acad Microbiol; fel Infectious Dis Soc; Am Soc Rickettsiology (pres 85-86). *Res:* Properties of the agent responsible for Legionnaire's disease; rickettsiae and rickettsial diseases. *Mailing Add:* Ctr Infectious Dis Ctr Dis Control Atlanta GA 30333

MCDANIEL, BENJAMIN THOMAS, b Pickens, SC, Feb 15, 35. ANIMAL GENETICS, ANIMAL BREEDING. *Educ:* Clemson Col, BS, 57; Univ Md, MS, 60; NC State Univ, PhD(animal breeding), 64. *Prof Exp:* Dairy husb, Animal Husb Res Div, USDA, 57-60; fel animal breeding, NC State Univ, 63-64; res dairy scientist, Animal Sci Res Div, USDA, 64-72; PROF ANIMAL SCI & GENETICS, NC STATE UNIV, 72- *Honors & Awards:* Res Award, Nat Asn Animal Breeders, 84; J L Lush Animal Breeding Award, 85. *Mem:* Am Dairy Sci Asn; Am Soc Animal Sci. *Res:* Genetics and breeding of domestic animals, with emphasis on dairy cattle. *Mailing Add:* Dept Animal Sci NC State Univ Box 7621 Raleigh NC 27695-7621

MCDANIEL, BOYCE DAWKINS, b Brevard, NC, June 11, 17; m 41; c 2. EXPERIMENTAL HIGH ENERGY PHYSICS. *Educ:* Ohio Wesleyan Univ, BS, 38; Case Sch Appl Sci, MS, 40; Cornell Univ, PhD(physics), 43. *Prof Exp:* Asst physics, Case Sch Appl Sci, 38-40; asst & res assoc, Cornell Univ, 40-42; mem staff, Radiation Lab, Mass Inst Technol, 43; physicist, Los Alamos Sci Lab, NMex, 43-45; from asst prof to prof, 45-87, assoc dir, 60-67, dir lab nuclear studies, 67-85, EMER PROF PHYSICS, CORNELL UNIV, 87- *Concurrent Pos:* Fulbright res award, Australian Nat Univ, 53; Guggenheim & Fulbright award, Univ Rome & Synchrotron Lab, Frascati, Italy, 59; trustee, Assoc Univs, Inc, 63-75; res collabr, Brookhaven Nat Lab, 66; trustee, Univ Res Asn Inc, 71-77; vis distinguished prof, Ariz State Univ, 88; chmn, Bd Overseers Superconducting Supercollider, 84-89. *Mem:* Nat Acad Sci; fel Am Phys Soc; fel AAAS. *Res:* Electron diffraction; slow neutron and gamma ray spectroscopy; high energy particle physics; accelerator design, construction and operation. *Mailing Add:* Newman Lab Nuclear Studies Cornell Univ Ithaca NY 14850

MCDANIEL, BURRUSS, JR, b Ft Smith, Ark, Apr 18, 27; m 58; c 2. ENTOMOLOGY. *Educ:* Univ Alaska, BA, 53; Tex A&M Univ, MS, 61, PhD(entom), 65. *Prof Exp:* Consult entomologist, Insect Control & Res, Inc, Md, 57-59; asst prof biol, Tex Col Arts & Indust, 61-66; assoc prof entom, 66-71, PROF ENTOM, SDAK STATE UNIV, 71- *Concurrent Pos:* Consult, Rockefeller Found, Mex, 57-59. *Mem:* Soc Study Evolution; Entom Soc Am; Am Ornith Union. *Res:* Systematic and ecological studies of soil microarthropods; investigation on Entomophaga Grylli, a pathogen of grasshoppers. *Mailing Add:* Dept Plant Sci SDak State Univ Box 2207A Brookings SD 57007

MCDANIEL, CARL NIMITZ, b Englewood, NJ, June 9, 42; m 67; c 2. DEVELOPMENTAL BIOLOGY, PLANT CELL CULTURE. *Educ:* Oberlin Col, BA, 64; Wesleyan Univ, MS, 66, PhD(biol), 73. *Prof Exp:* Instr biol & chem, US Naval Acad, 69; fel biol, Yale Univ, 73-75; asst prof, 75-80, assoc prof, 81-86, PROF BIOL, RENSSELAER POLYTECH INST, 87- *Concurrent Pos:* Vis assoc res prof, Univ Calif, Davis, 81-82, Univ Pa, Philadelphia, 87; McMaster fel, Commonwealth Sci & Indust Res Orgn, Canberra, Australia, 89; Benedict distinguished vis prof, Carleton Col, Northfield, Minn, 91. *Mem:* Bot Soc Am. *Res:* Development of shoot apical meristems and regulation of flowering. *Mailing Add:* Dept Biol-MRC Rensselaer Polytech Inst Troy NY 12180-3590

MCDANIEL, CARL VANCE, b Grafton, WVa, Dec 4, 29; m 52; c 2. PHYSICAL CHEMISTRY. *Educ:* Univ Pittsburgh, BS, 57; Mass Inst Technol, PhD(phys chem), 62. *Prof Exp:* Res technologist, US Steel Corp, 52-58; asst phys chem, Mass Inst Technol, 58-62; res chemist, 62-65, RES SUPVR PHYS CHEM, W R GRACE & CO, COLUMBIA, 65- *Mem:* Am Chem Soc; AAAS; Catalysis Soc. *Res:* Chemical metallurgy; physicochemical properties of ion exchange systems; physicochemical properties of molecular sieves; hydrocarbon catalysis and catalysts; coal gasification and liquefaction; synthetic fuels. *Mailing Add:* 10725 Crestview Lane Laurel MD 20707

MCDANIEL, EARL WADSWORTH, b Macon, Ga, Apr 15, 26; m 48; c 2. PHYSICS. *Educ:* Ga Inst Technol, BA, 48; Univ Mich, MS, 50, PhD, 54. *Prof Exp:* Asst physics, Univ Mich, 48-54; res physicist, 54-55, from asst prof to prof, 55-70, REGENTS' PROF PHYSICS, GA INST TECHNOL, 70- *Concurrent Pos:* Instr, Ga Power Co, 57-; consult, Union Carbide Nuclear Co, 59-, US Army Missile Command, 70- & United Technol Res Ctr, 76-; Guggenheim fel & Fulbright sr res scholar, Univ Durham, Eng, 66-67. *Mem:* Am Phys Soc. *Res:* Atomic collisions; gaseous electronics; plasma physics. *Mailing Add:* Dept of Physics Ga Inst of Technol Atlanta GA 30332

MCDANIEL, EDGAR LAMAR, JR, b Augusta, Ga, Dec 28, 31; m 54; c 2. INDUSTRIAL ORGANIC CHEMISTRY. *Educ:* Univ Tenn, BS, 53, MS, 54, PhD(chem), 56. *Prof Exp:* Asst chem, Univ Tenn, 52-56; res chemist, 56-59, sr res chemist, 60-68, res assoc, 68-74, div head, Eng Res Div, 74-77, staff asst gen mgt, Tenn Eastman Co, 77-80, coordr appl res, 80-81, ASST DIR DEVELOP, EASTMAN CHEM DIV, EASTMAN KODAK CO, 81- *Mem:* Am Chem Soc; Am Inst Chem. *Res:* Catalytic reactions; petrochemistry; chemical kinetics; chemical reaction engineering. *Mailing Add:* 4506 Mitchell Rd Kingsport TN 37664

MCDANIEL, FLOYD DELBERT, SR, b Memphis, Tenn, Aug 27, 42; m; c 5. MATERIALS CHARACTERIZATION, SEMICONDUCTOR PHYSICS. *Educ:* Memphis State Univ, BS, 66; Univ Ga, MS, 68, PhD(physics), 71. *Prof Exp:* Res assoc & vis asst prof nuclear physics, Univ Ky, 71-73; vis res prof atomic physics, 74-75, from asst prof to assoc prof, 75-84, PROF MAT SCI, UNIV NTEX, 84-, CO-DIR INDUST, UNIV COOP RES CTR, 90- *Concurrent Pos:* Referee, Nuclear Instruments & Methods, 77-, Phys Rev, Phys Rev Lett, 79-, NSF, 88-; consult, Southwest Res Inst, San Antonio, 80-, Tex Instruments Inc, Dallas, 80-, Tex Utilities Elec Co, Dallas, 88-, Inter Digital Modeling Corp, 89-; vis scientist atomic physics, Hahn Meitner Inst, Berlin, FRG, 89. *Mem:* Sigma Xi; Am Phys Soc; Mat Res Soc; AAAS. *Res:* Experimental and theoretical atomic and nuclear physics; materials modification and characterization; trace element analysis in semiconductor materials and metals by accelerator mass spectrometry and nuclear reaction analysis; author of over 100 publications. *Mailing Add:* Dept Physics Univ NTex Denton TX 76203

MCDANIEL, GAYNER RAIFORD, b Milport, Ala, Feb 4, 29; m 48; c 2. GENETICS, PHYSIOLOGY. *Educ:* Auburn Univ, BS, 53, MS, 54; Kans State Univ, PhD(genetics), 60. *Prof Exp:* Exten poultryman, Kans State Univ, 59-61; dir res & develop turkeys, Ralston Purina Co, 61-68; assoc prof reproductive physiol, 68-81, PROF POULTRY SCI, AUBURN UNIV, 81- *Mem:* Artificial insemination of broiler breeders in cages; reproductive performance of breeders; environmental factors related to hatchability. *Mailing Add:* Dept Poultry Sci Auburn Univ Auburn AL 36849

MCDANIEL, IVAN NOEL, b Martinsville, Ill, Feb 13, 28; m 57; c 3. MEDICAL ENTOMOLOGY. *Educ:* Eastern Ill Univ, BS, 51; Univ Ill, MS, 52, PhD(entom), 58. *Prof Exp:* Asst biol, Univ Ill, 53-57; asst prof, 57-63, ASSOC PROF ENTOM, UNIV MAINE, ORONO, 63- *Concurrent Pos:* Mem sci adv panel, World Health Orgn, 75-80. *Mem:* Sigma Xi; Entom Soc Am; Entom Soc Can; Am Mosquito Control Asn. *Res:* Identification of attractants and stimulants that influence oviposition of mosquitoes; biological control of mosquitoes and black flies. *Mailing Add:* Dept of Entom 303 Deering Hall Univ of Maine Orono ME 04469

MCDANIEL, LLOYD EVERETT, b Michigan Valley, Kans, Apr 4, 14; m 44; c 2. MICROBIOLOGY. *Educ:* Kans State Univ, BS, 35, MS, 37; Univ Wis, PhD(bact), 41. *Prof Exp:* Res microbiologist, Merck Sharp & Dohme Res Labs Div, Merck & Co, Inc, 40-46, head develop sect, 46-51, mgr microbiol develop, 51-56, asst dir microbiol res, 56-61; from assoc prof to prof microbiol, Waksman Inst Microbiol, Rutgers Univ, 61-84; RETIRED. *Mem:* emer mem Am Soc Microbiol. *Res:* Fermentation research and technology; polyene macrolide antibiotics; production of chemicals by microorganisms. *Mailing Add:* 1055 Sleepy Hollow Lane Plainfield NJ 07060

MCDANIEL, MAX PAUL, b Ft Worth, Tex, May 11, 47; m 69. SURFACE CHEMISTRY. *Educ:* Southern Ill Univ, BA, 69; Northwestern Univ, MS, 70, PhD(phys chem), 74. *Prof Exp:* Res assoc catalysis, Nat Ctr Sci Res, France, 74; res chemist, 75-88, BR MGR POLYOLEFIN CATALYST & RESIN DEVELOP, PHILLIPS PETROL CO, 88- *Mem:* Am Chem Soc; Soc Plastics Engrs. *Res:* Modifying and testing ethylene polymerization catalysts, seeking to improve polymer performance through catalyst innovations; better understanding of the mechanism of ethylene polymerization. *Mailing Add:* 1601 Melmart Dr Bartlesville OK 74003

MCDANIEL, MICHAEL LYNN, INTERCELLUAR CALCIUM, ARACHIDONIC METABOLISM. *Educ:* St Louis Univ, PhD(biochem), 70. *Prof Exp:* ASSOC PROF PATH, SCH MED, WASH UNIV, 70- *Mailing Add:* Dept Path Sch Med Wash Univ 660 S Euclid Ave Box 8118 St Louis MO 63110

MCDANIEL, MILTON EDWARD, b Elk City, Okla, Jan 17, 38; m 60; c 1. PLANT BREEDING, GENETICS. *Educ:* Okla State Univ, BS, 60; Va Polytech Inst, PhD(plant breeding), 65. *Prof Exp:* Asst prof small grains breeding, 65-77, ASSOC PROF AGRON & GENETICS, TEX A&M UNIV, 77- *Mem:* Am Soc Agron. *Res:* Oat and wheat breeding and genetic research. *Mailing Add:* Dept Soil & Crop Sci Rm 427A Tex A&M Univ College Station TX 77843

MCDANIEL, ROBERT GENE, b Wash, DC, Aug 2, 41; m 76; c 4. GENETICS, PLANT PHYSIOLOGY. *Educ:* Univ WVa, AB, 63, PhD(genetics), 67. *Prof Exp:* Asst agronomist, 67-71, assoc prof agron & plant genetics, 71-75, PROF PLANT SCI, UNIV ARIZ, 75 - *Concurrent Pos:* Sabbatical, Calif Inst Technol, 74. *Mem:* AAAS; Am Soc Agron; Crop Sci Soc Am; Am Soc Plant Physiol; Int Soc Plant Molecular Biol; Am Genetic Asn; Sigma Xi. *Res:* Biochemical studies of mitochondrial heterosis in eukaryotes; biochemistry of plant hormones; histones and gene regulation; genetic engineering; plant secondary metabolites. *Mailing Add:* Dept Plant Sci Univ Ariz Tucson AZ 85721

MCDANIEL, ROGER LANIER, JR, b Suffolk, Va, Dec 29, 52; m; c 2. PESTICIDES. *Educ:* Campbell Univ, BS, 75; NC State Univ, PhD(org chem), 81. *Prof Exp:* Sr chemist, Union Carbide Agr Prod Co, 81-88; CHIEF, ENVIRON SCI, NC STATE LAB PUB HEALTH, 89- *Mem:* Am Chem Soc. *Res:* Synthesis and reactions of spirocyclic enones; synthesis of strained bridgehead olefins; herbicides and plant growth regulators. *Mailing Add:* 3700 Avert Ferry Rd Raleigh NC 27606

MCDANIEL, SUSAN GRIFFITH, b Kansas City, Kans, Dec 3, 38; wid. ZOOLOGY. *Educ:* Kans State Teachers Col, BS, 59, MS, 62; Univ Okla, PhD(zool), 66. *Prof Exp:* Asst prof biol, 67-79, asst provost, 73-77, from asst vchancellor to assoc vchancellor acad affairs, 77-84, ASSOC PROF BIOL, EAST CAROLINA UNIV, 79- *Mem:* Am Mus Natural Hist; Sigma Xi. *Res:* Animal ecology and behavior. *Mailing Add:* Dept Biol East Carolina Univ Greenville NC 27858-4353

MCDANIEL, TERRY WAYNE, magnetic recording physics, for more information see previous edition

MCDANIEL, VAN RICK, b San Antonio, Tex, Oct 29, 45; m 66; c 5. MAMMALOGY, HERPETOLOGY. *Educ:* Tex A&I Univ, BS, 67, MS, 69; Tex Tech Univ, PhD(zool), 73. *Prof Exp:* PROF & CHMN DEPT BIOL, ARK STATE UNIV, 72- *Mem:* Am Soc Mammalogists; Am Inst Biol Sci; Sigma Xi; Nat Speleol Soc. *Res:* Taxonomy and natural history of mammals; natural history of Ozark cave communities. *Mailing Add:* Dept Biol Sci Ark State Univ State University AR 72467

MCDANIEL, WILLARD RICH, b Hammond, Ind, June 9, 34; div; c 2. METEOROLOGY, GEOLOGY. *Educ:* Miami Univ, AB, 56, MS, 57; Tex A&M Univ, PhD(meteorol), 67. *Prof Exp:* Geologist, Lion Oil Co, 57-58; assoc prof, 59-74, PROF & HEAD DEPT EARTH SCI, E TEX STATE UNIV, 74- *Mem:* Am Meteorol Soc. *Res:* Applied climate; air pollution. *Mailing Add:* Dept Physics East Tex State Univ Commerce TX 75428

MCDANIEL, WILLIAM FRANKLIN, b Washington, DC, Nov 26, 51; m 75; c 1. BIOPSYCHOLOGY, FUNCTIONAL NEUROANATOMY. *Educ:* Duke Univ, BS, 73; Appalachian State Univ, MA, 74; Univ Ga, PhD(psychol), 77. *Prof Exp:* From asst prof to assoc prof, 77-88, PROF PSYCHOL, GA COL, 88-; NEUROPSYCHOL CONSULT, CENT GA REHAB HOSP, 90- *Concurrent Pos:* Res grants, Fac Res Fund, Ga Col, 77-91, Am Philos Soc, 78- 79 & Nat Sci Found, 80-81. *Honors & Awards:* Zimmer Award in Res, 77. *Mem:* Soc Neurosci; Southern Soc Philos & Psychol; Southeastern Psychol Asn; Int Brain Res Orgn. *Res:* Functions of the neocortex; recovery from brain damage; neocortical mechanisms of vision and spatial behaviors; neural tissue transplantation; neuropeptides and behavior. *Mailing Add:* Dept Psychol Ga Col Milledgeville GA 31061

MCDANIEL, WILLIE L(EE), JR, b Montgomery, Ala, Sept 19, 32. ELECTRICAL ENGINEERING, CONTROL SYSTEMS. *Educ:* Ala Polytech Inst, BEE, 57; Miss State Univ, MSEE, 61; Auburn Univ, PhD(elec eng), 65. *Prof Exp:* Design engr, Bell Tel Labs, Inc, 57-59; asst prof elec eng, Miss State Univ, 59-62; res asst flight dynamics, Boeing Co, 63; instr elec eng, Auburn Univ, 63-65; assoc prof, 65-68, assoc dean eng, 71-78, PROF ELEC ENG, MISS STATE UNIV, 68-, DEAN ENG, 78- *Mem:* Inst Elec & Electronics Engrs; Am Soc Eng Educ. *Res:* Optimization techniques for large scale systems; development of design tool aids using the digital computer in an adaptive mode. *Mailing Add:* Col of Eng Miss State Univ PO Box 5328 Mississippi State MS 39762

MCDANIELS, DAVID K, b Hoquiam, Wash, May 21, 29; m 53, 66; c 3. NUCLEAR PHYSICS, SOLAR ENERGY. *Educ:* Wash State Univ, BS, 51; Univ Wash, Seattle, MS, 58, PhD(physics), 60. *Prof Exp:* Physicist, Hanford Atomic Prod Oper, Gen Elec Co, 51-54; res instr physics, Univ Wash, Seattle, 60-61; NSF fel, Nuclear Res Ctr, Saclay, France, 62-63; from asst prof to assoc prof, 63-69, PROF PHYSICS, UNIV ORE, 69-, DEPT HEAD, 86- *Concurrent Pos:* Vis staff mem, Los Alamos Sci Lab, 70-84, Uppsola Univ, Sweden, 79. *Mem:* Int Solar Energy Soc; Am Phys Soc; Am Asn Physics Teachers; Sigma Xi; AAAS. *Res:* Fast neutron radiative capture; inelastic proton scattering at medium energies; solar radiation; solar energy applications to flat-plate collector systems; giant multipole resonances; Electron intensity in radiation belts. *Mailing Add:* Dept Physics Univ Ore Eugene OR 97403

MCDAVID, JAMES MICHAEL, b Kingsport, Tenn, Dec 8, 45. MATERIALS SCIENCE. *Educ:* Univ Tenn, Knoxville, BSEPhys, 68; Univ Wash, MS, 69, PhD(physics), 75. *Prof Exp:* Process & metall engr, Northwest Div, Davis Walker Corp, 76-77; res assoc & mem fac elec eng, Univ Wash, 77-81; mem tech staff & prog engr, 81-89, sematech assignee, 89-91, SR MEM TECH STAFF, TEX INSTRUMENTS, 91- *Mem:* Am Phys Soc. *Res:* Advanced processes for very-large-scale integrated circuits; high-reliability materials; electron devices. *Mailing Add:* 5550 Spring Valley Unit #F-24 Dallas TX 75240

MCDERMED, JOHN DALE, b Alva, Okla, Dec 20, 41; m 85. MEDICINAL CHEMISTRY. *Educ:* Okla State Univ, BS, 63; Univ Tex, Austin, PhD(org chem), 68. *Prof Exp:* NIH fel chem, Ind Univ, 68-69; sr res scientist med chem, 70-79, PRIN SCIENTIST, WELLCOME RES LABS, 79- *Concurrent Pos:* Lectr chem, Eve Col, Univ NC, 74-76. *Mem:* Am Chem Soc; AAAS. *Res:* Neurotransmitters; dopamine receptors, agonists and antagonists; peptides; cardiovascular drugs. *Mailing Add:* Div Org Chem Burroughs Welcome Co 3030 Cornwallis Rd Research Triangle Park NC 27709

MCDERMOTT, DANA PAUL, b Huntington, WVa, Nov 16, 48; m 78; c 1. MOLECULAR SPECTROSCOPY, VIBRATIONAL ANALYSIS. *Educ:* Centre Col, AB, 70; Univ Calif, Berkeley, PhD(phys chem), 74. *Prof Exp:* Res fel nuclear inorg chem, Argonne Nat Lab, 75; prof phys chem, Univ Mali, 75-77; asst prof gen chem, Ind Univ E, 78-81; ASST PROF GEN & PHYS CHEM, LAFAYETTE COL, 81- *Mem:* Am Chem Soc; Soc Appl Spectros. *Res:* Studies of vibrational motions of cyclic amides and peptides by infrared and Raman spectroscopy; determination of internal rotational barriers about bonds in amide groups by far-infrared spectroscopy. *Mailing Add:* Lees Col Jackson KY 41339-1196

MCDERMOTT, DANIEL J, MEDICAL AFFAIRS. *Educ:* Georgetown Univ, Washington, DC, BS, 58; Univ Minn, Minneapolis, MS, 65; Marquette Univ, Milwaukee, Wis, PhD(physiol), 69. *Prof Exp:* Res assoc, Med Col Wis, Milwaukee, 64-72, asst prof physiol, 72-73, asst clin prof, 73-76; asst dir cardiovasc clin res, G D Searle & Co, Skokie, Ill, 73-75, assoc dir, 75-80; vpres & dir med res, Smith Labs, Northbrook, Ill, 80-86; assoc dir clin & tech affairs, Boots Co, Inc, Lincolnshire, Ill, 86-88; DIR CLIN RES, GENSIA PHARMACEUT, INC, 88- *Concurrent Pos:* Mem, Comt Allergy Chymopapain, Nat Inst Allergy & Infectious Dis. *Mem:* Fel Royal Soc Med; assoc Am Physiol Soc; Am Soc Clin Pharmacol & Exp Therapeut; NY Acad Sci; Am Heart Asn; Reticuloendothelial Soc; AAAS; Sigma Xi. *Res:* Author of various publications. *Mailing Add:* Dept Med Affairs Gensia Pharmaceut Inc 11025 Roselle St San Diego CA 92121-1207

MCDERMOTT, JOHN FRANCIS, b Hartford, Conn, Dec 12, 29; m; c 2. PSYCHIATRY, CHILD PSYCHIATRY. *Educ:* Cornell Univ, AB, 51; NY Med Col, MD, 55; Am Bd Psychiat, cert psychiat, 62, cert child psychiat, 65. *Prof Exp:* Intern, Henry Ford Hosp, 55-56; resident psychiat, Med Ctr, Univ Mich, 56-58, child psychiatry, 60-62, from instr to assoc prof, 62-69; chmn, Comt Cert Child Psychiat, 73-78, dir, 83-90, emer dir, Am Bd Psychiat & Neurol, Inc, 91-; PROF PSYCHIAT & CHMN DEPT, SCH MED, UNIV HAWAII, 69- *Honors & Awards:* Scholar-in-Residence, Rockefeller Found, Bellagio, Italy, 85. *Mem:* Fel Am Psychiat Asn; Am Orthopsychiat Asn; Am Col Psychiatrists; Am Acad Child Psychiat. *Res:* Male-female adolescent personality development; assessment of medical and psychiatric competence; cross-cultural study of family functioning. *Mailing Add:* Univ Hawaii Sch Med 1356 Lusitana St Honolulu HI 96813

MCDERMOTT, JOHN JOSEPH, b Newark, NJ, May 31, 27; m 54; c 2. PARASITOLOGY, MARINE BIOLOGY. *Educ:* Seton Hall, BS, 49; Rutgers Univ, MS, 51, PhD(zool), 54. *Prof Exp:* Asst biol, genetics & parasitol, Rutgers Univ, 51-53, parasitol, Bur Biol Res, 53-54; res assoc oyster invests, Agr Exp Sta, Oyster Res Lab, 54-56, asst res specialist, 56-58; from asst prof to assoc prof, 58-69, chmn dept, 63-73, PROF BIOL, FRANKLIN & MARSHALL COL, 69- *Concurrent Pos:* Vis prof marine sci, Va Inst Marine Sci, 68-74; dir summer course, Bermuda Biol Sta, 77-78; Fulbright Scholar, Denmark, 89. *Mem:* AAAS; Am Soc Parasitol; Ecol Soc Am; Am Soc Zool; Crustacean Soc; Estuarine Res Fed. *Res:* Marine biology and ecology; biology of marine symbiotic relationships; nemertean feeding. *Mailing Add:* Dept of Biol Franklin & Marshall Col Box 3003 Lancaster PA 17604-3003

MCDERMOTT, KEVIN J, b Teaneck, NJ, Nov 21, 35; m 59; c 4. INDUSTRIAL ROBOT RESEARCH, FLEXIBLE MANUFACTURING SYSTEMS. *Educ:* NJ Inst Technol, BS, 65; Columbia Univ, MS, 70; Farleigh Dickerson Univ, DEd(educ), 75. *Prof Exp:* ASSOC PROF, INDUST ENG & DIR, COMPUT-AIDED DESIGN/COMPUT-AIDED ROBOTICS CONSORTIUM, NJ INST TECHNOL, 83- *Concurrent Pos:* Chmn eng dept, NJ Inst Technol, 83-; IBM fel, 87. *Mem:* Inst Indust Engrs; fel Soc Mech Engrs; fel Inst Elec & Electronic Engrs. *Res:* Industrial robot work cells; manufacturing systems; expert systems; analysis of industrial of Robotics, Flexible Manufacturing Systems; expert & vision systems in computer aided design and manufacturing; author of over 30 technical papers. *Mailing Add:* NJ Inst Technol King Blvd Newark NJ 07102

MCDERMOTT, LILLIAN CHRISTIE, b New York, NY, Feb 9, 31; m 54; c 3. PHYSICS, SCIENCE EDUCATION. *Educ:* Vassar Col, BA, 52; Columbia Univ, MA, 56, PhD(physics), 59. *Hon Degrees:* DSc, Ripon Col, 91. *Prof Exp:* Instr physics, City Univ New York, 61-62; lectr physics, Seattle Univ, 65-69; from lectr to assoc prof, 67-80, PROF PHYSICS, UNIV WASH, 81- *Concurrent Pos:* Dir, Physics Educ Group, Univ Wash. *Honors & Awards:* Millikan Lect Award, Am Asn Physics Teachers, 90. *Mem:* Sigma Xi; fel Am Phys Soc; Am Asn Physics Teachers; Nat Sci Teachers Asn; fel AAAS. *Res:* Physics education; preparing teachers to teach physics and physical science, investigating conceptual difficulties in physics, developing curriculum for special groups of students. *Mailing Add:* Dept Physics Univ Wash Seattle WA 98195

MCDERMOTT, MARK NORDMAN, b Yakima, Wash, Feb 6, 30; m 54; c 3. OPTICS. *Educ:* Whitman Col, BA, 52; Columbia Univ, MA, 56, PhD(physics), 59. *Prof Exp:* Asst physics, Columbia Univ, 53-59, instr, 60-62; res assoc, Univ Ill, 59-60; from asst prof to assoc prof, 62-74, PROF PHYSICS, UNIV WASH, 74-, CHMN DEPT, 84- *Mem:* Fel Am Phys Soc; AAAS; Am Asn Phys Teachers. *Res:* Radiofrequency spectroscopy; atomic beams magnetic resonance; optical pumping. *Mailing Add:* Dept Physics FM-15 Univ of Wash Seattle WA 98195

MCDERMOTT, MARK RUNDLE, INFECTIOUS DISEASE, VACCINE. *Educ:* McMaster Univ, PhD(immunol), 79. *Prof Exp:* ASST PROF IMMUNOL, MCMASTER UNIV, 82- *Mailing Add:* Dept Path McMaster Univ 1200 Main St W Rm 3N7 Hamilton ON L8N 3Z5 Can

MACDERMOTT, RICHARD J, JR, EXPERIMENTAL BIOLOGY. *Educ:* Oberlin Col, Ohio, BA, 65; Ohio State Univ, Columbus, MD, 69; Am Bd Internal Med, cert, 75; Am Bd Immunol & Allergy, cert, 77; Am Bd Gastroenterol, cert, 79. *Prof Exp:* Intern med, Peter Bent Brigham Hosp, Harvard Med Sch, Boston, Mass, 69-70, resident, 70-71, fel tumor & human cellular immunol, Dana-Farber Cancer Inst, 73-74, NIH sr fel, Dept Immunol, Brigham & Womens Hosp, 84-85; fel gastroenterol, Univ Hosp, Med Sch, Boston Univ, Mass, 71-73; from asst prof to prof med, Wash Univ, Barnes Hosp, St Louis, Mo, 77-90; T GRIER MILLER PROF MED, GASTROINTESTINAL SECT, SCH MED, UNIV PA, PHILADELPHIA, 89-, CHIEF, 89- *Concurrent Pos:* Mem, Grants Rev Comt, Nat Found Iteitis & Colitis, 80-89, chmn, 82-89 & Grants Coun, 90-; mem, Gastroenterol Grants Rev Comt, NIH, 82-85, Mucosal Immunol Prog Proj Site Visit Team, 83, chmn, 84, mem, Gen Internal Med A2 Study Sect, 85-89, Transplantation & Mucosal Immunol Site Unit Team, 88; mem, Gastroenterol Res Group Steering Comt, Am Gastroenterol Asn, 84-90, Res Comt, 85-89, chmn, Immunol & Microbiol Sect, 89- 92; mem, Intestinal Dis Res Visit Site Team, Can Found Iteitis & Colitis, 87-89, Am Asn Study Liver Dis, 90; prin investr, NIH grant, 90-95. *Honors & Awards:* Watman Achievement Award, 69. *Mem:* Am Fedn Clin Res; Am Asn Immunologists; Am Gastroenterol Asn; Am Soc Clin Invest; Am Soc Microbiol; Nat Coalition Immune Syst Disorders; Nat Found Ileitis & Colitis. *Res:* Author of various publications. *Mailing Add:* Sch Med Clin Res Bldg Suite 600 Univ Pa 422 Curie Blvd Philadelphia PA 19104-6144

MCDERMOTT, RICHARD P, b Weston, WVa, Jan 2, 28; m 54. SPEECH PATHOLOGY, AUDIOLOGY. *Educ:* Univ Mo, BA, 52; Western Mich Univ, BS, 53; Ohio State Univ, MA, 55; Univ Iowa, PhD(speech path, audiol), 62. *Prof Exp:* Instr speech path, Univ Iowa, 58-60; asst prof, St Cloud State Col, 60-62, assoc prof speech path & dir speech & hearing serv, 62-64; asst prof speech path, 64-67, assoc prof, 67-77, PROF SPEECH PATH & AUDIOL, UNIV MINN, MINNEAPOLIS, 77-, ASST DIR SPEECH & HEARING CLIN, 64- *Mem:* Am Speech & Hearing Asn; Sigma Xi. *Res:* Speech physiology; articulation skills. *Mailing Add:* Speech Sci 110 Shevlin Univ Minn Minneapolis MN 55455

MCDERMOTT, ROBERT EMMET, b Maywood, Ill, Oct 5, 20; m 43; c 2. FORESTRY. *Educ:* Iowa State Col, BS, 43, MS, 47; Duke Univ, PhD(bot, forestry), 52. *Prof Exp:* Jr forester, Cook County Forest, Ill, 46; instr bot, Iowa State Col, 46-47; forestry, Univ Mo, 48-49; asst bot, Duke Univ, 50-52; from asst prof to assoc prof forestry, Univ Mo, 52-59; prof forestry & head, Dept Forestry & Wildlife, Pa State Univ, 59-69, assoc dean, Grad Sch, 66-69, asst dir, Sch Forest Resources, 59-65; dean, Grad Sch, Univ Ark, Fayetteville, 69-72; provost, Pa State Univ, 72-81, prof forestry, 81-83; RETIRED. *Mem:* AAAS; Ecol Soc Am; Soc Am Foresters. *Res:* Forest ecology and physiology; rural land use. *Mailing Add:* 4305 Long Dr Harrisburg PA 17112

MCDERMOTT, WILLIAM VINCENT, JR, b Salem, Mass, Mar 7, 17; wid; c 3. SURGERY. *Educ:* Harvard Univ, AB, 38, MD, 42. *Prof Exp:* Intern surg, Mass Gen Hosp, Boston, 42-43, from asst resident surgeon to resident surgeon, 46-49; instr, 51-54, clin assoc, 54-57, asst clin prof, 57-63, DAVID CHEEVER PROF SURG, HARVARD MED SCH, 63-; CHMN, DEPT SURG, NEW ENG DEACONESS HOSP, 80- *Concurrent Pos:* NIH res fel, 49-50; fel surg, Mass Gen Hosp, 50-51; consult staff & asst surg, Mass Gen Hosp, 51-54, asst surgeon, 54-57, assoc vis, 57-63, vis, 63-; dir, Fifth Surg Serv & Sears Surg Lab, Boston City Hosp, 63-73; dir, Harvard Surg Serv & Sci Dir, Cancer Res Inst, 73-80. *Mem:* Am Surg Asn; Am Acad Arts & Sci; Soc Univ Surg; Am Col Surg; Am Fedn Clin Res. *Res:* Diseases of the liver and portal circulation; endocrine system; gastrointestinal physiology; cancer. *Mailing Add:* Dept Surg New England Deaconess Hosp 110 Francis St Boston MA 02215

MCDEVITT, DANIEL BERNARD, b Pocatello, Idaho, Apr 14, 27; m 52. COMMUNICATION-COMPUTER-CONTROL SYSTEMS, APPLICATION ENGINEERING. *Educ:* Univ Idaho, BS, 50. *Hon Degrees:* DSc, Univ Karachi, 66 & Univ Riyadh, 77. *Prof Exp:* Systs engr value eng control & distrib systs, Gen Elec Co, 50-54, systs specialist first com appln, UNIVAC I, 54-56, regional mgr power, control & commun systs, 56-59; mgr mkt control, commun syst networks, Nelson Elec Mfg Co, 59-62; pres systs appln, Progress Eng & Consult Enterprises, 65-67; pres, I-C Computer Corp, 67-69; OWNER COMPUTER COMMUN & CONTROL SYSTS, DAN B MCDEVITT & ASSOC, 62-; CHMN & PRES INFO COMMUN MGT, ICM COMPUTER CORP, 70-; PRES & CHIEF EXEC OFFICER, PROGRESS ENG, INC, 75- *Concurrent Pos:* Lectr, Univ Okla & Tulsa Univ, 62-64; consult, US Agency Int Develop, 63-66; pres, Res & Develop Inst US, 63-, Saudi-Am Group, 76-82, Ethanol Mkt & Refining Group, 79-82; chmn, First Select Acad, Govt & Bus Conf, 65-66 & US Coalition for Clear Air, 70-82; adv, US Small Bus Admin, 65-67; comn mem, King Faisal Air Acad Develop Comn, Kingdom Saudi Arabia, 74-77. *Honors & Awards:* Cert Appreciation, Kingdom Iran, Ministry Elect & Water, 66; Medal, Dept Defense, Kingdom Saudi Arabia, 75. *Mem:* AAAS; sr mem Inst Elec & Electronics Engrs; Nat Soc Prof Engrs. *Res:* Application engineering-interactive, distributed, communication-computer and control systems and networks; automation systems engineering; dynamic models. *Mailing Add:* PO Drawer 700270 Tulsa OK 74170-0270

MCDEVITT, DAVID STEPHEN, b Philadelphia, Pa, Nov 15, 40; m 63; c 2. DEVELOPMENTAL BIOLOGY. *Educ:* Villanova Univ, BS, 62; Bryn Mawr Col, MA, 63, PhD(biol), 66. *Hon Degrees:* MA, Univ Pa. *Prof Exp:* Investr, Biol Div, Oak Ridge Nat Lab, 66-68; from asst prof to assoc prof, 68-83, PROF ANIMAL BIOL, SCH VET MED, UNIV PA, 83- *Concurrent Pos:* Consult, Biol Div, Oak Ridge Nat Lab, 69-; sr res fel, Cancer Res Campaign, UK, 74-75; guest res scientist, NIH, 83-84. *Mem:* AAAS; Int Soc Develop Biologists; Am Soc Cell Biol; Asn Res Vision & Ophthal. *Res:* Lens-specific proteins in lens development and lens regeneration as studied by biochemical and immunological techniques; structure of crystallin genes. *Mailing Add:* Dept of Animal Biol Univ of Pa Sch of Vet Med Philadelphia PA 19104

MCDEVITT, HUGH O'NEILL, b Cincinnati, Ohio, Aug 26, 30. IMMUNOLOGY. *Educ:* Stanford Univ, AB, 52; Harvard Univ, MD, 55; Am Bd Internal Med, dipl, 68. *Hon Degrees:* Dr, Univ Paris, 90. *Prof Exp:* Intern, Peter Bent Brigham Hosp, Boston, 55-56, sr asst resident med, 61-62; asst resident, Bell, 56-57; res fel, dept bact & immunol, Harvard Univ, 59-61; USPHS spec fel, Nat Inst Med Res, Mill Hill, London, Eng, 62-64; from asst prof to assoc prof med, Stanford Univ, 66-78, chief, Div Immunol, Dept Med, 70-76, dir, Clin Immunol Lab, 71-78, prof med microbiol & med, 78-88, Joseph D Grant prof med, 80-89, chmn dept, 86-90, PROF MICROBIOL, IMMUNOL & MED, SCH MED, STANFORD UNIV, 88-, BURT & MARION AVERY PROF IMMUNOL, 89- *Concurrent Pos:* Physician, Stanford Univ Hosp, 66-; consult physician, Vet Admin Hosp, Palo Alto, Calif, 68-; mem, var adv comts, NIH, 68-; sr investr, Arthritis Found, 68, Russell Cecil fel, 72-73; assoc ed, J Immunol, 71-75; Nat Cancer Inst outstanding investr award, 89-; K P Chang vis prof, Univ Hong Kong, 91. *Honors & Awards:* Dyer Lectr, NIH, 75; Borden Award, 77; Alena Lengerova Mem Lectr, 80; Passano Award, 81; Albion O Bernstein Award, 83; Lita Annenberg Hazen Award, 85; Kinyoun Lectr, NIH, 87; Kroc Lectr, Brigham & Women's Hosp, 88; Joseph T Bunim Lectr, Am Rheumatism Asn, 88. *Mem:* Nat Acad Sci; Inst Med-Nat Acad Sci; Am Fedn Clin Res; Am Soc Clin Invest; Am Asn Immunologists (pres, 81-82); Asn Am Physicians; Am Acad Arts & Sci; AAAS. *Mailing Add:* Dept Med Microbiol Stanford Univ Hosp D-345 Fairchild Stanford CA 94305

MACDIARMID, ALAN GRAHAM, b Masterton, NZ, Apr 14, 27; m 54; c 4. INORGANIC CHEMISTRY, SOLID STATE CHEMISTRY. *Educ:* Univ NZ, BSc, 48, MSc, 50; Univ Wis, MS, 52, PhD(chem), 53; Cambridge Univ, PhD(chem), 55. *Prof Exp:* Asst lectr chem, St Andrews Univ, 55; from instr to assoc prof, 55-64, PROF CHEM, UNIV PA, 64- *Concurrent Pos:* Sloan fel, Univ Pa, 59-63. *Mem:* Am Chem Soc; Royal Soc Chem. *Res:* Preparation and characterization of organosilicon compounds, derivatives of sulfur nitrides and quasi one-dimensional semiconducting and metallic covalent polymers such as polyacetylene and its derivatives. *Mailing Add:* Dept Chem Univ Pa Philadelphia PA 19104-6323

MCDIARMID, DONALD RALPH, b Vancouver, BC, Apr 6, 37; m 71; c 2. IONOSPHERIC & MAGNETOSPHERIC PHYSICS. *Educ:* Univ BC, BASc, 60, MASc, 61, PhD(elec eng), 65. *Prof Exp:* Asst res officer, 65-71, assoc res officer, 71-80, SR RES OFFICER, IONOSPHERIC & MAGNETOSPHERIC PHYSICS, SOLAR TERRESTRIAL PHYSICS SECT, HERZBERG INST ASTROPHYSICS, NAT RES COUN CAN, 80- *Mem:* Am Geophys Union; Can Asn Physicists; Inst Elec & Electronics Engrs. *Res:* Study of radio aurora; mechanisms and relationships to other auroral and magnetospheric phenomena; theoretical and observational study of geomagnetic pulsations. *Mailing Add:* Solar Terrestrial Physics Sect Nat Res Coun Ottawa ON K1A 0R6 Can

MCDIARMID, IAN BERTRAND, b Carleton Place, Ont, Oct 1, 28; m 51; c 2. SPACE PHYSICS. *Educ:* Queen's Univ, Can, BA, 50, MA, 51; Univ Manchester, PhD(physics), 54. *Prof Exp:* Head cosmic ray sect, 55-69, asst dir div physics, 69-75, asst dir, Herzberg Inst Astrophysics, 75-80, DIR CAN CTR SPACE SCI, NAT RES COUN, CAN, 80- *Concurrent Pos:* Sessional lectr, Univ Ottawa, 56- *Mem:* Royal Soc Can; Am Geophys Union; Can Asn Physicists. *Res:* Cosmic rays; high energy particle physics; space research; rocket borne particle detectors. *Mailing Add:* 1106-60 Mc Leod Ottawa ON K2P 2G1 Can

MCDIARMID, ROY WALLACE, b Santa Monica, Calif, Feb 18, 40; m 87. HERPETOLOGY, MUSEUM CURATOR. *Educ:* Univ Southern Calif, AB, 61, MS, 66, PhD(biol), 69. *Prof Exp:* Instr biol, Univ Chicago, 68-69; asst prof biol, Univ SFla, 69-77, assoc prof, 77-79; RES ZOOLOGIST, NAT MUS NATURAL HIST, US FISH & WILDLIFE SERV, 79- *Concurrent Pos:* Res assoc Los Angeles County Mus Natural Hist, 69-; coordr advan biol course, Orgn Trop Studies, 71; zool coordr, Cerro de la Neblinn Expedition, 83-86; assoc ed herpetol, Am Nat Mus, 86- *Mem:* Am Soc Naturalists; Herpetologists' League (pres, 80-82); Am Soc Ichthyologists & Herpetologists; Asn Trop Biol; Ecol Soc Am. *Res:* Major patterns of evolution in anuran amphibians; systematics of neotropical amphibians and reptiles; biogeography and ecology of neotropical vertebrates; biosystematics of neotropical frogs of families Bufonidae and Centrolenidae. *Mailing Add:* Biol Survey Sect Nat Mus Natural Hist US Fish & Wildlife Serv Washington DC 20560

MCDIARMID, RUTH, b New York, NY, Feb 3, 38; c 2. BIOCHEMISTRY, QUANTUM CHEMISTRY. *Educ:* Cornell Univ, BA, 60; Radcliffe Col, MA, 62; Harvard Univ, PhD(biochem), 65. *Prof Exp:* Postdoctoral fel, 65-71, RES CHEMIST, NAT INST DIABETES, DIGESTIVE & KIDNEY DIS, NIH, 71- *Concurrent Pos:* Prog mgr, Chem Div, NSF, 86-87. *Mem:* Am Chem Soc; Am Physics Soc; AAAS. *Res:* Electronic excited state spectroscopy and dynamics; intramolecular energy transfer assessment of quantum mechanical models and calculation of electronic excited states; properties of electronic excited states. *Mailing Add:* NIH 9000 Rockville Pike Bldg 2 Rm B1-07 Bethesda MD 20892

MACDIARMID, WILLIAM DONALD, b Arcola, Sask, June 22, 26; m 53; c 4. INTERNAL MEDICINE, HUMAN GENETICS. *Educ:* Univ Sask, BA, 47; Univ Toronto, MD, 49; Am Bd Internal Med, dipl, 67; FRCPS(C) & cert, 67 & 77. *Prof Exp:* Instr internal med, Col Med, Univ Utah, 64-66, from asst prof to assoc prof, 66-69; prof med, fac med, Univ Man, 69-75; prof & chmn med, fac med, Mem Univ Nfld, 75-79; chmn dept, Univ Man, 79-85, prof med, 79-91, prof genetics, 84-91; RETIRED. *Concurrent Pos:* NIH training grant, Univ Utah, 60-62; Neuromuscular Found fel, Univ London, 62-64; consult, Winnipeg Children's Hosp, 69-75, 84-91 & Winnipeg Gen Hosp, 70-91; physician-in-chief, St Boniface Gen Hosp; sr consult, Janeway Child Health Ctr, St Clare's Mercy Hosp, 75-79; active staff, Health Sci Ctr, Winnipeg, 79-91, physician in chief, 79-85; prof pediat, Univ Man, 86-91, chmn Bd, Man Cancer Found, 87- 88. *Mem:* fel Am Col Physicians; fel Can Col Med Geneticists; Genetics Soc Can; Am Soc Human Genetics; fel Royal Soc Med; Soc Med Bioethics Can. *Res:* Genetics; bioethics. *Mailing Add:* 4142 Cortez Pl Victoria BC V8N 4R5 Can

MCDIFFETT, WAYNE FRANCIS, b Uniontown, Pa, Sept 2, 39; m 65; c 1. ECOLOGY. *Educ:* WVa Univ, AB, 61, MS, 64; Univ Ga, PhD(zool), 70. *Prof Exp:* Instr biol, Frostburg State Col, 63-65; asst prof, 69-74, ASSOC PROF BIOL, BUCKNELL UNIV, 74- *Mem:* Ecol Soc Am; Am Soc Limnol & Oceanog. *Res:* Fresh-water ecology; productivity and energy relationships. *Mailing Add:* Dept of Biol Bucknell Univ Lewisburg PA 17837

MCDIVITT, MAXINE ESTELLE, b Rosedale, Ind, Aug 16, 12. FOOD SCIENCE, NUTRITION. *Educ:* Univ Ill, BS, 40, MS, 41; Univ Wis, PhD(foods & nutrit), 52. *Prof Exp:* Instr home econ, Univ Mo, 41-44, asst prof, 45-47; instr, Univ Iowa, 47-49; from assoc prof to prof, Univ Wis-Madison, 52-67; prof, 67-77, EMER PROF HOME ECON, UNIV WIS-MILWAUKEE, 77- *Concurrent Pos:* On leave, UN Food & Agr Orgn, India, 64-66; USDA grant, Univ Wis-Milwaukee, 67-77; Danforth Assoc. *Honors & Awards:* Lita Bane Lect, Univ Ill, 68. *Mem:* Am Home Econ Asn; Am Dietetics Asn. *Res:* Bacteriological aspects of food handling procedures; heat resistance of food poisoning organisms; heat transfer in foods; food habits in urban communities; consumer acceptance of dairy products; studies of milk flavor. *Mailing Add:* 3916 N Oakland Ave #124 Shorewood WI 53211

MCDIVITT, ROBERT WILLIAM, b West Sunbury, Pa, Mar 30, 31; m 61; c 2. PATHOLOGY. *Educ:* Harvard Univ, BA, 53; Yale Univ, MD, 56. *Prof Exp:* Attend pathologist, Mem-Sloan Kettering Cancer Ctr, 64-71; assoc prof surg, Med Sch, Cornell Univ, 68-71; prof path, Med Sch, Univ Utah, 71-; prof, Dept Path, 79-87, DIR ANAT PATH, WASH UNIV SCH MED, 87- *Concurrent Pos:* Chmn, Southwest Oncol Group, 75-; chmn, DCCR Path Working Group, Nat Cancer Inst, 77-78, DCT Path Working Group, 78 &

DCT Path Task Force, 78-; consult, Ctr for Dis Control, Atlanta, 78; Surg Pathologist in chief, Barnes Hosp, St Louis, 87- *Honors & Awards:* Am Soc Clin Oncol. *Mem:* Soc Surg Oncol; Int Acad Path; Am Soc Clin Path; Arthur Purdy Stout Soc; Sigma Xi. *Res:* Etiology and epidemiology of human breast cancer; breast cancer kinetics and oncogene expression. *Mailing Add:* Dept Path Wash Univ Sch Med 660 S Euclid Ave PO Box 8118 St Louis MO 63110

MCDOLE, ROBERT E, b Eugene, Ore, Oct 7, 30; c 3. SOIL FERTILITY, SOIL GENESIS. *Educ:* Ore State Univ, BS, 52; Univ Idaho, MS, 68, PhD(soils), 69. *Prof Exp:* Soil scientist, Bur Indian Affairs, US Dept Interior, Idaho, Ore & Wash, 52-65; from asst res prof to assoc res prof soils, Aberdeen Br Sta, 69-77, EXTEN SOIL SPECIALIST, DEPT PLANT & SOIL SCI, UNIV IDAHO, 77- *Concurrent Pos:* Vis prof, Wash State Univ, 74-75. *Mem:* Soc Agron; Soil Sci Soc Am; Soil Conserv Soc Am; Sigma Xi. *Res:* Soil surveys of Northwest Indian reservations; study of loess soil materials in Southern Idaho; soil fertility of potato production; soil fertility in small grains and pulse crops. *Mailing Add:* Dept of Plant & Soil Sci Univ of Idaho Moscow ID 83843

MCDONAGH, JAN M, b Wilmington, NC, Nov 9, 42; m 68, 73; c 1. BIOCHEMISTRY, PATHOLOGY. *Educ:* Wake Forest Univ, BS, 64; Univ NC, PhD(biochem), 68. *Prof Exp:* NIH fels, Univ NC, Chapel Hill, 68-69, Med Univ Klinik, Basel, Switz, 69-70 & Karolinska Inst, 70-71; asst prof path, Med Sch, Univ NC, Chapel Hill, 71-76, asst prof biochem, 74-76, assoc prof path & biochem, 76-82; ASSOC PROF PATH, HARVARD MED SCH, BOSTON, 82- *Concurrent Pos:* Mem coun thrombosis, Am Heart Asn, 72-, estab investr, 77-78; mem, NIH Study Sect, Hemat. *Mem:* Int Soc Thrombosis & Haemostasis; Am Chem Soc; Am Soc Path; Am Soc Biol Chem; Am Soc Hemat; Japan Hematol Soc. *Res:* Biochemistry of blood coagulation and fibrinolysis; pathogenesis of thrombosis. *Mailing Add:* Dept Path Harvard Med Sch Beth Isreal Hosp Boston MA 02215-5491

MCDONALD, ALAN T(AYLOR), b Los Angeles, Calif, Oct 11, 38; m 58, 82; c 6. MECHANICAL ENGINEERING. *Educ:* Purdue Univ, BS, 60, MS, 62, PhD(mech eng), 65. *Prof Exp:* Asst prof mech eng, Univ Calif, Davis, 64-67; assoc prof, 67-74, PROF MECH ENG, SCH MECH ENG, PURDUE UNIV, WEST LAFAYETTE, 74- *Mem:* Am Soc Mech Engrs; Soc Automotive Engrs; Am Inst Aeronaut & Astronaut; Am Soc Eng Educ; Sigma Xi. *Res:* Fluid mechanics. *Mailing Add:* Sch Mech Eng Purdue Univ West Lafayette IN 47907

MACDONALD, ALEX BRUCE, b Anaconda, Mont, Mar 7, 34; m 59; c 2. IMMUNOLOGY, BIOCHEMISTRY. *Educ:* Carroll Col, Mont, AB, 56; Mich State Univ, PhD(biochem), 67. *Prof Exp:* Chemist, Anaconda Aluminum Co, Mont, 56-57; biochemist, Abbott Labs, 59-63; asst biochem, Mich State Univ, 63-67; res assoc, 69-70, asst prof microbiol, 70-75, assoc prof immunol, Sch Public Health, Harvard Univ, 75-78; head dept microbiol, 78-84, PROF MICROBIOL, UNIV MASS, 84- *Concurrent Pos:* NIH fel immunol, Med Sch, Univ Ill, 67-69; vis lectr microbiol, Harvard Univ, 78- *Mem:* Fedn Am Socs Exp Biol; Am Asn Immunol. *Res:* Immunological response to infectious agents, including chlamylia and virus studies on possible competition between substances in immune response leading to detrimental effects in the host. *Mailing Add:* Dept Microbiol Univ Mass Amherst MA 01003

MACDONALD, ALEXANDER, JR, b Quincy, Mass, Oct 29, 36; m 62; c 1. ANALYTICAL CHEMISTRY. *Educ:* Northeastern Univ, BS, 59; Univ Iowa, MS & PhD(chem), 63. *Prof Exp:* Asst prof chem, Mich State Univ, 65-67; sr chemist, 67-71, res fel, 71-74, res group chief, 74-75, ASST DIR ANIMAL HEALTH RES, HOFFMAN-LA ROCHE, INC, 76- *Mem:* Am Chem Soc. *Res:* Chromatography; aerosols; automated analysis; residue analysis; veterinary drug metabolism. *Mailing Add:* 16 Cypress Ave North Caldwell NJ 07006

MACDONALD, ALEXANDER DANIEL, b Sydney, NS, Apr 8, 23; m 46; c 4. PHYSICS. *Educ:* Dalhousie Univ, BSc, 45, MSc, 47; Mass Inst Technol, PhD(physics), 49. *Prof Exp:* Res assoc physics, Mass Inst Technol, 48-49; from asst prof to prof, Dalhousie Univ, 49-60; sr specialist, Microwave Physics Lab, Gen Tel & Electronics Lab, Inc, 60-62; prof appl math & head div, Dalhousie Univ, 62-65; sr mem res labs, Lockheed Missiles & Space Co, 65-73, dir, Electronic & Commun Sci Lab, Palo Alto Labs, 73-78, chief scientist, Space Systs Div, 78-80; CONSULT, 81- *Concurrent Pos:* Res scientist, Defence Res Bd, Can, 49-52; mem nat comn electronics, Int Union Radio Sci, 52-; specialist, Sylvania Elec Prod Co, 56-57. *Mem:* Fel Am Phys Soc. *Res:* Physical electronics; ultrasonics; theoretical physics; computer science. *Mailing Add:* 3056 Greer Rd Palo Alto CA 94303

MCDONALD, ALISON DUNSTAN, epidemiology, for more information see previous edition

MACDONALD, ALLAN HUGH, b Antigonish, NS Can, Dec 1, 51; m 74; c 2. STRONG-CORRELATIONS, ELECTRONIC PROPERTIES. *Educ:* St Francis Xavier Univ, BSc, 73; Univ Toronto, MSc, 74, PhD(physics), 78. *Prof Exp:* Res assoc, Nat Res Coun Can, 78-79; res scientist, Nat Res Coun Can, 80-87; PROF IND UNIV, 87- *Concurrent Pos:* Vis prof, ETH-Hongerburg, Zurich, Switzerland, 82-83; deputy chmn, Condenser Mater Div, chmn, 87 & post chmn Can Asn Physicists, 88; vis scientist, Max Plank Inst Festkorperforshung, Stutgart, WGer, 88. *Honors & Awards:* Herzberg Medal, Can Asn Physicists, 87. *Mem:* Can Asn Physicist; Am Phys Soc. *Res:* Theory condensed matter especially electronic properties in superconductors & two-dimensional systems. *Mailing Add:* Dept Physics Ind Univ Bloomington IN 47401

MCDONALD, ARTHUR BRUCE, b Sydney, NS, Aug 29, 43; m 66; c 4. PHYSICS. *Educ:* Dalhousie Univ, BSc, 64, MSc, 65; Calif Inst Technol, PhD(physics), 70. *Prof Exp:* Nat Res Coun Can & Rutherford Mem fels, 70-71; res physicist, Chalk River Nuclear Labs, Atomic Energy Can, Ltd, 71-82; PROF PHYSICS, PRINCETON UNIV, 82- *Concurrent Pos:* Vis scientist, Univ Wash, 77 & Los Alamos Nat Lab, 81. *Mem:* Fel Am Phys Soc; Can Asn Physicists. *Res:* Nuclear physics, especially investigations of structure of nuclei using particle accelerators; weak interaction measurements. *Mailing Add:* Physics Dept Queens Univ Kingston ON K7I 3N6 Can

MCDONALD, BARBARA BROWN, b Ray, Ariz, Feb 14, 24; m 52. CYTOLOGY. *Educ:* Simmons Col, BS, 48; Columbia Univ, MA, 55, PhD, 57. *Prof Exp:* From instr to assoc prof, 56-73, PROF BIOL, DICKINSON COL, 73- *Mem:* AAAS; Am Soc Cell Biol; Histochem Soc; Genetics Soc Am; Soc Protozool. *Res:* Cytochemistry; protozoology; nucleic acids in Tetrahymena. *Mailing Add:* 1152 Rock Ledge Dr Carlisle PA 17013

MCDONALD, BERNARD ROBERT, b Kansas City, Kans, Nov 17, 40; m 63; c 2. MATHEMATICS. *Educ:* Park Col, BA, 62; Kans State Univ, MA, 64; Mich State Univ, PhD(math), 68. *Prof Exp:* From asst prof to prof math, Univ Okla, 68-85, chmn dept, 81-84; prog dir algebra & number theory prog, 83-86, PROG DIR & HEAD OFF SPEC PROJS, DIV MATH SCI, NSF, WASHINGTON, DC, 86-, DEP DIV DIR, 88- *Concurrent Pos:* Vis prof, Pa State Univ, 74-75, Queen's Univ, 79 & Univ Calif, Santa Barbara, 80-81. *Mem:* Am Math Soc; Math Asn Am; Soc Indust & Appl Math. *Res:* Algebra; matrix theory; commutative rings; finite rings; combinatorial theory. *Mailing Add:* Div Math Sci Rm 339 Nat Sci Found 1800G St NW Washington DC 20550

MCDONALD, BRUCE EUGENE, b Mannville, Alta, Apr 30, 33; m 62; c 5. NUTRITION, FOOD SERVICE MANAGEMENT. *Educ:* Univ Alta, BSc, 58, MSc, 60; Univ Wis, PhD(biochem, poultry nutrit), 63. *Prof Exp:* Res assoc, Univ Ill, 63-64; asst prof animal sci, Macdonald Col, McGill Univ, 64-68; from assoc prof to prof nutrit, 68-77, dean, fac human ecol, 77-85, PROF FOOD & NUTRIT, UNIV MAN, 85- *Concurrent Pos:* Vis prof, Inst Nutrit, Univ Uppsala, Sweden, 76-77, Sch Hotel & Food Admin, Univ Guelph, 86-87. *Mem:* Nutrit Soc Can; Am Inst Nutrit; Can Inst Food Sci & Technol; Am Oil Chemists Soc; Can Dietetic Asn. *Res:* Effect of dietary fat on lipid metabolism in the human; biochemistry of lipids. *Mailing Add:* Dept Foods & Nutrit Fac Human Ecol Univ Man Winnipeg MB R3T 2N2 Can

MACDONALD, CAROLYN TROTT, b Iowa City, Iowa, June 23, 41; c 3. APPLIED MATHEMATICS, MATHEMATICS EDUCATION. *Educ:* Univ Minn, BS, 61; Univ Ore, MA, 63 & 65; Brown Univ, PhD(chem), 68. *Prof Exp:* Asst prof chem, Moorhead State Col, 68-71; asst prof phys sci & chmn dept, Univ Mo-Kansas City, 72-76; assoc prof math & physics & chairperson dept, Baptist Col, Charleston, SC, 76-79; SDIP coordr, Rockhurst Col, Kansas City, Mo, 79-81; LECTR BUS OPERS & ANAL, UNIV MO-KANSAS CITY, 83-; INSTR DATA PROCESSING, PENN VALLEY COMMUNITY COL, 83-; SELF EMPLOYED. *Concurrent Pos:* Proj Dir, NSF grants, 74-75, 78-79 & 81; consult, 81- *Mem:* Int Coun Comput Educ; Asn Women Math; Math Asn Am. *Res:* Mathematical modeling. *Mailing Add:* 5412 Locust St Kansas City MO 64110

MCDONALD, CHARLES JOSEPH, b Cambridge, Mass, Feb 20, 41. POLYMER COLLOIDS, POLYMER REACTIONS. *Educ:* Boston Col, BS, 63; Holy Cross Col, MS, 64; Clarkson Col Technol, PhD(polymer chem), 70. *Prof Exp:* Res chemist, Immont Corp, 64-66; fel, Uppsala Univ, Sweden, 70-73; ASSOC SCIENTIST, DOW CHEM CORP, 73- *Mem:* Am Chem Soc; Sigma Xi; AAAS. *Res:* Polymer solution properties; emulsion technology. *Mailing Add:* 2915 Oakhaven Ct Midland MI 48640-8847

MCDONALD, CLARENCE EUGENE, b McPherson, Kans, Oct 10, 1926; m 55; c 3. CEREAL CHEMISTRY. *Educ:* McPherson Col, AB, 50; Kans State Univ, MS, 53; Purdue Univ, PhD(biochem), 57. *Prof Exp:* Asst, Kans State Univ, 52-53; asst, Purdue Univ, 53-56, res fel, 57; chemist, Western Utilization Res & Develop Div, Agr Res Serv, USDA, 57-64; assoc prof, 64-77, PROF CEREAL SCI & FOOD TECHNOL, NDAK STATE UNIV, 77- *Mem:* Am Asn Cereal Chem; Inst Food Technol. *Res:* Isolation and characterization of wheat lipids, proteins and enzymes; analysis of cereals (analytical chem) and their products. *Mailing Add:* Dept Cereal Sci & Food Technol N Dak State Univ Fargo ND 58105

MCDONALD, DANIEL JAMES, b New York, NY, Jan 27, 25; m 52. GENETICS. *Educ:* Siena Col, BS, 50; Columbia Univ, MA, 52, PhD(zool), 55. *Prof Exp:* Lectr zool, Columbia Univ, 53-55, instr, 55-56; from asst prof to prof biol, 56-83, EMER PROF BIOL, DICKINSON COL, 83- *Concurrent Pos:* Instr, Long Island Univ, 55-56. *Mem:* AAAS; Am Soc Nat; Ecol Soc Am; Genetics Soc Am; Soc Study Evolution. *Res:* Population genetics; ecology of the flour beetles Tribolium confusum and Tribolium castaneum. *Mailing Add:* Dept Biol Dickinson Col Carlisle PA 17013

MCDONALD, DANIEL PATRICK, b Los Angeles, Calif, Oct 30, 46; m 66; c 2. RESEARCH ADMINISTRATION. *Educ:* Univ Calif, BS, 69; Univ Colorado, PhD(inorganic chem), 74. *Prof Exp:* Sr res chemist, Williams Co, 75-77; res mgr, Mi Chem Corp, 77-86; RES MGR, FMC LITHIUM DIV, 86- *Mem:* Am Chem Soc; Am Ceramic Soc; Sigma Xi. *Res:* Inorganic chemical process; mineral deposits. *Mailing Add:* 501 Shadowview Gastonia NC 28054

MACDONALD, DAVID HOWARD, b Cleveland, Ohio, Sept 23, 34; m 56; c 3. PLANT NEMATOLOGY, EDUCATION. *Educ:* Purdue Univ, BS, 56; Cornell Univ, MS, 62, PhD(pomol), 66. *Prof Exp:* From asst prof to assoc prof plant nematol, 66-74, assoc prof plant path, 74-77, PROF PLANT PATH, UNIV MINN, ST PAUL, 77- *Mem:* Am Phytopath Soc; Soc Nematol. *Res:* Biological, chemical and physical factors affecting plant parasitic nematodes. *Mailing Add:* Dept Plant Path Univ Minn 1991 Upper Buford Cr St Paul MN 55108

MACDONALD, DAVID J, b San Diego, Calif, May 14, 32; m 62; c 2. PHYSICAL CHEMISTRY, INORGANIC CHEMISTRY. *Educ:* Calif Inst Technol, BS, 53, MS, 54; Univ Calif, Los Angeles, PhD(phys chem), 60. *Prof Exp:* Chem engr, Apache Powder Co, 54-55; res asst, Univ Calif, Los Angeles,

60; asst prof chem, Univ Nev, Reno, 63-68; RES CHEMIST, RENO RES CTR, US BUR MINES, 68- *Res:* Kinetics of substitution reactions in coordination compounds; extractive metallurgy; liquid-liquid extraction. *Mailing Add:* Reno Res Ctr US Bur Mines 1605 Evans Ave Reno NV 89512-2295

MCDONALD, DAVID WILLIAM, b Shreveport, La, Aug 4, 23; m 48; c 4. ORGANIC CHEMISTRY. *Educ:* Southwestern La Univ, BS, 43; Univ Tex, PhD(chem), 51. *Prof Exp:* Res chemist, Monsanto Chem Co, Mo, 51-54, res group leader, 56-59, res sect leader, 59-64, mgr polyolefins res, Tex, 64-67, prod adminr polyolefins, 67-69, dir res & develop, Plastic Div, 69-74, dir technol, 74-80, dir technol plans corp res & develop staff, 80-82; sr assoc, Technol Mgt Group, 83-89, DEP DIR, PUGH-ROBERTS ASSOC, CAMBRIDGE, MASS, 89- *Concurrent Pos:* Affil prof technol mgt, Wash Univ, St Louis, Mo, 83-88; vis scholar & lectr, Stanford Univ, 89. *Mem:* Am Chem Soc; AAAS; Am Inst Chemists. *Res:* Petrochemicals; monomers; polymer synthesis, properties and applications; polymer combustion properties; safe uses of plastics; management of technology; management of research and development. *Mailing Add:* 611 Capistrano Way San Mateo CA 94402

MCDONALD, DONALD, b Montgomery, Ala, Oct 16, 30; c 3. CIVIL ENGINEERING, STRUCTURAL MECHANICS. *Educ:* Auburn Univ, BCE, 52; Univ Ill, MS, 57, PhD(civil eng), 59. *Prof Exp:* Sr res engr, Lockheed Missiles & Space Co, 59-61, res specialist, 61-62; from asst prof to assoc prof civil eng, NC State Univ, 62-67; mgr struct & mech dept, Lockheed Missiles & Space Co, 67-73; assoc head, Civil Eng Dept, Tex A&M Univ, 74-79, interim dean eng & assoc dep chancellor, 83-84, prof civil eng, 73-90, head civil eng dept, 79-86, provost & vpres acad affairs, 86-90; PRES, THE AM UNIV CAIRO, 90-, PROF ENG, 90- *Mem:* Am Soc Civil Engrs; Am Inst Aeronaut & Astronaut; Am Soc Eng Educ; Nat Soc Prof Engrs. *Res:* Structural vibrations; numerical analysis. *Mailing Add:* Am Univ Cairo 866 UN Plaza Suite 517 New York NY 10017

MCDONALD, DONALD BURT, b Salt Lake City, Utah, Mar 5, 32; m 77; c 3. LIMNOLOGY, TOXICOLOGY. *Educ:* Univ Utah, BS, 54, MS, 56, PhD(limnol), 62. *Prof Exp:* Proj leader fisheries, Utah Fish & Game Dept, 58-60; instr biol, Carbon Col, 60-62; PROF ENG, UNIV IOWA, 62- *Concurrent Pos:* Consult, Iowa Elec Co, 71-; dir, D B McDonald Res, 73- *Mem:* Am Waterworks Asn; Water Pollution Control Fedn; Am Soc Testing & Mat. *Res:* Environmental toxicology and hazardous waste assessment; applied limnology. *Mailing Add:* Dept Environ Eng Univ Iowa Iowa City IA 52242

MCDONALD, DONALD C, b Norwood, Ohio, Dec 25, 19; c 2. ENGINEERING. *Educ:* Mass Inst Technol, SB, 41, SM, 42. *Prof Exp:* Res asst, Servo Lab, Mass Inst Technol, 42-45; proj engr, Doelcam Corp, Mass, 46-47; div supvr, Controls & Instrumentation Div, Willow Run Res Ctr, Univ Mich, 48-49; asst dir, Res Labs Div, Cook Elec Co, Ill, 50-56; dir eng, Friez Instrument Div, Bendix Aviation Corp, Md, 56-57; vpres in charge eng, Cook Elec Co, 58-59 & Sola Elec Co, 60-64; dir eng, Int Register Co, 65-67; vpres eng, Electronics Corp Am, 67-70; VPRES ENG, O S WALKER CO, INC, 70- *Mem:* Inst Elec & Electronics Engrs. *Res:* Electronics; electromechanical components; subsystems and systems in the instrumentation and control fields; magnetics. *Mailing Add:* O S Walker Co Inc Rockdale St Worcester MA 01606

MACDONALD, DONALD LAURIE, b Toronto, Ont , Nov 16, 22; nat US; m 47; c 3. BIOCHEMISTRY. *Educ:* Univ Toronto, BA, 44, MA, 46, PhD(chem), 48. *Prof Exp:* Asst, Univ Calif, 48-50, instr, 50-52, asst prof biochem, 52-57; vis scientist, Nat Inst Arthritis & Metab Dis, 57-60, chemist, 60-61; USPHS fel, Heidelberg, 61-62; prof biochem, 62-67, actg chmn dept biochem & biophysics, 76-78 & 84-85, prof biochem & biophysics, Ore State Univ, 67-87; RETIRED. *Concurrent Pos:* USPHS career develop award, 62-67; fel, Harvard Med Sch, 70-71; vis investr, Fred Hutchinson Cancer Res Ctr, Univ Wash, Seattle, 78-79. *Mem:* Am Chem Soc; Am Soc Biol Chem. *Res:* Degradation of monosaccharides; sugar phosphates; glycoproteins. *Mailing Add:* Dept of Biochem & Biophysics Ore State Univ Corvallis OR 97331-6503

MACDONALD, DONALD MACKENZIE, cellulose chemistry; deceased, see previous edition for last biography

MACDONALD, DOUGLAS GORDON, b Dryden, Ont, May 13, 39; m 64; c 3. CHEMICAL ENGINEERING, BIOCHEMICAL ENGINEERING. *Educ:* Queen's Univ, Ont, BSc, 62, MSc, 65, PhD(chem eng), 68. *Prof Exp:* PROF CHEM ENG, UNIV SASK, 68- *Mem:* Chem Inst Can; Asn Prof Eng Sask; Can Soc Chem Eng. *Res:* Bioconversion of cellulosic materials to protein; acid hydrolysis of biomass; pyrolysis of biomass; gasohol fermentation. *Mailing Add:* Dept Chem & Chem Eng Univ Sask Saskatoon SK S7N 0W0 Can

MACDONALD, EVE LAPEYROUSE, b Baton Rouge, La, Jan 2, 29; m 50; c 1. DEVELOPMENTAL BIOLOGY, CELL BIOLOGY. *Educ:* Wellesley Col, AB, 50; Bryn Mawr Col, MA, 65, PhD(develop biol), 67. *Prof Exp:* Teaching asst, Bryn Mawr Col, 64-66; from asst prof to assoc prof biol, Wilson Col, 67-76, dir inst electron micros, 75-76; res scientist, Burroughs Wellcome Co, 76-80; mem staff, Brennan Assocs, 80-83; PRES, ELMAC SCI TECH, INC, 83- *Concurrent Pos:* NSF fel, 66; co-instr develop biol, Marine Biol Lab, Woods Hole, Mass, 76; prog coordr, Marine lab, Electron Microscope Inst, Duke Univ, 76-79. *Mem:* AAAS; Am Soc Zool; Sigma Xi; Soc Develop Biol; Electron Microscope Soc Am. *Res:* Development of the amphibian eye at different temperatures; cellular differentiation in Rana pipiens and Xenopus laevis; ultrastructure of differentiating cells; polymerization of microtubules. *Mailing Add:* 13703 Exotica Lane West Palm Beach FL 33414

MCDONALD, FRANCIS RAYMOND, b Phila, Pa, Sept 28, 24; m 55; c 1. SPECTROSCOPY, ORGANIC CHEMISTRY. *Educ:* Phila Col Textiles & Sci, BS, 54. *Prof Exp:* Asst chief chemist, Niagara Falls Div, Int Minerals & Chem Corp, NY, 55-57; chemist, 57-61, res chemist, 61-75, proj leader, Laramie energy res ctr, US Energy Res & Develop Admin, 75-77, SECT SUPVR, GEN ANAL SECT, DIV RES SUPPORT, US DEPT ENERGY, LARAMIE ENERGY TECHNOL CTR, 77- *Mem:* Soc Appl Spectros; Coblentz Soc. *Res:* Molecular structure of petrochemical compounds found in oil shale and shale oil using nuclear magnetic resonance, infrared and ultraviolet spectroscopy; mass spectroscopy, environmental monitoring; air pollution; stack and process monitoring of in situ retorting of oil shale, tar sands and coal gasification. *Mailing Add:* 4511 Bluebird Ln Laramie WY 82070

MCDONALD, FRANK ALAN, b Dallas, Tex, Jan 11, 37; m 64; c 2. THEORETICAL PHYSICS, ACOUSTICS. *Educ:* Southern Methodist Univ, BS & BA, 58; Yale Univ, MS, 59, PhD(physics), 65. *Prof Exp:* Asst prof physics, Tex A&M Univ, 64-69; from assoc prof to prof physics, Southern Methodist Univ, 69-85; RES SCIENTIST, IBM, 85- *Mem:* Am Phys Soc; Am Asn Physics Teachers; Sigma Xi. *Res:* Photo acoustic spectroscopy; photoacoustics in condensed matter. *Mailing Add:* IBM T J Watson Res Ctr PO Box 218 Yorktown Heights NY 10598

MCDONALD, FRANK BETHUNE, b Columbus, Ga, May 28, 25; m 51, 87; c 3. PHYSICS. *Educ:* Duke Univ, BS, 48; Univ Minn, MS, 51, PhD, 55. *Prof Exp:* Asst physics, Duke Univ, 47-48; asst, Univ Minn, 48-51; asst prof, Univ Iowa, 53-59; head, Fields & Particles Br, NASA Goddard Space Flight Ctr, 59-70, proj scientist, Explorer Satellites & High Energy Astron Observ, 64-89, chief, Lab High Energy Astrophysics, 70-82, chief scientist, 82-87, assoc dir & chief scientist, 87; PROF, DEPT PHYSICS & ASTROPHYSICS, UNIV MD, 89- *Concurrent Pos:* Prof, Univ Md, 64- *Mem:* Nat Acad Sci; Am Geophys Union; Am Phys Soc. *Res:* Study of primary cosmic radiation at high altitudes by means of rockets, balloons and satellites. *Mailing Add:* Dept Physics-Astrophysics Univ Md College Park MD 20742

MCDONALD, GEORGE GORDON, b Chicago, Ill, Feb 20, 44. BIOCHEMISTRY, BIOPHYSICS. *Educ:* Loyola Univ Chicago, BS, 66; Johns Hopkins Univ, PhD(biochem), 70. *Prof Exp:* Fel biochem, 70-74, asst prof, 74-80, RES ASSOC PROF, DEPT BIOCHEM & BIOPHYSICS, UNIV PA, 80- *Concurrent Pos:* Career investr fel, Am Heart Asn, 71-73; dir mid Atlantic Nuclear Magnetic Resonance Facil, 73- *Mem:* AAAS; Am Chem Soc; Am Soc Biol Chemists; Acad Sci. *Res:* Relationship between the structure and function of macromolecules; nuclear magnetic resonance. *Mailing Add:* 329 Ripplewood Dr Mesquite TX 75150

MCDONALD, GERAL IRVING, b Wallowa, Ore, Dec 31, 35; m 62; c 3. PLANT PATHOLOGY. *Educ:* Wash State Univ, BS, 63, PhD(plant path), 69. *Prof Exp:* Res plant geneticist, 66-68, RES PLANT PATHOLOGIST, INTERMT FOREST & RANGE EXP STA, US FOREST SERV, 68- *Concurrent Pos:* Affiliate prof, Univ Idaho, 68- *Mem:* Am Phytopath Soc; Sigma Xi. *Res:* Genetics, computer simulation and evolution of host-pest interaction in forest trees. *Mailing Add:* Intermt Forest & Range Exp Sta Forestry Sci Lab US Forest Serv 1221 S Main Moscow ID 83843

MCDONALD, GERALD O, b Wichita, Kans, Nov 23, 22. ONCOLOGY, SURGERY. *Educ:* Northwestern Univ, MD, 47. *Prof Exp:* Prof surg, Univ Ill, 64-75; prof, Chicago Med Sch, Univ Health Sci, 75-84; DEP DIR SURG SERV, VET ADMIN, 84- *Mem:* Am Soc Cell Biol; Am Asn Clin Res; Am Fedn Clin Res; Am Col Surgeons; Am Col Gastroenterol; Am Col Nutrit. *Res:* Cell culture. *Mailing Add:* Vet Admin Cent Off PO Box 995 Mclean VA 22101

MACDONALD, GORDON J, b Staten Island, NY, May 18, 34; m 56; c 2. PHYSIOLOGY, ANATOMY. *Educ:* Rutgers Univ, BS, 55, MS, 58, PhD, 61. *Prof Exp:* Res asst dairy sci, Rutgers Univ, 57-58; trainee endocrinol, Univ Wis, 61-63; res fel physiol, Sch Dent Med, Harvard Univ, 63-64, res asst, 64-68; res assoc anat, New Eng Regional Primate Res Ctr, 68-71, asst prof anat, Lab Human Reproduction & Reproductive Biol, Harvard Med Sch, 69-73; assoc prof, 73-79, PROF ANAT, ROBERT WOOD JOHNSON RUTGERS MED SCH, UNIV MED & DENT NJ, 79- *Mem:* Am Physiol Soc; Endocrine Soc; Soc Study Reproduction; NY Acad Sci; Int Soc Res Reprod. *Res:* Role of pituitary gonadotropins in ovarian function; hormonal control of mammary gland growth and lactation; endocrinology reproduction. *Mailing Add:* Dept Neurosci & Cell Biol UMDNJ Robert Wood Johnson Med Sch Piscataway NJ 08854-5635

MACDONALD, GORDON JAMES FRASER, b Mexico City, July 30, 29; US citizen. MATHEMATICAL STATISTICS, THEORETICAL PHYSICS. *Educ:* Harvard Univ, AB, 50, AM, 52, PhD(geophysics), 54. *Prof Exp:* From vpres to exec vpres, Inst Defense Analysis, 66-68; prof physics & geophysics, Univ Calif, Santa Barbara, 68-70; mem, Coun Environ Qual, Exec Off Pres, Washington, DC, 70-72; Henry R Luce Third Century prof & dir, Environ Studies, Dartmouth Col, 72-79; chief scientist, Mitre Corp, 79-83, vpres & chief scientist, 83-90; PROF INT RELS & DIR ENVIRON STUDIES PROG, UNIV CALIF, SAN DIEGO, 90- *Concurrent Pos:* mem, Bd Trustees, Inst Defense Anal, 66-70 & MITRE Corp, 68-70, 72-77; vchancellor, Res & Grad Affairs, Univ Calif, Santa Barbara, 68-70; mem, Bd Dir, Inst Congress, 74-78, Environ Law Inst, 75-84 & Adv Bd, Gas Res Inst, 76-79; distinguished vis scholar, MITRE Corp, 77-79; adj prof, Environ Studies, Dartmouth Col, 79-84. *Honors & Awards:* James B Macelwane Award, Am Geophys Union, 65. *Mem:* Nat Acad Sci; Am Acad Arts & Sci; Am Philos Soc; fel Am Geophys Union; fel Royal Astron Soc. *Res:* Earth's interior to the upper atmosphere. *Mailing Add:* Univ Calif San Diego 9500 Gilman Dr La Jolla CA 92093-0518

MCDONALD, H(ENRY) S(TANTON), b Pa, Oct 28, 27; m 54; c 2. COMPUTER SCIENCE, ELECTRICAL ENGINEERING. *Educ:* Cath Univ, BEE, 50; Johns Hopkins Univ, MS, 53, DE, 55. *Prof Exp:* Instr elec eng, Johns Hopkins Univ, 53-55, res staff asst, 54-55; mem tech staff, 55-60, head signal processing res dept, 60-67, asst dir commun principles res ctr, 67-72, asst dir, Electronic & Comput Systs Res Lab, 72-76, CONSULT, SYSTS ARCHIT RES, AT&T BELL LABS, 76- *Concurrent Pos:* Bell Labs fel, 88. *Honors & Awards:* Inst Elec & Electronic Engrs ASSP Sol Award,77. *Mem:* Fel Inst Elec & Electronics Engrs; Asn Comput Mach; Sigma Xi; AAAS. *Res:* Perception and encoding of visual and acoustic signals; communication theory; computer design and applications; computer graphics; communications system design. *Mailing Add:* AT&T Bell Labs Whippany Rd Rm 14J-411 Whippany NJ 07981

MACDONALD, HAROLD CARLETON, b Englishtown, NS, Sept 18, 30; m 55; c 5. GEOLOGY. *Educ:* State Univ NY, Binghamton, BA, 60; Univ Kans, MS, 62, PhD(geol), 69. *Prof Exp:* Explor geologist, Sinclair Oil & Gas Co, 62-65; res assoc geol, Ctr Res, Univ Kans, 69-70; asst prof, Ga Southern Col, 70-71; assoc prof geol, Univ Ark, Fayetteville, 71-73; prof geol, 76-88, UNIV PROF, UNIV ARK, FAYETTEVILLE, 88- *Mem:* Geol Soc Am; Am Asn Petrol Geologists; Sigma Xi; Am Soc Photogram; Am Geophys Union. *Res:* Geoscience evaluation of side-looking airborne imaging radars; geological remote sensing. *Mailing Add:* Dept Geol Univ Ark Fayetteville AR 72701

MCDONALD, HARRY SAWYER, b New Orleans, La, Sept 24, 30; m 56; c 2. COMPARATIVE PHYSIOLOGY, HERPETOLOGY. *Educ:* Loyola Univ, La, BS, 54, Univ Notre Dame, PhD(zool), 58. *Prof Exp:* Nat Heart Inst fel zool, Univ Calif, Los Angeles, 58-59, cardiovasc trainee, 59-60; from asst prof to assoc prof biol, St John's Univ, NY, 60-65; from asst prof to assoc prof, 65-68, PROF BIOL, STEPHEN F AUSTIN STATE UNIV, 68- *Mem:* Am Soc Zool; Am Soc Ichthyol & Herpet; Soc Study Amphibians & Reptiles. *Res:* Reptile electrocardiography; breathing in snakes; thermal acclimation; snake anatomy and organ topography. *Mailing Add:* Dept Biol Stephen F Austin State Univ Box 13003 Nacogdoches TX 75962

MCDONALD, HECTOR O, b Casper, Wyo, Sept 4, 30; m 52; c 2. PHYSICAL INORGANIC CHEMISTRY. *Educ:* Cent Methodist Col, AB, 52; Ala Polytech Inst, MS, 54; Univ Ark, PhD(chem), 60. *Prof Exp:* Control chemist, Southern Cotton Oil Co, Ill, 54-55; res asst chem, Univ Ark, 55-59; assoc prof, West Tex State Univ, 59-63; assoc prof, 63-86, PROF CHEM, UNIV MO-ROLLA, 86- *Concurrent Pos:* Res chemist, US Bureau Mines, 72- *Mem:* Am Chem Soc; Royal Soc Chem. *Res:* Chemistry of inorganic complexes; chemical kinetics in aqueous media, non-aqueous solvent systems and mass spectrometry. *Mailing Add:* Dept Chem Univ Mo Rolla MO 65401

MCDONALD, HENRY C, b Long Beach, Calif, July 2, 28; m 55; c 3. ENGINEERING. *Educ:* Univ Calif, AB, 52. *Prof Exp:* Proj engr, Lawrence Livermore Nat Lab, 60-69, leader, Physics Systs Div, 63-69, dep dept head, 69-72, head, Dept Elec Eng, 72-74, assoc dir eng, 74-86; RETIRED. *Mem:* Inst Elec & Electronics Engrs. *Res:* Nuclear engineering. *Mailing Add:* 201 Midland Way Danville CA 94526

MACDONALD, HUBERT C, JR, b Detroit, Mich, Aug 3, 41; m 67; c 4. ANALYTICAL CHEMISTRY, ENVIRONMENTAL SCIENCES. *Educ:* Wheeling Col, BS, 63; Univ Mich, MS, 65, PhD(chem), 69. *Prof Exp:* Res scientist, Koppers Co, Inc, 69-78; dir, Senate Environ Testing Lab, Inc, 78-82; SR ENGR & TECH DIR, E & ISD WESTINGHOUSE ELEC CORP, 82- *Concurrent Pos:* Consult water & waste water industs, 78-82. *Mem:* Am Chem Soc; Spectros Soc; Am Soc Testing & Mat. *Res:* Electrochemistry; x-ray spectroscopy; inorganic chemistry; catalyst testing; gas chromatography; atomic absorption spectroscopy; analysis of drinking water by atomic absorption and gas chromatographic methods; analysis of industrial waste streams and other watewater streams; application of analytical chemistry in the electric equipment and electric power generation industries; analysis of PCBs in oil and environmental samples. *Mailing Add:* Eng Serv Div Lab Westinghouse Elec Corp 517 Pkwy View Dr Pittsburgh PA 15205-1410

MCDONALD, HUGH JOSEPH, b Glen Nevis, Ont, July 27, 13; nat US; m; c 3. BIOCHEMISTRY, PHYSICAL CHEMISTRY. *Educ:* McGill Univ, BSc, 35; Carnegie-Mellon Univ, MS, 36, DSc(phys chem), 39; Am Bd Clin Chem, dipl, 52. *Hon Degrees:* Grad, Catholic Univ Rio de Janeiro, 62. *Prof Exp:* Asst chem, Carnegie-Mellon Univ, 36-38, instr, 38-39; from instr to prof, Ill Inst Technol, 39-48; from prof & chmn dept, 48-79 to EMER PROF BIOCHEM & BIOPHYS, STRITCH SCH MED, LOYOLA UNIV, CHICAGO, 79- *Concurrent Pos:* Res scientist, Manhattan Proj, Columbia Univ, 43-44; dir corrosion res lab, Ill Inst Technol, 44-48; ed theoret sect, Corrosion & Mat Protection, 45-48; consult, Argonne Nat Lab, 46-79, State Ill Dept Pub Health, 67-81; assoc ed, Clin Chem, 55-60, Anal Biochem, 60-70; prof chem, Kendall Col, Evanston, Ill, 81-86. *Honors & Awards:* Nat Award Educ & Training, Am Asn Clin Chem, 77. *Mem:* Fel AAAS; Soc Exp Biol & Med; Am Soc Biochem & Molecular Biol; Biophys Soc; Am Asn Clin Chem (vpres, 52, pres, 53); Nat Acad Clin Biochem; Am Chem Soc; Sigma Xi. *Res:* Biochemical aspects of diabetes and atherosclerosis; separation processes; electrophoresis; lipoproteins; nutrition. *Mailing Add:* 5344 Cleveland St Skokie IL 60077-2414

MCDONALD, IAN CAMERON CRAWFORD, b Flint, Mich, Feb 20, 39; m 64; c 2. ENTOMOLOGY, GENETICS. *Educ:* Southern Methodist Univ, BS, 62, MS, 65; Va Polytech Inst, PhD(entom), 68. *Prof Exp:* Asst, Va Polytech Inst, 64-67; entomologist, Metab & Radiation Res Lab, Entom Res Div, 68-81, GENETICIST, BIOSCI RES LAB, AGR RES SERV, USDA, 81- *Mem:* Entom Soc Am; Am Genetic Asn. *Res:* Biochemical genetics of insects, hybridoma research on insect proteins. *Mailing Add:* Biosci Res Lab Agr Res Serv USDA State Univ Sta Box 5674 Fargo ND 58105

MACDONALD, IAN FRANCIS, b Montreal, Que, Feb 8, 42; m 63; c 2. FLOW IN POROUS MEDIA, ENHANCED OIL RECOVERY. *Educ:* NS Tech Col, BE, 63; Univ Wis-Madison, PhD(chem eng), 68. *Prof Exp:* Nat Res Coun Can overseas fel, Inst Sci & Technol, Univ Manchester, 68-69; res engr, Exp Sta, E I du Pont de Nemours & Co, Inc, 69-70; from asst prof to assoc prof, 70-88, PROF CHEM ENG, UNIV WATERLOO, 88- *Concurrent Pos:* Vis assoc prof chem eng, Univ Wash, 78-79. *Mem:* Soc Rheol; Can Soc Chem Engrs. *Res:* Flow in and structure of porous media; computer reconstruction of porous media. *Mailing Add:* Dept of Chem Eng Univ of Waterloo Waterloo ON N2L 3G1 Can

MCDONALD, IAN MACLAREN, b Regina, Sask, May 20, 28; m 53; c 5. PSYCHIATRY. *Educ:* Univ Man, MD, 53; Royal Col Physicians & Surgeons Can, dipl psychiat, 58. *Prof Exp:* Psychiatrist, Crease Clin Psychol Med, 53-54 & Psychiat Serv, Prov Sask, 54-56; fel neurol, Col Med, Univ Sask, 56; fel psychiat, Med Ctr, Univ Colo, 56-58; lectr, 58, from asst prof to assoc prof, 59-67, head Univ Dept Psychiat, Head Psychiat Univ Hosp, 71-83, PROF PSYCHIAT, UNIV SASK, 67-, DEAN, 83- *Concurrent Pos:* Mem Sask Bd of Review. *Mem:* Am Psychiat Asn; Can Psychiat Asn. *Res:* Social psychiatry. *Mailing Add:* Col Med Univ Sask Saskatoon SK S7N 0W0 Can

MCDONALD, JACK RAYMOND, b Birmingham, Ala, May 20, 44; m 71; c 3. ELECTROSTATIC PRECIPITATION. *Educ:* Samford Univ, BS, 66; Auburn Univ, MS, 68, PhD(physics), 77. *Prof Exp:* Instr physics & math, Jefferson State Jr Col, 72-74; assoc & res physicist, Southern Res Inst, 74, sect head, 77-82; vpres, Paul & McDonald Assocs, Inc, 82-83; mgr res & develop & actg vpres, Crestmont Assocs, Inc, 84-85; PRES, JACK R MCDONALD, INC, 85- *Concurrent Pos:* Consult & expert witness. *Mem:* Air Pollution Control Asn; Am Soc Mech Engrs. *Res:* Fundamental mechanisms in the separation of particles from gas streams by electrostatic precipitation, filtration, and scrubbing; investigate applications to stationary and mobile sources of gas streams; coal utilization & impact on boilers & precipitators. *Mailing Add:* Jack R McDonald Inc Rte 1 Box 491 Trussville AL 35173

MACDONALD, JAMES, b Cheltenham, Eng, Sept 19, 52; m 81; c 2. STELLAR EVOLUTION. *Educ:* Cambridge Univ, BA, 73, PhD(astrophys), 79. *Prof Exp:* Postdoctoral, Astron Ctr, Univ Sussex, 78-81 & Dept Astron, Univ Ill, 81-84; vis asst prof, Ariz State Univ, 84-85; asst prof, 85-91, ASSOC PROF PHYSICS & ASTRON, UNIV DEL, 91- *Mem:* Int Astron Union; Am Astron Soc. *Res:* Structure and evolution of the stars. *Mailing Add:* Dept Physics & Astron Univ Del Newark DE 19716

MACDONALD, JAMES CAMERON, biochemistry; deceased, see previous edition for last biography

MCDONALD, JAMES CLIFTON, mycology; deceased, see previous edition for last biography

MACDONALD, JAMES DOUGLAS, b Portland, Ore, July 14, 48; m 67; c 2. PLANT PATHOLOGY, SOIL MICROBIOLOGY. *Educ:* Univ Calif, BS, 73, MS, 75, PhD(plant path), 77. *Prof Exp:* ASST PROF PLANT PATH, UNIV CALIF, DAVIS, 78- *Mem:* Am Phytopath Soc; Am Soc Hort Sci. *Res:* Diseases of ornamental plants; abiotic diseases of plants; environmental stresses as factors in plant disease. *Mailing Add:* Dept Plant Path Univ Calif Davis CA 95616

MCDONALD, JAMES FREDERICK, b Detroit, Mich, Sept 26, 39; m 64. OPTIMIZATION AND APPLIED PROBABILITY. *Educ:* Wayne State Univ, BS, 61, PhD(physics), 67. *Prof Exp:* From asst prof to assoc prof, 67-76, PROF MATH, UNIV WINDSOR, 76- *Concurrent Pos:* Nat Res Coun Can grant, 67-, Dept Univ Affairs grant, Univ Windsor, 69; NATO grant, 90. *Mem:* Am Phys Soc; Am Math Soc; Math Asn Am; Sigma Xi; Can Math Soc; Math Prog Soc. *Res:* Optimization and applied probability. *Mailing Add:* Dept Math Univ Windsor Windsor ON N9B 3P4 Can

MCDONALD, JAMES LEE, JR, b LaGrange, Ky, May 21, 39; m 63; c 3. ORAL BIOLOGY, NUTRITION. *Educ:* Ind Univ, AB, 62, PhD(dent sci), 68. *Prof Exp:* From asst prof to assoc prof, 68-82, PROF PREV DENT, SCH DENT, IND UNIV, INDIANAPOLIS, 83- *Mem:* AAAS; assoc Am Dent Asn; Int Asn Dent Res. *Res:* Smoking cessation and nicotine addiction; nutritional control of dental caries. *Mailing Add:* Dept Prev & Comm Dent Ind Univ Purdue Univ Med 1121 W Mich St Indianapolis IN 46202

MACDONALD, JAMES REID, b St Helena, Calif, Aug 22, 18; m 41, 57, 69, 83; c 3. VERTEBRATE PALEONTOLOGY, GEOLOGY. *Educ:* Univ Calif, BA, 40, MS, 47, PhD(paleont), 49. *Prof Exp:* Lab asst mus paleont, Univ Calif, 45-46, lab technician, 46-49; from asst prof to assoc prof geol, SDak Sch Mines & Technol, 49-57, cur, 49-57; occup analyst, Idaho Dept Hwy, 57-58; res assoc, Am Mus Natural Hist, 58-60; spec projs technician, Idaho Dept Hwy, 60-61; dep dir admin, State of Idaho, 61-62; assoc prof geol, Univ Idaho, 62; sr cur vert paleont, Los Angeles County Mus Natural Hist, 62-69; instr geog, Calif State Polytech Col, San Luis Obispo, 70; prof geol, Foothill Col, 70-80; CONSULT RES ASSOC, MUS GEOL, SDAK SCH MINES & TECHNOL, RAPID CITY, 80- *Concurrent Pos:* Adj prof, Univ Southern Calif, 65-69. *Mem:* Soc Vert Paleont. *Res:* Pliocene faunas of Nevada, California and South Dakota; Oligocene and Miocene mammals of South Dakota and adjacent regions. *Mailing Add:* 1840 Tranquil Trail Rapid City SD 57701

MACDONALD, JAMES ROSS, b Savannah, Ga, Feb 27, 23; m 46; c 3. SOLID STATE PHYSICS. *Educ:* Williams Col, BA, 44; Mass Inst Technol, SB, 44, SM, 47; Oxford Univ, DPhil(physics), 50, DSc, 67. *Prof Exp:* Asst, Radar & Electronics Lab, Mass Inst Technol, 43-44, asst elec eng, 46-47; physicist, Armour Res Found, Ill Inst Technol, 50-52; assoc physicist, Argonne Nat Lab, 52-53; res physicist semiconductor & solid state physics, Tex Instruments Inc, 53-55, dir solid state physics res, 55-61, dir physics res lab, 61-63, dir cent res labs, 63-72, asst vpres corp res & eng, 67, vpres, 68-74, vpres corp res & develop, 73-74; W R Kenan Jr prof, 74-91, EMER W R

KENAN JR PROF PHYSICS, DEPT PHYSICS & ASTRON, UNIV NC, CHAPEL HILL, 91- *Concurrent Pos:* Adj assoc prof, Southwestern Med Sch, Univ Tex, 54-71, adj prof, 71-74; chmn, Numerical Data Adv Bd, Nat Acad Sci, 70-74, mem, Comt Motor Vehicle Emissions, 71-73, chmn, 73-74; mem coun, Nat Acad Eng, 71-74; dir, Simmonds Precision Prod, Inc, 79-83. *Honors & Awards:* Achievement Award, Inst Elec & Electronics Engrs, 62, Merit Serv Award, 74; Achievement Award, Inst Radio Engrs; George E Pake Prize, Am Phys Soc, 85; Edison Gold Medal, Inst Elec & Electronic Engrs, 88. *Mem:* Nat Acad Sci; Nat Acad Eng; fel Am Phys Soc; fel Inst Elec & Electronic Engrs; fel AAAS; Sigma Xi. *Res:* Impedance spectroscopy; data analysis; electrolyte double layer; space-charge. *Mailing Add:* 308 Laurel Hill Rd Chapel Hill NC 27514

MACDONALD, JAMES SCOTT, b Appleton, Wis, July 4, 48. ENVIRONMENTAL TOXICOLOGY, BIOCHEMISTRY. *Educ:* DePauw Univ BA, 70; Univ Cincinnati, PhD(toxicol), 75; Am Bd Toxicol, dipl, 80. *Prof Exp:* Fel toxicol, Ctr Toxicol, Vanderbilt Univ, 75-77; sr res toxicologist, 77-79, ASSOC DIR TOXICOL, MERCK INST THERAPEUT RES, 79- *Mem:* NY Acad Sci; AAAS. *Res:* Mechanisms of toxicity; carcinogenesis; promotion of tumorigenesis; drug metabolism. *Mailing Add:* Dept Safety Assessment Merck Sharp & Dohme Res Lab West Point PA 19486

MCDONALD, JAY M, EXPERIMENTAL BIOLOGY. *Prof Exp:* PROF & CHMN DEPT PATH, UNIV ALA, BIRMINGHAM, 90- *Mailing Add:* Dept Path Univ Ala Birmingham Sta Birmingham AL 35294

MCDONALD, JIMMIE REED, b Austin, Tex, Aug 20, 42; m 63; c 2. PHYSICAL CHEMISTRY, MOLECULAR SPECTROSCOPY. *Educ:* Southwestern Univ, Tex, BS, 64; La State Univ, Baton Rouge, PhD(phys chem), 68. *Prof Exp:* Vis asst prof, La State Univ, Baton Rouge, 68-70; RES CHEMIST, NAVAL RES LAB, 70- *Mem:* Am Chem Soc; Am Phys Soc; Optical Soc Am; AAAS. *Res:* Molecular electronic spectroscopy; photochemistry; chemical kinetics. *Mailing Add:* 35 Herrington Dr Upper Marlboro MD 20772

MCDONALD, JOHN ALEXANDER, PULMONARY DISEASE. *Educ:* Rice Univ, PhD(biochem), 70; Duke Univ, MD, 73. *Prof Exp:* ASSOC PROF PULMONARY MED & DIR, PULMONARY DIS DIV, SCH MED, WASH UNIV, 85- *Mailing Add:* Dept Pulmonary Wash Univ 660 S Euclid PO Box 8052 St Louis MO 63110

MACDONALD, JOHN BARFOOT, b Toronto, Ont, Feb 23, 18; m 67; c 5. MICROBIOLOGY. *Educ:* Univ Toronto, DDS, 42; Univ Ill, MS, 48; Columbia Univ, PhD(bact), 53. *Hon Degrees:* AM, Harvard Univ, 56; LLD, Univ Man, 62, Simon Fraser, 65, Wilfred Laurier Univ, 76, Brock Univ, 76 & Univ Western Ont, 77; DSc, Univ BC, 67 & Univ Windsor, 77. *Prof Exp:* Lectr prev dent, Univ Toronto, 42-44, instr bact, 46-47; res asst, Univ Ill, 47-48; asst prof, Univ Toronto, 49-53, assoc prof bact & chmn div dent res, 53-56, prof, 56; prof microbiol, Sch Dent Med, Harvard Univ, 56-62, dir postdoctoral studies, 60-62; pres, Univ BC, 62-67; prof higher educ, Univ Toronto, 68-83; exec dir, Coun Ont Univs, 68-76; pres, 76-81, Addiction Res Found, chmn, 81-87. *Concurrent Pos:* Charles Tomes lectr, Royal Col Surgeons, Eng, 62; dir, Forsyth Dent Infirmary, 56-62, consult, 62-; consult, Univ BC, 55-56; consult, Dent Med Sect, Corp Res Div, Colgate-Palmolive Co, 58-62; mem bd, Banff Sch Advan Mgt, 62-67, chmn, 66-67; consult, Donwood Found, Toronto, 67-68, Sci Coun Can & Can Coun Support Res in Can Univs, 67-69 & Addiction Res Found, Toronto, 68; mem dent study sect, NIH, 61-65; mem, Ont Coun Health, 81-84. *Mem:* AAAS; Am Soc Microbiol; NY Acad Sci; Can Ment Health Asn; Int Asn Dent Res (pres, 68). *Res:* Ecology of mucous membranes; mixed anaerobic infections. *Mailing Add:* 1137 Royal York Rd No 1008 Etobicoke ON M9A 4A7 Can

MCDONALD, JOHN C, b Baldwin, Miss; c 3. SURGERY, IMMUNOLOGY. *Educ:* Miss Col, BS, 51; Tulane Univ, La, MD, 55. *Prof Exp:* Intern, Confederate Mem Med Ctr, Shreveport, La, 55-56; resident, E J Meyer Mem Hosp, Buffalo, NY, 58-63; Buswell res fel & instr surg, State Univ NY Buffalo, 63-65; assoc dir surg res lab, E J Meyer Mem Hosp, Buffalo, 65-68, head sect transplantation, 66-68; assoc prof surg, Sch Med, Tulane Univ, La, 68-72, dir surg labs, 68-77, prof surg & assoc prof microbiol, 72-77; PROF & CHMN DEPT SURG, SCH MED, LA STATE UNIV, 77-, SURGEON-IN-CHIEF, LA STATE MED CTR, 77- *Concurrent Pos:* Head sect transplantation, State Univ NY Buffalo, 66; attend surgeon, E J Meyer Mem Hosp, Buffalo, 67; attend surgeon & head sect transplantation, Deaconess Hosp, Buffalo, 68; consult, Roswell Park Mem Hosp, 68 & Masten Park Rehab Ctr, 68; dir transplantation, Tulane Univ La & Charity Hosp, New Orleans; consult surgeon, Lallie Kemp Charity Hosp, Keesler AFB, Biloxi, Miss, Pineville Vet Admin Hosp & Huey P Long Charity Hosp; clin asst surgeon, Touro Infirm, Med staff, Southern Baptist Hosp & assoc mem, dept surg, Hotel Dieu Hosp, New Orleans; dir, Tulane Univ Med Ctr Histocompatibility Testing Lab, La Organ Procurement Prog; consult, Northwest La Emergency Med Serv. *Mem:* Am Col Surgeons; Am Surg Asn; Soc Univ Surgeons; Transplantation Soc; United Network Organ Sharing; Am Soc Transplant Surgeons (treas, 84-86, pres,87-88). *Res:* Transplantation. *Mailing Add:* Dept of Surg La State Univ Sch Med Shreveport LA 71130

MACDONALD, JOHN CAMPBELL FORRESTER, physics, for more information see previous edition

MCDONALD, JOHN CHARLES, b San Bernardino, Calif, Jan 23, 36; m 57; c 3. ELECTRICAL ENGINEERING. *Educ:* Stanford Univ, BS, 57, MS, 59 & PhD (elec eng), 64. *Prof Exp:* Sr engr, GTE Sylvania, 59-63; vpres, Vidar Corp, 63-80; pres, MBX, Inc, 80-82; EXEC VPRES, CONTEL CORP, 82- *Concurrent Pos:* Mem adv bd, Manhattan Col & Univ Md, 85-; mem bd dirs, TelWatch, Inc, 85-, Contel Corp, 87-; chmn, Nat Acad Sci Comt Telecommun, 86-; pres, Inst Elec Electronics Commun Soc, 88-89. *Mem:* Inst Elec Electronics Soc (pres, 88-). *Res:* Digital telecommunications including digital transmission and switching. *Mailing Add:* 54 Comstock Hill Rd New Canaan CT 06840

MACDONALD, JOHN CHISHOLM, b Boston, Mass, Mar 17, 33; m 64; c 2. ANALYTICAL CHEMISTRY. *Educ:* Boston Col, BS, 55, MS, 57; Univ Va, PhD(chem), 62. *Prof Exp:* AEC fel & res assoc chem, Pa State Univ, 62-63; sr res chemist, Monsanto Res Corp, Mass, 63-66; from asst prof to assoc prof, 66-75, PROF CHEM, FAIRFIELD UNIV, 75- *Concurrent Pos:* NIH spec res fel, Med Lab, Sch of Med, Yale Univ, 72-73; vis prof, Arrhenius Lab, Univ Stockholm, 79. *Mem:* AAAS; Am Chem Soc; Sigma Xi. *Res:* Chemical analysis and instrumentation; computers in chemistry; science for society; chemometrics; orthomolecular medicine. *Mailing Add:* Dept Chem Fairfield Univ Fairfield CT 06430

MCDONALD, JOHN F(RANCIS), b Narberth, Pa, Jan 14, 42. ELECTRICAL ENGINEERING, COMPUTER ENGINEERING. *Educ:* Mass Inst Technol, BSEE, 63; Yale Univ, MEng, 64, PhD(elec eng), 69. *Prof Exp:* Mem tech staff, Bell Telephone Labs, 65; from lectr to asst prof eng & comput sci, Yale Univ, 69-74; ASSOC PROF, DEPT ELEC & SYSTS ENG, RENSSELAER POLYTECH INST, 74- *Concurrent Pos:* Contract co-supvr, US Navy Underwater Sound Lab, Conn, 69-71; consult, CTMS Inc, NY, Argonne Nat Lab, 75, Westinghouse Hanford Eng, Develop Lab, 76-77, Gen Elec Corp Res & Develop Ctr, 78, TV Data Corp, 78, Teledyne Gurley Corp, 79. *Mem:* Sigma Xi; Inst Elec & Electronics Engrs; Asn Comput Mach; Am Inst Physics. *Res:* Computer design; communication; detection estimation systems research; control system studies; information theory and coding; digital test set generation; phased array design; microprocessor systems; medical instrumentation. *Mailing Add:* Six Twilight Dr Clifton Park NY 12065

MACDONALD, JOHN JAMES, b New Glasgow, NS, Oct 31, 25; m 52; c 7. ELECTROCHEMISTRY. *Educ:* St Francis Xavier Univ, Can, BSc, 45; Univ Toronto, MA, 47, PhD(chem), 51. *Prof Exp:* assoc prof chem, 49-60, dean sci, 60-70, acad vpres, 70-78, EXEC VPRES, ST FRANCIS XAVIER UNIV, 78- *Concurrent Pos:* Researcher, Ottawa Univ, 59-60; mem, Can Nat Coun, UNESCO, 68-; chmn bd gov, Atlantic Inst Educ, 70-74; mem, Maritime Prov Higher Educ Comn, 74-84 & Can Coun, 75-76; vis prof, Univ Toronto, 76-77; mem, Sci Coun Can, 77-83, chmn transp study comt, 81; mem, Atlantic Region Adv Comt, Nat Res Coun & Res Develop Adv Comt, Nat Sci & Eng Res Coun, 81-85; dir, Mar Tel & Tel, 73-, Force Ten Inc, 83-86 & Adminet Inc, 83-85; mem, NS Coun of Appl Sci & Tech, 87-90; chair, Expo 86 Symposia on Transp; founding chair, World Union of Transp; chmn bd gov, St Martha's Regional Hosp, 85-; exec, NS Asn Health Orgn, 91- *Honors & Awards:* Centennial Medal, 67. *Mem:* Chem Inst Can; Can Soc Study Higher Educ. *Res:* Electrochemical kinetics, higher education. *Mailing Add:* St Francis Xavier Univ Antigonish NS B2G 1C0 Can

MCDONALD, JOHN KENNELY, b Vancouver, Can, Oct 4, 30; nat US; m 53; c 3. BIOCHEMISTRY. *Educ:* Univ BC, BSA, 53; Purdue Univ, MS, 56; Ore State Univ, PhD(chem-biochem), 60. *Prof Exp:* Teaching asst bact, Purdue Univ, West Lafayette, Ind, 53-56; grad res fel biochem, Sci Res Inst, Ore State Univ, Corvallis, 56-60; protein chemist, Protein Res Sect, Cutter Labs, Berkeley, Calif, 60-62; res scientist, Nat Aeronaut & Space Admin, Ames Res Ctr, Moffett Field, Calif, 62-75; PROF, DEPT BIOCHEM, MED UNIV SC, CHARLESTON, 75- *Concurrent Pos:* Vis prof, Dept Biochem, Med Univ SC, 74-75. *Mem:* Am Soc Biol Chemists; Endocrine Soc; AAAS; Sigma Xi. *Res:* Author of numerous books. *Mailing Add:* Dept Biochem Med Univ SC 171 Ashley Ave Charleston SC 29425

MACDONALD, JOHN LAUCHLIN, b Woodstock, NB, Oct 7, 38; US citizen; m 63; c 4. CATEGORY THEORY, PROGRAMMING LANGUAGES. *Educ:* Harvard Univ, AB, 59; Univ Chicago, MS, 61, PhD(math), 65. *Prof Exp:* Humboldt fel math, Math Seminar, Frankfurt, 65-66; asst prof, 66-74, ASSOC PROF MATH, UNIV BC, 74- *Concurrent Pos:* Nat Res Coun grant, 66-88; vis prof math, Res Inst, Swiss Fed Inst Technol, 70-71; Can Coun fel, Swiss Fed Inst Technol & Florence, Italy, 77-78. *Mem:* Am Math Soc; Can Math Cong. *Res:* Category theory; algebraic topology; universal algebra; computer science. *Mailing Add:* Dept Math Univ BC Vancouver BC V6T 1W5 Can

MACDONALD, JOHN MARSHALL, b Dunedin, NZ, Nov 9, 20; m 52; c 3. PSYCHIATRY, CRIMINOLOGY. *Educ:* Univ Otago, MD, 46; Univ London, dipl, 50. *Prof Exp:* House physician, New Plymouth Gen Hosp, 46; asst physician, Belmont Hosp, London, Eng, 48-49; asst physician, Royal Edinburgh Hosp, 49-51; from instr to prof psychiat, 51-88, dir forensic psychiat, 60-88, EMER PROF PSYCHIAT, MED CTR, UNIV COLO, 88-; from instr to prof psychiat, 51-88, dir forensic psychiat, Med Ctr, 60-88, EMER PROF PSYCHIAT, SCH MED, UNIV COLO MED CTR, DENVER, 88- *Concurrent Pos:* Med consult, Colo State Hosp, 62- & Fitzsimons Gen Hosp, US Army, 64-77; med adv, Social Security Admin, 64-74. *Mem:* Fel Am Psychiat Asn. *Res:* Crime. *Mailing Add:* 2205 E Dartmouth Circle Englewood CO 80110

MCDONALD, JOHN N, b Bayonne, NJ, Apr 26, 42; c 2. MATHEMATICS. *Educ:* King's Col, AB, 64; Rutgers Univ, MS, 66, PhD(math), 69. *Prof Exp:* Vis assoc prof, Univ Conn, 75-76; asst prof, 69-75, ASSOC PROF MATH, ARIZ STATE UNIV, 76- *Concurrent Pos:* Grant, Ariz State Univ, 78. *Res:* Convex sets; spaces of analytic functions; stochastic processes. *Mailing Add:* Dept Math Ariz State Univ Tempe AZ 85287

MACDONALD, JOHN ROBERT, b Detroit, Mich, Oct 5, 54. BIOCHEMICAL TOXICOLOGY, IN VITRO TOXICOLOGY. *Educ:* Univ Mich, BS, 77; Univ Ariz, MS, 81, PhD(pharmacol & toxicol), 84. *Prof Exp:* Vis toxicol scholar & postdoctoral, Dept Path, Col Med, Univ Calif, San Francisco, 84-85, res toxicologist, 85-87; sr scientist, 87-90, RES ASSOC, DEPT PATH & EXP TOXICOL, PARKE-DAVIS PHARMACEUT RES DIV, WARNER-LAMBERT CO, 90- *Mem:* Soc Toxicol; Am Asn Pathologists; AAAS. *Res:* Safety assessment of new therapeutic agents; biochemical mechanisms of target organ toxicity and use of in vitro models to elucidate toxic mechanisms. *Mailing Add:* Park-Davis Pharmaceut Res Div Warner-Lambert Co 2800 Plymouth Rd Ann Arbor MI 48105

MCDONALD, JOHN STONER, b Mt Hope, Wash, Sept 20, 32; m 63; c 1. VETERINARY MEDICINE, MICROBIOLOGY. *Educ:* Wash State Univ, BA, 54, DVM, 56; Univ Idaho, MS, 58; Iowa State Univ, PhD(vet microbiol), 67. *Prof Exp:* Sta vet, Univ Idaho, 56-58; pvt pract, 58-61; res vet, Nat Animal Dis Ctr, Agr Res Serv, USDA, 61-84; PROF, WASHINGTON STATE UNIV, 84- *Mem:* Am Vet Med Asn; Am Col Vet Microbiol. *Res:* Pathogenesis of udder infection and mastitis. *Mailing Add:* Col Vet Med Washington State Univ Pullman WA 99164-6610

MCDONALD, JOHN WILLIAM, b Decatur, Ill, Mar 29, 45; m 72; c 2. BIOINORGANIC CHEMISTRY. *Educ:* Grinnell Col, BA, 67; Northwestern Univ, PhD(inorg chem), 71. *Prof Exp:* Sr res assoc, 71-73, staff scientist, 73-78, investr, Charles F Kettering res lab, 78-84, prin res scientist, Battelle-Kettering Lab, 84-87, PRIN RES SCIENTIST, BATTELLE COLUMBUS LAB, 87- *Mem:* Am Chem Soc. *Res:* Synthesis and characterization of metal complexes as probes for the mechanism of action of metalloenzymes. *Mailing Add:* 606 Sharon Dr Fairborn OH 45324-5829

MCDONALD, JOHN WILLIAM DAVID, b Chatham, Ont, Jan 7, 38; m 70; c 2. HEMATOLOGY, GASTROENTEROLOGY. *Educ:* Univ Western Ont, MD, 61, PhD(biochem), 66; FRCP(C), 69. *Prof Exp:* Intern med, Victoria Hosp, London, Ont, 61-62; resident, Montreal Gen Hosp, 62-63; res & clin fel med & hemat, Royal Victoria Hosp, Montreal, 67-70; asst prof med, Victoria Hosp, 70-72, from asst prof to prof med, Univ Hosp, 72-78, asst prof biochem, 72-78, asst dean res, Fac Med, 75-78, HON LECTR, DEPT BIOCHEM & PROF MED, UNIV WESTERN ONT, 78-, CHMN, DEPT MED, 85-, CHIEF MED, UNIV HOSP, 85- *Concurrent Pos:* Dir biochem-radioisotope-hemat lab & physician, Dept Med, Victoria Hosp, 70-72; physician & dir biochem-radioisotope-hemat, Hemat Serv, Hosp, Univ Western Ont, 72-; physician, Hemat-Oncol Serv, Dept Med, Univ Hosp, 72- & physician, Gastroenterol Serv, Dept Med, 78- *Mem:* Royal Col Physicians & Surgeons Can; Can Soc Hemat; Can Soc Clin Invest; Am Soc Hemat; Am Fedn Clin Res; Am Gastroenterol Asn. *Res:* Effect of drugs on platelet prostaglandin synthesis and platelet function; clinical trials in ulcer disease; Crohn disease. *Mailing Add:* Dept Med Univ Hosp PO Box 5339 A London ON N6A 5A5 Can

MCDONALD, KEITH LEON, b Murray City, Utah, Apr 20, 23; m 56. THEORETICAL PHYSICS, ASTROPHYSICS. *Educ:* Univ Utah, BS, 50, MS, 51, PhD(physics), 56. *Prof Exp:* Mem staff, Los Alamos Sci Lab, Calif, 56-57; physicist, Dugway Proving Ground, 57-60; physics fac mem, Brigham Young Univ, 60-62 & Univ Utah, 62-63; res physicist, Nat Bur Standards, 63-64; physics fac mem, Idaho State Univ, 65; theoret geophysicist, Inst Earth Sci, Environ Sci Serv Admin-Inst Environ Res, 66-69; consult, Environ Res Labs, Environ Sci Serv Admin, Colo, 69-70 & Nat Oceanic & Atmospheric Admin, 71; CONSULT, 71- *Mem:* Am Phys Soc; Am Astron Soc; Am Geophys Union; Am Asn Physics Teachers; Sigma Xi. *Res:* Electromagnetic theory and hydromagnetism; theoretical optics; kinetic theory; statistical mechanics; solar-terrestrial relations; the role of the breach in the sun's upper toroid as a source of high speed streams; geomagnetism; solar atmosphere and interior; planetary interiors, geomagnetic field reversals, description of earth's velocity and magnetic vector modes; cosmic hydromagnetic dynamos; continued precession of Earth's rotation axis over its surface towards its geomagnetic axis. *Mailing Add:* PO Box 2433 Salt Lake City UT 84110

MACDONALD, KENNETH CRAIG, b San Francisco, Calif, Oct 14, 47; m. MARINE GEOPHYSICS. *Educ:* Univ Calif, Berkeley, BS, 70; Mass Inst Technol, PhD(marine geophyics), 75. *Prof Exp:* Green scholar geophysics, Inst Geophysics & Planetary Physics, 75-76; res geophysicist marine geophysics, Scripps Inst Oceanog, 76-79; assoc prof, 79-83, PROF MARINE GEOPHYSICS, UNIV CALIF, SANTA BARBARA, 83- *Concurrent Pos:* Ed, J Geophys Res, 78-81, assoc ed, Earth & Planetary Sci Letter, 78-88; mem, Ocean Sci Bd, Nat Res Acad, 79-82, Woods Hole Oceanog Inst Corp, 79-85, 86-92; res geophysicist, Scripps Inst Oceanog, 79-82; chief scientist on over 20 sea-going expeditions; ocean sci bd, Nat Acad Sci, 80-83. *Honors & Awards:* Newcomb-Cleveland Prize, AAAS, 80. *Mem:* Am Geophys Union; Geol Soc Am. *Res:* Marine magnetics; marine seismology; tectonics of spreading centers and transform faults; fine scale structural geology of submarine plate boundaries. *Mailing Add:* Dept Geol Univ Calif Santa Barbara CA 93106

MCDONALD, KIRK THOMAS, b Vallejo, Calif, Oct 20, 45. EXPERIMENTAL HIGH ENERGY PHYSICS. *Educ:* Univ Ariz, BS, 66; Calif Inst Technol, PhD(physics), 72. *Prof Exp:* Res assoc, Europ Coun Nuclear Res, Geneva, 72-74 & Enrico Fermi Inst, Univ Chicago, 75-76; from asst prof to assoc prof, 76-85, PROF HIGH ENERGY PHYSICS, PRINCETON UNIV, 85- *Res:* Experimental research into the structure of hadronic matter, as probed by high transverse momentum jets and by massive drell-yan muon pairs; studies of non-linear quantum electrodynamics using terawatt lasers. *Mailing Add:* Box 708 Princeton NJ 08544

MCDONALD, LARRY WILLIAM, b Louisville, Nebr, May 25, 28; m 55; c 3. PATHOLOGY, NEUROPATHOLOGY. *Educ:* Univ Calif, Berkeley, AB, 50; Northwestern Univ, MD, 55. *Prof Exp:* Res pathologist, Pondville State Hosp, Norfolk, Mass, 59-61; res assoc, Univ Calif, Berkeley, 62-68; assoc prof path, Sch Med, Univ Calif, Davis, 68-75; prof path, sch med, Wright State Univ, 75-77; PROF NEUROPATH, UNIV ILL MED CTR, CHICAGO, 78- *Concurrent Pos:* Damon Runyon grant, 60-62; USPHS trainee radiobiol, Cancer Res Inst, New Eng Deaconess Hosp, Boston, 61-62; consult, Vet Admin Hosp, Martinez, Calif, 66-73, chief electron micros in path, 67; consult, Lawrence Berkeley Lab, Univ Calif, 68-69 & Sacramento County Coroners Off, 71-75; chief lab serv, Vet Admin Ctr, Dayton, Ohio, 75-76. *Mem:* AAAS; Am Asn Neuropath; AMA; Am Asn Pathologists. *Res:* Effects of radiation on the nervous system, particularly the mechanism of production of the delayed effects; vascular disease in the nervous system; mechanisms of induction of tumors in the central nervous system; vestibular function aberrations; effects of metabolic disease on nervous system. *Mailing Add:* Univ Ill Med Ctr Neuropath Dept M/C 799 PO Box 6998 Chicago IL 60680

MCDONALD, LEE J, b Park Rapids, Minn, Mar 20, 62. BIOCHEMISTRY, ENZYMOLOGY. *Educ:* St Johns Univ, BS, 84; Wrights Univ, PhD(biomed sci), 89. *Prof Exp:* NAT RES COUN RES ASSOC, LAB CELL METAB, NIH, 89- *Mem:* Am Soc Cell Biol; AAAS. *Mailing Add:* Lab Cell Metab NIH Bldg 10 Rm 5N307 Bethesda MD 20892

MCDONALD, LESLIE ERNEST, b Middletown, Mo, Oct 14, 23; m 46; c 3. PHYSIOLOGY. *Educ:* Mich State Univ, BS, 48, DVM, 49; Univ Wis, MS, 51, PhD(physiol), 52. *Prof Exp:* Practicing vet, 49-50; instr physiol, Univ Ill, 52-53, assoc prof, 53-54; prof & head dept, Okla State Univ, 54-69; assoc dean, Col Vet Med, Univ Ga, 69-71; dean, Col Vet Med, Ohio State Univ, 71-72; PROF PHYSIOL, COL VET MED, UNIV GA, 72- *Concurrent Pos:* Consult, USPHS, 64-; mem, Comt Vet Drug Efficacy, Nat Res Coun, 66-; coun res, Am Vet Med Asn, 74-84. *Mem:* Soc Exp Biol & Med; Am Physiol Soc; Am Vet Med Asn; Brit Soc Study Fertil; distinguished fel Am Acad Pharm & Ther. *Res:* Endocrinology; reproductive physiology. *Mailing Add:* 4917 Crestview Dr Stillwater OK 74074

MCDONALD, LYNN DALE, b Dumas, Tex, Oct 4, 42; m 65; c 2. RESEARCH ADMINISTRATION, AGRONOMY. *Educ:* Tex Technol Col, BS, 65; Univ Ariz, MS, 71; Tex Tech Univ, PhD (agron), 85. *Prof Exp:* Asst breeder, cotton, Cokers Pedigreed Seed Co, Tex, SC, 71-73, Miss, 73-76, Cotton breeder, 76-83; res dir cotton, Stoneville Pedigreed Seed Co, 85-90; DIR RES CORN, TRIUMPH SEED CO, 90- *Concurrent Pos:* Hybreed Prod Specialist, Wash State Univ, Sudan, 84. *Mem:* Am Soc Agron. *Res:* Development of corn hybrids. *Mailing Add:* 4406 58th St Lubbock TX 79414

MCDONALD, MALCOLM EDWIN, b Ann Arbor, Mich, May 29, 15; m 41; c 2. PARASITOLOGY, ECOLOGY. *Educ:* Parsons Col, BS, 37; Univ Iowa, MS, 39; Univ Mich, PhD(wildlife mgt), 51. *Prof Exp:* Asst prof biol, Parsons Col, 40-42; instr, Beloit Col, 50-51; asst prof, Union Col, NY, 51-56 & Univ Nev, 56-57; wildlife res biologist, US Fish & Wildlife Serv, US Dept Interior, 57-81; RETIRED. *Concurrent Pos:* Parasitologist, Malaria Surv Unit, Sanit Corps, 42-46. *Mem:* Ecol Soc Am; Am Soc Parasitol; Wildlife Soc; Wilson Ornith Soc; Wildlife Dis Asn; Sigma Xi. *Res:* Animal parasitology and ecology; parasites of waterfowl; wildlife conservation. *Mailing Add:* Box 9291 Madison WI 57315-0291

MCDONALD, MARGUERITE BRIDGET, b Chicago, Ill, Oct 14, 50. OPHTHALMOLOGY, CORNEAS. *Educ:* Manhattanville Col, BA, 72; Columbia Univ, MD, 76. *Prof Exp:* Med intern, Lenox Hill Hosp, 76-77; ophthal resident, Manhattan Eye & Ear Hosp, 77-80; corneal fel 80-81, asst prof, 81-85, ASSOC PROF OPHTHAL, LA STATE UNIV EYE CTR, NEW ORLEANS, 85-, RES AFFIL, 82- *Concurrent Pos:* Dir, cornea serv, La State Univ Eye Ctr, 87-, refractive surg res group, 84-; dir refractive surg res, Am Med Optics, 84-, Cooper Vision Laser Div, 86-; adv bd, In Vision Inst, Bausch & Lomb, 87- *Mem:* Nat Asn Residents & Interns; AMA; Am Med Womens Asn; Asn Res Vision & Ophthal; Am Acad Ophthal. *Res:* Refractive surgery using the excimer laser to increase visual acuity of myopia, hyperopia and astigmatism in human patients. *Mailing Add:* LSU Eye Ctr 2020 Gravier St New Orleans LA 70112

MACDONALD, MARNIE L, BIOCHEMICAL GENETICS. *Educ:* Am Bd Med Genetics, dipl, PhD(med genetics), 90. *Prof Exp:* DIR BIOCHEM GENETICS, SOUTHWEST BIOMED RES INST, 88- *Mem:* Am Soc Biochem & Molecular Biol; Am Soc Human Genetics; NY Acad Sci. *Res:* Author of various publications. *Mailing Add:* Dept Biochem Genetics Southwest Biomed Res Inst 6401 E Thomas Rd Scottsdale AZ 85251

MACDONALD, MICHAEL J, PEDIATRIC ENDROCROCRINOLOGY. *Educ:* Univ Notre Dame, BS, 66; Wash Univ, MD, 70. *Prof Exp:* Intern pediat, Children's Hosp Buffalo, State Univ NY, 70-71; resident pediat, Mass Gen Hosp, Harvard Med Sch, Boston, Mass, 71-72; resident pediat & res fel, Elliott P Joslin Res Lab, Harvard Med Sch & Peter Brigham Hosp, Boston, Mass, 72-74; from asst prof to assoc prof, Dept Pediat, 76-87, dir, Pediat Diabetes Prog, 80-87, PROF PEDIAT, HEAD, PEDIAT DIABETES & ENDOCRINOL DIV & DIR, PEDIAT DIABETES PROG, UNIV WIS MED SCH, 87- *Concurrent Pos:* Fel, pediat endocrinol, Children's Hosp Med Ctr, Harvard Med Sch, 73-74; fel, Inst Enzyme Res, Univ Wis, Madison, 74-76, assoc prof, 80-83. *Mem:* Am Soc Biol Chemist; Europ Asn Study Diabetes; Can Diabetes Asn; Lawson Wilkins Pediat Endocrine Soc. *Res:* Biochemistry of insulin release; genetics of insulin - dependent diabetes mellitus; diabetes mellitus in children. *Mailing Add:* Dept Pediat H4 4Csc Univ Wis Sch Clin Ctr 600 Highland Ave Madison WI 53792

MACDONALD, NOEL C, b San Francisco, Calif, Dec 31, 40. NANODYNAMICS, MICROACTUATORS & SENSORS. *Educ:* Univ Calif, Berkeley, BS, 63, MS, 65, PhD(elec eng); Harvard Bus Sch, PMD, 80. *Prof Exp:* Actg asst prof, Univ Calif Berkeley, 67-68; mem tech staff, Rockwell Int Sci Ctr, 68-69; vpres eng, Phys Electronics Indust, Inc, 69-77; dir mktg, Perkin Elmer Corp, 77-79, gen mgr, Physical Elec div, 79-80, Group dir mktg, Semiconductor Equip Group, 80-82; PROF ELEC ENG, CORNELL UNIV, 84-, DIR SCH ELEC ENG, 89- *Honors & Awards:* Victor Macres Award, Electron Probe Anal Soc Am, 73. *Mem:* Inst Elec & Electronic Engrs; Am Phys Soc; Am Vacuum Soc; Electrochem Soc; Sigma Xi; AAAS. *Res:* Microdynamics and microactuators; microelectronics; Nanostructure science; electron beam instrumentation for lithography; characterization and metrology. *Mailing Add:* Dept Elec Eng Cornell Univ 224 Phillips Hall Ithaca NY 14853-5401

MACDONALD, NORMAN SCOTT, b Boston, Mass, Jan 16, 17; m 51; c 2. RADIOACTIVE TRACERS. *Educ:* Western Reserve Univ, AB, 38; Ohio State Univ, MSc, 40, PhD(org chem), 42. *Prof Exp:* Asst chem, Ohio State Univ, 38-42; res assoc, Mass Inst Technol, 42-43; asst prof, Occidental Col, 46-48; assoc prof biophys, Univ Calif, Los Angeles, 49-67, prof radiol, Sch Med, 67-, dir, Biomed & Cyclotron Facil, 71-, EMER PROF RADIOL, UNIV CALIF, LOS ANGELES. *Mem:* Am Chem Soc; Am Soc Biol Chem; Am Soc Nuclear Med. *Res:* Cyclotron production of radionuclides and labeling of compounds for applications in nuclear medicine; measurement of radioactivity in human body. *Mailing Add:* Radiol Sci Ar 259 Chs Univ Calif 405 Hilgard Ave Los Angeles CA 90024

MCDONALD, P(ATRICK) H(ILL), JR, b Carthage, NC, Dec 25, 24; m 51; c 3. MECHANICAL ENGINEERING & ENGINEERING MECHANICS. *Educ:* NC State Univ, BS, 47; Northwestern Univ, MS, 51, PhD(mech), 53. *Prof Exp:* Instr mech eng, NC State Univ, 47-48; instr mech & hydraul, Clemson Col, 48-50; asst, Northwestern Univ, 51-52; from res assoc prof to res prof mech eng, 53-58, prof & grad adminstr, 58-60, prof & head dept, 60-65, head dept, 65-76, HARRELSON PROF ENG MECH, NC STATE UNIV, 65- *Concurrent Pos:* Exec chmn, 5th Southeastern Conf Theoret & Appl Mech, 68-70. *Honors & Awards:* Pi Tau SIgma Gold Medal Award, Am Soc Mech Engrs, 57; Award, Am Soc Eng Educ, 59. *Mem:* Am Soc Mech Engrs; Am Soc Eng Educ; Soc Exp Mech. *Res:* Applied and continuum mechanics; stress analysis; vibration; shock and wave motion; engineering cybernetics. *Mailing Add:* 3120 Tanager St Raleigh NC 27606

MACDONALD, PAUL CLOREN, REPRODUCTIVE BIOLOGY. *Educ:* Univ Tex, MD, 55. *Prof Exp:* PROF OBSTET/GYNEC & BIOCHEM, CECIL H & IDA, CECIL GREEN CTR REPRODUCTIVE BIOL SCI, 74-, DIR ENDOCRINOL, 74- *Mailing Add:* Green Ctr Reproductive Biol 5323 Harry Hines Blvd Dallas TX 75235

MCDONALD, PAUL THOMAS, genetics, entomology, for more information see previous edition

MCDONALD, PERRY FRANK, b Garner, Tex, July 26, 33; m 59; c 2. SOLID STATE PHYSICS. *Educ:* Tex Christian Univ, BA, 54; NTex State Univ, MA, 60; Univ Ala, Tuscaloosa, PhD(physics), 68. *Prof Exp:* Res asst, Socony Mobil Field Res Lab, Tex, 60-61; res physicist, Brown Eng CoInc, Ala, 61-66 & US Army Missile Command, Redstone Arsenal, 66-68; ASSOC PROF PHYSICS, SAM HOUSTON STATE UNIV, 68- *Mem:* Am Phys Soc; Sigma Xi. *Res:* Acoustic paramagnetic resonance; electron paramagnetic resonance; spin lattice interactions. *Mailing Add:* Dept Physics Sam Houston State Univ Huntsville TX 77340

MACDONALD, PETER MOORE, b Toronto, Ont, Dec 29, 53; m 76. DEUTERIUM NMR, SURFACE ELECTROSTATICS. *Educ:* Univ Guelph, BSc, 76; Univ Western Ont, MSc, 78; Univ Alta, PhD(biochem), 84. *Prof Exp:* MRC postdoctoral fel biophys chem, Dept Biophys Chem, Bioctr Univ Basel, Switz, 85-87; postdoctoral fel radiol, Dept Radiol, Deaconess Hosp, Harvard Med Sch, 87-88; ASST PROF CHEM, DEPT CHEM, UNIV TORONTO, 88- *Concurrent Pos:* Med Res Coun Can postdoctoral fel, 85-87; Nat Sci & Eng Res Coun Can univ res fel, 88-93; dir phys & biol res, Ctr Magnetic Resonance Imaging, Toronto Western Hosp, 90- *Mem:* Chem Inst Can. *Res:* Solid state nuclear magnetic resonance spectroscopy of biomembrane proteins and lipids, polymers and zeolites; particular emphasis on Deuterium NMR studies of surface electrostatics in biomembranes and latex particles. *Mailing Add:* Dept Chem Univ Toronto 3359 Mississauga Rd Mississauga ON L5L 1C6

MCDONALD, PHILIP MICHAEL, b Seattle, Wash, Feb 5, 36; m 60; c 3. FORESTRY, ECOLOGY. *Educ:* Wash State Univ, BS, 60; Duke Univ, MF, 61; Ore State Univ, PhD(forest sci), 78. *Prof Exp:* RES FORESTER, PAC SOUTHWEST FOREST & RANGE EXP STA, 61- *Mem:* Ecol Soc Am. *Res:* Silviculture-ecology; cutting methods; seed production; regeneration; growth, including evaluation of succession; dynamics; biomass of conifers, hardwoods, woody shrubs and lesser vegetation. *Mailing Add:* Pac SW Forest & Range Exp Sta 2400 Washington Plaza Redding CA 96001

MACDONALD, R NEIL, b Calgary, Alta, Jan 6, 35; m 62; c 4. MEDICINE. *Educ:* Univ Toronto, BA, 55; McGill Univ, MD, 59. *Prof Exp:* Assoc dean, fac med, McGill Univ, 67-70; prof med, Univ Alta, 71-80, dir, Cross Cancer Inst, 75-87, dir div oncol, 81-87, ALTA CANCER BD PROF-PALLIATIVE MED, UNIV ALTA, 87- *Concurrent Pos:* From assoc dir to dir, Oncol Day Ctr, Royal Victoria Hosp, 67-71, assoc dir, Palliative Care Ctr, 80-81; chmn Serv Patients Comt, Can Cancer Soc, 81-84, mem, Nat Bd, 86-89; mem, Transfusion Adv Comt, Red Cross, 84-88; med oncol examr, Royal Col Physicians & Surgeons, 85-86, chief examr, 87-90; expert comt cancer, WHO, 88- *Honors & Awards:* Queen's Jubilee Award, 77. *Mem:* Can Oncol Soc (pres, 77); fel Royal Col Physicians Can; Am Soc Clin Oncol (secy-treas, 79-82); Can Cancer Soc (vpres, 86-); fel Am Col Physicians. *Res:* Co-analgesic agents; analgesic administration; asthenic syndrome in patients with advanced cancer. *Mailing Add:* Cross Cancer Inst 11560 University Ave Edmonton AB T6G 1Z2 Can

MCDONALD, RALPH EARL, b Indianapolis, Ind, May 12, 20; m 42; c 3. DENTISTRY. *Educ:* Ind Univ, BS, 42, DDS, 44, MS, 51; Am Bd Pedodontics, dipl, 52. *Prof Exp:* From instr to assoc prof pedodontics, 46-57, chmn dept, 53-68, asst dean & secy grad dent educ, 64-68, dean, 69-85, PROF PEDODONTICS, SCH DENT, IND UNIV, INDIANAPOLIS, 57-, EMER DEAN, 85- *Concurrent Pos:* Consult, USPHS, 50-; mem exam bd, Am Bd Pedodontics, 55-62, chmn bd, 60-62; ed, J Pediat Dent, 82- *Honors & Awards:* Distinguished Serv Award, Am Soc Dent for Children; Hon mem Brazilian Acad Dent & Am Dent Soc Ireland. *Mem:* Am Soc Dent for Children (secy-treas, 57-59, vpres, 59, pres, 62); fel Am Col Dent; Am Acad Pedodont (pres, 67); hon mem Brazilian Acad Dent & Am Dent Soc Ireland; fel Int Col Dent. *Res:* Clinical dental caries control; pathology of the dental pulp. *Mailing Add:* Emer Dean Ind Univ Sch Dent 1121 W Mich St Indianapolis IN 46202

MCDONALD, RAY LOCKE, b Chula Vista, Calif, Oct 5, 31; m 60; c 2. PHYSICAL CHEMISTRY. *Educ:* San Diego State Col, AB, 55; Ore State Univ, PhD(chem), 60. *Prof Exp:* Res assoc chem, Mass Inst Technol, 60-61; asst prof, NDak State Univ, 61-65; from asst prof to assoc prof, 65-74, assoc chmn dept, 71-74, prof & dept chmn, 74-77, assoc dean, Col Arts & Sci, 77-81, asst vchancellor acad prog, 82-86, PROF CHEM & ASSOC CHMN DEPT, UNIV HAWAII, 86- *Mem:* Am Chem Soc. *Res:* Ionic interactions in nonaqueous solvents; solvent extraction of ions; atmospheric chemistry. *Mailing Add:* Dept Chem Univ Hawaii Honolulu HI 96822

MACDONALD, RICHARD ANNIS, b Manistee, Mich, July 23, 28; m 54; c 3. MEDICINE, PATHOLOGY. *Educ:* Albion Col, AB, 51; Boston Univ, MD, 54. *Prof Exp:* Intern med, Boston City Hosp, Mass, 54-55, resident path, Mallory Inst Path, 56-59; instr, Harvard Med Sch, 59-60, assoc, 60-62, asst prof, 62-65; prof, Sch Med, Univ Colo, 66-69; prof, Sch Med, Boston Univ, 69-71; prof & chmn dept, Sch Med, Univ Mass, 71-73; prof path, Sch Med, Boston Univ, 73-85; PROF PATH, SCH MED, UNIV SOUTH FLA, 85- *Concurrent Pos:* Fel path, Harvard Med Sch, 57-59; Am Heart Asn sr res fel, Mallory Inst Path, Boston City Hosp, 58-59; chief path dept, Norwood Hosp, 73-83; pres, Mass Soc Path, 81-83; dir, Walwood Path, 84; chief, Lab Serv, Bay Pines Vet Admin Med Ctr, 85-89. *Mem:* AMA; Am Asn Path & Bact; Soc Exp Biol & Med; Am Soc Exp Path; Int Acad Path. *Res:* Liver and hematological diseases; nutrition; computers in medicine. *Mailing Add:* 551 Sandy Hook Rd Treasure Island FL 33706

MACDONALD, RICHARD G, b Detroit, Mich, Sept 28, 53; m; c 2. BIOCHEMISTRY. *Educ:* Univ Conn, Storrs, BA, 75, MS, 76; Univ Vt, Burlington, PhD(biochem), 81. *Prof Exp:* Res assoc, Worcester Found Exp Biol, Shrewsbury, Mass, 80- 82, sr res assoc, 82-83; instr biochem, Med Ctr, Univ Mass, Worcester, 83-88; ASST PROF BIOCHEM, MED CTR, UNIV NEBR, OMAHA, 88- *Concurrent Pos:* NIH grant, 84-87; Seed Res Grant Prog, Med Ctr, Univ Nebr, 88-89 & 90-91; Inst grant, Am Cancer Soc, 88-89; State Cancer grant, Nebr Dept Health, 89-90. *Mem:* Endocrine Soc; Am Soc Biochem & Molecular Biol. *Res:* Author of various publications. *Mailing Add:* Dept Biochem Univ Nebr Med Ctr 600 S 42nd St Omaha NE 68198-4525

MCDONALD, RICHARD NORMAN, b Detroit, Mich, Feb 26, 31; m 56; c 2. ORGANIC CHEMISTRY. *Educ:* Wayne State Univ, BS, 54, MS, 55; Univ Wash, PhD(org chem), 57. *Prof Exp:* Res chemist, Pioneering Res Div, Textile Fibers Dept, E I du Pont de Nemours & Co, 57-60; from asst prof to assoc prof, 60-68, PROF CHEM, KANS STATE UNIV, 68- *Concurrent Pos:* Hon lectr, Mid Am State Univ Asn, 78-79; mem comt disposal hazardous indust wastes, Nat Res Coun, 81-82. *Mem:* Am Chem Soc; Royal Soc Chem. *Res:* Synthesis and chemistry of non-benzenoid aromatic compounds, small ring compounds and strained polycyclic structures; chemistry of hypovalent radicals; gas phase ion-molecule reactions; structures of gas phase ions; reaction mechanisms. *Mailing Add:* Chem Dept Kans State Univ Manhattan KS 66506-3701

MCDONALD, ROBERT H, JR, b Cincinnati, Ohio, Nov 15, 33; m 58; c 2. CLINICAL PHARMACOLOGY. *Educ:* Xavier Univ, BS, 55; Stritch Sch Med, MD, 59; Univ Pittsburgh, MBA. *Prof Exp:* Intern, Health Ctr Hosps, Univ Pittsburgh, 59-60; jr resident internal med, Univ Pittsburgh, 60-61; NIH fel clin pharmacol, Emory Univ, 61-63; assoc surgeon, Nat Heart Inst, 63-65; USPHS hon res asst physiol, Univ Col, London, 65-66; from asst prof to assoc prof, 66-77, PROF MED & PHARMACOL, SCH MED, UNIV PITTSBURGH, 77-, CHIEF DIV CLIN PHARM & HYPERTENSION, PROF EPIDEMOL, GRAD SCH PUB HEALTH, 81- *Concurrent Pos:* mem, Coun High Blood Pressure & Coun Circulation, Am Heart Asn. *Mem:* Am Fedn Clin Res; Am Soc Pharmacol & Exp Therapeut; Cent Soc Clin Res; Am Physiol Soc; Am Heart Asn; Am Med Asn; Am Soc Clin Pharmacol & Therapeut. *Res:* Hypertension; adrenergic mechanisms; drug abuse; biobehavior modification of cardiovascular function. *Mailing Add:* Dept Med Univ Pitt 488 Scaife Hall Pittsburgh PA 15261

MACDONALD, ROBERT NEAL, b Mansfield, Ohio, Nov 2, 16; m 41; c 3. ORGANIC CHEMISTRY. *Educ:* Oberlin Col, AB, 38; Yale Univ, PhD(org chem), 41. *Prof Exp:* chemist, E I Du Pont de Nemours & Co, Inc, 41-81; CONSULT, 81- *Mem:* Sigma Xi. *Res:* Synthesis of high polymers. *Mailing Add:* 5 Crestfield Rd Wilmington DE 19810-1401

MCDONALD, ROBERT SKILLINGS, b Pittsburgh, Pa, July 19, 18; m 43; c 3. PHYSICAL CHEMISTRY, SPECTROSCOPY. *Educ:* Univ Maine, BS, 41; Mass Inst Technol, PhD(phys chem), 52. *Prof Exp:* Jr engr, Hygrade Sylvania Co, Mass, 51; res physicist, Am Cyanamid Co, Conn, 42-46; res assoc, Mass Inst Technol, 46-50; res assoc, Res & Develop Ctr, Gen Elect Co, 51-83, INFRARED SPECTROS, COMPUT APPLN, 83- *Concurrent Pos:* Joint Comt Atomic & Molecular Phys Data (JCamp), JCamp-DX. *Mem:* Am Chem Soc; Am Phys Soc; Coblentz Soc; Soc Appl Spectros. *Res:* Infrared spectroscopy; surface chemistry; computer applications in analytical chemistry; mass spectroscopy. *Mailing Add:* Nine Woodside Dr Burnt Hills NY 12027

MACDONALD, RODERICK, JR, b Charleston, SC, Oct 16, 26; m 51; c 5. OPHTHALMOLOGY. *Educ:* Davidson Col, BS, 47; Med Col SC, MD, 50. *Prof Exp:* Resident ophthal, New Orleans Eye, Ear, Nose & Throat Hosp, La, 52-53 & 55-56; instr ophthal, Sch Med, Tulane Univ, 56-57; exec dir, Dept Ophthal, Sch Med, Univ Louisville, 57- 65, from asst prof to prof ophthal, 57-73, actg chmn dept, 65, chmn dept, 65-73, assoc pharmacol, 69, assoc dean, 69-70, vdean, 70-73; prof ophthal & chmn dept, Med Col Va, 73-76; dean, 76-83, PROF OPHTHAL, SCH MED, UNIV SC, 76- *Concurrent Pos:* Mem vision res training comt, NIH, 68-71. *Mem:* AAAS; Am Col Surg; Am Ophthal Soc; Royal Soc Med; Am Acad Ophthal. *Res:* External diseases of the eye; ocular immunology. *Mailing Add:* 2015 Monument Ave Richmond VA 23220

MACDONALD, RODERICK PATTERSON, b Detroit, Mich, Nov 9, 24. BIOCHEMISTRY. *Educ:* Mich State Univ, BS, 47; Univ Detroit, MS, 49; Wayne State Univ, PhD(physiol chem), 52; Am Bd Clin Chem, dipl; Can Bd Clin Chem, dipl. *Prof Exp:* Asst chem, Univ Detroit, 47-49; spec instr, Mortuary Sch, Wayne State Univ, 49-51; supvr chem lab, 52-59, res assoc, 52-75, dir clin chem, 59-75, CLIN CHEMIST, HARPER HOSP, 75- *Concurrent Pos:* Asst prof path, Wayne State Univ, 64- *Honors & Awards:* McLean Award, 52. *Mem:* Am Chem Soc; Am Asn Clin Chem (secy, 65-68); Can Soc Clin Chem. *Res:* Metabolism and physiology of bone; thyroid diseases and hypometabolism; ultramicro clinical methods; general clinical chemistry; clinical enzymology. *Mailing Add:* 31356 Churchill Rd Birmingham MI 48009-4302

MACDONALD, ROSEMARY A, b Leamington Spa, Eng, Oct 7, 30; m 65; c 2. SOLID STATE PHYSICS. *Educ:* St Andrews Univ, BSc, 54; Oxford Univ, DPhil(solid state physics), 59. *Prof Exp:* Lectr physics, Somerville Col, Oxford Univ, 57-59; res assoc, Univ Md, 59-61; NATO res fel, Bristol Univ, 61-62; lectr, Sheffield Univ, 62-64; PHYSICIST, NAT INST STANDARDS & TECHNOL, 64- *Mem:* Am Phys Soc; Brit Inst Physics; Sigma Xi. *Res:* Lattice dynamics; point defects; phase transitions; molecular dynamical calculations of nonequilibrium behavior of solids and liquids; modeling transport in porous media. *Mailing Add:* Thermophysics Div Nat Inst Stand & Technol Phys A105 Gaithersburg MD 20899

MACDONALD, RUSSELL EARL, b NB, Can, Feb 18, 28; m 59; c 3. BACTERIOLOGY. *Educ:* Acadia Univ, BA, 50, MA, 52; Univ Mich, PhD(bact), 57. *Prof Exp:* Asst bact, Acadia Univ, 50-52; asst, Univ Mich, 53-55, instr & res assoc, 56-57; from asst prof to assoc prof bact, 57-, EMER PROF, CORNELL UNIV. *Concurrent Pos:* Jane Coffin Childs Mem Fund Med Res fel, 64-65; vis scientist, Ames Res Ctr, 74-75. *Mem:* Am Soc Microbiol. *Res:* Microbial ecology; molecular biology; control mechanisms; ribosomes; membrane transport. *Mailing Add:* c/o Liberman 1316 E Tenth St Brooklyn NY 11230

MACDONALD, RUTH S, FOOD SCIENCE. *Prof Exp:* ASST PROF, FOOD SCI & NUTRIT DEPT, UNIV MO, 87- *Mailing Add:* Food Sci & Nutrit Dept Univ Mo 122 Eckles Hall Columbia MO 65211

MCDONALD, T(HOMAS) W(ILLIAM), b Winnipeg, Man, June 24, 29; m 53; c 3. HEAT TRANSFER, THERMAL SYSTEMS ANALYSIS. *Educ:* Queen's Univ, BSc, 52, MSc, 55; Purdue Univ, PhD(convective heat transfer), 65. *Prof Exp:* Univ Sask, 54-68; head dept, 72-82, PROF MECH ENG, UNIV WINDSOR, 68- *Mem:* Can Soc Mech Engrs; Am Soc Heating, Refrig & Air-Conditioning Engrs; Int Solar Energy Soc; Solar Energy Soc Can. *Res:* Waste heat recovery systems; simulation of thermal and flow systems; two phase flow thermal systems; hysteresis phenomenon associated with incipient nucleate boiling. *Mailing Add:* Dept Mech Eng Univ Windsor Windsor ON N9B 3P4 Can

MCDONALD, TED PAINTER, b Loudon, Tenn, Nov 23, 30; m 55; c 3. BIOCHEMISTRY. *Educ:* Univ Tenn, BS, 55, MS, 58, PhD(animal sci, biochem), 65. *Prof Exp:* Radiobiologist, Oak Ridge Nat Lab, 58-64; res asst blood platelet & mucopolysaccharide chem, Res Lab, AEC, 64-65; res assoc, 65, from res instr to prof exp hemat, Mem Res Ctr, 65-83, PROF HEMAT, COL VET MED, UNIV TENN, KNOXVILLE, TENN, 83- *Concurrent Pos:* Mem coun thrombosis, Am Heart Asn. *Mem:* Am Soc Hemat; Int Soc Hemat; Soc Exp Biol & Med; Int Soc Exp Hemat; Radiation Res Soc. *Res:* Blood platelet physiology; mucopolysaccharide chemistry; experimental hematology. *Mailing Add:* Dept Animal Sci Univ Tenn Knoxville TN 37996

MCDONALD, TERENCE FRANCIS, b Dublin, Ireland, Jan 28, 43; Can citizen. MEMBRANE IONIC CHANNELS, HEART DYSFUNCTION. *Educ:* Univ Alta, Edmonton, BSc, 64; Dalhousie Univ, Halifax, PhD(physiol & biophysics), 71; Univ London, DIC, 76. *Prof Exp:* From asst prof to assoc prof, 75-83, PROF PHYSIOL, DALHOUSIE UNIV, 83-, HEAD, 87- *Concurrent Pos:* Vis prof, II Physiol Inst Univ Saarlandes, Homburg, Ger, 76-90; mem, Sci Rev Exec, Can Heart Found, 82-86; assoc ed, Can J Physiol & Pharmacol, 87- *Mem:* Can Physiol Soc; Can Fedn Biol Sci; Brit Physiol Soc; Biophys Soc Can. *Mailing Add:* Dept Physiol & Biophys Dalhousie Univ Halifax NS B3H 4H7 Can

MACDONALD, THOMAS THORNTON, b Glasgow, Scotland, Jan 9, 51. IMMUNOLOGY. *Educ:* Univ Glasgow, BSc Hons, 73, PhD(immunol), 77. *Prof Exp:* Res asst, Univ Glasgow, 73-76; res assoc, Trudeau Inst, Saranac Lake, NY, 76-78; ASST PROF MICROBIOL, THOMAS JEFFERSON UNIV, 78-; DEPT PEDIAT GASTROENTEROL, ST BARTHOLOMEW CTR CLIN RES, LONDON, ENG. *Concurrent Pos:* Fels, Nat Found Ileitis & Colitis Inc, 76-78 & res grant, 78-79; Asn Gnotobiotics fel, 78. *Mem:* Brit Soc Immunol; Am Soc Microbiol; Asn Gnotobiotics. *Res:* The role of cell-mediated immunity in host defenses at mucous surfaces. *Mailing Add:* Dept Pediat Gastroenterol St Bartholomew Ctr Clin Res Bartholomew Close London 0166067721 England

MACDONALD, TIMOTHY LEE, b Long Beach, Calif, Mar 12, 48; m 71; c 2. ORGANIC CHEMISTRY, BIO-ORGANIC CHEMISTRY. *Educ:* Univ Calif, Los Angeles, BSc, 71; Columbia Univ, PhD(chem), 75. *Prof Exp:* NIH fel chem, Stanford Univ, 75-77; asst prof chem, Vanderbilt Univ, 77-82; ASSOC PROF CHEM, UNIV VA, 88- *Concurrent Pos:* A P Sloan Found res fel, 81-83. *Honors & Awards:* Lederle Award, 87. *Mem:* Am Chem Soc. *Res:* Synthesis of natural products, development of new synthetic methods, determination of mechanisms of enzymes with toxicological relevance; enzymes in DNA processing and studies of aluminum neurotoxicity. *Mailing Add:* Univ Va McCormick Rd Charlottesville VA 22901

MCDONALD, TIMOTHY SCOTT, b Albuquerque, NMex, Feb 20, 46; m 71; c 2. COMPUTER SCIENCE. *Educ:* Univ NMex, BS, 70, MS, 72, PhD(elec eng), 77. *Prof Exp:* From res asst to res engr, 67-77, PROG MGR MINICOMPUT SYSTS & DATA COMMUN, DIKEWOOD INDUSTS, INC, 77- *Concurrent Pos:* Consult, Equ-A-Ex Corp, 78- *Res:* Digital signal analysis; minicomputer system software development; data communications; computerized mass spectrometry. *Mailing Add:* 2801 Charleston NE Albuquerque NM 87110

MCDONALD, W JOHN, b Lethbridge, Alta, Sept 29, 36; m 61; c 3. SUBATOMIC PHYSICS. *Educ:* Univ Sask, BSc, 59, MSc, 61; Univ Ottawa, Ont, PhD(physics), 64. *Prof Exp:* From asst prof to assoc pro, 65-75, PROF PHYSICS, UNIV ALTA, 75-, DEAN SCI, 81- *Mem:* Am Asn Physics Teachers; Am Phys Soc; Can Asn Physicists. *Res:* Subatomic physics research using accelerators; intermediate energy proton and electron scattering and reactions. *Mailing Add:* Fac Sci Univ Alta Edmonton AB T6G 2E9 Can

MACDONALD, WILLIAM, b Salem, Ohio, Nov 25, 27; m 65; c 2. NUCLEAR PHYSICS, EDUCATIONAL SOFTWARE. *Educ:* Univ Pittsburgh, BS, 50; Princeton Univ, PhD(physics), 55. *Prof Exp:* Asst, Princeton Univ, 50-54; theoret physicist, Radiation Lab, Univ Calif, 54-55; vis prof physics, Univ Wis, 55-56; from asst prof to assoc prof, 56-63, PROF PHYSICS, UNIV MD, COLLEGE PARK, 63- *Concurrent Pos:* NATO sr fel, 62-63; vis scientist, Atomic Energy Res Estab, Harwell, Eng, 69-70, Nat Bur Standards, 77; prog dir theoret physics, NSF, 87-88. *Mem:* Fel Am Phys Soc; Sigma Xi; Am Asn Physics Teachers. *Res:* Structure of light nuclei; nuclear reaction theory; plasma physics; space physics; educational software. *Mailing Add:* Dept Physics & Astron Univ Md College Park MD 20740

MCDONALD, WILLIAM CHARLES, b Chicago, Ill, Feb 14, 33; m 57; c 4. MICROBIOLOGY. *Educ:* Univ Okla, BS, 55; Univ Tex, PhD(bact), 59. *Prof Exp:* Res assoc biochem genetics, Children's Hosp, Buffalo, NY, 59-60; asst prof bact, Wash State Univ, 62-65; from assoc prof to prof biol, Tulane Univ, 65-73; chmn dept, 73-83, PROF BIOL, UNIV TEX, ARLINGTON, 73- *Concurrent Pos:* State of Wash initiative measure grants, 63-66; NSF res grants, 63-69. *Mem:* fel Am Acad Microbiol; Am Soc Microbiol; Sigma Xi. *Res:* Effects of radiation on bacteria; microbial genetics; temperature sensitivity and resistance in microorganisms; biology of bacteriophages. *Mailing Add:* Dept Biol Univ Tex Arlington TX 76019

MACDONALD, WILLIAM DAVID, b Chatham, Ont, Feb 28, 37; m 70; c 1. TECTONICS OF LATIN AMERICA-CARIBBEAN. *Educ:* Univ Western Ont, BSc, 59; Princeton Univ, PhD(geol), 65. *Prof Exp:* Asst prof geol, Villanova Univ, 64-65; from asst prof to assoc prof, 65-78, PROF GEOL, STATE UNIV NY BINGHAMTON, 78- *Concurrent Pos:* Vis prof & researcher, Univ Sao Paulo, 71, Cambridge Univ, 72 & Stanford Univ, 78-79. *Mem:* Geol Soc Am; Am Geophys Union; Geol Asn Can. *Res:* Structural geology, tectonics and paleomagnetism of Latin America and Caribbean regions; computer applications in geology and geophysics; orientation analysis. *Mailing Add:* Dept Geol Sci State Univ NY Binghamton NY 13902-6000

MCDONALD, WILLIAM E, JR, b Columbus, Ohio, Nov 21, 16; m 40. PHARMACOLOGY, TOXICOLOGY. *Educ:* Emory Univ, AB, 39; Univ Fla, MS, 51; Univ Miami, PhD(pharmacol), 61; Am Acad Indust Hyg, dipl; Am Bd Toxicol, dipl. *Prof Exp:* Formulator, Biscayne Chem Co, 39-40; spec officer, Seaboard Air Line RR, 40-41; chemist, Fla State Bd Health, 41-46, indust hyg chemist, 46-55; instr, Sch Med, Univ Miami, 55-63, asst prof pharmacol & toxicol, 63-75, assoc dir, Res & Teaching Ctr Toxicol, 73-75; dir sci div, 75-80, PRIN TOXICOLOGIST, TRACOR JITCO, INC, 75-, SCI ADV TO PRES, 81- *Concurrent Pos:* Am Cancer Soc instnl grants, 64-66. *Mem:* Soc Toxicol; Am Chem Soc; Am Conf Govt Indust Hygienists; Am Indust Hyg Asn; Sigma Xi. *Res:* Pharmacology, toxicology and industrial hygiene of silicones, nitroolefins, benzene, food additives including drugs, pesticides and anticorrosives; bladder carcinogens; subliminal toxicology of pesticides; carcinogenesis bioassay. *Mailing Add:* Tracor Jitco Inc 1601 Research Blvd Rockville MD 20850-3191

MCDONALD, WILLIAM TRUE, b Bellingham, Wash, July 18, 35; m 61; c 3. AEROSPACE SCIENCES. *Educ:* Calif Inst Technol, BS, 57, MS, 58; Mass Inst Technol, ScD(instrumentation), 68. *Prof Exp:* Res engr, Jet Propulsion Lab, 60; supvr, Autonetics Div, NAm Aviation, Inc, 60-62; staff engr, Instrumentation Lab, Mass Inst Technol, 62-64, staff engr, Exp Astron Lab, 64-68; mem tech staff, 68-80, DIV CHIEF SCIENTIST, ROCKWELL INT CORP, 80- *Concurrent Pos:* Founder & dir, Dover Instrument Corp, Mass, 64-68; consult, Rockwell Int Corp, 66-67, Med Systs Tech Serv, Inc, 78-79 & Asst Secy Defense, Health Affairs, 79-; consult & auth in field of small arms ballistics, 71-; lectr Col Eng, W Coast Univ, Los Angeles, 80- *Honors & Awards:* Engr of Year, Rockwell Int Corp, 76. *Res:* Missile and spacecraft guidance, navigation and control systems engineering; visible and infrared sensor systems. *Mailing Add:* Autonetics Rockwell Int Corp 3370 Miraloma Ave Anaheim CA 92803

MCDONEL, EVERETT TIMOTHY, b Lima, Ohio, Oct 6, 33; m 55; c 6. POLYMER CHEMISTRY, TIRE ENGINEERING. *Educ:* Case Inst Technol, BS, 55, MS, 57, PhD(org chem), 60. *Prof Exp:* Res asst polymer chem, Case Inst Technol, 57-59; res chemist, B F GOODRICH, 59-61, sr res chemist tire mat, 61-64, group mgr, Tire Res, Tire Div, 72-82, group mgr int tech liaison, 82-86, SECT LEADER TIRE MAT, RES CTR, B F GOODRICH, 64-, SR TIRE TECH CONSULT, 86- *Concurrent Pos:* Lectr, Ctr Prof Advan. *Mem:* Am Chem Soc. *Res:* Oxidation and vulcanization of elastomers; relation of polymer structure to performance; new elastomers for tires. *Mailing Add:* Uniroyal Goodrich Tire Co 600 South Main Akron OH 44397-0001

MCDONELL, WILLIAM ROBERT, b New Rockford, NDak, Mar 8, 25; m 55; c 2. RADIATION EFFECTS ON MATERIALS. *Educ:* Univ Mich, BS, 47, MS, 48; Univ Calif, PhD(chem), 51. *Prof Exp:* Chemist, Radiation Lab, Univ Calif, 48-51 & Argonne Nat Lab, 51-53; chemist, Savannah River Lab, E I Du Pont de Nemours & Co, Inc, 53-60, res supvr, 60-70, sr res scientist, 69-70, res assoc, 70-89; ADV SCIENTIST, WESTINGHOUSE SAVANNAH RIVER CO, 89- *Concurrent Pos:* Mem, High Temperature Fuels Comt, AEC, 65-69; mem, Tech Sub-comt Fuel Performance Integral Fast Reactor, Univ Chicago, 85- *Mem:* Am Chem Soc; Am Nuclear Soc; AAAS. *Res:* Nuclear and radiation chemistry; effects of radiation on metals and ceramics; reactor fuel behavior; development of radioisotopic heat and radiation sources; surface effects of radiation on materials; technology and costs of nuclear waste disposal; heavy water reactor development. *Mailing Add:* 1318 Evans Rd Aiken SC 29803

MCDONNEL, GERALD M, b Salt Lake City, Utah, Feb 13, 19; m 41; c 7. RADIOLOGY, RADIOBIOLOGY. *Educ:* Univ Utah, BA, 40; Temple Univ, MD, 43. *Prof Exp:* Intern, Temple Univ Hosp, 43, resident internal med, 44; assoc prof radiol, ctr health sci, Univ Calif, Los Angeles, 59-67; chancellor, Am Col Radiol, 66-72. *Concurrent Pos:* Mem sci adv bd, US Air Force, 60-76

& Defense Sci Bd, 64-68; mem staff dept radiol, Hosp of the Good Samaritan, Los Angeles, 68-85. *Mem:* AMA; Am Col Radiol; Radiol Soc NAm. *Res:* Nuclear weapons effects; radiation effects and space radiation effects in humans; megavoltage diagnostic radiology. *Mailing Add:* 4766 Bryn Mawr Rd Los Angeles CA 90027-2258

MCDONNELL, ARCHIE JOSEPH, b New York, NY, June 3, 36; m 66; c 2. WATER QUALITY ASSESSMENT, ENVIRONMENT. *Educ:* Manhattan Col, BS, 58; Pa State Univ, MS, 60, PhD(civil eng), 62. *Prof Exp:* From asst prof to assoc prof, 63-73, PROF ENVIRON ENG, PA STATE UNIV, 73-, DIR, ENVIRON RESOURCES RES INST, 86- *Concurrent Pos:* Dir, Water Resources Res Inst, Pa State Univ, 69-82, Inst Res Land & Water Resources, 82-86. *Honors & Awards:* James R Croes Medal, Am Soc Civil Engrs, 76; Karl M Mason Medal, Asn Environ Professionals, 91. *Mem:* Am Soc Civil Engrs; Water Pollution Control Asn; Int Asn Water Pollution Res; Am Soc Limnol & Oceanog; Am Water Resources Asn. *Res:* Direction of funded research projects in the principal areas of stream eutrophication; characterization of physical and biological reactions in natural waters; statistical methods in water quality data analysis. *Mailing Add:* Environ Resources Res Inst Pa State Univ University Park PA 16802

MCDONNELL, FRANCIS NICHOLAS, b Ottawa, Ont, Sept 28, 40; m 67; c 3. NUCLEAR ENGINEERING, PHYSICS. *Educ:* Royal Mil Col, BEng, 63; Univ Toronto, MASc, 66; Univ Manchester, PhD, 70. *Prof Exp:* Res officer, Appl Math Br, Atomic Energy Can Ltd, 70-79, sr adv to vpres res, 79-81, exec asst to pres, 82-83, dir bus initiatives, 84-86, res & develop mgr, 87-89, GEN MGR, ATOMIC ENERGY CAN LTD, 90- *Mem:* nuclear fuel waste management; Can Nuclear Soc. *Res:* Reactor physics; measurement and analysis; reactor dynamics; code development; research and development management. *Mailing Add:* Atomic Energy Can Ltd Whiteshell Labs Pinawa MB R0E 1L0 Can

MACDONNELL, JOHN JOSEPH, b Springfield, Mass, Mar 28, 27. MATHEMATICS. *Educ:* Boston Col, AB, 50, MA, 51; Cath Univ Am, PhD(math), 57. *Prof Exp:* From instr to assoc prof, 60-69, ASSOC PROF MATH, COL OF THE HOLY CROSS, 69- *Mem:* Math Asn Am; Am Math Soc; Sigma Xi; Am Asn Univ Professors. *Res:* Convergence theorems for Dirichlet type series. *Mailing Add:* Dept Math Col the Holy Cross Worcester MA 01610-2395

MACDONNELL, JOSEPH FRANCIS, b Springfield, Mass, May 4, 29. MATHEMATICS. *Educ:* Boston Col, BA, 54, MA, 59 & 62; Fordham Univ, MS, 64; Columbia Univ, EdD, 72. *Prof Exp:* Asst prof, Al Hikma Univ, Baghdad, 64-69; from asst prof to assoc prof, 69-85, PROF MATH, FAIRFIELD UNIV, 85- *Mem:* Am Math Asn; Sigma Xi. *Res:* Ruled surfaces; moebius surfaces of n half twists. *Mailing Add:* Dept Math Fairfield Univ Fairfield CT 06430

MCDONNELL, LEO F(RANCIS), b Edmonton, Alta, May 10, 26; nat US; m 57; c 5. CHEMICAL ENGINEERING. *Educ:* Univ Alta, BSc, 48; Lawrence Col, MS, 55, PhD(chem eng), 59. *Prof Exp:* Res engr, Minn & Ont Paper Co, 48-53; tech dir, Cup Div, Scott Paper Co, 58-61, tech dir, Marinette Plant, 61-65, brand tech mgr, staff tech servs, 65-73, int develop mgr, 73-86; RETIRED. *Concurrent Pos:* Consult, 87- *Mem:* Tech Asn Pulp & Paper Indust. *Res:* Pulp, paper and fiber insulation board manufacture and conversion; organic protective and decorative coating adhesives. *Mailing Add:* 1999 Kimberwick Rd Media PA 19063

MCDONNELL, MARK JEFFERY, b New London, Conn, June 13, 53; m 77; c 3. COMMUNITY ECOLOGY, LANDSCAPE ECOLOGY. *Educ:* Conn Col, BA, 75; Univ NH, MS, 79, Rutgers Univ, PhD(ecol), 83. *Prof Exp:* Postdoctoral assoc, 83-84, asst scientist, 84-91, ASSOC SCIENTIST, INST ECOSYST STUDIES, 91- *Mem:* Ecol Soc Am; Int Asn Landscape Ecol (treas, 89-92); Brit Ecol Soc; Torrey Bot Club; New Eng Bot Club. *Res:* Vegetation dynamics, especially seed dispersal; urban-rural gradient ecology; ecology of forests in urban environments. *Mailing Add:* Inst Ecosyst Studies PO Box AB Millbrook NY 12545

MCDONNELL, SANFORD, b Little Rock, Ark, Oct 12, 22; m; c 2. MECHANICAL ENGINEERING. *Educ:* Princeton Univ, BA, 45; Univ Colo, BS, 48; Wash Univ, MS, 54. *Hon Degrees:* LLD, Wash Univ, 85, Univ Mo, 85, Maryville Col, 88, Okla Christian Col, 89, Lehigh Univ, 89; DSc, St Louis Univ, 85, Univ Colo, 89. *Prof Exp:* Training prog, McDonnell Douglas Corp, 48, stress engr, 49-51, aerodynamicist, 51, design engr, 52-55, F-101 group leader, 55, F-101 asst proj engr, 56, F-101 proj mgr, 57-59, vpres proj mgt, 59-61, F4H vpres-gen mgr, 61, dir of the co & vpres gen mgr, 62, exec comt, 63-65, vpres-aircraft gen mgr, 65, pres, McDonnell Aircraft Co, 66, dir, 67-71, pres, 71, chief exec officer, 72-75, chmn, exec comt, 75-80, chmn & chief exec officer, 80-88, emer chmn, 88; RETIRED. *Mem:* Nat Acad Eng; fel Am Inst Aeronaut & Astronaut. *Res:* Science administration. *Mailing Add:* McDonnell Douglas Corp PO Box 516 St Louis MO 63166-0516

MACDONNELL, WILFRED DONALD, b Boston, Mass, Nov 3, 11; m 37, 71; c 3. METALLURGY. *Educ:* Mass Inst Technol, BS, 34. *Hon Degrees:* LLD, Lawrence Inst Technol, 61. *Prof Exp:* Trainee to gen supt, Bethlehem Steel Co, NY, 34-49, asst gen mgr, Pa, 49-57; vpres opers, Great Lakes Steel Co Div, Nat Steel Corp, Mich, 57-59, pres, 59-62; vpres, Kelsey-Hayes Co, 62, pres, 62-69, pres & chief exec officer, 69-76, chmn & chief exec officer, 76-77; RETIRED. *Concurrent Pos:* Consult, Kelsey-Hayes Co, 78-86. *Mem:* Nat Acad Engrs; Soc Automotive Engrs. *Res:* Steel industry management; management of supplier to transportation industry. *Mailing Add:* 2121 View Point Dr Naples FL 33963-7936

MCDONNELL, WILLIAM VINCENT, b Carbondale, Pa, Oct 28, 22; m 49; c 2. PATHOLOGY. *Educ:* Univ Scranton, AB, 43; Jefferson Med Col, MD, 47; Am Bd Path, dipl. *Hon Degrees:* LHD, Penna Col Pediat Med. *Prof Exp:* Vpres med affairs, West Jersey Health Syst, 72-90; from asst prof to assoc prof, 52-69, PROF PATH, JEFFERSON MED COL, 69- *Concurrent Pos:* Asst dir clin lab, Methodist Hosp & Jefferson Med Col, 52-61; coordr cancer teaching, Jefferson Med Col, 55-60; dir clin lab, West Jersey Hosp, 61-79; med dir, West Jersey Health Syst, 69-79. *Mem:* Am Soc Clin Pathologists; AMA; Col Am Pathologists; Am Acad Med Dis; Am Col Physician Execs; fel Am Col Phys. *Res:* Human pathology. *Mailing Add:* W Jersey Health Syst Mt Ephraim & Atlantic Ave Camden NJ 08104

MCDONOUGH, EUGENE STOWELL, b Abingdon, Ill, Apr 16, 05; m 30, 75. MYCOPATHOLOGY, GENETICS. *Educ:* Marquette Univ, BS, 28, MS, 31; Iowa State Col, PhD(bot), 36. *Prof Exp:* Asst bot, Marquette Univ, 28-29; instr, Mich State Col, 29-30; from instr to prof, 30-70, EMER PROF BOT, MARQUETTE UNIV, 70- *Concurrent Pos:* Co-dir, Holton & Hunkel Res Award, 44-50; researcher, Mycol Unit, USPHS grant, Commun Dis Ctr, Ga, 58-59. *Mem:* Fel AAAS; Mycol Soc Am; Int Soc Human & Animal Mycol; Med Mycol Soc Americas. *Res:* Cytology and genetics of fungi; polyploidy; human heredity; host-parasite relations of plant diseases; diseases of flowering plants; chromosome structure; antimycotics; phytopathology; medical mycology; soil mycolysis; epidemiology; epidemiological studies on blastomycosis. *Mailing Add:* Dept of Biol Marquette Univ Milwaukee WI 53233

MCDONOUGH, EVERETT GOODRICH, organic chemistry, for more information see previous edition

MCDONOUGH, JAMES EDWARD, b Stillacoom, Wash, Oct 8, 59. EXPERIMENTAL HIGH ENERGY PHYSICS. *Educ:* Temple Univ, BA, 82, MA, 84, PhD(physics), 87. *Prof Exp:* Res physicist, Yale Univ, 87-89; POSTDOCTORAL FEL PHYSICS, UNIV TEX, AUSTIN, 90- *Honors & Awards:* Louis Rosen Prize, Los Alamos Meson Physics Facil Users Group, 87. *Mem:* Am Physics Soc. *Res:* Interests are in rare and forbidden decays that probe fundamental symmetries. *Mailing Add:* Physics Dept Univ Tex Austin TX 78712

MCDONOUGH, JAMES FRANCIS, b Boston, Mass, June 7, 39; c 7. STRUCTURES, COMPUTER APPLICATIONS. *Educ:* Northeastern Univ, BSCE, 62, MSCE, 64; Univ Cincinnati, PhD(civil eng), 68, MBA, 81. *Prof Exp:* Asst prof, 68-74, assoc prof, 74-78, head & prof Civil Eng, 78-87, ASSOC DEAN, UNIV CINCINNATI, 86- *Concurrent Pos:* Prof civil eng, Univ Kabul, Afghanistan, 69-71; vis prof, NC State Univ, 71; consult, Huskey Prod & various firms, 78-; William Thomas Prof Civil Eng, 78. *Honors & Awards:* Western Elec Award, Am Soc Eng Educ, 75 & Dow Award, 77. *Mem:* Am Soc Civil Eng; Am Soc Eng Educ; Nat Soc Prof Engrs. *Mailing Add:* Univ Cincinnati Mail Location 18 Cincinnati OH 45221

MCDONOUGH, KATHLEEN H, b St Charles, Mo, Sept 16, 47. CARDIOVASCULAR PHYSIOLOGY. *Educ:* Univ Mo, PhD(physiol), 77. *Prof Exp:* Res asst prof, 79-82, asst prof, 82-86, ASSOC PROF PHYSIOL, LA STATE UNIV, 86- *Mem:* Am Physiol Soc; Am Heart Asn; Shock Soc. *Mailing Add:* Dept Physiol La State Univ Med 1901 Perdido St New Orleans LA 70112

MCDONOUGH, LESLIE MARVIN, b Tacoma, Wash, July 2, 30; c 2. INSECT SEX PHEROMONES, INSECT SEMIOCHEMICALS. *Educ:* Univ Puget Sound, BS, 54; Univ Wash, PhD(chem), 60. *Prof Exp:* Res chemist, E I du Pont de Nemours & Co, 60-64; assoc res chemist, Midwest Res Inst, 64-65; res leader, 65-88, LEAD SCIENTIST, AGR RES SERV, USDA, 88- *Mem:* Entom Soc Am; Am Chem Soc; Int Soc Chem Ecol; Controlled Release Soc. *Res:* Identification and synthesis of insect sex pheromones; insect behavioral response to sex pheromones; controlled release of sex pheromones; development of applications of sex pheromones for control of insect pests. *Mailing Add:* 308 N 58th Ave Yakima WA 98908

MCDONOUGH, ROBERT NEWTON, b Huntingdon, Pa, Feb 3, 35; m 65; c 2. ELECTRICAL ENGINEERING. *Educ:* Johns Hopkins Univ, BEngS, 56, DE(elec eng), 63. *Prof Exp:* Asst prof elec eng, Univ Del, 62-65; mem tech staff, Bell Tel Labs, 65-69; assoc prof elec eng & asst dean, Univ Del, 69-80; PRIN ENGR, APPL PHYSICS LAB, JOHNS HOPKINS UNIV, 80- *Concurrent Pos:* Ed, Scripta Publ Corp, 68- *Mem:* Inst Elec & Electronic Engrs. *Res:* Acoustic array processing; system modeling; estimation and detection theory. *Mailing Add:* Appl Physics Lab John Hopkins Univ John Hopkins Rd Laurel MD 20723

MCDONOUGH, THOMAS JOSEPH, b Toronto, Ont, Sept 16, 40. CHEMICAL ENGINEERING, ORGANIC CHEMISTRY. *Educ:* Univ Toronto, BASc, 62, PhD(chem eng), 72. *Prof Exp:* Res engr, Can Int Paper Co, 62-66 & Can Industs Ltd, 72-78; from asst prof to assoc prof, 78-83, PRIN RES ENGR, INST PAPER CHEM, 78-, PROF, 84- *Concurrent Pos:* Erco res fel, Univ Toronto, 85. *Honors & Awards:* C Howard Smith Medal, Can Pulp & Paper Asn, 71. *Mem:* Can Pulp & Paper Asn; Tech Asn Pulp & Paper Indust. *Res:* Pulping and bleaching technology; chemistry and kinetics of wood pulping and bleaching reactions. *Mailing Add:* Inst Paper Sci & Technol 575 14th St NW Atlanta GA 30318

MCDONOUGH, THOMAS REDMOND, b Boston, Mass. JOVIAN ASTROPHYSICS. *Educ:* Mass Inst Technol, SB, 66; Cornell Univ, PhD(astrophys), 73. *Prof Exp:* Res assoc astrophys, Cornell Univ, 73-75; resident res assoc, Jet Propulsion Lab, Nat Res Coun, NASA, 76-77; LECTR ENG, CALIF INST TECHNOL, 79- *Concurrent Pos:* Consult, Jet Propulsion Lab, 78-81 & Comt for the Sci Invest of Claims of the Paranormal, 90-; coordr search for extraterrestrial intel, Planetary Soc, 81-; tech adv, Avco Embassy Pictures, 81-82 & New World Pictures, 82-84. *Honors & Awards:* Spec Citation, NASA, 79. *Mem:* Int Astron Union; fel Brit Interplanetary Soc; Am Astron Soc; Am Phys Soc; Am Inst Aeronaut & Astronaut; Am Geophys Union. *Res:* Astrophysics of Jovian planets; Search for extraterrestrial intelligence; popular science, especially space science. *Mailing Add:* 138-78 Calif Inst Technol Pasadena CA 91125

MACDORAN, PETER FRANK, b Los Angeles, Calif, Jan 2, 41; m 64; c 2. GEODESY. *Educ:* Calif State Univ, Northridge, BS, 64; Univ Calif, Santa Barbara, MS, 66. *Prof Exp:* Aries Proj mgr radio geod, Jet Propulsion Lab, Calif Inst Technol, 68-; PRES, ISTAC INC. *Honors & Awards:* NASA Except Sci Achievement Award, 70. *Mem:* Am Geophys Union; Inst Elec & Electronics Engrs. *Res:* Research and development of radio interferometry for applications to problems of geodesy and earth physics. *Mailing Add:* ISTAC Inc 444 N Altadena Dr Suite 101 Pasadena CA 91107

MCDOUGAL, DAVID BLEAN, JR, b Chicago, Ill, Jan 31, 23; m 47; c 3. PHARMACOLOGY. *Educ:* Princeton Univ, BA, 45; Univ Chicago, MD, 47. *Prof Exp:* Instr anat, Johns Hopkins Univ, 50-53; from instr to assoc prof, 57-71, PROF PHARMACOL, SCH MED, WASH UNIV, 71- *Res:* Neurochemistry. *Mailing Add:* Rte 1 Box 85A Labadie MO 63055

MCDOUGAL, ROBERT NELSON, b Breckenridge, Minn, Mar 21, 20; m 45; c 3. MECHANICAL ENGINEERING, MATHEMATICS. *Educ:* NDak State Univ, BS, 61, MS, 64; Univ Nebr, Lincoln, PhD(eng mech), 71. *Prof Exp:* From instr to assoc prof mech eng, 63-77, asst dean sch eng, NDak State Univ, 71-77; asst prof mech eng, 77-81, ASSOC PROF ENG MECH, UNIV NEBR, 81- *Concurrent Pos:* Vpres, Forensic Engrs, Ltd, 70- *Mem:* AM Soc Mech Eng; Am Soc Eng Educ; Soc Exp Stress Anal; Am Soc Testing & Mat. *Res:* Heat transfer through composite materials; stress analysis in structural and machine elements. *Mailing Add:* 4830 Fleetwood Circle Lincoln NE 68516

MCDOUGALD, LARRY ROBERT, b Broken Bow, Okla, Dec 31, 41; m 62; c 1. PARASITOLOGY. *Educ:* Southeastern State Col, BS, 62; Kans State Univ, MS, 66, PhD(parasitol), 69. *Prof Exp:* Teacher pub schs, Kans, 62-64; instr biol, Mt St Scholastica Col, 67-68; instr zool, Kans State Univ, 68-69; fel poultry sci, Univ Ga, 69-71; sr parasitologist, Parasitol Res Dept, Eli Lilly & Co, 71-77; assoc prof, 77-89, PROF, POULTRY SCI DEPT, UNIV GA, 89- *Honors & Awards:* D W Brooks Award. *Mem:* Am Soc Parasitol; Am Micros Soc; Poultry Sci Asn; Am Asn Avian Pathologists. *Res:* Physiological and chemical aspects of host-parasite relationships; chemotherapy of protozoan parasites. *Mailing Add:* Poultry Sci Dept Univ Ga Athens GA 30602

MACDOUGALL, EDWARD BRUCE, b Ontario, Can, Aug 17, 39; m 60; c 3. GEOGRAPHICAL SCIENCES, COMPUTER ENVIRONMENTAL ASSESSMENT. *Educ:* Univ Toronto, BScF, 61, MScF, 62, PhD(geog), 67. *Hon Degrees:* MA, Univ Penn, 72. *Prof Exp:* Asst prof geog, Univ Toronto, 67-69; asst prof & assoc prof, Univ Penn, 67-75; assoc prof forestry & geog, Univ Toronto, 75-77, prof planning, Univ Mass, 77-84; dean, 84-89, PROF PLANNING, UNIV MASS, 89- *Honors & Awards:* Gold Medal, Can Inst Forestry, 61. *Mem:* Hon mem Am Soc Landscape Architect; Asn Am Geograhers. *Res:* Applications of computers in environmental assessment; land use planning; landscape design; author of two books. *Mailing Add:* 109 Hills N Univ Mass Amherst MA 01003

MCDOUGALL, JAMES K, EXPERIMENTAL BIOLOGY. *Prof Exp:* RES PROF PATH, UNIV WASH, 78- *Mailing Add:* Fred Hutchinson Cancer Res Ctr 1124 Columbia St Seattle WA 98104

MACDOUGALL, JOHN DOUGLAS, b Toronto, Ont, Mar 9, 44; m 68; c 2. GEOCHEMISTRY, METEORITICS. *Educ:* Univ Toronto, BSc, 67; McMaster Univ, MSc, 68; Univ Calif, San Diego, PhD(earth sci), 72. *Prof Exp:* Asst res geologist, Univ Calif, Berkeley, 72-74; from asst prof to assoc prof, 74-84, chmn, Geol Res Div, 85-89, PROF EARTH SCI, SCRIPPS INST OCEANOG, UNIV CALIF, SAN DIEGO, 84- *Mem:* AAAS; Am Geophys Union; Meteoritical Soc; Geochem Soc. *Res:* Isotope geology and geochronology; evolution of earth's mantle; geochemistry of volcanic rocks; evolution of meteorites. *Mailing Add:* Geol Res Div Scripps Inst Oceanog La Jolla CA 92093-0220

MCDOUGALL, KENNETH J, b Ashland, Wis, Aug 31, 35; div; c 2. MICROBIAL GENETICS. *Educ:* Northland Col, BS, 57; Marquette Univ, MS, 59; Kans State Univ, PhD(genetics), 64. *Prof Exp:* Instr zool, Northland Col, 59-60; res assoc genetics, Rice Univ, 63-64, NIH fel, 64-65, res assoc, 65-66; from asst prof to assoc prof, 66-77, PROF BIOL, UNIV DAYTON, 77-, CHMN DEPT, 85- *Mem:* AAAS; Genetics Soc Am; Am Soc Microbiol; Environ Mutagen Soc. *Res:* Mutagenesis and antimutagenesis. *Mailing Add:* Dept of Biol Univ of Dayton Dayton OH 45469

MACDOUGALL, ROBERT DOUGLAS, b McVille, NDak, Jan 2, 22; m 61; c 2. PETROLEUM GEOLOGY. *Educ:* Univ Mont, BA, 49; Univ Minn, Minneapolis, MS, 52. *Prof Exp:* Field geologist, Arabian Am Oil Co, Saudi Arabia, 52-56, subsurface geologist, 56-59; consult, 59-62; geologist, US Geol Surv, 62-82; geologist, Minerals Mgt Serv, 82-86; RETIRED. *Mem:* Am Asn Petrol Geol; fel Royal Geog Soc; Mensa. *Res:* Subsurface geology. *Mailing Add:* 646 Oak St Mandeville LA 70448

MCDOUGALL, ROBERT I, b Ft Wayne, Ind, Apr 26, 29; m 71; c 2. MATERIALS TESTING, PLASTICS ENGINEERING. *Educ:* Ind Univ, AB, 51; Carnegie-Mellon Univ, PhD(chem), 57. *Prof Exp:* Sr chemist, Exxon Res & Eng, 56-71; sr develop chemist, Signal Chem Co, 71-72 & Van Lear Plastics, 73-74; QUAL CONTROL SUPRV, MGR & SR LAB ENGR, IGLOO CORP, 77- *Mem:* Am Chem Soc; Soc Plastics Engrs. *Mailing Add:* Igloo Corp PO Box 19322 Houston TX 77224-9322

MCDOUGLE, PAUL E, b Peru, Ind, May 25, 28; div; c 3. MATHEMATICS. *Educ:* Purdue Univ, MS, 49; Univ Va, PhD(math), 58. *Prof Exp:* Asst prof, 58-67, assoc chmn dept,70-84 ASSOC PROF MATH,UNIV MIAMI, 67-, assoc chmn dept, 70-84. *Mem:* Math Asn Am. *Res:* Topology; study of decomposition spaces. *Mailing Add:* Dept Math Univ Miami Univ Station Coral Gables FL 33124

MCDOW, JOHN J(ETT), b Covington, Tenn, Jan 6, 25; m 46; c 2. AGRICULTURAL ENGINEERING. *Educ:* Univ Tenn, BS, 48; Mich State Univ, MS, 49, PhD, 57. *Prof Exp:* Instr, Mich State Univ, 49; instr & asst prof, Okla State Univ, 49-51; prof agr eng & head dept, La Polytech Inst, 51-62; head agr eng dept, 62-73, PROF AGR ENG, UNIV TENN, KNOXVILLE, 62-, DEAN ADMIS & RECORDS, 73- *Concurrent Pos:* Consult, Collaborator, Agr Res Serv, US Dept Agr, 70-76. *Mem:* Am Soc Agr Engrs. *Res:* Chemical and mechanical brush control; solar energy use in agriculture; utilization of solar energy in agriculture; silo structures; electric treatment of plant materials. *Mailing Add:* Dept Agr Eng Knoxville TN 37996

MCDOWELL, CHARLES ALEXANDER, b Belfast, Ireland, Aug 29, 18; m 45; c 3. PHYSICAL CHEMISTRY. *Educ:* Queen's Univ, Belfast, BSc, 41-42, DSc, 55. *Prof Exp:* Asst lectr chem, Queen's Univ, Belfast, 41-42; lectr inorg & phys chem, Univ Liverpool, 45-55; prof chem & head dept, 55-81, UNIV PROF CHEM, UNIV BC, 81- *Concurrent Pos:* Spec lectr & spec sci medal, Univ Liege, 55; Nat Res Coun Can sr res fel, Cambridge Univ, 63-64; vis prof, Kyoto Univ, 65 & 69; Killiam sr fel, 69-70; distinguished vis prof, Univ Fla, Gainesville, 74; fac sci distinguished lectr, Univ Calgary, 78. *Honors & Awards:* Letts Gold Medal Theoret Chem, 41; Centennial Medal, Govt Can, 67; Chem Inst Can Award, 69; Frontiers of Chem lectr, Wayne State Univ, 78; Queen Elizabeth's Silver Jubilee Medal, 78; Montreal Medal, Chem Inst Can, 82. *Mem:* Am Phys Soc; Am Chem Soc; fel Royal Soc Can; fel Chem Inst Can (pres, 78-79); fel Am Inst Physics. *Res:* Mass spectrometry; chemical kinetics; electron and nuclear spin resonance spectroscopy; molecular structure; electronic structures of molecules; photoelectron spectrometry. *Mailing Add:* Univ Brit Col 2075 Westbrook Mall Vancouver BC V6T 1Y6 Can

MCDOWELL, DAWSON CLAYBORN, b Chicago, Ill, July 19, 13; m 36; c 1. METEOROLOGY. *Educ:* Univ Chicago, BS, 35, MS, 42. *Prof Exp:* Instr, Univ Chicago, 42-45; PROF METEOROL & DIR INST TROP METEOROL, UNIV PR, 45- *Concurrent Pos:* Mem, Nat Adv Comt Educ. *Mem:* Am Meteorol Soc; Am Geophys Union; AAAS. *Res:* Tropical meteorology. *Mailing Add:* Dept Meteorol Univ PR PO Box 22931 Rio Piedros PR 00931

MACDOWELL, DENIS W H, b Belfast, Northern Ireland, Jan 26, 24; US citizen; m 49; c 4. ORGANIC CHEMISTRY. *Educ:* Queen's Univ, Belfast, BSc, 45, MSc, 47; Mass Inst Technol, PhD(org chem), 55. *Prof Exp:* Res chemist, Linen Indust Res Asn, 45-48; instr, Univ Toronto, 48-51; NIH fel org chem, Ohio State Univ, 55-57; lectr, Univ Toronto, 57-59; PROF CHEM, WVA UNIV, 59- *Mem:* Am Chem Soc. *Res:* Polycyclic aromatic hydrocarbon derivatives; thiophene chemistry. *Mailing Add:* Dept of Chem WVa Univ Morgantown WV 26506

MCDOWELL, ELIZABETH MARY, b Kew Gardens, Eng, Mar 30, 40. PATHOLOGY. *Educ:* Univ London, BVetMed, 64; Univ Cambridge, PhD(path), 71, MA, 72. *Prof Exp:* Vet gen pract, 64-66; from instr to assoc prof, 71-80, PROF PATH, UNIV MD, BALTIMORE, 80- *Mem:* Royal Col Vet Surg; Am Asn Pathologists; Int Acad Path. *Res:* Cellular and subcellular pathology; pulmonary pathology. *Mailing Add:* Dept Path Univ of Md Med Ten S Pine St Baltimore MD 21201

MCDOWELL, FLETCHER HUGHES, b Denver, Colo, Aug 5, 23; m 58; c 3. NEUROLOGY. *Educ:* Dartmouth Col, AB, 43; Cornell Univ, MD, 47. *Prof Exp:* From instr to assoc prof med, 52-68, PROF NEUROL, MED COL, CORNELL UNIV, 68-, ASSOC DEAN MED COL, 70-; MED DIR, BURKE REHAB HOSP, WHITE PLAINS, 73- *Mem:* Am Acad Neurol; Am Neurol Asn; Am Fedn Clin Res. *Res:* Cerebrovascular disease. *Mailing Add:* Dept Neurol & Rehab Med Cornell Univ Med Col 1300 York Ave New York NY 10021

MCDOWELL, FRED WALLACE, b Abington, Pa, Sept 30, 39; m 64; c 3. GEOCHEMISTRY. *Educ:* Lafayette Col, AB, 61; Columbia Univ, PhD(geochem), 66. *Prof Exp:* Researcher geochronology, Swiss Fed Inst Technol, 66-69; RES SCIENTIST, UNIV TEX, AUSTIN, 69- *Mem:* Fel Geol Soc Am; Sigma Xi. *Res:* Application of isotopic age determination methods to geologic problems, including volcanism, orogenesis and metallization. *Mailing Add:* 1704 Cannonwood Lane Univ of Tex Austin TX 78745

MCDOWELL, HARDING KEITH, b High Point, NC, Feb 5, 44; m 75; c 2. CHEMICAL PHYSICS. *Educ:* Wake Forest Univ, BS, 66; Harvard Univ, PhD(chem physics), 72. *Prof Exp:* Res assoc chem, State Univ NY Stony Brook, 72-74; asst prof chem, Clemson Univ, 74-78, assoc prof, 78-; AT LOS ALAMOS NAT LAB. *Concurrent Pos:* Alfred P Sloan Res Fel, 78-80. *Mem:* Am Phys Soc; Am Chem Soc; AAAS. *Res:* Study of many-body effects in chemical systems. *Mailing Add:* Los Alamos Nat Lab CLS-2 MS K765 Los Alamos NM 87545

MCDOWELL, HERSHEL, b Gaffney, SC, Aug 9, 30; m 56; c 2. PHYSICAL CHEMISTRY. *Educ:* Morgan State Col, BS, 57; Howard Univ, PhD(chem), 67. *Prof Exp:* Res asst immunol, Johns Hopkins Univ, 57-58; res asst, Charlotte Mem Hosp, NC, 58-62; res assoc chem, Am Dent Asn, Nat Bur Standards, DC, 65-69; prof chem & chmn dept, Fed City Col, DC, 69-78, PROF CHEM & ACTG CHMN DEPT, UNIV DC, 78- *Mem:* Am Chem Soc. *Res:* Solubility and thermodynamic properties of the calcium orthophosphates; stability of complexes between calcium and phosphate ions and organic ligands. *Mailing Add:* 7703 Whittier Blvd Bethesda MD 20817-6643

MACDOWELL, JOHN FRASER, b Oct 10, 32; US citizen; m 57; c 6. GLASS CERAMICS, COATINGS. *Educ:* Univ Mich, BS, 58, MS, 59. *Prof Exp:* Chemist, Corning Glass Works, 59-61, sr chemist, 61-64, res mgr glass ceramics, 64-66, dir chem res, 66-78, dir glass res, 78-82, RES FEL, CORNING INC, 82- *Concurrent Pos:* Mem panel ceramic processing, Mat Adv Bd, Nat Acad Sci-Nat Res Coun, 65- *Mem:* Am Chem Soc; fel Am

Ceramic Soc; Mineral Soc Am; Brit Soc Glass Technol; Sigma Xi. *Res:* Devitrification of glass and formation of ceramic materials; glass-ceramic coatings for metals and intermetallics; 25 US patents. *Mailing Add:* Corning Inc Sullivan Park FR-5 Corning NY 14831

MCDOWELL, JOHN PARMELEE, b Ridgewood, NJ, Feb 13, 31; m 55; c 2. SEDIMENTARY PETROLOGY. *Educ:* Yale Univ, BS, 53; Dartmouth Col, MA, 55; Johns Hopkins Univ, PhD, 63. *Prof Exp:* Geologist, US Geol Surv, 54-58; instr, 58-63, asst dean, 64-67, assoc dean, Col Arts & Sci, 67-83, ASST PROF GEOL, TULANE UNIV LA, 63-, CHMN, DEPT GEOL, 83- *Concurrent Pos:* Spec geologist, Ont Dept Mines, Can, 56-57; fel acad admin, Am Coun Educ, 66-67. *Mem:* Am Asn Petrol Geol; fel Geol Soc Am; Soc Econ Paleont & Mineral; AAAS. *Res:* Sedimentary petrology. *Mailing Add:* Dept Geol Tulane Univ 6823 St Charles Ave New Orleans LA 70118

MCDOWELL, JOHN WILLIS, b Honolulu, Hawaii, Dec 12, 21; m 50; c 2. PARASITOLOGY. *Educ:* Colo State Univ, BS, 47, MS, 48; Univ NC, MPH, 50; Okla State Univ, PhD(zool), 53. *Prof Exp:* Adv parasitol, USPHS, Int Coop Admin, Cambodia, Laos & Vietnam, 51-53, co-dir malaria control, Govt Iran, 54-56, tech adv malaria eradication, US Opers Mission, Philippines, 56-60, regional adv, Western Pac Region, AID, 60-61, spec asst to chief training br, Commun Dis Ctr, 61-62, asst chief vector borne dis sect, 62-64, chief eval unit, Aedes Aegypti Eradication Br, 64-66, asst chief eval sect, Malaria Eradication Br, 66-69; assoc prof, 69-72, PROF BIOL, BERRY COL, 72- *Concurrent Pos:* Mem independent malaria assessment team, US-WHO, Thailand & Iran, 63 & Vietnam, 64; consult, 11th Exper Comt Malaria, WHO, 64; US del, Int Cong Trop Med & Malaria, Lisbon, 58, Asian Malaria Conf, New Delhi, 58, Anti-Malaria Coord Bd, Southeast Asia, 59 & Borneo Conf Malaria Semarang, 60; Lilly Found fel environ educ, Ohio State Univ, 74-75; mem joint WHO/US Task Force on Malaria Training in Asia, 78. *Mem:* Am Soc Parasitol; Am Soc Trop Med & Hyg; Royal Soc Trop Med & Hyg; Am Mosquito Control Asn. *Res:* Field investigation and control of parasitic diseases, especially those which are arthropod borne; eradication and control of vector-borne disease or disease vectors, particularly malaria, filariasis and yellow fever; technical, organizational and administrative evaluation of such programs. *Mailing Add:* Dept Biol Berry Col Mt Berry GA 30149

MCDOWELL, LEE RUSSELL, b Sodus, NY, Apr 11, 41; m 71; c 3. ANIMAL NUTRITION. *Educ:* Univ Ga, BS, 64, MS, 65; Wash State Univ, PhD(animal nutrit), 71. *Prof Exp:* Res asst animal nutrit, Univ Ga, 64-65; agr vol, Peace Corps, Bolivia, 65-67; res asst animal nutrit & teaching asst animal prod, Wash State Univ, 68-71; from asst prof to assoc prof, 71-81, PROF ANIMAL NUTRIT, UNIV FLA, 81- *Concurrent Pos:* Mineral res consult, 74. *Honors & Awards:* Gustav Bohstedt Award, 84; Animal Sci Award, Int Agr, 88; Moorman Travel Award, 90. *Mem:* Am Soc Animal Sci; Am Dairy Sci Asn; Latin Am Asn Animal Prod. *Res:* Problems dealing with international animal nutrition and production; chemical composition and feeding value of tropical feeds; determining the location of mineral deficiencies or toxicities that are inhibiting livestock production in tropical countries. *Mailing Add:* Dept Animal Sci Univ Fla Animal Sci Bldg Rm 125c Gainesville FL 32611

MCDOWELL, MARION EDWARD, b Torrington, Wyo, Nov 4, 21; m 44; c 2. MEDICINE, NUTRITION. *Educ:* Univ Wyo, BS, 42; Univ Rochester, MD, 45; Am Bd Internal Med, dipl, 53, recert, 74. *Prof Exp:* Instr med, Sch Med, Univ Rochester, 45-46 & 48-50, from intern to assoc resident, Sch Med, Univ Rochester & Strong Mem Hosp, 45-50; med res, Walter Reed Army Med Ctr, 50-54, dir med div, Walter Reed Army Inst Res, 54-56, from asst chief to chief dept med, Tokyo Army Hosp, 56-58, chief outpatient dept, US Army Med Hosp, Camp Zama, Japan, 58-59, dep comdr & chief metab div, US Army Med Res & Nutrit Lab, 59-60, cmndg officer, 60-64; dir med educ, St Joseph Hosp, Denver, 64-84; ASSOC CLIN PROF MED, SCH MED, UNIV COLO, 80- *Concurrent Pos:* Fel renal dis, Peter Bent Brigham Hosp, 50; Army liaison rep, Food & Nutrit Bd, Nat Acad Sci-Nat Res Coun & Nutrit Study Sect, NIH, 61-64; mem subcomt feeding & nutrit space, Nat Res Coun, 62-64; asst clin prof med, Sch Med, Univ Colo, 64-80. *Mem:* Fel Am Col Physicians; Am Soc Internal Med. *Res:* Internal medicine and clinical research; renal disease; fluid and electrolyte metabolism; hemodialysis in renal failure; clinical nutrition; military medicine; medical education. *Mailing Add:* 7865 E Mississippi Ave Apt 1407 Denver CO 80231

MCDOWELL, MAURICE JAMES, b Dannevirke, NZ, Dec 6, 22; nat US; m 57; c 4. AUTOMOTIVE COATINGS, COLLOIDS. *Educ:* Univ Otago, NZ, BSc, 43, MSc, 44; Brown Univ, PhD(chem), 50. *Prof Exp:* Lectr, Univ Otago, NZ, 44-47; chemist, E I du Pont de Nemours & Co, Inc, 50-52; res assoc, Univ Montreal, 52; chem assoc, 53-60, res supvr, 60-63, RES ASSOC, FABRICS & FINISHES DIV, E I DU PONT DE NEMOURS & CO, INC, 64- *Mem:* Am Chem Soc; Fedn Soc Coating Technol. *Res:* Polymer research; colloid chemistry; research on coatings for automobiles. *Mailing Add:* 407 Meetinghouse Lane Media PA 19063

MCDOWELL, ROBERT CARTER, b Glens Falls, NY, Feb 15, 35; m 63; c 3. GEOLOGY. *Educ:* Va Polytech Inst, BS, 56, MS, 64, PhD(geol), 68. *Prof Exp:* Instr geol, Va Polytech Inst, 65-66; asst prof, Wis State Univ, River Falls, 66-67; GEOLOGIST, US GEOL SURV, 67- *Mem:* Geol Soc Am; Am Geophys Union; Am Asn Petrol Geologists. *Res:* Structural geology and petrography of the Arvonia slate quarries, Virginia; structural geology and stratigraphy of lower Cambrian quartzites, Virginia; field geology of northeastern Kentucky; structural geology of the Southern and Central Appalachians; field geology of the Newark Basin. *Mailing Add:* 928 Nat Ctr US Geol Surv Reston VA 22092

MCDOWELL, ROBERT E, JR, b Charlotte, NC, June 27, 21; m 45; c 3. ANIMAL SCIENCE, ANIMAL PHYSIOLOGY. *Educ:* NC State Col, BS, 42; Univ Md, MS, 49, PhD(dairy husb), 55. *Prof Exp:* Dist supvr, Vet Training Prog, NC, 46; dairy husbandman, Dairy Husb Res Br, Agr Res Serv, USDA, 46-58, sr res scientist, Dairy Cattle Res Br, Animal Husb Res Div, 58-67; prof, 67-86, EMER PROF INT ANIMAL SCI, CORNELL UNIV, 86- *Concurrent Pos:* Instr, Johns Hopkins Univ, 53-57; consult to various US agencies & foreign govt, 56-; vis prof, Cornell Univ, 66; prof, Univ PR, 69-80; vis prof, Fac de Agron, Cent Univ, Venezuela, 74, Nat Univ Pedro Henriquez Urena, Dominican Repub, 76, NC State Univ, 86-; chmn bd trustees, Int Livestock Ctr, Africa, 79-85. *Honors & Awards:* Superior Serv Award, USDA, 62; Int Animal Agr Award, Am Soc Animal Sci, 79, Int Dairy Prod Award, 89. *Mem:* Am Soc Animal Sci; Am Dairy Sci Asn; AAAS. *Res:* Means of improving contribution of livestock to man's need in developing countries; use of crossbreeding as mating system in animal production; farming systems in developing countries. *Mailing Add:* Dept of Animal Sci NC State Univ Raleigh NC 27695

MCDOWELL, ROBERT HULL, b Chicago, Ill, Dec 1, 27; m 56; c 3. MATHEMATICS. *Educ:* Univ Chicago, PhB, 47, BS & MS, 50; Purdue Univ, PhD, 59. *Prof Exp:* Asst mathematician, Argonne Nat Lab, 53-54; asst math, Purdue Univ, 54-55; instr & asst prof, Rutgers Univ, 59; asst prof, Purdue Univ, 59-60; from asst prof to assoc prof, 60-72, PROF MATH, WASH UNIV, 72-, CHMN DEPT, 74- *Concurrent Pos:* Dir comt undergrad prog math, NSF, 64-66, mem panel teacher training, mem adv group commun, mem comt teaching undergrad math; Nat Acad Sci interacad sci exchange, Czech, 81, 83. *Mem:* Am Math Soc; Math Asn Am. *Res:* Set theoretic topology and algebra and their interrelationships; mathematical structures. *Mailing Add:* Dept of Math Wash Univ St Louis MO 63130

MACDOWELL, ROBERT W, b Detroit, Mich, Dec 11, 24; m 44; c 3. MATHEMATICS. *Educ:* Oberlin Col, AB, 48; Univ Mich, AM, 49, PhD(math), 53. *Prof Exp:* From instr to asst prof math, Univ Rochester, 51-57; res assoc, Cornell Univ, 57-58; from assoc prof to prof, Antioch Col, 58-70; vpres & dean, 74-77, PROF MATH, HIRAM COL, 70- *Concurrent Pos:* Sci fac fel, 57-58. *Mem:* Am Math Soc; Math Asn Am; Asn Symbolic Logic. *Res:* Model theory in mathematical logic; logic of quantum mechanics. *Mailing Add:* 11129 Limeridge Rd Hiram OH 44234

MCDOWELL, ROBIN SCOTT, b Greenwich, Conn, Nov 14, 34; m 63; c 2. MOLECULAR SPECTROSCOPY. *Educ:* Haverford Col, BA, 56; Mass Inst Technol, PhD(phys chem), 60. *Prof Exp:* Asst chem, Mass Inst Technol, 56-60; staff mem phys chem, Los Alamos Nat Lab, 60-81, asst group leader, 81-82, res fel, 83-91; SR CHIEF SCIENTIST, CHEM STRUCT & DYNAMICS PROG, MOLECULAR SCI RES CTR, BATTELLE PAC NORTHWEST LABS, 91- *Mem:* AAAS; Optical Soc Am; Coblentz Soc (pres, 87-89); Soc Appl Spectros. *Res:* High-resolution infrared fourier-transform and laser spectroscopy; vibrational spectra of inorganic and isotopically-substituted molecules, especially fluorides; molecular dynamics and force fields; partition functions and thermodynamic properties of ideal gases; infrared analytical methods; atmospheric optics. *Mailing Add:* Battelle Pac Northwest Labs Richland WA 99352

MCDOWELL, SAM BOOKER, b New York, NY, Sept 13, 28; m 52; c 2. HERPETOLOGY. *Educ:* Columbia Univ, AB, 47, PhD(zool), 57. *Prof Exp:* From asst instr to assoc prof biol, 56-68, PROF ZOOL, RUTGERS UNIV, NEWARK, 68- *Mem:* Am Soc Ichthyol & Herpet; Soc Study Evolution; Zool Soc London; Soc Study Amphibians & Reptiles; Linnean Soc London. *Res:* Vertebrate paleontology; teleost and reptilian anatomy; origin of higher taxonomic categories; New Guinea snakes; primitive teleosts. *Mailing Add:* 900 W 190th St Apt B6 New York NY 10040

MACDOWELL, SAMUEL WALLACE, b Recife, Brazil, Mar 24, 29; m 53; c 3. PHYSICS. *Educ:* Univ Recife, BS, 51; Univ Birmingham, PhD(physics), 58. *Prof Exp:* Instr physics, Brazilian Ctr Phys Res, 54-56; instr, Princeton Univ, 59-60; from assoc prof to prof, Brazilian Ctr Phys Res, 60-63; mem, Inst Advan Study, 63-65; assoc prof, 65-67, PROF PHYSICS, YALE UNIV, 67- *Concurrent Pos:* Prof, Cath Univ Rio de Janiero, 62-63, vis prof, 69, 74-75, 82 & 88. *Mem:* Fel Am Phys Soc; Brazilian Acad Sci; NY Acad Sci. *Res:* Strong and weak interactions of elementary particles; s-matrix theory; analytic properties of scattering amplitudes and form factors; group theory and symmetries of interactions; gauge theories; spontaneous symmetry breaking; renormalization group; theory of gravitation and supergravity. *Mailing Add:* Dept of Physics Yale Univ 468 Sloane Physics Lab New Haven CT 06511

MCDOWELL, WILBUR BENEDICT, b Omaha, Nebr, Feb 27, 20; m 47; c 3. ORGANIC CHEMISTRY, PHARMACEUTICAL CHEMISTRY. *Educ:* Ohio State Univ, BSc, 41, MSc, 42, PhD(chem), 44. *Prof Exp:* Asst chem, Ohio State Univ, 42-43; res assoc, Squibb Inst Med Res, 44-52; head synthetic org develop sect, Olin Mathieson Chem Corp, 53-59, sr res chemist, 59-66; assoc mgr, New York, 66-69, sci mgr prof serv, E R Squibb & Sons, 69-84, archivist, Squibb Corp, Princeton, 85-89; ARCHIVIST, BRISTOL-MYERS SQUIBB CO, PRINCETON, 90- *Mem:* fel AAAS; Am Chem Soc; Fel Am Inst Chem; Soc Nuclear Med; NY Acad Sci. *Res:* Synthesis and isolation of antibiotics; synthetic organic process development. *Mailing Add:* Four Fairview Ave East Brunswick NJ 08816

MCDOWELL, WILLIAM H, b Philadelphia, Pa, Apr 26, 53. AQUATIC ECOLOGY. *Educ:* Amherst Col, BA, 75; Cornell Univ, PhD(aquatic ecol), 82. *Prof Exp:* Res assoc, Ecosysts Res Ctr, Cornell Univ, 81-82; sr scientist, Univ PR, 82-85; SR STAFF ASSOC, STATE UNIV NY OSWEGO, 85- *Mem:* Am Soc Limnol & Oceanog; AAAS; Ecol Soc Am; Sigma Xi. *Res:* Biochemistry of dissolved organic carbon in terrestrial and aquatic ecosystems; mechanisms controlling stream water chemistry; nutrient cycling tropical ecosystems. *Mailing Add:* James Hall Forest Resources Univ NH Durham NH 03824

MCDOWELL, WILLIAM JACKSON, b McMinnville, Tenn, July 14, 25; m 51; c 2. SOLVENT EXTRACTION CHEMISTRY. *Educ:* Tenn Technol Univ, BS, 51; Univ Tenn, MS, 54. *Prof Exp:* Jr chemist, Oak Ridge Nat Lab, 51-52, res chemist, 54-84, res group leader, 75-89, SR RES STAFF MEM, MARTIN MARIETTA ENERGY SYSTS, INC, OAK RIDGE NAT LAB,

84- *Concurrent Pos:* Tech Dir & Chief Exec Officer, ENTRAC, Inc, 89-; consult, Oak Ridge Nat Lab, Chem Div, 89- *Honors & Awards:* Indust Res-100 Award, Indust Res, 81. *Mem:* Am Chem Soc; Am Nuclear Soc; Sigma Xi; AAAS. *Res:* Inorganic separations chemistry, primarily liquid-liquid extraction and ion exchange, relating to hydrometallurgical processes and analytical procedures; alpha counting and spectrometry using liquid scintillation methods especially photon electron-rejecting alpha liquid scintillation. *Mailing Add:* 10903 Melton View Lane Knoxville TN 37931

MCDUFF, CHARLES ROBERT, b Austin, Tex, Nov 17, 29; m 55; c 2. HAZARDOUS WASTE MANAGEMENT. *Educ:* Univ Miss, BS, 52; Univ Wis, MS, 61. *Prof Exp:* Captain, US Army Chem Corps, 55-63; microbiologist res, Economics Lab, Inc, 63-65, mgr microbiol, 65-69, dir biol res, 69-77; DIR GOVT TECH AFFAIRS, ECOLAB INC, 77- *Concurrent Pos:* Mem, bd dir, Independent Sch Dist 197, W St Paul, Minn, 74- 80, chmn, 75-80. *Mem:* Am Soc Microbiol; Am Pub Health Asn; Nat Environ Health Asn; Soc Indust Microbiol; Int Asn Milk, Food & Environ Sanitarians. *Res:* Development of new disinfectants and sanitizers; development of compliance programs for environmental laws and regulations. *Mailing Add:* Ecolab Inc Ecolab Ctr St Paul MN 55102

MCDUFF, DUSA MARGARET, b London, UK, Oct 18, 45; c 2. SYMPLECTIC GEOMETRY, DIFFEOMORPHISM GROUPS. *Educ:* Univ Edinburgh, UK, BSc, 67; Univ Cambridge, UK, PhD(math), 71. *Prof Exp:* Sci Res Coun fel math, Univ Cambridge, UK, 70-72; lectr, Univ York, UK, 72-76 & Univ Warwick, UK, 76-78; from asst prof to assoc prof, 78-84, PROF MATH, STATE UNIV NY, STONY BROOK, 84- *Concurrent Pos:* Vis asst prof, Mass Inst Technol, 74-75; visitor, Inst Advan Study, Princeton, 76-77; prin investr, NSF grants, 78- *Honors & Awards:* Ruth Lyttle Satter Prize, Am Math Soc, 91. *Mem:* Am Math Soc; Asn Women Sci. *Res:* Study of groups of diffeomorphisms, especially those which preserve symplectic or volume forms; global symplectic geometry; foliations. *Mailing Add:* Dept Math State Univ NY Stony Brook NY 11794

MCDUFF, ODIS P(ELHAM), b Gordo, Ala, May 16, 31; m 53; c 4. ELECTRICAL ENGINEERING. *Educ:* Univ Ala, BS, 52; Mass Inst Technol, SM, 53; Stanford Univ, PhD(quantum electronics), 66. *Prof Exp:* Test engr, Gen Elec Co, 52, circuit develop engr, 56-59, supv engr, 59; mem tech staff elec eng, Bell Tel Lab, 53-54; asst prof microwaves, 59-66, assoc prof quantum electronics, Dept Elec Eng, 66-68, prof & head dept, Dept Elec Eng, 68-83, PROF ELEC ENG, UNIV ALA, TUSCALOOSA, 83- *Concurrent Pos:* NSF res grant, 67-68; mem laser technol comt, Army Missile Command, Ala, 67-70. *Mem:* Sr mem Inst Elec & Electronics Engrs; Am Soc Eng Educ; Opt Soc Am; Soc Photo Opt Inst Engrs. *Res:* Quantum electronics; nonlinear optics; microwaves; laser mode coupling; laser modulation. *Mailing Add:* Dept Elec Eng PO Box 870286 Tuscaloosa AL 35487

MCDUFFIE, BRUCE, b Atlanta, Ga, Aug 25, 21; m 50; c 3. CHEMISTRY. *Educ:* Princeton Univ, AB, 42, MA, 46, PhD(chem), 47. *Prof Exp:* Asst chem, Cornell Univ, 42; asst Manhattan Proj, Princeton Univ, 42-47; instr, Emory Univ, 47-51; assoc prof, Washington & Jefferson Col, 51-58; from assoc prof to prof chem, State Univ NY, Binghamton, Harper Col, 58-88; RETIRED. *Honors & Awards:* Spec Award Merit, US Environ Protection Agency, 76. *Mem:* AAAS; Am Chem Soc; Scientists Inst Pub Info; Sigma Xi. *Res:* Electroanalytical chemistry; trace methods and environmental analysis for toxic metals and organic pollutants; river pollution and sludge disposal studies. *Mailing Add:* 1601 Edgewood Circle Chattanooga TN 37405

MCDUFFIE, FREDERIC CLEMENT, b Lawrence, Mass, Apr 27, 24; m 52; c 4. MEDICINE. *Educ:* Harvard Med Sch, MD, 51. *Prof Exp:* From intern to jr asst resident, Peter Bent Brigham Hosp, Boston, 51-53; fel phys chem, Harvard Univ, 53-54; fel microbiol, Col Physicians & Surgeons, Columbia Univ, 54-56; sr asst resident, Peter Bent Brigham Hosp, 56-57; from asst prof to assoc prof med & microbiol, Sch Med, Univ Miss, 57-65; asst prof, med & microbiol, Mayo Med Sch, Univ Minn, 65-69, assoc prof, Mayo Grad Sch Med, 69-74, prof, 74-79, consult, Mayo Clin, 65-79; SR VPRES MED AFFAIRS, ARTHRITIS FOUND, 79-; PROF MED, EMORY UNIV, 79- *Concurrent Pos:* Vis investr, Ctr Dis Control, 79- *Mem:* Am Asn Immunologists; Am Rheumatism Asn; Am Fedn Clin Res; Am Col Physicians; Soc Exp Biol & Med. *Res:* Complement system in connective tissue diseases; antibody production in chickens; autoimmune phenomena in rheumatic diseases. *Mailing Add:* Piedmont Hosp Arthritis Ctr 2001 Peachtree Rd NE 200 Atlanta GA 30309

MCDUFFIE, GEORGE E(ADDY), JR, b Washington, DC, Apr 4, 25; m 50; c 6. ELECTRICAL ENGINEERING. *Educ:* Cath Univ Am, BEE, 49, MEE, 52; Univ Md, PhD, 62. *Prof Exp:* From instr to assoc prof elec eng, 49-63, chmn dept, 69-70, actg dean, 70-71, PROF ELEC ENG, CATH UNIV AM, 63-, DEAN, 76- *Concurrent Pos:* NSF grant, 58-60; physicist, Naval Ord Lab, Md, 58- *Mem:* Am Soc Eng Educ; Inst Elec & Electronics Engrs. *Res:* Properties of liquid dielectrics and their measurement under varying temperature, pressure and frequency; modern network theory; pattern recognition. *Mailing Add:* Dept of Elec Eng Cath Univ Am 620 Michigan Ave NE Washington DC 20064

MCDUFFIE, NORTON G(RAHAM), JR, b Beaumont, Tex, Nov 26, 30; m 61; c 2. BIOCHEMISTRY, CHEMICAL ENGINEERING. *Educ:* Tex A&M Univ, BS, 52; Univ Tex, PhD(chem), 62. *Prof Exp:* Process engr, Mobil Oil Co, 52-57; fel biochem, Clayton Found Biochem Inst, 62-63; asst prof chem, Lamar Univ, 63-64; asst prof biochem, Sch Med, Univ Okla, 64-69; ASSOC PROF CHEM ENG, UNIV CALGARY, 69- *Concurrent Pos:* Assoc, Okla Med Res Found, 64-69; vis lectr, Univ Tex, Austin, 67-68; consult, Mobil Chem Co, 68-70. *Mem:* Am Chem Soc; Am Inst Chem Engrs; Am Soc Microbiol; Sigma Xi. *Res:* Cancer; chemical engineering design; viruses; electron microscopy. *Mailing Add:* 1745 SW Whiteside Dr Corvallis OR 97333

MACE, ARNETT C, JR, b Hackers Valley, WVa, Nov 18, 37; m 62; c 2. FOREST HYDROLOGY. *Educ:* WVa Univ, BS, 60; Univ Ariz, MS, 62, PhD(forest hydrol), 68. *Prof Exp:* Res forester forest hydrol, Res Br, US Forest Serv, 62-64; instr, Univ Ariz, 64-67; from asst prof to assoc prof, Univ Minn, 67-74, head forest biol dept, 72-74, prof & head forest resources dept, 74-78; DIR, SCH FOREST RESOURCES & CONSERV, UNIV FLA, 78-, DIR, CTR NATURAL RESOURCES. *Mem:* Soc Am Foresters; Am Geophys Union; Am Water Resources Asn. *Res:* Applied and basic research related to forest land management practices and policies, with particular emphasis on water resources problems related to forest land management. *Mailing Add:* Sch Forest Resources & Conserv Univ Fla 118 Newins Ziegler Hall Gainesville FL 32611

MACE, JOHN WELDON, b Buena Vista, Va, July 9, 38; m 62; c 3. PEDIATRIC ENDOCRINOLOGY, METABOLISM. *Educ:* Columbia Union Col, BA, 60; Loma Linda Univ, MD, 66. *Prof Exp:* Attend physician pediat, Naval Hosp, San Diego, 68-70; fel endocrinol, Univ Colo Med Ctr, Denver, 70-72; instr pediat, 70-72, asst prof, 72-75, PROF & CHMN, DEPT PEDIAT, SCH MED, LOMA LINDA UNIV, 75-, MED DIR, CHILDREN'S HOSP, 90- *Concurrent Pos:* Consult, Calif Newborn Screening Comt, 78-86, Calif Med Asn Pediat Adv Comt, 75-; chmn, Calif Children's Serv Adv Comt, 88- *Mem:* Am Acad Pediat; Lawson Wilkens Pediat Endocrine Soc; Asn Med Sch Pediat Dept Chmn; AAAS; NY Acad Sci; Sigma Xi. *Res:* Behavioral factors as related to juvenile diabetes patients and diabetic control; growth promoting effects of oxandrolowe on short stature. *Mailing Add:* 20 W South Ave Redlands CA 92373

MACE, KENNETH DEAN, b Pekin, Ill, Jan 29, 26; c 1. MICROBIOLOGY, CELL PHYSIOLOGY. *Educ:* Univ Ark, BS, 51, MS, 57, PhD(plant physiol), 64. *Prof Exp:* Bacteriologist, Pepsi Cola Co, NY, 51-54; asst epidemiol, Univ Ark, 54-55 & bact, 55-57; bact chemist, Pepsi Cola Co, NY, 57-59; asst bot & bact, Univ Ark, 59-64; from asst prof to assoc prof, 64-71, PROF BIOL, STEPHEN F AUSTIN STATE UNIV, 71- *Concurrent Pos:* Dept Health, Educ & Welfare grants, 71-73. *Mem:* AAAS; Sigma Xi; NY Acad Sci; Am Soc Microbiol. *Res:* Cyclic adenosine monophosphate relationships early in infection by New Castle disease virus; humoral and secretory immune response to whole cell and ribosomal preparation of Candida albicans; cyclic adenosine monophosphate levels in liver cells in response to ethyl alcohol intake at two levels of intake; development of certain monoclonal antibodies by the hybridoma technique; immunology; fatty acid assimilation and turnover process; arachidorie acid; cholerterol. *Mailing Add:* Dept Biol Box 13003 Stephen F Austin State Univ Nacogdoches TX 75962

MCEACHRAN, JOHN D, b Iron Mountain, Mich, Nov 1, 41; m 72; c 1. ICHTHYOLOGY, TAXONOMY. *Educ:* Mich State Univ, BA, 65; Col William & Mary, MA, 68, PhD(marine biol), 73. *Prof Exp:* Asst prof, 73-78, ASSOC PROF WILDLIFE & FISHERIES SCI, TEX A&M UNIV, 78- *Concurrent Pos:* Prin investr, NSF grant, 78-81. *Mem:* Am Soc Ichthyologists & Herpetologists; Soc for Study Evolution; Soc Syst Zool; AAAS. *Res:* Systematics; evolutionary biology and zoogeography of the flattened elasmobranchs; systematic and ecological interrelationships of fishes. *Mailing Add:* Dept Oceanog Tex A&M Univ College Station TX 77843

MACEK, ANDREJ, b Zagreb, Yugoslavia, Oct 24, 26; nat US; m 56; c 2. CHEMICAL ENGINEERING. *Educ:* Georgetown Univ, BS, 50; Cath Univ Am, MS, 51, PhD(phys chem), 53. *Prof Exp:* Asst, Cath Univ Am, 51-53, res assoc, 53-54; asst prof, Lafayette Col, 54-55; res assoc, US Naval Ord Lab, 55-60; phys chemist, Atlantic Res Corp, 60-69, chief kinetics & combustion group, 69-74; proj leader flame & combustion res, Nat Bur Stand, 74-75; chief, Power & Combustion Br, Energy Res & Develop Admin, Dept Energy, 75-78; COMBUSTION SCIENTIST, NAT INST STANDARDS & TECHNOL, 78- *Concurrent Pos:* Adj prof, Am Univ, 58-72. *Mem:* Am Chem Soc; Combustion Inst. *Res:* Combustion; chemical kinetics and thermodynamics. *Mailing Add:* 859 Golden Arrow St Great Falls VA 22066

MACEK, JOSEPH, b Rapid City, SDak, July 4, 37; m 64; c 3. PHYSICS. *Educ:* SDak State Col, BS, 60; Rensselaer Polytech Inst, PhD(physics), 64. *Prof Exp:* Nat Res Coun fel, Nat Bur Stand, 64-66; Oxford-Harwell fel, Atomic Energy Res Estab, UK, 66-68; from asst prof to assoc prof, 68-74, PROF PHYSICS, UNIV NEBR, LINCOLN, 74- *Mem:* Am Phys Soc; Sigma Xi. *Res:* Atomic physics. *Mailing Add:* 200 S College Univ Tenn Knoxville TN 37996-1501

MACEK, ROBERT JAMES, b Rapid City, SDak, July 14, 36; m 62; c 2. ELEMENTARY PARTICLE PHYSICS, PARTICLE ACCELERATOR PHYSICS. *Educ:* SDak State Univ, BS, 58; Calif Inst Technol, PhD(physics), 65. *Prof Exp:* Res assoc physics, Univ Pa, 66-69; GROUP LEADER & MEM STAFF MEDIUM ENERGY PHYSICS DIV, LOS ALAMOS SCI LAB, 69-; group leader & mem staff, 69-87, PROG MGR, MEDIUM ENERGY PHYSICS, LOS ALAMOS NAT LAB, 87- *Mem:* AAAS; Am Phys Soc; Sigma Xi. *Res:* Weak interactions and rare decay modes of mesons; pion production and absorption reactions on nuclei; nuclear instrumentation; particle beam optics and beam instrumentation; storage ring beam dynamics; accelerator technology, development and operations. *Mailing Add:* Los Alamos Nat Lab MP-D0 MS-848 Los Alamos NM 87545

MCELGUNN, JAMES DOUGLAS, b Vancouver, BC, Mar 22, 39; m 61; c 3. FORAGE CROP PRODUCTION, CROP PHYSIOLOGY. *Educ:* Mont State Col, BS, 62, MS, 64; Mich State Univ, PhD(crop prod), 67. *Prof Exp:* Res scientist forage, Agr Can Res Br, Swift Current, SK, 67-80; dir, Kamloops Range Res Sta, 80-85; DIR, BEAVERLODGE RES STA, 85- *Concurrent Pos:* Can chmn, Lucerne Weather Data Acquisition, World Meteorol Orgn, UN, 75-; assoc ed, Can J Plant Sci, 75-81; consult, Can-China deleg, 81, 82 & 83. *Res:* Forage crop agronomy under dryland and irrigated conditions including establishment, management and harvesting. *Mailing Add:* Agr Can Res Sta PO Box 1204 Beaverlodge AB T0H 0C0 Can

MCELHANEY, JAMES HARRY, b Philadelphia, Pa, Oct 27, 33; m 55; c 3. MECHANICAL ENGINEERING, BIOMECHANICS. *Educ:* Villanova Univ, BSME, 55; Pa State Univ, MSME, 60; WVa Univ, PhD(biomech), 64. *Prof Exp:* Asst prof mech eng, Villanova Univ, 55-61; from assoc prof to prof theoret & appl mech, WVa Univ, 61-69; assoc prof biomech & head dept, Hwy Safety Res Inst, Univ Mich, Ann Arbor, 69-77; PROF BIOMED ENGR, DUKE UNIV, 77- *Res:* Biomechanics of head and neck injuries; protective systems; sensory feedback & protheses. *Mailing Add:* Dept Bioeng Duke Univ 136 Engr Bldg Durham NC 27706

MCELHANEY, RONALD NELSON, b Youngstown, Ohio, Jan 5, 42; m 65, 79; c 3. BIOCHEMISTRY, BIOPHYSICS. *Educ:* Wash & Jefferson Col, BA, 64; Univ Conn, PhD(biochem), 69. *Prof Exp:* NIH fel biochem, State Univ Utrecht, 69-70; from asst prof to assoc prof, 70-79, PROF BIOCHEM, UNIV ALTA, 79- *Concurrent Pos:* Med Res Coun Can res grant, Univ Alta, 70- *Mem:* Biophys Soc; Can Biochem Soc; Am Soc Microbiol; Am Soc Biol Chemists; Am Oil Chem Soc. *Res:* Biological membrane structure and function; role of membrane polar lipids and cholesterol in passive and mediated membrane transport and in membrane enzyme activity; calorimetric fourier-transform infrared and nuclear magnetic resonance studies of model and biomembranes. *Mailing Add:* Dept Biochem Univ Alta Edmonton AB T6G 2H7 Can

MCELHINNEY, JOHN, b Philadelphia, Pa, Mar 25, 21; m 42; c 2. PHYSICS. *Educ:* Ursinus Col, BS, 42; Univ Ill, MS, 43, PhD(physics), 47. *Prof Exp:* Asst physics, Univ Ill, 42-44 & Manhattan Dist, 44-45, spec res assoc, 47-48; assoc scientist, Los Alamos Sci Lab, 48-49; physicist, Nat Bur Standards, 49-55; head, Nuclear Interactions Br, Naval Res Lab, 55-62, assoc supt, Nucleonics Div, 57-66, head, Linac Br, 62-66, supt, Nuclear Sci Div, 66-74, Supt, Radiation Tech Div, 74-80, consult, Condensed Matter & Rad Sci Div, 80-83; RETIRED. *Concurrent Pos:* Partic sabbatical study & res prog, Stanford Univ-Lawrence Livermore Lab, 69-70. *Mem:* Am Phys Soc; Am Nuclear Soc; Inst Elec & Electronics Engrs. *Res:* Nuclear physics with betatron, synchrotron, electron linac, Van de Graaff, cyclotron; nuclear and x-ray instrumentation; application of nuclear technology; radiation dosimetry. *Mailing Add:* 11601 Stephen Rd Silver Spring MD 20904

MCELHINNEY, MARGARET M (COCKLIN), b Grandview, Iowa; div; c 2. BIOLOGY. *Educ:* Iowa Wesleyan Col, BS, 45; Univ Northern Colo, MS, 60; Ball State Univ, EdD(sci ed), 66. *Prof Exp:* Teacher pub schs, Iowa, seven years; assoc prof, 61-74, prof biol, Ball State Univ, 74-88; RETIRED. *Concurrent Pos:* Nat Sci Teachers Asn; Am Asn Biol Teachers. *Res:* Museum technology; developing instructional materials and preparing exhibits for direct experience learning. *Mailing Add:* 9281 Kanawha Tucson AZ 85741

MCELHOE, FORREST LESTER, JR, b Carnation, Wash, Feb 15, 23; m 53; c 2. PHYSICAL GEOGRAPHY, ECONOMIC GEOGRAPHY. *Educ:* Univ Wash, BA, 48, MA, 50; Ohio State Univ, PhD(geog), 55. *Prof Exp:* Instr geog, Ohio State Univ, 55-56; asst prof, Univ Ky, 56-68; asst prof, 68-72, ASSOC PROF GEOG, CALIF STATE POLYTECH UNIV, POMONA, 72- *Mem:* Asn Am Geog; AAAS; Nat Coun Geog Educ; Sigma Xi. *Res:* Climate. *Mailing Add:* 6353 Moonstone Ave Alta Loma CA 91701

MCELIGOT, DONALD M(ARINUS), b Passaic, NJ, Mar 9, 31; m 57; c 3. FLUID MECHANICS, THERMAL SCIENCE. *Educ:* Yale Univ, BEME, 52; Univ Wash, MSE, 58; Stanford Univ, PhD(thermo sci eng), 63. *Prof Exp:* Engr, Gen Elec Co, 55-57 & Knolls Atomic Power Lab, 57; thermalhydromech scientist, Gould Ocean Systs Div, Westinghouse Elec Corp, 84-88, mgr hydromech res & technol, Naval Systs Div, 87-89; from assoc prof to prof mech eng, 63-85, EMER PROF AEROSPACE & MECH ENG, UNIV ARIZ, 85- *Concurrent Pos:* Consult, Los Alamos Sci Lab, 67-90; mem vis staff, Imp Col, Sci Technol, Univ London, 69-70, Heat Transfer Comt, Am Soc Mech Eng, Gas Turbine Div, 76-, gen papers Comt, 79-82, chmn, 81-82; guest prof, Univ Karlsruhe, 75-76, 79 & Max Planck Inst for Fluid Flow Res, WGer, 82-83 & 84; capt, Aeronaut Eng Duty, US Naval Reserve, 80-; sr Fulbright res award, WGer, 82-83; adj prof, Univ RI, 86; assoc tech ed, J Heat Transfer, Am Soc Mech Engrs, 86-; prin investr, NSF Heat Trans Prog, Off Naval Res Power, Appl Hydrodynamics, Westinghouse. *Mem:* fel Am Soc Mech Engrs; Am Phys Soc; Sigma Xi; US Naval Inst. *Res:* Turbulent and transitional flow; heat transfer; experiments and numerical analyses. *Mailing Add:* PO Box 4008 Middletown RI 02840

MCELLIGOTT, JAMES GEORGE, b New York, NY, June 20, 38; m 67; c 2. NEUROSCIENCE, NEUROPHARMACOLOGY. *Educ:* Fordham Univ, BS, 60; Columbia Univ, MA, 63; McGill Univ, PhD(psychol), 66. *Prof Exp:* NIH fel neuroanat, Univ Calif, Los Angeles, 67-68, asst res anatomist, Sch Med, 68-71; from asst prof to assoc prof, 71-87, PROF PHARMACOL, SCH MED, TEMPLE UNIV, 87. *Mem:* AAAS; Soc Neurosci; Sigma Xi. *Res:* Neurophysiology; bioengineering; computer analysis of neuroscientific data. *Mailing Add:* Dept Pharmacol Temple Univ Sch Med 3400 N Broad St Philadelphia PA 19140

MCELLIGOTT, MARY ANN, b Brooklyn, NY, 1953. PHYSIOLOGY. *Educ:* State Univ NY Stony Brook, BS, 75; Fordham Univ, Bronx, NY, MS, 77; Rutgers Univ, New Brunswick, NJ, PhD(physiol), 80. *Prof Exp:* Teaching asst introductory biol, Fordham Univ, 75-76 & introductory physiol, Gen Physiol & Grad Mammalian Physiol, Rutgers Univ, 76-79; postdoctoral fel, Rutgers Univ, New Brunswick, NJ, 80-81; sr res scientist, Dept Drug Discovery, Growth Physiol & Biochem, Merck Sharp & Dohme, Inc, Rahway, NJ, 84-88; dir biol res, 88, PROJ DIR CLIN RES, ROBERTS PHARMACEUT CORP, EATONTOWN, NJ, 88- *Concurrent Pos:* Paul Dudley White fel, Am Heart Asn; postdoctoral fel, Muscular Dystrophy Asn. *Mem:* Sigma Xi; AAAS; Am Physiol Soc; NY Acad Sci. *Res:* Author of various publications. *Mailing Add:* Roberts Pharmaceut Corp 15 Remington Circle Cranbury NJ 08512

MCELLIGOTT, PETER EDWARD, b New York, NY, Apr 4, 35; m 62; c 2. SURFACE CHEMISTRY, TRIBOLOGY. *Educ:* NC State Col, BS, 57; Rensselaer Polytech Inst, MS, 61, PhD(mat sci), 66. *Prof Exp:* Physicist, Alco Prod, Inc, NY, 57-59; physicist, Knolls Atomic Power Lab, 59-62, phys chemist, 62-73, liasion scientist, 73-77, PROJ DEVELOP MGR, RES & DEVELOP CTR, GEN ELEC CO, 77- *Mem:* Fel Am Inst Chem; Am Soc Lubrication Eng. *Res:* Friction, lubrication and wear of materials; electrical contact phenomena; surface chemistry of metals and semiconductors. *Mailing Add:* Res & Develop Ctr Gen Elec Co PO Box 8 Schenectady NY 12301

MCELLISTREM, MARCUS THOMAS, b St Paul, Minn, Apr 19, 26; m 57; c 6. NUCLEAR PHYSICS. *Educ:* Col St Thomas, BA, 50; Univ Wis, MS, 51, PhD, 56. *Prof Exp:* Asst physics, Univ Wis, 50-55; res assoc, Ind Univ, 55-57; from asst prof to assoc prof, 57-65, PROF PHYSICSS, UNIV KY, 65- *Concurrent Pos:* Pres, Adena Corp, 71-74; prog officer nuclear sci, NSF, 81-82; chair-elect, Univ Senate, 90-91, chair, 91-92; univ res prof, Univ Ky, 78-79, distinguished prof arts & sci, 81-82. *Mem:* AAAS; fel Am Phys Soc. *Res:* Analysis of nuclear reactions; measurements of differential cross sections of nuclear reactions; fast neutron physics, neutron induced reaction cross sections; radioisotopes for nuclear medicine. *Mailing Add:* Dept Physics & Astron Univ Ky Lexington KY 40506

MCELRATH, GAYLE W(ILLIAM), b Randolph, Minn, Dec 15, 15; m 46; c 5. QUALITY CONTROL ENGINEERING. *Educ:* Univ Minn, BS, 41, MS, 46. *Prof Exp:* From instr to asst prof math & mech eng, Univ Minn, 43-53, from assoc prof to prof mech eng, 53-70, head indust eng, 55-70; vpres, Bayer & McElrath, Inc, 60-78, PRES, MCELRATH & ASSOCS, INC, 78- *Concurrent Pos:* Consult indust & govt, 50-; dir, Coun Int Progress Mgt. *Honors & Awards:* E L Grant Award, Am Soc Qual Control, 69. *Mem:* Fel AAAS; fel Am Soc Qual Control (vpres, 61-65); Opers Res Soc Am; Am Statist Asn; Sigma Xi. *Res:* Industrial statistics; design and analysis of experiments; statistical quality control. *Mailing Add:* McElrath & Assocs Inc Southdale Off Ctr 6750 France Ave S Suite 240 Minneapolis MN 55435

MCELREE, HELEN, b Waxahachie, Tex, Nov 22, 25. IMMUNOBIOLOGY. *Educ:* Col Ozarks, BS, 47; Univ Okla, MS, 54; Univ Kans, PhD, 59. *Prof Exp:* Asst prof bact, Univ Kans, 59-62; assoc prof, 62-68, PROF BIOL, EMPORIA KANS STATE COL, 68- *Mem:* AAAS; Am Soc Microbiol. *Res:* Role of reticulo-endothelial system in viral infections. *Mailing Add:* Dept of Biol Emporia Kans State Col 1200 Coml St Emporia KS 66801

MCELROY, ALBERT DEAN, b Quinter, Kans, Feb 25, 22; m 55; c 2. INORGANIC CHEMISTRY. *Educ:* Sterling Col, BS, 47; Univ Kans, PhD(inorg chem), 51. *Prof Exp:* Res asst, Univ Ill, Urbana, 51-53; res chemist, Callery Chem Co, Pa, 53-64; res chemist supvr, Midwest Res Inst, Kansas City, 64-87; RETIRED. *Concurrent Pos:* Consult, 87- *Mem:* Am Chem Soc; Sigma Xi. *Res:* Synthesis and product research in boron hybride derivatives, perchloro and perfluoramino compounds; water and wastewater treatment processes; corrosion; deicing chemical research. *Mailing Add:* 9913 Juniper Overland Park KS 66207

MCELROY, DAVID L(OUIS), b Fairfield, Ala, June 16, 30; m 54; c 4. METALLURGY. *Educ:* Univ Ala, BS, 51, MS, 53; Univ Tenn, PhD(metall), 57. *Prof Exp:* Res engr, Eng Exp Sta, Univ Tenn, 56-59; metallurgist, Oak Ridge Nat Lab, 59-90; RETIRED. *Concurrent Pos:* Nat Acad Sci adv panel mem, Heat Div, Nat Bur Standards, 69-72 & Off Standard Reference Data, 71-73; Ford Found lectr, Univ Tenn, 71-72; foreign assignment, AERE Harwell, UK, 74-75; mem, Dept Commerce Criteria Comt Lab Accreditation, 78; chmn, Int Thermal Conductivity Conf Bd, 81-85. *Honors & Awards:* Annual Thermal Conductivity Conf Award, 71. *Mem:* Fel Am Soc Metals; Am Soc Testing & Mat. *Res:* Temperature measurement at high temperatures; thermocouple research at 2000 degrees Fahrenheit; adiabatic calorimetry to 950 degrees centigrade on iron-carbon alloys; thermal conductivity of nuclear refractory materials to 2400 degrees centigrade; steel transformation; heat capacity Americium, Pu-C 4 to 300 Kelvin; thermal insulation properties. *Mailing Add:* 5823 Westover Dr Knoxville TN 37919-4145

MCELROY, MARY KIERAN, inorganic chemistry, analytical chemistry, for more information see previous edition

MCELROY, MICHAEL BRENDAN, b Shercock Co Cavan, Ireland, May 18, 39; m 63. APPLIED MATHEMATICS, PHYSICS. *Educ:* Queen's Univ, Belfast, BA, 60, PhD(math), 62. *Prof Exp:* Proj assoc, Theoret Chem Inst, Univ Wis, 62-63; from asst physicist to physicist, Kitt Peak Nat Observ, 63-71; PHYSICIST, HARVARD CTR FOR EARTH & PLANETARY PHYSICS, 71- *Concurrent Pos:* Mem Mars panel, Lunar & Planetary Missions Bd, NASA, 68-69, mem, Stratospheric Res Adv Comt, Space & Terrestrial Appl Adv Comt; mem, Comt Atmospheric Sci, Nat Acad Sci, Space Sci Bd, chmn, Comt Planetary & Lunar Explor. *Honors & Awards:* James B Macelwane Award, Am Geophys Union, 68; Newcomb Cleveland Prize, AAAS, 77; Pub Serv Medal, NASA, 78. *Mem:* AAAS; Am Astron Soc; Am Geophys Union. *Res:* Physics and chemistry of planetary atmospheres. *Mailing Add:* Ctr for Earth & Planetary Physics Harvard Univ Cambridge MA 02138

MCELROY, PAUL TUCKER, b Boston, Mass, Sept 25, 31; m 63; c 4. UNDERWATER ACOUSTICS. *Educ:* Harvard Univ, AB, 53, MA, 60, PhD(solid state physics), 68. *Prof Exp:* Asst scientist underwater acoustics, Woods Hole Oceanog Inst, 68-75; SR SCIENTIST, BOLT BERANEK & NEWMAN, INC, 75- *Mem:* AAAS; Sigma Xi. *Res:* Structure-borne vibrations; low noise; sonar arrays; scattering of sound from marine organisms; statistical analysis of relationship of acoustic scattering and biological targets; signal processing; medical ultrasonics; computer control of real-time processes. *Mailing Add:* Eight Valley Rd Acton MA 01720

MCELROY, WILBUR RENFREW, b Fayetteville, Pa, Aug 20, 14; m 39; c 1. POLYMER CHEMISTRY, POLYMER ENGINEERING. *Educ:* Gettysburg Col, AB, 36; Pa State Univ, MS, 38; Purdue Univ, PhD(chem), 43. *Prof Exp:* Res chemist, Texaco Co, NY, 39-40 & 43-45; proj leader, Jefferson Chem Co, 45-49; chem eng consult & owner, Wayne Labs, Pa, 49-78; PRES, ACTION PROD, INC, 69- *Concurrent Pos:* Dir chem technol, Cent Res Labs, Westinghouse Air Brake Co, Va, 52-55; sr group leader, Mobay Chem Co, 55-64; exec vpres, Conap, Inc, 64-69. *Mem:* Am Chem Soc; fel Am Inst Chem; Am Soc Test & Mat; NY Acad Sci; Nat Soc Prof Eng. *Res:* Process and product development in chemical intermediates and urethanes; orthopedic pads, radiation bolus and other health care products; industrial elastomers, sealants, coatings. *Mailing Add:* RFD 4 Box 240 Smithsburg MD 21783-9034

MCELROY, WILLIAM DAVID, b Rogers, Tex, Jan 22, 17; m 40, 67; c 5. BIOLOGY, BIOCHEMISTRY. *Educ:* Stanford Univ, BA, 39; Reed Col, MA, 41; Princeton Univ, PhD(biol), 43. *Hon Degrees:* DSc, Univ Buffalo, 62, Mich State Univ, Loyola Univ, Chicago & Providence Col, 70, Del State Col & Univ Pittsburgh, 71 & Notre Dame Univ, 75; DPS, Providence Col, 70; LLD, Univ Pittsburgh, 71, Univ Calif, San Diego, 72 & Fla State Univ, 73, Johns Hopkins Univ & Pasadena City Col, 77, Calif Sch Prof Psychol, 78. *Prof Exp:* Asst, Reed Col, 39-41; asst, Princeton Univ, 43-44, res assoc, 44-45; Nat Res Coun fel, Stanford Univ, 45; from instr to prof biol, Johns Hopkins Univ, 45-69, chmn dept, 56-69, dir, McCollum-Pratt Inst, 49-69; dir, NSF, 69-72; chancellor, 72-80, prof, 80-88, EMER PROF BIOL, UNIV CALIF, SAN DIEGO, 88- *Concurrent Pos:* Trustee, Brookhaven Nat Lab, 54; ex officio mem, Nat Sci Bd, 69-72; co-ed, Anal Biochem. *Honors & Awards:* Barnett Cohen Award, Am Soc Microbiol, 58; Andrew White Medal, Loyola Col, Md, 71; Howard N Potts Medal, Franklin Inst, 71; Rumford Award, Am Acad Arts & Sci. *Mem:* Nat Acad Sci; AAAS (pres, 75-76); Am Chem Soc; Am Acad Arts & Sci; Am Soc Microbiol. *Res:* Bioluminescence; bacterial mutations; biochemical genetics; mechanism of inhibitor action; bacterial and mold metabolism. *Mailing Add:* Univ Calif San Diego La Jolla CA 92093

MCELROY, WILLIAM NORDELL, b Minneapolis, Minn, Nov 28, 26; m 50; c 4. REACTOR PHYSICS. *Educ:* Univ Southern Calif, BA, 51; Ill Inst Technol, MA, 60, PhD(physics), 65. *Prof Exp:* Sr engr, Atomics Inst, 51-57, sr tech specialist, 65-67; mgr reactor oper, Ill Inst Technol, 57-65; res assoc, Battelle-Northwest, 67-70; mgr & fel scientist irradiation characterization & anal, Hanford Engr Develop Lab, Westinghouse Hanford Co, 70-87. *Concurrent Pos:* Mem, Nat Acad Sci-Nat Acad Engr-Nat Res Coun Panel For Nat Bur Standards Ctr For Radiation Res, 73- *Mem:* Am Nuclear Soc; AAAS. *Res:* Reactor environmental characterization and fuels and materials data correlation and damage analysis for light water, fast reactor and controlled thermonuclear reactors. *Mailing Add:* 113 Thayer Rd Richland WA 99352

MCELROY, WILLIAM TYNDELL, JR, physiology; deceased, see previous edition for last biography

MCELWEE, EDGAR WARREN, ornamental horticulture; deceased, see previous edition for last biography

MCELWEE, ROBERT L, b Elkins, WVa, Oct 4, 27; m 51; c 3. FOREST MANAGEMENT, GENETICS. *Educ:* WVa Univ, BSF, 51; NC State Col, MS, 60, NC State Univ, PhD, 70. *Prof Exp:* Mgt forester, Gaylord Container Corp, 51-56; liaison geneticist, NC State Univ, 56-63, dir hardwood res prog, 63-70; proj leader, Exten Forestry, Wildlife & Outdoor Recreation, Va Polytech Inst & State Univ, 71-89; RETIRED. *Concurrent Pos:* Assoc prof, Univ Maine, 70-71. *Mem:* Soc Am Foresters. *Mailing Add:* 3736 Millstone Ridge Rd Blacksburg VA 24060

MCENALLY, TERENCE ERNEST, JR, b Richmond, Va, Apr 21, 27; m 64; c 2. PHYSICS, ELECTRONICS ENGINEERING. *Educ:* Va Polytech Inst, BS, 50, MS, 55; Mass Inst Technol, PhD(physics), 66. *Prof Exp:* Instr physics, Va Polytech Inst, 52-55; instr elec eng, NC State Univ, 61-63; instr physics, Mass Inst Technol, 66-67; ASSOC PROF PHYSICS, E CAROLINA UNIV, 67- *Mem:* Am Phys Soc. *Res:* Electron spin resonance of transition metal ions in crystals; exchange coupling of paramagnetic ions in crystals; theory of nuclear magnetic resonance of fluid molecules. *Mailing Add:* Dept Physics E Carolina Univ Greenville NC 27858

MCENTEE, KENNETH, b Oakfield, NY, Mar 30, 21; m 52; c 2. VETERINARY PATHOLOGY. *Educ:* State Univ NY, DVM, 44. *Hon Degrees:* Dr, Royal Vet Col, Sweden, 75. *Prof Exp:* Asst path, NY State Vet Col, Cornell Univ, 47-48, from asst prof to assoc prof vet path, 48-55, chmn dept large animal med, obstet & surg, 65-69, assoc dean clin studies, 69-73, prof vet path, 55-; PROF REPRODUCTIVE PATHOL, DEPT PATHOBIOL, COL VET MED, URBANA, IL. *Concurrent Pos:* Commonwealth Sci & Indust Res Orgn sr res fel, Univ Melbourne, 65-66; vis lectr & researcher, Royal Vet Col, Sweden, 73, 75 & 77; vis prof, Vet Col, Belo Horizonte, Brazil, 73-74. *Honors & Awards:* Borden Award, Am Vet Med Asn, 71. *Mem:* Am Vet Med Asn; Am Col Vet Path; Int Fertil Asn; Int Acad Path; Am Asn Pathologists. *Res:* Pathology of reproduction in dairy cattle. *Mailing Add:* 24 Townhill Dr Eustis FL 32726

MCENTEE, THOMAS EDWIN, b San Mateo, Calif, Feb 27, 43; m 63, 74; c 3. ORGANIC CHEMISTRY. *Educ:* Univ Vt, BA, 66; Univ Colo, PhD(chem), 72. *Prof Exp:* Res chemist, Arapahoe Chem, Inc, Div Syntex Corp, 73-78, sr res chemist, 78-79, mkt tech analyst, 79-81; com develop mgr, Fine Chem Div, Upjohn Co, 81-85; com develop mgr, Biesterfeld, USA, 85; ENVIRON SYSTS SCIENTIST, MITRE CORP, 86- *Mem:* Am Chem Soc. *Res:* Synthesis of organic medicinals; industrial methods for fine organic chemicals synthesis. *Mailing Add:* MITRE Corp 7525 Colshire Dr McLean VA 22102-3481

MCENTIRE, RICHARD WILLIAN, b Miami, Fla, Sept 10, 42; m 66; c 2. SPACE PHYSICS. *Educ:* Mass Inst Technol, BS, 64; Univ Minn, PhD(physics), 72. *Prof Exp:* Res asst physics, Univ Minn, 65-72; SR STAFF PHYSICIST, APPL PHYSICS LAB, JOHNS HOPKINS UNIV, 72- *Mem:* Am Phys Soc; Am Geophys Union; AAAS. *Res:* Planetary magnetospheric physics; interplanetary phenomena; solar and galactic cosmic rays. *Mailing Add:* Appl Physics Lab Johns Hopkins Univ Laurel MD 20723-6099

MCENTYRE, JOHN G(ERALD), b Topeka, Kans, Nov 3, 20; m 48; c 3. CIVIL ENGINEERING. *Educ:* Kans State Univ, BS, 42, MS, 48; Cornell Univ, PhD(surv, mapping), 54. *Prof Exp:* From instr to prof civil eng, Kans State Univ, 46-65; from assoc prof to prof civil eng technol, 65-71, prof land surv, 71-86, EMER PROF, PURDUE UNIV, 86- *Concurrent Pos:* Consult, Cadastral Surv, Afghanistan, 63-65; vis prof, Kabul Univ, Afghanistan, 65-67. *Honors & Awards:* Earl J Fennell Award, Am Cong Surv & Mapping, 84; Surveying Excellence Award, Nat Soc Prof Surveyors, 80. *Mem:* Fel Am Soc Civil Engrs; Nat Soc Prof Engrs; Am Soc Photogram; fel Am Cong Surv & Mapping. *Res:* Land surveying and precise surveying problems. *Mailing Add:* 160 Pathway Lane West Lafayette IN 47906

MACERO, DANIEL JOSEPH, b Revere, Mass, Nov 19, 28; m 52; c 2. ANALYTICAL CHEMISTRY. *Educ:* Mass Inst Technol, BS, 51; Univ Vt, MS, 53; Univ Mich, PhD(chem), 58. *Prof Exp:* Instr chem, Univ Mich, 56-57; asst prof, 57-64, actg chmn, 70-71, assoc prof anal chem, 64-77, vchmn chem, 70-79, PROF ANALYTICAL CHEM, SYRACUSE UNIV, 77- *Concurrent Pos:* Res assoc, Univ NC, Chapel Hill, 67-68; acad dir, Summer Inst Microcomput, Am Chem Soc, 80-85; treas, Div Comput Chem, Am Chem Soc, 84- *Mem:* Am Chem Soc. *Res:* Kinetics of electron transfer; electrochemistry of transition metal complexes; computer assisted instruction; intro of microcomputer technology into the chemistry curriculum; development of analytical instrumentation; the study of electrode processes; development of new electroanalytical probes; use of computer method for experiment control and data enhancement. *Mailing Add:* Dept of Chem Syracuse Univ 1-014 Ctr Sci & Technol Syracuse NY 13244

MCEVILLY, THOMAS V, b East St Louis, Ill, Sept 2, 34; m 55, 70; c 6. GEOPHYSICS, SEISMOLOGY. *Educ:* St Louis Univ, BS, 56, PhD(geophys), 64. *Prof Exp:* Geophysicist, Calif Co, 57-60; res asst, St Louis Univ, 60-64; asst prof seismol, 64-68, assoc prof, 68-74, chmn dept, 76-80, PROF, DEPT GEOL & GEOPHYS, UNIV CALIF, BERKELEY, 74-, ASST DIR, SEISMOL STA, 68-, ASSOC DIR & HEAD, EARTH SCI DIV, LAWRENCE BERKELEY LAB, 82-, CHMN BD DIRS, INCORP RES INST SEISMOL, 84- *Concurrent Pos:* Vpres eng, W F Sprengnether Instrument Co, 63-68. *Mem:* AAAS; Seismol Soc Am; Am Geophys Union; Soc Explor Geophys; Royal Astron Soc. *Res:* Structure of the earth as revealed by seismic surface and body wave propagation; nature of earthquake sequences; seismic instrumentation. *Mailing Add:* Dept of Geol & Geophys Univ of Calif 2120 Oxford St Berkeley CA 94720

MCEVILY, ARTHUR JOSEPH, JR, b New York, NY, Dec 20, 24; m 53; c 4. METALLURGY. *Educ:* Columbia Univ, BS, 45, MS, 49, DEngSc(metall), 59. *Prof Exp:* Res scientist, NASA, 49-61; sr res scientist, Sci Lab, Ford Motor Co, 61-67; head dept metall, 67-78, PROF METALL, UNIV CONN, 67- *Concurrent Pos:* Consult, several indust orgn, 67- *Honors & Awards:* Howe Medal, Am Soc Metals, 64. *Mem:* Am Soc Metals; Am Soc Mech Engrs; Am Soc Testing & Mat; Am Inst Mining, Metall & Petrol Engrs. *Res:* Fatigue and fracture of metals; stress corrosion; alloy development; mechanical metallurgy. *Mailing Add:* Univ Conn U-136 97 No Eaglevil Univ Conn Storrs CT 06268

MCEVOY, FRANCIS JOSEPH, chemistry, for more information see previous edition

MCEVOY, JAMES EDWARD, b London, Eng, Aug 5, 20; US citizen; m 41; c 2. GENERAL CHEMISTRY. *Educ:* Temple Univ, BA, 55. *Prof Exp:* Res chemist, Houdry Process & Chem Co, 55-64, proj dir catalyst res, 64-66, sect head explor process res, 66-69, asst dir chem res, Houdry Labs, 69-71, asst dir res & develop, Houdry Div, 71-72, dir contract res & develop, 72-78, mgr govt relations, 76-78, mgr bus develop, 78-80; dir, Admin Sci Ctr, Air Prod & Chem Inc, 80-82; EXEC DIR, COUN CHEM RES, INC, 82- *Concurrent Pos:* Mem, Inst Cong Catalysis. *Honors & Awards:* Joseph Stewart Award, Am Chem Soc. *Mem:* Am Chem Soc; Coun Chem Res. *Res:* Applied catalysis; catalytic cracking; hydrogenation; desulfurization; fuel cell catalysts; catalytic oxidation of environmental pollutants; auto emissions control catalysts; synthetic fuels from coal; research administration; university/industry collaboration in chemical sciences and engineering. *Mailing Add:* 1875 Quarter Mile Rd Bethlehem PA 18015-9323

MCEWAN, ALAN THOMAS, b London, Eng, May 2, 40; m 64; c 4. COMPUTER SCIENCE. *Educ:* Univ Hull, BSc, 61. *Prof Exp:* Res mathematician, English Elec, Luton, Eng, 61-62; data processing mgr, Assoc Newspapers, London Ltd, 62; res asst comput sci, Univ Hull, 62-64; lect numerical anal, Cripps Comput Ctr, Univ Nottingham, 64-66; assoc prof comput sci & dir, Comput Ctr, Lakehead Univ, 66-76; head, comput serv, Bedford Inst Oceanog, 76-80; DIR, COMPUT CTR, ACADIA UNIV, 80- *Mem:* Asn Comput Mach; Brit Comput Soc. *Res:* Compiler writing techniques and operating system organization; image processing. *Mailing Add:* Dept Comp Ctr Acadia Univ Wolfville NS B0P 1X0 Can

MACEWAN, DOUGLAS W, b Ottawa, Ont, Nov 11, 24; c 4. RADIOLOGY. *Educ:* McGill Univ, BSc, 48, MD, CM, 52; FRCPS(C), 58. *Prof Exp:* Asst radiologist, Montreal Children's Hosp, 58-63; asst prof radiol, McGill Univ, 58-65; chief radiologist, Health Sci Centre, 66-82; PROF RADIOL, UNIV MAN, 66-; PROF COMMUNITY HEALTH SCI, 90- *Concurrent Pos:* Radiologist, Montreal Gen 63-65; mem bd chancellors, Am Col Radiol, 85-88; mem bd chancellors, Am Col Radiol, 85-91. *Honors & Awards:* Queen's Silver Jubilee Medal, 77. *Mem:* Can Med Asn; Can Asn Radiol; fel Am Col Radiol; Radiol Soc NAm. *Res:* Renal disease; standards, administration and costs of radiology; public policy and economics of radiology. *Mailing Add:* Health Sci Centre 820 Sherbrook St Winnipeg MB R3A 1R9 Can

MCEWAN, IAN HUGH, polymer chemistry, for more information see previous edition

MCEWEN, BRUCE F, b June 3, 45. EXPERIMENTAL BIOLOGY. *Educ:* Cornell Univ, BS, 68, PhD(biochem), 82. *Prof Exp:* Res, Dept Biochem & Chem, Mass Inst Technol, Cambridge, 68-72; res asst I & doctoral res, Dept Biochem, Cornell Univ, Ithaca, NY, 75-82, postdoctoral fel, Dept Nutrit Sci, 82-85; RES SCIENTIST, WADSWORTH CTR LABS & RES, NY STATE DEPT HEALTH, ALBANY, 85- *Mem:* Am Soc Biochem & Molecular Biol; Electron Micros Soc Am; Am Soc Cell Biol. *Res:* Author of numerous publications. *Mailing Add:* Wadsworth Ctr Labs & Res NY State Dept Health Empire State Plaza PO Box 509 Albany NY 12201-0509

MCEWEN, BRUCE SHERMAN, b Ft Collins, Colo, Jan 17, 38; m 60; c 2. NEUROBIOLOGY. *Educ:* Oberlin Col, AB, 59; Rockefeller Univ, PhD(biol), 64. *Prof Exp:* USPHS fel, Inst Neurobiol, Gothenburg Univ, Sweden, 64-65; asst prof zool, Univ Minn, Minneapolis, 66; from asst prof to assoc prof, 66-81, PROF NEUROSCI, ROCKEFELLER UNIV, 81 - *Mem:* Soc Gen Physiol; Soc Neurosci; Am Soc Neurochem; Endocrine Soc. *Res:* Gene activity in nervous tissue, focusing on steroid hormone action. *Mailing Add:* Lab Neuroendocrinol Rockefeller Univ New York NY 10021

MCEWEN, C(ASSIUS) RICHARD, b Missoula, Mont, Dec 22, 25; m 50; c 2. INSTRUMENTATION. *Educ:* Calif Inst Technol, BS, 46; Mont State Univ, BA, 48; Stanford Univ, MS, 50, PhD(chem eng), 52. *Prof Exp:* Sect leader, Res Dept, Union Oil Co Calif, 52-64; res dir, Spinco Div, Beckman Instruments, Inc, Palo Alto, 64-83; RETIRED. *Mem:* AAAS; Am Chem Soc; Sigma Xi. *Mailing Add:* 753 Garland Dr Palo Alto CA 94303

MCEWEN, CHARLES MILTON, JR, INTERNAL MEDICINE. *Educ:* Harvard Univ, MD, 59. *Prof Exp:* Asst chmn, dept med, 75-86, DIR QUAL ASSESSMENT, SINAI HOSP, 86-; assoc prof med, Wayne State Univ, 71-88; MED DIR UTILIZATION REV, ST JOSEPH'S HOSP, MT CLEMENS, MICH, 88- *Mailing Add:* St Joseph's Hosp 15855 19 Mile Mt Clemens MI 48044

MCEWEN, CHARLES NEHEMIAH, b Matoaca, Va, Feb 9, 42; m 67; c 2. MASS SPECTROMETRY. *Educ:* Col William & Mary, BS, 65; Atlanta Univ, MS, 70; Univ Va, PhD(chem), 73. *Prof Exp:* Teacher chem, Hermitage High Sch, Henrico County, Va, 65-68; RES SUPVR SEPARATION SCI & MASS SPECTROMETRY, CENT RES & DEVELOP DEPT, E I DU PONT DE NEMOURS & CO, INC, 73- *Concurrent Pos:* Chmn, Anal Topical Group, Am Chem Soc, Del Valley, 82; mem bd dirs, Am Soc Mass Spectrometry, 85. *Mem:* Am Chem Soc; Am Soc Mass Spectrometry; AAAS. *Res:* Mass spectrometry of biological materials with emphasis on biopolymers; emphasis on special techniques and new applications such as electrospray ionization. *Mailing Add:* E I du Pont de Nemours & Co PO Box 80228 Wilmington DE 19880-0228

MCEWEN, CURRIER, b Newark, NJ, Apr 1, 02; m 30; c 4. MEDICINE. *Educ:* Wesleyan Univ, BS, 23; NY Univ, MD, 26. *Hon Degrees:* DSc, Wesleyan Univ, 50; DSc, Marietta Col, 52. *Prof Exp:* Intern, Bellevue Hosp, 26-27, asst res, 27-28; asst med & asst resident physiciatn, Rockefeller Inst, 28-30, assoc med, 30-32; from asst prof to prof, 33-70, asst dean & secy, 32-37, dean, 37-55, EMER PROF MED, SCH MED, NY UNIV, 70- *Concurrent Pos:* Consult physician, Bellevue Hosp Ctr, 32-; consult physician, Goldwater Mem Hosp, 46-70; vis physician, Univ Hosp, 49-70; consult, Vet Admin Hosps, New York, 55-70 & Vet Admin Cent Off, 59-70; consult physician, Maine Med Ctr, Regional Mem Hosp, Cent Maine Med Ctr, Togus Vet Admin Hosps, Mid-Maine Med Ctr, Northern Maine Med Ctr, Cary Med Ctr. *Mem:* Am Soc Clin Invest; Master Am Rheumatism Asn (pres, 52-53); Asn Am Physicians; master Am Col Physicians. *Res:* Rheumatic and collagen diseases. *Mailing Add:* South Harpswell ME 04079

MCEWEN, DAVID JOHN, b Winnipeg, Man, July 23, 30; US citizen; m 54; c 4. ANALYTICAL CHEMISTRY. *Educ:* Acadia Univ, BSc, 52; Purdue Univ, MS, 55, PhD(chem), 57. *Prof Exp:* Res chemist, Cent Res Lab, Can Industs, Ltd, Que, 57-60; sr res chemist, Res Labs, 60-80, staff res scientist, 80-86, SENIOR STAFF RES SCIENTIST, GEN MOTORS CORP, WARREN, 86- *Mem:* Soc Appl Spectros; Coblentz Soc. *Res:* Gas chromatography; mass spectrometry; analysis of vehicle emissions, polymers, solvents, synthesized chemicals; fourier-transform infrared spectroscopy. *Mailing Add:* 3477 Newgate Rd Troy MI 48084

MCEWEN, EVERETT E(DWIN), b Providence, RI, July 15, 32; m 56; c 2. CIVIL ENGINEERING. *Educ:* Univ RI, BS, 54; Univ Ill, MS, 56; Rensselaer Polytech Inst, DEng, 64. *Prof Exp:* Asst struct eng, Univ Ill, 54-56; instr civil eng, 56-57; from instr to asst prof, Rensselaer Polytech Inst, 57-62; asst prof, 62-65, ASSOC PROF CIVIL ENG, UNIV RI, 65- *Mem:* Am Soc Civil Engrs; Am Soc Eng Educ; Am Concrete Inst. *Res:* Structural dynamics; numerical methods; offshore towers; reinforced concrete. *Mailing Add:* Dept of Civil Eng Univ of RI Kingston RI 02881

MCEWEN, FREEMAN LESTER, b Bristol, PEI, Nov 11, 26; m 47; c 3. ENTOMOLOGY. *Educ:* McGill Univ, BS, 50; Univ Wis, MS, 52, PhD(entom), 54. *Hon Degrees:* LLD, Univ Prince Edward Island, 86. *Prof Exp:* Asst, Univ Wis, 50-54; agr res off, Sci Serv Can, 54; from asst prof to prof entom & head dept, NY State Agr Exp Sta, Cornell Univ, 54-69; prof zool, 69-71, prof environ biol & chmn dept, Univ Guelph, 71-83. *Mem:* AAAS; fel Entom Soc Can (pres, 78-79); Sigma Xi; Am Inst Biol Sci; Entom Soc Am. *Res:* Vegetable insect control; transmission of plant virus diseases by insects; microbial control of insects. *Mailing Add:* Dean OAC Univ Guelph Guelph ON N1G 2W1 Can

MCEWEN, GERALD NOAH, JR, b Washington, DC, Jan 1, 43; m 66; c 2. SCIENCE ADMINISTRATION. *Educ:* Ind State Univ, Terre Haute, BS, 66, MA, 68; Univ Ill, Urbana, PhD(physiol), 73. *Prof Exp:* Asst prof biol, Vincennes Univ, 68-69; instr, Univ Ill, 71-72, vis lectr physiol, 73-74; fel, Nat Res Coun, NASA, 74-76; sr bioscientist, Stanford Res Inst, 76-80; dir tech affairs, 80-86, VPRES SCI DEPT, COSMETIC, TOILETRY, FRAGRANCE ASN, 86- *Concurrent Pos:* Consult, 74-80. *Mem:* Am Physiol Soc; Aerospace Med Asn; Dermal Clin Eval Soc. *Res:* Evaluation of information on health effects of potentially hazardous substances. *Mailing Add:* 1101 17th St NW Suite 300 Washington DC 20036

MACEWEN, JAMES DOUGLAS, b Detroit, Mich, July 31, 26; m 47; c 1. TOXICOLOGY. *Educ:* Wayne State Univ, BS, 49, PhD(pharmacol, physiol), 62; Univ Mich, Ann Arbor, MPH, 57; Am Bd Indust Hyg, dipl, 64; Am Bd Toxicol, dipl, 80. *Prof Exp:* Chemist, Water Bd, City of Detroit, Mich, 49, asst indust hyg, Bur Indust Hyg, 51-56; chemist, USPHS Hosp, Detroit, 50-51; asst prof & res assoc toxicol, Wayne State Univ, 56-63; dir environ health, Toxic Hazards Res Unit, Systemed Corp, 63-72; DIR ENVIRON HEALTH, TOXIC HAZARDS RES LAB, DAYTON, OHIO, 72- *Mem:* Soc Toxicol; Am Indust Hyg Asn; Am Conf Govt Indust Hyg; NY Acad Sci. *Res:* Inhalation toxicology; environmental analysis and control. *Mailing Add:* Horse Shoe Bend Rd Rte 1 Box 2897 Spring City TN 37381

MCEWEN, JOAN ELIZABETH, b Lawrence, Kansas July, 3, 52; m 74; c 2. MOLECULAR GENETICS, MEMBRANE BIOGENESIS. *Educ:* Tufs Univ, BS, 74; Albert Einstein Col Med, PhD, 80. *Prof Exp:* ASST PROF, MICROBIOL, UNIV CALIF, 84- *Mem:* Am Soc Microbiol; Genetics Soc Am. *Res:* Saccharonyces cerevisiae; genetic analysis of nuclear-encoded protines are targeted & translocated into mitochondria. *Mailing Add:* Dept Microbiol UCLA 405 Hilgard Ave Los Angeles CA 90024-1489

MCEWEN, ROBERT B, b Fall River, Mass, July 29, 34; m 58; c 3. REMOTE SENSING, GEOGRAPHIC INFORMATION SYSTEMS. *Educ:* Univ NH, BS, 56, MS, 65; Cornell Univ, PhD(civil eng), 68. *Prof Exp:* Instr civil eng, Univ NH, 62-65; sci fac fel, Cornell Univ, 67-68; sr scientist, Autometric/Raytheon, 68-69; PHYS SCIENTIST, US GEOL SURV, 69- *Concurrent Pos:* Adj prof, George Washington Univ, 70-76 & Va Polytech Inst & State Univ, 76-; bd dirs, Am Soc Photogrammetry & Remote Sensing, 76; sr scientist & res prog mgr, US Geol Surv Nat Mapping Div. *Honors & Awards:* Alan Gordan Award, Am Soc Photogrammetry, 77; Meritorious Serv, Dept Interior, 81. *Mem:* Am Soc Photogrammetry & Remote Sensing; Am Soc Civil Engrs; Sigma Xi. *Res:* Satellite remote sensing and geographic information systems. *Mailing Add:* 3512 Wilson St Fairfax VA 22030

MCEWEN, WILLIAM EDWIN, b Oaxaca, Mex, Jan 13, 22; m 45; c 3. ORGANIC CHEMISTRY. *Educ:* Columbia Univ, AB, 43, MA, 45, PhD(chem), 47. *Prof Exp:* Asst chem, Columbia Univ, 43-45; tech engr, Carbide & Carbon Chem Corp, Oak Ridge, 45-46; from asst prof to prof chem, Univ Kans, 47-62; prof chem & head dept, 62-76, COMMONWEALTH PROF, UNIV MASS, AMHERST, 77- *Concurrent Pos:* Res collab, Brookhaven Nat Lab. *Mem:* Fel NY Acad Sci; fel Am Chem Soc. *Res:* Mechanisms of organic reactions; stereochemistry of organophosphorus compounds; heterocyclic, organosulfur and organoantimony chemistry; organometallic compounds. *Mailing Add:* Dept of Chem Univ of Mass Amherst MA 01003

MACEY, ROBERT IRWIN, b Minneapolis, Minn, Sept 22, 26; m 56; c 2. PHYSIOLOGY, BIOPHYSICS. *Educ:* Univ Minn, BA, 47; Univ Chicago, PhD(math biol), 54. *Prof Exp:* Instr chem & physiol, George Williams Col, 49-51; instr math, Ill Inst Technol, 53-54; res assoc physiol, Aeromed Lab, Univ Ill, 55-57, asst prof, Col Med, 57-60; from asst prof to assoc prof, 60-69, PROF PHYSIOL, UNIV CALIF, BERKELEY, 69-, CHMN DEPT, 84- *Concurrent Pos:* Asst, Univ Chicago, 53-54; consult, Rand Corp, 63 & NIH, 65. *Mem:* Biophys Soc; Soc Math Biol; Am Physiol Soc. *Res:* Theoretical biophysics; membrane transport; kidney; blood. *Mailing Add:* Dept Physiol Univ Calif 2549 Life Sci Bldg Berkeley CA 94720

MACEY, WADE THOMAS, b Mt Airy, NC, Jan 13, 36; m 58; c 3. MATHEMATICS. *Educ:* Guilford Col, BS, 60; Fla State Univ, MS, 62, PhD(math educ), 70. *Prof Exp:* Instr math, Oxford Col, Emory Univ, 62-65; prof & head, Dept Math, Pfeiffer Col, 67-82; PROF DEPT MATH SCI, APPALACHIAN STATE UNIV, 82- *Mem:* Math Asn Am. *Res:* Effect of prior instruction of selected topics of logic on the understanding of the limit of a sequence. *Mailing Add:* Dept Math Sci Appalachian State Univ Boone NC 28608

MCFADDEN, BRUCE ALDEN, b La Grande, Ore, Sept 23, 30; m 58; c 3. BIOCHEMISTRY, MICROBIOLOGY. *Educ:* Whitman Col, AB, 52; Univ Calif, Los Angeles, PhD(chem), 56. *Hon Degrees:* DSc, Whitman Col, 78. *Prof Exp:* Asst chem, Univ Calif, Los Angeles, 52-54; from instr to assoc prof, Wash State Univ, 56-66, dir develop, Div Sci, 74-78, prof chem, 66-74, chmn, dept biochem & biophys, 78-84, PROF BIOCHEM, WASH STATE UNIV, 74- *Concurrent Pos:* NIH res career develop award, 63-69; vis prof, Univ Ill, Urbana, 66-67, Tech Univ, Munchen, 80-81, Univ Florence, 80; Guggenheim fel & vis prof, Univ Leicester, 72-73; NIH spec fel, 73, mem, NIH ad hoc rev comt, 78-79, 82; ed, Arch Microbiol, 77-83; Humboldt Distinguished Sr Scientist Award, US, 80-81; prin investr, DOE, NIH, NSF, NASA, Res Corp, Frasch Found, 58-, mem rev comt, Frasch Found, 82- *Mem:* Fel AAAS; Am Soc Biol Chem & Molecular Biol; Am Soc Microbiol; Am Chem Soc; Am Soc Plant Physiol; Protein Soc. *Res:* Microbial assimilation of one-carbon and two-carbon compounds; plant biochemistry; genetic and chemical modification of enzymes. *Mailing Add:* Dept Biochem & Biophys Wash State Univ Pullman WA 99164-4660

MCFADDEN, DAVID LEE, b Orange, Calif, Oct 18, 45; m 70; c 2. CHEMICAL PHYSICS. *Educ:* Occidental Col, AB, 67; Mass Inst Technol, PhD(phys chem), 72. *Prof Exp:* Res fel chem, Harvard Univ, 72-73; from asst prof to assoc prof, 73-89, chmn, 89-92, PROF CHEM BOSTON COL, 89- *Mem:* Am Phys Soc; Am Chem Soc; AAAS. *Res:* Gas phase reaction dynamics and kinetics. *Mailing Add:* Dept Chem Boston Col Chestnut Hill MA 02167

MACFADDEN, DONALD LEE, b Port Deposit, Md, Dec 12, 26; m 53; c 4. PHYSIOLOGY, BIOLOGY. *Educ:* Univ Del, BS, 53, MS, 55; Univ Kans, PhD(physiol), 59. *Prof Exp:* Asst prof nutrit, Univ Mass, 60-63; prof physiol, 63-73, PROF BIOL, KING COL, 73-, HEAD DEPT, 63- *Concurrent Pos:* Sterling-Winthrop Res Corp res grant, 61-64; res consult, Chevron Res Co, 74-80. *Mem:* AAAS; Am Inst Biol Sci. *Res:* Mode of action of antibiotics in growth stimulation; studies on histological and immunological mechanisms involved in homograft rejections with intent of reducing severity of this response. *Mailing Add:* Dept Natural Sci King Col Bristol TN 37620

MCFADDEN, EDWARD REGIS, PULMONARY, ASTHMA. *Educ:* Univ Pittsburgh, MD, 63. *Prof Exp:* PROF MED, UNIV HOSP, CASE WESTERN RESERVE UNIV, 84- *Res:* Airway disease. *Mailing Add:* Dept Med Case Western Reserve Univ 2074 Abington Rd Cleveland OH 44106

MCFADDEN, GEOFFREY BEY, b Stillwater, Minn, June 18, 53; m 77; c 3. HYDRODYNAMIC STABILITY THEORY. *Educ:* Rice Univ, BA, 75; NY Univ, MS, 77, PhD(math), 79. *Prof Exp:* NSF res fel math sci, Courant Inst Math Sci, NY Univ, 79-80, res scientist appl math, 80-82; MATHEMATICIAN, CTR COMPUT & APPL MATH, NAT INST STANDARDS & TECHNOL, 82- *Concurrent Pos:* Vis scientist, Inst Theoret Physics, Univ Calif, Santa Barbara, 84, Inst Math Appl, Univ Minn, 90. *Honors & Awards:* Silver Medal, Dept Com, 85. *Mem:* Sigma Xi; Soc Indust & Appl Math; Am Phys Soc. *Res:* Convection and interface stability during crystal growth. *Mailing Add:* A151 Technol Bldg Nat Inst Standards & Technol Gaithersburg MD 20899

MCFADDEN, HARRY WEBBER, JR, b Greenwood, Nebr, Dec 9, 19; m 45; c 2. MEDICAL MICROBIOLOGY, INFECTIOUS DISEASES. *Educ:* Univ Nebr, AB, 41, MD, 43; Am Bd Path, dipl, 51 & 65. *Prof Exp:* From intern to resident path, 43-45, resident clin path & bact, 47-49, fel & instr path & bact, 49-52, asst prof path & microbiol, 52-55, interim chancellor, Med Ctr, 72 & 76-77, interim assoc dean grad studies, Med Ctr, 72-79, PROF MED MICROBIOL & CHMN DEPT, COL MED, UNIV NEBR, OMAHA, 55-, PROF PATH, 68- *Concurrent Pos:* Trustee, Am Bd Path, 70- *Mem:* Col Am Pathologists; Am Soc Microbiol; Am Soc Clin Pathologists; NY Acad Sci; Am Fedn Clin Res; Sigma Xi. *Res:* Immunology, microbiology and pathology. *Mailing Add:* 10710 Poppleton Ave Omaha NE 68124

MCFADDEN, JAMES DOUGLAS, b Winchester, Va, Feb 19, 34; m 61; c 4. METEOROLOGY, OCEANOGRAPHY. *Educ:* Va Polytech Inst, BS, 56; Univ Wis-Madison, PhD(meteorol), 65. *Prof Exp:* Res meteorologist, NOAA, 65-73, dir, res flight facil 73-75, dir, flights oper, res facil ctr, nat oceanic & atmospheric admin, 75-85, MGR, AIRBORNE SCI PROGS, NOAA-OAO 85-,. *Mem:* Am Meteorol Sco; Am Geophys Union; Am Geog Soc; Sigma Xi. *Res:* Airborne meteorological and oceanographic research; cloud physics; weather modification; remote sensing of the environment. *Mailing Add:* Nat Oceanic & Atmospheric Admin/Off Aircraft Opers Ctr Box 020197 Miami FL 33102-0197

MACFADDEN, KENNETH ORVILLE, b Philadelphia, Pa, Sept 5, 45; m 68; c 2. ANALYTICAL CHEMISTRY. *Educ:* Juniata Col, BS, 66; Georgetown Univ, PhD(phys chem), 72. *Prof Exp:* Fel kinetics, Univ Calgary, 71-73; asst prof chem, Stockton State Col, 73-75; analytical chemist, Air Prod & Chem Co, 75-77, group leader, 77-78, sect mgr, 79-81, mgr analysis serv, 81-82, mgr res, 82-84; DIR, ANALYTICAL RES, W R GRACE, 84- *Mem:* Am Chem Soc; AAAS; Sigma Xi. *Res:* Mass spectrometric analysis and methods development. *Mailing Add:* 6575 River Clyde Dr Highland MD 20777

MCFADDEN, LORNE AUSTIN, b Lower Stewiacke, NS, Oct 12, 26; nat US; m 53; c 3. PLANT PATHOLOGY. *Educ:* McGill Univ, BSc, 49; Cornell Univ, MS, 54, PhD(plant path), 56. *Prof Exp:* Plant pathologist, NS Agr Col, 49-50; asst, Cornell Univ, 50-56; asst plant pathologist, Subtrop Exp Sta, Univ Fla, 56-63; assoc prof plant sci, Univ NH, 63-69, prof plant path, 69-72; PROF BIOL & HEAD DEPT, NS AGR COL, 72- *Concurrent Pos:* Exten horticulturist, Agr Exp Sta, Univ NH, 63-69. *Mem:* Am Phytopath Soc; Can Phytopath Soc; Agr Inst Can. *Res:* Bacterial plant pathogens; diseases of ornamental and other crop plants. *Mailing Add:* Dept of Biol NS Agr Col Truro NS B2N 5E3 Can

MCFADDEN, LUCY-ANN ADAMS, b New York, NY, May 23, 52; m 82; c 2. PLANETARY SCIENCE, REMOTE SENSING. *Educ:* Hampshire Co, BA, 74; Mass Inst Technol, MS, 77; Univ Hawaii, PhD(geol & geophys), 83. *Prof Exp:* Res assoc, Inst Astron, Univ Hawaii, 77-78; res asst, planetary geosci div, Hawaii Inst Geophys, 78-83,; res assoc, dept geol, Univ Md & Goddard Space Flight Ctr, NASA, 83-84, res assoc astron prog, Univ Md, 84-86, asst res scientist astron prog, Univ Md, 86-87; ASST RES SCIENTIST, CALIF SPACE INST, UNIV CALIF SAN DIEGO, 87- *Concurrent Pos:* Prin investr, planetary geol prog, NASA, 84-, planetary astron prog, 85; mem, Comt Data Mgt & Comput, Nat Acad Sci, 85-88. *Mem:* AAAS; Am Astron Soc; Am Geophys Union; Meteoritical Soc. *Res:* Determination of the surface composition of planet-crossing asteroids to understand their nature, source and evolution; relationship between asteroids and comets based on composition of solid components; studies of reflectance properties of meteorites as a means of developing interpretive techniques of remotely measured spectra of asteroids and comets; ultra-violet spectroscopy of composition of comets; applications of automation techniques for space sciences. *Mailing Add:* Calif Space Inst 0216 Univ Calif San Diego 2265 Sverdrup Hall La Jolla CA 92093-0216

MCFADDEN, PETER W(ILLIAM), b Stamford, Conn, Aug 2, 32; div; c 3. MECHANICAL ENGINEERING. *Educ:* Univ Conn, BS, 54, MSME, 57; Purdue Univ, PhD(mech eng), 59. *Prof Exp:* Asst instr graphics, Univ Conn, 54-56; instr thermodyn, Purdue Univ, 56-59, from asst prof to prof heat transfer & head sch mech eng, 59-71; dean sch eng, 71-85, PROF MECH ENG, UNIV CONN, 71- *Concurrent Pos:* Res fel, Swiss Fed Inst Technol, 60-61. *Mem:* Am Soc Mech Engrs; Am Soc Eng Educ. *Res:* Cryogenic heat transfer; combined heat and mass convection; optical investigations of transport phenomena; interaction of thermal radiation and convection; boiling studies; convection problems. *Mailing Add:* Dept Mech Engr Univ Conn Storrs CT 06268

MCFADDEN, ROBERT B, b Belfast, Northern Ireland, Oct 7, 34; m 57; c 2. MATHEMATICS. *Educ:* Queen's Univ Belfast, BA, 55, MA, 59, PhD(math), 62. *Prof Exp:* Lectr math, Queen's Univ Belfast, 62-68; prof, Northern Ill Univ, 68-89; CHAIR, UNIV LOUISVILLE, 89- *Concurrent Pos:* Asst prof, Ind Univ, 64-65 & La State Univ, 65-66; NSF res grants, Northern Ill Univ, 69-73; vis prof, Monash Univ, 73-74 & St Andrews Univ, 80-81. *Mem:* Math Asn Am; Am Math Soc. *Res:* Theory of partially ordered semigroups; algebraic theory of semigroups; automated reasoning. *Mailing Add:* Dept Math Univ Louisville Louisville KY 40292

MCFADYEN, JOHN ARCHIBALD, JR, b Scranton, Pa, July 10, 22; m 46; c 3. GEOLOGY. *Educ:* Williams Col, BA, 48; Lehigh Univ, MS, 50; Columbia Univ, PhD(geol), 62. *Prof Exp:* Asst geol, Lehigh Univ, 48-50 & Columbia Univ, 50-52; from instr to prof geol, 52-74, Edward Brust prof geol & mineral, 74-75, chmn dept, 68-83, EDNA McCONNELL CLARK PROF GEOL, WILLIAMS COL, 75-, EMER PROF, 83- *Concurrent Pos:* Geologist, Vt Geol Surv, 51-54; res assoc, Mineral Inst, Univ Oslo, 65-66; vis prof geol, Univ Aarhus, 72-73; consult struct geol; vis scientist, Imperial Col London, 79-80. *Mem:* Geol Soc Am. *Res:* Structural geology; deformation and flow of solids; genesis of mylonites and cataclasitese. *Mailing Add:* 98 Water St Stonington CT 06378

MCFALL, ELIZABETH, b San Diego, Calif, Oct 28, 28; m 80. BIOCHEMISTRY. *Educ:* San Diego State Col, BS, 50; Univ Calif, MA, 54, PhD(biochem), 57. *Prof Exp:* Asst resident biochemist, Univ Calif, 57; res assoc microbiol, Mass Inst Technol, 60-61, 62-63; from asst prof to assoc prof, 63-72, PROF MICROBIOL, SCH MED, NY UNIV, 72- *Concurrent Pos:* Res fel bact, Harvard Med Sch, 57-60; Nat Inst Med Res fel, London, 61-62; USPHS career develop award, 63-73. *Mem:* Am Soc Biol Chemists; Am Soc Microbiol. *Res:* Molecular genetics; regulatory mechanisms in microorganisms. *Mailing Add:* Dept Microbiol NY Univ Sch Med New York NY 10016

MCFARLAN, EDWARD, JR, b Brooklyn, NY, Mar 24, 21; m 49; c 2. PETROLEUM GEOLOGY, SEDIMENTOLOGY. *Educ:* Williams Col, BA, 43; Univ Tex, MA, 48. *Prof Exp:* Res geologist, Humble Oil & Ref Co, 48-59, area stratigrapher, La, 59-64, area explor geologist, 64-65; div stratigrapher, STex, 65-66, Mgr basin geol div, Esso Prod Res Co, 66-69, mgr stratig geol div, Exxon Prod Res Co, 69-73, explor adv, 73-77, geol scientist, 77-81, SR GEOL SCIENTIST, EXPLORATION DEPT, HQ, EXXON CO USA, 81- *Honors & Awards:* Cert Merit, Am Asn Petrol Geol, 75, 76, 79 & 80. *Mem:* Soc Econ Paleont & Mineral; fel Geol Soc Am; Am Inst Prof Geol; Am Asn Petrol Geol; Sigma Xi. *Res:* Recent and Pleistocene geology; sedimentation; photogeology; geomorphology; Cenozoic and Mesozoic structure and stratigraphy of the Gulf Coast and adjacent offshore areas; Regional petroleum geology of the Cenozoic and Mesozoic sequences in the US Gulf Coast from Florida to Mexico. *Mailing Add:* 10631 Gawain St Houston TX 77024

MCFARLAND, CHARLES ELWOOD, b Kirkwood, Mo, June 20, 27; m 50; c 2. PHYSICS. *Educ:* Mo Sch Mines, BS, 49; Wash Univ, PhD(physics), 55. *Prof Exp:* Sr scientist, Bettis Atomic Power Div, Westinghouse Elec Corp, 55-56; nuclear physicist, Internuclear Co, 56-60; ASSOC PROF PHYSICS, UNIV MO-ROLLA, 60- *Mem:* Am Phys Soc; Sigma Xi. *Res:* Solid state physics; nuclear physics. *Mailing Add:* Dept of Physics Univ of Mo Rolla MO 65401

MCFARLAND, CHARLES MANTER, b Dayton, Ohio, Jan 11, 20; m 42; c 3. CHEMICAL ENGINEERING, EXTRACTIVE METALLURGY. *Educ:* Univ Dayton, BChE, 41. *Prof Exp:* Asst investr metall, NJ Zinc Co, Pa, 41-48, investr, 48-60; mem res staff mat appl & eval, Res Lab, Gen Elec Co, 61-65, group liaison scientist, 65-71, mgr separation technol, Phys Chem Lab, 77-79, mem staff process develop, Metall Lab, Corp Res & Develop, 71-83; RETIRED. *Concurrent Pos:* Consult, 83- *Mem:* Am Chem Soc; Am Inst Mining, Metall & Petrol Engrs. *Res:* Extractive metallurgy of titanium, copper, zinc, rare earths and other metals. *Mailing Add:* 37 Knollwood Dr Yarmouth Port MA 02675

MCFARLAND, CHARLES R, b Columbus, Ohio, Aug 26, 27; m 50. MICROBIOLOGY. *Educ:* Otterbein Col, BS, 49; Ohio State Univ, MS, 50; WVa Univ, PhD(microbiol), 68. *Prof Exp:* Clin microbiologist, Clifton Springs Sanitarium & Clin, NY, 55-57, Aultman Hosp, Canton, Ohio, 57-60 & Good Samaritan Hosp, Dayton, 60-65; asst prof, 67-74, ASSOC PROF MICROBIOL, WRIGHT STATE UNIV, 74- *Concurrent Pos:* Consult, Barney's Children's Med Ctr, Dayton, Ohio, 69- *Mem:* AAAS; Am Soc Microbiol. *Res:* Chemistry, metabolism and biological significance of microbial lipids; bacterial endogenous carbon and energy reserves. *Mailing Add:* 2760 Nantucket Rd Venia OH 45385

MCFARLAND, CHARLES WARREN, b Schenectady, NY, Jan 24, 42; m 64; c 1. ANALYTICAL CHEMISTRY, INFORMATION MANAGEMENT. *Educ:* Oberlin Col, AB, 64; Case Western Reserve Univ, PhD(org chem), 71. *Prof Exp:* Res fel, Case Western Reserve Univ, 71-73; res chemist, 73-75, res mgr, 75-77, tech asst to res dir, 77-88, QUANTUM MECH, MCGEAN-ROHCO, INC, 88- *Concurrent Pos:* Res assoc, Cuyahoga County Coroner's Off, 72-73. *Mem:* Am Electroplaters & Surface Finishers Soc; Am Chem Soc; fel Am Inst Chemists; AAAS; Sigma Xi. *Res:* Applications of advanced analytical instrumentation and computer modeling to metal finishing problems; technical information retrieval and management. *Mailing Add:* 12699 Cedar Rd Cleveland Heights OH 44106

MACFARLAND, CRAIG GEORGE, b Great Falls, Mont, July 17, 43; m 88; c 5. WILDLANDS MANAGEMENT. *Educ:* Austin Col, Tex, BA, 65; Univ Wis-Madison, MA, 69. *Hon Degrees:* DSc, Austin Col, 78. *Prof Exp:* Dir, Charles Darwin Res Sta, Galapagos Islands, Ecuador, 74-78; head, Wildlands Mgt Prog Cent Am, Trop Agr Res & Training Ctr, 78-85; PRES, CHARLES DARWIN FOUND, GALAPAGOS ISLES, 85- *Concurrent Pos:* Pvt consult, 85- *Honors & Awards:* Int Conserv Medal, Zool Soc, 78 & 83. *Mem:*

Ecol Soc Am; Sigma Xi; Int Soc Trop Foresters; Asn Trop Biol; Int Soc Ecol Econ. *Res:* Development of methods and techniques for planning and management of protected areas and wildlife in the tropics; land use. *Mailing Add:* Charles Darwin Found for the Galapagos Isles 836 Mabelle Moscow ID 83843

MACFARLAND, HAROLD NOBLE, industrial hygiene, toxicology, for more information see previous edition

MCFARLAND, HENRY F, MEDICINE, MUSCULAR & NEUROLOGICAL DISEASES. *Educ:* Univ Colo, MD, 66. *Prof Exp:* DEP CHIEF NEUROL BR, NIH, 82- *Res:* Multiple sclerosis. *Mailing Add:* Dept Neurology NINDS NIH Bldg 10 Rm 5B16 9000 Rockville Pike Bethesda MD 20892

MCFARLAND, JAMES THOMAS, biochemistry; deceased, see previous edition for last biography

MCFARLAND, JAMES WILLIAM, b Sacramento, Calif, Nov 16, 31; m 61; c 3. MEDICINAL CHEMISTRY. *Educ:* Chico State Col, BA, 54; Univ Calif, Berkeley, PhD(org chem), 57. *Prof Exp:* Fulbright fel, Univ Munich, 57-58; res fel org chem, Univ Calif, Berkeley, 58-59; res chemist, 60-68, proj leader, 68-87, PRIN RES INVESTR, PFIZER, INC, 87- *Mem:* Am Chem Soc. *Res:* Heterocyclic syntheses; quantitative and graphical structure-activity correlations; pesticides; anthelmintic, antibacterial and anticoccidial agents. *Mailing Add:* Cent Res Div Pfizer Inc Groton CT 06340

MCFARLAND, JOHN WILLIAM, b Elkton, Tenn, Aug 16, 23; m 47; c 4. ORGANIC CHEMISTRY. *Educ:* DePauw Univ, AB, 49; Vanderbilt Univ, PhD(chem), 53. *Prof Exp:* Fel, Mass Inst Technol, 53-55; res chemist, E I du Pont de Nemours & Co, Inc, Del, 55-61; assoc prof, 61-67, PROF CHEM, DEPAUW UNIV, 67- *Concurrent Pos:* Fel, Univ Groningen, 71. *Mem:* Am Chem Soc. *Res:* Organic syntheses and reaction mechanisms; polymerization. *Mailing Add:* Dept of Chem DePauw Univ Locust St Greencastle IN 46135

MCFARLAND, KAY FLOWERS, b Daytona Beach, Fla, Jan 27, 42; m 63; c 4. INTERNAL MEDICINE, ENDOCRINOLOGY. *Educ:* Wake Forest Col, BS, 63; Bowman Gray Sch Med, MD, 66. *Prof Exp:* From instr to assoc prof internal med & endocrinol, Med Col Ga, 71-77; PROF OBSTET & GYNEC, UNIV SC SCH MED, 77- *Concurrent Pos:* Endocrinologist, pvt pract, 72-75. *Mem:* Am Diabetes Asn; fel Am Col Physicians. *Res:* Clinical research on diabetes. *Mailing Add:* Dept Obstet Sch Med Univ SC 6425 Westshore Rd Columbia SC 29206

MCFARLAND, MACK, b Houston, Tex, Sept 9, 47; m 68; c 2. ATMOSPHERIC CHEMISTRY. *Educ:* Univ Tex, Austin, BS, 70; Univ Colo, PhD(chem physics), 73. *Prof Exp:* Intern surg, Univ Calif Hosp, Univ Calif, San Francisco, 59-60, asst resident surg,; proj scientist atomospheric measurements, York Univ, 74-75; chemist, Aeronomy Lab, Nat Oceanic & Atomosperic Admin, US Dept Com, 75-83; RES CHEMIST, E I DU PONT DE NEMOURS & CO, INC, 83- *Mem:* Am Geophys Union. *Res:* Atmospheric Sciences. *Mailing Add:* 1385 Pakerville Rd Kennet Square PA 19348

MCFARLAND, RICHARD HERBERT, b Cleveland, Ohio, Jan 20, 29; m 50; c 3. ELECTRICAL ENGINEERING. *Educ:* Univ Ohio, BS, 50; Ohio State Univ, MS, 57, PhD(elec eng), 61. *Prof Exp:* From instr to asst prof elec eng, Ohio State Univ, 57-61; assoc prof, 62-69, PROF ELEC ENG & DIR AVIONICS RES, OHIO UNIV, 69- *Mem:* Inst Elec & Electronics Engrs; Am Soc Eng Educ; Am Inst Navig; Sigma Xi. *Res:* Air navigation, including systems for permitting aircraft to land under conditions of low ceiling and visibility. *Mailing Add:* Avionics Eng Ctr-Elec Eng 205 Clippinger Labs Athens OH 45701

MCFARLAND, ROBERT HAROLD, b Severy, Kans, Jan 10, 18; m 40; c 2. PHYSICS. *Educ:* Kans State Teachers Col Emporia, BS & BA, 40; Univ Wis, PhM, 43, PhD(physics), 47. *Prof Exp:* Instr high sch, Kans, 40-41 & Univ Wis, 43-44; sr engr, Sylvania Elec Prod, Inc, 44-46; from asst prof to prof physics, Kans State Univ, 47-60, dir nuclear lab, 58-60; physicist, Lawrence Livermore Lab, Univ Calif, 60-69; vpres acad affairs, Univ Mo, 74-75; dean grad sch, Univ Mo-Rolla, 69-79, prof, 69-85, dir Inst Anal & Planning, 79-82, prog dir, Off Fusion Energy, 82-84, EMER PROF PHYSICS, UNIV MO-ROLLA, 85- *Concurrent Pos:* Consult, Well Survs, Inc, Okla, 53-54; mem, Grad Rec Exam Bd, 71-75; Vis prof, Univ Calif, Berkeley, 80-81; Mendenhall fel, Univ Wis-Madison, 43; numerous grants related to research. *Mem:* Fel Am Phys Soc; Am Asn Physics Teachers; fel AAAS; Sigma Xi. *Res:* Atomic spectra; gaseous electronics; nuclear physics; effect of humidity on low pressure discharges; fluorescent and discharge spectra of mercury with zinc, thallium and indium metals; use of tracers; electron impact ionization of alkali metals; threshold polarization of helium radiation; resonance transfer and excitation; charge exchange. *Mailing Add:* Physics Dept Univ Mo Rolla MO 65401

MCFARLAND, WILLIAM D, b Sedalia, Mo, July 4, 45; m 68; c 3. DIGITAL IMAGE ANALYSIS, COMPUTER ENGINEERING. *Educ:* Univ Mo, BSEE, 68, MS, 71, PhD(elec eng), 73. *Prof Exp:* asst prof bioeng, 73-75, assoc prof elec eng, 75-81, ASSOC PROF RADIOL, MED SCH, UNIV MO, 75-, PROF ELEC ENG, 81- *Concurrent Pos:* Coordr, Elec Eng Dept, Comput Eng Div, Univ Mo-Columbia, 79-81; consult, McDonnell-Douglas Astronaut Corp, 81- *Mem:* Inst Elec & Electronic Engrs; Soc Photo-Optical Instrumentation Engrs. *Res:* Digital image analysis specifically in medical imaging and remote sensing; computerized tomography imaging; digital radiography; automated diagnosis from x-ray films; applied image analysis to landsat and digitized aerial photography. *Mailing Add:* Elec Eng 303 Univ Mo 5100 Rockhill Rd Columbia MO 65211

MCFARLAND, WILLIAM NORMAN, b Toronto, Ont, Sept 11, 25; nat US; wid; c 3. ZOOLOGY. *Educ:* Univ Calif, Los Angeles, BA, 51, MA, 53, PhD(zool), 59. *Prof Exp:* Chemist & biologist, Marineland of Pac, Calif, 54-57; marine biologist, State Game & Fish Comn, Univ Tex, 58; res scientist, Inst Marine Sci & lectr zool, Univ Tex, 58-61; from asst prof to assoc prof, 61-73, chmn, sect ecol & systs, 67-69, PROF ZOOL, CORNELL UNIV, 73-, CHMN, SECT ECOL & SYSTS, 83- *Mem:* AAAS; Am Soc Ichthyol & Herpet; Am Soc Zool; Am Fisheries Soc; Ecol Soc Am; Int Soc Reef Studies; Sigma Xi. *Res:* Fish physiology; comparative physiology and ecology. *Mailing Add:* 19522 Sierra Raton Irvine CA 92715

MACFARLANE, DONALD ROBERT, b Oshkosh, Wis, July 10, 30; m 62; c 2. NUCLEAR ENGINEERING, CHEMICAL ENGINEERING. *Educ:* Ill Inst Technol, BS, 52; Purdue Univ, MS, 57, PhD(nuclear eng), 66. *Prof Exp:* Chem engr, Sinclair Res Labs, Ill, 52-54; assoc chem engr, Argonne Nat Lab, 57-73; sr engr, Commonwealth Edison, Co, 73-74; vpres, Eta, Inc, 74-89; STAFF MEM, LOS ALAMOS NAT LAB, 89- *Concurrent Pos:* Prof, Midwest Col Eng, 71-; vis prof, Purdue Univ, 86-87. *Mem:* Am Nuclear Soc; Nat Soc Prof Engrs; Am Inst Chem Engrs. *Res:* Environmental impacts of nuclear power; nuclear reactor safety problems; numerical analysis. *Mailing Add:* 3205 Calle Celestial Santa Fe NM 87501

MACFARLANE, DUNCAN LEO, b Morristown, NJ, Nov 13, 62. LASERS. *Educ:* Brown Univ, ScB, 84, ScM, 85; Portland State Univ, PhD(elec eng), 89. *Prof Exp:* Sr scientist, W J Shafer Assocs, 85-86; lectr eng & physics, Portland State Univ, 87-89; ASST PROF ENG, UNIV TEX DALLAS, 89- *Mem:* Sigma Xi; Inst Elec & Electronic Engrs; Optical Soc Am. *Res:* Mode-locked lasers; laser amplifiers; laser instabilities; moving beams of light around in space and time. *Mailing Add:* Univ Tex Dallas BE 28 PO Box 830688 Richardson TX 75083

MCFARLANE, ELLEN SANDRA, b Halifax, NS, May 19, 38. ONCOLOGY, HERPES VIRUS GROUP. *Educ:* Dalhousie Univ, BSc, 59, MSc, 61, PhD(biochem), 63. *Prof Exp:* From asst prof to assoc prof & lectr biochem, Dalhousie Univ, 71-84, prof microbiol, 84-; RETIRED. *Concurrent Pos:* Nat Cancer Inst Can fel, Univ Man, 63-64; Med Res Coun Can fel, Harvard Univ, 64 & Dalhousie Univ, 65; Med Res Coun Can scholar microbiol, Dalhousie Univ, 66-71. *Mem:* Can Biochem Soc; Brit Biochem Soc; Can Microbiol Soc. *Res:* The study of viral oncology with particular reference to human tumors; herpes virus infections. *Mailing Add:* 5770 Ogilvie St Halifax NS B3H 1C2 Can

MCFARLANE, FINLEY EUGENE, b Atkins, Va, Nov 24, 40; m 64; c 2. POLYMER CHEMISTRY. *Educ:* King Col, AB, 63; Univ NC, PhD(chem), 68. *Prof Exp:* Sr res chemist, 67-75, RES ASSOC, TENN EASTMAN CO, 75- *Mem:* Am Chem Soc. *Res:* Polyester chemistry; catalysis, kinetics and mechanisms of polymerization and degradation reactions; polymer synthesis; preparation and properties of liquid-crystalline polyesters. *Mailing Add:* 1249 Eastbrook Dr Kingsport TN 37663

MCFARLANE, HAROLD FINLEY, b Hagerstown, Md, Apr 23, 45; m 68; c 1. NUCLEAR ENGINEERING, NEUTRON PHYSICS. *Educ:* Univ Tex, Austin, BS, 67; Calif Inst Technol, MS, 68, PhD(eng sci), 71. *Prof Exp:* Asst prof nuclear eng, NY Univ, 71-72; NUCLEAR ENGR, 72-84, ZPR PROG MGR, ARGONNE NAT LAB, 84- *Concurrent Pos:* Staff, Energy Res Adv Bd Nuclear Power Reactor Eval Panel, 88; Tech Prog Chmn, Int Reactor Phys Conf, 88. *Mem:* Am Nuclear Soc. *Res:* Physics of fast breeder reactors. *Mailing Add:* 3545 Sun Circle Idaho Falls ID 83404

MCFARLANE, JOANNA, b Montreal, Que, May 12, 61; m 85. PHYSICAL INORGANIC CHEMISTRY, CHEMICAL KINETICS. *Educ:* McGill Univ, BSc, 83; Univ Toronto, PhD(chem), 90. *Prof Exp:* RES SCIENTIST, AECL RES, WHITESHELL LABS, 89- *Mem:* Chem Inst Can; Can Soc Mass Spectros. *Res:* Study thermodynamics of chemical systems at high temperatures using mass spectrometric analysis of volatiles from a Knudsen cell; model kinetics of chemical reactions occurring in an electric discharge. *Mailing Add:* AECL Research Whiteshell Labs Pinawa MB R0E 1L0 Can

MCFARLANE, JOHN ELWOOD, b Tisdale, Sask, Aug 1, 29; m 60; c 3. INSECT PHYSIOLOGY. *Educ:* Univ Sask, BA, 49, MA, 51; Univ Ill, PhD(entom), 55. *Prof Exp:* From asst prof to assoc prof, 55-72, actg chmn, Dept Entom, 80-81, PROF ENTOM, 73-, CHMN, DEPT ENTOM, MCGILL UNIV, 86- *Mem:* fel Entom Soc Can; Sigma Xi. *Res:* Physiology of development and aggregation in insects. *Mailing Add:* Fac Agr Macdonald Campus 21 111 Lakeshore Rd Ste Anne de Bellevue PQ H9X 1C0 Can

MACFARLANE, JOHN O'DONNELL, medical microbiology, oncology; deceased, see previous edition for last biography

MCFARLANE, JOHN SPENCER, b Lkothair, Mont, Aug 26, 15; m 47. AGRONOMY, PHYTOPATHOLOGY. *Educ:* Mont State Col, BS, 38; Univ Wis, PhD(genetics), 43. *Prof Exp:* Asst, Univ Wis, 38-43, instr genetics & entom, 43-44; asst olericulturist, Univ Hawaii, 44-46; assoc geneticist, Curly Top Resistance Breeding Comt, Utah, 46-47; geneticist, Bur Plant Indust Soils & Agr Eng, USDA, Calif, 47-53, supt agr res sta, 64-72, leader sugar beet invests, 69-72, geneticist, Agr Res Serv, 53-82, location leader, Agr Res Sta & res leader sugar beet prod, 72-82; RETIRED. *Concurrent Pos:* Consult, 82-85. *Mem:* Am Phytopath Soc; Am Soc Sugar Beet Technol. *Res:* Genetics of sugar beet; breeding for disease resistance; breeding hybrid sugar beet varieties. *Mailing Add:* 37 Santa Ana Dr Salinas CA 93901

MACFARLANE, JOHN T, b Hamilton, Ont, Nov 23, 23; m 46; c 5. PHYSICS, THERMAL PHYSICS. *Educ:* McMaster Univ, BA, 44; Univ Montreal, MSc, 53. *Prof Exp:* Asst prof physics, Sir George Williams Univ, 45-51; lectr physics & math, Univ Col, Ethiopia, 51-54; res scientist, Can Armament Res & Develop Estab, Que, 54-58; assoc prof physics, Univ Col, Haile Sellassie, 58-63, dean sci, 59-63, prin, 61-63; prof physics, Univ Libya, 63-64; head dept, Univ Col Sci Educ, Ghana, 64-65; assoc prof, Mt Allison

Univ, 65-69; vrector, Nat Univ Rwanda, 69-72; assoc prof, 72-75, PROF PHYSICS, MT ALLISON UNIV, 75- *Concurrent Pos:* Asst lectr, Univ Montreal, 47-49; mem, Nat Comts Educ in Ethiopia, 60-63. *Mem:* Inst Elec & Electronics Engrs; Am Asn Physics Teachers. *Res:* Systems analysis; gas laser physics. *Mailing Add:* Dept Physics Mt Allison Univ Sackville NB E0A 3C0 Can

MCFARLANE, KENNETH WALTER, b Glasgow, Scotland, Mar 7, 37; div; c 2. PHYSICS, RESEARCH ADMINISTRATION. *Educ:* Glasgow Univ, BSc, 58; Univ Birmingham, PhD(high energy physics), 64. *Prof Exp:* Res investr high energy physics, Univ Pa, 63-66, asst prof, 66-69; assoc prof, 69-75, chmn dept, 70-75, PROF PHYSICS, TEMPLE UNIV, 75- *Concurrent Pos:* Prin investr, Acad Elec Contracting, Energy Res Develop Admin & Dept of Energy contracts & grants, 71- *Mem:* Am Phys Soc; Am Asn Physics Teachers; Am Asn Univ Profs. *Res:* Nucleon scattering and meson production; proton-antiproton interactions; K meson decays; pi meson decays; neutron interactions. *Mailing Add:* Dept Physics Barton Hall Temple Univ Philadelphia PA 19122

MACFARLANE, MALCOLM DAVID, b Cambridge, Mass, Sept 26, 40; m 75. CLINICAL PHARMACOLOGY, MEDICAL RESEARCH. *Educ:* New Eng Col Pharm, BS, 62; Georgetown Univ, PhD(pharmacol), 67. *Prof Exp:* Instr pharmacol, Kirksville Col, 67-69; asst prof, Univ Southern Calif, 69-74; dir res, Inst Res, Meyer Labs, 74-79; vpres prof serv, Glaxo Inc, NC, 79-81, dir regulatory affairs, 81-86, vpres, 86-89; VPRES REGULATORY AFFAIRS, GENENTECH INC, 89- *Concurrent Pos:* Consult, Kirksville Osteop Hosp & Still-Hildreth Hosp, 67-69, Rom-Amer Pharmaceuts, Ltd, LAC/USC Med Ctr, 69-74, Calif Dept Consumer Affairs, 72-74 & Superior Court of Calif, 72-74. *Mem:* Fel Am Col Clin Pharmacol; Am Pharmaceut Asn; Am Soc Pharmacol & Exp Therapeut; Sigma Xi; Am Soc Clin Pharmacol & Therapeut; Drug Info Soc; Pharmaceut Mfg Assoc. *Res:* Pharmacology of the autonomic nervous system; cardiovascular and renal pharmacology; microvascular renal physiology; narcotic analgesics and treatment of addiction; neuropsychopharmacology; clinical nutrition; clinical and geriatric pharmacology; regulation affairs. *Mailing Add:* Genentech Inc 460 Pt San Bruno Blvd South San Francisco CA 94080

MACFARLANE, MALCOLM HARRIS, b Brechin, Scotland, May 22, 33; m 57; c 4. NUCLEAR PHYSICS, THEORETICAL PHYSICS. *Educ:* Edinburgh Univ, MA, 55; Univ Rochester, PhD(physics), 59. *Prof Exp:* Res assoc physics, Argonne Nat Lab, 59-60, assoc scientist, 61-68, sr scientist physics, 68-80; PROF PHYS, IND UNIV, 80- *Concurrent Pos:* Fel, John Simon Guggenheim Found, 66-67; prof physics, Univ Chicago, 69-80; assoc ed, Phys Review Letters, 72-77. *Mem:* Fel Am Phys Soc. *Res:* Theory of nuclear reactions and nuclear structure. *Mailing Add:* Phys Dept Ind Univ Bloomington IN 47401

MACFARLANE, ROBERT, JR, b Brooklyn, NY, Aug 26, 30; m 52; c 5. PHYSICAL CHEMISTRY. *Educ:* Brown Univ, ScB, 52; Yale Univ, PhD(phys chem), 56. *Prof Exp:* Sr res chemist, Chem Div, US Rubber Co, 56-65; from res chemist to sr res chemist, Imp Oil Enterprises, Ltd, Esso Res & Eng Co, 65-74; group leader, 74-77, res supvr, 76-77, MGR QUAL ASSURANCE, ALLIED CHEM CORP, 77- *Concurrent Pos:* Lectr grad sch, Brooklyn Polytech Inst, 69-70. *Mem:* Am Chem Soc; Soc Rheol; NY Acad Sci; Soc Plastics Engrs; Sigma Xi. *Res:* Emulsion stability; controlled aglomeration of colloids; mechanics of liquid-liquid mixing; molecular structure versus physical properties of polymers; polymerization kinetics and mechanisms; thermally stable polymers; techniques of polymer characterization. *Mailing Add:* 25 Linda Pl Fanwood NJ 07023

MACFARLANE, ROGER MORTON, b Dunedin, NZ, Oct 25, 38; m 59; c 2. SOLID STATE PHYSICS. *Educ:* Univ Canterbury, BSc, 59, PhD(solid state physics), 64. *Prof Exp:* Asst lectr physics, Univ Canterbury, 64-65; res assoc, Stanford Univ, 65-68; STAFF MEM, IBM RES LAB, 68- *Concurrent Pos:* Sci Res Coun sr vis fel, Oxford Univ, 74. *Mem:* Am Phys Soc. *Res:* Laser spectroscopy and studies of phase transitions in organic and magnetic solids; biophysics. *Mailing Add:* IBM Res Lab Dept K32/802 650 Harry Rd San Jose CA 95120

MACFARLANE, RONALD DUNCAN, b Buffalo, NY, Feb 21, 33; m 56; c 2. BIOPHYSICAL CHEMISTRY. *Educ:* Univ Buffalo, BA, 54; Carnegie Inst Technol, MS, 57, PhD(chem), 59. *Prof Exp:* Res fel nuclear chem, Lawrence Radiation Lab, Univ Calif, Berkeley, 59-62; asst prof chem, McMaster Univ, 62-64, assoc prof, 65-67; PROF CHEM, TEX A&M UNIV, 67- *Concurrent Pos:* Guggenheim fel, 69-70. *Mem:* Am Chem Soc; Am Phys Soc. *Res:* Mass spectroscopy of biomolecules; pattern recognition. *Mailing Add:* Dept Chem Tex A&M Univ College Station TX 77840

MCFARLANE, ROSS ALEXANDER, b Toronto, Ont, Can, June 10, 31; m 60; c 2. PHYSICS. *Educ:* McMaster Univ, BSc, 53; McGill Univ, MSc, 55, PhD, 59. *Prof Exp:* Asst physics, Eaton Electronics Lab, McGill Univ, 59; mem staff, Res Lab Electronics, Mass Inst Technol, 59-61; mem staff, Bell Tel Labs, Inc, 61-69; assoc prof, Cornell Univ, 69-74, prof elec eng, 74-79; HEAD, QUANTUM ELECTRONICS SECT, HUGHES RES LABS, 79- *Mem:* fel Optical Soc Am; Inst Elec & Electronics Eng. *Res:* Microwave physics and frequency standards; noise and signal studies of electron beam type microwave devices; paramagnetic resonance; chemical lasers; reaction kinetics; atomic and molecular spectroscopy; nonlinear optics; photoacoustics; carrier dynamics in semiconductors. *Mailing Add:* Hughes Res Labs 3011 Malibu Canyon Rd Malibu CA 90265

MCFARLIN, DALE ELROY, MEDICINE, MUSCULAR & NEUROLOGICAL DISEASES. *Educ:* Vanderbilt Univ, MD, 61. *Prof Exp:* CHIEF NEUROL BR, NIH, 75- *Concurrent Pos:* Prof Neurol, George Wash Univ. *Res:* Multiple sclerosis. *Mailing Add:* NIH Bldg 10 Rm 5B16 9000 Rockville Pike Bethesda MD 20892

MCFARLIN, RICHARD FRANCIS, b Oklahoma City, Okla, Oct 12, 29; m 53; c 4. INORGANIC CHEMISTRY. *Educ:* Va Mil Inst, BS, 51; Purdue Univ, MS, 53, PhD, 56. *Prof Exp:* Sr res chemist, Monsanto Chem Co, Mo, 56-60; supvr inorg chem, Int Minerals & Chem Corp, 60-62; mgr Atlanta Res Ctr, Armour Agr Chem Co, 62-64, tech dir, 64-65; vpres & tech dir, Armour Agr Chem Co, 65-68; vpres com develop, 68-69 & develop & admin, 69-74, VPRES OPERS, USS AGRI-CHEM, INC, US STEEL CORP, 74-, VPRES PLANNING & ADMIN, 82- *Concurrent Pos:* Mem adv mgt prog, Columbia Univ, 67. *Mem:* AAAS; Am Chem Soc. *Res:* Nitrogen and industrial chemicals; industrial explosives; rocket oxidants; fertilizer materials; metal hydrides; inorganic phosphates. *Mailing Add:* 6611 Sweetbrian Lane Lakeland FL 33813-3598

MCFARREN, EARL FRANCIS, b Akron, Ohio, June 30, 19; m 45; c 3. ANALYTICAL CHEMISTRY. *Educ:* Bowling Green State Univ, BA, 41. *Prof Exp:* Asst chem, Bowling Green State Univ, 41-42; chemist, Nat Dairy Res Labs, Inc, Div Nat Dairy Prod Corp, 46-52 & Robert A Taft Sanit Eng Ctr, USPHS, 52-64; with training prog, Environ Control Admin, Environ Protection Agency, 64-69, chief anal qual control, Water Supply Prog Div, 69-72, chief water supply prog support activ, 72-76, chief distrib qual sect, Drinking Water Res Div, 76-78; RETIRED. *Concurrent Pos:* Consult, Dept Interior, Marshall Islands, 57. *Honors & Awards:* USPHS Awards, 58 & 63; US Environ Protection Agency Award, 78; Am Water Works Asn Award, 80. *Mem:* Am Chem Soc; Am Water Works Asn; Am Soc Testing & Mat; Sigma Xi. *Res:* Paper chromatography of amino acids and sugars; enzymatic digestion of casein; reactivation of normal alkaline milk phosphatase; bioassay and chemical assay of paralytic shellfish poison and poisonous fishes; gas chromatography of pesticides; chemical analysis of water; statistics; electron microscopy of asbestos in drinking water. *Mailing Add:* 304 Forestland Ct West Columbia SC 29169

MCFATE, KENNETH L(EVERNE), b LeClaire, Iowa, Feb 5, 24; m 54; c 3. AGRICULTURAL ENGINEERING, COMMUNICATIONS. *Educ:* Iowa State Univ, BS, 50; Univ Mo, Columbia, MS, 59. *Prof Exp:* Assoc agr engr, Iowa State Univ, 51-53, exten agr engr, 53-56; from asst prof to prof, 56-86, EMER PROF AGR ENG, UNIV MO-COLUMBIA, 86- *Concurrent Pos:* Dir, Mo Farm Electrification Coun, 56-76; vpres, Penreico, Inc, 69-76 & treas, 76-78; dir bd, Am Soc Agr Engrs, 70-72, mem, tech comt, pres & rep, Int Cong Agr Eng, Piacenza, Italy, 71; interim mgr, Nat Farm Elec Coun, 75, exec mgr, 76-77; exec mgr, Nat Food & Energy Coun, 76-86, pres, 86-; coun mem, Agr Sci & Technol, 77; mem, food indust adv comt, US Dept Energy, 79; deleg leader, People-to-People Deleg, People's Repub China, 83; dir bd, Int Comn Agr Engrs, 89. *Honors & Awards:* George Kable Electrification Award, 74; Safety Award, Nat Safety Coun, 75; Citation for Outstanding Serv to 4-H, Nat 4-H Coun, 82. *Mem:* Fel Am Soc Agr Engrs. *Mailing Add:* 409 Vandiver W Suite 202 Nat Food & Energy Coun Inc Columbia MO 65202

MCFEAT, TOM FARRAR SCOTT, b Montreal, Que, Feb 5, 19; m 47; c 2. ETHNOLOGY. *Educ:* McGill Univ, BA, 50; Harvard Univ, AM, 54, PhD, 57. *Prof Exp:* Assoc prof anthrop, Univ NB, 54-59; sr ethnologist, Nat Mus Can, 59-63; sr ethnologist, Carleton Univ, 63-64; chmn dept, Univ Toronto, 64-74, prof anthrop, 64-84. *Mem:* Fel Am Anthrop Asn. *Res:* Culture process, particularly concepts of growth, evolution and pattern in diachronic analysis; small group culture, especially influence of information on structure of n-generation groups; Canadian Indian and other ethnic communities; certain aspects of mass media analysis. *Mailing Add:* Dept Anthrop Univ Toronto Scarborough Col Scarborough ON M1C 1A4 Can

MCFEDRIES, ROBERT, JR, b Chicago, Ill, Nov 11, 30; m 52; c 2. CHEMICAL ENGINEERING, POLYMER SCIENCE. *Educ:* Purdue Univ, BS, 52, MS, 56. *Prof Exp:* Develop engr, 55-58, sect head, Plastics & Packaging Div, 58-62, div leader packaging res, 62-65, lab dir flexible packaging, 65-67, tech mgr, 67-72, dir res & develop, Packaging Dept, 72-73, exec vpres, Dow Chem Invest & Finance Co, 73-76, dir res & develop, Designed Prod Dept, 79-81, dir licensing, 81-83, DIR MERGERS & ACQUISTIONS, DOW CHEM CORP, 83-, PORTFOLIO INVESTMENTS, 85-, INVESTOR RELATIONS, 86-, VPRES EXEC DEPT, 87- *Concurrent Pos:* Bd mem, Liana Ltd, 87-, Dorinco Reinsurganie, 86-, United Ageiseeds, 87-, Magma Power, 87- *Mem:* Am Chem Soc; Am Soc Soc Testing & Mat; Soc Plastics Engrs; Am Inst Chem Engrs. *Res:* Epoxy modification of alkyd resins; effects of radiation on polymers; styrene foam systems; ion exchange resins. *Mailing Add:* 651 Bering Dr No 1605 Houston TX 77057-2133

MCFEE, ALFRED FRANK, b Knoxville, Tenn, Aug 19, 31; div; c 2. CYTOGENETICS, RADIOBIOLOGY. *Educ:* Univ Tenn, BS, 53, MS, 59; Cornell Univ, PhD(animal breeding), 63. *Prof Exp:* From asst prof to assoc prof, Agr Res Lab, Univ Tenn-AEC, 63-75, prof, comp animal res lab, Univ Tenn-Dept Energy, 75-81; SCIENTIST, OAK RIDGE ASSOC UNIVS, 81-; PRIN INVESTR, NAT TOXICOL PROG. *Mem:* Radiation Res Soc Am; Environ Mutagen Soc. *Res:* Adverse effects of energy-related pollutants on chromosome structure and cell behavior as they relate to embryonic survival; genotoxic evaluation of chemicals. *Mailing Add:* Med Health Sci Div Oak Ridge Asn Univ PO Box 117 Oak Ridge TN 37831

MCFEE, ARTHUR STORER, b Portland, Maine, May 1, 32; m 67. SURGERY. *Educ:* Harvard Univ, BA, 53, MD, 57; Univ Minn, MS, 66, PhD(surg), 67; Am Bd Surg, dipl, 67. *Prof Exp:* Intern surg, Univ Minn Hosp, 57-58, spec fel surg, 58-65; from asst prof to assoc prof, 67-73, PROF SURG, UNIV TEX HEALTH SCI CTR, SAN ANTONIO, 74- *Concurrent Pos:* Co-dir surg intensive care unit, Bexar County Hosp, San Antonio, 68- *Mem:* Asn Acad Surgeons; Asn Hist Med; AMA; Am Col Surgeons; Soc Surg Alimentary Tract. *Res:* Local gastric hypothermia; gastric physiology; prevention of hepatic metastatic diseases; emergency medicine and transportation. *Mailing Add:* Dept Surg Health Sci Ctr Univ Tex 7703 Floyd Curl Dr San Antonio TX 78284

MCFEE, DONALD RAY, b Union Co, Ind, July 4, 29; m 54; c 3. INDUSTRIAL HYGIENE & ENVIRONMENTAL HEALTH & SAFETY, TECHNOLOGY & MANAGEMENT. *Educ:* Purdue Univ, BSc, 51; Univ Cincinnati, MSc, 60, DrSc(indust health), 62; Am Bd Indust Hyg, cert comprehensive pract indust hyg, 64; Bd Cert Safety Prof, cert, 71; Bd Cert Prod Safety Mgrs, cert, 81. *Prof Exp:* Tool engr, Boeing Airplane Co, Wash, 51-53, facil engr, 53-54, sr facil engr, Pilotless Aircraft Div, 58; assoc indust hygienist, Indust Hyg & Safety Div, Argonne Nat Lab, 61-72, supvr indust hyg group, 71-72; VPRES, OCCUSAFE, INC, WHEELING, 72- *Concurrent Pos:* Lectr, Nat Safety Coun, 63-71. *Mem:* Am Indust Hyg Asn (vpres, 77, pres elect, 78, pres, 79); Am Acad Indust Hyg; Am Indust Hyg Found (treas, 80, vpres, 81, pres, 82); Am Soc Safety Engrs. *Res:* Industrial and environmental health and safety engineering and management, including industrial hygiene; air pollution; ventilation, hazard analysis and engineered control; government regulations; product safety; dusts and fume characteristics; mineral and man made fibers; toxicology of organic solvents; noise, air sampling; air cleaning and filtration; combustible gas detection systems. *Mailing Add:* 25 W 210 Highview Dr Naperville IL 60540

MCFEE, RAYMOND HERBERT, b Somerville, Mass, Mar 1, 16; m 38; c 2. SOLID STATE PHYSICS. *Educ:* Mass Inst Technol, SB, 37, SM, 38, PhD(physics), 43. *Prof Exp:* Tech asst, Geophys Res Corp, Okla, 38-40; asst physics, Mass Inst Technol, 41-42, physicist, Div Indust Coop, 42-43; chief physicist, White Res Assocs, Boston, 43-45; physicist, Cambridge Thermionic Corp, 45-46; sr elec engr, Submarine Signal Co, 46; res physicist, Electronics Corp of Am, 46-53, dir eng, 53-54, res, 54-56; dir res avionics div, Aerojet-Gen Corp, 56-60 & Azusa Plant, 60-64; sect mgr, Jet Propulsion Lab, Calif Inst Technol, 64-67; assoc dir, Douglas Advan Res Labs, McDonnell Douglas Corp, Calif, 67-70, sr staff engr, McDonnell Douglas Astronaut Co, 70-73, prin engr/scientist, 73-81; CONSULT, 81- *Mem:* AAAS; Am Phys Soc; fel Optical Soc Am; assoc fel Am Inst Aeronaut & Astronaut. *Res:* Solid state physics; semiconductors; spectroscopy; electronics; optical system design; purification of materials by recrystallization from the melt; infrared systems and techniques; space science; optics of solar power systems. *Mailing Add:* 5163 Belmez Laguna Hills CA 92653-1810

MCFEE, RICHARD, b Pittsburgh, Pa, Jan 24, 25. ELECTRICAL ENGINEERING. *Educ:* Yale Univ, BE, 47; Syracuse Univ, MS, 49; Univ Mich, PhD(elec eng), 55. *Prof Exp:* Instr elec eng, Syracuse Univ, 48-49; res assoc, Hosp, Univ Mich, 49-51; mem tech staff, Bell Tel Labs, Inc, 52-57; from assoc prof to prof elec eng, Syracuse Univ, 57-82; INDEPENDENT RESEACHER, 82- *Mem:* Inst Elec & Electronic Engrs; AAAS; Sigma Xi. *Res:* Biophysics; linear and nonlinear systems; device development; cryogenics. *Mailing Add:* PO Box 989 Kauai HI 96755

MCFEE, WILLIAM WARREN, b Concord, Tenn, Jan 8, 35; m 57; c 3. FOREST SOILS, SOIL FERTILITY. *Educ:* Univ Tenn, BS, 57; Cornell Univ, MS, 63, PhD(soils), 66. *Prof Exp:* Asst soils, Cornell Univ, 61-65; from asst prof to assoc prof, 65-73, PROF AGRON, PURDUE UNIV, WEST LAFAYETTE, 73-, DIR NATURAL RESOURCE & ENVIRON SCI PROG, 75- *Concurrent Pos:* Vis prof, Univ Fla, 86-87. *Honors & Awards:* Educ Award, Soil Sci Soc Am, 87. *Mem:* Fel Soil Sci Am; fel Am Soc Agron; Sigma Xi; Soc Am Foresters; Int Soil Sci Soc; Soil Sci Soc Am (pres, 91-92). *Res:* Relationship of soils to plant nutrition; mechanisms of ion uptake; forest tree-site relationships; mined land reclamation; effects of atmospheric deposition on soils; forest tree species; common agricultural plants. *Mailing Add:* Dept Agron Purdue Univ West Lafayette IN 47906

MCFEELEY, JAMES CALVIN, b Altoona, Pa, Aug 6, 40; m 63; c 2. BOTANY. *Educ:* Otterbein Col, BS, 65; Ohio Univ, MS, 68; Ohio State Univ, PhD(plant path), 71. *Prof Exp:* Teacher, High Sch, Ohio, 65-66; from asst to assoc prof biol, East Tex State Univ, 72-89, asst dean grad sch, 81-89; RETIRED. *Concurrent Pos:* Res assoc, Purdue Univ, 71-72; consult, Tulsa Dist, US Corps Engrs, 73 & Tex Hwy Dept, 73- *Mem:* Am Phytopath Soc; Am Bot Soc; Mycol Soc Am; Sigma Xi. *Res:* Study of the effect of strong 60-hertz electric fields on biological systems. *Mailing Add:* 1116 Briarwood Dr Commerce TX 75428

MCFEELY, RICHARD AUBREY, b Trenton, NJ, Dec 3, 33; div; c 3. VETERINARY MEDICINE, CYTOGENETICS. *Educ:* Pa State Univ, BS, 55; Univ Pa, VMD, 61, MS, 67; Am Col Theriogenologists, dipl, 73. *Prof Exp:* Pvt pract, 61-62; res asst comp cardiol, Sch Vet Med, 62, USPHS fels reprod physiol, 62-63, grad div, Sch Med, 63-65 & cytogenetics, Sch Vet Med, 65-66, from asst prof to assoc prof clin reprod, 66-75, chief, Sect Reprod, 68-73, chief of staff, 73-76, PROF CLIN REPROD, SCH VET MED, UNIV PA, 75-, ASSOC DEAN, 76- *Concurrent Pos:* NIH res grant, 64-; Lalor fel award, 71-72. *Mem:* Am Vet Med Asn; Am Asn Vet Clinicians; Soc Theriogenol. *Res:* Chromosome abnormalities in domestic mammals; especially sex determination and altered reproductive function. *Mailing Add:* 428 Bartram Rd Kennett Square PA 19348

MCFERON, D(EAN) E(ARL), b Portland, Ore, Dec 24, 23; m 45; c 4. MECHANICAL ENGINEERING, HEAT TRANSFER. *Educ:* Univ Colo, BSME, 45, MSME, 48; Univ Ill, PhD(mech eng), 56. *Prof Exp:* Instr mech eng, Univ Colo, 46-48; from instr to assoc prof, Univ Ill, 48-58; prof, 58-82, EMER PROF MECH ENG, UNIV WASH, 82- *Concurrent Pos:* Res fel, NSF-Atomic Energy Comn, Argonne Nat Lab, 56, res assoc, 57-58; NSF fac fel, 67-68. *Honors & Awards:* Am Soc Heating, Refrig & Air Conditioning; Soc Mfg Engrs. *Mem:* Am Soc Mech Engrs; Am Soc Eng Educ; Sigma Xi (pres, 78). *Res:* Heat transfer and fluid flow; power plant cooling systems; thermodynamics. *Mailing Add:* Dept Mech Eng FU-10 Univ Wash Seattle WA 98195

MCFETERS, GORDON ALWYN, b Ayer, Mar 28, 39; m 63; c 2. MICROBIAL PHYSIOLOGY, WATER MICROBIOLOGY. *Educ:* Andrews Univ, BA, 61; Loma Linda Univ, MS, 63; Ore State Univ, PhD(microbiol), 67. *Prof Exp:* Asst prof, 67-72, assoc prof, 72-78, PROF MICROBIOL, MONT STATE UNIV, 78- *Concurrent Pos:* Sigma Xi fac res award, Mont State Univ. *Honors & Awards:* Wiley Award, Am Water Works Asn, 84. *Mem:* Am Soc Microbiol; fel Am Acad Microbiol. *Res:* Water microbiology; environmental microbiology. *Mailing Add:* Dept Microbiol Mont State Univ Bozeman MT 59717

MCGAHREN, WILLIAM JAMES, b Ballyshannon, Ireland, Feb 16, 24; US citizen; m 58; c 4. ORGANIC CHEMISTRY. *Educ:* Chelsea Polytech Col, BSc, 53; Brooklyn Col, MA, 57; Brooklyn Polytech Inst, PhD(org chem), 66. *Prof Exp:* Chemist, Charles Pfizer Co, 53-63; res fel chem, Brooklyn Polytech Inst, 64-66; PRIN SCIENTIST NATURAL PROD CHEM, LEDERLE LABS, AM CYANAMID CO, PEARL RIVER, NY, 66- *Mem:* Am Chem Soc; Sigma Xi; AAAS. *Res:* Isolation from microbial sources of novel products that exhibit a specific biological activity; elucidation of structure and sterochemistry of these materials; microbial enzyme transformation of substrates of commercial interest; protein chemistry. *Mailing Add:* 64 Glenwood Ave Demarest NJ 07627

MCGANDY, EDWARD LEWIS, physical chemistry, computer and control systems; deceased, see previous edition for last biography

MCGANITY, WILLIAM JAMES, b Kitchener, Ont, Sept 21, 23; nat US; m 48; c 3. OBSTETRICS & GYNECOLOGY. *Educ:* Univ Toronto, MD, 46; FRCS(C), 53; Am Bd Nutrit, dipl. *Prof Exp:* Intern, Toronto Gen Hosp, Ont, Can, 46-47; resident obstet & gynec, Univ Toronto & Toronto Gen Hosp, 49-52; from instr to assoc prof, Sch Med, Vanderbilt Univ, 52-59; PROF OBSTET & GYNEC & CHMN DEPT, UNIV TEX MED BR, GALVESTON, 60- *Concurrent Pos:* Res fel nutrit, Univ Toronto, 47-48; res fel, Vanderbilt Univ, 48-49; Lowell M Palmer sr fel, 54-56; part time lectr, Univ Toronto, 49-52; consult, Interdept Comt Nutrit Nat Defense, DC, 56-73, Dept Air Force, 60-, Dept Army, 60-67 & Comt Consider Folic Acid, Food & Drug Admin, 60-63; mem, Comts Dietary Allowances & Int Nutrit, Food & Nutrit Bd, Nat Res Coun, 59-63; dean fac, Univ Tex Med Br, Galveston, 64-67; mem comt maternal nutrit, 66-70; chmn, Comt Maternal Health & Sci Activ, Tex Med Asn, 61-67, Coun Foods & Nutrit, AMA, 63-66, Nutrit Study Sect, NIH, 63-67 & Res Panel Maternal & Child Health, Children's Bur, 64-66; panel mem, White House Conf Nutrit & Health, 69; co-chmn panel nutrit & health, US Senate Select Comt Nutrit & Human Needs, 74; chmn, Family Self-Support Serv Br Adv Coun, Tex Dept Human Resources; chmn, Nutrit & Fitness Adv Coun, State of Tex; consult, Diag & Therapeut Technol Assessment Prog, AMA, 86-; chmn, Task Force on Nutrit, Am Col Obstet & Gynec. *Honors & Awards:* Hendry Prize, Univ Toronto, 47; Agnes Higgins Award, Outstanding Achievement field of Maternal & Fetal Nutrit, March of Dimes Birth Defects Found, 81. *Mem:* Am Col Obstet & Gynec; Am Inst Nutritnt; AMA; Am Soc Clin Nutrit (past pres); Asn Prof Gynec & Obstet. *Res:* Nutrition in reproduction; nutrition among underdeveloped populations; physiology of human reproduction. *Mailing Add:* Dept of Obstet & Gynec Univ of Tex Med Br Galveston TX 77550

MCGANN, LOCKSLEY EARL, b Kingston, Jamaica, Aug 11, 46; Can citizen; m 69; c 3. CRYOBIOLOGY, BIOPHYSICS. *Educ:* Univ Waterloo, BSc, 69, MSc, 70, PhD(physics), 73. *Prof Exp:* Med Res Coun Can fel, Div Cryobiol, Clin Res Ctr, Harrow, Eng, 73-75; from asst prof to assoc prof biomed eng, 75-80, assoc prof path, 80-88, PROF PATH, UNIV ALTA, 88- *Concurrent Pos:* Consult comput applications. *Mem:* Soc Cryobiol; Soc Anal Cytol. *Res:* Interactions of living systems with the environment during cooling to and warming from low temperatures; development of methods for the frozen preservation of cells, tissues and organs; osmotic properties of cells and tissues; osmotic water and solute movements in cells and tissues. *Mailing Add:* Dept Path Univ Alta Edmonton AB T6G 2R7 Can

MCGARITY, ARTHUR EDWIN, b Chicago, Ill, Apr 2, 51; m 77; c 1. SOLAR ENERGY, OPERATIONS RESEARCH. *Educ:* Trinity Univ, BS, 73; John Hopkins Univ, MSE & PhD(environ eng), 78. *Prof Exp:* Elec engr instrumentation, San Antonio City Pub Serv Bd, 73-74; ASSOC PROF ENG, SWARTHMORE COL, 78- *Concurrent Pos:* Consult, NSF, 75-76; res assoc, Donnovan, Hammester, Rattien, Inc, Washington, DC, 77-78; vis asst prof, G W C Whiting Sch Eng, Johns Hopkins Univ, 81; scientist in residence, Solar Energy Group, Argonne Nat Lab, 81-82; vis fel, Ctr Energy & Environ Studies, Princeton Univ. *Mem:* Int Solar Energy Soc; Inst Mgt Sci; Sigma Xi. *Res:* Solar energy engineering and economics; mathematical programing applications on arms control; operations research applied to public sector problems involving the supply of energy, water and other natural resources. *Mailing Add:* Dept Eng Swarthmore Col 135 Rutgers Ave Swarthmore PA 19081

MCGARITY, WILLIAM CECIL, b Jersey, Ga, Oct 5, 21; m 50; c 3. SURGERY. *Educ:* Emory Univ, BA, 42, MD, 45. *Prof Exp:* Instr, 51-54, assoc, 54-59, from asst prof to assoc prof, 59-76, PROF SURG, SCH MED, EMORY UNIV, 76- *Concurrent Pos:* Mem, Am Bd Surg. *Mem:* AMA; Am Heart Asn; fel Am Col Surg. *Res:* Burns treated with steroids; transminase values in biliary tract and liver pathology; prognosis and course of dogs following litigation of common duct; arterial emboli. *Mailing Add:* Dept Surg Emory Univ Med Sch 1365 Clifton Rd NE Atlanta GA 30322

MCGARR, ARTHUR, b San Francisco, Calif, May 24, 40; m 71; c 1. GEOPHYSICS. *Educ:* Calif Inst Technol, BS, 62, MS, 63; Columbia Univ, PhD(geophys), 68. *Prof Exp:* Res asst, Lamont Geol Observ, Columbia Univ, 63-68; sr res fel, Univ Witwatersrand, 68-70, sr res officer geophys, Bernard Price Inst Geophys Res, 70-78; GEOPHYSICIST, US GEOL SURV, 78- *Concurrent Pos:* Assoc ed, j Geog Res, 82-85; vis prof, Univ Paris, 81. *Mem:* AAAS; Am Geophys Union; Seismol Soc Am. *Res:* Seismology; tectonophysics. *Mailing Add:* 3666 La Calle Ct Palo Alto CA 94306

MCGARRITY, GERARD JOHN, b Brooklyn, NY, Oct 10, 40; m 64; c 2. MICROBIOLOGY, CELL BIOLOGY. *Educ:* St Joseph Col, Pa, BS, 62; Jefferson Med Col, MS, 65, PhD(microbiol), 70. *Prof Exp:* Res assoc cell biol, 65-71, vpres sci affairs, 84-85, HEAD DEPT MICROBIOL, INST MED

RES, 71-, ACTG PRES, 85-, PRES, 86- *Concurrent Pos:* Instr Univ Pa, 71-75; exchange visitor, Czechoslovak Acad Sci, 75; adj prof, Thomas Jefferson Univ, Philadelphia, Pa, 84-; chair, Recombinant Adv Comt, NIH, 84-85 & 87-91; team leader, Int Res Prog Comp Mycoplasmology, Int Org Mycoplasmology, 84-; bd dirs, Vitro Cell Biol & Biotechnology Prog, State Univ NY, Plattsburgh, 85-, Thomas Jefferson Univ, Philadelphia, 89-, Inst Coop Environ Mgt, 90- *Mem:* AAAS; Am Soc Microbiol; Tissue Culture Asn (vpres, 80-82, pres, 82-84); Int Org Mycoplasmology; Int Asn Cell Culture (pres, 87); Am Asn Cancer Res. *Res:* Mycoplasma infection of cell cultures; detection of environmental mutagens; mutagenicity testing; cell culture facilities. *Mailing Add:* Coriell Inst Med Res 401 Haddon Ave Camden NJ 08103

MCGARRY, FREDERICK J(EROME), b Rutland, Vt, Aug 22, 27; m 50; c 6. MATERIALS ENGINEERING, POLYMER SCIENCE. *Educ:* Middlebury Col, AB, 50; Mass Inst Technol, SB, 50, SM, 53. *Prof Exp:* Res asst civil eng, 50-51, from instr to assoc prof, 51-65, PROF CIVIL ENG, MASS INST TECHNOL, 65- *Concurrent Pos:* Consult, numerous indust orgn. *Mem:* AAAS; Am Soc Testing & Mat; Am Concrete Inst; Soc Rheol; Am Chem Soc. *Res:* Relationship between the mechanical properties and the structure and composition of polymers; reinforced polymers; adhesion. *Mailing Add:* Rm 8-211 Dept Mat Sci & Eng Mass Inst Technol Cambridge MA 02139

MCGARRY, JOHN DENIS, b Dec 6, 40; m 67; c 3. BIOCHEMISTRY. *Educ:* Univ Manchester, BSc, 62, PhD(biochem), 66. *Prof Exp:* From asst prof to assoc prof, 69-77, PROF INTERNAL MED & BIOCHEM, UNIV TEX HEALTH SCI CTR DALLAS, 77- *Concurrent Pos:* Fels, Univs Liverpool & Wales, 65-67; fel, Univ Tex Southwestern Med Sch Dallas, 68-69; sub-ed, J Lipid Res, 73-77; mem, Comt Res, Am Diabetes Asn, 77-79; mem, Nat Med Sci Res & Adv Comt, Juv Diabetes Fedn, 79-84; spec reviewer, Metab Study Sect, NIH, 79-81, mem, 81-85, chmn, 84-85; assoc ed, Diabetes, 79-83 & Am J Physiol, 82-88; overseas lectr to Australian Biochem Soc, 81; consult-reviewer, NIH progs & Vet Admin progs; Metab-Nutrit Study Sect, Med Res Coun Can, 90- *Honors & Awards:* Lilly Award, Am Diabetes Asn, 78; Jacobaeus Lectr, Nordisk Insulin Found, 86; David Rumbough Sci Award, Juv Diabetes Found, 87; Joslin Medal, 87. *Mem:* Am Diabetes Asn; Brit Biochem Soc; Am Soc Biol Chem. *Res:* Mechanism of control hepatic ketogenesis-ketosis of starvation and uncontrolled diabetes; hormonal and metabolic derangements in diabetic ketoacidosis; regulation of hepatic carbohydrate and lipid metabolism in various nutritional states. *Mailing Add:* Dept of Internal Med Univ of Tex Health Sci Ctr 5323 Harry Hines Blvd Dallas TX 75235-8858

MCGARRY, MARGARET, b Boston, Mass, Apr 11, 28. INORGANIC CHEMISTRY, ANALYTICAL CHEMISTRY. *Educ:* Regis Col, Mass, AB, 57; Univ Pa, PhD(chem), 64. *Prof Exp:* From instr to asst prof, 64-70, ASSOC PROF CHEM, REGIS COL, MASS, 70- *Mem:* Am Chem Soc; Sigma Xi. *Res:* Liability of aqueous solutions of alkali silicates; spectral properties of dyes in colloidal systems; flocculation of colloids; electron microscopy of colloidal substances; ion exchange separations; gas chromatography. *Mailing Add:* 235 Wellesley St Weston MA 02193

MCGARRY, MICHAEL P, b Buffalo, NY, Oct 30, 42; m 64; c 7. EXPERIMENTAL HEMATOLOGY, ANIMAL LAB RESOURCES. *Educ:* Canisius Col, Buffalo, NY, BS, 64; Purdue Univ, PhD(develop biol), 69. *Prof Exp:* Sr cancer res scientist, Dept Biol Resources, Lab Exp Hemat & Leukemogenesis, Springville Labs, 70-75, assoc cancer res scientist, Large Animal Resource Facil, 75-77, cancer res scientist V, 77-82, cancer res scientist VI, 82-87; from asst res prof to assoc res prof, 71-78, RES PROF, PHYSIOL DEPT, ROSWELL PARK DIV, STATE UNIV NY, BUFFALO, 78-, HEAD, DEPT CANCER RES, 85- *Concurrent Pos:* Lab teaching asst, Dept Biol Sci, Purdue Univ, 64-66; guest lectr, Div Exp Biol, Baylor Col Med, 68-70; clin instr, Dept Med Technol, Sch Health Related Prof, State Univ NY, Buffalo, 72-75, chmn, Dept PhD Preliminary Exam Comt, Physiol Dept, Roswell Park Div, 73-85, dir, Grad Studies, 74-78, rep, Grad Sch Div Comt, 75-79; adj asst prof, Biol Dept, Bowling Green State Univ, 73-75; admin asst, Dept Viral Leukemia, Roswell Park Cancer Inst, 73-75, mem, Biohazard Control Comt, 76-81, asst dir, Animal Serv, 80-85; mem, Inst Animal Care & Use Comt, 76-, chmn, 80-88; assoc ed, Exp Hemat, 79-; spec consult, Am Asn Accreditation of Lab Animal Care, 83-87, Vet Admin Spec Grant Rev, 87, 88 & 89; external consult, Prog Proj, Div Hemat & Oncol, Dept Med, Univ Va; dir, Dept Lab Animal Resources Planning Comt, Vivarium Component, Inst Major Modernization, 85- *Mem:* Int Soc Exp Hemat; AAAS; Am Asn Cancer Res; Am Soc Hemat; Am Asn Lab Animal Sci; Am Asn Immunologists; NY Acad Sci; Cell Kinetics Soc; Am Asn Exp Biol. *Res:* Biochemical defects in a series of single gene pigment mutations; megakaryocyte-platelet granules in storage pool disease; differentiation and the rate of proliferation of eosinophil granulocyte progenitors; spontaneous murine models for human malignant disease; role of eosinophil granulocytes in cell-mediated immune destruction of tumors; author of various publications. *Mailing Add:* Lab Animal Resources Roswell Park Cancer Inst Carleton & Elm St Buffalo NY 14263

MCGARRY, PAUL ANTHONY, b Warren, Pa, Nov 13, 28; m 55; c 3. PATHOLOGY, NEUROPATHOLOGY. *Educ:* Pa State Univ, BS, 50; Temple Univ, MD, 54; Am Bd Path, dipl & cert anat & clin path, 63, cert neuropath, 68, cert forensic path, 83. *Prof Exp:* From instr to assoc prof, 63-72, PROF PATH, SCH MED, LA STATE UNIV, 72- *Concurrent Pos:* Asst vis pathologist, Charity Hosp, New Orleans, La, 63-65, vis pathologist, 65-; vis pathologist, New Orleans Vet Admin, La, 68- *Mem:* Am Asn Neuropath. *Res:* Cerebrovascular disease; cervical spondylotic myelopathy; cerebrospinal fluid cytology; pathology of head, neck, and back injuries. *Mailing Add:* 43 Stilt St New Orleans LA 70124

MCGARVEY, BRUCE RITCHIE, b Springfield, Mo, Mar 10, 28; m 54; c 3. PHYSICAL CHEMISTRY. *Educ:* Carleton Col, BA, 50; Univ Ill, MA, 51, PhD(chem), 53. *Prof Exp:* From instr to asst prof chem, Univ Calif, 53-57; from asst prof to assoc prof, Kalamazoo Col, 57-62; from assoc prof to prof, Polytech Inst Brooklyn, 62-72, actg head dept, 71-72; PROF CHEM, UNIV WINDSOR, 72- *Concurrent Pos:* Guggenheim fel, Imp Col, Univ London, 67-68. *Mem:* Am Chem Soc; Am Phys Soc; fel Chem Inst Can. *Res:* Nuclear magnetic and electron spin resonance. *Mailing Add:* Dept Chem & Biochem Univ Windsor Windsor ON N9B 3P4 Can

MCGARVEY, FRANCIS X(AVIER), b Kingston, NY, Mar 16, 19; m 41; c 4. CHEMICAL ENGINEERING. *Educ:* Univ Pa, BS, 41, MS, 44. *Prof Exp:* Engr, US Navy, NJ, 41-44; phys chemist, Manhattan Proj, Los Alamos Sci Lab, 44-45; sr chemist res dept, Rohm & Haas Co, 45-58, tech consult, Foreign Opers Div, 58-65; dir res & develop, Barnstead Still & Sterilizer Co, 65-66; vpres, Hartung Assocs, 66-68; pres, Puricons Inc, 68-76; MGR TECH CTR, SYBRON CHEM CO, 76- *Mem:* Am Chem Soc; Electrochem Soc; Am Water Works Asn; Am Inst Chem Engrs. *Res:* Technical development of the use of ion exchange materials in industry, study of their role in nuclear reactor development, uranium recovery, sugar processing, desalination, water treatment and radioactivity waste disposal. *Mailing Add:* Sybron Chem Co Birmingham NJ 08011

MCGARY, CARL T, b Pittsburgh, Pa, Oct 24, 61. PATHOLOGY, CELL BIOLOGY. *Educ:* Univ Pa, BA, 83; Pa State Univ, MD, 87; Univ Tex, PhD(cell biol), 90. *Prof Exp:* RES PHYSICIAN PATH, DEPT PATH, MILTON S HERSHEY MED CTR, COL MED, PA STATE UNIV, 90- *Mem:* Am Soc Cell Biol. *Mailing Add:* Dept Path Milton S Hershey Med Ctr Col Med Pa State Univ Hershey PA 17033

MCGARY, CHARLES WESLEY, JR, b New Castle, Pa, Dec 12, 29; m 49; c 5. BIODEGRADABLE STARCH POLYMERS. *Educ:* Westminster Col, Pa, BS, 51; Purdue Univ, PhD(phys org chem), 55. *Prof Exp:* Res chemist, Chem Div, Union Carbide Corp, 54-61, group leader, 61-66, asst dir chem & plastics div, 66-73, prod mgr, 73-77, int mkt mgr, 77-80; dir res & develop, Riverain Corp, 80-82; dir res & develop, Deseret Polymer Res Div, Warner-Lambert Co,82-86; vpres res & develop, Riverain Corp, 86-90; DIR RES & DEVELOP, NOVON PRODS DIV, WARNER-LAMBERT CO, 90- *Mem:* Am Chem Soc; Soc Plastics Eng. *Res:* Plastics intermediates; epoxy resin; polyesters; plasticizers; vinyl resins; latex paints; coatings; water soluble polymers; polyurethane; biomaterials; sealants and adhesives; degradable starch polymers. *Mailing Add:* Warner Lambert Co 182 Tabor Rd Morris Plains NJ 07950

MCGAUGH, JAMES L, b Long Beach, Calif, Dec 17, 31; m 52; c 3. PSYCHOBIOLOGY, NEUROBIOLOGY. *Educ:* San Jose State Univ, BA, 53; Univ Calif, Berkeley, PhD(psychol), 59. *Prof Exp:* From asst prof to assoc prof, San Jose State Univ, 57-61; Nat Res Coun sr fel physiol psychol, Adv Inst Health, Italy, 61-62; assoc prof psychol, Univ Ore, 62-64; chmn dept psychobiol, Univ Calif, Irvine, 64-67, 71-74 & 86-89, dean, Sch Biol Sci, 67-70, vchancellor acad affairs, 75-77, exec vchancellor, 78-82, PROF PSYCHOBIOL, UNIV CALIF, IRVINE, 66-, DIR, CTR NEUROBIOL LEARNING & MEMORY, 83-- *Concurrent Pos:* Mem biol sci training rev comt, Nat Inst Ment Health, chmn, 71-72, mem preclin psychopharmacol res rev comt, 75-78, mem, Nat Acad Sci Comt, aging, 78-79, W Clement Stone Found, 79-86; consult, Vet Admin; ed, Behav & Neural Biol, 72-; mem, External Adv Rev Comt, Beckman Inst Advan Sci & Technol, Univ Ill, 90-94; William James fel, Am Psychol Soc. *Honors & Awards:* Distinguished Sci Contrib Award, Am Psychol Asn, 81. *Mem:* Nat Acad Sci; Int Brain Res Orgn; Soc Neurosci; Am Psychol Soc (pres, 89-91); Am Col Neuropsychopharmacol; Sigma Xi; fel AAAS; Psychonomic Soc. *Res:* Biological bases of behavior; neurobiology of learning and memory. *Mailing Add:* Ctr Neurobiol of Learning & Memory Univ of Calif Irvine CA 92717

MCGAUGH, JOHN WESLEY, b Garden City, Kans, June 26, 38; m 64; c 3. ANIMAL SCIENCE. *Educ:* Colo State Univ, BS, 66; Univ Ky, MS, 68, PhD(reprod physiol), 71. *Prof Exp:* Res asst, Univ Ky, 66-71; dist sales mgr, Am Breeders Serv Inc, 71-75; asst prof, 75-80, ASSOC PROF AGR, FT HAYS STATE UNIV, 80- *Mem:* Am Soc Animal Sci; Sigma Xi. *Mailing Add:* PO Box 828 Chadron NE 69337

MCGAUGHAN, HENRY S(TOCKWELL), b Philadelphia, Pa, Nov 5, 17; m 41; c 3. ELECTRICAL ENGINEERING. *Educ:* Univ Mich, BS, 41; Cornell Univ, MEE, 49. *Prof Exp:* From jr engr to sub-sect chief, US Naval Ord Lab, 41-47; from instr to prof, 47-83, EMER PROF ELEC ENG, CORNELL UNIV, 83- *Concurrent Pos:* Consult, US Naval Ord Lab, 55-56; vis prof, Cornell Aeronaut Lab, Cornell Univ, 57-58; tech assistance expert, Int Telecommun Union, Inst Electronics, Taiwan, 62-63. *Mem:* Inst Elec & Electronics Engrs. *Res:* Underwater sound measurements and analysis; electronic instrumentation in radio astronomy; statistical studies of integrated defense systems; electronic countermeasures in communications and radar. *Mailing Add:* Dept Elec Eng Cornell Univ Phillips Hall Ithaca NY 14850

MCGAUGHEY, CHARLES GILBERT, b San Diego, Calif, Sept 8, 25. BIOCHEMISTRY, ORAL DISEASES RESEARCH. *Educ:* Univ Calif, BA, 50; Univ Southern Calif, MA, 52. *Prof Exp:* Asst biochem, Univ Southern Calif, 51-52; radiol biochemist, US Naval Radiol Defense Lab, 52; res biochemist med, Vet Admin Hosp, Long Beach, Calif, 52-63, res biochemist, Oral Dis Res Lab, 63-81; RETIRED. *Mem:* AAAS. *Res:* Mechanism of formation of dental plaque and calculus; biochemistry and physiology of cancer cell; mechanisms of tumor promotion; effects of polyphosphates on parameters related to dental caries and calculus; mammalian bioassay for tumor initiation and promotion. *Mailing Add:* 337 Winnipeg Pl Long Beach CA 90814

MCGAVIN, MATTHEW DONALD, b Goondiwindi, Australia, July 25, 30; m 61; c 3. VETERINARY PATHOLOGY. *Educ:* Univ Queensland, BVSc, 52, MVSc, 62; Mich State Univ, PhD(path), 64; Am Col Vet Path, dipl, 63. *Prof Exp:* Vet off, Animal Res Inst, Yeerongpilly, Australia, 52-59, sr histopathologist, 59-61 & 64-68; from assoc prof to prof path, Kans State Univ, 68-76; PROF PATH, UNIV TENN, 76- *Concurrent Pos:* Assoc ed, Vet Path, 86-88, ed, 88- *Mem:* Am Vet Med Asn; Am Col Vet Path; Australian Vet Asn. *Res:* Comparative myopathies. *Mailing Add:* Dept Path Univ Tenn Knoxville TN 37901

MCGAVIN, RAYMOND E, physics; deceased, see previous edition for last biography

MCGAVOCK, WALTER DONALD, b Nashville, Tenn, Apr 18, 33; m 79; c 2. ECOLOGICAL PARISITOLOGY. *Educ:* Mid Tenn State Univ, BS, 56, MA, 58; Univ Tenn, PhD(zool), 67. *Prof Exp:* Actg chmn dept biol, Limestone Col, 58-62; asst prof, Wofford Col, 62-64; instr & res asst zool, Univ Tenn, 66-67; assoc prof biol, ETenn State Univ, 67-70; prof, Tusculum Col, 70-75; assoc prof, ETenn State Univ, 75-80; CHMN, DIV NATURAL SCI, TUSCULUM COL, 81- *Concurrent Pos:* Lectr, Spartanburg Gen Hosp, SC, 62-64; AEC res fel, 66. *Mem:* Am Soc Parasitol; Sigma Xi. *Res:* Host-parasite relationships in fish; life cycles, cytology and taxonomy of caryophyllidea; effects of yomesan on bile duct tapeworm, H microstoma; effects of ultraviolet radiation on H diminuta eggs and cysticercoids. *Mailing Add:* Dept Biol Tusculum Col PO Box 5048 Greeneville TN 37743

MCGAVRAN, MALCOLM HOWARD, b Harda, India, Oct 18, 29; US citizen; c 5. PATHOLOGY. *Educ:* Bethany Col, WVa, AB, 51; Wash Univ, MD, 54; Am Bd Path, dipl, 59. *Prof Exp:* From instr to assoc prof path, Med Sch, Wash Univ, 57-70; prof path, Hershey Col Med, Pa State Univ & dir anat path, Hershey Med Ctr Hosp, 70-75; PROF PATH & DIR DIV ANAT PATH, BAYLOR COL MED, 75-; CHIEF ANAT PATH, THE METHODIST HOSP, 75- *Concurrent Pos:* Consult, Vet Admin Hosp, Houston. *Mem:* AAAS; Am Asn Path; Int Acad Path. *Res:* Surgical pathology; oncology; electron microscopy. *Mailing Add:* 1515 Halcombe Houston TX 77030

MCGEACHIN, ROBERT LORIMER, b Pasadena, Calif, May 13, 17; m 47; c 2. BIOCHEMISTRY. *Educ:* Univ Nebr, BS, 39, MS, 40; Wash Univ, PhD(org chem), 42. *Prof Exp:* Asst chem, Univ Nebr, 39-40; asst, Univ Ill, 46-47; from asst prof to assoc prof, 47-66, prof, 66-85, EMER PROF BIOCHEM, SCH MED, UNIV LOUISVILLE, 85- *Concurrent Pos:* USPHS fel, Univ Ill, 46-47; chmn dept biochem, Sch Med, Univ Louisville, 72-76. *Mem:* Am Chem Soc (treas, Div Biol Chem, 63-66); Soc Exp Biol & Med; Am Soc Biol Chem. *Res:* Biochemistry and physiology of mammlian amylases. *Mailing Add:* Univ of Louisville Sch of Med Louisville KY 40292

MCGEADY, LEON JOSEPH, b Freemansburg, Pa, July 5, 21; m 49; c 6. PHYSICAL METALLURGY. *Educ:* Lehigh Univ, BS, 43, MS, 46, PhD, 50. *Prof Exp:* Sci investr, Lehigh Univ, 43-49; from asst prof to assoc prof, 49-57, dir eng, 75-86, PROF METALL ENG, LAFAYETTE COL, 57- *Concurrent Pos:* consult, Naval Res Lab, Washington, DC, 64-74. *Mem:* Am Soc Metals; Am Welding Soc; Am Soc Eng Educ; Am Inst Mining, Metall & Petrol Engrs. *Res:* Welding; internal friction in metals; fracture and plastic deformation. *Mailing Add:* Dept of Metall Eng Dana Hall of Eng Lafayette Col Easton PA 18042

MCGEAN, THOMAS J, b New York, NY, Apr 8, 37; m 62; c 2. INNOVATIVE TRANSIT SYSTEMS & EQUIPMENT, AUTOMATED PEOPLE MOVERS. *Educ:* New York Univ, BS, 59; Calif Inst Technol, MS, 60. *Prof Exp:* Hughes fel, Hughes Aircraft Co, 59-60; mem tech staff, Bell Telephone Labs, 60-66, Computer Sci Corp, 66-69, Mitre Corp, 69-74; dir technol, Deleuw Cather & Co, 74-75; partner, Lea Elliott McGean, 81-87; vpres, ND Lea & Assoc, 84-85; PRIN, THOMAS J MCGEAN PE, 75- *Concurrent Pos:* Assoc prof lectr, George Washington Univ, 72-75; vis prof, Howard Univ, 73-74; panel mem, Off Technol Assessment, Automated Guideway Transit Assessment, 74-75. *Honors & Awards:* David Orr Prize in Mech Eng, 59; William R Bryan Medal in Eng, 59. *Mem:* Am Soc Civil Engrs; fel Inst Transp Engrs; Am Soc Mech Engrs; Am Pub Transit Asn; Transp Res Bd. *Res:* Planning and introduction into revenue service of innovative transit systems and equipment; first linear induction motor transit system at Duke Univ; first automated people mover in major downtown area-Miami Metromover. *Mailing Add:* 3711 Spicewood Dr Annandale VA 22003

MCGEARY, DAVID F R, b Bellefonte, Pa, Dec 23, 40; m 67; c 2. GEOLOGY, OCEANOGRAPHY. *Educ:* Williams Col, BA, 62; Univ Ill, Urbana, 64; Scripps Inst Oceanog, PhD(oceanog), 69. *Prof Exp:* From asst prof to assoc prof, 69-79, PROF GEOL, CALIF STATE UNIV, SACRAMENTO, 79- *Mem:* Sigma Xi; Geol Soc Am; Soc Econ Paleontologists & Mineralogists; Am Asn Petrol Geologists. *Res:* Marine geology. *Mailing Add:* 1840 Arroyo Vista Way Folsom CA 95630

MCGEE, CHARLES E, b Baylor Co, Tex, July 29, 35; m 60; c 1. ANALYTICAL CHEMISTRY, HEALTH PHYSICS. *Educ:* ETex State Univ, BS, 62; Purdue Univ, MS, 66, PhD(bionucleonics), 68. *Prof Exp:* Res asst physiol, Southwestern Med Sch, Univ Tex, 60-62; from asst prof to assoc prof radiochem, Sch Pharm, Temple Univ, 67-74, proj dir, Radiol Health Specialist Training Proj, 68-74; mgr health physics, 74-78, mgr consult & lab serv, 78-80, dir mkt, 80-81, DIR LABS, RADIATION SAFETY OFFICER, TECH AUDITS, RADIATION MGT CORP, 81- *Concurrent Pos:* Qual assurance, Instrument Calibration-Radiation. *Mem:* Am Chem Soc; Health Physics Soc. Environmental health; health physics, environmental radiochemistry. *Res:* Radiobioassay; analytical chemistry section. *Mailing Add:* Radiation Mgt Consultants 5301 Tacony St Box 208 Philadelphia PA 19137

MCGEE, HENRY A(LEXANDER), JR, b Atlanta, Ga, Sept 12, 29; m 51; c 3. PHYSICAL CHEMISTRY. *Educ:* Ga Inst Technol, BChE, 51, PhD(chem eng), 55. *Prof Exp:* Res asst, Theoret Chem Lab, & Sch Chem Eng, Univ Wis, 55; instr, Huntsville Ctr, Univ Ala, 56-59; from assoc prof to prof chem eng, Ga Inst Technol, 59-71; head dept, 71-81, PROF CHEM ENG, VA POLYTECH INST & STATE UNIV, 71- *Concurrent Pos:* lectr, US & Europe; res chemist, Army Rocket & Guided Missile Agency, Ala, 56-58; phys scientist, Army Ballistic Missile Agency, 59 & George C Marshall Space Flight, Ctr, NASA, 59. *Mem:* Fel AAAS; Am Chem Soc; fel Am Inst Chem Engrs; Sigma Xi. *Res:* Cryogenics, especially low temperature chemistry and related techniques; experimental and theoretical kinetics and thermodynamics; chemical and physical properties of matter at extremes of temperature and pressure; laser applications in chemical processing. *Mailing Add:* Va Polytech Inst & State Univ Blacksburg VA 24061

MCGEE, JAMES PATRICK, b New York, NY, Dec 27, 41; m 66; c 2. ECONOMIC SYSTEMS AS FACTORS OF PRODUCTION, INFORMATION INTEGRATING ECONOMIC SYSTEMS. *Educ:* City Univ New York, BBA, 65, MBA, 72; Polytech NY, MS, 78; Southwestern Univ, PhD(computer sci), 82; Walden Univ, PhD(info sci), 84. *Prof Exp:* Programmer, US Army, 65-68; sr programmer, Shell Oil Co, 68-70; staff mem, I/S CHQ, IBM Corp, 70-78, mgr I/S advan technol, 78-82, mgr systs anal, IS&TC, 83-85, res staff mem, T J Watson Res Ctr, 86-90, SR SYSTS ANALYST, IBM ENTERPRISE SYSTS, 90- *Concurrent Pos:* Adj lectr, Grad Div, Baruch Col, City Univ New York, 74-76; mem bd dirs, White Pond Community Ctr, 75-76, Int Technol Inst, 85-88 & Technol Transfer Soc, 86-88; mem, X3 DBSSG, Am Nat Standards Inst, 85- *Mem:* Inst Elec & Electronic Engrs; Asn Comput Mach; Am Asn Artificial Intel; Am Soc Qual Control; Data Processing Mgt Asn; Int Cert Computer Prof. *Res:* Inter-disciplinary approach to representation and using large scale enterprise system models in terms of knowledge engineering, software engineering, system theory and economics. *Mailing Add:* IBM Enterprise Systs RHQ IS & TC Rte 100 Somers NY 10589

MACGEE, JOSEPH, b Edinburgh, Scotland, Nov 24, 25; nat US; m 47, 81; c 1. BIOCHEMISTRY, ANALYTICAL CHEMISTRY. *Educ:* Univ Calif, PhD(comp biochem), 54. *Prof Exp:* Lab technician res & develop, Merck & Co, Inc, 44-50; asst bact, Univ Calif, 52-53; res analysis biochemist, Procter & Gamble Co, 55-61; from asst prof to assoc prof exp biol, Col Med, Univ Cincinnati, 61-77, asst prof biol chem, 61-70, assoc prof biol chem, 70-88 prof exp med, 77-88; res biochemist, Vet Admin Hosp, 61-88; RETIRED. *Concurrent Pos:* USPHS fel, Univ Ill, 54-55. *Res:* Analytical biochemistry. *Mailing Add:* 9469 Treetop Lane Cincinnati OH 45247-2331

MCGEE, LAWRENCE RAY, b Salt Lake City, Utah, Oct 29, 52; m 77; c 3. CHEMISTRY. *Educ:* Univ Utah, BA, 74; Calif Inst Technol, PhD(chem), 82. *Prof Exp:* Res chemist, Cent Res & Develop, E I Du Pont de Nemours & Co, Inc, 81-89; SCIENTIST, BIOORGANIC GROUP, GENENTECH, 89- *Concurrent Pos:* Moderator, Nat FidoNet Sci Conf. *Mem:* Am Chem Soc. *Res:* Natural product synthesis with emphasis on pharmacologically active, novel structural types; development of new synthetic methods; use of organometallic centers for stereo control in organic synthesis. *Mailing Add:* One Crater Lake St Pacifica CA 94044-4441

MCGEE, THOMAS DONALD, b Tripoli, Iowa, June 9, 25; m 48, 81; c 4. BIOMEDICAL ENGINEERING. *Educ:* Iowa State Univ, BSc, 48, MSc, 58, PhD(ceramic eng, metall), 61. *Prof Exp:* Res engr, A P Green Refractories Co, Mo, 48-54, res eng supvr, 54-56; from asst prof to assoc prof, 61-65, PROF CERAMIC ENG, IOWA STATE UNIV, 65- *Honors & Awards:* Greaves-Walker Award, Nat Inst Ceramic Eng. *Mem:* Fel Am Ceramic Soc; Nat Inst Ceramic Eng; Am Soc Eng Educ; Brit Soc Glass Technol. *Res:* Refractories; glass; ceramic bone implants. *Mailing Add:* Dept Mats Sci & End Iowa State Univ Ames IA 50011

MCGEE, THOMAS HOWARD, b New York, NY, Dec 19, 41; m 66; c 3. CHEMICAL KINETICS, PHOTOCHEMISTRY. *Educ:* St John's Univ, NY, BS, 62; Univ Conn, PhD(phys chem), 66. *Prof Exp:* Fel chem, Univ Tex, Austin, 66-67; from asst prof to assoc prof chem, 74-82, chmn dept natural sci, 76-79, PROF CHEM, YORK COL, NY, 83- *Concurrent Pos:* NSF acad year award, 68-69; res collabr, Chem Dept, Brookhaven Nat Lab, 75-77 & 83-84. *Mem:* Am Chem Soc. *Res:* Chemical kinetics; photochemistry. *Mailing Add:* Glenby Lane Brookville NY 11545

MCGEE, WILLIAM F, b Toronto, Ont, Jan 16, 37; m 62; c 5. ELECTRICAL ENGINEERING. *Educ:* Univ Toronto, BASc, 59, MA, 60; Univ Ill, PhD(elec eng), 62. *Prof Exp:* Asst prof elec eng, Univ Waterloo, 63-66; MGR, BELL NORTHERN RES, 66- *Mem:* Inst Elec & Electronics Engrs. *Res:* Approximation problems in electric network synthesis; digital transmission system design. *Mailing Add:* Dept Elec Eng Univ Ottawa 770 King Edward Ave Ottawa ON K1N 6N5 Can

MCGEE, WILLIAM WALTER, b Toledo, Ohio, June 27, 39; m 63; c 1. ANALYTICAL CHEMISTRY. *Educ:* Univ Toledo, BS, 61, MS, 63; Univ Fla, PhD(chem), 66. *Prof Exp:* Group leader gas chromatography lab, B F Goodrich Co, 66-68; from asst prof to assoc prof anal chem, 68-77, ASSOC PROF FORENSIC SCI, UNIV CENT FLA, 77- *Mem:* Am Chem Soc. *Res:* Fundamental processes which occur in the flame, as used in flame spectroscopy. *Mailing Add:* Forensic Sci Teaching Lab Univ Cent Fla PO Box 25000 Orlando FL 32816

MCGEER, EDITH GRAEF, b New York, NY, Nov 18, 23; m 54; c 3. ORGANIC CHEMISTRY. *Educ:* Swarthmore Col, BA, 44; Univ Va, PhD(org chem), 46. *Hon Degrees:* DSc, Univ Victoria, BC, 87. *Prof Exp:* Technician, Squibb Inst Med Res, E R Squibb & Sons, 43; res chemist, Exp Sta, E I du Pont de Nemours & Co, Inc, 46-54; res assoc, 54-74, from assoc prof to prof, 74-90, head, 83-89, EMER PROF, KINSMEN LAB NEUROL RES, UNIV BC, 90- *Mem:* Can Biochem Soc; Int Neurochem Soc; Am Neurochem Soc; Soc Neurosci; Can Fedn Biol Sci; Sigma Xi. *Res:* Synaptic transmission; neurochemistry and anatomy; neurochemisty and pathology. *Mailing Add:* Kinsmen Lab Neurol Res Univ BC Vancouver BC V6T 1W5 Can

MCGEER, JAMES PETER, b Vancouver, BC, May 14, 22; m 48; c 4. PHYSICAL CHEMISTRY, EXTRACTIVE METALLURGY. *Educ:* Univ BC, BA, 44, MA, 46; Princeton Univ, MA, 48, PhD(chem), 49. *Prof Exp:* Res phys chemist & group leader, Aluminum Labs, Ltd, 49-68; tech supt, Aluminum Co Can Ltd, 68-70, asst mgr reduction div, Arvida Works & mgr reduction technol, Quebec Smelters, Alcan, 70-73, mgr reduction technol, Alcan Smelters Serv, 73-77, dir res, 77-83, dir, Kingston Res & Develop Ctr, Alcan Int Ltd, 83-87; MANAGING DIR, ONTARIO CTR MAT RES, 88- *Concurrent Pos:* Lectr, Can Coun Am Soc Metals, 85-86; vchmn, adv comt, Mat Res Inst Can; chmn, Can Univ Indust Coun Advan Ceramics; distinguished lectr, Can Inst Mining & Metall, 87, Can Res Mgt Asn, 91. *Mem:* Metall Soc; Chem Inst Can; Can Inst Mining & Metall; Can Res Mgt Asn. *Res:* Electrometallurgy of aluminum; carbon; smelting furnace operation; hydrometallurgy of bauxite. *Mailing Add:* 1149 Front Rd Kingston ON K7M 4M2 Can

MCGEER, PATRICK L, b Vancouver, BC, Can, June 29, 27; m 54; c 3. BIOCHEMISTRY, IMMUNOLOGY. *Educ:* Univ BC, BA, 48, MD, 58; Princeton Univ, PhD(phys chem), 51; FRCP(C). *Hon Degrees:* LLD, British Columbia Open Univ. *Prof Exp:* Res chemist, Polychem Dept, E I du Pont de Nemours & Co, 51-54; res assoc, 56-58, from asst prof to assoc prof, 59-74, PROF, KINSMEN LAB NEUROL RES, UNIV BC, 86- *Concurrent Pos:* Intern Vancouver Gen Hosp, 58-59; dir, BC Hydro, BC Petrol Corp, Discovery Found, Discovery Park11Inc & BC Res Coun; minister educ, BC, 74-77, sci & technol, 77-79, univ sci & commun, 79-86 & Int Trade & Investr including Sci & Commun. *Honors & Awards:* Nat Res Award, Can Mental Health Assoc, 60. *Mem:* Fel AAAS; Soc Neurosci; Can Biochem Soc; Am Neurochem Soc; Int Brain Res Orgn; Int Soc Neurochem; Sigma Xi. *Res:* Medical biochemistry; neurochemistry; neurophysiology; immunology. *Mailing Add:* Kinsmen Lab Neurol Res Univ BC Vancouver BC V6T 1W5 Can

MCGEE-RUSSELL, SAMUEL M, b Sutton, Eng, Aug 24, 27; m 55; c 3. BIOLOGY, ELECTRON MICROSCOPY. *Educ:* Oxford Univ, BA, 51, MA, 54, DPhil(zool), 55. *Prof Exp:* Asst lectr zool, Birkbeck Col, Univ London, 54-56; vis scientist fel, NIH, 56-57; lectr zool, Birkbeck Col, Univ London, 57-61; lab dir & staff scientist, Virus Res Unit, Med Res Coun, 62-68; sr res assoc & lectr cell biol, 68-72, PROF CELL BIOL & ELECTRON MICROS, STATE UNIV NY ALBANY, 72- *Concurrent Pos:* Dir, Co of Biologists, 67- *Mem:* Fel Zool Soc London; fel Royal Micros Soc; Brit Soc Exp Biol (secy, 62-66); Am Soc Cell Biol; Electron Micros Soc Am. *Res:* Light microscopy; microtechnique and microtomy; cell biology; histochemistry; protozoology; invertebrate zoology especially of mollusca; cell motility. *Mailing Add:* Dept Biol Sci State Univ NY 1400 Washington Ave Albany NY 12222

MCGEHEE, OSCAR CARRUTH, b Baton Rouge, La, Nov 29, 39; m 89; c 1. MATHEMATICS. *Educ:* Rice Univ, BA, 61; Yale Univ, MA, 63, PhD(math), 66. *Prof Exp:* From instr to asst prof math, Univ Calif, Berkeley, 65-71; assoc prof, 71-79, chmn, 79-84, dean, Div Acad Serv, 86-90, PROF MATH, LA STATE UNIV, BATON ROUGE, 79- *Concurrent Pos:* NATO fel, Fac Sci, Orsay, France, 67-68. *Mem:* Am Math Soc; Math Asn Am; Math Soc France; Sigma Xi; AAAS. *Res:* Commutative harmonic analysis; functional analysis. *Mailing Add:* Dept Math La State Univ 301 Lockett Hall Baton Rouge LA 70803

MCGEHEE, RALPH MARSHALL, b Magnolia, Miss, Apr 23, 21; m 45; c 1. APPLIED MATHEMATICS, NUMERICAL ANALYSIS. *Educ:* La Col, BA, 42; NC State Col, BEE, 49, MS, 50, PhD(elec eng), 53. *Prof Exp:* Staff mem, Sandia Lab, US AEC, 53-61; mem tech staff, Tucson Eng Lab, Hughes Aircraft Co, 61-62; assoc prof math, 62-66, head dept comput sci, 66-67, from assoc prof to prof, 66-84, EMER PROF COMPUT SCI, NMEX INST MINING & TECHNOL, 84- *Mem:* Inst Elec & Electronics Eng; Asn Comput Mach; Soc Indust & Appl Math; Opers Res Soc Am; Am Meteorol Soc. *Res:* Applied analysis; numerical analysis of modeling; atmospheric dynamics on thunderstorm scales. *Mailing Add:* Dept of Comput Sci NMex Inst of Mining & Technol Socorro NM 87801

MCGEHEE, RICHARD PAUL, b San Diego, Calif, Sept 20, 43; m 67, 81; c 2. MATHEMATICS. *Educ:* Calif Inst Technol, BS, 64; Univ Wis-Madison, MS, 65, PhD(math), 69. *Prof Exp:* Vis mem, Courant Inst Math Sci, NY Univ, 69-70; from asst prof to assoc prof, 70-79, PROF MATH, UNIV MINN, MINNEAPOLIS, 79-; HEAD OF DEPT. *Mem:* Am Math Soc; Math Asn Am. *Res:* Dynamical systems. *Mailing Add:* Sch Math Univ Minn Minneapolis MN 55455

MCGEHEE, RICHARD VERNON, b Tyler, Tex, Aug 1, 34; m 58; c 2. GEOLOGY. *Educ:* Univ Tex, BS, 55, PhD(geol), 63; Yale Univ, MS, 56, Tex A&M, MS, 78. *Prof Exp:* Petrol geologist, Phillips Petrol Co, 56-57; instr geol, Univ Kans, 63; asst prof, Western Mich Univ, 63-66 & SDak Sch Mines & Technol, 66-67; assoc prof, Western Mich Univ, 67-72; vis prof geol, Inst Geol, Nat Univ Mex, 72-74; from assoc prof to prof geol, Div Earth & Phys Sci, Univ Tex, San Antonio,74-79; assoc prof, 78-81, PROF HEALTH & PHYS EDUC SOUTHEASTERN LA UNIV, 81- *Concurrent Pos:* Fulbright prof geol & phys educ, Univ Liberia, 82-83, Fulbright prof phys educ, Nat Teachers Col, Honduras, 88. *Mem:* Am Alliance Health,. *Res:* History of sport in Latin America. *Mailing Add:* Box 677 Southeastern La Univ Hammond LA 70402

MCGEORGE, A(RTHUR), JR, b Wilmington, Del, July 6, 21; m 52; c 3. CHEMICAL ENGINEERING. *Educ:* Univ Pa, BS, 42, PhD(chem eng), 51. *Prof Exp:* Jr chemist, E I du Pont de Nemours & Co, 42-44; develop engr, B F Goodrich Chem Co, 44-47; asst instr chem eng, Univ Pa, 47-50; res engr nylon res, 51-52, res supvr, 53-57, sr res supvr, Dacron Res Lab, 57-58, tech supvr, 58-65, TECH MGR DEVELOP DEPT, NEWPORT LAB, E I DU PONT DE NEMOURS & CO, INC, 65- *Mem:* Am Chem Soc; Soc Rheol; Sigma Xi; Am Inst Chem Engrs. *Res:* Fluid flow; rheology; heat transfer; molecular structure of polymers; design of experiments. *Mailing Add:* 28 Brandywine Falls Rd Wilmington DE 19806-1002

MCGERITY, JOSEPH LOEHR, b New York, NY, Sept 16, 28; m 65; c 2. ALLERGY, IMMUNOLOGY. *Educ:* Duke Univ, BS, 50, MD, 54. *Prof Exp:* Resident internal med, Letterman Gen Hosp, 59-62; fel allergy-immunol, Univ Buffalo, 65-67; assoc prof med & dir allergy serv, Chest Div, Dept Med, Univ Calif, San Francisco, 75-83. *Concurrent Pos:* Med mem, US Army Phys Disability Agency, 73-; assoc clin prof med, 74-88, clin prof med, Stanford Univ, 88- *Mem:* Am Acad Allergy; Asn Mil Allergists (pres, 77). *Res:* Detection of clinical allergens by botanical and immunologic investigation; influence of allergic disease and therapy thereof on physiology of pregnancy. *Mailing Add:* Allergy Div Chest Clin 536-A 400 Parnasus Ave San Francisco CA 94143

MCGERVEY, JOHN DONALD, b Pittsburgh, Pa, Aug 9, 31; m 57; c 3. EXPERIMENTAL SOLID STATE PHYSICS. *Educ:* Univ Pittsburgh, BS, 52; Carnegie Inst Technol, MS, 55, PhD(physics), 61. *Prof Exp:* Instr math, Carnegie Inst Technol, 57-60; from asst prof to assoc prof, 60-78, PROF PHYSICS, CASE WESTERN RESERVE UNIV, 78- *Concurrent Pos:* Vis scientist, Inst Solid State Physics & Nuclear Res, W Ger, 72-73; Sci Res Coun vis fel, Univ East Anglia, Norwich, England, 78-79; vis scientist, Univ der Bundeswehr, Munich, WGer, 88; bd mem, Cleveland Collab Sci Educ. *Mem:* AAAS; Am Phys Soc; Am Asn Physics Teachers; Am Asn Univ Prof; Sigma Xi; Fedn Am Scientists. *Res:* Study of materials by positron annihilation, including studies of fatigue and radiation damage in metals, and physical aging in polymers. *Mailing Add:* Dept Physics Case Western Reserve Univ Cleveland OH 44106

MCGHAN, WILLIAM FREDERICK, b Sacramento, Calif, July 6, 46; m; c 4. PHARMACY ADMINISTRATION, HEALTH SERVICES & ECONOMICS. *Educ:* Univ Calif, PharmD, 70; Univ Minn, Minneapolis, PhD(pharm admin), 79. *Prof Exp:* Resident, Med Ctr, Univ Calif, San Francisco, 70-71; pharm coordr, Appalachian Student Health Proj, 69-70; staff dir, Student Am Pharm Asn, 71-74 & staff dir, Acad Pharmaceut Sci, 74-76; instr pharm, Univ Minn, 76-78; Asst Prof Pharm, Sch Pharm, Univ Southern Calif, 78-82; from assoc prof to prof pharmaceut pract, Univ Ariz, 82-89; PROF & CHMN, DEPT PHARM PRACT & ADMIN, PHILADELPHIA COL PHARMACEUT & SCI, 89-, EXEC DIR, INST PHARMACEUT ECON, 89- *Concurrent Pos:* Dir, Nat Student Drug Educ grant, 71-73; fel, Am Found Pharmaceut Educ, 76-78; consult, Nat Comn Protection Human Subjects, 76-78 & Vet Admin San Diego Pharm Servs, 78-; mem comt, Nat Drug Educ Asn, 74-75; pres, Acad Pharm Res & Sci, 88. *Honors & Awards:* Archambault Award, Am Soc Consult Pharmacists, 87; Lyman Award, Am Asn Col Pharm, 90. *Mem:* Am Pharmaceut Asn; Am Asn Pharmaceut Sci; Am Pub Health Asn; Am Soc Hosp Pharmacists; Sigma Xi. *Res:* Administrative and social pharmacy; evaluation of health services; drug utilization review; health services research; cost effectiveness analysis; decision support systems. *Mailing Add:* Philadelphia Col Pharm & Sci Woodland Ave at 43rd St Philadelphia PA 19104-4495

MCGHEE, CHARLES ROBERT, b Chattanooga, Tenn, July 17, 34; m 64; c 3. SYSTEMATIC ZOOLOGY. *Educ:* Mid Tenn State Univ, BS, 61, MA, 62; Va Polytech Inst, PhD(zool), 70. *Prof Exp:* From asst prof to assoc prof, 68-79, PROF BIOL,MID TENN STATE UNIV, 79- *Concurrent Pos:* Instr entom & invertebrate zool, Upper Cumberland Biol Sta. *Mem:* Am Arachnol Soc; Sigma Xi. *Res:* Systematics of phalangid genus Leiobunum; collection identification and systematic revision of the genus leiobunum in the eastern United States; new species have been described and phyletic groups have been established. *Mailing Add:* Dept Biol Mid Tenn State Univ PO Box 280 Murfreesboro TN 37130

MCGHEE, GEORGE RUFUS, JR, b Henderson, NC, Sept 25, 51; m 71. PALEOECOLOGY, EVOLUTIONARY MORPHOLOGY. *Educ:* NC State Univ, BSc, 73; Univ NC, Chapel Hill, MSc, 75; Univ Rochester, PhD(geol), 78. *Prof Exp:* asst prof, 78-83, ASSOC PROF GEOL & ECOL, RUTGERS UNIV, 83- *Concurrent Pos:* Sci asst, Univ Tübingen, 77; vis scientist, Field Mus Natural Hist, 81; prin investr, NSF Grant, Earthsci Div, 81-85, Petrol Res Fund, 82-86; res assoc, Am Mus Natural Hist, 82-85; Gastdozent, Univ Tübingen, 82 & 83, guest prof, 84; invited mem, Inst Marine & Coastal Sci, Rutgers Univ, 90-; US chmn, Int Geol Correlation Prog 216: Global Biol Events in Earth Hist, 84-92; assoc ed, Paleobiol, 82-89, adj ed, 90-; pres, Paleont Soc, Northeast Sect, 90. *Mem:* Int Paleont Asn; Paleont Soc; Paläontologische Ges; Brit Paleont Asn; AAAS; Soc Syst Zool; Geol Soc Am. *Res:* Marine community paleoecology; evolution of ecosystems; theorectical morphology; evolutionary theory; phenotypic evolution; animal form and function in nature. *Mailing Add:* Dept Geol Sci Wright Geol Lab Rutgers Univ New Brunswick NJ 08903

MCGHEE, JERRY ROGER, b Knoxville, Tenn, June 25, 41; m 61; c 2. MICROBIOLOGY, IMMUNOLOGY. *Educ:* Univ Tenn, Knoxville, BS, 64; Univ Tenn, Memphis, PhD(microbiol), 69. *Prof Exp:* Res asst prof, Univ Tenn, Memphis, 69; instr microbiol & dent res, 69-71, asst prof microbiol, 72-75, ASSOC PROF MICROBIOL & SCIENTIST INST DENT RES, UNIV ALA, BIRMINGHAM, 75- *Concurrent Pos:* Teaching fel microbiol, Med Units, Univ Tenn, Memphis, 64-66; NIH fel microbiol, Univ Chicago, 71-72; sabbatical, Lab Micro Immunol, Nat Inst Dent Res, NIH, 77-78. *Mem:* Am Soc Microbiol; Am Asn Immunol; Soc Exp Biol & Med; Reticuloendothelial Soc; Tissue Cult Asn. *Res:* Host-parasite interrelationships; cellular rates of synthesis; effective immunity to dental cavities and enteric diseases; local immunity; LPS effects on lymphoid cells. *Mailing Add:* Dept Microbiol Univ Ala Birmingham BHS Rm 392 UAB Sta Birmingham AL 35294

MCGHEE, ROBERT B, b Detroit, Mich, June 6, 29; m 52; c 3. ELECTRICAL ENGINEERING. *Educ:* Univ Mich, BS, 52; Univ Southern Calif, MS, 57, PhD(elec eng), 63. *Prof Exp:* Mem tech staff, Hughes Aircraft Co, 55-63; from asst prof to assoc prof elec eng, Univ Southern Calif, 63-68; prof elec eng, Ohio State Univ, 68-88; PROF, DEPT COMPUT SCI, NAVAL POSTGRAD SCH, 88- *Mem:* Inst Elec & Electronics Engrs. *Res:* Control and switching theory; non-linear system identification; statistical analysis; computer design and applications; prosthetics; digital computers; biomedical engineering. *Mailing Add:* Dept Comput Sci Naval Postgrad Sch Monterey CA 93940

MCGHEE, TERENCE JOSEPH, b Summit, NJ, July 5, 36; m 69; c 1. SANITARY ENGINEERING. *Educ:* Newark Col Eng, BS, 59; Va Polytech Inst, MS, 63; Univ Kans, PhD(civil eng), 68. *Prof Exp:* Ciivl engr, US Forest Serv, 59-62; res engr, Boeing Co, 62-64 & Lord Mfg Co, 64-65; asst prof civil eng, Univ Louisville, 68-70; from asst prof to assoc prof, Univ Nebr, 70-75; assoc prof, 75-79, PROF CIVIL ENG, TULANE UNIV, 79- *Concurrent Pos:* Sr proj adv, NY Assocs, Consult Engrs, 76-79 & United Res Serv, 79-85. *Mem:* Water Pollution Control Fedn; Am Water Works Asn; Am Soc Civil Engrs. *Res:* Water quality; water pollution; wastewater and water treatment. *Mailing Add:* Dept of Civil Eng Tulane Univ 6823 St Charles Ave New Orleans LA 70118

MCGIFF, JOHN C, b New York, NY, Aug 6, 27; m 58; c 5. PHARMACOLOGY, INTERNAL MEDICINE. *Educ:* Georgetown Univ, BS, 47; Columbia Univ, MD, 51; Am Bd Internal Med, dipl, 64. *Prof Exp:* Res assoc, Sch Med, St Louis Univ, 58-59; from asst instr to instr, Univ Pa, 60-62; assoc med & pharmacol, 62-64, asst prof, 64-65; assoc prof internal med & chief cardiovasc sect, Sch Med, St Louis Univ, 65-70, prof med, 70-71; prof pharmacol & med & dir clin pharm sect, Med Col Wis, 71-75; prof & chmn dept pharmacol, Univ Tenn, Ctr Health Sci, 75-80; MEM FAC, NEW YORK MED COL, 80- *Concurrent Pos:* Fel med, Col Physicians & Surgeons, Columbia Univ, 57-58; Am Heart Asn res fel, 57-59; fel & co-recipient inst grant, USPHS, 63-65; Burroughs Wellcome Fund scholar; estab investr, Am Heart Asn, 64-69; vis prof, Tulane Univ, 69-; mem, Cardiovasc B Study Sect, NIH; mem med adv bd, Coun High Blood Pressure Res. *Mem:* Am Fedn Clin Res; Am Physiol Soc; NY Acad Sci; Am Soc Pharmacol; Am Soc Clin Invest. *Res:* Cardiorenal pharmacology and physiology; clinical pharmacology. *Mailing Add:* Pharmacol Dept Basic Sci Bldg NY Med Col Valhalla NY 10595

MCGILL, DAVID A, b Albany, NY, Sept 18, 30. OCEANOGRAPHY. *Educ:* Bucknell Univ, BSc, 52; Columbia Univ, MA, 56; Yale Univ, PhD(zool), 63. *Prof Exp:* Technician chem oceanog, Woods Hole Oceanog Inst, 56-59, fel, 59-62, asst scientist, 62-66; asst prof, Southeastern Mass Tech Inst, 66-68; PROF OCEAN SCI, US COAST GUARD ACAD, 68- *Concurrent Pos:* Mem, Int Geophys Year Surv Atlantic Ocean, 57-58; Sci Comt Oceanic Res-UNESCO grants, Hawaii, 61 & Australia, 62; mem, Int Indian Ocean Expeds, 62-65. *Mem:* AAAS; Soc Study Evolution; Am Soc Limnol & Oceanog; Marine Biol Asn UK; Am Geophys Union. *Res:* Distribution of nutrient elements in sea water, chiefly inorganic and total phosphorus. *Mailing Add:* Dept Sci US Coast Guard Acad New London CT 06320

MCGILL, DAVID JOHN, b New Orleans, La, Oct 9, 39; m 61; c 3. ENGINEERING MECHANICS. *Educ:* La State Univ, BS, 61, MS, 63; Univ Kans, PhD(eng mech), 66. *Prof Exp:* asst prof, 66-80, PROF ENG MECH, GA INST TECHNOL, 80- *Mem:* Sigma Xi. *Res:* Dynamics. *Mailing Add:* 3753 Gladney Dr Ga Inst Technol 225 North Ave NW Atlanta GA 30341

MCGILL, DAVID PARK, b Waverly, Nebr, Sept 3, 19; m 44; c 1. AGRONOMY. *Educ:* Univ Nebr, BS, 41, MS, 49; Iowa State Col, PhD(agron), 54. *Prof Exp:* Asst agron, 46-51, from asst agronomist to assoc agronomist, 51-62, PROF AGRON, UNIV NEBR, LINCOLN, 62- *Mem:* Am Soc Agron. *Res:* Genetics teaching. *Mailing Add:* Dept Agron Univ Nebr Lincoln NE 68503

MCGILL, DOUGLAS B, b New York, NY, Aug 14, 29; m 53; c 4. GASTROENTEROLOGY. *Educ:* Yale Univ, BA, 51; Tufts Univ, MS, 55; Univ Minn, MS, 61. *Prof Exp:* Intern med, Boston City Hosp, 55-56; resident, Mayo Grad Sch Med, Univ Minn, 58-61; assoc prof med, Med Sch, 73-77, CONSULT MED & GASTROENTEROL, MAYO CLIN, 61-, PROF MED, MAYO MED SCH, 77- *Mem:* AMA; fel Am Col Physicians; Am Gastroenterol Asn; Am Asn Study Liver Dis (pres, 86-87); Am Fedn Clin Res; Brit Soc Gastroenterol. *Res:* Liver disease; colon cancer detection. *Mailing Add:* 303 Sixth Ave SW Rochester MN 55901

MCGILL, GEORGE EMMERT, b Des Moines, Iowa, June 10, 31; m 55; c 3. STRUCTURAL GEOLOGY, PLANETARY SCIENCE. *Educ:* Carleton Col, BA, 53; Univ Minn, MS, 55; Princeton Univ, PhD(geol), 58. *Prof Exp:* From asst prof to assoc prof, 58-73, PROF GEOL, UNIV MASS, AMHERST, 73-, HEAD DEPT GEOL & GEOG, 77- *Mem:* Am Geophys Union; fel Geol Soc Am; Sigma Xi; AAAS. *Res:* Structural and regional geology; astrogeology. *Mailing Add:* Dept of Geol & Geog Univ of Mass Amherst MA 01003

MCGILL, HENRY COLEMAN, JR, b Nashville, Tenn, Oct 1, 21; m 45; c 3. PATHOLOGY. *Educ:* Vanderbilt Univ, BA, 43, MD, 46. *Prof Exp:* Intern, Vanderbilt Univ, 46-47; asst, Sch Med, La State Univ, 47-48, from asst prof to prof , 50-66, head dept, 61-66,; chmn dept, 66-72, prof path, Univ Tex Health Sci Ctr, San Antonio, 72-91; SCI DIR, SOUTHWEST FOUND BIOMED RES, SAN ANTONIO, TEX, 79- *Concurrent Pos:* Mem Coun Arteriosclerosis & Epidemiol, Am Heart Asn. *Mem:* Am Asn Path; Int Acad Path; Sigma Xi. *Res:* Arteriosclerotic heart disease. *Mailing Add:* Southwest Found Biomed Res San Antonio TX 78228-0147

MCGILL, JULIAN EDWARD, b Blacksburg, SC, Oct 22, 32; m 59; c 2. ANALYTICAL CHEMISTRY, SCIENCE ADMINISTRATION. *Educ:* Erskine Col, BA, 55; Clemson Univ, MS, 68, PhD(chem), 71. *Prof Exp:* Qual control engr, Celanese Fibers Co, 59-61; asst prof, 70-76, ASSOC PROF PHARM CHEM, COL PHARM, MED UNIV SC, 76-, DIR CHEM AFFAIRS, PHARMACEUT DEVELOP CTR, 90- *Mem:* Am Chem Soc; Am Asn Pharmaceut Scientists. *Res:* Spectrophotometric analysis; pharmaceutical analysis, high performance liquid chromatography. *Mailing Add:* Col of Pharm Med Univ of SC 171 Ashley Ave Charleston SC 29425-2303

MCGILL, LAWRENCE DAVID, b Lincoln, Nebr, Mar 24, 44; m 66. VETERINARY PATHOLOGY. *Educ:* Okla State Univ, BS, 66, DVM, 68; Tex A&M Univ, PhD(vet path), 72; Am Col Vet Pathologists, dipl. *Prof Exp:* NIH fel, Tex A&M Univ, 68-71, teaching asst vet path, 71; asst prof, Univ Minn, St Paul, 71-72; asst prof vet path, Univ Nebr, Lincoln, 72-77; vet pathologist, 77-81, chief path, Salt Lake City, 81-85, MED DIR, VET REF LAB, SAN LEANDRO, CALIF, 85- *Mem:* Int Acad Path; Am Vet Med Asn; Am Col Vet Path; AAAS. *Res:* Immunopathology; pathogenesis of viral infections; ultrastructural pathology; host-virus relationships. *Mailing Add:* Animal REF Path PO Box 30095 Salt Lake City UT 84130

MCGILL, ROBERT MAYO, b Marianna, Ark, Nov 29, 25; wid. PHYSICAL INORGANIC CHEMISTRY. *Educ:* Univ Ark, BS, 45, MA, 48; Univ Tenn, PhD(phys chem), 55. *Prof Exp:* Chemist, 48-78, sr staff consult, Res Lab, K-25 Plant, 78- 84, CONSULT, UNION CARBIDE CORP, 84- *Concurrent Pos:* Consult, isotope separations. *Mem:* Am Chem Soc; AAAS; fel Am Inst Chemists; Sigma Xi. *Res:* Molecular weight determination; adsorption of fatty acids on metals; chemistry of uranium and fluorine; kinetics of precipitation; isotope separations. *Mailing Add:* Rte 2, Box 254 Ten Mile TN 37880

MCGILL, SUZANNE, b Port Arthur, Tex, Mar 10, 44. MATHEMATICS. *Educ:* Univ St Thomas, BA, 67; Tex Christian Univ, MS, 70, PhD(math), 72. *Prof Exp:* Instr, Pub Sch, Tex, 67-68; instr math, Tex Christian Univ, 72; PROF MATH, UNIV S ALA, 72- *Mem:* Math Asn Am. *Res:* Abstract algebra. *Mailing Add:* Dept Math Univ SAla 307 University Blvd Mobile AL 36688

MCGILL, THOMAS CONLEY, JR, b Port Arthur, Tex, Mar 20, 42; m 66. PHYSICS, ELECTRICAL ENGINEERING. *Educ:* Lamar State Col, BS, 63 & 64; Calif Inst Technol, MS, 65, PhD(elec eng, physics), 69. *Prof Exp:* Asst prof, 71-74, assoc prof, 74-77, PROF APPL PHYSICS, CALIF INST TECHNOL, 77-; CONSULT, ARCO SOLAR, 81- *Concurrent Pos:* Mem tech staff, Hughes Res Labs, 67-73, consult, 73-; NATO fel theoret physics, Bristol Univ, 69-70; Air Force Nat Res Coun fel, Princeton Univ, 70-71; Alfred P Sloan Found fel, 74-76; mem adv res proj agency, Mat Res Coun. *Mem:* AAAS; Am Inst Physics; Inst Elec & Electronics Eng; Sigma Xi. *Res:* Solid state physics, particularly semiconductors and insulators. *Mailing Add:* 380 Laguna Rd Pasadena CA 91105

MCGILL, WILLIAM BRUCE, b Morden, Man, Oct 20, 45; m 66; c 3. SOIL BIOLOGY, BIOCHEMISTRY. *Educ:* Univ Man, BSA Hons, 67, MSc, 69; Univ Sask, PhD(soil biol & biochem), 72. *Prof Exp:* From asst prof to assoc prof soil biochem, 71-81, chmn Dept Soil Sci, 79-89, PROF SOIL BIOCHEM, UNIV ALTA, 81- *Concurrent Pos:* Sr res fel, Commonwealth Sci & Indust Res Orgn, 84-85. *Mem:* Can Soc Soil Sci; Can Soc Microbiol; Soil Sci Soc Am; Int Soc Soil Sci. *Res:* Soil carbon, nitrogen and phosphorous cycling; modelling of organic matter turnover in soil; soil enzymology; land reclamation. *Mailing Add:* Dept Soil Sci Univ Alta Edmonton AB T6G 2E3 Can

MCGILLEM, C(LARE) D(UANE), b Clinton, Mich, Oct 9, 23; m 47; c 1. ELECTRICAL ENGINEERING, BIOENGINEERING & BIOMEDICAL ENGINEERING. *Educ:* Univ Mich, BSEE, 47; Purdue Univ, MSE, 49, PhD(elec eng), 55. *Prof Exp:* Res engr, Diamond Chain Co, Inc, Ind, 47-51; proj engr mil electronics, US Naval Avionics Facility, 51-53, div head radar, 53-56; dept head mil & electronic eng, AC Spark Plug Div, Gen Motors Corp, Mich, 56-59, dir electronics & appl res, Wis, 59-60, prog mgr lunar explor, Defense Res Labs, Calif, 60-63; assoc prof elec eng, 63-67, dir electronic systs res lab, 65-68, assoc dean eng & dir eng exp sta, 68-72, PROF ELEC ENG, SCH ELEC ENG, PURDUE UNIV, 67- *Concurrent Pos:* Pres & mem bd dirs, Technol Assocs, Inc, 76-, pres, Vetronics, Inc, W Lafayette, 82-87 & mem bd dirs, 82- *Honors & Awards:* Centennial Medal, Inst Elec & Electronics Engrs, 84, J Fred Peoples Award, 88. *Mem:* Fel Inst Elec & Electronics Engrs; Geosci Electronics Soc (pres, 76). *Res:* Monopulse radar; automatic tracking and firecontrol systems; radar resolution and scattering; microwave radiometry; signal detection and estimation; random signal analysis; bioelectric signal processing; statistical communications theory; information processing. *Mailing Add:* Sch Elec Eng Purdue Univ W Lafayette IN 47907

MCGILLIARD, A DARE, b Stillwater, Okla, Oct 15, 26; m 51; c 3. ANIMAL NUTRITION. *Educ:* Okla State Univ, BS, 51, MS, 52; Mich State Univ, PhD(animal nutrit), 61. *Prof Exp:* From asst prof to prof animal sci, Iowa State Univ, 57-87; RETIRED. *Concurrent Pos:* Chmn physiol panel, Rumen Function Conf, 69- *Honors & Awards:* Am Feed Mfrs Award, 72. *Mem:* Am Dairy Sci Asn; Am Inst Nutrit; Am Soc Animal Sci. *Res:* Nutritional physiology and biochemistry; experimental surgery of digestive tract; cardiovascular and lymphatic systems. *Mailing Add:* Dept Animal Sci 313 Kid EE Iowa State Univ 22 Norwood Dr Balairsta AR 72714

MCGILLIARD, LON DEE, b Manhattan, Kans, Aug 9, 21; m 46, 75; c 4. ANIMAL BREEDING, DAIRY SCIENCE. *Educ:* Okla State Univ, BS, 42; Mich State Univ, MS, 47; Iowa State Univ, PhD(animal breeding, dairy husb), 52. *Prof Exp:* Asst dairy prod, Mich State Univ, 46-47; asst animal breeding, Iowa State Univ, 48-49, from res assoc to assoc prof animal breeding & dairy husb, 49-55; assoc prof, 55-62, PROF DAIRY CATTLE BREEDING, MICH STATE UNIV, 62- *Concurrent Pos:* Assoc ed, J Dairy Sci, 69-72, ed, 73-85. *Honors & Awards:* Award of Hon, Am Dairy Sci Asn, 86. *Mem:* Am Dairy Sci Asn; Sigma Xi; Coun Biol Ed. *Res:* Improving dairy cattle through breeding; quantitative genetics. *Mailing Add:* Dept Animal Sci 121 A Anthony Hall Mich State Univ East Lansing MI 48824

MCGILLIARD, MICHAEL LON, b Lansing, Mich, Sept 29, 47; m 70. DAIRY SCIENCE, ANIMAL GENETICS. *Educ:* Mich State Univ, BS, 69; Iowa State Univ, MS, 70, PhD(animal breeding), 74. *Prof Exp:* Exten specialist dairy sci, 75-80, DAIRY MGT RES EXP DESIGN, VA POLYTECH INST & STATE UNIV, 81- *Mem:* Am Dairy Sci Asn. *Res:* Determining the influence of management strategies on quantitative measures of dairy herd success. *Mailing Add:* Dept Dairy Sci Va Polytech Inst & State Univ Blacksburg VA 24061-0315

MACGILLIVRAY, ARCHIBALD DEAN, b Vancouver, BC, Dec 28, 29; m 57; c 3. APPLIED MATHEMATICS. *Educ:* Univ BC, BASc, 55; Calif Inst Technol, MS, 57, PhD(aeronaut), 60. *Prof Exp:* Instr math, Calif Inst Technol, 60-61; mathematician, Lincoln Lab, Mass Inst Technol, 61-62, instr math, 62-64; from asst prof to assoc prof, 64-84, PROF MATH, STATE UNIV NY, BUFFALO, 84- *Concurrent Pos:* Sr scientist, Jet Propulsion Labs, Calif Inst Technol, 60-61; consult, Lincoln Lab, 62-64. *Honors & Awards:* Royal Inst Scholar, 52. *Mem:* Am Math Soc; Am Phys Soc. *Res:* Asymptotic expansions of solutions of differential equations. *Mailing Add:* Dept of Math State Univ of NY Buffalo NY 14214

MACGILLIVRAY, JEFFREY CHARLES, b Washington, DC, Apr 14, 52; m 87. DIGITAL IMAGE, MATHEMATICAL PHYSICS. *Educ:* Mass Inst Technol, SB(elec eng) & SB(physics), 73, ScD(physics), 78. *Prof Exp:* staff electro-optic scientist, Optical Systs Div, Itek Corp, 78-82; proj leader, Imagitex Inc, 82-83; eng, Apollo Comput Inc, 83-84; CONSULT, 85- *Res:* Digital image processing; mathematical modeling of physical systems. *Mailing Add:* 652 Timbertop Rd New Ipswich NH 03071

MACGILLIVRAY, M ELLEN, b Fredericton, NB, Nov 26, 25; m 48. ENTOMOLOGY. *Educ:* Univ NB, BA, 47; Univ Mich, MSc, 51; State Univ Leiden, DSc, 58. *Hon Degrees:* DSc, Univ NB, 81. *Prof Exp:* Sr agr asst, Div Entom, Can Dept Agr, 47-48, tech officer, 48-54, res officer, 54-59, res scientist, Res Br, 59-80; RETIRED. *Mem:* Entom Soc Can (pres, 76-77); Can Soc Zool; Entom Soc Am; hon mem Acadian Entom Soc. *Res:* Systematics and biology of aphids of the Atlantic provinces of Canada and the New England states; potato pest management. *Mailing Add:* 156 Glengarey Place Fredericton NB E3B 5Z9 Can

MACGILLIVRAY, MARGARET HILDA, b San Fernando, Trinidad, WI, Aug 30, 30; Can citizen; m 57; c 3. ENDOCRINOLOGY, IMMUNOLOGY. *Educ:* Univ Toronto, MD, 56. *Prof Exp:* From lectr to assoc prof pediat endocrinol, 64-76, PROF PEDIAT & CO-DIR DIV ENDOCRINOL, CHILDREN'S HOSP & MED SCH, STATE UNIV NY BUFFALO, 76- *Concurrent Pos:* AEC grant biol, Calif Inst Technol, 60-61; USPHS grants endocrinol & metab, Mass Gen Hosp, 61-64 & juvenile diabetes, 76- *Mem:* Soc Pediat Res; Endocrine Soc; Andrology Soc. *Res:* The role of growth hormone in dwarfism of childhood; hormonal control of carbohydrate metabolism in hypopituitary children and juvenile onset diabetes. *Mailing Add:* Childrens Hosp 219 Bryant St Buffalo NY 14222

MCGILVERY, ROBERT WARREN, b Coquille, Ore, Aug 25, 20; m 43; c 4. BIOCHEMISTRY. *Educ:* Ore State Col, BS, 41; Univ Wis, PhD(physiol chem), 47. *Prof Exp:* From asst prof to assoc prof physiol chem, Univ Wis, 48-57; from assoc prof to prof, 57-85, EMER PROF BIOCHEM, UNIV VA, 85- *Concurrent Pos:* USPHS sr res fel, Univ Wis, 47-48; vis prof biochem, State Univ NY Buffalo, 71. *Mem:* AAAS; Am Soc Biol Chemists. *Res:* Metabolic economy. *Mailing Add:* 1090 W Leigh Dr Charlottesville VA 22901

MCGINN, CLIFFORD, b New York, NY, Aug 22, 22; m 54; c 4. ORGANIC CHEMISTRY. *Educ:* Fordham Univ, BS, 43, MS, 46; Syracuse Univ, PhD(chem), 57. *Prof Exp:* From instr to assoc prof, 49-70, PROF CHEM, LEMOYNE COL, NY, 70- *Mem:* Am Chem Soc. *Res:* Theoretical organic chemistry; chemical bonding. *Mailing Add:* Dept of Chem LeMoyne Col LeMoyne Heights Syracuse NY 13214

MCGINNES, BURD SHELDON, b Pittsburgh, Pa, Aug 10, 21; m 46; c 1. WILDLIFE RESEARCH. *Educ:* Pa State Univ, BS, 48, MS, 49; Va Polytech Inst & State Univ, PhD(biol), 58. *Prof Exp:* Proj leader, Del Game & Fish Comn, 49-51 & 52-55; leader wild turkey invest, Pa Game Comn, 58; leader coop wildlife res unit, VA Polytech Inst & State Univ, 58-82; RETIRED. *Mem:* Wildlife Soc (vpres, 75-76); Sigma Xi; Wildlife Dis Asn. *Res:* Animal ecology; wildlife diseases and management techniques. *Mailing Add:* 1417 Palmer Dr Blacksburg VA 24060-5632

MCGINNES, EDGAR ALLAN, JR, b Chestertown, Md, Feb 15, 26; m 51; c 3. WOOD CHEMISTRY, WOOD TECHNOLOGY. *Educ:* Pa State Univ, BS, 50, MF, 51; State Univ NY Col Forestry, Syracuse Univ, PhD(wood technol), 55. *Prof Exp:* Jr res chemist, Am Viscose Corp, 51-52, res chemist, 55-58, res specialist, 59-60; assoc prof, 60-63, PROF FORESTRY, UNIV MO-COLUMBIA, 63- *Concurrent Pos:* Consult, wood failure. *Honors & Awards:* Soc Wood Sci & Tech (past pres). *Mem:* Fel AAAS; fel Am Inst Chemists; Am Chem Soc; Soc Am Foresters; Forest Prod Res Soc. *Res:* Wood anatomy; influence of environmental stress on wood structure. *Mailing Add:* 900 Bourn Ave Univ Mo Columbia MO 65203

MCGINNESS, JAMES DONALD, b Evansville, Ind, June 23, 30; m 50; c 5. COATING PLASTICS, URETHANE COATINGS. *Educ:* Univ Evansville, AB, 52. *Prof Exp:* Chemist, Sherwin-Williams Co, 52-55, chemist & group leader spectros, 55-61, dir anal res dept, Ill, 61-70, mgr reliability, Ohio, 70-74, group mgr automotive chem coatings, 74-77; mgr high solids & bus mach coatings, 77-83, MGR EXTERIOR AUTOMOTIVE COATINGS RES & DEVELOP, RED SPOT PAINT & VARNISH CO, 83- *Mem:* Am Chem Soc; Soc Plastics Engrs; Soc Automotive Engrs. *Res:* Infrared spectroscopy; gas chromatography; x-ray emission; urethane technology; coatings technology; color science; polymer characterization; nuclear magnetic resonance. *Mailing Add:* RR 2 Box 476 Sledd Gilbertsville KY 42044

MCGINNESS, WILLIAM GEORGE, III, b Lock Haven, Pa, Apr 9, 48; c 2. ORGANIC CHEMISTRY, OPTICAL SCANNING. *Educ:* Lock Haven State Col, BS, 70. *Prof Exp:* Dir mfg, Novamont Corp, 76-80; vpres tech serv, Natmar Inc, 80-83; PRES, ANGSTROM TECHNOL, 83- *Mem:* Am Chem Soc; Robotics Int; Soc Mech Engrs; Am Inst Chemists. *Res:* Quantum physics; organic chemistry; electro-optical scanning of robotic vision; automatic identification. *Mailing Add:* 2251 Clarkson Rd Union KY 41091

MCGINNIES, WILLIAM GROVENOR, b Steamboat Springs, Colo, Aug 14, 99; m 25; c 1. PLANT ECOLOGY. *Educ:* Univ Ariz, BSA, 22; Univ Chicago, PhD(plant ecol), 32. *Prof Exp:* Asst range examr, US Forest Serv, 24-26; asst prof animal husb, univ & range specialist, Exp Sta, Univ Ariz, 27-29, assoc prof & range ecologist, 30-32, actg head dept bot, 32-35; mgr & dir land mgt, Navajo Dist, Soil Conserv Serv, USDA, 35-38, chief div range res, Southwest Forest & Range Exp Sta, US Forest Serv, 38-42; chief div surv & invest, Guayule Emergency Rubber Proj, 42-44; dir, Rocky Mountain Forest & Range Exp Sta, US Forest Serv, 45-53 & Cent States Forest Exp Sta, 54-60; dir lab tree-ring res & coordr arid lands res prog, 60-65, prof & proj mgr, Off Arid Lands Res, 65-70, EMER PROF DENDROCHRONOL & EMER DIR ARID LANDS STUDIES, UNIV ARIZ, 70- *Mem:* Ecol Soc Am; Soc Range Mgt; Tree Ring Soc; Sigma Xi. *Res:* ecology of desert plants; frequency and abundance of plants; water requirement of Xerophytes. *Mailing Add:* 2001 W Rudasill Rd #5311 Tucson AZ 85704

MCGINNIES, WILLIAM JOSEPH, b Tucson, Ariz, Jan 2, 27; m 49; c 3. RANGE SCIENCE, PLANT ECOLOGY. *Educ:* Colo State Univ, BS, 48, PhD, 67; Univ Wis, MS, 52. *Prof Exp:* Forester watershed mgt, Forest Serv, Idaho, USDA, 48, range conservationist grazing mgt & range reseeding, Utah, 49-53, range conservationist range reseeding, Agr Res Serv, Utah, 54-56, Colo, 56-66, range scientist range reseeding, 66-75, range scientist & strip mine revegetation, short grass ecology, agr res serv, 76-87, collaborator, range & strip mine revegetation, short grass ecology, 76-87; RETIRED. *Concurrent Pos:* Fac affil & mem grad fac, Colo State Univ, 66- *Honors & Awards:* Trail Grass Award, Soc for Range Mgt, 87. *Mem:* Fel AAAS; Am Soc Surface Mine Reclamation. *Res:* Adaptability of species for range seeding; methods of establishing seeded grasses on rangeland; techniques for measuring range herbage and environmental requirements of range plants; rehabilitation of strip mines and other disturbed areas; seeding and management of saline (Solonetz or Natrustoll) soils; fertilization of rangeland; grazing management of rangelands; ecology of shortgrass range; repeat photography of vegetation. *Mailing Add:* 1909 Navajo Dr Ft Collins CO 80525

MCGINNIS, CHARLES HENRY, JR, b Newark, NJ, Nov 24, 34; m 56; c 2. AVIAN PHYSIOLOGY. *Educ:* Rutgers Univ, New Brunswick, BS, 56; Purdue Univ, MS, 61; Mich State Univ, PhD(poultry physiol), 65. *Prof Exp:* Asst prof physiol, Sch Vet Med, Univ Minn, St Paul, 65-67; avian physiologist, 67-69, head poultry res sect, 69-76, MGR NUTRIT SERV, RHONE POULENC INC, 77- *Mem:* Poultry Sci Asn; World Poultry Sci Asn; Am Physiol Soc; Sigma Xi; NY Acad Sci. *Res:* Nutrition in poultry. *Mailing Add:* Feed Additives Div Phone, Poulenc Inc, Suite 620 500 Northridge Rd Atlanta GA 30350

MCGINNIS, DAVID FRANKLIN, JR, b Baltimore, Md, Nov 10, 44; m 68; c 2. SATELLITE REMOTE SENSING. *Educ:* Univ Del, BS, 65; Pa State Univ, MS, 67, PhD(civil eng), 71. *Prof Exp:* HYDROLOGIST RES, LAND SCI BR, NAT ENVIRON SATELLITE DATA & INFO SERV, NAT OCEANIC & ATMOSPHERIC ADMIN, US DEPT COMMERCE, 71- *Mem:* Am Meteorol Soc; Am Soc Photogrammetry & Remote Sensing; Am Soc Civil Engrs. *Res:* Use of satellite data to study water resources; agricultural science; anthropogenic effects. *Mailing Add:* FB4 Rm 3010 Nat Oceanic & Atmospheric Admin Washington DC 20233

MCGINNIS, EUGENE A, b Jessup, Pa, May 15, 21; m 41; c 5. PHYSICS. *Educ:* Univ Scranton, BS, 48; NY Univ, MS, 53; Fordham Univ, PhD(physics), 60. *Prof Exp:* Instr, Clarks Summit Sch, 46-48; from instr to assoc prof, 48-63, chmn dept, 67-76, PROF PHYSICS, UNIV SCRANTON, 63- *Mem:* Am Asn Physics Teachers; Soc Appl Spectros. *Res:* Emission and molecular spectroscopy, particularly Raman spectorscopy of gases. *Mailing Add:* 1218 Clay Ave Dunmore PA 18510

MCGINNIS, GARY DAVID, b Everett, Wash, Oct 1, 40; m 64; c 3. CARBOHYDRATE CHEMISTRY. *Educ:* Pac Lutheran Univ, BS, 62; Univ Wash, MS, 68; Univ Mont, PhD(org chem), 70. *Prof Exp:* Prod chemist, Am Cyanamid Co, 64-67; fel, Univ Mont, 70-71; asst prof wood chem, Forest Prod Lab, 71-77, ASSOC PROF WOOD SCI & TECHNOL & CHEM & ASSOC TECHNOLOGIST, FOREST PROD UTILIZATION LAB, MISS STATE UNIV, 77- *Mem:* Am Chem Soc; Forest Prod Res Soc; Sigma Xi. *Res:* Thermal decomposition of wood; high pressure liquid chromatography of natural products; chemical composition of wood barks. *Mailing Add:* 910 Howard Starkville MS 39159

MCGINNIS, JAMES, b Cliffside, NC, Apr 10, 18; m 52; c 1. NUTRITION. *Educ:* NC State Col, BS, 40; Cornell Univ, PhD(nutrit), 44. *Prof Exp:* Asst nutrit, Cornell Univ, 40-43, res assoc, 43-44; from asst prof to assoc prof poultry sci, 44-50, PROF POULTRY SCI, POULTRY SCIENTIST & EXTEN POULTRY SCI SPECIALIST, WASH STATE UNIV, 50-, PROF ANIMAL SCI. *Concurrent Pos:* Asst dir agr develop dept, Chas Pfizer & Co, 52-53. *Mem:* AAAS; Am Chem Soc; Soc Exp Biol & Med; Poultry Sci Asn; Am Inst Nutrit. *Res:* Nutritional requirements of poultry; biochemistry of proteins; nutritional and endocrinological interrelationships; Browning reaction in foods; antibiotics and nutrition. *Mailing Add:* Dept Animal Sci Wash State Univ Pullman WA 99164

MCGINNIS, JAMES F, b Troy, NY, Dec 15, 41. MOLECULAR NEUROBIOLOGY, PSYCHIATRY. *Educ:* Siena Col, BS, 63; State Univ NY, PhD(cell & molecular biol), 70. *Prof Exp:* PROF, DEPT ANAT & CELL BIOL, UNIV CALIF, LOS ANGELES, 87- *Mem:* Asn Res Vision & Ophthal; Am Soc Biochem & Molecular Biol; AAAS; Am Soc Neurochem. *Mailing Add:* Dept Anat & Cell Biol Sch Med Univ Calif 760 Westwood Plaza Los Angeles CA 90024

MCGINNIS, JAMES LEE, b Jacksonville, Ill, Nov 28, 45; m 77; c 1. ORGANIC CHEMISTRY, CATALYSIS. *Educ:* MacMurray Col, BA, 67; Columbia Univ, PhD(chem), 76. *Prof Exp:* Res assoc chem, Mass Inst Technol, 76-77; res chemist, 77-81, sr res chemist, 81-83, RES SUPVR, HOECHST-CELANESE RES DIV, 83- *Mem:* Am Chem Soc. *Res:* Homogeneous catalysis; reaction mechanisms; physical organic chemistry; organometallics; polymer chemistry; heterogeneous catalysis. *Mailing Add:* Hoechst-Celanese Res Div 86 Morris Ave Summit NJ 07901

MCGINNIS, JOHN THURLOW, ecology, biostatistics; deceased, see previous edition for last biography

MCGINNIS, LYLE DAVID, b Appleton, Wis, Mar 5, 31; m 59; c 5. GEOPHYSICS. *Educ:* St Norbert Col, BSc, 54; St Louis Univ, MSc, 60; Univ Ill, PhD(geol), 65. *Prof Exp:* Geophys trainee, Carter Oil Co, Okla, 54-55; IGY geophysicist, Antarctica, 57-59; assoc geophysicist, Ill State Geol Surv, 59-66; tech expert, UN Develop Prog, Afghanistan, 66-67; from assoc prof to prof geol, Northern Ill Univ, 67-83, chmn dept, 80-83; chmn dept geol, La State Univ, 83-85; SR SCIENTIST, ARGONNE NAT LAB, 85- *Concurrent Pos:* Consult geophysicist, Desert Res Inst, Univ Nev, 61-62; consult, Minn Geol Surv, 70-; NSF coordr, Antarctic Drilling Prog, 71-; geophysicist, US Geol Surv, Woods Hole, Mass, 75-76; adj prof geol, Northern Ill Univ, 85- *Mem:* AAAS; fel Geol Soc Am; Soc Explor Geophys; Am Geophys Union; Seismol Soc Am. *Res:* Geology and geophysics of continental interiors. *Mailing Add:* EES-362 Eng Geosci Argonne Nat Lab Argonne IL 60439

MCGINNIS, MICHAEL RANDY, b Hayward, Calif, Oct 22, 42; m 75; c 2. MYCOLOGY, MICROBIOLOGY. *Educ:* Calif State Polytech Col, BS, 61; Iowa State Univ, PhD(mycol), 69. *Prof Exp:* Fel, Ctr Dis Control, USPHS, 71-73; supvr, Mycol Lab, SC Dept Health & Environ Control, 74-75; assoc dir, Clin Microbiol Lab, NC Mem Hosp, Univ NC, 75-88; PROF, DEPT PATHOL, UNIV TEX MED BR, 88-, VCHMN, DEPT PATH, 89- *Concurrent Pos:* Vpres, Int Soc Human & Animal Mycol; ed, jour Clin Microbiol; Ed-in-chief, Current Topics in Mycol; Diplomat, Am Bd Med Microbiol, chair, Mycol Div, Am Soc Microbiol. *Honors & Awards:* Meridian Award, 88. *Mem:* Mycol Soc Am; Am Soc Microbiol; Int Soc Human & Animal Mycol; fel Am Acad Microbiol; fel Infectious Dis Soc Am. *Res:* Taxonomic relations of the human pathogenic dematiaceous hyphomycetes. *Mailing Add:* Dept Path Univ Tex Med Br Galveston TX 77550

MCGINNIS, ROBERT CAMERON, b Edmonton, Alta, Aug 18, 25; m 51; c 3. AGRONOMY. *Educ:* Univ Alta, BSc, 49, MSc, 51; Univ Man, PhD(cytogenetics), 54. *Prof Exp:* Asst, Univ Alta, 49-51; cytologist, Can Dept Agr, 51-60; assoc prof plant sci, Univ Man, 60-65, prof & Head dept, 65-75; assoc dir, Int Crops Res Inst for Semi-Arid Tropics, 75-80; DEAN FAC AGR, UNIV MAN, 80- *Concurrent Pos:* Dir, Plant Breeding Sta, Njoro, Kenya, 73-75. *Honors & Awards:* Centennial Medal, Can Govt, 67. *Mem:* Sigma Xi; NY Acad Sci; Genetics Soc Can; fel Agr Inst Can; hon mem Can Seed Growers Asn. *Res:* Improving crops and farming systems for the semi-arid and humid tropics. *Mailing Add:* Univ Man Winnepeg MB R3T 2N2 Can

MCGINNIS, WILLIAM JOSEPH, b St Louis, Mo, Apr 30, 23; m 51; c 4. INORGANIC CHEMISTRY, PHYSICAL CHEMISTRY. *Educ:* Colo Col, BA, 48; Iowa State Col, MS, 51. *Prof Exp:* Chemist, Los Alamos Sci Lab, Univ Calif, 48-49; res asst, Inst Atomic Res, Iowa State Col, 49-51; chemist, Int Minerals & Chem Corp, 51-56; chemist, Pigments Dept, E I du Pont de Nemours & Co, Inc, 56-63, sr res chemist, 63, tech supvr, Tenn, 63-69, prod specialist, Del, 69-70, tech serv supvr, 70-74, tech serv consult, Chem & Pigments Dept, 74-85; RETIRED. *Concurrent Pos:* Consult pigmentation problems in paper indust, 85-90. *Mem:* Am Chem Soc; Tech Asn Pulp & Paper Indust. *Res:* Rare earth metals; potassium and magnesium compounds; phosphate chemistry; titanium chemistry and compounds, particularly titanium dioxide pigments. *Mailing Add:* 2407 Landon Dr Wilmington DE 19810

MCGINNISS, VINCENT DANIEL, b Philadelphia, Pa, Feb 9, 42; m 68; c 3. POLYMER CHEMISTRY. *Educ:* Univ Fla, BS, 63; Univ Ariz, PhD(phys org chem), 70. *Prof Exp:* Analytical chemist agr, Agr Dept, Univ Fla, 62-63; res chemist polymers, Peninsular Chem Res, 64-65; teaching asst org chem, Univ Ariz, 66-68; sr res chemist, Glidden-Durkee, 70-72, scientist polymers, 72-77; SR RES SCIENTIST POLYMERS & COATINGS, COLUMBUS LABS, BATTELLE MEM INST, 77- *Honors & Awards:* Roon Award, Fedn Socs Paint Technol, 73 & 77. *Mem:* Am Chem Soc. *Res:* Kinetics and mechanisms of photopolymerization reactions; photoinitiation and network formation as applied to coatings and films; solar energy applications of polymers and coatings, electrical properties of polymers and coatings, prediction of the chemical, electrical, physical and mechanical properties of polymers and coatings; polymer synthesis and applications. *Mailing Add:* 1315 N State Rt 3 PO Box 702 Sunbury OH 43074

MCGIRK, RICHARD HEATH, b Boonville, Mo, Jan 25, 45; m 66; c 2. ORGANIC POLYMER CHEMISTRY. *Educ:* Tex Christian Univ, BS, 67; Univ Colo, PhD(chem), 71. *Prof Exp:* Fel, Univ Col, Univ London, 71-72 & Johns Hopkins Univ, 72-73; SR RES CHEMIST ORG POLYMER CHEM, POLYMER PROD DEPT, E I DU PONT DE NEMOURS & CO, INC, BEAUMONT, 73- *Mem:* Sigma Xi; Am Chem Soc. *Res:* Polymer, vulcanization of elastomers, isocyanate and polyurethane chemistry; polyester elastomers; ethylene-propylene-diene monomer elastomers. *Mailing Add:* E I du Pont de Nemours & Co PO Box 3269 Beaumont TX 77704-3269

MCGLAMERY, MARSHAL DEAN, b Mooreland, Okla, July 29, 32; m 57; c 2. AGRONOMY, WEED SCIENCE. *Educ:* Okla Agr & Mech Col, BS, 56; Okla State Univ, MS, 58; Univ Ill, Urbana, PhD, 65. *Prof Exp:* Instr soils, Panhandle Agr & Mech Col, 58-60; agronomist, Agr Bus Co, 60-61; instr soils, 61-63, from res assoc to assoc prof weed control, 64-75, PROF WEED CONTROL, UNIV ILL, URBANA, 75- *Concurrent Pos:* AID weed control consult, India, 67-68, Jamaica, 85; vis assoc prof, Univ Minn, 71-72; vis prof, NC State Univ, 81; CICP-AID consult, Grenada, 86. *Mem:* Weed Sci Soc Am; Coun Agr Sci & Technol; Am Soc Agron. *Res:* Herbicide residue; weed taxonomy; herbicide selectivity. *Mailing Add:* Agron Dept Univ Ill 1102 S Goodwin Urbana IL 61871

MACGLASHAN, DONALD W, JR, b Wichita Falls, Tex, Oct 21, 53. IMMUNOLOGY. *Educ:* Calif Inst Technol, BS, 75; Johns Hopkins Univ, MD, 79, PhD(immunol), 81. *Prof Exp:* From instr to asst prof, 83-90, ASSOC PROF IMMUNOL, SCH MED, JOHNS HOPKINS UNIV, 90- *Mem:* Am Asn Immunol. *Mailing Add:* Dept Med Johns Hopkins Asthma & Allergy Ctr 301 Bayview Rd Baltimore MD 21224

MCGLASSON, ALVIN GARNETT, b Boone Co, Ky, July 27, 25; m 49; c 3. MATHEMATICS. *Educ:* Eastern Ky State Col, BS, 49; Univ Ky, MS, 52. *Prof Exp:* From instr to prof, 50-83, EMER PROF MATH, EASTERN KY UNIV, 83- *Concurrent Pos:* NSF fel, Univ Kans, 59. *Mem:* Math Asn Am; Nat Coun Teachers of Math. *Res:* Geometry and preparation of high school teachers of mathematics; computer science. *Mailing Add:* 124 Buckwood Dr Richmond KY 40475-2222

MCGLAUCHLIN, LAURENCE D(ONALD), electronics, optics, for more information see previous edition

MCGLINN, WILLIAM DAVID, b Leavenworth, Kans, Feb 15, 30; m 53; c 5. THEORETICAL PHYSICS. *Educ:* Univ Kans, BS, 52, PhD(physics), 59. *Prof Exp:* From instr to asst prof physics, Northwestern Univ, 58-63; physicist, Argonne Nat Lab, 63-65; assoc prof, 65-68, PROF PHYSICS, UNIV NOTRE DAME, 68- *Mem:* Am Phys Soc. *Res:* Particle physics; strong and weak interaction theory. *Mailing Add:* Dept of Physics Univ of Notre Dame Notre Dame IN 46556

MCGLOIN, PAUL ARTHUR, b Woburn, Mass, Jan 15, 23; m 53; c 3. MATHEMATICS, COMPUTER SCIENCE. *Educ:* Boston Univ, AB & AM, 49; Rensselaer Polytech Inst, PhD(math), 68. *Prof Exp:* Instr math, Univ Conn, 49-51 & 53-55; asst prof, 55-65, ASSOC PROF MATH, RENSSELAER POLYTECH INST, 65- *Mem:* Asn Comput Mach. *Res:* Graph theory, non-Hamiltonian graphs. *Mailing Add:* 2001 15th St Troy NY 12180

MCGLYNN, SEAN PATRICK, b Ireland, Mar 8, 31; nat US; m 55; c 5. PHYSICAL CHEMISTRY. *Educ:* Nat Univ Ireland, BSc, 51, MSc, 52; Fla State Univ, PhD(phys chem), 56. *Prof Exp:* Fel chem, Fla State Univ, 56 & Univ Wash, 56-57; from asst prof to prof, 57-68, dean, Grad Sch, 80-81, BOYD PROF CHEM, LA STATE UNIV, BATON ROUGE, 68-, VCHANCELLOR RES, 81- *Concurrent Pos:* Assoc prof, Yale Univ, 61; NSF lectr, 63 & 64; consult biophys prog, Mich State Univ, 63-65; consult, Am Optical Co & Am Instrument Co, 63- & Bell Tel Labs, 65; Alfred E Sloan Found fel, 64-; US Sr scientist award, Alexander von Humboldt Found, 79. *Honors & Awards:* Southwestern US Award, Am Chem Soc, 67 & Fla Award, 70. *Mem:* AAAS; Am Chem Soc; Am Phys Soc. *Res:* Biophysics; excitons; spectroscopy; fluorescence and phosphorescence; energy transfer; electronic structure of inorganic anions; charge transfer; conductivity of aromatics; spin-orbital perturbation; rotation barriers; vacuum ultraviolet. *Mailing Add:* Dept of Chem La State Univ Baton Rouge LA 70803-1804

MCGONIGAL, WILLIAM E, b Decatur, Ga, Aug 20, 39; m 59; c 1. ORGANIC CHEMISTRY. *Educ:* Ga Inst Technol, BS, 61, PhD(org chem), 65. *Prof Exp:* Res chemist, Indust & Biochem Dept, E I du Pont de Nemours & Co, Inc, Del, 65-67, sr res chemist, 67-68, res supvr, 68-69, admin asst, 69-73, tech area supvr, Biochem Dept, Tex, 73-75, biochem prod supt, Biochem Dept, Belle WVa, 75-76, lab dir, Biochem Dept, 76-79, prin consult, Corp Plans Dept, 79-80, mgr new bus develop, Clin Systs Div, 80-82, mkt mgr, 82-84, DIR TECHNOL, DIAG RES & DEVELOP DIV, MED PRODS DEPT, E I DU PONT DE NEMOURS & CO INC, 84- *Concurrent Pos:* Mem chem abstracts adv comt, 77-80; mem res dirs coun, Nat Agr Chem Asn, 79-80; NIH fel, 61-65; mem, Budget & Finance Comt, Am Chem Soc, 81-89. *Honors & Awards:* Nat Sci Found Res Award, 60; M A Ferst Res Award, Sigma Xi, 65. *Mem:* Am Chem Soc; Sigma Xi. *Res:* Synthetic, organic, carbohydrate, heterocyclic, agricultural, pharmaceutical and diagnostic chemistry. *Mailing Add:* 204 Haystack Lane Wilmington DE 19807

MCGOOKEY, DONALD PAUL, b Sandusky, Ohio, Sept 19, 28; m 51; c 6. PETROLEUM GEOLOGY. *Educ:* Bowling Green State Univ, BS, 51; Univ Wyo, MA, 52; Ohio State Univ, PhD(geol), 58. *Prof Exp:* Div stratigr, Texaco, Inc, Denver, 52-69, exec producing comnr, New York, Texaco Inc, 69-71, chief geologist, Houston, 71-78, asst div mgr explor, Midland Tex, 78-79; sr vpres explor, Omni-Explor Inc, 79-80; CONSULT GEOLOGIST, MIDLAND TEX, 80-; PARTNER, DM EXPLOR CO, 84- *Concurrent Pos:* Bownocker fel, Ohio State Univ, 57-58. *Mem:* Soc Independent Prof Earth Scientists; Geol Soc Am; Am Asn Petrol Geologists; Soc Explor Paleontologists & Mineralogists. *Res:* Investigation of sedimentary basins of North America by detailed study of the chronology of structural development and related sedimentary deposits. *Mailing Add:* 310 W Illinois Suite 314 Midland TX 79701

MCGOUGH, WILLIAM EDWARD, psychiatry; deceased, see previous edition for last biography

MCGOVERN, JOHN JOSEPH, b Pittsburgh, Pa, June 21, 20; m 87; c 6. CHEMISTRY, ENVIRONMENTAL SCIENCES. *Educ:* Carnegie Inst Technol, BS, 42, MS, 44, ScD(chem), 46. *Prof Exp:* Instr, Carnegie Inst Technol, 44-45; fel spectros, Mellon Inst, 45-48, sr fel spectros & chem anal groups, 48-50; asst chief chemist, Kobuta Plant Chem Div, Koppers Co, 50-56, chief chemist, 56-58; head res serv, Mellon Inst, 58-71, dir educ & res serv, Carnegie-Mellon Univ, 71-73; asst dir, Carnegie-Mellon Inst Res, 73-76; info scientist, Univ Pittsburgh, 76-79; educ serv mgr, Air Pollution Control Asn, 79-84, mem serv mgr, 84-87; CONSULT, 88- *Concurrent Pos:* Pres, Int Pittsburgh Conf Anal Chem & Appl Spectros, 51. *Mem:* Am Chem Soc; Air & Waste Mgt Asn; Sigma Xi. *Res:* Administration of technical activities; chemistry and analytical chemistry. *Mailing Add:* 1606 Parkline Dr Pittsburgh PA 15227-1645

MCGOVERN, JOHN PHILLIP, b Washington, DC, June 2, 21. ALLERGY. *Educ:* Duke Univ, MD, 45; Am Bd Pediat, cert, 52, cert allergy, 56; Am Bd Allergy & Immunol, cert, 72. *Hon Degrees:* Twenty-three from US cols & univs, 71-87. *Prof Exp:* Intern & asst pediat, Yale Med Sch, Yale-New Haven Med Ctr, 45-46; asst resident & assoc pediat, Duke Univ Sch Med & Hosp, 48; asst chief resident, Children's Hosp, Washington, DC, 49, chief resident, 50; assoc, Sch Med, George Washington Univ, 50-51, asst prof, 51-54; assoc prof, Sch Med, Tulane Univ, La, 54-56; clin prof allergy, 56-70, prof & chmn dept hist med, 70-81, PROF HIST & PHILOS OF BIOMED SCI, GRAD SCH BIOMED SCI, UNIV TEX, HOUSTON, 81- *Concurrent Pos:* Asst pediat, Sch Med, Yale Univ, 45-46; chief, Out-Patient Dept, Children's Hosp, Washington, DC, 50-51, assoc attend staff, 51-54, attend, Allergy Clin, 53-54; assoc pediat, George Washington Univ Hosp, 50-54; Markle scholar med sci, 50-55; assoc pediat, Sch Med, George Washington Univ, 50-51, asst prof & chief, Pediat Div, 51-54; attend physician, Doctors Hosp, 52-54; chief, Tulane Pediat Allergy Clin, Charity Hosp, New Orleans, 54-55 & vis physician, Tulane Pediat Serv, 54-56; consult, Huey Long Charity Hosp, La, 54-56, US Pub Health Hosp, New Orleans, 54-56, Wilford Hall USAF Hosp, Tex, 59-, Hermann Hosp, Houston, 62- & St Luke's Episcopal Hosp, Houston, 75-; assoc staff pediat, Tex Children's Hosp, Houston, 56-59, chief allergy serv, 57-74, dir, Allergy Clin, Jr League Diag Clins, 57-68, active staff pediat, 60-68, consult staff, Med Serv, Allergy Sect, 68-; dir, McGovern Allergy Clin, Houston; clin prof pediat, Baylor Col Med, 56-, chief, Allergy Sect, Dept Pediat, 56-78, dir, Allergy Fel Training Prog, Allergy Res Lab, Dept Microbiol, 57-70, clin assoc prof microbiol, 57-70, adj prof immunol, Dept Microbiol & Immunol, 69-; regional consult, Children's Asthma Res Inst & Hosp, Denver, 60-78, Nat Found Asthmatic Children, Tucson, 60-70, Asthmatic Children's Found, 63-; distinguished adj prof health & safety educ, Kent State Univ, 72-; adj prof physiol & health sci, Ball State Univ, 76-; adj prof, Grad Prog Biomed Commun, Univ Tex, 76-77, prof, 77-, clin prof immunol & allergy, MD Anderson Hosp & Tumor Inst, 76-, consult, Res Med Libr, 76-, adj prof environ sci, Sch Pub Health, 78-, clin prof internal med, Med Sch, Houston, 78-, adj prof hist & philos of med, Inst Med Humanities, Med Br, Galveston, 82-, clin res prof, Sch Nursing & Dent Sci Inst, Health Sci Ctr, 85-; fel, Green Col, Oxford Univ, 82- *Honors & Awards:* Clemens von Pirquet lectr & Award, Med Ctr, Georgetown Univ, 75; William A Howe Award, Am Sch Health Asn, 83. *Mem:* Am Osler Soc (secy, 70-71, second vpres, 71-72, vpres, 72-73, pres, 73-74); AMA; Am Asn Cert Allergists (pres, 72-73); fel AAAS; fel Am Acad Pediat; Am Asn Hist Med; fel Royal Col Physicians; fel Am Col Allergists (pres, 68-69); Sigma Xi; Asn Res Nervous & Ment Dis; fel Am Acad Allergy; Am Asn Study Headache (pres, 63-64); Pan Am Med Asn; Am Asn Immunologists; Soc Exp Biol & Med; fel Am Med Writers Asn; fel Am Col Physicians; fel Am Col Chest Physicians; Asthma Care Asn Am (second vpres, 80-). *Res:* Clinical allergy; immunology of hypersensitivity. *Mailing Add:* McGovern Allergy Clin 6969 Bromton Rd Houston TX 77025

MCGOVERN, TERENCE JOSEPH, b New York, NY, Aug 9, 42; m 68; c 3. COMPUTER SYSTEMS. *Educ:* Cooper Union, BE, 65; Northwestern Univ, MS, 68, PhD(chem eng), 70. *Prof Exp:* Proj engr, Chem & Plastics Div, 70-76, SR PROGRAMMER ANALYST, UNION CARBIDE CORP, 76- *Concurrent Pos:* Adj fac mem, WVa Col Grad Studies. *Mem:* Am Inst Chem Engrs; Am Chem Soc. *Res:* Reaction engineering; simulation; modeling; polymerization; kinetics; computer systems for medical records, work history and environmental impact analysis. *Mailing Add:* 2029 Huber Rd Charleston WV 25314

MCGOVERN, TERRENCE PHILLIP, organic chemistry; deceased, see previous edition for last biography

MCGOVERN, WAYNE ERNEST, b Orange, NJ, June 3, 37; m 63; c 2. METEOROLOGY. *Educ:* Newark Col Eng, BSME, 59, MS, 64; NY Univ, PhD(meteorol), 67. *Prof Exp:* Aerospace scientist, NASA, 67-68; meteorologist, Nat Meteorol Ctr, 68-69; assoc prof meteorol, Univ Ariz, 69-72; meteorologist, res & develop, Nat Oceanic & Atmospheric Admin, 72-75; US sci coordr, Global Weather Exp, 76-82; MESOSCALE BR CHIEF, NAT WEATHER SERV, 82- *Concurrent Pos:* Dept Com sci fel, 75-76. *Mem:* Am Meteorol Soc; Royal Numismatic Soc; Am Numismatic Soc. *Res:* Meteorological operational research. *Mailing Add:* Nat Weather Serv 1325 East-West Hwy Silver Spring MD 20910

MCGOVREN, JAMES PATRICK, b Washington, Ind, June 12, 47; m 68; c 2. PHARMACOLOGY, PHARMACEUTICS. *Educ:* Purdue Univ, BS, 70; Univ Ky, PhD(pharmaceut sci), 75. *Prof Exp:* Scientist cancer res, 75-83, ASSOC DIR, CANCER & INFECTIOUS DIS RES, UPJOHN CO, 83- *Mem:* Am Asn Cancer Res. *Res:* Pharmacology and experimental therapeutics of antitumor agents; drug disposition, drug metabolism and pharmacokinetics; analytical methods for drugs in biofluids; molecular pharmacology of DNA-interactive drugs. *Mailing Add:* Cancer Res 7252-267-4 UpJohn Co Kalamazoo MI 49001

MCGOWAN, BLAINE, JR, veterinary medicine; deceased, see previous edition for last biography

MCGOWAN, ELEANOR BROOKENS, b Santa Barbara, Calif, Oct 7, 44; m 67; c 1. BIOCHEMISTRY. *Educ:* Grinnell Col, BA, 65; Univ Iowa, PhD(biochem), 69. *Prof Exp:* Res assoc biochem, Brandeis Univ, 69-71 & Harvard Med Sch, 71-72; res assoc, 72-74, sr res assoc, 74-77, ASST PROF BIOCHEM, STATE UNIV NY DOWNSTATE MED SCH, 77- *Concurrent Pos:* Muscular Dystrophy Asn fel, 74-75; Young investr, Nat Heart, Lung & Blood Inst, NIH, 77-80. *Mem:* Biophys Soc; Am Soc Cell Biol; NY Acad Sci; Harvey Soc; Sigma Xi. *Res:* Protein structure, function and breakdown; calcium-binding proteins; muscle proteins; chromophores in proteins; muscular dystrophy; platelet function. *Mailing Add:* Dept Biochem SUNY Downstate Med Ctr PO Box 8 450 Clarkson Ave Brooklyn NY 11203

MCGOWAN, FRANCIS KEITH, b Baileyville, Kans, May 2, 21; m 44. PHYSICS. *Educ:* Kans State Teachers Col, Emporia, AB, 42; Univ Wis, PhM, 44; Univ Tenn, PhD(physics), 51. *Prof Exp:* Asst physics, Univ Wis, 42-44; res engr, Sylvania Elec Prod, Inc, NY, 44-46; PHYSICIST, OAK RIDGE NAT LAB, 46- *Mem:* Fel Am Phys Soc. *Res:* Short-lived nuclear states; short-lived isomers of nuclei; gamma-ray spectroscopy; coulomb excitation of nuclei; cross sections for charged-particle induced reactions. *Mailing Add:* Oak Ridge Nat Lab PO Box 2008 Bldg 6000 Oak Ridge TN 37831

MCGOWAN, JAMES WILLIAM, b Pittsburgh, Pa, July 5, 31; m 58; c 6. ATOMIC PHYSICS, MOLECULAR PHYSICS. *Educ:* St Francis Xavier Univ, BSc, 53; Carnegie-Mellon Univ, MS, 58; Laval Univ, DSc(physics), 61. *Prof Exp:* Instr physics, St Francis Xavier Univ, 55-56; res asst, Westinghouse Res Labs, 57-58; instr physics, St Lawrence Col, Laval Univ, 58-59; staff mem, Gen Atomic Div, Gen Dynamics Corp, 62-69; chmn, Dept Physics, Univ Western Ont, 69-72, chmn, Founding Ctr Interdisciplinary Studies Chem Physics, 73-76, prof physics, 69-84; dir, Nat Mus Sci & Technol, Ottawa, 84-88; DIR & CHMN, IMAGES IN TIME & SPACE, 89- *Concurrent Pos:* Fel, Joint Inst Lab Astrophys, Univ Colo, Boulder, 66; adv, NSF, Nat Bur Standards, AEC, Advan Res Proj Agency, US Dept Defense, Defense Atomic Support Agency & Defense Res Bd Can; consult, Vacuum Electronics Corp, 59-61, Bach-Simpson Ltd, 71-85 & Med Physics Group, Lawrence Berkeley Labs, 78-82; vis prof, Centro Atomico, Arg, 71, Fac Univ Namur, Belg, 76-77 & Univ Cath Louvain, Belg, 78-88; mem, Comt Sci & Technol in Developing Nations, India, 77; mem, Can Pugwash Group, 78-88; sr indust fel, Bell Northern Res Co, Ottawa, 81-82; founding chmn, Very Large Scale Integration Study Group, 81 & coordr, Can Very Large Scale Integration Implementation Group, 81-84; secy comn physics develop, Int Union Pure & Appl Physics, 81-84; mem bd dir, Ont Sci Ctr, 80-87; found chmn, Sci Tech & You, London, 78-82; pres, Asn Advan Sci in Canada, 85-; mem bd, Sci for Peace & Ctr Bras D'Or; chmn, Assoc Sci & Technol Inc, 88- *Mem:* Fel AAAS; fel Am Phys Soc; Can Asn Physicists. *Res:* Experimental atomic and molecular collision physics; energy deposition studies; positrons and ions in solids; charge transfer lasers; laser interaction with eyes; soft x-ray studies of biologically significant molecules using synchrotron radiation; very large scale integration related studies of polymers; x-ray lithography and surfaces; science policy related to regional and international development. *Mailing Add:* 330 E Cordova St Pasadena CA 91101

MCGOWAN, JOAN A, b New York, NY, Dec 4, 44; m 66; c 2. CELL BIOLOGY, TISSUE CULTURE. *Educ:* Marymount Manhattan Col, BS, 66; Cornell Univ, MNS, 68; Brown Univ, PhD(biochem), 78. *Prof Exp:* Assoc 78-81, instr, 81-83, ASST PROF BIOCHEM, SCH MED, HARVARD UNIV, 83-; BIOLOGIST, MASS GEN HOSP, BOSTON, 78-; STAFF SCIENTIST, SHRINERS BURNS INST, BOSTON, 81- *Mem:* AAAS; Am Soc Cell Biol; Tissue Cult Asn. *Res:* Regulation of liver growth and differentiation using isolated liver cells in culture as a model system. *Mailing Add:* 301 Brookline St Cambridge MA 02139

MCGOWAN, JOHN ARTHUR, b Oshkosh, Wis, Aug 22, 24; m 64; c 2. BIOLOGICAL OCEANOGRAPHY. *Educ:* Ore State Univ, BS, 50, MS, 51; Univ Calif, San Diego, PhD(oceanog), 60. *Prof Exp:* Marine biologist, Trust Territory of Pac Islands, 56-58; asst res biologist, 60-62, from asst prof to assoc prof, 62-72, PROF OCEANOG, SCRIPPS INST OCEANOG, UNIV CALIF, SAN DIEGO, 72- *Concurrent Pos:* Mem UNESCO consult comt, Indian Ocean Biol Ctr, 63-66; ed, J Plank Res, 80-87; mem AAAS Chile-Am panel cooperative res, 88- *Mem:* fel AAAS; Am Soc Limnol & Oceanog; Marine Biol Asn UK. *Res:* Ecology and zoogeography of oceanic plankton; climate and oceanic ecosystems, time series and coastal zone ecology; biology of squid. *Mailing Add:* Dept Oceanog A-028 Univ Cal San Diego La Jolla CA 92093

MCGOWAN, JOHN JOSEPH, b Mobile, Ala, Aug 31, 51; m 77. AETROVIRUSES. *Educ:* Univ S Ala, BS, 73; Univ Miss, MS, Med Ctr, PhD(virol), 80. *Prof Exp:* Fel virol, Univ Va, 80-82; asst prof microbiol, Uniformed Serv Univ Health Sci, 82-86; proj officer, AIDS prog, Treatment Br, 86-87, CHIEF AIDS PROG, DEVELOP TREATMENT BR, NAT INST ALLERGY & INFECTIOUS DIS, NIH, 88- *Concurrent Pos:* Adj prof, dept microbiol, Uniformed Serv Univ Health Sci, 86- *Honors & Awards:* McClasky Award, Am Soc Microbiol, 78, Mahaffey Award, 79. *Mem:* Sigma Xi; Am Soc Virol; Am Soc Microbiol. *Res:* Viral animal models. *Mailing Add:* 7605 Maryknoll Ave Bethesda MD 20817

MCGOWAN, JON GERALD, b Lockport, NY, May 3, 39; m 65. MECHANICAL ENGINEERING. *Educ:* Carnegie Inst Technol, BS, 61, PhD(mech eng), 65; Stanford Univ, MS, 62. *Prof Exp:* Develop engr, E I du Pont de Nemours & Co, Inc, 65-67; from asst prof to assoc prof mech & aerospace eng, 67-76, PROF MECH ENG, UNIV MASS, AMHERST, 76- *Concurrent Pos:* Consult, Combustion Eng, Inc, Conn, 70- *Mem:* AAAS; Am Soc Mech Engrs; Air Pollution Control Asn; Int Solar Energy Soc. *Res:* Solar and wind energy conversion; thermodynamics; combustion; air pollution control; heat transfer; fluid mechanics. *Mailing Add:* 134 Main St Northfield MA 01360

MCGOWAN, MICHAEL JAMES, b Chicago, Ill, Apr 22, 52; m 80; c 3. MEDICAL ENTOMOLOGY, ACAROLOGY. *Educ:* Univ Kans, BA, 74, MA, 78; Okla State Univ, PhD(entom), 80. *Prof Exp:* Fel entom, Okla State Univ, 80; res assoc, Univ Fla, 80; entomologist IV, West Fla Anthropod Res Lab, Agr Res Serv, USDA, 80-82; sr entomologist, 82-87, prod regist mgr, 87-88, MGR, ANIMAL SCI REGULATORY AFFAIRS, LILLY RES LAB, DIV ELI LILLY CO, GREENFIELD, IND, 88- *Mem:* AAAS; Entom Soc Am. *Res:* Biology, ecology and physiology of insects of medical and veterinary importance, specifically flies and diseases which flies transmit. *Mailing Add:* Lilly Res Lab-Div Eli Lilly Co PO Box 708 Bldg 202 Greenfield IN 46140-0708

MCGOWAN, WILLIAM COURTNEY, JR, b Mobile, Ala, Aug 8, 37; m 65; c 2. SOLID STATE PHYSICS. *Educ:* Spring Hill Col, BS, 59; Univ NC, PhD(physics), 65. *Prof Exp:* Electronic engr, White Sands Missile Range, 65-66, asst chief anti-tank div & res engr, 66; chmn dept, 66-80, from assoc prof to prof physics, Western Carolina Univ, 76-82; PROF PHYSICS, CLARION UNIV PA, 82-, CHMN DEPT, 88- *Mem:* Am Asn Physics Teachers. *Res:* Imperfections in ionic crystals; motion and charge on dislocations in silver chloride; range and velocity measurement accuracy as affected by radar signal design. *Mailing Add:* Dept Physics Clarion Univ Pa Clarion PA 16241

MCGOWN, EVELYN L, b Council, Idaho, Oct 23, 44. ANALYTICAL CHEMISTRY. *Educ:* Univ Idaho, BS, 66; Univ Minn, PhD(nutrit & biochem), 71. *Prof Exp:* RES CHEMIST, DEPT BLOOD RES, LETTERMAN ARMY RES INST, 76- *Mem:* Am Asn Clin Chem; Am Inst Nutrit; Am Chem Soc. *Mailing Add:* Chem Br Letterman Army Inst Res Presidio San Francisco CA 94129

MCGRADY, ANGELE VIAL, b New Rochelle, NY, Dec 31, 41; m 68; c 2. PSYCHOPHYSIOLOGY. *Educ:* Chestnut Hill Col, BS, 63; Mich State Univ, MS, 66; Univ Toledo, PhD(physiol), 72, MEd, 84. *Prof Exp:* Instr vet physiol, Wash State Univ, 65-66; res assoc physiol Kresge Eye Inst, Wayne State Univ, 66-68; from instr to asst prof, 68-80, ASSOC PROF PHYSIOL, MED COL OHIO, 80-, ASSOC PROF PSYCHIAT, 82- *Concurrent Pos:* Lic prof clin counr, Ohio, 86; exec bd, Asn Appl Psychophysiol & Biofeedback, 89-92. *Honors & Awards:* Res Recognition Award, Biofeedback Soc Am. *Mem:* Soc Neurosci; Biofeedback Soc Am; Am Physiol Soc. *Res:* neurosciences, biofeedback for control of hypertension and migraine headache; stress physiology; psychophysiology. *Mailing Add:* Dept Physiol C S 10008 Toledo OH 43699

MCGRAIL, DAVID WAYNE, oceanography, sedimentology; deceased, see previous edition for last biography

MCGRATH, ARTHUR KEVIN, b New York, NY. TECHNICAL MANAGEMENT, GRAPHIC USER INTERFACES. *Educ:* Dowling Col, BA, 70. *Prof Exp:* Engr, WSMW-TV, 75-82; engr, Video Tape Assocs, 82-83; ed, Proj Mgt, Learning Tree Int, 86-89; PROGRAMMER, CONTRACT ENGRS, 83-; INSTR ADVAN C, TECHNOL EXCHANGE, 90- *Concurrent Pos:* Consult, Satellite Data Telecommun, 83-84 & Satellite Reservation Systs, 84-86; ed, Qual Assurance, Technol Exchange, 90-; instr, OS/2 kernel & OS/2 pres & mgr, Productivity Solutions, 90- *Mem:* Int Neural Network Soc; Am Soc Training & Development. *Res:* Editor and writer on technical management, neural network, operating system, communications and programming language courses. *Mailing Add:* PO Box 128 Barboursville VA 22923

MCGRATH, CHARLES MORRIS, b Seattle, Wash, Sept 17, 43. ONCOLOGY. *Educ:* Univ Portland, BS, 65; Univ Calif, Berkeley, PhD(virol), 70. *Prof Exp:* Cell biologist, Univ Calif, Berkeley, 70-71; viral oncologist, Inst Jules Bordet, Brussels, Belg, 71-72; res scientist viral oncol, 72-83, CHIEF TUMOR BIOL LAB VIRAL ONCOL, MICH CANCER FOUND, 73- *Concurrent Pos:* NIH res grant, Nat Cancer Inst, 75. *Mem:* Tissue Cult Asn; Am Asn Cancer Res. *Res:* Viral and hormonal involvement in malignant conversion of breast epithelial cells. *Mailing Add:* Dept Bio Sci Oakland Univ Rochester MI 48063

MCGRATH, EUGENE R, ELECTRIC POWER SYSTEMS. *Prof Exp:* CHMN BD, CONSOL EDISON CO, NY, 90- *Mem:* Nat Acad Eng. *Mailing Add:* Consol Edison Co NY Off Chmn 8414 Irving Pl Rm 1610 New York NY 10003

MCGRATH, JAMES EDWARD, b Easton, NY, July 11, 34; m 59; c 6. ORGANIC POLYMER CHEMISTRY, POLYMER SCIENCE. *Educ:* Siena Col, BS, 56; Univ Akron, MS, 64, PhD(polymer sci), 67. *Prof Exp:* Res chemist, ITT-Rayonier, 56-59 & Goodyear Tire & Rubber Co, 59-65; fel polymer sci, Univ Akron, 67; res scientist, Union Carbide Corp, 67-75; assoc prof, 75-78, PROF POLYMER SCI, VA POLYTECH INST & STATE UNIV, 78-, CO-DIR, POLYMERIC MAT & INTERFACES LAB, 78- *Concurrent Pos:* Polymer consult; chmn, Gordon Res Conf on Elastomers, 79. Corp, 75- *Mem:* Am Chem Soc; NY Acad Sci. *Res:* Anionic polymerization; block and graft copolymers; reactions of polymers. *Mailing Add:* Dept Chem Va Polytech Inst & State Univ Blacksburg VA 24060

MCGRATH, JAMES J, b Brooklyn, NY, Oct 30, 31. PLANT ANATOMY. *Educ:* Univ Notre Dame, AB, 55; Univ Calif, Davis, MA, 64, PhD(bot), 66. *Prof Exp:* Asst prof biol & rector, 65-71, ASSOC PROF BIOL, UNIV NOTRE DAME, 71-, ASST CHMN DEPT, 74- *Mem:* Am Forestry Asn; Bot Soc Am; Int Soc Plant Morphol. *Res:* Developmental plant anatomy; seasonal changes in secondary phloem of angiosperms. *Mailing Add:* Dept Biol Univ Notre Dame Notre Dame IN 46556

MCGRATH, JAMES JOSEPH, m; c 4. INHALATION TOXICOLOGY, AIR POLLUTION. *Educ:* Ind Univ, PhD(physiol), 68. *Prof Exp:* PROF PHYSIOL, SCH MED, TEX TECH UNIV, 79- *Res:* Inhalation toxicology; high altitude. *Mailing Add:* Physiol Dept Tex Tech Univ Lubbock TX 79430

MCGRATH, JAMES W, b Kirkland, Ill, May 17, 12. CHEMICAL PHYSICS, X-RAYS. *Educ:* Ft Hays Kans State Univ, BA, 33; Univ Kans, MA, 34; Univ Iowa, PhD(physics), 39. *Prof Exp:* From instr to assoc prof physics, Mich State Univ, 39-46; prof physics, Kent State Univ, 46-75, dean, Grad Sch & Res, 68-73, assoc provost, 72-73, vpres grad studies & res, 73-75, EMER PROF PHYSICS, KENT STATE UNIV, 75- *Concurrent Pos:* Assoc ed, Am J Physics, 50-52; consult, Goodyear Aircraft Corp, 52-56; prin investr, Air Force Off Sci Res & NSF, 57-69. *Mem:* Fel Am Phys Soc. *Res:* Nuclear magnetic resonance in solids; x-ray magnetic absorption in gold and lead. *Mailing Add:* 1544 Morris Rd Kent OH 44240-4530

MCGRATH, JOHN F, b New York, NY, Dec 30, 27; m 50; c 3. PHYSICAL CHEMISTRY, INORGANIC CHEMISTRY. *Educ:* Siena Col, NY, BS, 49; State Univ NY, MS, 51; Rensselaer Polytech Inst, MS, 60. *Prof Exp:* Teacher, Colonie Cent High Sch, 53-60; from asst prof to assoc prof phys sci, 61-74, chmn div natural sci, 73-76, PROF PHYS SCI, COL ST ROSE, 74- *Concurrent Pos:* Partic, NSF Summer Inst, Rensselaer Polytech Inst, 55-59; consult geol Lake Superior region res, Mich Technol Univ, 65; partic, Inst Sci, Ithaca Col, 68 & Eastern Regional Inst, 68-71; ed, Bull Sci Teachers Asn NY State, 71-74. *Mem:* Am Chem Soc. *Res:* Instrumental analysis; determination of stability constants of series of esters of some complexes of tripositive cobalt. *Mailing Add:* Seven Lawnridge Ave Albany NY 12208-3111

MCGRATH, JOHN JOSEPH, b Randolph, Vt, Apr 25, 49; m 79; c 3. BIOLOGICAL & MODEL MEMBRANES. *Educ:* Stanford Univ, BS, 71; Mass Inst Technol, MS, 74, PhD(mech eng), 77. *Prof Exp:* Res asst mech eng, Mass Inst Technol, 72-77, res assoc toxicol, 77-78; asst prof, 78-82, PROF MECH ENG, MICH STATE UNIV, 82-, ACTG CHAIRPERSON MECH ENG, 90- *Concurrent Pos:* Vis scientist, dept surg, Cambridge Univ, 79; ed, Cryoletters, 83- & Cryobiol, 85-; consult, Nat Environ Res Coun, UK, 84-; Fulbright scholar, Cambridge Eng, 86-87. *Mem:* Soc Cryobiol (secy, 85-86); Sigma Xi; Am Soc Mech Engrs; Biophys Soc; NY Acad Sci; Am Soc Eng Educ. *Res:* Biophysics of biological and model membranes; mechanisms of chilling and freezing injury of biomaterials; biomedical instrumentation; advanced thermal diagnostics. *Mailing Add:* A106 Res Complex-Eng Mich State Univ East Lansing MI 48823-1226

MCGRATH, JOHN THOMAS, b Philadelphia, Pa, Sept 27, 18; m 47; c 3. PATHOLOGY. *Educ:* Univ Pa, VMD, 43. *Prof Exp:* Instr path, Vet Sch, 47-48, assoc, 48-50, from asst prof to assoc prof, 50-58, PROF PATH, GRAD SCH MED, UNIV PA, 58- *Mem:* Am Vet Med Asn; Am Asn Neuropath; Am Col Vet Path; Am Acad Neurol. *Res:* Nervous disorders of animals; pathology of central nervous system diseases of animals. *Mailing Add:* 463 Forrest Ave Drexel Hill PA 19026

MCGRATH, MICHAEL GLENNON, b St Louis, Mo, Oct 12, 41; m 67, 90. ORGANIC CHEMISTRY. *Educ:* Col Holy Cross, BS, 63; Mass Inst Technol, PhD(chem), 67. *Prof Exp:* Asst prof, 67-72, ASSOC PROF CHEM, COL OF THE HOLY CROSS, 72- *Mem:* Am Chem Soc; Nat Asn Adv Health Prof. *Res:* Synthesis and reactions of small ring carbon compounds; carbene chemistry; synthesis of non-benzenoid aromatic compounds. *Mailing Add:* Dept of Chem Col of the Holy Cross Worcester MA 01610

MCGRATH, ROBERT L, b Iowa City, Iowa, Nov 2, 38; m 60; c 2. NUCLEAR PHYSICS. *Educ:* Oberlin Col, BA, 60; Univ Iowa, MS, 62, PhD(physics), 65. *Prof Exp:* Fel physics, Univ Iowa, 65; res assoc, Lawrence Radiation Lab, Univ Calif, 66-68; from asst prof to assoc prof, 68-82, PROF, STATE UNIV NY STONYBROOK, 83- *Concurrent Pos:* Alexander von Humboldt Found sr US scientist award, 74-75; nuclear physics prog officer, NSF, 76-77; vis physicist, GANIL, Caen, France, 86. *Mem:* Fel Am Phys Soc. *Res:* Low and intermediate energy; heavy ion reactions. *Mailing Add:* Dept Physics State Univ NY Stony Brook NY 11794

MCGRATH, THOMAS FREDERICK, b Braddock, Pa, Jan 4, 29; m 52; c 4. ORGANIC CHEMISTRY. *Educ:* Franklin & Marshall Col, BS, 50; Univ Pittsburgh, PhD(org chem), 55. *Prof Exp:* Res chemist, Reilly Tar & Chem Corp, 54-55 & Am Cyanamid Co, 57-63; assoc prof, 63-69, PROF CHEM, SUSQUEHANNA UNIV, 70- *Mem:* Am Acad Arts & Sci; Sigma Xi; Am Chem Soc. *Res:* Organometallic and organofluorine chemistry. *Mailing Add:* 613 N Orange St Selinsgrove PA 17870-1633

MCGRATH, W PATRICK, clinical chemistry, for more information see previous edition

MCGRATH, WILLIAM ROBERT, b Weymouth, Mass, Sept 25, 33; m 57, 74; c 3. REGULATORY TOXICOLOGY, RISK ASSESSMENT. *Educ:* Mass Col Pharm, BS, 55, MS, 57; Univ Wash, PhD(pharmacol), 60. *Prof Exp:* Head, Pharmacol Dept, William S Merrell Co, Richardson-Merrell, Inc, 60-67; dir labs, Enzomedic Labs, Eversharp, Inc, 67-69; vpres chem & biol res, USV Pharmaceut Corp, Revlon, Inc, 69-75, div mgr, Nat Health Labs, 75-76 & 77-78; regional vpres, Damon Corp, 76-77; PRES, BOGSIDE AVIATION SERV, 78-; DIR PROD SAFETY, KIMBERLY-CLARK CORP, 81- *Concurrent Pos:* Lectr, Dept Pharmacol & Toxicol, Med Br, Univ Tex, 64-67; affil assoc prof, Dept Pharmacol, Sch Med, Univ Wash, 67-69. *Res:* General pharmacology and toxicology; risk assessments; product, environmental and workplace hazards. *Mailing Add:* Kimberly Clark Corp 1400 Holcomb Bridge Rd Roswell GA 30076-2199

MCGRATH, WILLIAM THOMAS, b Lincoln, Nebr, Feb 27, 33; m 67; c 2. FOREST PATHOLOGY, FOREST FIRE. *Educ:* Mont State Univ, BSF, 60; Univ Wis, PhD(plant path), 67. *Prof Exp:* Tech asst, Australian Forestry & Timber Bur, 61-62; ASSOC PROF FORESTRY, STEPHEN F AUSTIN STATE UNIV, 68- *Concurrent Pos:* US Forest Serv jumper, 57-60, 75-85. *Mem:* Am Phytopath Soc; Am Inst Biol Sci; Sigma Xi; Int Soc Arboricult. *Res:* Pine rusts; identification of Western Gulf Coast wood decay fungi; American mistletoe host specificity. *Mailing Add:* Sch of Forestry Stephen F Austin State Univ Nacogdoches TX 75962

MCGRAW, CHARLES PATRICK, b Sherman, Tex, Feb 17, 42; m 67; c 4. NEUROPHYSIOLOGY. *Educ:* Belmont Abbey Col, BCh, 64; ETex State Univ, MS, 67; Baylor Univ, cert biomed eng, 68; Tex A&M Univ, PhD(neurophysiol), 69. *Prof Exp:* Mem fac, ETex State Univ, 65-67; mem fac, Tex A&M Univ, 67-69; mem fac & neurophysiologist, Div Neurosurg & Dept Physiol, Tex Med Br, Galveston, 69-73; mem fac & assoc prof, Dept Neurol, Anat & Physiol, Bowman Gray Sch Med, 73-80; MEM FAC, PROF & ASST DIR NEUROSCI PROGS, DIV NEUROSURG, MED SCH, UNIV LOUISVILLE, 80- *Concurrent Pos:* HEW grant. *Mem:* AAAS; Am Inst Biol Sci; Am Heart Asn; Am Physiol Soc; Am Soc Clin Hypnosis. *Res:* Tritium labeled testosterone locating centers in the rat brain with radioaudiography; neural mechanisms involved in the immobility reflex from phylogenetic ontogenetic, pharmacological and surgical approaches; head injury and stroke; neurological and neurosurgical research; cerebral blood flow; autoregulation; patient monitoring; monitoring of intracranial pressure; biomedical instrumentation. *Mailing Add:* Dept Surg Neurosurg Univ Louisville Sch Med Louisville KY 40292

MCGRAW, DELFORD ARMSTRONG, b Keyrock, WVa, May 13, 17; m 41; c 2. PHYSICS. *Educ:* Concord Col, AB, 37; WVa Univ, MS, 39. *Prof Exp:* Supvr, Ballistics Lab, Hercules Powder Co, 41-45; res physicist, Owens-Ill Inc, 46-55, sr physicist, 55-66, mgr res instrumentation, 66-69, assoc dir res, 69-70, dir eng res, 70-73, mgr process systs, 73-78, chief engr, eng tech serv,

78-81; RETIRED. *Honors & Awards:* Forrest Award, Am Ceramic Soc, 52 & 59. *Mem:* Fel Am Ceramic Soc; fel Brit Soc Glass Technol. *Res:* Velocity of sound in gases; ballistics of small arms and rocket powders; mechanical and thermal properties of glass and plastics. *Mailing Add:* 3410 Chapel Dr Toledo OH 43615

MCGRAW, GARY EARL, b Wellsville, NY, Sept 26, 40; m 63; c 3. PHYSICAL CHEMISTRY. *Educ:* Univ Mich, BS, 62; Pa State Univ, PhD(phys chem), 65. *Prof Exp:* US Dept Health, Educ & Welfare fel, Div Air Pollution, Univ Calif, Berkeley, 66-67; res chemist, Tenn Eastman Co, 67-68, sr res chemist, 69-74, res assoc, 74-75, actg div head, Phys & Anal Chem Div, 75-76, actg div dir, Anal Sci Div, 76-77, staff asst to exec vpres develop, 77, asst works mgr, 78, asst div supvr, Polymers Div, 78-80, dir, Develop Div, Tex Eastman Co, 80-82, staff asst, Eastman Chem Div, 82-83, dir, Tech Serv & Develop Div, 83-85, asst dir res, Eastman Chem Div, Eastman Kodak Co, 85-88, VPRES DEVELOP, EASTMAN CHEM CO, 88- *Mem:* AAAS; Am Chem Soc; Sigma Xi. *Res:* Spectroscopy; polymer morphology. *Mailing Add:* Eastman Chem Co PO Box 1972 Kingsport TN 37662

MCGRAW, GERALD WAYNE, b Tampa, Fla, Apr 30, 43; m 67; c 3. WOOD CHEMISTRY. *Educ:* Ouachita Baptist Univ, BS, 65; Fla State Univ, PhD(org chem), 71. *Prof Exp:* From asst prof to assoc prof, 70-80, PROF CHEM, LA COL, 80- *Concurrent Pos:* Consult, USDA Forest Serv. *Mem:* Am Chem Soc. *Res:* Chemistry of condensed tannins; utilization of tannins in PF resins. *Mailing Add:* Dept Chem La Col Col Sta Box 501 1140 Coll Dr Pineville LA 71360

MCGRAW, JAMES CARMICHAEL, b Martins Ferry, Ohio, Mar 20, 28; m 51; c 3. ZOOLOGY, PARASITOLOGY. *Educ:* Oberlin Col, AB, 51; Ohio State Univ, MS, 57, PhD(zool, parasitol), 68. *Prof Exp:* ASST PROF BIOL, STATE UNIV NY COL PLATTSBURGH, 64- *Mem:* AAAS; Am Soc Parasitol; Am Inst Biol Sci; Sigma Xi. *Res:* Nematodes parasitic in amphibians, especially the genus Cosmocercoides. *Mailing Add:* Dept of Biol State Univ of NY Col Plattsburgh NY 12901

MCGRAW, JOHN LEON, JR, b Port Arthur, Tex, June 16, 40; m 57; c 4. CELL BIOLOGY, ELECTRON MICROSCOPY. *Educ:* Lamar State Col, BS, 62; Tex A&M Univ, MS, 64, PhD(biol), 68. *Prof Exp:* Res grants, Res Ctr, 67-74, from asst prof to assoc prof, 67-77, PROF BIOL, LAMAR UNIV, 78- *Honors & Awards:* Regents' Merit Award, Lamar Univ, 72; Edwin S Hays Sci Teacher Award, 76. *Mem:* Sigma Xi; Electron Micros Soc Am. *Res:* Parasites of fishes; morphology and ecological distribution of Myxomycetes; Tardigrada-taxonomy; ultrastructural studies and toxicology of dimethyl sulfoxide and fluorocarbons; ultrastructural studies related to vertebral subluxations, nerve damage, and acupuncture; computerized teaching techniques. *Mailing Add:* Dept Biol Box 10037 Beaumont TX 77710

MCGRAW, LESLIE DANIEL, b Hutchinson, Minn, Dec 19, 20; m 47; c 2. CHEMISTRY. *Educ:* Col St Thomas, BS, 42; Carnegie Inst Technol, MS, 43, DSc(phys chem), 46. *Prof Exp:* Instr chem, Pa Col Women, 45-46; asst prof, Webster Col, 46-48; from assoc consult chemist to asst div chief, Battelle Mem Inst, 48-63; dir res phys chem, North Star Res & Develop Inst, 63-65; div chief chem processes & res serv, Nat Steel Corp, 65-86; RETIRED. *Mem:* Am Chem Soc; Electrochem Soc; Am Electroplaters Soc. *Res:* Analytical chemistry; electro-mechanical research; operations research; information services; corrosion; electrochemical processes. *Mailing Add:* 526 Briar Lane NE Grand Rapids MI 49503-2174

MCGRAW, ROBERT LEONARD, b Phillipsburg, NJ, Jan 18, 49. ATMOSPHERIC SCIENCES. *Educ:* Drexel Univ, BS, 72; Univ Chicago, SM, 74, PhD(phys chem), 79. *Prof Exp:* Res assoc, Univ Calif, Los Angeles, 77-80; ASST SCIENTIST, BROOKHAVEN NAT LAB, 81- *Mem:* Am Chem Soc. *Res:* Physical chemistry including the statistical mechanics of phase transitions and nucleation; mechanisms of gas-to-particle conversion in combustion streams and polluted air. *Mailing Add:* Dept Agron 138 Mumford Hall Univ Mo Columbia MO 65211

MCGRAW, TIMOTHY E, PATHOLOGY. *Prof Exp:* ASSOC PROF, PATH DEPT, COL PHYSICIANS & SURGEONS, COLUMBIA UNIV, 89- *Mailing Add:* Path Dept Columbia Univ Col P & S 630 W 168th St Rm 1428 New York NY 10032

MCGRAY, ROBERT JAMES, entomology, public health, for more information see previous edition

MACGREGOR, ALEXANDER WILLIAM, b Strathpeffer, Scotland, Jan 29, 38; Can citizen; m 67. CEREAL CHEMISTRY, PLANT BIOCHEMISTRY. *Educ:* Univ Edinburgh, BSc Hons, 60, PhD(chem), 64. *Prof Exp:* Asst prof agr biochem, Edinburgh Sch Agr, Univ Edinburgh, 64-68; RES SCIENTIST CEREAL CHEM, GRAIN RES LAB, CAN GRAIN COMN, 68- *Mem:* Royal Soc Chem; Am Asn Cereal Chem; Am Soc Plant Physiol. *Res:* Chemistry and biochemistry of cereal grains during growth, maturation and germination; malt quality evaluation. *Mailing Add:* Grain Res Lab Can Grain Comn 1404-303 Main St Winnipeg MB R3C 3G8 Can

MCGREGOR, BONNIE A, b Fitchburg, Mass, June 13, 42; m 75. OCEANOGRAPHY. *Educ:* Tufts Univ, BS, 64; Univ RI, MS, 67; Univ Miami, PhD(marine sci), 75. *Prof Exp:* Res asst oceanog, Univ RI, 65-70, res assoc, 70-72; oceanogr, Atlantic Oceanog & Meteorol Labs, Nat Oceanog & Atmospheric Admin, 72-78; res assoc, Tex A&M Univ, 78-79; GEOLOGIST, US GEOL SURVEY, 79- *Mem:* Am Geophys Union; Asn Women Geoscientists; Soc Econ Paleontologists & Mineralogists; Am Asn Petrol Geologists; Sigma Xi. *Res:* Processes responsible for the evolution of the sea floor; sedimentary framework and the processes which determine the sediment stability and evolution of continental margins. *Mailing Add:* 915 Nat Ctr Reston VA 22092

MACGREGOR, C(HARLES) W(INTERS), b Dayton, Ohio, May 25, 08; m 51; c 1. MECHANICAL ENGINEERING. *Educ:* Univ Mich, BS, 29; Univ Pittsburgh, MS, 32, PhD(mech), 34. *Prof Exp:* Res engr, Westinghouse Elec & Mfg Co, Pa, 29-34; asst engr, US Bur Reclamation, 34; from instr to prof mech eng, Mass Inst Technol, 34-52; vpres in chg eng & sci studies, Univ Pa, 52-54; private consult, 54-60; eng consult & mgr adv tech, Systs Develop Div Develop Lab, IBM Corp, 60-68, eng consult, 68-71; PRIVATE CONSULT, 71- *Honors & Awards:* Levy Medal, Franklin Inst, 41; Dudley Medal, Am Soc Testing & Mat, 41; Ord Develop Award, US Navy, 45. *Mem:* Am Soc Mech Engrs; Am Soc Metals; Am Soc Testing & Mat; fel Franklin Inst; Am Inst Mining, Metall & Petrol Engrs; fel Am Acad Arts & Sci. *Res:* Plasticity and brittle fracture of materials; elasticity; stress analysis; testing of materials; rolling of metals; applied mechanics. *Mailing Add:* 112 Jerusalem Rd Cohasset MA 02025

MCGREGOR, DENNIS NICHOLAS, b Oelwein, Iowa, Nov 6, 43; m 65; c 2. COMMUNICATIONS, SYSTEMS ENGINEERING. *Educ:* Cath Univ Am, BEE, 65, MEE, 66. *Prof Exp:* Mem tech staff commun syst, Comput Sci Corp, 66-70, sr mem tech staff, 70-71; prin staff, commun systs div, ORI, Inc, 71-73, assoc prog dir, 73-75, sr scientist, 75-79, prin scientist, 79-81; head commun network sect, 81-87, HEAD NETWORK RES & SIMULATION SECT, NAVAL RES LAB, 81- *Mem:* Sr mem Inst Elec & Electronics Engrs; assoc Sigma Xi. *Res:* Analysis and design of communication networks and satellite and terrestrial communications systems; advanced networking techniques; modulation and coding; anti-jam techniques; network simulation; surveillance and navigation techniques; propagation and interference effects. *Mailing Add:* 5912 Oakland Park Dr Burke VA 22015

MACGREGOR, DONALD MAX, b UK, Oct 2, 49. ELECTRONICS ENGINEERING, APPLIED MATHEMATICS. *Educ:* St Catharine's Col, Cambridge, UK, BA, 70; Univ Col N Wales, PhD(electronic eng), 73. *Prof Exp:* LEAD ENGR, ELECTROCON INT, INC, 73- *Res:* Computer-aided simulation and design of electron tubes including cathode-ray tubes, klystrons, traveling-wave tubes, crossed-field tubes and gyroresonant devices; semiconductor device modeling; fault analysis and production costing for electric utilities. *Mailing Add:* Electrocon Int, Inc 611 Church St Ann Arbor MI 48104-3098

MCGREGOR, DONALD NEIL, b Cameron, Tex, Mar 27, 36; m 57; c 1. ORGANIC CHEMISTRY. *Educ:* Rice Univ, BA, 57; Mass Inst Technol, PhD(org chem), 61. *Prof Exp:* Res asst chem, Lever Bros Co, 56 & 57; teaching asst org chem, Mass Inst Technol, 57-58; sr res scientist, Bristol Labs, 61-69, asst dir org chem res, 69-73, dir antibiotic chem res, 73-77, assoc dir med chem, 77-82, assoc dir anti-infective chem, 82-85, ref fel, 85-89, sr res fel, 90-91, ASSOC DIR PHARMACOL DEVELOP PLANNING, BRISTOL-MYERS SQUIBB, 91- *Concurrent Pos:* Vis scientist, McGill Univ, 85-86; vis scholar, Wesleyan Univ, 86-89. *Mem:* Am Chem Soc; Am Soc Microbiol; NY Acad Sci. *Res:* Chemistry and biochemistry of antiinfective agents. *Mailing Add:* 11 Oak Hill Dr Clinton CT 06413-2428

MACGREGOR, DOUGLAS, b Fresno, Calif, Dec 5, 25; m 56; c 4. ENGINEERING. *Prof Exp:* Petrol engr, Arabian Am Oil Co, 46-48; engr plant construct, CWI, Stockholm, 49-50; proj engr agr, Pineapple Res Inst, Hawaii, 50-53; proj engr automation, HMRN, Samoa, 53-55; asst plant engr, Kennecott Copper Corp, 55-57; staff engr, US Steel Corp, 57-58; proj engr, Kaiser Eng, 59-61; hazards engr, Exp Gas-Cooled Reactor Proj, Oak Ridge, Tenn, 61-63; vpres res & develop, Terralab Engrs, 63-; AT INST SCI RES, SALT LAKE CITY. *Concurrent Pos:* Pres, Energy Bioneers Inc; dir, Unilink Corp; chmn, Intergalactic Corp; dir, Edenglo Corp. *Mem:* Am Soc Testing & Mat; Am Nuclear Soc; AAAS; Am Inst Chem Engrs; Am Chem Soc. *Res:* Agriculture; hydroelectrometallurgy; test methods and procedures; application engineering; safety analysis; consumer safety procedures; failure analysis; nutrition; recipe development; photo engineering; acoustics; mechanics. *Mailing Add:* Terralab Engrs PO Box 7025 Salt Lake City UT 84157

MCGREGOR, DOUGLAS D, b Hamilton, Ont, Mar 5, 32; m 63; c 1. EXPERIMENTAL PATHOLOGY. *Educ:* Univ Western Ont, BA, 54, MD, 56; Oxford Univ, DPhil(path), 63. *Prof Exp:* Asst prof path, Case Western Reserve Univ, 63-68; assoc prof, Univ Conn, 68-69; mem, Trudeau Inst, 69-76, PROF IMMUNOL & DIR, JAMES A BAKER INST FOR ANIMAL HEALTH, CORNELL UNIV, 76- *Res:* Development and immunological activity of lymphocytes. *Mailing Add:* 3108 Winchester Plano TX 75075

MCGREGOR, DOUGLAS H, b Temple, Tex, Aug 28, 39; m 69; c 2. ANATOMIC PATHOLOGY, CLINICAL PATHOLOGY. *Educ:* Duke Univ, BA, 61, MD, 66. *Prof Exp:* Res trainee exp path, USPHS, Sch Med, Duke Univ, 63-66; intern, resident II & chief resident path, Ctr Health Sci, Univ Calif, Los Angeles, 66-68, res trainee immunopath, 67-68; pathologist, Atomic Bomb Casualty Comn, Japan, 68-71; chief resident path, Queen's Med Ctr, 71-73; from asst prof to assoc prof, 73-82, PROF, DEPT PATH & ONCOL, MED CTR, UNIV KANS, 82- *Concurrent Pos:* Auditor, Res Training Prog, Sch Med, Duke Univ, 64-65; investr radiation res & epidemiol path, Atomic Bomb Casualty Comn, USPHS, 68-71. *Mem:* Fel Am Soc Clin Pathologists; Int Acad Path; Am Asn Pathologists; fel Col Am Pathologists; Soc Exp Biol & Med; AAAS; NY Acad Sci; Am Asn Univ Professors. *Res:* Biology and pathology of parathyroid hormone secretion; platelet-leukocyte aggregation; ultrastructure and pathobiology of neoplasms; radiation carcinogenesis; morphogenesis of arteriosclerosis; morphology and histochemistry of lung carcinoma; platelet satellitism; non-occlusive mesenteric infarction; liposarcoma; Avitene granulomas; diabetic nephropathy; parathormone-secreting carcinoma; gastric hyalinization; basal and squamous cell carcinomas of skin; breast and gynecologic radiation carcinogenesis; morphological correlates of parathormone biosynthesis; morphogenesis of calcification in atherosclerosis; author of various publications. *Mailing Add:* 9400 Lee Blvd Leawood KS 66206

MCGREGOR, DOUGLAS IAN, b Ottawa, Ont, June 22, 42; m 74; c 2. PLANT BIOCHEMISTRY, PLANT PHYSIOLOGY. *Educ:* Carleton Univ, BSc, 64; Purdue Univ, PhD(biochem, physiol), 69. *Prof Exp:* RES SCIENTIST OIL SEED CROPS, AGR CAN, 70- *Concurrent Pos:* Chmn Can assoc comt, Int Standards Orgn, 77-; adj prof dept biochem, Univ Sask, Saskatoon. *Mem:* Asn Off Anal Chem. *Res:* Improvement of the efficiency of production, adaptability and quality of rape seed, mustard and their products through physiological biochemical research. *Mailing Add:* Agr Can Res Sta 107 Sci Crescent Sask SK S7N 0X2 Can

MCGREGOR, DUNCAN J, b St Joseph, Mo, Jan 3, 21; m 46; c 3. GEOLOGY. *Educ:* Univ Kans, BS, 43, MS, 48; Univ Mich, PhD(geol), 53. *Prof Exp:* Geologist indust minerals, State Geol Surv, Kans, 47; asst dist geologist petrol geol, Sinclair Oil & Gas Co, 48; asst prof geol & head indust minerals sect, Geol Surv, Ind Univ, 53-63; PROF GEOL & STATE GEOLOGIST, STATE GEOL SURV, UNIV SDAK, 63- *Mem:* AAAS; Asn Am State Geologists; Geol Soc Am; Soc Econ Geologists; Soc Econ Paleontologists & Mineralogists; Am Asn Petrol Geologists; Sigma Xi. *Res:* Industrial minerals; economic geology. *Mailing Add:* SDak Geol Surv Ackley Sci Ctr Univ SDak Vermillion SD 57069

MACGREGOR, ELIZABETH ANN, b Edinburgh, UK, 41; Brit & Can citizen; m 67. ENZYME STRUCTURE & FUNCTION RELATIONSHIPS, POLYMER DEGRADATION. *Educ:* Univ Edinburgh, UK, BSc, 63, PhD(chem), 66. *Prof Exp:* Postdoctoral fel, Univ Edinburgh, UK, 66-67; postdoctoral fel, 68-70, lectr, 70-81, INSTR CHEM, UNIV MAN, CAN, 81- *Mem:* Royal Soc Chem. *Res:* Relationship between structure and function of starch- degrading enzymes, particularly alpha-amylases; computer modelling of active sites and action patterns is carried out. *Mailing Add:* Dept Chem Univ Man Winnipeg MB R3T 2N2 Can

MACGREGOR, IAN DUNCAN, b Calcutta, India, Jan 5, 35; Can citizen; m 56; c 4. PETROLOGY, GEOCHEMISTRY. *Educ:* Univ Aberdeen, BSc, 57; Queen's Univ, Ont, MSc, 60; Princeton Univ, PhD(geol), 64. *Prof Exp:* Sr asst, Geol Surv Can, Ottawa, 57, party chief, 58-59; fel exp petrol, Geophys Lab, Washington, DC, 64-65; assoc prof high pressure exp petrol, Southwest Ctr Advan Studies, Tex, 65-69; chmn dept, 69-74, PROF GEOL, UNIV CALIF, DAVIS, 69-; AT DIV EARTH SCIS, NAT SCI FOUND. *Concurrent Pos:* Chmn petrol panel, Joint Oceanog Insts Deep Earth Sampling, 66- *Mem:* Am Geophys Union; Mineral Soc Am. *Res:* Field geology; experimental petrology. *Mailing Add:* 5005 River Hill Rd Bethesda MD 20816

MACGREGOR, JAMES GRIERSON, b Vegreville, Alta, Feb 14, 34; m 56; c 3. STRUCTURAL ENGINEERING. *Educ:* Univ Alta, BSc, 56; Univ Ill, Urbana, MS, 58, PhD(civil eng), 60. *Prof Exp:* Res assoc, Univ Ill, 59-60; from asst prof to prof, 60-89, UNIV PROF CIVIL ENG, UNIV ALTA, 89-; PRES, MKM ENGR CONSULT & VPRES, AM CONCRETE INST, 90- *Concurrent Pos:* Vpres, European Comt Concrete. *Honors & Awards:* State of Art in Civil Eng Award, Am Soc Civil Eng, 68 & 74, Reese Res Prize, 76 & 79, Can-Am Amity Award, 79 & Norman Medal, 83; Wason Medal, Am Concrete Inst, 72, Reese Medal, 72 & 87, Bloem Award, 74, Kelly Award, 86. *Mem:* Am Soc Civil Eng; Am Concrete Inst; Royal Soc Can; Can Acad Eng. *Res:* Strength and stability of reinforced concrete; shear strength of reinforced concrete; safety of concrete structures. *Mailing Add:* Dept Civil Engr Univ Alta Edmonton AB T6G 2G7 Can

MACGREGOR, MALCOLM HERBERT, b Detroit, Mich, Apr 24, 26; m 49; c 3. PHYSICS. *Educ:* Univ Mich, BA, 49, MS, 50, PhD(physics), 54. *Prof Exp:* PHYSICIST, LAWRENCE LIVERMORE NAT LAB, UNIV CALIF, 53- *Concurrent Pos:* NATO fel, Inst Theoret Physics, Denmark, 60-61; lectr, Univ Calif, 60, 62 & 63; consult, Appl Radiation Corp, 56-73, Gen Atomic Div, Gen Dynamics Corp, 62-69 & Sci Appln, Inc, 69-73. *Mem:* Fel Am Phys Soc. *Res:* Neutron scattering; two-nucleon problem; elementary particle structure. *Mailing Add:* Lawrence Livermore Nat Lab PO Box 808 Livermore CA 94550

MCGREGOR, MAURICE, MEDICINE. *Prof Exp:* PRES, EVAL COUN HEALTH TECHNOL ASSESSMENT, 88- *Mailing Add:* Eval Coun Health Technol Assessment 800 Victoria Sq PO Box 215 Rm 4205 Montreal PQ A4Z 1E3 Can

MACGREGOR, RONAL ROY, b Hayward, Calif, July 4, 39; m 69; c 2. BIOCHEMISTRY. *Educ:* Calif State Col, Long Beach, BS, 64; Ind Univ, PhD(biochem), 68. *Prof Exp:* From res assoc to co-prin investr, Calcium Res Lab, Kans City VA Med Ctr, Kansas City, Mo, 68-86; PROF, DEPT ANAT & CELL BIOL, UNIV KANS MED CTR, 86- *Concurrent Pos:* From instr to asst prof biochem Univ Mo, 71-80; adj asst prof, 77-81, adj/res assoc prof, 81-85, assoc prof, Dept Biochem, Univ Kans, Med Ctr, 85-87. *Mem:* Am Soc Biol Chemists; Endocrine Soc; Am Soc Cell Biol; Am Soc Bone & Mineral Res. *Res:* The packaging and secretion of parathormone; the mechanisms of these processes and their control. *Mailing Add:* Dept Anat & Cell Biol Univ Kans Med Sch 39 St Rainbow Blvd Kansas KS 66103

MACGREGOR, RONALD JOHN, b South Bend, Ind, Nov 30, 38; m 60; c 3. ENGINEERING SCIENCES, NEUROPHYSIOLOGY. *Educ:* Purdue Univ, BS, 62, MS, 64, PhD(eng sci), 67. *Prof Exp:* Res asst, Purdue Univ, 62-66; engr bioeng, Rand Corp, Calif, 66-71; asst prof chem eng & eng design, 71-78, ASSOC PROF CHEM ENG, UNIV COLO, 78- *Concurrent Pos:* Consult bioeng, Rand Corp, 65-66; prin investr, NSF grant, 72-77 & NIH grant, 74-77. *Mem:* Am Inst Chem Engrs. *Res:* Neural modeling particular reference to theoretical models for sustained activity in neuronal networks. *Mailing Add:* Aerospace Engr Box 429 Univ Colo Boulder CO 80309

MCGREGOR, RONALD LEIGHTON, b Manhattan, Kans, Apr 4, 19; m 42. BOTANY. *Educ:* Univ Kans, AB, 41, MA, 47, PhD, 54. *Prof Exp:* Asst instr, State Biol Surv, 41-42 & 46, from instr to assoc prof, 47-60, chmn dept, 61-69, chmn div biol sci, 69-76, prof bot, 61-89, dir herbarium & dir, State Biol Surv, 73-89; RETIRED. *Mem:* Am Bryol Soc; Bot Soc Am; Am Fern Soc; Am Soc Plant Taxonomists; Brit Bryol Soc. *Res:* Systematic botany; ecology; flora of Kansas and the southwest United States. *Mailing Add:* Dept Biol Herbarium West Campus Univ Kans 10 Abotr Lawrence KS 66045

MCGREGOR, STANLEY DANE, b Endicott, NY, Nov 11, 38; m 60; c 2. ORGANIC CHEMISTRY. *Educ:* WVa Wesleyan Col, BS, 60; Univ Wis, MS, 61, PhD(org chem), 66. *Prof Exp:* Instr chem, Davis & Elkins Col, 61-63; fel, Univ Fla, 66-67; res chemist, E C Britton Res Lab, 68-71, RES CHEMIST, AGR SYNTHESIS LAB, DOW CHEM USA, 71- *Mem:* Am Chem Soc; AAAS. *Res:* Synthesis of biologically active organic compounds; heterocyclic synthesis. *Mailing Add:* 5901 Sommerset Midland MI 48640

MACGREGOR, THOMAS HAROLD, b Jersey City, NJ, June 21, 33; m 56; c 1. MATHEMATICS. *Educ:* Lafayette Col, AB, 54; Univ Pa, MA, 56, PhD(math), 61. *Prof Exp:* Instr math, Col South Jersey, Rutgers Univ, 58-59; prof, Lafayette Col, 59-67; chmn dept, 75-78, PROF MATH, STATE UNIV NY ALBANY, 67- *Concurrent Pos:* NSF grants, 62-64, 65-67 & 68-71. *Mem:* Am Math Soc; Math Asn Am. *Res:* Complex analysis; geometric function theory. *Mailing Add:* Dept of Math State Univ of NY Albany NY 12222

MCGREGOR, WALTER, b Ukraine, Nov 2, 37; US citizen; m 65; c 2. SURGICAL INSTRUMENTATION, SURGICAL TECHNIQUE IMPROVEMENTS THROUGH IMPROVED SUPPORT SYSTEMS. *Educ:* Fairleigh Dickinson Univ, BS, 73, MBA, 75. *Prof Exp:* Mech designer eng, Mack Motors, 57-62; prod develop eng prod develop, Magnus Inc, 62-63; design eng prod develop, Ethicon Inc, 63-65, sect leader equip develop, 65-68, supvr equip develop, 68-76, MGR PROD DEVELOP & DIR PROD DEVELOP & MAT SCI, ETHICON INC, 76- *Concurrent Pos:* Consult surg, Wilmer Inst, Johns Hopkins, 65-70; needle & inst adv, Surg Panel, 68-; lectr, Ethicon Resident Surgeons Meeting, 71- & Univ Va Med Sch Plastic Surg, 86-; tech presenter, Soc Contemp Opthal, 82; founding mem, Am Asn Med Systs & Informatics, 82- *Mem:* Fel Soc Advan Med Syst; Am Asn Med Systs & Informatics. *Res:* Development of surgical instrumentation made from high technology materials as well as processes for all surgical specialties and the new surgical field of endoscopy; nine publications and six patents. *Mailing Add:* 104 Hoffman Rd Flemington NJ 08822

MCGREGOR, WHEELER KESEY, JR, b Akron, Ohio, Apr 20, 29; m 48. PHYSICS. *Educ:* Univ Tenn, BS, 51, MS, 61, PhD(physics), 69. *Prof Exp:* Instrument engr controls anal, Arnold Eng Develop Ctr, ARO, Inc, Engine Testing Facil, Instrument Br, 51-56, supvr controls sect, 56-58, res engr measurement tech, Res Br, 58-68, supvr physics sect res, 68-70, res engr, T-Proj Br, 70-75, supvr physics sect res, Technol Appl Br, 75-77, staff scientist, Advan Diag Br, Aro, Inc, 78-80; phys scientist, US Air Force Rocket Propulsion Lab, Edwards Air Force Base, 77-78; SR TECH SPECIALIST, SVERDRUP TECHNOL INC, 81- *Concurrent Pos:* Prof physics & adj fac, Univ Tenn Space Inst, Tullahoma, 76- *Honors & Awards:* Gen H H Arnold Award, Am Inst Aeronaut & Astronaut, 62. *Mem:* Am Phys Soc; assoc fel Am Inst Aeronaut & Astronaut; Sigma Xi. *Res:* Spectroscopy and radiative transfer in gases (combustion flames and plasmas). *Mailing Add:* Stillwood Dr Rte 8 Box 8070 Manchester TN 37355

MCGREGOR, WILLIAM HENRY DAVIS, b SC, Mar 25, 27; m 50; c 2. PLANT PHYSIOLOGY, FORESTRY. *Educ:* Clemson Col, BS, 51; Univ Mich, BSF & MF, 53; Duke Univ, PhD(physiol), 58. *Prof Exp:* Res forester, Forest Serv, USDA, 53-57, plant physiologist, 57-60; from assoc prof to prof forestry, 60-69, head dept, 69-70, dean, Col Forest & Recreation Resources, 70-78, PROF FORESTRY, CLEMSON UNIV, 78- *Concurrent Pos:* Asst, Duke Univ, 55-57. *Mem:* Sigma Xi; fel Soc Am Foresters. *Res:* Forest influences, forest ecology and tree physiology, particularly vegetation analysis, succession and hydrology. *Mailing Add:* 210 Thomas St Clemson SC 29631

MCGREW, ELIZABETH ANNE, b Faribault, Minn, Aug 30, 16. PATHOLOGY. *Educ:* Carleton Col, AB, 38; Univ Minn, MB, 44, MD, 45. *Prof Exp:* From instr to asst prof path, Col Med, 46-47, from asst pathologist to assoc pathologist, Res & Educ Hosps, 47-62, PROF PATH, UNIV ILL COL MED, 62-, PATHOLOGIST, UNIV HOSPS, 62- *Mem:* Am Soc Clin Path; Col Am Path; Am Asn Path & Bact; Am Asn Cancer Res; Am Soc Cytol; Sigma Xi. *Res:* Cytology and cancer. *Mailing Add:* 548 Judson Ave Evanston IL 60602

MCGREW, LEROY ALBERT, b Galva, Ill, Nov 1, 38; m 63; c 3. ORGANIC CHEMISTRY. *Educ:* Knox Col, BA, 60; Univ Iowa, MS, 63, PhD(chem), 64. *Prof Exp:* Prof chem, Ball State Univ, 64-77; PROF CHEM & HEAD DEPT, UNIV NORTHERN IOWA, 77- *Mem:* Am Chem Soc. *Res:* Mechanisms of organic reactions, especially reactions of isocyanates; use of audio-visual media in chemical education; the place of chemistry and biochemistry in the liberal arts. *Mailing Add:* Dept Chem Univ Northern Iowa Cedar Falls IA 50613

MCGRIFF, RICHARD BERNARD, b St Petersburg, Fla, July 15, 35; m 69; c 2. ORGANIC CHEMISTRY, ANALYTICAL CHEMISTRY. *Educ:* Fla A&M Univ, BS, 55; Calif Inst Technol, MS, 59; Univ Wis, PhD(chem), 67. *Prof Exp:* Med chemist, Riker Labs, Inc, 59-61; res chemist org chem, Organics Dept, Hercules, Inc, 67-73; tech specialist, Special Mat Area, 73-75, SR CHEMIST ANALALYTICAL CHEM, MAT ANALYSIS AREA, XEROX CORP, 75- *Mem:* Am Chem Soc. *Res:* Thermal analysis of polymers; structure-property relationships; gas chromatography-mass spectroscopy; liquid chromatography-mass spectroscopy; reprographic imaging materials; oxidation and electrical conduction in polymers. *Mailing Add:* 886 Independence Dr Webster NY 14580

MCGROARTY, ESTELLE JOSEPHINE, b Lafayette, Ind, Sept 14, 45; m 67; c 2. MEMBRANE BIOPHYSICS, MEMBRANE BIOCHEMISTRY. *Educ:* Purdue Univ, BS, 67, PhD(molecular biol), 71. *Prof Exp:* Lectr, Dept Biol, Purdue Univ, 71-72; res assoc biochem, 73-73, asst prof biophysics, 73-

78, actg chmn biophys, 75-76, assoc prof, 78-81, assoc prof, 81-87, PROF BIOCHEM, MICHIGAN STATE UNIV, 87-, DIR UNDERGRAD PROG, 90- *Mem:* Biophys Soc; Am Soc Microbiol; Int Endotoxin Soc. *Res:* Membrane biophysics and biochemistry; interactions of surface active compounds with erythrocyte membranes; structure, composition and physical properties of bacterial membranes; biophysical and biochemical changes in cell membranes induced upon transformation. *Mailing Add:* Dept Biochem Mich State Univ East Lansing MI 48824-1319

MCGRODDY, JAMES CLEARY, b New York, NY, Apr 6, 37; c 4. SOLID STATE PHYSICS. *Educ:* St Joseph's Col, Pa, BS, 58; Univ Md, PhD(physics), 64. *Prof Exp:* Res assoc physics, Univ Md, 64-65; res scientist, 65-77, dir, Semiconductor Sci & Technol, 77-81, VPRES LOGIC & MEMORY, IBM RES, IBM CORP, 81- *Concurrent Pos:* Vis prof, Tech Univ Denmark, 70-71. *Mem:* Nat Acad Eng; fel Inst Elec & Electronics Engrs; fel Am Phys Soc. *Res:* Semiconductor technology; experimental semiconductor and surface physics. *Mailing Add:* 796 Long Hill Rd West Briarcliff Manor NY 10510

MCGRORY, JOSEPH BENNETT, b Philadelphia, Pa, Feb 23, 34; m 57; c 2. NUCLEAR PHYSICS. *Educ:* Univ of the South, BA, 55; Vanderbilt Univ, MS, 57, PhD(physics), 64. *Prof Exp:* Sanitarian health physics, USPHS, 57-60; nuclear physicist nuclear theory, 63-76, nuclear theory group leader, 76-90, ASST DIR, PHYSICS DIV, OAK RIDGE NAT LAB, 90- *Mem:* AAAS; fel Am Phys Soc. *Res:* Microscopic theory of nuclear structure; nuclear shell model. *Mailing Add:* Physics Div Bldg 6003 Oak Ridge Nat Lab Oak Ridge TN 37830-6373

MCGUIGAN, FRANK JOSEPH, b Oklahoma City, Okla, Dec 7, 24. PSYCHOLOGY. *Educ:* Univ Calif, Los Angeles, BA, 45, MA, 49; Univ Southern Calif, PhD, 50. *Prof Exp:* Instr, Pepperdine Col, 49-50; asst prof, Univ Nev, 50-51; res assoc, Psychol Corp, 50-51; res scientist, sr res scientist, actg dir res, Human Resources Res Off, George Wash Univ, 51-55; prof psychol & chmn dept, Hollins Col, Roanoke, Va, 55-76; res prof Grad Sch, prof dept psychol, Dept Psychiat & Behav Sci, Sch Med & dir Performance Res Lab, Inst Advan Study, Univ Louisville, 76-83; PROF PSYCHOL & DIR, INST STRESS MGT, US INT, UNIV SAN DIEGO, 83- *Concurrent Pos:* Vis prof, Univ Hawaii, 65 & Univ Calif, Santa Barbara, 66; adj res prof, NC State Univ, 70-72; vis scientist, Nat Acad Sci, Hungary, 75 & Bulgaria, 87; adj prof psychiat, Sch Med, Univ Louisville, 86- *Honors & Awards:* Award for Outstanding Contrib in Psychol, 73; Honor Medal for Sci Accomplishment, Scientists of Bulgaria, 81; Medal of Honor, Tours, France, 81; Medal of Sechenov, USSR Acad Med Sci, 82 & Medal of Anokhin, 84. *Mem:* Am Physiol Soc; Biofeedback Soc Am; Inter-Am Soc Psychol; Int Cong Appl Psychol; Psychonomic Soc; Soc Psychophysiol Res; Sigma Xi; Am Psychol Asn; Am Psychol Soc. *Res:* Experimental methodology; cognitive psychology; stress management; progressive relaxation. *Mailing Add:* US Int Univ 10455 Pomerado Rd San Diego CA 92131

MCGUIGAN, JAMES E, b Paterson, NJ, Aug 20, 31; m 56; c 3. GASTROENTEROLOGY, IMMUNOLOGY. *Educ:* Seattle Univ, BS, 52; St Louis Univ, MD, 56; Am Bd Gastroenterol, dipl. *Prof Exp:* Intern med, Pa Hosp, Philadelphia, 56-57; resident internal med, Sch Med, Univ Wash, 60-62; asst prof med, Wash Univ, 66-69; chief div gastroenterol, 69-78, PROF MED, COL MED, UNIV FLA, 69-, CHMN DEPT MED, 76- *Concurrent Pos:* NIH fel gastroenterol, 62-64; Nat Inst Allergy & Infectious Dis spec fel immunol, Sch Med, Wash Univ, 64-66; NIH res career develop award, 66-69. *Mem:* Asn Am Physicians; Asn Prof Med; AAAS; Am Soc Clin Invest. *Res:* Immunological phenomena associated with gastrointestinal diseases. *Mailing Add:* J Hills Miller Ctr Gainesville FL 32610

MCGUIGAN, ROBERT ALISTER, JR, b Evanston, Ill, July 21, 42; div; c 2. MATHEMATICS. *Educ:* Carleton Col, BA, 64; Univ Md, College Park, PhD(math), 68. *Prof Exp:* Asst prof math, Univ Mass, Amherst, 68-74; from asst prof to assoc prof math, 74-84, chmn dept, 75-81, chmn comput sci dept, 85-86, PROF MATH & COMPUT SCI, WESTFIELD STATE COL, 84- *Concurrent Pos:* Res anal, Nat Hwy Traffic Safety Admin, Washington, DC, 79-80; consult, Abt Assocs, Inc, 81. *Mem:* Am Math Soc; Math Asn Am. *Res:* Banach spaces; spaces of continuous functions; non-standard analysis; history of mathematics; theoretical linguistics; graph theory; computer science; data analyses, structures and the theory of computation. *Mailing Add:* Dept of Math Westfield State Col Westfield MA 01085

MCGUINNESS, DEBORAH LOUISE, b Drexel Hill, Pa, Mar 28, 58. KNOWLEDGE REPRESENTATION. *Educ:* Duke Univ, BS, 80; Univ Calif, Berkeley, MS, 81. *Prof Exp:* Res scientist videotex, artificial intel & databases, Home Info Systs, 80-84, res scientist, Artificial Intel & Comput Environ Res, 84-86, RES SCIENTIST KNOWLEDGE REPRESENTATION, ARTIFICIAL INTEL RES DEPT, AT&T BELL LABS, 86- *Mem:* Am Asn Artificial Intel. *Res:* Knowledge representation system design; explanation in systems based on description logic or terminologic logics. *Mailing Add:* AT&T Bell Labs Rm 2B-439 600 Mountain Ave Murray Hill NJ 07974-2070

MCGUINNESS, EUGENE T, b Newark, NJ, Feb 2, 27; m 59; c 2. BIOCHEMISTRY. *Educ:* St Peter's Col, BS, 49; Fordham Univ, MS, 54; Rutgers Univ, PhD(biochem), 61. *Prof Exp:* Anal chemist, Wallace & Tiernan, Inc, 49-50; res asst, Fordham Univ, 51-52; asst lab control supvr, P Ballantine & Sons, 52-55; from instr to assoc prof, 55-79, PROF CHEM, SETON HALL UNIV, 79- *Mem:* AAAS; Am Chem Soc; Am Soc Biol Chemists; Sigma Xi. *Res:* Mechanisms of enzyme action; biological oxidation; evolution of enzyme function. *Mailing Add:* 59 Monroe Ave Roseland NJ 07068

MCGUINNESS, JAMES ANTHONY, b Staten Island, NY, Nov 4, 41. ORGANIC CHEMISTRY. *Educ:* St Peter's Col, NJ, BS, 63; Columbia Univ, MA, 64, PhD(chem), 68. *Prof Exp:* RES CHEMIST RES & DEVELOP, UNIROYAL INC, 67- *Mem:* Am Chem Soc; Royal Soc Chem. *Res:* Polymer degradation, especially photochemical; polymer modification by organic reactions; synthesis of organic chemicals as agricultural products. *Mailing Add:* 38 Highland Ave Naugatuck CT 06770

MCGUINNESS, OWEN P, b Bronx, NY, Mar 16, 56. PHYSIOLOGY. *Educ:* State Univ NY, BS, 78; La State Univ, PhD, 84. *Prof Exp:* Postdoctoral fel, 85-88, ASST PROF PHYSIOL, SCH MED, VANDERBILT UNIV, 88- *Mem:* Am Diabetes Asn; Am Physiol Soc. *Mailing Add:* Molecular Physiol & Biophys Dept Sch Med Vanderbilt Univ 702 Light Hall Nashville TN 37232

MCGUIRE, CHARLES FRANCIS, b Heber City, Utah, May 13, 29; m 52; c 5. AGRONOMY, CEREAL CHEMISTRY. *Educ:* Brigham Young Univ, BS, 54; Utah State Univ, MS, 65; NDak State Univ, PhD(agron), 68. *Prof Exp:* Lab technician & mgr qual control, Pillsbury Co, 55-63; from asst prof to assoc prof, 68-77, PROF AGRON & CEREAL TECHNOL, MONT STATE UNIV, 77- *Mem:* Am Asn Cereal Chemists; Am Soc Agron. *Res:* Improvement in wheat and barley quality through improved breeding and cultural practices. *Mailing Add:* Dept Plant & Soil Sci Mont State Univ Bozeman MT 59717

MCGUIRE, CHRISTINE H, b Jenkins, Ky, Aug 1, 18; m 43. EVALUATION OF PROFESSIONAL COMPETENCE, LICENSURE & CERTIFICATION. *Educ:* Muskingum Col, BA, 37; Ohio State Univ, MA, 38. *Prof Exp:* Res asst economics, Univ Chicago, 38-41, assoc prof, 41-61; prof med educ, Col Med, Univ Ill, 61-89, assoc dir, 66-89; CONSULT, WHO, 66- *Concurrent Pos:* Consult, var med schs & prof asns, nat & int, 62-; nat chair, Group on Med Educ, Asn Am Med Cols, 71-72. *Honors & Awards:* John Hubbard Award, Nat Bd Med Examiners, 87; Distinguished Career Award, Am Educ Res Asn, 87; Merrel Flair Award, Asn Am Med Cols, 87. *Mem:* Nat Coun Measurement Educ (pres, 67-68); Am Educ Res Asn (vpres, 69-70). *Res:* Development and validation of more effective methods of evaluation of professional competence of physicians and medical students, with special emphasis on the use of simulation techniques to assess professionals. *Mailing Add:* 2231 E 67th St Apt 14D Chicago IL 60649

MCGUIRE, DAVID KELTY, b Pittsburgh, Pa, Dec 18, 34; div; c 3. SCIENCE POLICY. *Educ:* St Vincent Col, BS, 57; Univ Pittsburgh, PhD(anal chem), 64. *Prof Exp:* Res assoc, Brookhaven Nat Lab, 64-65; asst prof chem, Rider Col, 65-67; asst prof, 67-70, assoc prof chem, 70-83, chmn dept, 71-76, dir, Title III Prog, Upsala Col, 79-83; dir of licensing, Apollo Lasers Inc, subsid of Patlex Corp Inc, 83-85, bd dir, 84-86, pres, 85-86, asst to chmn, Patlex Corp Inc, 87; dir of opers, Environ Div, 88-89, PRIN CHEMIST & SR PROJ MGR, LOUIS BERGER & ASSOC INC, 88- *Mem:* AAAS; Am Chem Soc. *Res:* Science, technology, and public policy relationships, with special attention to the energy situation. *Mailing Add:* 51A James St Montclair NJ 07042

MCGUIRE, EUGENE J, b New York, NY, May 15, 38; div; c 3. ATOMIC PHYSICS. *Educ:* Manhattan Col, BEE, 59; Cornell Univ, PhD, 65. *Prof Exp:* Mem tech staff physics res, 65-74, supvr laser theory div, Sandia Labs, 74-82. *Mem:* Am Phys Soc. *Res:* High power gas laser theory; x-ray lasers; auger transitions and electron spectroscopy; gas discharges; inelastic atomic processes. *Mailing Add:* Sandia Labs Box 5800 Albuquerque NM 87185

MCGUIRE, FRANCIS JOSEPH, b Baltimore, Md, Aug 29, 32; m 63. ORGANIC CHEMISTRY, ANALYTICAL CHEMISTRY. *Educ:* Loyola Col, Md, BS, 54; Johns Hopkins Univ, MA, 56, PhD(org chem), 61. *Prof Exp:* Res chemist, E I du Pont de Nemours & Co, 61-63; from instr to asst prof, 63-67, chmn dept, 65-67, dean studies, 67-77, ASSOC PROF CHEM, LOYOLA COL, MD, 67-, DEAN UNDERGRAD STUDIES & ACAD REC, 77- *Concurrent Pos:* Acad internship prog fel, Am Coun on Educ, 69-70. *Mem:* Am Chem Soc. *Res:* Structure determination and mechanism of cyclization reactions of natural products; synthesis of polymers; problems in chemical education. *Mailing Add:* Off of Dean Studies Day Div Loyola Col 4501 N Charles St Baltimore MD 21210

MCGUIRE, GEORGE, b Edinburgh, Scotland, Apr 25, 40; m 64; c 2. INORGANIC CHEMISTRY. *Educ:* Heriot Watt Univ, BSc, 62; Edinburgh Univ, PhD(inorg chem), 65; FRIC, 73. *Prof Exp:* Lab supvr, Que Iron & Titanium Corp, 65-69; sr chemist, Johnson Matthey Chem, 69-71, develop supt, 71-73; develop supt refining, Matthey Rustenburg Refiners, 73-74; res mgr, Matthey Bishop Inc, 74-78, VPRES RES, JOHNSON MATTHEY INC, 78- *Mem:* Royal Inst Chem; Inst Works Mgrs; AAAS. *Res:* Chemistry and metallurgy of platinum group metals; applications to catalysis, drugs, refining of precious metals, powders and organometallic complexes. *Mailing Add:* Johnson Mathey Tech Ctr Blounts Ct Sonning Common Reading RG4-9NH England

MCGUIRE, JAMES HORTON, b Canandaigua, NY, June 7, 42; c 4. ATOMIC PHYSICS. *Educ:* Rensselaer Polytech Inst, BS, 64; Northeastern Univ, MS, 66, PhD(physics), 69. *Prof Exp:* Asst prof physics, Tex A&M Univ, 69-72; from asst prof to assoc prof, 72-82, PROF PHYSICS, KANS STATE UNIV, 82- *Concurrent Pos:* US Dept Energy grant; consult, Lawrence Livermore Lab, 78; gen comt, Int Conf Phys Elec Atom Collisions, 85-; secy-treas, div atomic & molecular optical physics, Am Phys Soc, 90- *Mem:* Fel Am Phys Soc; Am Chem Soc. *Res:* Atomic scattering theory; excitation; ionization; charge transfer; electron systems; multi electron excitation by charge particle impact. *Mailing Add:* Dept of Physics Kans State Univ Manhattan KS 66506

MCGUIRE, JAMES MARCUS, b Gassville, Ark, July 30, 35; m 56; c 4. PLANT PATHOLOGY. *Educ:* Univ Ark, BS, 56, MS, 57; NC State Univ, PhD(plant path), 61. *Prof Exp:* Asst prof plant path, SDak State Univ, 61-63; res assoc, Univ Ark, Fayetteville, 63-65, assoc prof plant path, 65-70, prof, 70-; HEAD, DEPT PLANT PATH & WEED SCI, MISS STATE UNIV. *Mem:* Am Phytopath Soc; Soc Nematol; Weed Sci Soc Am. *Res:* Nematode transmission of viruses; interactions between nematodes and other plant pathogens; virus diseases of horticultural plants. *Mailing Add:* Dept Plant Path Miss State Univ PO Drawer PG Miss State MS 39762

MCGUIRE, JOHN ALBERT, b Banner, Miss, July 31, 31; m 55; c 1. ANIMAL BREEDING, ANIMAL GENETICS. *Educ:* Miss State Univ, BS, 51, MS, 57; Auburn Univ, PhD(animal breeding), 69. *Prof Exp:* Instr animal sci, Miss State Univ, 55-56, supt, Natchez Br Exp Sta, 56-65; from asst prof to assoc prof, 68-84, prof, 84-87, DEPT HEAD & PROF BIOSTATISTICS, AUBURN UNIV, 87- *Mem:* Biomet Soc. *Res:* Animal genetics; improvement of important production traits in beef cattle and sheep by selection for genetically superior animals. *Mailing Add:* Res Data Anal Auburn Univ Auburn AL 36849-5402

MCGUIRE, JOHN J, b Mar 19, 30; US citizen; m 54; c 6. ORNAMENTAL HORTICULTURE. *Educ:* Rutgers Univ, BS, 58; Univ RI, MS, 61, PhD(biol sci), 68. *Prof Exp:* Instr hort, Va Polytech Inst & State Univ, 61-62; from instr to prof hort, 62-85, chmn, Dept Plant Sci, 78-85, EMER PROF, UNIV RI, 91- *Honors & Awards:* L C Chadwick Award, Am Asn Nurserymen, 84. *Mem:* Am Soc Hort Sci; Int Plant Propagators Soc. *Res:* Winter hardiness of woody ornamental plants; marketing technology of container grown ornamental plants; propagation of woody plants; solar heated propagation beds. *Mailing Add:* 2071 Ministerial Rd Wakefield RI 02879

MCGUIRE, JOHN JOSEPH, b San Luis Obispo, Calif, Nov 18, 49. BIOCHEMICAL PHARMACOLOGY. *Educ:* Univ Calif, Los Angeles, BS, 71; Univ Calif, Berkeley, PhD(biochem), 76. *Prof Exp:* Postdoctoral res scientist pharmacol, Yale Univ Sch Med, 76-81, 83-86, spec fel, Leukemia Soc Am, Yale Univ Sch Med, 81-83; cancer res scientist IV, 86-89, CANCER RES SCIENTIST V, ROSWELL PARK CANCER INST, 89-; ASST RES PROF PHARMACOL, STATE UNIV NY BUFFALO, 86- *Concurrent Pos:* Scholar, Leukemia Soc Am, Yale Univ Sch Med, Roswell Park Cancer Inst, 84-89. *Mem:* Sigma Xi; Am Chem Soc; Am Asn Cancer Res; AAAS. *Res:* Folic acid and one-carbon metabolism and biochemistry; biochemical pharmacology of antineoplastic agents; mechanisms of drug resistance. *Mailing Add:* Roswell Park Cancer Inst 666 Elm St Buffalo NY 14263

MCGUIRE, JOHN L, b Kittanning, Pa, Nov 3, 42; m 69; c 2. PHARMACOLOGY. *Educ:* Butler Univ, BS, 65; Princeton Univ, MS, 68; Columbia Univ, PhD, 69. *Prof Exp:* Pop Coun fel, 69; pharmacologist, 69-72, section head molecular biol, 72-75, exec dir res, 75-80, VPRES PRECLIN RES & DEVELOP LABS, ORTHO PHARMACEUT CORP, RARITAN, NJ, 80- *Concurrent Pos:* Adj assoc prof, Dept Med, M S Hershey Sch Med, Pa State Univ, 78-; adj prof, Dept Animal Sci, Rutgers Univ, 83-; consult to NASA, 85; adj prof, Eastern Va Sch Med, 87- *Mem:* Am Soc Pharmacol & Exp Therapeut; Endocrine Soc; Soc Exp Biol & Med; Am Physiol Soc; Am Soc Clin Pharmacol & Therapeut; Royal Soc Med. *Res:* Endocrine and gastrointestinal pharmacology; toxicology; endocrinology. *Mailing Add:* R W Johnson Pharmaceut Res Inst (J&J) PO Box 300 Raritan NJ 08869-0602

MCGUIRE, JOHN MURRAY, b New Bedford, Mass, May 15, 29; m 54; c 5. ANALYTICAL CHEMISTRY. *Educ:* Univ Miami, BS, 48, MS, 51; Univ Fla, PhD(phys chem), 55. *Prof Exp:* Prod chemist, Gen Elec Co, 55-57; sr res chemist, Wash Res Ctr, W R Grace & Co, 57-60; sr chemist, Cathode Ray Tube Dept, Gen Elec Co, 60-63, supvr fluid develop eng, TV Receiver Dept, 63-68, supvr fluids eval, Advan Eng Proj Oper, Visual Commun Dept, 68-70, advan mat engr, Audio Prod Dept, 70-71; res chemist, Contaminants Characterization Prog, Southeast Water Lab, 71-73, chief org analysis sect, Analytical Chem Br, Environ Res Lab, 73-86, SUPVRY RES CHEMIST, MULTISPECTRAL ANALYSIS PROJ, MEASUREMENTS BR, ENVIRON RES LAB, ENVIRON PROTECTION AGENCY, 86- *Concurrent Pos:* Ed adv, Biomed Mass Spectrometry, 74-84. *Honors & Awards:* Silver Medal, Environ Protection Agency. *Mem:* Am Chem Soc; Am Soc Mass Spectrometry. *Res:* Mass spectrometric and gas chromatographic analysis of water contaminants; computer control of analytical instrumentation; computerized spectrum matching. *Mailing Add:* Athens Environ Res Lab Col Sta Rd Athens GA 30613

MCGUIRE, JOSEPH CLIVE, b Columbus, Ohio, Oct 14, 20; m 43; c 4. PHYSICAL CHEMISTRY, METALLURGY. *Educ:* Franklin Col, AB, 43; Univ NMex, MS, 63. *Prof Exp:* Chemist, Continental Can Co, 43-44, Manhattan Eng Dist, 44-46, Sherwin Williams Co, 46-47; staff mem, Argonne Nat Lab, 47-53; consult, 53-54; staff mem, Los Alamos Sci Lab, 54-66; sr lab scientist, Donald W Douglas Labs, McDonnell Douglas Astronaut Co, Richland, 66-73; prin engr, 73-80, fel engr, Westinghouse-Hanford Co, 80-85; INDEPENDENT CONSULT, 86- *Concurrent Pos:* Guest scientist, Los Alamos Nat Lab, 82-83. *Honors & Awards:* Chemist of the Year, Am Chem Soc, 84. *Mem:* Am Chem Soc; Sigma Xi; fel Am Inst Chemists. *Res:* Gas-metal reactions of refractory and transition metals; high temperature vacuum metallurgy; tritium and helium three production; radioisotope transport in liquid metals; heat pipes; helium venting from radioisotope capsules; radionuclide traps; sodium technology; tritium permeation; nuclear waste disposal. *Mailing Add:* 1637 Mowry Sq Richland WA 99352

MCGUIRE, JOSEPH SMITH, JR, b Logan, WVa, Apr 19, 31; div; c 7. MEDICINE. *Educ:* WVa Univ, AB, 52; Yale Univ, MD, 55. *Prof Exp:* Intern med, Yale-New Haven Med Ctr, 55-56; clin assoc, Nat Inst Arthritis & Metab Dis, 56-58, investr, 58-59; from asst prof to assoc prof med, 61-72, PROF DERMAT, SCH MED, YALE UNIV, 72- *Concurrent Pos:* USPHS spec fel med, Yale Univ, 59-61, USPHS award, Sch Med, 61-63, dir clin res training prog, Yale Univ, 64-69, mem gen med A study sect, 75-79; mem bd scientific couns, Nat Cancer Inst, NIH, 84-88. *Mem:* AAAS; Soc Invest Dermat (pres, 87-88); Am Dermat Asn; Am Soc Cell Biol; Am Acad Dermat; Am Soc Clin Invest. *Res:* Control of cell division in cultivated malignant and non-malignant cells; control of cell division, differentiation of keratinocytes and structure of keratins; production of cytokines by keratinocytes. *Mailing Add:* 800 Howard Ave New Haven CT 06510

MCGUIRE, ODELL, b Knoxville, Tenn, Apr 19, 27; m 57; c 3. GEOLOGY. *Educ:* Univ Tulsa, BS, 56; Columbia Univ, MA, 58; Univ Ill, PhD(geol), 62. *Prof Exp:* Assoc prof, 62-71, PROF GEOL, WASHINGTON & LEE UNIV, 71- *Concurrent Pos:* Geologist, Va Div Mining Resources. *Res:* Fossil population studies; geology of Appalachian region. *Mailing Add:* Dept of Geol Washington & Lee Univ Lexington VA 24450

MCGUIRE, ROBERT FRANK, b Greeneville, Tenn, Oct 8, 37; m 62. PHYCOLOGY. *Educ:* Union Col, Ky, BA, 60; Univ Tenn, Knoxville, MS, 64, PhD(bot), 71. *Prof Exp:* Teacher, Pub Schs, Ky, 60-62; instr math & sci, Eastern Ky Univ, 64-67; from teaching asst to instr bot, Univ Tenn, Knoxville, 67-71; from asst prof to assoc prof, 71-83, PROF BIOL, UNIV MONTEVALLO, 83- *Concurrent Pos:* NDEA fel, Univ Tenn, Knoxville, 70 & mem; Mellon workshop on advan biol, Vanderbilt Univ, 82; workshop comput bioeduc, Notre Dame Univ, 84; Deleg to China, Citizen Ambassador Prog & Nat Asn Biol Teachers, 88; teacher, Upward Bound Prog, Univ Montevallo, 87-91. *Mem:* Nat Asn Biol Teachers; Sigma Xi; Am Inst Biol Sci; Int Phycol Soc; Phycol Soc Am. *Res:* Application of the principles of numerical taxonomy to the Cyanophyta and Chlorophyta; these procedures have been used with species of Chlorococcum, clones of Chara and selected blue-green algae. *Mailing Add:* Dept Biol Sta 6463 Univ Montevallo Montevallo AL 35115-3605

MCGUIRE, STEPHEN CRAIG, b New Orleans, La, Sept 17, 48; m 71; c 2. NUCLEAR STRUCTURE. *Educ:* Southern Univ & A&M Col, BS, 70; Univ Rochester, MS, 74; Cornell Univ, PhD(nuclear sci), 79. *Prof Exp:* Teaching asst & res asst physics, Univ Rochester, 70-74; res asst, Ward Lab, Cornell Univ, 75-78; develop assoc, Chem Technol Div, Oak Ridge Nat Lab, 78-82; asst prof, 82-86, ASSOC PROF, DEPT PHYSICS, ALABAMA A&M UNIV, 86- *Concurrent Pos:* Lectr physics, Stanford Linear Accelerator Ctr, 76; physicist, Lawrence Livermore Nat Lab, 83 & 84; fac fel, Nat Aeronaut & Space Admin, Am Soc Eng Educ, Marshall Space Flight Ctr, 87. *Honors & Awards:* Utilization Award for Aerospace-Related Innovation, NASA Off Technol, 87. *Mem:* Am Phys Soc; Am Nuclear Soc; Am Asn Physics Teachers; Sigma Xi; Nat Soc Black Physicists (past-pres). *Res:* Relativistic Nucleus - Nucleus Collisions in emulsion detectors; heavy-ion-induced single-event-upset mechanisms in semiconductor circuits. *Mailing Add:* 118 Homestead Circle Ithaca NY 14850

MCGUIRE, STEPHEN EDWARD, b Excelsior Springs, Mo, Mar 25, 42; m 63; c 2. INDUSTRIAL CHEMISTRY. *Educ:* Lamar State Col Technol, BS, 63; Univ Tex, Austin, PhD(org chem), 67. *Prof Exp:* Res chemist, 67-70, sr res chemist, Continental Oil Co, 70-73, RES GROUP LEADER, CONOCO INC, 73- *Mem:* Am Oil Chemists' Soc. *Res:* Friedel-Crafts alkylations; carbonium ions; detergent intermediates. *Mailing Add:* Ten Hillcrest Ponca City OK 74604

MCGUIRE, TERRY RUSSELL, b Wadsworth, Ohio, June 12, 50; m 83; c 1. BEHAVIOR GENETICS, BIOMEDICAL GENETICS. *Educ:* Ohio State Univ, BS, 71; Univ Ill, Urbana-Champaign, PhD(genetics), 78. *Prof Exp:* Postdoctoral fel biochem genetics, Univ Rochester, 78-79; asst prof, 79-85, ASSOC PROF, DEPT BIOL SCI, RUTGERS UNIV, 85- *Mem:* Behav Genetics Asn; Genetic Soc Am; Soc Study Evolution; Am Genetic Asn. *Res:* Behavioral potential is encoded in genome and expressed to influence individual differences in behavior; model systems-geotaxis in drosophila melanogaster, community ecology of mycophagus drosophila; behavior of entomopathogenic nematodes. *Mailing Add:* Dept Biol Sci Nelson Biol Labs Rutgers Univ PO Box 1059 Piscataway NJ 08855

MCGUIRE, TRAVIS CLINTON, IMMUNOLOGY, BIOCHEMISTRY. *Educ:* Wash State Univ, PhD(path), 68. *Prof Exp:* PROF IMMUNOPATH, WASH STATE UNIV, 68- *Mailing Add:* Dept Vet Microbiol/Path Wash State Univ Pullman WA 99164-7040

MCGUIRE, WILLIAM, b Staten Island, NY, Dec 17, 20; m 44; c 2. CIVIL ENGINEERING. *Educ:* Bucknell Univ, BS, 42; Cornell Univ, MCE, 47. *Prof Exp:* Struct engr, Jackson & Moreland, 47-49; from asst prof to assoc prof civil eng, 49-60, PROF CIVIL ENG, CORNELL UNIV, 60- *Concurrent Pos:* Consult engr, 51-; design engr, Pittsburgh-Des Moines Steel Co, 54-56; vis prof, Asian Inst Technol, 68-70, Univ Strathclyde, 86; Gledden sr fel, Univ Western Australia, 73; vis scholar, Univ Tokyo, 79. *Honors & Awards:* Norman Medal, Am Soc Civil Engrs, 62. *Mem:* Fel Am Soc Civil Engrs; Int Asn Bridge & Struct Engrs; Am Concrete Inst. *Res:* Structural engineering; computer aided design. *Mailing Add:* Col of Eng Hollister Hall Cornell Univ Ithaca NY 14850

MCGUIRE, WILLIAM L, b Detroit, Mich, Oct 22, 37. MEDICINE. *Educ:* Western Reserve Univ, AB, 58, MD, 64; Am Bd Internal Med, dipl. *Prof Exp:* Intern med, Univ Hosps, Cleveland, Ohio, 64-65, resident med, 65-66; clin assoc, Endocrinol Br, Nat Cancer Inst, Bethesda, Md, 66-69; from asst prof to assoc prof, 69-75, PROF MED & CHIEF, DIV MED ONCOL, HEALTH SCI CTR, UNIV TEX, SAN ANTONIO, 75- *Concurrent Pos:* Mem, Treatment Comt, Breast Cancer Task Force, Nat Cancer Inst, 73-77, chmn, 75-77, chmn, Task Force, 80-81; mem bd govs, Bexar County Unit, Am Cancer Soc, 75-79; mem, Cancer Coord Comt, Health Sci Ctr, Univ Tex, 75-, chmn, 77-; counr, Southern Sect, Am Fedn Clin Res, 75-78; mem, US-France Coop Breast Cancer Res Prog, Nat Cancer Inst, 75-79 & Biochem & Chem Carcinogenesis Res Comt, Am Cancer Soc, 80-82; ed-in-chief, Breast Cancer Res & Treatment, 81-; Gen Motors vis Harvard prof clin invest, Dana Farber Cancer Inst; clin res prof, Am Cancer Soc; Bloomingdale award breast cancer res, 90. *Mem:* Asn Am Physicians; Am Soc Clin Invest; Am Soc Biol Chemists; Am Soc Cell Biol; Am Physiol Soc; Am Asn Cancer Res; Am Soc Clin Oncol; Am Fedn Clin Res; Endocrine Soc. *Res:* Author of various publications. *Mailing Add:* Dept Med Oncol Univ Tex Health Sci Ctr 7703 Floyd Curl Dr San Antonio TX 78284-7884

MCGUIRK, WILLIAM JOSEPH, b Terre Haute, Ind, Sept 19, 44; m 73; c 3. SOLAR ENERGY, TECHNICAL COMMUNICATIONS. *Educ:* Purdue Univ, BS, 66, MS, 67. *Prof Exp:* Proj engr, Monsanto Co, 67-71, Talley Indust Ariz, 71-73 & Bechtel Inc, 73-75; sr mech engr, Ariz Pub Serv Co, 75-77; eng supvr, Talley Indust Ariz, 77-78; sr res proj engr, 78-83, MGR R&D, ARIZ PUB SERV CO, 83- *Concurrent Pos:* Mem, State Ariz Solar Energy Adv Coun. *Mem:* Am Soc Mech Engrs; Asn Energy Engrs. *Res:* Monitors broad perspective of research and development for the electric utility industry; administer selected projects in solar energy and electric vehicles. *Mailing Add:* 10202 N 58th Pl Scottsdale AZ 85253

MCGURK, DONALD J, b Wichita, Kans, June 2, 40. ORGANIC CHEMISTRY. *Educ:* Univ Nebr, BS, 62; Okla State Univ, PhD(org chem), 68. *Prof Exp:* From asst prof to assoc prof, 68-84, PROF CHEM, SOUTHWESTERN STATE COL, OKLA, 85- *Mem:* AAAS; Am Chem Soc; Asn Comput Mach. *Res:* Chemistry of natural products; structure of molecules in oil of catnip; identity of volatile compounds produced by ants. *Mailing Add:* Dept Chem & Comput Sci Southwestern Okla State Col 100 Campus Dr Weatherford OK 73096

MACH, GEORGE ROBERT, b Cedar Falls, Iowa, July 23, 28; m 52; c 3. MATHEMATICS. *Educ:* Univ Northern Iowa, BA, 50; Univ Iowa, MS, 51; Purdue Univ, PhD(math), 63. *Prof Exp:* From instr to assoc prof, 54-67, PROF MATH, CALIF POLYTECH STATE UNIV, SAN LUIS OBISPO, 67- *Mem:* Math Asn Am. *Res:* Mathematics education. *Mailing Add:* Dept Math Calif Polytech State Univ San Luis Obispo CA 93407

MACH, MARTIN HENRY, b New York, NY, Feb 10, 40; div; c 1. ENVIRONMENTAL ANALYSIS, PHYSICAL ORGANIC CHEMISTRY. *Educ:* City Col New York, BS, 61; Clark Univ, MA, 65; Univ Calif, Santa Cruz, PhD(chem), 73. *Prof Exp:* Assoc scientist chem, Polaroid Corp, Mass, 65-69; mem tech staff chem, Aerospace Corp, 73-81; SR STAFF SCIENTIST, TRW CORP, 81- *Concurrent Pos:* Consult, Aerospace Corp, 81-82. *Mem:* Sigma Xi. *Res:* Analytical organic chemistry using interfaced, computerized gas chromatography-mass spectrometry, including environmental analysis, forensic science, fuel and synfuel synthesis and analysis, and lubrication phenomena. *Mailing Add:* TRW 0-1/2030 1 Space Park Dr Redondo Beach CA 90278

MACH, WILLIAM HOWARD, meteorology; deceased, see previous edition for last biography

MACHA, MILO, b Motycin, Czech, July 18, 18; US citizen; m 50; c 2. SOLID STATE ELECTRONICS, CERAMICS. *Educ:* Prague Tech Univ, dipl eng, 39, PhD(ceramics), 47. *Prof Exp:* Asst to Prof R Barta, Prague Tech Univ, 45-47; mgr, Ceramic Res Inst Gouda, Neth, 47-50; consult ceramics, N V Naga Hidjau, Indonesia, 50-52 & Georgian China Ltd, Can, 52-53; res scientist, Gen Foods Corp, Calif, 54-55; mgr solar cells res, Int Rectifier Corp, 55-60; sr res scientist, Acoustica Assoc Inc, 60-61; proj mgr, Lear Siegler Inc, 61-63; sr res scientist, Librascope Div, Gen Precision Inc, 63-64 & Douglas Aircraft Co, 64-66; head physics sect, Librascope Div, Gen Precision Inc, 66-71; sr res engr, Teledyne Systs Co, 71-75; PRES/OWNER, SOLLOS, INC, 75- *Concurrent Pos:* Translr & abstractor, Am Chem Soc. *Mem:* Electrochem Soc; sr mem Inst Elec & Electronics Engrs; Am Ceramic Soc; Acoust Soc Am. *Res:* Electrooptical devices, including preparation of single crystal materials in bulk form as well as thin films, device construction and application; research and development of photovoltaic arrays based on proprietory microcell structure. *Mailing Add:* 1519 Comstock Ave Los Angeles CA 90024

MACHAC, JOSEF, b Prerov, Czech, Jan 13, 54; m 82; c 2. INTERNAL MEDICINE, NUCLEAR MEDICINE. *Educ:* Brown Univ, ScB, 75, MD, 78. *Prof Exp:* Resident, med, Mt Sinai Med Ctr, 78-81, nuclear med, 81-83, fel, nuclear cardiol, 83-84, cardiol, 85-87, PHYSICIAN, NUCLEAR CARDIOL, MT SINAI MED CTR, 84- *Concurrent Pos:* Res collaborator, Brookhaven Nat Lab, 87- *Mem:* Am Col Physicians; Sigma Xi; Am Fedn Clin Res. *Res:* Computer analysis of functional images of radio-nuclide imaging of the heart. *Mailing Add:* 428 Jefferson Ave Haworth NJ 07641

MACHACEK, MARIE ESTHER, b Cedar Rapids, Iowa, Sept 12, 47; m 67; c 3. THEORETICAL HIGH ENERGY PHYSICS. *Educ:* Coe Col, BA, 69; Univ Mich, MS, 70; Univ Iowa, PhD(physics), 73. *Prof Exp:* Teaching asst physics, Univ Iowa, 70-71; hon fel physics, Univ Wis-Madison, 73-74; jr fel, Mich Soc Fellows, Univ Mich, 74-77; hon res fel, Harvard Univ, 76-77, lectr physics, 77-79; asst prof, 79-83, ASSOC PROF PHYSICS, NORTHEASTERN UNIV, 83- *Concurrent Pos:* Lectr physics, Univ Wis-Madison, 74; prin investr, NSF res grants, 79-86; Zone 1 counr, Soc Physics Students, 84-87. *Mem:* Am Phys Soc. *Res:* Unified theories of weak, electromagnetic and strong interactions; nonabelian gauge theories, quark models and applications to new narrow resonance phenomena; perturbative quantum chromodynamics, grand unified models, proton decay, renormalization group analysis of fermion masses, supersymmetry, hadron collider physics and other predictions of unified theories. *Mailing Add:* 250 Seven Bridge Rd Lancaster MA 01523

MACHACEK, MILOS, b Prague, Czech, Sept 11, 32; US citizen; m 52; c 2. ATOMIC PHYSICS. *Educ:* Univ Tex, BS, 58, MA, 60, PhD(physics), 64. *Prof Exp:* Staff res engr, AC Electronics Div, Mass, 64-69, STAFF RES ENGR, DELCO ELECTRONICS, GEN MOTORS CORP, 69- *Concurrent Pos:* Russ physics translr, Consult Bur, Inc, 67- *Mem:* Am Phys Soc. *Res:* Computational atomic and molecular physics; magnetic resonance; quantum electronics; underwater acoustics; electromagnetic wave propagation. *Mailing Add:* 5230 Paso Cameo Santa Barbara CA 93111

MACHACEK, OLDRICH, b Nachod, Czech, Arp 3, 30; US citizen; m 54; c 3. CHEMISTRY OF EXPLOSIVES, PROPELLANTS. *Educ:* Univ Chem Technol, Prague, MS, 53, PhD(chem of propellants), 61. *Prof Exp:* Proj leader propellants, Res Inst Indust Chem, Pardubice, Czech, 54-69; res chemist org chem, Universal Oil Prod, 70-72; res scientist explosives, Atlas Powder Co, 72-78; DIR RES & DEVELOP EXPLOSIVES, THERMET ENERGY CORP, 78- *Mem:* Am Chem Soc. *Res:* Chemistry of modern commerical explosives and propellants; watergels, emulsions, aluminised and permissible explosives; rocket propellants and double based smokeless powders. *Mailing Add:* 6841 Greenwich Dallas TX 75230

MACHADO, EMILIO ALFREDO, experimental pathology; deceased, see previous edition for last biography

MCHALE, EDWARD THOMAS, b Hazleton, Pa, Dec 10, 32; m 57; c 3. CHEMISTRY. *Educ:* King's Col, Pa, BS, 54; Pa State Univ, PhD(fuel sci), 64. *Prof Exp:* Res chemist, Gen Chem Div, Allied Chem Corp, 54-58; chemist, Reaction Motors Div, Thiokol Chem Corp, 58-60; prin scientist, Atlantis Res Corp, 64-72, res mgr, 72-90; STAFF SCIENTIST, EOS TECHNOLOGIES, 90- *Mem:* Am Chem Soc; Combustion Inst; Inst Energy. *Res:* Combustion science and engineering; physical and analytical chemistry; coal-water fuels; energy technology. *Mailing Add:* 8634 Braeburn Dr Annandale VA 22003

MCHALE, JOHN T, b New York, NY, Nov 2, 33; div; c 1. BOTANY, CYTOLOGY. *Educ:* Iona Col, BS, 55; Univ Tex, PhD(bot), 65. *Prof Exp:* Asst prof, Loyola Univ, La, 65-69, assoc prof biol, 69-74; BIOL WRITER, 74- *Concurrent Pos:* Sigma Xi grant-in-aid-of res, 67; consult cytol & cytogenetics, Gulf South Res Inst, La, 68-70 & 72-74; NIH training grant, Univ Calif, Berkeley, 70-71, res assoc, 71-72; res physiologist consult, Vet Admin Hosp, Martinez, Calif, 70-72. *Mem:* Am Physiol Soc. *Res:* Mechanisms of aging on the cellular level; bioenergetics of subcellular organelles; ultrastructure of plant cells; plant morphogenesis; cytology and cytogenetics. *Mailing Add:* 725 18th Ave San Francisco CA 94121

MACHALEK, MILTON DAVID, b Oenaville, Tex, July 30, 41; m 63; c 1. PLASMA PHYSICS, FUSION ENERGY. *Educ:* Harvard Univ, AB, 63; Univ Chicago, MS, 68; Univ Tex, Austin, PhD(physics), 72. *Prof Exp:* Res asst nuclear physics, Argonne Nat Lab, 65-68; res scientist plasma physics, Univ Tex, Austin, 72-74; staff scientist plasma physics, Los Alamos Sci Lab, Controlled Thermonuclear Res Div, 74-78; assoc group leader accelerator design, Accelerator Technol Div, Los Alamos Nat Lab, 78-80; div head, 80-90, HEAD, OFF TECHNOL TRANSFER, PRINCETON PLASMA PHYSICS LAB, 90- *Mem:* Am Phys Soc; sr mem Inst Elec & Electronic Engrs; fel Int Technol Inst; Fusion Power Assocs. *Res:* Plasma physics; controlled thermonuclear research; high current particle accelerators; superconductivity. *Mailing Add:* Princeton Plasma Physics Lab PO Box 451 Princeton NJ 08543

MCHARDY, GEORGE GORDON, b New Orleans, La, Mar 7, 10; c 3. MEDICINE. *Educ:* Spring Hill Col, BA, 32; Tulane Univ, MD, 36; Am Bd Internal Med, dipl & cert gastroenterol, 43. *Prof Exp:* Asst prof med, Tulane Univ La, 38-51; chief div gastroenterol, 60-71, assoc prof, 51-58, prof, 59-81, EMER PROF MED, SCH MED, LA STATE UNIV, NEW ORLEANS, 81- *Concurrent Pos:* Mem, World Cong Gastroenterol, 58-62; lectr, Univ Brazil, 60; sr vis physician, Charity Hosp. *Honors & Awards:* Rudolf Schindler Mem Award, Am Soc Gastrointestinal Endoscopy, 70. *Mem:* AMA; Am Gastroenterol Asn (treas, 53-59, vpres, 59-62, pres, 62); Am Soc Gastrointestinal Endoscopy (pres, 65); Am Col Physicians; Bockus Inst Soc Gastroenterol. *Res:* Gastroenterology. *Mailing Add:* Med Ctr of New Orleans 3638 St Charles Ave New Orleans LA 70115

MCHARGUE, CARL J(ACK), b Corbin, Ky, Jan 30, 26; m 48, 60; c 3. PHYSICAL METALLURGY. *Educ:* Univ Ky, BS, 49, MS, 51, DrEng(phys metall), 53. *Prof Exp:* Res assoc, Univ Ky, 49-52; res metallurgist, Ky Res Found, 52-53; metallurgist, Oak Ridge Nat Lab, 53-59, mgr res sect, 59-79, mgr mat sci, Metall & Ceramics Div, 79-86, sr res staff, 86-90; PROF CHEM & METALL ENG, UNIV TENN, 59- *Concurrent Pos:* Lectr, Univ Ky, 50-53 & Univ Tenn, 56-59; bd dir, TMS/AIME, 87-; vis prof, Univ Newcastle upon Tyne, UK, 87; adj prof mat sci & eng, Vanderbilt Univ, 88- *Honors & Awards:* Mat Sci Award, US Dept Energy, 82. *Mem:* Fel Am Soc Metals; fel Metall Soc; Sigma Xi; Am Inst Mining, Metall & Petrol Engrs; Mat Res Soc. *Res:* Plastic deformation of crystalline solids; radiation damage; ion implantation ceramics. *Mailing Add:* Dept Mat Sci & Eng Univ Tenn 434 Dougherty Eng Bldg Knoxville TN 37996-2200

MCHARRIS, WILLIAM CHARLES, b Knoxville, Tenn, Sept 12, 37; m 60; c 1. NUCLEAR CHEMISTRY. *Educ:* Oberlin Col, BA, 59; Univ Calif, Berkeley, PhD(nuclear chem), 65. *Prof Exp:* From asst prof to assoc prof, 65-71, PROF CHEM & PHYSICS, COL NATURAL SCI, MICH STATE UNIV, 71- *Concurrent Pos:* Consult, Heavy Elements Group, Argonne Nat Lab, 66-; sabbatical, Lawrence Berkeley Lab, Univ Calif, 71-72; Sloan fel, 72-76. *Mem:* AAAS; Am Phys Soc; Am Chem Soc; Sigma Xi. *Res:* Nuclear spectroscopy and reactions in actinides, lead region and deformed rare earths; on-line spectroscopy with cyclotron; beta-decay theory. *Mailing Add:* Dept of Chem Mich State Univ East Lansing MI 48823

MACHATTIE, LESLIE BLAKE, meteorology, for more information see previous edition

MACHATTIE, LORNE ALLISTER, bacterial viruses, plasmids, for more information see previous edition

MACHELL, GREVILLE, b Blackburn, Eng, Nov 19, 29; m 52; c 4. ORGANIC CHEMISTRY, TEXTILE CHEMISTRYN. *Educ:* Univ London, BSc, 48, Hons, 50, PhD(org chem), 52. *Prof Exp:* Res chemist, Brit Rayon Res Asn, 55-59 & Brit Celanese Ltd, 60-61; res chemist, Deering Milliken Res Corp, 61-64, dept mgr radiation chem, 64-67, mgr tech opers div, 67-69, gen mgr decorative fabrics div, 69-75, dir develop, Decorative Fabrics Div, Deering Milliken Inc, Ga, 75-76, res assoc, 76-83, SR TECHNOLOGIST, MILLIKEN RES CORP, 84- *Mem:* Am Chem Soc. *Res:* Alkaline degradation of carbohydrates; chemical modification of textile fibers; specialty chemicals; radiation-initiated graft polymerization; polymer degradation and stabilization. *Mailing Add:* Milliken Res Corp 920 Milliken Rd Spartanburg SC 29304

MACHEMEHL, JERRY LEE, b Bryan, Tex, Jan 8, 38; m 59; c 3. COASTAL ENGINEERING, OCEAN ENGINEERING. *Educ:* Tex A&M Univ, BSCE, 62, MSCE, 68, BSARCO, 70, PhD(civil eng), 70. *Prof Exp:* Eng res assoc, Tex A&M Univ, 65-68, instr civil eng, 69-70; civil engr, Coastal Eng Res Ctr, 68-69; from asst prof to assoc prof civil eng, 70-78, ASSOC PROF MARINE SCI & ENG, NC STATE UNIV, 78- *Concurrent Pos:* Prin investr, Sea Grant Prog, Nat Oceanic & Atmospheric Admin, 73-74 & 78-80, Univ NC Water Resources Inst, 73-74. *Honors & Awards:* Exten Serv Award, NC State Univ, 73. *Mem:* Am Soc Civil Engrs; Sigma Xi. *Mailing Add:* Dept Civil Eng Tex A&M Univ College TX 77843

MACHEMER, PAUL EWERS, b Romney, WVa, Jan 30, 19; m 41; c 3. ANALYTICAL CHEMISTRY. *Educ:* Princeton Univ, AB, 40; Univ Pa, MS, 43, PhD(chem), 49. *Prof Exp:* From chief chemist, Analysis Lab, Warner Co, 41-44; from jr chemist to sr shift supvr, Manhattan Proj, Carbide & Carbon Chem Co, 44-45; from asst prof to assoc prof, Villanova Univ, 49-55; from asst prof to prof, 55-84, EMER PROF CHEM, COLBY COL, 84- *Concurrent Pos:* Vis prof, Rosemont Col, 51-52. *Mem:* AAAS; Am Chem Soc. *Res:* Inorganic analytical chemistry; instrumental analysis. *Mailing Add:* PO Box 219 Port Clyde ME 04855

MACHEMER, ROBERT, b Muenster, Ger, Mar 16, 33; US citizen; c 1. VITREORETINAL DISEASES & SURGERY. *Educ:* Freiburg, Ger, MD, 59; Univ Goettingen, MD, 68. *Prof Exp:* Sci asst, der Univ Augenklinik, Ger, 62-68; instr ophthal, Bascom Palmer Eye Inst, Miami, Fla, 68-70; chief ophthal sec, Vet Admin Hosp, Miami, Fla, 69-71; from asst prof to prof ophthal, Univ Miami Sch Med, Fla, 70-78; PROF & CHMN OPHTHAL, DUKE UNIV EYE CTR, DURHAM, NC, 78- *Concurrent Pos:* Mem study sect, Nat Eye Inst, 82-84; Helena Rubinstein Found prof, 83. *Honors & Awards:* Trustees Award, Outstanding Ophthal Achievement, Am Acad Ophthal, 78; Award of Merit in Retina Res, Retina Soc, 80; V Graefe Prize, Ger Ophthal Soc, 81; Jackson Mem Lectr, Am Acad Ophthal, 84; Donders Lectr, Neth Ophthal Soc, 88; Proctor Lectr, Arvo, 88. *Mem:* Am Acad Ophthal; Am Ophthal Soc; Retina Soc; Asn Res Vision & Ophthal; Ger Ophthal Soc; Pan Am Asn Ophthal. *Res:* Development of experimental model of retinal detachment and studies on pathogenesis of proliferative vitreoretinopathy; development of vitreous surgery and instruments for this surgery; experimental and clinical studies on the treatment of proliferative vitreoretinopathy. *Mailing Add:* Duke Univ Eye Ctr PO Box 3802 Durham NC 27710

MCHENRY, CHARLES S, b Jan 1, 48; US citizen. NUCLEIC ACID ENZYMOLOGY, MOLECULAR VIROLOGY. *Educ:* Purdue Univ, BS, 70; Univ Calif, San Francisco & Santa Barbara, PhD(org chem-biochem), 74. *Prof Exp:* Postdoctoral fel nucleic acids-enzymol, Stanford Univ, 74-76; from asst prof to assoc prof biochem & molecular biol, Med Sch, Univ Tex, 76-84; PROF BIOCHEM, BIOPHYS & GENETICS, HEALTH SCI CTR, UNIV COLO, 85-, DIR, PROG MOLECULAR BIOL, 85- *Concurrent Pos:* Vis prof, Biozentrum, Univ Basel, Switz, 84-85; mem, Nucleic Acids & Proteins Adv Comt, Am Cancer Soc, 86-90, comt chair, 90; mem adv bd, Max Planck Inst Molecular Genetics, Berlin. *Mem:* AAAS; Am Chem Soc; Am Soc Biochem & Molecular Biol; Am Soc Microbiol. *Res:* Enzymology structure of multienzyme complexes; regulation of gene expression; mechanism and regulation of DNA synthesis; molecular virology. *Mailing Add:* Health Sci Ctr Univ Colo 4200 E Ninth Ave B121 Denver CO 80262

MCHENRY, HENRY MALCOLM, b Los Angeles, Calif, May 19, 44; m 66; c 2. BIOLOGICAL ANTHROPOLOGY. *Educ:* Univ Calif, Davis, BA, 66, MA, 67; Harvard Univ, PhD(anthrop), 72. *Prof Exp:* From asst prof to assoc prof, 71-81, PROF ANTHROP, UNIV CALIF, DAVIS, 81- *Mem:* AAAS; Sigma Xi; Am Asn Phys Anthrop; Am Anthrop Asn; Human Brit Coun; Soc Vert Paleont. *Res:* Paleoanthropology; australopithecine postcranial anatomy. *Mailing Add:* Dept of Anthrop Univ of Calif Davis CA 95616

MCHENRY, HUGH LANSDEN, b Baxter, Tenn, Aug 19, 37; m 63; c 2. STATISTICS, MATHEMATICS. *Educ:* Tenn Technol Univ, BS, 60; George Peabody Col, MA, 61, PhD(math), 70. *Prof Exp:* Instr math, Okla Christian Col, 61-63, asst prof, 65-68; from asst prof to assoc prof, 68-78, PROF MATH, MEMPHIS STATE UNIV, 78- *Mem:* Am Statist Asn; Math Asn Am. *Res:* Mathematics and statistics education. *Mailing Add:* 2429 Hawkhurst Memphis TN 38119

MCHENRY, K(EITH) W(ELLES), JR, b Champaign, Ill, Apr 6, 28; m 52; c 2. CHEMICAL ENGINEERING. *Educ:* Univ Ill, BS, 51; Princeton Univ, PhD(chem eng), 58. *Prof Exp:* Chem engr, Whiting Res Labs, Standard Oil Co, Ind, 55-58, group leader, 58-62, proj mgr, Res & Develop Dept, Am Oil Co, 62-66, res assoc, 66-67, asst dir, 67-70, dir process & anal res, Whiting Res Labs, Amoco Oil Co, 70-74, mgr process res, Naperville, Ill, 74-75, vpres res & develop, 75-89, SR VPRES TECHNOL, AMOCO CORP, CHICAGO, ILL, 89- *Concurrent Pos:* Mem adv coun, Chem Eng Dept, Univ Ill, Urbana, Beckman Inst & Catalysis Prog, Univ Calif, Berkeley. *Honors & Awards:* Charles D Hurd lectr, Northwestern Univ, 81; Thiele Fuels Eng lectr, Univ Utah, 83; Gerster Mem Lectr, Univ Del, 87. *Mem:* Nat Acad Eng; Am Chem Soc; Am Inst Chem Engrs; AAAS; Indust Res Inst. *Res:* Heterogeneous catalysis; cracking; reforming; engineering kinetics; reactor design; research management; recipient of four patents and contributor to several technical and research management publications. *Mailing Add:* Amoco Corp 200 E Randolph Dr Mail Code 4904A Chicago IL 60601

MCHENRY, KELLY DAVID, b Williamsport, Pa, Feb 24, 52. FERROELECTRICS, COMPOSITES. *Educ:* Pa State Univ, BS, 74, MS, 76, PhD(ceramic sci), 78. *Prof Exp:* Prin res scientist, Honeywell Corp Technol Ctr, 78-81, PRIN DEVELOP ENGR, HONEYWELL CERAMICS CTR, 81- *Mem:* Am Ceramic Soc. *Res:* Processing and development of ferroelectric ceramics & lightweight composite ceramics to enhance mechanical dielectric and optical properties; novel surface finishing techniques for polishing of ceramics. *Mailing Add:* Alliant Tech Systs MM48 3700 7225 Northland Dr Brooklyn Park MN 55428

MCHENRY, WILLIAM EARL, b Camden, Ark, July 22, 50; m 76. ORGANIC CHEMISTRY, PESTICIDE CHEMISTRY. *Educ:* Southern Ark Univ, BS, 72; Miss State Univ, PhD(org chem), 77. *Prof Exp:* Teaching asst, Miss State Univ, 72-77, res asst, 75-76, asst prof, 77-80, ASSOC PROF ORG CHEM, MISS STATE UNIV, 80-, ASSOC DEAN, 88-; PROG DIR, NSF, 91- *Concurrent Pos:* Prin investr alternative toxicants contract, Miss Dept Agr & Com, 78-79. *Mem:* Am Chem Soc; AAAS. *Res:* Synthesis of organophosphorous compounds; chemical and physical properties of small ring heterocycles. *Mailing Add:* Dept of Org Synthesis Drawer CH & Box 5328 Miss State Univ Mississippi State MS 39762

MACHEREY, ROBERT E, b Evanston, Ill, Jan 5, 18. NUCLEAR ENGINEERING. *Educ:* Purdue Univ, BS, 41; Ill Inst Technol, MS, 50. *Prof Exp:* Supt shops, Argonne Nat Lab, 70-86; RETIRED. *Mem:* Am Soc Metals Int. *Mailing Add:* 737 S Cornell Ave Villa Park IL 60181

MACHETTE, MICHAEL N, b San Jose, Calif, Nov 21, 49; m 72; c 1. QUATERNARY GEOLOGY, NEO- TECTONICS. *Educ:* San Jose State Univ, BS, 72; Univ Colo, Boulder, MS, 75. *Prof Exp:* RES GEOLOGIST, US GEOL SURVEY, DENVER, 72- *Mem:* Fel Geol Soc Am; Am Geophys Union. *Res:* Quaternary geology. *Mailing Add:* US Geol Surv Mail Stop 966 Fed Ctr Box 25046 Denver CO 80225

MACHIELE, DELWYN EARL, b Zeeland, Mich, Dec 30, 38; m 62; c 2. ORGANIC CHEMISTRY. *Educ:* Hope Col, AB, 60; Univ Ill, Urbana, PhD(org chem), 64. *Prof Exp:* RES ASSOC, EASTMAN KODAK CO, 64- *Mem:* Am Chem Soc. *Mailing Add:* 163 Bunker Hill Dr Rochester NY 14625

MACHIN, J, b Herne Bay, Eng, June 29, 37; m 61; c 3. COMPARATIVE PHYSIOLOGY. *Educ:* Univ London, BSc, 59, PhD(zool), 62. *Hon Degrees:* DSc, London Univ, 83. *Prof Exp:* From asst prof to assoc prof, 62-78, PROF ZOOL, UNIV TORONTO, 78- *Concurrent Pos:* Vis assoc prof, Univ Wash; vis prof, Ariz State, 83. *Honors & Awards:* T H Huxley Award, Zool Soc London, 62. *Mem:* Am Physiol Soc; Marine Biol Asn UK; Brit Soc Exp Biol; Zool Soc London. *Res:* Comparative physiology of terrestrial moist skinned animals, particularly physics and physiology of evaporation; water transport in insects; osmotic regulation in cells. *Mailing Add:* Dept Zool Univ Toronto Toronto ON M5S 1A1 Can

MACHLEDER, WARREN HARVEY, b New York, NY, Aug 2, 43; m 67; c 2. ORGANIC CHEMISTRY. *Educ:* NY Univ, BA, 64; Ind Univ, Bloomington, PhD(chem), 68. *Prof Exp:* res chemist, 68-74, proj leader, 74-78, mem staff mkt res, 78-80, mkt planning mgr, 80-87, RES MGR, PETROL CHEM, ROHM & HAAS CO, 87- *Mem:* Am Chem Soc; Soc Auto Engrs. *Res:* Allene oxidations; petroleum additives, antiwear agents and carburetor detergents; viscosity index improvers. *Mailing Add:* 1098 Boxwood Lane Blue Bell PA 19422

MACHLIN, E(UGENE) S(OLOMON), b New York, NY, Dec 29, 20; m 43, 60; c 3. MATERIALS SCIENCE. *Educ:* City Col New York, BME, 42; Case Inst Technol, MS, 48; Mass Inst Technol, ScD, 51. *Prof Exp:* Aeronaut res scientist, Nat Adv Comt Aeronaut, 42-48; res assoc metall, Mass Inst Technol, 48-50, asst prof, 51; from asst prof to assoc prof, 51-58, prof metall, 58-89, Howe Prof Metall, 89-90, EMER PROF, COLUMBIA UNIV, 91- *Concurrent Pos:* Guggenheim fel and Fulbright lectr, Italy, 65-66. *Honors & Awards:* Mathewson Gold Medal, Am Inst Mining, Metall & Petrol Engrs, 54. *Mem:* Am Soc Metals; fel Am Inst Mining, Metall & Petrol Engrs. *Res:* Synthesis of diamond films; high Tc,Jc superconductors; thin films. *Mailing Add:* 1106 SW Mudd Bldg Columbia Univ 520 W 120th St New York NY 10027

MACHLIN, IRVING, b Brooklyn, NY, Mar 10, 19; m 48; c 2. HIGH TEMPERATURE STRUCTURAL MATERIALS, AIRCRAFT ENGINE MATERIALS. *Educ:* Cooper Union Inst Technol, BChE, 40. *Prof Exp:* Mat engr mat, testing, specif, appln, res & develop, US Naval Gun Factory, Bur Ord, Naval Weapons & Naval Air Systs Command, Dept Navy, 40-80; RETIRED. *Concurrent Pos:* Tech chmn, Refractory Metals Comt, Am Soc Metals Int. *Mem:* Fel Am Soc Metals Int. *Res:* Directed successful high temperature materials development programs for naval aircraft. *Mailing Add:* 1903 Ladd St Silver Spring MD 20902

MACHLIN, LAWRENCE JUDAH, b New York, NY, June 24, 27; m 53; c 3. NUTRITIONAL BIOCHEMISTRY. *Educ:* Cornell Univ, BS, 48, MNS, 49; Georgetown Univ, PhD(biochem), 54. *Prof Exp:* Poultry nutritionist, USDA, 49-50, biochemist, 50-53, biochemist, Agr Res Serv, 53-56; biochemist, Monsanto Co, 56-73; sr res group chief, 73-84, DIR DEPT CLIN NUTRIT, HOFFMANN-LA ROCHE, 84- *Concurrent Pos:* Lectr, Washington Univ, 69-72; adj prof nutrit, NY Univ, 77-85 & Cornell Univ Med Col, 79- *Mem:* Am Inst Nutrit; Am Col Nutrit; Am Soc Clin Nutrit; Soc Exp Biol & Med; NY Acad Sci; Soc Free Radicals Res. *Res:* Biochemical function and nutritional role of vitamin E, vitamin A, ascorbic acid, carotene, antioxidants and fatty acids and amino acids. *Mailing Add:* 18 Locust Pl Livingston NJ 07039

MACHLOWITZ, ROY ALAN, b Brooklyn, NY, Apr 16, 21; m 49; c 2. ANALYTICAL BIOCHEMISTRY, VIROLOGY. *Educ:* Brooklyn Col, AB, 41. *Prof Exp:* Jr chemist, US Naval Boiler & Turbine Lab, 42-45, chemist, Naval Air Exp Sta, 45-51; res assoc antibiotics, 51-56 & virus & tissue cult res, 56-64, sr res biochemist, Merck Sharp & Dohme Res Labs, 64-83; RETIRED. *Mem:* Am Chem Soc. *Res:* Purification of antibiotics; purification and assay of viruses. *Mailing Add:* 520 Laverock Rd Glenside PA 19038

MACHLUP, STEFAN, b Vienna, Austria, July 1, 27; nat US; m 61, 71; c 2. BIOPHYSICS. *Educ:* Swarthmore Col, BA, 47; Yale Univ, MS, 49, PhD, 52. *Prof Exp:* Asst, Yale Univ, 49-51; mem tech staff, Bell Tel Labs, Inc, 52-53; res assoc physics, Univ Ill, 53-55; mem sci staff, Van der Waals Lab, Amsterdam, 55-56; asst prof, 56-61, ASSOC PROF PHYSICS, CASE WESTERN RESERVE UNIV, 61- *Concurrent Pos:* Consult, Res Ctr, Clevite Corp, 57-62, Educ Serv Inc, Auckland, NZ, 61, Ibadan, Nigeria, 62 & Watertown, Mass, 64-65; NSF sci fac fel, Univ Liverpool, 62-63; consult, Mass Inst Technol, 68, 69; consult-evaluator, N Cent Asn Cols & Secondary Schs; mem, Comn Scholars, State Ill Bd Higher Educ. *Mem:* AAAS; Am Phys Soc; Am Asn Physics Teachers; Biophys Soc. *Res:* Theory of solids; imperfections in metals; polaron mobility; fluctuations and irreversible processes; noise in semiconductors; oscillatory chemical reactions; underwater sound scattering; ion transport in membranes. *Mailing Add:* Dept Physics Case Western Reserve Univ Cleveland OH 44106

MACHNE, XENIA, b Trieste, Italy, Sept 28, 21. PHYSIOLOGY. *Educ:* Univ Padua, MD, 46. *Prof Exp:* Asst physiol, Univ Parma, 47-48; asst, Univ Bologna, 48-51; res assoc physiol, Univ Ore, 56-58; asst prof pharmacol, Univ Ill, 58-62; assoc prof physiol, Sch Med, Tulane Univ, 62-71; prof pharmacol, Univ Minn, Minneapolis, 71-76; RETIRED. *Concurrent Pos:* Brit Coun scholar, Univ Col, Univ London, 51-53; Fulbright fel, Univ Calif, Los Angeles, 53-55. *Mem:* Am Physiol Soc. *Res:* Neurophysiology; neuropharmacology; electrophysiological methods; microelectrode techniques. *Mailing Add:* Via Celesti 13 Desenzano del Garda Italy

MACHOL, ROBERT E, b New York, NY, Oct 16, 17; m 46; c 2. SYSTEMS ANALYSIS, AERONAUTICAL ENGINEERING. *Educ:* Harvard Univ, BA, 38; Univ Mich, MS, 53, PhD(chem), 57. *Prof Exp:* Syst engr, Willow Run Res Ctr, Univ Mich, 51-58; prof elec eng, Purdue Univ, 58-61; vpres, Conductron Corp, 61-64; prof systs eng & head dept, Univ Ill, Chicago, 64-67; prof systs, Grad Sch Mgt, Northwestern Univ, Evanston, 67-86; asst adminr, 86-87, CHIEF SCIENTIST, SCI & ADVAN TECHNOL, FED AVIATION ADMIN, 87- *Concurrent Pos:* Vis prof, Waseda Univ, Japan, 57, Europ Inst Bus Admin, France, 71, Stanford Univ, 74-75, Univ Sydney, 78 & others; consult, indust, govt & educ, 58-86; liaison scientist, US Off Naval Res, London, 78-81; ed-in-chief J Inst Mgt Sci, Studies in Mgt Sci, 77-85. *Mem:* Fel AAAS; sr mem Inst Elec & Electronics Engrs; Opers Res Soc Am (pres, 71-72); Inst Mgt Sci. *Res:* Design, analysis and evaluation of future aviation systems. *Mailing Add:* Fed Aviation Admin ASD-4 800 Independence Ave SW Washington DC 20591

MACHOVER, MAURICE, b New York, NY, Dec 5, 31; m 64. MATHEMATICAL ANALYSIS. *Educ:* Brooklyn Col, BS, 56; Columbia Univ, MA, 58; NY Univ, MS, 60, PhD(math), 63. *Prof Exp:* Lectr math & physics, Brooklyn Col, 56-58; asst math, NY Univ, 58-63, asst res scientist, 63; asst prof math, Fairleigh Dickinson Univ, 64-65, assoc prof & chmn dept, 65-67; ASSOC PROF MATH, ST JOHN'S UNIV, NY, 67- *Concurrent Pos:* Asst physics, Columbia Univ, 56-58. *Mem:* Am Math Soc; Math Asn Am; Asn Symbolic Logic; Soc Indust & Appl Math; Sigma Xi. *Res:* Eigenfunction expansions as applied to self adjoint differential and integral equations and their extentions to non-self adjoint problems. *Mailing Add:* Dept Math & Comput Sci St John's Univ Grand Cent Utopia Pkwy Jamaica NY 11439

MACHT, MARTIN BENZYL, b Baltimore, Md, Aug 31, 18; m 39. PHYSIOLOGY. *Educ:* Johns Hopkins Univ, AB, 39, PhD(neurophysiol), 42, MD, 45. *Prof Exp:* Asst instr psychol, Johns Hopkins Univ, 39-42, instr physiol, Med Sch, 42-45; intern & asst resident med, Jewish Hosp, Cincinnati, Ohio, 45-46; physiologist & staff surgeon, Climatic Res Lab, Mass, 46-48; instr pharmacol, 48-54, asst clin prof med, 54-77, ASST PROF PHYSIOL, COL MED, UNIV CINCINNATI, 54-, PROF MED, 77- *Concurrent Pos:* Chief res, Jewish Hosp, 48-49, dir med educ, 49-50; co-dir internal med, Rollman Receiving Hosp. *Mem:* Am Physiol Soc; assoc Am Psychol Asn; AMA; Am Diabetes Asn; Am Geriat Soc. *Res:* Neurophysiology; localization of function in central nervous system; temperature regulation; peripheral blood flow; frost bite; neural basis of emotion; decerebrate preparations in the chronic state; psychopharmacology; psychosomatic medicine; renal disease. *Mailing Add:* Univ Cincinnati Col Med Cincinnati OH 45219

MACHTA, JONATHAN LEE, b Washington, DC, May 19, 51; m; c 2. PHYSICS. *Educ:* Univ Mich, BSc, 75; Univ Ill, MSc, 76; Mass Inst Technol, PhD(physics), 80. *Prof Exp:* Teaching res fel physics, Inst Physics, Sci & Technol, Univ Md, College Park, 80-82; asst prof, 82-87, ASSOC PROF PHYSICS, UNIV MASS, AMHERST, 87- *Mem:* Am Phys Soc. *Res:* Theoretical condensed matter physics; transport in disordered systems. *Mailing Add:* Dept Physics Univ Mass Amherst MA 01003

MACHTINGER, LAWRENCE ARNOLD, b St Louis, Mo, Mar 11, 36; m 64; c 2. MATHEMATICS. *Educ:* Wash Univ, BSChE, 59, AM, 63, PhD(math), 65. *Prof Exp:* Asst prof math, Webster Col, 64-65; asst prof, St Louis Univ, 65-66; asst prof, Ill Inst Technol, 66-72; dir sci & math proj, ASSOC PROF MATH, PURDUE UNIV, N CENT CAMPUS, 72- *Concurrent Pos:* Dir, Curric Improv & Teacher Training Inst Proj; consult, high sch math; consult, Madison Proj, Syracuse Univ & Webster Col, 64-72; mem math adv coun, State of Ind, 74-76. *Mem:* Am Math Soc; Math Asn Am; Nat Coun Teachers Math. *Res:* Group theory and ordered rings. *Mailing Add:* Dept Math Purdue Univ Hwy 421 Westville IN 46391

MCHUGH, ALEXANDER E(DWARD), metallurgy; deceased, see previous edition for last biography

MCHUGH, JAMES ANTHONY, JR, b Stockton, Calif, Oct 7, 37; m 57; c 4. SURFACE ANALYSIS. *Educ:* Univ Pac, BS, 59; Univ Calif, Berkeley, PhD(chem), 63. *Prof Exp:* Phys chemist, 63-74, mgr, Mass Spectrometry Res & Develop, 74-75, MGR, CHEM LAB, KNOLLS ATOMIC POWER LAB, GEN ELEC CO, 75- *Mem:* Am Soc Mass Spectrometry; Am Chem Soc; Am Inst Chemists. *Res:* Mass spectrometry; systematics of nuclear fission; medium energy ion-surface interactions and secondary positive ion emission; ionization phenomena; surface analysis; secondary ion mass spectrometry; ion microprobe; microbeam analysis. *Mailing Add:* Knolls Atomic Power PO Box 1072 Schenectady NY 12302

MCHUGH, JOHN LAURENCE, b Vancouver, BC, Nov 24, 11; nat US; m 41, 79; c 3. FISHERIES MANAGEMENT. *Educ:* Univ BC, BA, 36, MA, 38; Univ Calif, Los Angeles, PhD(zool), 50. *Prof Exp:* Asst, Univ BC, 36-38; fishery biologist, Pac Biol Sta, BC, 38-41; asst ichthyol, Scripps Inst, Univ Calif, 47-48, res assoc, 48-51; dir, Va Fisheries Lab, 51-59; chief div biol res, Bur Commercial Fisheries, US Fish & Wildlife Serv, 59-63, asst dir biol res, 63-66, dept dir bur, 66-68; actg dir, Off Marine Resources, US Dept Interior, 68-70; head, Off Int Decade Ocean Explor, NSF, 70; prof, 70-82, EMER PROF MARINE RESOURCES, MARINE SCI RES CTR, STATE UNIV NY STONY BROOK, 82- *Concurrent Pos:* Prof marine biol, Col William & Mary, 51-59; mem, Nat Res Coun, 65-70; adv comn marine resources to dir gen, Food & Agr Orgn, 66-70; mem US nat sect, Int Biol Prog, 67-70; comt int marine sci affairs policy, Nat Acad Sci, 70-74; US comnr, Inter-Am Trop Tuna Comn, 60-70; vchmn, Int Whaling Comn, 68-71, chmn, 71-72; mem hard clam adv comt, Nassau-Suffolk Regional Marine Resources Coun, 73-82; consult, Town of Islip, NY & Islip Town Shellfish Mgt Comn, 74-78; mem, Mid Atlantic Fishery Mgt Coun, 76-79; mem, Sci & Statist Comt Coun, 79-84; fel, Woodrow Wilson Int Ctr for Scholars; J L McHugh fel, Marine Sci Res Ctr. *Honors & Awards:* Award of Excellence, Am Fisheries Soc. *Mem:* AAAS; Am Inst Biol Sci; Am Fisheries Soc; Inst Fishery Res Biol; hon mem Am Inst Fishery Research Biologists; hon mem Estuarine Res Soc; hon mem Nat Shellfish Asn. *Res:* Oceanography; fishery biology and management; resolution of social-political impediments to application of scientific knowledge in fishery utilization and management. *Mailing Add:* Marine Sci Res Ctr State Univ of NY Stony Brook NY 11794-5000

MCHUGH, KENNETH LAURENCE, b Brooklyn, NY, Mar 22, 27; m 53; c 5. ORGANIC CHEMISTRY. *Educ:* Hofstra Col, BA, 51; Univ Conn, PhD(chem), 59. *Prof Exp:* Jr res chemist, Evans Res & Develop Corp, 51-53; res chemist, Conn Hard Rubber Co, 53-56; asst instr chem, Univ Conn, 56-58; sr res chemist, 58-60, res group leader, Spec Proj Dept, 60-61 & Org Chem Div, 61-64, proj mgr org develop dept, 64-66, sr proj mgr, 66-68, proj mgr commercial develop, Chemstrand Res Ctr, New Enterprises Div, 68-71, mgr commercial develop, New Enterprises Div, 71-77, MGR NEW VENTURES DEVELOP, CORP RES & DEVELOP, MONSANTO CO, 77- *Mem:* Am Chem Soc; Am Soc Lubrication Engrs; Coord Res Coun. *Res:* High temperature stable fluids, principally those derived from organometallic, organophosphorus and polyaromatic ether chemistry for use as turbine lubricants, lubricant additives, power transmission fluids and thermodynamic fluids for Rankine and Brayton power cycles; air pollution control by catalytic processes. *Mailing Add:* 414 Gray Ave St Louis MO 63117

MCHUGH, PAUL RODNEY, b Lawrence, Mass, May 21, 31; m 59; c 3. NEUROLOGY, PSYCHIATRY. *Educ:* Harvard Univ, AB, 52, MD, 56. *Prof Exp:* Intern med, Peter Bent Brigham Hosp, 56-57; res neurol, Mass Gen Hosp, 57-60; clin asst psychiat, Maudsley Hosp, London, Eng, 60-61; res asst neuroendocrinol, Walter Reed Army Inst Res, 61-64; asst prof neurol & psychiat, Med Col, Cornell Univ, 64-71, prof psychiat, 71-73; prof psychiat & chmn dept, Med Sch, Univ Ore, 73-75; CHMN & PSYCHIATRIST-IN-CHIEF, DEPT PSYCHIAT & BEHAV SCI, JOHNS HOPKINS UNIV SCH MED, 75- *Concurrent Pos:* Clin dir & supvr psychiat educ, Westchester Div, New York Hosp-Cornell Med Ctr, 69-73. *Mem:* Am Psychiat Asn; Am Neurol Asn; Am Physiol Soc; Harvey Soc; Am Psychopath Asn; Sigma Xi. *Res:* Neural mechanisms of visceral, endocrine and behavioral control. *Mailing Add:* Dept of Psychiat/Behav Sci Johns Hopkins Hosp 600 N Wolfe St Baltimore MD 21205

MCHUGH, RICHARD B, b Villard, Minn, Oct 25, 23; m 51; c 5. BIOMETRICS, BIOSTATISTICS. *Educ:* Univ Minn, PhD, 55. *Prof Exp:* Assoc prof psychol statist, Iowa State Univ, 50-56; from assoc prof to prof, 56-86, dir, biomet,68-71, EMER PROF BIOMET, UNIV MINN, MINNEAPOLIS, 86- *Concurrent Pos:* Nat Inst Gen Med Sci fel, Univ Calif, Berkeley, 64-65; vis scholar, Univ London, 71-72; consult, Dept Med & Surg, Vet Admin; mem pancreatic cancer study sect, Nat Cancer Inst, 74-78; mem oncol adv comt, Food & Drug Admin, 77-; Int Union Against Cancer res fel, Univ Oxford, 78-79; ed, The Am J Epidemiol, 80- *Mem:* Hon fel AAAS; Pop Asn Am; hon fel Am Statist Asn; hon fel Am Pub Health Asn; Biomet Soc; Int Statist Inst. *Res:* Research design in the health and life sciences; mathematical demography; epidemetrics; biomathematical models in assay; biostatistics in health services research; cancer clinical trials and surveys. *Mailing Add:* Sch Pub Health Univ Minn Mayo Box 197 Minneapolis MN 55455

MCHUGH, STUART LAWRENCE, b San Francisco, Calif, Nov 7, 49. MATERIALS SCIENCE, GEOPHYSICS. *Educ:* Univ Nev, Reno, BS, 71, BS, 72; Stanford Univ, MS, 74, MS, 76, PhD(geophys), 77. *Prof Exp:* Geophysicist, US Geol Surv, 73-77; Geophysicist, fracture mech, SRI Int, 77-81; RES SCIENTIST, LOCKHEED CORP, 81- *Concurrent Pos:* Consult, US Geol Surv, 77-, SRI Int, 81-, NSF traineeship, 72-73. *Mem:* Am Geophys Union; Am Phys Soc; Int Asn Math & Comput Modeling; AAAS. *Res:* Fracture mechanics; deformation and fracture of materials; mathematical models of the constitutive behavior of materials. *Mailing Add:* Lockheed 0/ 9330 Bldg 204 3251 Hanover St Palo Alto CA 94304

MCHUGH, WILLIAM DENNIS, b Berwick, Eng, May 8, 29; US citizen; m 58; c 2. DENTISTRY. *Educ:* Univ St Andrews, Scotland, LDS, 50, BDS, 54, DDSc(dent), 59; Royal Col Surgeons, Edinburgh, FDS, 70. *Prof Exp:* Prof & chair, Univ St Andrews, UK, 65-70; DIR, EASTMAN DENT CTR, 70-; PROF & ASSSOC DEAN, UNIV ROCHESTER, NY, 70- *Concurrent Pos:* Ed, Proc Brit Soc Periodont, 60-62; assoc ed, J Periodont Res, 66-71 & J Am Col Dentists, 87-; chmn, Adv Comt Periodont Dis, Nat Inst Dent Res, 74-75, Periodont Dis Concensus Panel, 81 & Adv Panel, NIH, Nat Inst Dent Res, Vet Admin, 84-85; mem, NIH Dent Res Coun, 87-91. *Mem:* Fel Am Col Dentists; fel AAAS; fel Int Col Dentists; Int Asn Dent Res (pres, 88-89); Am Asn Dent Res (pres, 83-84); Am Asn Dent Schs (vpres, 80-83). *Res:* Periodontal tissues and diseases affecting them; dental plaque development, prevention and control; dental health improvement. *Mailing Add:* 625 Elmwood Ave Rochester NY 14620

MCHURON, CLARK ERNEST, ENGINEERING GEOLOGY. *Educ:* Syracuse Univ, AB, 40. *Prof Exp:* Asst geologist, US Bur Reclamation, Denver, 46-47, proj geologist Wyo, 47-50, chief geologist, Eklutna, Alaska, 50-51, Alaska Dist Geologist, US Bur Reclamation, 51-53; CONSULT, ENGR GEOLOGIST, 53- *Concurrent Pos:* Geol Soc Am; Asn Engr Geologists. *Res:* Engineering geology. *Mailing Add:* 274 Mocking Bird Circle Santa Rosa CA 95405

MACHUSKO, ANDREW JOSEPH, JR, b Hiller, Pa, Dec 31, 37; m 62; c 2. MATHEMATICS. *Educ:* Calif State Col, Pa, BS, 59; Univ Ga, MA, 64, PhD(math), 68. *Prof Exp:* Teacher high sch, Ohio, 59-62; instr math, Univ Ga, 67-68; asst prof, Univ Tenn, Knoxville, 68-70; assoc prof, 70-71, PROF MATH, CALIF STATE COL, PA, 71- *Mem:* Math Asn Am. *Res:* Topology; algebra. *Mailing Add:* 618 Front St Brownsville PA 15417

MACIAG, THOMAS EDWARD, b Bayonne, NJ, Nov 19, 46; m 74; c 1. CELL BIOLOGY, EXPERIMENTAL PATHOLOGY. *Educ:* Rutgers Univ, BA, 68; Univ Pa, PhD(molecular biol), 75. *Prof Exp:* Fel biochem & biophys, Sch Med, Univ Pa, 75-76; sr res investr, Collab Res Inc, 76-79; res fel med, 79-80, asst prof path, Harvard Med Sch, 80-83; assoc prof biochem, George Washington Univ, Sch Med, 84-86; DIR, BIOTECHNOL RES CTR, 83- *Concurrent Pos:* Adj prof biol & chem, Regis Col, Mass, 78- *Mem:* Am Soc Biol Chemists; Am Soc Cell Biol; Tissue Culture Asn. *Res:* Mechanisms of action of growth factors in the blood vessel wall; biochemistry of angiogenesis; developmental biology of neovascularization and the neural crest; role of proteases and extracellular matrix in developmental biology and homeostasis; diseases involving cell proliferation, atheriosclerosis and cancer; biochemical events during wound healing. *Mailing Add:* Dept Molecular Biol Am Red Cross 15601 Crabbs Branch Way Rockville MD 20855 Inc 4 Res Ct

MACIAG, WILLIAM JOHN, JR, b Rome, NY, May 28, 36; m 58; c 2. MICROBIOLOGY. *Educ:* Univ Buffalo, BS, 57; Syracuse Univ, MS, 59, PhD(microbiol), 63. *Prof Exp:* Asst instr microbiol, Syracuse Univ, 59-63; res microbiologist, 63-72, sr res microbiologist, 72-79, head, animal care, 79-83, supvr, safety & occup health, 83-85, COORDR, REGULATORY COMPLIANCE & ENVIRON AFFAIRS, STINE-HASKELL LABS, E I DU PONT DE NEMOURS & CO, INC, 85- *Concurrent Pos:* mem Comn F, Int Sci Racho Union. *Mem:* AAAS; Soc Indust Microbiol; Am Soc Microbiol. *Res:* Pharmaceutical drug research; intermediary metabolism; autotrophic mechanisms; laboratory safety and occupational health. *Mailing Add:* Sypherd Dr Newark DE 19711

MACIAS, EDWARD S, b Milwaukee, Wis, Feb 21, 44; m 67; c 2. NUCLEAR CHEMISTRY, RADIOCHEMISTRY. *Educ:* Colgate Univ, AB, 66; Mass Inst Technol, PhD(nuclear chem), 70. *Prof Exp:* Asst prof, 70-76, ASSOC PROF CHEM, WASHINGTON UNIV, 76- *Concurrent Pos:* Consult, Argonne Nat Lab, 67, US Dept Transp, 75-76 & Meteorol Res Inc, 75-; vis scientist, Lawrence Livermore Lab, Univ Calif, 71; vis prof, Calif Inst Technol, 78-79. *Mem:* AAAS; Sigma Xi; Am Phys Soc; Am Chem Soc; Air Pollution Control Asn. *Res:* Aerosol chemistry and physics; nuclear structure studies; atomic structure studies via x-ray spectroscopy; atmospheric chemistry; effects of air pollution; visibility and atmospheric optics; air pollution. *Mailing Add:* Dept Chem Wash Univ 1134 Lindell-Skinker Blvd St Louis MO 63130

MACIEJKO, ROMAN, b Ger, Jan 22, 46; Can citizen; m 77; c 2. OPTOELECTRONICS, FIBER OPTICS. *Educ:* Univ Laval, Que, BA, 66, BS, 70; State Univ NY, Stony Brook, MS, 71, PhD(physics), 75. *Prof Exp:* Postdoctoral physics, Tech Univ Aachen, Ger, 75-78; scientist physics, Bell-Northern Res, Ottawa, 78-82, mgr fiber optics, 82-85; asst prof, 85-87, ASSOC PROF ENG PHYSICS, ECOLE POLYTECHNIQUE, MONTREAL, 87- *Concurrent Pos:* Consult, Optoelectronics, 85-; mem, Physics & Astron Comt, Natural Sci Eng Res Coun, Can, 85-87, Elec Eng Grant Select Comt, 90- *Mem:* Inst Elec & Electronics Engrs; Am Phys Soc; Optical Soc Am; Can Asn Physicists; Sigma Xi. *Res:* Optoelectronics; fiber optics; integrated optics; modelling, fabrication and characterization of semiconductor lasers; optical properties of materials, both linear and non-linear. *Mailing Add:* Dept Eng Physics Ecole Polytechnique Box 6079 Sta A Montreal PQ H3C 3A7 Can

MACIEL, GARY EMMET, b Niles, Calif, Jan 18, 35; m 56; c 2. PHYSICAL CHEMISTRY, ANALYTICAL CHEMISTRY. *Educ:* Univ Calif, Berkeley, BS, 56; Mass Inst Technol, PhD(chem), 60. *Prof Exp:* Res asst chem, Mass Inst Technol, 59-60, NSF fel, 60-61; from asst prof to prof, Univ Calif, Davis, 61-70; PROF CHEM, COLO STATE UNIV, 71- *Mem:* AAAS; Am Chem Soc; Sigma Xi. *Res:* Nuclear magnetic resonance and its application to chemical problems, particularly involving solid samples and/or using less common nuclei; relationships between high-resolution nuclear magnetic resonance parameters; fossil fuels, polymers, ceramics, surfaces. *Mailing Add:* Dept of Chem Colo State Univ Ft Collins CO 80523

MCILHENNY, HUGH M, b Gettysburg, Pa, Sept 25, 38; m 62; c 2. DRUG REGULATORY AFFAIRS, DRUG METABOLISM. *Educ:* Pa State Univ, BS, 60; Univ Mich, MS, 64, PhD(pharm, chem), 66. *Prof Exp:* Res asst antibiotics, Parke-Davis & Co, Inc, 60-62; anal chemist, 66-75, res investr, 75-79, REGULATORY AFFAIRS LIAISON, PFIZER, INC, 79- *Mem:* Am Soc Pharmacol & Exp Therapeut. *Res:* Fate of foreign substances in biological systems; factors influencing drug metabolism; drug bioavailability; development of assay methods for the measurement of drugs and their metabolites. *Mailing Add:* 16 Village Dr East Lyme CT 06333

MCILRATH, THOMAS JAMES, b Dowagiac, Mich, May 10, 38; m 62; c 2. VACUUM ULTRAVIOLET RADIATION PHYSICS. *Educ:* Mich State Univ, BS, 60; Princeton Univ, PhD(physics), 66. *Prof Exp:* Res fel astrophysics, Harvard Col Observ, 67-70, res assoc, 70-73; from vis assoc prof to assoc prof, 73-81, PROF, INST PHYS SCI & TECHNOL, UNIV MD, 81- *Concurrent Pos:* Lectr, Astron Dept, Harvard Univ, 70-73; staff physicist, Nat Bur Standards, Gaithersberg, Md, 74-; consult, Plasma Lab, Princeton Univ, 84- & AT&T Bell Tel Labs, 85-86. *Honors & Awards:* Indust Res-100 Award, Indust Res Mag Inc, 80; Silver Medal, US Dept Com, 80. *Mem:* Fel Optical Soc Am; fel Am Phys Soc. *Res:* Atomic and molecular structure; high resolution laser spectroscopy; vacuum ultraviolet spectroscopy; non-linear mixing in gases; generation of coherent vacuum ultraviolet radiation; photochemistry and spectroscopy of small molecules; atmospheric remote sensing; optical techniques for study of atomic structure; laser-matter interactions; molecular fluorescense. *Mailing Add:* Inst Phys Sci & Technol Univ Md College Park MD 20742

MCILRATH, WAYNE JACKSON, b Laurel, Iowa, Oct 18, 21; m 42; c 3. PLANT PHYSIOLOGY. *Educ:* Iowa State Teachers Col, BA, 43; Univ Iowa, MS, 47, PhD(plant physiol), 49. *Prof Exp:* From asst to instr bot, Univ Iowa, 46-49; asst prof plant physiol, Agr & Mech Col Tex & Agr Exp Sta, 49-51; from asst prof to prof bot, Univ Chicago, 51-64; dean grad sch, 64-73, PROF BIOL SCI, NORTHERN ILL UNIV, 64- *Concurrent Pos:* Consult, Argonne Nat Lab, 56-; OEEC sr vis fel, Sorbonne, 61; res assoc, Univ Chicago, 64-66; chmn coun, Cent States Univs, Inc, 67-68, bd dirs, 69-80, tres, 71-80; mem comn scholars, Ill Bd Educ, 68-71; mem coun & gov bd, Quad-Cities Grad Study Ctr, 70-73. *Mem:* AAAS; Am Soc Plant Physiologists; Bot Soc Am; Am Inst Biol Sci; Scand Soc Plant Physiol; Sigma Xi. *Res:* Mineral nutrition of plants; physiology of growth and development. *Mailing Add:* Dept of Biol Sci Northern Ill Univ De Kalb IL 60115

MCILRATH, WILLIAM OLIVER, b Coulterville, Ill, Aug 30, 36; m 63; c 3. PLANT BREEDING, GENETICS. *Educ:* Univ Ill, BS, 59, MS, 64; Okla State Univ, PhD(agron), 67. *Prof Exp:* From asst prof to assoc prof agron, La State Univ, 67-73; RES AGRONOMIST, AGR RES SERV, USDA, 73- *Mem:* Am Soc Agron; Crop Sci Soc Am; Am Genetic Asn; Soc Advan Breeding Res in Asia & Oceania; Sigma Xi. *Res:* Effects of artificial shading on development and morphology of corn; heterosis; combining ability and quantitative genetics of hexaploid wheat; breeding, genetics of rice. *Mailing Add:* Rte 1 Box 560 Beaumont TX 77706

MCILREATH, FRED J, b Amsterdam, NY, Apr 1, 29; m 52; c 5. PHYSIOLOGY, PHARMACOLOGY. *Educ:* Siena Col, BS, 51; Univ Ky, MS, 55; McGill Univ, PhD(physiol), 59. *Prof Exp:* Sr pharmacologist, Strasenburgh Labs, 59-62; from asst dir to assoc dir clin invest, Riker Labs, 62-71; asst dir, G D Searle & Co, 71-73, dir regulatory affairs, 73-78; vpres clin & regulatory affairs, 78-80, VPRES RES & DEVELOP, REED & CARNRICK LABS, 80- *Concurrent Pos:* Head, Pulmonary Dis Sect, Riker Labs, 68-71. *Mem:* AAAS; Am Physiol Soc; Am Soc Clin Pharmacol & Therapeut; NY Acad Sci. *Res:* Clinical pharmacology and drug development; effects of drugs on air flow dynamics. *Mailing Add:* Reed & Carnrick Labs One New England Ave Piscataway NJ 08855

MCILROY, M DOUGLAS, b Newburgh, NY, Apr 24, 32; m 65; c 2. COMPUTER PROGRAMMING & SYSTEMS. *Educ:* Cornell Univ, BEP, 54; Mass Inst Technol, PhD(math), 59. *Prof Exp:* Mem tech staff, 58-86, dept head, 65-86, DISTINGUISHED MEM TECH STAFF COMPUTER SCI, AT&T BELL LABS, 86- *Concurrent Pos:* Vis lectr, Oxford Univ, 67-68. *Mem:* AAAS; Asn Comput Mach. *Res:* Systems programming, especially computer languages and text processing; graphic and geometric algorithms; computer cartography; synthetic speech; computer security. *Mailing Add:* AT&T Bell Labs Rm 2C-526 600 Mountain Ave Murray Hill NJ 07974-2070

MCILVAIN, JESS HALL, b Ponder, Tex, Mar 29, 33; m 59; c 2. CONSTRUCTION FORENSICS, COEFFICIENT OF FRICTION OF WALKING SURFACES. *Educ:* Tex Tech Univ, BArch, 59. *Prof Exp:* Archit dir, Tile Coun Am, 69-80; OWNER & CONSULT, JESS MCILVAIN & ASSOC, 80- *Concurrent Pos:* Owner, Jess McIlvain Archit, 69-80; consult, Handbook Ceramic Tile Installation, 70-91; lectr, Tile Coun Am, 70-80, Int Tile Expos, 87-91, Int Ceramics Symp, Bologna & Int Tile Fair, Bologna, 89; secy, A108 Comt, Ceramic Tile, Am Nat Standards Inst, 74-80, A137-1 Comt, 76-80. *Mem:* Am Soc Testing & Mat. *Res:* Causes of installation failures of ceramic tile, marble and stone, waterproofing, terrazzo, and masonry; avoiding tile and stone installation problems; author of various publications. *Mailing Add:* 6012 Woodacres Dr Bethesda MD 20816

MCILWAIN, CARL EDWIN, b Houston, Tex, Mar 26, 31; m 52; c 2. PHYSICS. *Educ:* NTex State Col, BME, 53; Univ Iowa, MS, 56, PhD(physics), 60. *Prof Exp:* Asst prof physics, Univ Iowa, 60-62; assoc prof, 62-66, PROF PHYSICS, UNIV CALIF, SAN DIEGO, 66- *Concurrent Pos:* Guggenheim fels, Eng, 67-68 & Eng, Ger & Sweden, 72; Alexander von Humboldt US Sr scientist award. *Honors & Awards:* Space Sci Award, Am Inst Aeronaut & Astronaut, 70; John a Fleming Medal, 75. *Mem:* Am Phys Soc; Am Geophys Union; Am Astron Soc. *Res:* Energetic particles in solar system. *Mailing Add:* CASS/0111 Univ of Calif at San Diego La Jolla CA 92093-0111

MCILWAIN, DAVID LEE, b Memphis, Tenn, Jan 7, 38; m 84; c 2. NEUROCHEMISTRY. *Educ:* Vanderbilt Univ, BA, 60; Wash Univ, MD, 64. *Prof Exp:* Fel biochem, Univ Calif, Berkeley, 64-66; fel neurochem, Col Physicians & Surgeons, Columbia Univ, 68-72; from asst prof to assoc prof, 72-84, PROF PHYSIOL, UNIV NC, CHAPEL HILL, 84- *Honors & Awards:* Javits Neurosci Investr Award. *Mem:* Soc Neurosci; Am Soc Neurochem. *Res:* Chemistry of spinal motoneurons. *Mailing Add:* Dept Physiol Sch Med Univ NC CB 7545 Chapel Hill NC 27514

MCILWAIN, JAMES TERRELL, b June 16, 36; m. VISION, OCULOMOTOR CONTROL. *Educ:* Tulane Univ, BS, 54, MD & MS, 61. *Prof Exp:* Intern, Montreal Gen Hosp, 61-62; Space Biol Lab & Dept Anat, Univ Calif, US Pub Health Serv, 62-64; Exp Neurophysiol Div, Max Planck Inst Psychiat, 65-66; fel NSF, Comp Lab Neurophysiol, fac sci, Univ Paris, France, 66-67; asst prof physiol, Col Med, Univ Cincinnati, 67-68; chief, Behav Radiol Lab, Walter Reed Army Inst Res, Wash, DC, 68-71; from asst prof to assoc prof med sci, 71-79, chmn, Sect Neurobiol, 79-89, PROF MED SCI, BROWN UNIV, PROVIDENCE, RI, 79-, DOROTHEA DOCTORS FOX PROF VISUAL SCI, 88- *Concurrent Pos:* consult to dir, Walter Reed Army Inst Res, 71-78; consult, Neurology Serv, Veterans Admin Med Ctr, 76-85; mem, Comt Man & Radiation, Inst Elec & Electronic Engrs, 71-77, Biopsychology Study Sect, NIH, 77-82; assoc ed, Visual Neurosci, 87- *Mem:* Am Physiol Soc; Soc Neurosci; Medieval Acad Am. *Res:* Anatomy and physiology of the vertebrate visual system; brainstem mechanisms of oculo motor control and visual orienting behavior. *Mailing Add:* Brown Univ Box G M416 Providence RI 02912

MCILWAIN, ROBERT LESLIE, JR, b Lancaster, SC, Apr 16, 29; m 54; c 3. HIGH ENERGY PHYSICS. *Educ:* Carnegie Inst Technol, BS, 53, PhD, 60. *Prof Exp:* Instr physics, Princeton Univ, 59-62; from asst prof to assoc prof, 62-76, PROF PHYSICS, PURDUE UNIV, WEST LAFAYETTE, 76- *Mem:* Am Phys Soc. *Res:* High energy nuclear physics and elementary particles. *Mailing Add:* Dept Physics Purdue Univ West Lafayette IN 47907

MCINDOE, DARRELL W, b Wilkinsburg, Pa, Sept 28, 30; c 5. ENDOCRINOLOGY, NUCLEAR MEDICINE. *Educ:* Allegheny Col, BS, 52; Temple Univ, MD, 56, MSD, 60. *Prof Exp:* Chief internal med & hosp serv, 7520th US Air Force Hosp, 64-68; Air Force Inst Technol fel, Royal Postgrad Med Sch, London, 68-69; chief & chmn, Endocrinol Serv, Nuclear Med, Keesler AFB, 69-75; dep dir, Armed Forces Radiobiol Inst, Defense Nuclear Agency, 75-77, dir, 77-79; asst prof, Dept Radiol & Nuclear Med, Uniformed Serv Univ Health Sci & staff physician, Dept Radiol, Nuclear Med Br, Nat Naval Med Ctr, 79-82; STAFF PHYSICIAN, NUCLEAR MED, ST JOSEPH HOSP, TOWSON, MD, 82- *Concurrent Pos:* Med adv to mgr, Nevada oper, Dept Energy, 80- *Mem:* Uniformed Serv Nuclear Med Asn (pres, 75); fel Am Col Nuclear Physicians; Soc Nuclear Med; Air Force Soc Physicians; fel Royal Soc Med; AMA; Health Physics Soc. *Res:* Nuclear weapons effects; basic science of endocrinology and nuclear medicine; diagnosis and management of thyroid carcinoma. *Mailing Add:* 12405 Borges Ave Silver Spring MD 20904

MCINERNEY, EUGENE F, b New York, NY, Apr 11, 38; m 65; c 4. PHOTOPOLYMERS, ORGANOMETALLIC CHEMISTRY. *Educ:* Fordham Univ, BS, 59; St John's Univ , MS, 63, PhD(organometallic chem), 67. *Prof Exp:* Jr chemist, Hoffman-La Roche, 59-61; res chemist, E I du Pont de Nemours & Co, Inc, 67-68; res specialist, Photo Prods Div, GAF Corp, 68-72, sr staff specialist, 72-74; mgr imaging systs, Horizons Res Inc, 74-76, mgr new ventures, 77-79; supvr, mkt develop, 80-84, proj leader, Dept Res & Develop, 84-86, indust analyst/licensing assoc, 87-89, MGR BUS DEVELOP/AEROSPACE, HERCULES INC, 90- *Mem:* Am Chem Soc; Soc Photog Scientists & Engrs; Soc Photo-Optical Instrumentation Engrs; Soc Plastic Engrs; Licensing Execs Soc; Mat Res Soc; Sigma Xi. *Res:* Organo tin/lead complexes; photopolymers/photoresists; semiconductor fabrication/materials; gas purification; advanced composites; nonlinear optos; technology licensing; commercial development. *Mailing Add:* Hercules Inc Hercules Plaza Wilmington DE 19894

MCINERNEY, JOHN GERARD, b Cork, Ireland, May 9, 59; m 87. SEMICONDUCTOR LASER PHYSICS, DYNAMICS OF NONLINEAR OPTICAL SYSTEMS. *Educ:* Nat Univ Ireland, Cork, Ireland, BSc, 80; Trinity Col, Dublin, Ireland, PhD(physics), 85. *Hon Degrees:* MA, Cambridge Univ, 86. *Prof Exp:* Res engr, Standard Telecommun Labs, Eng, 81; res asst physics, Trinity Col, Dublin, Ireland, 82-84; res fel physics & eng, Robinson Col & Cavendish Lab, Cambridge Univ, UK, 84-86; asst prof elec eng & physics, 86-90, REGENTS' LECTR OPTOELECTRONICS, UNIV NMEX, 90-, ASSOC PROF ELEC ENG & PHYSICS, 91- *Concurrent Pos:* Instr, Dublin City Univ, Ireland, 82-84; fel, Robinson Col, Cambridge Univ, UK, 84-86; prin investr, Sandia Nat Labs, Los Alamos Nat Lab, USAF Rome Air Develop Ctr, NSF, USAF Phillips Lab, 86-; consult, Eastman Kodak Corp, Alcoa Inc, McDonnell-Douglas Corp, Mission Res Corp, RDA Logicon, 87- *Mem:* Optical Soc Am; Inst Elec & Electronics Engrs; Inst Elec Engrs UK; Am Inst Physics. *Res:* Semiconductor laser physics and engineering; ultrafast optical processes in semiconductors; optical communication; photonic logic and switching; optical bistability; instabilities and chaos; engineering and biomedical applications of lasers; author of over 100 publications. *Mailing Add:* Ctr High Technol Mat Univ NMex Albuquerque NM 87131

MACINNES, DAVID FENTON, JR, b Abington, Pa, Mar 19, 43; m; c 3. INORGANIC & POLYMER CHEMISTRY, SCIENCE EDUCATION. *Educ:* Earlham Col, BA, 65; Princeton Univ, MA, 70, PhD(chem), 72. *Prof Exp:* Teacher chem, Westtown Sch, Pa, 70-73; asst prof, 73-80, ASSOC PROF CHEM, GUILFORD COL, 81- *Concurrent Pos:* Vis assoc prof, Simon Fraser Univ, 87-88; Lindberg Grant recipient, 83. *Mem:* Am Chem Soc; Sigma Xi. *Res:* Transition metal complexes; x-ray crystallography; chemical education; computers and chemistry; conducting polymers; organic batteries. *Mailing Add:* Dept of Chem Guilford Col Greensboro NC 27410

MACINNIS, AUSTIN J, b Virginia, Minn, Mar 15, 31; m 57; c 3. PARASITOLOGY, BIOCHEMISTRY. *Educ:* Concordia Col, Moorhead, Minn, BA, 57; Fla State Univ, MS, 59, PhD(parasitol), 63. *Prof Exp:* Asst, Fla State Univ, 57-59; NIH scholar parasitol, Rice Univ, 63-65; from asst prof to prof zool, 65-77, PROF CELL BIOL, UNIV CALIF, 77- *Concurrent Pos:* NSF foreign travel award, 65-; ed, J Parasitol, 79-83, series co-ed, Human Parasitic Dis. *Mem:* Fel AAAS; Am Soc Parasitol; Am Soc Trop Med & Hyg; fel Royal Soc Trop Med & Hyg. *Res:* Behavior, molecular and immunoparasitology. *Mailing Add:* Dept Biol 2203 Life Sci Univ Calif 405 Hilgard Ave Los Angeles CA 90024

MACINNIS, CAMERON, b West Bay, NS, Mar 25, 26. CIVIL ENGINEERING. *Educ:* Dalhousie Univ, BSc, 46; NS Tech Col, BE, 48; Durham Univ, PhD(concrete), 62. *Prof Exp:* Res engr, Hydro Elec Power Comn Ont, 48-59; res assoc, Durham Univ, 59-62; assoc prof civil eng, 63-68, dean eng, 79-89, PROF CIVIL ENG, UNIV WINDSOR, 68-, HEAD DEPT, 76- *Honors & Awards:* Wason Medal, Am Concrete Inst, 75. *Mem:* Eng Inst Can; Am Concrete Inst; Am Soc Testing & Mat; Can Soc Civil Eng. *Res:* Concrete technology, especially frost durability, high-strength, drying, shrinkage and creep. *Mailing Add:* Dept Civil Eng Univ Windsor Windsor ON N9B 3P4 Can

MACINNIS, MARTIN BENEDICT, b Big Pond, NS, Aug 16, 25; m 53; c 4. CHEMISTRY, PHYSICS. *Educ:* St Francis Xavier Univ, BSc, 46; Col of the Holy Cross, MS, 53. *Prof Exp:* Prof chem, Loyola Col Montreal, 46-52; from engr to sr engr, Sylvania Elec Prod Inc, 53-60, develop engr, 60-63, advan develop engr, GTE Sylvania Inc, 63-74, head dept chem develop, 69-74, sect head chem develop & process eng, 74-80, eng mgr chem, GTE Prods Corp, 74-86; DIR NORTHEASTERN PA CONSULTS, INC, 86- *Honors & Awards:* Leslie H Warner Tech Achievement Awards. *Mem:* Am Chem Soc. *Res:* Chemistry of tungsten, cobalt, rhenium, molybdenum, silicon nitride, rare earths, tantalum and niobium; solvent extraction; pyrometallurgy; hydrometallurgy; chemical vapor deposition; germanium and silicon; sugar chemistry; organic phosphors and fire retardants. *Mailing Add:* 507 Main St Northeast Pa Consults Towanda PA 18848

MCINROY, ELMER EASTWOOD, b Ont, Can, Nov 2, 21; m 46; c 3. AGRICULTURAL CHEMISTRY. *Educ:* Ont Agr Col, BSA, 44. *Prof Exp:* Res chemist, Defense Industs, Ltd, Can, 43; metall chemist, Deloro Smelting & Refining Co, 44-46; res biochemist, Cent Res Labs, Gen Foods Corp, NJ, 46-50, lab supvr, Gaines Div, 50-53; dir animal nutrit, Foxbilt, Inc, Iowa, 53-56; dir animal nutrit, 56-65, vpres nutrit & tech serv, 65, DIR, ARBIE MINERAL FEED CO, INC, 65- *Mem:* AAAS; Am Chem Soc; Am Soc Animal Sci; Am Registry Prof Animal Scientists; Am Feed Indust Asn. *Res:* Animal, human and dog nutrition; food technology; audio-visual training; technical sales; technical service; public speaking. *Mailing Add:* Arbie Mineral Feed Co Inc 1905 Knollwood Dr Marshalltown IA 50158

MCINTIRE, CHARLES DAVID, b St Louis, Mo, Sept 20, 32; m 65; c 3. AQUATIC ECOLOGY. *Educ:* Southern Methodist Univ, BBA, 54; Ore State Univ, BS, 58, MS, 60, PhD(bot), 64. *Prof Exp:* Res asst fisheries, 58-60, res asst bot, 61-63, asst prof, 64-69, assoc prof, 69-76, PROF BOT, ORE STATE UNIV, 76- *Concurrent Pos:* Acad Natural Sci Philadelphia McHenry Fund grant, 67; NSF sci fac fel, Ctr Quant Sci, Univ Wash, 70-71. *Mem:* Ecol Soc Am; Am Soc Limnol & Oceanog; Phycol Soc Am; Sigma Xi. *Res:* Physiological ecology of marine and freshwater algae; trophic ecology of aquatic ecosystems; mathematical ecology and systems analysis; diatom systematics. *Mailing Add:* 2025 NW Elder St Corvallis OR 97330

MCINTIRE, JUNIUS MERLIN, b Price, Utah, Jan 27, 18; m 47; c 2. BIOCHEMISTRY. *Educ:* Brigham Young Univ, AB, 40; Univ Wis, MS, 42, PhD(biochem), 44. *Prof Exp:* Res biochemist, Western Condensing Co, Wis, 44-50; chief dairy oil & fat div, Qm Food & Container Inst, 50-55; asst dir res, 55-67, GEN MGR RES, CARNATION CO, 67- *Mem:* AAAS; Am Chem Soc; Sigma Xi. *Res:* Animal nutrition; vitamin assays; protein chemistry of milk; food processing of milk products; distribution and nutritional significance of certain members of vitamin B complex. *Mailing Add:* 6939 Columbus Ave Van Nuys CA 91405

MCINTIRE, KENNETH ROBERT, b Portland, Ore, Mar 31, 33; m 54; c 5. BIOLOGY, IMMUNOLOGY. *Educ:* Univ VA, BA, 55, MD, 59. *Prof Exp:* From intern to resident internal med, Univ Hosps of Cleveland, Western Reserve Univ, 59-61; res assoc, 61-63, staff scientist, Lab Biol, 63-70, staff scientist, Lab Cell Biol, 70-75, sr investr, Lab Immunodiag, 75-78, CHIEF DIAG BR, NAT CANCER INST, 78- *Mem:* Am Asn Immunologists; Int Soc Oncodevelop Biol & Med; AAAS; Am Asn Cancer Res. *Res:* Identification and study of circulating tumor associated antigens for the early diagnosis of cancer and for monitoring the effectiveness of therapy. *Mailing Add:* 15 W Main St Falmouth MA 02540

MCINTIRE, LARRY V(ERN), b St Paul, Minn, June 28, 43; m 69. BIOPHYSICS, IMMUNOLOGY. *Educ:* Cornell Univ, BChE & MS, 66; Princeton Univ, MA, 68, PhD(chem eng), 70. *Prof Exp:* From asst prof to prof chem eng, 70-83, chmn dept chem eng, 82-89, E D BUTCHER PROF CHEM & BIOMED ENG, RICE UNIV, 83-, CHMN, INST BIOSCI & BIOENG, 90- *Concurrent Pos:* NATO-NSF sr postdoctoral fel, Imp Col, London, 76-77; adj prof med, Sch of Med, Univ Tex, Houston, 77- , Baylor Col Med, Houston, 82- *Mem:* AAAS; Soc Rheol; Am Inst Chem Engrs; Am Heart Asn; NY Acad Sci; Tissue Cult Asn; Biomed Eng Soc. *Res:* Rheology of nonlinear materials; fluid mechanics; red and white blood cell deformability studies; biorheology; biomedical engineering; kinetics of protein polymerizations; tissue culture reactors; tissue engineering. *Mailing Add:* Dept Chem Eng Rice Univ Houston TX 77251-1892

MCINTIRE, LOUIS V(ICTOR), b Lafayette, La, Jan 6, 25; c 5. CHEMICAL ENGINEERING. *Educ:* Southwestern La Inst, BS, 44; Ohio State Univ, MSc, 48, PhD(chem eng), 51. *Prof Exp:* Instr chem eng, Southwestern La Inst, 46-47, assoc prof, 53-56; res asst, Mathieson Proj, Res Found, Ohio State Univ, 48-51; chem engr, Mathieson Chem Corp, 51-53; chem engr, E I du Pont de Nemours & Co, Inc, 56-58, supvr, 58-59, sr res engr, 59-72; CONSULT, 72- *Mailing Add:* 1732 Greenbriar Orange TX 77630

MCINTIRE, MATILDA S, b Brooklyn, NY, July 15, 20; c 1. PEDIATRICS. *Educ:* Mt Holyoke Col, BA, 42; Albany Med Col, MD, 46; Am Bd Pediat, dipl. *Hon Degrees:* DSc, Mt Holyoke Col, 89. *Prof Exp:* Intern, Flower & Fifth Ave Hosps, New York, 46-47; pediatrician, 8th Army Hq, Japan, 48-49; resident pediat, St Louis Univ Med Sch, 52-53 & Univ Nebr Hosp, 53-54; pvt pract, 54-56; instr pediat, Sch Med, Creighton Univ, 55-61; pediat consult & asst res prof, Col Med, Univ Nebr, 61-66; from assoc clin prof to clin prof pediat, 66-73, prof pediat & dir community health, 73-90, CLIN PROF PUB HEALTH & PREV MED, SCH MED, CREIGHTON UNIV, 68- *Concurrent Pos:* Children's Mem Hosp trainee, 73; dir div prev dis control, Omaha-Douglas County Health Dept, 56-61, dir div maternal & child health, 66-; asst prof pediat, Col Med, Univ Nebr, 61-72, assoc prof, 72-, asst prof food & nutrit, 70-75; consult, co-investr & prin investr, USPHS grants, 66-74; consult toxicol info prog, Nat Libr Med; consult fac, Inst Clin Toxicol, Houston, Tex; mem infant stand comt, Nebr Dept Pub Welfare; mem comt accident prev, Am Acad Pediat, 74-; med consult, Omaha Douglas County Health Dept, 79-90; med dir, Nebr Reg Poison Control Ctr, Childrens Mem Hosp, Omaha, Nebr, 79. *Mem:* AAAS; Am Col Prev Med; AMA; fel Am Pub Health Asn; fel Am Acad Pediat; fel Am Acad Clin Toxicol. *Mailing Add:* 1510 S 80th St Omaha NE 68124

MCINTIRE, SUMNER HARMON, b Essex, Mass, July 7, 12; m 36; c 2. PHYSICS, ACADEMIC ADMINISTRATION. *Educ:* Bowdoin Col, AB, 33; NY Univ, MA, 37; Boston Univ, MS, 54. *Hon Degrees:* DSc, 64, DEd, 77, Norwich Univ. *Prof Exp:* Instr, Westbrook Jr Col, 34-36 & Thornton Acad, 37-38; instr chem, 36-37, from instr to assoc prof physics, 38-57, chmn dept, 62-72, prof, 57-77, dean univ, 72-77, EMER PROF PHYSICS, NORWICH UNIV, 77-, EMER DEAN UNIV, 77- *Concurrent Pos:* State exam consult, Vt, 44-46; res worker, Bur Indust Res, Northfield, 47; consult, Vt Dept Pub Safety, 54-; sci adv to Gov, 71-76; mem, State Bd Radiologic Technol, 78-, chmn, 85- *Mem:* Nat Asn Res Sci Teaching; Am Asn Physics Teachers; Am Asn Higher Educ; Am Conf Acad Deans. *Res:* Wood-machining practices in Vermont. *Mailing Add:* 51 Central Northfield VT 05663

MCINTIRE, WILLIAM S, b Chicago, Ill, Apr 12, 49. BIOCHEMISTRY. *Educ:* Univ Ill, BS, 71; Univ Calif, Berkeley, MS, 76, PhD(comp biochem), 83. *Prof Exp:* Phys sci technician, 76-79, chemist, 79-83, RES CHEMIST, VET ADMIN MED CTR, SAN FRANCISCO, 83-; ASSOC RES BIOCHEM, UNIV CALIF, SAN FRANCISCO, 88- *Concurrent Pos:* Asst res biochem, Univ Calif, San Francisco, 87-88; NSF grant, 87-90; prog proj grant, NIH, 89-94; merit rev grant, Vet Admin, 91- *Mem:* AAAS; Am Chem Soc; Am Inst Chem; Am Soc Biochem & Molecular Biol; Int Union Pure & Appl Chem; NY Acad Sci. *Mailing Add:* Molecular Biol Div Vet Admin Med Ctr 4150 Clement St San Francisco CA 94121

MCINTOSH, ALAN WILLIAM, Harvey, Ill, Sept 5, 44; m 66; c 2. AQUATIC TOXICOLOGY, BIOMONITORING. *Educ:* Univ of Ill, BS, 66, MS, 68; Mich State Univ, PhD(fisheries & limnol), 72. *Prof Exp:* Asst prof environ health, Purdue Univ, 72-77; ASSOC PROF ENVIRON SCI, RUTGERS UNIV, 77-; DIR, VT WATER RESOURCES & LAKE STUDIES CTR, SCH NATURAL RESOURCES, UNIV VT. *Concurrent Pos:* Dir, Div Water Resources, Ctr for Coastal & Environ Studies, Rutgers Univ, 77-; vis prof, Waikato Univ, Hamilton NZ, 83; consult, Univ Vt, 84- *Mem:* Soc Environ Toxicol & Chem. *Res:* Fate of trace elements in surface waters; effects of acidification on trace element accumulation by biota. *Mailing Add:* Univ Vt Sch Natural Resources Aiken Ctr Burlington VT 05405

MCINTOSH, ALEXANDER OMAR, b Acton, Ont, Oct 27, 13. INDUSTRIAL CHEMISTRY. *Educ:* Univ Toronto, BA, 39; Univ Minn, MS, 40; Univ Glasgow, PhD(phys chem), 51. *Prof Exp:* Asst geol, Univ Minn, 39-40; chemist, Defence Industs, Ltd, 41-44; res chemist, Can Indust, Ltd, 44-48; demonstr, Univ Glasgow, 48-50; res chemist, Can Indust Ltd, 51-56, group leader, Cent Res Lab, 56-62, patent asst, Legal Dept, 62-65, patent agt, Legal Dept, 65-76; consult, 76-81; RETIRED. *Mem:* Fel Chem Inst Can. *Res:* Protection of chemical inventions. *Mailing Add:* 404-120 Edinburgh Rd S Guelph ON N1H 5P7 Can

MCINTOSH, ARTHUR HERBERT CRANSTOUN, b St Thomas, VI, Apr 2, 34; m 64; c 3. MICROBIOLOGY. *Educ:* McMaster Univ, BA, 59; Univ Guelph, MS, 62; Harvard Univ, DSc(microbiol), 69. *Prof Exp:* NIH fel, Stanford Res Inst, 69-71; staff researcher microbiol, Boyce Thompson Inst, NY, 71-74; asst res prof, Waksman Inst Microbiol, Rutgers Univ, New Brunswick, 74-79; RES MICROBIOLOGIST, CONTROL INSECTS RES LAB, USDA, 79- *Concurrent Pos:* NSF grant, 75. *Mem:* Sigma Xi; Soc Invert Path; Tissue Culture Asn. *Res:* Invertebrate and vertebrate tissue culture and virology; in vitro safety testing of viral insecticides; transfection of lepidopteran cells with baculovirus DNA. *Mailing Add:* Biol Control Insects Res Lab PO Box 7629 Columbia MO 65205

MCINTOSH, BRUCE ANDREW, b Walkerton, Ont, Oct 30, 29; m 54; c 4. PHYSICS, ELECTRONICS. *Educ:* Western Ont Univ, BSc, 52, MSc, 53; McGill Univ, PhD(electron beams), 58. *Prof Exp:* Asst res officer, Nat Res Coun Can, 53-55, assoc res officer, 58-70, secy assoc comt meteorites, 60, sr res officer, 70-90; RETIRED. *Honors & Awards:* Hon Gold Medal, Czech Acad Sci, 89. *Mem:* Can Asn Physicists; Int Astron Union; Can Astron Soc. *Res:* Meteoritics; upper atmosphere physics. *Mailing Add:* 25 Seguin St Gloucester ON K1A 0R6 Can

MCINTOSH, ELAINE NELSON, b Webster, SDak, Jan 30, 24; m 55; c 3. PHYSIOLOGICAL BACTERIOLOGY, NUTRITION. *Educ:* Augustana Col, SDak, AB, 45; Univ SDak, Vermillion, MA, 49; Iowa State Univ, PhD(phys bact & biochem), 54. *Prof Exp:* From instr to asst prof chem, Sioux Falls Col, SDak, 45-48; res fel biochem, Univ SDak, Vermillion, 48-49; instr bact, Iowa State Univ, 49-54; res assoc dairy sci, Univ Ill, Urbana, 54-55; res assoc home econ res, Iowa State Univ, 55-62; from asst prof to prof human biol, Univ Wis-Green Bay, 68-90, asst to chancellor, 74-75, spec asst to vchancellor, 75-76, chmn human biol, 75-80, leader, Human Nutrit/Dietetics Prog, 75-90, EMER PROF HUMAN BIOL, UNIV WIS-GREEN BAY, 90- *Concurrent Pos:* Sigma Xi res grant, 66; pres, Wis Nutrit Coun, 75. *Honors & Awards:* Chancellor's Res Award, Univ Wis-Green Bay, 69. *Mem:* Sigma Xi; Inst Food Technologists; Am Dietetic Asn. *Res:* Nutrition education; community health. *Mailing Add:* Environ Sci 301 Univ Wis Green Bay WI 54311-7001

MACINTOSH, FRANK CAMPBELL, b Baddeck, NS, Dec 24, 09; m 38; c 5. PHYSIOLOGY. *Educ:* Dalhousie Univ, BA, 30, MA, 32; McGill Univ, PhD(physiol), 37. *Hon Degrees:* LLD, Univ Alta, 64, Queen's Univ, Can, 65, Dalhousie Univ, 76; MD, Univ Ottawa, 74; DSc, McGill Univ, 80, St Francis Xavier Univ, 85. *Prof Exp:* Instr biol, Dalhousie Univ, 30-31, biochem, 31-32 & pharmacol, 32-33; demonstr physiol, McGill Univ, 36-37; mem res staff, Med Res Coun Gt Brit, 38-49; Joseph Morley Drake prof, 49-78, EMER PROF PHYSIOL, MCGILL UNIV, 80- *Concurrent Pos:* Treas, Int Union Physiol Sci, 62-68; mem, Sci Coun Can, 66-71; ed, Can J Physiol & Pharmacol, 68-72; pres, Thirtieth Int Congress Physiol Scis, 86. *Mem:* Am Physiol Soc; Royal Soc London; Royal Soc Can; Can Physiol Soc (pres, 60-61); Brit Physiol Soc; Int Soc Neurochem; Pharmacol Soc Can; Brit Pharmacol Soc. *Res:* Acetylcholine metabolism. *Mailing Add:* Dept of Physiol McGill Univ 3655 Drummond St Montreal PQ H3G 1Y6 Can

MCINTOSH, HAROLD LEROY, b Fairfax, Mo, Dec 25, 31; m 53; c 3. PHYSICS. *Educ:* Tarkio Col, BA, 53; Univ Colo, MS, 61; Ore State Univ, PhD(physics, chem), 68. *Prof Exp:* Pub sch teacher, Mo, 56-61; acad dean, 74-80, PROF PHYSICS, TARKIO COL, 62-, PRES, 80- *Mem:* Am Asn Physics Teachers. *Res:* Chemistry; mathematics. *Mailing Add:* 705 College Tarkio MO 64491

MCINTOSH, HENRY DEANE, b Gainesville, Fla, July 19, 21; m 45; c 3. INTERNAL MEDICINE. *Educ:* Davidson Col, BS, 43; Univ Pa, MD, 50; Am Bd Internal Med, dipl, 57; Am Bd Cardiovasc Dis, dipl, 64. *Hon Degrees:* DSc, Univ Francisco Marroquin, Guatemala City, Guatemala. *Prof Exp:* Intern med, Duke Univ Hosp, 50-51; asst res, Lawson Vet Admin Hosp, 51-52; from instr to assoc med, Sch Med, Duke Univ, 54-58, from asst prof to prof, 58-70; prof med & chmn dept, Baylor Col Med, 70-77; MEM STAFF, CARDIOL SECT, WATSON CLIN, LAKELAND, 77-; CLIN PROF MED, UNIV FLA, 77- *Concurrent Pos:* Am Heart Asn fel cardiol, Duke Univ Hosp, 52-54; consult, Vet Admin Hosps, Durham, NC, 56-70 & Fayetteville, 57-70, Womack Army Hosp, Ft Bragg, 57-70, Watts Hosp, Durham, 57-70 & Portsmouth Naval Hosp, Va, 57-70; dir cardiovasc lab, Med Ctr, Duke Univ, 56-70, dir cardiovasc div, 66-70; asst ed, Mod Concepts Cardiovasc Dis, 65-68; mem cardiovasc study sect, Nat Heart Inst, 65-69; mem, Subspecialty Bd Cardiovasc Dis, Am Bd Internal Med, 68-76, & Am Bd Emergency Med, 79; chief med serv, Methodist Hosp, Houston, 70-77, Ben Taub Gen Hosp, Jefferson Davis Hosp & Houston Vet Admin Hosp; consult, St Luke's Hosp; ed consult, Am J Cardiol & Circulation, 70-76; chmn coun clin cardiol, Am Heart Asn, 74-76; adj prof med, Baylor Col Med, 77-; chmn, Comt Prevention Cardiovasc Disease, Am Col Cardiol, 81-; N Am Soc Pacing & Electrophysiology Comt, Continuing Med Educ; coun exec comt, Geriatric Cardiol. *Honors & Awards:* Distinguished Achievement Award, Coun on Clin Cardiol, Am Heart Asn. *Mem:* Am Soc Internal Med; Am Heart Asn; Am Col Physicians; Asn Am Physicians; distinguished fel Am Col Cardiol (pres, 73-74). *Res:* Cardiovascular hemodynamics, especially factors controlling cardiac output. *Mailing Add:* Watson Clin PO Box 95000 Lakeland FL 33804-5000

MCINTOSH, JOHN MCLENNAN, b Galt, Ont, Apr 16, 40; m 65; c 2. SYNTHETIC ORGANIC CHEMISTRY. *Educ:* Queen's Univ, Ont, BSc, 62; Mass Inst Technol, PhD(org chem), 66. *Prof Exp:* Fel org chem, Nat Res Coun Can, 66-68; teaching fel, Univ Waterloo, 68; asst prof, 68-73, assoc prof, 73-80, PROF ORG CHEM, UNIV WINDSOR, 80- *Mem:* Am Chem Soc; fel Chem Inst Can. *Res:* Synthesis of complex molecules; new methods of organic synthesis, especially those utilizing organo-sulfur and organo-phosphorus compounds; phase-transfer catalysis; synthesis of chiral compounds; asymmetric alkylations. *Mailing Add:* Dept Chem Univ Windsor Windsor ON N9B 3P4 Can

MCINTOSH, JOHN RICHARD, b New York, NY, Sept 25, 39; m 61; c 3. CELL BIOLOGY. *Educ:* Harvard Univ, BA, 61, PhD(biophys), 67. *Prof Exp:* Sch teacher, Mass, 61-63; from instr to asst prof biol, Harvard Univ, 67-70; asst prof, 71-73, assoc prof, 73-77, PROF BIOL, UNIV COLO, BOULDER, 77- *Concurrent Pos:* Consult, Educ Serv Inc, 62-63; NIH fel, Biol Labs, Harvard Univ, 68; ed, J Cell Biol, 77-81; mem, NSF Cell Biol Panel, 77-78; mem, Coun Am Soc Cell Biol, 78-81; mem, Cell & Develop Panel, Am Cancer Soc, 82-86. *Honors & Awards:* K R Porter lectr, Am Soc Cell Biol, 83. *Mem:* Am Soc Cell Biologists; Biophys Soc; Soc Develop Biologists. *Res:* Mitosis and cell motion; control of cell form; structure and biochemistry of the mitotic apparatus in an effort to understand the mechanisms of chromosome movement. *Mailing Add:* 870 Willowbrook Rd Boulder CO 80302

MCINTOSH, JOHN STANTON, b Ford City, Pa, Jan 6, 23. THEORETICAL PHYSICS. *Educ:* Yale Univ, BS, 48, MS, 49, PhD(physics), 52. *Prof Exp:* Res assoc physics, Proj Matterhorn, Princeton Univ, 52-53, instr, Univ, 53-56; from instr to asst prof, Yale Univ, 56-63; assoc prof, 63-65, PROF PHYSICS, WESLEYAN UNIV, 65- *Concurrent Pos:* Consult, Los Alamos Sci Lab & Rand Corp, 53; res assoc, Peabody Mus Natural Hist, Yale Univ, 65-71. *Mem:* Am Phys Soc; Sigma Xi. *Res:* Low energy nuclear and nuclear scattering theories; heavy ion scattering and reactions. *Mailing Add:* Sci Ctr Tower Rm 253 Wesleyan Univ Middletown CT 06457

MCINTOSH, LEE, b Wichita, Kans, June 29, 49; m. PLANT GENES, PHOTOSYNTHESIS. *Educ:* Univ Calif, Irvine, BSc, 72; Univ Wash, Seattle, PhD(develop bot), 77. *Prof Exp:* Res fel, Biol Lab, Harvard Univ, 77-81; from asst prof to assoc prof biochem, 81-89, PROF BIOCHEM, DEPT ENERGY, PLANT RES LAB, MICH STATE UNIV, 90- *Mem:* Am Soc Plant Physiol; Int Soc Plant Molecular Biol. *Res:* Molecular basis controlling the development of photosynthetic competence in higher plants; developmental genetics of plant mitochondria; plant biochemistry. *Mailing Add:* Plant Res Lab Mich State Univ East Lansing MI 48824

MCINTOSH, ROBERT EDWARD, b Hartford, Conn, Jan 19, 40; m 62; c 5. ELECTRICAL ENGINEERING, APPLIED PHYSICS. *Educ:* Worcester Polytech Inst, BSEE, 62; Harvard Univ, SM, 64; Univ Iowa, PhD(elec eng), 67. *Prof Exp:* Mem tech staff microwave eng, Bell Tel Labs, 62-65; teaching asst elec eng, Univ Iowa, 65-66; from asst prof to assoc prof eng, 67-72, coordr Microwave Electronics Group, 81-87, acting head, 83-84, PROF ENG, UNIV AMHERST, 72-, DIR MICROWAVE REMOTE SENSING LAB, 81- *Concurrent Pos:* Guest prof physics, Cath Univ Nijmegen, 73-74; ed, Antennas & Propagation Soc Newsletter, 79-; assoc ed, Transaction Antennas & Propagation, Inst Elec & Electronic Engrs, 83-85, ed, 88-, assoc ed, Radio Sci, guest ed, Transactions on Educ & Geosci & Remote Sensing, Inst Elec & Electronic Engrs; vis scientist, NASA Langley Res Ctr, 80-81; pres, Quadrant Eng Inc, 81- *Honors & Awards:* Centennial Medal, Inst Elec & Electronic Engrs, 84, Distinguished Serv Award, 85. *Mem:* fel Inst Elec & Electronics Engrs; Int Radio Sci Union; Am Phys Soc; Optical Soc Am; fel IEEE; Antennas & Propagation Soc (secy-treas, 81-83, vpres, 84, pres, 85); Remote Sensing Soc (secy-treas, 81-83, pres, 84); Am Geophys Union. *Res:* Microwave systems; remote sensing; electromagnetics. *Mailing Add:* Dept of Elec & Comput Eng Univ of Mass Amherst MA 01002

MCINTOSH, ROBERT PATRICK, b Milwaukee, Wis, Sept 24, 20; m 47; c 2. PLANT ECOLOGY. *Educ:* Lawrence Col, BS, 42; Univ Wis, MS, 48, PhD, 50. *Prof Exp:* Asst bot, Univ Wis, 46-50; from instr to asst prof, Middlebury Col, 50-53; asst prof, Vassar Col, 53-58; from asst prof to prof, 58-87, EMER PROF BIOL, UNIV NOTRE DAME, 87- *Concurrent Pos:* Ed, Am Midland Naturalist, 70-; prog dir, Nat Sci Found, 77-78. *Mem:* AAAS; Ecol Soc Am; Brit Ecol Soc; Soc Am Naturalists. *Res:* Forest ecology; history of ecology. *Mailing Add:* Dept Biol Univ Notre Dame Notre Dame IN 46556

MCINTOSH, THOMAS HENRY, b Ames, Iowa, July 3, 30; m 55; c 3. ENVIRONMENTAL SCIENCES. *Educ:* Iowa State Univ, BS, 56, MS, 58, PhD(soil microbiol), 62. *Prof Exp:* Instr soils, Iowa State Univ, 60-62; from asst prof to assoc prof, Univ Ariz, 62-68; asst dean col environ sci, 68, assoc prof sci & environ change, 68-71, asst dean cols, 70-73, assoc dean, 73-75, asst chancellor student & admin serv, 75-76, PROF SCI & ENVIRON CHANGE, UNIV WIS-GREEN BAY, 71-, ASST TO CHANCELLOR, 76- *Concurrent Pos:* Vis prof, Iowa State Univ, 67. *Mem:* Am Soc Agron; Soil Sci Soc Am; Am Soc Microbiol; Am Chem Soc; Soil Conserv Soc Am; Am Soc Photogram & Remote Sensing. *Res:* Nitrogen transformations in soil and aquatic systems; biogeochemistry. *Mailing Add:* 148 Rose Lane Green Bay WI 54302

MCINTOSH, THOMAS JAMES, b Geneva, NY, Feb 4, 47; m 68; c 3. BIOPHYSICS. *Educ:* Univ Rochester, BS, 69; Carnegie-Mellon Univ, MS, 71, PhD(physics), 73. *Prof Exp:* Res assoc, 74-77, from asst prof to assoc prof anat, 77-90, PROF CELL BIOL, DUKE UNIV MED CTR, 90- *Mem:* Biophys Soc. *Res:* Structure of biological membranes; interactions between membrane surfaces. *Mailing Add:* Dept Cell Biol Duke Univ Med Ctr Durham NC 27710

MCINTOSH, WILLIAM DAVID, b Pryor, Okla, Oct 14, 36; m 74. MATHEMATICS. *Educ:* Southwestern Col, Kans, BA, 58; Univ Kans, MA, 60, PhD(math), 65. *Prof Exp:* Asst instr math, Univ Kans, 58-63, asst, 63-64, asst instr, 64-65; asst prof, Univ Mo-Columbia, 65-70; prof math & chmn dept, 70-80, PROF MATH, CENT METHODIST COL, 70- *Mem:* Am Math Soc; Math Asn Am; Sigma Xi. *Res:* Theory of retracts. *Mailing Add:* 600 Green Acres Dr Fayette MO 65248

MCINTURFF, ALFRED D, b Clinton, Okla, Feb 24, 37; m 60, 64; c 3. HIGH ENERGY PHYSICS, CRYOGENICS. *Educ:* Okla State Univ, BS, 59; Vanderbilt Univ, MS, 60, PhD(physics), 64. *Prof Exp:* Res assoc, Synchrofron Lab, Calif Inst Technol, 62-65; res assoc, Vanderbilt Univ, 64-66, sr res engr, Atomics Int, 66-67; mem staff, Advan Accelerator Design Div, Brookhaven Nat Lab, 67-71, physicist, Accelerator Dept, 71-80; PHYSICIST, CRYOGENIC RES, FERMI NAT ACCELERATOR LAB, 81- *Concurrent Pos:* Physicist, Rutherford Lab, Chilton, Didcot, UK, 72-73, Attaché Scientifique CERN Lab, Geneva, Switzerland, 88-89. *Mem:* Am Phys Soc; Sigma Xi. *Res:* High field super conductors; photostar research; photosigma production; sigma magnetic moment studies; accelerator magnet design and testing. *Mailing Add:* Fermi Nat Accelerator Lab PO Box 500 Batavia IL 60510

MCINTYRE, ADELBERT, b Providence, RI, Jan 1, 29; div; c 5. ATMOSPHERIC SCIENCES, SPACE SCIENCES. *Educ:* Univ RI, BSc, 58. *Prof Exp:* Instr physics, Univ RI, 58-61; spec scientist space physics, Off Aerospace Res, US Air Force, 61-63; spec scientist space physics & optics, Air Force Cambridge Res Labs, 63-74, res physicist optics & Tech dir missile surveillance technol, Air Force Geophys Lab, 74-83, supvry res physicist optics, chief atmospheric backgrounds br & dep dir IR Tech Div, Air Force Geophysics Lab, Air Force Systs Command, 83-86; pres & tech dir, Infratech, Inc, 86-88; SCIENTIST, AEROJET ELECTRONIC SYSTS DIV, 88- *Concurrent Pos:* Teaching fel, Univ RI, 58-59; prin investr & res physicist, Navy Underwater Ord Lab, 59-61; consult, US Comt Space Res, Nat Acad Sci-Res Coun, 65-74, consult & mem, Ballistic Missile Defense Working Group for Churchill Rocket Range, 69-74 & Off Undersecy Defense for Res & Eng, 74-83; mem & consult, Missile Plume Technol Working Group, US Air Force, 74-83; tech dir, field widened interferometer, US Air Force, 82-84; tech consutl, Defense Technologies Study Team (Fletcher comt), Dept Defense, 83-84 & Strategic Defense Initiative Orgn, 85- *Mem:* Sigma Xi; AAAS; Int Platform Asn; Am Inst Aeronaut & Astronaut. *Res:* Rocket satellite and balloon-borne measurements of infrared targets and backgrounds phenomenology; first in situ measurements of a high altitude rocket engine plume using a mother-daughter rocket payload; tech dir for first rocketborne field-widened interferometer measurements of an aurora; tech dir for first balloon-borne spectral measurements of a ground release of a gas from 90,000 ft altitude. *Mailing Add:* 1729 Orangewoood Ave Uplands CA 91786

MCINTYRE, ANDREW, b Weehawken, NJ, Sept 17, 31; m 67, 78. MARINE GEOLOGY, PALEOCLIMATOLOGY. *Educ:* Columbia Univ, BA, 55, PhD(marine biol), 67. *Prof Exp:* Lectr geol, Barnard Col, Columbia Univ, 61-62; instr, City Col New York, 63-64; res assoc, 62-67, SR VIS RES ASSOC OCEANOG, LAMONT GEOL OBSERV, COLUMBIA UNIV, 67- *Concurrent Pos:* NSF res grants, 62-; from asst prof to assoc prof geol, Queens Col, NY, 67-77, prof, 77- *Mem:* AAAS; Geol Soc Am; NY Acad Sci; Am Geophys Union; Am Quaternary Asn. *Res:* Biogeography and ecology of Coccolithophorida, Pleistocene paleoclimatology and paleooceanography; nanoplankton of the oceans; skeletal ultramicrostructure of nanoplankton and microplankton. *Mailing Add:* Lamont-Doherty Geol Observ Columbia Univ Palisades NY 10964

MACINTYRE, BRUCE ALEXANDER, b Oak Park, Ill, Sept 10, 42; m 65; c 2. PHYSIOLOGY. *Educ:* Carroll Col, Wis, BS, 63; Ind Univ, Bloomington, PhD(physiol), 68. *Prof Exp:* Asst prof, 68-73, ASSOC PROF BIOL, CARROLL COL, WIS, 73- *Mem:* AAAS; Am Inst Biol Sci. *Res:* Human thermoregulatory control mechanisms; cellular enzymology, including carbohydrate metabolism of choriocarcinoma cells. *Mailing Add:* Dept Biol Carroll Col 100 N East Ave Waukesha WI 53186

MCINTYRE, DONALD, b Detroit, Mich, Sept 8, 28; m 57; c 4. POLYMER CHEMISTRY. *Educ:* Lafayette Col, BA, 49; Cornell Univ, PhD(chem), 54. *Prof Exp:* Chemist, Monsanto Chem Co, Mass, 53-54; chemist, Nat Bur Standards, 56-62, sect chief macromolecules sect, 62-66; PROF POLYMER SCI & CHEM, UNIV AKRON, 66- *Mem:* Am Chem Soc; Am Phys Soc. *Res:* Physical chemistry of polymers; solution properties; kinetics; phase equilibria of polymers; characterization of polymers; cycloparaffins; structure of polymers. *Mailing Add:* 1365 Hillandale Dr Akron OH 44313-1823

MCINTYRE, DONALD B, b Edinburgh, Scotland, Aug 15, 23; m 57; c 1. GEOLOGY. *Educ:* Univ Edinburgh, BSc, 45, PhD, 47. *Hon Degrees:* DSc, Univ Edinburgh, 51. *Prof Exp:* Lectr econ geol, Univ Edinburgh, 48-52, lectr petrol, 52-54; from assoc prof to prof, Pomona Col, 54-75, chmn dept, 55-84, Seaver prof sci, 75-89, EMER PROF GEOL, POMONA COL, 89- *Concurrent Pos:* Res assoc, Univ Calif, 52; Guggenheim Found fel, 69-70; consult, IBM Sci Ctr. *Honors & Awards:* Pigeon Award, Geol Soc London, 52. *Mem:* Geol Soc Am; Geol Asn London; NY Acad Sci. *Res:* Structural geology of deformed rocks; sampling methods; computer applications; history of geology. *Mailing Add:* Luachmhor Church Rd Kinfauns Perth Ph2-7Ld Scotland

MCINTYRE, GARY A, b Portland, Ore, July 16, 38; m 63; c 2. PLANT PATHOLOGY. *Educ:* Ore State Univ, BS, 60, PhD(plant path), 64. *Prof Exp:* Nat Defense Educ Act fel, Ore State Univ, 60-64; from asst prof to prof plant path, Univ Maine, Orono, 64-75, actg head, dept bot & plant path, 68-69, chmn dept, 69-75; chmn, dept bot & plant path, 75-84, HEAD DEPT PLANT PATH & WEED SCI, COLO STATE UNIV, 84- *Concurrent Pos:* Dir, Biol Core Curriculum, Colo State Univ, 76-81; coordr, Western Regional Integrated Pest Mgt Prog, 78- *Mem:* Am Phytopath Soc; Potato Asn Am; Sigma Xi; AAAS. *Res:* Potato diseases; physiology of parasitism. *Mailing Add:* Dept Plant Path & Weed Sci Colo State Univ Ft Collins CO 80523

MACINTYRE, GILES T, b Bridgeton, NJ, Oct 6, 26; div; c 3. VERTEBRATE ZOOLOGY & PALEONTOLOGY. *Educ:* Columbia Univ, BS, 55, MA, 57, PhD(zool), 64. *Prof Exp:* Asst zool, Columbia Univ, 56-57, lectr, 57-59; lectr biol, 62-64, instr, 64-65, asst prof, 66-68, ASSOC PROF BIOL, QUEENS COL, NY, 69- *Concurrent Pos:* Higgins & NIH fel. *Mem:* Am Soc Mammal; Soc Syst Zool; Soc Study Evolution; Soc Vert Paleont; Sigma Xi. *Res:* Mesozoic mammals; quasi-mammals and therapsid reptiles; carnivore evolution and systematics; basicranial osteology and associated anatomy; functional anatomy of teeth and jaws; adaptive radiation. *Mailing Add:* Dept Biol Queens Col Flushing NY 11367

MCINTYRE, JAMES DOUGLASS EDMONSON, b Toronto, Ont, Feb 16, 34; m 60; c 3. PHYSICAL CHEMISTRY. *Educ:* Univ Toronto, BA, 56, MA, 58; Rensselaer Polytech Inst, PhD(phys chem), 61. *Prof Exp:* Res assoc chem, Princeton Univ, 60-62; mem tech staff chem res, Bell Labs, 62-87; prog officer, NSF, 88-90; RES PROF, NAT COLD FUSION INST, UNIV UTAH, 91- *Mem:* Am Chem Soc; Electrochem Soc. *Res:* Electrode kinetics and adsorption; electrodeposition; chemistry and physics of surfaces; catalysis; optical properties of thin films; modulation spectroscopy. *Mailing Add:* Chem Dept Univ Utah Salt Lake City UT 84102

MCINTYRE, JOHN A, b Beloit, Wis, July 1, 42; m; c 2. TRANSPLANT, IMMUNOLOGY. *Educ:* Rockford Col, Ill, AB, 66; Wake Forest Univ, PhD, 71; MRCPath, 82. *Prof Exp:* Fel, Dept Path, Harvard Med Sch, Boston, Mass, 71-74; asst prof, Dept Med & Dept Path & dir, Tissue Typing Lab Med, Univ SC, Charleston, 74-79; dir, Div Immunobiol, Blond McIndoe Ctr Transplantation Biol, Queen Victoria Hosp, East Grinstead, Sussex, Eng, 79-82; vis assoc prof, Div Maternal & Fetal Med, 82-83, assoc prof, 83-86, PROF, DEPT OBSTET & GYNEC, SCH MED, SOUTHERN ILL UNIV, 86-; DIR RES, METHODIST CTR REPRODUCTION & TRANSPLANTATION IMMUNOL, METHODIST HOSP IND, INC, INDIANAPOLIS, 86- *Concurrent Pos:* Adj asst prof, Dept Med, Med Univ SC, Charleston, 79-82; assoc prof, Dept Med Microbiol & Immunol, Sch Med, Southern Ill Univ, Springfield, 83-86, Dept Microbiol, Carbondale, Ill, 83-86, adj prof, Dept Med Microbiol & Immunol, 86-; dir, Histocompatibility Lab, Methodist Hosp Ind, Indianapolis, 86-; counr, Am Soc Immunol Reproduction, 86-87; adj prof, Dept Biol, Ind Univ-Purdue Univ, Indianapolis, 87- *Mem:* Sigma Xi; Am Soc Immunol Reproduction (pres-elect, 87-88, pres, 89-90); AAAS; Soc Leucocyte Biol; Am Soc Cell Biol; Transplantation Soc; Am Soc Histocompatibility & Immunogenetics; Am Asn Immunologists; Int Soc Develop & Comp Immunol; Am Soc Reproductive Immunol. *Res:* Author of various publications. *Mailing Add:* Dept Transplant-Immunol Methodist Hosp Ind 1701 N Senate Blvd Indianapolis IN 46202

MCINTYRE, JOHN ARMIN, b Seattle, Wash, June 2, 20; m 47; c 1. POSITRON EMISSION TOMOGRAPHY. *Educ:* Univ Wash, BS, 43; Princeton Univ, MA, 48, PhD(physics), 50. *Prof Exp:* Instr elec eng, Carnegie Inst Technol, 43-44; radio engr, Westinghouse Elec Corp, 44-45; res assoc, Stanford Univ, 50-57; from asst prof to assoc prof physics, Yale Univ, 57-63; assoc dir cyclotron inst, 65-70, PROF PHYSICS, TEX A&M UNIV, 63- *Concurrent Pos:* Consult, Varian Assocs, 55; vis scientist, NSF, 60-65; Oak Ridge Inst Nuclear Studies res partic grant, 63-67; Welch Found grants, 64-66 & 67-70; councilor, Oak Ridge Assoc Univs, 65-71; NIH grant, 78-79; Am Cancer Soc grant, 80-82. *Mem:* AAAS; fel Am Phys Soc; Soc Nuclear Med; Am Sci Affil. *Res:* Scintillation counters; Compton effect; high energy electron scattering; heavy ion scattering and transfer reactions; gamma ray spectroscopy; three-nucleon scattering; nuclear instrumentation for medicine and technology. *Mailing Add:* Dept Physics Tex A&M Univ College Station TX 77840

MACINTYRE, JOHN R(ICHARD), b Eng, June 29, 08; nat US; m 34; c 3. ELECTRICAL ENGINEERING. *Educ:* Univ Mich, BSEE, 34. *Prof Exp:* Elec instrument design engr, Gen Elec Co, 36-40, aircraft instrument design engr, 40-44, adv develop engr, Electromech Devices, 44-54, adv develop eng mgr, 54-58, consult engr, 58-64, instrumentation & adv systs engr, 64-73; RETIRED. *Mem:* Soc Prof Engrs; fel Inst Elec & Electronics Engrs; fel Instrument Soc Am. *Res:* Development and design of instrument and control systems and mechanisms. *Mailing Add:* 231 Queensbury Dr Apt 4 Huntsville AL 35802

MCINTYRE, JUDITH WATLAND, b Kansas City, Mo, Aug 19, 30; m 57; c 3. ORNITHOLOGY, ETHOLOGY. *Educ:* Carleton Col, BA, 52; Univ Minn, MA, 70, PhD(zool), 75. *Prof Exp:* Asst prof, 77-80, assoc prof biol, 80-86, PROF BIOL, UTICA COL, SYRACUSE UNIV, 86- *Concurrent Pos:*

Dir, Oikos Res Found, 76-; coordr, Conf on Common Loon Res & Mgt, 77-79; res grants, Dept Environ Conserv, Endangered Species Div, NY, 78-79, Nat Geog Soc, 79-80 & 89, NSF, 81-82, NAm Loon Fund, 81-88 & Exxon Corp, 89-90; vpres, NAm Loon Fund. *Mem:* Am Ornithologists Union; AAAS; Cooper Ornith Soc; Wilson Soc; Animal Behav Soc; Asn Field Ornithologists; Sigma Xi. *Res:* Behavior of common and yellow-billed loons; sound transmission over water; vocal communication; reservoirs as loon habitat. *Mailing Add:* Dept Biol Utica Col Syracuse Univ Utica NY 13502

MCINTYRE, LAURENCE COOK, JR, b Knoxville, Tenn, July 9, 34; m 57; c 4. ATOMIC PHYSICS. *Educ:* Stanford Univ, BS, 57; Univ Wis, MS, 61, PhD(physics), 65. *Prof Exp:* Instr physics, Princeton Univ, 65-66; from asst prof to assoc prof, 66-78, PROF PHYSICS, UNIV ARIZ, 78- *Mem:* Am Phys Soc. *Res:* Atomic and molecular physics; beam foil spectroscopy; molecular dissociation. *Mailing Add:* Dept of Physics Univ of Ariz Tucson AZ 85721

MACINTYRE, MALCOLM NEIL, b Honolulu, Hawaii, Aug 12, 19; m 41; c 4. CYTOGENETICS. *Educ:* Univ Mich, BA, 48, MA, 50, PhD(zool), 55. *Prof Exp:* From instr to prof anat, 54-84, EMER PROF, CASE WESTERN RESERVE UNIV, 84- *Concurrent Pos:* Mem anat sci training comt, NIH, 62-, chmn, 65- *Mem:* AAAS; Teratol Soc; Am Asn Anat; Am Soc Zool; Soc Human Genetics. *Res:* Human reproduction and development; human reproductive failure; association of chromosomal abnormalities with congenital malformations; amniotic fluid cultures; prenatal genetic evaluation; genetic counseling. *Mailing Add:* 2573 Wellington Rd Cleveland OH 44118

MCINTYRE, MARIE CLAIRE, b Orange, Tex, Mar 17, 61. ENGINEERING. *Educ:* Pomona Col, BA, 82; Scripps Inst Oceanog, Univ Calif San Diego, PhD(oceanog), 89. *Prof Exp:* Res asst, Scripps Inst Oceanog, 82-89; ENGR, DEEP OCEAN ENG, 89- *Concurrent Pos:* Vis lectr, Naval Postgrad Sch, 88; lectr, Univ Calif, Berkeley, 91. *Honors & Awards:* Chapman-Scaaefer Award, Marine Technol Soc, 87. *Mem:* Marine Technol Soc; Am Geophys Union. *Res:* Design, development, repair and operation of remotely operated vehicles in the underwater environment. *Mailing Add:* 1431 Doolittle Dr San Leandro CA 94577

MCINTYRE, OSWALD ROSS, b Chicago, Ill, Feb 13, 32; m 57; c 3. HEMATOLOGY. *Educ:* Dartmouth Col, AB, 53; Harvard Univ, MD, 57. *Prof Exp:* Intern, Hosp Univ Pa, 57-58; resident med, Dartmouth Affiliated Hosps, 58-60; trainee hemat, Dartmouth Med Sch, 60-61; asst chief clin res sect, Pakistan SEATO Cholera Res Lab, 61-62; mem bact prof, Div Biol Standards, NIH, 62-63; trainee hemat, 63-64, from asst prof med to assoc prof med, 66-76, JAMES J CARROL PROF ONCOL, DARTMOUTH MED SCH, 76-, CHMN, CANCER & LEUKEMIA GROUP B, 90- *Concurrent Pos:* Markle scholar acad med, 65; counsult, Hitchcock Clin, 64; attend physician, Vet Admin Hosp, White River Junction, Vt, 64; dir, Norris Cotton Cancer Ctr, Hanover, NH. *Mem:* AAAS; Am Fedn Clin Res; Am Soc Hemat; NY Acad Sci; Am Asn Cancer Res; Am Soc Clin Oncol. *Res:* Cancer clinical trials in myeloma leukemia and lymphoma. *Mailing Add:* Norris Cotton Cancer Ctr & Dartmouth Dept Med Hinman Box 7920 Hanover NH 03756

MCINTYRE, PATRICIA ANN, b Christopher, Ill, Sept 1, 26. MEDICINE. *Educ:* Kalamazoo Col, AB, 48; Johns Hopkins Univ, MD, 52; Am Bd Internal Med, dipl, 62; Am Bd Nuclear Med, dipl, 72. *Prof Exp:* Intern, Mass Gen Hosp, 52-53; asst resident internal med, Johns Hopkins Hosp, 53-55, instr med, 57-58, clin dir dept med, Mem Med Ctr, 58-64; res assoc hemat div, Sch Med, 64-65, instr med, 65-67, instr, Univ, 66-67, asst prof med & radiol sci, 67-71, assoc prof radiol sci, Sch Hyg, 71-73, ASSOC PROF MED & RADIOL, SCH MED, JOHNS HOPKINS UNIV, 71-, ASSOC PROF RADIOL SCI, SCH MED & ASSOC PROF ENVIRON HEALTH, SCH HYG, 73- *Concurrent Pos:* Fels hemat, Sch Med, Johns Hopkins Univ, 55-57; dir med educ, Hosp for Women, 57-58; chmn, panel int comt standardization hemat. *Mem:* Am Fedn Clin Res; Am Soc Hemat; Soc Nuclear Med; fel Am Col Physicians. *Res:* Hematology; nuclear medicine. *Mailing Add:* 615 N Wolfe St Baltimore MD 21205

MCINTYRE, ROBERT GERALD, b Cleveland, Okla, Mar 26, 24; m 59; c 3. APPLIED MATHEMATICS, THEORETICAL PHYSICS. *Educ:* US Naval Acad, BS, 45; Univ Okla, PhD(physics), 59. *Prof Exp:* Res mathematician, Sohio Petrol Co, 56-59; asst prof math & physics, Grad Inst Technol, Univ Ark, 57-60; mem tech staff & theoret physicist, Tex Instruments, Inc, 60-63; asst prof math, Okla State Univ, 63-65; from assoc prof to prof physics, 65-69, prof math, 69-74, PROF PHYSICS, UNIV TEX, EL PASO, 74- *Mem:* Am Math Soc; Am Phys Soc. *Mailing Add:* Dept Physics Univ Tex El Paso TX 79968

MCINTYRE, ROBERT JOHN, b Bathurst, NB, Dec 19, 28; m 55; c 2. SOLID STATE PHYSICS. *Educ:* St Francis Xavier Univ, Can, BSc, 50; Dalhousie Univ, MSc, 53; Univ Va, PhD(physics), 56. *Prof Exp:* Sr mem sci staff, RCA Victor Co, Ltd, 56-67; dir semiconductor electronics lab, RCA Inc, 67-76, mgr, electrooptics res & develop, 76-90; DIR, DETECTOR RES & DEVELOP, EG&G CAN, INC, 91- *Honors & Awards:* David Sarnoff Award for Outstanding Technical Achievement, 79. *Mem:* Fel Inst Elec & Electronics Engrs; Can Asn Physicists. *Res:* Solid state surface, semiconductor and device physics. *Mailing Add:* EG&G Can Inc Box 900 Vaudreuil PQ J7V 8P7 Can

MCINTYRE, RUSSELL THEODORE, b Alexis, Ill, Mar 20, 25; m 48; c 3. BIOCHEMISTRY. *Educ:* Monmouth Col, BS, 49; Kans State Col, MS, 50, PhD(chem), 52. *Prof Exp:* Asst prof chem, La State Univ, 52-53; asst biochemist agr chem & biochem, 53-56; biochemist, Haynie Prod, Inc, NJ, 56-61; dir qual control, 61-65, dir spec prof develop, indust prod group, Stokely Van Camp, Inc, 65-84, DIR MKT, SPECIALTY PROD, CAPITAL CITY PROD CO, INC, COLUMBUS, 84- *Mem:* Am Chem Soc; Inst Food Technologists; Am Oil Chemists Soc. *Res:* Electrophoresis of plant proteins; biochemistry of fishes; lipid chemistry; polyglycerols and their derivatives. *Mailing Add:* 3780 Smiley Rd Hilliard OH 43026

MACINTYRE, STEPHEN S, b Honolulu, Hawaii, June 25, 47. CELL BIOLOGY, RHEUMATOLOGY. *Educ:* Yale Univ, BS, 69; Case Western Res Univ, PhD(molecular biol), 75, MD, 76. *Prof Exp:* ASST PROF MED, CASE WESTERN RESERVE UNIV, 81- *Concurrent Pos:* Mem staff, Cleveland Metrop Hosp, 76- *Mem:* Am Soc Cell Biol; Am Fedn Clin Res; AAAS. *Mailing Add:* Dept Med Case Western Reserve Univ 3395 Scranton Rd Cleveland OH 44109

MCINTYRE, THOMAS WILLIAM, b Chicago, Ill, Apr 22, 41; m 60; c 3. ORGANIC CHEMISTRY, RESEARCH ADMINISTRATION. *Educ:* DePauw Univ, BA, 62; Univ Iowa, PhD(org chem), 67. *Prof Exp:* Sr org chemist, Eli Lilly & Co, 67-71, res scientist, 71-73, mgr prod introd & chmn cardiovasc proj teams, 73-77, mgr prod licensing, 77-78; vpres & gen mgr, 78-87, EXEC VPRES, KINGSWOOD LABS, INC, 87- *Concurrent Pos:* Consult, chem & pharmaceut res & develop, 78-; mem bd dirs, Kingswood Labs, Telesis Inc, Calida, Inc. *Mem:* AAAS; Am Chem Soc; Sigma Xi. *Res:* Electrochemical reactions of organic compounds; synthesis, modification and purification of cephalosporin antibiotics; process research on plant fungicides and herbicides; synthetic saliva. *Mailing Add:* 336 Heather Dr Carmel IN 46032

MCINTYRE, WILLIAM ERNEST, JR, b Alvy, WVa, Dec 19, 25; m 50; c 1. CHEMISTRY, ORGANIC CHEMISTRY. *Educ:* Salem Col, WVa, BS, 48; Carnegie Inst Technol, MS, 51, PhD(org chem), 53. *Prof Exp:* Lab instr chem, Carnegie Inst Technol, 48-51; res chemist, E I du Pont de Nemours & Co, 52-66, staff scientist, Film Dept, 66-77, sr res chemist, Polymer Prod Dept, 77-88, SR CHEMIST, ELECTRONICS DEPT, E I DU PONT DE NEMOURS & CO, 88- *Mem:* AAAS; Am Chem Soc; NY Acad Sci. *Res:* Polymer synthesis, evaluation and chemistry; plastic film manufacture. *Mailing Add:* Dupont Co Inc PO Box 89 Circleville OH 43113

MACINTYRE, WILLIAM JAMES, b Canaan, Conn, Nov 26, 20; m 47; c 2. NUCLEAR MEDICINE, MEDICAL PHYSICS. *Educ:* Western Reserve Univ, BS, 43, MA, 47; Yale Univ, MS, 48, PhD(physics), 50. *Prof Exp:* Res asst, Atomic Energy Med Res Proj, 47, res assoc, 49-51, sect chief radiation physics, 51-58, sr instr radiol, 50-52, asst prof, 52-63, from asst prof to assoc prof biophys, 58-71, PROF BIOPHYS, SCH MED, CASE WESTERN RESERVE UNIV, 71-; STAFF PHYSICIST, CLEVELAND CLIN, 72- *Concurrent Pos:* Dir, 2nd Regional Training Course Med Applns Radiosotopes, Cairo, Egypt, 61; lectr, Mid East Regional Ctr Arab Countries, Egypt, 64; mem US nat comt med physics, Int Atomic Energy Agency, 66-72; mem, Coun Cardiovasc Radiol, Am Heart Asn, Diag Radiol Study Sect, NIH, 82-86; chmn sci comt 18B, Nat Coun Radiation Protection, 76-; chmn, Fed Coun Nuclear Med Orgn, 77-82. *Honors & Awards:* Paul C Aebersold Award, Outstanding Achievement in Basic Sci Appl to Nuclear Med, Soc Nuclear Med, 87. *Mem:* Radiol Soc NAm; Am Asn Physicists in Med; Soc Nuclear Med (pres, 76-77); Am Phys Soc; Soc Magnetic Imaging in Med; Soc Magnetic Resonance Imaging; Am Roentgen Ray Soc; Biophys Soc. *Res:* techniques of radionuclide and functional imaging; emission computed tomography, cardiovascular nuclear medicine; nuclear magnetic resonance imaging, gated cardiac NMRI; measurement and analysis of electrical parameters of the brain. *Mailing Add:* 3108 Huntington Rd Shaker Heights OH 44120-2410

MACIOLEK, JOHN A, b Milwaukee, Wis, Nov 2, 28; m 56; c 5. LIMNOLOGY, TROPICAL STREAM ECOLOGY. *Educ:* Ore State Univ, BS, 50; Univ Calif, Berkeley, MA, 54; Cornell Univ, PhD(limnol), 61. *Prof Exp:* Fishery biologist, East Fish Nutrit Lab, NY, 56-61 & Sierra Nevada Aquatic Res Lab, Calif, 61-65, leader, Hawaii Coop Fishery Res Unit, 66-77, sr adv, Hawaii Coop Fishery Res Unit, 77-79, Chief Aquatic Ecol Res, Seattle Nat Fishery Res Ctr, US Fish & Wildlife Serv, 79-84. *Concurrent Pos:* NSF-Am Soc Limnol & Oceanog travel grant, XIV Cong, Int Soc Limnol, Austria, 59 & XVI Cong, Poland, 65; assoc prof, Univ Hawaii, 67-77; assoc zoologist, Hawaii Inst Marine Biol, 69-79; actg chmn, Hawaii State Natural Area Reserves Syst Comn, 73-76; affil grad fac, Univ Wash, 75-79; affil fac, Univ Guam, 80-85; res assoc, B P Bishop Mus, 81-; adj prof biol, Mont Stoke Univ, 84-86; freelance res biologist, 84- *Mem:* Western Soc Naturalists; Int Soc Limnol. *Res:* Insular estuary and stream ecology; diadromous fauna of Oceania; insular aquatic ecosystem classification and inventory; anchialine pool fauna; aquatic zoogeography. *Mailing Add:* PO Box 7117 Mammoth Lakes CA 93546

MACIOLEK, RALPH BARTHOLOMEW, b Milwaukee, Wis, July 30, 39; m 74; c 2. METALLURGICAL ENGINEERING, PHYSICS. *Educ:* Univ Wis, BS, 61, MS, 62, PhD(metall eng), 66; Univ Minn, MBA, 83. *Prof Exp:* Res scientist mat sci, Corp Mat Sci Ctr, Honeywell Inc, 65-78; res specialist mat sci, 3M Co, 78-84; PROG MGR, HONEYWELL INC, 84- *Mem:* Am Soc Metals. *Res:* Crystal growth; solidification; hybrid microelectronics; metallurgy; thin films; soft and hard ferrites. *Mailing Add:* 10031 James Circle Bloomington MN 55431

MACIOR, LAZARUS WALTER, b Yonkers, NY, Aug 26, 26. BOTANY. *Educ:* Columbia Univ, AB, 48, MA, 50; Univ Wis, PhD(bot, zool), 59. *Prof Exp:* Asst bot, Columbia Univ, 48-50; instr biol, St Francis Col, Wis, 60-62; instr, Marquette Univ, 62-64; asst prof biol, Loras Col, 65-67; from asst prof to assoc prof, 67-72, PROF BIOL, UNIV AKRON, 72- *Concurrent Pos:* Res assoc, Inst Arctic & Alpine Res, Univ Colo, 66-67 & Inst Polar Studies, Ohio State Univ, 71-73; res fac assoc entom, Univ Calif-Davis, 84. *Mem:* Arctic Inst N Am; Soc Study Evolution; Int Asn Plant Taxon; Am Inst Biol Sci; Am Bryological & Lichenological Soc; fel AAAS; Bot Soc Am; Ecol Soc Am; Am Soc Plant Taxon; Int Orgn Plant Biosyst. *Res:* insect-flower pollination relationships; evolutionary systematics. *Mailing Add:* Dept Biol Univ Akron Akron OH 44325-3908

MCIRVINE, EDWARD CHARLES, b Winnipeg, Man, Dec 19, 33; US citizen; m 54; c 2. RESEARCH ADMINISTRATION, SOLID STATE PHYSICS. *Educ:* Univ Minn, BS, 54; Cornell Univ, PhD(theoret physics), 59. *Prof Exp:* Res scientist, Gen Atomic Div, Gen Dynamics Corp, 58-60; res scientist, Sci Lab, Ford Motor Co, 60-66, prog mgr, 66-69; var mgt positions, Res & Develop, Xerox Corp, 69-87; dean, Col Graphic Arts & Photography, Rochester Inst Technol, 87-91; CONSULT, RES & DEVELOP MGT, 91- *Concurrent Pos:* Dir, Detection Systs Inc, 81- *Mem:* Fel Am Phys Soc; Am Inst Physics; Soc Imaging Sci & Technol. *Res:* Imaging science; research and development management; condensed matter physics. *Mailing Add:* Ten Pond View Dr Pittsford NY 14534

MCISAAC, PAUL R(OWLEY), b Brooklyn, NY, Apr 20, 26; m 49; c 4. ELECTRICAL ENGINEERING. *Educ:* Cornell Univ, BEE, 49; Univ Mich, MSE, 50, PhD(elec eng), 54. *Prof Exp:* Engr, Sperry Gyroscope Co, 54-59; assoc prof elec eng, 59-65, assoc dean engr, 75-80, PROF ELEC ENG, CORNELL UNIV, 65- *Concurrent Pos:* Vis prof, Chalmers Univ Technol, Sweden, 65-66, Royal Inst Technol, Sweden, 87-88. *Mem:* Inst Elec & Electronic Engrs; Sigma Xi. *Res:* Electromagnetic theory; microwave theory; symmetry analysis. *Mailing Add:* 107 Forest Home Dr Ithaca NY 14850

MCISAAC, ROBERT JAMES, b Brooklyn, NY, Jan 9, 23; m 76; c 3. PHARMACOLOGY. *Educ:* Univ Buffalo, BS, 49, PhD, 54. *Prof Exp:* Assoc pharmacol, Sch Med, Univ Buffalo, 54-55; from asst prof to prof, 58-85, asst vpres res, 81-88, EMER PROF PHARMACOL, SCH MED, STATE UNIV NY, BUFFALO, 85- *Concurrent Pos:* Res fel, Grad Sch Med, Univ Pa, 56-58; spec res fel, Univ Lund, Sweden, 66-67. *Mem:* AAAS; Am Soc Pharmacol & Exp Therapeut; NY Acad Sci; Soc Neurosci. *Res:* Synaptic transmission; neuroeffector release; distribution and transport of drugs. *Mailing Add:* 3071 Sweet Home Rd Amherst NY 14228

MACIVER, DONALD STUART, b Cambridge, Mass, Oct 3, 27; m 47; c 2. INDUSTRIAL CHEMISTRY. *Educ:* Boston Univ, AB, 52; Univ Pittsburgh, PhD(phys chem), 57. *Prof Exp:* Sect head catalysis, Gulf Res & Develop Co, 57-64; mgr chem res, Eastern Res Ctr, NY, 64-67, dir, Western Res Ctr, Richmond, Calif, 67-74, dir, St Gabriel Plant, La, 72-74, vpres & gen mgr, Wyo, 74-76, vpres, Agr Chem Div, 76-81, VPRES & GEN MGR, AGR & DRUG INTERMEDIATES DIV, STAUFFER CHEM CO, 81- *Mem:* Am Chem Soc. *Res:* Adsorption; catalysis; resonance spectroscopy; fuel cell technology; petroleum processing; petrochemicals; chemical processes; agricultural chemicals; caustic-chlorine manufacture. *Mailing Add:* Digital Equip Corp Mail Stop WJ01-1/CS 5 Carlisle Rd Westford MA 01886

MCIVER, NORMAN L, b Abbey, Sask, Sept 7, 31; m 58; c 3. GEOLOGY. *Educ:* Univ Sask, BA, 56; Johns Hopkins Univ, PhD(geol), 61. *Prof Exp:* Res geologist, Shell Develop Co, Tex, 60-68, staff geologist, Shell Oil Co, La, 68-71, staff geologist, Shell Can Ltd, 71, sr staff geologist, 71-77, dist mgr, 77-81, MGR GEOL, SHELL OIL CO, 81- *Mem:* Soc Econ Paleontologists & Mineralogists; Am Asn Petrol Geologists. *Res:* Stratigraphy and sedimentation of clastic sediments. *Mailing Add:* Shell Oil Co Box 991 Houston TX 77001

MCIVER, ROBERT THOMAS, JR, b Macon, Ga, June 19, 45; m 69; c 4. SPECTROMETRY, PHYSICAL CHEMISTRY. *Educ:* Univ Kans, BS, 67; Stanford Univ, PhD(chem), 71. *Prof Exp:* Asst prof, 71-75, assoc prof, 75-77, PROF CHEM, UNIV CALIF, IRVINE, 77- *Concurrent Pos:* Prin investr, NSF & Petrol Res Fund, 72-, NIH, 75-; Alfred P Sloan fel, Sloan Found, 73; vis fel, Joint Inst Lab Astrophysics, Univ Colo, 79. *Mem:* Am Chem Soc; Am Soc Mass Spectrometry; Am Phys Soc; Sigma Xi. *Res:* Ion-molecule reactions in the gas phase; effects of solvation on chemical reactions; development of analytical applications of Fourier transform mass spectrometry. *Mailing Add:* Dept Chem Univ Calif Irvine CA 92717

MCIVER, SUSAN BERTHA, b Hutchinson, Kans, Nov 6, 40. ENTOMOLOGY. *Educ:* Univ Calif, Riverside, BA, 62; Wash State Univ, MS, 64, PhD(entom), 67. *Prof Exp:* From asst prof to prof zool, Univ Toronto, 67-84; prof & chmn, Dept Environ Biol, Univ Guelph, 84-89; CONSULT, 89- *Honors & Awards:* C Gordon Hewitt Award, Entom Soc Can, 78. *Mem:* Entom Soc Am; Can Soc Zoologists; fel Entom Soc Can. *Res:* Medical entomology; behavior and sensory physiology and morphology of blood feeding arthropods. *Mailing Add:* C95 Reynolds Rd RR 1 Fulford Harbour BC V0S 1C0 Can

MCIVOR, KEITH L, IMMUNOLOGY REGULATION. *Educ:* Univ Wash, Seattle, PhD(microbiol), 69. *Prof Exp:* ASSOC PROF IMMUNOL, WASH STATE UNIV, 76- *Res:* Role of the macrophage. *Mailing Add:* Dept Microbiol Wash State Univ SW 1125 Alvar Dr Pullman WA 99163

MACK, ALEXANDER ROSS, b Oberon, Man, May 6, 27; c 4. AGRONOMY, REMOTE SENSING. *Educ:* Univ Man, BSA, 49; Iowa State Univ, MS, 52; Purdue Univ, PhD(soil), 59. *Prof Exp:* Agronomist plant & soil, 49-54, res scientist soil environ, 55-70, RES SCIENTIST REMOTE SENSING, CAN DEPT AGR, 71- *Concurrent Pos:* Consult, Food & Agr Orgn, 66; chmn, Ont Soil Fertil Comt, 67-69; chmn, Climate & Soil Classification, Nat Soil Survey comt, 70-71; secy, Agr & Geog Working Group Prog Planning for Resource Satellites, 71-; chmn, Agr Working Group Remote Sensing, 72-79; reg Agr Can, Can Adv Comt Remote Sensing, 72- *Mem:* Soil Sci Soc Am; Can Soc Soil Sci (secy, 66-69); Int Soc Soil Sci; Am Soc Photogram; Can Soc Remote Sensing (secy, 72-75). *Res:* Effect of environment on crop production in various soil climatic zones from satellite imagery. *Mailing Add:* 1978 Gorfield Ottawa ON K2C 0W8 Can

MACK, CHARLES EDWARD, JR, b Freeport, NY, July 31, 12; m 43; c 2. AERODYNAMICS. *Educ:* Mass Inst Technol, BS, 34; Stevens Inst Technol, MS, 44; NY Univ, ScD(aerodyn), 48. *Prof Exp:* Instr math, Stanton Prep Acad, 40-41; asst, Iowa State Univ, 41-42; res engr appl math & head theoret sect, Res Dept, Grumman Aircraft Eng Corp, 42-48, dir res, Res Dept, 48-78, CONSULT, GRUMMAN AEROSPACE CORP, 78- *Concurrent Pos:* Lectr, NY Univ, 48- & Adelphi Col, 52- *Mem:* Sigma Xi; Inst Aeronaut & Space; AAAS. *Res:* Flutter and vibration of aircraft; transonic and supersonic aerodynamics; dynamic analysis; linearized treatment of supersonic flow through and around ducted bodies of narrow cross section. *Mailing Add:* Midline South Syosset NY 11791

MACK, CHARLES LAWRENCE, JR, b Cleveland, Ohio, July 20, 26; m 50; c 4. APPLIED PHYSICS. *Educ:* Harvard Univ, BSc, 48; Univ Pa, MSc, 53. *Prof Exp:* Instr physics, Univ Pa, 49-50, instr med, Sch Med, 52-53; mem staff physics, Lincoln Lab, Mass Inst Technol, 53-57; tech consult, Supreme Hq, Allied Powers Europe, Paris, 57-59; mem staff, Mitre Corp, Bedford, Mass, 59; mem staff physics, Lincoln Lab, Mass Inst Technol, 59-85; CONSULT, FOREIGN TECHNOL, 85- *Concurrent Pos:* Science writing. *Res:* Conversion of solar to electrical energy; energy storage; development of science. *Mailing Add:* 7 Parker St Lexington MA 02173

MACK, DICK A, b Santa Cruz, Calif, Nov 22, 21; m 45; c 2. ELECTRONICS ENGINEERING. *Educ:* Univ Calif, Berkeley, BS, 43. *Prof Exp:* Radio engr, Naval Res Lab, DC, 43-46; electronics engr, Lawrence Radiation Lab, Univ Calif, 46-51; res engr, Stanford Res Inst, 51-52; Electronics engr bevatron elec eng, 52-55, electronics engr nuclear instrumentation, 55-69, head, Dept Electronics Eng, 69-77,EMER DEPT HEAD, DEPT ELECTRONICS ENG, LAWRENCE BERKELEY LAB, 77- *Concurrent Pos:* chmn, US Camac Mech & Power Supply working group, 71-76 & Fast Syst Design Group, Dept Energy, 78-; prin investr, Surv Instrumentation for Environ Qual Monitoring, NSF, 71-77; consult control systs & nuclear instrumentation, 77-; assoc ed, Inst Elec & Electronic Engrs Transactions on Nuclear Sci, 78-85, ed, 86-; ed, Inst Elec & Electronic Engrs Trans on Nuclear Sci, 86- *Mem:* sr mem Inst Elec & Electronics Engrs; Sigma Xi. *Mailing Add:* 600 Lockewood Ln Santa Cruz CA 95066

MACK, DONALD R(OY), b Seattle, Wash, Mar 14, 25; m 54; c 2. ELECTRICAL ENGINEERING. *Educ:* Univ Wash, BSEE, 48; Polytech Inst Brooklyn, MS, 67, PhD(syst sci), 69. *Prof Exp:* Engr, Gen Elec Co, 48-56, supvr proj eng, Mat Lab, Large Steam Turbine Generator Dept, 56-61, consult eng educ, 61-70, prog mgr advan course eng, 70-78, mgr tech educ, 78-87; RETIRED. *Concurrent Pos:* Adj prof, Rensselaer Polytech Inst, 69-78. *Mem:* Inst Elec & Electronics Engrs; Am Soc Eng Educ. *Res:* Industrially oriented graduate education in electrical and mechanical engineering; system engineering. *Mailing Add:* 404 A Montauk Lane Stratford CT 06497

MACK, FREDERICK K, b Watertown, NY, July 5, 24; c 1. GROUND WATER HYDROLOGY, GEOLOGY. *Educ:* Syracuse Univ, BA, 49, MS, 50. *Prof Exp:* Geologist, US Geol Survey, Albany, NY, 51-59, hydrologist, Annapolis, Md, 59-80; HYDROGEOLOGIST, MD GEOL SURV, ANNAPOLIS, 80- *Mem:* AAAS; Am Geophys Union; fel Geol Soc Am; Intl Asn Hydrogeol; Asn Groundwater Sci Engrs. *Mailing Add:* 19 Second St Greenwood Acres Annapolis MD 21401

MACK, GARY W, b Vallejo, Calif, Mar 13, 54. PHYSIOLOGY. *Educ:* Univ Calif, Davis, MA, 77; Univ Hawaii, PhD(physiol), 85. *Prof Exp:* ASST FEL, JOHN B PIERCE FOUND LAB, 88- *Mem:* Am Col Sports Med; Am Physiol Soc. *Mailing Add:* John B Pierce Found Lab 290 Congress Ave New Haven CT 06519

MACK, HARRY JOHN, b Gatesville, Tex, Mar 18, 26; m 55; c 4. HORTICULTURE. *Educ:* Tex A&M Univ, BS, 50, MS, 52; Ore State Univ, PhD(hort), 55. *Prof Exp:* Instr hort, Tex A&M Univ, 50-51; from instr to assoc prof, 55-69, PROF HORT, ORE STATE UNIV, 69- *Mem:* Am Soc Hort Sci; Am Sci Affil. *Res:* Vegetable crops physiology; population density; mineral nutrition; irrigation and water relations. *Mailing Add:* Dept Hort Ore State Univ Corvallis OR 97331-2911

MACK, JAMES PATRICK, b Newark, NJ, Dec 9, 39; m 68; c 1. CELL BIOLOGY. *Educ:* Monmouth Col, NJ, BS, 62; William Paterson Col NJ, MA, 66; Columbia Univ, EdD(cell biol), 71. *Prof Exp:* Teacher biol, Shore Regional High Sch, NJ, 62-66; asst prof cell physiol, Monmouth Col, NJ, 68-69; teacher biol chem, Lakewood High Sch, NJ, 71-74; res scientist cell biol, Lamont Doherty Geol Observ, Columbia Univ, 69-71; ASSOC PROF BIOL & CHMN DEPT, MONMOUTH COL, NJ, 74- *Concurrent Pos:* Teacher biol, Jersey City State Col, 70-73 & biol & chem, Ocean County Col, NJ, 71-; res scientist cell biol, Creedmore Inst, Queens Village, NY, 74-75; NSF grant. *Mem:* Sigma Xi; Am Chem Soc; Am Inst Biol Sci. *Res:* Determining the role of vitamin A, outside the visual cycle, in cellular membranes; antimicrobial activity in lichens; effect of vitamin E and hydrocortisone on aging cells. *Mailing Add:* Dept Biol Monmouth Col W Long Branch NJ 07764

MACK, JULIUS L, b Gadsden, SC, June 14, 30; m 58; c 3. CHEMICAL PHYSICS. *Educ:* SC State Col, BS, 52; Howard Univ, MS, 57, PhD(phys chem), 65. *Prof Exp:* Res chemist, Naval Ord Sta, 56-71; chmn div natural sci, 71-73, dean sch natural, appl & health sci, 72-78, PROF CHEM, UNIV DC, 71- *Mem:* AAAS; Am Chem Soc; Am Phys Soc; NY Acad Sci; Sigma Xi. *Res:* Study of structural and thermodynamic properties of high temperature molecules; infrared spectroscopy of matrix isolated species and mass and infrared spectra of hot vapors; vibrational spectra of unstable species; laser studies of molecular dynamics; gas-surface interactions; kinetics of decompostions. *Mailing Add:* 7030 Oregon Ave NW Washington DC 20015

MACK, LAWRENCE LLOYD, b Springfield, Mo, Dec 10, 42; m 66; c 2. PHYSICAL CHEMISTRY, BIOCHEMISTRY. *Educ:* Middlebury Col, AB, 65; Northwestern Univ, PhD(phys chem), 69. *Prof Exp:* Fel, Univ Calif, Berkeley, 69-70, instr chem & NIH fel, 70; asst prof chem, St Lawrence Univ, 70-72; ASSOC PROF CHEM, BLOOMSBURG STATE COL, 72- *Mem:* Am Chem Soc; Sigma Xi. *Res:* Development and employment of physical methods to study solution behavior of biopolymers; biophysical chemistry and biophysics; computer assisted instruction. *Mailing Add:* Dept of Chem Bloomsburg State Col Bloomsburg PA 17815

MACK, LAWRENCE R(IEDLING), b Detroit, Mich, Oct 9, 32; m 54; c 4. FLUID MECHANICS. *Educ:* Univ Mich, BSE(eng mech) & BSE(math), 54, MSE, 55, PhD(eng mech), 58. *Prof Exp:* Asst prof mech & hydraulics & res engr, Inst Hydraulic Res, Univ Iowa, 58-60; res engr, 60-62, from asst prof to assoc prof eng mech, 60-68, from asst prof to assoc prof civil eng, 60-73, head Eng Mech Div, 71-77, ASSOC PROF AEROSPACE ENG & ENG MECH, UNIV TEX, AUSTIN, 68-, UNDERGRAD ADV, AEROSPACE ENG, 81- *Concurrent Pos:* NSF sci fac fel, Case Western Reserve Univ, 68. *Mem:* Am Soc Eng Educ; Soc Natural Philos; Am Acad Mech. *Res:* Nonlinear oscillations; gravity waves; flow through porous media; slow viscous flow; convective heat transfer; municipal annexations in Germany. *Mailing Add:* 5824 Trailridge Dr Austin TX 78731

MACK, LESLIE EUGENE, b Winchester, Mass, June 24, 29; m 70; c 4. GROUNDWATER GEOLOGY. *Educ:* Duke Univ, BA, 51; Univ Kans, MA, 57, PhD(geol), 59; Univ Ark, JD, 64. *Prof Exp:* Geologist, US Geol Surv, 56-59; water mgt consult, Rockwin Fund, Inc, 59-70; vpres, Resources, Inc, 70-73; sr hydrologist, Dames & Moore, 73-75; asst dir, Am Petrol Inst, 75-80; DIR, ARK WATER RESOURCES RES CTR, UNIV ARK, 80- *Concurrent Pos:* Consult geol, 59-; asst prof, Grad Inst Technol, Univ Ark, 60-61, prof groundwater geol, 81-84; water mgt consult, Nat Water Comn, 70-71. *Mem:* Am Water Resources Asn; Nat Water Well Asn; Int Water Resources Asn; Am Asn Petrol Geologists; Am Inst Hydrol; Am Water Works Asn. *Res:* Hydrogeological studies in the development of springs for the bottled water industry; location and sealing leaky dams and impoundments; groundwater contamination problems; groundwater supplies; expert witness. *Mailing Add:* 2735 Valencia Fayetteville AR 72703

MACK, MARK PHILIP, b Buffalo, NY, Jan 14, 50; m. POLYMER CHEMISTRY, CATALYSIS. *Educ:* State Univ NY, Col Buffalo, BA, 71; State Univ NY Buffalo, PhD(chem), 76. *Prof Exp:* Res assoc, Duke Univ, 75-77; res scientist chem, Continental Oil Co, Okla, 77-80, group leader, Conoco, Inc, 80-85; SR SUPVR, POLYMER PROD DEPT, E I DU PONT DE NEMOURS & CO INC, WILMINGTON, DEL, 85- *Concurrent Pos:* Samuel B Silbert fel, State Univ NY, Buffalo, 74-75. *Mem:* Am Chem Soc; Sigma Xi; Soc Plastic Engrs. *Res:* Synthesis and processing of alpha-olefin polymers; drag reducers; flow improvers; high density polyethylene; engineering plastics. *Mailing Add:* Alathon Polymers Div PO Box 2917 Alvin TX 77512

MACK, MICHAEL E, b Poughkeepsie, NY, May 28, 39; m 63; c 4. ION BEAM-TARGET INTERACTION, ION BEAM OPTICS. *Educ:* Mass Inst Technol, BS, 61, PhD(physics), 67. *Prof Exp:* Prin scientist, United Technologies Res Lab, 67-73; dir laser systs, Avco Everett Res Lab, 73-81; tech dir, Eaton Ion Beam Systs Div, 81-86; eng mgr, Varian Extrion, 86 & Eaton Semiconductor Equip Div, 87-89; VPRES ENG, GENUS ION TECHNOL DIV, 89- *Concurrent Pos:* Lectr, Int Implant Conf Sch, 86-; mem, Int Ion Implant Conf Comt, 90- *Mem:* Am Phys Soc; Inst Elec & Electronics Engrs. *Res:* Ion implantation of semiconductors; minimizing ion beam induced device charging and reducing particulate contamination; author of 55 publications. *Mailing Add:* Seven Hidden Ledge Rd Manchester MA 01944

MACK, MICHAEL J, b Viola, Ill, July 3, 25; m 49; c 8. TRANSMISSION DESIGN, VEHICLE DESIGN. *Educ:* Univ Notre Dame, BS, 46; Univ Iowa, MS, 49. *Prof Exp:* Engr prod eng, John Deere, 46-47 & 49-56, proj engr, 56-69, div engr, 69-71, dir Prod Eng Ctr, 71-87; RETIRED. *Concurrent Pos:* Consult, 89-91. *Mem:* Fel Soc Automotive Engrs. *Res:* Power shifting transmissions in agricultural tractors and industrial tractors; design and research of entire vehicle; application of ultra sonics to agricultural activities. *Mailing Add:* 1433 Olympic Dr Waterloo IA 50701

MACK, RICHARD BRUCE, b South Paris, Maine, Sept 18, 28; m 54. ELECTROMAGNETIC MEASUREMENT TECHNIQUES, ANTENNAS. *Educ:* Colby Col, BA, 51; Harvard Univ, MS, 57, PhD, 64. *Prof Exp:* Physicist electromagnetic, Air Force Cambridge Res Labs, 51-77; CONSULT ELECTROMAGNETIC MEASUREMENT TECH, 77- *Mem:* Inst Elec & Electronics Engrs; Antenna & Propagation Soc; Microwave Theory & Tech Soc; Antenna Measurement Tech Asn; AAAS. *Res:* Experimental techniques for measurement of electromagnetic scattered fields; radar scattering control; measurement techniques for evaluating microwave properties of materials. *Mailing Add:* 35 Kenwin Rd Winchester MA 01890

MACK, RICHARD NORTON, b Providence, RI, July 31, 45; m 72; c 1. PLANT ECOLOGY. *Educ:* Western State Col Colo, BA, 67; Wash State Univ, PhD(bot), 71. *Prof Exp:* Instr bot, Wash State Univ, 70; asst prof biol sci, Kent State Univ, 71-75; from asst prof to assoc prof, 75-83, PROF BOT & BIOL, 83-, CHMN DEPT, WASH STATE UNIV, 86- *Concurrent Pos:* Res assoc, Univ Col N Wales, Bangor, 73-74; mem, sub-panel ecol, NSF, 82-85. *Mem:* Ecol Soc Am; Brit Ecol Soc; Am Quaternary Asn; Bot Soc Am. *Res:* Plant population biology; community ecology; Holocene vegetation history of Pacific Northwest. *Mailing Add:* Dept of Bot Wash State Univ Pullman WA 99164

MACK, ROBERT EMMET, b Morris, Ill, Mar 26, 24; m 51; c 5. INTERNAL MEDICINE. *Educ:* Univ Notre Dame, BS, 46; St Louis Univ, MD, 48. *Prof Exp:* From intern to resident internal med, St Louis Univ, 48-52, from instr to asst prof, 53-61; chief med serv, Hutzel Hosp, 61-66, med dir, 65-66, dir, 66-70, pres, 70-80; asst prof, 61-67, PROF INTERNAL MED, SCH MED, WAYNE STATE UNIV, 67-; VPRES, DETROIT MED CTR CORP, 80- *Concurrent Pos:* Chief, Med Serv, Vet Admin Hosp, 56-61; consult, St Louis City Hosp. *Mem:* Am Thyroid Asn; Am Physiol Soc; Endocrine Soc; Am Col Physicians; Am Fedn Clin Res. *Res:* Thyroid gland; cardiac output and coronary blood flow; radioisotopes. *Mailing Add:* Detroit Med Ctr Corp 4201 St Antoine Detroit MI 48201

MACK, RUSSEL TRAVIS, b San Diego, Calif, Apr 27, 45; m; c 2. INFRARED THERMAL IMAGING, SURFACE HEAT FLOW MEASUREMENT. *Educ:* Trinity Univ, San Antonio, Tex, BS, 81. *Prof Exp:* Res engr, D H Titanium Co, Dow Chem USA, 81-82, mech engr, Mech & Mat Technol, 82-84, sr mech engr, 84-87, mech specialist, 87-91, SR NDT SPECIALIST, INFRARED MECH & MAT TECHNOL, DOW CHEM USA, 91- *Concurrent Pos:* Rep, C-16 Comt Thermal Insulation, Am Soc Testing & Mat, 86-, chmn, Subcomt C-16 30 Sect 7, In-Situ Test Methods, 88-; US rep, TC 163 Working Group, Indust Heat Flow Calculations, Int Standards Orgn, 87-; chmn, Infrared Thermal Personnel Qualifications Comt, Am Soc Nondestructive Testing, 90- *Mem:* Am Soc Mech Engrs; Am Soc Testing & Mat; Am Soc Nondestructive Testing. *Res:* Dow's method for heat flow mapping; thermodynamic process analysis; error analysis on heat flow transducers; original postulates in heat-flow mathematics and finite element method; granted 6 patents. *Mailing Add:* 304 Balsam Lake Jackson TX 77566

MACK, SEYMOUR, b New York, NY, March 4, 22. IGNEOUS METAMORPHIC PETROLOGY, HYDROGEOLOGY. *Educ:* City Univ New York, BS, 47; Syracuse Univ, MS, 50, PhD(geol), 57. *Prof Exp:* Geologist, US Geol Surv, 52-57; PROF GROUND WASTE GEOL, CALIF STATE UNIV, FRESNO, 57- *Concurrent Pos:* Distinguished lectr, Fresno State Univ, Calif, 65. *Mem:* Geol Soc Am. *Res:* Groundwater hydrology of Upton Kalamath base in California. *Mailing Add:* Dept Geol Cal State Univ Fresno CA 93740

MACK, THOMAS MCCULLOCH, b Reno, Nev, Apr 20, 36; m 58; c 3. CHRONIC DISEASE EPIDEMIOLOGY, TWIN RESEARCH. *Educ:* Carleton Col, BA, 57; Columbia Univ, MD, 61; Harvard Univ, MPH, 69. *Prof Exp:* Intern med, Univ Colo Hosp, 61-62; resident, Univ Wash Hosp, 62-63, fel infectious dis, 63-64; med epidemiologist, Ctrs Dis Control, 64-65; dir epidemiol, Pakistan Med Res Inst, 65-68; fel, Sch Pub Health, Harvard Univ, 68-70, asst prof, 70-74; assoc prof, 74-81, PROF PREV MED, SCH MED, UNIV SOUTHERN CALIF, 81- *Concurrent Pos:* Mem cancer immunodiag comt, Nat Cancer Inst, NIH, 73-76; dir, Cancer Surveillance Syst, Los Angeles County, 74-86; adj prof, Univ Calif, Los Angeles, 75-; mem, epidemiol & dis control study sect, NIH, 76-80; mem, toxicity data elements comt, Nat Acad Sci, 80-84; mem, sci working group malaria, WHO, 81-87; mem adv panel, Nat Death Index, 82-; mem, sci adv panel, Air Resources Bd, Calif, 83-88; co-ed, Cancer on Five Continents, Int Aging Res, 84-; prin investr, Int Twin Study. *Mem:* Soc Epidemiol Res; Am Epidemiol Soc; Int Epidemiol Soc; Am Pub Health Asn. *Res:* Etiology of chronic diseases. *Mailing Add:* PMB-B105 2025 Zonal Ave Los Angeles CA 90033

MACK, TIMOTHY PATRICK, b Poughkeepsie, NY, Mar 17, 53; m 76; c 5. INTEGRATED PEST MANAGEMENT. *Educ:* Colgate Univ, BA, 75; Pa State Univ, MS, 79, PhD(entom), 81. *Prof Exp:* Res technologist statist, Pa State Univ, 81; asst prof, Dept Entom, 81-86, ASSOC PROF ENTOM, AUBURN UNIV, 86- *Concurrent Pos:* Res award, Sigma Xi, 81, Dir's Award, 86. *Honors & Awards:* Elco Award, Entom Soc Pa, 80; Award, Potato Asn Am, 80. *Mem:* Entom Soc Am; Am Peanut Res & Educ Soc. *Res:* Integrated pest management systems for soybean and peanut insects; population dynamics, modelling and statistical analysis of insect populations; predator-prey theory. *Mailing Add:* Entom Dept Auburn Univ 301 Funchess Hall Auburn AL 36849-5413

MACK, WALTER NOEL, POLIO, ENTERIC VIRUSES. *Educ:* Univ Calif, San Francisco, PhD(path), 47. *Prof Exp:* Prof microbiol, Inst Water Res, 73-77; RETIRED. *Mailing Add:* Giltner Hall Mich State Univ East Lansing MI 48823

MCKAGUE, ALLAN BRUCE, b Weston, Ont, Oct 21, 40; m 63; c 2. ORGANIC CHEMISTRY. *Educ:* McMaster Univ, BSc, 62; Univ BC, PhD(chem), 67. *Prof Exp:* Chemist I, Toronto Food & Drug Directorate, 62-63; Nat Res Coun fel, Australian Nat Univ, 67-69; res chemist & res dir indust org chem, Orchem Res Co, 69-71; SUPVR ORG ANALYSIS, BC RES COUN, 71- *Mem:* Chem Inst Can. *Res:* Applied organic synthesis; environmental chemistry; organic analytical chemistry. *Mailing Add:* Univ Toronto 200 College St Toronto ON M5F 1A4 Can

MCKAGUE, HERBERT LAWRENCE, b Altoona, Pa, June 24, 35; m 63; c 2. GEOLOGY, GEOCHEMISTRY. *Educ:* Franklin & Marshall Col, BS, 57; Wash State Univ, MS, 60; Pa State Univ, PhD(mineral, petrol), 64. *Prof Exp:* Fel gem minerals, Gemol Inst Am, 64-66; asst prof geol, Rutgers Univ, New Brunswick, 66-72; sr geologist, Lawrence Livermore Nat Lab, Univ Calif, 72-84, containment prog leader, 79-81; CONSULT, 81- *Concurrent Pos:* Instr geol, Chabot Community Col, 81-89, Los Positas Community Col, 89- *Mem:* Geol Soc Am; Am Geophys Union; Minerals & Geotech Logging Soc. *Res:* Genesis, mineralogy, petrology and geochemistry of alpine ultramafic rocks; mineralogy of gem minerals; geologic and geophysical criteria for containment of explosions, geologic interpretation of borehole gravimetry; seismo-tectonic assessment of high level waste repository sites. *Mailing Add:* Lawrence Livermore Lab L-222 Univ Calif Livermore CA 94550

MACKAL, ROY PAUL, b Milwaukee, Wis, Aug 1, 25; div; c 1. ZOOLOGY, BIOCHEMISTRY. *Educ:* Univ Chicago, BS, 49, PhD(biochem), 53. *Hon Degrees:* Wilmington Col, DSc, 82. *Prof Exp:* Res assoc biol, Univ Chicago, 53-84, from instr to assoc prof biochem, 53-91, univ safety & energy coordr, 74-91; RETIRED. *Mem:* Am Soc Biol Chemists; Int Soc Cryptozool. *Res:* Virus; bacteriophage; chemical physical anthropology; synthesis of DNA; energy conservation; vertebrate zoology; safety and environmental health. *Mailing Add:* 9027 S Oakley Ave Chicago IL 60620

MACKANESS, GEORGE BELLAMY, b Sydney, Australia, Aug 20, 22; m 45; c 1. IMMUNOLOGY. *Educ:* Univ Sydney, MB & BS, 45; Univ London, dipl clin path, 48; Oxford Univ, MA, 49, DPhil(path), 53. *Prof Exp:* Resident med, Sydney Hosp, 45-46; resident path, Kanematsu Inst, Univ Sydney, 46-47; demonstr, Oxford Univ, 48-53; asst prof, Australian Nat Univ, 54-62; prof microbiol, Univ Adelaide, 62-65; dir, Trudeau Inst, 65-76; pres, Squibb

Inst Med Res, 76-88; RETIRED. *Concurrent Pos:* USPHS traveling fel, Rockefeller Univ, 59-60; USPHS res grant, Trudeau Inst, 65-; consult, USPHS, 66-; adj prof, Sch Med, NY Univ, 68-; consult, Nat Acad Med, 69-71 & US Armed Force Epidemiol Br, 69-; Mem, Bd Sci Consult, Mem Sloan-Kettering Cancer Ctr, 81-, Assembly Life Sci, Nat Res Coun, 81-, Bd Trustees, Josiah Macy Found, 81- *Honors & Awards:* Paul Ehrlich Prize, 75. *Mem:* AAAS; Am Soc Microbiol; Am Thoracic Soc; Am Asn Immunol. *Res:* Cellular aspects of immunology; resistance to infectious disease; antitumor immunity. *Mailing Add:* 2783 Little Creek Rd Johns Island SC 29455

MCKANNAN, EUGENE CHARLES, b Philadelphia, Pa, Apr 16, 28; m 52; c 5. MATERIALS SCIENCE ENGINEERING. *Educ:* West Chester Univ, BS, 49; Univ Ala, Huntsville, MS, 68. *Prof Exp:* Engr mat, US Army Guided Missile Ctr, 51-52; plastics engr extrusions, Plastics Dept, E I du Pont de Nemours & Co, Inc, 52-61; br chief eng physics, Marshall Space Flight Ctr, NASA, 61-68, div chief metall, 68-75, prog mgr mat processing, 75-82, dep lab dir mat, 82-88; mgr mat, Martin-Marietta Corp, 88-89; LEAD ENGR MAT, BOEING CO, 89- *Concurrent Pos:* Chmn, Rogers Comn Space Shuttle Accident, Subcomt Orbiter Payloads, 86. *Mem:* Am Inst Physics. *Res:* Industrial computed tomography for non-invasive testing; solid rocket motor materials; turbine blade alloy and its solidification process; space flight experiments on materials in low gravity; extrusion of cable insulation. *Mailing Add:* 2512 Vista Dr Huntsville AL 35803

MACKAUER, MANFRED, b Wiesbaden, Ger, June 3, 32; m 59; c 1. PEST MANAGEMENT, BIOLOGICAL CONTROL. *Educ:* Univ Frankfurt, Drphilnat, 59. *Prof Exp:* Res asst parasitol, Univ Frankfurt, 59-61; res scientist, Res Inst Belleville, Can Dept Agr, 61-67; chmn, dept, 76-81, PROF BIOL SCI, SIMON FRASER UNIV, 67-, DIR, CTR FOR PEST MGT, 82- *Concurrent Pos:* Chmn, Biol Control of Myzus Persicae Proj, Int Biol Prog, 68-74; vis prof entom, Univ Calif, Berkeley, 78; ed, Entomologia Exp & Applicata, 85- *Honors & Awards:* Gold Medal, Entomol Soc Can, 89. *Mem:* Entom Soc Am; fel Entom Soc Can. *Res:* Bionomics, phylogeny and taxonomy of parasitic Hymenoptera; host specificity of hymenopterous parasites of aphids; biological controls of pest insects, especially aphids; insect parasitology; quality control of mass-produced insects; agricultural pest management. *Mailing Add:* Dept of Biol Sci Simon Fraser Univ Burnaby BC V5A 1S6 Can

MCKAVENEY, JAMES P, b Pittsburgh, Pa, Sept 26, 25; m 52; c 2. ANALYTICAL CHEMISTRY. *Educ:* Univ Pittsburgh, BS, 49, MS, 51, PhD(anal chem), 57. *Prof Exp:* Asst technologist chem, Res Div, US Steel Corp, 51-56; assoc chemist, Res Div, Crucible Steel Co Am, 56-58, res engr, 58-59, supvr analytical chem, 59-64, mgr, 64-67, assoc dir chem & melting, 67-69; mgr analytical serv, Occidental Res Corp, 69-76, sr scientist, 76-80, technol rev scientist, 80-82. *Concurrent Pos:* Pres, Pittsburgh Conf Analytical Chem & Appl Spectros, 65-66; consult, 82- *Mem:* Am Chem Soc; Am Inst Chemists; Am Soc Testing & Mat; NY Acad Sci; Electrochem Soc. *Res:* Spectroscopy of steel and refractory metals; analysis of industrial acids; semiconductor electrodes as analytical sensors; separation of mercury and other heavy metals from water; protective coatings for metals; analysis of coal and non-ferrous minerals; electrically conductive coatings and articles; corrosion resistant metal coatings. *Mailing Add:* 940 Fenn Ct Claremont CA 91711

MACKAY, COLIN FRANCIS, b Waterbury, Conn, Sept 21, 26. PHYSICAL CHEMISTRY. *Educ:* Univ Notre Dame, BS, 50; Univ Chicago, PhD(chem), 56. *Prof Exp:* From asst prof to assoc prof, 56-68, PROF CHEM, HAVERFORD COL, 68- *Mem:* AAAS; Am Chem Soc. *Res:* Atomic reactions; chemistry of highly reactive species; laser photochemistry. *Mailing Add:* Dept Chem Haverford Col Haverford PA 19041-1382

MCKAY, DALE ROBERT, b Oakland, Calif, Oct 26, 46; m 77; c 1. NUCLEAR MAGNETIC RESONANCE. *Educ:* Calif State Univ, Hayward, BA, 68; Univ Wyo, PhD(physics), 80. *Prof Exp:* res assoc, Colo State Univ, 79-83; opers mgr, Chemagnetics, Ft Collins, Colo, 83-87; SR SCIENTIST, QUANTUM MAGNETICS, SAN DIEGO, CALIF, 88- *Mem:* AAAS. *Res:* Solid state nuclear magnetic resonance, applied especially to security systems. *Mailing Add:* Quantum Magnetics 11578 Sorrento Valley Rd, #30 San Diego CA 92121

MACKAY, DONALD ALEXANDER MORGAN, b Gt Brit, Feb 8, 26; nat US; m 52; c 4. FOOD CHEMISTRY. *Educ:* Oxford Univ, BA, 48; Yale Univ, PhD(chem), 54. *Prof Exp:* Proj leader, Evans Res & Develop Corp, NY, 53-56, from assoc dir res to dir res, 56-61; mgr corp res, Gen Foods Corp, 61-64; asst to vpres technol, Coca-Cola Co, 64-66; vpres res & develop, Evans Res & Develop Corp, 66-69; vpres res & develop, Life Savers, Inc, Squibb Corp, 73-82; vpres, Gillette Res Inst, 82-84; vpres, MacKay Assocs, 85-87, PRES, APPL MICROBIOL, 87- *Mem:* Am Chem Soc; Int Asn Dent Res; NY Acad Sci. *Res:* Chromatography; trace analysis; odor measurement; flavor, sulfur and keratin; reaction mechanisms; biogenesis of natural products; product development; consumer studies; new product planning; research organization; dental research; nutrition; food science; candy technology; applications of bioengineered proteins. *Mailing Add:* 135 Deerfield Lane Pleasantville NY 10570

MACKAY, DONALD DOUGLAS, b Lorne, NS, Apr 29, 08; m 36; c 2. CHEMISTRY. *Educ:* Acadia Univ, BSc, 29. *Prof Exp:* Chemist, Aluminum Co Can, Ltd, 29-31; chemist, McColl-Frontenac Oil Co, 31-35; chemist, Aluminum Co Can, Ltd, 35-36, head tech control, Alumina Plant, 36-37, chief chemist, Demerara Bauxite Co, Ltd Div, 38, from supvr to asst supt, Chem Div, Aluminum Co, 39-42, chem engr, 42-45, chief alumina div, Raw Mat Dept, Aluminium Labs, Ltd Div, 45-49, head, 49-66, vpres div, 59-66, dir, 65-66; managing dir, Indian Aluminium Co, Ltd, Calcutta, 66-69; vpres raw mat, Alcan Aluminium Ltd & pres, Alcan Ore Ltd, Montreal, Que, 69-73; RETIRED. *Concurrent Pos:* Dir & vpres, Fluoresqueda, SA, Mex, 57-66; dir, Nfld Fluorspar, Ltd, Nfld & Southeast Asia Bauxites, Ltd, Singapore, 58-66; dir, Alcan Queensland Pty, Ltd, 64-66; aluminum indust adv & consult, 73- *Mem:* Am Chem Soc; Chem Inst Can. *Res:* Investigation and acquisition of raw material deposits throughout the world for the aluminum industry, especially bauxite and fluorspar; appraisal of economics of producing alumina and fluoride materials from these deposits. *Mailing Add:* 415 Greenview Ave #1107 Ottawa ON K2B 8G5 Can

MCKAY, DONALD EDWARD, b Washington, DC, Aug 21, 38; m 63; c 4. ORGANIC CHEMISTRY, SCIENCE EDUCATION. *Educ:* Univ Calif, BS, 60; Univ Ill, PhD(org chem), 66. *Prof Exp:* Asst, Univ Ill, 60-66; from res chemist to sr res chemist, Org Chem Div, Am Cyanamid Co, 74-82; INSTR CHEM & PHYSICS, TIMOTHY CHRISTIAN HIGH SCH, PISCATAWAY, NJ, 83- *Mem:* Am Chem Soc. *Res:* Synthetic organic chemistry; small ring compounds; dyestuffs; chemiluminescent reactions; process development. *Mailing Add:* 1715 Rosewood Dr Piscataway NJ 08854

MCKAY, DONALD GEORGE, pathology; deceased, see previous edition for last biography

MCKAY, DOUGLAS WILLIAM, b Howland, Maine, Mar 10, 27; c 3. ORTHOPEDIC SURGERY. *Educ:* Univ Maine, BA, 51; Tufts Univ, MD, 55. *Prof Exp:* Intern, Maine Gen Hosp, Bangor, 55-56; resident adult orthop surg & trauma, Vet Admin Prog, McKinney & Dallas, Tex, 56-57; children's orthop surg, Newington Hosp Crippled Children, Conn, 59-60; pvt pract orthop surg, Covington, Ky, 60-61; chief surgeon, Carrie Tingley Hosp Crippled Children Truth or Consequences, NMex, 61-67; chief surgeon, Shriner's Hosp Crippled Children, 67-72; PROF ORTHOP SURG & CHILD HEALTH & DEVELOP, MED SCH, GEORGE WASHINGTON UNIV, 72-; CHIEF PEDIAT ORTHOP SURG, CHILDREN'S HOSP NAT MED CTR, 72-, CHMN DEPT, 78- *Concurrent Pos:* Clin instr, Univ Colo, 61-67; consult, William Beaumont Gen Hosp, US Army, El Paso, Tex, 64-67; adj prof, Univ NMex, 66-67; consult, US Air Force Hosp, Lackland AFB, 67- & Vet Admin Hosp, Shreveport; dir orthop residency prog, Confederate Mem Med Ctr, 68-74; prof orthop & head dept, Sch Med, La State Univ, Shreveport, 69-74. *Mem:* Pediat Orthop Soc (secy, 71-72). *Res:* Orthopedic pathology. *Mailing Add:* AAOS Ped ORS PO Drawer 670 Ville Platte LA 70586

MCKAY, EDWARD DONALD, III, b Robinson, Ill, June 30, 49; m 69; c 3. QUATERNARY GEOLOGY, ENGINEERING GEOLOGY. *Educ:* Hanover Col, BA, 71; Univ Ill, Urbana, MS, 75, PhD(geol), 77. *Prof Exp:* Geologist, Ill State Geol Surv, 76-80; geologist, CGS Inc, 80-83; HEAD, COMPUT RES SECT, ILL STATE GEOL SURV, 83- *Mem:* Am Quaternary Asn; Geol Soc Am. *Res:* Pleistocene glacial history; computer applications in geology; geologic disposal of radioactive wastes; geologic site characterization; instrumentation to measure rock mass displacements around excavations; rock and soil mechanics. *Mailing Add:* 2610 Coppertree Rd Champaign IL 61821

MACKAY, FRANCIS PATRICK, b Waterbury, Conn, July 12, 29. ORGANIC CHEMISTRY. *Educ:* Univ Notre Dame, BS, 51; Col Holy Cross, MS, 52; Pa State Univ, PhD(chem). *Prof Exp:* Res chemist, E I Du Pont de Nemours & Co, Inc, 56-58; asst prof, 58-70, ASSOC PROF CHEM, PROVIDENCE COL, 70-, CHMN DEPT, 73- *Mem:* Am Chem Soc; Sigma Xi. *Mailing Add:* Dept Chem Providence Col River Ave & Eaton St Providence RI 02918-0002

MCKAY, JACK ALEXANDER, b Alhambra, Calif, Apr 3, 42; m 65. PHYSICS. *Educ:* Stanford Univ, BS, 63, MS, 67; Carnegie-Mellon Univ, PhD(physics), 74. *Prof Exp:* res physicist, Naval Res Lab, 74-84; RES SCIENTIST, PHYS SCI INC, 84- *Mem:* Am Phys Soc; Inst Elec & Electronics Engrs; AAAS; Optical Soc Am; Am Geophys Union. *Res:* Interaction of high-energy pulsed laser radiation with materials; laser probes. *Mailing Add:* 3200 19th St NW Washington DC 20010

MCKAY, JAMES BRIAN, b Uniontown, Pa, Sept 2, 40; m 65; c 3. ANALYTICAL CHEMISTRY. *Educ:* Phila Col Pharm, BSc, 62; Wayne State Univ, MS, 64, PhD(anal chem), 66. *Prof Exp:* Asst prof chem, Univ Conn, 66-72; CHMN DEPT CHEM, EDINBORO STATE COL, 72-, ASSOC PROF CHEM, 76- *Mem:* Am Chem Soc. *Res:* Solvent extraction; column chromatography; organic reagents; radiochemistry; ion exchange. *Mailing Add:* Dept of Chem Edinboro State Col Edinboro PA 16444

MCKAY, JAMES HAROLD, b Seattle, Wash, July 23, 28; m 47; c 4. MATHEMATICS. *Educ:* Univ Seattle, BS, 48; Univ Wash, MS, 50, PhD(math), 53. *Prof Exp:* Assoc math, Univ Wash, 51-53, instr, 53-54; instr, Mich State Univ, 54-56, asst prof, 56-57; asst prof, Univ Seattle, 57-59; assoc prof, 59-63, PROF MATH, OAKLAND UNIV, 63- *Concurrent Pos:* Assoc dean sci & eng, Oakland Univ, 61-65, chmn dept math, 63-65; NSF fac fel, Univ Calif, Berkeley, 66-67; mem, Nat Coun, Am Asn Univ Profs, 75-78. *Mem:* Am Math Soc; Math Asn Am. *Res:* Algebra. *Mailing Add:* Dept Math Oakland Univ Rochester MI 48309-4401

MCKAY, JERRY BRUCE, b Wyandotte, Mich, Dec 20, 35; m 58; c 2. ORGANIC CHEMISTRY, POLYMER CHEMISTRY. *Educ:* Mich State Univ, BS, 58; Stanford Univ, MS, 60; Univ Ohio, PhD(org chem), 66. *Prof Exp:* Chemist, Stanford Res Inst, 59-62; res chemist, Dacron Res Lab, 65-68, res supvr, Textile Fibers Dept, 68-71, res & develop supvr, Kinston Plant, 71-73, process supvr, 73-78, develop assoc, 78-81, res assoc, Dacron Res & Develop, 81-87, SR RES ASSOC, DACRON TECH, KINSTON PLANT, E I DU PONT DE NEMOURS & CO, INC, 87- *Mem:* Am Chem Soc; Sigma Xi; Soc Plastic Eng; Textile Inst. *Res:* Structure of synthetic fibers, polymer structure and property analysis; monomer and polymer synthesis; organic synthesis; polymer process development; resource management; recovery of polymer wastes; process hazard analysis; textile fiber technology; polyester composition. *Mailing Add:* E I du Pont de Nemours & Co Box 800 Kinston NC 28501

MACKAY, JOHN WARWICK, b Can, June 1, 23; m 48; c 4. SOLID STATE PHYSICS. *Educ:* Univ Sask, BSc, 45; Purdue Univ, MS, 49, PhD(physics), 53. *Prof Exp:* Asst physics, Nat Res Coun Can, 45-46; asst, Purdue Univ, 46-53, from asst prof to prof physics, 66-88; RETIRED. *Concurrent Pos:* Consult, NASA, 63-68. *Mem:* Am Phys Soc. *Res:* Microwave electron accelerators; radiation damage in semiconductors. *Mailing Add:* RR 2 Box 249 Delphi IN 46923

MACKAY, KENNETH DONALD, b Detroit, Mich, July 18, 42; m 64; c 2. ORGANIC CHEMISTRY. *Educ:* Univ Mich, BS, 64; Univ Minn, PhD(chem), 69. *Prof Exp:* Sr res chemist, Gen Mills, Inc, 68-74, group leader, 74-75, res assoc, 75-77; tech mgr, 77-80, assoc dir res & develop, 80-82, vpres & dir res, 82-86, PRES & DIR RES, HENKEL RES CORP, 86- *Mem:* Am Chem Soc; Am Inst Mining, Metall & Petrol Engrs; Indust Res Inst. *Res:* Physical organic chemistry; organic synthesis; solvent extraction; polymer chemistry. *Mailing Add:* 1467 White Oak Dr Santa Rosa CA 95405

MCKAY, KENNETH GARDINER, b Montreal, Que, Apr 8, 17; m 42; c 2. TELECOMMUNICATIONS. *Educ:* McGill Univ, BSc, 38, MSc, 39; Mass Inst Technol, ScD(physics), 41. *Hon Degrees:* DEng, Stevens Inst Technol. *Prof Exp:* Demonstr physics, McGill Univ, 38-39; jr radio res engr, Nat Res Coun Can, 41-46; res physicist, Bell Tel Labs, 46-52, head phys electronics res, 52-54, physics of solids res, 54-57, dir solid state device develop, 57-59, vpres systs eng, 59-62, exec vpres, 62-66, vpres eng, Am Tel & Tel Co, 66-73; exec vpres, Bell Tel Labs Inc, 73-80; chmn bd, Charles Stark Draper Lab Inc, Cambridge, Mass, 82-87; SCI & TECHNOL ADV GROUP, TAIWAN, 82- *Concurrent Pos:* Mem tech adv bd, Dept Com, 70-72; counr, Nat Acad Eng, 68-, Nat Acad Scis, 76- *Mem:* Nat Acad Sci; Nat Acad Eng; fel Am Phys Soc; fel Inst Elec & Electronics Engrs; NY Acad Sci. *Res:* Secondary electron emission; electron bombardment conductivity; electrical breakdown; light emitting diodes; semiconducting nuclear detectors. *Mailing Add:* 200 E 66th St New York NY 10021

MACKAY, KENNETH PIERCE, JR, b Detroit, Mich, Feb 7, 39; m 63; c 2. METEOROLOGY, AIR POLLUTION. *Educ:* Univ Mich, BSE, 61, MS, 65; Univ Wis, PhD(meteorol), 70. *Prof Exp:* Asst res meteorologist, Univ Mich, 61-65; res asst meteorol, Univ Wis, 65-69; asst prof, 69-75, assoc prof, 75-80, PROF METEOROL, SAN JOSE STATE UNIV, 80-, ASSOC DEAN EDUC EQUITY, 89- *Concurrent Pos:* Vis scientist, Univ Autonoma Metropolitana, Mex, 78, Environ Protection Agency, 80. *Mem:* Am Meteorol Soc; Air Pollution Control Asn. *Res:* Statistical analysis applied to wind power potential; performance evaluation of air quality models; boundary layer wind dynamics and turbulence. *Mailing Add:* Dept Meteorol San Jose State Univ Washington Sq San Jose CA 95192

MCKAY, LARRY LEE, b Oregon City, Ore, June 3, 43; m 64; c 2. PLASMID BIOLOGY, BIOTECHNOLOGY. *Educ:* Univ Mont, BA, 65; Ore State Univ, PhD(microbiol), 69. *Prof Exp:* Fel microbiol, Mich State Univ, 69-70; from asst prof to assoc prof, 70-78, PROF FOOD MICROBIOL, UNIV MINN, ST PAUL, 78- *Concurrent Pos:* Vis prof, Latin Am Prog, ASM, 81; Sci lectr, Inst Food Technologists, 84-87; chair food sci, Kraft-General Foods, Inst Food Technol, 89-94. *Honors & Awards:* Pfizer Award, Am Dairy Sci Asn, 76, Dairy Res Found Award, 82; Res Award, Am Cultured Dairy Prod Inst, 85; Fisher Sci Co Award, Am Soc Microbiol, 87; Borden Award, Am Dairy Sci Asn, 90. *Mem:* Am Dairy Sci Asn; Am Soc Microbiol; Inst Food Technol; Am Cultured Dairy Prod Inst. *Res:* Dairy and food starter culture; food fermentations; plasmid biology and genetics of lactic acid bacteria; biotechnology as applied to lactococci used in milk fermentations; bacteriophages of lactococci. *Mailing Add:* Dept Food Sci & Nutrit Univ Minn 1334 Eckles Ave St Paul MN 55108

MACKAY, LOTTIE ELIZABETH BOHM, b Vienna, Austria, June 7, 27; US citizen; m 52; c 4. SCIENCE WRITING, SCIENCE EDUCATION. *Educ:* Vassar Col, AB, 47; Yale Univ, PhD(org chem), 52. *Prof Exp:* Sr res chemist, Standard Oil Develop Corp, 52-53 & Burroughs-Wellcome Co, 53-55; sci consult, Hudson Inst, 67; sci & math ed AV educ mat, Educ AV Corp, 68-70; ed, 70-77, EXEC ED SCI & MATH, AV EDUC MAT & COMPUT SOFTWARE, PRENTICE-HALL MEDIA, INC, 77- *Mem:* Nat Sci Teachers Asn; Nat Coun Teachers Math; Am Chem Soc; Am Asn Physics Teachers; Am Nat Metric Coun. *Mailing Add:* 135 Deerfield Lane Pleasantville NY 10570

MCKAY, MICHAEL DARRELL, b Temple, Tex, Jan 5, 44; m 66; c 3. STATISTICS. *Educ:* Univ Tex, BA, 66; Tex A&M Univ, PhD(statist), 72. *Prof Exp:* Res mathematician, Southwest Res Inst, 66-68; asst prof statist, Tex A&M Univ, 71-73; STAFF MEM, LOS ALAMOS NAT LAB, 73- *Mem:* Am Statist Asn. *Res:* Model validation; mathematical modeling; computer applications in statistics. *Mailing Add:* Los Alamos Nat Lab MS600 Los Alamos NM 87545

MACKAY, RALPH STUART, b San Francisco, Calif, Jan 3, 24; m 60, 84. BIOPHYSICS, ELECTRONICS. *Educ:* Univ Calif, Berkeley, AB, 44, PhD(physics), 51. *Prof Exp:* Asst physics, Univ Calif, Berkeley, 44-48, elec eng, 47-49, lectr, 49-52, electron microscopist, 46-51, asst prof elec eng, 52-57, dir res & develop lab, Med Ctr, San Francisco, 54-58, assoc res biophysicist, 54-57, lectr, 55-60, assoc clin prof exp radiol & assoc res physicist, 58-60, res biophysicist, Berkeley, 60-67, assoc clin prof optom, 60-62, clin prof, 62-64, biophysicist, Space Sci Lab, 63, lectr med physics, 63-67; prof surg, Med Ctr, 67-74; prof, 67-88, EMER PROF BIOL, BOSTON UNIV, 87-; EMER PROF SCI, SAN FRANCISCO STATE UNIV, 90- *Concurrent Pos:* Guggenheim fel, Karolinska Inst, Sweden, 56 & 57; vchmn, Inst Elec & Electronics Engrs, 56-58, mem admin comt eng med & biol, 72-74; vis prof med physics, Alcohol Res Inst, Karolinska Inst, Sweden, 59; Fulbright fel, Cairo Univ, Egypt, 60, vis prof, 60-61; sr scientist, Galapagos Int Sci Proj, 64; US ed, Ultrasonics, 65-; mem res comt, Franklin Park & Stoneham Zoos, Boston, 67-70; Erskine fel, Univ Canterbury, NZ, 69 & 80; consult, Dep Proj Off, US Naval Radiol Defense Lab, 56; chronic uremia consult, Nat Inst Arthritis & Metab Dis, 66-69; mem bd dir, Biotronics, Inc, 60-67, mine adv comt, Nat Acad Sci-Nat Res Coun, 61-74, bio-instrumentation adv coun, Am Inst Biol Sci, 65-71 & comt emergency med serv, Nat Res Coun, 68-70; lectr biomed telemetry, Univ Calif, San Francisco, 66, Smithsonian Inst, 66, Boston Univ, Am Inst Biol Sci, 67, Int Inst Med Electronics & Biol Eng, London, 68, Am Mus Nat Hist, NY, 70, Commonwealth Sci & Indust Res Orgn, Australia, 70, Marine Lab, Woods Hole, 72; distinguished visitor, US Antarctic Prog, 70; mem bd gov, Int Inst Med Electronics & Biol Eng, Paris, 70-73; vis prof elec eng & comput sci, Univ Calif, Berkeley, 73-74; vis prof radiol, Univ Calif, Davis, 78; dean prof sci, San Francisco State Univ, 86-90. *Honors & Awards:* Apollo Award, Am Optom Asn, 62; Fel, Inst Elec & Electronics Engrs, 62; Guggenheim Fel, 56-57; Fullbright Fel, 60; Erskine Fel, 69, 80; Career Achievement Award, Inst Elec & Electronic Engrs, 88. *Mem:* Sigma Xi; Fel Inst Elec & Electronics Eng; Am Inst Biol Sci; Undersea Med Soc; Biomed Eng Soc; Acoust Soc Am. *Res:* Medical engineering; biology; marine mammals. *Mailing Add:* 2083 16th Ave San Francisco CA 94116

MACKAY, RAYMOND ARTHUR, b New York, NY, Oct 30, 39; m 66; c 2. PHYSICAL CHEMISTRY. *Educ:* Rensselaer Polytech Inst, BS, 61; State Univ NY Stony Brook, PhD(phys chem), 66. *Prof Exp:* Guest res assoc, Nuclear Eng Dept, Brookhaven Nat Lab, 63-64, res assoc chem, 66; res chemist Phys Res Lab, US Army Edgewood Arsenal, Md, 67-69; from asst prof to assoc prof, Drexel Univ, 69-80, prof chem, 80-; chief, Res Directorate, Chem Div, 83-85, actg res dir, Chem Res Develop & Eng Ctr, 88-89, CHIEF, DETECTION DIRECTORATE, CHEM RES DEVELOP & ENG CTR, ABERDEEN PROVING GROUND, MD, 85-88 & 89- *Concurrent Pos:* Asst prof exten div, Univ Del, 67-68; consult, Edgewood Arsenal, Md, 70-80; assoc ed, J Am Oil Chemists Soc, 81-; sr res assoc, Nat Res Coun, 82; mem bd dirs, Sigma Xi, 86-93. *Honors & Awards:* Res & Develop Achievement Award, US Army, 80. *Mem:* AAAS; Am Chem Soc; Sigma Xi; NY Acad Sci; fel Am Inst Chemists. *Res:* Gas-aerosol reactions; photochemistry; liquid crystals; reactions in microemulsions. *Mailing Add:* Detection Directorate Chem Res Develop & Eng Ctr Aberdeen Proving Ground MD 21010-5423

MCKAY, RICHARD A(LAN), b Salt Lake City, Utah, May 19, 27. CHEMICAL & ELECTRICAL ENGINEERING. *Educ:* Calif Inst Technol, BS, 49, MS, 50, PhD(chem eng), 59. *Prof Exp:* Res engr, Dow Chem Co, 50-52; mgr, Nevin H McKay & Co, 52-54; MEM TECH STAFF, JET PROPULSION LAB, CALIF INST TECHNOL, 59- *Mem:* Am Chem Soc. *Res:* Solid propellant engineering; chemical thermodynamics; geothermal process engineering. *Mailing Add:* Jet Propulsion Lab 601- 101 4800 Oak Grove Dr Pasadena CA 91109

MCKAY, ROBERT HARVEY, b Cordova, Alaska, June 12, 27; m 58; c 3. BIOCHEMISTRY. *Educ:* Univ Wash, BS, 53; Univ Calif, PhD(biochem), 59. *Prof Exp:* Asst biochem, Univ Calif, 53-58; fel, Brandeis Univ, 58-61; Am Cancer Soc fel, Harvard Med Sch, 61-63; asst prof, 63-68, ASSOC PROF BIOCHEM, UNIV HAWAII, 68- *Concurrent Pos:* NIH fel, Univ Hawaii, 64-73. *Mem:* AAAS; Am Chem Soc; Am Soc Biol Chem; Sigma Xi. *Res:* Enzymology; protein chemistry; optical and fluorescent properties of proteins; enzyme-coenzyme interactions; iron metabolism; iron storage proteins. *Mailing Add:* Dept of Biochem & Biophys Univ Hawaii 1960 East-West Rd Honolulu HI 96822

MCKAY, ROBERT JAMES, JR, b New York, NY, Oct 8, 17; m 43; c 4. PEDIATRICS. *Educ:* Princeton Univ, AB, 39; Harvard Univ, MD, 43. *Prof Exp:* From asst prof to assoc prof, 50-55, chmn dept, 50-83, PROF PEDIAT, UNIV VT, 55- *Concurrent Pos:* Markle scholar med sci, 50-55; Fulbright lectr, State Univ Groningen, 60. *Mem:* Soc Pediat Res; Am Pediat Soc; Am Soc Human Genetics; Can Pediat Soc; fel Am Acad Pediat. *Res:* Rheumatoid arthritis; chromosomes; genetics. *Mailing Add:* Dept of Pediat Univ Vt Burlington VT 05405

MACKAY, ROSEMARY JOAN, b Eng, July 18, 36; m 62. FRESHWATER ECOLOGY, SCIENTIFIC JOURNAL EDITING. *Educ:* Univ London, BSc, 57, Dipl, 58; McGill Univ, MSc, 68, PhD(biol), 72. *Prof Exp:* High sch teacher biol, Govt Kenya, 58-61; high sch teacher biol, St Mary's Acad, Winnipeg, 61-62; engr water anal, Winnipeg Water Works, 62-63; technician med, Montreal Gen Hosp, 63-65; res fel entom, Royal Ont Mus, 73-74; from asst prof to assoc prof, 74-88, PROF ZOOL, UNIV TORONTO, 88- *Concurrent Pos:* Fel, McGill Univ, 72-73; Nat Res Coun Can fel, 73-74; assoc ed, Am Midland Naturalist, 77-82; managing ed & ed-in-chief, J NAm Benthological Soc, 85- *Mem:* NAm Benthol Soc (pres, 81-82); Ecol Soc Am; British Ecol Soc; Freshwater Biol Asn; AAAS; Coun Biol Ed. *Res:* Ecology of aquatic insects, especially Trichoptera, with emphasis on resource-partitioning among closely related species; investigating ecology of sympatric Hydropsychidae in streams and rivers; behavioral interactions recorded by video. *Mailing Add:* Dept Zool Univ Toronto Toronto ON M5S 1A1 Can

MCKAY, SANDRA J, b Philadelphia, Pa, Sept 6, 47. SYNTHETIC ORGANIC CHEMISTRY. *Educ:* Dickinson Col, BS, 69; Princeton Univ, MA, 71, PhD(chem), 73. *Prof Exp:* Res chemist, Jackson Lab, 73-78, res chemist, Elastomer Chem Dept, 78-81, sr lab supvr, Polymer Prod Dept, 81-82, AREA SUPT, TECH DEPT, E I DU PONT DE NEMOURS & CO, INC, 82- *Mem:* Am Chem Soc. *Res:* Management of manufacturing technology. *Mailing Add:* Polymer Prod PO Box 80310 E I Du Pont de Nemours & Co Inc Wilmington DE 19880-0310

MCKAY, SUSAN RICHARDS, b Plainfield, NJ; c 2. CONDENSED MATTER PHYSICS, STATISTICAL MECHANICS. *Educ:* Princeton Univ, AB, 75; Univ Maine, MS, 79; Mass Inst Technol, PhD(physics), 87. *Prof Exp:* Engr new prod res & develop, Gillette Advan Technol Lab, 75-77; res assoc, 79-80, ASST PROF PHYSICS, DEPT PHYSICS, UNIV MAINE, 86- *Mem:* Am Phys Soc; Sigma Xi; Mat Res Soc. *Res:* Theoretical condensed matter physics with over twenty publications in the areas of phase transitions and critical phenomena, spin glasses and amorphous magnetism, chaos and nonlinear systems and surface and interface physics. *Mailing Add:* Dept Physics Univ Maine Orno ME 04469

MACKAY, VIVIAN LOUISE, b Columbus, Ohio, Jan 8, 47. MOLECULAR BIOLOGY, BIOCHEMICAL GENETICS. *Educ:* Capital Univ, BS, 68; Case Western Reserve Univ, PhD(microbiol), 72. *Prof Exp:* Fel biochem, Univ Calif, Berkeley, 72-74; asst res prof, 74-79, assoc prof microbiol, Waksman Inst Microbiol, Rutgers Univ, 79-84; SR SCIENTIST, ZYMO GENETICS, INC, 82- *Mem:* AAAS. *Res:* Biochemical and genetic investigations of cellular regulatory system in yeast that controls sexual conjugation, meiosis, genetic recombination, and DNA repair; expression of foreign proteins in yeast. *Mailing Add:* Zymo Genetics Inc 4225 Roosevelt Way NE Seattle WA 98105

MACKAY, W(ILLIAM) B(RYDON) F(RASER), b Winnipeg, Man, May 21, 14; m 41; c 3. INDUSTRIAL & MANUFACTURING ENGINEERING. *Educ:* Univ Man, BSc, 38; Univ Minn, BMetE, 40, MS, 47, PhD(phys metall), 53. *Prof Exp:* Instr metall eng, Univ Minn, 46-53, asst prof, 53-56; prod develop engr, Atlas Steels Ltd, 56-57, res & develop engr, 58, chief metall engr, 59, actg mgr metall, 60, mgr res & develop, 60-61, dir res & technol, Atlas Titanium Ltd, 62-63, mgr appl res, Atlas Steels Co, 63-66; prof head dept, 66-77, actg dean, Fac Appl Sci, 76-77, assoc dean, 78-81, prof, 81-84, EMER PROF METALL ENG, QUEEN'S UNIV, ONT, 84- *Concurrent Pos:* Wing commander, Royal Canadian Air Force, 40-46; Prof mech eng, Royal Mil Col, Can, 80-; consult, 46-56 & 66-; Am Soc for Metals, 64-67, Chmn Documentation Comm, 67-68, Publication Coun, 69-71, mem, Nat Nominating Comm, 79; teaching/post-retirement appt, 84- *Mem:* Fel Am Soc Metals; fel Eng Inst Can; fel Inst Metals Brit; Can Soc Mech Engrs; Asn Prof Eng Ont. *Res:* Alloy development; tool, stainless and high strength steels; metal failures; metal processing. *Mailing Add:* Ravensview RR1 Kingston ON K7L 4V1 Can

MACKAY, WILLIAM CHARLES, b Innisfail, Alta, Nov 29, 39; div; c 3. ENVIRONMENTAL PHYSIOLOGY. *Educ:* Univ Alta, BSc, 61, BEd, 65, MSc, 67; Case Western Reserve Univ, PhD(biol), 71. *Prof Exp:* Res assoc physiol, Yale Univ, 70-71; from asst prof to assoc prof, 71-84, PROF ZOOL, UNIV ALTA, 84- *Mem:* AAAS; Freshwater Biol Asn, UK; Am Soc Zool; Can Soc Zool; Am Fisheries Soc. *Res:* Comparative and environmental physiology of aquatic animals. *Mailing Add:* Dept Zool Univ Alta Edmonton AB T6G 2M7 Can

MCKAYE, KENNETH ROBERT, b Camp Lejeune, NC, July 12, 47. ECOLOGY, BEHAVIORAL BIOLOGY. *Educ:* Univ Calif, Berkeley, AB, 70, MA, 72, PhD(zool), 75. *Prof Exp:* Asst prof biol, Yale Univ, 75-81; res assoc prof, Duke Univ, 81-83; ASSOC PROF, UNIV MD, 83- *Mem:* Ecol Soc Am; Soc Study Evolution; Animal Behav Soc; Am Soc Ichthyologists & Herpetologists. *Res:* Evolution of social behavior and the role behavior plays in determining the community structure of animals; study of the behavioral ecology of cichlid fishes; use of indigenous fish in aquaculture. *Mailing Add:* Univ Md Frostburg MD 21532

MACKE, GERALD FRED, b Elbing, Ger, Mar 6, 39; US citizen; m 69; c 2. PLASTICS ENGINEERING, INFRARED SPECTROSCOPY. *Educ:* Univ Utah, BS, 64. *Prof Exp:* Chemist anal, Hercules, Inc, 61-68; prod develop mgr plastics, Gen Elec Co, 68-81; proj mgr plastics, Hoover Universal, Inc, 81-83; QUAL ASSURANCE MGR PLASTICS, PAWNEE INDUSTS, INC, 83- *Concurrent Pos:* Bd dirs, Soc Plastic Engrs, 75-78. *Mem:* Am Chem Soc. *Res:* Product development involving plastic compositions and modifications; reinforced thermoplastic materials for injection molding, blow molding, and extrusion. *Mailing Add:* PO Box 12775 Wichita KS 67277

MACKE, H(ARRY) JERRY, b Newport, Ky, Aug 26, 22; m 48; c 2. EXPERIMENTAL STRESS, PHOTOELASTICITY. *Educ:* Univ Ky, BS, 47; Harvard Univ, SM, 48, ScD(appl mech), 51. *Prof Exp:* Mech anal specialist, Flight Propulsion Div, Gen Elec Co, 51-63, sr engr, 63-68, mgr appl mech, Aircraft Engine Group, 68-71, consult engr appl mech, Group Eng Div, 71-76, mgr appl exp stress, Aircraft Engine Bus Group, Design Eng Oper, 77-87, CONSULT, GEN ELEC AIRCRAFT ENGINES, GEN ELEC CO, 88- *Concurrent Pos:* Adj assoc prof aerospace eng, Univ Cincinnati, 61-67. *Mem:* Am Soc Mech Engrs; Soc Exp Mech. *Res:* Experimental stress analysis; mechanical vibrations; mechanical analysis of aircraft gas turbines. *Mailing Add:* 7305 Drake Rd Indian Hill Cincinnati OH 45243-1419

MCKEAGUE, JUSTIN ALEXANDER, b Sibbald, Alta, Nov 7, 24; m 52; c 5. SOIL GENESIS, SOIL CLASSIFICATION. *Educ:* Univ BC, BA, 47, BSA, 55; Univ Alta, MSc, 58; Cornell Univ, PhD(soils), 61. *Prof Exp:* Res officer, Alta Soil Surv, Can Agr, Edmonton, 55-59, res scientist soil genesis, Soil Res Inst, Can Agr, Ottawa, 62-71, 72-78; res scientist, Land Resource Res Inst, 78-89; RETIRED. *Concurrent Pos:* Res scientist, Res Sta, Ste Foy, Que, 71-72; mem, Can Soil Surv Comt, 65-75; vis res, Rothamstad Exp Sta Eng, 82-83; field dir res, Tanzania-Can Wheat Proj, Arusha, Tanzania, 86-89; ed, Geoderma, 89-; lectr, Carleton Univ, 90. *Mem:* Int Soc Soil Sci; fel Can Soc Soil Sci. *Res:* Diagnostic criteria of soil classification; genesis of soils; soil micromorphology & macromorphology. *Mailing Add:* 1289 Amesbrooke Dr Ottawa ON K2C 2E7 Can

MCKEAN, DAVID JESSE, b Indiana, Pa, Jan 21, 46. BIOCHEMISTRY, GENETICS. *Educ:* Johns Hopkins Univ, PhD(biol), 73- *Prof Exp:* PROF IMMUNOL, MAYO CLIN, 76-, DEAN, MAYO GRAD SCH, 87- *Mem:* Am Asn Immunologists. *Res:* Structure-function analysis class II (Ia) membrane proteins; biochemical events that regulate T lymphocyte activation. *Mailing Add:* Dept Immunol Guggenheim Bldg Mayo Med Sch First St SW Rochester MN 55901

MCKEAN, HENRY P, b Wenham, Mass, Dec 14, 30. MATHEMATICS. *Educ:* Dartmouth Col, AB, 52; Princeton Univ, PhD, 55. *Prof Exp:* Instr, Princeton Univ, 55-57; instr, Mass Inst Technol, 58-63, prof, 64-66; prof, Rockefeller Univ, 66-70; dep dir & chmn, Dept Mat, 84-88, PROF, COURANT INST MATH SCI, NY UNIV, 70-, DIR, 88- *Concurrent Pos:* Vis prof, Kyoto Univ, 57-58 & Rockefeller Univ, 63-64; George Eastman prof, Balliol Col, Oxford Univ, 79-80. *Mem:* Nat Acad Sci; Am Acad Arts & Sci. *Res:* Author of numerous mathematical publications. *Mailing Add:* Courant Inst Math Sci 251 Mercer St New York NY 10012

MCKEAN, JOSEPH WALTER, JR, b Sewickley, Pa, June 11, 44; m 64; c 2. STATISTICS. *Educ:* Geneva Col, Pa, BS, 66; Univ Ariz, MS, 68; Pa State Univ, PhD(statist), 75. *Prof Exp:* Instr math, Waynesburg Col, Pa, 68-72; asst prof math, Univ Tex, Dallas, 75-78; ASST PROF MATH, WESTERN MICH UNIV, KALAMAZOO, 78- *Mem:* Inst Math Statist; Am Statist Asn; Biometric Soc; Math Asn Am. *Res:* Non-parametric statistics; particularly robust statistical methods for linear models based on ranks. *Mailing Add:* Dept Math Western Mich Univ Kalamazoo MI 49008

MCKEAN, THOMAS ARTHUR, b Boise, Idaho, Jan 27, 41. PHYSIOLOGY. *Educ:* Whitman Col, AB, 63; Univ Ore, PhD(physiol), 68. *Prof Exp:* Fel physiol, Sch Med, Univ Minn, 68-69; asst dir sci mkt, Hoechst Pharmaceut Co, 69-70; asst prof zool & physiol, Univ Wyo, 70-74; assoc prof, 74-82, PROF ZOOL, UNIV IDAHO, 82- *Concurrent Pos:* Affil prof physiol & biophys, Sch Med, Univ Wash, 84-; actg dir, Wash, Alaska, Mont & Idaho Med Prog, 77-78 & 86-88. *Mem:* AAAS; Am Physiol Soc; Am Heart Asn. *Res:* Oxygen transport to tissues; physiology of diving mammals; comparative cardiac physiology. *Mailing Add:* Dept of Biol Sci Univ of Idaho Moscow ID 83843

MCKEAN, WILLIAM THOMAS, JR, b Littleton, Colo, Jan 23, 38; m 67; c 1. CHEMICAL ENGINEERING. *Educ:* Univ Colo, BS, 60; Univ Wash, PhD(chem eng), 68. *Prof Exp:* Sr develop engr, Battelle-Northwest, 67-70; assoc prof wood & paper sci & chem eng, NC State Univ, 70-76; MEM STAFF, WEYERHAEUSER CO, WASH, 76- *Concurrent Pos:* Consult, Battelle-Northwest, 71-72. *Mem:* Am Chem Soc; Tech Asn Pulp & Paper Indust; Am Inst Chem Eng. *Res:* Oxidative pulping and odor reduction; slow release agents from pulping waste; kinetics of formation and destruction of malodorous sulfur compounds from pulping systems; recovery of heavy metal from industrial waste water. *Mailing Add:* 11017 Alton Ave NE Seattle WA 98195-5823

MCKEARN, THOMAS JOSEPH, b Rockford, Ill, Oct 4, 48. ANTIBODIES. *Educ:* Ind Univ, BA, 71; Univ Chicago, PhD(immunol), 74, MD, 76. *Prof Exp:* Instr to asst prof path, Univ Pa, 77-81, Head, Immunoprotein Lab, Univ Hosp, 78-81; VPRES RES & DEVELOP, CYTOGEN CORP, 81- *Concurrent Pos:* Adj asst prof path, Univ Pa, 81- *Mem:* Am Asn Immunol; AAAS; Am Asn Pathologists; Transplantation Soc. *Res:* Site-selective covalent modification of monoclonal antibodies; design and synthesis of specialized antibody linker systems; biological effects of monoclonal antibodies and monoclonal antibody conjugates. *Mailing Add:* Cytogen Corp 201 College Rd E Princeton NJ 08540

MCKEARNEY, JAMES WILLIAM, b Bayshore, NY, Apr 4, 38; div; c 2. PSYCHOPHARMACOLOGY. *Educ:* C W Post Col, LI Univ, BA, 62; Univ Pittsburgh, MS, 65, PhD(psychol, 66. *Prof Exp:* Res fel pharmacol, Harvard Med Sch, 66-68, instr psychobiol, 68-69; staff scientist, 70-72, SR SCIENTIST, WORCESTER FOUND FOR EXP BIOL, 72- *Concurrent Pos:* Prin investr, Nat Inst Mental Health & Nat Inst on Drug Abuse res grants, 70-; adj prof psychol, Boston Univ, 77- *Mem:* Am Soc Pharmacol & Exp Therapeut; Am Psychol Asn; Behav Pharmacol Soc; Soc Neurosci. *Res:* Control of human and animal behavior by external environmental events; the determinants of the effects of drugs on behavior; neuropharmacology. *Mailing Add:* Worcester Found for Exp Biol 222 Maple Ave Shrewsbury MA 01545

MCKEE, CHRISTOPHER FULTON, b Washington, DC, Sept 6, 42; m 65; c 3. ASTROPHYSICS. *Educ:* Harvard Univ, AB, 63; Univ Calif, Berkeley, PhD(physics), 70. *Prof Exp:* Physicist, Lawrence Livermore Lab, Univ Calif, 69-70; res fel astrophys, Calif Inst Technol, 70-71; asst prof astron, Harvard Univ, 71-74; from asst prof to assoc prof, 74-78, PROF PHYSICS & ASTRON, UNIV CALIF, BERKELEY, 78- *Concurrent Pos:* Dir, Space Sci Lab, 85- *Honors & Awards:* Fel, Am Phys Soc. *Mem:* Am Astron Soc; Am Phys Soc; Int Astron Union. *Res:* Theoretical astrophysics; theory of interstellar medium; astrophysical fluid dynamics; active galactic nuclei. *Mailing Add:* Dept Physics Univ Calif Berkeley CA 94720

MCKEE, CLAUDE GIBBONS, b Md, June 30, 30; m 58; c 2. AGRONOMY. *Educ:* Univ Md, BS, 51, MS, 55, PhD(agron), 59. *Prof Exp:* Asst, Univ Md, 51-54, exten instr, 54-56; exec secy, Md Tobacco Improv Found, Inc, 57-62; EXTEN TOBACCO SPECIALIST, UNIV MD, 62- *Res:* Production of Maryland tobacco, especially quality constituents. *Mailing Add:* 2005 Largo Rd Upper Marlboro MD 20772

MCKEE, DAVID EDWARD, b Pittsburgh, Pa, July 18, 38; m 61; c 2. MECHANICAL ENGINEERING. *Educ:* Lehigh Univ, BSME, 60; WVa Univ, MSE, 63, PhD(mech eng), 67. *Prof Exp:* Mech design engr, Rust Eng, Pa, 60-62; from instr to asst prof mech eng, WVa Univ, 63-73; eng consult, 77-79, consult supvr eng, 79-85, MGR RES & DEVELOP, E I DU PONT DE NEMOURS & CO INC, 85- *Concurrent Pos:* Consult, Morgantown Energy Res Ctr, US Bur Mines, 71-73. *Mem:* Am Soc Mech Eng; Am Soc Eng Educ; Sigma Xi. *Res:* Fluid mechanics; two phase flow; irreversible thermodynamics; heat transfer. *Mailing Add:* Dept Eng E I Du Pont De Nemours 101 Beach St Wilmington DE 19898

MCKEE, DAVID JOHN, b Detroit, Mich, Jan 11, 47; m 65; c 2. BIOCHEMISTRY. *Educ:* Univ S Fla, BA, 68; Fla State Univ, MS, 71, PhD(sci & human affairs), 76; Duke Univ, MBA, 80. *Prof Exp:* Chemist, Oak Ridge Nat Lab, 65-67; biochem trainee res, Fla State Univ, 68-72; chem lectr gen med chem, Univ S Pac, Suva, Fiji, 73-74; lab dir & environ chemist environ regulation, Tallahassee, Fla, 75-77; chemist & proj mgr criteria doc & health assessments, Environ Criteria & Assessment Off, 77-80, PHYSICAL SCIENTIST & NAT PROG MGR, OFF AIR QUAL PLANNING & STANDARDS, US ENVIRON PROTECTION AGENCY, RES TRIANGLE PARK, NC, 80- *Concurrent Pos:* Mem environ comt, Fla Elec Power Coord Group, 75-77; chmn carbon monoxide criteria doc task force, Environ Criteria & Assessment Off, US Environ Protection Agency, 77-80, co-chmn sulfur oxides/particulate matter criteria doc task force, 78-; prog mgr, Nat Ambient Air Qual Standards, sulfur dioxide, nitrogen dioxide, hydrocarbons & ozone, Off Air Qual Planning & Standards/US Environ Protection Agency, 80-; distinguished lectr med sci, Am Med Asn, 87; special achievement award, 79-80 & 84-90. *Honors & Awards:* Bronze Medal, 79; Sustained Super Performance Award, 86 & 89. *Mem:* Air Pollution Control Asn; Soc Risk Anal. *Res:* Biochemistry of blood proteins; science and human affairs; organic synthesis; health effects of air and water pollution. *Mailing Add:* Mail Drop 12 Off Air Qual Planning & Standards US Environ Protection Agency Research Triangle Park NC 27711

MCKEE, DOUGLAS WILLIAM, b Toronto, Ont, Can, Oct 6, 30; m 59; c 4. PHYSICAL CHEMISTRY. *Educ:* Univ London, BSc, 51, PhD(chem), 54, DSc, 82. *Prof Exp:* Res fel, Nat Res Coun Can, 54-55; Welch fel, Rice Univ, 55-56; res chemist, Linde Co div, Union Carbide Corp, 56-60; RES CHEMIST, RES LABS, GEN ELEC CO, 60- *Concurrent Pos:* Lectr, Canisius Col, 58-60. *Honors & Awards:* Graffin Lectr, Am Carbon Soc, 87. *Mem:* Am Carbon Soc (chmn); Royal Soc Chem; Catalysis Soc; Am Chem Soc; Mat Res Soc. *Res:* Surface chemistry; adsorption; diffusion; ion exchange; catalysis; surface chemistry of carbon; carbon fibers; colloid chemistry and surface activity; fuel cells; corrosion; high temperature materials; coal reactivity. *Mailing Add:* Gen Elec Co PO Box 1088 Schenectady NY 12305

MCKEE, EDITH MERRITT, b Oak Park, Ill, Oct 9, 18. COASTAL ZONE USE, OFFSHORE BOTTOM MAPPING. *Educ:* Northwestern Univ, BS, 46. *Prof Exp:* Jr geologist, US Geol Survey, 43-46; asst geologist, Res Unit Shell Oil Co, 47-49; geologist, Arabian Am Oil Co, 49-54; geologist, Underground Gas Storage Co, 56-58; CONSULT GEOLOGIST, 58- *Concurrent Pos:* Prin investr, Mapping Lake Mich Basin, 67-69 & Mapping Lake Superior basin, 69-70; appointee, Fed EnergyAdmin Environ Comt, 73-75; pres appointee, Nat Adv Comt Oceans & Atmosphere, 75-76. *Mem:* Fel Geol Soc Am; Am Inst Prof Geologists; fel Marine Technol Soc; AAAS. *Res:* Detailed three dimensional mapping of surface, subsurface and submarine topography and geology; world-wide mineral resource exploration and economic development; shore stabilization; bathymetric navigation; search retrieval off shore; containing off shore oil spills. *Mailing Add:* PO Box 3 Good Hart MI 49737

MCKEE, GUY WILLIAM, b Renovo, Pa, Apr 14, 19; m 45; c 5. AGRONOMY, SEED SCIENCE. *Educ:* Pa State Univ, BS, 52, MS, 54, PhD(agron), 59. *Prof Exp:* Soil conservationist, Soil Conserv Serv, USDA, 49-51 & 54; from instr to prof agron, Pa State Univ, 54-87, emer prof, 87-; RETIRED. *Honors & Awards:* Pugh Medal, 52. *Mem:* Fel AAAS; Soil Conservation Soc Am; Am Soc Plant Physiol; Am Meteorol Soc; Ecol Soc Am; Am Soc Agron; Sigma Xi. *Res:* Crop ecology and physiology; plant growth modeling; seed technology and physiology; improvement of quality of the environment; intra-specific taxonomy of crop species; revegetation of disturbed sites. *Mailing Add:* Briarcrest Gardens 23 Westmont Hershey PA 17033

MCKEE, HERBERT C(HARLES), b San Antonio, Tex, Feb 26, 20; m 48; c 2. RESEARCH ADMINISTRATION. *Educ:* Muskingum Col, BSc, 42; Ohio State Univ, MSc, 47, PhD(chem eng), 49. *Prof Exp:* Res assoc chem eng, Res Found, Ohio State Univ, 48-50; chem engr, Jefferson Chem Co, 50-53; from engr to dir, Southwest Res Inst, 53-76; ASST HEALTH DIR ENVIRON CONTROL, CITY OF HOUSTON HEALTH DEPT, 76- *Concurrent Pos:* Chmn, Tex Air Control Bd, 66-73; mem, Tex Energy Adv Comt, 73-79; mem, Tex Nuclear Adv Comt, 77-80. *Mem:* Am Inst Chem Engrs; Air Pollution Control Asn; Am Acad Environ Engrs; Am Chem Soc; Am Soc Testing & Mat. *Res:* Environmental control; occupational health, noise and radiation; air and water pollution. *Mailing Add:* 8010 Neff Houston TX 77036-6408

MCKEE, J(EWEL) CHESTER, JR, b Madison, Wis, Nov 4, 23; m 69; c 5. RESEARCH ADMINISTRATION. *Educ:* Miss State Univ, BS, 44; Univ Wis, MS, 49, PhD(elec eng), 52. *Prof Exp:* Instr math & elec eng, Miss State Univ, 46-47; part-time instr elec eng, Univ Wis, 47-48; from asst prof to prof, Miss State Univ, 49-71, head dept elec eng, 57-61, asst dean grad sch, 61-62, coordr res, 62-71, dean, grad sch, 62-69, vpres res and grad studies, 69-79; dir associateships & fellowships, Nat Res Coun, 79-89, emer prof & vpres, 79-; RETIRED. *Concurrent Pos:* Exec dir, Gov Emergency Coun for Recovery & Redevelop Planning Following Hurricane Camille, Miss, 69-70; pres, Conf Southern Grad Schs, 73-74; chmn, Coun Grad Schs US, 77-78; prog consult, Nat Res Coun, 89- *Mem:* AAAS; Sigma Xi; Inst Elec & Electronic Engrs; Nat Soc Prof Engrs; Am Soc Eng Educ. *Res:* Application of analog computers to system analysis; application of digital computers to network synthesis using linear programming techniques. *Mailing Add:* Seven Ridgewood Dr Starkville MS 39759-9158

MCKEE, JAMES STANLEY COLTON, b Belfast, Northern Ireland, June 6, 30; m 62; c 2. PHYSICS. *Educ:* Queens Univ Belfast, BSc, 52, PhD(theoret physics), 56; Univ Birmingham, DSc, 68. *Prof Exp:* Asst lectr physics, Queens Univ Belfast, 54-56; lectr, Univ Birmingham, 56-64, sr lectr, 64-74; vis prof, Lawrence Radiation Lab, 66-67 & 72; PROF PHYSICS & DIR ACCELERATOR CENTRE, UNIV MAN, 74- *Concurrent Pos:* Dir, Solar Physics Lab, 84-; pres, Can Asn Physicists, 86-87; ed, Physics in Can, 90- *Mem:* Fel Inst Physics London; Can Asn Physicists. *Res:* Few body problems, environmental physics, solar physics, space materials and analytical techniques. *Mailing Add:* Dept Physics Univ Man Winnipeg MB R3T 2N2 Can

MCKEE, JAMES W, b Lawrenceburg, Ky, Sept 25, 32; div; c 3. PALEONTOLOGY, SEDIMENTOLOGY. *Educ:* La State Univ, BS, 60, MS, 64, PhD(geol), 67. *Prof Exp:* From instr to prof geol, Univ Wis-Oshkosh, 65. *Res:* Sediment-fossil relationships; paleolimnology; stratigraphy and tectonics, northern Mexico. *Mailing Add:* Geol Dept Univ Wis 800 Algoma Blvd Oshkosh WI 54901

MCKEE, KEITH EARL, b Chicago, Ill, Sept 9, 28; m; c 2. MANUFACTURING SYSTEMS, ROBOTICS. *Educ:* Ill Inst Technol, BS, 50, MS, 56, PhD(eng), 62. *Prof Exp:* Engr, Swift & Co, 53-54; mgr struct res, Ill Inst Technol Res Inst, 54-62; dir, Mech Design & Prod Assurance, Andrew Corp, 62-76; asst dir, 67-68, dir eng mech, 68-80, DIR, MFG PROD CTR, IIT RES INST, 77- *Concurrent Pos:* Adj prof, Ill Inst Technol, 80- *Honors & Awards:* Gold Medal, Soc Mfg Engrs, 91. *Mem:* Am Soc Civil Engrs; Soc Mfg Engrs; Inst Indust Eng; Nat Asn Comput Graphics. *Res:* Application of technology to improve productivity and automation; robotics; manufacturing technology; manufacturing management; information technology; higher order robotic languages. *Mailing Add:* 18519 Clyde Homewood IL 60430

MCKEE, MICHAEL GEOFFREY, b Santa Monica, Calif, May 11, 31; m 56; c 4. PSYCHOPHYSIOLOGY. *Educ:* Univ Calif, Berkeley, AB, 53, PhD(clin psychol), 60. *Prof Exp:* Assessment psychologist, chief res & actg chief of staff, psychol servs staff, Cent Intel Agency, 60-68; STAFF MEM, CLEVELAND CLIN FOUND, 69-, & HEAD, BIOFEEDBACK SECT, 76- *Concurrent Pos:* Consult psychologist, Edward Glaser & Assoc, 66-69; res assoc, Human Interaction Res Inst, 66-83; lectr psychol, Univ Calif, Berkeley, 68-69, Cleveland State Univ Grad Prog, 73-76, grad prog, Case Western Reserve Univ, 78-83; clin instr pastoral psychol, Case Western Reserve Univ, 74-78; consult, Vet Admin Hosps, 80- *Mem:* Am Psychol Asn; Asn Appl Psychophysiol & Biofeedback; AAAS; Asn Advan Behav Ther; Sigma Xi. *Res:* The interaction of personality and physiologic variables contributing to psychophysiologic disorders; the relationship of sympathetic nervous systems arousal to measures of immunocompetence. *Mailing Add:* Psychol Sect-P57 One Clinic Ctr Cleveland Clinic Found 9500 Euclid Ave Cleveland OH 44195-5189

MCKEE, MICHAEL LELAND, b Hampton, Va, Dec 21, 49; m 82. POTENTIAL SURFACES, ELECTRON-DEFICIENT SYSTEMS. *Educ:* Lamar Univ, BS, 71; Univ Tex, Austin, PhD(chem), 77. *Prof Exp:* ASSOC PROF CHEM, AUBURN UNIV, 81- *Mem:* Am Chem Soc. *Res:* Quantum chemistry; potential surfaces; electron-deficient systems. *Mailing Add:* Dept Chem Auburn Univ Auburn AL 36849-5312

MCKEE, PATRICK ALLEN, b Tulsa, Okla, Apr 30, 37; m 63; c 5. MEDICAL RESEARCH, HEMATOLOGY. *Educ:* Univ Okla, MD, 62. *Prof Exp:* From intern to resident internal med, Med Ctr, Duke Univ, 62-65; clin assoc epidemiol heart dis, Nat Heart Inst, 65-67; chief resident & fel internal med & hemat, Med Ctr, Univ Okla, 67-69; from assoc med to assoc prof internal med & hemat, 69-75, ASST PROF BIOCHEM, MED CTR, DUKE UNIV, 71-, PROF INTERNAL MED, 75-, CHIEF DIV GEN MED, 76- *Concurrent Pos:* Assoc ed, Circulation, 73-; mem hemat study sect, NIH, 73-; fel coun thrombosis, Am Heart Asn. *Mem:* Am Fedn Clin Res; Am Heart Asn; Am Soc Hemat; Am Soc Biol Chemists; Am Soc Clin Invest; Sigma Xi. *Res:* Biochemistry of structure-function relationships of human blood coagulation proteins, particularly with respect to mechanisms of thromboses in human disease. *Mailing Add:* Dept of Med OK Univ Health Sci Ctr PO Box 26901 Oklahoma City OK 73190

MCKEE, RALPH WENDELL, b Boynton, Okla, Nov 13, 12; m 38; c 3. BIOLOGICAL CHEMISTRY. *Educ:* Kalamazoo Col, AB, 34, MS, 35; St Louis Univ, PhD(biochem), 40. *Prof Exp:* Asst, St Louis Univ, 35-39, sr asst, 39-40; instr indust hyg, Sch Pub Health, Harvard Univ, 40-45, assoc biochem, Harvard Med Sch, 45-47, asst prof, 47-52; prof, 53-81, EMER PROF BIOL CHEM & EMER ASST DEAN, UNIV CALIF, LOS ANGELES, 81- *Concurrent Pos:* Head biochem, Cancer Res Inst, New Eng Deaconess Hosp, 50-52. *Mem:* Fel AAAS; Am Chem Soc; Am Soc Biol Chemists; Soc Exp Biol & Med; Sigma Xi. *Res:* Isolation and chemistry of vitamin K; physiology and toxicology of carbon disulfide; biochemical studies of lewisite and mustard gas; growth, metabolism and nutrition of malarial parasites; interrelationship of ascorbic acid and cortical hormones; cytochemistry; chemistry and metabolism of cancer cells; methycation of niacinamide. *Mailing Add:* 858 Oreo Pl Pacific Palisades CA 90272

MCKEE, ROBERT B(RUCE), JR, b Kalispell, Mont, Jan 15, 24; m 49; c 3. MECHANICAL ENGINEERING. *Educ:* Mont State Col, BSc, 48; Univ Wash, MSc, 52; Univ Calif, Los Angeles, PhD(eng), 67. *Prof Exp:* Test engr, Pratt & Whitney Aircraft Div, United Aircraft Corp, 48-50; mech engr, Dow Chem Co, 52-56; from asst prof to assoc prof, 57-72, PROF MECH ENG, UNIV NEV, RENO, 72- *Mem:* Am Soc Mech Engrs; Nat Soc Prof Engrs; Am Soc Heating, Ventilating & Refrig Engrs. *Res:* Plastics fabrication and physical testing; energy conservation; solar energy. *Mailing Add:* Dept of Mech Eng Univ of Nev Reno NV 89557

MCKEE, RODNEY ALLEN, b Freeport, Tex, Oct 1, 47; m 69; c 1. SOLID STATE KINETICS. *Educ:* Lamar Univ, BS, 70; Univ Tex, PhD(mat sci), 75. *Prof Exp:* Metallurgist, Nat Bur Standards, 74-75; MEM STAFF, METALS & CERAMICS DIV, OAK RIDGE NAT LAB, 75- *Concurrent Pos:* Nat Acad Sci-Nat Res Coun assoc, Nat Bur Standards, 74-75. *Mem:* Am Phys Soc; Am Crystallog Asn. *Res:* Kinetics of diffusion processes in solids; thermomigration; oxidation of metals and alloys. *Mailing Add:* Metals & Ceramics Div Oak Ridge Nat Lab PO Box X Oak Ridge TN 37830

MCKEE, THOMAS BENJAMIN, b New Castle, Pa, Dec 14, 35; m 59; c 2. ATMOSPHERIC SCIENCE. *Educ:* Univ NC, BS, 58; Col William & Mary, MA, 63; Colo State Univ, PhD(atmospheric sci), 72. *Prof Exp:* Res engr physics, NASA Langley Res Ctr, 58-72; fel atmospheric sci, Colo State Univ, 72, asst prof, 72-73; asst prof environ sci, Univ Va, 73-74; asst prof, 74-77, assoc prof, 77-81, PROF ATMOSPHERIC SCI, COLO STATE UNIV, 81-, HEAD DEPT, 84- *Concurrent Pos:* Colo State climatologist, Colo State Univ, 74- *Mem:* Am Meteorol Soc; Am Asn State Climatologists. *Res:* Remote sensing in the atmosphere; transfer of solar radiation in clouds; temperature inversions in mountain valleys; climate. *Mailing Add:* Dept of Atmospheric Sci Colo State Univ Ft Collins CO 80523

MCKEE, W(ILLIAM) DEAN, JR, b Sigourney, Iowa, May 23, 20; m 42; c 3. CERAMICS ENGINEERING. *Educ:* Tarkio Col, AB, 41; Univ Mo-Rolla, BSc, 51, MSc, 52, PhD(ceramic eng), 55. *Prof Exp:* Develop engr, Receiving Tube Sub-Dept, Gen Elec Co, Ky, 53-56; prin ceramist, Ceramic Div, Battelle Mem Inst, 56-57; sr ceramist, Res & Develop Div, Carborundum Co, NY, 57-60, sr res assoc, 60-62; res specialist, Autonetics Div, NAm Aviation, Inc, 62-69, supvr thin film process eng, Rockwell Int, 69-73; HEAD, HYBRID MICROCIRCUIT BR, NAVAL OCEAN SYSTS CTR, 73-, SR CONSULT, SOLID STATE ELEC DIV. *Mem:* Am Ceramic Soc; Inst Elec & Electronics Engrs; Int Soc Hybrid Microelectronics. *Res:* Materials and processes for microelectronics; electronic ceramics; high temperature chemistry and crystal growth; refractory materials. *Mailing Add:* 1279 Bangor St San Diego CA 92106

MCKEEHAN, CHARLES WAYNE, b Greencastle, Ind, Nov 16, 29; m 53; c 4. PHARMACEUTICAL CHEMISTRY. *Educ:* Purdue Univ, BS, 51, MS, 53, PhD(pharmaceut chem), 57. *Prof Exp:* Sr pharmaceut chemist, Eli Lilly & Co, 57-62, corp trainee, 62-64, proj coordr new prod, 64-68, head liquid-ointment parenteral prod pilot plants, 68-69, head pharmaceut res, 69-71, head liquid-ointment prod develop, 72-78, head parenteral liquid-ointment prod develop, 79-87, HEAD DRY PROD DEVELOP, ELI LILLY & CO, 87- *Mem:* Am Chem Soc; Am Pharmaceut Asn; Sigma Xi. *Res:* Planning and coordination of new product development; physical pharmacy; medical and health sciences; research and development administration in the area of new drug dosage forms. *Mailing Add:* 399 Exeter Ct Greenwood IN 46143

MCKEEHAN, WALLACE LEE, b Texarkana, Tex, Jan 22, 44; m; c 1. CELL BIOLOGY, BIOCHEMISTRY. *Educ:* Univ Fla, BS, 65; Univ Tex, Austin, PhD, 69. *Prof Exp:* Res scientist assoc III, Univ Tex, Austin, 69-70; res scientist, Basel Inst Immunol, Switzerland, 71-73; res assoc molecular, cellular & develop biol, Univ Colo, Boulder, 74-78; sr scientist, 78-87, DEP DIR, W ALTON JONES SCI CTR, 87- *Concurrent Pos:* Assoc ed, Invitro Cell Develop Biol, 87-; mem exec bd, Tissue Cult Asn, 84-88. *Mem:* Tissue Cult Asn; AAAS; Am Chem Soc. *Res:* Regulation of cell growth and function. *Mailing Add:* W Alton Jones Cell Sci Ctr Old Barn Rd Lake Placid NY 12946

MCKEEMAN, WILLIAM MARSHALL, b Pasadena, Calif, Aug 20, 34; c 7. SOFTWARE SYSTEMS. *Educ:* Univ Calif, BA, 56; George Washington Univ, MS, 61; Stanford Univ, PhD(comput sci), 66. *Prof Exp:* Instr physics, US Naval Acad, 59-61; asst prof comput sci, Stanford Univ, 66-68; prof info sci, Univ Calif, Santa Cruz, 68-78; staff scientist, Palo Alto Res Ctr, Xerox Corp, 79-80; chair fac, 80-83, PROF INFO TECHNOL, WANG INST GRAD STUDIES, MASS, 80- *Concurrent Pos:* Indust consult, 61- *Mem:* Sigma Xi; Inst Elec & Electronics Engrs; Am Asn Univ Professors. *Res:* N-dimensional geometry; design of algorithms; mechanical translators and computing machines; software engineering. *Mailing Add:* 62 Truell Rd Hollis NH 03049

MCKEEN, COLIN DOUGLAS, b Strathroy, Ont, June 23, 16; m 42; c 3. PLANT PATHOLOGY. *Educ:* Univ Western Ont, BA, 38; Univ Toronto, MA, 40, PhD(plant path), 42. *Prof Exp:* Class asst bot, Univ Toronto, 38-42; mem, Harrow Res Sta, Can Dept Agr, 46-62, Vegatable Path, head fruit & vegatable path, 62-73, nat res coord plant path, 73-78, nat leader crop protection coordr, 75-78; Founding ed-in-chief, Can J Plant Path, 78-81; CHMN, ONT CHESTNUT COUN, 81- *Mem:* AAAS; Am Phytopath Soc; fel Can Phytopath Soc; Can Microbiol Soc; Australian Phytopathol Soc. *Res:* Crop protection in greenhouse and field crops; soil-borne fungal pathogens and chestnut blight. *Mailing Add:* 3 Keppler Cresant Ottawa ON K2H 5Y1 Can

MCKEEN, WILBERT EZEKIEL, b Strathroy, Ont, Can, Feb 20, 22; m 50; c 6. PLANT PATHOLOGY. *Educ:* Univ Western Ont, BSc, 45, MSc, 46; Univ Toronto, PhD, 49. *Prof Exp:* Plant pathologist, Sci Serv Ont, 49-51, BC, 51-57; assoc prof, 57-60, PROF PLANT SCI, UNIV WESTERN ONT, 60- *Honors & Awards:* Gold Medal, Univ Western Ont, 45; Wintercorbyn Award, Univ Toronto. *Mem:* AAAS; Am Phytopath Soc; Can Soc Phytopath; Can Soc Microbiol; Mycol Soc Am. *Res:* Tobacco blue mold and root rot diseases. *Mailing Add:* Dept of Plant Sci Univ of Western Ont London ON N6A 5B7 Can

MCKEEVER, CHARLES H, b Gibson City, Ill, June 6, 12. ORGANIC CHEMISTRY. *Educ:* Ill Wesleyan Univ, BS, 36; Univ Ill, PhD(chem), 40. *Prof Exp:* Org res chemist, Rohm & Haas Co, 40-60, res supvr, 60-72, process & employee health adv, 72-78; RETIRED. *Mem:* Am Chem Soc. *Res:* Process research. *Mailing Add:* 1406 Holcomb Rd Box 3202 Meadowbrook PA 19046-6702

MCKEEVER, L DENNIS, b Pittsburgh, Pa, Jan 23, 41; m 63; c 3. POLYMER CHEMISTRY, PHYSICAL CHEMISTRY. *Educ:* Univ Pittsburgh, BS, 62; Univ Calif, Irvine, PhD(phys chem), 66. *Prof Exp:* Lab dir, Dow Chem Co, 66-73, dir, Cent Res Plastics Lab, Mich, 73-77, MGR CHEM PROD DEPT, DOW CHEM LATIN AM, 77- *Mem:* Am Chem Soc. *Res:* Organo-electrochemistry; organo-alkali metal chemistry; anionic and free radical polymerization; thermoplastics; styrene based polymers; plastic foams. *Mailing Add:* US Area Res & Develop Dow Chem Co 2040 Dow Ctr Midland MI 48674

MCKEEVER, PAUL EDWARD, b Pasadena, Calif, Dec 3, 46; m 71; c 3. NEUROPATHOLOGY, IMMUNOLOGY. *Educ:* Brown Univ, BS, 68; Univ Calif, Davis, MD, 72; Med Univ SC, PhD(path), 76. *Prof Exp:* Intern anat path, Univ Calif, San Diego, 72-73; resident neuropath, Med Univ SC, 73-76; res assoc immunol, Nat Inst Allergy & Infectious Dis, 76-79, NIH, dir neuropath, Surg Neurol Br, Nat Inst Neurol Disorders & Stroke, 79-; CHIEF, SECT NEUROPATH, DEPT PATH, SCH MED, UNIV MICH. *Concurrent Pos:* USPHS trainee cardiopulmonary med, Univ Calif, San Diego, 72-73; Southern Med Asn guest scientist, Path Inst, Med Univ SC, 75-76; consult neuropath, Lab Path, Nat Cancer Inst, 76-; clin assoc prof, Uniformed Serv Univ Health Sci, 78- *Mem:* Int Acad Path; Am Asn Neuropathologists; Reticuloendothelial Soc; Am Col Physicians; AAAS. *Res:* Membrane and protein receptor cell biology; neoplasms as models of normal neural and glial cell structure. *Mailing Add:* Dept Path Med Sch Univ Mich Box 0602 MSI Ann Arbor MI 48109-0602

MCKEEVER, STEPHEN WILLIAM SPENCER, b Widnes, UK, Sept 22, 50; UK citizen; m 74; c 2. RADIATION DOSIMETRY, THERMOLUMINESCENCE DATING. *Educ:* Univ Col North Wales, Bangor, BS, 72, MS, 73, PhD(materials sci), 75. *Prof Exp:* Postdoctoral fel physics, Univ Birmingham, UK, 75-80, Univ Sussex, UK, 80-82; from asst prof to assoc prof, 82-90, PROF PHYSICS, OKLAHOMA STATE UNIV, 90- *Concurrent Pos:* Vis prof, Naval Surface Warfare Ctr, Md, 83-85, Naval Res Lab, 89-90; nobel res fel, Oklahoma State Univ, 87- *Mem:* Inst Physics; Am Inst Physics. *Res:* Solid state physics; materials science, identification and characterization of defects in insulating and semiconducting materials, using thermoleminescence, ionic thermocurrents, optical techniques and thermally stimulated conductivity. *Mailing Add:* Dept Physics Oklahoma State Univ Stillwater OK 74078

MCKEEVER, STURGIS, b Renick, WVa, Sept 6, 21; m 46. ECOLOGY, MAMMALOGY. *Educ:* NC State Univ, BS, 48, MS, 49, PhD(animal ecol), 55. *Prof Exp:* Proj leader biol res, WVa Conserv Comn, 49-51; biologist, Commun Dis Ctr, USPHS, 55-57; asst zoologist, Agr Exp Sta, Univ Calif, Davis, 57-63; from asst prof to prof biol, Ga Southern Col, 63-88; RETIRED. *Mem:* Sigma Xi. *Res:* Animal ecology; mammalian reproduction; wildlife diseases and pathology; parasitology of mammals. *Mailing Add:* 101 S Edgewood Dr Statesboro GA 30458

MCKEITH, FLOYD KENNETH, b Billings, Mont, July 3, 55; m 81. MEAT SCIENCE, MUSCLE BIOLOGY. *Educ:* Wash State Univ, BS, 77; Tex A&M Univ, MS, 79, PhD(animal sci), 82. *Prof Exp:* Teaching asst animal sci, Tex A&M Univ, 77-79, res asst, 79-81; ASST PROF ANIMAL SCI, UNIV ILL, URBANA, 81- *Mem:* Am Meat Sci Asn; Am Soc Animal Scientist; Inst Food Technologists. *Res:* Palatability and cutability characteristics of beef, pork, and lamb. *Mailing Add:* Dept Animal Sci Univ Ill 1301 W Gregory Dr Urbana IL 61801

MCKELL, CYRUS MILO, b Payson, Utah, Mar 19, 26; m 47; c 3. RANGE MANAGEMENT, RESEARCH ADMINISTRATION. *Educ:* Univ Utah, BS, 49, MS, 50; Ore State Univ, PhD(bot), 56. *Prof Exp:* Asst plant ecol, Univ Utah, 49-50; prin high sch, Utah, 52-53; asst bot, Ore State Univ, 53-55, instr, 55-56; agr res scientist plant physiol, Agr Res Serv, USDA, Univ Calif, Davis, 56-61, assoc prof agron & vchmn dept, Univ Calif, Davis & Univ Calif, Riverside, 61-66, chmn dept, Univ Calif, Riverside, 66-69; head range dept, Utah State Univ, 69-71, dir environ & man prog, 71-76, prof range sci, 69-81, dir inst land rehab, 76-81; vpres res, Native Plants Inc, Salt Lake City, Utah, 81-88; DEAN, SCH NATURAL SCI, WEBER STATE UNIV, OGDEN, UTAH, 88- *Concurrent Pos:* Consult, Ford Found Mex, 65 & US AID, Bolivia & Ford Found, Agr, 71; Fulbright res fel, Spain, 67-68; panelist, Nat Acad Sci, Brazil, 74 & 76; consult, Food & Agr Orgn, UN, 75 & 78; technol innovation, Nat Acad Sci Comt, 79-87; off technol assessment, US Cong, 81, 82, & 84-85; chair, Utah Gov Adv coun Sci & Technol, 90- *Mem:* Am Soc Agron; Soc Range Mgt; Soil Conserv Soc Am; Sigma Xi; fel AAAS. *Res:* Environmental physiology of range plants; rehabilitation of disturbed arid lands; land use planning; application of plant biotechnology to agricultural production. *Mailing Add:* 2248 E 4000 South Holaday UT 84124

MACKELLAR, ALAN DOUGLAS, b Detroit, Mich, Sept 3, 36; m 87; c 3. PHYSICS. *Educ:* Univ Mich, BSE(physics) & BSE(math), 58; Tex A&M Univ, PhD(physics), 66. *Prof Exp:* Nuclear engr, Oak Ridge Nat Lab, 60-61; instr physics, Tex A&M Univ, 61-63; fel, Oak Ridge Nat Lab, 63-65; instr, Mass Inst Technol, 65-67; mem fac, Dept Physics, Rice Univ, 67-68; from asst prof to assoc prof, 68-78, PROF PHYSICS, UNIV KY, 78-, CHMN, 85- *Mem:* Am Phys Soc. *Res:* Theoretical nuclear physics, especially scattering theory; nuclear many body problems and relativistic effects in nuclei; theoretical atomic physics, including ion-atom collisions. *Mailing Add:* Dept Physics & Astron Univ of Ky Lexington KY 40506

MCKELLAR, BRUCE HAROLD JOHN, b Forbes, NSW Australia, May 7, 41; m 63; c 2. THEORETICAL PHYSICS. *Educ:* Univ Sydney, BSc, 62, PhD, 66; Univ Melbourne, DSc, 76. *Prof Exp:* Lectr physics, Univ Sydney, 65-68, sr lectr, 69-72; PROF THEORET PHYSICS, UNIV MELBOURNE, 72-, DEAN, FAC SCI, 91. *Concurrent Pos:* Australian Am Educ Found Fel, Sch Natural Sci, Inst Adv Study, 66-68, sr fel, Cyclotron Lab, Mich State Univ, 71; Fulbright grants, Australia-Am Educ Found, 66, 71 & 82; actg chmn, Sch Physics, Univ Melbourne, 73-74, chmn, 77-79; foreign collabr, Theoret Physics Serv, Ctr Nuclear Study, Saclay, France, 75-76; res medal, Royal Soc Victoria, 77; consult, Los Alamos Sci Lab, 78-, vis staff mem, 82; vis, TRIUMF Lab, 84-85 & 86-87; mem, Australian Res Grants Comt, 85-87, Australian Res Coun, 88-, Res Grants Comt & Chair Math, Phys & Chem Sci Panel, 89; vis prof, Dept Physics, Univ Wash, 88; chair, Math Phys & Chem Sci Adv Comt, Australian Res Coun, 88. *Honors & Awards:* Pawsey Medal, Australian Acad Sci, 73, Lyle Medal, 91. *Mem:* Inst Physics; Am Phys Soc; Optical Soc Am. *Res:* Author of over 150 publications in scientific journals on theoretical physics. *Mailing Add:* Univ Melbourne Parkville Victoria 3052 Australia

MCKELLAR, HENRY NORTHINGTON, JR, b Lumberton, NC, Feb 4, 47; m 71; c 3. ENVIRONMENTAL SCIENCES, AQUATIC ECOLOGY. *Educ:* Univ NC, BS, 69, MS, 71; Univ Fla, PhD(environ eng sci), 75. *Prof Exp:* Guest researcher plankton ecol, Asko Lab, Univ Stockholm, 75-76; fel ecol models, Baruch Inst, 76-77, asst prof, 77-83, ASSOC PROF ENVIRON HEALTH SCI & MARINE SCI, UNIV SC, 83- *Concurrent Pos:* Baruch res assoc, Belle Baruch Inst Marine Biol & Coastal Res, 77-; co-prin investr, NSF res grant, 77-81; vis scientist, Smithsonian Environ Res Ctr, 84-85. *Mem:* Am Soc Limnol & Oceanog; Estuarine Res Fedn; Ecol Soc Am; Soc Wetland Sci. *Res:* Systems ecology; ecological modeling; productivity and nutrient cycling in aquatic and wetland environments; eutrophication; environmental impact analysis; environmental planning. *Mailing Add:* Dept of Environ Health Sci Univ of SC Columbia SC 29208

MACKELLAR, WILLIAM JOHN, b Detroit, Mich, July 14, 35; m 62; c 4. ANALYTICAL CHEMISTRY. *Educ:* Concordia Col, Moorhead, Minn, BA, 65; Wayne State Univ, PhD(chem), 70. *Prof Exp:* From asst prof to assoc prof, 69-85, PROF CHEM, CONCORDIA COL, MOORHEAD, MINN, 85- *Concurrent Pos:* Chem analyst, Ctr Environ Studies, Tri Col Univ, 74 & Lower Sheyenne River Basin Study, NDak Water Resources Res Inst, 75-78; res assoc, NDak State Univ, 76-78; chem analyst, Fish & Wildlife Serv, 78-79; res assoc, USDA, 79-80; mem staff, microprocessor controlled instrumentation develop, 79-; mgr, Buffalo-Red River Watershed Bd, Minn. *Mem:* Am Chem Soc. *Res:* Coordination complex formation reactions and mechanisms; chemical instrumentation; microprocessor based systems; electroanalytical determination of chemical pollutants. *Mailing Add:* Dept Chem Concordia Col Moorhead MN 56560

MCKELLIPS, TERRAL LANE, b Terlton, Okla, Dec 2, 38; m 58; c 2. MATHEMATICS. *Educ:* Southwestern State Col, BSEd, 61; Okla State Univ, MS, 63, EdD(math), 68. *Prof Exp:* Asst prof math, Southwestern State Col, 62-66; instr math & educ, Okla State Univ, 67-68; prof math & dean, 68-89, VPRES ACAD AFFAIRS, CAMERON UNIV, 89- *Concurrent Pos:* Vis assoc prof, Okla State Univ, 72-; consult, Consult Bur, Math Asn Am, 73-89. *Mem:* Math Asn Am; Am Math Soc. *Res:* Point set topology. *Mailing Add:* Sch Math & Appl Sci Cameron Univ Lawton OK 73505

MCKELVEY, DONALD RICHARD, b Indiana, Pa, June 19, 38; m 60. PHYSICAL ORGANIC CHEMISTRY. *Educ:* NMex Inst Mining & Technol, BS, 60; Carnegie Inst Technol, PhD(chem), 64. *Prof Exp:* Res assoc & instr chem, Univ Pittsburgh, 64-66; asst prof, Cornell Univ, 66-69; assoc prof, 69-73, PROF CHEM, INDIANA UNIV PA, 73- *Concurrent Pos:* Mem staff, Mellon Inst, 64-66. *Mem:* Am Chem Soc; Royal Soc Chem; Sigma Xi. *Res:* Reaction mechanisms and solvent effects. *Mailing Add:* Dept of Chem Indiana Univ of Pa RD 1 Box 1398 Indiana PA 15701

MCKELVEY, EUGENE MOWRY, b Greensburg, Pa, June 13, 34; m 60; c 2. INTERNAL MEDICINE, HEMATOLOGY. *Educ:* Yale Univ, BS, 56; Johns Hopkins Univ, MD, 60. *Prof Exp:* From Intern to resident, Boston City Hosp, 60-62; clin assoc, Med Br, Nat Cancer Inst, 62-64; assoc staff physician, Cleveland Clin, 66-67; asst prof med, Northwestern Univ, Chicago, 67-73; assoc prof & assoc internist, M D Anderson Hosp & Tumor Inst, 73-80, prof med & internist, 80-; ASSOC VPRES RES, M D ANDERSON CANCER CTR, UNIV TEX, HOUSTON, 83- *Concurrent Pos:* Fel med, Cleveland Clin, 64-66; hematologist, Vet Admin Res Hosp, Chicago, 67-73. *Mem:* Am Hemat Soc; Am Soc Clin Oncol; Am Asn Cancer Res; AMA; AAAS. *Res:* Oncology; investigational chemotherapy; combined modality treatment; antibody and antigen detection in malignancy. *Mailing Add:* 1515 Holcombe Blvd Houston TX 77030

MCKELVEY, JAMES M(ORGAN), b St Louis, Mo, Aug 22, 25; m 57; c 2. CHEMICAL ENGINEERING. *Educ:* Mo Sch Mines, BS, 45; Wash Univ, MS, 47, PhD(chem eng), 50. *Prof Exp:* Instr chem eng, Wash Univ, 46-50; res engr, Exp Sta, E I du Pont de Nemours & Co, 50-54; asst prof chem eng, Johns Hopkins Univ, 54-57; from assoc prof to prof, 57-64, DEAN SCH ENG, WASH UNIV, 64- *Mem:* Am Chem Soc; Am Inst Chem Engrs; Soc Rheol; Soc Plastics Engrs; Am Soc Eng Educ. *Res:* Flow of non-Newtonian fluids; plastics extrusion; polymer processing. *Mailing Add:* Sch of Eng Wash Univ St Louis MO 63130

MCKELVEY, JOHN MURRAY, b Stanley, NC, Nov 1, 37; m 73; c 1. THEORETICAL CHEMISTRY, PHOTOCHEMISTRY. *Educ:* Mercer Univ, AB, 61; Univ of Ga, MS, 65; Ga Inst Technol, PhD(phys org chem), 71. *Prof Exp:* Fel theoret chem, Chem Lab, Advan Normal Sch for Women, Paris, France, 71-73 & Dept of Chem, Univ Calif, Berkeley, 73-75; RES SCIENTIST, EASTMAN KODAK CO, 75- *Mem:* Am Chem Soc. *Res:* Applied quantum mechanics in photochemistry and mechanistic and synthetic organic chemistry. *Mailing Add:* Dept Info & Comput Tech Bldg 83 Floor 2 Eastman Kodak Co Rochester NY 14650-2216

MCKELVEY, JOHN PHILIP, b Ellwood City, Pa, Nov 9, 26; m 50; c 2. SOLID STATE PHYSICS. *Educ:* Pa State Univ, BS, 49, MS, 50; Univ Pittsburgh, PhD(physics), 57. *Prof Exp:* Asst physics, Pa State Univ, 49-50, instr math, 50-51; res physicist, Res Labs, Westinghouse Elec Corp, 51-59, supvry physicist, 59-62; from assoc prof to prof physics, Pa State Univ, 62-74; head, Dept Physics & Astron, 74-82, prof, 74-87, EMER PROF PHYSICS, CLEMSON UNIV, 87- *Concurrent Pos:* Asst dean col sci, Pa State Univ, 69-73, adj prof physics, 89-; assoc ed, Am J Physics, 85-88; vis prof physics, Va Polytech Inst & State Univ, 87-88. *Mem:* Am Asn Physics Teachers; fel Am Phys Soc. *Res:* Semiconductor physics; solid state theory; statistical physics. *Mailing Add:* State Col 410 S Gill St State College PA 16801

MCKELVEY, ROBERT WILLIAM, b Ligonier, Pa, Apr 27, 29; m 52; c 2. MATHEMATICS. *Educ:* Carnegie Inst Technol, BS, 50; Univ Wis, MS, 52, PhD(math), 54. *Prof Exp:* Instr math, Purdue Univ, 54; res fel, Inst Fluid Dynamics & Appl Math, Univ Md, 54-56; from asst prof to prof, Univ Colo, 56-70; PROF MATH, UNIV MONT, 70- *Concurrent Pos:* Fac fel, Univ Colo, 60, chmn dept math, 65-66; mem, Math Res Ctr, Madison, Wis, 64-65; vis prof, Univ Utah, 66-67, Humboldt State, 78, Oregon State, 81 & Univ BC, 80 & 81-82; exec dir, Rocky Mountain Math Consortium, 67-75; ed, Natural Resource Modeling, 85-; exec ed, Natural Resource Modeling. *Mem:* Am Math Soc; Math Asn Am; Soc Indust & Appl Math; AAAS; Asn Environ & Resource Economists; Resource Modeling Asn. *Res:* Asymptotic theory of differential equations; differential boundary value problems; linear operations in Hilbert space; functional analysis; mathematical models in ecology and natural resource management. *Mailing Add:* Dept Math Univ Mont Missoula MT 59801

MCKELVEY, RONALD DEANE, b Battle Creek, Mich, June 24, 44; m; c 3. PHYSICAL ORGANIC CHEMISTRY. *Educ:* Western Mich Univ, BS, 66; Univ Wis-Madison, PhD(org chem), 71. *Prof Exp:* Fel org chem, Univ Calif, Berkeley, 72; assoc prof org chem, Inst Paper Chem, 72-78; LECTR DEPT CHEM, UNIV WIS-LA CROSSE, 78- *Concurrent Pos:* Fel, Japan Soc Prom Sci, Inst Molecular Sci, Okazaki, Japan, 83, 87; Fulbright fel, Pune Univ, India, 87. *Mem:* Am Chem Soc. *Res:* Free radical reactions; nuclear magnetic resonance; anomeric effect. *Mailing Add:* Dept Chem Univ Wis La Crosse 1725 St St La Crosse WI 54601

MCKELVIE, DOUGLAS H, b Collbran, Colo, Mar 15, 27; m 50; c 3. VETERINARY PATHOLOGY. *Educ:* Colo Agr & Mech Col, BS, 50, DVM, 52; Univ Calif, Davis, PhD(comp path), 68; Am Col Lab Animal Med, dipl. *Prof Exp:* Pvt pract, 52-60; res vet, Radiobiol Lab, Univ Calif, 60-68; prof clin sci & dir lab animal med, Col Vet Med & Biomed Sci, Colo State Univ, 68-74; dir div animal resources & assoc prof path, 74-85, clin vet & assoc prof path, Col Med, Univ Ariz, 85-88; spec asst to dir, Univ Animal Care, 87-; RETIRED. *Concurrent Pos:* Consult, Hill Found dog breeding grant, Med Sch, Univ Ore, 66-; consult, Am Asn Accreditation Lab Animal Care; assoc ed, Lab Animal Sci, 69-76; deleg from Am Soc Lab Animal Practr to house deleg Am Vet Med Asn, 78-86. *Mem:* Am Vet Med Asn; Am Asn Lab Animal Sci; Am Soc Lab Animal Practr (pres-elect, 77-78, pres, 78). *Res:* Production and care of laboratory dogs; serum chemistry of the dog; effects of internal and external irradiation on maturing and adult bone; laboratory animal medicine and biology. *Mailing Add:* Univ Animal Care 1501 N Campbell Ave Tucson AZ 85724

MCKELVIE, NEIL, b Welwyn Garden City, Eng, Dec 9, 30; US citizen; m 59; c 1. ORGANO PHOSPHORUS CHEMISTRY. *Educ:* Cambridge Univ, BA, 53; Columbia Univ, PhD(chem), 61. *Prof Exp:* Chemist, Fairey Aviation Co, Eng, 53-54; res chemist, Am Cyanamid Co, Conn, 59-61; res asst, Yale Univ, 61-62; asst prof, 62-70, ASSOC PROF CHEM, CITY COL NEW YORK, 70- *Mem:* Am Chem Soc; Royal Soc Chem; Sigma Xi; NY Acad Sci. *Res:* Organophosphorus chemistry; chemistry of other elements in groups IV, V and VI. *Mailing Add:* 109-23 71st Rd Forest Hills NY 11375

MCKELVY, JEFFREY F, NEUROSCIENCE. *Educ:* Univ Akron, Ohio, BSc, 63; Case Western Reserve Univ, Cleveland, Ohio, MD, 63, PhD(physiol), 64; Johns Hopkins Univ, Baltimore, Md, PhD(biochem), 68. *Prof Exp:* Jane Coffin Child postdoctoral fel, Dept Biophys, Weizmann Inst Sci, Rehovoth, Israel, 68-69; postdoctoral fel, Dept Physiol Chem, Sect Neurochem, Roche Inst Molecular Biol, Nutley, NJ, 69-71; asst prof anat, Health Ctr, Univ Conn, Farmington, 71-76; asst prof biochem, Health Sci Ctr, Univ Tex, Dallas, 76-79, assoc prof psychiat, 78-79; prof psychiat & biochem & dir, Neuroendocrine Prog, Dept Psychiat & Western Psychiat Inst & Clin, Sch Med, Univ Pittsburgh, Pa, 79-81; prof neurobiol & behav, State Univ NY, Stony Brook, 81-87, prof psychiat, Sch Med, Health Sci Ctr, 83-87 & prof molecular microbiol, 81-87; area head, 87-90, HEAD, NEUROSCI CTR, NEUROSCI RES DIV, ABBOTT LABS, ABBOTT PARK, ILL, 90- *Concurrent Pos:* Vis scientist, INSERM, Paris, France, 75; res career develop award, Nat Inst Arthritis & Metab Dis, 77-82; mem, Neurol B Study Sect, Nat Inst Neurol & Commun Dis & Stroke, 78-82; mem, Task Force Eval Res Needs Endocrinol & Metab Dis, NIH, 79-80; ed, Current Methods Cellular Neurobiol, 83; mem, Cellular & Molecular Neuropharmacol Panel, Pharmacol Sect Eval Group, Neurosci Res Br, NIMH, 83, Cellular Neurobiol & Psychopharmacol Study Sect, 86-88, chmn, 89-; adj prof, Dept Biochem, Robert Wood Johnson Med Sch, Piscataway, NJ, 87-90 & neurobiol & physiol, Northwestern Univ, Evanston, Ill, 87-; chair, NSF Master Grant Comt, State Univ NY & Dept Neurochem Facil. *Mem:* AAAS; Am Col Psychiatrists; Am Col Neuropsychopharmacol; Am Soc Biochem & Molecular Biol; Am Soc Neurochem; Collegium Int Neuropsychopharmacol; Endocrine Soc; Int Soc Neuroendocrinol; Int Brain Res Orgn; Soc Neurosci. *Res:* Clinical neuropharmacology and psychopharmacology; molecular neurobiology, neuropharmacology and neuroendocrinology; author of various publications. *Mailing Add:* Neurosci Res Div Abbott Labs One Abbott Park Rd Abbott Park IL 60064

MCKENNA, CHARLES EDWARD, b Long Beach, Calif, May 9, 44; m 74. BIO-ORGANIC CHEMISTRY. *Educ:* Oakland Univ, BA, 66; Univ Calif, San Diego, PhD(chem), 71. *Prof Exp:* Res assoc chem, Univ Calif, San Diego, 71-72; NIH fel, Harvard Univ, 72-73; Nat Acad Sci exchange scholar, Bakh Inst, Moscow, 73; from asst prof to assoc prof, 73-88, PROF CHEM, UNIV SOUTHERN CALIF, 89- *Concurrent Pos:* Frasch awards comt, Am Chem Soc. *Mem:* Am Chem Soc; Am Soc Biochem & Molecular Biol; Int Soc Antiviral Res; Inst Genetic Med. *Res:* Mechanism of biological dinitrogen fixation; molybdoenzymes; biologically important organophosphorus compounds; anti-viral agents, including anti-HIV (AIDS) agents. *Mailing Add:* Dept of Chem Univ of Southern Calif Los Angeles CA 90089-0744

MCKENNA, EDWARD J, b Boston, Mass, Dec 30, 58. PHARMACOLOGY. *Educ:* Carnegie Mellon Univ, BS, 81; Univ RI, MA, 83; Univ Cincinnati, PhD(pharmacol), 89. *Prof Exp:* POSTDOCTORAL, HOWARD HUGHES MED INST, UNIV CALIF, LOS ANGELES, 89- *Mem:* Biophys Soc; NY Acad Sci; Am Soc Pharmacol & Exp Therapeut. *Mailing Add:* Dept Physiol Molecular Biol Inst Univ Calif Los Angeles CA 90024

MCKENNA, JACK F(ONTAINE), b White Plains, NY, Dec 24, 55; m 77; c 2. KINETICS. *Educ:* Clemson Univ, BS, 77, PhD(chem), 82. *Prof Exp:* Vis instr chem, Clemson Univ, 81-82; from asst prof to assoc prof, 82-90, PROF CHEM, ST CLOUD STATE UNIV, 90- *Mem:* Am Inst Chemists; Nat Sci Teachers Asn; Sigma Xi; Am Chem Soc. *Res:* Liquid and gaseous phase kinetics; thin films; polymers. *Mailing Add:* MS-363 St Cloud State Univ 720 Fourth Ave S St Cloud MN 56301-4498

MCKENNA, JAMES, b Canton, Ohio, June 18, 29; m 71; c 2. THEORY & APPLICATION OF QUEUEING NETWORKS, RESEARCH MANAGEMENT. *Educ:* Mass Inst Technol, BS, 51, MS, 54; Princeton Univ, PhD(math), 61. *Prof Exp:* Dept head, AT&T Bell Labs, 60-89; DIV MGR, BELLCORE, 89- *Concurrent Pos:* Chmn, bd trustees, Soc Indust & Appl Math, 89- *Mem:* Soc Indust & Appl Math; AAAS; Inst Elec & Electronics Engrs; Asn Comput Mach. *Res:* Mathematical physics; quantum field theory; optics; elasticity; statistical mechanics; queueing theory; performance analysis. *Mailing Add:* Bell Commun Res 445 South St Rm 2L-309 Box 1910 Morristown NJ 07960-1910

MCKENNA, JOHN DENNIS, b New York, NY, Apr 1, 40; m 64; c 2. RESEARCH ADMINISTRATION, TECHNICAL MANAGEMENT. *Educ:* Manhattan Col, Riverdale, NY, BS, 61; Newark Col, Riverdale, NY, MS, 68; Rider Col, Lawrenceville, NJ, MBA, 74; Walden Univ, PhD(mgt), 91. *Prof Exp:* Tech asst, Eldib Eng & Res, 64-67; proj leader, Princeton Chem Res, 67-68; proj dir, Res Cottrell, 68-72; vpres, Enviro-Systs & Res, 72-78; PRES, ETS, INC, 79- *Concurrent Pos:* Div chmn & tech coun, Air & Waste Mgt Tech Coun; bd mem, Va State Adv Bd Air Pollution. *Mem:* Air & Waste Mgt Asn; Am Inst Chem Engrs. *Res:* Air pollution control equipment; fabric filtration; flue gas desulfurization. *Mailing Add:* 1401 Municipal Rd Roanoke VA 24012

MCKENNA, JOHN MORGAN, b Providence, RI, Oct 11, 27; m 54; c 6. IMMUNOLOGY. *Educ:* Providence Col, BS, 50, MS, 52; Lehigh Univ, PhD(biol), 59; Am Bd Microbiol, dipl. *Prof Exp:* Res assoc, Sharp & Dohme Res Labs div, Merck & Co, Inc, 55-59; assoc, Harrison Dept Surg Res, Sch Med, Univ Pa, 59-66; from assoc prof to prof microbiol, Sch Med, Univ Mo, 66-72, res assoc, Space Sci Res Ctr, 66-72; chmn dept, 72-78, PROF MICROBIOL, SCH MED, TEX TECH UNIV, 72- *Mem:* AAAS; fel Am Acad Microbiol; Am Soc Microbiol; NY Acad Sci; Am Asn Immunologists; Sigma Xi. *Res:* Antibody formation; virology; tumor immunity. *Mailing Add:* Dept Microbiol PO Box 4569 Tex Tech Univ Sch Med Lubbock TX 79430

MCKENNA, MALCOLM CARNEGIE, b Pomona, Calif, July 21, 30; m 52; c 4. VERTEBRATE PALEONTOLOGY. *Educ:* Univ Calif, AB, 54, PhD(paleont), 58. *Prof Exp:* Instr paleont, Univ Calif, 58-59; from asst prof to assoc prof, 60-72, PROF GEOL SCI, COLUMBIA UNIV, 72-; FRICK CUR, DEPT VERT PALEONT, AM MUS NATURAL HIST, 68- *Concurrent Pos:* From asst cur to assoc cur, Dept Vert Paleont, Am Mus Natural Hist, 60-65, Frick assoc cur, 65-68. *Mem:* AAAS; Soc Vert Paleont; Am Soc Mammalogists; Soc Study Evolution; Geol Soc Am. *Res:* Evolution of the Mammalia during the late Mesozoic and Cenozoic eras; stratigraphic paleontology of Mesozoic and Cenozoic continental sediments; mammalian order Insectivora and its close allies; plate tectonics; biogeography. *Mailing Add:* Am Mus Natural Hist Cent Pk W 79th St New York NY 10024

MCKENNA, MARY CATHERINE, b Bethesda, Md, Dec 17, 45; m 74; c 1. NUTRITION, BIOSTATISTICS. *Educ:* Univ Md, BA, 68, PhD(nutrit biochem), 79. *Prof Exp:* Staff Fel, Nutrit Biochem Sect, Lab Nutrit & Endocrinol, NIH, 79-82, Guest Worker, Pediat Metab Br & Lab Cellular & Develop Biol, 82; AFFIL ASST PROF NUTRIT, UNIV MD, 83-, ASST PROF PEDIAT, SCH MED, 82- *Honors & Awards:* Res Excellence Award, Sigma Xi, 77. *Mem:* Am Inst Nutrit; Am Chem Soc; Sigma Xi; Am Soc Neurochemistry; AAAS; NY Acad Sci. *Res:* Nutrition and brain development; metabolism of ketone bodies, malate and lactate in individual brain cell types; metabolic trafficking in brain. *Mailing Add:* 13088 Williamfield Dr Ellicott City MD 21043

MCKENNA, MICHAEL JOSEPH, b Buffalo, NY, Oct 9, 46; m 69; c 2. TOXICOLOGY, PHARMACOLOGY. *Educ:* St John Fisher Col, BA, 68; Univ Rochester, PhD(toxicol), 75; Am Bd Toxicol, dipl. *Prof Exp:* Res mgr, Toxicol Res Lab, Health & Environ Res, Dow Chem USA, 75-88; assoc dir, Path Exp Toxicol Dept, Warner Lambert-Parke Davis, Ann Arbor, 88-89, VPRES, DRUG DEVELOP, PARKE-DAVIS PHARMACEUT RES, ANN ARBOR, 89- *Mem:* AAAS; Am Indust Hyg Asn; Sigma Xi; Soc Toxicol. *Res:* Inhalation toxicology, pharmacokinetics, drug toxicity. *Mailing Add:* Drug-Develop Warner Lambert-Parke Davis 2800 Plymouth Rd Ann Arbor MI 48105

MCKENNA, OLIVIA CLEVELAND, b Pittsburgh, Pa, Mar 6, 39; m 74. NEUROBIOLOGY, NEUROCYTOLOGY. *Educ:* Mt Holyoke Col, AB, 61; Boston Univ, PhD(physiol), 68. *Prof Exp:* From asst res scientist to assoc res scientist neurocytol, Med Ctr, NY Univ, 68-72, asst prof physiol, 72-74; lectr biol & res assoc, Columbia Univ, 74-75; from asst prof to assoc prof, 75-86, PROF BIOL, CITY COL NEW YORK, 87- *Concurrent Pos:* NIH fel, 68-70; adj asst prof physiol, Hunter Col, 74-75. *Mem:* Soc Neurosci; Am Asn Anatomists; NY Acad Sci. *Res:* Identification and cytological characterization of brain areas that process visual information necessary for stabilizing eye movements. *Mailing Add:* Dept of Biol City Col NY 138th St & Convent Ave New York NY 10031

MCKENNA, ROBERT WILSON, b Graceville, Minn, Oct 30, 40; m 64; c 4. PATHOLOGY. *Educ:* Col St Thomas, Minn, BS, 62; Univ Minn, MD, 66. *Prof Exp:* Med intern, Univ Calif, San Diego, 66-67; fel, 69-73, from instr to asst prof path, 73-77, ASSOC PROF PATH, UNIV MINN, MINNEAPOLIS, 77- *Mem:* Am Soc Hemat; Am Asn Cancer Res; Acad Clin Lab Physicians & Scientists; Am Soc Clin Path. *Res:* Morphology, cytochemistry immunology and ultrastructure of hematologic malignancies. *Mailing Add:* Dept Path Univ Tex Health Sci Ctr 5323 Harry Hines Blvd Dallas TX 75235-9072

MCKENNA, THOMAS MICHAEL, b Brooklyn, NY, July 26, 47. METABOLISM. *Educ:* Colo State Univ, BS, 70, MS, 72; Univ Okla, PhD(physiol), 79. *Prof Exp:* Teaching asst, Dept Zool & Entom, Colo State Univ, 70-72; res-teaching asst, Dept Zool, Univ Okla, 72-80, res assoc, 80-82; instr & fel, Endocrine-Metab Div, Med Ctr, Univ Rochester, 82-84; PRIN INVESTR, SEPTIC SHOCK RES PROG, NAVAL MED RES INST, BETHESDA, MD, 85- *Concurrent Pos:* Reviewer, Circulatory Shock, Am Rev Respiratory Dis, Am J Med Sci, J Clin Invest, Chest, Crit Care Med & Swiss Nat Sci Found; res scientist, Casualty Care Res Dept, Naval Med Res Inst. *Mem:* Am Physiol Soc; Shock Soc; AAAS. *Res:* Integrative and regulatory processes in normal and stressed physiological systems; endocrine, receptor and second messenger actions in pituitary; cardiovascular, immune and renal systems in normal and disordered states; author of numerous publications. *Mailing Add:* Septic Shock Res Prog Naval Med Res Inst MS 42 Bethesda MD 20814-5055

MCKENNA, WILLIAM GILLIES, b Kilmarnock, Scotland, Sept 18, 49; m 74. MEDICAL SCIENCES. *Educ:* Univ Edinburgh, BSc, 72; Albert Einstein Col Med, MD & PhD(cell biol), 81. *Prof Exp:* Intern med, 81-82, ASST RADIATION ONCOL, JOHNS HOPKINS HOSP, 82-, RES ASSOC, DEPT PHARMACOL & EXP THERAPEUTS, SCH MED, 82- *Res:* Use of monoclonal antibodies in cancer therapy. *Mailing Add:* Univ Pa Hosp 2114 Green St Philadelphia PA 19130

MCKENNEY, CHARLES LYNN, JR, b Riverdale, Md, Oct 20, 45; m 66; c 2. MARINE ENVIRONMENTAL PHYSIOLOGY, CRUSTACEAN REPRODUCTIVE BIOLOGY. *Educ:* Univ SC, BS, 70, MS, 73; Tex A&M Univ, PhD(zool), 79. *Prof Exp:* Teaching asst biol & zool, Dept Biol, Univ SC, 70-73; res assoc, Duke Univ Marine Lab, 73-75; teaching asst invert physiol, Dept Biol, Tex A&M Univ, 75-76, res fel, 76-79; RES AQUATIC BIOLOGIST, ENVIRON RES LAB, US ENVIRON PROTECTION AGENCY, GULF BREEZE, FLA, 79- *Concurrent Pos:* Fac assoc, Dept Biol, Univ WFla, 82-83, lectr, 89-90. *Mem:* Am Soc Zool; Crustacean Soc; Estuarine Res Fedn. *Res:* Marine environmental physiology; marine toxicology; physiological mechanism of pollutant toxicity; reproductive and larval biology of marine crustaceans; energy metabolism of marine organisms. *Mailing Add:* Environ Res Lab US Environ Protection Agency Gulf Breeze FL 32561

MCKENNEY, DEAN BRINTON, b Newton, Mass, Mar 1, 40; m 62; c 3. OPTICS. *Educ:* Bowdoin Col, AB, 62; Univ Rochester, MS, 65; Univ Ariz, PhD(optics), 69. *Prof Exp:* Res assoc optics, Univ Ariz, 67-69, asst prof, 69-71; pres, Helio Assocs, Inc, 71-80; PROJ MGR, HUGHES AIRCRAFT CO, 80-, SR SCIENTIST, 90- *Mem:* Optical Soc Am. *Res:* Physical and thin film optics; optical properties of solids. *Mailing Add:* 6101 N Camino Esquina Tucson AZ 85718

MCKENNEY, DONALD JOSEPH, b Eganville, Ont, May 3, 33; m 59; c 3. PHYSICAL CHEMISTRY. *Educ:* Univ Western Ont, BSc, 57, MSc, 58; Univ Ottawa, PhD(phys chem), 63. *Prof Exp:* Lectr gen chem, Royal Mil Col Can, 58-60 & Exten Div, Univ Ottawa, 62-63; Nat Res Coun Can fel, Cambridge Univ, 63-64; from asst prof to assoc prof gen & phys chem, 64-84, PROF PHYS CHEM, UNIV WINDSOR, 84- *Concurrent Pos:* Vis prof, Queen Mary Col, Univ London, 71-72; vis scientist, Nat Res Coun Can, Ottawa, 84, Commonwealth Sci & Indust Res Orgn, Canberra, Australia, 85. *Mem:* Chem Inst Can; Soil Sci Soc Am. *Res:* Chemical kinetics; sources and sinks of atmospheric trace gases; dentrification. *Mailing Add:* Dept Chem & Biochem Univ Windsor Windsor ON N9B 3P4 Can

MCKENNEY, KEITH HOLLIS, b Yankton, SDak. GENE EXPRESSION & REGULATION, MOLECULAR GENETICS. *Educ:* Am Univ, BS, 74; Johns Hopkins Univ, PhD(biochem), 82. *Prof Exp:* Staff fel res, Nat Inst Neurol & Commun Dis & Stroke, NIH, 84-86; RES CHEMIST RES, CTR ADVAN RES BIOTECHNOL, NAT INST STANDARDS & TECHNOL, GAITHERSBURG, MD, 87- *Concurrent Pos:* Adj prof, Sch Med & Health Sci, George Washington Univ, 85- & Grad Sch Arts & Sci, 87-; postdoctoral fel res, Med Res Coun Ctr, Cambridge, Eng. *Res:* Understanding protein and nucleic acid function at the molecular level to elucidate the basic mechanisms regulating two fundamental biologicl processes, transcription and replication. *Mailing Add:* Ctr Advan Res Biotechnol Nat Inst Standards & Technol 9600 Gudelsky Dr Rockville MD 20850

MACKENSIE, ALAN P, b Ilford, Eng, July 14, 32. CRYOBIOLOGY. *Educ:* Univ London, PhD(chem), 57. *Prof Exp:* RES ASSOC PROF, CTR BIOENG, UNIV WASH, 80- *Mailing Add:* Ctr Bioeng WD-12 Univ Wash Seattle WA 98195

MCKENTLY, ALEXANDRA H, b Wilmington, Del, Oct 17, 56. PLANT BIOTECHNOLOGY, SCIENCE COMMUNICATIONS & EDUCATION. *Educ:* Pa State Univ, BS, 78; Univ Fla, MS, 81, PhD(plant biotechnol), 89. *Prof Exp:* Hort mgr, 81-87, SR RES SCIENTIST & COMMUN SPECIALIST, LAND, EPCOT CTR, 88- *Concurrent Pos:* Consult, Kraft Gen Foods, Glenview, Ill, 88-; adj prof, Interdisciplinary Ctr Biotechnol Res, Univ Fla, Gainesville, 88- *Mem:* AAAS; Tissue Cult Asn; Inst Food Technologies; Am Soc Agron; Am Soc Hort Sci. *Res:* Application of biotechnology in plant science, specifically technologies of recombinant DNA, tissue culture, and plant regeneration; science education and communication; increasing the public's understanding of the science of biotechnology in agriculture and food development. *Mailing Add:* Land Epcot Ctr PO Box 10000 Lake Buena Vista FL 32830

MCKENZIE, ALLISTER ROY, b Moose Jaw, Sask. PHYTOPATHOLOGY. *Educ:* Alberta Univ, BSc, 59, MSc, 62; Univ Adelaide, PhD(fungal genetics), 67. *Prof Exp:* RES SCIENTIST, RES BR, AGR CAN, 66- *Mem:* Agr Inst Can; Can Phytopath Soc; Potato Asn Am. *Res:* Tuber-borne potato diseases; disease-indexed stem cuttings. *Mailing Add:* Plant Protection Div Agr Can Code W Neatby Bldg Ottawa ON K1A 0C6 Can

MACKENZIE, ANGUS FINLEY, b Elrose, Sask, Can Oct 2, 32; m 55; c 3. SOIL CHEMISTRY. *Educ:* Univ Sask, BSA, 54, MSc, 57; Cornell Univ, PhD(soil chem), 59. *Prof Exp:* Asst prof soil chem, Ont Agr Col, 59-62; from asst prof to assoc prof, 62-72, head dept, 66-77, PROF SOIL SCI, MACDONALD COL, MCGILL UNIV, 72- *Concurrent Pos:* Assoc dean res, MacDonald Col, McGill Univ, 80-85. *Mem:* Am Soc Agron; fel Can Soc Soil Sci; Int Soc Soil Sci. *Res:* Chemistry of the major plant nutrients in the soil; fertilizer technology; organic matter transformations. *Mailing Add:* Dept Renewable Resources Macdonald Col 21111 Lake Shore Rd Ste Anne de Bellevue PQ H9X 1C0 Can

MCKENZIE, BASIL EVERARD, b Jamaica, WI, Sept 14, 35; m 66; c 1. VETERINARY PATHOLOGY. *Educ:* Jamaica Sch Agr, dipl agr, 56; Tuskegee Univ, DVM, 66; Cornell Univ, MSc, 69; Univ Wis, PhD(vet path), 71. *Prof Exp:* Artificial insemination officer, Ministry of Agr & Lands, Jamaica, 56-60; res asst reprod path, Cornell Univ, 66-68; exp pathologist,

Lederle Labs, Pearl River, NY, 71-73; from asst prof to assoc prof path, Tuskegee Univ, 73-76; head path, Drug Safety Eval Div, Ortho Pharmaceut Corp, 76-77, dir, 77-90; EXEC DIR, DRUG SAFETY EVAL DIV WORLDWIDE, RWJ, PHARMACEUT RES INST, 90- *Concurrent Pos:* Adj prof toxicol, Philadelphia Col Pharm, 81-; expert toxicol & pharmacol, French Ministry Health. *Mem:* Am Vet Med Asn; Sigma Xi; Soc Toxicol Pathologists; Am Col Vet Pathologists; Int Acad Path. *Res:* Viral myocarditis; comparative cardiovascular pathology; pathology of the male and female reproductive system. *Mailing Add:* Drug Safety Eval Div RWJ Pharm Res Inst Raritan NJ 08869

MACKENZIE, CHARLES WESTLAKE, III, b New York, NY, Jan 25, 46; m 69; c 2. BIOCHEMISTRY. *Educ:* Univ Pac, BS, 68; Univ Southern Calif, PhD(biochem), 74. *Prof Exp:* Res chemist, Curtis Nuclear Corp, Los Angeles, 69-70; from teaching asst to res asst biochem, Univ Southern Calif, 71-74; fel pharmacol, Univ Minn, 74-76, res specialist med, 76-78; res assoc pharmacol, 78-79, instr, 79-81, ASST PROF PHARMACOL, UNIV NEBR MED CTR, 81- *Concurrent Pos:* Fel, USPHS, NIH, 75-77. *Res:* Mechanism of action and roles of cyclic nucleotides; biochemical control mechanisms in animal cells; mechanism of action of hormones. *Mailing Add:* DataChip Corp 5624 Pierce St Omaha NE 68106-1647

MACKENZIE, CORTLANDT JOHN GORDON, b Toronto, Ont, Sept 6, 20; m 45; c 3. ENVIRONMENTAL HEALTH. *Educ:* Queens Univ, Ont, MD, CM, 51; Univ Toronto, DPH, 56; FRCP(C), 61. *Prof Exp:* Dir health units, Peace River, 54-55, West Kootenay & Selkirk, 56-59 & Cent Vancouver Island, 59-63; from asst prof to assoc prof prev med, 63-71, prof & dept head, 71-81, prof, 82-86, EMER PROF HEALTH CARE & EPIDEMIOL, UNIV BC, 87- *Concurrent Pos:* Res fel, Univ BC, 61-62; mem main bd exam, Med Coun Can, 66, chmn prev med exam comt, 72-; vpres, Family Planning Fedn Can, 71-72; consult family planning & birth control, Dept Nat Health & Welfare, 71-72, mem med adv group health of immigrants, 72-; Can deleg, Int Planned Parenthood Western Hemisphere, 72; chmn adv comt pub health option, Brit Col Inst Technol; mem, Pollution Control Bd BC & Traffic Injury Res Found Can; chmn, Royal Comn Herbicides & Pesticides; mem test comt, Med Coun Can; vis prof, Univ Papua, New Guinea, 76 & 81. *Honors & Awards:* Defries Medal Award, Can Public Health Assoc, 85. *Mem:* Fel Royal Soc Health; Can Med Asn; Can Pub Health Asn; Can Asn Teachers Social & Prev Med (secy). *Res:* Tuberculosis survey effectiveness; pediatric hospitalization; birth control acceptance; health effects environmental, arsenic, mercury, brocroles, vibration white finger disease in forest workers and miners. *Mailing Add:* Dept Health Care & Epidemiol Univ BC Vancouver BC V6T 1W5 Can

MACKENZIE, COSMO GLENN, b Baltimore, Md, May 22, 07; m 36; c 2. NUTRITION. *Educ:* Johns Hopkins Univ, AB, 32. *Hon Degrees:* ScD, John Hopkins Univ, 36. *Prof Exp:* Asst prof biochem, Sch Hyg & Pub Health, Johns Hopkins Univ, 38-42; from asst prof to assoc prof, Med Col, Cornell Univ, 46-50; chmn dept, 50-73, prof, 50-75, EMER PROF BIOCHEM, SCH MED, UNIV COLO, 75- *Concurrent Pos:* USPHS fel, Johns Hopkins Univ, 36-38; assoc investr, Webb-Waring Lung Inst, 75-80. *Mem:* Am Soc Biol Chem; Soc Exp Biol & Med; fel NY Acad Sci; Am Inst Nutrit; fel AAAS. *Res:* Vitamin E; antioxidants; antithyroid action of thioureas and sulfonamides; thyroid-pituitary axis; biochemistry of one-carbon compounds; s-amino acids and enzymes; regulation of lipid metabolism, lipid accumulation and lipid-rich particles in cultured mammalian cells. *Mailing Add:* Dept Path Univ Colo Med Sch 4200 E Ninth Ave Box 216 Denver CO 80262

MACKENZIE, DAVID BRINDLEY, b Victoria, BC, May 1, 27; nat US; m 54; c 4. GEOLOGY. *Educ:* Calif Inst Technol, BS, 50; Princeton Univ, PhD(geol), 54. *Prof Exp:* Geologist, Am Overseas Petrol Ltd, 53-57; res geologist, 57-63, mgr geol res, 63-77, mgr regional explor, 77-80, mgr minerals explor, Denver Res Ctr, 80-83, mgr explor, North Sea, 83-86, CONSULT, MARATHON OIL CO, 86- *Mem:* Fel Geol Soc Am; Am Asn Petrol Geol. *Res:* Sedimentology; economic geology; petroleum geology. *Mailing Add:* 1000 Ridge Rd Littleton CO 80120

MCKENZIE, DAVID BRUCE, b Saskatoon, Sask, May 12, 54; m 81; c 1. FORAGE MANAGEMENT, SOIL PRODUCTIVITY. *Educ:* Univ Man, BSA, 77, MSc, 80; Tex A&M Univ, PhD(soil chem & plant nutrit), 84. *Prof Exp:* Area soil conservationist, 84-88, RES SCIENTIST, AGR CAN, 88- *Concurrent Pos:* Adj prof, Mem Univ Nfld, 89- *Mem:* Am Soc Agron; Soil Sci Soc Am; Crop Sci Soc Am; Sigma Xi; Can Soc Agron; Can Soc Soil Sci. *Res:* Development of forage management methods for increased crop quality and production; diversification of forage species in production; on-farm applied research across a range of soil and climate environments. *Mailing Add:* PO Box 7098 St John's NF A1E 3Y3 Can

MACKENZIE, DAVID ROBERT, b Beverly. Mass, Oct 19, 41; m 63; c 3. PLANT PATHOLOGY, PLANT BREEDING. *Educ:* Univ NH, BS, 64; Pa State Univ, MS, 67, PhD(plant path), 70. *Prof Exp:* Res asst plant path, Pa State Univ, 64-70; mem field staff, Rockefeller Found, 70-73; plant breeder, Asian Veg Res & Develop Ctr, 73-74; asst prof plant path, Pa State Univ, University Park, 74-; AT DEPT PLANT PATH & CROP PHYSIOL, LA STATE UNIV, BATON ROUGE. *Mem:* Am Phytopath Soc. *Res:* Plant breeding and pathology; basic epidemiological investigations through genetic modeling or pastforecasting. *Mailing Add:* Dept Plant Path & Crop Physiol La State Univ 302 Life Sci Bldg Baton Rouge LA 70803-1720

MCKENZIE, DONALD EDWARD, b Rivers, Man, Aug 9, 24; nat US citizen; m 50; c 3. PHYSICAL CHEMISTRY. *Educ:* Univ Man, BSc, 45, MSc, 47; Univ Southern Calif, PhD, 50. *Prof Exp:* Res officer, Atomic Energy Can, Ltd, Chalk River, Ont, 50-57; MGR CHEM TECHNOL, ENERGY SYSTS GROUP, ROCKWELL INT CORP, 57- *Mem:* Am Chem Soc; Sigma Xi; Air Pollution Control Asn. *Res:* High temperature chemistry; fused salts; electrochemistry; air and water pollution; management of research and development for coal gasification and liquefaction; chemical reactions in fused salts; high temperature batteries; disposal of combustible waste; gas scrubbing. *Mailing Add:* Rockwell Int Na 04 Rocketdyne Div 6633 Canoga Ave Canoga Park CA 91304

MACKENZIE, DONALD ROBERTSON, b Man, Can, Dec 9, 21; m 49; c 3. PHYSICAL CHEMISTRY. *Educ:* Queen's Univ, Ont, MA, 44; Univ Toronto, PhD, 50. *Prof Exp:* Asst res officer, Atomic Energy Can, Ltd, 50-58; chemist, Dept Appl Sci, Brookhaven Nat Lab, 58-78, chemist, Dept Nuclear Energy, 78-89, sr proj engr, Safety & Environ Protection Div, 89-91; RETIRED. *Concurrent Pos:* Consult nuclear waste mgt. *Mem:* Am Chem Soc; NY Acad Sci; Fedn Am Sci; AAAS. *Res:* Radiation, fluorine and nuclear chemistry; chemical kinetics of high temperature reactions in coal hydrogenation; nuclear waste management. *Mailing Add:* Four George Ct Bellport NY 11713

MACKENZIE, FREDERICK THEODORE, b Garwood, NJ, Mar 17, 34; m 87; c 2. GEOCHEMISTRY, SEDIMENTOLOGY. *Educ:* Upsala Col, BS, 55; Lehigh Univ, MS, 59, PhD(geol), 62. *Prof Exp:* Geologist, Shell Oil Co, 62-63; staff geochemist & mem corp, Bermuda Biol Sta Res, 63-65; res fel geol, Harvard Univ, 65; vis scholar, Northwestern Univ, Evanston, 65-66, asst prof geol, 67-69, from assoc prof to prof geol sci, 70-81, chmn dept, 70-77; PROF OCEANOG & HEAD MARINE GEOL & GEOCHEM, UNIV HAWAII, HONOLULU, 81- *Concurrent Pos:* Vis lectr, Lehigh Univ, 63-65; comt mem, Nat Acad Sci, 72-; vis scientist, Am Geophys Union, Upsala Col, 72; res fel, Univ Brussels, 74, Franqui int chair appl sci, 88-89; assoc ed, J Sedimentry Petrol, 77-80, Geochem, 79- & Geol, 89-; adj prof geol sci, Northwestern Univ, 81-85; Merrill W Haas distinguished prof geol, Univ Kans, 84; trustee, Bermuda Biol Sta Res, 89-; eminent scholars lectr, Univ S Fla, 89. *Honors & Awards:* W A Tarr Award Earth Sci, Am Geophys Union. *Mem:* Int Asn Cosmochem & Geochem; Am Asn Petrol Geologists; fel AAAS; Soc Econ Paleont & Mineral (vpres); fel Geol Soc Am; fel Mineral Soc Am; Sigma Xi; Int Asn Sedimentologists; Geochem Soc; Am Geophys Union. *Res:* Control of the chemical composition of seawater; history of earth's surface environment from a chemical and sedimentologic approach; chemical cycles of the elements and man's contributions; diagenesis of carbonates and clastics. *Mailing Add:* Dept Oceanog Sch Ocean Sci & Technol Univ Hawaii Honolulu HI 96822

MCKENZIE, GARRY DONALD, b Niagara Falls, Ont, June 8, 41; m 65; c 1. GLACIAL GEOLOGY. *Educ:* Univ Western Ont, BSc, 63, MSc, 64; Ohio State Univ, PhD(geol), 68. *Prof Exp:* Asst to dir, Inst Polar Studies, 68-69, exec officer dept, 69-72, asst prof, 69-75, ASSOC PROF GEOL, OHIO STATE UNIV, 72- *Concurrent Pos:* Glacial res, Alaska, Antarctica, Ontario, 65-; prog assoc, Polar Prog Div, NSF, 80-82; prog mgr, Ohio Bd Regents, 85- *Mem:* Geol Soc Am; Sigma Xi; Geol Asn Can; Am Geophys Union. *Res:* Quaternary stratigraphy, glacial geomorphology erosion and sedimentation in disturbed landscapes; environmental geology; Antarctic resources; metageology; cryospeleology. *Mailing Add:* Dept Geol Sci Ohio State Univ Columbus OH 43210-1398

MACKENZIE, GEORGE HENRY, b Bishop Auckland, Eng, Feb 11, 40; Can & British citizen; m 70; c 2. ACCELERATOR PHYSICS, APPLIED PHYSICS. *Educ:* Univ Birmingham, BSc, 61, PhD(physics), 65. *Prof Exp:* Res assoc physics, Cyclotron Lab, Mich State Univ, 65-68; res assoc physics, 68-75, res physicist, 75-83, SR RES SCIENTIST, TRI UNIV MESON FACIL, UNIV BC, 83- *Concurrent Pos:* Consult, Superconducting Cyclotron Proj, Chalk River, 75-76, Maria Proj, 80, Review Comt Cheer Proj, 81. *Mem:* Can Asn Physicists; Am Phys Soc. *Res:* Accelerator beam dynamics and diagnostics; accelerator physics including beam dynamics, beam instrumentation, machine commissioning and operation and general facility design. *Mailing Add:* Tri Univ Meson Facil 4004 Wesbrook Mall Vancouver BC V6T 2A3 Can

MCKENZIE, GERALD MALCOLM, PARKINSONS DISEASE, SCHIZOPHRENIA. *Educ:* Dalhousie Univ, PhD(physiol), 67. *Prof Exp:* ASSOC PROF PHARMACOL, DALHOUSIE UNIV, 75- *Res:* Pharmacology of the basal ganglia. *Mailing Add:* Dept Pharmacol Dalhousie Univ Sir Chas Tupper Bldg Halifax NS B3H 4H7 Can

MCKENZIE, INNES KEITH, b Stornoway, Scotland, Nov 16, 22; Can citizen; m 46; c 5. PHYSICS. *Educ:* Univ Western Ont, BSc, 48, MSc, 49; Univ BC, PhD(physics), 53. *Prof Exp:* Sci officer physics, Can Defence Res Bd, 53-60; from assoc prof to prof, Dalhousie Univ, 60-67; chmn dept, 67-70, prof, 67-88, EMER PROF PHYSICS, UNIV GUELPH, 88- *Concurrent Pos:* Res grants, 61-; mem, Can Assoc Comt Space Res, 61- *Res:* Positron annihilation; metal defects; Compton profiles; x-ray fluorescence. *Mailing Add:* Dept Physics Univ Guelph Guelph ON N1G 2W1 Can

MCKENZIE, JAMES MONTGOMERY, b Ashburton, NZ, Nov 10, 25; US citizen; m 51; c 4. RADIATION PHYSICS, ELECTRONICS. *Educ:* Univ Canterbury, BSc, 49, MSc, 51; Univ Otago, PhD(physics), 57. *Prof Exp:* Res physicist, Atomic Energy Can, 56-60; mgr, Amperex Electronic Corp, 60-63; tech dir, Harshaw Chem Co, 63-65; MEM TECH STAFF, BELL LABS, 65-; DISTINGUISHED MEM TECH STAFF, SANDIA LABS. *Mem:* Brit Inst Elec Engrs. *Res:* Use of unfolding codes with quantitative measurements to determine dependent parameters; differential neutron spectra by foil activation; location and source strength of spent-fuel from time-variant radiation measurements; surveillance and containment safeguards for nuclear fuel ultrasonic seals; radiation detection. *Mailing Add:* Orgn 5210 Sandia Labs, Bldg 821, Rm 1178 Albuquerque NM 87185

MACKENZIE, JAMES W, b Cleveland, Ohio, Oct 17, 25; m 50; c 1. THORACIC SURGERY. *Educ:* Univ Mich, BS, 48, MD, 51. *Prof Exp:* Resident surg, Univ Mich, 52-53, 55-58, resident thoracic surg, 58-60, instr surg, 60-62; from asst prof to prof, Univ Mo, 62-69; dean, 71-75, PROF SURG & CHMN DEPT, RUTGERS MED SCH, COL MED & DENT NJ, 69- *Concurrent Pos:* Chief sect thoracic & cardiovasc surg, Sch Med, Univ Mo, 62-69; consult, Ellis Fischel State Cancer Hosp, 64-69. *Mem:* Am Col Surg; Sigma Xi. *Res:* Cardiac surgery. *Mailing Add:* Dept Surg Univ Med & Dent NJ Robert Wood Johnson Med Sch CN19 New Brunswick NJ 08903

MCKENZIE, JESS MACK, b Woodsboro, Tex, Sept 2, 32; m 55; c 2. PHYSIOLOGY. *Educ:* NTex State Col, BA, 54, MA, 56; Univ Tex, PhD(physiol), 60. *Prof Exp:* Physiologist, Dept Space Med, US Air Force Sch Aerospace Med, 58-59; chief hemat sect, Physiol Labs, Civil Aeromed Inst, 59-85; chief, Progs Div, Dugway Proving Ground, 85-89; SCI SAFETY OFFICER, WEBER STATE UNIV, 89- *Concurrent Pos:* from adj asst prof zool to asst prof, Sch Med, Univ Okla, 59-74; adj prof zool, Univ Okla, 73- *Mem:* Am Physiol Soc; Soc Exp Biol & Med; Sigma Xi; Am Chem Soc. *Res:* Blood physiology; automated and computer-aided measurement of stress in large populations; effects of stress on health; biological-chemical-radiological safety. *Mailing Add:* 1588 Apache Way Ogden UT 84403-4408

MACKENZIE, JOHN D(OUGLAS), b Hong Kong, Feb 18, 26; nat US; m 54; c 3. MATERIALS SCIENCE, PHYSICAL CHEMISTRY. *Educ:* Univ London, BSc, 52, PhD(phys chem), 54; Royal Inst Chem, FRIC. *Prof Exp:* Asst chem, Princeton Univ, 54-56; Imperial Chem Industs fel, Cambridge Univ, 56-57; phys chemist, Res Labs, Gen Elec Co, 57-63; prof mat sci, Rensselaer Polytech Inst, 63-69; PROF MAT SCI, UNIV CALIF, LOS ANGELES, 69- *Concurrent Pos:* Lectr, Princeton Univ, 55-56; guest lectr, Nat Bur Standards, 59; ed, Physiochem Measurements at High Temperatures, 59; ed, Modern Aspects of the Vitreous State, 60, 62 & 64; ed-in-chief, J Non-Crystalline Solids. *Honors & Awards:* Lebeau Medal, High Temperature Soc France; Meyer Award, Am Ceramic Soc, 64, Toledo Award, 69; N F Mott Award 87; Sr Res Award, Am Soc Eng Educ, 88. *Mem:* Nat Acad Eng; Am Chem Soc; Electrochem Soc; Am Phys Soc; Am Soc Testing & Mat. *Res:* Structure of glasses and liquids; high pressure and high temperature physical chemistry; electronic ceramics; solid waste recycling; biomaterials; glass technology; sol-gel science. *Mailing Add:* Mat Sci Dept Sch Eng & Appl Sci Univ Calif Los Angeles CA 90024

MCKENZIE, JOHN MAXWELL, b Glasgow, Scotland, Nov 13, 27; c 4. INTERNAL MEDICINE, ENDOCRINOLOGY. *Educ:* Univ St Andrews, MB, ChB, 50, MD, 58. *Prof Exp:* House surgeon, Dundee Royal Infirmary, Scotland, 50-51; house physician, Therapeut Unit, Maryfield Hosp, 51; asst, dept pharmacol & therapeut, Univ St Andrews, 53-55; registr, Maryfield Hosp, 55-56; asst med, Sch Med, Tufts Univ, 56-57; registr, Maryfield Hosp, 57-58; instr med, Sch Med, Tufts Univ, 58-59; lect med, McGill Univ, 60-61, from asst prof to prof med & clin med, 61-81; PROF MED & CHMN DEPT, SCH MED, UNIV MIAMI, 81- *Concurrent Pos:* Res fel endocrinol, New Eng Ctr Hosp, 56-57, res assoc endocrinol, 58-59; clin asst, Royal Victoria Hosp, Montreal, Que, 59-63, from assoc physician to physician, 67-81; physician-in-chief, Jackson Mem Hosp, Miami, Fla, 81- *Honors & Awards:* Ayerst Award, Endocrine Soc, 61; Killam Award, Can Coun, 80; Park Davis distinguished lectr, Am Thyroid Asn, 81. *Mem:* AAAS; Endocrine Soc; Am Fedn Clin Res; Am Soc Clin Invest; Am Thyroid Asn (pres, 83-84). *Res:* Pathogenesis of Graves' disease, particularly the role of the thyroid stimulating antibody in that syndrome; investigations of mode of action of thyrotropin in the thyroid gland. *Mailing Add:* Dept Med Med Sch Univ Miami PO Box 016760 Miami FL 33101

MCKENZIE, JOHN WARD, b Dillon, SC, Sept 11, 18; m 49; c 1. HISTOLOGY, EMBRYOLOGY. *Educ:* The Citadel, BS, 40; Univ SC, MS, 50; Univ NC, PhD(zool), 54. *Prof Exp:* Asst gen biol, Univ SC, 49-50; asst gen zool, Univ NC, 51-54; from asst prof to assoc prof micros anat, 54-74, PROF ANAT, MED COL GA, 74- *Concurrent Pos:* Wilson scholar, Marine Biol Lab, Woods Hole, 52; mem staff, Eugene Talmadge Mem Hosp. *Mem:* AAAS; Electron Micros Soc Am. *Res:* Neurophysiology in primates. *Mailing Add:* Dept Anat 331 Heath Dr Augusta GA 30909

MCKENZIE, JOSEPH ADDISON, b Trinidad, WI, Nov 6, 30; Can citizen; m 57; c 2. ETHOLOGY, BIOCHEMICAL SYSTEMATICS. *Educ:* McMaster Univ, BA, 57; Univ Toronto, MA, 60; Univ Western Ont, PhD(ethology), 64. *Prof Exp:* Fisheries officer, Govt Trinidad & Tobago, 60-61; asst prof zool, Laurentian Univ, 64-67; asst prof zool, 67-69, assoc prof animal behav, 69-76, PROF ANIMAL BEHAV, UNIV NB, 76- *Concurrent Pos:* Nat Res Coun Can operating grants, 64- *Mem:* Can Soc Zoologists; Am Fisheries Soc. *Res:* Comparative behavior of stickleback family; biochemical systematics of fish; ichthyology. *Mailing Add:* Dept Biol Univ NB College Hill Rd Box 4400 Fredericton NB E3B 5A3 Can

MACKENZIE, KENNETH VICTOR, b Brandon, Man, Aug 29, 11; nat US; m 33, 68; c 4. PHYSICS, ACOUSTICS. *Educ:* Univ Wash, Seattle, BS, 34, MS, 36; Willamette Univ, DSc, 68. *Hon Degrees:* DSc, Willamette Univ, 83. *Prof Exp:* Physicist, Ore State Hwy Dept, 36-41; head physicist, Puget Sound Magnetic Degaussing Range, US Navy, 41-44; assoc physicist, Appl Physics Lab, Univ Wash, Seattle, 44-46; physics group leader deep & shallow water propagation, US Navy Electronics Lab, San Diego, 46-51, sect head, Oceanog Br, 51-55, head shallow water acoust processes sect, 55-61; exchange scientist from US Off Naval Res to Her Majesty's Underwater Weapons Estab, Eng, 61-62; head deep submergence group, Naval Undersea Ctr, 62-67, supvry physicist, Acoust Propagation Div, Ocean Sci Dept, 67-73; sr staff physicist & primary adv sci & eng directorate, US Naval Oceanog Off, Washington, DC, 73-76; sr staff physicist, Naval Ocean Res & Develop Activ, 76-79; PRES, MACKENZIE MARINE SCI CONSULT, SAN DIEGO, CALIF, 79- *Concurrent Pos:* Consult allied govts & defense indust; hon comt mem, Int Ocean Develop Confs, Tokyo, 71-; consult & lectr, Japanese Oceanog Inst, 62-, Chinese Oceanog Inst & Univ, 80-,. *Honors & Awards:* Cent Merit, Off Sci Res, 45; Navy Electronics Lab Award, 60. *Mem:* Fel Acoust Soc Am; Am Geophys Union; fel Marine Technol Soc; Sigma Xi; Optical Soc Am; Am Phys Soc; Instrument Soc; Int Navigation; Navy League. *Res:* Deep dives aboard manned deep sub vehicles; bathyscaphs Trieste I and II; deepstar-4000, Alvin Aluminaut, USN smallest and deepest nuclear research submarine NR-1; author of over 185 scientific papers. *Mailing Add:* Marine Sci Consults Midway Sta PO Box 80715 San Diego CA 92138

MCKENZIE, MALCOLM ARTHUR, b Providence, RI, Apr 21, 03. FOREST PATHOLOGY. *Educ:* Brown Univ, PhB & MA, 26, PhD(forest path), 35. *Prof Exp:* Field asst forest path, Forest Prod Lab, US Forest Serv, 26-27; instr biol, Univ NC, 27-29; agent forest path, Bur Plant Indust, USDA, 29-35; pathologist shade tree dis, 35-36, from asst res prof to res prof bot, 36-50, prof plant path & dir shade tree labs, 50-73, EMER PROF PLANT PATH, UNIV MASS, AMHERST, 73-; CONSULT SHADE TREE MGT, 73- *Concurrent Pos:* Lectr, Brown Univ, 29-35; actg head dept entom & plant path, Univ Mass, Amherst, 65-68. *Honors & Awards:* Hon life award, Int Shade Tree Conf, 75; Award of Merit, 85. *Mem:* Am Phytopath Soc; Mycol Soc Am; Soc Indust Microbiol; Int Soc Arboriculture. *Res:* Tree pests; Dutch Elm disease; wood decay; tree hazards in public utility work; municipal tree maintenance; continuing education programs in tree workshops and environmental pollution control; urban forestry. *Mailing Add:* PO Box 651 North Amherst MA 01059

MACKENZIE, MALCOLM R, b Oakland, Calif, Jan 15, 35; m 59; c 3. INTERNAL MEDICINE, IMMUNOLOGY. *Educ:* Univ Calif, Berkeley, AB, 56; Univ Calif, San Francisco, MD, 59. *Prof Exp:* Intern & resident, Univ Calif Hosps, San Francisco, 59-62; asst res physician, Div Hemat, Univ Calif, 66-67; asst prof med, Univ Calif, San Francisco, 67-68; asst prof, Univ Cincinnati, 68-70; assoc prof, 70-76, PROF MED, SCH MED, UNIV CALIF, DAVIS, 76- *Concurrent Pos:* USPHS fel physiol chem, Univ Wis, 62-64; Am Cancer Soc fel, Div Hemat, Univ Calif, 65-66; Am Cancer Soc scholar, 66-68. *Mem:* Am Fedn Clin Res; Am Rheumatism Asn; Am Soc Hemat; Am Asn Immunol; Am Soc Clin Oncol. *Res:* Immuno deficiency diseases; origin of human lymphocyte malignancies; human multiple myeloma. *Mailing Add:* Dept Internal Med Div Hemat & Oncol Univ Calif Sch Med 4301 X St Sacramento CA 95817

MCKENZIE, RALPH NELSON, b Cisco, Tex, Oct 20, 41; m 61; c 2. MATHEMATICS. *Educ:* Univ Colo, Boulder, BA, 63, PhD(math), 66. *Prof Exp:* From asst prof to assoc prof, 67-77, PROF MATH, UNIV CALIF, BERKELEY, 77- *Concurrent Pos:* NSF fel, Inst Advan Study, 71-72; Alfred P Sloan fel, 72-74; vis res mathematician ETH, Zurich, 78; vis prof Univ of Hawaii, 83 & Siena, Italy, 84. *Mem:* Am Math Soc. *Res:* Direct products of relational systems; equational varieties of algebras; algorithmicity in algebra. *Mailing Add:* Dept Math Univ Calif Berkeley CA 94720

MACKENZIE, RICHARD STANLEY, b Detroit, Mich, Dec 28, 33; m 57; c 2. DENTISTRY. *Educ:* Univ Mich, DDS, 58, MS, 60; Univ Pittsburgh, PhD(psychol), 65. *Prof Exp:* Assoc prof dent behav sci & head dept, Univ Pittsburgh, 65-68, dir educ res, Sch Dent, 66-69, prof higher educ, Sch Dent Med, 68-69; dir, Off Dent Educ, 69-74, PROF DENT EDUC, COL DENT & COL EDUC, UNIV FLA, 69-, CHMN DEPT, 75- *Concurrent Pos:* Mem USPHS adv comt, Dent Health Res & Educ, 68-72; mem oral biol training comt, Vet Admin Hosps, Washington, DC, 69-72, mem res serv merit rev bd, 72-75; USPHS career develop award, 70-75; consult, WHO-Pan Am Health Orgn, 71-76 & Am Dent Asn, 72-; chmn adv comt, Educ Testing Serv, 73-78; deleg, Coun Fac, Am Asn Dental Sch, 78-88, officer, Educ Res Develop & Curric, 82-86, officer, Coun Fac, 86-90; mem, Spec Rev Comt, Oral Health Behav Res Ctrs, Nat Inst Dent Res, 84-85. *Mem:* AAAS; Am Dent Asn; Am Educ Res Asn; Sigma Xi. *Res:* Analysis and evaluation of clinical judgment. *Mailing Add:* 3715 SE 37th St Gainesville FL 32601

MACKENZIE, ROBERT DOUGLAS, b Chicago, Ill, Aug 18, 28; m 52; c 4. BIOCHEMISTRY. *Educ:* Univ Cincinnati, BS, 52; Mich State Univ, MS, 54, PhD(biochem), 57. *Prof Exp:* Asst, Mich State Univ, 53-57, res assoc, 57; res biochemist, Merell-Nat Labs Div, Richards-Merrell, Inc, 57-63, head hemat sect, pharmacol dept, 63-76, sect head, drug metab dept, 76-81, MGR, RES SAFETY, MERRELL DOW PHARMACEUT INC, DOW CHEM CO, 81- *Concurrent Pos:* Adj assoc prof, Univ Cincinnati, 67-69; mem coun thrombosis, Am Heart Asn. *Mem:* AAAS; Am Chem Soc; Soc Exp Biol & Med; Am Soc Pharmacol & Exp Therapeut; Sigma Xi; Am Heart Asn. *Res:* Animal and lipid metabolism; blood coagulation; radiobiochemistry; toxicology; nutrition. *Mailing Add:* 5532 Morrow-Blackhawk Rd Morrow OH 45152

MACKENZIE, ROBERT EARL, b Calif, Mar 17, 20; m 50; c 4. MATHEMATICS. *Educ:* Calif Inst Technol, BS, 42; Princeton Univ, MA, 48, PhD(math), 50. *Prof Exp:* Physicist, US Naval Ord Lab, 42-45; from instr to assoc prof math, Ind Univ, Bloomington, 50-90, asst chmn dept, 62-67; RETIRED. *Concurrent Pos:* Ed, J Math & Mech, 56-62 & 71-76. *Mem:* Am Math Soc. *Res:* Modern algebra and algebraic number theory. *Mailing Add:* 6695 E State Rd 46 Bloomington IN 47401

MACKENZIE, ROBERT EARL, b Oct 12, 41; m; c 2. TRANSFORMED CELLS, FOLATE. *Educ:* Cornell Univ, PhD(biochem), 69. *Prof Exp:* From asst prof to assoc prof, 71-82, PROF BIOCHEM, MCGILL UNIV, 83- *Mem:* Am Soc Biochem & Molecular Biol; Can Biochem Soc; Protein Soc. *Mailing Add:* Dept Biochem McGill Univ 3655 Drummond St Rm 816 Montreal PQ H3G 1Y6 Can

MCKENZIE, ROBERT LAWRENCE, US citizen. LASER SPECTROSCOPY, FLUID MECHANICS. *Educ:* Univ Cincinnati, BS, 59; Stanford Univ, MS, 67; York Univ, PhD(physics), 76. *Prof Exp:* RES SCIENTIST, AMES RES CTR, NASA, 59- *Concurrent Pos:* Consult laser applications & fluid mech. *Honors & Awards:* H J Allen Award, NASA, 74. *Mem:* Am Phys Soc; Am Inst Aeronaut & Astronaut. *Res:* Diagnostic applications to turbulent flows. *Mailing Add:* 825 Cathedral Dr Sunnyvale CA 94087

MCKENZIE, RONALD IAN HECTOR, b Saskatoon, Sask, Oct 8, 30; m 51; c 4. PLANT BREEDING, PLANT GENETICS. *Educ:* Univ Sask, BSA, 51, MSc, 54; Univ Minn, PhD(plant genetics), 57. *Prof Exp:* RES SCIENTIST PLANT BREEDING, AGR CAN RES STA, RES BR, 56- *Concurrent Pos:* Hon prof plant sci, Univ Man, 65- *Mem:* Agr Inst Can; Genetics Soc Can; Am Soc Agron. *Res:* Development of high yielding, semi-dwarf, disease resistant wheat of medium quality for Manitoba. *Mailing Add:* 817 KilkennyCan 817 Kilkenny Dr Winnipeg MB R3T 4Y5 Can

MACKENZIE, SCOTT, JR, b Sedalia, Mo, Mar 10, 20; m 47; c 3. CHEMISTRY. *Educ:* Univ Pa, BS, 42; Univ Ill, MS, 44, PhD(org chem), 47. *Prof Exp:* Res chemist, E I du Pont de Nemours & Co, Va, 44-46; fel, Gen Mills Co, Univ Minn, 47-48; instr chem, Columbia Univ, 48-51; from asst prof to assoc prof, 51-66, PROF CHEM, UNIV RI, 66- *Mem:* Am Chem Soc. *Res:* Organic chemistry; carbanions; ultraviolet spectroscopy. *Mailing Add:* 35 Little Rest Rd Kingston RI 02881

MCKENZIE, WENDELL HERBERT, b Wykoff, Minn, Nov 23, 42; m 64; c 2. HUMAN GENETICS. *Educ:* Westmar Col, BA, 64; NC State Univ, MS, 69, PhD(genetics), 73. *Prof Exp:* Teacher, Pub Schs, Iowa, 64-67; res asst genetics, NC State Univ, 68-69, univ instr human genetics, 69-71; res cytogeneticist, Univ Colo Med Ctr, Denver, 72; univ instr human genetics, 73, asst prof, 73-77, assoc prof human genetics, 77-86, PROF GENETICS, 86-, ALUMNI DISTINGUISHED PROF, NC STATE UNIV, 87- *Concurrent Pos:* travel grant, Genetics Soc Am, 78. *Mem:* Genetics Soc Am; Am Soc Human Genetics; Am Inst Biol Sci; AAAS; Sigma Xi. *Res:* Human cytogenetics, including chromosome structure and variation, methodologies of chromosome banding techniques and mutagenic effects of environmental agents as pollutants. *Mailing Add:* Dept Genetics Box 7614 NC State Univ Raleigh NC 27695-7614

MCKENZIE, WILLIAM F, b Dallas, TX, Aug 17, 41; m 81. THEORETICAL GEOCHEMISTRY, STABLE ISOTOPE GEOCHEMISTRY. *Educ:* Univ Okla, BS, 63; Colo Sch Mines, MS, 70; Univ Calif, Berkeley, MA, 78, PhD(geol), 80. *Prof Exp:* Res chemist, Pan Am Petrol Corp Res Ctr, 63-64, Marathon Oil Co Res Ctr, 64-67; instr, Colo Sch Mines, 71-72; vis chemist res, Lab Nuclear Geol, 73-74; geochemist res, US Geol Surv, 74-76; geologist explor, Chevron Resources, 80-84; lectr, Univ Calif, Berkeley, 84-85; RES CHEMIST, LAWRENCE LIVERMORE NAT LAB, 85- *Mem:* Fel Geol Soc Am; Am geophys union; fel Soc Econ Geologists; Mineral Soc Am; Int Asn Geochem & Cosmochem; Geochem Soc. *Res:* Water-rock interactions as related to weathering diagenesis, metamorphism and ore deposition. *Mailing Add:* Lawrence Livermore Lab PO Box 808 Livermore CA 94550

MCKEON, CATHERINE, b Babylon, NY, March 23, 53; m 84. GENE REGULATION. *Educ:* State Univ NY, Buffalo, BA, 75; Med Col Va, PhD(human genetics), 80. *Prof Exp:* fel, Lab Molecular Biol, 80-84, SR STAFF FEL, PEDIAT BR, NAT CANCER INST, NIH, 84- *Concurrent Pos:* Fel, NIH, 80-84. *Mem:* Am Soc Human Genetics; Sigma Xi. *Res:* Differentiation of tissues arising from the embryonic neural crest; determining alterations in gene expression in normal tissues and tumor tissues of neural crest origin. *Mailing Add:* Lab Molecular Biology Rm 2D-27 NIH Nat Cancer Inst Bldg 37 Bethesda MD 20205

MCKEON, JAMES EDWARD, b Derby, Conn, June 25, 30; m 52; c 4. ORGANIC CHEMISTRY. *Educ:* Wesleyan Univ, BA, 51, MA, 53; Yale Univ, PhD(org chem), 60. *Prof Exp:* Res chemist, Chem Div, 59-63, res scientist & group leader, 63-69, sr res scientist, 69-73, assoc dir res, 73-77, dir res, Chem & Plastics Div, 77-84, VPRES TECHNOL, UNION CARBIDE CORP, 84- *Mem:* Am Chem Soc; Sigma Xi; Ind Res Inst. *Res:* Reactions of molecules coordinated with metals; homogeneous catalysis; oxidation processes. *Mailing Add:* 4 Carriage Lane New Fairfield CT 06812

MCKEON, MARY GERTRUDE, b New Haven, Conn, June 8, 26. ELECTROCHEMISTRY. *Educ:* Albertus Magnus Col, BA, 47; Yale Univ, MS, 52, PhD(chem), 53. *Prof Exp:* Lab asst, Yale Univ, 49-52; from instr to assoc prof chem, Conn Col, 52-71, dean sophomores, 63-69, chmn dept, 72-79, Margaret W Kelly prof, 74-77, PROF CHEM, CONN COL, 71- *Concurrent Pos:* NSF sci fac fel, Harvard Univ, 59-60; vis assoc prof, Wesleyan Univ, 70-71. *Mem:* Am Chem Soc; Sigma Xi. *Res:* Synthetic organic; organic polarography; electroanalytical chemistry. *Mailing Add:* Dept Chem Conn Col New London CT 06320

MCKEOWN, JAMES JOHN, b Albert Lea, Minn, Oct 29, 30; m 58; c 3. PHYSICAL CHEMISTRY. *Educ:* St John's Univ, Minn, BS, 53; Iowa State Univ, PhD(chem), 58. *Prof Exp:* Asst chem, Iowa State Univ, 58; res chemist, Procter & Gamble Co, 58-60; sr chemist, 60-64, supvr, 64-66, mgr appl res, 66-69, tech dir dielec mat & systs lab, 69-73, tech dir Electric Prods Lab, 73-83, vpres res & develop, Indust & Com Sector, 83-84, VPRES INDUST SPECIALITIES DIV, MINN MINING & MFG CO, 84- *Mem:* Sigma Xi. *Res:* Ceramics; electronics; engineering management; materials science; microelectronics; polymers. *Mailing Add:* 3M Ctr Bldg 220-7E-01 St Paul MN 55144-1000

MCKEOWN, JAMES PRESTON, b Vicksburg, Miss, Mar 2, 37; m 62; c 1. ETHOLOGY. *Educ:* Univ of the South, BS, 59; Univ Miss, MS, 62; Miss State Univ, PhD, 68. *Prof Exp:* From instr to assoc prof, 62-75, PROF BIOL & CHMN DEPT, MILLSAPS 75- *Mem:* AAAS; Am Soc Ichthyol & Herpet. *Res:* Celestial navigation by amphibians; experimental embryology. *Mailing Add:* Dept of Biol Millsaps Col Jackson MS 39210

MCKEOWN-LONGO, PAULA JEAN, EXTRACELLULAR MATRIX ASSEMBLY, CELL BIOLOGY. *Educ:* Univ Conn, PhD(cell biol), 81. *Prof Exp:* RES ASST PROF CELL BIOL, STATE UNIV NY, ALBANY, 84- *Mailing Add:* Dept Biol State Univ NY 47 New Scotland Ave Albany NY 12208

MACKERELL, ALEXANDER DONALD, JR, b Chester, Pa, Mar 22, 59. BIOCHEMISTRY. *Educ:* Gloucester County Col, AS, 79; Univ Hawaii, BS, 81; Rutgers Univ, PhD(biochem), 85. *Prof Exp:* Lab asst, Hawaii Inst Geophys, 79-81; biochem res asst, Rutgers Univ, 81-85, postdoctoral fel, Ctr Alcohol Studies, 85; postdoctoral fel, Dept Med Biophys, Karolinska Inst, 86-88; postdoctoral fel, 88-91, RES ASSOC, DEPT CHEM, HARVARD UNIV, 91- *Concurrent Pos:* Consult, Polygen Corp, 90- *Mem:* Am Soc Biochem & Molecular Biol. *Res:* Biological molecules via empirical energy functions along with the interpretation of biochemical experimental results via theoretical approaches. *Mailing Add:* Dept Chem Harvard Univ 12 Oxford St Box 198 Cambridge MA 02138

MACKERER, CARL ROBERT, b Jersey City, NJ, Sept 2, 40; m 66; c 3. BIOCHEMICAL PHARMACOLOGY & TOXICOLOGY. *Educ:* Rutgers Univ, BA, 63; Univ Nebr, PhD(med biochem), 71. *Prof Exp:* Cardiovasc pharmacologist, Hoffman-La Roche, Inc, 63-67; res investr, G D Searle, Inc, 71-73, sr res investr, 73-75, res scientist, 75-77, sr res scientist, 78; MGR BIOCHEM TOXICOL, MOBIL OIL CORP, 78- *Concurrent Pos:* Instr biochem, Sch Med, Univ Nebr, 71; assoc ed, Toxicol & Indust Health, 84- *Mem:* Am Col Toxicol; Am Soc Pharmacol & Exp Therapeut; Biochem Soc; Am Chem Soc. *Res:* Management of research in analytical chemistry, genetic toxicology, pharmacokinetics, clinical chemistry and pharmacology; determination and explanation of toxic effects of pure chemicals and mineral oils. *Mailing Add:* Five Blue Spruce Dr Pennington NJ 08534-2110

MCKERNS, KENNETH (WILSHIRE), b Hong Kong, Mar 5, 19; nat US; m 43; c 2. BIOCHEMISTRY. *Educ:* Univ Alta, BSc, 42, MSc, 46; McGill Univ, PhD(biochem), 51. *Prof Exp:* Demonstr, Univ Alta, 40-42; sr demonstr & lectr, McGill Univ, 46-51; chief biochemist, Can Packers, Ltd, Toronto, 51-55; lectr, Univ St Andrews, 55-56; sr res scientist & group leader, Lederle Labs Div, Am Cyanamid Co, NY, 56-60; from assoc prof to prof obstet & gynec, Col Med, Univ Fla, 60-78; PRES, INT FOUND BIOCHEM ENDOCRINOL, 78 -; PRES & CHIEF EXEC OFFICER, BIOMOL, INC, 81 -; PRES AGRO-STIM, INC, 85 - *Concurrent Pos:* Fel, McGill-Montreal Gen Hosp Res Inst, 50-51; NIH spec res fel, 69-70; vis lectr, Harvard Med Sch, 69-70; consult, AID, 81 - *Mem:* AAAS; Am Soc Biol Chem. *Res:* Endocrine regulation of intermediate metabolism; protein hormone action and purification; effect of hormones on growth, metabolism and disease states; biologically active peptides, synthesis and mechanism action; growth stimulation farm animals. *Mailing Add:* Bayside Rd Ellsworth ME 04605

MCKETTA, JOHN J, JR, b Wyano, Pa, Oct 17, 15; m; c 4. CHEMICAL ENGINEERING. *Educ:* Tri-State Col, BS, 37; Univ Mich, BSE, 43, MS, 44, PhD(chem eng), 46. *Hon Degrees:* DrEng, Tri-State Col, 67, Drexel Univ, 77; DrSc, Univ Toledo, 73. *Prof Exp:* Group leader, Tech Dept, Wyandotte Chems Corp, Mich, 37-40, asst supt, Caustic Soda Div, 40-41; chem dir, C B Schneible Co, 41-42; instr chem eng, Univ Mich, 44-45; from asst prof to prof, Univ Tex, 46-52; ed dir, Gulf Pub Co, 52-54; grad prof chem eng, 54-63, chmn dept chem eng, 50-52, 55-63, dean col eng, 63-69, E P Schoch prof chem eng, 70-82, JOE C WALTER PROF CHEM ENG, UNIV TEX, AUSTIN, 82- *Concurrent Pos:* Asst dir, Tex Petrol Res Comt, 51-52, 54-55 & 58-60; mem bd dirs, Vulcan Mat Co, 66-85 & Eng Joint Coun, 67-74; exec vchancellor acad affairs, Univ Tex Syst, 69-70; mem bd dirs, Houston Oil & Mineral Corp, 72-80, Marley Corp, 77-82, Big Three Corp, 77-80, Howell Corp, 78-, Dresser Indust, 80-87 & Tesoro Petrol, 78- *Honors & Awards:* Warren K Lewis Excellence Chem Eng, Am Inst Chem Engrs, 69, Fuels & Petrochem Div Award, 83, F J Van Atwerpen Award Outstanding Contrib Field of Chem Eng, 85; Charles M Schwab Mem Award, Am Iron & Steel Inst, 73; Triple E Award, Environ Develop Assn, 76; Boris Pregel Award in Sci & Technol, NY Acad Sci, 78; J Donald Lindsay Lectr, Tex A&M, 82; Herbert Hoover Award, 89. *Mem:* Nat Acad Eng; fel Am Inst Chem Engrs (pres, 62); hon fel Soc Tech Commun. *Res:* Solubility of hydrocarbon systems at high pressure; vapor-liquid-liquid equilibrium in hydrocarbon-water systems; energy supply and demand. *Mailing Add:* Dept Chem Eng Univ Tex Austin TX 78712

MACKEVETT, EDWARD M, b New York, NY, Sept 29, 18. MINERAL DEPOSITS. *Educ:* Univ Calif, Los Angeles, BA, 47; Calif Technol, MS, 50. *Prof Exp:* Staff, US Geol Survy, Mineral Deposits Br, Calif, 47-55, Alaskan Geol Br, 56-79; Consult, 80-86; RETIRED. *Mailing Add:* 927 Lobelia Lane San Luis Obispo CA 93401

MACKEY, BRUCE ERNEST, b Akron, Ohio, Feb 9, 39; m 61; c 2. BIOMETRICS, PLANT BREEDING. *Educ:* Univ Akron, BS, 61; Cornell Univ, MS, 64, PhD(plant breeding), 66. *Prof Exp:* Statistician, Biomet Serv, Md, 66-69, biometrician, Biomet Serv, Calif, 69-71, BIOMETRICIAN, WESTERN REGIONAL RES LAB, AGR RES SERV, USDA, 71- *Mem:* Am Stat Asn. *Res:* Applications of quantitative genetics to plant breeding research; design and analysis of agricultural research data. *Mailing Add:* Western Regional Res Lab 800 Buchanan St Berkeley CA 94710

MACKEY, GEORGE W, b St Louis, Mo, Feb 1, 16; m 60; c 1. MATHEMATICS. *Educ:* Univ Rice, BA, 39; Harvard Univ, AM, 39, PhD, 42; Oxford Univ, MA, 66. *Prof Exp:* Instr math, Ill Inst Technol, 42-43; fac instr math, Harvard Univ, 43-46, from asst prof to prof, 46-56, chmn dept, 63-66, Landon T Clay prof, 69-85, LANDON T CLAY EMER PROF MATH & THEORET SCI, HARVARD UNIV, 85- *Concurrent Pos:* Fel John Simon Guggenheim Found, France, 49-50, 60-61 & 70-71, Italy & Ger, 54, Eng & Russia, 61-62; George Eastman vis prof, Oxford Univ, 66-67; prof assoc, Univ Paris, 78; consult, Sch Physic Theory, 70-; vis prof, Inst Des Hautes Etudes Scientifigu Iltes, Bur Sup Yvette, France, Tata Inst, Bumbay, India, 70-71, Max Planck Inst, Bunn, Ger, 85-86. *Honors & Awards:* Steele Prize, Am Math Soc. *Mem:* Nat Acad Sci; AAAS; Am Math Soc (vpres, 64-65); Am Philos Soc. *Res:* Abstract analysis; infinite dimensional representations of locally compact groups and applications to quantum mechanics and other branches of mathematics. *Mailing Add:* Dept Math Harvard Univ One Oxford St Cambridge MA 02138

MACKEY, HENRY JAMES, b Vicksburg, Miss, Nov 25, 35; m 59; c 4. SOLID STATE PHYSICS. *Educ:* La State Univ, BS, 57, MS, 59, PhD(physics), 63. *Prof Exp:* Res assoc physics, La State Univ, 63-64; from asst prof to assoc prof, 64-69, PROF PHYSICS, NTEX STATE UNIV, 69- *Mem:* Am Phys Soc. *Res:* Electron, phonon transport phenomena in metals; Fermi surface mappings; surface scattering contribution to electrical resistivity. *Mailing Add:* Dept of Physics NTex State Univ Box 5368 Denton TX 76203

MACKEY, JAMES E, b Tupelo, Miss, Feb 4, 40; m 65; c 4. ENERGY CONSERVATION. *Educ:* Tulane Univ, BS, 62; Univ Miss, MS, 65, PhD(physics), 69. *Prof Exp:* From asst prof to assoc prof, 68-78, PROF PHYSICS, HARDING UNIV, 78- *Concurrent Pos:* Physicist, Westinghouse-Hanford Co, Richland, Wash, 81. *Mem:* Am Asn Physics Teachers. *Res:* Residential energy conservation; image restoration and enhancement. *Mailing Add:* Dept Physics Harding Univ Box 582 Searcy AR 72143

MACKEY, JAMES P, b Akron, Ohio, Feb 28, 30; m 50; c 1. BIOLOGY. *Educ:* Univ Akron, BS, 51; Ohio State Univ, MS, 54; Univ Ore, PhD(biol), 57. *Prof Exp:* From asst prof to assoc prof, 57-74, PROF BIOL, SAN FRANCISCO STATE UNIV, 74- *Mem:* Ecol Soc Am; Soc Study Evolution; Am Soc Ichthyol & Herpet. *Res:* Vertebrate ecology; herpetology; intraspecific variation of tree frogs. *Mailing Add:* Dept Biol San Francisco State Univ 1600 Holloway Ave San Francisco CA 94132

MACKEY, MICHAEL CHARLES, b Kansas City, Kans, Nov 16, 42; m 77; c 5. APPLIED MATHEMATICS. *Educ:* Univ Kans, BA, 63; Univ Wash, PhD(physiol), 68. *Prof Exp:* Res assoc electrocardiol, Sch Med, Univ Okla, 63-64; biophysicist, Phys Sci Lab, Div Comput Res & Technol, NIH, 69-71; from asst prof to assoc prof, 71-81, PROF PHYSIOL, MCGILL UNIV, 81-, DIR, CTR NONLINEAR DYNAMICS, 88-, PROF PHYSICS, 89-, PROF MATH, 90- *Concurrent Pos:* Vis res prof, Silesian Univ, Katowice, Poland, 82-, Univ Marie-Curie, Lublin, Poland, 88; res dir, E N Huyck Biol Preserve, Rensselaerville, NY, 84; vis prof, Oxford Univ, 86, Univ Bremen, Ger 87. *Mem:* Am Math Soc; Cell Kinetic Soc; Soc Indust & Appl Math; Soc Math Biol; Can Soc Theoret Biol (pres, 85-87); Math Asn Am; Am Phys Soc. *Res:* Mechanisms involved in the generation of periodic and chaotic behavior in biological systems, and mathematical models of these systems; neural networks; control of the cell cycle; genetic evolution; tissue growth and homeostasis; biochemical oscillators; alternate formulations of quantum mechanics; foundations of thermodynamics. *Mailing Add:* Rm 1124 Dept Physiol McGill Univ 3655 Drummond Montreal PQ H3G 1Y6 Can

MCKHANN, CHARLES FREMONT, b Boston, Mass, Jan 29, 30; m 54; c 3. SURGERY, MICROBIOLOGY. *Educ:* Harvard Univ, BA, 51; Univ Pa, MD, 55. *Prof Exp:* From instr to asst prof surg, Harvard Med Sch, 64-67; prof surg & microbiol, Med Ctr, Univ Minn, Minneapolis, 68-88; PROF SURG, SCH MED, YALE UNIV, NEW HAVEN, CT, 88- *Concurrent Pos:* Nat Cancer Inst spec res fel tumor biol, Karolinska Inst, Sweden, 61-62; Am Cancer Soc clin fel, Mass Gen Hosp, Boston, 63-64; Andres Soriano investr oncol, Mass Gen Hosp, 64-67. *Mem:* Am Asn Cancer Res; Transplantation Soc; Am Surg Soc; Am Asn Immunol; Am Col Surg. *Res:* Tumor immunology. *Mailing Add:* Dept Surg Hosp Yale Univ 333 Cedar St New Haven CT 06510

MCKHANN, GUY MEAD, b Boston, Mass, Mar 20, 32; m 57; c 5. NEUROLOGY, NEUROCHEMISTRY. *Educ:* Yale Univ, MD, 55. *Prof Exp:* Intern med, NY Hosp, 55-56; resident pediatrics, Johns Hopkins Hosp, 56-57; res assoc neurochem, Nat Inst Neurol Dis & Blindness, 57-60; resident neurol, Mass Gen Hosp, 60-63; from asst prof to assoc prof pediat & neurol, Sch Med, Stanford Univ, 63-69; PROF NEUROL & EXEC HEAD DEPT, SCH MED, JOHNS HOPKINS UNIV, 69- *Concurrent Pos:* Joseph P Kennedy, Jr scholar, 63-66; John & Mary R Markle scholar acad med, 64-69. *Mem:* Inst Med-Nat Acad Sci; Am Neurol Asn; Am Neurochem Soc; Soc Neurosci; Am Acad Neurol. *Res:* Lipid metabolism in the developing nervous system; metabolism of myelin; cellular neurophathology; neuroimmunology. *Mailing Add:* Dept Neurol Johns Hopkins Univ Sch Med 720 Rutland Ave Baltimore MD 21205

MACKI, JACK W, b Mullan, Idaho, June 16, 39; m 62; c 3. MATHEMATICS. *Educ:* Univ Idaho, BS, 60; Calif Inst Technol, PhD(math), 64. *Prof Exp:* Staff mem, Los Alamos Sci Lab, 64-65; from asst prof to assoc prof, 66-75, PROF MATH, UNIV ALTA, 75- *Concurrent Pos:* Chmn, Rocky Mountain Math Consortium. *Mem:* Soc Indust & Appl Math; Math Asn Am; Am Math Soc; Can Math Soc. *Res:* Ordinary differential equations; control theory. *Mailing Add:* Dept Math Univ Alta Edmonton AB T6G 2G1 Can

MCKIBBEN, GERALD HOPKINS, b Baldwyn, Miss, Aug 23, 39; m 61; c 2. ENTOMOLOGY. *Educ:* Miss State Univ, BSc, 62. *Prof Exp:* RES ENTOMOLOGIST, USDA, 62- *Mem:* Am Chem Soc; Entom Soc Am. *Res:* Insect pheromones; controlled release formulations; computer simulation modeling. *Mailing Add:* Harned Res Lab PO Box 5367 Mississippi State MS 39762-5367

MCKIBBEN, JOHN SCOTT, b Toledo, Ohio, Jan 25, 37. VETERINARY ANATOMY. *Educ:* Purdue Univ, BS, 59, DVM, 63; Iowa State Univ, MS, 66, PhD(vet anat), 69. *Prof Exp:* Clinician vet med, Rowley Mem Animal Hosp, Soc Prev Cruelty Animals, Mass, 63-64; from instr to asst prof vet anat, Iowa State Univ, 64-69; from assoc prof to prof anat & histol, Auburn Univ, 69-88; PVT PRACT VET MED, ROME CITY, IND, 88- *Honors & Awards:* Am Vet Med Asn Award, 75, 76. *Mem:* Am Vet Med Asn; Am Asn Vet Anat; World Asn Vet Anat; Am Asn Anatomists. *Res:* Neurology, cardiology, and teaching methods; cardiac autonomic innervation and denervation; multimedia programming of teaching materials; parrot anatomy; sharpei dogs. *Mailing Add:* Sylvan Vet Hosp Rd Nine S 875 Kelly St Exten Rome City IN 46784

MCKIBBEN, JOSEPH L, b Auxvasse, Mo, Oct 22, 12; m 42; c 4. PARTICLE PHYSICS. *Educ:* Park Col, BS, 35; Univ Wis, PhD(physics), 40. *Prof Exp:* Fel & physicist, Univ Wis, 40-43; group leader, Los Alamos Lab, 43-68, staff mem, 68-80; RETIRED. *Mem:* Am Phys Soc; AAAS. *Res:* Pressurized electrostatic accelerator. *Mailing Add:* 113 Aztec Ave White Rock Los Alamos NM 87544-3456

MCKIBBEN, ROBERT BRUCE, b Cincinnati, Ohio, Sept 1, 43. COSMIC RAY PHYSICS. *Educ:* Harvard Col, BA, 65; Univ Chicago, MS, 67, PhD(physics), 72. *Prof Exp:* Res assoc physics, 72-74, sr res assoc, 75-88, SR SCIENTIST PHYSICS, ENRICO FERMI INST, UNIV CHICAGO, 88- *Mem:* Am Geophys Union; Sigma Xi; Am Physical Soc. *Res:* Distribution of solar and galactic cosmic rays within the solar system and studies of energetic charged particles in planetary magnetospheres. *Mailing Add:* Enrico Fermi Inst Physics LASR 933 E 56th St Univ of Chicago Chicago IL 60637

MCKIBBIN, JOHN MEAD, b Tucson, Ariz, Nov 15, 15; m 44; c 4. BIOCHEMISTRY. *Educ:* Mich State Univ, BS, 38; Univ Wis, MS, 40, PhD(biochem), 42. *Prof Exp:* Instr nutrit, Harvard Med Sch & Sch Pub Health, Harvard Univ, 42-45; from asst prof to prof biochem, Col Med, Syracuse Univ & State Univ NY Med Ctr, 45-61; prof, 61-82, EMER PROF BIOCHEM DEPT, MED CTR, UNIV ALA, BIRMINGHAM, 86- *Mem:* Am Chem Soc; Soc Exp Biol & Med; Am Soc Biol Chem; Soc Complex Carbohydrates. *Res:* Nutrition, chemistry and metabolism of phospholipids and glycolipids. *Mailing Add:* 3747 Brookwood Rd Birmingham AL 35223-1538

MCKIBBINS, SAMUEL WAYNE, b West Allis, Wis, Nov 27, 31; m 53; c 3. CHEMICAL ENGINEERING. *Educ:* Univ Wis-Madison, BS, 53, MS, 55, PhD(chem eng), 58. *Prof Exp:* Chem engr, Kimberly-Clark Corp, 57-62; group leader process res, Union-Camp Corp, 62-64; assoc dir paper res, res planning & environ control, Continental Can Co, 64-70; dir res & develop, Am Can Co, 70-76; dir, Chem, Energy & Effluent Technol, 76-80, dir process engr & develop, 80-90, TECH DIR, CHAMPION INT, 90- *Concurrent Pos:* Mem pulp indust liaison comt, US Dept Health Educ & Welfare, 69. *Mem:* Am Inst Chem Engrs; Tech Asn Pulp & Paper Indust. *Res:* Environmental control; chemical kinetics; mass transfer; pulp and paper; process and systems analysis. *Mailing Add:* 42 Meadow View Rd Westport CT 06880

MACKICHAN, BARRY BRUCE, b Danville, Pa, May 15, 44; m 69; c 4. MATHEMATICS. *Educ:* Harvard Col, AB, 65; Stanford Univ, MS, 67, PhD(math), 68. *Prof Exp:* C L E Moore instr math, Mass Inst Technol, 69-70; asst prof, Duke Univ, 70-75; assoc prof math, NMex State Univ, 76-84; pres, TCI Software Res Inc, 81-88; SOFTWARE DESIGN ENGR, MICROSOFT CORP, 88- *Concurrent Pos:* Mem, Inst Advan Study, Princeton, NJ, 68-69 & 75-76; develop T3 sci word processing syst. *Res:* Overdetermined systems of partial differential equations; theory of functions of several complex variables; computer systems for technical text processing; computer aided instruction in mathematics; object oriented software development. *Mailing Add:* Microsoft Corp One Microsoft Way Redmond WA 98052-6399

MACKICHAN, JANIS JEAN, b Detroit, Mich, Aug 23, 51; m 86. PHARMACOKINETICS. *Educ:* Univ Mich, BS, 75, PharmD, 77. *Prof Exp:* Fel pharmacokinetics, Sch Pharm, State Univ NY Buffalo, 77-78, res instr neurol, Sch Med, 78-79; asst prof pharm, 79-85, ASSOC PROF PHARM, DIV PHARM PRACTICE, COL PHARM, OHIO STATE UNIV, 85- *Concurrent Pos:* Clin asst prof, dept pediat, Col Med, Ohio State Univ, 79-81; Young Investr Award, Am Heart Asn, 80-82. *Honors & Awards:* William H Rorer Award, Am Col Gastroenterol, 80. *Mem:* Am Asn Pharmaceut Scientists; Am Soc Clin Pharmacol & Therapeut; Am Col Clin Pharmacol; Am Col Clin Pharm; Am Soc Hosp Pharmacists; Am Assoc Cols Pharm. *Res:* Quantitation of the time-course of drugs in the human body and correlation of pharmacokinetics with physiologic processes to predict and avoid drug-drug interactions and to design improved drug dosage regimens in patients. *Mailing Add:* Col Pharm 217 Lloyd M Parks Hall Ohio State Univ 500 W 12th Ave Columbus OH 43210

MACKIE, GEORGE ALEXANDER, b Winnipeg, Man, Nov 11, 45; m 78. MOLECULAR BIOLOGY, BIOCHEMISTRY. *Educ:* Univ Toronto, BSc, 67; Cornell Univ, PhD(biochem & molecular biol), 71. *Prof Exp:* Med Res Coun fel, dept molecular biol, Univ Geneva, Switz, 71-73, researcher, 73-74; from asst prof to assoc prof, 74-88, PROF BIOCHEM, UNIV WESTERN ONT, 88- *Mem:* Am Soc Biochem & Molecular Biol. *Res:* Regulation of gene expression; structure of the ribosome; plant virology. *Mailing Add:* Dept Biochem Univ Western Ont London ON N6A 5C1 Can

MACKIE, GEORGE OWEN, b Louth, Eng, Oct 20, 29; m 56; c 5. ZOOLOGY. *Educ:* Oxford Univ, BA, 53, MA & DPhil, 56. *Prof Exp:* Lectr zool, Univ Alta, 56-58, from asst prof to prof, 58-68; chmn dept biol, 71-74, PROF BIOL, UNIV VICTORIA, BC, 68- *Concurrent Pos:* Nat Res Coun Can overseas fel, 63-64; ed, Can J Zool, 81-89, assoc ed, Acta Zool, 74-77 & 79-,; vis prof biol, Univ Calif, Los Angeles, 78 & Stanford Univ, 81-82 & 90; Killam res fel, 86-88. *Honors & Awards:* Fry Medal, Can Soc Zoologists. *Mem:* Am Soc Cell Biol; Am Soc Zool; Soc Exp Biol & Med; Can Soc Zool; Brit Soc Exp Biol; fel Royal Soc Can. *Res:* Neurobiology, especially coelenterata; Tunicata. *Mailing Add:* Dept Biol Univ Victoria Victoria BC V8W 2Y2 Can

MACKIE, RICHARD JOHN, b Foster City, Mich, July 6, 33; m 57; c 4. WILDLIFE MANAGEMENT, ECOLOGY. *Educ:* Mich State Univ, BS, 58; Wash State Univ, MS, 60; Mont State Univ, PhD(wildlife mgt), 65. *Prof Exp:* Res biologist, Mont Fish & Game Dept, 60-65, res coordr, 65-66; from asst prof to assoc prof entom, fisheries & wildlife, Univ Minn, St Paul, 66-70; assoc prof, 70-76, PROF WILDLIFE MGT, MONT STATE UNIV, 76- *Concurrent Pos:* Coordr statewide deer & habitat res, Mont Dept Fish, Wildlife & Parks, 70- *Mem:* Wildlife Soc (pres, 90-92); Soc Range Mgt. *Res:* Reproductive cycle of chukar; range ecology and relations of mule deer, elk and cattle; big game habitat relationships; big game range survey techniques; deer population ecology; browse plant ecology; interspecific relations. *Mailing Add:* Dept Biol Mont State Univ Bozeman MT 59717-0346

MCKIERNAN, MICHEL AMEDEE, b Chicago, Ill, Feb 17, 30; m 58. MATHEMATICS. *Educ:* Loyola Univ, Ill, BS, 51, MA, 52; Ill Inst Technol, PhD(math), 56. *Prof Exp:* Mathematician, Armour Res Found, 55-56; from instr to asst prof math, Ill Inst Technol, 56-61; mathematician, Inst Air Weapons Res, Univ Chicago, 61-62; assoc prof, 62-68, PROF MATH, UNIV WATERLOO, 68- *Mem:* Am Math Soc; Math Asn Am. *Res:* Functional equations. *Mailing Add:* Dept of Math Univ of Waterloo Waterloo ON N2L 3G1 Can

MACKIEWICZ, JOHN STANLEY, b Waterbury, Conn, July 12, 30; m 57; c 1. PARASITOLOGY. *Educ:* Cornell Univ, BS, 53, MS, 54, PhD(parasitol), 60. *Prof Exp:* Asst med entom & parasitol, Cornell Univ, 54-57, 59-60, instr, 57-59; NIH fel parasitol, Switz, 60-61; from asst prof to assoc prof biol, 61-68, prof biol sci, 68-73, DISTINGUISHED TEACHING PROF BIOL SCI, STATE UNIV NY ALBANY, 73- *Concurrent Pos:* Vis assoc prof, Univ Tenn, 67-68; vis scientist, India, 79. *Mem:* AAAS; Am Soc Parasitol; Am Soc Trop Med & Hyg; Soc Syst Zool; Am Micros Soc; Sigma Xi. *Res:* Parasites of freshwater fish; Cestoidea; Caryophylidea; conservation. *Mailing Add:* Dept Biol State Univ of NY Albany NY 12222

MCKILLOP, ALLAN A, b Calif, July 24, 25; m 54. MECHANICAL ENGINEERING. *Educ:* Univ Calif, BS, 50, PhD, 62; Mass Inst Technol, ME, 59. *Prof Exp:* Lectr agr eng, 51-53, from instr to assoc prof, 53-70, PROF AGR ENG, UNIV CALIF, DAVIS, 70- *Mem:* Am Soc Eng Educ; Am Soc Mech Engrs; Sigma Xi. *Res:* Experimental and numerical analysis in fluid mechanics and heat transfer. *Mailing Add:* Dept Eng Univ of Calif Davis CA 95616

MCKILLOP, J(OHN) H, b Detroit, Mich, Sept 21, 27; Can citizen; m 54; c 2. GEOLOGY. *Educ:* St Francis Xavier Univ, BSc, 51; Mem Univ Nfld, MSc, 61. *Prof Exp:* Geologist, Geol Surv Can, 51; asst govt geologist, Govt Nfld, 51-61, chief geologist, 61-62, dir mineral resources, Nlfd Dept Mines, Agr & Resources, 62-72, dep minister, Dept Mines & Energy, Govt Nfld & Labrador, 72-87; channing sr fel mem, Univ Nfld, 87-89; RETIRED. *Mem:* Can Inst Mining & Metall (vpres, 73-74); Geol Asn Can; Prospectors & Develop Asn Can. *Res:* Indian rocks and minerals in Newfoundland and Labrador. *Mailing Add:* 17 Dublin Rd St John's NF A1B 2E7 Can

MCKILLOP, LUCILLE MARY, b Chicago, Ill, Sept 28, 24. MATHEMATICS. *Educ:* St Xavier Col, Ill, BS, 51; Univ Notre Dame, MS, 59; Univ Wis, PhD(math educ), 65. *Prof Exp:* Elem teacher, St Patrick Acad, Ill, 47-51, St Xavier Acad, 51-52, Siena High Sch, 52-57 & Marquette High Sch, 57-58; from instr to prof math, St Xavier Col, Ill, 58-73; PRES, SALVE REGINA-NEWPORT COL, 73- *Concurrent Pos:* Consult, Archdioceasan Sch Bd Prog Math, 65-66; chmn div liberal arts & humanities, St Xavier Col, 66-73; secy, Comt Math Prep Teachers Elem Sch Math, 67-72. *Res:* Evolution of concepts in mathematics, particularly the evolution of concepts in finite geometries; investigations of collineations in projective planes with coordinates in a Galois field; self-generative quality of historical studies of mathematical creativity. *Mailing Add:* Salve Regina Col 100 Ochre Point Ave Newport RI 02840

MCKILLOP, WILLIAM L M, b Aberdeen, Scotland, June 3, 33; m 58; c 4. FOREST ECONOMICS. *Educ:* Univ Aberdeen, BSc, 54; Univ NB, MSc, 59; Univ Calif, Berkeley, MA & PhD(agr econ), 65. *Prof Exp:* Res officer forestry, Can Dept Forestry, 58-59, forest economist, 59-61; asst prof forestry & forest economist, 64-69, assoc prof, 69-75, PROF FORESTRY, UNIV CALIF, BERKELEY, 75- *Mem:* Am Econ Asn; Am Agr Econ Asn; Soc Am Foresters. *Res:* Econometrics; economic theory; forest economics and statistics. *Mailing Add:* Dept Forestry & Resource Mgt 145 Mulford Hall Univ Calif Berkeley CA 94720

MCKIM, HARLAN L, b Gothenburg, Nebr, Sept 28, 37; m 61; c 1. GEOLOGY, SOIL SCIENCE. *Educ:* Univ Nebr, BS, 62, MS, 67; Iowa State Univ, PhD(soil sci), 72. *Prof Exp:* Soil scientist, US Army Cold Regions Res & Eng Lab, 62-64, Soil Conserv Serv, 64-66 & Soil Surv Lab, 66-68; phys scientist, Water Resources Support Ctr, 79-81, SOIL SCIENTIST, COLD REGIONS RES & ENG LAB, US ARMY, 81- *Concurrent Pos:* Res assoc soil sci, Iowa State Univ, 68-72. *Mem:* Soil Conserv Soc Am; Soil Sci Soc Am; Am Quaternary Asn; Am Soc Agron; Am Soc Photogram. *Res:* Influence of soil moisture on runoff and utilization of remote sensing on water resources; field experiments and performance and management requirements on new data acquisition systems; use of remote sensing in Corps of Engineers programs; satellite sensor projects. *Mailing Add:* Meriden Rd Lebanon NH 03766

MCKIMMY, MILFORD D, b Beaverton, Mich, Dec 22, 23; m 54; c 3. WOOD TECHNOLOGY. *Educ:* Mich State Univ, BS, 49; Ore State Col, MS, 51; State Univ NY, PhD(wood technol), 55. *Prof Exp:* From instr to prof forest prod, Ore State Univ, 53-; RETIRED. *Concurrent Pos:* Charles Bullard forest res fel, Harvard Univ, 66-67. *Mem:* Soc Am Foresters; Soc Wood Sci & Technol; Forest Prod Res Soc; Tech Asn Pulp & Paper Indust. *Res:* Growth quality relationships of wood; application of genetics to wood quality. *Mailing Add:* Dept Forest Prods Forest Res Lab 105 Ore State Univ Corvallis OR 97331-5709

MACKIN, ROBERT BRIAN, b Huntington, NY, June 3, 60; m 82; c 2. PROPEPTIDE PROCESSING, PROTEIN CHEMISTRY. *Educ:* Carleton Col, BA, 82; Emory Univ, PhD(cell biol), 87. *Prof Exp:* Postdoctoral fel, Clayton Found Labs Peptide Biol, Salk Inst Biol Studies, La Jolla, Calif, 87-88; staff scientist, Dept Molecular Neuroendocrinol, Max Planck Inst Exp Med, Göttingen, Ger, 88-91; RES ASST PROF CELL BIOL, DEPT ANAT & CELL BIOL, SCH MED, EMORY UNIV, ATLANTA, 91- *Mem:* Sigma Xi; Am Soc Cell Biol; Endocrine Soc; Protein Soc. *Res:* Cellular and biochemical mechanisms involved in the synthesis and post-translational processing of precursors of biologically active peptides. *Mailing Add:* Dept Anat & Cell Biol Emory Univ Sch Med Atlanta GA 30322

MACKIN, ROBERT JAMES, JR, b Little Rock, Ark, Dec 4, 25; c 6. PHYSICS. *Educ:* Yale Univ, BE, 49; Calif Inst Technol, MS, 51, PhD(physics), 53. *Prof Exp:* Res assoc, Calif Inst Technol, 53-54 & Nuclear Physics Br, Off Naval Res, 54-56; res assoc, Thermonuclear Exp Div, Oak Ridge Nat Lab, 56-59, group leader, 59-62; mgr physics sect, Space Sci Div, 62-67, mgr, Lunar & Planetary Sci Sect, 67-69, mgr, Space Sci Div, 69-78, mgr energy technol develop, 79-83, asst dir, Arroyo Ctr, 83-85, dep mgr, 85-87, mgr, Army Progs, 87-88, PROG DIR TECHNOL, JET PROPULSION LAB, CALIF INST TECHNOL, 88- *Concurrent Pos:* Traveling lectr, Oak Ridge Inst Nuclear Studies, 58-60; vis mem staff, Off Energy Res, Dept Energy, 78-79; mem bd dir, Huntington Med Res Inst, 73- *Mem:* AAAS; Am Phys Soc; Sigma Xi. *Res:* Plasma and interplanetary physics; controlled fusion; planetary science; military technology. *Mailing Add:* 2626 N Holliston Ave Altadena CA 91001

MACKIN, WILLIAM MICHAEL, IMMUNOLOGY, CELL BIOLOGY. *Educ:* Univ Ill, PhD(immunol-microbiol), 80. *Prof Exp:* PRIN INVESTR, EI DUPONT DE NEMOURS & CO, 84- *Mailing Add:* Biomed Prod Dept Exp Sta E400-2269 EI Dupont De Nemours & Co Wilmington DE 19880-0400

MCKINLEY, CAROLYN MAY, b Lima, Ohio, May 13, 45. GENETICS, CYTOGENETICS. *Educ:* Ohio Northern Univ, BA, 72; Univ Toledo, PhD(biol), 77. *Prof Exp:* investr, Oak Ridge Nat Lab-Univ Tenn Grad Sch Biomed Sci, 77-79, res assoc comp mutagenesis, 79-81; asst prof, Widener Univ, 81-87; CONSULT, 87- *Mem:* Genetics Soc Am; Environ Mutagen Soc. *Res:* Genetics of recombination; mutagenesis and sister chromatid exchange. *Mailing Add:* 509 S Orange St Media PA 19063

MCKINLEY, CLYDE, b Mongo, Ind, Apr 19, 17; m 74; c 4. CHEMICAL ENGINEERING. *Educ:* Tri-State Col, BS, 37; Univ Mich, MS, 41, ScD(chem eng), 43. *Prof Exp:* Chem engr, Gas Corp Mich, 38 & Ohio Gas Light & Coke Co, 39; consult, Eng Div, Dow Chem Co, 40-42; chem engr, sect leader & supt, Spec Prod Dept, Gen Aniline & Film Corp, 43-53; dir res & develop, Air Prod & Chem Inc, 53-69, dir, Cryogenic Systs Div Res, 69-77, dir, Corp Res Serv, 77-83; RETIRED. *Mem:* AAAS; Am Chem Soc; Am Inst Chem Engrs; Sigma Xi. *Res:* Cryogenic processes and new product research and development. *Mailing Add:* 202 N 27th St Allentown PA 18104

MCKINLEY, JOHN MCKEEN, b Wichita, Kans, Feb 1, 30; m 53; c 3. ASTROPHYSICS. *Educ:* Univ Kans, BS, 51; Univ Ill, PhD(physics), 62. *Prof Exp:* Asst prof physics, Kans State Univ, 60-66; assoc prof, 66-71, PROF PHYSICS, OAKLAND UNIV, 71- *Concurrent Pos:* Assoc ed, Am J Physics, 79-82; res assoc, Goddard Space Flight Ctr, NASA, 80-81 & 82-83. *Mem:* Am Phys Soc; AAAS; Am Asn Physics Teachers. *Res:* Theoretical astrophysics. *Mailing Add:* Dept Physics Oakland Univ Rochester MI 48309-4401

MCKINLEY, MARVIN DYAL, b Ocala, Fla, Mar 3, 37; m 58; c 3. CHEMICAL ENGINEERING. *Educ:* Univ Fla, BChE, 59, MSE, 60, PhD(chem eng), 63. *Prof Exp:* Engr, E I du Pont de Nemours & Co, Tex, 63-65; from asst prof to assoc prof, Univ Ala, 56-76, prof chem eng, 76-, actg head, Dept Chem & Metall Eng, 81-83, head Chem & Metall Eng, 83-86, head, Chem Eng Dept, 86-88, Reichold-Shumaker Prof Chem Eng, 87-88, 90-91. *Concurrent Pos:* Vis assoc prof, Busan Nat Univ, Korea, 72-73. *Mem:* Am Inst Chem Engrs; Am Chem Soc; Am Soc Eng Educ; Air & Waste Mgr Asn; Nat Fire Protection Asn. *Res:* Phase equilibria; pyralysis, separation process, incineration, mass transfer. *Mailing Add:* Dept Chem Eng Univ Ala Box 870203 Tuscaloosa AL 35487-0203

MCKINLEY, MICHAEL P, b Waterloo, Iowa, Sept 19, 46. CELL BIOLOGY, NEUROLOGY. *Educ:* Ariz State Univ, PhD(cell biol), 78. *Prof Exp:* Fel neurol, 78-81, res asst, 82, ASST PROF NEUROL & ANAT, UNIV CALIF, SAN FRANCISCO, 82- *Mem:* AAAS; Am Soc Cell Biol; Electron Micros Soc Am. *Mailing Add:* Dept Neurol HSE 781 Univ Calif San Francisco CA 94143

MCKINLEY, VICKY L, b Dayton, Ohio, Feb 13, 57. MICROBIAL ECOLOGY. *Educ:* Wright State Univ, BS, 79; Univ Cincinnati, MS, 81, PhD(biol), 85. *Prof Exp:* Qual control chemist, Nat Distillers & Chem Co, 79-80; Alta Oil Sands Tech & Res Authority postdoctoral fel, Univ Calgary, 85-86; ASST PROF, ROOSEVELT UNIV, 87- *Mem:* Am Soc Microbiol; AAAS; Sigma Xi. *Res:* Microbial ecology; microbial heterotrophic decomposition of plant litter, composts, organic compounds, lignin, cellulose, PAHs and toxics; microbial communities of lakes, streams, soils, groundwater and composts; starvation survival of pseudomonas and other bacteria in freshwaters. *Mailing Add:* Dept Biol Roosevelt Univ 430 S Michigan Ave Chicago IL 60605

MCKINLEY, WILLIAM ALBERT, b Dallas, Tex, Aug 23, 17; m 40; c 2. PHYSICS. *Educ:* Univ Tex, BA, 39; Mass Inst Technol, PhD(physics), 47. *Prof Exp:* Mem staff, Radiation Lab, Mass Inst Technol, 44-46, res assoc, Instrumentation Lab, 46-47; from asst prof to assoc prof, 47-54, prof physics, 54-85, EMER PROF, RENSSELAER POLYTECH INST, 85- *Mem:* AAAS; Am Phys Soc; Am Asn Physics Teachers. *Res:* Quantum field theory; theory of atomic and nuclear collisions. *Mailing Add:* Dept of Physics Rensselaer Polytech Inst Troy NY 12181

MCKINNELL, ROBERT GILMORE, b Springfield, Mo, Aug 9, 26; m 64; c 3. DEVELOPMENTAL BIOLOGY. *Educ:* Univ Mo, AB, 48; Drury Col, BS, 49; Univ Minn, PhD(zool), 59. *Prof Exp:* Res assoc embryol, Inst Cancer Res, 58-61; asst prof zool, Tulane Univ, 61-65, assoc prof biol, 65-69, prof, 69-70; prof zool, Univ Minn, Minneapolis, 70-76; PROF GENETICS & CELL BIOL, UNIV MINN, ST PAUL, 76- *Concurrent Pos:* Instr, Univ Minn, 58; sr sci fel, NATO, St Andrews Univ, Scotland, 74; mem, Adv Coun, Inst Lab Animal Resources, 74-77; vis scientist biomed res, Dow Chem USA, Tex Div, Freeport, 76; royal soc guest res fel, Nuffield Dept Path, Oxford Univ, 81-82; panel mem, Genetic & Cellular Resources Prog, Nat Inst Aging. *Mem:* AAAS; fel Linnean Soc London; Soc Develop Biol; Int Soc Differentiation (secy-treas, 75-); Sigma Xi. *Res:* Transplantation of nuclei from normal, neoplastic and aging anuran cells; use of mutant genes as nuclear markers; invasion and metastasis of Lucke renal adenocarcinoma. *Mailing Add:* Dept Genetics & Cell Biol Univ Minn 1445 Gortner Ave St Paul MN 55108

MCKINNELL, W(ILLIAM) P(ARKS), JR, b Springfield, Mo, Dec 24, 24; m 50; c 4. METALLURGY. *Educ:* Univ Minn, BChE, 45; Univ Mo, BS, 47; Ohio State Univ, MS, 54, PhD(metall), 56. *Prof Exp:* Metallurgist, Chrysler Corp, 47-50; asst prof metall, Va Polytech Inst, 50-51; advan res engr, Denver Res Ctr, 56-60, mgr, Analysis Dept, 60-62, Chem Eng Dept, 62-63, Eng &

Chem Dept, 63-67 & Commercial Develop Div, 67-72, RES DIR, DENVER RES CTR, MARATHON OIL CO, 73- *Concurrent Pos:* Adj assoc prof, Univ Denver, 63-67. *Mem:* Am Soc Metals; Nat Asn Corrosion Engrs; Am Inst Chem Engrs; Am Inst Mining, Metall & Petrol Engrs; Sigma Xi. *Res:* Corrosion of and brittle failure of metals. *Mailing Add:* 730 Front Range Rd Littleton CO 80120

MCKINNEY, ALFRED LEE, b Houston, Tex, Aug 19, 37; m 60; c 1. MATHEMATICAL ANALYSIS, NUMERICAL ANALYSIS. *Educ:* La Tech Univ, BS, 59, MS, 61; Univ Okla, PhD(math), 72. *Prof Exp:* Res mathematician, United Gas Res Lab, 61-65; chmn math dept, Okla Col Lib Arts, 68-72, chmn math & sci div, 72-74; assoc prof, 74-81, PROF MATH, LA STATE UNIV, SHREVEPORT, 81- *Mem:* Asn Comput Mach; Data Processing Mgt Asn; Math Asn Am. *Res:* Numerical approximations, particularly computer-oriented approaches to solutions of calculus of variations or control problems. *Mailing Add:* Dept of Math La State Univ 8515 Youree Dr Shreveport LA 71115

MACKINNEY, ARCHIE ALLEN, JR, b St Paul, Minn, Aug 16, 29; m 55; c 3. HEMATOLOGY. *Educ:* Wheaton Col, Ill, BA, 51; Univ Rochester, MD, 55; Am Bd Internal Med, dipl, 62. *Prof Exp:* Resident med, Univ Wis Hosps, 55-59; clin assoc hemat, Nat Inst Arthritis & Metab Dis, 59-61; clin investr, 61-64, from asst prof to assoc prof med, 64- 74, PROF MED, SCH MED, UNIV WIS-MADISON, 74-; CHIEF HEMAT, VET ADMIN HOSP, 64- *Concurrent Pos:* Vet admin res grants, 61-; NIH res grants, 64-67 & 70-75; chief nuclear med, Vet Admin Hosp, 64-74. *Mem:* Am Fedn Clin Res; Cent Soc Clin Res; Am Soc Hemat; Sigma Xi. *Res:* Cell proliferation in vitro; lymphomagenesis. *Mailing Add:* 190 N Prospect Ave Madison WI 57305

MACKINNEY, ARLAND LEE, b Hendersonville, NC, Nov 29, 31; m 55; c 3. NUCLEAR PHYSICS, TECHNICAL MANAGEMENT. *Educ:* NC State Col, BS, 53; Ind Univ, MS, 55; Mass Inst Technol, SM, 67. *Prof Exp:* Physicist, Knolls Atomic Power Lab, AEC, 55-58; physicist, Babcock & Wilcox Co, 58-60, supvr, 60-64, sect chief, 64-67, tech adv to mgr physics lab, 67-68, from asst mgr to mgr, Qual Control Dept, 68-71, spec asst nuclear opers to div vpres, Mount Vernon Plant, Ind, 71-72, plant mgr, 73-75, qual assurance mgr, 76-79, 87-89, mgr gen serv, Nuclear Power Generation Div, 79-82, mgr, Nuclear Parts Ctr, 82-83, mgr, Comput Serv, 83-87, SR PROJ MGR, NUCLEAR MGT & RESOURCE COUN, BABCOCK & WILCOX CO, 89- *Mem:* Am Nuclear Soc; Am Soc Qual Control. *Mailing Add:* 1505 Crystal Dr Apt 605 Arlington VA 22202-4118

MCKINNEY, CHARLES DANA, JR, physical chemistry; deceased, see previous edition for last biography

MCKINNEY, CHESTER MEEK, b Cooper, Tex, Jan 29, 20; m 48; c 2. PHYSICS, UNDERWATER ACOUSTICS. *Educ:* ETex State Teachers Col, BS, 41; Univ Tex, MA, 47, PhD(physics), 50. *Prof Exp:* Res physicist, Univ Tex, 45-65 & 80-85, dir, Appl Res Lab, 65-80; RETIRED. *Concurrent Pos:* Assoc prof, Tex Tech Col, 50-53; mem naval res adv comt, Lab Adv Bd Naval Ships, 75-77; mem, US Navy Underwater Sound Adv Group, 62-64 & 75-77, chmn, 71-73; mem, mine adv comt, Nat Res Coun, 59-72 & Naval Studies Bd, 79-91; liason scientist, US Off Naval Res, London, 83-84. *Honors & Awards:* David Bushnell Award, Am Defense Preparedness Asn, 85; Distinguished Tech Achievement Award, Inst Elec & Electronic Engrs Oceanic Eng Soc, 88. *Mem:* Acoust Soc Am (vpres, 84-85, pres, 87-88); Inst Elec & Electronic Engrs; hon fel Brit Inst Acoust. *Res:* Underwater acoustics; electronics; microwaves; dielectric waveguides and antennae. *Mailing Add:* 4305 Farhills Dr Austin TX 78731

MCKINNEY, DAVID SCROGGS, physical chemistry; deceased, see previous edition for last biography

MCKINNEY, EARL H, b Wilkinsburgh, Pa, May 24, 29; m 52; c 3. NUMERICAL ANALYSIS. *Educ:* Washington & Jefferson Col, AB, 51; Univ Pittsburgh, MS, 56, PhD(math), 61. *Prof Exp:* Instr math, Univ Pittsburgh, 55-59; asst prof, Northern Ill Univ, 59-62; prof math & head dept, 62-70, PROF, BALL STATE UNIV, 70- *Concurrent Pos:* NSF res grant appl math, Argonne Nat Lab, 68-69. *Mem:* Am Math Soc; Math Asn Am; Soc Indust & Appl Math. *Res:* Analysis; applied mathematics; numerical analysis; interpolation and numerical integration. *Mailing Add:* Dept Computer Sci Ball State Univ Muncie IN 47306

MCKINNEY, FRANK KENNETH, b Birmingham, Ala, Apr 13, 43; m 64; c 4. INVERTEBRATE PALEONTOLOGY. *Educ:* Old Dom Col, BS, 64; Univ NC, Chapel Hill, MS, 67, PhD(paleont), 70. *Prof Exp:* Asst prof, 68-76, PROF GEOL, APPALACHIAN STATE UNIV, 76- *Concurrent Pos:* Fel, Smithsonian Inst, 72-73; vis prof, Univ Durham, 78; exchange scientist, USSR, 78, Czechoslovakia, 81, 83 & Yugoslavia, 87-90; res assoc, Field Mus of Natural Hist, Chicago & Am Mus Natural Hist, NY. *Mem:* Paleont Soc; Int Palaeont Asn; Soc Econ Paleont & Mineral; Int Bryozool Asn; Am Soc Zoologists; AAAS. *Res:* Bryozoans, evolution and functional morphology particularly Fenestrata; biostratigraphy. *Mailing Add:* Dept Geol Appalachian State Univ Boone NC 28608

MCKINNEY, GORDON R, b Indianapolis, Ind, Oct 14, 23; m 47; c 1. PHARMACOLOGY, MEDICAL INFORMATION. *Educ:* DePauw Univ, AB, 46; Univ Notre Dame, MS, 48; Duke Univ, PhD, 51. *Prof Exp:* Lectr pharmacol, Duke Univ, 52-53; from asst prof to assoc prof, Sch Med, WVa Univ, 53-59; res pharmacologist, Pharmaceut Div, Mead Johnson & Co, 59-61, sr res fel, 61-68, dir pharmacol, 68-75, dir biol res,75-78, assoc dir med serv, 78-80, dir med commun, 80-86; dir med commun, 86-87, DIR REGULATORY AFFAIRS, BRISTOL-MYERS US PHARMACEUT GROUP, 87- *Concurrent Pos:* Am Cancer Soc res fel med, Duke Univ, 51-53; Lederle med fac award, WVa Univ, 55-57. *Mem:* Drug Info Asn; Am Soc Pharmacol & Exp Therapeut; Am Med Writers Asn; Soc Exp Biol & Med; Endocrine Soc; Sigma Xi; AAAS; Soc Study Reproduction. *Res:* Adrenergic, biochemical and endocrine pharmacology; medical/drug information service. *Mailing Add:* Mead Johnson & Co Evansville IN 47721-0001

MACKINNEY, HERBERT WILLIAM, b London, Eng, Feb 22, 07; nat US; m 44; c 1. POLYMER CHEMISTRY, ENVIRONMENTAL CHEMISTRY. *Educ:* Swiss Fed Inst Technol, dipl, 28; McGill Univ, MSc, 33, PhD(cellulose chem), 35. *Prof Exp:* Chemist, Lever Bros Ltd, Eng, 28-31; res chemist, Can Int Paper Co, Ont, 31-32; res assoc, Macdonald Col, McGill Univ, 36; res chemist, Kendall Co, 36-39, Sylvania Indust Corp, 39-40 & Bakelite Corp Div, Union Carbide Corp, 41-58; staff chemist, Int Bus Mach Corp, 58-61, adv chemist, 61-66, sr chemist, IBM Corp, 66-72; RETIRED. *Concurrent Pos:* Consult, adhesives, microencapsulation, econometrics of precious metals. *Honors & Awards:* Honor Scroll, Am Inst Chemists, 57. *Mem:* Am Chem Soc; fel Am Inst Chemists; Soc Plastics Engrs. *Res:* Polymers and plastics; adhesives; environmental deterioration of materials; plastic laminates. *Mailing Add:* 740 S Alton Way Denver CO 80231-1603

MCKINNEY, JAMES DAVID, b Gainesville, Ga, Dec 28, 41; m 70; c 3. ENVIRONMENTAL CHEMISTRY. *Educ:* Univ Ga, BS, 63, PhD(org chem), 68. *Prof Exp:* Pub health scientist, Pesticide Toxicol Lab, Food & Drug Admin, Ga, 67-69; res scientist environ chem, 69-74, head chem sect, 74-78, chief, Environ Chem Br, 78-79, chief, lab environ chem, 80-83, SR RES SCIENTIST, LAB MOLECULAR BIOPHYS, NAT INST ENVIRON HEALTH, 83- *Mem:* Am Chem Soc; Am Inst Chemists. *Res:* Environmental health chemistry; structure-activity relationships as predictive tools in chemical toxicology; analytical chemistry and residue analysis of environmental/biological samples; bioorganic chemistry and biomechanism elucidation; synthetic organic chemistry; medicinal chemistry. *Mailing Add:* Nat Inst Environ Health Sci PO Box 12233 Research Triangle Park NC 27709-2233

MCKINNEY, JAMES T, b Detroit, Mich, May 28, 38; m 61; c 2. SURFACE PHYSICS. *Educ:* Univ Detroit, BS, 60; Univ Wis, MS, 62, PhD(physics), 66. *Prof Exp:* Scientist, Fundamental Res Lab, US Steel Res Ctr, 66-71; sr physicist, 72-76, RES SCIENTIST, 3M CO, 77- *Mem:* Am Phys Soc. *Res:* Ion scattering; secondary ion mass spectroscopy; interaction of charged particles with solid surfaces; surface analytical instrumentation. *Mailing Add:* Bldg-236-GD-18 3M Ctr St Paul MN 55144-1000

MCKINNEY, JOHN EDWARD, b Altoona, Pa, Apr 6, 25; m 58. THERMODYNAMICS, RHEOLOGY. *Educ:* Pa State Univ, BS, 50. *Prof Exp:* Physicist, Polymers Div, 50-89, GUEST SCIENTIST, NAT INST STANDARDS & TECHNOL, 89- *Concurrent Pos:* Guest worker, Nat Physics Lab, Teddington, Eng, 64. *Honors & Awards:* Bronz Medal, Nat Bur Standards, 83. *Mem:* Rheology Soc; Int Asn Dent Res. *Res:* Experimental rheology, acoustics, thermodynamics, dynamic mechanical, dielectric, piezoelectric and pyroelectric properties of polymers and glasses; physical properties including wear and fatigue of dental restorative materials; related theoretical development of liquid and glassy states; development of related instrumentation. *Mailing Add:* Polymer Sci & Standards Div Nat Inst Standards & Technol Washington DC 20234

MCKINNEY, MAX TERRAL, b Esto, Fla, Sept 25, 35; m 53; c 2. MATHEMATICS. *Educ:* Troy State Univ, BS, 56; Auburn Univ, MEd, 62, DEd, 64. *Prof Exp:* Proj mathematician, Vitro Corp Am, 56-57; high sch teacher, Ga, 57-61; asst prof, 64-66, assoc prof, 66-81, PROF MATH, GA SOUTHWESTERN COL, 81-, CHMN DEPT, 66- *Res:* Statistics and algebraic fields. *Mailing Add:* Dept of Math Ga Southwestern Col Wheatley St Americus GA 31709

MCKINNEY, MICHAEL, b Bucyrus, Mo, June 14, 50. NEUROPHARMACOLOGY, MOLECULAR BIOLOGY. *Educ:* US Naval Acad, BS, 72; Johns Hopkins Univ, PhD, 82. *Prof Exp:* Postdoctoral fel, Mayo Clin, 82-84, assoc consult & asst prof, 85-86; group leader, Abbott Labs, 86-89; ASSOC PROF & CONSULT, MAYO CLIN, 89- *Mem:* Am Soc Pharmacol & Exp Therapeut; AAAS; Am Soc Neurochem; Soc Neurosci. *Mailing Add:* Dept Pharmacol Mayo Clin 4500 San Pablo Rd Jacksonville FL 32224

MCKINNEY, PAUL CAYLOR, b Otterbein, Ind, Aug 21, 30; m 66; c 2. PHYSICAL CHEMISTRY. *Educ:* Wabash Col, AB, 52; Northwestern Univ, PhD, 58. *Prof Exp:* from asst prof to assoc prof, 58-76, chmn dept, 78-81, PROF CHEM, WABASH COL, 77-, DEAN COL, 82- *Mem:* Am Chem Soc; AAAS; NY Acad of Sci; Ind Acad of Sci. *Res:* Molecular mechanics. *Mailing Add:* Wabash Col Crawfordsville IN 47933

MCKINNEY, PETER, b Baltimore, Md, Nov 2, 34; m 73; c 2. PLASTIC SURGERY. *Educ:* Harvard Univ, AB, 56; McGill Univ, MD, CM, 60. *Prof Exp:* Intern, Montreal Gen Hosp & resident, New York City Hosp, 60-61; asst, Bellevue-Jacobi Hosp & teacher gen surg, Albert Einstein Col Med, 61-64; resident plastic surg, New York Hosp, Med Ctr, Cornell Univ, 64-67, chief resident, 66-67; instr & assoc surg, 67-70, asst prof, 70-74, ASSOC PROF CLIN SURG, SCH MED, NORTHWESTERN UNIV, CHICAGO, 74- *Concurrent Pos:* Instr surg, Sch Med, Cornell Univ, 64-67. *Mem:* Am Soc Aesthet Plastic Surg; Plastic Surg Res Coun; Am Soc Plastic & Reconstruct Surgeons; Am Asn Plastic Surgeons; Am Col Surgeons. *Res:* Experimental closure of perforation in dog septums. *Mailing Add:* Northwestern Univ 707 N Fairbanks Suite 1207 Chicago IL 60611

MCKINNEY, RALPH VINCENT, JR, b Columbus, Ohio, Jan 9, 33; m 55; c 4. PATHOLOGY, CELL BIOLOGY. *Educ:* Bowling Green State Univ, BS, 54; Ohio State Univ, DDS, 61; Univ Rochester, PhD(path), 71. *Prof Exp:* Asst instr dent hyg, Ohio State Univ, 60-61; clin asst prof oper dent, Case Western Reserve Univ, 61-65; PROF ORAL PATH & ORAL BIOL, GRAD FAC, MED COL GA, 70-, CHMN, 79- *Concurrent Pos:* NIH fel path, Univ Rochester, 65-70; NIH grants, 72-74, 72-78 & 89-94; Nat Inst Dent Res contract, 73-77; pvt dent pract, Ohio, 61-65; mem active staff, Med Col Ga Hosp & Clin, Augusta, 70-; oral path diag serv, Med Col Ga, 73-; pres dent found, 74-76; chmn, Med Col Ga Fac, 79; consult, Vet Admin Hosps, Augusta, 79-; secy bd of dir, Med Col Ga Res Inst, Inc, 80-; consult, Eisenhower Army Med Ctr, Ft Gordon, GA, 81-; consult, Scripps Clin & Res

Found, La Jolla, 86- *Honors & Awards:* Isiah Lew Award for Distinguished Res in Implantology, Am Acad Implant Dent Res Found. *Mem:* Int Acad Path; Am Acad Oral Path; Int Asn Dent Res (bd dirs, 80-81); Sigma Xi; Soc Biomaterials; Am Acad Implant Dent; fel Am Col Dentists; Int Cong Oral Implant (treas, 88-). *Res:* Dental implants, emphasis on biological tissue interface; microcirculation; wound healing; peripheral interests, inflammation, connective tissue. *Mailing Add:* Dept Oral Path Med Col of Ga Augusta GA 30912-1110

MCKINNEY, RICHARD LEROY, b Altoona, Pa, May 23, 28; m 56; c 3. MATHEMATICS. *Educ:* Syracuse Univ, AB, 51, MA, 52; Univ Wash, PhD(math), 58. *Prof Exp:* Asst, Univ Wash, 53-58; from instr to asst prof math, Univ Calif, Riverside, 58-62; asst prof, 62-67, ASSOC PROF MATH, UNIV ALTA, 67- *Concurrent Pos:* Hon res assoc, Univ Col, Univ London, 68-69. *Mem:* Am Math Soc; Can Math Soc; Math Asn Am. *Res:* Linear spaces; convex sets; functional analysis; topology. *Mailing Add:* Dept Math Univ Alta Edmonton AB T6G 2E2 Can

MCKINNEY, ROBERT WESLEY, b East St Louis, Ill, Dec 11, 31; m 61; c 2. ANALYTICAL CHEMISTRY. *Educ:* Southern Ill Univ, BA, 53; Univ Kans, PhD(anal chem), 57. *Prof Exp:* Analytical chemist, Celanese Corp Am, 57-60; analytical chemist, 60-64, group leader analytical chem, 64-72, MGR ANALYTICAL CHEM, W R GRACE & CO, 72- *Mem:* Am Chem Soc. *Res:* Instrumental and wet analytical chemistry; gas and liquid chromatography. *Mailing Add:* Washington Res Ctr W R Grace & Co 7379 Rte 32 Columbia MD 21044

MCKINNEY, ROGER MINOR, b Deerbrook, Wis, May 31, 26; m 52; c 3. ORGANIC CHEMISTRY, IMMUNOCHEMISTRY. *Educ:* Wis State Col, River Falls, BS, 50; St Louis Univ, MS, 56, PhD(chem), 58. *Prof Exp:* Chemist, Lambert Pharmacal Co, Mo, 50-52; chemist, Universal Match Corp, 52-55; res chemist, 58-66, Aedes Aegypti Eradication Prog, 66-68, Tech Develop Labs, 68-71, res chemist, Tech Develop Labs, 71-72, RES CHEMIST, CTR INFECTIOUS DIS, CTR DIS CONTROL, DEPT HEALTH & HUMAN SERVS, USPHS, 72- *Mem:* AAAS; Am Chem Soc. *Res:* Synthesis of radioactive isotope labeled insecticides; technical aspects of immunofluorescent staining; basic immunochemistry studies; diagnostic reagents through hybridoma technology. *Mailing Add:* 4872 Cambridge Dr CDC Bldg 5 Rm 308 Atlanta GA 30333

MCKINNEY, RONALD JAMES, b Altadena, Calif, Mar 14, 49; m 70; c 2. HOMOGENEOUS CATALYSIS. *Educ:* Anderson Col, BA, 70; Univ Calif, Los Angeles, PhD(inorg chem), 74. *Prof Exp:* Teaching fel, Sch Chem, Univ Bristol, UK, 74-76; res chem, 76-89, PROJ LEADER, CENT RES & DEVELOP DEPT, E I DU PONT DE NEMOURS, 89- *Concurrent Pos:* Co-chmn, Nat Organometallic Chem Workshop, 84-87. *Mem:* Am Chem Soc. *Res:* Synthesis of organometallic compounds; homogeneous catalysis of organic reactons involving carbon-carbon bond formation; olefin dimerization and hydrocyanation; theoretical studies on organometallic compounds and reactions (MO). *Mailing Add:* Cent Res & Develop Dept E I du Pont de Nemours & Co Exp Sta B328 Wilmington DE 19880-0328

MCKINNEY, ROSS E(RWIN), b San Antonio, Tex, Aug 2, 26; m 52; c 4. SANITARY ENGINEERING. *Educ:* Southern Methodist Univ, BS & BA, 48; Mass Inst Technol, SM, 49, ScD(sanit eng), 51. *Prof Exp:* Asst sanit chem, Mass Inst Technol, 49-51; actg head, Div Sanit Sci, Southwest Found Res & Educ, 51-53; asst prof sanit eng, Mass Inst Technol, 53-59, assoc prof, 59-60; prof civil eng, Univ Kans, 60-66, chmn dept, 63-66, Parker prof civil eng, 66-76, N T VEATCH PROF ENVIRON ENG, UNIV KANS, 76- *Concurrent Pos:* Adv Prof, Tongji Univ, Shanghai, People's Repub China, 85. *Honors & Awards:* Harrison Prescott Eddy Award, Water Pollution Control Fedn, 62; Rudolph Hering Award, Am Soc Civil Engrs, 62; Presidential Commendation, 71; Environ Qual Award, Environ Protection Agency, 79; Thomas R Camp Medal, Water Pollution Control Fedn, 82. *Mem:* Nat Acad Engrs; Am Soc Microbiol; Am Soc Civil Engrs; Am Water Works Asn; Am Chem Soc; Am Soc Eng Educ; Am Asn Univ Prof; Am Acad Environ Eng; Am Pub Works Asn; AAAS. *Res:* Application of fundamental microbiology to design of liquid waste treatment systems. *Mailing Add:* Dept Civil Eng Univ Kans Lawrence KS 66044

MCKINNEY, TED MEREDITH, b Huntsville, Ala, Apr 18, 38. ANALYTICAL CHEMISTRY. *Educ:* Harvard Univ, AB, 60; Cornell Univ, PhD(chem), 65. *Prof Exp:* Res assoc chem, Cornell Univ, 64-66; asst prof, Univ Calif, Riverside, 66-71; WRITER & ED, 71-; CONSULT, ROCKWELL INT SCI CTR, THOUSAND OAKS, CALIF, 81- *Concurrent Pos:* Consult, Beckman Instruments, Inc, Calif, 69; Photog instr, Riverside Community Col, 79- *Mem:* Am Chem Soc. *Res:* Magnetic resonance; electroanalytical chemistry; optical spectroscopy; explosives. *Mailing Add:* 5156 Colina Way Riverside CA 92507

MCKINNEY, THURMAN DWIGHT, b Bowling Green, Ky, June 23, 47; m 69; c 2. NEPHROLOGY. *Educ:* Western Ky Univ, BS, 70; Vanderbilt Univ, MD, 73. *Prof Exp:* Med intern & resident, Univ Calif, San Francisco, 73-75; res assoc renal physiol, NIH, 75-77; asst prof med, Vanderbilt Univ, 77-83; staff nephrologist, Nashville Vet Admin Med Ctr, 77-83; at dept med, Health Sci Ctr, Univ Tex, San Antonio, 83-87; PROF MED, DIR, NEPHROLOGY SECT, IND UNIV MED CTR, INDIANAPOLIS, 87- *Mem:* Am Soc Nephrology; Am Fedn Clin Res; Am Soc Clin Invest; Int Soc Nephrology; Am Physiol Soc; Southern Soc Clin Invest. *Res:* Renal physiology with emphasis on renal acidification and organic base transport by renal tubules in vitro and tissue culture, membrance vesicles. *Mailing Add:* Dir Nephrology Sect Ind Univ Med Ctr 1120 South Dr Fesler Hall 108 Indianapolis IN 46202-5116

MCKINNEY, WILLIAM ALAN, b Omaha, Nebr, Dec 18, 27; m 52; c 2. METALLURGICAL ENGINEERING. *Educ:* Univ Ariz, BS, 51. *Prof Exp:* Process engr, Grand Cent Aircraft Co, 51-52; metallurgist, 52-68, supvry metallurgist, 68-75, RES DIR, US BUR MINES, 75- *Concurrent Pos:* Adj prof metall eng, Univ Utah, 76- *Mem:* Am Inst Mining, Metall & Petrol Engrs. *Res:* Minerals beneficiation, hydrometallurgy and pyrometallurgy of copper; chemical processing of copper and by product molybdenite concentrates; flue gas desulfurization. *Mailing Add:* 3936 Sunny Dale Dr Salt Lake City UT 84124

MCKINNEY, WILLIAM MARK, b Spring Valley, NY, Dec 26, 23; m 51; c 1. PLANETARY SCIENCES, EARTH SCIENCES. *Educ:* Lang Col, BA, 48; Univ Fla, PhD(geog), 58. *Prof Exp:* Consult pub health, Ga Dept Pub Health, 53-58; from instr to asst prof geog, Southern Ore Col, 58-63; from asst prof to prof geog & geol, Univ Wis-Stevens Point, 63-88; RETIRED. *Concurrent Pos:* Guest investr, Lowell Observ, 73-78. *Mem:* Geol Soc Am; Nat Coun Geog Educ. *Res:* Analysis and mapping of atmospheric phenomena of Mars as photographed in various wavelengths of light; study and development of instrumentation for demonstrating principles of astronomical geography. *Mailing Add:* 1540 NW Kings Blvd Corvallis OR 97330

MCKINNEY, WILLIAM MARKLEY, b Roanoke, Va, June 6, 30; m 52; c 3. NEUROLOGY. *Educ:* Univ NC, Chapel Hill, BA, 51; Univ Va, MD, 59. *Prof Exp:* From instr to assoc prof neurol, 63-76, PROF NEUROL, BOWMAN GRAY SCH MED, 76-, RES ASSOC RADIOL, 67-, FAC CHMN, POSTGRAD COURSE MED SONICS, 75- *Concurrent Pos:* Dir sonic lab, Bowman Gray Sch Med, 63-76, mem subcomt on stroke, Regional Med Prog, 67-; consult, Vet Admin Hosp, Salisbury, NC; chairperson, Adv Comt, Ultrasonic Tissue Signature Characterization, Nat Bur Standards, NIH, NSF, Gaithersburg, 77- *Mem:* Am Acad Neurol; Am Fedn Clin Res; Asn Res Nerv & Ment Dis; fel Am Inst Ultrasound in Med (secy, 67, pres, 74-76); Sigma Xi. *Res:* Diagnostic ultrasound in medicine; the application of ultrasound to medicine; cerebrovascular disease; crystallography; urinary lithiasis. *Mailing Add:* Dept of Neurol Bowman Gray Sch of Med Winston-Salem NC 27103

MCKINNIS, CHARLES LESLIE, b Cape Girardeau, Mo, July 10, 23; m 44; c 1. CHEMISTRY. *Educ:* Southeast Mo State Col, BS, 46; Mo Sch of Mines, BS, 47, MS, 48; Ohio State Univ, PhD(glass sci), 54. *Prof Exp:* Res engr, Res Lab, Pittsburgh Plate Glass Co, 48-50; res assoc, Res Found, Ohio State Univ, 50-54; res chemist, Midwest Res Inst, 54-55; sr res scientist, Owens Corning Fiberglas Corp, 55-75, res assoc, glass res & develop, Tech Ctr, 75-82; RETIRED. *Honors & Awards:* Frank Forrest Award, Am Ceramic Soc (Credible paper in Glass Sci), 59. *Mem:* Sr & fel Am Ceramic Soc; Nat Inst Ceramic Engrs; Sigma Xi. *Res:* Glass science; glass structure, fiber strength and volume surface properties affecting it; heat transfer. *Mailing Add:* 109 Mt Parnassus Dr Box 46 Granville OH 43023

MCKINNON, DAVID M, b Scotland, Aug 11, 38; m 63; c 2. ORGANIC CHEMISTRY. *Educ:* Univ Edinburgh, BSc, 60, PhD(chem), 63. *Prof Exp:* Fel chem, Dalhousie Univ, 63-65; from asst prof to assoc prof, 65-76, PROF CHEM, UNIV MAN, 76- *Mem:* Chem Inst Can. *Res:* Chemistry of heterocyclic sulphur and nitrogen compounds. *Mailing Add:* Dept Chem Univ Man Winnipeg MB R3T 2N2 Can

MCKINNON, WILLIAM BEALL, b Montreal, Que, Aug 14, 54; US citizen. PLANETARY SCIENCE. *Educ:* Mass Inst Technol, SB, 76; Calif Inst Technol, MS, 79, PhD(planetary sci & geophysics), 80. *Prof Exp:* Res assoc, Lunar & Planetary Lab, Univ Ariz, 80-82, sr res assoc, 82; asst prof, 82-88, ASSOC PROF, DEPT EARTH & PLANETARY SCI, WASH UNIV, 88- *Concurrent Pos:* Vis res assoc, Dept Earth & Space Sci, State Univ NY, Stony Brook, 81; mem, Comt Lunar Planetary Explor, 85-88, Lunar Planetary Geosci Rev Panel, 86-88, Planetary Geol Geophys Working Group, 86-91, Cassini Panel, 90, Origins of Solar Systs Rev Panel, 91-92 & Planetary Res Anal Study Group, 91; ed, EOS Planetology, 86-88. *Mem:* Am Geophys Union; Am Astron Soc; AAAS; Sigma Xi; Meteoritical Soc; Planetary Soc. *Res:* Cratering and tectonics, including physical gemorphology, planetary interiors and radio astronomy. *Mailing Add:* Dept Earth & Planetary Sci Wash Univ One Brookings Dr St Louis MO 63130-4899

MCKINSEY, RICHARD DAVIS, b New York, NY, May 20, 21; m 44; c 3. BOTANY. *Educ:* Ill Inst Technol, BS, 48; Stanford Univ, MA, 53, PhD(biol), 58. *Prof Exp:* From instr to assoc prof biol, Fac Arts & Sci, Univ Va, 57-; RETIRED. *Res:* Intermediate metabolism of fungi. *Mailing Add:* Dept Biol Univ Va Charlottesville VA 22903

MCKINSTRIE, COLIN J(OHN), b Glasgow, Scotland, June 30, 60; m 86; c 3. NONLINEAR OPTICS OF INERTIAL-CONFINEMENT-FUSION PLASMAS, NONLINEAR WAVES & INSTABILITIES. *Educ:* Univ Glasgow, BSc, 81; Univ Rochester, MS, 82, PhD(mech & aerospace sci), 86. *Prof Exp:* Postdoctoral fel, Los Alamos Nat Lab, 85-88; ASST PROF MECH ENG & SCIENTIST LAB LASER ENERGETICS, UNIV ROCHESTER, 88- *Concurrent Pos:* Vis scientist, Rutherford Appleton Lab, Eng, 87; NSF presidential young investr award, 90. *Res:* Nonlinear optics of inertial-confinement-fusion plasmas and related media; nonlinear waves and instabilities. *Mailing Add:* Dept Mech Eng Univ Rochester Rochester NY 14627

MCKINSTRY, DONALD MICHAEL, b Lancaster, Pa, June 10, 39; m 66; c 2. TOXINOLOGY. *Educ:* Univ Md, BS, 64, MS, 65, PhD(dairy sci), 71. *Prof Exp:* Lab technician dairy sci, Univ Md, 67-70; asst prof biol, 70-75, ASSOC PROF BIOL, BEHREND COL, PA STATE UNIV, 75- *Mem:* Sigma Xi. *Res:* Toxicity of venemous animals. *Mailing Add:* Behrend Col Pa State Univ Station Rd Erie PA 16563

MCKINSTRY, DORIS NAOMI, b McVeytown, Pa, Sept 8, 36. CLINICAL PHARMACOLOGY. *Educ:* Pa State Univ, BS, 58; Univ Pa, PhD(pharmacol), 65. *Prof Exp:* Res assoc pharmacol, Merck Sharp & Dohme Res Labs, 58-61; sr scientist, McNeil Labs, Inc, 66-69; sr res toxicologist, Merck Sharp & Dohme Res Labs, 69-70; sr res investr, 71-75, assoc clin pharmacol dir, 75-79, DIR CLIN PHARMACOL, SQUIBB INST MED RES, 79- *Concurrent Pos:* Fel pharmacol, Univ Pa, 65-66. *Mem:* Am Soc Pharmacol & Exp Therapeut; Am Soc Clin Pharmacol & Therapeut; Am Col

Clin Pharmacol; NY Acad Sci; AAAS. *Res:* Cardiovascular pharmacology and physiology; radiocontrast agents; anti-inflammatory agents; ancitiotics. *Mailing Add:* Squibb Inst for Med Res Clin Pharmacol Div PO Box 4000 Princeton NJ 08540

MCKINSTRY, HERBERT ALDEN, b Rochester, NY, Apr 22, 25; m 45; c 4. SOLID STATE PHYSICS. *Educ:* Alfred Univ, BS, 47; Pa State Univ, MS, 50, PhD, 60. *Prof Exp:* Res asst, 47-60, res assoc, 60-64, asst prof, 64-69, ASSOC PROF SOLID STATE SCI, PA STATE UNIV, UNIV PARK, 69- *Res:* X-ray diffraction; x-ray fluorescence; computer modeling of material properties; computer generated movies for instruction; ceramics. *Mailing Add:* 522 S Pugh St State College PA 16801

MCKINSTRY, KARL ALEXANDER, b Phoenix, Ariz, Oct 10, 43; m 68; c 3. CHEMICAL ENGINEERING. *Educ:* Univ Mich, Ann Arbor, BSE, 66; Col Sch Mines, MS, 69, PhD(chem eng), 70. *Prof Exp:* Group leader res & develop, 70-80, MGR PROCESS ENG & RES ASSOC PHARMACEUT & PROCESS DESIGN, DOW CHEM CO, 80- *Mem:* Am Chem Soc; Am Inst Chem Eng. *Res:* Process engineering; process development; reaction engineering; heat transfer; evaporation; pharmaceuticals. *Mailing Add:* 2205 Parkwood Midland MI 48640-3240

MCKISSON, R(ALEIGH) L(EWELLYN), b Stockton, Calif, Feb 10, 22; m 44; c 2. CHEMICAL ENGINEERING. *Educ:* Univ Calif, BS, 47, MS, 48, PhD(chem eng), 50. *Prof Exp:* Chemist, Radiation Lab, Univ Calif, 50; chem engr, Calif Res & Develop Co, 50-53; proj leader, Food Mach & Chem Corp, 53-55; sect leader, NAm Aviation, Inc, Rockwell Int, 55-71, mgr, Energy Systs Group, 71-84; RETIRED. *Mem:* Am Chem Soc; Sigma Xi. *Res:* High temperature chemistry; liquid metal chemistry; flue gas desulfurization. *Mailing Add:* PO BOx 1508 Blue Jay CA 92317-1508

MCKITRICK, MARY CAROLINE, b Chicago, Ill, Nov 17, 55. EVOLUTION, PHYLOGENETIC SYSTEMATICS. *Educ:* Princeton Univ, AB, 78; Univ Ariz, MS, 81; Univ Pittsburgh, PhD(biol), 84. *Prof Exp:* Frank M Chapman postdoctoral fel ornith, Am Mus Nat Hist, 85-86; ASST PROF BIOL, UNIV MICH, 86- *Concurrent Pos:* Counr, Soc Syst Biol, 85-90. *Mem:* Am Ornithologists' Union; Soc Study Evolution. *Res:* Avian systematics; patterns of evolution in archosaur limb musculature using phylogenetic approach; ontogenetic patterns in archosaur hindlimb musculature; evolution of behavior. *Mailing Add:* Mus Zool Univ Mich Ann Arbor MI 48109

MACKIW, VLADIMIR NICOLAUS, b Stanislawiw, Western Ukraine, Sept 4, 23; nat Can; m 51; c 3. INORGANIC CHEMISTRY, PHYSICAL CHEMISTRY. *Educ:* Univ Breslau, dipl, 46; Univ Erlangen, MS, 46. *Hon Degrees:* DSc, Univ Alta, 76. *Prof Exp:* Chemist, Lingman Lake Gold Mines, 48 & Prov Bur Mines, Man, 49; res chemist, Sherritt Gordon Ltd, 49-52, dir res, 52-55, dir Res & Develop Div, 55-64, dir, Sherritt Gordon Mines Ltd, 64-90, vpres 67-68, vpres Technol & Corp Develop, 68-72, exec vpres, 72- 88, CONSULT, SHERRITT GORDON LTD, 88- *Concurrent Pos:* Mem, Nat Res Coun Can, 71-77; mem, Nat Adv Comt Mining & Metall Res, 72-79, co-chmn, 75-79; chmn, Nickel Develop Inst, 84-; adv, govt technol missions, Belgium, USSR, & People's Repub China; invited lectr, Toronto, BC, McMaster, Waterloo, Windsor & Alta Univs. *Honors & Awards:* Jules Garnier Prize, Fr Metall Soc, 66; Int Nickel Co Medal & Inco Platinum Medal, Can Inst Mining & Metall, 66, Airey Award, Metall Soc, 72; R S Jane Mem Lect Award, Chem Inst Can, 67; Gold Medal, Inst Mining & Metall Engrs, UK, 77; James Douglas Gold Medal Award, Am Inst Mining, Metall & Petrol Engr, 91. *Mem:* Fel Chem Inst Can; Can Inst Mining & Metall; Am Inst Mining, Metall & Petrol Engrs; Am Powder Metall Inst; Soc Chem Indust Can; Shevchenco Sci Soc (past pres). *Res:* Hydrometallurgical processes for recovery of nickel, cobalt, copper, zinc, lead and other non-ferrous metals from their ores and concentrates; commercial production of wrought forms of nickel, colbalt and their alloys from metallic powders; inorganic chemicals; kinetics and thermodynamics of inorganic reactions; powder metallurgy; processes for production of synthetic fertilizers; author and co-author of over fifty publications; holder of over forty-five patents. *Mailing Add:* 9 Blair Athol Crescent Etobicoke ON M9A 1X6 Can

MACKLEM, PETER TIFFANY, b Kingston, Ont, Oct 4, 31; m 54; c 5. PULMONARY PHYSIOLOGY, EXPERIMENTAL MEDICINE. *Educ:* Queen's Univ, Ont, BA, 52; McGill Univ, MD, CM, 56; FRCPS(C), 63. *Prof Exp:* Fel, Royal Victoria Hosp, Montreal, Que, 60-61, res fel, 61-63; Meakins Mem fel, 63-64; McLaughlin traveling res fel, Sch Pub Health, Harvard Univ, 64-65; from asst prof to assoc prof, 65-71, chmn, dept med, 80-85, PROF EXP MED, MCGILL UNIV, 71-; PHYSICIAN-IN-CHIEF, ROYAL VICTORIA HOSP, 79- *Concurrent Pos:* Watson scholar, McGill Univ, 61-62; asst physician, Royal Victoria Hosp, 67-71, sr physician, 72-, dir, Meakins-Christie Labs, 72-79. *Mem:* Am Physiol Soc; Am Soc Clin Invest; Am Thoracic Soc; Can Soc Clin Invest; Asn Am Physicians. *Res:* Mechanical properties of lungs; relationship between lung structure and function; airway dynamics. *Mailing Add:* Dept Physiol & Med Montreal Chest Hosp 3650 St Urboun St Montreal PQ H2X 2P4 Can

MACKLER, BRUCE, b Philadelphia, Pa, May 23, 20; m 49; c 2. PEDIATRICS. *Educ:* Temple Univ, MD, 43. *Prof Exp:* Intern, Temple Univ Hosp, 43-44; resident physician pediat, Willard Parker Hosp, New York, 44; resident physician, Univ Iowa, 46-47; resident physician, Children's Hosp, Univ Cincinnati, 47-48, res assoc, 50-53, asst prof pediat, Univ, 53-54; asst prof enzyme chem, Univ Wis, 55-57; assoc prof, 57-61, PROF PEDIAT, SCH MED, UNIV WASH, 61- *Concurrent Pos:* USPHS fel res found, Children's Hosp, Univ Cincinnati, 48-50; fel, Inst Enzyme Res, Univ Wis, 53-55; estab investr, Am Heart Asn, 55-60. *Honors & Awards:* Borden Award, 68. *Mem:* Am Soc Biol Chem; Am Soc Pediat Res. *Res:* Carbohydrate metabolism; metalloflavoproteins; electron transport systems. *Mailing Add:* Dept Pediat Univ Wash Seattle WA 98195

MACKLER, BRUCE F, b Philadelphia, Pa, Mar 11, 42; m; c 2. IMMUNOLOGY, CHEMISTRY. *Educ:* Temple Univ, Philadelphia, Pa, BS, 64; Pa State Univ, State Col, MS, 65; Univ Ore, Portland, PhD(immunol & microbiol), 70. *Hon Degrees:* JD, STex Col Law, Houston, 79. *Prof Exp:* Fel, Kennedy Inst Rheumatology, London, UK, 70-72; vis scientist, NIH, Nat Inst Dent Res & Litton Bionetics, Inc, 72-74; assoc prof, Dent Br, Dent Sci Inst, Univ Tex, Houston, 74-79; assoc, Weitzmann & Royal, Washington, DC, 79-80 & Leighton, Conklin, Lemov, Jacobs & BuCkley, Washington, DC, 80-81; of coun, William J Skinner, Rockville, Md, 81-83 & Meyer, Faller, Wiseman & Greenberg, PC, Washington, DC, 83-84; partner, Mackler & Gibbs, PC, Washington, DC, 84-90; GEN COUN, ASN BIOTECHNOL CO, WASHINGTON, DC, 83-; PARTNER, BAKER & HOSTETLER, WASHINGTON, DC, 91- *Concurrent Pos:* Vis assoc prof, Dept Pediat, Baylor Col Med, Houston, Tex, 74-79; assoc mem, Grad Sch Biomed Sci, Univ Tex, Houston, 74-79; assoc mem, Grad Sch, Univ Md, Baltimore, 79-83 & clin assoc prof, Dept Oral Path, Dent Sch, 79-82; co-prin investr, NIH, 82-85. *Mem:* AAAS. *Res:* Author of various publications. *Mailing Add:* 9124 Wandering Trail Dr Potomac MD 20854

MACKLER, SAUL ALLEN, b New York, NY, Dec 9, 13; m 40; c 3. SURGERY. *Educ:* Columbia Univ, BS, 33; Univ Chicago, MD, 37. *Prof Exp:* Intern, Michael Reese Hosp, 38-39, resident surg, 40; assoc prof thoracic surg, Med Sch, Univ Chicago, 47-; prof thoracic surg, Cook County Grad Sch Med, 46-; AT DEPT SURG, UNIV HEALTH SCI-CHICAGO MED. *Concurrent Pos:* Fel thoracic surg, Barnes Hosp, St Louis, 41 & 42; attend, assoc & consult thoracic surgeon var hosps, 46-; chmn dept surg, Michael Reese Hosp, 62-64. *Mem:* Fel AMA; Soc Thoracic Surg; fel Am Col Surg; fel Int Col Surg; fel Am Col Chest Physicians; Am Asn Thoracic Surg. *Res:* Physiology and disease of the esophagus and mediastinum; cancer of the esophagus; disease of the heart and great vessels; injuries of the chest. *Mailing Add:* 111 Woodly Rd Winnetka IL 60093

MACKLES, LEONARD, b New York, NY, Jan 17, 29; m 54; c 2. COSMETIC & PHARMACEUTICAL CHEMISTRY. *Educ:* Long Island Univ, BS, 51. *Prof Exp:* Org chemist, Colloids Inc, 53-56, head org chemist, Arlen Chem Corp, 56-58; tech dir, Chemclean Corp, 58-61; head chemist, Schenley Res Inst, 61-63, asst dir res, 63-65; sr res scientist prod develop, Bristol Myers Co, 65-73, prin res investr concept develop, Prod Div, 73-84; PRIN, LINK LABS, 84- *Mem:* Fel AAAS; Am Chem Soc; fel Am Inst Chemists; NY Acad Sci; Soc Cosmetic Chemists. *Res:* Development of consumer products in the fields of pharmaceuticals, toiletries and household specialties. *Mailing Add:* 311 E 23rd St New York NY 10010

MACKLIN, JOHN WELTON, b Ft Worth, Tex, Dec 11, 39; c 1. INORGANIC CHEMISTRY, SPECTROSCOPY. *Educ:* Linfield Col, BA, 62; Cornell Univ, PhD(inorg chem), 68. *Prof Exp:* ASSOC PROF CHEM, UNIV WASH, 68- *Mem:* Am Chem Soc; Microbeam Anal Soc; Nat Orgn Black Chemists & Chem Engrs; Int Soc Study Origin Life. *Res:* Spectroscopic measurements, particularly Raman, applied to elucidation of structural characteristics of inorganic solids, liquids and solutions. *Mailing Add:* Dept of Chem Univ of Wash Seattle WA 98195

MACKLIN, MARTIN, b Raleigh, NC, Aug 27, 34; m 79; c 4. PSYCHOPHARMOCOLOGY, ALCOHOLISM. *Educ:* Cornell Univ, BME, 57, MIE, 58; Case Western Reserve Univ, PhD(biomed eng), 67, MD, 77; Am Bd Psychiat & Neurol, dipl, 83. *Prof Exp:* Instr mech eng, Cornell Univ, 56-58; sr engr, Hamilton Standard Div, United Aircraft Corp, 58-61; prod planning specialist, Moog Servocontrols, Inc, 61-62; staff specialist, Thompson-Ramo-Wooldridge, Inc, 62-65; asst prof biomed eng, Case Western Reserve Univ, 67-72, assoc prof, 72-83, asst prof psychiat, 81-83; ADMIN DIR, MENTAL HEALTH SERV, ASHTABULA COUNTY MED CTR, 83- *Concurrent Pos:* Established investr, Am Heart Asn, 69-74; resident psychiat, Case Western Reserve Univ Hosps; admin dir, Ment Health Serv, Ashtabula County Med Ctr, 83- *Mem:* Am Psychiat Asn; Am Med Soc Alcoholism & Drug Dependencies; Soc Gen Physiologists. *Res:* Nature and causes of alcoholism and drug dependence; transport of ions in epithelia and intestinal villi; fetal electrocardiography. *Mailing Add:* 345 Rogers Pl Ashtabula OH 44004

MACKLIN, PHILIP ALAN, b Richmond Hill, NY, Apr 13, 25; m 53; c 3. QUANTUM MECHANICS, MATHEMATICAL PHYSICS. *Educ:* Yale Univ, BS, 44; Columbia Univ, MA, 49, PhD(physics), 56. *Prof Exp:* Physicist, Carbide & Carbon Chem Corp, Tenn, 46-47; res scientist, AEC, Columbia Univ, 49-51; instr physics, Middlebury Col, 51-54, actg chmn dept, 53-54; from asst prof to assoc prof, 54-61, chmn dept, 72-85, PROF PHYSICS, MIAMI UNIV, 61- *Concurrent Pos:* Vis prof, Univ NMex, 57-68; physicist, Los Alamos Sci Labs, 60-62; vis prof, Boston Univ, 85-86; vis scientist, Mass Inst Technol, 85-86. *Mem:* AAAS; Am Phys Soc; Am Asn Physics Teachers; Sigma Xi. *Res:* Beta and gamma spectroscopy; interpretation of quantum mechanics. *Mailing Add:* 211 Oakhill Dr Oxford OH 45056

MACKLIN, RICHARD LAWRENCE, b Jamaica, NY, Dec 24, 20; m 45; c 4. NUCLEAR PHYSICS. *Educ:* Yale Univ, BS, 41, PhD(org chem), 44. *Prof Exp:* Lab asst, Yale Univ, 41-44; chemist, Indust Labs, Carbide & Carbon Chem Co, 44-48, sr physicist, 48-52; sr physicist, Oak Ridge Nat Lab, 52-90; RETIRED. *Mem:* Fel AAAS; fel Am Phys Soc; Int Asn Geochem & Cosmochem; Sigma Xi. *Res:* Radioactivity; nuclear data; neutron capture experiments; nuclear physics instrumentation; astrophysics and cosmology experiments; nuclear safety. *Mailing Add:* 225 Outer Dr Oak Ridge TN 37830-3810

MACKLIN, RUTH, b Newark, NJ, Mar 27, 38; m 57; c 2. BIOMEDICAL ETHICS, HEALTH POLICY. *Educ:* Cornell Univ, BA, 58; Case Western Reserve Univ, MA, 66, PhD(philos), 68. *Prof Exp:* From asst prof to assoc prof philos, Case Western Reserve Univ, 68-76; assoc behav studies, Hastings Ctr, 76-80; assoc prof, 80-84, PROF BIOETHICS, ALBERT EINSTEIN COL MED, 84- *Concurrent Pos:* Fel, Hastings Ctr, 81; mem, Inst Med, Nat Acad Sci, 89; consult, Centers Dis Control, 89-90; appointee, Sci & Ethical

Rev Group, WHO, 89- & Nat Biotechnol Policy Bd, 91-; consult, Human Genome Initiative, NIH, 90- *Mem:* Inst Med-Nat Acad Sci; Am Soc Law & Med; Am Philos Asn. *Res:* Ethical and legal issues in clinical medical practice and health policy, including human subjects research, AIDS research and policies; new reproductive technologies; hospital ethics committees; decisions surrounding life-sustaining therapy. *Mailing Add:* Albert Einstein Col Med 1300 Morris Park Ave Bronx NY 10461

MCKLVEEN, JOHN WILLIAM, b Washington, DC, May 31, 43; m 66; c 2. NUCLEAR ENGINEERING, ENVIRONMENTAL ENGINEERING. *Educ:* US Naval Acad, BS, 65; Univ Va, ME, 71, PhD(nuclear environ eng), 74. *Prof Exp:* RADIATION SAFETY OFFICER & MEM FAC ELEC ENG, ARIZ STATE UNIV, 74- *Concurrent Pos:* Consult, low-level radiation measurements, environ monitoring, radioactive waste disposal, fast neutorn activation analysis & energy educ, several nat labs, utilities, indust & mining orgns, 74-; pres, Radiation & Environ Monitoring, Inc, 81-; radiation safety officer, Ariz State Univ, 74-80; adv radiation & hazardous mat, Ariz Senate, 79-; chmn, Environ Sci Div, Am Nuclear Soc, 83-84. *Honors & Awards:* Nat Pub Commun Award, Am Nuclear Soc, 86. *Mem:* Am Nuclear Soc; Health Physics Soc; Sigma Xi. *Res:* Uranium exploration; mining and milling; assay of natural radioactivity; uranium and thorium by liquid scintillation techniques; fast neutron activation analysis applications; energy education. *Mailing Add:* Col Eng & Appl Sci Ariz State Univ Tempe AZ 85287

MACKMAN, NIGEL, b Rustington, Eng, Dec 18, 59. MOLECULAR BIOLOGY. *Educ:* Univ Leicester, BS, 81, PhD(genetics), 84. *Prof Exp:* Postdoctoral fel, Univ Leicester, 84-87; postdoctoral fel, 87-89, ASST MEM, RES INST SCRIPPS CLIN, 89- *Concurrent Pos:* Mem, Thrombosis Coun, Am Heart Asn. *Mem:* Am Heart Asn; Am Soc Microbiol; Am Soc Biochem & Molecular Biol. *Mailing Add:* Dept Immunol Res Inst Scripps Clin 10666 N Torrey Pines Rd La Jolla CA 92037

MCKNEALLY, MARTIN F, EXPERIMENTAL BIOLOGY. *Prof Exp:* THORACIC SURGEON, TORONTO GEN HOSP, 90- *Mailing Add:* Div Thoracic Surg Toronto Gen Hosp Eaton N 10-226 200 Elizabeth St Toronto ON M5G 2C4 Can

MACKNIGHT, FRANKLIN COLLESTER, paleontology, science education, for more information see previous edition

MCKNIGHT, JAMES POPE, b Arlington, Tenn, Sept 19, 21; m 49; c 4. DENTISTRY. *Educ:* Memphis State Univ, BS, 48; Univ Tenn, DDS, 51, cert, 52; Ind Univ, MSD, 64. *Prof Exp:* Pvt pract, 52-56; from instr to assoc prof, Col Dent, Univ Tenn, Memphis, 56-69, prof pedodont & chmn dept, 69-86, consult, 87-90; RETIRED. *Mem:* Am Dent Asn; Am Soc Dent for Children. *Res:* Treatment of the dental pulp; dental care of handicapped children. *Mailing Add:* Dept Pediat Dent Univ of Tenn Col of Dent Memphis TN 38103

MCKNIGHT, JOHN LACY, b Monroe, Mich, Sept 13, 31; m 64; c 1. THEORETICAL PHYSICS, HISTORY OF SCIENCE. *Educ:* Univ Mich, AB, 53; Yale Univ, MS, 54, PhD(physics), 57. *Prof Exp:* Asst prof, 57-59, assoc prof, 59-68, PROF PHYSICS, COL WILLIAM & MARY, 68- *Mem:* Am Phys Soc; Soc Hist Technol; Philos Sci Asn; Hist Sci Soc; Sigma Xi. *Res:* Logical foundations of quantum mechanics; 18th century physics and scientific apparatus; history of scientific ideas. *Mailing Add:* Dept Physics Col William & Mary Williamsburg VA 23185

MCKNIGHT, LEE GRAVES, b Washington, DC, Sept 7, 33; m 55; c 2. CHEMICAL PHYSICS. *Educ:* Va Mil Inst, BS, 55; Univ Mich, MS & PhD(chem), 61. *Prof Exp:* NATO fel, Univ Col, Univ London, 61-62; lectr chem, Univ Mich, 63; mem tech staff physics, Bell Labs, Whippany, NJ, 63-73, MEM TECH STAFF, BELL LABS, MURRAY HILL, NJ, 73- *Mem:* Am Chem Soc; Am Phys Soc. *Res:* Ion-molecule interactions; carbon arcs; computer-process monitor-control. *Mailing Add:* Bell Labs Rm 30E 033 Murray Hill NJ 07974

MCKNIGHT, LEE WARREN, b Victoria, Tex, Nov 19, 56; m 87. HIGH RESOLUTION SYSTEMS, BROADBAND NETWORKS. *Educ:* Tufts Univ, BA, 78; John Hopkins Univ, MA, 81; Mass Inst Technol, PhD(polit sci), 89. *Prof Exp:* Consult, Kalba Bowen Assoc, 84-85; res asst, Dept Polit Sci, 82-84 & 85-88, consult, Media Lab, 89, POSTDOCTORAL ASSOC, CTR TECHNOL POLICY & INDUST DEVELOP, MASS INST TECHNOL, 89- *Concurrent Pos:* Max Planck Found res fel, 85; mem, Open High Resolution Systs; contrib, Promethee Inst, Paris, 87-; postdoctoral fel, Mass Inst Technol, 89. *Honors & Awards:* Stein Rokkan Award, Int Pol Sci Asn, 85. *Mem:* Inst Elec & Electronic Engrs; Am Polit Sci Asn; Acad Polit Sci; Int Polit Sci Asn; Int Commun Asn. *Res:* International technology policy for information and communication industries; cross-industry and cross-technology dialog with government and academics on critical issues; design audience research studies on new communication technologies. *Mailing Add:* Ctr Technol Policy & Indust Develop Mass Inst Technol Cambridge MA 02139

MCKNIGHT, RANDY SHERWOOD, b Los Angeles, Calif, June 18, 43. APPLIED MATHEMATICS. *Educ:* Univ Calif, BS, 66; Rice Univ, MS, 69, PhD(math sci), 72. *Prof Exp:* Sci programmer, IBM Corp, 66; res scientist geophys & reservoir modeling, 71-78, MGR GEOPHYS RES, DENVER RES CTR, MARATHON OIL CO, 78- *Mem:* Soc Indust & Appl Math; Inst Elec & Electronics Engrs; Sigma Xi. *Res:* Application of optimization theory and numerical analysis to direct and inverse problems in exploration geophysics and petroleum engineering; system theory to processing and interpretation of seismic data. *Mailing Add:* Denver Res Ctr Marathon Oil Co PO Box 269 Littleton CO 80160

MCKNIGHT, RICHARD D, b Cincinnati, Ohio, June 30, 44; m 78. REACTOR PHYSICS. *Educ:* Univ Cincinnati, BS, 67, MS, 69, PhD(nuclear eng), 73. *Prof Exp:* NUCLEAR ENGR, APPL PHYSICS DIV, ARGONNE NAT LAB, 73- *Mem:* Am Nuclear Soc. *Res:* Zero power reactor critical assembly theory and analysis; measurement and calculation of integral reactor parameters; reactor analysis methods development and validation; nuclear data testing. *Mailing Add:* Reactor Anal Div Argonne Nat Lab 9700 S Cass Ave Argonne IL 60439

MCKNIGHT, STEVEN L, b El Paso, Tex, Aug 27, 49; m 78. EMBRYOLOGY. *Educ:* Univ Tex, BA, 74; Univ Va, PhD(biol), 77. *Prof Exp:* Asst staff mem, Develop Biol Prog, Fred Hutchinson Cancer Res Ctr, 81-84; staff assoc, 79-81, STAFF MEM, DEPT EMBRYOL, CARNEGIE INST WASH, 84-, INVESTR, HOWARD HUGHES MED INST, 88- *Concurrent Pos:* Andrew Fleming Award biol res, Univ Va, 77; Helen Hay Whitney Found fel, 77-80; res grant, NIH, 80-88; Basil O'Connor res grant, 82-84; ed, Molecular & Cellular Biol, 88-; mem, Study Sect Molecular Cytol, Div Res Grants, NIH, 85-88, chmn, 88-90. *Honors & Awards:* Dai Nakada Mem Lectr, Univ Pittsburgh, 85; DeWitt Stetten Lectr, NIH, 87; Eli Lilly Award, 89; Newcomb Cleveland Award, AAAS, 89; Steinburg/Wylie Lectr, Univ Md, 90; Monsanto Award, Nat Acad Sci, 91; Bernard Cohen Lectr, Univ Pa, 91; Gerhard Schmidt Lectr, Tufts Univ, 91. *Res:* Author of various publications. *Mailing Add:* Dept Embryol Carnegie Inst Washington 115 W University Pkwy Baltimore MD 21210

MCKNIGHT, THOMAS JOHN, b Marietta, Ohio, Nov 5, 06; c 3. PARASITOLOGY. *Educ:* Okla Agr & Mech Col, BS, 25; Univ Okla, MS, 47, PhD, 59. *Prof Exp:* Instr high sch, 29-30; supt schs, 31-41; instr zool, Univ Okla, 46; prof biol, East Cent State Col, 47-76, chmn dept, 58-76, emer prof & consult, 76-; RETIRED. *Concurrent Pos:* Consult, Environ Survs, Kerr Magee Corp, Okla, 74-75 & US Corps Engrs, 73-74. *Mem:* AAAS; Am Soc Parasitologists; Nat Sci Teachers Asn; fel Royal Soc Health. *Res:* Bacteriology; animal parasitology and microbiology; taxonomy and physiology of parasites; taxonomy of parasites of reptiles, soil bacteria and Actinomycetes. *Mailing Add:* 1020 S Stockton Ada OK 74821

MCKNIGHT, WILLIAM BALDWIN, b Macon, Ga, July 4, 23; m 55; c 2. LASERS. *Educ:* Purdue Univ, BS, 50; Oxford Univ, PhD(physics), 68. *Prof Exp:* Physicist, Navy Underwater Sound Reference Lab, 52-53; test engr, Ord Missile Labs, Redstone Arsenal, 53-56, chief, Infrared Br, 56-58, Electro-Optical Br, 58-62 & Appl Physics Br, Res & Develop Directorate, US Missile Command, 62-74; RES PROF, UNIV ALA, HUNTSVILLE, 74-; PRES, TECHNOL RES ASSOC, 84- *Mem:* Am Phys Soc; Optical Soc Am; Inst Elec & Electronics Engrs. *Res:* Infrared radiation and detection; rocketry; missile guidance systems; solid state, molecular and x-ray lasers. *Mailing Add:* RR 1 Box 141A Gordo AL 35466-9728

MAC KNIGHT, WILLIAM JOHN, b New York, NY, May 5, 36; m 67. PHYSICAL CHEMISTRY. *Educ:* Univ Rochester, BS, 58; Princeton Univ, MA, 63, PhD(phys chem), 64. *Prof Exp:* Res assoc, Princeton Univ, 64-65; from asst prof to assoc prof, 65-74, head, Dept Polymer Sci & Eng, 76-85, PROF CHEM, UNIV MASS, AMHERST, 74-, PROF POLYMER SCI & ENG, 85-, HEAD DEPT, 89- *Concurrent Pos:* Guggenheim fel, 85. *Honors & Awards:* Am Phys Soc Prize High Polymer Physics, 84. *Mem:* Am Chem Soc; fel Am Phys Soc; fel AAAS. *Res:* Physical chemistry of high polymers; sulfur chemistry. *Mailing Add:* Polymer Sci & Eng Univ of Mass Amherst MA 01003

MACKO, DOUGLAS JOHN, b Tarrytown, NY, Jan 27, 43; m 69; c 2. DENTISTRY, PEDIATRICS. *Educ:* Bates Col, BS, 65; Univ Pa, DMD, 69; Univ Conn, Pediat Dent, 75. *Prof Exp:* ASST PROF PEDIAT DENT, UNIV CONN HEALTH CTR, 75-, ASST PROF PEDIAT, 76- *Concurrent Pos:* Prin investr, Nat Inst Dent Res, 78- *Mem:* Int Soc Dent Res; Am Acad Pediat Dent. *Res:* Basic and applied research in dental caries and its prevention; salivary research related to glucose metabolism; histochemical studies of materials on the dental pulp. *Mailing Add:* Two White Oak Rd Farmington CT 06032

MACKO, STEPHEN ALEXANDER, b Mobile, Ala, Sept 21, 51; m 77; c 3. ORGANIC GEOCHEMISTRY, ISOTOPE GEOCHEMISTRY. *Educ:* Carnegie-Mellon Univ, BS & BA, 73; Univ Maine, Orono, MS, 76; Univ Tex, Austin, PhD(chem), 81. *Prof Exp:* Fel geochem, Geophys Lab, Wash, 81-83; from asst prof to assoc prof earth sci, Mem Univ Nfld, 83-89; ASSOC PROF ENVIRON SCI, UNIV VA, 89- *Concurrent Pos:* Res collabr, Brookhaven Nat Lab, 82-85, guest assoc scientist, 86-88; assoc ed, Can Bull Petrol Geol, 88-90, Geochimica et Cosmochimica Acta, 90-; invited guest researcher, Geotop, Univ Que, Montreal, 89-90; panel mem, Ocean Drilling Proj, SGPP, 89-90; vis investr, Geophys Lab, Wash, 90. *Honors & Awards:* President's Medal for Outstanding Res, Mem Univ, 87. *Mem:* Geochem Soc (secy-treas, 88-); Europ Asn Org Geochem; Am Geophys Union; Sigma Xi. *Res:* Marine nitrogen cycle; diagenesis; fossils; extraterrestrial organic matter. *Mailing Add:* Dept Environ Sci Univ Va Charlottesville VA 22903

MCKONE, THOMAS EDWARD, b Cresco, Iowa, Apr 14, 51; m 87; c 1. ENVIRONMENTAL FATE & EXPOSURE. *Educ:* St Thomas Col, BA, 74; Univ Calif, Los Angeles, MS, 77, PhD(eng), 81. *Prof Exp:* Mem tech staff, TRW Energy Systs Group, 79-81; fel adv comt reactor safeguards, US Nuclear Regulatory Comn, 81-83; RES ENGR, LAWRENCE LIVERMORE NAT LAB, UNIV CALIF, 83- *Concurrent Pos:* Fel, Southern Calif Edison, 78 & Adv Comt Reactor Safeguards, 80; chair, Risk Assessment/Mgt Comt, Air Pollution Control Asn, 84-; nat lectr, NAm Asn Environ Educ, 84-85; vis scientist, Harvard Univ, 87-88; ed bd, Risk Analysis. *Mem:* Air & Waste Mgt Asn; AAAS; Soc Risk Analysis; Int Soc Ecol Econ; Int Soc Exposure Analysis. *Res:* Chemical transport and accumulation of toxic chemicals in multiple environmental media (air, water, soil); developing multimedia compartment models that can be used in quantitative risk assessments; human exposure and health risk assessment. *Mailing Add:* Lawrence Livermore Nat Lab L-453 PO Box 5507 Livermore CA 94550

MACKOWIAK, ELAINE DECUSATIS, b Hazleton, Pa, Apr 28, 40; wid; c 2. PHARMACOLOGY, PHARMACEUTICAL CHEMISTRY. *Educ:* Temple Univ, BS, 62, MS, 65; Thomas Jefferson Univ, PhD(pharmacol), 74. *Prof Exp:* Asst chief pharmacist, Holy Redeemer Hosp, Meadowbrook, Pa, 62-63; lectr radiol health, Sch Dist Philadelphia, 64-68; from instr to assoc prof, 64-86, actg chmn, 74-75, CHMN, DEPT PHARM PRACTICE, 85-, PROF PHARMACEUT CHEM, TEMPLE UNIV, 86- *Mem:* Am Pharmaceut Asn; Health Physics Soc; Sigma Xi; Am Asn Col Pharm. *Res:* Melanin formation; human autopsied samples for research; enzyme purification; pharmacy manpower, especially women and pharmaceutical education; trace metal analysis; radiopharmaceuticals; patient compliance; over the counter drugs. *Mailing Add:* Temple Univ Sch of Pharm 3307 N Broad St Philadelphia PA 19140

MACKOWIAK, ROBERT CARL, internal medicine, cardiology, for more information see previous edition

MCKOWN, CORA F, b Atoka, Okla, Aug 21, 43. RESOURCE MANAGEMENT, RESEARCH ADMINISTRATION. *Educ:* Southeastern State, Okla, BS, 64; Okla State, MS, 68; Univ Mo, PhD(int design & housing com devel), 72. *Prof Exp:* Asst assoc prof housing & design, Univ Ark, 72-77; prof housing & design, Tex Tech Univ, 77-87; PRES DESIGN FIRM & CONSULT ENERGY & DESIGN, 87- *Mem:* Am Asn Housing Educators; Environ Design Res Asn; Am Home Econ Asn. *Res:* Energy efficient housing and interior components; specification of interior materials. *Mailing Add:* 547 Sunny Acres Glenwood Springs CO 81601

MCKOY, BASIL VINCENT, b Trinidad, BWI, Mar 25, 38; US citizen; m 67; c 1. THEORETICAL CHEMISTRY. *Educ:* NS Tech Univ, BE, 60; Yale Univ, PhD(chem), 64. *Prof Exp:* From instr to assoc prof chem, 64-73, PROF THEORET CHEM, CALIF INST TECHNOL, 73- *Concurrent Pos:* Sloan Found fel, 69-71; Guggenheim fel, 73; consult, Lawrence Livermore Lab, Univ Calif, 75- & Inst Defense Analyses, 84- *Mem:* Fel Am Phys Soc. *Res:* Studies of single and multiphoton ionization processes in molecules; energy transfer in electron-molecule collisions in gases and adsorbed systems. *Mailing Add:* Dept of Chem Calif Inst of Technol Pasadena CA 91109

MCKOY, JAMES BENJAMIN, JR, b Americus, Ga, Oct 31, 27; m 47; c 5. POLYMER SCIENCE & TECHNOLOGY. *Educ:* Ga Inst Technol, BS, 48. *Prof Exp:* Engr chem synthesis, Union Carbide Corp, 48-52; tech supvr tire yarn, Chemstrand, Monsanto Textiles, US, 52-57, tech supvr textile yarns, 56-61 sect head nylon develop, 61-65, mgr acrilan develop, Monsanto Co, 65-66, prod tech dir, Monsanto Textiles, Europe, 67-70, sr res specialist monsanto textiles, US, 70-82; RETIRED. *Mem:* Am Inst Chem Engrs. *Res:* Polymer and fiber science, with particular interest in melt spinning dynamics and properties of fibers; fluid-bed catalysis, synthesis of organic chemicals; process control theory and practice. *Mailing Add:* 408 Tudor Dr No 1B Cape Coral FL 33904-9474

MACKSEY, HARRY MICHAEL, b Detroit, Mich, Feb 20, 47; m 68. SEMICONDUCTORS. *Educ:* Univ Mich, Ann Arbor, BS, 68; Univ Ill, Urbana, MS, 70, PhD(physics), 72. *Prof Exp:* Fel physics, Univ Ill, Urbana, 72-73; MEM TECH STAFF, TEX INSTRUMENTS, INC, 73- *Mem:* Am Phys Soc; Inst Elec & Electronics Engrs. *Res:* Development of high power gallium arsenide field-effect transistors for microwave amplification. *Mailing Add:* Star Rte Box 182 Diga WA 98279

MACKSON, CHESTER JOHN, b Crystal Falls, Mich, July 14, 19; m 46; c 6. AGRICULTURAL ENGINEERING, PACKAGING. *Educ:* Mich State Univ, BS, 43, MA, 49, MS, 55; Cornell Univ, PhD(agr eng), 62. *Prof Exp:* High sch teacher, Mich, 49-51; from asst prof to assoc prof agr eng, 54-68, PROF AGR ENG, MICH STATE UNIV, 68-, DIR SCH PACKAGING, 77- *Concurrent Pos:* Ed, Packaging Sci & Technol. *Mem:* Am Soc Agr Engrs; Am Soc Eng Educ; Inst Food Technol. *Res:* Farm power and machinery; technical training. *Mailing Add:* Sch Packaging 142 Packaging Mich State Univ East Lansing MI 48824

MACKULAK, GERALD THOMAS, b Gary, Ind, Mar 19, 52; m 80. SIMULATION. *Educ:* Purdue Univ, BS, 74, MS, 75, PhD(indust eng), 79. *Prof Exp:* Systs analyst info systs, Burroughs Corp, 75-76; systs engr simulation, Pritsker & Assoc, 79-80; asst prof, 80-85, ASSOC PROF INDUST ENG, ARIZ STATE UNIV, 85- *Concurrent Pos:* Consult. *Res:* Application of simulation techniques to areas of production control, computer graphics and computer aided manufacturing; IDEF modeling methods. *Mailing Add:* Dept Indust-Mgt Systs Eng Ariz State Univ Tempe AZ 85287

MCKUSICK, VICTOR ALMON, b Parkman, Maine, Oct 21, 21; m 49; c 3. MEDICAL GENETICS. *Educ:* Johns Hopkins Univ, MD, 46; Am Bd Internal Med, dipl, 54. *Hon Degrees:* Numerous from US & foreign univs, 74-91. *Prof Exp:* Intern clin med, Johns Hopkins Univ/USPHS, 46-52, from instr to prof med, Johns Hopkins Sch Med, 52-85, chief div med genetics, Dept Med, 57-73, prof epidemiology & biol, 69-78, William Osler prof med, 78-85, chmn dept med, 73-85; physician-in-chief, 73-85, chief div med genetics, 85-89, PROF MED GENETICS, JOHNS HOPKINS HOSP, 85- *Concurrent Pos:* Exec chief cardiovasc unit, Baltimore Marine Hosp, 48-50; resident, Osler Med Clin, 51-52, physician, 53-; med adv bd, Howard Hughes Med Inst, 67-83; comt mapping & sequencing human genome, Nat Acad Sci, 86-88, chmn compt DNA technol, 89-91; pres, Int Med Cong, Ltd, 72-78, 8th Int Cong Human Genetics, 91; mem, Nat Adv Res Resources Coun, 70-74, bd sci adv, Roche Inst Molecular Biol, 67-71, human genome adv comt, NIH 88-; trustee, Jackson Lab, 79-; founding mem, Am Bd Med Genetics, 79-82; founder & pres, Human Genome Orgn, 88-90; co-chmn, Centennial, Johns Hopkins Med, 89-90. *Honors & Awards:* John Phillips Award, Am Col Physicians, 72; William A Allan Award, Soc Human Genetics, 77; Gairdner Int Award, 77; Kober Medal, Asn Am Physicians, 90. *Mem:* Nat Acad Sci; hon fel Am Acad Orthop Surg; Am Soc Clin Invest; Am Soc Human Genetics; fel Am Col Physicians; Royal Col Physicians London; AAAS; Am Philos Soc; Asn Am Physicians. *Res:* Medical genetics. *Mailing Add:* Dept Med Johns Hopkins Hosp Baltimore MD 21205

MACKWAY-GIRARDI, ANN MARIE, b Philadelphia, Pa, May 9, 43; m 88; c 2. MEDICAL INFORMATION ANALYSIS, CANCER PREVENTION-CARCINOGENESIS & CHEMOTHERAPY. *Educ:* Univ Del, BS, 65; Univ Pittsburgh, PhD(pharmacol), 84. *Prof Exp:* Tech specialist polymers, Patent Sect, Film Dept, DuPont, 65-68; fel diabetes res, Sch Med, Univ Pittsburgh, 84-86, lectr pharmacol, 86-88; MED INFO, CANCER, INFO VENTURES, INC, 89- *Mem:* Am Soc Pharmacol & Exp Therapeut. *Res:* Medical information organization and analysis-cancer chemoprevention, carcinogenesis and cancer chemotherapy; diabetic autonomic neuropathy. *Mailing Add:* 14 E Wayne Terr Collingswood NJ 08108

MCKYES, EDWARD, b Annapolis Royal, NS, Nov 24, 44; m 75; c 3. TILLAGE & CROP PRODUCTION. *Educ:* McGill Univ, BEng, 66, MEng, 67, PhD(soil mech), 69. *Prof Exp:* Res assoc soil mech, dept civil eng, McGill Univ, 69-72, lectr, 71-72, from asst prof to assoc prof, 72-82, chmn dept, 77-87, PROF SOIL MECH & AGR ENG, MCGILL UNIV, 82-, ASSOC DIR CTR DRAINAGE STUDIES, 89- *Concurrent Pos:* Vis lectr, dept civil eng, Univ Ariz, Tucson, 71; external examr fac agr, Univ WI, Trinidad & Tobago, 78-87; examr, Order Engrs Que, Montreal, 79-; ed, Can Agr Eng J & Can Soc Agr Eng, 79-, Soil & Tillage Res J & Int Orgn Soil Tillage, 80-; chmn, Nat Educ Comt, Can Soc Agr Eng, 80-81 & Can Expert Comt Mechanization, 81-84; secy, Can Comt Agr Eng Serv, 81-84; mem, Comt New Agr Eng Prog, Univ Zimbabwe, 87-92. *Honors & Awards:* Micro photog Award, Am Soc Testing & Mat, 71. *Mem:* Can Soc Agr Eng; Can Soc Terrain-Vehicle Systs; Int Soc Terrain-Vehicle Systs; Int Soil Tillage Res Orgn; Am Soc Agr Engrs. *Res:* Interactions among the physical and mechanical properties of soil; machinery traffic and soil manipulation; soil cutting and tillage; water movement and plant growth; machinery design, treatment of organic food production and processing waste materials. *Mailing Add:* Dept Agr Eng Box 950 MacDonald Col Ste-Anne-de Bellevue PQ H9X 1C0 Can

MCLACHLAN, DAN, JR, chemical physics; deceased, see previous edition for last biography

MACLACHLAN, GORDON ALISTAIR, b Saskatoon, Sask, June 30, 30; m 59; c 2. PLANT BIOCHEMISTRY. *Educ:* Univ Sask, MA, 54; Univ Man, PhD(biochem), 56. *Prof Exp:* Nat Res Coun res fel plant physiol, Imp Col, Univ London, 56-58, sci officer, Res Inst Plant Physiol, 58-59; asst prof bot, Univ Alta, 59-62; assoc prof, McGill Univ, 62-69, chmn dept, 70-75, dean grad studies & vprin res, 80-90, PROF BIOL, MCGILL UNIV, 70- *Concurrent Pos:* Assoc ed, Can J Biochem, 72-75, Can J Botany, 79-88, Plant Physiol, 80-85 & ed, Plant Mol Biol, 88-90; vis commonwealth prof, Australia, 75-76. *Honors & Awards:* Gold Medal, Can Soc Plant Physiol. *Mem:* Can Soc Plant Physiol (pres, 74); Am Soc Phytochem; Am Soc Plant Physiol; Am Soc Cell Biol; Can Soc Cell Biol (pres, 81); fel Royal Soc Can. *Res:* Metabolism of growing plants; biosynthesis and hydrolysis of cellulose and xyloglucan. *Mailing Add:* Dept Biol McGill Univ 1205 Penfield Ave Montreal PQ H3A 1B1 Can

MCLACHLAN, JACK (LAMONT), b Huron, SDak, Apr 1, 30; m 51; c 2. PHYCOLOGY. *Educ:* Ore State Univ, BSc, 53, MA, 54, PhD, 57. *Prof Exp:* Asst, Ore State Univ, 55-57; NIH res fel, Woods Hole Oceanog Inst, 57-59; NIH res fel appl biol, 60-61, from asst res officer to assoc res officer, 61-74, SR RES OFFICER, DEPT MARINE PLANTS, ATLANTIC REGIONAL LAB, NAT RES COUN CAN, 74- *Mem:* Phycol Soc Am; Am Soc Limnol & Oceanog; Brit Phycol Soc; Int Phycol Soc. *Res:* Marine algae. *Mailing Add:* 942 Brussels Halifax NS B3H 2T1 Can

MACLACHLAN, JAMES ANGELL, b Cambridge, Mass, May 18, 38; m 60; c 2. HIGH ENERGY PHYSICS, ACCELERATOR PHYSICS. *Educ:* Univ Mich, AB, 59; Yale Univ, MS, 62, PhD(physics), 68. *Prof Exp:* Consult programmer, Yale Comput Ctr, 69; PHYSICIST, FERMI NAT ACCELERATOR LAB, 69- *Concurrent Pos:* Dir, Endura Plastic Inc, Kirtland, Ohio. *Mem:* Am Phys Soc. *Res:* Accelerator and storage ring design-proton/anti proton-proton. *Mailing Add:* Fermi Nat Accelerator Lab PO Box 500 Batavia IL 60510

MACLACHLAN, JAMES CRAWFORD, b Detroit, Mich, Jan 13, 23; m 50. GEOLOGY. *Educ:* Wayne State Univ, AB, 48; Princeton Univ, MA, 51, PhD(geol), 52. *Prof Exp:* Consult geologist, Ministerio de Minas e Hidrocarburos, Caracas, Venezuela, 49-51; geologist, Mineral Deposits Br, US Geol Surv, 52-53 & Fuels Br, 53-56; mem explor projs group, Phillips Petrol Co, 57-62; consult, Shallow Well Explor Co, 62-67, secy-treas, 65-67; chmn dept earth sci, 70-74, from asst prof to prof geol, 67-87, EMER PROF, METROP STATE COL, 87- *Concurrent Pos:* Consult geologist, 62- *Mem:* Geol Soc Am; Am Asn Petrol Geol. *Res:* Stratigraphy and sedimentation of Paleozoic rocks of Eastern Colorado and Wyoming; Precambrian basement; mineralogy and petrology. *Mailing Add:* Dept Geol Metrop State Col 1006 11th St Denver CO 80204

MCLACHLAN, JOHN ALAN, b Pittsburgh, Pa, July 8, 43; m 68; c 3. DEVELOPMENTAL ENDOCRINOLOGY. *Educ:* Johns Hopkins Univ, BA, 65; George Washington Univ, PhD(pharmacol), 71. *Prof Exp:* Res assoc pharmacol, training prog, Nat Cancer Inst, 71-73, res assoc, lab reproductive develop toxicol, Nat Inst Environ Health Sci, 73-76, SECT HEAD, LAB REPRODUCTIVE DEVELOP TOXICOL, NAT INST ENVIRON HEALTH SCI, NIH, 76-, LAB CHIEF, 83- *Concurrent Pos:* Mem, Cancer Res Ctr, Univ NC, 78- *Honors & Awards:* Commendation Medal, USPHS, 80. *Mem:* AAAS; Am Asn Cancer Res; Am Soc Pharmacol & Exp Therapeut; Endocrine Soc. *Res:* Estrogen-induced differentiation defects in the reproductive tract of mammals, including development of hormone responsiveness, mechanisms of sex differentiation and cell transformation by estrogens in vivo and in vitro; transplacental carcinogenesis. *Mailing Add:* Lab Reproductive Develop Toxicol Nat Inst Environ Health Sci PO Box 12233 Research Triangle Park NC 27709

MCLACHLAN, RICHARD SCOTT, b London, Ont, June 6, 46; m 73; c 2. NEUROPHYSIOLOGY, EPILEPSY. *Educ:* Univ Western Ont, BSc, 68, MD, 72, FRCP(C), 78. *Prof Exp:* Teaching fel EEG, Epilepsy Unit, Univ Western Ont, 78-79 & MRC fel neurophysiol, Montreal Neurol Inst, McGill Univ, 81-83; ASST PROF CLIN NEUROSCI, PHYSIOL & MED & DIR, CLIN NEUROPHYSIOL LAB, EPILEPSY UNIT, UNIV WESTERN ONT, 83- *Concurrent Pos:* Career scientist, Ont Ministry Health, 83- *Honors & Awards:* Jasper Prize, Can Soc Clin Neurophysiologists, 83. *Mem:* Can Soc Clin Neurophysiologists; Can Neurol Soc; Am Epilepsy Soc; Int League Against Epilepsy; Soc Neurosci. *Res:* Neurophysiology of epilepsy both in vitro and in vivo in experimental models of epilepsy in animals and in humans undergoing surgery for treatment of their epilepsy. *Mailing Add:* 3 Linksgate Three Linksgate London ON N6G 2A6 Can

MCLACHLIN, JEANNE RUTH, GENE TRANSFER, RETROVIROL VECTORS. *Educ:* Univ Western Ont, London, Can, PhD(develop biol), 84. *Prof Exp:* VIS FEL, LAB MOLECULAR HEMAT, NIH, 84- *Mailing Add:* Lab Molecular Hemat NIH Bldg Ten Rm 7D05 Bethesda MD 20892

MCLAEN, DONALD FRANCIS, b Butte, Mont, Sept 22, 42; m 64; c 3. ORGANIC CHEMISTRY, PHOTOGRAPHIC CHEMISTRY. *Educ:* Carroll Col, Mont, BA, 64; Univ Nebr, Lincoln, PhD(org chem), 68. *Prof Exp:* NIH fel org chem, Univ Ill, Urbana, 68-69; RES ASSOC, EASTMAN KODAK CO, 69- *Mem:* Am Chem Soc; Soc Photog Sci & Eng. *Res:* Organic chemical research related to photographic chemistry and processing. *Mailing Add:* Prof Photography Div Eastman Kodak Co 8-69 Kodak Park Rochester NY 14650-1912

MCLAFFERTY, FRED WARREN, b Evanston, Ill, May 11, 23; m 48; c 5. ANALYTICAL CHEMISTRY. *Educ:* Univ Nebr, BS, 43, MS, 47; Cornell Univ, PhD(org chem), 50. *Hon Degrees:* DSc, Univ Nebr, 83 & Univ Liege, 89. *Prof Exp:* Fel, Univ Iowa, 50; chemist, Dow Chem Co, 50-52, div leader, Mass Spectrometry Sect, Spectros Lab, 52-56, dir, Eastern Res Lab, Framingham, Mass, 56-64; prof chem, Purdue Univ, 64-68; PROF CHEM, CORNELL UNIV, 68- *Concurrent Pos:* Chmn chem sect, AAAS, 84-85; mem, Chem Sci & Technol Bd, Nat Res Coun, 84-86; co-chmn, Chem Chinese Devolop Proj, World Bank, 85-86; mem, panel chem res, Bd Army Sci & Technol, Nat Res Coun, 86-89, Num Data Adv Bd, 87-90 & Bd Radioactive Waste Mgt, 90- *Honors & Awards:* Chem Instrumentation Award, Am Chem Soc, 71, Anal Chem Award, 81, Nichols Medal, 84, Oesper Award, 85 & S C Lind Award, 86; Sir J J Thomson Medal, Intl Mass Spectrometry, 85. *Mem:* Nat Acad Sci; AAAS; Am Chem Soc; Am Acad Arts & Sci; hon mem Ital Chem Soc; Am Chem Soc (chmn, Anal Chem Div). *Res:* Mass spectrometry; molecular structure determination; on-line computers. *Mailing Add:* Dept Chem Baker Lab Cornell Univ Ithaca NY 14853-1301

MCLAFFERTY, GEORGE H(OAGLAND), JR, b Newport, RI, June 11, 26; m 50; c 2. AERONAUTICAL ENGINEERING. *Educ:* Mass Inst Technol, BS, 47, MS, 48. *Prof Exp:* Res engr, Res Labs, United Aircraft Corp, 48-55, supvr inlet group, 55-59, asst to chief res engr, 59-62, sr prog mgr, 62-76; mgr chem, 76-80, mgr Nozzle Prog, Laser Develop, Pratt & Whitney Aircraft, 80-87; PRES, MCLAFFERTY CONSULT, INC, 87- *Mem:* Am Inst Aeronaut & Astronaut; AAAS. *Res:* Fluid dynamics, particularly supersonic and hypersonic inlets; all phases of advanced nuclear rockets; high power lasers; jet engine nozzles. *Mailing Add:* 642 Riverside Rd North Palm Beach FL 33408

MCLAFFERTY, JOHN J, JR, b Carbondale, Ill, July 6, 29; m 60; c 1. ANALYTICAL CHEMISTRY. *Educ:* Southern Ill Univ, BA, 51; Loyola Univ, Ill, MS, 64, PhD(anal chem), 66. *Prof Exp:* Lab supvr, Fansteel Metall Corp, 51-62; staff chemist, Union Carbide Corp, 66-70; chief chemist, Stellite Div, Cabot Corp, 70-87; CHIEF CHEMIST, HAYNES INT, 87- *Mem:* Soc Appl Spectros (pres, 75); Am Chem Soc. *Res:* X-ray spectroscopy; emission spectroscopy. *Mailing Add:* 5423 West 80th South Kokomo IN 46901-9785

MCLAIN, DAVID KENNETH, b Marietta, Ga, Aug 23, 37; m 64; c 4. MATHEMATICAL ANALYSIS. *Educ:* Ga Inst Technol, BS, 59, MS, 61; Carnegie Inst Technol, MS, 64, PhD(math), 67. *Prof Exp:* Assoc scientist, Bettis Atomic Power Lab, 60-62, sr mathematician, 66-73, FEL MATHEMATICIAN, RES LABS, WESTINGHOUSE ELEC CORP, 73- *Mem:* AAAS; Am Math Soc; Soc Indust & Appl Math. *Res:* Calculus of variations; differential and integral equations; linear algebra; electromagnetic fields. *Mailing Add:* Dept of Math Westinghouse Res Labs Pittsburgh PA 15235

MCLAIN, DOUGLAS ROBERT, b Detroit, Mich, Apr 2, 38; m 62; c 2. PHYSICAL OCEANOGRAPHY, FISHERIES OCEANOGRAPHY. *Educ:* Univ Mich, BSE, 61, MS, 63, PhD(oceanog), 69. *Prof Exp:* Oceanographer, Bur com Fisheries, US Fish & Wildlife Serv, Auke Bay, Alaska, 63-68, Bur Com Fisheries, Ann Arbor, Mich, 68-89; oceanographer, Nat Marine Fisheries Serv, 70-85, OCEANOGRAPHER, NAT OCEAN SERV, NAT OCEANIC & ATMOSPHERIC ADMIN, MONTERREY, CALIF, 85- *Concurrent Pos:* Chmn group of experts on opers & tech appln, Integrated Global Ocean Serv Syst, WMO/IOC, 86-; vis prof, Dept Oceanog, Naval Postgrad Sch, Monterey, Calif, 88- *Mem:* Oceanog Soc. *Res:* Ocean monitoring; synoptic oceanography; fisheries oceanography; use of microcomputers in oceanography. *Mailing Add:* 13 Wyndemere Vale Monterey CA 93940

MCLAIN, STEPHAN JAMES, b Tacoma, Wash, Sept 5, 53; m 75; c 1. OLEFIN METATHESIS, ORGANOFLUORINE CHEMISTRY. *Educ:* Iowa State Univ, BS, 75; Mass Inst Technol, PhD(inorg chem), 79. *Prof Exp:* Miller fel, Univ Calif, Berkeley, 79-80; RES CHEMIST, E I DUPONT DE NEMOURS & CO, INC, 80- *Mem:* Am Chem Soc; Sigma Xi. *Res:* Synthesis of new polymers and new metal catalysts for polymerization. *Mailing Add:* PO Box 80328 E I DuPont de Nemours & Co Inc Wilmington DE 19880-0328

MCLAIN, WILLIAM HARVEY, b Chicago, Ill; c 6. CHEMICAL ENGINEERING. *Educ:* Univ Chicago, BA, 50; Univ Denver, BSCh, 52, PhD(chem eng), 69; Univ Wash, MS, 60. *Prof Exp:* Mem staff combustion, Aeronaut Res Lab, Wright-Patterson AFB, 52-56; mem staff chem kinetics, Dept Chem, Univ Wash, 56-58; res chemist systs eng, Boeing Sci Res Labs, 58-61; scientist propulsion technol, Denver Div, Martin Marietta Corp, 61-63; chemist, Denver Res Inst, 63-73; asst mgr & mgr, Fire Res & Technol Sect, Div Struct Res & Ocean Eng, Southwest Res Inst, 74-76, inst scientist fire sci & prin investr, 76-79; tech dir, Stan Chem Inc, 79-83; SR FIRE PROTECTION ENG, US COAST GUARD, 83- *Concurrent Pos:* Sr physicist high temperature processes, Ger Res Inst Aeronaut & Rocket Propulsion, 71-72; mem life safety comt & fire test comt, Nat Fire Protection Asn; mem code admin comt, Nat Conf States Bldg Codes & Standards. *Mem:* Am Inst Astronaut & Aeronaut; Am Chem Soc; Nat Fire Protection Asn; Sigma Xi; Soc Fire Protection Engrs. *Res:* Combustion; fire science and technology; chemical kinetics; thermodynamics. *Mailing Add:* 178 Savage Hill Rd Berlin CT 06037

MCLAMORE, WILLIAM MERRILL, b Shreveport, La, Mar 15, 21; m 49; c 2. DRUG RESEARCH. *Educ:* Rice Inst, BA, 41, MA, 43; Harvard Univ, PhD(org chem), 49. *Prof Exp:* Jr chemist, Shell Develop Co, 43-45; res chemist, Pfizer Inc, 50-58, res assoc, 58-61, sect mgr, 61-68, res adminr, 68-76, sci liaison dir, 76-86; RETIRED. *Mem:* Am Chem Soc. *Res:* Constituents of bone oil; polymers and resins; synthesis of substituted benzoquinones; synthesis of alkaloids; total synthesis of steroids; structure and synthesis of antibiotics; synthesis of organic medicinals. *Mailing Add:* 22 S Glenwoods Rd Gales Ferry CT 06335-0632

MCLANE, GEORGE FRANCIS, b Philadelphia, Pa, Nov 19, 38. SOLID STATE ELECTRONICS. *Educ:* Villanova Univ, Pa, BA, 60; Univ Pa, Philadelphia, PhD(elec eng), 71. *Prof Exp:* Elec engr, Sperry Univac, 60-62 & RCA Serv Co, 62-64; NASA fel, Univ Pa, 65-68, res asst, 68-71; ELECTRONICS ENGR, NAVAL RES LAB, 71- *Mem:* Inst Elec & Electronics Engrs; Am Phys Soc. *Res:* Solid state electronics, including fabrication and characterization of lead salt hetero junction diode lasers and determination of the laser vulnerability of satellite solar cell arrays. *Mailing Add:* 207 Baltimore Ave Point Pleasant Beach NJ 08742

MCLANE, PETER JOHN, b Vancouver, BC, July 6, 41; m 67; c 2. COMMUNICATION SYSTEMS. *Educ:* Univ BC, BASc, 65; Univ Pa, MSEE, 66; Univ Toronto, PhD(elec eng). *Prof Exp:* Jr res officer, Dept Mech Eng, Nat Res Coun, 66-67; from asst prof to assoc prof, 69-78, CHMN GRAD STUDIES,QUEENS UNIV, 78-, PROF DEPT ELEC ENG, 78-, CHMN COMMUN GROUP, 84- *Concurrent Pos:* Vis assoc prof, Dept Elec Eng, Univ BC, 77-78; consult, Can Fed Dept Commun, 72- & AT&T Bell Labs, 84-88; mem, Scholarship Comt, Nat Sci & Eng Res Coun, 79-82, chmn, 81-82; assoc ed, Commun Mag, Inst Elec & Electronics Engrs, 80-84 & Trans on Commun, 82- *Mem:* Inst Elec & Electronics Engrs; Can Asn Univ Teachers. *Res:* Digital signal processing in communication and radar systems; application of microprocessors and very large scale integration technologies to communications receivers; author of over 100 publications. *Mailing Add:* Dept Elec Eng Queen's Univ Kingston ON K7L 3N6 Can

MCLANE, ROBERT CLAYTON, b Hinsdale, Ill, May 12, 24; m 50; c 3. SYSTEMS DESIGN & SYSTEMS SCIENCE. *Educ:* Univ Ill, BSEE, 49; Univ Wis, MSEE, 50. *Prof Exp:* Engr, Eng Labs, Firestone Tire & Rubber Co, 50-51; res assoc air defense systs, Eng Res Inst-Willow Run Res Ctr, Univ Mich, 51-53; res engr, Aeronaut Div, Minneapolis-Honeywell Regulator Co, 53-58, sr develop engr, 58-62, sect head eng dept, 67-68, staff engr res dept & prin systs engr, Systs & Res Ctr, 69-73, automatic test equip engr, 73-76, reliability & automatic test engr, Defense Syst Div, 76-79, sr training engr, 79-85, sr logistics engr, Avionics Div, Honeywell Inc, 85-86; RETIRED. *Concurrent Pos:* Mentor elementary, intermediate, jr & sr high sch & Univ Minn via Community Resource Pool & Mgt Assistance Proj, 88- *Mem:* Inst Elec & Electronic Engrs. *Res:* Study of control and display relationships for manned systems; analysis and synthesis of control and display systems; hybrid computer simulation techniques for man-in-the-loop system experiments. *Mailing Add:* 4527 Arden Ave Edina MN 55424

MACLANE, SAUNDERS, b Norwich, Conn, Aug 4, 09; m 33, 86; c 2. MATHEMATICS. *Educ:* Yale Univ, PhB,30; Univ Chicago, MA, 31; Univ Gottingen, PhD(math), 34. *Hon Degrees:* MA, Harvard Univ, 42; DSc, Purdue Univ, 65, Yale Univ, 69, Coe Col, 74 , Univ Pa, 77, Union Col, 90; LLD, Glasgow Univ, 71. *Prof Exp:* Pierce instr math, Harvard Univ, 34-36; instr, Cornell Univ, 36-37; instr, Univ Chicago, 37-38; from asst prof to prof, Harvard Univ, 38-47; prof, 47-63, Max Mason Distinguished Serv prof, 63-82, EMER PROF MATH, UNIV CHICAGO, 82- *Concurrent Pos:* Dir, Appl Math Group, Columbia Univ, 44-45; Guggenheim fel, Swiss Fed Inst Technol & Columbia Univ, 47-48 & 72-73; mem exec comt, Int Math Union, 54-58; vis prof, Univ Heidelberg, 58, 65 & 76, Univ Frankfurt, 60 & Tulane Univ, 69; mem coun, Nat Acad Sci, 59-62 & 69-72 & Am Acad Arts & Sci, 81-85; Fulbright fel, Australian Nat Univ, 69; mem, Nat Sci Bd, 74-80; US Sr Scientist Alexander von Humboldt prize, Ger, 82-83. *Honors & Awards:* Chauvenet Prize, Math Asn Am, 41; Distinguished Serv Award, Math Asn Am, 75; Nat Medal of Sci, 89. *Mem:* Nat Acad Sci (vpres, 73-81); Am Math Soc (vpres, 46-48, pres, 73-74); Am Philos Soc (vpres, 68-71); Math Asn Am (vpres, 48, pres, 50); Am Acad Arts & Sci. *Res:* Algebra; topology; algebraic topology; logic; category theory; philosophy of mathematics. *Mailing Add:* Dept Math 5734 University Ave Chicago IL 60637

MCLANE, VICTORIA, b New York, NY, Sept 23, 39; m 89; c 1. PHYSICS. *Educ:* Adelphi Univ, BA, 61. *Prof Exp:* SR PHYSICS ASSOC, BROOKHAVEN NAT LAB, 62- *Mem:* Am Phys Soc; Asn Women in Sci. *Res:* Neutron reactions and charged-particle reactions. *Mailing Add:* Brookhaven Nat Lab Bldg 197D Upton NY 11973

MCLAREN, DIGBY JOHNS, b Carrickfergus, Northern Ireland, Dec 11, 19; m 42; c 3. EVENT STRATIGRAPHY, RESOURCES & DEVELOPMENT. *Educ:* Univ Cambridge, BA, 41, MA, 46; Univ Mich, PhD(geol), 51. *Hon Degrees:* DSc, Univ Ottawa, 80. *Prof Exp:* Mem staff, Geol Serv Can, 48-59, chief paleontologist, 59-67, dir, Inst Sedimentary & Petrol Geol, 67-73, dir, Geol Serv Can, 73-80, asst dep minister sci & technol, 81, sr sci adv, Dept Energy, Mines & Resources, 81-85; VIS PROF, UNIV OTTAWA, 81- *Concurrent Pos:* Chmn, Comn on Stratig, Int Union Geol Sci, 72-76; chmn, Bd Int Geol Correlation Prog, UNESCO-Int Union Geol Sci, 76-80; organizer, Dahlem Workshops on Resources & Develop, Berlin, 86; chmn second int symp, Devonian Syst, Calgary, 87. *Honors & Awards:* Leopold von Buch Medal, Geol Soc Germany, 82; Edward Fitzgerald Coke Medal, Geol Soc London, 85; Sir William Logan Medal, Geol Soc Can, 87. *Mem:* Foreign assoc Nat Acad Sci; fel Am Geol Soc (pres, 82); fel Royal Soc Can (pres, 87-90); Can Soc Petrol Geol (pres, 71); Am Paleont Soc (pres, 69); foreign assoc Geol Soc France; fel Europ Union Geosci. *Res:* Devonian paleontology and stratigraphy of Western Canada; science in government; world resource futures; biological extinctions in the stratigraphic record and their significance in evolutionary theory; contribution to Canadian global change program. *Mailing Add:* 248 Marilyn Ave Ottawa ON K1V 7E5 Can

MCLAREN, EUGENE HERBERT, b Troy, NY, Aug 3, 24; m 57; c 3. ATMOSPHERIC CHEMISTRY. *Educ:* NY State Col Teachers Albany, BA, 48, MA, 49; Washington Univ, PhD(phys chem), 55. *Prof Exp:* Instr chem, State Univ NY Col Teachers Albany, 50-52, assoc prof, 55-57; sr res chemist, Pan-Am Petrol Corp, 57-60; prof sci & math, 60-69, prof chem & sr res assoc Atmospheric Sci Res Ctr, 69-89, EMER PROF, STATE UNIV NY ALBANY, 89- *Concurrent Pos:* Assoc dean, State Univ NY, Albany, 68-69, chmn div sci & math, 61-68; univ fel & res scientist, Max Planck Inst Chem, 69-70. *Mem:* AAAS; Am Chem Soc; Am Meteorol Soc. *Res:* Physical chemistry of atmospheric particulates and gases; geochemistry of carbonates; x-ray diffraction and spectroscopy. *Mailing Add:* Dept Chem State Univ NY Albany NY 12222

MCLAREN, IAN ALEXANDER, b Montreal, Que, Jan 11, 31; m 56; c 3. BIOLOGY. *Educ:* McGill Univ, BSc, 52, MSc, 55; Yale Univ, PhD(zool), 61. *Prof Exp:* Asst scientist, Fisheries Res Bd Can, 55-63; asst prof biol, Marine Sci Ctr, McGill Univ, 63-66; assoc prof, 66-69, PROF BIOL, DALHOUSIE UNIV, 69- *Concurrent Pos:* Can Coun fel, McGill Univ, 64-66. *Mem:* Can Soc Zoologists; Am Soc Naturalists; Ecol Soc Am. *Res:* Population; evolutionary ecology; birds; sea mammals; zooplankton. *Mailing Add:* Dept Biol Dalhousie Univ Halifax NS B3H 4J1 Can

MCLAREN, J PHILIP, b Los Angeles, Calif, May 17, 41; div; c 3. INVERTEBRATE ZOOLOGY, MARINE BIOLOGY. *Educ:* Bethel Col, Ind, BA, 64; Western Mich Univ, MA, 68, PhD(sci educ biol & earth sci), 76. *Prof Exp:* Asst prof natural sci, Bethel Col, Ind, 66-76; adj prof biol, Barrington Col, 70-79, Carribean Theological Serminary, 79 & Point Loma Nazarene Col, 80-82; assoc prof, 76-87, PROF BIOL & EDUC, EAST NAZARENE COL, 87- *Concurrent Pos:* Mem bd, Nat Marine Educrs Asn, 83-; pres, Mass Marine Educ, 83-87; Consult, Boston Globe Found Bowdoin Proj, 84- New Eng Aquarium Educ Comt, 85- *Mem:* Nat Marine Educrs Asn; AAAS; Nat Asn Biol Teachers; Nat Sci Teachers Asn. *Res:* Coral reef ecology in Belize; salp distribution in the Bahamas; inshore scallop distribution vasing; science education. *Mailing Add:* Eastern Nazarene Col 23 E Elm Ave Wollaston MA 02170

MCLAREN, LEROY CLARENCE, b Bishop, Calif, Jan 18, 24; m 45; c 3. MICROBIOLOGY, VIROLOGY. *Educ:* San Jose State Col, AB, 49; Univ Calif, Los Angeles, MA, 51, PhD(microbiol), 53; Am Bd Med Microbiol, dipl. *Prof Exp:* Asst microbiol, Univ Calif, Los Angeles, 49-50, asst bact, 50-52, instr infectious dis, Sch Med, 53-55; from instr to assoc prof bact, Sch Med, Univ Minn, 55-64; prof microbiol, Sch Med, Univ NMex, 64-90, chmn dept, 64-76, dir clin virol lab, Med Ctr, 78-90, EMER PROF MICROBIOL, UNIV NMEX, 90- *Concurrent Pos:* USPHS career res award, 62-64; mem, Microbiol Fel Rev Panel, NIH, 64-68; consult, Lilly Res Labs, 64-68 & Los Alamos Nat Lab, 85; mem, comt personnel for res, Am Cancer Soc, 73-77; mem, Bd Educ & Training, Am Soc Microbiol, 72-76. *Mem:* Fel AAAS; fel Am Acad Microbiol; Am Soc Microbiol; Am Asn Immunol; Tissue Cult Asn. *Res:* Tissue culture of animal viruses; stability of viruses; animal virus multiplication; mechanisms of viral susceptibility and resistance; infectious agents in inflammatory bowel disease; clinical virology. *Mailing Add:* Dept Microbiol Univ NMex Sch Med Albuquerque NM 87131

MACLAREN, MALCOLM DONALD, b Tarrytown, NY, Aug 5, 36; m 68. PROGRAMMING LANGUAGES, COMPILERS. *Educ:* Harvard Univ, AB, 58, MA, 60, PhD(math), 62. *Prof Exp:* Mem staff math, Boeing Sci Res Labs, Wash, 60-64; from asst mathematician to assoc mathematician, Argonne Nat Lab, 64-72; mgr advan lang systs develop, Cambridge Info Systs Lab, Honeywell, Inc, 74-76; SR CONSULT ENGR, DIGITAL EQUIP CORP, 76- *Concurrent Pos:* Vchmn comt X3J1, Am Nat Standards Inst, 71-72. *Mem:* Am Math Soc; Asn Comput Mach; Soc Indust & Appl Math. *Res:* Computation and programming, especially design, definition and implementation of languages, operating systems and similar software; formal theory of languages and programming. *Mailing Add:* 4910 E Mercer Way Mercer Island WA 98040

MCLAREN, MALCOLM G(RANT), b Denver, Colo, July 22, 28; m 50; c 4. CERAMICS. *Educ:* Rutgers Univ, BS, 50, MS, 51, PhD, 62. *Prof Exp:* Res asst, Rutgers Univ, 50-54; res scientist, Wright Air Develop Ctr, Ohio, 55-57; assoc prof ceramics, 62-69, PROF CERAMICS & CHMN DEPT, RUTGERS UNIV, NEW BRUNSWICK, 69- *Concurrent Pos:* Chief ceramist, Paper Makers Importing Co, Inc, 54-55, vpres, 57-62; indust consult ceramics; govt appt, US Dept Energy Technol Panel, 78. *Mem:* Am Ceramic Soc; Ceramic Educ Coun (pres, 71-72); Can Ceramic Asn; Sigma Xi; Am Soc Eng Educ; Int Ceramic Fedn (pres, 90-91). *Res:* Whitewares; refractories; electronic ceramics. *Mailing Add:* Dept Ceramic Sci & Eng Rutgers Univ Piscataway NJ 08855-0909

MACLAREN, NOEL K, b New Zealand, July 28, 39; m 88; c 4. DIABETES RESEARCH. *Educ:* Otago Univ, NZ, MD, 63. *Prof Exp:* PROF PATH, UNIV FLA, 78-, CHMN DEPT, 81- *Mem:* Am Soc Path; Soc Pediat Res; Am Diabetes Asn; Am Endocrinol Soc; Soc Clin Immunologists. *Res:* Pathogenesis, genetic predispositions and natural histories of the autoimmune endocrinopathies is the primary thrust. *Mailing Add:* Dept Path Box J-275 Univ Fla J Hillis Miller Health Ctr Gainesville FL 32610

MACLAREN, RICHARD OLIVER, b Missoula, Mont, Sept 4, 24; m 48; c 5. PHYSICAL CHEMISTRY. *Educ:* Univ Ore, BA, 49, MA, 50; Univ Wash, PhD(chem), 54. *Prof Exp:* Res chemist, Olin Mathieson Chem Corp, 54-55, pilot plant supvr, 55-56, group leader, 56-57, head thermodyn sect, 57-60; chief chem sect, United Tech Ctr, United Aircraft Corp, 60-63, mgr combustion res & develop br, 63-84, mgr projectile tech, Chem Systs Div, United Technol Corp, 84-89; RETIRED. *Mem:* Am Chem Soc. *Res:* Physical and inorganic chemistry involving high temperature processes; electrochemistry; solid propellant combustion, ignition and propellant chemistry; chemistry of fluorine compounds; chemical propulsion concepts. *Mailing Add:* 1031 Pinenut Ct Sunnyvale CA 94087

MCLAREN, ROBERT ALEXANDER, b Hamilton, Ont, Mar 29, 46; m 69; c 3. INFRARED ASTRONOMY. *Educ:* Univ Toronto, BSc, 68, MSc, 69, PhD(physics), 73. *Prof Exp:* NATO fel astron, Univ Calif, Berkeley, 73-75; from asst prof to assoc prof astron, Dept Astron, Univ Toronto, 75-90; Can-France-Hawaii Telescope, 82-90; ASSOC DIR, INST ASTRON, MAUNA KEA DIV, 90- *Concurrent Pos:* Resident astronomer, Canada-France-Hawaii Telescope Corp, 83-84, assoc dir, 84-87, dir, 87-90. *Mem:* Can Asn Physicists; Am Astron Soc; Can Astron Soc; Int Astron Union. *Res:* Infrared photometry and spectrophotometry of astronomical objects; determination of distances to nearby galaxies using infrared photometry; astronomical instrumentation. *Mailing Add:* Inst Astron Univ Hawaii 2680 Woodlawn Dr Honolulu HI 96822

MCLAREN, ROBERT WAYNE, b Chicago, Ill, Aug 31, 36; m 69; c 3. ELECTRICAL ENGINEERING. *Educ:* Univ Ill, Urbana, BScEE, 59, MScEE, 60; Purdue Univ, PhD(elec eng), 66. *Prof Exp:* Res engr, Autonetics Div, NAm Aviation, Inc, 61-62; instr elec eng, Purdue Univ, 62-66; asst prof, 66-69, eng exp sta res grant, 67-68, assoc prof elec eng, 69-77, PROF ELEC & COMPUT ENG, UNIV MO-COLUMBIA, 78- *Concurrent Pos:* Various summer prog, 69-81. *Mem:* Inst Elec & Electronics Engrs; Sigma Xi. *Res:* Artificial intelligence; pattern recognition; learning and control systems; automata theory; image analysis; robotics. *Mailing Add:* Dept of Elec Eng Univ of Mo Columbia MO 65211

MACLATCHY, CYRUS SHANTZ, b Galt, Ont, May 24, 41; m 63; c 3. PLASMA PHYSICS, COMBUSTION. *Educ:* Acadia Univ, BS, 64; Univ BC, MS, 66, PhD(physics), 70. *Prof Exp:* Asst prof, 70-76, assoc prof, 76-81, PROF PHYSICS, ACADIA UNIV, 81- *Res:* Use of Langmuir probes in diagnostics of combustion plasmas; edge physics in Tohamak plasmas. *Mailing Add:* Dept Physics Acadia Univ Wolfville NS B0P 1X0 Can

MCLAUGHLIN, ALAN CHARLES, b Vancouver, BC, Aug 17, 45. CEREBRAL PHYSIOLOGY, MEMBRANE BIOPHYSICS. *Educ:* Univ BC, BSc, 67; Univ Pa, PhD(biophys), 73. *Prof Exp:* Postdoctoral fel biochem, Oxford Univ, Eng, 74-77; biophysicist, Brookhaven Nat Lab, 77-83; lectr biophys, Univ Pa, 83-87; SECT CHIEF PHYS CHEM, NAT INST ALCOHOL ABUSE & ALCOHOLISM, 87- *Concurrent Pos:* Adj assoc prof, Univ Pa, 88- *Mem:* Soc Magnetic Resonance Med; Am Soc Biochem & Molecular Biol. *Res:* Development and use of non-invasive nuclear magnetic resonance imaging techniques for the study of cerebral metabolism and physiology. *Mailing Add:* LMMB NIAAA 12501 Washington Ave Rockville MD 20852

MCLAUGHLIN, BARBARA JEAN, b Miami, Fla, Nov 3, 41; c 2. VISION CELL BIOLOGY. *Educ:* Univ Fla, BS, 63; Stanford Univ, PhD(anat), 71. *Prof Exp:* From asst prof to prof anat & ophthal, Univ Tenn, Memphis, 82-87; PROF OPHTHAL & ANAT SCI & NEURO BIOL, UNIV LOUISVILLE, 87- *Concurrent Pos:* Agr Res Coun, Eng Underwood Fund grant zool, Univ Cambridge, 71-73; fel neurosci, City of Hope Med Ctr, Duarte, Calif, 73-74; mem comt to study status of women in anat, Am Asn Anatomists, 78-80; mem, proj rev A comt, neurol disorders prog, NIH, 78-82, behav neurosci study sect, Nat Inst Neurol & Commun Disorders & Stroke, 84-85, visual sci A-1 study sect, 84 & visual sci A-2 study sect, 85-89; vis scientist ophthalmol & anat sci & neurobiol, Univ Louisville, 87- *Mem:* Soc Neurosci; Am Asn Anat; Am Soc Cell Biol; Asn Res Vision & Ophthal. *Res:* Corneal and retinal cell biology; cytochemical and immunocytochemical localization of various molecular components in adult and developing retinal tissue; pigment epithelial-photoreceptor interactions; retinal degeneration; immuno-and cytochemistry of developing retinal synapses; cell biology of corneal endothelial dysfunction; blood retinal-barrier changes in retinal dystrophy. *Mailing Add:* Dept Ophthal Ky Lions Eye Res Inst Univ Louisville 301 E Muhammad Ali Blvd Louisville KY 40292

MCLAUGHLIN, CALVIN STURGIS, b St Joseph, Mo, May 29, 36; m 60; c 3. BIOCHEMISTRY, GENETICS. *Educ:* King Col, BS, 58; Mass Inst Technol, PhD(biochem), 64. *Prof Exp:* From asst prof to assoc prof, 66-72, vchmn, Dept Biochem, 79-83, PROF BIOCHEM, UNIV CALIF, IRVINE, 72-,; dir, Cancer Res Inst, 81-83. *Concurrent Pos:* Am Cancer Soc fel, Inst Phys Chem Biol, Paris, 64-66; dir, Cancer Res Inst, 81-83. *Mem:* Genetics Soc Am; Am Soc Microbiol; Am Soc Biol Chem. *Res:* Biochemistry and biochemical genetics of protein and RNA synthesis; mechanism of action of antibiotics; regulation of protein and RNA synthesis; mycology. *Mailing Add:* Dept of Biol Chem Univ of Calif-Calif Col of Med Irvine CA 92717

MCLAUGHLIN, CAROL LYNN, GROWTH OF ANIMALS, CONTROL OF FOOD INTAKE. *Educ:* Univ Pa, PhD(anat), 81. *Prof Exp:* RES SPECIALIST, MONSANTO CHEM CO, 85- *Mailing Add:* Monsanto Co BB3G 700 Chesterfield Village Pkwy Chesterfield MO 63198

MCLAUGHLIN, CHARLES ALBERT, b Chatham, Ill, Nov 12, 26; m 49; c 2. ZOOLOGY. *Educ:* Univ Ill, BS, 49, MS, 51, PhD(zool), 58. *Prof Exp:* Asst, Mus Natural Hist, Univ Ill, 50-51, exhibit preparator, 56, sci artist zool, Univ 51-55, asst, 55-56; assoc curator ornith & mammal, Los Angeles County Mus Natural Hist, 57-62, curator, 62-65, sr curator mammal, 65-67; assoc prof zool & curator mammal, Univ Wyo, 67-71; dir educ & gen cur, San Diego Zoo, 71-80; EXEC DIR, NATURAL HIST MUS, SAN DIEGO, 80- *Mem:* Am Soc Mammal; Am Asn Zool Parks & Aquariums; Soc Syst Zool; Am Asn Mus; Am Soc Zool. *Res:* Mammalogy, especially taxonomy, ecology and distribution, particularly of rodents and bats. *Mailing Add:* 6641 Jackson Dr San Diego CA 92119

MCLAUGHLIN, CHARLES WILLIAM, JR, b Washington, Iowa, Feb 3, 06; m 39; c 1. SURGERY. *Educ:* Univ Iowa, BS, 27; Wash Univ, MD, 29; Am Bd Surg, dipl, 40. *Prof Exp:* From instr to assoc prof, 35-55, prof surg, Col Med, 55-77, mem fac, Sch Med & sr consult surgeon, Univ Nebr, Omaha, 77-85, SR CONSULT SURG, UNIV NEB MED CTR, 77- *Concurrent Pos:* Res fel surg, Univ Pa, 31-34; nat consult, Gen Surgeon, US Air Force; assoc dir, Nebr Methodist Hosp Found, 51- *Mem:* Fel Am Col Surg; AMA; Asn Mil Surg US; Am Surg Asn. *Res:* Abdominal and pediatric surgery; author 100 publications. *Mailing Add:* 10085 Fieldcrest Dr Omaha NE 68114

MCLAUGHLIN, DAVID, zoology; deceased, see previous edition for last biography

MCLAUGHLIN, DONALD REED, b Los Angeles, Calif, Oct 6, 38; m 64; c 6. CHEMICAL PHYSICS. *Educ:* Univ Calif, Los Angeles, BS, 60; Univ Utah, PhD(chem), 65. *Prof Exp:* Asst prof, 65-72, dir, Ctr Grad Studies, Los Alamos, 81-87, ASSOC PROF CHEM, UNIV NMEX, 72- *Concurrent Pos:* Assoc, Rocky Mt Univs, Inc, Fac Orientation & Training Summers Fels, 65-69; vis staff mem, Los Alamos Nat Lab, 71-84. *Res:* Theoretical chemistry, especially quantum mechanics, statistical mechanics, chemical kinetics and thermodynamics. *Mailing Add:* Dept Chem Univ NMex Albuquerque NM 87131

MACLAUGHLIN, DOUGLAS EARL, b Indiana, Pa, Nov 18, 38; m 67; c 1. SOLID STATE PHYSICS. *Educ:* Amherst Col, BA, 60; Univ Calif, Berkeley, PhD(physics), 66. *Prof Exp:* NATO fel physics, Atomic Energy Res Estab, Harwell, Eng, 66-67; res assoc, Lab Physique Solides, Fac Sci, Orsay, France, 67-69; from asst prof to assoc prof, 69-78, PROF PHYSICS, UNIV CALIF, RIVERSIDE, 78- *Concurrent Pos:* Vchmn Physics Dept, Univ Calif, Riverside, 76-81, 87-; chmn, Muon Spin Relaxation working group, Los Alamos Meson Physics Facil, 80-81; vis fac mem, Univ Amsterdam, 80 & 83; vis affil, Los Alamos Nat Lab, 80-; mem steering comt, INCOR Prog, High-Temperature Superconductivity, Univ Calif, Los Alamos, 88-; chair, Workshop Superconductivity, Univ Calif Riverside, 89. *Mem:* Am Phys Soc; AAAS; Sigma Xi. *Res:* Use of magnetic resonance and relaxation to investigate electronic structure of metals, superconductors, spin glasses, rare earth compounds, heavy-fermion materials and high-temperature superconductors. *Mailing Add:* Dept of Physics Univ of Calif Riverside CA 92521-0413

MCLAUGHLIN, EDWARD, b Ballymena, Ireland, Oct 16, 28; m 56; c 4. CHEMICAL ENGINEERING, PHYSICAL CHEMISTRY. *Educ:* Queen's Univ, Belfast, BSc, 53, MSc, 54; Univ London, PhD(phys chem), 56, DSc, 74, Imp Col, dipl, 57. *Prof Exp:* Asst lectr, Imp Col, Univ London, 56-58, from lectr to sr lectr chem physics, 58-66, reader, 66-70; asst dir dept chem eng, 61-70; chmn, 79-87, PROF CHEM ENG, LA STATE UNIV, BATON ROUGE, 70-, DEAN ENG, 87- *Concurrent Pos:* Sr foreign scientist, NSF, 67-68. *Mem:* Am Soc Eng Educ; Sigma Xi; Am Inst Chem Engrs. *Res:* Equilibrium and transport properties of fluids and their mixtures; statistical mechanical theory of fluids; light initiated detonation; chemical reaction induced pressure pulses. *Mailing Add:* Col Eng CEBA La State Univ Baton Rouge LA 70803

MCLAUGHLIN, ELLEN WINNIE, b Roosevelt, NY, Aug 17, 37. EXPERIMENTAL EMBRYOLOGY. *Educ:* State Univ NY Albany, BS, 58; Univ NC, Chapel Hill, MA, 62; Emory Univ, PhD(biol), 67. *Prof Exp:* Instr biol, Converse Col, 60-63; from asst prof to assoc prof, 67-75, PROF BIOL, SAMFORD UNIV, 75- *Concurrent Pos:* Res grants, Samford Univ, 68-69, & 84-87. *Mem:* AAAS; Am Soc Zool; Am Sci Affiliation. *Res:* Effects of heavy metals and pesticides on aquatic vertebrate and invertebrate embryos, including amphibians, fish and snails. *Mailing Add:* Dept Biol Samford Univ Birmingham AL 35229

MCLAUGHLIN, FOIL WILLIAM, b NC, Dec 9, 23; m 48; c 4. AGRONOMY. *Educ:* NC State Univ, BS, 49, MS, 53. *Prof Exp:* Res asst prof field crops, 53-63, exten assoc prof, 63-68, DIR, NC CROP IMPROV ASN, NC STATE UNIV, 62-, EXTEN PROF CROP SCI, 68- *Concurrent Pos:* Exec vpres, Asn Off Seed Certifying Agencies in US & Can, 81- *Honors & Awards:* Hon NC Seedsman Award, 65; Hon Seedsman, Am Seed Trade Asn, 81. *Mem:* Hon mem Asn Off Seed Certifying Agencies. *Res:* Crop and seed improvement, especially breeding and quality control in seed development and production; seed certification. *Mailing Add:* 804 Runnymede Rd Raleigh NC 27607

MCLAUGHLIN, FRANCIS X(AVIER), b Philadelphia, Pa, Sept 2, 30. DATA PROCESSING, STATISTICS. *Educ:* Villanova Univ, BS, 52, MS, 59. *Prof Exp:* Jr chemist, Waste Control Lab, Atlantic Refining Co, 52-54, asst chemist, 54-55, asst chemist, Res & Develop Dept, 55-56; res engr, Franklin Inst, Pa, 56-58, methods analyst, 58, sr methods analyst, 58-60; specialist mgt sci, Missile & Space Div, Gen Elec Co, 60-64, mgr, 65-69; vpres mgt & comput sci, Sci Resources Corp, 69-70; consult data processing, 70-71; vpres mgt & comput sci, Mauchly Mgt Serv, Inc, 71-73; mgr statist, Union Fidelity Ins Corp, 73-76; CONSULT, 76- *Concurrent Pos:* Instr, Villanova Univ, 86-87. *Mem:* Am Soc Qual Control; Sigma Xi; Biomet Soc; The Inst Mgt Sci; Am Statist Asn. *Res:* Operations research; systems analysis and design; computer applications and programming; time-shared and remote computer processing systems. *Mailing Add:* PO Box 616 Drexel Hill PA 19026-0616

MCLAUGHLIN, GERALD WAYNE, b Nashville, Tenn, Aug 16, 42; m 65; c 2. RESEARCH MANAGEMENT, APPLIED STATISTICS. *Educ:* Univ Tenn, BS, 64, MS, 65, PhD(orgn psychol), 69. *Prof Exp:* Res asst indust mgt, Univ Tenn, 64-65, res asst econ, 65-66; asst dir instnl res, US Mil Acad, 69-71; asst prof, 71-81, ASSOC PROF OFF INSTNL RES, VA POLYTECH INST & STATE UNIV, 81- *Mem:* Asn Instnl Res; Psychomet Soc. *Res:* Administration and conducting applied research; testing and measurement; role analysis in education, industry and other fields. *Mailing Add:* 308 Ardmore St Blacksburg VA 24060

MCLAUGHLIN, HARRY WRIGHT, b Highland Park, Mich, June 28, 37; m 61; c 2. MATHEMATICS, COMPUTER GRAPHICS. *Educ:* DePauw Univ, BS, 59; Kans State Univ, MS, 61; Univ Md, PhD(math), 66. *Prof Exp:* Assoc mathmatician, Applied Physics Lab, Johns Hopkins Univ, 61-63; asst prof math, Univ Calif, Riverside, 66-67; asst prof, 67-72, assoc prof, 72-77, PROF MATH & COMPUT SCI, RENSSELAER POLYTECH INST, 77- *Mem:* Math Asn Am; Am Math Soc; Soc Indust & Applied Math. *Res:* Mathematical models for computer aided design; algorithms for optimal knot location in spline theory; approximation techniques. *Mailing Add:* Dept Math Sci Rensselaer Polytech Inst Troy NY 12180

MCLAUGHLIN, J(OHN) F(RANCIS), b New York, NY, Sept 21, 27; m 50; c 4. CIVIL ENGINEERING. *Educ:* Syracuse Univ, BCE, 50; Purdue Univ, MSCE, 53, PhD(civil eng), 57. *Prof Exp:* From instr to prof civil eng, 50-68, head, Sch Civil Eng, 68-78, asst dean eng, 78-80, ASSOC DEAN ENG, PURDUE UNIV, 80- *Mem:* Am Soc Civil Engrs; Nat Soc Prof Engrs; hon mem Am Concrete Inst (vpres, 77-79, pres, 79-80). *Res:* Mineral aggregates for construction purposes; properties of portland cement concrete. *Mailing Add:* Sch Civil Eng Purdue Univ Lafayette IN 47907

MCLAUGHLIN, JACK A, EYE RESEARCH. *Prof Exp:* ASSOC DIR EXTRAMURAL & COLLABORATIVE PROG, NAT EYE INST, NIH, 87- *Mailing Add:* NIH Nat Eye Inst Extramural & Collaborative Prog Bldg 31 Rm 6A04 Bethesda MD 20892

MCLAUGHLIN, JACK ENLOE, b St Maries, Idaho, Aug 17, 23; m 49. MATHEMATICS. *Educ:* Univ Idaho, BS, 44; Calif Inst Technol, PhD(math), 50. *Prof Exp:* From instr to assoc prof, 50-62, PROF MATH, UNIV MICH, ANN ARBOR, 62- *Concurrent Pos:* Res fel, Harvard Univ, 60. *Mem:* Am Math Soc; Math Asn Am. *Res:* Group theory. *Mailing Add:* 408 Wesley Ann Arbor MI 48103

MCLAUGHLIN, JAMES L, b Detroit, Mich, July 16, 42; m 64; c 1. PLANT PATHOLOGY. *Educ:* Eastern Mich Univ, BA, 64; Univ Ill, MS, 66, PhD(plant path), 69. *Prof Exp:* Asst prof biol, Univ Wis-Superior, 69-72; asst prof, 72-73, chmn dept, 73-77, ASSOC PROF BIOL, COL ST SCHOLASTICA, 73- *Mem:* Am Inst Biol Sci; Am Phytopath Soc. *Res:* Mycology; plant physiology; bacteriology. *Mailing Add:* Dept Natural Sci Col State Scholastica Duluth MN 55811

MCLAUGHLIN, JERRY LOREN, b Coldwater, Mich, Oct 14, 39; m 60, 81; c 2. PHARMACOGNOSY. *Educ:* Univ Mich, BS, 61, MS, 63, PhD(pharmacog), 65. *Prof Exp:* Asst prof pharmacog, Univ Mich, 65-66 & Univ Mo-Kansas City, 66-67; mem fac, Col Pharm, Univ Wash, 67, from asst prof to assoc prof pharmacog, 67-71; assoc prof, 71-75, exec asst to dean, 75-80, PROF PHARMACOG, SCH PHARM & PHARMACAL SCI, PURDUE UNIV, 75- *Concurrent Pos:* Nat Inst Ment Health res grants, 66-67, 69-72; NIH res grants, 74-77 & 84-93; Nat Sci Found res grants, 74, 75 & 79, AID res grant, 82-85. *Honors & Awards:* Cancer Res Award, Purdue Univ, 87. *Mem:* AAAS; Am Pharmaceut Asn; Am Soc Pharmacog (vpres, 81-82, pres, 82-83); Soc Econ Bot; Acad Pharmaceut Sci; fel, Acad Pharm Sci. *Res:* Cactus alkaloids, their isolation and biosynthesis; active constituents of antitumor and poisonous plants; natural insecticides. *Mailing Add:* 2940 St Rd 26W West Lafayette IN 47906

MCLAUGHLIN, JOHN J A(NTHOANY), microbiology; deceased, see previous edition for last biography

MCLAUGHLIN, JOHN ROSS, b Clayton, NMex, June 4, 39; m 64; c 2. INSECT BEHAVIOR, CHEMICAL ECOLOGY. *Educ:* Colo State Univ, BS, 66, MS, 67; Univ Calif, Riverside, PhD(entom), 72. *Prof Exp:* Res assoc insect behav, Dept Entom, Univ Calif, Riverside, 71-72, res assoc insect ecol, Dept Plant Sci, 72-73; asst prof insect behavior, Dept Entom & Nematol, Univ Fla, 73-74; RES ENTOMOLOGIST INSECT ATTRACTANTS, BEHAV & BASIC BIOL RES LAB, AGR RES SERV, USDA, 74- *Concurrent Pos:* Assoc ed, Fla Entomologist, 78-84, ed, 84-; chmn, Entom Soc Am, 79-80 & 88. *Mem:* Int Soc Chem Ecol; Entom Soc Am. *Res:* Insect behavior and ecology, particularly the study of mating behavior, pheromones and insect-host relationships mediated by semiochemicals. *Mailing Add:* Insect Attractants PO Box 14565 Gainesville FL 32604

MCLAUGHLIN, JOYCE ROGERS, b Milwaukee, Wis; c 2. DIFFERENTIAL EQUATIONS. *Educ:* Kans State Univ, BS; Univ Md, MA; Univ Calif, Riverside, PhD(math), 68. *Prof Exp:* asst prof, 78-87, ASSOC PROF, RENSSELAER POLYTECH INST, 87- *Mem:* Am Math Soc; Soc Indust & Appl Math; Math Asn Am; Asn Am Univ Prof. *Res:* Inverse boundary value problems; algorithms for constructing materials parameters from spectral data as well as existence, uniqueness and continuous dependence questions for this nonlinear problem; stability of nonlinear feedback control problems. *Mailing Add:* Dept Math Sci Rensselaer Polytech Inst Troy NY 12181

MCLAUGHLIN, KENNETH PHELPS, geology, for more information see previous edition

MCLAUGHLIN, MICHAEL RAY, b Carroll, Iowa, June 20, 49; m 70; c 3. PLANT VIROLOGY. *Educ:* Iowa State Univ Sci & Technol, BS, 71, MS, 74; Univ Ill, PhD(plant path), 78. *Prof Exp:* Vis asst prof plant path, Clemson Univ, 78-79; asst prof plant path, Univ Tenn, 79-82; RES PLANT PATHOLOGIST, USDA-ARS, 82- *Concurrent Pos:* Adj asst prof, Miss State Univ, 82-87, adj prof, 88- *Mem:* Am Phytopath Soc; AAAS; Int Working Group Legume Virologists; Sigma Xi. *Res:* Plant virology, especially legume viruses; serology. *Mailing Add:* USDA-ARS Crop Sci Res Lab PO Box 5367 Mississippi State MS 39762-5367

MCLAUGHLIN, PATRICIA J, b Harrisburg, Pa. DEVELOPMENTAL NEUROBIOLOGY, DEVELOPMENTAL CELLULAR NEUROBIOLOGY. *Educ:* Lebanon Valley Col, BS, 74; Shippensburg Univ, MS, 76. *Prof Exp:* Res asst anat, 76-81, res support assoc anat, 83-88, SR RES SUPPORT ASSOC, ANAT, MILTON S HERSHEY MED CTR, 88- *Mem:* AAAS; Soc Neurosci; Int Brain Res Org. *Res:* The isolation, identification and characterization of endogenous opioid receptors and peptides associated with growth of developing normal and abnormal tissues; identification of opioid peptide precursors and RNA and DNA components of those peptides related to growth regulation. *Mailing Add:* Dept Anat Milton S Hershey Med Ctr 500 University Dr Hershey PA 17033

MCLAUGHLIN, PATSY ANN, b Seattle, Wash, May 27, 32; m 70. SYSTEMATIC ZOOLOGY, ENVIRONMENTAL BIOLOGY. *Educ:* Univ Wash, BA, 57; George Washington Univ, MPh, 69, PhD(zool), 72. *Prof Exp:* Fishery biologist res, US Bur Com Fisheries, 57-60; zoologist, Dept Oceanog, Univ Wash, 60-65; supvr, Smithsonian Oceanog Sorting Ctr, 65-68; from res asst to asst prof, Rosenstiel Sch Marine & Atmospheric Sci, Univ Miami, 69-75; courtesy assoc prof, 75-77, prof, 75-87, COURTESY PROF RES TEACHING, DEPT BIOL SCI, FLA INT UNIV, 90- *Concurrent Pos:* contrib specialist, Norwegian Res Coun Sci & Humanities, 73-; prin investr, NSF, 73-75, 76-78 & 90-91; prin investr, NSF, 73-75 & 76-78; prin investr, Fla State Univ Syst Sea Grant Prog, 77-78; dir, Appl Marine Ecol Serv, Inc, 75-76; res scientist & adj prof, Shannon Point Marine Ctr, 87- *Mem:* Willi Hennig Soc; Soc Syst Zool; Crustacean Soc. *Res:* Systematics larval development and comparative morphology of crustaceans, with particular emphasis on pagurids, cirripeds and cephalocarids; marine environmental assessment, with emphasis on benthic communities. *Mailing Add:* Shannon Point Marine Ctr 1900 Shannon Point Rd Anacortes WA 98221-4042

MCLAUGHLIN, PETER, b New Haven, Conn, Nov 23, 48; m; c 2. ONCOLOGY. *Educ:* Harvard Col, BA, 70; Tufts Univ Sch Med, MD, 74. *Prof Exp:* Intern internal med, Hartford Hosp, 74-75, resident intern med, 75-77; fel med, 78-80, asst internist & asst prof 80-86, ASSOC INTERNIST & ASSOC PROF MED, MD ANDERSON CANCER CTR, 86- *Mem:* Am Col Physicians; Am Soc Clin Oncol; Am Asn Cancer Res; AAAS. *Res:* Treatment of lymphomas and Hodgkin's disease. *Mailing Add:* Dept Hemat Univ Tex M D Anderson Cancer Ctr 1515 Holcombe Box 68 Houston TX 77030

MCLAUGHLIN, PHILIP V(AN DOREN), JR, b Elizabeth, NJ, Nov 10, 39; m 61; c 3. SOLID MECHANICS, COMPOSITE MATERIALS. *Educ:* Univ Pa, BS, 61, MS, 64, PhD(eng mech), 69. *Prof Exp:* Assoc engr, Vertol Div, Boeing Co, 62-63, engr, 63; res engr, Tech Ctr, Scott Paper Co, 63-65, res proj engr, 65-69, sr res proj engr, 69; asst prof theoret & appl mech, Univ Ill, Urbana, 69-73, asst dean, Col Eng, 71-72; proj mgr, Mat Sci Corp, Pa, 73-76; assoc prof, 76-81, PROF MECH ENG, VILLANOVA UNIV, 81- *Concurrent Pos:* NSF grant, Univ Ill, Urbana, 70-72, Naval Air Eng Ctr grants, 78- 83, Lawrence Livermore Nat Lab grants, 79-81 & RCA Corp grants, 85-87; fac res grant, Villanova Univ, 78 & 90; consult, US Naval Air Eng, 77-79, US Steel Corp, 80-81, Coal Tech Corp, 88 & Air Prod Corp, 89; chmn, EMD Comt, Inelastic Behavior, Am Soc Civil Engrs, 77-79, comt mem, 74-86, Aerospace Serv Div, Aerospace Struct & Mat, 86-; assoc ed, J Eng, Mech Div, Am Soc Civil Engrs, 77-79; vis prof, Dept Eng, Univ Cambridge, Eng, 90-91; sr res assoc, Nat Res Coun, 83-84. *Mem:* Am Acad Mechanics; Am Soc Eng Educ; Sigma Xi; Am Soc Civil Engrs; Am Soc Mech Engrs. *Res:* Plasticity; composite materials; materials, flow and fracture; nondestructive testing, design; numerous papers, presentations and reports. *Mailing Add:* Dept Mech Eng Villanova Univ Villanova PA 19085

MCLAUGHLIN, RENATE, b Ger; US citizen; m 64; c 2. USE OF GRAPHING CALCULATORS IN TEACHING MATHEMATICS, USE OF COMPUTERS IN TEACHING MATHEMATICS. *Educ:* Univ Mich, AM, 65, PhD(math), 68. *Prof Exp:* From asst prof to assoc prof, 68-75, PROF MATH, UNIV MICH, FLINT, 75- *Concurrent Pos:* Ed, Mich Math J, 68-75; lectr math, Tech Univ, Berlin, Ger, 74 & Univ Salzburg, Austria, 75; vchair, Math Asn Am, Mich Sect, 90-91, chair, 91-92. *Mem:* Am Math Soc; Nat Coun Teachers Math; Math Asn Am. *Res:* Classical complex analysis; role of technology in the teaching of mathematics. *Mailing Add:* 1432 Duffield Rd Lennon MI 48449

MCLAUGHLIN, ROBERT EVERETT, b Aurora, Ind, Nov 30, 19; m 45; c 3. PALEONTOLOGY. *Educ:* Tulane Univ, BS, 51, MS, 52; Univ Tenn, Knoxville, PhD(paleont), 57. *Prof Exp:* Asst bot, Tulane Univ, 50-51, asst geol, 52; from asst to instr bot, 52-54, from instr to prof geol, 54-85, EMER PROF GEOL, UNIV TENN, KNOXVILLE, 85- *Concurrent Pos:* Southern Fels Fund award, Harvard Univ, 55-56; consult, GEO MAC Enterprises, 85- *Mem:* Paleont Asn; Paleont Soc; Geol Soc Am; Am Asn Stratig Palynologists; Soc Econ Paleontologists & Mineralogists. *Res:* Palynology of tertiary and older deposits; Paleozoic paleontology; coal paleobotany; history of ideas-roots of science. *Mailing Add:* Dept Geol Sci Univ Tenn Knoxville TN 37996-1410

MCLAUGHLIN, ROBERT LAWRENCE, b Beaver, Pa, Sept 17, 23; m 46; c 2. ORGANIC CHEMISTRY. *Educ:* Pa State Univ, BS, 44, MS, 46, PhD(org chem), 49. *Prof Exp:* Mem, Am Petrol Inst Proj, Pa State Univ, 44-46, asst, 44-49; res scientist, NASA, Ohio, 49-50, group leader, 50-51; fel, Mellon Inst, 51-55, sr fel, 55-57; sr res chemist, Res & Develop Lab, Mobil Chem Co, 57-60, group leader, 60-62, sect leader, 62-64; coordr res, Velsicol Chem Co, 64-65, dir res, Resin Prod Div, 65-69; res mgr, Armour Dial Inc, 69-73; tech dir, De Soto, Inc, 73-84; RETIRED. *Mem:* Am Chem Soc; Am Oil Chemists Soc. *Res:* Hydrocarbon analysis, synthesis, purifications, separations, adducts, processing and properties; organic and petroleum chemicals; monomers and polymeric materials, synthesis, processing and evaluations; soaps and detergents; personal care products; cleaning products. *Mailing Add:* 2333 Schiller Ave Wilmette IL 60091-2329

MCLAUGHLIN, STUART GRAYDON ARTHUR, b NVancouver, BC, Dec 15, 42; m 80. MEMBRANE BIOPHYSICS. *Educ:* Univ BC, BSc, 64; PhD(biophysics), 68. *Prof Exp:* Fel biophysics, Univ Dundee, Scotland, 68, Univ Calif, Los Angeles & Univ Chicago, 69; res physiologist, Univ Calif, Los Angeles, 70; res fel, Calif Inst Technol, 71; from asst prof to assoc prof, 72-81, PROF PHYSIOL & BIOPHYSICS, STATE UNIV NY, STONY BROOK, 81- *Concurrent Pos:* Mem, Physiol Study Sect, NIH, 79-82; prin investr, NIH & NSF grants. *Mem:* Biophys Soc (pres, 86); Soc Gen Physiologists; AAAS. *Res:* Electrostatic potentials at membrane-solution interfaces; role of phosphatidylinositol 4,5 bisphosphate protein kinase C and phospholipase C in intracellular signaling. *Mailing Add:* Physiol & Biophys Health Sci Ctr State Univ NY Stony Brook NY 11794

MCLAUGHLIN, THAD G, b Wichita, Kans, Oct 16, 13; m 42. HYDROGEOLOGY. *Educ:* Wichita State Univ, BA, 35; Univ Kans, PhD(geol), 39. *Prof Exp:* Br area chief ground water, US Geol Surv, Denver 59-65, regional hydrologist, 66-72, asst dir, 72-75; RETIRED. *Concurrent Pos:* Consult, 75-85. *Mem:* Am Water Resource Asn (pres, 77); Am Asn Petrol Geol Soc; Geol Soc Am. *Mailing Add:* 7355 W Maple Dr Lakewood CO 80226

MCLAUGHLIN, THOMAS G, b McIntosh, SDak, Nov 26, 33; m 69; c 1. MATHEMATICS. *Educ:* Univ Calif, Los Angeles, BA, 59, MA, 62, PhD(math), 63. *Prof Exp:* From instr to asst prof math, Univ Ill, Urbana, 63-66; vis asst prof, Cornell Univ, 66-67; assoc prof math, Univ Ill, Urbana, 67-74; vis assoc prof, 73-74, assoc prof, 74-75, PROF MATH, TEX TECH UNIV, 75- *Concurrent Pos:* Vis lectr math, Rutgers Univ, 79. *Mem:* Am Math Soc. *Res:* Recursive function theory. *Mailing Add:* Dept of Math Tex Tech Univ Lubbock TX 79409

MCLAUGHLIN, WALLACE ALVIN, b Calgary, Alta, May 5, 27; m 51; c 5. CIVIL ENGINEERING. *Educ:* Univ Sask, BSc, 51; Purdue Univ, MSCE, 58, PhD(civil eng), 65. *Prof Exp:* Proj engr, Sask Dept Hwy, 51-54, div engr, 54-57, asst planning engr, 58-61; from asst prof to assoc prof civil eng, 61-66, assoc chmn dept, 65-66, chmn, dept civil eng, 66-72, dean eng, 74-82, PROF CIVIL ENG & DIR CONSTRUCT PROG, UNIV WATERLOO, 69- *Concurrent Pos:* Dept Trans fel, Inst Transp & Traffic Eng, Univ Calif, Berkeley, 59-70. *Mem:* Am Asn Cost Engrs. *Res:* Construction engineering and planning. *Mailing Add:* Dept Civil Eng Univ Waterloo Waterloo ON N2L 3G1 Can

MCLAUGHLIN, WILLIAM IRVING, b Oak Park, Ill, Mar 6, 35; m 60; c 4. EPISTEMOLOGY. *Educ:* Univ Calif, Berkeley, BS, 63, MA, 66, PhD(math), 68. *Prof Exp:* Mem tech staff celestial mech, Bellcomm, Inc, 68-71; mem tech staff celestial mech, 71-81, supvr, inner Planets Trajectory & Mission Design Group, 81-83, mission design mgr, Infrared Astron Satellite, 82-83, mgr flight eng off, Voyager Uranus Proj, 83-86, MGR, MISSION PROFILE & SEQUENCING SECT, JET PROPULSION LAB, CALIF INST TECHNOL, 86- *Honors & Awards:* Apollo Achievement Award, Nat Aeronautics & Space Admin, 69, Pioneer 10 Mission Analysis Award, 74, Viking Flight Award, 77,; Exceptional Serv Medal, NASA, 84, Outstanding Leadership Medal, 86. *Mem:* Fel Brit Interplanetary Soc; Int Acad Astronaut. *Res:* Scientific epistemology. *Mailing Add:* Jet Propulsion Lab 4800 Oak Grove Dr Pasadena CA 91109

MCLAUGHLIN, WILLIAM LOWNDES, b Stony Point, Tenn, Mar 30, 28; m 51; c 2. RADIATION PHYSICS, FIBER OPTICS. *Educ:* Hampden-Sydney Col, BS, 49; George Washington Univ, MS, 63. *Prof Exp:* Physicist, Radiation Phys Div, 51-54, 56-64, proj leader, 64-69, PROJ LEADER, CTR RADIATION RES, NAT BUR STANDARDS, 69- *Concurrent Pos:* Ed, Int J Appl Radiation & Isotopes, 73-88, Radiation Physics & Chem, 81-; consult, Risoe Nat Lab, Denmark, 73-; mem adv panel, Int Atomic Energy Agency, 77-87; ed-in-chief N Am, Appl Radiation & Isotopes, 88- *Honors & Awards:* Silver Medal, Dept Com, 69; Gold Medal, Dept Com, 79; Technol Transfer Award, Fed Lab Consortium, 84; Appl Res Award, Nat Bureau Standards, 85; Radiation Sci & Technol Award, Am Nuclear Soc, 87; Res & Develop 100 Awards, 88, 90. *Mem:* Am Nuclear Soc; Am Phys Soc; Optical Soc Am; Health Physics Soc; Radiation Res Soc; Soc Photograph Scientists & Engrs. *Res:* Measurement of ionizing radiation; x-ray, gamma-ray and electron spectrometry, absorption and scattering measurement and computation; radiation chemistry; photographic processes; industrial radiation processing; fiber optics sensors; medical physics. *Mailing Add:* Radiation Physics C214 Nat Inst Standards & Technol Gaithersburg MD 20899

MCLAUGHLIN-TAYLOR, ELIZABETH, b Hamilton, Scotland, June 11, 56. IMMUNOLOGY. *Educ:* Univ Glasgow, BS, 79; Univ Southern Calif, PhD(immunol), 83. *Prof Exp:* Postdoctoral fel, Dept Molecular Genetics, Beckman Res Inst, City of Hope, 83-87, asst res scientist, Dept Hemat & Bone Marrow Transplant, 87-91; SR SCIENTIST, FENWALL DIV, BAXTER HEALTH CARE CORP, 91- *Mem:* Am Asn Immunologists; AAAS. *Mailing Add:* Fenwall Div Baxter Health Care Corp 3015 Daimler St Santa Ana CA 92705

MCLAURIN, JAMES WALTER, b Natchez, Miss, July 1, 10; m 35. MEDICINE, SURGERY. *Educ:* Univ Ark, MD, 34. *Prof Exp:* Instr, Eye, Ear, Nose & Throat Clins, Univ Ark, 34; intern, Mary's Help Hosp, San Francisco, Calif, 34-35; resident, Los Angeles Children's Hosp, 35-36; instr otolaryngol, Sch Med, La State Univ, 46-49; chmn dept otolaryngol, 49-58, med dir, Speech & Hearing Ctr, 52-63, Otto Joachiem prof otolaryngol, 49-69, clin prof, 69-80, EMER CLIN PROF OTOL, SCH MED, TULANE UNIV, 80- *Concurrent Pos:* Mem, Nat Comt Deafness Res Found; pvt pract otolaryngology, Baton Rouge, 36- *Mem:* Am Laryngol, Rhinol & Otol Soc; Am Otol Soc; AMA; fel Am Col Surg; Am Acad Otolaryngol. *Res:* Otitis externa and otitis media; speech and hearing problems. *Mailing Add:* 2856 Reymond Ave Baton Rouge LA 70808

MCLAURIN, ROBERT L, b Dallas, Tex, Jan 5, 22; m 46; c 5. NEUROSURGERY. *Educ:* Rice Inst, BA, 44; Harvard Med Sch, MD, 44. *Prof Exp:* Asst surg, Harvard Med Sch, 51-53; from instr to assoc prof surg, 53-60, actg dir div neurosurg, 54-55, PROF SURG, COL MED, UNIV CINCINNATI, 53-, DIR DIV NEUROSURG, 55- *Concurrent Pos:* Res fel neurosurg, Children's Hosp & Peter Bent Brigham Hosp, Boston, Mass, 51-53. *Mem:* Cong Neurol Surg; Am Asn Neurol Surg; Am Acad Neurol Surg (secy-treas, 58-63, pres, 71-72); Soc Brit Neurol Surg; Soc Neurol Surg (treas, 70-). *Res:* Clinical and experimental aspects of intracranial trauma and hemorrhage. *Mailing Add:* 415 Bond Place Apt 9 D Cincinnati OH 45206

MCLAURIN, WAYNE JEFFERSON, b Hattiesburg, Miss, July 1, 42; m 66; c 2. HORTICULTURE. *Educ:* Southeastern La Univ, BS, 71; La State Univ, MS, 74, PhD(hort), 79. *Prof Exp:* County agent, Ark Coop Extension Serv, 74-75; ASST PROF VEGETABLE RES, UNIV TENN INST AGR, 79- *Mem:* Am Soc Hort Sci. *Res:* Environmental manipulation of vegetable crops; breeding of sweet potatoes; breeding and adaptability of heat tolerance in tomatoes; nitrogen fixiation in legume crops; plant stress due to soil fertility levels. *Mailing Add:* Exten Hort Univ Ga Athens GA 30602

MACLAURY, MICHAEL RISLEY, b Hays, Kans, Dec 28, 43; m 66; c 1. MATERIAL SCIENCE. *Educ:* Antioch Col, BS, 67; Stanford Univ, PhD(inorg chem), 74. *Prof Exp:* Res asst nuclear med, Stanford Med Ctr, 70-72; staff chemist, Gen Elec Res & Develop Ctr, 74-81, tech coordr chem reactions & systs, Polymer Physics & Eng Br, 81-85, mgr planning & resources, 85-89, MGR POLYMER FLAMMABILITY & STABILIZATION PROG, CHEM LABS, GEN ELEC RES & DEVELOP CTR, 89- *Concurrent Pos:* Fel chem catalysis, NSF US/USSR Exchange Prog, 74; Woodrow Wilson Fel, 68. *Mem:* Am Chem Soc; Am Inst Chemists. *Res:* Mechanisms of organometallic reactions as they relate to possible catalytic reactions; the role of inorganic materials in flame retarding polymers. *Mailing Add:* Gen Elec Res & Develop Ctr PO Box 8 Bldg K-1 Rm 4A38 Schenectady NY 12301

MACLAY, CHARLES WYLIE, b Fannettsburg, Pa, Oct 17, 29; m 50; c 2. MATHEMATICS. *Educ:* Shippensburg State Col, BS, 56; Univ Va, MEd, 59, EdD(math educ), 68. *Prof Exp:* Teacher math, Seaford Spec Schs, Del, 56-58; instr, Shippensburg State Col, 60-62; teacher, Abington Schs, 62-69; PROF MATH, EAST STROUDSBURG STATE COL, 69- *Concurrent Pos:* Consult, Elem/Soc Sch Math. *Mem:* Math Asn Am; Nat Coun Teachers Math. *Res:* Learning sequences and sequencing in math; curriculum constructing and testing. *Mailing Add:* 1006 Lindberg Ave East Stroudsburg PA 18360

MCLAY, DAVID BOYD, b Toronto, Ont, Feb 29, 28; m 53. MOLECULAR PHYSICS. *Educ:* McMaster Univ, BSc, 50, MSc, 51; Univ BC, PhD(physics), 56. *Prof Exp:* Lectr physics, Victoria Col, 54-56; asst prof, Univ New Brunswick, 56-62; from asst prof to assoc prof, 62-77, assoc dean studies, 76-85, PROF PHYSICS, QUEEN'S UNIV, ONT, 76- *Mem:* Can Asn Physicist; Am Asn Physics Teachers; Can Astron Soc. *Res:* Microwave spectroscopy and paramagnetic spectroscopy of gases; dielectric and nuclear paramagnetic relaxation; infrared laser spectroscopy. *Mailing Add:* Dept Physics Queen's Univ Kingston ON K7L 3N6 Can

MACLAY, G JORDAN, b Washington, DC, May 17, 42; m 87; c 1. SOLID STATE ELECTRONICS, MICROELECTRONIC SENSORS. *Educ:* Queens Col, City Univ New York, MA, 65; Yale Univ, MPh, 68, PhD(physics), 72. *Prof Exp:* Fel res asst physics, High Energy Physics Div, Argonne Nat Lab, 72; lectr, Roosevelt Univ, 74-77; consult eng, Innovator Assocs, 77-81; ASSOC PROF ELEC ENG, DEPT ELEC ENG & COMPUT SCI & DIR, MICROELECTRONICS LAB, UNIV ILL, CHICAGO, 81- *Concurrent Pos:* Physicist, Control Instrument Div, Warner & Swasey Co, Flushing, Long Island, 61-65; vis asst prof elec eng, Univ Ill, Chicago, 79-81; vis scientist, Fraunhofer Inst Microelectronics, Duisburg, 88. *Mem:* Am Phys Soc; Am Vacuum Soc; Inst Elec & Electronics Engrs; Sigma Xi. *Res:* Microelectronic sensors; chemical and gas sensors; micromachining and microfabrication technology; quantum phenomena; solid state devices and materials; signal analysis and pattern recognition. *Mailing Add:* Dept Elec Eng & Comput Sci Univ Ill Box 4348 Chicago IL 60680

MACLAY, WILLIAM NEVIN, b Belleville, Pa, Dec 30, 24; m 49; c 5. PHYSICAL CHEMISTRY, RESEARCH ADMINISTRATION. *Educ:* Juniata Col, BS, 47; Yale Univ, PhD(phys chem), 50. *Prof Exp:* Assoc prof chem, Davis & Elkins Col, 50-51; res chemist, B F Goodrich Co, 51-59; mgr latices res, Koppers Co Inc, 59-65, mgr polystyrene res, 62-65, asst mgr plastics res, 65-67, mgr com develop, 67-68, asst mgr res, 68, vpres & dir res, 68-85; RETIRED. *Mem:* Am Chem Soc. *Res:* Surface and colloid chemistry; high polymer latices; polystyrene molding resins; copolymerization; graft polymers. *Mailing Add:* 539 Greenleaf Dr Monroeville PA 15146-1201

MACLEAN, BONNIE KUSESKE, b St Cloud, Minn, Jan 23, 42; m 64; c 2. ENTOMOLOGY. *Educ:* Gustavus Adolphus Col, BA, 63; Purdue Univ, MS, 65, PhD(entom), 72. *Prof Exp:* Instr biol, Youngstown State Univ, 68-69; RES & WRITING, 69-; ASST PROF BIOL, THIEL COL, 80- *Concurrent Pos:* instr biol, Thiel Col, 77-80. *Mem:* Sigma Xi; Entom Soc Am; AAAS; Am Inst Biol Sci; Ecol Soc Am. *Res:* Trichoptera of Ohio. *Mailing Add:* 280 EW Reserve Rd Poland OH 44514

MACLEAN, DAVID ANDREW, b Fredericton, NB, Feb 27, 52; m 82; c 1. FOREST ECOLOGY. *Educ:* Univ NB, BSc Hons, 73, PhD(ecol), 78. *Prof Exp:* RES SCIENTIST FOREST ECOL, MARITIMES FOREST RES CTR, CAN FORESTRY SERV, 78- *Concurrent Pos:* Postgrad scholar, Nat Res Coun Can, 75-78; postdoctoral fel, Nat Res Coun Can, 78; hon res assoc, Univ NB, 82- *Mem:* Ecol Soc Am; Can Inst Forestry. *Res:* Impact of spruce budworm defoliation on tree growth and productivity, mortality and stand deterioration; modelling of forest growth and effects of defoliation on growth. *Mailing Add:* Forestry Can - Maritimes Box 4000 Fredericton NB E3B 5P7 Can

MACLEAN, DAVID BAILEY, b Summerside, PEI, July 15, 23; m 45; c 7. ORGANIC CHEMISTRY. *Educ:* Acadia Univ, BSc, 42; McGill Univ, PhD(chem), 46. *Prof Exp:* Res chemist, Dom Rubber Co, 46-49; assoc prof indust chem, NS Tech Col, 49-54; from assoc prof to prof, 54-89, EMER PROF CHEM, MCMASTER UNIV, 90- *Mem:* Am Chem Soc; Chem Inst Can; Royal Soc Can; Am Soc Mass Spectrometry. *Res:* Isolation, structure and synthesis of alkaloids and related heterocyclic systems; mass spectrometry of organic compounds. *Mailing Add:* Dept Chem McMaster Univ Hamilton ON L8S 4M1 Can

MACLEAN, DAVID BELMONT, b Cleveland, Ohio, Sept 15, 41; m 64; c 2. INSECT ECOLOGY & BEHAVIOR. *Educ:* Heidelberg Col, BS, 63; Purdue Univ, MS, PhD (entom, quant ecol), 69. *Prof Exp:* From asst prof to assoc prof, 68-85, PROF BIOL, YOUNGSTOWN STATE UNIV, 86- *Concurrent Pos:* Mem exec coun, Ohio Biol Survey. *Mem:* Sigma Xi; Entom Soc Am; Entom Soc Can; Ecol Soc Am; Am Inst Biol Sci. *Res:* Interrelationships of insect and plant communities; ecology and biogeography of trichoptera; multivariate methods in the study of ecology, behavior and evolution. *Mailing Add:* Dept Biol Youngstown State Univ Youngstown OH 44555

MACLEAN, DAVID CAMERON, b New Rochelle, NY, Dec 8, 33; m 56; c 3. PLANT PHYSIOLOGY. *Educ:* Univ Conn, BS, 60; Mich State Univ, MS, 62, PhD(plant physiol), 65. *Prof Exp:* Res assoc plant physiol, Mich State Univ, 61-65; PLANT PHYSIOLOGIST, BOYCE THOMPSON INST PLANT RES, 65- *Concurrent Pos:* Consult air pollution, 65-; mem panel on med & biol effects of environ pollutants, Nat Res Coun-Nat Acad Sci, 73-77; pres, Environ Strategies Inc, 77- *Honors & Awards:* Dow Award, Am Soc Hort Sci, 70. *Mem:* AAAS; Am Soc Plant Physiol; Am Soc Hort Sci; NY Acad Sci; Sigma Xi. *Res:* Climatic stresses and photosynthesis; effects of air pollutants on vegetation; plant growth and development. *Mailing Add:* Boyce Thompson Inst Plant Res Tower Rd Ithaca NY 14853

MACLEAN, DONALD ISADORE, b Norwalk, Conn, Nov 25, 29. PHYSICAL CHEMISTRY. *Educ:* Boston Col, AB, 53; Catholic Univ, PhD(phys chem), 58. *Prof Exp:* Humboldt res fel, Univ Gottingen, 62-65; from asst prof to assoc prof phys chem, Boston Col, 66-73; vpres acad affairs, Creighton Univ, 73-76; PRES, ST JOSEPH'S UNIV, 76- *Concurrent Pos:* Adj prof, Boston Col, 73-75; mem bd trustees, St Louis Univ, 75-81, bd dirs, Am Coun Educ, 78-, Philadelphia Urban Coalition, 76- *Mem:* AAAS; Sigma Xi. *Res:* Fast reaction kinetics; combustion chemistry; mass, electron spin resonance and optical spectroscopy. *Mailing Add:* Springs Hill Col 4000 Sauphin St Mobile AL 36608

MCLEAN, DONALD LEWIS, b Norwood, Mass, Oct 2, 28; m 52; c 3. ENTOMOLOGY. *Educ:* Tufts Univ, BS, 53; Univ Mass, MS, 55; Univ Calif, PhD(entom), 58. *Prof Exp:* Prof enton, Univ Calif, Davis, 58-87, chmn dept, 74-79, dean biol sci, 79-85; DEAN COL AGR & LIFE SCI, UNIV VT, 87- *Mem:* Entom Soc Am; AAAS; Am Inst Biol Sci. *Res:* Insect transmission of plant viruses, especially aphid vectors; biological studies dealing with aphid feeding; culturing and chemical studies with aphid cells and symbiotes. *Mailing Add:* Dean's Off Col Agr & Life Sci Univ Vt Burlington VT 05405-0106

MCLEAN, DONALD MILLIS, b Melbourne, Australia, July 26, 26; Can citizen; m 76. VIROLOGY, MEDICAL MICROBIOLOGY. *Educ:* Univ Melbourne, BSc, 47, MD, 50; FRCP(C), 67. *Prof Exp:* Dir virol, Hosp Sick Children, Toronto, 58-67; head, Div Med Microbiol, Univ BC, 67-80, prof, 67-91; RETIRED. *Concurrent Pos:* Consult microbiol, Univ Hosp-Univ BC Site, Vancouver, 68-91. *Mem:* Am Soc Trop Med & Hyg; Am Soc Microbiol; Am Epidemiol Soc; Can Med Asn; Am Soc Virol; Am Mosquito Control Asn. *Res:* Arbovirus vectors and reservoirs; same-day virus diagnostic tests. *Mailing Add:* 6-5885 Yew St Vancouver BC V6M 3Y5 Can

MCLEAN, EDGAR ALEXANDER, b Gastonia, NC, July 25, 27; m 51; c 5. PLASMA PHYSICS. *Educ:* Univ NC, Chapel Hill, BS, 49; Univ Del, MS, 51. *Prof Exp:* RES PHYSICIST, NAVAL RES LAB, 51- *Concurrent Pos:* Res consult, Space Sci Dept, Cath Univ Am, 64-69 & Univ Western Ont, 73- *Honors & Awards:* Res Publ Award, Naval Res Lab, 71, 74 & 77. *Mem:* Inst Elec & Electronics Engrs; AAAS; Am Phys Soc; Sigma Xi. *Res:* Optical diagnostics in the field of plasma physics, including spectroscopic, interferometric, laser-scattering, and high-speed photographic measurements on shock tubes, laser-produced plasmas, and various controlled fusion devices. *Mailing Add:* 19 Mel Mara Dr Oxon Hill MD 20745

MCLEAN, EDWARD BRUCE, b Washington Court House, Ohio, Jan 10, 37; m 86; c 3. ORNITHOLOGY, MAMMALOGY. *Educ:* Ohio State Univ, BSc, 58, MSc, 63, PhD(zool), 68. *Prof Exp:* Asst prof biol, Southern Univ, Baton Rouge, 68-70; from asst prof to assoc prof, 70-78, chmn dept, 81-88, PROF BIOL, JOHN CARROLL UNIV, 78-,. *Concurrent Pos:* Ecol consult, 72- *Mem:* Am Ornith Union; Wilson Ornith Soc; Cooper Ornith Soc; Asn Field Ornith. *Res:* Bioacoustics; vertebrate ecology; ethology. *Mailing Add:* Dept of Biol John Carroll Univ Cleveland OH 44118

MCLEAN, EUGENE OTIS, soil chemistry, for more information see previous edition

MCLEAN, FLYNN BARRY, b Ft Bragg, NC, Nov 15, 41; m 66; c 2. SOLID STATE PHYSICS, SOLID STATE ELECTRONICS. *Educ:* Rensselaer Polytech Inst, ScB, 63; Brown Univ, ScM, 66, PhD(physics), 68. *Prof Exp:* Physicist, Harry Diamond Labs, 68-69; res assoc solid state physics, Brookhaven Nat Lab, 70-71; PHYSICIST, HARRY DIAMOND LABS, 71- *Mem:* Am Phys Soc; fel Inst Elec & Electronic Engrs. *Res:* Radiation effects in solids; electrical transport in amorphous insulators; physics of metal-oxide-semiconductor systems; electron tunneling through thin insulating films; modeling of cosmic ray induced upset in integrated circuits; physics of thin film ferroelectrics. *Mailing Add:* Radiation Effects Physics Br SLCHD-NW-RP Harry Diamond Labs 2800 Powder Mill Rd Adelphi MD 20783-1197

MCLEAN, FRANCIS GLEN, b Central City, Nebr, Sept 5, 38; m 64; c 2. GEOTECHNICAL & STRUCTURAL ENGINEERING, EARTHQUAKE ENGINEERING. *Educ:* Univ Nebr, Omaha, BS(IE), 62, BS(CE), 63; Univ Nebr, Lincoln, MS(CE), 65; Northwestern Univ, PHD(civil eng), 70. *Prof Exp:* Prin soil mech engr & assoc, Westenhoff & Novick Consult Engrs, Chicago, Ill, 70-72; res civil engr, US Army Corps Engrs, Waterways Exp Sta, Vicksburg, Miss, 72-75, chief, earthquake eng & geophys div, 75- 77; chief, geotech & mat br, US Army Corps Engrs, S Pac Div, San Francisco, Calif, 77-82; chief, res & lab serv div, 82-88, CHIEF GEOTECH ENG BR, US BUR RECLAMATION, DENVER, COLO, 88- *Concurrent Pos:* Asst prof, Univ Nebr, Omaha, 63-67; lectr, Univ Calif, San Jose, 80-81; mem, US Comt Large Dams & US Nat Comt Soil Mech & Found Engrs. *Mem:* Am Soc Civil Engrs; Earthquake Eng Res Inst. *Res:* Various studies in geotechnical & earthquake engineering fields for water resource projects, ie, large clams, primarily related to analysis techniques, design & construction methods, including centrifugal testing and new materials. *Mailing Add:* PO Box 280564 Lakewood CO 80228-0564

MACLEAN, GRAEME STANLEY, b Melbourne, Australia, Feb 11, 50. PHYSIOLOGY, INTERNAL MEDICINE. *Educ:* Monash Univ, Australia, BSc, 72; Univ Mich, Ann Arbor, MS, 74, PhD(biol sci), 77; Univ Tex, Southwestern Med Sch, MD, 86. *Prof Exp:* Lectr, Univ Mich, 77; asst prof Physiol, Univ Tex, Arlington, 77-82; PHYSICIAN, PVT PRACT, 88- *Mem:* Am Soc Zoologists; Am Soc Mammalogists; Am Physiol Soc; Am Med Asn. *Res:* Environmental physiology of vertebrates; respiratory and cardiovascular physiology of reptiles, hibernating mammals, and fossorial mammals. *Mailing Add:* 302 W Ninth St Suite E Dallas TX 75208

MCLEAN, HUGH, b El Centro, Calif, Nov 8, 39; m 82; c 3. GEOLOGY, SEDIMENTARY PETROLOGY. *Educ:* Calif Maritime Acad, BS, 60, San Diego State Col, BA, 66; Univ Wash, MS, 68, PhD(geol), 70. *Prof Exp:* Geologist petrol explor, Texaco, Inc, 70-74; GEOLOGIST, US GEOL SURV, 74- *Mem:* Geol Soc Am; Am Asn Petrol Geologists. *Res:* Reservoir properties of clastic rocks; depositional environments and sedimentary processes as a reflection of tectonic setting; geologic mapping in Alaska, California, and Baja California Sur, Mexico. *Mailing Add:* US Geol Surv MS 999 345 Middlefield Rd Menlo Park CA 94025

MCLEAN, IAN WILLIAM, b Durham, NC, Sept 21, 43; m 86; c 3. PATHOLOGY, OPHTHALMOLOGY. *Educ:* Univ Mich, BS, 65, MD, 69; Univ Colo Med Ctr, cert path, 73; Am Bd Path, cert anat path, 74. *Prof Exp:* Fel ophthal path, Armed Forces Inst Path, 71-72; vis scientist, Inst Biol Sci, Oakland Univ, 72; STAFF PATHOLOGIST, OPHTHAL DIV, ARMED FORCES INST PATH, 73- *Mem:* Asn Res Vision & Ophthalmol. *Res:* Evaluation of factors relating to mortality in patients with intraocular malignant melanoma and retinoblastoma. *Mailing Add:* Ophthal Div Armed Forces Inst of Path Washington DC 20306

MCLEAN, JAMES AMOS, b Flint, Mich, Dec 2, 21; m 54; c 2. INTERNAL MEDICINE, ALLERGY. *Educ:* Univ Mich, BS & MD, 46; Baylor Univ, MS, 52. *Prof Exp:* From asst prof to assoc prof, 56-67, asst allergy, Health Serv, 65-77, prof, 67-86, EMER PROF INTERNAL MED, MED CTR, UNIV MICH, ANN ARBOR, 86- *Mem:* AMA; Am Acad Allergy & Immunol; Am Fedn Clin Res; Am Thoracic Soc. *Res:* Clinical allergy. *Mailing Add:* Dept Int Med TC 3918/0380 Ann Arbor MI 48109

MCLEAN, JAMES DENNIS, b Bay City, Mich, Nov 23, 40; m 68. ANALYTICAL CHEMISTRY, POLAROGRAPHY. *Educ:* Univ Mich, Ann Arbor, BS, 62; Mich State Univ, PhD(anal chem), 67. *Prof Exp:* Anal specialist, 67-79, RES ASSOC, DOW CHEM CO, 80- *Honors & Awards:* Vernon A Stenger Award, 74 & 82; John C Vaaler Award, 74 & 80; A O Beckmann Award, Instrument Soc Am, 75; I R 100 Award, 85. *Mem:* Am Chem Soc; NY Acad Sci; Sigma Xi. *Res:* Polarographic analysis; anodic stripping voltammetry; electrochemical flo-thru detectors and liquid chromatography detectors; square-wave voltammetry. *Mailing Add:* Dow Chemical Co 1897 Bldg Midland MI 48667

MCLEAN, JAMES DOUGLAS, b Regina, Sask, Feb 12, 20; m 50; c 4. DENTISTRY. *Educ:* Univ Toronto, DDS, 42. *Hon Degrees:* FRCD(C). *Prof Exp:* From lectr to asst prof dent, Univ Alta, 47-53; dean, Fac Dent, 54-75, PROF DENT, DALHOUSIE UNIV, 53- *Mem:* Fel Int Col Dent; fel Am Col Dent. *Mailing Add:* 21 Meadowbrook Dr Apt 105 Bedford NS B4A 1P7 Can

MCLEAN, JAMES H, b Detroit, Mich, June 17, 36. INVERTEBRATE ZOOLOGY. *Educ:* Wesleyan Univ, BA, 58; Stanford Univ, PhD(biol), 66. *Prof Exp:* CUR MOLLUSKS, LOS ANGELES COUNTY MUS NATURAL HIST, 64- *Mem:* Am Malac Union(pres, 89); Soc Syst Zool. *Res:* Systematics of marine mollusks, especially prosobranch gastropods of the eastern Pacific. *Mailing Add:* Invert Zool 900 Exposition Blvd Los Angeles CA 90007

MCLEAN, JEFFERY THOMAS, b Memphis, Tenn, July 5, 44; m 74. MATHEMATICS. *Educ:* Hendrix Col, BA, 66; Ohio State Univ, MS, 67, PhD(math), 73. *Prof Exp:* Asst prof math, Ohio Northern Univ, 73-74 & Ohio State Univ, Lima, 74-75; assoc prof & chmn dept math, Ohio Northern Univ, 75-78; assoc prof math & chmn dept, 78-, ASSOC PROF, DEPT MATH, UNIV ST THOMAS. *Mem:* Am Math Soc; Math Asn Am. *Res:* Finite projective planes; mathematical perspective in art; mathematical limitations of technology. *Mailing Add:* Dept Math Univ St Thomas 2115 Summit Ave Main No 4304 St Paul MN 55105

MCLEAN, JOHN A, JR, b Chapel Hill, Tenn, Nov 8, 26; m 58; c 3. METAL CONTAINING POLYMERS. *Educ:* Tenn State Univ, BS, 48; Univ Ill, MS, 56, PhD(chem), 59. *Prof Exp:* High sch teacher, Ill, 48-53; asst chmn undergrad progs, 74-84, PROF CHEM, UNIV DETROIT, 59-, CHMN DEPT, 84- *Concurrent Pos:* Bk rev ed, J Appl Spectros, 75-78. *Mem:* Am Chem Soc; Soc Appl Spectros. *Res:* Kinetics, structure, mechanisms, synthesis and catalytic properties of transition metal complexes; transition metal catalysts on polystyrene-divinylbenzene supports; thermoplastics containing metal carboxylate groups. *Mailing Add:* 9260 W Outer Dr Detroit MI 48219

MCLEAN, JOHN ALEXANDER, b Takapuna, NZ, Mar 31, 43; m 68. FOREST ENTOMOLOGY. *Educ:* Univ Auckland, New Zealand, BSc, 65, MSc, 68; Simon Fraser Univ, Can, PhD(forest entomol), 76. *Prof Exp:* Sec sch teacher sci, Lytton High Sch, New Zealand, 68-69; lectr biol, Univ S Pacific, Fiji, 70-73; from asst prof to assoc prof, 77-86, PROF FOREST ENTOMOL, UNIV BC, 86- *Concurrent Pos:* Consult forest pest mgt, Integrated Intensive Forest Mgt Proj, People's Repub China-Canadian Int Develop Agency, 85, 86 & 87; Hurricane Gilbert assessment damage to Jamacian forests, UN Develop Prog, 88. *Mem:* Entomol Soc Am; Sigma Xi. *Res:* Management of forest insects in western North America with special emphasis on the use of semiochemicals for surveying and mass trapping ambrosia beeches and the western spruce budworm; marking insects. *Mailing Add:* Fac Forestry Univ Brit Columbia 2357 Main Mall Vancouver BC V6T 1W5 Can

MCLEAN, JOHN DICKSON, b Washington, DC, Mar 16, 52; m 81; c 3. COMPUTER SECURITY, SOFTWARE ENGINEERING. *Educ:* Oberlin Col, BA, 74; Univ NC, MA, 76, MS, 80, PhD(philos), 80. *Prof Exp:* Instr philos, Univ NC, 74-79, res asst computer sci, 79-80; computer scientist, 80-88, SUPVRY COMPUTER SCIENTIST, NAVAL RES LAB, 88- *Concurrent Pos:* Mem, Verification Working Group, Nat Computer Security Ctr, 87-; adj prof, Computer Sci Dept, Univ Md, 88 & Nat Cryptographic Sch, 91-; assoc ed, J Computer Security, 90-; prog chair, Computer Security Found Workshop, Inst Elec & Electronic Engrs, 90, prog co-chair, Symp Security & Privacy, 91-92. *Mem:* Inst Elec & Electronic Engrs Computer Soc; Asn Comput Mach. *Res:* Formal methods in computer security including security models and information theory; software engineering; software specification and verification. *Mailing Add:* Naval Res Lab Code 5543 Washington DC 20375-5000

MCLEAN, JOHN ROBERT, b St Thomas, Ont, Apr 15, 26; m 51; c 2. BIOCHEMISTRY. *Educ:* Queen's Univ, Ont, BSc, 50, PhD(biochem), 54. *Prof Exp:* Instr biochem, Yale Univ, 54-56; assoc res chemist, 56-60, sr res biochemist, 60-72, sect dir neuropharmacol, 72-77, assoc dir nuerochem sect, 77-82, ASST DIR CLIN SCI, PARKE DAVIS & CO, 83- *Mem:* Am Soc Pharmacol & Exp Therapeut. *Res:* Drugs acting on the central nervous system. *Mailing Add:* 1708 Corington Dr Ann Arbor MI 48103

MCLEAN, KATHARINE WEIDMAN, b Camden, NJ, Oct 18, 27; m 51; c 3. BIOCHEMISTRY. *Educ:* Cornell Univ, AB, 48; Univ Ill, MS, 49, PhD(biochem), 51. *Prof Exp:* Org res chemist, Rohm and Haas Co, 51-52; anal chemist, Ariz Testing Labs, 52-57; TEACHER CHEM, PHOENIX COL, 57- *Mem:* Am Chem Soc; Am Inst Chemists. *Res:* Vitamin BT; explosives. *Mailing Add:* Dept Chem Phoenix Col 1202 W Thomas Rd Phoenix AZ 85013

MCLEAN, LARRY RAYMOND, b Detroit, Mich, Aug 21, 54; m 77; c 3. PHYSICAL BIOCHEMISTRY, LIPID-PEPTIDE INTERACTIONS. *Educ:* Mich State Univ, BS, 76; Med Col Pa, PhD(biochem), 81. *Prof Exp:* Postdoctoral fel biochem, Royal Free Hosp Sch Med, 81-82; postdoctoral fel biochem, Col Med, Univ Cincinnati, 82-85; SR RES BIOCHEMIST, MARION MERRELL DOW RES INST, 85- *Mem:* Biophys Soc; Am Soc Biochem & Molecular Biol. *Res:* Interaction of soluble peptides with lipids; lung surfactants; peptide-peptide interactions, especially amphipathic alpha-helical peptides; physical properties of lipids. *Mailing Add:* Dept Chem Marion Merrell Dow Res Inst 2100 E Galbraith Rd Cincinnati OH 45215-6300

MACLEAN, LLOYD DOUGLAS, b Calgary, Alta, June 15, 24; m 54; c 5. SURGERY. *Educ:* Univ Alta, BSc, 43, MD, 49; Univ Minn, PhD(surg), 57; FRCPS(C). *Prof Exp:* From instr to assoc prof, Univ Minn, 56-65; PROF SURG, McGILL UNIV, 65-, CHMN DEPT, 68- *Concurrent Pos:* Chief surg serv, Ancker Hosp, St Paul, 57-65; surgeon-in-chief, Royal Victoria Hosp, Montreal, 65-; officer, Order of Can, 85. *Mem:* Am Physiol Soc; Soc Exp Biol & Med; Soc Univ Surgeons; AMA; Am Surg Asn. *Res:* Blood flow to heart; intestinal blood flow in shock due to hemorrhage and endotoxin and cardiac excitability; shock and transplantation. *Mailing Add:* Dept Surg McGill Univ Royal Victoria Hosp 687 Pine Ave W Montreal PQ H3A 1A1 Can

MCLEAN, MARK PHILIP, b Dekalb, Ill, Feb 20, 57; m 83; c 3. ENDOCRINOLOGY, BIOCHEMISTRY. *Educ:* Northern Ill Univ, BS, 79, MS, 81; Univ Ill, PhD(physiol), 86. *Prof Exp:* Teaching asst physiol, Northern Ill Univ, 79-81, Univ Ill, 81-82; res asst physiol, Univ Ill, 82-86, res assoc endocrinol, 86-90; ASST PROF, UNIV SFLA, 90- *Concurrent Pos:* Res Assoc, Univ Ill Chicago, 86-; Nat Res Serv Award, NIH, 86-89. *Honors & Awards:* New Investr Award, Endocrine Soc, 89; Serono Young Investr Award, 89. *Mem:* Soc Study Reproduction; Electron Micros Soc Am; Endocrine Soc; Soc Exp Biol & Med; Am Physiol Soc; Am Soc Cell Biol. *Res:* Mechanism of estradiol action in the corpus luteum; estradiol was shown to promote luteal growth, vascularization and steriodogenesis; direct effect of estradiol and cholesterol biosynthesis, mobilization and transport; current investigation of specific proteins produced in response to estradiol. *Mailing Add:* Dept Obstet/Gynec Univ SFla Tampa FL 33606

MCLEAN, NORMAN, JR, b San Diego, Calif, May 8, 26; m 63; c 1. INVERTEBRATE ZOOLOGY. *Educ:* Univ Calif, Berkeley, BS, 51, PhD(zool), 65. *Prof Exp:* Asst prof to assoc prof, 65-74, PROF ZOOL, SAN DIEGO STATE UNIV, 74- *Res:* Functional morphology of the molluscan digestive tract. *Mailing Add:* Dept Biol San Diego State Univ 5300 Campanile Dr San Diego CA 92182

MACLEAN, PAUL DONALD, b Phelps, NY, May 1, 13; m 42; c 5. NEUROPHYSIOLOGY. *Educ:* Yale Univ, BA, 35, MD, 40. *Hon Degrees:* DSc, State Univ NY, Binghamton, 86. *Prof Exp:* Intern, Johns Hopkins Hosp, 40-41; asst resident med, New Haven Hosp, Conn, 41-42; res asst path, Sch Med, Yale Univ, 42; clin instr med, Med Sch, Washington Univ, 46-47; USPHS res fel psychiat, Harvard Med Sch & Mass Gen Hosp, 47-49; Asst prof physiol, Sch Med, Yale Univ, 49-51, from asst prof to assoc prof psychiat, physiol & neurol, 51-56, assoc prof physiol, 56; chief limbic integration & behav sect, 57-71, chief lab brain evolution & behav, 71-85, SR RES SCIENTIST, NIMH, NIH, 85- *Concurrent Pos:* Dir EEG lab, New Haven Hosp, Conn, 51-52; attend physician, Grace-New Haven Hosp, 53-56; NSF sr postdoctoral fel, Switz, 56-57. *Honors & Awards:* Distinguished Res Award, Asn Res Nerv & Ment Dis, 64; Salmon Medal Distinguished Res in Psychiat, 66; Salmon Lectr, NY Acad Med, 66; Superior Serv Award US Dept Health, Educ & Welfare, 67; Hincks Lectr, Ont Ment Health Found, 69; Spec award, Am Psychopath Asn, 71; Karl Spenser Lashley Award, Am Philos Soc, 72; Mider Lectr, NIH, 72; Adolph Meyer Lectr, Am Psychiat Asn, 82. *Mem:* Am Electroencephalog Soc; Am Neurol Asn; Am Asn Anatomists; Soc Neurosci; Am Physiol Soc; Am Asn Neurol Surgeons. *Res:* Forebrain mechanisms of species-typical and emotional behavior. *Mailing Add:* 9916 Logan Dr Potomac MD 20854

MCLEAN, R T, b Westerville, Ohio, July 18, 22; m 53. ALGEBRA. *Educ:* Otterbein Col, BS, 46; Bowling Green State Univ, MA, 50; Univ Pittsburgh, PhD(group theory), 61. *Prof Exp:* High sch teacher math, Ohio, 45-49, 50-51; asst chem, Bowling Green State Univ, 49-50; from asst prof to prof math, Univ Steubenville, 52-67, head dept, 53-67, chmn div natural sci, 63-67; chmn dept math sci, 67-73, dir, NSF Math Teacher Develop Prog, 77-80, chmn dept math, 78-82, PROF MATH, LOYOLA UNIV, LA, 67- *Concurrent Pos:* Nat Defense Act Workshops Elem & Secondary Math Teacher Improvement, 59-67; NSF In-Serv Insts, 62-67; HEW Experienced Teacher Fel Prog strengthening elem sch sci teaching, 68-70; Loyola Univ fac res grants, 73-77 & 78-81; innovation & expansion grant, Radio for Blind & Print Handicapped, 77-80, pres & chmn bd; nat chmn, Affil Leadership League of & for the Blind, vpres, Am Coun of the Blind & Light House for the Blind; mem bd dirs, Nat Accredidation Coun & Nat Industs for the Blind; pres & chmn bd, Lighthouse for the Blind, New Orleans; dir, Asn Radio Reading Serv. *Mem:* Am Math Soc; Nat Coun Teachers Math; Sigma Xi; Math Asn Am; Nat Asn Blind Teachers. *Res:* Theory of groups; higher education program design; careers through mathematics; group generators; braille large print and voice output for computers and all other digital information devices; institutional research. *Mailing Add:* Dept of Math Sci Loyola Univ St Charles Ave New Orleans LA 70118

MCLEAN, RICHARD BEA, b Raleigh, NC, Aug 27, 46; m 68; c 1. MARINE BIOLOGY. *Educ:* Fla State Univ, BA, 68, PhD(marine biol), 75. *Prof Exp:* Res asst spiny lobsters, Fla State Univ, 69-70, instr biol, 70-73; res assoc marine environ impact, 74-76, RES STAFF FISH POP DYNAMICS-RESERVOIRS, CAUSE & EFFECTS IMPINGEMENT, OAK RIDGE NAT LAB, 76- *Concurrent Pos:* NASA fel, 72. *Mem:* Am Inst Biol Sci; Am Soc Zool; Animal Behav Soc; Sigma Xi. *Res:* Behavioral ecology of marine benthic communities, emphasizing interactions such as predation, symbiosis and competition; population dynamics and bioenergetics of reservoir fishes. *Mailing Add:* Equatic Ecol Sect Bldg 1505 Oak Ridge Nat Lab Oak Ridge TN 37830

MCLEAN, ROBERT GEORGE, b Warren, Ohio, Jan 10, 38; div; c 2. EPIZOOTIOLOGY, VERTEBRATE ECOLOGY. *Educ:* Bowling Green State Univ, BSE & BS, 61, MA, 63; Pa State Univ, PhD(zool), 66. *Prof Exp:* Chief parasitol br, Third US Army Med Lab, Ft McPherson, Ga, 66-68; chief rabies ecol subunit, Ctr Dis Control, Ft Collins, Colo, 68-69, mem rabies control unit, Viral Zoonoses Sect, Viral Dis Br, Epidemiol Prog, 69-73, mem, Vert Ecol Br & Vector-Borne Dis Br, 73-79, asst chief, Arbovirus Ecol Br, 79-82, res ecologist, Med Entom Res & Training Unit, Guatemala, 82-84, res ecologist, Arbovirus Ecol Br, 84-90, CHIEF VERT ECOL SECT, MED ENTOMOL & ECOL BR, CTR DIS CONTROL, FT COLLINS, COLO, 90- *Concurrent Pos:* Part-time instr biol, Ga State Univ, 67-68; fac affil zool, fishery & wildlife, Colo State Univ, 73-; asst ed, J Wildlife Dis, 84- *Mem:* Wildlife Dis Asn; Am Soc Trop Med & Hygiene; Ecol Soc Am; Am Soc Mammalogist; AAAS. *Res:* Homing ability and courtship behavior of pigeons; population control of vertebrate pest animals with chemosterilants; ecological studies of zoonotic diseases in birds and mammals; ecology of Colorado tick fever in northern Colorado and St Louis encephalitis in western Tennessee; vertebrate hosts of leishmaniasis in Central America. *Mailing Add:* Arbovirus Ecol Br Ctr Dis Control PO Box 2087 Ft Collins CO 80522

MCLEAN, ROBERT J, b New Haven, Conn, Aug 15, 40; m 63; c 4. CELL BIOLOGY. *Educ:* Univ Conn, BA, 62, MS, 64, PhD(phycol), 67. *Prof Exp:* NIH fel bot, Univ Tex, 67-68; from asst prof to assoc prof, 68-75, PROF BIOL SCI, STATE UNIV NY COL BROCKPORT, 75- *Mem:* AAAS; Am Inst Biol Sci; Phycol Soc Am; Am Soc Cell Biol; Int Phycol Soc. *Res:* Cell recognition and membrane biology. *Mailing Add:* 27 Lancet Way Brockport NY 14420

MACLEAN, STEPHEN FREDERICK, JR, b Los Angeles, Calif, Jan 18, 43; m 67; c 1. ECOLOGY. *Educ:* Univ Calif, Santa Barbara, BA, 64; Univ Calif, Berkeley, PhD(ecol), 69. *Prof Exp:* Actg asst prof zool, Univ Mont, 69-70; asst prof, Univ Ill, 70-71; NSF res grant Arctic Alaska, 71-72, asst prof, 71-74, assoc prof biol sci, 74-76, ASSOC PROF BIOL SCI, BIOL & WILDLIFE, UNIV ALASKA, 77- *Concurrent Pos:* NSF res grant Arctic Alaska, Univ Mon, 70-71; prog integrator, US Int Biol Prog Tundra Biome, 71- *Mem:* Ecol Soc Am; Brit Ecol Soc; Arctic Inst NAm; Am Ornithologists Union; Cooper Ornith Soc. *Res:* Population ecology and energetics of arctic birds, mammals and insects; systems analysis of tundra ecosystem. *Mailing Add:* Dept Biol & Wildlife Univ Alaska 202 Bunnell Fairbanks AK 99701

MCLEAN, STEWART, b Moascar, Egypt, Nov 19, 31; m 57, 81; c 5. ORGANIC CHEMISTRY. *Educ:* Glasgow Univ, BSc, 54; Cornell Univ, PhD(org chem), 58. *Prof Exp:* Fel org chem, Univ Wis, 57-58; fel, Nat Res Coun Can, 58-60; from asst prof to assoc prof, PROF ORG CHEM, UNIV TORONTO, 70- *Mem:* Am Chem Soc; fel Chem Inst Can; Royal Soc Chem. *Res:* Structural and synthetic organic chemistry; mechanistic studies. *Mailing Add:* Dept of Chem Univ of Toronto 80 St George St Toronto ON M5S 1A1 Can

MACLEAN, WALLACE H, b PEI, Jan 10, 31; m 60; c 2. GEOLOGY, GEOCHEMISTRY. *Educ:* Colo Sch Mines, GeolE, 55; McGill Univ, MSc, 64, PhD(geol), 68. *Prof Exp:* Mine geologist, United Keno Hill Mines, Can, 55-57; econ geologist, Ministry Petrol & Mineral Resources, Saudi Arabia, 57-62; prof assoc, 67-70, from asst prof to assoc prof, 70-84, PROF GEOL SCI, MCGILL UNIV, 84- *Honors & Awards:* Waldemar Lindgren Citation Award, 69; Barlow Mem Medal, Can Inst Mining & Metall, 82. *Mem:* Geol Asn Can; Mineral Asn Can; Can Inst Mineral Metall; Soc Econ Geologists; Soc Geol Appl Mineral Deposits. *Res:* Genesis of mineral deposits; phase equilibria of sulfide-silicate liquid systems pertaining to magmatic ores; field and hydrothermal alteration studies on Archean massive sulfides. *Mailing Add:* Dept Geol McGill Univ 3450 University St Montreal PQ H3A 2A7 Can

MACLEAN, WALTER M, b Modesto, Calif, Mar 15, 24; m 54; c 2. NAVAL, MARINE & OFFSHORE ENGINEERING. *Educ:* Merchant Marine Acad, grad, 45; Univ Calif, Berkeley, BS, 56, ME, 57, DEng(naval archit), 67. *Prof Exp:* Asst engr, US Lines, Inc, 45-46 & Am Pres Lines, Inc, 46-52; from draftsman to naval architect, Morris Guralnick, Naval Architect, 55-59; jr engr, Univ Calif, Berkeley, 59-65; prof eng, Webb Inst Naval Archit, 65-72; head eng, US Merchant Marine Acad, 72-75; Actg dir, 80-87, MGR, REQ DEVELOP LAB, NAT MARITIME RES CTR, 75-; PROF ENG, US MERCHANT MARINE ACAD, 87- *Concurrent Pos:* Mem, comt 8, slamming & impact, Int Ship & Offshore Struct Cong, 70 & 73, US Observ, 82 & 85 & comt II 2, 88 & 91; mem, TC-7, Inter Coop Marine Eng Systs, 83- & US Rep to Standing Comt, 84-88; mem, Hull Structure Comt, Soc Naval Architects & Marine Engrs, Tech & Res Comt & Ship Tech Oper Comt. *Mem:* Am Soc Mech Engrs; Am Soc Naval Engrs; Soc Naval Archit & Marine Engrs; NY Acad Sci. *Res:* Naval architecture; structural design; structural seaworthiness of ships, particularly ship slamming and the resulting structural damage and means of obviating such damages; ship instrumentation for operations, test and evaluation. *Mailing Add:* 24 Harbor Way Sea Cliff NY 11579-2114

MACLEAN, WILLIAM C, JR, b Chicago, Ill, Oct 1, 40; m; c 2. PEDIATRIC NUTRITION. *Educ:* Princeton Univ, AB, 62; McGill Univ, MD, 66; Am Bd Pediat, cert, 72. *Prof Exp:* Intern pediat, Johns Hopkins Hosp, 66-67, asst resident, 67-68, sr asst resident, 70-71, resident, Outpatient Dept, 71-72, chief resident, Dept Pediat, 72-73, from asst prof to assoc prof, Dept Pediat, 73-82, from asst prof to assoc prof, Dept Int Health, 73-82; instr, Dept Med & Surg, Med Field Serv Sch, Ft Sam Houston, San Antonio, Tex, 69-70; clin assoc prof, 82-86, CLIN PROF, DEPT PEDIAT, OHIO STATE UNIV, 86-; RES SCIENTIST & MED DIR PEDIAT NUTRIT, ROSS LABS, COLUMBUS, OHIO, 82- *Concurrent Pos:* Field dir, Inst Nutrit Invest, Nutrit Res Inst, Miraflores, Lima, Peru, 73-74, assoc dir res, 74-82; active med staff, Dept Pediat, Johns Hopkins Hosp, Baltimore, Md, 73-82, actg dir neonatology, 74-75 & head, Pediat Gastroenterol & Nutrit Unit, 76-82; postdoctoral fel, USPHS-NIH, 73-75; assoc dir, Nutrit Prog, Sch Hyg & Pub Health, Johns Hopkins Univ, 76-82; consult, Nutrit Eval Lab, Manila, Philippines, 77-82; actg chief, Sect Nutrit, Children's Hosp, Columbus, 85-88; mem, Nutrit Study Sect, NIH, 83-87, chmn, 85-87; consult, Task Force Clin Testing Infant Formula, Comt Nutrit, Am Acad Pediat, 88. *Mem:* Am Acad Pediat; Am Soc Clin Nutrit; Soc Pediat Res; Am Pediat Soc; Am Inst Nutrit; NAm Soc Pediat Gastroenterol; Soc Int Nutrit. *Res:* Author of various publications. *Mailing Add:* Med Dept Ross Labs 625 Cleveland Ave Columbus OH 43215

MACLEAN, WILLIAM PLANNETTE, III, b Bainbridge, Md, Sept 20, 43; m 66. BIOLOGY, MARINE SCIENCES. *Educ:* Princeton Univ, BA, 65; Univ Chicago, PhD(evolutionary biol), 69. *Prof Exp:* From asst prof to assoc prof, Col Virgin Islands, 69-81, chmn, Div Sci & Math, 78-82, dir Marine Sci Ctr, 81-86, vpres acad affairs, 86-91, PROF BIOL, COL VIRGIN ISLANDS, 81-, DIR MARINE SCI PROG, 91- *Mem:* Ecol Soc Am; Am Soc Ichthyologists & Herpetologists; Soc Study Evolution. *Res:* Island biogeography; co-evolution; environmental impacts. *Mailing Add:* Univ Virgin Islands St Thomas VI 00802

MACLEAY, RONALD E, b Buffalo, NY, Dec 3, 35; m 60; c 4. SYNTHETIC ORGANIC CHEMISTRY. *Educ:* St Bonaventure Univ, BS, 57, MS, 59; Univ Buffalo, PhD(org chem), 65. *Prof Exp:* Res & develop chemist, 59-61, sr res chemist, 64-66, res group leader nitrogen chem, 66-75, sr group leader res, 75-79, res fel, 79-80, mgr process develop, Lucidol Div, 80-82, RES FEL, PENNWALT CORP, 82- *Mem:* Am Chem Soc. *Res:* Reduction of carbonyl compounds with lithium tetrakis-aluminate; synthesis of azo compounds for free radical initiation of polymerization. *Mailing Add:* Ten Mahogany Dr Williamsville NY 14221-6695

MCLEISH, KENNETH R, b Noblesville, Ind, Oct 27, 47; m; c 1. MEDICINE. *Educ:* Miami Univ, Oxford, Ohio, BA, 68; Ind Univ, Indianapolis, MD, 72; Am Bd Internal Med, dipl, 75 & 76. *Prof Exp:* Med intern, Med Ctr, Ind Univ, Indianapolis, 72-73, med resident, 73-74; res fel, Immunol Sect, Renal Div, Peter Bent Brigham Hosp, Boston, Mass, 74-75; clin fel, Renal Div, Med Ctr, Ind Univ, Indianapolis, 75-76, res fel, 76-77, asst prof med, Vet Admin Hosp, 77; asst prof med, Med Col Ohio, Toledo, 77-80; from asst prof to assoc prof, 80-89, PROF MED, HEALTH SCI CTR, UNIV LOUISVILLE, KY, 89-, ASSOC PROF BIOCHEM, 90- *Concurrent Pos:* Prin investr, fel grant, Kidney Found Ind, 76-77, Biomed Res Support grant, 79-81, Kidney Found Ohio, 79-80, Am Heart Asn, 80-81 & 85-86, Univ Louisville Med Sch Res Fund, 81-82, Am Cancer Soc, 81-82, Nat Inst Arthritis, Diabetes & Digestive & Kidney Dis, NIH, 82-84 & Vet Admin Merit Rev, 91-; mem, Tissue Comt, Univ Louisville, 81-83, dir, Med Intensive

Care Unit, 83-84, mem, Humana Hosp Univ Spec Care Unit Comt, 83-84, dir, Res Labs, Div Nephrology, Dept Med, 88-, mem, Res Comt, 89-, chmn, 90-; co-prin investr, Am Diabetes Asn, 83-85; assoc, Dept Microbiol & Immunol, Health Sci Ctr, Univ Louisville, Ky, 83-, mem, Grad Sch Fac, 83-, assoc, Dept Biochem, 88-90; actg chief, Nephrology Div, Vet Admin Hosp, 84-87, chief, Renal Sect, 86-87; med dir, Organ Procurement Agency, Univ Louisville, 86-87; mem bd dirs, Ky Organ Donor Affil, 87; res fel, Alexander von Humboldt Found, Dept Pharmacol, Univ Heidelberg, WGer, 87-89, sabbatical, 87-88; co-investr, NASA, 90-, collab investr, 90-; mem, Coun Kidney Cardiovasc Dis, Am Heart Asn. *Mem:* Fel Am Col Physicians; Am Fedn Clin Res; Am Soc Nephrology; Int Soc Nephrology; Am Asn Immunologists. *Res:* Author of various publications. *Mailing Add:* Kidney Dis Prog Univ Louisville 500 S Floyd St Louisville KY 40292

MCLELLAN, REX B, b Leeds, Eng, Nov 21, 35; m 58; c 1. METALLURGY. *Educ:* Univ Sheffield, BMet, 57; Univ Leeds, PhD(metall), 62. *Prof Exp:* Sci officer metall, UK Atomic Energy Comn, 57-59; sr res fel, Univ Leeds, 62-63; res assoc, Univ Ill, Urbana, 63-64; from asst prof to assoc prof, 64-69, PROF MAT SCI, RICE UNIV, 77- *Honors & Awards:* Mappin Medal, 57. *Mem:* Metall Soc; Am Inst Mining, Metall & Petrol Engrs; Am Soc Metals. *Res:* Thermodynamics and statistical mechanics of solid solutions; diffusion and relaxation phenomena in solid solutions. *Mailing Add:* Dept Mech Eng Rice Univ Houston TX 77251

MCLEMORE, BENJAMIN HENRY, JR, b Memphis, Tenn, Oct 4, 24; c 2. MATHEMATICAL STATISTICS. *Educ:* Dillard Univ, AB, 44; Univ Ill, MS, 52, AM, 55, PhD(statist), 59. *Prof Exp:* High sch instr, Tex, 44-45; instr math, Dillard Univ, 45-47; instr, Jackson State Col, 47-53, head dept, 56-64; ASSOC PROF MATH, CLEVELAND STATE UNIV, 64- *Mem:* Am Statist Asn; Math Asn Am; Inst Math Statist. *Mailing Add:* Dept Math Cleveland State Univ Euclid Ave & E 24th St Cleveland OH 44115

MCLEMORE, BOBBIE FRANK, b Jasper, Tex, May 22, 32; m 50; c 2. PLANT PHYSIOLOGY. *Educ:* Tex A&M Univ, BS, 53; La State Univ, MS, 57, PhD(forestry), 67. *Prof Exp:* Lab asst bot, Tex A&M Univ, 50-53; res asst agron, Tex Agr Exp Sta, 55; silviculturist, Southern Forest Exp Sta, US Forest Serv, 57-89; RETIRED. *Mem:* Soc Am Foresters. *Res:* Storage, processing, testing and dormancy of southern pine seed; pine seed, cone and conelet physiology; herbicides, uneven-aged management of loblolly-shortleaf pines. *Mailing Add:* Rte 4 Box 424 Jasper TX 75951

MCLENDON, GEORGE L, b Fortworth, Tex, June 6, 52; m 72; c 2. BIOINORGANIC CHEMISTRY. *Educ:* Univ Tex, El Paso, BS, 72; Tex A&M Univ, PhD(chem), 76. *Prof Exp:* Asst prof, 76-81, assoc prof, 81-84, PROF CHEM, UNIV ROCHESTER, 84- *Concurrent Pos:* Henry & Camille Dreyfuss teacher scholar award, 79-85; A P Sloan fel, 80-84; Guggenheim fel, 90. *Honors & Awards:* Pure Chem Award, Am Chem Soc, 87, Eli Lilly Award, 90. *Mem:* Am Chem Soc; Mat Res Soc; Am Phys Soc. *Res:* Inorganic chemistry and protein chemistry; electron transfer reactions; structure and function of heme proteins; photocatalysis; quantisation in semiconductors. *Mailing Add:* Dept Chem Univ Rochester Rochester NY 14627

MCLENDON, WILLIAM WOODARD, b Durham, NC, Oct 29, 30; m 52; c 3. PATHOLOGY, LABORATORY MEDICINE. *Educ:* Univ NC, BA, 53, MD, 56. *Prof Exp:* Intern & resident path, Columbia-Presby Med Ctr, New York, 56-58; resident, Univ NC, Chapel Hill, 58-61; asst chief path serv, US Army Hosp, Landstuhl, Ger, 61-63; assoc pathologist, Moses Cone Hosp, Greensboro, NC, 63-69, dir labs, 69-73; PROF PATH, SCH MED, UNIV NC, CHAPEL HILL, 73-, CHMN DEPT HOSP LABS, NC MEM HOSP, 73- *Concurrent Pos:* Asst chief ed, Arch Path, 74-; assoc ed, Yearbook Pathol, 81- *Mem:* AAAS; Soc Comput Med; Acad Clin Lab Physicians & Sci (pres, 81-82); Col Am Path; Am Soc Clin Path. *Res:* Endocrine pathology; automation and computerization of the clinical laboratory. *Mailing Add:* Dept Path Hosp Labs Univ NC Mem Hosp Chapel Hill NC 27514

MCLENNAN, BARRY DEAN, b Bracken, Sask, Apr 15, 40; m 63; c 3. EDUCATION ADMINISTRATION, RESEARCH ADMINISTRATION. *Educ:* Brandon Univ, BSc, 60; Univ Sask, MSc, 63; Univ Alta, PhD(biochem), 66. *Prof Exp:* Instr chem, Brandon Univ, 60-61; from asst prof to assoc prof, 69-73, assoc dean, Col Grad Studies & Res, 84-90, PROF BIOCHEM, UNIV SASK, 78-, DEAN, COL GRAD STUDIES & RES, 90- *Concurrent Pos:* Inst fel biochem, Roswell Park Mem Inst, 66-67, Nat Cancer Inst Can fel, 66-68; fel McMaster Univ, 67-69; Med Res Coun Can grant, Univ Sask, 69-88, vis scientist award, 79-80; MEM, Commonwealth Scholar Comt, 87-; chmn, Med Res Coun Can Studentship Comt, 87. *Mem:* AAAS; Can Biochem Soc (secy, 80-83); Can Fedn Biol Soc(pres, 85-86). *Res:* Structure and chemistry of nucleic acids; function of modified nucleosides in transfer RNA and bacterial pathogenicity. *Mailing Add:* Dean Col Grad Studies & Res Univ Sask Saskatoon SK S7N 0W0 Can

MACLENNAN, CAROL G, b Plainfield, NJ, Oct 5, 38; c 1. MAGNETOSPHERIC PHYSICS, PLANETARY SCIENCE. *Educ:* Brown Univ, AB, 60. *Prof Exp:* Sr tech aide, Bell Tel Labs, 60-63, assoc mem tech staff, 64-82; MEM TECH STAFF, AT&T BELL LABS, 82- *Mem:* Am Geophys Union. *Res:* Data analysis associated with studies of the earths magnetosphere via ground-based magnetometers and long cables; the study of planetary magnetospheres using Voyager, Galileo and Ulysses spacecraft. *Mailing Add:* AT&T Bell Labs Rm 1E-436 Murray Hill NJ 07974-2070

MCLENNAN, CHARLES EWART, obstetrics & gynecology; deceased, see previous edition for last biography

MACLENNAN, DAVID HERMAN, b Swan River, Man, July 3, 37; m 65; c 2. MEMBRANE TRANSPORT, MEMBRANE SYNTHESIS. *Educ:* Univ Man, BSA, 59; Purdue Univ, MS, 61, PhD(biol), 63. *Hon Degrees:* Royal Soc Can, FRSC, 85. *Prof Exp:* Fel, Enzyme Inst, 63-64; asst prof, Univ Wis, 64-69; assoc prof, 69-74, chmn dept, 78-90, PROF, BANTING & BEST DEPT MED RES, UNIV TORONTO, 74- *Concurrent Pos:* Prof, dept biochem, Univ Toronto, 80-, J W Billes prof med res, 87- *Honors & Awards:* Gold Medal, Univ Man, 59; Ayerst Award, Can Biochem Soc, 74. *Mem:* Can Biochem Soc; Am Soc Biol Chem; Royal Soc Can; Biophys Soc. *Res:* Proteins involved with calcium transport, sequestration and release in the sarcoplasmic reticulum of muscle cells. *Mailing Add:* Banting & Best Dept Med Res Univ Toronto 112 College St Toronto ON M5G 1L6 Can

MACLENNAN, DONALD ALLAN, b San Turse, PR, Mar 27, 36; m 67; c 3. ATOMIC PHYSICS. *Educ:* Univ Calif, Berkeley, BS, 59, PhD(physics), 66; Case Western Reserve Univ, MBA, 72. *Prof Exp:* Engr, Atomic Power & Equip Div, 59, 60, res scientist, Res & Develop Ctr, 66-68, group leader discharge eng, 68-70, mgr fluorescent lamp eng, Lamp Div, Gen Elec Co, 70-75; mgr flash tube eng, EG & G, Inc, 75-78, mgr flash sources & lamp systs bus, Electro-optics Div, 79-85; vpres, Canrad Hanovia, 85-88; mfg vpres, Agr Int, 88-89; CONSULT, 88- *Mem:* Am Phys Soc; Inst Elec & Electronic Engrs. *Res:* Experimental atomic and molecular physics; gaseous electronics; surface physics. *Mailing Add:* 107 Rittswood Dr Butler PA 16001-2138

MCLENNAN, DONALD ELMORE, b London, Ont, Dec 5, 19; m 43, 66; c 7. ELECTRODYNAMICS. *Educ:* Univ Western Ont, BA, 41; Univ Toronto, PhD(physics), 50. *Prof Exp:* Res scientist ballistics, Can Armament Res & Develop Estab, 50-59; prof physics, Col William & Mary, 59-67; prof physics & astron, 67-90, EMER PROF PHYSICS & ASTRON, YOUNGSTOWN STATE UNIV, 90- *Mem:* Am Phys Soc; AAAS. *Res:* Unified field theory. *Mailing Add:* Dept Physics & Astron Youngstown State Univ Youngstown OH 44555

MCLENNAN, HUGH, b Montreal, Que, Oct 22, 27; m 49; c 2. PHYSIOLOGY. *Educ:* McGill Univ, BSc, 47, MSc, 49, PhD(biochem), 51. *Prof Exp:* Asst lectr biophys, Univ Col, Univ London, 52-53; res fel, Montreal Neurol Inst, 53-55; asst prof physiol, Dalhousie Univ, 55-57; from asst prof to assoc prof, 57-65, PROF PHYSIOL, UNIV BC, 65- *Mem:* Am Physiol Soc; Can Physiol Soc (secy, 65-); Brit Biochem Soc; Brit Physiol Soc. *Res:* Neurophysiology. *Mailing Add:* Dept Physiol Univ BC 2194 Health Sci Mall Vancouver BC V6T 1Z2 Can

MCLENNAN, JAMES ALAN, JR, b Atlanta, Ga, Nov 24, 24; m 52; c 2. STATISTICAL MECHANICS. *Educ:* Harvard Univ, AB, 48; Lehigh Univ, MS, 50, PhD(physics), 52. *Prof Exp:* Tech engr, Gen Elec Co, 52-53; from instr to assoc prof, 53-62, chmn dept, 68-78, PROF PHYSICS, LEHIGH UNIV, 62- *Concurrent Pos:* Nat Sci Found fel, Lehigh Univ, 60-61. *Mem:* Fel Am Phys Soc. *Res:* Quantum theory of elementary particles; nonequilibrium statistical mechanics and kinetic theory. *Mailing Add:* Dept of Physics Lehigh Univ Bethlehem PA 18015

MCLENNAN, MILES A, b Bay City, Mich, Nov 11, 02. ELECTRICAL ENGINEERING. *Educ:* Univ Mich, BSEE, 28. *Prof Exp:* Res & develop staff, RCA, 31-36; RETIRED. *Mem:* Fel Inst Elec & Electronic Engrs. *Mailing Add:* 3775-A41 Modoc Rd Santa Barbara CA 93105

MCLENNAN, SCOTT MELLIN, b London, Ont, Apr 20, 52; m 85. CRUSTAL EVOLUTION, SEDIMENTATION & TECTONICS. *Educ:* Univ Western Ont, London, BSc Hons, 75, MSc, 77; Australian Nat Univ, PhD(geochem), 81. *Prof Exp:* Res fel geochem, Res Sch Earth Sci, Australian Nat Univ, 81-86; asst prof, 87-89, ASSOC PROF GEOL, DEPT EARTH & SPACE SCI, STATE UNIV NY, STONY BROOK, 89- *Concurrent Pos:* NSF presidential young investr award, 89-; assoc ed, Geochim Cosmochim Acta, 89-; vis fel, Res Sch Earth Sci, Australian Nat Univ, 89. *Mem:* Geochem Soc; Am Geophys Soc. *Res:* Examining the chemical and radiogenic isotopic composition of sedimentary rocks in order to evaluate models of crustal evolution and plate tectonic history. *Mailing Add:* Dept Earth & Space Sci State Univ NY Stony Brook NY 11794

MACLEOD, CAROL LOUISE, b San Francisco, Calif; m; c 2. TUMOR CELL HETEROGENEITY. *Educ:* Univ Calif, San Diego, PhD(molecular biol), 79. *Prof Exp:* ASSOC PROF MED, UNIV CALIF, SAN DIEGO, 83- *Concurrent Pos:* Salk Inst Biol Res, 79-83. *Mem:* Am Asn Cancer Res; Am Microbiol Soc; Asn Res Leukemia & Related Dis; Asn Women Sci; Am Soc Cell Biol. *Res:* The regulation of gene expression in cancer cells and the role of differentation in the generation of tumor cell heterogeneity and tumor progression using molecular biological approaches. *Mailing Add:* Cancer Ctr 0812 Univ Calif La Jolla CA 92093-0812

MACLEOD, CHARLES FRANKLYN, ecology, for more information see previous edition

MACLEOD, DONALD IAIN ARCHIBALD, b Glasgow, Scotland, Oct 2, 45; m 74. VISION. *Educ:* Univ Glasgow, MA, 67; Cambridge Univ, PhD(exp psychol), 74. *Prof Exp:* Res assoc vision, Inst Perception, Soesterberg, 67-68; res assoc zool, Cambridge Univ, 72; vis asst prof psychobiol, Fla State Univ, 72-74; asst prof, 74-78, ASSOC PROF PSYCHOL, UNIV CALIF, SAN DIEGO, 78- *Honors & Awards:* WAH Rushton Mem Lectr, 81. *Mem:* Psychonomic Soc; Asn Res Vision & Ophthal; fel Optical Soc Am. *Res:* Retinal mechanisms in human vision; human color vision. *Mailing Add:* Dept Psychol G009 Univ Calif San Diego Box 109 La Jolla CA 92093-0109

MCLEOD, DONALD WINGROVE, b Rochester, NY, Feb 15, 35; m 58; c 3. HIGH ENERGY PHYSICS. *Educ:* Univ Rochester, BS, 56; Cornell Univ, PhD(exp physics), 62. *Prof Exp:* Res assoc, Argonne Nat Lab, 62-64, asst scientist, 64-66; from instr to assoc prof, 64-77, PROF EXP HIGH ENERGY PHYSICS, UNIV ILL, CHICAGO CIRCLE, 77- *Mem:* Am Phys Soc. *Res:* Experimental high energy physics, using counter and wire chamber techniques, emphasis on strong interactions. *Mailing Add:* Dept Physics MC 273 Univ Ill Box 4348 Chicago IL 60680

MCLEOD, EDWARD BLAKE, b Los Angeles, Calif, July 25, 24. MATHEMATICS. *Educ:* Occidental Col, BA, 47, MS, 49; Stanford Univ, PhD(math), 54. *Prof Exp:* Mathematician, NAm Aviation, Inc, 47-48; asst, Stanford Univ, 51-53; instr math, Univ Colo, 53-55; asst prof, Ore State Col, 55-63; sr mathematician, Dynamics Sci Corp, 63-64; assoc prof, 64-72, PROF MATH, CALIF STATE UNIV, LONG BEACH, 72- *Mem:* Am Math Soc; Math Asn Am; Soc Indust & Appl Math; Am Inst Aeronaut & Astronaut. *Res:* Complex variables; fluid dynamics; time series forecasting and applications of mathematics to biology. *Mailing Add:* Dept Math Calif State Univ 1250 Bellflower Blvd Long Beach CA 90840

MACLEOD, ELLIS GILMORE, b Washington, DC, Sept 3, 28. EVOLUTIONARY BIOLOGY, ENTOMOLOGY. *Educ:* Univ Md, BS, 55, MS, 60; Harvard Univ, PhD(biol), 64. *Prof Exp:* Fel evolutionary biol, Harvard Univ, 64-66; asst prof, 66-69, assoc prof entom, 69-77, ASSOC PROF GENETICS & DEVELOP, UNIV ILL, URBANA, 78- *Concurrent Pos:* NSF grant syst biol, 69- *Mem:* Royal Entom Soc London; AAAS. *Res:* Speciation and higher levels of evolution of insects, including studies of phylogeny deduced from the fossil record, and such comparative studies of contemporary species as behavior, ecology and chromosome cytology. *Mailing Add:* Dept Entomol Univ Ill Urbana 5055 Goodwin Ave Urbana IL 61801

MCLEOD, GUY COLLINGWOOD, plant physiology; deceased, see previous edition for last biography

MCLEOD, HARRY O'NEAL, JR, b Shreveport, La, Feb 26, 32; m 59; c 2. PETROLEUM ENGINEERING. *Educ:* Colo Sch Mines, BPeEng, 53; Univ Okla, MPeEng, 63, PhD(eng sci), 65. *Prof Exp:* Petrol engr, Phillips Petrol Co, 53-54 & 56-58; res engr, Jersey Prod Res Co, 63-64; sr res engr, Dowell Div, Dow Chem Co, 65-69; asst prof petrol eng & dir info serv, Univ Tulsa, 69-75; ENGR PROF, CONOCO INC, 75- *Honors & Awards:* Prod Eng Award, Soc Petrol Engrs, 89. *Mem:* Soc Petrol Engrs; Sigma Xi. *Res:* Fluid flow through porous media; oil and gas well stimulation; pressure transient testing of oil and gas wells. *Mailing Add:* 2006 Southwick Houston TX 77080-6315

MACLEOD, HUGH ANGUS, b Glasgow, UK, June 20, 33; m 57; c 5. THIN FILM OPTICS. *Educ:* Univ Glasgow, BSc, 54; Inst Physics, UK, FInstP, 69; Coun Nat Acad, DTech, 79. *Prof Exp:* Engr, Sperry Gyroscope Co, Ltd, 54-60; chief develop engr, Williamson Mfg Co, Ltd, 60-62; sr physicist, Mervyn Instruments Ltd, 62-63; thin films mgr, Sir Howard Grubb Parsons & Co, Ltd, 63-71; reader thin film physics, Newcastle Upon Tyne Polytech, 71-79; PROF OPTICAL SCI, UNIV ARIZ, 79- *Honors & Awards:* Gold Medal, Soc Photo-Optical Instrumentation Engrs, 87. *Mem:* Inst Physics; Optical Soc Am; Am Vacuum Soc; Soc Photo-Optical Instrumentation Engrs; Soc Vacuum Coaters. *Res:* Optical thin films, coatings and filters, properties, processes, design, manufacture and measurement. *Mailing Add:* Optical Sci Ctr Univ Ariz Tucson AZ 85721

MCLEOD, JOHN, b Mayertown, NMex, Dec 23, 32. PHYSICS. *Educ:* Princeton Univ, PhD(physics), 62. *Prof Exp:* RES SCIENTIST, LOS ALAMOS NAT LAB, 60- *Mailing Add:* Five Maya Lane Los Alamos NM 87544

MACLEOD, JOHN ALEXANDER, b Brantford, Can, Mar 15, 45; m 67; c 2. SOIL SCIENCE, AGRONOMY. *Educ:* Macdonald Col, BSc, 66; McGill Univ, MSc, 68; Cornell Univ, PhD(soil fertil), 71. *Prof Exp:* RES SCIENTIST SOIL FERTIL, AGR CAN, RES STA, 71- *Mem:* Am Soc Agron; Soil Sci Soc Am; Agr Inst Can; Can Soc Soil Sci; Can Soc Agron; Sigma Xi. *Res:* Nutrient sources and methods of application for forages and cereals; nitrogen nutrition, disease and lodging interaction of cereals; ammonia losses from fertilizer. *Mailing Add:* Res Sta-Box 1210 Agr Can Charlottetown PE C1A 7M8 Can

MCLEOD, JOHN HUGH, JR, b Hattiesburg, Miss, Feb 27, 11; m 51; c 2. MECHANICAL & ELECTRICAL ENGINEERING. *Educ:* Tulane Univ, BS, 33. *Prof Exp:* Engr var indust orgns, 33-39; field engr, Taylor Instrument Co, NY, 40-42; res & develop engr, Leeds & Northrup Co, Pa, 43-47; sect head guid systs & guided missiles, US Naval Air Missile Test Ctr, Calif, 47-56; design specialist, Gen Dynamics/Astronaut, 56-63, consult, 63-64; INDEPENDENT CONSULT, 64- *Concurrent Pos:* Ed & publ, Simulation Coun Newsletter, 52-55, ed, Simulation, 63-74, chmn Simulation Coun, 52; ed & publ, Simulation in the Serv of Soc, 71-; assoc ed, Behav Sci, 73-; grantee, for the study of professional ethics for simulationists, Nat Acad Sci/Nat Endowment Humanities, 83. *Honors & Awards:* Sr Sci Simulation Award, Electronic Assocs, Inc, 65; Inst Mgt Sci Award, 86; John McLeod Simulation Award, Soc Comput Simulation, 87. *Mem:* Inst Elec & Electronics Engrs; Soc Comput Simulation; AAAS. *Res:* Application of computer modeling and simulation technology, especially for study, analysis and prediction of response of physiological, societal and global systems to therapy or corrective measures. *Mailing Add:* Simulation in Serv of Soc 8484 La Jolla Shores Dr La Jolla CA 92037

MCLEOD, JOHN MALCOLM, entomology, for more information see previous edition

MACLEOD, JOHN MUNROE, b Vermilion, Alta, Sept 3, 37; m 59; c 3. RADIO ASTRONOMY. *Educ:* Univ Alta, BSc, 59; Univ Ill, MS, 60, PhD(elec eng), 64. *Prof Exp:* RES OFFICER, RADIO ASTRON SECT, HERZBERG INST ASTROPHYS, NAT RES COUN CAN, 64- *Concurrent Pos:* Pres, Can Astron Soc, 84-86. *Mem:* Can Astron Soc; Am Astron Soc. *Res:* Interstellar molecules; extragalactic variable sources; recombination lines. *Mailing Add:* 29 W Park Dr Gloucester ON K1B 3G6 Can

MCLEOD, KENNETH WILLIAM, b Miami, Okla, Oct 14, 47; m 77. PLANT ECOLOGY. *Educ:* Okla State Univ, BS, 69, MS, 71; Mich State Univ, PhD(plant ecol), 74. *Prof Exp:* res assoc plant ecol, 74-80, ASSOC RES ECOL, SAVANNAH RIVER ECOL LAB, UNIV GA, 80- *Mem:* Ecol Soc Am; Am Inst Biol Sci. *Res:* Factors that govern the establishment and distribution of plant populations, especially ecophysiology; elemental cyling in forests; radioecology. *Mailing Add:* Savannah River Ecol Lab Drawer E Aiken SC 29801

MCLEOD, LIONEL EVERETT, b Wainwright, Alta, Aug 9, 27; m 52; c 4. MEDICAL EDUCATION, MEDICAL ADMINISTRATION. *Educ:* Univ Alta, BSc, 49, MD, 51; McGill Univ, MSc, 56; FRCP, 57; FRCPS(C). *Hon Degrees:* LLD, Univ Alta, 88. *Prof Exp:* Intern, Univ Alta Hosp, 51-52, lectr med & biochem, 57-60, mem staff, 57-68, from asst prof to assoc prof, 60-68; resident internal med, Univ Minn, Minneapolis, 53-55; clin fel endocrinol, Royal Victoria Hosp, 55-57; PROF MED, UNIV CALGARY, FOOT HILLS PROV GEN HOSP, 68-; PRES, ALTA HERITAGE FOUND MED RES, 68- *Concurrent Pos:* Nat Res Coun Can Med Div fel, dept invest med, McGill Univ, 55-57; dir endocrinol & metab, Univ Alta Hosp, 60-68 & Renal Dial Unit, 62-68; prof med, Univ Calgary; prof health sci admin & community med, Univ Alta, hon prof med; mem coun, Royal Col Physicians & Surgeons Can, 72-80 & 81-85, chmn, comt credentials, 74-76, comt accreditation, 78-80 & Task Force Can Inst Health, 84-87, mem comt health & pub policy, 84-; rep to liaison comt med educ in US, Asn Can Med Col, 78-81, co-chmn, comt accreditation Can Med Sch, 79-81, chmn, Task Force Restructuring, 79-80; mem coun, Can Soc Clin Invest, 68-72, chmn educ comt, 69-71; mem coun, Can Coun Hosp Accreditation, 75-78, chmn prog & standard comt, 75-76 & coun, 76-77; mem bd, Nat Inst Nutrit, 84-87; policy & priorities comt, Med Res Coun Can, 85-; mem adv bd, Can Inst Advan Res, 85- & adv comt pop health prog, 87-; res adv comt, Gov Prov Alta, 83-85, Task Force Res & Sci, 85-; mem bd dirs, Chembiomed Ltd, 86-, compensation comt, 86-87 & equity partic comt, 86-87, chmn equity partic comt, 86-87; consult, Health & Welfare Can, Nat Nutrit Surv, Univ Toronto & Kates-Peat-Marwick, Prov Ont, Queen's Univ & Kingston Gen Hosp, Alta Hosp & Med Care & Alta Comt Hosp Utilization; ext adv, prin adv comt dean med, Queen's Univ, 86; bd dirs, Alta Div & chmn med adv comt, Can Heart Found, 70-76 & Can Cancer Soc, 75-80. *Honors & Awards:* Rankin Prize, 48. *Mem:* Am Soc Artificial Internal Organs; Can Soc Clin Invest; Can Fedn Biol Soc; fel Royal Col Physicians & Surgeons Can; fel Am Col Physicians; Can Soc Nephrol; Soc Health & Human Values; Am Fedn Clin Res; Soc Res Adminr. *Res:* Endocrinology and metabolism; application of intermittent hemodialysis in chronic renal failure; author of over 10 publications. *Mailing Add:* Alta Heritage Found Med Res 3125 Manulife Pl 10180-101 St Edmonton AB T5J 3S4 Can

MACLEOD, LLOYD BECK, b Apr 27, 30; Can citizen; m 55; c 3. AGRONOMY. *Educ:* McGill Univ, BSc, 52, MSc, 53; Cornell Univ, PhD(soil sci), 62. *Prof Exp:* Res officer agron, Res Br, Can Dept Agr, 53-59, head soils & plant nutrit sect, 62-65, res scientist, PEI, 65-67, head soils & plant nutrit sect, 67-70, chief liaison officer PEI, 78-83, dir res sta, Agr Can, 70-90; RETIRED. *Honors & Awards:* Fel, Can Soc Soil Sci, 77. *Mem:* Am Soc Agron; fel Can Soc Soil Sci; Can Soc Agron. *Res:* Soil fertility and plant nutrition of forage species; nutrient competition; aluminum tolerance; effect of potassium on utilization of ammonium and nitrate sources of nitrogen by forage and cereal crops; research management and administration. *Mailing Add:* PO Box 1323 Charlottetown PE C1A 7N1 Can

MACLEOD, MICHAEL CHRISTOPHER, b Allentown, Pa, Feb 7, 47; m 69; c 2. CHEMICAL CARCINOGENESIS. *Educ:* Calif Inst Technol, BS, 69; Univ Ore, PhD(molecular biol), 74. *Prof Exp:* Fel biol, Univ Ore, 74-75; fel gene regulation, biol div, Oak Ridge Nat Lab, 75-77, res assoc chem carcinogenesis, 77-82; asst biochemist & asst prof biochem, 82-86, ASSOC DIR, RES DIV, SYST CANCER CTR, UNIV TEX, SMITHVILLE,85-, ASSOC BIOCHEMIST & ASSOC PROF BIOCHEM, 86- *Concurrent Pos:* Mem, Nat Adv Comt Biochem & Chem Carcinogenesis, Am Cancer Soc, 88-90, chair, 91; assoc ed, Molecular Carcinogenesis, 88. *Mem:* Am Soc Biol Chem; Am Asn Cancer Res; AAAS. *Res:* Interaction of polycyclic aromatichydrocarbons with cellular macromolecules; DNA adducts, DNA repair and mutagenesis; prevention of carcinogenesis by chemical intervention (chemoprevention). *Mailing Add:* Sci Park Res Div Univ Tex M D Anderson Cancer Ctr PO Drawer 389 Smithville TX 78957

MCLEOD, MICHAEL JOHN, b Quantico, Va, Jan 24, 48; m 72; c 1. POPULATION GENETICS, FRESHWATER ECOLOGY. *Educ:* Lincoln Mem Univ, BS, 69; E Tenn State Univ, MS, 73; Miami Univ, Ohio, PhD(zool), 77. *Prof Exp:* Teaching fel zool, Miami Univ, Ohio, 74-77, asst prof, 77-78; from asst prof to assoc prof, 78-86, PROF BIOL, BELMONT ABBEY COL, 86-, ASST VPRES ACAD AFFAIRS, 88- *Concurrent Pos:* Vis prof, Miami Univ, Ohio, 80 & 81; adj prof, Sacred Heart Col, 81-85; consult, Gaston County Pub Schs, 84-; Am coun educ fel, 87-88. *Mem:* Sigma Xi; Am Soc Zoologists; Am Malacological Union. *Res:* Genetic structure of populations of colonizing species; genetics of the domestication process in plants and the population genetics of freshwater molluscs. *Mailing Add:* Dept Biol Belmont Abbey Col Belmont NC 28012

MCLEOD, NORMAN BARRIE, b Regina, Sask, Apr 7, 32; m 57; c 3. NUCLEAR ENGINEERING, PHYSICS. *Educ:* Univ Toronto, BASc, 55; Univ Mich, MSE, 57; Mass Inst Technol, PhD(nuclear eng), 62. *Prof Exp:* Instrumentation engr jet engines, Orenda Engines, Malton, Ont, 55-57; teaching asst nuclear reactor, Mass Inst Technol, 57-61; staff consult nuclear fuel, Nus Corp, 61-72, vpres, South East Opers, 73-77, vpres & gen mgr energy syst div, 77-80, vpres, nuclear waste proj, 80-84; VPRES, E R JOHNSON ASSOCS, INC, 84- *Mem:* Sigma Xi; Am Nuclear Soc. *Res:* Resources, production capability, costs and prices of major fuels, with emphasis on uranium and coal; nuclear waste disposal. *Mailing Add:* 8010 Bradley Blvd Bethesda MD 20817-1908

MCLEOD, NORMAN WILLIAM, civil engineering, chemical engineering; deceased, see previous edition for last biography

MCLEOD, RAYMOND, JR, b Cameron, Tex, Aug 19, 32. COMPUTER SCIENCE, MANAGEMENT INFORMATION SYSTEMS. *Educ:* Baylor Univ, BBA, 54; Tex Christian Univ, MBA, 57; Univ Colo, DBA, 75. *Prof Exp:* ASSOC PROF BUS COMPUTER SYSTS, COL BUS, TEX A&M UNIV, 80- *Concurrent Pos:* Chair, Systs Info Group Personnel Res, Asn Comput Mach, 89- *Mem:* Asn Comput Mach; Info Resources Mgt Asn; Decision Sci Inst. *Mailing Add:* Dept Bus Anal & Res Col Bus Tex A&M Univ College Station TX 77843

MCLEOD, RIMA W, b Berkeley, Calif, Sept 14, 45. IMMUNOLOGY, INFECTIOUS DISEASES. *Educ:* Univ Calif, San Francisco, MD, 71. *Prof Exp:* ATTEND PHYSICIAN, DEPT MED, DIV INFECTIOUS DIS, MICHAEL REESE MED CTR, 79-; ASSOC PROF MED, UNIV CHICAGO, 85- *Mailing Add:* Dept Med/Infectious Dis Michael Reese Hosp & Med Ctr Lake Shore Dr at 31st St Chicago IL 60616

MACLEOD, ROBERT ANGUS, b Athabasca, Alta, July 13, 21; m 48; c 6. MICROBIOLOGY. *Educ:* Univ BC, BA, 43, MA, 45; Univ Wis, PhD(biochem), 49. *Prof Exp:* Instr chem, Univ BC, 45-46; asst prof biochem, Queen's Univ, Ont, 49-52; biochemist, Fisheries Res Bd, Can, 52-60; assoc prof agr bact, 60-64, chmn dept, 68-70 & 74-79, prof, 64-86, EMER PROF MICROBIOL, MCGILL UNIV, MACDONALD CAMPUS, 86- *Honors & Awards:* Harrison Prize, Royal Soc Can, 60; Award, Can Soc Microbiol, 73. *Mem:* Fel Royal Soc Can; Am Soc Biol Chem; Am Soc Microbiol; Can Soc Microbiol (pres, 76-77). *Res:* Nutrition and metabolism of marine bacteria; function of inorganic ions in bacterial metabolism; microbial biochemistry. *Mailing Add:* Dept Microbiol MacDonald Col McGill Univ 21111 Lakeshore Rd Ste Anne de Bellevue PQ H9X 1C0 Can

MACLEOD, ROBERT M, PROTEIN CHEMISTRY, HEMOGLOBIN RESEARCH. *Educ:* Univ Ill, PhD(biophysics), 66. *Prof Exp:* ASSOC PROF BIOCHEM, UNIV TENN, 74- *Res:* Serum proteins. *Mailing Add:* Dept Biochem Univ Tenn Memphis TN 38163

MCLEOD, ROBERT MELVIN, b Newco, Miss, June 19, 29; m 65; c 2. MATHEMATICAL ANALYSIS. *Educ:* Miss State Univ, BS, 50; Rice Univ, MA, 53, PhD(math), 55. *Prof Exp:* Instr math, Duke Univ, 55-58, asst prof, 58-61; assoc prof, Am Univ Beirut, 61-65; assoc prof, Univ Tenn, 65-66; ASSOC PROF MATH, KENYON COL, 66- *Mem:* Am Math Soc; Math Asn Am. *Res:* Function theory. *Mailing Add:* Dept Math Kenyon Col Gambier OH 43022

MACLEOD, ROBERT MEREDITH, b Newark, NJ, May 14, 29; m 51; c 4. BIOCHEMISTRY. *Educ:* Seton Hall Univ, BS, 52; NY Univ, MS, 56; Duke Univ, PhD(biochem), 59. *Prof Exp:* Res biochemist, Schering Corp, NJ, 48-56; instr biochem, Sch Med, Duke Univ, 59-60; asst prof biochem in internal med, 60-66, chmn div biomed eng, 64-65, dir div cancer studies, 69-72, assoc prof, 66-73, PROF INTERNAL MED, SCH MED, UNIV VA, 73-, DIR, ENDOCRINOL LAB, 78- *Concurrent Pos:* Am Heart Asn res fel, 59-60; USPHS res grant, 64-, career develop award, 65-71; cancer travel fel, WHO, 68. *Mem:* AAAS; Am Physiol Soc; Am Fedn Clin Res; Am Asn Cancer Res; Int Soc Neuroendocrinol. *Res:* Hormonal control of biochemical mechanisms which regulate normal and neoplastic growth. *Mailing Add:* Dept Immunol Univ Va Health Sci Ctr Box 135 Charlottesville VA 22908

MCLEOD, ROBIN JAMES YOUNG, b Arbroath, Scotland. NUMERICAL ANALYSIS. *Educ:* St Andrews Univ, BSc, 66; Dundee Univ, PhD(math), 73. *Prof Exp:* Res fel math, Dundee Univ, 73-74; lectr, Univ Man, 74-75; asst prof, Univ Calgary, 75-76; scientist, Inst Comput Appln Sci & Eng, 76-77; asst prof, Rensselaer Polytech Inst, 77-78; assoc prof math, NMex State Univ, 78-86; comput graphics mgr, Tektronix, Inc, 86-90. *Mem:* Inst Math. *Res:* Numerical analysis; curved finite elements; parametric curve and surface inter-polation; techniques in computer aided design and difference methods for trajectomy problems in ordinary differential equations; application of geometry in numerical analysis; computer graphics. *Mailing Add:* Rte 4 Box 689 Strawberry Hill Hillsboro OR 97123

MCLEOD, SAMUEL ALBERT, b Tampa, Fla, Nov 13, 52. ADAPTIVE FUNCTIONAL MORPHOLOGY. *Educ:* Univ Calif, Berkeley, AB, 74, PhC, 77, PhD(paleont), 81. *Prof Exp:* Curatorial asst, 81, ASST CUR, LOS ANGELES CO MUS NATURAL HIST, 81- *Mem:* Soc Vert Paleont; Am Soc Mammalogists; Am Soc Zoologists; AAAS; Soc Marine Mammal. *Res:* Evolutionary biology, paleobiology, including functional morphology, systematics and paleocology of selected marine vertebrates (primarily cetaceans and chelonions and elasmobranchs). *Mailing Add:* Los Angeles County Mus Natural Hist 900 Exposition Blvd Los Angeles CA 90007

MACLEOD, STUART MAXWELL, b Toronto, Ont, Can, June 20, 43. PEDIATRIC PHARMACOLOGY. *Educ:* Univ Toronto, MD, 67; McGill Univ, PhD(pharmacol), 72. *Prof Exp:* DIR & PROF, DIV PHARMACOL, HOSP FOR SICK CHILDREN, TORONTO, CAN, 79- *Mem:* Can Soc Clin Invest; Soc Pediat Res; Am Fedn Clin Res; Am Soc Pharmacol & Exp Therapeut. *Res:* Anti-natal drug treatment. *Mailing Add:* Dept Pharmacol McMaster Univ 1200 Main St W Hamilton Ontario ON L8N 3Z5 Can

MCLEOD, WILLIAM D, b Toronto, Ont, Nov 16, 30; m 56; c 1. BIOMEDICAL ENGINEERING. *Educ:* Univ Toronto, BASc, 58, MAS, 61; Case Western Reserve Univ, PhD(mech eng), 65. *Prof Exp:* Asst dir cybernetic systs group, Case Western Reserve Univ, 64-66; asst prof mech eng & res assoc bioeng inst, Univ NB, 66-68; dir bioeng res, Insts Achievement Human Potential, Pa, 68-69; asst prof phys med, Emory Univ, 69-77, clin asst prof rehab med, 77-80; MEM STAFF, HUSHOTON ARTHOPEDIC CLIN, COLUMBUS, GA, 80- *Mem:* Inst Elec & Electronics Eng; Int Soc Electromyographic Kinesiology. *Res:* Information processing from bio-electric signals, primarily myo-electric signals; electroencephalographic signals; human operator modelling with handicapped people using electromyographic and electroencephalographic signals. *Mailing Add:* Richards Med 1450 Brooks Rd Memphis TN 38116

MCLERAN, JAMES HERBERT, b Audubon, Iowa, Apr 9, 31; m 57, 79; c 1. DENTISTRY, ORAL SURGERY. *Educ:* Simpson Col, BS & BA, 53; Univ Iowa, DDS, 57, MS, 62; Am Bd Oral Surg, dipl, 64. *Hon Degrees:* DSc, Simpson Col. *Prof Exp:* Instr oral surg, Univ Iowa, 59-60; resident, Univ Hosps, Iowa City, Iowa, 60-62; pvt pract, Calif, 62-63; from asst prof to assoc prof oral surg, Univ Iowa, 63-69; prof & chmn dept, Sch Dent, Univ NC, Chapel Hill, 69-72; assoc dean, 72-74, PROF ORAL SURG, COL DENT, UNIV IOWA, 72-, DEAN COL, 74- *Mem:* Am Asn Dent Schs (pres, 78-79; Int Asn Dent Res; Am Soc Oral & Maxillofacial Surg; Am Dent Asn. *Res:* Pain control in dentistry; temporomandibular joint disfunction; bacteremia following oral surgical procedures. *Mailing Add:* Off of the Dean Univ Iowa City IA 52242

MCLEROY, EDWARD GLENN, b Atlanta, Ga, June 23, 26; c 2. PHYSICS. *Educ:* Emory Univ, BA, 49, MS, 51. *Prof Exp:* Asst math, Ga Inst Technol, 49; physicist, US Navy Mine Defense Lab, 51-53, 54-56; asst physics, Brown Univ, 53-54; asst prof, Marine Lab, Univ Miami, 57; head acoustics sect, US Navy Mine Defense Lab, 57-72; physicist, Naval Coastal Systs Lab, 72-84; CONSULT, TECH MARINE SERV, 84- *Concurrent Pos:* Consult, Tech Marine Serv Inc, 84- *Mem:* AAAS; Acoust Soc Am; Am Asn Physics Teachers; Am Geophys Union; Soc Explor Geophys. *Res:* Marine physics, particularly marine acoustics. *Mailing Add:* Box 4647 Panama City FL 32401

MACLIN, ERNEST, b New York, NY, Jan 25, 31; m 56; c 3. LABORATORY MEDICINE. *Educ:* City Col NY, BME, 52, MME, 69. *Prof Exp:* Res engr, Ford Instr Corp, 57-58; proj des engr, Singer, Kearfott Div, 58-68; proj engr, Technicon, 68-69; dir res & develop, Electro Nucleonics Inc, 69-83, vpres res & develop, 83-90; PRES, ERNEST MACLIN & ASSOCS INC CONSULT FIRM, 90- *Concurrent Pos:* Design engr architect eng, A Wilson Assoc, 56-57. *Mem:* Am Asn Clin Chem; Am Soc Mech Engr; Nat Comt Clin Lab Standards. *Mailing Add:* 659 Rutgers Pl Paramus NJ 07652

MCLINDEN, LYNN, b 43; US citizen. NONLINEAR ANALYSIS, OPTIMIZATION THEORY. *Educ:* Princeton Univ, AB, 65; Univ Wash, PhD(math), 71. *Prof Exp:* Vis asst prof math, Math Res Ctr, Univ Wis-Madison, 71-73; from asst prof to assoc prof, 73-85, PROF MATH, UNIV ILL, URBANA, 85- *Concurrent Pos:* NSF res grant, 75-, res fel, Ctr Opers Res & Econometrics, Université Catholique de Louvain (Louvain-la-Neuve, Belg), 78-79; vis mem, Math Res Ctr, Univ Wis-Madison, 83-85. *Mem:* Math Prog Soc; Opers Res Soc Am; Am Math Soc; Soc Indust & Appl Math. *Res:* Convex, nonsmooth and nonlinear analysis; optimization theory, including nonlinear programming, minimax, complementarity, variational inequality and equilibrium problems. *Mailing Add:* 1409 W Green St Univ Ill Urbana-Champaign Urbana IL 61801

MCLOUGHLIN, DANIEL JOSEPH, b Columbus, Ohio, Aug 3, 47; m 86; c 2. ENZYMOLOGY, PROTEIN CHEMISTRY. *Educ:* Wayne State Univ, BS, 69; John Carroll Univ, MS, 71; Univ Mo, Kansas City, PhD(chem), 76. *Prof Exp:* Res assoc chem, Univ Ore, Eugene, 76-78; res assoc biochem, Sch Med, Washington Univ, 78-81; asst prof chem, 81, ASSOC PROF CHEM, XAVIER UNIV, 88- *Concurrent Pos:* Consult, Ralston-Purina, St Louis, Mo, 80-81; Fries-Fries, Cincinnati, Ohio, 85-86. *Mem:* Am Chem Soc; AAAS. *Res:* Enzymology; study of protein-lipid interactions; fluorescent labeling of proteins; synthesis of enzyme active site inhibitors. *Mailing Add:* Dept Chem Xavier Univ Cincinnati OH 45207-1096

MCLUCAS, JOHN L(UTHER), b Fayetteville, NC, Aug 22, 20; m 46, 81; c 4. ELECTRONICS, PHYSICS. *Educ:* Davidson Col, BS, 41; Tulane Univ, MS, 43; Pa State Univ, PhD(physics), 50. *Hon Degrees:* DSc, Davidson Col, 76. *Prof Exp:* Physicist servomechanism design, Air Force Cambridge Res Ctr, 46-47; proj engr, Radio Countermeasures, HRB-Singer Co, Pa, 48-50, vpres & tech dir, 50-57, pres, 57-62; dep dir defense res & eng, Off Secy of Defense, Pentagon, 62-64, asst secy gen for sci affairs, NATO, France, 64-66; pres, Mitre Corp, Mass, 66-69; from undersecy to secy, US Air Force, 69-75; adminr, Fed Aviation Admin, 75-77; pres, Comsat Gen Corp, 77-80, pres, Comsat World Systs, 80-85; CONSULT AEROSPACE, 85- *Concurrent Pos:* Mem, US Air Force Sci Adv Bd, 67-69, 77-83; consult aerospace, 85- *Mem:* Nat Acad Eng; fel Am Inst Aeronaut & Astronaut; fel Inst Elec & Electronics Engrs; Am Phys Soc. *Res:* Management. *Mailing Add:* 1213 Villamay Blvd Alexandria VA 22307

MCMACKEN, ROGER, b Spokane, Wash, July 24, 43; m 68; c 3. BIOCHEMISTRY, GENETICS. *Educ:* Univ Wash, BS, 65; Univ Wis-Madison, PhD(biochem), 70. *Prof Exp:* Res assoc biochem, Sch Med, Yale Univ, 70-71; NIH fel biochem, Univ Fla, 72-74, Stanford Univ, 74-76; from asst prof to prof, 77-90, E V MCCOLLUM PROF BIOCHEM & CHMN, DEPT BIOCHEM, JOHNS HOPKINS UNIV, 90- *Mem:* AAAS; Am Soc Biochem & Molec Biol; Am Soc Microbiol; The Protein Soc. *Res:* Enzymology and regulation of DNA replication; nucleoprotein complexes; heat shock proteins. *Mailing Add:* Dept Biochem Johns Hopkins Univ 615 N Wolfe St Baltimore MD 21205

MCMAHAN, ELIZABETH ANNE, b Davie Co, NC, May 5, 24. ZOOLOGY. *Educ:* Duke Univ, AB, 46, MA, 48; Univ Hawaii, PhD(entom), 60. *Prof Exp:* Res asst, Parapsychol Lab, Duke Univ, 43-54; Am Asn Univ Women fel, Univ Chicago, 60-61; from asst prof to prof, 61-87, EMER PROF ZOOL, UNIV NC, CHAPEL HILL, 87- *Concurrent Pos:* Dept Nat Sci, Col Agr, Port Antonio, Jamaica, WI, 87-88. *Mem:* AAAS; Animal Behav Soc; Am Inst Biol Sci; Am Soc Zool; Entom Soc Am; Int Union for the Study of Social Insects. *Res:* Termite colony development; termite feeding behavior and temporal polyethism; biology of dragonflies. *Mailing Add:* Dept Biol Univ NC Chapel Hill NC 27599-3280

MACMAHAN, HORACE ARTHUR, JR, b Freeport, Maine, Aug 13, 28; div; c 1. EARTH SCIENCE EDUCATION. *Educ:* Univ Maine, BA, 54; Univ Utah, MSEd, 63; Univ Colo, EdD(sci educ), 67. *Prof Exp:* Staff asst, Earth Sci Curric Proj, Univ Colo, 63-64; assoc prof earth sci, State Univ NY Col Oneonta, 67-68; assoc prof sci educ, Weber State Col, 68-69; PROF GEOG

& GEOL, EASTERN MICH UNIV, 69- *Mem:* Nat Earth Sci Teachers Asn; Nat Sci Teachers Asn. *Res:* Developed composite paleogeographic maps of North America for each geologic time period; determination of the most effective mode of presenting map concepts in geology. *Mailing Add:* Dept of Geog & Geol Eastern Mich Univ Ypsilanti MI 48197

MCMAHAN, UEL JACKSON, II, b Kansas City, Mo, July 22, 38; m 60; c 4. NEUROBIOLOGY. *Educ:* Westminster Col, BA, 60; Univ Tenn, PhD(anat), 64. *Prof Exp:* Instr anat, Sch Med, Yale Univ, 65-67; instr, 67-72, asst prof, 72-75, assoc prof neurobiol, Harvard Med Sch, 72-77; PROF NEUROBIOL, SCH MED, STANFORD UNIV, 77- *Concurrent Pos:* NIH Career Develop Award. *Honors & Awards:* Jacob Javitz Neurosci Investr Award, NIH. *Mem:* Am Soc Neuroscience. *Res:* Structure and function of synapses. *Mailing Add:* Dept of Neurobiol Stanford Univ Sch Med Stanford CA 94305

MCMAHAN, WILLIAM H, b Sylacauga, Ala, Apr 19, 37; m 61. INORGANIC CHEMISTRY. *Educ:* Auburn Univ, BS, 59, MS, 61; Univ Kans, PhD(chem), 65. *Prof Exp:* Assoc prof chem, 65-77, PROF CHEM, MISS STATE UNIV, 77-, COORDR FAC DEVELOP CTR, 85 - *Mem:* Am Chem Soc. *Res:* Solution chemistry of low dielectric nonaqueous solvents; coordination chemistry of hydroxamic acids; chemical education. *Mailing Add:* Dept Chem Miss State Univ Box 741 Mississippi State MS 39762-0741

MACMAHON, BRIAN, b Sheffield, Eng, Aug 12, 23; US citizen; m 48; c 4. CANCER. *Educ:* Univ Birmingham, Eng, MRCS & LRCP, 46, MB & ChB, 49, DPH, 49, PhD(soc med), 52; Harvard Univ, SM, 53. *Hon Degrees:* DMedSc, Univ Athens, 76; DSc State Univ NY, Buffalo, 86. *Prof Exp:* Ship surgeon, Alfred Holt & Co, Liverpool, Eng, 46-48; univ res fel, Dept Social Med, Univ Birmingham, Eng 49-50, res fel, Med Res Coun, 50-52; Rockefeller Found fel, Dept Epidemiol, Harvard Sch Pub Health, 52-53; vis lectr, environ med, State Univ NY, Downstate Med Ctr, 53-54; lectr social med, Univ Birmingham, 54-55; from assoc prof to prof environ health & community health, State Univ NY, Downstate Med Ctr, 55-58; prof & head dept, 58-89, Henry Pickering Walcott prof, 76- 89, EMER PROF, SCH PUB HEALTH, HARVARD UNIV, 89- *Concurrent Pos:* Consult, Damon Corp, 74-81, Brush Wellman Inc, 77-79, Brookhaven Nat Lab, 78, E I du Pont de Nemours, 78-; adv comt, NIH, Epidemiol & Biometry, Div Res Grants, 61-65, 64, 65; mem comt, Radiation Exposure of Women, 66-70; mem, Breast Cancer Task Force, NIH, 71-73; assoc dean acad affairs, Harvard Univ, Sch Pub Health, 77-78; adv coun, Elec Power Res Inst, 85-88; mem adv bd, Int Life Sci Inst, 85-; ed, Cancer-Causes & Control, 89- *Honors & Awards:* Lucy Wortham James Clin Res Award, 78; John Snow Award, Am Pub Health Asn, 80; Helen Schmerker Mem Lectr, Johns Hopkins Univ, 81. *Mem:* Fel Am Pub Health Asn. *Res:* Epidemiology. *Mailing Add:* 89 Warren St Needham MA 02192

MCMAHON, BRIAN ROBERT, b Harrow, Eng, May 27, 36; m 86; c 4. COMPARATIVE ANIMAL PHYSIOLOGY. *Educ:* Univ Southampton, BSc, 64; Bristol Univ, PhD(zool), 68. *Prof Exp:* PROF ZOOL, UNIV CALGARY, 68- *Concurrent Pos:* Mem, Grant Selection Comt, Nat Sci Eng & Res Coun, 80-83; secy, Western Can Marine Biol Soc, 85- *Mem:* Brit Soc Exp Biol; Am Soc Zoologists; Am Physiol Soc; Sigma Xi. *Res:* Physiological compensation to environmental stress; control of respiration in invertebrates; evolution of respiratory and circulatory mechanisms; gas exchange dynamics across gill surfaces; toxicology; respiratory pigment function. *Mailing Add:* Dept Biol Sci Univ Calgary 2500 Univ Dr Calgary AB T2N 1N4 Can

MCMAHON, CHARLES J, JR, b Philadelphia, Pa, July 10, 33; m 59; c 5. METALLURGY, MATERIALS SCIENCE. *Educ:* Univ Pa, BS, 55; Mass Inst Technol, ScD(metall), 63. *Prof Exp:* Instr & res asst phys metall, Mass Inst Technol, 58-63; fel, Sch Metall & Mat Sci, 63-64, from asst prof to assoc prof, 64-74, PROF MAT SCI & ENG, UNIV PA, 74-, CHMN DEPT, 88- *Concurrent Pos:* Ford Found resident eng pract, Gen Elec Co, 68-69; overseas fel, Churchill Col, Univ Cambridge, 73-74; chmn, Gordon Conf Phys Metall, 75; Alexander von Humboldt Sr US scientist award, 83; vis prof, Inst Metall Physics, Univ Gottingen, 83-84. *Honors & Awards:* Champion H Mathewson Gold Medal, Am Inst Mech Engrs, 75; Henry Marion Howe Medal, Am Soc Metals, 75, Albert Sauver Achievement Award, 81; Sauver Award, Am Soc Metals, 81. *Mem:* Nat Acad Eng; fel Am Soc Metals; fel Inst Metals London; AAAS; fel Am Inst Mining, Metall & Petrol Engrs; Sigma Xi; fel Metall Soc. *Res:* Deformation and fracture of solids; segregation to surfaces and interfaces. *Mailing Add:* Dept Mat Sci & Eng Univ Pa 3231 Walnut St Philadelphia PA 19104-6272

MCMAHON, DANIEL STANTON, biochemistry, theoretical biology; deceased, see previous edition for last biography

MCMAHON, DAVID HAROLD, b Troy, NY, Apr 27, 42; c 1. ANALYTICAL CHEMISTRY. *Educ:* Col of the Holy Cross, BS, 63; Univ NH, PhD(chem), 68. *Prof Exp:* Res chemist, Esso Res & Eng Co, NJ, 67-70; res chemist, 70-74, group leader, 74-78, SR SCIENTIST, RES & DEVELOP DIV, UNION CAMP CORP, 78- *Concurrent Pos:* Assoc ed, J Chromatographic Sci; vchmn, E-19 comt chromatography, Am Soc Testing & Mat. *Mem:* Am Chem Soc; Am Soc Testing & Mat; NAm Thermal Anal Soc. *Res:* Chromatographic analysis of natural products; polynuclear aromatic hydrocarbon pollution analyses; characterization of natural products and polymers; flavor and fragrances characterization. *Mailing Add:* Union Camp Technol Ctr PO Box 3301 Princeton NJ 08543-3301

MCMAHON, DONALD HOWLAND, b Buffalo, NY, Apr 18, 34; m 54. OPTICAL PHYSICS. *Educ:* Univ Buffalo, BA, 57; Cornell Univ, PhD(exp physics), 64. *Prof Exp:* Staff scientist, Sperry Rand Res Ctr, 63-73, dept mgr optics, 73-; staff, Polaroid Corp, Cambridge, Mass, 83-86; tech mgr, Digital Equip Corp, 86-89; RETIRED. *Mem:* Am Phys Soc; Inst Elec & Electronics Engrs; Optical Soc Am. *Res:* Fiber Optics; optical pattern recognition; modern optics; optical information processing; holography; quantum electronics; nonlinear optics; paramagnetic resonance. *Mailing Add:* One La Fond Dr Gansevoort NY 12831

MCMAHON, DOUGLAS CHARLES, mathematics; deceased, see previous edition for last biography

MCMAHON, E(DWARD) LAWRENCE, b Peoria, Ill, Nov 19, 31; m 54; c 4. ELECTRICAL ENGINEERING. *Educ:* Fournier Inst Technol, BS, 52; Univ Ill, MS, 53, PhD(elec eng), 55. *Prof Exp:* Res engr, Hughes Aircraft Co, 55-56; asst prof elec eng, Mich State Univ, 56-59; assoc prof elec eng, 59-77, fac consult, Radiation Lab, 65-71, consult, electronic defense group, Res Inst, ASSOC PROF ELEC ENG & COMPUT SCI, UNIV MICH, ANN ARBOR, 77- *Mem:* Inst Elec & Electronics Engrs. *Res:* Network theory and synthesis. *Mailing Add:* 1715 Hermitage Rd Ann Arbor MI 48104

MCMAHON, FRANCIS GILBERT, b Kalamazoo, Mich Sept 10, 23; m 54; c 4. INTERNAL MEDICINE. *Educ:* Univ Notre Dame, BS, 45; Univ Mich, MS, 51, MD, 53; Am Bd Internal Med, dipl, 59. *Prof Exp:* Intern, Univ Wis Hosps, 56-58; vis physician, Charity Hosp & Med Sch, La State Univ, 58-60, clin asst prof med, 59-60; dir med res, Upjohn Co, Mich, 60-64; vpres-in-charge med res, Ciba Pharmaceut Co, 64-67; exec dir, Merck Sharp & Dohme, 67-68; prof med & head therapeut & dir clin pharmacol, 68-77, CLIN PROF MED, MED SCH, TULANE UNIV, 77- *Honors & Awards:* Univ Notre Dame Sci Award, 64. *Mem:* Int Soc Clin Pharmacol (vpres); Am Soc Clin Pharmacol & Therapeut; fel Am Col Physicians; Endocrine Soc; AMA. *Res:* Hypertension; diabetes; clinical pharmacology; endocrinology; bioavailability of drugs and drug metabolism. *Mailing Add:* Dept of Med Clin Res Ctr 147 S Liberty St New Orleans LA 70112

MCMAHON, GARFIELD WALTER, b Man, Feb 25, 32. ACOUSTICS. *Educ:* Univ Man, BSc, 52; Univ BC, MSc, 55. *Prof Exp:* Sci officer, 55-70, TRANSDUCER GROUP LEADER, DEFENSE RES ESTAB ATLANTIC, 70- *Mem:* Fel Acoust Soc Am. *Res:* Underwater acoustics; transducer calibration and design; properties of piezo-electric ceramics; vibrations of solid cylinders. *Mailing Add:* Dept Econ Laurentian Univ Ramsey Lake Rd Sundbury ON P3E 2C6 Can

MACMAHON, HAROLD EDWARD, b Aylmer, Ont, Mar 30, 01; US citizen; m 34; c 4. PATHOLOGY, BACTERIOLOGY. *Educ:* Univ Western Ont, MD, 25; Am Bd Path, dipl & cert path anat. *Hon Degrees:* BA, Univ Western Ont, 22, ScD, 48. *Prof Exp:* Intern, Montreal Gen Hosp, 25-26; asst, Boston City Hosp, 26-29; asst path, Univ Hamburg, 29-30; prof path & chmn dept, 30-71, EMER PROF PATH, SCH MED & SCH DENT MED, TUFTS UNIV, 71- *Concurrent Pos:* Instr, Harvard Med Sch, 28-29; asst, Univ Berlin, 31-32; pathologist-in-chief, Tufts-New Eng Med Ctr Hosps; consult pathologist, Mt Auburn Hosp, New Eng Med Ctr Hosps, Carney Hosp, Lynn Hosp, Leonard Morse Hosp, Malden Hosp & Cape Cod Hosp; consult, Armed Forces Inst Path, Boston Vet Hosp & USPHS Hosp, Boston; vis prof, Med Sch, Univ Mass, 71-; vis prof pathol, Med Sch, Univ Mass, Worcester, 80- *Mem:* Hon fel Royal Col Physicians; Am Asn Path & Bact; Am Med Asn; Int Acad Path; Ger Path Soc. *Res:* Pathology of the heart, lungs, liver, kidneys and blood vessels. *Mailing Add:* 19 Hubbard Park Rd Cambridge MA 02138

MACMAHON, JAMES A, b Dayton, Ohio, Apr 7, 39. ECOLOGY, VERTEBRATE ZOOLOGY. *Educ:* Mich State Univ, BS, 60; Univ Notre Dame, PhD(biol), 64. *Prof Exp:* Asst dir biol, Dayton Mus Natural Hist, 63-64; from asst prof to assoc prof, Univ Dayton, 64-71; assoc prof zool, 71-74, asst dir US Int Biol Prog-Desert Biomed, 71-79, PROF BIOL, UTAH STATE UNIV, 74-, HEAD DEPT, 85- *Mem:* Soc Study Evolution; Soc Syst Zool; Ecol Soc Am; Am Soc Zoologists; Am Soc Ichthyologists & Herpetologists. *Res:* Theory of community organization; community ecology of deserts; biology of desert perennials; energy exchange in plant and animal populations; biology of reptiles and amphibians; biology of arachnids. *Mailing Add:* Dept Biol Utah State Univ UMC 5305 Logan UT 84321-5305

MCMAHON, JOHN ALEXANDER, b Monongahela, Pa, July 31, 21; m 77; c 4. HEALTH ADMINISTRATION. *Educ:* Duke Univ, AB, 42; Harvard Univ, JD, 48. *Hon Degrees:* LLD, Wake Forest Univ, 78. *Prof Exp:* Prof pub law & govt & asst dir, Inst Govt, Univ NC, 48-59; gen coun & secy-treas, NC Asn County Comnr, 59-65; pres, NC Blue Cross & Blue Shield Inc, 68-72; pres, Am Hosp Asn, Chicago, 72-86; CHMN, DEPT HEALTH ADMIN, DUKE UNIV, 86- *Mem:* Inst Med-Nat Acad Sci. *Res:* County and state government; hospital administration. *Mailing Add:* 181 Montrose Dr Durham NC 27707

MCMAHON, JOHN MARTIN, b Buffalo, NY, Dec 24, 15; m 42; c 6. MEDICINE. *Educ:* Georgetown Univ, BS, 36, MD, 40; Univ Minn, MS, 50. *Prof Exp:* Instr psychosom med, Sch Med, Tulane Univ, 50-52; clin prof med, Sch Med, 52-85, DIR ARTHRITIS CLIN, MED CTR, UNIV ALA, BIRMINGHAM, 66-; CHIEF MED & ASSOC DIR MED EDUC, BIRMINGHAM BAPTIST HOSP, 54-, PROG DIR, INTERNAL MED RESIDENCY, 85- *Concurrent Pos:* Attend consult, Vet Admin Hosp, Birmingham. *Honors & Awards:* Benemerenti Medal by Pope Paul VI, 67. *Mem:* AMA; Am Rheumatism Asn; Am Heart Asn; fel Am Col Physicians; Am Col Gastroenterol (past pres). *Res:* Internal medicine; gastroenterology; arthritis. *Mailing Add:* 801 Princeton Ave SW Suite 430 Birmingham AL 35211

MCMAHON, JOHN MICHAEL, b St Paul, Minn, May 13, 41; m 65; c 2. ATOMIC PHYSICS, QUANTUM ELECTRONICS. *Educ:* Boston Col, BS, 63; Dartmouth Col, MA, 65. *Prof Exp:* Physicist, 65-74, supvry res physicist, Laser Res, 74-81, assoc supt, 81-88, CHIEF SCIENTIST, OPTICAL SCI DIV, NAVAL RES LAB, 88- *Concurrent Pos:* Consult, Div Laser Fusion, Dept of Energy, 76- & var US Navy commands, 68-; assoc ed, Microwave & Optical Technol Letters, 88- *Mem:* Am Phys Soc; Sigma Xi. *Res:* Solid state lasers; non-linear optics; laser produced plasma; laser fusion; optical diagnostics of plasmas; laser applications. *Mailing Add:* Optical Sci Div Code 6501 Naval Res Lab Washington DC 20375

MCMAHON, KENNETH JAMES, b Flandreau, SDak, July 9, 22; m 47; c 2. BACTERIOLOGY. *Educ:* SDak State Univ, BS, 47; Okla State Univ, MS, 49; Kans State Univ, PhD(bact), 54. *Prof Exp:* Asst bact, Okla State Univ, 47-48, instr, 48-49; from instr to prof, Kans State Univ, 49-70, actg head dept, 61-62, 63-64; actg chmn, dept vet sci, 80-81, prof & chmn dept, 70-87, EMER

PROF BACT, NDAK STATE UNIV, 87- *Mem:* Am Soc Microbiol; fel Am Acad Microbiol. *Res:* Bacteriology of animal diseases and insect pathogens; serology of brucellosis. *Mailing Add:* Dept Microbiol NDak State Univ Fargo ND 58105

MCMAHON, M MOLLY, MEDICINE. *Prof Exp:* ASST PROF MED, MAYO CLIN, 89- *Mailing Add:* Dept Med Mayo Clin W 18 200 First St SW Rochester MN 55905

MCMAHON, PAUL E, b Burlington, Vt, July 2, 31; m 82; c 1. PHYSICAL CHEMISTRY. *Educ:* St Michael's Col, Vt, BS, 54; Univ Vt, MS, 56; Univ Ill, PhD(phys chem), 61; Rutgers Univ, MBA, 76. *Prof Exp:* Spectroscopist, Ill State Geol Surv, 56-57, & Univ Ill, 57-61; res chemist, Chemstrand Div, Monsanto Co, NC, 61-68; MGR, CELANESE RES CO, 68- *Res:* Molecular structure and motion in small molecules and polymers; structure-property relations of polymers and composites; characterization and evaluation of fiber reinforced composites. *Mailing Add:* Hoechst Celanese Div 86 Morris Ave Summit NJ 07901

MCMAHON, RITA MARY, b New York, NY, Mar 5, 22. CYTOLOGY. *Educ:* Fordham Univ, BS, 49, MS, 51, PhD(biol), 53. *Prof Exp:* From instr to assoc prof sci, Sch Educ, Fordham Univ, 54-67, chmn dept, 57-67; assoc prof, 67-77, prof, 77-81, EMER PROF BIOL, WESTERN CONN STATE UNIV, 81- *Concurrent Pos:* Adj prof, Grad Sch, New Eng Inst, 69- *Mem:* AAAS; Bot Soc Am; Am Inst Biol Sci; Environ Mutagen Soc; Genetics Soc Am; Sigma Xi. *Res:* Plant tissue and cell culture; polyploidy in development. *Mailing Add:* Box 350 Rural Rte 2 Pound Ridge NY 10576

MCMAHON, ROBERT FRANCIS, III, b Syracuse, NY, June 17, 44; m 80. FRESHWATER & MARINE BIOLOGY. *Educ:* Cornell Univ Sch Arts & Sci, BA, 66; Syracuse Univ, PhD(zool), 72. *Prof Exp:* From teaching asst biol to res asst, Syracuse Univ, 67-72; from asst prof to assoc prof, 72-84, PROF BIOL, UNIV TEX, ARLINGTON, 84- *Concurrent Pos:* Fulbright-Hayes res fel, Trinity Col, Dublin, Ireland, 78-79; adj prof, La State Univ, 85- *Mem:* Marine Biol Lab; Malacol Soc London; Marine Biol Asn UK; Sigma Xi; Am Zool Soc; Scottish Marine Biol Asn. *Res:* Physiological ecology of freshwater, estuarine and marine invertebrate organisms; study of growth and reproduction, bioenergetic and physiological variation in freshwater animals and the physiological basis for intertidal zonation in marine animals; ecophenotypic morphological and physiological interpopulation variation. *Mailing Add:* Dept Biol Univ Tex Box 19498 Arlington TX 76019

MCMAHON, THOMAS ARTHUR, b Dayton, Ohio, Apr 21, 43; m 65; c 2. BIOMECHANICS, BIOENGINEERING. *Educ:* Cornell Univ, BS, 65; Mass Inst Technol, SM, 67, PhD(fluid mech), 70. *Prof Exp:* Res fel, 69-71, lectr, 70-71, from asst prof to assoc prof, 71-77, PROF BIOMECH, DIV ENG & APPL PHYSICS, HARVARD UNIV, 77- *Honors & Awards:* Rosenthal Prize, 88; Storer lectr, Univ Calif, Davis, 91. *Mem:* Am Physiol Soc; Biomed Eng Soc. *Res:* The theory of models applied to living systems; mechanics of locomotion; cardiac mechanics; muscle mechanics and biophysics. *Mailing Add:* Div Eng & Appl Physics Pierce Hall Harvard Univ Cambridge MA 02138

MCMAHON, THOMAS JOSEPH, b Rahway, NJ, 1943; m 66; c 2. SEMICONDUCTORS, ELECTROOPTICS. *Educ:* Univ Ill, Urbana, BS, 65, MS, 66; Univ Mo-Rolla, PhD(physics), 69. *Prof Exp:* Res physicist, Naval Weapons Ctr, China Lake, 69-80; SR SCIENTIST, SOLAR ENERGY RES INST, GOLDEN, COLO, 80- *Mem:* Am Phys Soc. *Res:* Glow discharge amorphous silicon and vapor phase epitaxial semiconductors are grown for application in the areas of infrared and visible photo detection, photo thermal and photovoltaic solar energy conversion and electrooptic devices. *Mailing Add:* 13400 W Seventh Golden CO 80401

MCMAHON, TIMOTHY F, b Washington, DC, July 16, 60; m. TOXICOLOGY. *Educ:* Mt St Mary's Col, BS, 83; Univ Md, PhD(pharmacol & toxicol), 88. *Prof Exp:* Lab res asst, Dept Pharmacol & Toxicol, Sch Pharm, Univ Md, 84-85, grad teaching asst, 85-86, Am Found Pharmaceut Educ fel, Sch Pharm, 86-88; IRTA fel, Nat Inst Environ Health Sci, 88-90; sr scientist, NSI Technol Serv Corp, NC, 90; TOXICOLOGIST, HEALTH EFFECTS DIV, ENVIRON PROTECTION AGENCY, WASHINGTON, DC, 90- *Mem:* AAAS; Am Soc Pharmacol & Exp Therapeut; Soc Toxicol. *Res:* Author of numerous publications and abstracts. *Mailing Add:* Health Effects Div H7509C Environ Protection Agency 401 M St SW Washington DC 20460

MCMANAMON, PETER MICHAEL, b Chicago, Ill, June 15, 37; m 68; c 2. TELECOMMUNICATIONS. *Educ:* Ill Inst Technol, BS, 59, MS, 62, PhD(elec eng), 70. *Prof Exp:* Staff engr, IIT Res Inst, 59-66, sect mgr commun, 66-71; group chief, Off Telecommun, 71-77, group chief telecommun, Nat Telecommun & Info Admin, 77-78, assoc dir, Inst Telecommun Sci, US Dept Com, 79-; PRES, CYBERLINK CORP. *Honors & Awards:* Gold Medal Award, US Dept Com, 78. *Mem:* Inst Elec & Electronics Engrs. *Res:* Statistical communication theory; information theory; telecommunications. *Mailing Add:* Cyberlink Corp 1790 30th St Suite 300 Boulder CO 80301

MCMANIGAL, PAUL GABRIEL MOULIN, b Los Angeles, Calif, Apr 15, 36; m 59; c 2. INVESTMENT MANAGEMENT, PHYSICS. *Educ:* Pomona Col, BA, 58; Univ Calif, Berkeley, PhD(physics), 63. *Prof Exp:* Mgr, Missile Systs Eng, Aeronutronic Div, Philco Ford Corp, 63-71; dir advan sensors, Defense Advan Res Proj Agency, Dept of Defense, 71-73; dir develop planning, Aeronutronic Div, Ford Aerospace & Commun Corp, 73-81, vpres tech affairs, 81-84, mgr, Advan Sensors & Fire Control Activ, Aeronutronic Div, 84-87; pres, Growth Endowment Mgt, 87-88; VPRES & GEN MGR, VEMCO CORP, 88- *Res:* Applied research in optics, photography and atmospheric physics; development in tactical missiles; advanced sensors; elementary particles. *Mailing Add:* 16 Inverness Lane Newport Beach CA 92660

MCMANNIS, JOHN D, b Orange, NJ, Dec 18, 55. IMMUNOLOGY. *Educ:* Univ Notre Dame, BS, 78; Rush Presby St Lukes Med Ctr, PhD(immunol), 84. *Prof Exp:* Postdoctoral, Dept Hemat, Univ Chicago, 84-85; asst prof, Dept Med, Loyola Univ Med Ctr, 85-89; THER SCIENTIST, BLOOD COMPONENT TECH, COBE BCT, INC, 89- *Mem:* Fed Am Socs Exp Biol; Am Diabetes Asn; Am Soc Apheresis. *Mailing Add:* Cobe BCT Inc 1201 Oak St Lakewood CO 80215

MCMANUS, DEAN ALVIS, b Dallas, Tex, July 8, 34; m 72. GEOLOGY, OCEANOGRAPHY. *Educ:* Southern Methodist Univ, BS, 54; Univ Kans, MS, 56, PhD(geol), 59. *Prof Exp:* From res assoc to res asst prof, 59-65, from asst prof to assoc prof, 65-71, PROF OCEANOG, UNIV WASH, 71- *Concurrent Pos:* Mem, Joint Oceanog Inst for Deep Earth Sampling & Pac Ocean Adv Panel, 65-71, asst chmn planning comt, 76-78; adj prof, marine studies, Univ Wash, 73-89 & Quaternary Res Ctr, 76-82; ed-in-chief, Marine Geol, 73- *Mem:* AAAS; Am Asn Petrol Geologists; Soc Econ Paleontologists & Mineralogists; Geol Soc Am; Am Geophys Union. *Res:* continental shelf topography and sediments; Arctic shelf sedimentation; holocene stratigraphy on continental shelf of Chukchi and Bering Sea. *Mailing Add:* Sch Oceanog WB-10 Univ Wash Seattle WA 98195

MCMANUS, EDWARD CLAYTON, b McIntosh, Minn, Aug 19, 18; m 66. PARASITOLOGY, PHARMACOLOGY. *Educ:* Univ Minn, BS, 40, PhD(pharmacol), 50; Iowa State Univ, DVM, 44. *Prof Exp:* Asst pharmacol, Univ Minn, 44-48; res assoc, Sharp & Dohme, Inc, 48-53, res assoc, Merck Sharp & Dohme Res Labs, 53-55, pathologist, 55-58, res assoc & res fel basic animal sci res, 58-83; RETIRED. *Mem:* Am Vet Med Asn; Am Soc Pharmacol & Exp Therapeut; Am Soc Vet Physiol & Pharmacol; Am Asn Vet Parasitol. *Res:* Gastrointestinal pharmacology; pathology; toxicology; parasitologic chemotherapy; effect of ivermectin on canine heartworm microfilarial infection. *Mailing Add:* 634 Baron De Kalb Rd Wayne PA 19087

MCMANUS, HUGH, b West Bromwich, Eng, May 10, 18; m 53; c 3. NUCLEAR THEORY, COMPUTATIONAL PHYSICS. *Educ:* Univ Birmingham, BSc, 39, PhD(math-physics), 47. *Prof Exp:* Res officer ionosphere & commun, Brit Admiralty, London, 40-42, res officer opers res, 42-43, res officer tech intel, 43-44; res fel theoret physics, Univ Birmingham, 47-49, lectr, 49-51; assoc res officer, Theoret Physics Div, Atomic Energy Can, Ltd, 51-60; prof, 60-88, EMER PROF PHYSICS & ASTRON, MICH STATE UNIV, 88- *Concurrent Pos:* Res assoc, Mass Inst Technol, 57-58, 70 & 79; Guggenheim fel, Nordic Inst Theoret Atomic Physics, 63-64; Sci Res Coun fel, Oxford Univ, 69-70; staff mem, Ctr Nuclear Studies, Saclay, France, 77; consult, Triumf, Vancouver, BC, 80-; lectr, 3e cycle, Univ Neuchatel, 82; Fulbright fel, KFA Jeulich, 87-88; prof dir, Theo Phys, NSF, 88-89. *Mem:* Fel Am Phys Soc. *Res:* Nuclear structure; scattering of light hadrons from nuclei; interface between nuclear and particle physics; historical evolution of ideas in physics. *Mailing Add:* Dept Physics Mich State Univ East Lansing MI 48824-1116

MCMANUS, IVY ROSABELLE, biochemistry; deceased, see previous edition for last biography

MCMANUS, JAMES MICHAEL, b Brooklyn, NY, May 22, 30; m 55; c 3. ORGANIC CHEMISTRY. *Educ:* Col Holy Cross, BS, 52; Niagara Univ, MS, 54; Mich State Univ, PhD(chem), 58. *Prof Exp:* Res chemist, 58-69, PATENT CHEMIST, PFIZER, INC, 69- *Concurrent Pos:* US patent agent, 77- *Mem:* Am Chem Soc. *Res:* Chemistry of tetrazoles, indoles and benzimidazoles; preparation and pharmacology of diuretics and sulfonylureas. *Mailing Add:* Pfizer Inc Eastern Point Rd Groton CT 06340

MCMANUS, LINDA MARIE, PULMONARY IMMUNOPATHOLOGY. *Educ:* Univ Colo, PhD(path), 78. *Prof Exp:* ASSOC PROF PATH, HEALTH SCI CTR, UNIV TEX, 82- *Res:* Hypersensitivity. *Mailing Add:* Dept Path Univ Tex Health Sci Ctr 7703 Floyd Curl Dr San Antonio TX 78284

MCMANUS, MYRA JEAN, b St Micheldes Saints, Que, Sept 7, 25; m 54; c 3. ENDOCRINOLOGY, HISTOLOGY. *Educ:* McGill Univ, MS, 48; Univ Western Ont, PhD(endocrinol), 52. *Prof Exp:* Res officer, Atomic Energy Can, Ltd, 52-56; res asst, Zool Dept, 70-75, RES ASSOC, ANAT DEPT, MICH STATE UNIV, 75- *Mem:* Am Asn Cancer Res; AAAS. *Res:* Distribution and function of zinc in the male reproductive system (rats) and the effects of zinc deficiency (dietary) on this system; hormones, steroid and polypeptide involved in development, differentiation and carcinogenesis of human breast epithelium. *Mailing Add:* 526 Kedzie Dr East Lansing MI 48823

MCMANUS, SAMUEL PLYLER, b Edgemoor, SC, Oct 29, 38; m 59; c 2. ORGANIC CHEMISTRY, POLYMER CHEMISTRY. *Educ:* The Citadel, BS, 60; Clemson Univ, MS, 62, PhD(chem), 64. *Prof Exp:* Res chemist, Marshall Lab, E I du Pont de Nemours & Co, 64; from asst prof to assoc prof, 66-73, chmn dept, 70-72 & 77-78, dir, Mat Sci Grad Prog, 88-89, PROF CHEM, UNIV ALA, HUNTSVILLE, 73-, DEAN GRAD SCH, 89-, ASSOC VPRES ACAD AFFAIRS, 90- *Concurrent Pos:* Consult, US Army Res Off, Durham, 68-73, 83-; indust consult; vis prof, Univ SC, 74-75; fel Am Soc Eng Educ, NASA, 81, 82. *Honors & Awards:* Army Commendation Medal; Nat Defense Serv Medal. *Mem:* Am Chem Soc; fel Am Inst Chem; Sigma Xi. *Res:* Neighboring group participation; polymer chemistry; medium effects; gravitational effects. *Mailing Add:* Dept Chem Univ Ala Huntsville AL 35899

MCMANUS, THOMAS (JOSEPH), b Omaha, Nebr, Feb 5, 25; m 51; c 5. CELL PHYSIOLOGY, HEMATOLOGY. *Educ:* Antioch Col, BS, 51; Boston Univ, MD, 55. *Prof Exp:* Asst hemat, Sloan-Kettering Inst, NY, 49-51; res fel med, Harvard Med Sch, 55-58, asst med, 58-59, res assoc, 59-61; from asst prof to assoc prof physiol & pharmacol, 61-88, dir grad studies, 72-75, 80-85, DIR, LAB CELL MEMBRANE PHYSIOL, SCH MED, DUKE UNIV, 68-, PROF CELL BIOL, 88- *Concurrent Pos:* Asst physician, Peter Bent Brigham Hosp, Boston, Mass, 55-58; vis prof, NC Col, 63; mem sci staff, Res Vessel Alpha Helix Amazon Exped, Brazil, 67; Sloan vis scholar, Antioch

Col, 70; counr, Soc Gen Physiol, 76-79; vis prof, Univ Melbourne, 81. *Mem:* Fel AAAS; Am Physiol Soc; Biophys Soc; Soc Gen Physiol; Am Soc Zool; Sigma Xi; Archeol Inst Am. *Res:* Cell membrane transport; cell volume regulation; comparative physiology of red cells. *Mailing Add:* Div Physiol Dept Cell Biol Duke Univ Med Ctr Box 3709 Durham NC 27710

MCMASTER, MARVIN CLAYTON, JR, b Gering, Nebr, June 27, 38; m 62; c 1. BIOCHEMISTRY. *Educ:* SDak Sch Mines & Technol, BS, 60; Univ Nebr, PhD(org chem), 66. *Prof Exp:* Fel polypeptides, Inst Molecular Biophys, Fla State Univ, 70-71; Nat Heart & Lung Inst spec fel biochem, Webb-Waring Lung Inat, Univ Colo Med Ctr, Denver, 71-73; scholar biochem lipid storage dis, Ment Health Res Inst, Univ Mich, Ann Arbor, 73-75; sr develop chemist pesticide prod, Ciba-Geigy Corp, 75-78; TECH SPECIALIST, WATERS ASSOCS, ST LOUIS OFF, 78- *Concurrent Pos:* Res chemist, Indust & Biochem Dept, E I du Pont de Nemours & Co, Inc, 65-68; sr scientist I, Indust Chem Div, Kraftco Corp, 68-70. *Mem:* Am Chem Soc. *Res:* Small ring heterocyclic compounds; polypeptide synthesis; biochemistry of lung metabolism; high pressure liquid chromatography; pesticides. *Mailing Add:* 2070 Cordoba Dr Florissant MO 63033

MCMASTER, PAUL D, b Norwich, Conn, Feb 24, 32; m 63; c 3. ORGANIC CHEMISTRY, MEDICINAL CHEMISTRY. *Educ:* Col Holy Cross, BS, 54; Clark Univ, PhD(chem), 61. *Prof Exp:* From asst prof to assoc prof, 61-80, PROF CHEM, COL HOLY CROSS, 80- *Concurrent Pos:* Consult drug design, Astra Pharmaceut Prod, Inc, Worcester, Mass, 73-80; chmn dept, Col Holy Cross, 72-79 & 85. *Mem:* Am Chem Soc. *Res:* Conformational analysis; drug design. *Mailing Add:* Dept Chem Col of the Holy Cross Worcester MA 01610

MCMASTER, PHILIP ROBERT BACHE, b Cambridge, Mass, Feb 19, 30; m 58; c 2. IMMUNOLOGY, EXPERIMENTAL PATHOLOGY. *Educ:* Princeton Univ, AB, 52; Johns Hopkins Univ, MD, 56. *Prof Exp:* Intern med, New York Hosp-Cornell Med Ctr, 56-57; sr asst surgeon, Lab Immunol, Nat Inst Allergy & Infectious Dis, 57-60; USPHS surgeon, Pasteur Inst Paris, 60-62; surgeon, Lab Immunol, 62-64, sr surgeon, Lab Germfree Animal Res, Nat Inst Allergy & Infectious Dis, 64-; PROF SURGERY, SCH MED, BROWN UNIV. *Mem:* Fedn Am Soc Exp Biol. *Res:* Immunopathology related to immune process and tissue destruction; physiology of ocular pressure in normal and abnormal states. *Mailing Add:* 135 Lloyd Ave Providence RI 02906

MCMASTER, ROBERT CHARLES, electrical engineering; deceased, see previous edition for last biography

MCMASTER, ROBERT H, b Flint, Mich, Feb 26, 16; m 39; c 2. MEDICAL RESEARCH. *Educ:* Ohio Univ, AB, 38; Univ Cincinnati, MD, 50. *Prof Exp:* Physician, Gallipolis Clin, Ohio, 51-56; physician, Chas Pfizer & Co, 56; assoc dir med res, Wm S Merrell Co, 56-67, dir sci training, 67-74, dir, employee health serv, Merrell Dow Pharmaceut, Inc, 74-89; CONSULT, HILL TOP PHARMATEST, INC, CINCINNATI, OHIO, 89- *Concurrent Pos:* Asst, Col Med, Univ Cincinnati, 57-66, from instr to assoc prof, 57-82, emer assoc clin prof med, 82-; clinician, Cincinnati Gen Hosp, 64-; past-pres, SW Ohio Chap & Ohio affil, Am Heart Asn. *Mem:* Am Soc Clin Pharm & Therapeut; AMA; Am Heart Asn; fel Royal Soc Health. *Res:* Employee health and safety; training of nonprofessionals in medical subjects. *Mailing Add:* 8040 Shawnee Run Rd Cincinnati OH 45243

MCMASTER, ROBERT LUSCHER, b Mobile, Ala, Apr 16, 20; m 48; c 1. OCEANOGRAPHY. *Educ:* Columbia Univ, BA, 43; Rutgers Univ, MS, 49 & PhD(geol), 53. *Prof Exp:* Res marine geologist, from asst prof to assoc prof, 55-90, EMER PROF OCEANOG, UNIV RI, 90- *Concurrent Pos:* Consult, coastal environ. *Mem:* Geol Soc Am; Am Asn Petrol Geol; Soc Econ Paleontologist & Mineralogists; Int Asn Sedimentol; Am Geophy Union; Sigma Xi. *Res:* Marine geological processes and developmental history of beaches, estuaries and continental shelves, slopes and rises. *Mailing Add:* Grad Sch Oceanog Univ RI Kingston RI 02881

MCMASTER, WILLIAM H, b Ft Lewis, Wash, Apr 16, 26; m 51; c 1. HYDRODYNAMICS, FLUID STRUCTURE INTERACTIONS. *Educ:* US Mil Acad, BS, 46; Univ Va, MS, 52, PhD(physics), 54. *Prof Exp:* Sr physicist, 55-62, group leader, 62-74, staff physicist, 74-84 Lawrence Livermore Nat Lab, Univ Calif, 74-84; CONSULT, 86- *Honors & Awards:* Sigma Xi Res Prize. *Mem:* Am Phys Soc; NY Acad Sci. *Res:* Polarization of radiation; nuclear test diagnostics; computational methods; x-ray physics; equation of state-solids; calculational methods in hydrodynamics; incompressible fluids. *Mailing Add:* 351 Polk Way Livermore CA 94550

MCMASTERS, ALAN WAYNE, b Ottawa, Kans, Dec 19, 34; m 57; c 4. OPERATIONS RESEARCH. *Educ:* Univ Calif, Berkeley, BS, 57, MS, 62, PhD(indust eng, opers res), 66. *Prof Exp:* Res engr, Pac Southwest Forest & Range Exp Sta, US Forest Serv, 55-61, opers analyst, 61-65; from asst prof opers res to assoc prof opers res & admin sci, 65-88, PROF OPERS RES & ADMIN SCI, NAVAL POSTGRAD SCH, 88- *Concurrent Pos:* Consult, Mellonics Systs Div, Litton Indust, Monterey, 69-70, BDM, Ford Ord, 73, Pro-Log, Monterey, 81, Test Int, San Jose, 82 & Opers Res Groups, Navy, 80- *Mem:* Opers Res Soc Am; Sigma Xi; Inst Mgt Sci; Soc Logistics Engrs. *Res:* Locational problems in distribution networks; Navy inventory models. *Mailing Add:* 1227 Josselyn Canyon Rd Monterey CA 93940

MCMASTERS, DENNIS WAYNE, b Chickasha, Okla, July 9, 40; m 68; c 2. PLANT PHYSIOLOGY. *Educ:* Okla Christian Col, BSE, 62; Univ Okla, MNS, 68; Univ Ark, PhD(bot), 73. *Prof Exp:* Teacher biol & chem, McAlester Pub Schs, 62-68; teacher, Tulsa Pub Schs, 68-69; instr bot, 72-80, from assoc prof to prof, 80-90, CHMN, DEPT BIOL, HENDERSON STATE UNIV, 90- *Res:* Gibberellins; algal physiology. *Mailing Add:* Dept Biol Sci Henderson State Univ 1100 Henderson St Arkadelphia AR 71923

MCMASTERS, DONALD L, b Crawfordsville, Ind, July 14, 31. ANALYTICAL CHEMISTRY. *Educ:* Wabash Col, AB, 53; Univ NDak, MS, 55; Ind Univ, PhD(anal chem), 59. *Prof Exp:* Asst prof chem, Beloit Col, 59-63; lectr, Univ Ill, 63-64; lectr chem, 64-75, CHEMIST & SAFETY CONSULT, DEPT ENVIRON HEALTH & SAFETY, IND UNIV, BLOOMINGTON, 75- *Mem:* Am Chem Soc; Sigma Xi. *Res:* Polarography in aqueous and non-aqueous solvents. *Mailing Add:* Dept Environ Health & Safety Ind Univ 840 St Rd 46 Bypass Bloomington IN 47408-2694

MCMEEKIN, DOROTHY, b Boston, Mass, Feb 24, 32. PLANT PATHOLOGY. *Educ:* Wilson Col, AB, 53; Wellesley Col, MA, 55; Cornell Univ, PhD(plant path), 59. *Prof Exp:* Asst prof biol, Upsala Col, 59-64, BOWLING GREEN STATE UNIV, 66-; prof natural sci, 66-89, PROF BOT & PLANT PATH, MICH STATE UNIV, 89- *Mem:* Am Phytopath Soc; Bot Soc Am; Mycol Soc Am. *Res:* Physiology of plant disease; science and art. *Mailing Add:* Ctr Interdisciplinary Studies Mich State Univ 335 N Kedzie East Lansing MI 48824

MCMENAMIN, JOHN WILLIAM, b Tacoma, Wash, Apr 1, 17; m 42; c 2. DEVELOPMENTAL BIOLOGY, MICROANATOMY. *Educ:* Occidental Col, AB, 40; Univ Calif, Los Angeles, MA, 46, PhD(zool), 49. *Prof Exp:* High sch & jr col teacher, Calif, 42-45; spec appt, Occidental Col, 46-47, from instr to assoc prof biol, 47-57, prof, 57-82; RETIRED. *Mem:* AAAS; Am Soc Zoologists; Sigma Xi. *Res:* Role of the thyroid gland in the development of the chick embryo; lipid and porphyrin metabolism in the chick embryo as influenced by hormones; histology and reproduction of fishes. *Mailing Add:* PO Box 190 Lakebay WA 98349-0190

MCMENAMIN, MARK ALLAN, b Portland, Ore, Feb 4, 58; m 81; c 2. COMPUTER SCIENCE, CELL BIOLOGY. *Educ:* Stanford Univ, BS, 79; Univ Calif, Santa Barbara, PhD(geol), 84. *Prof Exp:* Lab scientist org geochem, US Geol Surv, 79-84; PROF GEOL, MT HOLYOKE COL, 84- *Concurrent Pos:* Ed, Scripta Technica, 87-; NSF presidential young investr, 88-93; nat lectr, Sigma Xi, 91-93. *Mem:* Paleont Soc; Sigma Xi. *Res:* Biospheric evolution, especially as concerns major transitions and transformations in earth history such as the PreCambrian-Cambrian transition; emergence of animals; Proterozoic super continents; atmospheric history; computer applications in the study of biospheric change. *Mailing Add:* Dept Geol Mount Holyoke Col South Hadley MA 01075-1484

MCMICHAEL, FRANCIS CLAY, b Philadelphia, Pa, Aug 8, 37; m 69; c 2. HYDRAULIC & ENVIRONMENTAL ENGINEERING. *Educ:* Lehigh Univ, BS, 58; Calif Inst Technol, MS, 59, PhD(civil eng), 63. *Prof Exp:* Res fel civil eng, Calif Inst Technol, 63-65; asst prof, Princeton Univ, 65-67; Am Iron & Steel Inst fel, 67-69, sr fel, 69-73, head dept, 75-79, PROF CIVIL ENG, ENG & PUB POLICY, CARNEGIE-MELLON UNIV, 75-, BLENKO PROF ENVIRON ENG, 81- *Concurrent Pos:* Mem sci adv bd, US Environ Protection Agency, 81- *Mem:* Am Soc Civil Engrs; Am Geophys Union; Water Pollution Control Fedn; Int Asn Hydraul Res; Am Inst Chem Engrs; Soc Risk Analysis. *Res:* Groundwater; hydrology; applied statistics; risk analysis; solid and hazardous waste management. *Mailing Add:* Dept Civil Eng Carnegie-Mellon Univ Pittsburgh PA 15213-3890

MCMICHAEL, KIRK DUGALD, b Schenectady, NY, July 13, 35; m 58; c 2. ORGANIC CHEMISTRY, SCIENCE EDUCATION. *Educ:* Shimer Col, AB, 53; Univ Chicago, MS, 56, PhD(chem), 60. *Prof Exp:* Res assoc chem, Univ Wis, 60-62; asst prof, 62-68, asst to chmn dept, 68-71, ASSOC PROF, 68- UNDERGRAD COORDR CHEM, WASH STATE UNIV, 81-, ASSOC CHAIR CHEM, 86- *Concurrent Pos:* petrol Res Fund grant, 64-66; SRC sr res assoc, Univ Glasgow, Scotland, 75-76. *Mem:* Fel AAAS; Am Chem Soc; Sigma Xi. *Res:* Characterization of allylic rearrangement pathways using deuterium isotope effects; long range allylic rearrangements in steroidal systems; thermodynamic isotope effects and their relationship to theory. *Mailing Add:* Dept Chem Wash State Univ Pullman WA 99164-4630

MCMICKING, JAMES H(ARVEY), b Detroit, Mich, Aug 5, 29; m 55; c 4. CHEMICAL ENGINEERING. *Educ:* Wayne State Univ, BS, 53, MS, 55; Ohio State Univ, PhD, 61. *Prof Exp:* ASSOC PROF CHEM ENG & ASSOC CHMN DEPT CHEM, WAYNE STATE UNIV, 58- *Mem:* Am Inst Chem Engrs; Sigma Xi. *Res:* Thermodynamic properties; process dynamics. *Mailing Add:* Dept Chem Eng Wayne State Univ Detroit MI 48202

MCMICKLE, ROBERT HAWLEY, b Paterson, NJ, July 30, 24; m 49; c 5. PHYSICS. *Educ:* Oberlin Col, BA, 47; Univ Ill, MS, 48; Pa State Univ, PhD(physics), 52. *Prof Exp:* Physicist, Res Ctr, B F Goodrich Co, 52-59; assoc prof, Robert Col, Istanbul, 59-63, prof & head dept, 63-71; prof & head dept physics, Bogazici Univ, Turkey, 71-76, vis prof, 76-79; prof physics, Schreiner Col, Kerrville, Tex, 79-80; prof physics, Luther Col, Decorah, Iowa, 80-81; sr lectr, Ctr Nuclear Studies, Memphis State Univ, 81-83; ACCREDITATION COORDR, NH YANKEE DIV, PUB SERV CO, 83- *Mem:* Am Phys Soc; Am Asn Physics Teachers; Sigma Xi; Am Nuclear Soc. *Res:* Solid state phenomena. *Mailing Add:* 36 Granite Dr North Hampton NH 03862

MCMILLAN, ALAN F, b Ont, Can, Feb 11, 25; m 59; c 2. PHYSICAL CHEMISTRY, CHEMICAL ENGINEERING. *Educ:* Queen's Univ, Ont, BSc, 48, MSc, 49; Mass Inst Technol, PhD(phys chem), 53. *Prof Exp:* Res chemist, Naval Res Estab, NS, 53-64; from assoc prof to prof chem eng, 64-86, head dept, 79-86, EMER PROF, TECH UNIV NS, 86- *Mem:* Fel Chem Inst Can; Can Soc Chem Eng. *Res:* Liquid-vapor equilibrium; corrosion; modelling of dispersion of pollutants in estuaries. *Mailing Add:* Dept Chem Eng Tech Univ NS Box 1000 Halifax NS B3J 2X4 Can

MCMILLAN, BROCKWAY, b Minneapolis, Minn, Mar 30, 15; m 42; c 3. RANDOM PROCESSES. *Educ:* Mass Inst Technol, BS, 36, PhD(math), 39. *Prof Exp:* H B Fine instr math, Princeton Univ, 40-41, res assoc fire control, 41-43; res mathematician, Bell Labs, 46-55, exec officer, 55-61; asst secy res & develop, Dept Air Force, 61-63, under secy, 63-65; exec dir mil res, Bell

Labs, 65-69, vpres mil syst, 69-80; RETIRED. *Concurrent Pos:* Mem staff, Pres Sci Adv Off, 58-59; chmn, Conf Bd Math Sci, 81-82. *Mem:* Nat Acad Eng; Inst Math Statist; Math Asn Am; Soc Indust & Appl Math (pres, 60-61); fel Inst Elec & Electronics Engrs. *Res:* Applications of random processes to communications and to statistical mechanics; physical realizability of electrical networks. *Mailing Add:* PO Box 27 Sedgwick ME 04676

MCMILLAN, CALVIN, b Murray, Utah, Feb 20, 22; m 50; c 4. PLANT ECOLOGY. *Educ:* Univ Utah, BS, 47, MS, 48; Univ Calif, Berkeley, PhD(bot), 52. *Prof Exp:* From asst prof to assoc prof bot, Univ Nebr, 52-58; assoc prof, 58-65, PROF BOT, UNIV TEX, AUSTIN, 65- *Concurrent Pos:* Vis assoc prof & actg dir, Bot Garden, Univ Calif, Berkeley, 64-65. *Honors & Awards:* Mercer Award, Ecol Soc Am, 60. *Mem:* AAAS; Bot Soc Am; Ecol Soc Am. *Res:* Ecotypes and ecosystem functions; ecology of seagrasses and mangroves. *Mailing Add:* Dept Bot Univ of Tex Austin TX 78713-7640

MCMILLAN, CAMPBELL WHITE, b Soochow, China, Jan 10, 27; US citizen; m 55; c 6. PEDIATRICS, HEMATOLOGY. *Educ:* Wake Forest Col, BS, 48, Bowman-Gray Sch Med, MD, 52. *Prof Exp:* Intern, Boston City Hosp, 52-53; asst resident pediat, Children's Hosp Med Ctr, 53-55; pediat registr, St Mary's Hosp Med Sch, London, Eng, 55-56; asst med, Nemazee Hosp, Shiraz, Iran, 56-58; fel pediat hemat, Harvard Med Sch, 58-61; from asst prof to assoc prof, 63-72, assoc dir clin res unit, 66-79, PROF PEDIAT, MED SCH, UNIV NC, CHAPEL HILL, 72- *Mem:* Soc Pediat Res; Am Pediat Soc. *Res:* Coagulation. *Mailing Add:* Dept of Pediat Univ of NC Sch of Med CB#7220 Chapel Hill NC 27599-7220

MCMILLAN, DANIEL RUSSELL, JR, b Atlanta, Ga, Feb 26; 35; m 65; c 1. TOPOLOGY. *Educ:* Emory Univ, BA, 56; Univ Wis, MA, 57, PhD(math), 60. *Prof Exp:* Res instr math, La State Univ, 60-61; asst prof, Fla State Univ, 61-62; vis mem, Inst Advan Study, 62-64; actg assoc prof, Univ Va, 64-65, assoc prof, 65-66; assoc prof, 66-69, PROF MATH, UNIV WIS-MADISON, 69- *Concurrent Pos:* Nat Sci Found fel, 62-63; Inst Adv Study fel, 63-64, 69-70; Sloan fel, 65-67. *Mem:* Math Asn Am; Am Math Asn. *Res:* Topology of combinatorial manifolds; local homotopy properties of topological embeddings; cellular sets in combinatorial manifolds; geometric topology of mappings, three-manifolds. *Mailing Add:* Dept Math Univ Wis 480 Lincoln Dr Madison WI 53706

MCMILLAN, DONALD BURLEY, b Toronto, Ont, Apr 25, 29; m 85. ZOOLOGY, COMPARATIVE HISTOLOGY. *Educ:* Univ Western Ont, BSc, 51, MSc, 53; Univ Toronto, PhD(zool), 58. *Prof Exp:* Demonstr zool, Univ Toronto, 53-56; instr, 56-59, lectr, 59-60, from asst prof to assoc prof, 60-78, PROF ZOOL, UNIV WESTERN ONT, 78- *Mem:* Am Soc Zoologists; Can Asn Anat; Can Soc Zool; Can Soc Cell Biol; Am Asn Anat. *Res:* Comparative histology of fish (Agnatha, Chondrichthyes, Osteichthyes) including ultrastructure. *Mailing Add:* Dept Zool Univ Western Ont London ON N6A 5B7 Can

MCMILLAN, DONALD EDGAR, b Butler, Pa, Sept 23, 37; m 61; c 2. PHARMACOLOGY, TOXICOLOGY. *Educ:* Grove City Col, BS, 59; Univ Pittsburgh, MS, 62, PhD(psychol), 65. *Prof Exp:* From instr to asst prof pharmacol, State Univ NY Downstate Med Ctr, 67-69; from asst prof to prof pharmacol, Univ NC, Chapel Hill, 69-78, asst prof psychol, 70-72, from clin assoc prof to clin prof psychol, 72-78; PROF PHARMACOL & TOXICOL & CHMN DEPT, UNIV ARK MED SCI CAMPUS, LITTLE ROCK, 78- *Concurrent Pos:* NIH training grant, Harvard Med Sch, 65-66; mem, Neurobiol Rev Bd & Vet Admin Rev Bd, NSF; mem, rev comt, Nat Inst Drug Abuse; mem, comt toxicity data elements, Nat Res Coun; mem, subcomt on halogenated org, Sci Adv Bd, Environ Protection Agency, chmn, neurotoxicol res rev comt; mem, Comt on Problems of Drug Dependence; mem, superfund grant review comt, Nat Inst Environ Health Sci; consult psychiat, Vet Admin Hosp, Little Rock, Ark; consult, Health Effects Inst; Wilbur Mill's chair, Alcoholism & Drug Abuse, 91- *Mem:* AAAS; NY Acad Sci; Soc Toxicol; Behav Pharmacol Soc (pres, 82-84); Am Soc Pharmacol & Exp Therapeut; Sigma Xi; Behav Toxicol Soc (pres, 88-90); Pavlovian Soc. *Res:* Behavioral pharmacology; mechanisms of drug tolerance and drug dependence; behavioral toxicology; drug abuse and alcoholism; occupational medicine; clinical toxicology; medical education. *Mailing Add:* Dept Pharmacol 4301 W Markham Little Rock AR 72201

MCMILLAN, DONALD ERNEST, b San Francisco, Calif, Dec 13, 31; m 56; c 6. ENDOCRINOLOGY, METABOLISM. *Educ:* Stanford Univ, AB, 53, MD, 57. *Prof Exp:* Intern med, USPHS, San Francisco, Calif, 57-58, dir metab serv & assoc chief med & endocrinol, 57-68, resident med, 61-63; chief endocrine & diabetes clin, Santa Barbara Gen Hosp, 68-82; DIR RES, HAL B WALLIS RES FACIL, EISENHOWER MED CTR, 82- *Concurrent Pos:* Resident med, USPHS, New Orleans, 59-60; fel endocrinol & metab, Univ Calif, San Francisco, 63-65; at USPHS Hosp, San Francisco, 68-79; dir diabetes res, Sansum Med Res Found, 68-82, lab dir, 72-82; adv, Lawson-Wilkins Pediat Endocrine Soc, 76; mem, res comt, Am Diabetes Asn, 80-83; spec reviewer, spec study sect, NIH, 80-81; clin prof med, Univ Southern Calif, 80-; mem, sci adv comt, Juv Diabetes Found, 82-85. *Honors & Awards:* Res Prize, USPHS Clin Soc, 63; Mary Jane Kugel Award, 82. *Mem:* Fel Am Col Physicians; Sigma Xi; Am Diabetes Asn (pres, 82-83); Europ Asn Study Diabetes; Am Physiol Soc; Endocrine Soc; Microcirculatory Soc; Biorheology Soc. *Res:* Hemorheology; atherogenesis; complications among long-standing diabetics. *Mailing Add:* 12901 N Bruce B Downs Blvd Box 45 Tampa FL 33612

MACMILLAN, DOUGLAS CLARK, b Dedham, Mass, July 15, 12; m 39; c 2. NAVAL ARCHITECTURE. *Educ:* Mass Inst Technol, BS, 34. *Prof Exp:* Engr, Fed Shipbuilding & Dry Dock Co, Kearny, NJ, 34-41; from chief marine engr to tech mgr, George G Sharp, 41-51, pres, 51-69, chmn bd, 69-70; RETIRED. *Concurrent Pos:* Mem bd dirs, Atomic Indust Forum, 65-67; consult naval archit, 70-; asst to gen mgr & consult, Quincy Shipbuilding Div, Gen Dynamics Corp, 72-77. *Honors & Awards:* Elmer A Sperry Medal, Am Soc Mech Engrs & David W Taylor Gold Medal, Soc Naval Architects & Marine Engrs, 69. *Mem:* Nat Acad Eng; fel Soc Naval Architects & Marine Engrs (hon vpres, vpres, 57-72); Am Soc Naval Engrs. *Mailing Add:* Box 834 Colony Dr East Orleans MA 02643

MCMILLAN, EDWIN MATTISON, nuclear physics, particle accelerators; deceased, see previous edition for last biography

MCMILLAN, GARNETT RAMSAY, b Madison, SC, June 11, 32; m 58. PHOTOCHEMISTRY. *Educ:* Univ Ga, BS, 53; Univ Rochester, PhD(chem), 58. *Prof Exp:* Res chemist, Celanese Corp Am, 57-62; res assoc, Ohio State Univ, 62-64; from asst prof to assoc prof, 64-73, prof chem, 73-77, PROF PHOTOCHEM, CASE WESTERN RESERVE UNIV, 77- *Res:* Reaction kinetics. *Mailing Add:* Dept of Chem Case Western Reserve Univ 2040 Adelbert Rd Cleveland OH 44106

MCMILLAN, GREGORY KEITH, b Brooklyn, NY. CONTROL SYSTEM DESIGN, DYNAMIC SIMULATION. *Educ:* Univ Kans, BS, 69; Univ Mo, Rolla, MS, 76. *Prof Exp:* Engr, Monsanto Chem Co, 69-74, sr engr, 74-76, eng specialist, 76-78, prin engr, 78-81, fel, 81-87, FEL, MONSANTO CHEM CO, 87- *Concurrent Pos:* Educ chmn, Instrument Soc Am, St Louis, 83-84, nat lectr, 86-; consult, Don H Munger & Co, 87. *Mem:* Instrument Soc Am (pres, 85-86). *Res:* Technical books, humorous books and articles on central loop performance, pH control, compressor control, biochemical control, how to become an instrument engineer. *Mailing Add:* Monsanto Chem Co F2W6 800 N Lindbergh St Louis MO 63167

MCMILLAN, HARLAN L, b Cabot, Ark, Sept 28, 26; m 48; c 4. BIOLOGY. *Educ:* Col Ozarks, BS, 50; Univ Ark, MS, 55; Purdue Univ, PhD(entom), 60. *Prof Exp:* Biol aide, Ark State Bd Health, 50-52, malaria control supvr, 52-53; med entomologist, Tech Develop Labs, Commun Dis Ctr, USPHS, Ga, 60-61; assoc prof biol, Col Ozarks, 61-64, prof & dean of men, 64-68, head dept biol, 61-69, chmn div natural sci & math, 68-69; assoc prof biol, 69-72, chmn dept, 72-73, dean sch arts & sci, 72-78, PROF BIOL, ARK TECH UNIV, 78- *Mem:* AAAS; Am Soc Allied Health Prof. *Res:* Culture of chick embryos in artificial media; electrophoresis studies of proteins in cultured chick embryo cells. *Mailing Add:* Dept of Biol Sci Ark Tech Univ Russellville AR 72801

MCMILLAN, HARRY KING, b Columbia, SC, Jan 27, 30; m 51, 81; c 3. MECHANICAL ENGINEERING. *Educ:* Univ SC, BS, 51; NC State Col, MS, 58; Purdue Univ, PhD(two phase flow), 63. *Prof Exp:* Instr mech eng, NC State Col, 54-59; asst, Purdue Univ, 59-62; assoc prof, 62-77, PROF MECH ENG, UNIV SC, 77- *Mem:* Sigma Xi. *Res:* Radiant and convective heat transfer. *Mailing Add:* 8009 Exeter Ln Columbia SC 29223-2523

MACMILLAN, J(OHN) H(ENRY), b Providence, RI, Dec 9, 28; m 51; c 3. NUCLEAR ENGINEERING. *Educ:* Mass Inst Technol, BS, 50; Univ London, DIC, 51. *Prof Exp:* Res engr, Boeing Airplane Co, 51-57; asst sect chief reactor eng, Atomic Energy Div, Babcock & Wilcox Co, 57-60, chief, Preliminary Design Sect, 60-61, asst mgr, Savannah Nuclear Power, 61-62, asst mgr, Appl Develop Dept, 62-65, mgr reactor eng, 65-66, eng mgr, Nuclear Power Generation Dept, 66-71, gen mgr, Reactor Dept, 71-73, vpres, Nuclear Power Generation Dept, 75-80, sr vpres & group exec, Nuclear Power Group, 80-90; DIR, LOW-LEVEL RADIOACTIVE WASTE MGT AUTHORITY, NC, 90- *Mem:* Am Nuclear Soc; Atomic Indust Forum; Am Nuclear Energy Coun. *Res:* Nuclear reactor technology, especially fluid flow, heat transfer, nuclear physics and transient reactor characteristics. *Mailing Add:* 116 W Jones St Raleigh NC 27603-8003

MCMILLAN, JAMES ALEXANDER, b Atascadero, Calif, Dec 18, 41; m 65; c 1. NEUROPHYSIOLOGY. *Educ:* Univ Calif, Davis, BS(vet sci), 63, BS(animal husb), 65, MS, 70, PhD(physiol), 72. *Prof Exp:* from asst prof to assoc prof, 73-88, PROF PHYSIOL, MONT STATE UNIV, 88- *Concurrent Pos:* NIH training grant, Sch Med, Univ Wash, 72-73; hon fel, Univ Edinburgh, 80-81. *Mem:* AAAS; Soc Neurosci; Am Physiol Soc. *Res:* Spinal cord and brain stem integration of motor and sensory functions; pain perception in humans. *Mailing Add:* Dept Biol Mont State Univ Bozeman MT 59717

MACMILLAN, JAMES G, b Bellingham, Wash, Feb 16, 42; m 64; c 2. ORGANIC CHEMISTRY. *Educ:* Western Wash State Col, BA, 64; Ohio State Univ, PhD(org chem), 69. *Prof Exp:* Fel org chem, Univ Calif, Berkeley, 69-70; lectr chem, Western Wash State Col, 70-71; lectr org chem, Ohio State Univ, 71-72; asst prof, 72-77, ASSOC PROF ORG CHEM, UNIV NORTHERN IOWA, 77- *Concurrent Pos:* Grants, Univ Northern Iowa, 73-78 & Res Corp, 77- *Mem:* Am Chem Soc; AAAS; Sigma Xi. *Res:* The synthesis of natural products of medicinal interest; the total synthesis of fungal metabolites. *Mailing Add:* 1419 Laurel Cr Cedar Falls IA 50613-3451

MACMILLAN, JOSEPH EDWARD, b Richmond, VA, Mar 3, 30; m 58; c 3. PHYSICAL CHEMISTRY, BIOCHEMISTRY. *Educ:* The Citadel, BS, 51; Univ SC, MS, 54; Cornell Univ, PhD(phys chem), 69. *Prof Exp:* Chemist, E I du Pont de Nemours, Inc, 53-54 & Allied Chem Corp, 54-56; LAB DIR CHEM, MacMILLAN RES, 56- *Mem:* Am Chem Soc; Am Oil Chemists' Soc; Nat Asn Corrosion Engrs; Cereal Chemists Asn; Am Soc Testing & Mat. *Res:* Oil chemistry; chemical corrosion research; spectrographic studies (emission, AA, IR, & UV). *Mailing Add:* 1221 Barclay Circle Marietta GA 30060-2903

MCMILLAN, JOSEPH PATRICK, b San Diego, Calif, Jan 17, 45. ECOLOGY, TOXICOLOGY. *Educ:* St Norbert Col, BSc, 67; Univ Ga, PhD(zool), 71. *Prof Exp:* NIH fel zool, Univ Tex, Austin, 71-72, asst prof, 73-75; asst prof, 75-80, ASSOC PROF, UNIV OF VI, 80- *Concurrent Pos:* Dir, Ciguatera Res Proj, 77- *Mem:* AAAS; Am Soc Zool; Am Inst Biol Sci; Int Soc Toxinology. *Res:* Toxicology and ecological physiology; ecology, toxicology and epidemiology of ciguatera fish poisoning, a circumtropical public health problem. *Mailing Add:* Div Sci & Math Univ of the VI St Thomas VI 00802

MCMILLAN, MALCOLM, b Victoria, BC, Aug 10, 36; m 57; c 3. THEORETICAL PHYSICS. *Educ:* Univ BC, BSc, 58, MSc, 59; McGill Univ, PhD(theoret nuclear physics), 61. *Prof Exp:* Nat Res Coun Can fel physics, Univ Turin, 61-62 & Cambridge Univ, 62-63; from asst prof to assoc prof, 63-

74, asst dean sci, 71, 85, PROF PHYSICS, UNIV BC, 74- *Mem:* Am Phys Soc; Can Asn Physicists. *Res:* Direct-interaction theories of relativistic particles; construction of Hamiltonians for systems of pions and nucleons at intermediate energies, pion scattering, production and absorption by nuclei at intermediate energies; unification of the classifications of triton state components; construction and application of approximate triton wave functions; Regge poles in potential scattering. *Mailing Add:* Dept Physics 6224 Agricultural Rd Vancouver BC V6T 1Z1 Can

MCMILLAN, MICHAEL LATHROP, b Detroit, Mich, Feb 1, 42; m 64; c 3. CHEMICAL ENGINEERING. *Educ:* Univ Mich, Ann Arbor, BSChE, 64, MSChE, 65; Ohio State Univ, PhD(chem eng), 70. *Prof Exp:* Res engr, 65-67, sr res engr, 70-77, SR STAFF RES ENGR, FUELS & LUBRICANTS DEPT, GEN MOTORS RES LABS, 78- *Mem:* Am Inst Chem Engrs; Am Chem Soc; Am Soc Testing & Mat; Soc Automotive Engrs; Sigma Xi. *Res:* Lubrication; rheology; non-Newtonian fluid flow; mass transfer; viscoelasticity; viscosity behavior of fluids; drag reduction; diesel fuels; wax formation; combustion. *Mailing Add:* Fuels & Lubricants Dept 12 Mile & Mound Rd Warren MI 48090

MCMILLAN, NEIL JOHN, b Souris, Man, Nov 11, 25; div; c 3. GEOLOGY, SOIL SCIENCE. *Educ:* Univ Man, BSc, 48; Univ Sask, MSc, 51; Univ Kans, PhD(geol), 55. *Prof Exp:* Geologist, Int Nickel Co, Can, summers, 48-50 & Chevron Standard, 51-52; field geologist, Gulf Oil Co, 53, Geol Surv Kans, 54 & Geol Surv Can, 55-56; sr geologist, Bur Mining Resource, Australia, 56-58; res geologist, Tenneco Oil Co, Houston, 59-71; sr geologist, Aquitane Co Can Ltd, 71-82; SR PETROL RES GEOLOGIST, GEOL SURV CAN, 82- *Concurrent Pos:* World Petrol Cong invited deleg, 79; Can deleg, Coop Offshore Resources Pac, Quangchow, China, 85. *Honors & Awards:* Medal of Merit, Can Soc Petrol Geol, 74. *Mem:* Geol Soc Australia; Geol Soc Can; Can Soc Petrol Geologists (pres, 77). *Res:* Contintental break-up as it relates to basin development and oil exploration; comparison of sediments on each side of the North Atlantic; provenance of clastics in the Labrador Sea; sedimentary basin classification and resources; fossil forests and fossil soils of the tertiary rocks of Canadian Arctic. *Mailing Add:* 211 Scarboro Ave SW Calgary AB T3C 2H4 Can

MACMILLAN, NORMAN HILLAS, b Montrose, Scotland, June 20, 41. MATERIALS SCIENCE. *Educ:* Cambridge Univ, BA, 63, MA, 67, PhD(metall), 69; Harvard Univ, MA, 64. *Prof Exp:* Res scientist, Res Inst Advan Studies, Martin-Marietta Corp, 69-74; asst prof metall, 75-78, assoc prof metall & mat res, 78-80, SR RES ASSOC, PA STATE UNIV, 80- *Concurrent Pos:* Lectr physics, Univ Aberdeen, Scotland, 74-75. *Res:* Structure and properties of defects in crystals; mechanical properties; crystallography; interatomic bonding; erosion and wear. *Mailing Add:* Mat Res Lab Pa State Univ University Park PA 16802

MCMILLAN, ODEN J, b Atlantic City, NJ, Mar 8, 44; m 66; c 3. MECHANICAL ENGINEERING. *Educ:* Ga Inst Technol, BME, 65; Stanford Univ, MS, 66, PhD(mech eng), 70. *Prof Exp:* Assoc sr res engr, Gas Turbine Res Dept, Gen Motors Res Labs, 70-75; res engr, 75-80, PROG MGR, NIELSEN ENG & RES, INC, 80- *Mem:* Am Soc Mech Engrs; Am Inst Aeronaut & Astronaut. *Res:* Turbulence; internal flow; aerodynamics. *Mailing Add:* 1062 Amarillo Ave Palo Alto CA 94303

MCMILLAN, PAUL FRANCIS, b Edinburgh, Scotland, June 3, 55. MATERIALS CHEMISTRY, MINERAL PHYSICS. *Educ:* Univ Edinburgh, BSc Hons, 77; Ariz State Univ, PhD(chem), 81. *Prof Exp:* Fac assoc, 81-83, asst prof, 83-89, ASSOC PROF CHEM, ARIZ STATE UNIV, 89- *Concurrent Pos:* Vis dir res, Inst Mat, Univ Nantes, France, 90. *Mem:* Mineral Soc Am; Am Geophys Union; Mat Res Soc. *Res:* Vibrational spectroscopy applied to problems in mineralogy and geochemistry; structure and dynamics of silicate glasses and melts; synthesis of new materials at high pressure. *Mailing Add:* Dept Chem Ariz State Univ Tempe AZ 85287

MCMILLAN, PAUL JUNIOR, b Atlanta, Ga, Sept 13, 30; m 55; c 4. BIOCHEMISTRY, HISTOCHEMISTRY. *Educ:* Southern Missionary Col, BA, 51; Loma Linda Univ, MS, 57, PhD(biochem), 60. *Prof Exp:* Instr, 60-61, from asst prof to assoc prof, 63-79, PROF ANAT, LOMA LINDA UNIV, 79- *Concurrent Pos:* USPHS fel histochem, NIH, 61-63; res assoc, Univ Mainz, 80-81. *Mem:* AAAS; Am Asn Anat; Int Soc Stereology; Histochem Soc; Sigma Xi; Soc Quant Morphol; Am Soc Bone & Min Res. *Res:* Interactions of form and function; metabolic and endocrine interrelations at the cellular level, with reference to the morphological design of tissues; quantitative morphology; bone cell biology. *Mailing Add:* Dept Anat Sch Med Loma Linda Univ Loma Linda CA 92350

MCMILLAN, PAUL N, b Charleston, WVa, June 22, 44. CELL BIOLOGY. *Educ:* Marshall Univ, BS, 66; Duke Univ, PhD(microbiol), 71. *Prof Exp:* Res assoc exp biol, Worcester Found, 73-76; dir med electron micros, RI Hosp, 76-89; asst prof, 82-91, ASSOC PROF PATH, BROWN UNIV, 91- *Mem:* Am Soc Cell Biol; NY Acad Sci; Electron Micros Soc Am. *Mailing Add:* Cent Res Labs RI Hosp 593 Eddy St Providence RI 02903

MCMILLAN, R BRUCE, b Springfield, Mo, Dec 3, 37; m 61; c 3. ARCHAEOLOGY, BIOGEOGRAPHY. *Educ:* Southwest Mo State Col, BS, 60; Univ Mo-Columbia, MA, 63; Univ Colo, PhD(anthrop), 71. *Prof Exp:* Res assoc archaeol, Univ Mo, 64-66; from assoc cur to cur anthrop, 69-73, asst mus dir, 73-77, MUS DIR, ILL STATE MUS, 77- *Concurrent Pos:* NSF fels, Ozark Pleistocene Springs, Ill State Mus, 71-72 & 72-73; lectr, Northwestern Univ, 72-73; consult, Midwest Res Inst, 72- *Mem:* Fel AAAS; fel Am Anthrop Asn; Soc Am Archaeol; Am Asn Mus (vpres). *Res:* Late Pleistocene and post-Pleistocene environments, especially eastern North America; man's adaptation to environmental change during the early Holocene in the eastern North American Praire Peninsula. *Mailing Add:* Ill State Mus Mus Dir Ill State Mus Springfield IL 62706

MCMILLAN, ROBERT, b Pittsburgh, Pa, Nov 19, 34; m 66; c 2. IMMUNOHEMATOLOGY. *Educ:* Pa State Univ, BS, 56; Univ Pa, MD, 60. *Prof Exp:* Assoc hemat, 68-74, ASSOC MEM HEMAT & ONCOL, SCRIPPS CLIN RES FOUND, 74-, DIR, WEINGART BONE MARROW TRANSPLANTATION CTR, 80- *Concurrent Pos:* Adj asst prof hemat, Univ Calif, San Diego, 68-75, assoc adj prof, 75-; consult, Vet Admin Hosp, San Diego, 74- *Mem:* Am Soc Clin Invest; Am Soc Hemat; Soc Exp Biol & Med. *Res:* Evaluation of antiplatelet antibodies in human disease, their synthesis, site of origin and antigens; study of platelet surface proteins; synthesis of human immunoglobulins. *Mailing Add:* Scripps Clin Res Found 10666 N Torrey Pines Rd La Jolla CA 92037

MCMILLAN, ROBERT MCKEE, b Morris, Ill, Dec 18, 24; m 51; c 5. MICROSCOPY, RUBBER CHEMISTRY. *Educ:* NDak State Normal & Indust Col, BS, 48; Univ Colo, MS, 57. *Prof Exp:* Instr chem, NDak State Sch Sci, 48-53; assoc prof, Northern State Teachers Col, SDak, 53-55 & Wis State Col, Superior, 55-57; res chemist micros, 57-59, staff microscopist, 59-64, sr microscopist, 64-80, SECT HEAD, GOODYEAR TIRE & RUBBER CO, 80- *Mem:* AAAS. *Res:* The microscopy of rubber, rubber chemicals, rubber products, plastics and fibers. *Mailing Add:* c/o Barbara Judy Goodyear Tire 142 Goodyear Blvd Akron OH 44305

MACMILLAN, ROBERT S(MITH), b Los Angeles, Calif, Aug 28, 24; m 62; c 1. ELECTRICAL ENGINEERING. *Educ:* Calif Inst Technol, BS, 48, MS, 49, PhD(elec eng, physics), 54. *Prof Exp:* Res engr, Jet Propulsion Lab, Calif Inst Technol, 53-55, asst prof elec eng, 55-58; assoc prof, Univ Southern Calif, 58-70; mem sr tech staff, Litton Systs, Inc, 69-79, assoc dir eng, 79-84; dir eng, Litton Data Command Systs, 84-89; PRES, MACMILLAN GROUP, 89- *Concurrent Pos:* Consult, Space Technol Labs, Thompson Ramo Wooldridge, Inc, 56-59 & US Air Force, 57-76. *Mem:* Am Phys Soc; Sigma Xi; Inst Elec & Electronic Engrs. *Res:* Ionosphere and the earth's upper atmosphere; ionospheric radio-wave propagation of very-low-frequency radio waves; quantum electronics; communications antennas; electromagnetic wave propagation. *Mailing Add:* 350 Starlight Crest Dr La Canada CA 91011

MCMILLAN, ROBERT THOMAS, JR, b Miami, Fla, June 1, 34; m 54; c 2. PLANT PATHOLOGY. *Educ:* Univ Miami, BS, 61, MS, 64; Wash State Univ, PhD(plant path), 68. *Prof Exp:* Asst plant pathologist, 67-74, ASSOC PROF PLANT PATH & ASSOC PLANT PATHOLOGIST, AGR RES & EDUC CTR, UNIV FLA, 67- *Mem:* Am Phytopath Soc; Am Soc Agron; Crop Sci Soc Am; Sigma Xi. *Res:* Tissue culture, physiology, etiology and control of fruit and vegetable diseases. *Mailing Add:* Agr Res & Educ Ctr Univ Fla 18905 SW 280th St Homestead FL 33031

MCMILLAN, ROBERT WALKER, b Sylacauga, Ala, Apr 18, 35; m 55; c 3. ATMOSPHERIC EFFECTS, MILLIMETER WAVE SCIENCE & TECHNOLOGY. *Educ:* Auburn Univ, BS, 57; Rollins Col, MS, 66; Univ Fla, PhD(physics), 74. *Prof Exp:* Asst engr, Westinghouse Elec Corp, 57-61; staff engr, Martin Marietta Aerospace, Orlando, 61-76; PRIN RES SCIENTIST, GA TECH RES INST, 76- *Mem:* Optical Soc Am; sr mem Inst Elec & Electronics Engrs. *Res:* Propagation of electromagnetic radiation including attenuation and turbulence effects; millimeter wave quasi-optics, sources of radiation, measurements, radar and radiometry; millimeter wave, microwave and optical spectroscopy; infrared and visible lasers and detectors. *Mailing Add:* 6332 Queen View Lane Stone Mountain GA 30087

MCMILLAN, WILLIAM GEORGE, b Montebello, Calif, Oct 19, 19; m 46; c 3. CHEMICAL PHYSICS. *Educ:* Univ Calif, Los Angeles, BA, 41; Columbia Univ, MA, 43, PhD(chem), 45. *Prof Exp:* Teaching asst chem, Columbia Univ, 41-44; Guggenheim fel, Inst Nuclear Studies, Chicago, 46-47; from asst prof to assoc prof, 47-58, chmn dept, 59-65, PROF CHEM, UNIV CALIF, LOS ANGELES, 58- *Concurrent Pos:* Chmn ad hoc group radiation effects, US Air Force, US Navy & Dir, Defense Res & Eng, 63-66, mem, 66-; chmn advan res projs agency defense sci sem, Univ Calif, Los Angeles, 64-66; chmn, Joint Chiefs-of-Staff Panel on Nuclear Test Ban, 65-66; vchmn sci adv comt, Defense Intel Agency, 65-71, mem, 71-; sci adv to Comdr, US Mil Assistance Command, Vietnam, 66-68; mem, Army Sci Adv Panel, 69- & Oak Ridge Nat Lab Adv Group on Civil Defense, 71-; chmn, Nat Acad Army Countermine Study Group, 71-72. *Honors & Awards:* Distinguished Civilian Serv Award, Dept of Army, 68; Distinguished Pub Serv Award, Dept of Defense, 69; Knight, Nat Order of Vietnam, 69. *Mem:* AAAS; Am Phys Soc; Am Chem Soc; Sigma Xi. *Res:* Statistical and quantum mechanics of small molecules; adsorption; equation of state; spectroscopy at high pressure; electrolytes. *Mailing Add:* Dept of Chem Univ of Calif Los Angeles CA 90024

MACMILLAN, WILLIAM HOOPER, b Boston, Mass, Oct 21, 23; m 48; c 3. PHARMACOLOGY. *Educ:* McGill Univ, BA, 48; Yale Univ, PhD, 54. *Prof Exp:* From instr to prof pharmacol, Col Med, Univ Vt, 54-76, chmn dept, 62-63, dean grad col, 63-69 & 71-76; PROF BIOL & DEAN GRAD SCH, UNIV ALA, 76- *Concurrent Pos:* USPHS fel, Oxford Univ, 58-59; Ford Found sci adv, Haile Selassie I Univ, Addis Ababa, Ethiopia, 69-71; consult, African Am Inst, 71-; educ consult, World Bank, 72-74; mem bd gov, Univ Press New Eng, 72-76; consult, New Eng Asn Schs & Cols, 72-76, Southern Asn Cols & Schs, 77-, Univ Ala Press, 81-; pres, Conf Southern Grad Schs, 82-83; bd dir, Coun Grad Sch, 86-; bd dir, Oak Ridge Assoc Univ, 88- *Mem:* Sigma Xi; AAAS; Am Soc Pharmacol & Exp Therapeut; NY Acad Sci. *Res:* Autonomic pharmacology; mechanism of drug action; graduate studies in biomedical sciences. *Mailing Add:* Grad Sch Univ of Ala Tuscaloosa AL 35487-0118

MCMILLEN, JANIS KAY, b El Dorado, Kans, Oct 21, 37. VIROLOGY. *Educ:* Trinity Univ, Tex, BS, 59; Univ Kans, PhD(microbiol), 71. *Prof Exp:* Res technician virol, Pitman-Moore Div, Dow Chem Co, 59-60 & Dept Pediat, Univ Kans Med Ctr, 60-66; Nat Inst Allergy & Infectious Dis fel, Div Biol, Kans State Univ, 71-74, res assoc, 74-76; sr res virologist, Jensen-Salsbery Labs Div, Richardson-Merrell, Inc, 76-80; head, Virol Res, Wellcome Res Labs, Burroughs-Wellcome Co, 80-82, dir, Biol Res &

Develop, 82-85; VPRES, PROD DEVELOP & REGIST, SYNTROVET INC, 85- *Mem:* NY Acad Sci; Sigma Xi; Am Soc Microbiol; US Animal Health Asn. *Res:* Viral pathogenesis; molecular mechanisms of viral infections. *Mailing Add:* 10104 Hemlock Dr Overland Park KS 66212

MACMILLEN, RICHARD EDWARD, b Upland, Calif, Apr 19, 32; m 80; c 1. PHYSIOLOGICAL ECOLOGY, MAMMALOGY. *Educ:* Pomona Col, BA, 54; Univ Mich, MS, 56; Univ Calif, Los Angeles, PhD(zool), 61. *Prof Exp:* Res zoologist, Univ Calif, Los Angeles, 59-60; from instr to assoc prof zool, Pomona Col, 60-68; assoc prof biol, 68-72, chmn, Dept Ecol & Evolutionary Biol, 72-74, 84-90, PROF BIOL, UNIV CALIF, IRVINE, 72- *Concurrent Pos:* NSF res grants, 61-79; Fulbright res scholar, Monash Univ, Australia, 66-67; vis prof zool, Univ New South Wales, Australia, 74-75; mem, Panel Pop Biol & Physiol Ecol, NSF, 76-79; assoc ed, J Mammal, 78-80; vis cur ecol, Australian Mus, 83; vis res scholar, Flinder Univ S Australia, 85, 89. *Mem:* Fel AAAS; Am Ornith Union; Ecol Soc Am; Am Soc Mammalogists; Australian Ecol Soc. *Res:* Assesments of the ways that higher vertebrates acquire and utilize limiting resources (primarily water and energy) with maximal efficiency, particularly nectarivorous birds and granivorous birds and rodents. *Mailing Add:* Dept of Ecol & Evolutionary Biol Univ of Calif Irvine CA 92717

MACMILLIAN, STUART A, b Edmonton, Alta, Oct 19, 51; m 83; c 2. SOFTWARE SYSTEMS, GENERAL COMPUTER SCIENCES. *Educ:* Univ Alta, BS, 74; Stanford Univ, MS, 81, PhD(math rec res), 84. *Prof Exp:* Sr staff engr, Corp Technol Ctr, FMC, 84-87; mgr, Knowledge Systs, 87-88, mgr, Prog Environ, 88-90, MGR BUS DEVELOP, SUN MICROSYSTS, 90- *Concurrent Pos:* Bd mem, Sun Collab Res Bd, 90- *Res:* Development of software systems that serve as personal assistants and advanced software development environments. *Mailing Add:* 341 O'Connor St Menlo Park CA 94025

MCMILLIN, CARL RICHARD, b Warren, Ohio, Aug 4, 46; m 70; c 3. ORTHOPEDIC RESEARCH, BIOCOMPATIBILITY. *Educ:* Gen Motors Inst, BME, 69; Case Western Reserve Univ, MS, 71, PhD(macromolecular sci), 74. *Prof Exp:* Prod engr, Packard Elec Div, Gen Motors Corp, 68-69; res assoc polymer stress cracking, Queen Mary Col, Univ London, 71-72; NIH res fel, Case Western Reserve Univ & vis scientist, Artificial Organs Div, Cleveland Clin, 74-75; sr res chemist, 75-78, contract mgr, Monsanto Res Corp, 78-83; assoc prof, dept biomed eng & dir Cardiovasc Lab, Inst Biomed Eng, Univ Akron, Ohio, 83-89; sr scientist, Inst Biomed Eng, Univ Akron, Ohio, 89-90, DIR, R & D, ACROMED CORP, 90- *Concurrent Pos:* Consult, adj staff, Artificial Organs Div, Cleveland Clin Found, 84- *Mem:* Am Chem Soc; Sigma Xi; Soc Biomat; Soc Advan Mat Process Eng. *Res:* Biomaterials, biocompatibility and design of prosthetic devices with emphasis on spinal orthopedics; rubber fatigue; materials development, selection, characterization and evaluation for in vivo applications. *Mailing Add:* 6099 Warblers Roost Brecksville OH 44141-1781

MCMILLIN, CHARLES W, b Indianapolis, Ind, Aug 22, 32. WOOD SCIENCE & TECHNOLOGY. *Educ:* Purdue Univ, BS, 54; Univ Mich, MWT, 57, PhD(wood sci), 69. *Prof Exp:* Assoc mech engr, Am Mach & Foundry Co, 57-64; res coordr, Southern Pine Asn, 64-65; PRIN WOOD SCIENTIST, SOUTHERN FOREST EXP STA, FOREST SERV, USDA, 65- *Concurrent Pos:* Chmn, Pulp & Paper Tech Comt, Forest Prod Res Soc, 71, 72, chmn-elect, Div C Processes, 71, chmn, 72, chmn-elect, Mid-South Sect, 73, chmn, 73. *Honors & Awards:* Wood Award, Wood & Wood Prod Mag, 57. *Mem:* Forest Prod Res Soc; Sigma Xi; Soc Wood Sci & Technol; Int Asn Wood Anatomists; Tech Asn Pulp & Paper Indust. *Res:* Wood machining; mechanical pulping; modification of properties; characterization; applications of scanning electron microscopy to wood science. *Mailing Add:* Forest Prod Res S Forest Exp Sta 2500 Shreveport Hwy Pineville LA 71360

MCMILLIN, DAVID ROBERT, b East St Louis, Ill, Jan 1, 48; m 74; c 2. INORGANIC CHEMISTRY. *Educ:* Knox Col, BA, 69; Univ Ill, PhD(inorg chem), 73. *Prof Exp:* Fel, Calif Inst Technol, 73-74; from asst prof to assoc prof, 75-85, PROF INORG CHEM, PURDUE UNIV, 85- *Concurrent Pos:* NIH fel, Calif Inst Technol, 74. *Mem:* Am Chem Soc; Inter-Am Photochem Soc; Sigma Xi. *Res:* Chemical and physical studies of copper proteins and related model systems; transition metal chemistry; photochemistry; photoluminescence spectroscopy; electron paramagnetic resonance spectroscopy. *Mailing Add:* Dept Chem Purdue Univ West Lafayette IN 47907

MCMILLIN, JEANIE, b Spartanburg, SC, Sept 26, 39; m 87; c 2. BIOCHEMISTRY, CHEMISTRY. *Educ:* Converse Col, BA, 61; Univ NC, Chapel Hill, PhD(biochem), 67. *Hon Degrees:* DSc, Converse Col, 88. *Prof Exp:* Instr biochem, Univ NC, Chapel Hill, 67-68; res assoc, Dept Med, Cornell Univ & Inst Muscle Dis, Inc, 68-69; res assoc, Baylor Col Med, 69-70, from res instr to instr, 70-72, res asst prof myocardial biol, 72-73, asst prof cell biophys, 73-77, from asst prof to assoc prof med & biochem, 77-85, prof med, 85-86, prof med & cell biol, 86-89, head, Lab Cardic Biochem, Univ Ala, 86-89; PROF PATH, UNIV TEX, HOUSTON, 89- *Concurrent Pos:* USPHS grant, Univ NC, Chapel Hill, 67-68; Musclar Dystrophy Asn Am grant, Cornell Univ & Inst Muscle Dis, Inc, 68-69, 80-; Tex Med Ctr grant myocardial biol, Baylor Col Med, 69-70; Tex Heart Asn grant, 70-72; NIH grants, 75-79, 78-81, 81-86, 85-89 & 87-91; mem, Int Study Group Res Cardiac Metab; consult, NIH, 79-82; mem, Cardiovasc Pulmonary Study Sect, NIH, 82-86; mem Cardiol Adv Comn, NIH, 87-91, chair, basic sci subcomt, 90-91; guest ed, Circulation Suppl, ed bds, Am J Physiol, 86-88, J Molecular Cell Cardiol, 87- *Mem:* Am Soc Biol Chemists; NY Acad Sci; Sigma Xi; Biophys Soc; Am Physiol Soc; fel Cardivasc Soc. *Res:* Effect of myocardial ischemia on the mitochondrial functions of energy production and fatty acid transport and oxidation; carnitine palmityltransferase system; carnitine acylcarnitine translocase. *Mailing Add:* Univ Texas Health Sci Ctr Houston TX 77030

MCMILLION, LESLIE GLEN, SR, b Nallen, WVa, June 24, 30; m 54; c 2. HYDROLOGY. *Educ:* Marshall Univ, BA, 52; Mich State Univ, MS, 57. *Prof Exp:* Geologist, US Geol Surv, 54-56; geologist, Tex Water Comn, 56-59, div dir, Tex Water Develop Bd, 59-65; res hydrologist, US Environ Protection Agency, 66-; RETIRED. *Concurrent Pos:* Prin partic, Tex-US Study Comn, 59-60; mem, Tex Govr's Water Pollution Adv Coun, 60-61, Working Group Preparation Regulations Safe Drinking Water Act, 75-76 & chmn, Nat Aquifer Protection Comn, 67-70. *Mem:* Nat Water Well Asn; Am Water Resources Asn; Geol Soc Am; Am Water Works Asn. *Mailing Add:* 7650 Buck Rd Las Vegas NV 89123

MCMINN, CURTIS J, b Lexington, Tenn, Oct 25, 29; m 54; c 3. CHEMICAL ENGINEERING. *Educ:* Middle Tenn State Univ, BS, 51; Vanderbilt Univ, MS, 55. *Prof Exp:* Design engr, Redstone Arsenal, 51-52; proj engr, 55-60, proj dir, 60-75, res dir, reduction res div, 75-84, SR SCIENTIST, MFG TECH LAB, REYNOLDS METALS CO, 84- *Mem:* Am Inst Chem Engrs; Metall Soc; Sigma Xi. *Res:* Oxidation of P-Xylene to terephthalic acid; reduction of alumina electrochemically; design of reduction cells; use of new materials in reduction cells; new processes for production of primary aluminum. *Mailing Add:* 2125 Chickasaw Dr Florence AL 35630

MCMINN, TREVOR JAMES, b Salt Lake City, Utah, Jan 23, 21. MATHEMATICS. *Educ:* Univ Utah, BA, 42; Univ Calif, Berkeley, PhD(math), 55. *Prof Exp:* Instr math, Univ Rochester, 50, Univ Calif, Riverside, 54-55 & Univ Calif, Berkeley, 55-56; asst prof, Univ Wash, 56-63; assoc prof, 63-69, PROF MATH, UNIV NEV, RENO, 69- *Mem:* Am Math Soc. *Res:* Measure theory; real analysis. *Mailing Add:* Dept of Math Univ of Nev Reno NV 89557

MCMORRIS, F ARTHUR, b Lawton, Okla, Sept 17, 44; m 85. NEUROCHEMISTRY, CELL BIOLOGY. *Educ:* Brown Univ, BA, 66; Yale Univ, PhD(biol), 72. *Prof Exp:* Teaching asst, Brown Univ, 66 & Yale Univ, 67-70; res assoc biol, Mass Inst Technol, 72-74; asst prof, 74-75, ASSOC PROF BIOL, WISTAR INST, UNIV PA, 85-, ASSOC DIR WISTAR PROG, MULTIPLE SCLEROSIS RES CTR, 86- *Concurrent Pos:* asst prof human genetics, Sch Med, Univ Pa, 74-75, assoc prof grad groups molecular biol, genetics, neurosci & cell biol, 75-, asst prof neurol, 84, assoc prof neurol, 85-; fel,NIH, 72-74; Huntington Dis Fel, Hereditary Dis Found, 75-76; Mary Jennifer Selznick Fel, 78; mem prog comt, Am Soc Neurochem, 86-88, Winter Conf Brain Res, 89-91; mem publ & educ comt, Am Soc Neurochem, 87-89; ad hoc mem, Neurol Sci 1 Study Sect, NIH, mem, 88-91. *Honors & Awards:* Sigma Xi Award, 73; Tarbox Distinguished Neuroscientist Award, 80. *Mem:* Am Soc Cell Biol; Am Soc Neurochem; Soc Develop Biol; Endocrine Soc; Soc Neurosci; NY Acad Sci; AAAS; Int Soc Neurochem. *Res:* Regulation of gene activity and differentiated cell function in the nervous system; chemistry and development of oligodendrocytes and myelin; multiple sclerosis. *Mailing Add:* The Wistar Inst 36th St at Spruce Philadelphia PA 19104

MCMORRIS, FRED RAYMOND, b Gary, Ind, Aug 28, 43; m 65; c 2. MATHEMATICS, BIOMATHEMATICS. *Educ:* Beloit Col, BS, 65; Univ Calif, Riverside, MA, 66; Univ Wis-Milwaukee, PhD(math), 69. *Prof Exp:* Asst prof, 69-74, assoc prof, 74-79, PROF MATH, BOWLING GREEN STATE UNIV, 79- *Concurrent Pos:* Nat Inst Gen Med Sci fel, Biomath Prog, NC State Univ, 71-73; vis assoc prof, Dept Math Sci, Clemson Univ, 78; vis prof dept math, Univ SC, 81. *Mem:* Am Math Soc; Math Asn Am; Soc Indust & Appl Math; Soc Math Biol. *Res:* Discrete mathematics; mathematical biology; numerical taxonomy. *Mailing Add:* Dept of Math Univ Louisville Louisville KY 40292

MCMULLEN, BRYCE H, b Tarkio, Mo, Apr 18, 21; m 44; c 2. CHEMICAL ENGINEERING. *Educ:* Centre Col, AB, 43; Ohio State Univ, MSc, 46, PhD(chem eng), 49. *Prof Exp:* Res chemist, Titanium Div, Nat Lead Co, 50-52, from develop engr to sr develop engr, 52-57, staff asst, 57-62, supvr develop & eng dept, 62-69, chief engr feed mat dept, 70, mgr, 71-77; SR DEVELOP ENGR, GLIDDEN PIGMENTS, 77- *Mem:* Sigma Xi; Am Chem Soc. *Res:* Catalyst for olefin polymerization; process for recovery of waste sulfuric acid; chloride process for manufacture of titanium dioxide. *Mailing Add:* 3901 Greenway Baltimore MD 21218

MACMULLEN, CLINTON WILLIAM, organic chemistry; deceased, see previous edition for last biography

MCMULLEN, JAMES CLINTON, b Alton, Ill, July 6, 42; m 63; c 3. CHEMISTRY. *Educ:* Wis State Univ-Superior, BS, 65; Univ SDak, PhD(chem), 69. *Prof Exp:* Asst prof, 69-76, assoc prof, 76-81, PROF CHEM, ST CLOUD STATE UNIV, 81- *Mem:* Am Chem Soc. *Res:* Analytical instrumentation; computer applications to chemistry. *Mailing Add:* 2532 De Soto St Cloud MN 56301-5991

MCMULLEN, JAMES ROBERT, b Clinton, Ind, May 22, 42; div; c 1. MICROBIOLOGY, BIOCHEMISTRY. *Educ:* Ind State Univ, BS, 64; Univ Wis-Madison, MS, 66. *Prof Exp:* Res microbiologist, Com Solvents Corp, 66-75; res microbiologist, Int Minerals & Chem Corp, 75-87; PRES, PITMAN-MOORE, INC, 87- *Mem:* Am Soc Microbiol; Sigma Xi; Soc Indust Microbiol. *Res:* Industrial fermentations; fungal metabolism; enzymes; cellulose utilization; rumen microbiology; microbiology of silage; fermentation with genetically engineered micoorganisms. *Mailing Add:* 845 Barton Ave Terre Haute IN 47803

MCMULLEN, ROBERT DAVID, b Medicine Hat, Alta, Feb 25, 30; m 53; c 2. ENTOMOLOGY, ECOLOGY. *Educ:* Univ Alta, BSc, 53; Wash State Univ, MSc, 60; Univ Calif, Berkeley, PhD(entom), 64. *Prof Exp:* Tech officer entom, Defence Res Bd Can, Suffield Exp Sta, Ralston, Alta, 53-56; res scientist entom, Res Sta, Agr Can, Harrow, Ont, 56-64; res scientist entom, Res Sta, Agr Can, Summerland, BC, 64-89; RETIRED. *Concurrent Pos:* Adj prof, Simon Fraser Univ, 74-; assoc ed, Can Entomologist, Entom Soc Can, 78- *Mem:* fel Entom Soc Can; Entom Soc Am. *Res:* Integrated pest management of insects and mites on deciduous fruits; vectors of diseases of deciduous fruit crops. *Mailing Add:* RR 4 Site 92-C1 Summerland BC V0H 1Z0 Can

MCMULLEN, WARREN ANTHONY, b Faulkton, SDak, Oct 22, 07; m 29; c 3. ORGANIC CHEMISTRY. *Educ:* Greenville Col, BS, 28; Univ Nebr, MA, 36; NY Univ, PhD(sci ed), 60. *Prof Exp:* Teacher high sch, Nebr, 28-43; teacher chem & physics, Cent Col, Kans, 43-45; from asst prof to prof, 45-73, EMER PROF CHEM, GREENVILLE COL, 73- *Concurrent Pos:* prin, 75-77, sci teacher, Oakdale Christian High School, Ky, 75-80. *Mem:* Am Chem Soc. *Mailing Add:* 5892 Moriah Pl Lakeland FL 33809

MCMURCHY, ROBERT CONNELL, geology; deceased, see previous edition for last biography

MCMURDIE, HOWARD FRANCIS, b Detroit, Mich, Feb 5, 05; m 28; c 3. CHEMISTRY. *Educ:* Northwestern Univ, BS, 28. *Prof Exp:* Jr chemist, Bur of Standards, Washington, DC, 28-33, Calif, 33-35, jr petrographer, DC, 35-44, gen phys scientist, 44-48, chief sect crystallog, 48-65, RES FEL, INT CTR DIFFRACTION DATA, NAT INST STANDARDS & TECHNOL, 66- *Concurrent Pos:* Consult, Am Ceramic Soc. *Mem:* Fel Am Ceramic Soc; fel Mineral Soc Am; Am Crystallog Asn. *Res:* Crystal chemistry; phase equilibrium of refractory oxides; high temperature x-ray diffraction; chemical analysis by x-ray diffraction; effect of heat on crystals; data compilation. *Mailing Add:* Nat Inst Standards & Technol Gaithersburg MD 20899

MCMURPHY, WILFRED E, b Lamont, Okla, Aug 3, 34; m 59; c 2. AGRONOMY. *Educ:* Okla State Univ, BS, 56, MS, 59; Univ Kans, PhD(agron), 63. *Prof Exp:* Res asst agron, Kans State Univ, 59-62; asst prof, SDak State Univ, 62-64; asst prof, 64-70, assoc prof, 70-76, PROF AGRON, OKLA STATE UNIV, 76- *Mem:* Am Soc Agron; Crop Sci Soc Am; Soc Range Mgt; Am Forage & Grassland Coun. *Res:* Range and pasture management; pasture production and ecology. *Mailing Add:* Dept of Agron Okla State Univ Stillwater OK 74074

MCMURRAY, BIRCH LEE, b Polk Co, NC, Oct 18, 31; m 56; c 3. VETERINARY MEDICINE, AGRICULTURE. *Educ:* NC State Univ, BS, 54; Univ Ga, DVM, 57. *Prof Exp:* Dir lab, Fla Dept Agr, 58-60; vet, Dr Robert E Lee Vet Hosp, 60-62; res vet, 62-64, mgr vet res, 64-72, dir int feed res, 72-78, PRIN VETERINARIAN POULTRY FIELD RES & TECH SERV, CENT SOYA CO, INC, 78- *Mem:* Am Vet Med Asn; Am Asn Avian Pathologists. *Res:* Programmed preventive veterinary medicine; preventive programs for coccidiosis, fowl cholera, Marek's disease, respiratory diseases of cattle, enteridides of swine, mastitis-metritis complex; interrelations of disease, nutrition and management. *Mailing Add:* 115 Clifton Dr Athens GA 30606

MCMURRAY, DAVID CLAUDE, b Greenville, Tex, May 3, 27; m 52; c 1. MECHANICAL ENGINEERING. *Educ:* Tex Tech Col, BSME, 50; Univ Wis, PhD(mech eng), 66. *Prof Exp:* Installation engr, Nordberg Mfg Co, 50-52, prod designer, 52-55; designer, W C Heath Assocs Inc, 55-57; mech engr, A O Smith Corp, 57-60, head eng anal, 60-62, consult engr, Res Div, 62-82, mgr, Consult Serv, Data Systs Div, 82-87; RETIRED. *Concurrent Pos:* Lectr, Marquette Univ, 66. *Res:* Industrial research and development in the areas of heat transfer, fluid flow, machinery and processes. *Mailing Add:* 4050 River Edge Circle Milwaukee WI 53209

MCMURRAY, DAVID N, MICROBIOLOGY, NUTRITION. *Educ:* Univ Wis, PhD(med microbiol), 72. *Prof Exp:* ASSOC PROF MED MICROBIOL & IMMUNOL, COL MED, TEX A&M UNIV, 82- *Mailing Add:* Microbiol Dept Tex A&M Univ Col Med College Station TX 77843

MCMURRAY, LOREN ROBERT, b Topeka, Kans, June 15, 31; m 58; c 2. MATHEMATICS, COMPUTER SIMULATION. *Educ:* Washburn Univ, BS, 53; Iowa State Univ, MS, 55; Univ Ill, MS, 57, PhD(math), 58. *Prof Exp:* Mem tech staff inertial guidance, 58-61, supvr, 61-62, MEM TECH STAFF ELECTRONIC DESIGN ANALYSIS AUTONETICS, ROCKWELL INT, 62- *Mem:* Inst Elec & Electronics Engrs. *Res:* Adaptive digital filters; modeling electronic components; algorithms for computers. *Mailing Add:* Rockwell Int Strategic Systs Div 3370 Miraloma Ave Anaheim CA 92803

MCMURRAY, WALTER JOSEPH, b Montague, Mass, Aug 22, 35; m 60; c 4. BIOCHEMISTRY. *Educ:* Amherst Col, AB, 58; Univ Ill, PhD, 62; Univ New Haven, MBA, 76; Univ Conn, JD, 81. *Prof Exp:* Res assoc chem, Mass Inst Technol, 63-64; instr med, 65-66, asst prof pharmacol, 66-67, from asst prof to assoc prof health sci resources, 67-75, res assoc, 75-81, sr res assoc, Sch Med, 81-83, RES SCIENTIST, LAB MED, SCH MED, YALE UNIV, 84- *Mem:* Am Soc Mass Spectrometry; Am Chem Soc; AAAS. *Res:* Application of mass spectronomy to problems in biomedical research; use of computers in chemistry. *Mailing Add:* Cancer Ctr Sch Med Yale Univ New Haven CT 06510

MCMURRAY, WILLIAM COLIN CAMPBELL, b Bangor, Northern Ireland, Mar 16, 31; nat Can; m 53; c 2. BIOCHEMISTRY. *Educ:* Univ Western Ont, BSc, 53, PhD(biochem), 56. *Prof Exp:* Proj assoc, Inst Enzyme Res, Univ Wis, 56-58; asst prof cancer res & lectr biochem, Univ Sask, 58-59; asst prof biochem & Jr Red Cross Res Prof ment retardation, 59-65, assoc prof biochem, 65-70, PROF BIOCHEM, UNIV WESTERN ONT, 70-, CHMN DEPT, 83- *Concurrent Pos:* Vis scientist, Inst Animal Physiol, Agr Res Coun, Cambridge, Eng, 67-68, 86-87. *Mem:* Am Soc Biol Chem; Can Biochem Soc. *Res:* Biochemistry of the brain; inborn errors of metabolism; metabolism of differentiation; lipid metabolism; mitochondrial enzymes and oxidative phosphorylation; membrane synthesis; mitochrondrial biogenesis; hormone interactions in cultured cells. *Mailing Add:* Dept Biochem Univ Western Ont Fac Med London ON N6A 5C1 Can

MCMURRY, JOHN EDWARD, b New York, NY, July 27, 42; m 64; c 3. SYNTHETIC ORGANIC CHEMISTRY. *Educ:* Harvard Univ, BA, 64; Columbia Univ, MA & PhD(chem), 67. *Prof Exp:* From asst prof to assoc prof chem, Univ Calif, Santa Cruz, 67-75, prof, 75-80; PROF CHEM, BAKER LAB, CORNELL UNIV, 80- *Honors & Awards:* A von Humboldt Sr Scientist Award, 87. *Mem:* Fel AAAS; Am Chem Soc; Royal Soc Chem. *Res:* Natural product synthesis; new synthetic reactions. *Mailing Add:* Dept Chem Baker Lab Cornell Univ Ithaca NY 14853

MCMURRY, WILLIAM, b Los Angeles, Calif, Aug 15, 29; m 55; c 4. POWER ELECTRONICS, MAGNETIC DEVICES. *Educ:* Battersea Polytechic, BSc, 50; Union Col, MS, 56. *Prof Exp:* Test Engr, General Elect Co, 50-51; PFC, US Army Signal Corp, 51-53; elec engr, Gen Elec Co, 53-89; CONSULT ENGR, 89- *Concurrent Pos:* Lectr, Inst Elec & Electronics Engrs, 88- *Honors & Awards:* William & Newell Award, Centennial Medal, Lamme Medal, Inst Elec & Electronics Engrs, 84. *Mem:* Inst Elec & Electronics Engrs. *Res:* Electrical power converters. *Mailing Add:* PO Box 741 Schenectady NY 12301

MCMURTREY, LAWRENCE J, b Ririe, Idaho, Mar 13, 24; m 47; c 6. SPACE VEHICLE DESIGN, SPACE PROPULSION. *Educ:* Univ Colo, BS, 44. *Prof Exp:* Eng officer commun, US Navy, 43-46; instr thermodyn, Univ Utah, 49-51; propulsion engr, Boeing Co, 46-49 & 51-56, chief flight tech propulsion, 56-58, proj mgr res, 58-71, navigation mgr, Lunar Rover Proj, 71-72, proj mgr new bus develop, 72-77; PRES, STRATAH CO & MCMURTREY ASSOC, 77- *Mem:* Am Rocket Soc; assoc fel Am Inst Aeronaut & Astronaut. *Res:* Space propulsion; liquid and solid rocket propulsion; attitude control propulsion; space vehicles; new applications using a complete space rocket stage to put pay loads into orbit; supersonic aerodynamics of propulsion inlets. *Mailing Add:* 12122 196th NE Redmond WA 98053

MCMURTRY, CARL HEWES, b Wellsville, NY, Dec 6, 31; m 55; c 3. CERAMICS, INORGANIC CHEMISTRY. *Educ:* Alfred Univ, BS, 53, MS, 58. *Prof Exp:* Res assoc ceramics, Alfred Univ, 53-57, instr mineral, 57-58; MGR ENG, RES & DEVELOP DIV, CARBORUNDUM CO, 58- *Mem:* Fel Am Ceramic Soc; Nat Inst Ceramic Engrs; Am Inst Mining, Metall & Petrol Engrs; Am Soc Testing & Mat; Can Ceramic Soc. *Res:* Refractories; semiconductors; nuclear ceramics, structural ceramics; carbides, nitrides, borides and silicides; carbon and graphite; ceramic matrix composites. *Mailing Add:* Technol Div Carborundum Co PO Box 832 Niagara Falls NY 14302

MCMURTRY, GEORGE JAMES, b Crawfordsville, Ind, July 19, 32; m 61; c 6. ELECTRICAL ENGINEERING. *Educ:* US Naval Acad, BS, 55; Univ Notre Dame, MS, 61; Purdue Univ, Lafayette, PhD(elec eng), 65. *Prof Exp:* Instr elec eng, Purdue Univ, 61-65; engr, Gen Elec Co, 65-67; from asst prof to assoc prof elec, Pa State Univ, 67-76, eng Ord Res Lab, 67-69, co-dir off remote sensing earth resources, 70-80, assoc dean instr, Col Eng, 80-84, assoc dean admin Col Eng, 84-87, acting dean Col Eng, 87-88, PROF ELEC ENG, PA STATE UNIV, UNIVERSITY PARK, 76- ASSOC DEAN ADMIN, COL ENG, 88- *Mem:* Inst Elec & Electronics Engrs; Am Soc Eng Educ. *Res:* Adaptive and learning control systems; remote sensing of earth resources by aircraft and spacecraft; pattern recognition; optimization procedures. *Mailing Add:* Pa State Univ 101 Hammond University Park PA 16802

MCMURTRY, IVAN FREDRICK, PULMONARY VASCULAR CONTROL. *Educ:* Colo State Univ, PhD(physiol), 73. *Prof Exp:* ASSOC PROF, DEPT MED, UNIV COLO, 76- *Res:* Baso constriction. *Mailing Add:* Dept Med B-133 Univ Colo Health Sci Ctr 4200 E Ninth Ave Denver CO 80220

MCMURTRY, JAMES A, b Lodi, Calif, Sept 21, 32; m 54; c 2. ENTOMOLOGY. *Educ:* San Jose State Col, AB, 54; Univ Calif, Davis, PhD(entom), 60. *Prof Exp:* Res assoc entom, Univ Calif, Davis, 56-60; asst entomologist, 60-66, ASSOC ENTOMOLOGIST & LECTR BIOL CONTROL, UNIV CALIF, RIVERSIDE, 66-, PROF ENTOM, 81- *Concurrent Pos:* Guggenheim fel, 68. *Mem:* Entom Soc Am; Entom Soc Can. *Res:* Biological control and population ecology of phytophagous mites. *Mailing Add:* Dept Entom Univ Calif 900 University Ave Riverside CA 92521

MCMURTY, BURTON J, b Houston, Tex; c 2. QUANTUM ELECTRONICS. *Educ:* Rice Univ, BA, 56, BSEE, 57; Stanford Univ, MSEE, 59, PhD(elec eng), 62. *Prof Exp:* Engr mgt, GTE Sylvania, 57-69; assoc, Jack Melchor, 69-70; pres, Palo Alto Investment Co, 70-73; GEN PARTNER, INST VENTURE ASSOC, 73-; GEN PARTNER, TECHNOL VENTURE INVESTORS, 80- *Concurrent Pos:* Comt Int Quantum Elec Conf, 68-70; comt Dept Defense Classified Laser Symposium. *Honors & Awards:* Alfred Nobel Prize, 64. *Mem:* Inst Elec Electronic Engrs. *Res:* start up investments in electronic technology and biotechnology; author of over 20 journals and books. *Mailing Add:* Technol Venture Investors 3000 Sand Hill Rd Bldg 4 Suite 210 Menlo Park CA 94025

MACNAB, ROBERT MARSHALL, b Barnsley, Eng, Feb 3, 40. BIOPHYSICS. *Educ:* Univ St Andrews, Scotland, BSc, 62; Univ Calif, Berkeley, PhD(chem), 69. *Prof Exp:* Technologist petrochem, Brit Petrol Co, 62-65; res assoc biochem, Univ Calif, Berkeley, 70-73; from asst prof to assoc prof, 73-83, PROF BIOPHYS, YALE UNIV, 83- *Res:* Bacterial flagellar motor; bacterial chemotaxis as a model for chemo-mechanical and sensory transduction. *Mailing Add:* Molecular Biophys & Biochem 61 Middle Rd Hamden CT 06517

MCNABB, CLARENCE DUNCAN, JR, b Beloit, Wis, July 7, 28; m 53; c 7. LIMNOLOGY, AQUACULTURE. *Educ:* Loras Col, BA, 51; Univ Wis, MS, 57, PhD(algal ecol), 60. *Prof Exp:* Pub health biologist, Wis State Bd Health, 57-59; from asst prof to assoc prof biol, Wis State Univ, Whitewater, 59-63; assoc prof, 68-72, PROF LIMNOL, MICH STATE UNIV, 72-, DIR USDA N CENT REG AQUACULT CTR, 88- *Concurrent Pos:* Dir, Field Ecol Progs, Wis State Univ Pigeon Lake Field Sta, 62-63; dir, Hydrobiol Sta, St Marys Col, Minn, 64-68; prog leader, Limnol Lab, Mich State Univ, 75-; vis prof, dept bot, Univ Auckland, NZ, 70 & Lab Ecol, Gadjah Mada Univ, Yogyakarta, Indonesia, 85; NATO sr scientist fel, Max Planck Inst, WGer, 75; grad fac external rev scientist, Univ Auckland, NZ, 72-74 & Agr Univ, Serdang, Malaysia, 87-; consult, Ministry Educ & Cult, Brasilia, Brazil, 77 & Jakarta, Indonesia, 80; vchmn, Environ Biol Grant Rev Bd, US EPA, 80-82, mem, 85-; prin investr, Pond Aquacult Collab Res Support Prog, US AID,

Bogor, Indonesia, 82-87 & Bangkok, Thailand, 87-; coop scientist, Lake Biwa Res Ctr, Shiga Univ, Otsu, Japan, 86-; mem, Nat Coun Aquacult Develop, USDA, 88- *Mem:* Am Soc Limnol & Oceanog; Ecol Soc Am; Int Asn Great Lakes Res; Asian Fisheries Soc. *Res:* Production of fish food resources in coastal wetlands of North American Laurentian Great Lakes; application of limnology to management of fish yield from tropical and temperate aquaculture ponds. *Mailing Add:* Dept Fisheries & Wildlife Mich State Univ 13 Nat Res Bldg E Lansing MI 48824-1222

MCNABB, F M ANNE, b Edmonton, Alta, Jan 17, 39; m 87; c 1. COMPARATIVE PHYSIOLOGY. *Educ:* Univ Alta, BEd, 60, BSc, 61; Univ Calif, Los Angeles, MA, 65, PhD(zool), 68. *Prof Exp:* Asst prof biol & allied health, Quinnipiac Col, 68-69; adj asst prof, 69-75, from asst prof to assoc prof, 75-88, PROF BIOL, VA POLYTECH INST & STATE UNIV, 88- *Concurrent Pos:* Fel zool, Yale Univ, 68-69; mem exec comt, Am Soc Zoologists, 86-88. *Mem:* Fel AAAS; Am Soc Zoologists; Sigma Xi; Endocrine Soc. *Res:* Development of thyroid function and its control in precocial and altricial birds; peripheral thyroid function in growth and development of birds genetically selected for high and low body weight. *Mailing Add:* Dept Biol Va Polytech Inst & State Univ Blacksburg VA 24061

MCNABB, HAROLD SANDERSON, JR, b Lincoln, Nebr, Nov 20, 27; m 49; c 2. FOREST PATHOLOGY. *Educ:* Univ Nebr, BSc, 49; Yale Univ, MS, 51, PhD(plant sci), 54. *Prof Exp:* Asst bot, Univ Nebr, 46-49; asst plant sci, Yale Univ, 49-52; asst prof bot, 53-54, asst prof bot & forestry, 54-56, assoc prof, 56-64, assoc inst atomic res, 54-76, PROF FORESTRY & PLANT PATH, IOWA STATE UNIV, 64- *Concurrent Pos:* Sci adv, Tree Res Inst; coordr forestry res, Iowa State Conserv Comn, 53-59; mem, Int Bot Cong, France, 54, Can, 59 & Gt Brit, 64; ed, Iowa Acad Sci, 64; vis res scientist, UK Forestry Comn, 73-; dep leader resistance to dis & insects, Int Union of Forestry Res Orgn, 81-; fel, Int Agr Club, Dutch Mining Agr, 83; fel, DeDorsch Kamp Res Sta, Netherlands, 83; fel, Inst Nat Agron Res & Grown Forest Indust, Nancy, France, 83. *Mem:* AAAS; Soc Am Foresters; Am Phytopath Soc; Mycol Soc Am; Sigma Xi; Bot Soc Am; Int Soc Arbor; fel Soc Am Foresters. *Res:* Diseases of forest and shade trees; deterioration of wood; transformation of Populus species. *Mailing Add:* Dept Forestry Iowa State Univ Ames IA 50011

MCNABB, ROGER ALLEN, b Moose Jaw, Sask, May 21, 38; m 91; c 1. COMPARATIVE PHYSIOLOGY. *Educ:* Univ Alta, BSc, 61, MSc, 63; Univ Calif, Los Angeles, PhD(zool), 68. *Prof Exp:* Res staff biologist, Yale Univ, 68-69; asst prof, 69-75, ASSOC PROF BIOL, VA POLYTECH INST & STATE UNIV, 75- *Concurrent Pos:* Assoc ed, Am Zoologist, 78-82. *Mem:* Am Soc Zool; Am Physiol Soc; Am Inst Biol Sci. *Res:* Physiology of environmental adaptations in lower vertebrate animals and invertebrates. *Mailing Add:* Dept Biol Va Polytech Inst & State Univ Blacksburg VA 24061-0406

MCNAIR, DENNIS M, b Dayton, Wash, June 7, 45; m 68. ANIMAL PHYSIOLOGY. *Educ:* Whitman Col, AB, 67; Southern Ill Univ Carbondale, MA, 73, PhD(zool), 80. *Prof Exp:* Asst prof invert, Western Ill Univ, 78-79; asst prof zool, Wright State Univ, 79-80; asst prof, 80-86, ASSOC PROF BIOL, UNIV PITTSBURGH, JOHNSTOWN, 86- *Concurrent Pos:* Consult, Pa Elec Co, 82-; dir, Cambia-Somerset Coun, Educ Health Prof, Inc, 87- *Mem:* AAAS; Sigma Xi; Am Soc Parasitologists; Entom Soc Am. *Res:* Aquatic insects as indicators of water quality; neuroanatomy of parasitic nematodes; effects of parasites on host behavior; invertebrate behavior and natural history; ectoparasites of birds and mammals. *Mailing Add:* Dept Biol Univ Pittsburgh Johnstown PA 15904

MCNAIR, DOUGLAS MCINTOSH, b Rockingham, NC, July 19, 27; div; c 1. PSYCHOPHARMACOLOGY, PSYCHOLOGY. *Educ:* Univ NC, Chapel Hill, AB, 48, PhD(clin psychol), 54. *Prof Exp:* Sr clin psychologist, Guildford County Ment Health Clin, 52-55; psychologist, Col Infirmary & asst prof psychol, Univ NC, Greensboro, 55-56; res psychologist & asst chief outpatient psychiat res lab, Vet Admin, Washington, DC, 56-64; asst prof, 64-65, assoc prof psychiat, 65-70, PROF PSYCHIAT, SCH MED, BOSTON UNIV, 70-, dir, Clin Psychopharm Lab, 80-87, PROF PSYCHOL & DIR, CLIN PSYCHOL TRAINING, 80- *Concurrent Pos:* Consult, 72-; mem psychopharmacol agents adv comt, Food & Drug Admin, 77- *Mem:* Psychomet Soc; fel Am Col Neuropsychopharmacol; Am Psychol Asn; fel Am Psychopathological Asn. *Res:* Drugs and behavior; psychotherapy; experimental design. *Mailing Add:* Psychol Dept Boston Univ 64 Cummington St Boston MA 02215

MCNAIR, HAROLD MONROE, b Miami, Ariz, May 31, 33; m 60; c 3. ANALYTICAL CHEMISTRY. *Educ:* Univ Ariz, BS, 55; Purdue Univ, MS, 57, PhD(anal chem), 59. *Prof Exp:* Fulbright & univ fels, Eindhoven Technol Univ, 59-60, Perkin Elmer res fel, 60; res chemist, Esso Res & Eng, NJ, 60-61; tech dir, Europe Div, F&M Sci Corp, Amsterdam, 61-63, gen mgr, 63-64; dir int opers, Varian Aerograph, Switz, 64-66, dir mkt, Calif, 66-68; assoc prof, 68-71, PROF ANALYTICAL CHEM, VA POLYTECH INST & STATE UNIV, 71- *Concurrent Pos:* Consult, Instrument Group, Varian Assocs, 68-; adj prof, Nat Univ Mex; Air Pollution Control Off, Res Triangle, NC; vis prof, Eindhoven Tech Univ, Holland, 81. *Mem:* Am Chem Soc; Am Soc Test & Mat; NY Acad Sci; Sigma Xi. *Res:* Gas chromatography; quantitative analysis of ionization detectors and temperature programming; trace gas analysis by ionization detectors; selection of selective liquid phases; theory of chromatography; liquid chromatography; quantitative analysis by capillary column gas chromatography; influence of injection systems; multidimensional chromatography. *Mailing Add:* Dept Chem Va Polytech Inst & State Univ Blacksburg VA 24061

MCNAIR, IRVING M, JR, b Herndon, Va, Dec 18, 32; m 59; c 2. ELECTRICAL ENGINEERING. *Educ:* Pa State Univ, BS, 54; Stevens Inst Technol, MS, 61. *Prof Exp:* Mem tech staff, Bell Labs, 54-61, supvr test sets & methods, 61-64, supvr outside plant & signaling group, 64-71, subscriber loop multiplexer group, 71-74, supvr test access automated distributing, 74-76, supvr spec serv, 76-87, supvr narrow channel field test cellular radio, 88-89; RETIRED. *Mem:* Inst Elec & Electronics Engrs. *Res:* Low power logic circuits; low power transistor circuits for use in portable test equipment; system design for switching and carrier systems for telephone loop application; circuit design for special services. *Mailing Add:* Six Standish Dr Morristown NJ 07960

MACNAIR, RICHARD NELSON, b Newton, Mass, Oct 19, 29; m 60; c 1. BEGONIA CULTURE, ORGANIC CHEMISTRY. *Educ:* Middlebury Col, AB, 52; Univ Del, PhD(org chem), 60. *Prof Exp:* Chemist, Arthur D Little, Inc, 60-63; sr chemist, Tracerlab Div, Lab for Electronics, Inc, 64; res chemist, US Army Natick Res, Develop & Eng Ctr, 64-69, supvry res chemist, 69-88; VOL CUR BEGONIAS, NORTHEASTERN UNIV BOT RES STA, 88- *Concurrent Pos:* Mem, Int Oceanog Found; comt chmn 52nd nat meeting printing & advertising, Am Inst Chemists, 75; pres, New Eng Inst Chemists, 90-92. *Mem:* Am Chem Soc; Am Carbon Soc; Sigma Xi; Am Inst Chemists. *Res:* Heterocyclic chemistry; graft polymerization; chemical protective clothing; activated carbon sorption-desorption; carbon fibers and fabrics. *Mailing Add:* 177 Hancock St Cambridge MA 02139

MCNAIR, ROBERT J(OHN), JR, b Owego, NY, Mar 24, 18; m 51; c 2. ELECTRICAL ENGINEERING. *Educ:* Univ Mich, BS, 50; Univ Wis, MS, 55, PhD(elec eng), 58. *Prof Exp:* Electronics engr, Bell Aircraft Corp, NY, 51-53, staff engr, 56-58; prin staff engr, Electronics Div, Avco Corp, 58-73, sci consult, 74-84; sci consult, Textron, 85-87; RETIRED. *Concurrent Pos:* Patent agent, US Patent & Trademark Off. *Mem:* Inst Elec & Electronics Engrs. *Res:* Radio communications theory and practice; turbine engine performance evaluation; solid state circuitry and devices. *Mailing Add:* 2920 Blue Haven Terr Cincinnati OH 45238

MCNAIR, RUTH DAVIS, b Flint, Mich, Mar 18, 21; m 50; c 2. CLINICAL CHEMISTRY. *Educ:* Univ Mich, BS, 41; Univ Cincinnati, BS, 42; Wayne State Univ, PhD(biochem), 48; Am Bd Clin Chem, dipl. *Prof Exp:* Dir biochem, Providence Hosp, 48-83; RETIRED. *Concurrent Pos:* Pres & bd dir, Am Bd Clin Chemists. *Mem:* AAAS; Sigma Xi; fel Am Inst Chemists; Am Chem Soc; fel Am Asn Clin Chemists. *Res:* Protein, nutrition and metabolism. *Mailing Add:* 6199 Thorncrest Birmingham MI 48010

MCNAIRN, ROBERT BLACKWOOD, b San Francisco, Calif, Mar 13, 40; m 62; c 2. PLANT PHYSIOLOGY. *Educ:* Univ Calif, Berkeley, BS, 62; Univ Calif, Davis, MS, 64, PhD(plant physiol), 67. *Prof Exp:* From asst prof to assoc prof, 67-76, PROF BOT, CALIF STATE UNIV, CHICO, 76- *Concurrent Pos:* Plant physiologist, Thornton Wholesale Florist & Flower Grower, 74-75. *Mem:* Am Soc Plant Physiol. *Res:* Phloem translocation; flowering; stress physiology. *Mailing Add:* 2815 San Verbena Way Chico CA 95926-0934

MCNAIRY, SIDNEY A, JR, b Memphis, Tenn, Oct 16, 37. MINORITY RESEARCH. *Educ:* LeMoyne Col, BS, 59; Purdue Univ, MS, 62, PhD(animal physiol-org chem), 65. *Prof Exp:* Prof, Dept Chem, Southern Univ, Baton Rouge, 65-75, dir, Health Res Ctr, 71-75; intergovt personnel act assignment, Div Res Resources, 75-76, health scientist-adminr, Minority Biomed Res Support Prog, 76-85, DIR, RES CTR MINORITY INST PROG, NAT CTR RES RESOURCES, NIH, 85- *Mem:* Soc Res Admin; AAAS. *Res:* Numerous publications; separation and identification of Triiodothyronine and Thyroxine; development of a synthetic medium for determining the biodegradability and toxicity of synthetic detergents. *Mailing Add:* NIH Div Res Resources Res Ctr Minority Inst Prog Westwood Bldg Rm 10A10 5333 Westbard Ave Bethesda MD 20892

MCNALL, LESTER R, b Gaylord, Kans, Oct 28, 27. PLANT NUTRITION. *Educ:* Univ Wis, BS, 50; Univ Calif, Los Angeles, PhD(org chem), 55. *Prof Exp:* Res chemist, Esso Res & Eng Co, 55-56; res chemist, Paper Mate Mfg Co, Gillette Co, 56-58, head chem res, 58-62, res div, 62-64; tech dir, Leffingwell Chem Co, 66-77, gen mgr Leffingwell Div, Thompson-Hayward Chem Co, 77-82; consult, 82-85; PRES, NUTRIENT TECHNOLOGIES, INC, 86- *Mem:* Am Chem Soc; Am Soc Hort Sci. *Res:* Chemistry of natural products; agricultural chemistry; minor elements in plant nutrition; foliar feeding. *Mailing Add:* 311 E Country Hills Dr La Habra CA 90631

MCNALLY, ELIZABETH MARY, b Chicago, Ill. MEDICINE. *Educ:* Columbia Univ, AB, 83; Albert Einstein Col Med, MS, 87, MD & PhD(microbiol & immunol), 90. *Prof Exp:* Intern-resident internal med, Brigham & Women's Hosp, Howard Med Sch, 90- *Mem:* Am Soc Cell Biol. *Res:* How structure relates to function in the major muscle protein myosin; muscle, both skeletal and cardiac, development and cell biology. *Mailing Add:* Dept Med Brigham & Women's Hosp Boston MA 02115

MCNALLY, JAMES HENRY, b Orange, NJ, Dec 18, 36; m 76. NUCLEAR WEAPON PHYSICS, VERIFICATION TECHNOLOGY. *Educ:* Cornell Univ, BEng Phys, 59; Calif Inst Technol, PhD(physics), 66. *Prof Exp:* Physicist & prog mgr, Los Alamos Nat Lab, 65-74; asst dir lasers & isotope separation, Energy Res & Develop Admin, 74-75; assoc div leader, Laser Fusion & Dep Inertial Fusion, Los Alamos Nat Lab, 76-81, asst nat security issues, 81-86, dep asst dir, verification & intel, US Arms Control & Disarmament Agency, 86-88, dir off staff, 88-90; CONSULT & WASH INST STAFF, 90- *Concurrent Pos:* Mem US delegation, Conf on Disarmament, Geneva, 69, 73, 74, threshold test ban treaty talks, Moscow, 74, Nuclear Testing Talks, Geneva, 86-88. *Mem:* Int Inst Strategic Study; AAAS; Am Phys Soc. *Res:* Nuclear weapons design; inertial fusion; laser isotope separation; national security issues. *Mailing Add:* 550 Rim Rd Los Alamos NM 87544-2931

MCNALLY, JAMES RAND, JR, b Boston, Mass, Nov 10, 17; m 42; c 7. PLASMA PHYSICS. *Educ:* Boston Col, BS, 39; Mass Inst Technol, SM, 41; PhD(physics), 43. *Prof Exp:* Asst spectros, Mass Inst Technol, 39-41, spectros & physics, 41-44, instr physics, 44-48; physicist, Stable Isotopes Div, Oak Ridge Nat Lab, 48-55, assoc dir div, 55-57, sr physicist, Physics Div, 57-60, sr staff physicist, Fusion Energy Div, 60-81; CONSULT, 82- *Concurrent Pos:* Mem advan fuels adv comt, Elec Power Res Inst; fusion energy consult. *Mem:* Fel AAAS; Am Phys Soc; fel Optical Soc Am; Soc Appl Spectros; Am Asn Physics Teachers. *Res:* Atomic physics; atomic spectroscopy; fusion physics; ion-layer; advanced fusion fuels; fusion chain reactions, grasers; gamma ray amplification by stimulated emission of radiation; nuclear dynamos. *Mailing Add:* 103 Norman Lane Oak Ridge TN 37830-5361

MCNALLY, JOHN G, b Brooklyn, NY, Mar 5, 32; m 64. ANALYTICAL CHEMISTRY. *Educ:* Polytech Inst Brooklyn, BS, 55, MS, 63. *Prof Exp:* Technician, Chas Pfizer & Co, 49-55, chemist, 55-57; chemist, Escambia Chem Corp, 57-61, head anal dept, 61-68; proj leader, 68-70, head chromatography group, 70-80, group leader chromatography & separations, 80-85, PRIN RES CHEMIST, AM CYANAMID CO, STAMFORD, 85- *Mem:* Am Chem Soc. *Res:* high performance liquid chromatography; size exclusion chromatography; polymer characterization; computerization; gas chromatography. *Mailing Add:* Am Cyanamid Co 1937 W Main St Stamford CT 06904

MCNALLY, KAREN COOK, b Clovis, Calif. GEOPHYSICS. *Educ:* Univ Calif, Berkeley, AB, 71, MA, 73, PhD(geophysics), 76. *Prof Exp:* Seismologist res asst geophysics, Univ Calif, Berkeley, 71-75; seismologist, Woodward-Clyde Consults, San Francisco, 75-76; fel, Calif Inst Technol, 76-78, sr res fel geophys, 78-81; assoc prof, 82-86, PROF GEOPHYS, UNIV CALIF, SANTA CRUZ, 87-, DIR, CHARLES F RICHTER SEISMOL LAB, 82- *Concurrent Pos:* Consult seismologist, Woodward-Clyde Consults, 76-; res assoc, Sierra Geophysics, Inc, 77-; dir, Inst Tectonics, Univ Calif, Santa Cruz, 87-; mem Nat Acad Sci comt seismol, 79-82. *Honors & Awards:* Ernest C Watson Lectr, Calif Inst Technol, 79; Richtmeyer Lectr, Am Phys Soc & Am Asn Physics Teachers, 82. *Mem:* Seismol Soc Am; Am Geophys Union; AAAS. *Res:* Regional tectonics; earthquake statistics; fracture mechanics; earthquake source mechanism; crust and upper mantle structure; author of numerous technical journal articles. *Mailing Add:* Dept Earth Sci Univ Calif Santa Cruz CA 95064

MCNAMARA, ALLEN GARNET, b Regina, Sask, Feb 28, 26; m 52; c 1. AERONOMY, SPACE SCIENCE. *Educ:* Univ Sask, BE, 47, MSc, 49, PhD(physics), 54; Univ Mich, MA, 51. *Prof Exp:* From res officer to assoc res officer, 51-63, SR RES OFFICER, NAT RES COUN CAN, 63-, HEAD PLANETARY SCI SECT, HERZBERG INST ASTROPHYS, 75-, PRIN RES OFFICER, 79- *Concurrent Pos:* Mem, Can Nat Comt Radio Sci, Int Sci Radio Union. *Mem:* Can Asn Physicists; Inst Elec & Electronics Engrs; Am Geophys Union. *Res:* Physics of upper atmosphere; scattering of radio waves by meteor ionization and aurora; rocket investigations of the aurora and upper atmosphere; rocket and satellite instrumentation for plasma studies. *Mailing Add:* Nat Res Coun Ottawa ON K1A 0R6 Can

MCNAMARA, DAN GOODRICH, b Waco, Tex, Oct 19, 22; m 49; c 5. PEDIATRICS. *Educ:* Baylor Univ, BS, 43, MD, 46. *Prof Exp:* Resident, Hermann Hosp, Houston, Tex, 49-50; from asst prof to assoc prof, 53-69, PROF PEDIAT, BAYLOR COL MED, 69- *Concurrent Pos:* Fel pediat cardiol, Cardiac Clin, Harriet Lane Home, Johns Hopkins Hosp, 51-53; dir cardiac clin, Tex Children's Hosp, Houston, Tex, 53- *Mem:* Am Acad Pediat. *Res:* Pediatric cardiology, especially secondary pulmonary hypertension and malfunctions of heart in infancy. *Mailing Add:* Dept Pediat Baylor Col Med One Baylor Plaza Houston TX 77030

MCNAMARA, DELBERT HAROLD, b Salt Lake City, Utah, June 28, 23; m 45; c 3. ASTROPHYSICS, ASTRONOMY. *Educ:* Univ Calif, BA, 47, PhD(astron), 50. *Prof Exp:* Asst astronomer, Univ Calif, Berkeley, 50-52, assoc res astronomer, 52-55; asst prof astron, 55-57, assoc prof, 57-62, PROF PHYSICS & ASTRON, BRIGHAM YOUNG UNIV, 62- *Concurrent Pos:* Prin scientist space sci lab, NAm Aviation, Inc; ed, Astron Soc Pac Publ. *Mem:* Am Astron Soc; Int Astron Union. *Res:* Stellar spectroscopy and photometry; variable stars; eclipsing binaries. *Mailing Add:* Dept Physics & Astron Esc 408 Brigham Young Univ Provo UT 84602

MCNAMARA, DENNIS B, b Shreveport, La, Feb 7, 42. ARACHIDONIC ACID METABOLISM, SIGNAL TRANSDUCTION. *Educ:* Univ Man, PhD(physiol), 73. *Prof Exp:* From asst prof to assoc prof, 81-90, PROF PHARMACOL, SCH MED, TULANE UNIV, 90- *Concurrent Pos:* Mem cardiopulmonary coun, Am Heart Asn. *Mem:* Am Heart Asn; Am Soc Pharmacol & Exp Therapeut; Int Soc Heart Res; NY Acad Sci. *Res:* Arachidonic acid metabolism; signal transduction. *Mailing Add:* Dept Pharmacol Sch Med Tulane Univ 1430 Tulane Ave New Orleans LA 70112

MCNAMARA, DONALD J, b Cleveland, Ohio, Jan 31, 44. NUTRITION & FOOD SCIENCE. *Educ:* Col Steubenville, BA, 66; Purdue Univ, PhD(biochem), 72. *Prof Exp:* Res assoc, Dept Physiol Chem, Med Col, Ohio State Univ, 72-74; from asst prof to assoc prof lipid metab, Rockefeller Univ, 74-85; PROF NUTRIT & BIOCHEM, DEPT NUTRIT & FOOD SCI, UNIV ARIZ, 85- *Concurrent Pos:* Postdoctoral fel, Nat Cancer Inst, 73-75. *Mem:* Am Inst Nutrit; Am Heart Asn; Am Coun Sci & Health; Am Soc Clin Nutrit; Am Soc Biochem & Molecular Biol. *Mailing Add:* Dept Nutrit & Food Sci Univ Ariz 309 Shantz Tucson AZ 85721

MCNAMARA, EDWARD P(AUL), b Troy, NY, Sept 27, 10; m 33; c 3. CERAMICS, PHYSICAL CHEMISTRY. *Educ:* Alfred Univ, BS, 35; Pa State Col, MS, 36. *Prof Exp:* Asst, Pa State Col, 35-36, head ceramic exten dept, 36-42; dir ceramics dept & NJ ceramic res sta, Rutgers Univ, 42-45; pres & gen mgr, Pfaltzgraff Pottery Co, 45-46; dir ceramics, Shenango Pottery Co, 46-53; tech dir, Cambridge Tile Mfg Co, 53-63, vpres res & develop, Ohio, 63-69; vpres res & develop, Marshall Tiles Inc, 69-71; pres, Monarch Tile Mfg Inc, 72-77, sr consult, 77-80. *Concurrent Pos:* Vol exec, Int Exec Serv Corps, Mex, 80-81, Singapore, 82, Portugal, 83. *Mem:* Fel Am Ceramic Soc; Am Soc Testing & Mat; Am Soc Qual Control; Nat Inst Ceramic Engrs (pres, 71-72). *Res:* Elastico-viscous properties of glasses; properties of borate glasses; reaction rates of limestone calcination; high temperature solution of limestones in slags; glassy and crystalline states of ceramic dielectrics; rheology of clays. *Mailing Add:* 5907 Pinto Path San Angelo TX 76901

MCNAMARA, JAMES ALYN, JR, b San Francisco, Calif, June 11, 43; m 70; c 2. ORTHODONTICS, ANATOMY. *Educ:* Univ Calif, Berkeley, AB, 64; Univ Calif, San Francisco, BS, DDS & cert orthod spec, 68; Univ Mich, Ann Arbor, MS, 69, PhD(anat), 72. *Prof Exp:* Res assoc, Ctr Human Growth & Develop, Univ Mich, Ann Aror, 70-72, assoc res scientist & prog dir, Exp Craniofacial Res, 72-77, asst prof anat, 72-77, int chmn, 87-91, ASSOC PROF ANAT, UNIV MICH, ANN ARBOR, 77-, RES SCIENTIST, CTR HUMAN GROWTH & DEVELOP, 79-, PROF ANAT & CELL BIOL, 83-, PROF, DEPT ORTHOD & PEDIAT DENT, 84- *Honors & Awards:* Milo Hellman Res Award, Am Asn Orthodontists, 73; E Sheldon Friel Mem Award, Europ Orthod Soc, 79; Chalmers J Lyons Mem lectr, Am Asn Oral & Maxillofacial Surgeons, 81; Northcroft Lectr, Brit Soc Study Orthod, 91. *Mem:* Am Asn Orthod; Int Asn Dent Res; Am Asn Anat; Am Dent Asn; Europ Orthod Soc. *Res:* Experimental studies of musculoskeletal interaction; craniofacial growth in man and non-human primates; cephalometric studies of treatment response; relationship of upper respiratory obstruction to craniofacial growth. *Mailing Add:* Dept Orthod & Pediat Dent Univ Mich Ann Arbor MI 48109

MCNAMARA, JAMES O'CONNELL, b Portage, Wis, Sept 25, 42; m 64; c 5. EPILEPSY RESEARCH. *Educ:* Marquette Univ, AB, 64; Univ Mich, MD, 68. *Prof Exp:* Dir, Epilepsy Ctr, Durham Vet Admin Med Ctr, 76-86, assoc prof med & neurol, 80-85, asst prof pharmacol, 81-86, STAFF NEUROLOGIST, DURHAM VET ADMIN MED CTR, 86-, PROF NEUROBIOL, 90-; DIR, DUKE CTR ADVAN STUDY EPILEPSY, 82-, PROF MED & NEUROL, 85-, ASSOC PROF PHARMACOL, 86- *Concurrent Pos:* Mem, Ethics Comt, Am Acad Neurol, 74; mem, Rules Comt, Am Epilepsy Soc, 83-84, Ad Hoc Comt Basic Sci, 84-, chmn, Long Range Planning Comt, 87-, couns, 87-89, 2nd vpres, 90-91; assoc ed, Epilepsy, Advances, 85-; Javits neurosci investr award, NIH, 87; bd dirs, Epilepsy Found Am, 87-, secy, Prof Adv Bd, 87-; William N Creasy vis prof clin pharmacol, 88. *Mem:* Am Acad Neurol; AAAS; Am Epilepsy Soc; Epilepsy Found Am; Am Soc Pharmacol & Exp Therapeut; Soc Neurosci (pres, 89). *Res:* Cellular and molecular mechanisms of the kindling model. *Mailing Add:* 401 Bryan Res Bldg Duke Univ Med Ctr Duke Box 3676 Durham NC 27710

MCNAMARA, JOHN REGIS, b Binghamton, NY, May 27, 41; m 72; c 2. CLINICAL PSYCHOLOGY, BEHAVIOR THERAPY. *Educ:* Univ Notre Dame, BA, 63; Xavier Univ, MA, 67; Univ Ga, PhD(psychol), 72. *Prof Exp:* Asst prof, clin psychol, Univ Mo, Kansas City, 71-72; PROF CLIN PSYCHOL, OHIO UNIV, 72- *Concurrent Pos:* Behav mgt consult, US Gen Acct Off, 77-82; assoc dir, Inst Health & Behav Sci, Ohio Univ, 80-91; clin consult, New Horizons Coun Ctr, 86-91. *Mem:* Am Psychol Asn; Asn Adv Behav Ther; Sigma Xi; Behav Ther & Res Soc. *Res:* Ethics of professional and scientific conduct; compliance problems in health psychology; broad spectrum research in behavior therapy. *Mailing Add:* Psychol Dept Ohio Univ Athens OH 45701-2979

MCNAMARA, JOSEPH JUDSON, b Oakland, Calif, Sept 12, 36; c 4. SURGERY. *Educ:* Wash Univ, MD, 61; cert, Am Bd Surg, 67, cert thoracic surg, 68. *Prof Exp:* Surgeon, Peter Bent Brigham & Mass Gen Hosps, 61-67; thoracic & cardiovasc surgeon, Baylor Univ Hosp, Dallas, 67-68; actg dir div surg, Walter Reed Army Inst Res, 69-70; PROF SURG, SCH MED, UNIV HAWAII, MANOA, 70- *Concurrent Pos:* Dir, cardiovasc res lab, Queen's Med Ctr, 70-; actg chmn, dept surg, John A Burns Sch Med, Univ Hawaii, 91-; interim prog dir, Univ Hawaii, Integrated Surg Residency Prog, 91- *Honors & Awards:* Sheard-Sanford Award, Am Soc Clin Path, 61. *Mem:* Fel Am Col Surg; fel Am Col Cardiol; Am Asn Thoracic Surg; Pac Coast Surg Asn; Am Surg Asn; Sigma Xi. *Res:* Oxygen free radicals & reperfusion injury. *Mailing Add:* Dept Surg Queens Med Ctr Harkness Pavilion 1301 Punchbowl St Honolulu HI 96813

MCNAMARA, MARY COLLEEN, b Albuquerque, NMex, Apr 5, 47. NEUROBIOLOGY, NEUROCHEMISTRY. *Educ:* Univ NMex, BS, 71, MS, 72; Univ NC, Chapel Hill, PhD(neurobiol), 75. *Prof Exp:* Teaching asst, Univ NMex, 71-72; instr introd psychol, Sch Med, Univ NC, 74-76, fel physiol, 75-77; asst prof physiol & psychol, Miami Univ, Oxford, 77-78; fel, Duke Univ Med Ctr, 78-80; RES ANALYST, DEPT PEDIAT, SCH MED, UNIV NC, CHAPEL HILL, 80- *Concurrent Pos:* Adj asst prof, Dept Psychol, Univ NC, 78-80. *Mem:* Sigma Xi; Soc Neurosci; Geront Soc; Am Physiol Soc. *Res:* Delineating age related changes in central neurotransmitters in response to stress. *Mailing Add:* 6309 Sage Hill Ctr Albuquerque NM 87120

MCNAMARA, MICHAEL JOSEPH, b New York, NY, May 16, 29; m 57; c 3. COMMUNITY HEALTH, PREVENTIVE MEDICINE. *Educ:* Fordham Univ, AB, 51; NY Univ, MD, 55; Am Bd Prev Med, dipl. *Prof Exp:* Intern med, Bellevue Hosp, NY Univ, 55-56, asst resident internal med, 56-58; from asst prof to assoc prof community med, Col Med, Univ Ky, 61-70; prof & chmn dept, 70-78, assoc dean student affairs, 79-84, PROF MED & CHIEF, DIV COMMUNITY MED, MED COL OHIO, 78- *Concurrent Pos:* Nat Found fel virus res, Col Med, Cornell Univ, 58-59. *Mem:* Soc Epidemiol Res; Asn Teachers Prev Med; Am Pub Health Asn; fel Am Col Prev Med. *Res:* Epidemiology of viral agents in human diseases; epidemiology of hospital acquired infection; development and evaluation of measurement of quality of medical care and programs for risk management in hospitals; teaching of community medicine to medical students and residents in general preventive medicine. *Mailing Add:* Dept Med Med Col Ohio CS No 10008 Toledo OH 43699

MCNAMARA, PAMELA DEE, b Jackson Heights, NY, July 13, 43; m 67. BIOLOGICAL MEMBRANES, AMINO ACID TRANSPORT. *Educ:* Hunter Col, City Univ New York, BA, 65; Univ Ill, PhD(microbiol), 69. *Prof Exp:* Lectr biol & cell physiol, Bryn Mawr Col, 69-70; res assoc, 70-72, RES STAFF, CHILDREN'S HOSP PHILADELPHIA, 72-, STAFF MEM, RES TRAINING PROG METAB DISEASES, 74- *Concurrent Pos:* Asst prof pediat res, Med Sch, Univ Pa, 76-80, res assoc prof, 80- *Mem:* Am Inst Nutrit; NY Acad Sci; Biochem Soc. *Res:* Metabolite transport in renal and intestinal tissues, especially as related to inherited and acquired metabolic disorders of transport; isolation and characterization of transporting epithelial membrane fractions. *Mailing Add:* Dept Metab One Children's Ctr Rm 7074 Children's Hosp Philadelphia Philadelphia PA 19104

MACNAMARA, THOMAS E, b Airdrie, Scotland, May 23, 29; US citizen; c 5. ANESTHESIOLOGY. *Educ:* Glasgow Univ, MB, ChB, 52. *Prof Exp:* Instr anesthesia, Med Ctr, Georgetown Univ, 57-60; asst, Mass Gen Hosp & Harvard Med Sch, 60-62; PROF ANESTHESIA & CHMN DEPT, MED CTR, GEORGETOWN UNIV, 63- *Concurrent Pos:* Lectr, Bethesda Naval Med Ctr, 63-75, consult, 63-; vpres fac senate, Georgetown Univ, 67-71, pres, 73-75; chief anesthesia dept, NIH Clin Ctr, Bethesda, Md, 75-; consult, DC Gen & Vet Admin Hosps & Charles Town Gen Hosp, Ranson, WVa. *Mem:* Am Soc Anesthesiol; Brit Med Asn; Royal Soc Med; Int Anesthesia Res Soc. *Mailing Add:* Dept Anesthesia Georgetown Univ 3900 Reservoir NW Washington DC 20007

MCNAMARA, THOMAS FRANCIS, b Brooklyn, NY, Jan 26, 28; m 53; c 5. MICROBIOLOGY. *Educ:* Manhattan Col, BS, 49; Hofstra Univ, MA, 50; Cath Univ Am, PhD (microbiol & immunol), 59. *Prof Exp:* Scientist, Eaton Labs, Norwich Pharmacol Co, 57-59; sr scientist, Lederle Labs, Am Cyanamid Co, 59-61; res dir, Consumer Prod Div, Warner Lambert Co, 61-73; PROF MICROBIOL, IMMUNOL, CLIN PHARMACOL, ORAL DIAGNOSTICS, SCH DENT MED, STATE UNIV NY, STONY BROOK, 73- *Concurrent Pos:* Grad fel, Cath Univ Am, 55-57; consult, contract res, Nat Inst Dent Res, 74-75, Imp Chem Indust, 78-80, Sigma Xi, 79-, Johnson & Johnson Prod Inc & Am Cyanamid Co, 81-; mem bd dirs, Sigma Xi, 85-, interim exec dir, 87. *Mem:* Int Asn Dental Res; AAAS; Sigma Xi. *Res:* Microbiology of oral diseases, such as dental caries and periodontal disease, and evaluation of chemotherapeutic as well as immunological approaches to control of these two oral diseases. *Mailing Add:* PO Box 44 Port Jefferson NY 11777

MCNAMEE, BERNARD M, b Philadelphia, Pa, Sept 13, 30; m 57; c 5. CIVIL ENGINEERING. *Educ:* Drexel Inst, BS, 53, MBA, 60; Univ Pa, MS, 63; Lehigh Univ, PhD(civil eng), 67. *Prof Exp:* Civil engr, Pa RR, 53; from instr to assoc prof, 55-73, chmn dept, 76-84, PROF CIVIL ENG, DREXEL UNIV, 74- *Concurrent Pos:* Dir, Archit Eng Prog, Drexel Univ, 84-88. *Mem:* Am Soc Civil Engrs; Am Soc Eng Educ; Am Concrete Inst; Sigma Xi. *Res:* Inelastic frame buckling; fatigue study; composite materials. *Mailing Add:* 108 Fox Hall Lane Narberth PA 19072-2156

MCNAMEE, JAMES EMERSON, b Englewood, NJ, Sept 17, 46; m 69; c 2. PHYSIOLOGY, BIOMEDICAL ENGINEERING. *Educ:* Rutgers Univ, BS, 69; Univ Southern Calif, MS, 71, PhD(biomed eng), 74. *Prof Exp:* Fel physiol, Cardiovascular Res Inst, Univ Calif, San Francisco, 74-76; ASST PROF PHYSIOL, UNIV SC, 76- *Concurrent Pos:* NIH prin investr, Cardiovascular Res Inst, Univ Calif, San Francisco, 74-76. *Mem:* Microcirculatory Soc; Inst Elec & Electronics Engrs; Am Physiol Soc. *Res:* Pulmonary edema; microvascular transport phenomena. *Mailing Add:* Dept Physiol Univ SC Sch Med Columbia SC 29208

MACNAMEE, JAMES K, b Philadelphia, Pa, May 15, 16; m 50. COMPARATIVE PATHOLOGY, MICROBIOLOGY. *Educ:* Auburn Univ, BS, 36, DVM, 37, MS, 42. *Prof Exp:* Prof & chmn dept path & parasitol, Col Vet Med, Auburn Univ, 42-47; prof path, Col Vet Med, Univ Ga, 47-49; comp pathologist, Med Res Labs, Edgewood Arsenal, Md, 49-55 & Biol Res Labs, Ft Detrick, Md, 55-58; dir diag lab, Hawaiian Med Lab, Hawaii, 58-62; dir vet sect, Sixth Army Med Lab, Ft Baker, Calif, 62-65; health scientist adminr, NIH, 65-76; CONSULT & LECTR DIS EXOTIC BIRDS, 76- *Concurrent Pos:* Med admin, virol. *Mem:* Sci Res Soc Am; Am Asn Zool Parks; Avicult Soc Am; Am Asn Vet Toxicologists; Am Vet Med Asn; Am Fedn Avicult. *Res:* Diseases of exotic birds; Pacheo disease; comparative pathology. *Mailing Add:* 133 Caribe Isle Bel Marin Keys Novato CA 94949-5316

MCNAMEE, LAWRENCE PAUL, b Pittsburgh, Pa, Sept 12, 34; m 57; c 7. ELECTRICAL ENGINEERING, COMPUTER SCIENCE. *Educ:* Univ Pittsburgh, BSEE, 56, MSEE, 58, PhD(elec eng), 64. *Prof Exp:* Teaching asst elec eng, Univ Pittsburgh, 56-57, from instr to asst prof, 58-66; from asst to assoc prof comput sci, 66-76, ASSOC PROF ENG & APPL SCI, UNIV CALIF, LOS ANGELES, 76- *Concurrent Pos:* Consult, Knowledge Availability Ctr, NASA, Pa, 65-66, TRW Systs, Inc, Calif, 68- & Hughes Aircraft Co, Calif, 71- *Mem:* Inst Elec & Electronics Engrs; Asn Comput Mach; Simulation Coun; Opers Res Soc Am; Am Math Soc. *Res:* Computer aided design; simulation; digital filtering. *Mailing Add:* Dept Comput Sci Univ Calif 3531 Boelter Hall Los Angeles CA 90024

MCNAMEE, MARK G, b Dec 13, 46. BIOCHEMISTRY, BIOPHYSICS. *Educ:* Mass Inst Technol, Cambridge, BS, 68; Stanford Univ, Calif, PhD(chem), 73. *Prof Exp:* Postdoctoral assoc, Col Physicians & Surgeons, Columbia Univ, NY, 73-75; from asst prof to assoc prof, 75-85, PROF BIOCHEM, DEPT BIOCHEM & BIOPHYS, UNIV CALIF, DAVIS, 85-, CHAIR, 90- *Concurrent Pos:* Chmn, Subcomt Cell Biol-Organ Systs Biol Curric Comt Biochem Grad Group, Univ Calif, 77, chair, Educ Policy Comt, 79-80, rep- at-large, Assembly Acad Senate, 80-81, mem, Biol Sci Curric Comt, 80-85, Acad Planning Coun, 82-85 & Neurobiol Ctr Planning Comt, 89-90; mem bd dirs, Davis Sci Ctr, 83-87, pres, 83-84; ad hoc mem, Study Sect, NIH, 85, mem, Study Sect, Neurol Sci I, 88- *Mem:* Soc Neurosci; Biophys Soc; Am Soc Biochem & Molecular Biol; Sigma Xi. *Res:* Author of various publications. *Mailing Add:* Dept Biochem & Biophys Univ Calif Davis CA 95616

MCNARY, ROBERT REED, b Dayton, Ohio, Oct 9, 03; m 48. BIOCHEMISTRY. *Educ:* Univ Cincinnati, Chem E, 26, PhD(biochem), 36; Antioch Col, AM, 33. *Prof Exp:* Res chemist, Thomas & Hochwalt Labs, Ohio, 26-28, Frigidaire Corp, 28-32, Kettering Lab, Col Med, Univ Cincinnati, 36-43 & Citrus Exp Sta, Fla Citrus Comn, 46-59; consult chemist, 59-74; RETIRED. *Res:* By-product development of citrus fruits; methods of treating citrus cannery waste water; industrial hygiene chemistry; chlorophyll decompositions; freon refrigerants. *Mailing Add:* 101 N Riverside Dr Apt 202 New Smyrna Beach FL 32168

MCNAUGHT, DONALD CURTIS, b Detroit, Mich, May 1, 34; m 62; c 4. ZOOPLANKTON ECOLOGY, AQUATIC TOXICOLOGY. *Educ:* Univ Mich, BS, 56, MS, 57; Univ Wis, PhD(zool), 65. *Prof Exp:* Asst prof limnol, Mich State Univ, 65-68; assoc prof limnol, State Univ NY, Albany, 68-80; PROF ECOL, UNIV MINN, 80- *Concurrent Pos:* Dir, Cranberry Lake Biol Sta, 68-73; mem bd, Int Asn Lake Res, 74-75 & 83-85. *Mem:* Am Soc Limnol & Oceanog; Int Asn Great Lakes Res; Int Asn Theoret & Appl Limnol; Ecol Soc Am; Am Fisheries Soc; Am Inst Biol Sci. *Res:* Zooplankton ecology-functional dynamics of grazing, excretion; community structure related perturbation; aquatic ecotoxicology involving grazing, behaviors; behaviors of crustaceans, as vertical migrations; use of multiple high-frequency acoustics to sample plankton. *Mailing Add:* Dept Ecol Evolution & Behav Univ Minn St Paul MN 55108

MACNAUGHTAN, DONALD, JR, b Keene, NH, Apr 13, 39; m 68; c 3. ANALYTICAL CHEMISTRY. *Educ:* Univ NH, BS, 67; Purdue Univ, PhD(anal chem), 72. *Prof Exp:* Asst prof, Univ Ill, 71-74 & Loras Col, 74-75; res specialist, 75-80, group leader, 80-85, SECT MGR, MOBAY CHEM CORP, 85- *Res:* The application and development of analytical methodology for the solution of industrial problems in production and applications. *Mailing Add:* 1234 Robin Dr New Martinsville WV 26055

MCNAUGHTON, DUNCAN A, b Cornwall, Ont, Dec 7, 10. GEOLOGY. *Educ:* Univ Southern Calif, BA, 30, PhD(geol), 51; Calif Inst Technol, MS, 34. *Prof Exp:* Consult geologist, Myer Achtischin, 54-58; partner, Bedmar & McNaughton, 58-60; PETROL CONSULT, 60- *Mem:* Am Asn Petrol Geologist; Geol Soc Am. *Mailing Add:* 102 Eagle Cove St Austin TX 78734

MACNAUGHTON, EARL BRUCE, b Maple, Ont, Aug 29, 19; m 43; c 2. PHYSICS. *Educ:* Univ Toronto, BA, 41, MA, 46, PhD(physics), 48. *Prof Exp:* Instr physics, Univ Toronto, 41-44; prof, Ont Agr Col, 48-56, head dept, 56-65; prof & head dept, Wellington Col, 65-70, assoc dean sci, 66-70, dean col phys sci, 70-81, prof physics, Univ Guelph, 82-85; RETIRED. *Mem:* Am Phys Soc; Am Asn Physics Teachers; Can Asn Physicists. *Res:* Molecular spectroscopy; electronics and instrumentation. *Mailing Add:* 11 Harcourt Dr Guelph ON N1G 1J7 Can

MCNAUGHTON, JAMES LARRY, b Morton, Miss, Sept 22, 48; m 77. NUTRITION. *Educ:* Miss State Univ, BS, 70, MS, 72, PhD(nutrit), 75. *Prof Exp:* Nutritionist, Miss State Univ, 70-75; NUTRITIONIST, POULTRY RES LAB, AGR RES STA, USDA, 75- *Mem:* Am Inst Nutrit; US Poultry Sci Asn; World Poultry Sci Asn; Sigma Xi. *Res:* Increase poultry productivity; improve quality of feeds and feed components of farm animals; develop processing techniques for feeds. *Mailing Add:* 16 Southdale Dr USDA-Sci & Educ Admin-Agr Res PO Box 5367 Starkville MS 39759

MCNAUGHTON, MICHAEL WALFORD, b Durban, SAfrica, Mar 2, 43; US citizen; m 69; c 2. EXPERIMENTAL NUCLEAR PHYSICS. *Educ:* Univ London, BSc, 62, PhD(physics), 72; Oxford Univ, MA, 66. *Prof Exp:* Physicist, Crocker Nuclear Lab, Univ Calif, Davis, 72-75; sr res assoc nuclear physics, Case Western Reserve Univ, 75-78; STAFF MEM, LOS ALAMOS NAT LAB, 78- *Concurrent Pos:* Adj prof physics, Univ NMex, Los Alamos, 87- *Mem:* Fel Am Phys Soc. *Res:* Polarized few nucleon nuclear physics, neutron physics, polarized proton beams and targets; proton polarimeters; elementary particle physics. *Mailing Add:* MP1o MS H841 Los Alamos Sci Lab Los Alamos NM 87545

MCNAUGHTON, ROBERT, b Brooklyn, NY, Mar 13, 24; m 74; c 2. MATHEMATICS, COMPUTER SCIENCES. *Educ:* Columbia Univ, BA, 48; Harvard Univ, PhD(philos), 51. *Prof Exp:* Asst prof philos, Stanford Univ, 54-57; from asst prof to assoc prof elec eng, Univ Pa, 57-64; vis assoc prof elec eng & mem staff, Proj MAC, Mass Inst Technol, 64-66; PROF MATH & COMPUT SCI, RENSSELAER POLYTECH INST, 66- *Honors & Awards:* Levy Medal, 56. *Mem:* Asn Comput Mach; Asn Symbolic Logic. *Res:* Theory of automata, applications of symbolic logic and combinatorics of words. *Mailing Add:* 2511 15th St Troy NY 12180

MCNAUGHTON, SAMUEL J, b Takoma Park, Md, Aug 10, 39; m 59; c 2. PLANT ECOLOGY. *Educ:* Northwest Mo State Col, BS, 61; Univ Tex, PhD(bot), 64. *Prof Exp:* Asst prof biol, Portland State Col, 64-65; USPHS trainee, Stanford Univ, 65-66; from asst prof to assoc prof, 66-73, PROF BOT, SYRACUSE UNIV, 73- *Concurrent Pos:* Res scientist, Serengeti Res Inst, Tanzania, 74-; adj prof, Univ Dar es Salaam, 74-75. *Mem:* AAAS; Bot Soc Am; Ecol Soc Am; Brit Ecol Soc. *Res:* Ecology of grasslands and grazing ecosystems. *Mailing Add:* Biol Res Labs Syracuse Univ Syracuse NY 13244-1220

MCNAY, JOHN LEEPER, CLINICAL PHARMACOLOGY. *Educ:* Harvard Univ, MD, 57. *Prof Exp:* PROF PHARMACOL & INTERNAL MED, SCH MED, IND UNIV, 82- *Mailing Add:* Lilly Labs for Clin Res Wishard Mem Hosp 1001 W 10th St Indianapolis IN 46202

MCNEAL, BRIAN LESTER, b Cascade, Idaho, Jan 27, 38; m 58; c 4. SOIL CHEMISTRY, WATER CHEMISTRY. *Educ:* Ore State Univ, BS, 60, MS, 62; Univ Calif, Riverside, PhD(soil chem), 65. *Prof Exp:* Lab asst soil chem, Ore State Univ, 58-59; student trainee, Agr Res Serv, USDA, 59-60, soil scientist, 60-61, res soil scientist, US Salinity Lab, Calif, 61-70; assoc prof, Wash State Univ, 70-76, prof soils, 76-83; prof & chmn, 83-90, PROF SOIL SCI, UNIV FLA, 90- *Concurrent Pos:* Vis prof, Colo State Univ, 80; chmn, Soil Chem Div, Soil Sci Soc Am, 77; mem bd dirs, Am Soc Agron, 86-89; pres, Western Soc Soil Sci, 78. *Mem:* Fel Soil Sci Soc Am; fel Am Soc Agron; Int Soc Soil Sci. *Res:* Pollution chemistry; chemistry of salt-affected soils; soil physical chemistry; modeling of soil chemical processes; land application of wastes; soil-supplied nutrition of vegetables and citrus. *Mailing Add:* Dept Soil Sci Univ Fla Gainesville FL 32611

MCNEAL, DALE WILLIAM, JR, b Kansas City, Kans, Nov 23, 39; m 66. BOTANY. *Educ:* Colo Col, AB, 62; State Univ NY Col Forestry, Syracuse Univ, MS, 65; Wash State Univ, PhD(bot), 69. *Prof Exp:* From asst prof to assoc prof, 69-79, chmn, Dept Biol Sci, 78-84, PROF BIOL, UNIV OF THE PAC, 79- *Mem:* AAAS; Am Bot Soc; Am Inst Biol Sci; Am Soc Plant Taxon; Int Soc Plant Taxon; Sigma Xi. *Res:* Biosystematics of Allium; floristics of unusual environments. *Mailing Add:* Dept Biol Sci Univ the Pac Stockton CA 95211

MCNEAL, FRANCIS H, b Bartlett, Ore, Dec 9, 20; m 47; c 2. PLANT BREEDING, PLANT GENETICS. *Educ:* Ore State Col, BS, 43, MS, 48; Univ Minn, PhD(genetics), 53. *Prof Exp:* res agronomist, Agr Res Serv, USDA, 47-80; RETIRED. *Concurrent Pos:* Mem, Hard Red Spring Wheat Regional Comt, 49-; secy & tech adv, Western Wheat Improv Comt, 56-; mem, Nat Wheat Improv Comt, 59- *Mem:* Fel Am Soc Agron; fel Am Soc Crop Sci. *Res:* Development of improved wheat varieties for western states; genetic and related studies of the wheat plant. *Mailing Add:* Dept of Plant & Soil Sci Mont State Univ Bozeman MT 59715

MCNEAL, ROBERT JOSEPH, b Knoxville, Tenn, Dec 23, 37; m 62; c 2. CHEMICAL PHYSICS, SPACE PHYSICS. *Educ:* Univ Calif, Berkeley, BS, 59; Columbia Univ, MA, 61, PhD(chem), 63. *Prof Exp:* NSF fel, Harvard Univ, 63-64; head lab aeronomy dept, Space Physics Lab, Los Angeles, 64-70, asst dir, Chem & Physics Lab, Aerospace Corp, 72-78; mgr policy analysis div, Environ Res & Tech, 79-80; MGR TROP CHEM PROG, NASA HQ, WASHINGTON, DC, 80- *Concurrent Pos:* Prog dir atmospheric chem, NSF, 78-79. *Mem:* Fel Am Phys Soc; Am Geophys Union. *Res:* Atomic and molecular physics; chemical kinetics; aeronomy; atmospheric chemistry and physics. *Mailing Add:* NASA Hq Code SEU8 Washington DC 20546

MACNEE, ALAN B(RECK), b New York, NY, Sept 19, 20; m 46; c 4. ELECTRICAL ENGINEERING. *Educ:* Mass Inst Technol, SB & SM, 43, ScD(elec eng), 48. *Prof Exp:* Asst engr, Western Elec Co, NY, 41; asst, Bell Tel Labs, Inc, 41; mem staff, Radiation Lab, Mass Inst Technol, 43-45; group leader, Continental TV Co, Mass, 46; asst elec commun, Mass Inst Technol, 46-48, res assoc, 48-49 & Chalmers Tech Sweden, 49-50; from asst prof to prof, 50-89, EMER PROF ELEC ENG, UNIV MICH, ANN ARBOR, 89- *Concurrent Pos:* Vis prof, Chalmers Tech Sweden, 61-62; NASA sr assoc, Goddard Space Flight Ctr, 71-72; mem tech staff, Sandia Labs, 80-81. *Mem:* Fel Inst Elec & Electronic Engrs; for mem Swedish Royal Soc Lit & Sci; fel AAAS. *Res:* Design of wideband amplifiers and receivers; electronic computing devices; semiconductor modeling; semiconductor and distributed circuit design; computer aided circuit analysis and optimization; network synthesis. *Mailing Add:* Dept Elec Eng & Comput Sci Univ Mich Ann Arbor MI 48109

MCNEELY, JAMES BRADEN, b Paducah, Ky, July 31, 33; m 55; c 3. DESIGN & PREP MULTIJUNCTION SOLAR PHOTOVOLTAIC CELLS, DEVELOPMENT OF III-V OPTICAL INTERCONNECTS. *Educ:* Purdue Univ, BS, 55; Mass Inst Technol, SM, 57; Wash Univ, St Louis, Mo, MA, 68. *Prof Exp:* Asst dir chem eng pract, Mass Inst Technol Sch Chem Eng Pract, 56-57; develop engr polyolefin pilot, Res Div, Celanese, 57-59; group leader advan mat, New Enterprise Div, Monsanto, 59-71; sect head emitter mat, Tex Instruments, Inc, 71; dir res & develop LED's/LCD's Develop, Litronix, 71-77; mgr crystal prod, Mat Res, Allied Chem, 77-78; dir tech mkt semicon lasers, Laser Diode, Inc, 78-83; DIR MAT DEVELOP SOLAR CELLS, ASTRO POWER, INC, 83- *Concurrent Pos:* Prin investr, Astro Power Inc, 84-90. *Mem:* Inst Elec & Electronic Engrs Device Soc; Inst Elec & Electronic Engrs Lasers & Electro-Optics Soc; Mat Res Soc; Am Asn Crystal Growth (pres, 77-78); Int Soc Optical Eng. *Res:* Commercial manufacturing processes; compound semiconductor devices for semiconductor lasers; light emitting diodes and displays; high efficiency solar cells; microwave devices and integrated circuits. *Mailing Add:* Astro Power Inc Solar Park Newark DE 19716-2000

MCNEELY, ROBERT LEWIS, b Morganton, NC, June 5, 38. ANALYTICAL CHEMISTRY, INSTRUMENTAL ANALYSIS. *Educ:* Duke Univ, BS, 60; Univ NC, Chapel Hill, PhD(chem), 69. *Prof Exp:* Teaching asst chem, Calif Inst Technol, 60-62; Peace Corps teacher chem, Govt Col Nigeria, 63-65; res asst, Univ NC, Chapel Hill, 65-69; from asst prof to prof chem, Univ Tenn, Chattanooga, 69-86, asst dean, 84-90, actg assoc provost, 87-89, GROTE PROF CHEM, UNIV TENN, CHATTANOOGA, 86-, DEPT HEAD, 90- *Mem:* Am Chem Soc; Am Asn Univ Professors. *Res:* Instrumental analysis. *Mailing Add:* 900 Mountain Creek Rd Apt P-222 Chattanooga TN 37405-4505

MCNEES, ROGER WAYNE, b San Antonio, Tex, May 26, 44; m 70. HUMAN FACTORS ENGINEERING. *Educ:* St Mary's Univ, Tex, BBA, 66, MA, 68; Tex A&M Univ, MBA, 71, PhD(indust eng), 76. *Prof Exp:* Acct, Southwest Res Inst, 67-68; res asst, Tex Transp Inst, 71-76, res assoc human factors, 77-79, eng res assoc, 79-86; AT TEX ENG EXTEN SERV, TEX A&M UNIV, 86- *Concurrent Pos:* Lectr, Civil Eng Dept, Tex A&M Univ, 81-82. *Mem:* Human Factors Soc; Am Inst Indust Engrs; Inst Transp Engrs. *Res:* Simulation development for traffic systems; static and real-time information systems; perception-decision-reaction time systems; safety and human factors engineering; accident investigation analysis. *Mailing Add:* Tex Eng Exten Serv Transport Div Tex A&M Univ Col Sta TX 77843-8000

MCNEESE, LEONARD EUGENE, b Round Rock, Tex, May 11, 35; m 54; c 3. CHEMICAL ENGINEERING. *Educ:* Tex Tech Univ, BS, 57; Univ Tenn, MS, 63. *Prof Exp:* Develop engr chem eng, 57-68, proj mgr molten salt reactor processing, 68-73, sect chief unit opers, Chem Technol Div, 73, dir molten salt reactor prog, 74-76, sect head eng coord & anal, Chem Technol Div, 76, assoc dir, Chem Technol Div, 76, dir fossil energy prog, 77-85, MGR ENVIRON RESTORATON & FACIL UPGRADE PROG, OAK RIDGE NAT LAB, 85- *Concurrent Pos:* Consult to commissioners, Generic Environ Statement on Mixed Oxides, Nuclear Regulatory Agency, 76. *Honors & Awards:* Engr of the Yr, Am Inst Chem Engrs, 76. *Mem:* Am Inst Chem Engrs; Am Nuclear Soc; Am Soc Mech Engrs; Sigma Xi. *Res:* Fossil energy research and development; coal liquefaction, gasification, and combustion; nuclear fuel reprocessing; nonaqueous high temperature molten salt systems; uranium hexafluoride-sodium fluoride sorption reactions. *Mailing Add:* Bldg 3047 Mailstop 6023 Oak Ridge Nat Lab PO Box 2008 Oak Ridge TN 37831-6023

MCNEICE, GREGORY MALCOLM, b Bracebridge, Ont, Aug 11, 39; m 64; c 4. BIOENGINEERING, CIVIL ENGINEERING. *Educ:* Univ Waterloo, BASc, 64; Univ London, PhD(struct eng), 68. *Prof Exp:* Res asst prof, 68-69, from asst prof to assoc prof, 69-76, PROF CIVIL ENG, UNIV WATERLOO, 76- *Concurrent Pos:* Univ Waterloo fel, 71-72; vis asst prof, Brown Univ, 71-72. *Mem:* Am Acad Mech; Int Soc Biomech; Am Soc Biomech. *Res:* Finite element research and orthopaedic bioengineering; applications of engineering mechanics and materials science to areas of orthopaedic bioengineering; design and implementation of spinal implants for non-fusion paediatrics; theoretical and experimental design of adult total joint replacement implants; full scale testing and analysis of spinal stability, hip joints, knee joints and various musculo/skeletal joints; stress analysis, including finite element techniques of various engineering structures including total joint replacements; analytical modelling and full scale testing of various engineering structures, including automotive parts, truck frames and tractor-trailer experimentation; mathematical modelling and experimental testing of impact dynamics on the human body; test facilities include the Hybrid III anthropomorphic dummy and supporting electronics; studies of head/neck injury in sports and the work place. *Mailing Add:* Dept Civil Eng Univ Waterloo Waterloo ON N2L 3G1 Can

MCNEIL, BARBARA JOYCE, b Cambridge, Mass, Feb 11, 41. RADIOLOGY, NUCLEAR MEDICINE. *Educ:* Harvard Univ, MD, 66, PhD(biol chem), 72. *Prof Exp:* Intern pediat, Mass Gen Hosp, 66-67; instr radiol, Harvard Med Sch, 74-75, from asst prof to assoc prof radiol, 75-83, STAFF RADIOLOGIST NUCLEAR MED, JOINT PROG NUCLEAR MED, HARVARD MED SCH, 74-, PROF RADIOL & CLIN EPIDEMIOL, 82-, RIDLEY WATTS PROF, HEALTH CARE POLICY, 90- *Concurrent Pos:* Consult, Mass Radiol Soc Comt Nuclear Med, 74-; mem, Spec Comt Pub Health & Efficacy, Soc Nuclear Med, 75-; mem, Radiopharmaceut Adv Comt, Food & Drug Admin, HEW, 76-80; res career develop award, NIH, 76-81; Kieckhefer lectr, Harvard-Mass Inst Technol Div Health Sci & Technol, 78-79; comnr, Prospective Payment Assessment Comn, 83-; coun, Inst Med, 90- *Mem:* Inst Med-Nat Acad Sci; Am Chem Soc; Soc Nuclear Med; Asn Univ Radiologists; AAAS; Fleischner Soc. *Res:* Cost effectiveness of diagnostic and therapeutic medicine; cost containment in hospitals; health policy. *Mailing Add:* Dept Health Care Policy 25 Shattuck St Parcel B Boston MA 02115

MCNEIL, JEREMY NICHOL, b Tonbridge, Eng, Nov 20, 44; Can citizen; m; c 2. ENTOMOLOGY, ECOLOGY. *Educ:* Univ Western Ont, BSc, 69; NC State Univ, PhD(entom, ecol), 72. *Prof Exp:* From asst prof to assoc prof, 78-82, PROF ENTOM, LAVAL UNIV, 82- *Concurrent Pos:* Assoc ed, Can Entomologist, 76- *Honors & Awards:* C Gordon Hewitt Award, Entom Soc Can, 79; Leon Provancher Prize, 82; Gold Medal, Entom Soc Can, 87. *Mem:* Entom Soc Am; fel Entom Soc Can (pres, 89-90); Can Zool Soc; Int Soc Chem Ecol; Royal Entom Soc; Can Soc Zool. *Res:* Ecological and behavioral aspects of pheromone-mediated communication in insects; insect migration in response to predictable and unpredictable habitat deterioration. *Mailing Add:* Dept Biol Laval Univ Quebec PQ G1K 7P4 Can

MACNEIL, JOSEPH H, b Sydney, NS, Apr 13, 31; m 57; c 5. FOOD SCIENCE. *Educ:* McGill Univ, BSA, 55; Mich State Univ, MS, 58, PhD(food sci), 61. *Prof Exp:* Asst agr rep, NS Dept Agr, 52-53; field serv rep, Maritime Coop Serv, 55-57; res asst poultry mkt, Mich State Univ, 57-61; asst prof poultry sci, Univ Conn, 61-62; sr res assoc food res, Lever Bros Co, NJ, 62-64; asst prof poultry sci, 64-71, assoc prof food sci, 71-73, PROF FOOD SCI, PA STATE UNIV, UNIVERSITY PARK, 73- *Concurrent Pos:* Fulbright scholar, 74-75; vis scientist, Univ Peridenyia, Sri Lanka, 84; distinguished vis prof, Univ Zimbabwe, 84-86. *Mem:* Inst Food Technol; Poultry Sci Asn; Sigma Xi. *Res:* Processing technology of poultry meat and fish quality maintenance, sensory evaluation and product development, flavor changes increased utilization of underutilized food materials. *Mailing Add:* Dept of Food Sci Pa State Univ University Park PA 16802

MCNEIL, KENNETH MARTIN, b Edinburgh, Scotland, Oct 20, 41. REACTOR DESIGN, SEPARATIONS PROCESSES. *Educ:* Univ Edinburgh, BSc, 62; Cambridge Univ, PhD(chem eng), 65. *Prof Exp:* Proj chem engr, Amoco Chem Corp, 65-67, sr proj chem engr, 67-70; asst prof chem eng, Drexel Univ, 70-76; CONSULT CHEM ENG, KENNETH M McNEIL ASSOCS, 77-; ASSOC PROF ENG, WIDENER UNIV, 82-, CHMN CHEM ENG, 83- *Concurrent Pos:* Adj prof continuing educ, Drexel Univ, 72-78. *Mem:* Am Inst Chem Engrs; Brit Inst Chem Engrs; Am Chem Soc; Am Soc Eng Educ. *Res:* Fluid-fluid reactions; catalytic fluid-solid reactions; gas absorption; pollution control systems; fluidization. *Mailing Add:* 969 E 20th St Chester PA 19013-5615

MCNEIL, LAURIE ELIZABETH, b Ann Arbor, Mich, Aug 19, 56; m 82. LIGHT SCATTERING SEMICONDUCTORS & INSULATORS, DISORDERED MATERIALS. *Educ:* Harvard Univ, AB, 77, MS, 77; Univ Ill, MS, 79, PhD(physics), 82. *Prof Exp:* Postdoctoral fel mat sci, Mass Inst Technol, 82-84; asst prof, 84-91, ASSOC PROF PHYSICS, UNIV NC, CHAPEL HILL, 91- *Concurrent Pos:* Fac res partic, Argonne Nat Lab, 90. *Mem:* Am Phys Soc; Mat Res Soc; Sigma Xi. *Res:* Experimental studies of structure-property relations in semiconductors and insulators; raman, m06ssbauer and photoluminescence spectroscopies; amorphous and disordered semiconductors and insulators. *Mailing Add:* Dept Physics & Astron Univ NC CB No 3255 Chapel Hill NC 27599-3255h

MACNEIL, MICHAEL, b Warsaw, NY, Aug 7, 52; m 80; c 2. SIMULATION, SYSTEMS ANALYSIS. *Educ:* Cornell Univ, BS, 74; Mont State Univ, MS, 77; SDak State Univ, PhD. *Prof Exp:* Instr, animal breeding, SDak State Univ, 79-80; RES ANIMAL SCI, USDA-AGR RES SERV, 80- *Concurrent Pos:* Adj asst prof, Univ Nebr-Lincoln, 86- *Mem:* Am Soc Animal Sci; Am Asn Artificial Intel. *Res:* The use of research tools of computer sciences mathematics and systems analysis to identify ways for increasing the efficiency of beef production through improved management of resources. *Mailing Add:* USDA-ARS-NPA Ft Keogh Livestock & Range Res Lab Rte 1 Box 2021 Miles City MT 59301-9202

MCNEIL, MICHAEL BREWER, b Houston, Tex, July 26, 38; m 60; c 1. METALLURGY. *Educ:* Rice Univ, BA, 59, MA, 62; Univ Mo, Rolla, PhD(metall eng), 66. *Prof Exp:* Mem tech staff, Tex Instruments Inc, 62-64; lectr metall, Univ Mo, Rolla, 66; res asst physics, Univ Bristol, 66; assoc physicist, Midwest Res Inst, Mo, 66-68; assoc prof ceramic & metall eng, Miss State Univ, 68-72; metallurgist, Nat Bur Standards, 72-77, US Dept Energy, 77-81, US Nuclear Regulatory Comn, 81-87, Naval Coastal Systs Ctr, 87-91; METALLURGIST, OFF RES, US NUCLEAR REGULATORY COMN, 91- *Concurrent Pos:* Nat Acad Sci vis lectr, Univs Bucuresti & Cluj, Romania, 71. *Mem:* Fel Brit Inst Metallurgists. *Res:* Electrometallurgy; corrosion; microbiological materials science. *Mailing Add:* Off Res Nuclear Regulatory Comn Washington DC 20555

MCNEIL, PAUL L, GENERAL BIOLOGY. *Prof Exp:* ASSOC PROF, DEPT ANAT & CELL BIOL, HARVARD MED SCH, 91- *Mailing Add:* Dept Anat & Cell Biol Harvard Med Sch 220 Longwood Ave Boston MA 02115

MCNEIL, PHILLIP EUGENE, b Cincinnati, Ohio, May 13, 41; m 66; c 3. ALGEBRA, COMPUTER SCIENCE EDUCATION. *Educ:* Ohio Univ, BS, 63; Pa State Univ, MA, 65, PhD(math), 68. *Prof Exp:* Asst prof math, Xavier Univ, 68- 70 & Univ Cincinnati, 70-73; prog assoc math, Inst Serv Educ, 71-73; assoc prof, 73-77, head dept, 76-80, PROF MATH, NORFOLK STATE UNIV, 77-, HEAD DEPT, 89-, DIR INSTRNL RESOURCE CTR, 90- *Concurrent Pos:* Consult, Educ Develop Prog, Univ Cincinnati, 70-71, Minorities Comt, Nat Res Coun, 72, Norfolk Pub Schs, Va, 73, Inst Serv Educ, 73-80 & Pre-Freshman Eng Prog, Dept Energy, Washington, DC, 81; consult & lectr, SEEK Proj, Hunter Col, 72-73 & Racine Pub Schs, Wis, 73; dir, Minority Inst Sci Improv Prog, 74-78; consult-lectr, Minority Access to Res Prog, 83 & Spec Progs Inst, State Univ NY, Albany, 85; mem, Adv Coun Regional Asn Coop in Sci & Math, Dept Energy, 85; pres, Belleville Sr Housing, Inc, 83-86; consult, calculus reform comt, Phoenix, 90. *Honors & Awards:* Inst Serv Educ Plaque, 73; Woodrow Wilson fel, 62. *Mem:* Am Math Soc; Math Asn Am; Nat Asn Mathematicians; Asn Educ Data Systs. *Res:* Development of structure theorems in the area of algebraic semigroups; curriculum development in undergraduate mathematics; curriculum development in computer science education. *Mailing Add:* Dept Math & Comput Sci Norfolk State Univ 2401 Corprew Ave Norfolk VA 23504

MCNEIL, RAYMOND, b St Fabien de Panet, Que, Nov 30, 36; m 63; c 2. ORNITHOLOGY. *Educ:* Laval Univ, BA, 59; Univ Montreal, BSc, 62, MSc, 64, PhD(ornith), 68. *Prof Exp:* Teacher ecol & ornith, Oriente, Venezuela, 65-67; PROF BIOL SCI, UNIV MONTREAL, 67- *Mem:* Am Ornith Union; Cooper Ornith Soc; Wilson Ornith Soc; Brit Ornith Union. *Res:* Population ecology of birds; natural history of birds; fat deposition in migratory birds and its relationships with flyways and the phenomenon of summering in southern latitudes; ecological relationship of migratory and sedentary birds in tropical environments; nocturnality or nocturnal behavior and activities of water birds. *Mailing Add:* Dept Biol Sci Univ Montreal Montreal PQ H3C 3J7 Can

MCNEIL, RICHARD JEROME, b Marquette, Mich, Dec 22, 32; m 60; c 3. INTERNATIONAL DEVELOPMENT POLICY. *Educ:* Mich State Univ, BS, 54, MS, 57; Univ Mich, PhD(natural resources), 63. *Prof Exp:* Biologist, Mich Dept Conserv, 57-59, res biologist deer, 60-64; asst prof, 64-69, ASSOC PROF CONSERV, CORNELL UNIV, 70-, ASSOC PROF NATURAL RESOURCES, 81- *Concurrent Pos:* Fulbright scholar, NZ, 62-63; vis scientist, UN Environ Prog, 75. *Mem:* Nat Asn Environ Prof. *Res:* International natural resource problems; resource policy and planning; development planning; environmental education; social surveys in environmental affairs. *Mailing Add:* 112 Fernow Hall Col Agr Cornell Univ Ithaca NY 14853

MACNEILL, IAN B, b Regina, Sask, Dec 12, 31; m 52; c 3. MATHEMATICS, STATISTICS. *Educ:* Univ Sask, BA, 62; Queen's Univ, Ont, MA, 65; Stanford Univ, PhD(statist), 69. *Prof Exp:* Asst prof math, Univ Toronto, 66-71; assoc prof appl math, 71-80, PROF & CHMN STATIST & ACTUARIAL SCI, UNIV WESTERN ONT, 80-, DIR STATIST LAB, 77- *Concurrent Pos:* Dir, Ont Qual Assurance Ctr, 85- *Mem:* Fel Am Statist Asn. *Res:* Time series analysis. *Mailing Add:* Dept of Statist & Actuarial Sci Univ of Western Ont London ON N6A 5B9 Can

MCNEILL, JOHN, b Edinburgh, Scotland, Sept 15, 33; m 61, 90; c 2. BOTANY, TAXONOMY. *Educ:* Univ Edinburgh, BSc, 55, PhD(plant taxon), 60. *Prof Exp:* From asst lectr to lectr agr bot, Univ Reading, 57-61; lectr bot, Univ Liverpool, 61-69; actg assoc prof pop & environ biol, Univ Calif, Irvine, 69; sect chief taxon & econ bot, Plant Res Inst, Agr Can, 69-73; res scientist, Agr Can, 73-77, sr res scientist, Biosyst Res Inst, 77-81; prof & chmn, Dept Biol, Univ Ottawa, 81-87; Regius Keeper, Royal Bot Garden, Edinburgh, 87-89; assoc dir, curatorial, 89, actg dir, 90, DIR ROYAL ONT MUS, 90- *Concurrent Pos:* Vis assoc prof, Univ Wash, 69; adj prof, Carleton Univ, Ottawa, 74-80; adj prof, Univ Ottawa, 87-; hon prof, Univ Edinburgh, 88-89; prof, Univ Toronto, 90- *Mem:* Am Soc Naturalists; Am Soc Plant Taxonomists (counr, 81-87); Can Bot Asn; Int Asn Plant Taxon; Soc Syst Zool; Biol Coun Can (vpres, 84-86, pres, 86-87); Can Coun Univ Biol (chmn, vpres, 83-84, pres, 84-85, past pres, 85-86); Int Union Biol Sci. *Res:* Taxonomy and biosystematics of vascular plants, especially weeds; applications of numerical taxonomy to plant classification. *Mailing Add:* Royal Ont Mus 100 Queen's Park Toronto ON M5S 2C6 Can

MCNEILL, JOHN HUGH, b Chicago, Ill, Dec 5, 38; Can citizen; m 63; c 2. PHARMACOLOGY. *Educ:* Univ Alta, BSc, 60, MSc, 62; Univ Mich, PhD(pharmacol), 67. *Prof Exp:* Lab asst pharm, Univ Alta, 59-62, lectr, 63; lectr, Dalhousie Univ, 62-63; asst instr pharmacol, Mich State Univ, 66-65, asst prof, 67-71; assoc prof pharmacol, 71-75, prof & chmn div pharmacol & toxicol, 75-81, from asst dean to assoc dean, 78-85, DEAN, PHARMACEUT SCI, UNIV BC, 85- *Concurrent Pos:* Teaching fel pharmacol, Univ Mich, 63-66; res prof, Med Res Coun, 81-82. *Honors & Awards:* Upjohn Award, 83; McNeil Award, 83; Jacob Biely Prize, 85. *Mem:* AAAS; Am Fedn Clin Res; Pharmacol Soc Can; Am Soc Pharmacol & Exp Therapeut; NY Acad Sci; Int Soc Heart Res; Can Cardiovasc Soc; Am Pharmaceut Asn. *Res:* Drug interactions with the adrenergic amines on cardiac cyclic AMP; role of cyclic AMP in the cardiac actions of drugs; diabetes; induced cardiac changes. *Mailing Add:* Fac of Pharmaceut Sci Univ of BC Vancouver BC V6T 1W5 Can

MCNEILL, JOHN J, b Washington, DC, Dec 4, 22; m 65; c 2. MICROBIOLOGY. *Educ:* Univ Md, BS, 51, MS, 53, PhD(bact), 57. *Prof Exp:* Asst bact, Univ Md, 51-55; from asst prof to assoc prof animal sci & microbiol, 56-, EMER PROF ANIMAL SCI & MICROBIOL, NC STATE UNIV. *Mem:* AAAS; Am Soc Microbiol; fel Am Acad Microbiol; Brit Soc Gen Microbiol. *Res:* Bacterial lipid metabolism; rumen microbiology. *Mailing Add:* 305 Forest Rd Raleigh NC 27605

MCNEILL, KENNETH GORDON, b Appleton, Eng, Dec 21, 26; m 59; c 1. NUCLEAR PHYSICS, NUCLEAR MEDICINE. *Educ:* Oxford Univ, BA, 47, MA & PhD(physics), 50. *Prof Exp:* Fel nuclear physics, Yale Univ, 50-51; fel, Glasgow Univ, 51-52, lectr physics, 52-57; assoc prof, 57-63, PROF PHYSICS, UNIV TORONTO, 63-, PROF MED, 69- *Concurrent Pos:* Spec staff mem, Toronto Gen Hosp, 74-; tech adv nuclear planning Solicitor-Gen Ont; Sir Thomas Lyle fel, Univ Melbourne, 84. *Mem:* Can Radiation Protection Asn; Can Asn Physicists; fel Inst Nuclear Eng. *Res:* Low energy nuclear physics; photodisintegration; applications of nuclear physics to medicine. *Mailing Add:* Dept Physics Univ Toronto Toronto ON M5S 1A7 Can

MCNEILL, KENNY EARL, b Louisville, Miss, June 28, 46; m 77; c 2. BOTANY, PLANT PATHOLOGY. *Educ:* Miss State Univ, BS, 68, MS, 70, PhD(plant path), 75. *Prof Exp:* Plant sci rep, Lilly Res Labs, 75-78, regulatory assoc, 78-80, mgr, Int Regulatory Serv, 80-82, prod regist mgr, 82-87, SR CLIN RES ADMINR, LILLY RES LABS, 87- *Mailing Add:* Lilly Corp Ctr Eli Lilly & Co Indianapolis IN 46285

MCNEILL, MICHAEL JOHN, b Algona, Iowa, Sept 12, 42; m 67; c 3. GENETICS, PLANT BREEDING. *Educ:* Iowa State Univ, BS, 64, MS, 67, PhD(plant breeding), 69. *Prof Exp:* Res plant pathologist plant path div, US Biol Res Lab, 69-71; RES GENETICIST, FUNK SEEDS INT, INC, 71- *Mem:* Am Phytopath Soc; Am Soc Agron; Crop Sci Soc Am. *Res:* Plant breeding and pathology dealing mainly with cereal crops. *Mailing Add:* 10 Oakridge Dr Algona IA 50511

MCNEILL, ROBERT BRADLEY, b Martinsburg, WVa, June 20, 41; c 2. MATHEMATICS. *Educ:* Univ WVa, AB, 63; Pa State Univ, MA, 65, PhD(math), 68. *Prof Exp:* From asst prof to assoc prof, 68-79, PROF MATH, NORTHERN MICH UNIV, 79- *Mem:* AAAS; Am Math Soc; Math Asn Am; Soc Indust & Appl Math. *Res:* Qualitative behavior of solutions of differential equations and differential systems. *Mailing Add:* Dept Math Northern Mich Univ Marquette MI 49855

MCNELIS, EDWARD JOSEPH, b Philadelphia, Pa, Aug 17, 30; m 56; c 2. ORGANIC CHEMISTRY. *Educ:* Villanova Univ, BS, 53; Columbia Univ, PhD(chem), 60. *Prof Exp:* Res chemist, Sun Oil Co, Pa, 60-67; assoc prof, 67-76, chmn dept, 70-73 & 77-87, PROF CHEM, WASH SQ COL, NY UNIV, 77- *Concurrent Pos:* Vis assoc prof, Haverford Col, 66-67; vis prof, Athens Univ, 73-74. *Mem:* Am Chem Soc; AAAS; New York Acad Sci. *Res:* Olefin metathesis; oxidations with oxides of iodine. *Mailing Add:* Dept Chem NY Univ New York NY 10003

MCNERNEY, JAMES MURTHA, b Pittsburgh, Pa, Apr 3, 27; m 56; c 3. TOXICOLOGY, ENVIRONMENTAL HEALTH. *Educ:* Univ Pittsburgh, BS, 51, ML, 56, MPH, 57. *Prof Exp:* Res asst, Indust Hyg Found, Mellon Inst, 51-53, res assoc, 53-55, res toxicologist, 55-57, chief toxicologist, 57-64; assoc dir, Toxic Hazard Res Unit, Aerojet-Gen Corp, 64-66; chief animal toxicol, Environ Health Lab, Am Cyanamid Co, 66-68, toxicol group leader, 68-69; dir, Inhalation Toxicol Dept, TRW Hazleton Labs, 69-70; staff toxicologist, Am Petrol Inst, 70-77; dir toxicol, Cosmetic, Toiletry & Fragrance Asn Inc, 77-81, vpres, 81-88, consult, 88-89; RETIRED. *Mem:* Int Soc Regulatory Toxicol & Pharmacol; Am Indust Hyg Asn; Soc Toxicol. *Res:* Occupational and environmental toxicology; industrial hygiene. *Mailing Add:* 9513 Leemay St Vienna VA 22182

MCNESBY, JAMES ROBERT, b Bayonne, NJ, Apr 16, 22; m 49; c 3. PHYSICAL CHEMISTRY. *Educ:* Univ Ohio, Athens, BS, 43; NY Univ, PhD(chem), 52. *Prof Exp:* Res chemist, Interchem Corp, 45-49; phys chemist, US Naval Ord Test Sta, Calif, 51-56; phys chemist, Nat Bur Standards, 57-62, chief photochem sect, 62-67, chief phys chem div, 67-74, chief off air & water measurement, 74-76; PROF & CHMN CHEM DEPT, UNIV MD, 76- *Concurrent Pos:* Rockefeller pub serv fel, Univ Leeds, 58-59. *Mem:* Am Chem Soc; The Chem Soc. *Res:* Kinetics of free radical reactions; photochemistry. *Mailing Add:* 13308 Valley Dr Rockville MD 20850

MCNEVIN, SUSAN CLARDY, b Washington, DC, Dec 8, 47; m; c 2. SURFACE CHEMISTRY. *Educ:* Brown Univ, BS, 69; Cornell Univ, MS, 71, PhD(chem), 74. *Prof Exp:* MEM TECH STAFF, BELL LABS, 74- *Mem:* Am Vacuum Soc. *Res:* Surface chemistry and electronic materials processing, specifically problems involved with plasma etching and plasma deposition. *Mailing Add:* 1584 Springfield Ave New Providence NJ 07974

MACNICHOL, EDWARD FORD, JR, b Toledo, Ohio, Oct 24, 18; m 40; c 2. BIOPHYSICS, BIOENGINEERING & BIOMEDICAL ENGINEERING. *Educ:* Princeton Univ, AB, 41; Johns Hopkins Univ, PhD, 52. *Prof Exp:* Mem staff, Radiation Lab, Mass Inst Technol, 41-43, assoc group leader, 43-44, group leader, 45-46; asst, Johnson Found, Univ Pa, 46-48; asst biophys, Johns Hopkins Univ, 49-51, from instr to prof, 52-68; dir, Nat Inst Neurol Dis & Stroke, 68-73; asst dir, Marine Biol Lab, Woods Hole, 73-76, dir, Lab Sensory Physiol, 73-85; RETIRED. *Concurrent Pos:* Vis prof, Venezuelan Inst Sci Res, 57; guest scientist, US Naval Med Res Inst, 57-60; actg dir, Nat Eye Inst, 68-69; mem bd dirs, Deafness Res Found, 73-83; prof physiol, Sch Med, Boston Univ, 73-; mem exec coun, Comt Vision, Nat Res Coun, 75-79, chmn, 78; co-ed, Sensory Processes, 78-81. *Honors & Awards:* Morlock Award, Inst Elec & Electronic Engrs, 56, Centennial Medal, 84. *Mem:* Am Phys Soc; Biophys Soc; Am Physiol Soc; fel Inst Elec & Electronics Engrs; NY Acad Sci; hon mem Asn Res Vision & Ophthalmol. *Res:* Neurophysiology of retina and other sensory systems; instrumentation for biological research. *Mailing Add:* 120 Racing Ave Boston MA 02540

MCNICHOLAS, JAMES J, b Brooklyn, NY, July 18, 34. CHROMATOGRAPHY EQUIPMENT, SEPARATION PRODUCTS & EQUIPMENT. *Educ:* Col City NY, BS, 57. *Prof Exp:* Nuclear engr, Brookhaven Nat Lab, 60-69; prin investr, Nat Lead Co, 70-73; res group leader, Wallace & Tiernan, 73-80; res mgr, MGI Int, 80-83; MGR ENG, WHATMAN SPECIALTY PROD CO, 83- *Mem:* Am Inst Chem Engrs; Am Chem Soc. *Res:* Technical products and apparatus; high tech inventions such as chromatography apparatus fermentation systems and equipment, and air pollution control products. *Mailing Add:* 31 Stearns Rd East Brunswick NJ 08816

MCNICHOLAS, JOHN VINCENT, b Youngstown, Ohio, Sept 18, 36; m 63; c 3. ACOUSTICS. *Educ:* John Carroll Univ, BS, 58, MS, 59; Cath Univ Am, PhD(physics), 68. *Prof Exp:* Physicist, Naval Ship Res & Develop Ctr, 61-68, Hydrospace Res Corp, 68-71; PHYSICIST, APPL HYDRO ACOUST RES, INC, 71- *Mem:* Acoust Soc Am; Sigma Xi. *Res:* Truck noise measurements; quiet truck development. *Mailing Add:* 9700 Delaware Ct Rockville MD 20850

MCNICHOLAS, LAURA T, DRUG ABUSE, NEUROTRANSMITTERS. *Educ:* Univ Ky, PhD(pharmacol), 80. *Prof Exp:* INSTR PHARMACOL, UNIV KY, 83- *Res:* Neuropsychopharmacology. *Mailing Add:* 377 Plainview Rd Lexington KY 40517

MCNICHOLS, GERALD ROBERT, b Cleveland, Ohio, Nov 21, 43; m 64; c 3. ENGINEERING, OPERATIONS RESEARCH. *Educ:* Case Inst Technol, BS, 65; Univ Pa, MS, 66; George Washington Univ, ScD(eng), 76. *Prof Exp:* Sr analyst, Off Secy Defense, 70-76; vpres, J Watson Noah Assocs Inc, 76-78; pres, 78-87, CHMN & CHIEF EXEC OFFICER, MGT CONSULT & RES, INC, 87- *Concurrent Pos:* Adj prof, Am Univ, 67-72, George Washington Univ, 69-, Southeastern Univ, 74-75; dir, Mil Opers Res Soc, 85-88. *Honors & Awards:* Frieman Award, Int Soc Parametric Analysts, 90. *Mem:* Opers Res Soc Am; Inst Mgt Sci; Int Soc Parametric Analysts; Soc Logistics Engrs; Inst Cost Anal (pres, 85-88); Mil Opers Res Soc (vpres, 87-88, treas, 86-87); Sigma Xi. *Res:* Operations research, cost estimating, cost analysis, risk assessment, statistical methods useful to solve military and government analytical and economic problems. *Mailing Add:* 8133 Rondelay Lane Fairfax Station VA 22039

MCNICHOLS, ROGER J(EFFREY), b Columbus, Ohio, Sept 25, 38; m 67; c 2. ENGINEERING. *Educ:* Ohio State Univ, BIE, 62, MS, 64, PhD(indust eng), 66. *Prof Exp:* Asst mgr, Summit Hardware Co, 50-63; instr math, Ohio State Univ, 62-66; from asst prof to prof indust eng, Tex A&M Univ, 66-76; assoc dean eng, 77-80, PROF INDUST ENG, UNIV TOLEDO, 76-, CHMN INDUST ENG, 90- *Concurrent Pos:* Mem staff, Ohio Malleable Div, Dayton Malleable Iron Co, 56-60; steering comt, Annual Reliability & Maintainability Conf, 69-71; consult, Frankford Arsenal, 71-; mem bd dirs, Ann Reliability & Maintainability Symp, 71-77; pres, McNichols, Street & Assocs, Consult Engrs, 69-76; head grad eng exten progs, Red River Army Depot, Tex A&M Univ, 74-76. *Honors & Awards:* Reliability & Maintainability Award, Soc Logistics Engrs, 77; AT&T Found Award, Am Soc Eng Educ, 88. *Mem:* Inst Indust Engrs; Math Asn Am; Am Soc Qual Control; Am Soc Eng Educ; Inst Elec & Electronic Engrs. *Res:* Operations research; reliablity; groundwater statistics maintainability; automatic control systems; applied mathematics; engineering economics. *Mailing Add:* Col Eng Univ Toledo Toledo OH 43606

MCNICOL, LORE ANNE, b La Jolla, Calif; m 67. MICROBIOLOGY, MOLECULAR BIOLOGY. *Educ:* Univ Mont, BA, 65; Boston Univ, PhD(med sci), 68. *Prof Exp:* NIH fel microbiol & molecular biol, Sch Med, Tufts Univ, 68-70, res fel, 70-71; assoc microbiol, Sch Med, Univ Pa, 71-75, asst prof, 75-76; sr res assoc biol, Calif Inst Technol, 76-77; fel, 77-78, ASST PROF MICROBIOL, UNIV MD, 78- *Concurrent Pos:* Guest lectr infectious dis, Med Sch, Univ Mass, 71-72; Pa plan scholar human genetics, Sch Med, Univ Pa, 72-74; assoc, Inst Cancer Res, 74-75, NIH fel, 75-76. *Mem:* AAAS; Am Soc Microbiol. *Res:* Restriction and modification of T-even phage DNA; plasmids in estuarine bacteria. *Mailing Add:* 6901 Pine Way University Park MD 20782

MCNIEL, JAMES S(AMUEL), JR, b Dallas, Tex, July 4, 21; m 45; c 3. CHEMICAL ENGINEERING, RESEARCH MANAGEMENT. *Educ:* Univ Tex, BS, 43, MS, 46, PhD(chem eng), 50. *Prof Exp:* Sr res technologist, Field Res Lab, Magnolia Petrol Co Div, Socony Mobil Oil Co, Inc, 49-57, res assoc, 57-58, res sect supvr, 58-64, mgr explor & prod res div, Mobil Res & Develop Corp, 64-74, pres, Mobil Tyco Solar Energy Corp, 74-80, sr res adv, Mobil Res & Develop Corp, 80-81; vpres, Res & Tech Serv, Core Labs, Inc, 82-86; PRES, MCNIEL ASSOCS, 86- *Concurrent Pos:* Distinguished Lectr, Soc Petrol Engrs, 76-77. *Mem:* Am Inst Chem Engrs; Am Inst Mining, Metall & Petrol Engrs; Sigma Xi. *Res:* Secondary recovery of petroleum. *Mailing Add:* 6239 Twin Oaks Dallas TX 75240

MCNIFF, EDWARD J, JR, b Danvers, Mass, Sept 26, 35; m 58; c 4. EXPERIMENTAL SOLID STATE PHYSICS. *Educ:* Boston Col, BS, 57; Northeastern Univ, MS, 61. *Prof Exp:* Electronics engr missile systs div, Sylvania Elec Prods, Inc, 57-61; staff physicist, Arthur D Little Co, 61-64; STAFF PHYSICIST, NAT MAGNET LAB, MASS INST TECHNOL, 64- *Mem:* Am Phys Soc; AAAS. *Res:* High temperature, high field superconductors; magnetic effects in metals and dilute alloys. *Mailing Add:* Nat Magnet Lab Bldg NW 14 Mass Inst Technol Cambridge MA 02139

MACNINTCH, JOHN EDWIN, b Moncton, NB, Nov 7, 35; m 58; c 2. BIOCHEMISTRY. *Educ:* McGill Univ, BScAgr, 58; Purdue Univ, MS, 63, PhD(biochem), 65. *Prof Exp:* Nat Heart Asn fel cardiovasc training prog, Bowman Gray Sch Med, 65-66; sr res biochemist, Biochem Dept, Bristol Labs Inc, 66-72, sr res biochemist, Pharmacol Dept, Fibrinolytic Res & Develop, 72-74, asst dir res planning, 74-79, dir sci info serv, 79-82, dir sci info serv, Pharmaceut Res & Develop Div, Bristol-Meyers Co 82-88, DIR SCI INFO SERV, BRISTOL-MYERS SQUIBB PHARMACEUT RES INST, WALLINGFORD, 88- *Concurrent Pos:* Chmn, Sci Info Subsection Steering Comt, Pharmaceut Mfrs Asn, 85-86, Steering Comt, 82-86. *Mem:* Am Chem Soc; Pharmaceut Mfrs Asn; Sigma Xi. *Res:* Atherosclerosis and general cardiovascular disease biochemistry; lipid biochemistry; biochemistry of drug addiction; administration and scientific information services; drug evaluations; research planning. *Mailing Add:* Pharmaceut Res & Develop Div Bristol Meyers Corp Five Research Parkway PO Box 5100 Wallingford CT 06492-7660

MCNITT, JAMES R, b Chicago, Ill, May 10, 32; m 63; c 4. GEOTHERMAL RESOURCES EXPLORATION. *Educ:* Univ Notre Dame, BS, 53; Univ Ill, MS, 54; Univ Calif, PhD(geol), 61. *Prof Exp:* Geologist, Calif Div Mines & Geol, 58-65; tech adv geothermal energy, UN, 65-70, proj mgr, Geothermal Explor Proj, Kenya, 70-74, sr tech adv geothermal energy, 74-80; VPRES, GEOTHERMAL, INC, 80- *Mem:* Fel Geol Soc Am; Soc Petrol Eng; Am Geophys Union. *Res:* Exploration and development of geothermal energy. *Mailing Add:* 1101 Ivy Court El Cerrito CA 94530

MCNITT, RICHARD PAUL, b Reedsville, Pa, Aug 1, 35; m 57; c 4. ENGINEERING SCIENCE & MECHANICS. *Educ:* Pa State Univ, BS, 57, MS, 60; Purdue Univ, PhD(eng sci), 65. *Prof Exp:* Instr eng mech, Pa State Univ, 57-59; instr eng sci, Purdue Univ, 59-65; from asst prof to assoc prof eng mech, Va Polytech Inst & State Univ, 65-74, prof eng sci & mech, 74-81; PROF & HEAD ENG SCI & MECH, PA STATE UNIV, 81- *Mem:* Fel Soc Eng Sci (secy, 74-80, pres, 81); Am Soc Eng Educ; Soc Exp Stress Anal; Am Acad Mech (treas, 77-81). *Res:* Fatigue; hydrogen embrittlement; continuum mechanics; fracture; environmental degradation of materials. *Mailing Add:* Dept Eng Sci & Mech Pa State Univ University Park PA 16802

MCNIVEN, HUGH D(ONALD), b Toronto, Ont, Aug 6, 22; nat US; m 59; c 1. MECHANICS. *Educ:* Univ Toronto, BASc, 44; Cornell Univ, MCE, 47; Columbia Univ, PhD(pure sci), 58. *Prof Exp:* Assoc prof, 57-64, PROF ENG SCI, UNIV CALIF, BERKELEY, 64- *Concurrent Pos:* NSF fel, 63-64. *Mem:* Am Soc Civil Engrs; Am Soc Mech Engrs; fel Acoust Soc Am; Am Math Soc. *Res:* Applied mechanics; high frequency vibrations of elastic solids; wave propagation in deformable media; earthquake engineering; mathematical modeling of earthquake structure. *Mailing Add:* Dept Civil Univ Calif Berkeley CA 94720

MCNOWN, JOHN S, b Jan 15, 16; US citizen; m 73; c 4. CIVIL ENGINEERING. *Educ:* Kans Univ, BS, 36, Iowa Univ, MS, 37; Minn Univ, PhD(hydraul eng), 42; Univ Grenoble, DSc, 51. *Prof Exp:* Instr math & mech, Univ Minn, 37-42; from asst prof to prof mech & hydraul, res engr & assoc dir, Inst Hydraul Res, Univ Iowa, 43-54; prof eng mech, Univ Mich, 54-57; prof & dean eng & archit, Univ Kans, 57-65, dir, Ctr Res, 60-65, Albert P Learned prof civil eng, 65-86; RETIRED. *Concurrent Pos:* Res assoc, Div War Res, Univ Calif, 42-43; Fulbright res scholar, Grenoble, France, 50-51; mem adv panel, NSF, 57-60; mem, Eng Educ Comt, Engrs Coun Prof Develop, 60-65; mem, Comn Eng Educ, 61-65; mem overseas liaison comt, Am Coun Educ, 67-74; tech educ specialist, World Bank, Wash, 72-73; consult, NSF, Sandia Corp, US Engr Corps, Ford Found, World Bank, Unesco, Bur Pub Roads, Bel Telephone Labs & Swed Power Bd; vis prof, Chalmers Tech Univ, Göteborg & Royal Tech Univ, Stockholm, 77-78. *Honors & Awards:* J C Stevens Prize, Am Soc Civil Engrs, 46, J Jas R Croes Medal, 55. *Mem:* Nat Acad Eng; fel Am Acad Mech. *Res:* Basic studies of the principles and applications of fluid flow at junctions. *Mailing Add:* Royal Inst Technol S-10044 Stockholm Sweden

MCNULTY, CHARLES LEE, JR, b Dallas, Tex, Feb 4, 18; m 42; c 3. PALEONTOLOGY. *Educ:* Southern Methodist Univ, BS, 40; Syracuse Univ, MS, 48; Univ Okla, PhD, 55. *Prof Exp:* Asst geol, Syracuse Univ, 40-42; asst prof Arlington State Col, 46-48; instr, Univ Okla, 48-49; assoc prof, Arlington State Col, 50-51; geologist, Concho Petrol Co, 51-53 & Continental Oil Co, 53-57; PROF GEOL, UNIV TEX, ARLINGTON, 57- *Mem:* Fel Geol Soc Am; Soc Econ Paleont & Mineral; Am Asn Petrol Geol; Swiss Geol Soc. *Res:* Micropaleontology, mainly small foraminifera; stratigraphy of Texas. *Mailing Add:* Dept Geol Box 19049 Univ Tex Arlington TX 76019

MCNULTY, GEORGE FRANK, b Palo Alto, Calif, June 18, 45; m 81; c 2. EQUATIONAL LOGIC, UNIVERSAL ALGEBRA. *Educ:* Harvey Mudd Col, BS, 67; Univ Calif, Berkeley, MS, 69, PhD(math), 72. *Prof Exp:* Nat Res Coun fel, Univ Man, 72-73; res instr, Dartmouth Col, 73-75; from asst prof to assoc prof, 75-86, PROF MATH, UNIV SC, 86- *Concurrent Pos:* Vis assoc prof, Univ Calif, San Diego, 79, Univ Hawaii, 82 & Univ Colo, 85; Fulbright fel, 82-83 & Alexander von Humboldt res fel, 83. *Mem:* Am Math Soc; Asn Symbolic Logic; Math Asn Am; Soc Indust & Appl Math. *Res:* Foundations of mathematics and general theory of algebraic structures, especially on the connections between model theory, set theory and the theory of equational classes. *Mailing Add:* Dept Math Univ SC Columbia SC 29208

MCNULTY, IRVING BAZIL, b Salt Lake City, Utah, Jan 6, 18; m 43; c 3. PLANT PHYSIOLOGY. *Educ:* Univ Utah, BS, 42, MS, 47; Ohio State Univ, PhD(plant physiol), 52. *Prof Exp:* Instr biol, bot & plant physiol, 47-53, from asst prof to assoc prof, 53-65, PROF BIOL, UNIV UTAH, 65- *Concurrent Pos:* Head dept bot, Univ Utah, 60-69. *Mem:* AAAS; Bot Soc Am; Am Soc Plant Physiol. *Res:* Physiology of mineral nutrition of halophytes. *Mailing Add:* Dept Biol Univ Utah Salt Lake City UT 84112

MCNULTY, JOHN ALEXANDER, b Bogota, Colombia, July 14, 46; US citizen; m; c 2. COMPARATIVE MORPHOLOGY, CELL BIOLOGY. *Educ:* Univ of the Pac, BA, 68; Univ Southern Calif, PhD(biol), 76. *Prof Exp:* Res asst biol, Univ Southern Calif, 70; from asst prof, 76-81 to assoc prof, 81-88, PROF ANAT, LOYOLA UNIV, CHICAGO, 88- *Mem:* Am Asn Anatomists; Int Soc Cell Biol; Am Soc Ichthyologists & Herpetologists; Am Soc Zoologists. *Res:* Neuropendocrinology of the pineal gland. *Mailing Add:* Dept of Anat Loyola Univ 2160 S First Ave Maywood IL 60153

MCNULTY, PETER J, b New York, NY, Aug 2, 41; m 66; c 2. BIOPHYSICS, RADIATION PHYSICS. *Educ:* Fordham Univ, BS, 62; State Univ NY Buffalo, PhD(physics), 65. *Prof Exp:* Asst physics, State Univ NY Buffalo, 62-65, fel, 65-66; from asst prof to prof physics, Clarkson Univ, 66-88; PROF & HEAD PHYSICS & ASTRON, CLEMSON UNIV, 88- *Concurrent Pos:* Nat Acad Sci-Nat Res Coun sr resident res assoc, 70-71 & 79-80; vis assoc scientist, Brookhaven Nat Lab, 72-73, res collabr, Med Dept, 71- *Mem:* AAAS; Am Phys Soc; Inst Elec & Electronic Engrs; Radiation Res Soc. *Res:* Biological effects of radiation; radiation dosimetry; soft errors in microelectronics; fluorescent and Raman scattering by microstructures. *Mailing Add:* Dept Physics & Astron Clemson Univ 117 Kinard Lab Physics Clemson SC 29634-1911

MCNULTY, RICHARD PAUL, b Scranton, Pa, Apr 23, 46. METEOROLOGY, THUNDERSTORMS. *Educ:* New York Univ, BS, 68, MS, 72, PhD(meteorol), 74. *Prof Exp:* Weather officer, US Navy, 69-72; acad assoc meteorol, Polytech Inst NY, 74-76; res meteorologist, Nat Severe Storms Forecast Ctr, Kansas City, Mo, 76-80; dep meterologist in-chg, Nat Weather Serv Forecast Off, Kans, 80-90, CHIEF, HYDROMETEROLOGY & MGT DIV, NAT WEATHER SERV TRAINING CTR, 91- *Concurrent Pos:* Assoc ed, Nat Weather Digest, 80-; adj instr, Washburn Univ, 82-88. *Mem:* Am Meteorol Soc; Am Geophys Union; Nat Weather Asn. *Res:* Techniques for forecasting severe thunderstorms and tornadoes, specifically the use of the jet stream, differential advection and statistical techniques; conceptual approach to thunderstorm forecasting. *Mailing Add:* Nat Weather Serv Training Ctr 617 Hardesty Kansas City MO 64124

MCNULTY, WILBUR PALMER, b Iowa City, Iowa, Sept 23, 25; m 59; c 5. TOXICOLOGY, VIROLOGY. *Educ:* Yale Univ, BS, 47, MD, 52. *Prof Exp:* Asst prof path, Yale Med Sch, 59-63; ASSOC PROF PATH, ORE HEALTH SCI UNIV, 63-; head, Div Primate Med, 63-87, SCIENTIST, ORE REGIONAL PRIMATE RES CTR, 87- *Res:* Pathogenesis of poisoning of primates by halogenated aromatic compounds; retroviral infections in primates. *Mailing Add:* Ore Regional Primate Res Ctr 505 NW 185th Ave Beaverton OR 97006

MCNUTT, CHARLES HARRISON, b Denver, Colo, Dec 11, 28; m 55; c 2. ARCHEOLOGY, ANTHROPOLOGY. *Educ:* Univ S, Sewanee, BS, 50; Univ NMex, MA, 54; Univ Mich, Ann Arbor, PhD(anthropol). *Prof Exp:* Seismic mapping, Century Geophys Corp, 52-53; tech adv, Nat Mus Can, 56; archeologist, Smithsonian Inst, 57-60; asst prof anthropol, Univ Tenn, Knoxville, 60-62 & Ariz State Col, 62-64; assoc prof, 64-68, PROF ANTHROPOL, MEMPHIS STATE UNIV, 68- *Concurrent Pos:* Prin investr, Tenn Valley Authority, Nat Park Serv, 60-61, Nat Sci Found, 65, Cumberland River, 80-83 & 86-87, Little Bear Creek, 84-85 & Jackson County, Tenn Dept Transp, 88-89; Sigma Xi lectr, 78. *Mem:* Fel Am Anthropol Asn; Soc Am Archeol. *Res:* Quantitative data analysis in archeology, general anthropological analytical methods and archeology of the eastern United States. *Mailing Add:* Dept Anthropol Memphis State Univ Memphis TN 38152

MCNUTT, CLARENCE WALLACE, b Ozan, Ark, Aug 5, 13; m 39; c 4. GENETICS. *Educ:* Henderson State Col, AB, 35; La State Univ, MS, 38; Brown Univ, PhD(biol, genetics), 41. *Prof Exp:* Instr gross anat, Univ Wis, 46-50; assoc prof anat, Univ Tex Med Br Galveston, 50-67; PROF ANAT, UNIV TEX HEALTH SCI CTR, SAN ANTONIO, 67- *Concurrent Pos:* Muellhaupt fel, Ohio State Univ, 41-42; vis staff dept surg, Brooke Gen Hosp, Ft Sam Houston, Tex. *Mem:* Fel AAAS; Soc Exp Biol & Med; Genetics Soc Am; Am Soc Human Genetics; Am Asn Anat; Sigma Xi. *Res:* Mammalian developmental genetics; human genetics; neurological conditions in mice and man. *Mailing Add:* 15711 NW Military Hwy San Antonio TX 78231

MCNUTT, DOUGLAS P, b Rome, Ga, Apr 24, 35; m 59; c 4. PHYSICS, COMPUTER SCIENCES. *Educ:* Wesleyan Univ, BA, 56; Univ Wis, MS, 57, PhD(physics), 62. *Prof Exp:* Proj assoc interference spectros, Univ Wis, 62-63; res physicist, US Naval Res Lab, 63-83; CONSULT, 83- *Mem:* AAAS; Optical Soc Am; Sigma Xi; Inst Elec & Electronic Engrs. *Res:* Spectroscopic determination of atmospheric sodium; x-ray infrared and microwave rocket astronomy and aeronomy. *Mailing Add:* 5918 Veranda Dr Springfield VA 22152-1416

MCNUTT, KRISTEN W, b Nashville, Tenn, Nov 17, 41. NUTRITION. *Educ:* Duke Univ, BA, 63; Columbia Univ, MD, 65; Vanderbilt Univ, PhD(biochem), 70. *Prof Exp:* Asst prof nutrit, Univ Ill Med Ctr, 81-82; assoc dir sci & pub affairs, Good Housekeeping Inst, 82-85; pres consumer & sci affairs, Kraft Inc, 85-88; PRES, CONSUMER CHOICES UNLTD, 88- *Mem:* Soc Nutrit Educ; Am Inst Nutrit. *Mailing Add:* Consumer Choices Unltd 1742 Asbury Ave Evanston IL 60201

MCNUTT, MARCIA KEMPER, b Minneapolis, Minn, Feb 19, 52; m 78; c 3. TECTONOPHYSICS. *Educ:* Colo Col, BA, 73; Scripps Inst Oceanog, PhD(earth sci), 78. *Prof Exp:* Vis asst prof, Univ Minn, 78-79; geophysicist, US Geol Surv, 79-82; ASST PROF, 82-, ASSOC PROF GEOPHYSICS, MASS INST TECHNOL, 86- *Concurrent Pos:* Mem, NASA Sci Steering Group Geopotential Res Mission, 78-, Comt Geodesy Nat Res Coun, 82-84 Geodynamics Comt, 84-87 & Tectonics Ed Search Comt, 83; assoc ed, J Geophys Res, 80-83, guest ed, 83; ed bd, Tectonophysics, 82-; mem comt earth sci, Nat Res Coun, 87- *Honors & Awards:* Macelevane Award, Am Geophys Union, 88. *Mem:* Am Geophys Union; John Muir Geophys Soc (secy, 79-83). *Res:* Studies of long-term rheology of the earth's crust and upper mantle using gravity and topography data, isotasy, paleomagnetism of seamounts, thermal modeling of lithosphere. *Mailing Add:* Dept Earth & Planetary Sci Mass Inst Technol Cambridge MA 02139

MCNUTT, MICHAEL JOHN, b Rochester, NY, April 22, 47. SILICON INTEGRATED CIRCUITS. *Educ:* Mass Inst Technol, SB, 69; Univ Ill, MS, 70, PhD(elec eng), 74; Calif State Univ, MBA, 85. *Prof Exp:* Res assoc, Univ Ill, 74-75, asst prof, 75-77; mem tech staff, Rockwell Int, 77-81, mgr process characterization & reliability, 81-83; vpres technol, Holt Inc, 83-86; PRIN ENGR, LORAL AEROSPACE CORP, 86- *Mem:* Soc Photo-Optical Instrumentation Engrs; Electrochem Soc; Inst Elec & Electronic Engrs. *Res:* Process reliability and parameter characterization for complementary metal-oxide semiconductor silicon-on-sapphire; characterization for complementary metal-oxide semiconductor bulk silicon; metal-oxide semiconductor device physics; high frequency integrated circuit design and functional test; VLSI reliability and failure analysis; infrared focal plane arrays. *Mailing Add:* 21242 Calle Olivia El Toro CA 92630-2151

MCNUTT, RALPH LEROY, JR, b Fort Worth, Tex, Oct 29, 53; m 80; c 2. SPACE PHYSICS, PLANETARY SCIENCE. *Educ:* Tex A&M Univ, BS, 75; Mass Inst Technol, PhD(physics), 80. *Prof Exp:* Mem tech staff, Sandia Nat Lab, Albuquerque, NMex, 80-81; res scientist, Ctr Space Res, 81-82, asst prof, 82-86, ASSOC PROF PHYSICS, MASS INST TECHNOL, 86- *Concurrent Pos:* Consult, Sandia Nat Lab, Alburquerque, NMex, 81-; prin invest, Voyager Uranus Data Anal Prog, 87-90. *Mem:* Am Geophys Union; British Interplanetary Soc; Sigma Xi; Planetary Soc. *Res:* Analysis and interpretation of plasma data taken inside of the Jovian and Saturnian magnetospheres; theory of magnetospheric equilibria. *Mailing Add:* 213 Hunnewell St Needham MA 02194

MCNUTT, ROBERT HAROLD, b Moncton, NB, July 4, 37; m 64; c 3. GEOCHEMISTRY. *Educ:* Univ NB, BSc, 59; Mass Inst Technol, PhD(geol), 65. *Prof Exp:* From asst prof to assoc prof, 65-81, chmn, Dept Geol, 84-88, PROF GEOL, MCMASTER UNIV, 81-, DEAN SCI, 89- *Concurrent Pos:* Ed, Geosci Can, 78-82. *Mem:* Am Geophys Union; Geol Asn Can; Geochem Soc. *Res:* Geochemistry and strontium Neodymium Lead and Osmium isotopic studies of Archean and Grenville gneissic terrains; isotoic geochemistry of Precambrian shield and sedimentary basin brines. *Mailing Add:* Dept of Geol McMaster Univ, 1280 Main St W Hamilton ON L8S 4L8 Can

MCNUTT, RONALD CLAY, b Birmingham, Ala, Oct 29, 29; m 54; c 1. ANALYTICAL CHEMISTRY, INORGANIC CHEMISTRY. *Educ:* Athens Col, BS, 59; Vanderbilt Univ, MS, 61, PhD(chem), 66. *Prof Exp:* Res technician chem, Chemstrand Corp, Ala, 53-59, res chemist, Chemstrand Res Ctr, NC, 61-62; assoc prof, Athens Col, 66-68, PROF CHEM & CHMN DEPT, 68- *Concurrent Pos:* Sabbatical, nondestructive evaluation, Marshall Space Flight Ctr, NASA, 81-82. *Mem:* Am Chem Soc; Am Soc Nondestructive Testing. *Res:* Analysis of polymers related to textile and tire industry; reactivity of coordinated ligands; analysis of liquid rocket fuels; analysis of contaminants on spacecraft surfaces; ultrasonic testing of solid propellants. *Mailing Add:* Dept of Chem Athens State Col Athens AL 35611-3589

MACOMBER, HILLIARD KENT, b Catania, Sicily, Italy, Dec 23, 33; US citizen; m 61; c 1. COMPUTATIONAL PHYSICS, NONLINEAR DYNAMICS. *Educ:* Univ Calif, Berkeley, BS, 55, MS, 57; Harvard Univ, PhD(appl physics), 65. *Prof Exp:* Sr scientist, Technol Div, GCA Corp, 65-67; asst prof, Robert Col, Turkey, 67-70; vis scientist & asst prof, Middle East Tech Univ, Turkey, 70-75; from asst prof to assoc prof, 76-86, PROF, UNIV NORTHERN IOWA, 86- *Concurrent Pos:* Fac res partic & consult, Energy & Environ Systs Div, Argonne Nat Lab, 78-80. *Mem:* Am Phys Soc; Am Asn Physics Teachers. *Res:* Plasma propulsion; kinetic theory; atomic collision theory; applied analysis in physics. *Mailing Add:* Dept Physics Univ Northern Iowa Cedar Falls IA 50614-0150

MACOMBER, RICHARD WILTZ, b Chicago, Ill, June 6, 32; m 57. PALEONTOLOGY, STRATIGRAPHY. *Educ:* Northwestern Univ, BS, 54, MS, 59; Harvard Univ, AM, 63; Univ Iowa, PhD(paleont), 68. *Prof Exp:* Geologist, Bear Creek Mining Co, 57-58; geologist, US Geol Surv, 62-63; cur geol, Northwestern Univ, 65-66; from instr to assoc prof, 66-76, PROF EARTH SCI, LONG ISLAND UNIV, 76-, CHMN PHYSICS, 85-88. *Mem:* Sigma Xi. *Res:* Ordovician brachiopod paleontology; Ordovician stratigraphy of North America and Europe. *Mailing Add:* Dept of Physics Long Island Univ Brooklyn NY 11201

MACOMBER, THOMAS WESSON, b Bakersfield, Calif, Nov 1, 12; m 46; c 4. STRUCTURAL DESIGN, HEAT TRANSMISSION. *Educ:* Stanford Univ, AB, 34, ME, 38. *Prof Exp:* Res engr, Ray Oil Burner Co, Calif, 34-35; design engr, F Jaden Mfg Co, Nebr, 38; designer, Cent Nebr Pub Power & Irrig Dist, 38; asst proj engr, Soil Conserv Serv, USDA, Nebr, 39; stress anal engr, NAm Aviation, Inc, Calif, 39-40; asst engr, Ames Aeronaut Lab, Nat Adv Comt Aeronaut, Calif, 40-43; mech res engr, Radiation Lab, Univ Calif, 43-45; design engr, Westinghouse Elec Corp, Calif, 45-55 & Ampex Corp, Calif, 55; test facility design engr, Atomic Power Equip Dept, Gen Elec Co, Calif, 55-57; design specialist, Lockheed Missile & Space Co, 57-70; consult, 71-72; sr engr, Western Consult Engrs, Gen Elec Co, San Jose, 73-80;

RETIRED. *Honors & Awards:* Silver Prize, AEC. *Mem:* Am Soc Mech Engrs; Audio Eng Soc. *Res:* Heat transfer; patent, hydraulic control device for aircraft; analysis of systems; analytical design of mechanism and structure; design to resist severe environments; computerized analysis and design of steam piping systems and of fuel storage units for nuclear power plants. *Mailing Add:* 216 Marich Way Los Altos CA 94022-1402

MACON, NATHANIEL, b Durham, NC, Nov 15, 26; m 53; c 4. SOFTWARE SYSTEMS, APPLIED MATHEMATICS. *Educ:* Univ NC, BA, 46, MA, 48, PhD(math), 51. *Prof Exp:* From asst prof to prof math & computer sci, Auburn Univ, 52-66; scientist computer sci, Inst Defense Anal, 66-68; prof math & computer sci, Am Univ, 68-80; software eng prof computer sci, Philips, Neth, 80-85; consult computer sci, Trans Atlantic Software Asn, 85-87; PROG DIR COMPUTER SCI, NSF, 87- *Concurrent Pos:* Postdoctoral student, Free Univ Amsterdam, Neth, 51-52; consult, var orgn, 53-87; vpres, exec secy & gov, Int Coun Computer Commun, 72- *Mem:* Int Coun Computer Commun; Asn Computer Mach; Math Asn Am; Am Math Soc; Sigma Xi. *Res:* Software development. *Mailing Add:* 6104 Namakagan Rd Bethesda MD 20816

MCOSKER, CHARLES C, b Ithaca, NY, Dec 31, 54. ANTI-INFECTIVE RESEARCH. *Educ:* Hope Col, BA, 76; Cornell Univ, PhD(biochem), 82. *Prof Exp:* Postdoctoral fel muscular dystrophy, Cornell Univ, 82-83; staff scientist, 83-91, GROUP LEADER, ANTI-INFECTIVE RES DEPT, NORWICH EATON PHARMACEUT, 91- *Mem:* Am Soc Cell Biol; Am Soc Microbiol. *Mailing Add:* Norwich Eaton Pharmaceut Box 191 Norwich NY 13815

MACOSKO, CHRISTOPHER WARD, b Bridgeport, Conn, June 14, 44; m 67; c 4. POLYMER SCIENCE, RHEOLOGY. *Educ:* Carnegie-Mellon Univ, BS, 66; Univ London, MSc, 67, Imp Col, dipl, 67; Princeton Univ, PhD(chem eng), 70. *Prof Exp:* Mem res staff, Western Elec Eng Res Ctr, 68-70; from asst prof to assoc prof, 70-79, PROF CHEM ENG & MAT SCI, UNIV MINN, MINNEAPOLIS, 79- *Concurrent Pos:* Consult, Rheometrics, Inc, 70-; vis prof, Univ Louis Pasteur, Strasbourg, France. *Honors & Awards:* Int Res Award, Soc Plastics Engrs, 86; Charles Stein Award in Mat, Am Inst Chem Engrs, 88. *Mem:* Am Inst Chem Engrs; Am Chem Soc; Soc Plastics Engrs; Soc Rheology; Brit Soc Rheology. *Res:* Polymer rheology; polymer processing; model networks, thermosets and polymer reaction molding; dynamic mechanical properties; suspension rheology coating floros. *Mailing Add:* Dept of Chem Eng & Mat Sci Univ of Minn Minneapolis MN 55455

MACOVSKI, ALBERT, b New York, NY, May 2, 29; m 50; c 2. ELECTRICAL ENGINEERING, MEDICAL INSTRUMENTS. *Educ:* City Col New York, BEE, 50; Polytech Inst Brooklyn, MEE, 53; Stanford Univ, PhD, 68. *Prof Exp:* Engr, RCA Labs, Inc, 50-57; from asst prof to assoc prof elec eng, Polytech Inst Brooklyn, 57-60; res engr, SRI Inst, 60-61, sr res engr, 61-68, staff scientist, 68-72; PROF, STANFORD UNIV, 72- *Concurrent Pos:* Consult, RCA Labs, Inc, NJ, 57-60; NIH spec res fel diag radiol, Univ Calif, San Francisco, 71-72. *Honors & Awards:* Award, Inst Elec & Electronics Engrs, 57, Zworykin Award, 73. *Mem:* Nat Acad Eng; fel Optical Soc Am; Am Asn Physicists in Med; Soc Mag Res Med; Sigma Xi; fel Inst Elec & Electronic Engrs. *Res:* Optical devices and systems; medical imaging systems. *Mailing Add:* Dept Elec Eng Stanford Univ Stanford CA 94305

MCOWEN, ROBERT C, b Jan 12, 51; m. PARTIAL DIFFERENTIAL EQUATIONS. *Educ:* Univ Calif, Berkeley, PhD(math), 78. *Prof Exp:* ASSOC PROF, DEPT MATH, NORTHEASTERN UNIV, 81- *Concurrent Pos:* Am Math Soc postdoctoral fel, 78. *Mem:* Am Math Soc. *Res:* Elliptic equations arising in differential geometry. *Mailing Add:* Dept Math Northeastern Univ Boston MA 02115

MACPEEK, DONALD LESTER, b Andover, NJ, Apr 4, 28; m 50; c 3. INDUSTRIAL ORGANIC CHEMISTRY. *Educ:* Rensselaer Polytech Inst, BS, 49, MS, 51, PhD(chem), 52. *Prof Exp:* Asst, Rensselaer Polytech Inst, 49-52; res chemist, Union Carbide Chem Co, Union Carbide Corp, 52-67, tech mgr oxidation prod opers group, Chem & Plastics Develop Div, 67-69, technol mgr aldehydes, alcohols & plasticizer intermediates opers group, 69-71 & acrolein, acrylic acid & acrylate esters opers group, 72-74, assoc dir res & develop dept, Chem & Plastics Div, 74-82; RETIRED. *Concurrent Pos:* Consult, 83-86. *Mem:* Am Chem Soc. *Res:* Oxidation of organic compounds, especially hydrocarbons and aldehydes; hydroformylation; specialty organic chemicals; process development; management of research and development. *Mailing Add:* 1518 Village Dr South Charleston WV 25309

MCPETERS, ARNOLD LAWRENCE, b Sept 13, 25; m 51; c 4. POLYMER SCIENCE. *Educ:* Univ NC, BS, 50, PhD(chem), 54. *Prof Exp:* Res chemist, Am Enka Corp, 53-58, head develop sect, Rayon Res Dept, 58-60; chemist, Chemstrand Corp, Monsanto Co, 60-61, from group leader to sr group leader, Monsanto Textiles Co, 62-85; RES ASSOC, NC STATE UNIV, 85- *Mem:* Am Chem Soc; Fiber Soc. *Res:* Improved acrylic and modacrylic fibers polymer structure; fiber morphology; fiber production processes; textile performance; photochemistry. *Mailing Add:* 6829 Perkins Dr Raleigh NC 27612

MCPETERS, RICHARD DOUGLAS, b Florence, Ala, July 3, 47. ATMOSPHERIC PHYSICS, OZONE MEASUREMENTS. *Educ:* Mass Inst Technol, BS, 69; Univ Fla, PhD(physics), 75. *Prof Exp:* assoc physics, Univ Fla, 75-76; staff scientist, Systs & Appl Sci Corp, 76-78, SPACE SCIENTIST, NASA GODDARD SPACE FLIGHT CTR, 78- *Mem:* Sigma Xi; Optical Soc Am; Am Meteorol Soc; Am Geophys Union. *Res:* Atmospheric optics of the solar aureole; total ozone and ozone profile determination from satellite measurements of backscattered ultraviolet. *Mailing Add:* Kirwanns Landing Chester MD 21619

MCPHAIL, ANDREW TENNENT, b Glasgow, Scotland, Sept 23, 37; m 61; c 2. CHEMISTRY. *Educ:* Glasgow Univ, BSc, 59, PhD(chem), 63. *Prof Exp:* Asst lectr chem, Glasgow Univ, 61-64; res assoc, Univ Ill, Urbana, 64-66; lectr, Univ Sussex, 66-68; assoc prof, 68-73, PROF CHEM, DUKE UNIV, 73- *Mem:* Royal Soc Chem; Am Crystallog Asn. *Res:* X-ray crystal structure analysis of organic molecules, particularly biologically active compounds; molecular conformations; studies of structure and bonding in transition metal complexes and in organometallic compounds. *Mailing Add:* Paul M Gross Chem Lab Duke Univ Durham NC 27706

MCPHAIL, JASPER LEWIS, b Slate Spring, Miss, Dec 30, 30; m 57; c 3. ENDOCRINOLOGIC SURGERY. *Educ:* Miss Col, Clinton, BS, 52; Baylor Med Col, Houston, Tex, MD, 56, Cert, 67; Oral Roberts Univ, Tulsa, Okla, MBA, 86. *Prof Exp:* Reader cardiac surg, Christian Med Col, Vellore, India, 62-66; assoc dean & assoc prof thoracic surg, Med Ctr, Univ Ark, Little Rock, 67-69; dir physiol & cardiac surg, Sch Health Sci, Univ Cent Ark, 69-75; dean, vpres planning & prof cardiac surg, Am Indian Sch Med, Phoenix, Ariz, 75-80; chmn, dept surg, 81-84, prof surg, Sch Med, Oral Roberts Univ, 81-87; dir cardiothoracic surg, City Faith Med & Res Ctr, Tulsa, OK, 81-87; med dir, St Elizabeth Hosp Med Ctr, Youngstown, Ohio, 87-89; SURG TEACHING FAC, ST PAUL'S HOSP, DALLAS, TX, 89- *Mem:* Am Asn Thoracic Surg; Soc Thoracic Surgeons; fel Am Col Surgeons; fel Am Col Cardiol; AMA; Int Soc Heart Transplantation. *Res:* Myocardial preservation; cardiac transplantation; mechanical assist devices. *Mailing Add:* 707 Parkway Blvd Cappell TX 75019-2704

MACPHAIL, MORAY ST JOHN, b Kingston, Ont, May 27, 12; m 39; c 1. MATHEMATICS. *Educ:* Queen's Univ Ont, BA, 33; McGill Univ, MA, 34; Oxford Univ, DPhil, 36. *Hon Degrees:* DSc, Carleton Univ, 78. *Prof Exp:* From instr to prof math, Acadia Univ, 37-47; vis lectr, Queen's Univ Ont, 47-48; from assoc prof to prof, 48-77, dir sch grad studies, 60-63, dean fac grad studies, 63-69, EMER PROF MATH, CARLETON UNIV, 77- *Concurrent Pos:* Instr, Princeton Univ, 41-42; vis prof, Univ Toronto, 47-48. *Mem:* Am Math Soc; Math Asn Am; fel Royal Soc Can; Can Math Soc. *Res:* Analysis; theory of series; sequence spaces by methods of functional analysis. *Mailing Add:* Dept Math & Statist Carleton Univ Ottawa ON K1S 5B6 Can

MACPHAIL, RICHARD ALLYN, b Midland, Mich, Jan 27, 53; m. VIBRATIONAL SPECTROSCOPY, DYNAMIC LIGHT SCATTERING. *Educ:* Oberlin Col, AB, 77; Univ Calif, Berkeley, PhD(chem), 81. *Prof Exp:* Postdoctoral scholar chem, Univ Calif, Los Angeles, 81-84; ASST PROF CHEM, DUKE UNIV, 84- *Mem:* Am Phys Soc; Am Chem Soc; AAAS. *Res:* Physical chemistry; vibrational spectroscopy and vibrational relaxation of molecules in condensed phases; liquids, solids and glasses; dynamic light scattering from viscous liquids and glasses; viscoelasticity. *Mailing Add:* Dept Chem Duke Univ Durham NC 27706

MACPHAIL, ROBERT C, b Buffalo, NY, Mar 31, 45. NEUROTOXICOLOGY. *Educ:* Wash & Jefferson Univ, BA, 67; Univ Md, MS, 71, PhD(psychol), 73. *Prof Exp:* Res asst, NASA Space Res Lab, Univ Md, 67-68, teaching & res asst, Dept Psychol, 68-73; instr, Dept Psychol, Roosevelt Univ, Chicago, 73-77; postdoctoral fel, Dept Pharmacol, Univ Chicago, 73-75, res assoc, Dept Pharmacol & Physiol Sci, 75-77; staff fel, Neurosci Prog, Div Molecular Biol, Nat Ctr Toxicol Res, 77-78; res psychologist, Neurotoxicol Div, Health Effects Res Lab, US Environ Protection Agency, 79-82, chief, Behav Toxicol Br, 82-88; res assoc prof, 83-88, RES PROF, DEPT PSYCHOL & NEUROBIOL CURRIC, UNIV NC, CHAPEL HILL, 89-; CHIEF, BEHAV & NEUROCHEM BR, NEUROTOXICOL DIV, HEALTH EFFECTS RES LAB, US ENVIRON PROTECTION AGENCY, 89- *Concurrent Pos:* Adj asst prof, Dept Pharmacol, Univ Ark Med Sci, 78 & Dept Psychol & Neurobiol Curric, Univ NC, Chapel Hill, 80-83; mem, Comt Neurobiol Principles Neurotoxicol, Nat Res Coun, 88-89; actg div dir, Neurotoxicol Div, Health Effects Res Lab, US Environ Protection Agency, 89. *Mem:* Behav Pharmacol Soc; Behav Toxicol Soc; Am Soc Pharmacol & Exp Therapeut; Soc Neurosci; Soc Toxicol; Int Neurotoxicol Asn. *Res:* Effects of chemicals on naturally occurring and acquired behavior; use of operant conditioning techniques in neurotoxicology and neuroscience; drug-toxicant behavior interactions; behavioral effects of pharmacological and environmental challenges; neurobehavioral screening techniques; author of various publications. *Mailing Add:* Neurotoxicol Div US Environ Protection Agency Health Effects Res Lab Research Triangle Park NC 27711

MACPHAIL, STUART, b Dec 27, 47. MURINE IMMUNOGENETICS, CELLULAR IMMUNOLOGY. *Educ:* Cambridge Univ, Eng, PhD(immunol), 76. *Prof Exp:* Res assoc, 79-85, LAB ASST MEMBER, MEM SLOAN-KETTERING CANCER RES, 85- *Concurrent Pos:* Scholar, Leukemia Soc Am, 85. *Mem:* Am Asn Immunol. *Res:* cellular immunology, characterize the product of a murine histocompatibility locus known as MIS, investigating the heterogeneity of cytotoxic Tcell subpopulations. *Mailing Add:* Sloan-Kettering Inst 1275 York Ave New York NY 10021

MACPHEE, KENNETH ERSKINE, organic chemistry, polymer chemistry, for more information see previous edition

MCPHEETERS, KENNETH DALE, b Mt Vernon, Ill, Dec 1, 52. INSTRUCTIONAL MATERIALS DEVELOPMENT, TISSUE CULTURE. *Educ:* Southern Ill Univ, BS, 74; Univ Ill, MS, 81, PhD(hort), 85. *Prof Exp:* Asst supt bldg & grounds, Mitchell Art Mus, Mt Vernon, Ill, 76-79; res asst hort, Dept Hort, 79-81, INSTR MAT SPECIALIST HORT & AGRON, VOC AGR SERV, UNIV ILL, 81- *Mem:* Sigma Xi. *Res:* Application of tissue culture techniques to improvement of horticultural crops, particularly small fruits; using somaclonal variation for accessing unique genes; agronomy; co-author of one textbook. *Mailing Add:* Univ Ill 1401 S Maryland Dr Urbana IL 61801

MCPHERRON, ROBERT LLOYD, b Chelan, Wash, Jan 14, 37; m 58; c 2. GEOPHYSICS, SPACE PHYSICS. *Educ:* Univ Wash, BS, 59; Univ Southern Calif, MS, 61; Univ Calif, Berkeley, PhD(physics), 68. *Prof Exp:* Res physicist space sci lab, Univ Calif, Berkeley, 66-68; res geophysicist, Inst Geophys & Planetary Physics, 68-69, asst prof space physics, 69-73, asst prof, 73-77, PROF GEOPHYS & SPACE PHYSICS, UNIV CALIF, LOS ANGELES, 77- *Mem:* AAAS; Am Geophys Union. *Res:* Magnetic field variations within the magnetosphere, including both macroscopic currents and wave phenomena and the part they play in magnetic storms and substorms; particles and fields; auroral phenomena. *Mailing Add:* Dept Earth & Space Sci Univ Calif 405 Hilgard Ave Los Angeles CA 90024-1567

MCPHERSON, ALEXANDER, b Columbus, Ohio, Feb 28, 44. BIOLOGICAL STRUCTURE. *Educ:* Duke Univ, BS, 66; Purdue Univ, West Lafayette, PhD(biol), 70. *Prof Exp:* Damon Runyon res fel biol, Mass Inst Technol, 70-71, Am Cancer Soc res fel, 71-73, res assoc, 74-75; assoc prof biol chem, Hershey Med Ctr, Pa State Univ, 75-; AT DEPT BIOCHEM, UNIV CALIF, RIVERSIDE. *Mem:* Am Crystallog Asn; Am Soc Biol Chemists. *Res:* Analysis and determination of the atomic structures of biological macromolecules by x-ray diffraction techniques and their correlation with mechanistic properties. *Mailing Add:* Dept Biochem Univ Calif Webber Hall E Rm 2466 Riverside CA 92521

MACPHERSON, ALISTAIR KENNETH, b Sydney, Australia, Jan 29, 36; m 59; c 4. MOLECULAR DYNAMICS. *Educ:* Univ Sydney, BE, 57, MEngSc, 66, PhD(mech eng), 68. *Prof Exp:* Res asst fire res, Commonwealth Exp Bldg Sta, 59-60; asst proj engr, Qantas Airways, 60-66; fel, Inst Aerospace Studies, Univ Toronto, 68-69; asst prof mech eng, Univ Man, 69-71; assoc prof, 71-74, PROF MECH ENG, LEHIGH UNIV, 74- *Concurrent Pos:* Adj prof, Univ Man, 71-72. *Mem:* Am Phys Soc; Am Meteorol Soc; Royal Meteorol Soc; Combustion Inst; Am Geophys Union. *Res:* Atmospheric flows; fluid flow at the molecular level; plasma flows; shock and detonation waves; statistical mechanics. *Mailing Add:* Dept of Mech Eng & Mech Lehigh Univ Bethlehem PA 18015

MCPHERSON, ALVADUS BRADLEY, b West Frankfort, Ill, July 10, 37; m 59; c 3. BIOLOGY. *Educ:* Southeastern La Col, BS, 60; La State Univ, MS, 67; Southern Ill Univ, PhD(biol), 71. *Prof Exp:* WATERS PROF BIOL & CHMN DEPT, CENTENARY COL LA, 71- *Concurrent Pos:* Fel trop med, La State Univ Med Sch, 66, 69-70; NSF fel marine ecol, Duke Univ, 68; alumni grant, Centenary Col La, 75 & 84, WHO, 83. *Mem:* Sigma Xi; Soc Vert Paleont; Southwestern Asn Naturalist. *Res:* Mammalian ecology, taxonomy, distribution, and parasitology. *Mailing Add:* Dept Biol Centenary Col Shreveport LA 71104

MACPHERSON, ANDREW HALL, b London, Eng, June 2, 32; Can citizen; m 57; c 3. ECOLOGY, ZOOGEOGRAPHY. *Educ:* Carleton Univ, BS, 54; McGill Univ, MSc, 57, PhD, 67. *Prof Exp:* Asst cur ornith, Nat Mus Can, 57-58; res wildlife biologist, Can Wildlife Serv, 58-64, regional supvr wildlife res, 64-67; sci adv, Sci Secretariat, Privy Coun Off, 67-68 & Sci Coun Can, 68-69; regional supvr wildlife res, 69-70, dir Western Region, 70-74, dir-gen Environ Mgt Serv, 74-79, regional dir-gen, Western & Nothern Region, Environ Can, 79-86; reg dir-gen, Northern Affairs, 86-88; CONSULT, 88- *Concurrent Pos:* Mem, Tech Comt for Caribou Preservation, 64-69; chmn, Polar Bear Group and mem, Survival Serv Comn, Int Union Conserv of Nature & Natural Resources, 70-72; mem, Mackenzie R Basin Comt, 74-88, Grants Comt World Wildlife Fund, Can, 78-84. *Honors & Awards:* Centennial Medal, 67. *Mem:* Fel Arctic Inst NAm (gov, 72-76). *Res:* Taxonomy of Laridae; ecology and population dynamics of Alopex; zoogeography of Arctic mammals; ecology and population processes of Rangifer; management of arctic wildlife resources; ice fishing. *Mailing Add:* 9619 96A St Edmonton AB T6C 3Z8 CAN

MCPHERSON, CHARLES WILLIAM, b Rugby, NDak,; m 56; c 2. LABORATORY ANIMAL MEDICINE. *Educ:* Univ Minn, BS, 54, DVM, 56; Univ Calif, Berkeley, MPH, 64; Am Col Lab Animal Med, dipl. *Prof Exp:* Vet, Animal Hosp Sect, NIH, 56-57, head, Primate Unit, 57-58, vet microbiologist, Comp Path Sect, 58-60, chief, Animal Prod Sect, Lab Aids Br, 60-64, head, Pathogen Free Unit & asst to chief, 64-66, chief, Lab Animal Med & Vivarium Sci Sect, Animal Resources Br, Div Res Resources, 66-70, chief br, 71-80; DIR ANIMAL RES, COL VET MED, NC STATE UNIV, 80- *Honors & Awards:* Co-recipient Res Award, Am Asn Lab Animal Sci, 63; Griffin Award, Am Asn Lab Animal Sci, 80. *Mem:* Am Vet Med Asn; Am Asn Lab Animal Sci; Am Soc Lab Animal Practitioners (pres, 82-83). *Res:* Diseases of laboratory animals; production of microbiologically defined laboratory animals. *Mailing Add:* Col Vet Med NC State Univ Raleigh NC 27606

MCPHERSON, CLARA, b Roscoe, Tex, Mar 10, 22; m 43; c 3. NUTRITION, FOODS. *Educ:* Tex Tech Col, BS, 43, MS, 47. *Prof Exp:* Instr food & nutrit, 47-48 & 55-60, asst prof, 61-68, ASSOC PROF FOOD & NUTRIT, TEX TECH UNIV, 68- *Mem:* Am Dietetic Asn; Am Home Econ Asn; Inst Food Technol; Soc Nutrit Educ; Sigma Xi. *Res:* Dietary studies of college students; frozen foods; development of high protein foods using cottonseed and soy protein; determination of quality of pork fed various rations. *Mailing Add:* 2131 56th St Lubbock TX 79412

MCPHERSON, CLINTON MARSUD, b Gainesville, Tex, Oct 6, 18; m 43; c 3. CHEMISTRY. *Educ:* Tex Tech Col, BS, 47, MEd, 52, DEd(psychol), 59. *Prof Exp:* Teacher pub schs, Tex, 50-56; instr chem, Tex Tech Univ, 58-59, asst prof, 60-74, assoc prof food & nutrit, 74-77, from asst prof to assoc prof chem, 77-84, EMER ASSOC PROF, TEX TECH UNIV, 84- *Res:* Inorganic chemistry; use of audio-visual materials. *Mailing Add:* 2131 56th St Lubbock TX 79412

MACPHERSON, COLIN ROBERTSON, b Aberdeen, Scotland, Sept 2, 25; m 49; c 4. PATHOLOGY. *Educ:* Univ Cape Town, MB, ChB, 46, MMed & MD, 54. *Prof Exp:* Asst lectr path, Univ Cape Town, 48-50, lectr, 50-55; asst renal physiol, Post-Grad Sch Med, Univ London, 55-56; from asst prof to prof bact & path, Ohio State Univ, 56-75, vchmn dept path, 60-72, actg chmn dept, 72-75; DIR, DIV LAB MED, UNIV CINCINNATI, 75-, VCHMN, DEPT PATHOL & LAB MED, 79- *Concurrent Pos:* Consult, Vet Admin Hosp, Cincinnati, Ohio, 65- *Res:* Immuno-hematology; laboratory screening procedures. *Mailing Add:* Med Ctr Univ Cincinnati Cincinnati OH 45267-0055

MACPHERSON, CULLEN H, b San Mateo, Calif, Dec 6, 27; m 51; c 3. BIOPHYSICS. *Educ:* San Jose State Col, AB, 49; Stanford Univ, MA, 51. *Prof Exp:* Mgr reproducing components div, Electro-Voice Inc, 54-56; biophysicist, Tektronix, Inc, 56-61; pres & chmn bd, Argonaut Assocs, Inc, 59- 78; chief res engr, Advan Develop Group, Temperature Controls Div, Cutler-Hammer, Inc, Beaverton, 78-80, Eaton Corp, 80-81; tech dir, NAm Automation Div, Kockums Indust, 81-83; vpres res & develop, LSI Inc, 83-85; SR VPRES & CHIEF OPER OFFICER, MODERATOR INC, 85- *Concurrent Pos:* Scientist, Dept Neurophysiol, Ore Regional Primate Res Ctr, 64-78, chmn, Dept Biophys, 70-78. *Mem:* AAAS; Am Asn Physics Teachers; Acoust Soc Am; Audio Eng Soc; Inst Elec & Electronics Eng. *Res:* Limited energy measurements in biological systems; constant current neuronal stimulation; stereo vector electrocardiography. *Mailing Add:* 2677 NW Westover Rd Portland OR 97210-3130

MCPHERSON, DONALD J(AMES), b Columbus, Ohio, Nov 18, 21; m 45; c 2. PHYSICAL METALLURGY. *Educ:* Ohio State Univ, BMetE, 43, MSc, 47, PhD(metall), 49. *Hon Degrees:* DSc, Ohio State Univ, 75. *Prof Exp:* Metall inspector, Carnegie-Ill Steel Corp, 43-44; res engr, Battelle Mem Inst, 46; assoc, Res Found, Ohio State Univ, 47-49; assoc metallurgist, Argonne Nat Lab, 49-50; from supvr phys metall & alloy develop to dir metals res, Ill Inst Technol Res Inst, 50-63, vpres, 63-69; vpres & dir technol, Kaiser Aluminum Tech Serv, Inc, 69-81, pres, Kaiser Aluminum & Chem Corp & vpres, Tech Serv, 81-; RETIRED. *Concurrent Pos:* Chmn, Nat Mat Adv Bd, 82- *Honors & Awards:* Citation, Bur Aeronaut, 55; Campbell Mem Lectr, Am Soc Metals, 74. *Mem:* AAAS; hon mem Am Soc Metals; Am Inst Mining Metall & Petrol Engrs; Am Ceramic Soc; Brit Inst Metals. *Res:* Phase diagrams; alloy development; research management. *Mailing Add:* 9369 Via Montoya Scottsdale AZ 85255

MCPHERSON, GEORGE, JR, b Westfield, NY, July 16, 21. ELECTRICAL MACHINERY. *Educ:* Ohio State Univ, BEE, 43, BSc, 48, MSc, 49. *Prof Exp:* Proj engr, Sound Div, Naval Res Lab, 43-44, 45-47; instr elec eng, Ohio State Univ, 48-49; teacher pub schs, Ohio, 49-50; prin elec engr, Battelle Mem Inst, 50-54; asst prof elec eng, Univ Ky, 54-56; prof elec eng, 56-85, EMER PROF, UNIV MO-ROLLA, 85- *Concurrent Pos:* Consult, elec motors and power lab design. *Mem:* Fel Inst Elec & Electronic Engrs; Am Asn Univ Prof; Sigma Xi. *Res:* Electromagnetic machines. *Mailing Add:* 18 Green Acres Rolla MO 65401

MCPHERSON, GUY RANDALL, b Wallace, Idaho, Feb 29, 60; m 83. TERRESTRIAL PLANT ECOLOGY, DISTURBANCE ECOLOGY. *Educ:* Univ Idaho, BS, 82; Tex Tech Univ, MS, 84, PhD(range sci), 87. *Prof Exp:* Res asst range sci, Tex Tech Univ, 83-87; postdoctoral res assoc ecol, Inst Ecol, Univ Ga, 87-88; vis asst prof rangeland ecol, Tex A&M Univ, 88-89; ASST PROF ECOL, UNIV ARIZ, 89- *Concurrent Pos:* Consult, Nature Conservancy, 89- *Mem:* Ecol Soc Am; Int Asn Landscape Ecol; Soc Am Foresters; Natural Areas Asn. *Res:* Terrestrial plant ecologist with primary interest in stability, resilience and disturbance regimes of semi-arid ecosystems; fundamental ecological processes in semi-arid savannas, grasslands and woodlands. *Mailing Add:* Forest Watershed Div Sch Renewable Natural Resources Univ Ariz Tucson AZ 85721

MCPHERSON, HAROLD JAMES, b Newry, NIreland, May 28, 39; Can citizen. PHYSICAL GEOGRAPHY, ENVIRONMENTAL HAZARDS. *Educ:* Queen's Univ, Ont, BA, 61; Univ Alta, MSc, 63; McGill Univ, PhD(geog), 67. *Prof Exp:* Asst prof geog, Queen's Univ, Ont, 66-70; from asst prof to assoc prof, 70-78, PROF GEOG, UNIV ALTA, 78- *Concurrent Pos:* Vis scholar, Univ Ariz, 74-75; Can Coun leave fel, Can Coun, 74. *Mem:* Asn Am Geog. *Res:* Water resources in developing countries; natural hazards; environmental quality; perception of environment; mountain geomorphology. *Mailing Add:* Dept of Geog Univ of Alta Edmonton AB T6G 2M7 Can

MACPHERSON, HERBERT GRENFELL, b Victorville, Calif, Nov 2, 11; m 37; c 2. NUCLEAR SCIENCE, NUCLEAR ENGINEERING. *Educ:* Univ Calif, AB, 32, PhD(physics), 37. *Prof Exp:* Jr meteorologist, USDA, 36-37; res physicist, Nat Carbon Co, 37-50, asst dir res, 50-56; with Oak Ridge Nat Lab, 56-60, assoc dir reactor prog, 60-63, asst lab dir, 63-64, dep dir, 64-70; prof nuclear eng, Univ Tenn, 70-76; CONSULT, 88- *Concurrent Pos:* Consult, Oak Ridge Nat Lab, 70-81 & US AEC, 72-74; actg dir, Inst Energy Analysis, 74-75, consult, 75-88. *Mem:* Nat Acad Eng; Am Nuclear Soc; Am Phys Soc. *Res:* Fundamentals of the carbon arc; high temperature properties of carbon and graphite; heavy particles in cosmic radiation; nuclear reactor technology; safety of nuclear reactors; energy policy. *Mailing Add:* 102 Orchard Circle Oak Ridge TN 37830

MCPHERSON, JAMES C, JR, b Hamilton, Tex, Dec 27, 26; m 45; c 4. BIOCHEMISTRY, MEDICINE. *Educ:* NTex State Col, BS, 46; Univ Tex, MA, 55, MD, 60. *Prof Exp:* Res scientist, Univ Tex Southwestern Med Sch Dallas, 60-61, from instr to asst prof biochem, 61-63; asst res prof biochem, 63-70, ASSOC PROF SURG, CELL & MOLECULAR BIOL, MED COL GA, 70- *Mem:* AAAS; Am Chem Soc; Am Oil Chem Soc; Am Asn Clin Chem; Soc Exp Biol & Med. *Res:* Lipid absorption and metabolism. *Mailing Add:* 3125 Ramsgate Rd Augusta GA 30904

MCPHERSON, JAMES KING, b Tucson, Ariz, Nov 11, 37; m 62; c 2. PLANT ECOLOGY. *Educ:* Univ Idaho, BS, 59; Univ Calif, Santa Barbara, MA, 66, PhD(bot), 68. *Prof Exp:* From asst prof to assoc prof, 68-82, PROF BOT, OKLA STATE UNIV, 82- *Mem:* AAAS; Bot Soc Am; Ecol Soc Am. *Res:* Ecological aspects of forest tree water relations; allelopathy and competition among plants. *Mailing Add:* Dept Bot Okla State Univ Stillwater OK 74078

MCPHERSON, JAMES LOUIS, b Chattanooga, Tenn, June 25, 22; m 48; c 3. POLYMER CHEMISTRY. *Educ:* Ga Inst Technol, BS, 44; Univ Tex, MA, 49; Ohio State Univ, PhD(org chem), 53. *Prof Exp:* Asst lab, Univ Tex, 47-48; res chemist explor sect, Plastics Dept, Exp Sta, E I du Pont de Nemours & Co, 53-58; proj leader, Cent Res Lab, Gen Aniline & Film Corp, 59-61; lab dir basic polymer res, Cent Res & Eng Div, Continental Can Co, 61-64; sr fel & prof, Mellon Inst, 64-67; mgr chem activ div, DeBell & Richardson, Inc, Mass, 67-69; prof chem, Lee Col, Tenn, 69-88; RETIRED. *Mem:* Am Chem Soc; The Royal Soc Chem; Sigma Xi. *Res:* Synthesis, properties and applications of polymers. *Mailing Add:* 7231 Short Tail Springs Rd Harrison TN 37341

MCPHERSON, JOHN EDWIN, b San Diego, Calif, June 8, 41; m 66; c 2. ENTOMOLOGY. *Educ:* San Diego State Univ, BS, 63, MS, 64; Mich State Univ, PhD(entom), 68. *Prof Exp:* From asst prof to assoc prof, 69-78, PROF ZOOL, SOUTHERN ILL UNIV, CARBONDALE, 78- *Mem:* Entom Soc Am; Entom Soc Can; Sigma Xi. *Res:* Bionomics and taxonomy of North American Pentatomoidea and aquatic Hemiptera of the group Insecta. *Mailing Add:* Dept Zool Southern Ill Univ Carbondale IL 62901

MCPHERSON, JOHN G(ORDON), b Nelson, NZ, Nov 18, 47; m 83; c 2. FLUVIAL SEDIMENTOLOGY. *Educ:* Victoria Univ, Wellington, BSc, 70, BSc hons, 71, PhD(geol), 75. *Prof Exp:* Lectr geol, Univ Cape Town, 76-80; assoc prof geol, Univ Tex, Arlington, 81-85; ASSOC, MOBIL RES & DEVELOP CORP, DALLAS, 85- *Mem:* Int Asn Sedimentologists; Am Asn Petrol Geologists; Geol Soc Am; Soc Econ Paleontologists & Mineralogists. *Res:* Terrigenour clastic depositional systems specializing in fluvial and alluvial-fan sedimentation; sandstone petrography, paleopedology and reservoir characterization. *Mailing Add:* Mobil Res & Develop Corp PO Box 819047 Dallas TX 75381

MCPHERSON, JOHN M, b Antioch, Calif, Aug 21, 48. EXPERIMENTAL BIOLOGY. *Educ:* Univ Calif, BA, 70, MA, 71, PhD(biochem), 79. *Prof Exp:* Staff res assoc II, Dept Cardiol, Univ Calif, Davis, 72-74; postdoctoral fel, Dept Biochem, Univ Wash, Seattle, 79-81; res biochemist, Biochem Sect, Collagen Corp, Palo Alto, Calif, 81-82, mgr, 82-88, res proj leader, Dermal Wound Repair, 85-88; dir, Dept Protein Chem, Integrated Genetics, Inc, Framingham, Mass, 88-89; VPRES, RECOMBINANT PROTEIN DEVELOP, GENZYME CORP, FRAMINGHAM, MASS, 89- *Mem:* Am Soc Cell Biol; AAAS. *Res:* Development of stable transfected mammalian cell lines which express recombinant proteins at predetermined target levels; development of purification and characterization methods; identifying novel proteins or engineered forms of proteins; author of various publications; granted 4 patents. *Mailing Add:* Genzyme Corp One Mountain Rd Framingham MA 01701

MACPHERSON, ROBERT DUNCAN, b Lakewood, Ohio, May 25, 44; m 69. TOPOLOGY. *Educ:* Swarthmore Col, BA, 66; Harvard Univ, MA, 69, PhD(math), 70. *Prof Exp:* Instr, Brown Univ, 70-72, from asst prof to assoc prof, 72-76; vis prof, Univ Paris, 76-77; PROF MATH, BROWN UNIV, 77- *Concurrent Pos:* Res fel fel, Oak Ridge Nat Lab, 64-65; mathematician, US Brazil exchange prog, 73; vis mem, Inst Advan Sci Studies, 74-75; fel, Nat Sci Res Ctr, France, 75. *Mem:* Am Math Soc. *Res:* Differential topology; topology of algebraic varieties; singularities. *Mailing Add:* MIT Rm 2-246 Cambridge MA 02139

MCPHERSON, ROBERT MERRILL, b Enid, Okla, Nov 12, 48; m 70; c 3. SOYBEAN PEST MANAGEMENT, TOBACCO PEST MANAGEMENT. *Educ:* Sam Houston State Univ, BS, 71; La State Univ, MS, 75, PhD(entom), 78. *Prof Exp:* Lab instr biol, Sam Houston State Univ, 70-71; res asst entom, La State Univ, 73-75, res assoc, 75-78; from asst prof to assoc prof entom, Va Polytech Inst & State Univ, 78-87; ASSOC PROF, UNIV GA, 87- *Concurrent Pos:* Co-chmn, Educ Comn, Va Agr Chem Asn, 80-81 & chmn, Steering & Budget Planning, 81-; tech comn mem, Southern Regional Res Proj, 81-; mem, Comt Insect Detection, Eval & Prediction, Entom Soc Am, 82-84, Prog Comt, Pub Relations Comt & Local Arrangements Comt, eastern br, 84-85; mem, Soybean & Corn Pesticide Impact & Assessment Team, Coop Exten Serv, USDA, 82-84. *Honors & Awards:* Researcher Award, Am Soybean Asn, 85. *Mem:* Entom Soc Am; Sigma Xi. *Res:* Develop and help implement insect pest management programs in soybeans and tobacco; ecology and population dynamics of arthropods in these crops, sampling techniques, economic threshold levels of pest species, and biological and chemical control techniques. *Mailing Add:* Entomology Dept Coastal Plain Exp Sta PO Box 748 Tifton GA 31793

MCPHERSON, ROBERT W, ANESTHESIOLOGY. *Prof Exp:* ASSOC PROF, DEPT ANESTHESIOL, JOHNS HOPKINS HOSP, 85- *Mailing Add:* Dept Anesthesiol Johns Hopkins Hosp Meyer 8-138 600 N Wolfe St Baltimore MD 21205

MACPHERSON, RODERICK IAN, b St Thomas, Ont, Feb 22, 35; m 57; c 5. RADIOLOGY, PEDIATRICS. *Educ:* Univ Man, BSc & MD, 58; Royal Col Physicians Can, cert radiol, 63; FRCP(C), 64. *Prof Exp:* Asst radiologist, Montreal Children's Hosp, Que, 64-65; asst radiologist, Royal Victoria Hosp, Montreal, 65; asst radiologist, Shaughnessy Hosp, Vancouver, BC, 65-67; assoc radiologist, Children's Hosp Winnipeg, 67-69; asst prof radiol, Univ Man, 67-69, assoc prof, 69-, assoc prof pediat, 71-; dir dept radiol, Children's Hosp Winnipeg, 69-; AT DEPT RADIOL-PEDIAT, MED UNIV SC. *Concurrent Pos:* Instr, McGill Univ, 65 & Univ BC, 65-67. *Mem:* Can Asn Radiol; Am Roentgen Ray Soc; Soc Pediat Radiol. *Res:* Diagnostic radiology; clinical, pathologic and radiologic aspects of pediatric chest diseases, renal diseases and skeletal diseases. *Mailing Add:* Dept Radiol-Pediat Med Univ SC 171 Ashley Ave Charleston SC 29425

MCPHERSON, ROSS, b Buffalo, NY, May 30, 34; m 57; c 3. PHYSICS. *Educ:* Queen's Univ, Ont, BSc, 59, MSc, 61; McGill Univ, PhD(physics), 64. *Prof Exp:* Res assoc, Brookhaven Nat Lab, 64-66; asst prof physics, Cornell Univ, 66-72; assoc prof physics, Univ Guelph, 72-79, MEM STAFF, BELL LABS, 79- *Mem:* AAAS; Am Phys Soc; Can Asn Physicists; Sigma Xi. *Res:* Nuclear physics; nuclear and digital instrumentation; nuclear engineering. *Mailing Add:* 14H 423 Bell Labs Whippany NJ 07981

MCPHERSON, THOMAS ALEXANDER, b Calgary, Alta, Mar 1, 39; m 67; c 4. IMMUNOLOGY. *Educ:* Univ Alta, MD, 62; Univ Melbourne, PhD(med, immunol), 69. *Prof Exp:* Sr resident med officer, Royal Adelaide Hosps, SAustralia, 63-64, med registr, 64-65, sr med registr, Renal Unit, 65; asst physician, Clin Res Univ, Walter & Eliza Hall Inst Med Res, Melbourne, 66-68; asst prof, 69-70, assoc prof, 70-77, PROF MED, UNIV ALTA, 77-DIR DEPT MED, CROSS INST, 73-, ASST DEAN MED, 81- *Concurrent Pos:* R S McLaughlin Res Found traveling fel, Southeast Asia & Europe, 68-69. *Mem:* Australasian Soc Med Res; Can Soc Immunol; Can Soc Clin Invest. *Res:* Induction and inhibition of experimental allergic encephalomyelitis using human encephalitogenic basic protein and synthetic polypeptides; the carcinoembryonic antigen in the human colon; trial of anti-thymocyte globulin in acute relapses of multiple sclerosis. *Mailing Add:* 307 Campus Tower 8625 112th St Edmonton AB T6G 0Y1 Can

MCPHERSON, THOMAS C(OATSWORTH), b Atlanta, Ga, May 12, 22; m 78; c 4. ALLERGY, IMMUNOLOGY. *Educ:* Emory Univ, AB, 43; Univ Md, Sch Med, MD, 46. *Prof Exp:* Admin asst to the dir of res, The Children's Cancer Res Found, 69-71; assoc med dir, Fisons Corp, Bedford, Mass, 72-77; clin res consult serv, Nashua, NH, 78; assoc dir, Domestic Clin Res, 3M Ctr, Minn, 78-80, ASSOC DIR MED SERV, RIKER LABS, INC, CA, 80-; assoc dir, Domestic Clin Res, 3M Ctr, Minn, 78-80, ASSOC DIR, MED SERV, RIKER LABS, INC, 3M CO, CALIF, 80- *Concurrent Pos:* Consult pediat, Martin Army Hosp, Ft Benning, Ga, 57-61; sr fel, Dept Biochem, Brandeis Univ, 66-67; consult, The Children's Cancer Res Found, Boston, Mass, 72-77. *Mem:* Am Acad Pediat; Drug Info Asn; Am Soc Clin Pharmacol & Therapeut; Am Acad Allergy. *Res:* Clinical research in allergy and hematology. *Mailing Add:* Riker Labs Inc 225-IN-07 3M Ctr St Paul MN 55144

MCPHIE, PETER, b Leeds, Eng, Oct 3, 42; US citizen; m 66; c 3. BIOPHYSICAL CHEMISTRY, ENZYMOLOGY. *Educ:* Univ Durham, BSc, 63; Univ London, PhD(biophys), 66. *Prof Exp:* Fel, Stanford Univ, 66-68; vis scientist, 68-74, RES CHEMIST BIOCHEM, NIH, 74- *Concurrent Pos:* adj prof, Georgetown Med Sch, 86- *Res:* Structure and function of proteins and nucleic acids. *Mailing Add:* Lab Biochem & Metab NIDDKD 9000 Rockville Pike Bethesda MD 20814

MCQUADE, HENRY ALONZO, b St Louis, Mo, Nov 1, 15. CYTOLOGY, CYTOGENETICS. *Educ:* Wash Univ, AB, 38, PhD, 49; Univ Mo, MA, 40. *Prof Exp:* Asst prof biol, Harris Teachers Col, 49-54; res assoc, Mallinckrodt Inst Radiol, Sch Med, Wash Univ, 54-56; res assoc, Radiation Res Lab, Col Med, Univ Iowa, 56-57; assoc prof radiobiol, Sch Med, Univ Mo-Columbia, 57-64, dir radioisotope lab, Med Ctr, 57-70, prof radiobiol, Sch Med, 64-81, chief radiol sci sect, Med Ctr, 70-81, EMER PROF, UNIV MO-COLUMBIA, 81- *Mem:* Genetics Soc Can; Genetics Soc Am; Bot Soc Am; Sigma Xi. *Res:* Electron microscopy of cells including neiocytes of wheat. *Mailing Add:* 1106 Maplewood Dr Columbia MO 65211

MCQUAID, RICHARD WILLIAM, b Woodland, Calif, Jan 5, 23; m 44; c 3. FUEL SCIENCE & TECHNOLOGY. *Educ:* Univ Calif, AB, 43; Johns Hopkins Univ, AM, 50, PhD(chem), 51. *Prof Exp:* Jr instr phys chem, Johns Hopkins Univ, 50-51; res chemist, Mutual Chem Co, 51-55; res chemist, Catalyst Res Corp, 55-57, mgr res & develop, 57-61; prin staff scientist, Aircraft Armaments Inc, 61-65; sr proj eng, David Taylor Naval Ship Res & Develop Ctr65-74, lubricants & hydraul fluids specialist, Annapolis Lab, 74-83; RETIRED. *Concurrent Pos:* Dept Defense Liason Rep, Comt Indust Hazards, Nat Acad Sci/Nat Market Adv Bd, 79-83. *Mem:* AAAS; Am Chem Soc; Sigma Xi. *Res:* Thermodynamics and structural inorganic chemistry; properties of highly desiccated silica and alumina gels; chemistry of chromium compounds; crystal optics; electrochemistry; corrosion; pyrotechnics; fuels and lubricants chemistry and technology. *Mailing Add:* 1501 Harris Mill Rd Parkton MD 21120

MCQUARRIE, BRUCE CALE, b Easton, Pa, Jan 6, 29; div; c 3. ALGEBRA. *Educ:* Lafayette Col, AB, 51; Univ NH, MA, 56; Boston Univ, PhD(math), 71. *Prof Exp:* From instr to prof, 60-90, EMER PROF MATH, WORCESTER POLYTECH INST, 90- *Concurrent Pos:* Vis instr, Tex A&M Univ, 69-70 & Trent Polytech, Nottingham, Eng, 79-80. *Res:* Near rings; endomorphisms of nonabelian groups. *Mailing Add:* Ten Eames Dr Auburn MA 01501

MCQUARRIE, DONALD ALLAN, b Lowell, Mass, May 20, 37; m 59; c 2. THEORETICAL CHEMISTRY. *Educ:* Lowell Technol Inst, BS, 58; Johns Hopkins Univ, MA, 60; Univ Ore, PhD(chem), 62. *Prof Exp:* Asst prof chem, Mich State Univ, 62-64; mem tech staff, NAm Aviation Sci Ctr, Calif, 64-68; prof chem, Ind Univ, Bloomington, 68-78; PROF CHEM, UNIV CALIF, DAVIS, 78- *Concurrent Pos:* Guggenheim fel, 75-76; vis prof, Univ Andes, Venezuela, 76, Melbourne Univ, 88; Royal Soc vis fel, Oxford Univ, 86. *Mem:* Am Chem Soc; Sigma Xi. *Res:* Statistical thermodynamics; stochastic processes; biophysics. *Mailing Add:* Dept of Chem Univ of Calif Davis CA 95616

MCQUARRIE, DONALD G, b Richfield, Utah, Apr 17, 31; m 56; c 2. SURGERY, COMPUTER SCIENCES. *Educ:* Univ Utah, BS, 53, MD, 56; Univ Minn, Minneapolis, PhD(surg), 65. *Prof Exp:* From instr to assoc prof, 65-72, PROF SURG, MED SCH, UNIV MINN, MINNEAPOLIS, 72-; STAFF SURGEON & ASSOC CHIEF SURG SERV, MINNEAPOLIS VET AFFAIRS MED CTR, 64-; GOV, AM COL SURG, 90- *Concurrent Pos:* USPHS fel, Univ Minn, Minneapolis, 62-65; Navy liaison mem, Div Med Sci, Nat Res Coun, 59-61; chmn surg partic comt, Vet Admin Ctr Off; dir,

Biophys Comput Proj. *Mem:* Asn Acad Surg; Soc Univ Surg; Am Surg Asn; Soc Head & Neck Surg; Soc Surg Oncol. *Res:* Tissue immunology; respiratory physiology; computer sciences; reconstruction after cancer treatment. *Mailing Add:* Minneapolis Vet Affairs Med Ctr One Veteran's Dr Minneapolis MN 55417

MACQUARRIE, IAN GREGOR, b Hampton, PEI, July 6, 33; m 57; c 4. BOTANY, PLANT PHYSIOLOGY. *Educ:* Dalhousie Univ, BSc, 57, MSc, 58; Univ London, PhD(plant physiol), 61. *Prof Exp:* Asst prof biol, Dalhousie Univ, 61-65 & St Dunstan's Univ, 65-67; asst prof, 67-74, assoc prof, 74-, PROF BIOL, UNIV PEI. *Mem:* Can Biol Asn; Can Soc Wildlife & Fishery Biol. *Res:* Upland game; fragile habitats; sand dune systems; hedgerows. *Mailing Add:* Dept Biol Sci Univ PEI Charlottestown PE C1A 4P3

MCQUARRIE, IRVINE GRAY, b Ogden, Utah, June 27, 39; m 80; c 4. EXPERIMENTAL NEUROLOGY, NEUROSURGERY. *Educ:* Univ Utah, BS, 61; Cornell Univ, MD, 65, PhD, 77; Am Bd Neurol Surg, cert, 79. *Prof Exp:* Chief neurosurg, Naval Hosp, Boston, 73-74; intern & asst surgeon gen surg, Cornell-New York Hosp Med Ctr, 65-68, asst surgeon neurosurg, 68-69, surgeon, 69-71 & 72-73; res fel physiol, Col Med, Cornell Univ, 70-72 and 74-76, asst prof physiol, 76-81, asst prof neurosurg, 77-80; asst prof neurosurg & anat, 81-84, assoc prof develop genetics & anat, 85-88, ASSOC PROF NEUROSURG, CASE WESTERN RESERVE UNIV, 85-, ASSOC PROF NEUROSCI, 88- *Concurrent Pos:* Extramural res fel, Nat Inst Neurol & Commun Disorders & Strokes, 71-72 & 74-76; Andrew W Mellon teacher-scientist, 77-79; asst attend neurosurg, NY Hosp, 77-79; vis asst prof anat, Case Western Reserve Univ, 79-81; consult, Dept Physiol, Med Col, Cornell Univ, 81-82; asst attend neurosurg, Univ Hosp Cleveland, 81-; clin investr, Vet Admin Med Ctr, Cleveland, Ohio, 81-84, med investr, 84- *Mem:* AAAS; Cong Neurol Surgeons; Soc Neurosci; Am Soc Cell Biol; Am Asn Neurol Surg. *Res:* Nerve regeneration in mammalian and amphibian peripheral nerves, studied by means of organ culture, histochemistry, axonal transport of radioactive proteins, gel electrophoresis and immunoblotting and insitu RNA hybridization. *Mailing Add:* 2040 Adelbert Rd Case Western Res Univ 2119 Abingon Rd Cleveland OH 44106

MACQUARRIE, RONALD ANTHONY, b Oakland, Calif, Jan 30, 43; m 83; c 3. ENZYMOLOGY, METABOLISM. *Educ:* Univ Calif, Berkeley, BS, 65; Univ Ore, PhD(chem), 69. *Prof Exp:* Teaching fel biochem, Cornell Univ, 70-72 & Univ Calif, Los Angeles, 72-73; from asst prof to assoc prof biochem, 73-83, chmn chem, 82-83, interim dean life sci, 85-86, PROF BIOCHEM, UNIV MO-KANSAS CITY, 83-, DIR MOLECULAR BIOL, 85- *Mem:* Am Chem Soc; AAAS; Sigma Xi; Am Soc Biol Chemists. *Res:* Structure and functions of enzymes; regulation of enzyme activity; kinetic properties of enzymes; structure of the active site. *Mailing Add:* Sch Basic Life Sci Univ Mo Kansas City MO 64110-2499

MCQUATE, JOHN TRUMAN, b Upper Sandusky, Ohio, Aug 28, 21; m 46; c 3. GENETICS, BIOCHEMISTRY. *Educ:* Heidelberg Col, BS, 43; Ind Univ, PhD(zool), 51. *Prof Exp:* From asst prof to prof, 51-84, EMER PROF ZOOL, OHIO UNIV, 84- *Concurrent Pos:* Am Cancer Soc res fel, Case Western Reserve Univ, 56-57, USPHS res fel, 57-59. *Mem:* Genetics Soc Am; Am Soc Human Genetics. *Res:* Radiation and chemical mutagenesis in Drosophila; human genetics and cytogenetics. *Mailing Add:* 24 Canterbury Dr Athens OH 45701

MCQUATE, ROBERT SAMUEL, b Lebanon, Pa, Sept 4, 47; m 70; c 2. BIO-INORGANIC CHEMISTRY, PHARMACOLOGY. *Educ:* Lebanon Valley Col, BS, 69; Ohio State Univ, PhD(chem), 73. *Prof Exp:* Res fel, NMex State Univ, 73-74; asst prof chem, Willamette Univ, 74-77; consumer safety officer, Regulatory Affairs, Bur Foods, Food & Drug Admin, 77-80; group leader, Regulatory Affairs & Nutrit, Armour-Dial, Inc, 80-83; sci dir, Soft Drink Asn, 83-86; EXEC DIR, ADVAN SCI & TECHNOL INST, 86-; FOUNDER, RS MCQUATE & ASSOCS, 88- *Concurrent Pos:* Sigma Xi grant-in-aid, 75; Petrol Res Fund grant-in-aid, 75; consult, Food Safety & Regulatory Affairs; expert witness, chem & pharmaceut safety. *Honors & Awards:* Merit Award, Food & Drug Admin. *Mem:* Am Chem Soc; Licensing Execs Soc; Inst Food Technologists; Soc Soft Drink Technologists; Regulatory Affairs Professionals Soc. *Res:* Risk analysis and risk communication; food safety; regulatory affairs pertaining to foods, drugs, cosmetics and household products. *Mailing Add:* Advan Sci & Technol Inst 318 Hendricks Hall Univ Oregon Eugene OR 97403

MCQUEEN, CHARLENE A, b Newark, NJ, Apr 15, 47. BIOCHEMICAL TOXICOLOGY, BIOCHEMICAL PHARMACOLOGY. *Educ:* Univ Mich, PhD(human genetics), 78. *Prof Exp:* Assoc, Div Path, Am Health Found, 80-84, head sect Biochem Toxicol, 84-90; ASSOC PROF DEPT PHARMACOL TOXICOL, UNIV ARIZ, 90- *Mem:* Soc Toxicol; Am Asn Cancer Res; Am Soc Pharmacol & Exp Therapeut; Int Soc Study Xenobiotics. *Res:* Pharmacogenetics; role of genetic differences in metabolism in susceptibility to chemical toxicity; genetic toxicology; chemical carcinogenesis. *Mailing Add:* Dept Pharmacol Toxicol Univ Ariz Col Pharm Tucson AZ 85715

MCQUEEN, DONALD JAMES, b Vancouver, BC, Sept 8, 43; m 65; c 2. LIMNOLOGY. *Educ:* Univ BC, BSc, 66, MSc, 68, PhD(ecol), 70. *Prof Exp:* From asst prof to assoc prof biol, 70-82, PROF BIOL, YORK UNIV, 82-, DIR BIOL GRAD PROG, 90- *Concurrent Pos:* Erskine fel, Univ Canterbury, New Zealand, 82 & 89; chair, Pop Biol Grants Comt, Nat Sci & Eng Res Coun, Can, 87, continuous funding, 70-90. *Mem:* Can Soc Zool; Int Asn Theoret & Appl Limnol; Am Soc Naturalists; Brit Ecol Asn; Ecol Soc Am; Am Soc Limnol & Oceanog. *Res:* Lake restoration; manipulation of fish, zooplankton and phytoplankton populations. *Mailing Add:* Dept Biol York Univ 4700 Keele St Toronto ON M3J 1B3 Can

MCQUEEN, HUGH J, b Alloa, Scotland, Sept 29, 33; Can citizen; m 59; c 6. MECHANICAL METALLURGY, SCIENCE POLICY. *Educ:* Loyola Col Montreal, BSc, 54; McGill Univ, BEng, 56; Univ Notre Dame, MS, 58, PhD, 61. *Prof Exp:* Assoc prof metall, Ecole Polytech, Montreal, 61-66; res scientist, Phys Metall Div, Dept Energy, Mines & Resources, Can, 66-68; assoc prof mech eng, Sir George Williams Univ, 68-72, chmn dept, 71-72, PROF MECH ENG, CONCORDIA UNIV, 73- *Honors & Awards:* Western Elec Fund Award, Am Soc Eng Educ, 76. *Mem:* Fel Am Soc Metals; Am Inst Mining, Metall & Petrol Engrs; Inst Metals Eng; Can Inst Mining & Metall; Am Soc Eng Educ. *Res:* Behavior of metals in hot working dynamic recovery; dynamic recrystallization; constitutive equations; hot ductility; mechanical properties of metals, especially dislocation, substructures, strengthening mechanisms, transmission electron microscopy. *Mailing Add:* Dept Mech Eng Concordia Univ 1455 Demaisonneuve Blvd W Montreal PQ H3G 1M8 Can

MCQUEEN, JOHN DONALD, medicine; deceased, see previous edition for last biography

MCQUEEN, RALPH EDWARD, Can citizen; m 82. RUMINANT NUTRITION, BIOCHEMISTRY. *Educ:* Univ BC, BSA, 66, MSc, 69; Cornell Univ, PhD(nutrit), 73. *Prof Exp:* RES SCIENTIST RUMINANT NUTRIT, AGR CAN, 74- *Concurrent Pos:* Assoc ed, Can J Animal Sci, 84-90. *Honors & Awards:* James Robb Award, 85. *Mem:* Am Soc Animal Sci; Am Dairy Sci Asn; Can Soc Sci Nutrit; Can Soc Animal Sci; Agr Inst Can. *Res:* Fermentation of fiber by ruminal microorganisms; chemical and biochemical methods for evaluating nutritive quality of feedstuffs. *Mailing Add:* Res Sta Agr Can PO Box 20280 Fredericton NB E3B 4Z7 Can

MACQUEEN, ROBERT MOFFAT, b Memphis, Tenn, Mar 28, 38; m 60; c 2. SOLAR PHYSICS. *Educ:* Rhodes Col, BS, 60; Johns Hopkins Univ, PhD(atmospheric sci), 68. *Prof Exp:* Actg asst prof physics, Rhodes Col, 61-63; instr astron, Goucher Col, 64-66; consult, Appl Physics Lab, Johns Hopkins Univ, 66; sr scientist, Nat Ctr Atmospheric Res, 67, dir, High Altitude Observ, 79-86, assoc & actg dir, 86-89; PROF PHYSICS, RHODES COL, 90- *Concurrent Pos:* Lectr, Dept Astrogeophys, Univ Colo, Boulder, 69-79, adj prof, 79-90; prin investr, NASA, Skylab, Apollo, Solar Maximum Mission, Int Solar Polar Mission; mem & chmn bd, Asn Univ Res Astron; mem vis comt, Bartol Res Inst 84-90. *Honors & Awards:* Exceptional Sci Achievement Medal, NASA, 74. *Mem:* Am Asn Physics Teachers; Am Geophys Union; fel Optical Soc Am; Am Astron Soc; Int Astron Union. *Res:* Structure and evolution of the solar electron corona; infrared thermal emission of the interplanetary medium. *Mailing Add:* Rhodes Col Memphis TN 38112

MACQUEEN, ROGER WEBB, b Toronto, Ont, Nov 5, 35; m 59; c 4. GEOLOGY. *Educ:* Univ Toronto, BA, 58, MA, 60; Princeton Univ, MA & PhD(geol), 65. *Prof Exp:* Geologist, V C Illing & Partners, Eng, 60-62; res scientist, Inst Sedimentary & Petrol Geol, Geol Surv Can, 65-76; assoc prof, 76-79, head petrol geol, 79-85, INST SEDIMENTARY & PETROL GEOL, GEOL SURV CAN, 85- *Concurrent Pos:* Spec lectr, Dept Geol, Univ Calgary, 67-68; vis assoc prof, Erindale Col, Univ Toronto, 71-72; chmn, Am Comn Stratig Nomenclature, 74-75; pres, Geol Asn Can, 77-78. *Mem:* AAAS; Geol Soc Am; fel Geol Asn Can (pres, 77-78); Can Soc Petrol Geologists. *Res:* Regional geology; organic geochemistry and sedimentology; base metals in sedimentary rocks. *Mailing Add:* Dept Earth Sci Univ Waterloo Waterloo ON N2L 3G1 Can

MCQUIGG, ROBERT DUNCAN, b Wooster, Ohio, Apr 7, 36; m 59; c 4. PHYSICAL CHEMISTRY, ENVIRONMENTAL CHEMISTRY. *Educ:* Muskingum Col, BS, 58; Ohio State Univ, PhD(phys chem), 64. *Prof Exp:* Asst prof chem, Univ Toledo, 64-65; from asst prof to assoc prof, 65-75, chmn dept, 72-75 & 78-82, PROF CHEM, OHIO WESLEYAN UNIV, 75-, CHMN DEPT, 88- *Concurrent Pos:* Woodrow Wilson fel, 58-59; vis res assoc, Ohio State Univ, 68, 73, 74, 76, 77, 81; vis scientist, Nat Ctr Atmospheric Res, Colo, 69-70; dir & res supvr, Undergrad Res Participation, NSF, Ohio Wesleyan Univ, 69, 80 & 81, chmn, Dept Chem, 72-75 & 78-82; vis sr chemist, Dept Appl Sci, Brookhaven Nat Lab, 75; Fulbright res scholar, Air Chem Div, Max Planck Inst Chem, Fed Repub Ger, 82-83. *Mem:* Am Chem Soc; AAAS. *Res:* Atmospheric chemistry, including photochemical kinetics, computer modelling, sampling and analysis; photochemistry of inorganic coordination compounds. *Mailing Add:* Dept Chem Ohio Wesleyan Univ Delaware OH 43015

MACQUILLAN, ANTHONY M, b London, Eng, Feb 3, 28; Can citizen; m 53; c 4. MICROBIAL PHYSIOLOGY, BIOCHEMISTRY. *Educ:* Univ BC, BSA, 56, MSc, 58; Univ Wis, PhD(bact), 62. *Prof Exp:* Nat Res Coun Can fel, 61-63; asst prof, 63-72, ASSOC PROF MICROBIOL, UNIV MD, COLLEGE PARK, 72- *Mem:* Am Soc Microbiol. *Res:* Physiology, genetics and halotolerant properties of the marine bacterium Vibrio parahaemolyticus; lactose metabolism in Vibrio; starvation physiology in yeast. *Mailing Add:* Dept Microbiol Univ Md College Park MD 20742

MCQUILLEN, HOWARD RAYMOND, b Dubuque, Iowa, May 6, 24; m 56; c 3. ELECTRONIC & OPTICAL ENGINEERING. *Educ:* Iowa State Col, BSEE, 50; Univ Ariz, MSEE, 60; Ariz State Univ, PhD(elec eng), 67. *Prof Exp:* Res engr, Radio Corp Am, 50-55 & Admiral Corp, 55-56; sr eng specialist, Goodyear Aerospace Corp, 56-67; mgr systs res & develop dept, GTE-Sylvania, 67-71; STAFF ENGR, ELECTROMAGNETIC SYSTS DIV, RAYTHEON CORP, 71- *Concurrent Pos:* Designer & conductor training courses practicing electronic engrs, 81- *Mem:* Sr mem Inst Elec & Electronics Engrs. *Res:* Electronic systems research and development using microwave, analog and digital circuits and optical processing; applications to electronic warfare, television, radar, communications, telemetry, navigation and other sensors and radiators. *Mailing Add:* Electromagnetic Systs Div Rayteon Corp Dept 9222 6380 Hollister Ave Goleta CA 93017

MCQUILLEN, MICHAEL PAUL, b New York, NY, Sept 9, 32; m 57; c 4. NEUROLOGY, BIOMEDICAL ETHICS. *Educ:* Georgetown Univ, BA, 53, MD, 57. *Prof Exp:* Intern, Royal Victoria Hosp, Montreal, 57-58; resident neurol, Georgetown Univ, 58-60; fel med, Johns Hopkins Univ, 60-62, instr, 62-65; from asst prof to prof neurol, Univ Ky, 65-74; PROF NEUROL &

CHMN, MED COL WIS, 74- *Concurrent Pos:* Vis scientist, Univ Copenhagen, 71-72; chmn, Myasthenia Gravis Found, 81-83; vis prof, Royal Col Surgeons, Ireland, 83. *Mem:* Fel Am Acad Neurol; Am Neurol Asn; Multiple Sclerosis Soc; Am Osteop Soc. *Res:* Neuromuscular disease, especially myasthenia gravis, and multiple sclerosis. *Mailing Add:* Neurol Med Ctr Univ Ky Lexington KY 40536-0084

MCQUISTAN, RICHMOND BECKETT, b West New York, NJ, June 12, 27; m 51; c 2. PHYSICS, STATISTICAL MECHANICS. *Educ:* Purdue Univ, BS, 50, MS, 52, PhD(physics, heat transfer), 54. *Prof Exp:* Instr, Purdue Univ, 52-54; head solid state sect, Farnsworth Electronics Co, 54-56; res supvr phys electronics, Honeywell Res Ctr, 56-61; assoc prof elec eng, Univ Minn, Minneapolis, 61-63; staff physicist, Honeywell Res Ctr, 63-66; chmn dept physics, 68-69, assoc dean col lett & sci, 69-71, asst to vchancellor, 65-66; dean grad sch, 72-75, PROF PHYSICS, UNIV WIS-MILWAUKEE, 66-, DIR LAB SURFACE STUDIES, 78- *Concurrent Pos:* Vis assoc prof, Univ Minn, Minneapolis, 63-64; vis prof, Univ Liverpool, 73-76; sr res fel, Brit Sci Res Coun, 75-76. *Mem:* Am Optical Soc; Am Phys Soc; Am Vacuum Soc; Orthop Res Soc. *Res:* Infrared detectors and detection theory; electrical and optical properties of vacuum deposited thin films; radiative heat transfer and thermodynamics; infrared gonimetric reflectance measurements; resistance-temperature characteristics of semiconductors; infrared modulation by free carrier absorption; statistical mechanics of lattice spaces. *Mailing Add:* Dept Physics Univ Wis Milwaukee WI 53201

MCQUISTON, FAYE C(LEM), b Perry, Okla, Jan 23, 28; m 55; c 3. MECHANICAL ENGINEERING, THERMAL SCIENCES. *Educ:* Okla State Univ, BS, 58, MS, 59; Purdue Univ, Lafayette, PhD(mech eng), 70. *Prof Exp:* Aerothermodyn engr, Gen Dynamics Corp, 59-61; res scientist gas dynamics, Ling-Temco-Vought Res Ctr, 61-62; from asst prof to assoc prof thermal environ eng, 62-75, PROF THERMAL ENVIRON ENG, OKLA STATE UNIV, 75- *Concurrent Pos:* Consult, Peerless Boiler Co, Oklahoma City, 63, C A Mathey Mach Works, Tulsa, 64, Evans & Throup, Ponca City, 66, Brewer Corp, Ft Worth, 70, Lapco, Inc, Tulsa, 71-, Collins-Soter Eng, Oklahoma City, 71-, Russell Coil Co, Los Angeles, 77- & Thermal Corp, Houston, 78-; NSF sci fac fel, 67-69. *Honors & Awards:* fel Am Soc Heating Refrig Air-conditioning Engrs, 86. *Mem:* Am Soc Mech Engrs; Am Soc Heating, Refrig & Air-Conditioning Engrs. *Res:* Compact heat exchanger design and optimization; condensation from binary gas mixtures; general heat transfer and fluid dynamics problems of air-conditioning industry; building thermal systems simulation. *Mailing Add:* Sch Mech & Aerospace Eng Okla State Univ Stillwater OK 74078

MACQUOWN, WILLIAM CHARLES, JR, b Pittsburgh, Pa, Sept 8, 15; m 39; c 2. GEOLOGY. *Educ:* Univ Rochester, BS, 38, MS, 40; Cornell Univ, PhD(struct geol), 43. *Prof Exp:* Asst geol, Univ Rochester, 38-40 & Cornell Univ, 40-43; petrol geologist, Magnolia Petrol Co, 43-46; asst prof geol, Univ Ky, 47-48; dist geologist, Coop Ref Asn, 48-50; geologist, Denver Area, Deep Rock Oil Corp, 50-54; chief geologist, Sohio Petrol Co, Okla, 54-57, mgr, Calgary, 57-60; PROF GEOL, UNIV KY, 61-; PETROL CONSULT, 60- *Mem:* AAAS; fel Geol Soc Am; Am Inst Prof Geologists; Sigma Xi; Soc Econ Paleont & Mineral. *Res:* Stratigraphy and carbonate petrology. *Mailing Add:* 641 Beth Lane Lexington KY 40503

MAC RAE, ALFRED URQUHART, b New York, NY, Apr 14, 32; m 67; c 2. PHYSICS. *Educ:* Syracuse Univ, BS, 54, PhD(physics), 60. *Prof Exp:* Head, Explor Semiconductor Technol Dept, 60-80, dir, Integrated Circuits Lab, 78-83, DIR, SATELLITES, BELL LABS, 83- *Concurrent Pos:* Dir, Electron Devices Soc, Inst Elec & Electronics Engrs, 86-87. *Mem:* Fel Am Phys Soc; Bohmische Phys Soc; fel Inst Elec & Electronics Engrs. *Res:* Low energy electron diffraction; surfaces; infrared photoconductivity; fluctuations in solids; ion implantation; silicon integrated circuits. *Mailing Add:* Bell Labs Rm 2F-335 Holmdel NJ 07733

MACRAE, ALFRED URQUHART, b Brooklyn, NY, April 14, 32; m 67; c 2. ELECTRONICS ENGINEERING, SOLID STATE PHYSICS. *Educ:* Syracuse Univ, BS, 54, PhD(physics), 60. *Prof Exp:* Mem tech staff basic res, Bell Labs, 60-67, supvr ion implantation, 67-69, dept head integrated circuits, 69-79, dir digital integrated circuits, Bell Labs, 79-81, dir network systs planning, 81-83, DIR SATELLITE TRANSMISSION, AT&T BELL LABS, 83- *Concurrent Pos:* Chmn, Inst Elec & Electronic Engrs Field Awards, 91-92. *Mem:* Fel Am Phys Soc; Fel Inst Elec & Electronic Engrs, Electron Devices Soc (pres, 87-88). *Res:* Infrared detection, electrical noise, surface effects, low energy electron diffraction and ion interactions with solids; electron devices, integrated circuits, SIC fabrication, high voltage circuits; microprocessors, satellites, video compression and wideband packet technology. *Mailing Add:* AT&T Bell Labs Crawfords Corner Rd Rm 2F-335 Holmdel NJ 07733

MACRAE, ANDREW RICHARD, b Cleveland, Ohio, Sept 24, 46; Can citizen; m 91. CLINICAL BIOCHEMISTRY, PRENATAL SCREENING. *Educ:* Univ Toronto, BSc, 65; Univ Guelph, MSc, 71; Univ London, PhD(clin biochem), 74; Univ Toronto, DCC, 76; Can Acad Clin Biochem, FCACB, 87. *Prof Exp:* Asst dir biochem, St Boniface Hosp, Winnipeg, Man, 77-83; DIR CLIN BIOCHEM, OSHAWA GEN HOSP, ONT, 83- *Concurrent Pos:* Lectr asst, Guys Hosp Med Sch, London, 72-74; asst prof, Fac Med, Univ Man, 77-82; pres, Man Soc Clin Chemists, 78-80, Ont Soc Clin Chemists, 87-89; chmn, Can Soc Clin Chemists, Task Force Prenatal Screening, 89- *Mem:* Can Soc Clin Chemists; Am Soc Human Genetics. *Res:* Detection and diagnosis of disease based on measurement of plasma constituents; prenatal screening for neural tube defects and chromosomal disorders; interpretive reporting of clinical tests; expert systems; nutritional biochemistry; continuing medical education. *Mailing Add:* Dept Lab Med Oshawa Gen Hosp 24 Alma St Oshawa ON L1G 2B9 Can

MCRAE, DANIEL GEORGE, b Jordan, Mont, Nov 25, 38; m 63; c 2. MATHEMATICS, OPERATIONS RESEARCH. *Educ:* Univ Mont, BA, 60, MA, 61; Univ Wash, PhD(math), 67. *Prof Exp:* Instr math, Univ Mont, 61-62; asst prof, Univ Ill, Urbana, 67-70; from asst prof to assoc prof, 70-79, PROF MATH, UNIV MONT, 79- *Concurrent Pos:* Vis lectr, Mat Asn Am, 72-88, bd gov, 85-88; vis assoc prof math, Ga Inst Technol, 76-77. *Mem:* Am Math Soc; Math Asn Am; Opers Res Soc Am; Decision Sci Inst; Sigma Xi. *Res:* Ring theory; operations research; homological algebra; microcomputers in education; categorical algebra. *Mailing Add:* 295 Horseshoe Ln Rte 5 Missoula MT 59803-9704

MACRAE, DONALD ALEXANDER, b Halifax, NS, Feb 19, 16; m 39; c 3. ASTRONOMY. *Educ:* Univ Toronto, BA, 37; Harvard Univ, PhD(astron), 43. *Prof Exp:* Asst, Univ Toronto, 37-38; res astronr, Univ Pa, 41-42; instr, Cornell Univ, 42-44; res physicist, Manhattan Proj, Tenn, 45-46; asst prof astron, Case Inst Technol, 46-53; from assoc prof to prof, 53-81, chmn dept & dir, David Dunlap Observ, 65-78, EMER DIR & PROF ASTRON, UNIV TORONTO, 81- *Concurrent Pos:* Mem bd trustees, Univs Space Res Asn, 69-76; mem bd dirs, Can-France-Hawaii Telescope Corp, 73-79. *Mem:* Fel Royal Soc Can; fel Royal Astron Soc; Can Astron Soc; Am Astron Soc; Royal Astron Soc Can. *Res:* Galactic and radio astronomy; stellar spectroscopy. *Mailing Add:* David Dunlap Observ Univ Toronto Richmond Hill ON L4C 4Y6 Can

MACRAE, DONALD RICHARD, b Poughkeepsie, NY, July 14, 34; c 1. PROCESS DEVELOPMENT, PLASMA PROCESSING. *Educ:* Syracuse Univ, BSCHE, 56; Princeton Univ, 58; Royal Inst Technol, Stockholm, Sweden, PhD(chem eng), 61. *Prof Exp:* Res assoc, dept ferrous metall, Royal Inst Technol, Sweden, 62-67; res engr, Bethlehem Steel Corp, 67-69, res supvr, 69-82, sr scientist, 82-85. *Mem:* Am Inst Metall Engrs. *Res:* Fundamental phenomena of high temperature extractive metallurgical processes including development of plasma reactors for reduction of iron ore directly to steel and for ferroalloy production; treatment and recycle of steelmaking dusts. *Mailing Add:* 418 High St Bethlehem PA 18018

MACRAE, EDITH KRUGELIS, b Waterbury, Conn, Jan 24, 19; m 50; c 1. ANATOMY. *Educ:* Bates Col, BS, 40; Columbia Univ, MA, 41, PhD(zool), 46. *Prof Exp:* Instr, Vassar Col, 45-47; Donner Found fel, Carlsberg Lab, Denmark, 47-48, Am Cancer Soc fel, 48-49; res assoc, Yale Univ, 49-51; instr biol, Mass Inst Technol, 51-53; asst res zoologist, Univ Calif, Berkeley, 54-56; from asst prof to prof anat, Univ Ill Med Ctr, 57-72, actg head dept, 69-70; PROF ANAT, MED SCH, UNIV NC, CHAPEL HILL, 72- *Concurrent Pos:* Guggenheim fel, 64-65. *Mem:* AAAS; Am Soc Anatomists; Am Soc Cell Biol; Electron Micros Soc Am. *Res:* Histology; cytochemistry; fine structure. *Mailing Add:* Anat Swing Bldg 217-B Univ NC Med Sch Chapel Hill NC 27514

MCRAE, EION GRANT, b Tambellup, Australia, Dec 25, 30; m 54; c 2. PHYSICS. *Educ:* Univ Western Australia, BSc, 52, MSc, 54; Fla State Univ, PhD(phys chem), 57. *Prof Exp:* Fel chem, Ind Univ, 57-58; sr res officer, Commonwealth Sci & Indust Res Orgn, Australia, 58-63; MEM TECH STAFF, BELL LABS, 63- *Mem:* Fel Am Phys Soc. *Res:* Theoretical and experimental work on the development of low-energy electron scattering techniques for studying the surface structures of crystals. *Mailing Add:* Room 1C320 Bell Labs Murray Hill NJ 07911

MACRAE, HERBERT F, b Middle River, NS, Mar 30, 26; m 55; c 4. BIOCHEMISTRY, ANIMAL SCIENCE. *Educ:* McGill Univ, BSc, 54, MSc, 56, PhD(biochem), 60. *Hon Degrees:* DSc, McGill Univ, 87. *Prof Exp:* Chemist, Can Dept Nat Health & Welfare, 60-61; asst prof biochem, MacDonald Col, McGill Univ, 61-67, assoc prof, 67-70, prof animal sci, 70-72, chmn dept, 67-72; prin, NS Agr Col, 72-89; RETIRED. *Concurrent Pos:* Chmn, Can Agr Res Coun, 86-93. *Mem:* Fel Agr Inst Can; Agr Inst Can; Can Soc Animal Sci. *Res:* Estrogens of avian species; milk proteins; organophosphorus insecticides; skeletal muscle enzymes and proteins. *Mailing Add:* Seven Hickman Dr Truro NS B2N 2Z2 Can

MCRAE, LORIN POST, b Tucson, Ariz, Feb 20, 36; m 56; c 7. BIOMEDICAL ENGINEERING, ELECTRONICS. *Educ:* Univ Ariz, BS, 61, PhD(elec eng), 68; NY Univ, MEE, 63. *Prof Exp:* Mem tech staff, Bell Tel Labs, 61-63; instr elec eng, Univ Ariz, 63-68; asst prof, Univ Wyo, 68-70; dir med electronics, 70-89, dir vascular lab, 74-89, CLIN ENGR SPECIALIST, TUCSON MED CTR, 89- *Concurrent Pos:* Chmn elec safety comt, Southern Ariz Hosp Coun, 71-72; mem bd examr, Biomed Equip Technicians, Asn Advan Med Instrumentation, 72-76, & Clin Eng Cert, Inst Clin Cert; pres, Med Electronic Devices, Inc, Tucson, Ariz, 75-77. *Honors & Awards:* Gerard B Lambert Award, Ocular Pulse Clin. *Mem:* Asn Advan Med Instrumentation; Inst Elec & Electronic Engrs; hon mem Soc Noninvasive Vascular Technol. *Res:* Epidemology of strokes; noninvasive detection of stroke risk patients; llrodynamics. *Mailing Add:* 3540 N Bonanza Ave Tucson AZ 85749

MACRAE, NEIL D, b Scotstown, Que, Oct 9, 35; c 1. GEOLOGY. *Educ:* Queen's Univ, Ont, BSc, 61; McMaster Univ, MSc, 63, PhD(geol), 66. *Prof Exp:* Res assoc petrol, Univ Minn, 65-66; fel, 66-68, asst prof, 68-72, ASSOC PROF GEOL, UNIV WESTERN ONT, 72- *Mem:* Soc Econ Geologists; Mineral Asn Can. *Res:* Geochemistry and petrology of silicate-oxide-sulfide relations in mafic and ultramafic intrusions; element fractionation modeling using secondary ion mass spectrometric analysis. *Mailing Add:* Dept of Geol Univ of Western Ont & Middlesex Col London ON N6A 5B7 Can

MACRAE, PATRICK DANIEL, b Calgary, Alta, Apr 26, 28; m 58. DENTISTRY. *Educ:* Univ Alta, DDS, 60, cert pediat dent, 67. *Prof Exp:* Assoc prof, 60-73, chmn, Div Pediat Dent, 78-87, PROF PEDIAT DENT, UNIV ALTA, 73-,. *Concurrent Pos:* Dir, Dent Serv, Glenrose Hosp, Edmonton, Alta, 66- *Mem:* Can Dent Asn. *Res:* Sociology of health care. *Mailing Add:* Dept Pediat Dent Univ Alta Fac Dent Edmonton AB T6E 2N8 Can

MACRAE, ROBERT ALEXANDER, b Charlotte, NC, Dec 8, 35; m 71; c 1. PHYSICS. *Educ:* Davidson Col, BS, 58; Vanderbilt Univ, MS, 60. *Prof Exp:* Instr physics, Davidson Col, 60-62; asst prof, Univ NC, Charlotte, 62-65; prof, Cent Piedmont Community Col, 65-67; ASST PROF PHYSICS, JACKSONVILLE STATE UNIV, 67- *Concurrent Pos:* Consult physics, Oak Ridge Nat Lab, 61-77; pres, Air Brake Consult, 75- *Mem:* Am Phys Soc; Optical Soc Am; Air Brake Asn; Sigma Xi; Soc Automotive Engrs; Am Soc Mech Engrs. *Res:* Optical properties of materials; vacuum ultraviolet spectroscopy; railroad train stopping performance. *Mailing Add:* PO Box 905 Jacksonville AL 36265-0905

MACRAE, ROBERT E, b Detroit, Mich, May 1, 34; m 57; c 1. MATHEMATICS. *Educ:* Univ Chicago, AB, 53, SM, 56, PhD(math), 61. *Prof Exp:* Staff mem, Los Alamos Sci Lab, 57-59; NSF fel, 61-62; Ritt instr math, Columbia Univ, 62-65; asst prof, Univ Mich, 65-67; assoc prof, 67-72, PROF MATH, UNIV COLO, BOULDER, 72- *Concurrent Pos:* Consult, Inst Defense Analysis, 65 & Math Rev, 65- *Res:* Commutative ring theory; homological algebra; algebraic curves. *Mailing Add:* Dept Math Univ Colo Box 426 Boulder CO 80309-0426

MCRAE, ROBERT JAMES, b Jordan, Mont, Aug 16, 31; m 56; c 13. HISTORY OF PHILOSOPHY OF SCIENCE. *Educ:* Univ Mont, BA, 54, MA, 57; Univ Wis, Madison, PhD(hist sci), 69. *Prof Exp:* Asst prof, 58-60, assoc prof physics, 60-62 & 66-69, dean Sch Liberal Arts, 78-81, acad vpres, 81-83, PROF PHYSICS, EASTERN MONT COL, 69-, DIR GRAD STUDIES & RES, 83- *Mem:* Hist Sci Soc; Am Asn Physics Teachers; Nat Sci Teachers Asn. *Res:* History of nineteenth and early twentieth century physics. *Mailing Add:* Off Grad Studies & Res Eastern Mont Col Billings MT 59101-2800

MACRAE, THOMAS HENRY, b Halifax, NS, Can, May 31, 48; m 69; c 2. CELL & MOLECULAR BIOLOGY, DEVELOPMENTAL BIOLOGY. *Educ:* Mount Allison Univ, BSc, 70; Univ Windsor, MSc, 73, PhD(biol), 76. *Prof Exp:* Res assoc, Univ Sherbrooke Med Ctr, 76-77, Univ Miss Med Ctr, 77-79 & Univ Ottawa, 79-80; asst prof, 80-87, ASSOC PROF BIOL, DALHOUSIE UNIV, 87- *Concurrent Pos:* Vis res fel, Univ Kent, 87. *Mem:* Am Soc Cell Biol; Can Soc Cell Mol Biol; Am Soc Biochem & Molecular Biol; Can Biochem Soc; Biochem Soc. *Res:* Examination of tubulin gene number; organization and expression in developing embroyos; regulation of microtubule assembly and organization; evolutionary and functional aspects of tubulin structure; effects of heavy metals on brine shrimp development. *Mailing Add:* Biol Dept Dalhousie Univ Halifax NS B3H 4J1 Can

MCRAE, VINCENT VERNON, b Columbia, SC, Sept 2, 18; m 41; c 2. SCIENCE POLICY, INFORMATION SCIENCE. *Educ:* Miner Teachers Col, BS, 40; Cath Univ, MS, 44, PhD(math), 55. *Prof Exp:* Teacher high sch, Washington, DC, 41-52; analyst opers res off, Johns Hopkins Univ, 52-60, chmn Stratspiel group & chief strategic div, 60-61; mem sr staff, Res Analysis Corp, 61-64; tech asst, Off Sci & Technol, White House, 64-74; DIR STRATEGIC PLANNING, FED SYSTS DIV, IBM CORP, 74- *Concurrent Pos:* Mem tech staff, Gaither Comt, 57; adv to del & mem US del, Surprise Attack Conf, Geneva, 58; res assoc, Coolidge Comt, 59; consult, Nat Security Coun, Dept of Defense, 73-78 & Off Sci & Technol Policy, White House; mem Defense Sci Bd, 73-80 & Naval Studies Bd. *Mem:* Fel AAAS; NY Acad Sci; Am Math Soc; Opers Res Soc Am; Sigma Xi. *Res:* Operations research; military operations research. *Mailing Add:* 5901 Montreal Rd Apt N305 Rockville MD 20852

MCRAE, WAYNE A, b Chicago, Ill, Aug 8, 25; m 79; c 2. MEMBRANE TECHNOLOGY, ELECTROCHEMICAL ENGINEERING. *Educ:* Harvard Univ, AB, 46, MA, 48. *Prof Exp:* Chemist, Ionics Inc, 49-55, asst tech dir, 55-61, vpres res & develop, 61-86, corp consult, 78-86; INDEPENDENT CONSULT, 86- *Mem:* Fel Am Inst Chemists; Europ Soc Membrane Sci & Technol; NAm Membrane Soc. *Res:* Ion-exchange membranes; electrodialysis for demineralization of water and membrane chloralkali cells. *Mailing Add:* PO Box 192 Zurich 8053 Switzerland

MCRAE, WAYNE ALAN, physical chemistry, for more information see previous edition

MACRANDER, ALBERT TIEMEN, b Hilversum, Holland, Apr 23, 50; US citizen; m 79. III-V SEMICONDUCTORS. *Educ:* Univ Ill, Urbana, BS, 72, MS, 73, PhD(physics), 77. *Prof Exp:* Assoc fel, Dept Mat Sci & Eng, Cornell Univ, 77-80; mem tech staff, Bell Labs, 80-90; PHYSICIST, ARGONNE NAT LAB, 90- *Mem:* Am Phys Soc; Inst Electric & Electronic Engrs; Electrochem Soc. *Res:* Point defects in noble gas solids, III-V semiconductors; x-ray optics; synchrotron radiation; ultra-precise x-ray lattice parameter determinations; high resolution x-ray diffraction. *Mailing Add:* Advan Photon Source Bldg 360 Argonne Nat Lab 9700 S Cass Ave Argonne IL 60439-4814

MCREE, DONALD IKERD, b Newton, NC, Dec 22, 34; m 59; c 3. BIOELECTROMAGNETICS, NEUROPHYSIOLOGY. *Educ:* Davidson Col, BS, 57; Col William & Mary, MS, 64; NC State Univ, PhD(eng sci & mech), 69. *Prof Exp:* Teaching asst physics, Duke Univ, 57-58; res engr, NASA, 59-64; sr res engr, Corning Glass Works, 64-69; res physicist, 69-89, HEALTH SCIENTIST ADMINR, NAT INST ENVIRON HEALTH SCI, 89- *Concurrent Pos:* Lieutenant, US Army, 58; from adj asst prof to adj assoc prof, NC State Univ, 70-85, adj prof, 85- *Honors & Awards:* NIH Dir Award, 78. *Mem:* Bioelectromagnetics Soc (vpres, 81-82, pres, 82-83); AAAS. *Res:* Biological effects of non-ionizing radiation, primarily microwave radiation; the interaction of microwaves with the nervous system; biophysics of the function of excitable membrane, including ionic conduction and gating mechanisms. *Mailing Add:* Nat Ist Environ Health Sci PO Box 12233 Research Triangle Park NC 27709

MCREE, G(RIFFITH) J(OHN), JR, b Blackstone, Va, Sept 21, 34; m 85; c 3. ELECTRICAL ENGINEERING. *Educ:* US Mil Acad, BS, 56; Univ Ariz, MS, 62; Univ Va, PhD(elec eng), 70. *Prof Exp:* Instr math, US Mil Acad, 65-67; teaching asst elec eng, Univ Va, 67-68, res asst control systs, 68-70; from asst prof to assoc prof, Dept Elec Eng, 70-87, ASSOC DEAN RES & GRAD STUDIES, OLD DOMINION UNIV, 87- *Concurrent Pos:* Pres, University Eng, LTD, PC, Norfolk, Va; prin investr, NASA contracts, 70- *Mem:* Inst Elec & Electronics Eng. *Res:* Automatic control of systems; simulation and modeling of systems to include large, complex socio-economic systems; studies in the improvement of engineering curricula. *Mailing Add:* 525 Virginia Dare Dr Virginia Beach VA 23451

MCREYNOLDS, LARRY AUSTIN, b Eugene, Ore, May 27, 46; m 77; c 2. PARASITOLOGY, MOLECULAR BIOLOGY. *Educ:* Ore State Univ, BSc, 68; Mass Inst Technol, PhD(biol), 74. *Prof Exp:* NIH fel, Baylor Col Med, 74-77; vis scientist DNA sequencing, Med Res Coun Lab Molecular Biol, Eng, 77-78; asst prof biochem, Col Med, Univ Ariz, 78-82; SR SCIENTIST MOLECULAR PARASITOL, NEW ENG BIOLABS, BEVERLY, MASS, 83- *Concurrent Pos:* NIH grant, gene transfer techniques, 78-82; vis lectr, WHO Special Prog Res & Training Trop Dis, WHO/TDR, Kamraj Univ, India, 84; grant to test DNA diag probes for filariasis, WHO/TDR, 87-88, mem, filariasis steering comt, 87-, sci working group filariasis, Woods Hole, Mass, 87, contract prod cloned filarial diag antigens, 90; organizer, workshop DNA diag & Filariasis, Jakarta, Indonesia, WHO/TDR & New Eng Biolabs, 89; mem special rev comt, trop dis res centers, NIAID, 90. *Mem:* Am Soc Trop Med & Hyg; Am Heartworm Soc. *Res:* Development of DNA diagnostic probes for human filarial parasites; use of repeated DNA's to study parasite evolution; cloning of antigens to develop vaccines for human and animal filarial parasites. *Mailing Add:* New Eng Biolabs 32 Tozer Rd Beverly MA 01915

MCREYNOLDS, RICHARD A, b Washington, DC, Sept 11, 44. IMMUNOPATHOLOGY, IMMUNOLOGY. *Educ:* George Washington Univ, BS, 66; Univ Va, MS, 68, MD, 71. *Prof Exp:* Intern, Peter Bent Brigham Hosp, 71-72, resident, 72-73; investr & staff investr path, NIH, 73-75; asst prof path, Univ Ala, Birmingham, 75-77; NIH res fel pulmonary immunopath, Univ Conn, 77-79; head path sect, Div Med Affairs, Allied Corp, 78-82; prin res pathologist, Lederle Labs, 82-85; ASST PROF PATH, SCH MED, E CAROLINA UNIV, 85- *Concurrent Pos:* Consult anat path, Vet Admin Hosp, Birmingham, 75-77; adj asst prof path, Col Physicians & Surgeons, Columbia Univ, 79-87; clin asst prof path, Univ Med & Dent, NJ-Rutgers Med Sch, 79-87, clin asst prof anat, 83-87. *Mem:* Am Asn Pathologists; Fedn Am Soc Exp Biol; Am Rheumatism Asn; Int Acad Path; Am Thoracic Soc; Sigma Xi. *Res:* Immunopathology; T-lymphocytes; experimental arthritis. *Mailing Add:* Dept Path E Carolina Univ Sch Med Greenville NC 27858-4354

MACRI, FRANK JOHN, b Portland, Maine, Jan 12, 23; m 48; c 4. PHARMACOLOGY. *Educ:* Univ Maine, BA, 43; George Washington Univ, BS, 50; Georgetown Univ, PhD(pharmacol), 53. *Prof Exp:* Analytical chemist, Naval Powder Factory, 43-45, org chemist, 46-48; biochemist, NIH, 48-50; pharmacologist, Irwin, Neisler & Co, 51-53; chief pharmacologist, Eaton Labs, Inc, 53-55; head, Sect Pharmacol, Nat Eye Inst, Bethesda, 55-79; PROF OPHTHAL & PHARMACOL & DIR OCULAR PHARMACOL, MED CTR, GEORGETOWN UNIV, 79- *Mem:* Am Soc Pharmacol & Exp Therapeut; Asn Res Vision & Ophthal; NY Acad Sci. *Res:* Pharmacodynamics of autonomic nervous system agents; effect of hormones on vascular physiology and pharmacology; structure-activity relationships in biological processes; pharmacology of the eye. *Mailing Add:* 3602 Janet Rd Silver Spring MD 20906

MACRIS, NICHOLAS T, b Brooklyn, NY, Oct 27, 31; m 60; c 3. INTERNAL MEDICINE. *Educ:* Columbia Col, AB, 53; State Univ NY, MD, 58. *Prof Exp:* CHIEF ALLERGY & IMMUNOL, DEPT MED, LENOX HILL HOSP, 70-, ATTEND PHYSICIAN MED, 74- & PATH, 89- *Concurrent Pos:* Attend physician med, NY Hosp, Med Ctr, Cornell Univ, 88-, co-dir, Allergy-Immunol Training Prog, Dept Med, 89- & clin prof med, Med Col, 89- *Mem:* Fel Am Acad Allergy & Immunol; Clin Immunol Soc; Am Asn Immunologists. *Res:* Immunodeficiency and autoimmunity. *Mailing Add:* Lenox Hill Hosp 100 E 77th St New York NY 10021

MACRISS, ROBERT A, b Athens, Greece, Nov 8, 30; US citizen; m 60; c 2. CHEMICAL ENGINEERING. *Educ:* Nat Tech Univ Athens, Ing Dipl, 55; Univ Alta, MSc, 59. *Prof Exp:* Pilot plant engr, Can Chem Co, Edmonton, 59-60; asst chem engr, Inst Gas Technol, 60-62, assoc chem engr, 62-64, chem engr air conditioning res & develop, 64-67, res supvr, 67-72, res supvr environ res & develop, 72-73, res mgr, 73-75, asst res dir, 75-77, assoc res dir energy conserv, 77-86, res dir energy utilization, 86-89. *Mem:* Am Soc Heating, Air Conditioning & Refrig Engrs. *Res:* Indoor pollution; environmental control; air conditioning; absorption cooling; thermodynamics; properties of fluid mixtures, gases and chemicals; gas appliances; solar heating/cooling; energy modeling; solar applications; energy conservation; air infiltration; energy management heat pumps, weatherization, advanced furnaces and retrofits. *Mailing Add:* 1603 Oak Terr St Joseph MI 49085

MCROBERTS, J WILLIAM, b Rochester, Minn, Dec 6, 32; m 59; c 2. UROLOGY. *Educ:* Princeton Univ, BA, 55; Cornell Univ, MD, 59. *Prof Exp:* Intern surg, New York Hosp & Cornell Univ Med Col, 59-60; resident surg, Mayo Clin, 60-61, resident urol, 61-64; from instr to assoc prof urol, Sch Med, Univ Wash, 67-72; assoc prof, 72-74, PROF SURG, COL MED, UNIV KY, 74-, CHIEF DIV UROL, 72- *Mem:* Am Col Surg; Am Urol Asn; Am Acad Pediat; Soc Pediat Urol; Soc Univ Urol. *Res:* Renal transplantation; pediatric urology, intersex. *Mailing Add:* Div of Urol Univ of Ky Col of Med Lexington KY 40536

MCROBERTS, KEITH L, b Clinton, Iowa, July 16, 31; m 52; c 4. INDUSTRIAL ENGINEERING, OPERATIONS RESEARCH. *Educ:* Iowa State Univ, BS, 53, MS, 59, PhD(eng valuation, opers res), 66. *Prof Exp:* Methods analyst, E I du Pont de Nemours & Co, 53-57; from instr to asst prof indust eng, Iowa State Univ, 57-61, asst prof indust eng & opers res, 62-67; opers analyst, Off Opers Analysis, US Air Force, 61-62; vis lectr mgt sci, Univ Bradford, 67-68; assoc prof, 68-72, PROF INDUST ENG & OPERS RES, IOWA STATE UNIV, 72-, CHMN DEPT INDUST ENG, 74- *Concurrent Pos:* Analyst consult, US Air Force, 58-, dir, Opers Anal Standby Unit, 66-70; consult to indust, 60- *Mem:* Fel Am Inst Indust Engrs; Opers Res Soc Am; Am Soc Eng Educ; Brit Oper Res Soc; Sigma Xi. *Res:* Search theory; simulation; material control. *Mailing Add:* Dept of Indust Eng 212 Marston Hall Iowa State Univ Ames IA 50011

MCROBERTS, MILTON R, biochemistry, for more information see previous edition

MCRORIE, ROBERT ANDERSON, b Statesville, NC, May 7, 24; m 46; c 3. BIOCHEMISTRY. *Educ:* NC State Col, BS, 49, MS, 51; Univ Tex, PhD(biochem), 54. *Prof Exp:* From asst prof to prof biochem, Univ Ga, 53-83; RETIRED. *Concurrent Pos:* Res partic, Oak Ridge Inst Nuclear Study, 57 & 58; assoc dean grad sch & dir gen res, Univ Ga, 59-71, asst vpres res, 68-71. *Mem:* AAAS; Am Chem Soc; Am Soc Microbiol; Am Soc Biol Chem; Sigma Xi. *Res:* Microbial nutrition; intermediary metabolism; enzymology; sperm enzymes; reproduction. *Mailing Add:* 305 Sherwood Dr Athens GA 30606

MCROWE, ARTHUR WATKINS, b San Francisco, Calif, Aug 16, 37; m 65; c 3. ORGANIC CHEMISTRY. *Educ:* Univ Calif, Berkeley, BS, 60; Univ Wis-Madison, PhD(org chem), 66. *Prof Exp:* Res chemist, B F Goodrich Res Ctr, 66-71, sr res chemist, 71-83; SR RES CHEMIST, DOVER CHEM CORP, 83- *Mem:* Am Chem Soc; Sigma Xi. *Res:* Physical organic chemistry; carbonium ion rearrangements; monomer processes; metallo-organic chemistry; polymer flammability; organic phosphorus chemistry. *Mailing Add:* 365 Springcrest Dr Akron OH 44313

MCROY, C PETER, b East Chicago, Ind, Jan 17, 41; m 87. BIOLOGICAL OCEANOGRAPHY. *Educ:* Mich State Univ, BS, 63; Univ Wash, MS, 66; Univ Alaska, Fairbanks, PhD(marine sci), 70. *Prof Exp:* From asst prof to assoc prof, 69-79, PROF MARINE SCI, INST MARINE SCI, UNIV ALASKA, FAIRBANKS, 79- *Concurrent Pos:* Fel, Univ Ga, 71-72; dir, int progs, Univ Alaska, 84-87; vis prof, Aarhus Univ, Denmark, 84 & Tokyo Univ, Japan, 88; consult, Baie James Ecol Soc, Que, 87-88, Ak Dep Environ Conserv Oil Spill, 90-91. *Honors & Awards:* Diamond Award, Bot Soc Am, 75. *Mem:* AAAS; Am Soc Limnol & Oceanog; Ecol Soc Am; Arctic Inst NAm; Int Asn Aquatic Vascular Plant Biologists; Sigma Xi; Int Asn Ecol. *Res:* Ecology of seagrass communities; productivity and ecology of northern seas; ecosystem analysis. *Mailing Add:* Inst Marine Sci Univ Alaska Fairbanks AK 99701

MCRUER, DUANE TORRANCE, b Bakersfield, Calif, Oct 25, 25; m 55; c 1. CONTROL ENGINEERING. *Educ:* Calif Inst Technol, BS, 45, MS, 48. *Prof Exp:* Control engr, Northrop Aircraft Inc, 48, engr-in-charge servo develop, supvr, Servomechanisms Sect asst chief flight control, 48-54; pres, Control Specialists Inc, 53-58; PRES, SYSTS TECHNOL, 58- *Concurrent Pos:* Chmn, Subcomt Avionics, 78-81; mem, Aeronaut Adv Comn, 80-90; mem, Aero/Astro Vis Comn, Mass Inst Technol, 85-88; mem, NRC Aeronaut & Space Eng Bd, 88-, chmn, 90. *Honors & Awards:* Louis E Levy Award, Franklin Inst, 60; Alexander C Williams Sr Award, Human Factors Soc, 76. *Mem:* Nat Acad Eng; fel Inst Elec & Electronic Engrs; fel Am Inst Aeronaut & Astronaut; fel Human Factors Soc; fel Soc Automotive Engrs. *Res:* Control systems engineering; manual and automatic flight control and guidance for manual aerospace and land vehicles; systems analysis techniques to the study of man-machine systems and dynamics of the human operations. *Mailing Add:* Systems Technol Inc 13766 S Hawthorne Blvd Hawthorne CA 90250-7083

MCSHANE, EDWARD JAMES, mathematics; deceased, see previous edition for last biography

MCSHARRY, JAMES JOHN, b Newark, NJ, May 28, 42; m 67; c 2. MICROBIOLOGY, VIROLOGY. *Educ:* Manhattan Col, BS, 65; Univ Va, MS, 67, PhD(microbiol), 70. *Prof Exp:* From asst prof to assocprof, 73-83, PROF MICROBIOL & IMMUNOL, ALBANY MED COL, 83- *Concurrent Pos:* Nat Inst Allergy & Infectious Dis fel, Rockefeller Univ, 70-73. *Mem:* AAAS; Am Soc Microbiol; Harvey Soc; Sigma Xi. *Res:* Studies on the structure and function of the membrane proteins and surface glycoproteins of enveloped RNA viruses; mode of action of antiviral agents; rapid techniques in viral diagnosis. *Mailing Add:* Dept Microbiol & Immunol Albany Med Col Albany NY 12208

MCSHARRY, WILLIAM OWEN, b Hammels, NY, Oct 14, 39; m 70; c 1. ANALYTICAL CHEMISTRY. *Educ:* Fordham Univ, BS, 61, MS, 67, PhD(chem), 69. *Prof Exp:* Sr scientist anal chem, Hoffmann-La Roche Inc, 69-75, group leader, 75-79, mgr anal chem, 79-85; MGR MAT CHARACTERIZATION, ENGELHARD CORP, 85- *Mem:* Am Chem Soc; Sigma Xi. *Res:* Characterization of precious metal powders and catalysts. *Mailing Add:* 29 Mountainview Dr Cedar Grove NJ 07009

MCSHEFFERTY, JOHN, b Akron, Ohio, Mar 14, 29; m 59; c 2. PHARMACEUTICAL CHEMISTRY. *Educ:* Univ Glasgow, BSc, 53, PhD(medicinal chem), 57. *Prof Exp:* Asst lectr pharm & chem, Univ Strathclyde, 53-54; res assoc, Sterling-Winthrop Res Inst, NY, 57-62; sr scientist, 62-63, dir pharmaceut develop, Ortho Pharmaceut Corp, Raritan, 63-80; PRES, GILLETTE RES INST, 80- *Concurrent Pos:* Chmn bd dirs, Dirs Indust Res. *Mem:* AAAS; Am Chem Soc; Am Pharmaceut Asn; NY Acad Sci; Am Acad Pharmaceut Sci; Sigma Xi; Indust Res Inst. *Res:* Toiletries; pharmaceutical development. *Mailing Add:* Gillette Res Inst 401 Professional Dr Gaithersburg MD 20879

MCSHERRY, CHARLES K, b New York, NY, Nov 22, 31; m 57; c 1. SURGERY. *Educ:* Fordham Univ, BS, 53; Cornell Univ, MD, 57; Am Bd Surg, dipl, 65; Bd Thoracic Surg, dipl, 66. *Prof Exp:* Asst surg, Med Col, Cornell Univ, 58-63, instr, 63-64, from clin instr to clin asst prof, 65-71, assoc prof, 71-77; PROF SURG, MT SINAI SCH MED, 77-; DIR SURG, BETH ISRAEL MED CTR, 77- *Concurrent Pos:* Asst attend surgeon, New York Hosp, 65-72, assoc attend surgeon, 72-77. *Mem:* Am Col Surg; Am Gastroenterol Asn; Asn Study Liver Dis; Soc Univ Surg; Am Surg Asn. *Res:* Gastrointestinal malignancy; gallstone formation. *Mailing Add:* Beth Israel Med Ctr 10 Nathan D Perlman Pl New York NY 10003

MCSHERRY, DIANA HARTRIDGE, b Long Branch, NJ; m 77; c 1. MEDICAL PHYSICS, COMPUTER SCIENCE. *Educ:* Harvard Univ, BA, 65; Rice Univ, MA, 67, PhD(nuclear physics), 69. *Prof Exp:* Fel nuclear physics, Rice Univ, 69; res physicist ultrasonics, Digicon, Inc, 69-74, exec vpres med ultrasound, 74-77, pres cardiol anal systs, Digisonics, Inc, 77-82; chmn bd, Info Prod Systs, Houston, 82-86; vpres, Digicon, Inc, 80-87; CHIEF OPERATING OFFICER, COGNISEIS DEVELOP, INC, 87- *Mem:* Inst Elec & Electronic Engrs; Am Inst Ultrasound Med; Am Phys Soc; Am Heart Asn. *Res:* Development of computer based cardiology analysis systems in the specific areas of echocardiology, ventriculography and hemodynamics. *Mailing Add:* 3034 Underwood Houston TX 77025

MCSORLEY, ROBERT, b Cincinnati, Ohio, July 23, 49; m 76. NEMATOLOGY. *Educ:* Univ Cincinnati, BS, 71, MS, 74; Purdue Univ, PhD(entom), 78. *Prof Exp:* From asst prof to assoc prof, 78-88, PROF NEMATOL, UNIV FLA, 88- *Mem:* Soc Nematologists; Ecol Soc Am; Org Trop Am Nematologists; Europ Soc Nematologists. *Res:* Pest management; population dynamics; ecology of plant-parasitic and free-living nematodes. *Mailing Add:* Dept Entom & Nematol Univ Fla Bldg 940 Gainesville FL 32611

MCSPADDEN, WILLIAM ROBERT, electrical engineering, for more information see previous edition

MCSWAIN, RICHARD HORACE, b Greenville, Ala, Sept 27, 49; m 72; c 3. FAILURE ANALYSIS, ACCIDENT INVESTIGATION. *Educ:* Auburn Univ, BS, 72, MS, 74; Univ Fla, PhD(mat eng), 85. *Prof Exp:* Res & teaching asst mat eng, Auburn Univ, 72-73; metallurgist, Southern Res Inst, 73-76; mat engr, Naval Aviation Depot, Pensacola, 77-88, head metall & mat engr, 88-90; PRES, MCSWAIN ENG, INC, 91- *Concurrent Pos:* Consult mat engr, 82-90; adj instr, Univ WFla, Pensacola, 88. *Mem:* Am Soc Metals Int; Soc Automotive Engrs; Nat Asn Corrosion Engrs; Am Soc Testing & Mat; Int Soc Air Safety Investigators; Electron Micros Soc Am. *Res:* Mechanism of solidification of gray and ductile iron and nickel-carbon alloys; mechanism of graphite growth in carbon-carbon composites; computerized fracture surface feature characterization. *Mailing Add:* 4021 Middlebury Dr Pensacola FL 32514-8047

MACSWAN, IAIN CHRISTIE, b Ocean Falls, BC, Apr 15, 21; US citizen; m 43; c 3. PLANT PATHOLOGY. *Educ:* Univ BC, BSA, 42, MSA, 62. *Prof Exp:* Prin res asst, Dom Lab Plant Path, Univ BC, 46-47; asst prof & plant pathologist, BC Dept Agr, 47-55; from asst prof to prof plant path, 55-87, EMER PROF PLANT PATH, ORE STATE UNIV, 87- *Mem:* Am Phytopath Soc; Can Phytopath Soc. *Res:* Extension pathology; diseases of horticultural crops. *Mailing Add:* 1629 NW 14th Corvallis OR 97330

MCSWEEN, HARRY Y, JR, b Charlotte, NC, Sept 29, 45; c 1. METEORITICS. *Educ:* Citadel, BS, 67; Univ Ga, MS, 69; Harvard Univ, PhD(geol), 77. *Prof Exp:* From asst to assoc prof, 77-87, PROF & DEPT HEAD GEOL, UNIV TENN, KNOXVILLE, 87- *Concurrent Pos:* Prin investr, NASA, 78; assoc ed, Geochem & Cosmochem Act, 85. *Mem:* fel Geol Soc Am; Meteoretical Soc (secy, 87-); fel Mineral Soc Am; Am Geophys Union. *Res:* Petrology and geochemistry of meteorites; petrogenesis of igneous and metamorphic rocks of Southern Appalachian origin. *Mailing Add:* Dept Geological Sci Univ Tennessee Knoxville TN 37996

MACSWEEN, JOSEPH MICHAEL, b Antigonish, NS, Mar 1, 33; m 57; c 4. IMMUNOLOGY. *Educ:* St Francis Xavier Univ, BSc, 52; Dalhousie Univ, MD, 57, MSc, 69; FRCP(C), 70. *Prof Exp:* Resident med, Dalhousie Univ, 66-69; fel clin immunol, Montreal Gen Hosp, 69-70; Med Res Coun res fel immunol, Walter & Eliza Hall Inst Med Res, Australia, 71-72; lectr, 72-74, from asst prof to assoc prof, 74-82, PROF MED, DALHOUSIE UNIV, 82- *Concurrent Pos:* Regional coordr myeloma chemother nat trial, Nat Cancer Inst Can, 74-76. *Mem:* Can Soc Clin Invest; Can Soc Immunol. *Res:* Lymphokine responses in renal transplantation; functional changes in mononuclear cell subpopulations in chronic lymphocytic leukemia. *Mailing Add:* Dept Med-Immunol Camphill Hosp 1763 Robi St Halifax NS B3H 3G2 Can

MCSWEENEY, FRANCES KAYE, b Rochester, NY, Feb 6, 48. PSYCHOLOGY. *Educ:* Smith Col, BA, 69; Harvard Univ, MA, 72, PhD(psychol), 74. *Prof Exp:* lectr psychol, McMaster Univ, 73-74; from asst prof to assoc prof, 74-83, PROF PSYCHOL, WASH STATE UNIV, 83-, CHMN DEPT, 86- *Concurrent Pos:* Basic behav processes rev panel, NIMH, 86-88, psychobiol & behav rev panel, 88-90; Woodrow Wilson fel. *Mem:* Sigma Xi; fel Am Psychol Asn; Asn Behav Anal; Soc Advan Behav Anal; Am Psychol Soc. *Res:* basic processes of animal learning and behavior in areas of operant and classical conditioning; behavioral contrast and quantitative theories. *Mailing Add:* Dept Psychol Wash State Univ Pullman WA 99164-4820

MCSWEENY, EDWARD SHEARMAN, b Tucson, Ariz, Oct 5, 34; m 68; c 2. AQUACULTURE, INVERTEBRATE SYSTEMATICS. *Educ:* Univ Ariz, BS, 64; Univ Miami, MS, 68, PhD(biol oceanog), 71. *Prof Exp:* Assoc res dir aquacult, Ocean Protein Corp, Dania, Fla, 70-75; mgr aquacult sci div, Syntex Corp, Palo Alto, Calif, 75-77; prawn res dir corp res & develop aquacult, Weyerhaeuser Co, 77-81, consult, 81-91; INDEPENDENT CONSULT, 91- *Mem:* AAAS; Am Fisheries Soc; World Maricult Soc. *Res:* Crustacean aquaculture; systematics of peracarid crustaceans and cephalopods. *Mailing Add:* 23700 SW 142nd Ave Homestead FL 33032

MCTAGGART, KENNETH C, b Vancouver, BC, Can, Aug 10, 19; m 51; c 2. PETROLOGY. *Educ:* Univ BC, BASc, 43; Queen's Univ, Ont, MSc, 46; Yale Univ, PhD(geol), 48. *Prof Exp:* Geologist, Geol Surv Can, 48-50; prof geol, Univ BC, 50-85; RETIRED. *Mem:* Geol Asn Can; fel Geol Soc. *Res:* Petrology and structure. *Mailing Add:* Dept Geol Univ BC 2075 Wesbrook Pl Vancouver BC V6T 1Z2 Can

MCTAGGART-COWAN, PATRICK DUNCAN, b Edinburgh, Scotland, May 31, 12; m 39; c 2. METEOROLOGY, SCIENCE POLICY. *Educ:* Univ BC, BA, 33; Oxford Univ, BA, 36. *Hon Degrees:* DSc, Univ BC, 60, McGill Univ, 74, Lakehead Univ, 74 & Univ NB, 76, Univ Guelph, 81; LLD, St Francis Xavier Univ, 70 & Simon Fraser Univ, 72. *Prof Exp:* Instr physics, Univ BC, 33-34; meteorologist in training, British Meteorol Off, 36; officer in charge, Meteorol Off, Botwood & Nfld, 37-42; chief meteorol officer, Royal Air Force Ferry Command, Que, 42-45; secy air navig, Provisional Int Civil Aviation Orgn, 45; asst dir & chief, Forecast Div, Meteorol Br, Can Dept Transport, 46-57, from assoc dir to dir, 58-64; pres, Simon Fraser Univ, 63-68; exec dir, Sci Coun Can, 68-75; dir, John Wiley & Sons Co, Ltd, 76-89; RETIRED. *Concurrent Pos:* Head task force, Oper Oil, 70-72. *Honors & Awards:* Losey Award, Inst Aeronaut Sci, 59; Paterson Medal, Meteorol Serv Can, 65; Charles F Brookes Award, Am Meteorol Soc, 65, Cleveland Abbe Award, 76; Can Centennial Medal, 67. *Mem:* Am Meteorol Soc (vpres, 60); Can Meteorol & Oceanog Soc; AAAS; Am Acad Arts & Sci; Am Geophys Union; Can Asn Physicists; Arctic Inst NAm; Asn Advan Sci Can. *Res:* Environment; climatic change. *Mailing Add:* High Falls Rd RR2 Bracebridge ON P1L 1W9 Can

MCTAGUE, JOHN PAUL, b Jersey City, NJ, Nov 28, 38; m 61; c 4. CHEMICAL PHYSICS. *Educ:* Georgetown Univ, BS, 60; Brown Univ, PhD(phys chem), 65. *Prof Exp:* Mem tech staff sci ctr, NAm Aviation Inc, 64-70; from asst prof to assoc prof chem, Univ Calif, Los Angeles, 70-74, prof, 74-82; dir, Nat Synchrotron Light Source, Brookhaven Nat Lab, 82-83; adj prof chem, Columbia Univ, 82-83; dep dir, 83-85, ACTG SCI ADV TO THE PRES & ACT DIR, OFF SCI & TECH POLICY, EXEC OFF PRES, 86- *Concurrent Pos:* Alfred P Sloan res fel, 71-74; John Simon Guggenheim Mem fel, 75-76. *Mem:* Am Phys Soc; Am Chem Soc. *Res:* Spectroscopic studies of molecular interactions; x-ray and neutron scattering; dynamics of condensed phases; surface physics; quantum solids and liquids. *Mailing Add:* Ford Motor Co VP-Tech Affairs The American Rd Dearborn MI 48121-1899

MACTAVISH, JOHN N, b Detroit, Mich, May 23, 39; c 1. ENVIRONMENTAL SCIENCES, GEOLOGY. *Educ:* Bowling Green Univ, BS, 61, MA, 63; Case Western Reserve Univ, PhD(geol), 71. *Prof Exp:* Asst prof geol, Col Arts & Sci, William James Col, Grand Valley State Cols, 68-71, asst prof environ studies, asst to the dean & dir environ studies prog, 71- *Mem:* Geol Soc Am; Paleont Soc; Nat Asn Geol Teachers. *Res:* Invertebrate paleontology. *Mailing Add:* 10925 Buchanan Grand Haven MI 49417

MCTERNAN, EDMUND J, b Hollis, NY, June 15, 30; m 52; c 6. PUBLIC HEALTH ADMINISTRATION. *Educ:* New Eng Col, BS, 56; Columbia Univ, MS, 58; Univ NC, Chapel Hill, MPH, 63; Boston Univ, EdD, 74. *Hon Degrees:* DSc, Thomas Jefferson Univ, 89. *Prof Exp:* Asst adminr, Emerson Hosp, Concord, Mass, 58-60; adminr, Mem Hosp, North Conway, NH, 60-62; adminr, Boston Dispensary & Rehab Inst, Mass, 63-65; assoc prof & dean div health sci, Northeastern Univ, 65-69; PROF HEALTH SCI & DEAN, SCH ALLIED HEALTH PROFESSIONS, HEALTH SCI CTR, STATE UNIV NY, STONY BROOK, 69- *Concurrent Pos:* USPHS fel core curric allied health educ, Northeastern Univ, 65-67; Bruner Found fel physician assoc educ, State Univ NY Stony Brook, 70-; mem nat adv allied health professions coun, NIH, 67-70; consult, col proficiency exam prof health educ, State Dept Educ, NY, 71- & USPHS, Kuwait Ministry Health, 83 & Ministry Health, Peru, 85; vis prof, Western Australian Inst Technol, 76, 78 & 80; consult, USPHS, Kuwait Ministry Health, 83 & Ministry Health, Peru, 85; mem, Nat Adv Coun Health Professions Educ, 88-91. *Mem:* Fel Asn Schs Allied Health Professions (pres, 84-86). *Res:* Education patterns and role relationships of allied health profession; continuing education needs in allied health professions. *Mailing Add:* Sch Allied Health Professions State Univ NY Health Sci Ctr Stony Brook NY 11790

MCTIGUE, DAVID FRANCIS, b Wilmington, Del, June 4, 52; m 85. EARTH SCIENCES. *Educ:* Williams Col, BA, 74; Stanford Univ, MS, 77, PhD(geol), 79. *Prof Exp:* Assoc, Mass Inst Technol, 79-81; SR MEM TECH STAFF, SANDIA NAT LABS, 81- *Mem:* Am Geophys Union; Soc Rheology. *Res:* Mechanics of geological materials; porous media; multiphase flow; crustal deformation. *Mailing Add:* Orgn 1511 Sandia Nat Labs Bldg 634 Rm 32 Albuquerque NM 87185

MCTIGUE, FRANK HENRY, b Holyoke, Mass, Dec 19, 19; m 43; c 2. PLASTICS CHEMISTRY. *Educ:* Williams Col, BA, 41; Yale Univ, PhD(org chem), 49. *Prof Exp:* Res chemist, Res Ctr, Plastic Tech Ctr, Hercules, Inc, 49 -69, res supvr, 69-75, mgr prod technol, 75-78, res scientist, 78-82; PRES/ CONSULT, MCTIGUE ASSOCS INC, 83- *Mem:* Am Chem Soc; Soc Plastics Eng. *Res:* Polymers, plastics and fibers; plastics formulation, stabilization, degradation. *Mailing Add:* 2551 Deepwood Dr Wilmington DE 19810

MACUR, GEORGE J, b Chicago, Ill, Feb 22, 33; m 60; c 2. PHYSICAL CHEMISTRY. *Educ:* DePaul Univ, BS, 54, MS, 57; Ill Inst Technol, PhD(chem), 65. *Prof Exp:* Chemist, IIT Res Inst, 60-65 & Argonne Nat Lab, 65-68; chemist, IBM Corp, 68-86; res assoc, State Univ NY, Binghamton, 87-88; CONSULT, 88- *Res:* Surface chemistry and technology; ozone technology; ozone, ultraviolet and hydrogen peroxide in water purification. *Mailing Add:* 3441 Alexander Pl Smyrna GA 30082

MACURDA, DONALD BRADFORD, JR, b Boston, Mass, Aug 8, 36; m 81; c 2. STRATIGRAPHY SEDIMENTATION, PALEONTOLOGY. *Educ:* Univ Wis, BS, 56, PhD(geol), 63. *Prof Exp:* From instr to prof geol & mineral, Univ Mich, Ann Arbor, 63-78, assoc cur invert paleont, Mus Paleont, 63-77, cur, 77-78; res specialist, Exxon Prod Res Co, 78-81; VPRES EXPLOR SERV ENERGISTS, 81- *Concurrent Pos:* NSF grants, Univ Mich. *Mem:* Am Asn Petrol Geologists; Geol Soc Am; Int Paleont Asn; Paleont Soc; Soc Econ Paleontologists & Mineralogists; Am Geophys Union; Soc Explor Geophysicists. *Res:* Seismic stratigraphy; study of the functional morphology and ecology of ancient crinoids, invertebrate. *Mailing Add:* Energists 10260 Westheimer Suite 300 Houston TX 77042

MCVAUGH, ROGERS, b Brooklyn, NY, May 30, 09; wid; c 2. BOTANY. *Educ:* Swarthmore Col, AB, 31; Univ Pa, PhD(bot), 35. *Prof Exp:* Asst inst bot, Univ Pa, 33-35; from instr to asst prof, Univ Ga, 35-38; assoc botanist, Div Plant Explor & Introd, Bur Plant Indust, USDA, 38-43, Div Soils & Agr Eng, 43-46; from assoc prof to prof bot, Univ Mich, Ann Arbor, 46-74, dir, Univ Herbarium, 72-75, Harley Harris Bartlett Prof Bot, 74-79, Cur Phanerogams, Herbarium, 46-55, vascular plants, 56-79; RES PROF BOT, UNIV NC, CHAPEL HILL, 80- *Concurrent Pos:* Prog dir syst biol, NSF, 55-56; adj res scientist, Hunt Inst Bot Doc, Carnegie Mellon Univ, 81- *Honors & Awards:* Merit Award, Bot Soc Am, 77; Asa Gray Award, Am Soc Plant Taxon, 84; Henry Allan Gleason Award, 84. *Mem:* Am Soc Plant Taxon (pres, 56); Int Asn Plant Taxon (vpres, 69-72, pres, 72-75); hon mem Soc Bot Mex. *Res:* Taxonomy of flowering plants; history of botany; flora of Mexico. *Mailing Add:* Dept Biol Coker Hall Univ NC Chapel Hill NC 27599-3280

MCVAY, FRANCIS EDWARD, applied statistics; deceased, see previous edition for last biography

MCVEAN, DUNCAN EDWARD, b Pontiac, Mich, June 8, 36; m 60; c 2. PHARMACEUTICAL CHEMISTRY. *Educ:* Univ Mich, BS, 58, MS, 60, PhD(pharmaceut chem), 63; Xavier Univ, MBA, 78. *Prof Exp:* Pharmaceut res chemist, William S Merrell Co, 65-71, head methods develop sect, Qual Control Dept, 71-74, proj asst to vpres qual opers, 74-75, head chem control sect, 75-78, dir qual control, Merrel-Nat Labs, 78-79; vpres Gentek Corp, 79-81; mgr pharmaceut develop, Adria Labs, 81-82, dir mfr, 82-84; VPRES, BEN VENUE LABS INC, 84- *Mem:* Sigma Xi; Am Acad Pharmaceut Sci; Asn Pharm Scientists; Am Hosp Pharm Asn; Parenteral Drug Asn. *Res:* Analytical chemistry; physical pharmacy; biopharmaceutics; pharmaceutical product development. *Mailing Add:* 6122 Liberty Rd Solon OH 44139

MCVEY, EUGENE STEVEN, b Wayne, WVa, Dec 6, 27; m 49; c 5. ELECTRICAL ENGINEERING. *Educ:* Univ Louisville, BS, 50; Purdue Univ, MS, 55, PhD(elec eng), 60. *Prof Exp:* Mem staff, Naval Ord, Ind, 50-56; group leader elec eng, Fransworth Electronics, 56-57; from instr to asst prof elec eng, Purdue Univ, 57-61; assoc prof, 61-66, PROF ELEC ENG, UNIV VA, 66- *Mem:* Fel Inst Elec & Electronics Engrs. *Res:* Automatic control systems; electronics; instrumentation; image processing; advanced automation. *Mailing Add:* 836 Broomley Rd Charlottesville VA 22901

MCVEY, JAMES PAUL, b Evansville, Ind, Aug 5, 43; m 70; c 2. SHRIMP HATCHERY & GROWOUT, MARINE FISH HATCHERY & GROWOUT. *Educ:* Univ Miami, Coral Gables, BS, 65; Univ Hawaii, MS, 67, PhD(zool), 70. *Prof Exp:* Fishery aid, Bur Commercial Fisheries, 63-65; res asst, Univ Hawaii, 65-70; biologist, Trust Territory of Pacific, 70-77; div chief, Nat Marine Fisheries Serv, 77-82; biologist, AID, Jakarta, Indonesia, 82-84; PROG DIR, NAT SEA GRANT PROG, 84- *Concurrent Pos:* Rep, Joint Comt Agr Res & Develop, 85-88; prog chmn, US, Japan Nat Resource Comt, 86-; vol, Int Exec Serv Corp, Egypt & Panama; partic, Worldnet Tech Exchange on Aquacult; mem, Joint Sub-Comt of Aquacult. *Honors & Awards:* Outstanding Performance Award, Nat Oceanic & Atmospheric Admin. *Mem:* World Aquaculture Assoc; Western Soc Naturalists; Nat Shellfish Assoc; Caribbean Aquaculture Assoc; Nat Aquacult Asn. *Res:* Primarily involved with the development of research and production techniques for shrimp, algae, molluscan and marine fish aquaculture. *Mailing Add:* Nat Sea Grant Col Prog 1335 East-West Hwy Rm 5492 Silver Spring MD 20910

MCVEY, JEFFREY KING, b Chicago, Ill, May 15, 50. CHEMICAL PHYSICS. *Educ:* Northwestern Univ, BS & MS, 72; Univ Chicago, PhD(chem), 77. *Prof Exp:* Fel chem, Columbia Univ, 77-78; asst prof chem, Princeton Univ, 78-; AT ROHM & HAAS CO. *Concurrent Pos:* Vis researcher, Exxon, 78; consult, Union Camp, 81; vis res, Dow Chem, 81, consult, 80- *Mem:* Am Chem Soc; Am Inst Physics. *Res:* Chemical physics, specifically exploring detailed mechanisms of unimolecular and biomolecular elementary processes in the gas phase; developing new techniques in molecular synthesis employing lasers to prepare reactants or facilitate the chemical reaction. *Mailing Add:* 79 Clinton St Labertville NJ 08530-1912

MCVICAR, JOHN WEST, b Rochester, NY, Oct 15, 28; m 56; c 4. VETERINARY MEDICINE, VIROLOGY. *Educ:* State Univ NY Vet Col, Cornell Univ, DVM, 52. *Prof Exp:* Partner, Loudoun Animal Hosp, 56-66; res vet, Sci & Educ Admin-Agr Res, USDA, 66-90; RETIRED. *Mem:* US Animal Health Asn; Infectious Dis Soc; NY Acad Sci; Am Vet Med Asn. *Res:* Transmission, early growth and pathogenesis of foot-and-mouth disease virus in cattle, sheep and goats and African swine fever virus in pigs; methods of virus detection and detection of infection. *Mailing Add:* Midlothian VA 24501

MCVICAR, KENNETH E(ARL), b Detroit, Mich, Nov 30, 20; m 45; c 3. ELECTRICAL ENGINEERING. *Educ:* Antioch Col, BS, 44; Mass Inst Technol, SM, 50. *Prof Exp:* Engr, Naval Res Lab, 44-46; owner, McVicar Radio Lab, 46-48; engr, Servomechanisms Lab, Mass Inst Technol, 48-51, staff mem, Digital Comput Lab, 51-54, sect leader, Lincoln Lab, 54-56, assoc group leader comput prog, 56-58, group leader syst design, 58-59; assoc tech dir syst res & planning, 59-62, tech dir syst planning, 62-67, tech dir strategic & range systs, 67-70, asst vpres, 70-72, VPRES, BEDFORD OPERS, MITRE

CORP, 72-, GEN MGR, 75- *Mem:* Inst Elec & Electronics Engrs; Sigma Xi. *Res:* Military systems research, planning, engineering and test; Air Defense, Command and Control data processing systems design. *Mailing Add:* 29 Clearwater Rd Winchester MA 01890

MACVICAR, MARGARET LOVE AGNES, solid state physics, materials science; deceased, see previous edition for last biography

MACVICAR, ROBERT WILLIAM, b Princeton, Minn, Sept 28, 18; m 48; c 2. CHEMISTRY. *Educ:* Univ Wyo, BA, 39; Okla State Univ, MS, 40; Univ Wis, PhD(biochem), 46. *Hon Degrees:* LLD, Univ Wyo, 77; DrSci, Dankook Univ, Korea, 80. *Prof Exp:* Assoc prof biochem, Okla State Univ, 46-49, prof & head dept, 49-57, dean grad sch, 53-64, vpres, 57-64; vpres acad affairs, Southern Ill Univ Syst, 64-68; chancellor, Southern Ill Univ, Carbondale, 68-70; pres, 70-84, EMER PRES, ORE STATE UNIV, 84- *Mem:* Am Chem Soc; Am Soc Biol Chemists; Am Inst Nutrit. *Res:* Nitrogen metabolism of plants and animals; nutrition and metabolism of large animals. *Mailing Add:* A-100 Radiation Ctr Ore State Univ Corvallis OR 97331-5903

MACVICAR-WHELAN, PATRICK JAMES, b St John's, Nfld; US citizen; m 71; c 1. ROBOTICS, MECHATRONICS SYSTEMS. *Educ:* St Francis Xavier Univ, BSc, 59; Dalhousie Univ, MSc, 61; Univ BC, PhD(physics), 65. *Prof Exp:* Asst prof physics, St Francis Xavier Univ, 64-65; asst prof math, St Mary's Univ NS, 65-66; Off Naval Res grant, Univ Calif, Berkeley, 66-70; assoc prof physics, Grand Valley State Col, 70-80; prin engr, Boeing Aerospcae Co, 81-82, PRIN SCIENTIST, BOEING ARTIFICIAL INTEL CTR, 82- *Concurrent Pos:* Res assoc, Nat Ctr Sci Res, France, 75-76 & 77-78; vis, Artificial Intel Lab, Stanford Univ, 79-81. *Mem:* Soc Mfg Engr; Comput-Aided Circuit Anal. *Res:* Machine vision; robotics; mechatronics systems; fuzzy logic and applications; pattern representation by tesselation atomic collisions; electrical breakdown; acoustics; low temperature physics; applied mathematics. *Mailing Add:* PO Box 936 Seahurst WA 98062

MACVITTIE, THOMAS JOSEPH, b Buffalo, NY, Aug 21, 41; m 64; c 3. EXPERIMENTAL HEMATOLOGY, INFECTIOUS DISEASE. *Educ:* Alfred Univ, BA, 64; Southern Univ NY, Buffalo, MS, 66, PhD(radiation biol), 70. *Prof Exp:* US Army, radiobiologist, Med Res Lab, Ft Knox, 70-72, RES PHYSIOLOGIST, ARMED FORCES RADIOBIOL RES INST, 72- *Mem:* Int Sci Exp Hematol. *Res:* Hematoporetic stem cell physiology in response to radiation and infectious disease; methods for enhancing recovery of stem cells and white blood cell production as well as increased non-specific resistance to gram negative bacteria. *Mailing Add:* 8840 N Westland Dr Gaithersburg MD 20877

MCVOY, KIRK WARREN, b Minneapolis, Minn, Feb 22, 28; m 53; c 3. THEORETICAL NUCLEAR PHYSICS. *Educ:* Carleton Col, BA, 50; Oxford Univ, BA, 52; Univ Gottingen, dipl, 53; Cornell Univ, PhD(physics), 56. *Prof Exp:* Res assoc physics, Brookhaven Nat Lab, 56-58; asst prof, Brandeis Univ, 58-62; from asst prof to assoc prof, 62-67, PROF PHYSICS, UNIV WIS-MADISON, 67- *Concurrent Pos:* Fulbright res grant, Univ Utrecht, 60-61; vis distinguished prof, Brooklyn Col, 70-71; vis prof, Ind Univ, 71-72 & Univ Groningen, Holland, 81; Alexander von Humboldt Sr Scientist Award, Heidelberg, Ger, 80-81. *Mem:* Fel Am Phys Soc. *Res:* Nuclear structure; scattering theory; nuclear reaction theory. *Mailing Add:* Dept Physics Univ Wis Madison WI 53706

MCWHAN, DENIS B, b New York, NY, Dec 10, 35; m 59; c 3. SOLID STATE PHYSICS. *Educ:* Yale Univ, BS, 57; Univ Calif, Berkeley, PhD(phys chem), 61. *Prof Exp:* Royal Inst Technol, Sweden, 61-62, CNRS, Grenoble, France, 73-74; distinguished mem tech staff, Phys Res Lab, AT&T Bell Labs, 62-90; CHMN, NAT SYNCHROTRON LIGHT SOURCE, BROOKHAVEN NAT LAB, 90- *Mem:* AAAS; Am Phys Soc. *Res:* Study of phase transitions using synchrotron x-ray and neutron scattering techniques including magnetic x-ray and time resolved scattering; magnetic and electronic properties of new materials. *Mailing Add:* NSLS Brookhaven Nat Lab Bldg 725 Upton NY 11973

MCWHINNEY, IAN RENWICK, b Burnley, Eng, Oct 11, 26; m 55; c 2. MEDICINE. *Educ:* Cambridge Univ, BA, 46, MB, ChB, 49, MD, 59. *Prof Exp:* Pvt pract, Eng, 55-68; PROF DEPT FAMILY MED, UNIV WESTERN ONT, 68-; MED DIR, PALLIATIVE CARE UNIT, PARKWOOD HOSP, LONDON, ONT, 86- *Concurrent Pos:* Nuffield travelling fel, US, 64-65; Fel Royal Col Physicians (London), 86. *Honors & Awards:* Foreign assoc mem, Inst Med Nat Acad Sci, 88. *Mem:* Fel Royal Col Gen Practitioners; Col Family Physicians Can; Soc Teachers Family Med. *Res:* Diagnostic process; social and behavioral aspects of medical practice. *Mailing Add:* Dept Family Med Univ Western Ont London ON N6A 5C1 Can

MCWHINNIE, DOLORES J, b Elmhurst, Ill, Sept 13, 33; m 84. HORMONES IN DEVELOPING SYSTEMS, NEUROENDOCRINOLOGY. *Educ:* DePaul Univ, BS, 55, MS, 58; Marquette Univ, PhD(biol sci), 65. *Prof Exp:* Res assoc radiation biol, Argonne Nat Lab, 59-60; ASSOC PROF DEVELOP BIOL & ENDOCRINOL, DEPAUL UNIV, 65-, DIR MED TECHNOL EDUC, 66- *Mem:* AAAS; Am Soc Zoologists; Am Inst Biol Sci. *Res:* Effects of hormones on mineral metabolism; metabolic aspects of bone growth and differentiation; metabolism of amphibian bone. *Mailing Add:* Dept Biol Sci DePaul Univ 1036 W Belden Ave Chicago IL 60614

MCWHIRTER, JAMES HERMAN, b Mercer, Pa, July 4, 24; m 52; c 5. MATHEMATICAL PROGRAMMING. *Educ:* Columbia Univ, BS, 45; Carnegie Inst Technol, MS, 48. *Prof Exp:* Lt-jg, USNR, 45-46; develop engr, Westinghouse Elec Transformer Div, 48-65, res engr, Westinghouse Res & Develop Ctr, 65-88; PRES, OPTIMIZATION, LTD, 88- *Concurrent Pos:* Cert financial planner, 89- *Mem:* Fel Inst Elec & Electronic Engrs. *Res:* Electrical engineering analysis with applications of mathematics and computers; electromagnetics; design optimization; semiconductor device simulation; electromagnetic numerical analysis. *Mailing Add:* 3660 Forbes Trial Dr Murrysville PA 15668

MCWHORTER, ALAN L, b Crowley, La, Aug 25, 30. SEMICONDUCTOR DEVICES, QUANTUM ELECTRONICS. *Educ:* Univ Ill, BS, 51; ScD, Mass Inst Technol, 55. *Prof Exp:* Staff mem, Lincoln Lab, 55-59, from asst prof to assoc prof, 59-65, HEAD, SOLID STATE DIV, LINCOLN LAB,MASS INST TECHNOL, 65-, PROF ELEC ENG, 66- *Honors & Awards:* David Sarnoff Award, Inst Elec & Electronics Engrs, 71. *Mem:* Nat Acad Eng; fel Am Phys Soc; fel Inst Elec & Electronics Engrs. *Mailing Add:* Lincoln Lab Mass Inst Technol PO Box 73 Lexington MA 02173-9108

MCWHORTER, CHESTER GRAY, b Brandon, Miss, May 3, 27; m 52; c 2. PLANT PHYSIOLOGY, WEED SCIENCE. *Educ:* Miss State Univ, BS, 50, MS, 52; La State Univ, PhD(bot), 58. *Prof Exp:* Agronomist, Agr Exp Sta, Miss State Univ, 52-56; plant physiologist, 58-75, DIR, SOUTHERN WEED SCI LAB, AGR RES, USDA, 75- *Concurrent Pos:* Plant physiologist, Agr Exp Sta, Miss State Univ, 58-, adj assoc prof weed control, 73-; mem, Nat Herbicide Assessment Team, 77-80; mem bd dirs, Coun Agr Sci & Technol, 78-80. *Honors & Awards:* Superior Serv Award, USDA, 78; Res Award, Weed Sci Soc Am, 77 & Am Soybean Asn, 78; 00078918g. *Mem:* AAAS; Am Soc Plant Physiologists; fel Weed Sci Soc Am (pres, 83-84); NY Acad Sci; Am Soybean Asn (pres, 83-84). *Res:* General weed control; Johnson grass control; weed control in soybeans. *Mailing Add:* Delta Brn Exp Sta Stoneville MS 38776

MCWHORTER, CLARENCE AUSTIN, medicine; deceased, see previous edition for last biography

MCWHORTER, EARL JAMES, b Argyle, NY, Sept 12, 29. ORGANIC CHEMISTRY. *Educ:* Rensselaer Polytech, BS, 50; Cornell Univ, PhD(org chem), 55. *Prof Exp:* Instr chem, 54-56, asst prof, 56-69, ASSOC PROF CHEM, UNIV MASS, AMHERST, 69- *Concurrent Pos:* Res assoc, New York Bot Garden, 63-64. *Mem:* Am Chem Soc; Sigma xi. *Res:* Polynuclear aromatic hydrocarbons; fungal polyacetylenes. *Mailing Add:* Dept of Chem Univ of Mass Amherst MA 01003

MCWHORTER, MALCOLM M(YERS), b Norfolk, Va, Jan 8, 26; m 51; c 1. ELECTRONICS. *Educ:* Ore State Col, BS, 49; Stanford Univ, MS, 50, PhD, 52. *Prof Exp:* Radio engr, KOAC, Ore State Col, 46-49; asst electronics, 51-52, res assoc nuclear induction, 52-53, wide band amplifiers, 53-54, from asst prof to assoc prof electronics, 54-69, PROF ELECTRONICS, STANFORD UNIV, 69- *Concurrent Pos:* Vpres, Vidar Corp, 58-69. *Mem:* Sigma Xi. *Res:* Transistor circuits; electronic measurement techniques. *Mailing Add:* 16 Portola Green Circle Portola Valley CA 94025

MACWILLIAMS, DALTON CARSON, b Winnipeg, Man, Mar 22, 28; US citizen; m 54; c 4. POLYMER CHEMISTRY. *Educ:* Univ Alta, BSc, 50; Univ Minn, PhD(anal phys chem), 55. *Prof Exp:* From chemist to sr res chemist, 55-75, SR RES SPECIALIST, WESTERN DIV RES, 72-, PROJ MGR, 75-, ASSOC SCIENTIST, CENT RES, DOW CHEM, USA, 84- *Concurrent Pos:* Semi Standards Comt, 84- *Mem:* Am Chem Soc. *Res:* Synthesis, rheology, interfacial properties and applications of water soluble polymers; physical organic chemistry; crystal growth; secondary oil recovery; wet end paper chemistry; polymers for electronics applications. *Mailing Add:* 213 Angela Ave Alamo CA 94507-1994

MCWILLIAMS, EDWARD LACAZE, b Shreveport, La, May 22, 41; m 65; c 2. ORNAMENTAL HORTICULTURE. *Educ:* Univ Southwestern La, BS, 63; Iowa State Univ, MS, 65, PhD(hort plant ecol), 66. *Prof Exp:* Hort botanist, Bot Gardens & asst prof bot, Univ Mich, Ann Arbor, 67-72; assoc prof, 72-78, PROF HORT, TEX A&M UNIV, 78- *Mem:* AAAS; Am Soc Plant Taxon; Ecol Soc Am; Am Asn Bot Gardens & Arboretums; Am Soc Hort Sci. *Res:* Horticultural taxonomy; introduction and propagation of ornamental plants; ecology and evolution of Amaranthus and Billbergia. *Mailing Add:* Dept of Hort Sci Tex A&M Univ College Station TX 77843

MCWILLIAMS, JAMES CYRUS, b Oklahoma City, Okla, Aug 22, 46; c 1. PHYSICAL OCEANOGRAPHY, ATMOSPHERIC DYNAMICS. *Educ:* Calif Inst Technol, BS, 68; Harvard Univ, MS, 69, PhD(appl math), 71. *Prof Exp:* Fel, Harvard Univ, 71-74; RES SCIENTIST OCEANOG, NAT CTR ATMOSPHERIC RES, 74- *Concurrent Pos:* Vis prof, Dept Oceanog, Univ Wash, 79; mem, Comt Atmospheric Sci, Nat Acad Sci, 80- *Mem:* Am Geophys Union. *Res:* General circulations of atmospheres and oceans; geostrophic turbulence; statistical estimation. *Mailing Add:* Oceanog Sect PO Box 3000 Boulder CO 80307

MCWILLIAMS, MARGARET ANN EDGAR, b Osage, Iowa, May 26, 29; c 2. NUTRITION, FOOD. *Educ:* Iowa State Univ, BS, 51, MS, 53; Ore State Univ, PhD(food, nutrit), 68. *Prof Exp:* From asst prof to assoc prof food & nutrit, 61-68, chmn dept, 68-76, PROF HOME ECON, CALIF STATE UNIV, LOS ANGELES, 68-, COORDR HEALTH RELATED PROGS, 85- *Concurrent Pos:* Mem, Nat Coun Home Econ Adminr, 68-; mem nat coord comt, Col Teachers Food & Nutrit, 69-; mem nat steering comt plan local prof involvement, White House Conf Food, Nutrit & Health, 69- *Mem:* Inst Food Technol; Am Dietetic Asn; Soc Nutrit Educ. *Res:* Experimental foods; nutrition education; organic chemistry; psychology; consumerism and relation to food and nutrition; author of thirteen text books. *Mailing Add:* Dept FSCS Calif State Univ 5151 State University Dr Los Angeles CA 90032

MCWILLIAMS, RALPH DAVID, b Ft Myers, Fla, Nov 5, 30; m 59; c 2. MATHEMATICAL ANALYSIS. *Educ:* Fla State Univ, BS, 51, MS, 53; Univ Tenn, PhD(math), 57. *Prof Exp:* Instr math, Univ Tenn, 56-57; instr, Princeton Univ, 57-59; asst prof math, 59-69, assoc chmn dept, 69-84, chmn dept, 84-90, PROF MATH, FLA STATE UNIV, 69- *Mem:* AAAS; Am Math Soc; Math Asn Am. *Res:* Functional analysis; weak topologies in Banach spaces. *Mailing Add:* Dept Math Fla State Univ Tallahassee FL 32306-3027

MCWILLIAMS, ROBERT GENE, b Junction City, Ore, Dec 26, 39; m 66; c 2. STRATIGRAPHY. *Educ:* Stanford Univ, BS, 62; Univ Wash, MS, 65, PhD(geol), 68. *Prof Exp:* Consult micropaleontologist, Phillips Petrol Co, 66; asst prof geol, 68-73, ASSOC PROF GEOL, MIAMI UNIV, 73- *Concurrent Pos:* Penrose grant, Geol Soc Am, 72; coordr sci & math, Miami Univ-Hamilton, 74-78; exchange prof, Mt Union Col, 75-76; vis scholar, Miami Univ Europ Ctr, Luxembourg, 78; dir, Miami Univ Geol Fld Sta, Dubois, Wyo. *Mem:* Fel Geol Soc Am. *Res:* Biostratigraphy and paleoecology of West Coast Tertiary Foraminifera and Mollusca; plate tectonics of United States Cordillera. *Mailing Add:* Dept Geol Miami Univ Oxford OH 45056

MCWILLIAMS, ROGER DEAN, b Ames, Iowa, Aug 18, 54; m 85; c 1. PLASMA PHYSICS. *Educ:* Univ Calif, BA, 75; Princeton Univ, PhD(astrophysics), 80. *Prof Exp:* From asst prof to assoc prof, 80-91, PROF PHYSICS, UNIV CALIF, IRVINE 91- *Concurrent Pos:* Consult, expert witness for physics & law, & educ, 83-, physics in sports 88- *Mem:* Am Phys Soc; Am Geophys Union. *Res:* Experimental plasma physics, including radio frequency waves in plasma; electron current generation; earth's magnetosphere. *Mailing Add:* Dept Physics Univ Calif Irvine CA 92717

MCWRIGHT, CORNELIUS GLEN, b Sebree, Ky, Aug 3, 29; m 57; c 3. IMMUNOLOGY, MICROBIOLOGY. *Educ:* Univ Evansville, BA, 52; George Washington Univ, MS, 65, PhD(microbiol), 70. *Prof Exp:* Spec agent, FBI, 55-56, supvry spec agent forensic biol, 56-73, chief biol sci res, 73-78; chief res, Fed Bur Invest Lab, 78-79; consult forensic med, Nat Inst Justice, 79-90; CONSULT, RAND CORP, 87- *Concurrent Pos:* Adj assoc prof biol sci, George Washington Univ, 70-75, res consult immunochem & forensic biol, Grad Sch, 72-73, immunochemist, Lab Virus & Cancer Res, Sch Med, 72-73, adj prof biol sci & forensic sci, 75-; res consult physiol fluids, Law Enforcement Assistance Admin, 75-79; asst prof forensic sci, Univ Va, 76-79; sr fel sci, Ctr Strategic & Int Studies, Georgetown Univ, 82-85; mem, Biotechnol Tech Adv Comt, US Dept Com, 90- *Mem:* Fel Am Acad Forensic Sci; AAAS; Am Soc Microbiol; Sigma Xi; Am Chem Soc. *Res:* Molecular biology and genetics of cell membranes and plasma proteins; immunochemistry; immunohematology; immunogenetics. *Mailing Add:* 7409 Estaban Pl Springfield VA 22151

MCWRIGHT, GLEN MARTIN, b Washington, DC, July 6, 58. ELECTRO-OPTICS, INTEGRATED OPTICS. *Educ:* Univ Va, BS, 80, MEng, 81, PhD(elec eng), 83; Univ Calif, Berkeley, MBA, 89. *Prof Exp:* Res asst elec eng, Univ Va, 80-83, teaching asst, 83; proj engr, Lawrence Livermore Nat Lab, 83-90; CONSULT, 90- *Concurrent Pos:* Vis scientist, AT&T Bell Labs, 85-86. *Mem:* Inst Elec & Electronics Engrs; Optical Soc Am. *Res:* Semiconductor devices; laser systems; high speed diagnostics; optical bistability. *Mailing Add:* 5629 Heming Ave Springfield VA 22151

MACY, RALPH WILLIAM, b McMinnville, Ore, July 6, 05; m 31, 81; c 1. ENTOMOLOGY. *Educ:* Linfield Col, BA, 29; Univ Minn, MA, 31, PhD(zool), 34. *Hon Degrees:* DSc, Linfield Col, 80. *Prof Exp:* Asst zool, Univ Minn, 29-33; from instr to prof biol, Col St Thomas, 34-42; prof biol, Reed Col, 42-55, head dept, 43-55; prof, 55-72, exec off dept, 58-65, EMER PROF BIOL, PORTLAND STATE UNIV, 72- *Concurrent Pos:* Lectr, Med Sch, Univ Ore, 43-45, prof exten div, 44-45; USPHS grants, 49-50 & 64-67; chief investr res proj, Off Naval Res, 50-59; trustee, Ore Mus Sci & Indust, 53-59 & Northwestern Sci Asn, 56-59; NSF grants, 60-64 & 66-72; del, Int Cong Trop Med & Malaria, Lisbon, 58, Int Cong Parasitol, Rome, 64 & DC, 70; guest investr, Univ Helsinki, 61, US Naval Med Res Unit, Cairo, 61-62 & Inst Trop Med, Lisbon, 62 & 69. *Mem:* Fel AAAS; fel Am Acad Microbiol; Am Micros Soc; Am Soc Parasitol. *Res:* Biology of helminths and trematodes; entomology. *Mailing Add:* Dept Biol Portland State Univ Portland OR 97207

MACY, WILLIAM WRAY, JR, b Urbana, Ill, Nov 10, 44; m 71; c 3. PLANETARY SCIENCES, ASTRONOMY. *Educ:* Pomona Col, BA, 67; Univ Ill, Urbana, MS, 68; Princeton Univ, PhD(astrophys), 73. *Prof Exp:* Physicist, Corona Naval Weapons Lab, 68-70; res assoc astron, Univ Tex, 73-76; astronomer, Univ Hawaii, 76-80; RES SCIENTIST, LOCKHEED RES LAB, 80- *Concurrent Pos:* Peyton fel, Princeton Univ, 70-72. *Mem:* Am Astron Soc; Sigma Xi; Int Astron Union. *Res:* Planetary atmospheres; optics. *Mailing Add:* 151 Melville Ave Palo Alto CA 94301

MACZURA, GEORGE, b Granite City, Ill, Jan 8, 30; m 52; c 10. CERAMICS ENGINEERING. *Educ:* Univ Mo-Rolla, BS, 52. *Hon Degrees:* CerEngr, Univ Mo-Rolla, 72. *Prof Exp:* Res engr, 52-64, sr res engr, 64-67, leader ceramic group, 67-71, SCI ASSOC, ALCOA LABS, ALUMINUM CO AM, 71- *Mem:* Fel Am Ceramic Soc; Nat Inst Ceramic Engrs; Sigma Xi; Am Soc Testing & Mat; Am Concrete Inst; Sigma Xi. *Res:* Manufacture of alumina for abrasives, ceramics, glass and refractory uses; refractory cements and castables; alumina fusion processes; high temperature measurement and processing; microscopy and particle size analysis. *Mailing Add:* 725 Orchard Hill Dr Pittsburgh PA 15238

MADAIO, MICHAEL P, MEDICINE. *Educ:* Fairfield Univ, BS, 70; Albany Med Col, MD, 74; Am Bd Internal Med, cert, 77. *Prof Exp:* Intern med, Med Col Va, Richmond, 74-75, resident med, 75-77, chief resident, 77-78; clin fel nephrology, Boston Univ, 78-79, res fel, 79-81; res fel immunol, Tufts Univ, Boston, 81-82, from instr to assoc prof med, Sch Med, 81-90; ASSOC PROF MED, SCH MED, UNIV PA, 90- *Concurrent Pos:* Asst physician, New Eng Med Ctr, Boston, 81-87, physician, 88-90; reviewer, var jour; chmn, Nat Kidney Found Mass, 87; ad hoc mem, Henry Ford Found, 88 & 89; mem, Biomed Res Support Grant Comt, New Eng Med Ctr, 89-90; chair basic immunol & path, Am Soc Nephrology, 90; prin investr var grants. *Mem:* Int Soc Nephrology; Am Soc Nephrology; NY Acad Sci; AAAS; Am Fedn Clin Res; Am Heart Asn; Am Soc Transplant Physicians; Am Coun Transplantation; Am Asn Immunologists. *Res:* Author of various publications. *Mailing Add:* 700 Clin Res Bldg Univ Pa 422 Curie Blvd Philadelphia PA 19104-6144

MADAN, MAHENDRA PRATAP, b Dehradun, India, Dec 24, 26; Can citizen; m 54; c 2. MOLECULAR PHYSICS, SOLID STATE PHYSICS. *Educ:* Univ Punjab, India, BA, 46; Univ Lucknow, India, MSc, 49, PhD(physics), 53, DSc, 58. *Prof Exp:* Asst prof physics, Univ Lucknow, India, 53-55, 58-61, 64-67, Bishop's Univ, Lennoxville, Can, 63-64; res staff mem, Mass Inst Technol, 55-58; fel physics, McGill Univ, Can, 61-63; assoc prof, St Dunstan's Univ, Can, 67-69; assoc prof, 69-75, PROF PHYSICS, UNIV PEI, 75-, CHMN DEPT, 83- *Concurrent Pos:* Mem, Coun Can Asn Physicists, 83-85. *Honors & Awards:* Bonarji Res Prize, Univ Lucknow, 58. *Mem:* Can Asn Physicists; fel Chem Inst Can. *Res:* Intermolecular forces and properties of gases; microwave gaseous electronics; solid state physics; microwave absorbtion and molecular relaxation processes; thermal properties. *Mailing Add:* Dept Physics Univ PEI Charlottetown PE C1A 4P3 Can

MADAN, RABINDER NATH, b New Delhi, India, Mar 11, 35; m 69; c 1. VARIATIONAL PRINCIPLES, FILTERING TECHNIQUES. *Educ:* Univ Delhi, BSc, 54, MSc, 56; Princeton Univ, MA, 64, PhD(physics), 67. *Prof Exp:* Lectr physics & math, Ramjas Col, Univ Delhi, 56-57; sr sci asst, Nat Phys Lab, Univ New Delhi, 57-62; teaching & res assoc physics, Princeton Univ, 62-67; res assoc theoret physics, Univ Calif, Santa Barbara, 67-68; res assoc & lectr, Univ Mass, Amherst, 68-69; assoc prof physics, NC State Univ, 69-77; mgr systs anal, Electronic Systs Div, Bunker-Ramo Corp, 77-79; sr scientist, syst anal, Hughes Aircraft Co, 79-84; PROG MGR, SYSTS & EM THEORY, OFF NAVAL RES, 84- *Concurrent Pos:* Fulbright travel award, Inst Int Educ, 62; prin investr, NASA res grant, 70-74, NSF res grant theoret physics, 73-75. *Mem:* Am Phys Soc; sen mem, Inst Elec & Electronic Engrs; fel Inst Math Appln. *Res:* Detection and estimation theories; system analysis of radars, sonars and sensor arrays; digital signal processing, variational and optimization principles; spectral line satellites in alkalis; Eikonal-Glauber amplitudes; nonlinear phenomenon. *Mailing Add:* Electronics Div Code 1114 Off Naval Res Arlington VA 22217-5000

MADAN, RAM CHAND, b Shaharsultan, India, Oct 8, 39; m 69; c 2. ENGINEERING MECHANICS. *Educ:* Panjab Univ, India, BSc, 65, Indian Inst Technol, India, M(tech), 68; Univ Iowa, MS, 73, PhD(mech & hydraul), 74. *Prof Exp:* Res asst, Nat Aeronaut Lab, India, 65-66; sr lectr appl mech, Engr Col, India, 68-70; lectr dynamics & solid mech, Univ Iowa, 72-73; sr engr, Rockwell Int, 74-77; lectr advan math, Iniv NC, Charlotte, 80-81; PRIN ENGR & SCIENTIST, McDONNEL DOUGLAS AIRCRAFT COMP, LONG BEACH, CALIF, 84- *Concurrent Pos:* Consult, Nuclear Power Serv, NJ, 77-84. *Mem:* Am Soc Mech Engrs. *Res:* Applications of optimization techniques; finite element methods; analytical and experimental stress analysis non-destructive testing, engineering mechanics; computer aided engineering designs; stress analysis in nuclear industry; damage tolerance in composites; composite structures for aerospace; composite component design; fracture mechanics; impact dynamics of aircraft, structure. *Mailing Add:* 13516 Palm Place Cerritos CA 90701

MADAN, STANLEY KRISHEN, b Lahore, Pakistan, May 1, 22; US citizen; m 58; c 1. INORGANIC CHEMISTRY. *Educ:* Forman Christian Col, Lahore, Pakistan, BSc, 45; Punjab Univ, Pakistan, MA, 50, MSc, 54; Univ Ill, Urbana, MS, 57, PhD(inorg chem), 60. *Prof Exp:* Res assoc irrig, Punjab Irrig Res Inst, Lahore, Pakistan, 45-48; demonstr physics, Forman Christian Col, 48-55; asst chem, Univ Ill, Urbana, 57-60; from asst prof to prof, 60-88, EMER PROF CHEM, STATE UNIV NY, BINGHAMTON, 88- *Mem:* Am Chem Soc. *Res:* Synthesis and structure determination of new coordination compounds; kinetics and substitution reactions of triaminotriethylamine metallic complexes. *Mailing Add:* Dept of Chem State Univ of NY Binghamton NY 13902-6000

MADAN, VED P, b Karachi, June 28, 42; Can citizen; m 68; c 2. ENGINEERING MECHANICS, APPLIED MATHEMATICS. *Educ:* Delhi Univ, BSc, 61, MSc, 63; Univ Toronto, PhD(appl math), 68; Univ Calgary, MSc, 89. *Prof Exp:* Sr res asst math, Indian Inst Technol, 63-65; fel, Univ Alberta, Edmonton, 68-70; coordr math, Red Deer Col, Alberta, 70-90; ASSOC PROF MATH, IND UNIV EAST, 90- *Concurrent Pos:* Vis prof, Univ Western Ont, 75-76; bd dir, Red Deer Col Press, 78-82; assoc ed, J Math & Its Applns, 84-; telephone tutor, Atabasca Univ, 85-; Sloan Found fel, Am Math Asn Two Yr Cols. *Mem:* Indian Math Soc; Am Math Soc; Can Math Cong; Am Math Asn Two Yr Cols. *Res:* Viscoelasticity; fluid-mechanics; biomathematics; applied mathematics; general computer science. *Mailing Add:* Dept Math Ind Univ E 2325 Chester Blvd Richmond IN 47374

MADANSKY, ALBERT, b Chicago, Ill, May 16, 34; m 56; c 4. STATISTICS. *Educ:* Univ Chicago, AB, 52, MS, 55, PhD(statist), 58. *Prof Exp:* Mathematician, Rand Corp, Calif, 57-65; vpres, Interpub Group of Co, Inc, NY, 65-68; pres, Dataplan, Inc, NY, 68-71; prof comput sci & chmn dept, City Col New York, 71-74; PROF BUS ADMIN & ASSOC DEAN, GRAD SCH BUS, UNIV CHICAGO, 74- *Concurrent Pos:* Fel, Ctr Advan Study Behav Sci, 61-62; Fulbright-Hayes lectr, 81. *Mem:* Fel Am Statist Asn; fel Inst Math Statist; fel Economet Soc. *Res:* Multivariate analysis; mathematical models in the social sciences. *Mailing Add:* Grad Sch of Bus Univ of Chicago Chicago IL 60637

MADANSKY, LEON, b Brooklyn, NY, Jan 11, 23; m 47; c 2. ELEMENTARY PARTICLE PHYSICS. *Educ:* Univ Mich, BS, 42, MS, 44, PhD(physics), 48. *Prof Exp:* From asst prof to assoc prof, 48-58, chmn dept, 65-68, prof physics, 58-77, DECKER PROF IN SCI & ENG, JOHNS HOPKINS UNIV, 78- *Concurrent Pos:* Res physicist, Brookhaven Nat Lab, 52-53; NSF sr fel, 61, 69; Guggenheim Mem Fel, 74. *Mem:* Fel Am Phys Soc. *Res:* Atomic properties of elementary particles; radiative effects in elementary particles; particle counters; nuclear spectroscopy; radiative effects in beta-decay; meson-nuclear scattering; electron positron collider physics; relativistic heavy ion collisions. *Mailing Add:* Dept of Physics Johns Hopkins Univ Baltimore MD 21218

MADARAS, RONALD JOHN, b Summit, NJ, Dec 18, 42; m 65; c 2. EXPERIMENTAL HIGH ENERGY PHYSICS. *Educ:* Cornell Univ, BEP, 65, MS, 65; Harvard Univ, PhD(physics), 73. *Prof Exp:* Res physicist, Lab Accelerateur Lineaire, Orsay, France, 72-75; RES PHYSICIST, LAWRENCE BERKELEY LAB, UNIV CALIF, 75- *Mem:* Am Phys Soc. *Res:* Electron-positron colliding beam physics; Hadron collider physics. *Mailing Add:* Lawrence Berkeley Lab 50B/5239 Univ Calif Berkeley CA 94720

MADAY, CLARENCE JOSEPH, b Chicago, Ill, Sept 18, 29; m 60; c 1. ENGINEERING MECHANICS. *Educ:* Ill Inst Technol, BS, 51, MS, 54; Northwestern Univ, PhD(lubrication), 60. *Prof Exp:* Jr engr, Hotpoint Co, 51-52, res engr, IIT Res Inst, 52-54 & Borg-Warner Res Ctr, 56-59; res scientist, Martin Co, 60-62; staff engr, Cook Elec Co, 62-64; asst prof, 64-65, ASSOC PROF ENG MECH, NC STATE UNIV, 65- *Concurrent Pos:* Consult, Hayes Int Corp, 65- *Mem:* Am Inst Aeronaut & Astronaut; Am Astronaut Soc. *Res:* Approximation methods in astrodynamics; variational principles in lubrication; dynamics; acoustics. *Mailing Add:* Dept Mech & Aero Eng NC State Univ Box 7910 Raleigh NC 27650

MADDAIAH, VADDANAHALLY THIMMAIAH, b Mysore, India, Nov 10, 29; m 53; c 2. BIOCHEMISTRY. *Educ:* Univ Mysore, BSc, 53, MSc, 54; Univ Ariz, PhD(biochem), 63. *Prof Exp:* Lectr chem, Univ Mysore, 54-60; NIH fel biol, Univ Calif, San Diego, 63-64; NIH fel biochem, Univ Alta, 64-66; sr res biochemist, Res & Develop Labs, Can Packers Ltd, 66-67; res assoc physiol & biophys, Sch Med, Univ Louisville, 67-69; RES BIOCHEMIST, NASSAU COUNTY MED CTR, 69-; ASSOC PROF PEDIAT, HEALTH SCI CTR, STATE UNIV NY STONY BROOK, 72- *Mem:* Am Physiol Soc. *Res:* Enzymology; mitochondrial metabolism; polypeptide hormones; membrane composition-function. *Mailing Add:* Dept of Pediat Nassau County Med Ctr East Meadow NY 11554

MADDEN, DAVID LARRY, b Willow Hill, Ill, Aug 10, 32; m 53; c 3. VETERINARY MICROBIOLOGY. *Educ:* Kans State Univ, BS, 56, DVM & MS, 58; Purdue Univ, PhD(vet microbiol), 63. *Prof Exp:* Instr vet microbiol, Sch Vet Sci & Med, Purdue Univ, 58-63; resident microbiol, Lab Bact Dis, Nat Inst Allergy & Infectious Dis, 63-68, off assoc dir, Nat Inst Child Health & Human Develop, 68-77, res microbiologist, Infectious Dis Br, Nat Inst Neurol Dis & Stroke, 70-77, head sect immunochem & clin invest, Nat Inst Neurol & Commun Dis & Stroke, 77-87; spec asst to chief animal res resources, Div Res Resources, Nat Inst Health, Bethesda, Md, 87-88; SCI ATTACHE, AM EMBASSY NEW DELHI INDIA, DEPT STATE, WASHINGTON, DC, 88- *Concurrent Pos:* Instr NIH Grad Sch Prog, 67-87. *Mem:* AAAS; Am Vet Med Asn; Am Asn Avian Path. *Res:* Etiology of neurological diseases; concerned with isolation of etiological agent and immunological response of the host. *Mailing Add:* Sci Attache Am Embassy-New Delhi India Dept State Washington DC 20520

MADDEN, HANNIBAL HAMLIN, JR, b New York, NY, Oct 5, 31; m 64, 84; c 2. PHYSICS. *Educ:* Williams Col, Mass, BA, 52, MA, 54; Brown Univ, PhD(physics), 59. *Prof Exp:* Asst physics, Williams Col, 52-54; asst, Brown Univ, 54-57, assoc, 58-59, instr, 59-60; asst prof, 60-67, assoc prof physics, Wayne State Univ, 67-75; mem tech staff, Sandia Labs, 74-87; res physicist, MPI/Stromungsf-Goettingen, 87-89, RES PHYSICIST, FRITZ-HABER INST, 89- *Concurrent Pos:* Fulbright travel grant, Hannover Tech Univ, 71-72. *Honors & Awards:* Fulbright fel, TH Darmstadt, 57-58. *Mem:* Am Phys Soc; Am Vacuum Soc. *Res:* Solid state and low temperature physics; solid state surface phenomena; auger electron spectroscopy; photoelectron spectroscopy; He-atom scattering from crystal surfaces; low energy electron diffraction; transport properties of liquid helium II. *Mailing Add:* Fritz-Haber Inst d Max-Planck Gesellschaft Faradayweg 4-6 Berlin W1000-33 Germany

MADDEN, JAMES H(OWARD), b Flint, Mich, Sept 2, 24; m 47; c 6. SYSTEMS ENGINEERING, HUMAN ENGINEERING. *Educ:* Univ Notre Dame, BS, 49, MS, 52. *Prof Exp:* Instr eng mech, Univ Notre Dame, 50-51; engr, Missile Div, Bendix Aviation Corp, 51-52; head reliability sect, Aviation Ord Dept, US Naval Ord Test Sta, 52-55, head, Weapon Support Br, 55-58; mgr, reliability dept, Aerojet-Gen Corp, 58-59, Reliability Div, 59-63, Reliability & Qual Assurance Div, 63-64, advan tactical weapons, 64-65, Apollo reliability, 65-66 & Apollo tech prog control, 66-70, mgr systs eval, Surface Effects Ships Div, 70-74; RETIRED. *Mem:* Assoc fel Am Inst Aeronaut & Astronaut. *Res:* Systems optimization engineering and analysis; reliability, maintainability and safety engineering, analysis and evaluation; physiological and psychological noise, vibration and acceleration habitability design and limit analysis. *Mailing Add:* 9725 Harborview Pl Gig Harbor WA 98332-1018

MADDEN, JOHN JOSEPH, b New York, NY, Oct 4, 43. DNA REPAIR, OPIATE RECEPTORS. *Educ:* Manhattan Col, BS, 64; Emory Univ, PhD(biochem), 68. *Prof Exp:* Fel crystalog, Univ Pittsburgh, 68-70 & Princeton Univ, 70-72; res assoc biophys, Univ Tex, Dallas, 72-76; asst prof, 79-83, ASSOC PROF, DEPT PSYCHIAT BIOCHEM & NEUROSCI PROG, EMORY UNIV, 83- *Concurrent Pos:* Robert A Welch Found res fel, 72-76; Nat Inst Drug Abuse Res Grantee, mem Study Sect on AIDS. *Mem:* Sigma Xi; Am Soc Photobiol; Huntington Dis Soc Am. *Res:* Relationship between DNA repair and mutagenesis in humans; effects of opiates on immune system; mechanisms of light-activation of proteins. *Mailing Add:* Human Genetics Lab Ga Mental Health Inst 1256 Briarcliff Rd Atlanta GA 30306

MADDEN, KEITH PATRICK, b Buffalo, NY, Jan 7, 53; m 84; c 1. RADIATION CHEMISTRY, FREE RADICAL CHEMISTRY. *Educ:* Cornell Univ, BA, 74; Univ Rochester, MS, 77, PhD(biophys), 79. *Prof Exp:* Res assoc chem, Univ Ill, Chicago Circle, 79-80; res assoc chem, 80-82, asst prof specialist 82-88, ASSOC PROF SPECIALIST, RADIATION LAB, UNIV NOTRE DAME, 88- *Concurrent Pos:* Vis res assoc, Wayne State Univ, 79-80. *Mem:* Am Phys Soc; Am Chem Soc; Sigma Xi. *Res:* Characterization of transient intermediates produced in chemical and biological systems by radiolysis and photolysis; free radical molecular dynamics and reactions using time domain magnetic resonance spectroscopy; kinetic studies of spin trapping reactions. *Mailing Add:* Radiation Lab Univ Notre Dame Notre Dame IN 46556

MADDEN, L V, PLANT PATHOLOGY. *Prof Exp:* FAC MEM, DEPT PLANT PATH, OHIO AGR RES & DEVELOP CTR, OHIO STATE UNIV. *Mailing Add:* Dept Plant Path Ohio Agr Res & Develop Ctr Ohio State Univ Wooster OH 44691

MADDEN, MICHAEL PRESTON, b Tulsa, Okla, Oct 26, 48; m 76. THERMAL ENHANCED OIL RECOVERY. *Educ:* Okla State Univ, BS, 70; Univ Okla, MS, 80, PhD(chem eng), 83. *Prof Exp:* Limnologist, Okla Water Resouces Bd, 75-78; instr chem eng, Univ Okla, 82-83; res eng, 83-85, mgr reservoir eng, Energy Prod Res Dept, 85-89, MGR GEOTECHNOL, NAT INST PETROL & ENERGY RES, 89- *Concurrent Pos:* Vis scientist, Inst Chem Physics, Moscow, 80; consult, SRE, Inc, 81-82. *Mem:* Am Chem Soc; Am Inst Chem Engrs; Soc Petrol Engrs. *Res:* Thermal enhanced oil recovery (streamflooding and in situ combustion); enhanced oil recovery computer simulation; data base development; subsurface waste disposal; reservoir engineering. *Mailing Add:* 415 Yale Dr Bartlesville OK 74006

MADDEN, RICHARD M, b Waubay, SDak, Apr 1, 28; m 54; c 1. DENTISTRY. *Educ:* Northern State Col, BS, 53; Univ Minn, DDS, 57; Univ Iowa, MS, 62; Am Bd Endodont, dipl, 65. *Prof Exp:* Pvt pract, Minn, 57-60; assoc prof dent, Univ Iowa, 62-70; assoc prof, 70-74, PROF ENDODONT & CHMN DEPT, UNIV TEX DENT BR, HOUSTON, 74- *Concurrent Pos:* Consult, M D Anderson Tumor Hosp, Tex Childrens Hosp, Vet Admin Hosp & USPHS. *Mem:* Am Asn Endodont; Int Asn Dent Res; Am Dent Asn; Am Asn Dent Schs; Sigma Xi; fel Int Col Dent. *Res:* Endodontics; aerosols. *Mailing Add:* Univ Tex PO Box 20068 Houston TX 77025

MADDEN, ROBERT E, b Oak Park, Ill, Sept 16, 25; c 4. THORACIC SURGERY, CARDIOVASCULAR SURGERY. *Educ:* Univ Ill, BS, 50, MS & MD, 52. *Prof Exp:* Am Cancer Soc fel, Hammersmith Hosp, London, 58-59; from asst prof to assoc prof, 61-71, PROF SURG, NY MED COL, 71- *Concurrent Pos:* Sr surgeon, Nat Cancer Inst, Bethesda, 59-60; mem, NY State Health Res Coun, 76-; ed-in-chief, NY Med Quartery, 79-89. *Honors & Awards:* Borden Res Award, Univ Ill, 52. *Mem:* Am Col Surgeons; Am Soc Clin Oncol; Am Asn Cancer Educ (pres, 79-80); Am Asn Cancer Res; Soc Surg Oncol; Int Soc Cardiovasc Surg; Societe Internationale de Chirurgie. *Res:* Surgical oncology; clinical chemotherapy studies; non invasive vascular testing. *Mailing Add:* New York Med Col Munger Pavilion Valhalla NY 10595

MADDEN, ROBERT PHYFE, b Schenectady, NY, Dec 20, 28; m 50; c 3. PHYSICS. *Educ:* Univ Rochester, BS, 50; Johns Hopkins Univ, PhD(physics), 56. *Prof Exp:* Physicist, Lab Astrophys & Phys Meteorol, Johns Hopkins Univ, 53-58; physicist & sect chief, US Army Eng Res & Develop Labs, Ft Belvoir, Va, 58-61; CHIEF FAR ULTRAVIOLET PHYSICS SECT, OPTICAL PHYSICS DIV, NAT BUR STANDARDS, 61- *Mem:* Fel Am Phys Soc; fel Optical Soc Am. *Res:* Atomic and molecular spectroscopy; optical instrumentation; surface physics; thin films. *Mailing Add:* A257 Physics Bldg Nat Bur Standards Washington DC 20234

MADDEN, SIDNEY CLARENCE, b Fresno, Calif, Oct 27, 07; m 33; c 4. PATHOLOGY. *Educ:* Stanford Univ, AB, 30, MD, 34. *Prof Exp:* Intern, Johns Hopkins Hosp, 33-34; asst resident path, Strong Mem Hosp, Univ Rochester, 34-40; asst path, Sch Med & Dent, Univ Rochester, 34-37, from instr to assoc prof, 37-45; prof & chmn dept, Sch Med, Emory Univ, 45-48; head div, Brookhaven Nat Lab, 48-51, sr physician, Lab Hosp, 49-51; chmn dept, 51-70, prof path, 51-75, EMER PROF PATH, SCH MED, UNIV CALIF, LOS ANGELES, 75- *Concurrent Pos:* Dir labs, Park Ave Hosp, 40-45; pathologist, Emory Hosp, 45-48 & Eggleston Hosp for Children, 45-48; consult, Vet Admin Ctr, Los Angeles, 51-70, Los Angeles County Harbor Gen Hosp, 52-70 & Armed Forces Inst Path, 55-65; sr adv, Atomic Bomb Casualty Comn, 58-66. *Honors & Awards:* Theobald Smith Award, AAAS, 43. *Mem:* AAAS; Am Soc Exp Path (pres, 52); Soc Exp Biol & Med; fel AMA; Am Asn Path & Bact (pres, 63). *Res:* Nucleic acid and protein metabolism after tissue injury; chemical carcinogenesis; immunology of neoplasia; molecular pathology. *Mailing Add:* Dept Path Univ Calif Sch Med Los Angeles CA 90024-1732

MADDEN, STEPHEN JAMES, JR, b Newton, Mass, June 8, 36; m 58; c 2. COMPUTER SCIENCE, GEODESY. *Educ:* Mass Inst Technol, BS, 59, MS, 62, PhD(appl math), 66. *Prof Exp:* Staff mathematician, Instrumentation Lab, 59-61, instr math, 65-66, assoc dir, Measurement Systs Lab, 66-74, SECT CHIEF, C S DRAPER LAB, MASS INST TECHNOL, 74- LECTR, DEPT AERONAUT & ASTRONAUT, 68- *Mem:* Am Math Soc; Soc Indust & Appl Math; Inst Elec & Electronics Engrs; Math Asn Am. *Res:* Electromagnetic propagation; stochastic processes; gradiometry. *Mailing Add:* Dept Aero & Astronautics Mass Inst Technol 77 Mass Ave Cambridge MA 02139

MADDEN, THEODORE RICHARD, b Boston, Mass, Mar 14, 25; m 74; c 2. GEOPHYSICS. *Educ:* Mass Inst Technol, BS, 49, PhD(geophys), 61. *Prof Exp:* From lectr to assoc prof, 52-64, PROF GEOPHYS, MASS INST TECHNOL, 64- *Mem:* Fel Am Geophys Union. *Res:* Geoelectricity and geomagnetism, inversion theory, rock physics, atmospheric gravity waves. *Mailing Add:* Dept Earth & Planetary Sci Mass Inst Technol 77 Mass Ave Cambridge MA 02139

MADDEN, THOMAS LEE, b Bellefontaine, Ohio, May 21, 42; m 65; c 4. ELECTRONICS ENGINEERING, ELECTROMAGNETIC THEORY. *Educ:* Ohio Univ, BS, 65; Ohio State Univ, MS, 73. *Prof Exp:* Proj engr electronic countermeasures, Systs Eng Group, 65-68; proj engr, 68-78, TASK MGR ELECTRONIC COUNTERMEASURES, AIR FORCE AVIONICS LAB, 78- *Honors & Awards:* Lett Commendation, Electronic Countermeasures Advan Develop Br, Air Force Avionics Lab, 78. *Mem:* Inst

Elec & Electronics Engrs. *Res:* Development of passive and active countermeasures to devices utilizing electromagnetic energy for the control of weapon systems to prevent them from destroying United States Air Force aircraft. *Mailing Add:* USAF Afwal/Aawd WL/AAWD-1 Wright Patterson AFB OH 45433-6543

MADDEX, PHILLIP J(OSEPH), chemical engineering; deceased, see previous edition for last biography

MADDIN, CHARLES MILFORD, b Vernon, Tex, Sept 7, 27; m 48; c 1. ANALYTICAL CHEMISTRY. *Educ:* Univ Tex, BA, 50, PhD(chem), 53. *Prof Exp:* Res scientist, Univ Tex, 51-53; lab group leader, 53-64, sect supvr, 64-73, res mgr, Dowell Div, Dow Chem USA, 73-84; environ tech specialist, Dowell Schlumberger, 84-86; CONSULT, 86- *Concurrent Pos:* Lectr, Univ Tulsa, 64-65. *Mem:* Am Chem Soc; Am Inst Chemists. *Res:* Instrumental analysis; chemical separations. *Mailing Add:* 7705 E 25th Pl Tulsa OK 74129

MADDIN, ROBERT, b Hartford, Conn, Oct 20, 18; m 45; c 2. METALLURGY, MATERIALS SCIENCE. *Educ:* Purdue Univ, BS, 42; Yale Univ, DrEng(metall), 48. *Hon Degrees:* MA, Univ Pa, 71. *Prof Exp:* Instr, New Haven Jr Col, 46-48; res assoc & fel, Yale Univ, 48-49; from asst prof to assoc prof phys metall, Johns Hopkins Univ, 49-55; from actg dir to dir sch metall & mat sci, 55-72, prof, 55-84, EMER PROF METALL, UNIV PA, 84- *Concurrent Pos:* Vis prof, Univ Birmingham, 54 & Oxford Univ, 70; ed-in-chief, Mat Sci & Eng; vis prof anthrop, Harvard Univ, 84-; dir, Ctr Archeol Res & Develop, Peabody Mus; hon prof Iron & steel technol, Beijing Univ; hon mem, Japan Inst Metals; vis fel, Wolfson Col, Oxford; hon cur archaeol sci, Peabody Mus, Harvard Univ; Humboldt Found Prize, 89-90. *Mem:* Fel Am Soc Metals; Am Inst Mining, Metall & Petrol Engrs; Brit Inst Metals; fel Metall Soc. *Res:* Deformation of materials; high temperature metals; diffusion; x-ray diffraction; structure and properties of amorphous alloys; history of metallurgy. *Mailing Add:* PO Box 568 Mashpee MA 02649

MADDISON, SHIRLEY EUNICE, immunology, parasitology, for more information see previous edition

MADDOCK, MARSHALL, b Glendale, Calif, Sept 2, 28; m; c 3. GEOLOGY. *Educ:* Univ Calif, AB, 49, PhD, 55. *Prof Exp:* PROF GEOL, SAN JOSE STATE UNIV, 55- *Mem:* Geol Soc Am; Nat Asn Geol Teachers; Sigma Xi. *Res:* Mineralogy; petrology; field geology. *Mailing Add:* Dept Geol San Jose State Col San Jose CA 95192

MADDOCK, THOMAS, JR, b Williams, Ariz, Apr 6, 07; m 35; c 1. CIVIL ENGINEERING. *Educ:* Univ Ariz, BS, 28. *Hon Degrees:* DSc, Univ Ariz, 71. *Prof Exp:* Dir mission, Food Supply Div, Inst Inter-Am Affairs, 43-46; head sedimentation sect, Bur Reclamation, 46-49, head flood sect, 49-50, asst chief hydrol br, 50-52, chief irrig opers br, 52-54; chief irrig analyst, Hoover Comn, 54-55; hydraul engr, 56-57; staff scientist, US Geol Surv, 57-59, chief gen hydrol br, 60-62, sr scientist, 63-71, sr scientist, Water Resources Div, 71-74; CONSULT, 76- *Concurrent Pos:* Consult, Orgn Am States, 74-, US Geol Surv, 75- & UNESCO, 76-80; vis scientist, Univ Ariz, 78-; mem, Indian Conferences on Ganges Flood Control & Village Water Supplies, Nat Acad Sci, 71 & 81. *Honors & Awards:* Bryan Award, Geol Soc Am, 58; Stevens Award, Am Soc Civil Engrs, 63. *Mem:* Fel Am Soc Civil Engrs; Am Geophys Union; hon mem Am Water Resources Asn. *Res:* Hydrology; fluvial morphology; effect of discharge and sediment load on channel characteristics; analysis of effectiveness of different types of flood control projects. *Mailing Add:* 2453 Avenida de Posada Tucson AZ 85718

MADDOCKS, ROSALIE FRANCES, b Lewiston, Maine, Aug 27, 38. GEOLOGY, MICROPALEONTOLOGY. *Educ:* Univ Maine, Orono, BA, 59; Univ Kans, MA, 62, PhD(geol), 65. *Prof Exp:* Res assoc, Smithsonian Inst, 65-67; PROF GEOL, UNIV HOUSTON, 67- *Mem:* Fel AAAS; Paleont Soc; Soc Syst Zool; Int Paleont Union; NAm Micropaleont Soc; Sigma Xi. *Res:* Systematics, ecology and evolution of living and fossil Ostracoda. *Mailing Add:* Dept of Geol Univ of Houston & 4800 Calhoun Rd Houston TX 77004

MADDOX, BILLY HOYTE, b Pahokee, Fla, July 2, 32; m 83; c 4. MATHEMATICS. *Educ:* Troy State Univ, BS, 53; Univ Fla, MEd, 57; Univ SC, PhD(math), 65. *Prof Exp:* Instr math, Troy State Univ, 57-60; prof & chmn dept, Presby Col SC, 64-66; assoc prof, 66-71, PROF MATH, ECKERD COL, 71- *Concurrent Pos:* Reviewer, Zentralblatt fur Mathematics, 66-81; partic, NSF Summer Inst Comput-Oriented Calculus, Fla State Univ, 69, Inst for Retraining in Computer Sci, 83-84. *Honors & Awards:* Nat Defense Serv Medal; Europ Occup Medal. *Mem:* Math Asn Am. *Res:* Algebra; ring and module theory, specifically absolutely pure modules; application of mathematics. *Mailing Add:* Dept of Math Eckerd Col St Petersburg FL 33733

MADDOX, JOSEPH VERNARD, b Montgomery, Ala, Apr 14, 38; m 71. INSECT PATHOLOGY. *Educ:* Auburn Univ, BS, 59, MS, 61; Univ Ill, PhD(entom), 66. *Prof Exp:* ASST PROF AGR ENTOM & ASSOC ENTOMOLOGIST, ILL NATURAL HIST SURV, UNIV ILL, URBANA, 66- *Mem:* AAAS; Entom Soc Am; Soc Invert Path. *Res:* Insect pathology, especially microsporidian diseases of insects. *Mailing Add:* Dept Entom Univ Ill 505 S Goodwin Ave Urbana IL 61801

MADDOX, LARRY A(LLEN), b Bryan, Tex, July 2, 43; m 65; c 1. CHEMICAL ENGINEERING. *Educ:* Tex A&M Univ, BS, 65; Univ Tex, Austin, PhD(chem eng), 70. *Prof Exp:* Chem engr, Res & Tech Lab, Mobil Chem Co, 65; res engr, Tech Ctr, Celanese Chem Co, 69-75, unit supvr, 75-76, process eng group leader, 76-78, PLANNING ASSOC, CELANESE CHEM CO, 78- *Mem:* AAAS; Am Inst Chem Engrs. *Res:* Heterogeneous and homogeneous catalysis; industrial crystallization; process design and development. *Mailing Add:* 1825 Lea Pampa TX 79065

MADDOX, R(OBERT) N(OTT), b Winslow, Ark, Sept 29, 25; m 51; c 2. NATURAL GAS CONDITIONING & PROCESSING, COMPUTER APPLICATIONS & PROCESS SIMULATION. *Educ:* Univ Ark, BS, 48; Univ Okla, MChE, 50; Okla State Univ, PhD(chem eng), 55. *Prof Exp:* Instr, Sch Chem Eng, Okla State Univ, 50-51; design engr, Process Div, Black, Sivalls & Bryson, Inc, 51-52; asst & instr, Okla State Univ, 52-53, from asst prof to prof, 53-76, head, Sch Chem Eng, 58-77, Leonard F Sheerar prof chem eng, 76-86, dir, Phys Properties Lab, 77-86; RETIRED. *Concurrent Pos:* Consult, Cities Serv Oil Co, Black, Sivalls & Bryson, Inc, Humble Oil & Refining Co, Pace Co & Iraq Petrol Corp; vis prof, Univ Cincinnati, 67; tech dir, Fluid Properties Res, Inc, 73-85. *Honors & Awards:* Joseph Stewart Award, Am Chem Soc, 71; Hanlon Award, Gas Processors Asn, 85; Phillips Lectr Eng Educ, Okla State Univ, 89. *Mem:* Fel Am Inst Chem Engrs; Am Chem Soc; Am Soc Eng Educ; Am Inst Mining, Metall & Petrol Engrs; fel Am Inst Chemists. *Res:* Computer applications; stagewise operations; physical properties; transport properties. *Mailing Add:* 1710 Davinbrook Lane Stillwater OK 74074

MADDOX, V HAROLD, JR, b New York, NY, Sept 25, 23; m 49; c 3. ORGANIC CHEMISTRY. *Educ:* Carnegie Inst Technol, BS, 49; Rutgers Univ, MS, 52, PhD(chem), 53. *Prof Exp:* Lab technician, Reichhold Chem, 41-43; asst chem, Rutgers Univ, 49-53; from assoc res chemist to res chemist, Parke, Davis & Co, 53-60, res leader org chem, 60-62, dir lab, 62-70, sect dir org chem, 70-80; mem staff, Pharm Res Div, Warner-Lambert Co, 80-85; CONSULT PROCESS DESIGN, 85- *Mem:* Am Chem Soc; NY Acad Sci; Royal Soc Chem. *Res:* Synthesis in the chrysene series; synthesis of sernyl; synthetic organic medicinals; process research. *Mailing Add:* 68 W 39th St Holland MI 49423

MADDOX, WILLIAM EUGENE, b Owensboro, Ky, Aug 3, 37; m 63. NUCLEAR PHYSICS. *Educ:* Murray State Univ, BS, 62; Ind Univ, MS, 64, PhD(physics), 68. *Prof Exp:* Asst prof, 67-69, assoc prof, 69-80, PROF PHYSICS, MURRAY STATE UNIV, 80- *Mem:* Am Phys Soc. *Res:* Experimental nuclear reaction physics. *Mailing Add:* Dept Physics Murray State Univ Murray KY 42071

MADDOX, YVONNE T, EXPERIMENTAL BIOLOGY. *Educ:* Va Union Univ, Richmond, BS, 65; Georgetown Univ, Washington, DC, PhD(physiol), 81. *Prof Exp:* Lab specialist, Dept Med, Med Col Va, 65-68; res assoc & group leader, Dept Inhalation Toxicol, Hazleton Labs, Va, 68-70; res assoc, Dept Ophthal, Wash Hosp Ctr, 70-73; teaching asst, Dept Biol, Am Univ, 73-75; instr, Exp Med Serv, Georgetown Univ, 76, res asst, Dept Physiol & Biophysics, Georgetown Univ Med Ctr, 76-81, NIH postdoctoral fel, 81-83, res assoc, 83-84, res asst prof, 83-85; HEALTH SCIENTIST ADMINR, NAT INST GEN MED SCI, NIH, 85-, DEP DIR, BIOPHYSICS & PHYSIOL SCI PROG, 91- *Concurrent Pos:* Co-prin investr, NIH Prog Proj grant, 82; prin investr, Am Heart Asn grant, 83, co-prin investr, 83; vis scientist, French, AEC, Saclay, France, 84; chair bd dirs, Ctr Develop & Pop Activ; mem, Coord Comt, Digestive Dis Interagency, Nutrit Coord Comt, PHS/NIH & Task Force Nursing Res. *Honors & Awards:* Mamie Doud Eisenhower Award, Am Heart Asn, 83 & 84. *Mem:* Am Physiol Soc; Soc Exp Biol & Med; Am Fedn Clin Res. *Res:* Author of various publications. *Mailing Add:* NIH Nat Inst Gen Med Sci 5333 Westbard Ave Rm 905 Bethesda MD 20892

MADDY, KEITH THOMAS, b Knoxville, Iowa, Oct 28, 23; m 46; c 6. VETERINARY MEDICINE, PUBLIC HEALTH. *Educ:* Iowa State Col, DVM, 45; Univ Calif, MPH, 54; Am Col Vet Prev Med, Am Col Vet Microbiol, Am Acad Vet & Comp Toxicol & Am Col Toxicol, dipl. *Prof Exp:* Asst prof, Univ Nev, 45-47; vet pathologist, USDA, 48-51; epidemiologist, Nat Commun Dis Ctr, 54-60, scientist, Lab Med & Biol Sci, Div Air Pollution, 60-62; scientist adminstr, Nat Inst Allergy & Infectious Dis, 62-64 & Nat Heart Inst, 64-69, chief pulmonary dis br, Nat Heart & Lung Inst, 69-71; STAFF TOXICOLOGIST PESTICIDE HEALTH & SAFETY BR, CALIF DEPT FOOD & AGR, 71-, BR CHIEF, 80- *Mem:* AAAS; Am Vet Med Asn; Am Pub Health Asn; Soc Occup & Environ Health; Soc Epidemiol Res. *Res:* Fungal infections occurring in man and animals; coccidioidomycosis; toxicology of pesticides. *Mailing Add:* 1413 Notre Dame Dr Davis CA 95616

MADDY, KENNETH HILTON, b Cleveland, Ohio, May 31, 23; m 46; c 3. BIOCHEMISTRY. *Educ:* Pa State Univ, BS, 44; Univ Wis, MS, 48; Pa State Univ, PhD(biochem), 52. *Prof Exp:* Res biochemist, Monsanto Co, Mo, 52-54, develop biochemist, 54-58, proj mgr, 58-60, mgr feed chem develop, 60-66, mgr new proj develop, 66-68, mgr life sci, New Enterprise Div, 68-69, Dir, Computerized Tech Dept, 69-71; pres, Maddy Assocs Inc, 71-75; mgr prod develop, Lonza Inc, 75-76; mgr tech serv, BASF Wyandotte Corp, 76-80; MEM FAC AGR, ARIZ STATE UNIV, 80- *Mem:* Am Chem Soc; Am Soc Animal Sci; Inst Food Technologists; Am Feed Mfrs Asn; fel Am Inst Chemists. *Res:* Animal and human nutrition, specifically amino acids; dietary energy and vitamins; computer program design and development for animal production and human nutrition; food & drug regulations. *Mailing Add:* Sch Agribus Ariz State Univ Tempe AZ 85287-3306

MADER, CHARLES LAVERN, b Dewey, Okla, Aug 8, 30; m 60; c 1. PHYSICAL CHEMISTRY. *Educ:* Okla State Univ, BS, 52, MS, 54; Pacific Western Univ, PhD, 80. *Prof Exp:* Mem staff, Los Alamos Nat Lab, 55-74, asst group leader, 74-76, group leader, 76-81, fel, 82-86; PRES, MADER CONSULT CO, 87- *Concurrent Pos:* Vis scientist, Hawaii Inst Geophys, Univ Hawaii, 72-73, vis colleague, Dept Oceanog, 87-89; affiliate prof marine Sci, Hawaii Loa Col, 89-; adj prof chem, Univ NMex, 80-; sr fel, Joint Inst Marine & Atmospheric Res, Univ Hawaii, 89- *Mem:* Am Chem Soc; fel Am Inst Chemists; Am Phys Soc; Combustion Inst; Sigma Xi; Marine Technol Soc. *Res:* Explosives; thermodynamics; hydrodynamics; equation of state; numerical modeling of detonation chemistry and physics; chemically reactive fluid dynamics; water waves and tsunamis; oceanography. *Mailing Add:* 1049 Kamehame Dr Honolulu HI 96825-2860

MADER, DONALD LEWIS, forest soils, forest hydrology; deceased, see previous edition for last biography

MADER, IVAN JOHN, b Iowa, Dec 29, 23; m 46; c 4. MEDICINE. *Educ:* Cornell Col, 41-43; Wayne State Univ, MS, 49, MD, 51; Am Bd Internal Med, dipl, 61. *Prof Exp:* Chief med & dir med educ, 62-68, chief of staff, 68-76, exec vpres & med dir, William Beaumont Hosp, 76-83. *Concurrent Pos:* Jr assoc, Detroit Gen Hosp, Mich, 56-; consult, Vet Admin, 57-; adj asst prof, Col Med, Wayne State Univ, 58- *Mem:* AMA; fel Am Col Physicians; Am Fedn Clin Res; Sigma Xi. *Res:* Renal disease. *Mailing Add:* 3535 W Thirteen Mile Royal Oak MI 48072

MADERA-ORSINI, FRANK, b Mayaguez, PR, Sept 29, 16; m 45. ENDOCRINOLOGY, IMMUNOLOGY. *Educ:* Ohio Univ, BS, 39; Univ Mo, MS, 56, PhD(biochem), 59. *Prof Exp:* Biochemist clin chem, St Joseph Mercy Hosp, 58-64; dir clin biochem & res, West Suburban Hosp, 64-87; RETIRED. *Concurrent Pos:* Consult, Ment Health State Hosp, Ill; consult, Peoples Community Hosps, Wayne, Mich, 58-64. *Mem:* Am Chem Soc; Am Asn Clin Chem; Am Soc Cell Biol; NY Acad Sci; Am Acad Clin Biochem. *Res:* Clinical and cell chemistry; aldosterone, antagonists and immunology; iron metabolism. *Mailing Add:* 3023 W 31st St Oak Park IL 60521

MADERSON, PAUL F A, b Kent, Eng, Dec 19, 38; m 61; c 1. DEVELOPMENTAL ANATOMY. *Educ:* Univ London, BSc, 60, PhD(zool), 62, DSc(zool), 72. *Prof Exp:* Asst lectr zool, Univ Hong Kong, 62-65; lectr, Univ Calif, Riverside, 65-66; NIH res assoc dermat, Mass Gen Hosp, Harvard Med Sch, 66-68; from asst prof to assoc prof, 68-73, PROF BIOL, BROOKLYN COL, 73- *Concurrent Pos:* Asst Prof, Boston Univ, 67-68; prof path, Univ Ark Med Ctr, 75-82; Assoc Ed, J Morphol, 76- *Mem:* Fel AAAS; Zool Soc London; Anat Soc Gt Brit & Ireland; Am Soc Zoologists; Am Soc Develop Biol; Am Asn Anat. *Res:* Anatomy and cell dynamics squamate epidermis; evolutionary morphology; evolution and development; development avian skin. *Mailing Add:* Dept Biol Brooklyn Col Brooklyn NY 11210

MADERSPACH, VICTOR, b Alsobarbatyen, Hungary, Oct 6, 18; m 44, 60; c 3. ENGINEERING MECHANICS. *Educ:* Hungarian Tech Mil Acad, BS, 39; Va Polytech Inst, MS, 62; Vienna Tech Univ, Dr Tech(solid mech), 64. *Prof Exp:* From asst prof to assoc prof eng mech, 57-87, EMER ASSOC PROF, VA POLYTECH INST & STATE UNIV, 87- *Mem:* Am Soc Eng Educ. *Res:* Plates; shells; thermal stresses; plasticity. *Mailing Add:* 102 Smithfield Dr Blacksburg VA 24060

MADEY, RICHARD, b Brooklyn, NY, Feb 23, 22; m 51; c 6. NUCLEAR PHYSICS. *Educ:* Rensselaer Polytech Inst, BEE, 42; Univ Calif, Berkeley, PhD(physics), 52. *Prof Exp:* Elec engr, Allen B DuMont Labs, 43-44; physicist, Lawrence Radiation Lab, Univ Calif, 47-53; assoc physicist, Brookhaven Nat Lab, 53-56; from scientist to sr scientist, Repub Aviation Corp, 56-61, chief staff scientist mod physics, 61-62, chief appl physics res, 63-64; prof physics, Clarkson Col Technol, 65-71, chmn dept, 65; chmn dept, 71-83, prof, 71-89, UNIV PROF PHYSICS, KENT STATE UNIV, 89- *Concurrent Pos:* Guest scientist, Nevis Cyclotron Lab, Columbia Univ, 55, 76 & 77, Brookhaven Nat Lab, 61 & 74, Foster Radiation Lab, McGill Univ, 67 & 68, Nat Res Coun Can, 68-70, Nuclear Struct Lab, Univ Rochester, 70 & Lawrence Berkeley Lab, Univ Calif, 71, 72, 77, 78, 79, 81, 84 & 88, Univ Md Cyclotron Lab, 73-75 & 78, quest scientist, Ohio Univ Accelerator Lab, 75, Ind Univ Cyclotron Facil, Bloomington, 76-77, 79-86 & 88-89, Space Radiation Effects Lab, Newport News, 77, Los Alamos Nat Lab, 83-85, Tri-Univ Meson Facil, Vancouver, Can, 86-87, Max Planck Inst Nuclear Physics, Heidelberg, WGer, 86, Mass Inst Technol Bates Linear Accelerator Ctr, 88-91; consult, Ross Radio Corp, 52-53, WCoast Electronics Lab, Willys Motors Co, 53-55, Kaiser Aircraft & Electronics Corp, 55-56, US AEC, 65-75, US Energy Res & Develop Admin, 75-77, US Dept Energy, 78-79, Lawrence Berkeley Lab, 78, Ecol Energy Systs, 79-80 & Life Systs, Inc, 81-84, Am Inst Biol Sci, 79, 83 & 84; prin investr contracts & grants, US Air Force, US AEC, NSF, Energy Res & Develop Admin, Dept of Energy, NASA, NIH; mem, Am Inst Biol Sci Radiation Adv Panel, NASA Off Life Sci, 79, Site Visit Team & Rev Panel, Nat Cancer Inst, 79, Am Inst Biol Sci Rev Panel, NASA Life Sci Flight Experts on Shuttle/Spacelab, 84, Site Visit Team & Rev Panel, Space Radiation Res, NASA Langley Res Ctr, 85. *Honors & Awards:* Army-Navy E Award, 43; Letter of Commendation, British Admiralty, 46; Naval Ordn Develop Award, Naval Ordn Lab, 46. *Mem:* Fel AAAS; Am Nuclear Soc; fel Am Phys Soc; sr mem Inst Elec & Electronic Engrs; fel NY Acad Sci; Sigma Xi; fel Am Inst Chemists; Am Geophys Union; Am Asn Univ Prof. *Res:* Nuclear, space and environmental sciences; interaction of radiation with matter and with biological systems; transport of gases through porous media; nuclear instrumentation. *Mailing Add:* Kent State Univ Kent OH 44242

MADEY, ROBERT W, b Norwalk, Conn, May 2, 33; m 60; c 3. NUCLEAR PHYSICS, ENGINEERING. *Educ:* Mass Inst Technol, BS, 55; Univ Md, PhD(nuclear eng), 63. *Prof Exp:* Consult, West Coast Electronics Lab, Willys Motors Co, 54 & Kaiser Aircraft & Electronics Corp, 55; instr nuclear eng, Univ Md, 57-60; staff scientist, Nat Bur Standards, 60-62, sci & eng asst to chief of reactor div, 62-63; head nuclear physics & eng group, 63-69, dir, Nuclear & Space Sci Progs, Grumman Aerospace Corp, 69-73, dir, Res & Develop, 73-76, vpres develop, Grumman Energy Systs, Inc, 76-84, DIR, CORP DEVELOP, GRUMMAN CORP, 84- *Concurrent Pos:* Adj prof, NY Inst Technol, 69. *Mem:* AAAS; Am Nuclear Soc; Am Inst Aeronaut & Astronaut; Int Solar Energy Soc; Am Gas Asn; Solar Energy Indust Asn; Sigma Xi; NY Acad Sci. *Res:* High energy physics and astronomy; space radiation; radiation detection and measurement techniques; radiation effects; nuclear reactor design; reactor instrumentation; shielding; Cerenkov radiation; high vacuum technology, thermionic emission; electrostatic deflection systems; uranium enrichment; photovoltaics; alternate energy sources. *Mailing Add:* 7 Milford Lane Huntington Station NY 11746

MADEY, THEODORE EUGENE, b Wilmington, Del, Oct 24, 37; m 60; c 4. SURFACE SCIENCE, CHEMICAL PHYSICS. *Educ:* Loyola Col, Baltimore, BS, 59; Univ Notre Dame, PhD(physics), 63. *Prof Exp:* Staff physicist, Nat Inst Standards & Technol, 65-81, res fel, 83-88, group leader, surface struct & kinetics, 81-88; DIR, LAB SURFACE MODIFICATION & STATE NJ PROF SURFACE SCI, DEPT PHYSICS, RUTGERS UNIV, 88- *Concurrent Pos:* Vis scientist, Inst Phys Chem, Tech Univ Munich, 73, Sandia Nat Labs, Albuquerque, 77 & Fritz Haber Inst of Max Planck, Gessellschaft, Berlin, 82; Chevron vis prof chem eng, Calif Inst Technol, 81. *Honors & Awards:* Gold Medal Award, US Dept Com, 81; M W Welch Award, Am Vacuum Soc, 85. *Mem:* Int Union Vacuum Sci Tech & Applications (secy gen, 86-89, pres-elect, 89-92); Am Inst Physics; Am Vacuum Soc (pres, 81). *Res:* Physics and chemistry of surfaces, including electron emission and electron spectroscopy, physisorption, chemisorption, surface reactions, structure and reactivity of ultrathin films, electron- and photon-induced radiation damage at surfaces, and the properties of metal/oxide interfaces. *Mailing Add:* Serin Physics Lab Rutgers Univ Piscataway NJ 08855-0849

MADHAV, RAM, b Bangalore, India, Jan 14, 38; m 67; c 2. CLINICAL CHEMISTRY. *Educ:* Univ Madras, BS, 57; Univ Delhi, MS, 60, PhD(chem), 65. *Prof Exp:* Res asst agr chem, Agr Res Inst, India, 57-58; res assoc chem, Univ Delhi, 65-68; res chemist, Carnegie-Mellon Univ, 68-71, res chemist, Mellon Inst, 71-75; res biochemist microbiol, Sch Med, Univ Pittsburgh, 75-80; dir, Kody Clin Ctr, Madras, 80-82; chemist, Naval Med Ctr, San Diego, Calif, 83-84; CHEMIST, VA MED CTR, PITTSBURGH, PA, 85- *Concurrent Pos:* USDA fel, Univ Delhi, 65-68; NIH fel, Carnegie-Mellon Univ, 68-71, Walter Reed Army Hosp fel, Mellon Inst, 71-75; co-investr, Army Inst Proj & NIH Res Proj; prin investr, Sigma Xi Soc Res; reviewer for clin chem J. *Mem:* Am Chem Soc; Sigma Xi; Am Asn Clin Chem; Am Soc Clin Path. *Res:* Natural products chemistry; oxygen heterocycles; synthesis of nitrogen heterocycles; synthesis of fatty acids; synthesis of spin labeled compounds, analogs of coenzymes; preparation of substrates on polymer supports for studying enzyme properties; kidney cell inhibitor on nucleoside kinases; serum factor with growth promoting property; clinical chemistry - study of isoenzymes; therapeutic drug monitoring. *Mailing Add:* 5036 Impala Dr Murrysville PA 15239

MADHAVAN, KORNATH, GENETICS, BIOCHEMISTRY. *Educ:* Annamalai of India, PhD(zool), 64. *Prof Exp:* ASSOC PROF BIOL, COL OF THE HOLY CROSS, 76- *Mailing Add:* Biol Dept Holy Cross Col Worcester MA 01610

MADHAVAN, KUNCHITHAPATHAM, b Mayuram, India, Feb 9, 45; m 71; c 1. CIVIL ENGINEERING, FOUNDATION ENGINEERING. *Educ:* Annamalai Univ, Madras, India, BS, 67; Indian Inst Technol, Kanpur, MS, 69; Univ Miss, PhD(civil eng), 75. *Prof Exp:* Sci teacher, Calhoun County Sch Syst, 72-74; design engr civil eng, Brice, Petrides & Assoc, Inc, Iowa, 75-76; from asst prof to assoc prof, Christian Bros Col, 76-89, PROF CIVIL ENG, CHRISTIAN BROS UNIV, 89-, DEPT HEAD, 85- *Concurrent Pos:* Found analyst, Test Inc, Consult Engrs, Memphis, 77-80; res grant & proj mgr water quality control, Off Planning & Develop, Memphis & Shelby County Govt, Tenn, 78; proj consult, US Testing Co, Memphis, 85- *Mem:* Am Soc Civil Engrs; Asn Groundwater Scientists & Engrs. *Res:* Structural mechanics; foundation engineering and wastewater treatment and water quality control. *Mailing Add:* Dept Civil Eng Christian Bros Univ Memphis TN 38104-5581

MADHU, SWAMINATHAN, b Madras, India, Jan 29, 31; US citizen; m 59; c 3. ELECTRICAL ENGINEERING. *Educ:* Madras Univ, MA, 51; Indian Inst Sci, Bangalore, dipl, 54; Univ Tenn, MS, 57; Univ Wash, PhD(elec eng), 64. *Prof Exp:* Instr elec eng, Univ Wash, 57-63; sr engr, Gen Dynamics/Electronics, 63-65; asst prof elec eng, Rutgers Univ, 65-68; assoc prof, 68-72, assoc dean grad studies, 80-84, actg dean eng, 82-83, PROF ELEC ENG, ROCHESTER INST TECHNOL, 72-, HEAD DEPT, 84- *Concurrent Pos:* Consult, Naval Res Lab, Washington, DC, 70- & Eastman Kodak Co, 73-75; educ consult, Nigerian Army Signal Corps, 76-79 & 90- *Mem:* Inst Elec & Electronics Engrs. *Res:* Automatic language translation; target classification and pattern recognition studies; application of lasers to communications; switching circuits. *Mailing Add:* Dept of Elec Eng 1 Lomb Mem Dr Rochester NY 14623

MADIA, WILLIAM J, m 70; c 3. NUCLEAR FUEL CYCLE. *Educ:* Indiana Univ Pa, BS, 69, MS, 70; Va Polytech Inst, PhD(nuclear chem), 75. *Prof Exp:* Res chemist, Battelle, 75-77, assoc sect mgr, 77-78, sect mgr, 78-80, dept mgr, 80-82, sr prog mgr, 82-84, gen mgr, 84-86, pres, Columbus Div, 86-89, CORP SR VPRES, BATTELLE, 89- *Concurrent Pos:* Bd mem, Franklin Country Children Serv, 88-; mem dean's adv bd, Col Eng, Ohio State Univ, 88-; mem bd trustees, Franklin Univ, 88-; mem external adv comt, Ga Tech Res Inst, 89- *Honors & Awards:* Sigma Xi Res Award in Chem, 76. *Res:* Nuclear fuel cycle technology with emphasis on plutonium reprocessing and high level waste disposal; theoretical studies associated with the acceleration of the radioactive decay process. *Mailing Add:* 469 Delegate Dr Worthington OH 43235

MADIAS, NICOLAOS E, b Samos, Greece, July 27, 44; US citizen; m; c 2. NEPHROLOGY. *Educ:* Athens Univ, MD 68; Am Bd Internal Med, cert, 73 & 76. *Prof Exp:* Intern, Waltham Hosp, Mass, 70-71; resident internal med, St Elizabeth's Hosp, Boston, Mass, 71-73; res fel nephrology, New Eng Med Ctr, Boston, Mass, 73-76; res fel med, 73-76, from asst prof to assoc prof, 76-88, PROF MED, SCH MED, TUFTS UNIV, 88-; CHIEF, DIV NEPHROLOGY, NEW ENG MED CTR, BOSTON, MASS, 82- *Concurrent Pos:* Postdoctoral res fel award, NIH, 74-76; asst med, Div Nephrol, New Eng Med Ctr, Boston, Mass, 76-78, dir, Hypertension Clin, 77-82, asst physician, Div Nephrology, 78-80, physician, 80-, dir res, 81-82; estab investr, Am Heart Asn, 81-86, prin investr, Grant-in-Aid, 84-87, mem, Kidney Coun; mem, Coun Clin Nephrology, Dialysis & Transplantation, Nat Kidney Found & Coun Hypertension; prog dir, NIMH training grant, 82-85; prin investr, NIMH training grant, 82-83; mem, Nephrology Subcomt, Am Col Physicians, 87-88 & Subspecialty Bd Nephrology, Am Bd Internal Med, 88- *Mem:* Am Fedn Clin Res; Am Soc Nephrology; Int Soc Nephrology; Europ Soc Clin Invest; Am Heart Asn; AMA; Am Soc Renal Biochem & Metab; AAAS; NY Acad Sci; fel Am Col Physicians. *Res:* Author of various publications. *Mailing Add:* Div Nephrology New Eng Med Ctr Hosps 750 Washington St Box 172 Boston MA 02111

MADIGOSKY, WALTER MYRON, b Derby, Conn, Dec 15, 33; m 62; c 2. ACOUSTICS. *Educ:* Fairfield Univ, BS, 55; Univ Del, MS, 57; Catholic Univ, PhD(physics), 63. *Prof Exp:* DISTINGUISHED SCIENTIST, US NAVAL SURFACE WEAPONS CTR, 57- *Concurrent Pos:* Sci officer phys acoust, Off Naval Res, 75-76. *Honors & Awards:* Independent Explor Develop Excellence Award, Navy Sci Technol Award, 86. *Mem:* Fel Acoust Soc Am; Inst Elec & Electronic Engrs; Am Chem Soc Rubber Div. *Res:* Materials physical acoustics; liquid state; ultrasonics; underwater acoustics; author of over 100 publications. *Mailing Add:* US Naval Surface Warfare Ctr 10901 New Hampshire Ave Silver Spring MD 20903-5000

MADIN, STEWART HARVEY, b Sheffield, Eng, Apr 3, 18; nat US; m 43; c 3. ANIMAL PATHOLOGY, VIROLOGY. *Educ:* Univ Calif, AB, 40, PhD, 60; Agr & Mech Col Tex, DVM, 43. *Prof Exp:* Asst histopath, Agr & Mech Col Tex, 42-43; assoc, Exp Sta, Off Naval Res, 43-44, assoc vet sci, 46-49, prin pathologist, 49-51, res pathologist, Naval Biol Lab, 51-57, asst sci dir, 57-60, from actg sci dir to dir, 60-68; actg dean, Sch Pub Health, 80-81, prof, 61-86, EMER PROF PUB HEALTH, EXP PATH & MED MICROBIOL, UNIV CALIF, BERKELEY, 86-; prof epidemiol & prev med, 67-86, EMER PROF EPIDEMIOL & PREV MED, SCH VET MED, UNIV CALIF, DAVIS, 86- *Concurrent Pos:* Lectr, Univ Calif, Berkeley, 50-61; mem res comt foot-and-mouth dis & chmn adv comt, Nat Acad Sci-Nat Res Coun, 67-; consult, Pan-Am Health Orgn, 67-; mem cell cult comt & consult-at-large, Nat Cancer Inst; chmn consults, Agr Res Serv; mem, Navy Sr Scientist's Coun; actg provost, Univ Calif, Berkeley, 82; mem, US Foot-and-Mouth Dis Comn II; mem, Comt Vesicular Dis & Animal Virus Classification; vol vet, Steinhardt Aquarium, San Francisco. *Mem:* Am Vet Med Asn; US Animal Health Asn; NY Acad Sci. *Res:* Pathology of infectious diseases; vesicular viruses of domestic animals; pathogenesis of viral diseases; experimental pathology; tissue culture; comparative medicine of terrestrial and marine animals and oncology. *Mailing Add:* Sch Pub Health Univ Calif Berkeley CA 94720

MADISON, BERNARD L, b Rocky Hill, Ky, Aug 1, 41. MATHEMATICS. *Educ:* Western Ky Univ, BS, 62; Univ Ky, MS, 64, PhD(math), 66. *Prof Exp:* From asst prof to prof math, La State Univ, Baton Rouge, 66-80; PROF & CHMN MATH, UNIV ARK, FAYETTEVILLE, 79- *Concurrent Pos:* Proj dir MS 2000, Nat Res Coun, 86- *Mem:* Am Math Soc; Math Asn Am; Sigma Xi. *Res:* Topology and algebraic semigroups. *Mailing Add:* 573 Rockcliff Fayetteville AR 72701

MADISON, CAROLINE RABB, zoology, genetics, for more information see previous edition

MADISON, DALE MARTIN, b Urbana, Ill, Oct 20, 42; m 66; c 2. RODENT SOCIOBIOLOGY. *Educ:* Univ Md, BSc, 65, MSc, 68, PhD(ethology), 71. *Prof Exp:* Res specialist, Univ Wis, 70-71, res assoc, 71-73; asst prof, McGill Univ, 73-77; asst prof, 77-81, ASSOC PROF BIOL, STATE UNIV NY BINGHAMTON, 82- *Honors & Awards:* Stoye Herpetol Award, Sigma Xi. *Mem:* AAAS; Animal Behav Soc; Am Soc Mammalogists. *Res:* Space utilization and social organization in vertebrates, especially small mammals; chemical communication and scent marking in vertebrates. *Mailing Add:* Dept Biol State Univ NY Binghamton NY 13901

MADISON, DON HARVEY, b Pierre, SDak, Jan 4, 45; m 66; c 2. ATOMIC PHYSICS. *Educ:* Sioux Falls Col, BA, 67; Fla State Univ, MS, 70, PhD(physics), 72. *Prof Exp:* Instr physics, Fla State Univ, 72; res assoc, Univ NC, 72-74; from asst prof to prof, 74-84, ELLIS & NELLE LEVITT PROF PHYSICS, DRAKE UNIV, 84- *Mem:* Am Phys Soc; Am Asn Physics Teachers. *Res:* Theoretical aspects of collisions between charged particles and atoms. *Mailing Add:* Dept Physics Univ Mo Rolla MO 65401

MADISON, JAMES ALLEN, b Woodstock, Ill, Jan 12, 28; m 54; c 4. GEOLOGY. *Educ:* Univ NC, BS, 51, MS, 55; Wash Univ, PhD(earth sci), 68. *Prof Exp:* assoc prof, 54-76, PROF EARTH SCI & CHMN DEPT, DEPAUW UNIV, 78- *Concurrent Pos:* Lectr, Jr Engrs & Scientists Summer Prog & NSF Summer Inst Jr High Sch Teachers, 59-70, 72 & 78; vis lectr high schs, Ind. *Mem:* Nat Asn Geol Teachers; Sigma Xi; Asn Eng Geologists; Am Inst Prof Geologists. *Res:* Mineralogy; geophysics; engineering geology; alpine ultramafics. *Mailing Add:* Dept of Earth Sci DePauw Univ Greencastle IN 46135

MADISON, JAMES THOMAS, b Adrian, Minn, Oct 23, 33; m 58; c 1. BIOLOGICAL CHEMISTRY. *Educ:* Colo State Univ, BS, 55; Univ Utah, PhD(biol chem), 62. *Prof Exp:* Res assoc biochem, Cornell Univ, 62-64; RES CHEMIST, PLANT, SOIL & NUTRIT LAB, AGR RES SERV, USDA, 64- *Concurrent Pos:* NIH fel, 62-64. *Mem:* AAAS; Am Chem Soc; Am Soc Biol Chemists; Am Soc Plant Physiol. *Res:* Regulation of amino acid and protein synthesis in plants. *Mailing Add:* US Plant Soil & Nutrit Lab Tower Rd Ithaca NY 14853

MADISON, JOHN HERBERT, JR, b Burlington, Iowa, Jan 16, 18; m 42; c 4. HORTICULTURAL SOILS & IRRIGATION, NATIVE PLANT MANAGEMENT. *Educ:* Oberlin Col, BA, 42; Cornell Univ, PhD(plant physiol), 53. *Prof Exp:* Instr, Cornell Univ, 52-53; from asst prof landscape hort & asst horticulturist to assoc prof & assoc horticulturist, 53-71, prof landscape hort & horticulturist, 71-80, EMER PROF ENVIRON HORT, UNIV CALIF, DAVIS, 80- *Concurrent Pos:* Fel, Univ Edinburgh, 69-70; res assoc, Univ BC, 75. *Honors & Awards:* Distinguished Serv Award, Golf Course Supt Asn Am, 80. *Res:* Monocotyledon biology, soil-water-plant relations of amended soils. *Mailing Add:* 32500 S Hwy One Gualala CA 95445

MADISON, LEONARD LINCOLN, b New York, NY, Feb 11, 20; div; c 2. ENDOCRINOLOGY, METABOLISM. *Educ:* Ohio State Univ, BA, 41; Long Island Col Med, MD, 44. *Prof Exp:* From asst prof to assoc prof, 52-64, PROF INTERNAL MED, UNIV TEX HEALTH SCI CTR, DALLAS, 64- *Concurrent Pos:* Mem gen med study sect, NIH, 62-66. *Mem:* Am Soc Clin Invest; Asn Am Physicians; Endocrine Soc; Am Diabetes Asn. *Res:* Cerebral, carbohydrate and hepatic metabolism; diabetes mellitus. *Mailing Add:* Dept of Internal Med Univ of Tex Health Sci Ctr & 5323 Harry Hines Blvd Dallas TX 75235

MADISON, ROBERTA SOLOMON, b Brooklyn, NY, Feb 10, 32; div; c 2. PUBLIC HEALTH, EPIDEMIOLOGY. *Educ:* Univ Calif, Los Angeles, AB, 66, MA, 69, MSPH, 72, DrPH(environ health, behav sci), 74. *Prof Exp:* Chief epidemiol analyst, County Los Angeles Occup Health Serv Scheduling Supvr, 72-75; asst prof, 75-79, PROF BIOSTATIST & EPIDEMIOL, CALIF STATE UNIV NORTHRIDGE, 79- *Concurrent Pos:* Epidemiologist, Dept Occup Health, City Hope Nat Med Ctr, 77-85. *Mem:* AAAS; Soc Occup & Environ Health; Am Pub Health Asn; fel Am Col Epidemiol; Am Stat Assoc. *Res:* Effect of occupational and environmental exposure on coke oven workers; alpha-1 antitrypsin deficiency as a risk factor in development of chronic obstructive lung disease; anti-oxidants and lung disease; hip fracture incidence in California, indoor pollutants. *Mailing Add:* Dept of Health Sci 18111 Nordhoff Northridge CA 91330

MADISON, VINCENT STEWART, b Adrian, Minn, Feb 10, 43; m 70; c 2. BIOPHYSICAL CHEMISTRY. *Educ:* Univ Minn, Minneapolis, BChem, 65; Univ Ore, PhD(chem), 69. *Prof Exp:* Instr chem, Nat Univ Trujillo, Peru, 70; fel biochem, Univ Calif Med Ctr, 70-71, Sch Med, Harvard, 71-74; from asst prof to assoc prof med chem, Univ Ill Med Ctr, 78-82; RES INVESTR TO RES LEADER, HOFFMAN-LA ROCHE INC, 82- *Concurrent Pos:* NIH fel, 71-73; vis scholar, Univ Calif, San Diego, 81-82. *Mem:* Am Chem Soc. *Res:* Molecular forces which determine peptide conformation and function; determination of conformation by circular dichroism and nuclear magnetic resonance spectroscopy; molecular graphics; structure-function relationships for pharmaceuticals. *Mailing Add:* Hoffmann-LaRoche Inc Nutley NJ 07110

MADISSOO, HARRY, b Paide, Estonia, July 4, 24; US citizen; m 64; c 2. VETERINARY TOXICOLOGY. *Educ:* Hannover Vet Col, DVM, 60. *Prof Exp:* Res asst physiol, Med Col, Cornell Univ, 52-57; from res scientist to sr res scientist toxicol, Squibb Inst Med Res, 60-64; res scientist, Bristol-Myers Co, 64-65, dir toxicol Bristol Labs, 65-82, dir toxicol, 82-90, SR RES ADV, BRISTOL-MYERS SQUIBB PHARMACEUT RES INST, 90- *Mem:* AAAS; Am Vet Med Asn; NY Acad Sci; Soc Toxicol. *Res:* Kidney physiology; veterinary pharmacology; toxicology of drugs, natural products and cosmetics. *Mailing Add:* Dept Toxicol Bristol-Myers Squibb Co Syracuse NY 13220-4755

MADIX, ROBERT JAMES, b Beech Grove, Ind, June 22, 38; c 4. CHEMISTRY, CHEMICAL ENGINEERING. *Educ:* Univ Ill, BS, 61; Univ Calif, PhD(chem eng), 63. *Prof Exp:* From asst prof to assoc prof, 65-77, fel, 69-72, PROF CHEM ENG, STANFORD UNIV, 77-, PROF CHEM, 81-, CHMN DEPT CHEM ENG, 83- *Concurrent Pos:* NSF fel, Max Planck Inst Phys Chem, 64-65; von Humboldt sr scientist, 78; consult, Monsanto Chem Intermediates Co, 78-84, Shell Develop Co, 80-88. *Honors & Awards:* Paul Emmett Award, Catalysis Soc NAm , 82; Debye Lectr, Cornell Univ, 85. *Mem:* Am Inst Chem Engrs; Am Chem Soc; Am Vacuum Soc. *Res:* Reactive gas-solid collisions on clean surfaces, molecular beam techniques employed to elucidate the reaction dynamics; reaction mechanisms and kinetics on metal surfaces. *Mailing Add:* Dept Chem Eng Stanford Univ & Stauffer 111-213 Stanford CA 94305

MADIYALAKAN, RAGUPATHY, b Madras, India, Jan 4, 52. ENZYMOLOGY, ONCOLOGY. *Educ:* Madras Univ, BSc, 72, MSc, 74, PhD(biochem), 78. *Prof Exp:* Sr res fel protein chem, Cent Leather Res Inst, Madras, India, 79-80; fel protein chem, State Univ NY Buffalo, 80-81; res assoc microbiol, Cornell Univ Ithaca, 81-83; res affil biochem, Roswell Park Mem Inst, 83-87; SR SCIENTIST IMMUNOL, BIOMIRA INC, EDMONTON, 87- *Concurrent Pos:* Enzymologist, Diag Clin Lab, Madras, 78-80. *Res:* Glycoprotein biosynthesis with a special reference to cancer and the possible utility of tumor associated carbohydrate antigens for the immunodiagnosis and immunotherapy of cancer. *Mailing Add:* 3304 117 St Edmonton AB T6J 3J4 Can

MADL, RONALD, b Lawrence, Kans, Nov 23, 44; m 69; c 1. BIOCHEMISTRY. *Educ:* Baker Univ, BS, 66; Kans State Univ, MS, 69, PhD(biochem), 73. *Prof Exp:* Food chemist protein chem, 73-76, mgr proj develop, 76-78, dir prod qual, 78-81, DIR TECH PLANNING & PROD QUAL, PROTEIN DIV, RALSTON PURINA CO, 81- *Mem:* Am Chem Soc; Am Asn Cereal Chemists; Inst Food Technologists. *Res:* Characterization of soy protein; study effects of processing parameters on functional characteristics; define tests to insure consistent processing. *Mailing Add:* Ralston Purina Co 900 Checkerboard Sq Plaza St Louis MO 63164-0002

MADOC-JONES, HYWEL, b Cardiff, Wales, Nov 7, 38; m 67; c 2. RADIOTHERAPY. *Educ:* Oxford Univ, BA, 60, MA, 66; Univ London, PhD(biochem), 65; Univ Chicago, MD, 73. *Prof Exp:* Fel cell biol, Ont Cancer Inst & Dept Med Biophys, Univ Toronto, 65-67; res asst prof radiation & cell biol, Mallinckrodt Inst Radiol, Sch Med, Wash Univ, St Louis, 67-68, asst prof, 68-69; res assoc radiol, Univ Chicago, 69-73, intern, Univ Chicago Hosps, 73-74; fel, M D Anderson Hosp & Tumor Inst, 74-77; assoc prof radiol, Mallinckrodt Inst Radiol, Sch Med, Wash Univ, St Louis, 77-80; PROF & CHMN, SCH MED, TUFTS UNIV, BOSTON, 80. *Mem:* fel Am Col Radiol; Radiol Soc NAm; Am Med Asn. *Res:* Radiosensitivity of tumor cells in tissue culture; mechanisms of action of cytotoxic drugs; control of cell division. *Mailing Add:* Dept Radiation Oncol New Eng Med Ctr 750 Washington St Boston MA 02111

MADOFF, MORTON A, b Clinton, Mass, Oct 20, 27; m 53; c 3. INFECTIOUS DISEASES, IMMUNOLOGY. *Educ:* Tulane Univ, BA, 51, MD, 55; Harvard Sch Pub Health, MPH, 73; Am Bd Prev Med, 74. *Prof Exp:* Asst med, New Eng Ctr Hosp, 58-60, asst physician, 60-64; chief infectious dis serv, Lemuel Shattuck Hosp, 64-69; dir biol lab, Mass Dept Pub Health,

67-70, dir, Bur Adult & Maternal Health Servs, 70-72, supt, State Lab Inst, 72-77; PROF, COMMUNITY HEALTH, SCH MED, TUFTS UNIV, 77-, CHMN, 81- *Concurrent Pos:* Nat Found res fel, 58-60; Nat Inst Allergy & Infectious Dis res career develop award, 62-67; prof, Sch Med, Tufts Univ. *Mem:* AAAS; Am Asn Immunologists; Infectious Dis Soc Am; Soc Exp Biol & Med; Soc Teachers Prev Med. *Res:* Cell membranes; microbiol exotoxins; public health and epidemiology. *Mailing Add:* Dept Community Health Tufts Med Sch 136 Harrison Ave Boston MA 02111

MADOLE, RICHARD, b Kirtland, Ohio, July 26, 36. QUATERNARY GEOLOGY & GEOMORPHOLOGY. *Educ:* Case Western Reserve, BS, 58, Ohio State Univ, MS, 60, Phd(geol), 63. *Prof Exp:* Explor geologist, Chevron, 63-65; sr geologist, Tex Instruments, 65-67; chmn, Dept Earth & Sci, Adrian Col, 67-71; res geologist & adj prof geol, Univ Colo, 72-74; GEOLOGIST, US GEOL SURV, NAT EARTHQUAKE CTR, 74- *Concurrent Pos:* Fel, Inst Polar Studies, Ohio State Univ, 62-63 & Univ Colo, 71-72; secy, Quaternary Geol & Geomorphol Div, 81-88, vchmn & chmn, 88-91. *Mem:* Geol Soc Am; Am Quaternary Asn. *Res:* Quaternary dating techniques, paleosersmology, paleoclimatology and recurrence intervals of natural hazards. *Mailing Add:* 737 Glenhaven Ct Boulder CO 80303

MADOR, IRVING LESTER, b Chicago, Ill, July 4, 21; m 43; c 2. CATALYSIS. *Educ:* Univ Md, BS, 41, PhD(phys chem), 49; Univ Chicago, MS, 48. *Prof Exp:* Res chemist, Nat Bur Standards, 42-45; sr chemist, Applied Physics Lab, Johns Hopkins Univ, 49-54; chemist, Res Div, Nat Distillers & Chem Corp, 54-58, asst mgr, 58-61, mgr, Explor Res, 61-74; assoc dir corp res, Allied Signal Corp, 74-80, Sr Staff Scientist, 81-87; RETIRED. *Concurrent Pos:* Consult, Free Radicals, Nat Bur Standards, 55-57. *Mem:* Am Chem Soc; AAAS; Sigma Xi. *Res:* Free radicals; organometallics; transition metal complexes; catalysis. *Mailing Add:* No 405 Bldg 1 7380 Oride Blvd Delray Beach FL 33446-3504

MADORE, BERNADETTE, b Barnston, Que, Jan 24, 18; US citizen. BOTANY, BACTERIOLOGY. *Educ:* Univ Montreal, AB, 42, BEd, 43; Catholic Univ, MS, 49, PhD(ecol, bot), 51. *Prof Exp:* Instr math, Can, 43-44, dean, 52-75, vpres, 75-78, PROF BIOL, ANNA MARIA COL, 49-, PRES, 78- *Mem:* AAAS; Am Soc Microbiol; Nat Asn Biol Teachers. *Res:* Ecology. *Mailing Add:* Anna Maria Col Paxton MA 01612

MADORE, MONICA AGNES, b Edmonton, Alta, May 5, 55. PHOTOSYNTHESIS, PHLOEM TRANSPORT. *Educ:* Carleton Univ, Ottawa, BSc Hons, 77, MSc, 80; Univ Guelph, PhD(hort), 85. *Prof Exp:* ASST PROF BOT & PLANT SCI, UNIV CALIF, RIVERSIDE, 86- *Mem:* Am Soc Plant Physiologists; Sigma Xi. *Res:* Photosynthetic carbon and nitrogen metabolism in plants with emphasis on phloem transport. *Mailing Add:* Dept Bot & Plant Sci Univ Calif Riverside CA 92521

MADOW, LEO, b Cleveland, Ohio, Oct 18, 15; m 42; c 2. PSYCHIATRY, NEUROLOGY. *Educ:* Western Reserve Univ, BA, 37, MD, 42; Ohio State Univ, MA, 38. *Prof Exp:* Prof neurol, Med Col Pa, 56-64, prof psychiat & chmn dept, 64-82; RETIRED. *Concurrent Pos:* Training analyst, Philadelphia Psychoanal Inst. *Mem:* Am Psychoanal Asn; Am Neurol Asn; Am Col Physicians; Am Col Phychiatrists. *Mailing Add:* 111 N 49th St Philadelphia PA 19139

MADRAS, BERTHA K, b Montreal, Que, Dec 9, 42; m 64; c 2. COCAINE ABUSE, PARKINSONS DISEASE & SCHIZOPHRENIA. *Educ:* McGill Univ, BSc, 63, PhD(biochem), 67. *Prof Exp:* Postdoctoral biochem, Tufts Univ, 66-67, Mass Inst Technol, 67-69; res assoc neurochem, Mass Inst Technol, 72-74; lectr pharmacol, Univ Toronto, 78-79; ASST PROF PHARMACOL, HARVARD MED SCH, 79-, ASSOC PROF PSYCHOBIOL, DEPT PSYCHIAT, 90- *Concurrent Pos:* Asst vis prof pharmacol, Univ Toronto, 84-; mem, Radiation Safety Comt, Harvard Univ, 87-, Substance Abuse Planning Group, 91; chmn fel comt, Ont Mental Health Found, 88-90. *Mem:* Soc Neurosci. *Res:* Molecular targets of cocaine and distribution of cocaine receptors in brain; antipsychotic and antiparkinsonian drugs; imaging probes for cocaine receptors, Parkinson's disease and for Huntington's disease; neuro and psychopharmacology with focus on stimulant drugs of abuse and Parkinson's disease. *Mailing Add:* New Eng Regional Primate Res Ctr Harvard Med Sch One Pine Hill Dr Southborough MA 01772-9102

MADRAZO, ALFONSO A, b Tabasco, Mex, Mar 6, 31; US citizen; m 60; c 2. PATHOLOGY. *Educ:* Nat Univ Mex, MD, 54; Am Bd Path, dipl, 65. *Prof Exp:* From instr to asst prof path, Col Med, Seton Hall Univ, 63-67; attend pathologist, St Vincents Hosp & Med Ctr, New York, 67-73; dir labs, Christ Hosp, Jersey City, NJ, 73-76; ASSOC DIR LABS, ST VINCENT'S HOSP & MED CTR, 76-; chief electron-micros lab, St Vincent's Hosp, 76-; LAB DIR, PALISADES GEN HOSP, NORTH BERGEN, NJ. *Concurrent Pos:* Res assoc path, Mt Sinai Sch Med, New York, 69-, assoc clin prof path, 77-; clin assoc prof path, NJ Col Med, 73-76; consult & attend pathologist, St Vincents Hosp & Med Ctr, 75- *Mem:* Fel Int Acad Path; fel NY Acad Med; fel Am Soc Cell Biol; fel Am Soc Clin Path; fel Col Am Pathologists. *Res:* Effects of radiation on human tissues, especially the kidney parenchyma and blood vessels. *Mailing Add:* Palisades Gen Hosp 7600 River Rd North Bergen NJ 07047

MADRI, JOSEPH ANTHONY, ANGIOGENESIS, CELL MATRIX INTERACTION. *Educ:* Ind Univ, PhD(chem), 73, MD, 75. *Prof Exp:* ASSOC PROF PATH, SCH MED, YALE UNIV, 85. *Mailing Add:* Dept Path Sch Med L111 Ln Yale Univ 310 Cedar St New Haven CT 06510

MADSEN, DAVID CHRISTY, b Chelsea, Mass, May 3, 43; m 67; c 2. PHYSIOLOGY, BIOCHEMISTRY. *Educ:* Merrimack Col, BA, 64; Univ Mass, Amherst, MA, 67, PhD(zool), 72. *Prof Exp:* Fel microbiol, Univ Notre Dame, 72-73, asst fac fel, Lobund Lab, 73-79; SR TECH DEVELOP MGR, CLINICAL NUTRIT, TRAVENOL LABS, INC, DEERFIELD, IL, 79- *Mem:* AAAS; NY Acad Sci; Asn Gnotobiotics; Am Soc Pareneral & External Nutrit. *Res:* Bile acid and cholesterol metabolism; role of dietary factors and of microflora of the intestine. *Mailing Add:* Backster Health Care Corp Three Park Lane N Deerfield IL 60015

MADSEN, DONALD H(OWARD), b Council Bluffs, Iowa, Oct 7, 22; m 49; c 2. MECHANICAL ENGINEERING. *Educ:* Iowa State Univ, BS, 44; Purdue Univ, MS, 48, PhD(mech eng), 53. *Prof Exp:* Asst engr, Eastman Kodak Co, 44-45; asst instr, Purdue Univ, 45-48; instr, Univ Kans, 49-51; res engr, Armour Res Found, Ill Inst Technol, 53; asst prof mech eng, SDak State Col, 53-54; assoc dean, Col Eng, 67-74, prof, 54-90, EMER PROF, UNIV IOWA, 90- *Concurrent Pos:* Consult engr, 70- *Honors & Awards:* Naval Ord Develop Award, 46. *Mem:* Am Soc Mech Engrs; Am Soc Eng Educ. *Res:* Heat transfer and thermodynamics. *Mailing Add:* 1315 Whiting Ave Court Iowa City IA 52240

MADSEN, FRED CHRISTIAN, b Rocksprings, Wyo, Dec 9, 45; m 73; c 1. ANIMAL NUTRITION. *Educ:* Austin Peay State Univ, BS, 68; Univ Tenn, Knoxville, MS, 72, PhD(animal nutrit), 74. *Prof Exp:* Res assoc nutrit & physiol, Comp Animal Res Lab, 74-76; tech serv coordr, Syntex Agribus, Inc, 76-78; ruminant nutritionist & consult, 78-81, DIR NUTRIT & RES, CONAGRA, INC, 81- *Mem:* Am Inst Nutrit; Am Soc Animal Sci. *Res:* Factors affecting insulin secretion and metabolism in the ruminant animal; energy supplementation and the relationship to mineral utilization and physical characteristics in the gastrointestinal tract. *Mailing Add:* 53 Memorial Ct Highland IL 62249

MADSEN, JAMES HENRY, JR, b Salt Lake City, Utah, July 28, 32; m 56; c 2. GEOLOGY, VERTEBRATE PALEONTOLOGY. *Educ:* Univ Utah, BS, 57, MS, 59. *Prof Exp:* Cur, Univ Utah, dinosaur lab, 59-77, asst res prof geol & geophys sci, 69-77, mem staff, Utah Mus Natural Hist, 71-77; paleontologist, Utah State, 77-87; PALEONT CONSULT, 87- *Concurrent Pos:* Adj cur paleont, Utah Mus Natural Hist, 77-87; adj prof geol, Brigham Young Univ & Ft Hays State Univ, 85- *Mem:* Paleont Soc; Soc Vert Paleont; Sigma Xi. *Res:* Stratigraphic and field geology; vertebrate paleontology, especially Late Jurassic and Early Cretaceous dinosaurs. *Mailing Add:* 4120 S 3340 E Salt Lake City UT 84124

MADSEN, KENNETH OLAF, b Lavoye, Wyo, May 30, 26; m 50; c 3. NUTRITIONAL BIOCHEMISTRY. *Educ:* Univ Wyo, BS, 50; Univ Wis, MS, 53, PhD(biochem), 58. *Prof Exp:* From instr to assoc prof, 58-70, PROF BIOCHEM, DENT BR, UNIV TEX, HOUSTON, 70- *Mem:* AAAS; Soc Exp Biol & Med; Int Asn Dent Res; Sigma Xi; Soc Nutrit Ed. *Res:* Bull semen metabolism; fluorine metabolism and toxicity; trace mineral nutrition; oral biology and dental research, especially dental caries in rats; biology of the cotton rat; dental nutrition. *Mailing Add:* PO Box 20068 Univ of Tex Dental Houston TX 77225

MADSEN, NEIL BERNARD, b Grande Prairie, Alta, Feb 8, 28; m 52; c 2. BIOCHEMISTRY. *Educ:* Univ Alta, BSc, 50, MSc, 52; Wash Univ, PhD, 55. *Prof Exp:* Instr biol chem, Wash Univ, 55-56; fel biochem, Oxford Univ, 56-57; res off, Microbiol Res Inst, Can Dept Agr, 57-62; assoc prof, 62-69, PROF BIOCHEM, UNIV ALTA, 69- *Concurrent Pos:* Med Res Coun Can vis scientist, Oxford Univ, Eng, 72-73; chmn sci policy comt, 73-75, chmn bd, Can Fedn Biol Soc, 77-78. *Honors & Awards:* G Malcolm Brown Mem Lectr. *Mem:* Am Soc Biol Chemists; Can Biochem Soc (past pres); fel Royal Soc Can. *Res:* Enzyme chemistry; structure-function relationships in glycogen phosphorylase and other enzymes related to glycogen metabolism as determined by protein chemistry, kinetics and x-ray crystallography; biological control of glycogen metabolism. *Mailing Add:* Dept Biochem Univ Alta Edmonton AB T6G 2H7 Can

MADSEN, OLE SECHER, b Aarhus, Denmark, Feb 24, 41; m 69; c 1. HYDRODYNAMICS, CIVIL ENGINEERING. *Educ:* Tech Univ Denmark, MSc, 64; Mass Inst Technol, ScD(hydrodynamics), 70. *Prof Exp:* Oceanogr, US Army Coastal Eng Res Ctr, Washington, DC, 69-72; assoc prof, 72-84, PROF CIVIL ENG, MASS INST TECHNOL, 84- *Res:* Transformation of water waves; the interaction between water waves and the coastal environment. *Mailing Add:* Ralph M Parsons Lab Mass Inst of Technol & 77 Mass Ave Cambridge MA 02139

MADSEN, PAUL O, b Denmark, July 25, 27; US citizen; m 55; c 3. UROLOGY. *Educ:* Copenhagen Univ, MD, 52, DrMed, 76; Univ Heidelberg, DrMed, 58. *Prof Exp:* Intern surg, Genesee Hosp, Rochester, NY, 53-54; resident surg & urol, Gen Hosp, Buffalo, 55-57; resident, Buffalo Gen Hosp & Roswell Park Mem Inst, 59-61; vis urologist, Univ Homburg, 61-62; assoc prof, 68-71, PROF UROL, UNIV WIS-MADISON, 71-; CHIEF UROL, VET ADMIN HOSP, 62- *Mem:* Int Soc Urol; Int Soc Surg; Am Urol Asn; Soc Univ Urologists. *Res:* Treatment of urinary tract infections; experimental pyelonephritis and hydronephrosis; prostatic tissue cultures; prostatitis. *Mailing Add:* Urol Vet Admin Hosp Univ Wis 600 Highland Ave Madison WI 53705

MADSEN, RICHARD ALFRED, b Sheffield, Ill, Feb 15, 33; m 57; c 4. HEAT TRANSFER, FLUID MECHANICS. *Educ:* Iowa State Univ, BS, 55; Purdue Univ, MS, 62, PhD(mech eng), 65. *Prof Exp:* Trainee, Caterpillar Tractor Co, 55-56, res design engr, 58-59; res asst, Jet Propulsion Ctr, Purdue Univ, 59-62, res asst, Heat Transfer Lab, 62-65; eng specialist, Ai res Mfg Co, Ariz, 65-66; sr eng scientist, McDonnell Douglas Corp, 66-78; sr res specialist, Aires Los Angeles Div, 78-89; SUBCONTRACT TECHNICIAN MGR, MCDONNELL DOUGLAS CORP, 89- *Mem:* Am Soc Mech Engrs. *Res:* Heat transfer; thermodynamics; fluid mechanics; cryogenics; low-gravity phenomena; isotope separation; spacecraft and environmental control. *Mailing Add:* McDonnell Douglas Corp 5301 Bolsa Ave Huntington Beach CA 92647

MADSEN, VICTOR ARVIEL, b Idaho Falls, Idaho, Feb 14, 31; m 55, 85; c 2. THEORETICAL NUCLEAR PHYSICS. *Educ:* Univ of Wash, BS, 53, PhD(physics), 61. *Prof Exp:* Res assoc physics, Case Inst Technol, 61-63; asst prof, 63-66, assoc prof, 66-71, PROF PHYSICS, ORE STATE UNIV, 71- *Concurrent Pos:* Physicist, Niels Bohr Inst, Copenhagen & Atomic Energy Res Estab, Eng, 69-70; consult, Lawrence Livermore Lab, 64-; guest scientist, KFA Julich, W Ger, 76-77, 84. *Mem:* Am Phys Soc. *Res:* Theory of nuclear inelastic scattering; core-polarization effects; charge-exchange reactions, isospin effects; theory of imaginary optical potential and inelastic form factor. *Mailing Add:* Dept Physics Ore State Univ Corvallis OR 97331

MAECK, JOHN VAN SICKLEN, obstetrics & gynecology; deceased, see previous edition for last biography

MAEDER, P(AUL) F(RITZ), b Basel, Switz, June 29, 23; nat US; m 46; c 2. ENGINEERING. *Educ:* Swiss Fed Inst Technol, Dipl Ing, 46; Brown Univ, PhD, 51. *Prof Exp:* Res assoc, Swiss Fed Inst Technol, 46-47; res assoc, Brown Univ, 47-51, assoc prof eng, 51-54, chmn exec comt div eng, 62-68, assoc provost, 68-77, vpres finance & opers, 72-77, prof eng, 54-87; RETIRED. *Concurrent Pos:* Consult, Aro, Inc, Tenn, Res Dept, United Aircraft Corp, Conn, Indust Div, Speidel Corp, RI, 56-63 & Agard, France, 58; mem transonic subcomt, Arnold Eng Develop Ctr, Tenn, 52-55; vis prof, Swiss Fed Inst Technol, 58; vis prof, Stanford Univ, Calif, 78-79; mem energy comt, Brown Univ, 79-; mem Gov Task Force Energy, 81-; consult, Elmwood Sensors, Inc, Fed Prod Corp, State Univ NY, Binghampton, RI Sch Design. *Mem:* Am Soc Mech Engrs; Am Inst Aeronaut & Astronaut; Am Soc Eng Educ; Nat Soc Prof Engrs; AAAS; Sigma Xi. *Res:* Engineering sciences, especially fluid mechanics, aerodynamics, magnetohydrodynamics and instrumentation. *Mailing Add:* The Landings Apt A304 790 Andrews Ave Delray Beach FL 33444

MAEHR, HUBERT, b Schlins, Austria, Feb 25, 35. NATURAL PRODUCTS CHEMISTRY. *Educ:* State Univ Agr & Forestry, Austria, Dipl Ing, 60; Rutgers Univ, PhD(chem microbial prod), 64; Vienna, Techn Dr, 77. *Prof Exp:* Res assoc chem antibiotics, Inst Microbiol, Rutgers Univ, 64-65; RES CHEMIST DIV CHEM RES, HOFFMANN-LA ROCHE INC, 66- *Concurrent Pos:* Univ lectr, Dept Org & Phamaceut Chem, Univ Innsbruck, 79- *Mem:* Am Chem Soc. *Res:* Separations, structure structure, elucidations and syntheses of natural products. *Mailing Add:* Chem Res Div Hoffmann-La Roche Inc Nutley NJ 07110-1199

MAEKS, JOEL, b Jamaica, NY, Apr 27, 34; div. COMMUNICATIONS, COMPUTER SCIENCE. *Educ:* Cornell Univ, BEP, 57; Mass Inst Technol, SM, 59, PhD(commun), 67; Imp Col, Univ London, dipl, 60. *Prof Exp:* Staff assoc, Lincoln Lab, Mass Inst Technol, 57-58, staff mem, 58-64 & 67-69, staff assoc, 64-67; mem tech staff, GTE Labs, Waltham, 69-73; MEM TECH STAFF, MITRE CORP, BEDFORD, 73- *Mem:* Inst Elec & Electronics Engrs. *Res:* Information theory; communication theory and systems; switching system control. *Mailing Add:* 9 Tyler Rd Lexington MA 02173-2403

MAENCHEN, GEORGE, b Moscow, Russia, Apr 27, 28; nat US; m 47; c 3. PHYSICS. *Educ:* Univ Calif, AB, 50, PhD(physics), 57. *Prof Exp:* PHYSICIST, LAWRENCE LIVERMORE LAB, 57- *Mem:* Am Phys Soc. *Res:* Nuclear and general physics. *Mailing Add:* 2347 Chateau Way Livermore CA 94550

MAENDER, OTTO WILLIAM, b Brooklyn, NY, Jan 18, 40; m 63; c 2. ORGANIC CHEMISTRY. *Educ:* Rochester Inst Technol, BS, 63; Iowa State Univ, MS, 65; Univ Ga, PhD(org chem), 69. *Prof Exp:* Sr res chemist, 69-79, res specialist, 79-88, SR RES SPECIALIST, MONSANTO CO, 88- *Mem:* Am Chem Soc. *Res:* Electron spin resonance study of oxidative processes; antidegradants; vulcanization. *Mailing Add:* Monsanto Co 260 Springside Dr Akron OH 44333

MAENGWYN-DAVIES, GERTRUDE DIANE, pharmacology; deceased, see previous edition for last biography

MAENZA, RONALD MORTON, b New York, NY, May 8, 36; m 60; c 3. PATHOLOGY. *Educ:* Columbia Univ, BA, 57, MD, 61; Am Bd Path, dipl, 66. *Prof Exp:* Intern med, Bronx Munic Hosp, 61-62; resident path, Columbia Presby Hosp, NY, 62-65; resident, Englewood Hosp, NJ, 65-66; asst prof, 68-73, ASSOC PROF PATH, SCHS MED & DENT MED, UNIV CONN, 73- *Concurrent Pos:* Consult, Vet Admin Hosp, Newington, Conn, 68-; teaching affil, Hartford Hosp, 68-; pathologist, New Britain Gen Hosp, 78- *Mem:* Int Acad Path; Am Asn Path & Bact. *Res:* Chemical carcinogenesis. *Mailing Add:* Hahnemann Univ Broad & Vine St Philadelphia PA 19102

MAERKER, GERHARD, b Bernburg, Ger, Nov, 4, 23; nat US; m 51; c 2. ORGANIC CHEMISTRY, LIPID CHEMISTRY. *Educ:* Philadelphia Col Pharm, BS, 51; Temple Univ, MA, 52, PhD(chem), 57. *Prof Exp:* Res chemist, Allied Chem Corp, 52-58; res chemist, USDA, 58-71, chief animal fat lab, 71-80, res leader food additives, 80-85, LEAD SCIENTIST FOOD SAFETY, EASTERN REGIONAL RES CTR, USDA, 85- *Mem:* AAAS; Am Chem Soc; Am Oil Chem Soc; Inst Food Technologists; Sigma Xi. *Res:* Synthesis, properties and reactions of chemical derivatives of fatty acids, particularly small ring heterocyclic derivatives such as epoxides and aziridines; lipid oxidation; cholesterol oxidation; radiolytic effects on lipids. *Mailing Add:* 606 Haws Lane Oreland PA 19075

MAERKER, JOHN MALCOLM, b Philadelphia, Pa, July 20, 45; div; c 1. ENHANCED OIL RECOVERY, RHEOLOGY. *Educ:* Univ Del, BChE, 67; Princeton Univ, MA, 69, PhD(chem eng), 71. *Prof Exp:* Sr res engr, chem oil recovery, 71-76, res specialist, 76-78, sr res specialist, 78-81, res assoc, 81-84, SR RES ASSOC, ENHANCED OIL RECOVERY, EXXON PROD RES CO, 84- *Concurrent Pos:* Fel, Nat Sci Found, 67-71. *Honors & Awards:* Cedrik K Ferguson Award, Soc Petrol Engrs, 80. *Mem:* Soc Rheology; Soc Petrol Engrs. *Res:* Flow of polymer solutions and microemulsions through porous media; mechanics of oil displacement and recovery from porous media through interfacial tension reduction; relationships between molecular and microscopic structure and bulk properties of microemulsions and polymer solutions. *Mailing Add:* Exxon Prod Res Co Box 2189 Houston TX 77252-2189

MAERKER, RICHARD ERWIN, b Jackson, Mich, July 9, 28; m 59; c 1. REACTOR SHIELDING & DOSIMETRY ANALYSIS. *Educ:* Univ Tenn, BS, 49, MS, 51, PhD(physics), 53. *Prof Exp:* Res physicist solid state, Kodak, Rochester, NY, 53-54; res assoc rocketry, Univ Tenn, 54-55; res assoc, Tulane Univ, 55-59; PHYSICIST REACTOR SHIELDING, OAK RIDGE NAT LAB, 59- *Concurrent Pos:* Mem, Cross Sect Eval Working Group, 69-85. *Mem:* Am Nuclear Soc; NY Acad Sci; Sigma Xi; Am Soc Testing & Mat. *Res:* Shielding of neutrons and photons from reactors; analysis of integral experiments on shielding; reactor dosimetry. *Mailing Add:* Oak Ridge Nat Lab Bldg 6025 PO Box 2008 Oak Ridge TN 37831-6364

MAESTRE, MARCOS FRANCISCO, b San Juan, PR, June 20, 32; US citizen; m 58; c 2. BIOPHYSICS. *Educ:* Univ Mich, BS, 54; Wayne State Univ, MS, 58; Yale Univ, PhD(biophysics), 63. *Prof Exp:* Res assoc, 63-65, ASSOC RES CHEMIST, SPACE SCI LABS, UNIV CALIF, BERKELEY, 65-, LECTR MED PHYSICS, 73- *Concurrent Pos:* USPHS fel, 63-65. *Mem:* Biophys Soc; Sigma Xi. *Res:* X-ray studies on virus structure; theory of the birefringence of nucleic adids; electric birefringence of bacteriophage; optical rotatory dispersion of viruses; viral nucleic acids and protein components. *Mailing Add:* 1222 Josephine Berkeley CA 94703

MAESTRELLO, LUCIO, b Legnago, Italy, Mar 4, 28; US citizen; m 58; c 4. ACOUSTICS. *Educ:* Galileo-Ferraris Inst, Dipl eng, 50; Univ Southampton, DPhil(acoustics), 76. *Prof Exp:* Res asst aerodynamics, Imp Col, Univ London, 53-54; res assoc acoustics, Inst Aerophys, Univ Toronto, 56-57; res engr, Boeing Co, 58-70; HEAD AEROACOUSTICS SECT, NASA-LANGLEY RES CTR, 70- *Mem:* Am Phys Soc. *Res:* Jet and boundary layer noise. *Mailing Add:* 310 Williamsburg Ct Newport VA 23606

MAESTRONE, GIANPAOLO, b Urgnano, Italy, Jan 31, 30; US citizen; m 56; c 3. VETERINARY MEDICINE, MICROBIOLOGY. *Educ:* Univ Milan, DVM, 51; Am Col Vet Microbiol, dipl, 67. *Prof Exp:* Asst prof infectious dis, Vet Sch, Univ Milan, 51-56; res assoc microbiol, Animal Med Ctr, NY, 57-61; sr res microbiologist, Squibb Inst Med Res, NJ, 61-66; sr res microbiologist, 66-75, asst group chief res, 76-78, clin vet, 73-86, res leader, 86-87, Hoffmann-La Roche, Inc, NJ; SCI & MED AFFAIRS DIR, AM PARKINSON DIS ASN NY, 87- *Mem:* Am Vet Med Asn; Am Asn Indust Vets; assoc fel NY Acad Med; Sigma Xi. *Res:* Diagnosis of infectious diseases; chemotherapy of infectious diseases and parasites; immunofluorescence applied to diagnosis of leptospiral and viral diseases; antibiotic sensitivity of clinical isolates; epidemiology of leptospirosis; experimental infections; fish and wildlife sciences. *Mailing Add:* Four Sophia Lane Staten Island NY 10304

MAEWAL, AKHILESH, b India, July 1, 49; US citizen. ENGINEERING MECHANICS, STRUCTURAL DYNAMICS. *Educ:* Indian Inst Technol, BS, 70; Univ Calif, San Diego, PhD(eng sci), 76. *Prof Exp:* Res engr, Univ Calif, San Diego, 76-77; staff scientist, Systs, Sci & Software, La Jolla, Calif, 77-79; asst prof appl mech, 79-83, ASSOC PROF MECH ENG, YALE UNIV, 83- *Concurrent Pos:* Banco Found award, Indian Inst Technol, 70. *Mem:* Am Soc Mech Engrs; Am Acad Mech. *Res:* Mechanical behavior of composite materials; nonlinear structural mechanics; computational mechanics. *Mailing Add:* Yale Univ 215 Becton Ctr PO Box 2157 Yale Sta New Haven CT 06520

MAEZ, ALBERT R, b Cuervo, NMex, Dec 19, 39; m 65; c 3. INDUSTRIAL TECHNOLOGY. *Educ:* NMex Highlands Univ, BA, 72, MA, 74. *Prof Exp:* Asst prof electronics technol, 72-77, ASST PROF ELEC ENG TECHNOL, NMEX HIGHLANDS UNIV, 77- *Concurrent Pos:* Vis prof, Los Alamos Nat Lab, 69, 70 & 89, Sandia Nat Labs, 90. *Mem:* Am Soc Eng Educ; Inst Elec & Electronics Engrs. *Res:* Printed circuit board production techniques. *Mailing Add:* PO Box 1243 Las Vegas NM 87701

MAFEE, MAHMOOD FOROOTAN, b Ghom, Iran, Dec 15, 41; m 72; c 3. HEAD & NECK RADIOLOGY. *Educ:* Tehran Univ, Iran, MD, 69. *Prof Exp:* Intern, Huron Rd Hosp, Cleveland, 72-73; resident radiol, Albert Einstein Col Med NY, 73-74; resident radiol, 74-76, from asst prof to assoc prof, 76-85, PROF RADIOL, UNIV ILL, CHICAGO, 85-, DIR RADIOL, EYE & EAR INFIRMARY, 85-, dir, magnetic resonance ctr, 87- *Concurrent Pos:* Vis Prof, dept radiol, Harvard MedSch, 85, lectr, 84-88; lectr, 5th Int Orbit Meeting, Amsterdam, Neth, 85 & AM Soc Head & Neck Radiol, 83-87; consult, Archives ophthalmol, 88 & AM Jour Neuro radiology. *Mem:* Am Med Assoc; Am Col Radiol; Radiol Soc N Am; Int Soc Magnetic Resonance; Int Soc Bronchoesophagology; Am Soc Head & Neck Radiol; Am Soc Neuroradiology. *Res:* Development and improving clinical application of CT scan and MR imaging in the diagnosis of head and neck pathology; diagnostic imaging of ocular tumors and diseases of the temporal bone and neurotalogical disorders; MR imaging and in-vivo proton spectroscopy of intra ocular tumors. *Mailing Add:* Univ Ill Hosp Rad/MR1 Ctr 840 S Wood St (M/C 711) Chicago IL 60612

MAFFETT, ANDREW L, b Port Royal, Pa, Oct 23, 21; m 43. MATHEMATICS. *Educ:* Gettysburg Col, AB, 43; Univ Mich, MA, 48. *Prof Exp:* From instr to asst prof math, Gettysburg Col, 47-54; res assoc, Univ Mich, 51-52, res engr, Radiation Lab, 54-56, asst head lab, 56-57; sr mathematician, Bendix Aviation Corp, 58-59; res mathematician, Inst Sci & Technol, Univ Mich, 60-61; assoc head sr anal staff, Conductron Corp, 61-67; sr scientist, KMS Indust, Inc, 67-69; adj prof math, Univ Mich-Dearborn, 69-84; ED, J APPL COMPUTATIONAL ELECTROMAGNETICS SOC, 88- *Concurrent Pos:* Consult, 69- *Mem:* Sigma Xi. *Res:* Electromagnetic boundary value problems arising from both the radiation and scattering of energy from various geometric shapes and configurations. *Mailing Add:* 2250 N Zeeb Rd R D One Dexter MI 48130-9714

MAFFLY, LEROY HERRICK, b Berkeley, Calif, Nov 26, 27; m 52; c 3. INTERNAL MEDICINE, NEPHROLOGY. *Educ:* Univ Calif, AB, 49, MD, 52. *Prof Exp:* Intern med, Univ Calif Hosp, 52-53, asst resident, 53-54; resident, Herrick Mem Hosp, 54-55; res fel, Harvard Med Sch & Mass Gen Hosp, 57-59; res fel, Sch Med, Univ Calif, San Francisco, 59-61; chief renal serv, Vet Admin Hosp, Palo Alto, Calif, 68-83; from asst prof to assoc prof, 61-70, chmn, Dept Physiol, 86-88, PROF MED, SCH MED, STANFORD UNIV, 70-, ASSOC DEAN STUDENT AFFAIRS, 83- *Concurrent Pos:* NSF fel, 57-58; USPHS fel, 58-60; Am Heart Asn fel, 60-61; estab investr,

Am Heart Asn, 61-66, mem exec comt, Coun Kidney in Cardiovasc Dis, 68-82, chmn comt, 71-73, mem cardiovasc A res study comt, 73-82, res comt, 76-82, mem comt teaching scholars, 76-81; mem adv comt renal dialysis ctrs, State of Calif, 66-70; mem gen med B study sect, NIH, 67-71; mem sci adv bd, Nat Kidney Found, 70-77; med test comt, Nat Bd Med Examrs, 81-88, chmn, 84-88. *Mem:* Am Soc Nephrology; Am Physiol Soc; Am Soc Clin Invest; Soc Med Decision Making; Asn Am Med Col. *Res:* Transport processes and permeability of biological membranes; nephrology; computer-assisted education. *Mailing Add:* Dept Med Stanford Univ Sch Med Stanford CA 94305

MAGA, JOSEPH ANDREW, b New Kensington, Pa, Dec 25, 40; m 64; c 2. FOOD SCIENCE, BIOCHEMISTRY. *Educ:* Pa State Univ, BS, 62, MS, 64; Kans State Univ, PhD(food sci), 70. *Prof Exp:* Proj leader dairy prod, Borden Foods Co, 64-66; group leader simulated dairy prod, Cent Soya Co, 66-68; from asst prof to assoc prof, 70-74, PROF FOOD SCI, COLO STATE UNIV, 78- *Mem:* Am Dairy Sci Asn; Am Asn Cereal Chemists; Am Chem Soc. *Res:* Flavor aspects of foods, especially high protein foods, including composition, chemistry and preferences. *Mailing Add:* Dept Food Sci & Human Nutrit Colo State Univ Ft Collins CO 80523

MAGAARD, LORENZ CARL, b Wallsbuell, Germany, May 21, 34; m 61; c 1. PHYSICAL OCEANOGRAPHY. *Educ:* Univ Kiel, Ger, BS, 58, MS, 61, PhD(math & phys oceanog), 63. *Prof Exp:* Asst scientist, Univ Kiel, Ger, 61-64, from asst prof to assoc prof, 64-75; PROF OCEANOG, UNIV HAWAII, HONOLULU, 75- *Concurrent Pos:* Vis investr, Woods Hole Oceanog Inst, Mass, 71; vis assoc prof, Univ Hawaii, Honolulu, 74-75. *Mem:* Am Geophys Union; Am Meteorol Soc; Marine Technol Soc. *Res:* Physical oceanography, especially internal waves and oceanic turbulence. *Mailing Add:* Dept Oceanog 1000 Pope Rd Honolulu HI 96822-9998

MAGARGAL, LARRY ELLIOT, b Bethel, Conn, June 14, 41; m 66; c 3. RETINA VASCULAR DISEASES OF THE EYE. *Educ:* Temple Univ, AB, 65, MD, 69. *Prof Exp:* CO-DIR RETINA VASCULAR UNIT, WILLS EYE HOSP, 77-, ASSOC SURGEON OPHTHAL, 84-; ASSOC SURGEON OPHTHAL, THOMAS JEFFERSON UNIV, 84- *Concurrent Pos:* Lectr, Am Acad Ophthal, 77-, Wills Eye Hosp Ann Conf, 77- & Am Soc Contemp Ophthal, 85-; consult, Holy Redeemer Hosp, 84-, Pa Hosp, 84-, Hay Pavilion Easton, 85- & Halleton Med Bldg, 88- *Honors & Awards:* Shumaker Award, Pa Acad Ophthal, 79. *Mem:* Fel Am Col Surgeons; fel Int Col Surgeons; Am Acad Ophthal; Am Soc Contemp Ophthal; Asn Ocular Surgeons. *Res:* Vascular diseases of the eye and treatment using laser photo coagulation; laser surgery of the eye. *Mailing Add:* Wills Eye Hosp Philadelphia PA 19107

MAGARGAL, WELLS WRISLEY, II, b Northampton, Mass; m 78; c 2. CELL BIOLOGY, CYTOSKELETON. *Educ:* Pa State Univ, BA, 72; Univ Mass, Amherst, MS, 76, PhD(biochem), 79. *Prof Exp:* Res assoc, 82-84, RES ASST PROF, UNIV ALA, BIRMINGHAM, 84- *Mem:* Biophys Soc; Tissue Cult Asn; Soc Exp Biol & Med. *Res:* Hypertension; role of ion transport in vascular smooth muscle; tone and hypertrophy and hyperplasia; role of actin binding proteins in smooth muscle growth and contractility. *Mailing Add:* 1029 Zeigler Res Univ Ala Birmingham AL 35294

MAGARIAN, CHARLES ARAM, b Boston, Mass, Dec 18, 27; m 51; c 3. CHEMICAL ENGINEERING. *Educ:* Mass Inst Technol, BS, 50. *Prof Exp:* Sect leader, Prod Control Lab, Monsanto Co, 50-52, prod supvr, 52-53, res engr, Plastics Res & Develop Div, 53-57, res group leader, 57-62, res mgr, 62-65, mgr res, 65-76, mgr com develop, Resin Prod Div, Monsanto Co, 76-82, mgr bus develop, 82-84, consult, 84; RETIRED. *Mem:* Am Inst Chem Engrs; Forest Prod Res Soc. *Res:* Organic and polymer chemistry; polymer synthesis and process development; thermosetting polymers; coating technology; wood products; synthetic adhesives; industrial polymer applications; research and development management. *Mailing Add:* 15 Ruth Dr Wilbraham MA 01095

MAGARIAN, EDWARD O, b East St Louis, Ill, Oct 3, 35; m 59; c 3. PHARMACEUTICAL CHEMISTRY. *Educ:* Univ Miss, BA, 58, PhD(pharmaceut chem), 64. *Prof Exp:* Asst prof, Univ RI, 64-67; assoc prof, Col Pharm, Univ Ky, 67-73; ASSOC PROF PHARMACEUT SCI, COL PHARM, NDAK STATE UNIV, 73 - *Concurrent Pos:* Consult, Pharmacol Res & Clin Studies Inst, Fargo. *Mem:* Am Chem Soc; Am Pharmaceut Asn; Am Heart Asn; Am Asn Cols Pharm. *Res:* Effects of drugs and dietary factors on blood lipids. *Mailing Add:* Dept Pharmaceut Sci NDak State Univ Fargo ND 58105

MAGARIAN, ELIZABETH ANN, b Orlando, Fla, July 13, 40. ALGEBRA. *Educ:* Asbury Col, AB, 60; Fla State Univ, MS, 61, PhD(math), 68. *Prof Exp:* Instr math, La State Univ, New Orleans, 61-64; asst prof, 68-73, ASSOC PROF MATH, STETSON UNIV, 73- *Mem:* Asn Women in Math; Math Asn Am. *Res:* Commutative ring theory; leadership styles of mathematics; using computers in teaching mathematics and statistics. *Mailing Add:* Dept of Math & Comput Sci Stetson Univ Box 8364 De Land FL 32720-3757

MAGARIAN, ROBERT ARMEN, b East St Louis, Ill, July 27, 30; m 50; c 4. MEDICINAL CHEMISTRY, ORGANIC CHEMISTRY. *Educ:* Univ Miss, BS, 56, BSPh, 60, PhD, 66. *Prof Exp:* NIH fel, Col Pharm, Univ Kans, 66-67; asst prof med chem, St Louis Col Pharm, 67-70; assoc prof, 70-78, PROF MED CHEM, COL PHARM, UNIV OKLA, 78- *Concurrent Pos:* NSF grant, St Louis Col Pharm, 69-70; sabbatical, Okla Med Res Foun, 82; grant, NIH, 87- *Honors & Awards:* Baldwin award, 78. *Mem:* Sigma Xi; Am Chem Soc. *Res:* Synthetic organic medicinal chemistry; relation of molecular structure to biological activity; non-steroidal anti-estrogens in breast cancer; college culture; drug design. *Mailing Add:* Col Pharm Univ Okla HSC Oklahoma City OK 73190

MAGASANIK, BORIS, b Kharkoff, Russia, Dec 19, 19; nat US; m 49. GENE REGULATION. *Educ:* City Col NY, BS, 41; Columbia Univ, PhD(biochem), 48. *Hon Degrees:* MA, Harvard Univ, 58. *Prof Exp:* Res asst biochem, Columbia Univ, 48-49; Ernst fel bact & immunol, Harvard Univ, 49-51, assoc, 51-53, from asst prof to assoc prof, 53-60; prof microbiol, 60-77, head dept biol, 67-77, JACQUES MONOD PROF MICROBIOL, MASS INST TECHNOL, 77- *Concurrent Pos:* Markle scholar, Harvard Univ, 51-56; Guggenheim fel, Pasteur Inst, Paris, 59; bd tutors, biochem sci, Harvard Univ, 51- *Mem:* Nat Acad Sci; Am Soc Biol Chem; Am Soc Microbiol; Am Acad Arts & Sci. *Res:* Microbial physiology and biochemistry. *Mailing Add:* Dept Biol Mass Inst Technol Cambridge MA 02139

MAGASI, LASZLO P, b Szekesfehervar, Hungary, June 30, 35; Can citizen; m 62; c 2. MYCOLOGY, FORESTRY. *Educ:* Univ BC, BScF, 59, MSc, 63; State Univ NY Col Environ Sci & Forestry, PhD(mycol), 72. *Prof Exp:* Res officer mycol, 63-73, identification officer, 73-75, SURV HEAD INSECTS & DIS, CAN FORESTRY SERV, 75- *Concurrent Pos:* Lectr, Univ NB, Fredericton, 78. *Mem:* Can Bot Asn; Can Inst Forestry. *Res:* Forest insect and disease surveys; scleroderris canker. *Mailing Add:* Maritimes Forest Res Ctr PO Box 4000 Fredericton NB E3P 5P7 Can

MAGAT, EUGENE EDWARD, b Kharkov, Russia, July 8, 19; nat US; m 45; c 3. CHEMISTRY. *Educ:* Mass Inst Technol, BS, 43, PhD(org chem), 45. *Prof Exp:* Res chemist, Carothers Res Lab, E I Du Pont de Nemours & Co, Inc, 45-50, res assoc, 50-52, res supvr, 52-62, res fel, 62-64, res mgr, Pioneering Res Lab, Textile Fibers Dept, 64-79; vis prof & prog dir, Ctr Indust Res Polymers, Univ Mass, 79-83; adj prof textile sci, NC State Univ & Duke Univ, 84-88; VIS RES SCIENTIST, DUKE UNIV, 88- *Concurrent Pos:* Adj prof, Duke Univ, 83-87. *Mem:* Am Chem Soc. *Res:* Condensation polymers; textile chemistry; radiation chemistry; fiber technology; university-industry interaction. *Mailing Add:* 109 Forest Ridge Dr Chapel Hill NC 27514

MAGAZINE, MICHAEL JAY, b New York, NY, Apr 29, 43; m 65; c 2. OPERATIONS RESEARCH, MANAGEMENT SCIENCE. *Educ:* City Col New York, BS, 64; NY Univ, MS, 66; Univ Fla, MEng & PhD(opers res), 69. *Prof Exp:* Res assoc opers res, Univ Fla, 67-69; asst prof indust eng, NC State Univ, 69-75; assoc prof, 75-82, PROF & CHMN MGT SCI, UNIV WATERLOO, 82- *Concurrent Pos:* NSF res grant, 70-71; vis Orgn Am State Prof opers res, Cath Univ Rio de Janeiro, 72; grant, Nat Res Coun, Can, 76-; consult; assoc ed three journals; vis scientist, Inst Nat Rech Info & Automatique, 82; NSERC Strategic Grant, 84-87; mem, Grant Selection Comt, Nat Sci & Eng Res Coun Can, 89-91; chair, Lancaster Prize Comt, 89 & 91. *Mem:* Opers Res Soc Am; Sigma Xi; Can Oper Res Soc; Inst Mgt Sci; Math Prog Soc; Opers Res Soc Am Coun. *Res:* Development of efficient approximation algorithms for production, scheduling, routing and other combinatorial problems; development of problem solving studio, an innovative teaching workshop for technical students, and manufacturing systems. *Mailing Add:* Dept Mgt Sci Univ Waterloo Waterloo ON N2L 3G1 Can

MAGAZINER, HENRY JONAS, b Philadelphia, Pa, Sept 13, 11; m 38; c 2. HISTORICAL ARCHITECTURE, ARBITRATION. *Educ:* Univ Pa, BArch, 36. *Prof Exp:* Chief, Architect's Squad, Iowa Ord Plant, 39-41; proj architect, Albert Kahn, FAIA, 41-42; test equip designer, Wright Aeronaut Corp, 42-45; apprentice architect, Louis Magaziner, AIA, 45-48; architect pvt pract, 48-72; regional hist architect & archit historian, US Dept Interior, Nat Park Serv, 72-87; CONSULT HIST ARCHIT, 87- *Concurrent Pos:* Mem, Nat Panel, Am Arbit Asn, 71- & Comt Bldg Performance, Am Soc Testing & Mat, 86- *Honors & Awards:* Presidential Award Excellence Design for Govt, Pres of US, 88. *Mem:* Fel Am Inst Architects; Am Arbit Asn; Soc Archit Historians; Am Soc Testing & Mat. *Res:* Historic structures; historical background of the period of a building, evolution and its physical condition and recommendations; stresses imposed on historic structures turned into museums; co-author of US Public Buildings Cooperative Use Act of 1976; originator of historic preservation clauses found in the major national building codes. *Mailing Add:* 2036 Rittenhouse Sq Philadelphia PA 19103-5621

MAGDE, DOUGLAS, b Rochester, NY, Feb 12, 42. CHEMICAL PHYSICS, LASER TECHNOLOGY. *Educ:* Boston Col, BS, 63; Cornell Univ, MS, 68, PhD(physics), 70. *Prof Exp:* Res assoc biophys, Cornell Univ, 70-72; res assoc chem physics, Wash State Univ, 72-74; from asst prof to assoc prof, 74-86, PROF CHEM, UNIV CALIF, SAN DIEGO, 86- *Mem:* Am Asn Physics Teachers; Am Chem Soc; Am Soc Photobiol; Am Photochem Soc. *Res:* Picosecond flash photolysis and spectroscopy; photosynthesis; photochemistry of coordination compounds; mechanism of protein reactions. *Mailing Add:* Dept of Chem B-014 Univ of Calif at San Diego La Jolla CA 92093

MAGDER, JULES, b Toronto, July 17, 34; m 63; c 2. PHYSICAL CHEMISTRY, INORGANIC CHEMISTRY. *Educ:* Univ Toronto, BA, 58, PhD(inorg chem), 61. *Prof Exp:* Proj supvr, Horizons, Inc, 61-64; proj mgr pigments & chem div, Glidden Co, 64-65; dir res, Princeton Chem Res, Inc, 65-71; vpres, 72-73, PRES, PRINCETON ORGANICS, INC, 73- *Mem:* Am Chem Soc; Am Ceramic Soc. *Res:* Heterogeneous catalysis; inorganic polymers; polymer technology; composites; inorganic foams; refractory and structural clay materials; cement and concrete materials. *Mailing Add:* 385 Walnut La Princeton NJ 08540

MAGDOFF, FREDERICK ROBIN, b Washington, DC, Apr 5, 42; c 1. SOIL CHEMISTRY, ENVIRONMENTAL SCIENCE. *Educ:* Oberlin Col, BA, 63; Cornell Univ, MS, 65, PhD(soil sci), 69. *Prof Exp:* Res scientist soils, Soil Conserv Div, Israeli Ministry Agr, 69-71; fel, Univ Wis, 72-73; asst prof, 73-79, PROF SOILS, UNIV VT, 79- *Mem:* Am Soc Agron; Soil Sci Soc Am. *Res:* Land disposal of human and agricultural wastes; soil fertility. *Mailing Add:* Dept Plant & Soil Sci Univ Vt Burlington VT 05405

MAGDOFF-FAIRCHILD, BEATRICE, MACROMOLECULES. *Educ:* Bryn Mawr Col, PhD(physics & math), 48. *Prof Exp:* SR RES SCIENTIST, ST LUKE'S ROOSEVELT HOSP MED CTR, 72-; SR RES SCIENTIST, DEPT MED, COLUMBIA UNIV, 83- *Res:* Structure of sickle cell hemoglobin fibers. *Mailing Add:* Dept Med St Luke's Hosp Amsterdam Ave & 114th St New York NY 10025

MAGE, MICHAEL GORDON, b New York, Aug 17, 34; m 55; c 3. IMMUNOLOGY. *Educ:* Cornell Univ, AB, 55; Columbia Univ, DDS, 60. *Prof Exp:* USPHS fel microbiol, Columbia Univ, 60-62; res immunochemist, Nat Inst Dent Res, 62-66, RES IMMUNOCHEMIST, NAT CANCER INST, 66- *Concurrent Pos:* Vis scientist, Nat Inst Allergy & Infectious Dis, 90-91. *Honors & Awards:* Commendation Medal, USPHS, 89. *Mem:* AAAS; Am Chem Soc; Am Asn Immunologists; Am Pub Health Asn; Sigma Xi. *Res:* Protein chemistry; mammalian cell separation; lymphocyte differentiation; immunochemical engineering. *Mailing Add:* 7008 Wilson Lane Bethesda MD 20817

MAGE, ROSE G, b New York, NY, Oct 8, 35; m 55; c 3. IMMUNOCHEMISTRY, MOLECULAR BIOLOGY. *Educ:* Columbia Univ, PhD(microbiol), 63. *Prof Exp:* Sr investr, Lab Immunol, 65-87, HEAD MOLECULAR IMMUNOGENETICS, LAB IMMUNOL, NAT INST ALLERGY & INFECTIOUS DIS, 88- *Concurrent Pos:* Adj prof genetics, George Washington Univ; NSF fel, 76; lectr immunol, WHO, Lausanne, 83-86. *Mem:* AAAS; Am Asn Immunologists; Can Soc Immunol; Am Soc Microbiol; NY Acad Sci. *Res:* Define in molecular terms the genetics and organization of rabbit immunoglobulin genes and analogous genes encoding antigen-specific receptors on T-cells. *Mailing Add:* Nat Inst Allergy & Infectious Dis NIH Bldg 10 11 N 311 Bethesda MD 20892

MAGEAU, RICHARD PAUL, b Flushing, NY, Oct 24, 41; m 64; c 3. MICROBIOLOGY, IMMUNOLOGY. *Educ:* Univ Conn, BA, 63; Univ Md, MS, 66, PhD(microbiol), 68. *Prof Exp:* Res fel & asst microbiol, Univ Md, 63-68; from asst prof to assoc prof, Kans State Col Pittsburgh, 68-74; res microbiologist, Sci Microbiol Div, Med Microbiol Group, Animal, Plant & Health Inspection Serv, 74-77, MICROBIOLOGIST-IN-CHG, IMMUNOL LAB, SCI MICROBIOL DIV, FOOD, SAFETY & QUALITY SERV, USDA, 77- *Mem:* Am Soc Microbiol; NY Acad Sci; AAAS. *Res:* Radioimmunoassay and other immunological methods of staphylococcal enterotoxin quantitation and detection; laboratory methods development. *Mailing Add:* 4500 Oaklyn Lane Bowie MD 20715

MAGEE, ADEN COMBS, III, b Dimmitt, Tex, Dec 8, 30; m 56; c 2. NUTRITION. *Educ:* Tex A&M Univ, BS, 53; NC State Univ, MS, 57, PhD(animal nutrit), 60. *Prof Exp:* Asst animal nutrit, NC State Univ, 55-60; from asst prof to assoc prof, 60-68, PROF NUTRIT, UNIV NC, GREENSBORO, 68- *Mem:* AAAS; Am Chem Soc; Biomet Soc; Am Inst Nutrit; Am Dietetic Asn. *Res:* Mineral metabolism and toxicities; mineral interrelationships. *Mailing Add:* Sch of Human Environ Sci Univ of N C Greensboro NC 27412-5001

MAGEE, DONAL FRANCIS, b Aberdeen, Scotland, June 4, 24; nat US; m 50; c 5. PHYSIOLOGY, PHARMACOLOGY. *Educ:* Oxford Univ, BA, 44, MA, BM & BCh, 48, DM, 72; Univ Ill, PhD(physiol), 52. *Prof Exp:* Instr & res assoc clin sci, Univ Ill, 48-51; from asst prof to prof pharmacol, Sch Med, Univ Wash, 51-65; prof physiol & chmn dept physiol & pharmacol, 65-76, PROF & CHMN DEPT PHYSIOL, CREIGHTON UNIV, 76- *Concurrent Pos:* Guggenheim Mem Found fel, Rowett Inst, Scotland, 59-60; Fogarty Inst fel, Inst Nat Sante Exper Rech Medicale, Marseille, France, 78-79. *Mem:* Soc Exp Biol & Med; Am Physiol Soc; Am Soc Pharmacol & Exp Therapeut; Am Gastroenterol Asn; Brit Med Asn; Phys Soc. *Res:* Gastrointestinal tract, especially physiology and pharmacology of pancreas and biliary canal and stomach. *Mailing Add:* Dept of Physiol Creighton Univ Omaha NE 68178

MAGEE, JOHN FRANCIS, b Bangor, Maine, Dec 3, 26; m 49; c 3. MATHEMATICS. *Educ:* Bowdoin Col, AB, 46; Harvard Univ, MBA, 48; Univ Maine, MA, 52. *Prof Exp:* Mem staff financial anal, Johns-Manville Co, 49-50; dir res, Opers Res Group, 50-59, head, 59-62, vpres mgt serv div, 63-68, mem corp tech staff, 68-69, exec vpres & dir, 69, chief operating officer, 71, pres, 72-86, chief exec officer, 74-88, CHMN, ARTHUR D LITTLE INC, 86- *Concurrent Pos:* Pres, Inst Mgt Sci, 71-72. *Honors & Awards:* Kimball Medal, Opers Res Soc Am. *Mem:* Opers Res Soc Am (pres, 66-67). *Res:* Operations research; marketing; financial planning and policy; managerial controls. *Mailing Add:* Arthur D Little Inc 25 Acorn Park Cambridge MA 02140

MAGEE, JOHN LAFAYETTE, b Franklinton, La, Oct 28, 14; m 48; c 3. PHYSICAL CHEMISTRY. *Educ:* Miss Col, AB, 35; Vanderbilt Univ, MS, 36; Univ Wis, PhD(chem), 39. *Prof Exp:* Nat Res Coun fel, Princeton Univ, 39-40, res assoc, 40-41; res physicist, B F Goodrich Co, 41-43; group leader, Los Alamos Sci Lab, 43-45 & Naval Ord Testing Sta, Calif, 45-46; sr scientist, Argonne Nat Lab, 46-48; from asst prof to assoc prof chem, Univ Notre Dame, 48-53, prof, 53-77, head, Dept Chem, 67-71, assoc dir, Radiation Lab, 54-71, dir, 71-75; vis sr staff mem, 76-77, staff sr scientist, Biol & Med Div, 77-86, CONSULT, CELLULAR & MOLECULAR BIOL DIV, LAWRENCE BERKELEY LAB, 87- *Concurrent Pos:* Observer, Proj Bikini, 46; mem weapons systs eval group, US Dept Defense, 55-57. *Mem:* AAAS; Am Chem Soc; Am Phys Soc; Radiation Res Soc; The Chem Soc. *Res:* Photochemistry; chemical reaction rate theory; radiation chemistry; application of statistical mechanics and quantum mechanics to chemistry. *Mailing Add:* Cellular & Molecular Biol Div Lawrence Berkeley Lab Berkeley CA 94720

MAGEE, JOHN ROBERT, b Bristol, RI, Aug 27, 16; m 49; c 5. TEXTILE CHEMISTRY. *Educ:* Brown Univ, AB, 39. *Prof Exp:* Chem supt, Nat Dairy Prod Corp, 39-48; gen supt, Va-Carolina Chem Corp, 48-58; asst mgr spec prod, Charles Pfizer & Co, Inc, 58; staff assoc res & develop, Monsanto Textiles Co, 58-60, sect head nylon develop, 60-61, mgr develop opers & tech serv, 62-85; RETIRED. *Concurrent Pos:* Consult, Charles Pfizer & Co, Inc & Textile Res Inst, NJ, 58. *Mem:* Am Chem Soc; Am Asn Textile Technol. *Res:* Melt and wet spun man-made fibers, especially nylon, polyester, acrylics, protein and copolymer blends. *Mailing Add:* 3844 Dunwoody Dr Pensacola FL 32503

MAGEE, JOHN STOREY, JR, b Baltimore, Md, Mar 23, 31; m 53; c 3. INORGANIC CHEMISTRY, PHYSICAL CHEMISTRY. *Educ:* Loyola Col, Md, BS, 53; Univ Del, PhD(inorg chem), 61. *Prof Exp:* Chemist, Davison Chem Corp Div, W R Grace & Co, 53-55, res chemist, Res Div, 61-63, res chemist, Davison Chem Corp Div, 63-71, res dir, 71-78, tech adminr, Petrol Catalyst Dept, 79-81; DIR RES & DEVELOP, KATALISTIKS, INC, 82- *Mem:* Am Chem Soc. *Res:* Preparation of catalysts for heterogeneous reactions; study of compounds containing metals in unusual oxidation states; preparation and evaluation of petroleum catalysts. *Mailing Add:* 12205 Mt Albert Rd Ellicott City MD 21043-1336

MAGEE, LYMAN ABBOTT, b Bogalusa, La, Apr 10, 26; m 57; c 2. MICROBIOLOGY. *Educ:* La Col, BS, 46; La State Univ, MS, 54, PhD(microbiol), 58. *Prof Exp:* Sr analyst, Cities Serv Refining Corp, La, 47-48; asst prof chem, La Col, 48-51, dean men, 51; Hite fel cancer res, M D Anderson Hosp & Tumor Inst, Univ Tex, 58-59; Nat Inst Allergy & Infectious Dis, fel & instr microbiol, 60-61, assoc prof biol, 61- 65, assoc prof microbiol, 63-77, chmn dept, 71-77, asst dean, 78- 84, assoc dean 84-87, PROF BIOL, UNIV MISS, 65- *Concurrent Pos:* Am Soc Microbiol Pres fel virol, Commun Dis Ctr, USPHS & Southern Res Inst, Ga, 65. *Mem:* Am Soc Microbiol. *Res:* Cancer viruses; metabolism of exanthem viruses; microbial decomposition of pesticides. *Mailing Add:* Dept of Biol Univ of Miss University MS 38677

MAGEE, MICHAEL JACK, b Coleman, Tex, Feb 16, 46; m 69; c 2. COMPUTER VISION, SPATIAL REASONING. *Educ:* Univ Tex, BA, 68, MA, 72, PhD(comput sci), 75. *Prof Exp:* Assoc programmer, Int Bus Mach Corp, 68-71; PROF COMPUT SCI, UNIV WYO, 75- *Concurrent Pos:* Consult, Sandia Nat Lab, Martin Marietta, 84-; prin investr, NSF, 86- *Mem:* Inst Elec & Electronic Engr Comput Soc. *Res:* Computer vision, image understanding and spatial reasoning. *Mailing Add:* Dept Comput Sci Univ Wyo Box 3682 Laramie WY 82071

MAGEE, PAUL TERRY, b Los Angeles, Calif, Oct 26, 37; m 64; c 2. GENETICS, MOLECULAR BIOLOGY. *Educ:* Yale Univ, BS, 59; Univ Calif, Berkeley, PhD(biochem), 64. *Prof Exp:* Am Cancer Soc fel, Lab Enzymol, Nat Ctr Sci Res, Gif-sur-Yvette, France, 64-66; from asst prof to assoc prof microbiol, 66-74, assoc prof human genetics, Sch Med, Yale Univ, 74-77; prof & chmn, Dept Microbiol & Pub Health, Mich State Univ, 77-87; PROF & DEAN, COL BIOL SCI, UNIV MINN, 87- *Concurrent Pos:* Mem adv bd, Genetic Biol Prog, NSF, 78-81; mem, Microbial Physiol Study Sect, NIH, 84-88. *Mem:* AAAS; Am Soc Microbiol; Genetic Soc Am; Am Soc Biol Chem. *Res:* Regulation of gene expression; developmental biology; the genetics and molecular biology of meiosis and ascospore formation in Saccharomy; the establishment of a genetic system and the genetics of virulence of Candida albicans; genetics and molecular biology of fungi. *Mailing Add:* Col Biol Sci Univ Minn 1475 Gortner Ave St Paul MN 55108

MAGEE, RICHARD STEPHEN, b East Orange, NJ, Mar 26, 41; m 69; c 1. MECHANICAL ENGINEERING. *Educ:* Stevens Inst Technol, BEng, 63, MS, 64, DSc(appl mech), 68. *Prof Exp:* Res engr, Combustion Lab, 66-68, from asst prof to assoc prof mech eng, 68-77, NSF & NASA grants, 69-70, PROF MECH ENG, STEVENS INST TECHNOL, 77- *Concurrent Pos:* Consult, Photochem, Inc, 70- *Mem:* Am Soc Mech Engrs; Am Soc Eng Educ; Combustion Inst. *Res:* Flammability characteristics of combustible materials, including ignition, flame spread and burning characteristics; pyrolysis characteristics of polymeric materials; basic incinerator design parameters. *Mailing Add:* 15 Valtee Dr Florham Park NJ 07932

MAGEE, THOMAS ALEXANDER, b Brookhaven, Miss, Apr 19, 30; m 61; c 1. ORGANIC CHEMISTRY. *Educ:* Tulane Univ, BS, 52, MS, 55; Univ Tenn, PhD(chem), 57. *Prof Exp:* Chemist, 57-76, res assoc 76-78, SR RES ASSOC, T R EVANS RES CTR, DIAMOND SHAMROCK CORP, 78-, GROUP LEADER, 80- *Mem:* Am Chem Soc. *Res:* Organometallic chemistry of the transition elements; pesticide chemistry. *Mailing Add:* 7301 Case Ave Mentor OH 44060

MAGEE, WAYNE EDWARD, b Big Rapids, Mich, Apr 11, 29; m 51; c 3. VIROLOGY, IMMUNOLOGY. *Educ:* Kalamazoo Col, BA, 51; Univ Wis, MS, 53, PhD(biochem), 55. *Prof Exp:* Res scientist microbiol, Upjohn Co, 55-60, proj leader biochem, 60-63, proj leader virol res, 63-66, sr res scientist, 67-71, sr scientist exp biol, 71; prof life sci, Ind State Univ, Terre Haute, 71-74; prof biol, Div Allied Health & Life Sci, Univ Tex, San Antonio, 75-81, dir, 75-80, prof biochem, Univ Tex Health Sci Ctr, 75-81; prof biochem & head dept bacteriol & biochem, Univ Idaho, 81-85; PROF BIOL SCI, HEAD DEPT BIOSCI & BIOTECHNOL & DIR, DIV LIFE SCI, DREXEL UNIV, PHILADELPHIA, 85- *Concurrent Pos:* Mem adj staff, Dept Biol, Western Mich Univ, 68-70, adj prof, 71; adj prof microbiol, Terre Haute Ctr Med Educ, Sch Med, Ind Univ, 72-74; adj found scientist, Southwest Found Res Educ, San Antonio, Tex, 78-81. *Mem:* Fel AAAS; Am Chem Soc; Am Soc Biol Chem; Am Soc Microbiol; NY Acad Sci; Soc Indus Micro; Sigma Xi. *Res:* Nitrogen fixation; nucleic acid biochemistry; biochemistry of virus infection; mechanism of action of therapeutic agents and interferon; phospholipid vesicles as carriers for drugs and nucleic acids; hybridoma technology. *Mailing Add:* Dept Biosci & Biotechnol Drexel Univ 32nd & Chestnut Sts Philadelphia PA 19104

MAGEE, WILLIAM LOVEL, b Ft William, Ont, Mar 24, 29; m 55; c 2. BIOCHEMISTRY. *Educ:* Univ Western Ont, BSc, 52, MSc, 54, PhD(biochem), 57. *Prof Exp:* Nat Multiple Sclerosis Soc overseas fel biochem, Guy's Hosp Med Sch, London, Eng, 58-61; res asst, 61-62, lectr, 62-63, asst prof, 63-68, ASSOC PROF BIOCHEM, HEALTH SCI CTR, UNIV WESTERN ONT, 68- *Mem:* AAAS; Can Biochem Soc. *Res:* Biochemistry of brain and peripheral nerve; biochemistry of demyelination; phospholipid chemistry and metabolism. *Mailing Add:* Dept Biochem Univ Western Ont Med Sch London ON N6A 5C1 Can

MAGEE, WILLIAM THOMAS, b San Antonio, Tex, Apr 22, 23; m 48; c 3. ANIMAL BREEDING. *Educ:* Tex A&M Univ, BS, 47; Iowa State Univ, MS, 48, PhD(animal husb), 51. *Prof Exp:* Asst animal husbandman, Tex A&M Univ, 53-55; from asst prof to assoc prof, 55-65, PROF ANIMAL BREEDING, MICH STATE UNIV, 65- *Mem:* Am Soc Animal Sci; Am Genetic Asn; Sigma Xi. *Res:* Evaluation of the effects of selection and mating systems on performance traits in beef cattle and swine. *Mailing Add:* Dept of Animal Husb Mich State Univ East Lansing MI 48823

MAGEL, KENNETH I, b Richmond, VA, Dec 11, 50; m 77; c 2. SOFTWARE ENGINEERING. *Educ:* Brown Univ, BSc, 72, MSc, 74, PhD(comput sci), 77. *Prof Exp:* Asst prof comput sci, Wichita State Univ, 76-78 & Univ Miss, Rolla, 78-82; assoc prof, Univ Tex, San Antonio, 82-83; PROF COMPUT SCI, NDAK STATE UNIV, 83- *Concurrent Pos:* Prin rep, Asn Comput Mach, 85-; comp sci chair, NDak State Univ, 88- *Mem:* Asn Comput Mach; Comput Soc Inst Elec & Electronic Engrs. *Res:* Software complexity metrics and human computer interfaces; development of flexible complexity metrics parameterized on theory and of a software system supporting tuning of an interface to human experiences. *Mailing Add:* NDak State Univ PO Box 5075 Fargo ND 58105-5075

MAGEN, MYRON S, b Brooklyn, NY, Mar 1, 36; m 52; c 3. COMMUNITY MEDICINE, MEDICAL EDUCATION. *Educ:* Col Osteop Med & Surg, Des Moines, DO, 51. *Hon Degrees:* ScD, Univ Osteop Med & Health Sci, Des Moines, 81. *Prof Exp:* Resident, Col Hosp, Des Moines, 54; chmn dept pediat, Still Col Hosp, 58-62 & Riverside Osteop Hosp, 62-68; assoc dean, Mich Col Osteop Med, 66-67, actg dean, 67-68, dean, 68-70; PROF & DEAN, COL OSTEOP MED, MICH STATE UNIV, 70- *Concurrent Pos:* Assoc prof & chair, Col Osteop Med & Surg, Des Moines, 58-62; deleg, White House Conf Children & Youth, 60; consult, Dept Health, Educ & Welfare, 69-72 & NY Dept Educ, 75; mem health adv coun, State of Mich, 70; mem adv comt, Nat Acad Sci, 72-76 & USPHS, 78-80; mem special med adv group, Vet Admin, Washington, DC, 73-77. *Mem:* Inst Med-Nat Acad Sci; NY Acad Sci; Nat Coun Int Health; Am Osteop Asn; Am Col Osteop Pediatricians (secy-treas, 61-63, vpres, 64, pres, 65); Asn Am Med Cols; Am Col Pediat; AAAS. *Res:* General pediatrics; health care policy; medical education; community medicine. *Mailing Add:* Col Osteop Med 308 E Fee Hall Mich State Univ East Lansing MI 48824

MAGER, ARTUR, b Nieglowice, Poland, Sept 21, 19; nat US; m 42; c 1. PHYSICS. *Educ:* Univ Mich, BS, 43; Case Inst Technol, MS, 51; Calif Inst Technol, PhD(aeronaut, physics) 53. *Prof Exp:* Test engr, Continental Aviation & Eng Co, 43-44; aeronaut res scientist, Nat Adv Comt Aeronaut, 46-53; res engr & consult, Naval Ord Test Sta, 53-54; res consult, Astro-Marquardt Corp, 54-60; dir, Nat Eng Sci Co, 60-61; dir spacecraft sci, Aerospace Corp, 61-64, gen mgr, Appl Mech Div, 64-68, vpres & gen mgr eng sci opers, 68-78, group vpres eng, 78-82; CONSULT, 82- *Concurrent Pos:* Res fel, Calif Inst Technol, 53-54; mem ballistic syst div reentry panel, Air Force Systs Command, 61-62; mem res adv comt space vehicle aerodyn, NASA, 63-65; mem bd counr, Sch Eng, Univ Southern Calif; mem Aeronaut Space Eng Bd, Nat Res Coun, 82-87; mem NASA adv coun, 83-86; chmn NASA Space Appln Adv Comt, 83-86. *Mem:* Nat Acad Eng; fel Am Inst Aeronaut & Astronaut (pres, 80-81); fel AAAS. *Res:* Boundary layers; heat transfer; gas dynamics; cycle analysis; chemical and electrical propulsion; magnetohydrodynamics. *Mailing Add:* 1353 Woodruff Ave Los Angeles CA 90024

MAGER, MILTON, b New York, NY, Dec 13, 20; m 47; c 3. BIOCHEMISTRY. *Educ:* NY Univ, BA, 43; Rensselaer Polytech Univ, MS, 49; Boston Univ, PhD(biochem), 54. *Prof Exp:* Instr chem, Albany Col Pharm, 46-49; biochemist, Qm Res & Eng Ctr, 53-56, chief biochem sect physiol br, 56-61; chief biochem lab, 61-67, dir biochem & pharmacol lab, 67-75, dir heat res div, US Army Res Inst Environ Med, 75-83; VIS LECTR, BIOL, FRAMINGHAM STATE COL, 83- *Concurrent Pos:* Asst, Sch Med, Boston Univ, 52-55, instr, 55-59, from asst res prof to assoc res prof, 59-83. *Honors & Awards:* Meritorious Civilian Serv Medal, Dept Army, 83. *Mem:* AAAS; Am Chem Soc; Am Asn Clin Chemists; Am Physiol Soc. *Res:* Biochemical and physiological responses of man and animals to environmental stress, primarily heat; temperature regulation; blood and tissue enzymes; methodology. *Mailing Add:* 19 Fenno Rd Newton Ctr Boston MA 02159

MAGERLEIN, BARNEY JOHN, b Columbus, Ohio, Nov 11, 19; m 44; c 4. MEDICINAL CHEMISTRY. *Educ:* Capital Univ, BS, 41; Ohio State Univ, PhD(org chem), 46. *Hon Degrees:* DSci Capital Univ, 78. *Prof Exp:* Asst org chem, Ohio State Univ, 41-44, Off Sci Res & Develop contract, Univ Res Found, 44-45; chemist, Upjohn Co, 46-71, distinguished res scientist, 71-88; RETIRED. *Concurrent Pos:* Vis scholar, Univ Calif, Los Angeles, 60-61. *Mem:* Am Chem Soc; Sigma Xi. *Res:* Organic synthesis; synthetic studies of morphine series; steroids and pteridines; antibiotics. *Mailing Add:* 1618 Timberlane Dr Upjohn Co Kalamazoo MI 49008

MAGERLEIN, JAMES MICHAEL, b Kalamazoo, Mich, Aug 25, 50. OPERATIONS RESEARCH, INDUSTRIAL ENGINEERING. *Educ:* Kalamazoo Col, BA, 72; Univ Mich, AM, 73, PhD(indust & opers eng), 78. *Prof Exp:* Res asst, Univ Mich, 74-78; mathematician oper res, 78-81, SR MATHEMATICIAN, UPJOHN CO, 81- *Mem:* Inst Mgt Sci. *Res:* Hospital systems; methodology for scheduling elective surgery; modelling, design and simulation of production systems. *Mailing Add:* Upjohn Co MS 9902-243-116 7000 Portage Rd Kalamazoo MI 49001

MAGGARD, SAMUEL P, b Whitesburg, Ky, June 19, 33; m 54; c 4. CIVIL ENGINEERING, STRUCTURAL ENGINEERING. *Educ:* Univ Ky, BS, 56, MS, 57; Purdue Univ, PhD(struct eng), 63. *Prof Exp:* Design engr, Ky Dept Hwy, 55-57; instr struct design, Univ Ky, 57-58, asst prof, 58-59; res assoc, Purdue Univ, 59-63; from asst prof to assoc prof, 63-67, PROF STRUCT DESIGN, N MEX STATE UNIV, 67-, HEAD DEPT CIVIL ENG, 66- *Concurrent Pos:* Consult nuclear effects br, Army Missile Test & Eval Directorate, White Sands Missile Range, 64- *Mem:* Am Soc Civil Engrs; Am Soc Eng Educ; Nat Soc Prof Engrs; Sigma Xi. *Res:* Structural analysis and design; economy and adequacy of structural materials and structures. *Mailing Add:* Dept of Civil Eng NMex State Univ University Park NM 88003

MAGGENTI, ARMAND RICHARD, b San Jose, Calif, Feb 15, 33; m 63; c 2. PLANT NEMATOLOGY. *Educ:* Univ Calif, Berkeley, BS, 54, PhD, 59. *Prof Exp:* From asst nematologist to nematologist, Univ Calif, Davis, 58-74, lectr nematol & chmn dept, 74-79. *Concurrent Pos:* Fulbright fels, Pakistan, 65, Iraq, 65-66. *Honors & Awards:* ASO distinguished lectr, 80. *Mem:* Soc Nematol; Soc Syst Zool; Sigma Xi. *Res:* Taxonomy and morphology of soil and freshwater nematodes; nematode parasites of fish; nematode parasites of birds. *Mailing Add:* Dept Nematol Univ Calif Davis CA 95616

MAGGIO, BRUNO, b Torino, Italy, July 20, 44; Arg citizen; m 72. MEMBRANE BIOPHYSICS, MODEL MEMBRANE SYSTEMS. *Educ:* Univ Nat Cordoba, Arg, Biochemist, 68, PhD(biochem), 72. *Prof Exp:* From asst prof to prof biophys chem, Fac C Quim, Univ Nat Cordoba, Arg, 73-88; PROF BIOPHYS CHEM, DEPT BIOCHEM & MOLECULAR BIOPHYS, MED COL VA, VA COMMONWEALTH UNIV, 89- *Concurrent Pos:* Res fel, Brit Coun, UK, 73-76; honor res assoc, dept biochem & chem, Univ London, UK, 73; assoc career investr, Nat Res Coun, Arg, 76-81, independent career investr, 81-87, prin career investr, 87-88; sr res fel, Dept Neurol, Sch Med, Yale Univ, 83-84, vis res scientist, 87. *Honors & Awards:* Outstanding Young Scientist, Cordoba, Arg, 79; B Houssay Sr Scientist Award, Nat Res Coun, Arg, 87. *Mem:* Am Soc Neurochem; Am Soc Biochem & Molecular Biol. *Res:* Biophysical chemistry of biomembranes; molecular properties and organization of glycolipids; lipid-lipid and lipid-protein interactions; artificial lipid-protein monolayers and bilayers; surface recognition, reactivity and modulation of membranes and membrane-active enzymes. *Mailing Add:* Dept Biochem & Molecular Biophys Med Col Va Va Commonwealth Univ Richmond VA 23298-0614

MAGGIO, EDWARD THOMAS, b Brooklyn, NY, Mar 28, 47. PROTEIN CHEMISTRY, NON-ISOTOPIC IMMUNOASSAYS. *Educ:* Polytech Inst Brooklyn, BS, 68; Univ Mich, MS, 69, PhD(biol chem), 73. *Prof Exp:* Fel pharmaceut chem, Univ Calif, San Francisco, 73-74, NIH fel pharmaceut chem, 74-75; group leader enzym, Int Diag Technol, Syva Res Inst, 75, mgr, Biochem Sect, 75-77, dir res & develop, 77-79; vpres res & develop, Scripps-Miles Inc, 80-82; PRES, SYNBIOTICS, INC, 82- *Concurrent Pos:* Adj assoc mem path dept, Scripps Clinic & Res Found, 80-82. *Mem:* Am Chem Soc; Am Soc Microbiol; Am Asn Clin Chem; NY Acad Sci. *Res:* Enzymology in clinical and pharmaceutical chemistry; enzyme kinetics; chemical modification of proteins; physical biochemistry; non-isotopic immunoassay methodology; anti-idiotypic monoclonal antibodies; infectious disease detection; therapeutic drug monitoring. *Mailing Add:* Synbiotics 11011 Via Frontera San Diego CA 92127

MAGGIORA, GERALD M, b Oakland, Calif, Aug 11, 38; m 63; c 2. MOLECULAR BIOPHYSICS. *Educ:* Univ Calif, Davis, BS, 64, PhD(biophys), 68. *Prof Exp:* Res assoc, Univ Kans, 68-69, USPHS fel, 69-70, from asst prof to assoc prof biochem, 70-85; MEM STAFF, UPJOHN RES LAB, 85- *Mem:* Am Chem Soc; Am Phys Soc; Int Soc Quantum Biol. *Res:* Molecular quantum mechanics; quantum biochemistry; photosynthesis; vision; electromagnetic properties of large molecules; interaction of light with matter. *Mailing Add:* Comput Chem Upjohn Res Lab 301 Henrietta St Kalamazoo MI 49001

MAGGIORE, CARL JEROME, b Grand Island, Nebr, June 28, 43; m 77; c 2. NUCLEAR PHYSICS, SOLID STATE PHYSICS. *Educ:* Creighton Univ, BS, 65; Mich State Univ, PhD(physics), 72. *Prof Exp:* Res assoc environ sci, Mt Sinai Med Sch, 70-74; mgr x-ray, Princeton Gamma Technol, 74-76; asst group leader solid state electronics, 76-81, STAFF MEM ELECTRONICS RES & DEVELOP, LOS ALAMOS NAT LAB, UNIV CALIF, 81- *Mem:* Am Phys Soc; AAAS; NY Acad Sci. *Res:* Ion beam analysis of the near surfaces of materials using backscattering, channeling, and ion-induced x-rays; compound semiconductor interfaces and device physics; electrochemistry and catalysis. *Mailing Add:* MS K765 Los Alamos Nat Lab PO Box 1663 Los Alamos NM 87545

MAGID, ANDY ROY, b St Paul, Minn, May 4, 44; m 66; c 2. MATHEMATICS. *Educ:* Univ Calif, Berkeley, BA, 66; Northwestern Univ, PhD(math), 69. *Prof Exp:* J F Ritt asst prof math, Columbia Univ, 69-72; assoc prof, 72-77, PROF MATH, UNIV OKLA, 77-, CHMN, 89- *Concurrent Pos:* Vis assoc prof math, Univ Ill, 75-76, Univ Calif, Berkeley, 80, Hebrew Univ, 84. *Mem:* Am Math Soc; Math Asn Am. *Res:* Commutative algebra; Galois theory; algebraic geometry. *Mailing Add:* Dept Math Univ Okla Norman OK 73069

MAGID, LINDA LEE JENNY, b Omaha, Nebr, Dec 13, 46; m 69. PHYSICAL ORGANIC CHEMISTRY. *Educ:* Rice Univ, BA, 69; Univ Tenn, Knoxville, PhD(chem), 73. *Prof Exp:* From instr to assoc prof chem, 73-88, assoc dean, Col Lib Arts, 87-90, PROF CHEM, UNIV TENN, KNOXVILLE, 88-, EXEC ASST TO CHANCELLOR, 90- *Mem:* Am Chem Soc; Nat Coun Univ Res Adminrs; Am Women Sci. *Res:* Elucidation of micellar and microemulsion structure using small-angle scattering, FTNMR, conductimetry, etc. *Mailing Add:* Dept Chem Univ Tenn Knoxville TN 37996-1600

MAGID, RONALD, b Brooklyn, NY, Dec 19, 38; m 60, 69; c 2. ORGANIC CHEMISTRY. *Educ:* Yale Univ, BS, 59, MS, 60, PhD(chem), 64. *Prof Exp:* Asst prof, Rice Univ, 64-70; from asst prof to assoc prof, 70-81, PROF CHEM, UNIV TENN, KNOXVILLE, 81- *Mem:* Am Chem Soc; Royal Soc Chem. *Res:* Orbital symmetry rules; mechanisms of organolithium reactions; mechanisms of reactions of allylic compounds; synthesis of strained compounds. *Mailing Add:* Dept of Chem Univ of Tenn Knoxville TN 37996

MAGIE, ALLAN RUPERT, b Umatilla, Fla, July 21, 36; m 61; c 3. PUBLIC HEALTH. *Educ:* Univ Calif, BA, 58, PhD(physiol), 63; Loma Linda Univ, MPH, 71. *Prof Exp:* Assoc prof biol, Pac Union Col 63-65; prof & dean sci & technol, Mountain View Col, Philippines, 65-70; asst prof environ health, 71-73, coordr, Doctor Health Sci Prog, 72-78, assoc prof, 73-77, PROF & CHMN ENVIRON & TROP HEALTH, SCH HEALTH, LOMA LINDA UNIV, 77- *Concurrent Pos:* Dir external MPH prog, Western Consortium Schs Pub Health, 73- *Mem:* Am Pub Health Asn; Sigma Xi. *Res:* Human health effects of ambient air pollutants, including fetal development, birth anomalies, respiratory disease and hospitalization; animal diseases transferrable to humans. *Mailing Add:* Dept Phys Educ Riverside Community Col 4800 Magnolia Ave Riverside CA 92506

MAGIE, ROBERT OGDEN, b Madison, NJ, July 30, 06; m 33; c 2. PLANT PATHOLOGY. *Educ:* Rutgers Univ, BS, 29; Univ Wis, MS, 30, PhD(plant path), 34. *Prof Exp:* Agt, Dutch elm dis invest, Bur Plant Indust, USDA, 35; Crop Protection Inst fel, NY Agr Exp Sta, Geneva, 35-36, res assoc, 36-40; asst prof plant path, Cornell Univ, 40-45; prof, 45-77, EMER PROF PLANT PATH, INST FOOD & AGR SCI, AGR RES & EDUC CTR, UNIV FLA, 77- *Concurrent Pos:* Consult, Israel, SAfrica, Brazil, NZ, Australia & France; ed & publ, GladioGrams, quart newslett. *Mem:* Am Phytopath Soc. *Res:* Diseases of cut flowers crops; gladiolus flower and corm production; control of Botrytis and Fusarium diseases; caladium tuber production. *Mailing Add:* Agr Res & Educ Ctr Inst of Food & Agr Sci Univ Fla Bradenton FL 34203

MAGILL, CLINT WILLIAM, b Washington, DC, Sept 15, 41; m 65; c 1. GENETICS, MICROBIOLOGY. *Educ:* Univ Ill, BS, 63; Cornell Univ, PhD(genetics), 69. *Prof Exp:* NIH fel biochem genetics, Univ Minn, 67-69; from asst prof to assoc prof, 69-90, PROF GENETICS, TEX A&M UNIV, 90- *Mem:* Genetics Soc Am. *Res:* Fungal variability; restriction fragment length polymorphism maps; host response. *Mailing Add:* Dept Plant Path & Microbiol Tex A&M Univ College Station TX 77843

MAGILL, DAVID THOMAS, b Evanston, Ill, May 14, 35; m 57; c 4. COMMUNICATIONS ENGINEERING. *Educ:* Princeton Univ, BSE, 57; Stanford Univ, MS, 60, PhD(elec eng), 64. *Prof Exp:* Engr, RS Electronics Corp, 57-58; asst ionospheric res, Stanford Univ, 58-60; res scientist commun & control syst theory, Lockheed Missiles & Space Co, 60-64; staff scientist commun theory & design & dept mgr, Philco-Ford Corp, 64-70; sr res engr, SRI Int, 70-; DEPT MGR, STANFORD TELECOMMUN. *Concurrent Pos:* Lectr, Exten Prog, Univ Calif, Berkeley, 64-65 & Santa Barbara, 70- *Mem:* Inst Elec & Electronics Engrs. *Res:* Satellite communication; modulation-demodulation and estimation-detection theories. *Mailing Add:* Stanford Telecommun 2421 Mission Col Blvd Santa Clara CA 95050

MAGILL, JANE MARY (OAKES), b Hamilton, Ont, Sept 30, 40; m 65; c 1. BIOCHEMISTRY, GENETICS. *Educ:* Univ Western Ont, BSc, 63; Cornell Univ, PhD(genetics), 68. *Prof Exp:* Asst scientist genetics, Univ Minn, 68-69; from instr, to asst prof, 70-82, ASSOC PROF BIOCHEM, TEX A&M UNIV, 82- *Mem:* Genetics Soc Am; Am Soc Biol Chemists. *Res:* DNA methylation; gene expression in fungi. *Mailing Add:* Dept of Biochem Tex A&M Univ College Station TX 77843

MAGILL, JOSEPH HENRY, b Drumnabreeze, Northern Ireland, Dec 16, 28; US citizen; m 56; c 2. PHYSICAL ORGANIC CHEMISTRY. *Educ:* Queen's Univ, Belfast, BSc, 52, PhD(chem), 56, DSc, 90; Univ London, DIC, 57; FRSC, 62 & 65. *Prof Exp:* Chem technologist polymer mfg, Chemstrand Ltd, Northern Ireland, 57-58; sr tech officer polymers, Brit Nylon Spinners Ltd, UK, 59-65; res fel polymer sci, Mellon Inst, 65-68; assoc prof mat eng, 68-75, PROF MAT ENG, UNIV PITTSBURGH, 75-, PROF CHEM-PETROL ENG & ADJ PROF CHEM, 80- *Concurrent Pos:* Asst tech officer, Imperial Chem Indust, UK, 54; fel, Nat Coal Bd, Imperial Col, 56-57; asst lectr, Northampton Polytech, London, 57-58; eve lectr, Newport Col Technol, UK, 61-62; vis fel, Mellon Inst, 62-64; sr res fel, Sci Res Coun, Univ Bristol, 75-76; Alexander von Humboldt Sr Award, Max Planck Inst & Univ Hamburg-Harburg, 84-85; vis prof, Univ Yamagata, Japan, 88 & 89. *Mem:* Fel Am Inst Physics; Am Chem Soc. *Res:* Physical properties and structural aspects of polymers and small molecules; flammability, thermal stability and toxicity of polymers; materials engineering. *Mailing Add:* Dept Metall & Mat Eng Univ Pittsburgh & 4200 Fifth Ave Pittsburgh PA 15260

MAGILL, KENNETH DERWOOD, JR, b Duncansville, Pa, Oct 21, 33; m 52, 74; c 3. MATHEMATICS. *Educ:* Shippensburg State Col, BS, 56; Pa State Univ, MA, 60, PhD(math), 63. *Prof Exp:* Teacher, Central Cove Schs, 56-57; instr math, Pa State Univ, 58-63; from asst prof to assoc prof, 63-67, chmn dept, 67-70, PROF MATH, STATE UNIV NY BUFFALO, 67- *Concurrent Pos:* Vis prof, Univ Leeds, 68, Univ Fla, Monash Univ, 85; vis mem, Inst Advan Studies, Australian Nat Univ, 70. *Mem:* Math Asn Am; Am Math Soc. *Res:* Topology; near-rings of continuous functions; transformation semigroups. *Mailing Add:* Dept Math SUNY Health Sci Ctr 334 Diefendor Hall 3425 Main St Buffalo NY 14214

MAGILL, ROBERT EARLE, b Ft Worth, Tex, May 8, 47; m 68; c 3. BRYOLOGY, TAXONOMY. *Educ:* Sul Ross State Univ, BS, 69, MS, 71; Tex A&M Univ, PhD(bot), 75. *Prof Exp:* Curatorial trainee, Mo Bot Garden, 75-76, cur cryptogams, Bot Res Inst, Pretoria, 76-81, from asst cur to assoc cur, 82-88, CUR BRYOPHYTES & DEPT HEAD, MO BOT GARDEN, 89- *Mem:* Am Bryol & Lichenological Soc; Brit Bryol Soc; Int Asn Plant Taxon. *Res:* Taxonomy and phytogeography of bryophytes; African mosses; tropicos computerized botanical data base. *Mailing Add:* PO Box 299 Mo Bot Garden St Louis MO 63166

MAGILL, THOMAS PLEINES, b Philadelphia, Pa, May 24, 03; m 43. MICROBIOLOGY, IMMUNOLOGY. *Educ:* Johns Hopkins Univ, AB, 25, MD, 30. *Prof Exp:* Instr med, Sch Med, Johns Hopkins Univ, 33-35; asst, Rockefeller Inst, NY, 35-36; mem staff, Rockefeller Found, 36-38; from asst prof to assoc prof bact & immunol, Med Col, Cornell Univ, 38-48; prof microbiol & immunol & chmn dept, Long Island Col Med, 48-50; chmn dept, 50-70, prof, 50-73, EMER PROF MICROBIOL & IMMUNOL, COL MED, STATE UNIV NY DOWNSTATE MED CTR, 73- *Mem:* Am Soc Immunologists (pres, 53); Soc Exp Biol & Med; Am Soc Microbiol; Harvey Soc. *Res:* Infectious diseases; filterable viruses and rickettsia; variation of influenza virus. *Mailing Add:* 140 Main St East Hampton NY 11937

MAGIN, RALPH WALTER, b Belleville, Ill, Oct 22, 37; m 61; c 2. ORGANIC CHEMISTRY, POLYMER CHEMISTRY. *Educ:* Univ Ill, BS, 59; Mass Inst Technol, PhD(org chem), 63. *Prof Exp:* Assignment, Univ Ariz, 63-65; sr res chemist, Org Chem Div, 66-68 & Monsanto Indust Chem Co, Mo, 68-75, res specialist, Monsanto Polymers & Petrochem, 75-78, sr res specialist, Monsanto Agr Prod Co, 78-80, group leader, 80-82, mgr, Formulations, 82-86, MGR, PROCESS DEVELOP, Monsanto Co, 86- *Mem:* Am Chem Soc. *Res:* Polyelectrolytes, both bioactive and watersoluble, and their effect on environment; polyester condensation polymers; encapsulation of agricultural products; agric formulations; process development. *Mailing Add:* Ethyl Corp PO Box 14799 Baton Rouge LA 70820

MAGINNES, EDWARD ALEXANDER, b Ottawa, Ont, Can, Apr 19, 33; m 64; c 3. HORTICULTURE. *Educ:* McGill Univ, BS, 56; Cornell Univ, MS, 60, PhD(floricult), 64. *Prof Exp:* Res officer horticult, Can Dept Agr, 56-60; from asst prof to assoc prof, 64-77, EXTEN SPECIALIST & PROF HORT SCI, UNIV SASK, 77- *Mem:* Am Soc Hort Sci. *Res:* Culture of horticultural crops; influence of photoperiod and temperature on flowering, light quality and plant growth; nutrition of floriculture crops; growth retardants; moisture stress and plant growth; protected cultivation using waste heat. *Mailing Add:* Dept of Hort Univ of Sask Saskatoon SK S7N 0W0 Can

MAGISON, ERNEST CARROLL, b Ft Lauderdale, Fla, Oct 15, 26; m 83; c 6. ELECTRICAL ENGINEERING. *Educ:* Tufts Univ, BS, 48. *Prof Exp:* Prod engr, 48-52, supvr prod develop, 52-74, MGR REGULATORY AFFAIRS, PROCESS CONTROL DIV, HONEYWELL INC, 74- *Concurrent Pos:* Adj prof dept elec eng, Eve Col, Drexel Univ, 52-; comt hazardous mat, Nat Res Coun, 68-74, comt eval indust hazards, 75- *Honors & Awards:* Standards & Practices Award, Instrument Soc Am, 74. *Mem:* Fel Instrument Soc Am (former vpres). *Res:* Electrical safety and standards; radiation pyrometry. *Mailing Add:* Honeywell Inc 1100 Virginia Dr Ft Washington PA 19034

MAGLACAS, A MANGAY, NURSING. *Educ:* Vanderbilt Univ, BSN; Univ Minn, MPH; Johns Hopkins Univ, DPH. *Hon Degrees:* DSc, Univ Ill, 87. *Prof Exp:* CONSULT & TEACHING PROF, 89- *Concurrent Pos:* Rockefeller fel, 64-67; Fulbright-Smith-Mundt scholar, 52-54; adj prof, Col Nursing, Univ Ill, Chicago; mem bd dirs, Int Nursing Found, Japan; mem bd dirs, Int Coun Nurses, 89-; various consult, adv serv & vis prof, US, Philippines, Thailand, Papua, New Guinea, Japan, USSR & Australia. *Mem:* Foreign assoc Inst Med-Nat Acad Sci. *Mailing Add:* 59 Chemin de Planta Cologny 1223 Switzerland

MAGLEBY, KARL LEGRANDE, b Provo, Utah. NEUROPHYSIOLOGY, BIOPHYSICS. *Educ:* Univ Utah, BS, 66; Univ Wash, PhD(physiol, biophys), 70. *Prof Exp:* NIH neurophys training grant, Univ Wash, 70-71; from asst prof to assoc prof, 71-80, PROF PHYSIOL & BIOPHYS, SCH MED, UNIV MIAMI, 80- *Concurrent Pos:* NIH res grants, Univ Miami, 72- *Mem:* Biophysical Soc; Soc Neurosci; Soc Gen Physiologists; Fedn Am Soc Exp Biol. *Res:* Synaptic transmission; mechanism of transmitter release; transmitter-receptor interaction; ion channels. *Mailing Add:* Dept Physiol & Biophys Univ Miami Sch Med Miami FL 33101

MAGLICH, BOGDAN, b Sombor, Yugoslavia, Aug 5, 28, US citizen; m; c 5. EXPERIMENTAL NUCLEAR PHYSICS, PARTICLE & PARTICLE BEAM PHYSICS. *Educ:* Univ Belgrade, dipl physics, 50; Univ Liverpool, MSc, 55; Mass Inst Technol, PhD(physics), 59. *Prof Exp:* UNESCO fel, Mass Inst Technol, 55-56; res assoc, Lawrence Radiation Lab, Univ Calif, Berkeley, 59-62; sr staff mem, European Orgn Nuclear Res, Geneva, 62-67; vis prof physics, Univ Pa, 67-68; prof & prin investr high energy physics, Rutgers Univ, 69-74; chmn & chief scientist, United Sci, Inc, 74-76; chmn & chief scientist, Aneutromic Energy Lab, Inc, 77-88; CHMN & CHIEF SCIENTIST, ADVAN PHYSICS CORP, 89- *Concurrent Pos:* Dir, Nat Comput Analysts, 72-; vis prof, dept elec eng, Polytech Inst New York, 81-82; sci proj dir, King Abdulaziz Univ Res Ctr, Jeddah Saudi Arabia, 81-83; prin investr, Air Force Weapons Lab, 85-86, USAF Space Tech Ctr, 87-88; chmn, TESCA Found, 86- *Honors & Awards:* White House Citation for discovery of omega meson, President US, 61. *Mem:* Fel Am Phys Soc; Sigma Xi; foreign mem Serbian Acad Sci & Art, Yugoslavia. *Res:* Aneutronic fusion; non-radioactive nuclear process for production of electrical energy; author or coauthor of over 100 publications; pi-meson, nucleon and antinucleon scattering and polarization; experimental searches short lived mesons; missing-mass spectrometry and heavy bosons; self-colliding beams, self collider (migma cell), high energy (Mev) fusion and aneutronic power; numerous discoveries, inventions, research books, and educational books and videos. *Mailing Add:* 1050 University Tower 4199 Campus Dr Irvine CA 92715

MAGLIOLA-ZOCH, DORIS, b Hoboken, NJ; m 78; c 2. PRODUCTIVITY IMPROVEMENT, BUSINESS CONSULTANT. *Educ:* Montclair State, BA, 76; Fairleigh Dickinson Univ, MS, 79. *Prof Exp:* Mgr methods eng, Prentice Hall, 77-80; mgr, Weiner & Co, CPA's, 80-85; PRES, ZOCH & ZOCH PRODUCTIVITY CONSULT, 85- *Concurrent Pos:* Mem nat comt, Am Inst Cert Pub Accountants, 81-84; adj prof indust eng, Fairleigh Dickinson Univ, 82-84; lectr, numerous nat & int orgn, 82-; chmn bd gov, Inst Indust Engrs, 89-90. *Honors & Awards:* Int Outstanding Indust Engr Award, Inst Indust Engrs, 87, Outstanding Leadership Award, 87, Distinguished Serv Award, 88, Cert Appreciation Award, 89. *Mem:* Inst Indust Engrs; Inst Mgt Consult. *Res:* Industrial engineering techniques to improve productivity and efficiency for white collar and knowledge workers; techniques which traditionally are used in a manufacturing environment are applied in offices which are information and paper intensive to streamline and improve efficiency. *Mailing Add:* 710 Rte 46 E Fairfield NJ 07004

MAGLIULO, ANTHONY RUDOLPH, b Brooklyn, NY, July 23, 31; m 67; c 7. NUTRITION, BIOCHEMISTRY. *Educ:* Brooklyn Col, BS, 55; Long Island Univ, MS, 61; St John's Univ, PhD(microbiol), 68. *Prof Exp:* Lab technician, Maimonides Hosp, Brooklyn, 55-56; res microbiol, Armed Forces Inst Path, 56-58; res asst, New York Hosp, 58-60; microbiologist chg clin microbiol, Misericordia Hosp, 60-61 & 62-63; lectr microbiol, Queens Col, NY, 61-62; teaching asst biol, St John's Univ, 63-65; lectr & instr, 65-68, asst prof, 68-73, ASSOC PROF BIOCHEM, JOHN JAY COL CRIMINAL JUSTICE, 73- *Concurrent Pos:* Instr nursing sci, Roosevelt Hosp, 68-69; coordr nursing chem & lectr nursing sci, Hunter Col, 68-69; consult nutrit, 79- *Mem:* AAAS; fel Am Inst Chemists; Am Chem Soc; Nat Sci Teachers Asn. *Res:* Isolation of new and unique microbial lipids; nutrition and behavior. *Mailing Add:* Dept Sci John Jay Col 445 W 59th St New York NY 10019

MAGLIVERAS, SPYROS SIMOS, b Athens, Greece, Sept 6, 38; US citizen; m 62; c 2. COMPUTER SCIENCE, MATHEMATICS. *Educ:* Univ Fla, BEE, 61, MA, 63; Univ Birmingham, PhD(math), 70. *Prof Exp:* Teaching asst math, Univ Fla, 61-62, interim instr, 62-63; instr, Fla Presby Col, 63-64; systs analyst, Inst Social Res, Univ Mich, Ann Arbor, 65-68; res fel math, Univ Birmingham, 68-70; from asst prof to prof math, State Univ NY Col Oswego, 70-78; PROF MATH & COMPUT SCI, UNIV NEBR-LINCOLN, 78- *Concurrent Pos:* Consult, Nat Broadcasting Co, 67-68; consult, Ctr Human Growth & Develop, Univ Mich, Ann Arbor, 68, res assoc, 71; vis scholar, Univ Mich, 72; vis assoc prof math, State Univ NY, Binghamton, 76; vis prof math & Eng Res Coun sr res fel, Univ Birmingham, 84-85. *Mem:* Am Math Soc; Math Asn Am; Edinburgh Math Soc; London Math Soc; Asn Comput Mach. *Res:* Combinatorial designs, permutation groups, finite geometries; data encryption, data security systems in computers and communications. *Mailing Add:* Dept Math Univ Nebraska Lincoln NE 68588

MAGNANI, JOHN LOUIS, b New York, NY, Oct 20, 52; m; c 2. BIOCHEMISTRY. *Educ:* Wash Univ, St Louis, AB, 74; Princeton Univ, MA, 76, PhD, 79. *Prof Exp:* Staff fel, Sect Biochem, Lab Biochem Pharmacol, Nat Inst Arthritis, Diabetes & Digestive Kidney Dis, NIH, 79-82, sr staff fel, 82-85 & Lab Struct Biol, 85-86, res chemist, 86-88; sr scientist, 88-89, proj leader cancer bus area, 89-90, VPRES RES & DEVELOP BIOCHEM, BIOCARB INC, 90- *Concurrent Pos:* Lectr, Imp Cancer Res Fund, London, 83, Univ Goteborg, Sweden, 83 & 91, Tokyo Soda Chem Co, Tokyo Univ, Tokyo Inst Technol & Yamanashi Med Sch, Japan, 86, Inst Nazionale Per Lo Studio E La Cura Dei Tumori, Milan, Italy & Solavo SpA, Sienna, Italy, 87; Am Soc Biol Chemists & Fedn Am Socs Exp Biol young investr travel grant, 85; chairperson, Session Glycoconjugates, Soc Complex Carbohydrates, 87. *Mem:* Soc Complex Carbohydrates; Fedn Am Soc Exp Biol. *Res:* Author of more than 50 technical publications. *Mailing Add:* BioCarb Inc 300 Professional Dr Suite 100 Gaithersburg MD 20879

MAGNANI, NICHOLAS J, b Mason City, Iowa, Apr 11, 42; m 63; c 2. METALLURGY. *Educ:* Iowa State Univ, BS, 64, PhD(metall), 68. *Prof Exp:* Mem tech staff, 68-73, supvr, Chem Metall Div, 73-79, MGR CHEM & CERAMICS, SANDIA NAT LABS, 79- *Mem:* Nat Asn Corrosion Engrs; Am Soc Metals; Am Ceramics Soc; Mat Res Soc. *Res:* Corrosion and stress corrosion cracking emphasizing uranium and uranium alloys; corrosion resistant materials. *Mailing Add:* Orgn 2520 907 Warm Sand Dr S E Albuquerque NM 87123

MAGNANTI, THOMAS L, b Omaha, Nebr, Nov 7, 45; m 67; c 1. OPTIMIZATION. *Educ:* Syracuse Univ, BS, 67; Stanford Univ, MS, 69, MS, 71, PhD(opers res), 72. *Prof Exp:* From asst prof to prof opers res, 71-85, head mgt sci, 82-88, GEORGE EASTMAN PROF MGT SCI, SLOAN SCH MGT, MASS INST TECHNOL, 85-, CO-DIR OPERS RES CTR, 86- *Concurrent Pos:* Prin investr, Dept Transp, 75-79, NSF, 80-82 & United Parcel Serv, 83-86; fel, Ctr Opers Res & Econometrics, 76-77; assoc ed, J Appl Math, J Algebraic & Discreet Methods & Mgt Sci, 76-83; co-ed, Math Prog, 81-83; ed, Opers Res, 83-88; codir, Leaders for Mfg Prog, Mass Inst Technol, 88- *Mem:* Nat Acad Eng; Inst Mgt Sci; Math Prog Soc; Opers Res Soc Am (vpres, 87-88, pres, 88-). *Res:* Optimization of large scale systems; linear, integer and nonlinear programming; combinatorial network optimization; network design; communication systems; logistics; manufacturing; scheduling; transportation planning. *Mailing Add:* Rm E53-351 Sloan Sch Mgt 50 Memorial Dr Cambridge MA 02139

MAGNARELLI, LOUIS ANTHONY, b Syracuse, NY, Mar 27, 45; m 69. ENTOMOLOGY. *Educ:* State Univ NY Col Oswego, BS, 67; Univ Mich, MS, 68; Cornell Univ, PhD(med entom), 75. *Prof Exp:* Teacher biol, WGenesee Cent Sch Dist, NY, 68-71; res asst, Cornell Univ, 71-75, experimentalist, 75; asst scientist, 75-78, assoc scientist entom, 78-81, agr scientist, 81-87, CHIEF SCIENTIST, CONN AGR EXP STA, 88- *Mem:* Entom Soc Am; Am Mosquito Control Asn; Am Soc Trop Med & Hyg; Am Soc Rickettsiologists; Am Soc Microbiol. *Res:* Blood feeding, sugar feeding, and ovarian studies of mosquitoes, deer flies and horse flies; Rocky Mountain Spotted Fever; rickettsiae in ticks and antibodies in mammals; ecology and epidemiology of Lyme disease. *Mailing Add:* Dept of Entom Conn Agr Exp Sta PO Box 1106 New Haven CT 06504-1106

MAGNELL, KENNETH ROBERT, b Detroit, Mich, July 27, 38; c 2. INORGANIC CHEMISTRY. *Educ:* Wayne State Univ, BS, 62, MS, 66; Univ Minn, Minneapolis, PhD(inorg chem), 70. *Prof Exp:* ASST PROF CHEM, CENT MICH UNIV, 70- *Mem:* Am Chem Soc. *Res:* Non-aqueous solvent equilibrium studies. *Mailing Add:* Dept of Chem Cent Mich Univ Mt Pleasant MI 48858

MAGNESS, T(OM) A(LAN), b San Diego, Calif, Oct 22, 27; m 61, 67; c 5. ASTRONAUTICAL ENGINEERING, SYSTEMS DESIGN. *Educ:* Univ Calif, Los Angeles, AB, 49, MA, 51, PhD(math), 56. *Prof Exp:* Asst math, Exten Div, Univ Calif, Los Angeles, 49-51; asst physics, Scripps Inst, Univ Calif, 51-53; mem tech staff, Ramo-Wooldridge Corp, 56-60, head sect guide anal, Space Tech Labs, 60-63, assoc mgr, Space Guid Dept, 63-64, mgr, Space Guid Dept, TRW Systs Group, Tex, 64-66, asst mgr, Systs Anal Lab, TRW Systs, 66-69, sr staff engr, 69-73, PROJ MGR, TRW SYSTS, 73- *Mem:* Inst Elec & Electronics Engrs. *Res:* Stochastic processes and noise theory applied to underwater sound; statistical analysis of geophysical observations; guidance and navigation of spacecraft in presence of noise; computer sciences and software systems. *Mailing Add:* TRW Electronics & Defense One Space Park R 2/1020 Redondo Beach CA 90278

MAGNIEN, ERNEST, b Ger, Mar 28, 25; nat US; m 49; c 2. ORGANIC CHEMISTRY. *Educ:* City Col New York, BS, 49; Polytech Inst Brooklyn, MS, 58. *Prof Exp:* Org chemist, Gane & Ingram, 49-53; res chemist, Burroughs-Wellcome, Inc, 53-60; sr res chemist, USV Pharmaceut Corp, 60-80, asst res fel, Revlon Health Care Group, Revlon Inc, 80-; RETIRED. *Mem:* Am Chem Soc. *Res:* Isolation of alkaloids and natural products; pharmaceutical compounds; organic synthesis; heterocyclic chemistry; nuclear chemistry. *Mailing Add:* PO Box 1045 Norwich VT 05055

MAGNIN, ETIENNE NICOLAS, ichthyology; deceased, see previous edition for last biography

MAGNO, MICHAEL GREGORY, b Newark, NJ, Aug 9, 42; m 65; c 3. PHYSIOLOGY. *Educ:* Rutgers Univ, New Brunswick, BS, 64, PhD(physiol), 69. *Prof Exp:* Instr physiol, Albany Med Col, 68-73; res fel, cardiovasc-pulmonary div, Hosp Univ Pa, 73-75, res assoc, 75-; ASST PROF, DEPT PHYSIOL & BIOPHYS, HAHNEMANN UNIV, PHILADELPHIA. *Mem:* Am Physiol Soc. *Res:* Capillary exchange in the lung; cardio-pulmonary adjustments to hypoxic in birds and mammals; bronchial circulation. *Mailing Add:* Dept Surg Thomas Jefferson Univ Col Bldg Rm 607 Philadelphia PA 19107

MAGNO, RICHARD, b Newark, NJ, May 5, 44; m 68; c 1. SOLID STATE PHYSICS. *Educ:* Stevens Inst Technol, BS, 66; Rutgers Univ, PhD(physics), 74. *Prof Exp:* fel physics, Univ Alta, 74-79; RES PHYSICIST, NAVAL RES LABS, 79- *Mem:* Sigma Xi; Am Phys Soc. *Res:* Inelastic electron tunneling is being used to study metal-insulator-metal junctions whose barriers contain organic molecules or were formed in an organic vapor glow discharge. *Mailing Add:* 4111 Marbourne Dr Ft Washington MD 20744

MAGNUS, ARNE, b Oslo, Norway, Aug 17, 22; nat US; m 50; c 3. MATHEMATICS. *Educ:* Univ Oslo, Cand Real, 52; Washington Univ, PhD(math), 53. *Prof Exp:* Instr math, Univ Kans, 52-54; asst prof, Univ Nebr, 54-56; asst prof, Univ Colo, 56-65, prof, 65-66; PROF MATH, COLO STATE UNIV, 66- *Mem:* Math Asn Am; Am Math Soc. *Res:* Analytic function of one and several variables; analytic theory of continued fractions; iteration. *Mailing Add:* Dept Math Colo State Univ Ft Collins CO 80523

MAGNUS, GEORGE, b Ganister, Pa, Jan 16, 30; m 54; c 4. PLASTICS CHEMISTRY. *Educ:* Franklin & Marshall Col, BS, 52; Univ Pittsburgh, PhD(org chem), 56. *Prof Exp:* Chemist, Union Carbide Chem Co, 56-71; chemist, 71-74, technol mgr urethanes, 74-81, SR RES CHEMIST, STEPAN CO, 81- *Mem:* Am Chem Soc. *Res:* Applied research and product development in rigid urethane foams; solid and microcellular urethane elastomers; applied research in spandex fibers; development of new poly-e-caprolactone and polyadipate polyols for solid and microcellular urethane elastomers; development of class I and class II rated rigid urethane foams; development of new polyols for class I and class II rated rigid urethane foams. *Mailing Add:* Edens & Winnetka Stepan Co Northfield IL 60093

MAGNUSON, EUGENE ROBERT, b Emerson, Nebr, Dec 5, 33; m 60. ORGANIC CHEMISTRY. *Educ:* Univ Nebr, BS, 55, MS, 58; Kans State Univ, PhD(org chem), 66. *Prof Exp:* Res chemist, Standard Oil Co Div, Am Oil Co, Ind, 60 & Allis-Chalmers Mfg Co, Wis, 63-69; PROF CHEM, MILWAUKEE SCH ENG, 69- *Mem:* Am Chem Soc; Soc Plastics Eng. *Res:* Insulation of electrical components as applied to motors, generators and transformers by encapsulation using various thermal and electrical resin systems. *Mailing Add:* 4725 N 147th St Brookfield WI 53005-1610

MAGNUSON, GUSTAV DONALD, b Chicago, Ill, Aug 22, 26; m 50; c 4. SOLID STATE PHYSICS, ATOMIC PHYSICS. *Educ:* Univ Chicago, PhB, 49, BS, 50; Univ Ill, MS, 52, PhD(physics), 57. *Prof Exp:* Asst, Univ Ill, 53-57; sr staff scientist, Gen Dynamics-Convair, Calif, 57-66; res assoc prof aerospace eng & eng physics, Univ Va, 66-69; res scientist, Atomic Physics Lab, Gulf Energy & Environ Systs, 69-73; res physicist, IRT Corp, 73-76; staff scientist, Gen Dynamics-Convair, 77-87; VPRES RES & DEVELOP, ISM TECHNOL,INC, 87- *Concurrent Pos:* Asst, Anderson Phys Lab, Ill, 53-55; instr, San Diego Community Col, 72- *Mem:* AAAS; Am Phys Soc; Am Asn Physics Teachers. *Res:* Molecular physics; radiation damage; surface-particle interactions; atomic collisions; large superconducting magnet systems for fusion research; radar cross-section calculations. *Mailing Add:* 1755 Catalina Blvd San Diego CA 92107

MAGNUSON, HAROLD JOSEPH, b Halstead, Kans, Mar 31, 13; m 35; c 2. OCCUPATIONAL MEDICINE. *Educ:* Univ Southern Calif, AB, 34, MD, 38; Johns Hopkins Univ, MPH, 42; Am Bd Prev Med, dipl. *Prof Exp:* Intern med & surg, Los Angeles County Hosp, 37-39; instr internal med, Sch Med, Univ Southern Calif, 39-41; asst surgeon, USPHS, 41-44, sr asst surgeon, 44-46, surgeon, 47-49, sr surgeon, 49-53, med dir, 53-62, chief oper res sect venereal dis prog, 55-56, chief div occup health, 56-62; dir inst indust health & chmn dept, Sch Pub Health, 62-69, prof indust health, Sch Pub Health & Prof Internal Med, Med Sch, 62-76, assoc dean sch pub health, 69-76, EMER PROF ENVIRON & INDUST HEALTH, SCH PUB HEALTH, UNIV MICH, 76- *Concurrent Pos:* Spec consult, USPHS, 40-41; instr, Johns Hopkins Univ, 43-45; res prof, Univ NC, 45-55; spec lectr, George Wash Univ, 59-62; vpres, Permanent Comn & Int Asn Occup Health; vchmn occup med, Am Bd Prev Med. *Honors & Awards:* Bronze Hektoen Medal, AMA, 55; William S Knudsen Award, Indust Med Asn, 70. *Mem:* Fel AAAS; fel Am Col Physicians; fel Am Pub Health Asn; fel AMA. *Res:* Public health; internal medicine. *Mailing Add:* 12305 Fernando Dr San Diego CA 92128

MAGNUSON, JAMES ANDREW, biochemistry, biophysics; deceased, see previous edition for last biography

MAGNUSON, JOHN JOSEPH, b Evanston, Ill, Mar 8, 34; m 59; c 2. HYDROBIOLOGY, MARINE SCIENCES. *Educ:* Univ Minn, BS, 56, MS, 58; Univ BC, PhD(zool), 61. *Prof Exp:* Chief tuna behav prog, Biol Lab, Bur Commercial Fisheries, US Fish & Wildlife Serv, 61-67; from asst prof to assoc prof, 68-74, chmn oceanog & limnol grad prog, 78-83 & 86, PROF ZOOL & DIR TROUT LAKE BIOL STA, UNIV WIS-MADISON, 74-, CTR LIMNOL DIR, 82- *Concurrent Pos:* Mem affil grad fac, Univ Hawaii, 63-67; prog dir ecol, NSF, 75-76; mem, Adv Comt on Marine Resources Res, Fleet Admin Off, 81-85, Ocean Policy Comt, Nat Acad Sci, 79, Fisheries Task Group, Chmn, 81-83, Sci Adv Comt to Great Lakes Fishery Comn, 76-81, bd tech exports chmn, 81. *Mem:* Am Fisheries Soc (pres, 81-82); Am Soc Ichthyol & Herpet; Animal Behav Soc; Ecol Soc Am; Am Inst Biol Sci. *Res:* Behavioral ecology of fishes; locomotion of scombrids; distributional ecology of fishes and macroinvertebrates in thermal fronts or gradients; comparative studies of factors determining community structure in lakes; ecology of Great Lakes; long term ecological research on northern lake ecosystems; ecology; fish and wildlife sciences. *Mailing Add:* Dept Zool Ctr Limnol Univ Wis Madison WI 53706

MAGNUSON, NANCY SUSANNE, b Chicago, Ill, Dec 17, 42; m 64. IMMUNE REGULATION. *Educ:* Univ Calif, Los Angeles, BS, 65; Wash State Univ, PhD(immunol), 78. *Prof Exp:* Res assoc, 78-81, asst prof, 81-86, ASSOC PROF IMMUNOL, WASH STATE UNIV, 86- *Mem:* Am Asn Immunol; Am Asn Vet Immunol; Am Soc Biol Chem; Am Soc Microbiol; Soc Intestinal Microbiol Ecol & Dis. *Res:* Regulation of cell-mediated immune response. *Mailing Add:* Rte 1 Box 431A Pullman WA 99163

MAGNUSON, VINCENT RICHARD, b Laurel, Nebr, May 5, 42; m 62. INORGANIC CHEMISTRY, CRYSTALLOGRAPHY. *Educ:* Univ Nebr, BS, 63; Univ Wis, MS, 65; Univ Ill, PhD(chem), 68. *Prof Exp:* ASSOC PROF CHEM, UNIV MINN, DULUTH, 68- *Mem:* Am Chem Soc; Am Crystallog Asn; Sigma Xi. *Res:* Structural properties of organometallic complexes of Group II and III metals; x-ray crystallography; computer modeling; chemical information storage and retrieval. *Mailing Add:* 1751 Carver Duluth MN 55812

MAGNUSON, WINIFRED LANE, b Brady, Tex, Oct 12, 35; m 58. INORGANIC CHEMISTRY. *Educ:* McMurry Col, AB, 59; Univ Kans, PhD(chem), 63. *Prof Exp:* Asst prof chem, McMurry Col, 63-69; CHMN DEPT CHEM, KY WESLEYAN COL, 69-, PROF CHEM, 80- *Concurrent Pos:* Petrol Res Fund grant, 63-64; Res Corp grant, 65-66; Robert A Welch Found grant, 66-69. *Mem:* Am Chem Soc. *Res:* Inorganic reactions in molten salts; reaction of metal carbonyls. *Mailing Add:* Dept of Chem Ky Wesleyan Col 3000 Frederica St Owensboro KY 42301

MAGNUSSON, LAWRENCE BERSELL, physical inorganic chemistry, for more information see previous edition

MAGNUSSON, PHILIP C(OOPER), b Seattle, Wash, Jan 18, 17; m 45; c 3. ELECTRICAL ENGINEERING. *Educ:* Univ Wash, BS, 37, EE, 47; Univ Calif, MS, 38; Mass Inst Technol, ScD(elec eng), 41. *Prof Exp:* Asst elec eng, Univ Calif, 37-38; asst elec engr, US Naval Ord Lab, Wash, 41-43 & Bonneville Power Admin, Ore, 46; from asst prof to prof, 46-83, EMER PROF ELEC ENG, ORE STATE UNIV, 83- *Concurrent Pos:* Elec engr, Bonneville Power Admin, 65-81. *Mem:* Inst Elec & Electronics Engrs; Am Soc Eng Educ; Sigma Xi. *Res:* Transient stability of electric power systems; numerical solution of differential equations and field problems; traveling waves on transmission systems. *Mailing Add:* Dept of Elec Eng Ore State Univ Corvallis OR 97331

MAGO, GYULA ANTAL, b Budapest, Hungary, Aug 16, 38. COMPUTER SCIENCE. *Educ:* Budapest Tech Univ, dipl elec eng, 62; Cambridge Univ, PhD(comput sci), 70. *Prof Exp:* Asst prof control eng, Budapest Tech Univ, 62-68; asst prof, 70-80, PROF COMPUT SCI, UNIV NC, CHAPEL HILL, 80- *Mem:* Asn Comput Mach. *Res:* Computer architecture; parallel computation. *Mailing Add:* Dept Comput Sci Univ NC 213 New West 035a Chapel Hill NC 27514

MAGOON, LESLIE BLAKE, III, b San Jose, Calif, Aug 15, 41; div; c 1. GEOLOGY, ORGANIC GEOCHEMISTRY. *Educ:* Univ Ore, BS, 64, MS, 66. *Prof Exp:* Geologist explor, Shell Oil Co, 66-74; GEOLOGIST, US GEOL SURV, 74- *Concurrent Pos:* Distinguished lectr, Am Asn Petrol Geologists, 89-90. *Mem:* Sigma Xi; Am Asn Petrol Geologists; Europ Asn Org & Geochemists. *Res:* Origin, migration, and accumulation of oil and gas; research stresses Alaskan geology and organic geochemistry of petroleum. *Mailing Add:* US Geol Surv 345 Middlefield Rd MS 999 Menlo Park CA 94025

MAGORIAN, THOMAS R, b Charleston, WVa, Nov 26, 28; m 50; c 3. GEOPHYSICS. *Educ:* Univ Chicago, PhD(geol), 52. *Prof Exp:* Res geologist, Ohio River Div Labs, US Corps Engrs, 50-52, Pure Oil Co, 52-53 & Shell Develop Co, 56-62; consult petrol explor, Tex, 62-63; prin geophysicist, Cornell Aeronaut Lab, 63-73; CHIEF GEOPHYSICIST, ECOL & ENVIRON, INC, 73- *Concurrent Pos:* Mem joint environ effects prog, Environ Characterization Comt, US Dept Defense, 65-; consult, Ltd War Lab & Ballistics Res Lab, US Army, 65-; mem air staff, US Air Force, 66-; dir earth sci prog, Rosary Hill Col, 70-, prof earth sci, 71-74. *Mem:* Fel AAAS; Am Asn Petrol Geol; Am Geophys Union; Geol Soc Am; Ecol Soc Am. *Res:* Geomorphology; biophysics; stratigraphic seismology; hydrology; plant ecology and paleocology; operations research; origin of oil; geotechnical analysis of energy systems. *Mailing Add:* 133 South Dr Eggertsville NY 14226

MAGOSS, IMRE V, b Nagykata, Hungary, July 19, 19; US citizen; m 44; c 2. UROLOGY. *Educ:* Pazmany Peter Univ, Hungary, MD, 43; Am Bd Urol, dipl, 60. *Prof Exp:* Assoc cancer res urologist, Roswell Park Mem Inst, 55-60; from asst prof to assoc prof urol surg, 60-71, prof surg, 71-76, PROF UROL, STATE UNIV NY BUFFALO, 76- *Mem:* AMA; Am Urol Asn; fel Am Col Surg. *Res:* Histochemical changes of prostatic carcinoma during progression. *Mailing Add:* Dept Urol 462 Grider St BB State Univ NY Health Sci Ctr 3435 Main St Buffalo NY 14214

MAGOUN, HORACE WINCHELL, neuroanatomy; deceased, see previous edition for last biography

MAGOVERN, GEORGE J, BIOMEDICAL ENGINEERING. *Educ:* Marquette Univ, MD, 47. *Hon Degrees:* Dr, Med Col Wis, 87, Duquesne Univ, 90. *Prof Exp:* CLIN ASSOC PROF MED, SCH MED, UNIV PITTSBURGH, 69-; CHMN & PROF SURG, MED COL PA, ALLEGHENY CAMPUS. *Concurrent Pos:* Sr staff mem, Allegheny Gen Hosp, chmn, Dept Surg; emer staff Presby-Univ Hosp; courtesy staff, St Margaret Mem Hosp; consult staff, Forbes Health Syst, Shadyside Hosp & St Francis Gen Hosp; mem, Evarts P Graham Mem Traveling fel comt, Am Asn Thoracic Surg, 68-71, postgrad educ comt, Soc Thoracic Surgeons, 74-77, house of deleg, AMA, 79-83, Standards & Ethics Comt, Am Asn Thoracic Surg, 80-83, Coordr Comt Perfusion Affairs, 81-83; bd dirs, Blue Cross of Western Pa, 76-82; adv bd, Am Soc Extracorporeal Technol, Soc Thoracic Surgeons Coun, Am Asn Thoracic Surg, 78-82; dir, Am Bd Thoracic Surg, AMA, 84-90; bd govs, Am Col Surgeons, 90, AIDS comt, 90. *Mem:* AAAS; Am Asn Thoracic Surg; Am Col Angiol; Am Col Cardiol; Am Col Chest Physicians; Am Col Physicians; AMA; Am Soc Artificial Internal Organs; Am Soc Extra-Corporeal Technol; Am Surg Asn. *Res:* Thoracic and cardiovascular surgery. *Mailing Add:* Dept Surg Allegheny Gen Hosp Pittsburgh PA 15212

MAGRATH, IAN, b Jurby, UK, Oct 31, 44; m 70; c 3. PEDIATRIC ONCOLOGY, NON-HODGKINS LYMPHOMA. *Educ:* Univ London, MBBS, 67, MRCP, 70, MRCPath, 85, FRCP, 86, FRCPath, 87. *Prof Exp:* Sr house officer, Hammersmith Hosp, London, 68-69; res fel, Kennedy Inst, London, 69-70; dir, Lymphoma Treatment Ctr, Kampala, Uganda, 70-73; SR INVESTR, PEDIAT BR, NAT CANCER INST, NIH, 74-, HEAD, SECT LYMPHOMA BIOL, 87- *Concurrent Pos:* Vis prof, Univ Madras, India, 84- *Honors & Awards:* Gov's Clin Gold Medal, 67. *Mem:* Am Soc Clin Oncol; Am Asn Cancer Res. *Res:* Clinical studies in non-Hodgkin's lymphomas in young people; investigation of molecular pathogenesis of Burkitt's lymphoma; molecular epidemiology of non-Hodgkin's lymphomas; oncology in developing countries. *Mailing Add:* Div Cancer Treatment Nat Cancer Inst NIH Bldg 10 Rm 13N240 Bethesda MD 20892

MAGRUDER, WILLIS JACKSON, b Lentner, Mo, Aug 7, 35; m 54; c 3. CHEMISTRY, SCIENCE EDUCATION. *Educ:* Northeast Mo State Col, BS, 57; State Col Iowa, MA, 61; Colo State Col, EdD(chem, sci educ), 66. *Prof Exp:* High sch teacher, Mo, 58-60; instr gen & org chem, Fullerton Jr Col, 61-64; from asst prof to assoc prof sci educ, 64-67, PROF SCI & SCI EDUC, NORTHEAST MO STATE COL, 67- *Concurrent Pos:* Mem sci curric comt, Sec Sch Chem & Phys, State Dept Educ, 65-67; Danforth Assoc, 68-; dir, Northeast Mo Regional Sci Fair, 69; mem, Adv Comt Kirksville R-III Bd Educ, 77-78; pres, Sci Teachers Mo, 78-80. *Mem:* Am Chem Soc; Nat Sci Teachers Asn. *Res:* Organolithium chemistry. *Mailing Add:* Dept of Sci Northeast Mo State Univ Kirksville MO 63501

MAGUDER, THEODORE LEO, JR, b Meriden, Conn, Oct 14, 39; m 64; c 4. PLANT ECOLOGY, FRESHWATER BIOLOGY. *Educ:* Fairfield Univ, BS, 61; St John's Univ, MS, 63; State Univ NY Col Forestry, Syracuse Univ, PhD(forest zool), 68. *Prof Exp:* Educ dir Environ Ctr, Great Mt Forest, 73-79; ASSOC PROF BIOL, UNIV HARTFORD, 68-, ASST CHMN, DEPT BIOL, 87- *Concurrent Pos:* NSF Col Sci Improv Prog grant prin investr, Woodcock Study, Conn, 70 & US Fish & Wildlife Serv, 71-73. *Mem:* Wildlife Soc; Sigma Xi; Am Inst Biol Sci. *Res:* Wildlife biology including the study of woodchuck, woodcock, crow and seagull population dynamics; development of environmental education programs in ecology and field biology; succession in riparian habitats of Wind Cave National Park, Hot Springs South Dakota; effects of snowmelt upon pH of pond and stream ecosytems after winter of acid precipitation in central Connecticut; vegetation of Fort Phil, Kearny, Wyoming; sand dune and kettle pond ecology, Cape Cod; American Western History. *Mailing Add:* 3 Orchard Lane Windsor CT 06095-3402

MAGUE, JOEL TABOR, b New Haven, Conn, Nov 23, 40; m 64; c 2. INORGANIC CHEMISTRY, ORGANOMETALLIC CHEMISTRY. *Educ:* Amherst Col, BA, 61; Mass Inst Technol, PhD(chem), 65. *Prof Exp:* NIH fel, Imp Col, Univ London, 65-66; from asst prof to assoc prof, 66-81, PROF CHEM, TULANE UNIV, 81- *Mem:* Am Chem Soc; Royal Soc Chem. *Res:* Organometallic complexes of the platinum metals; synthetic and structural studies. *Mailing Add:* Dept of Chem Tulane Univ New Orleans LA 70118

MAGUIRE, BASSETT, systematic botany; deceased, see previous edition for last biography

MAGUIRE, BASSETT, JR, b Birmingham, Ala, Aug 30, 27; m 50; c 2. ECOLOGY. *Educ:* Cornell Univ, AB, 53, PhD(zool), 57. *Prof Exp:* From instr to assoc prof, 57-78, PROF ZOOL, UNIV TEX, AUSTIN, 78- *Concurrent Pos:* Ed, Ecol Modeling & Eng, 74- *Mem:* AAAS; Ecol Soc Am; Am Soc Limnol & Oceanog; Am Micros Soc; Am Soc Naturalists. *Res:* Mechanisms of community structure determination; internal niche structure; ecological and ecosystem analysis; physiological ecology; ecology of closed systems. *Mailing Add:* Dept of Zool Univ of Tex Austin TX 78712

MAGUIRE, GERALD QUENTIN, JR, b Indiana, Pa, Jan 6, 55; m 87. SPECIAL PURPOSE PROCESSORS, DISTRIBUTED COMPUTING. *Educ:* Indiana Univ, Pa, BA, 75; Univ Utah, MS, 81, PhD (comput sci), 83. *Prof Exp:* asst prof, 83-89, ASSOC PROF COMPUT SCI, COLUMBIA UNIV, 89- *Concurrent Pos:* Vis comput scientist, NY Univ Med Ctr, 76-; res collabr, Brookhaven Nat Lab, 81-87; invited lectr, Leiden State Univ, Neth, 86; US govt & Swed Fulbright-Hayes Comn, 83-84. *Mem:* Inst Elec & Electronic Engrs; Asn Comput Mach; Am Physics Soc. *Res:* Development of tools and environments for non-textual data processing (e.g. images, graphics, symbolic computation); particularly picture archiving, communication systems and portable standard LISP. *Mailing Add:* Dept Comput Sci Columbia Univ 450 Comput Sci Bldg New York NY 10027-7031

MAGUIRE, HELEN M, BLOOD FLOW REGULATION IN PLACENTA. *Educ:* Univ New South Wales, Sidney, Australia, PhD(biol org chem), 57. *Prof Exp:* PROF PHARMACOL, COL HEALTH SCI, UNIV KANS, 83- *Mailing Add:* Dept Pharmacol Univ Kans 39th & Rainbow Blvd Kansas City KS 66103

MAGUIRE, HENRY C, JR, b New York, NY, May 4, 28; m 53; c 3. IMMUNOLOGY, DERMATOLOGY. *Educ:* Princeton Univ, BA, 49; Columbia Univ, 49-50; Univ Chicago, MD, 54; Univ Pa, dipl, 61. *Prof Exp:* Asst instr dermat, Sch Med, Univ Pa, 58-61, instr, 61, assoc, 61-64, asst prof, 64-67, asst instr, Div Grad Med, 58-61, instr, 61-63, assoc, 63-65, asst prof, 65-67; assoc prof med, 67-75, assoc prof microbiol, 69-75, PROF MICROBIOL, IMMUNOL & MED, HAHNEMANN MED COL, 75- *Concurrent Pos:* Guest investr, Rockefeller Univ, 66-67; chief investr, NIH res grant, 68-; investr, Inst Cancer Res, 72-75. *Mem:* AAAS; Am Fedn Clin Res; fel Am Col Physicians; Am Asn Immunologists; Am Acad Dermat. *Res:* Delayed hypersensitivity; tumor immunology; immunological adjuvants; hair growth. *Mailing Add:* Thomas Jefferson Univ Hosp 111 S 11th St Suite 4250 Philadelphia PA 19107

MAGUIRE, JAMES DALE, b Chelan, Wash, Sept 16, 30; m 55; c 4. AGRONOMY. *Educ:* Wash State Univ, 52; Iowa State Univ, MS, 57; Ore State Univ, PhD(crops), 68. *Prof Exp:* Settler assistance agt irrig eng, US Bur Reclamation, 54-56; instr seed technol, Wash State Univ, 57-66; res assoc seed physiol, Ore State Univ, 66-67; assoc prof, 67-77, PROF AGRON, WASH STATE UNIV, 77- *Concurrent Pos:* Wash Crop Improv Asn fel, Wash State Univ, 68-70; vis prof, Univ Nottingham, 71-, Lisbon & Nat Ctr Sci Res, Paris, 79- & Warsaw, 81; consult, Indian Agr Res Conf, 79- *Mem:* Sigma Xi; Am Soc Agron; Asn Off Seed Analysts. *Res:* Seed physiology studies dealing with seed vigor evaluation and determination of metabolic processes involved in dormancy, germination and seedling growth. *Mailing Add:* 291 Johnson Hall Dept of Agron Wash State Univ Pullman WA 99164

MAGUIRE, MARJORIE PAQUETTE, b Pearl River, NY, Sept 2, 25; m 50; c 2. CYTOGENETICS. *Educ:* Cornell Univ, BS, 47, PhD(cytol), 52. *Prof Exp:* Instr bot, Cornell Univ, 52-53; res assoc, Genetics Found, 57-60, res scientist, 60-75, assoc prof, 75-81, PROF ZOOL, UNIV TEX, AUSTIN, 81- *Concurrent Pos:* NIH career develop award, 65 & 70. *Mem:* Genetics Soc Am; Am Genetic Asn; Bot Soc Am; Am Soc Cell Biol; Int Soc Plant Molecular Biol. *Res:* Meiotic mechanisms of homologue pairing; crossing over and disjunction. *Mailing Add:* Dept Zool Univ Tex Austin TX 78712

MAGUIRE, MILDRED MAY, b Leetsdale, Pa, May 7, 33. SYNTHETIC INORGANIC & ORGANOMETALLIC CHEMISTRY. *Educ:* Carnegie-Mellon Univ, BS, 55; Univ Wis, MS, 60; Pa State Univ, PhD(chem), 67. *Prof Exp:* Chemist, Koppers Co, 55-58 & Am Cyanamid Co, 60-63; PROF CHEM, WAYNESBURG COL, 67- *Concurrent Pos:* Leverhulme vis res fel, Univ Leicester, Eng, 80-81; Oak Ridge fac res partic, US Dept Energy, Pittsburgh, 79-80 & 82-85; Leverhulme res fel & AAUW Curie fel, UK, 80-81; vis prof, Univ Leicester, Eng, 89. *Mem:* Am Chem Soc; Am Asn Univ Profs; Am Asn Univ Women. *Res:* Electron spin resonance of transition metal complexes and charge transfer complexes, magnetic resonance of coal liquids, heteroatom nuclear magnetic resonance and solid state nuclear magnetic resonance, radiolysis and photolysis of transition metal complexes. *Mailing Add:* Dept Chem Waynesburg Col Waynesburg PA 15370

MAGUIRE, ROBERT JAMES, b Saskatoon, Sask, Sept 13, 46; m 71; c 1. PHYSICAL CHEMISTRY, BIOCHEMISTRY. *Educ:* Univ Ottawa, BSc, 68; Univ Alta, PhD(chem), 72. *Prof Exp:* Fel, Univ Ottawa, 72-73; RES SCIENTIST CHEM, NAT WATER RES INST, CAN DEPT ENVIRON, 73- *Mem:* Chem Inst Can. *Res:* Environmental modeling of toxic substance distribution and transformation in aquatic systems; macromolecule and small molecule interactions. *Mailing Add:* 3274 Northgate Dr Burlington ON L7N 2N5 Can

MAGUIRE, YU PING, b Hanko, China, May 27, 47; m; c 1. CHEMOSENSITIVITY TEST FOR TUMOR PATIENTS, IMMUNOTHERAPY. *Educ:* Paterson State Col, NJ, BS, 74; Rutgers Univ, New Brunswick, NJ, MS, 76, PhD(food sci), 78. *Prof Exp:* Postdoctoral fel path, Univ Wash, Seattle, 78-79, hemat, Med Sch, 80-83; vis scientist med path, Univ Naples, Italy, 79-80; sr scientist oncol, Swedish Hosp, Tumor Inst, Seattle, Wash, 84-88; DIR TUMOR DIAG ONCOL, BARTELS DIV, BAXTER HEALTHCARE, ISSAQUAH, WASH, 88- *Concurrent Pos:* Consult, Oncogen, Div Bristol Meyers, Seattle, Wash, 85-88. *Mem:* AAAS; Am Asn Cancer Res; Women Cancer Res. *Res:* Generate an in vitro system and/or diagnostic test which serves as an indicator of patients with neoplastic diseases to conventional and investigational therapy; tumor biology; anti-neoplastic reagents. *Mailing Add:* Bartels Div Baxter Healthcare 2005 NW Sammamish Rd Issaquah WA 98027

MAGWIRE, CRAIG A, b Meadow Grove, Nebr, July 9, 22; m 56; c 4. MATHEMATICAL STATISTICS. *Educ:* Nebr State Col, BA, 43; Univ Mich, MS, 47; Stanford Univ, PhD(math statist), 53. *Prof Exp:* Appl sci rep, Int Bus Mach, 53-55; assoc prof math, US Naval Postgrad Sch, 55-59; res mathematician, Aerojet Gen Corp, 59-65; dir comput ctr, Desert Res Inst, Univ Nev, 65-69; dir comput ctr, 69-73, prof math, 73-86, EMER PROF MATH, PORTLAND STATE UNIV, 86- *Concurrent Pos:* Mem bd dir, Hazelnut Growers of Oregon, 83- *Res:* Sequential decision theory. *Mailing Add:* 1560 Cherry Crest Ave Lake Oswego OR 97034

MAGYAR, ELAINE STEDMAN, b York, Maine, Feb 21, 46; m 72; c 3. ORGANIC CHEMISTRY. *Educ:* Mt Holyoke Col, AB, 68; Northwestern Univ, PhD(chem), 72. *Prof Exp:* Instr chem, Northwestern Univ, 72-73; lectr, Univ Mass, Amherst, 73-75; adj asst prof, Brandeis Univ, 77-78; asst prof, 78-81, ASSOC PROF CHEM, RI COL, 81- *Concurrent Pos:* Fel, Univ Mass, 73-75. *Mem:* Am Chem Soc; AAAS; Sigma Xi. *Res:* Organic reaction mechanisms; nuclear magnetic resonance spectroscopy. *Mailing Add:* 51 Forbes St Providence RI 02908

MAGYAR, JAMES GEORGE, b New Milford, Conn, Aug 29, 47; m 72; c 3. ORGANIC CHEMISTRY, PHOTOCHEMISTRY. *Educ:* Dartmouth Col, AB, 69; Northwestern Univ, PhD(chem), 74. *Prof Exp:* Res & teaching fel chem, Mt Holyoke Col, 73-75; asst prof, Hamilton Col, 75-77; adj asst prof, Brandeis Univ, 77-78; lectr, 78-80, asst prof, 80-88, ASSOC PROF CHEM, RI COL, 88- *Mem:* Am Chem Soc. *Res:* Mechanistic organic photochemistry; nuclear magnetic resonance spectroscopy. *Mailing Add:* Dept Phys Sci RI Col Providence RI 02908-1991

MAH, RICHARD S(ZE) H(AO), b Shanghai, China, Dec 16, 34; m 62; c 1. CHEMICAL ENGINEERING. *Educ:* Univ Birmingham, BSc, 57; Univ London, PhD(chem eng), 61; Imp Col, dipl, 61. *Prof Exp:* Jr chem engr, APV Co Ltd, Eng, 57-58; res fel chem eng, Univ Minn, 61-63; res engr, Tech Ctr, Union Carbide Corp, 63-67; group head & sr proj analyst, Esso Math & Systs, Inc, 67-72; assoc prof chem eng, 72-77, PROF CHEM ENG, NORTHWESTERN UNIV, EVANSTON, 77- *Concurrent Pos:* Consult, Argonne Nat Lab, 75-78, E I Du Pont de Nemours & Co, Inc, 81-89 & Shell Develop Co, 89-; secy, Cache Corp, 78-80, vpres, 82-84, pres, 84-86, trustee, 74- *Honors & Awards:* Comput in Chem Eng Award, Am Inst Chem Engrs, 81; Youden Prize, Am Soc Qual Control, 86; Thiele Award, Am Inst Chem Engrs, 90. *Mem:* fel Am Inst Chem Engrs; Am Chem Soc; Am Soc Eng Educ. *Res:* Computer-aided process planning, design and analysis. *Mailing Add:* Dept Chem Eng Northwestern Univ Evanston IL 60208-3120

MAH, ROBERT A, b Fresno, Calif, Oct 28, 32. MICROBIOLOGY. *Educ:* Univ Calif, Davis, AB, 57, MA, 58, PhD(microbiol), 63. *Prof Exp:* Asst prof biol, San Fernando Valley State Col, 62-64; from asst prof to assoc prof microbiol, Univ NC, Chapel Hill, 64-71; PROF ENVIRON MICROBIOL, UNIV CALIF, LOS ANGELES, 71- *Mem:* AAAS; Am Soc Microbiol; Soc Protozoologists. *Res:* Physiology and metabolism of anaerobic bacteria and protozoa; microbial physiology and ecology of anaerobic habitats, especially methane producing bacteria; cycling of matter by microbes. *Mailing Add:* Div Environ Health Sci Univ of Calif Sch of Pub Health Los Angeles CA 90024-1772

MAHADEVA, MADHU NARAYAN, b Kallidaikurichi, India, Apr 27, 30; US citizen; m 57; c 2. ZOOLOGY. *Educ:* Univ Madras, MA, 51, MS, 52; Univ Calif, Los Angeles, PhD(zool), 56. *Prof Exp:* Lectr biol & head dept, Harward Col, Ceylon, 50-51; lect & head dept, Dharmaraja, Kandy & Zahira Cols, Ceylon, 52-53; res officer, Fisheries Res Sta, Govt of Ceylon, 53-61; head dept biol, Jamshedpur Coop Col, India, 61-67, Govt of India Univ Grants Comn fac res grant, 66-67; actg assoc prof zool & Acad Senate res grant, Univ Calif, Los Angeles, 67-68; assoc prof, 68-76, PROF BIOL, UNIV WIS-OSHKOSH, 76- *Concurrent Pos:* Fulbright Smith-Mundt, US Govt, 53-56; teaching asst, Univ Calif, Los Angeles, 54-55; hon head dept biol, Navalar Hall, Tamil Univ Movement, Ceylon, 56-61; NSF instnl grant, Univ Wis, 70-71, fac res grants from pres & chancellor, 72-73 & 74-75; Danforth assoc, 75-81; curric develop grants, Univ Wis, 78-; hon vis prof zool, Univ Madras, India, 79; distinguished teacher, Univ Wis, 83-84; mem, deleg biol teacher to People's Repub China, 88. *Honors & Awards:* Mandala Award, Col Educ, Univ Wis-Oshkosh. *Mem:* China Asn Sci & Technol; Am Fisheries Soc; Nat Asn Biol Teachers; Sigma Xi; Am Fisheries Soc. *Res:* 25 publications in the areas of ichthyology, genetics and evolutionary biology and philosophy and teaching of biology. *Mailing Add:* Dept Biol Univ Wis 800 Algoma Blvd Oshkosh WI 54901

MAHADEVAN, PARAMESWAR, b India, Apr 23, 26; m 58; c 2. ATOMIC PHYSICS, MOLECULAR PHYSICS. *Educ:* Univ Kerala, BSc, 44, MSc, 46; Univ London, PhD(physics), 58. *Prof Exp:* Lectr physics, Univ Kerala, 47-54; fel elec eng, Univ Fla, 59-60; consult physics, Gen Dynamics/Convair, 60-66; res scientist, Douglas Advan Res Labs, 67-74; STAFF SCIENTIST, AEROSPACE CORP, EL SEGUNDO, 74- *Mem:* AAAS; Am Phys Soc; fel Brit Inst Physics. *Res:* Atomic and molecular collision processes; measurement of interaction crossections; phenomena accompanying the impact of ions and atoms on clean metallic surfaces, such as sputtering, secondary electron emission and ion reflection; chemistry and physics of the earth's ionosphere; space science. *Mailing Add:* 2556 Crown Way Fullerton CA 92633

MAHADEVAN, SELVAKUMARAN, b Madras, India, Sept 29, 48; c 3. BIOLOGICAL OCEANOGRAPHY, BENTHIC ECOLOGY. *Educ:* Univ Madras, BSc, 67; Annamalai Univ, MSc, 71; Fla State Univ, PhD(oceanog), 77. *Prof Exp:* Environ scientist, Conserv Consult Inc, Palmetto, Fla, 75-78; staff scientist marine biol, 78-80, dir environ assessment, 80-83, co-dir, 84-85, EXEC DIR, MOTE MARINE LAB, SARASOTA, FLA, 86- *Concurrent Pos:* Exec comt, Coun Ocean Affairs, Southern Asn Marine Labs, Nat Asn Marine Labs, 89-91. *Mem:* Nat Asn Marine Labs (pres, 90); Estuarine Res Fedn; Deep-Sea Biol Soc; Asn Marine Lab Caribbean. *Res:* Marine benthic ecology; thermal pollution; environmental assessment and prediction; biological data analysis; ecological theories and applications to benthic communities; deep-sea ecology; recolonization patterns of benthos. *Mailing Add:* Mote Marine Lab 1600 Ken Thompson Pkwy Sarasota FL 34236

MAHADEVIAH, INALLY, nuclear chemistry, inorganic chemisty; deceased, see previous edition for last biography

MAHADIK, SAHEBARAO P, b India, June 1, 41. BIOCHEMISTRY. *Educ:* PhD(biochem), 69. *Prof Exp:* PROF EXP SCI, COLUMBIA UNIV, 69- *Mem:* Am Soc Neurochem; Int Soc Develop Neurosci. *Mailing Add:* Div Neurosci NY State Psychiat Inst Box 64 722 W 168th St New York NY 10032

MAHAFFEY, KATHRYN ROSE, b Johnstown, Pa, Dec 24, 43; m 77; c 2. NUTRITION, EXPERIMENTAL PATHOLOGY. *Educ:* Pa State Univ, BS, 64; Rutgers Univ, MS, 66, PhD(nutrit), 68. *Prof Exp:* NIH fel endocrine pharmacol, Sch Med, Univ NC, Chapel Hill, 67-69, res assoc, 69-70, asst prof path, 70-72; proj mgr lead contamination of foods, US Food & Drug Admin, 72-83, asst to dir, nutrit div, 75-78, res biochemist, 78-83; chief, priorities res anal br, Nat Inst Occup Safety & HHS, Cincinnati, 83-88; SCI ADV, OFF

DIR, NAT INST ENVIRON HEALTH SCI, CINCINNATI, 88- *Concurrent Pos:* Lectr, Sch Home Econ, Univ NC, Greensboro, 68-71; prog coordr, Univs Assoc Res & Educ in Path, Inc, Bethesda, Md, 71-72; adj asst prof community med, Sch Med, Georgetown Univ, 72-74; adj assoc prof internal med, Univ Cincinnati Med Col; mem adv comt, Priorities Biol & Med Res to Evaluate Etiology, Prevalence & Toxicity of Lead Exposure, HUD, 75-76; mem, sci rev group, Health Effects of Airborne Lead, Environ Protection Agency, 82-83, subcomt benefits anal, Clean Air Sci Adv Comt, Sci Adv Bd, 87, Carcinogen Assessment Group, 86 & 89, sci adv bd, Airborne Lead Criterion Document, 84-89; consult, adv comt childhood lead screening prog, Ctr Dis Control, 84-85, adv comt prev pediat lead poisoning, 90-; mem, HEW Interagency Comt Hyperkinesis & Food Additives, 75-76, Nat Inst Occup Safety & Health Comt Health Hazards to Women in the Workplace, 76; mem, Comt Lead in Human Environ, Bd Environ & Life Sci, Nat Acad Sci-Nat Res Coun, 78-80, Comt Susceptibility Critical Populations to Lead, 88-, Comt Monitoring Human Tissues, 88-; consult, Food Chem Codex Comt, Food & Nutrit Bd, Inst Med, 88-, Comt Eval Seafood Safety, 89-; rep, Task Force Environ Cancer, Heart & Lung Dis, NIEHS, 90- , Nat Ocean Pollution Policy Bd, Nat Oceanic & Atmospheric Admin,90-; adv problems of chem pollution, Ministry Health, Govt Mex, 90- *Mem:* Am Inst Nutrit; Am Soc Exp Path; Am Pub Health Asn; Soc Toxicol. *Res:* Metal toxicity, especially lead toxicity; factors affecting susceptibility to disease (e.g., age, diet, nutritional status); human epidemiology as related to differential diagnosis of chemically induced disease; environmental health research and the legislative process. *Mailing Add:* Nat Inst Environ Health Sci Univ Cincinnati Med Ctr Kettering Lab 3223 Eden Ave Cincinnati OH 45221

MAHAFFY, JOHN HARLAN, b Kansas City, Mo. COMPUTATIONAL PHYSICS, PHYSICS. *Educ:* Univ Nebr, BS, 70; Univ Colo, PhD(astrophys), 74. *Prof Exp:* Staff mem nuclear reactor safety, Los Alamos Nat Lab, 76-81, assoc group leader, Safety Code Develop Group, Energy Div, 81-85; PROF, PA STATE UNIV, 85- *Res:* Numerical modeling of multi-phase fluid flow and application of these models on multiprocessor computers. *Mailing Add:* Appl Res Lab Pa State Univ PO Box 30 State College PA 16804

MAHAJAN, DAMODAR K, b Pilode, Maharastra; m 54; c 2. BIOCHEMISTRY, ENDOCRINOLOGY. *Educ:* Univ Poona, BSc(Hons), 51, MSc, 53, PhD(biochem), 64; Univ Utah, dipl steroid biochem, 61. *Prof Exp:* Pop Coun New York fel, Inst Steroid Biochem, Univ Utah, 60-62, Am Cancer Soc fel, Dept Biochem, 62-63; sr res asst steroid biochem, Cancer Res Inst, Bombay, India, 63-65, res officer, 65-69; instr steroid hormones, Dept Med & dir, Intermountain Regional Lab, Univ Utah, 69-72; asst prof hormones, Dept Obstet & Gynec, Pa State Univ, 72-75; ASST PROF HORMONES, DEPT REPRODUCTIVE BIOL, CASE WESTERN RESERVE UNIV, 75- *Concurrent Pos:* Teacher biochem, Univ Bombay, 68; vis prof, Univ del Salvador, Inst Latino Am de Fisiologia de la Reproduction, Buenos Aires, Argentina, 72. *Mem:* AAAS; Endocrine Soc; Soc Study Reproduction. *Res:* Metabolism of steroid hormones in health and endocrine disorders; inhibitory effect of progesterone, steroidogenic ability of different cells of the ovary; bioconversion of steroids by reproductive organs; radioimmunoassay of steroids; steroid binding proteins; isolation and physiological effects of ovarian peptides. *Mailing Add:* Dept Obstet & Gynec La State Univ Sch Med New Orleans LA 70112

MAHAJAN, OM PRAKASH, b Sujanpurtira, India, July 13, 36; m 66; c 2. FUEL CHEMISTRY. *Educ:* Panjab Univ, BSc Hons, 57, MS, 59, PhD(chem), 65. *Prof Exp:* Lectr chem, Panjab Univ, 61-66; res assoc, Pa State Univ, 66-69; lectr, Panjab Univ, 69-73; sr res assoc, Pa State Univ, 73-79; SR RES CHEMIST, AMOCO RES CTR, 79- *Mem:* Am Chem Soc. *Res:* Coal: characterization and reactivity, liquifaction, gasification, and catalysis; activated carbons: adsorption characteristics, nature of surface and surface complexes; adsorption: physical adsorption, pore structure, surface area, and molecular sieves; author or coauthor of over 70 publications. *Mailing Add:* Amoco Res Ctr Rm 2067 MS H-2 PO Box 3011 Naperville IL 60566-7011

MAHAJAN, ROOP LAL, b Jassar, Punjab, India, Mar 15, 43; m 78; c 3. HEAT & MASS TRANSFER, MANUFACTURING SCIENCE & ENGINEERING. *Educ:* Punjab Univ, Chandigarh, India, BSME, 64, MSME, 69; Cornell Univ, PhD(mech eng), 76. *Prof Exp:* Mem, Air Conditioning & Refrigeration, Am Refrigerator Co, Pvt, Ltd, Calcutta, India, 64-65; asst prof mech eng, Design & Heat Transfer, Punjab Eng Col, Chandigash, 65-72; mem tech staff, thermal eng, AT&T Bell Labs, 76-79, supvr thermal eng, Eng Res Ctr, 79-; PROF MECH ENG, UNIV COLO, 91- *Concurrent Pos:* UN Develop programme vis scientist, Indian Inst Sci, Bangalore, India, 88; adj prof mech eng, Mech Eng Dept, Univ Pa, 89-91; Bell Lab fel, 89; adj prof mech eng, Mech Eng Dept, Drexel Univ, 90. *Mem:* Am Soc Mech Engrs. *Res:* Advancing the manufacturing science of the thermally-based processes in semiconductor materials, device fabrication and packaging; several publications on chemical vapor deposition, vapor phase soldering and buoyancy induced flows. *Mailing Add:* Dept Mech Eng Univ Colo Boulder CO 80309-0427

MAHAJAN, SATISH CHANDER, b Chandigarh, India, Dec 6, 35; m 62; c 2. REPRODUCTIVE PHYSIOLOGY, ENDOCRINOLOGY. *Educ:* Punjab Univ, India, BVSc, 57; Agra Univ, MVSc, 60; Rutgers Univ, PhD(animal sci), 65. *Prof Exp:* Exten officer vet sci, Block Develop Area, 57-58; res asst animal sci, Indian Vet Res Inst Izatnagar, 60-61; PROF BIOL, LANE COL, 65- *Concurrent Pos:* Ford Found fel, Univ Wis-Madison, 65-66; prog dir & grants, MARC, Minority Biomed Res Support Prog-NIH & Multinat Staged Improvement Prog-Dept Educ, 89- *Mem:* AAAS; Am Soc Animal Sci. *Res:* Study of factors which affect fertilization and survival of embryo in various animals. *Mailing Add:* Dept Biol Lane Col Jackson TN 38301

MAHAJAN, SUBHASH, b Gurdaspur, India, Oct 4, 39; m 65; c 3. PHYSICAL METALLURGY. *Educ:* Panjab Univ, Chandigarh, BSc, 59; Indian Inst Sci, BE, 61; Univ Calif, Berkeley, PhD(mat sci), 65. *Prof Exp:* Res asst mat sci, Univ Calif, Berkeley, 61-65; res metallurgist , Univ Denver, 65-68; Harwell fel mat sci, Ukaea, Harwell, Eng, 68-71; mem tech staff phys metall, Bell Labs, Murray Hill, 71-83, group supvr, 81-83; PROF METALL ENG & MAT SCI, CARNEGIE-MELLON UNIV, 83- *Concurrent Pos:* Daad fel, Univ Gottingen, WGer, 76; mem bd dirs, Metall Soc, 86-; author & indust consult. *Mem:* Sigma Xi; Mat Res Soc; Am Inst Mining Metall & Petroleum Engrs; Electrochem Soc. *Res:* Correlations between structure and properties of materials encompassing magnetic mats, metals, ceramics and semiconductors. *Mailing Add:* Dept Metall Eng & Mat Sci Carnegie-Mellon Univ Pittsburgh PA 15213-9998

MAHAJAN, SUDESH K, b New Delhi, India, Jan 20, 48; US citizen; c 3. INTERNAL MEDICINE, NEPHROLOGY. *Educ:* Maulana Azad Med Col, MD, 70. *Prof Exp:* Fel nephrology, Univ Chicago, 76; Instr, 76-78, asst prof, 79-83, ASSOC PROF MED, SCH MED, WAYNE STATE UNIV, 83-; CHIEF, NEPHROLOGY SECT, VET ADMIN MED CTR, ALLEN PARK, MICH, 76- *Concurrent Pos:* Reviewer manuscripts, Mineral & Electrolyte J, 84; mem, Adv Panel Renal Electrolyte & Nutrit, US Pharmacopeial Conv, Inc, 85-; consult, Int J Artificial Organisms. *Mem:* Fel Am Col Physicians; Am Fedn Clin Res; fel Am Col Nutrit; Am Soc Nephrology; Int Soc Nephrology; Am Inst Clin Nutrit; Am Soc Clin Nutrit; Am Soc for Artificial Internal Organs; Int Soc Artificial Organs. *Res:* Metabolism of zinc and other trace elements in patients with chronic renal disease; chronic renal failure; dialyzed uremic patients and renal transplant recipients. *Mailing Add:* Nephrology Sect Med Serv Vet Admin Med Ctr Southfield & Outer Dr Allen Park MI 48101

MAHAJAN, SWADESH MITTER, b Doda, Punjab, India, May 25, 44. PLASMA CONFINEMENT & STABILITY. *Educ:* Delhi Univ, India, BSc, 63, MSc, 66; Univ Md, College Park, PhD(physics), 73. *Prof Exp:* Lectr physics, Delhi Univ, 66-68; fel, Plasma Physics Lab, Princeton Univ, 73-75; res scientist, Bhabha Atomic Res Ctr, Bombay, 75-76; SR RES SCIENTIST, INST FUSION STUDIES, UNIV TEX, AUSTIN, 77- *Concurrent Pos:* Lectr, Int Centre Theoret Physics, 79, dir, 88 & Inst Math Sci, India, 87-; vis prof, Phys Res Lab, Ahmedobed, 85 & 86; head, Plasma Physics Group, Int Centre Theoret Physics, Trieste. *Mem:* Fel Am Phys Soc. *Res:* Plasma confinement, auxiliary heating of plasmas in laboratories and in astrophysical situations; solutions of nonlinear, field theories including Yang-Mills fields; foundations of quantum mechanics and relativity theory; plasma heating and anomalous transport. *Mailing Add:* Dept Physics-IFS Univ Tex Austin TX 78712

MAHALINGAM, LALGUDI MUTHUSWAMY, b Madras, India, May 19, 46. SOLID STATE PHYSICS, SOLAR ENERGY. *Educ:* Univ Madras, BSc, 66; Carnegie-Mellon Univ, MS, 72, PhD(solid state physics), 76. *Prof Exp:* Sci officer nuclear reactor physics, Bhabha Atomic Res Ctr, India, 67-70; fel & res physicist Solar Energy, Carnegie-Mellon Univ, 76-79; SR MEM TECH STAFF & MGR PACKAGE DESIGN TECHNOL, MOTOROLA, INC, 89- *Mem:* Assoc mem Sigma Xi. *Res:* Solar energy, both thermal and direct conversion processes; materials research particularly optical, thermal, electric, magnetic and surface-applied properties; applied magnetism. *Mailing Add:* Motorola Inc 5005 E McDowell Rd MS-B136 Phoenix AZ 85008

MAHALINGAM, R, US citizen. AIR POLLUTION, HAZARDOUS WASTES. *Educ:* Bombay Univ, BS, 57; Indian Inst Sci, Dip II Sc, 59; Purdue Univ, MS, 63; Univ Newcastle-Upon-Tyne, PhD(chem eng), 66. *Prof Exp:* Chem engr, Century Rayon Corp, 59-61 & Monsanto Chemicals Ltd, 66-67; develop & design engr, Div Halcon Int, Sci Design Co, 67-69; from asst prof to assoc prof, 69-78, PROF CHEM ENG, WASH STATE UNIV, 78- *Concurrent Pos:* Head chem energy & processing res, Wash State Univ, 75-80, prof mat sci & eng, 72-89, prof environ eng, 87-; vis prof chem eng, Indian Inst Technol, 85- *Mem:* Fel Am Inst Chem Engrs (secy, 85); Am Chem Soc. *Res:* Aerosol reactors; fluid-bed electropolymerization reactors; three-phase foam slurry reactors; non-newtonian fluids; coal conversion; air pollution/air toxics; hazardous wastes remediation; polymer thin films; plasma polymerization; polymeric microencapsulation. *Mailing Add:* Dept Chem Eng Wash State Univ Pullman WA 99164-2710

MAHALL, BRUCE ELLIOTT, b Springfield, Mass, Apr 21, 46. PLANT ECOLOGY. *Educ:* Dartmouth Col, AB, 68; Univ Calif, Berkeley, PhD(bot), 74. *Prof Exp:* Fel physiol plant ecol, Carnegie Inst Washington, 74-75; asst prof physiol plant ecol, 75-82, ASSOC PROF DEPT BIOL SCI, UNIV CALIF, SANTA BARBARA, 82- *Mem:* Ecol Soc Am; Brit Ecol Soc; Am Soc Plant Physiologists; AAAS; Bot Soc Am. *Res:* Environmental and physiological restrictions of plant distributions and dimensions of plant niches. *Mailing Add:* Dept Biol Sci Univ Calif Santa Barbara CA 93106

MAHAN, DONALD C, b E Chicago,Ind,May 28, 38; m 62; c 3. ANIMAL NUTRITION. *Educ:* Purdue Univ, W Lafayette, BS, 60, MS, 65; Univ Ill, Urbana, PhD(nutrit), 69. *Prof Exp:* From asst prof to assoc prof, 69-81, PROF ANIMAL SCI, OHIO STATE UNIV & OHIO AGRRES DEVELOP CTR, 81- *Honors & Awards:* Gustav Bohstedt Award, 83; Feed Indust Nutrit Res Award, 87. *Mem:* Am Soc Animal Sci; Am Inst Nutrit; Sigma Xi. *Res:* Protein nutrition in reproducing swine; selenium and vitamin E nutrition in swine; calcium and phosphorus nutrition in swine; management and nutrition interrelationship with weanling swine. *Mailing Add:* Dept of Animal Sci Ohio State Univ Columbus OH 43210-1095

MAHAN, GERALD DENNIS, b Portland, Ore, Nov 24, 37; m 65; c 3. THEORETICAL SOLID STATE PHYSICS. *Educ:* Harvard Univ, AB, 59; Univ Calif, Berkeley, PhD(physics), 64. *Prof Exp:* Physicist, Gen Elec Res & Develop Ctr, 64-67; assoc prof physics, Univ Ore, 67-73; prof physics, Ind Univ, Bloomington, 73-84; PROF PHYSICS, DEPT PHYSICS & ASTRON, UNIV TENN, 84- *Concurrent Pos:* Alfred P Sloan res fel, 68-70. *Mem:* Fel Am Phys Soc. *Res:* Theory of optical and transport phenomena in solids. *Mailing Add:* Dept Physics & Astron Univ Tenn Knoxville TN 37996

MAHAN, HAROLD DEAN, ornithology, ecology, for more information see previous edition

MAHAN, KENT IRA, b Springfield, Mo, June 7, 42; m 65; c 1. ANALYTICAL CHEMISTRY, PHYSICAL CHEMISTRY. *Educ:* Southwest Mo State Col, BS, 64; Columbia Univ, PhD(chem), 69. *Prof Exp:* From asst prof to assoc prof, 69-77, PROF CHEM, UNIV SOUTHERN COLO, 77- *Concurrent Pos:* Fel, Univ Denver, 81. *Mem:* Sigma Xi; Am Chem Soc. *Res:* Environmental chemistry; applied atomic spectroscopy; trace metal. *Mailing Add:* 4 Sandcastle Ct Pueblo CO 81001

MAHAN, SUMAN MULAKHRAJ, b Nyeri, Kenya, Apr 20, 55; m; c 2. MOLECULAR BIOLOGY, IMMUNOLOGY. *Educ:* Univ Nairobi, BVM, 78; Univ Birmingham, UK, MSc, 80, PhD(immunol), 84. *Prof Exp:* Postdoctoral assoc, Int Lab Res Animal Dis, 85-88; postdoctoral assoc, 88-90, ASST SCIENTIST, DEPT INFECTIOUS DIS, UNIV FLA, 90- *Res:* Applied immunology to diseases of livestock; molecular biology of heartwater-developing ambient based diagnostic tests-serological and nucleic acid as well as subunit vaccines. *Mailing Add:* Dept Infectious Dis Univ Fla 1FAS 0633 Gainesville FL 32611-0633

MAHANTHAPPA, KALYANA T, b Hirehalli, India, Oct 29, 34; m 61; c 3. ELEMENTARY PARTICLE PHYSICS. *Educ:* Univ Mysore, BSc, 54; Univ Delhi, MSc, 56; Harvard Univ, PhD(physics), 61. *Prof Exp:* Res assoc physics, Univ Calif, Los Angeles, 61-63; asst prof, Univ Pa, 63-66; assoc prof, 66-69, fac res fel, 70-71, 76-77, 83-84, PROF PHYSICS, UNIV COLO, BOULDER, 69- *Concurrent Pos:* Mem, Inst Advan Study, 64-65; dir, Boulder Summer Inst Theoret Physics, 68-69; vis scientist, Int Ctr Theoret Physics, Trieste, 70-71; dir, NATO Advan Study Inst, 79, NATO Workshop Superstrings, 87; sr vis fel, Imp Col, London, 83-84; gen dir, Elem Particle Physics, Theoret Advan Inst, 89-; sr vis fel, Sci & Eng Res Coun, UK, Imp Col, London, 83-84. *Mem:* AAAS; fel Am Phys Soc. *Res:* Quantum field theory; elementary particle physics; quantum electro-dynamics; gauge theories; supersymmetry; supergravity. *Mailing Add:* 2865 Darley Ave Boulder CO 80303

MAHANTI, SUBHENDRA DEB, b Cuttack, India, Sept 24, 45; m 72; c 2. THEORETICAL CONDENSED MATTER PHYSICS. *Educ:* Utkal Univ. India, BSc, 61; Univ Allahabad, MSc, 63; Univ Calif, Riverside, PhD(physics), 68. *Prof Exp:* Res asst physics, Univ Allahabad, 63-64 & Univ Calif, Riverside, 64-68; mem tech staff physics, Bell Tel Lab, Murray Hill, NJ, 68-70; from asst prof to assoc prof, 70-82, PROF PHYSICS, MICH STATE UNIV, 82- *Honors & Awards:* Fulbright Lectr, India. *Mem:* Am Phys Soc; Am Biophys Soc. *Res:* Structural, electronic and magnetic phase transitions in solids; electron nuclear interactions; energy and charge transport in biological systems; physics of intercalation compounds. *Mailing Add:* Dept Physics Mich State Univ East Lansing MI 48824

MAHAR, THOMAS J, b Albany, NY, July 29, 49; m 76; c 1. DIFFERENTIAL EQUATIONS, BIFURCATION THEORY. *Educ:* Rensselaer Polytech Inst, BSc, 71, PhD(math), 75. *Prof Exp:* Vis mem res, Courant Inst Math Sci, NY Univ, 75-77; asst prof, Math Dept, Utah State Univ, 77-78; asst prof appl math, Dept Eng Sci & Appl Math, Northwestern Univ, 78-; AT DEPT MATH, US NAVAL ACAD. *Concurrent Pos:* Vis prof, Rensselaer Polytech Inst, 75 & 79. *Mem:* Soc Indust & Appl Math; Am Math Soc. *Res:* Bifurcation phenomena in nonlinear problems; analysis of discontinuous boundary value problems; numerical models in mechanics and population biology. *Mailing Add:* Dept Math US Naval Acad Annapolis MD 21402

MAHARD, RICHARD H, b Van Buren Co, Mich, July 5, 15. GEOLOGY. *Educ:* Eastern Mich Univ, AB, 35; Columbia Univ, MA, 41, Phd(geol), 49. *Prof Exp:* prof, 41-80, EMER PROF GEOL & GEOG, DENISON UNIV, OHIO, 80- *Mem:* Geological Soc Am. *Mailing Add:* 115 W Elm St Box 342 Granville OH 43023

MAHAVIER, WILLIAM S, b Houston, Tex, July 30, 30; m; c 2. MATHEMATICS. *Educ:* Univ Tex, BS, 51, PhD(math), 57. *Prof Exp:* Physicist, US Air Missile Test Ctr, Calif, 51-52; mathematician, Defense Res Lab, Univ Tex, 54-57; instr math, Ill Inst Technol, 57-59; asst prof, Univ Tenn, 59-64; from asst prof to assoc prof, 64-70, PROF MATH, EMORY UNIV, 70- *Mem:* Am Math Soc; Math Asn Am. *Mailing Add:* Dept of Math Emory Univ 1364 Clifton Rd NE Atlanta GA 30322

MAHBOUBI, EZZAT, b Dargaz, Iran, Apr 21, 29; US citizen; m 57; c 2. CANCER EPIDEMIOLOGY, PREVENTIVE ONCOLOGY. *Educ:* Univ Teheran, Iran, MD, 61; Univ Calif, Los Angeles, MPH, 64. *Prof Exp:* Assoc prof prev med & epidemiol, Sch Med & Sch Pub Health, Univ Teheran, Iran, 65-73; prof cancer epidemiol and head dept epidemiol, Eppley Cancer Inst, Univ Nebr Med Ctr, Omaha, 73-79, prof epidemiol, 79-86; actg chmn, Dept Biol & Med Res, King Faisal Specialist Hosp, Riyadh, Saudi Arabia, 83-87, cancer epidemiologist, 83-88; PVT PRACT, 90- *Concurrent Pos:* Dir prev med & epidemiol, Med Res Sta, Babol, Iran, 67-73; adv, Int Agency Res Cancer, WHO, Jamaica & Signapore, 68 & 71; head Res Ctr, Int Agency Res Cancer, Teheran, 71-73; consult, Reserve Mining Co, Duluth, Minn, 73-75; fel grad fac, Univ Nebr Systs, 76-89; mem, Expert Adv Panel on Food Additives & Contaminants, WHO, 77-82; vis prof, Col Med, Univ Baghdad, Iraq, 79; actg chief epidemiol, Am Health Found, 80-81. *Mem:* Am Soc Prev Oncol; Am Asn Cancer Res; Int Epidemiol Asn; NY Acad Sci; Am Med Asn. *Res:* Environmental factors and specific anatomic cancer sites; cancers of esophagus and oral cavity; carcinogens in the environment; design and development of Epidemiological studies focusing on the etiology of different malignancies in Saudi Arabia and identification of most common cancers in the country; study of possible ways and means for modification of carcinogenesis and search for applicable methods of cancer control; dietary and life-style factors among different population strata; initiation of case-control studies of breast cancer and oval and nasopharynz. *Mailing Add:* 8380 Greensboro Dr 910 Mclean VA 22102

MAHDY, MOHAMED SABET, b Heliopolis, Egypt, Oct 29, 30; Can citizen; m 51; c 2. VIROLOGY, IMMUNOLOGY. *Educ:* Cairo Univ, BSc, 50; Univ Pittsburgh, MPH, 60, DScHyg(virol, microbiol), 63. *Prof Exp:* Res asst tissue cult & virol, US Naval Med Res Unit 3, 52-59; investr virol & head sect, Microbiol Unit, Med Res Inst, Egypt, 63-65; head lab virus res, 65-71, res scientist, 65-68, asst to chief virologist, 71-75, HEAD ARBOVIRUSES & SPEC PATHOGENS, LAB SERV BR, ONT MINISTRY HEALTH, 76- *Concurrent Pos:* Guest investr, US Naval Med Res Unit 3, 63-64; lectr, Sch Hyg, Univ Toronto, 68-71, asst prof, 71-72; consult microbiol labs, Can Ctr Inland Waters, 72-77; asst prof, Univ Toronto & Erindale Col, 78-, chmn comt surveillance of anthropod-borne dis, 78-; consult contingency planning comt, Dangerous Exotic Dis, 78- *Mem:* Am Soc Microbiol; Can Soc Microbiol; Brit Soc Gen Microbiol; Can Pub Health Asn; Am Soc Trop Med Hyg. *Res:* Immune and non-specific responses to viral infections; assay of immunoglobulins to identify viral infections; health hazards of environmental pollution with viruses; arboviruses; high-hazard viruses and containment. *Mailing Add:* 50 Ruddington Dr Apt 411 Willowdale ON M2K 2J8 Can

MAHENDROO, PREM P, b Indore, India, July 4, 31; m 62; c 2. SOLID STATE PHYSICS, CHEMICAL PHYSICS. *Educ:* Agra Univ, BSc, 50; Panjab Univ, MA, 55; Univ Tex, Austin, PhD(physics), 60. *Prof Exp:* Physicist, Div Acoust, Nat Phys Lab, India, 51-56; from asst prof to prof physics, Tex Christian Univ, 60-81; dir spectros, Alcon Labs Inc, 81-82, dir physics, 82-87, new technol & instrumentation, Res & Develop Div, 87-89, DIR PHYS SCI RES, ALCON LABS INC, 89- *Concurrent Pos:* Consult, Gen Dynamics Corp, Tex, 61- & Alcon Labs, 65-66; vis prof, Univ Nottingham, 73-74; cancer sci fel, Nat Cancer Inst, 76-78. *Honors & Awards:* Fulbright Award, 56. *Mem:* Fel AAAS; Am Phys Soc; Soc Magnetic Res Med; Biophys Soc; Asn Res Vision & Ophthal. *Res:* Biophysics and biochemstry of the eye; technology for biomedical research. *Mailing Add:* Alcon Labs Inc PO Box 6600 Ft Worth TX 76115

MAHER, F(RANCIS) J(OSEPH), b Brooklyn, NY, June 11, 15; m 39; c 2. ENGINEERING MECHANICS. *Educ:* Manhattan Col, BS, 36; Va Polytech Inst, MS, 37. *Prof Exp:* Instr appl mech, Va Polytech Inst, 37-41, asst prof, 41-44; contract employee, Ballistics Sect, Naval Res Lab, 44-45; assoc prof eng mech, 45-52, prof, 52-79, EMER PROF ENG SCI & MECH, VA POLYTECH INST & STATE UNIV, 79- *Concurrent Pos:* Instr, non-resident, Mass Inst Technol, 50; res grants, Res Corp, 52-55, US Army Res Off, 63-65, NSF, 73-78 & NASA, 65-66; consult, Thompson Ramo Wooldridge, Inc, Bethlehem Steel Co, Greiner Eng Sci, Inc & Steinman, Boynton, Gronquist & Birdsall, MAR Assocs, Williams, Worrell, Kelly & Greer, Jackson Durkee, Williams Realty Corp. *Honors & Awards:* Western Elec Award, AM Soc Eng Educ. *Mem:* Am Soc Eng Educ; assoc fel Am Inst Aeronaut & Astronaut; Nat Soc Prof Engrs; Sigma Xi. *Res:* Aerodynamic properties of suspension bridges and other structures; structural vibrations, dynamics and properties of materials; aerodynamic interference of adjacent structures. *Mailing Add:* Box 817 Blacksburg VA 24063

MAHER, GALEB HAMID, b Lebanon, Dec 2, 37; US citizen; m 65; c 3. MATERIALS SCIENCE, PHYSICAL CHEMISTRY. *Educ:* North Adams State Col, BS, 62; Williams Col, MS, 68; Rensselaer Polytech Inst, PhD(mat sci), 72. *Prof Exp:* Ceramic engr, 62-68, ceramic sr engr res & develop, 68-72, SR SCIENTIST MAT RES & DEVELOP, SPRAGUE ELEC CO, 72- *Mem:* Am Ceramic Soc. *Res:* Electrical properties of ceramic dielectrics; physical and electrical properties of thin film dielectrics. *Mailing Add:* 116 Autumn Dr North Adams MA 02147

MAHER, GEORGE GARRISON, b Washington, DC, Apr 17, 19; m 43; c 5. BIO-ORGANIC CHEMISTRY. *Educ:* NDak Agr Col, BSc, 41, MSc, 49; Ohio State Univ, PhD(org chem), 54. *Prof Exp:* Chemist, Weldon Springs Ord Works, 41-42; chief chemist, Ky Ord Works, 43-44; chemist, Tenn Eastman Corp, 44-45; asst agr chemist, NDak Agr Col, 45-49; res assoc, Res Found, Ohio State Univ, 49-54; sr res chemist, Clinton Corn Processing Co, 54-64; prin res chemist, Northern Regional Res Lab, USDA, 64-86; RETIRED. *Mem:* AAAS; Am Chem Soc; NY Acad Sci; Sigma Xi. *Res:* Carbohydrate chemistry; starch derivatives; sugars and syrups; physical characterization; enzymology; propellant and explosive chemistry; chemical research and development of agriculturally derived commodities. *Mailing Add:* 4101 W Simpson Dr Dunlap IL 61525

MAHER, JAMES VINCENT, b New York, NY, Aug 25, 42; m 66; c 2. PHYSICS. *Educ:* Univ Notre Dame, BS, 64; Yale Univ, MS, 65, PhD(physics), 69. *Prof Exp:* Fel, Argonne Nat Lab, 68-70; asst prof physics, 70-74, assoc prof, 74-80, PROF PHYSICS, UNIV PITTSBURGH, 80- *Concurrent Pos:* chair, Dept Physics & Astron, Univ Pittsburgh, 91- *Mem:* Fel Am Phys Soc; Sigma Xi; fel AAAS; Am Asn Crystal Growth. *Res:* Fluctuation phenomena in fluids; instabilities at liquid interfaces. *Mailing Add:* Dept Physics Univ Pittsburgh Pittsburgh PA 15260

MAHER, JOHN FRANCIS, b Hempstead, NY, Aug 3, 29; m 53; c 5. MEDICINE, NEPHROLOGY. *Educ:* Georgetown Univ, BS, 49, MD, 53. *Prof Exp:* Intern med, Boston City Hosp, 53-54; resident, Georgetown Univ Hosp, 56-58, Nat Inst Arthritis & Metab Dis fel nephrol, Univ & Hosp, 58-60, from instr to assoc prof med, Sch Med, 60-69; prof med & dir div nephrol & clin res ctr, Univ Mo-Columbia, 69-74; prof med & dir, Div Nephrol, Univ Conn, Farmington, 74-79; PROF MED & DIR, DIV NEPHROL, UNIFORMED SERV UNIV HEALTH SCI, 79- *Concurrent Pos:* Ed, Clin Res, 67-69. *Mem:* Am Fedn Clin Res; Am Soc Artificial Internal Organs; Am Col Physicians; Am Soc Nephrol; Am Heart Asn; fel Royal Col Physicians Ireland; Sigma Xi. *Res:* Hemodialysis, kinetics; toxic nephropathy; dialysis of poisons; fluid and electrolyte homeostasis; transplantation immunology; uremia; renal pathology. *Mailing Add:* Dept Med Uniformed Serv Univ Health Sci Bethesda MD 20814-4799

MAHER, JOHN PHILIP, b Pittsfield, Mass, Oct 2, 42; m 65; c 2. PHYSICS, ELECTRICAL ENGINEERING. *Educ:* Northeastern Univ, BS, 65; Williams Col, MS, 69. *Prof Exp:* Engr, Sprague Elec Co, 65-69, sr engr, 69-73; sr scientist mat sci, Sprague Elec Div of Gen Cable Corp, 73-85; mgr thin film res & develop, Englehard Corp, 85-88; gen mgr, 88-90, PRES, CERMET MAT, 90- *Honors & Awards:* IR 100, Indust Res Mag, 71. *Mem:* Inst Elec & Electronics Engrs; Int Soc Hybrid Microcircuits. *Res:* Electronic properties of materials; high frequency measurements; thin and thick films; ceramics; electronic and electromechanical devices. *Mailing Add:* Cermet Material Six Meco Dr Wilmington DE 19804

MAHER, LOUIS JAMES, JR, b Iowa City, Iowa, Dec 18, 33; m 56; c 3. PALYNOLOGY, QUATERNARY GEOLOGY. *Educ:* Univ Iowa, BA, 55, MS, 59; Univ Minn, Minneapolis, PhD(geol), 61. *Prof Exp:* NATO fel, Cambridge Univ, 61-62; from asst prof to assoc prof, 62-70, chmn dept, 80-84, PROF GEOL, UNIV WIS-MADISON, 70- *Mem:* AAAS; Geol Soc Am; Ecol Soc Am; Am Quaternary Asn. *Res:* Quaternary palynology in central and western United States; microcomputer data analysis. *Mailing Add:* Dept of Geol & Geophys Univ of Wis Madison WI 53706

MAHER, MICHAEL JOHN, b Napa, Calif, Oct 28, 28; m 54; c 2. COMPARATIVE ENDOCRINOLOGY. *Educ:* Univ Calif, Los Angeles, AB, 52, PhD(zool), 58. *Prof Exp:* Asst, Univ Calif, Los Angeles, 53-57; res scientist zool, Barnard Col, Columbia Univ, 58-59; res fel anat, Albert Einstein Col Med, 59-60, instr, 60-62; from asst prof to assoc prof zool, 62-69, ASSOC PROF PHYSIOL & CELL BIOL, UNIV KANS, 69- *Mem:* AAAS; Am Soc Zool. *Res:* Reptilian physiology; vertebrates; role of the thyroid gland in metabolism, especially cold blooded vertebrates; role of pancreatic and adrenal hormones in carbohydrate metabolism of reptiles. *Mailing Add:* Dept Biol Univ of Kans Lawrence KS 66045

MAHER, PHILIP KENERICK, b Catonsville, Md, Dec 13, 30; m 56; c 4. PHYSICAL CHEMISTRY. *Educ:* Randolph-Macon Col, BS, 52; Cath Univ Am, MS, 55, PhD(phys chem), 56. *Prof Exp:* Sr res chemist, W R Grace & Co, Md, 56-60, res supvr, 60-62, mgr, 62-68, dir, 68-75; PRES, CATALYST RECOVERY INC, 75-, PRES, KATALISTIKS INST, 81- *Mem:* Am Chem Soc; Am Inst Chem Eng; Am Inst Chemists; Am Mgt Asn; Catalysis Soc. *Res:* Physical and surface chemistry of inorganic materials, including catalysis, adsorption and desorption phenomena; thermal and hydrothermal reactions of inorganic materials such as silicates, aluminate and phosphates; catalyst preparation, regeneration and rejuvenation. *Mailing Add:* 2515 Merrymans Mill Rd Phoenix MD 21131

MAHER, STUART WILDER, b Knoxville, Tenn, Aug 11, 18; m 43; c 2. GEOLOGY. *Educ:* Univ Tenn, AB, 46, MS, 48. *Prof Exp:* Asst geol, Univ Tenn, 45-46, instr, 46-48; geologist, US Geol Surv, 48-51; res assoc, Univ Tenn, 51-54; sr geologist, Tenn State Div Geol, 54-57, prin geologist, 57-66, chief geologist, 66-81; PRIN GEOLOGIST, KENWILL INC, 81- *Concurrent Pos:* Asst prof, Univ Tenn, Knoxville, 71. *Mem:* Am Inst Prof Geologists; Int Asn Genesis Ore Deposits; Soc Econ Geol. *Res:* Economic geology. *Mailing Add:* Kenwill Inc 505 E Broadway Maryville TN 37801

MAHER, TIMOTHY JOHN, b Boston, Mass, Nov 24, 53; m 77; c 3. NEUROPHARMACOLOGY, BIOCHEMICAL PHARMACOLOGY. *Educ:* Boston State Col, BS, 76; Mass Col Pharm, PhD(pharmacol), 80. *Prof Exp:* From asst prof to assoc prof, 80-88, CHMN DEPT, 85-, MASS COL PHARM, 85-, PROF PHARMACOL, 88- *Concurrent Pos:* Res fel neurosci, Mass Inst Technol, 80-84, res assoc, 84- 87, lectr, 87-; mem, bd dirs, Mass Soc Med Res, 83-, bd trustees, Ctr Brain Sci & Metab, 84- *Mem:* Am Soc Pharmacol Exp Therapeut; Soc Neurosci; Int Soc Neurochem; NY Acad Sci; Int Brain Res Orgn; Soc Study Ingestive Behaviors. *Res:* Effects of nutrients on central neurotransmitter synthesis and function; amino acid metabolism. *Mailing Add:* Mass Col Pharm 179 Longwood Ave Boston MA 02115

MAHER, VERONICA MARY, b Detroit, Mich, Feb 20, 31. CANCER, MOLECULAR BIOLOGY. *Educ:* Marygrove Col, BS, 51; Univ Mich, MS, 58; Univ Wis, PhD(molecular biol), 68. *Prof Exp:* Res assoc radiol, Sch Med, Yale Univ, 68-69; asst prof biol, Marygrove Col, 69-70; res scientist & chief carcinogenesis lab, Mich Cancer Found, 70-76; assoc prof, 76-80, PROF MICROBIOL, PUB HEALTH & BIOCHEM, MICH STATE UNIV, 80-, CO-DIR CARCINOGENESIS LAB, 76-, ASSOC DEAN GRAD STUDIES, COL OSTEOP MED, 87- *Concurrent Pos:* Mem carcinogenesis contract comt, Nat Cancer Inst, NIH, 75-77, sci adv, Carcinogenesis Prog, Div Cancer Cause & Prev, 75-76; sci adv, Nat Ctr Toxicol Res, 78-80; mem, Spec Cancer Proj Adv Bd, Nat Cancer Inst, 81- *Mem:* Am Asn Cancer Res; Am Soc Microbiol; Environ Mutagen Soc; Tissue Cult Asn; Genetics Soc Am. *Res:* Mutagenic and carcinogenic action of physical and of chemical carcinogenic agents in human cells in culture and the effect of DNA repair on this interaction. *Mailing Add:* Carcinogenesis Lab Fee Hall Mich State Univ East Lansing MI 48824

MAHER, WILLIAM J, b Brooklyn, NY, Mar 10, 27; m 88. VERTEBRATE ECOLOGY, ORNITHOLOGY. *Educ:* Purdue Univ, BS, 51; Univ Mich, MS, 53; Univ Calif, Berkeley, PhD(zool), 61. *Prof Exp:* Lectr biol, San Francisco State Col, 61-62; actg asst prof, Univ Calif, Santa Barbara, 62-63; from asst prof to assoc prof, 63-74, PROF BIOL, UNIV SASK, 74- *Honors & Awards:* Edward Prize, Wilson Ornith Soc, 71. *Mem:* AAAS; Am Ornithologists Union; Wilson Ornith Soc; Cooper Ornith Soc. *Res:* Predator-prey relationships; competition between closely related predators; adaptability of developmental rates to arctic and tropical environment. *Mailing Add:* Dept Biol Univ Sask Saskatoon SK S7N 0W0 Can

MAHESH, VIRENDRA B, b Khanki Punjab, India, Apr 25, 32; m 55; c 3. ENDOCRINOLOGY, ORGANIC CHEMISTRY. *Educ:* Patna Univ, BSc, 51; Univ Delhi, MSc, 53, PhD(org chem), 55; Oxford Univ, DPhil(biol sci), 58. *Prof Exp:* J H Brown Mem fel physiol, Sch Med, Yale Univ, 58-59; from asst res prof to assoc res prof, 59-66, prof, 66-70, chmn dept endocrinol, 72-86, REGENTS PROF ENDOCRINOL, MED COL GA, 70-, DIR, CTR POP STUDIES, 71-, CHMN DEPT PHYSIOL & ENDOCRINOL, 86- *Concurrent Pos:* Mem Reproductive Biol Study Sect, NIH, 77-81, Human Embryol & Develop Study Sect, 82-86, 90-93. *Honors & Awards:* Rubin Award, Am Soc Study Sterility, 63; Billings Silver Medal, AMA, 65; World Decoration Excellence, Am Biog Inst, 89; Distinguished Scientist Award, Asn Scientists Indian Origin Am, 89. *Mem:* Am Fertil Soc; Sigma Xi; Endocrine Soc; Am Physiol Soc; Soc Gynec Invest; Soc Study Reprod; Int Soc Reproductive Med (pres, 80-82); Am Soc Biochem & Molecular Biol; NY Acad Sci; Int Soc Neuroendocrinology; Royal Soc Chem. *Res:* Isolation, secretion, biosynthesis and metabolism of various steroid hormones; mechanism of hormone action; control of gonadotropin secretion; ovulation; reproductive physiology; biochemistry. *Mailing Add:* Dept Physiol & Endocrinol Med Col of Ga Augusta GA 30912-3000

MAHESHWARI, KEWAL KRISHNAN, b India. DNA CLONING & SEQUENCING, PCR & RFLP ANALYSIS & GENE EXPRESSION. *Educ:* Punjab Agr Univ, India, BSc, 74, MSc, 76; Monash Univ, Australia, PhD(biochem), 83. *Prof Exp:* Postdoctoral fel molecular biol, Dept Biochem, McMaster Univ, Ont, 83-84; postdoctoral fel molecular biol, Dept Anat & Cell Biol, Health Sci Ctr, Brooklyn, State Univ NY, 84-85, res assoc Dept Path, 85-89; INSTR MOLECULAR BIOL, DEPT PEDIAT, MED UNIV SC, CHARLESTON, 89- *Mem:* Am Soc Biochem & Molecular Biol; Am Asn Clin Chem. *Res:* Peroxisomal disorders in humans of biochemical molecular biology level; cloning, sequencing and expression studies of the fatty acyl CoA synthetase gene gene in rats and humans. *Mailing Add:* Pediat Dept CSB-316 Med Univ SC 171 Ashley Ave Charleston SC 29425

MAHESWARI, SHYAM P, b Mithi, India, Mar 22, 47; US citizen; m 67; c 3. ELECTROCHEMISTRY, SURFACE CHEMISTRY. *Educ:* Gujarat Univ, India, BSc, 70; Detroit Inst Technol, BS, 72; Cent Mich Univ, MA, 82. *Prof Exp:* Chemist plating, Int Tool Co Ltd, Windsor, Can, 72-75; qual control chemist pharmaceut, Conaught Labs, Toronto, Can, 75-77; PROJ SCIENTIST ELECTROCHEM, GEN MOTORS TECH CTR, 77- *Mem:* Am Chem Soc. *Res:* Surface chemistry of metals and lubricants using FT infrared techniques. *Mailing Add:* 850 Hampton Cir Rochester Hills MI 48307

MAHGOUB, AHMED, pharmacology, for more information see previous edition

MAHIG, JOSEPH, b New York, NY, Aug 25, 30; m 56; c 2. ACOUSTICS, AUTOMATIC CONTROL SYSTEMS. *Educ:* City Col New York, BME, 51; Ohio State Univ, MS, 59, PhD(mech eng), 62. *Prof Exp:* Engr, Gen Elec Co, Ohio, 56-57; instr mech, Ohio State Univ, 57-62, asst prof, 62-66; assoc prof, 66-73, PROF MECH ENG, UNIV FLA, 73- *Concurrent Pos:* NSF fel. *Mem:* Am Soc Eng Educ; Am Soc Mech Engrs; Am Inst Aeronaut & Astronaut; Acoust Soc Am. *Res:* Noise and vibration transmission; nonlinear multivariable stability analyses; experimental impedance of human brain; hydrofoil flutter theory and development. *Mailing Add:* 535 Riverdale Dr Merritt Island FL 32953

MAHILUM, BENJAMIN COMAWAS, b Calatrava, Philippines, Sept 9, 31; m 56; c 6. SOIL SCIENCE, CROP SCIENCE. *Educ:* Cent Mindanao Univ, Bukidnon, Philippines, BS, 57; Univ Hawaii, MS, 66; Okla State Univ, PhD(soil sci), 71. *Prof Exp:* Asst prof soil sci, Mindanao Inst Technol, Kabacan, Philippines, 71-73; assoc prof & res coordr, Univ Eastern Philippines, 73-74; assoc prof & dept chmn agron & soils, Visayas State Col Agr, Philippines, 74-77; vis asst prof agron, Okla State Univ, 77-79; asst prof soil sci, Univ Hawaii, 79-86; CONSULT CROPPING & SOILS, TROP-AG HAWAII, 86- *Concurrent Pos:* Dir, Regional Coconut Res Ctr, VISCA, Baybay, Philippines, 74-77; mem nat res comts, Philippine Coun Agr & Resources Res & Develop, Los Banos Philippines, 73-77. *Mem:* Int Soil Sci Soc; AAAS; Soil Sci Soc Am; Am Soc Agron; Sigma Xi. *Res:* Sewage and industrial waste utilization for crop production and nutrient recycling; utilization of hilly lands for crop and animal production and reforestation; revegetation of eroded lands, mine spoils and roadsides; fertilizer efficiency studies, cropping systems and soil and water management. *Mailing Add:* PO Box 1213 Honokaa HI 96730

MAHL, MEARL CARL, b Pipestone, Minn, Sept 13, 38; m; c 3. MICROBIOLOGY, BIOCHEMISTRY. *Educ:* SDak State Univ, BS, 60, MS, 62; Univ Wis-Madison, PhD(microbiol), 66. *Prof Exp:* Res microbiologist, S C Johnson & Son, Inc, 66-75, sr res microbiologist, 75-88, res assoc, 88-89, MICROBIOL SECT MGR, S C JOHNSON & SON, INC, 90- *Concurrent Pos:* Chmn, Bd Wastewater Standards, Racine, Wis, 68-; test collabr, Am Soc Testing Mat, 76- *Mem:* Am Soc Microbiol; Am Soc Testing Mat. *Res:* Virucidal effects of disinfectant chemicals; epidemiology of viruses and their control in the environment; disinfectants and antiseptics. *Mailing Add:* S C Johnson & Son Inc 1525 Howe St Racine WI 53403

MAHLBERG, PAUL GORDON, b Milwaukee, Wis, Aug 1, 28; m 54; c 2. PLANT ANATOMY, CELL BIOLOGY. *Educ:* Univ Wis, BS, 50, MS, 51; Univ Calif, Berkeley, PhD, 58. *Prof Exp:* Drug salesman, Pitman-Moore Co, Wis, 53-54; asst, Univ Calif, 55-58; instr bot, Univ Pittsburgh, 58-65; PROF BOT, IND UNIV, BLOOMINGTON, 77- *Mem:* AAAS; Bot Soc Am. *Res:* Laticifer ontogeny and physiology in normal and abnormal tissues; localization and biosynthesis of alkaloids, terpenoids and hydrocarbons in plant cells. *Mailing Add:* Dept of Biol Ind Univ Bloomington IN 47401

MAHLE, CHRISTOPH E, b Stuttgart, Ger, Mar 7, 38; Austrian citizen; m; c 2. SATELLITE COMMUNICATIONS, MICROWAVE ENGINEERING. *Educ:* Swiss Fed Inst Technol, dipl elec eng, 61, Dr Sc Techn(microwave technol), 66. *Prof Exp:* Res asst microwave eng, Swiss Fed Inst Technol, 61-67; Mem tech staff, microwave res & develop, Comsat Labs, 68-71, sect head, 71-73, dept mgr, 73-81, dir tech mgt, Microwave Com Lab, 81-83, exec dir, Microelectronics Div, 83-84 & Microwave Tech Div, 83-89, EXEC DIR, SATELLITE TECH DIV, COMSAT LABS, 89- *Mem:* Fel Inst Elec & Electronic Engrs. *Res:* Spacecraft bus technologies; microwave circuits for satellites and earth stations; analysis and verification of satellite performance; communication satellite systems and radio wave propagation. *Mailing Add:* Comsat Corp 22300 Comsat Dr Clarksburg MD 20871

MAHLE, NELS H, b Highland Park, Mich, June 23, 43. ANALYTICAL CHEMISTRY. *Educ:* Eastern Mich Univ, BS, 66; Northern Ill Univ, PhD(anal chem), 74. *Prof Exp:* Res assoc analytical chem, Purdue Univ, 72-74; sr analytical chemist, Dow Chem Co, 74-77, res specialist, 77-79, proj leader, Analysis Labs, 79-90, SR ENVIRON SPECIALIST, ENVIRON ANALYSIS RES LAB, DOW CHEM CO, 90- *Mem:* Am Soc Mass Spectrometry; Am Chem Soc. *Res:* Mass spectrometry; polymer characterization; separation science; computer analysis of spectrometry data; environmental analysis. *Mailing Add:* 1411 Glendale Midland MI 48640

MAHLER, DAVID S, b Jersey City, NJ, Oct 18, 46; m 72; c 1. MACHINE VISION, PARTICLE MEASUREMENT. *Educ:* Stevens Inst Technol, BE, 68, Fairleigh Dickinson Univ, MS, 70; Univ Del, PhD(physics), 77. *Prof Exp:* Fel, Univ Guelph, 76-78; HEAD INSTRUMENT DEPT, KLD ASSOC, INC, 78- *Mem:* Am Phys Soc; Am Soc Testing & Mat; Fine Particle Soc. *Res:* Develop hot wire instrumentation to measure particle size distribution in liquid sprays; develop video based image processing system to measure railroad track wear; characteristics of roadway surfaces. *Mailing Add:* Decilog Inc 555 Broadhollow Rd Melville NY 11747

MAHLER, HALFDAN, b Vivild, Denmark, Apr 21. 23; m; c 2. MEDICINE. *Educ:* Univ Copenhagen, MD, 48. *Hon Degrees:* LLD, Univ Nottingham, Eng, 75, McMaster Univ, 89, Univ Exeter, 90, Univ Toronto, 90; MD, Karolinska Inst, Sweden, 77, Warsaw Med Acad, 80, Charles Univ, Czech, 82, Mahidol Univ, Thailand, 82, Univ Newcastle-upon-Tyne, 90; DPH, Seoul Nat Univ, Korea, 79; DSc, Univ Lagos, Nigeria, 79, State Univ NY, 90; Dr, Nat Fed Univ Villarreal, 80, Univ Gand, Belg, 83, Nat Univ Nicaragua, 83, Semmelweis Univ Med, Hungary, 87, Aarhus Univ, Denmark, 88, Univ Copenhagen, 88, Emory Univ, 89, Aga Khan Univ, Pakistan, 89; LHD, City Univ NY, 89. *Prof Exp:* Planning officer tuberculosis, Ecuador, 50-51; sr officer, Nat Tuberculosis Prog, WHO, India, 51-61, chief, Tuberculosis Unit, Geneva, 62-69, dir, Proj Systs Anal, 69-73, asst dir-gen, Div Strengthening Health Serv & Div Family Health, 70-73, dir-gen, WHO, 73-88, EMER DIR-GEN, WHO, 88-; SECY-GEN, INT PLANNED PARENTHOOD FEDN, 89- *Concurrent Pos:* Secy, Expert Adv Panel Tuberculosis, WHO, 62-69; hon prof, Nat Univ Mayor San Marcos, Lima, Peru, 80, Fac Med, Univ Chile, 82, Beijing Med Col, China, 83, Shanghai Med Univ, 86; hon academician, Nat Acad Med, Mex, 88, Nat Acad Med, Buenos Aires, 88. *Honors & Awards:* Jana Evangelisty Purkyne Medal, Prague, 74; Carlo Forlanini Gold Medal, 75; Ernst Carlsens Found Prize, Copenhagen, 80; Georg Barfred-Pedersen Prize, Copenhagen, 82; Hagedorn Medal & Prize, Denmark, 86; Freedom from Want Medal, Roosevelt Inst, 88. *Mem:* Assoc mem Belg Soc Trop Med; hon fel Royal Col Physicians; hon fel Royal Soc Med; hon mem Med Asn Arg; hon mem Latin-Am Med Asn; hon mem Ital Soc Trop Med; hon mem Am Pub Health Asn; hon mem Swed Soc Med; hon foreign corresp mem Brit Med Asn; hon mem Med Soc Geneve. *Res:* Epidemiology and control of tuberculosis; political, social, economic and technological priority in the health care sector; application of systems analysis to health care problems. *Mailing Add:* Int Planned Parenthood Fedn Regent's Col Inner Circle Regent's Park London NW1 4MS England

MAHLER, RICHARD JOSEPH, b New York, NY, Mar 4, 34; m 60; c 2. INTERNAL MEDICINE, ENDOCRINOLOGY. *Educ:* NY Univ, BA, 55; New York Med Col, MD, 59. *Prof Exp:* Am Diabetes Asn metabolic res fel, New York Med Col, 62-63; NY Acad Med Glorney-Raisbeck traveling fel, Univ Durham, 63-64; from instr to assoc prof med, New York Med Col, 64-71; assoc dir, Dept Metab & Endocrinol, City of Hope Med Ctr, Duarte, Calif, 71-72; dir, Metab & Endocrinol, Eisenhower Med Ctr, 72-88; CLIN ASSOC PROF MED, CORNELL UNIV MED COL, 88- *Concurrent Pos:* Am Diabetes Asn res & develop award, 66-68; NSF res grant, 66-69; New York City Health Res Coun res grants, 66-70, career scientist award, 68-71. *Mem:* Am Fedn Clin Res; Am Diabetes Asn; Asn Am Med Cols; Endocrine Soc; Am Physiol Soc. *Res:* Hormonal and non-hormonal influences on insulin action, the influence of these factors on its mechanism and its relationship to diabetes mellitus. *Mailing Add:* 791 Park Ave New York NY 10021

MAHLER, ROBERT JOHN, b Los Angeles, Calif, Mar 17, 32; m 59; c 2. SOLID STATE PHYSICS. *Educ:* Univ Calif, Los Angeles, BA, 55; Univ Colo, PhD(physics), 63, MBA, 81. *Prof Exp:* Oceanogr, US Navy Hydrographic Off, 57-58; asst nuclear magnetic resonance, Univ Colo, 58-63; physicist, Nat Bur Standards, 63-68, chief solid state electronics sect, 68-71, res physicist, 71-74, chief off prog develop, 74-80; actg dep dir, 83-85, DIR OFF PROG, ENVIRON RES LABS, NAT OCEANIC & ATMOSPHERIC ADMIN, 80-, DEP DIR, 85- *Concurrent Pos:* Res assoc, Univ Colo, Boulder, 63-64, lectr, 64- *Mem:* Am Phys Soc; Am Geophys Union. *Res:* Investigations of phonon-nuclear spin system interactions using pulsed nuclear magnetic resonance techniques; infrared detectors; measurement techniques. *Mailing Add:* 2100 Kohler Dr Boulder CO 80303

MAHLER, WALTER, b Vienna, Austria, May 7, 29; nat US; m 65; c 1. MATERIALS SCIENCE. *Educ:* Monmouth Col, BS, 50; Univ Southern Calif, PhD(inorg chem), 58. *Prof Exp:* Res chemist, Am Dent Asn, Ill, 54; RES CHEMIST, CENT RES & DEVELOP, EXP STA, DUPONT, 58- *Mem:* Am Chem Soc. *Res:* Amorphous and crystalline solids; polar and nonpolar liquids; reactive and inent gases; hot and cold plasmas; catalysts; inorganic fibers. *Mailing Add:* DuPont Exp Sta PO Box 80328 Wilmington DE 19880-0328

MAHLER, WILLIAM FRED, b Iowa Park, Tex, Aug 30, 30; m 55; c 2. PLANT TAXONOMY. *Educ:* Midwestern State Univ, BS, 55; Okla State Univ, MS, 60; Univ Tenn, Knoxville, PhD(bot), 68. *Prof Exp:* From instr to asst prof bot, Hardin-Simmons Univ, 60-66; from asst prof to assoc prof bot, 68-88, cur herbarium, 71- 88, EMER PROF BOT, SOUTHERN METHODIST UNIV, 88-; DIR, BOT RES INST TEX, 88- *Concurrent Pos:* Ed, Sida, Contrib to Bot, 71-83, publ, 71-; Harold Beaty Award, Tex Orgn Endangered Species, 88. *Mem:* Int Asn Plant Taxon; Am Soc Plant Taxon; Am Bryol & Lichenological Soc. *Res:* Floristic studies and pollen morphology in relation to taxonomic concepts; nomenclature. *Mailing Add:* PO Box 947025 Bot Res Inst Tex Ft Worth TX 76147-9025

MAHLMAN, BERT H, b Bismarck, NDak, Nov 2, 22; m 48; c 3. POLYMER CHEMISTRY, MATERIALS SCIENCE. *Educ:* Univ Minn, BChE, 49. *Prof Exp:* Chemist, Hercules Powder Co, 49-61; res chemist, Hercules, Inc, 61-72, res scientist, 77-82; RETIRED. *Mem:* Am Chem Soc. *Res:* Polymer and coatings research and development; radiation chemistry. *Mailing Add:* 576 W Lafayette Dr W Chester PA 19382-6849

MAHLMAN, HARVEY ARTHUR, b La Crosse, Wis, Aug 8, 23; m 45; c 2. ANALYTICAL CHEMISTRY, CHEMICAL ENGINEERING. *Educ:* Univ Minn, BChem & BBA, 49; Univ Tenn, PhD, 56. *Prof Exp:* Chemist, Clinton Lab, Tenn, 44-46; asst, Gen Elec Co, 49-52; res chemist, Oak Ridge Nat Lab, 52-73; nuclear staff power resources specialist, Fla Power & Light, Miami, 73-74; STAFF SPECIALIST, MECH ENG BR, TENN VALLEY AUTH, 74- *Concurrent Pos:* NIH spec fel, Univ Paris, France, 63-64. *Mem:* NY Acad Sci; Sigma Xi; Am Chem Soc. *Res:* Radiochemical and neutron activation analysis; radiation chemistry; nuclear stream generator specialization. *Mailing Add:* 12007 S Fox Den Dr Knoxville TN 37922

MAHLMAN, JERRY DAVID, b Crawford, Nebr, Feb 21, 40; m 62; c 2. NUMERICAL MODELING, STRATOSPHERE. *Educ:* Chadron State Col, AB, 62; Colo State Univ, MS, 64, PhD(atmospheric sci), 67. *Prof Exp:* Instr atmospheric sci, Colo State Univ, 64-67; from asst prof to assoc prof, US Naval Postgrad Sch, 67-70; res meteorologist, 70-84, DIR, METEOROL, GEOPHYS FLUID DYNAMICS LAB, NAT OCEANIC & ATMOSPHERIC ADMIN, 84- *Concurrent Pos:* Vis lectr & assoc prof, Princeton Univ, 77-80, prof, 80-; sci adv comt, High Altitude Pollution Prog, Fed Aviation Agency, 78-82; mem panel, Middle Atmosphere Prog, Nat Acad Sci, 78-84, chmn, 81-84, mem comt solar terrestrial res, 81-84; assoc ed, J Atmospheric Sci, 79-85; mem, Int Comn Meteorol Upper Atmosphere, 80-, Nat Acad Sci Climate Res Comt, 86-89, NASA- WMO Panel Ozone Trend Detection, 86-88, Nat Acad Sci Panel Dynamical Extended Range Forecasting, 87-88, US-USSR Joint Nat Acad Sci Comt on Global Ecol, 89-, Joint Sci Comt, World Climate Res Prog, 91- *Honors & Awards:* Gold Medal, US Dept Com, 86. *Mem:* Sigma Xi; fel Am Meteorol Soc; fel Am Geophys Union. *Res:* Modeling and diagnosis of the general circulation of the atmosphere, including the transport and chemistry of atmospheric trace constituents. *Mailing Add:* Geophys Fluid Dynamics Lab NOAA Princeton Univ Princeton NJ 08540

MAHLO, EDWIN K(URT), b Berlin, Ger, Dec 24, 23; US citizen; m 48; c 2. CHEMICAL ENGINEERING, ORGANIC CHEMISTRY. *Educ:* Chemotechnicum, Berlin, BS, 48. *Prof Exp:* Indust chemist, Dr S Goetze & Co, Ger, 48-50, prod mgr, 50-53; mfg supt petrochem, Thiochem, Inc, Tex, 53-55; chem engr, Monsanto Co, 55-59, sr engr, 59-61, group leader, Plastics Div, 61-65, SR RES GROUP LEADER, MONSANTO PLASTICS & RESINS CO, 65- *Mem:* Am Chem Soc. *Res:* Chemical process research and development in plastics and organic chemicals. *Mailing Add:* 180 Parker St East Longmeadow MA 01028

MAHLSTEDE, JOHN PETER, b Cleveland, Ohio, June 5, 23; m 48, 85; c 4. ORNAMENTAL HORTICULTURE. *Educ:* Miami Univ, Ohio, BS, 47; Mich State Univ, MS, 48, PhD(ornamental hort), 51. *Prof Exp:* From asst prof to assoc prof, 51-57, asst dir, 65-66, actg assoc dir, 66-67, head dept hort, 61-65, prof hort, 57-87, assoc dir, Agr & Home Econ Exp Sta, 67-87, assoc dean, Col Agr, 75-87, EMER PROF HORT, IOWA STATE UNIV, 87- *Concurrent Pos:* Vpres, Agr Res Inst, 74-75; chmn, Exp Sta Comt Policy, Nat Asn State Univ & Land Grant Col, 75-76, mem, Environ Quality Comt, Div Agr, 77; mem, Joint Coun & Exec Comt, Food & Agr Sci, Sci & Educ Admin, US Dept Agr, 77-79; Adv Coun, US Nat Arboretum, 70-80. *Honors & Awards:* Colman Award, Am Asn Nurserymen, 58; Int Plant Propagators' Soc Award, 67; Spec Res Recognition Award, Am Soybean Asn, 89. *Mem:* Fel AAAS; Sigma Xi; fel Am Soc Hort Sci (pres, 71-72); Int Plant Propagators Soc (pres, 65). *Res:* Nursery management; plant propagation; morphological subjects; packaging and propagation. *Mailing Add:* 2858 Torrey Pines Rd Ames IA 50010

MAHLUM, DANIEL DENNIS, b Wolf Point, Mont, May 5, 33; m 56; c 2. COMPLEX MIXTURES, NON-IONIZING RADIATION. *Educ:* Whitworth Col, Wash, BS, 55; Univ Idaho, MS, 58; Univ Wis, PhD(biochem), 62. *Prof Exp:* Res technician agr chem, Univ Idaho, 55-57, asst agr chemist, 57-58; biol scientist, Hanford Labs, Gen Elec Co, 61-63, sr scientist, 63-65; sr scientist, Energy Res & Develop Admin, 65-75, radiobiologist, 75-76; staff scientist, Battelle-Northwest, 76-83, mgr, Develop Toxicol Sect, 83-91; SR PROG OFFICER, NAT ACAD SCI, 91- *Concurrent Pos:* Co-chmn, Conf Early Nutrit & Environ Influences upon Behav Develop, 71; co-chmn, Develop Toxicol Energy Related Pollutants, 77, Coal Conversion & Environ, 80 & chmn, Nat Coun Radiation Protection & Measurements Comt, Biol Effects Magnetic Fields. *Mem:* Radiation Res Soc; Sigma Xi; Am Inst Nutrit; Soc Toxicol. *Res:* Radionuclide metabolism and toxicity; bioeffects of electromagnetic fields; carcinogenesis; toxicology of complex mixtures. *Mailing Add:* Biol & Chem Lab Battelle Pac NW Labs Richland WA 99352

MAHMOODI, PARVIZ, b Tehran, Iran, Sept 1, 31; m 58; c 2. ENGINEERING, PHYSICS. *Educ:* Abadan Inst Technol, Iran, BSc, 55; Univ Minn, MS, 56, PhD(mech eng), 59. *Prof Exp:* Asst, Univ Minn, 56-58, instr mech eng, 58-59; sr res engr, Minn Mining & Mfg Co, 59-60; res fel, Harvard Univ, 60-61; prof physics & head dept, Abadan Inst Technol, 61-62; res specialist, 62-74, CORP SCIENTIST, 3M CO, 74- *Concurrent Pos:* Lectr, Boston Univ, 60-61; dean Sch Allied Health Sci, Univ Med Sci Iran, 78-79. *Honors & Awards:* Sci Gold Medal. *Mem:* AAAS; Am Soc Testing & Mat; Acoust Soc Am. *Res:* Heat transfer; elasticity; viscoelasticity; vibration and acoustics. *Mailing Add:* Cent Res Labs 3M Ctr 3M Co 201-BS-08 St Paul MN 55144

MAHMOUD, ADEL A F, b Cairo, Egypt, Aug 24, 41. INFECTIOUS DISEASES, TROPICAL MEDICINE. *Educ:* Cairo Univ, Egypt, MD, 63; Ain Shams Univ, Egypt, MPH, 67; Univ London, PhD (tropical med), 71. *Prof Exp:* Chief, Div Geog, 77-87, PROF MED, CASE WESTERN

RESERVE UNIV, 80-, PROF & CHMN MED, 87-; PHYSICIAN INCHARGE, UNIV HOSP, CLEVELAND, 87- *Concurrent Pos:* Ed, Trop & Geog Med, 84- *Honors & Awards:* B K Ashford Award, Am Soc Tropical Med Hygiene, 83; Squibb Award, Infect Dis Soc Am 84. *Mem:* Inst Med-Nat Acad; Am Asn Physicians; Am Soc Clin Investigation; Am Asn Immunologists; Int Soc Infect Dis. *Res:* Studies of mechanisms of disease and resistance in schistosomiasis; techniques of immunology, molecular biology and biochemistry; production of an antiparasitic vaccine. *Mailing Add:* Dept Med Univ Hosp Cleveland Cleveland OH 44106

MAHMOUD, ALY AHMED, b Cairo, Egypt, Jan 25, 35; m 62; c 2. ELECTRICAL ENGINEERING, POWER SYSTEMS. *Educ:* Ain Shams Univ, Cairo, BSc, 58; Purdue Univ, MS, 61, PhD, 64. *Prof Exp:* Instr elec eng, Ain Shams Univ, 58-60; asst prof, Univ NB, 64 & Univ Assiut, 64-66; from asst prof to prof, Univ Mo-Columbia, 66-76; prof elec eng & dir power affil prog, Iowa State Univ, 76-, dir Iowa Test & Eval Facil, 80-; AT COL ENG, UNIV NEW ORLEANS. *Concurrent Pos:* Vis prof, Ain Shams Univ, Cairo, 64-66; consult, Nat Res Ctr, Cairo, 64-66; sr res elec engr, Naval Civil Eng Lab, 68-69; prog mgr, NSF, 74-76. *Mem:* Inst Elec & Electronics Engrs; Am Phys Soc; Am Soc Eng Educ. *Res:* Power system analysis, transmission and distribution systems; load modeling and management; power system operation and control. *Mailing Add:* Dept Eng Purdue Univ 2101 Coliseum E Ft Wayne IN 46805

MAHMOUD, HORMOZ MASSOUD, b Teheran, Iran, May 2, 18; nat US; m 54. THEORETICAL PHYSICS. *Educ:* Univ Tehran, EE, 40; Ind Univ, MS, 49, PhD(physics), 53. *Prof Exp:* Res assoc theoret physics, Ind Univ, 53-54 & Cornell Univ, 54-56; asst prof physics & assoc physicist, Ames Lab, Iowa State Univ, 56-60; assoc prof, 60-70, PROF PHYSICS, UNIV ARIZ, 70- *Mem:* Am Phys Soc; Sigma Xi. *Res:* Field theory; elementary particles. *Mailing Add:* Dept Physics Bldg 81 Univ of Ariz Tucson AZ 85721

MAHMOUD, IBRAHIM YOUNIS, b Baghdad, Iraq, Sept 9, 33; m 65; c 4. HERPETOLOGY. *Educ:* Ark Agr & Mech Col, BS, 53; Univ Ark, MA, 55; Univ Okla, PhD(zool), 60. *Prof Exp:* Prof biol, Northland Col, 60-63; assoc prof, 63-68, PROF BIOL, UNIV WIS-OSHKOSH, 68- *Concurrent Pos:* Bd regents res grants, Univ Wis, 63-65, 67-69 & 70-71; CAS grant, 75-81. *Mem:* AAAS; Am Soc Zool; Am Soc Ichthyol & Herpet; NY Acad Sci; Sigma Xi. *Res:* Reptilian physiology with special emphasis on steroid metabolism in reptiles. *Mailing Add:* 5322 Wild Rose Lane Oshkosh WI 54901

MAHON, HAROLD P, b Los Angeles, Calif, Jan 21, 31. SOLID STATE PHYSICS. *Educ:* Ore State Univ, BA, 53, MS, 54; Univ Wash, PhD(physics), 61. *Prof Exp:* Instr physics, One State Univ, 53-54; res assoc, Univ Wash, 54-61; instr & res assoc, physics, Univ Colo, 61-63; ASSOC PROF, UNIV MASS, 67- *Concurrent Pos:* Vis guest, Univ Zurich, 63-67; res assoc, Air Force Res Labs, 69-74. *Mem:* Am Phys Soc; Am Asn Teachers. *Res:* Co-author of various books. *Mailing Add:* Dept Physics Univ Mass-Boston Harbor Campus Boston MA 02125

MAHON, RITA, b Eng, Sept 23, 49; m 75; c 2. OPTICS, ATOMIC & MOLECULAR PHYSICS. *Educ:* Univ London, BSc, 70, PhD(physics), 73. *Prof Exp:* Nat Res Coun Can fel, Ctr Res Exp Space Sci, York Univ, 73-75; res assoc physics, Inst Physics & Astron, 80-85, RES ASSOC PHYSICS, INST PHYS SCI & TECHNOL, UNIV MD, 75-80, 86-; SCIENTIST, JAYCOR INC, US NAVAL RES LAB, WASHINGTON, DC, 85- *Concurrent Pos:* Consult, Los Alamos, 79, Plasma Fusion Ctr, Mass Inst Technol, 85-86. *Mem:* Am Phys Soc; Optical Soc Am. *Res:* Quantum optics; laser spectroscopy; atomic and plasma physics. *Mailing Add:* Laser Physics Br Code 6540 Naval Res Lab Washington DC 20375

MAHON, WILLIAM A, b Airdrie, Scotland, Aug 20, 29; m 56; c 4. CLINICAL PHARMACOLOGY. *Educ:* Harvard Univ, SM, 62; FRCP(C), 60. *Prof Exp:* Sr resident med, St Joseph's Hosp, London, Ont, 56-57; teaching fel path, Univ Western Ont, 57-58; asst resident, Lemuel Shattuck Hosp, 58-59, chief resident, 59-60; assoc, Sch Med, Tufts Univ, 60-62; from asst prof to assoc prof pharmacol & med, Univ Alta, 62-66; asst prof therapeut & assoc med, 66-68, assoc prof pharmacol, 66-85, assoc prof med, 68-85, PROF PHARMACOL & MED, UNIV TORONTO, 85-; VPRES MED AFFAIRS, TORONTO GEN HOSP, 89- *Mem:* Am Fedn Clin Res; Can Soc Pharmacol. *Res:* Cardiovascular pharmacology in man and animals; distribution of antibiotics in normal man and in man with renal failure; methods for studying drug action in man. *Mailing Add:* Toronto Gen Hosp Toronto ON M5G 2C4 Can

MAHONEY, ARTHUR W, b Canton, Maine, Aug 11, 39; m 61; c 5. NUTRITIONAL EPIDEMIOLOGY, TOXICOLOGY. *Educ:* Univ Maine, BS, 61, PhD(animal nutrit), 65. *Prof Exp:* Fel, Harvard Sch Pub Health, 65-67; sr scientist nutrit, Mead Johnson Co, 67-69; assoc prof, 69-74, PROF HUMAN NUTRIT, UTAH STATE UNIV, LOGAN, 74- *Mem:* Am Inst Nutrit; Sigma Xi; fel Am Col Nutrit. *Res:* Effects of food processing on bioavailability of iron, calcium and protein; mechanism of seizure response in magnesium deficiency; diet in the etiology of colon cancer; food composition. *Mailing Add:* Dept Nutrit & Food Sci Utah State Univ Logan UT 84322-8700

MAHONEY, BERNARD LAUNCELOT, JR, b Boston, Mass, Nov 1, 36; m 65; c 3. ANALYTICAL CHEMISTRY, PHYSICAL CHEMISTRY. *Educ:* Boston Col, BS, 58, MS, 60; Univ NH, PhD(phys chem), 67. *Prof Exp:* From asst prof to assoc prof, Mary Wash Col, 65-74, chmn dept, 71-84, prof chem, 74-87, DISTINGUISHED PROF CHEM, MARY WASHINGTON COL, 88- *Mem:* Am Chem Soc; Soc Appl Spectros; fel Am Inst Chem. *Res:* Spectroscopy; photochemistry; chemical kinetics. *Mailing Add:* Dept Chem-Geol Mary Washington Col Fredericksburg VA 22401

MAHONEY, CHARLES LINDBERGH, b Geneva, NY, Mar 18, 28; m 57; c 2. ENVIRONMENTAL SCIENCES. *Educ:* Colo State Univ, BS, 53; State Univ NY Col Environ Sci & Forestry, MS, 55, PhD(ecol, forest zool), 65. *Prof Exp:* From asst prof to prof biol, State Univ NY Col Geneseo, 55-68; dir, Pingree Park Campus, 68-74, PROF RESOURCE CONSERV, COL FORESTRY & NATURAL RESOURCES, COLO STATE UNIV, 68- *Concurrent Pos:* Mem Yellowstone field res exped, Atmospheric Sci Res Ctr, State Univ NY Albany, 63-64; res scientist, Bahamas Marine Surv, Off Naval Res, 67; staff mem, Coop Sci Prog, Univ Sopron, Hungary, 74; consult atmospheric monitoring, Stillwater Platinum Mines, Mont, 76; res scientist, Nowcast Weather Satellite Prog, NASA-Colo State Univ, 76; fel explor San Francisco, Fund Improv Secondary Educ, HEW, 80; consult ed, J Environ Educ, 81- *Mem:* AAAS; Explorers Club; Nat Wildlife Fedn; Sigma Xi; Conserv Educ Asn. *Res:* Soil insects as indicators of use patterns in recreation areas. *Mailing Add:* Col Forest/Nat Res Colo State Univ Ft Collins CO 80523

MAHONEY, FRANCIS JOSEPH, b Boston, Mass, Mar 18, 36; m 70; c 3. PHYSICS, BIOLOGY. *Educ:* Col Holy Cross, BS, 57; Univ Rochester, MS, 58; Harvard Univ, MS, 60; Mass Inst Technol, PhD(nuclear eng), 68. *Prof Exp:* Asst radiation physics, Harvard Univ, 58-60, asst radiation physics & biol, 60-61; physicist, Cambridge Nuclear Corp, 61-64; res asst nuclear eng, Mass Inst Technol, 64-67; res physicist, US Army Natick Labs, 67-71; grants assoc, NIH, 71-72; PROG DIR FOR RADIATION, NAT CANCER INST, 72- *Concurrent Pos:* Radiation physicist, Avco Corp, 66-67. *Mem:* AAAS; Am Phys Soc; Radiation Res Soc. *Res:* Radiation physics and biology; nuclear physics and engineering. *Mailing Add:* 1978 Lancashire Dr Rockville MD 20854

MAHONEY, JOAN MUNROE, b Providence, RI, May 13, 38; m 71; c 2. BIOCHEMISTRY. *Educ:* Chatham Col, BS, 60; St Lawrence Univ, MS, 62; State Univ NY Upstate Med Ctr, PhD(biochem), 69. *Prof Exp:* Instr chem labs, St Lawrence Univ, 60-62; USPHS training grant, Inst Enzyme Res, Univ Wis, 68-69; res assoc mitochondrial metab, Sch Med, Ind Univ, Indianapolis, 69-71; instr, 71-74, asst prof biochem, Terre Haute Ctr Med Educ, Ind State Univ, 74-78; from asst prof to assoc prof, 78-89, PROF BIOCHEM, UNIV OSTEOP MED & HEALTH SCI, 89- *Concurrent Pos:* Curric coordinator, Univ Osteop Med & Health Sci. *Mem:* AAAS; Am Chem Soc; Biophys Soc. *Res:* Mitochondrial metabolism; drugs and liposomes; membranes and their control of processes; enzyme regulation. *Mailing Add:* Univ Osteop Med & Health Sci 3200 Grand Ave Des Moines IA 50312

MAHONEY, MARGARET E, b Nashville, Tenn, Oct 24, 24. SCIENCE POLICY. *Educ:* Vanderbilt Univ, BS, 46. *Hon Degrees:* LHD, Meharry Med Col, 77, Univ Fla, 80, Williams Col, 83; LLD, Med Col Pa, 82, Smith Col, 85, Beaver Col, 85. *Prof Exp:* Foreign affairs officer, US Dept State, Washington, DC, 46- 53; exec assoc & assoc secy, Carnegie Corp, NY, 53-72; vpres, Robert Wood Johnson Found, Princeton, NJ, 72-80; PRES, COMMONWEALTH FUND, 80- *Concurrent Pos:* Mem adv bd, Off Chief Med Examr, NY, 87-, Duke Univ Med Ctr, 87-, Barnard Col Inst Med Res, 86-; mem, Mayor's Comn Sci & Technol, NY, 86-; adv comt, Synergos Inst, 87-; mem vis comt coun, div biol sci, Pritzker Sch Med, Univ Chicago, 84-; mem corp vis comt med dept, Mass Inst Technol, 75- *Honors & Awards:* Frank H Lahey Mem Award, 84. *Mem:* Inst Med-Nat Acad Sci; fel AAAS; NY Acad Med; fel Am Acad Arts & Sci. *Res:* National public policy and international policy issues; organizational theory and management. *Mailing Add:* Commonwealth Fund One E 75th St New York NY 10021

MAHONEY, MICHAEL S, b New York, NY, June 30, 39. COMPUTER SCIENCE, HISTORY OF COMPUTERS. *Educ:* Harvard Univ, AB, 60; Princeton Univ, PhD(hist sci), 67. *Prof Exp:* From instr to assoc prof, 65-80, PROF HIST, PRINCETON UNIV, 80- *Concurrent Pos:* Mem conf bd, Asn Comput Mach, 87- *Mem:* Asn Comput Mach; Inst Elec & Electronics Engrs Computer Soc; Hist Sci Soc. *Mailing Add:* Prog Hist Sci Princeton Univ 303 Dickinson Hall Princeton NJ 08544

MAHONEY, ROBERT PATRICK, b Poughkeepsie, NY, Dec 27, 34; m 61; c 5. MICROBIAL PHYSIOLOGY, ELECTRON MICROSCOPY. *Educ:* State Univ NY Co, New Paltz, BS, 58; Syracuse Univ, MS, 61, PhD(microbiol), 64. *Prof Exp:* Asst prof, 64-68, chmn dept, 77-80, ASSOC PROF BIOL, SKIDMORE COL, 68- *Concurrent Pos:* NSF fac fel & guest investr, Biol Dept, Woods Hole Oceanog Inst, 70-71. *Mem:* AAAS; Am Soc Microbiol; Electron Micros Soc Am; NY Acad Sci. *Res:* Physiology and electron microscopy of autotrophic bacteria; iron and sulfur oxidizing species. *Mailing Add:* Dept of Biol Skidmore Col Saratoga Springs NY 12866

MAHONEY, WALTER C, b Seattle, Wash, Sept 12, 51; m; c 2. BIOCHEMISTRY. *Educ:* Univ Wash, BS, 74; Purdue Univ, PhD(biochem), 80. *Prof Exp:* Res technician, Dept Med, Div Med Genetics, Univ Wash, Seattle, 74-76, sr fel, 81-82; dir protein & nucleic acid chem, Immuno Nuclear Corp, 82-84, dir res, 84-85; ADJ ASST PROF & ASSOC MEM, DEPT GENETICS & CELL BIOL, UNIV MINN, 84-; DIR ASSAY DEVELOP, BIOTOPE, INC, 89- *Concurrent Pos:* Consult, Microchem Facil, Univ Minn, 85-87; peptide-protein chemist, Instrument Develop Group, Kallestad Diagnostics, Erbamount, 85, mgr new technol develop, 85-86, assoc dir res & develop, 86-87, dir, 87-89; co-investr, Nat Inst Diabetes & Digestive & Kidney Dis, NIH, 87-90. *Mem:* Am Soc Biochem & Molecular Biol. *Res:* Author of numerous technical publications. *Mailing Add:* Biotope Inc 12277 134th Ct NE Redmond WA 98052

MAHONEY, WILLIAM C, b New York, NY, Jan 6, 27; m 53; c 4. GEODESY, PHOTOGRAMMETRY. *Educ:* Syracuse Univ, BS, 51; Ohio State Univ, MS, 55, PhD(photogrammetry), 61. *Prof Exp:* Proj engr, Engineer Res & Develop Lab, Ohio State Univ, 51-53, res asst photogram, Univ Res Found, 53-55, res assoc, 55-59; supvry cartographer, Missile Support Div, Aeronaut Chart & Info Ctr, US Air Force, 59-64, phys scientist & chief tech develop, 64-69, chief, Missile Support Div, 69-71, tech adv to chief P&D plant, Sci & Tech Off, 71-72; chief, Advan Technol Div, 72-78, DEP DIR SYSTS & TECH, AEROSPACE CTR, DEFENSE MAPPING AGENCY, 78- *Mem:* Am Soc Photogram; Am Geophys Union; Am Cong Surv & Mapping. *Res:* Analytical photogrammetry; block adjustment; map compilation, celestial mechanics; computer technology. *Mailing Add:* Dept Art N Adam St Col Church St North Adams MA 01247

MAHONY, DAVID EDWARD, b St John, NB. MICROBIOLOGY. *Educ:* Acadia Univ, BSc, 62; Dalhousie Univ, MSc, 64; McGill Univ, PhD(bact), 67. *Prof Exp:* Lectr, 67-69, asst prof, 69-76, assoc prof, 76-81, PROF MICROBIOL, FAC MED, DALHOUSIE UNIV, 81- *Mem:* Can Soc Microbiol; Am Soc Microbiol; Can Col Microbiologists. *Res:* Bacteriocins, toxins and plasmids of Clostridium perfringens and difficile; typing methodologies. *Mailing Add:* Dept of Microbiol Fac Med Dalhousie Univ Halifax NS B3H 4H7 Can

MAHONY, JOHN DANIEL, b New York, NY, Jan 15, 31. NUCLEAR CHEMISTRY, AQUATIC CHEMISTRY. *Educ:* St John's Univ, NY, 51; Univ Conn, MS, 53; Univ Calif, Berkeley, PhD(nuclear chem), 65. *Prof Exp:* Asst prof chem, Marquette Univ, 65-67; from asst prof to assoc prof, 67-81, PROF CHEM, MANHATTAN COL, 81-, PROF ENVIRON ENG, 90- *Concurrent Pos:* Vis res collabr, State Univ NY Stony Brook, 67-72. *Mem:* Am Chem Soc; Am Microchem Soc; Sigma Xi. *Res:* Sediment metal chemistry and toxicology; environmental applications of chemisorption; radiochemical tracers; nuclear reactions and nuclear fission. *Mailing Add:* Dept Chem Manhattan Col Bronx NY 10471

MAHOOD, GAIL ANN, b Oakland, Calif, June 27, 51; m 82. VOLCANOLOGY, IGNEOUS PETROLOGY. *Educ:* Univ Calif, Berkeley, AB, 74, MA, 76, PhD(geol), 80. *Prof Exp:* Geologist, Copan Archeol Proj, Honduras, 78; ASST PROF GEOL, STANFORD UNIV, 79- *Concurrent Pos:* Prin investr, NSF, 80-, Petrol Res Fund, Am Chem Soc, 81-83; vis asst prof, dept geosci, Pa State Univ, 83; mem Int Asn Volcanology & Chem of Earth's Interior subcomt, US Nat Comt, Int Union Geol & Geophys, 85-; mem Geochem Subcomt, Off Int Affairs, Nat Res Coun, 85- *Mem:* Geol Soc Am; Am Geophys Union. *Res:* Compositional zonation in silicic to intermediate magma chambers; crystal-melt partitioning; histories of caldera complexes; triggering mechanisms for explosive volcanism; physical volcanology of pyroclastic rocks; potassium-aryl dating of young volcanic rocks. *Mailing Add:* Dept Geol Stanford Univ Stanford CA 94305

MAHOWALD, ANTHONY P, b Albany, Minn, Nov 24, 32; m 71; c 3. DEVELOPMENTAL GENETICS. *Educ:* Spring Hill Col, BS, 58; Johns Hopkins Univ, PhD(biol), 62. *Prof Exp:* Res fel biol, Johns Hopkins Univ, 62-66; asst prof, Marquette Univ, 66-70; asst mem & vis scientist, Inst Cancer Res, 70-72; from assoc prof to prof biol, Ind Univ, Bloomington, 76-82; Henry Willson Payne prof & chmn dept, Develop Genetics & Anat, Case Western Reserve Univ, Cleveland, 82-88, chmn dept genetics, 88-90; LOUIS BLOCK PROF & CHMN, DEPT MOLECULAR GENETICS & CELL BIOL, UNIV CHICAGO, 90- *Concurrent Pos:* Mem, Soc Scholars, Johns Hopkins Univ. *Mem:* Genetics Soc Am (secy, 86-88); AAAS; Am Soc Cell Biologists; Soc Develop Biol; Am Soc Human Genetics; Soc Develop Biol (pres, 89); fel AAAS. *Res:* Developmental and molecular genetics of drosophila; vitellogenesis; molecular analysis of maternal effect mutations affecting both germ cell formation and gastrulation; developmental biology; sex determination of germ cells. *Mailing Add:* Dept Molecular Genetics & Cell Biol Univ Chicago 920 E 58th St Chicago IL 60637

MAHOWALD, MARK EDWARD, b Albany, Minn, Dec 1, 31; m 54; c 5. MATHEMATICS. *Educ:* Univ Minn, BA, 53, MA, 54, PhD(math), 55. *Prof Exp:* Sr engr, Gen Elec Co, 56-57; asst prof math, Xavier Univ, Ohio, 57-59 & Syracuse Univ, 59-63; chmn dept, 72-75, PROF MATH, NORTHWESTERN UNIV, EVANSTON, 63- *Concurrent Pos:* Sloan fel, 65-67. *Mem:* Am Math Soc; Math Asn Am; Sigma Xi. *Res:* Homotopy theory; algebraic topology; topological groups. *Mailing Add:* 2109B Sherman Ave Evanston IL 60201

MAHOWALD, THEODORE AUGUSTUS, b St Cloud, Minn, June 22, 30; m 54; c 7. BIOCHEMISTRY. *Educ:* St John's Univ, Minn, BA, 52; St Louis Univ, PhD(biochem), 57. *Prof Exp:* USPHS fel, Enzyme Inst, Univ Wis, 57-60; mem staff, Scripps Clin & Res Found, Univ Calif, 60-62; asst prof biochem, Med Col, Cornell Univ, 62-69; ASSOC PROF BIOCHEM, MED COL, UNIV NEBR AT OMAHA, 69- *Mem:* AAAS; Am Soc Biol Chemists; Am Chem Soc. *Res:* Metabolism of bile acids; mechanism and sites of enzyme action; metab of methionine. *Mailing Add:* Dept Biochem Univ Nev Med Ctr 600 S 42nd St Omaha NE 68198-4525

MAHROUS, HAROUN, b Fayum, Egypt, Apr 15, 21; m 51; c 1. ELECTRICAL ENGINEERING. *Educ:* Cairo Univ, BSEE, 43; Swiss Fed Inst Technol, DSc(elec eng), 51. *Prof Exp:* Asst physics, Cairo Univ, 43-45; sr lectr elec eng, Ibrahim Univ, Egypt, 51-54; assoc prof, Pratt Inst, 54-56; assoc prof, Univ Ill, 56-58; PROF ELEC ENG & CHMN DEPT, PRATT INST, 58- *Concurrent Pos:* Consult, Solid State Memory Div, Int Bus Mach Corp, 63 & Bell Tel Labs, 66-; prog dir, Eng Div, NSF, 68. *Mem:* Sr mem Inst Elec & Electronics Engrs; NY Acad Sci. *Res:* Computer systems; memory; microwaves and electromagnetic theory; electrical power field. *Mailing Add:* Dept of Elec Eng Pratt Inst 200 Willoughby Ave Brooklyn NY 11205

MAHRT, JEROME L, b Colon, Nebr, Dec 20, 37; m 60; c 3. PARASITOLOGY, ZOOLOGY. *Educ:* Utah State Univ, BS, 60, MS, 63; Univ Ill, PhD(vet parasitol), 66. *Prof Exp:* From asst prof to assoc prof, 66-83, assoc dean sci, 80-81, PROF ZOOL, UNIV ALTA, 83- *Mem:* Can Soc Zoologists; Am Soc Parasitologists; Soc Protozoologists. *Res:* Protozoan parasitology with specialization in the Coccidia and blood parasites of reptiles and birds. *Mailing Add:* Dept of Zool Univ of Alta Edmonton AB T6G 2E2 Can

MAHTAB, M ASHRAF, b Ambala, India, Oct 15, 35; Can citizen; c 2. ROCK MECHANICS, NUMERICAL ANALYSIS. *Educ:* Mont Univ, Butte, BS, 59; McGill Univ, MEng, 65; Univ Calif, Berkeley, PhD(civil eng), 70. *Prof Exp:* Lectr mining eng, Eng Univ, Pakistan, 59-63; res asst mining eng, McGill Univ, 63-65; teaching fel geol eng, Univ Calif, Berkeley, 65-69; mining engr, US Bur Mines, Colo, 70-75; rock mech engr, Acres Consult Serv Ltd, Can, 75-79; ASSOC PROF MINING, COLUMBIA UNIV, 80- *Concurrent Pos:* Sr mining engr, Iron Ore Co, Can, 64-65; consult, United Nations, GeoMech Inc & Engrs Int, 82- *Mem:* Am Inst Mining Engrs Soc Mining Eng; Asn Eng Geologists; Int Soc Rock Mech. *Res:* Characterization of jointed rock mass; support of tunnels in loosening rock; rock bursts and gas outbursts; progressive failure and directional strength of pillars; thermal-mechanical analysis of radioactive waste repositories. *Mailing Add:* RR 1 160D Accord NY 12404

MAI, KLAUS L(UDWIG), b Changsha, China, Mar 7, 30; US citizen; m 57; c 4. CHEMICAL ENGINEERING. *Educ:* Gonzaga Univ, BS, 51; Univ Wash, MS, 52, PhD(chem eng), 54. *Prof Exp:* Develop engr, Procter & Gamble Co, 55-57; technologist process develop, Shell Chem Co, 57-59, asst mgr opers, 59-60, chief technologist, 60-61, mgr lab tech, 61-63, res dir tech serv & appln, 64, div mgr opers, 65-66, plant mgr, 66-68, mgr mfg, 68-69, gen mgr, Indust Chem Div, 69-72, gen mgr, Polymers Div, 72-74, overseas assignment, 74-76, vpres transp & supplies, Shell Oil Co, 76-77, vpres, 77-81, PRES, SHELL DEVELOP CO, 81- *Concurrent Pos:* Dir, Am Inst Chem Engrs, chmn, Coun Chem Res; mem, Indust Adv Bd, Am Chem Soc. *Mem:* Fel Am Inst Chem Engrs; Am Chem Soc; Soc Chem Indust; Indust Res Inst; Sigma Xi. *Res:* Business, manufacturing, marketing and research administration. *Mailing Add:* PO Box 2463 Houston TX 77024

MAI, WILLIAM FREDERICK, b Greenwood, Del, July 23, 16; m 41; c 3. PLANT PATHOLOGY. *Educ:* Univ Del, BS, 39; Cornell Univ, PhD(plant path), 45. *Prof Exp:* Asst prof, 46-52, prof, 52-81, Liberty Hyde Bailey Prof, 81-83, EMER PROF PLANT PATH, CORNELL UNIV, 83- *Honors & Awards:* Adventurer in Res Award, IX Int Cong Plant Protection, 79; Hon Mem, Soc Nematologists, 80. *Mem:* Fel Am Phytopath Soc; Soc Nematol (vpres, 68, pres, 69); Am Inst Biol Sci; Potato Asn Am; Soc Europ Nematol; Sigma Xi. *Res:* Plant pathogenic and soil inhabiting nematodes; diseases of plants caused by nematodes. *Mailing Add:* 613 E Shore Dr Ithaca NY 14850

MAIBACH, HOWARD I, b New York, NY, July 18, 29; c 3. DERMATOLOGY. *Educ:* Tulane Univ, AB, 50, MD, 55; Am Bd Dermat, dipl, 61; Univ Paris, PhD. *Prof Exp:* Asst instr, Sch Med, Univ Pa, 58-61; from asst prof to assoc prof, 61-71, PROF DERMAT, SCH MED, UNIV CALIF, SAN FRANCISCO, 71-, VCHMN DIV DERMAT, 62- *Concurrent Pos:* USPHS fel, Pa Hosp, 59-61; lectr, Grad Sch Med, Univ Pa, 60-61; pvt pract, 61-; mem staff, Herbert C Moffitt Hosps, Univ Calif, 61-; consult, Sonoma State Hosp, Eldridge, 62-64, Calif State Pub Health Serv, Stanford Res Inst, Calif Med Facil, Vacaville, Letterman & San Francisco Gen Hosps, 62- & Vet Admin Hosp, San Francisco, 63- *Mem:* AAAS; Soc Invest Dermat; Am Acad Dermat; fel Am Col Physicians; Am Fedn Clin Res. *Res:* Dermatology. *Mailing Add:* Dept Dermat Univ Calif Sch Med San Francisco CA 94143

MAIBENCO, HELEN CRAIG, b Scotland, June 9, 17; nat US; m 57; c 2. ANATOMY. *Educ:* Wheaton Col, Ill, BS, 48; De Paul Univ, MS, 50; Univ Ill, PhD, 56. *Prof Exp:* Instr pharmacol, Presby Hosp Sch Nursing, 50-54; asst, Univ Ill Col Med, 54-56, from instr to prof anat, 56-73; PROF ANAT, RUSH MED COL, 73- *Mem:* AAAS; Endocrine Soc; Am Asn Anatomists; Sigma Xi. *Res:* Physiology of connective tissues; endocrine relationships; morphological changes associated with aging; connective tissues of the female reproductive system. *Mailing Add:* 600 S Paulina Chicago IL 60612

MAICKEL, ROGER PHILIP, b Floral Park, NY, Sept 8, 33; m 56; c 2. PHARMACOLOGY, SCIENCE ADMINISTRATION. *Educ:* Manhattan Col, BS, 54; Georgetown Univ, MS, 57, PhD(chem), 60. *Prof Exp:* Chemist, Lab Chem Pharmacol, Nat Heart Inst, 55-60, biochemist, 60-63, sect head biochem function, 63-65; from assoc prof to prof pharmacol, Ind Univ, Bloomington, 65-77, chief, med sci sect, Inst Res Pub Safety, 70-75, dir pharmacol, 71-77; head dept, 77-83, PROF PHARMACOL & TOXICOL, SCH PHARM & PHARMACOL SCI, PURDUE UNIV, WEST LAFAYETTE, 77- , DIR, LAB ANIMAL PROG, 88- *Concurrent Pos:* Exec ed, Life Sci, 65-69; Am Chem Soc tour lectr, 65- *Mem:* Fel AAAS; Am Chem Soc; Am Soc Pharmacol & Exp Therapeut; fel Am Inst Chemists; NY Acad Sci; fel Royal Soc Chem. *Res:* Toxicology. *Mailing Add:* Dept Pharmacol & Toxicol Purdue Univ West Lafayette IN 47907

MAIELLO, JOHN MICHAEL, b New York, NY, Oct 5, 42. MYCOLOGY, ENVIRONMENTAL BIOLOGY. *Educ:* Hunter Col, BA, 65; Rutgers Univ, PhD(mycol), 72. *Prof Exp:* Asst prof, 72-80, ASSOC PROF MYCOL, RUTGERS UNIV, NEWARK, 80- *Concurrent Pos:* Dir & cur, mycol cult collection, Rutgers Univ, 72- *Mem:* Sigma Xi; Mycol Soc Am; AAAS. *Res:* Morphology and physiology of pycnidial development in the sphaeropsidales; pathology of economically important plants; air pollutants injurious to vegetation in New Jersey; deep level optical spectroscopy in semiconductors. *Mailing Add:* Dept Bot Rutgers Univ Newark NJ 07102

MAIENSCHEIN, FRED (CONRAD), b Belleville, Ill, Oct 28, 25; m 48; c 2. REACTOR PHYSICS & SHIELDING. *Educ:* Rose Polytech Inst, BS, 45; Ind Univ, MS, 48, PhD(physics), 49. *Prof Exp:* Sr physicist, Nuclear Energy Propulsion for aircraft div, Fairchild Engine & Airplane Corp, 49-51; sr physicist, Oak Ridge Nat Lab, 51-66, dir eng physics & maths div, 66-90; RETIRED. *Mem:* Am Phys Soc; Sigma Xi; fel Am Nuclear Soc. *Res:* Neutron and reactor physics; shielding; engineering physics. *Mailing Add:* 838 W Outer Dr Oak Ridge TN 37830

MAIENSCHEIN, JANE ANN, b Oak Ridge, Tenn, Sept 23, 50; m 82. HISTORY & PHILOSOPHY OF EMBRYOLOGY, GENETICS. *Educ:* Yale Univ, BA, 72; Ind Univ, MA, 75, PhD(hist & philos sci), 78. *Prof Exp:* Asst prof hist sci, Dickinson Col, 78-80; asst prof philos, 81-86, assoc prof, 86-90, PROF PHILOS & ZOOL, ARIZ STATE UNIV, 90-, CHAIR PHILOS, 91- *Concurrent Pos:* Vis scholar, Harvard Univ, 83-84; vis assoc prof, Stanford Univ, 87. *Mem:* Int Soc Hist Philos & Social Studies Biol (pres, 89-91); AAAS; Am Soc Zoologists; Sigma Xi. *Res:* History and philosophy of embryology, genetics and cytology in order to explore the nature of the biological sciences generally and the factors that cause changes through time and place. *Mailing Add:* Dept Philos Ariz State Univ Tempe AZ 85287-2004

MAIER, CHARLES ROBERT, b Leoti, Kans, Oct 9, 28; m 56; c 3. BOTANY. *Educ:* Emporia State Univ, BS, 53, MS, 55; Ore State Univ, PhD(plant path), 59. *Prof Exp:* Asst, Emporia State Univ, 54-55; from asst prof to assoc prof plant path, NMex State Univ, 58-68; PROF BIOL, WAYNE STATE COL, 68-, CUR ARBORETUM, 77- *Concurrent Pos:* Res prof plant path, Univ Nebr-Lincoln, 72-78. *Honors & Awards:* LAL, hon Phys Sci; BBB, hon Biol. *Mem:* Nat Asn Biol Teachers; Nat Sci Teachers Asn. *Res:* Soil microbiology; control of diseases of cotton; antibiotic chemotherapy of hop downy mildew; organismal pollution of irrigation water. *Mailing Add:* Dept Biol Math Sci Div Wayne State Col Wayne NE 68787-1488

MAIER, CHRIS THOMAS, b Pontiac, Mich, Jan 14, 49; m 81; c 2. INSECT ECOLOGY, INSECT BEHAVIOR. *Educ:* Univ Mich, Ann Arbor, BS, 71; Univ Ill, Urbana, MS, 73, PhD(entom), 77. *Prof Exp:* Asst entomologist, 77-82, ASSOC ENTOMOLOGIST, CONN AGR EXP STA, 82- *Mem:* Xerces Soc; Entom Soc Am; Entom Soc Can. *Res:* Host-specificity of leafminers; distribution of rare and endangered insects; ecology of insects inhabiting peatlands; biological control of apple leafminers; behavior of flower flies. *Mailing Add:* Dept Entom Conn Agr Exp Sta 123 Huntington St New Haven CT 06504-1106

MAIER, EUGENE ALFRED, b Tillamook, Ore, May 7, 29; m 52; c 4. MATHEMATICS. *Educ:* Univ Ore, BA, 50, MA, 51, PhD(math), 54. *Prof Exp:* From asst prof to assoc prof & chmn dept, 55-61; assoc prof, 61-70, PROF MATH, UNIV ORE, 70- *Mem:* Am Math Soc; Math Asn Am. *Res:* Number theory; mathematics education. *Mailing Add:* 2021 Skyview Ct Milwaukee OR 97222

MAIER, EUGENE JACOB RUDOLPH, b Washington, DC, Sept 1, 31; m 59; c 3. GEOPHYSICS. *Educ:* Mass Inst Technol, BS, 53; Carnegie Inst Technol, MS, 59, PhD(meson physics), 62. *Prof Exp:* Physicist, Appl Physics Lab, Johns Hopkins Univ, 56; physicist, 62-87, HEAD, SOLAR RADIATION OFF, GODDARD SPACE FLIGHT CTR, NASA, GREENBELT, MD. *Mem:* AAAS; Am Phys Soc; Am Geophys Union. *Res:* Structure and direct measurements of ionosphere and atmosphere; measurements of interplanetary medium; spacecraft experiment instrumentation; high energy meson physics; energetic particle experiments; solar radiation and solar activity. *Mailing Add:* 1173 River Bay Rd Annapolis MD 21401

MAIER, GEORGE D, b Chicago, Ill, July 24, 30. BIOCHEMISTRY. *Educ:* Cornell Col, BA, 53; Iowa State Univ, MS, 56, PhD(biochem), 62. *Prof Exp:* Asst biochemist, Am Meat Inst Found, Chicago, 57-59; fel, Mass Gen Hosp, Boston, 62-63; fel, Sch Med, Western Reserve Univ, 63-65; asst prof chem, 65-74, ASSOC PROF CHEM, COLBY COL, 74- *Res:* Protein primordial synthesis; thiamine and renin chemistry. *Mailing Add:* 1355 W Estes Chicago IL 60626

MAIER, HERBERT NATHANIEL, b Baltimore, Md, Oct 15, 43; m 76; c 1. PHYSICAL CHEMISTRY, ANALYTICAL CHEMISTRY. *Educ:* Wilkes Col, BS, 65, MS, 67; Pa State Univ, PhD(phys chem), 71. *Prof Exp:* Dir labs, Precision Gas Prods Inc & Burdox Inc, 75-77; dir labs, Ideal Gas Prods, 78-80, plant mgr, 80-84; plant mgr, Alphagaz, 84-85; DIR RES & DEVELOP, CRYOGENIC RARE GAS LABS, 85- *Mem:* Am Chem Soc; Optical Soc Am; Soc Photo-optical Instrumentation Engrs. *Res:* Obtaining response factors; obtaining difficult separations; increasing sensitivity on gas chromatographic analysis of gas mixtures; perfecting elemental fluorine analysis; perfecting hydrogen chloride analysis. *Mailing Add:* 1683 Seignidus Dr Charleston SC 29407-8232

MAIER, ROBERT HAWTHORNE, b New York, NY, Oct 26, 27; m 52; c 3. TRACE ELEMENTS, PLANT PHYSIOLOGY. *Educ:* Univ Miami, BS(chem) & BS(bot), 51; Univ Ill, MS, 52, PhD(soil & plant chem), 54. *Prof Exp:* Asst agron, Univ Ill, 51-54; from asst prof to prof agr chem & soils, Univ Ariz, 56-67, asst dean grad col, 66-67; asst chancellor & prof chem, Univ Wis-Green Bay, 67-70, vchancellor & prof environ sci, 70-75, prof sci, Environ Change & Pub & Environ Admin, 75-79; prof biol & polit sci & vchancellor acad affairs, 79-83, PROF EXP SURG, BIOL & POLIT SCI, E CAROLINA UNIV, 83- *Concurrent Pos:* Vis sci, Am Soc Argon, Liberal Arts Col, 63-64; Am Coun Educ fel acad admin, Univ NC, 65-66; mem, NC Gov's Comt on the Year 2000, 81-83. *Mem:* Fel AAAS; fel Am Inst Chemists; fel Am Soc Agron; fel Soil Sci Soc Am; Am Chem Soc; NY Acad Sci; Am Inst Biol Sci. *Res:* Cellular physiology and biochemistry of metals, metal chelates and chelating agents; analytical chemistry of biological material; chemistry of soil-plant-human relationships. *Mailing Add:* Dept Surg E Carolina Univ Greenville NC 27858-4354

MAIER, ROBERT J, b Detroit, Mich, July 1, 51. MICROBIOLOGY. *Educ:* Univ Wis, MS, 75, PhD(bacteriol), 77. *Prof Exp:* Res assoc bacteriol, Univ Wis, 77; postdoctoral fel, Lab Nitrogen Fixation Res, Ore State Univ, 77-78, postdoctoral res assoc, 78-79; from asst prof to assoc prof, 79-88, PROF BIOL, MCCOLLUM-PRATT INST, JOHNS HOPKINS UNIV, 88-; SR RES SCIENTIST, CTR MARINE BIOTECHNOL, UNIV MD, 89- *Concurrent Pos:* Ed, Appl & Environ Microbiol. *Mem:* Am Soc Biol Chem & Molecular Biol; Am Soc Microbiol. *Res:* Microbiology. *Mailing Add:* Biol Dept Johns Hopkins Univ Charles & 34th Sts Baltimore MD 21218

MAIER, ROBERT S, b Santa Monica, Calif, Oct 27, 57. APPLIED MATHEMATICS. *Educ:* Calif Inst Technol, MS, 80; Rutgers, State Univ NJ, PhD(math), 83. *Prof Exp:* Instr math, Univ Tex, Austin, 83-86; vis asst prof, 86-88, ASST PROF MATH, UNIV ARIZ, 88- *Mem:* Am Math Soc; Asn Comput Mach; Soc Indust Appl Math. *Res:* Applied probability; stochastic modelling, including the performance analysis of computer systems. *Mailing Add:* Dept Math Univ Ariz Tucson AZ 85721

MAIER, SIEGFRIED, b Stuttgart, Ger, Apr 22, 30; US citizen; m 54; c 3. MICROBIOLOGY. *Educ:* Capital Univ, BS, 58; Ohio State Univ, MSc, 60, PhD(microbiol), 63. *Prof Exp:* Asst prof bact, 63-68, assoc prof microbiol, 68-77, PROF MICROBIOL, OHIO UNIV, 77- *Concurrent Pos:* AEC res partic, Argonne Nat Lab, 69-70; vis prof, Univ Kiel, Ger, 76; taxonomist, Comt Approved Bact Names, Int Comt Syst Bact, 75-80. *Mem:* Am Soc Microbiol; Can Soc Microbiol; Sigma Xi. *Res:* Bacteriophages; physiology and structure of Beggiatoaceae; anaerobic sporulation in bacillus. *Mailing Add:* Dept Zool & Biomed Sci Ohio Univ Athens OH 45701

MAIER, V(INCENT) P(AUL), b Carteret, NJ, Dec 11, 30; m 53; c 2. BIOCHEMISTRY, PHYTOHORMONES. *Educ:* Univ Calif, Los Angeles, BS, 52; Univ Calif, Davis, MS, 54, PhD(agr chem), 57. *Prof Exp:* Chemist, USDA, 57-58, prin chemist, 59-61, invest head, 62-69, lab chief, 70-77, dir, Fruit & Veg Chem Lab, Agr Res Serv, 77-89; RETIRED. *Mem:* AAAS; Am Soc Plant Physiologists; Am Chem Soc; Inst Food Technol; Plant Growth Regulator Soc Am. *Res:* Sub-zero enzyme kinetics; heme compound catalyzed lipid oxidation; chemistry of dates; biochemistry and enzymes of citrus fruits; biosynthesis and regulation of flavonoid and limonoid bitter principles; bioregulation of plant hormones. *Mailing Add:* 400 Lugonia St New Port Beach CA 92663

MAIER, WILLIAM BRYAN, II, b Oklahoma City, Okla, Jan 30, 37; m 58; c 4. ATOMIC & MOLECULAR COLLISIONS. *Educ:* Univ Chicago, BS, 59, MS, 60, PhD(physics), 65. *Prof Exp:* STAFF MEM, LOS ALAMOS NAT LAB, 64- *Concurrent Pos:* Fel, NSF, 61-64; sci res coun vis fel, Gt Brit, 79-80. *Mem:* Am Phys Soc. *Res:* Ion-molecule reactions; electron-impact excitation of molecules; upper atmospheric chemistry; long-lived ionic states of small molecules; optical spectroscopy; photochemistry; synthesis of unstable compounds in cold solutions; isotope separation; spectra of solid solutions; properties of UF_6; classical field theory; theory of transient electric discharges. *Mailing Add:* 430 Camino Encantado Los Alamos NM 87545

MAIERHOFER, CHARLES RICHARD, civil engineering; deceased, see previous edition for last biography

MAILEN, JAMES CLIFFORD, b Colorado Springs, Colo, Aug 19, 37; m 63; c 1. CHEMICAL ENGINEERING. *Educ:* Kans State Univ, BS, 59; Univ Fla, PhD(chem eng), 64. *Prof Exp:* Develop engr, Oak Ridge Nat Lab, Martin Marietta Energy Systs, Inc, 63-73; GROUP LEADER, OAK RIDGE NAT LAB, 73- *Mem:* AAAS; Am Chem Soc; fel Am Inst Chem Engrs; Am Nuclear Soc; Sigma Xi. *Res:* Nuclear reactor safety; separations technology; high temperature reactions. *Mailing Add:* 134 Cumberland View Dr Oak Ridge TN 37830

MAILLIE, HUGH DAVID, b Chester, Pa, Nov 2, 32; m 58; c 3. RADIOBIOLOGY, HEALTH PHYSICS. *Educ:* La Salle Col, BA, 54; Univ Rochester, MS, 56, PhD(radiation biol), 63. *Prof Exp:* From instr to asst prof, 62-70, ASSOC PROF RADIATION BIOL, UNIV ROCHESTER, 70-, DIR HEALTH PHYSICS DIV, 72- *Mem:* AAAS; Health Physics Soc; Sigma Xi. *Res:* Radiation dosimetry and its application to the understanding of biological effects in man and laboratory animals. *Mailing Add:* Dept Rad Biol & Biophys Univ Rochester Rochester NY 14642

MAILLOT, PATRICK GILLES, b Paris, France, June 27, 58; m 83; c 2. COMPUTER GRAPHICS, VISUALIZATION & MULTIMEDIA. *Educ:* Univ Lyon, France, BS, 79, MS, 82, PhD(computer sci graphics), 86. *Prof Exp:* Engr, Secapa, France, 83-87; mgr, Thomson, France, 87-88; proj leader, 88-91, SOFTWARE MGR, SUN MICROSYSTS, 91- *Concurrent Pos:* Lectr, Univ Lyon, France, 83-86. *Mem:* Asn Comput Mach. *Res:* Graphics systems architectures. *Mailing Add:* Sun Microsysts Inc MTV 21-04 2550 Garcia Ave Mountain View CA 94043

MAILLOUX, GERARD, b Montreal, Que, Jan 17, 45. BIOLOGY, ENTOMOLOGY. *Educ:* Univ Sherbrooke, BS, 68; Univ Montreal, MSc, 71; Rutgers Univ, PhD(entom), 76. *Prof Exp:* Researcher, St Jean Res Sta, 76-79, RESEARCHER ENTOM, AGR QUEBEC, L'ASSOMPTION RES STA, CAN DEPT AGR, 79- *Mem:* Entom Soc Can; Japanese Soc Pop Ecol. *Res:* Economic entomology; population pest management; sampling and forecasting insect pests populations. *Mailing Add:* Ministere de l'Agriculture des Pescheries et de l'Alimentation 335 Chemin des 25 Est St Bruno de Montarville PQ J3V 4P6 Can

MAILLOUX, ROBERT JOSEPH, b Lynn, Mass, June 20, 38; m 67; c 3. ELECTROMAGNETISM, APPLIED PHYSICS. *Educ:* Northeastern Univ, BS, 61; Harvard Univ, SM, 62, PhD(appl physics), 65. *Prof Exp:* Res engr microwave technol, NASA Electronics Res Ctr, 65-70; res physicist, Air Force Cambridge Res Lab, 70-76; RES PHYSICIST ANTENNAS, ELECTROMAGNETIC SCI DIV, ROME AIR DEVELOP CTR, 76- *Concurrent Pos:* Res fel, Harvard Univ, 68; tech coop prog deleg, US, Brit, Can & Australian Exchange Agreement, 77-82; Inst Elec & Electronics Engrs fel, 78; nat lectr, Antenna & Propagation Soc, 81; lectr, Tufts Univ, 85-88; adj prof, Univ Mass, 86-88, Northeastern Univ, 88- *Honors & Awards:* Marcus O'Day Award, Air Force Cambridge Res Lab, 71. *Mem:* Inst Elec & Electronics Engrs; Sigma Xi; Antenna & Propagation Soc (vpres, 82, pres, 83). *Res:* Antennas; periodic structures; electromagnetic scattering; diffraction; publications and 5 textbook chapters on phased arrays, conformed antennas and limited field of view scanning systems. *Mailing Add:* 98 Concord Rd Wayland MA 01778

MAILMAN, DAVID SHERWIN, b Chicago, Ill, June 29, 38; div; c 4. PHYSIOLOGY. *Educ:* Univ Chicago, BS, 58; Univ Ill, PhD(physiol), 64. *Prof Exp:* Fel biophys, Univ Md, 62-64; asst prof, 64-70, assoc prof, 70-75, PROF BIOL, UNIV HOUSTON, 75- *Concurrent Pos:* Adj assoc prof physiol, Univ Tex Med Sch Houston, 73-83. *Mem:* Am Physiol Soc; Am Soc Zoologists; Soc Exp Biol & Med; Sigma Xi. *Res:* Salt and water transport regulation by physical forces and hormones; membrane physiology; intestinal regulation. *Mailing Add:* Dept of Biol Univ of Houston Houston TX 77204

MAILMAN, RICHARD BERNARD, b New York, NY, Feb 6, 45; m 79; c 1. NEUROPHARMACOLOGY, NEUROTOXICOLOGY. *Educ:* Rutgers Univ, New Brunswick, BS, 68; NC State Univ, MS, 72, PhD(physiol, toxicol), 74. *Prof Exp:* Res assoc toxicol, NC State Univ, 74-75; fel neurobiol, 76-77, from res asst prof to res assoc prof pyschiat & chief neurotoxicol sect, 80-87, assoc prof, 87-88, PROF PSYCHIAT & PHARMACOL, BRAIN & DEV RES CTR, UNIV NC, CHAPEL HILL, 88- *Concurrent Pos:* Burroughs Wellcome Toxicol Scholar Award, 87-92. *Mem:* Am Soc Pharmacol & Exp Therapeut; Int Soc Neurochem; Am Soc Neurochem; Soc Toxicol; AAAS; fel Am Col Neuropsychopharmacology. *Res:* Pharmacological and toxicological mechanisms of drugs and toxicols on the central nervous system. *Mailing Add:* Univ NC Biol Res Ctr Chapel Hill NC 27599-7250

MAIMAN, THEODORE HAROLD, b Los Angeles, Calif, July 11, 27. PHYSICS, ADVANCED TECHNOLOGY. *Educ:* Univ Colo, BS, 49; Stanford Univ, MS, 51, PhD(physics), 55. *Prof Exp:* Sect head, Hughes Res Labs, 55-61; pres & founder, Korad Corp, Santa Monica, Calif, 61-68, Maiman Assoc, Los Angeles, 68-76; vpres & founder, Laser Video Corp, Los Angeles, 72-75; vpres technol, TRW Electronics Co, Los Angeles, 75-83; CONSULT, 83- *Honors & Awards:* Fannie & John Hertz Found Award, 66; Ballantine Award, Franklin Inst, 62. *Mem:* Nat Acad Engrs; fel Am Phys Soc; Optical Soc Am; Inst Elec & Electronics Engrs; Soc Motion Picture & TV Engrs; Sigma Xi. *Res:* Lasers. *Mailing Add:* One Rubio Rd Santa Barbara CA 93103-2134

MAIMONI, ARTURO, b Bogota, Colombia, Jan 24, 27; nat US; m 52; c 3. CHEMICAL ENGINEERING, NUCLEAR ENGINEERING. *Educ:* Calif Inst Technol, MSChE, 49; Univ Calif, PhD(chem eng), 56. *Prof Exp:* Asst to chief chemist, Cementos Diamante, Colombia, 45-46; jr res engr, Fluor Corp, 49; chemist, 52-55, chem engr, 55-64, dep div leader process & mat develop, 64-68, asst div leader, Gen Chem Div, 68-70, div leader, Inorg Mat Div, 70-76, proj leader nuclear mat safeguards, 76-79, proj leader, Int Energy Technol Assessment, 79-80, mem staff Long Range Planning, 81-83, PROJ LEADER ALUMINUM AIR BATTERY, LAWRENCE LIVERMORE NAT LAB, 84- *Concurrent Pos:* Fulbright prof, Univ Seville, 63-64. *Mem:* Am Chem Soc; Am Inst Chem Engrs; Electrochem Soc. *Res:* Crystallization; sedimentation; high temperature materials; nuclear safeguards and technology; coal gasification; materials selection for complex engineering systems. *Mailing Add:* 134 Crestview Dr Orinda CA 94563

MAIN, ALEXANDER RUSSELL, biochemistry, enzymology; deceased, see previous edition for last biography

MAIN, ANDREW JAMES, b Taunton, Mass, May 13, 42; m 64; c 3. MEDICAL ENTOMOLOGY, EPIDEMIOLOGY. *Educ:* Univ Mass, BS, 64, MS, 70; Yale Univ, MPH, 72, DrPh, 76. *Prof Exp:* Biologist, Encephalitis Field Sta, Mass Dept Pub Health, 60-70; instr med entom, 76-78, asst prof med entom, Dept Epidemiol & Pub Health, 78-83, ASSOC PROF EPIDEMIOL, YALE UNIV, 83 - *Mem:* Wildlife Dis Asn; Am Mosquito Control Asn. *Res:* Arbovirology; disease ecology; diseases of wildlife; arthropods of medical and veterinary importance. *Mailing Add:* Med Zool Dept US Naval Med Res Unit No 3 FPO NY 09527-1600

MAIN, CHARLES EDWARD, b Triadelphia, WVa, July 25, 33; m 54; c 2. PHYTOPATHOLOGY, EPIDEMIOLOGY. *Educ:* WVa Univ, BS, 59, MS, 61; Univ Wis, PhD(plant path), 64. *Prof Exp:* Fel plant path, Univ Wis, 64; asst prof, 64-76, PROF PLANT PATH, NC STATE UNIV, 76-; RES PLANT PATHOLOGIST, AGR RES SERV, USDA, 64- *Mem:* AAAS; Am Phytopath Soc. *Res:* Botanical epidemiology; disease forecasting and management; tobacco diseases; estimation and quantification of crop losses in agricultural cropping systems. *Mailing Add:* Dept Plant Path NC State Univ Box 7616 Raleigh NC 27695-7616

MAIN, JAMES HAMILTON PRENTICE, b Biggar, Scotland, June 7, 33; m 61; c 2. ORAL PATHOLOGY. *Educ:* Univ Edinburgh, BDS, 55, PhD(path), 64. *Prof Exp:* Lectr oral path, Univ Edinburgh, 61-66, sr lectr & consult, 66-69; PROF ORAL PATH, UNIV TORONTO, 69- *Concurrent Pos:* USPHS int res fel, NIH, 64-65; USPHS res grant, Univ Edinburgh, 66-69; Nat Cancer Inst Can res grant, Univ Toronto, 70-; mem, Nat Cancer Inst Can, 86-; head dent, Sunnybrook Hosp, 71- *Honors & Awards:* Colgate Prize, Int Asn Dent Res, 66; Clarke Prize Cancer Res, 68. *Mem:* Am Acad Oral Path; Royal Col Path; Int Asn Dent Res; Royal Col Dentists Can. *Res:* Induction of neoplasia; epithelio-mesenchymal interactions; developmental aspects of oncogenesis; biological testing of dental materials; salivary gland tumors. *Mailing Add:* Dept Oral Path Univ Toronto Fac Dent Toronto ON M5C 1G6 Can

MAIN, ROBERT ANDREW, b Billings, Mont, Sept 23, 23; div. ZOOLOGY, LIMNOLOGY. *Educ:* Univ Calif, Santa Barbara, AB, 48; Univ Wash, MS, 53; Univ Mich, PhD(zool), 61. *Prof Exp:* Asst zool, Univ Mich, 53-57, instr, 61; instr, Univ NH, 58-61; asst prof biol, Va Polytech, 61-62, Western Ill Univ, 62-64 & Tex Woman's Univ, 64-66; assoc prof, 66-70, PROF BIOL, CALIF STATE UNIV, HAYWARD, 70- *Mem:* Am Soc Limnol & Oceanog; Int Asn Theoret & Applied Limnol; Western Soc Naturalists. *Res:* Freshwater zooplankton and bottom fauna; life histories of planktonic copepods; biological illustration. *Mailing Add:* Dept Biol Sci Calif State Univ Hayward CA 94542

MAIN, STEPHEN PAUL, b Iowa City, Iowa, Aug 26, 40; m 63; c 1. BOTANY, AQUATIC ECOLOGY. *Educ:* Valparaiso Univ, BS, 62, MALS, 65; Ore State Univ, PhD(bot), 72. *Prof Exp:* Teacher biol, Crescent-Iroquois Community High Sch, Ill, 62-63; lab instr, Valparaiso Univ, 63-65; teacher, Santiam High Sch, Ore, 65-69; asst prof, 71-79, assoc prof biol, 79-90, PROF BIOL, WARTBURG COL, 90- *Mem:* Sigma Xi; Am Inst Biol Sci; Phycol Soc Am; Ecol Soc Am; Int Soc Diatom Res; Nat Asn Biol Teachers. *Res:* Ecology and taxonomy of diatoms in rivers, wetlands, and marine shoreline systems. *Mailing Add:* Dept Biol Wartburg Col Waverly IA 50677

MAIN, WILLIAM FRANCIS, b Fresno, Calif, July 2, 21; m 43; c 3. SYSTEMS DESIGN, SYSTEMS SCIENCE. *Educ:* Fresno State Col, AB, 43; Harvard Bus Sch, AMP, 64. *Prof Exp:* Electronic scientist, US Naval Res Lab, 43-55; mgr radar & data link dept, Missile & Space Div, Lockheed Aircraft Corp, 55-56, mgr electronic res, 56-65; dir electronics, Lockheed Missiles & Space Co, Calif, 65-66, asst gen mgr electronics, 66-69, asst chief engr space systs, 69-72, asst prog mgr, 72-74; pres, Pan Data Corp, 74-76; pres, Wackenhut Electronics, 76-79, tech dir, 79-84; CONSULT, W F MAIN ASSOCS, 84- *Mem:* Am Phys Soc; Sigma Xi. *Res:* Radar; communications and data processing systems. *Mailing Add:* 1326 Crane St Menlo Park CA 94025

MAINEN, EUGENE LOUIS, organic chemistry, for more information see previous edition

MAINES, MAHIN D, b Arak, Iran, July 31, 41; m 62; c 2. PHARMACOLOGY. *Educ:* Ball State Univ, BS, 64, MA, 67; Univ Mo, PhD(pharmacol), 70. *Prof Exp:* Res assoc pharmacol, Univ Mo, 70-71; NIH fel, Univ Minn, 71-73; res assoc, 73-75, from asst prof to assoc prof pharmacol, Rockefeller Univ, 75-78; from assoc prof to prof pharmacol, Med Ctr, Univ Ill, 82-85; DEAN'S PROF TOXICOL, DEPT BIOPHYSICS, MED CTR, UNIV ROCHESTER, 85- *Concurrent Pos:* Irma T Hirschl Trust career scientist award, 76-80; mem, Toxicol Study Sect, Nat Sci Adv Comt, NIH, 80-84, Med Adv Bd, Leukemia Res Found, 84-88, NIH Reviewers Reserve, 89-92; Burroughs-Wellcome Toxicol Scholar Award, 90; prin investr, NIH grants, 78- *Mem:* Am Soc Pharmacol & Exp Therapeut; Soc Toxicol; Am Soc Biochem & Molecular Biol; Am Soc Cell Biol. *Res:* Regulation of biosynthesis and degradation of cellular heme and hemoproteins, with emphasis on the elucidation of the mechanisms of regulatory activities of metal ions. *Mailing Add:* Dept Biophysics Med Ctr Univ Rochester Rochester NY 14642

MAINIER, ROBERT, b Pittsburgh, Pa, Oct 27, 25; m 50; c 6. ANALYTICAL CHEMISTRY. *Educ:* Univ Pittsburgh, BS, 49. *Prof Exp:* Chemist, Barrett Co, Allied Chem & Dye, 49-50; fel anal chem, Mellon Inst, 50-53; LAB GROUP MGR ABSORPTION SPECTROS, RES DEPT, KOPPERS CO INC, 53- *Mem:* Am Chem Soc; Soc Appl Spectros; Coblentz Soc. *Res:* Utilization of infrared, ultraviolet and nuclear magnetic resonance spectroscopy for characterization and analyses of commercial products. *Mailing Add:* 143 Cresent Garden Dr Pittsburgh PA 15235-3539

MAINIGI, KUMAR D, TOXICOLOGY. *Educ:* Aligarh Univ, India, BSc, 60, MSc, 62; MS Univ, Baroda, India, PhD(biochem), 67; Am Bd Toxicol, dipl. *Prof Exp:* Res assoc, Cancer Res Ctr, Pharmacol Dept, Baylor Col Med, 67-69; postdoctoral assoc, Inst Cancer Res, Foxchase Comprehensive Cancer Res Ctr, 74-77; sr res assoc & Nat Res Serv Awardee Toxicol, Div Nutrit Sci, Cornell Univ, 78-82; res instr toxicol, Albany Med Col, Ctr Exp Path & Toxicol, 80-81; dir, Life Sci Lab, Am Standards Testing Bur, Inc, 82-83, sci & consult, Environ Health Res, 83-85; sr staff toxicologist, Dynamac Int Corp, 85-87; TOXICOLOGIST, FOOD & DRUG ADMIN, 87- *Mem:* Soc Toxicol; Asn Govt Toxicologists; Am Asn Cancer Res; Am Soc Pharmacol & Exp Therapeut. *Res:* Modulatory effects of environmental pollutants nutrients, drugs, and sex hormones on initiation, promotion, progression and inhibition of carcinogenesis and toxicity; cellular mechanisms of carcinogen transport; enzymes in animal and human neoplasms; author of various publications. *Mailing Add:* Food & Drug Admin 5600 Fishers Lane Suite 12020 Rockville MD 20857

MAINLAND, GORDON BRUCE, b Elmhurst, Ill, May 22, 45; m 67; c 2. THEORETICAL HIGH ENERGY PHYSICS. *Educ:* Cornell Univ, BS, 67; Univ Tex, PhD(theoret high energy physics), 71. *Prof Exp:* Fel theoret high energy physics, Univ Tex, 72; scholar, Sch Theoret Physics, Dublin Inst Advan Studies, 72-74; fel, 74-75, from asst prof to assoc prof, 75-87, PROF THEORET HIGH ENERGY PHYSICS, OHIO STATE UNIV, 87- *Res:* Field theory, gauge theory, relativistic bound states, symmetries. *Mailing Add:* Dept Physics Ohio State Univ 174 W 18th Ave Columbus OH 43210

MAINO, VERNON CARTER, BIOCHEMISTRY, MEMBRANES. *Educ:* Univ Rochester, PhD(microbiol), 72. *Prof Exp:* SR RES SCIENTIST, BECTON DICKINSON MONOCLONAL CTR, 82- *Concurrent Pos:* Adj prof anthropology, Calif State Univ, Hayward, 86- *Mailing Add:* 380 Winsor Blvd Cambria CA 93428

MAINS, GILBERT JOSEPH, b Clairton, Pa, Apr 20, 29; m 51; c 2. PHYSICAL CHEMISTRY. *Educ:* Duquesne Univ, BS, 51; Univ Calif, PhD(chem), 54. *Prof Exp:* Fulbright fel, Cambridge Univ, 54-55; from asst prof to assoc prof chem, Carnegie Inst Technol, 55-65; prof, Univ Detroit, 65-68, Poetker prof physics & chem, 68-71, chmn dept chem, 65-68; prof chem & head dept chem, 71-77, PROF CHEM, OKLA STATE UNIV, 77- *Concurrent Pos:* Fel, Lawrence Radiation Lab, Univ Calif, 59-60; vis prof, Univ Tex, 86-87. *Mem:* AAAS; Am Chem Soc; Am Phys Soc; Radiation Res Soc. *Res:* Photochemistry; solar energy; fossil fuel chemistry; chemical kinetics; elementary reactions in photolysis, radiolysis and pyrolysis; reactions of free radicals; excited molecules and ions; quantum chemistry. *Mailing Add:* 301 E Redbud Dr Stillwater OK 74075

MAINS, RICHARD E, b Mar 21, 47. NEUROSCIENCE. *Educ:* Brown Univ, Providence, RI, ScB & MS, 86; Harvard Univ, Cambridge, Mass, PhD(neurobiol), 73. *Prof Exp:* From asst prof to assoc prof, Physiol Dept, Health Sci Ctr, Univ Colo, 76-83; PROF, NEUROSCI DEPT, SCH MED, JOHNS HOPKINS UNIV, 83- *Concurrent Pos:* Co-dir grad prog, Neurosci Dept, Sch Med, Johns Hopkins Univ, 83-90; DABR2 Study Sect, Nat Inst Drug Abuse, 88- *Honors & Awards:* Vincent Du Vigneaud Award, 84; Ernst Oppenheimer Award, Endocrine Soc, 86. *Mem:* Am Soc Cell Biol; Soc Neurosci; Endocrine Soc. *Res:* Biosynthesis of neuropeptides; plasticity of peptide biosynthesis in corticotropes; author of various publications. *Mailing Add:* Neurosci Dept Sch Med Johns Hopkins Univ 725 N Wolfe St Baltimore MD 21205

MAINS, ROBERT M(ARVIN), b Denver, Colo, Jan 18, 18; m 40; c 2. STRUCTURAL DYNAMICS. *Educ:* Univ Colo, BS, 38; Univ Ill, MS, 40; Lehigh Univ, PhD(civil eng), 46. *Prof Exp:* Asst, Univ Ill, 38-40; instr mech, Mo Sch Mines, 40-41; engr tests, Eng Lab, Lehigh Univ, 41-42, asst dir, 42-44; engr & proj supvr, Appl Physics Lab, Johns Hopkins Univ, 44-46, engr & asst group supvr, 50-55; from asst prof to assoc prof civil eng, Cornell Univ, 46-50; specialist shock & vibration anal, Submarine Adv Reactor Eng, Knolls Atomic Power Lab, Gen Elec Co, 55-57, consult engr, Proj D1G, 57-60, sr mech engr, Gen Eng Lab, 60-65; chmn dept, 65-69, prof, 65-84, EMER PROF CIVIL ENG, WASH UNIV, 84- *Concurrent Pos:* Consult on Struct Probs & Expert Witness. *Honors & Awards:* Irwin Vigness Award, Inst Environ Sci. *Mem:* Fel Am Soc Civil Engrs; Am Soc Mech Engrs; Am Soc Eng Educ; Soc Exp Stress Anal; fel Inst Environ Sci; NY Acad Sci. *Res:* Properties of materials; static and dynamic behavior of structures; bi-axial stress effects; bond stresses in concrete; applications of models; vibration testing machine for heavy equipment; strain gage wind tunnel balance; process for steptapering aircraft tubing; galvanized sheet bend tester; methods of controlling vibration response by increased structural damping. *Mailing Add:* 420 S Kirkwood Rd Kirkwood MO 63122

MAINSTER, MARTIN ARON, b Toronto, Ont, June 30, 42; US citizen; m 86; c 5. OPHTHALMOLOGY, PHYSICS. *Educ:* NC State Univ, BS, 63, PhD(physics), 69; Univ Tex Med Br Galveston, MD, 75. *Prof Exp:* Sr res scientist, Life Sci Div, Technol Inc, 68-70, mgr biomath anal, 70-71; resident ophthal, Scott & White Mem Inst, 76-79; assoc scientist, Retina Found, 79-85, dir, Clin Res Ctr, 81-85; assoc scientist, Inst Ophthal, Harvard Univ, 80-85; assoc, Retina Assocs, Boston, 81-85; PROF, DEPT OPHTHAL, KANSAS UNIV, 85- *Concurrent Pos:* Ford Found fel; Nat Sci Found fel; fel retinal dis, Retina Found & Mass Eye & Ear Infirmary, 81. *Honors & Awards:* Honor Award, Am Acad Ophthal, 86. *Mem:* Am Acad Ophthal; Am Phys Soc; Asn Res Vision & Ophthal; Optical Soc Am; AMA. *Res:* Macular and retinal vascular disease; radiational eye hazards; ophthalmic laser systems; mathematical biophysics. *Mailing Add:* Kansas Univ Med Ctr 39th & Rainbow Blvd Kansas City KS 66103

MAIO, DOMENIC ANTHONY, b Washington, DC, June 22, 35; m 58; c 3. PHYSIOLOGY, AEROSPACE MEDICINE. *Educ:* Georgetown Univ, BS, 56; George Washington Univ, MS, 57; Tex A&M Univ, PhD(physiol), 68. *Prof Exp:* US Air Force, 58-, res scientist, US Air Force Sch Aerospace Med, 63-71, staff officer, Aerospace Biotechnol, Off Dep Chief Staff Res & Develop, Hq US Air Force, 71-74, spec asst int res & develop, off asst secy Air Force for res & develop, Pentagon, Washington, DC, 74-76; prog mgr, med & biosci, Air Force Off of Sci Res, 76-78; DIR BIOTECHNOL, AIRFORCE SYSTS COMMAND, WASHINGTON, DC, 78- *Concurrent Pos:* Liaison rep to appl physiol study group, NIH. *Mem:* Assoc fel Aerospace Med Asn. *Res:* Altitude and hyperbaric physiology; biotechnology. *Mailing Add:* Stafford Court Dr Sterling VA 22170

MAIO, JOSEPH JAMES, b Priest River, Idaho, July 29, 29. MOLECULAR BIOLOGY, MICROBIOLOGY. *Educ:* Univ Wash, BS, 55, MS, 57, PhD(microbiol), 61. *Prof Exp:* NIH fel tissue cult, Univ Pavia, 61-63; NIH trainee, 64-66, fel cell biol, 66-67, from asst prof, to assoc prof, 67-78, PROF CELL BIOL, ALBERT EINSTEIN COL MED, 78- *Concurrent Pos:* NIH career develop award, Albert Einstein Col Med, 69-74. *Mem:* AAAS; Am Soc Cell Biol; NY Acad Sci; Fedn Am Soc Exp Biol. *Res:* Host-induced modification in bacteriophage; predatory fungi; enzymology and active transport processes of tissue culture cells; mammalian cytogenetics; nucleic acids of mammalian cells; human retroviruses. *Mailing Add:* Dept of Cell Biol Albert Einstein Col of Med Bronx NY 10461

MAIONE, THEODORE E, b Orange, NJ, Oct 21, 56. CELL BIOLOGY. *Educ:* Mich Univ, BA, 78; Cornell Univ, PhD(plant physiol-biol), 83. *Prof Exp:* Postdoctoral res asst, Brandeis Univ, 83-85; PRIN INVESTR, REPLIGEN CORP, 85- *Mem:* Am Soc Biochem & Molecular Biol; Am Soc Photobiol; NY Acad Sci; Am Chem Soc. *Mailing Add:* Repligen Corp One Kendall Sq Bldg 700 Cambridge MA 02139

MAIORANA, VIRGINIA CATHERINE, b Hagerstown, Md, Aug 16, 47; m 74. EVOLUTIONARY BIOLOGY. *Educ:* Univ Md, BS, 69; Univ Calif, Berkeley, MA, 71, PhD(zool), 74. *Prof Exp:* Actg asst prof biol, Univ Calif, Berkeley, 74; asst prof, 78-90, RES ASSOC BIOL, UNIV CHICAGO, 74- *Concurrent Pos:* Asst prof biol, Mundelein Col, 75-76; postdoc fel, NIMH, 76-78. *Mem:* Soc Study Evolution; Am Soc Zoologists; Animal Behav Soc; fel Linnean Soc London. *Res:* Empirical and theoretical investigations on the evolution of life history and behavior within and among phylogenetic groups and their relations to community structure and function; herbivore-plant interactions; human origins and evolution; body size and its influence on behavior and community structure. *Mailing Add:* Dept Ecol & Evolution Univ Chicago 1101 E 57th St Chicago IL 60637

MAIR, ROBERT DIXON, b Tide Head, NB, Feb 11, 21; nat US; m 43; c 4. OLEFIN POLYMERIZATION CATALYSTS, SENSORY EVALUATION OF MATERIAL AND PRODUCTS. *Educ:* Univ NB, BSc, 41; Brown Univ, ScM, 43, PhD(chem), 49. *Prof Exp:* Chemist, Polymer Corp, 43-46; res chemist anal methods develop, Hercules Inc, 48-58, sr res chemist, 58-71, res scientist, 71-78, res assoc polymer characterization & catalyst develop, 78-82; RETIRED. *Concurrent Pos:* Mem, Am Soc Testing & Mat Comt Sensory Eval, 58-78; consult, Wiltech Assoc, 85-87; plastics recycling, 85-87. *Mem:* Am Chem Soc; Sigma Xi. *Res:* Infrared spectroscopy; molecular structure of benzene; catalysis; organic peroxide analysis; polymer fractionation; thermal analysis; molecular characterization of polyolefins. *Mailing Add:* 215 N Garfield St Kennett Square PA 19348

MAIRHUBER, JOHN CARL, b Rochester, NY, Dec 14, 22; m 46; c 2. MATHEMATICS. *Educ:* Univ Rochester, BS, 42, MS, 50; Univ Pa, PhD(math), 59. *Prof Exp:* Instr math, Univ Rochester, 56-58; from asst prof to assoc prof, Univ NH, 58-64; prof, Univ Richmond, 64-68; head dept, 68-77, PROF MATH, UNIV MAINE, ORONO, 68- *Mem:* Am Math Soc; Math Asn Am. *Res:* Complex variables; theory of numbers and approximations. *Mailing Add:* Dept of Math Univ of Maine Orono ME 04469

MAIS, DALE E, b Mar 24, 52; m 76; c 2. CARDIOVASCULAR. *Educ:* Ind Univ, Bloomington, BS, 74, MS, 77, PhD(pharmacol), 83. *Prof Exp:* Teaching asst chem, Indian Univ, Bloomington, 73-77; mgr org synthesis, Lafayette Pharmacal Inc, Ind, 77-79; res asst, Ind Univ, 79-83; postdoctoral fel, Med Univ SC, Charleston, 83-86, asst prof, 86-89; SR PHARMACOLOGIST, DEPT CARDIOVASCULAR PHARMACOL, ELI LILLY & CO, 89- *Concurrent Pos:* Nat Res Serv Award, 83-86; expert analyst, Org Chem Ed Chemtracts, 90-; adj assoc prof, Dept Pharmacol & Toxicol, Sch Med, Ind Univ, 90- *Mem:* Am Soc Pharmacol & Exp Therapeut; Am Chem Soc; Sigma Xi. *Res:* Toxicology; author of various publications; granted 1 patent. *Mailing Add:* Dept Cardiovascu Eli Lilly & Co Lilly Corp Ctr Indianapolis IN 46285

MAISCH, WELDON FREDERICK, b Pana, Ill, Jan 19, 35; m 60; c 2. PRODUCT RESEARCH, INDUSTRIAL MICROBIOLOGY. *Educ:* Ill Wesleyan Univ, BS, 57; Univ Ill, MS, 60, PhD(microbiol), 67. *Prof Exp:* Asst plant bacteriologist, A E Staley Mfg Co, 59-61; sr scientist, Mead Johnson & Co Div, Bristol-Meyers Co, 66-68; assoc res scientist, Hiram Walker & Sons, Inc, 68-71, sr res scientist, 71-74, dir res, 74-80; mgr microbiol group, Archer Daniels Midland Co, 80-83; sr res scientist, Joseph E Seagram, 83-85; sr prod & develop supvr, 85-88, MGR PROD & DEVELOP, BROWN FORMAN CORP, 88- *Mem:* Am Soc Microbiol; Am Chem Soc; Am Cereal Chem; Inst Food Technologists. *Res:* Beverage alcohol products and production; food microbiology; biotechnology and fermentation science. *Mailing Add:* Brown Forman Corp PO Box 1080 Louisville KY 40201

MAISCH, WILLIAM GEORGE, b Philadelphia, Pa, Feb 15, 29. PHYSICAL CHEMISTRY. *Educ:* Univ Pa, BS, 51; Brown Univ, PhD(chem), 56. *Prof Exp:* Res assoc phys chem, Eng Exp Sta, Univ Ill, 55-57; asst prof, Inst Molecular Physics, Univ Md, 57-63; RES CHEMIST, US NAVAL RES LAB, 63- *Mem:* Am Chem Soc; Am Phys Soc; Sigma Xi. *Res:* Optical properties of magnetic materials; molecular and crystal structure; high pressure spectroscopy. *Mailing Add:* Code 4610 US Naval Res Lab Washington DC 20375

MAISEL, HERBERT, b Brooklyn, NY, Sept 22, 30; m 57; c 2. COMPUTER SCIENCE, STATISTICS. *Educ:* City Col New York, BS, 51; NY Univ, MS, 52; Cath Univ Am, PhD(math), 64. *Prof Exp:* Anal statistician, Develop & Proof Serv, Aberdeen Proving Ground, 52-56, chief statist sect, 56-58; chief methodology & reliability div, Off Naval Inspector Ord, Washington, DC, 58-59; mathematician, Oper Math Br, Off Qm Gen, 59-62; tech chief modeling div, US Army Strategy & Tactics Group, 62-63; dir comput ctr, Georgetown Univ, 63-76; systs adv, Social Security Admin, 76-84; from asst prof to assoc prof, 63-76, PROF COMPUT SCI, COMPUT & SCI DEPT, GEORGETOWN UNIV, 76- *Concurrent Pos:* Consult, Social Security Admin, 66-73, Nat Bur Standards, 68-72 & Baltimore Housing Authority, 72-73; app to spec study group for suppl security income prog, Dept Health, Educ & Welfare, 75-76. *Honors & Awards:* Outstanding Contribution Award, Asn Comput Mech. *Mem:* AAAS; Asn Comput Mach; Fedn Am Scientists; Comt Concerned Scientists. *Res:* Simulation and other stochastic applications of computers; application of numerical and statistical methods to problems in physical and life sciences; teaching of computer science. *Mailing Add:* 9432 Curran Rd Silver Spring MD 20901

MAISSEL, LEON I, b Cape Town, SAfrica, May 31, 30; US citizen; m 56; c 2. PHYSICS, COMPUTER SCIENCES. *Educ:* Cape Town Univ, BSc, 49, MSc, 51; Univ London, PhD(physics), 55. *Prof Exp:* Res physicist, Philco Corp, Pa, 56-60; SR TECH STAFF MEM, IBM CORP, 60- *Mem:* Fel Inst Elec & Electronic Engrs. *Res:* Thin films, particularly cathodic sputtering, and their application; computer aided design, particularly array logic. *Mailing Add:* Dept G57 Bldg 708 Data Syst Div IBM Corp PO Box 950 Poughkeepsie NY 12602

MAITRA, UMADAS, b Jalpaiguri, India. BIOCHEMISTRY, MOLECULAR BIOLOGY. *Educ:* Univ Calcutta, BSc, 56, MSc, 58, DSc, 78; Univ Mich, Ann Arbor, PhD(biol chem), 63. *Prof Exp:* Jane Coffin Childs Found Med Res fel, 63-65, from instr to asst prof, 65-72, assoc prof, 72-77, PROF DEVELOP BIOL & CANCER, ALBERT EINSTEIN COL MED, 77- *Concurrent Pos:* Am Heart Asn estab investr, Albert Einstein Col Med, 67-72, Am Cancer Soc fac res award, 72-77. *Mem:* Am Soc Biol Chemists. *Res:* Molecular mechanisms of transcription and translation. *Mailing Add:* Dept of Develop Biol Albert Einstein Col of Med 1300 Morris Park Ave Bronx NY 10461

MAIZELL, ROBERT EDWARD, b Baltimore, Md, Aug 3, 24; m 54; c 2. INDUSTRIAL CHEMISTRY. *Educ:* Loyola Col, BS, 45; Columbia Univ, BS, 47, MS, 49, DLS, 57. *Prof Exp:* Chemist, Manhattan Proj, 45; reference asst, Sci & Technol Div, New York Pub Library, 47-48; teaching asst sci lit, Columbia Univ, 50; chg tech info serv, Olin Mathieson Chem Corp, NY, 50-58; NSF doc res proj dir, Am Inst Physics, NY, 58-60; supvr tech info serv, Olin Corp, 60-65, mgr, 65-72, tech mgr, 72-74, mgr bus & sci intel serv, 74-82, consult scientist new technol, 82-85; DIR, TECH INFO CONSULT, 85- *Concurrent Pos:* Chmn continuing educ comt, Olin Corp, 69-75; mem adv coun, Smithsonian Sci Info Exchange, 74-82; chmn, Subcomt On-line Serv, Mfg Chemists Asn, 75-78; mem, Data Comt & Public Liaison Subcomt, US Interagency Toxic Substances Info Network, 81-83; mem, Tech Adv Bd, New Haven Technol Investment Fund, 83-; mem adv bd, Index Chemicus, 83-; rep, Sci Park's Off, US Patent Off, 87- *Mem:* Am Chem Soc; fel Am Inst Chemists. *Res:* New uses, processes, and products; industrial processes for chemicals; technical and marketing intelligence; technological forecasting; new methods for disseminating chemical information; materials science; technology transfer; patent searching. *Mailing Add:* Dir, Tech Info Consult 5 Sci Park New Haven CT 06511

MAJAK, WALTER, b Montreal, Que, Mar 7, 42; m 68; c 2. PLANT & ANIMAL BIOCHEMISTRY, NATURAL PRODUCTS. *Educ:* McGill Univ, BSc, 63; Dalhousie Univ, MSc, 65; Univ BC, PhD(plant biochem), 72. *Prof Exp:* RES SCIENTIST TOXIC PLANTS, AGR CAN, 72- *Mem:* NAm Phytochem Soc. *Res:* Identification and quantitative determination of toxic constituents in cultivated and native forages; animal and microbiol metabolism of toxic compounds; detoxification of poisonous compounds by rumen microbes; etiology of pasture bloat. *Mailing Add:* Agr Can Res Sta 3015 Ord Rd Kamloops BC V2B 8A9 Can

MAJARAKIS, JAMES DEMETRIOS, surgery; deceased, see previous edition for last biography

MAJCHROWICZ, EDWARD, b Stryj, Poland, Mar 18, 20; m 56; c 2. BIOCHEMISTRY, BIOCHEMICAL PHARMACOLOGY. *Educ:* Univ Birmingham, BSc, 48; McGill Univ, PhD(biochem), 59. *Prof Exp:* Asst chemist, A Guinness, Son & Co, Eng, 49-56; res assoc biochem, McGill Univ & McGill Montreal Gen Hosp Res Inst, 56-59, fel neurochem, 59-60; res assoc biochem, Med Sch, Univ Va, Charlottesville, 61-62, asst prof, 62-63; asst prof, Sch Med, Univ NC, Chapel Hill, 63-67; sr res scientist, Squibb Inst Med Res, 67-68; res scientist, Nat Ctr Prev & Control Alcoholism, NIMH, 68-71, head biochem prog, Nat Inst Alcohol Abuse & Alcoholism, 71-76, dir, Nat Inst Alcohol Abuse, Alcohol, Drug Abuse & Ment Health Admin, 76-88; RETIRED. *Mem:* AAAS; Am Chem Soc; NY Acad Sci; Am Soc Pharmacol & Exp Therapeut; Am Soc Neurochem. *Res:* Neurochemistry, metabolism and metabolic effects of aliphatic alcohols, aldehydes and fatty acids; biogenic amine metabolism; biological aspects of mental diseases; biochemistry and microbiology of fermentation processes. *Mailing Add:* 1001 Montezuma Dr Ft Washington MD 20744

MAJERUS, PHILIP W, b Chicago, Ill, July 10, 36; m; c 4. HEMATOLOGY, MEDICINE. *Educ:* Notre Dame Univ, BS, 58; Wash Univ, MD, 61. *Prof Exp:* Intern & asst resident med, Mass Gen Hosp, 61-63; res assoc, Lab Biochem, Nat Heart Inst, Bethesda, Md, 63-66; asst prof biochem, 66-75, from asst prof to assoc prof med, 66-71, PROF MED, SCH MED, WASH UNIV, ST LOUIS, MO, 71-, CO-DIR, DIV HEMAT-ONCOL, 73-, PROF BIOCHEM, 76- *Concurrent Pos:* Mem, Biochem Fel Rev Comt, NIH, 69-70, Hemat Study Sect, 74-78; prin investr & dir, Specialized Ctr Res Thrombosis, Wash Univ, 74-; ed, J Clin Invest, 77-82; comt mem, Am Heart Asn, 88-; chmn, Prog Molecular Med in Cancer Res, James S McDonnell Found, 88-, Sect 41, Med Genetics, Hemat & Oncol, Nat Acad Sci, 89-, Searle Scholars Prog, 89- & bd sci counselors, Nat Heart, Lung & Blood Inst, 89- *Honors & Awards:* Am Cancer Soc Awards, 66 & 75; Dameshek Medal & Prize, Am Soc Hemat, 81; Distinguished Career Award, Int Soc Thrombosis & Hemostasis, 85; Lewis A Connor Lectr, Am Heart Asn, 90; Kober Lect Award, Asn Am Physicians, 91. *Mem:* Nat Acad Sci; Inst Med-Nat Acad Sci; Am Soc Biol Chemists; Am Soc Clin Invest (pres, 81); Am Fedn Clin Res; Sigma Xi; Am Soc Hemat; Asn Am Physicians; fel Am Col Physicians; fel Am Acad Arts & Sci. *Res:* Structure and function of human blood platelets; holder of one US patent. *Mailing Add:* Div Hemat-Oncol Wash Univ Sch Med 660 S Euclid Box 8125 St Louis MO 63110

MAJEWSKI, ROBERT FRANCIS, b Chicago, Ill, Oct 1, 27; m 61; c 3. ORGANIC CHEMISTRY. *Educ:* Univ Ill, BS, 51; Univ Notre Dame, PhD(org chem), 55. *Prof Exp:* Chemist, Armour & Co labs, 51; sr chemist res & develop, 54-60, group leader, 60-68, sect leader chem res, 68-72, SR PRIN INVESTR, MEAD JOHNSON RES CTR, 72- *Mem:* AAAS; Am Chem Soc; Sigma Xi. *Res:* Unsaturated lactones; medicinal chemistry in endocrine, central nervous and cardiovascular systems; development of organic chemical processes. *Mailing Add:* 401 Kings Valley Rd Evansville IN 47711

MAJEWSKI, THEODORE E, b Boonton, NJ, July 5, 25; m 53; c 6. ORGANIC CHEMISTRY. *Educ:* Syracuse Univ, BA, 51; Univ Del, MS, 53, PhD(org chem), 60. *Prof Exp:* Res chemist, Dow Chem Co, 57-64, proj leader benzene chem, 64-66, org process titled specialist, 66-69; RES SCIENTIST, PHILIP MORRIS, INC, 69- *Concurrent Pos:* Consult, Herald Pharmacal. *Mem:* Am Chem Soc; AAAS. *Res:* flavor technology on cigaret; taste research on cigaret; menthol, tobacco pyrolysis, cellulose pyrolysis, filter and adhesive research; rheology; water soluble natural products for use in adhesives and binders for tobacco. *Mailing Add:* Philip Morris USA Res & Develop PO Box 26583 Richmond VA 23261

MAJITHIA, JAYANTI, b Dar Es Salaam, Tanzania, June 12, 38; Can citizen; m 64; c 2. HARDWARE SYSTEMS, ELECTRICAL ENGINEERING. *Educ:* Univ London, Eng, BSc, 60; McMaster Univ, Can, MSc, 68, PhD(elec eng), 71. *Prof Exp:* Res engr, Cable & Wireless, UK, 60-62; lectr elec eng, Univ Khartoum, Sudan, 62-66; prof, Univ Waterloo, 71-81; PROF COMPUT SCI & CHMN DEPT COMPUT & INFO SCI, COL PHYS SCI, UNIV GUELPH, 81- *Mem:* Sr mem Inst Elec & Electronics Engrs. *Res:* Computer networks; computer architecture; hardware design languages for very large scale integrated design. *Mailing Add:* Dept Comput & Info Sci Univ Guelph Guelph ON N1G 2W1 Can

MAJKRZAK, CHARLES FRANCIS, East Orange, NJ, Mar 5, 50; m 73; c 2. NEUTRON SCATTERING. *Educ:* Montclair State Col, NJ, BA, 72; Univ RI, PhD(physics), 78. *Prof Exp:* Teaching asst physics, Univ RI, 72-78; res assoc, 78-80, asst physicist, 80-82, assoc physicist, 82-84, PHYSICIST NEUTRON SCATTERING, BROOKHAVEN NAT LAB, 84- *Concurrent Pos:* Vis scientist, Inst Solid State Nuclear Res Ctr, Julich, WGermany, 83. *Mem:* Am Phys Soc. *Res:* Neutron scattering studies of magnetic materials; development of polarized neutron scattering methods and instrumentation; thin film multilayer deposition techniques; properties of synthetic superlattices. *Mailing Add:* Nat Bur Standards Bldg 235 Gaithersburg MD 20899

MAJMUDAR, HARIT, electrical engineering; deceased, see previous edition for last biography

MAJMUNDAR, HASMUKHRAI HIRALAL, b Baroda, India, Nov 19, 32; US citizen; m 62; c 2. LANDSLIDE HAZARD MAPPER, APPLIED GEOCHEMIST. *Educ:* MS Univ Baroda, India, BSc Hons, 55; Banaras Hindu Univ, India, MSc, 57; Univ Nancy, France, DSc(geol), 61. *Prof Exp:* Assoc prof geol, MS Univ Baroda, India, 57-64; Nat Acad Sci-Nat Res Coun resident res assoc geochem, Goddard Space Flight Ctr, NASA, 64-66; spec lectr, head Geochem Sect & Nat Res Coun fel, Dalhousie Univ, Halifax, Can, 66-68; prof geol, Appalachian State Univ, Boone, NC, 68-70; mgr, Geochem Sect, 70-84, GEOLOGIST, DIV MINES & GEOL, STATE CALIF, 84- *Concurrent Pos:* US cand-chief tech adv expert appl chem, UN Develop Prog, 76-81. *Res:* Landslide hazard maps; compiling geological maps and classifying areas with relative landslide and debris-flow susceptibilities; existing slope failures inventory by field mapping and analysis of aerial photographs. *Mailing Add:* 376 Borica Dr Danville CA 94526

MAJOR, CHARLES WALTER, b Framingham, Mass, Jan 31, 26; m 51; c 3. PHYSIOLOGY. *Educ:* Dartmouth Col, AB, 48; Univ Tenn, MS, 54, PhD(zool), 57. *Prof Exp:* Res fel, Nat Cancer Inst, 57; instr physiol, Sch Med, Univ Rochester, 57-59; from asst prof to assoc prof, 59-71, PROF ZOOL, UNIV MAINE, 71- *Res:* Comparative physiology and toxicology in marine ecosystems. *Mailing Add:* Dept Zool Murray Hall Univ ME Orono ME 04469

MAJOR, FOUAD GEORGE, b Lebanon, Feb 7, 29; US citizen; m 55; c 2. ATOMIC PHYSICS, SPECTROSCOPY. *Educ:* Victoria Univ, Wellington, NZ, BSc, 49, MSc, 52; Univ Wash, PhD(physics), 62. *Prof Exp:* Guest prof physics, Univ Bonn, 65-67; physicist, NASA Goddard Space Flight Ctr, 67-73; vis scientist, Univ Paris-Sud, Orsay, 73-74; prof math, Univ DC, 74-77; prof physics, Kuwait, 77-88; VIS PROF, CATHOLIC UNIV AM, 88- *Concurrent Pos:* Consult, Litton Guid & Control Div, 76-77. *Honors & Awards:* Apollo Achievement Award, NASA, 69. *Mem:* Am Phys Soc. *Res:* Laser spectroscopy; laser scattering. *Mailing Add:* 6702-41st Ave University Park MD 20782

MAJOR, JACK, b Salt Lake City, Utah, Mar 15, 17; m 47; c 3. PLANT ECOLOGY. *Educ:* Utah State Agr Col, BS, 42; Univ Calif, PhD(soil sci), 53. *Prof Exp:* Range researcher, US Forest Serv, Utah, 42-49; range weed control researcher, 53-60, from asst prof to assoc prof bot, 55-71, PROF BOT, UNIV CALIF, DAVIS, 71- *Mem:* Ecol Soc Am; Am Bryol & Lichenol Soc; Brit Ecol Soc; Brit Soc Soil Sci. *Res:* Plant community and soil relationships; California vegetation; vegetation near Atlin Lake, British Columbia; vegetation of Teton and Gros Ventre Ranges, Wyoming. *Mailing Add:* 623 E St Davis CA 95616

MAJOR, JOHN KEENE, b Kansas City, Mo, Aug 3, 24; m 70; c 3. SCIENCE ADMINISTRATION. *Educ:* Yale Univ, BS, 43, MS, 47, Fac Sci Paris, DSc, 51. *Prof Exp:* Instr physics, Yale Univ, 52-55; assoc prof, Western Reserve Univ, 55-57, chmn dept, 55-64, Perkins prof, 57-66; staff assoc, NSF, 64-68; prof & dean Grad Sch, Univ Cincinnati, 68-71 & NY Univ, 71-; vis scholar, Sloan Sch Mgt, Mass Inst Technol, 73-74; prof & vpres acad affairs, Northeastern Ill Univ, 74-77; gen mgr, WONO, Syracuse, NY, 77; dir res & mkt, WFMT, Chicago, Ill, 78-81; Gen Mgr, 81-88, PRES & CHMN BD, KCMA, TULSA, OKLA, 81- *Concurrent Pos:* Consult, Sonar Anal Group, 46-47 & NSF, 68-69. *Mem:* Sigma Xi; Audio Eng Soc. *Res:* Experimental nuclear and solid-state physics; Mossbauer effect; resonant and non resonant scattering of gamma rays; architectural acoustics. *Mailing Add:* 126 E 26 Pl Tulsa OK 74114-2422

MAJOR, ROBERT WAYNE, b Newark, Ohio, Sept 5, 37; m 63; c 2. PHYSICS. *Educ:* Denison Univ, BS, 58; Iowa State Univ, MS, 60; Va Polytech Inst, PhD(physics), 66. *Prof Exp:* Instr physics, Denison Univ, 60-61; asst prof, The Citadel, 61-62; from asst prof to assoc prof, 66-74, prof physics, 74-80, chmn dept, 79-81, R E LOVING PROF PHYSICS, UNIV RICHMOND, 80- *Concurrent Pos:* Oak Ridge Assoc Univs fac res fel, 74, 81 & 88; mem bd, Southeast Univs Res Asn, 84- *Mem:* AAAS; Am Asn Physics Teachers; Sigma Xi. *Res:* Laser-modulated optical absorption in II-VI crystals; photoacoustic kinetics in solids; varistor physics. *Mailing Add:* 9626 Hitchin Dr Univ Richmond Box 218 Richmond VA 23233

MAJOR, SCHWAB SAMUEL, JR, b Windsor, Mo, July 2, 24; m 51; c 3. PHYSICS. *Educ:* Wichita State Univ, BA, 49; Kans State Univ, MS, 53, PhD(physics), 67. *Prof Exp:* Elec engr, Derby Oil Refinery, Kans, 51; staff engr, Boeing Airplane Co, 51-53; instr physics, Southwestern Col, Kans, 53-55; asst prof, Midland Col, 55-59; assoc prof physics, Univ Mo-Kansas City, 59-81; adminr eng staff develop, Kansas City Div, Allied Signal Aerospace Co, 81-90; CONSULT TECHNOL MGT, 90- *Concurrent Pos:* Res assoc, Midwest Res Inst, Mo, 62-64. *Mem:* Am Phys Soc; Am Asn Physics Teachers; Optical Soc Am; Am Inst Aeronaut & Astronaut. *Res:* Applied quantum statistical mechanics; microwave theory; effect of stress and fatigue on pilot operation and error; nonlinear dynamics; methods and techniques for industrial training of scientists and engineers; technology management. *Mailing Add:* 8 W 70th St Kansas City MO 64113-2565

MAJORS, RONALD E, b Ellwood City, Pa, Apr 10, 41; m 71. ANALYTICAL CHEMISTRY, CHROMATOGRAPHY. *Educ:* Fresno State Col, BS, 63; Purdue Univ, PhD(anal chem), 68. *Prof Exp:* Res asst, Purdue Univ, 63-68; res chemist, Celanese Res Co, NJ, 68-71; res chemist, Varian Aerograph, 71-73, prod mgr, Varian European Opers, 73-75, appln mgr, Varian Assocs, 75-77, Liquid Chromatography Appln & Column Develop, 77-81, mkt mgr, 81-85, LIFE SCI MKT MGR, VARIAN INSTRUMENT DIV, VARIAN ASSOCS, 85- *Mem:* Am Chem Soc; Chromatographic Soc; Soc Analytical Chem. *Res:* Liquid and gas chromatography applications; liquid chromatography column technology. *Mailing Add:* Hewlett Packard Rte 41 PO Box 900 Avondale PA 19311-0900

MAJUMDAR, DALIM K, civil engineering; deceased, see previous edition for last biography

MAJUMDAR, DEBAPRASAD (DEBU), b Calcutta, India, Dec 10, 41; US citizen; m 71; c 2. NUCLEAR ENGINEERING, PHYSICS. *Educ:* Univ Calcutta, BS, 61, MS, 63; Univ Pa, MS, 66; Univ Mich, Ann Arbor, MS, 73; State Univ NY Stony Brook, PhD(physics), 69. *Prof Exp:* Res assoc physics, Syracuse Univ, 69-71; res assoc & lectr, physics Univ Mich, Ann Arbor, 71-73, assoc res scientist nuclear eng, Phoenix Mem Lab, 73-74; nuclear engr, Brookhaven Nat Lab, 74-80; prog mgr, 80-86, chief reactor res & technol div, 86-89, MGR, UNIV & SCI EDUC PROG, IDAHO OPERS OFF, US DEPT ENERGY, 89- *Mem:* Am Phys Soc; Am Nuclear Soc. *Res:* Light water reactor safety; liquid metal fast breeder reactor safety; reactor operation, nuclear waste management; production and use of hydrogen as energy; thermal hydraulic and core disruptive computer code developments and applications. *Mailing Add:* 1749 Delmar Dr Idaho Falls ID 83404

MAJUMDAR, S K, b India, Apr 12, 38; US citizen; m 62; c 4. GENETIC TOXICOLOGY, CELL BIOLOGY. *Educ:* Calcutta Univ, BS; Univ Ky, PhD(biol), 68. *Prof Exp:* Asst prof, Ark Polytech Col, 69-75; assoc prof, 76-83, PROF BIOL, LAFAYETTE COL, 84- *Concurrent Pos:* Vis assoc prof, Rutgers Univ, 79; ed, J Pa Acad Sci, 88-, pres, 86-88. *Honors & Awards:* Distinguished Serv Award, Nat Asn Acad Sci. *Mem:* Fel AAAS; Nat Asn Acad Sci (secy, 85-87, treas, 87-89, pres, 91-); Sigma Xi; Am Cell Biol Soc; Am Tissue Culture Asn; Am Soc Metals. *Res:* Genotoxicity, electron microscopy and cell biology; author of one book. *Mailing Add:* Dept Biol Lafayette Col Easton PA 18042

MAJUMDAR, SAMIR RANJAN, b Chittagong, India, Nov 26, 36; m 62; c 2. APPLIED MATHEMATICS. *Educ:* Univ Calcutta, BA, 56, MA, 58; Jadavpur Univ, India, PhD(fluid mech), 63; Univ London, PhD(mech of continuous medium), 65. *Prof Exp:* Lectr math, Jadavpur Univ, 59-62; asst prof, Univ Ariz, 65-69; assoc prof, 69-76, PROF MATH, UNIV CALGARY, 76- *Concurrent Pos:* Assoc fel, Inst Math & Appln, UK, 65. *Mem:* Am Math Soc; Soc Natural Philos; Calcutta Math Soc; Soc Ind Appl Math; Inst Math Appln. *Res:* Hydrodynamics, especially slow motion of viscous liquids. *Mailing Add:* Dept Math Statist & Comput Sci Univ Calgary 2500 University Dr NW Calgary AB T2N 1N4 Can

MAJUMDAR, SHYAMAL K, b Jessore, India, Apr 12, 38; US citizen; m 62; c 4. CELL BIOLOGY, ELECTRON MICROSCOPY. *Educ:* Calcutta Univ, India, BS, 59; Univ Ky, MS, 65, PhD(biol), 68. *Prof Exp:* Asst prof genetics & bot, Russellville Col, Ark, 68-69; vis assoc prof biol-res, Rutgers Univ, New Brunswick, 78-79; from asst prof to assoc prof genetics-cell biol, 69-82, PROF BIOL GENETICS & MICROBIOL, LAFAYETTE COL, EASTON, PA, 82- *Concurrent Pos:* Distinguished Award, Nat Asn Acad Sci, 86-87; pres, Pa Acad Sci, 86-88; ed, J Pac Acad Sci, 88- *Mem:* Fel AAAS; Nat Asn Acad Sci (pres, 91-92); Am Genetic Asn; Am Soc Cell Biol; Am Soc Microbiol. *Res:* Determine genotoxic properties of environmental chemicals and drugs using tissue culture; cell biology; genetics and electron microscopy techniques; acid rain effects in aquatic and terrestrial systems utilizing microbiological techniques. *Mailing Add:* Dept Biol Lafayette Col Easton PA 18042

MAK, KING KUEN, b Canton, China, Jan 16, 46; US citizen; m 74; c 2. HIGHWAY SAFETY, TRANSPORTATION ENGINEERING. *Educ:* Univ Hong Kong, BSc, 68; Ga Inst Technol, MSCE, 71, MSOR, 73. *Prof Exp:* Res Eng, Ga Inst Technol, 70-73, asst res engr, Ga Dept Transp, 73-74; civil engr, Ins Inst Hwy Safety, 74-76; sr res engr, Southwest Res Inst, 76-83; RES ENG, TEX TRANS INST, 83- *Concurrent Pos:* Mem, Comt Trans-Planning Prog Eval, Transp Res Bd, 74-76, Comt Methodol Eval Hwy Improv, 85- & Comt Traffic & Hwy Safety, Am Soc Civil Engrs, 76-85; chmn, Environ Factors Sect, Asn Advan Automotive Med, 88- & Comt Traffic Control Locations Adverse Geomet, Inst Transp Engrs, 90- *Mem:* Am Soc Civil Engrs; Asn Advan Automotive Med; Transp Res Bd; Inst Transp Engrs. *Res:* Highway safety evaluation; statistical analysis of traffic accident data; accident investigation and reconstruction; traffic engineering. *Mailing Add:* Nine Cotswold Lane San Antonio TX 78257-1221

MAK, LINDA LOUISE, b Oakland, Calif, Sept 16, 47; m 77; c 2. CELL BIOLOGY, EMBRYOLOGY. *Educ:* Univ Calif, Berkeley, AB, 69, PhD(zool), 75, MBA, 83; Ariz State Univ, MS, 71. *Prof Exp:* res biologist path, Vet Admin Hosp, San Francisco, 75-81; exec vpres, 85-87, SR VPRES, HEALTH/MED INVESTMENTS, SAN MATEO, 83-, CONSULT, HEALTH/MED VENTURES, INC, 87- *Concurrent Pos:* Scholar anat, Univ Calif, San Francisco, 76-79. *Mem:* Am Soc Cell Biol; Soc Develop Biol; Sigma Xi; AAAS. *Res:* Cell adhesion and movement; morphology and function of cardiac tissue; vertebrate neurulation. *Mailing Add:* 2041 Greenwood San Carlos CA 94070

MAK, SIOE THO, b Medan, Indonesia, Sept 3, 32; US citizen; m 59; c 2. ELECTRICAL ENGINEERING, ENGINEERING PHYSICS. *Educ:* Univ Indonesia, Engr, 58; Ill Inst Technol, MSc, 63, PhD(elec eng), 70. *Prof Exp:* Sr lectr, Bandung Inst Technol, 61-67; sr scientist, Res Div, Joslyn Mfg & Supply Co, 70-76, res proj mgr, 76-78; sr staff engr, Electronics & Space Div, Emerson Elec Co, 78-87, sr staff scientist, Cance Load Mgt Systs Div, 87-90; SR STAFF SCIENTIST DISTRIB CONTROL SYSTS, INC, ESCO ELECTRONICS CORP, 90- *Concurrent Pos:* Mem tech working group, Inst Elec & Electronic Engrs, Power Eng Soc, 76-78; instr, Ill Inst Technol, 70-78; adj prof, Wash Univ, 85. *Mem:* Inst Elec & Electronic Engrs; AAAS. *Res:* Applications of dielectrics to extra and ultra high voltage power transmission; two-way communication on power lines for load management and automated distribution systems; distribution automation; systems design & system science. *Mailing Add:* Distrib Control Systs Inc Esco Electronics Corp 5657 Campus Pkwy Hazelwood MO 63042

MAK, STANLEY, b Canton, China, Sept 20, 33; Can citizen; m; c 1. VIROLOGY. *Educ:* Univ Sask, BSc, 58, MSc, 59; Univ Toronto, PhD(biophys), 62. *Prof Exp:* Asst prof biol, Queen's Univ, Ont, 62-68; assoc prof, 68-74, PROF BIOL, MCMASTER UNIV, 74- *Concurrent Pos:* Mem, Cancer Res Unit, Nat Cancer Inst Can, 76- *Mem:* Can Asn Physicists; Can Soc Cell Biol; Am Soc Microbiol. *Res:* Molecular biology of animal virus infection; transcription cellular transformation of oncogenic viruses. *Mailing Add:* Dept of Biol McMaster Univ 1280 Main St W Hamilton ON L8S 4L8 Can

MAK, WILLIAM WAI-NAM, b Hong Kong, Aug 5, 49; Can citizen; m 75; c 2. GROWTH FACTOR, PROTEIN ENGINEERING. *Educ:* Chinese Univ Hong Kong, BSc Hons, 71; Univ Toronto, Can, MSc, 74, PhD(biochem), 77. *Prof Exp:* Postdoctoral fel cancer, McArdle Lab Cancer Res, Univ Wis, 77-80; res scientist, Connaught Res Inst, 81-82; res assoc biochem, Banting & Best Dept Med Res, Univ Toronto, 83-86; PROF BIOTECHNOL, DEPT BIOL PHARM SCI & APPL CHEM, SENECA COL, 87- *Concurrent Pos:* Res consult, Queen Elizabeth Hosp Res Inst, Toronto, 83-; acad coordr, Dept Biol Pharm Sci & Appl Chem, Seneca Col, 88- *Mem:* AAAS. *Res:* Treatment of osteoporosis; regulators of bone synthesis and degradation, such as parathyroid hormone. *Mailing Add:* Seneca Col 1750 Finch Ave E North York ON M2J 2X5 Can

MAKAR, ADEEB BASSILI, b Alexandria, Egypt; m 68; c 2. BIOCHEMICAL PHARMACOLOGY, TOXICOLOGY. *Educ:* Alexandria Univ, Egypt, BS, 58; Univ Minn, Minneapolis, PhD(pharmacol), 66. *Prof Exp:* Instr pharmacol, Alexandria Univ, Egypt, 58-62, from lectr to prof, 66-80; ASSOC PROF PHARMACOL, UNIV OSTEOP MED & HEALTH SCI, DES MOINES, IOWA, 83- *Concurrent Pos:* Res consult, Alexandria Drug Co, Egypt, 72-74; int res fel, Univ Iowa, Iowa City, 74-76; prin investr, NIH grant, 77-80. *Mem:* Am Soc Pharmacol & Exp Therapeut; Int Soc Environ Toxicol & Cancer; Am Col Toxicol; NY Acad Sci; Am Chem Soc. *Res:* Biochemical pharmacology and toxicology of methanol, formate, folate and other aspects of the one-carbon pool metabolism; methanol metabolism and toxicity. *Mailing Add:* Univ Osteop Med & Health Sci 3200 Grand Ave Des Moines IA 50312

MAKAR, BOSHRA HALIM, b Sohag, Egypt, Sept 23, 28; m 60; c 2. PURE MATHEMATICS, MATHEMATICAL ANALYSIS. *Educ:* Univ Cairo, BSc, 47, MSc, 52, PhD(math anal), 55. *Prof Exp:* Lectr math, Univ Cairo, 48-55, from asst prof to assoc prof, 55-65; vis assoc prof, Am Univ, Beirut, 66; assoc prof, Mich Tech Univ, 66-67; PROF MATH, ST PETER'S COL, NJ, 67- *Concurrent Pos:* Egyptian Govt sci exchange mission, Moscow State Univ, 63-64. *Mem:* AAAS; Am Math Soc. *Res:* Functions of a complex variable; functional analysis, cryptology. *Mailing Add:* 410 Fairmont Ave Jersey City NJ 07306

MAKAREM, ANIS H, b Rasel-Metn, Lebanon, Dec 21, 33; US citizen; m 66; c 2. BIOCHEMISTRY. *Educ:* Concord Col, BS, 57; Univ Calif, San Francisco, MS, 64, PhD(biochem), 65. *Prof Exp:* Fel nutrit, Univ Calif, Berkeley, 65-66; res training fel clin chem, Med Sch, Yale Univ, 66-67, instr, 67-68; asst dir endocrinol, 68-69, ASST DIR CHEM, BIOSCI LABS, VAN NUYS, 69-; EXEC DIR, INST LAB MED, NORTHRIDGE, CALIF, 74- *Mem:* Am Asn Clin Chem. *Res:* Radio immunoassays for quantitation of hormones and vitamins; disc and agarose gel electrophoresis for serum proteins; lipoproteins and hemoglobins. *Mailing Add:* 31551 Saddle Tree Dr West Lake Village CA 91361

MAKAREWICZ, JOSEPH CHESTER, b Attleboro, Mass, Aug 5, 47; m 71. LIMNOLOGY. *Educ:* Southeastern Mass Univ, BS, 69; Cornell Univ, PhD(aquatic ecol), 75. *Prof Exp:* Res asst ecol, Cornell Univ, 69-71; instr pop biol, Southeastern Mass Univ, 71-72; instr biol, Bristol Community Col, 72; res asst ecol, Hubbard Brook Ecosyst Study, 72-74; ASSOC PROF BIOL, STATE UNIV NY COL BROCKPORT, 74- *Mem:* AAAS; Am Soc Limnol & Oceanog; Int Asn Theoret & Appl Limnol; Sigma Xi. *Res:* Zooplankton and phytoplankton dynamics; production biology of zooplankton; pesticide dynamics especially mirex. *Mailing Add:* Dept Biol Sci SUNY Col at Brockport 218 Lennon Hall Brockport NY 14420

MAKAREWICZ, PETER JAMES, b Buffalo, NY, Feb 15, 52; m 75. CHEMICAL ENGINEERING. *Educ:* Univ Notre Dame, BS, 73; Princeton Univ, MA, 75, PhD(chem eng), 77. *Prof Exp:* res chemist, 77-80, SR RES SCIENTIST CHEM ENG, RES LABS, EASTMAN KODAK CO, 80- *Concurrent Pos:* NSF grad fel, 73-76; fel, Textile Res Inst, 75-77. *Mem:* Assoc mem Am Inst Chem Engrs; Am Chem Soc. *Res:* Polymer structure-property relationships; separation phenomena and membrane fabrication; bioengineering; gelatin and gelation phenomena. *Mailing Add:* 24 Brook Valley Dr Rochester NY 14624-5348

MAKDANI, DHIREN D, EXPERIMENTAL BIOLOGY. *Educ:* Gujarat Univ, India, BS, 52, MS, 54; Mich State Univ, PhD(nutrit), 70. *Prof Exp:* Teacher, Dept Nutrit, Gujarat Univ, India, 54-66 & 70-72; res specialist, Dept Med & Dept Food Sci & Nutrit, Mich State Univ, 72-78, asst prof, Dept Food Sci & Human Nutrit, 79-80; res nutritionist, Dept Pediat, William Beaumont Hosp, Royal Oak, Mich, 78-80; assoc prof, 80-85, PROF HUMAN NUTRIT, LINCOLN UNIV, 85- *Concurrent Pos:* Prin investr, Human Nutrit Res Prog, Lincoln Univ, Jefferson City, Mo, 85- *Mem:* Am Soc Clin Nutrit; Am Inst Nutrit; Nutrit Today Soc; AAAS. *Res:* Assessment of vitamin A status of children; impression cytology for assessment of vitamin A status; vitamin A supplementation of children; long term effects of dietary fiber on trace mineral utilization; influence of dietary fiber on lipid metabolism. *Mailing Add:* Lincoln Univ 820 Chestnut St Jefferson City MO 65101

MAKELA, LLOYD EDWARD, b Duluth, Minn, June 11, 36; m 62; c 3. PAPER SCIENCE. *Educ:* Univ Wis-Madison, BS, 58, MS, 59. *Prof Exp:* Process engr, Kimberly Clark Corp, 59-63 & Am Can Co, 63-65; process engr, 65, supvr new prod develop, 65-69, mgr prod develop, 69-73, DIR RES & DEVELOP, MOSINEE PAPER CORP, 74- *Mem:* Tech Asn Pulp & Paper Indust. *Res:* Flame-resistant, grease-resistant, solvent silicone coating; ink-jet printing; creped tape backing papers. *Mailing Add:* Mosinee Paper Corp 100 Main St Mosinee WI 54455

MAKEMSON, JOHN CHRISTOPHER, b San Francisco, Calif, Sept 20, 42. MARINE MICROBIOLOGY. *Educ:* San Francisco State Col, BA, 64, MA, 66; Wash State Univ, PhD(bacteriol), 70. *Prof Exp:* Asst prof biol, Am Univ Beirut, 70-75; special fel, Harvard Univ, 75-77; PROF, FLA INT UNIV, 78- *Concurrent Pos:* Vis lectr biol, San Francisco State Univ, 77-78; vis lectr biochem, Univ Ill, 78. *Mem:* AAAS; Am Soc Microbiol. *Res:* Bioluminescence and cellular regulation; coral tumor associated algae. *Mailing Add:* Dept Biol Sci Fla Int Univ Miami FL 33199

MAKEPEACE, GERSHOM REYNOLDS, b Ludlow, Mass, Oct 8, 19; m 40; c 2. AEROSPACE ENGINEERING, WEAPONS TECHNOLOGY. *Educ:* Calif Inst Technol, BS, 42. *Prof Exp:* Mat lab supvr, Menasco Mfg Co, Calif, 42-47; mgr propulsion dept, Naval Ord Test Sta, 47-55; pres, Sandshell Corp, 55-58; propulsion prog mgr, Lockheed Aircraft Corp, 58-59, vpres & tech dir, 59-69; DIR ENG TECH & RES, US DEPT DEFENSE, OFF UNDER SECY, 69- *Mem:* Am Inst Aeronaut & Astronaut; Am Defense Preparedness Asn; Sci Res Soc Am; Sigma Xi. *Res:* Rocket propulsion engineering. *Mailing Add:* 507 Angelita Dr Prescott AZ 86301

MAKER, PAUL DONNE, b Detroit, Mich, Dec 7, 34; m 56; c 2. SPECTROSCOPY. *Educ:* Univ Mich, BSE, 56, MS, 57, PhD(physics), 61. *Prof Exp:* Assoc res scientist, Univ Mich Willow Run Lab, 60-61; prin res scientist, Physics Dept, Eng & Res Staff, Ford Motor Co, 61-87; MEM TECH STAFF, JET PROPULSION LAB, 87- *Concurrent Pos:* Assoc ed, Optics Letters, 81-84. *Mem:* Am Phys Soc. *Res:* Fourier transform infrared spectroscopy; hyper-raman scattering; nonlinear optics. *Mailing Add:* 4800 Oak Grove Pasadena CA 91109

MAKHIJA, SURAJ PARKASH, b Campbellpur, Pakistan, Apr 8, 36; US citizen; m 64; c 3. INORGANIC CHEMISTRY, ANALYTICAL CHEMISTRY. *Educ:* Agra Univ, BSc, 57; Sagar Univ, MSc, 59; Ind Univ, Bloomington, PhD(chem), 67. *Prof Exp:* Instr chem, DAV Col, Jullundur, India, 59-60; asst prof, JV Col, India, 60-61; grad res asst, Ind Univ, 61-66; assoc prof, 66-69, PROF CHEM, ALA STATE UNIV, 69- *Mem:* Am Chem Soc. *Res:* Electrochemistry in non-aqueous solvents, coordination compounds. *Mailing Add:* 3206 Capstone Ct Montgomery AL 36106

MAKHLOUF, GABRIEL MICHEL, b Haifa, Israel, June 11, 29; m 60; c 3. MEDICINE, PHYSIOLOGY. *Educ:* Univ Liverpool, MB, ChB, 53; Univ Edinburgh, PhD(med), 65; FRCP, 72. *Prof Exp:* Sr res asst gastroenterol, Univ Edinburgh, 62-65; asst prof med, Tufts Univ, 66-68; assoc prof, Med Col, Univ Ala, Birmingham, 68-70; assoc prof, 70-72, PROF MED, MED COL VA, 72-, DIR GASTROENTEROL RES, 70- *Concurrent Pos:* Consult, Lemuel Shattuck Hosp, Boston, 66-68 & Med Ctr, Univ Ala, 68- *Mem:* AAAS; Am Gastroenterol Asn; Am Fedn Clin Res; Am Physiol Soc; Biophys Soc. *Res:* Exocrine physiology, particularly gastric physiology; membrane transport; gut peptides. *Mailing Add:* Dept Internal Med Va Commonwealth Univ Box 711 MCV Richmond VA 23298-0711

MAKHOUL, JOHN IBRAHIM, b Lebanon, Sept 19, 42; US citizen. ELECTRICAL ENGINEERING, COMPUTER SCIENCE. *Educ:* Am Univ Beirut, BE, 64; Ohio State Univ, MSc, 65; Mass Inst Technol, PhD(elec eng), 70. *Prof Exp:* sr scientist, Speech Commun, 70-80, prin scientist & mgr, 80-89, CHIEF SCIENTIST, SPEECH & SIGNAL PROCESSING DEPT, SPEECH COMMUN, BOLT BERANEK & NEWMAN INC, 89- *Concurrent Pos:* Res affil, Res Lab Electronics, Mass Inst Technol, 70-; adj prof, Northeastern Univ, 82- *Honors & Awards:* Sr Award, Inst Elec & Electronic Engrs Signal Processing Soc, 78, Tech Achievement Award, 82, Soc Award, 88. *Mem:* Fel Inst Elec & Electronics Engrs; Fel Acoust Soc Am. *Res:* Speech communication (speech analysis, synthesis, compression, recognition, enhancement); digital signal processing (spectral analysis and modeling, linear prediction, lattice methods, adaptive filtering); neural networks. *Mailing Add:* Bolt Beranek & Newman Inc 10 Moulton St Cambridge MA 02138

MAKI, ARTHUR GEORGE, JR, b Portland, Ore, Nov 24, 30; m 66, 80; c 1. SPECTROSCOPY, MOLECULAR STRUCTURE. *Educ:* Univ Wash, Seattle, BS, 53; Ore State Col, PhD(phys chem), 60. *Prof Exp:* PHYSICIST, NAT BUR STANDARDS, 58- *Concurrent Pos:* Lectr, Louvain Univ, Belgium, 82. *Mem:* AAAS; Am Phys Soc; Optical Soc Am; Sigma Xi. *Res:* Molecular spectroscopy studies relating to frequency standards and calibration, the measurement of atmospheric pollutants, and molecular structure determination; infrared studies of unstable molecules and high temperature species; studies related to molecular lasers. *Mailing Add:* 15012 24th Ave SE Mill Creek WA 98012

MAKI, AUGUST HAROLD, b Brooklyn, NY, Mar 18, 30; m 52; c 4. BIOPHYSICAL CHEMISTRY. *Educ:* Columbia Univ, AB, 52; Univ Calif, PhD(chem), 57. *Prof Exp:* Instr chem, Harvard Univ, 57-60, asst prof, 60-64; from assoc prof to prof, Univ Calif, Riverside, 64-74, PROF CHEM, UNIV CALIF, DAVIS, 74- *Concurrent Pos:* Guggenheim fel, 70-71; assoc ed, Photochemistry & Photobiology, 75-79. *Mem:* AAAS; Am Soc Photobiology; Am Chem Soc; Sigma Xi. *Res:* Studies of molecular paramagnetism, principally by optical detection of electron paramagnetic resonance in excited triplet states; aromatics applications to macromolecules and macromolecular assemblies. *Mailing Add:* Dept of Chem Univ of Calif Davis CA 95616

MAKI, KAZUMI, b Takamatsu, Japan, Jan 27, 36; m 69. THEORETICAL PHYSICS. *Educ:* Kyoto Univ, BS, 59, MS, 61, PhD(physics), 64. *Prof Exp:* Res assoc physics, Res Inst Math Sci, Kyoto Univ, 64 & Enrico Fermi Inst, Univ Chicago, 64-65; from asst prof to assoc prof, Univ Calif, San Diego, 65-67; prof, Tohoku Univ, 67-74; PROF PHYSICS, UNIV SOUTHERN CALIF, 74- *Concurrent Pos:* Fulbright travel grants, 64-66; vis prof, Univ Paris-Sud, 69-70, Max-Planck Inst, Stuttgart, 86-87, Univ Paris-VII, 90; assoc ed, J Low Temperature Physics, 69-; sabbatical leave, vis scientist, Laue-Langevin Inst, France, 79 & vis prof, Univ Paris-Sud, 80-; Guggenheim fel, 79. *Honors & Awards:* Nishina Found Prize, 72; Alexander von Humboldt US Sr Scientist Prize, 86. *Mem:* AAAS; Am Phys Soc; Phys Soc Japan. *Res:* Theory of condensed matter physics; theory of superconductivity, superfluidity and charge density wave and spin density wave. *Mailing Add:* Dept of Physics Univ of Southern Calif Univ Park Los Angeles CA 90089-0484

MAKI, LEROY ROBERT, b Astoria, Ore, May 27, 27; m 51; c 5. MICROBIOLOGY, ECOLOGY. *Educ:* State Col Wash, BS, 51; Univ Wis, PhD(bact), 55. *Prof Exp:* From asst prof to assoc prof microbiol, Univ Wyo, 55-65, prof, 65-91; RETIRED. *Mem:* Fel Am Acad Microbiol; AAAS; Am Soc Microbiol. *Res:* Aerobiology; taxonomy of air bacteria; bacterially induced ice nucleation; sheep diseases. *Mailing Add:* Dept Molecular Biology Univ Wyo Laramie WY 82071

MAKIELSKI, SARAH KIMBALL, b Ft Defiance, Ariz, Nov 23, 38; m 63. POPULATION BIOLOGY, ENVIRONMENTAL HEALTH. *Educ:* Columbia Univ, BA, 60, MA, 61, PhD(zool), 65; Univ Va, MUP, 68. *Prof Exp:* Asst prof biol, Loyola Univ, La, 70-72; health planner, New Orleans Area Health Planning Coun, 72-75; spec asst to dir urban energy studies, Inst Human Rels, Loyola Univ, LA, 75-79, ASST PROF, DEPT POLIT SCI, 88- *Concurrent Pos:* Fac res grant, Loyola Univ, La, 71-72; adj asst prof, Dept Biol Sci, Loyola Univ, 83- *Mem:* AAAS; Am Soc Cell Biol; Lepidop Soc; Pop Asn Am; Am Soc Planning Offs. *Res:* Biological approach to the study of urban systems. *Mailing Add:* Dept Polit Sci Loyola Univ New Orleans LA 70118

MAKIN, EARLE CLEMENT, JR, chemistry; deceased, see previous edition for last biography

MAKINEN, KAUKO K, b Turku, Finland, July 18, 37; m 63; c 2. ORAL BIOLOGY. *Educ:* Univ Turku, Finland, BA, 64, MA, 65; Univ Helsinki, Finland, PhD(biochem), 69. *Prof Exp:* Instr anat & biochem, Univ Oulu, Finland, 64-65; teacher microbiol chem, Turku Trade Sch, Finland, 65; chemist & lab head, Dent Sch, Univ Turku, Finland, 65-69, assoc prof dent biochem, 70-74; prof clin biochem, Univ Kuopio, Finland, 73; PROF BIOCHEM, SCH DENT, UNIV MICH, ANN ARBOR, 84- *Concurrent Pos:* Chmn, Biochem Soc Turku, 68-70, mem bd, 67-75; chmn, Sect Proteolytic Enzymes, Regulation of Enzyme Activity, Fed Europ Biochem Socs; corresp, Finnish Dent Soc, 70-; sr lectr enzymol, Univ Helsinki, Finland, 70-75; vis scientist, Nat Inst Dent Res, Bethesda, Md, 75-76; sci adv, Europ Orgn Caries Res, 75-78 & 78-81; consult, Fed Am Socs Exp Biol, 77-78; co-pres, Orgn Europ Caries Res, 77-78, chmn, Sessions Caries Res, pres, Organizing Comt 25th Annual Cong, 78; chmn, Sessions Cariology, Int Asn Dent Res; mem bd, Europ Res Group Oral Biol, 78-80; chmn, Sessions Sweeteners, Fed Dent Int; expert dent & biochem, State Med Bd, Helsinki, Finland, 78-83; sr res fel, Acad Finland, 79; vis prof biochem & biophysics, Texas A&M Univ, College Station, Tex, 79-80. *Honors & Awards:* Gustave Komppa Prize, Finnish Chem Soc, 70; 2nd Eero Tammisalo Silver Medal, Finnish Dent Asn, 75; Geraldo Halsfeld Medal, 81. *Mailing Add:* Oral Biol Dept Dent Sch Dent Bldg Rm 4208 Univ Mich 1011 N University Ave Ann Arbor MI 48109

MAKINEN, MARVIN WILLIAM, b Chassell, Mich, Aug 19, 39; m 66; c 2. MOLECULAR BIOPHYSICS. *Educ:* Univ Pa, AB, 61, MD, 68; Univ Oxford, DPhil, 76. *Prof Exp:* Intern, Columbia-Presbyterian Med Ctr NY, 68-69; res assoc, USPHS, Nat Inst Health, 69-71; NIH fel, Lab Molecular Biophys, Univ Oxford, 71-74; from asst prof to assoc prof biophys & theoret biol, 74-84, assoc prof, 84-86, PROF BIOCHEM & MOLECULAR BIOL, UNIV CHICAGO, 86-, CHMN, 88- *Concurrent Pos:* Estab investr, Am Heart Asn, 75-80; John E Fogarty sr int fel & Europ Molecular Biol Orgn sr fel, Lab Phys Chem, Univ Groningen, Neth, 84-85. *Mem:* Am Chem Soc; Biophys Soc; Am Soc Biochem & Molecular Biol; fel Am Inst Chemists; Protein Soc; AAAS. *Res:* Electron spin resonance, electron nuclear double resonance; structural basis of enzyme function; stereochemical and electronic basis of catalytic action of enzymes; structure-function relationships of biological macromolecules; molecular dynamics. *Mailing Add:* Dept Biochem & Molecular Biol Cummings Life Sci Ctr Univ Chicago 920 E 58th St Chicago IL 60637

MAKINODAN, TAKASHI, b Hilo, Hawaii, Jan 19, 25; m 54. IMMUNOLOGY. *Educ:* Univ Hawaii, BS, 48; Univ Wis, MS, 50, PhD(zool, biochem), 53. *Prof Exp:* Asst serol, Univ Wis, 50-53; res assoc immunohemat, Mt Sinai Med Res Found, Ill, 53-54; NIH fel, 54-55; assoc biologist, Biol Div, Oak Ridge Nat Lab, 55-56, biologist, 56-57, head immunol group, 57-72; chief cellular & comp physiol br, Geront Res Ctr, Baltimore City Hosps, 72-76; DIR, GERIAT RES, EDUC & CLIN CTR, VET ADMIN, WADSWORTH MED CTR, 76- *Concurrent Pos:* NSF sr fel, 61-62; mem microbiol fel rev comt, NIH, 67-70; prof, Grad Sch Biomed Sci, Univ Tenn, 68-72; dir training prog, Nat Inst Child Health & Human Develop, 68-72; mem adv panel regulatory biol prog, NSF, 71-73; mem adv panel, Lobund Inst, Notre Dame, 71-73; mem pub info comt, Fedn Am Socs Exp Biol, 74- *Mem:* AAAS; Int Soc Hemat; Geront Soc (vpres, 74-75); Am Soc Microbiol; Am Asn Immunologists. *Res:* Radiation immunology; mechanism of antibody formation; aging of the immune system. *Mailing Add:* Geriat Res Educ Clin Ctr 691/11G Vet Admin Wadsworth Med Ctr Wilshire & Sawtelle Blvds Los Angeles CA 90073

MAKITA, TAKASHI, b Kyoto, Japan, Sept 21, 35; m 66; c 2. ENZYME CYTOCHEMISTRY, COMPARATIVE ANATOMY. *Educ:* Univ Tokyo, Bachelor, 61, Master, 63, Dr, 66. *Prof Exp:* From asst prof to assoc prof, 66-85, PROF VET ANAT, YAMAGUCHI UNIV, 85- *Concurrent Pos:* Dean, Fac Agr, 88- *Mem:* Am Asn Anatomists; Am Asn Cell Biol; Electron Micros Soc Am; Royal Micros Soc; Am Asn Histochem & Cytochem; Electron Micros Soc Japan. *Res:* Gross anatomy of wine and Japanese monkey; in situ observation with wet-scanning electron microscopy and environmental-scanning electron microscopy as well as localization of enzymatic activities and elements at transmission electron microscopic level. *Mailing Add:* Dept Vet Anat Yamaguchi Univ 1677-1 Yoshida Yamaguchi 753 Japan

MAKKER, SUDESH PAUL, b Naushehra Punjab, India, June 8, 41; US citizen; m 69; c 2. IMMUNOPATHOLOGY, PEDIATRIC NEPHROLOGY. *Educ:* India Inst Med Sci, New Delhi, MD, 64; Am Bd Pediat, dipl pediat, 71, dipl pediat nephrol, 74. *Prof Exp:* Intern internal med, India Inst Med Sci, New Delhi, 65, resident, 66; intern rotating, Queen's Gen Hosp, NY, 66-67; resident pediat, Univ Chicago, 67-69; fel pediat nephrol, Case Western Reserve Univ, Cleveland, 69-71, from instr to sr instr pediat, 71-73, from asst prof to assoc prof, 76-83; prof & div head Pediat & Nephrol, Health Sci Ctr, Univ Tex, San Anntonio, 83-91; PROF PEDIAT, DIR PEDIAT NEPHROL, UNIV CALIF, DAVIS, 91- *Concurrent Pos:* Prin investr, NIH grants 73-75, 74-75, 79-82, 83-86, Environ Protection Agency grant, 84-86. *Mem:* Soc Pediat Res; Am Asn Immunologists; Soc Exp Biol & Med; Am Soc Nephrol; Sigma Xi; Am Acad Pediat; Am Pediat Soc. *Res:* Immune mechanisms in glomerulonephritis and autoimmune glomerulonephritis; mechanism of heavy metal environmental toxins induced autoimmunity; biochemistry and molecular biology of glomerular antigens. *Mailing Add:* Univ Calif Davis Med Ctr 2516 Stockton Blvd Sacramento CA 95817

MAKMAN, MAYNARD HARLAN, b Cleveland, Ohio, Oct 6, 33; m 59; c 2. PHARMACOLOGY, BIOCHEMISTRY. *Educ:* Cornell Univ, BA, 55; Case Western Reserve Univ, MD & PhD(pharmacol), 62. *Prof Exp:* Asst prof, 64-70, assoc prof, 70-79, PROF BIOCHEM & MOLECULAR PHARMACOL, ALBERT EINSTEIN COL MED, 79- *Concurrent Pos:* NIH spec fel, 65-66, career develop award, 66-71. *Mem:* AAAS; Am Soc Biol Chemists; Am Soc Pharmacol & Exp Therapeut. *Res:* Hormone action; neurotransmitter action; control of hormone receptors in normal and malignant cells; biochemical influence of hormones, neurotransmitters, cyclic adenosine monophosphate and related drugs on cultured cells, brain and retina. *Mailing Add:* Dept of Biochem Albert Einstein Col of Med 1300 Morris Park Ave Bronx NY 10461

MAKOFSKE, WILLIAM JOSEPH, b Brooklyn, NY, July 27, 43; m 67; c 2. THERMAL MODELING OF BUILDINGS. *Educ:* Pratt Inst, BS, 64; Rutgers Univ, MS, 66, PhD(physics), 68. *Prof Exp:* Fel & instr physics, Rutgers Univ, 68; res assoc & lectr, Univ Minn, Minneapolis, 68-71 & Columbia Univ, 71-74; from asst prof to assoc prof physics, Ramapa Col, 74-76, dir, Inst Environ Studies, 85-88, prog dir, dept math-physics, 85-90, PROF PHYSICS, RAMAPO COL, 79- *Concurrent Pos:* Consult energy, 76-; vis res physicist, Building Res Establishment, England, 83-84; consult radon, 88- *Mem:* Am Asn Physics Teachers; Am Phys Soc; Sigma Xi. *Res:* Theoretical and experimental nuclear structure physics; solar and renewable energy technologies; thermal behavior of greenhouses and buildings; environmental radon. *Mailing Add:* Dept Physics Ramapo Col 505 Ramapo Valley Rd Mahwah NJ 07430

MAKOFSKI, ROBERT ANTHONY, b Newport Township, Pa, Dec 27, 30; m 53; c 4. TRANSPORTATION ENGINEERING. *Educ:* Pa State Univ, BS, 52; Univ Va, MS, 56. *Prof Exp:* Aeronaut res scientist, Nat Adv Comt Aeronaut, 52-56; sr engr, 57-68, prog mgr Urban Transp Systs, 68-82, head tech serv, 82-86, ASST DIR, APPL PHYSICS LAB, JOHNS HOPKINS UNIV, 86- *Concurrent Pos:* Parsons vis prof, Johns Hopkins Univ, 80-81. *Res:* Internal and unsteady flow; helicopters; hypersonic fluid mechanics; boundary layers; research facility development; transportation system design and evaluation; transportation system operation and control; system costs and economics. *Mailing Add:* 1905 Cedar Circle Dr Catonsville MD 21228

MAKOMASKI, ANDRZEJ HENRYK, b Chojnice, Poland, July 15, 32; Can citizen; m 62; c 3. MECHANICAL ENGINEERING. *Educ:* Univ London, BSc, 54, PhD(mech eng), 57. *Prof Exp:* Tech asst, De Havilland Engine Co Ltd, 57-58; lectr, Dept Mech Eng, Univ Col, London, 58-62; Nat Res Coun Can res assoc, Inst Aerospace Studies, Univ Toronto, 62-64; assoc res officer, 64-72, SR RES OFFICER, NAT RES COUN CAN, 73- *Mem:* Am Phys Soc. *Res:* Fluid mechanics; combustion. *Mailing Add:* 2257 Prospect Ave Ottawa ON K1H 7G2 Can

MAKOUS, WALTER L, b Milwaukee, Wis, Nov 22, 34; m 82; c 3. VISION. *Educ:* Univ Wis, BS, 58; Brown Univ, PhD(psychol), 64. *Prof Exp:* mem staff, Res Div, Int Bus Mach Corp, 63-66; from asst prof to prof psychol, Univ Wash, 66-79; PROF PSYCHOL, VISUAL SCI & OPHTHAL, UNIV ROCHESTER, 79- *Concurrent Pos:* Mem, NSF Panel, 77-82; topical ed, J Optical Soc Am, 83-87. *Mem:* Fel AAAS; fel Optical Soc Am; Asn Res Vision & Ophthal; Soc Neurosci; Human Factors Soc; Psychonomic Soc; Am Psychol Soc. *Res:* Psychophysical investigation of the organization and function of the visual system. *Mailing Add:* Ctr Visual Sci Univ Rochester Rochester NY 14627

MAKOWKA, LEONARD, b Toronto, Ont, 1953; m; c 3. LIVER TRANSPLANTATION, HEPATOLOGY. *Educ:* Univ Toronto, MD, 77, MS, 79, PhD(path), 82. *Prof Exp:* Res assoc, Dept Surg-Path, Univ Toronto, Ont, 82, asst prof, 82-85; from asst prof to assoc prof, Dept Surg, Univ Pittsburgh, 86-89; DIR SURG, DEPT SURG, CEDARS-SINAI MED CTR, LOS ANGELES, CALIF, 89-, DIR TRANSPLANTATION SERV, 89- *Concurrent Pos:* Vis prof, Dept Surg, Univ Pittsburgh, Pa, 85 & 86-87. *Honors & Awards:* Medalist in Surg, Royal Col Physicians & Surgeons Can, 82. *Mem:* Am Col Surgeons; Soc Univ Surgeons; Am Soc Transplant Phyhsicians; Am Soc Transplant Surgeons; Int Hepato-Biliary-Pancreatic Asn. *Res:* Immunosuppressive agents, other than cyclosporine, to decrease the activity of the immune system of patients who have organ transplants, primarily liver, kidney, heart and lung. *Mailing Add:* Dept Surg Cedars-Sinai Med Ctr 8700 Beverly Blvd Rm 8215 Los Angeles CA 90048

MAKOWSKI, EDGAR LEONARD, b Milwaukee, Wis, Oct 27, 27; m 52; c 6. OBSTETRICS & GYNECOLOGY. *Educ:* Marquette Univ, BS, 51, MD, 54. *Prof Exp:* Intern, Evangelical Deaconess Hosp, Milwaukee, Wis, 54-55; resident obstet & gynec, Univ Minn, 55-59, from instr to assoc prof, 59-66; assoc prof, 66-69, chmn dept, 76-88, PROF OBSTET & GYNEC, MED CTR, UNIV COLO, DENVER, 69- *Concurrent Pos:* Fel physiol, Sch Med, Yale Univ, 63-64. *Res:* Reproductive and fetal physiology. *Mailing Add:* Dept of Obstet & Gynec Univ Colo Med Ctr 4200 E 9th Ave Denver CO 80262-0001

MAKOWSKI, LEE, b Providence, RI, Nov 4, 49; m; c 2. STRUCTURAL BIOLOGY, X-RAY DIFFRACTION. *Educ:* Brown Univ, BS, 71; Mass Inst Technol, MS, 73, PhD(elec eng), 76. *Prof Exp:* Fel, Structural Biol Lab, Brandeis Univ, 76-78, res assoc, 78-80; Irma T Hirschl Career Scientist, 80-85, asst prof biochem, Dept Biochem, Columbia Univ, 80-87; PROF PHYSICS, DEPT PHYSICS, BOSTON UNIV, 87- *Concurrent Pos:* Alfred P Sloan Found fel, 78-82; assoc ed, Biophys J; guest asst scientist, Brookhaven Nat Lab; mem coun & exec bd, Biophys Soc. *Mem:* Biophys Soc; Am Crystallog Asn; AAAS; Am Soc Cell Biol. *Res:* X-ray diffraction and electron microscopy to determine the structure of macromolecular assemblies; image and data processing strategies are being developed for extracting information from data collected using non-crystalline specimens. *Mailing Add:* Physics Dept Boston Univ 590 Commonwealth Ave Boston MA 02215

MAKOWSKI, MIECZYSLAW PAUL, b Warsaw, Poland, Jan 15, 22; US citizen; m 45; c 2. ELECTROCHEMISTRY, PHYSICAL CHEMISTRY. *Educ:* Western Reserve Univ, BA, 57, MS, 61, PhD(electrochem), 64. *Prof Exp:* Asst mgr plastics technol, Smith-Phoenix Mfg Co, Ohio, 50-55; res chemist, Clevite Corp, 55-61; res asst electrochem, Western Reserve Univ, 61-62; sr res chemist, Gould Inc, 62-64 mgr chem & polymers sect, 64-74, assoc dir explor develop, 74-76, dir, Gould Labs, 76-80, vpres tech admin, 80-81, vpres sci affairs, 81-87; PRES, TECTRA CONSULT, INC, 87- *Concurrent Pos:* Mem, Frontiers in Chem Lecture Series Comt, Case Western Reserve Univ, 72-80; mem, Ill Gov Comn Sci & Technol, 87. *Honors & Awards:* Heise Award, Cleveland Sect, Electrochem Soc. *Mem:* Am Chem Soc; Electrochem Soc; AAAS. *Res:* Electrode kinetics; hydrogen electrode; fuel cell electrode structure; electrodeposition; electroless deposition of metals; surface area studies; applied polymer research. *Mailing Add:* 305 Shady Dr Palatine IL 60067-7551

MAKOWSKI, ROBERTE MARIE DENISE, b Winnipeg, Man. BIOLOGICAL CONTROL, WEED SCIENCE. *Educ:* Univ Ottawa, BSc, 77, MSc, 81; Univ Sask, PhD(agr & crop sci), 87. *Prof Exp:* Res asst underwater archeol, Nat Hist Sites, Parks, Can, 77-78; teaching asst gen biol & invert zool, Univ Ottawa, 78-81; biologist biocontrol, RPS, Res Br, Ottawa, 81-83, scientist training, Regina, 83-87, RES SCIENTIST BIOCONTROL WEEDS, REGINA RES STA, AGR CAN, 87-, SECT HEAD, 90- *Concurrent Pos:* Res asst ichthyol, Nat Mus Natural Sci, Can, 77-78; chair, Resolutions Comt, Can Phytopath Soc, 89-90; agr Can rep, Biotechnol Biocontrol working group, USDA, 90- *Mem:* Can Phytopath Soc; Entom Soc Can; Weed Sci Soc Am. *Res:* Biological control of weeds using plant pathogens; plant diseases are evaluated and developed for their potential as biological herbicides against important weeds to Canadian agriculture; co-responsible for first biological herbicide submitted for registration in Canada. *Mailing Add:* Res Sta Agr Can PO Box 440 Regina SK S4P 3A2 Can

MAKRIYANNIS, ALEXANDROS, b Cairo, Egypt, Sept 3, 39; US citizen. MEDICINAL CHEMISTRY. *Educ:* Univ Cairo, BPharmChem, 60; Univ Kans, PhD(med chem), 67. *Prof Exp:* Fel org chem, Univ Calif, Berkeley, 67-69; sr res chemist drug synthesis, Smith Kline & French Labs, Pa, 69-70; res assoc biochem & pharmacol, Tufts Med Sch, Mass, 71-72, asst prof, 72-74; from asst prof to assoc prof, 74-85, PROF MED CHEM, UNIV CONN, 85- *Concurrent Pos:* Mem, Inst Mat Sci, Univ Conn, 74-, polymer sci prog, 80-; vis scientist, Mass Inst Tech, 83. *Mem:* Am Chem Soc; Sigma Xi. *Res:* Drug design and synthesis; drug-membrane interactions; nuclear magnetic resonance spectroscopy; molecular pharmacology; cannabinoids, steroids, antineoplastic lipids. *Mailing Add:* Univ of Conn U-92 Storrs CT 06268

MAKSOUDIAN, Y LEON, b Beirut, Lebanon, Oct 30, 33; US citizen; c 3. MATHEMATICS. *Educ:* Calif State Polytech Col, BS, 57; Univ Minn, Minneapolis, MS, 61, PhD, 70. *Prof Exp:* Instr math, Westmont Col, 57-58, Northwestern Col, Minn, 58-62 & Univ Minn, Minneapolis, 62-63; from asst prof to assoc prof, 63-72, PROF STATIST, CALIF STATE POLYTECH UNIV, SAN LUIS OBISPO, 72- *Mem:* Am Statist Asn; Math Asn Am. *Res:* Probability and statistics. *Mailing Add:* Dept of Comput Sci & Statist Calif State Polytech Univ San Luis Obispo CA 93407

MAKSUD, MICHAEL GEORGE, b Chicago, Ill, Mar 26, 32; m 59; c 2. EXERCISE PHYSIOLOGY. *Educ:* Univ Ill, Urbana, BS, 55; Syracuse Univ, MA, 57; Mich State Univ, PhD(phys educ, physiol), 65. *Prof Exp:* Instr phys educ, Univ Ill, Chicago, 59-63; from asst prof to assoc prof phys educ, Univ Wis-Milwaukee, 65-72, prof, 72-80, dir exercise physiol lab, 65-80, assoc dean, Grad Sch, 76-80; DEAN, COL HEALTH & HUMAN PERFORMANCE, ORE STATE UNIV, 80- *Concurrent Pos:* Clin assoc, Med Col Wis, 66-80; consult physiol, Res Serv, Wood Vet Admin Ctr, 67-80; mem, Human Biol Coun, Am Asn Health, Phys Educ & Recreation. *Mem:* Am Physiol Soc; Fel Am Col Sports Med; Am Asn Health, Phys Educ & Recreation; Fel Am Acad Phys Educ. *Res:* Physiological basis of performance; nutritional effects of biochemical adaptation. *Mailing Add:* Col Health & Human Performance Ore State Univ Corvallis OR 97331-6802

MAKSYMOWYCH, ROMAN, b Kaminka, W Ukraine, Oct 15, 24; nat US; m 51; c 4. BOTANY, BIOMETRY OF GROWTH. *Educ:* Univ Pa, MS, 56, PhD(bot), 59. *Prof Exp:* Instr biol, Univ Pa, 59-60; from asst prof to assoc prof, 61-65, chmn dept, 76-79, PROF BIOL, VILLANOVA UNIV, 66- *Concurrent Pos:* NSF res grants, 59, 61, 65, 67; vis prof, Oxford Univ, 68; vis researcher, Brookhaven Nat Lab, 81 & 82. *Mem:* Bot Soc Am; Shevchenko Sci Soc (vpres, 78-79). *Res:* Plant growth and development; cell division and DNA biosynthesis in leaves and roots of Xanthium plants; hormonal regulation of plant development; analysis of stem growth in terms of relative elemental rates. *Mailing Add:* Dept Biol Villanova Univ Villanova PA 19085

MAL, AJIT KUMAR, b West Bengal, India, Oct 2, 37; m; c 1. ENGINEERING MECHANICS, GEOPHYSICS. *Educ:* Univ Calcutta, MSc, 59, DPhil(appl math/mech), 64. *Prof Exp:* Lectr math, Bengal Eng Col, India, 61-63; geophysicist, Univ Calif, Los Angeles, 64-66; asst res engr & lectr appl mech, Univ Calif, Berkeley, 66-67; from asst prof to assoc prof mech, 67-74, PROF MECH ENG, UNIV CALIF, LOS ANGELES, 74- *Concurrent Pos:* NSF grants earthquake eng & mech; consult, 80-; sr Fulbright fel, WGer, 84-85. *Mem:* Am Soc Mech Engrs; Am Acad Mech. *Res:* Wave propagation in solids; earthquake engineering; applied mathematics. *Mailing Add:* Dept Mech Aerospace & Nuclear Eng Univ Calif Sch Eng & Appl Sci Los Angeles CA 90024

MALACARA, DANIEL, b Leon, Mex, June 7, 37; m 64; c 4. OPTICS. *Educ:* Univ Mex, BSc, 61; Univ Rochester, MSc, 63, PhD(optics), 65. *Prof Exp:* Asst prof astron, Tonantzinla Observ, 64-66; from assoc prof to prof optics, Univ Mex, 66-72; PROF & TECH DIR, NAT INST ASTROPHYS OPTICS & ELECTRONICS, 72- *Concurrent Pos:* Mem admis comn, Mex Acad Sci & acad judgement comn, Physics Inst, Univ Mex, 69-71. *Honors & Awards:* Sci Instrumentation Award, Mex Acad Sci, 68. *Mem:* Optical Soc Am; Int Astron Union; Int Comn Optics; Mex Acad Sci. *Res:* Optical testing and design of instruments and components; interferometry. *Mailing Add:* Ctr Investigacion-Optica Ac Apartado PO 948 Leon GTO Mexico

MALACINKSI, GEORGE M, b Norwood, Mass, Nov 25, 40; m 65; c 2. DEVELOPMENTAL BIOLOGY, BIOCHEMISTRY. *Educ:* Boston Univ, AB, 62; Univ Ind, MA, 64, PhD(microbiol), 66. *Prof Exp:* USPHS fel biochem & develop biol, Univ Wash, 66-68; asst prof, 68-74, assoc prof zool, 74-80, PROF BIOL DEPT, IND UNIV, BLOOMINGTON, 80- *Concurrent Pos:* Res assoc, Univ Zurich, Switz, 74-75. *Mem:* AAAS; Soc Develop Biol. *Res:* Biochemical and molecular basis of the regulatory mechanisms which control the ordered sequence of events which characterize the various stages in the developmental cycle of various animals. *Mailing Add:* Dept Zool Ind Univ Bloomington IN 47401

MALACZYNSKI, GERARD W, b Stanislawow, Poland, May 26, 42; m 65; c 2. QUANTUM ELECTRONICS. *Educ:* Gdansk Tech Univ, Poland, MSc, 65, PhD(electronics), 68. *Prof Exp:* Mem sci staff, Inst Fluid Flow Mach, Polish Acad Sci, 65-68, asst prof, 69-76; chief & mgt engr, Radio & TV Broadcasting Ctr, Gdansk, Poland, 76-81; sr res engr, 83-84, MEM STAFF RES ENG, ELEC ENG DEPT, GEN MOTORS RES LABS, 84- *Concurrent Pos:* Vis scientist, Elec Eng Dept, Mass Inst Technol, 69-70, res engr, High Voltage Lab, 83; serv engr, RCA Jersey Island Ltd, 81. *Honors & Awards:* Sci Award, Polish Acad Sci, 71. *Mem:* Inst Elec & Electronics Engrs; Electrostatics Soc Am. *Res:* The interaction of electromagnetic fields with matter; continuum electromechanics; physics of ionized gases; complex heat exchanges in solid bodies. *Mailing Add:* 1183 Ashover Dr Bloomsfield MI 48013

MALAH, DAVID, b Poland, Mar 31, 43; Israeli citizen; m 66; c 1. DIGITAL SIGNAL PROCESSING OF SPEECH & IMAGES. *Educ:* Technion-Israel Inst Technol, BSc, 64, MSc, 67; Univ Minn, PhD(elec eng), 71. *Prof Exp:* Lectr elec eng, Univ NB, Can, 70-71, asst prof, 71-72; lectr, Technion-Israel Inst Technol, 72-77, sr lectr, 77-81, assoc prof, 81-90, PROF ELEC ENG, TECHNION-ISRAEL INST TECHNOL, 90- *Concurrent Pos:* Head, Signal Processing Lab, Technion, 75-; consult, AT&T Bell Labs, 79-80 & 88-89, mem tech staff, 80-81. *Mem:* Fel Inst Elec & Electronic Engrs Acoust Speech & Signal Processing Soc; Europ Asn Signal Processing. *Res:* Image and speech communication with emphasis on image and speech coding, image and speech enhancement and digital signal processing techniques. *Mailing Add:* Elec Eng Dept Technion Technion City Haifa 32000 Israel

MALAHOFF, ALEXANDER, b Moscow, USSR, Feb 7, 39; US citizen; m 62; c 2. GEOLOGY, GEOPHYSICS. *Educ:* Univ NZ, BS, 60; Victoria Univ, NZ, MSc, 62; Univ Hawaii, PhD(geophys), 65. *Prof Exp:* Sci officer, Dept Sci & Indust Res, NZ, 59-60; asst geophys, Univ Wis, 63-64; asst geophys, Univ Hawaii, 64-65, asst geophysicist & asst prof geosci, 65-69, asst prof oceanog, 66-69, assoc prof geosci & oceanog, 69-71; prog dir marine geol & geophys prog, Off Naval Res, 71-75; chief scientist, Nat Ocean Surv, Dept Com, Nat Oceanic & Atmospheric Admin, 76-84; PROF OCEANOG & DIR, HAWAII UNDERSEA RES LAB, UNIV HAWAII, 84- *Concurrent Pos:* Mem, US-USSR Bilateral Sci Comt, 72- & US-Japan Natural Resources Comt, 76-; mem staff, deep submergence studies, US Navy, 72-76 & aeromagnetic studies world ocean, US Naval Res Lab, 76-; mem staff, ocean ridge res, Nat Ocean Surv, Nat Oceanic & Atmospheric Admin, 76- & shipboard multibeam studies, 79- *Honors & Awards:* Silver Medal, US Dept Com, 81. *Mem:* Am Geophys Union; Soc Explor Geophys; Royal Soc NZ; Geol Soc NZ; Geol Soc Am. *Res:* Solid earth geophysics; marine geophysical studies of the Pacific Ocean crust; tectonics of the ocean ridge system; minerals of the ocean floor; submarine hydrothermal systems; ocean technology. *Mailing Add:* 1484 Humuwili Place Kailua Oahu HI 96734-3714

MALAIYA, YASHWANT KUMAR, b Sagour, India, May 19, 51; m 79. RELIABILITY & FAULT TOLERANCE. *Educ:* Univ Saugor, India, MS, 71; Birla Inst Technol & Sci, MS, 74; Utah State Univ, PhD(elec eng), 79. *Prof Exp:* Asst prof comput sci, State Univ NY, 78-82; ASSOC PROF COMPUT SCI, COLO STATE UNIV, 82- *Concurrent Pos:* Co-prin investr, Proj Intermittent Faults, NSF, 79-83, Proj Testability Very Large Scale Integration, Rome Air Develop Ctr, 81-82 & Proj Functional Testing, US Army, 82-84. *Mem:* Inst Elec & Electronics Engrs. *Res:* Modeling, testing and reliability analysis of systems with intermittent faults; testing and testability of complementary metal-oxide semiconductor very large scale integration; design automation; multiple processor systems. *Mailing Add:* Dept Comput Sci Colo State Univ Ft Collins CO 80523

MALAMED, SASHA, b New York, NY, May 6, 28; m 56; c 1. CELLULAR ENDOCRINOLOGY. *Educ:* Univ Pa, BA, 48, MS, 50; Columbia Univ, PhD(zool), 55. *Prof Exp:* Asst zool, Univ Pa, 49-50; asst, Columbia Univ, 50-54; res assoc, Univ Iowa, 54-55 & Columbia Univ, 55-56; USPHS res fel physiol, Western Reserve Univ, 56-58; instr anat, Albert Einstein Col Med, 58-59, asst prof, 59-67; assoc prof, 67-74, PROF ANAT, ROBERT WOOD JOHNSON MED SCH, UNIV MED & DENT NJ, 74- *Concurrent Pos:* Vis lectr, Cornell Col, 55; instr, Hunter Col, 55; vis prof, St Bartholomew's Hosp Med Sch, London, 86-87. *Mem:* Am Soc Cell Biol; Am Asn Anatomists; Biophys Soc; Am Soc Zoologists; Am Physiol Soc; Endocrine Soc. *Res:* Ultrastructural and steroidogenic relationships of adrenocortical cells; somatotroph development of structure and function; innervation of endocrine glands. *Mailing Add:* Robert Wood Johnson Med Sch Univ Med & Dent NJ Piscataway NJ 08854

MALAMUD, DANIEL F, b Detroit, Mich, June 5, 39; m 61; c 2. CELL BIOLOGY, BIOCHEMISTRY. *Educ:* Univ Mich, BS, 61; Western Mich Univ, MA, 62; Univ Cincinnati, PhD(zool), 65. *Prof Exp:* Instr biol, Univ Cincinnati, 65-66; asst prof path, Temple Univ, 68-69; asst prof path, Harvard Med Sch, 70-77; assoc prof, 77-84, PROF BIOCHEM, SCH DENT MED, UNIV PA, 84-, CHMN DEPT, 85- *Concurrent Pos:* USPHS res fel, Fels Res Inst, Sch Med, Temple Univ, 66-68; asst biologist, Mass Gen Hosp, 69-77; res career develop award, 72-77; Fulbright Scholar, Univ Philippines, 75. *Mem:* Am Soc Biol Chemists; Am Soc Cell Biol; Soc Develop Biol; Sigma Xi; NY Acad Sci; Am Soc Microbiol. *Res:* Salivary gland growth and secretion; human salivary proteins and anti-microbial proteins, anti-human immunodeficiency virus activity. *Mailing Add:* Dept Biochem Univ Pa Sch Dent Med Philadelphia PA 19104-6002

MALAMUD, ERNEST I(LYA), b New York, NY, May 8, 32. COLLIDER EXPERIMENTS, ACCELERATOR PHYSICS. *Educ:* Univ Calif, Berkeley, AB, 54; Cornell Univ, PhD(physics), 59. *Prof Exp:* Res assoc, Cornell Univ, 59-60; privat docent, Univ Lausanne, 61-62; Ford fel, 63; guest prof, Univ Heidelberg, 64; from asst prof to assoc prof physics, Univ Ariz, 64-66; vis assoc prof, Univ Calif, Los Angeles, 66-67, assoc prof in residence, 67-68; head, Main Ring Accelerator Sect, 70-71, head internal target sect, 72-73, ombudsman, 76, head, Meson Dept, 78-81, PHYSICIST, FERMI NAT ACCELERATOR LAB, 68- *Concurrent Pos:* Group leader, Tevatron I Sect, 83-85; vis prof, Univ Lausanne, 85-86; staff mem, Exploratorium, San Francisco, 82; exec dir, Sci & Technol Interactive Ctr, 88- *Mem:* Inst Elec & Electronic Engrs; Am Phys Soc; Sigma Xi. *Res:* High energy physics; designing, constructing and commissioning of Fermilab 400 GEV main accelerator; initiating Soviet-American joint experiment on pp scattering; investigations of hadron jets; installation, controls coordination and commissioning of the Autiproton Source; design and construction of D0 Muon Detector; luminosity upgrade of the Fermilab Tevatron. *Mailing Add:* Fermi Nat Accelerator Lab PO Box 500 Batavia IL 60510

MALAMUD, HERBERT, b New York, NY, June 28, 25; m 51; c 3. MEDICAL PHYSICS. *Educ:* City Col, New York, BS, 49; Univ Md, MS, 52; NY Univ, PhD(physics), 57; Long Island Univ, MS, 76; Am Bd Sci Nuclear Med, dipl, 79. *Prof Exp:* Sr engr, Physics Labs, Sylvania Elec Prod Co, 57-59; specialist res engr, Repub Aviation Corp, 59-64; res sect head, Sperry Gyroscope Co, 64-65; dir physics res, Radiation Res Corp, 65-67; vpres, Plasma Physics Corp, 67-70; sr physicist, Dept Nuclear Med, Queen's Hosp Ctr, 70-79; CONSULT PHYSICS, 76-; TECH DIR NUCLEAR ASSOC, VICTOREEN, 79- *Mem:* AAAS; Am Phys Soc; Soc Nuclear Med; Am Asn Physicists in Med. *Res:* Medical radiation physics; plasma and atomic physics. *Mailing Add:* 30 Wedgewood Dr Westbury NY 11590

MALAMUD, NATHAN, b Kishinev, Russia, Jan 28, 03; nat US; m 30; c 2. NEUROPATHOLOGY. *Educ:* McGill Univ, MD, 30. *Prof Exp:* Asst neuropathologist & instr psychiat, Med Sch, Univ Mich, 34-45; prof, 46-71, neuropathologist, Langley Porter Neuropsychiat Inst, 46-75, EMER PROF NEUROPATH IN RESIDENCE, SCH MED, UNIV CALIF, SAN FRANCISCO, 71-, NEUROPATHOLOGIST, DEPT PATH, SCH MED, 75- *Concurrent Pos:* Consult, Armed Forces Inst Path, 44-, Letterman Army Med Ctr, 46-, Oakland Naval Hosp, 51-, Nat Inst Neurol Dis & Stroke, 55- & Vet Admin Hosps, Martinez & San Francisco, 60-75. *Mem:* Am Asn Neuropath (vpres, 58-59); Am Psychiat Asn; Am Acad Neurol. *Res:* Cerebral palsy; mental retardation; geriatric disorders; chronic alcoholism; encephalitis; radiation; epilepsy; heredodegenerative disorders. *Mailing Add:* 320 Silver Ave San Francisco CA 94112

MALAMY, MICHAEL HOWARD, b Brooklyn, NY, Apr 20, 38; m 58; c 2. MOLECULAR BIOLOGY, BACTERIAL GENETICS. *Educ:* NY Univ, BA, 58, PhD(microbiol), 63. *Prof Exp:* From asst prof to assoc prof, 66-76, PROF MOLECULAR BIOL, TUFTS UNIV, 76- *Mem:* Am Soc Biol Chemists; Am Soc Microbiol. *Res:* Regulation of gene expression in bacteria, bacterial viruses and plasmids. *Mailing Add:* Dept Molecular Biol Sch Med Tufts Univ Boston MA 02111

MALAN, RODWICK LAPUR, b Du Quoin, Ill, Sept 7, 16; m 43; c 6. ORGANIC CHEMISTRY. *Educ:* Univ Ariz, BS & MS, 41; Univ Colo, PhD(org chem), 46. *Prof Exp:* Lab asst, Univ Ariz, 40-41; asst, Univ Colo, 41-44; chemist, Res Labs, Eastman Kodak Co, 44-49, Film Emulsion Div, 50-56, tech assoc, 57-82; RETIRED. *Concurrent Pos:* Lectr eve div, Rochester Inst Technol, 48-68. *Mem:* Am Chem Soc. *Res:* Hemicelluloses and pectic materials in corn leaves; pyridine chemistry; photographic chemicals for the color processes; chemical emulsions. *Mailing Add:* 6 Countryside Rd Fairport NY 14450

MALANGA, CARL JOSEPH, b New York, NY, Aug 26, 39; m 66; c 1. CELL PHYSIOLOGY, PHARMACOLOGY. *Educ:* Fordham Univ, BS, 61, MS, 67, PhD(biol sci), 70. *Prof Exp:* Instr biol sci labs, Col Pharm, Fordham Univ, 64-67, instr anat, physiol & pharmaceut, 67-70; from asst prof to assoc prof therapeut, 70-76, chmn basic pharm sci, 80-91, PROF BIOPHARM, SCH PHARM, WVA UNIV, 78-, ASSOC DEAN, ACAD AFFAIRS & ADMIN, 91- *Concurrent Pos:* Chmn Curric Comt, Sch Pharm Exec Comn & Grad Sch Exec Comn, WVa Univ. *Mem:* AAAS; Am Pharmaceut Asn; Am Asn Pharmaceut Scientists; Am Soc Zoologists; Am Soc Pharmacol & Exp Therapeut. *Res:* Effects of serotonin, catecholamines and drugs on ciliary activity, energy metabolism and mucus synthesis and secretion. *Mailing Add:* WVa Univ Sch Pharm-Health Sci Ctr N Morgantown WV 26506

MALANIFY, JOHN JOSEPH, b Troy, NY, Apr 26, 34; m 56; c 2. NUCLEAR PHYSICS. *Educ:* Rensselaer Polytech Inst, BS, 55, PhD(physics), 64. *Prof Exp:* Staff mem nuclear physics, Los Alamos Sci Lab, Univ Calif, 64-66 & Oak Ridge Nat Lab, 66-69; STAFF MEM, LOS ALAMOS NAT LAB, 69- *Mem:* Am Nuclear Soc; Inst Nuclear Mat Mgt; Am Phys Soc. *Res:* Direct nuclear reaction mechanism and the nucleon-nucleon problem, especially polarization; delayed neutrons and gamma rays from fission; nuclear accountability; x-ray fluorescence; muonic atoms; neutron time of flight. *Mailing Add:* Los Alamos Nat Lab Group Q2 MS J562 Los Alamos NM 87545

MALARKEY, EDWARD CORNELIUS, b Girardville, Pa, Dec 7, 36; m 65; c 1. OPTICAL SIGNAL PROCESSING, INTEGRATED OPTICS. *Educ:* La Salle Col, AB, 58; Mass Inst Technol, PhD(phys chem), 63. *Prof Exp:* Physicist, 63-81, ADV PHYSICIST, APPL SCI GROUP, ADVANCED TECHNOL DIV, WESTINGHOUSE ELEC CORP, 81- *Concurrent Pos:* Assoc prof, Anne Arundel Community Col, Md, 67. *Mem:* NY Acad Sci; Optical Soc Am; Soc Photo-Optical Instrumentation Engrs; Sigma Xi. *Res:* Optical computer development; acousto-optical system design, analysis and development; integrated optics system design, component analysis and device development; laser and laser resonator design and development; computer modeling; optical signal processing. *Mailing Add:* 353 White Cedar Lane Severna Park MD 21146

MALASHOCK, EDWARD MARVIN, b Omaha, Nebr, Mar 27, 23; m 44; c 3. UROLOGY. *Educ:* Univ Nebr, BA, 43, MD, 46; Am Bd Urol, dipl, 56. *Prof Exp:* Resident urol surg, Beth Israel Hosp, New York, 50-53; clin asst, 53-55, assoc, 55-57, from asst prof to assoc prof, 57-76, CLIN PROF UROL, COL MED, UNIV NEBR, OMAHA, 76-, ASSOC, PHYS MED & REHAB, 58- *Concurrent Pos:* Chief urol sect, Tenth Gen Hosp, Manila, 48-49; pres med staff, Bishop Clarkson Mem Hosp, 76-78; pres, Metropolitan Omaha Med Soc, 79-80. *Mem:* AMA; Am Urol Asn; Am Col Surgeons. *Res:* Urological surgery; neurogenic bladder dysfunction; new drugs as related to urological problems. *Mailing Add:* 8602 Hickory St Omaha NE 68124

MALASPINA, ALEX, b Athens, Greece, Jan 4, 31; nat US; m 54; c 4. NUTRITION, FOOD REGULATION. *Educ:* Mass Inst Technol, BS, 52, SM, 53, PhD(food tech), 55. *Prof Exp:* Asst, Mass Inst Technol, 53-55; coordr, New Prod Dept, Chas Pfizer & Co, 55-61; mgr qual control dept, 61-69, vpres qual control & develop dept, Coca-Cola Export Corp, 69-77, vpres external tech affairs dept, 78-86, SR VPRES, COCA-COLA CO, 86- *Concurrent Pos:* Pres, Int Tech Caramel Asn, 77 & Int Life Sci Inst, 78; vpres, Toxicol Forum, 79- *Mem:* AAAS; Am Chem Soc; Am Inst Chemists; Inst Food Technologists; NY Acad Sci. *Res:* Quality control and new product development on carbonated beverages and protein drinks; ingredient safety, food regulations, nutrition and health. *Mailing Add:* Coca-Cola Co PO Drawer 1734 Atlanta GA 30301

MALATHI, PARAMESWARA, MEMBRANE TRANSPORT, BIOCHEMISRY. *Educ:* Indian Inst Sci, Bangalore, India, PhD(biochem), 66. *Prof Exp:* ASSOC PROF PHYSIOL & BIOPHYS, ROBERT WOOD JOHNSON SCH MED, UNIV MED & DENT NJ, 82- *Res:* Absorption from small intestine and re-absorption from the kidney. *Mailing Add:* Univ Med & Dent NJ Robert Wood Johnson Med Sch Piscataway NJ 08854

MALAVIYA, BIMAL K, b Allahabad, India, Oct 28, 35. ENERGY TECHNOLOGY. *Educ:* Banaras Univ, BSc, 54, MSc, 56; Harvard Univ, AM, 59, PhD, 64. *Prof Exp:* Lectr, Banaras Univ, 56-58; DSR res fel, Mass Inst Technol, 61-65; from asst prof to assoc prof nuclear eng & sci, 65-76, PROF NUCLEAR ENG & DIR SPEC PROGS IN NUCLEAR TECHNOL, RENSSELAER POLYTECH INST, 76- *Mem:* AAAS; Am Nuclear Soc; NY Acad Sci; Am Soc Eng Educ. *Res:* Energy technology; energy planning and policy; nuclear power; reactor physics and engineering. *Mailing Add:* 1532 Bouton Rd Troy NY 12180

MALAWISTA, STEPHEN E, b New York, NY, Apr 4, 34; m 69. INTERNAL MEDICINE, RHEUMATOLOGY. *Educ:* Harvard Univ, AB, 54; Columbia Univ, MD, 58; Am Bd Internal Med, dipl, 65; Am Bd Dermat, dipl, 66. *Prof Exp:* Intern internal med, Yale-New Haven Med Ctr, 58-59, asst resident, 59-60; clin assoc, Nat Inst Arthritis & Metab Dis, 60-62; asst resident, Yale-New Haven Med Ctr, 62-63, NIH spec fel, 63-66; from asst prof to assoc prof med, 66-75, chief rheumatol, 67-88, PROF MED, SCH MED, YALE UNIV, 75- *Concurrent Pos:* Asst attend physician, Yale-New Haven Med Ctr, 66-69, attend physician, 69-; attend physician, Vet Admin Hosp, West Haven, 66-69, consult rheumatology, 69-; sr investr, Arthritis Found, 66-70; consult, Gaylord Hosp, 68-; Nat Inst Arthritis & Metab Dis career res develop award, 70-75; vis scientist, Cent Med Genetics, CNRS, Gif-Sur-Yvette, 72-73, Inst Cell Path, Nat Inst Health & Med Res, Le Kremlin-Bicetre, 79-80, dept Molecular Genetics, Pasteur Inst, Paris, France, 87-88. *Honors & Awards:* CIBA-Geigy Ilar Rheumatology Prize, Int League Against Rheumatism, 85. *Mem:* AAAS; fel Am Col Physicians; Am Soc Invest; Am Col Rheumatology (pres elect, 90-91, pres, 91-92); Am Soc Cell Biol; Asn Am Physicians. *Res:* Rheumatic diseases; gout; Lyme disease; inflammation; phagocytosis. *Mailing Add:* Dept of Internal Med Yale Univ Sch Med 333 Cedar New Haven CT 06510

MALBICA, JOSEPH ORAZIO, b Brooklyn, NY, Apr 6, 25; m 47; c 4. BIOCHEMISTRY. *Educ:* Brooklyn Col, BS, 49; Fordham Univ, MS, 54; Rutgers Univ, PhD(biochem), 67. *Prof Exp:* Chemist, Hoffman-La Roche Inc, NJ, 54-65; instr physiol & biochem Rutgers Univ, 66-67; res biochemist, Hess & Clark Div, Richardson-Merrell, Inc, 67-69; supvr biochem-pharmacol, Stuart Pharmaceut, Div ICI Americas Inc, 69-80, sect mgr, drug metab, 80-87; RETIRED. *Concurrent Pos:* Mem adv bd, Cecil Community Col, Northeast, Md, 86-87. *Mem:* Am Chem Soc; Am Soc Pharmacol & Exp Therapeut; NY Acad Sci; Sigma Xi. *Res:* Biosynthesis of natural and unnatural products of pharmacology; fermentation; isolation and purification of natural products; drug metabolism; anti-inflammatories; muscle relaxants; drug-related studies with cellular organelles; ion-transport; drug kinetics; bioanalytical methods development. *Mailing Add:* 19020 SW Christensen Rd McMinnville OR 97128

MALBON, CRAIG CURTIS, b Providence, RI, June 1, 50; m 72; c 2. BIOCHEMISTRY, PHARMACOLOGY. *Educ:* Worcester State Col, BA, 72; Case Western Reserve Univ, PhD(biochem), 76. *Prof Exp:* NIH fel, Div Biol & Med, Brown Univ, 76-78; from asst prof to prof pharmacol sci, 78-89, VCHMN, SCH MED HEALTH SCI CTR, STATE UNIV NY, STONY BROOK, 88-, ASSOC DEAN, BIOMED SCI, 89- *Concurrent Pos:* Res career develop award, NIH, 81 & mem, Celluar Biol & Physiol Study Sect, 81-85; dir, Diabetes & Metabolic Res Prog, 80-; vis prof biochem, Univ Mass Med Ctr, Worcester, Mass, 86; dir, Inst Nat Res Serv Awards Prog, NIH, 86- *Honors & Awards:* Samuel A Talbot Award, Biophys Soc, 75; Univ Res Award, Sigma Xi, 75; Res career develop award, NIH, 81. *Mem:* Biophys Soc; NY Acad Sci; Sigma Xi; Am Soc Biol Chemists; Am Physiol Soc. *Res:* Hormone action and regulation of metabolism; molecular and cellular pharmacology; molecular and cell biology of transmembrane signaling; membrane biochemistry. *Mailing Add:* Dept Pharmacol Sci-HSC State Univ NY Stony Brook NY 11794-8651

MALBON, WENDELL ENDICOTT, b Norfolk, Va, July 18, 18; m 42; c 1. MATHEMATICS. *Educ:* Univ Va, BCh, 41, MA, 52, PhD(math), 55. *Prof Exp:* Asst math, Univ Va, 50-54, from instr to assoc prof, 54-69; prof math, Old Dominon Univ, 69-; RETIRED. *Mem:* Am Math Soc; Math Asn Am. *Res:* Point set topology; quasi-compact mappings; application of topology to theory of functions of a complex variable. *Mailing Add:* 3300 Ocean Shore Ave Virginia Beach VA 23451

MALBROCK, JANE C, b NJ. MATHEMATICAL ANALYSIS. *Educ:* Montclair State Col, BA, 64; Pa State Univ, MA, 66, PhD(math anal), 71, Fairleigh Dickinson Univ, MS, 82. *Prof Exp:* From asst prof to assoc prof, 71-85, PROF MATH & COMPUT SCI, KEAN COL NJ, 85- *Mem:* Am Math Soc; Math Asn Am; Asn Comput Mach. *Res:* Approximation theory. *Mailing Add:* Dept of Math & Comput Sci Kean Col of NJ Morris Ave Union NJ 07083

MALCHICK, SHERWIN PAUL, b St Paul, Minn, Aug 13, 29; m 51, 80; c 3. MANAGEMENT, ORGANIC CHEMISTRY. *Educ:* Univ Minn, BA, 49; Univ Rochester, PhD(chem), 52. *Prof Exp:* Lab asst, Univ Rochester, 49-51, asst, 51-52; chemist, Stand Oil Co, Ind, 52-59, group leader explor res, 59-60; group leader, Nalco Chem Co, 60-66, sect head corp res, 66-67, tech mgr corp res, 67-69, tech dir com develop, 69-71, mgr textile chem, 71-73; dir res, Pigments Div, Chemetron Corp, 73-77, dir technol, 77-79; gen mgr, Pigments Div, BASF Wyandotte Corp, 80-81, vpres, Pigments & Performance Chem Group, 81-85; CONSULT, 86- *Mem:* Sigma Xi; Am Chem Soc. *Res:* Organic synthesis; polymerization; polymer application; additives; coordination complexes; paper chemicals; colloid chemistry; textile chemistry; surface chemistry; organic pigments and dispersions. *Mailing Add:* 1705 Hillsboro SE Grand Rapids MI 49546-9754

MALCOLM, ALEXANDER RUSSELL, b Providence, RI, June 28, 36; m 64. BIOCHEMICAL GENETICS, GENETIC TOXICOLOGY. *Educ:* Univ RI, BS, 64, MS, 70, PhD(biophys), 76. *Prof Exp:* Chemist, Elec Boat Div, Gen Dynamics Corp, 64-67 & USPHS, 67-69; chemist, 69-72, RES CHEMIST, ENVIRON RES LAB, US ENVIRON PROTECTION AGENCY, 72- *Concurrent Pos:* Panel mem subcomt environ mutagenesis, Comt Coord Toxicol & Related Progs, Dept Health, Educ & Welfare, 74-; adj asst prof, Dept Pharmacol & Toxicol, Col Pharm, Univ RI, Kingston, 79- *Mem:* Am Chem Soc; Genetic Toxicol Asn; Environ Mutagen Soc; Sigma Xi; Tissue Cult Asn; Soc Risk Anal. *Res:* Development and application of in vitro mammalian cell methods in cellular toxicology; development, validation and application of multiple end point, in vitro, cell culture techniques to identify and assess broad classes of toxic substances for regulatory purposes. *Mailing Add:* Hundred Acre Pond Rd PO Box 93 West Kingston RI 02892

MALCOLM, DAVID ROBERT, b Green Lake, Wis, Jan 11, 26; m 49. ENTOMOLOGY. *Educ:* Minn State Teachers Col, Winona, BS, 49; State Col Wash, MS, 51, PhD(zool), 54. *Prof Exp:* Asst zool, State Col Wash, 49-53, res assoc entom, 50-51; instr, Iowa State Col, 54; from instr to prof, Portland State Col, 54-69; prof biol & chmn div sci, 69-76, DEAN COL ARTS & SCI, PAC UNIV, 76- *Concurrent Pos:* Asst dean grad studies, Portland State Col, 67-69. *Mem:* Sigma Xi; Soc Syst Zool; Am Inst Biol Sci. *Res:* Biology and taxonomy of Chelonethida. *Mailing Add:* Dept Biol Pac Univ Forest Grove OR 97116

MALCOLM, EARL WALTER, b Menominee, Mich, July 17, 37; m 59; c 2. PULP & PAPER CHEMISTRY. *Educ:* Western Mich, BS, 59; Lawrence Inst Paper Chem, MS, 61, PhD(chem), 64. *Prof Exp:* Chemist, Dow Chem, 63-65; res mgr, Dexter Corp, 65-76; DIV DIR CHEM & BIOL SCI, INST PAPER CHEM, 76- *Mem:* Tech Asn Pulp & Paper Indust; Am Chem Soc. *Res:* Pulp and paper technology with emphasis on reactor kinetics. *Mailing Add:* Inst Paper Sci & Technol 575 14th St NW Atlanta GA 30318

MALCOLM, JANET MAY, b Bronx, NY, Mar 25, 25; m 46; c 3. OPERATIONS RESEARCH, RESOURCE MANAGEMENT. *Educ:* Rutgers Univ, New Brunswick, BS, 45; Northwestern Univ, MS, 46; Columbia Univ, PhD(phys chem), 51. *Prof Exp:* Lectr, Hunter Col, 48-49; res instr phys chem, Univ Miami, 51-53, Sch Med, 57-59; prof chem, Univ El Salvador, 62; asst prof opers res & info sci & dir opers res prog, Am Univ, 70-75; PVT CONSULT, 75- *Mem:* Fel AAAS; Asn Comput Mach; Opers Res Soc Am; Inst Mgt Sci; Am Chem Soc; Sigma Xi. *Res:* Rank sum statistics; physical chemistry of blood serum; refrigerant desiccants; kinetics; fluid dynamics analogs to linear, quadratic and separable convex programming; indigenous marketing system in Ghana and Nigeria; socioeconomic system modelling. *Mailing Add:* 1607 Kirby Rd McLean VA 22101

MALCOLM, JOHN LOWRIE, b Westfield, NJ, July 30, 20; m 46; c 3. SOILS. *Educ:* Rutgers Univ, BSc, 43, MSc, 45, PhD(soils), 48. *Prof Exp:* Assoc soil chemist, Subtrop Exp Sta, Univ Fla, 48-59; soils adv, US Opers Mission, Int Coop Admin, El Salvador, 59-63, USAID, India, 63-69; soils specialist, Tech Off, East Asia Bur, AID, Washington, DC, 70, proj mgr fertilizer specialist, Off Agr, Develop & Support Bur, 70-81, SOIL & FERTILIZER SPECIALIST, BUR SCI & TECHNOL, AID, 81- *Concurrent Pos:* Proj mgr, Food & Agr Orgn, UN, Ghana, 69. *Mem:* AAAS; Am Chem Soc; Soil Sci Soc Am. *Res:* Soil and analytical chemistry; agronomy; vegetable and sub-tropical fruit production. *Mailing Add:* 1607 Kirby Rd McLean VA 22101

MALCOLM, MICHAEL ALEXANDER, software systems, for more information see previous edition

MALCOM, SHIRLEY MAHALEY, b Birmingham, Ala, Sept 6, 46; m 75; c 2. SCIENCE ADMINISTRATION. *Educ:* Univ Wash, BS, 67; Univ Calif, Los Angeles, MA, 68; Pa State Univ, PhD(ecol), 74. *Hon Degrees:* LHD, Col St Catherine, 90, DSc, NJ Inst Tech. *Prof Exp:* Asst prof biol, Univ NC, Wilmington, 74-75; res asst, staff assoc & proj dir, Off Opportunities Sci, AAAS, 75-77; prog mgr, Sci Educ Directorate, NSF, 77-79; prog head off opportunities sci, 79-89, HEAD, DIRECTORATE EDUC & HUMAN RESOURCES PROGS, AAAS, 89- *Concurrent Pos:* Mem, equal opportunities sci & technol comt, NSF, 83-86, chair, 84-86; technol & women's employ panel, Comn Behav & Social Sci & Educ, Nat Res Coun, 84-86 & panel to evaluate Nat Ctr Educ Statist, Comt Nat Statist, 84-; adv coun, Carnegie Forum Educ & Econ, 84-88, task force teaching profession, 85-87; bd mem, Nat Ctr Educ & Ecom, 89-; mem, Smithsonian Adv Coun, 90- *Mem:* AAAS; Sigma Xi. *Res:* Advocacy, research and program development to increase the participation of women, minorities and disabled persons in science and engineering. *Mailing Add:* Educ & Human Resources Am Asn Advan Sci 1333 H St NW Washington DC 20005

MALCUIT, ROBERT JOSEPH, b Fredericksburg, Ohio, Feb 11, 36. PETROLOGY, ASTROGEOLOGY. *Educ:* Kent State Univ, BS, 68, MS, 70; Mich State Univ, PhD(geol), 73. *Prof Exp:* From asst prof to assoc prof, 72-84, PROF GEOL, DENISON UNIV, 84- *Mem:* Geol Soc Am; Am Geophys Union; AAAS; Sigma Xi. *Res:* Igneous and metamorphic petrology; zircons as petrogenetic indicators; geologic evidence relating to origin and evolution of Earth-Moon system; origin of massif-type anorthosite; origin of continents. *Mailing Add:* Dept Geol & Geog Denison Univ Granville OH 43023

MALDACKER, THOMAS ANTON, b New York, NY, Apr 3, 46. ANALYTICAL CHEMISTRY. *Educ:* Fordham Univ, BS, 67; Purdue Univ, MS, 69, PhD(anal chem), 73. *Prof Exp:* SR SCIENTIST RES & DEVELOP, SANDOZ INC, 73- *Mem:* Am Chem Soc. *Res:* Development of liquid and gas chromatographic techniques for the analysis of drug substance and dosage, degradation and by-products. *Mailing Add:* 30 Country Wood Dr Morris Plains NJ 07950

MALDE, HAROLD EDWIN, b Reedsport, Ore, July 9, 23; m 54; c 2. GEOLOGY, ARCHAEOLOGY. *Educ:* Willamette Univ, AB, 47. *Prof Exp:* Geologist, US Geol Surv, 51-87; RETIRED. *Concurrent Pos:* Affil prof, Univ Idaho, 68-77 & 87-88; mem Colo consult comt, Nat Register Hist Places, 72-80; mem Nat Acad Sci comts, Potential Rehab Lands Surface Mined for Coal Western US, 73, Surface Mining & Reclamation, 78-79 & Alaskan Surface Coal Mining, 80, mem, US Dept Interior Oil Shale Environ Panel, 76-80; mem, Nat Acad Sci Deleg Paleoanthrop, People's Repub China, 75; assoc ed, Bulletin, Geol Soc Am, 82-88. *Honors & Awards:* Kirk Bryan Award, Geol Soc Am, 70; Meritorious Serv Award, Us Dept Interior, 79. *Mem:* Fel Geol Soc Am; fel AAAS; Am Quaternary Asn; Sigma Xi. *Res:* Cenozoic and Quaternary geology; geomorphology; environmental geology; stratigraphy and paleo-geomorphology applied to early man; technical photography. *Mailing Add:* 842 Grant Place Boulder CO 80302

MALDONADO, C(LIFFORD) DANIEL, b Honolulu, Hawaii, Apr 21, 28; m 51; c 3. ELECTRICAL ENGINEERING, PHYSICS. *Educ:* Univ Wis, BS, 53, MS, 55, PhD(elec eng), 58. *Prof Exp:* Asst elec eng, Univ Wis, 54-58, fel, 58-59; res engr, Gen Tel & Electronics Corp, 59-62; sr scientist, Northrop Space Labs, 62-63, mem res staff, 63-67, sr mem res staff, Northrop Corp Labs, 67-69; SR MEM TECH STAFF, AUTONETICS DIV, ROCKWELL INT CORP, 69- *Mem:* AAAS; Inst Elec & Electronics Engrs; Am Phys Soc; Sigma Xi. *Res:* Mathematical and theoretical physics; electromagnetic theory; transport phenomena in semiconductors; collective behavior studies in gaseous and solid state plasmas. *Mailing Add:* 27111 Mission Hills Dr San Juan Capistrano CA 92675

MALDONADO, JUAN RAMON, b Holguin, Cuba, May 6, 38; US citizen; m 62; c 3. APPLIED PHYSICS. *Educ:* Univ Havana, PhD(phys math sci), 61; Univ Md, College Park, PhD(exp solid state physics), 68. *Prof Exp:* Elec engr, CMQ TV, Havana, 57-61; instr physics-math, Univ Havana, 60-61; supvr eletronics,62-65, res asst solid state physics, Univ Md, College Park, 65-68; mem tech staff physics, Bell Tel Labs, 68-80; mgr, X-Ray Lithography Processes, T J Watson Res Ctr, 80-88, SR SCIENTIST, ADVAN TECHNOL CTR, HOPEWELL JUNCTION, 88- *Mem:* Am Inst Physics; Inst Elec & Electronic Engrs; AAAS; Sigma Xi. *Res:* Research and development in x-ray lithography for the fabrication of sub half micron size integrated circuits. *Mailing Add:* IBM Advan Tech Ctr GTD Rte 52 Hopewell Junction NY 12533

MALE, JAMES WILLIAM, b Schenectady, NY, Dec 8, 45. SYSTEMS ANALYSIS. *Educ:* Union Col, BS, 68; Johns Hopkins Univ, PhD(environ eng), 73. *Prof Exp:* Asst prof enivron eng, Ill Inst Technol, 73-77; ASSOC PROF CIVIL ENG, UNIV MASS, 77-, PROG COORDR, ENVIRON ENG, 81- *Mem:* Asn Environ Eng Professionals; Opers Res Soc Am; Am Water Resources Asn; Am Soc Civil Eng; Am Geophys Union; Sigma Xi. *Res:* Water resource systems analysis; solid waste management; transportation systems; environmental engineering. *Mailing Add:* Dept Civil Eng Univ Mass Amherst MA 01003

MALEADY, N(OEL) R(ICHARD), b Pittsfield, Mass, Dec 25, 16; m 46; c 5. CHEMICAL ENGINEERING. *Educ:* Worcester Polytech Inst, BS, 40. *Prof Exp:* Group leader, Gen Elec Co, Mass, 41-52, mgr eng lab, Ga, 53-61, mgr mat design lab, oper, Ind, 62-68, mgr advan mfg eng, Neutron Devices Dept, 68-71; CONSULT INDUST PROCESSES, MAT RES & DEVELOP & MGT ORGN, 71- *Concurrent Pos:* Mem, Eckerd Col; chmn, Atlanta Chap, Am Inst Chem Engrs, 58-60. *Honors & Awards:* Coffin Award, Gen Elec Co, 50. *Mem:* Am Chem Soc; Am Inst Chem Engrs. *Res:* Vacuum processing; drying; laboratory administration. *Mailing Add:* 105 Island Way No 112 Clearwater FL 34630-2218

MALECH, HARRY LEWIS, b Carlstadt, NJ, Nov 10, 46; m 72; c 3. INFECTIOUS DISEASES, IMMUNOLOGY. *Educ:* Brandeis Univ, BA, 68, Yale Univ, MD, 72. *Prof Exp:* Resident, Hosp Univ Pa, 72-74; res assoc, Nat Cancer Inst, 74-76; fel infectious dis, Yale Univ, 76-78, from asst prof to assoc prof med, 78-86; SECT HEAD BACT DIS, NAT INST ALLERGY & INFECTIOUS DIS, 86- *Concurrent Pos:* Assoc ed, J Immunol, 84- *Mem:* Am Soc Clin Invest; Am Soc Cell Biol; Am Asn Immunologists; Infectious Dis Am; AAAS. *Res:* Structure and function of human phagocytic cells; diagnosis and treatment of disorders of host defense leading to recurrent infections. *Mailing Add:* Nat Inst Allergy & Infectious Dis Lab Clin Invest Nat Inst Allergy & Infectious Dis Bldg 10 Rm 11N112 Bethesda MD 20892

MALECHA, SPENCER R, b Chicago, Ill, Nov 13, 43; m 71. GENETICS. *Educ:* Loyola Univ, Ill, BS, 65; Univ Hawaii, MS, 68, PhD(genetics), 71. *Prof Exp:* Asst zool, 66-68, asst genetics, 68-69, asst prof genetics, 71-80, ASSOC PROF ANIMAL SCI, SCH MED, UNIV HAWAII, MANOA, 80- *Concurrent Pos:* Ford Found fel, Univ Chicago, 72-73. *Mem:* AAAS; Soc Study Evolution; Am Soc Nat; Nat Asn Biol Teachers. *Res:* Ecological genetics; genetic variation in natural populations. *Mailing Add:* Hawaii Aquacult Co PO Box 61970 Honolulu HI 96830-1970

MALECHEK, JOHN CHARLES, b San Angelo, Tex, Aug 6, 42. RANGE SCIENCE, ECOLOGY. *Educ:* Tex Tech Univ, BS, 64; Colo State Univ, MS, 66; Tex A&M Univ, PhD(range sci), 70. *Prof Exp:* From asst prof to assoc prof, 70-82, PROF RANGE SCI, UTAH STATE UNIV, 82- *Mem:* Am Soc Animal Sci; Soc Range Mgt. *Res:* Nutrition and behavior of free-grazing livestock; agro-ecology of tropics; projects in Utah and in dry tropics of Brazil. *Mailing Add:* Dept Range Sci Utah State Univ UMC 5230 Logan UT 84322

MALECKAR, JAMES R, b Salt Lake City, Utah, July 28, 54. IMMUNOLOGY. *Educ:* Ariz Univ, BA, 76; Tulane Univ, MS, 79; Mich Univ, PhD(microbiol), 83. *Prof Exp:* Postdoctoral fel, Scripps Clin & Res Found, 83-87; staff scientist, Biotherapeutics, 87-89; SR SCIENTIST, DON & SYBIL HARRINGTON CANCER CTR, 89-; ASST PROF MICROBIOL, TEX TECH UNIV, 90- *Mem:* Am Asn Immunologists. *Mailing Add:* Don & Sybil Harrington Cancer Ctr 1500 Wallace Blvd Amarillo TX 79106

MALECKI, GEORGE JERZY, life sciences, metallurgy, for more information see previous edition

MALEENY, ROBERT TIMOTHY, b Staten Island, NY, Jan 1, 31; m 54; c 2. CHEMISTRY. *Educ:* Wagner Col, BS, 52; St John's Univ, MS, 54. *Prof Exp:* NSF res asst chem, St John's Univ, 52-53; asst tech dir, Dodge & Olcott, 53-66; tech dir, Globe Extracts, 66-68; vpres tech dir, Major Prod, 68-70 & Aromatics, Int, 70-74; dir res, Monsanto Flavor/Essence, Monsanto Co, 74-79; gen mgr, Synfleur, Div Nestle Enterprises, 79-83; PRES, FLAVOR & FRAGRANCE SPECIALTIES, INC, 83- *Mem:* Am Chem Soc; AAAS; Soc Flavor Chemists; Chem Sources Asn. *Res:* Development of flavors, fragrances and unique aromatic chemicals, including malodor counteractants. *Mailing Add:* 16 Mohawk Dr Ramsey NJ 07446

MALEK, EMILE ABDEL, b El Mansura, Egypt, Aug 22, 22; US citizen; m 54; c 3. MEDICAL PARASITOLOGY. *Educ:* Cairo Univ, BSc, 43, MSc, 47; Univ Mich, Ann Arbor, PhD(parasitol), 52. *Prof Exp:* Teaching asst zool, Cairo Univ, 43-47; scientist schistosomiasis control, Ministry of Health, Egypt, 52-53; from lectr to reader parasitol, Univ Khartoum, 53-59; from asst prof to assoc prof, 59-73, PROF PARASITOL, MED SCH & SCH PUB HEALTH, TULANE UNIV, 74- *Concurrent Pos:* NIH res career award, 62-; consult, WHO & Pan Am Health Orgn, 56, 61-64, 66 & 74, Peace Corps, 67 & USPHS, 74-; mem expert adv panel, WHO, 64-, scientist parasitic dis, 67-69. *Mem:* Am Soc Parasitol; Am Malacol Soc; Am Soc Trop Med & Hyg; Royal Soc Trop Med & Hyg. *Res:* Epidemiology and control of schistosomiasis; medical malacology; snail-transmitted helminthiases. *Mailing Add:* Dept Pub Health Tulane Univ Law New Orleans LA 70118

MALEK, MIROSLAW, b Wroclaw, Poland, Apr 9, 47; US citizen; m 83. RELIABLE & PARALELL COMPUTING. *Educ:* Technical Univ Wroclaw, MS, 70, PhD(CSc), 75. *Prof Exp:* From res sci to asst prof, Technical Univ Wroclaw, 70-77; from asst prof to prof, 77-89, BETTIE MARGARET SMITH PROF, UNIV TEX, 89- *Mem:* Inst Elec & Electronic Engrs; Asn Comput Mach. *Res:* Parellel computing; real-time systems; fault-tolerant computing. *Mailing Add:* Dept Elect & Comp Engr Univ Texas Austin TX 78712

MALEK, RICHARD BARRY, b Westfield, NJ, July 6, 36; m 63; c 3. PLANT NEMATOLOGY. *Educ:* Univ Maine, BS, 58; Rutgers Univ, MS, 60, PhD(plant nematol), 64. *Prof Exp:* Res asst entom, Rutgers Univ, 58-64; asst prof plant path, SDak State Univ, 64-68; asst prof, 68-75, ASSOC PROF NEMATOL, UNIV ILL, URBANA, 75- *Mem:* Soc Nematol; Helminthol Soc Wash; Europ Soc Nematologists; Orgn Trop Am Nematologists; Am Phytopathological Soc. *Res:* Nematode diseases of plants and their control. *Mailing Add:* Dept Plant Path Univ Ill N-519 Turner Hall 1102 S Goodwin Urbana IL 61801

MALEMUD, CHARLES J, b Brooklyn, NY, Nov 5, 45; m 86; c 3. CELL BIOLOGY, CELL PHYSIOLOGY. *Educ:* Long Island Univ, BSc, 66; George Washington Univ, PhD(exp path), 73. *Prof Exp:* Res asst, Coney Island Hosp, Brooklyn, NY, 66-68; biologist, Nat Inst Arthritis & Metab Dis, 68-73; lectr path, State Univ NY, Stony Brook, 73-74, instr, 74-77; from asst prof to assoc prof, 78-89, PROF MED, CASE WESTERN RESERVE UNIV, 89- *Concurrent Pos:* Consult, CIBA-Geigy Corp, 74-76; investr, Woods Hole Marine Biol Lab, 81; chmn, Cartilage Study Group, Am Rheumatism Asn, 81-83, Gen Med Study Sect, NIH, 87-90. *Mem:* Am Soc Cell Biol; Am Rheumatism Asn; Orthop Res Soc; NY Acad Sci; Soc Exp Biol & Med. *Res:* Relationship between aging and degenerative joint disease (osteoarthritis); comparing extracellular matrix biosynthesis and catabolism of rabbit and human cartilage from samples of different ages and as tissue or cells undergo senescence in culture. *Mailing Add:* Dept Med Wearn Bldg Rm 549 Case Western Reserve Univ Cleveland OH 44106

MALENCIK, DEAN A, MUSCLE PROTEINS, PROTEIN PHOSPHOLATION. *Educ:* Ore State Univ, PhD(biochem), 72. *Prof Exp:* RES ASSOC, DEPT BIOCHEM, ORE STATE UNIV, 80- *Mailing Add:* Dept Biochem Ore State Univ Corvallis OR 97331

MALENFANT, ARTHUR LEWIS, b Wakefield, RI, May 17, 37; m 57; c 3. ANALYTICAL CHEMISTRY. *Educ:* Univ RI, BS, 60; Mass Inst Technol, PhD(anal chem), 67. *Prof Exp:* Chemist, Corning Glass Works, NY, 60-63; res asst anal chem, Mass Inst Technol, 63-67; dir tech serv, 67-74, vpres res & develop, 74-79, VPRES INT DIV, INSTRUMENTATION LAB, INC, 79- *Mem:* Am Chem Soc; Soc Appl Spectros; Asn Advan Med Instrumentation; Am Asn Clin Chemists; Sigma Xi. *Res:* Analytical techniques and methods to promote new approaches in biomedical instrumentation. *Mailing Add:* 75 Henry St Cambridge MA 02139

MALENKA, BERTRAM JULIAN, b New York, NY, June 8, 23; m 48; c 2. THEORETICAL HIGH ENERGY PHYSICS. *Educ:* Columbia Univ, AB, 47; Harvard Univ, AM, 49, PhD(physics), 51. *Prof Exp:* Res fel, Harvard Univ, 51-54; asst prof physics, Wash Univ, 54-56; assoc prof, Tufts Univ, 56-60; assoc prof, 60-62, PROF PHYSICS, NORTHEASTERN UNIV, 62- *Concurrent Pos:* Adv, Harvard Univ & Mass Inst Technol, 54-; consult, Arthur D Little Inc, 59- & Am Sci & Eng, 59- *Mem:* Am Phys Soc; NY Acad Sci; Ital Phys Soc. *Res:* Theoretical nuclear physics; scattering theory at high energies; elementary particles; accelerator theory and design. *Mailing Add:* Dept Physics Northeastern Univ Boston MA 02115

MALER, GEORGE J(OSEPH), b Denver, Colo, Aug 12, 24; m 45; c 2. ELECTRICAL ENGINEERING. *Educ:* Univ Colo, BS, 45, MS, 57. *Prof Exp:* Instr eng, Univ Colo, 46-51; engr, Mountain States Tel & Tel Co, 54-56; actg asst prof eng, Univ Colo, Boulder, 56-57, from asst prof to assoc prof, 57-66, prof & assoc dean, 66-88; RETIRED. *Mem:* Inst Elec & Electronics Engrs; Am Soc Eng Educ; Nat Soc Prof Engrs. *Res:* Circuit theory. *Mailing Add:* Old Post Circle Boulder CO 80301

MALERICH, CHARLES, b Rochester, Minn, Sept 25, 44; m 73; c 1. PHYSICAL CHEMISTRY. *Educ:* St John's Univ, BS, 66; Yale Univ, PhD(phys chem), 71. *Prof Exp:* Res assoc, Univ Calif, Irvine, 70-71; instr chem, Univ Utah, 71-73; PROF CHEM, BARUCH COL, 73- *Concurrent Pos:* NSF indust res fel, Grumman Aerospace Corp, 81. *Mem:* Am Chem Soc; Am Asn Univ Professors. *Res:* Dependence of the physical chemistry of metal parphyrins on molecular structures; develop model systems for heme proteins. *Mailing Add:* Dept Natural Sci Chem Baruch Col 17 Lexington Ave New York NY 10010

MALES, JAMES ROBERT, b Noblesville, Ind, Sept 28, 45; m 72; c 2. BEEF CATTLE. *Educ:* Pa State Univ, BS, 67; Mich State Univ, MS, 69; Ohio State Univ, PhD(ruminant nutrit), 73. *Prof Exp:* Res assoc ruminant nutrit, Okla State Univ, 73-74; asst prof animal sci, Southern Ill Univ, Carbondale, 74-78; asst prof & asst animal scientist, Wash State Univ, Pullman, 78-82, assoc prof & assoc animal scientist, 82-87,prof & animal scientist, 87-88; PROF & HEAD, DEPT ANIMAL & RANGE SCI, SDAK STATE UNIV, BROOKINGS, 88- *Mem:* Am Soc Animal Sci; Am Inst Nutrit; AAAS; Coun Agr Sci & Technol; Soc Exp Biol & Med. *Res:* Maximizing meat production from low quality roughages with minimum protein supplementation; beef cattle performance as it is related to maximum efficiency of beef production. *Mailing Add:* Dept Animal & Range Sci SDak State Univ Box 2170 Brookings SD 57007-0392

MALETSKY, EVAN M, b Pompton Lakes, NJ, June 9, 32; m 54; c 4. MATHEMATICS. *Educ:* Montclair State Col, BA, 53, MA, 54; NY Univ, PhD(math educ), 61. *Prof Exp:* PROF MATH, MONTCLAIR STATE COL, 57- *Mem:* Math Asn Am; Nat Coun Teachers Math. *Res:* Training of mathematics teachers and mathematics curriculum changes in the junior and senior high school. *Mailing Add:* 34 Pequannock Ave Pompton Lakes NJ 07442

MALETTE, WILLIAM GRAHAM, b Springfield, Mo, Mar 27, 22; m 45; c 2. SURGERY. *Educ:* Drury Col, 40-42; Wash Univ, MD, 53. *Prof Exp:* Physician, USAF, 53-58, chief exp surg, USAF Sch Aviation Med, 58-61, chief unit II surg & chief vascular surg serv, Air Force Hosp, Tex, 61-63; from asst prof to assoc prof surg, Med Ctr, Univ Ky, 63-75, assoc dean Vet Admin affairs, 71-75; dir emergency med serv, Kern Med Ctr, Bakersfield, Calif, 75-77; PROF SURG, COL MED, UNIV NEBR & SCH MED, CREIGHTON UNIV, 77- *Concurrent Pos:* Chief surg serv, Vet Admin Hosp, Lexington, Ky, 63-75, chief staff, Univ Div, 71-75; participating nat surg consult & mem surg res comt, Vet Admin, Washington, DC; consult, USPHS Hosp, Lexington; chief surg serv, Vet Admin Med Ctr, Omaha, Nebr, 77-86; pres, Triton-Chito Inc, 86- *Mem:* Fel Am Col Chest Physicians; fel Am Col Cardiol; Aerospace Med Asn; Am Soc Artificial Internal Organs; fel Am Col Surgeons. *Res:* Cardiovascular surgery; tissue transplantation. *Mailing Add:* 16461 Delate Rd NE Poulsbo WA 98370

MALEWICZ, BARBARA MARIA, b Lipno, Poland, Apr 1, 42; m 67; c 1. MOLECULAR BIOLOGY. *Educ:* Tech Univ, Gdansk, MS, 65, PhD(biochem), 74. *Prof Exp:* Res asst chem, Polish Acad Sci, 66-68, res assoc biochem, 68-77; fel, cell biol, 77-79, RES ASSOC BIOL, HORMEL INST, UNIV MINN, 79- *Concurrent Pos:* Instr, Univ Gdansk, 75-77. *Mem:* Am Soc Microbiol; AAAS; Am Oil Chemists Soc; Tissue Cult Asn. *Res:* Phospholipid metabolism in mammalian cells; membrane regulation by lysophospholipids; cell membrane transport; LDL endocytosis; human endothelial cells; nuclear magnetic resonance spectroscopy. *Mailing Add:* Hormel Inst 801 16th Ave NE Austin MN 55912

MALEWITZ, THOMAS DONALD, b Holland, Mich, Apr 13, 29; m 57; c 3. ANATOMY. *Educ:* Hope Col, AB, 51; Univ Kans, MA, 53; Mich State Univ, PhD(anat, histol), 56. *Prof Exp:* Asst instr biol, Univ Kans, 52-53; instr anat, Mich State Univ, 56-57; asst prof anat & physiol, Col Pharm, Univ Fla, 57-61 & anat, Woman's Med Col Pa, 61-66; asst prof, 66-67, ASSOC PROF BIOL, VILLANOVA UNIV, 67-, HEALTH RELATED SCI ADV, 77- *Mem:* Am Asn Anat; Am Soc Cell Biol. *Res:* Human and animal histology; animal pathology; parasitology. *Mailing Add:* Dept Biol Villanova Univ Villanova PA 19085

MALEY, FRANK, b Brooklyn, NY, July 26, 29; m 53; c 3. BIOCHEMISTRY. *Educ:* Brooklyn Col, BS, 52; Univ Wis, MS, 53, PhD, 56. *Prof Exp:* USPHS res fel, Sch Med, NY Univ, 56-58; from asst prof to assoc prof, Albany Med Col, 58-61, prof, 70-81; prof, Sch Pub Health, State Univ NY; sr res scientist, 58-61, assoc res scientist, 61-69, DIR LAB OF BIOCHEM, WADSWORTH CTR OF LAB & RES, NY STATE DEPT HEALTH, 69- *Concurrent Pos:* Mem, Exp Therapeut Study Sect, USPHS, 80-83. *Mem:* Am Soc Biol Chem; Am Chem Soc; Soc Complex Carbohydrates. *Res:* Nucleotide interconversions and nucleic acid metabolism; one carbon and hexosamine metabolism; chemical synthesis of hexosamine derivatives; glycoprotein structure function and biosynthesis; regulation of enzyme activity and synthesis; isolation of phage induced enzymes; site specific mutagenesis structure; function of thymidylate synthase and deoxycytidylate deaminase, action of introns and exons. *Mailing Add:* Div of Lab & Res NY State Dept of Health Albany NY 12201-0509

MALEY, GLADYS FELDOTT, b Aurora, Ill, June 7, 26; m 53; c 3. BIOCHEMISTRY. *Educ:* NCent Col, Ill, BA, 48; Univ Wis, MS, 50 & PhD, 53. *Prof Exp:* Am Heart Asn res fel, Sch Med, NY Univ, 56-58; adv res fel, 58-60, Am Heart Asn estab investr, 60-65, USPHS res career develop award, 65-70, assoc res scientist, 70-83; PRIN RES SCIENTIST, NY STATE DEPT HEALTH, 83- *Concurrent Pos:* Asst prof, Albany Med Col, Union Univ, NY, 59-65; adj prof, Rensselaer Polytech Inst, 81-; prof, Sch Pub Health Sci, SUNY, Albany, NY, 85- *Mem:* AAAS; Am Soc Biol Chem; Am Soc Microbiol. *Res:* Nucleotide interconversions; nucleic acid metabolism; structure and function of regulatory proteins. *Mailing Add:* Wadsworth Ctr Labs & Res NY State Dept Health Albany NY 12201-0509

MALEY, MARTIN PAUL, b Seattle, Wash, Aug 27, 36; m 66. PHYSICS. *Educ:* Yale Univ, BS, 56; Rice Univ, MA, 63, PhD(physics), 65. *Prof Exp:* Fel, Los Alamos Sci Lab, 65-68; asst prof physics, Rensselaer Polytech Inst, 68-74; STAFF MEM, LOS ALAMOS NAT LAB, 74- *Res:* Low temperature physics; magnetism; ferromagnetic resonance; microwave ultrasonics; superconductivity. *Mailing Add:* Los Alamos Nat Lab MS K764 Los Alamos NM 87545

MALEY, S(AMUEL) W(AYNE), b Sidney, Nebr, Mar 1, 28; m 63; c 2. ELECTRICAL ENGINEERING, APPLIED MATHEMATICS. *Educ:* Univ Colo, BS, 52, MS, 57, PhD(elec eng), 59. *Prof Exp:* Elec engr, Beech Aircraft Corp, 52-53 & Land-Air Div, Dynalectron Corp, 53-56; res assoc elec eng, 58-60, vis lectr, 60-61, from asst prof to assoc prof, 61-67, PROF ELEC ENG, UNIV COLO, BOULDER, 67- *Concurrent Pos:* Consult, Automation Industs, Inc, 58-68, Nat Ctr Atmospheric Res, 63-66 & IBM Corp. *Mem:* AAAS; Inst Elec & Electronics Engrs; Soc Indust & Appl Math; Int Union Radio Sci. *Res:* Electromagnetic propagation in the trophosphere, ionosphere and space; electromagnetic waves in plasmas; communication theory, including information theory, stochastic processes and coding theory; theory of automata; computer logic. *Mailing Add:* Dept Elec Eng Univ Colo Boulder CO 80309

MALEY, WAYNE A, b Stanley, Iowa, Mar 9, 27; m 59; c 3. ASSOCIATION MANAGEMENT, TECHNICAL INTERPRETATIVE WRITING. *Educ:* Iowa State Univ, 49. *Prof Exp:* Power appln engr, Southwestern Elec, 49-53; field engr, Am Zinc Inst, 53-60; mkt develop engr, US Steel Corp, 60-71, prod develop engr, 71-76; air qual & energy consult, Taylor Assocs, 77-81; dir mem & educ, 81-91, DIR SPEC PROJS, AM SOC AGR ENGRS, 91- *Concurrent Pos:* Pres, Educ Concerns Health Orgn, 79-83; mem, bd dirs, Am Soc Agr Engrs, 79-81. *Mem:* Fel Am Soc Agr Engrs; Sigma Xi; Am Agr Ed Asn. *Res:* Farm equipment and material handling devices; computer technology to develop tractor rollover protection; structures and vibrations-noise reduction. *Mailing Add:* 2592 Stratford Dr St Joseph MI 49085

MALGHAN, SUBHASCHANDRA GANGAPPA, b Gadag, India, Mar 5, 47; US citizen; m 70; c 2. CERAMIC POWDERS, APPLIED SURFACE CHEMISTRY OF FINE POWDERS. *Educ:* Karnatak Univ, Dharwar, BSc, 67; Indian Inst Sci, Bangalore, BE, 70; Univ Nev, Reno, MS, 71; Univ Calif, Berkeley, PhD(mat sci), 74. *Prof Exp:* Mineral engr, NC State Univ, Raleigh, 74-77; prog leader mineral eng, Kennecott Copper Corp, Utah, 77-79; sr supervisor, Exxon Res, NJ, 79-86; sr scientist, Norton Co, Mass, 86-88; GROUP LEADER POWDER SCI, NAT INST STANDARDS & TECHNOL, 88- *Concurrent Pos:* Lectr, var univs, 79- & Japan Fine Ceramic Soc, 91; prin reviewer, Minerals & Metall J, 85-; chmn, US Tech Group, Int Energy Agency, 88- & Subtask C-21 Comt Particle Size Gravity Sedimentation, 90-; adj prof, Clemson Univ, 91- *Mem:* Am Ceramic Soc; Am Soc Metals; Am Soc Testing & Mat; Soc Mining & Metall Engrs. *Res:* Underlying physical and chemical interactions in the synthesis and processing of advanced ceramic powders; internationally accepted standards. *Mailing Add:* Nat Inst Standards & Technol A-258-223 Gaithersburg MD 20899

MALHERBE, ROGER F, b Lausanne, Switz, Jan 5, 46; m 77. ORGANIC CHEMISTRY. *Educ:* Univ Lausanne, Switz, Dipl, 68, PhD(org chem), 72. *Prof Exp:* Res asst org chem, Yale Univ, 73-75; sr scientist, 75-81, STAFF SCIENTIST ORG CHEM, CHEM INDUST CIBA-GEIGY CORP, 81- *Mem:* Swiss Chemists Soc; Am Chem Soc. *Res:* Synthesis of additives for polymers, in particular antioxidants and light stabilizers for plastics and coatings; new synthetic methods; polyimides, photopolymers. *Mailing Add:* Bleichestr Seven Basel 4058 Switzerland

MALHIOT, ROBERT, b Chicago, Ill, Nov 21, 26; m 53; c 3. PHYSICS. *Educ:* Denver Univ, BS, 49; Ill Inst Technol, MS, 54, PhD(physics), 57. *Prof Exp:* From asst prof to assoc prof, 57-70, PROF PHYSICS, ILL INST TECHNOL, 70- *Concurrent Pos:* Actg chmn dept physics, Ill Inst Technol, 62-68, chmn dept, 68-70. *Mem:* AAAS; Am Phys Soc; Fedn Am Scientists; Am Asn Physics Teachers; Sigma Xi. *Res:* Gravitation and general relativity. *Mailing Add:* 3048 Crescent Beach Rd Manistee MI 49660

MALHOTRA, ASHWANI, b Lahore, WPakistan, July 6, 43; m 71; c 1. ENZYMOLOGY, CARDIOVASCULAR PHYSIOLOGY. *Educ:* Univ Delhi, BS, 62, MS, 64, PhD(org chem), 69. *Prof Exp:* Res asst chem, Univ Conn, 69-70; Fogarty Int vis fel cancer res, Nat Cancer Inst, 70-72; fel, New Eng Inst, 72-73; RES ASSOC CARDIOL, MONTEFIORE HOSP & MED CTR, ALBERT EINSTEIN COL MED, 73- *Mem:* Am Chem Soc. *Res:* Cardiovascular physiology and biochemistry with special emphasis on excitation-contraction coupling of contractile proteins and protein synthesis. *Mailing Add:* 27 Overlook Ardsley NY 10502

MALHOTRA, OM P, WARFARIN INDUCED PROTHROMBIN. *Educ:* Punjab Univ, India, DVM, 49; Univ Ill, PhD(vet physiol), 62. *Prof Exp:* RES CHEMIST, VET ADMIN MED CTR, CLEVELAND, 67-; ASSOC PROF EXP PATH, SCH MED, CASE WESTERN RESERVE UNIV, 74- *Mailing Add:* Vet Admin Hosp Cleveland OH 44106

MALHOTRA, SUDARSHAN KUMAR, b Bhera, India, June 20, 33; m 63; c 2. NEURAL CELL BIOLOGY. *Educ:* Oxford Univ, DrPhil(biol sci), 60, MA, 61, DSc, 85. *Prof Exp:* Sr studentship, Royal Comn for Exhib of 1851, 60-62; res fel biol, Calif Inst Technol, 63-65, sr res fel, 66-67; dean life sci, Jawaharlal Nehru Univ, New Delhi, 71-72; PROF CELL BIOL & DIR BIOL SCI ELECTRON MICROS & PROF, DEPT ZOOL, UNIV ALTA, 67-, HON PROF PATH, 83- *Concurrent Pos:* Res fel, New Col, Oxford Univ, 61-63; Del E Webb vis assoc, Calif Inst Technol, 80-81; McCalla prof, Univ Alta, 87-88. *Mem:* AAAS; Brit Cell Biol Soc; Royal Micros Soc; Can Soc Cell Biol; Int Soc Develop Neurosci; Soc Neurosci. *Res:* Cytology of neural cells; nerve-muscle interaction; structure, function and biogenesis of cellular membranes and membranous organelle; cell biology of glial cells; editor two books. *Mailing Add:* 12916-63rd Ave Edmonton AB T6H 1S1 Can

MALICK, JEFFREY BEVAN, b Brooklyn, NY, Nov 14, 42; m 62; c 5. NEUROPHARMACOLOGY, PSYCHIATRY. *Educ:* Rutgers Univ, BA, 65, MS, 68; NY Univ, PhD(psychobiol & pharmacol), 73. *Prof Exp:* Res asst neuropharmacol & biochem, Lederle Labs, Div Am Cyanamid, 62-67; supvr neurophysiol, Union Carbide Corp, 67-69; sr res scientist neuropharmacol, Schering Corp, 69-74; sr res pharmacologist, ICI Americas, Inc, 74-76, mgr cent nervous syst pharmacol, 76-87, sr proj mgr, 87-89, DIR, CENT NERVOUS SYST & ANESTHESIA CLIN RES, ICI PHARMACEUT, 90- *Concurrent Pos:* Adj assoc prof neurophysiol, Fairleigh Dickinson Univ, 74-75; adj assoc prof neurochem, Univ Del, 80-86. *Mem:* Am Soc Pharmacol & Exp Therapeut; Soc Neurosci; Col Int Neuropsychopharmacol; Soc Exp Biol & Med; Am Psychol Asn. *Res:* Psychopharmacology; neurochemical and neuroanatomical substrates of behavior and drug action; aggressive behavior; depression; alcoholism and drug abuse; neuropharmacology; neurophysiology; anxiety and antipsychotic research. *Mailing Add:* Clin Res Dept ICI Pharmaceut Group Div ICI Americas Inc Wilmington DE 19897

MALIK, ASRAR B, b Lahore, Pakistan, Dec 1, 45; US & Can citizen; m 75; c 2. CARDIOVASCULAR & PULMONARY PHYSIOLOGY. *Educ:* Univ Western Ont, BSc(Hons), 68; Univ Toronto, MSc, 69, PhD(physiol), 71. *Prof Exp:* Demonstr physiol, Univ Toronto, 68-71, demonstr histol, 69-70; instr physiol, Wash Univ, 71-73; from asst prof to assoc prof, 73-80, PROF PHYSIOL & CELL BIOL, ALBANY MED CO, 80-; PROF BIOMED ENG, RENSSELAER POYLTECH INST, 83-; PROF BIOMED SCI, STATE UNIV NY-ALBANY, 87- *Concurrent Pos:* Assoc staff surg, Jewish Hosp St Louis, 72-73; res career develop award, 77-82; consult, NIH, Am Heart Asn & Am Lung Asn; exec dir, NY State Lung Res Inst, Albany. *Honors & Awards:* Merit Award, Nat Heart, Lung, Blood Inst, NIH. *Mem:* Am Physiol Soc; Biophys Soc; Am Heart Asn; Am Thoracic Soc. *Res:* Lung fluid exchange; regulation of pulmonary circulation; endothelial cell biology; tissue inflammation; mechanisms of vascular injury with special emphasis on cell-cell interactions and role of mediators. *Mailing Add:* Dept Physiol & Cell Biol Albany Med Col 43 New Scotland Ave Albany NY 12208

MALIK, DAVID JOSEPH, b Pittsburg, Calif, July 24, 45; m 85; c 2. INTERMOLECULAR ENERGY TRANSFER. *Educ:* Calif State Univ, Hayward, BS, 68, MS, 69; Univ Calif, San Diego, PhD(chem), 76. *Prof Exp:* Res fel chem, Univ Calif, San Diego, 76-77, lectr phys chem, 76; res fel, Univ Ill, Urbana-Champaign, 77-80; asst prof, 80-86, actg chmn, 89-90, pres, Sch Sci, 89-91, ASSOC PROF CHEM, IND UNIV-PURDUE UNIV, INDIANAPOLIS, 86- *Concurrent Pos:* Prin investr, Res Corp grant, 81-83 & Petrol Res Fund, Am Chem Soc, 84-86, 88-90. *Mem:* Am Chem Soc; Am Phys Soc; Sigma Xi. *Res:* Investigation of atomic and molecular interactions and energy transfer; effects of electric fields and field gradients on vibrational properties, including their role in determination of potential energy surfaces; scattering theory and computational methods. *Mailing Add:* Dept Chem Ind Univ-Purdue Univ Indianapolis IN 46205-3825

MALIK, FAZLEY BARY, b Bankura, India, Aug 16, 34. ATOMIC PHYSICS, NUCLEAR PHYSICS. *Educ:* Calcutta Univ, India, BS, 53; Dhaka Univ, Bangladesh, MSc 55; Gottingen Univ, Fed Repub Germany, PhD, 58. *Prof Exp:* Res assoc, Max Planck Inst, Munich, Fed Repub Germany, 58-60; res assoc, Princeton Univ, NJ, 60-63; asst prof physics, Yale Univ, New Haven, 64-68; from assoc prof to prof, Ind Univ Bloomington, 68-80; chmn dept, 80-85, PROF PHYSICS, SOUTHERN ILL UNIV, CARBONDALE, 80- *Concurrent Pos:* Vis prof, Tech Univ Karlsruhe, 65, Tech Univ Darmstadt, 67, Univ Geneva, Univ Lausanne & Univ Neuchatel, 71-72, Univ Frankfurt, 71, Univ Helsinki & Univ Jyvaskyla, 83 & Univ Tubingen, 85-86; mem, Gov Task Force Fusion, Ill, 83-; hon prof, Northwest Norman Univ, Chang Chun, China, 84; ed, Condensed Matter Theories, 85; adv, Bangladesh Planning Comn, 72; Univ of Chicago Select Comt, Nat Sychrotron Source, 85-87, Counr, Oakridge Assoc Univ, 85-87; Fulbright Scholar, Finland, 87. *Mem:* Am Phys Soc; Sigma Xi. *Res:* Theory of atomic, molecular and nuclear structure; reactions and many body aspects; fission. *Mailing Add:* Dept Physics Southern Ill Univ Carbondale IL 62901

MALIK, JIM GORDEN, b Elyria, Ohio, Oct 5, 28; m 53; c 3. PHYSICAL CHEMISTRY, INORGANIC CHEMISTRY. *Educ:* Wabash Col, AB, 50; Mich State Univ, PhD, 54. *Prof Exp:* Instr, Univ Minn, Duluth, 54-55, asst prof, 55-56; asst prof chem, Knox Col, 56-57; from asst prof to assoc prof, San Diego State Col, 57-64; prof, Sonoma State Col, 64-65; PROF CHEM, SAN DIEGO STATE UNIV, 65- *Res:* Preparation of inorganic coordination compounds; atomic and molecular structure; interferometric measurements; chemical education. *Mailing Add:* Dept Chem San Diego State Univ 5300 Campanile Dr San Diego CA 92182

MALIK, JOHN S, b Geddes, SDak, Sept 3, 20; m 54; c 1. EXPERIMENTAL PHYSICS. *Educ:* Kans State Teachers Col, AB, 42; Univ Mich, MS, 47, PhD, 50. *Prof Exp:* PHYSICIST, LOS ALAMOS NAT LAB, 50- *Concurrent Pos:* Staff mem & sci adv, Opers Off, Dept Energy, Nev, 69- *Mem:* Fel AAAS; Am Phys Soc; Sigma Xi. *Res:* Nuclear weapon testing and phenomenology; diagnostics; gamma rays; electro-magnetic pulse; nuclear weapons test hazards evaluation. *Mailing Add:* 810 46th St Los Alamos NM 87544

MALIK, JOSEPH MARTIN, b Monroe, Mich, Dec 2, 44; m 71; c 2. CHEMICAL PROCESS DEVELOPMENT, ENVIRONMENTAL FATE TESTING. *Educ:* Mich State Univ, BS, 66; Univ Wis, Madison, PhD(org chem), 72. *Prof Exp:* Res assoc, Univ Ill, Urbana, 72-73; sr res chemist, 73-77, sr group leader, 77-81, res mgr metab, Monsanto Agr Prod Co, 81-87, mgr residue chem, 87-89, MGR PROCESS CHEM, MONSANTO, 89- *Mem:* Am Chem Soc; Asn Off Anal Chemists. *Res:* Pesticide metabolism in soil, plants and animals; high pressure liquid chromatography; environmental fate determination; enzyme kinetics; pesticide residue analysis; pesticide chemical process development; pilot plant synthesis. *Mailing Add:* Monsanto Agr Co 03C 800 N Lindbergh Blvd St Louis MO 63167

MALIK, MAZHAR ALI KHAN, industrial engineering, statistics, for more information see previous edition

MALIK, MAZHAR N, b Pakistan, Aug 19, 40. DEVELOPMENTAL DISABILITIES. *Educ:* Punjab Col, MS, 64; Univ Glasgow, Scotland, PhD(biochem), 69. *Prof Exp:* Postdoctoral fel, Med Sch, Univ St Louis, 69-71; asst prof biol, 71-77, ASSOC DIR CLIN LAB, MED CTR, STATE UNIV NY, 83-; RES SCIENTIST, NY STATE INST DISABILITIES, 77- *Concurrent Pos:* Investr, Am Health Asn, 75-80; vis prof, Univ NY, 77-79. *Mem:* Am Soc Biochem & Molecular Biol; NY Acad Sci. *Mailing Add:* Basic Res in Develop Disabilities NY State Inst 1050 Forest Hill Rd Staten Island NY 10314

MALIK, NORBERT RICHARD, b Cedar Rapids, Iowa, June 6, 36; m 65; c 3. ELECTRICAL ENGINEERING. *Educ:* Univ Iowa, BS, 59, MS, 60; Iowa State Univ, PhD(elec eng), 64. *Prof Exp:* Instr elec eng, Univ Iowa, 59-60; asst prof, Kans State Univ, 64-67; asst prof, 67-69, assoc prof, 69-80, PROF ELEC & COMPUT ENG, UNIV IOWA, 80- *Concurrent Pos:* Consult, Autonetics Div, NAm Aviation, Teleglobe, Booze-Hamilton-Jordan, Rockwell Collins, Intext Inc & Phillips Oil Co. *Mem:* Inst Elec & Electronics Engrs. *Res:* Digital signal processing; speech processing. *Mailing Add:* Dept Elec & Comput Eng Univ of Iowa Iowa City IA 52242

MALIK, OM PARKASH, b Sargodha, Pakistan, Apr 20, 32; m 68; c 3. ELECTRICAL MACHINES, DIGITAL CONTROL & POWER SYSTEMS PROTECTION. *Educ:* Delhi Polytech Inst, NDEE, 52; Roorkee Univ, ME, 62; Univ London, PhD(elec eng) & dipl, Imp Col, 65. *Prof Exp:* Asst engr, Punjab State Elec Bd, India, 53-61; elec engr, English Elec Co, Eng, 65-66; asst prof elec eng, Univ Windsor, 66-68; assoc prof, 68-74, assoc dean, Acad Fac Eng, 79-90, PROF ELEC ENG, UNIV CALGARY, 74- *Concurrent Pos:* Chmn, student activ, Can Region, 78, 79 & 80. *Honors & Awards:* Centennial Medal, Inst Elec & Electronic Engrs, 84, Merit Award, Western Can Coun, 86. *Mem:* Am Soc Eng Educ; Brit Inst Elec Engrs; Inst Elec & Electronic Engrs; Can Elec Asn. *Res:* Mathematical models for simulation and analysis of electrical machines and power systems; optimal and adaptive control of synchronous machines; on-line digital control of generating units; digital protection of power systems; expert systems applications in power systems. *Mailing Add:* Dept Elec Eng Univ Calgary 2500 Univ Dr NW Calgary AB T2N 1N4 Can

MALIK, VEDPAL SINGH, b Sunna, India, Jan 1, 42; US citizen; m 72, 81; c 3. MEDICAL RESEARCH. *Educ:* Indian Agr Res Inst, New Delhi, MSc, 64; Dalhousie Univ, PhD(biol), 70. *Prof Exp:* Teaching asst biol, Dalhousie Univ, 65-68; res assoc microbiol, Sherbrooke Med Ctr, Can, 70, Tex Med Ctr, Houston, 70-71 & Mass Inst Technol, 71-72; RES SCIENTIST MICROBIOL, UPJOHN CO, 72-,; sr scientist & proj leader biotechnol, 81-89, TECHNOL LEADER BIOTECHNOL, PHILIP MORRIS, RICHMOND, 89- *Concurrent Pos:* Res assoc, Nat Cancer Inst, 72; adj prof microbiol, Western Mich Univ, Kalamazoo, 78, vis scientist, cell biol, Rockefeller Univ, 85-86; consult, United Nations Orgn Biotechnol, 86-87, Govt Saudia Arabia, biotechnol, 87. *Mem:* Am Soc Microbiol. *Res:* Physiology and genetics of industrial organisms; novel microbial metabolites; molecular biology of plants; biotechnology. *Mailing Add:* 2714 Teaberry Dr Richmond VA 23236

MALIN, DAVID HERBERT, b Washington DC, June 9, 44; m 70. BIOPSYCHOLOGY, PSYCHOPHARMACOLOGY. *Educ:* Harvard Univ, BA, 66; Univ Mich, PhD(psychobiol), 72. *Prof Exp:* Res assoc neuropharmacol, Baylor Col Med, 72-74; asst prof, 74-77, ASSOC PROF PSYCHOL, UNIV HOUSTON-CLEAR LAKE, 77- *Concurrent Pos:* Adj asst prof, Baylor Col Med, 74-76. *Mem:* Soc Neurosci; Am Psychol Asn; AAAS; Psychonomic Soc. *Res:* Neurochemical mechanisms of opiate dependence; the role of endorphins and catecholamines in memory, appetite, respiratory control and drug dependence; behavioral pharmacology of opiate antagonists. *Mailing Add:* Dept Human Sci Univ Houston Clear Lake Houston TX 77058

MALIN, EDYTH THERESA LASKY, b Fitchburg, Mass, May 21, 26; m 47; c 3. BIOCHEMISTRY, BIOPHYSICAL CHEMISTRY. *Educ:* Vanderbilt Univ, BA, 47; Am Univ, MS, 69; Bryn Mawr Col, PhD(chem), 75. *Prof Exp:* Chemist, Nat Bur Standards, 50-51; fel, dept biochem, Sch Med, Univ Pa, 75-77; Nat Res Coun assoc, 77-79, res chemist, 77-85, LEAD SCIENTIST, EASTERN REGIONAL RES CTR, USDA, 85- *Concurrent Pos:* Ed asst, Chem & Eng News, 51-53; manuscript ed, Am Chem Soc Appl Pubs, 53-64; bibliogr, Life Sci Res Off, Fedn Am Soc Exper Biol, 64-67; NIH trainee, dept biochem, Sch Med, Univ Pa, 75-76; lectr biochem, chem dept, St Joseph's Univ, 80, Spring Garden Col, 82 & Eastern Regional Res Ctr, USDA, 83, Beaver Col, 89. *Mem:* Am Chem Soc; Sigma Xi; Protein Soc; AAAS; Int Asn Women Biochemists; Asn Women Sci. *Res:* Chemical modification; membrane proteins; protein structure and dynamics; enzyme mechanisms; molecular modeling. *Mailing Add:* Eastern Regional Res Ctr USDA 600 E Mermaid Lane Philadelphia PA 19118

MALIN, HOWARD GERALD, b Providence, RI, Dec 2, 41. PODIATRIC MEDICINE, PODIATRIC DERMATOLOGY. *Educ:* Univ RI, Kingston, AB, 64; Brigham Young Univ, MA, 69; Calif Col Podiatric Med, BSc, 69, DPM(podiatric med), 72; Am Bd Podiatric Orthopedics, dipl, 78; Pepperdine Univ, MSc, 78; Am Bd Podiatric Pub Health, dipl, 89. *Prof Exp:* Asst instr basic sci, Calif Col Podiatric Med, 69-72, res asst, 70-72; instr, NY Col Podiatric Med, 73-74; staff podiatrist, Prospect Hosp, NY, 74-77; staff podiatrist & chief, Podiatry Serv, David Grant US Air Force Med Ctr, Travis AFB, Calif, 77-80; CHIEF, PODIATRIC SECT, VET ADMIN MED CTR, WVA, 80-, STAFF PODIATRIST, 80- *Concurrent Pos:* Practitioner, 74-77; ed, Futura Publ Co & David Grant US Air Force Med Ctr, 79-80; staff reserve podiatrist, Malcolm Grow US Air Force Med Ctr, 80-; consult podiatric sports med, Travis AFB, 77-80; ed, Current Podiatric Medicine, 84-; clin prof, Alderson-Broaddus Col, 86, assoc clin prof, Univ Osteop Med & Health Sci; adj prof Barry Univ, Col Podiatric Med; exten dir, Podiatric Med, Pa Col Podiatric Med; instr Podiatric Med, Ohio Col Podiatric Med. *Honors & Awards:* Basic Sci Award, Calif Col Podiatric Med, 72. *Mem:* Am Podiatry Asn; NY Acad Sci; AAAS; Asn Military Surgeons US; fel Am Col Foot Orthopedics; fel Am Soc Podiatric Med; fel Am Soc Podiatric Dermat; fel Am Asn Hosp Podiatrists; fel Am Col Podiatric Radiol; Am Col Foot Surgeons; Am Acad Podiatric Med. *Res:* Specialties of podiatric medicine and podiatric dermatology. *Mailing Add:* Podiatry Dept Vets Admin Med Ctr Martinsburg WV 25401

MALIN, JOHN MICHAEL, b Cleveland, Ohio, July 9, 42; m 67; c 2. INORGANIC CHEMISTRY, PHOTOCHEMISTRY. *Educ:* Univ Calif, Berkeley, BS, 63; Univ Calif, Davis, PhD(chem), 68. *Prof Exp:* NIH fel inorg chem, Stanford Univ, 68-70; Nat Acad Sci fel, Inst Chem, Univ Sao Paulo, Brazil, 70-73; vis asst prof inorg chem, Univ Mo, Columbia, 73-74, asst prof, 74-77, assoc prof, 77-81; prog officer, Chem Dynamics, NSF, 79-81; ASST PROG ADMINR, PETROL RES FUND-AM CHEM SOC, WASHINGTON, DC, 81- *Mem:* Sigma Xi; Am Chem Soc; AAAS. *Res:* Studies of the synthesis, spectra, structures and reactivity, both thermal and photoinduced, of transition metal complexes. *Mailing Add:* 705 N Edison St Arlington VA 22203

MALIN, MICHAEL CHARLES, b Burbank, Calif, May 10, 50. PLANETARY GEOLOGY, GEOMORPHOLOGY. *Educ:* Univ Calif, Berkeley, AB, 71; Calif Inst Technol, PhD(planetary sci), 76. *Prof Exp:* Sr scientist, Jet Propulsion Lab, Calif Inst Technol, 75-78, mem tech staff planetary geol, 78-79; from asst prof to assoc prof, 79-87, PROF GEOL, ARIZ STATE UNIV, 87- *Concurrent Pos:* Prin investr, Mars Observer Camera; John D & Catherine T MacArthur fel, 87-92. *Mem:* Am Geophys Union; Am Astron Soc; Asn Computing Mach. *Res:* Evolution of landforms; volcanology; remote sensing; specific areas; Mars, heavily cratered terrains, channels and other erosional landforms; polar regions; Venus, tectonic and volcanic features. *Mailing Add:* 1011 Camino Del Rio S Suite 210 San Diego CA 92108

MALIN, MURRAY EDWARD, b New York, NY, June 25, 27; m 52; c 2. PHYSICAL CHEMISTRY. *Educ:* City Col New York, BS, 48; Harvard Univ, AM & PhD(phys chem), 51. *Prof Exp:* Mem staff, Los Alamos Sci Lab, 51-56; vpres res & technol, Res & Develop Div, Avco Corp, 56-57, tech opers, Avco Space Systs Div, 57-68; mgr spec systs, 68-71, prog mgr, 71-75, DIV MGR, POLAROID CORP, 75- *Concurrent Pos:* Mem, Adv Comt Fluid Mech, NASA, 65- *Mem:* Am Phys Soc; Am Chem Soc; fel Am Inst Aeronaut & Astronaut. *Res:* High temperature; thermodynamics; theory and structure of detonation waves; instant color transparencies; space and missile physics. *Mailing Add:* Polaroid Corp 1 Upland Rd Norwood MA 02062

MALIN, SHIMON, b Ramat-Gan, Israel, July 2, 37; m 60; c 3. THEORETICAL PHYSICS. *Educ:* Hebrew Univ, Jerusalem, MSc, 61; Univ Colo, Boulder, PhD(physics), 68. *Prof Exp:* From asst prof to assoc prof, 68-84, PROF PHYSICS, COLGATE UNIV, 84- *Mem:* Am Phys Soc. *Res:* Group theory and its applications to general relativity; foundations of quantum mechanics; cosmology. *Mailing Add:* Dept Physics & Astron Colgate Univ Hamilton NY 13346

MALINA, JOSEPH F(RANCIS), JR, b Brooklyn, NY, Aug 24, 35; m 65; c 3. ENVIRONMENTAL & CIVIL ENGINEERING. *Educ:* Manhattan Col, BCE, 53; Univ Wis, MSSE, 59, PhD(saüit eng), 61; Acad of Environ Engrs, Dipl. *Prof Exp:* Instr civil eng, Ufiv Wis, 57-58, asst sanit eng, 58-61; from asst prof to assoc prof environ health eng, Univ Tex, Austin, 61-70, dir environ health eng res lab, 70-76, chmn dept, civil eng, 76-88, PROF CIVIL ENG, UNIV TEX, AUSTIN, 70-, CW COOK PROF ENVIRON ENG, 80- *Concurrent Pos:* Res grant, 62-65; Eng Found award, Univ Tex, Austin, 62; regist prof eng, Tex. *Honors & Awards:* Gordon M Fair Award, Am Acad Environ Engrs; Arthur Sidney Bedell Award, Water Pollution Control Fedn; Joe J King Prof Achievement Award. *Mem:* Am Soc Civil Engrs; Am Water Works Asn; Water Pollution Control Fedn; Am Soc Eng Educ; Inter-Am Asn Sanit Engrs; Sigma Xi; Int Asn Water Pollution Res & Control. *Res:* Water and wastewater treatment, renovation and reuse; hazardous waste; sludge management; solid wastes disposal. *Mailing Add:* Dept Civil Eng ECJ 8-6 Univ Tex Austin TX 78712

MALINA, MARSHALL ALBERT, b Chicago, Ill, July 11, 28; m 55; c 2. RESOURCE MANAGEMENT. *Educ:* Ill Inst Technol, BS, 49; Northwestern Univ, MBA, 69. *Prof Exp:* Qual control chemist, Capitol Chem Co, 49-50; prod mgr, Hamilton Industs, 50-52; mgr, Gen Anal Sect, Velsicol Chem Corp, 52-68, mgr qual control, 68-69, dir qual control & tech serv, 69-70, dir res, 71-74, mgr corp develop, 74-76, dir corp develop, 76-79, dir capital expenditures, 80-86; MGR DIR, MALINA & ASSOCS, 86- *Mem:* Am Chem Soc; Am Soc Testing & Mat; fel Asn Off Anal Chemists; Com Develop Asn; Soc Plastics Eng. *Res:* Agricultural pesticides; polymers; polymer additives; plasticizers; benzoic acid and derivatives; pollution control; statistics; licensing, acquisitions, capital budgeting and planning, corporate planning. *Mailing Add:* 16012 Glen Haven Dr 1 Tampa FL 33618

MALINA, ROBERT MARION, b Brooklyn, NY, Sept 19, 37. PHYSICAL ANTHROPOLOGY. *Educ:* Manhattan Col, BS, 59; Univ Wis, MS, 60, PhD(phys educ), 63; Univ Pa, PhD(anthrop), 68. *Prof Exp:* Asst prof, 67-71, assoc prof, 71-77, PROF ANTHROP, UNIV TEX, AUSTIN, 77- *Concurrent Pos:* Vis prof, Cath Univ Leuven, Belgium, 81. *Mem:* Am Asn Health, Phys Educ & Recreation; Am Asn Phys Anthrop; Soc Study Human Biol; Human Biol Coun; Soc Res Child Develop. *Res:* Human growth and development; motor development; growth and nutrition in Mexico and Central America; growth and athletic performance. *Mailing Add:* Dept Anthrop Univ of Tex Austin TX 78712

MALINAUSKAS, ANTHONY PETER, b Ashley, Pa, Mar 24, 35; m 57; c 6. PHYSICAL CHEMISTRY. *Educ:* Kings Col, Pa, BS, 56; Boston Col, MS, 58; Mass Inst Technol, PhD(phys chem), 62. *Prof Exp:* Mem res staff, Oak Ridge Nat Lab, 62-73, head, Chem Develop Sect, 73-83, dir, Nuclear Regulatory Comn Progs, 83-89, DIR, WASTE RES & DEVELOP PROGS, OAK RIDGE NAT LAB, 89- *Honors & Awards:* Award, Am Nuclear Soc, 81; E O Lawrence Award, 85. *Mem:* Sigma Xi; Am Chem Soc; Am Nuclear Soc. *Res:* Thermal transpiration; transport properties of gases; nuclear safety; fission product chemistry and transport; nuclear fuel reprocessing; environmental restoration; waste management. *Mailing Add:* 107 Newton Lane Oak Ridge TN 37830

MALINDZAK, GEORGE S, JR, b Cleveland, Ohio, Jan 3, 33; m 59; c 4. MEDICAL SCIENCES, ENVIRONMENTAL HEALTH. *Educ:* Western Reserve Univ, AB, 53-56; Ohio State Univ, MSc, 57-58, PhD(physiol & biophys), 58-61. *Prof Exp:* Phys metallurgist, Thompson Co, Cleveland, Ohio, 56-57; res asst, Dept Physiol, Ohio State Univ, Columbus, 57-60, NIH fel, Dept Physiol, 60-61, res assoc, 61-62; instr, Dept Physiol, Bowman Gray Sch Med, Winston-Salem, NC, 62-63, from asst prof to assoc prof, 63-73, dir, Computer Ctr, 64-65; res physiologist, Clin Studies Div, Environ Protection Agency, Univ NC, Chapel Hill, 73-76; chmn, Dept Physiol, Northeastern Ohio Univs Col Med, Rootstown, 76-84, prof, 76-85; prof, Div Biol Sci, Kent State Univ, Ohio, 77-85; head, Dept Biomed Eng, La Tech Univ, Ruston, 85-87, prof, 85-88, coordr physiol eng, Ctr Rehab Sci & Biomed Eng, 86-88; HEALTH SCI ADMINR, NAT INST ENVIRON HEALTH SCI, NIH, RESEARCH TRIANGLE PARK, NC, 88- *Concurrent Pos:* Sr res investr, NC Heart Asn, 65-70 & 70-73, mem res comt, 68-69 & 71-73; consult, Biol Appln Team, Res Triangle Inst, NASA, 67-74, Simulators Inc, 68-69, Peer Rev Corp, Region 6, Health Serv Admin, Akron, Ohio, 76-82, Tech Adv Serv for Attorneys, 79- & Int Chelation Res Found, 82-; State bd dirs, NC Chap, Nat Cystic Fibrosis Found, 68-71; mem, Eng Med & Biol Group, Winston-Salem Chap, Inst Elec & Electronic Engrs, 68-73; bd dirs, High Rock Lake Asn, 72-77 & Portage County Heart Asn, Ohio, 77-85; mem, Spec Interest Group Biomed Comput, Asn Comput Mach, 73-77, Spec Comt Eng Educ, NC State Univ, 73-75 & res rev comt, Ohio Affil, Am Heart Asn, 77-85; mem, Prof Educ Comt, Akron Chap, Am Heart Asn, 77-82, res comt, 77-84 & bd trustees, Dist Chap, 79-84; consult, grad med res, develop & educ, Col Med, Northeastern Ohio Univs, 84-85; mem, Peer Res Rev Comt, Am Heart Asn La, 85-87 & Steering comt, Southern Biomed Eng Conf, 85-89. *Mem:* Am Physiol Soc; Am Soc Pharmacol & Exp Therapeut; Asn Chairmen Depts Physiol; Biomed Eng Soc; Sigma Xi; Am Heart Asn; Asn Comput Mach; Am Asn Univ Professors; Biophys Soc; Acoust Soc Am. *Res:* Cardiopulmonary toxicology; coronary and cerebral vascular reactivity and spasm; coronary and cerebral ischemia and hypoxia and vascular control; carbon monoxide toxicity; autonomic and receptor physiology and pharmacology; control of blood flow in health and disease; microcirculation of the heart; cardiopulmonary function and environmental toxicology; hypoxia, ischemia, and circulator (coronary and cerebral) function; pathophysiology of coronary and peripheral vascular atherosclerosis; echocardiography and ventricular function; spinal cord trauma; cardiac function and rehabilitation of alcoholics; peripheral vascular disease; medical electronics and medical engineering; cardiovascular modeling; indicator-dilution techniques and analyses; mathematical and computer analyses of biological systems. *Mailing Add:* Div Extramural Res & Training NIEHS NIH, PO Box 12233 Research Triangle Park NC 27709

MALING, GEORGE CROSWELL, JR, b Boston, Mass, Feb 24, 31; m 60; c 3. PHYSICS, NOISE CONTROL ENGINEERING. *Educ:* Bowdoin Col, AB, 54; Mass Inst Technol, SB & SM, 54, EE, 58, PhD(physics), 63. *Prof Exp:* Mem res staff, Mass Inst Technol, 63-65; adv physicist, 65-70, SR PHYSICIST, IBM CORP, 70- *Concurrent Pos:* Consult pvt pract, 59-65 & Indust Acoustics Co, 63-65. *Mem:* Fel Inst Elec & Electronics Engrs; fel Acoust Soc Am; Inst Noise Control Eng (secy, 71-74, pres, 75); fel AAAS; fel Audio Eng Soc. *Res:* Physical acoustics; large amplitude wave propagation; instabilities in inhomogeneous media; radiation phenomena; noise control. *Mailing Add:* Dept CT1A IBM Corp PO Box 950 Bldg 704 Poughkeepsie NY 12602

MALININ, GEORGE I, b Krasnodar, USSR, Sept 19, 29; US citizen; m 64. ANALYTICAL HISTOCHEMISTRY, CRYOBIOLOGY. *Educ:* Concord Col, BS, 58; George Washington Univ, MS, 60; Catholic Univ Am, PhD(cell biol), 72. *Prof Exp:* Res scientist, Am Found Biol Res, 68-70; guest scientist, Naval Med Res Inst, 71-72; res assoc, Dept Biol, Georgetown Univ, 72-74; res assoc, 74-76, chief, Biochem Res Lab, Children's Hosp, 76-77; assoc res prof, 77-79, RES PROF BIOPHYSICS, GEORGETOWN UNIV, 79- *Concurrent Pos:* Consult, Washington Hosp Ctr, 77-86. *Mem:* AAAS; Soc Cryobiol; Histochem Soc; Am Soc Photobiol. *Res:* Cell differentiation induction by membrane active compounds; effects of trace elements on defferentiation and on cell membranes; interaction of external electric and acustic fields on cells and their components at ambient and ultra low temperatures. *Mailing Add:* Dept Physics Georgetown Univ Washington DC 20057

MALININ, THEODORE I, b Krasnodar, USSR, Sept 13, 33; US citizen; m 60; c 4. PATHOLOGY, EXPERIMENTAL SURGERY. *Educ:* Concord Col, BS, 55; Univ Va, MS, 58, MD, 60. *Prof Exp:* Fel path, Johns Hopkins Univ, 60-61; pathologist, Nat Cancer Inst, 61-64; asst prof path, Sch Med,

Georgetown Univ, 64-69; clin assoc prof, 69-70, PROF SURG & PATH, SCH MED, UNIV MIAMI, 70- *Concurrent Pos:* Guest scientist, Tissue Bank, Naval Med Res Inst, 64-70; ed, Cryobiol, 64-70; consult, Bur Health Manpower, USPHS, 66-67; mem staff, Vet Admin Hosp, Miami, 70-80. *Mem:* AAAS; Tissue Cult Asn; Soc Cryobiol; Am Asn Path; Path Soc Gt Brit & Ireland; Orthopedic Res Soc; Am Acad Orthop Surgeons. *Res:* Organ perfusion and preservation; experimental myocardial infarction; hemorrhagic shock; tissue banking; bone and cartilage transplantation; biological behavior of neoplastic tissue. *Mailing Add:* Dept of Surg Univ of Miami Sch of Med Miami FL 33101

MALINOW, MANUEL R, b Buenos Aires, Arg, Feb 27, 20; m 52; c 3. CARDIOLOGY. *Educ:* Univ Buenos Aires, MD, 45. *Hon Degrees:* DHC, Univ Buenos Aires. *Prof Exp:* Res fel cardiovasc med, Michael Reese Hosp, Chicago, 45-46; dir res dept & electrocardiologist, Hosp Ramos Mejia, Buenos Aires, 47-57; chief res physiol, Buenos Aires Med Sch, 56-63; PROF MED, MED SCH, UNIV ORE, 64-; SECT HEAD CARDIOVASC DIS, ORE REGIONAL PRIMATE RES CTR, 64- *Concurrent Pos:* Chief cardiol serv, Munic Inst Radiol & Physiother, Arg, 50-53; chief sect atherosclerosis, Nat Acad Med, Buenos Aires, 60-63, physician, Hosp, 62-63. *Honors & Awards:* Paul D White Prize, Arg Soc Cardiol, 54 & Gold Medal, 60; Ciba Found Award, 59; Rafael M Bullrich Prize, Nat Acad Med, Buenos Aires, 59; Gold medal, Inter-Am Cong Cardiol, Brazil, 60; Sesquicentenary Prize, Arg Med Asn. *Mem:* AAAS; Royal Soc Med; NY Acad Sci; Am Heart Asn. *Res:* Experimental cardiology; atherosclerosis; blood cholesterol; cardiovascular diseases; exercise and electrocardiography. *Mailing Add:* Ore Regional Primate Res Ctr 505 NW 185th Beaverton OR 97006

MALINOWSKI, EDMUND R, b Mahanoy City, Pa, Oct 16, 32; m 58; c 2. PHYSICAL & ANALYTICAL CHEMISTRY, SPECTROSCOPY. *Educ:* Pa State Univ, BS, 54; Stevens Inst Technol, MS, 56, PhD(phys chem), 61. *Prof Exp:* Res assoc, Nuclear Magnetic Resonance Lab, 60-63, from asst prof to assoc prof, 63-70, PROF CHEM, STEVENS INST TECHNOL, 70- *Mem:* Am Chem Soc; Int Chemometrics Soc. *Res:* Applications of factor analysis to chemistry; chemometrics; spectroscopy. *Mailing Add:* Dept Chem & Chem Eng Stevens Inst Technol Hoboken NJ 07030

MALINS, DONALD CLIVE, b Lima, Peru, May 19, 31; US citizen; m 62; c 3. TOXICOLOGY. *Educ:* Univ Wash, BA, 53; Seattle Univ, BS, 56; Univ Aberdeen, PhD(biochem), 67, DSc, 76. *Prof Exp:* Org chemist, Tech Lab, US Bur Com Fisheries, 56-62, res chemist, 62-66, Food Sci Pioneer Res Lab, 66-71, prog dir, Pioneer Res Unit, Northwest Fisheries Ctr, 71-74, dir, Environ Conserv Div, Northwest Fisheries Ctr, Nat Marine Fisheries Serv, 74-87; HEAD, ENVIRON BIOCHEM PROG, PAC NORTHWEST RES FOUND, 87- *Concurrent Pos:* Lectr, Univ Wash, 68-74, affil assoc prof, 74-78, affil prof, 78-; res prof, Seattle Univ, 71-; ed, Biochem & Biophys Perspectives in Marine Biol, 74-79, Effects Petrol on Arctic & Subarctic Marine Environ, 77; ed-in-chief, Aquatic Toxicol, 79-; consult pollution marine environ, UN Environ Prog, 80-; sci consult, Nat Oceanic & Atmospheric Admin, 89-; mem, bd dir, Am Oceans Campaign, 89-; sr sci consult, US Dept Justice, 90-; adv bd, Int Joint Comn, 90- *Honors & Awards:* US Dept Interior Achievement Awards, 60; Spec Achievement Award, US Dept Com, 75 & 80, Gold Medal, 82. *Mem:* Am Chem Soc; Am Soc Biochem & Molecular Biol. *Res:* Synthesis and metabolism of lipids; chemistry of bioacoustics; chemistry of aromatic compounds; biochemistry of aromatic compounds and pathologic and other responses from chemical exposures of marine organisms. *Mailing Add:* Pacific Northwest Research Foundation 720 Broadway Seattle WA 98122

MALIS, LEONARD I, b Philadelphia, Pa, Nov 23, 19; m; c 2. NEUROSURGERY. *Educ:* Univ Va, MD, 43. *Prof Exp:* Attend neurosurgeon, 51-70, NEUROSURGEON IN CHIEF & DIR DEPT, MT SINAI HOSP, 70-; PROF NEUROSURG & CHMN DEPT, MT SINAI SCH MED, 70- *Concurrent Pos:* Res collabr, Med Dept, Brookhaven Nat Lab, 56-70; consult neurosurgeon, Beth Israel Med Ctr, NY. *Mem:* Am Acad Neurol Surg; Soc Neurol Surg; Am Asn Neurol Surg; Am Physiol Soc; Cong Neurol Surg. *Res:* Neurophysiology. *Mailing Add:* 1176 Fifth Ave New York NY 10029

MALISZEWSKI, CHARLES R, b Lowell, Mass, Oct 30, 54; m. IMMUNOLOGY. *Educ:* Boston Col, Chestnut Hill, Mass, 76; Case Western Reserve Univ, Cleveland, Ohio, PhD(microbiol), 82. *Prof Exp:* Postdoctoral fel, Microbiol Lab, Dartmouth Med Sch, 82-85; assoc staff scientist, 85-86, staff scientist, 86-90, SR STAFF SCIENTIST, DEPT IMMUNOL, IMMUNEX CORP, 91- *Concurrent Pos:* Postdoctoral fel, NIH, 82-85; Small Bus Innovation Res Prog grant, US Dept Agr, 88; ad hoc reviewer, J Immunol, Vet Immunol & Immunopath, Infection & Immunity. *Mem:* Am Asn Immunologists. *Res:* Immunology and immunopathology; author of various publications. *Mailing Add:* Dept Immunol Immunex Corp 51 University St Seattle WA 98101

MALITSON, HARRIET HUTZLER, b Richmond, Va, June 30, 26; m 51; c 2. ASTRONOMY. *Educ:* Goucher Col, AB, 47; Univ Mich, MS, 51. *Prof Exp:* Jr physicist, Nat Bur Standards, 47-49 & 51-52 & US Naval Res Lab, 52-57; astronomer, Goddard Space Flight Ctr, NASA, 60-82; RETIRED. *Concurrent Pos:* Mem, Comn J, Int Union Radio Sci & Comn 10, Int Astron Union. *Mem:* Am Astron Soc. *Res:* Solar physics; solar-terrestial relationships; radio and space astronomy. *Mailing Add:* 13315 Magellan Ave Rockville MD 20853

MALITZ, SIDNEY, b Brooklyn, NY, Apr 20, 23; m 45; c 2. PSYCHIATRY, PSYCHOPHARMACOLOGY. *Educ:* Univ Chicago, MD, 46; Columbia Univ, cert psychoanal med, 59; Am Bd Psychiat & Neurol, dipl, 53. *Prof Exp:* Resident psychiatrist, NY State Psychiat Inst, 48-51, sr resident psychiatrist, 51-52; asst, Columbia Univ, 55-57, assoc, 57-59, asst clin prof, 60-65, assoc prof, 65-69, vchmn, Dept Psychiat, 72-75 & 76-78, actg chmn dept, 75-76 & 81-84, PROF CLIN PSYCHIAT, COL PHYSICIANS & SURGEONS, COLUMBIA UNIV, 70- *Concurrent Pos:* Sr res psychiatrist, NY State Psychiat Inst, 54-56, actg chief psychiat res, 56-65, chief dept biol psychiat, 65-72 & 84-, dep dir, Inst, 72-75 & 77-78; asst vis psychiatrist, Francis Delafield Hosp, 54-60; asst attend psychiatrist, Vanderbilt Clin & Presby Hosp, 56-58, assoc attend psychiatrist, 58-71, attend psychiatrist, Presby Hosp, 71-, actg dir psychiat serv, 75-76 & 81-84; actg dir, NY State Psychiat Inst, 75-76 & 81-84, dep dir & asst dir, Psychiat Serv, 76-78; asst examr, Am Bd Psychiat & Neurol, 57-; archivist & historian, Am Col Psychiatrists, 78-; pres, NY Psychiat Soc, 91, Vidonian Soc, 91, Benjamin Rush Soc, 91- *Mem:* fel Am Col Psychoanalysts; fel Am Psychiat Asn; fel Am Col Psychiatrists; Am Psychopath Asn; fel Royal Col Psychiatrists; Group Advan Psychiat; fel Am Col Neuropsychopharmacol. *Res:* Psychopharmacology and lateralization of brain function. *Mailing Add:* NY State Psychiat Inst 161 Ft Washington Ave New York NY 10032

MALKANI, MOHAN J, b Hyderabad, Pakistan, Sept 17, 33; m 66; c 2. CONTROL ENGINEERING, AI EXPERT SYSTEM. *Educ:* Univ Baroda, BS, 53, MS, 55; Miss State Univ, MS, 64; Vanderbilt Univ, PhD, 80. *Prof Exp:* Instr elec eng, Tuskegee Inst, 63, asst prof, 64-67; prof elec eng & head dept, 67-79, ASSOC DEAN, SCH ENG & TECHNOL, TENN STATE UNIV, 79- *Concurrent Pos:* Vis scientist, Lincoln Lab, Mass Inst Technol, 70-73, 75-76, 78, 79 & 80-81. *Mem:* Sr mem Inst Elec & Electronics Engrs; Am Soc Energy Engrs; Sr Mem Soc Mfr Engrs; Am Soc Engr Educ. *Res:* Neural networks; control systems; digital filter design; energy; artifical intelligence; signal processoring. *Mailing Add:* Sch Eng & Technol Tenn State Univ Nashville TN 37209-1561

MALKASIAN, GEORGE D, JR, b Springfield, Mass, Oct 26, 27; m 54; c 3. MEDICINE. *Educ:* Yale Univ, BA, 50; Boston Univ, MD, 54; Univ Minn, MS, 63. *Prof Exp:* PROF OBSTET-GYNEC, MAYO MED SCH GRAD SCH MED, 61- *Concurrent Pos:* Chmn, Dept Ob- Gyn, Mayo Clin, Rochester, 76-86; mem, Cancer Chemotherapy Investigational Rev Comt, Nat Cancer Inst, 83-87; pres, Cent Asn Obstetricians & Gynecologists, 84-85. *Mem:* Am Col Obstetricians & Gynecologists; Soc Gynecologic Oncologists; Am Col Surgeons. *Res:* Chemotherapy; clinical evaluation of drug treatment for ovarian carcinoma, cervical carcinoma and; endometrial carcinoma; clinical evaluation of tumor markers for ovarian cancer. *Mailing Add:* 1750 11th Ave NE Rochester MN 55904

MALKEVITCH, JOSEPH, b Brooklyn, NY, May 24, 42; m 72; c 1. MATHEMATICS. *Educ:* Queens Col, NY, BS, 63; Univ Wis-Madison, MS, 65, PhD(math), 69. *Prof Exp:* Teaching asst, Univ Wis-Madison, 63-68; from asst prof to assoc prof, 68-80, PROF MATH, YORK COL, NY, 81- *Mem:* AAAS; Am Math Soc; Math Asn Am; AAAS; fel NY Acad Sci; Sigma Xi. *Res:* Convex polytopes; graph theory; arrangements of curves; euclidean geometry. *Mailing Add:* Dept Math York Col Jamaica NY 11451

MALKIEL, SAUL, b Boston, Mass, Dec 28, 12; m 45, 84; c 5. ALLERGY, IMMUNOCHEMISTRY. *Educ:* Clark Univ, AB, 34; Boston Univ, MA, 36, PhD(chem), 42, MD, 44. *Prof Exp:* Asst med sci, Sch Med, Boston Univ, 35-37, asst chem, 37-44; asst path, Yale Univ, 44-45; vis investr, Rockefeller Inst, 45-48; asst prof med, Med Sch, Northwestern Univ, 48-54; assoc path, Harvard Med Sch, 54-83, res assoc, Sidney Farber Cancer Inst Inc, 63-83; ASSOC PROF MED, MED SCH, UNIV MASS, 71- *Concurrent Pos:* Fel, Sch Med, Boston Univ, 45; Nat Res Coun fel med sci, Rockefeller Inst, 45-46, Am Cancer Soc sr fel, 46-48; lectr, Univ Pa, 48; assoc, Peter Bent Brigham Hosp, 54-63; res assoc, Children's Hosp, 63-83; mem corp, Marine Biol Lab, Woods Hole; pvt pract allergy, 54- *Mem:* AAAS; Am Asn Immunologists; Soc Exp Biol & Med; Am Acad Allergy (pres, 70-71); Soc Leukocyte Biol; Sigma Xi. *Res:* Immunochemistry of hemocyanins, hemoglobin, viruses and allergic reactions; effects of stress on experimental asthma; clinical and experimental studies with the antihistamines; effect of cortisone on antibody production; fractionation and adjuvant studies on Bordetella pertussis; allergens from ragweed pollen; immuno-competence of the leukemic cell; tumor viruses; immune response to malarial infection; clinical allergy. *Mailing Add:* 130 Lincoln St Worcester MA 01605

MALKIN, AARON, b Winnipeg, Man, Mar 23, 26; m 53; c 2. CLINICAL BIOCHEMISTRY, ONCOLOGY. *Educ:* Univ Man, MD, 49; McGill Univ, PhD(biochem), 56,. *Prof Exp:* HEAD, DEPT CLIN BIOCHEM, SUNNYBROOK MED CTR, UNIV TORONTO, 61-, PROF BIOCHEM, 73-, PROF DEPT MED, 79- *Mem:* Am Asn Can Res; Am Soc Clin Oncol; Biochem Soc; Int Soc Oncodevelop Biol & Med; fel Royal Col Physicians & Surgeons Can. *Res:* Evaluation of the utility of tumor markers; pathogenesis of AIDS. *Mailing Add:* 38 Snowshoe Millway Willowdale ON M2L 1T5 Can

MALKIN, HAROLD MARSHALL, b San Francisco, Calif, Oct 9, 23; m 49; c 5. BIOCHEMISTRY. *Educ:* Univ Calif, AB, 47, MA, 49; Univ Chicago, MD, 51. *Prof Exp:* Nat Found Infantile Paralysis res fel, Univ Calif, 51-53; Am Cancer Soc fel, Univ Brussels, 53-54; dir, Malkin Med Lab, Malkin Med Lab/Solano Labs, 54-72, pres & med dir, 72-78; MED DIR, OSLERWELCH LABS, 78-; CONSULT, SMITHKLINE BEECHAM LAB, 86- *Concurrent Pos:* Asst, Univ Calif, 51-; clin instr histol, Sch Med, Stanford Univ, 56-62; intern path, Sch Med, Stanford Univ, 61-62. *Mem:* AAAS; Am Chem Soc; Am Asn Clin Chemists. *Res:* Cellular physiology as related to structure; nucleic acid and protein metabolism; cancer; clinical chemistry; medical history. *Mailing Add:* 250 The Uplands Berkeley CA 94705

MALKIN, IRVING, b Cleveland, Ohio, Dec 28, 25; m 47; c 3. INORGANIC CHEMISTRY, ELECTROCHEMISTRY. *Educ:* Western Reserve Univ, BS, 54; Case Inst Technol, MS, 60. *Prof Exp:* Chief chemist, I Schumann & Co, 50-56; res & process control inorg chem, Precision Metalsmiths Inc, 56-59; Mgr Electrode Technol, Electrolytic Systs Div, Diamond Shamrock Corp, 59-82; RETIRED. *Concurrent Pos:* Consult. *Mem:* Am Chem Soc; Electrochem Soc. *Res:* Catalytically active electrode surfaces; electrochemical processes; heterogeneous catalysis; corrosion inhibiting metal coatings; controlled release technology. *Mailing Add:* 4161 Silsby Rd University Heights OH 44118

MALKIN, LEONARD ISADORE, b New York, NY, Dec 17, 36; m 59; c 2. BIOCHEMISTRY. *Educ:* NY Univ, AB, 57; Univ Calif, San Francisco, PhD(biochem), 62. *Prof Exp:* Resident res assoc biochem, Western Regional Res Lab, USDA, Calif, 62-63; NIH fel, Brown Univ, 63-65 & Mass Inst Technol, 65-68; asst prof biochem, Dartmouth Med Sch, 68-73; ASSOC PROF BIOCHEM, SCH MED, WAYNE STATE UNIV, 73- *Mem:* AAAS; Am Soc Biol Chemists. *Res:* Protein synthesis in animal cells; production and utilization of messenger RNA, both endogenous and viral, in animal systems. *Mailing Add:* Dept Biochem Wayne State Univ 4374 Scott 540 E Canfield Detroit MI 48201

MALKIN, MARTIN F, b Newark, NJ, June 28, 37; m 60; c 1. RESEARCH ADMINISTRATION. *Educ:* Univ Mich, BS, 59; Brooklyn Col, MA, 65; NY Univ, PhD(biol), 68. *Prof Exp:* Res assoc biochem, Rockefeller Univ, 67-69; sr res scientist, Merck Inst Therapeut Res, 69-72, res fel, Dept Basic Animal Sci Res, Merck & Co, 72-74, proj coordr, Dept Proj Planning & Mgt, 74-76, dir, Dept Proj Planning & Mgt, Nippon Merk Banyu, 76-78, dir, Proj Planning & Mgt, 78-81, dir, 81-85, EXEC DIR, PLANNING & MGT, MERCK & CO, 86- *Mem:* AAAS; Proj Mgt Inst; Proj Mgt Asn. *Res:* Management of development research projects in animal and human health areas; drug development in Japan. *Mailing Add:* Merck & Co 126 Lincoln Ave Rahway NJ 07065

MALKIN, RICHARD, b Chicago, Ill, Mar 25, 40; m 60; c 3. BIOCHEMISTRY. *Educ:* Antioch Col, BS, 62; Univ Calif, Berkeley, PhD(biochem), 67. *Prof Exp:* Fel biochem, Univ Goteborg, 67-69; from asst biochemist & lectr to assoc biochemist & lectr cell physiol, Univ Calif, 69-79, from assoc prof to prof cell physiol, 79-81, prof molecular plant biol, 82-89, chmn, Div Molecular Plant Biol, 88-89, CHMN, DEPT PLANT BIOL, UNIV CALIF, 87-, PROF PLANT BIOL, 89- *Concurrent Pos:* NATO fel, 67-68; Am Cancer Soc fel; Guggenheim Found fel, 85-86; vis prof, Dept Bot, Imp Col, London, 77-78. *Mem:* Am Soc Plant Physiol; Am Soc Biol Chemists; Biophys Soc. *Res:* Photosynthesis; investigations of electron transport; biological applications of electron paramagnetic resonance spectroscopy. *Mailing Add:* Dept Plant Biol Plant Biol Bldg Univ Calif 111 Genetics Berkeley CA 94720

MALKIN, STEPHEN, b Boston, Mass, June 20, 41; m 64. MECHANICAL ENGINEERING. *Educ:* Mass Inst Technol, SB, 63, SM, 65, ScD(mech eng), 68. *Prof Exp:* Asst prof mech eng, Univ Tex, Austin, 68-74; assoc prof, State Univ NY, Buffalo, 74-76; assoc prof mech eng, Technion, Israel Inst Technol, 76-86; PROF MECH ENG & DIR MFG ENG, UNIV MA, AMHERST, 88- *Mem:* Am Soc Mech Engrs; Sigma Xi. *Res:* Materials processing; metal cutting; friction and wear of materials; mechanical behavior of materials. *Mailing Add:* Mech Eng Dept Eng Lab Bldg Univ Ma Amherst MA 01003

MALKINSON, ALVIN MAYNARD, b Buffalo, NY, Jan 5, 41; m 67; c 2. BIOCHEMISTRY, PHARMACOLOGY. *Educ:* Univ Buffalo, BA, 62; Johns Hopkins Univ, PhD(biol), 68. *Prof Exp:* Lectr biochem, Univ Nairobi, 69-71; fel, Leicester Univ, 71-72; fel pharmacol, Yale Med Sch, 72-74; asst prof psychiat, Sch Med, Univ Minn, 74-78; from asst prof to assoc prof, 78-90, PROF PHARM, UNIV COLO, 90- *Concurrent Pos:* NIH fel, 72-74; asst prof, Nat Found Basil O'Connor starter grant, 75-78, Nat Inst Gen Med Sci grant, 76-78, Am Cancer Soc Instnl grant, 78, Colo Heart Asn Grant, 79, Nat Inst Environ Health grant, 81-94, Nat Cancer Inst, 82-91, Am Cancer Soc, 87, Nat Inst Heart Lung Blood, 86-89. *Mem:* Am Asn Cancer Res; Am Soc Pharmacol & Exp Therapeut; Environ Mutagen Soc; Soc Toxicol. *Res:* Role of protein phosphorylation in the control of normal and neoplastic lung growth; pulmonary effects of butylated hydroxytoluene; genetics of neoplasia. *Mailing Add:* Dept Pharm EKLP Univ Colo W122 Boulder CO 80309

MALKINSON, FREDERICK DAVID, b Hartford, Conn, Feb 26, 24; m 79; c 3. MEDICINE. *Educ:* Harvard Univ, DMD, 47, MD, 49. *Prof Exp:* From instr to assoc prof dermat, Sch Med, Univ Chicago, 54-68; PROF & CHMN, DEPT DERMAT, RUSH-PRESBY-ST LUKE'S MED CTR, 68-, CLARK W FINNERUD PROF DERMAT, RUSH MED SCH, 81- *Concurrent Pos:* Mem bd dirs, Soc Invest Dermat, 63-68, Am Acad Dermat, 64-67 & 87-89, Dermat Found, 79-; assoc prof oral med, Zoller Dent Clin, Univ Chicago, 67-68, res assoc, Sect Dermat, Dept Med, 68-70; prof med & dermat, Univ Ill, 68-70; ed, Yearbk of Dermat, 71-78; chief ed, AMA Archives Dermat, 79-83; mem, Comt Cutaneous Radiobiol, Nat Coun on Radiation Protection and Measurement, 84- *Mem:* AAAS; Soc Invest Dermat (vpres, 79); Am Acad Dermat (vpres, 88-89); Am Dermat Asn; Radiation Res Soc; Am Fedn Clin Res; Dermat Found (pres, 83-85). *Res:* Percutaneous absorption; adrenal steroid effects on skin; radiation effects on skin. *Mailing Add:* Rush-Presby-St Luke's Med Ctr 1653 W Congress Pkwy Chicago IL 60612

MALKUS, DAVID STARR, b Chicago, Ill, June 30, 45; m 77; c 4. COMPUTATIONAL MECHANICS, FINITE ELEMENTS. *Educ:* Yale Univ, BA, 68; Boston Univ, MA, 75, PhD(math), 76. *Prof Exp:* Nat Res Coun fel & mathematician, US Nat Bur Standards, 75-77; from asst prof to assoc prof math, Ill Inst Technol, 77-84, fac res fel, 78-79; assoc prof, 84-86, PROF ENG MECH & MATH SCI, UNIV WIS, 86- *Concurrent Pos:* chair prof, Nanjing Aero-Inst, Peoples Repub China, 85. *Mem:* Soc Indust & Appl Math; Soc Rheology; Am Acad Mech. *Res:* Finite element analysis; computational methods in fluid dynamics; computational methods in elasticity and structural dynamics; numerical methods for non-Newtonian flows. *Mailing Add:* Eng Mech & Ctr Math Sci Univ Wis Madison WI 53706

MALKUS, WILLEM VAN RENSSELAER, b Brooklyn, NY, Nov 19, 23; m 48; c 4. PHYSICS. *Educ:* Univ Chicago, PhD(physics), 50. *Prof Exp:* Asst prof natural sci, Univ Chicago, 50-51; phys oceanogr, Woods Hole Oceanog Inst, 51-60; prof geophys, Univ Calif, Los Angeles, 60-67, prof geophys & math, 67-69; PROF APPL MATH, MASS INST TECHNOL, 69- *Concurrent Pos:* Prof appl math, Mass Inst Technol, 59-60; Guggenheim fel, Cambridge Univ & Univ Stockholm, 71-72. *Mem:* Nat Acad Sci; Am Phys Soc; fel Am Acad Arts & Sci. *Res:* Fluid dynamics. *Mailing Add:* Dept Math Mass Inst Technol Cambridge MA 02139

MALL, SHANKAR, b Varanasi, India, June 10, 43; m 65; c 1. FRACTURE MECHANICS, STRESS ANALYSIS. *Educ:* Banaras Hindu Univ, India, BS, 64, MS, 66; Univ Wash, PhD(mech eng), 77. *Prof Exp:* Lectr mech eng, Banaras Hindu Univ, India, 67-74; res asst, Univ Wash, 74-77, res assoc, 77-78; from asst prof to assoc prof mech eng, Univ Maine, Orono, 78-83; Dept Eng Mech, Univ Mo, Rolla, 83-86; HEAD & PROF, AIR FORCE INST TECHNOL, WRIGHT-PATTERSON AIR FORCE BASE, OHIO, 86- *Mem:* Am Soc Mech Engrs; Am Soc Aeronaut & Astronaut; Sigma Xi. *Res:* Finite element method; fracture and fatigue; composite materials; fatigue; author of 80 technical papers. *Mailing Add:* Aeronaut & Astronaut Air Force Inst Technol Wright-Patterson AFB OH 45433

MALLAMS, ALAN KEITH, b Johannesburg, SAfrica, June 11, 40. NATURAL PRODUCTS CHEMISTRY. *Educ:* Univ Witwatersrand, BSc, 62, PhD(chem), 64; Univ London, PhD(chem), 67. *Prof Exp:* Res off org chem, African Exlosives & Chem Industs, SAfrica, 64; Exhibit 1851 fel, Queen Mary Col, Univ London, 64-66, res asst, 66-67; sr scientist, Med Res Div, 67-70, prin scientist, 70-72, SECT HEAD, ANTIBIOTICS & ANTIINFECTIVES CHEM RES, SCHERING CORP, 72- *Mem:* Am Chem Soc; Royal Soc Chem. *Res:* Synthesis and structural elucidation of carotenoids, antibiotics, carbohydrates and natural products of medicinal interest; antibiotics research. *Mailing Add:* Res Div Schering Corp 60 Orange St Bloomfield NJ 07003

MALLAMS, JOHN THOMAS, b Ashland, Pa, Aug 29, 23; m 45; c 3. MEDICINE, RADIOLOGY. *Educ:* Temple Univ, MD, 46; Am Bd Radiol, dipl. *Prof Exp:* Intern, US Naval Hosp, Philadelphia, 46-47, resident radiol, 47-48, radiologist, Parris Island, SC, 48-49, chief radiol, Beaufort, 49-50; asst prof radiol, Baylor Col Med, 52-54, mem attend staff & dir irradiation ther & tumor clins & dir, Sammons Res Div, Med Ctr, 54-68, prof radiother, Col Dent, 66-68; prof clin radiol, Med Sch, Yale Univ, 68-70; assoc dean patient serv, 71-72, PROF RADIOL & CHMN DEPT, MARTLAND HOSP UNIT, COL MED & DENT NJ, 70- *Concurrent Pos:* Fel, Robert Packer Hosp & Guthrie Clin, Sayre, Pa, 50-51; Am Cancer Soc fel clin radiation ther, Frances Delafield Hosp, New York, 51-52; assoc radiologist, Jefferson Davis Hosp, Houston & attend radiologist, Vet Admin Hosp, 52-54; prof, Univ Tex Southwest Med Sch, 66-68; vis prof, Med Sch, Yale Univ, 67-68; Am Col Radiol consult, State of NJ, 70; consult, Conn Hosp Planning Comn, Branford, 71. *Mem:* Fel Am Col Radiol; AMA; Am Radium Soc; Am Roentgen Ray Soc; Radiol Soc NAm; Sigma Xi. *Res:* Radiotherapy. *Mailing Add:* 1860 N Atlantic Ave 503B c/o RYL TW Cocoa Beach FL 32931-2960

MALLER, JAMES LEIGHTON, PROTEIN PHOSPHOLATION, MITOSIS. *Educ:* Univ Calif, Berkeley, PhD(molecular biol), 74. *Prof Exp:* ASSOC PROF CELL REGULATION, CANCER CHEMOTHERAPY & ANESTHETICS, UNIV COLO, DENVER, 78- *Res:* Cancer. *Mailing Add:* Dept Pharmacol Univ Colo Med Ctr 4200 E Ninth Ave Denver CO 80220

MALLER, OWEN, psychophysiology, health psychology, for more information see previous edition

MALLERY, CHARLES HENRY, b Southampton, NY, June 3, 43; m 66; c 2. DEVELOPMENTAL PHYSIOLOGY, BIOCHEMISTRY. *Educ:* Univ Ga, BS, 65, PhD(bot, biochem), 70. *Prof Exp:* Res fel, Lab Quant Biol, 70-72, asst prof biol, 72-78, assoc chmn, Biol Dept, 80-83, ASSOC PROF BIOL, UNIV MIAMI, 78-, ASSOC DEAN, COL ARTS & SCI, 85- *Concurrent Pos:* NSF grants, 74, 77-79 & 81-83; consult, Dade County Dept Educ & State Atty Off, 74- *Mem:* Am Soc Plant Physiologists; Am Soc Zoologists; Bot Soc Am. *Res:* Biochemical development in the establishment of a mature eucaryotic plant cell; ion regulating mechanisms in plants and animals; sodium-potassium-AtPase and bicarbonate AtPases in teleosts. *Mailing Add:* Dept of Biol Univ Miami Coral Gables FL 33124

MALLET, VICTORIN NOEL, b Shippegan, NB, Dec 12, 44; m 66; c 3. ANALYTICAL CHEMISTRY, PESTICIDE CHEMISTRY. *Educ:* Univ Moncton, BSc, 66, MSc, 70; Dalhousie Univ, PhD(anal chem), 71. *Prof Exp:* Lab supvr chem, Can Celanese Co, 66-67; chmn, chem dept, 72-78, PROF CHEM, UNIV MONCTON, 71- *Honors & Awards:* Caledon Award, 84. *Mem:* Chem Inst Can; Asn Can Fr Advan Sci. *Res:* Analytical chemistry of pesticides and derivatives thereof; gas chromatography; thin-layer chromatography; high performance liquid chromatography; fluorescence spectroscopy. *Mailing Add:* Universitair de Moncton Dept of Chem/ Biochem, 98 Chapman Moncton NB E1A 3E9 Can

MALLETT, GORDON EDWARD, b Lafayette, Ind, Nov 30, 27; m 50; c 4. MICROBIOLOGY. *Educ:* Purdue Univ, BS, 49, MS, 52, PhD(bact), 56. *Prof Exp:* Bacteriologist, US Army Res & Develop Labs, Md, 56-57; sr microbiologist, Eli Lilly & Co, 57-69, head fermentation prod res dept, 66-69, dir res, Lilly Res Ctr, Ltd, Eng, 69-75, DIR CORP QUAL ASSURANCE, ELI LILLY & CO, IND, 75- *Mem:* Sigma Xi; Am Chem Soc. *Res:* Bacterial cell structure; microbiological conversions of steroids, alkaloids, antibiotics; microbiological conversion mechanisms. *Mailing Add:* 60 Raintree Dr Zionsville IN 46077-2008

MALLETT, RUSSELL LLOYD, b Seattle, Wash, Nov 2, 35; m 61; c 2. APPLIED MECHANICS, DYNAMICS. *Educ:* Mass Inst Technol, BS, 58, PhD(appl math), 70; Stanford Univ, MS, 66. *Prof Exp:* Res engr, Boeing Co, 58-65; asst prof, 70-75, sr res assoc appl mech, Stanford Univ, 75-81; assoc prof, Rensselaer Polytechnic Inst, 81-85; SCI ASSOC, ALCOA LAB, 85. *Mem:* Am Acad Mech; Am Soc Mech Engrs; Soc Indust & Appl Math; Sigma Xi. *Res:* Rigid body, structural and continuum dynamics, shell theory and nonlinear continuum mechanics; efficient modeling, computational techniques and finite element analysis; cable dynamics and metal forming analysis. *Mailing Add:* ATC-D-AMCT Alcoa Ctr Apollo PA 15069

MALLETT, WILLIAM ROBERT, b Painesville, Ohio, Sept 12, 32; m 57; c 2. PETROLEUM CHEMISTRY, ANALYTICAL CHEMISTRY. *Educ:* Miami Univ, BA, 61, MS, 63; Rensselaer Polytech Inst, PhD(energy transfer), 66. *Prof Exp:* Res chemist, 66-68, sr res chemist, 68-81, res assoc, 81-84, supvr prod eval, 84, supvr fuels res, 84-90, STAFF CONSULT, UNOCAL CORP, 90- *Concurrent Pos:* Vis res chemist, Maruzen Oil Co, Japan, 73-75. *Mem:* Am Chem Soc; Soc Automotive Engrs; Am Soc Testing & Mat. *Res:* Fuels research. *Mailing Add:* 1273 Genoa Pl Placentia CA 92670

MALLETTE, JOHN M, b Houston, Tex, Aug 6, 32; m 59; c 3. ENDOCRINOLOGY, EXPERIMENTAL EMBRYOLOGY. *Educ:* Xavier Univ, BS, 54; Tex Southern Univ, MS, 58; Pa State Univ, PhD(zool), 62. *Prof Exp:* Res technician anat, Dent Br, Univ Tex, 57-58; instr biol, Tex Southern Univ, 58-59; asst zool, Pa State Univ, 59-62; assoc prof biol, Tenn State Univ, 62-64, prof biol sci & chmn grad curric, 64-74; vchancellor acad affairs, Univ Tenn, Nashville, 74-79; ASSOC VPRES RES & DEVELOP, TENN STATE UNIV, NASHVILLE, 79- *Concurrent Pos:* Dir, Undergrad Res Participation Prog, NSF, 64-66; grants assoc, NIH, 67; dir allied health, Meharry Med Col & Tenn State Univ. *Mem:* Fel AAAS; Am Soc Zool; Nat Inst Sci; fel NY Acad Sci. *Res:* Growth of trypsin; dissociated glands in vitro and vivo; teratogenic effects of drugs in avian embryos. *Mailing Add:* Dept Biol Sci Tenn State Univ 3500 Centennial Blvd Nashville TN 37203

MALLEY, ARTHUR, b Chicago, Ill, Jan 7, 31; m 61. IMMUNOLOGY. *Educ:* San Francisco State Col, BA, 53, BS, 57; Ore State Univ, PhD(biochem), 61. *Prof Exp:* Fel immunochem, Calif Inst Technol, 61-63; from asst prof to assoc prof, 63-71, PROF BACT, ORE REGIONAL PRIMATE RES CTR, 71- *Mem:* AAAS; Am Chem Soc; Transplantation Soc; Am Asn Immunologists. *Res:* Isolation of antigens and antibodies involved in various allergic diseases, and the regulation of antibody formation. *Mailing Add:* Ore Regional Primate Res Ctr 505 NW 185th Ave Beaverton OR 97005

MALLEY, LINDA ANGEVINE, b Kensington, MD, May 13, 55; m 82; c 1. PULMONARY TOXICOLOGY, INHALATION TOXICOLOGY. *Educ:* Col William & Mary, BS, 77; Univ Miss Med Ctr, PhD(toxicol), 81. *Prof Exp:* Res toxicologist, Shell Develop Co, 81-86; SR RES TOXICOLOGIST, E I DU PONT DE NEMOURS & CO, 86- *Mem:* Sigma Xi. *Res:* Stimulating property of pyrethroid insecticides on the sensory nerve fibers in the skin; toxicity evaluation of various agricultural chemicals and alternative chlorofluoro carbons. *Mailing Add:* 17 Boyds Valley Rd Newark DE 19711

MALLI, GULZARI LAL, b Lehlian, India, Feb 12, 38; Can citizen; m 74; c 2. QUANTUM CHEMISTRY, CHEMICAL PHYSICS. *Educ:* Univ Delhi, BSc, 58; McMaster Univ, MSc, 60; Univ Chicago, MS, 63, PhD(chem physics), 64. *Prof Exp:* Mem res staff physics, Yale Univ, 64-65; asst prof theoret chem, Univ Alta, 65-66; asst prof, 66-69, assoc prof, 69-75, PROF CHEM, SIMON FRASER UNIV, 75- *Concurrent Pos:* Vis fel, Mellon Inst, 67-68; vis assoc prof, Univ Houston, 72; Alexander von Humboldt fel, 78, 82; vis prof, IBM Corp, Kingston, NY, 89-90. *Mem:* Sigma Xi; Am Phys Soc. *Res:* Quantum mechanics of atoms and molecules; relativistic quantum chemistry; relativistic many-electron atomic and molecular self-consistent field theory; electron correlation in many-electron systems. *Mailing Add:* Dept of Chem Simon Fraser Univ Burnaby BC V5A 1S6 Can

MALLIA, ANANTHA KRISHNA, b Kerala, India, May 22, 41; m 71; c 1. PROTEIN CHEMISTRY, AFFINITY CHROMATOGRAPHY. *Educ:* Kerala Univ, India, BSc, 62; Banaras Hindu Univ, MSc, 64; Indian Inst Sci, PhD(biochem), 70. *Prof Exp:* Res assoc biochem, Columbia Univ, 71-74, neurochem, Univ Mich, Ann Arbor, 74-77; MGR RES BIOCHEM, PIERCE CHEM CO, ROCKFORD, 77- *Mem:* Am Chem Soc; AAAS. *Res:* Isolation, purification and characterization of enzymes and proteins; application of affinity chromatography for the purification of enzymes, lectins and other proteins; use of affinity chromatography to develop new procedures for clinical tests. *Mailing Add:* Pierce Chem Co PO Box 117 Rockford IL 61105-0117

MALLIK, ARUP KUMAR, b Calcutta, India, Nov 14, 43; US citizen; m 70; c 1. INDUSTRIAL ENGINEERING, OPERATIONS RESEARCH. *Educ:* Jadav Pur Univ, BS, 64; NC State Univ, Raleigh, MS, 70, PhD(indust eng), 72. *Prof Exp:* Asst prof indust eng, WVa Univ, 71-77, assoc prof, 77-; CHMN, DEPT INDUST ENG, NC AGR & TECH STATE UNIV, GREENSBORO. *Mem:* Am Soc Indust Engrs; Oper Res Soc Am; Am Soc Eng Educ; Inst Mgt; Sigma Xi. *Res:* Economic analysis and production scheduling. *Mailing Add:* Dept Indust Eng N Car A&T St Univ 1601 E Market St Greenboro NC 27411

MALLIN, MORTON LEWIS, b Feb 13, 26; m 60; c 2. MICROBIOLOGY, BIOCHEMISTRY. *Educ:* Phila Col Pharm, BS, 50; Hahnemann Med Col, MS, 52; Cornell Univ, PhD(bact), 56. *Prof Exp:* Res fel biochem, McCollum-Pratt Inst, Johns Hopkins Univ, 56-57; NIH res fel, Brandeis Univ, 57-59; res assoc, May Inst Med Res, Jewish Hosp, Cincinnati, 59-64; assoc prof, 64-71, chmn, 66-74, PROF MICROBIOL, OHIO NORTHERN UNIV, 71- *Mem:* Am Soc Microbiol; Sigma Xi. *Res:* Biochemical problems related to hypertension; oxidative phosphorylation in animal mitochondria and bacterial particles; phosphorous metabolism in anaerobic bacteria. *Mailing Add:* 2904 Hanover Dr Lima OH 45805

MALLING, GERALD F, b Evanston, Ill, Apr 27, 38; m 65; c 4. CHEMICAL ENGINEERING, APPLIED MATHEMATICS. *Educ:* Northwestern Univ, BS, 61, PhD(chem eng), 66; Univ Ill, Urbana, MS, 63. *Prof Exp:* SR ANALYST, OAK RIDGE GASEOUS DIFFUSION PLANT, ENERGY SYSTS DIV, MARTIN MARIETTA CORP, 66- *Mem:* Am Phys Soc; Sigma Xi. *Res:* Rarefied and continuum gas flow and diffusion phenomena. *Mailing Add:* 902 W Outer Dr Oak Ridge TN 37830

MALLING, HEINRICH VALDEMAR, b Copenhagen, Denmark, Apr 21, 31; m 55, 68; c 7. GENETICS, MICROBIOLOGY. *Educ:* Univ Copenhagen, BSc, 51, MSc, 53, PhD(genetics), 57. *Prof Exp:* Lectr genetics, Univ Copenhagen, 57; res staff mem, Leo Pharmaceut Prod, Denmark, 53-58; fel genetics, Univ Copenhagen, 58-61, lectr, 61-63; mem res staff, Biol Div, Oak Ridge Nat Lab, 63-72; MEM LAB GENETICS, NAT INST ENVIRON HEALTH SCI, 72- *Honors & Awards:* Environ Mutagen Soc Award, 80. *Mem:* Genetics Soc Am; Environ Mutagen Soc. *Res:* Botany; cytology; mutation induction; mutagenesis in mammals; induction of cancer. *Mailing Add:* Nat Inst Environ Health Sci PO Box 12233 Research Triangle Park NC 27709

MALLINSON, GEORGE GREISEN, b Troy, NY, July 4, 17; m 43, 54; c 4. SCIENCE ADMINISTRATION, SCIENCE EDUCATION. *Educ:* NY State Col Teachers Albany, BA, 38, MA, 41; Univ Mich, PhD(sci educ, statist), 47. *Prof Exp:* Teacher high schs, NY, 38-42; dir sci educ, Iowa State Teachers Col, 47-48; from assoc prof to prof exp psychol & statist, 48-53, prof sci educ & res methodology, 54-77, dean grad studies, 56-77, distinguished prof, 77-88, DISTINGUISHED EMER PROF SCI EDUC, WESTERN MICH UNIV, 88- *Concurrent Pos:* Burke Aaron Hinsdale scholar, Univ Mich, 47-48; actg dir, Grad Div, Western Mich Univ, 54-55, dir grad studies, 55-56; dir, NSF Summer Insts, In-serv Insts, Sec Sci Training Progs, 57-74; Nat Defense Educ Act grant, 63-66; ed, Sch Sci & Math, 57-81; mem, Coun & Coop Comt, Teaching Sci & Math, AAAS, 54-73; chmn coun, Cent States Univs, Inc, 65-66, pres, Bd Dirs, 70-73 & 83-84. *Honors & Awards:* C Warren Bledsoe Award, Am Asn Workers Blind, 82. *Mem:* AAAS; Nat Asn Res Sci Teaching (pres, 53-54); Asn Educ & Rehab Blind & Visually Impaired. *Res:* Scientific manpower; factors related to achievement in science; basic hydrothermal methods for amelioration of taconite ore; environmental sciences. *Mailing Add:* Sangren Hall Rm 3409 Western Mich Univ Kalamazoo MI 49008-3899

MALLIS, ARNOLD, entomology; deceased, see previous edition for last biography

MALLISON, GEORGE FRANKLIN, b Suffolk Co, NY, May 31, 28; m 53, 69; c 3. ENVIRONMENTAL HEALTH, PUBLIC HEALTH. *Educ:* Cornell Univ, BCE, 51; Univ Calif, MPH, 57. *Prof Exp:* Asst sanit engr, Cornell Univ, 51; from jr asst to asst sanit engr & asst to chief tech develop labs, Tech Br, USPHS, 51-53, sr asst sanit engr, Phoenix Field Sta, 54-57, sanit engr & asst to chief, Tech Br, 57-59, asst chief & sr sanit engr, 59-63, chief microbiol control sect, Ctr Dis Control, 63-74, sanit engr dir, 66-81, asst dir bact div, 74-81; CONSULT ENVIRON & INFECTION CONTROL, 81- *Concurrent Pos:* Mem, Conf Fed Environ Engrs; consult environ epidemiol, US Environ Protection Agency, 71-79,. *Mem:* Am Hosp Asn; AAAS; Am Pub Health Asn; Am Indust Hyg Asn; Asn Practr Infection Control; Nat Environ Health Asn. *Res:* Control of nosocomial infections; hospital and environmental sanitation; water supply; epidemiologic aspects and control of environmental contamination; sanitary engineering; microbiology; solid and other waste disposal; vector control; control of legionellosis; published over seventy-five scientific and technical papers; investigation of disease outbreaks, including twenty-three cases of legionellosis. *Mailing Add:* 88 Kenmore Place Glen Rock NJ 07452

MALLMANN, ALEXANDER JAMES, b Sheboygan, Wis, Dec 12, 37; m 68; c 2. SOLAR ENERGY TECHNOLOGY. *Educ:* Univ Wis-Milwaukee, BS, 65, MS, 68; Marquette Univ, PhD(mat sci), 77. *Prof Exp:* PROF PHYSICS & R D PETERS PROF MAT SCI, MILWAUKEE SCH ENG, 68- *Concurrent Pos:* Mem tech prog comt meteorol optics, Optical Soc Am, 82-83; mem apparatus comt, Am Asn Physics Teachers, 90- *Honors & Awards:* Outstanding Teaching Award, Inland Steel-Ryerson Found, 83. *Mem:* Am Asn Physics Teachers; Optical Soc Am; Int Solar Energy Soc. *Res:* Meteorological optics; optical and thermodynamic problems in solar energy technology; coauthor introductory college physics textbooks. *Mailing Add:* Dept Physics Milwaukee Sch Eng PO Box 644 Milwaukee WI 53201-0644

MALLONEE, JAMES EDGAR, b Frederick, Md, June 20, 15; m 43; c 2. INDUSTRIAL ORGANIC CHEMISTRY. *Educ:* Col William & Mary, BS, 35; Univ Va, PhD(org chem), 40. *Prof Exp:* Control chemist, Solvay Process Co, Va, 36-37; res chemist, Jackson Lab, E I du Pont de Nemours & Co, 40-43, chemist, Louisville Works, 43-47, res chemist, Fine Chem Div, 47-52, sr res chemist, Chambers Works, 52-79; RETIRED. *Mem:* Am Chem Soc. *Res:* Morphine chemistry; neoprene intermediates; industrial organic chemical research and development. *Mailing Add:* Coffee Run Condo No E3E Hockessin DE 19707-1618

MALLORY, CHARLES WILLIAM, b Brewster, Kans, Sept 17, 25; m 50; c 1. NUCLEAR & CIVIL ENGINEERING. *Educ:* Univ Colo, BSME, 46; Rensselaer Polytech Inst, BCE, 50. *Prof Exp:* Proj officer, Bur Yards & Docks Proving Ground, US Navy, Calif, 46-47, staff construct officer, Opers Highjump & Windmill, Antarctica, 46-47, 47-48, div dir automatic tel commun syst, Island Pub Works Dept, Guam, 49, supt pub works dept, Portsmouth Naval Shipyard, NH, 50-52; spec asst, Schenectady Opers Off, Atomic Energy Comn, 52-54, proj officer, 54-58, chief water syst proj br, Div Reactor Develop, 58-62; dir nuclear power div, Bur Yards & Docks, US Navy, DC, 62-64; mgr eng dept, Hittman Assoc Inc, 65-71, vpres & dir environ sci, 71-77, vpres eng, 77-83, vpres mkt develop & vpres eng, Westinghouse-Hittman, 83-87, sr tech adv, Westinghouse Environ Tech, 87-88, regional mgr, 88-89; PVT ENG & MGT CONSULT, 89- *Mem:* Am Soc Mech Engrs; Am Nuclear Soc; Water Pollution Control Fedn. *Res:* Urban hydrology; wastewater treatment; energy utilization; power systems; water resources, reuse and desalination; nuclear and hazardous waste disposal; eleven US patents. *Mailing Add:* 536 Heavitree Hill Severna Park MD 21146

MALLORY, CLELIA WOOD, b Brooklyn, NY, Feb 9, 38; m 65. ORGANIC CHEMISTRY. *Educ:* Bryn Mawr Col, BA, 59, MA, 60, PhD(chem), 63. *Prof Exp:* Res assoc chem, Bryn Mawr Col, 63-77; lectr chem, Yale Univ, 77-80; lectr, 80-84, SR LECTR CHEM, UNIV PA, 84- *Concurrent Pos:* Vis fel, Bryn Mawr Col, 86- *Mem:* Am Chem Soc; Int-Am Photochem Soc; Int Union Pure & Appl Chem. *Res:* Photochemistry; nuclear magnetic resonance spectroscopy. *Mailing Add:* Dept Chem Univ Penn Philadelphia PA 19104-6323

MALLORY, FRANK BRYANT, b Omaha, Nebr, Mar 17, 33; m 51, 65; c 3. ORGANIC CHEMISTRY. *Educ:* Yale Univ, BS, 54; Calif Inst Technol, PhD(chem), 58. *Prof Exp:* From asst prof to assoc prof, 57-69, acad dep to pres, 78-81, PROF CHEM, BRYN MAWR COL, 69-, CHMN DEPT, 82- *Concurrent Pos:* Guggenheim fel, 63-64; Sloan res fel, 64-68; NSF sr fel, 70-71; vis assoc, Calif Inst Technol, 63-64; vis prof, State Univ NY, Albany, 67 & Yale Univ, 68; vis fel, Cornell Univ, 70-71; vis prof, Yale Univ, 77-79; mem exec comt, Organic Div, Am Chem Soc, 85-88; mem, Ed Adv Bd, J Organic Chem, 88-; mem, Sci & Arts Comt, Franklin Inst, 86-; Franklin Inst, 86-, vis scholar, Univ Pa, 88-89. *Honors & Awards:* Bond Award, Am Oil Chemists Asn. *Mem:* Am Chem Soc; affil Int Union Pure & Appl Chem; Int Am Photochemical Soc; Am Chem Soc. *Res:* Preparative and mechanistic aspects of organic photochemistry; nuclear magnetic resonance spectroscopy. *Mailing Add:* Dept of Chem Bryn Mawr Col Bryn Mawr PA 19010

MALLORY, FRANK FENSON, b Hamilton, Ont, Sept 25, 44; m 69; c 4. ELECTROPHORESIS, KARYOTYPING. *Educ:* Laurentian Univ, BSc, 69, MSc, 72; Univ Guelph, PhD(behav), 79. *Prof Exp:* Teacher sci, sec sch syst, Sudbury, 71-72; from instr to asst prof biol, Wilfred Laurier Univ, 72-83; ASST PROF BIOL, LAURENTIAN UNIV, 83- *Concurrent Pos:* Adj prof biol, Univ Waterloo, 80-83. *Mem:* Can Soc Zoologists; Am Soc Mammalogists; Soc Obstetricians & Gynecologists Can. *Res:* Meadow voles, varying lemmings, arctic fox, wolves and beluga whales; population dynamics of predator/prey systems using a holistic approach correlating genetics, physiology, anatomy and behavior; laboratory and field research. *Mailing Add:* Dept Biol Laurentian Univ Sudbury ON P3E 2C6 Can

MALLORY, HERBERT DEAN, physical chemistry; deceased, see previous edition for last biography

MALLORY, MERRIT LEE, b Lansing, Mich, May 21, 38; m 63; c 3. ACCELERATOR PHYSICS. *Educ:* Mich State Univ, BS, 60, MS, 61, PhD(physics), 66. *Prof Exp:* Fel accelerators, Oak Ridge Nat Lab, 66-68, staff physicist, 68-76; physicist accelerators, Mich State Univ, 76-90; PROF RADIATION PHYSICS, UNIV TEX, M D ANDERSON CANCER CTR, 90- *Mem:* Am Phys Soc; AAAS. *Res:* Cyclotron design and operation with main interest in ion sources; cryogenics and vacuums; development of radioactive ion beams. *Mailing Add:* 5038 Wigton Dr Houston TX 77096

MALLORY, THOMAS E, b Alhambra, Calif, July 2, 40. PLANT MORPHOLOGY, PLANT CYTOLOGY. *Educ:* Univ Redlands, BS, 62; Univ Calif, Davis, MS, 65, PhD, 68. *Prof Exp:* Asst prof, 68-72, assoc prof, 72-76, PROF BIOL, CALIF STATE UNIV, FRESNO, 76- *Mem:* Bot Soc Am; Sigma Xi. *Res:* Plant morphogenesis; morphogenesis of lateral roots; mechanisms of action of herbicides. *Mailing Add:* Fresno State Univ Fresno CA 93726

MALLORY, VIRGIL STANDISH, b Englewood, NJ, July 14, 19; m 46; c 4. GEOLOGY, MUSEUM CONSULTANT. *Educ:* Oberlin Col, AB, 46; Univ Calif, MA, 48, PhD(paleont), 52. *Prof Exp:* Lectr paleont, Univ Calif, 50-51, consult, Mus Paleont, 52; from asst prof to prof, 53-84, EMER PROF GEOL, UNIV WASH, 84-; cur Invert Paleont, 62-90, EMER CUR GEOL, THOMAS BURKE WASH STATE MUS, 90- *Concurrent Pos:* Consult, 54-; mem, Gov Comn Petrol Laws, Wash, 56-57; chmn, Geol & Paleont Div, Thomas Burke Wash State Mus, 61-84; geologist, US Geol Surv, 63-; ed invert paleont, Quaternary Res, 70-77; mem, Abstr Rev Comt, Geol Soc Am, 75-76 & Minerals Mus Adv Comt, 75-85. *Honors & Awards:* Agnes Anderson Award, 63. *Mem:* Fel AAAS; fel Geol Soc Am; Int Paleont Union; Soc Econ Paleont & Mineral; Am Asn Petrol Geol; Paleont Res Inst; Sigma Xi; Am Asn Mus; Paleontological Soc (secy, 56-58). *Res:* Biostratigraphy of west coast ranges, especially Lower Tertiary; west coast Lower Tertiary foraminifera; Lower Tertiary and Cretaceous molluscan paleontology; Pacific coast structural geology; wine consultant and lecturer to societies and restaurants. *Mailing Add:* Paleont Div Burke Wash State Mus Univ Wash Seattle WA 98195

MALLORY, WILLAM R, b Dudley, Mo; c 4. THEORETICAL PHYSICS, OPTICS. *Educ:* Univ Ill, Urbana-Champaign, BS, 59; Syracuse Univ, PhD(physics), 70. *Prof Exp:* Physicist, Gen Elec Co, 59-67; asst prof physics & astron, Univ Mont, 70-75; SR SCIENTIST, SYSTS RES LABS, INC, 75- *Mem:* Optical Soc Am; Soc Photo Instrumentation Engrs. *Res:* Theoretical optics; quantum and statistical optics; design of optical instrumentation. *Mailing Add:* Systs Res Labs Inc 2800 Indian Ripple Rd Dayton OH 45440

MALLORY, WILLIAM WYMAN, b New Rochelle, NY, Apr 19, 17; m 42; c 2. GEOLOGY. *Educ:* Columbia Univ, BA, 39, MA, 46, PhD(geol), 48. *Prof Exp:* Asst geol, Columbia Univ, 40-41; asst geologist, Phillips Petrol Co, Okla, 43-45, explor geologist, 46-55, supvr explor projs, 55-57; petrol consult, 58; mem, Paleotectonic Map Proj, US Geol Surv, 59-75, tech supv officer, Oil & Gas Br, 75-79; pres, Digital Data For Oil Independents, 79-81; VOL GEOLOGIST, US GEOL SURV, DENVER, COLO, 89- *Concurrent Pos:* Ed-in-chief, Geol Atlas Rocky Mountains Region, Rocky Mountain Asn Geologist, 72; consult geologist, petrol data syst, 81-83. *Mem:* Fel Geol Soc Am; Am Asn Petrol Geologists; Sigma Xi. *Res:* Stratigraphy and tectonics of Colorado and western United States; continental framework and petroleum exploration in western United States; petroleum data system (computerized oil and gas field records). *Mailing Add:* 1130 W 25th Pl Denver CO 80215

MALLOV, SAMUEL, b New York, NY, Apr 19, 19; m 43; c 2. PHARMACOLOGY. *Educ:* City Col New York, BS, 39; NY Univ, MS, 41; Syracuse Univ, PhD(biochem), 52. *Prof Exp:* Chemist, Labs, Westinghouse Elec Co, 43-44; org chemist, US Bur Mines, 44-45; res chemist, Coal Lab, Carnegie Inst Technol, 45-48; asst, 48-52, from instr to assoc prof, 53-70, PROF PHARMACOL, COL MED, STATE UNIV NY UPSTATE MED CTR, 71- *Concurrent Pos:* Am Heart Asn fel, State Univ NY Upstate Med Ctr, 52-53. *Mem:* AAAS; Am Soc Pharmacol & Exp Therapeut; Sigma Xi; Res Soc Alcoholism; Int Soc Heart Res. *Res:* Cardiovascular, alcohol. *Mailing Add:* Dept Pharmacol State Univ NY Upstate Med Ctr 766 Irving Ave Syracuse NY 13210

MALLOW, JEFFRY VICTOR, b New York, NY, June 28, 38; m 70. ATOMIC PHYSICS, SCIENCE ANXIETY. *Educ:* Columbia Univ, AB, 64; Northwestern Univ, MS, 66, PhD(physics astron), 70. *Prof Exp:* Fel physics, Hebrew Univ, Jerusalem, 70-71; res assoc, Northwestern Univ, 71-74; asst prof physics, Oakland Univ, 74-76; asst prof, 76-78, ASSOC PROF PHYSICS, LOYOLA UNIV CHICAGO, 78- *Concurrent Pos:* Res consult, Northwestern Univ, 74- *Mem:* AAAS; Am Phys Soc; Sigma Xi. *Res:* Atomic Hartee-Fock and Dirac-Fock theory; laser radar investigation of atmospheric constituents; theory of muonic atoms. *Mailing Add:* Dept Physics Loyola Univ Chicago 6525 N Sheridan Rd Chicago IL 60626

MALLOWS, COLIN LINGWOOD, b Great Sampford, Eng, Sept 10, 30; US citizen; m 56; c 3. DATA ANALYSIS. *Educ:* Univ London, BSc, 51, PhD(statist), 53. *Prof Exp:* Asst lectr statist, Univ Col London, 55-57, lectr, 58-60; res assoc statist tech res group, Princeton Univ, 57-58; mem tech staff, 60-69, dept head, 69-87, DISTINGUISHED MEM TECH STAFF, AT&T BELL LABS, 87- *Concurrent Pos:* Adj assoc prof, Columbia Univ, 60-64. *Mem:* Fel Inst Math Statist; fel Am Statist Asn; Math Asn Am; Int Statist Inst; Royal Statist Soc. *Res:* Data analysis, especially informal and graphical methods; algebraic coding theory. *Mailing Add:* AT&T Bell Labs Murray Hill NJ 07974

MALLOY, ALFRED MARCUS, b Pittsburgh, Pa, Oct 6, 03; div. PHYSICAL CHEMISTRY, ELECTROCHEMISTRY. *Educ:* Carnegie Inst Technol, BS, 25; Mich State Univ, MS, 27. *Prof Exp:* Asst electrochem, Mich State Univ, 25-27; foundry chemist, Cadillac Motorcar Co, Mich, 27-29; asst fel serol res, Univ Mich Hosp, 29-30; metall observer, Carnegie-Ill Steel Corp, Pa, 36-37; chemist, Bur Tests, Allegheny County, 37-40; chem engr, Navy Bur Aeronaut, DC, 40-59; head, Mat Protection Sect, Bur Naval Weapons, 59-72, proj officer, Naval Air Systs Command, 65-68, engr Mat Br, 72-73; RETIRED. *Concurrent Pos:* Naval officer, 42-46, reserve, 46-65; mem, Working Group Aircraft Camouflage, NATO, France, 52; vol cancer res lab, Georgetown Med Ctr, 83. *Mem:* Nat Asn Corrosion Engrs; fel Am Inst Chemists; Sigma Xi. *Res:* Environmental behavior of aircraft and missile material on land, sea and air; corrosion; thin free films; organic coatings; camouflage and visibility; adhesion; surface effects; biomedical techniques; air and water pollution control; invented electroplating of chromium process as pioneer; introduced polyurethane as coating for naval aircraft. *Mailing Add:* 1201 S Scott Arlington VA 22204

MALLOY, DONALD JON, b Rochester, NY, Dec 21, 50; m 72; c 2. NUCLEAR ENGINEERING, THERMAL HYDRAULIC ANALYSIS. *Educ:* Purdue Univ, BS, 72, MSNE, 74, PhD(nuclear eng), 76. *Prof Exp:* ENGR NUCLEAR REACTOR ANALYSIS SYSTS DESIGN, ARGONNE NAT LAB, 76- *Honors & Awards:* Argonne Pacesetter Award. *Res:* Reactor analysis; fuel cycle analysis; reactor physics and safety; reactor thermal-hydraulics analysis, advanced concepts. *Mailing Add:* 413 Assembly Dr Bolingbrook IL 60440

MALLOY, JOHN B, b Chicago Ill, May 6, 28; m 53; c 2. CHEMICAL ENGINEERING. *Educ:* Mass Inst Technol, BS, 50, MS, 51; Univ Chicago, MBA, 61. *Prof Exp:* Chem engr, Am Oil Co, 51-54, group leader, 54-64; specialist mgt sci, Amoco Chem Corp, 64-70, mgr planning & environ anal, 70-76; dir chem indust anal Standard Oil Ind, 76-88; RETIRED. *Mem:* Am Inst Chem Engrs; Am Chem Soc. *Res:* Application of scientific methods to business problems; analysis of economic environment for chemicals; author of one publication. *Mailing Add:* Seven Clearview Dr South Dennis MA 02660

MALLOY, MICHAEL H, NUTRITION, EPIDEMIOLOGY. *Educ:* Univ Tex, MD, 73. *Prof Exp:* DIR COMMUNITY PEDIAT, MED BRANCH, UNIV TEX, 80- *Mailing Add:* Peds/C3-T16 Child Health Ctr C33 Univ Tex Med Sch 301 Univ Blvd Galveston TX 77550

MALLOY, THOMAS BERNARD, JR, b El Campo, Tex, Aug 20, 41; m 68; c 2. MOLECULAR SPECTROSCOPY. *Educ:* Univ St Thomas, Tex, BA, 64; Tex A&M Univ, PhD(chem), 70. *Prof Exp:* Res assoc chem, Mass Inst Technol, 70-71; from asst prof to prof physics & chem, Miss State Univ, 71-79; SR RES CHEMIST, SHELL DEVELOP CO, 79- *Concurrent Pos:* Sabbatical, Univ Ga, 76-77. *Mem:* Am Phys Soc; Am Chem Soc; Coblenty Soc; Sigma Xi. *Res:* Analytical applications of infrared, ultraviolet-visible spectroscopy, Raman spectroscopy, and coherent Raman spectroscopy. *Mailing Add:* 6929 Westchester St Houston TX 77005

MALLOY, THOMAS PATRICK, b Chicago, Ill, July 28, 41; m 64; c 2. ORGANIC CHEMISTRY. *Educ:* Ill Inst Technol, BS, 65; Loyola Univ, PhD(org chem), 70. *Prof Exp:* Sr res scientist, De Soto, Inc, 70-71; consult org chem, Bernard Wolnak & Assocs, 71-74; group leader, Signal Res Ctr, 74-80, mgr, 80-84, assoc dir, Allied-Signal, 84-89; DIR, UOP, 89- *Mem:* Am Chem Soc. *Res:* Physical organic, mainly molecular, rearrangements; mechanisms of reaction and organic synthesis; catalysis of organic reactions. *Mailing Add:* 730 Red Bridge Lake Zurich IL 60047

MALLOZZI, PHILIP JAMES, b Norwalk, Conn, Feb 12, 37; m 61; c 4. PHYSICS. *Educ:* Harvard Univ, BA, 60; Yale Univ, MS, 61, PhD(physics), 64. *Prof Exp:* Instr physics, Yale Univ, 64-66; mem tech staff, Columbus Labs, Battelle Mem Inst, 66-70, dir, Laser Applications Ctr, 70-81; CONSULT LASER APPLICATIONS, 81- *Honors & Awards:* Indust Res 100 Award, 80. *Mem:* Am Phys Soc; AAAS; Sigma Xi. *Res:* Plasma physics; laser produced plasmas; laser physics; astrophysics; use of laser-generated plasmas for x-ray production; applications of laser-produced x-rays, especially to x-ray microscopy, exafs, and x-ray lithography. *Mailing Add:* 1783 Home Rd Delaware OH 43015

MALM, DONALD E G, b Tallant, Okla, June 3, 30; m 57; c 1. NUMBER THEORY, THEORETICAL COMPUTER SCIENCE. *Educ:* Northwestern Univ, BS, 52; Brown Univ, AM, 54, PhD(math), 59. *Prof Exp:* Instr math, Rutgers Univ, 57-59; vis lectr, Royal Holloway Col, Univ London, 59-60; asst prof, State Univ NY Stony Brook, 60-62; asst prof, 62-65, assoc prof, 65-76, PROF MATH, OAKLAND UNIV, 76- *Mem:* Asn Comput Mach; Am Asn Univ Prof; Am Math Soc; Math Asn Am. *Res:* number theory. *Mailing Add:* Dept of Math Oakland Univ Rochester MI 48063

MALM, NORMAN R, b Boulder, Colo, June 9, 31; m 55; c 3. PLANT BREEDING, AGRONOMY. *Educ:* Colo State Univ, BS, 54; Univ Ill, MS, 56, PhD(agron), 60. *Prof Exp:* Agronomist, NMex State Univ, 61-68 & Univ Nebr, Lincoln, 68-69; agronomist, 69-72, assoc prof, 72-77, PROF AGRON & COTTON BREEDER, NMEX STATE UNIV, 77- *Mem:* Am Soc Agron; Crop Sci Soc Am. *Res:* Cotton breeding research for high quality fiber, disease resistance and insect resistance. *Mailing Add:* Dept Agron NMex State Univ Las Cruces NM 88003

MALMBERG, JOHN HOLMES, b Gettysburg, Pa, July 5, 27; m 52; c 2. PLASMA PHYSICS. *Educ:* Ill State Univ, BS, 49; Univ Ill, MS, 51, PhD(physics), 57. *Prof Exp:* Mem staff plasma physics, Gen Atomic Div, Gen Dynamics Corp, 57-69; PROF PHYSICS, UNIV CALIF, SAN DIEGO, 67- *Honors & Awards:* Maxwell Prize Plasma Physics, Am Physics Soc. *Mem:* Fel Am Physics Soc. *Res:* Experimental plasma physics; non neutral plasmas; fundamental properties of waves; controlled thermonuclear research; development of seismic prospecting systems. *Mailing Add:* Dept Physics 0319 9500 Gilman Dr Univ Calif San Diego La Jolla CA 92093-0319

MALMBERG, MARJORIE SCHOOLEY, b Estherville, Iowa, Aug 20, 21; m 45; c 5. PHYSICAL CHEMISTRY. *Educ:* Wellesley Col, AB, 42; Univ Md, Col Park, PhD(phys chem), 67. *Prof Exp:* Jr chemist, Nat Bur Standards, 42-45, chemist, 48-55; US Army grant & res assoc, Univ Md, Col Park, 67-70; Nat Inst Gen Med Sci-Nat Inst Arthritis & Metab Dis spec res fel nuclear magnetic resonance, Nat Bur Standards, 70-72, guest worker, 72-74; eng analyst, NUS Corp, 74-75, sect leader, 75-78, consult, 78-81; RETIRED. *Mem:* Am Chem Soc; Health Phys Soc; Sigma Xi. *Res:* Molecular microdynamics in fluids; structure of liquids; relaxation phenomena in biological molecules and polymers; infrared, raman and nuclear magnetic resonance spectroscopy; light scattering; radiological environmental monitoring. *Mailing Add:* 3300 Gregg Rd Brookeville MD 20833

MALMBERG, PAUL ROVELSTAD, b New Haven, Conn, Apr 15, 23; m 44; c 5. SOLID STATE ELECTRONICS. *Educ:* Thiel Col, BS, 44. *Prof Exp:* Fel life preservers, Mellon Inst, 46-48; instr physics, Univ Pittsburgh, 50-51; physicist, Westinghouse Res Labs, 51-60; Int Atomic Energy Agency UN vis prof, Tsing Hua Univ, Taiwan, 60-61; mgr sci instrumentation, 61-66, fel engr, 66-68, mgr advan circuit fabrication technol, 68-70, mgr electron beam fabrication technol, 70-72, mgr electron imaging technol, Elec Sci Div, 72-75, fel engr, 75-77, MGR THIN FILM TECHNOL, WESTINGHOUSE RES & DEVELOP LABS, 77-; MGR DESIGN ENG, LITTON PANELVISION. *Mem:* Am Phys Soc; Inst Elec & Electronic Engrs; Electrochem Soc; Soc Info Display; Fedn Am Scientists. *Res:* Advanced integrated circuits and solid state devices made by electron and ion beam techniques; thin film devices and systems for information processing and display and for signal transduction. *Mailing Add:* Malmberg Assoc 130 Gordon St Pittsburgh PA 15218

MALMBERG, PHILIP RAY, b Norwood, Mass, Oct 13, 20; m 56; c 3. NUCLEAR PHYSICS, SOLID STATE PHYSICS. *Educ:* Ill State Norm Univ, BEd, 40; Univ Iowa, MS, 44, PhD(physics), 55. *Prof Exp:* Jr engr, Res Lab, Sylvania Elec Prod, Inc, 44-46; res assoc nuclear physics, Univ Iowa, 50-53; physicist, Naval Res Lab, 55-85; RETIRED. *Mem:* Sigma Xi; Am Phys Soc. *Res:* Radiation damage by charged particles; ion implantation; materials analysis; development of specialized equipment. *Mailing Add:* 6818 Farmer Dr Ft Washington MD 20744

MALME, CHARLES I(RVING), b Crookston, Minn, Aug 13, 31; m 61; c 2. ELECTRICAL ENGINEERING, ACOUSTICS. *Educ:* Univ Minn, BEE & BS, 54; Mass Inst Technol, SM, 58, EE, 59. *Prof Exp:* Res asst acoustics, Mass Inst Technol, 56-59; consult appl physics, 60-69, mgr phys sci lab, 69-75, SR SCIENTIST, BOLT BERANEK & NEWMAN, INC, 76- *Mem:* Acoust Soc Am; Marine Technol Soc; Sigma Xi. *Res:* Underwater sound propagation (marine environmental acoustics); sound propagation measurements; high intensity sound instrumentation; explosive effects research. *Mailing Add:* 25 Rockwood Rd Hingham MA 02043

MALMGREN, RICHARD AXEL, b St Paul, Minn, Dec 31, 21; m 46; c 2. PATHOLOGY. *Educ:* Wagner Col, BS, 42; Cornell Univ, MD, 45; Am Bd Path, dipl, 57. *Prof Exp:* Intern, Grasslands Hosp, 45-46; head serol unit, Biol Sect, Nat Cancer Inst, 48-53, med officer chg cancer invest unit, Tenn, 53-56, head cytopath serv, 56-72; prof & head cytopath serv, dept path, Med Ctr, George Washington Univ, 74-76; RETIRED. *Mem:* Am Soc Cytol (pres, 75). *Res:* Cancer immunology; cytology; pathologic physiology of cancer; demonstration of immuno depression of carcinogens and cancer chemotherapeutic agents; prevention of carcinogen induced cancer by vitamin A; field demonstration of Papanicolaou technique for mass screening. *Mailing Add:* 1686 N Harbor Ct Annapolis MD 21401

MALMSTADT, HOWARD VINCENT, b Marinette, Wis, Feb 17, 22; m 47; c 3. INSTRUMENTATION, SPECTROSCOPY. *Educ:* Univ Wis, BS, 43, MS, 48, PhD(chem), 50. *Prof Exp:* Res assoc chem, Univ Wis, 50-51; from instr to asst prof chem, 51-57, assoc prof anal chem, 57-61, prof, 61-78, EMER PROF CHEM, UNIV ILL, URBANA, 78-; PROVOST & VPRES, UNIV OF THE NATIONS, INT OFF, HAWAII, 89- *Concurrent Pos:* Industrial consultant, 54-78; Guggenheim fel, 60; dir, short courses electronics to sci, 60-74; chmn, Analysis Div, Am Chem Soc, 60- instr 78-88; prin investr NSF grants, 64-78 & NIH grants, 75-80; consult, UN Univ, Brazil, 78; Fulbright-Hayes distinguished prof, 78; vis prof, Japan, 84 & China, 86; provost & sr vpres, Pac & Asia Christian Univ, Hawaii, 79-89. *Honors & Awards:* Chem Instrumentation Award, Am Chem Soc, 63; Educ Award, Instrument Soc Am, 70; Analytical Chemistry Fisher Award, Am Chem Soc, 76; Analytical Chem Award, Fed Analytical Chem & Spectros Soc, 87. *Mem:* Am Chem Soc; Am Asn Clin Chemists; Soc Appl Spectros. *Res:* Science and technology; tertiary education for developing nations; community technology systems; spectrochemical methods; automatic titrations; molecular absorption spectrometry; scientific instrumentation; automation; reaction-rate methods. *Mailing Add:* 75-5851 Kuakini Hwy Kailua-Kona HI 96740

MALMUTH, NORMAN DAVID, b Brooklyn, NY, Jan 22, 31; m 60; c 1. FLUID MECHANICS, APPLIED MATHEMATICS. *Educ:* Univ Cincinnati, AE, 53; Polytech Inst Brooklyn, MAE, 56; Calif Inst Technol, PhD(aeronaut), 62. *Prof Exp:* Res engr, Grumman Aircraft Eng Corp, 53-56; mem tech staff, Los Angeles Div, NAm Rockwell Corp, 56-58, MEM TECH STAFF, SCI CTR, ROCKWELL INT CORP, 68- *Concurrent Pos:* Teaching asst, Calif Inst Technol, 61-62; lectr, Univ Calif, Los Angeles; vchmn pub affairs, AIAA, Ventura Pac sect; mem, NASA User Interface Group/Comt. *Mem:* Assoc fel Am Inst Aeronaut & Astronaut; Soc Indust & Appl Maths; Am Phys Soc; Am Soc Mech. *Res:* Transonic and hypersonic gas dynamics; perturbation theory; elastroplasticity; viscous flows over protuberances; nonlinear heat conduction; micrometeorology and chemical kinetics; hypersonic stability; chemical vapor deposition. *Mailing Add:* 182 Maple Rd Newbury Park CA 91320

MALO, JACQUES, b Montreal, Que, Can, Nov 5, 50. PULMONARY. *Educ:* Univ Montreal, MD, 74; Med Coun Can, LMCC, 75; Prof Corp Med Que, CSPQ, 79. *Prof Exp:* Res scholar, Univ Man, 79-81; med intern, Hosp Verdun, Univ Montreal, 74-75, resident, Hosp Sacre, 75-78, resident, Hosp Ste Justine, 78-79, resident, Hosp Montreal, 78-79, asst clin prof, 81-85, asst prof, 85-88, PNEMOLOGIST, HOSP SACRE-COEUR, UNIV MONTREAL, 81-, ASSOC PROF, 88-; HOSP SACRE-COEUR, 81- *Concurrent Pos:* Mem, Comt Eval Dept Med, Univ Montreal, 83, pres, 86. *Mem:* Am Physiol Soc; Am Thoracic Soc; Am Col Chest Physicians. *Mailing Add:* Div Pulmonary Hosp Sacre-Coeur 5400 Gouin Blvd W Montreal PQ H4J 1C5 Can

MALOCHA, DONALD C, b Chicago, Ill, Oct 17, 50; m 72; c 2. SURFACE ACOUSTIC WAVE TECHNOL, SOLID STATE DEVICES. *Educ:* Univ Ill, BS, 72, MS, 74, PhD(elec eng), 77. *Prof Exp:* Res assoc, Saw Technol Group, Univ Ill-Urbana, 77-78; mem tech staff, Corp Res Lab, Texas Instruments-Dallas, 78-80; mgr advan prod develop, Sawtek Inc-Orlando, 80-82; from asst prof to assoc prof, 82-88, PROF SURFACE ACOUST WAVE TECHNOL, UNIV CENT FLA, 82-, LEADER, SOLID STATE DEVICES & SYSTS LAB, 82- *Concurrent Pos:* Vis scholar, Swiss Fed Inst Technol, Inst Field Theory & High Frequencies, Zurich, 89; vis mem tech staff, Motorola, Advan Component Technol Group- Phoenix, Ariz, 89-90. *Honors & Awards:* Outstanding Serv Award, Inst Elec & Electronic Engrs, 91. *Mem:* Sr mem Inst Elec & Electronic Engrs; emer affil Elec Industs Asn; Int Electrotechnic Comn. *Res:* Surface acoustic wave technology and acoustic charge transport technology and their applications; design, analysis, synthesis and fabrication of devices and systems. *Mailing Add:* Elec Eng Dept Univ Cent Fla Orlando FL 32816

MALOFF, BRUCE L(ARRIE), b Syracuse, NY, Aug 26, 53; m 76. DIABETES, INFLAMMATION. *Educ:* Syracuse Univ, BS, 73; State Univ NY, Albany, PhD(cell physiol), 78. *Prof Exp:* Res assoc physiol, Sch Med, Upstate Med Ctr, State Univ NY, 77-78; instr & fel med & endocrinol, Sch Med, Univ Rochester, 78-81, sr instr, 81; res pharmacologist, E I Du Pont de Nemours & Co, Inc, Wilmington, Del, 81-86, sr res pharmacologist, Pharmaceut Div, 87-88, dir biochempharmacol, 88-90; DIR, PHARMACOL SERV, PANLABS INC, BOTHELL, WASH, 90- *Mem:* Inflammation Res Asn; AAAS; Endocrine Soc; Am Soc Pharmacol & Exp Ther; Soc Indust Microbiol. *Res:* Development of pharmacotherapeutic agents with novel cellular mechanisms of action; bioscience services. *Mailing Add:* Panlabs Inc 11804 North Creek Pkwy S Bothell WA 98011-8805

MALOFSKY, BERNARD MILES, b New York, NY, Oct 7, 37; m 64; c 2. ORGANIC POLYMER CHEMISTRY. *Educ:* Calif Inst Technol, BS, 59; Univ Wash, PhD(org chem), 64. *Prof Exp:* Res chemist, Textile Fibers Dept, E I du Pont de Nemours & Co, Inc, 64-70; from res & develop chemist to technol mgr, 72-74, mgr prod develop, 74-77, assoc dir chem technol, 77-82, vpres, Res & Develop, 82-85, vpres & dir, New Bus Develop, 85-90, VPRES & CHIEF CHEMIST, LOCTITE CORP, 90- *Mem:* Am Chem Soc; Soc Advan Mat & Processing Engr; Soc Mfg Engrs; Inst Polymer Chem; Sigma Xi. *Res:* Anaerobic adhesives and sealants, particularly thermal resistance, cure systems, structural adhesives of high peel and impact strength, primers, ultraviolet curing adhesives, powdered metal and casting impregnation; cyanoacrycate manufacture and products. *Mailing Add:* Loctite Corp 705 N Mountain Rd Newington CT 06111

MALONE, CREIGHTON PAUL, b Beaver City, Nebr, May 3, 33; m 57; c 2. PHYSICAL CHEMISTRY. *Educ:* Univ Colo, BA, 58, PhD(phys chem), 62. *Prof Exp:* Asst phys chem, Univ Colo, 58-62; res chemist, Eng Dept, 62-65, sr res chemist, 65-69, sr res chemist, 69-78, RES ASSOC, TEXTILE FIBERS DEPT, E I DU PONT DE NEMOURS & CO, INC, 78- *Mem:* Am Chem Soc. *Res:* Magnetic susceptibility of small particles; infrared adsorption and reflection spectroscopy; liquid chromatography; polymer physical chemistry; textile physical chemistry. *Mailing Add:* Textile Res Lab Chestnut Run E I du Pont de Nemours & Co Inc 19 Fountain Ct Wilmington DE 19808

MALONE, DENNIS P(HILIP), b Buffalo, NY, Sept 3, 32; m 54; c 2. PHYSICS, ELECTRICAL ENGINEERING. *Educ:* Univ Buffalo, BA, 54; Yale Univ, MSc, 55, PhD(physics), 60. *Prof Exp:* Res physicist, Cornell Aeronaut Lab, 59-60, head mod physics br, 60-65; assoc prof eng, 65-70, PROF ELEC ENG & CHMN DEPT, STATE UNIV NY BUFFALO, 70- *Mem:* Am Phys Soc; Am Nuclear Soc; Inst Elec & Electronics Engrs. *Res:* Quantum electronics; frequency multiplication techniques; coherent optics; electron polarization studies; ion molecule reactions. *Mailing Add:* Dept Elec Eng Bell Hall State Univ NY Buffalo NY 14260

MALONE, DIANA, b Chicago, Ill, May 6, 35. ANALYTICAL CHEMISTRY, CHEMICAL EDUCATION. *Educ:* Mundelein Col, BS, 56; Univ Notre Dame, MS, 65; Univ Iowa, PhD(sci educ, chem), 77. *Prof Exp:* Instr chem & math, Assumption High Sch, Davenport, Iowa, 59-69; ASSOC PROF CHEM, CLARKE COL, 69- *Mem:* Am Chem Soc. *Res:* Analytical and inorganic chemistry; coordination compounds preparation and characterization. *Mailing Add:* 1550 Clark Dr Dubuque IA 52001

MALONE, JAMES MICHAEL, b Berkeley, Calif, Sept 13, 46; m 83; c 2. VASCULAR SURGERY. *Educ:* Univ Calif, San Francisco, BMS, 68, MD, 71. *Prof Exp:* Intern surg, Univ Calif, San Francisco, 71-72, asst resident, 72-76, chief resident, 76-77; ASSOC PROF SURG & CHIEF VASCULAR SURG, TUCSON VET ADMIN MED CTR, 77- *Concurrent Pos:* Vet Admin career develop award, 77-80. *Mem:* Am Col Surgeons; Int Soc Cardiovasc Surg; Soc Vascular Surg; Clin Soc Vascular Surg; Asn Acad Surg. *Res:* Mechanisms of tumor metastasis; prosthetic vascular graft healing; amputation rehabilitation. *Mailing Add:* 1501 N Campbell Ave Tucson AZ 85719

MALONE, JAMES W(ILLIAM), b Arizona, La, Apr 11, 25; m 52; c 3. CHEMICAL ENGINEERING. *Educ:* La Polytech Inst, BS, 47; La State Univ, MS, 55. *Prof Exp:* Asst prof chem, 47-54, 55-56, assoc prof chem eng, 56-67, PROF CHEM ENG, LA TECH UNIV, 67- *Mem:* Am Inst Chem Engrs. *Res:* Heat transfer; concentration of native iron ore; liquid-liquid solubility. *Mailing Add:* 109 Woodhaven Rd E Ruston LA 71272

MALONE, JOHN IRVIN, b Altoona, Pa, Oct 10, 41; m; c 4. PEDIATRICS. *Educ:* Pa State Univ, BS, 63; Univ Pa, MD, 67. *Prof Exp:* Intern pediat, Children's Hosp, Philadelphia, 67-68, resident, 68-69; instr, Sch Med, Univ Pa, 69-72; asst prof, 72-76, assoc prof, 76-80, PROF PEDIAT, UNIV SOUTH FLA, 80-, CO-DIR, DIABETES CENTER, 79- *Concurrent Pos:* Res fel, Div Biochem Develop & Molecular Dis, Children's Hosp, Philadelphia, 69-71; chief resident pediat hosp, Univ Pa, 71-72; mem staff, Philadelphia Gen Hosp, 71, Tampa Gen Hosp, 72 & All Children's Hosp, St Petersburg, Fla, 73, Shriners Crippled Childrens Hosp, Tampa; co-dir, Fla's Camp Children & Youth Diabetes, 73-; dir, SunCoast Regional Diabetes Prog, Tampa, 76- *Mem:* AAAS; Am Diabetes Asn; Soc Pediat Res; Am Fedn Clin Res. *Res:* Pathogenesis of insulin dependent diabetes and long term complications. *Mailing Add:* Dept Pediat Univ South Fla 4202 Fowler Ave Tampa FL 33620

MALONE, JOSEPH JAMES, b St Louis, Mo, Sept 9, 32; m 60; c 4. ALGEBRA. *Educ:* St Louis Univ, BS, 54, MS, 58, PhD(math), 62. *Prof Exp:* Instr math, Rockhurst Col, 60-62; asst prof, Univ Houston, 62-67; from assoc prof to prof, Tex A&M Univ, 67-71; head dept, 71-78, PROF MATH, WORCESTER POLYTECH INST, 71- *Mem:* Am Math Soc; Math Asn Am; Soc Indust & Appl Math; Am Asn Univ Prof. *Res:* Abstract algebra; groups; near-rings; endomorphisms and automorphisms of groups; classifying those nonabelian groups whose morphisms generate a ring rather than a distributively generated near-ring. *Mailing Add:* 45 Adams St Westboro MA 01581

MALONE, LEO JACKSON, JR, b Wichita, Kans, July 24, 38; m 64; c 4. INORGANIC CHEMISTRY. *Educ:* Univ Wichita, BS, 60, MS, 62; Univ Mich, PhD(inorg chem), 64. *Prof Exp:* From asst prof to assoc prof, 64-73, PROF CHEM, ST LOUIS UNIV, 73- *Mem:* AAAS; Am Chem Soc; Sigma Xi. *Res:* Chemistry of carbon-monoxide-borane. *Mailing Add:* 46 Frontenac Estates St Louis MO 63131

MALONE, MARVIN HERBERT, b Fairbury, Nebr, Apr 2, 30; m 52; c 2. NATURAL PRODUCT PHARMACOLOGY. *Educ:* Univ Nebr, BS, 51, MS, 53, PhD(pharmacol, pharmaceut sci), 58. *Prof Exp:* Asst pharmacol, Univ Nebr, 51-53& 56-58; asst pharmacodyn, Squibb Inst Med Res, 53-56; asst prof pharmacol, Col Pharm, Univ NMex, 58-60; assoc prof, Sch Pharm, Univ Conn, 60-69; prof physiol & pharmacol, Univ Pac, 69-84, distinguished prof pharmacol & toxicol, 84-90, chmn dept, 87-90, EMER PROF, UNIV PAC, 90-; HEAD, WOODWORM ASSOCS, 90- *Concurrent Pos:* Consult, Drug Plant Lab, Univ Wash, 60-64, Res Path Assocs, Md, 67-70, Amazon Natural Drug Co, NJ, 67-70, Imp Chem Indust US Inc, Del, 68-78 & Sisa Inst Res Inc, Mass, 77-82; ed, Wormwood Rev, 61-, Pac Info Serv on St Drugs, 71-78, Am J Pharm Educ, 75-79, Pharmat, 84-87 & J Ethnopharmacol, 85-91; mem, Task Force on Plants for Fertility Regulation, Spec Prog Res, Develop & Res Training in Human Reproduction, WHO, Geneva, Switz, 82-87; mem, Med Therapeut & Drug Adv Comt, State Calif, 85-90. *Honors & Awards:* Mead Johnson Labs Award, 64. *Mem:* fel AAAS; Am Soc Pharmacol & Exp Therapeut; Am Soc Pharmacog; Int Soc Ethnopharmacol. *Res:* Screening and assay of natural products; pharmacodynamics of psychotropic and autonomic agents; biometrics; pharmacology of inflammation and antiinflammation. *Mailing Add:* Dept Physiol-Pharmacol Univ Pac Sch Pharm Stockton CA 95211-0197

MALONE, MICHAEL JOSEPH, b Portland, Maine, Apr 28, 30; m 57; c 1. NEUROLOGY, NEUROCHEMISTRY. *Educ:* Boston Col, AB, 51; Georgetown Univ, MD, 56. *Prof Exp:* Resident neurol, Boston Vet Admin Hosp, Boston Univ, 60-63; Nat Inst Neurol Dis & Stroke spec fel neurochem, Harvard Med Sch, 63-65; res assoc, Mass Gen Hosp, 65-69; lectr, Boston City Hosp, 69-70; prof, Med Sch, George Washington Univ, 70-75; prof neurol & pediat & dir neurol res, Boston City Hosp, 75-76; DIR, GRECC BEDFORD VET ADMIN MED CTR, 76-; PROF NEUROL & PSYCHIAT, BOSTON UNIV, 76- *Concurrent Pos:* Vet Admin clin investr, Harvard Med Sch, 65-68; asst prof, Sch Med, Boston Univ, 67-70; Nat Inst Neurol Dis & Stroke res grant, Boston Univ, 68-71, career develop award, 69; chief neurol, Children's Hosp, Washington, DC, 70-75; consult, Walter Reed Army Med Ctr, Washington, DC, 71-, US Naval Hosp, Bethesda, Md, 71- & NIH, 71- *Mem:* Am Acad Neurol; Am Soc Neurochem; Int Soc Neurochem; Soc Neurosci; NY Acad Sci. *Res:* Biochemistry of maturation; biochemical pathology of nervous tissue; gerontology. *Mailing Add:* Dept Neurol Med Univ SC 171 Ashley Ave Charleston SC 29425-2232

MALONE, PHILIP GARCIN, b Louisville, Ky, Jan 12, 41; m 75; c 2. GEOCHEMISTRY. *Educ:* Univ Louisville, BA, 62; Ind Univ, Bloomington, MA, 64; Case Western Reserve Univ, PhD(geol), 69. *Prof Exp:* Nat Acad Sci-Nat Res Coun res assoc, Smithsonian Inst, 69-70; from asst prof to assoc prof geol, Wright State Univ, 70-78; geologist, 78-87, GEOPHYSICIST, US ARMY ENG, WATERWAYS EXP STA, 87- *Mem:* Asn Eng Geologists; Sigma Xi; Am Asn Prof Geologists. *Res:* Containment, treatment and disposal of toxic and hazardous wastes. *Mailing Add:* 705 Santa Rosa Dr Vicksburg MS 39180-0631

MALONE, ROBERT CHARLES, b Wichita, Kans, Mar 14, 45; m 68. THEORETICAL PHYSICS. *Educ:* Wash Univ, BA, 67; Cornell Univ, MS, 70, PhD(theoret physics), 73. *Prof Exp:* STAFF PHYSICIST THEORET PHYSICS, LOS ALAMOS NAT LAB, 72- *Mem:* Am Phys Soc; Am Meteorol Soc. *Res:* Hydrodynamics, atmospheric dynamics. *Mailing Add:* Los Alamos Nat Lab Group C-3/MS B265 Los Alamos NM 87545

MALONE, STEPHEN D, b St Petersburg, Fla, July 3, 44; m 69; c 2. SEISMOLOGY, VOLCANOLOGY. *Educ:* Occidental Col, Los Angeles, Calif, BA, 66; Univ Nev, Reno, PhD(geophysics), 72. *Prof Exp:* Res asst seismol, seismol lab, Univ Nev, Reno, 66-72; res assoc geophysics, 72-75, sr res assoc, 75-84, RES PROF GEOPHYSICS, GEOPHYSICS PROG, UNIV WASH, SEATTLE, 84- *Concurrent Pos:* Travel assistantship, Nat Acad Sci, 70; fel, Inst Geophys Mining, 72; assoc ed, J Geophys Res, 85-86. *Mem:* Am Geophys Union; Seismol Soc Am; AAAS. *Res:* Earthquake and volcanic seismology; seismic network operation; earthquake and volcanic hazards; volcanic eruption processes; regional geophysical structure. *Mailing Add:* Geophysics AK-50 Univ Wash Seattle WA 98195

MALONE, THOMAS, REPRODUCTIVE BIOLOGY. *Educ:* Harvard Univ, PhD(biol), 52. *Prof Exp:* Dep dir, NIH, 77-86; ASSOC CHMN, ASN AM MED COLS, 86- *Mailing Add:* Asn Am Med Cols Univ Md Du Pont Circle Suite 200 Washington DC 20036

MALONE, THOMAS C, b Banana River, Fla, Sept 7, 43; m 84; c 2. BIOLOGICAL OCEANOGRAPHY, EUTROPHICATION. *Educ:* Colo Col, BA, 65; Univ Hawaii, MS, 67; Stanford Univ, PhD(biol), 71. *Prof Exp:* Asst prof, City Col, City Univ New York, 71-76; sr res assoc, Columbia Univ, 76-80; oceanogr, Brookhaven Nat Lab, 80-82; assoc prof, 82-83, actg dir, Ctr Environ & Estuarine Studies, 88-90, PROF, UNIV MD SYST, 83-, DIR, HORN PT ENVIRON LAB, 90- *Concurrent Pos:* Prin investr, Nat Oceanic & Atmospheric Admin, NSF & Dept Natural Resources, 73-; vchmn, Syst Adv Coun, Univ Nat Oceanog Lab, 85-89; chmn, Chesapeake Res Consortium, 88-90; mem, Gov Chesapeake Bay Work Group, 88-90. *Mem:* Am Soc Limnol & Oceanog; Am Geophys Union; Estuarine Res Fedn; Phycological Soc Am. *Res:* Dynamics of coastal ecosystems; phytoplankton ecology; eutrophication. *Mailing Add:* Horn Pt Environ Lab PO Box 775 Cambridge MD 21613

MALONE, THOMAS E, b Henderson, NC, June 3, 26. BIOMEDICAL RESEARCH. *Educ:* NC Cent Univ, BS, 48, MS, 49; Harvard Univ, PhD, 52. *Prof Exp:* Prof zool, NC Cent Univ, Durham, 52-58; resident res assoc, Argonne Nat Lab, 58-59; mem fac, Loyola Univ, Chicago, 59-62; asst chief res grants sect, Nat Inst Dent Res, NIH, 63-64, dep chief extramural progs, 64-66 chief periodont dis & soft tissue studies, 66-67; prof & chmn, Dept Biol, Am Univ Beirut, Lebanon, 67-69; assoc dir extramural progs, Nat Inst Dent Res, NIH, 69-72, assoc dir extramural res & training, dep dir, NIH, actg dir, 72- 86; assoc vchancellor res, Univ Md, Baltimore, 86-88; VPRES BIOMED RES, ASN AM MED COLS, 88. *Honors & Awards:* Super Serv Award, Dept Health, Educ & Welfare, 71, Distinguished Serv Award, 74; Cert Merit, Am Col Deentsts, 75; Sr Exec Serv Presidential Merit Award, 80, Distinguished Exec Rank Award, 83. *Mem:* Inst Med-Nat Acad Sci. *Res:* Dentistry; peridontal disease. *Mailing Add:* Asn Am Med Cols One Dupont Circle NW Suite 200 Washington DC 20036

MALONE, THOMAS FRANCIS, b Sioux City, Iowa, May 3, 17; m 42; c 6. ENVIRONMENT. *Educ:* SDak Sch Mines & Technol, BS, 40; Mass Inst Technol, ScD, 46. *Hon Degrees:* DEng, SDak Sch Mines & Technol, 62; DHL, St Joseph Col, 65, Bates Col, 88. *Prof Exp:* Asst, Mass Inst Technol, 41-42, from asst prof to assoc prof meteorol, 43-55; dir weather res ctr, Travelers Ins Co, Conn, 55-57, dir res, 57-64, second vpres, 64-66, vpres & dir res, 66-67, sr vpres, 68-70; dean grad sch, Univ Conn, 70-73; dir, Holcomb Res Inst, Butler Univ, 73-83; EXEC SCIENTIST, CONN ACAD SCI & ENG, 87-; DISTINGUISHED PROF, NC STATE UNIV, 90- *Concurrent Pos:* With Off Naval Res, 50-53; ed, Compendium Meteorol; mem adv panel sci & technol, Comt Sci & Astronaut, US House of Rep; mem geophys res bd & comt water, Nat Acad Sci, chmn comt atmospheric sci, 62-68, dep foreign secy, 68-73, chmn geophys res bd, 69-75, chmn bd int orgn & progs, 69-76, mem space applns bd, 72-77, foreign secy, 78-82; secy-gen comt atmospheric sci, Int Union Geod & Geophys, 65-68; chmn, Nat Motor Vehicle Safety Adv Coun, 67-69; secy-gen sci comt probs of environ, Int Coun Sci Unions, 64-68; mem, Nat Adv Comt Oceans & Atmosphere, 71-75; chmn bd trustees, Univ Corp Atmospheric Res, 73-74; mem, Weather Modification Adv Bd, Dept Commerce, 77-80; pres, Inst Ecol, 78-81; trustee, Int Found Sci; pres, Sigma Xi, 88-; scholar in residence, St Joseph Col, 83-; comt global change, Nat Acad Sci, 87-91. *Honors & Awards:* Losey Award, Int Aerospace Sci, 60; Brooks Award, Am Meteorol Soc, 64; Abbe Award, 68; Int Meteorol Orgn Prize, 85; Waldo E Smith Award, Am Geophys Union, 85. *Mem:* Nat Acad Sci; fel AAAS; fel & hon mem Am Meteorol Soc (secy, 57-60, pres, 60-62); fel Am Geophys Union (vpres, 60-61, pres, 61-64, secy int partic, 64-72); Int Coun Sci Unions (vpres, 70-72, treas, 78-); fel Am Acad Arts & Sci. *Res:* Applied meteorology; synoptic climatology; environment. *Mailing Add:* Five Bishop Rd Unit 203 West Hartford CT 06119

MALONE, WINFRED FRANCIS, b Revere, Mass, Feb 10, 35; m 75. PHARMACOLOGY, TOXICOLOGY. *Educ:* Univ Mass, Amherst, BS, 57, MS, 61; Rutgers Univ, New Brunswick, MS, 64; Univ Mich, Ann Arbor, PhD, 70. *Prof Exp:* Asst prof, Univ Lowell, 64-66; res & develop officer, NIH,

69-72; sci adv, Hazardous Mat Adv Comt, Environ Protection Agency, 72-73, staff dir, Sci Adv Bd, 73-74; prog dir, Prev Cancer Control Prog, 74-80, CHIEF CHEMOPREV BR, NAT CANCER INST, 80- *Mem:* Soc Clin Trials; Am Soc Prev Oncol; Royal Soc Health; NY Acad Sci; Am Pub Health Asn; AAAS. *Res:* Preventive clinical trials; chemoprevention. *Mailing Add:* Nat Cancer Inst Div Cancer Prev & Control Exec Plaza N 201 Bethesda MD 20892

MALONEY, CLIFFORD JOSEPH, b Wheelock, NDak, Mar 25, 10; m 42; c 1. STATISTICS. *Educ:* NDak Agr Col, BS, 34; Univ Minn, MA, 37; Iowa State Col, PhD(statist), 48. *Prof Exp:* Instr math, NDak Agr Col, 35-41; instr math, Iowa State Col, 41-42; statistician, Bur Agr Econ, USDA, 42-46; instr math, Iowa State Col, 46, res assoc statist method, 46-47; chief statist br, Chem Corps, Ft Detrick, Md, 47-58, Biol Labs, Biomath Div, 58-62; chief biomet sect, div biol standard, 61-78, CONSULT STATIST, NIH, 78- *Mem:* Biomet Soc; Math Asn Am; Am Statist Asn; Inst Math Statist. *Res:* Biometrics; computing; geometry. *Mailing Add:* 6021 Landon Ln Bethesda MD 20817

MALONEY, DANIEL EDWIN, b Jericho, Vt, Feb 9, 26; m 53; c 8. POLYMER CHEMISTRY. *Educ:* St Michael's Col, BS, 47; Univ Notre Dame, MS, 49, PhD(org chem), 51. *Prof Exp:* Res chemist, Plastics Dept, E I DuPont de Nemours & Co, 51-61, sr res chemist, 61-67, res assoc, 67-84, sr tech consult, 84-88; CONSULT, 88- *Mem:* Am Chem Soc; Electrochem Soc. *Res:* Plastics; polyolefins; ion exchange membranes. *Mailing Add:* 109 Highland Dr Hockessin DE 19707

MALONEY, J(AMES) O(HARA), b St Joseph, Mo, Apr 29, 15; m 40; c 3. CHEMICAL ENGINEERING. *Educ:* Univ Ill, BS, 36; Pa State Col, MS, 39, PhD(chem eng), 41. *Prof Exp:* Chem engr, Exp Sta, E I du Pont de Nemours & Co, 41-43, 45; sect chief, Plutonium Proj, Univ Chicago, 43-45; exec dir, Res Found, 45-61, chmn chem eng dept, Univ, 45-64, prof, 45-85, EMER PROF CHEM ENG, UNIV KANS, 85 - *Concurrent Pos:* Fulbright lectr, Univ Naples, 56-57; Smith-Mundt lectr, UAR, 60; Ford Found consult, Univ Alexandria, 66; Fulbright lectr, Nat Tech Univ Athens, 69; lectr, Chonnam Univ, Korea, 79. *Mem:* Am Chem Soc; fel Am Inst Chem Engrs; Am Asn Univ Prof; fel AAAS. *Res:* Absorption of gases in liquids; scale up of centrifugal equipment; development of laboratory experiments; structuring of technical information. *Mailing Add:* Dept Chem Eng Univ Kans Lawrence KS 66045

MALONEY, JAMES VINCENT, JR, b Rochester, NY, June 30, 25; m 57; c 4. SURGERY. *Educ:* Univ Rochester, MD, 47; Am Bd Surg, dipl, 55; Am Bd Thoracic Surg, dipl, 57. *Prof Exp:* Lectr biol, Sampson Col, 47; surg house officer, Johns Hopkins Univ Hosp, 47-48, from asst res to res surgeon, 50-55, instr, Sch Med, 54-55; res fel physiol, Sch Pub Health, Harvard Univ, 48-50; from asst prof to assoc prof, 55-65, chief div thoracic surg, 59-76, PROF SURG, CHMN DEPT, SCH MED, UNIV CALIF, LOS ANGELES, 76- *Concurrent Pos:* Consult, Vet Admin Hosp, Los Angeles & Harbor County Gen Hosp, Torrance, 57-; Markle scholar, 58- *Mem:* Soc Univ Surgeons; Soc Clin Surgeons; Am Physiol Soc; AMA; Am Col Surg; Sigma Xi. *Res:* Surgery and physiology, especially the cardiorespiratory system. *Mailing Add:* Dept Surg Sch Med Univ Calif Los Angeles CA 90024

MALONEY, JOHN F, b Waltham, Mass, Feb 14, 36; m 64; c 3. FOOD SCIENCE & TECHNOLOGY. *Educ:* Northeastern Univ, BS, 59; Mass Inst Technol, PhD(food sci & technol), 65. *Prof Exp:* Sr res food scientist, Gen Mills, Inc, Minneapolis, 65-69, develop leader, 69-71, head prod develop, 71-72; mgr & dir res, 72-76, VPRES RES & DEVELOP & TECH SERV, H P HOOD INC, 76- *Mem:* Inst Food Technologists; Am Dairy Sci Asn; Nat Restaurant Asn; Sigma Xi. *Res:* Technical management including research and development, quality assurance and engineering for food and dairy products; development of new food products. *Mailing Add:* 151 Hallet St Boston MA 02124

MALONEY, JOHN P, b Omaha, Nebr, Dec 9, 29. FUNCTIONAL ANALYSIS. *Educ:* Iowa State Univ, BS, 58; Georgetown Univ, MA, 62, PhD(math), 65. *Prof Exp:* Prod engr, Western Elec Co, 58-59; elec engr, US Govt, 58-63; instr math, Georgetown Univ, 63-65; asst prof, Univ Nebr, Lincoln, 65-67; from asst prof to assoc prof, 67-76, PROF MATH, UNIV NEBR, OMAHA, 76- *Mem:* Math Asn Am; Soc Indust & Appl Math. *Res:* Integral equations; solar energy. *Mailing Add:* Dept of Math & Comput Sci Univ of Nebr Omaha NE 68182

MALONEY, KENNETH LONG, b Wilkes-Barre, Pa, Oct 1, 45; m; c 3. FUEL SCIENCE, PHYSICAL CHEMISTRY. *Educ:* Wilkes Col, BS, 67; Pa State Univ, PhD(fuel sci), 71. *Prof Exp:* Res assoc, Drexel Univ, 71-72 & Princeton Univ, 72-73; sr scientist, Ultrasysts, Inc, 73-74; mgr, Fossil Fuel Systs Studies, KVB, Inc, 74-81; PRES, CALPENN ASSOCS INC, 81- *Concurrent Pos:* Instr, Drexel Univ, 71-73; adj prof, Univ Calif, Irvine, 74-76; vpres, Calpenn Assocs, 77- *Mem:* Am Chem Soc; Combustion Inst; Soc Photo-optical Engrs; Sigma Xi. *Res:* Combustion generated-pollutants; fossil fuel flame processes; chemistry of post flame reactions; one patent. *Mailing Add:* 433 Locust St Laguna Beach CA 92651

MALONEY, KENNETH MORGAN, b New Orleans, La, Oct 11, 41. MATERIALS SCIENCE, PROCESS DESIGN. *Educ:* Southern Univ, BS, 63, Univ Wash, PhD(phys chem), 68. *Prof Exp:* Res asst phys chem, Univ Wash, 63-68; sr res scientist reaction dynamics, Pac Northwest Labs, Battelle Mem Inst, 68-70; sr scientist & tech leader, Lamp Div, Gen Elec Co, 70-74, mgr advan eng mat sci, 74-76; mgr mat tech & mat eng, Xerox Corp, 76-81; assoc dir forward & contract res, Allied Corp, 81-; AT PHILLIP MORRIS RES & DEVELOP. *Concurrent Pos:* Mem, Nat Res Coun Adv Comt to US Army Res Off, 74-; fel, Am Inst Chemists, 75. *Mem:* Am Chem Soc; Am Inst Chemists; Sigma Xi. *Res:* Reaction dynamics of metal-oxygen combustion systems; properties of liquid and solid state photoconductive materials; process dynamics of controlling high temperature processes. *Mailing Add:* Phillip Morris Res & Develop PO Box 26583 Richmond VA 23261

MALONEY, MICHAEL STEPHEN, b Green Bay, Wis, Feb 27, 47; m 75; c 3. CELL BIOLOGY, DEVELOPMENTAL BIOLOGY. *Educ:* Ripon Col, BA, 69; Univ Kans, PhD(cell biol), 76. *Prof Exp:* Asst prof biol, Radford Univ, 76-82; asst prof, 82-84, ASSOC PROF BIOL, BUTLER UNIV, 85- *Mem:* Am Soc Cell Biol; Soc Develop Biol; Soc Protozoologists; Sigma Xi. *Res:* Cellular control mechanisms involved in oral regeneration in the ciliated protozoan Stentor Coeruleus, particularly those involving the cell surface and calcium ion fluxes. *Mailing Add:* Dept Biol Sci Butler Univ 4600 Sunset Ave Indianapolis IN 46208

MALONEY, PETER CHARLES, b Boston, Mass, Nov 5, 41. MEMBRANE BIOCHEMISTRY, PHYSIOLOGY. *Educ:* Swarthmore Col, BA, 63; Brown Univ, PhD(biol sci), 70. *Prof Exp:* Res fel physiol, Harvard Med Sch, 72-74, res assoc, 74-76; from asst prof to assoc prof, 76-88, PROF PHYSIOL, JOHNS HOPKINS MED SCH, 88- *Concurrent Pos:* Mellon scholar. *Mem:* Sigma Xi; Biophysics Soc; Am Soc Microbiol. *Res:* Physiology of ionic movements across cell membranes. *Mailing Add:* Dept Physiol Sch Med Johns Hopkins Univ 720 Rutland Ave Baltimore MD 21205

MALONEY, THOMAS EDWARD, b Niagara Falls, NY, Sept 7, 23; m 46; c 3. ENVIRONMENTAL SCIENCES. *Educ:* Univ Buffalo, BA, 49, MA, 53. *Prof Exp:* Res biologist, Robert A Taft Sanit Eng Ctr, USPHS, 51-65; chief plankton res sect, Nat Marine Water Qual Lab, Fed Water Pollution Control Admin, US Environ Protection Agency, 65-68, chief, Physiol Control Br, Nat Eutrophication Res Prog, 68-71, dep chief prog & chief, Physiol Control Br, Environ Protection Agency, 71-72, chief, Eutrophication & Lake Restoration Br, 72-78, actg dir, Assessment & Criteria Develop Div, Corvallis Environ Res Lab, US Environ Res Lab, 78-80; RETIRED. *Concurrent Pos:* Chmn, Plankton Subcomt, Biol Methods Comt, Environ Protection Agency, Biostimulation Joint Task Group & Phytoplankton Subcomt, Standard Methods for Examination Water & Wastewater, 73- *Mem:* AAAS; Am Soc Limnol & Oceanog; Water Pollution Control Fedn; Am Phycol Soc; Am Inst Biol Sci. *Res:* Culturing of algae; algal physiology; chemical control of algal growth; environmental requirements of planktonic organisms; coordination and review of research and development programs to provide for control of accelerated eutrophication and development of lake restoration technology. *Mailing Add:* 3630 NW Roosevelt Dr Corvallis OR 97330

MALONEY, THOMAS M, b Raymond, Wash, Feb 18, 31; m 60; c 3. WOOD COMPOSITES, COMPOSITION BOARD MATERIALS. *Educ:* Wash State Univ, BA, 56. *Prof Exp:* Jr wood technologist, 56-70, asst wood technologist, 60-72, head wood technol, 72-85, DIR WOOD ENG LAB, WASH STATE UNIV, 85- *Concurrent Pos:* Consult, Food & Agr Orgn, UN, 74, UN Indust Develop Orgn, 82- , many indust cos; contributing ed, Plywood & Panel World, 76-86; ed, Wood Indust Abstracts, 80-84; prin investr numerous res projs. *Mem:* Soc Wood Sci & Technol (pres, 76-77); Forest Prods Res Soc (pres, 81-82). *Res:* Wood composite and composition board materials. *Mailing Add:* Wood Eng Lab Wash State Univ Pullman WA 99164-3020

MALONEY, TIMOTHY JAMES, b Dayton, Ohio, Aug 11, 49. SEMICONDUCTOR DEVICES, INTEGRATED CIRCUIT RELIABILITY. *Educ:* Mass Inst Technol, SB, 71; Cornell Univ, MS, 73, PhD(elec eng), 76. *Prof Exp:* Postdoctoral assoc elec eng, Sch Elec Eng, Cornell Univ, 76-77; from engr to sr engr, Cent Res Lab, Varian Assocs Inc, 77-84; SR STAFF ENGR, INTEL CORP, 84- *Concurrent Pos:* Lectr, Univ Calif Los Angeles, 83, Univ Wis-Madison, 87-89, Univ Calif, Berkeley, 89-90; tech prog chmn, Elec Overstress/Electrostatic Discharge Symp, 89-90, vchmn, 90-91, gen chmn, 91-92. *Mem:* Sr mem Inst Elec & Electronics Engrs. *Res:* Electrical overstress and electronic discharge in semiconductor devices, especially MOS integrated circuits; design and layout of integrated circuits for protection against EOS/ESD; integrated circuit reliability and failure analysis; metal electromigration phenomena in integrated circuits. *Mailing Add:* Intel Corp (SC9-06) 2250 Mission College Blvd Santa Clara CA 95052-8125

MALONEY, WILLIAM THOMAS, b Warren, Ohio, Dec 12, 35; m 58; c 2. MAGNETISM, MAGNETIC RECORDING. *Educ:* Case Western Reserve Univ, BS, 57, MS, 58; Harvard Univ, AM, 61, PhD(appl physics), 64. *Prof Exp:* Asst appl physics, Harvard Univ, 60-63, lectr & res fel, 63-65; res staff mem, Sperry Res Ctr, 65-83,; RES STAFF MEM, POLAROID CORP, 83- *Mem:* Inst Elec & Electronics Engrs; Am Phys Soc; Magnetics Soc. *Res:* Plasma physics; optical signal processing; optical memories; magnetic materials; magnetic recording. *Mailing Add:* 119 Willis Rd Sudbury MA 01776

MALOOF, FARAHE, internal medicine, endocrinology; deceased, see previous edition for last biography

MALOOF, GILES WILSON, b San Bernardino, Calif, Jan 4, 32; m 58; c 2. MATHEMATICS, GEOPHYSICS. *Educ:* Univ Calif, Berkeley, BA, 53; Univ Ore, MA, 58; Ore State Univ, PhD(math), 62. *Prof Exp:* Engr, Creole Petrol Corp, 53-54; engr, US Navy Ord Res Labs, 58-59; instr math, Ore State Univ, 61-62, asst prof, 62-68, res assoc geophys oceanog, 63-68; head dept, 68-75, dean grad sch, 70-75, PROF MATH, BOISE STATE UNIV, 68- *Concurrent Pos:* Dir, Northwest Col & Univ Asn Sci, 73-, pres, 90-; dir, Northwest Sci Asn, 77-; vis prof, Ore State Univ, 77-78. *Mem:* Soc Indust & Appl Math; Am Math Soc; Math Asn Am. *Res:* Numerical filtering as applied in the interpretation of geophysical data; nonlinear functional analysis applied to integral equations. *Mailing Add:* Dept Math Boise State Univ 1910 University Dr Boise ID 83725

MALOTKY, LYLE OSCAR, b New London, Wis, Apr 14, 46; m 68; c 2. POLYMER CHEMISTRY. *Educ:* Augsburg Col, BA, 68; Univ Akron, PhD(polymer sci), 73; George Wash Univ, MEA, 80. *Prof Exp:* chemist, Naval Explosive Ord Disposal Facil, 73-81, head chem eng div, Naval Explosive Ord Disposal Technol Ctr, 81-84; prog mgr, Security Res & Develop Off, 84-90, SCI ADV, CIVIL AVIATION SECURITY, FED

AVIATION ADMIN, 90- *Mem:* Am Chem Soc. *Res:* Directs agencies security related research; participant in several national and international committees applying high technology to counter terrorism. *Mailing Add:* 13203 Chalfont Ave Ft Washington MD 20744

MALOUF, EMIL EDWARD, b Ogden, Utah, Mar 23, 16; m 42; c 4. HYDROMETALLURGY, INORGANIC CHEMISTRY. *Educ:* Univ Utah, BS, 39. *Prof Exp:* Chief exp mfg, Ogden Arsenal, US Army, 40-44; res chemist, Kennecott Copper Corp, 47-52, proj, develop engr, 52-65, chief sect hydrometall develop, 65-77; CONSULT, 77- *Concurrent Pos:* Adj prof, Univ Utah. *Honors & Awards:* Robert Earll McConnell Award, Am Inst Mining & Metall Engrs, 72. *Mem:* Am Inst Mining & Metall Engrs; Am Inst Mining, Metall & Petrol Engrs; Am Soc Metals; Sigma Xi. *Res:* Analytical procedures and processes for recovery of rhenium; copper leaching processes from mine waste; copper precipitation units for efficient high volume recovery of copper from solutions. *Mailing Add:* 132 Dorchester Dr Salt Lake City UT 84103

MALOUF, GEORGE M, b San Diego, Calif, Mar 21, 50; m 76; c 2. PROCESS CHEMISTRY, PRODUCT DEVELOPMENT. *Educ:* Univ Calif, San Diego, BA, 72; Santa Barbara, PhD(inorg chem), 77. *Prof Exp:* Fel surface chem, Aerospace Corp, Univ Southern Calif, 77-78; res chemist inorg chem, US Boras Res Corp, 78-79; ENVIRON MGR, US BORAX & CHEM CORP, 90- *Mem:* Am Chem Soc. *Res:* Photochemistry of transition metal complexes; surface chemistry of aluminum with emphasis on developing corrosion resistant coatings; solubility kinetics and thermodynamics of borate solutions; scale prediction in process equipment; fine particle analysis; corrosion of process equipment; pesticide and agricultural products research and development. *Mailing Add:* 19 Elderberry Irvine CA 92715-3703

MALOUF, ROBERT EDWARD, b Dallas, Tex, Aug 21, 46; m 66; c 2. MARINE & SHELLFISH BIOLOGY. *Educ:* Univ Mont, BA, 68; Ore State Univ, MS, 70, PhD(fisheries), 77. *Prof Exp:* Res biologist aquacult, Col Marine Studies, Univ Del, 70-72; res asst fisheries, Ore State Univ, 72-77; from asst prof to assoc prof Marine Biol, State Univ NY, Stony Brook, 77-91, dir NY Sea Grant Inst, 87-91; PROF FISHERIES & DIR ORE SEA GRANT, ORE STATE UNIV, 91- *Concurrent Pos:* Sea Grant prof, NY Sea Grant Inst, 77-80. *Mem:* Nat Shellfisheries Asn (secy/treas, 82-84, vpres, 84-85, pres elect, 85-86, pres, 86-87); World Aquacult Soc; Am Malacological Union; AAAS. *Res:* Biology of commercially important bivalve molluscs, particularly reproductive processes and growth energetics; application of findings to resource management and aquaculture. *Mailing Add:* Ore Sea Grant Adm A500G Ore State Univ Corvallis OR 97331-2131

MALOWANY, ALFRED STEPHEN, b Joliette, Que, Aug 16, 39; m 64; c 2. ELECTRICAL ENGINEERING. *Educ:* McGill Univ, BEng, 59, MEng, 62, PhD(control), 67. *Prof Exp:* Lectr, 60-65, asst prof, 67-77, ASSOC PROF ELEC ENG, MCGILL UNIV, 77- *Mem:* Can Pulp & Paper Asn; Inst Elec & Electronics Engrs; Sigma Xi. *Res:* Biological and industrial modelling; control and optimization; computer graphics. *Mailing Add:* Dept Elec Eng McGill Univ 3480 University St Montreal PQ H3A 2A7 Can

MALOY, JOHN OWEN, b Orange, NJ, Feb 7, 32; div; c 1. HIGH ENERGY PHYSICS, SPACE PHYSICS. *Educ:* Univ Ariz, BS, 54; Calif Inst Technol, PhD(physics), 61. *Prof Exp:* Group leader systs anal, Jet Propulsion Lab, Calif Inst Technol, 60-61, res fel physics, Synchrotron Lab, 61-63, sr res fel, 63-67; mgr Advan Develop Div, Analog Technol Corp, 67-71; chief scientist, Beckman Instruments Inc, 71-74; staff scientist, Ball Bros Res Corp, 74-77; res staff physicist, Univ Southern Calif, 74-77; PRES, MOUNTAIN INSTRUMENTS CORP, 77- *Concurrent Pos:* Consult, Electro-Optical Systs, Inc, Calif, 63-66, Beckman Instruments, Inc, 74-76, Jet Propulsion Lab, Calif Inst Technol, 76- & Melcon, 76-77. *Mem:* AAAS; Am Geophys Union; Sigma Xi. *Res:* Photoproduction of pi mesons; accelerator physics and technology; development of radio frequency acceleration system; solar and planetary science; space science instrumentation systems design; instrument program management. *Mailing Add:* PO Box 2083 Mammoth Lakes CA 93546-2083

MALOY, JOSEPH T, b Mt Pleasant, Pa, Apr 19, 39; m 70; c 1. ANALYTICAL CHEMISTRY, ELECTROCHEMISTRY. *Educ:* St Vincent Col, BA, 61; Univ Tex, Austin, MA, 67, PhD(chem), 70. *Prof Exp:* Teacher, Mt Pleasant area schs, Pa, 61-65; from asst prof to assoc prof, WVa Univ, 70-79; ASSOC PROF CHEM, SETON HALL UNIV, 79-; PRIN, ELECTROANAL CONSULT ASSOCS, 81- *Concurrent Pos:* Sr investr, Air Force Aero-Propulsion Lab, Wright-Patterson AFB, 78-81 & Res Inst, Univ Dayton, 81-84; vis prof, Univ Tex, 82 & 84; div ed, J Electrochem Soc, 83-; consult, AT&T Bell Labs, 84-85; mem bd dirs, Soc Electroanalytical Chem, 84-88; sr assoc, Nat Res Coun-Air Force Systs Command. *Mem:* Am Chem Soc; AAAS; Sigma Xi; Soc Electroanal Chem (pres, 85-86). *Res:* Electrochemistry; electroanalytical techniques; cybernetic instrumentation for electroanalysis; digital simulation of problems involving mass transport and chemical reactions; high energy density batteries; chromatography; computational chemistry; artificial intelligence. *Mailing Add:* Dept of Chem Seton Hall Univ South Orange NJ 07079

MALOY, OTIS CLEO, JR, b Coeur d'Alene, Idaho, Jan 19, 30; m 53. PLANT PATHOLOGY. *Educ:* Univ Idaho, BS, 51, MS, 55; Cornell Univ, PhD, 58. *Prof Exp:* Asst plant path, Cornell Univ, 55-58; forest pathologist, US Forest Serv, 58-59; res forester, Potlatch Forests, Inc, 59-63; exten plant pathologist, 63-76, PROF PLANT PATH, WASH STATE UNIV, 76- *Mem:* Am Phytopath Soc. *Res:* Ecology and physiology of soil microorganisms; root rots; diseases of forest trees. *Mailing Add:* 2293 Wallen Rd Moscow ID 83843

MALOY, W LEE, b Kansas City, Mo, Oct 6, 40. EXPERIMENTAL BIOLOGY. *Educ:* Univ Md, BS, 67; Case Western Reserve Univ, MS, 69, PhD(inorg chem), 76. *Prof Exp:* Res technician, Res Div, Cleveland Clin Found, Ohio, 71-73; res assoc, Dept Biochem, Case Western Reserve Univ, Cleveland, Ohio, 76-78; staff fel, Lab Immunogenetics, Nat Inst Allergy & Infectious Dis, NIH, Bethesda, Md, 78-81, sr staff fel, 81-82, expert, 82- 86, sr staff fel, Bone Res Br, Nat Inst Dent Res, 86-87, sect head, Synthetic Peptide Antigen Sect, Biol Resources Br, 87-89; DIR PEPTIDE CHEM, MAGAININ SCI INC, PLYMOUTH MEETING, PA, 89- *Mem:* Am Asn Immunologists; Regulatory Affairs Prof Soc. *Res:* Peptide and protein chemistry; synthesis of angiotensin analogs by the solid phase method; amino acid sequence of the biotinyl subunit of the enzyme transcarboxylase; structure-function relationships in the immune system; author of various publications. *Mailing Add:* Magainin Sci Inc 5110 Campus Dr Plymouth Meeting PA 19462

MALOZEMOFF, ALEXIS P, b Santa Rosa, Calif; m 81; c 3. MAGNETICS, SUPERCONDUCTIVITY. *Educ:* Harvard Univ, BA, 66; Stanford Univ, PhD(mat sci eng), 70. *Prof Exp:* Nat Sci Found fel res, Clarendon Lab, Oxford, eng, 70-71; staff mem mgr res, IBM Res, Yorktown, 71-87, div coordr high temperature superconductivity, 87-90; VPRES RES & DEVELOP, AM SUPERCONDUCTOR CORP, 91- *Concurrent Pos:* Max Planck fel res, Max Planck Inst Metals Res, 77-78; vis prof res, Univ Grenoble & Lab Louis Neel, 85, Univ Paris, 90. *Mem:* Sr mem Inst Elec & Electronic Engrs; fel Am Phys Soc. *Res:* Magnetic domain walls; amorphous magnets; spin glasses; high temperature superconductors. *Mailing Add:* Am Superconductor Corp 149 Grove St Watertown MA 02172

MALOZEMOFF, PLATO, b St Petersburg, Russia, Aug 26, 09; m 42; c 2. MINING ENGINEERING. *Educ:* Univ Calif, Berkeley, BS, 31; Mont Sch Mines, MS, 32. *Hon Degrees:* DSc, Colo Sch Mines, 57. *Prof Exp:* Metall researcher, Mont Sch Mines, 31-34 & Alaska Juneau Gold Mining Co, 34-35; lab & field metallurgist, Pan Am Eng Corp, Berkeley, Calif, 35-40; mgr, var pvt gold mining enterprises, SAm & Cent Am, 40-44; eng analyst, Off Price Admin, Washington, DC, 44-45; mining engr, 45-52, vpres, 52-54, pres, chmn & chief exec officer, 54-85, EMER CHMN, NEWMONT MINING CORP, 85- *Concurrent Pos:* Dir, Browning-Ferris Industs; vpres & trustee, Am Mus Natural Hist; mem coun, Woodrow Wilson Int Ctr Scholars; mem adv comt, John F Kennedy Sch Govt, Harvard Univ; mem, indust adv bd, Nat Acad Eng, 83-; dir, Newmont Mining Corp, 53-87, consult, 85-87. *Honors & Awards:* Charles F Rand Mem Gold Medal, Am Inst Mining, Metall & Petrol Engrs, 72; Gold Medal, Inst Mining & Metall, London, 74; Gold Medal Award, Mining & Metall Soc Am, 76. *Mem:* Nat Acad Eng; Am Inst Mining; Am Inst Mining, Metall & Petrol Engrs; Mining & Metall Soc Am; Inst Mining & Metall London. *Mailing Add:* 230 Park Ave Suite 1154 New York NY 10169

MALPASS, DENNIS B, b Biloxi, Miss. ZIEGLER-NATTA CATALYSIS. *Educ:* Tulane Univ, New Orleans, BS, Univ Tenn, Knoxville, PhD(chem), 70. *Prof Exp:* Tech mgr, Tex Alkyls Inc, 70-80; staff chemist, Exxon Chem Am, 80-82; mgr res & develop, Tex Alkyls Inc, 82-; AT AKZO CHEM INC. *Mem:* Am Chem Soc; Soc Plastics Engrs. *Res:* Synthesis and properties of organometallics; applications of organometallics to synthetic organic chemistry; Ziegler-Natta polymerization of olefins. *Mailing Add:* AKZO Chem Inc Livingston Ave Dobbs Ferry NY 10522-3401

MALSBERGER, RICHARD GRIFFITH, b Philadelphia, Pa, Jan 12, 23; m 44; c 1. BIOLOGY. *Educ:* Lehigh Univ, BA, 48, MS, 49, PhD(bact), 58. *Prof Exp:* Mem staff, Biol Prod Dept, Merck Sharp & Dohme, 50-53, res assoc virol, 53-57, mgr control, 57-59; from asst prof to prof biol, 59-, EMER PROF, LEHIGH UNIV. *Mem:* AAAS; Am Inst Biol Sci; Tissue Cult Asn; Am Soc Microbiol. *Res:* Viral vaccines, immunology and multiplication; viral diseases of freshwater fishes; immunology. *Mailing Add:* RD 4 34 Pleasant View Rd Bethlehem PA 18015

MALSKY, STANLEY JOSEPH, b New York, NY, July 15, 25; m 65; c 1. RADIOLOGICAL PHYSICS, MEDICAL PHYSICS. *Educ:* NY Univ, BSc, 46, MA, 50, MSc, 53, PhD, 63. *Prof Exp:* Nuclear physicist, US Navy, 50-54; asst chief radiother, Vet Admin Hosp, Bronx, 54-73; PRES, RADIOL PHYSICS ASSOCS, INC, 73- *Concurrent Pos:* Asst prof, NY Univ, 59-63; res collab, Med Div, Brookhaven Nat Lab, 60-67; co-dir & prof radiol sci, Manhattan Col, 63-74; res prof radiol, Sch Med, NY Univ, 74-76; res grants, Nat Cancer Inst & Bur Radiol Health. *Honors & Awards:* James Picker Award Res Dosimetry. *Mem:* Fel AAAS; fel Am Pub Health Asn; Royal Soc Health; Am Asn Physicists Med; Health Physics Soc; Sigma Xi. *Res:* Solid state dosimetry; medical and radiological physics. *Mailing Add:* PO Box 31 Elmsford Post Office Elmsford NY 10523

MALSTROM, ROBERT ARTHUR, b Syracuse, NY, Mar 24, 50; m 81; c 1. LASERS, FIBER OPTICS. *Educ:* Syracuse Univ, BS, 74, PhD(chem), 80. *Prof Exp:* res chemist, Savannah River Lab, E I du Pont de Nemours Co, 80-89; PRIN SCIENTIST, WESTINGHOUSE SAVANNAH RIVER CO, 89- *Mem:* Am Chem Soc; Am Phys Soc. *Res:* Development of fiber optic sensors; adaptation of laser and fiber optic technology for the analysis of chemical species; gas phase Raman studies; solution and gas phase photochemistry and spectroscopy. *Mailing Add:* Westinghouse Savannah River Co Savannah River Lab 773-A Aiken SC 29808

MALT, RONALD A, b Pittsburgh, Pa, 31; m 51; c 3. SURGERY, MOLECULAR BIOLOGY. *Educ:* Wash Univ, AB, 51; Harvard Univ, MD, 55; Am Bd Surg, dipl, 62; Bd Thoracic Surg, dipl, 63. *Prof Exp:* Intern surg, Mass Gen Hosp, 55-56, resident, 58-62; from asst to assoc, 62-68,from asst prof to assoc prof, 68-75, PROF SURG, HARVARD MED SCH, 75-; CHIEF GASTROENTEROL SURG, MASS GEN HOSP, 70- *Concurrent Pos:* USPHS spec res fel biol, Mass Inst Technol, 62-63, fel, Sch Advan Study, 63-64; Am Heart Asn estab investr, 63-68; res assoc, Mass Inst Technol, 62-64, from asst surgeon to surgeon, Mass Gen Hosp, 62-; assoc surgeon, Shriners Burns Inst, 67- *Mem:* Am Surg Asn; Am Soc Clin Invest; Am Soc Cell Biol; Am Physiol Soc. *Res:* Regeneration; molecular events in renal, hepatic, and enteric growth and neoplasia; liver, biliary and portal-system surgery; replantation of limbs. *Mailing Add:* Mass Gen Hosp Boston MA 02114

MALTBY, FREDERICK L(ATHROP), b Bradford, Pa, Dec 14, 17. ELECTRICAL ENGINEERING. *Educ:* Grove City Col, BS, 40; Univ Buffalo, MA, 43. *Prof Exp:* Fel, Radium Lab, State Inst Malignant Dis, NY, 40-42; instr physics, Univ Buffalo, 42-44; elec eng, Rudloph Wurlitzer Co, 44; from sr engr to tech dir measurement & control, Bristol Co, 44-52, tech dir, Fielden Instrument Div, 52-57; PRES, DREXELBROOK ENG CO, 57- *Mem:* Am Phys Soc; Instrument Soc Am; Inst Elec & Electronics Engrs; Sigma Xi. *Mailing Add:* 780 Glen Rd Jenkintown PA 19046

MALTENFORT, GEORGE GUNTHER, b Landsberg, Germany, Aug 13, 13; US citizen; m 46; c 1. CHEMISTRY. *Educ:* Northwestern Univ, BS, 34. *Prof Exp:* Chemist, Transparent Package Co, 34-42 & 45-46; chemist, Container Div Lab, Container Corp Am, 46-58, tech dir, 58-78; CONSULT, 78- *Concurrent Pos:* Dir, Res & Develop Assocs, 65-68 & 74-76; mem packaging comt, Nat Acad Sci-Nat Res Coun, 74-77. *Honors & Awards:* Container Div Medal, Tech Asn Pulp & Paper Indust, 67, Tappi Gold Medal, 79. *Mem:* Am Chem Soc; Am Soc Qual Control; Tech Asn Pulp & Paper Indust; Am Soc Test & Mat. *Res:* Packaging, sampling, statistics and development of test methods and instruments. *Mailing Add:* 3355 Capitol St Skokie IL 60076

MALTER, MARGARET QUINN, organic chemistry, for more information see previous edition

MALTESE, GEORGE J, b Middletown, Conn, June 24, 31; m 56; c 2. MATHEMATICS. *Educ:* Wesleyan Univ, BA, 53; Yale Univ, PhD(math), 60. *Prof Exp:* NATO fel, Univ Gottingen, 60-61; instr math, Mass Inst Technol, 61-63; from asst prof to prof, Univ Md, College Park, 63-74; PROF MATH, UNIV MUNSTER, GER, 74- *Concurrent Pos:* Vis prof, Univ Frankfurt, 66-67 & 70-71, Univ Palermo, 70-71, Univ Kuwait, 77, Univ Bari, 79 & Univ Bahrain, 88-89. *Mem:* Am Math Soc; Austrian Math Soc; Ital Math Soc; Ger Math Soc. *Res:* Functional analysis with emphasis on Banach algebra theory and the spectral theory of linear operators. *Mailing Add:* Math Inst Univ Munster Einstein Str 64 Munster 44 Germany

MALTHANER, W(ILLIAM) A(MOND), b Columbus, Ohio, July 9, 15; m 50; c 2. ELECTRICAL ENGINEERING, SYSTEMS RESEARCH & ADMINISTRATION. *Educ:* Rensselaer Polytech Inst, BEE, 37. *Prof Exp:* Mem tech staff, Bell Tel Labs, 37-58, systs res engr, 58-63, systs develop engr, 63-75; CONSULT COMMUN SYSTS, 75- *Mem:* AAAS; fel Inst Elec & Electronics Engrs. *Res:* Transmission, handling and storage of information; complex communication switching systems; electronic computing and control systems. *Mailing Add:* 3001 Seventh Ave W Bradenton FL 34205-4115

MALTZ, MARTIN SIDNEY, b Bronx, NY, Jan 6, 41; m 69; c 2. MODEL IMAGING SYSTEMS. *Educ:* Rensselaer Polytech Inst, BSEE, 62; Mass Inst Technol, MS, 63, PhD(solid state physics), 68. *Prof Exp:* SR SCIENTIST ELEC ENG & SOLID STATE PHYSICS, XEROX RES, 68- *Mem:* Inst Elec & Electronics Engrs. *Res:* Xerography; image science. *Mailing Add:* Xerox Corp 800 Phillips Rd Webster NY 14580

MALTZ, MICHAEL D, b Brooklyn, NY, Dec 18, 38; m 66, 84; c 3. OPERATIONS RESEARCH. *Educ:* Rensselaer Polytech Inst, BEE, 59; Stanford Univ, MSEE, 61, PhD(elec eng), 63. *Prof Exp:* Res assoc control systs, Tech Univ, Denmark, 63-64; staff assoc systs eng, Arthur D Little, Inc,, 64-69; opers res analyst, Nat Inst Law Enforcement & Criminal Justice, 69-72; assoc prof criminal justice & systs eng, Univ Ill, Chicago, 72-80, assoc prof criminal justice & quant methods, 80-84, dir, Res Law & Justice Ctr, 85-88, PROF CRIMINAL JUSTICE, INFO & DECISION SCI, UNIV ILL, CHICAGO, 84- *Concurrent Pos:* Lectr, Grad Sch Eng, Northeastern Univ, 65; vis lectr, Opers Res Soc Am & Inst Mgt Sci, 73-75; eval consult to numerous agencies & industs, 73-; mem bd dirs, Int Pub Prog Anal, St Louis, 74-80; consult, Task Force on Criminal Justice Res & Develop, Nat Adv Comn on Criminal Justice Standards & Goals, 75-76, Task Force on Bid-Rigging, Nat Dist Attys Asn, Econ Crime Proj, 77-78, Rand Corp, 75-82, Police Exec Res Forum, 76-; mem acad adv coun, 10th Cong Dist Ill, 75-80; mem bd dirs, Bus & Prof People for Pub Interest, Chicago, 76-, Ill Citizens Handgun Control, 80-83, Chicago Law Enforcement Study Group, 82-; consult ed, J Res Crime & Delinquency, 76-; mem working group on FBI Comput Criminal Hist Prog, Off Technol Assessment, Cong US, 77-78; assoc ed, Opers Res, 78-83. *Mem:* AAAS; Opers Res Soc Am; Inst Mgt Sci; Sigma Xi; Am Soc Criminol. *Res:* Criminal justice system research, operations research. *Mailing Add:* Dept Criminal Justice M/C 222 Box 4348 Univ Ill Chicago Chicago IL 60680

MALTZEFF, EUGENE M, b Khabarovsk, Russia, Oct 31, 12; US citizen; m 46; c 2. FISHERIES. *Educ:* U Wash, BS, 39. *Prof Exp:* Aquatic biologist, Fish & Wildlife Serv, US Bur Com Fisheries, 44-48, fishery res biologist, 42-57, fishery biologist, 58-68, foreign fisheries analyst, Pac Northwest Region I, 68-70, foreign fisheries analyst, Nat Marine Fisheries Serv, 70-74; CONSULT,75- *Concurrent Pos:* Consult marine affairs, 75-; interpreter, Can Drilling Res Asn, Can Dept of State, 76. *Mem:* Am Fisheries Soc; Am Inst Fishery Res Biol. *Res:* Pacific salmon; stream improvement; Indian fisheries; foreign fishing. *Mailing Add:* 4501 Stanford Ave NE Seattle WA 98105

MALUEG, KENNETH WILBUR, b Appleton, Wis, Apr 19, 38; m 64; c 2. LIMNOLOGY. *Educ:* Univ Wis, BS, 60, MS, 63, PhD(zool), 66. *Prof Exp:* res aquatic biologist, Corvallis Environ Res Lab, US Environ Protection Agency, 66-88; CONSULTANT, 88- *Mem:* Am Soc Limnol & Oceanog; Sigma Xi; Amer Inst Biol Sci; Int Asn Theoret & Appl Limnol; NAm Lake Mgt Soc. *Res:* Lake restoration; eutrophication control; nonpoint source pollution control; sediment; sediment-water exchange; sediment bioassays. *Mailing Add:* 3455 NW Roosevelt Dr Corvallis OR 97333

MALVEAUX, FLOYD J, b Opelousas, La, Jan 11, 40; m 65; c 4. MICROBIOLOGY, MICROBIAL PHYSIOLOGY. *Educ:* Creighton Univ, BS, 61; Loyola Univ, La, MS, 64; Mich State Univ, PhD(microbiol), 68. *Prof Exp:* Instr soil microbiol, Mich State Univ, 68; ASST PROF MICROBIOL, COL MED, HOWARD UNIV, 68- *Mem:* AAAS; Am Soc Microbiol. *Res:* Characterization of enzymes and extracellular proteins of pathogenic bacteria as these products relate to virulence; physiology of microorganisms associated with plaque formation and periodontal disease. *Mailing Add:* Col Med Howard Univ Washington DC 20001

MALVEN, PAUL VERNON, b Annapolis, Md, Oct 24, 38; m 63; c 2. NEUROENDOCRINOLOGY, REPRODUCTIVE PHYSIOLOGY. *Educ:* Univ Ill, Urbana, BS, 60; Cornell Univ, PhD(animal physiol), 64. *Prof Exp:* NIH fel, Univ Calif, Los Angeles, 64-65; from asst to assoc prof animal sci, 66-72, PROF ANIMAL SCI, PURDUE UNIV, 72- *Honors & Awards:* Physiol & Endocrinol Award, Am Soc Animal Sci; Upjohn Physiol Award, Am Dairy Sci Asn. *Mem:* Am Physiol Soc; Endocrine Soc; Soc Neurosci; Am Soc Animal Sci; Soc Study Reproduction; Int Soc Neuroendocrinol. *Res:* Neuroendocrinology of reproduction; lactational physiology. *Mailing Add:* Dept Animal Sci Purdue Univ West Lafayette IN 47906

MALVICK, ALLAN J(AMES), b Chicago, Ill, Oct 15, 35; m 63; c 3. ENGINEERING MECHANICS. *Educ:* Univ Notre Dame, BS, 57, MS, 59, ScD(eng sci), 61. *Prof Exp:* Asst prof eng sci, Univ Notre Dame, 61-65; assoc prof civil eng, 65-71, assoc prof optical sci, 67-71, PROF CIVIL ENG, UNIV ARIZ, 71- *Mem:* Am Soc Civil Engrs; Optical Soc Am; Sigma Xi. *Res:* Deformation of optical mirrors. *Mailing Add:* Dept of Civil Eng Univ of Ariz Tucson AZ 85721

MALVILLE, JOHN MCKIM, b San Francisco, Calif, Apr 24, 34; m 60; c 2. ASTROPHYSICS. *Educ:* Calif Inst Technol, BS, 56; Univ Colo, PhD(astrophys), 61. *Prof Exp:* Res assoc astron, Univ Mich, 62-63, asst prof, 63-65; mem sr staff solar physics, High Altitude Observ, 65-70; asst dean col arts & sci, 69-70, assoc prof, 70-73, PROF ASTRO GEOPHYS, UNIV COLO, BOULDER, 73-, CHMN DEPT, 77- *Concurrent Pos:* Am Astron Soc vis prof, 64- *Mem:* AAAS; Am Astron Soc; Int Astron Union. *Res:* Solar physics; radio astronomy; auroral, atomic and molecular physics; interstellar medium; philosophy of science; science education. *Mailing Add:* Dept Astro Univ Colo Boulder CO 80309

MALVIN, GARY M, b Brooklyn, NY, June 10, 54. RESEARCH. *Educ:* Univ Mich, BA, 76; Univ NMex Sch Med, PhD(physiol), 83. *Prof Exp:* Scientist, Dept Physiol, Max-Planck Inst Exp Med, Gottingen, Ger, 83-84; sr fel, Dept Physiol & Biophys, Sch Med, Univ Wash, 84-88; ASSOC SCIENTIST, RES DIV, LOVELACE MED FOUND, 88- *Concurrent Pos:* Lectr & lab instr human physiol, Univ NMex, 79-83, Univ Wash, 85-88, Univ NMex, 89-; vis scientist, Dept Physiol, Univ Mich, 81, Dept Zoophysiol, Univ Aarhus, Denmark, 82, Dept Physiol, Max-Planck Inst, Gottingen, Ger, 84; Individual Nat Res Serv award, NIH, 84-87, First Award, 88-; adj asst prof, Dept Biol, Univ NMex, 88- & Dept Physiol, 89- *Honors & Awards:* Scholander Award, Am Physiol Soc, 88. *Mem:* Am Physiol Soc. *Res:* Cellular mechanisms of smooth muscle function; effects of high frequency ventilation on gas exchange; adrenergic control of cardiovascular shunting. *Mailing Add:* Res Div Lovelace Med Found 2425 Ridgecrest Dr SE Albuquerque NM 87108

MALVIN, RICHARD L, b Aug 19, 27; US citizen; m 49; c 2. PHYSIOLOGY, HYPERTENSION. *Educ:* McGill Univ, BSc, 50; NY Univ, MS, 54; Univ Cincinnati, PhD(physiol), 56. *Prof Exp:* Res assoc, 56-57, from instr to assoc prof, 57-67, PROF PHYSIOL, MED SCH, UNIV MICH, 67- *Concurrent Pos:* Vis prof, Univ Atago, Christchurch, NZ. *Honors & Awards:* Gold Medal, Brit Med Asn, 68. *Mem:* AAAS; Am Physiol Soc; Am Soc Nephrology; Sigma Xi; Int Soc Hypertension. *Res:* Renal physiology; salt and water balance; control of secretion of renin and antidiuretic hormone; major research interest are in control of salt and water balance; the renin angiotensin system and hypertension. *Mailing Add:* Dept Physiol Univ Mich 7730 Med Sci II Ann Arbor MI 48109

MALY, EDWARD J, b Troy, NY, Nov 10, 42; m 67; c 2. ECOLOGY, EVOLUTION. *Educ:* Univ Rochester, BS, 64; Princeton Univ, PhD(biol), 68. *Prof Exp:* Asst prof biol, Tufts Univ, 68-75; from asst prof to assoc prof biol, 75-87, chmn dept, 85-88, PROF BIOL, CONCORDIA UNIV, 87- *Mem:* Sigma Xi; Ecol Soc Am; Am Soc Limnol & Oceanog. *Res:* Population dynamics; predator-prey interactions and population growth rates; life histories and diversity of fresh-water animals. *Mailing Add:* Dept Biol Concordia Univ 1455 De Maisonneuve Blvd W Montreal PQ H3G 1M8 Can

MALZAHN, DON EDWIN, b Perry, Okla, June 8, 45; m 78. INDUSTRIAL ENGINEERING, REHABILITATION ENGINEERING. *Educ:* Okla State Univ, BS, 68, MS, 69, PhD(indust eng), 75. *Prof Exp:* Industr engr, Charles Mach Works, 70-73; asst prof, 73-81, ASSOC PROF INDUST ENG, WICHITA STATE UNIV, 81- *Concurrent Pos:* Dir proj 1, Rehab Eng Ctr, Wichita State Univ, 76- *Mem:* Am Inst Indust Engrs; Am Soc Eng Educ; Human Factors Soc. *Res:* Development of job modification and adaptive device strategies for the employment of the severely handicapped in mainstream industry. *Mailing Add:* Dept Indust Eng Box 35 Wichita State Univ Wichita KS 67208

MALZAHN, RAY ANDREW, b Ft Madison, Iowa, July 8, 29; m 53; c 2. ORGANIC CHEMISTRY, ACADEMIC ADMINISTRATION. *Educ:* Gustavus Adolphus Col, BA, 51; Univ NDak, MS, 53; Univ Md, PhD(org chem), 62. *Prof Exp:* Assoc prof chem, WTex State Univ, 63-67, dean, Col Arts & Sci, 67-71, vpres acad affairs, 71-77, prof chem, 67-80; PROF CHEM & DEAN, SCH ARTS & SCI, MO SOUTHERN STATE COL, 80- *Concurrent Pos:* Fel, Univ Ariz, 61-63. *Mem:* Am Chem Soc; AAAS. *Res:* Pyrolysis of allyl and propargyl ethers; polymerization of monomers derived from natural products; synthesis of arylsilanes containing carboxyl groups. *Mailing Add:* Dean Sch Arts & Sci Mo Southern State Col Joplin MO 64801-1595

MALZER, GARY LEE, b Nebraska City, Nebr, Nov 8, 45; m 68; c 3. SOIL FERTILITY, CHEMISTRY. *Educ:* Univ Nebr, BS, 67, MS, 70; Purdue Univ, PhD(agron soil fertil), 73. *Prof Exp:* Soil conservationist, Soil Conservation Serv, USDA, 67-68; res assoc agron, Purdue Univ, 73-74; asst prof, 74-79,

ASSOC PROF SOIL SCI, UNIV MINN, ST PAUL, 79- *Mem:* Am Soc Agron; Soil Sci Soc Am; Soil Conserv Soc Am. *Res:* Soil fertility and plant nutrition, particularly methodology to characterize and improve fertilizer use efficiency for agricultural production. *Mailing Add:* Dept Soil Sci Univ Minn 1529 Gortner Ave St Paul MN 55108

MAMANDRAS, ANTONIOS H, b Athens, Greece, Sept 13, 49; m 76; c 2. CRANIOFACIAL GROWTH & COMPUTERIZED CEPHALOMETRICS. *Educ:* Univ Athens, DDS, 73; Univ Man, MSc, 80. *Prof Exp:* Clin instr, fac dent, Univ Man, 78-80; asst prof & dir orthod, fac dent, Univ Western Ont, 82-88, asst prof fac grad studies, 84-88, ASSOC PROF & DIR ORTHOD, FAC DENT, UNIV WESTERN ONT, 89-, ASSOC PROF, FAC GRAD STUDIES, 89- *Mem:* Am Asn Orthod; Can Asn Orthod; Can Asn Univ Teachers. *Res:* Clinical orthodontic research. *Mailing Add:* Dent Sci Bldg Rm 1013 Univ Western Ont London ON N6A 5C1 Can

MAMANTOV, GLEB, b Karsava, Latvia, Apr 10, 31; US citizen; m 56; c 3. ANALYTICAL & INORGANIC CHEMISTRY. *Educ:* La State Univ, BS, 53, MS, 54, PhD(chem), 57. *Prof Exp:* Res chemist electrochem dept, E I Du Pont de Nemours & Co, 57-58; instr & res assoc chem, Univ Wis, 60-61; from asst prof to prof chem, 61-86, HEAD DEPT, UNIV TENN, KNOXVILLE, 79-, DISTINGUISHED PROF CHEM, 86-; PRES, MOLTEN SALT TECHNOL, 87- *Concurrent Pos:* Consult, Oak Ridge Nat Lab, 62-; NATO sr fel, Ger, 71; NSF travel award, Int cong Polarography, 64, Inst Inorg Chem, Kiev, 74; div ed, Phys Electorchem Div, J Electrochem Soc, 80; invited lectr, several Gordon & Euchem conferences on molten salts, 29th IUPAC cong, Vancouver, 81; mem, Coun Chem Res, 84-86; dir, molten salt chem, NATO Advan Study Inst, Camerino, Italy, 86; chmn, Gordon Res Conf Molten Salts & Liquid Metals, 87, discussion leader, 89. *Honors & Awards:* Meggers Award, Soc Appl Spectros, 83; Charles H Stone Award, Am Chem Soc, 89. *Mem:* Fel AAAS; Am Chem Soc; Electrochem Soc; fel Am Inst Chemists; Sigma Xi; Soc Appl Spectors; Soc Electroanal Chem. *Res:* Electrochemistry and chemistry in molten salts; fluorine chemistry; electroanalytical chemistry; auth or co-auth of 220 publications. *Mailing Add:* Dept Chem Univ Tenn Knoxville TN 37996-1600

MAMELAK, JOSEPH SIMON, b Lodz, Poland, Dec 14, 23; US citizen; m 59; c 2. COMPUTER SCIENCE, APPLIED MATHEMATICS. *Educ:* McGill Univ, BS, 45, MS, 46; Univ Pittsburgh, PhD(math), 49. *Prof Exp:* Asst prof math, Univ WVa, 53-56; sr analyst opers res, Univac, Sperry Rand Corp, 56-59; mgr sci appln, RCA, 59-62, proj mgr automated design, 62-65; head dept, 65-71, PROF MATH, COMMUNITY COL PHILADELPHIA, 66- *Concurrent Pos:* Consult, City Philadelphia, 65-; mem, var med & educ insts, 66-; mem, Am Stand Inst, 64- *Mem:* Am Math Soc; Am Statist Asn; Asn Comput Mach; Math Asn Am; Can Math Cong. *Res:* Water pollution models using computer simulation; automated circuit design and integrated circuit layout using computers; computer utilization in schools. *Mailing Add:* 70 Knollwood Dr Cherry Hill NJ 08034

MAMET, BERNARD LEON, b Brussels, Belg, Feb 7, 37; m 63; c 1. GEOLOGY. *Educ:* Free Univ Brussels, LSc, 57, PhD(stratig), 62; French Petrol Inst, cert eng geol, 59; Univ Calif, Berkeley, MA, 60. *Prof Exp:* Researcher, Royal Inst Natural Sci, Belg, Brussels, 56-61; asst researcher, Nat Found Sci Res, Belg, 63-65; from asst prof to assoc prof, 65-76, PROF GEOL, UNIV MONTREAL, 76- *Mem:* Geol Soc Belg; Geol Soc France; Belg Soc Geol, Paleont & Hydrol; Am Asn Petrol Geologists; Soc Econ Paleontologists & Mineralogists. *Res:* Carboniferous microfacies. *Mailing Add:* Dept Geol Univ Montreal CP 6128 Succ A Montreal PQ H3C 3J7 Can

MAMIYA, RICHARD T, b Honolulu, Hawaii, Mar 8, 25; m 50; c 8. THORACIC SURGERY, CARDIOVASCULAR SURGERY. *Educ:* Univ Hawaii, BSc, 50; St Louis Univ, MD, 54. *Prof Exp:* From instr to sr instr surg, Sch Med, St Louis Univ, 59-61, dir sect surg, Cochran Vet Admin Hosp, 59-61; from assoc prof to prof surg, Sch Med, Univ Hawaii, Manoa, 67-74, chmn dept, 67-74. *Concurrent Pos:* Consult, US Army Tripler Gen Hosp, 68- *Mem:* AMA; Am Col Surg. *Mailing Add:* 1380 Lusitana St No 710 Honolulu HI 96813

MAMOLA, KARL CHARLES, b Greenport, NY, May 23, 42; m 63; c 1. SOLID STATE PHYSICS. *Educ:* State Univ NY, Stony Brook, BS, 63; Fla State Univ, MS, 65; Dartmouth Col, PhD(physics), 73. *Prof Exp:* From instr to asst prof physics, 65-69, asst prof, 72, ASSOC PROF PHYSICS, APPALACHIAN STATE UNIV, 72-, CHAIRPERSON DEPT, 76- *Mem:* Am Phys Soc; Am Asn Physics Teachers; Nat Sci Teachers Asn; Sigma Xi. *Res:* Point deflects in crystals at high pressures using optical absorption and magnetic resonance spectroscopy. *Mailing Add:* Dept of Physics Appalachian State Univ Boone NC 28608

MAMRACK, MARK DONOVAN, b Cleveland, Ohio, Nov 18, 51. CELLULAR BIOCHEMISTRY. *Educ:* Purdue Univ, BS, 72; Baylor Col Med, PhD(pharmacol), 78. *Prof Exp:* Teaching fel, biol div, Oak Ridge Nat Lab, 78-81; vis asst prof cell biol, biomed sci, Univ Tenn Grad Sch & biol div, Oak Ridge Nat Lab, 81-83; ASST PROF CELL BIOL, DEPT BIOL SCI, WRIGHT STATE UNIV, 83- *Mem:* Am Soc Cell Biol. *Res:* Signal transduction inside cells involving protein phosphorylation, particularly protein kinases that are independent of cyclic-nucleotide regulations; model systems including mouse epidermis, bovine heart and normal human fibroblasts. *Mailing Add:* Dept Biol Sci Wright State Univ Dayton OH 45435

MAMRAK, SANDRA ANN, b Cleveland, Ohio, Sept 8, 44; m 78; c 1. COMPUTER SCIENCES. *Educ:* Notre Dame Col, BS, 67; Univ Ill, Urbana-Champaign, MS, 73, PhD(comput sci), 75. *Prof Exp:* Comput scientist, Nat Bur Standards, 75-79; mem tech staff, Bell Labs, 80; fac appointee, Lawrence Livermore Nat Lab, 81-84; PROF COMPUT SCI, OHIO STATE UNIV, 75- *Concurrent Pos:* Sabbatical, Xerox Parc, 90-91. *Mem:* Asn Comput Mach. *Res:* Heterogeneity in computer systems; translation and exchange of electronic data. *Mailing Add:* Dept of Comput & Info Sci Ohio State Univ Columbus OH 43210

MAN, CHI-SING, b Hong Kong, Aug 23, 47; m 73; c 2. CONTINUUM MECHANICS & THERMODYNAMICS. *Educ:* Univ Hong Kong, BSc, 68, MPhil, 76; Johns Hopkins Univ PhD(mech), 80. *Prof Exp:* Tutor math & physics, Hong Kong Baptist Col, 70-72, asst lectr physics, 72-76; fel mech, Johns Hopkins Univ, 80-81; asst prof civil eng, Univ Man, 81-85; asst prof, 85-88, ASSOC PROF MATH, UNIV KY, 88- *Concurrent Pos:* Mem, Inst Math & Applns, Univ Minn, 84-85. *Mem:* Soc Natural Philos; Am Math Soc. *Res:* Ultrasonic measurement of residual stress in plastically deformed sheets; constitutive equations for creep of ice; stress waves in lungs; foundations of continuum thermodynamics and Gibbsian thermostatics; partial differential equations. *Mailing Add:* Dept Math Univ Ky Lexington KY 40506-0027

MAN, EUGENE H, b Scranton, Pa, Dec 14, 23; m 45, 75; c 6. NEUROCHEMISTRY. *Educ:* Oberlin Col, AB, 48; Duke Univ, PhD(chem), 52. *Prof Exp:* Res chemist chem dept, E I du Pont de Nemours & Co, 51-58, textile fibers dept, 58-60, supvr nylon tech div, 60-61, sr supvr, 61-62; coordr res, 62-66, dean res & sponsored progs, 66-79, PROF CHEM, UNIV MIAMI, 68- *Concurrent Pos:* Vis investr, Scripps Inst Oceanog, Univ Calif, San Diego, 71-72; Lady Davis vis fel, Technion-Israel Inst Technol, Haifa, Israel, 90. *Mem:* AAAS; Am Chem Soc; Sigma Xi; fel Am Inst Chem. *Res:* Implications of altered amino acids on aging of brain proteins; immunological dysfunctions; neurochemical dysfunctions-Alzheimer's disease. *Mailing Add:* Dept Chem Univ Miami Coral Gables FL 33124

MAN, EVELYN BROWER, b Lawrence, NY, Oct 7, 04. CLINICAL CHEMISTRY. *Educ:* Wellesley Col, AB, 25; Yale Univ, PhD(physiol chem), 32. *Prof Exp:* Instr chem, Conn Col, 25-27; technician sch med, Yale Univ, 28-29, asst, 29-30, Am Asn Univ Women fel, 33-34, from instr to asst prof psychiat, 34-50, res assoc med, 50-61; assoc mem, 61-71, EMER ASSOC MEM, INST HEALTH SCI, BROWN UNIV, 71- *Mem:* AAAS; Endocrine Soc; Am Thyroid Asn; Am Chem Soc; Am Soc Biochem & Molecular Biol. *Res:* Lipemia and iodemia in thyroid diseases; pregnancy and infancy; screening for neonatal hypothyroidism. *Mailing Add:* 275 Steele Rd Apt B 407 West Hartford CT 06117

MAN, SHU-FAN PAUL, b Hong Kong, Jan 13, 45; Can citizen; m 71; c 2. MEDICINE, ENVIRONMENTAL HEALTH. *Educ:* Univ Alta, BSc, 69, MD, 70; FRCP(C), 75. *Prof Exp:* From asst prof to assoc prof, 76-85, PROF MED, UNIV ALTA, 85- *Concurrent Pos:* Counr, Fedn Clin Res, 81-84. *Mem:* Fel AM Col Physicians; Am Physiol Soc; Am Thoracic Soc; fel Am Col Chest Physicians; Fedn Clin Res. *Res:* Pulmonary physiology related to defense mechanisms. *Mailing Add:* 2E433 WCMHSC Univ Alta Edmonton AB T6G 2B7 Can

MANABE, SYUKURO, b Japan, Sept 21, 31; US citizen; m 62; c 2. CLIMATE DYNAMICS, CLIMATE MODELING. *Educ:* Tokyo Univ, BA, 53, MA, 55, DSc, 58. *Prof Exp:* Res meteorologist, 58-79, MEM SR EXEC SERV, GEOPHYS FLUID DYNAMICS LAB, NAT OCEANIC & ATMOSPHERIC ADMIN, 79- *Concurrent Pos:* Adj prof, Dept Geol & Geophysics, Princeton Univ, 68- *Honors & Awards:* Meisinger Award, Am Meteorol Soc, 67 & Second Half Century Award, 77; Fujiwara Award, Japan Meteorol Soc, 66; Gold Medal Award, Dept Com, 70. *Mem:* Nat Acad Sci; fel Am Geophys Union; fel Am Meteorol Soc. *Res:* Study of climate variation by use of mathematical models of climate; climate change resulting from the future increase of atmospheric carbon dioxide. *Mailing Add:* Geophys Fluid Dynamics Lab/NOAA Princeton Univ PO Box 308 Princeton NJ 08542

MANAK, RITA C, BIOCHEMISTRY, MICROBIOLOGY. *Educ:* Univ Ill, PhD(immunochem), 75. *Prof Exp:* MGR, DEPT BIOTECHNOL, IRCERCA, 84- *Res:* Immunology. *Mailing Add:* Dept Technol Transfer Univ Ariz 1745 N Campbell Tucson AZ 85719

MANAKER, ROBERT ANTHONY, b Avenel, NJ, Feb 28, 18; m 53; c 1. MICROBIOLOGY. *Educ:* Rutgers Univ, BS, 50, PhD(microbiol), 53. *Prof Exp:* Merck-Waksman res fel, Rutgers Univ, 53-54, instr microbiol, Inst Microbiol, 54-55, asst prof virol, 55-56; res microbiologist, Nat Cancer Inst, 56-72, chief viral biol br, 72-76, chief, Lab of DNA Tumor Viruses, 76-80; RETIRED. *Mem:* AAAS; Am Asn Cancer Res; Am Soc Microbiol. *Res:* Virus-tumor relationships. *Mailing Add:* 5305 Balto Ave Chevy Chase MD 20815

MANAKKIL, THOMAS JOSEPH, b Gothuruthy, India, Dec 31, 33. PHYSICS. *Educ:* Univ Kerala, BSc, 53; Univ Saugar, MSc, 58; NMex State Univ, MS, 65, PhD(physics), 67. *Prof Exp:* Instr physics, Sacred Heart Col, India, 53-55 & 57-58; from asst prof to assoc prof, 67-77, PROF PHYSICS, MARSHALL UNIV, 77- *Concurrent Pos:* NSF. *Mem:* Am Phys Soc; Am Asn Physics Teachers; Inst Fundamental Studies Asn. *Res:* Magnetic resonance; crystal field studies; nuclear physics; interactions of gamma photons with fibers. *Mailing Add:* Dept of Physics Marshall Univ Huntington WV 25701

MANALIS, MELVYN S, b Los Angeles, Calif, Nov 16, 39; m 64; c 3. WIND ENERGY, GENERAL ENERGY. *Educ:* Calif State Univ, Northridge, BA, 61, Santa Barbara, PhD(physics), 70; Univ NH, MS, 64. *Prof Exp:* Scientist physics, Te Co, 70-72; res asst, Univ Calif, Santa Barbara, 65-70, lectr, Dept Physics, 75, res physicist, Quantum Inst, 72-79, ADJ LECTR ENVIRON STUDIES, UNIV CALIF, SANTA BARBARA, 75-; CONSULT, US DEPT COM, MD, 86- *Concurrent Pos:* Scientist II physics, Univ Colo, Boulder, 66; physicist, Nat Bur Standards, Washington, DC, 67; prin investr, Univ Calif, Santa Barbara, 79- *Mem:* Am Phys Soc; Am Asn Physics Teachers; AAAS; Sigma Xi. *Res:* Energy renewable technologies for energy end-use cost effectiveness; wind energy instigation with particular emphasis on desalination of salt water using wind energy. *Mailing Add:* Environ Studies Univ Calif Phelps 3206 Santa Barbara CA 93106-4160

MANALO, PACITA, b Manila, Phillipines, May 17, 32. PATHOLOGY. *Educ:* Univ Santo Thomas, Phillipines, MD, 55. *Prof Exp:* ASSOC PROF PATH, SCH MED, UNIV NEV, 76- *Mem:* Am Soc Clin Path; Am Asn Pathologists; Am Med Women's Asn. *Mailing Add:* Dept Path Univ Nev Sch Med Manville Med Sci Bldg Reno NV 89520

MANASEK, FRANCIS JOHN, b New York, NY, July 22, 40. CELL BIOLOGY, DEVELOPMENTAL BIOLOGY. *Educ:* NY Univ, AB, 61; Harvard Univ, DMD, 66. *Prof Exp:* Fel anat, Harvard Med Sch, 66-68; vis investr develop biol, Carnegie Inst Washington, 68-69; from instr to asst prof anat, Harvard Med Sch, 69-74; ASSOC PROF ANAT, UNIV CHICAGO, 74-, ASSOC PROF PEDIAT & COMT DEVELOP BIOL, 80- *Concurrent Pos:* Res assoc path & cardiol, Children's Hosp, 69-74; NIH res career develop award, 71; fel, Med found Inc, 69. *Mem:* Am Asn Anatomists; Am Soc Cell Biol; Biophys Soc; Int Soc Heart Res; Soc Develop Biol. *Res:* Cell biology of developing cardiac muscle particularly synthesis of structural macromolecules and their role in myocardial morphogenesis. *Mailing Add:* Dept Anat Dartmouth Med Sch Hanover NH 03756

MANASSAH, JAMAL TEWFEK, b Haifa, Palestine, Feb 23, 45; US citizen; m 79; c 2. OPTICS, SCIENCE POLICY. *Educ:* Am Univ Beirut, BSc, 66; Columbia Univ, MA, 68 PhD(theoret physics), 70. *Prof Exp:* Mem, Inst Advan Study, Princeton, 70-72; asst prof & secy bd grad, Am Univ Beirut, 72-75; chief sci adv, Kuwait Inst Sci Res 76-81; chief operating officer, Kuwait Found Advan Sci, 79-81; PROF ELECTROPHYSICS, DEPT ELEC ENG, CITY COL NY, 81- *Concurrent Pos:* Consult, Columbia Radiation Lab, Optical Sci Ctr, Univ Ariz, Ctr Educ Res & Develop, Beirut, Al-Hazen Res Ctr, Baghdad, Ford Found, NY, Beirut, Cairo & Khartom, NSF, Washington, DC, Welfare Asn Geneva, Khayatt & Co, Inc & mang int eng & investment co; mem adv comt int prog, NSF, 78-83, organizing comt, Chem Res Appl World Needs Two, 80-83; chmn, Int Conf Series Innovations Technol, 79-81; comr, Develop Lebanese Boys Scout Asn, 72-75; coordr, Task Force Excellence Welfare Asn, Switz, 84-85 & mem in many task forces & comn dedicated to develop countries technol & educ develop; Columbia fac fel; Pfister fel. *Honors & Awards:* ABI Key Award, 87. *Mem:* Optical Soc Am; AAAS; NY Acad Sci; Asn Mem Inst Advan; Int Platform Asn. *Res:* statistical quantum electrodynamics of two-level systems, collective Lamb and blackbody shifts, pion condensation, ultrafast phenomena and nonlinear optics; author and co- author of more than 100 monographs, articles & reports in theoretical physics, photonics, research management methodology and techno-economics. *Mailing Add:* Dept Elec Eng City Col NY Convent Ave & 140th St New York NY 10031

MANASSE, FRED KURT, b Frankfurt, Ger, July 27, 35; m 56; c 5. ELECTRICAL ENGINEERING, PHYSICS. *Educ:* City Col New York, BEE, 56, MEE, 58; Princeton Univ, AM, 59, PhD(physics), 62. *Prof Exp:* Mem tech staff, Bell Tel Labs, 56-63; asst prof elec eng, City Col New York, 63-65 & Princeton Univ, 65-68; assoc prof, Dartmouth Col, 68-72; prof, chmn dept elec eng, asst dean continuing educ & dir, Ctr Teaching Innovations, Drexel Univ, 72-75; vis prof & Nat Acad Sci fel, Bucharest Inst Physics, 73; vis prof elec eng, State Univ NY, Stony Brook, 76; prof elec eng, Univ NH, 76-80; pres & chief scientist, Aeta Corp, 80-85; div staff, Mitre Corp, 85-87; PRIN ENGR, MISSILE SYSTS DIV, RAYTHEON CO, 87- *Concurrent Pos:* Consult, Int Bus Mach Corp, NY, 59-60, Pitt Precision Prod, 60-61, Opinion Res Corp, 67-68 & US Energy Res & Develop Admin, 76-77; mem, Comt Gravitation & Relativity, 60-68; mem, Adv Comt Nonionizing Radiation, NJ Dept Health, 67-68; elec eng ed, Addison-Wesley Publ Corp, 67-78; ed-in-chief, T H E J; vpres, Sanders Assoc, 68-82, AETA Corp, 76-80, Frequency Sources, 81-82 & Ferrofluidics Corp, 85. *Mem:* AAAS; Am Soc Eng Educ. *Res:* Semiconductor device physics; switching circuits; memory devices; computer language and system design; general relativity; field interactions in solids; solid state devices; educational innovation; minority education; solar energy; small scale hydro; magnetics. *Mailing Add:* 225 Trapelo Rd Waltham MA 02154

MANASSE, ROGER, b New York, NY, Apr 9, 30; m 52; c 2. PHYSICS. *Educ:* Mass Inst Technol, BS, 50, PhD(physics), 55. *Prof Exp:* Mem staff instrumentation lab, Mass Inst Technol, 50-52, Lincoln Lab, 54-59; subdept head radar dept, Mitre Corp, 59-60, assoc dept head, 60-64, dept head, 64-67; staff mem, Gen Res Corp, 67-70; independent consult, 70-71; vpres & chief scientist, Spectra Res Systs, 72-73; independent consult, 74-83; pres, Swerling Manasse & Smith, Inc, 83-85; INDEPENDENT CONSULT, 86- *Concurrent Pos:* Mem, Air Force Sci Adv Bd, 69-78; mem numerous govt adv comts. *Mem:* Inst Elec & Electronics Eng; Sigma Xi. *Res:* Radar systems analysis and measurement theory; develop advanced radar concepts; technical direction of research programs. *Mailing Add:* 234 Canon Dr Santa Barbara CA 93105

MANASTER, ALFRED B, b Chicago, Ill, May 25, 38; m 60; c 3. MATHEMATICAL LOGIC. *Educ:* Univ Chicago, BS, 60; Cornell Univ, PhD(math), 65. *Prof Exp:* Res assoc math, Cornell Univ, 65; instr, Mass Inst Technol, 65-67; asst prof, 67-71, ASSOC PROF MATH, UNIV CALIF, SAN DIEGO, 71- *Mem:* Am Math Soc; Math Asn Am; Asn Symbolic Logic. *Res:* Recursive function theory. *Mailing Add:* Dept Math C-012 Univ Calif Box 109 La Jolla CA 92093

MANATT, STANLEY L, b Glendale, Calif, July 13, 33; m 58; c 4. ORGANIC CHEMISTRY. *Educ:* Calif Inst Technol, BS, 55, PhD(chem, physics), 59. *Prof Exp:* Wis Alumni Res Found fel, Univ Wis, 58-59; sr scientist, Jet Propulsion Lab, Calif Inst Technol, 59-64, res specialist, 65-66, mem tech staff, 66-70, asst mgr biosci & planetology sect, 70-73, staff scientist, Sci Data Analysis Sect, 73-78, STAFF SCIENTIST, INFO SYSTS RES SECT, JET PROPULSION LAB, CALIF INST TECHNOL, 78- *Concurrent Pos:* Vis prof, Inst Org Chem, Univ Cologne, 74-75 & Univ Siegen, Ger, 81; Am ed, Org Magnetic Resonance, 68-; Alexander von Humboldt Award, 74-75; Int Cancer Res Technol Transfer travel grant, Univ Siegen, 81. *Mem:* AAAS; Am Chem Soc. *Res:* Nuclear magnetic resonance; polymers; propellant and fluorocarbon chemistry; gas chromatography techniques; spacecraft material problems; extraterrestrial life detection; analytical chemistry; small-ring compounds; theoretical calculations on aromatic molecules; polypeptide synthesis; steric effects in organic molecules. *Mailing Add:* 5447 La Forest Dr La Canada CA 91011

MANCALL, ELLIOTT L, b Hartford, Conn, July 31, 27; c 2. NEUROPATHOLOGY, NEUROLOGY. *Educ:* Trinity Col, BS, 48; Univ Pa, MD, 52; Am Bd Psychiat & Neurol, dipl & cert neurol, 59. *Prof Exp:* Asst prof neurol, Jefferson Med Col, 53-64, assoc prof, 64-65; prof med & neurol, 65-75, CHMN, DEPT NEUROLOGY, HAHNEMANN MED COL, 75- *Concurrent Pos:* Fulbright fel, Nat Hosp Neurol Dis, London, 54-55; teaching fel neuropath, Harvard Med Sch, 56-57; mem vis fac, Sch Med, Emory Univ, 64; vis lectr, US Naval Hosp, 64-; consult, Valley Forge Gen Army Hosp, 67-74 & Pennhurst State Sch & Hosp, 68-73; chief neurol serv, Philadelphia Gen Hosp, 69-74. Mem; dir, Am Bd Psychiat & Neurol, 82- *Mem:* fel Am Acad Neurol; Am Asn Neuropath; Am Neurological Asn; Soc Neurosci; Asn Univ Prof Neurol. *Res:* Neurology and neuropathology of metabolic diseases of the nervous system. *Mailing Add:* Dept of Neurol Hahnemann Med Col 85230 N Broad St Philadelphia PA 19102

MANCE, ANDREW MARK, b Braddock, Pa, Jan 21, 52; m 74; c 2. CHEMISTRY, THIN FILM CERAMICS. *Educ:* Thiel Col, BA, 73; Carnegie-Mellon Univ, PhD(chem), 79. *Prof Exp:* SR RES SCIENTIST, GEN MOTORS RES LABS, 78- *Mem:* AAAS; Am Chem Soc; Am Electroplaters Soc; Am Ceramic Soc. *Res:* Metal deposition from electroless paths and in vacuum and surface chemistry; materials and material processing for electronic applications. *Mailing Add:* Electrical & Electronics Eng Dept Gen Motors Res Labs 30500 Mound Rd Warren MI 48090-9055

MANCERA, OCTAVIO, organic chemistry, for more information see previous edition

MANCHE, EMANUEL PETER, b New York, NY, Apr 30, 31; m 61; c 3. CHEMISTRY. *Educ:* City Col New York, BS, 56; Brooklyn Col, MA, 59; Rutgers Univ, NB, PhD(chem), 65. *Prof Exp:* Res chemist, Am Chicle Co, 56-57; lectr chem, Brooklyn Col, 58-62, sch gen studies, 58-59; teaching asst, Rutgers Univ, 62-64, instr, 65; advan res engr, Gen Tel & Electronics Res Labs, 65-68; from asst prof to assoc prof, 68-80, PROF CHEM, YORK COL, CITY UNIV NEW YORK, 81- *Concurrent Pos:* Instr univ col, Rutgers Univ, 64-73, prof, 73-74; invited lectr, United Hosps, Newark, 72-73. *Mem:* Am Chem Soc; Sigma Xi. *Res:* Chemical instrumentation; thermal methods of analysis including thermogravimetry and differential thermal analysis; thermoluminescent dating. *Mailing Add:* Dept of Natural Sci York Col City Univ New York Jamaica NY 11451-0999

MANCHEE, ERIC BEST, b Toronto, Oct 16, 18; m 45; c 3. GEOPHYSICS, SCIENCE ADMINISTRATION. *Educ:* Univ Toronto, BASc, 49, MA, 51. *Prof Exp:* Geophysicist, Calif Standard Co, Alta, 51-59, dist geophysicist, 59-62; head array seismol sect, Dept Energy, Mines & Resources, 62-74, head spec proj, 74-75, br prog officer, Earth Physics Br, 75-83; RETIRED. *Mem:* Can Asn Physicists; Can Geophys Union. *Res:* Exploration and array seismology; earthquake-explosion differentiation, administration science programs. *Mailing Add:* 2420 Rector Ave Ottawa ON K2C 1M3 Can

MANCHESKI, FREDERICK J, AUTOMOTIVE AFTERMARKET DESIGN. *Prof Exp:* CHMN BD, ECHLIN INC, 63- *Mem:* Nat Acad Eng. *Mailing Add:* Echlin Inc 100 Double Beach Rd Branford CT 06405

MANCHESTER, KENNETH EDWARD, b Winona, Minn, Mar 22, 25; m 46; c 2. SURFACE CHEMISTRY. *Educ:* San Jose State Col, AB, 49; Stanford Univ, MS, 50, PhD(thermochem), 55. *Prof Exp:* Fel chem, Stanford Univ, 52-55; chemist surface chem, Shell Develop Co, 55-62; sect head semiconductor chem, Sprague Elec Co, 62-63, dept head, 63-69, dir semiconductor res & develop, 69-76, chief scientist, 76-79, dir semiconductor qual assurance & reliability, 79-85, corp vpres, Res Develop & Eng, 85-89; CONSULT, ALLEGRO MICROSYSTS INC, 89- *Mem:* Am Chem Soc; Am Asn Contamination Control; Am Inst Mining, Metal & Petrol Eng; Sigma Xi. *Res:* Energetics of liquid-liquid and liquid-solid interfaces; interaction of energetic ions or electrons with solid substrates. *Mailing Add:* Allegro Microsysts Inc PO Box 15036 115 Northeast Cutoff Worcester MA 01615

MANCINELLI, ALBERTO L, b Rome, Italy, Nov 22, 31; m 62. PLANT PHYSIOLOGY. *Educ:* Univ Rome, Dr rer nat(bot), 54. *Prof Exp:* Vol asst prof bot, Univ Rome, 54-64; from asst prof to assoc prof plant physiol, 64-81, PROF BIOL SCI, COLUMBIA UNIV, 81- *Concurrent Pos:* Ital Nat Res Coun fel, 56-59; Ital Nat Comt Nuclear Energy fel, 59-62; NATO fel, 62-63; NSF res grant, 65-82. *Mem:* AAAS; Am Soc Plant Physiol; Am Inst Biol Sci; Japanese Soc Plant Physiol; Am Soc Photobiol; Sigma Xi. *Res:* Metabolism during seed germination; reactions controlling light responses in plant growth and development, particularly phytochrome controlled and high energy reaction controlled responses. *Mailing Add:* Dept of Biol Sci Box 22 Schermerhorn Hall Columbia Univ New York NY 10027

MANCINI, ERNEST ANTHONY, b Reading, Pa, Feb 27, 47; m 69; c 2. PETROLEUM GEOLOGY, MICROPALEONTOLOGY. *Educ:* Albright Col, BS, 69; Southern Ill Univ, MS, 72; Tex A&M Univ, PhD(geol), 74. *Prof Exp:* Explor geologist, Cities Serv Oil Co, 74-76; from asst prof to assoc prof, 76-84, PROF GEOL, UNIV ALA, 84- *Concurrent Pos:* Petrol geologist, Mineral Resources Inst, Univ Ala, 76-78; state geologist & oil & gas supvr, State of Ala, 82-; mem, Interstate Oil Compact comn & Outer Continental Shelf Policy adv bd, 83-; ed, Asn Am State Geologists J, 84-86. *Honors & Awards:* A I Levorsen Petrol Geol Award, Am Asn Petrol Geologists, 80; Pratt-Haas Distinguished Lectr, 87-88. *Mem:* Am Asn Petrol Geologists; Soc Econ Paleontologists & Mineralogists; Paleont Soc; Cushman Found Foraminiferal Res; NAm Micropaleont Soc. *Res:* Petroleum geology; stratigraphy; paleoenvironments; micropaleontology (foraminifera). *Mailing Add:* Geol Surv of Ala PO Box 0 Tuscaloosa AL 35486

MANCLARK, CHARLES ROBERT, b Rochester, NY, June 22, 28; m 53; c 2. MICROBIOLOGY, IMMUNOLOGY. *Educ:* Calif Polytech State Univ, BS, 53; Univ Calif, Los Angeles, PhD(bact), 63. *Prof Exp:* Asst prof microbiol, Calif State Univ, Long Beach, 61-64; asst res bacteriologist, Univ Calif, Los Angeles, 63-65; asst prof microbiol, Univ Calif, Irvine, Col Med, 65-67; res microbiologist & asst to chief, Lab Bact Prod, Div Biologics Stand, NIH, 67-72; RES MICROBIOLOGIST, CTR BIOLOGICS EVAL & RES, FOOD & DRUG ADMIN, 72- *Honors & Awards:* Health Inst Peru Medal, Lima, 80. *Mem:* AAAS; fel Am Soc Microbiol; assoc mem Mycol Soc Am; assoc mem Am Soc Indust Microbiol; Int Asn Biol Stand; Sigma Xi. *Res:* Host-parasite relationships in pertussis; immunity and the immune response; diagnostic bacteriology; microbial spoilage of marine fish; bacterial taxonomy; isolation and study of penicillin resistant neisseria gonorrhoeae; nutritional studies of haemophilus ducreyi; author of numerous scientific publications. *Mailing Add:* Ctr Biol Eval & Res 8800 Rockville Pike Bethesda MD 20892

MANCUSO, JOSEPH J, b Hibbing, Minn, Dec 9, 33; m 57; c 4. ECONOMIC GEOLOGY. *Educ:* Carleton Col, BA, 55; Univ Wis, MS, 57; Mich State Univ, PhD(geol), 60. *Prof Exp:* Instr geol, Mich State Univ, 58-60; asst prof, 60-71, chmn dept, 75-78, PROF GEOL, BOWLING GREEN STATE UNIV, 71- *Concurrent Pos:* Consult geologist. *Mem:* Am Inst Prof Geologists; Soc Econ Geol. *Res:* Economic geology, mineralogy and stratigraphy of the Lake Superior iron formations; Precambrian geology of Wisconsin, Michigan and Minnesota. *Mailing Add:* Dept of Geol Bowling Green State Univ Bowling Green OH 43403

MANCUSO, RICHARD VINCENT, b Rochester, NY, Nov 4, 38; m 64; c 2. NUCLEAR PHYSICS. *Educ:* St Bonaventure Univ, BS, 60; State Univ NY Buffalo, PhD(physics), 65. *Prof Exp:* Teaching asst physics, State Univ NY Buffalo, 61-65; Nat Acad Sci-Nat Res Coun res assoc nuclear physics, Van De Graaff Br, Naval Res Lab, Washington, DC, 67-69; asst prof, 69-74, ASSOC PROF PHYSICS, STATE UNIV NY COL BROCKPORT, 74- *Mem:* Am Phys Soc; Sigma Xi. *Res:* Gamma ray spectroscopy; level structures of medium weight nuclei; charged particle reactions and reaction mechanisms; application of nuclear techniques to non-nuclear problems. *Mailing Add:* Dept of Physics State Univ of NY Col Brockport NY 14420

MANDAL, ANIL KUMAR, b West Bengal, India, Nov 12, 35; m 64; c 2. CARDIOVASCULAR DISEASES. *Educ:* Univ Calcutta, MB, BS, 59; Am Bd Internal Med, dipl, 72. *Prof Exp:* Med officer, Inst Postgrad Med Educ & Res, Calcutta, 63-66; registr, R G Kar Med Col, Calcutta, 66-67; lectr path, Univ Edinburgh, 68-69; instr med, Univ Ill, Chicago Circle, 71-72; asst prof med, Col Med, Univ Okla, 72-75, assoc prof, 75-; PROF MED, WRIGHT STATE UNIV. *Concurrent Pos:* Consult physician, Vet Admin Hosp, Muskogee, Okla, 72-; asst physician, Okla Med Res Found, 72-; attend nephrologist, Vet Admin Hosp & Univ Hosp, Oklahoma City, 75- *Mem:* Fel Am Col Physicians; Am Fedn Clin Res; Am Soc Nephrol; Electron Micros Soc Am; Sigma Xi. *Res:* Pathological study by light, electron and fluorescence microscopy of kidney in experimental renal disease and hypertension. *Mailing Add:* Div Nephro-Prof Med Wright State Univ Vet Admin Med Ctr Campus Box 927 Dayton OH 45401

MANDARINO, JOSEPH ANTHONY, b Chicago, Ill, Apr 20, 29; m 56; c 4. MINERALOGY. *Educ:* Mich Col Mining & Technol, BS, 50, MS, 51; Univ Mich, PhD(mineral), 58. *Prof Exp:* Asst prof mineral, Mich Col Mining & Technol, 57-59; assoc cur, 59-65, CUR DEPT MINERAL, ROYAL ONT MUS, 65- *Concurrent Pos:* Mem, Joint Comt Powder Diffraction Standards; Nat Res Coun Can sr res fel, Fr Bur Geol Mines Res, 68-69. *Mem:* Mineral Soc Am; Mineral Asn Can; Mineral Soc Gt Brit & Ireland. *Res:* Crystal optics; crystallography; descriptive mineralogy. *Mailing Add:* Royal Ont Mus 100 Queens Pkwy Toronto ON M5S 2C6 Can

MANDAVA, NAGA BHUSHAN, b Bhushanagulla, India, Oct 14, 34; m 57; c 3. BIO-ORGANIC CHEMISTRY. *Educ:* Univ Andhra, India, BSc, 55; Banaras Hindu Univ, MSc, 57; Indian Inst Sci, Bangalore, PhD(chem), 62. *Prof Exp:* Res assoc, Okla State Univ, 63-65, State Univ NY Stony Brook, 65-66 & Laval Univ, 66-68; chemist, Plant Sci Res Div, 68, res chemist, Plant Physiol Inst, Sci & Educ Agr Res Serv, USDA, 68-82; sci adv, Environ Protection Agency, 82-86; SR PARTNER, TODHUNE, MANDAVA & ASSOCS, 86- *Concurrent Pos:* Nat Res Coun Can fel, 66-68. *Mem:* Fel Am Inst Chem; Am Chem Soc; Int Plant Growth Substances Asn; Sigma Xi; Soc Toxicol & Environ Chem; Soc Risk Analysis. *Res:* Organophosphorus compounds; plant hormones, lipids, steroids, carbohydrates, alkaloids, pesticides and heterocyclic compounds; application of spectroscopy and computers to structural and stereochemical problems; bioassays and tracer techniques. *Mailing Add:* 15404 Tindlay St Silver Spring MD 20904

MANDEL, ANDREA SUE, b New York, NY, Mar 25, 51; m 70; c 1. PACKAGING ENGINEERING & DEVELOPMENT, PLASTICS ENGINEERING. *Educ:* City Col NY, BEME, 73; Rutgers Univ, MS, 77. *Prof Exp:* Scientist, Johnson & Johnson, 73-78; chief packaging engr, Howmedica/Pfizer, 78-80; mgr packaging, Drake Bakeries/Borden, 80-86; Mgr package eng, Lehn & Fink/Sterling Drug/Kodak, 86-88; SR MGR PACKAGE DEVELOP, CHURCH & DWIGHT, INC, 88- *Concurrent Pos:* Adv bd, Modern Plastics Mag, 80; chairperson, Conf Tamper Evidence, Packaging Inst Am, 83-; chmn, NJ Chap Packaging Inst, 84-85. *Mem:* Inst Packaging Professionals; Am Soc Mech Engrs; Soc Plastics Engrs. *Res:* Technical design and development of consumer products and medical packaging. *Mailing Add:* 46 Ellsworth Dr Robbinsville NJ 08691

MANDEL, BENJAMIN, virology, for more information see previous edition

MANDEL, BENJAMIN J, b Poland, Sept 1, 12; US citizen; m 37; c 2. APPLIED STATISTICS. *Educ:* City Col New York, BS, 34; George Washington Univ, MA, 38; Goteborg Sch Econ & Bus Admin, Sweden, Ekonomie Licentiate, 68. *Prof Exp:* Jr & asst statistician, Social Security Admin, 38-44, chief statist div, 44-60; asst dir statist standards & opers, US Off Educ, 60-61; dir off statist prog, US Post Off Dept, 61-70; STATIST SCI CONSULT, 70-; CONSULT INSTR, GRAD SCH, USDA & OFF PERSONNEL MGT, 70- *Concurrent Pos:* Prof statist, Univ Baltimore, 46-70, chmn dept, 48-70, emer prof, 70-; vis prof mgt & statist, Dept Agr Grad Sch, 62-; lectr statist for mgt, Bur Training, US Civil Serv Comn, 65-; vis lectr statist, Am Statist Asn, NSF, 71-72; consult var agencies; spec lectr, George Washington Univ, 85-87. *Mem:* Am Statist Asn; fel Am Soc Qual Control; Am Asn Retired Teachers; Smithsonian Assocs. *Res:* Extension of application of statistical theory and techniques to new areas of management, administration, accounting, auditing, inspection and quality assurance. *Mailing Add:* 6101 16th St NW Washington DC 20011

MANDEL, HAROLD GEORGE, b Berlin, Ger, June 6, 24; nat US; m 53; c 2. PHARMACOLOGY. *Educ:* Yale Univ, BS, 44, PhD(org chem), 49. *Prof Exp:* Asst & lab instr chem, Yale Univ, 42-44, lab instr org chem, 47-49; res assoc, 49-50, asst res prof, 50-52, from assoc prof to prof, 52-58, CHMN DEPT PHARMACOL, SCH MED, GEORGE WASHINGTON UNIV, 60- *Concurrent Pos:* Advan Commonwealth Fund fel, Molteno Inst, Eng, 56 & Pasteur Inst, France, 57; travel award, Int Pharmacol Cong, Prague, 63, Helsinki, 75 & Paris, 78; Commonwealth Fund sabbatical leave, Univ Auckland & Univ Med Sci, Thailand, 64; Am Cancer Soc Eleanor Roosevelt Int fel, Chester Beatty Res Inst, London, 70-71; Am Cancer Soc scholar cancer res clin pharmacol, Univ Calif, San Francisco, 78-79; lectr, US Naval Dent Sch, 59-61, 71-75, Washington Hosp Ctr, 60-66, US Army Dent Sch, 72-75, Holy Cross Hosp, 72-74 & Food & Drug Admin, 67, 78, 79 & 81; consult, Fed Aviation Agency, 61-62; mem biochem comt, Cancer Chemother Nat Serv Ctr, 58-61, med adv comt, Therapeut Res Found, Inc, 62-70, pharmacol & exp therapeut B study sect, USPHS, 63-68, comt probs drug safety, Nat Acad Sci-Nat Res Coun, 65-71 & 72-76, mem drug metab workshop progs, NY Univ, 66, George Washington Univ, 67 & Univ Calif, 68; mem cancer chemother comt & workshops, Poland, 68, Curacao, 71, Ger, 73, Belg, 73, & Arg, 73, Int Union Against Cancer, 66-74, res comt, Children's Hosp, Washington, DC, 69-85 & sci adv comt, Registry Tissue Reactions to Drugs, 70-76; mem chemother comt, Am Cancer Soc, 69-73; mem cancer res training comt, Nat Cancer Inst, 71-73, mem cancer spec prog adv comt, 74-78, chmn, 76-78; consult, Roswell Park Inst, Buffalo, 72-74; consult toxicol & mem toxicol adv comt, Food & Drug Admin, 75-78; mem merit rev bd, Vet Admin, 75-78; mem comt Toxicol, Nat Acad Sci, Nat Res Coun, 78-82; vis prof pharmacol, Sch Med, Stanford Univ, 78, Univ Calif, San Francisco, 78-79, Mahidol Univ, Bangkok, Thailand, 81; consult, Environ Protection Agency, 78-82; mem, Kettering Award Selection Comt, Gen Motors Cancer Res Found, 79-81; mem, Nat Large Bowel Cancer Working Cadic, 80-84; panel mem, Comn Human Resources, Nat Acad Sci, 81-86, chmn subcomt, long term health effects, 82-86; vis scientist, Med Res Coun Toxicol Unit, Carshalton, Surrey, Eng, 86, 88 & 90, Int Agency Against Cancer, Lyon, France, 89. *Honors & Awards:* Abel Award, Am Soc Pharmacol & Exp Therapeut, 58. *Mem:* AAAS; Am Soc Pharmacol & Exp Therapeut (secy, 61-63, pres, 73-74); Am Chem Soc; Asn Med Sch Pharmacol (treas, 71-73, pres, 76-78); Am Asn Cancer Res; Sigma Xi; Am Soc Biochem & Molecular Biol; Int Soc Biochem Pharmacol. *Res:* Cancer chemotherapy and carcinogenesis; drug metabolism; mechanism of action of antimetabolites and other anti-cancer drugs; action of growth inhibitory drugs. *Mailing Add:* Dept Pharmacol Sch Med George Washington Univ Washington DC 20037

MANDEL, IRWIN D, b New York, NY, Apr 9, 22; m 44; c 3. PREVENTIVE DENTISTRY, ORAL BIOLOGY. *Educ:* City Col New York, BS, 42; Columbia Univ, DDS, 45. *Hon Degrees:* DSc, Col Med & Dent NJ, 81; DSc, Univ Goteborg, 84. *Prof Exp:* Res asst dent & oral surg, 46-50, instr, 50-57, from asst clin prof to clin prof, 57-69, dir, Lab Clin Res, 60-69, dir, div prev dent, 69-84, PROF DENT & ORAL SURG, COLUMBIA UNIV, 69-, DIR, CTR CLIN RES DENT, 84- *Concurrent Pos:* USPHS grant, 60-; Health Res Coun City of New York career scientist award, 69-72; chmn, Coun Dent Res, Am Dent Asn, 79-81; consult & chmn, Oral Biol & Med Study Sect, 72-76, Coun Dent Therapeut, 68- *Honors & Awards:* Gold Medal Award, Dent Res, Am Dent Asn, 85. *Mem:* Fel AAAS; Sigma Xi; Am Asn Dent Res (pres); Am Dent Asn; Int Asn Dent Res. *Res:* Plaque, calculus and periodontal disease; salivary composition and relation to oral and systemic disease; host factors in caries resistance. *Mailing Add:* 630 W 168th St New York NY 10032

MANDEL, JAMES A, b Pittsburgh, Pa, Dec 25, 34; m 59; c 2. CIVIL ENGINEERING. *Educ:* Carnegie-Mellon Univ, BS, 56, MS, 62; Syracuse Univ, PhD(civil eng), 67. *Prof Exp:* Design engr, Richardson, Gordon & Assocs, Pa, 56-61; sr stress engr, Goodyear Aerospace Corp, Ohio, 62-64; from asst prof to assoc prof civil eng, 67-78, PROF CIVIL ENG, SYRACUSE UNIV, 78- *Mem:* Am Concrete Inst. *Res:* Structural analysis; composite materials; shell structures; mechanics; finite element analysis. *Mailing Add:* Dept Civil Eng Syracuse Univ Syracuse NY 13244-1190

MANDEL, JOHN, b Antwerp, Belgium, July 12, 14; nat US; m 38; c 2. MATHEMATICAL STATISTICS. *Educ:* Univ Brussels, BS, 35, MS, 37; Eindhoven Technol Univ, PhD(appl statist), 65. *Prof Exp:* Res chemist, Soc Belge De Recherches, Belgium, 38-40; anal & develop chemist, Foster D Snell, Inc, 41-43; res chemist, B G Corp, 44-47; gen phys scientist, 47-48, anal statistician, 48-57, MATH STATISTICIAN, NAT BUR STANDARDS, 58- *Honors & Awards:* Gold Medal Award, US Dept Com, 73; Shewhart Medal, Am Soc Qual Control, 80, Deming Medal, 81. *Mem:* Fel Am Statist Asn; Inst Math Statist; fel Am Qual Control; fel Royal Statist Soc. *Res:* Statistical design of experiments; statistical analysis of data obtained in physical and chemical experimentation; development of statistical techniques for the physical sciences; interlaboratory testing; statistical evaluation of measuring processes. *Mailing Add:* Nat Inst Standards & Technol Chem Bldg A323 Gaithersburg MD 20899

MANDEL, JOHN HERBERT, b New York, NY, Mar 11, 25; m 50; c 3. MEDICAL MICROBIOLOGY. *Educ:* City Col New York, BS, 47; Univ Calif, Berkeley, MA, 49. *Prof Exp:* Chief diag serol sect & res bacteriologist, US Army Grad Sch, Walter Reed Army Med Ctr, 50-51; chief, Serol & Hemat Sect, FDR Vet Admin Hosp, Montrose, NY, 51-55; DIR, LEHIGH VALLEY LABS, INC, 55- *Concurrent Pos:* Dir, Am Bd Bioanalysts; instr microbiol, Pa State Univ, 69-74; pres, Allentown Bd Health, 80- *Mem:* Am Soc Microbiol; Am Asn Bioanalysts; Am Pub Health Asn; Int Soc Human & Animal Mycol; Am Soc Med Technol; Am Pract Infection Control. *Res:* Rapid isolation and identification of pathogenic microorganisms from biological materials, food and water; identification and significance of yeasts isolated from clinical specimens. *Mailing Add:* Microbiol Dept Lehigh Valley Labs 1740 Allen St Allentown PA 18104-5070

MANDEL, LAZARO J, b Lima, Peru, Oct 13, 40; US citizen; m 63. PHYSIOLOGY, BIOPHYSICS. *Educ:* Mass Inst Technol, BS, 61, MS, 62; Univ Pa, PhD(biomed eng), 69. *Prof Exp:* USPHS fel, Yale Univ Med Sch, 69-72; from asst prof to assoc prof, 72-83, PROF PHYSIOL, MED CTR, DUKE UNIV, 84- *Mem:* AAAS; Biophys Soc; Am Physiol Soc; Am Soc Nephrology. *Res:* Active and passive transport; biological energy conversion; properties of biological membranes; epithelial transport and metabolism. *Mailing Add:* Dept of Physiol Duke Univ Med Ctr Box 3709 Durham NC 27710

MANDEL, LEONARD, b May 9, 27; US citizen; m 53; c 2. QUANTUM OPTICS, LASERS. *Educ:* Univ London, BSc, 47, BSc, 48, PhD(physics), 51. *Prof Exp:* Tech officer, Imp Chem Indust Res Labs, 51-54; lectr & sr lectr physics, Imp Col, London, 54-64; PROF PHYSICS, UNIV ROCHESTER, 64- *Concurrent Pos:* Prof optics, Univ Rochester, 77-80; vis prof, Univ Tex, Austin, 84; mem bd, Optical Soc Am, 85-88. *Honors & Awards:* Max Born Prize, Optical Soc Am, 82; Marconi Medal, Ital Nat Res Coun, 87; Thomas Young Medal, Brit Inst Physics, 89. *Mem:* Fel Am Phys Soc; fel Optical Soc Am. *Res:* Optical coherence; quantum interactions of light; non-classical states of light; lasers; tests of locality in quantum mechanics; author or coauthor of over 245 publications, coeditor of six books. *Mailing Add:* Dept Physics & Astron Univ Rochester Rochester NY 14627

MANDEL, LEWIS RICHARD, b Brooklyn, NY, Nov 13, 36; m 60; c 3. BIOCHEMISTRY, RESEARCH ADMINISTRATION. *Educ:* Columbia Univ, BS, 58, PhD(biochem), 62. *Prof Exp:* Asst prof pharmacol, Col Pharm, Columbia Univ, 62-64; sr res biochemist, 64-67, sect head, 67-71, asst dir, 71-74, dir biochem, 74-76, dir univ rel, 76-79, sr dir sci & indust rel, 79-90, EXEC DIR, INDUST & ACAD REL, MERCK SHARP & DOHME RES LABS, 90- *Concurrent Pos:* NIH Res Grant, 63-64. *Mem:* Am Soc Biol Chem; Am Soc Pharmacol & Exp Therapeut. *Res:* Lipid metabolism; prostaglandins; transmethylation reactions. *Mailing Add:* Merck Sharp & Dohme Res Labs PO Box 2000 Rahway NJ 07065

MANDEL, MANLEY, b Philadelphia, Pa, July 10, 23; m 52; c 3. MICROBIOLOGY, MOLECULAR BIOLOGY. *Educ:* Brooklyn Col, BA, 43; Mich State Univ, MS, 47, PhD(bact), 52. *Prof Exp:* Asst biol, Brooklyn Col, 46; guest investr microbiol, Haskins Labs, 46 & 52; asst bact, Univ Calif, 48-50; instr, Univ Mass, 52-53, asst prof, 53-63; res assoc, Brandeis Univ, 63; assoc prof biol, Grad Sch Biomed Sci, Univ Tex, Houston, 63-66, assoc biologist, 63-66, prof, 66-81, chief, Sect Molecular Biol, Univ Tex M D Anderson Hosp & Tumor Inst, 63-79, biologist, 66-81; dir res, Immunopath Labs Int, 81-89; DIR EDUC, TEX INST REPRODUCTIVE MED & ENDOCRINOL, 90- *Concurrent Pos:* Mem molecular biol study sect, NIH, 66-70, sci review comt health related facil, 70-74; mem bd trustees, Am Type Cult Collection, 73-79, Nucleic Acid & Protein Biosynthesis Adv Comt, Am Cancer Soc, 70-75, Res Comt, 79-80; mem bd dir, Tex Found Res Reproductive Med, 90- *Mem:* Am Soc Microbiol; Biophys Soc; Genetics Soc Am; Brit Soc Gen Microbiol; fel Am Acad Microbiol. *Res:* Molecular and genetic relations of protists; role of modification in viral DNA-host interactions. *Mailing Add:* 4115 Falkirk Lane Houston TX 77025

MANDEL, MORTON, b Brooklyn, NY, July 6, 24; m 52; c 2. GENE CLONING, GENE TRANSFER. *Educ:* City Col New York, BCE, 44; Columbia Univ, MS, 49, PhD(physics), 57. *Prof Exp:* Instr civil eng, Stevens Inst Technol, 48; asst microwave components, Columbia Univ, 52-56; mem tech staff solid state devices, Bell Tel Labs, 56-57; res assoc paramagnetic resonance, Stanford Univ, 57-61, asst prof physics, 59-61; eng specialist, Gen Tel & Electronics Lab, 61-63; res assoc genetics, Sch Med, Stanford Univ, 63-64; USPHS fel microbial genetics, Karolinska Inst, Sweden, 64-66; assoc prof biophys, 66-69, chmn dept biochem & biophys, 71-72, PROF BIOPHYS, SCH MED, UNIV HAWAII, 69- *Concurrent Pos:* Consult, Fairchild Semiconductor Corp, 57-58, Hewlett-Packard Co, 58 & Rheem Semiconductor Corp, 59; vis prof, Worcester Found Exp Biol, 72-73, vis prof human genetics, Sch Med, Yale Univ, 79-80; prin investr, Am Cancer Soc, 79-86, Minority Biomed Res Support, NIH, 85-86; vis scientist, DuPont Molecular Biol Group, 86-87; consult, Lawrence Livermore Nat Lab, Human Genome Proj, 88; Am Cancer Soc Scholar, 79-80; Eleanor Roosevelt Int Cancer fel. *Mem:* Fel Am Phys Soc; AAAS; Sigma Xi. *Res:* Phage genetics; bacterial transformation and transfection; gene transfer in mammalian cells; cloning of sea urchin genes and study of their regulation. *Mailing Add:* Dept Biochem & Biophys Univ Hawaii Sch Med Honolulu HI 96822

MANDEL, NEIL, b Queens, NY, Jan 28, 47. MOLECULAR STRUCTURE. *Educ:* Univ Pa, PhD(chem), 71. *Prof Exp:* SR RES SCIENTIST, VET ADMIN MED CTR, 76-; CHMN DEPT BIOPHYSICS, MED COL WIS, 76- *Concurrent Pos:* Prof Med Biochem, 82- *Mailing Add:* Dept Med Med Col Wis Vet Admin Med Ctr Res Serv 151 5000 W National Ave Milwaukee WI 53295

MANDEL, ZOLTAN, b Czech, July 18, 24; US citizen; m 57; c 1. TEXTILE CHEMISTRY. *Educ:* Western Reserve Univ, BS, 51, MS, 52, PhD(org chem), 55. *Prof Exp:* Res chemist, Diamond Alkali Co, 54-55; res chemist, 55-63, sr res chemist, 63-65, res assoc, 65-72, SR RES CHEMIST, E I DU PONT DE NEMOURS & CO, INC, 72- *Mem:* Am Chem Soc. *Res:* Synthetic fibers. *Mailing Add:* 4013 Greenmount Rd Wilmington DE 19810

MANDELBAUM, HUGO, b Sommerhausen, Bavaria, Oct 18, 01; nat US; m 31; c 5. PHYSICAL OCEANOGRAPHY. *Educ:* Univ Hamburg, Dr rer nat(geophys), 34. *Prof Exp:* Teacher high sch, Hamburg, 30-38; prin, Jewish Day Sch, Mich, 40-48; prof geol, 48-71, EMER PROF GEOL, WAYNE STATE UNIV, 71- *Mem:* AAAS; Soc Explor Geophys; Seismol Soc Am; Am Geophys Union. *Res:* Tides and tidal currents; air-sea boundary problems; sedimentation; the astronomical foundation of the Jewish calendar. *Mailing Add:* Rechov Hapisgah 62 Jerusalem 96382 Israel

MANDELBAUM, ISIDORE, b New York, NY. SURGERY. *Educ:* NY Univ, AB, 48; State Univ NY Downstate Med Ctr, MD, 52. *Prof Exp:* From instr to assoc prof, 61-72, PROF SURG, SCH MED, IND UNIV, INDIANAPOLIS, 72- *Concurrent Pos:* Dazian fel path, Mt Sinai Hosp, New York, 54-55; Nat Heart Inst fel, Ind Univ, Indianapolis, 61-62; consult, Vet Admin Hosp, Indianapolis, 61- *Mem:* Am Asn Thoracic Surg; Soc Vascular Surg; Int Cardiovasc Soc; Am Soc Clin Invest; Am Col Surg. *Res:* Cardiothoracic surgery; cardiovascular and pulmonary research. *Mailing Add:* Dept of Surg Ind Univ Sch Med 1100 W Michigan St Indianapolis IN 46223

MANDELBERG, HIRSCH I, b Baltimore, Md, Apr 16, 34; m 58; c 2. PHYSICS. *Educ:* Johns Hopkins Univ, BE, 54, PhD(physics), 60. *Prof Exp:* Res physicist res inst adv study div, Martin Co, 56-60; res physicist, 60-71, LAB DIR, LAB PHYS SCI, US DEPT DEFENSE, 71- *Concurrent Pos:* Visitor, Univ Col, Univ London, 67-68; assoc prof, Univ Col, Univ Md, 61-66. *Mem:* Am Phys Soc. *Res:* Optics; optical propagation; non-linear optics; lasers; optical spectroscopy; laser applications; optical communications. *Mailing Add:* 6800 Pimlico Dr Baltimore MD 21209

MANDELBERG, MARTIN, b Bronx, NY, Oct 12, 46; m 79; c 4. DECISION SUPPORT SYSTEMS. *Educ:* Drexel Inst Technol, BS, 69; Univ Conn, MS, 73; US Naval War Col, MA, 76; Naval Postgrad Sch, PhD(elec eng), 82. *Prof Exp:* Assoc engr, Gen Signal Co, 65 & 66, Lockheed Missile & Space Co, 67, Lockheed Electronics Co, 68; res & develop engr, Elec Boat Div, Gen Dynamics Corp, 69-70; prog mgr & systs engr, US Naval Underwater Systs Ctr, 70-80; prog mgr & prin res engr, US Coast Guard Res & Develop Ctr, 80-82; prin res engr, Data Systs Div, Gen Dynamics Corp, 82-83; chief scientist, 83-85, head, Systs Anal Div, Defense Mobilization Systs Planning Activ, Off Sect Defense, 86-87; SR SCIENTIST, CHIEF SCIENTIST & ASST VPRES, SCI APPLICATIONS INT CORP, 87- *Concurrent Pos:* Mem, Res Rev Comt, Analog Test Prog Generation Comt & Joint Serv Electronics Prog, Off Naval Res, 78-82; adj lectr, Hartford Grad Ctr, 81-82; adj fac mem, Univ Conn, 82; adj assoc prof, Univ Md, Univ Col, 85- *Honors & Awards:* Secy Transp Silver Medal, 82. *Mem:* Sigma Xi; Inst Elec & Electronic Engrs; Asn Comput Mach; Nat Soc Prof Engrs; AAAS. *Res:* Advanced mathematical techniques for experimental modeling, design and simulation of complex engineering systems; advanced information management and decision support systems. *Mailing Add:* 5004 N 16th St Arlington VA 22205

MANDELBROT, BENOIT B, b Warsaw, Poland, Nov 20, 24; m 55; c 2. APPLIED MATHEMATICS. *Educ:* Polytech Sch, Paris, Engr, 47; Calif Inst Technol, MS, 48, Prof Eng, 49; Univ Paris, PhD(math), 52. *Hon Degrees:* Numerous from US & foreign univs, 86-88. *Prof Exp:* Mathematician, Philips Electronics, Paris, 50-53; mem staff sch math, Inst Adv Study, NJ, 53-54; assoc, Inst Henri Poincare, Paris, 54-55; asst prof math, Univ Geneva, 55-57; jr prof, Lille Univ & Polytech Sch, Paris, 57-58; res staff mem, 58-72, sci adv to dir res, 72-76, FEL, INT BUS MACH CORP, 74-; ABRAHAM ROBINSON PROF MATH SCI, YALE UNIV, 87- *Concurrent Pos:* Vis prof, Harvard Univ, 62-64 & 79-80; inst lectr, Mass Inst Technol, 64-68 & 74-76; staff mem, Nat Bur Econ Res, 69-78; Trumbull lectr & vis prof, Yale Univ, 70; vis prof, Albert Einstein Col Med, 71; lectr, Col France, 73, 74 & 77; vis prof, Downstate Med Ctr, State Univ NY, 74; prof pract math, Harvard Univ, 84-87; Charles M & Martha Hitchcock prof, Univ Calif, 91-92. *Honors & Awards:* F Barnard Medal for Meritorious Serv in Sci, Nat Acad Sci, 85; Franklin Medal, Signal & Eminent Serv Sci, Franklin Inst, 86; Charles Proteus Steinmetz Medal, 88; Alexander von Humboldt-Stiftung Sr Award, 88. *Mem:* Nat Acad Sci; fel Economet Soc; fel Am Statist Asn; fel Inst Elec & Electronics Eng; Am Math Soc; Sigma Xi; fel Inst Math Statist; Am Acad Arts & Sci; Europ Acad Arts Sci & Humanities; fel Am Phys Soc; fel Am Geophys Union. *Res:* Fractals and fractal geometry of nature; thermodynamics, noise and turbulence; natural languages; astronomy; geomorphology; commodity and security prices; self-similar or sporadic chance phenomena; computer art and graphics; theory of stochastic processes, especially its applications. *Mailing Add:* Math Dept Yale Univ 12 Hillhouse Ave New Haven CT 06520

MANDELCORN, LYON, b Montreal, Que, June 27, 26; nat US; m 55; c 3. PHYSICAL CHEMISTRY. *Educ:* NY Univ, BA, 47; McGill Univ, PhD(phys chem), 51. *Prof Exp:* Res fel photochem, Nat Res Coun Can, 51-53; res assoc microcalorimetry, Univ Montreal, 53-54; SECT MGR, WESTINGHOUSE ELEC CORP RES & DEVELOP CTR, 54- *Mem:* Am Chem Soc; Sigma Xi; Inst Elec & Electronics Engrs. *Res:* Electrochemistry; free radicals; clathrates; dielectrics and insulation; research on development and behavior and properties of dielectrics for high voltage power equipment. *Mailing Add:* Westinghouse Res & Develop Ctr 1310 Beulah Rd Churchill Borough Pittsburgh PA 15235

MANDELIS, ANDREAS, b Corfu, Kerkyra, Greece, June 22, 52; Can & Greek citizen; m 78; c 2. PHOTOTHERMAL SCIENCE, MATERIALS PHYSICS. *Educ:* Yale Univ, BS, 74; Princeton Univ, MA, 76, MSE, 77, PhD(appl physics) 79. *Prof Exp:* Mem sci staff, Bell Northern Res, 80-82; from asst prof to assoc prof 82-90, PROF MECH ENG, UNIV TORONTO, 90- *Concurrent Pos:* Prin investr, Ont Laser & Lightwave Res Ctr, 87-; vis prof, Ecole Polytechnique Federale de Lausanne, Switz, 88-89, Physics Dept, Catholic Univ, Leuven, Belg, 90. *Mem:* Sigma Xi; Am Phys Soc; Can Asn Physicists; Inst Elec & Electronic Engrs; Am Soc Mech Engrs. *Res:* Photothermal and photoacoustic spectroscopy of condensed phases; development of new photothermal NDE techniques; photopyroelectric spectroscopy and depth profiling/imaging; frequency multiplexed photothermal impulse response instrumentation and detection; photothermal solid state sensor devices; thermal-wave tomography. *Mailing Add:* Dept Mech Eng Univ Toronto Toronto ON M5S 1A4 Can

MANDELKERN, LEO, b New York, NY, Feb 23, 22; m 46; c 3. POLYMER CHEMISTRY, BIOPHYSICS. *Educ:* Cornell Univ, AB, 42, PhD(chem), 49. *Prof Exp:* Res assoc chem, Cornell Univ, 49-52; phys chemist, Nat Bur Standards, 52-62; PROF CHEM & BIOPHYS, FLA STATE UNIV, 62-, R O LANTON DISTINGUISHED PROF CHEM, 84. *Concurrent Pos:* Vis prof, Med Sch, Univ Miami, 63, Med Sch, Univ Calif, San Francisco, 64 & Cornell Univ, 67; consult, NIH, 70-73. *Honors & Awards:* Fleming Award, 59; Polymer Award, Am Chem Soc, 75 & Whitby Award, 88; Mettler Award, NAm Thermal Anal Soc, 84. *Mem:* Fel NY Acad Sci; fel AAAS; Am Chem Soc; fel Am Phys Soc; Biophys Soc. *Res:* Physical chemistry of high polymers; biophysics and macromolecules. *Mailing Add:* Fla State Univ Inst Molecular Biophys Tallahassee FL 32306-1096

MANDELKERN, MARK, b Milwaukee, Wis, July 18, 33; m; c 3. MATHEMATICS. *Educ:* Marquette Univ, BS, 55; Univ Rochester, PhD(math), 66. *Prof Exp:* Instr math, State Univ NY Stony Brook, 62-65; asst prof, Univ Kans, 66-69; from asst prof to assoc prof, 69-87, PROF MATH, NMEX STATE UNIV, 87- *Mem:* Math Asn Am; Am Math Soc. *Res:* Constructive mathematics. *Mailing Add:* Dept Math NMex State Univ Las Cruces NM 88003

MANDELKERN, MARK ALAN, b New York, NY, Jan 28, 43; m 80; c 3. ELEMENTARY PARTICLE PHYSICS, MEDICAL & BIOPHYSICS. *Educ:* Columbia Univ, AB, 63; Univ Calif, Berkeley, PhD(physics), 67; Univ Miami, MD, 75; dipl, Am Bd Nuclear Med, 84. *Prof Exp:* From asst prof to assoc prof, 68-82, PROF PHYSICS, UNIV CALIF, IRVINE, 82-; STAFF PHYSICIAN, VET ADMIN MED CTR, WADSWORTH, 84- *Concurrent Pos:* Researcher, Saclay Nuclear Res Ctr, France, 70-71, CERN, Switzerland, 77- & Brookhaven Nat Lab, 78-79; vis prof, Univ Brasilia, 75; fel nuclear med, Univ Calif, Los Angeles, 82-84, Fermilab, 86-; clin prof, Univ Calif Los Angeles, 88- *Res:* Strong interactions; physiology; positron emission tomography; nuclear medicine. *Mailing Add:* Dept of Physics Univ of Calif Irvine CA 92664-4546

MANDELL, ALAN, b New Bedford, Mass, Feb 26, 26; m 46; c 2. SCIENCE EDUCATION, BIOLOGY. *Educ:* Holy Cross Col, BS, 46; Univ Va, MEd, 56; Univ NC, DEd, 66. *Prof Exp:* Prof biol, Frederick Col, 61-65, chmn dept, 65-67; PROF SCI EDUC, OLD DOMINION UNIV, 67- *Concurrent Pos:* Consult surplus property div, US Off Educ, 56. *Mem:* Sci Teachers Asn. *Mailing Add:* Dept Curric & Instr Old Dominion Univ Hampton Blvd Norfolk VA 23508

MANDELL, ARNOLD J, b Chicago, Ill, July 21, 34; c 2. PSYCHIATRY, NEUROCHEMISTRY. *Educ:* Stanford Univ, BA, 54; Tulane Univ, MD, 58. *Prof Exp:* Resident psychiat, Sch Med, Univ Calif, Los Angeles, 59-62, chief resident, 62-63, from asst prof to assoc prof, 63-68; assoc prof psychiat, human behav & psychobiol, Univ Calif, Irvine, 68-69; chmn dept, 69-74, co-chmn, 75-77, PROF PSYCHIAT, UNIV CALIF, SAN DIEGO, 69- *Concurrent Pos:* NIMH career teacher award, Univ Calif, Los Angeles, 62-67; referee, Science, Psychopharmacologica, Community Behav Sci & Am J Psychiat, 64-; mem res comt, Calif Interagency Coun Drug Abuse, 69-; ctr study narcotic addiction & drug abuse, NIMH, 69-, ad hoc sci adv bd biol res, 71-, consult, Lab Clin Sci; mem adv bd, Jerusalem Ment Health Ctr, Israel, 71-; ad hoc sci adv bd, President's Spec Actg Off Drug Abuse Prev, 71-; staff psychiatrist, Vet Admin Hosp, San Diego, Calif, 72-; consult, Ill State Psychiat Res Inst. *Honors & Awards:* A E Bennett Award Res Biol Psychiat, 62. *Mem:* AAAS; Am Inst Chem; Am Psychiat Asn; Soc Biol Psychiat; Am Col Psychiat. *Res:* Neurochemical and biochemical correlates of behavior in animals and man. *Mailing Add:* Dept Psychiat M-003 Univ Calif San Diego Box 109 La Jolla CA 92093

MANDELL, LEON, b Bronx, NY, Nov 19, 27; m 59, 71; c 2. SYNTHETIC ORGANIC & NATURAL PRODUCTS CHEMISTRY. *Educ:* Polytech Inst Brooklyn, BS, 48; Harvard Univ, MA, 49, PhD(chem), 51. *Prof Exp:* Sr chemist, Merck & Co, Inc, 51-53; from asst prof to prof org chem, Emory Univ, 55-64, chmn dept, 68-84; dean, 84-90, PROF, COL NAT SCI, UNIV SFLA, 90- *Concurrent Pos:* Consult, Schering Corp, 58-88. *Mem:* Am Chem Soc. *Res:* Synthetic methods in organic chemistry; nuclear magnetic resonance spectroscopy; mechanisms of organic reactions; natural product chemistry. *Mailing Add:* Col Arts & Sci Univ S Fla Tampa FL 33620

MANDELL, ROBERT BURTON, b Alhambra, Calif, Nov 13, 33; m 59; c 2. OPTOMETRY, VISUAL PHYSIOLOGY. *Educ:* Los Angeles Col Optom, OD, 56; Ind Univ, MS, 58, PhD(physiol optics), 62. *Prof Exp:* From asst prof to assoc prof, 62-73, PROF OPTOM, UNIV CALIF, BERKELEY, 73- *Mem:* AAAS; Optical Soc Am; Am Acad Optom; Am Optom Asn; Sigma Xi. *Res:* Corneal contour; contact lenses. *Mailing Add:* Dept of Optom Univ of Calif 2120 Oxford St Berkeley CA 94720

MANDELS, MARY HICKOX, b Rutland, Vt, Sept 12, 17; m 42; c 2. MICROBIAL PHYSIOLOGY. *Educ:* Cornell Univ, BS, 39, PhD(plant physiol), 47. *Prof Exp:* Microbiologist Pioneering Res Div, Qm Res & Eng Ctr, 55-62, microbiologist, Food Lab, Natick Develop Ctr, 62-71, MICROBIOLOGIST ENVIRON SCI DIV, US ARMY RES & DEVELOP COMMAND, 71- *Mem:* Bot Soc Am; Am Soc Microbiol; Sigma Xi. *Res:* Fungal enzymes; cellulose saccharification; fermentation technology. *Mailing Add:* 106 Everett St Natick MA 01760

MANDELSTAM, PAUL, b Boston, Mass, Apr 18, 25. BIOCHEMISTRY, INTERNAL MEDICINE. *Educ:* Harvard Univ, AB, 44, AM, 46, MD, 50, PhD, 53; Am Bd Internal Med, dipl, 57. *Prof Exp:* Intern, Med Serv, Beth Israel Hosp, Boston, 50-51, asst resident, 52-53; from asst prof to assoc prof, 60-73, PROF MED, COL MED, UNIV KY, 73- *Concurrent Pos:* Res fel, Sch Med, Yale Univ, 55-57; Nat Found res fel, Sch Med, Wash Univ, 57-59; Nat Inst Neurol Dis & Blindness spec trainee, 59-60; asst physician, New Haven Hosp, 55-57. *Mem:* Fel Am Col Physicians; Am Soc Gastrointestinal Endoscopy; Cent Soc Clin Res; Am Gastroenterol Asn; Am Physiol Soc. *Res:* Active transport; gastroenterology. *Mailing Add:* Univ Ky Col Med Lexington KY 40536

MANDELSTAM, STANLEY, b Johannesburg, SAfrica, Dec 12, 28. THEORETICAL PHYSICS. *Educ:* Univ Witwatersrand, BSc, 51; Cambridge Univ, BA, 54; Univ Birmingham, PhD(math physics), 56. *Prof Exp:* Asst math physics, Univ Birmingham, 56-57; Boese fel physics, Columbia Univ, 57-58; asst res physics, Univ Calif, Berkeley, 58-60; prof math physics, Univ Birmingham, 60-63; PROF PHYSICS, UNIV CALIF, BERKELEY, 63- *Concurrent Pos:* Prin investr grant, NSF. *Mem:* Fel Royal Soc; Am Phys Soc. *Res:* Theoretical physics of elementary particles. *Mailing Add:* Dept Physics Univ Calif Berkeley CA 94720

MANDERS, KARL LEE, b Rochester, NY, Jan 21, 27; m 69; c 2. CHRONIC PAIN MANAGEMENT, HYPERBARIC MEDICINE. *Educ:* Univ Buffalo, MD, 50. *Prof Exp:* Coroner, Marion County Coroner, Indianapolis, 76-84, pres Hyperbaric Assoc Ind, 85-, resident, Midwest Pain Soc, 88-89, MED DIR COMMUNITY HOSP REHAB CTR CHRONIC PAIN, 73-, PRES, NECK & BACK INST IND, 85-; Coroner, Marion County Coroner, Indianapolis, 76-84; pres, Hyperbaric Assoc Ind & Neck & Back Inst Ind, 85-90, Midwest Pain Soc, 88-89; MED DIR, COMMUNITY HOSP REHAB CTR CHRONIC PAIN, 73- *Concurrent Pos:* Pres, private practice, neurosurgical Assoc of Ind, 56-, Ind Coroners Assoc, 79; prof adv coun, Am Bd Med Psychotherapists, 85-; mem ed rev bd, J Pain Mgt, 83. *Mem:* Am Assoc Neuro Sci; fel Int Col Surgeons; Am Acad Forensic Sci; Am Soc Stereotaxic & functional neurosurg. *Res:* Developed the first center in Indiana for treatment of chronic pain; hyperbaric oxygen therapy for head injuries; contemporary chronic pain therapy. *Mailing Add:* 7209 N Shadeland Ave Indianapolis IN 46250-2021

MANDERS, PETER WILLIAM, b London, UK, Apr 1, 53; Brit citizen. MATERIALS ENGINEERING. *Educ:* Univ Cambridge, BA, 74, MA, 76; Univ Surrey, PhD(composites), 79. *Prof Exp:* Res fel, Univ Surrey, 76-81; RES ASSOC, UNIV DEL, 81- *Mem:* Inst Metallurgists. *Res:* Strength and fracture of fiber reinforced composite materials; statistical aspects of failure, and the experimental and theoretical study of the micromechanisms of failure in composite materials; mechanics of composite materials. *Mailing Add:* Union Carbide PO Box 670 Bound Brook NJ 08805

MANDERSCHEID, LESTER VINCENT, b Oct 9, 30; m 53; c 4. PRICE ANALYSIS, APPLIED STATISTICS. *Educ:* Iowa State Univ, BS, 51, MS, 52; Stanford Univ, PhD(agr econ), 61. *Prof Exp:* Res asst agr econ, Iowa State Univ, 51-52 & Stanford Univ, 52-56; from asst prof to assoc prof agr econ, 56-70, assoc chair, 73-87, PROF AGR ECON, MICH STATE UNIV, 70-, CHAIR DEPT, 87- *Concurrent Pos:* Consult, Pfizer Co, 70-72, Dept Agr Econ, Tex A&M Univ, 89 & Ceisin, NASA, 90; dir, Am Agr Econ Asn, 82-85. *Mem:* Am Agr Econ Asn (pres, 87-90); Am Econ Asn; Am Statist Asn; Am Eval Asn; Am Asn Univ Professors. *Res:* Behavior of agricultural prices. *Mailing Add:* Agr Econ Dept Mich State Univ East Lansing MI 48824-1039

MANDEVILLE, CHARLES EARLE, b Dallas, Tex, Sept 3, 19; m 43; c 2. PHYSICS. *Educ:* Rice Univ, BA, 40, MA, 41, PhD(physics), 43. *Prof Exp:* Mem staff radiation lab, Mass Inst Technol, 43-45; instr physics, Rice Univ, 45-46; physicist, Bartol Res Found, 46-53, asst dir, 53-59; prof physics & head dept, Univ Ala, 59-61, res physicist, 61; prof physics, Kans State Univ, 61-67; prof, 67-84, head dept, 67-75, dir spec proj, comt energy res, 75-84, EMER PROF PHYSICS, MICH TECHNOL UNIV, 85- *Concurrent Pos:* Vis lectr, Philadelphia Col Osteop Med, 50-68; consult, US Naval Ord Test Sta, 54-60, Curtiss-Wright Corp, 56-60, res ctr, Babock & Wilcox Co, 58-60, US Army Rocket & Guided Missile Agency, Redstone Arsenal, 59-61, US Naval Radiol Defense Lab, 62-64, Kaman Nuclear Corp, 64-67 & Commonwealth-Edison Co, 73-74. *Mem:* Fel AAAS; fel Am Phys Soc; Sigma Xi. *Res:* Nuclear, experimental and solid state physics; primary disintegrations; energies of gamma rays; coincidence experiments; neutron scattering; luminescence; biophysics; theory of magnetism. *Mailing Add:* 3131 Quenby Rd Houston TX 77005-2337

MANDICS, PETER ALEXANDER, b Budapest, Hungary, May 29, 37; US citizen; m 68; c 2. ADVANCED WEATHER SERVICE SYSTEMS, ATMOSPHERIC PHYSICS. *Educ:* Univ Colo, Boulder, BS, 62; Mass Inst Technol, SM, 63; Stanford Univ, PhD(elec eng), 71. *Prof Exp:* Instr & res asst elec eng, Univ Colo, Boulder, 63-65; physicist, Wave Propagation Lab, 71-79, supvry physicist, Prog Regional Observing & Forecasting Serv, 79-81, chief, Explor Develop Facil, 81-88, DIR, PROG REGIONAL OBSERVING & FORECASTING SERV, ENVIRON RES LAB, NAT OCEANIC & ATMOSPHERIC ADMIN, US DEPT COM, 88- *Concurrent Pos:* Nat Res Coun res assoc atmospheric physics, Wave Propagation Lab, Nat Oceanic & Atmospheric Admin, US Dept Com, 71-73. *Mem:* Inst Elec & Electronic Engrs; Int Union of Radio Sci; Nat Weather Asn. *Res:* Development of advanced weather forecasting systems; computer systems development, computer networking and data communications; remote atmospheric sensing using acoustic, radio and optical wave propagation. *Mailing Add:* NOAA/ERL US Dept of Commerce 325 Broadway Boulder CO 80303

MANDIL, I HARRY, b Istanbul, Turkey, Dec 11, 19; US citizen; m 46; c 2. ELECTRICAL ENGINEERING. *Educ:* Univ London, BSc, 39; Mass Inst Technol, MS, 41; dipl, Oak Ridge Sch Reactor Technol, 50. *Hon Degrees:* DSc, Thiel Col, 60. *Prof Exp:* Field engr, Norcross Corp, 41-42, asst to pres, 46-49; chief reactor eng br, Naval Reactors, AEC, proj mgr, Shippingport Atomic Power Sta & dir reactor eng div, Naval Nuclear Propulsion Hq, Bur Ships, US Navy, 50-64; prin officer & dir, MPR Assoc Inc, 64-85; RETIRED. *Concurrent Pos:* Vis comt Nuclear Eng Dept, Mass Inst Technol. *Honors & Awards:* Prime Movers Award, Am Soc Mech Engrs, 56. *Mem:* Am Soc Mech Engrs; Am Nuclear Soc; NY Acad Sci. *Res:* Development of nuclear power for generation of electricity and for application to ships. *Mailing Add:* 701 Heathery Lane Naples FL 33963-8514

MANDL, ALEXANDER ERNST, b Vienna, Austria, May 18, 38; US citizen; m 60; c 2. CHEMICAL KINETICS, LASER PHYSICS. *Educ:* City Col New York, BS, 60; New York Univ, MS, 63, PhD(physics), 67. *Prof Exp:* Instr physics, New York Univ, 66-67; electron scattering physics, Nat Bur

Standards, 67-69; STAFF SCIENTIST PHYSICS, AVCO EVERETT RES LAB, 69- *Mem:* Am Phys Soc. *Res:* Basic shock tube studies on alkali halides; mercury halides; halide negative ions; rare gas halides; author of 50 major journal publications; laser research-excimers; co-author book on alkali halides. *Mailing Add:* Avco Everett Res Lab Everett MA 02149

MANDL, INES, b Vienna, Austria, Apr 19, 17; nat US; m 36. BIOCHEMISTRY. *Educ:* Nat Univ Ireland, dipl, 44; Polytech Inst Brooklyn, MS, 47, PhD(chem), 49. *Hon Degrees:* Dr, Univ Bordeaux, 83. *Prof Exp:* Res chemist, Res Labs, Interchem Corp, 45-49; res assoc, 49-55, assoc, 55-56, asst prof biochem, 56-72, assoc prof reprod biochem, 72-76, prof, Col Physicians & Surgeons, 76-86, EMER PROF REPROD BIOCHEM, COLUMBIA UNIV, 86- *Concurrent Pos:* Dir gynec labs, Delafield Hosp, 59-76; ed-in-chief, Connective Tissue Res-An Int J, 72-85. *Honors & Awards:* Neuberg Medal, 77; Garvan Medal, 83. *Mem:* AAAS; Am Soc Biol Chem; Am Chem Soc; Am Asn Cancer Res; fel NY Acad Sci; fel Geront Soc; Am Thoracic Soc. *Res:* Enzymes; proteins; amino acids; carbohydrates; proteolytic enzyme inhibitors; alpha-antitrypsin; emphysema; respiratory distress syndrome; enzymes of bacterial origin and their medical applications; microstructure of collagen and elastin. *Mailing Add:* 166 W 72nd St New York NY 10023

MANDL, PAUL, b Vienna, Austria, Feb 9, 17; nat Can; m 50. APPLIED MATHEMATICS. *Educ:* Univ Toronto, BA, 45, MA, 48, PhD(math), 51. *Prof Exp:* Jr res officer aerodyn, Nat Res Coun Can, 45-48, from asst to assoc res officer, 48-60, sr res officer, 60-66; prof math, Carleton Univ, 66-82; RETIRED. *Concurrent Pos:* Lectr univ exten dept, McGill Univ, 58-61 & Carleton Univ, 60-61, vis assoc prof math, 64-65, part-time lectr wing theory, Grad Div, 65-; consult lab unsteady aerodynamics, Nat Aeronaut Estab; vis prof mechanics of fluids, Univ Manchester, Eng, 74; vis prof, Inst Gasdynamics & Thermodynamics, Tech Univ Vienna, Austria, 81; adj prof, Carleton Univ, 82-87, chmn orgn comt, Eighth Can Symp on Fluid Dynamics, 88; chmn orgn comt, Eighth Can Symp on Fluid Dynamics, 88. *Honors & Awards:* F W Casey Baldwin Award, Can Aeronaut & Space Inst; W R Tumbull lectr, Can Aeronaut & Space Inst, Vancouver BC, 86. *Mem:* Can Aeronaut & Space Inst; Can Math Soc; Can Appl Math Soc. *Res:* Theoretical fluid mechanics; aerodynamics; rheology; theory of screw propellers in transonic flow. *Mailing Add:* 162 Camelia Ave Ottawa ON K1K 2X8 Can

MANDL, RICHARD H, b New York, NY, Oct 20, 34; m 56; c 4. BIOLOGY. *Educ:* NY Univ, BA, 66. *Prof Exp:* Res assoc environ biol, Boyce Thompson Inst Plant Res, 52-69, asst plant physiologist, 69-77, assoc environ biologist, 77-90, EMER ENVIRON BIOLOGIST, BOYCE THOMPSON INST PLANT RES, 90- *Mem:* AAAS. *Res:* Environmental biology; development of new food products; methods for protein, amino acid, iodo-amino acid, organic phosphate and fluoride analysis; effects of air pollutants on plants. *Mailing Add:* Boyce Thompson Inst Plant Res Tower Rd Ithaca NY 14853

MANDLE, ROBERT JOSEPH, b New York, NY, May 18, 19; m 43; c 4. MICROBIOLOGY. *Educ:* Lebanon Valley Col, BS, 42; Univ Pa, PhD, 51. *Prof Exp:* Asst microbiol, Rockefeller Inst, NJ, 45-50; instr, Univ Del, 50-51; asst prof, 51-65, PROF MICROBIOL, JEFFERSON MED COL, 65- *Concurrent Pos:* Fulbright-Hays grant, Sch Med Technol, Catholic Univ, Quito, Ecuador, 80-81. *Honors & Awards:* William H Rorer Award, Am Col Gastroenterol, 74. *Mem:* Am Soc Microbiol; Sigma Xi; Reticuloendothelial Soc; Mycol Soc Am; Mycol Soc NAm. *Res:* Physiology of micro-organisms; infections and resistance. *Mailing Add:* PO Box 1009 Patagonia AZ 85624-1009

MANDRA, YORK T, b New York, NY, Nov 24, 22; m 46. GEOLOGY. *Educ:* Univ Calif, AB, 47, MA, 49; Stanford Univ, PhD(geol), 58. *Prof Exp:* Asst paleont, Univ Calif, 49-50; from instr to assoc prof, 50-65, head sect & chmn dept, 60-67, PROF GEOL, SAN FRANCISCO, 65- *Concurrent Pos:* Danforth Found teaching fel, 56-57, NSF, teaching fel, 59-60; NSF fac fel & vis prof, Univ Glasgow & Univ Aix-Marseille, 59-60; vis prof, Syracuse Univ, 63, Univ Maine, 69 & Univ Calif, Santa Barbara, 72-; res assoc, Calif Acad Sci, 66-86; NSF res grants, 67-77; vis scientist, New Zealand Geol Surv, 70; fel, Calif Acad Sci, 73- *Honors & Awards:* Robert Wallace Webb Award, 77. *Mem:* Fel AAAS; fel Geol Soc Am; Paleont Soc; Nat Asn Geol Teachers (vpres, 72-73, pres, 73-74); Soc Econ Paleont & Mineral. *Res:* Micropaleontology, especially stratigraphic and paleoecologic aspects of Mesozoic and Cenozoic silicoflagellates; societal problems of energy. *Mailing Add:* 8 Bucareli Dr San Francisco CA 94132

MANDULA, JEFFREY ELLIS, b New York, NY, July 23, 41; m 63. PHYSICS. *Educ:* Columbia Col, NY, AB, 62; Harvard Univ, AM, 64, PhD(physics), 66. *Prof Exp:* NSF fel, Harvard Univ, 66-67; res fel physics, Calif Inst Technol, 67-69; mem natural sci, Inst Advan Study, 69-70; asst prof theoret physics, Calif Inst Technol, 70-74; assoc prof appl math, Mass Inst Technol, 74-79; prof math & physics, Washington Univ, 79-; AT DIV HIGH ENERGY & PHYSICS, US DEPT ENERGY. *Concurrent Pos:* Dir theoret physics prog, NSF, 80-81. *Mem:* AAAS; Am Phys Soc. *Res:* Theoretical elementary particle physics; quantum field theory. *Mailing Add:* ER-221 GTN Dir High Energy Physics US Dept Energy Washington DC 20545

MANDULEY, ILMA MORELL, b Holquin, Cuba, Sept 19, 29. MATHEMATICS. *Educ:* Univ Havan, PhD(math), 53. *Prof Exp:* From asst prof to prof math, Polytech Inst Holquin, Cuba, 55-61; instr, 61-64, ASST PROF MATH, GUILFORD COL, 64- *Concurrent Pos:* Instr, Friends Sch, Cuba, 52-61. *Mem:* AAAS; Math Asn Am; Am Asn Univ Professors. *Res:* Euclidean and projective geometry; differential analysis; ordinary and partial differentiation equations. *Mailing Add:* Dept Math Guilford Col 4213-B Edith Lane Greensboro NC 27410

MANDY, WILLIAM JOHN, b Lackawanna, NY, Mar 12, 33; m 59; c 3. IMMUNOBIOLOGY, IMMUNOGENETICS. *Educ:* Elmhurst Col, 58; Univ Ill, Urbana, PhD(microbiol), 63; Univ Houston, MS, 62. *Prof Exp:* From asst prof to assoc prof, 65-73, PROF IMMUNOL, UNIV TEX, AUSTIN, 73- *Concurrent Pos:* NIH fel immunochem, Sch Med, Univ Calif, 63-65; USPHS career develop award, 66- *Mem:* AAAS; Am Soc Microbiol; Am Soc Immunol. *Res:* Structure of the gamma globulin molecule and nature of naturally occurring antiglobulin factors. *Mailing Add:* Dept of Microbiol Univ of Tex Austin TX 78712

MANE, SATEESH RAMCHANDRA, b Malacca, Malaysia, June 26, 59. PARTICLE ACCELERATORS. *Educ:* Cambridge Univ, UK, BA, 81; Cornell Univ, MS, 85, PhD(physics), 87. *Prof Exp:* Res assoc physics, Univ Mich, Ann Arbor, 87; res assoc physics, Fermi Nat Accelerator Lab, 87-89; ASSOC SCIENTIST, BROOKHAVEN NAT LAB, 89- *Mem:* Am Phys Soc. *Res:* Dynamics of high energy particle accelerators, especially synchrotron radiation and polarization of particle beams. *Mailing Add:* Brookhaven Nat Lab Bldg 911 Upton NY 11973-9999

MANEATIS, GEORGE A, HIGH VOLTAGE SUBSTATION DESIGN. *Prof Exp:* PRES, PAC GAS & ELEC CO. *Mem:* Nat Acad Eng; fel Inst Elec & Electronics Engrs. *Mailing Add:* Pac Gas & Elec Co 77 Beale St 32nd Floor San Francisco CA 94106

MANEN, CAROL-ANN, b Newark, NJ, Feb 4, 43. ENVIRONMENTAL TOXICOLOGY, RESEARCH ADMINISTRATION. *Educ:* Gettysburg Col, AB, 64; Univ Ill, MSc, 66; Univ Maine, PhD(zool), 73. *Prof Exp:* Fel pharmacol, Med Sch, Univ Ariz, 73-77; res asst biochem, Roche Inst Molecular Biol, 77-78; asst prof biol, Univ Ala, 78-80; mem staff, 80-89, SR CHEMIST, NAT OCEANIC & ATMOSPHERIC ADMIN, 89- *Concurrent Pos:* NIH fel, Nat Inst Child Health Develop, 75-77. *Mem:* Am Soc Zoologists; Soc Develop Biol. *Res:* Assess the fate and effects of pollutants in Arctic and sub-Arctic marine ecosystems. *Mailing Add:* Nat Oceanic & Atmospheric Admin 6001 Executive Blvd WSC-1 Rm 323 Rockville MD 20852

MANERA, PAUL ALLEN, b Clovis, Calif, Nov 11, 30; m 59; c 2. HYDROGEOLOGY. *Educ:* Fresno State Col, BA, 55; Ari State Univ, MA, 63 & 71, PhD(geog), 82. *Prof Exp:* Mining geologist, Holly Minerals Corp, 55-56; geohydrologist, Samuel F Turner & Assocs, 57-62; CONSULT GEOHYDROLOGIST, MANERA INC, 62- *Concurrent Pos:* Lectr, Fresno State Col, geol, 55-56, Ari State Univ, geog, 61-62. *Mem:* Geol Soc Am; Am Geophys Union; Am Water Resources Asn; Asn Ground Water Scientists & Engrs; Europ Soc Explor Geophysicists. *Res:* Availability of groundwater resources and its impact on the population of the Cave Creek-Carefree Basin, Maricopa County, Arizona; physical factors limiting the agricultural development of Butler Valley, Yuma County, Arizona. *Mailing Add:* 5251 N 16th St Suite 302 Phoenix AZ 85016

MANERI, CARL C, b Cleveland, Ohio, Jan 25, 33; m 53; c 3. MATHEMATICS. *Educ:* Case Western Reserve Univ, BS, 54; Ohio State Univ, PhD(math), 59. *Prof Exp:* Res fel & instr math, Univ Chicago, 60-62; asst prof, Syracuse Univ, 62-65; chmn dept, 68-71, ASSOC PROF MATH, WRIGHT STATE UNIV, 65- *Mem:* Am Math Soc; Math Asn Am. *Res:* Algebra; combinatorics and finite mathematics; geometry. *Mailing Add:* Dept Math Wright State Univ Dayton OH 45435

MANES, ERNEST GENE, b New York, NY, 1943; m 63; c 2. MATHEMATICS. *Educ:* Harvey Mudd Col, BSc, 63; Wesleyan Univ, PhD(math), 67. *Prof Exp:* Asst prof math, Harvey Mudd Col, 67-68 & Univ Hawaii, 68-69; Killam fel, Dalhousie Univ, 69-71; asst prof, 71-75, ASSOC PROF MATH, UNIV MASS, AMHERST, 75- *Concurrent Pos:* NSF grants, 72-74, 75-77 & 77-79. *Mem:* Math Asn Am; Am Math Soc; Asn Comput Mach. *Res:* Applied category theory, semantics of computation, universal algebra. *Mailing Add:* Dept of Math & Statist Univ of Mass Amherst MA 01002

MANES, KENNNETH RENE, b Brooklyn, NY, Oct, 9, 42; m 69; c 3. INERTIAL CONFINEMENT FUSION, LASERS FOR MATERIALS PROCESSING. *Educ:* Purdue Univ, BS, 65; Stanford Univ, MS, 67, PhD(elec eng),70. *Prof Exp:* Killam fel teaching & res, Elec Eng Dept, Univ Alberta, Can, 70-72; ASSOC PROJ MGR, LASER FUSION RES & ADVAN APPLICATIONS/LASER MAT PROCESSING, LAWRENCE LIVERMORE NAT LAB, 72- *Mem:* Am Phys Soc. *Res:* High power solid-state laser; Raman convertor and free-electron laser development. *Mailing Add:* Lawrence Livermore Nat Lab Box 5508 Livermore CA 94550

MANES, MILTON, b New York, NY, Oct 14, 18; m 45; c 2. PHYSICAL CHEMISTRY. *Educ:* City Col New York, BS, 37; Duke Univ, PhD(phys chem), 47. *Prof Exp:* From lab asst to jr chemist, US Food & Drug Admin, 37-41; phys chemist, US Bur Mines, 47-52; sr chemist & mgr statist design group, Koppers Co, Inc, 52-58; supvr phys chem res, Pittsburgh Chem Co, 59-64; sr fel & head adsorption fel, Mellon Inst, 64-67; prof, 67-85, RES EMER PROF CHEM, KENT STATE UNIV, 85- *Concurrent Pos:* Vis prof, Cornell Univ, 64; adj prof, Indust /environ Health Sci, Grad Sch Pub Health, Univ pittsburgh, 86- *Mem:* AAAS; Am Chem Soc. *Res:* Adsorption; near equilibrium thermodynamics and kinetics; activated carbon. *Mailing Add:* Amberson Towers (412) 5 Bayard Rd Pittsburgh PA 15213

MANFREDI, ARTHUR FRANK, JR, b Neosho, Mo, Aug 8, 44; m 67; c 2. CONTROL THEORY. *Educ:* Cooper Union, BSME, 65, Polytech Inst Brooklyn, MSME, 68; George Washington Univ, DSc(eng), 75. *Prof Exp:* Missile analyst, Cent Intel Agency, 69-77, chief, Ballistic Missile Br, 77-80, dep chief, S&T Div, 80-82, policy analyst, 83- 85, asst nat intel officer, 86-89, SR ANALYST, CENT INTEL AGENCY, 89- *Concurrent Pos:* Prof lectr, George Washington Univ, 77-; chief space systs, Nat Security Coun, 82-83; sr analyst, Cong Res Serv, 85-86; adj fac mem, Univ Va, 86- *Mem:* Am Soc Mech Engrs; Am Defense Preparedness Asn. *Res:* Analysis of foreign weapon and space systems. *Mailing Add:* 1315 Round Oak Ct McLean VA 22101

MANGAN, GEORGE FRANCIS, JR, biochemistry, for more information see previous edition

MANGAN, JERROME, b Columbus, Ohio, Nov 18, 34; div; c 2. DEVELOPMENTAL BIOLOGY, GENETICS. *Educ:* Univ Cincinnati, BA, 60, MS, 63; Brown Univ, PhD(biol), 66. *Prof Exp:* NSF fel, Albert Einstein Col Med, 66-67; asst prof biol, Univ Chicago, 67-70; asst prof, 70-72, assoc prof, 72-76, PROF BIOL, CALIF STATE UNIV, FRESNO, 76-, CHMN DEPT, 78- *Concurrent Pos:* Am Cancer Soc & NIH grants, Univ Chicago, 67-68; NSF grant, Univ Chicago & Fresno State Col, 68-72. *Mem:* Genetics Soc Am; Am Soc Microbiol. *Res:* The nature of control of biochemical processes in bacteria and lower eukaryotic organisms. *Mailing Add:* Dept Biol Calif State Univ Fresno CA 93740

MANGAN, ROBERT LAWRENCE, b Washington, DC, Aug 6, 45; m 78; c 1. ECOLOGY, ENTOMOLOGY. *Educ:* Kent State Univ, BA, 67; Univ Ariz, PhD(ecol), 78. *Prof Exp:* Res asst ecol, US Int Biol Prog, Desert Biome Validation Prog, 74-77; res assoc ecol, US Pasture Res Lab, Pa State Univ, 78-81; res entomologist, screwworm proj, Tex A&M Univ, 82-83; res entomologist, 83-86, RES LEADER, SCREWWORM PROJ, USDA-ARS, 86- *Mem:* Entom Soc Am; Ecol Soc Am; Pan Pac Entom Soc; AAAS; Animal Behav Soc; Entom Soc Mex. *Res:* Theoretical and experimental studies of insect community ecology; reproductive behavior and population genetics; systematics of acalypterate diptera; agricultural applications of insect population and community ecology. *Mailing Add:* 1123 Kerria St Mcallen TX 78501

MANGANARO, JAMES LAWRENCE, b Brooklyn, NY, Aug 27, 39; m 74; c 2. PROCESS & NEW OPPORTUNITY DEVELOPMENT, PROTEIN PURIFICATION. *Educ:* Mass Inst Technol, SB, 61, SM, 62; Rensselaer Polytech Inst, PhD(chem eng), 65. *Prof Exp:* Res engr electrochem, Res & Develop Ctr, Gen Elec Co, 65-67; asst prof, Manhattan Col, 67-70; SR PRIN ENGR CHEM ENG, FMC CORP, 70- *Concurrent Pos:* Consult, Charles A Manganaro Engrs, 67-; prof engr, NJ, NY, 82. *Mem:* Am Inst Chem Engrs; Sigma Xi. *Res:* Process development for production of chlorinated isocyanurates, sodium tripolyphosphate, barium and strontium salts and organic phosphate esters; elemental phosphorous, heat, mass and momentum transfer; math modeling, computational fluid mechanics, venture analysis, economic evaluation, and protein chromatography. *Mailing Add:* FMC Corp PO Box 8 Princeton NJ 08543

MANGANELLI, RAYMOND M(ICHAEL), b Elizabeth, NJ. ENVIRONMENTAL SCIENCES, CHEMISTRY. *Educ:* Rutgers Univ, PhD(environ sci), 53. *Prof Exp:* PROF ENVIRON SCI, RUTGERS UNIV, 60- *Mem:* Am Indust Hyg Asn; Water Pollution Control Fedn; Air Pollution Control Asn; Am Chem Soc; Sigma Xi. *Res:* Chemistry, physics and biology of air and water pollution. *Mailing Add:* 60 Bauer Terr Hillside NJ 07205

MANGANELLO, S(AMUEL) J(OHN), b Johnstown, Pa, Jan 2, 30; m 57; c 4. PHYSICAL METALLURGY, PROCESS. *Educ:* Univ Pittsburgh, BS, 51, MS, 57. *Prof Exp:* Technologist, Res Lab, USSR Tech Ctr, USX Corp, 53-56, head forgings sect, 56-62, sr res engr ord prod, 63-66, group leader heavy prod, 68-80, assoc res consult, armor, rolls & forgings, 66-85, supvr, Sect Bar & Plate, 85-87, res mgr, 87-89, RES CONSULT, USS TECH CTR, USX CORP, 89- *Mem:* Am Soc Metals; Am Inst Mining, Metall & Petrol Engrs; Am Defense Preparedness Asn. *Res:* Steel products; patents; heavy forgings; alloy steels; ordnance products; armor; rolls; railroad products; steel processing; steel bars; steel plates. *Mailing Add:* USS Tech Ctr USX Corp Monroeville PA 15146

MANGANIELLO, EUGENE J(OSEPH), b New York, NY, June 8, 14; m 45; c 4. MECHANICAL ENGINEERING. *Educ:* City Col New York, BSE, 34, EE, 35. *Prof Exp:* Mech engr, Nat Adv Comt Aeronaut, Langley Field, Va, 36-42, head heat transfer sect, 42-45, chief thermodyn br, 45-49, asst chief res, Lewis Res Ctr, NASA, 49-52, from asst dir to dep dir, 61-73; RETIRED. *Mem:* AAAS; Soc Automotive Engrs; Am Inst Aeronaut & Astronaut. *Res:* Aircraft; missiles; spacecraft propulsion; power generation. *Mailing Add:* 329 Hillcrest Dr Leucadia CA 92024

MANGANIELLO, LOUIS O J, b Waterbury, Conn, June 6, 15; m 50; c 2. NEUROSURGERY. *Educ:* Harvard Univ, AB, 37; Univ Md, MD, 42; Augusta Law Sch, JD, 67; Am Bd Neurol Surg, dipl. *Prof Exp:* Fel neurosurg, Sch Med, Univ Md, 46-47; asst resident, Univ Md Hosp, 47-48, chief resident, 49-50; asst resident, Baltimore City Hosp, 48-49; instr neuroanat & neurosurg, Sch Med, Univ Md, 50-51; ASSOC PROF NEUROSURG, MED COL GA, 51- *Concurrent Pos:* Consult, Hosps, Ga. *Mem:* Am Asn Neurol Surg; Am Asn Cancer Res; AMA; Am Psychiat Asn; fel Am Col Surg. *Res:* Cancer detection and therapy; porphyrin metabolism. *Mailing Add:* 656 Milledge Rd Augusta GA 30904

MANGANIELLO, VINCENT CHARLES, b Jersey City, NJ, Mar 16, 39; m 64; c 3. HORMONE ACTION, CYCLIC NUCLEOTIDES. *Educ:* Johns Hopkins Univ, MD & PhD(physiol chem), 67. *Prof Exp:* CHIEF BIOCHEM PHYSIOL SECT, CELLULAR METAB LAB, NAT HEART LUNG BLOOD INST, NIH, 75- *Mem:* Am Soc Biochem & Molecular Biol; Am Fed Clin Res. *Mailing Add:* Cellular Metab Lab Bldg 10 Rm 5N323 Biochem Physiol Sect Nat Heart Lung Blood Inst NIH Bethesda MD 20892

MANGASARIAN, OLVI LEON, b Baghdad, Iraq, Jan 12, 34; US citizen; m 59; c 3. APPLIED MATHEMATICS. *Educ:* Princeton Univ, BSE, 54, MSE, 55; Harvard Univ, PhD(appl math), 59. *Prof Exp:* Mathematician, Shell Develop Co, 59-67; lectr math programming, Univ Calif, Berkeley, 65-67; assoc prof comput sci, 67-69, chmn dept, 70-73, PROF COMPUT SCI, UNIV WIS-MADISON, 69- *Mem:* Math Prog Soc; Opers Res Soc; Soc Indust & Appl Math; Asn Comput Mach. *Res:* Development and use of theory and computational methods of mathematical programming in various fields of applied mathematics such as operations research, optimal control theory and numerical analysis. *Mailing Add:* Dept Comput Sci 4247 Univ Wis 1210 W Dayton St Madison WI 53706

MANGAT, BALDEV SINGH, b Ludhiana, India, May 7, 35; m 60; c 2. ENTOMOLOGY, ZOOLOGY. *Educ:* Univ Punjab, India, MSc, 58; Univ Wis, PhD(entom), 65. *Prof Exp:* Instr entom, Punjab Agr Col & Res Inst, 58-60; assoc prof biol, Alcorn Agr & Mech Col, 65-66; assoc prof, 66-69, PROF BIOL, ALA A&M UNIV, 69- *Mem:* AAAS; Entom Soc Am. *Res:* Biology and physiology of corn earworm, Heliothis zea. *Mailing Add:* Dept Biol Concordia Univ 1455 De Maisonneuve Blvd W Montreal PQ H3G 1M8 Can

MANGE, ARTHUR P, b St Louis, Mo, Jan 28, 31; m 60; c 3. HUMAN GENETICS. *Educ:* Cornell Univ, BEngPhys, 54; Univ Wis, MS, 58, PhD(genetics), 63. *Prof Exp:* Instr biol, Case Western Reserve Univ, 62-64; from asst prof to assoc prof, 64-90, PROF ZOOL, UNIV MASS, AMHERST, 90- *Concurrent Pos:* Sr res assoc, Univ Wash, 71-72. *Mem:* AAAS; Am Soc Human Genetics; Genetics Soc Am. *Res:* Writer on human genetics. *Mailing Add:* Dept of Zool Univ of Mass Amherst MA 01003

MANGE, FRANKLIN EDWIN, b St Louis, Mo, Feb 12, 28; m 54; c 4. ORGANIC CHEMISTRY. *Educ:* Mass Inst Technol, SB, 48; Univ Ill, PhD(org chem), 51. *Prof Exp:* Res chemist, Petrolite Corp, 51-57, group leader, 57-63, sect mgr, 63-67, res dir, Tretolite Div, 67-82, vpres res & develop, 82-85, dir corp develop, 85-89, TECH DIR, INTEGRATED TECHNOL GROUP, PETROLITE CORP, 89- *Mem:* Am Chem Soc; Sigma Xi. *Res:* Polymer synthesis; surfactants; waxes; demulsification; flocculation; petroleum chemistry; water treatment. *Mailing Add:* 18 Granada Way St Louis MO 63124

MANGE, PHILLIP WARREN, b Kalamazoo, Mich, June 5, 25; m 51; c 2. PLANETARY ATMOSPHERES, SPACE PHYSICS. *Educ:* Kalamazoo Col, AB, 49; Pa State Univ, MS, 52, PhD(physics), 54. *Prof Exp:* Asst prof eng res, Ionosphere Res Lab, Pa State Univ, 54-55; admin asst to gen secy, Spec Comt, Int Geophys Year, Belgium, 55-57, prog officer, US IGY Nat Comt, Nat Acad Sci, DC, 57-59; physicist, 59-70, assoc supt, Space Sci Div, 70-86, SCI CONSULT TO DIR RES, US NAVAL RES LAB, 86- *Mem:* AAAS; Am Phys Soc; Am Geophys Union; Am Astron Soc; Am Inst Aeronaut & Astronaut. *Res:* Structure of high atmosphere; ultraviolet environment in the solar system. *Mailing Add:* Code 1004 US Naval Res Lab Washington DC 20375-5000

MANGELSDORF, CLARK P, b Bryan, Tex, Oct 28, 28; m 51, 83; c 3. CIVIL ENGINEERING. *Educ:* Swarthmore Col, BS, 53; Mass Inst Technol, MS, 54, ScD, 64. *Prof Exp:* Instr civil eng, Swarthmore Col, 54-56 & Univ Ill, 56-59; from asst prof to assoc prof, Swarthmore Col, 59-68; dir mining eng, 79-84, assoc prof, 68-91, EMER PROF CIVIL ENG, UNIV PITTSBURGH, 91- *Concurrent Pos:* NSF sci fac fel, 60-62; res assoc, Scott Paper Co, 62-69; sr sci officer, Tech Inst Delft, Neth, 66-67; res assoc, PPG Res, 69-71; design engr, Dravo Corp; structural engr, US Bur of Mines, 84- *Mem:* Am Soc Civil Engrs; Structural Stability Res Coun; Soc Mining Engrs. *Res:* Structural mechanics; stability of structures; mine roof support systems. *Mailing Add:* Dept Civil Eng Univ Pittsburgh Pittsburgh PA 15261

MANGELSDORF, PAUL CHRISTOPH, economic botany, genetics; deceased, see previous edition for last biography

MANGELSDORF, PAUL CHRISTOPH, JR, b New Haven, Conn, Jan 31, 25; m 49; c 4. MARINE GEOCHEMISTRY. *Educ:* Swarthmore Col, BA, 49; Harvard Univ, PhD(chem physics), 55. *Prof Exp:* Instr, Univ Chicago, 55-57, asst prof chem, 57-60; assoc prof, 61-69, PROF PHYSICS, SWARTHMORE CO., 69-, CHMN DEPT, 78-; RES ASSOC PHYS CHEM, WOODS HOLE OCEANOG INST, 60- *Mem:* Geochem Soc; Am Phys Soc; Am Soc Limnol & Oceanog; Am Geophys Union; Am Asn Physics Teachers. *Res:* Fluid dynamics; transport properties in liquids; thermodynamics of electrolyte solutions; chemistry of sea water. *Mailing Add:* 110 Cornell Ave Swarthmore PA 19081

MANGELSON, FARRIN LEON, b Levan, Utah, May 12, 12; m 36; c 9. BIOCHEMISTRY, NUTRITION. *Educ:* Univ Wash, BS, 38; Utah State Univ, MS, 50, PhD, 63. *Prof Exp:* Teacher high schs, Utah, 36-41; chemist, Remington Arms Co, Inc Div, E I du Pont de Nemours & Co, 42-43; asst state chemist, Utah, 44-47; asst chem, Exp Sta, Utah State Univ, 47-49, res instr, 49-51; from instr to prof chem, 51-80, chmn, Div Phys Sci & Math, 69-77, emer prof chem, 80-; RETIRED. *Concurrent Pos:* Mem res staff, Univ Calif, Berkeley, 64. *Mem:* AAAS; Am Chem Soc. *Res:* Kidney function in cattle as affected by ingestion of inorganic fluorides; effect of animal fats and proteins on blood serum cholesterol level in humans; molecular size of myosin; human nutrition. *Mailing Add:* 260 W 1500 N Orem UT 84057

MANGELSON, NOLAN FARRIN, b Nephi, Utah, Jan 17, 36; m 61; c 8. PHYSICAL CHEMISTRY, NUCLEAR PHYSICS. *Educ:* Utah State Univ, BS, 61; Brigham Young Univ, MS, 63; Univ Calif, Berkeley, PhD(chem), 68. *Prof Exp:* Res fel, Nuclear Physics Lab, AEC, Univ Wash, 67-69; from asst prof to assoc prof, 69-78, PROF CHEM, BRIGHAM YOUNG UNIV, 78- *Mem:* Am Chem Soc; Am Phys Soc; Sigma Xi. *Res:* Particle induced x-ray and gamma-ray emission methods are used for element analysis of air particulates, plant and animal samples, and water; EXAFS analysis is also used; nuclear reactions and spectroscopy. *Mailing Add:* Dept Chem Brigham Young Univ Provo UT 84602-1002

MANGEN, LAWRENCE RAYMOND, b Minneapolis, Minn, Sept 12, 27; m 50; c 3. GEOLOGY, PETROLEUM ENGINEERING. *Educ:* Univ Minn, BS, 55, MS, 57. *Prof Exp:* Jr engr, Shell Oil Co, 57, field engr, 57-59, work over engr, 59-61, prod geologist, 61-62; engr, US Fed Power Comn, 62-68, supvr eng, 68, asst sect head eng, 68-77; br chief, Fed Energy Admin, 77-78; sr staff engr, Energy Info Admin, 78-79, sr tech advr, 79-81, SR STAFF ENGR, OIL & GAS RESERVES & PROD EVAL, US DEPT ENERGY, 81- *Mem:* Fel AAAS; Soc Petrol Eng. *Mailing Add:* 2724 Keystone Lane Bowie MD 20715

MANGER, MARTIN C, b Bethlehem, Pa, Sept 20, 37; m 63. ORGANIC CHEMISTRY, ENVIRONMENTAL SCIENCES. *Educ:* Muhlenberg Col, BS, 59; St Lawrence Univ, MS, 61; Rutgers Univ, MS, 65; Sheffield Univ, PhD(chem), 68. *Prof Exp:* Res chemist, E R Squibb & Sons, Inc, 64-65; assoc prof, 68-73, PROF CHEM, ALA A&M UNIV, 73- *Concurrent Pos:* Sci consult, Asn Educ & Prof Opportunities Found prog for gifted children, 68- *Mem:* Am Chem Soc; Royal Soc Chem; Am Inst Chem. *Res:* Natural products; conformational inversion of bridged biphenyls; pollution technology; biological pigments. *Mailing Add:* 1525 Path Finder Way Lilburn GA 30247-2352

MANGER, WALTER LEROY, b Baltimore, Md, Sept 24, 44; m 67; c 1. BIOSTRATIGRAPHY, PALEONTOLOGY. *Educ:* Col Wooster, BA, 66; Univ Iowa, MS, 69, PhD(geol), 71. *Prof Exp:* Asst prof geol, Northeastern Univ, 71-72; from asst prof to assoc prof, 72-81, PROF GEOL, UNIV ARK, FAYETTEVILLE, 81-, CHMN, 84- *Concurrent Pos:* Cur geol, Univ Ark, Fayetteville, 72-; NSF grants Carboniferous ammonoids, 75-77, 77-79, 80-83 & 85-87; subcontractor, Dames & Moore, Houston, 76; consult, Northwest Minerals Inc, Fayetteville, 79-, Quintana Petrol, Houston, 80- & IRS, 88-; expert witness, Ark Oil & Gas Comn, 80- *Honors & Awards:* S Cent Fedn Award, Am Fedn Mineral Socs, 80. *Mem:* Soc Econ Paleontologists & Mineralogists; Paleont Soc; Brit Palaeont Asn; Paleont Res Inst; Res Soc NAm; fel Geol Soc Am. *Res:* Ammonoid biostratigraphy of the Carboniferous Period, particularly the Mississippian-Pennsylvanian boundary, on a worldwide basis; ammonoid phylogeny and taxonomy, and Carboniferous lithostratigraphy of North American midcontinent. *Mailing Add:* Dept Geol Univ Ark Fayetteville AR 72701

MANGER, WILLIAM MUIR, b Greenwich, Conn, Aug 13, 20; m 64; c 4. MEDICINE. *Educ:* Yale Univ, BS, 44; Columbia Univ, MD, 46; Univ Minn, PhD(med), 58; Am Bd Internal Med, dipl, 57. *Prof Exp:* Intern med, Columbia Presby Med Ctr, 46-47, resident, 49-50; instr, 57-66, assoc, 66-69, LECTR, COL PHYSICIANS & SURGEONS, COLUMBIA UNIV, 80-; CLIN PROF, MED CTR, NY UNIV, 83- *Concurrent Pos:* Dir, Manger Res Found, 58-77; chmn, Nat Hypertension Asn, 77-; asst physician, Presby Hosp, 57-66, asst attend, 66-70; clin asst vis physician, Columbia Div, Bellevue Hosp, 64-68; asst attend, Dept Med, Bellevue Hosp, 69-77, assoc attend, 77-83; attend, Dept Med, Bellevue Hosp, 83-; consult, Southampton Hosp, 71-; fel coun circulation & coun high blood pressure res, Am Heart Asn; ed, Am Lect in Endocrinol, 62-75; from asst clin prof to assoc clin prof, Med Ctr, NY Univ, 68-83. *Honors & Awards:* Meritorious Res Award, Mayo Found, 55. *Mem:* Am Physiol Soc; Soc Pharmacol & Exp Therapeut; fel Am Col Clin Pharmacol; fel Acad Psychosom Med; Soc Exp Biol & Med; Harvey Soc; fel Am Soc Clin Pharmacol & Therapeut; fel Am Col Physicians; fel Am Col Cardiol; fel Am Geriat Soc. *Res:* Chemical quantitation of epinephrine and norepinephrine in plasma and relationship of these pressor amines to hypertension, circulatory shock and mental disease; pheochromocytoma; endocrine aspects of tumor development; salt-induced hypertension. *Mailing Add:* Inst Rehab Med 400 E 34th St Rm 709 New York NY 10016

MANGHAM, JESSE ROGER, b Plains, Ga, Nov 18, 22; m 43; c 5. TOXIC CHEMICALS REGULATIONS, ORGANIC CHEMISTRY. *Educ:* Univ Ga, BS, 43; Ohio State Univ, MS, 46, PhD(org chem), 48. *Prof Exp:* Asst, Ohio State Univ, 43-44 & 46-48; res chemist, Va-Carolina Chem Corp, 49-51, group leader org chem, 51-53, sect leader, 53-54; proj leader, Ethyl Corp, 54-67, sr appln chemist, 67-76, appln res assoc, 70-77, sr environ health assoc, 77-87; CONSULT, TOXIC CHEMICALS REGULATIONS, 88- *Mem:* Am Chem Soc; Sigma Xi. *Res:* Organometallics of aluminum, magnesium and boron; organophosphorus chemistry; brominated chemicals; chlorinated solvents. *Mailing Add:* Consult Toxic Chemicals Regulations 263 Riverside Mall Suite 300 Baton Rouge LA 70801

MANGHNANI, MURLI HUKUMAL, b Karachi, Pakistan, Apr 4, 36; m 62; c 4. GEOPHYSICS, GEOCHEMISTRY. *Educ:* Jaswant Col, India, BS, 54; Indian Sch Mines & Appl Geol, Dhanbad, BS, 57; Bihar Univ, MS, 58; Mont State Univ, PhD(geochem, geol, geophys), 62. *Prof Exp:* Fel geophys & NSF res grant, Univ Wis, 62-63; from asst prof to assoc prof geophys & geochem, 64-76, PROF GEOPHYS, UNIV HAWAII, 76-; GEOPHYSICIST, HAWAII INST GEOPHYS, 74- *Concurrent Pos:* Asst geophysicist, Hawaii Inst Geophys, 63-74; prog dir, Exp & Theoret Geophys, Div Earth Sci, NSF, Washington, DC, 81-82. *Mem:* Am Geophys Union; Am Ceramic Soc; AAAS; Geol, Mining & Metall Soc India. *Res:* High pressure and temperature laboratory experimentation of rock materials believed to form the lower crust and upper mantle of the earth; gravity; seismology. *Mailing Add:* Dept Geophys Univ Hawaii Honolulu HI 96822

MANGLITZ, GEORGE RUDOLPH, b Washington, DC, Aug 26, 26; m 53; c 5. ENTOMOLOGY. *Educ:* Univ Md, BS, 51, MS, 52; Univ Nebr, PhD(entom), 62. *Prof Exp:* Asst entomologist, United Fruit Co, Guatemala, 51; assoc prof entom, USDA, Univ Nebr, Lincoln, 65-73, res entomologist, 52-88, prof entom, 73-88; RETIRED. *Mem:* Entom Soc Am; Am Inst Biol Sci. *Res:* Field crop insects, particularly plant resistance to insects. *Mailing Add:* Dept Entom Univ of Nebr Lincoln NE 68583-0816

MANGO, FRANK DONALD, b San Francisco, Calif, Dec 31, 32; m 59; c 2. ORGANIC CHEMISTRY, INORGANIC CHEMISTRY. *Educ:* San Jose State Col, BS, 59; Stanford Univ, PhD(chem), 63. *Prof Exp:* RES ASSOC, SHELL DEVELOP CO, 63- *Mem:* Am Chem Soc. *Res:* Geochemistry, catalysis. *Mailing Add:* 806 Soboda Houston TX 77079

MANGOLD, DONALD J, b Lacoste, Tex, July 1, 29; m 61; c 3. ORGANIC CHEMISTRY, POLYMER CHEMISTRY. *Educ:* St Edward's Univ, BS, 50; Univ Tex, Austin, PhD(chem), 54. *Prof Exp:* Group leader, Olin Corp, 54-60, sect chief, 60-63, tech field rep, 63-66, proj mgr, 66-69; dir res & develop, Flecto Co, 69-72; mgr floor prod div, West Chem Prod, Inc, 72-77; mgr org & polymer sect, 77-84, DIR, DEPT APPL CHEM & CHEM ENG, SOUTHWEST RES INST, 84- *Mem:* Am Chem Soc; Am Inst Chemists; Sigma Xi; Am Defense Prep Asn. *Res:* Microencapsulation; drug development. *Mailing Add:* 14711 Bold Venture San Antonio TX 78248

MANGULIS, VISVALDIS, b Tukums, Latvia, Nov 25, 30; US citizen; m 53; c 2. RADAR, COMMUNICATIONS. *Educ:* Brooklyn Col, BS, 56; NY Univ, MS, 58. *Prof Exp:* Scientist, TRG, Inc, Control Data Corp, 56-62, sect head, 62-68; mem tech staff, Gen Tel & Electronics Labs, 68; sr staff consult, Comput Applns, Inc, 68-70; vpres & treas, Questek, Inc, 70-75; prin mem eng staff, RCA Missile & Surface Radar Div, 75-78, mem tech staff, RCA Labs, 78-85, staff engr, 85-88; CONSULT ENGR, GE AEROSPACE, 88- *Mem:* Am Phys Soc; Inst Elec & Electronics Engrs. *Res:* Design of radar systems, satellite communications; aircraft collision avoidance; operations research; sonar, phased arrays; acoustics; wave radiation and propagation; hydrodynamics; nuclear reactor shielding. *Mailing Add:* 127 Ainsworth Ave East Brunswick NJ 08816

MANGUM, BILLY WILSON, b Mize, Miss, Dec 8, 31; m 63; c 1. LOW TEMPERATURE PHYSICS, SOLID STATE PHYSICS. *Educ:* Univ Southern Miss, BA, 53; Tulane Univ, MS, 55; Univ Chicago, PhD(phys chem), 60. *Prof Exp:* NSF fel physics, Clarendon Lab, Oxford, 60-61; actg sect chief low temperature physics, 67-68, PHYSICIST, NAT BUR STANDARDS, 61- *Mem:* Am Phys Soc. *Res:* Magnetism at very low temperatures; cooperative phenomena; electron-lattice interactions. *Mailing Add:* Nat Bur Standards Phys B128 Gaithersburg MD 20899

MANGUM, CHARLOTTE P, b Richmond, Va, May 19, 38. INVERTEBRATE ZOOLOGY, COMPARATIVE PHYSIOLOGY. *Educ:* Vassar Col, AB, 59; Yale Univ, MS, 61, PhD(biol), 63. *Prof Exp:* Res assoc biol, Yale Univ, 63; NIH res fel zool, Bedford Col, London, 63-64; from asst prof to assoc prof, 64-74, PROF BIOL, COL WILLIAM & MARY, 74- *Concurrent Pos:* Assoc, Va Inst Marine Sci, 64-; vis investr, Marine Biol Lab, Mass, 66, instr, 69-73 & 84; NSF res grants, 66-88; Col William & Mary fac res fel, 73; mem, Corp Marine Biol Lab, Woods Hole; lectr, Univ Aarhus, Denmark, 74. *Mem:* Fel AAAS; Am Soc Zool; Marine Biol Asn UK; Am Physiol Soc; Brit Soc Exp Biol. *Res:* Comparative physiology of respiratory pigments. *Mailing Add:* Dept Biol Col William & Mary Williamsburg VA 23185

MANGUM, FREDRICK ANTHONY, b Provo, Utah, Mar 3, 38; m 57; c 4. AQUATIC ECOSYSTEMS, FISHERIES. *Educ:* Calif State Col, Long Beach, BS, 60, MS, 62; Brigham Young Univ, PhD(aquatic ecol), 75. *Prof Exp:* Teacher biol sci, Orange Unified Sch Dist, 62-71; AQUATIC ECOLOGIST, FOREST SERV, USDA, 72- *Concurrent Pos:* Dir, Aquatic Ecosyst Anal Lab, 74- *Mem:* NAm Benthological Soc; Am Fisheries Soc. *Res:* Environmental profiles of aquatic invertebrate taxa; water quality. *Mailing Add:* Brigham Young Univ 105 Page Provo UT 84602

MANGUM, JEFFREY GARY, b Willows, Calif, Dec 29, 62; m 85; c 1. RADIO ASTRONOMY, MOLECULAR CLOUDS & STAR FORMATION. *Educ:* Univ Calif, Berkeley, BA, 85; Univ Va, MA, 88, PhD(astron), 90. *Prof Exp:* Jr res assoc, Nat Radio Astron Observ, 89-90; POSTDOCTORAL RES FEL, ASTRON DEPT, UNIV TEX, 90- *Mem:* Am Astron Soc. *Res:* Measuring the intensities and distributions of molecules in molecular clouds results in measurement of the density, temperature and composition of regions on the verge of forming stars. *Mailing Add:* Dept Astron Univ Tex Austin TX 78712

MANGUM, JOHN HARVEY, b Rexburg, Idaho, Apr 16, 33; div; c 3. BIOCHEMISTRY. *Educ:* Brigham Young Univ, BS, 57, MS, 59; Univ Wash, PhD(biochem), 63. *Prof Exp:* Res assoc biochem, Scripps Clin & Res Found, 62-63; from asst prof to assoc prof, 63-74, PROF CHEM, BRIGHAM YOUNG UNIV, 74-, DIR, BRIGHAM YOUNG UNIV CANCER RES CTR, 85- *Concurrent Pos:* Vis prof, NIH, 70-71; vis scholar, Univ Cincinnati, 83 & Univ Calif, San Diego, 84. *Mem:* Am Chem Soc; Sigma Xi. *Res:* Enzymology; one-carbon metabolism; methionine biosynthesis; virus-induced acquisition of metabolic function; folate mediated reactions in brain metabolism; purine metabolism. *Mailing Add:* Dept Chem Brigham Young Univ Provo UT 84601

MANGUS, MARVIN D, b Altoona, Pa, Sept 13, 24; m 50; c 2. GEOLOGY. *Educ:* Pa State Univ, BS, 45, MS, 46. *Prof Exp:* Geologist, US Geol Surv, 47-58; sr geologist, Guatemalan Atlantic Corp, 58-60; surface geologist, Atlantic Refining Co, Pa, 60-65, sr surface geologist, Alaska, 65-69; consult geologist, Calderwood & Mangus, 69-76; INDEPENDENT CONSULT GEOLOGIST, 76- *Mem:* Explorer's Club; Am Asn Petrol Geol; Can Soc Petrol Geol; fel Geol Soc Am; Am Inst Prof Geol. *Res:* Surface geologic mapping and regional geologic studies in Alaska, Central America, Canada and Bolivia; geological mapping in arctic Alaska, the British Mountains of northern Yukon Territory, the arctic islands of Canada and the Northwest Territories of Canada; geological well site; environmental studies of Alaska. *Mailing Add:* 1045 E 27th Ave Anchorage AK 99504

MANHART, JOSEPH HERITAGE, b Greencastle, Ind, Mar 26, 30; m 57; c 2. ORGANIC CHEMISTRY. *Educ:* DePauw Univ, AB, 52; Ohio State Univ, PhD(org chem), 60. *Prof Exp:* sr scientist, Alcoa Tech Ctr, Aluminum Co Am, 60-85; TECH DIR, LORIN INDUST, 85- *Mem:* Am Chem Soc; Sci Res Soc Am; Am Electroplaters' Soc. *Res:* Anodizing electrolytes; fire retardant fillers; lithographic printing plates; ultraviolet curable coatings. *Mailing Add:* 429 E Circle Dr North Muskegon MI 49445

MANHART, ROBERT (AUDLEY), b Charleston, Ill, Oct 5, 25; m 49; c 2. NANOSECOND PULSE TECHNIQUES, ELECTRO-OPTICAL SYSTEMS. *Educ:* Rose Polytech Inst, BS, 45; Univ Ill, MS, 47; Stanford Univ, PhD(elec eng), 61. *Prof Exp:* Design engr, Radio Corp Am, 45-46; develop engr, NAm Aviation, Inc, 47-48; head electronics dept, Res & Develop Div, NMex Sch Mines, 48-50; develop engr, Bell Aircraft Corp, 50-51; supvr electronic instruments group, Calif Res & Develop Co, Standard Oil Co Calif, 51-54, res engr, Calif Res Corp, 54-55; from asst to assoc prof elec eng, Univ Ariz, 55-58 & Ariz State Univ, 59-61; chmn dept, 61-77, prof elec eng, 61-, EMER PROF, UNIV NEV, RENO. *Concurrent Pos:* Consult, Wesix Elec Heater Co, 56-57, US Army Electronic Proving Ground, 57-58,

Sperry Phoenix Co, 59, Motorola, Inc, 60, Edgerton, Germeshausen & Grier, Inc, Nev, 61-65, Lawrence Livermore Lab, 71-73 & Los Alamos Nat Lab, 72- *Mem:* Sr mem Inst Elec & Electronics Engrs; Am Soc Eng Educ; Nat Soc Prof Engrs; Sigma Xi. *Res:* Theoretical and experimental semiconductor device research; circuit theory and network synthesis; nanosecond pulse systems including superconductors; fiber optic communication systems. *Mailing Add:* Dept Elec Eng Univ Nev Reno NV 89557

MANHAS, MAGHAR SINGH, b Kothe Manhasan, India, Aug 17, 22; m 53; c 5. ORGANIC CHEMISTRY. *Educ:* Punjab Univ, India, BSc, 43; Allahabad Univ, MSc, 45, DPhil, 50. *Hon Degrees:* MEng, Stevens Inst Technol, 74. *Prof Exp:* Asst prof chem, BR Col, Agra, India, 50-52 & Univ Saugar, 52-60; from asst prof to assoc prof, 61-70, PROF CHEM, STEVENS INST TECHNOL, 70- *Concurrent Pos:* Res assoc, Stevens Inst Technol, 60-61; Ottens res award, 68. *Honors & Awards:* Davis Award, 81. *Mem:* Am Chem Soc; fel Royal Soc Chem; Sigma Xi. *Res:* Heterocyclic chemistry; medicinal chemistry; stereochemistry. *Mailing Add:* Dept of Chem & Chem Eng Stevens Inst of Technol Hoboken NJ 07030

MANHEIM, FRANK T, b Leipzig, Ger, Oct 14, 30; US citizen; m 61; c 3. GEOCHEMISTRY. *Educ:* Harvard Univ, AB, 51; Univ Minn, MSc, 53; Univ Stockholm, Fil Lic, 61, DSc(geochem), 74. *Prof Exp:* Geochemist, Geol Surv, Sweden, 61-62; fel & res asst strontium isotopes, Yale Univ, 63; res geologist, US Geol Surv, 64-73; chmn dept marine sci, Univ S Fla, 74-76; OFF MARINE GEOL, US GEOL SURV, 77- *Concurrent Pos:* Mem, Nat Acad Sci Comt on USSR & Eastern Europe, 75-78; Deep Sea Drilling Proj; Adv Panel Inorg Geochem; vis fel, Nat Acad Sci, Bulgaria, 74. *Honors & Awards:* Olin Found Award, 88. *Mem:* AAAS; Am Geophys Union; Geochem Soc. *Res:* Geochemistry of recent and fossil sediments; marine hard mineral resources; chemistry of ground and natural waters; suspended matter in ocean waters; marine policy and scientific communications. *Mailing Add:* US Geol Surv Off Marine Geol Woods Hole MA 02543

MANHEIMER, WALLACE MILTON, b New York, NY, Feb 10, 42; m 65; c 3. PLASMA PHYSICS. *Educ:* Mass Inst Technol, BS, 63, PhD(physics), 67. *Prof Exp:* Prof physics, Mass Inst Technol, 68-70; PHYSICIST, NAVAL RES LABS, 70- *Mem:* Am Phys Soc. *Res:* Turbulence theory; laser plasma interaction; relativistic beams; controlled thermonuclear fusion. *Mailing Add:* Code 4740 Naval Res Labs Washington DC 20375

MANHOLD, JOHN HENRY, JR, b Rochester, NY, Aug 20, 19; m 71. PATHOLOGY. *Educ:* Univ Rochester, BA, 40; Harvard Univ, DMD, 44; Wash Univ, MA, 56. *Prof Exp:* Instr oral path, Med & Dent Schs, Tufts Univ, 47-48, dir cancer teaching prog, 48-50; asst prof gen & oral path, Sch Dent, Wash Univ, 55-56; assoc prof path, Univ Med & Dent NJ, 56-57, prof gen & oral path & dir dept, 57-87; MED DIR, WOOG INT, 87-; CONSULT PHARMACEUT INDUST, 87- *Concurrent Pos:* Ed, Clin Prev Dent. *Mem:* AAAS; Acad Psychosom Med (pres); Am Psychol Asn; AMA; fel Am Col Dent; Am Soc Clin Pathologists; Int Col Dent. *Res:* Psychosomatics; oral diagnosis; tissue metabolism; author of seven books. *Mailing Add:* PO Box 9150 Treasure Island FL 33740-9159

MANI, INDER, polymer chemistry, latex chemistry, for more information see previous edition

MANI, RAMA I, b Madras, India, Apr 10, 27. ORGANIC CHEMISTRY. *Educ:* Univ Bombay, BSc, 47, MSc, 51, PhD(chem), 61. *Prof Exp:* Res scientist chem, Indian Coun Med Res, India, 51-53; res scientist chem, Coun Sci & Indust Res, 57-60; res assoc, Stanford Univ, 60-62 & Univ Southern Calif, 62-63; res assoc, Vanderbilt Univ & Meharry Med Col, 63-65; ASSOC PROF CHEM, TENN STATE UNIV, 65- *Mem:* Am Chem Soc; Sigma Xi. *Res:* Synthetic organic chemistry; boron trifluoride in the synthesis of plant phenolics. *Mailing Add:* Dept Chem Tenn State Univ 3500 J A Merritt Blvd Nashville TN 37209

MANIAR, ATISH CHANDRA, b Unjha, India, Jan 21, 26; Can citizen; m 51; c 5. MICROBIOLOGY, PUBLIC HEALTH. *Educ:* Univ Bombay, BSc, 48, MSc, 52, PhD(microbiol), 56, DSc(microbiol), 69. *Prof Exp:* Bacteriologist, Caius Res Lab, Bombay, India, 48-55; res bacteriologist, Alembic Chem, India, 55; bacteriologist, Glaxo Labs, Bombay, 55-59; bacteriologist, Hindustan Antibiotics Ltd, India, 61-63; bacteriologist, Winnipeg Gen Hosp, 64-67; BACTERIOLOGIST, CADHAM PROVINCIAL LAB, MAN, 67- *Concurrent Pos:* Nat Res Coun Can fel antibiotics, Can Commun Dis Ctr, Ottawa, 59-61; Royal Soc Health fel, Med Col, Winnipeg, 71; lectr, St Xavier's Col, India, 48-52; mem bd studies & fac sci, Univ Bombay, 57-59; lectr, Univ Man, 64- *Mem:* Am Soc Microbiol; Can Pub Health Asn; Indian Asn Microbiol; Royal Soc Health; Can Soc Microbiol; Can Col Microbiol; Am Acad Microbiol; Can Asn Clin Microbiol & Infectious Dis; Sigma Xi. *Res:* Mode of action of antibiotics; aminoacidopathy in newborn babies; staphylococcal toxins; verotoxin and difficile toxin; detection of mutagenic agents; diagnoses of chlamydia infection. *Mailing Add:* 68 Viola St Winnipeg MB R2V 3B9 Can

MANIATIS, THOMAS PETER, b Denver, Colo, May 8, 43; m 68. MOLECULAR BIOLOGY. *Educ:* Univ Colo, Boulder, BA, 65, MA, 67; Vanderbilt Univ, PhD(molecular biol), 71. *Prof Exp:* NIH fel, Harvard Univ, 71-73; European molecuiar biol org res fel, Med Res Coun Molecular Biol, Cambridge, Eng, 73-74; res assoc biol, Harvard Univ, 74-75, asst prof biochem & molecular biol, 75-77; sr staff investr, Cold Spring Harbor Lab, 75-77; from assoc prof to prof biol, Calif Inst Technol, 77-81; chmn dept, 85-88, PROF BIOCHEM & MOLECULAR BIOL, HARVARD UNIV, 81- *Concurrent Pos:* Assoc ed, Cell, 80-; mem, molecular biol study sect, NIH, 81-84, chmn, 82-84; Burroughs-Wellcome vis prof cell biol, 84; mem adv comt, Searle Scholars Prog, 85-87, Jane Coffin Childs Bd Sci Adv & Sci Rev Bd, Howard Hughes Med Inst, 89- *Honors & Awards:* Career Develop Award, Rita Allen Found, 78; Eli Lilly Res Award, Am Soc Microbiol, 81; Carter-Wallace Lectr, Princeton Univ, 81; Christian A Herter Lectr, NY Univ, 83; William D McElroy Distinguished Lectr, Univ Calif, San Diego, 84; Smith, Kline & French Lectr, Univ Calif, San Francisco, 85, Cetus Lectr, Berkeley, 85; Richard Lounsbery Award, US & French Nat Acad Sci, 85. *Mem:* Nat Acad Sci; fel Am Acad Arts & Sci; fel AAAS. *Res:* Control of gene expression. *Mailing Add:* Dept Biochem & Molecular Biol Harvard Univ Cambridge MA 02138

MANICKAM, JANARDHAN, b Madras, India, May 12, 46; m 77; c 2. COMPUTATIONAL PHYSICS, MAGNETOHYDRODYNAMICS ANALYSIS. *Educ:* Osmania Univ, Hyderabad, India, BSc, 66; Andhra Univ Waltair AP, India, MSc, 68; Stevens Inst Technol, PhD(plasma physics), 75. *Prof Exp:* PRIN RES PHYSICIST, PLASMA PHYSICS LAB, PRINCETON UNIV, 75- *Concurrent Pos:* Adj prof, Rutgers Univ, 77 & 79; vis scientist, Japan Atomic Energy Res Inst, 86 & CRPP, Ecole Polytechnique, Lausanne, Switz, 88; vchmn, Int Sherwood Comt Theory of Controlled Nuclear Fusion, 91. *Mem:* Am Phys Soc. *Res:* Magnetohydrodynamics equilibrium and stability analysis of tokamaks; tokamak design studies and analysis of tokamak experiments; computational physics. *Mailing Add:* Plasma Physics Lab Princeton Univ PO Box 451 Princeton NJ 08543

MANIGLIA, ROSARIO, b Brooklyn, NY, Nov 18, 18. PATHOLOGY. *Educ:* Hahnemann Med Col, MD, 46. *Prof Exp:* LAB DIR, HOLY SPIRIT HOSP, CAMP HILL, PA, 63- *Mem:* AMA; Col Am Pathologists; Am Soc Clin Pathologists; Am Asn Pathologists; Am Asn Blood Banks. *Mailing Add:* Dept Path Holy Spirit Hosp Camp Hill PA 17011

MANILOFF, JACK, b Baltimore, Md, Nov 6, 38; m 60; c 2. BIOPHYSICS, MICROBIOLOGY. *Educ:* Johns Hopkins Univ, BA, 60; Yale Univ, MS, 64, PhD(biophys), 65. *Prof Exp:* Res assoc chem, Brown Univ, 64-66; from asst prof to assoc prof, 66-79, PROF MICROBIOL, UNIV ROCHESTER, 79- *Concurrent Pos:* NIH res career develop award, 70-75; Fogarty Sr Int fel, USPHS-NIH, 87-88; distinguished vis fel, Christ's Col, Cambridge Univ, Eng, 87-88; lectr, Am Soc Microbiol Found, 89-90. *Mem:* AAAS; Biophys Soc; Am Soc Microbiol; Sigma Xi. *Res:* Molecular and cellular biology of mycoplasma cells and their viruses; theoretical aspects of biological processes. *Mailing Add:* Dept of Microbiol Univ Rochester Med Ctr Box 672 Rochester NY 14642

MANINGER, RALPH CARROLL, b Harper, Kans, Dec 24, 18; m 42; c 3. PHYSICS. *Educ:* Calif Inst Technol, BS, 41. *Prof Exp:* Res engr & group leader, Off Sci Res & Develop Proj, Columbia, 41-45; contract physicist, Taylor Model Basin, US Dept Navy, DC, 45, sect head, US Navy Electron Lab, Calif, 45-48; physicist, Vitro Corp Am, 48-51, from asst dir to dir phys res & develop, 51-53; tech dir, Precision Tech, Inc, 53-57, br mgr librascope div, Gen Precision, Inc, 57-62; head, Eng Res Div, Electronics Dept, 62-68, dept head electronics, Eng Dept, 68-71, head environ studies, 71-74, head, Technol Appln Group, 75-78, sr staff consult, Lawrence Livermore Nat Lab, 79-85, CONSULT, UNIV CALIF, 86- *Concurrent Pos:* Mem adv comt, Statewide Air Pollution Res Ctr, Univ Calif, 71-73. *Mem:* AAAS; Acoust Soc Am; fel Inst Elec & Electronics Engrs; Nuclear & Plasma Sci Soc (pres, 81-82). *Res:* Under-water acoustics and electronics; non-linear vibrations; solid state radiation detectors and electron devices; high speed pulse circuitry; microwave generation and propagation; initiation of explosives; fast reactions in solids; quantum electronics; environmental systems research; radioactve waste management. *Mailing Add:* 146 Roan Dr Danville CA 94526

MANION, JAMES J, b Butte, Mont, May 17, 22; m 46; c 3. ECOLOGY. *Educ:* Univ Portland, BS, 48; Univ Notre Dame, MS, 50, PhD(zool), 52. *Prof Exp:* Asst prof biol & head dept, St Mary's Col, 52-58; chmn div natural sci & math, 58-77, from assoc prof to prof biol sci, 71-87, EMER PROF, CARROLL COL, 87- *Concurrent Pos:* Dean fac, Carroll Col, 54-71, acad vpres, 66-71. *Mem:* AAAS. *Res:* Ecology of amphibians. *Mailing Add:* 1002 Garfield Helena MT 59601

MANION, JERALD MONROE, b Beebe, Ark, Sept 24, 40; m 59; c 2. ORGANIC CHEMISTRY. *Educ:* Harding Col, BS, 62; Univ Miss, PhD(chem), 65. *Prof Exp:* PROF CHEM & CHMN DEPT, UNIV CENT ARK, 65- *Mem:* Am Chem Soc. *Res:* Gas phase kinetics of reverse Diels-Alder reactions. *Mailing Add:* Rte 2 Box 93 Conway AR 72032

MANIOTIS, JAMES, b Detroit, Mich, Aug 17, 29; m 55; c 1. MYCOLOGY. *Educ:* Wayne State Univ, AB, 52, MS, 57; Univ Iowa, PhD(bot), 60. *Prof Exp:* Res assoc virol, Child's Res Ctr, Mich, 56-57; asst bot, Univ Iowa, 57-60; instr bot, Univ Tex, 60-61; asst prof biol, Wayne State Univ, 61-65; assoc prof bot, 65-69, ASSOC PROF BIOL, WASH UNIV, 69- *Mem:* AAAS; Mycol Soc Am; Bot Soc Am; Am Inst Biol Sci. *Res:* Biochemical-genetical bases for pathogenicity in ringworm fungi; biology of membrane fusion in slime molds. *Mailing Add:* Dept Biol Washington Univ Lindell-Skinker Blvd St Louis MO 63130

MANIRE, GEORGE PHILIP, b Roanoke, Tex, Mar 25, 19; m 43; c 2. MICROBIOLOGY. *Educ:* NTex State Col, BS, 40, MS, 41; Univ Calif, Berkeley, PhD(bact), 49. *Prof Exp:* Instr bact, Univ Tex Southwestern Med Sch, 49-50; from asst prof to assoc prof, Univ NC, Chapel Hill, 50-59, asst vchancellor health sci, 65-66, chmn dept, 66-79, vchancellor & grad dean, 79-86, prof, 59-89, Kenan prof, 71-89, EMER PROF, UNIV NC CHAPEL HILL, 89- *Concurrent Pos:* Fulbright scholar, Serum Inst, Copenhagen, 56; China Med Bd Alan Gregg fel, Virus Inst, Kyoto, Japan, 63-64; USPHS spec fel, Lister Inst, London, 71-72; mem health sci advan award comt, NIH, 67-71 & chmn, 69-71; vis scientist, Japan Soc Promotion Sci, 79. *Mem:* Am Asn Immunol; fel Am Acad Microbiol; Infectious Dis Soc Am; Am Soc Microbiol; fel AAAS. *Res:* Mechanisms of pathogenesis of microorganisms, especially diseases of Chlamydia. *Mailing Add:* Dept Micro & Immunol Univ NC Sch Med Chapel Hill NC 27599-7155

MANIS, ARCHIE L, b Oklahoma City, Okla, Nov 1, 39; m 71; c 2. PLANT PATHOLOGY, MYCOLOGY. *Educ:* Abilene Christian Univ, BS, 61; Sam Houston State Univ, MEd, 66; Tex A&M Univ, PhD(plant path), 71. *Prof Exp:* Asst prof biol, David Lipscomb Col, 71-73; asst prof & chmn div math & sci, Jackson State Community Col, 73-76; assoc prof biol, Abilene Christian Univ, 76-; RETIRED. *Mem:* Int Soc Plant Pathologists; Am Phytopathological Soc. *Res:* Fungal physiology; antibiotics, fungicides and fungistats. *Mailing Add:* 2001 Shadow Ridge Harker Heights TX 76543

MANIS, MERLE E, b St Ignatius, Mont, Aug 20, 34; m 61; c 4. MATHEMATICS. *Educ:* Univ Mont, BA, 60, MA, 61; Univ Ore, PhD(math), 66. *Prof Exp:* From instr to asst prof, 62-73, assoc prof, 73-78, PROF MATH, UNIV MONT, 78- *Concurrent Pos:* NSF res contract, 67-68. *Mem:* Am Math Soc; Sigma Xi. *Res:* Ring theory; valuation theory; D K Harrison's theory of primes; pruffer rings; rings with several objects. *Mailing Add:* 231 South 5th E Missoula MT 59801

MANISCALCO, IGNATIUS ANTHONY, b New York, NY, June 25, 44; m 67; c 2. ORGANIC CHEMISTRY, BIOCHEMISTRY. *Educ:* Manhattan Col, BS, 65; Fordham Univ, PhD(org chem), 71. *Prof Exp:* Instr chem, Univ Va, 70-71; from asst prof to assoc prof, 71-85, PROF CHEM, SPRINGFIELD COL, 85- *Concurrent Pos:* Res fel, Univ Va, 70-71. *Mem:* Am Chem Soc. *Res:* Heterocyclic organic chemistry. *Mailing Add:* Dept of Chem Springfield Col Bemis Hall Springfield MA 01109-3788

MANJARREZ, VICTOR M, b Los Angeles, Calif, June 13, 33; m 66. MATHEMATICS. *Educ:* Spring Hill Col, BS, 57; Harvard Univ, MA, 58, PhD(math), 63. *Prof Exp:* Instr math, Cath Univ Am, 65-67; asst prof, Univ Houston, 67-71; assoc prof, 71-75, PROF MATH, CALIF STATE UNIV, HAYWARD, 75- *Mem:* Am Math Soc; Math Asn Am. *Res:* Interpolation and approximation in the complex domain; topological vector spaces. *Mailing Add:* Dept of Math Calif State Univ Hayward CA 94542

MANJOINE, MICHAEL J(OSEPH), b Muscatine, Iowa, Apr 29, 14; m 38; c 2. MECHANICAL METALLURGY. *Educ:* Iowa State Univ, BSME, 36, BSEE, 37; Univ Pittsburgh, MS, 39. *Prof Exp:* Res engr, Res Labs, Westinghouse Elec Corp, 40-61, mgr eng mech, 61-63, consult reactor eng, Astronuclear Lab, 63-69, consult engr, mech dept, 69-84; RETIRED. *Concurrent Pos:* Am chmn, Int Conf on Creep, 63. *Honors & Awards:* Dudley Medal, Am Soc Testing & Mat, 53; Nadai Award, Am Soc Mech Engrs, 80, Pressure Vessel & Piping Award, 88; Edgar C Bain Award, Am Soc Metals, 88. *Mem:* Fel Am Soc Mech Engrs; fel Am Soc Metals. *Res:* Mechanics of materials; elevated temperature properties of materials; high speed tensile testing; effect of notches; testing machine design; elevated temperature design. *Mailing Add:* 25 Lewin Lane Pittsburgh PA 15235

MANJULA, BELUR N, b Bangalore, India, Oct 6, 43. PROTEIN OF GROUP A STREPTOCOCCUS. *Educ:* Indian Inst Sci, PhD(biochem), 70. *Prof Exp:* Asst prof, 78-84, ASSOC PROF BACTERIOL & IMMUNOL, ROCKEFELLER UNIV, 84- *Mem:* Am Soc Biol Chemists; Am Asn Immunol; Harvey Soc; Am Heart Asn; Protein Soc; Am Soc Microbiol. *Mailing Add:* Dept Immunol Rockefeller Univ 1230 York Ave New York NY 10021

MANJUNATH, PUTTASWAMY, b June 23, 52; Can citizen; m. PROTEIN STRUCTURE-FUNCTION, REPRODUCTIVE BIOCHEMISTRY. *Educ:* Karnataka Univ, BSc, 71, MSc, 73; Univ Mysore, PhD(biochem), 79. *Prof Exp:* Res fel biochem, Coun Sci & Indust Res, New Delhi, India, 73-76, sci asst, 76-78; scientist S-1 biochem, Indian Coun Agr Res, New Delhi, 78-80; fel biochem, Clin Res Inst Montreal, 80-82, prof assoc, 82-84, Que Health Res Fund Scholar, 84-90; ASST PROF, UNIV MONTREAL, 90-, QUE HEALTH RES FUND SCHOLAR. *Mem:* Endocrine Soc; Can Biochem Soc; Soc Biol Chemists, India; Am Soc Biochem & Molecular Biol; NY Acad Sci; Soc Study Reproduction. *Res:* Isolation and physiochemical characterization of glycoprotein; structure and function of proteins; lipoprotein and lipid metabolism; radioreceptor assays, radioimmunoassays; high-performance liquid chromatography; fast protein and peptide liquid chromatography; gas-liquid chromatography; amino acid analysis. *Mailing Add:* 350 Prince Arthur w Montreal PQ H2X 3R4 Can

MANKA, CHARLES K, b Flemington, Mo, Sept 28, 38; m 61; c 1. PLASMA PHYSICS, SPECTROSCOPY. *Educ:* William Jewell Col, AB, 60; Univ Ark, MS, 64, PhD(plasma physics), 65. *Prof Exp:* Asst prof, Sam Houston State Univ, 65-68, dir dept, 68-75, prof physics, 75-83; RES PHYSICIST, SPACE PLASMA BR, 83- *Res:* Temperatures of exploding wires and other transient plasma formed in vacuum; applications to criminology and police science; forensic physics. *Mailing Add:* Naval Res Lab Code 4783 Washington DC 20276-2728

MANKA, DAN P, b Farrell, Pa, May 2, 14; m 42; c 4. PROCESS CONTROL. *Educ:* Valparaiso Univ, BS, 36; Carnegie Inst Technol, MS, 47. *Prof Exp:* Chemist, Wheeling Steel Corp, 36; res chemist, Koppers Co, 36-41; sr res chemist, Jones & Laughlin Steel Corp, 41-77; CONSULT, 77- *Concurrent Pos:* Consult, Kaiser Stell Corp, 78-79 & Mellon Inst Res, 78; ed, Acad Press, Inc, 80-83; guest speaker, Am Inst Chemists, 83-84, Peoples Repub China, US-China Sci Exchange, 84. *Mem:* Am Soc Testing & Mat; Am Chem Soc; fel Am Inst Chemists. *Res:* Organic chemicals in coke oven gas; development and installation of equipment for analysis of gases from blast furnaces, coke oven gas, and from basic oxygen furnaces; recovery of phenols from waste liquor; solids formation in coal tar. *Mailing Add:* 1109 Lancaster Ave Pittsburgh PA 15218-1012

MANKAU, REINHOLD, b Chicago, Ill, July 22, 28; m 54; c 2. NEMATOLOGY, SOIL BIOLOGY. *Educ:* Univ Ill, BS, 51, MS, 53, PhD(plant path), 56. *Prof Exp:* Res asst, Univ Ill, 54-56; Fulbright res fel, India, 56-57; asst nematologist, 58-63, assoc nematologist & assoc prof, 63-75, PROF NEMATOL, UNIV CALIF, RIVERSIDE, 75- *Concurrent Pos:* Fulbright res fel, India, 64-65. *Mem:* Am Phytopath Soc; Mycol Soc Am; Soc Nematol; Indian Phytopath Soc; Soc Europ Nematologists; Orgn Tropical Am Nematologists. *Res:* Biological control of plant-parasitic nematodes; soil biology and biochemistry; nematode-soil microbial relationships. *Mailing Add:* Dept Nematol Univ Calif Riverside CA 92521

MANKAU, SAROJAM KURUDAMANNIL, b Kottayam, India, June 5, 30; nat US; m 54; c 1. PARASITOLOGY, NEMATOLOGY. *Educ:* Univ Madras, BS, 49; Univ Ill, PhD(zool), 56. *Prof Exp:* Asst, Univ Ill, 53-56; instr biol, Univ Redlands, 58-59; res assoc plant nematol, Citrus Exp Sta, Univ Calif, Riverside, 59-60, asst prof zool, 60-63, res assoc nematol, 59-68; asst prof biol, 68-72, assoc prof, 72-81, PROF BIOL, CALIF STATE UNIV, SAN BERNARDINO, 81, HEAD DEPT. *Mem:* Am Soc Parasitol. *Res:* Helminthology; biology of soil nematodes and invertebrates. *Mailing Add:* Div Natural Sci Calif State Univ 5500 State College Pkwy San Bernadino CA 92407

MANKE, PHILLIP GORDON, b Pawhuska, Okla, Aug 27, 29; m 51; c 4. CIVIL ENGINEERING. *Educ:* Okla State Univ, BS, 56, MS, 57; Tex A&M Univ, PhD(civil eng), 65. *Prof Exp:* Civil engr, Humble Oil & Refining Co, 57-59; instr civil eng, Okla State Univ, 59-60, asst prof, 60-63; res asst pavement design, Tex A&M Univ, 64-65; from asst prof to assoc prof civil eng, 65-72, PROF CIVIL ENG, OKLA STATE UNIV, 72- *Concurrent Pos:* NSF res initiation grant, 66-67, sci res grant, 68-70; mem, Hwy Res Bd, Nat Acad Sci-Nat Res Coun; prin investr sponsored highway mat res projs, Okla Dept Transp, 72- *Mem:* Asn Asphalt Paving Technologists; Sigma Xi. *Res:* Soil-asphalt stabilization; asphaltic materials; asphalt paving mixtures; nondestructive testing of soils and highway materials. *Mailing Add:* Sch of Civil Eng Okla State Univ Stillwater OK 74078

MANKES, RUSSELL FRANCIS, b Schenectady, NY, May 21, 49; m 72; c 2. REPRODUCTIVE TOXICOLOGY, TOXICOLOGY. *Educ:* Albany Med Col, PhD(exp path & toxicol), 77. *Prof Exp:* ASSOC PROF PHARMACOL & TOXICOL, ALBANY MED COL, 80- *Mailing Add:* Dept Pharmacol & Toxicol A-136 Albany Med Col 43 New Scotland Ave Albany NY 12208

MANKIN, CHARLES JOHN, b Dallas, Tex, Jan 15, 32; m 53; c 3. GEOLOGY. *Educ:* Univ Tex, BS, 54, MA, 55, PhD(geol), 58. *Prof Exp:* Instr geol, Univ Tex, 56-57; asst prof, Calif Inst Technol, 58-59; asst prof, 59-63, dir Sch Geol & Geophys, 64-77, dir Energy Resources Inst, 78-87, PROF GEOL, UNIV OKLA, 66-; DIR OKLA GEOL SURV, 67- *Concurrent Pos:* Assoc prof, Univ Okla, 63-66, actg dir sch geol & geophys, 63-64; mem, Nat Petrol Coun-Exec Adv Coun Nat Gas Surv, 75; mem, Comm on Phys Sci, Math & Resources, Nat Acad Sci-Nat Res Coun, 77-88 & Bd on Mineral & Energy Resources, 76-88, vchmn, 77-79, chmn, 79-88; mem, US Nat Comt Geol, 77-80; counr, Geol Soc Am, 84-86. *Mem:* Asn Am State Geol (pres, 75-76); Am Asn Petrol Geologists; Am Inst Prof Geologists (vpres, 84, pres-elect, 86, pres, 87); Geol Soc Am; Soc Econ Paleont & Mineral. *Res:* Sedimentary petrology and geochemistry; clay mineralogy. *Mailing Add:* Okla Geol Surv 100 E Boyd Rm N-131 Norman OK 73019

MAN-KIN, MAK, b Hong Kong, China, Oct 17, 39; Brit citizen; m 66; c 2. ATMOSPHERIC SCIENCE. *Educ:* Univ Toronto, BASc, 63; Mass Inst Technol, MS, 66, PhD(meteorol), 68. *Prof Exp:* Res fel meteorol, Nat Res Coun Can, 68-69; vis assoc prof, Nat Taiwan Univ, China, 69-70; assoc prof, 70-80, PROF METEOROL, UNIV ILL, URBANA-CHAMPAIGN, 80- *Concurrent Pos:* Res grants, NSF, 72- *Mem:* Am Meteorol Soc. *Res:* Dynamics of global atmospheric waves; tropical meteorology; monsoonal circulation; general circulation of the atmosphere. *Mailing Add:* 2206 S Race Urbana IL 61801

MANKIN, RICHARD WENDELL, b Houston, Tex, Feb 14, 48. INSECT OLFACTORY PHYSIOLOGY, INSECT BEHAVIORAL PHYSIOLOGY. *Educ:* NMex State Univ, BS(physics) & BS(math), 70; Univ Fla, MS, 76, PhD(entom), 79. *Prof Exp:* Assoc, Univ Fla, 79-809-80; RES ENTOMOLOGIST, AGR RES SERV, USDA, 80- *Concurrent Pos:* Adj assoc prof, Univ Fla, 82-83. *Mem:* Asn Chemoreception Sci; Entom Soc Am. *Res:* Sex pheromone dispersal, detection and discrimination by insects; alternatives to pesticides by finding means to interfere with insect sexual communication. *Mailing Add:* USDA Agr Res Serv PO Box 14565 Gainesville FL 32604

MANKIN, WILLIAM GRAY, b Memphis, Tenn, Sept 2, 40; m 72; c 2. ATMOSPHERIC SCIENCE. *Educ:* Rhodes Col, BS, 62; Johns Hopkins Univ, PhD(physics), 69. *Prof Exp:* Res asst astron, Univ Mass, 67-69; sci visitor, 69-71, staff scientist, High Altitude Observ, 71-73, SCIENTIST, ATMOSPHERIC CHEM DIV, NAT CTR ATMOSPHERIC RES, 73- *Concurrent Pos:* Lectr, Univ Mass, Amherst & Smith & Mt Holyoke Cols, 68-69; vis scientist, Rutherford Appleton Lab, England, 81-82. *Honors & Awards:* Technol Advan Award, Nat Ctr Atmospheric Res, 87. *Mem:* Fel Optical Soc Am; Am Asn Physics Teachers; AAAS; Am Geophys Union. *Res:* Infrared spectroscopy and radiometry for atmospheric and astrophysical applications; stratospheric composition, polar zone; solar atmosphere; development of optical techniques for geophysics. *Mailing Add:* Nat Ctr Atmospheric Res PO Box 3000 Boulder CO 80307

MANLEY, AUDREY FORBES, b Jackson, Miss, Mar 25, 34. HEALTH RESOURCE ADMINISTRATION. *Educ:* Spleman Col, Atlanta, AB, 55; Meharry Med Col, Nashville, MD, 59; Johns Hopkins Univ, MPH, 87; Am Bd Pediat, cert, 66. *Prof Exp:* Clin instr, Pediat Midway Tech Allied Health Training Prog, 66-69; instr, pediat, Pritzker Sch Med, Univ Chicago, 67-69; clin asst prof, Univ Calif Med Sch, 69-70; asst prof, Dept Pediat, Gynecol & Obstet, 72-76; comt officer, USPHS & med dir chief, Sickle Cell/Genetic Dis Serv Br, Off Maternal & Child Health, Bur Community Health Serv, Health Serv Admin, Rockville, Md, 78-83, assoc admin clin affairs, Off Admin, Health Resources & Serv Admin, 83-85, med dir, Off Planning, Eval & Legis, 85-86, dir, Health Serv Corp, Bur Health Care Delivery & asst, Health Resources & Serv Admin, 87-89, DEP ASST SECY HEALTH, DEPT HEALTH & HUMAN SERV, USPHS, 89- *Concurrent Pos:* Clin asst prof,

Univ Calif, Med Sch, Dept Pediat, San Francisco 72-76 & Howard Univ, Sch Med, Wash, 81; consult, Am Acad Pediat, 69; asst prof, Dept Pediat Gynec & Obstet, Emory Univ, Grody Mem Hosp, Atlanta, 72-76; chief, Family Health & Prevent Servs, Div Clin Servs, Bur Community Health Servs, Health Serv Admin, Md, 76-78; Jesse Smith Noyes Found fel, 55-59; asst surgeon gen, USPHS, 88. *Mem:* Inst Med-Nat Acad Sci; Fel Am Acad Pediat; Nat Med Asn; fel Nat Inst Health; Am Pub Health Asn; NY Acad Sci; Am Asn Univ Women; AAAS; Am Soc Human Genetics; Nat Coun Negro Women. *Res:* author of various articles & journals. *Mailing Add:* Hubert H Humphrey Bldg Rm 7162 200 Independence Ave Washington DC 20201

MANLEY, CHARLES HOWLAND, b Acushnet, Mass, Feb 27, 43; m 65; c 1. FOOD CHEMISTRY. *Educ:* Southeastern Mass Univ, BS, 64; Univ Mass, Amherst, MS, 68, PhD(food sci), 69. *Prof Exp:* Sr food technologist, Nestle Co, 69-71; sr res chemist, Givaudan Corp, 71-74; group leader tea res, 74-76, mgr beverage prods develop, 76-77; mgr indust prods, Thomas J Lipton Inc, 77-81; mkt mgr, Nat Flavors & Seasonings Div, Nat Starch & Chem Corp, 81-83, dir, Flavor Opers, 83-85; vpres, Int Bus Develop, 85-87, VPRES & GEN MGR, FOOD & FLAVOR INGREDIENT DIV, QUEST INT, 87- *Mem:* Am Chem Soc; Inst Food Technologists; Sigma Xi; Am Inst Chemists. *Res:* Chemistry of flavor components in natural foods; isolation and characterization of volatile and non-volatile constituents of tea by the use of chromatographic and chemical techniques. *Mailing Add:* 88 Choclaw Trail Ringwood NJ 07456

MANLEY, DONALD GENE, b Monterey Park, Calif, Sept 15, 46; m 69; c 2. PEST MANAGEMENT, INSECT TAXONOMY. *Educ:* Univ Calif, Los Angeles, BA, 73; Calif State Univ, Long Beach, MA, 75; Univ Ariz, PhD(entom), 78. *Prof Exp:* From asst prof to assoc prof, 78-87, PROF ENTOMOL, CLEMSON UNIV, 87- *Concurrent Pos:* Assoc ed, J Agr Entom, 84- *Mem:* Entom Soc Am; Am Registry Prof Entomologists; Int Soc Hymenopterists. *Res:* Pest management of insects on field crops; development of economic thresholds for major pests, and scouting techniques; taxonomy and biology of mutillid wasps. *Mailing Add:* Pee Dee Res & Educ Ctr Rte 1, Box 531 Florence SC 29501-9603

MANLEY, EMMETT S, b Jackson, Tenn, Nov 6, 36; m 58; c 2. PHARMACOLOGY, PHYSIOLOGY. *Educ:* Univ Tenn, BS, 59, PhD(pharmacol, physiol), 63. *Prof Exp:* Asst physiol, Bowman Gray Sch Med, 64-65; from instr to asst prof, 65-72, assoc prof pharmacol, Med Units, Univ Tenn, Memphis, 72-75; prof & chmn, Dept of Pharmacol, Kirksville Col Osteopath Med, 75-85; ASST DEAN, UNIV TENN COL PHARM, 86- *Concurrent Pos:* USPHS trainee, 64. *Mem:* Am Soc Pharmacol & Exp Therapeut. *Res:* Cardiovascular pharmacology; catecholamines; hypercapnia; acid-base balance and drug response; blood flow determination; hemorrhagic shock; myocardial function. *Mailing Add:* Col Pharm Univ Tenn 847 Monroe Memphis TN 38163

MANLEY, HAROLD J(AMES), b Brooklyn, NY, July 7, 30; m 52; c 3. COMMUNICATIONS ENGINEERING, ENGINEERING MANAGEMENT. *Educ:* Worcester Polytech Inst, BS, 52, MS, 54; Stanford Univ, Eng, 68. *Prof Exp:* Physicist, Eng Labs, Signal Corps, 52; consult, David Clark Co, Mass, 53-54; electronic engr, Appl Res Lab, Sylvania Elec Prod Inc Div, 54-56, sr engr, 56-58, adv res engr, 58-59, eng specialist, 59-64, sr eng specialist, 64-68, mgr speech systs res dept, 68-70, mgr advan signal processing, 70-73, mgr systs eng, 73-77, asst dir eng, Sylvania Electronic Systs, Eastern Div, 77-81, TECH DIR, NAVY SYSTS BUS AREA, GEN TEL & ELECTRONICS CORP, 81-; ASST DIR ENG, EASTERN DIV, GTE SYLVANIA. *Honors & Awards:* Leslie H Warner Tech Achievement Award, Gen Tel & Electronics Inc, 76. *Mem:* Assoc mem Inst Elec & Electronics Engrs; Acoust Soc Am. *Res:* Digital signal processing; digital communications; modulation and demodulation systems; speech processing; speech analysis-synthesis. *Mailing Add:* GTE Sylvania Eastern Div 77 A St Needham MA 02194

MANLEY, JOHN HENRY, b Harvard, Ill, July 21, 07; m 35; c 2. NEUTRON PHYSICS, BIOPHYSICS. *Educ:* Univ Ill, BS, 29; Univ Mich, PhD(physics), 34. *Prof Exp:* Instr physics, Univ Mich, 31-33; lectr, Columbia Univ, 34-37; assoc, Univ Ill, 37-41, asst prof, 41-45; assoc prof, Wash Univ, 46-47; assoc dir, Los Alamos Sci Lab, 47-51; prof physics & exec officer, Univ Wash, 51-57, res adv, 57-72; CONSULT, LOS ALAMOS NAT LAB, 72- *Concurrent Pos:* Res assoc, Univ Chicago, 42-43; scientist, Los Alamos Sci Lab, 43-46; asst dir res, AEC, Washington, DC, 47; Guggenheim fel, 54; US State Dept fel, 58. *Mem:* Fel AAAS; fel Am Phys Soc; Sigma Xi. *Res:* Biophysics; nuclear physics; electron microscopy. *Mailing Add:* 1469 46th St Los Alamos NM 87544

MANLEY, KIM, b Chicago, Ill, Apr 28, 43. STRATIGRAPHY-SEDIMENTATION. *Educ:* Univ Colo, BA, 64 & PhD(geol), 76; Univ Tex, Med, 68. *Prof Exp:* Teaching asst anthropol, Univ Colo, 64; res asst anthropol, Mus NMex, 65; prof asst geol, Pan Am Petrol Corp, 66; teaching asst geol, Univ Colo, 69-70; asst prof geol, Western Wash State Col, 70-74; geologist, US Geol Surv, 74-82; GEOL CONSULT, 82-; PROF, UNIV NMEX, LOS ALAMOS, 86- *Concurrent Pos:* Chmn stratig session, Pen Rose Conf Rio Grande Graben, Western Wash State Col, 74; continuing educ adv comt, US Geol Surv, 75-78, chmn, Stratig Sect, GSA Rocky Mt Sect Meeting, 76 & br sem series, 75-77, br prom comt, 78, chmn, Denver Regional Comt Women Scientists Career Pattern Geol Div, 78-80; invited lectr, St Barbara's Day Celeb, Kans State Geol Surv, 77; co-leader 30th field conf, NMex Geol Soc, 79 & 35th field conf, 84; invited lectr, Geosci Div Lect Series, Los Alamos Nat Lab, 81; consult, Shannon & Wilson Inc, 74 & Los Alamos Nat Lab, 83-87; prof, Univ Colo, 86- *Mem:* Geol Soc Am. *Res:* History of rifting in northern Rio Grande, Rift, New Mexico; sauropod dinosaur gastroliths. *Mailing Add:* 4691 Ridgeway Dr Los Alamos NM 87544

MANLEY, ROCKLIFFE ST JOHN, b Kingston, Jamaica, Mar 26, 25; Can citizen; m 58; c 1. POLYMER CHEMISTRY. *Educ:* McGill Univ, BSc, 50, PhD(phys chem), 53; Uppsala Univ, DSc, 56. *Prof Exp:* Nat Res Coun Can fel, 53-55; RES ASSOC CHEM, McGILL UNIV, 58-; SR SCIENTIST, PULP & PAPER RES INST CAN, 58-, PRIN SCIENTIST, 78- *Concurrent Pos:* Secy-treas, Can High Polymer Forum, 67-69, prog chmn, 69-71, chmn forum, 71-73. *Mem:* AAAS; Fiber Soc; Am Phys Soc; Chem Inst Can; Tech Asn Pulp & Paper Indust. *Res:* Macromolecular science; flow properties of model disperse systems; polymer solution properties; morphology of crystalline polymers; polymer crystallization; molecular morphology and biosynthesis of cellulose. *Mailing Add:* Dept Chem McGill Univ 3420 University St Montreal PQ H3A 2A7 Can

MANLEY, THOMAS CLINTON, b Ithaca, NY, Feb 15, 11; m 40; c 2. ELECTROCHEMISTRY. *Educ:* Cornell Univ, BChem, 32, MChem, 33; Rutgers Univ, PhD(phys chem), 38. *Prof Exp:* Pilot plant engr, Welsbach Corp, 40, res engr, 41-47, asst dir res, 48-49, dir res, 49-76; CONSULT, 76- *Mem:* Am Chem Soc. *Res:* Electric discharges in gases; chemical reactions in electric discharges; ozone production properties; reactions of ozone; absorption of gases. *Mailing Add:* 3141 Maple Rd Huntingdon Valley PA 19006

MANLEY, THOMAS ROY, b McKeesport, Pa, June 15, 18; m 43; c 2. ENTOMOLOGY. *Educ:* Fairmont State Col, AB, 40; WVa Univ, MS, 46. *Prof Exp:* Horticulturist, Oglebay Inst, 46-48 & Case Western Reserve Univ, 49-52; mgr, Champlain View Gardens, 52-55; teacher, Selinsgrove Area High Sch, 56-64; prof, 64-81, EMER PROF, BLOOMSBURG STATE COL, 81- *Concurrent Pos:* John Hay fel, Yale Univ, 61-62, curatorial affil, Peabody Mus, 64-; NSF res grants, 64-65. *Honors & Awards:* NAm Gladiolus Coun Gold Medal, 48; Dean Herbert Medal, Plant Life Soc, 54; New Eng Gladiolus Soc Gold Medal, 54. *Mem:* Soc Study Evolution; Lepidopterist Soc; Entom Soc Am; Sigma Xi. *Res:* Evolution of Lepidopterous insects in suture zones of North America; genetics of genus Automeris and taxonomic status of subspecies in sympatry. *Mailing Add:* Rte 1 Box 269 Port Trevorton PA 17864

MANLY, DONALD G, b Cleveland, Ohio, Oct 7, 30; m 52; c 5. ORGANIC CHEMISTRY. *Educ:* Brown Univ, ScB, 52; Lehigh Univ, MS, 54, PhD(org chem), 56. *Prof Exp:* Proj leader chem, Quaker Oats Co, 56-57, group leader, 57-63, sect leader chem res, 63-65; res mgr chem, Glyco Chem Inc, 65-67, dir res, 67-69; assoc dir appln res, 69-71, Air Prod & Chem, Inc, dir corp res, 71-76, group dir res, 76-77; vpres res & develop, Abex Corp, 77-80; vpres technol, Anaconda Indust, 80-82; pres, Wisc Centrifugal, 82-85; PRES, ROLI INK CORP, 85- *Concurrent Pos:* Small bus & mgt consult. *Mem:* Am Chem Soc; Am Oil Chem Soc; Indust Res Inst. *Res:* Heterogeneous catalysis; heterocyclic compounds; fluorine chemistry; enzyme technology; pollution control; surface chemistry. *Mailing Add:* 122E E Sutton Pl Waukesha WI 53188

MANLY, JETHRO OATES, b NC, Jan 21, 14; m 41; c 3. PHYCOLOGY. *Educ:* Col William & Mary, BS, 37; Duke Univ, PhD, 53. *Prof Exp:* Instr biol, Col William & Mary, 46-49; instr zool, Duke Univ, 49-50, bot, 52-55; PROF BIOL, PFEIFFER COL, 55- *Mem:* Fel AAAS; Sigma Xi. *Res:* Taxonomy and distribution of marine diatoms. *Mailing Add:* Pfeiffer Col Misenheimer NC 28109

MANLY, KENNETH FRED, b Cincinnati, Ohio, July 12, 41; m 62; c 2. COMPUTING IN MOLECULAR BIOLOGY. *Educ:* Calif Inst Technol, BS, 64; Mass Inst Technol, PhD(microbiol), 69. *Prof Exp:* Sr cancer res scientist, 71-74, assoc cancer res scientist, 74-78, actg head, viral oncol dept, 78-79, CANCER RES SCIENTIST V, ROSWELL PARK MEM INST, 78- *Concurrent Pos:* Am Cancer Soc fel, Mass Inst Technol, 69-71; Nat Cancer Inst grant, 72-75, 78-81. *Mem:* AAAS; Am Soc Microbiol; Asn Comput Mach. *Res:* Computer applications in molecular biology and genetics. *Mailing Add:* Molecular & Cellular Biol Dept Roswell Park Mem Inst Buffalo NY 14263

MANLY, PHILIP JAMES, b Cincinnati, Ohio, Apr 12, 44; m 67; c 4. HEALTH PHYSICS. *Educ:* Mass Inst Technol, BS, 67; Rensselaer Polytech Inst, MS, 71; Am Bd Health Physics, cert. *Prof Exp:* Shift radcon dir, Gamma Corp, 71-72, head, Tech Div, 72-74, sr health physicist, 74, head, Training Div, Pearl Harbor Naval Shipyard, 74-78, PRES, GAMMA CORP, HEALTH PHYSICS CONSULTING, 78-; pres, Osteon Inc, Med Equip Mfg, 83-88; CONSULT, 88- *Mem:* Health Physics Soc; assoc mem Sigma Xi. *Res:* Practical application of radiation protection principles. *Mailing Add:* PO Box 430 Wahiawa HI 96786

MANLY, WILLIAM D, b Malta, Ohio, Jan 13, 23; m 49; c 4. HIGH TEMPERATURE ALLOYS, NUCLEAR MATERIALS. *Educ:* Univ Notre Dame, BS, 47, MS, 49. *Prof Exp:* Metallurgist/mgr, Oak Ridge Nat Lab, 49-60, dir, Gas Cooled Reactor Proj, 60-64; mgr, Union Carbide Corp, 64-65, gen mgr, 67-69, vpres, 69-70; sr vpres, Cabot Corp, 70-84, exec vpres, 85-86; CONSULT, 86- *Honors & Awards:* Merit Award, Am Nuclear Soc, 83; Medal Advancement of Sci, Am Soc Metals, 87. *Mem:* Nat Acad Eng; fel Am Soc Metals (pres, 72-73); fel Am Inst Mining & Metall Engrs; fel Am Nuclear Soc; Nat Asn Corrosion Engrs. *Res:* High temperature; corrosion; reactor materials; high temperature materials; materials research management; author of 75 publications and recipient of 5 patents. *Mailing Add:* Oak Ridge Nat Labs PO Box 2008 Bldg 4500-S Mail Stop 6118 Oak Ridge TN 37831-6118

MANN, ALAN EUGENE, b New York, NY, Sept 19, 39; m 83; c 2. PHYSICAL ANTHROPOLOGY, PRIMATOLOGY. *Educ:* Univ Pittsburgh, BA, 61; Univ Calif, Berkeley, MA & PhD(anthrop), 68. *Prof Exp:* Teaching asst phys anthrop, Columbia Univ, 63-64; asst, Univ Calif, Berkeley, 66-68, actg asst prof, 68-69; asst prof & asst cur mus, 69-75, ASSOC PROF ANTHROP & ASSOC CUR PHYS ANTHROP, UNIV PA MUS, 75- *Concurrent Pos:* Res assoc, Laborataire d'Anthropologie, Univ de Bordeaux I, France; Fulbright fel, 90. *Mem:* Am Asn Phys Anthrop; Royal Anthrop Inst Gt Brit & Ireland. *Res:* Analysis of hominid evolution, with emphasis on the reconstruction of behavior. *Mailing Add:* Dept Anthrop Univ Pa Philadelphia PA 19104

MANN, ALFRED KENNETH, b New York, NY, Sept 4, 20; m 46; c 4. ASTROPHYSICS. *Educ:* Univ Va, AB, 42, MS, 46, PhD(physics), 47. *Prof Exp:* Instr physics, Columbia Univ, 47-49; from asst prof to assoc prof, 49-57, PROF PHYSICS, UNIV PA, PHILADELPHIA, 57- *Concurrent Pos:* Fulbright fel, Australian Nat Univ, 55-56; NSF sr fel, Cern, 62-63; assoc dir, Princeton-Pa Acceleration, 66-67; phys prog eval comt, Regents of NY State, 74-75; mem bd trustees, Assoc Univ, Inc, 70-83; subpanel on new facil, High Energy Physics adv panel, Erda, 75; mem bd trustees, Univ Res Asn, Inc, 79-84; Guggenheim fel, 81-82; mem adv panel High Energy Physics, Dept Energy, 83; chmn exec comt, Div Particles & Fields, Am Phys Soc, 83; vis prof, Univ Calif, Los Angeles, 89 & 90. *Honors & Awards:* US Naval Ordnance Develop Award, 45; Asahi Prize, Japan, 87; Rossi Prize, Am Astron Soc, 89. *Mem:* Sigma Xi; fel Am Phys Soc. *Res:* Mass spectroscopy; molecular beams; elementary particle physics; structure of weak interactions through studies of K-meson decays and high energy neutrino interactions; neutrino astronomy. *Mailing Add:* Dept Physics Univ Pa Philadelphia PA 19104

MANN, BENJAMIN MICHAEL, b Philadelphia, Pa, Apr 17, 48; m 80; c 1. TOPOLOGY. *Educ:* Univ Calif, Los Angeles, BA, 70; Stanford Univ, MS, 71, PhD(math), 75. *Prof Exp:* lectr math, Rutgers Univ, 75-76; asst prof, Harvard Univ, 76-80; asst prof math, Bowdoin Col, 80-; AT UNIV NMEX. *Mem:* Am Math Soc. *Res:* Corbordism; infinite loop space theory; surgery; homotopy theory. *Mailing Add:* Dept Math & Statist Univ NMex Albuquerque NM 87131

MANN, CHARLES KENNETH, b Fairmont, WVa, Jan 2, 28; m 57; c 2. ANALYTICAL CHEMISTRY. *Educ:* George Washington Univ, BS, 50, MS, 52; Univ Va, PhD(chem), 55. *Prof Exp:* Analytical chemist, Nat Bur Standards, 50-52; instr chem, Univ Tex, 55-58; from asst prof to assoc prof, 58-68, PROF CHEM, FLA STATE UNIV, 68- *Mem:* Am Chem Soc; Soc Appl Spectros. *Res:* Organic electrochemistry; electroanalytical chemistry. *Mailing Add:* Dept of Chem Fla State Univ Tallahassee FL 32306

MANN, CHARLES ROY, b New York, NY, Mar 27, 41. MATHEMATICAL STATISTICS, APPLIED STATISTICS. *Educ:* Polytech Inst Brooklyn, BS, 61; Mich State Univ, MS, 63; Univ Mo, PhD(statist), 69. *Prof Exp:* Instr math, Univ Maine, 63-64; asst prof statist, George Washington Univ, 69-73; head statist div, Group Opers Inc, 73-77; PRES, CHARLES R MANN ASSOCS, INC, 77- *Concurrent Pos:* Consult, US Info Agency, 71-, Mack Trucks, 74-, Equal Employment Opportunity Comn & Dept Justice, 75-, Dean Witter Reynolds, Inc, 75-; prof lectr statist, George Wash Univ, 91- *Mem:* Inst Math Statist; Am Statist Asn. *Res:* Bayesian statistics; density estimation; data analysis; legal applications of statistics. *Mailing Add:* Suite 950 1828 L St NW Washington DC 20036-5104

MANN, CHRISTIAN JOHN, b Junction City, Kans, Oct 16, 31; m 61; c 4. MATHEMATICAL GEOLOGY, PETROLEUM GEOLOGY. *Educ:* Univ Kans, BS, 53, MS, 57; Univ Wis, PhD(geol), 61. *Prof Exp:* Geologist, Gulf Oil Corp, 53 & Calif Oil Co, 57-64; sr earth scientist, Hazleton Nuclear Sci Corp, 64-65; asst prof geol, 65-69, assoc prof, 69-78, PROF GEOL, UNIV ILL, URBANA, 81- *Concurrent Pos:* Dept ed, Math Geol, 80-84, ed, 84-89; mem, Nat Comt Math Geol, 81-82; coord, Lithostratigraphic Sect, Proj 148, Int Union Geol Sci 78-83, mem, 71-83; consult, Sandia Nat Lab, 84-, US Nuclear Regulatory Comn, 85; mem, Comt Publ, Int Asn Math Geologists, 85-; mem, Proj 261, Int Union Geol Sci, 88-; mem, Peer Rev Panel, Waste Isolation Pilot Plant (nuclear waste disposal), 88- *Mem:* Fel AAAS; fel Geol Soc Am; Int Asn Math Geol; Math Geologists of US (pres, 76-78); Am Asn Petrol Geologists. *Res:* Mesozoic and Paleozoic stratigraphy; quantitative lithostratigraphic correlation; quantitative analysis of cycles in geology; nature of geologic data; data enhancement; regional stratigraphic synthesis; archaeological geology; geologic repositories for high-level nuclear wastes. *Mailing Add:* Dept Geol Univ Ill 1301 West Green St Urbana IL 61801

MANN, DAVID EDWIN, JR, b Johnson City, Tenn, Feb 13, 22; m 50; c 3. PHARMACOLOGY. *Educ:* Harvard Univ, BS, 44; Purdue Univ, MS, 48, PhD(physiol), 51. *Prof Exp:* Asst prof physiol & pharmacol, Sch Pharm, 50-54, assoc prof pharmacol & chmn dept, 54-60, PROF PHARMACOL, SCH PHARM & SCH DENT, TEMPLE UNIV, 60- *Honors & Awards:* Lindback Award, 66. *Mem:* AAAS; Am Pharmaceut Asn. *Res:* Teratology; toxicology; carcinogenesis. *Mailing Add:* 321 Gribbel Rd Wyncote PA 19095

MANN, DAVID R, b Bellefonte, Pa, July 21, 44. PHYSIOLOGY. *Educ:* Juniata Col, BS, 66; Rutgers Univ, PhD(zool), 71. *Prof Exp:* Teaching asst, Dept Zool, Rutgers Univ, New Brunswick, NJ, 66-71; res assoc, Dept Physiol, Sch Med, Univ Md, Baltimore, 71-73; from asst prof to assoc prof biol, State Univ NY, Binghamton, 73-78; assoc prof, 79-84, PROF PHYSIOL, MOREHOUSE SCH MED, ATLANTA, GA, 84- *Concurrent Pos:* Henry Rutgers fel, Rutgers Univ, 66; postdoctoral fel reproductive endocrinol, Sch Med Univ Md, 71-73, VA Med Ctr, Augusta, Ga & Univ Mo, 90-; affil scientist, Yerkes Regional Primate Res Ctr, Emory Univ, Atlanta, Ga, 80-, adj prof, Dept Psychol, 91-; co-chmn, Reproductive Endocrinol Session, IXth Cong Int Primatological Soc, Atlanta, Ga, 82; adj prof, Dept Biol, Ga State Univ, Atlanta, 89; mem, Peer Rev Comt, Am Heart Asn, Ga Affil, 90-91. *Mem:* Soc Study Reproduction; Endocrine Soc; Am Physiol Soc; Soc Exp Biol & Med; Am Soc Andrology. *Res:* The role of the adrenal gland in reproduction; endogenous opioids and sexual development; neonatal testosterone and primate sexual development; author of various publications. *Mailing Add:* Dept Physiol Morehouse Sch Med 720 Westview Dr SW Atlanta GA 30310

MANN, FREDERICK MICHAEL, b San Francisco, Calif, May 8, 48. NUCLEAR PHYSICS. *Educ:* Stanford Univ, BS, 70; Calif Inst Technol, PhD(physics), 75. *Prof Exp:* Res fel physics, Calif Inst Technol, 75; advan engr, Westinghouse Hanford Co, 75-78, sr engr physics, 78-85, prin scientist, 85-90, FEL SCIENTIST, WESTINGHOUSE HANFORD CO, 90- *Concurrent Pos:* Sub-comt chmn, Cross Sect Eval Working Group, 80; basic nuclear data, Nat Acad Sci Panel, 82-88. *Mem:* Am Phys Soc; Am Nuclear Soc. *Res:* Nuclear data evaluation, nuclear computer code development, nuclear cross section measurement. *Mailing Add:* 240 Saint Ct Richland WA 99352-1912

MANN, GEORGE VERNON, b Lehigh, Iowa, Sept 15, 17; m 47. BIOCHEMISTRY, HUMAN NUTRITION. *Educ:* Cornell Col, BA, 39; Johns Hopkins Univ, DSc(biochem), 42, MD, 45. *Prof Exp:* Asst chemist, State Health Dept, Md, 40-41; intern med, Johns Hopkins Hosp, 44-45; intern, Peter Bent Brigham Hosp, 46; asst prof nutrit, Harvard Univ, 49-55, asst med, 50-58; from asst prof med to assoc prof med, Sch Med, Vanderbilt Univ, 58-87, assoc prof biochem, 58-87; SR CONSULT, WHO, 87- *Concurrent Pos:* Nutrit Found res fel, Sch Pub Health, Harvard Univ, 47-49; asst resident, New Eng Deaconess Hosp, 46; asst, Peter Bent Brigham Hosp, 47-48; estab investr, Am Heart Asn, 54-62; asst dir Framingham Heart Study, USPHS, 55-58, consult, 58-; career investr, Nat Heart Inst, 62- *Mem:* AAAS; Am Heart Asn; Am Inst Nutrit; NY Acad Sci; Sigma Xi. *Res:* Atherosclerosis-cardiovascular diseases; epidemiology; cardiovascular disease; human nutrition; water purification. *Mailing Add:* 3710 Westbrook Ave Nashville TN 37205

MANN, JAMES, b Paterson, NJ, Nov 29, 13; m 42; c 4. MEDICINE. *Educ:* Univ Ill, AB, 35; Wash Univ, MD, 40; Am Bd Psychiat & Neurol, dipl, 50. *Prof Exp:* From asst prof to assoc prof, 54-63, PROF PSYCHIAT, SCH MED, BOSTON UNIV, 63- *Concurrent Pos:* Dir psychiat, Briggs Clin, Boston State Hosp, 49-52, dir psychiat, 52-59; mem fac, Sch Social Work, Smith Col, 51-58; mem, Boston Psychoanal Inst, 53-, training analyst, 62-, dean, 71-73; vis prof, Hadassah Med Sch, Hebrew Univ Jerusalem, 55-56, Inst Living, Hartford, Conn, 58 & Grad Sch, Brandeis Univ, 59. *Mem:* fel Am Psychiat Asn; Am Psychoanal Asn. *Res:* Psychotherapy of schizophrenia; time limited psychotherapy; group psychotherapy; dynamics of teaching. *Mailing Add:* 20 Locke Rd Waban MA 02168

MANN, JAMES EDWARD, JR, b Bluefield, WVa, Nov 17, 36; m 62; c 4. APPLIED MATHEMATICS, APPLIED MECHANICS. *Educ:* Va Polytech Inst, BS, 59; Harvard Univ, SM, 60, PhD(eng), 64. *Prof Exp:* Res engr, Esso Prod Res Co, 63-65; from asst prof to assoc prof appl math, Univ Va, 65-82; PROF MATH, WHEATON COL, 82- *Mem:* Soc Indust & Appl Math; Math Asn Am. *Res:* Wave propagation phenomena; problems that arise in electrodynamics, acoustics and water waves. *Mailing Add:* Dept Math Wheaton Col Wheaton IL 60187

MANN, JOHN ALLEN, b NJ, Sept 28, 21; m 43; c 3. GENERAL GEOLOGY, EDUCATION. *Educ:* Princeton Univ, BA, 43, MA, 49, PhD(geol), 50. *Prof Exp:* Geologist, Stand Oil Co, Calif, 50-54, dist geologist, 54-64; mgr, geol res, Chevron Res Co, 64-68, sr staff geologist, Chevron Oil Field Res Co, 68, Div Geologist, Sotex Div, Chevron Oil Co, 68-71, sr staff geologist-schs coordr, Chevron Corp, 71-82; RETIRED. *Concurrent Pos:* Mem staff, Sec Sch Gen Geol Adults, Whittier, Calif, 59-68 & Midland, Tex, 69-71; consult, geol educ, 82-90. *Mem:* Fel Geol Soc Am; Sigma Xi; Am Asn Petrol Geologists; Nat Asn Geol Teachers. *Res:* Petroleum exploration; education. *Mailing Add:* 4769 Serrente Plaza Yorba Linda CA 92686

MANN, JOSEPH BIRD, (JR), b Kearny, NJ, Dec 1, 23; m 45; c 3. CHEMICAL PHYSICS. *Educ:* Union Univ, NY, BS, 44; Mass Inst Technol, PhD(phys chem), 50. *Prof Exp:* Asst photochem, Cabot Solar Energy Fund, Mass Inst Technol, 48-50; mem staff, 50-86, LAB ASSOC, LOS ALAMOS NAT LAB, 86- *Mem:* Fel AAAS; Am Inst Chem; Am Phys Soc. *Res:* Calculation of excitation and ionization cross sections of atoms and ions; relativistic Hartree-Fock calculations of electron structure of atoms. *Mailing Add:* 2551 35th St Los Alamos NM 87544

MANN, KENNETH GERARD, b Floral Park, NY, Jan 1, 41; m 64; c 4. BIOCHEMISTRY. *Educ:* Manhattan Col, BS, 63; Univ Iowa, PhD(biochem), 67. *Prof Exp:* Fel, Univ Iowa, 67-68; NIH fel, Duke Univ, 68-70; from asst prof to prof biochem, Univ Minn, 70-84; from assoc prof to prof biochem, Mayo Med Sch, Mayo Found, 74-84; PROF & CHMN BIOCHEM, UNIV VT, 84- *Concurrent Pos:* NIH res grant, Univ Minn, St Paul, 71-74, Dreyfus teacher grant, 71-76; estab investr, Am Heart Asn, 74-79. *Mem:* Am Soc Biol Chemists; Am Chem Soc; Sigma Xi; Am Soc Hematol; Am Heat Asn; Int Soc Thrombosis & Haemostasis. *Res:* Protein chemistry; blood clotting. *Mailing Add:* 2543 Oakhill Rd Shelburne VT 05482

MANN, KENNETH H, b Dovercourt, Eng, Aug 15, 23; m 46; c 3. ECOLOGY OF COASTAL WATERS, BIOLOGICAL PHYSICAL INTERACTIONS. *Educ:* Univ London, BSc, 49, DSc(zool), 66; Univ Reading, PhD(zool), 53. *Prof Exp:* Asst lectr zool, Univ Reading, 49-51, lectr, 51-64, reader, 64-67; sr biologist, Marine Ecol Lab, Bedford Inst, Fisheries Res Bd Can, 67-72; chmn dept, Dalhousie Univ, Halifax, NS, 72-78, prof biol, 72-80; dir, Marine Ecol Lab, 80-87, RES SCIENTIST, HABITAT ECOL DIV, DEPT FISHERIES & OCEANS, BEDFORD INST OCEANOG, DARTMOUTH, CAN, 87- *Concurrent Pos:* Mem productivity freshwater subcomt, Brit Nat Comt Int Biol Prog, 64-67; consult, London Anglers Asn, 58-64; ed, J Animal Ecol, 66-67; mem, Can Comt Man & Biosphere, 73-80; Can rep, Sci Comt Oceanic Res & chmn, Working Group Math Models & Biol Oceanog, 77-87; Queen Elizabeth II sr fel, Australia, 78; Killman sr fel, Can, 80. *Mem:* Am Soc Limnol & Oceanog; Brit Ecol Soc; fel Royal Soc Can. *Res:* Functioning of aquatic ecosystems; primary and secondary productivity in coastal, estuarine and fresh waters; dynamics of marine food chains; biological-physical interactions in the sea. *Mailing Add:* Habitat Ecol Div Bedford Inst Oceanog Box 1006 Dartmouth NS B2Y 4A2 Can

MANN, LARRY N, b Philadelphia, Pa, Aug 21, 34; m 59; c 3. TOPOLOGICAL TRANSFORMATION GROUPS, GROUPS OF ISOMETRIES. *Educ:* Univ Pa, BS, 55, MA, 56, PhD(math), 59. *Prof Exp:* Mathematician, Radio Corp Am, 58-60; lectr math, Univ Va, 60-61, asst prof, 61-63; mem, Inst Defense Analysis, 63-64; asst, Inst Advan Study, 64-65; assoc prof, 65-70, PROF MATH, UNIV MASS, AMHERST, 70-, DEPT HEAD MATH & STATIST, 82- *Concurrent Pos:* Assoc, Off Naval Res, 60-61. *Mem:* Am Math Soc. *Res:* topological and riemannian transformation groups. *Mailing Add:* Dept of Math Univ of Mass Amherst MA 01003

MANN, LAWRENCE, JR, b Baton Rouge, La, Feb 12, 26; m 52; c 2. INDUSTRIAL ENGINEERING. *Educ:* La State Univ, BS, 49; Purdue Univ, MS, 50, PhD(indust eng), 65. *Prof Exp:* Engr, Esso Standard Oil Co, 50-57; sr design engr, Lummus Co, 57-59; assoc prof, 59-68, PROF INDUST ENG, LA STATE UNIV, 68- *Concurrent Pos:* Bur Pub Rds res grants, 61-62, 65-66; mem res rev panel, Nat Acad Sci, 65-66. *Mem:* Am Inst Indust Engrs; Am Soc Mech Engrs; Am Soc Eng Educ; Nat Soc Prof Engrs. *Res:* Highway maintenance costs; highway accident causes; general industrial maintenance. *Mailing Add:* Dept Indust Eng La State Univ Baton Rouge LA 70803

MANN, LESLIE BERNARD, b Granger, Wash, Oct 19, 19; m 44; c 3. NEUROLOGY. *Educ:* La Sierra Col, BS, 44; Loma Linda Univ, MD, 45. *Prof Exp:* Intern, Los Angeles County Hosp, 44-45; intern, 50, DIR EEG LAB, WHITE MEM MED CTR, 50-; assoc prof neurol, Col Med, Loma Linda Univ, 58-; RETIRED. *Concurrent Pos:* Chief neuromed serv, White Mem Med Ctr, 52-62; assoc prof, Univ Calif, Irvine-Calif Col Med, 66-69; consult, Los Angeles County Gen Hosp, 52-62, Glendale Adventist Hosp, Rancho Los Amigos Hosp, 62-77 & Alhambra Community Hosp, 55-85. *Mem:* Am Acad Neurol; Am EEG Soc; Am Epilepsy Soc. *Res:* Pediatric neurology; electroencephalography; epilepsy. *Mailing Add:* 737 Michael Ct Redlands CA 92373

MANN, LEWIS THEODORE, JR, b New York, NY, Aug 5, 25; m 57; c 3. CLINICAL BIOCHEMISTRY, IMMUNOCHEMISTRY. *Educ:* Mass Inst Technol, SB, 46; Columbia Univ, AM & PhD(org chem), 51; Am Bd Clin Chem, dipl. *Prof Exp:* Lectr biochem, Harvard Med Sch, 58-66; asst prof immunol, 68-72, Sch Med, Univ Conn, 68-72, assoc prof radiol, 72-75; clin chemist, Fresno Community Hosp, 75-79; CLIN CHEMIST, VET ADMIN MED CTR, FRESNO, 79-; ASST PROF LAB MED, UNIV CALIF, SAN FRANCISCO, 79- *Concurrent Pos:* Res fel path, Harvard Med Sch, 56-58; Nat Inst Gen Med Sci spec fel, McIndoe Mem Res Inst, East Grinstead, Eng, 66-67 & Inst for Exp Immunol, Copenhagen, 67-68; vis scientist, Hartford Hosp, Conn, 74-75; adj assoc prof chem, Calif State Univ, Fresno, 75-, adj prof chem, 81- *Mem:* AAAS; Am Chem Soc; Am Asn Clin Chem. *Res:* Radioimmunoassay and non-radio immune assay techniques; chromatographic methods in clinical analyses. *Mailing Add:* 462 W Sample Ave Fresno CA 93704

MANN, LLOYD GODFREY, b Sterling, Mass, July 2, 22; m 59; c 3. EXPERIMENTAL NUCLEAR PHYSICS. *Educ:* Worcester Polytech, BS, 44; Univ Ill, MS, 47, PhD(physics), 50. *Prof Exp:* Mem staff radiation lab, Mass Inst Technol, 44-46; instr physics, Stanford Univ, 50-53; mem staff, 53-87, CONSULT, LAWRENCE LIVERMORE LAB, UNIV CALIF, 88- *Mem:* Am Phys Soc; AAAS. *Res:* Properties of nuclear energy levels and radiations; light-ion reactions. *Mailing Add:* Lawrence Livermore Lab Univ Calif Box 808 Livermore CA 94550

MANN, MARION, b Atlanta, Ga, Mar 29, 20; m 43; c 2. PATHOLOGY. *Educ:* Tuskegee Inst, BS, 40; Howard Univ, MD, 54; Georgetown Univ, PhD, 61; Nat Bd Med Examiners, dipl; Am Bd Path, dipl. *Hon Degrees:* DSc, Georgetown Univ, 79. *Prof Exp:* Intern, USPHS Hosp, Staten Island, NY, 54-55; resident, Univ Hosp, Georgetown Univ, 56-60, instr path, 60-61; from asst prof to prof, 61-70, dean, Col Med, 70-79, prof, 79-88, ASSOC VPRES RES, HOWARD UNIV, 88- *Concurrent Pos:* Pvt pract med, Washington, DC, 61-; prof lectr, Sch Med, Georgetown Univ, 70-73; mem bd dirs, Nat Med Fel, Inc. *Mem:* Inst Med-Nat Acad Sci; Nat Med Asn. *Mailing Add:* 1460 Locust Rd Washington DC 20072

MANN, MICHAEL DAVID, b Gold Beach, Ore, May 20, 44; m 66; c 2. NEUROPHYSIOLOGY. *Educ:* Univ Southern Calif, BA, 66; Cornell Univ, PhD(neurobiol & behav), 71. *Prof Exp:* Ford Found fel, Cornell Univ, 71; USPHS fel, Univ Wash, 71-73; asst prof 73-77, ASSOC PROF PHYSIOL & BIOPHYSICS, UNIV NEBR MED CTR, OMAHA, 77- *Mem:* Soc Neurosci; Cajal Club. *Res:* Somatosensory system, especially in the cerebral cortex, with a view to understanding the role of the system in controlling and modulating behavior; evolution and development of the central nervous system. *Mailing Add:* Dept of Physiol & Biophysics Univ of Nebr Med Ctr Omaha NE 68198-4575

MANN, NANCY ROBBINS, b Chillicothe, Ohio, May 6, 25; m 49; c 2. STATISTICAL ANALYSIS. *Educ:* Univ Calif, Los Angeles, BA, 48, MA, 49, PhD(biostatist), 65. *Prof Exp:* Mathematician, Inst Numerical Analysis, Nat Bur Standards, 49-50; sr physicist, NAm Rockwell Corp, 62-70, mem tech staff & proj develop engr, 70-74, sr scientist, 74-75, proj mgr reliability & statist, Rockwell Sci Ctr, Rockwell Int Corp, 75-78; vis res statistician, 78-80, RES BIOMATHEMATICIAN, DEPT BIOMATH & PSYCHIAT, UNIV CALIF, LOS ANGELES, 80- *Concurrent Pos:* Consult, US Army Mat Command & Missile Command; mem adv comt, US Census Bur, 72-74; mem, Comt Nat Statist, Nat Acad Sci, 78-80 & Bd Adv, US Naval Postgrad Sch, Monterey, Calif, 78-82. *Honors & Awards:* Reliability Soc Annual Award, Inst Elec & Electronics Engrs, 82. *Mem:* Fel Am Statist Asn (vpres, 82-84); Inst Math Statist; Int Statist Inst; Sigma Xi. *Res:* Point and interval estimation theory; order statistics; statistical methods in reliability. *Mailing Add:* 1870 Veteran Ave 103 Los Angeles CA 90025

MANN, PAUL, b Lewiston, Idaho, Sept 3, 17; m 44; c 5. ELECTRICAL ENGINEERING. *Educ:* Univ Idaho, BS, 38, MS, 51. *Prof Exp:* Power sta operator, Pac Power & Light Co, 38-39; elec tester & field serv engr, Westinghouse Elec Corp, 39-46; asst prof elec eng, Va Polytech Inst, 47-48; from asst prof to assoc prof, 48-65, PROF ELEC ENG, UNIV IDAHO, 65- *Mem:* Nat Soc Prof Engrs; Inst Elec & Electronics Engrs. *Res:* Electric power systems; energy sources. *Mailing Add:* 407 Spruce St Caldwell ID 83605

MANN, RALPH WILLARD, b Robinson, Ill, July 12, 16; m 44; c 2. PHYSICS. *Educ:* DePauw Univ, AB, 38; Wash Univ, St Louis, MS, 40. *Prof Exp:* Asst physics, Wash Univ, St Louis, 39-41; physicist, Naval Ord Lab, 41-45; sr res geophysicist, Humble Oil & Refining Co, 46-54, res specialist, 54-63, sr res specialist geophys, Esso Prod Res Co, Tex, 64-71; environ health specialist, Tex Air Pollution Control Serv, 71-73, mem staff, Tex Air Control Bd, 73-84; RETIRED. *Mem:* Inst Elec & Electronics Engrs; Sigma Xi. *Res:* Conductivity of liquid dielectrics; underwater ordnance; geophysical research and instrumentation; underwater gravity meter; seismic prospecting methods and apparatus; digital computer applications. *Mailing Add:* 5013 Westview Dr Austin TX 78731

MANN, RANVEER S, b Mirzapur, India, Mar 31, 24; Can citizen; m 44; c 2. CHEMICAL ENGINEERING. *Educ:* Univ Allahabad, BSc, 42, MSc, 44; Polytech Inst Brooklyn, MChE, 48; Univ Hull, PhD(eng chem), 58. *Prof Exp:* Res chemist, Dept Sci & Indust Res, Govt India, 44-46; asst dir indust, Govt Punjab, 48-53; res scientist, Univ Tex, 53-54; instr chem eng, Univ BC, 54-56; res scientist, Nat Res Coun Can, 56; sci officer, Bur Mines, Can, 58-60; asst prof, 60-64, ASSOC PROF CHEM ENG, UNIV OTTAWA, 64- *Mem:* Fel Royal Inst Chem; Am Inst Chem Engrs; Am Chem Soc; fel Brit Inst Petrol. *Res:* Kinetics and mechanism of vapor phase catalytic reactions and heat transfer in fluidized beds and two phase flow of fluids; reactions, including dehydrogenation of cyclohexane, hydrogenation of methylacetylene and allene; oxidation of hydrocarbons. *Mailing Add:* Dept of Chem Eng Univ of Ottawa Ottawa ON K1N 6N5 Can

MANN, RICHARD A(RNOLD), b Lake Geneva, Wis, Dec 15, 21; m 66; c 1. ENGINEERING MECHANICS, COMPUTER SCIENCE. *Educ:* Univ Wis, BS, 44, PhD(eng mech), 66; Northwestern Univ, MS, 48. *Prof Exp:* Engr, Douglas Aircraft Co, Inc, 44, 48-49 & Nat Adv Comt Aeronaut, 44-47; instr mech eng, Univ Wis, 49-51, instr eng mech, 51-65; asst prof eng, Wright State Campus, Miami-Ohio State Univ, 65-70; assoc prof civil eng, 70-77, prof civil eng, 77-79, PROF INDUST ENG & COMPUT SCI, UNIV NEW HAVEN, 79- *Concurrent Pos:* Consult, Forest Prod Lab, USDA, 57-65. *Mem:* Am Soc Eng Educ; Asn Comput Mach. *Res:* Elastic stability. *Mailing Add:* 300 Orange Ave West Haven CT 06516

MANN, ROBERT BRUCE, b Montreal, Que, Dec 11, 55; m 78; c 1. REGULARIZATION TECHNIQUES, UNIFIED FIELD THEORIES. *Educ:* McMaster Univ, BSc, 78; Univ Toronto, MSc, 79, PhD(physics), 82. *Prof Exp:* Fel physics, Harvard Univ, 82-84; asst prof, Univ Toronto, 84-87; ASSOC PROF PHYSICS, UNIV WATERLOO, 87- *Mem:* Can Asn Physicists; Can Sci & Christian Affil. *Res:* Superstrings; sigma models; anomalies; gravitation, two-dimensional models. *Mailing Add:* Dept Physics Univ Waterloo Waterloo ON N2L 3G1 Can

MANN, ROBERT W(ELLESLEY), b Brooklyn, NY, Oct 6, 24; m 50; c 2. REHABILITATION ENGINEERING, ENGINEERING DESIGN. *Educ:* Mass Inst Technol, SB, 50, SM, 51, ScD(mech eng), 57. *Prof Exp:* Draftsman, Bell Tel Labs, Inc, NY, 42-43, 46-47; res supvr dynamic anal & control lab, 51-57, from asst prof to prof mech eng, 53-63, head eng design div, 57-66, Germeshausen prof, 70-72, prof eng, 72-74, WHITAKER PROF BIOMED ENG, MASS INST TECHNOL, 74- *Concurrent Pos:* Engr, Bell Tel Labs, Inc, NY, 50; mem comt prosthetic res & develop, Nat Acad Sci-Nat Res Coun, 63-69, mem comt skeletal syst, Nat Res Coun, 69-71, mem comt life scis, Nat Res Coun, 84-, mem Bd Health Scis Policy, Inst Med, 82-86; found chmn, Ctr Sensory Aids Eval & Develop, Mass Inst Technol, 64-75, dir & co-dir, Harvard-MIT Rehab Engr Ctr, dir Biomech & Human Rehab, MIT, 75-; mem bd dirs, Carroll Ctr for the Blind, 67-74, pres, 68-74; consult eng sci, Mass Gen Hosp, 69-; eng lectr, Fac Med, Harvard Univ, 73-79; res assoc orthop surg, Children's Hosp Med Ctr, 73-; Sigma Xi nat lectr, 79-81; prin investr, prog proj grant, Nat Heart & Lung Inst, 72-77 & NIH grant, 74-80; mem, Sci Adv Bd, Becton, Dickson & Co, 75-81, Sci Merit Rev Bd, Vet Admin, 83-, vis comt, Dept Mech Eng, Lehigh Univ, 83-86 & numerous comts, govt & nat socs; assoc ed, Am Soc Mech Engrs J Biomech Eng, 76-82; prof Harvard-MIT Div, Health Sci & Tech, 79. *Honors & Awards:* Talbert Abrams Photogram Award, 62; IR-100 Innovation Award, 72; Goldenson Award for Outstanding Res Phys Handicapped, 76; Am Soc Mech Engrs Gold Medal, 77; H R Lissner Award for Outstanding Bioeng, 77. *Mem:* Nat Acad Sci; Inst Med-Nat Acad Sci; Nat Acad Eng; fel Am Acad Arts & Sci; fel Inst Elec & Electronics Engrs; fel AAAS; fel Am Soc Mech Engrs. *Res:* Engineering design; technology and human rehabilitation; human musculoskeletal biomechanics; synovial joint biomechanics and the mechanogenesis of osteoarthritis; application of technologies of computer-aided design and gate analysis to a system for computer-aided simulation of orthopedic surgical procedures. *Mailing Add:* Mass Inst Technol Rm 3-144 77 Massachusetts Ave Cambridge MA 02139

MANN, RONALD FRANCIS, b Winnipeg, Man, Aug 18, 31; m 55; c 3. CHEMICAL ENGINEERING, NUCLEAR ENGINEERING. *Educ:* Queen's Univ, Ont, BSc, 56, MSc, 60, PhD(chem eng), 66. *Prof Exp:* Lectr chem, Royal Mil Col, Ont, 58-60; develop engr, Atomic Power Dept, Can Gen Elec Co, 60-62; from asst prof to assoc prof, 62-75,PROF CHEM ENG, ROYAL MIL COL CAN, 75- *Concurrent Pos:* Res grant, Defence Res Bd, 65- & Dept Nat Defence, 77-; consult, Dept Nat Defence, 75- *Mem:* Fel Chem Inst Can; Can Soc Chem Eng; Am Soc Eng Educ; Can Nuclear Soc. *Res:* Chemical reaction engineering; catalysis; heat transfer; fuel cells. *Mailing Add:* 51 Mackenzie Crescent Kingston ON K7M 2S2 Can

MANN, STANLEY JOSEPH, b Worcester, Mass, Sept 18, 32; m 69; c 2. GENETICS, BIOLOGY. *Educ:* Clark Univ, AB, 53, MA, 58; Brown Univ, PhD(biol), 61. *Prof Exp:* Cancer res scientist, Springville Labs, Roswell Park Mem Inst, 61-63, sr cancer res scientist, 63-66; asst prof anat in dermat, 66-69, exec dir, Federated Med Resources, 79-84, ASSOC PROF ANIMAL GENETICS DERMAT, TEMPLE UNIV, 69-, DIR LAB ANIMAL RESOURCES, 77-, ASSOC DEAN, SCH MED, 86- *Concurrent Pos:* Asst res prof biol, State Univ NY Buffalo, 64-66. *Res:* Mammalian genetics; biology of the skin; phenogenetics of hair mutants in the house mouse; morphology and development of normal and abnormal mammalian hair follicles. *Mailing Add:* Temple Univ Sch Med 3400 N Broad St Philadelphia PA 19140

MANN, THURSTON (JEFFERSON), b Lake Landing, NC, June 22, 20; m 45; c 3. GENETICS. *Educ:* NC State Univ, BS, 41, MS, 47; Cornell Univ, PhD(genetics, plant breeding), 50. *Prof Exp:* From asst prof to assoc prof agron, 49-53, in charge agron teaching, 53-55, prof crop sci, 55-64, prof genetics & head dept, 64-73, prof genetics & crop sci, 73-76, ASST DIR RES, SCH OF AGR & LIFE SCI, NC STATE UNIV, 76- *Concurrent Pos:* Vis agronomist, Coop State Res Serv, USDA, 74. *Mem:* Am Soc Agron; Am Genetic Asn; Genetics Soc Am. *Res:* Tobacco genetics; interspecific hybridization and breeding procedures; inheritance of alkaloids in Nicotiana. *Mailing Add:* 100 Patterson Hall NC State Univ Raleigh NC 27650

MANN, WALLACE VERNON, JR, b Pembroke, Mass, Mar 17, 30; m 52; c 4. DENTISTRY. *Educ:* Williams Col, BA, 51; Tufts Univ, DMD, 55; Univ Ala, MS, 62. *Prof Exp:* From instr to prof dent, Sch Dent, Univ Ala, Birmingham, 62-74, chmn dept periodont, 65-74, asst dean sch, 66-74; DEAN DENT, UNIV MISS MED CTR, 74-, PROF ENDODONTICS & PERIODONTICS, 77-, DENTIST-IN-CHIEF, UNIV HOSP, 77- *Concurrent Pos:* Sr res trainee, Sch Dent, Univ Ala, Birmingham, 62-63; NIH career develop award, 63-66; dir clin res training grant for DMDþPhD prog, Sch Dent Univ Ala, Birmingham, 67-74. *Mem:* Am Dent Asn; Am Asn Dent Schs; fel Am Col Dentists; Int Asn Dent Res; Sigma Xi. *Res:* Physiology and biochemistry of periodontal tissues. *Mailing Add:* Univ Provost Univ Louisville Louisville KY 40292

MANN, WILFRID BASIL, b London, Eng, Aug 4, 08; m 38; c 3. PHYSICS OF IONIZING RADIATION. *Educ:* Univ London, BSc, 30, PhD(physics), 34,. *Hon Degrees:* DSc, Univ London, 51. *Prof Exp:* Lectr physics, Imp Col, Univ London, 33-46; Nat Res Coun Can atomic energy proj, 46-48; attache, Brit Embassy, DC, 48-51; chief, Radioactivity Sect, Nat Bur Standards, 51-80; RETIRED. *Concurrent Pos:* Sci liaison officer, Brit Commonwealth Sci Off, DC, 43-45; sci adv, UK Del, UN AEC, 46-51; adj prof, Am Univ, 61-68; emer NAm ed-in-chief, Int J Appl Radiation & Isotopes, 65-; mem, Fed Radiation Coun, 69-70; dep chief, Appl Radiation Div, Ctr Radiation Res, 74-78; chmn sci 18A, Nat Coun Radiation Protection & Measurements, 72-; ed, Int J Nuclear Med & Biol, 73-, Environ Int, 77-; pres, Int Comt Radionuclide Metrology, 78-80. *Honors & Awards:* Medal of Freedom, 48; Gold Medal, US Dept Com, 58; Edward Bennett Rosa Award, 77. *Mem:* Am Phys Soc; Brit Inst Physics & Phys Soc (ed, Reports Progress Physics, 41-46); hon mem Nat Coun Radiation Protection & Measurements. *Res:* Radioactivity standardization; microcalorimetry; author of four published books. *Mailing Add:* 5710 Warwick Pl Chevy Chase MD 20815

MANN, WILLIAM ANTHONY, b Trenton, NJ, Oct 12, 43; m 88. EXPERIMENTAL HIGH ENERGY PHYSICS. *Educ:* Yale Univ, BS, 65; Univ Mass, MS, 67, PhD(physics), 70. *Prof Exp:* Res assoc, Argonne Nat Lab, 70-72; res assoc, 72-74, from asst prof to assoc prof, 74-86, PROF PHYSICS, TUFTS UNIV, 86- *Concurrent Pos:* Vis prof physics, State Univ NY, Stony Brook, 80-81; guest physicist, Argonne Nat Lab, 83-84. *Mem:* Am Phys Soc; Sigma Xi; Fedn Am Scientists. *Res:* Experimental fundamental particle interactions; neutrino interactions, structure of nucleons; nucleon decay, monopoles, cosmic ray physics; dynamics of multiparticle production; neutral current reactions. *Mailing Add:* Dept Physics Tufts Univ Medford MA 02155

MANN, WILLIAM ROBERT, b Honea Path, SC, Sept 21, 20; m 47; c 3. APPLIED MATHEMATICS. *Educ:* Univ Rochester, AB, 41; Univ Calif, PhD(math), 49. *Prof Exp:* Instr, 49-50, from asst prof to assoc prof, 50-60, PROF MATH, UNIV NC, CHAPEL HILL, 60- *Mem:* Am Math Soc. *Res:* Nonlinear boundary value problems; iterative techniques. *Mailing Add:* Dept Math Univ NC 426 Whitehead Circle Chapel Hill NC 27514

MANNA, ZOHAR, b Haifa, Israel, Jan 17, 39; m; c 4. COMPUTER SCIENCE. *Educ:* Israel Inst Technion, BS, 61, MS, 65; Carnegie-Mellon Univ, PhD(comput sci), 68. *Prof Exp:* PROF COMPUT SCI, STANFORD UNIV, 68- *Concurrent Pos:* prof appl math, Weizmann Inst, Israel, 72-; Guggenheim fel. *Res:* Mathematical theory of computation. *Mailing Add:* Dept Comput Sci Stanford Univ Stanford CA 94305

MANNE, ALAN S, OPERATIONS RESEARCH. *Prof Exp:* PROF, STANFORD UNIV. *Mem:* Nat Acad Eng. *Mailing Add:* Dept Opers Res Terman Eng 432A Stanford Univ Stanford CA 94305-4022

MANNE, VEERASWAMY, b Achanta, Andhra Pradesh, June 1, 52; m 80; c 1. ENZYMOLOGY, PROTEIN CHARACTERIZATION. *Educ:* Andhra Univ, BSc, 72; Univ Myeore, MSc, 74; Indian Inst Sci, PhD(biochem), 78. *Prof Exp:* Res assoc biochem, Hoffmann LaRoche, Inc, 84-86 & Wistar Inst, 86-88; SR RES INVESTR BIOCHEM, BRISTOL MYERS SQUIBB PHARMACEUT RES INST, 89- *Mem:* AAAS; Am Soc Biochem & Molecular Biol. *Res:* Oncogene products and their biochemical role in transformation PI metabolism and signal transduction across the membrane; post-translational protein processing and G protein function. *Mailing Add:* Bristol-Myers Squibb Pharmaceut Res Inst Princeton NJ 08540

MANNELLA, GENE GORDON, b Niles, Ohio, Aug 23, 31; m 58; c 4. CHEMICAL ENGINEERING. *Educ:* Case Inst Technol, BSChE, 53; Rensselaer Polytech Inst, MChE, 55, PhD(chem eng), 56. *Prof Exp:* Dep asst adminr, Energy Res & Develop Adminr, 77; sr vpres, Inst Gas Technol, 77-79; vpres, Mech Technol Inc, 79-82; dir, Wash Off, Elec Power Res Inst, 82-89; managing dir, Solar Energy Res Inst, 89-90; VPRES, GAS RES INST, 90- *Honors & Awards:* Apollo Achievement Award, NASA, 69. *Mailing Add:* Gas Res Inst 8600 Bryn Mawr Ave Chicago IL 60631

MANNER, RICHARD JOHN, b Buffalo, NY, Mar 22, 20; m 43; c 4. MEDICAL RESEARCH. *Educ:* Rensselaer Polytech Inst, BS, 41; Univ Rochester, MD, 51. *Prof Exp:* Physicist, photomicrography dept, Eastman Kodak Co, 41-42, electronics dept engr, eng dept, 46-47, physicist, color control studio, 48, 49 & 50; intern med, Rochester Gen Hosp, 51-52, preceptorship internal med, 52-55, physician, 52-60; assoc dir clin res, Mead Johnson Res Ctr, Evansville, Ind, 60-65; staff physician, med dept, Wyeth Labs, 65-67; dir med affairs, Warren-Teed Pharmaceut Inc, Columbus, 67-75; dir pharmaceut res, Ross Labs Div, Abbott Labs, Columbus, 75-76; vpres res & develop, Dome Labs Div, Miles Lab Inc, 76-79, vpres res serv, Miles Pharmaceut Div, 79-82, Med Affairs, Hollister-Stier Lab Div, 82-89, res & develop, 85-89; PHARMACEUT CONSULT, 89- *Concurrent Pos:* Sch physician, Rochester Inst Technol, 53-54; indust physician, Stecher Traung Lithography Corp, 54-60; pvt pract, Rochester, NY, 55-60; physician, St Mary's Hosp, Evansville, Ind, 60-65; mem, bd dirs, Consol Biomed Labs, Columbus, Ohio, 69-75. *Mem:* AMA; Soc Clin Trials; Am Soc Clin Pharmacol & Therapeut; Sigma Xi; Am Col Gastroenterol; Am Col Allergists. *Res:* Allergy and immunology; pharmaceutical research and development. *Mailing Add:* 12 Cassway Rd Woodbridge CT 06525

MANNERING, GILBERT JAMES, b Racine, Wis, Mar 9, 17; m 39, 69; c 3. PHARMACOLOGY, BIOCHEMISTRY. *Educ:* Univ Wis, BS, 40, MS, 43, PhD(biochem), 44. *Prof Exp:* Sr biochemist, Parke, Davis & Co, 44-50; consult, Chem Dept, 406th Med Gen Lab, Tokyo, Japan, 50-54; from asst prof to assoc prof pharmacol & toxicol, Univ Wis, 54-62; PROF PHARMACOL, MED SCH, UNIV MINN, MINNEAPOLIS, 62- *Concurrent Pos:* Consult, Wis State Crime Lab, 54-62; spec consult, Interdept Comt Nutrit Nat Defense, NIH, Ethiopia, 58; mem toxicol study sect, USPHS, 62-65, mem pharmacol-toxicol rev comt, 65-67 & pharmacol study sect, 68-69; mem comt probs drug safety, Nat Acad Sci-Nat Res Coun, 65-71. *Mem:* Am Soc Pharmacol & Exp Therapeut; Sigma Xi. *Res:* Biochemical pharmacology; drug metabolism; toxicology. *Mailing Add:* Pharmacol 102 Millard H Univ Minn Minneapolis MN 55455

MANNERING, JERRY VINCENT, b Custer City, Okla, June 14, 29; m 53; c 3. AGRONOMY. *Educ:* Okla State Univ, BS, 51; Purdue Univ, MS, 56, PhD, 67. *Prof Exp:* Asst agronomist, Univ Idaho, 56-58; soil scientist, Agr Res Serv, USDA, 58-67; exten agronomist & prof agron, 67-90, EMER PROF AGRON, PURDUE UNIV, 90- *Concurrent Pos:* Consult, Food Agr Orgn, Bulgaria, 72, EMBRAPA, Brazil, 75, Sabbatical to Tillage Lab, Wageningen, The Netherlands, 78. *Mem:* Fel Am Soc Agron; fel Soil Sci Soc Am; Int Soil Sci Soc; fel Soil Conserv Soc Am; Int Soil Tillage Res Orgn. *Res:* Soil erosion; efficient water use; soil management for crop production. *Mailing Add:* Dept Agron Purdue Univ West Lafayette IN 47907

MANNEY, THOMAS RICHARD, b El Paso, Tex, Dec 20, 33; m 56; c 3. BIOPHYSICS. *Educ:* Western Wash State Col, BA, 57; Univ Calif, Berkeley, PhD(biophys), 64. *Prof Exp:* Teaching asst, Western Wash State Col, 56-58; res asst, Univ Calif, Berkeley, 58-59, biophysicist, Donner Lab, 59-60; staff biologist, Oak Ridge Nat Lab, 64-65; asst prof microbiol, Case Western Reserve Univ, 65-71; assoc prof biol, 71-77, PROF PHYSICS, KANS STATE UNIV, 71- *Concurrent Pos:* USPHS res career develop award, 67-70; vis assoc prof, Univ Calif, 70-71. *Mem:* AAAS; Genetics Soc Am; Am Soc Microbiol. *Mailing Add:* Dept Physics Kans State Univ Cardwell Hall Manhattan KS 66506

MANNIK, MART, b Tallinn, Estonia, Jan 21, 32; US citizen; m 76; c 2. RHEUMATOLOGY, IMMUNOLOGY. *Educ:* Ohio Northern Univ, AB, 55; Western Reserve Univ, MD, 59. *Prof Exp:* Intern & resident internal med, Mass Gen Hosp, 59-61; guest investr immunol, Rockefeller Univ, 61-63, asst prof immunol & rheumatology, 65-67; clin assoc rheumatology, Nat Inst Arthritis & Metab Dis, 63-65; assoc prof, 67-73, PROF RHEUMATOLOGY, UNIV WASH, 73-, HEAD DIV, 67- *Concurrent Pos:* Mem bd dirs, Arthritis Found, 69-75, bd trustees, 85; co-chmn, res work group, Nat Arthritis Comn, 75-76; consult, spec projs rev group A, Nat Inst Arthritis, Diabetes & Digestive & Kidney Dis, 81-84; mem, Nat Arthritis & Musculoskeletal & Skill Dis Adv Coun, 87-90. *Honors & Awards:* Lee Howley Prize for Res in Arthritis, 88. *Mem:* Am Rheumatism Asn; Am Soc Clin Invest (vpres, 77-78); Asn Am Physicians; Am Asn Immunologists. *Res:* Immunoglobulin structure and discovery of idiotypes; role of immune complexes in diseases and the significance of rheumatoid factors in rheumatoid arthritis; author of over 200 scientific publications. *Mailing Add:* Dept Med Sch Med RG 28 Univ Wash Seattle WA 98195

MANNING, BRENDA DALE, b Spartanburg, SC, Oct 6, 45. BIOLOGICAL CHEMISTRY. *Educ:* Antioch Col, BS, 67; Univ Mich, MS, 69, PhD(biol chem), 73. *Prof Exp:* Res assoc genetics, Univ Mich, 73-76; ASST PROF CHEM, EASTERN MICH UNIV, 76- *Concurrent Pos:* Fac res grant, Eastern Mich Univ, 77-78. *Mem:* Sigma Xi. *Res:* Steroid hormone control in the eukaryotic systems of mammals. *Mailing Add:* 158 Canton St North Easton MA 02356

MANNING, CHARLES RICHARD, JR, b Erie, Pa, Mar 12, 30; m 55; c 3. CERAMICS ENGINEERING. *Educ:* Fla State Univ, BS, 58; Va Polytech Inst, MS, 62; NC State Univ, PhD(ceramic eng), 67. *Prof Exp:* Mat res engr & group leader, NASA Langley Res Ctr, Va, 58-67; assoc prof, 67-72, PROF CERAMIC ENG, SCH ENG, NC STATE UNIV, 72-; PRES, ACCIDENT RECONSTRUCT ANALYSIS INC, 71- *Mem:* Am Ceramic Soc; Am Soc Metals; Sigma Xi. *Res:* Metal-ceramic bonding; high-temperature behavior of materials and stress corrosion. *Mailing Add:* Accident Reconstruct Analysis Inc 5801 Lease Lane Raleigh NC 27613

MANNING, CLEO WILLARD, b Woodhull, Ill, Oct 10, 15; m 38; c 5. PLANT BREEDING. *Educ:* Ill Wesleyan Univ, BS, 40; Agr & Mech Col, Univ Tex, MS, 42; Iowa State Univ, PhD, 54. *Prof Exp:* Agent, Bur Plant Indust, Soils & Agr Eng, USDA, 41-45; asst agronomist, Agr & Mech Col, Univ Tex, 41-45, agronomist, 45-48, botanist, 48, agronomist & assoc prof, 48-51; plant breeder in charge, 51-76, SR PLANT BREEDER, STONEVILLE PEDIGREED SEED CO, 76- *Mem:* AAAS; Am Inst Biol Sci; Am Genetic Asn; Crop Sci Soc Am; Cotton Improv Conf. *Res:* Genetics and breeding of cotton; soybeans; plant exploration; interspecific relationships in Gossypium. *Mailing Add:* PO Box 213 Stoneville MS 38776

MANNING, CRAIG EDWARD, b South Kingston, RI, May 7, 59. MINERALOGY-PETROLOGY. *Educ:* Univ Vt, BA, 82; Stanford Univ, MS, 85, PhD(geol), 89. *Prof Exp:* Res scientist, US Geol Surv, 89-91; ASST PROF GEOL & GEOCHEM, UNIV CALIF, LOS ANGELES, 91- *Mem:* AAAS; Am Geophys Union; Mineral Soc Am. *Res:* Metamorphic petrology; experimental geochemistry; heat and mass transport in magna-hydrothermal systems. *Mailing Add:* Dept Earth & Space Sci Univ Calif 405 Hilgard Ave Los Angeles CA 90024-1567

MANNING, DAVID TREADWAY, b Santa Monica, Calif, Sept 19, 28; m 56; c 2. BIO-ORGANIC CHEMISTRY, AGRICULTURAL CHEMISTRY. *Educ:* Calif Inst Technol, BS, 51, PhD(chem), 55. *Prof Exp:* Res proj chemist, Union Carbide Corp, 54-62, res scientist, 62-75, sr res scientist, 75-90; PRIN SCIENTIST II, RHÔNE-POULENC AGR CO, 90- *Mem:* AAAS; Am Chem Soc; NY Acad Sci; Plant Growth Regulator Working Group. *Res:* Nitrosation reactions; organic reactions of nitrosyl chloride; chemistry of oximes; nitrogen-containing heterocyclic compounds; pesticide chemistry; plant growth regulators; insecticides. *Mailing Add:* Rhône-Poulenc Agr Co PO Box 12014 T W Alexander Dr Research Triangle Park NC 27709

MANNING, DEAN DAVID, b Grand Junction, Colo, Oct 11, 40; m 66. IMMUNOLOGY. *Educ:* Colo State Univ, BS, 62, MS, 64; Mont State Univ, PhD(microbiol), 72. *Prof Exp:* Fel microbiol, Mont State Univ, 72-74; asst prof, 75-77, ASSOC PROF MED MICROBIOL, MED SCH, UNIV WISMADISON, 77- *Concurrent Pos:* NIH fel, 73-74. *Mem:* Am Asn Immunol. *Res:* Control of the immune reponse with particular reference to heavy chain isotype suppression. *Mailing Add:* Dept Med Microbiol Univ Wis Med Sch 470 N Charter St Madison WI 53706

MANNING, ERIC G(EORGE), b Windsor, Ont, Aug 4, 40; m 61; c 2. COMPUTER SCIENCE. *Educ:* Univ Waterloo, BSc, 61, MSc, 62; Univ Ill, PhD(elec eng), 65. *Prof Exp:* Asst elec eng, Univ Ill, 62-63, asst, Coord Sci Lab, 63-65; asst prof elec eng & Ford fel, Proj Mac, Mass Inst Technol, 65-66; mem tech staff, Bell Tel Labs, 66-68; assoc prof comput sci, 68-76, PROF COMPUT SCI, UNIV WATERLOO, 76-,. *Concurrent Pos:* Sci adv, Sci Coun Can, 70-71; dean eng, Univ Victoria, 86-; dir, Comput Commun Networks Group, 73-83; dir, inst CPIR Res, 83-86. *Mem:* Fel Inst Elec & Electronics Engrs; Can Info Processing Soc; Natural Sci & Org Res Coun Can. *Res:* Computer maintenance problems; computer networks. *Mailing Add:* Faculty of Eng Univ Victoria PO Box 1700 Victoria BC V8W 2Y2 Can

MANNING, FRANCIS S(COTT), b Barbados, WI, Sept 16, 33; nat US; m 60, 78; c 2. CHEMICAL ENGINEERING. *Educ:* McGill Univ, BEng, 55; Princeton Univ, MSE & AM, 57, PhD, 59. *Prof Exp:* From instr to assoc prof chem eng, Carnegie Inst Technol, 59-68; head dept, 68-76, PROF CHEM ENG, UNIV TULSA, 68-, DIR, PETROL & ENERGY RES INST, 76-, HEAD DEPT, 79- *Honors & Awards:* Robert W Hunt Award, Am Inst Mining, Metall & Petrol Engrs, 69. *Mem:* Am Inst Chem Engrs; Am Inst Mining, Metall & Petrol Engrs; Am Soc Eng Educ; Am Chem Soc. *Res:* Mixing; thermodynamics; kinetics; water pollution control. *Mailing Add:* Dept of Chem Eng Univ of Tulsa & 600 S Coll Ave Tulsa OK 74104

MANNING, GERALD STUART, b New York, NY, Dec 9, 40; m 64. BIOPHYSICAL CHEMISTRY. *Educ:* Rice Univ, BA, 62; Univ Calif, San Diego, PhD(phys chem), 65. *Prof Exp:* NATO fel, Univ Brussels, 65-66; Nat Sci Found fel, Rockefeller Univ, 66-67, asst prof chem, 67-69; assoc prof, 69-75, PROF CHEM, RUTGERS UNIV, 75- *Concurrent Pos:* Alfred P Sloan fel, 70-72. *Res:* Polyelectrolytes; biopolymer conformation; DNA folding. *Mailing Add:* Dept Chem Rutgers Univ New Brunswick NJ 08903

MANNING, HAROLD EDWIN, b Huntsville, Ala, Mar 18, 35; m 66, 87; c 4. HETEROGENEOUS CATALYSIS. *Educ:* Auburn Univ, BS, 58; Trinity Univ, MS, 62. *Prof Exp:* Chemist, Petro-Tex Chem Corp, 62-66, res chemist, 66-67, head res group, 67-84, SR RES CHEMIST, TEX PETROCHEM CORP, 84- *Mem:* Am Chem Soc; Catalysis Soc. *Res:* Heterogeneous catalysis; reaction mechanisms and surface chemistry. *Mailing Add:* Tex Petrochem Corp 8600 Park Pl Blvd Houston TX 77017

MANNING, IRWIN, b Brooklyn, NY, Mar 7, 29; m 64; c 2. THEORETICAL PHYSICS. *Educ:* Mass Inst Technol, BS, 51, PhD(physics), 55. *Prof Exp:* Res assoc & asst prof, Syracuse Univ, 55-57; res assoc, Univ Wis, 57-59; res physicist, 59-73, supvry res physicist, 73-78, RES PHYSICIST, US NAVAL RES LAB, 78- *Concurrent Pos:* Mem ad hoc panel on the use of accelerators to study irradiation effects, Nat Acad Sci, 74. *Mem:* Am Phys Soc; Sigma Xi. *Res:* Atomic scattering at high energies; phenomena associated with the penetration of matter by energetic particles such as ion implantation, sputtering, neutron radiation damage; scattering theory; quantum mechanics; thermodynamic and statistical mechanics of irreversible processes; solid state physics. *Mailing Add:* US Naval Res Lab Code 4691 Washington DC 20375

MANNING, JAMES HARVEY, b Hancock, Mich, Aug 13, 40; m 66; c 4. PAPER CHEMISTRY. *Educ:* Mich Technol Univ, BS, 62; Lawrence Univ, MS, 64, PhD(paper technol), 67. *Prof Exp:* Prod engr paper, Kimberly-Clark Corp, 61, res chem tissue, 62; qual control engr bd, Westvaco, 63; proj engr pulping, Int Paper Co, 64, res scientist cellulose, 67-72, group dir nonwovens, 73-77, tech dir, 77-79; assoc dir, Am Can Co, 79-83; ASSOC DIR NONWOVENS RES & DEVELOP, JAMES RIVER CORP, 83- *Concurrent Pos:* Secy, Dissolving Pulp Comt, Tech Asn Pulp & Paper Indust, 72-74. *Mem:* Tech Asn Pulp & Paper Indust; NY Acad Sci. *Res:* Direction of a group responsible for the development of new nonwoven products, with emphasis on wet-laid nonwoven products and processes. *Mailing Add:* James River Corp PO Box 899 Neenah WI 54956

MANNING, JAMES MATTHEW, b Boston, Mass, Jan 3, 39; m 64; c 2. BIOCHEMISTRY. *Educ:* Boston Col, BS, 60; Tufts Univ, PhD(biochem), 66. *Prof Exp:* Nat Sci Found fel biochem, Univ Rome, 66-67; res assoc, 67-69, asst prof, 69-72, ASSOC PROF BIOCHEM, ROCKEFELLER UNIV, 72- *Honors & Awards:* Merit Award, NIH. *Res:* Protein and synthetic peptide chemistry; mechanism of enzyme action, especially of pyridoxal phosphate enzymes; amino acid metabolism and methods for the determination of amino acids; collagen biosynthesis; chemical aspects of hemoglobinopathies; gene/enzyme relationships. *Mailing Add:* Dept Biochemistry Rockefeller Univ 1230 York Ave New York NY 10021

MANNING, JARUE STANLEY, b Indiana, Pa, Sept 25, 34; m 60; c 1. VIROLOGY, BIOPHYSICS. *Educ:* San Francisco State Col, BA, 62; Univ Calif, Berkeley, PhD(biophys), 69. *Prof Exp:* Nat Cancer Inst fel, Univ Calif, Berkeley, 69-70; asst prof microbiol, sch vet med, 71-73, asst res virologist, comp oncol lab, 71-73, from asst prof to assoc prof, 73-81, chmn bacteriol, 81-86, PROF MICROBIOL, UNIV CALIF, DAVIS, 80- *Honors & Awards:* Hektoen Gold Medal, Am Med Asn, 68. *Mem:* AAAS; Am Soc Microbiol; Biophys Soc. *Res:* Molecular virology of animal viruses including insect viruses; cell-virus interaction and virus-induced cytopathology; mechanisms of viral replication; characterization of viral components; viral immunity, including host defense mechanisms. *Mailing Add:* Dept Microbiol Univ of Calif Davis CA 95616

MANNING, JEROME EDWARD, b Minneapolis, Minn, Dec 31, 40; m 62; c 3. ACOUSTICS. *Educ:* Mass Inst Technol, SB, 62, SM, 63, ScD(mech eng), 65. *Prof Exp:* Sr scientist, Bolt Beranek & Newman, Inc, Mass, 65-68; PRES, CAMBRIDGE COLLAB, INC, 68- *Concurrent Pos:* Lectr, Mass Inst Technol, 67-69. *Mem:* Acoust Soc Am. *Res:* Sound induced vibration; noise; random vibrations. *Mailing Add:* Cambridge Collaborative Inc 689 Concord Ave Cambridge MA 02138

MANNING, JERRY EDSEL, b Redland, Calif, Oct 19, 44; m 67; c 2. BIOCHEMISTRY. *Educ:* Univ Utah, BS, 66, PhD(biochem), 71. *Prof Exp:* Res fel biol, Univ Utah, 71-72 & chem, Calif Inst Technol, 73-74; from asst prof to assoc prof, 75-85, PROF, MOLECULAR BIOL, UNIV CALIF, IRVINE, 80- *Concurrent Pos:* Jane Coffin Childs fel, 73; Petrol Res Fund res grant, 74; NIH & Res Corp res grants, 75. *Mem:* Am Soc Cell Biol. *Res:* Molecular mechanisms that govern the regulation of genetic activity in the eukaryotic genome. *Mailing Add:* Dept of Molecular Biol & Biochem Univ of Calif Irvine CA 92664

MANNING, JOHN CRAIGE, b Detroit, Mich, Jan 5, 20; m 46; c 2. GEOLOGY, HYDROLOGY & WATER RESOURCES. *Educ:* Univ Idaho, BS, 42; Stanford Univ, PhD, 51. *Prof Exp:* Geologist, US Geol Surv & US Army Corps Engrs, 46-47; field geologist, Amerada Petrol Corp, 49; instr & asst prof geol, Stanford Univ, 50-54; res geologist, Shell Oil Co, 54-56; chief geologist, Philadelphia Oil Co, 56; vpres & tech dir, Ranney Method West Corp, 56-57; pres, Hydro Develop, Inc, 61-70; prof, 70-80, EMER PROF EARTH SCI, CALIF STATE COL, BAKERSFIELD, 80- *Concurrent Pos:* Consult geologist, 57-; prof, Fresno State Col, 58-70. *Mem:* Am Asn Petrol Geol; Am Inst Mining, Metall & Petrol Eng; AAAS. *Res:* Economic geology, especially engineering ground water and mining geology; development of ground water and industrial mineral deposits. *Mailing Add:* PO Box 13470 Las Vegas NV 89112

MANNING, JOHN RANDOLPH, b Norristown, Pa, Aug 24, 32; m 60; c 2. SOLID STATE PHYSICS, METALLURGY. *Educ:* Ursinus Col, BS, 53; Univ Ill, MS, 54, PhD(physics), 58. *Prof Exp:* Asst physics, Univ Ill, 57, res assoc, 58; physicist, Nat Bur Standards, 58-67, chief, Metall Physics Sect, 67-75, Transformations & Kinetics Group, 75-79, CHIEF METALL PROCESSING GROUP, NAT INST STANDARDS & TECHNOL, 79- *Mem:* Am Phys Soc; Am Inst Mining, Metall & Petrol Engrs; Am Soc Metals. *Res:* Diffusion in solids; kinetic processes and defects in metals. *Mailing Add:* Metall Div Nat Inst Standards & Technol Gaithersburg MD 20899

MANNING, JOHN W, b New Orleans, La, Nov 14, 30; m 54; c 6. PHYSIOLOGY. *Educ:* Loyola Univ, La, BS, 51; Tulane Univ, MS, 55; Loyola Univ Chicago, PhD(physiol), 58. *Prof Exp:* Vis scientist, Karolinska Inst, Sweden, 63-64; from asst prof to assoc prof physiol, 64-70, assoc prof anat, 67-71, dir grad studies physiol, 70-74, actg chmn, 79-81, PROF PHYSIOL, EMORY UNIV, 70- *Concurrent Pos:* USPHS fel, Emory Univ, 58-61, Am Heart Asn Advan res fel, 61-65; guest referee, Am J Physiol, 70-; vis prof, Shinshu Univ, Matsumoto, Japan, 72. *Mem:* AAAS; Am Physiol Soc; Am Asn Anat; Soc Neurosci; Can Physiol Soc. *Res:* Central nervous system regulation of cardiovascular activity; central response and transmission small cutaneous afferents; energetics of cardiac muscle. *Mailing Add:* Dept of Physiol Emory Univ 1364 Clifton Rd NE Atlanta GA 30322

MANNING, LAURENCE A(LBERT), b Palo Alto, Calif, Apr 28, 23; m 54; c 3. ELECTRICAL ENGINEERING. *Educ:* Stanford Univ, AB, 44, MSc, 48, PhD, 49. *Prof Exp:* Asst elec eng, Stanford Univ, 43-44; res assoc, Radio Res Lab, Harvard Univ, 44-45; asst physics, 45-46, res assoc elec eng, 46-47, actg asst prof, 47-50, from asst prof to assoc prof, 51-59, PROF ELEC ENG, STANFORD UNIV, 60- *Mem:* Int Union Radio Sci; fel Inst Elec & Electronics Engrs; Am Geophys Union. *Res:* Ionospheric physics; radio communications; radar astronomy; electrical circuits; computer organization. *Mailing Add:* Dept Elec Eng Stanford Univ Durand 347 Stanford CA 94305

MANNING, M(ELVIN) L(ANE), b Miller, SDak, Nov 26, 00; m 41; c 2. POWER ENGINEERING. *Educ:* SDak State Col, BS, 27; Univ Pittsburgh, MS & cert, 36. *Hon Degrees:* DrEng, SDak State Univ, 78. *Prof Exp:* Motor design engr, Westinghouse Elec Corp, 28-32, res engr, High Voltage Lab, 36-42; instr math, Univ Pittsburgh, 32-36; assoc prof elec eng, Ill Inst Technol, 42-43 & Cornell Univ, 43-45; chief engr, Kuhlman Elec Co, 45-49; res engr, McGraw Edison Co, 49-59; dean eng, 59-66, prof, 59-72, EMER PROF ELEC ENG, S DAK STATE UNIV, 72-; CONSULT, CHASE-FOSTER DIV, KEENE CORP, RI, 71- *Concurrent Pos:* mem transformer comt, Inst Elec & Electronics Engrs, 73- *Mem:* Am Soc Eng Educ; Am Soc Testing & Mat; fel Inst Elec & Electronics Engrs; Sigma Xi. *Res:* Transformer insulation; electrical insulation, silicones; application to dry-type transformers; power engineering, transformers; electrical insulation. *Mailing Add:* Dept Elec Eng Harding Hall SDak State Univ Brookings SD 57007

MANNING, MAURICE, b Loughrea, Ireland, Apr 10, 37; m 65; c 3. BIOCHEMISTRY & MOLECULAR BIOLOGY, CHEMISTRY. *Educ:* Nat Univ Ireland, BSc, 57, MSc, 58, DSc, 74; Univ London, PhD(chem), 61. *Hon Degrees:* Dr, Univ Gdansk, Poland, 87. *Prof Exp:* Res assoc biochem, Med Col, Cornell Univ, 61-64; res assoc, Rockefeller Univ, 64-65; asst prof, McGill Univ, 65-69; assoc prof, 69-73, PROF BIOCHEM, MED COL OHIO, 73- *Concurrent Pos:* Fulbright travel grant, 61-64. *Mem:* Am Soc Biol Chem; AAAS; The Chem Soc; Am Chem Soc; NY Acad Sci; Sigma Xi. *Res:* Solid phase peptide synthesis; design of selective agonists and antagonists of oxytocin and vasopressin; investigation of the roles and mechanism of action of oxytocin and vasopressin with selective agonists and antagonists; peptide. *Mailing Add:* Dept Biochem & Molecular Biol Med Col Ohio CS 10008 Toledo OH 43699

MANNING, MONIS JOSEPH, b Allentown, Pa, Mar 7, 31; m 57; c 3. POLYMER CHEMISTRY, SPECTROSCOPY. *Educ:* Pa State Univ, BS, 53; Univ Cincinnati, MS, 58, PhD(org chem, molecular spectros), 60. *Prof Exp:* Chemist, Arthur D Little, Inc, 60-66; CHEMIST, POLAROID CORP, 66- *Mem:* AAAS; Am Chem Soc. *Res:* Application of chemical and physical science to plastics problem-solving; thermal analysis; analytical spectrophotometry; spectral color-matching in plastic; plastics molding; spectral properties of optical plastics. *Mailing Add:* Polaroid Corp 784 Memorial Dr Cambridge MA 02139

MANNING, PATRICK JAMES, b St Paul, Minn, Sept 4, 38; m 60; c 3. DISEASES OF LABORATORY ANIMALS, ANIMAL MODELS OF HUMAN DISEASES. *Educ:* Univ Minn, BS, 63, DVM, 65; Wake Forest Univ, MS, 70. *Prof Exp:* Intern vet med, Univ Calif, Davis, 65-66; teaching asst path & lab animal med, Bowman Gray Med Sch, Wake Forest Univ, 66-70; from asst prof to assoc prof comp path, Univ Mo-Columbia, 70-74; PROF LAB MED & PATH, MED SCH, UNIV MINN, MINNEAPOLIS, 74-; VET MED OFFICER, VET ADMIN MED CTR, MINNEAPOLIS, 80- *Concurrent Pos:* Mem, Inst Lab Animal Resources Coun, Nat Res Coun-Nat Acad Sci, 75-78; mem, animal resources rev comt, NIH, 81-85; mem, sci adv bd, Food & Drug Admin, 82-85. *Honors & Awards:* Griffin Award, Am Asn Lab Animal Sci, 85. *Mem:* Am Vet Med Asn; Am Soc Microbiol; Int Acad Path; Am Asn Pathologists; AAAS; Sigma Xi. *Res:* Comparative pathology of atherosclerosis; immunopathology of Pasteurellasis; somatic antigens of Pasteurella multocida. *Mailing Add:* Dept Lab Med & Path UMHC Univ Minn Box 351 Minneapolis MN 55455

MANNING, PHIL RICHARD, b Kansas City, Mo, May 14, 21; m 48; c 2. INTERNAL MEDICINE. *Educ:* Univ Southern Calif, AB, 45, MD, 48. *Prof Exp:* Intern, Los Angeles County Hosp, 47-48; resident internal med, Vet Admin Hosp, Van Nuys & Long Beach, Calif, 48-50; from instr to assoc prof, 54-64, dir postgrad div, 53-59, PROF MED, SCH MED, UNIV SOUTHERN CALIF, 64-, ASSOC DEAN POSTGRAD DIV, 59-, ASSOC VPRES HEALTH AFFAIRS, 79-, PAUL INGALLS HOAGLAND PROF CONTINUING MED EDUC & DIR, CONTINUING EDUC HEALTH PROF DEVELOP DEMONSTRATION CTR, 80- *Concurrent Pos:* Fel, Mayo Clin, 50-52. *Honors & Awards:* Laureate Award, Am Col Physicians, 90. *Mem:* Master Am Col Physicians. *Res:* Evaluation of teaching techniques in continuing in health sciences education; development of the community hospital and physician's office as an intramural teaching center. *Mailing Add:* KAM 317 Postgrad Div Univ Southern Calif Sch Med 1975 Zonal Ave Los Angeles CA 90033

MANNING, R DAVIS, JR, BODY FLUID VOLUMES, HYPERTENSION. *Educ:* Univ Miss, PhD(biomed eng-physiol), 73. *Prof Exp:* PROF PHYSIOL & BIOPHYSICS, MED CTR, UNIV MISS, 83- *Mailing Add:* Dept Physiol/Biophys Univ Miss Med Ctr 2500 N State St Jackson MS 39216

MANNING, R(OBERT) E(DWARD), b Williamstown, Mass, May 23, 27; m 47; c 4. CHEMICAL ENGINEERING. *Educ:* Pa State Univ, BS, 48, MS, 49, PhD(chem eng), 54. *Prof Exp:* PRES, CANNON INSTRUMENT CO, 54- *Mem:* Am Chem Soc; fel Am Soc Testing & Mat; Soc Rheology. *Res:* Distillation and viscosity measurement. *Mailing Add:* PO Box 16 State College PA 16804-0016

MANNING, RAYMOND B, b Brooklyn, NY, Oct 11, 34; m 57; c 3. INVERTEBRATE ZOOLOGY, MARINE BIOLOGY. *Educ:* Univ Miami, BS, 56, MS, 59, PhD(marine sci), 63. *Prof Exp:* Res instr, Inst Marine Sci, Univ Miami, 59-63; assoc curator Crustacea, Div Marine Invert, 63-65, curator in charge div Crustacea, 65-67, chmn dept invert zool, 67-71, CURATOR DIV CRUSTACEA, SMITHSONIAN INST, 71- *Mem:* Biol Soc Wash; Crustacean Soc. *Res:* Systematics and biology of decapod and stomatopod Crustacea. *Mailing Add:* IZ NHB 163 Smithsonian Inst Washington DC 20560

MANNING, ROBERT JOSEPH, b Kansas City, Kans, Jan 12, 20; m 49; c 4. PHYSICAL CHEMISTRY, ORGANIC CHEMISTRY. *Educ:* St Benedict's Col, Kans, BS, 43; Univ Kansas City, MS, 48. *Prof Exp:* Phys chemist, US Naval Ord Test Sta, 48-53; sr chemist, Beckman Instruments, Inc, 53-63, prod line mgr, 63-73, prin chemist, 73-86; RETIRED. *Mem:* Am Chem Soc; Optical Soc Am; Soc Appl Spectros (pres-elect, 76, pres, 77). *Res:* Ultraviolet and infrared absorption spectroscopy, especially the near infrared; reflectance spectroscopy. *Mailing Add:* PO Box 1627 Davidson NC 28036

MANNING, ROBERT M, b Perea, Ohio, Jan 20, 56. ATMOSPHERIC ELECTROMAGNETIC WAVE PROPAGATION. *Educ:* Case Western Reserve Univ, BS, 80, MS, 82, PhD(appl phys), 85. *Prof Exp:* Mem fac, phys & elec eng, Case Westrn Reserve Univ, 80-83; consult, Analex Corp, 83-84, sr scientist, 84-86; SR SCIENTIST, NASA LEWIS RES CTR, 86-; PRES, APPL RES TECHNOL, INC, 86- *Concurrent Pos:* Conf chmn, Int Soc Optical Eng, 86-87; sect chmn, Inst Elec Electronics Eng, 87-88. *Mem:* NY Acad Sci; Optical Soc Am; Inst Elec & Electronics Eng; Am Phys Soc; Int Soc Optical Eng. *Res:* Stochastic electromagnetic wave propagation in turbulent and turbid media; imaging through the atmosphere and ocean (statistical optics); adaptive optics; atmospheric thermal blooming of high-energy laser beams; microwave and optical communications link modeling and analysis; remote sensing of atmospheric propagation parameters; author of textbook, laser and image propagation in the turbid atmosphere. *Mailing Add:* 16993 Stag Thicket Lane Strongsville OH 44136

MANNING, ROBERT THOMAS, b Wichita, Kans, Oct 16, 27; m 49; c 3. MEDICINE, BIOCHEMISTRY. *Educ:* Univ Wichita, AB, 50; Univ Kans, MD, 54; Am Bd Internal Med, dipl, 62, recertified, 74. *Prof Exp:* Intern, Kans City Gen Hosp, 54-55; resident, Med Ctr, Univ Kans, 55-58, instr internal med, 58-59, assoc, 59-62, asst prof, 62-64, assoc prof internal med & biochem, 64-69, prof med & assoc dean, 69-71; prof internal med, Eastern Va Med Sch, 71-77, dean, 71-74, chmn dept, 74-77; prof med, Sch Med & dir internal med, Wesley Med Ctr, 77-90, assoc dean clin affairs, 82-88, CHMN INTERNAL MED, WESLEY MED CTR, UNIV KANSAS, WICHITA, 91- *Concurrent Pos:* Nat Inst Arthritis & Metab Dis fel, 56-58; physician Wichita Vet Admin Hosp, 77- nat consult, US Air Force, 73-78. *Mem:* Fel Am Col Physicians; Am Fedn Clin Res; Cent Soc Clin Res; Am Asn Study Liver Dis; AMA; Sigma Xi. *Res:* Liver disease; biometrics. *Mailing Add:* 1010 N Kansas Wichita KS 67214

MANNING, SHERRELL DANE, US citizen. STRUCTURAL ENGINEERING. *Educ:* Tex Tech Univ, BS, 57, PhD(eng mech), 69; Southern Methodist Univ, MS, 62. *Prof Exp:* Assoc engr, Gen Dynamics Corp, 57-58; struct engr, 58-62; lead struct engr, Vought Corp, 62-63; sr struct engr, Gen Dynamics Corp, 63-66; instr civil eng, Tex Tech Univ, 67-68; res engr, Struct Res Dept, Southwest Res Inst, 68-69; eng specialist, 69-80, SR ENG SPECIALIST, GEN DYNAMICS CORP, 80- *Mem:* Am Soc Testing & Mat. *Res:* Fatigue and fracture, structural reliability, advanced composites, computer applications, stress analysis. *Mailing Add:* Gen Dynamics Corp MZ 2846 PO Box 748 Ft Worth TX 76106

MANNING, WILLIAM JOSEPH, b Grand Rapids, Mich, June 13, 41; m 69; c 2. PHYTOPATHOLOGY BOTANY. *Educ:* Mich State Univ, BS, 63; Univ Del, MS, 65, PhD, 68. *Prof Exp:* From asst to assoc prof plant path, Suburban Exp Sta, Univ Mass, Waltham, 68-77; assoc prof, 77-81, PROF PLANT PATH, UNIV MASS, AMHERST, 81- *Mem:* AAAS; Am Phytopath Soc; Sigma Xi; Air Pollution Control Asn. *Res:* Ecology of soil-borne fungi that cause root diseases of plants; interactions between air pollutants and biological incitants of plant diseases; air pollution effects on economic plants; microbiology. *Mailing Add:* Dept Plant Path Univ Mass Amherst MA 01003

MANNINO, JOSEPH ROBERT, b Altoona, Pa, May 6, 41; m 78; c 1. ENDOCRINOLOGY, MEDICAL EDUCATION. *Educ:* Juniata Col, BS, 63; ECarolina Univ, MA, 65; Kansas City Col Osteop Med, DO, 71; Colo State Univ, PhD(endocrinol), 74. *Prof Exp:* Pvt pract, Denver, Colo, 72-77; prof family osteop med & dir med educ, Kans City Col Osteop Med, 77-80; DIR GEN PRACTICE RESIDENCY, DOCTORS HOSP, COLUMBUS, OHIO, 80- *Concurrent Pos:* Lectr, Searle & Co, 74-; dir med educ, Rocky Mt Hosp, Denver, 75-77; lectr, Sandoz Pharmaceut, 78-; prof family med, Col Osteopathic Med, Ohio Univ, 80- *Mem:* Sigma Xi; fel Am Col Gen Pract; NY Acad Sci; Endocrine Soc; fel Am Soc Colposcopy & Cervical Path. *Res:* Applied endocrinology, especially as it pertains to hypertension; metabolic disease models. *Mailing Add:* 590 Indian Mound Columbus OH 43213

MANNINO, RAPHAEL JAMES, b Brooklyn, NY, Jan 28, 47; m 74; c 3. CELL BIOLOGY, GENE TRANSFER. *Educ:* City Univ New York, BS, 67; Sch Med, Johns Hopkins Univ, PhD(biochem), 73. *Prof Exp:* Res assoc biochem, Univ Basel, Switz, 73-76; adj asst prof microbiol, Univ Med & Dent NJ, Med Sch, Rutgers Univ, 76-80; asst prof, 80-83, ASSOC PROF MICROBIOL & IMMUNOL, ALBANY MED COL, 83- *Concurrent Pos:* Sinsheimer Found scholar, 81. *Mem:* Am Soc Cell Biol; Am Soc Virol; AAAS. *Res:* Growth regulation of tumor cells; protein lipid vesicles as vaccines and high efficiency delivery vehicles to animal cells. *Mailing Add:* Dept Microbiol & Immunol Albany Med Col & 43 New Scotland Ave Albany NY 12208

MANNIS, FRED, b Boston, Mass, May 3, 37; m 61; c 2. PHYSICAL CHEMISTRY. *Educ:* Harvard Univ, AB, 58; Mass Inst Technol, PhD(phys chem), 63. *Prof Exp:* Nat Sci Found fel, Univ Col, N Wales, 63; chemist, Cent Res Dept, 63-73, SR RES CHEMIST, PLASTICS DEPT, RES & DEVELOP DIV, E I DU PONT DE NEMOURS & CO, INC, 73- *Res:* Mossbauer effect; nuclear magnetic resonance; heterogeneous catalysis; materials research; electrochemistry; cation exchange membranes; optical fibers. *Mailing Add:* 3204 Coachman Rd Wilmington DE 19803

MANNIX, EDWARD T, MEDICINE. *Educ:* Southern Ore State Col, Ashland, BS, 82; Ind Univ, Bloomington, MS, 84, PhD(human performance & exercise physiol), 87. *Prof Exp:* Assoc instr kineseology, 83-87, res assoc, 87-90, ASST SCIENTIST, PULMONARY DIV, SCH MED, IND UNIV, INDIANAPOLIS, 90-, ASST PROF, DEPT PHYSIOL & BIOPHYS, 91- *Concurrent Pos:* Res physiologist, Vet Admin Med Ctr, Indianapolis, 87-90; postdoctoral fel, DIV Pulmonary & Critical Care Med, Sch Med, Ind Univ, 87-90; co-investr, Vet Affairs Merit Rev Grant, 90-92; prin investr Biomed Res Grant, Ind Univ, 91-92. *Mem:* Am Col Sports Med; Am Fedn Clin Res; Am Physiol Soc. *Res:* Metabolism and muscle energetics in cold; effects of oral glucose-phosphate on cardiac output and oxygen deficit; author of various publications. *Mailing Add:* Dept Physiol & Biophys Sch Med Ind Univ Indianapolis IN 46202

MANNO, BARBARA REYNOLDS, b Columbus, Ohio, Mar 16, 36; m 68; c 2. TOXICOLOGY, PHARMACOLOGY. *Educ:* Otterbein Col, BS, 57; Ind Univ, Indianapolis, MS, 68, PhD(pharmacol), 70; dipl, Am Bd Forensic Toxicol, 77. *Prof Exp:* Asst prof pharmacol, Sch Pharm, Auburn Univ, 70-71; from asst prof to assoc prof, 71-80, pharmacologist, Vet Admin Hosp, 71-78, PROF PHARMACOL, MED SCH, LA STATE UNIV, SHREVEPORT, 80-, MEM GRAD FAC, 71-, DIR, CLIN TOXICOL LAB, 76 - *Concurrent Pos:* Assoc prof clin med technol, La Tech Univ, 74-79; exec dir, La Regional Poison Ctr, 77-84; consult, Schumpert Mem Hosp, 79 -; reviewer, J Anal,

Toxicol, 79 -; clin prof, Dept Med Technol, Sch Allied Health Sci, LSU Med Ctr, Shreveport, La; pres, Southwestern Asn Toxicologists, 88-89. *Mem:* AAAS; Soc Toxicol; Am Acad Forensic Sci; Am Asn Clin Chem; Am Acad Clin Toxicol; Am Soc Pharmacol & Exp Therapeut. *Res:* Cardiovascular actions of marijuana; analytical and experimental toxicology; environmental toxicology; forensic urine drug screening; human psychomotor performance and drugs. *Mailing Add:* 622 Cumberland Dr Shreveport LA 71106

MANNO, JOSEPH EUGENE, b Warren, Pa, May 5, 42; m 68; c 2. TOXICOLOGY, PHARMACOLOGY. *Educ:* Duquesne Univ, BS, 65, MS, 67; Ind Univ, Indianapolis, PhD(toxicol), 70; dipl, Am Bd Forensic Toxicol, 77. *Prof Exp:* Asst prof pharmacol & toxicol, Sch Pharm, Auburn Univ, 70-71; asst prof, 71-74, assoc prof, 74-78, PROF PHARMACOL & TOXICOL & CHIEF, SECT TOXICOL, SCH MED, LA STATE UNIV, SHREVEPORT, 78-, ASST DEAN GRAD STUDIES, 84- *Mem:* Soc Toxicol; Am Acad Forensic Sci; Am Chem Soc; Am Acad Clin Toxicol; Am Soc Pharmacol & Exp Therapeut. *Res:* Pharmacology, toxicology and pharmacokinetics of marijuana and the cannabinoids; performance pharmacology; forensic and analytical toxicology. *Mailing Add:* 622 Cumberland Dr Shreveport LA 71106

MANNWEILER, GORDON B(ANNATYNE), b Naugatuck, Conn, Dec 1, 16; m 82; c 6. METALLURGICAL ENGINEERING. *Educ:* Yale Univ, BS, 41. *Prof Exp:* Eng trainee, Eastern Co, 41-43, res engr, 47-53, dir res, 53-60 & 60-70, chief metallurgist, 70; RETIRED. *Concurrent Pos:* consult cast metals, 70-80. *Honors & Awards:* Silver Award, Am Soc Metals. *Mem:* Am Soc Metals; Am Foundrymen's Soc; fel Am Soc Testing & Mat; Rotary Int. *Res:* Production, development and control of cast aluminum, carbon and alloy steels, gray, malleable and pearlitic malleable irons; casting alloys and their processing. *Mailing Add:* 435 Hillside Ave Naugatuck CT 06770

MANNY, BRUCE ANDREW, b Dayton, Ohio, May 24, 44; m 75; c 4. LIMNOLOGY, AQUATIC ECOLOGY. *Educ:* Oberlin Col, AB, 66; Rutgers Univ, MS, 68; Mich State Univ, PhD(bot), 71. *Prof Exp:* Nat Sci Found res assoc limnol, W K Kellogg Biol Sta, Mich State Univ, 71-73; proj leader, 73-86, SR SCIENTIST, GREAT LAKES FISHERY LAB, US FISH & WILDLIFE SERV, 86- *Concurrent Pos:* Prin investr numerous res contracts, 78- *Honors & Awards:* Nat Sci Found Travel Award, Int Cong Limnol, Leningrad, 71. *Mem:* Ecol Soc Am; Am Soc Limnol & Oceanog; Int Asn Great Lakes Res; Int Asn Theoret & Appl Limnol; Am Fisheries Soc. *Res:* Ecological effects of nutrients and contaminants on animals and plants in lakes and streams; nitrogen cycle; ecosystem eutrophication; early life history of lake trout; lake trout reproduction and spawning habitat requirements in the Great Lakes. *Mailing Add:* Great Lakes Fishery Lab 1451 Green Rd Ann Arbor MI 48105

MANO, KOICHI, theoretical physics, for more information see previous edition

MANOCHA, MANMOHAN SINGH, b Sheikhupura, India, Feb 25, 35; Can citizen; m 63; c 3. MYCOLOGY, PLANT PATHOLOGY. *Educ:* Punjab Univ, India, BSc, 55, MSc, 57; Indian Agr Res Inst, New Delhi, PhD(mycol, plant path), 61. *Prof Exp:* Coun Sci & Indust Res fel, Indian Agr Res Inst, New Delhi, 61-63; Can Dept Agr grant, Univ Sask, 63-65; fel, Nat Res Coun Can, 65-66; from asst prof to assoc prof, 66-75, chmn biol, 80-85, PROF BIOL, BROCK UNIV, 75- *Concurrent Pos:* Alexander von Humboldt-Stiftung fel, Inst Plant Path, Univ Gottingen, WGermany, 71-72; Nat Sci & Eng Res Coun exchange scientist, Brazil, 78-79, Japan, 83 & WGermany, 85. *Mem:* Can Phytopath Soc; Indian Phytopath Soc; Mycol Soc Am. *Res:* Fine structure and physiology of microorganisms; study of host-parasite interaction at cellular and molecular level; cell-cell interaction; biochemistry of host specificity and recognition. *Mailing Add:* Dept Biol Sci Brock Univ St Catharines ON L2S 3A1 Can

MANOCHA, SOHAN LALL, b Sultan Pur Lodhi, India, Aug 12, 36; m 64; c 2. HISTOCHEMISTRY, NEUROANATOMY. *Educ:* Punjab Univ, India, BSc, 56, MSc, 57, PhD(biol), 61. *Prof Exp:* Lectr zool, Govt Col, Rupar, India, 61-62; Ont Cancer Res Found fel, Queen's Univ, Ont, 62-64; res assoc histochem, 64-67, from asst prof to assoc prof neurohistochem, 67-75, CHMN DIV NEUROHISTOCHEM, YERKES PRIMATE RES CTR, EMORY UNIV, 75- *Mem:* Histochem Soc; Am Asn Anat; Soc Neurosci; Int Primatol Soc. *Res:* Fields of cytology, cytogenetics, histology, histochemistry, neuroanatomy and experimental nutrition using biological material related to reproductive system, skin, nervous system and biology of malnutrition; alteration of the nervous system under the impact of experimental dietary deficiency of protein in the diets of pregnant female and multigenerational study of its impact. *Mailing Add:* 5825 Glen Ridge Dr Atlanta GA 30328

MANOGUE, WILLIAM H(ENRY), b Queens, NY, Nov 27, 26; m 51; c 5. CHEMICAL ENGINEERING. *Educ:* Cornell Univ, BChE, 49; Univ Del, PhD(chem eng), 57. *Prof Exp:* Chem engr, Chem Warfare Labs, US Army Chem Ctr, Md, 50-56; sr res chem engr, Eastern Lab, 57-69, mem sci staff, Cent Res & Develop Dept, Exp Sta, 69-86, sr res chem engr, Du Pont Polymers, 87-89, TECHNOL ASSOC, E I DU PONT DE NEMOURS & CO INC, 89- *Concurrent Pos:* Lectr, Univ Del, 57-67; vis lectr, Univ Colo, 67-68; adj prof, Univ Del, 71-; mem, Eng Found Bd, 72-84; mem bd dir, Accreditation Bd Eng & Technol, 84-87, Eng Found Projs Comt, 85- & Eng Accreditation Comt, 88- *Mem:* Am Chem Soc; fel Am Inst Chem Engrs; Catalysis Soc. *Res:* Heterogeneous catalysis; kinetics; process development. *Mailing Add:* 224 Beverly Rd Newark DE 19711

MANOHAR, MURLI, b Amritsar, India, Oct 3, 47. PHYSIOLOGY, CARDIOVASCULAR SURGERY. *Educ:* Panjab Agr Univ, BVSc, 68; Haryana Agr Univ, MVSc, 70; Univ Wis, PhD(physiol), 78. *Prof Exp:* Res assoc, Univ Wis, 78; from asst prof to assoc prof, 79-87, PROF CARDIOPULMONARY PHYSIOL, COL VET MED, UNIV ILL, 87- *Concurrent Pos:* Sr res fel, Indian Coun Agr Res, 70-73. *Honors & Awards:* Beecham Award Res Excellence, 86. *Mem:* Am Soc Vet Physiologists & Pharmacologists; Am Physiol Soc; Am Heart Asn; Sigma Xi. *Res:* Coronary physiology and pharmacology in large domestic animals; cardiopulmonary aspects of high altitude physiology; coronary circulation; exercise physiology of horse. *Mailing Add:* Dept Vet Biosci Col Vet Med Univ Ill 212 Large Animal Clin 1102 Hazelwood Dr Urbana IL 61801

MANOLSON, MORRIS FRANK, b Calgary, Alta, July 20, 59; m 85. MOLECULAR BIOLOGY. *Educ:* McGill Univ, BS, 81, PhD(cell & molecular biol), 88. *Prof Exp:* MED RES COUN CAN FEL, CARNEGIE MELLON UNIV, 88- *Mem:* Am Soc Cell Biol; Genetics Soc Am. *Res:* Genetic and biochemical analysis of yeast vacuoles. *Mailing Add:* Dept Biol Sci Carnegie Mellon Univ 4400 Fifth Ave Pittsburgh PA 15213-3890

MANOOCH, CHARLES SAMUEL, III, b Big Springs, Tex, Aug 18, 43; m 65; c 3. VERTEBRATE ZOOLOGY, FISH ECOLOGY. *Educ:* Campbell Col, BS, 66; NC State Univ, MS, 72, PhD(zool), 75. *Prof Exp:* Biologist, Fla Game & Fresh Water Fish Comn, 68-70; res asst fisheries, NC State Univ, 70-72; BIOLOGIST, NAT MARINE FISHERIES SERV, 72- *Concurrent Pos:* Adj asst prof zool, NC State Univ, 78- *Mem:* AAAS; Am Fisheries Soc; Am Soc Ichthyologists & Herpetologists. *Res:* Population physiology of fish: growth, mortality, reproduction, population dynamics; reproductive histology. *Mailing Add:* Box 164 Atlantic Beach NC 28512

MANOOGIAN, ARMEN, physics; deceased, see previous edition for last biography

MANOS, CONSTANTINE T, b White Plains, NY, Jan 2, 33; m 71; c 2. GEOLOGY, SEDIMENTOLOGY. *Educ:* City Col New York, BS, 58; Univ Ill, MS, 60, PhD(geol), 63. *Prof Exp:* Res asst, State Geol Surv, Ill, 58-63; asst prof, Plattsburgh, 63-64, New Paltz, 64-66, assoc prof, 66-70, chmn dept geol sci, 69-72, PROF GEOL, 70-, & CHMN DEPT GEOL SCI, STATE UNIV NY COL, NEW PALTZ, 86- *Concurrent Pos:* Grants-in-aid, State Univ NY Res Found, 65 & 68; NSF fels, Am Univ, 66, Hofstra Univ, N Ill Univ & Fairleigh Dickinson Univ, 73; consult, x-ray diffraction analysis of soils, clay minerals in soils. *Mem:* Fel Geol Soc Am; Sigma Xi; Soc Econ Paleontologists & Mineralogists; Nat Asn Geol Teachers. *Res:* sedimentation; stratigraphy; heavy mineral analysis; x-ray diffraction analysis of clay minerals. *Mailing Add:* Dept Geol Sci State Univ NY Col New Paltz NY 12561-1143

MANOS, PHILIP, b Thessaloniki, Greece, May 8, 28; US citizen; m 57; c 2. ORGANIC POLYMER CHEMISTRY. *Educ:* Univ Thessaloniki, BS, 51; Boston Univ, PhD(chem), 61. *Prof Exp:* USPHS fel, Boston Univ, 59-60; res chemist, 60-68, to sr res chemist, 60-75, res assoc petrochem dept, 75-84, SR RES ASSOC, POLYMER PROD DEPT, E I DU PONT DE NEMOURS & CO, INC, 84- *Mem:* Am Chem Soc; AAAS. *Res:* Synthetic and mechanistic studies in organic chemistry; polymer and membrane chemistry; petroleum additives; polymers for electronics (printed circuit boards). *Mailing Add:* 215 Waverly Rd Wilmington DE 19803

MANOS, WILLIAM P, b Milwaukee, Wis, May 2, 19; m 49; c 3. MECHANICAL ENGINEERING. *Educ:* Ill Inst Technol, BS, 50, MS, 54, PhD(mech eng), 62. *Prof Exp:* Res engr, Mech Eng, Pullman-Standard, Pullman, Inc, 50-52; res engr, Armour Res Found, Ill Inst Technol, 52-55; sr staff mem, Labs Appl Sci, Univ Chicago, 55-62; sr staff mem supv anal, Pullman-Standard, Pullman, Inc, 62-63, assoc dir res & develop, 63-65, eng mgr, 65-66, vpres res & develop, 68-78; CONSULT ENG, 78- *Honors & Awards:* Railroad Award, Am Soc Mech Engrs, 65; Arnold Stucki Award, Am Soc Mech Engrs, 90. *Mem:* Am Soc Mech Engrs; Sigma Xi. *Res:* Dynamic analysis of machines and ordnance weapons; guidance systems; missiles; rail vehicles. *Mailing Add:* 10350 S Longwood Dr Chicago IL 60643

MANOUGIAN, EDWARD, b Highland Park, Mich, Apr 11, 29; m 62; c 2. THEORETICAL BIOLOGY. *Educ:* Wayne State Univ, BS, 51; Univ Mich, MD, 55. *Prof Exp:* NIH fel biophys, Univ Calif, Berkeley, 60-62; res assoc biomed, Donner Lab, Univ Calif, Berkeley, 62-77; PVT RESEARCHER, 77- *Mem:* AAAS; Am Math Soc. *Res:* Biological rhythms. *Mailing Add:* 1517 Summit Rd Berkeley CA 94708

MANOUGIAN, MANOUG N, b Jerusalem, Palestine, Apr 29, 35; m 60; c 1. MATHEMATICS. *Educ:* Univ Tex, Austin, BA, 60, MA, 64, PhD(math), 68. *Prof Exp:* Instr math, Haigazian Col, Lebanon, 60-62, asst prof, 64-66; chmn dept, 74-84, PROF MATH, UNIV SOUTH FLA, 68- *Concurrent Pos:* Nat Sci Found grant, Univ S Fla, 71 & 77; educ grant, State Fla. *Mem:* Am Math Soc; Math Asn Am. *Res:* Analysis; differential and integral equations. *Mailing Add:* Dept of Math Univ of SFla 4202 Fowler Ave Tampa FL 33620

MANOUSAKIS, EFSTRATIOS, b Ithaca, Greece, July 11, 57. THEORETICAL CONDENSED MATTER PHYSICS, THEORETICAL LOW TEMPERATURE PHYSICS. *Educ:* Univ Athens, Greece, dipl physics, 81; Univ Ill, Urbana, MS, 83, PhD(physics), 85. *Prof Exp:* Res assoc physics, Mass Inst Technol, 85-87; res assoc, 87-88, ASSOC PROF PHYSICS, FLA STATE UNIV, 88- *Concurrent Pos:* Develop scholar award, Fla State Univ, 90. *Mem:* Am Phys Soc; Sigma Xi. *Res:* Theoretical condensed matter physics; many-body problem, superfluidity, superconductivity, magnetism and electronic correlations in solids; computational physics; many-body theory. *Mailing Add:* Dept Physics 318 Keen Bldg Fla State Univ Tallahassee FL 32306

MANOWITZ, BERNARD, b Jersey City, NJ, Mar 6, 22; m 51; c 2. CHEMICAL ENGINEERING. *Educ:* Newark Col Eng, BS, 43; Columbia Univ, MS, 47. *Hon Degrees:* DSc, Long Island Univ, 82. *Prof Exp:* Assoc chem engr reactor eng, Clinton Lab, Monsanto Chem Co, 44-46; chem engr waste processing & radiation chem, Brookhaven Nat Lab, 47-60, assoc head chem eng div, Nuclear Eng Dept, 60-62, head radiation div, 62-74, assoc chmn, 74-79, chmn dept appl sci, 79-89, SR CHEM ENGR, BROOKHAVEN NAT LAB, 89- *Concurrent Pos:* Ed, J Appl Radiation & Isotopes, 57-75. *Honors & Awards:* Radiation Indust Award, Am Nuclear Soc, 71. *Mem:* Fel Am Nuclear Soc; Am Inst Chem Engrs; Am Geophys Union. *Res:* Applied radiation; atmospheric chemistry; atmospheric physics; oceanography. *Mailing Add:* 216 Lakeview Ave E Brightwaters NY 11718

MANOWITZ, PAUL, b Monticello, NY, Dec 13, 40; m 68; c 2. BIOLOGICAL PSYCHIATRY. *Educ:* Cornell Univ, BA, 62; Brandeis Univ, PhD(biochem), 67. *Prof Exp:* Fel psychiat, NY Univ Sch Med, 67-70, instr exp psychiat, 70-72; asst prof, 72-78, ASSOC PROF PSYCHIAT, UNIV MED & DENT NJ, ROBERT WOOD JOHNSON MED SCH, 78- *Mem:* Am Soc Neurochem; Soc Biol Psychiat; Res Soc Alcoholism; Int Soc Biomed Res Alcoholism; Soc Neurosci; World Fedn Biol Psychiat. *Res:* Biochemical basis of normal and abnormal human behavior. *Mailing Add:* Dept Psychiat UMDNJ-Robert Wood Johnson Med Sch 675 Hoes Lane Piscataway NJ 08854-5635

MANRING, EDWARD RAYMOND, b Springfield, Ohio, Mar 21, 21; m 40; c 3. PHYSICS. *Educ:* Ohio Univ, BS, 44, MS, 48, PhD(physics), 52. *Prof Exp:* Res physicist, Monsanto Chem Co, 48-51; physicist, Geophys Res Directorate, Upper Air Observ, NMex, 52-60; head, Observational Physics Group, Geophys Corp Am, 60-66; prof physics, NC State Univ, 66-84; RETIRED. *Mem:* Am Phys Soc. *Res:* Nuclear physics; radio frequency spectroscopy; night sky intensity. *Mailing Add:* 1601 Dixie Trail Raleigh NC 27607

MANRIQUEZ, ROLANDO PAREDES, b Quezon City, Philippines, Apr 19, 51; US citizen. ELECTRICAL ENGINEERING. *Educ:* Am Univ, BS, 76; Cath Univ, MS, 78. *Prof Exp:* Physicist electromagnetic effects, Harry Diamond Labs, 74-82; sr elec engr, Energistics Corp, 83-84; elec engr prog mgt, Naval Sea Systs Command, 84-85; elec engr prog mgr, Defense Nuclear Agency, 90-91. *Mem:* Am Phys Soc; Inst Elec & Electronics Engrs. *Res:* Electromagnetic effects on equipment. *Mailing Add:* 6300 Stevenson Ave No 313 Alexandria VA 22304

MANS, RUSTY JAY, b Newark, NJ, Sept 30, 30; m 52; c 5. BIOCHEMISTRY, ENZYMOLOGY. *Educ:* Univ Fla, BS, 52, MS, 54, PhD(biochem), 59. *Prof Exp:* Res assoc, Enzyme Group, Biol Div, Oak Ridge Nat Lab, 59-61, biochemist, 61-64; assoc prof biochem genetics, Univ Md, 64-69; prof immunol, med microbiol & radiol, Radiation Biol Lab, 69-72, PROF BIOCHEM & MOLECULAR BIOL, UNIV FLA, 72- *Concurrent Pos:* AEC res contract, 65-; consult & dir, Univ Fla, Japan Exchange Prog. *Mem:* Am Soc Biol Chem; Am Soc Plant Physiol; Int Soc Plant Molecular Biol. *Res:* Biosynthesis of proteins and nucleic acids; mechanism of eukaryotic transcription; mode of regulatin of cytoplasmic male sterility im maize; RNA processing. *Mailing Add:* Dept Biochem & Molecular Biol Univ Fla J245 JHMHC Gainesville FL 32610

MANSBACH, CHARLES M, II, b Norfolk, Va, Aug 21, 37. EXPERIMENTAL BIOLOGY. *Educ:* Yale Univ, BA, 59; NY Univ, MD, 63. *Prof Exp:* Head, Div Gastroenterol, Naval Hosp, Portsmouth, Va, 68-70 & Vet Admin Med Ctr, Durham, NC, 70-83; assoc med, Duke Univ Med Ctr, 70-71, from asst prof to assoc prof med, 71-86, chief, Intestinal Serv, 83-86, assoc prof physiol, 84-86; PROF MED & PHYSIOL & CHIEF, DIV GASTROENTEROL, UNIV TENN, MEMPHIS, 86- *Concurrent Pos:* Res fel gastroenterol, NIH, 66-68; vis scientist, Ctr Biochem & Molecular Biol, Marseille, France & dir, Lab Dr Robert Verger, 78-79; prog specialist gastroenterol, Vet Admin, 80-83; mem, Training & Educ Comt, Am Gastroenterol Asn, 80-83; dir, Fel Training Prog Gastroenterol, Med Ctr Duke Univ, 81-86; prog chmn gastroenterol, Southern Soc Clin Invest, 82, 84, 87 & 89; mem, Spec Rev Comt, NIH, 84, 87 & 88, Spec Grants Rev Comt, 87-91 & Health Reviewers Reserve Comt, 90-94; ad hoc mem, Ment Rev Bd, Vet Admin, 86, 87 & 90; affil & mem, Grant Rev Comt, Am Heart Asn, 87-90; mem, Res Comt, Univ Tenn, Memphis, 87-90, Memphis Vet Admin Med Ctr, 87; mem, Abstract Selection Comt Nat Meeting, Am Gastroenterol Asn, 88-89. *Mem:* Am Soc Clin Invest; Am Fedn Clin Res; Am Gastroenterol Asn; AAAS; Am Soc Biol Chem & Molecular Biol; Am Physiol Soc; Am Asn Study Liver Dis; fel Am Col Gastroenterol. *Res:* Author of various publications. *Mailing Add:* Div Gastroenterol Med Group Rm 555D Univ Tenn 951 Court Ave Memphis TN 38163

MANSBERGER, ARLIE ROLAND, JR, b Turtle Creek, Pa, Oct 13, 22; m 46; c 3. SURGERY. *Educ:* Univ Md, MD, 47. *Prof Exp:* Res fel surg, Univ Md, Baltimore City, 49-50, asst, Sch Med, 53-56, instr, 56-59, from asst prof to prof surg, 59-74, dir clin res, Shock-Trauma Unit, 62-66, chief clin adv, 66-71, head div gen surg, Univ Md Hosp, 71-74; PROF SURG & CHMN DEPT, MED COL GA, 73- *Concurrent Pos:* Consult surgeon, Montebello State Hosp, Baltimore, 56-73; consult surgeon, Div Voc Rehab, State of Md, 58-73; actg chmn dept surg, Univ Md Hosp, 70-71; examr, Am Bd Surg, 73-; chief consult, Vet Admin Hosp, Augusta, Ga, 73- *Mem:* AMA; fel Am Col Surg. *Res:* Ammonia metabolism in surgical diseases; biochemical and metabolic factors in shock. *Mailing Add:* Dept of Surg Med Col of Ga 1120 15th St Augusta GA 30902

MANSELL, ROBERT SHIRLEY, b Roswell, GA, Apr 28, 38; m 65; c 3. SOIL PHYSICS. *Educ:* Univ Ga, BSA, 60, MS, 63; Iowa State Univ, PhD(agron), 68. *Prof Exp:* From asst prof to assoc prof, 68-79, PROF SOIL PHYSICS, UNIV FLA, 79- *Mem:* Soil Sci Soc Am; Am Soc Agron; Am Geophys Union. *Res:* Movement of chemicals and water through saturated-unsaturated soil; water movement in layered soils; soil and water pollution. *Mailing Add:* Dept Soil Sci Univ Fla G149 McCarty Hall Gainesville FL 32611

MANSFELD, FLORIAN BERTHOLD, b Leipzig, Ger, Mar 6, 38; m 63; c 2. ELECTROCHEMISTRY, CORROSION. *Educ:* Univ Munich, BS, 60, MS, 64, PhD(phys chem), 67. *Prof Exp:* Fel, Mass Inst Technol, 67-68; Nat Acad Sci fel, NASA, 68-69; group leader, Rockwell Int Sci Ctr, 81-85; PROF MAT SCI, UNIV SOUTHERN CALIF, 85- *Concurrent Pos:* US Sr Scientist Award, Humboldt Found, W Germany. *Honors & Awards:* Whitney Award, Nat Asn Corrosion Engrs; S Tour Award, Am Soc Testing & Mat. *Mem:* Electrochem Soc; Nat Asn Corrosion Engrs; Am Soc Testing & Mat; Int Soc Electrochem; Sigma Xi. *Res:* Theoretical and practical problems of electrochemistry and corrosion. *Mailing Add:* Dept Mat Sci VHE 714 Univ Southern Calif Los Angeles CA 90089-0241

MANSFIELD, ARTHUR WALTER, b London, Eng, Mar 29, 26; m 57; c 3. MARINE BIOLOGY. *Educ:* Cambridge Univ, BA, 47, MA, 51; McGill Univ, PhD, 58. *Prof Exp:* Meteorologist, Falkland Islands Dependencies Surv, South Georgia Island, 51, base leader & biologist, S Orkneys, 52-53; demonstr zool, McGill Univ, 54-56; scientist & dir, Arctic Biol Sta, Fisheries & Marine Serv, Environment Can, 56-80, dir, 80-87, SCIENTIST, ARCTIC BIOL STA, FISHERIES & OCEANS CAN, 80-, COORDR, 87- *Concurrent Pos:* Lectr, McGill Univ, 64-65. *Mem:* Fel Arctic Inst NAm. *Res:* Arctic marine biology, principally marine mammals. *Mailing Add:* Arctic Biol Sta Fisheries & Oceans Can 555 St Pierre Blvd Ste Anne de Bellevue PQ H9X 3R4 Can

MANSFIELD, CHARLES ROBERT, b Lewiston, Idaho, July 27, 38; m 65; c 1. PHYSICS, OPTICS. *Educ:* Ore State Univ, BS, 62; Univ Idaho, MS, 65, PhD(physics), 70. *Prof Exp:* Sr scientist, EG&G, Los Alamos Div, 73-76; STAFF MEM, LOS ALAMOS SCI LABS, UNIV CALIF, 76- *Concurrent Pos:* Fel, NASA-Nat Res Coun, Johnson Space Ctr, 69-71; res assoc, William Marsh Rice Univ, 71-73. *Res:* Applications of modern optics to problems in science and engineering; specializing in atomic and molecular physics; interferometry and metrology; integration and operation of large laser systems. *Mailing Add:* 498 Quartz St Los Alamos NM 87544

MANSFIELD, CLIFTON TYLER, b Sept 12, 36; US citizen; m 77; c 2. ANALYTICAL CHEMISTRY, ENVIRONMENTAL SCIENCE. *Educ:* Miss Col, BS, 59; Univ Fla, PhD(anal chem), 63. *Prof Exp:* Asst prof chem, Millsaps Col, 63-67; sr chemist, R J Reynolds Tobacco Co, 67-73, group leader, 73-77, sect head, 77-81, prog mgr filters & mat, 81-85, res & develop Planning, Res Dept, 85-87; SR GROUP LEADER, ANALYTICAL CHEM, TEXACO RES, 89- *Concurrent Pos:* Post doct fel, Univ Ala, Birmingham, 87-89. *Mem:* Am Soc Testing & Mat. *Res:* Analytical chemistry of sugars, high pressure liquid chromatography of natural products, environmental monitoring. *Mailing Add:* Texaco Res PO Box 1608 Port Arthur TX 77641

MANSFIELD, JOHN E, b Cleveland, Ohio, July 2, 38; m 68; c 2. TECHNICAL INTELLIGENCE, THEORETICAL PHYSICS. *Educ:* Univ Detroit, AB, 60; St Louis Univ, MS & PhL, 63; Harvard Univ, AM, 66, PhD(physics), 70. *Prof Exp:* Res fel physics, Univ Notre Dame, 68-71; res physicist, Sci Applns, Inc, 71-74; sr scientist, 75-76, chief nuclear energy & appl sci div, Defense Intelligence Agency, 76-82; asst dir theoret res, Defense Nuclear Agency, 82-84; prof staff mem, Comt Armed Serv, US House of Rep, 84-86; dir, strategic tech off, Defense Adv Res Proj Agency, 86-88, chief scientist, 88-89; prof staff mem, US Senate Armed Serv Comt, 88-91; CONSULT, 91- *Mem:* Am Phys Soc. *Res:* Internal symmetries; bootstraps; S-matrix theory; hydrodynamics; statistical physics, thermodynamics. *Mailing Add:* 1503 Brook Meade Pl Vienna VA 22182

MANSFIELD, JOHN MICHAEL, b Louisville, Ky, Nov 5, 45; m 66; c 1. IMMUNOLOGY, MICROBIOLOGY. *Educ:* Miami Univ, BA, 67, MA, 69; Ohio State Univ, PhD(microbiol), 71. *Prof Exp:* NSF trainee, 71; fel microbiol & immunol, Sch Med, Union Louisville, 71-73, asst prof microbiol & immunol,73-; AT DEPT VET SCI, LAB IMMUNOL, UNIV WIS. *Mem:* AAAS; Am Soc Microbiol; Am Soc Trop Med & Hyg. *Res:* Tumor immunobiology; immunopathology of experimental African trypanosomiasis; cellular immunology; cytogenetics. *Mailing Add:* Dept Vet Sci-Lab Immunol Univ Wis 1655 Linden Dr Madison WI 53706

MANSFIELD, JOSEPH VICTOR, b Chicago, Ill, Mar 9, 07; m 35; c 1. ORGANIC CHEMISTRY. *Educ:* Iowa State Col, BS, 31; Univ Chicago, PhD(org chem), 42. *Prof Exp:* Chemist, Chicago Steel Co, 28-29; instr physiol chem, Chicago Med Sch, 31-36; pres & res dir, Mansfield Photo Res Labs, 35-42; chief specialist photog, Air Ctr, Photog Sch, US Navy, Fla, 42-43, chief photog mate, Photo Sci Lab, Va, 43-45; dir educ, Chicago Sch Photog, 45-46; from asst prof to prof, 46-75, EMER PROF CHEM, UNIV ILL, CHICAGO CIRCLE, 75- *Mem:* AAAS; Am Chem Soc; Photog Soc Am; Am Inst Chem; Royal Photog Soc. *Res:* Photographic chemical products; analytical chemistry. *Mailing Add:* 6455 La Jolla Blvd Apt 339 La Jolla CA 92037-6644

MANSFIELD, KEVIN THOMAS, b Yonkers, NY, Mar 26, 40; m 62; c 3. CHEMICAL PATENT PRACTICE. *Educ:* Fordham Univ, BS, 62; Ohio State Univ, MS, 65, PhD(org chem), 67. *Prof Exp:* Res chemist, Silicones Res Ctr, Union Carbide Corp, NY, 67-69; prod develop chemist, Ciba-Geigy Corp, 69-71, process develop chemist, 71-74, develop chemist, Basel, Switzerland, 74-75, group leader plastics develop, 75-78, sr group leader process develop, Plastics & Additives Dir, 78-84, PATENT AGENT, CIBA-GEIGY CORP, 84- *Mem:* Am Chem Soc. *Res:* Process development; epoxies and other polymers of commercial interest; polymer antioxidants. *Mailing Add:* CIBA-GEIGY Corp Patent Corp Seven Skyline Dr Hawthorne NY 10532

MANSFIELD, LARRY EVERETT, b Seattle, Wash, Sept 8, 39; m 62, 84; c 1. MATHEMATICS. *Educ:* Whitman Col, BA, 61; Univ Wash, PhD(math), 65. *Prof Exp:* Asst prof math, Queens Col, NY, 65-74, assoc prof, 74-90; CONSULT, MACINTOSH COMPUTER, 90- *Mem:* Am Math Soc; Math Asn Am. *Res:* Differential geometry. *Mailing Add:* 71 Pacific Ave Franklin Square NY 11010

MANSFIELD, LOIS E, b Portland, Maine, Jan 2, 41. NUMERICAL ANALYSIS. *Educ:* Univ Mich, BS, 62; Univ Utah, MS, 66, PhD(math), 69. *Prof Exp:* Vis asst prof comput sci, Purdue Univ, 69-70; from asst prof to assoc prof, Univ Kans, 70-78; assoc prof math, NC State Univ, 78-79; assoc prof, 79-83, PROF APPL MATH, UNIV VA, 83- *Concurrent Pos:* Vis asst prof math, Univ Utah, 73-74; mem, Adv Panel, Comput Sci Sect, NSF, 75-78; vis scientist, Inst Comput Appln Sci & Eng, NASA Langley Res Ctr, Hampton, Va, 77, consult, 78. *Mem:* Am Math Soc; Soc Indust & Appl Math; Asn Comput Mech. *Res:* Numerical solution of partial differential equations; algorithms for parallel computers. *Mailing Add:* Dept Appl Math Univ Va Olsson Hall 111M Charlottesville VA 22903

MANSFIELD, MARC L, b Vernal, Utah, Mar 30, 55; m 77; c 3. THEORETICAL POLYMER PHYSICS, MATERIAL SCIENCE ENGINEERING. *Educ:* Univ Utah, BA, 77; Dartmouth Col PhD(chem), 81. *Prof Exp:* Postdoctoral fel chem, Colo State Univ, 81-83; asst prof mat sci, Univ Md, 83-85; SR SCIENTIST & ASSOC RES PROF CHEM, MICH MOLECULAR INST, 85- *Honors & Awards:* Presidential Young Investr Award, Nat Sci Found, 85; Award Initiatives in Res, Nat Acad Sci, 88. *Mem:* Am Chem Soc; Am Phys Soc. *Res:* Theoretical polymer sci; statistical mechanic; polymer crystallization. *Mailing Add:* 1910 W St Andrews Rd Midland MI 48640

MANSFIELD, MAYNARD JOSEPH, b Marietta, Ohio, Jan 28, 30; m 53, 75; c 2. TOPOLOGY. *Educ:* Marietta Col, BA, 52; Purdue Univ, MS, 54, PhD(math), 56. *Prof Exp:* Asst math, Purdue Univ, 52-54, statist, Statist Lab, 54-55, instr math, 56-57; asst prof, Washington & Jefferson Col, 57-60, assoc prof, 60-63; assoc prof, Ind-Purdue Univ, 63-65, asst dean, Grad Sch, 77-81, chmn dept math, 63-88, prof math, 65-88, PROF COMPUTER SCI,IND-PURDUE UNIV, FT WAYNE, 88-, DEAN, SCH ENG & TECHNOL, 90- *Mem:* Am Math Soc; Math Asn Am; Asn Comput Mach; Inst Elec & Electronic Engrs. *Res:* Environments for instruction in programming. *Mailing Add:* Sch Eng & Technol Ind-Purdue Univ Ft Wayne IN 46805-1499

MANSFIELD, RALPH, b Chicago, Ill, Aug 21, 12; m 40; c 4. ELECTRICAL ENGINEERING, MATHEMATICS. *Educ:* Univ Chicago, BS, 35, SM, 37. *Prof Exp:* Instr math, Lewis Inst Technol, 36; statistician, Cost of Living Surv, US Dept Commerce, 36; teacher, Chicago Teachers Col, 37-43; res engr, Joseph Weidenhoff, Inc, 43-45, asst chief engr, 45-46, actg chief engr, 46-47; chief engr & vpres, Auto-Test, Inc, 47-77; RETIRED. *Concurrent Pos:* Mem, Chicago Eclipse Exped, 32; instr, Armour Inst Technol, 37-40 & Ill Inst Technol, 40-; chmn dept math, Wright Jr Col, 62-65 & loop Col, 66-77. *Mem:* AAAS; Am Math Soc; Math Asn Am; Am Statist Asn; Inst Elec & Electronics Engrs; Sigma Xi. *Res:* Differential equations; population problems; internal combustion engine ignition; testing circuits; electrical measurements. *Mailing Add:* 1942 Creekside Rd Santa Rosa CA 95405

MANSFIELD, ROGER LEO, b Boston, Mass, Feb 18, 44; m 69, 87; c 1. SPACE ORNITHOLOGY. *Educ:* Univ Cinn, BS, 65; Univ Nebr, MA, 72. *Prof Exp:* Orbital analyst, Defense Meteorol Satellite Prog, 67-73; instr math, Dept Math, US Air Force Acad, 73-74; aerospace engr, Philco-Ford Corp, 74-75 & Data Dynamics Inc, 75-76; aerospace engr, 76-78, team leader, 78-84, prin engr, 84-86, SUPERVR ASTRODYNAMICS, 86- *Concurrent Pos:* Lectr astrodynamics, dept elec eng, Univ Colo, 84; owner, Astron Data Serv, 75- *Mem:* Am Astron Soc; Math Asn Am; Nat Space Soc; Int Planetarium Soc. *Res:* Publishes literature for astronomy and aerospace education. *Mailing Add:* Astron Data Serv 3922 Leisure Lane Colorado Springs CO 80917-3502

MANSFIELD, VICTOR NEIL, b Norwalk, Conn, Mar 7, 41; m 68; c 2. ASTROPHYSICS, COMPUTER SCIENCE. *Educ:* Dartmouth Col, BA, 63, MS, 64; Cornell Univ, PhD(astrophys), 72. *Prof Exp:* Res assoc astron, Cornell Univ, 71-73; from asst prof to assoc prof, 73-89, CHMN DEPT, PHYSICS & ASTRON , COLGATE UNIV, 84-, PROF, 89- *Concurrent Pos:* Vis asst prof astron, Cornell Univ & vis scientist, Nat Astron & Ionosphere Ctr, 75-76; sr systs analyst, comput sci, Digicomp Res, Ithaca, NY, 81; vpres, Odyssey Res Assocs, Ithaca, NY, 82-83; consult, numerical methods tools, Borland Int, Scotts Valley, Calif, 85-86. *Mem:* Int Astron Union; Am Astron Soc; Am Phys Soc. *Res:* Physics and philosophy, theoretical astrophysics, general relativity and cosmology, computer science. *Mailing Add:* Picnic Area Rd Burdelt NY 14818

MANSINHA, LALATENDU, b Orissa, India, July 2, 37; m 63. GEOPHYSICS, APPLIED MECHANICS. *Educ:* Indian Inst Technol, Kharagpur, BSc, 57, MTech, 59; Univ BC, PhD(geophys, physics), 63. *Prof Exp:* Fel geophys, Rice Univ, 62-65; vis lect, 65-66, from asst prof to assoc prof, 66-79, PROF GEOPHYS, UNIV WESTERN ONT, 79- *Concurrent Pos:* NASA-Nat Res Coun sr res assoc, Goddard Space Flight Ctr, Md, 70-71. *Mem:* Am Geophys Union; Seismol Soc Am; Soc Explor Geophys. *Res:* Time series analysis; electrical, seismic and gravity methods of exploration; Chandler wobble. *Mailing Add:* Dept Geophys Univ Western Ont & Middlesex Col London ON N6A 5B9 Can

MANSKE, WENDELL J(AMES), b Minneapolis, Minn, Apr 6, 24; m 47; c 3. CHEMICAL ENGINEERING. *Educ:* Univ Minn, BChemEng, 49; US Merchant Marine Acad, BS, 50. *Prof Exp:* Proj engr solvent extraction, Swift & Co, Tenn, 49-51; proj engr process develop, 3M Co, 51-54, group supvr tape develop, 54-57, group supvr chem eng res, 57-60, supvr prod develop, 60-64, proj supvr, 64-66, mgr new prod develop, Paper Prod Div, 66-73, sr new bus developer, New Bus Ventures Div, 73-78, sr prod develop specialist, Med Prod Div, Minn Mining & Mfg Co, 78-81, div scientist, Packaging Systs Div, 81-87; PRES, INTROTECH INC, 87- *Mem:* 15 United States patents. *Res:* Oxidation catalysis; combustion of lean gas mixtures; microencapsulation; paper based feedback systems for programmed education; temperature, time and humidity indicators. *Mailing Add:* 706 Birchwood Ave St Paul MN 55110

MANSKI, WLADYSLAW J, b Lwow, Poland, May 15, 15; US citizen; m 41; c 2. MICROBIOLOGY, IMMUNOCHEMISTRY. *Educ:* Univ Warsaw, PhM, 39; Univ Wroclaw, DSc, 51. *Prof Exp:* Instr anal chem, Inst Inorg & Phys Chem, Univ Warsaw, 36-39; instr, Inst Chem, Univ Lublin, 44-45; head chem lab, Inst Microbiol, Univ Wroclaw, 45-49; Rockefeller fel, US, Denmark & Sweden, 49-50; head immunochem, Inst Immunol & Exp Ther, Polish Acad Sci, 51-55, head macromolecular biochem, Inst Biochem & Biophys & head biochem lab, State Inst Hyg, 55-57; from asst prof to assoc prof, 58-75, PROF MICROBIOL, ASSIGNED TO OPHTHALMOL, COL PHYSICIANS & SURGEONS, COLUMBIA UNIV, 75- *Mem:* AAAS; Am Asn Immunol; Am Chem Soc; Harvey Soc; Brit Biochem Soc. *Mailing Add:* 10 Downing St New York NY 10014

MANSMANN, HERBERT CHARLES, JR, b Pittsburgh, Pa, Apr 11, 24; m 47; c 6. ALLERGY IMMUNOLOGY. *Educ:* Univ Pittsburgh, BS, 49; Jefferson Med Col, MD, 51; Am Bd Pediat, cert, 59, Bd Pediat Allergy, cert, 60, Am Bd Allergy & Immunol, cert, 71. *Prof Exp:* Rotating intern, St Francis Hosp, 51-52; resident pediat, Children's Hosp, Pittsburgh, 52-54; fel allergy immunol, Mass Gen Hosp, 54-55; fel microbiol, Col Med, NY Univ, 55-56; fel path, Sch Med, Univ Pittsburgh, 56-69, asst prof pediat, 62-68; ASSOC PROF MED, DIR ALLERGY & CLIN IMMUNOL & PROF PEDIAT, JEFFERSON MED COL, 68- *Concurrent Pos:* Dir pulmonary prog, Children's Heart Hosp, 70-; vis prof clin pharmacol, Children's Hosp Philadelphia, 75-76; mem & consult adv comt, Pulmonary-Allergy & Clin Immunol, Food & Drug Admin, 75-80. *Mem:* Am Acad Pediat; Am Acad Allergy; Am Col Allergy; Asn Care Asthma; Am Pediat Soc. *Res:* Management of the child with chronic and acute asthma; management of asthma with the use of therapeutic theophylline dosing. *Mailing Add:* Dept Pediat Jefferson Med Col 1025 Walnut St Philadelphia PA 19107

MANSON, ALLISON RAY, b Boonville, Mo, Jan 24, 39; m 61; c 4. CHEMICAL ENGINEERING. *Educ:* Va Polytech Inst, BS, 62, PhD(statist), 65. *Prof Exp:* Coop engr, Oak Ridge Gaseous Diffusion Plant, Tenn, 59-61; from asst prof to prof statist, NC State Univ, 65-88; EXEC DIR, KILKELLY ENVIRON ASSOC, 88- *Concurrent Pos:* VPres, Assessment Anal Assocs, Inc, 79-; treas, Kilkelly Environ Assocs, Inc, 81- *Mem:* Am Inst Chem Engrs; Am Chem Soc; Am Statist Asn; Inst Math Statist; Biomet Soc; Sigma Xi. *Res:* Design and analysis of experiments; engineering applications of statistics; response surface methodology and its application to engineering problems. *Mailing Add:* Kilkelly Environ Assoc PO Box 31265 Raleigh NC 27622

MANSON, DONALD JOSEPH, b Chewelah, Wash, Dec 10, 30; m 67; c 3. PHYSICS, ELECTRONICS. *Educ:* St Louis Univ, BA, 59, PhL, 60, MS, 63, PhD(physics), 67. *Prof Exp:* Fel physics, St Louis Univ, 66-67, asst prof, 67-69; asst prof radiol, Med Ctr, Univ Mo-Columbia, 69-71, assoc prof radiol sci & elec eng, 71-75; vpres res, Alpha Electronic Labs, 76-78, pres, 76-85; PRES/OWNER, COMPUQUICK, 86- *Concurrent Pos:* Res asst biomed comput labs, Washington Univ, 67-69; assoc investr, Space Sci Res Ctr, 70-; res coordr, Nat Inst Gen Med Sci grant, 70-; tech proj dir, USPHS grant, 71- *Mem:* Am Phys Soc; Am Asn Physics Teachers; Asn Comput Mach; Inst Elec & Electronics Engrs; Am Asn Physicists in Med. *Res:* Computer diagnosis in radiology; computer applications in medicine; image analysis; ecological systems analysis; computer modeling of biological systems; energy conservation; consumer electronics. *Mailing Add:* CompuQuick Rte 1 Box 348 New Bloomfield MO 65063-9722

MANSON, JEANNE MARIE, reproductive toxicology, teratology, for more information see previous edition

MANSON, JOSEPH RICHARD, b Petersburg, Va, Nov 6, 42; m 67; c 2. SURFACE PHYSICS. *Educ:* Univ Richmond, BS, 65; Univ Va, PhD(physics), 69. *Prof Exp:* PROF PHYSICS, CLEMSON UNIV, 81- *Honors & Awards:* Alexander von Humboldt US Sr Scientist Award, 90. *Mem:* Am Physiol Soc; Am Asn Physics Teachers. *Res:* Theoretical work in the field of scattering of low energy atoms by solid surfaces. *Mailing Add:* Dept Physics & Astron Clemson Univ 201 Kinard Hall Clemson SC 29631

MANSON, LIONEL ARNOLD, b Toronto, Ont, Dec 24, 23; nat US; m 45; c 3. IMMUNOBIOLOGY, MOLECULAR BIOLOGY. *Educ:* Univ Toronto, BA, 45, MA, 47; Wash Univ, PhD(biol chem), 49. *Prof Exp:* Nat Res Coun fel med sci, Western Reserve Univ, 49-50, from instr to sr instr microbiol, 50-54; res asst prof, Sch Med, Univ Pa, 54-66, assoc prof, 66-73; fel, Wistar Inst Anat & Biol, 54-57, assoc mem, 58-65, mem, 65-77, prof, 77-89; PROF MICROBIOL, SCH MED, UNIV PA, 74- *Concurrent Pos:* Mem grad group microbiol, Grad Fac, Univ Pa, 55-, molecular biol, 65-, immunology, 71-, chmn immunology, 74-81; sr Fulbright scholar, France, 63-64, genetics, 77-; fel, Nat Cancer Inst Israel, 71-72; sr fel biol, Sch Med, Univ PA, 87- *Mem:* Sigma Xi; Am Soc Microbiol; Am Asn Biol Chemists; Am Asn Immunol; Am Chem Soc. *Res:* Transplantation immunobiology; molecular biology and biochemical genetics of humoral and cell-mediated immunity; structure, function and biosynthesis of cellular membranes; cell membrane differentiation; interaction of sub-cellular organelles in macromolecular biosynthesis; molecular biology of the immune response induced by tumor-specific and transplantation antigens; biochemical analysis of subcellular organelles. *Mailing Add:* Dept Biol Univ Penn Philadelphia PA 19104

MANSON, NANCY HURT, CARDIOVASCULAR PHYSIOLOGY. *Educ:* Med Col Va, PhD(physiol), 80. *Prof Exp:* ASST PROF PHYSIOL, DEPT PHYSIOL, MED COL VA, 84- *Mailing Add:* 565 Cedar Run Rd Manakin Sabot VA 23103

MANSON, S(AMUEL) S(TANFORD), b Jerusalem, July 4, 19; nat US; m 46; c 5. ENGINEERING. *Educ:* Cooper Union Inst Technol, BS, 41; Univ Mich, MS, 42. *Prof Exp:* Head, Stress & Vibration Sect, NASA, 45-49, chief, Strength of Mat Br, 49-55, asst chief, Mat & Thermodyn Div, 55-56, chief, Mat & Struct Div, Lewis Res Ctr, 74; PROF MECH & AEROSPACE ENG, CASE WESTERN RESERVE UNIV, 74- *Concurrent Pos:* Lectr, Univ Mich, 52, Mass Inst Technol, 57, Pa State Univ, 58-, Univ Calif, 59, Wayne State Univ, 63-64, Ga Inst Technol, 63 & Carnegie Inst Technol, 64-65; mem comt vibration & flutter, NASA, 48-49, res adv comt mat, 56-65, res adv fatigue, 62-72, Nat Acad Sci fracture guid comt res prog ship steel fracture, Univ Ill, 63-64. *Honors & Awards:* Medal for Sci Achievement, NASA, 67; Von Karmon 2nd Prize, 74; Austen Lilligren Award, Am Die Casting Inst, 74. *Mem:* Soc Exp Anal(pres, 56); Am Soc Mech Engrs; Am Soc Test & Mat; hon mem Japan Soc Mat Sci; fel Royal Aeronaut Soc. *Mailing Add:* Dept Mech & Aerospace Eng Case Western Reserve Univ Cleveland OH 44106

MANSON, SIMON V(ERRIL), b Jerusalem, Palestine, Mar 9, 21; US citizen; m 54. MECHANICAL ENGINEERING, MATHEMATICS. *Educ:* Columbia Univ, BA, 41, BS, 43. *Prof Exp:* Aeronaut res scientist, Aircraft Powerplants, Lewis Lab, Nat Adv Comt Aeronaut, 43-53; head heat transfer & asst chief anal, Nuclear Powerplants, Res Div, Curtiss-Wright Corp, 53-59, head heat transfer & cooling, Aerospace Powerplants, Wood-Ridge Div, 59-60; chief engr, Rocket Res & Develop, Dynetics, Inc, 60-61; pres, S V Manson & Co, Inc, 61-66; head advan component technol, Space Nuclear Power, Off Advan Res & Technol, 66-70, mgr, Off Aeronaut & Space Technol Progs, Space Nuclear Systs Off, NASA Atomic Energy Comn, 70-73, chief Solar Energy Appln, Off of Appln, 73-74, prog mgr, Satellite Power Systs, Off Energy Progs, 74-81, mgr, Orbital Transfer Vehicle Engine Concept Studies & Space Platform Technol Progs, 81-84, mgr, Space Power & Energy Progs, Off Aeronaut & Space Technol, Hq, NASA, 84-85; PVT INDUST & COM CONSULT ENGR, 85- *Concurrent Pos:* Nat Adv Comt Aeronaut consult, Aircraft Reactor Exp, Oak Ridge Nat Lab, 50-51; mem, NASA-AEC Adv Rankine Ad Hoc Comt, 67-68; chmn, NASA-AEC-Dept Defense Organic Rankine Coord Comt, 67-68; mem, Proj Independence Energy Conserv Comt, 74; mem solar & mech group, Interagency Power Group, 75-85; consult, Argonne Nat Lab, 85-87. *Mem:* AAAS; Am Nuclear Soc; Am Inst Aeronaut & Astronaut; Nat Soc Prof Engrs; NY Acad Sci. *Res:* Heat transfer; fluid flow; conceptual design in aircraft, rocket and aerospace powerplants; technology of advanced components of aerospace nuclear power systems; technology of satellite power systems; orbital transfer vehicles; space platform systems; space power and energy; space environment. *Mailing Add:* 21 Squirrel Run Morris Township NJ 07960-6411

MANSON, STEVEN TRENT, b Brooklyn, NY, Dec 12, 40; m 68; c 2. ATOMIC PHYSICS, CHEMICAL PHYSICS. *Educ:* Rensselaer Polytech Inst, 61; Columbia Univ, MA, 63, PhD(physics), 66. *Prof Exp:* Nat Acad Sci-Nat Res Coun res assoc, Nat Bur Standards, 66-68; from asst prof to assoc prof, 68-75, PROF PHYSICS, GA STATE UNIV, 75- *Concurrent Pos:* Consult, Oak Ridge Nat Lab, 68-69 & 75-81; consult, Argonne Nat Lab & Pac Northwest Labs, Battelle Mem Inst, 73- & Lawrence Livermore Nat Lab, 85-; NSF Res Grants, 73- & US Army Res Off Grants, 74- *Mem:* Fel Am Phys Soc; Brit Inst Physics. *Res:* Theoretical atomic collisions; photoionization; ion-atom collisions; energy and angular distributions of ionized electrons; penetration of charged particles into matter. *Mailing Add:* Dept Physics Astron Ga State Univ Atlanta GA 30303

MANSON-HING, LINCOLN ROY, b Georgetown, Guyana, May 20, 27; US citizen; m 51; c 3. DENTISTRY. *Educ:* Tufts Univ, DMD, 48; Univ Ala, MS, 61. *Prof Exp:* From asst prof to assoc prof, 56-58, PROF DENT, SCH DENT, UNIV ALA, BIRMINGHAM, 68-, CHMN DEPT DENT RADIOL, 62- *Concurrent Pos:* USPHS grants, 58-72; consult, State Ala Cleft Palate Clins, 57-65, Vet Admin Hosp & US Air Force; Fulbright-Hays lectr, UAR, 64-65. *Mem:* AAAS; Int Asn Dent Res; Am Dent Asn; Am Acad Dent Radiol; Am Bd Oral Maxillofacial Radiol; Sigma Xi. *Res:* Dental radiology. *Mailing Add:* Univ Ala 1919 Seventh Ave Birmingham AL 35233

MANSOORI, G ALI, b Naragh, Iran, Oct 8, 40; div; c 1. CHEMICAL ENGINEERING. *Educ:* Univ Tehran, BS, 63; Univ Minn, Minneapolis, MS, 67; Univ Okla, PhD(chem eng), 69. *Prof Exp:* Postdoctoral fel, Rice Univ, 69-70; asst prof, 70-73, assoc prof, 73-77, PROF CHEM ENG, UNIV ILL, CHICAGO, 77- *Concurrent Pos:* Nat Inst Standards & Technol, 83-87, Petro Stat Labs 85-; ed-in-chief, Energy Sources Jour, Advan in Thermodyn Book Ser. *Mem:* Am Inst Chem Eng; Soc Petrol Engrs. *Res:* Thermodynamics and statistical mechanics; oil and gas recovery and processing. *Mailing Add:* Box 4348 Univ Ill Chicago M/C 110 Chicago IL 60680

MANSOUR, MOHAMED, b Dumyat, Egypt, Aug 30, 28; Swiss citizen; m 53; c 4. STABILITY THEORY, SYSTEM THEORY. *Educ:* Univ Alexandria, BSc, 51, MSc, 53; ETH, DrSc Techn, 65. *Prof Exp:* Asst elec eng, Alexandria Univ, 51-61; asst prof, Queen's Univ Can, 67-68; res asst, ETH, 61-65, lectr, 66-67, dean, 76-78, PROF AUTOMATIC CONTROL & HEAD INST AUTOMATIC CONTROL, ETH, 68- *Concurrent Pos:* Vis prof, Univ Fla, Univ Ill, Univ Calif, Berkeley & Australian Nat Univ, 74-89; chmn & co-chmn, int conferences, Int Fedn Automatic Control, 74-91, mem coun, 81-, deleg UN, 82-90; hon prof, Gansu Tech Univ, 87 & Guangxi Univ, 88; assoc fel, Third World Acad Sci, 89; mem coun, Swiss Fedn Automatic Control. *Honors & Awards:* Outstanding Serv Award, Int Fedn Automatic Control, 90. *Mem:* Fel Inst Elec & Electronic Engrs. *Res:* Stability theory; robust control; multidimensional systems; oscillations in nonlinear systems. *Mailing Add:* Eth-Zentrum Inst Fur Automatik Zurich CH 8092 Switzerland

MANSOUR, MOUSTAFA M, hepatic fibrosis, recombinant dna, for more information see previous edition

MANSOUR, TAG ELDIN, b Belkas, Egypt, Nov 6, 24; nat US; m 55; c 3. BIOCHEMISTRY, MOLECULAR BIOLOGY. *Educ:* Cairo Univ, BVSc, 46; Univ Birmingham, PhD(pharmacol), 49, DSc(biochem), 74. *Prof Exp:* Lectr, Cairo Univ, 50-51; Fulbright instr physiol, Sch Med, Howard Univ, 51-52; sr instr & res assoc pharmacol, Med Sch, Western Reserve Univ, 52-54; res assoc, Sch Med, La State Univ, 54-56, from asst prof to assoc prof, 56-61; from assoc prof to prof, 61-78, DONALD E & DELIA B BAXTER PROF PHARMACOL & CHMN DEPT, SCH MED, STANFORD UNIV, 78- *Concurrent Pos:* Res fel pharmacol, Univ Birmingham, 49-50; Commonwealth Fund fel, 65; vis prof, Univ Wis, 69-70; consult, WHO, 70; mem study sect pharmacol, USPHS, 72-75 & pharmacol sci rev comt, 85- *Honors & Awards:* Josiah Macy, Jr, Award, 81; Heath Clark lectr, London Sch Hygiene & Trop Med, 81. *Mem:* AAAS; Am Soc Pharmacol & Exp Therapeut; Am Soc Biol Chemists. *Res:* Molecular and biochemical pharmacology; enzyme regulation; action of drugs on enzyme systems; regulation of cellular metabolism; chemotherapy of helminthiasis; physiology; biochemistry and molecular biology of parasitic helminths. *Mailing Add:* Dept of Pharmacol Stanford Univ Sch of Med Stanford CA 94305

MANSPEIZER, WARREN, b New York, NY, July 16, 33; m 62; c 4. GEOLOGY. *Educ:* City Col New York, BS, 56; WVa Univ, MS, 58; Rutgers Univ, PhD(geol), 63. *Prof Exp:* Assoc prof, 69-79, assoc dean acad affairs, 70-71, chmn dept geol, 71-77, PROF GEOL, RUTGERS UNIV, NEWARK, 79- *Concurrent Pos:* Scientist, Nat Sci Found-Moroccan Study Group, 72-75; vis scientist, Israel Geol Surv, Jerusalem. *Mem:* Geol Soc Am; Am Asn Petrol Geol; Nat Asn Geol Teachers; Soc Econ Paleontol & Mineral. *Res:* Origin of the continental margins' opening of the Atlantic Ocean; stratigraphy of the Triassic basalts in Morocco and Eastern North America; Dead Sea; tectonism and sedimentation. *Mailing Add:* Dept of Geol Rutgers Univ Newark NJ 07102

MANSUR, CHARLES I(SAIAH), b Kansas City, Mo, Dec 22, 18; m 60; c 2. CIVIL ENGINEERING. *Educ:* Univ Mo, BS, 39; Harvard Univ, MS, 41. *Prof Exp:* Jr sanit engr, W K Kellogg Health Found, 39; asst sanit eng, Grad Sch Eng, Harvard Univ, 39-41; chief, Seepage Sect, Waterways Exp Sta, 41-43, chief, Design & Anal Sect, 46-54, asst chief, Embankment & Found Br, 54-57; chief, Geol, Soils & Mat Br, Miss River Comn, 57; chief engr, Luhr Bros, Inc, 58-59; supv engr, Fruin-Colnon Contracting Co, 59-61; vpres, Fruco & Assocs, 61-66, pres, 66-69; vpres, McClelland Engrs, Inc, 69-80, sr vpres, 80-84; CONSULT, 84- *Concurrent Pos:* In charge, Meramec stream pollution surv, State Dept Health, Mo, 40; vpres & chief engr, Independent Wellpoint Corp, 57-58. *Honors & Awards:* Thomas A Middlebrooks Award, Thomas Fitch Rowland Prize, James R Croes Medal, Am Soc Civil Engrs. *Mem:* Am Soc Civil Engrs. *Res:* Soil mechanics and foundation engineering; dewatering. *Mailing Add:* 1715 N Geyer Rd St Louis MO 63131

MANSUR, LOUIS KENNETH, b Lowell, Mass, Apr 18, 44; c 3. DEFECTS IN SOLIDS. *Educ:* Univ Lowell, BS, 66; Cornell Univ, MEng, 68, PhD(mat sci & nuclear eng), 74. *Prof Exp:* Reactor engr, Atomic Energy Comn, 66-67; reactor physicist, Gen Elec Co, 68-69; reactor engr, Atomic Energy Comn, 70-71; mat scientist, 74-78, GROUP LEADER, OAK RIDGE NAT LAB, 79- *Concurrent Pos:* Ed, J Nuclear Mat, 90- *Honors & Awards:* Tech Achievement Award, Martin Marietta Energy Systs, 85, 86 & 87; Mat Sci Award, Dept Energy, 83 & 90. *Mem:* Fel Am Nuclear Soc; fel Am Soc Metals; AAAS; Metall Soc; Sigma Xi. *Res:* Theory and mechanisms of radiation effects in metals and alloys; defect behavior in solids; ion beam modification; alloy design for demanding applications; diffusion in solids. *Mailing Add:* Bldg 5500 Oak Ridge Nat Lab Oak Ridge TN 37831-6376

MANTAI, KENNETH EDWARD, b Jamaica, NY, Oct 19, 42; m 62; c 2. AQUATIC BIOLOGY & PLANT PHYSIOLOGY. *Educ:* Univ Maine, BS, 64; Ore State Univ, PhD(plant physiol), 68. *Prof Exp:* Res fel plant physiol, Carnegie Inst, Washington, 68-69; res assoc, Brookhaven Nat Lab, 69-71; asst prof, 71-76, chmn dept, 80-84, PROF BIOL, STATE UNIV NY COL FREDONIA, 80- *Concurrent Pos:* NSF res grant, 76. *Mem:* Am Soc Plant Physiologists; Int Asn Great Lakes Res; Sigma Xi; Int Asn Theoret & Appl Limnol; AAAS. *Res:* Physiology of the green alga Cladophora glomerata in response to environmental conditions found in Lake Erie; physiology of aquatic plants. *Mailing Add:* Dept of Biol State Univ NY Col Fredonia NY 14063

MANTEI, ERWIN JOSEPH, b Benton Harbor, Mich, Nov 1, 38. GEOCHEMISTRY. *Educ:* St Joseph's Col, Ind, BS, 60; Univ Mo, MS, 62, PhD(geochem), 65. *Prof Exp:* Asst prof earth sci, 65-68, assoc prof geochem, 68-70, assoc prof, 70-72, PROF GEOL, SOUTHWEST MO STATE UNIV, 72- *Concurrent Pos:* Dr Carl Hasselman stipend, Mineral Inst, Univ Heidelberg, 71-72. *Mem:* AAAS; Geochem Soc; Mineral Soc Am. *Res:* Trace element distribution in minerals and rocks associated with ore deposits. *Mailing Add:* Dept Geol Southwest Mo State Univ - 901 S Nat Springfield MO 65802

MANTEI, KENNETH ALAN, b Los Angeles, Calif, Nov 22, 40; m 67; c 2. PHYSICAL CHEMISTRY. *Educ:* Pomona Col, BA, 62; Ind Univ, Bloomington, PhD(chem), 67. *Prof Exp:* Res scientist, Univ Calif, Los Angeles, 67-68; assoc prof, 68-80, chmn dept, 75-84, PROF CHEM, CALIF STATE COL, SAN BERNARDINO, 80- *Mem:* Am Chem Soc; Asn Comput Mach. *Res:* Kinetics and mechanism of gas-phase reactions; flash photolysis. *Mailing Add:* Dept Chem Calif State Col San Bernardino CA 92407

MANTEL, LINDA HABAS, b New York, NY, May 12, 39; m 66. COMPARATIVE PHYSIOLOGY. *Educ:* Swarthmore Col, BA, 60; Univ Ill, Urbana, MS, 62, PhD(physiol), 65. *Prof Exp:* Res fel living invert, Am Mus Natural Hist, 65-68; from asst prof to assoc prof, 68-85, asst provost res & grad studies, 82-87, PROF BIOL, CITY COL NEW YORK, 85-, CHMN, BIOL DEPT, 87- *Concurrent Pos:* Nat Inst Child Health & Human Develop fel, 65-66; res assoc, Am Mus Natural Hist, 68-; mem, Extra-Mural Assoc, NIH, 82; ed, Biol of Crustacea, Academic Press, 82-85 & Kaleidoscope, 84-88; mem publ comt, Columbia Univ Press, 84-88; consult, Educ Testing Serv, 85-, Dept Environ Protection, NJ, 87-; proj rev, Nat Sci Found; dir, Howard Hughes Proj Undergrad Educ in Biol Sci, 89-; vis prof, Autonomus Univ Santo Domingo, 91; chmn-elect Div Comp Physiol & Biochem, Am Soc Zoologists; Nat Counr & Nat Bd Mem, Asn Women in Sci, 91- *Honors & Awards:* Outstanding Woman Scientist Award, Asn Women in Sci, 88. *Mem:* Fel AAAS; fel NY Acad Sci; Am Soc Zool; Asn Women Sci; Crustacean Soc; Am Inst Biol Sci. *Res:* Adaptations of animals to their environment, particularly invertebrates; comparative physiology of salt and water balance, particularly in crustaceans; neuroendocrine control of adaptive mechanisms; effect of pollutants on physiology of crustaceans. *Mailing Add:* Dept Biol City Col New York New York NY 10031

MANTELL, CHARLES L(ETNAM), b Brooklyn, NY, Dec 9, 97; m 26; c 2. CHEMICAL ENGINEERING. *Educ:* City Col New York, BA & BS, 18; Columbia Univ, MA, 24, PhD(chem eng), 27. *Prof Exp:* Chem engr, Aluminum Co Am, 18-21; indust engr, Celluloid Corp, 21-22; prof chem eng, Pratt Inst, 22-37; tech dir, Wilbur B Driver Co, NJ, 34-39 & United Merchants & Mfrs, Inc, 40-47; PROF CHEM ENG, NEWARK COL ENG, 48- *Concurrent Pos:* Consult, 24-; dir res, Am Gum Importers Asn, 34-40; dir res, Netherlands Indies Labs, 36-50. *Mem:* Am Chem Soc; Electrochem Soc (vpres); fel Am Inst Chem Engrs; Am Inst Mining, Metall & Petrol Engrs. *Res:* Nonferrous metallurgy; electrochemical engineering. *Mailing Add:* 96 Powderhill Rd Montvale NJ 07645-1018

MANTELL, GERALD JEROME, b US, May 11, 23; m 48; c 4. ORGANIC POLYMER CHEMISTRY. *Educ:* Queen's Univ, Ont, BSc, 45; NY Univ, PhD(chem), 49. *Prof Exp:* Res chemist, E I du Pont de Nemours & Co, 50-58; mgr applns res, Spencer Chem Co, 58-63; mgr, Gulf Oil Corp, 63-66; dir polymer res, 66-69, dir polymer & applns res & develop, 69-71, dir res & appln develop, Plastics Div, 71-77, CHEM GROUP MGR PLANNING & COORD, AIR PROD & CHEM, INC, 77- *Mem:* Am Chem Soc; Tech Asn Pulp & Paper Indust; Am Asn Textile Chemists & Colorists; Soc Plastics Engrs. *Res:* Elastomers; free radical reactions; textile chemicals; polymers; adhesive and paper applications; polyvinyl chloride polymers and uses. *Mailing Add:* 2805 Liberty St PO Box 538 Allentown PA 18104-4735

MANTELL, M(URRAY) I(RWIN), b New York, NY, Sept 6, 17; m 44; c 4. CIVIL ENGINEERING. *Educ:* Univ Fla, BME, 40, ME, 48; Univ Southern Calif, MS, 45; Univ Tex, PhD, 52. *Prof Exp:* Pres & supt construct, Mantell Construct Co, 40-41, 46; struct engr, R A Belsham, 41; supvr sci sect, Charleston Navy Yard, 41-43; supvr alignment sect, Terminal Island Naval Shipyard, 43-45; chmn civil eng dept, 48-76 & 82-83, PROF CIVIL ENG, UNIV MIAMI, 46- *Concurrent Pos:* Vis prof, Univ Sheffield, 65-66. *Honors & Awards:* Western Elec Award, Am Soc Eng Educ. *Mem:* Am Soc Civil Engrs; Nat Soc Prof Engrs; Am Soc Eng Educ. *Res:* Structures; construction; naval architecture; engineering economics and administration; planning and zoning. *Mailing Add:* Dept Civil Eng Univ Miami Univ Station Coral Gables FL 33124

MANTEUFFEL, THOMAS ALBERT, b Woodstock, Ill, Nov 15, 48; m 72. NUMERICAL ANALYSIS. *Educ:* Univ Wis-Madison, BS, 70; Univ Ill, Urbana, MS, 72, PhD(math), 75. *Prof Exp:* Asst prof math, Emory Univ, 75-76; MEM TECH STAFF, SANDIA LABS, 76- *Mem:* Am Math Soc; Soc Indust & Appl Math. *Res:* Numerical linear algebra. *Mailing Add:* Univ Colo Denver CO 80202

MANTEY, PATRICK E(DWARD), b Ft Morgan, Colo, Dec 15, 38; m 86; c 2. COMPUTER SCIENCE, ELECTRICAL ENGINEERING. *Educ:* Univ Notre Dame, BS, 60; Univ Wis-Madison, MS, 61; Stanford Univ, PhD(elec eng), 65. *Prof Exp:* Res assoc elec eng, Stanford Univ, 65-67; mem res staff systs, IBM Res, 67-72, dept mgr comput sci, 72-84; PROF & CHMN DEPT COMPUTER ENG, UNIV CALIF, SANTA CRUZ, 84-, JACK BASKIN CHAIR, 89- *Concurrent Pos:* Lectr, Stanford Univ, 65-70 & Grad Sch Bus, 79, res assoc, dept anethesia, Med Sch, 65-67 & consult assoc prof elec eng, 78-79; consult, Philco-Ford Corp, 66-67 & IBM, 84-; mem bd dirs, Nat Ctr Geog Info & Anal, 88- *Mem:* Sr mem Inst Elec & Electronic Engrs; Sigma Xi. *Res:* Mathematical system theory; computer systems and applications; digital signal processing; computer graphics; image processing; text and document processing; computer aided engineering design; decision support systems; pictorial information systems. *Mailing Add:* Dept Comput Eng Univ Calif 391 A Appl Sci Santa Cruz CA 95064

MANTHEY, ARTHUR ADOLPH, b New York, NY, June 7, 35; m 62. PHYSIOLOGY. *Educ:* Dartmouth Col, BA, 57; Columbia Univ, PhD(physiol), 65. *Prof Exp:* Fel physiol, Col Physicians & Surgeons, Columbia Univ, 65; instr med, 66-70, asst prof, 70-74, ASSOC PROF PHYSIOL & BIOPHYS, CTR HEALTH SCI, UNIV TENN, MEMPHIS, 74- *Mem:* Sigma Xi. *Res:* Electrophysiology of muscle and nerve. *Mailing Add:* 3121 Beechrun Dr Memphis TN 38128

MANTHEY, JOHN AUGUST, b Akron, Ohio, Mar 22, 25; m 49; c 5. AGRICULTURAL BIOCHEMISTRY. *Educ:* Kent State Univ, BS, 49; Univ Wyo, MS, 52. *Prof Exp:* Bacteriologist, Cleveland City Hosp, Ohio, 50-51; anal res chemist, Strong, Cobb & Co, Inc, 51-54; biochemist, Res Labs, Eli Lilly & Co, Inc, 54-63, anal biochemist, 63-86; RETIRED. *Concurrent Pos:* Lectr, Butler Univ, 58-86. *Res:* Subcellular fractionation techniques in plant and animal tissues; mechanisms of metabolism of agrichemicals and therapeutic drug agents. *Mailing Add:* 1016 N Cumberland Rd Indianapolis IN 46229

MANTIL, JOSEPH CHACKO, b India, Apr 22, 37; US citizen; m 66; c 2. NUCLEAR MEDICINE, MAGNETIC RESONANCE. *Educ:* Poona Univ, India, BS, 56; Univ Detroit, MS, 60; Ind Univ, PhD(physics), 65; Wright State Univ, MS, 75; Univ Juarez, Mex, MD, 77. *Prof Exp:* Res physicist, Aerospace Res Labs, Wright Patterson AFB, Ohio, 64-75; resident med, Good Samaritan Hosp, 77-80; fel nuclear med, Univ Cincinnati, 80-82; staff physician, 82-85, DIR NUCLEAR MED, KETTERING MED CTR, 86- *Concurrent Pos:* Dir, Kettering-Scott Magnetic Resonance Lab, 84-; assoc prof, Dept Med, Wright State Univ, 87-, chief, Div Nuclear Med, 88- *Mem:* Am Phys Soc; Soc Nuclear Med; Soc Magnetic Resonance Med. *Res:* Clinical research in nuclear medicine including positron emission tomography, and in nuclear magnetic resonance spectroscopy; high technology assessment and implementation in medicine. *Mailing Add:* Dept Nuclear Med Kettering Med Ctr Kettering OH 45429

MANTLE, J(OHN) B(ERTRAM), b London, Eng, June 17, 19; m 42; c 2. EXPERIMENTAL MECHANICS. *Educ:* Univ Sask, BE, 41; Univ Ill, MSc, 47. *Prof Exp:* Instr mech eng, 45-47, from asst prof to prof, 47-66, head dept, 58-66, dean fac eng, 66-79, EMER PROF, UNIV REGINA, 79- *Concurrent Pos:* Vis prof, mech eng, Univ Melbourne, Australia, 64-65, Univ Queensland, 74-75. *Mem:* Am Soc Eng Educ; Soc Exp Stress Anal; fel Eng Inst Can (vpres, 63-65); Can Soc Mech Engrs (vpres, 71-73). *Res:* Photoelastic and experimental stress analysis. *Mailing Add:* PO Box 2892 Creston BC V0B 1G0 Can

MANTSCH, HENRY H, b Mediasch, Transylvania, July 30, 35; Can citizen; m 59; c 2. BIOPHYSICAL CHEMISTRY, MOLECULAR SPECTROSCOPY. *Educ:* Univ Cluj, BSc, 58, PhD(phys chem), 64. *Prof Exp:* Res scientist phys chem, Romanian Acad Sci, 58-65; res fel phys chem, Univ Munich, 66-68; res assoc molecular spectros, Nat Res Coun Can, 68-72; prof phys biochem, Babes Univ, Cluj, 73-74; Humboldt prof biophys chem, Justus Liebig Univ, Giessen, 75-77; SR RES OFFICER & HEAD MOLECULAR SPECTROS SECT, NAT RES COUN CAN, 77- *Concurrent Pos:* Adj prof biophy chem, Carleton Univ, Ottawa, 78- *Honors & Awards:* Herzberg Award, 84. *Mem:* Am Biophys Soc; fel Chem Inst Can; Can Spectros Soc; fel Royal Soc Can; Am Soc Appl Spectros. *Res:* Physical biochemistry; fourier transform infrared spectroscopy; raman spectroscopy; multinuclear magnetic resonance; use of molecular spectroscopic techniques for the study of complex biological systems. *Mailing Add:* Div Chem Nat Res Coun Can 100 Sussex Dr Ottawa ON K1A 0R6 Can

MANTSCH, PAUL MATTHEW, b Ravenna, Ohio, Apr 10, 41; m 70; c 3. CRYOGENICS, SUPERCONDUCTING MAGNETS. *Educ:* Case Inst Technol, BS, 63; Univ Ill, MS, 65, PhD(physics), 70. *Prof Exp:* Res fel, physics, German Electron Synchrotron, Hamburg, 70-73; assoc head, Res Serv Dept, 75-78, head, 78-80, dep head, Tech Support Sect, 80-84, PHYSICIST, FERMI NAT ACCELERATOR LAB, 73-, HEAD, TECH SUPPORT SECT, 84 - *Concurrent Pos:* Vis scientist, Mass Inst Technol, 73. *Mem:* Am Phys Soc. *Res:* High energy physics; hadron jets and photon interactions at 400 GeV. *Mailing Add:* Fermi Nat Accelerator Lab Box 500 Batavia IL 60510

MANUDHANE, KRISHNA SHANKAR, industrial pharmacy, for more information see previous edition

MANUEL, OLIVER K, b Wichita, Kans, Oct 13, 36; m 60; c 9. NUCLEAR CHEMISTRY, GEOCHEMISTRY. *Educ:* Kans State Col, Pittsburg, BS, 59; Univ Ark, MS, 62, PhD(chem), 64. *Prof Exp:* Nat Sci Found fel, Univ Calif, Berkeley, 64; from asst prof to assoc prof, 64-73, PROF CHEM, UNIV MO-ROLLA, 73-, CHMN DEPT, 82- *Concurrent Pos:* Res chemist, US Geol Surv, Denver, 79-80; Fulbright fel adv res abroad, Tata Inst Fundamental Res, Bombay, India, 83-84; Fulbright scholar, 83. *Honors & Awards:* Spec Recognition, Prin Investr Lunar Sci Prog, NASA, 79. *Mem:* AAAS; Am Chem Soc; Meteoritical Soc. *Res:* Cosmochemistry; noble gas mass spectrometry to study chronology of solar system; geochemistry of tellurium and the halogens; formation of the earth; element synthesis; environmental geochemistry. *Mailing Add:* Dept Chem Univ Mo Rolla MO 65401

MANUEL, THOMAS ASBURY, b Austin, Tex, Jan 3, 36; m 58, 84; c 2. SYNTHETIC INORGANIC & ORGANOMETALLIC CHEMISTRY, POLYMER CHEMISTRY. *Educ:* Ohio Wesleyan Univ, BA, 57; Harvard Univ, AM, 58, PhD(chem), 61. *Prof Exp:* Res chemist, Cent Basic Res Lab, Esso Res & Eng Co, 60-63, sr chemist, 63-66, proj leader, Enjay Polymer Labs, 66-67, res assoc & sect head, 67-73, mgr, European Elastomers Tech Serv, Esso Chem Europe, Inc, 74-75, mgr new elastomers, Elastomers Technol Div, Exxon Chem Co, 75-77; dir res & develop, Dept Polymer Chem, Air Prods & Chem, Inc, 77-79, gen sales mgr, Dept Polymer Chem, 80-84, dir polymer chem tech, 85-86, gen mgr, technol & comn develop, 87-89, GEN MGR, CORP SCI & TECHNOL CTR, AIR PROD & CHEM, INC, 89- *Mem:* Am Chem Soc; Soc Chem Indust; Tech Asn Pulp & Paper Indust. *Res:* Metal carbonyls; organometallic compounds; coordination compounds; polymers; elastomers; emulsions. *Mailing Add:* Air Prod & Chem Inc 7201 Hamilton Blvd Allentown PA 18195

MANUELIDIS, ELIAS EMMANUEL, b Constantinople, Turkey, Aug 15, 18; US citizen; m 66; c 2. PATHOLOGY, NEUROPATHOLOGY. *Educ:* Univ Munich, MD, 42. *Hon Degrees:* MA, Yale Univ, 64. *Prof Exp:* Sci asst path, Univ Munich, 43-46; lab dir, Ger Res Inst Psychiat, Max Planck Inst, 46-49; lab dir, Hosp Int Refugee Orgn, 49-50; neuropathologist, US Army Europ Command, 98th Gen Hosp, 50-51; from instr to assoc prof neuropath, 51-64, PROF PATH, SCH MED, YALE UNIV, 64-, PROF NEUROL, 72-, CUR BRAIN TUMOR REGISTRY, 58- *Concurrent Pos:* USPHS spec fel, 66-67; vis lectr, Harvard Med Sch, 66-67; consult, Fairfield, Norwich & Norwalk Hosps & NIH. *Mem:* AAAS; Am Acad Neurol; Am Asn Path & Bact; Am Asn Neuropath; NY Acad Sci; Sigma Xi. *Res:* Encephalitides and tumors; tissue cultures and transplantation of brain tumors; electron microscope utilizing transplanted tumors. *Mailing Add:* 585 Ellsworth Ave New Haven CT 06511

MANUWAL, DAVID ALLEN, b South Bend, Ind, Oct 13, 42; m 68; c 2. WILDLIFE ECOLOGY. *Educ:* Purdue Univ, BS, 66; Univ Mont, MS, 68; Univ Calif, Los Angeles, PhD(zool), 72. *Prof Exp:* Biologist, Pt Reyes Bird Observ, Bolinas, Calif, 71; from asst prof to assoc prof, 72-83, PROF WILDLIFE SCI, COL FOREST RESOURCES, UNIV WASH, 83- *Concurrent Pos:* Consult, Ctr Northern Studies, Wolcott, Vt, 73; Nat Wildlife Fedn fel, 73 & 75. *Mem:* Ecol Soc Am; Am Ornithologists Union; Wildlife Soc; Cooper Ornith Soc; Wilson Ornith Soc; Wildlife Soc; Soc Conserv Biol. *Res:* Timing and synchrony of reproduction and the social structure of seabird populations; impact of timber management on forest bird communities. *Mailing Add:* Col of Forest Resources Wildlife Sci Group Univ of Wash Seattle WA 98195

MANVEL, BENNET, b Springfield, Ill, Jan 23, 43; m 64; c 3. GRAPH THEORY, ALGORITHMS. *Educ:* Oberlin Col, BS, 64; Univ Mich, MS, 66, PhD(math), 70. *Prof Exp:* PROF MATH, COLO STATE UNIV, 70- *Concurrent Pos:* Vis scholar, Cambridge Univ, 76-77. *Mem:* Am Math Soc; Soc Indust & Appl Math; Math Asn Am. *Res:* Graph theory and computer applications, including graph reconstruction, connectivity and coloring algorithms. *Mailing Add:* Dept Math Colo State Univ Ft Collins CO 80523

MANVILLE, JOHN FIEVE, b Victoria, BC, Mar 18, 41; m 65; c 3. PLANT CHEMISTRY, INSECT CHEMISTRY. *Educ:* Univ BC, BSc, 64, PhD(chem), 68. *Prof Exp:* Res scientist, Forest Prod Lab, 68-79, RES SCIENTIST, PAC FOREST RES CTR, CAN FORESTRY SERV, 79- *Mem:* Chem Inst Can; Weed Sci Soc Am. *Res:* Extractives; herbicides. *Mailing Add:* 2220 Tashy Pl Victoria BC V8Z 4R6 Can

MANWILLER, FLOYD GEORGE, b Bailey, Iowa, May 8, 34; m 59; c 2. FOREST PRODUCTS, WOOD QUALITY. *Educ:* Iowa State Univ, BS, 61, PhD(wood tech, plant cytol), 66. *Prof Exp:* Res assoc wood anat, Iowa State Univ, 64-65; wood scientist, Southern Forest Exp Sta, US Forest Serv, 66-78; PROF FORESTRY, IOWA STATE UNIV, 78- *Concurrent Pos:* Affil mem grad fac, La State Univ, 74-79. *Honors & Awards:* Wood Award, Forest Prod Res Soc & Wood & Wood Prod Mag, 66. *Mem:* Forest Prod Res Soc; Int Asn Wood Anatomists; Soc Wood Sci & Technol; Nat Asn Cols & Teachers Agr. *Res:* Wood anatomy; physical, mechanical and chemical properties of wood. *Mailing Add:* Dept Forestry 251 Bessey Hall Iowa State Univ Ames IA 50011

MANYAN, DAVID RICHARD, b Providence, RI, Nov 9, 36; m 65; c 2. BIOCHEMISTRY. *Educ:* Bowdoin Col, AB, 58; Univ RI, MS, 65, PhD(biochem), 67. *Prof Exp:* Asst chemist, Metals & Controls Div, Tex Instruments, Inc, 59-60; NIH fel dermat, Sch Med, Univ Miami, 67-69, res scientist, 69-71, instr med, 71-72; dir, Am Heart Asn, 72-73; vis investr, Howard Hughes Med Inst, Univ Miami, 73-75; asst prof, St Frances Col, 75-78, assoc prof chem, 78-80; asst prof, 80-82, chmn dept , 82-85, ASSOC PROF BIOCHEM & NUTRIT, UNIV NEW ENG, COL OSTEOP MED, 82-, ASSOC DEAN BASIC SCI, 84- *Mem:* AAAS; Am Chem Soc; fel Am Inst Chem; Am Soc Clin Invest; NY Acad Sci; Sigma Xi. *Res:* Enzymology, protein synthesis and drug effects in mitochondria. *Mailing Add:* Dept Biochem Univ New England 11 Hills Beach Rd Biddeford ME 04005

MANYIK, ROBERT MICHAEL, b San Francisco, Calif, June 11, 28; m 52; c 2. PETROLEUM CHEMISTRY. *Educ:* Univ Calif, BS, 49; Duke Univ, PhD(org chem), 54. *Prof Exp:* res chemist, Union Carbide Corp, 53-86; RETIRED. *Concurrent Pos:* Asst prof, WVa State Col. *Mem:* Am Chem Soc. *Res:* Organo-metallic reagents; olefins polymerization; homogeneous catalysis. *Mailing Add:* 1146 Summit Dr St Albans WV 25177

MANZ, AUGUST FREDERICK, b Newark, NJ, Mar 7, 29; m 56; c 2. WELDING SAFETY & HEALTH ISSUES, INSTRUCTION LITERATURE. *Educ:* NJ Inst Technol, BSEE, 57, MSEE, 59. *Prof Exp:* Develop engr, Linde Div, Union Carbide Corp, 57-64, proj engr, 64-69, spec proj engr, 69-73, proj scientist, 73-76, asst mgr regulations technol, 76-82, mgr regulations technol, 82-85, sr engr, 85-86; PRES, A F MANZ ASSOCS, 86- *Concurrent Pos:* Chmn, NJ Sect, Am Welding Soc, 69-70, bd dirs, Nat Orgn, 80-89; adj prof, Kean Col NJ Indust Studies, 75-; dist dir, Am Welding Soc, 80-86, direct at large, 86-89; chmn, Am Welding Soc, Labeling & Safe Practices Comt, 80-, Am Nat Standards Z49.1 Comt, 84- & Nat Fire Protection Asn, 51B Comt, 84-; ed adv, Welding Design & Fabrication Magazine, 87- *Honors & Awards:* Nat Meritorious Award, Am Welding Soc, 88, Samuel Wylie Miller Mem Award, 89, William Irrgang Award, 90 & Airco Welding Award, 91. *Mem:* Am Welding Soc; Am Nat Standards Inst; Nat Fire Protection Asn. *Res:* Over 30 US patents on power supplies and other welding subjects; inventor of Hot Wire and One Knob welding process; author, co-author and contributor to over a dozen books related to welding. *Mailing Add:* 470 Whitewood Rd Union NJ 07083

MANZELLI, MANLIO ARTHUR, entomology; deceased, see previous edition for last biography

MANZER, FRANKLIN EDWARD, b Maine, Feb 28, 32; m 54; c 4. PLANT PATHOLOGY. *Educ:* Univ Maine, BS, 55; Iowa State Col, PhD(plant path), 58. *Prof Exp:* From asst prof to assoc prof, 58-66, PROF PLANT PATH, UNIV MAINE, ORONO, 66- *Mem:* Am Phytopath Soc; Potato Asn Am; Europ Asn Potato Res; Sigma Xi. *Res:* All phases of potato disease; remote sensing. *Mailing Add:* 307 Deering Hall Univ of Maine Orono ME 04469

MANZO, RENE PAUL, CELL BIOLOGY. *Educ:* NY Univ, MS, 86. *Prof Exp:* Res assoc, Dept Neuroradiol, 82-91, RES RADIOL, CORNELL MED COL, 91- *Res:* Cerebrospinal fluid physiology; nuclear magnetic resonance; hydrocephalus. *Mailing Add:* 140 W 71st New York NY 10023

MAO, CHUNG-LING, b Nanking, China, Apr 21, 36; m 66; c 2. ORGANIC CHEMISTRY, POLYMER CHEMISTRY. *Educ:* Cheng Kung Univ, Taiwan, BSc, 59; Tex Tech Univ, MS, 64; Va Polytech Inst, PhD(chem), 67. *Prof Exp:* NSF & NIH fels & res assoc org chem, Duke Univ, 67-69; res chemist org chem & polymerizations, Res Ctr, World Hq, Uniroyal Inc, 69-73, res scientist, Oxford Mgt & Res Ctr, 73-77, sr res scientist, 77-80; sr res assoc, Avery Int Res & Develop Ctr, 80-83; ACRYLIC EMULSION RES & DEVELOP MGR, AIR PROD & CHEM INC, 83- *Mem:* Am Chem Soc; Sci Res Soc Am. *Res:* Organic syntheses; carbanion chemistry; molecular rearrangements; high temperature resistant polymers; elastomers and polyurethane chemistry; polymer chemistry. *Mailing Add:* 4696 Sweetbriar Circle Emmaus PA 18049

MAO, HO-KWANG, b Shanghai, China, June 18, 41; m 68; c 3. HIGH PRESSURE GEOPHYSICS, GEOCHEMISTRY. *Educ:* Taiwan Nat Univ, BS, 63; Univ Rochester, MS, 66, PhD(geol sci), 68. *Prof Exp:* Res assoc geochem, Univ Rochester, 67-68; GEOPHYSICIST, GEOPHYS LAB, CARNEGIE INST, 70- *Concurrent Pos:* Res fel, Carnegie Inst, 68-70. *Honors & Awards:* Mineral Soc Am Award, 79; Bridgman Medal, 89; Day Award, 90. *Mem:* AAAS; Am Geophys Union; Mineral Soc Am; Am Phy Soc. *Res:* High-pressure geochemistry; geophysics and physics. *Mailing Add:* Carnegie Inst Geophys Lab 5251 Broad Branch Rd NW Washington DC 20015

MAO, IVAN LING, b Shanghai, China, Sept 16, 40; US citizen; m 67; c 2. BIOMETRICAL GENETICS, LIVESTOCK BREEDING. *Educ:* Nat Taiwan Univ, BSc, 64; Univ Guelph, MSc, 67; Cornell Univ, PhD(animal breeding), 70. *Prof Exp:* Res assoc, Cornell Univ, 70-71 & Univ Guelph, 71-72; from asst prof to assoc prof, 72-84, PROF ANIMAL SCI, MICH STATE UNIV, 84- *Concurrent Pos:* Vis prof, Sweden, 82; NATO vis prof, Italy, 84; hon prof, Repub China, 86. *Mem:* Am Dairy Sci Asn; Biomet Soc. *Res:* Livestock breeding and genetics; mating systems and application of biometrical procedures. *Mailing Add:* Dept Animal Sci Mich State Univ 102 Anthony Hall East Lansing MI 48824

MAO, JAMES CHIEH HSIA, b China, Apr 3, 28; m 58; c 1. BIOCHEMISTRY. *Educ:* Taiwan Nat Univ, BS, 52; Univ Wis, MS, 59, PhD(biochem), 63. *Prof Exp:* Sr biochemist, Abbott Labs, 63-66, assoc res fel, 66-89; RETIRED. *Mem:* AAAS; Am Chem Soc; Am Soc Biol Chemists; NY Acad Sci; Am Soc Microbiol. *Res:* Metabolism of erythromycin; mode of action of antibiotics; antiviral research; nucleic acid histochemistry; enzymology. *Mailing Add:* 134 Hamilton Rd Landenberg PA 19350

MAO, SHING, b Kiangsi, China, May 21, 35; m 60; c 5. SOLID STATE ELECTRONICS. *Educ:* Nat Taiwan Univ, BS, 57; Carnegie Inst Technol, MS, 59; Stanford Univ, PhD(microwave electronics), 63. *Prof Exp:* Asst instrumentation, inst fluid dynamics & appl math, Univ Md, 58; engr, Semiconductor Div, Westinghouse Elec Corp, 59; res asst microwave electronics, Electron Devices Res Labs, Stanford Univ, 59-62, instr physics, 61-62; sr res scientist, Raytheon Co, 62-66; sect mgr, Semiconductor Res & Develop Labs, Tex Instruments, Inc, 66-68, eng mgr, Tex Instruments Singapore (PTE) Ltd, 68-71, prog mgr, Tex Instruments Taiwan Ltd, 71-76, progs mgr, Semi Conductor Group, 76-77; vpres, UTL Corp, 77-84; chmn bd, Lite-On Corp, 84-88; CHMN BD, LITE-ON INC, 88-; DIR BD, DIODES INC, 90- *Mem:* Inst Elec & Electronics Engrs. *Res:* Transversewaves interaction in electron beam; field effect transistors; solid-state physics, particularly III-V compounds; bulk semiconductor effect; microwave solid-state devices and applications. *Mailing Add:* 1809 Weanne Dr Richardson TX 75082

MAO, SIMON JEN-TAN, b Taipei, Taiwan, Oct 10, 46; m 73; c 2. BIOCHEMISTRY. *Educ:* Fu Jen Univ, BS, 69; Univ Southwestern La, MS, 72; Baylor Col Med, PhD(biochem), 78. *Prof Exp:* Res assoc, Baylor Col Med, 78-79; asst prof biochem, Mayo Med Clin & Grad Sch, 80-; at div cardiol, Mt Sinai Sch Med, New York; AT MERRIL RES INST. *Concurrent Pos:* NIH career develop award, 80- *Res:* Structural and functional studies on human plasma lipoproteins; cause of atheriosclerosis. *Mailing Add:* Merrill Res Inst 2110 E Galbraith Rd Cincinnati OH 45215

MAPES, GENE KATHLEEN, b Oakland, Calif, Feb 21, 46; m 65; c 1. BOTANY, PALEONTOLOGY. *Educ:* Univ Ark, BS, 68; Univ Iowa, MS, 75, PhD(bot), 78. *Prof Exp:* ASSOC PROF, DEPT BOT, OHIO UNIV, ATHENS, 78- *Concurrent Pos:* Vis prof, Univ Mich, 84-85; ed, Ohio Geol Surv, Kans Geol Surv, 86-88. *Mem:* Sigma Xi; Bot Soc Am; Int Orgn Paleobotanists; AAAS; Geol Soc Am. *Res:* Earliest conifers; biology and systematics, paleoecology associated floras; Paleozoic gymnosperms. *Mailing Add:* Dept Bot Ohio Univ Athens OH 45701

MAPES, ROYAL HERBERT, b Manhattan, Kans, Nov 16, 42; m 65; c 1. PALEONTOLOGY. *Educ:* Univ Ark, BS, 66, MS, 68; Univ Iowa, PhD(geol), 77. *Prof Exp:* Geologist, Phillips Petrol Co, 72-74, explor geologist, 77-78; from asst prof to assoc prof, 78-90, PROF PALEONT, OHIO UNIV, 90- *Mem:* Soc Econ Paleontologists & Mineralogists; Paleont Res Inst; Geol Soc Am; Paleont Soc; Sigma Xi. *Res:* Paleobiology, paleoecology and biostratigraphy of Upper Paleozoic Cephalopoda; recent Nautilus. *Mailing Add:* Dept Geol Sci Ohio Univ Athens OH 45701

MAPES, WILLIAM HENRY, b Jonesboro, Ark, Dec 30, 39; wid; c 2. THERMODYNAMICS. *Educ:* Carnegie Inst Technol, BS, 61; Univ Wash, PhD(chem), 66; Univ Louisville, MS, 82. *Prof Exp:* Chemist, Gen Elec Co, 69-76, sr chemist, Major Appliance Lab, 76-80, sr chemist, 80-88, mgr eng chem, 88-89, staff scientist, Mat & Processes Lab, 89-90, STAFF ENGR, MAT & PROCESSES LAB, GEN ELEC CO, 90- *Mem:* Royal Soc Chem. *Res:* Halocarbon thermodynamics; solution thermodynamics; heat transfer. *Mailing Add:* Appl Sci & Technol Lab GE Co Appliance Park Louisville KY 40225

MAPLE, M BRIAN, b Chula Vista, Calif, Nov 20, 39; m 62. PHYSICS, PHYSICAL CHEMISTRY. *Educ:* San Diego State Col, BS & AB, 63; Univ Calif, San Diego, MS, 65, PhD, 69. *Prof Exp:* Asst res physicist, Univ Calif-San Diego, 69-75, actg assoc prof, 75-77, assoc prof, 77-81, PROF PHYSICS & RES PHYSICIST, UNIV CALIF, SAN DIEGO, 81-, DIR CTR INTERFACE & MAT SCI, 90- *Concurrent Pos:* Vis scientist, Univ Chile, Santiago, 71 & 73; asst adj prof physics, Univ Calif, San Diego, 73-75; vis prof, Inst de Fisica Jose Balseiro, Argentina, 74; assoc res physicist, Univ Calif, San Diego, 75-79, Inst Theoret Physics, Univ Calif, Santa Barbara, 80; fel, John Simon Guggenheim Mem Found, 84; vchmn, Div Condensed Matter Physics, Am Phys Soc, 86-87, chmn, 87-88; lectr, 76, 82 & 83; chmn various confs, 79-90; mem various comts, 77- *Mem:* Fel Am Phys Soc; AAAS; Am Vacuum Soc; Mat Res Soc. *Res:* Superconductivity; magnetism; properties of alloys; low temperature and high pressure physics; surface physics; catalysis; valence fluctuation and heavy Fermion. *Mailing Add:* Dept Physics 0319 Univ Calif-San Diego 9500 GilmanDr La Jolla CA 92093-0319

MAPLE, WILLIAM THOMAS, b Salem, Ohio, Aug 14, 42; m 66; c 2. VERTEBRATE ZOOLOGY, ECOLOGY. *Educ:* Miami Univ, AB, 64; Kent State Univ, MA, 68, PhD(ecol), 74. *Prof Exp:* Instr biol, Kent State Univ, 65-68; asst prof, 73-79, ASSOC PROF BIOL & DIR ECOL FIELD STA, BARD COL, 79- *Concurrent Pos:* Dir natural sci, Nantucket Maria Mitchell Asn, 84; mem bd dirs, Environ Res & Educ Orgn, Hudsonia, Ltd & Winnakee Land Trust. *Mem:* Am Soc Ichthyologists & Herpetologists; Soc Study Amphibians & Reptiles; Sigma Xi; Herpetologists League. *Res:* Vertebrate ecology and natural history; evolutionary ecology of amphibians and reptiles. *Mailing Add:* Dept Biol Bard Col Annandale-on-Hudson NY 12504

MAPLES, GLENNON, b Perkinston, Miss, Aug 24, 32; m 68. MECHANICAL ENGINEERING. *Educ:* Miss State Univ, BS, 55, MS, 61; Okla State Univ, PhD(mech eng), 67. *Prof Exp:* Mech engr, Shell Oil Co, 55-58; instr eng graphics, Miss State Univ, 58-61, asst prof mech eng, 61-62; res assoc, Univ Fla, 62-63; asst, Okla State Univ, 63-66; asst prof, 66-77, ASSOC PROF MECH ENG, AUBURN UNIV, 77- *Concurrent Pos:* NSF grant, 68-69. *Mem:* AAAS; Am Soc Mech Engrs; Am Inst Aeronaut & Astronaut. *Res:* Convection and radiation heat transfer; cavitation; water resources. *Mailing Add:* Dept Chem Eng Auburn Univ Auburn AL 36849

MAPLES, LOUIS CHARLES, underwater sound propagation; deceased, see previous edition for last biography

MAPLES, WILLIAM PAUL, b Jefferson City, Mo, Sept 1, 29; m 49; c 2. PARASITOLOGY. *Educ:* George Peabody Col Teachers, BS, 53, MA, 56; Univ Ga, PhD(parasitol), 66. *Prof Exp:* Soil engr, State Tenn Hwy Dept, 53-54; teacher, Davidson County Bd Educ, Tenn, 54-56; asst prof physics & zool, WGa Col, 56-59; res asst parasitol, Univ Ga, 59-63, res scholar, Southeastern Coop Wildlife Dis Study, 63-64, res assoc, 64-66, asst dir lab serv, 66-67; assoc prof, 67-70, PROF BIOL, W GA COL, 70- *Mem:* Am Soc Parasitol; Am Micros Soc. *Res:* Physiology, life history, embryology and taxonomy of helminths; pathobiology. *Mailing Add:* Dept Biol West Ga Col Carrollton GA 30118

MAPLES, WILLIAM ROSS, b Dallas, Tex, Aug 7, 37; m 58; c 2. PHYSICAL ANTHROPOLOGY, FORENSIC ANTHROPOLOGY. *Educ:* Univ Tex, Austin, BA, 59, MA, 62, PhD(anthrop), 67; Am Bd Forensic Anthrop, dipl, 78. *Prof Exp:* Mgr, Darajani Primate Res Sta, Kenya, 62-63; mgr, Southwest Primate Res Ctr, Nairobi, Kenya, 64-65; asst prof anthrop, Western Mich Univ, 66-68; from asst prof to assoc prof anthrop, Fla State Mus, Univ Fla, 68-78, chmn, Dept Soc Sci & assoc cur mus, 73-78, prof anthrop & cur phys anthrop, 78-86, CUR-IN-CHG, CA POUND HUMAN IDENTIFICATION LAB, FLA MUS NATURAL HIST, UNIV FLA, 86- *Concurrent Pos:* Western Mich Univ Fac Res Fund grant path of the Kenya baboon, 67-68, Univ Fla biomed sci grant, Kenya, 69-70, NSF grant, Kenya, 69-71, Univ Fla biomed sci grant, Kenya, 72-73. *Mem:* Am Asn Phys Anthrop; fel Am Anthrop Asn; fel Am Acad Forensic Sci. *Res:* Primate taxonomy and primate behavior, particularly as related to adaptation; forensic identification and trauma analysis of human skeletal remains. *Mailing Add:* CA Pound Human Identification Lab Fla Mus Natural Hist Univ Fla Gainesville FL 32611-2035

MAPOTHER, DILLON EDWARD, b Louisville, Ky, Aug 22, 21; m 46; c 3. ACADEMIC ADMINISTRATION, INTELLECTUAL PROPERTY RIGHTS. *Educ:* Univ Louisville, BS, 43; Carnegie-Mellon Univ, DSc(physics), 49. *Prof Exp:* Engr, Res Lab, Westinghouse Elec Co, 43-46; from instr to assoc prof, 49-59, dir off comput serv, 71-76, PROF PHYSICS, UNIV ILL, URBANA-CHAMPAIGN, 59-, ASSOC VCHANCELLOR RES, 76-, ASSOC DEAN, GRAD COL, 79- *Concurrent Pos:* Consult, govt labs, indust & univs, 54-; Sloan fel, 57-61, Guggenheim fel, 60-61; vis prof, Cornell Univ, 60-61. *Mem:* Fel Am Phys Soc; Am Asn Physics Teachers; AAAS; Sigma Xi. *Res:* Experimental physics of solids; low temperature physics; superconductivity; calorimetry; magnetic phase transitions; thermodynamics. *Mailing Add:* Swanlund Admin Bldg 601 E John St Champaign IL 61820

MAPP, FREDERICK EVERETT, b Atlanta, Ga, Oct 12, 10; m 63; c 2. ZOOLOGY. *Educ:* Morehouse Col, BS, 32; Atlanta Univ, MS, 34; Harvard Univ, MA, 42; Univ Chicago, PhD(zool), 50. *Prof Exp:* Instr high sch, Ga, 33-40; prof biol & head dept, Knoxville Col, 44-46; lectr, Roosevelt Col, 48-50; prof & head dept, Tenn Agr & Indust State Col, 51-52; prof, 52-73, chmn dept, 62-73, DAVID PACKARD PROF BIOL, MOREHOUSE COL, 73- *Mem:* AAAS; Am Soc Zool; Am Micros Soc; NY Acad Sci; Sigma Xi. *Res:* Regeneration; experimental morphology; transplantation; tissue culture. *Mailing Add:* 703 Waterford Rd NW Atlanta GA 30318

MAQUSI, MOHAMMAD, b Dawaimeh, Palestine, Aug 28, 44; Jordanian citizen; m 75. COMMUNICATION TRANSMISSION, CATV TRANSMISSION. *Educ:* NMex State Univ, BS, 69, MS, 71, MS & ScD(elec eng), 73. *Prof Exp:* Asst prof elec eng, Mosul Univ, Iraq, 73-76; vpres, Jordan Univ Sci, 87-90; prof elec eng, chmn dept & assoc dean eng, 76-87, VPRES, UNIV JORDAN, 90- *Concurrent Pos:* Vis assoc prof elec eng, Univ Tex, El Paso, 79-80; consult engr, Gen Tel Sylvania, 80-82 & Rogers Cables, Can, 84; vis prof elec eng, McMaster Univ, Can, 83-84. *Mem:* Fel Inst Elec & Electronic Engrs. *Res:* Modeling and performance evaluation of nonlinear communication channels; catv and satellite transmission systems; application of efficient modulation techniques; design and evaluation of digital data equalizers for nonlinear channels. *Mailing Add:* Elec Eng Dept Univ Jordan Amman Jordan

MAR, BRIAN W(AYNE), b Seattle, Wash, Aug 5, 33; m 55; c 5. SYSTEMS ENGINEERING. *Educ:* Univ Wash, BS, 55, MS, 56, PhD(chem eng), 58, MS, 67. *Prof Exp:* Asst chem eng, Univ Wash, 55-58, asst prof chem & nuclear eng, 58-59; res specialist, Boeing Co, 59-67; res assoc prof civil eng, 67-71, res prof, 71-73, assoc dean, Col Eng, 75-81, PROF CIVIL ENG, UNIV WASH, 73- *Mem:* Water Pollution Control Fedn; Am Soc Civil Engrs; Sigma Xi. *Res:* Water management; system engineering; environmental systems engineering and management; simulation gaming. *Mailing Add:* 10615 60th Ave Seattle WA 98178

MAR, HENRY Y B, ceramics engineering, for more information see previous edition

MAR, JAMES WAH, b Oakland, Calif, Mar 10, 20; m 42; c 4. AERONAUTICS. *Educ:* Mass Inst Technol, SB, 41, SM, 47, ScD(civil eng), 49. *Prof Exp:* Head struct test sect, Curtiss-Wright Corp, 41-44; engr, Struct Res Group, Mass Inst Technol, 49-50, from asst prof to prof aeronaut & astronaut, 49-90, dept head, 81-83; RETIRED. *Concurrent Pos:* Mem panel thermal protection syst, Mat Adv Bd, Nat Acad Sci-Nat Res Coun, 57-69; mem comt thermal protection aerospace vehicles, Mat Adv Bd, Nat Acad Sci, 63-65; consult, Comt Space Vehicle Struct, NASA, 63- & Res & Technol Adv Comt Mat & Struct; mem adv bd, Mat Comt Struct Design with Fibrous Composites, Nat Acad Sci-Nat Res Coun-Nat Acad Eng, 67-; chief scientist, US Air Force, 71-72, mem sci adv bd vehicles panel. *Mem:* Nat Acad Eng; Am Inst Aeronaut & Astronaut. *Res:* Structural design of flight vehicles; aerothermoelastic problems; use of new materials. *Mailing Add:* PO Box 51281 Pacific Grove CA 93950

MAR, RAYMOND W, b Seattle, Wash, Oct 22, 42; m 64; c 3. MATERIALS SCIENCE, CHEMISTRY. *Educ:* Univ Wash, BS, 64; Univ Calif, Berkeley, MS, 66, PhD(mat sci), 68. *Prof Exp:* Ministry of Defense res fel, Univ Leeds, 68-69; STAFF MEM, MAT DIV, SANDIA LABS, 70- *Mem:* Am Ceramic Soc; Sigma Xi. *Res:* Synthesis and thermochemical characterization of materials and compounds; equilibrium and kinetic studies of sublimation reactions; calorimetric studies; high temperature phase relationships. *Mailing Add:* 1190 San Rafael San Leandro CA 94577

MARA, MICHAEL KELLY, b Westport, NZ, July 17, 52; m 80; c 1. APPLIED STATISTICS. *Educ:* Univ Canterbury, BSc, 74, MSc, 76, PhD(math statist), 79. *Prof Exp:* Teaching fel math, Univ Canterbury, 77-78, scientist, Appl Math Div, Dept Sci & Indust Res, 79-80; asst prof statist, Univ Minn, 80-; AT MGT EDGE LTD, NEW ZEALAND. *Mem:* NZ Math Soc; NZ Statist Asn; Am Statist Asn. *Res:* Empirical Bayesian decision theory; nonparametric density estimation; biometrics, specifically capture-recapture methods. *Mailing Add:* Mgt Edge Ltd PO Box 12461 Thorndon Wellington New Zealand

MARA, RICHARD THOMAS, b New York, NY, Mar 18, 23; m 46; c 1. PHYSICS. *Educ:* Gettysburg Col, AB, 48; Univ Mich, MS, 50, PhD(physics), 53. *Prof Exp:* From asst prof to prof, 53-70, SAHM PROF PHYSICS, GETTYSBURG COL, 70-, CHMN DEPT, 58- *Mem:* Am Phys Soc. *Res:* Molecular structure; classical field theory. *Mailing Add:* 1550 Table Rock Rd Gettysburg PA 17325

MARABLE, NINA LOUISE, b Wilmington, NC, July 26, 39. PROTEIN & ENERGY METABOLISM, FOOD CHEMISTRY. *Educ:* Agnes Scott Col, BA, 61; Emory Univ, MS, 63; Mt Holyoke Col, PhD(chem), 67. *Prof Exp:* Instr chem, Mary Baldwin Col, 66-67; asst prof, Sweet Briar Col, 67-69; asst prof chem, Va Polytech Inst & State Univ, 69-72, asst prof human nutrit, 73-80; res chemist, Protein Nutrit Lab, USDA, 80-; ASSOC PROF, DEPT NUTRIT & FOOD SCI, UNIV TENN, KNOXVILLE. *Mem:* Am Chem Soc; Inst Food Technologists; Am Col Sports Med; Sigma Xi; Am Inst Nutrit. *Res:* Protein nutritive quality; processing effects on food quality; methods of protein quality evaluation; protein and energy metabolism in humans. *Mailing Add:* PO Box 2522 Cullowhee NC 28723

MARADUDIN, ALEXEI, b San Francisco, Calif, Dec 14, 31; m 54; c 2. THEORETICAL SOLID STATE PHYSICS. *Educ:* Stanford Univ, BS, 53, MS, 54; Bristol Univ, PhD(physics), 56. *Hon Degrees:* Prof, Univ Pierre Marie Curie, Paris, 87. *Prof Exp:* Res assoc physics, Univ Md, 56-57, res asst prof, 57-58, asst res prof, Inst Fluid Dynamics & Appl Math, 58-60; physicist, Westinghouse Res Labs, 60-65; chmn dept, 68-71, PROF PHYSICS, UNIV CALIF, IRVINE, 65- *Concurrent Pos:* Consult, Semiconductor Br, US Naval Res Lab, 58-60, Los Alamos Sci Lab, 65-67 & 83- & Gen Atomic Div, Gen Dynamics Corp, 65-71; Alexander von Humboldt US sr scientist award, 80-81. *Mem:* Fel Am Phys Soc; Sigma Xi; Optical Soc Am. *Res:* Lattice dynamics; electronic properties of solids; statistical mechanics; surface physics. *Mailing Add:* Dept Physics Univ Calif Irvine CA 92717

MARAGOUDAKIS, MICHAEL E, b Myrthios, Greece, Aug 4, 32; m 68; c 2. BIOCHEMICAL PHARMACOLOGY. *Educ:* Nat Univ Athens, BS, 56; Ore State Univ, MS, 61, PhD(biochem), 64. *Prof Exp:* Res asst biochem, Ore State Univ, 61-63, instr, 64; res assoc, Albert Einstein Med Ctr, 64-66; Ciba fel, Ciba Pharmaceut Co, 66-67, sr biochemist, Ciba-Geigy Corp, 67-69, head biochem pharmacol, 69-78, MGR BASIC BIOCHEM, CIBA- GEIGY CORP, 78-; CHMN DEPT PHARMACOL, MED SCH UNIV PATRAS, GREECE, 78- *Concurrent Pos:* Chmn bd, Nat Pharmaceut Indust, 83-87. *Mem:* Am Chem Soc; Am Soc Biol Chem; Am Soc Pharmacol & Exp Therapeut. *Res:* Intermediary metabolism and enzymology; mode of action of drugs at the molecular level; angiogenesis; basement membrane biosynthesis and degradation. *Mailing Add:* Med Sch Univ Patras Patras 26500 Greece

MARAMOROSCH, KARL, b Vienna, Austria, Jan 16, 15; nat US; m 38; c 1. VIROLOGY, ENTOMOLOGY. *Educ:* Warsaw Tech Univ, MA, 38; Columbia Univ, PhD(bot, plant path), 49. *Prof Exp:* Lectr biol & animal breeding, Agr Sch Rumania, 45-46; from asst to assoc, Rockefeller Inst, 49-60; entomologist, Boyce Thompson Inst Plant Res, 60-63, prog dir, 63-74; prof microbiol, 74-85, PROF ENTOM, RUTGERS UNIV, NEW BRUNSWICK, NJ, 85- *Concurrent Pos:* Vis prof, State Agr Univ, Wageningen, 53; del, Int Cong Microbiol, Italy, 53, Sweden, 58, Can, 62, USSR, 66, Mex, 70; vis prof, Cornell Univ, 57, Rutgers Univ, 67-68, Fordham Univ, 73, Univ Agr Sci, Bangalore, India, 79, Hokkaido Univ, Japan, 80, Fudan Univ, 82 & Justus-Liebig Univ, Giessen, Ger, 83; ed, Methods Virol, 67-, Advan Virus Res, 72-, Advan Cell Culture, 81-; virologist, Food & Agr Orgn, UN, Philippine Islands, 60, world-wide coconut dis surv, 63; coordr, US-Japan Virus-Vector Conf, Japan, 65; coordr, Invertebrate Tissue Culture, Tokyo, 74; mem exec comt, Int Comn Virus Nomenclature, 65-78; mem, Leopoldina Acad, 71-; food & fiber panel, Nat Acad Sci, 66; consult, US State Dept, Agency Int Develop, India, 67 & Int Rice Res Inst, Philippines, 67; consult, All India Cent Rice Improv Proj, Hyderabad, & Ford Found, Nigeria, 71; Fulbright distinguished prof, Yugoslavia, 72 & 78; pres, Int Conf Comp Virol, 69, 73, 77 & 82; mem trop med panel, NIH, 72-76; ed-in-chief, J NY Entom Soc, 72-83; pres, Hist Br, Tissue Culture Asn; fel, Nat Acad Sci, India, 78-; consult, Food & Agr Orgn, Sri Lanka, 81, 82 & 83, Mauritius, 85. *Honors & Awards:* Morrison Prize, NY Acad Sci, 51; Campbell Award, AAAS, 58; Ciba-Geigy Nat Award, Entomological Soc Am, 76; Waksman Award, Theobald Smith Soc, 78; Jurzykowski Award, 80; Wolf Prize, 80; Distinguished Serv Award, Am Inst Biol Sci, 83. *Mem:* Tissue Culture Asn; Harvey Soc; Soc Invertebrate Path; Am Phytopath Soc; NY Acad Sci (recording secy, 60-62, vpres, 62-63); Sigma Xi; Entom Soc; hon fel Indian Virol Soc; Int Asn Med Forest Plants (pres, 89-). *Res:* Plant pathology; insect transmission of viruses and mycoplasma-like agents; comparative virology; parasitology. *Mailing Add:* 17 Black Birch Lane Scarsdale NY 10583

MARAN, JANICE WENGERD, b Baltimore, Md, June 30, 42. ELECTROPHYSIOLOGY, NEUROPHARMACOLOGY. *Educ:* Juniata Col, BS, 64; Stanford Univ, PhD(physiol), 74. *Prof Exp:* Res asst, Stanford Univ, 64-66, res assoc, 66-69; NATO fel, Univ Bristol, 74-75; NIH fel, Med Sch, Johns Hopkins Univ, 76-77; res scientist, McNeil Labs, 77-78, sr scientist, McNeil Labs, 78-; mgr new prod planning & develop, Ayerst Lab, NY, 83-88; DIR BUS DEVELOP, PAREXEL INT, 88- *Mem:* Am Physiol Soc; Soc Neurosci; Biomed Eng Soc; Sigma Xi; NY Acad Sci; AAAS; Prof Mgt Inst. *Res:* Discovery of novel psychotropic drugs; elucidation of the mechanism of action of novel and existing central nervous system active drugs through electrophysiological techniques; techniques used include recording of single and multiple neurons, electroencephalogram, and evoked potentials. *Mailing Add:* Parexel Int One Alewife Pl Cambridge MA 02140

MARAN, STEPHEN PAUL, b Brooklyn, NY, Dec 25, 38; m 71; c 3. ASTROPHYSICS. *Educ:* Brooklyn Col, BS, 59; Univ Mich, MA, 61, PhD(astron), 64. *Prof Exp:* Astronomer-in-charge remotely controlled telescope, Kitt Peak Nat Observ, Ariz, 64-69; proj scientist for orbiting solar observs, 69, mgr, Oper Kohoutek, 73-74, head advan systs & ground observations br, 70-77, SR STAFF SCIENTIST, GODDARD SPACE FLIGHT CTR, NASA, 77- *Concurrent Pos:* Assoc ed, Earth & Extraterrestrial Sci, 69-79; ed, Astrophys Letters, 75-77, assoc ed, 77-85; sr lectr, Dept Astron, Univ Calif, Los Angeles, 76; co-investr, Hubble Space Telescope High Resolution Spectrograph, 77-; consult, Nat Geographic Soc, 79-; press officer, Am Astron Soc, 85-; contrib ed, Air & Space/Smithsonian Mag, 90- *Honors & Awards:* A Dixon Johnson Lectr in Sci Commun, Pa State Univ, 90. *Mem:* Am Astron Soc; Am Phys Soc; Royal Astron Soc; Int Astron Union; Am Geophys Union. *Res:* Comets; radio astronomy; infrared astronomy; space instrumentation; nebulae; ultraviolet astronomy. *Mailing Add:* NASA Goddard Space Flight Ctr Code 680 Greenbelt MD 20771

MARANGOS, PAUL JEROME, b Brooklyn, NY, July 2, 47; m 68; c 2. NEUROSCIENCE, NEUROPHARMACOLOGY. *Educ:* RI Col, BA, 69; Univ RI, PhD(biochem), 73. *Prof Exp:* Asst neurochem, Roche Inst Molecular Biol, 73-76; staff fel, NIMH, 76-79, biologist, 79-88; biologist, Geusia Pharm, 88-90; PRES, NEUROTHERAPEUT CORP, 90- *Concurrent Pos:* Adj prof, Dept Biochem, Sch Med, George Washington Univ, 79- *Honors & Awards:* Bennett Award, Soc Biol Psychiat, 80. *Mem:* Soc Neurosci; Am Soc Neurochem; AAAS. *Res:* Function of brain proteins, such as neuron specific enolase, and brain receptors, including adenosine receptors; development of new psychotherapeutic drugs, based on their interaction with brain receptor systems. *Mailing Add:* NeuroTherapeut Corp 11211 Sorrento Valley Rd Suite H San Diego CA 92121

MARANO, GERALD ALFRED, b Philadelphia, Pa, Feb 15, 44; m 79; c 2. TECHNICAL MANAGEMENT, TRACER STUDIES. *Educ:* LaSalle Col, BA, 66; Univ Pa, PhD(inorg anal), 71; Ga Inst Tech, PhD, 75. *Prof Exp:* Asst prof chem, Ky State Univ, 71-75; res & teaching fel, Ga Inst Technol, 75-76; sr develop chemist res & develop, Ciba-Geigy Corp, 76-80; SR RES ASSOC RES & DEVELOP, INT PAPER CO, 80- *Concurrent Pos:* Prin investr res projs, NASA, 73-75 & NSF, 74-75. *Mem:* Am Chem Soc; Int Food Tech Asn. *Res:* Hydrolic flow mechanisms of waste water lagoons and their efficiencies; heat flow calorimetry as related to process development and safety; synthesis of novel and interesting inorganic and organo metallic compounds; food package testing and shelf-life evaluations. *Mailing Add:* 7941 Terry Dr Mobile AL 36695-9562

MARANS, NELSON SAMUEL, b Washington, DC, June 5, 24; m 54; c 3. CHEMISTRY. *Educ:* George Washington Univ, BS, 44; Pa State Univ, MS, 47, PhD, 49. *Prof Exp:* Asst anal develop, Allegany Ballistics Lab, George Washington Univ, 44-45; fel & lectr, De Paul Univ, 49-50; res assoc chem invest, Allegany Ballistics Lab, Hercules Powder Co, 50-54; group leader, Org Group, Mineral Benefication Lab, Columbia Univ, 54-55; sr scientist, Westinghouse Elec Corp, 55-57; res supvr, Res Div, W R Grace & Co, 57-62, res assoc, 62-87; CONSULT, 87- *Honors & Awards:* IR 100 Award. *Mem:* Am Chem Soc; Sigma Xi. *Res:* Radiation chemistry; polymers; amino acids; organosilicon chemistry. *Mailing Add:* 12120 Kerwood Rd Silver Spring MD 20904-2816

MARANTZ, LAURENCE BOYD, b Los Angeles, Calif, July 8, 35; m 62; c 3. ANALYTICAL CHEMISTRY. *Educ:* Calif Inst Technol, BS, 57; Univ Calif, Los Angeles, PhD(org chem), 62. *Prof Exp:* Res chemist, Rocket Power Res Lab, Maremont Corp, 62-64, head org sect, 64-66; mat res & develop engr, Douglas Aircraft Missile & Space Systs Div, McDonnell Douglas Corp, 66-67; spec mem staff, Marquardt Corp, Calif, 67, spec mem advan tech staff, CCI-Marquardt Corp, 67-71, chief chemist, Med Systs Div, CCI Corp, Van Nuys, 71-73, dir chem, CCI Life Systs, 73-75, dir sci serv, 75-76, dir chem, Redy labs, 76-78; dir chem, Organon Teknika Corp, 78-81; MEM STAFF, MARANTZ CONSULT, 81- *Concurrent Pos:* Indust bldg mgt. *Mem:* AAAS; Am Chem Soc; Sigma Xi. *Res:* Organo fluorine compounds; phosphorous polymers; high temperature thermodynamics; explosives and rocket propellants; ion exchange systems; polymers; instrumental design and operation; zirconium compounds; electrochemistry; medical equipment; analytical chemistry. *Mailing Add:* 3447 Alana Dr Sherman Oaks CA 91403

MARANVILLE, JERRY WESLEY, b Hutchinson, Kans, Sept 21, 40; m 67; c 3. AGRONOMY. *Educ:* Colo State Univ, BS, 62, MS, 64; Kans State Univ, PhD(agron), 67. *Prof Exp:* From asst prof to assoc prof, 67-80, PROF AGRON, UNIV NEBR, LINCOLN, 80- *Mem:* Crop Sci Soc Am; Am Soc Agron. *Res:* Crop physiology and protein biochemistry; development of grain sorghum strains which are high in mineral uptake and utilization efficiency. *Mailing Add:* Dept Agron Univ Nebr Lincoln NE 68583

MARANVILLE, LAWRENCE FRANK, b Utica, Kans, Jan 6, 19; m 45; c 3. PULP CHEMISTRY. *Educ:* Phillips Univ, BA, 40; State Col Wash, MS, 42; Univ Chicago, PhD(phys chem), 49. *Prof Exp:* Anal chemist, Aluminum Co Am, 42-45, res chemist, 49-51; res chemist, Olympic Res Div, Rayonier Inc, 51-59, group leader, ITT Rayonier Inc, 60-72, asst to res supvr, 72-76, technol assessment specialist, 77-81; RETIRED. *Mem:* AAAS; Am Chem Soc; Tech Asn Pulp & Paper Indust. *Res:* Raman spectroscopy; medium strong electrolytes; electrical conductivity of molten salts; lignin and tannin chemistry; ultraviolet and infrared spectrophotometry; chemical cellulose; pulp chemistry. *Mailing Add:* 1128 Harvard Ave Shelton WA 98584

MARASCIA, FRANK JOSEPH, b New York, NY, Aug 6, 28; m 51; c 3. ORGANIC CHEMISTRY. *Educ:* Bucknell Univ, BS, 52; Univ Maine, MS, 54; Univ Del, PhD(chem), 58. *Prof Exp:* Control chemist, M W Kellogg Co, Pullman, Inc, 52; asst to sales & tech supvr, Southern Dist, Bound Brook Labs, Am Cyanamid Co, 54, res chemist, 54-55; res supvr org chem dept, Jackson Labs, NJ, 55-57, res chemist, 57-63, asst to sales, 64, tech supvr southern dist, 64-65, res supvr, Jackson Labs, 65-66, sales supvr southern district, NC, 66-67, asst mgr chem, Org Chem Div, 67-70, asst dir tech lab, 70-73, RES MGR, CHEM, DYES & PIGMENTS DEPT, SPEC CHEM & PRODS DIV, E I DU PONT DE NEMOURS & CO, INC, 73- *Mem:* AAAS; Am Asn Textile Chem & Colorists; Am Chem Soc; Tech Asn Pulp & Paper Indust. *Res:* Synthetic organic chemistry; sulfur and nitrogen compounds; textile chemicals in intermediates; paper chemicals. *Mailing Add:* 2342 Oakwood Way SE Smyrna GA 30080

MARATHAY, ARWIND SHANKAR, b Bombay, India, Dec 11, 33; m 63; c 1. OPTICAL PHYSICS. *Educ:* Univ Bombay, BSc, 54; Imp Col, Univ London, dipl, 56; Univ London, MSc, 57; Boston Univ, PhD(physics), 63. *Prof Exp:* Fel physics, Boston Univ, 63-64; sr scientist optical physics, Tech Opers, Mass, 64-69; assoc prof, 69-80, PROF OPTICAL SCI, UNIV ARIZ, 80- *Concurrent Pos:* Fel mech eng, Univ Pa, 66-67; consult, Lincoln Labs, Ariz & Tech Opers, Mass, 69-; vis prof, Indian Inst Sci, Bangalore, 75-76 & Sch Optom, Univ Ala, Birmingham, 78. *Mem:* Optical Soc Am; Am Asn Physics Teachers. *Res:* Physical optics; coherence theory; partial polarization; electro-optic light modulators and scanners; optical bistability; optical computing. *Mailing Add:* Dept Optical Sci Univ Ariz Tucson AZ 85721

MARAVETZ, LESTER L, b Cresco, Iowa, May 6, 37; m 62; c 4. AGRICULTURAL CHEMISTRY. *Educ:* Loras Col, BS, 59; Creighton Univ, MS, 61; Loyola Univ, PhD(org chem), 65. *Prof Exp:* Res chemist, Esso Res & Eng Co, 65-70; sr res chemist, Mobil Chem Co, 70-79, assoc chemist, 80, agr chem res, 70-81; RES ASSOC, HERBICIDE RES, FMC CORP, 81- *Mem:* Am Chem Soc. *Mailing Add:* 843 Carleton Rd Westfield NJ 07090

MARAVOLO, NICHOLAS CHARLES, b Chicago, Ill, Dec 4, 40. PLANT MORPHOGENESIS. *Educ:* Univ Chicago, BS, 62, MS, 64, PhD(bot), 66. *Prof Exp:* from asst prof to assoc prof, 66-83, PROF BOT, LAWRENCE UNIV, 83- *Concurrent Pos:* Consult, State Wis, 72-73 & East-Cent Wis Regional Planning Comt, 73-80. *Mem:* Bot Soc Am; Am Asn Plant Physiol; Am Bryol & Lichenological Soc. *Res:* Plant growth and development; biochemical changes associated with differentiation in lower green plants; hormonal physiology of development in bryophytes. *Mailing Add:* Dept of Biol Lawrence Univ Appleton WI 54912

MARBAN, EDUARDO, b Cuba, June 27, 54; US citizen. CARDIOLOGY. *Educ:* Wilkes Col, BS, 74; Yale Univ, MD, 80, PhD(physiol), 81. *Prof Exp:* Res fel physiol, Sch Med, Yale Univ, 80-81; intern med, Johns Hopkins Hosp, 81-84; res assoc physiol, Sch Med, Univ Md, 83-85; res fel cardiol, Johns Hopkins Hosp, 84-85; asst prof med, 85-88, ASSOC PROF MED, JOHNS HOPKINS UNIV, 88- *Concurrent Pos:* Ed, Yale J Biol & Med, 76-79; lectr, Yale Univ, 77-81, vis fel, 78 & co-instr, Sch Med, 79; attend physician, Johns Hopkins Univ Sch Med Basic Med Clerkship, 85-86 & Adv Med Clerkship, 87; mem, Cardiovasc & Pulmonary Study Sect, NIH, 87; prin investr, NIH, 86- *Mem:* Biophys Soc; Cardiac Electrophysiol Soc; Am Fed Clin Res; AAAS; Int Soc Heart Res; Sigma Xi; Am Heart Asn. *Res:* Calcium metabolism in heart; basic cardiac electrophysiology; calcium channels; cardiac arrhythmias; ischemia and reperfusion. *Mailing Add:* 6902 Chansory Lane University Park MD 20782-1408

MARBERRY, JAMES E(DWARD), b Carbondale, Ill, Jan 11, 31; m 62; c 2. CHEMICAL ENGINEERING. *Educ:* Purdue Univ, BSChE, 52, MSChE, 54; Univ Mich, PhD(chem eng), 60. *Prof Exp:* Res asst, Lilly Varnish Co, Ind, 53-54; sr res technologist, Mobil Oil Corp, 60-68, staff reservoir engr, Mobil Oil Can, Ltd, 68-71, sr res engr, Field Res Lab, Mobil Res & Develop Corp, 71-74, staff engr, Mobil Oil AG in Deutschland, 74-77, assoc engr, Mobil Res & Develop Corp, 77-85; RETIRED. *Mem:* Soc Petrol Engrs; Sigma Xi. *Res:* Chemical, miscible and thermal recovery of oil; simulation of laboratory and field experiments. *Mailing Add:* 4626 Dove Creek Way Dallas TX 75232

MARBLE, ALEXANDER, b Troy, Kans, Feb 2, 02; m 30; c 1. MEDICINE. *Educ:* Univ Kans, AB, 22, AM, 24; Harvard Univ, MD, 27. *Prof Exp:* Moseley traveling fel, Austria, Ger & Eng, 31-32; pres physician, Joslin Clin, Joslin Diabetes Ctr, Inc, 32-83, pres, Joslin Diabetes Found, 68-77, EMER PRES, JOSLIN DIABETES CTR, INC, 77; RETIRED. *Concurrent Pos:* Staff physician, New Eng Deaconess Hosp, 32-85; from asst clin prof to clin prof med, Harvard Med Sch, 55-68, emer prof, 68- *Honors & Awards:* Banting Medal, Am Diabetes Asn, 59. *Mem:* AAAS; Am Soc Clin Invest; emer mem Endocrine Soc; Am Diabetes Asn (pres, 58-59); Int Diabetes Fedn (hon pres, 77-). *Res:* Diabetes mellitus and carbohydrate metabolism. *Mailing Add:* 215 Badger Terr Bedford MA 01730

MARBLE, EARL R(OBERT), JR, metallurgy, for more information see previous edition

MARBLE, FRANK E(ARL), b Cleveland, Ohio, July 21, 18; m 43; c 2. PROPULSION, COMBUSTION. *Educ:* Case Inst Technol, BS, 40, MS, 42; Calif Inst Technol, AE, 47, PhD(aeronaut, math), 48. *Prof Exp:* Head, Heat Transfer & Engine Cooling Sect, Lewis Flight Propulsion Lab, Nat Adv Comt Aeronaut, Cleveland, Ohio, 42-44, Chief, Compressor & Turbine Res Br, 44-46; instr aeronaut, Calif Inst Technol, 48-49, from asst prof to prof jet propulsion & mech eng, 49-81, Richard L & Dorothy M Hayman prof mech eng & prof jet propulsion, 81-89, EMER RICHARD L & DOROTHY M HAYMAN PROF MECH ENG & JET PROPULSION, 89-; PROF & ENDOWED CHAIR MECH ENG & JET PROPULSION, RICHARD L & DOROTHY M HAYMAN FOUND, 80- *Concurrent Pos:* Instr fluid dynamics & gas turbines, Case Sch Appl Sci, 42-46; consult, 48-; chief combustion res, Jet Propulsion Lab, Calif Inst Technol, 49-57; vis prof, Cornell Univ, 56, Mass Inst Technol, 80-81, Chinese Acad Sci, 82; mem, var comts & bds, Nat Res Coun, Nat Acad Sci, 56- *Mem:* Nat Acad Eng; Am Phys Soc; fel Am Inst Aeronaut & Astronaut; Combustion Inst; Sigma Xi. *Res:* Fluid mechanics; gas dynamics; magnetohydrodynamics; propulsion. *Mailing Add:* Guggenheim Aeronaut Lab Calif Inst Technol Pasadena CA 91125

MARBLE, HOWARD BENNETT, JR, b Shelburne Falls, Mass, June 14, 23; m 48; c 5. DENTISTRY, ORAL SURGERY. *Educ:* Tufts Univ, DMD, 47; Am Bd Oral & Maxillofacial Surg, dipl, 62. *Prof Exp:* Assoc prof oral surg, Sch Dent, Med Col Ga, 69-78, prof, 78-; RETIRED. *Concurrent Pos:* Chief dent serv, Vet Admin Hosp, Augusta, Ga, 69- *Mem:* Am Soc Oral Surg; Int Asn Oral Surg; Int Asn Dent Res. *Res:* Bone graft substitutes; bone healing. *Mailing Add:* 4052 Fairlene Ct Augusta GA 30906

MARBOE, CHARLES CHOSTNER, PATHOLOGY, IMMUNOLOGY. *Educ:* Pa State Univ, MD, 76. *Prof Exp:* ASST PROF PATH, COLUMBIA UNIV, 76- *Mailing Add:* Dept Path Columbia Univ 630 W 168th St New York NY 10032

MARBURG, STEPHEN, b Frankfurt, Ger, July 16, 33; US citizen; m 56; c 2. ORGANIC CHEMISTRY. *Educ:* City Col New York, BS, 55; Harvard Univ, MA, 57, PhD(org chem), 60. *Prof Exp:* Chemist, Hoffman-La Roche, Inc, 55; NIH fel biochem, Brandeis Univ, 59-61; NIH fel phys org chem, Mass Inst Technol, 61-62; asst prof chem, Boston Univ, 62-65; sr res chemist, 65-74, res fel, 74-80, sr res fel, 80-88, SR INVESTR, MERCK SHARP & DOHME INC, 88- *Mem:* Am Chem Soc. *Res:* Synthetic organic and physical organic chemistry; organic fluorine chemistry; macromolecular chemistry; chemistry of vaccines. *Mailing Add:* 50 Concord Ave Metuchen NJ 08840

MARBURGER, RICHARD EUGENE, b Detroit, Mich, May 26, 28; m 50; c 2. PHYSICS. *Educ:* Wayne State Univ, BS, 50, MS, 52, PhD, 62. *Prof Exp:* Physicist, Res Labs, Gen Motors Corp, 52-53, res physicist, 55-58, sr res physicist, 58-69; mem staff, 69-70, dir, Sch Arts & Sci, 70-72, dean acad affairs, 72-75, PROF PHYSICS, LAWRENCE INST TECHNOL, 75-, VPRES ACAD AFFAIRS, 75-, PRES & CHIEF ADMIN OFFICER, 77- *Mem:* Am Phys Soc; Am Asn Physics Teachers. *Res:* Phase transformations in solids; residual stress analysis by x-ray diffraction; metal physics; x-ray diffraction techniques; Mossbauer effect. *Mailing Add:* Lawrence Inst of Technol 21000 W 10 Mile Rd Southfield MI 48075

MARCATILI, ENRIQUE A J, b Arg, 1925. ELECTRICAL ENGINEERING. *Prof Exp:* Elec & aeronaut engr, Cordoba Univ, Arg, 47-48, aircraft design engr & univ prov, 48-53; head, Transmission & Circuit Res Dept, Bell Labs, 53-91; RETIRED. *Honors & Awards:* Co-recipient W R G Baker Prize Award, 75. *Mem:* Nat Acad Eng; fel Inst Elec & Electronics Engrs. *Res:* Millimeter waveguide; optical communications. *Mailing Add:* Two Markwood Lane Rumson NJ 07760

MARC DE CHAZAL, L E, b St Denis, Reunion, Nov 23, 21; US citizen; m 51; c 2. CHEMICAL & NUCLEAR ENGINEERING. *Educ:* La State Univ, BS, 49, MS, 51; Okla State Univ, PhD(chem eng), 53. *Prof Exp:* Asst prof chem eng, 53-57, from assoc prof to prof chem & nuclear eng, 57-88, EMER PROF, UNIV MO-COLUMBIA, 88- *Concurrent Pos:* Vis res assoc, Atomic Energy Res Estab, Eng, 59-60; res assoc, United Kingdom Atomic Energy Authority, 66-67. *Mem:* AAAS; Am Inst Chem Engrs; Am Chem Soc; Am Soc Eng Educ. *Res:* Mass transfer; drop formation and interfacial tension of liquid-liquid systems; statistics of solid mixtures; fluid mechanics. *Mailing Add:* 1506 Wilson Ave Columbia MO 65201

MARCEAU, NORMAND LUC, b Lambton, Que, Nov 11, 42; m 65; c 3. BIOPHYSICS, CELL BIOLOGY & BIOCHEMISTRY. *Educ:* Laval Univ, BASc, 66; Univ Toronto, MSc, 68, PhD(biophys), 71. *Prof Exp:* From asst prof to assoc prof, 71-82, scientist biophys, Ctr Hosp, 71-81, PROF MED, LAVAL UNIV, 83-; RES SCIENTIST, CANCER CTR, HOTEL-DIEU HOSP, 81- *Concurrent Pos:* Main investr, Coun Res Health, Que grant, 72-73, Defense Res Bd grant, 73-76, Med Res Coun grant, 73-, Dept Educ, Que, grant, 76-80 & Nat Cancer Inst Can, 78- *Mem:* Biophys Soc; Tissue Culture Asn; Can Soc Biochem; Can Cell Biol; Am Asn Cancer Res; Am Cell Biol; Am Asn Cancer Res. *Res:* Cytokeratins and epithelial cell differentiation; Hepatocyte precursors and their role in hepatocarcinogenesis. *Mailing Add:* Ctr Rech L'Hotel-Dieu de Quebec 11 Cote de Palais Quebec PQ G1R 2J6 Can

MARCELIN, GEORGE, b Havana, Cuba, Oct 8, 48; US citizen; m 70. HETEROGENEOUS CATALYSIS, SOLID STATE NUCLEAR MAGNETIC RESONANCE. *Educ:* Univ Miami, BS, 69; Rice Univ, MA, 73, PhD(chem), 74. *Prof Exp:* Instr chem, Bolles Sch, 73-76; sect head, Org Chem Div, SCM Corp, 76-80; sr res chemist, Gulf Res & Develop Co, 80-85; RES PROF CHEM ENG, UNIV PITTSBURGH, 85- *Concurrent Pos:* Adj instr, Fla State Univ, 74-76. *Mem:* Am Chem Soc; NAm Catalysis Soc. *Res:* Heterogeneous catalysis over supported-metal catalysts; synthesis of new catalytic supports and their interaction with active metals; catalytic activation and reaction of small molecules. *Mailing Add:* Dept Chem & Petrol Eng Univ Pittsburgh Pittsburgh PA 15261

MARCELLI, JOSEPH F, b Schenectady, NY, Nov 22, 26; m 52. ORGANIC CHEMISTRY. *Educ:* Rensselaer Polytech Inst, BS, 48, MS, 50, PhD, 57. *Prof Exp:* Asst, Rensselaer Polytech Inst, 49-51, res assoc, 53-56; from instr to prof, 56-85, dean health & Phys sci, 71-84, CHMN DEPT CHEM, HUDSON VALLEY COMMUNITY COL, 58-, EMER PROF & EMER DEAN, 85- *Concurrent Pos:* Actg chmn, Dept Chem, Hudson Valley Community Col, 57-58, dir phys sci div, 67-70, dean sci & actg dean arts, 70-71. *Mem:* AAAS; fel Am Inst Chem; Am Pub Health Asn; Am Chem Soc; Health Physics Soc; Sigma Xi. *Mailing Add:* 28 Locust Ave Troy NY 12180

MARCELLINI, DALE LEROY, b Oakland, Calif, Mar 19, 37; m 58; c 3. HERPETOLOGY, ANIMAL BEHAVIOR. *Educ:* San Francisco State Univ, BA, 64, MA, 66; Univ Okla, PhD(zool), 70. *Prof Exp:* Instr biol, Calif State Univ, Hayward, 71-74; RES CUR AMPHIBIANS & REPTILES, NAT ZOOL PARK, SMITHSONIAN INST, 74-, CUR HERPOTOL, 78- *Mem:* Soc Study Amphibians & Reptiles; Animal Behav Soc; Am Soc Ichthyologists & Herpetologists; Am Asn Zool Parks & Aquariums; Sigma Xi. *Res:* Behavior and ecology of Gekkonid lizards, expecially acoustic behavior; breeding and captive maintenance of vertebrates. *Mailing Add:* 3401 N Vermont St Arlington VA 22207

MARCELLINO, GEORGE RAYMOND, b Brooklyn, NY, Apr 28, 49; m 69; c 2. AUDIOLOGY, PSYCHOACOUSTICS. *Educ:* City Univ New York, BA, 70, MSc, 72, PhD(hearing sci), 77. *Prof Exp:* Res assoc psychoacoust, NY Univ Med Ctr, 71-72; audiologist, Bellevue Hosp, New York, 71-72; clin res audiologist, Mt Sinai Hosp, New York, 72-73; asst prof audiol & hearing sci, Brooklyn Col, City Univ New York, 73-78; AUDITORY PROD MGR, TELESENSORY SYSTS, INC, 78- *Mem:* Acoust Soc Am; Am Speech & Hearing Asn; Soc Med Audiol. *Res:* Auditory time and frequency analysis; synthetic speech perception in sensorineural hearing-impairment; neonatal hearing detection. *Mailing Add:* 107 Wilder Ave Los Gatos CA 95030

MARCELO, CYNTHIA LUZ, b New York, NY, Aug 13, 45. CELL PHYSIOLOGY, INFLAMMATION. *Educ:* Col Mt St Vincent, BS, 67; State Univ NY, Buffalo, MS, 69, PhD(cell physiol), 73. *Prof Exp:* Teaching fel biochem, Univ Mich Med Sch, 73-75, from instr to asst prof physiol in dermat, 75-83; ASSOC RES SCIENTIST PHYSIOL & DIR RES DIV DERMAT, UNIV MICH, 83- *Concurrent Pos:* Prin investr, NIH, NAIMDD, 79-; mem, GMA Study Sect, 82-, sci comt, Dermat Found, 83-86 & sci & prog comt, Soc Invest Dermat, 84-89. *Mem:* Soc Invest Dermat; Am Acad Dermat; Am Soc Cell Biol; AAAS; Dermat Found. *Res:* Control of epidermal cell function; use of tissue culture technology to study growth factors; effects of hormones and vitamin A on growth and function; eicosanoid metabolism. *Mailing Add:* Med Sch Univ Mich KI 6558-0528 Ann Arbor MI 48109-0528

MARCH, BERYL ELIZABETH, b Port Hammond, BC, Aug 30, 20; m 46; c 1. POULTRY NUTRITION, FISH NUTRITION. *Educ:* Univ BC, BA, 42, MSA, 62. *Hon Degrees:* DSc, Univ BC. *Prof Exp:* Res asst, Can Fishing Co, 44-47; instr, 47-59, res assoc, 59-62, from asst prof to assoc prof, 62-70, PROF ANIMAL SCI, UNIV BC, 70- *Honors & Awards:* Nutrit Res Award, Am Feed Mfg Asn; Earle Willard McHenry Award, Can Soc Nutrit Sci. *Mem:* Am Inst Nutrit; fel Poultry Sci Asn; Can Soc Nutrit Sci; fel Agr Inst Can; fel Royal Soc Can. *Res:* Physiology and nutrition. *Mailing Add:* Dept Animal Sci Univ BC Vancouver BC V6T 2A2 Can

MARCH, JERRY, b Brooklyn, NY, Aug 1, 29; m 54; c 3. ORGANIC CHEMISTRY. *Educ:* City Col NY, BS, 51; Brooklyn Col, MA, 53; Pa State Univ, PhD(chem), 57. *Prof Exp:* From asst prof to assoc prof, 56-68, PROF CHEM, ADELPHI UNIV, 68- *Concurrent Pos:* Vis prof, Univ Strasbourg, 67-68 & Imp Col, Univ London, 78. *Mem:* Am Chem Soc. *Res:* Organic synthesis; organometallic compounds. *Mailing Add:* Dept of Chem Adelphi Univ Garden City NY 11530

MARCH, RALPH BURTON, b Oshkosh, Wis, Aug 5, 19; m 42; c 3. ENTOMOLOGY. *Educ:* Univ Ill, AB, 41, MA, 46, PhD(entom, chem), 48. *Prof Exp:* From jr entomologist to entomologist, Citrus Exp Sta, 48-83, dean grad div, Univ, 61-69, head div toxicol & physiol, Dept Entom, 69-72, prof, 61-83, chmn dept entom, 78-83, EMER PROF ENTOM, UNIV CALIF, RIVERSIDE, 83- *Mem:* AAAS; Entom Soc Am; Am Chem Soc. *Res:* Physiological, biochemical and toxicological studies on the mode of action of insecticides; relation of chemical structure to insecticidal activity; resistance of insects to insecticides. *Mailing Add:* 300 Deer Valley Rd No 2N San Rafael CA 94903-5514

MARCH, RAYMOND EVANS, b Newcastle upon Tyne, Eng, Mar 13, 34; m 58; c 3. PHYSICAL CHEMISTRY. *Educ:* Leeds Univ, BSc, 57; Univ Toronto, PhD(phys chem), 61. *Prof Exp:* Res chemist, Johnson & Johnson Ltd, Can, 61-62; res assoc & res fel, McGill Univ, 62-65; from asst prof to prof chem, 65-76, chmn dept, 79-83, MEM STAFF DEPT CHEM, TRENT UNIV, 76- *Concurrent Pos:* Adj prof chem, Queen's Univ, Kingston, Ont, 81- *Mem:* Fel Chem Inst Can; Am Soc Mass Spectrometry; Can Inst Phys; Spectrometry Soc Can. *Res:* Ion trap mass spectrometry; multiphoton absorption by gaseous ions; multi-sector mass spectrometry. *Mailing Add:* Dept Chem Trent Univ Peterborough ON K9J 4V5 Can

MARCH, RICHARD PELL, b Medford, Mass, May 1, 22; m 46; c 3. FOOD SCIENCE. *Educ:* Univ Mass, BS, 44; Cornell Univ, MS, 48. *Prof Exp:* From instr to prof, 48-77, EMER PROF DAIRY INDUST, CORNELL UNIV, 77- *Concurrent Pos:* Exec secy, NY State Asn Milk & Food Sanit, 56-79, ed, Ann Report, 64-79; exec bd, Int Asn Milk, Food & Environ Sanit, 70-78; chmn, NE Dairy Pract Coun, 70-76, exec secy, 77-90. *Res:* Dairy industry extension in the field of milk and milk handling on farms; processing in fluid milk plants. *Mailing Add:* 1260 Ellis Hollow Rd Ithaca NY 14850

MARCH, ROBERT HENRY, b Yarmouth, NS, Can, July 30, 37; m 63; c 3. PHYSICS. *Educ:* Dalhousie Univ, MSc, 60; Oxford Univ, PhD(physics), 65. *Prof Exp:* From asst prof to assoc prof, 65-77, chmn dept, 69-78, PROF PHYSICS, DALHOUSIE UNIV, 77- *Mem:* Can Asn Physicists. *Res:* Low temperature physics. *Mailing Add:* Dept Physics Dalhousie Univ Halifax NS B3H 4H6 Can

MARCH, ROBERT HERBERT, b Chicago, Ill, Feb 28, 34; m 53, 79; c 1. PHYSICS. *Educ:* Univ Chicago, AB, 52, SM, 55, PhD(physics), 60. *Prof Exp:* Lectr physics, Midwest Univ Res Asn, 60-61; from instr to assoc prof, 61-71, PROF PHYSICS, UNIV WIS-MADISON, 71- *Concurrent Pos:* Vis scientist, Europ Orgn Nuclear Res, 65 & 67, Fermi Nat Accelerator Lab, 71- & Stanford Linear Accelerator Ctr, 75-77; vis prof, Univ Athens, Greece, 81-;

chair, Integrated Lib Studies Prog, Univ Wis-Madison, 87-90. *Honors & Awards:* Sci Writing Award, Am Inst Physics-US Steel Found, 71 & 75. *Mem:* Am Phys Soc; Am Asn Univ Professors. *Res:* Experimental high-energy physics and astrophysics; science writing. *Mailing Add:* Dept Physics Univ Wis Madison WI 53706

MARCH, SALVATORE T, b Staten Island, NY, Oct 22, 50. INFORMATION SYSTEMS DESIGN. *Educ:* Cornell Univ, BS, 72, MS, 75, PhD(phys database design), 78. *Prof Exp:* From asst prof to assoc prof, 79-91, PROF INFO & DECISION SCI, CARLSON SCH MGT, UNIV MINN, 91- *Mem:* Asn Comput Mach. *Mailing Add:* Info & Decision Sci Dept Carlson Sch Mgt Univ Minn 217-19th Ave S Minneapolis MN 55455

MARCHALONIS, JOHN JACOB, b Scranton, Pa, July 22, 40; m 78; c 3. BIOCHEMISTRY, IMMUNOLOGY. *Educ:* Lafayette Col, AB, 62; Rockefeller Univ, PhD(biochem), 67. *Prof Exp:* Am Cancer Soc fel, Walter & Eliza Hall Inst Med Res, Melbourne, 67-68; asst prof med sci, Brown Univ, 69-70; sr lectr, Walter & Eliza Hall Inst Med Res, Melbourne, 70-73, assoc prof molecular immunol & head lab, 73-76; head, Cell Biol & Biochem Sect, Cancer Biol, Frederick Cancer Res Ctr, Nat Cancer Inst, 77-79; prof & chmn dept biochem & molecular biol, Med Univ SC, Charleston, 80-88; PROF & CHMN DEPT MICRO-IMMUNOL, UNIV ARIZ, TUCSON, 88- *Concurrent Pos:* Consult, Miriam Hosp, Providence, RI, 69-70; assoc, dept microbiol, Monash Univ Med Sch, Melbourne, 75-76; adj prof pathol, Med Sch, Univ Penn, 77-; F R Lillie fel, Marine Biol Lab; mem, Allergy & Immunol Study Sect, NIH, 79-; ed, Cancer Biol Rev, J Immunogenetics, Develop Comp Immunol, Quart Rev Biol, Continued Topics Immunobiol, Immunochem, Exp Parasitol, Molecular Immunol, Exp Clin Immunol Genetics, J Protein Chem; consult allied instrumentation, bd dir, Am Type Cult Collect. *Mem:* AAAS; Am Asn Immunol; Am Soc Biol Chemists; NY Acad Sci; Sigma Xi; fel Am Inst Chem. *Res:* Molecular and cellular basis of immunological specificity. *Mailing Add:* Dept Micro-Immunol Univ Ariz Tucson AZ 85724

MARCHAND, ALAN PHILIP, b Cleveland, Ohio, May 23, 40; m. PHYSICAL ORGANIC CHEMISTRY, POLYCYCLIC CAGE MOLECULES. *Educ:* Case Western Reserve Univ, BS, 61; Univ Chicago, PhD(org chem), 65. *Prof Exp:* Instr phys chem, Huston-Tillotson Col, 63-65; NIH fel phys org chem, Univ Calif, Berkeley, 65-66; from asst prof to prof org chem, Univ Okla, 66-82, adj prof pharm, 79-82; prof org chem, 82-86, DISTINGUISHED RES PROF CHEM, UNIV N TEX, 86- *Concurrent Pos:* Sr res fel, Fulbright-Hays grant, Inst Chem, Univ Liege, Belg, 72-73; consult ed, VCH Publ Inc, New York, 79-; Fulbright-Hays grant, Univ Hyderabad, India, 84; regent's fac lectr, Univ NTex, 89; prog officer, Div Chem, NSF, 90-91. *Honors & Awards:* J Clarence Karcher lectr, Univ Okla, 87. *Mem:* Am Chem Soc; Royal Soc Chem; Sigma Xi; fel Am Inst Chemists. *Res:* Nuclear magnetic resonance of rigid polycyclic systems; synthesis of new high energy polycyclic hydrocarbon fuels; organometallic chemistry; molecular recognition and inclusion phenomena; synthesis of polynitropolycyclic compounds. *Mailing Add:* Dept Chem Box 5068 Univ NTex Denton TX 76203-5068

MARCHAND, E ROGER, b Palo Alto, Calif, June 17, 36; m 59; c 2. ANATOMY. *Educ:* San Diego State Col, BS, 63; Univ Calif, Los Angeles, PhD(anat), 68. *Prof Exp:* ADJ ASSOC PROF NEUROSCI & ACAD ADMINR, OFF LEARNING RESOURCES, SCH MED, UNIV CALIF, SAN DIEGO, 68- *Mem:* AAAS; Soc Neurosci; Am Asn Anatomists; Am Soc Zoologists. *Res:* Neuroanatomy. *Mailing Add:* Dept Neurosci Univ Calif Sch Med San Diego 9500 Gilman Dr La Jolla CA 92093-0611

MARCHAND, ERICH WATKINSON, b Hartford, Conn, July 7, 14; m 41; c 3. OPTICS. *Educ:* Harvard Univ, AB, 36; Univ Wash, MS, 41; Univ Rochester, PhD(math), 52. *Prof Exp:* Instr math, Univ Rochester, 43-49; physicist optics, Res Labs, Eastman Kodak Co, 49-79; CONSULT, 79- *Concurrent Pos:* Univ of Rochester, 80- *Mem:* Fel Optical Soc Am; Math Asn Am. *Res:* Mathematics; optics. *Mailing Add:* 192 Seville Dr Rochester NY 14617

MARCHAND, JEAN-PAUL, b Murten, Switz, Mar 25, 33. MATHEMATICAL PHYSICS. *Educ:* Univ Bern, Dipl math, 58; Univ Geneva, DrSc(physics), 63. *Prof Exp:* Asst physics, Univ Bern, 58-60; asst, Univ Geneva, 60-63, res assoc, 63-67; asst prof physics & math, Univ Denver, 67-68, math, 68-69, assoc prof, 69-79, prof, 79-81. *Concurrent Pos:* Lectr, Univ Bern, 65-66; invited prof, Swiss Fed Inst Technol, Lausanne, 81, 83, 85 & 87. *Res:* Mathematical foundations of quantum mechanics and statistical mechanics. *Mailing Add:* Dept Math Univ Denver Denver CO 80208

MARCHAND, MARGARET O, b Shorncliffe, Man, Can, Oct 17, 25; US citizen; m 57; c 3. MATHEMATICS, STATISTICS. *Educ:* Univ Man, BA, 45; Univ Minn, MA, 48, PhD(math), 50. *Prof Exp:* Assoc prof math, Southwest Mo State Col, 50-52; statistician, Man Cancer Res Inst, Can, 52-56; asst prof math, Bemidji State Col, 56-57, 58-59, 65-66; asst prof, Univ Denver, 57-58; instr corresp dept, Univ Minn, 59-66; asst prof math, Lakehead Univ, 66-68; prof, Wis State Univ-Superior, 68-71; from assoc prof to prof math, Adrian Col, 71-90; RETIRED. *Mem:* Am Math Soc; Math Asn Am; Sigma Xi. *Mailing Add:* 2823 Elmwood Dr Adrian MI 49221

MARCHAND, NATHAN, b Shawinigan Falls, Que, June 20, 16; US citizen; m 38; c 4. ELECTRICAL ENGINEERING, ELECTRONICS. *Educ:* City Col New York, BS, 37; Columbia Univ, MS, 41. *Prof Exp:* Sr engr, Int Tel & Tel Corp Labs, 41-45; head sect electronic circuits, Sylvania Elec Prods, Inc, 49-51; ELECTRONIC CONSULT & PRES, MARCHAND ELECTRONIC LABS, INC, 51- *Concurrent Pos:* Lectr, Grad Sch Elec Eng, Columbia Univ, 45-47; consult & staff mem, NY Univ-Bellevue Med Ctr, 47-60; expert consult, US Air Force Atlantic Missile Range, 51-54; mem bd dirs, Panoramic Radio Prod Inc, 54-63; consult, Air Navig Develop Bd, Spec Asst to the President on Aviation Facilities, Airways Modernization Bd & Fed Aviation Agency, 55-64; consult, US Army Aviation Res & Develop Command, 76-81; mgr, Special Electronics Mission Aircraft Proj, US Army Troop Readiness and Support Command, 80-83; consult engr, 45-49. *Mem:* Fel AAAS; fel Inst Elec & Electronics Engrs. *Res:* Communications, navigation, data acquisition, transmission, radiation, electronic counter measures, data links and system analysis as accomplished by electronics. *Mailing Add:* 311 Riversville Rd Greenwich CT 06831

MARCHAND, RICHARD, b Montreal, Que, Mar 25, 52; m 79; c 3. KINETIC MODELLING, ATOMIC & RADIATION PROCESSES. *Educ:* Univ Montreal, BSc, 74; Univ Toronto, MSc, 75; Princeton Univ, MA, 77, PhD(plasma physics), 79. *Prof Exp:* Res assoc, Lab Plasma & Fusion Energy Studies, Univ Md, 79-81; res assoc plasma physics, Univ Alta, 81-83, univ res fel, 83-87; res assoc, 87-90, PROF PLASMA PHYSICS INRS-ENERGIE, UNIV QUE, 90. *Concurrent Pos:* Adv, Can Asn Physicists, 88-90. *Mem:* Am Phys Soc; Can Asn Physicists. *Res:* Theoretical and computer modelling of plasmas. *Mailing Add:* INRS-ENERGIE CP 1020 Univ Que Montreal PQ J3X 1S2 Can

MARCHANT, DAVID DENNIS, b Murray, Utah, May 1, 43; m 66; c 8. CERAMICS, METALLURGY. *Educ:* Univ Utah, BS, 68; Mass Inst Technol, ScD(ceramics), 74. *Prof Exp:* Systs analyst comput, Food & Drug Admin, 68-70; scientist mat, Argonne Nat Lab, 74-77; sr res scientist mat, Pac Northwest Lab, Battelle Mem Inst, 77-84; RES SCIENTIST, B P AM, CLEVELAND, 84- *Mem:* Am Ceramic Soc. *Res:* High temperature thermophysical properties of ceramics and metals; high temperature electrochemical interactions between ceramics, metal and coal slag; electro-optical materials; electronic ceramics; substrates. *Mailing Add:* 7077 Longview Dr Solon OH 44139

MARCHANT, DOUGLAS J, b Malden, Mass, Dec 31, 25; m 55; c 5. OBSTETRICS & GYNECOLOGY. *Educ:* Tufts Univ, BS, 47, MD, 51; Am Bd Obstet & Gynec, dipl, cert gynec oncol, 74. *Prof Exp:* Assoc prof, 65-75, dir, Off Cancer Control, 76-77, PROF OBSTET & GYNEC, SCH MED, TUFTS UNIV, 75-, PROF SURG, 83- *Concurrent Pos:* Consult, St Margaret's Hosp, Dorchester, Mass, 57- & Boston City Hosp, 58-; sr gynecologist, New Eng Med Ctr Hosp, 67-; consult, Sturdy Mem Hosp, Attleboro, Mass, Choate Hosp, Wouborn & Lemuel Shattuck Hosp, Boston; dir gynec oncol, Tufts New Eng Med Ctr, 74-77, 81-83, dir, Cancer Ctr, 77-83. *Mem:* AMA; Am Col Obstet & Gynec; Am Col Surg; Soc Gynec Oncol; Am Asn Obstet & Gynec; Soc Surg Oncol; Soc Pelvic Surgeons; Soc Study Breast Dis (pres, 79-81). *Res:* Transport of urine; ureteral activity; breast cancer--epidemiology and receptor studies; immunology of pregnancy; gynecologic oncology. *Mailing Add:* Dept of Obstet & Gynec New Eng Med Ctr 750 Washington St Box 319 Boston MA 02111

MARCHANT, GUILLAUME HENRI (WIM), JR, b Brisbane, Australia, Mar 8, 46; m 72; c 2. ENVIRONMENTAL SCIENCES. *Educ:* Univ Southern Miss, BS, 71. *Prof Exp:* Asst geologist, Southern Res Inst, 72-73; start up & testing engr, Pollution Control Div, Carborundum Co, 73-74; assoc engr, 74-76, supvr, Control Device Sect, 76-77, SECT HEAD, CONTROL DEVICE EVAL SECT, CONTROL DEVICE RES DIV, SOUTHERN RES INST, 77- *Res:* Preparation of proposals, plans and manage field test programs for evaluation of environmental control devices; data analysis; reporting; evaluation. *Mailing Add:* 204 Wind View Trace Birmingham AL 35210

MARCHANT, LELAND CONDO, b Columbus, NDak, May 17, 31; m 54; c 4. ENGINEERING MANAGEMENT, PETROLEUM ENGINEERING. *Educ:* Univ Okla, BS, 60. *Prof Exp:* Petrol res engr, Petrol Res Lab, US Bur Mines, WVa, 60-62 & Laramie Energy Technol Ctr, 62-76; proj mgr, Laramie Energy Technol Ctr, 76-83, RES MGR, WESTERN RES INST, LARAMIE, WYO, 83- *Mem:* Am Inst Mining, Metall & Petrol Engrs. *Res:* Thermal methods of oil recovery; general petroleum reservoir evaluation; well log analysis; petroleum reservoir engineering; tar sand research. *Mailing Add:* 1215 Reynolds Laramie WY 82070

MARCHASE, RICHARD BANFIELD, b Sayre, Pa, Mar 12, 48; m 79; c 2. GLYCOPROTEINS, INTRACELLULAR VESICLES. *Educ:* Cornell Univ, BS, 70; Johns Hopkins Univ, PhD(biophys), 76. *Prof Exp:* Fel anat, Med Sch, Duke Univ, 76-78, asst prof, 79-86; assoc prof cell biol & anat, 86-90, PROF CELL BIOL & DIR GRAD STUDIES, UNIV ALA, BIRMINGHAM, 90- *Concurrent Pos:* Adv panelist develop biol, Nat Sci Found, 83-87. *Honors & Awards:* Presidential Young Investr Award, NSF. *Mem:* Am Soc Cell Biol; Am Soc Zoologists; Int Soc Develop Neurosci; Sigma Xi. *Res:* Roles of oligosaccharide sequences in regulated exocytosis; oligosaccharides as determinants of growth control; intercellular adhesion and recognition. *Mailing Add:* Dept Cell Biol 668 BHSB Univ Ala Birmingham AL 35294

MARCHATERRE, JOHN FREDERICK, b Hermansville, Mich, Aug 28, 32; m 57; c 3. NUCLEAR ENGINEERING, RESEARCH ADMINISTRATION. *Educ:* Mich Technol Univ, BS, 50, MS, 54; Univ Chicago, MBA, 80. *Prof Exp:* Group leader & proj mgr, Reactor Anal & Safety Div, Argonne Nat Lab, 60-70, assoc div dir, 70-86, div dir, 86-90, DIV DIR, REACTOR ENG DIV, ARGONNE NAT LAB, 90- *Mem:* Am Nuclear Soc. *Res:* Nuclear reactor safety; safety of liquid metal reactors; heat transfer and hydrodynamics. *Mailing Add:* Reactor Eng Div Argonne Nat Lab 9700 S Cass Ave Argonne IL 60439

MARCHELLO, JOSEPH M(AURICE), b East Moline, Ill, Oct 6, 33; m; c 2. CHEMICAL ENGINEERING. *Educ:* Univ Ill, BS, 55; Carnegie Inst Technol, PhD(chem eng), 59. *Prof Exp:* Asst prof chem eng, Okla State Univ, 59-61; from assoc prof to prof, 62-78, head dept, 67-73, provost, Div Math & Phys Sci & Eng, Univ Md, College Park, 73-78; chancellor, Univ Mo-Rolla, 78-85; PRES, OLD DOMINION UNIV, 85- *Mem:* Am Chem Soc; Am Inst Chem Engrs; AAAS; Nat Soc Prof Engrs. *Res:* Transport phenomena; fluid mechanics; mass and heat transfer; reaction kinetics; applied mathematics and models. *Mailing Add:* Col Eng & Technol Old Dominion Univ Norfolk VA 23529

MARCHESE, FRANCIS THOMAS, b Brooklyn, NY, May 12, 49. THEORETICAL CHEMISTRY. *Educ:* Niagara Univ, BS, 71; Youngstown State Univ, MS, 73; Univ Cincinnati, PhD(chem), 79. *Prof Exp:* Fel res found, City Univ New York, 78-; AT DEPT COMPUT SCI, PACE UNIV. *Mem:* Am Chem Soc. *Res:* Statistical mechanical treatments of liquids and solutions; solvent effects on ground and excited state properties and structure; quantum theory of molecules; theoretical spectroscopy. *Mailing Add:* Dept Comput Sci Pace Univ New York NY 10038

MARCHESI, VINCENT T, b New York, NY, Sept 4, 35; m 59; c 3. BIOCHEMISTRY, PATHOLOGY. *Educ:* Yale Univ, BA, 57, MD, 63; Oxford Univ, PhD(exp path), 61. *Prof Exp:* From intern to resident path, Wash Univ, 63-65; res assoc cell biol, Rockefeller Univ, 65-66; staff assoc, Nat Cancer Inst, 66-68; chief sect chem path, Nat Inst Arthritis, Metab & Digestive Dis, 68-77; ANTHONY N BRADY PROF PATH, SCH MED, YALE UNIV, 77- *Mem:* Am Soc Cell Biol; Histochem Soc; NY Acad Sci. *Res:* Inflammation, blood vessel permeability and the biochemical properties of cell surfaces; physical and chemical properties of cell membranes. *Mailing Add:* Dept Path Brady Mem Lab Yale Univ Sch Med New Haven CT 06520

MARCHESSAULT, ROBERT HENRI, b Montreal, Que, Sept 16, 28; m 52; c 6. PHYSICAL CHEMISTRY, POLYMER CHEMISTRY. *Educ:* Univ Montreal, BSc, 50; McGill Univ, PhD(phys chem), 54. *Hon Degrees:* DSc, Concordia Univ, Montreal. *Prof Exp:* Res chemist, Am Viscose Corp, 56-59, res assoc, 59-61; from assoc prof to prof polymer & phys chem, State Univ NY Col Forestry, Syracuse Univ, 61-69; prof chem & dir dept, Univ Montreal, 69-78; MGR, XEROX RES CTR CAN, 78- *Concurrent Pos:* Fel, Univ Uppsala, 55; distinguished res fel, State Univ NY & vis prof, Univ Strasbourg, 67-68; vis prof, Univ Kyoto. *Honors & Awards:* Anselme Payen Award, Am Chem Soc, 76. *Mem:* AAAS; Am Chem Soc; Am Phys Soc; Tech Asn Pulp & Paper Indust; Chem Inst Can; fel Royal Soc Can. *Res:* Physical chemical studies on natural and synthetic polymers, especially solid state characterization by electromagnetic scattering techniques. *Mailing Add:* McGill Univ 3420 Univ St Montreal ON H3A 2A7 Can

MARCHETTA, FRANK CARMELO, b Utica, NY, Apr 28, 20; m 49; c 3. MEDICINE. *Educ:* Univ Buffalo, MD, 44; Am Bd Surg, dipl, 54. *Prof Exp:* Resident gen surg, Deaconess Hosp, Buffalo, 47-50; resident, 50-51, assoc surgeon, 51-54, ASSOC CHIEF CANCER RES HEAD & NECK SURG, ROSWELL PARK MEM INST, 54-; CLIN ASSOC PROF ORAL PATH, DENT SCH & RES ASSOC PROF SURG, MED SCH, STATE UNIV NY BUFFALO, 59- *Concurrent Pos:* Consult, head & neck oncol, Roswell Park Mem Inst, 76- & Buffalo Vet Admin Hosp, 77. *Mem:* Fel Am Col Surg; Soc Head & Neck Surgeons. *Res:* Cancer. *Mailing Add:* 2804 Main Buffalo NY 14214

MARCHETTE, NYVEN JOHN, b Murphys, Calif, June 26, 28; m 50; c 1. VIROLOGY, RICKETTSIAL DISEASES. *Educ:* Univ Calif, BA, 50, MA, 53; Univ Utah, PhD, 60. *Prof Exp:* Bacteriologist, Ecol Res Lab, Univ Utah, 55-61, res microbiologist, 60-61; asst res microbiologist, Hooper Found, Univ Calif, San Francisco, 61-69, from asst res prof rickettsiology to assoc res prof virol, Univ, 63-70, assoc res microbiologist, Med Ctr, 69-70; assoc prof, 70-74, PROF TROP MED, SCH MED, UNIV HAWAII, MANOA, 74- *Concurrent Pos:* Fel, Inst Ctr Med Res & Training, Med Ctr, Univ Calif, San Francisco, 61-63; fel, Inst Med Res, Kuala Lumpur, 61-63; chief arbovirus res lab, Fac Med, Univ Malaya, 63-65; Fogarty sr int fel, John Curtin Sch Med Res, Australian Nat Univ, Canberra, 77-78. *Mem:* AAAS; Am Soc Microbiol; Am Soc Trop Med. *Res:* Ecology infectious diseases, such as virology, rickettsiology; immunology; immunopathology; pathogenesis of virus infections; diagnostic virology. *Mailing Add:* Univ of Hawaii Sch of Med Leahi Hosp Honolulu HI 96816

MARCHETTI, ALFRED PAUL, b Bakersfield, Calif, Feb 16, 40; m 63; c 3. SOLID STATE PHYSICS. *Educ:* Univ Calif, Riverside, BA, 61, PhD(phys chem, spectros), 66; Univ Calif, Berkeley, MS, 63. *Prof Exp:* NIH fel, Univ Pa, 66-69; sr res chemist, 69-75, lab head, 86-90, RES ASSOC, EASTMAN KODAK CO, 75- *Concurrent Pos:* Adj prof, Univ Rochester, 81-82. *Mem:* Am Phys Soc; Sigma Xi. *Res:* Electronic spectroscopy of organic and inorganic molecules and crystals; Stark and Zeeman effect in molecules; exciton theory and energy transfer processes in organic and inorganic crystals; photophysical processes in silver halides. *Mailing Add:* 227 Henderson Dr Penfield NY 14526

MARCHETTI, MARCO ANTHONY, b New York, NY, Feb 15, 36; m 58; c 5. PLANT PATHOLOGY. *Educ:* Pa State Univ, BS, 57; Iowa State Univ, MS, 59, PhD(plant path), 62. *Prof Exp:* Res plant pathologist, US Army Biol Ctr, Ft Detrick, 62-71; res plant pathologist, Epiphytology Res Lab, 71-73, res plant pathologist, Plant Dis Res Lab, 73-74, RES PLANT PATHOLOGIST, RICE RES, TEX A&M UNIV AGR RES & EXTEN CTR, USDA, 74- *Concurrent Pos:* Assoc ed, Phytopathology, 85-87. *Mem:* Am Phytopath Soc; Mycol Soc Am; Sigma Xi; AAAS; Am Inst Biol Sci; NY Acad Sci. *Res:* Rice diseases; inheritance of resistance; disease loss assessment; epidemiology; rice varietal improvement. *Mailing Add:* Tex A&M Univ Agr Res & Exten Ctr Rte 7 Box 999 Beaumont TX 77713-8530

MARCHIN, GEORGE LEONARD, b Kansas City, Kans, July 12, 40; m 74; c 4. MOLECULAR BIOLOGY, PARASITOLOGY. *Educ:* Rockhurst Col, AB, 62; Univ Kans, PhD(microbiol), 67. *Prof Exp:* Res assoc microbiol, Purdue Univ, 67-68, NIH fel, 68-70; asst prof, 70-75, ASSOC PROF BIOL, KANS STATE UNIV, 75- *Concurrent Pos:* Comnr, Adv Lab Comn, Kans State Bd Health, 70-75; dir, Allied Health Prog, Div Biol, Kans State Univ, 70-75; NIH res grant, 72-75, 75-78; vis scientist, Microbiol Inst, Umea Univ, Umea, Sweden, 77-78. *Mem:* AAAS; Am Soc Microbiol; Sigma Xi; Am Soc Parasitologists. *Res:* Enzymology; bacterial physiology; analysis of the mechanism of pathogenesis of the parasite Giardia Lamblia and disinfection of water samples containing cysts of this organism; DNA cloning strategies to study vitamin C biosynthesis and T4 bacteriophage gene regulation. *Mailing Add:* Div Biol Kans State Univ Manhattan KS 66506

MARCHINTON, ROBERT LARRY, b New Smyrna Beach, Fla, Mar 3, 39; m 64; c 1. WILDLIFE ECOLOGY, ETHOLOGY. *Educ:* Univ Fla, BSF, 62, MS, 64; Auburn Univ, PhD(zool), 68. *Prof Exp:* Mgr, Loxahatchee Refuge, US Fish & Wildlife Serv, Fla, 62; wildlife biologist, Fla Game & Fresh Water Fish Comn, 64; from asst prof to assoc prof, 67-85, PROF WILDLIFE ECOL, SCH FOREST RESOURCES, UNIV GA, 85- *Concurrent Pos:* Ga Forest Res Coun & McIntire-Stennis grant, Univ Ga, 68-85, Southeastern Coop Wildlife Dis Study Group Contract, 71-72. *Mem:* Soc Am Foresters; Am Soc Mammal; Wildlife Soc. *Res:* Behavioral ecology of deer. *Mailing Add:* Sch of Forest Resources Univ of Ga Athens GA 30602

MARCHIORO, THOMAS LOUIS, b Spokane, Wash, Aug 1, 28; c 7. SURGERY. *Educ:* Gonzaga Univ, BS, 51; St Louis Univ, MD, 55. *Prof Exp:* Intern, St Mary's Group of Hosps, Mo, 55-56; asst in surg, Sch Med, Univ Colo, 59-60, from instr to assoc prof, 60-67; assoc prof, 67-69, PROF SURG, SCH MED, UNIV WASH, 69- *Concurrent Pos:* Clin investr, Denver Vet Admin Hosp, Colo, 62-65; consult, Children's Orthop, Vet Admin, USPHS & Harborview Hosps, Seattle, Wash, 67-; chmn, End Stage Renal Disease Network Coord Coun #2, 81-82. *Mem:* Soc Univ Surg; Asn Acad Surg (secy, 67-70, pres, 74); Soc Vasc Surg; Am Heart Asn; Am Soc Transplant Surgeons (pres-elect, 76, pres, 77). *Res:* Transplantation. *Mailing Add:* Dept Surg Univ Wash Seattle WA 98195

MARCHMAN, JAMES F(RANKLIN), III, b Lexington, Ky, May 15, 43; m 66. AEROSPACE ENGINEERING. *Educ:* NC State Univ, BS, 64, PhD(mech eng), 68. *Prof Exp:* Aerospace engr, US Army Aviation Test Activity, Edwards AFB, 66; asst mech eng, NC State Univ, 67-68; from asst prof aerospace eng to assoc prof aerospace & ocean eng, 72-87, PROF AEROSPACE ENG, VA POLYTECH INST & STATE UNIV, 87- *Mem:* Assoc fel Am Inst Aeronaut & Astronaut; Am Soc Eng Educ. *Res:* Subsonic aerodynamics; low RE Aerodynamics; flow visualization; wind tunnel aerodynamics. *Mailing Add:* Dept Aerospace & Ocean Eng Va Polytech Inst & State Univ Blacksburg VA 24061

MARCHOK, ANN CATHERINE, b McKeesport, Pa, May 17, 36. CANCER RESEARCH,CELL BIOLOGY. *Educ:* Seton Hill Col, BA, 58; La State Univ, MS, 60; Univ Conn, PhD(cell biol), 66. *Prof Exp:* Res assoc, Univ Conn, 65-67, instr physiol, 67-68, asst prof anat & cell biol, Health Ctr, 68-73; staff scientist, 73-78, GROUP LEADER, OAK RIDGE NAT LAB, 78- *Concurrent Pos:* Lectr, Grad Sch Biomed Sci, Univ Tenn, Oak Ridge, 75-88, adj prof, 88- *Mem:* Sigma Xi; Am Soc Cell Biol; NY Acad Sci; Am Asn Cancer Res; Tissue Cult Asn. *Res:* Cellular and biochemical mechanisms regulating the induction and progression of cancer, particularly during the development of lung cancer. *Mailing Add:* 131 Connors Dr Oak Ridge TN 37830-7663

MARCIANI, DANTE JUAN, protein chemistry, physical chemistry, for more information see previous edition

MARCIANO, WILLIAM JOSEPH, b New York, NY, Oct 11, 47; m 74; c 2. ELEMENTARY PARTICLE PHYSICS. *Educ:* New York Univ, BS, 69, MS, 71, PhD(physics), 74. *Prof Exp:* Res assoc physics, Rockefeller Univ, 74-78, asst prof, 78-80; assoc prof physics, Northwestern Univ, 80-81; physicist, 81-86, SR PHYSICIST, BROOKHAVEN NAT LAB, 86- *Concurrent Pos:* Assoc ed, Physics Rev Lett, 83-86. *Mem:* Am Phys Soc. *Res:* Gauge theories of weak electromagnetic and strong interactions. *Mailing Add:* Brookhaven Nat Lab Upton NY 11973

MARCINIAK, EWA J, MEDICINE. *Educ:* Med Sch Wroclaw, Poland, MD, 51, PhD, 56. *Prof Exp:* Fac mem, Med Sch Wroclaw, Poland, 52-64; from asst prof to assoc prof, 69-78, PROF MED, UNIV KY, 79- *Concurrent Pos:* Dir, Coagulation Lab, Univ Ky Med Ctr Hosp, Bone Marrow Transplantation Lab; ad hoc reviewer, Blood, J Lab & Clin Med, Brit J Haematology, Thrombosis & Haemostasis, Thrombosis & Res Sci. *Mem:* Am Physiol Soc; NY Acad Sci; Am Soc Hemat; Int Soc Thrombosis & Hemostasis; AAAS; Soc Exp Biol & Med; World Fedn Hemophilia. *Res:* Hemostasis and thrombosis; clinical, biochemical and physiological aspects of natural coagulation inhibitors; immunohematology; methods of bone marrow processing for allogeneic transplantation; numerous articles. *Mailing Add:* Dept Med Univ Ky Rm 403 800 Rose St Lexington KY 40536-0093

MARCINKOWSKI, M(ARION) J(OHN), b Baltimore, Md, Feb 27, 31; div; c 2. MATERIALS SCIENCE, PHYSICS. *Educ:* Univ Md, BS, 53; Univ Pa, MS, 55, PhD(metall eng), 59. *Prof Exp:* Supv scientist, Edgar C Bain Lab Fundamental Res, US Steel Corp, 56-63; assoc prof metall, univ assoc & metallurgist, Inst Atomic Res, Iowa State Univ, 63-68; PROF MAT SCI, DEPT MECH ENG, UNIV MD, COLLEGE PARK, 68- *Honors & Awards:* Alexander von Humboldt sr US scientist award. *Res:* Theory of the mechanical behavior of matter. *Mailing Add:* Dept Mech Engr Univ Md College Park MD 20742

MARCINKOWSKY, ARTHUR ERNEST, b Moosehorn, Man, Nov 8, 31; div; c 4. PHYSICAL CHEMISTRY. *Educ:* Univ Man, BSc, 55, MSc, 58; Rensselaer Polytech Inst, PhD(phys chem), 61. *Prof Exp:* Sci teacher, Foxwarren Collegiate, Man, 52-53; lectr phys chem, Royal Mil Col, Ont, 57-58; res assoc & fel, Rensselaer Polytech Inst, 58-61; develop chemist, Chem & Plastics Div, Union Carbide Can, Ltd, 61-63; res assoc, Chem Div, Oak Ridge Nat Lab, 64-67; res scientist, 66-77, SR RES SCIENTIST, TECH CTR, UNION CARBIDE CORP, 77- *Honors & Awards:* Sci Award, Am Chem Soc, 72. *Mem:* Am Chem Soc; Catalysis Soc; Sigma Xi; Am Inst Chem Engrs. *Res:* Physical chemistry of electrolyte solutions in aqueous, non-aqueous and mixed solvent media; polymerization of olefins via stereospecific catalysis; water purification and desalination; heterogeneous catalysis; chromatography; industrial separations; process development. *Mailing Add:* 809 Bauer Ave South Charleston WV 25302

MARCO, GINO JOSEPH, b Leechburg, Pa, Dec 19, 24; m 51; c 5. BIOCHEMISTRY, ORGANIC CHEMISTRY. *Educ:* Carnegie Inst Technol, BS, 50; Univ Pittsburgh, MS, 52, PhD(biochem), 56. *Prof Exp:* Res biochemist agr chem, Monsanto Chem Co, 56-60, proj leader animal nutrit & biochem, 60, group leader chem biol & animal feed res, Agr Chem Div, Res Dept, Monsanto Co, 60-66, group leader biochem of pesticide metab & residues, 66-69; mgr metab invests, Agr Biochem Dept, CIBA-Geigy Agr Chem Div, CIBA-Geigy Corp, 69-84, sr group leader residue anal, 84-86, sr group leader spec studies, 86-87; CONSULT, 88- *Concurrent Pos:* Fel, Pesticide Div, Am Chem Soc. *Mem:* Fel AAAS; Am Chem Soc; fel Am Inst Chemists. *Res:* Animal biochemistry; especially in ruminant metabolism, physiology and biochemistry; organic synthesis of agricultural and radioactive chemicals; process development; development of analytical methods and metabolic information in plants, animals, fish and environment systems for use in submitting pesticide petitions. *Mailing Add:* 1904 Tennyson Dr Greensboro NC 27410

MARCONI, GARY G, b Columbus, Ohio, Aug 31, 44; c 1. NATURAL PRODUCTS CHEMISTRY, LABORATORY AUTOMATION. *Educ:* Univ Dayton, BS, 66; Case Western Reserve Univ, PhD(chem), 70. *Prof Exp:* Sr biochemist, 70-81, RES SCIENTIST, ELI LILLY & CO, 81- *Mem:* Am Chem Soc; Am Soc Pharmacog; AAAS. *Res:* Isolation and purification of natural products produced in fermentations; automated chromatography; computer systems. *Mailing Add:* Lilly Corporate Ctr MC 539 Indianapolis IN 46285

MARCOTTE, BRIAN MICHAEL, b Lewiston, Maine, May 29, 49. DEVELOPMENTAL NEUROETHOLOGY. *Educ:* Stonehill Col, BS; Clark Univ, MA, 73; Dalhousie Univ, PhD(biol), 77. *Prof Exp:* Asst prof biol, Univ Victoria, Can, 77-80; oceanog, McGill Univ, Can, 80-87; DIR MARINE SCI, MAINE DEPT MARINE RESOURCES, 87- *Concurrent Pos:* Scholar, Bamfield Marine Sta, Can, 78 & 84, Huntsman Marine Lab, 81-84; chief exec officer, GAEA Consults, 80-87; vis lectr, Brit Museum Nat Hist, UK, 84. *Mem:* Fel Linnean Soc London; NY Acad Sci; AAAS; Am Fisheries Soc; Am Soc Zoologists; Int Asn Meiobenthologists. *Res:* Crustacean systematics and evolution; developmental neuroethology. *Mailing Add:* Bur Marine Sci Dept Marine Resources West Boothbay Harbor ME 04575-0008

MARCOTTE, PATRICK ALLEN, b New Orleans, La, July 26, 52; m 83; c 2. BIOCHEMISTRY. *Educ:* Mass Inst Technol, SB, 73, PhD(org chem), 77. *Prof Exp:* Fel, Dept Chem, Mass Inst Technol, 77-78 & Dept Pharmacol & Exp Therapeut, Johns Hopkins Univ Sch Med, 78-81; scientist, 81-84, sr scientist, Biochem Group & group leader, Renin Inhibitor Proj, 84-88, RES SCIENTIST, BASIC RES GROUP, PHARMACEUT PROD DIV, ABBOTT LABS, 88- *Mem:* Am Chem Soc; Sigma Xi; assoc mem Am Soc Biochem & Molecular Biol. *Res:* Syntheses of amino acid, steroid and antibiotic derivatives; characterization of enzymes and the evaluation of compounds as enzyme inhibitors or substrates; enzyme-activated pro-drugs as novel pharmaceuticals; numerous publications; recipient of one patent. *Mailing Add:* Abbott Labs ID-48R AP-9A Abbott Park IL 60064

MARCOTTE, RONALD EDWARD, b Taunton, Mass, Aug 27, 39; m 64; c 1. CHEMICAL KINETICS, MASS SPECTROMETRY. *Educ:* Univ Fla, BS, 62, PhD(phys chem), 68. *Prof Exp:* Ohio State Univ Res Found, vis res assoc, Aerospace Res Labs, Wright-Patterson AFB, 68-70; from asst prof to assoc prof, 70-80, PROF CHEM, TEX A&I UNIV, 80- *Mem:* Am Chem Soc; Am Soc Mass Spectrometry. *Res:* Formation and decay of reactive intermediates; design and construction of double mass spectrometer system for study of ion-molecule reaction energetics; special interest in halocarbons and atmospheric gases; critical point studies. *Mailing Add:* Dept Chem Campus Box 161 Tex A&I Univ Kingsville TX 78363

MARCOUX, FRANK W, b Melrose, Mass, Oct 6, 52; m; c 2. PHARMACOLOGY. *Educ:* Univ Mass, Amherst, 65-74; Univ Ala Birmingham, PhD(physiol & biophys), 79. *Prof Exp:* Neurosurg res asst, Stroke Lab, Mass Gen Hosp & Harvard Med Sch, Boston, 74-77; res asst, Stroke Res Ctr & Dept Physiol & Biophysics, Col Med, Univ Vt, Burlington, 79-82; sr scientist, Pharmacol Dept, Parke-Davis Res, Ann Arbor, Mich, 83-85, res assoc, 85-88, sr res assoc, 88-89, SECT DIR, NEUROL DISORDERS, PHARMACOL DEPT, PARKE-DAVIS RES, ANN ARBOR, MICH, 89- *Concurrent Pos:* Mem, Stroke Coun, Brain Study Sect, Am Heart Asn, 90-91. *Mem:* Am Physiol Soc; AAAS; Am Heart Asn. *Res:* Vascular diseases, especially cerebrovascular diseases; vascular, neurochemical and pathological aspects of cerebral ischemia; nerve cell injury in cell culture and the role of calcium; pharmacology of vascular head pain; brain edema and blood brain barrier; neurokinins and vascular permeability; hypertension and neural regulation of blood pressure; central nervous system recovery mechanism and plasticity; neurodegenerative disorders; drug discovery in cerebrovascular diseases; cerebrovascular physiology and pharmacology; cerebral blood flow and metabolism in animal models for stroke, including the nonhuman primate; pharmaco-mechanical study of isolated cerebral resistance arteries; pharmacological regulation of blood brain barrier permeability; sensory innervation of the cerebral circulation; models of cerebral edema; hypoxic nerve cell injury in cell culture and the roles of excitotoxicity and calcium; author of various publications; granted three patents. *Mailing Add:* Dept Pharmacol Parke-Davis Pharmaceut Res Div Warner-Lambert Co 2800 Plymouth Rd Ann Arbor MI 48105

MARCOUX, JULES E, b Charny, Que, Jan 26, 24; m 55; c 6. PHYSICS. *Educ:* Laval Univ, BA, 47, BASc, 52; Univ Toronto, MA, 54, PhD(physics), 56. *Prof Exp:* Nat Res Coun Can fel, 56-57; prof physics, Royal Mil Col, Que, 57-62; prof, Laval Univ, 62-64; prof physics, Royal Mil Col, Que, 64-90; RETIRED. *Mem:* Can Asn Physicists; Am Asn Physics Teachers; NY Acad Sci. *Res:* Physical constants of rare gases in the liquid and solid states. *Mailing Add:* 29 rue des Tilleuls St Luc PQ J0J 2A0 Can

MARCOVITZ, ALAN BERNARD, b Boston, Mass, July 4, 36; m 60; c 2. ELECTRICAL & COMPUTER ENGINEERING. *Educ:* Mass Inst Technol, SBEE & SMEE, 59; Columbia Univ, PhD(elec eng), 63. *Prof Exp:* Instr elec eng, Columbia Univ, 59-63; from asst prof to assoc prof, Univ Md, 63-70; chmn dept, Fla Atlantic Univ, 70-76, prof elec eng, 70-80, asst dean eng, 82-88, PROF COMPUTER ENG, FLA ATLANTIC UNIV, 89- *Concurrent Pos:* Reviewer, Acad Press, 67-; consult, US Army Logistics Mgt Ctr, 68-72. *Mem:* Inst Elec & Electronics Engrs; Am Asn Eng Educ. *Res:* Computer systems; time-sharing system design; computer-aided design; computer education. *Mailing Add:* Col Eng Fla Atlantic Univ Boca Raton FL 33431

MARCU, KENNETH BRIAN, b Brooklyn, NY, June 15, 50; c 1. MOLECULAR IMMUNOLOGY, MOLECULAR ONCOLOGY. *Educ:* State Univ NY, Stony Brook, BS, 72, PhD(molecular biol), 75. *Prof Exp:* Postdoctoral assoc, Inst Cancer Res, Philadelphia, 75-78; from asst prof to assoc prof, 78-88, PROF, STATE UNIV NY, STONY BROOK, 88- *Concurrent Pos:* Prin investr, NIH grants, 78-, 81-84 & 84-, res career develop award, 81-86; mem, Basel Inst Immunol, 83-84, Adv Comt Proteins & Nucleic Acids, Am Cancer Soc, 84-89 & Adv Comt Immunol & Immunother, Am Cancer Soc, 91-; ad hoc mem, Immunol Study Sect, NIH, 91-; assoc ed, J Immunol, 91-93. *Mem:* Am Soc Microbiol; AAAS; Am Asn Immunologists. *Res:* Regulation of genes in mammalian cells with particular emphasis on proto-oncogenes which are involved in normal cell growth control; regulation of lymphoid antigen receptor gene rearrangements; genes which transform lymphoid cells. *Mailing Add:* Dept Biochem & Cell Biol State Univ NY Stony Brook NY 11794

MARCUM, JAMES BENTON, b Cedar Co, Mo, June 25, 38; m 64; c 3. ANIMAL GENETICS, CYTOGENETICS. *Educ:* Univ Mo-Columbia, BSAgr, 60, PhD(animal genetics), 69; Cornell Univ, MS, 61; Midwestern Baptist Theol Sem, MDiv, 65. *Prof Exp:* Lectr animal breeding, Univ Libya, 69-71; from asst prof to assoc prof, 71-85, chmn dept vet & animal sci, 78-85, PROF ANIMAL GENETICS, UNIV MASS, AMHERST, 85-, ASSOC DEAN & DIR RESIDENT INSTR, 85 - *Mem:* Am Soc Animal Sci; Am Genetic Asn; Nat Asn Col & Teachers Agr; Am Asn Higher Ed; Sigma Xi. *Res:* Chromosomal identification and mapping in domestic animals; relationships between cytogenetics and reproductive biology; freemartin syndrome. *Mailing Add:* Col Food & Natural Resources Univ Mass Amherst MA 01003

MARCUS, AARON JACOB, b Brooklyn, NY, Nov 6, 25; m 55; c 3. INTERNAL MEDICINE, HEMATOLOGY. *Educ:* Univ Va, BA, 48; New York Med Col, MD, 53. *Prof Exp:* CHIEF HEMAT SECT, NEW YORK VET ADMIN HOSP, 58-; PROF MED, MED COL, CORNELL UNIV, 74- *Concurrent Pos:* NIH res fel, Montefiore Hosp, 56-58; attend physician, New York Hosp, 74- *Mem:* Am Soc Clin Invest; Asn Am Physicians; Am Physiol Soc; Am Soc Hemat. *Res:* Hemostasis, coagulation and thrombosis; biochemistry and physiology of blood platelets. *Mailing Add:* Dept Hematol-Oncol NY Vet Admin Med Ctr Cornell Med Col 408 First Ave 13 W New York NY 10010

MARCUS, ABRAHAM, b New York, NY, Oct 26, 30; m 55; c 4. BIOCHEMISTRY. *Educ:* Yeshiva Univ, BA, 50; Univ Buffalo, AM, 54, PhD, 56. *Prof Exp:* Asst, Univ Buffalo, 52-54; biochemist, Agr Mkt Serv, Plant Indust Sta, USDA, 58-67; mem staff biol div, 67, assoc mem, 67-71, SR MEM, INST CANCER RES, 71- *Concurrent Pos:* USPHS res fel, Univ Chicago, 56-58; mem staff biophys, Weizmann Inst Sci, 64-65; vis prof, Bar-Ilan Univ, Israel, 70-71. *Mem:* Am Chem Soc; Am Soc Biol Chem; Am Soc Plant Physiol. *Res:* Metabolic pathways as ascertained by enzymatic studies; metabolic control of growth and development. *Mailing Add:* Inst for Cancer Res 7701 Burholme Ave Philadelphia PA 19111

MARCUS, ALLAN H, b New York, NY, July 14, 39. STATISTICS, ENVIRONMENTAL SCIENCES. *Educ:* Case Western Reserve Univ, BS, 61; Univ Calif, Berkeley, MA, 63, PhD(statist), 65. *Prof Exp:* Asst prof math, Case Western Reserve Univ, 64-67; mem staff, Bellcomm, Inc, DC, 67-68; assoc prof statist & earth & planetary sci, Johns Hopkins Univ, 68-73; assoc prof math, Univ Md, Baltimore County, 73-77; assoc prof math, 77-79, prof math, Wash State Univ, 79-87, Statist Consult, Wash State Univ Comput Ctr, 79-87; SR RES SCI, BATTELLE-COLUMBUS LAB, 87- *Concurrent Pos:* Consult, Rand Corp, 63-65; fel, Statist Lab, Cambridge Univ, 65-66; exec secy, Power Plant Siting Adv Comt, State of Md, 73-77. *Mem:* Am Statist Asn; Biometrics Soc; Soc Toxicol; Soc Risk Anal. *Res:* Applied statistics; urban transportation and environmental sciences; environmental health; biomathematics; air pollution; toxicology and pharmacokinetics; risk assessment. *Mailing Add:* Battelle Res Triangle Park Off 100 Park Dr Suite 207 PO Box 13758 Research Triangle Park NC 27709-3758

MARCUS, ANTHONY MARTIN, b London, Eng, June 21, 29; Can citizen; div; c 1. PSYCHIATRY. *Educ:* Cambridge Univ, BA, 52, MA, 56 & LMS, 56; McGill Univ, dipl psychiat, 62; Royal Col Physicians & Surgeons Can, spec cert, 62; Am Bd Psychiat & Neurol, dipl, 63. *Prof Exp:* From instr to asst prof, 62-70, dir Div Forensic Psychiat, 67-84, actg head dept, 70-72, ASSOC PROF PSYCHIAT, UNIV BC, 70-; CONSULT, PSYCHIATRIST SCHIZOPHRENIC SERV, 85- *Concurrent Pos:* Res fel psychiat, Montreal Gen Hosp, Que, 61-62; Can Penitentiary Serv res grant study dangerous sexual offenders; vis staff, Vancouver Gen Hosp; mem Lt Gov BC Order in Coun Rev Bd. *Honors & Awards:* Bronze Medal for Bravery, Royal Can Humane Asn, 77. *Mem:* fel Am Psychiat Asn; Can Psychiat Asn; fel Am Col Psychiat. *Res:* Forensic psychiatry; clinical psychiatry and psychopathology; teaching. *Mailing Add:* Dept Psychiat Univ BC Vancouver BC V6T 1W5 Can

MARCUS, BRUCE DAVID, b New York, NY, May 26, 37; m 58; c 2. THERMAL SCIENCES, THERMO-MECHANICS. *Educ:* Cornell Univ, BME & MME, 59, PhD(thermal processes), 63. *Prof Exp:* Postdoctoral fel heat transfer, Delft Tech Univ, 63-64; mem tech staff, TRW Defense & Space Systs Group, 65-68, proj mgr, heat pipe projs, 69-74, solar elec propulsion syst, 78-80 & excimer laser proj, 82-84, PROJ MGR, TRW ELECTRONICS & DEFENSE, 69-, PROJ MGR, ORBITING SOLAR LAB, 88- *Concurrent Pos:* Mem, Comt Theory & Fundamental Res, Nat Heat Transfer Div, Am Soc Mech Engrs, 69-75; adj assoc prof, Dept Energy & Kinetics, Univ Calif, Los Angeles, 71-75; consult, McGraw-Hill, Inc, 75-78. *Mem:* Am Soc Mech Engrs; Am Inst Aeronaut & Astronaut. *Res:* Heat and mass transfer; heat pipes; thermosiphons; spacecraft thermal control; thermo-mechanical design of telecopes. *Mailing Add:* 1877 Comstock Ave Los Angeles CA 90025

MARCUS, BRYAN HARRY, b Los Angeles, Calif, Aug 29, 49. MATHEMATICS. *Educ:* Pomona Col, BA, 71; Univ Calif, Berkeley, MA, 72, PhD(math), 75. *Prof Exp:* Asst prof, 75-81, ASSOC PROF MATH, UNIV NC, 81- *Concurrent Pos:* IBM Corp fel, 76-77; NSF res grants, 76- *Res:* Erogodic theory; dynamical systems. *Mailing Add:* 650 Harry Rd San Jose CA 95120

MARCUS, CAROL JOYCE, b New York, NY, Aug 13, 43; c 1. BIOCHEMISTRY. *Educ:* Cornell Univ, BS, 65; Duke Univ, PhD(biochem), 72. *Prof Exp:* From instr to asst prof biochem, Univ Tenn Ctr Health Sci, Memphis, 72-78; SR STAFF FEL, EXP PATH LAB, NIH, 78- *Concurrent Pos:* Vis scientist, Biochem & Metab Lab, NIH, 76-78. Acad Sci. *Res:* Adeno associated virus; viral proteins; viral replication; Xenopus oocyte microinjection. *Mailing Add:* 5604 Vernon Pl Bethesda MD 20817

MARCUS, CAROL SILBER, b New York, NY, July 2, 39; m 58; c 2. RADIATION BIOLOGY, BIOPHYSICS. *Educ:* Cornell Univ, BS, 60, MS, 61, PhD(phys biol), 63; Univ Southern Calif, MD, 77. *Prof Exp:* Teaching fel, NY State Grad Col, 60-62; NIH fel, 62-63; vis res scientist, Lab Biol Med, Neth, 63-64; asst res biochemist, Lab Nuclear Med & Radiation Biol, Univ Calif, Los Angeles, 65-67; STAFF PHYSICIAN & DIR, NUCLEAR MED OUTPATIENT CLINIC, DIV NUCLEAR MED, LOS ANGELES COUNTY HARBOR, UCLA MED CTR, 82-, ASST PROF RADIOL, UNIV CALIF, LOS ANGELES, 82- *Concurrent Pos:* Instr, Santa Monica City Col Exten, 66-69; Pfeiffer Found fel, Sch Pharm, Univ Southern Calif, 69-70; consult, Gen Elec Co & Innotek, Inc, 70-72, XMI Assocs, 71-73, Lab Nuclear Med & Radiation Biol, Univ Calif, Los Angeles, 72-77, Cutter Labs, 82- & Vesta Res, 84-; adv & consult, Radiobiol for Nuclear Med Technol Training Prog, Los Angeles City Col, 71-73; consult & mem Radiopharmaceut Adv Comt, US Food & Drug Admin, 72-77 & 82-87; adj asst prof radiopharm, Sch Pharm, Univ Southern Calif, 73-77, adj assoc prof, 77-; intern internal med, Med Ctr, LAC-Univ Southern Calif, Los Angeles, 77-78, resident, 78-80; resident nuclear med, Wadsworth Med Ctr, Vet Admin, Los Angeles, Calif, 80-82; reviewer, Noninvasive Med Imaging, 84-85 & J Nuclear med, 85; consult, Cutter Lab, 82- & Vestar res, 48-86; adj fac, Dept clin Sci, calif state Univ, Dominque Hills, 85; consult, E I du Pont de Nemours Inc, 87-, Syncor Inc, 87- & Millinckrodt Inc, 87- *Mem:* AAAS; Soc Nuclear Med; Am Col Nuclear Physicians; NY Acad Sci; AMA. *Res:* Use of semiconductor microprobe radiation detectors and short lived radionuclides for biomedical applications; radiolabeling of cellular blood elements for diagnostic nuclear medicine techniques; radiobiological aspects of radiopharmaceutical research and development; regulatory aspects of nuclear medicine procedures. *Mailing Add:* Univ Calif Los Angeles Med Ctr Los Angeles County Harbor 1000 W Carson St Torrence CA 90509

MARCUS, DONALD M, b New York, NY, Dec 10, 30; m 58; c 3. INTERNAL MEDICINE, IMMUNOCHEMISTRY. *Educ:* Princeton Univ, BA, 51; Columbia Univ, MD, 55. *Prof Exp:* Intern internal med, Presby Hosp, New York, 55-56, asst resident, 56-57; assoc resident, Strong Mem Hosp, Rochester, 59-60; assoc med, Albert Einstein Col Med, 63-64, asst prof, 64-70, assoc prof med & microbiol, 70-75, prof med, microbiol & immunol, 75-80, dir, Div Rheumatology & Immunol, 73-80; PROF, DEPTS MICROBIOL & IMMUNOL & MED, BAYLOR COL MED, 80- *Concurrent Pos:* Helen Hay Whitney Found fel, 60-63; career scientist, Health Res Coun, New York, 63- *Honors & Awards:* Carl Landsteiner Mem Award, Am Asn Blood Banks, 80; Philip Levine Award, Am Soc Clin Pathologist, 85. *Mem:* Am Asn Immunol; Am Soc Biol Chem; Am Rheumatism Asn; Am Soc Clin Invest. *Res:* Immunochemistry of human blood group antigens; blood group and cell membrane antigens; glycosphingolipids; structure and regulation of anitbodies against carbohydrate determinants; mechanism of hapten-antibody interactions. *Mailing Add:* Primary Dept Med Baylor Col Med 1200 Moursund Ave Houston TX 77030

MARCUS, ELLIOT M, b Bridgeport, Conn, Sept 13, 32; m 62; c 1. NEUROLOGY, CLINICAL NEUROPHYSIOLOGY. *Educ:* Yale Univ, BA, 54; Tufts Univ, MD, 58. *Prof Exp:* Intern med, Yale New Haven Hosp, 58-59; clin fel neurol, New Eng Med Ctr, Tufts Univ, 59-60, chief resident, 61-62, NIH res fel neurophysiol, 62-63; vis fel neuropath, Columbia Presby, 60-61; chief neurol, Walson Army Hosp, Ft Dix, 63-65; instr physiol & neurol, 65-66, asst prof, 66-70, from assoc prof to prof neurol, 70-76, LECTR NEUROL, TUFTS UNIV, 76-; PROF NEUROL, MED SCH, UNIV MASS, 76- *Concurrent Pos:* Capt, Med Corp, Army US, 63-65; chair neurosci teaching, Tufts, 66-76; asst neurologist, New Eng Med Ctr, 65-70, neurologist, 70-75, sr neurologist, 75-76; chief, div neurol, St Vincent Hosp, 76-, dir, neurophysiol labs, 76 -; chmn, Comt Nat Needs Neurol, Am Acad Neurol; chmn, Dept Neurol, St Vincent Hosp, 89- *Mem:* Am Acad Neurol; Am Neurol Asn; Am Epilepsy Soc; Am EEG Soc; Asn Res Nerv & Ment Dis. *Res:* Development of experimental models of generalized epilepsy; corpus callusum in bilateral epileptic discharges; genetic, endocrine and age factors in seizures; anatomical physiological studies of the corpus callosum; prognosis in coma: use of evoked potentials relationship to age, etiology and brain stem auditory evoked potentials; integrated teaching of neurosciences. *Mailing Add:* Div Neurol St Vincent Hosp Worcester MA 01604

MARCUS, FRANK, b Berlin, Ger, July 27, 33; US citizen; m 59; c 3. PROTEIN CHEMISTRY, ENZYMOLOGY. *Educ:* Univ Chile, Santiago, PhC, 58. *Prof Exp:* Res assoc biochem, Univ Chile, 58-61; Rockefeller Found fel, Australian Nat Univ, 62-63, & Univ Wis-Madison, 64; asst prof biochem, Univ Chile, 65-68; prof & chmn, Inst Biochem, Southern Univ Chile, Valdivia, 68-74; vis prof, Univ Wis-Madison, 74-77; from assoc prof to prof biochem, Univ Health Sci, Chicago Med Sch, 77-88; DIR PROTEIN CHEM, CHIRON CORP, EMORYVILLE, CALIF, 88- *Concurrent Pos:* Consult, Photosynthesis Panel, US Dept Agr; Off Grants & Prog Systs, USDA, 85-86. *Mem:* Am Soc Biol Chemists; Am Chem Soc; Am Soc Plant Physiologists; Protein Soc. *Res:* Protein purification and characterization of proteins; gluconeogenesis and its regulation; structure, function and evolution of enzyme. *Mailing Add:* Dir Protein Chem Chiron Corp 4560 Horton St Emeryville CA 94608

MARCUS, FRANK I, b Haverstraw, NY, Mar 23, 28; m 57; c 3. CARDIOLOGY, INTERNAL MEDICINE. *Educ:* Columbia Univ, BA, 48; Tufts Univ, MS, 51; Boston Univ, MD, 53. *Prof Exp:* Intern med, Peter Bent Brigham Hosp, Boston, 53-54, asst resident, 56-57; clin fel, Georgetown Univ Hosp, 58-59, chief med resident, 59-60; from instr to assoc prof med, Georgetown Univ, 60-68; prof med & chief cardiol, 69-82, DIR, ELECTROPHYSIOL, ARIZ MED CTR, UNIV ARIZ, 82- *Concurrent Pos:* Mass Heart Asn res fel cardiol, Peter Bent Brigham Hosp, 57-58; Markle scholar, Georgetown Univ, 60-65, NIH career develop award, 65-68; chief cardiol, Georgetown Univ Med Serv Div, DC Gen Hosp, 60-68; fel coun clin cardiol, Am Heart Asn, 65-; consult, Vet Admin & Davis-Monthan AFB Hosps, Tucson, 69-; ed, Modern Concepts Cardiovas Dis, 82-85; distinguished prof internal med (cardiology), Endowed Chair principally supported by Am Heart Asn & Flinn Found, 82; gov, Am Col Cardiol, State of Ariz; mem, bd trustees & asst secy, Am Col Cardiol, 87-89; mem, adv bd, Int Cardiol Inst Therapeut Res, 88- & Diag & Therapeut Technol Assessment Panel Med Experts, AMA, 90- *Mem:* Am Fedn Clin Res; Am Heart Asn; fel Am Col Physicians; Asn Univ Cardiologists (vpres, 89-90, pres, 90-91); Am Soc Pharmacol & Exp Therapeut; fel Am Col Cardiology. *Res:* Cardiovascular research; pharmacology; digitalis; metabolism; clinical electrophysiology. *Mailing Add:* Ariz Med Ctr Cardiol Sect Univ Ariz Tucson AZ 85724

MARCUS, GAIL HALPERN, b New York, NY, Jan 28, 47; m 68. NUCLEAR ENGINEERING, PHYSICS. *Educ:* Mass Inst Technol, SB, SM, 68, ScD(nuclear eng), 71. *Prof Exp:* Physicist, US Army Electronics Command, Ft Monmouth, 68-71; res asst, Mass Inst Technol, 70-71; nuclear engr, Analysis Servs Inc, 72-77, dep mgr, support systs div, 77-80; asst chief, Sci Policy Res Div, Cong Res Serv, 80-85, dir, policy, planning & control staff, Off Nuclear Regulatory Res, 85-86, TECH ASST, OFF COMN, US NUCLEAR REGULATORY COMN, 87- *Concurrent Pos:* Instr, Grad Sch, US Dept Agr, 76; mem bd dirs, Educ Prospective Engrs Pub Policy, Am Soc Eng Educ, 82-, Am Nuclear Soc, 82-85; mem vis comt, Mass Inst Technol, 84-; NSF peer review comt Nuclear Regulatory Comn Probabilistic Risk Assessment Ref Document, 84-85; comnr, Sci Manpower Comn, 85-87; tech comt Thermal Reactor Safety Res, Int Atomic Energy Agency, 85-86; indust adv coun, Univ Lowell, 86; res & develop coun, Am Mgt Asn, 86-; Nat Res Coun comt Future Needs Nuclear Eng Educ, 89-90. *Mem:* Fel Am Nuclear Soc; AAAS. *Res:* Energy systems and policy; nuclear power; nuclear reactor safety research; risk assessment; radiation damage to materials; science policy. *Mailing Add:* US Nuclear Regulatory Comm Washington DC 20555

MARCUS, GEORGE JACOB, b Toronto, Ont, Mar 17, 33; m 56; c 2. REPRODUCTIVE BIOLOGY. *Educ:* Univ Toronto, BA, 56, PhD(biochem), 61. *Prof Exp:* Res assoc biodynamics, Weizmann Inst, 63-68; asst prof pop dynamics, Sch Hyg, Johns Hopkins Univ, 68-72; asst prof reproductive biol, Dept Anat & Lab Human Reproduction & Reproductive Biol, Harvard Med Sch, 72-74; RES SCIENTIST, ANIMAL RES INST, RES BR, CAN DEPT AGR, 75- *Concurrent Pos:* Pop Coun med fel, Weizmann Inst, 61-63; asst ed, Biol Reproduction, 70-74. *Res:* Biochemical aspects of nidation; decidual induction; preimplantation embryonic development; maternal recognition of pregnancy. *Mailing Add:* 271 Crocus Ave Ottawa ON K1H 6E7 Can

MARCUS, HARRIS L, b Ellenville, NY, July 5, 31; m 62; c 2. METALLURGY, MATERIALS SCIENCE. *Educ:* Purdue Univ, BS, 63; Northwestern Univ, PhD(mat sci), 66. *Prof Exp:* Shift supvr, Channel Master Corp, NY, 56-60; mem tech staff, metals & controls div, Tex Instruments, Inc, Mass, 66-68; group leader fracture & metal physics, Sci Ctr, N Am Rockwell Corp, 68-75; Harry L Kent Jr prof mech eng & mat sci & eng, 75-90, CULLEN PROF ENG, UNIV TEX, AUSTIN, 90- *Concurrent Pos:* Dir, Ctr Mat Sci & Eng, Univ Tex, Austin, 80- *Honors & Awards:* Von Karman Spec Award, 74. *Mem:* Am Inst Mining, Metall & Petrol Engrs; fel Am Soc Metals; Am Phys Soc; Am Soc Testing & Mat; Metall Soc. *Res:* Application of Auger and Mossbauer spectroscopy to metallurgy; fracture and fatigue of structural alloys;effect of metallurgical variables and environment; electronic packaging; magnetic materials; materials processing. *Mailing Add:* Dept Mech Eng Ctr for Mat Sci & Eng Univ of Tex Austin TX 78712

MARCUS, JOHN STANLEY, b Los Angeles, Calif. MATERIAL HARDNESS TEST EQUIPMENT, THERMOMETRY. *Educ:* Univ Calif, Los Angeles, BA, 79, MS, 81, PhD(physics, 83. *Prof Exp:* Dir res & develop, Amcor Indust Inc, 83-87; PRES & CHIEF EXEC OFFICER, PAC TRANSDUCER CORP, 87- *Concurrent Pos:* Res assoc physics, Univ Calif, Los Angeles, 82-83; consult, Rand Corp, 84-87. *Mem:* Am Phys Soc; Soc Mfg Engrs; Am Soc Metals; Am Standards Testing Mat. *Res:* Product research and development for thermometers, material hardness testers and other industrial test equipment. *Mailing Add:* Pac Transducer Corp 2301 Fed Ave Los Angeles CA 90064

MARCUS, JOSEPH, b Cleveland, Ohio, Feb 27, 28; c 2. CHILD PSYCHIATRY. *Educ:* Hadassah Med Sch, Hebrew Univ, MD, 58; Western Reserve Univ, BSc, 63. *Prof Exp:* Resident psychiat, Ministry of Health, Israel, 58-61; actg head child psychiat, Ness Ziona Rehab Ctr; sr psychiatrist, Lasker Dept Child Psychiat, Hadassah Hosp, 62-64, consult, Tel Hashomer Govt Hosp, 65-66; res assoc, Israel Inst Appl Social Res, 66-69; assoc dir, Jerusalem Infant & Child Develop Ctr, 69-70; head dept child psychiat, Eytanim Hosp, 70-72; dir child psychiat & develop, Jerusalem Ment Health Ctr, 72-75; prof & dir, 75-85, EMER PROF PSYCHIAT, UNIV CHICAGO & CO-DIR, UNIT RES CHILD PSYCHIAT & DEVELOP, 86- *Concurrent*

Pos: Vis res psychiatrist, Univ Calif, Los Angeles, 87- *Mem:* Am Acad Child Psychiat; Soc Res Child Develop. *Res:* Development of infants of parents with serious mental diseases, especially behavioral, neurological, physiological and biochemical aspects. *Mailing Add:* 910 Chelham Way Santa Barbara CA 93108

MARCUS, JULES ALEXANDER, b Coytesville, NJ, May 10, 19; div; c 4. PHYSICS. *Educ:* Yale Univ, BS, 40, MS, 44, PhD(physics), 47. *Prof Exp:* Instr physics, Yale Univ, 42-44; res physicist, Appl Physics Lab, Johns Hopkins Univ, 44-46; asst physics, Yale Univ, 46-47; fel, Inst Study Metals, Univ Chicago, 47-49; from asst prof to assoc prof, 49-61, PROF PHYSICS, NORTHWESTERN UNIV, 61- *Mem:* Fel Am Phys Soc; Am Asn Physics Teachers; AAAS. *Res:* Low temperature solid state physics; de Haas-van Alphen effect; experimental determination of Fermi surfaces; Overhauser spin-density-waves in chromium; galvanomagnetic effects in metals; first observation of quantum oscillations of magnetic properties of normal metals at low temperatures; first observation of magnetically induced two-fold symmetry in potassium and sodium metals. *Mailing Add:* 2801 Girard Ave Evanston IL 60201

MARCUS, LESLIE F, b Los Angeles, Calif, Oct 22, 30; m 58; c 1. BIOMETRY, PALEONTOLOGY. *Educ:* Univ Calif, Berkeley, BA, 51, MA, 59, PhD(paleont), 62. *Prof Exp:* From asst prof to assoc prof statist, Kans State Univ, 60-67; assoc prof, 67-70, PROF BIOL, QUEENS COL, NY, 70-; RES ASSOC, DEPT INVERT, AM MUS NATURAL HIST, 76- *Concurrent Pos:* Vis asst prof, Univ Kans, 63-64; NSF sci fac fel, Columbia Univ, 66-67. *Mem:* AAAS; Soc Study Evolution; Soc Syst Zool; Am Statist Asn; Biomet Soc; Sigma Xi. *Res:* Vertebrate paleontology; statistical methods application to study of natural selection in fossils; geographic variation; morphology; numerical classification; data base management of museum collections; multivariate statistics. *Mailing Add:* Invertebrates Dept Am Mus Natural Hist Cent Park W & 79th St New York NY 10024

MARCUS, MARK, b Des Moines, Iowa, Oct 7, 44; m 66; c 2. HAZARDOUS WASTE, QUALITY CONTROL. *Educ:* Univ Iowa, BS, 66; Kans State Univ, PhD(chem), 70. *Prof Exp:* Asst prof anal chem, Marquette Univ, 70-74; mgr environ anal chem, Midwest Res Inst, 74-79; mgr anal chem, Environ Health Ctr, Stauffer Chem Co, 79-83; SR DIR ANAL PROGS, CHEM WASTE MGT, WASTE MGT, INC, 83- *Mem:* Am Chem Soc; Asn Official Anal Chemists; Sigma Xi; Am Soc Testing Mats. *Res:* Analysis of hazardous waste necessary for disposal decisions; quality control and quality assurance policies. *Mailing Add:* 150 W 137th St Chicago IL 60627

MARCUS, MARVIN, b Albuquerque, NMex, July 31, 27; m 65; c 2. ALGEBRA. *Educ:* Univ Calif, Berkeley, BA, 50, PhD(math), 54. *Prof Exp:* Res assoc, Univ Calif, Berkeley, 53-54; assoc prof math, Univ BC, 54-60, 61-62; res mathematician, Numerical Anal Sect, Nat Bur Standards, 60-61; chmn dept, 63-68, PROF MATH, UNIV CALIF, SANTA BARBARA, 62-, DEAN RES DEVELOP, 78-, ASSOC VCHANCELLOR, RES & ACADEMIC DEVELOP, 78- *Concurrent Pos:* Fulbright grant, 54; consult, US Naval Test Sta, 55; Nat Res Coun fel, 56-57; NSF grant, 58-59, 63-66, 81-83; vis distinguished prof, Univ Islamabad, W Pakistan, 70; dir, Inst Interdisciplinary Applns of Algebra and Combinatorics, 73-79; Air Force grant, 79-80; Fund Improv Postsecondary Educ grant, 79-81. *Mem:* Am Math Soc; Math Asn Am; Soc Indust & Appl Math; Soc Tech Commun; Sigma Xi. *Res:* Linear and multilinear algebra. *Mailing Add:* 2937 Kenmore Pl Santa Barbara CA 93105

MARCUS, MELVIN GERALD, b Seattle, Wash, Apr 13, 29; m 53; c 4. PHYSICAL GEOGRAPHY. *Educ:* Univ Miami, BA, 56; Univ Colo, MS, 57; Univ Chicago, PhD(geog), 63. *Prof Exp:* Res asst climat, Lab Climat, 58-59; from instr to asst prof geog, Rutgers Univ, 60-64; from asst prof to prof, Univ Mich, Ann Arbor, 64-73, chmn dept, 67-71; dir, Ctr Environ Studies, 74-81, PROF GEOG, ARIZ STATE UNIV, TEMPE, 74- *Concurrent Pos:* Sr scientist, Icefield Ranges Res Proj, Arctic Inst NAm, 64-71; consult, High Sch Geog Proj, Boulder, 67; chmn, Comn Col Geog, 68-71; vis prof, Univ Canterbury, 72 & 81, US Mil Acad, 85-86. *Mem:* Asn Am Geog (pres, 77-78); Am Geog Soc (vpres, 88-); Glaciol Soc; Sigma Xi. *Res:* Glaciological and climatological work, particularly in Alpine regions; physical geography to include geographic education and urban environments; environmental education. *Mailing Add:* Geog Sci Ariz State Univ Tempe AZ 85287

MARCUS, MELVIN L, cardiology; deceased, see previous edition for last biography

MARCUS, MICHAEL ALAN, b New York City, NY, Nov 13, 52; m 72; c 3. POLYMER PHYSICS, FIBER OPTIC SENSORS. *Educ:* Rensselaer Polytech Inst, BS, 73; Cornell Univ, PhD(appl physics), 78. *Prof Exp:* NIH fel appl physics, Cornell Univ, 75-78; SR RES SCIENTIST, KODAK RES LABS, 78- *Concurrent Pos:* Lectr, Brockport State Univ, 79-, Univ Rochester, 80-81; pres, Beta Physics, 84-86; ferroelectrics comt, Inst Elec & Electronic Engrs. *Mem:* Am Phys Soc; Optical Soc Am; Am Vacuum Soc; Inst Elec & Electronic Engrs; Int Soc Optical Eng. *Res:* Electrical, thermal and mechanical properties of ferroelectric materials; polymeric ferroelectrics and device applications of these materials; applications for fiber optic sensors, process monitoring and control sensors. *Mailing Add:* Eastman Kodak Co B-23 Kodak Park Rochester NY 14652-4201

MARCUS, MICHAEL BARRY, b Brooklyn, NY, Mar 5, 36; m 64; c 3. MATHEMATICS. *Educ:* Princeton Univ, BSE, 57; Mass Inst Technol, MS, 58, PhD(math), 65. *Prof Exp:* Staff mem math & electronics, Rand Corp, 58-67; from asst prof to prof math, Northwestern Univ, 67-86; prof math, Tex A&M Univ; PROF, DEPT MATH, CITY COL, CITY UNIV NEW YORK GRAD CTR, 86- *Concurrent Pos:* Asst, Mass Inst Technol, 62-65; NSF vis prof, Westfield Col, Univ London, 70-71; vis mem, Courant Inst, NY Univ, 73-74; guest prof, Aarhus Univ Denmark, 77-78. *Mem:* Fel Inst Math Statist; Am Math Soc. *Res:* Probability theory; analysis. *Mailing Add:* Dept Math City Col City Univ New York New York NY 10031

MARCUS, MICHAEL JAY, b Boston, Mass, May 13, 46. TELECOMMUNICATIONS POLICY, COMMUNICATIONS ARCHITECTURE. *Educ:* Mass Inst Technol, SB, 68, ScD(elec eng), 72. *Prof Exp:* Proj officer, US Air Force, 72-75; res staff mem, Inst Defense Anal, 75-79; spec asst to chief scientist, 79-81, div chief, 81-87, ASST BUR CHIEF, Fed Commun Comn, 88-,. *Concurrent Pos:* Prof lectr, George Washington Univ, 77-81; vis assoc prof, MIT, 86. *Mem:* Inst Elec & Electronics Engrs. *Res:* Technology assessment in support of telecommunication's policy development and review of Federal Communications Commission technical rules to remove implicit and explicit barriers to innovation. *Mailing Add:* Fed Commun Comn Rm 734 1919 M St NW Washington DC 20554

MARCUS, NANCY HELEN, b New York, NY, May 17, 50. POPULATION GENETICS. *Educ:* Goucher Col, BA, 72; Yale Univ, MPhil, 75, PhD(biol), 76. *Prof Exp:* Scholar, biol oceanog, Woods Hole Oceanog Inst, 76-77, investr, 77-78, from asst scientist to assoc scientist, 78-87; ASSOC PROF, DEPT OCEANOG, FLA STATE UNIV, 87- *Concurrent Pos:* Instr develop biol, Marine Biol Lab, 78- *Honors & Awards:* Fel, AAAS, 89. *Mem:* Am Soc Zoologists (secy div of ecol, 85-87); Soc Develop Biol; Soc Study Evolution; Am Soc Limnol & Oceanog; AAAS. *Res:* Zooplankton ecology and population biology; dormancy. *Mailing Add:* Dept Oceanog Fla State Univ Tallahassee FL 32306

MARCUS, PAUL MALCOLM, b New York, NY, Feb 4, 21. MATHEMATICAL PHYSICS. *Educ:* Columbia Univ, BA, 40; Harvard Univ, MA, 42, PhD(chem physics), 43. *Prof Exp:* Mem staff, Radiation Lab, Mass Inst Technol, 43-46; fel, Nat Res Coun, 46-47, res assoc physics, 47-48; sci liaison officer, Off Naval Res, US Govt, Eng, 49-50; res asst prof physics, Univ Ill, 50-52; lectr, Carnegie Inst Technol, 52-53, res physicist, 53-58, asst prof physics, 58-59; PHYSICIST, T J WATSON RES CTR, IBM CORP, 59- *Mem:* Fel Am Phys Soc. *Res:* Low temperature and solid state physics; radiation theory. *Mailing Add:* IBM Res Ctr Yorktown Heights NY 10598

MARCUS, PHILIP IRVING, b Springfield, Mass, June 3, 27; m 54; c 3. VIROLOGY. *Educ:* Univ Southern Calif, BS, 50; Univ Chicago, MS, 53; Univ Colo, PhD(microbiol, biophys), 57. *Prof Exp:* Lab asst infrared studies bacteria, Univ Chicago, 51-52, lab asst med & gen microbiol & microbiologist, 52-53, asst steroid enzyme induction, 53-54; asst biophys, Med Ctr, Univ Colo, 54-57, instr, 57-59, asst prof, 59-60; asst prof microbiol & immunol, Albert Einstein Col Med, 60-62, assoc prof, 62-66, prof, 66-69; head microbiol sect, 69-74, prof biol, 69-85, PROF, DEPT MOLECULAR & CELL BIOL & DIR, BIOTECHNOL CTR, UNIV CONN, 85- *Concurrent Pos:* USPHS sr res fel, 60-65, res career develop awardee, 65-69; on leave from Albert Einstein Col Med to Salk Inst, 67-68; mem sci bd, Damon Runyon Mem Fund Cancer Res, 69-73; ed, J Cellular Physiol, 69-, J Interferon Res, 80-83, ed-in-chief-, 84-; mem, sci adv comt res grants, Am Cancer Soc, 85-88; sci adv comt, Am Found AIDS Res, 89- *Mem:* AAAS; British Soc Microbiol; Am Soc Microbiol; Am Soc Cell Biol; NY Acad Sci; Int Soc Interferon Res; Harvey Soc; Am Soc Virol. *Res:* Single-cell cloning techniques for mammalian cells; host-cell animal virus interactions; mechanism of cell-killing by viruses; interferon induction and viral inhibition; cell surfaces; viral hemadsorption; viral interference; interferon action; persistent infection. *Mailing Add:* Dept Molecular & Cell Biol U-44 Univ Conn Storrs CT 06268

MARCUS, PHILIP SELMAR, b New York, NY, Jan 30, 36; m 66, 84; c 2. MATHEMATICS. *Educ:* Univ Chicago, AB, 56, BS, 58, MS, 59; Ill Inst Technol, PhD(math), 68. *Prof Exp:* Instr math, De Paul Univ, 62-66; mem fac, Shimer Col, 66-70, dir, Shimer-in-Oxford Prog, 68-69, chmn, Div Nat Sci, 67-70; asst prof math, Ind Univ, South Bend, 70-76; assoc prof math, Christian Bros Col, Memphis, 76-79; assoc prof, 79-88, PROF MATH, EUREKA COL, 88- *Mem:* Math Asn Am; Am Math Soc. *Res:* Probability; geometry; mathematics education. *Mailing Add:* Dept Math Eureka Col 300 E Col Ave Eureka IL 61530

MARCUS, PHILIP STEPHEN, b Philadelphia, Pa, Sept 7, 51; m 86. COMPUTATION FLUID DYNAMICS, NONLINEAR STABILITY FLUIDS. *Educ:* Caltech Univ, BS, 73; Princeton Univ, PhD(physics), 78. *Prof Exp:* Res assoc astron, Ctr Radiophysics & Space Res, Cornell Univ, 77-80; asst prof appl math, dept math, Mass Inst Technol, 80-83; assoc prof appl math & astron, dept astron, Div Appl Sci, Harvard Univ, 83-86; PROF FLUID DYNAMICS, UNIV CALIF BERKELEY, 86- *Concurrent Pos:* Consult, Lawrence Livermore Lab, 84-; assoc ed, J Computational Physics, 85- *Mem:* Fel Am Phys Soc. *Res:* Fluid dynamics with an emphasis on the supercomputers and with applications in nonlinear physics; the onset of chaotic behavior in fluid systems including turbulence and applications in astronomy. *Mailing Add:* Dept Mech Eng Univ Calif Berkeley Berkeley CA 94501

MARCUS, RICHARD, b Rochester, NY, Mar 6, 46; m 68; c 2. PSYCHOPHARMACOLOGY. *Educ:* Rensselaer Polytech Inst, BS, 67; Boston Univ, MA, 70, PhD(exp psychol), 74. *Prof Exp:* EXEC SECY, RES REV COMT TREATMENT, DEVELOP & ASSESSMENT, NIMH, 80- *Concurrent Pos:* Adminr, Am Col Neuropsychopharmacol. *Mem:* Am Psychol Asn; Am Soc Pharmacol & Exp Therapeut; Am Col Neuropsychopharmacol. *Mailing Add:* NIMH 5600 Fishers Lane Rockville MD 20857

MARCUS, ROBERT BORIS, b Chicago, Ill, Nov 26, 34; m 57; c 2. MATERIALS SCIENCE. *Educ:* Univ Chicago, BS, 56, SM, 58; Univ Mich, PhD(phys chem), 62. *Prof Exp:* Fel phys chem, Univ Mich, 61-62, instr, 62-63; mem tech staff, 63-67, supvr struct anal group, Bell Labs, 67-84, DIST RES MGR, BELL COMMUN RES, INC, 84- *Mem:* Electron Micros Soc; Am Vacuum Soc. *Res:* Microstructure and electrical properties of materials; laser probing of semiconductors; field emission. *Mailing Add:* 133 Colchester Rd New Providence Rd NJ 07974

MARCUS, ROBERT BROWN, b Phila, Pa, Dec 1, 18; m 42; c 2. PHYSICAL GEOGRAPHY. *Educ:* West Chester Univ, Pa, BS, 40; Univ Fla, MA, 53, EdD(geog), 56. *Prof Exp:* Teacher high sch, NC, 41-42; head sci dept & master chem & physics, Pennington Sch, NJ, 46-51; from instr to prof phys sci & geog, 54-76, prof geog, 76-85, chmn dept, 79-82, EMER PROF GEOG, UNIV FLA, 85- *Mem:* Asn Am Geog; Nat Coun Geog Educ; Int Geog Union. *Res:* Utilization of water and natural resources. *Mailing Add:* Dept Geog Univ Fla Gainesville FL 32611

MARCUS, ROBERT TROY, b Brookline, Mass, Dec 18, 49; m 74; c 4. COLOR SCIENCE & TECHNOLOGY. *Educ:* Rensselaer Polytechnic Inst, BS, 68, PhD(chem), 74. *Prof Exp:* Sr res physicist, PP6 Indust, Inc, 74-78, res assoc, 78-80; mgr sci systs, Mobay Chem Corp, 80-82; sect mgr, Travenol Labs, 82-84; res scientist, UniRoyal, Inc, 84-86; dir, Munsell Lab, MacBeth Div, Kollmorgen Instruments Corp, 86-88; SR RES ASSOC, PPG INDUST, INC, 88- *Concurrent Pos:* Mem, US Nat Comt, Comn Int D'Ellairange & Coun Optical Radiation Measurement. *Mem:* Am Soc Testing & Mat; Inter-Soc Color Coun. *Res:* Industrial color measurment, color difference evaluation and computer color control; color science and technology. *Mailing Add:* 3860 Dolphin Dr Allison Park PA 15101

MARCUS, RUDOLPH ARTHUR, b Montreal, Que, July 21, 23; nat US; m 49; c 3. PHYSICAL CHEMISTRY. *Educ:* McGill Univ, BSc, 43, PhD(phys chem), 46. *Hon Degrees:* DSc, Univ Chicago, 83; DSc, Polytechnic Univ, 86; Filda, Goteburg Univ, 87; DSc, McGill Univ, 88. *Prof Exp:* Jr res officer photochem, Nat Res Coun Can, 46-49; res assoc theoret chem, Univ NC, 49-51; asst prof phys chem, Polytech Inst Brooklyn, 51-54, from assoc prof to prof, 54-58; prof phys chem, Univ Ill, Urbana, 64-78; A A NOYES PROF CHEM, CALIF INST TECHNOL, PASADENA, 78- *Concurrent Pos:* Temp mem, Courant Insp Math Sci, 60-61; NSF sr fel, 60-61; Sloan fel, 60-63; vis sr scientist, Brookhaven Nat Lab, 62-64; coun mem, Gordon Res C/nf, 65-68, chmn bd trustees, 68-69; mem adv coun chem dept, Princeton Univ, 72-78, Calif Inst Technol, 77-78 & Polytech Inst NY, 77-80; Fulbright-Hays sr scholar, 72 & 73, Alexander von Humboldt sr US scientist award, 76; mem, Nat Res Coun-Nat Acad Sci Climatic Impact Comt, Panel Atmospheric Chem, 75-78; chmn, Nat Res Coun-Nat Acad Sci Comt, Kinetics of Chem Reactions, 75-77 & Sci Comt, Chem Sci, 77-78; mem, Review Comt, Radiation Lab, Univ Notre Dame, 75-80; vis prof theoret chem, Oxford Univ, 75-76; prof fel, Univ Col, Oxford, UK, 75-76. *Honors & Awards:* Irving Langmuir Award, Am Chem Soc, 78; Robinson Medal, Faraday Div, Royal Soc Chem, 82; Wolf Prize, 85; Peter Debye Award, Am Chem Soc, 88 & Willard Gibbs Medal, 88; Centenary Medal, Faraday Div, Royal Soc Chem, 88; Nat Medal of Sci, 89; Theodore William Richards Medal, 90. *Mem:* Nat Acad Sci; Am Chem Soc; Am Phys Soc; Am Acad Arts & Sci; foreign mem Royal Soc London; Int Acad Quantum Molecular Sci; Am Philos Soc. *Res:* Theoretical chemical kinetics; electron transfer, electrode and unimolecular reactions; semiclassical theory of reactive and nonreactive collisions of bound states and of spectra. *Mailing Add:* Div Chem & Chem Eng Calif Inst Technol Pasadena CA 91125

MARCUS, RUDOLPH JULIUS, b Frankfurt, Ger, Mar 30, 26; nat US; m 78. TECHNOLOGY TRANSFER. *Educ:* Wayne State Univ, BS, 48; Univ Utah, PhD, 54. *Prof Exp:* Chemist, Sun Oil Co, 48-49; phys chemist, Stanford Res Inst, 54-64; CHEMIST, OFF NAVAL RES, PASADENA, CALIF, 64-; CONSULT, 84- *Concurrent Pos:* Sci dir, Off Naval Res, Tokyo, 79-80; books adv bd, Am Chem Soc, 84-87, chmn, Div Comput Chem, 84. *Mem:* Fel AAAS; Am Chem Soc (secy, Div Comput Chem, 74-79). *Res:* Statistical thermodynamics; research methodology in the orient; use of alphanumeric data bases in research; structure - activity relationships; experimental design; interactive computer applications; structure-activity relationships. *Mailing Add:* 605 Cavedale Rd Sonoma CA 95476-3040

MARCUS, SANFORD M, b New York, NY, Mar 18, 32; m 59; c 2. PHYSICS. *Educ:* Brooklyn Col, BS, 54; Columbia Univ, MS, 57; Univ Pa, PhD(physics), 64. *Prof Exp:* Engr, Radio Corp Am, 57-59; PHYSICIST, E I DU PONT DE NEMOURS & CO, 64- *Mem:* Am Phys Soc. *Res:* Superconductivity, especially experimental work by means of tunneling; thick film materials for display, microcircuit and hybrid applications; solid state devices; transport properties of metals; printing technology. *Mailing Add:* Photo Prod E352/170 E I du Pont de Nemours & Co PO Box 80352 Wilmington DE 19898-0352

MARCUS, STANLEY, b New York, NY, Jan 20, 16; m 39; c 2. MICROBIOLOGY, IMMUNOLOGY. *Educ:* City Col New York, BA, 37; Univ Mich, MS, 39, PhD(microbiol), 42. *Prof Exp:* PROF MICROBIOL, COL MED, UNIV UTAH, 49- *Concurrent Pos:* Nat Inst Allergy & Infectious Dis res career award, 61. *Mem:* AAAS; Am Soc Microbiol; Am Asn Immunol; Soc Exp Biol & Med; Reticuloendothelial Soc; Sigma Xi. *Res:* Mechanisms of specific and nonspecific resistance to infectious and neoplastic disease; theory of testing; pyrogen tests; nontoxic enteric vaccines; standardization of mycotic sensitins; proficiency testing as basis of evaluation surveys. *Mailing Add:* 1400 Fed Way Salt Lake City UT 84102

MARCUS, STANLEY RAYMOND, b Providence, RI, Feb 29, 16; m 42; c 1. ENGINEERING ADMINISTRATION, RESEARCH MANAGEMENT. *Educ:* Univ RI, BS, 38; George Washington Univ, MS, 58. *Prof Exp:* Mech engr, Naval Torpedo Sta, Newport, 40-45; chief engr, Div War Res, Columbia Univ, 45-46; self employed, 46-51; proj engr, Bur Ord, US Navy, 51-56; coordr underwater ord, Off Naval Res, 56-59; coordr antisubmarine warfare weapons systs, Dept Chief Naval Opers, 59-60, asst tech dir systs planning, 60-63; dir, Res Div, Bur Naval Weapons, 63-66; dep commander & chief scientist res & technol, Naval Ord Systs Command, 66-74; asst dep commander & tech dir res & technol, 77-80, DIR, OFF RES & TECHNOL, NAVAL SEA SYST COMMAND, 80- *Concurrent Pos:* Consult, 80- *Honors & Awards:* Cindy Award, Int Film Producers Am, 72. *Mem:* Am Mgt Asn; AAAS; Nat Soc Prof Engrs; Acoust Soc Am; Opers Res Soc Am. *Res:* Management, planning, appraisal and development of research and development efforts for a broad area of technology applications for ordinance and ship systems. *Mailing Add:* 2111 Jeff Davis Hwy Apt 809N Arlington VA 22202-3131

MARCUS, STEPHEN, b New York, NY, Dec 27, 39; m 70; c 3. LASERS. *Educ:* Rensselaer Polytech Inst, BS, 61; Columbia Univ, MA, 63, PhD(physics), 68. *Prof Exp:* Res assoc elec eng, Cornell Univ, 67-69; staff scientist laser physics, United Aircraft Res Labs, 69-70; MEM STAFF, LINCOLN LAB, MASS INST TECHNOL, 70- *Mem:* Am Phys Soc; Sigma Xi. *Res:* Gas lasers, primarily pulsed and continuous wave carbon dioxide lasers and their applications to laser radar systems. *Mailing Add:* Mass Inst Technol Lincoln Lab 244 Wood St Lexington MA 02173-9108

MARCUS, STEVEN IRL, b St Louis, Mo, Apr 2, 49; m 78; c 2. CONTROL THEORY, SYSTEMS. *Educ:* Rice Univ, BA, 71; Mass Inst Technol, MS, 72, PhD(elec eng), 75. *Prof Exp:* From asst prof to assoc prof, 75-84, assoc chmn, 84-89, PROF ELEC ENG, UNIV TEX, AUSTIN, 84- *Concurrent Pos:* Assoc ed, Trans Automatic Control, Inst Elec & Electronic Engrs, 80-81; co-managing ed, Acta Appl Math, 83-; assoc ed, Math Control, Signals & Systs, 87-, Inst Elec & Electronic Engrs Trans Info Theory, 90-, J Discrete Event Dynamic Systs, 90-, Soc Indust & Appl Math J Control & Optimization, 90- *Mem:* Fel Inst Elec & Electronic Engrs; Soc Indust Appl Math; Am Math Soc; Opers Res Soc Am. *Res:* Analysis and optimization techniques for nonlinear stochastic systems; stochastic control theory; the application of geometric techniques to the design of control systems and the application of the control techniques to manufacturing. *Mailing Add:* Dept Elec & Comput Eng Univ Tex Austin TX 78712

MARCUSE, DIETRICH, b Koenigsberg, Ger, Feb 27, 29; m 59; c 2. ELECTRICAL ENGINEERING. *Educ:* Free Univ Berlin, Dipl phys, 54; Karlsruhe Tech Univ, DrIng, 62. *Prof Exp:* Mem tech staff, Siemens & Halske, Ger, 54-57; DISTINGUISHED MEM TECH STAFF, AT&T BELL LABS, 57- *Honors & Awards:* Quantum Electronics Award, Inst Elec & Electronic Engrs, 81; Max Born Award, Optical Soc Am, 89. *Mem:* Fel Inst Elec & Electronic Eng; fel Optical Soc Am. *Res:* Circular electric waveguide; microwave masers; light communications. *Mailing Add:* Crawford Hill Lab Rm HOH-L-117 Bell Labs Box 400 Holmdel NJ 07733

MARCUVITZ, NATHAN, b Brooklyn, NY, Dec 29, 13; m 46; c 2. MATHEMATICAL PHYSICS. *Educ:* Polytech Inst Brooklyn, BEE, 35, MEE, 41, DEE, 47. *Prof Exp:* Develop engr, Radio Corp Am, 35-40; mem staff, Radiation Lab, Mass Inst Tech, 42-46; from asst prof to prof elec eng, Polytech Inst Brooklyn, 46-65, dir microwave res inst, 57-61, vpres res & actg dean, Grad Ctr, 61-63, prof electrophys, 61-65, dean res & grad ctr, 64-65, inst prof, 65-66; prof appl physics, NY Univ, 66-73; prof, 73-78, INST PROF APPL PHYSICS, POLYTECH UNIV, 78 - *Concurrent Pos:* Asst dir res, Defense Res & Eng, Dept Defense, DC, 63-64; Gordon MacKay vis prof, Harvard Univ, 71. *Honors & Awards:* Microwave Career Award, Inst Elec & Electronics Engrs, 85, Henry Hertz Medal, 88. *Mem:* Nat Acad Eng; Am Phys Soc; fel Inst Elec & Electronics Engrs. *Res:* Electromagnetics; plasma dynamics; nonlinear and turbulent wave phenomena. *Mailing Add:* Dept Elect Eng & Electrophys Polytech Inst New York Rte 110 Farmingdale NY 11735

MARCY, JOSEPH EDWIN, b Bristol, Va; m 73; c 2. FOOD SCIENCE. *Educ:* Univ Tenn, Knoxville, BS, 74, MS, 76; NC State Univ, PhD(food sci), 80. *Prof Exp:* Res asst, Univ Tenn, 74-76; res asst, NC State Univ, 76-80; asst prof food sci, Univ Fla, 80-84; sr food scientist, 84-85, MGR BUS DEVELOP, RAMPART PACKAGING, INC, 85-; ASSOC PROF, VA TECH UNIV, 88- *Mem:* Am Chem Soc; Inst Food Technologists. *Res:* Citrus juice processing including analysis of chemical constituents and evaluation of processing procedures; aseptic packaging; plastic food containers. *Mailing Add:* 118 FST Bldg Va Tech Blacksburg VA 24061-0418

MARCY, WILLARD, b Newton, Mass, Sept 27, 16; m 38; c 2. ORGANIC CHEMISTRY, CHEMICAL ENGINEERING. *Educ:* Mass Inst Technol, SB, 37, PhD(org chem), 49. *Prof Exp:* Asst supt, Am Sugar Ref Co, 37-42; res assoc org chem, Mass Inst Technol, 46-49; chem engr, Res & Develop Div, Am Sugar Ref Co, 49-58, head process develop, 58-64; dir patent progs, Invention Admin Res Corp, 64-67, vpres patents, 67-80, vpres, 80-83; pres, Ardus, Inc, 83-85; CONSULT, 85- *Concurrent Pos:* US Chem Corp, plants div, 42-46. *Mem:* AAAS; Am Chem Soc; fel Am Inst Chem (pres, 84-85); NY Acad Sci. *Res:* Carbohydrates; war gases; sugar refining; sugar by-products; university invention administration. *Mailing Add:* 621 Caminito Del Sol Santa Fe NM 87505-4913

MARCZYNSKA, BARBARA MARY, b Cracow, Poland; US citizen; m 56; c 1. IMMUNOLOGY, GENETICS. *Educ:* Acad Med Cracow, MS, 56, PhD(immunol, genetics), 62. *Prof Exp:* From instr to assoc prof genetics & embryol, Med Sch, Cracow, 56-64; res asst, 65-72, ASST PROF VIROL, RUSH-PRESBY-ST LUKE'S MED CTR, 72- *Mem:* Am Soc Microbiol; Sigma Xi. *Res:* Virological and immunological aspects of virus-induced oncogenic transformation in non-human primates. *Mailing Add:* 1217 Hinman Evanston IL 60201

MARCZYNSKI, THADDEUS JOHN, b Poznan, Poland, Nov 30, 20; m 56; c 2. PHARMACOLOGY, NEUROPHYSIOLOGY. *Educ:* Cracow Acad Med, MD, 51, DMSc, 59. *Prof Exp:* Res asst pharmacol, Cracow Acad Med, 54-59, asst prof, 62-64; from asst prof to assoc prof, 64-73, PROF PHARMACOL, UNIV ILL COL MED, 73-, STAFF MEM, INTERCAMPUS BIOENG DEPT, 78-, PROF, DEPT PSYCHIAT, 78- *Concurrent Pos:* Rockefeller Found & Brain Res Inst fel, Univ Calif, Los Angeles, 61-62; NIH res grant, Univ Ill Col Med, 66-72. *Mem:* AAAS; Soc Neurosci; Am Soc Pharmacol & Exp Therapeut. *Res:* Pharmacology and electrophysiology of the central nervous system; analysis of information and its transmission coding in neuronal pathways; positive reinforcement and sensory input. *Mailing Add:* Dept Pharmacol Univ Ill Col Med PO Box 6998 Chicago IL 60680

MARDELLIS, ANTHONY, b Neuville-sur-Saone, France, July 17, 20; wid. MATHEMATICS. *Educ:* Univ Calif, Berkeley, BA, 50, MA, 52. *Prof Exp:* Asst math, Univ Calif, Berkeley, 51-55; instr, Loyola, Calif, 55-56; from asst prof to assoc prof, 56-70, chmn dept, 63-67, PROF MATH, CALIF STATE UNIV, LONG BEACH, 70- *Mem:* Am Math Soc; Math Asn Am; Math Soc France; Sigma Xi. *Res:* Picard-Vessiot theory; differential algebra. *Mailing Add:* Dept Math & Comput Sci Calif State Univ Long Beach CA 90840

MARDEN, JOHN IGLEHART, b Chicago, Ill, Sept 24, 51. STATISTICS. *Educ:* Univ Chicago, AB, 73, PhD(statist), 78. *Prof Exp:* Vis lectr, 77-78, ASST PROF MATH, UNIV ILL, URBANA-CHAMPAIGN, 78- *Mem:* Am Statist Asn; Inst Math Statist. *Res:* Hypothesis testing in multivariate analysis, combining independent tests of significance. *Mailing Add:* Dept Statist Univ Ill 725 S Wright St Champaing IL 61820

MARDEN, MORRIS, b Boston, Mass, Feb 12, 05; m 32; c 2. MATHEMATICS. *Educ:* Harvard Univ, AB, 25, AM, 27, PhD(math), 28. *Prof Exp:* Instr math, Harvard Univ, 25-27; Nat Res Coun fel, Uni- Wis, Princeton Univ, Univ Zurich & Univ Paris, 28-30; from asst prof to prof, 30-64, distinguished prof, 64-75, chmn dept, 57-61, 63-64, EMER DISTINGUISHED PROF MATH, UNIV WIS-MILWAUKEE, 75- *Concurrent Pos:* Invited lectr, Math Inst, Polish Acad Sci, 58-62, Math Insts, Univs Jerusalem, Haifa & Tel Aviv, 62, 68 & 73, Greece, 62, Japan, India, Spain & Eng, 64, Mex, Peru, Chile, Argentina, Uruguay & Brazil, 65, Budapest, Belgrade, Goteburg, 67, Montreal, 67 & 70, Finland & NZ, 70 & Australia, 71; consult, Allis-Chalmers Co, 48-60; asst ed, Bull, Am Math Soc, 42-45; vis distinguished prof math, Calif Poly State Univ, San Luis Obispo, 75-77. *Mem:* Fel AAAS; Am Math Soc; Math Asn Am. *Res:* Zeros of polynomials; entire and potential function; functions of a complex variable. *Mailing Add:* 7100 N Barnett Lane Milwaukee WI 53217

MARDER, A R, b Oct 24, 40; m 62; c 3. HIGH TEMPERATURE MATERIALS, COATINGS. *Educ:* Polytech Inst, Brooklyn, BS, 62, MS, 65; Lehigh Univ, PhD, 68. *Prof Exp:* Engr, Curtiss-Wright Corp, Woodridge, NJ, 62-65; res dept, Bethlehem Steel Corp, PA, 65-86; assoc dir, Energy Liaison Prog, 86-87, ASSOC DIR, ENERGY RES CTR, LEHIGH UNIV, 87- *Concurrent Pos:* Adj prof mat sci & eng, Lehigh Univ, 82- *Honors & Awards:* Taylor Welding Award, Am Welding Soc, 62; Joseph Villela Award, Am Soc Testing & Mat, 74. *Mem:* Fel Am Soc Metals Int; Am Inst Mech Engrs; Am Welding Soc; Am Soc Testing & Mat; Soc Automotive Engrs; Nat Asn Civil Engrs. *Res:* Processing, structure and property relationships in zinc and zinc alloy coatings and high temperature properties and coating for energy related materials. *Mailing Add:* Energy Res Ctr Lehigh Univ Bethleham PA 18015

MARDER, EVE ESTHER, b New York, NY, May 30, 48. NEUROBIOLOGY. *Educ:* Brandeis Univ, AB, 69; Univ Calif, San Diego, PhD(biol), 74. *Prof Exp:* Res fel, Univ Ore, 75 & Sch Higher Educ, France, 76-78; from asst prof to assoc prof biol, 84-90, PROF BIOL, BRANDEIS UNIV, 90- *Concurrent Pos:* Panel mem, Neurol Sci Study Sect, NIH, 83-86; Sloan fel, 80. *Honors & Awards:* Javits Award in Neurosci, 87. *Mem:* AAAS; Biophys Soc; Soc Neurosci. *Res:* Mechanisms by which neurotransmitters and neuropeptides modulate neural circuits. *Mailing Add:* Biol Dept Brandeis Univ Waltham MA 02254

MARDER, HERMAN LOWELL, b New York, NY, Mar 3, 31; m 55; c 4. ORGANIC POLYMER CHEMISTRY, RESEARCH ADMINISTRATION. *Educ:* State Univ NY, BS, 54, MS, 57, PhD(chem), 59. *Prof Exp:* Res chemist, E I du Pont de Nemours & Co, Inc, 58-61; sect head, Colgate Palmolive Co, 61-66; dir res & develop, Boyle Midway Div, Am Home Prod Corp, 66-69; vpres res & develop, Int Playtex Co, 69-78; vpres res & develop, 78-81, VPRES OPERS & TECHNOL, CHURCH & DWIGHT CO, INC, 81- *Mem:* Am Chem Soc. *Mailing Add:* 29 Constitution Hill W Princeton NJ 08540-6753

MARDER, STANLEY, b Philadelphia, Pa, Aug 21, 26; m 53; c 3. INFORMATION SCIENCE, SYSTEMS ANALYSIS. *Educ:* Univ Pa, BA, 50; Columbia Univ, PhD(physics), 58. *Prof Exp:* Res physicist, Carnegie Inst Technol, 56-60; staff mem, Inst Defense Analysis, 60-73; res physicist, 73-74, DIR, WASHINGTON OFF, ENVIRON RES INST MICH, ARLINGTON, VA, 74- *Mem:* Inst Elec & Electronic Engrs; Am Econ Asn; Am Phys Soc. *Res:* Signal processing; image evaluation; radar system analysis. *Mailing Add:* Dept Data Processing San Diego Mesa Col 7250 Masa Col Dr San Diego CA 92111

MARDIAN, JAMES K W, b Pasadena, Calif, July 1, 46; m 72; c 2. BIOCHEMISTRY. *Educ:* Cornell Univ, BS, 68; Calif Inst Technol, MS, 70; Pac Sch Relig, MA, 73; Ore State Univ, PhD(biophys), 78. *Prof Exp:* res assoc biochem, Grad Sch Biomed Sci, Univ Tenn-Oak Ridge, 78-80; engr aerodyn, Garrett Turbine Engine Co, Phoenix, 80-83; Motorola Gort Electronics Group, 83-84; Glen-Mar Door, 84-89; OWNER, MARDIAN & ASSOC, 89- *Mem:* Am Soc Cell Biol. *Res:* Compressor aerodynamics; yeast chromatin; structure of active chromatin; chromosomal proteins. *Mailing Add:* 5826 N 70th Pl Paradise Valley AZ 85253

MARDINEY, MICHAEL RALPH, JR, b Brooklyn, NY, Dec 16, 34; c 4. IMMUNOLOGY, INTERNAL MEDICINE. *Educ:* Hamilton Col, AB, 56; Seton Hall Col Med & Dent, MD, 60; Am Bd Allergy & Immunol, dipl, 74. *Prof Exp:* Intern med, Kings County Hosp Ctr, Brooklyn, 60-61; resident med, Ctr Med, Baylor Univ, 61-62; clin assoc, Immunol Br, Nat Cancer Inst, 65-67, head immunol & cell biol sect, Baltimore Cancer Res Ctr, Nat Cancer Inst, 67-77; pres, Immunodiag & Immunotherapeut, Inc, 78-84; PRES, ALLERGY CONTROL, INC, 84- *Concurrent Pos:* Res fel, Exp Path Div, Scripps Clin & Res Found, Univ Calif, 62-65; instr, Sch Med, Johns Hopkins Univ & physician, Allergy & Infectious Dis Clin, Johns Hopkins Hosp; physician to med staff, Good Samaritan Hosp; staff physician, Howard County Gen Hosp & South Baltimore Gen Hosp, 70-; consult med, Greater Baltimore Med Ctr. *Mem:* Transplantation Soc; Am Soc Exp Path; Am Asn Immunol; Am Asn Cancer Res; Am Acad Allergy. *Res:* Immunopathology; tumor immunology; allergy. *Mailing Add:* Allergy Control Inc 9380 Balitmore Nat Pike Ellicott City MD 21043

MARDIX, SHMUEL, b Lodz, Poland, May 22, 31; Israel & US citizen; m 57. SOLID STATE PHYSICS. *Educ:* Hebrew Univ Jerusalem, MSc, 66, PhD(physics), 69. *Prof Exp:* Res assoc x-ray topog, Bristol Univ, 69-70; res assoc x-ray imaging, 71-72, assoc prof solid state physics, 73-77, PROF SOLID STATE PHYSICS, UNIV RI, 78- *Res:* Photoelectronic effects in materials, x-ray imaging and recording, x-ray crystallography, and photographic processes. *Mailing Add:* Dept of Elec Eng Univ of RI Kingston RI 02881

MARE, CORNELIUS JOHN, b Middleburg, SAfrica, Aug 27, 34; m 60; c 4. VETERINARY MICROBIOLOGY, TROPICAL DISEASES. *Educ:* Pretoria Univ, BVSc, 57; Iowa State Univ, PhD(vet microbiol), 65. *Prof Exp:* Private practice, 57-58; vet diagnostician, Allerton Diag Lab, SAfrica, 58-59; res virologist, Onderstepoort Vet Res Inst, 59-62, sr res virologist, 65-67; res assoc microbiol, Vet Med Res Inst Iowa, 62-65; assoc prof virol, Iowa State Univ, 67-72, prof vet microbiol, 72-76; prof vet sci & head dept,Univ Ariz, 76-84, head, Vet Diag Lab, 76-84, dir, Int Prog, 84-86; adv to dir Agr Resources, Lesotho, 86-88; PROF VET SCI, UNIV ARIZ, 88- *Concurrent Pos:* Fulbright scholar, 62-65; vis prof,Plum Island Animal Dis Ctr, NY, 73 & EAfrican Vet Res Inst, Nairobi, Kenya, 74; vis scientist, Upper Volta, 79, Egypt, 81. *Mem:* Conf Res Workers Animal Dis; US Animal Health Asn; Am Vet Med Asn; Wildlife Dis Asn. *Res:* Viruses, mycoplasma and chlamydia of domestic animals and man, especially viruses of the herpes virus group; bovine malignant Catarrhal fever; disease ecology in arid lands. *Mailing Add:* Dept Vet Sci Univ of Ariz Tucson AZ 85721

MAREK, CECIL JOHN, b Chicago, Ill, Mar 12, 40; m 61; c 4. CHEMICAL ENGINEERING. *Educ:* Ill Inst Technol, BS, 61, PhD(chem eng), 67. *Prof Exp:* Instr chem eng, Ill Inst Technol, 64-67; asst prof, Drexel Inst, 67-70; AEROSPACE ENGR, NASA LEWIS RES CTR, 70- *Mem:* Am Inst Chem Engrs; Combustion Inst. *Res:* Combustion modeling; radiant heat transfer; film cooling; turbulent mixing; particle-gas interaction; icing research; experimental combustion shear layer. *Mailing Add:* NASA Lewis Res Ctr 21000 Brookpark Rd Cleveland OH 44135

MAREK, CHARLES R(OBERT), b Chicago, Ill, May 8, 40; m 62; c 3. ENGINEERING, MATERIALS SCIENCE. *Educ:* Univ Ill, Urbana, BS, 63, MS, 64, PhD(civil eng), 67. *Prof Exp:* Asst civil eng, Univ Ill, Urbana, 62-66, from instr to asst prof, 66-72; construct mat engr, 72-80, sr mat engr, 80-84, TECH DIR, VULCAN MAT CO, BIRMINGHAM, 84- *Concurrent Pos:* Mem comt, Transp Res Bd, Nat Acad Sci-Nat Res Coun; fourth vchmn, Comt D-4 Rd & Paving Mat, Am Soc Testing & Mat; chmn tech comt, Nat Stone Asn. *Mem:* Am Soc Civil Engrs; Asn Asphalt Paving Technol; Am Soc Testing & Mat; Nat Stone Asn; Transp Res Bd; Nat Mgt Asn. *Res:* Highway and airfield pavement constituent material properties and behavior and pavement design; quality control of aggregate production. *Mailing Add:* 1404 Cosmos Circle Vestavia Hills AL 35216

MAREK, EDMUND ANTHONY, b Enid, Okla, June 29, 48; m 69; c 2. SCIENCE EDUCATION, GENERAL BIOLOGY. *Educ:* Univ Okla, BS, 70, MEd, 72, PhD(sci educ), 77. *Prof Exp:* Teaching biol, Norman High Sch, Okla, 71-73; chmn, dept biol, W Mid-High Sch, Norman, 73-77; sci curric coordr, Pub Schs, Norman, Okla, 77-78; asst prof biol, Southwest Tex State Univ, 78-82, asst dir, Edwards Aquifer Res Ctr, 79-82; asst prof, 82-85, ASSOC PROF SCI EDUC, UNIV OKLA, 85- *Concurrent Pos:* Mem, Sci Cert Comt, Univ Okla, 73-78, adj asst prof, 77-78 & 82-84; prog comt, Nat Asn Res Sci Teaching, 79-81, 84-86, 82-84 & consult, Health Sci Enrichment Inst, 85; reviewer, grant proposal, NSF, Washington, 81 & Res in Sci Teaching, 82. *Mem:* Nat Asn Res in Sci Teaching; Nat Sci Teachers Asn; Sch Sci Math Asn; Sigma Xi. *Res:* Science education; curriculum and teaching methodology and its effects on learning, intellectual development and attitudinal qualities; concept understanding and misunderstanding. *Mailing Add:* Dept Educ Univ Okla Main Campus Norman OK 73019

MAREN, THOMAS HARTLEY, b New York, NY, May 26, 18; m 41, 80; c 3. PHARMACOLOGY. *Educ:* Princeton Univ, AB, 38; Johns Hopkins Univ, MD, 51. *Hon Degrees:* MD Honoris Causa, Uppsala Univ, Sweden, 77. *Prof Exp:* Res chemist, Wallace Labs, Carter Prods, Inc, NJ, 38-40, group leader, 41-44; chemist, Sch Hyg & Pub Health, Johns Hopkins Univ, 44-46, instr pharmacol, Med Sch, 46-51; pharmacologist, Chemother Dept, Res Div, Am Cyanamid Co, 51-54, group leader, 54-55; prof pharmacol & therapeut & chmn dept, Col Med, 55-78, GRAD RES PROF, UNIV FLA, 78- *Concurrent Pos:* Investr, Mt Desert Island Biol Lab, 53- *Honors & Awards:* Werckey Award Res, Am Soc Pharmacol & Exp Therapeut, 78. *Mem:* Asn Res Vis & Ophthal; Sigma Xi; AAAS; NY Acad Sci. *Res:* Renal ocular, cerebrospinal and electrolyte pharmacology and physiology; carbonic anhydrase and its inhibitors; comparative physiology. *Mailing Add:* Univ Fla Col Med Box J267 Gainesville FL 32610

MARENGO, NORMAN PAYSON, b New York, NY, Feb 21, 13; m 39; c 2. BOTANY. *Educ:* NY Univ, BS, 36, MA, 39, MS, 42, PhD(biol), 49. *Prof Exp:* Asst ed, NY Univ, 36-39, instr, 39-43, biol, 46-48; instr biol, Lafayette Col, 48-49, asst prof, 49-50; asst prof biol, Hofstra Col, 50-54; teacher sci, Cent High Sch, Merrick, 54-55; asst prof biol & gen sci, C W Post Col, Long Island Univ, 55-57, assoc prof, 57-60, dir div sci, 61-67, chmn dept biol, 63-67, prof, 61-81, emer prof biol, 81-; RETIRED. *Mem:* Bot Soc Am; Am Fern Soc. *Res:* Developmental genetics; botanical cytology; microscopical technique. *Mailing Add:* Dept Biol C W Post Col Long Island Univ Greenvale NY 11548

MARENUS, KENNETH D, BASIC RESEARCH. *Educ:* Univ Calif, Los Angeles, BA, 74, PhD(biol), 80. *Prof Exp:* Postdoctoral fel biophys, Johns Hopkins Univ, Baltimore, Md, 80-82; sr scientist, Mary Kay Cosmetics, Dallas, Tex, 82-83, group leader, 84-86; mgr appln res, Electro Scar Inc, Topsfield, Mass, 86; mgr, 86-88, DIR BIOL RES, ESTEE LAUDER CO, MELVILLE, NY, 88- *Concurrent Pos:* Adj asst prof dermat, Southwestern Med Sch, Univ Tex, Dallas, 83-86, biol, C W Post Campus, Long Island Univ, Brookville, NY, 86- *Mem:* Dermal Clin Eval Soc (pres, 89-); Am Soc Cell Biol; Soc Investigative Dermat; Electron Micros Soc Am; Soc Cosmetic

Chemists; NY Acad Sci; Int Soc Bioeng & Skin; Am Soc Photobiol. *Res:* Skin biology; biochemistry; cell biology; tissue culture; photobiology; molecular biology; histology; electron microscopy; author of various publications; granted two patents. *Mailing Add:* Estee Lauder Res & Develop 125 Pine Lawn Rd Melville NY 11747

MARES, FRANK, b Czech, Nov 1, 32; US citizen; m 58; c 2. ORGANOMETALLIC CHEMISTRY, POLYMER CHEMISTRY. *Educ:* Prague Tech Univ, MS, 57; Czech Acad Sci, PhD(org chem), 60. *Prof Exp:* Staff mem, Czech Acad Sci, 60-65; fel org chem, Univ Calif, Berkeley, 65-67; res group leader, Czech Acad Sci, 67; res assoc, Univ Tubingen, 68; res assoc, Univ Calif, Berkeley, 69, lectr, 69-70; sr res chemist, 70-74, res group leader ganometallic chem, 76-77, tech res supvr ganometallic chem, 77-80 scientist, 80-83, MGR POLYMER & ORGANOMETALLIC CHEM, ALLIED CHEM CORP, 83- *Concurrent Pos:* Assoc ed, Collection Czech Chem Commun, 61-68. *Mem:* Am Chem Soc. *Res:* Transition metals in organic reactions; kinetics and mechanism of organic reactions; organic synthesis; polymer synthesis; eng plastics; bioresorbable polymers. *Mailing Add:* Allied-Signal Corp PO Box 1021101 Columbia Rd Morristown NJ 07962

MARES, MICHAEL ALLEN, b Albuquerque, NMex, Mar 11, 45; m 66; c 2. MAMMALIAN BIOGEOGRAPHY. *Educ:* Univ NMex, BS, 68; Ft Hays Kans State Univ, MS, 69; Univ Tex, Austin, PhD(zool), 73. *Prof Exp:* From asst prof to assoc prof biol, Univ Pittsburgh, 73-81; assoc prof & cur mammals, 81-85, DIR, OKLA MUS NAT HIST, UNIV OKLA, 83-, PROF ZOOL, 85- *Concurrent Pos:* Adj prof ecol, Univ Nac de Cordoba, Argentina, 71-72 & Univ Nac de Tucuman, 72; Fulbright res fel, ecol, Salta, Argentina, 74; Nat Chicano Coun fel, Mus Northern Ariz, Flagstaff, 78; Ford Found res fel, Univ Ariz, Tucson, 80-81; vis prof, Univ Nac de Tucuman, 74; scientist fel, Univ Ariz, Tucson, 80-81; ecol consult, NUS Corp, 80-81; consult, Arg Nat Sci Found, 83, World Wildlife Found, Brazil, 86. *Mem:* Ecol Soc Am; Am Soc Mammalogists; AAAS; Soc Study Evolution; Interam Asn Adv Sci; Asn Trop Ecol; Soc Conserv Biol; Am Soc Naturalists. *Res:* Examination of convergent evolution, adaptation and community organization of desert rodents of the world; ecology, conservation, evolution and systematics of South American mammals; spatial organization in vertebrates; island biogeographic patterns of birds; tropical ecology. *Mailing Add:* Okla Mus Nat Hist Univ Okla Norman OK 73019

MARET, S MELISSA, b San Antonio, Tex, Aug 27, 53. CLINICAL MICROBIOLOGY. *Educ:* Univ NC, Chapel Hill, PhD(immunol), 80. *Prof Exp:* Contract supvr, Nat Cancer Inst, NIH, 82-83; SR SCIENTIST, BECTON-DICKINSON, INC, 83- *Mem:* Am Asn Immunologists; Am Soc Microbiol. *Mailing Add:* R&D Dept Becton Dickenson Advanced Diagnostics 1400 Coppermine Terr Baltimore MD 21209

MARETZKI, ANDREW, b Berlin, Germany, Feb 23, 26; nat US; m 57; c 2. BIOCHEMISTRY. *Educ:* Univ Cincinnati, BS, 52; Pa State Univ, MS, 58, PhD, 60. *Prof Exp:* Asst res biochemist, Parke, Davis & Co, 52-53; res asst biochem, Pa State Univ, 55-60; res biochemist, Kitchawan Res Labs, 60-61; assoc scientist, Nuclear Ctr & Sch Med, Univ Puerto Rico, 61-65; assoc biochemist, 66-69, BIOCHEMIST & LEADER, CELL BIOL GROUP, EXP STA, HAWAIIAN SUGAR PLANTERS ASN, 69- *Mem:* Am Soc Plant Physiol; Sigma Xi. *Res:* Structure of antibiotics; toxins; plant enzyme systems; membrane transport; sugar assimilation in plants. *Mailing Add:* c/o Hspa 99-193 Aiea Hts Dr Aiea HI 96701

MAREZIO, MASSIMO, crystallography, for more information see previous edition

MAREZIO, MASSIMO, b Rome, Italy, Aug 25, 30; US citizen. STRUCTURE & TRANSPORT PROPERTIES, X-RAY DIFFRACTION. *Educ:* Univ Rome, Laurea, 54, Libera Docenza, 64. *Prof Exp:* Res assoc, Physics Dept, Univ Chicago, 59-63; MEM TECH STAFF, AT&T BELL LABS, 63-73 & 86-; DIR RES, NAT CTR SCI RES, GRENOBLE, FRANCE, 73- *Mem:* Am Crystallog Asn. *Res:* Determination of structures by diffraction techniques; structure and physical properties; transport properties; superconductivity, high Tc superconductors. *Mailing Add:* AT&T Bell Labs Rm 1C-211 600 Mountain Ave Murray Hill NJ 07974-2070

MARFAT, ANTHONY, b Zadar, Yugoslavia, Sept 8, 51; m 77; c 2. MEDICINAL CHEMISTRY, DESIGN & SYNTHESIS OF NEW DRUGS FOR TREATMENT OF INFLAMMATORY DISEASES. *Educ:* Hunter Col, BA, 74; State Univ NY, Stony Brook, MS, 77, PhD(org chem), 78. *Prof Exp:* Postdoctoral fel chem, Harvard Univ, 78-81; res scientist chem, Pfizer Inc, 81-83, sr res scientist, 83-86, sr res investr, 86-90, PRIN RES INVESTR CHEM, PFIZER INC, 90- *Mem:* Am Chem Soc; Sigma Xi. *Res:* Pulmonary and inflammatory diseases, in particular, chemistry and biology of eicosanoids (leukotrienes, PAF, lipoxins, prostaglandins, etc) and their relationship or roles in various inflammatory diseases. *Mailing Add:* 160 Lantern Hill Rd Mystic CT 06355

MARFEY, SVIATOPOLK PETER, b Kobaki, Poland, June 1, 25; nat US; m 64; c 2. ORGANIC CHEMISTRY, BIOCHEMISTRY. *Educ:* Wayne State Univ, BS, 49, MS, 53, PhD(chem), 55. *Prof Exp:* Res asst chem, Princeton Univ, 55-56; res assoc, Rockefeller Univ, 56-59; res assoc, Harvard Univ, 59-67; ASSOC PROF BIOL SCI, STATE UNIV NY ALBANY, 67- *Concurrent Pos:* Dir, Walker Biochem Res Lab, Mass Eye & Ear Infirmary, 59-67; chmn, Eastern NY Sect, Am Chem Soc, 85-87. *Mem:* Am Soc Biochem & Molecular Biol; Am Chem Soc; Sigma Xi. *Res:* Chemistry of proteins and nucleic acids; synthesis and biotechnological applications of left-handed B-DNA containing 2-deoxy-L-ribose. *Mailing Add:* Dept Biol Sci State Univ NY Albany NY 12203

MARG, ELWIN, b San Francisco, Calif, Mar 23, 18; m 42; c 1. VISION, NEUROSCIENCES. *Educ:* Univ Calif, AB, 40, PhD(physiol optics), 50. *Prof Exp:* Instr optom, 50-51, from asst prof to assoc prof, 51-56, Miller res prof, 67-68, PROF OPTOM & PHYSIOL OPTICS, UNIV CALIF, BERKELEY, 62- *Concurrent Pos:* NSF sr fel, Nobel Inst Neurophysiol, Karolinska Inst, Sweden, 57; Guggenheim fel, Madrid, 64; res assoc neurosci, Mt Zion Hosp & Med Ctr, San Francisco, 69- *Honors & Awards:* Apollo Award, Am Optom Asn, 62; Chas F Prentice Award, Am Acad Optom, 81. *Mem:* Am Physiol Soc; Optical Soc Am; Asn Res Vision & Ophthal; Am Acad Optom; Soc Neurosci; Inst Elec & Electronic Engrs; Soc Cerebral Blood Flow & Metab. *Res:* Neurophysiology of visual system and brain; automated eye examination; phosphene visual prosthesis; diagnosis and prognosis by single neuron responses from the brain in neurosurgery; visual acuity and development in infants; visual evoked potentials; functional brain imaging (positron emission tomography, nuclear magnetic resonance); magnetic stimulation of the human visual system. *Mailing Add:* Sch Optom Univ Calif Berkeley CA 94720

MARGACH, CHARLES BOYD, optometry; deceased, see previous edition for last biography

MARGANIAN, VAHE MARDIROS, b Jlala, Lebanon, May 28, 38; US citizen; m 62; c 3. INORGANIC CHEMISTRY. *Educ:* San Francisco State Col, BS, 60; Clemson Univ, MS, 64, PhD(inorg chem), 66. *Prof Exp:* Teaching & res fel chem, Clemson Univ, 62-66; NSF res fel inorg chem, Univ Mass, Amherst, 66-67; from asst prof to assoc prof, 67-74, PROF CHEM, BRIDGEWATER STATE COL, 74- *Mem:* Am Chem Soc; Sigma Xi. *Res:* Synthesis and structural studies of oxo-compounds with tellurium IV halides; characterization of the products of cadmium II halides with N-bases; P-NMR of platinum II hydride systems. *Mailing Add:* Dept of Chem Bridgewater State Col Bridgewater MA 02324

MARGARETTEN, WILLIAM, b Brooklyn, NY, Sept 19, 29. PATHOLOGY. *Educ:* NY Univ, AB, 50; Northwestern Univ, MS, 51; State Univ NY, MD, 55; Am Bd Path, dipl, 67. *Prof Exp:* Asst prof path, Columbia Univ, 66-67; from asst prof to assoc prof, 67-75, PROF PATH, UNIV CALIF, SAN FRANCISCO, 75-, ASSOC DEAN ACAD AFFAIRS, 88- *Mem:* AAAS; Am Asn Path & Bact; Am Soc Exp Path. *Res:* Coagulation; endotoxin; inflammation. *Mailing Add:* Dept of Path Univ of Calif San Francisco CA 94143

MARGARITONDO, GIORGIO, b Rome, Italy, Aug 24, 46; m 71; c 2. SOLID STATE PHYSICS, SURFACE SCIENCE. *Educ:* Univ Rome, PhD(physics), 69. *Prof Exp:* Fel physics, Ital Nat Res Coun, 69-71, mem res staff, 71-78; from asst prof to assoc prof, 78-83, PROF PHYSICS, UNIV WISMADISON, 83-, ASSOC DIR, RES SYNCHROTRON RADIATION CTR, 84- *Concurrent Pos:* Elected mem scientific coun, Nat Group Struct Matter, Ital Nat Res Coun, 74-75; consult, Bell Labs, Murray Hill, NJ, 75-77. *Honors & Awards:* Romnes Award, 83. *Mem:* Am Phys Soc; Am Vacuum Soc; Ital Phys Soc; Europ Phys Soc. *Res:* Photoemission spectroscopy with synchrotron radiation and electron spectroscopy in general on clean surfaces, interfaces, bulk states and superconductors. *Mailing Add:* Synchrotron Radiation Ctr 3731 Schneider Dr Stoughton WI 53589

MARGAZIOTIS, DEMETRIUS JOHN, b Athens, Greece, Oct 14, 38; m 67; c 2. NUCLEAR PHYSICS. *Educ:* Univ Calif, Los Angeles, BA, 59, MA, 61, PhD(physics), 66. *Prof Exp:* From instr to assoc prof, 64-73, chmn dept, 73-79 & 83-89, PROF PHYSICS, CALIF STATE UNIV, LOS ANGELES, 73- *Mem:* Am Phys Soc. *Res:* Few nucleon problem; nuclear structure. *Mailing Add:* Dept of Physics Calif State Univ Los Angeles CA 90032

MARGEN, SHELDON, b Chicago, Ill, May 9, 19; m 44; c 4. HUMAN NUTRITION. *Educ:* Univ Calif, AB, 38, MA, 39, MD, 43. *Prof Exp:* USPHS sr res fel, 47-48; res assoc, US Metab Unit, Univ Calif, Berkeley, 47-50, clin instr med, Sch Med, 48-56, lectr soc res, Sch Soc Welfare, 56-62, assoc res biochemist, 52-60, res biochemist, 60-62, nutritionist, Agr Exp Sta, 62-70, chmn Dept Nutrit Sci, 70-74, prof human nutrit, 62-79, prof, 79-89, EMER PROF HEALTH NUTRIT, UNIV CALIF, BERKELEY, 89- *Concurrent Pos:* Schering fel, 48-49; Damon Runyon fel, Nat Res Coun, 49-51; res assoc, Inst Metab Res, Alameda, 50-52. *Mem:* Endocrine Soc; Am Fedn Clin Res; Am Med Asn; Am Inst Nutrit; Am Soc Clin Nutrit; Sigma Xi. *Res:* Energy and general protein metabolism; protein turnover; Human nutrition, experimental and programmatic; international nutrition; public health nutrition; nutrition and food policy. *Mailing Add:* Dept SA HS Sch Pub Health Univ Calif Berkeley CA 94720

MARGERISON, RICHARD BENNETT, b Philadelphia, Pa, Feb 24, 32; m 53; c 4. ORGANIC CHEMISTRY, PHARMACEUTICAL CHEMISTRY. *Educ:* Lehigh Univ, BS, 53, MS, 55; Univ Va, PhD(chem), 57. *Prof Exp:* Asst, Lehigh Univ, 53-55; res chemist med chem, Wallace & Tiernan, Inc, 57-58; sr chemist develop res, Ciba Pharmaceut Prods Inc, 58-67, mgr process res & develop, Ciba Agrochem Co, 67-70, mgr chem mfg, 70-74, dir chem mfg, Pharmaceut Div, 74-82, exec dir chem opers, 82-87, VPRES PHARMACEUT OPERS, CIBA GEIGY CORP, 88- *Concurrent Pos:* Chmn, bulk pharmaceut chem comt, Pharmaceut Mfrs Asn. *Mem:* NY Acad Sci. *Res:* Preparation of nitrogen and sulfur aliphatic, aromatic and heterocyclic compounds as medicinal agents; substituted piperazines, diphenyl sulfides, gem-diphenyl compounds; medium size heterocyclic rings; sulfonamides; ureas. *Mailing Add:* 556 Morris Ave Summit NJ 07901

MARGERUM, DALE WILLIAM, b St Louis, Mo, Oct 20, 29; m 53; c 3. INORGANIC CHEMISTRY, ANALYTICAL CHEMISTRY. *Educ:* Southeast Mo State Col, BA, 50; Iowa State Univ, PhD, 55. *Prof Exp:* Chemist, Ames Lab, Iowa State Univ, 52-53; from instr to assoc prof, 54-65, PROF CHEM, PURDUE UNIV, 65- *Concurrent Pos:* NSF sr fel, Max Planck Inst Phys Chem, Gottingen, 63-64; vis prof, Univ Kent, Canterbury, 70; adv bd, Res Corp, 73-78, NIH Med Chem Study Sect, 65-69, Air Force Office Sci Res, 78-82; consult, Eastman Kodak, 81-, Great Lakes Chem, 87-, W R Grace, 88- *Mem:* AAAS; Am Chem Soc. *Res:* Coordination chemistry; bio-inorganic; kinetics; fast reactions in solution; analytical applications of kinetics; inorganic-analytical studies of environmental solution chemistry. *Mailing Add:* Dept of Chem Purdue Univ West Lafayette IN 47907

MARGERUM, DONALD L(EE), b St Louis, Mo, Mar 29, 26; m 49; c 3. ELECTRONICS ENGINEERING. *Educ:* Univ Mo, BSEE, 49; Northwestern Univ, MSEE, 50. *Prof Exp:* Res engr antennas, Aerophys Lab, NAm Aviation, Inc, 50-52; proj engr, Microwave Eng Co, 52-53; sr engr, Stanford Res Inst, 53-56; dir electronics, Systs Labs Corp, 56-59; asst gen mgr electronic warfare systs, Electronic Specialty Co, 59-66; vpres, Raven Electronics, Inc, 66-70; sr staff engr, Aerospace Corp, 70-71; chief engr, Rantec Div, Emerson Elec, 71-73; mem tech mgt, Ventura Div, Northrop Corp, 73-78; PRIN ENGR, ELECTROMAGNETICS SYSTS DIV, RAYTHEON, 78-, CONSULT ENGR. *Mem:* Inst Elec & Electronics Engrs; Sigma Xi. *Res:* Design, development and analysis of electronic warfare systems. *Mailing Add:* Raytheon Corp 6380 Hollister Goleta CA 93017

MARGERUM, JOHN DAVID, b St Louis, Mo, Oct 20, 29; m 54; c 3. PHYSICAL CHEMISTRY, LIQUID CRYSTALS. *Educ:* Southeast Mo State Col, AB, 50; Northwestern Univ, PhD(phys chem), 56. *Prof Exp:* Res chemist spectros, Wood River Res Lab, Shell Oil Co, 54-55; mem staff, Qm Res Eng Ctr, US Army, 55-57, sect chief, 57-59; res specialist photochem, Turbo Div, Sunstrand Corp, 59-62; mem tech staff chem, 62-67, SECT HEAD CHEM, RES LABS, HUGHES AIRCRAFT CO, 67-, SR SCIENTIST, 78-, ASST DEPT MGR, 89- *Honors & Awards:* Holley Medal, Am Soc Mech Engrs, 77. *Mem:* Fel AAAS; Am Chem Soc; Inter-Am Photochem Soc; Electrochem Soc; Sigma Xi; Soc Info Display. *Res:* Photochemistry on polymers, dyes, lasers, photochromic materials, photogalvanic and fuel cells; liquid crystal materials and electrooptical devices; electrochemical studies on secondary batteries, display technology; thin film silicon services. *Mailing Add:* Hughes Res Labs RL 70 3011 Malibu Canyon Rd Malibu CA 90265

MARGETTS, EDWARD LAMBERT, b Vancouver, BC, Mar 8, 20; m 41; c 2. MEDICINE, PSYCHIATRY. *Educ:* Univ BC, BA, 41; McGill Univ, MD, CM, 44; FRCP(C); FRC Psych; FRAI. *Prof Exp:* Psychiatrist, Royal Victoria Hosp, Montreal, 49-55; specialist psychiatrist, Kenya Govt & med supt, Mathari Hosp, Nairobi, 55-59; head, dept psychiat, Vancouver Gen Hosp, 72-84; prof psychiat & lectr hist med, 60-85, EMER PROF PSYCHIAT, UNIV BC, 85- *Concurrent Pos:* Asst to dir, Allan Mem Inst Psychiat, 49-51; chief serv, Shaughnessy Vet Hosp, Vancouver, BC, 64-70; Ment Health Unit, WHO, Switz, 70-72; mem, Expert Adv Coun Ment Health, WHO. *Mem:* Am Psychiat Asn; Am Asn Hist Med; Am Anthrop Asn; Can Psychiat Asn; Royal Micros Soc. *Res:* Ethnic, cultural and international psychiatry; history of medicine; archaeology and anthropology applied to medicine. *Mailing Add:* 6171 Collingwood St Vancouver BC V6N 1T5 Can

MARGO, CURTIS EDWARD, b Los Angeles, Calif, May 15, 48; m 77; c 1. OPHTHALMOLOGY, ANATOMIC PATHOLOGY. *Educ:* Univ Southern Calif, BS, 70; Emory Univ Sch Med, MD, 74. *Prof Exp:* Intern, Grady Mem Hosp, 74-75; resident internal med, Univ Va, Charlottesville, 77-78; resident ophthal, Univ Fla, Gainesville, 78-81; Fel path, Armed Forces Inst Path, Washington, DC, 81-83; asst prof ophthal, La State Univ Sch Med, 82-83; ASST PROF OPHTHAL/PATH, UNIV SFLA COL MED, 84- *Mem:* Am Acad Ophthal. *Res:* Light microscopic and ultrastructural features of diseases of the eye and orbit. *Mailing Add:* Col Med Box 21 Univ SFla 12901 N 30 St Tampa FL 33612

MARGOLIASH, EMANUEL, b Cairo, Egypt, Feb 10, 20; m 44; c 2. MOLECULAR BIOLOGY. *Educ:* American Univ, Beirut, BA, 40, MA, 42, MD, 45. *Prof Exp:* Res fel exp path, Hebrew Univ, Israel, 45-48, sr asst, 49-51, lectr & actg head, Cancer Res Labs, Hadassah Med Sch, 54-58; res assoc biochem, Molteno Inst, Cambridge Univ, 51-53; res assoc, Nobel Inst, Sweden, 58, Univ Utah, 58-60 & Montreal Res Inst, McGill Univ, 60-62; head, Protein Sect, Abbott Labs, 62-71; prof, 71-90, EMER PROF BIOCHEM & MOLECULAR BIOL, NORTHWESTERN UNIV, 90-; PROF BIOL SCI, UNIV ILL, CHICAGO, 89- *Concurrent Pos:* Prof lectr, Univ Chicago, 64-71; Rudi Lemberg fel, 81; Guggenheim fel, 83; Owen L Coon prof molecular biol, Northwestern Univ, 88-90. *Mem:* Nat Acad Sci; Am Acad Arts & Sci; Am Soc Biol Chem; Am Chem Soc; Brit Biochem Soc; Can Biochem Soc. *Res:* Structure-function relations of heme proteins; molecular evolution and immunology; energy conservation mechanisms. *Mailing Add:* LMB-Biol Sci 4297A SEL M/C 067 Univ Ill Box 4348 Chicago IL 60680

MARGOLIES, MICHAEL N, b New York, NY, May 17, 38; m 67; c 2. PROTEIN CHEMISTRY. *Educ:* Columbia Univ Sch Med, MD, 62. *Prof Exp:* ASSOC PROF SURG, HARVARD MED SCH, 78- *Mem:* Am Soc Biol Chemists; Am Asn Immunologists; Soc Univ Surgeons; fel Am Col Surgeons; Protein Soc. *Res:* Immunology; protein chemistry. *Mailing Add:* Dept Surg Mass Gen Hosp 15 Parkman St Boston MA 02114

MARGOLIN, BARRY HERBERT, b New York, NY, Jan 8, 43; m 69; c 1. MATHEMATICAL STATISTICS, APPLIED STATISTICS. *Educ:* City Col New York, BS, 63; Harvard Univ, MA, 64, PhD(statist), 67. *Prof Exp:* Instr educ statist, Harvard Univ, 66-67; asst prof statist, Yale Univ, 67-72, assoc prof statist, 72-77; math statistician, Nat Inst Environ Health Sci, 77-87; PROF & CHMN, BIOSTATISTICS, UNIV N CAROLINA, CHAPEL HILL, 87- *Concurrent Pos:* Consult, Consumers Union, 67-77 & IBM Co, 69-70. *Honors & Awards:* Shewall Award, Chem Div, Am Soc Quality Control, 77; George W Snedecor Award, Am Statist Asn, 81. *Mem:* Fel Am Statist Asn; Int Statist Inst; Inst Math Statist; Environ Mutagen Soc. *Res:* Data analysis; design and analysis of experiments; categorical data; contingency tables; genetic toxicology. *Mailing Add:* Sch Pub Health Dept Biostatist CB No 7400 McGavran-Greenberg Hall The Univ of N Carolina Chapel Hill NC 27599-7400

MARGOLIN, ESAR GORDON, b Omaha, Nebr, Mar 17, 24; m 56; c 2. INTERNAL MEDICINE. *Educ:* Univ Nebr, BA, 45, MD, 47. *Prof Exp:* From asst prof to assoc prof, 58-69, clin prof, 69-72, PROF MED, COL MED, UNIV CINCINNATI, 72-; dir, Dept Internal Med, 59-86, DIR GERIATRIC PROG, DEPT INTERNA MED, JEWISH HOSP, 89- *Concurrent Pos:* Fel med, Harvard Univ, 53-55. *Mem:* Am Col Physicians; Am Soc Geriatrics. *Res:* Kidney and electrolytes; geriatrics. *Mailing Add:* Univ Cincinnati Col Med Cincinnati OH 45267-0535

MARGOLIN, HAROLD, b Hartford, Conn, July 12, 22; m 46; c 3. PHYSICAL & MECHANICAL METALLURGY. *Educ:* Yale Univ, BE, 43, MEng, 47, DEng(metall), 50. *Prof Exp:* Res assoc, Res Div, NY Univ, 49-55, eng scientist, 55-56, assoc prof metall, 56-63, prof metall eng, 63-73; PROF PHYS & ENG METALL, POLYTECH UNIV, 73- *Concurrent Pos:* Theodore Krengel vis prof, Dept Mat Eng Tech, Israel. *Mem:* Fel Am Soc Metals; Am Inst Mining, Metall & Petrol Engrs; Mat Res Soc; Sigma Xi; AAAS. *Res:* Titanium metallurgy; plastic flow and fracture; grain boundary strengthening; Bauschinger behavior fatigue. *Mailing Add:* Polytech Inst New York 333 Jay St Brooklyn NY 11201

MARGOLIN, JEROME, b Brooklyn, NY, Nov 13, 27; m 65; c 3. ELECTRICAL ENGINEERING. *Educ:* Univ Mich, BSE(math) & BSE(elec eng), 51, MSE, 52. *Prof Exp:* Staff mem radar, Lincoln Lab, Mass Inst Technol, 52-59 & Shape Air Defense Tech Ctr, 59; sub dept head, Mitre Corp, 59-61; SR STAFF MEM, LINCOLN LAB, MASS INST TECHNOL, 61- *Concurrent Pos:* Consult, Weapons Systs Eval Group, Inst Defense Analysis, 59-61. *Mem:* Inst Elec & Electronics Engrs. *Res:* Radar system synthesis and signal processing. *Mailing Add:* Mass Inst Technol Lincoln Lab 244 Wood St PO Box 73 Lexington MA 02173

MARGOLIN, PAUL, genetics, molecular biology; deceased, see previous edition for last biography

MARGOLIN, SOLOMON B, b Philadelphia, Pa, May 16, 20; m 87; c 4. EXPERIMENTAL THERAPEUTICS, PHARMACODYNAMICS. *Educ:* Rutgers Univ, BSc, 41, MS, 43, PhD(physiol & biochem), 45. *Prof Exp:* Res asst physiol, Rutgers Univ, 43-45; res biologist, Silmo Chem Co, 47-48; pharmacologist, Schering Corp, 48-52, dir pharmacol res, 52-54; chief pharmacologist, Maltbie Div, Wallace & Tiernan, Inc, 54-56; chief pharmacologist, Wallace Labs Div, Carter-Wallace, Inc, 56-60, dir biol res, 60-64, vpres res, 64-68; pres, AMR Biol Res, Inc, 68-78; prof & chair, 78-89, EMER PROF, PHARMACOL DEPT, SCH MED, ST GEORGE'S UNIV, 89-; PRES, MARNAC, INC, 90- *Concurrent Pos:* Assoc dean basic sci, Sch Med, St George's Univ, 87-89. *Mem:* AAAS; Endocrine Soc; Am Chem Soc; Soc Exp Biol Med; Am Soc Pharmacol & Exp Therapeut; NY Acad Sci. *Res:* Antihistamines; anticholinergics; sedative-hypnotics; tranquilizers; muscle relaxants; adrenal hormones; cardiovascular agents; anti-inflammatory agents; anti-fibrotic agent; release of neurohumors from central nervous system by morphine and opiates. *Mailing Add:* Marnac Inc 6723 Desco Dr Dallas TX 75225

MARGOLIN, SYDNEY GERALD, psychiatry; deceased, see previous edition for last biography

MARGOLIS, ASHER J(ACOB), b New York, NY, Nov 22, 14; m 39; c 2. CHEMICAL ENGINEERING. *Educ:* Columbia Univ, AB, 35, BS, 36, ChE, 37. *Prof Exp:* Chemist, Am Smelting & Ref Co, Utah, 37-38; chief chemist & chem engr, Sweets Co Am, NJ, 38-41; chief chemist, Nutrine Candy Co, 41-43; chem engr, Emulsol Corp, 43; assoc chem engr, metall lab, Univ Chicago, 43-45; sr chem engr, Sherwin-Williams Co, 45-51; mgr, pilot plant, Simoniz Co, 51-66; group leader process develop, Armour-Dial, 66-70; dir test & rentals, Eimco Process Mach Div, Envirotech Corp, 70-72; civil engr, Metrop Sanit Dist Greater Chicago, 72-79; RETIRED. *Mem:* AAAS; Am Chem Soc; Am Inst Chem Engrs. *Res:* Emulsion technology; polishes and waxes; polyurethane foams; aerosol technology; soap and detergents; filtration and other liquid-solids separations. *Mailing Add:* 1338 E Madison Park Chicago IL 60615

MARGOLIS, BERNARD, b Montreal, Que, Aug 15, 26; m 54. PHYSICS. *Educ:* McGill Univ, BSc, 47, MSc, 49; Mass Inst Technol, PhD, 52. *Prof Exp:* Instr physics, Mass Inst Technol, 53-54; instr, Columbia Univ, 54-57, res physicist, 57-59; assoc prof physics, Ohio State Univ, 59-61; assoc prof math physics, 61-63, PROF PHYSICS, MCGILL UNIV, 63- *Mem:* Am Phys Soc. *Res:* Theoretical physics. *Mailing Add:* Dept of Physics McGill Univ 8045 3600 University St Montreal PQ H3A 2T8 Can

MARGOLIS, FRANK L, b Brooklyn, NY, Jan 21, 38; m 61; c 3. NEUROCHEMISTRY, NEUROSCIENCES. *Educ:* Antioch Col, BS, 59; Columbia Univ, PhD(biochem), 64. *Prof Exp:* USPHS trainee biochem, Columbia Univ, 64-65; fel, Lab Comp Physiol, Univ Paris, 65-66; asst res microbiologist, Sch Med, Univ Calif, Los Angeles, 66-69; res assoc, 69-71, asst mem, 71-74, assoc mem, 74-81, chmn, Dept Neurosci, 87-88, FULL MEM, ROCHE INST MOLECULAR BIOL, 81-, LAB HEAD, 88- *Concurrent Pos:* Adj prof, City Univ New York, 71-; mem, Panel Sensory Physiol & Perception, NSF, 79-81. *Honors & Awards:* Freeman Award, Innovative Res in Olfaction and Taste, 85; Philips Mem Lectr in Biol, Haverford Col. *Mem:* Am Soc Pharmacol Exp Therapeut; Am Soc Neurochem; Int Soc Neurochem; Am Soc Biochem & Molecular Biol; Soc Neurosci; Sigma Xi; Asn Chemoreception Sci. *Res:* Regulation of mammalian gene expression; biochemistry and molecular biology of taste and olfaction; neuronal plasticity. *Mailing Add:* Roche Inst Molecular Biol Nutley NJ 07110

MARGOLIS, HAROLD STEPHEN, b Tucson, Ariz, Feb 27, 46; m 71; c 3. PEDIATRICS, INTERNATIONAL HEALTH. *Educ:* Univ Ariz, BS, 68, MD, 72. *Prof Exp:* Dep Chief, 83-87, CHIEF, HEPATITIS BR, DIV VIRAL RICKETTSIAL DIS, CTRS DIS CONTROL, 87-; DIR, COLLABORATING CTR & REF & RES VIRAL HEPATITIS, WHO, 87- *Mem:* Fel Am Acad Pediat; Infectious Dis Soc Am; Am Soc Virol; Sigma Xi. *Res:* Prevention and control of infectious disease, particularly viral hepatitis; vaccines for viral hepatitis; immunization strategies; defining molecular basis for pathogenesis of hepatitis virus mediated liver injury. *Mailing Add:* 1600 Clifton Rd Ctr for Infectious Dis 1600 Clifton Rd NE Atlanta GA 30333

MARGOLIS, JACK SELIG, b Los Angeles, Calif, Mar 9, 32. CHEMICAL PHYSICS, SPECTROSCOPY. *Educ:* Univ Calif, Los Angeles, AB, 54, PhD(physics), 60. *Prof Exp:* Engr, Collins Radio Co, 54-55; asst physics, Univ Calif, Los Angeles, 55-60; mem tech staff, Sci Ctr, NAm Aviation, Inc, 60-64; lectr physics, Univ Calif, Santa Barbara, 64-65, asst prof, 65-66; MEM TECH STAFF, JET PROPULSION LAB, CALIF INST TECHNOL, 66- *Concurrent Pos:* Asst, Scripps Inst, Univ Calif, 55-56; prof dept earth & planetary sci, Washington Univ, 78-79. *Mem:* Am Phys Soc; Am Astron Soc. *Res:* Spectroscopy of the earth's atmosphere; theoretical rare earth and molecular spectroscopy; induced Raman effect; charge transfer complexes; atmospheric radiation. *Mailing Add:* Jet Propulsion Lab 4800 Grove Dr Pasadena CA 91103

MARGOLIS, LEO, b Montreal, Que, Dec 18, 27. PARASITOLOGY, FISH HEALTH. *Educ:* McGill Univ, BSc, 48, MSc, 50, PhD(parasitol), 52. *Prof Exp:* Asst parasitol, McGill Univ, 49-52; asst zool, Macdonald Col, 50-51; from assoc scientist to prin scientist, Fisheries Res Bd Can, 52-67, head var res div & sect, 67-90, SR SCIENTIST, PAC BIOL STA, 90- *Concurrent Pos:* Co-chmn, Can Comt Fish Dis, 70-73; assoc ed, Can J Zool, 71-81; mem, Comt Biol & Res, Int North Pac Fisheries Comn, 71-; mem ed comt, Bull Int N Pac Fisheries Comn, 76-84; chmn, Parasitol Sect, Can Soc Zool, 77-78; mem, Sci Subvention Comt, Dept Fisheries & Oceans, Can, 78-81, Adv Bd Sci Info, 79-83; consult, Int Develop Res Ctr, Southeast Asian Fish Dis, 86; assoc ed, J World Aquacult Soc, 86-; bd trustees, Zool Educ Trust, 88-92, chmn, 90-91; bd dir, Can Fed Biol Soc, 90-91; app officer, Order Can, 90. *Honors & Awards:* RA Wardle Invitiational Lect Award, 82. *Mem:* Am Soc Parasitol; Wildlife Dis Asn; Can Soc Zool (pres, 90-91); fel Royal Soc Can; Am Fisheries Soc; Aquacult Asn Can; World Aquacult Soc; Asian Fisheries Soc. *Res:* Parasites of fish and marine mammals; diseases of fish. *Mailing Add:* Pac Biol Sta Dept Fisheries & Oceans Nanaimo BC V9R 5K6 Can

MARGOLIS, PHILIP MARCUS, b Lima, Ohio, July 7, 25; m 59; c 4. PSYCHIATRY. *Educ:* Univ Minn, BA, 46, BS, 47, BM, 48, MD, 49; Am Bd Psychiat & Neurol, dipl. *Prof Exp:* Harvard fel psychiat, Univ Minn, 49-53, instr, Med Sch, Univ Minn, 53-56; from asst prof to assoc prof, Sch Med, Univ Chicago, 56-66; dir med educ, 78-81, assoc chief, Clin Affairs, Univ Mich Hosp, 81-85, PROF PSYCHIAT, MED SCH, UNIV MICH, ANN ARBOR, 66-, PROF COMMUNITY MENT HEALTH, SCH PUB HEALTH, 66- *Concurrent Pos:* Clin fel, Mass Gen Hosp, Boston, 52-53; consult, Vet Admin Hosp, Minneapolis, Minn, 49-52 & Family Serv Agency, St Paul, 54-56; chief psychiat in-patient serv, Billings Hosp, Univ Chicago Clins, 56-66; consult, Child & Family Serv, Chicago, 56-60 & State Psychiat Inst, 60-66; sr psychiat consult, Peace Corps, 61-66; dir, Washtenaw County Ment Health Clin, 66-72; consult, Geriatric Clin, 78-84. *Mem:* Fel Am Psychiat Asn (secy, 89-91); AMA; fel Am Orthopsychiat Asn; World Fedn Ment Health; Int Asn Social Psychiat. *Res:* Social and community psychiatry; preventive psychiatry; crisis and brief therapy; suicide studies; consultation process; inpatient psychosocial issues. *Mailing Add:* 900 Wall St Univ Mich Ann Arbor MI 48105

MARGOLIS, RENEE KLEIMANN, b Paris, France, Oct 31, 38; US citizen; m 59. PHARMACOLOGY, NEUROCHEMISTRY. *Educ:* Univ Chicago, BS, 60, PhD(pharmacol), 66. *Prof Exp:* Res scientist, NY State Res Inst Neurochem, 66-68; instr pharmacol, Mt Sinai Sch Med, 68-70; from asst prof to assoc prof, 75-81, PROF PHARMACOL, STATE UNIV NY HEALTH SCI CTR AT BROOKLYN, 81- *Concurrent Pos:* mem, Behav & Neurosci Study Sect, NIH, 80-85. *Mem:* Am Soc Neurochem; Int Soc Neurochem; Soc Complex Carbohydrates; Soc Neurosci; Am Soc Biol Chemists. *Res:* Neurobiology of glycoconjugates. *Mailing Add:* Dept Pharmacol State Univ NY Health Sci Ctr Brooklyn NY 11203

MARGOLIS, RICHARD URDANGEN, b Pittsburgh, Pa, Sept 7, 37; m 59. PHARMACOLOGY, BIOCHEMISTRY. *Educ:* Univ Chicago, BS, 59, PhD(pharmacol), 63, MD, 66. *Prof Exp:* Res assoc pharmacol, Univ Chicago, 63-66; from instr to assoc prof, 66-77, PROF PHARMACOL, SCH MED, NY UNIV, 77- *Concurrent Pos:* mem neurol sci study sect, Div Res Grants, NIH, 80-84. *Honors & Awards:* Javits Neurosci Investigator Award. *Mem:* Soc Neurosci; Am Soc Biol Chemists; Int Soc Neurochem; Am Soc Pharmacol & Exp Therapeut; Am Soc Neurochem. *Res:* Neurobiology of glycoconjugates. *Mailing Add:* Dept Pharmacol NY Univ Sch Med New York NY 10016

MARGOLIS, ROBERT LEWIS, b Newark, NJ, Feb 11, 46. CYTOSKELETON, CELL MOTILITY. *Educ:* Rutgers State Univ, NJ, BA, 69; Wesleyan Univ, Middletown, Conn, PhD(biochem), 75. *Prof Exp:* Fel, pharmacol dept, Sch Med, Stanford Univ, 75-76; fel, biochem dept, Univ Calif, Santa Barbara, 76-79; asst mem, 79-84, ASSOC MEM, BASIC SCI DIV, FRED HUTCHINSON CANCER RES CTR, SEATTLE, WASH, 84- *Concurrent Pos:* Adj assoc prof, dept pathol, Univ Wash, 82-, dept biochem, 84-; res fel, Inserm U244, Grenoble, France, 83- *Mem:* Am Soc Cell Biol; Am Soc Biol Chemists. *Res:* Mechanisms of microtubule assembly and the basis of microtubule associated motility; functions of microtubule associated proteins; analysis of the centromere; mitotic mechanisms. *Mailing Add:* Fred Hutchinson Cancer Res Ctr 1124 Columbia St Seattle WA 98104

MARGOLIS, RONALD NEIL, b Brooklyn, NY, Oct 12, 50; m 76; c 2. METABOLISM, CELLULAR ENDOCRINOLOGY. *Educ:* State Univ NY, Albany, BS, 71; Upstate Med Ctr, Syracuse Univ, NY, PhD(anat), 76. *Prof Exp:* Fel cell biol, Univ Va, Charlottesville, 76-79, fel metab, Diabetes Res Training Ctr, 79-80; ASSOC PROF CELL BIOL & HIST, COL MED & ASSOC PROF, DEPT ONCOL, CANCER RES CTR, HOWARD UNIV, 80- *Mem:* Am Asn Anatomists; Am Soc Cell Biol; AAAS; Endocrine Soc; Am Diabetes Asn. *Res:* Hormonal and metabolic regulation of hepatic glycogen metabolism and the effects of diabetes mellitus on the regulation. *Mailing Add:* NIDDK NIH Westwood Bldg Rm 605 Bethesda MD 20892

MARGOLIS, SAM AARON, b Cambridge, Mass, Nov 17, 33; m 60; c 2. BIOCHEMISTRY, MOLECULAR BIOLOGY. *Educ:* Boston Univ, AB, 55, PhD(biochem), 63; Univ RI, MS, 57. *Prof Exp:* Staff scientist, Worcester Found Exp Biol, 63-64; pharmacologist, Food & Drug Admin, 66-68; from staff fel to sr staff fel biochem, Nat Inst Allergy & Infectious Dis, 68-70; sr staff fel biochem, Nat Cancer Inst, 70-72; RES CHEMIST, NAT BUR STANDARDS, 72- *Concurrent Pos:* Fel, Inst Enzyme Res, Univ Wis, 64-66; instr, Sch Med, Boston Univ, 63-64. *Mem:* AAAS; Am Chem Soc; NY Acad Sci. *Res:* Protein hormones and antihormones, characterization and isolation; association of metabolic pathways with biological membranes; modification of viral growth and reproduction by natural and synthetic substances; standards for clinical chemistry and food science. *Mailing Add:* 5902 Roosevelt St Bethesda MD 20817

MARGOLIS, SIMEON, b Johnstown, Pa, Mar 29, 31; m 54; c 3. BIOCHEMISTRY. *Educ:* Johns Hopkins Univ, BA, 53, MD, 57, PhD(lipoprotein struct), 64. *Prof Exp:* From intern to asst resident, Johns Hopkins Hosp, 57-59; res assoc biochem, Nat Heart Inst, 59-61; resident med, Johns Hopkins Hosp, 64-65; asst prof med & physiol chem, Johns Hopkins Univ, 65-68, assoc prof med, 68-77, assoc prof biol chem, 75-81, assoc dean acad affairs, 84-90, PROF BIOL CHEM, SCH MED, JOHNS HOPKINS UNIV, 81-, PROF MED, 77- *Concurrent Pos:* Fel biochem, Sch Med, Johns Hopkins Univ, 61-64, assoc dean fac affairs, 90-; Nat Heart Inst res grant, 65-; mem metab study sect, USPHS, mem gen clin res ctr study sect; mem coun on arteriosclerosis, Am Heart Asn; investr, Howard Hughes Med Inst, 76-81; assoc ed, Am J Clin Nutrit, 81-; mem, Nat Diabetes Adv Bd, 85-88. *Mem:* Am Diabetes Asn; Endocrine Soc; Am Soc Clin Invest; Am Soc Biol Chem; Am Chem Soc. *Res:* Lipid biochemistry; regulation of lipid biosynthesis; metabolism of isolated hepatocytes; structure of human serum lipoproteins; role of serum lipoproteins in atherosclerosis; regulation of hepatic cholesterol metabolism. *Mailing Add:* Dept Med Johns Hopkins Univ Sch Med Baltimore MD 21205

MARGOLIS, STEPHEN BARRY, b Newport News, Va, Mar 23, 50; m 86; c 1. COMBUSTION THEORY, BIFURCATION THEORY. *Educ:* Col William & Mary, BS, 72; Brown Univ, ScM, 73, PhD(appl math), 76. *Prof Exp:* Fel appl math, Brown Univ, 72-73, teaching & res asst, 73-76; SR MEM TECH STAFF APPL MATH & COMBUSTION THEORY, SANDIA NAT LABS, 76- *Concurrent Pos:* Vis scholar, dept eng sci & appl math, Northwestern Univ, 85-87. *Mem:* Soc Indust & Appl Math; Am Phys Soc; Combustion Inst; Sigma Xi. *Res:* Combustion theory, fluid dynamics, heat transfer; theory of flame propagation in gaseous and condensed premixed combustion stability of solid and liquid propellants; nonlinear stability and bifurcation theory; asymptotic and perturbation methods; nonlinear ordinary and partial differential equations. *Mailing Add:* Combustion Res Facil Sandia Nat Labs Livermore CA 94551-0969

MARGOLIS, STEPHEN G(OODFRIEND), b Philadelphia, Pa, Dec 15, 31; m 55; c 2. ELECTRICAL & NUCLEAR ENGINEERING. *Educ:* Univ Pa, BS, 53; Mass Inst Technol, SM, 55; Univ Pittsburgh, PhD(elec eng), 62. *Prof Exp:* Res engr, Jet Propulsion Lab, Univ Calif, 55-56; engr, Bettis Atomic Power Lab, Westinghouse Elec Corp, 56-60, sr engr, 60-63, fel engr, 63-66; assoc prof, Div Interdisciplinary Studies & Res, 66-71, PROF ELEC ENG & ENG SCI, SCH ENG, STATE UNIV NY BUFFALO, 71- *Mem:* Inst Elec & Electronics Engrs. *Res:* Dynamics, stability and control of nuclear reactors and power plants. *Mailing Add:* Dept of Elec & Comput Eng State Univ NY N Campus 201 Bell Hall Buffalo NY 14260

MARGOLIUS, HARRY STEPHEN, b Albany, NY, Jan 29, 38; m 64; c 2. CLINICAL PHARMACOLOGY, MOLECULAR PHARMACOLOGY. *Educ:* Union Univ, BS, 59; Albany Med Col, PhD(pharmacol), 63; Univ Cincinnati, MD, 68. *Prof Exp:* From intern to resident med, Harvard Med Serv II & IV, Boston City Hosp, 68-70; res assoc pharmacol, Exp Therapeut Br, Nat Heart & Lung Inst, 70-72, sr clin investr hypertension res, Hypertension-Endocrine Br, 72-74; prog dir, Gen Clin Res Ctr, Med Univ, SC, 74-84, from asst prof to assoc prof med, 74-80, assoc prof pharmacol, 74-77, PROF PHARMACOL, MED UNIV SC, 77-, PROF MED, 80-, CHMN, 90- *Concurrent Pos:* Attend physician, Clin Ctr, NIH, 70-74; attend physician, Med Univ SC Hosp, Charleston Mem Hosp & Vet Admin Hosp, 74-; mem, Hypertension Task Force, Nat Heart & Lung Inst, 75-77; Nat Heart, Lung & Blood Inst res grants, 75-; mem, Cardiovasc & Renal Study Sect, Nat Heart & Lung Inst, 76-80, Cardiovasc & Renal Drugs Adv Comm, Food & Drug Admin, 82-86; Burroughs-Wellcome scholar clin pharmacol, 76; mem med adv bd, Coun High Blood Pressure Res, Am Heart Asn; vis scholar, Univ Cambridge, 80-81; consult, Monsanto Co & Searle Co; chmn, Gordon Conf Kallikreins & Kinins, 86, Nat Bd Med Examrs, 87-91, Pharmacol comt, 90- *Mem:* Am Soc Clin Invest; Am Fedn Clin Res; Am Soc Pharmacol & Exp Therapeut; Am Heart Asn; Am Soc Clin Pharmacol & Therapeut; Brit Pharmacol Soc. *Res:* Studies of tissue kallikrein and kinins and their roles in cellular function, ion transport and cardiovascular and renal diseases using isolated cells, tissues, whole animals and clinical investigation. *Mailing Add:* Dept Pharmacol Med Univ SC 171 Ashley Ave Charleston SC 29425

MARGOLSKEE, ROBERT F, b Boston, Mass, Oct 28, 54; m 77; c 3. MOLECULAR BIOLOGY SENSORY TRANSDUCTION IN TASTE CELLS. *Educ:* Harvard Col, AB, 76; Johns Hopkins Univ MD, 83, PhD(molecular biol), 83. *Prof Exp:* Postdoctoral fel biochem, Sch Med, Stanford Univ, 83-87; ASST MEM NEUROSCI, ROCHE INST MOLECULAR BIOL, 87- *Concurrent Pos:* Fel, Jane Coffin Childs Mem Fund, 83-86; adj prof, Dept Biol, Columbia Univ, 90- *Mem:* AAAS; Am Soc Microbiol; Soc Neurosci; Am Soc Biochem & Molecular Biol; Am Soc Pharmacol & Exp Therapeut; Soc Analysis Cytol. *Res:* Molecular basis of sensory transduction in mammalian taste cells; identified proteins common to signal transduction in taste, olfaction and vision. *Mailing Add:* Dept Neurosci Roche Inst 340 Kingsland St Nutley NJ 07110

MARGON, BRUCE HENRY, b New York, NY, Jan 7, 48; m 76; c 1. ASTROPHYSICS. *Educ:* Columbia Univ, AB, 68; Univ Calif, Berkeley, MA, 71, PhD(astron), 73. *Prof Exp:* NATO fel astron, Univ Col, Univ London, 73-74; asst res astronomer, Univ Calif, Berkeley, 74-76; from asst prof to assoc prof astron, Univ Calif, Los Angeles, 76-80; PROF & CHMN ASTRON, UNIV WASH, 81-87 & 90- *Concurrent Pos:* Alfred P Sloan Fel, 79-83. *Honors & Awards:* Pierce Prize, Am Astron Soc, 81. *Mem:* Am Astron Soc; Royal Astron Soc; Int Astron Union; Astron Soc Pac. *Res:* Extrasolar x-ray and ultraviolet astronomy; optical observations of x-ray sources. *Mailing Add:* Dept Astron FM-20 Univ Wash Seattle WA 98195

MARGOSHES, MARVIN, b New York, NY, May 23, 25; m 55; c 4. TECHNICAL MANAGEMENT, TECHNOLOGY TRANSFER. *Educ:* Polytech Inst Brooklyn, BS, 51; Iowa State Col, PhD(phys chem), 53. *Prof Exp:* Asst, Inst Atomic Res, Iowa State Col, 50-53; res fel med, Harvard Med Sch, 54-56, res assoc, 56-57; res assoc spectrochem anal sect, Nat Bur Standards, 57-69; proj dir, Digilab Div, Block Eng, Inc, 69-70; tech dir, Technicon Instrument Corp, 71-89; PRES, TECH TRANSFER SERV INC, 90- *Concurrent Pos:* Ed, Atomic Spectra Sect, Spectrochimica Acta, 66-73. *Honors & Awards:* Award, Soc Appl Spectros, 76. *Mem:* Am Chem Soc; Soc Appl Spectros (pres, 74); NY Acad Sci; Sigma Xi; Licensing Exec Soc. *Res:* Analytical spectroscopy; clinical chemistry. *Mailing Add:* Tech Transfer Serv 69 Midland Ave Tarrytown NY 10591-4317

MARGOSSIAN, SARKIS S, b Beirut, Lebanon, June 20, 40; m 88. MUSCLE CONTRACTION. *Educ:* Pa State Univ, PhD(biochem), 69. *Prof Exp:* ASSOC PROF BIOCHEM, ALBERT EINSTEIN COL MED, 80- *Concurrent Pos:* Estab investr, Am Heart Asn, 74-79. *Mem:* Am Soc Biochem & Molecular Biol; Biophys Soc; NY Acad Sci; AAAS; Am Heart Asn. *Res:* Molecular basis of cardiac contraetility. *Mailing Add:* Div Cardiology Montefiore Hosp Med Ctr 111 E 210 St Bronx NY 10467

MARGRAVE, JOHN LEE, b Kansas City, Kans, Apr 13, 24; m 50; c 2. HIGH TEMPERATURE CHEMISTRY, FLUORINE CHEMISTRY. *Educ:* Univ Kans, BS, 48, PhD(chem), 50. *Prof Exp:* Atomic Energy Comn fel, Univ Calif, 51-52; from instr to prof chem, Univ Wis, 52-63; chmn, Dept Chem, Rice Univ, 67-72, dean advan studies & res, 72-80, vpres, 80-86, PROF CHEM, RICE UNIV, 63-, E D BUTCHER PROF CHEM, 86-; vpres res, 86-89, DIR, MAT RES CTR, HOUSTON AREA RES CTR, 85-, CHIEF SCI OFFICER, 89- *Concurrent Pos:* Sloan res fel, 57-58; Guggenheim fel, 61; pres, Marchem, Inc, 70-; mem, bd trustees, Ctr Res, Inc, Univ Kans, 71-75; consult, Nat Bur Standards, Argonne Nat Lab, Lawrence Radiation Lab, Oak Ridge Nat Lab, NASA & private indust; vpres, bd dirs, Rice Ctr for Community Design & Res, 72-86; mem, dir bd of dirs, Gulf Univs Res Consortium, 74-80, Houston Area Res Ctr, 82-86, Coun Chem Res, 85-; ed, High Temp Sci, 69-; consult, var govt agencies & pvt indust. *Honors & Awards:* Inorg Chem Award, Am Chem Soc, 67, Fluorine Chem Award, 81; Reilly Lectr, Notre Dame Univ, 68; IR-100 Award, 70 & 86; Seydel-Wooley Lectr, Ga Inst Technol, 70; DuPont Lectr, Univ SC, 71; Abbott Lectr, Univ NDak, 72; Robert A Welch Distinguished Lectr, 85. *Mem:* Nat Acad Sci; fel AAAS; fel Am Phys Soc; Am Ceramic Soc; Am Chem Soc; Chem Soc, London; Mat Res Soc; fel Am Inst Chemists. *Res:* High temperature chemistry and thermodynamics; fluorine chemistry; optical and mass spectroscopy; synthetic inorganic, plasma and high pressure chemistry; ESCA; levitation calorimetry, microscale laser methods and matrix isolation spectroscopy; untraviolet-visable infrared, raman and ESR studies. *Mailing Add:* Dept Chem Rice Univ Houston TX 77251

MARGRAVE, THOMAS EWING, JR, b Langley Field, Va, Nov 15, 38; div; c 4. ASTRONOMY. *Educ:* Univ Notre Dame, BS, 61; Rensselaer Polytech Inst, MS, 63; Univ Ariz, PhD(astron), 67. *Prof Exp:* Physicist, US Naval Avionics Facil, 61; aerospace technologist, Manned Spacecraft Ctr, NASA, 63; asst prof astron, Georgetown Univ, 67-69; asst prof, 69-73, assoc prof, 73-80, PROF ASTRON, UNIV MONT, 80- *Concurrent Pos:* NSF sci equip grants, Univ Mont, 70-72 & 75-77; Univ Mont Found res grants, 71-72, 73-75 & 78-81. *Mem:* Am Astron Soc; Astron Soc Pacific. *Res:* Photoelectric photometry of variable stars, ephemerides of eclipsing binary stars; frequency analysis of Delta Scuti variable stars. *Mailing Add:* 400 Johnson St Vienna VA 22180

MARGULIES, DAVID HARVEY, b Passaic, NJ, Dec 8, 49; m 79; c 2. MOLECULAR IMMUNOLOGY. *Educ:* Columbia Univ, New York, AB, 71; Albert Einstein Col Med, Bronx, MD & PhD(cell biol), 78. *Prof Exp:* Intern, Columbia Presby Med Ctr, 78-79, resident internal med, 79-80; res assoc molecular biol, Lab Molecular Genetics, Nat Inst Child Health & Human Develop, 80-83, investr, 83-87, SR INVESTR MOLECULAR IMMUNOL, LAB IMMUNOL, NAT INST ALLERGY & INFECTIOUS DIS, NIH, 87- *Concurrent Pos:* Mem, Immunol Adv Comt, Am Cancer Soc 88-91, Fel Subcomt, Arthritis Found, 89-90 & NIH Res Scholars Prog Comt, Howard Hughes Med Inst, 91-; ed, Current Protocols Immunol, 90-; assoc ed, J Immunol, 91-93. *Mem:* Am Asn Immunologists; AAAS; Am Soc Clin Invest. *Res:* Structure, function, biochemistry, and genetics of cell surface molecules involved in cell-cell interactions in the immune response; molecules encoded by the major histocompatibility complex. *Mailing Add:* Lab Immunol NIAID NIH Bldg 10 Rm 11N311 Bethesda MD 20892

MARGULIES, MAURICE, b Brooklyn, NY, Feb 9, 31; m 67; c 3. BIOCHEMISTRY, PLANT PHYSIOLOGY. *Educ:* Brooklyn Col, BA, 52; Yale Univ, MS, 53, PhD(microbiol), 57. *Prof Exp:* Res assoc biol, Haverford Col, 57; McCollum Pratt fel, Johns Hopkins Univ, 57-59; biochemist, Radiation Biol Lab, Smithsonian Inst, 59-86; COLLABORATING SCIENTIST, AGR RES SERV, USDA, 86- *Concurrent Pos:* Lectr, George Washington Univ, 64-67; res fel, Harvard Univ, 69-70; vis scientist, Weizmann Inst, Rehovot, Israel, 87-88; biotech consult, 89- *Mem:* AAAS; Am Soc Plant Physiol; Am Soc Biol Chem; Am Chem Soc. *Res:* Chloroplast biochemistry-protein synthesis, synthesis of chloroplast membranes, photosynthesis, electron transport; membrane protein structure. *Mailing Add:* Climate Stress Lab USDA, Agr Res Serv 046A-BARC West Beltsville MD 20705

MARGULIES, SEYMOUR, b Jaslo, Poland, Oct 3, 33; US citizen; m 59; c 2. EXPERIMENTAL HIGH-ENERGY PHYSICS. *Educ:* Cooper Union, BEE, 55; Univ Ill, MS, 56, PhD(physics), 62. *Prof Exp:* Nat Acad Sci-Nat Res Coun res fel, Max Planck Inst Nuclear Physics, Ger, 61-63; res assoc nuclear & high energy physics, Nevis Labs, Columbia Univ, 63-65; from asst prof to assoc prof, 65-87, PROF HIGH-ENERGY PHYSICS, UNIV ILL, CHICAGO, 87- *Concurrent Pos:* Res grant, co-prin investr, NSF, 73- *Mem:* Am Phys Soc; Sigma Xi. *Res:* Mossbauer effect; nuclear spectroscopy; nuclear disintegrations following capture of negative pi-mesons; strong interactions of elementary particles, particularly multiparticle production; hadron jets and particle constituents; high transverse-momemtum reactions; study of heavy quarks, high energy physics instrumentation; development of inner tracker for superconducting super collider. *Mailing Add:* Dept Physics M-C 273 Univ Ill Chicago Chicago IL 60680

MARGULIES, WILLIAM GEORGE, b New York, NY, Oct 31, 40; m 64; c 2. MATHEMATICS. *Educ:* State Univ NY Col Long Island, BS, 62; Brandeis Univ, MS, 64, PhD(math), 67. *Prof Exp:* Asst prof math, Wash Univ, 66-69; asst prof, 69-76, PROF MATH, CALIF STATE UNIV, LONG BEACH, 76- *Concurrent Pos:* NSF grant, 70-72. *Mem:* Math Asn Am; Am Math Soc; Soc Indust & Appl Math. *Res:* Analysis, partial differential equations; least action principle. *Mailing Add:* Dept of Math Calif State Univ 1250 Bellflower Blvd Long Beach CA 90840

MARGULIS, ALEXANDER RAFAILO, b Belgrade, Yugoslavia, Mar 31, 21; nat US; m 46. RADIOLOGY. *Educ:* Harvard Med Sch, MD, 50. *Prof Exp:* Intern, Henry Ford Hosp, Detroit, 50-51; resident radiol, Univ Mich Hosps, 51-53; clin instr, Univ Mich, 53-54; from instr to asst prof radiol, Univ Minn, 54-57; from asst prof to prof, Mallinckrodt Inst Radiol, Sch Med, Wash Univ, 59-63; prof radiol & chmn dept, 63-89, ASSOC CHANCELLOR, SPEC PROJS, UNIV CALIF, SAN FRANCISCO, 89- *Concurrent Pos:* Mem comt radiol, Nat Acad Sci-Nat Res Coun, 64-; consult, Off Surgeon Gen, 67-71, Vet Admin Hosp, Ft Miley & Letterman Gen Hosp, San Francisco & Oak Knoll Naval Hosp, Oakland. *Honors & Awards:* Medaille Antoine Beclere, France, 78. *Mem:* AMA; fel Am Col Radiol; Am Roentgen Ray Soc; Asn Univ Radiol (past pres); Soc Gastrointestinal Radiol (pres, 72); Soc Magnetic Resonance Med (past pres). *Res:* Gastroenterology and arteriography. *Mailing Add:* 500 Parnussus Ave Univ of Calif San Francisco CA 94143-0292

MARGULIS, LYNN, b Chicago, Ill, Mar 5, 38; m 57, 67; c 4. CELL BIOLOGY, MICROBIAL EVOLUTION & ECOLOGY. *Educ:* Univ Chicago, AB, 57; Univ Wis, MS, 60; Univ Calif, Berkeley, PhD(genetics), 65. *Hon Degrees:* DSc, Southeastern Mass Univ, 89, Westfield State Col, 89. *Prof Exp:* From adj asst prof to prof, 66-86, DISTINGUISHED UNIV PROF BIOL, DEPT BIOL, BOSTON UNIV, 86- *Concurrent Pos:* Sherman Fairchild Distinguished scholar, Calif Inst Technol, 76-77; Guggenheim fel, 79; mem space sci bd, Nat Acad Sci Comt Lunar & Planetary Studies, 75-77; assoc ed, Pre cambrian Res, 79-; mem, Workshop on Global Habitability, NASA, 82, adv coun mem, 82-86; assoc managing ed, BioSysts, 83-; vis prof, dept microbiol, Univ Barcelona, 85, 86 & 88, dept marine biol, Scripps Inst Oceanog, 80; mem exec coun, Int Soc Study Origin Life, 89-92. *Honors & Awards:* Diamond Award, Int Bot Cong, 75; Miescher-Ishida Award, Int Soc Endocytobiol, 86; Distinguished Serv Award, Nat Asn Biol Teachers, 88. *Mem:* Nat Acad Sci; Catalan Soc Biol; Int Soc Study Origin Life; Sigma Xi; Soc Evolutionary Protistol; fel AAAS. *Res:* Origin and evolution of cells; cytoplasmic genetics; microtubules and kinetosomes; evolution of biochemical pathways; morphogenesis in protists; spirochetes of termites; numerous publications. *Mailing Add:* Dept Bot Univ Mass Amherst MA 01003

MARGULIS, THOMAS N, b New York, NY, Sept 7, 37. STRUCTURAL CHEMISTRY. *Educ:* Mass Inst Technol, BS, 59; Univ Calif, Berkeley, PhD(chem), 62. *Prof Exp:* Asst prof chem, Brandeis Univ, 62-67; assoc prof, 67-75, PROF CHEM, UNIV MASS, BOSTON, 75- *Mem:* Am Crystallog Asn; Am Chem Soc. *Res:* Crystal and molecular structure by x-ray diffraction; small ring compounds; structural chemistry of drugs. *Mailing Add:* Dept of Chem Univ of Mass Boston MA 02125

MARIA, NARENDRA LAL, b Chamba, India, Apr 22, 28; m 57; c 1. APPLIED MATHEMATICS. *Educ:* Panjab Univ, India, BA, 48, MA, 49; Univ Calif, Berkeley, PhD(appl Math), 68. *Prof Exp:* Lectr math, Panjab Univ, India, 50-51, sr lectr, 51-59, asst prof, 59-65; teaching assoc, Univ Calif, Berkeley, 65-67; vis lectr, Stanislaus State Col, 67-68, assoc prof, 68-70, chmn dept, 70-90, PROF MATH, STANISLAUS STATE COL, 70- *Mem:* Am Math Soc. *Res:* Partial differential equations; analysis. *Mailing Add:* Dept Math Cal St Univ Stanislaus 801 W Monte Vista Ave Turlock CA 95380

MARIANELLI, ROBERT SILVIO, b Wilmington, Del, Dec 17, 41; m 61; c 2. INORGANIC CHEMISTRY. *Educ:* Univ Del, BA, 63; Univ Calif, Berkeley, PhD(chem), 66. *Prof Exp:* Asst prof chem, Univ Nebr, Lincoln, 66-71, assoc prof, 71-80; MEM STAFF, DIV CHEM SCI, OFF BASIC ENERGY, DEPT ENERGY, 80- *Concurrent Pos:* Staff mem, Div Chem Sci, Dept Energy, 77-79. *Mem:* AAAS; Am Chem Soc; Royal Soc Chem; Sigma Xi. *Res:* The chemistry of metalloporphyrins and related compounds. *Mailing Add:* Div Chem Sci Off Basic Energy Dept Energy Washington DC 20545

MARIANI, HENRY A, b Medford, Mass, Sept 13, 24. BIOCHEMISTRY, PHYSICAL CHEMISTRY. *Educ:* Boston Col, AB, 47, Tufts Univ, MS, 49; Boston Col, PhD, 81. *Prof Exp:* Instr chem, St Anselm's Col, 49-50; asst prof org chem & biochem, Merrimack Col, 52-60; chmn dept sci, Medford Pub Schs, Mass, 60-62; assoc prof biochem & phys chem, Boston State Col, 62-81; PROF GEN CHEM, BIOCHEM & NUTRIT, UNIV MASS, BOSTON, 81- *Mem:* AAAS; Am Chem Soc; Sigma Xi. *Res:* Cell membranes and transport-photosynthesis; electro-organic fluorination of aromatic polycyclics, application to chemical carcinogenesis. *Mailing Add:* 216 Fulton St Medford MA 02115

MARIANI, TONI NINETTA, EXPERIMENTAL PATHOLOGY. *Educ:* Univ Mich, PhD(physiol), 67. *Prof Exp:* ASSOC PROF PATHOBIOL, UNIV MINN, 67- *Res:* Tumor immunobiology. *Mailing Add:* 1924 E River Terr Minneapolis MN 55414

MARIANO, PATRICK S, b Passaic, NJ, Aug 31, 42. CHEMISTRY. *Educ:* Fairleigh Dickinson Univ, BSc, 64; Univ Wis, PhD(chem), 69. *Prof Exp:* NIH fel, Yale Univ, 68-70; asst prof chem, Tex A&M Univ, 70-77, assoc prof, 77-; PROF, DEPT CHEM, UNIV MD, COL PARK. *Mem:* Am Chem Soc; The Chem Soc. *Res:* Organic chemistry; photochemistry; synthetic chemistry. *Mailing Add:* 16311 Cambridge Ct Mitchellville MD 20716

MARIANOWSKI, LEONARD GEORGE, b Hammond, Ind, Oct 31, 35; m 59; c 4. CHEMICAL ENGINEERING, ELECTROCHEMISTRY. *Educ:* Purdue Univ, BS, 57, MS, 59. *Prof Exp:* Process engr, Commercial Solvents Corp, 59-60, chem engr, 60-65, supvr, 65-70, mgr, 70-76, asst dir, 76-78; assoc dir, 78-87, DIR ENERGY CONVERSION & STORAGE RES, INST GAS TECHNOL, 87- *Mem:* Am Chem Soc; Am Inst Chem Engrs; Electrochem Soc. *Res:* Direct energy conversion processes for electricity production; fuel cells, using molten carbonate electrolytes. *Mailing Add:* Inst Gas Technol 3424 S State St Chicago IL 60616

MARIANS, KENNETH J, MOLECULAR BIOLOGY. *Educ:* Polytech Inst Brooklyn, NY, BS, 72; Cornell Univ, Ithaca, NY, PhD(biochem), 76. *Prof Exp:* Postdoctoral fel, Dept Develop Biol & Cancer, Albert Einstein Col Med, Bronx, NY, 76-78, from asst prof to assoc prof, 78-84; assoc prof & dir, 84-88, PROF, GRAD PROG MOLECULAR BIOL, GRAD SCH MED SCI, CORNELL UNIV, NEW YORK, NY, 88-; CHMN, PROG MOLECULAR BIOL, MEM SLOAN-KETTERING CANCER CTR, NEW YORK, NY, 91- *Concurrent Pos:* Postdoctoral fel, Am Cancer Soc, 76; assoc mem, Prog Molecular Biol, Mem Sloan-Kettering Cancer Ctr, New York, NY, 84-88, mem, 88-; mem, NJ State Cancer Comn Study Sect, 87-90 & Biochem Study Sect, NIH, 89- *Honors & Awards:* Sinsheimer Found Award, 82; Irma T Hirschl Career Scientist Award, 85. *Res:* Cancer research; author of various publications. *Mailing Add:* Dept Molecular Biol Mem Sloan-Kettering Cancer Ctr 1275 York Ave Box 97 New York NY 10021

MARICICH, TOM JOHN, b Anacortes, Wash, Dec 20, 38; m 64; c 3. ORGANIC CHEMISTRY. *Educ:* Univ Wash, BS, 61; Yale Univ, MS, 63, PhD(chem), 65. *Prof Exp:* Chemist, Shell Develop Co, Calif, 65-67; asst prof org chem, N Dak State Univ, 67-70, assoc prof, 70-75; from asst prof to assoc prof, 75-84, PROF CHEM, CALIF STATE UNIV, LONG BEACH, 84-, ACTG ASSOC DEAN, SCH NATURAL SCI, 88- *Concurrent Pos:* Consult, Niklor Chem Co, 82-88, Impex Co, 86-90; expert witness, chem litigations, 86- *Mem:* Am Chem Soc. *Res:* Reactive organic intermediates, nitrenes; sulfur-nitrogen functional groups and heterocycles; synthesis of antitumor agents; chemistry of sulfonimidates. *Mailing Add:* Dept Chem Calif State Univ Long Beach CA 90840

MARICONDI, CHRIS, b Oct 13, 41; US citizen; m 70. INORGANIC CHEMISTRY. *Educ:* WVa Univ, AB, 64; Univ Pittsburgh, PhD(chem), 69. *Prof Exp:* Asst prof, 69-75, ASSOC PROF CHEM, PA STATE UNIV, McKEESPORT, 75- *Mem:* Am Chem Soc. *Res:* Molecular structure. *Mailing Add:* Dept Chem Pa State Univ University Dr McKeesport PA 15132-7647

MARICQ, HILDEGARD RAND, b Rakvere, Estonia, Apr 23, 25; US citizen; m 48; c 3. MEDICINE, HEALTH SCIENCES. *Educ:* Free Univ Brussels, Cand, 49, MD, 53. *Prof Exp:* Intern, Jersey City Med Ctr, 55-56; resident psychiat, Essex County Overbrook Hosp, Cedar Grove, NJ, 57-61; resident, Vet Admin Hosp, Lyons, NJ, 61-62, res assoc, 62-63, clin investr, 63-65, dir, Microcirc Lab, 67-69, sr psychiatrist, 67-73, dir Schizophrenia Res Sect, 69-73; res assoc, dept med, Col Physicians & Surgeons, Columbia Univ, 73-75; assoc prof, 75-81, PROF RES MED, MED UNIV SC, 81- *Concurrent Pos:* Res fel psychiat, Col Physicians & Surgeons, Columbia Univ, 65-67; res assoc, dept psychiat, Rutgers Med Sch, 67-71, res asst prof, 71-73. *Mem:* AAAS; Am Rheumatism Asn; Am Physiol Soc; Microcirc Soc; Soc Psychophysiol Res; Soc Biol Psychiat. *Res:* Somatic research in schizophrenia; microcirculation; human genetics; psychophysiology; microcirculation in connective tissue diseases; peripheral circulation; epidemiology of Raynaud phenomenon and scleroderma spectrum disorders. *Mailing Add:* Dept of Med Med Univ of SC Charleston SC 29425

MARICQ, JOHN, b Anderlecht, Belg, Sept 14, 22; US citizen; m 48; c 3. ORGANIC CHEMISTRY, MEDICAL ELECTRONICS. *Educ:* Free Univ Brussels, Lic en Sc, 48, Dr en Sc, 51. *Prof Exp:* Res chemist, Pharmaceut Div, Belgian Union Chem, 50-54; sr chemist, Hoffmann-La Roche, Inc, 54-74, tech fel, Tech Develop Dept, 74-77; RETIRED. *Concurrent Pos:* Adj asst prof med, Med Univ SC, 80- *Mem:* Am Chem Soc. *Res:* Synthetic organic chemistry; research and development of new drugs, vitamins, carotenoids and aromatics; medical electronics. *Mailing Add:* 728 Jim Isle Dr Battery Point Charleston SC 29412

MARIEB, ELAINE NICPON, b Northhampton, Mass, Apr 5, 36; m 58; c 2. ANATOMY, PHYSIOLOGY. *Educ:* Westfield State Col, BSEd, 64; Mt Holyoke Col, MA, 66; Univ Mass, Amherst, PhD(cell biol), 69; Fitchburg State Col, BS, 84; Univ Mass, Amherst, MS(nursing), 85. *Prof Exp:* Instr zool, anat, physiol & embryol, Springfield Col, 66-67; from asst prof to assoc prof, 69-78, PROF BOT, ANAT & PHYSIOL, HOLYOKE COMMUNITY COL, 78-, PROF BIOL, 81- *Mem:* AAAS; Am Soc Zool; Sigma Xi. *Res:* Kinetic studies on the synthesis of sRNA in yeast; species and tissue variations in transfer RNA populations; college textbook publications in the field of human anatomy and physiology. *Mailing Add:* Dept Biol Holyoke Community Col 303 Homestead Ave Holyoke MA 01040

MARIELLA, RAYMOND PEEL, b Philadelphia, Pa, Sept 5, 19; m 43; c 4. ORGANIC CHEMISTRY. *Educ:* Univ Pa, BS, 41; Carnegie Inst Technol, MS, 42, DSc(org chem), 45. *Prof Exp:* Asst, Carnegie Inst Technol, 41-44, instr, 44, res chemist, 44-45; postdoctoral fel, Eli Lilly & Co, Univ Wis, 45-46; from instr to asst prof chem, Northwestern Univ, 46-51; from assoc prof to prof chem, Loyola Univ Chicago, 51-77, chmn dept, 51-70, dean grad sch, 69-77; exec dir, Am Chem Soc, 77-82; CONSULT, 83- *Concurrent Pos:* Mem, Gov Sci Adv Coun, Ill; exec comt, Coun Grad Schs, 71-74; exec comt mem, Midwestern Asn Grad Schs, 72-77; ed, annual Proc, Midwestern Asn Grad Schs, 72-77; assoc vpres res, Loyola Univ, Chicago, 74-77; consult, 83- *Honors & Awards:* McCormack-Freud Hon Lectr in chem, Ill Inst Technol, 61; mems & fels lectr, Am Inst Chemists, 77. *Mem:* AAAS; Am Chem Soc; Sigma Xi. *Res:* Synthesis of new pyridine compounds; hyperconjugations; ultraviolet absorption spectra; small ring synthesis; synthesis of carcinolytic substances. *Mailing Add:* 21215 123rd Dr Sun City West AZ 85375-1944

MARIEN, DANIEL, b New York, NY, Aug 19, 25; m 59; c 3. GENETICS, ZOOLOGY. *Educ:* Cornell Univ, BS, 49; Columbia Univ, MA, 51, PhD, 56. *Prof Exp:* From instr to assoc prof, 53-70, PROF BIOL, QUEENS COL, NY, 70- *Mem:* Sigma Xi. *Res:* Population genetics and evolution. *Mailing Add:* 64-18 136 St Flushing NY 11367

MARIK, JAN, b Ungvar, USSR, Nov 12, 20; m 48; c 1. MATHEMATICAL ANALYSIS. *Educ:* Univ Prague, RNDr(math), 49. *Prof Exp:* Asst math, Prague Tech Univ, 48-50; grant, Czech Acad Sci, 50-52, sci worker, 52-53; asst, Prague Univ, 53-56, docent, 56-60, prof, 60-69; vis prof, Mich State Univ, 69-70, prof math, 70-90; RETIRED. *Res:* Surface integral and non-absolute convergent integrals in Euclidean spaces; representation of functionals by integrals; oscillatory properties of differential equations of second order. *Mailing Add:* Dept of Math Mich State Univ East Lansing MI 48824

MARIKOVSKY, YEHUDA, b Nov 21, 24. BIOLOGY OF AGING, BIOLOGICAL ULTRASTRUCTURE. *Educ:* Weizmann Inst Sci, Israel, PhD(biol), 73. *Prof Exp:* Head, Clin Lab, Med Corp Israel Air Force, 49-54; RES FEL BIOL, WEIZMANN INST SCI, 55- *Concurrent Pos:* Vis assoc prof, Med Sch, Tufts Univ, Boston, Mass, 74-75; actg head, Dept Biol Ultrastruct, Weizmann Inst Sci, 77-78; vis prof, Med Col, Rush Univ, Chicago, Ill, 79-90. *Mem:* NY Acad Sci; Am Soc Cell Biol; Europ Asn Cancer Res. *Res:* Erythrocyte aging; membrane sialic acid; band 3; cryptic antigens and the complement system; immune-recognition process of senescent erythrocytes; membrane alteration in malignant cell transformation. *Mailing Add:* Dept Membrane Res & Biophys Weizmann Inst Sci Rehovot 76100 Israel

MARIMONT, ROSALIND BROWNSTONE, b New York, NY, Feb 3, 21; m 51; c 2. APPLIED MATHEMATICS. *Educ:* Hunter Col, BA, 42. *Prof Exp:* Physicist electronics, Nat Bur Stand, 42-51, electronic scientist digital comput design, 51-60; mathematician, NIH, 60-79; RETIRED. *Mem:* AAAS. *Res:* Applications of linear algebra to biological problems including compartmental analysis and classification schemes; mathematical modeling of biological systems, particularly human visual and auditory systems; applications of high level computer language to development of mathematical models of biological systems; statistical analysis of employment with regard to sex or race discrimination. *Mailing Add:* 11512 Yates St Silver Spring MD 20902

MARIN, MATTHEW GRUEN, PHYSIOLOGY. *Educ:* Univ Rochester, MD, 69. *Prof Exp:* CO-DIRECTOR PULMONARY & CRITICAL CARE UNIT, SCH MED & DENT, UNIV ROCHESTER, 76- *Mailing Add:* Dept Med & Radiation Biol Sch Med & Dent Univ Rochester Rochester NY 14642

MARIN, MIGUEL ANGEL, b Seville, Spain, Dec 26, 38; m 79; c 5. COMPUTER ENGINEERING, DATA PROCESSING. *Educ:* Univ Madrid, Licenciado, 63, DSc(phys sci), 69; Univ Calif, Los Angeles, PhD(eng), 68. *Prof Exp:* Asst prof elec eng, McGill Univ, 68-72; vpres, Assyst Assocs Ltd, 69-71; MGR, HARDWARE & SOFTWARE, EDP PLANNING, HYDRO-QUEBEC, 77- *Concurrent Pos:* Nat Res Coun Can fel, McGill Univ, 68-71; design engr, Cent Dynamics, Que, 69; comput scientist, Philips Data Systs, Holland, 71-72; mgr sci appln, Hydro-Quebec, 72-77; auxiliary prof, McGill Univ, 72-; adj prof, Concordia Univ, 78-; vis researcher, Electrosc Lab, Ohio State Univ. *Mem:* Inst Elec & Electronics Engrs; Asn Comput Mach; Sigma Xi. *Res:* Computer systems engineering; logic design of digital systems; computer applications to power utilities, instruction and to computer design. *Mailing Add:* Inst de Recherceh D'Hydro 1800 Montee Ste-Julie Varennes PQ J0L 2P0 Can

MARINACCIO, PAUL J, b Bridgeport, Conn, May 30, 37; m 59; c 3. POLYMER CHEMISTRY, PHYSICAL CHEMISTRY. *Educ:* Fairfield Univ, BS, 59; Purdue Univ, MS, 61. *Prof Exp:* Res chemist, Rexall Chem Co, 61-64, supvr additives, 64-67; sr chemist, AMF Inc, 67-70, mgr plastics eng, 70-78, dir plastics eng, 78-85; dir, Foster Miller Inc, 85- *Mem:* Am Chem Soc; Am Mgt Asn; Indust Res Inst; Soc Plastics Engrs; Soc Advan Mat & Process Eng; NY Acad Sci. *Res:* Membranes; stereospecific polymer catalysts; polymer additives; composites. *Mailing Add:* 47 River Rd PO Box 1328 East Orleans MA 02643

MARINCHAK, ROGER ALAN, b Milwaukee, Wis, Nov 27, 52; div; c 1. CARDIOLOGY. *Educ:* Med Col Pa, MD, 77. *Prof Exp:* Resident internal med, Med Col Pa, 77-80; staff physician, Southbridge Med Adv Coun Inc, 80-83; instr med, 83-85, fel cardiol, 83-86, clin instr emergency med, 85-86, ASST PROF MED, MED COL PA, 85-, MEM MED STAFF, HOSP MED COL PA, 85- *Concurrent Pos:* Asst, sect internal med, St Francis Hosp, Wilmington, Del, 80-84 & Wilmington Med Ctr, Del, 80-85; mem, med staff, Jefferson Park Hosp, Philadephia, 86-; dir, Heart Station, Hosp Med Col Pa, 87- *Mem:* Fel Am Col Cardiol; Fel Am Col Physicians; N Am Soc Pacing & Electrophysiol; Am Fed Clin Res. *Res:* Cardiac arrythmias; anti-arrythmic drug investigations. *Mailing Add:* Lankenay Med Off Bldg E Suite 556 Winnwood PA 19096

MARINE, WILLIAM MURPHY, b Cleveland, Ohio, Oct 21, 32; c 4. PREVENTIVE MEDICINE, INTERNAL MEDICINE. *Educ:* Emory Univ, BA, 53, MD, 57; Univ Mich, MPH, 63; Am Bd Internal Med, dipl, 65. *Prof Exp:* From intern to resident med, NY Hosp-Cornell Med Ctr, 57-59; mem staff, Epidemic Intel Serv Kansas City Field Sta, 59-61; resident med, Grady Mem Hosp, Atlanta, Ga, 61-62; trainee epidemiol, Univ Mich, 62-64; from asst prof to assoc prof prev med, Sch Med, Emory Univ, 64-70, prof prev med

& community health, 70-75; PROF & CHMN DEPT PREV MED & COMPREHENSIVE HEALTH CARE, UNIV COLO MED CTR, 75- *Concurrent Pos:* Milbank Mem Fund fac fel, 65; med consult, Southeastern Region, Job Corps, 73- *Mem:* Am Epidemiol Soc; AMA; Am Fedn Clin Res; Am Pub Health Asn; Asn Teachers Prev Med (secy-treas, 74). *Res:* Epidemiology and immunology of respiratory virus infections, especially influenza; evaluation of health care delivery. *Mailing Add:* Dept Prev Med Univ Colo Med Ctr 4200 E Ninth Ave C245 Denver CO 80220

MARINENKO, GEORGE, b Voronezh, USSR, Sept 16, 35; US citizen; m 74; c 5. ELECTROANALYTICAL CHEMISTRY. *Educ:* Am Univ, BS, 59, MS, 61, PhD(physical chem), 72. *Prof Exp:* Scientific Translator chem, Cyrillic Prog, Libr Congress, 58-59, scientific lexicographer chem, Am Info Div, 59-60; res chemist, Nat Bur Standards, 60-90; SYSTS SCIENTIST, MITRE CORP, 90- *Concurrent Pos:* Adj prof chem, Am Univ, 74-78; US delegate to USSR, Environ Agreement, 76; consult, office solid waste, EPA, 87-88. *Honors & Awards:* Silver Medal, Dept Com, 71. *Mem:* Electrochem Soc; Am Chem Soc; Am Soc Toxic Mat; Sigma Xi. *Res:* Developed high precision methods of analysis, redetermined a number of physical constants; conducted electrolyte measurements in biological fluids; conducted coulometric methods of analysis. *Mailing Add:* 1701 Siever Ct Germantown MD 20874

MARINETTI, GUIDO V, b Rochester, NY, June 26, 18; m 42; c 2. BIOCHEMISTRY. *Educ:* Univ Rochester, BS, 50, PhD(biochem), 53. *Prof Exp:* Res biochemist, West Regional Res Lab, USDA, 53-54; from instr to assoc prof, 54-66, PROF BIOCHEM, SCH MED & DENT, UNIV ROCHESTER, 66- *Concurrent Pos:* Lederle med fac award, 55-56; Nat Heart Inst & Nat Sci Found grants, 55- *Mem:* AAAS; Am Chem Soc; Sigma Xi; Am Soc Biol Chem. *Res:* biosynthesis of phosphatides and neutral glycerides and regulatory or control mechanisms in this process; the topology and function of phospholinids in cellular membranes; catecholamine receptors on cell membranes. *Mailing Add:* Med Ctr Univ Rochester 601 Elmwood Ave Box 607 Rochester NY 14642

MARINI, MARIO A, b Ascoli Piceno, Italy, Oct 18, 25; nat US. BLOOD RESEARCH. *Educ:* St Michael's Col, Winouski Park, Vt, BS, 49; Wayne Univ, Detroit, Mich, MS, 52, PhD(biol chem), 55. *Prof Exp:* Teaching asst, Wayne Univ, 51-53, res asst, 53-55; res fel, Univ Minn, 55-58; postdoctoral fel, NIH, Cornell Univ, 58-60; from asst prof to assoc prof biochem, Med Sch, Northwestern Univ, 60-86; RES CHEMIST, BRD, LETTERMAN INST RES, 86- *Concurrent Pos:* Vis prof, Div Sci, Inst Regina Elena, Rome, Italy, 69, Univ Calif, San Diego, 72 & Univ Rome, 82; int exchange scientist, NSF, 75-77; Fulbright-Hayes fel, Univ Rome, 75-76. *Mem:* Am Soc Biol Chemists; NY Acad Sci; Am Chem Soc; Biophys Soc; Sigma Xi; Int Cong Biochem; Int Calorimetry Conf; Int Thermodynamic Conf. *Res:* Produced a highly purified single adult human hemoglobin; evaluating novel modification reagents using HbAo for a blood substitute; author of numerous publications. *Mailing Add:* Div Blood Res Letterman Army Inst Res Presidio San Francisco San Francisco CA 94129

MARINI, ROBERT C, ENVIRONMENTAL CONTROL. *Educ:* Northeastern Univ, BS, 54; Harvard Univ, MS, 55. *Prof Exp:* Jr engr, Camp Dresser & McKee, 55-56, proj engr, 58-64, assoc, 64-66, sr vpres, 70-77, pres, Environ Eng Div, 77-82, exec vpres, 82-84, PARTNER, CAMP DRESSER & MCKEE, 67-, PRES, 84-, CHIEF EXEC OFFICER & CHMN BD, 89- *Concurrent Pos:* Instr hydraul, Lincoln Inst, Northeastern Univ, 61-64. *Mem:* Nat Acad Eng; fel Am Consult Eng Coun; fel Am Soc Civil Engrs; Am Pub Works Asn; Am Water Resources Asn; Am Water Works Asn; Int Asn Water Pollution Control & Res; Water Pollution Control Fedn. *Res:* Author of various publications. *Mailing Add:* Off Pres Camp Dresser & McKee Inc One Cambridge Ctr Cambridge MA 02142-1403

MARINO, A A, b Philadelphia, Pa, Jan 12, 41; m 62; c 4. ELECTROMAGNETISM. *Educ:* St Joseph's Col, BS, 62; Syracuse Univ, PhD(biophys), 68. *Prof Exp:* Biophysicist, Vet Admin Med Ctr, Syracuse, 64-81; PROF ORTHOP SURG, LA STATE UNIV MED CTR, 81- *Concurrent Pos:* Ed, J Bioelec, 82- *Res:* Biological effects of electromagnetic energy. *Mailing Add:* Dept Orthop Surg La State Univ Med Ctr Shreveport LA 71130

MARINO, JOSEPH PAUL, b Hazleton, Pa, Apr 20, 42; m 67; c 3. ORGANIC CHEMISTRY. *Educ:* Pa State Univ, BS, 63; Harvard Univ, AM, 65, PhD(chem), 67. *Prof Exp:* NIH fel, Harvard Univ, 67-69; from asst prof to assoc prof, 69-82, PROF CHEM, UNIV MICH, ANN ARBOR, 82-, PROF MED CHEM, 89- *Concurrent Pos:* Fulbright scholar, 85. *Mem:* Am Chem Soc. *Res:* Sulfur chemistry; ylides; synthesis of natural products; organometallic chemistry; new synthetic methods. *Mailing Add:* Dept Chem Univ Mich Ann Arbor MI 48109

MARINO, LAWRENCE LOUIS, hydrodynamics, nuclear physics, for more information see previous edition

MARINO, PAMELA A, b Milford, Conn, Feb 28, 51. CARCINOGENETICS. *Educ:* Univ Conn, BA, 73, PhD(biomed sci-molecular biol-biochem), 86. *Prof Exp:* Asst res, Dept Pulmonary Med, Yale Med Sch, 73-76, assoc res, Pulmonary Biochem Res Lab, Dept Pediat, 76-78, lab supvr, 78-80; fel, Lab Molecular Biol, 86-89, IRTA fel, Lab Exp Carcinogenesis, 89-90, SR STAFF FEL, LAB EXP CARCINOGENESIS, NAT CANCER INST, NIH, BETHESDA, 90- *Concurrent Pos:* Am Cancer Soc fel, 86-89; NIH intramural res training award, 89. *Mem:* Assoc mem Am Soc Biol Chemists & Molecular Biologists. *Res:* Gene regulation; transcriptional and post-transcriptional control mechanisms; protein/membrane biochemistry. *Mailing Add:* Nat Cancer Inst NIH Bldg 37 Rm 3C25 Bethesda MD 20892

MARINO, RICHARD MATTHEW, b Cleveland, Ohio, Sept 22, 57. LASER RADAR RESEARCH, SENSOR SYSTEMS ENGINEERING. *Educ:* Cleveland State Univ, BS, 79; Case Western Reserve Univ, MS, 82, PhD(physics), 85. *Prof Exp:* Res fel physics, Case Western Reserve Univ, 79-85; STAFF MEM, LASER RADAR, MASS INST TECHNOL/LINCOLN LAB, 85-, PROJ LEADER, SENSOR SYSTS ENG, 87- *Concurrent Pos:* Ed, Mass Inst Technol/Lincoln Lab Sensor Systs Eng, 88-, mem comt, sci & eng support group, 89- *Mem:* Am Phys Soc; AAAS; Sigma Xi. *Res:* New applications and measurements of laser radar remote sensors; elementary particle physics experiments namely, anti-proton/proton interactions. *Mailing Add:* 13 Dudley St Cambridge MA 02140

MARINO, ROBERT ANTHONY, b Positano, Italy, Feb 19, 43; US citizen; m 67; c 2. PHYSICS. *Educ:* City Col New York, BS, 64; Brown Univ, PhD(physics), 69. *Prof Exp:* Res assoc physics, Brown Univ, 69-70; from asst prof to assoc prof, 70-81, PROF PHYSICS, HUNTER COL, CITY UNIV NY, 81- *Concurrent Pos:* Consult, US Army Res Off, 70-; vis sr scientist, Block Eng, Cambridge, Mass, 76-77; vis prof, Univ Geneva, 85-86. *Mem:* Am Phys Soc; Am Asn Physics Teachers. *Res:* Nuclear quadrupole resonance; nuclear magnetic resonance. *Mailing Add:* Dept Physics Hunter Col 695 Park Ave New York NY 10021

MARINOS, PETE NICK, b Sparta, Greece, July 9, 35; US citizen; m 68. COMPUTER SCIENCE, SYSTEMS ENGINEERING. *Educ:* Clemson Univ, BSEE, 59, MSEE, 61; NC State Univ, PhD(elec eng), 64. *Prof Exp:* Electronic engr, Lockheed-Georgia Co, 59; instr elec eng, Clemson Univ, 59-61, asst prof, 64-66; instr, NC State Univ, 61-64; assoc prof, Univ Ala, Huntsville, 66-68; assoc prof, 68-72, PROF ELEC ENG, DUKE UNIV, 72- *Concurrent Pos:* Consult, Southern Bell Tel Co, 64-66, US Army Missile Command, 67-68, Chrysler Corp, 68 & US Naval Res Lab, 71- *Mem:* Inst Elec & Electronics Engrs; Simulation Coun; Sigma Xi. *Res:* Switching, automata and systems theory. *Mailing Add:* Dept Elec Eng Duke Univ Durham NC 27706

MARIN-PADILLA, MIGUEL, b Jumilla, Spain, July 9, 30; nat US; m 58; c 2. PATHOLOGY, PEDIATRIC PATHOLOGY. *Educ:* Univ Granada, BS, 49, MD, 55; Educ Coun Foreign Med Grads, cert, 60; Am Bd Path, dipl & cert anat path, 65. *Prof Exp:* Teaching fel path, Sch Med, Boston Univ, 60-62 & Harvard Med Sch, 61-62; from instr to assoc prof, 62-75, PROF PATH, DARTMOUTH MED SCH, 75-, PROF MATERNAL & CHILD HEALTH, 83- *Concurrent Pos:* Consult, Vet Admin Hosp, White River Junction, Vt, 64- *Mem:* Teratol Soc; Am Asn Anat; Soc Neurosci. *Res:* Development pathology; neurohistology; human and experimental teratology. *Mailing Add:* Two Maynard Hanover NH 03755

MARINUS, MARTIN GERARD, b Amsterdam, Neth, June 22, 44; m 70; c 3. MICROBIAL GENETICS. *Educ:* Univ Otago, NZ, BSc, 65, PhD(microbiol), 68. *Prof Exp:* Fel microbiol, Yale Univ, 68-70; vis fel microbiol, Free Univ, Amsterdam, Neth, 70-71; instr pharmacol, Col Med & Dent NJ, Rutgers Med Sch, 71-74; from asst prof to assoc prof, 74-86, PROF PHARMACOL, MED SCH, UNIV MASS, 86- *Concurrent Pos:* Fac res award, Am Cancer Soc, 76-81. *Mem:* Am Soc Microbiol; Genetics Soc Am. *Res:* Function of methylated bases in nucleic acids. *Mailing Add:* Dept Pharmacol Med Sch Univ Mass 55 Lake Ave N Worcester MA 01655

MARION, ALEXANDER PETER, b New York, NY, Apr 24, 15; m 43. PHYSICAL CHEMISTRY. *Educ:* City Col New York, BS, 36, MS, 39; NY Univ, PhD(chem), 44. *Prof Exp:* Lectr asst, 37-41, tutor, 41-43, from instr to prof, 43-76, EMER PROF CHEM, QUEENS COL, NY, 76- *Concurrent Pos:* Designer, Microchem Serv, 42-48. *Mem:* AAAS; Am Chem Soc; Sigma Xi. *Res:* Chemical kinetics; teaching aids; electronic laboratory apparatus. *Mailing Add:* 475B Heritage Hills Somers NY 10589-1920

MARION, C(HARLES) P(ARKER), b Montclair, NJ, Jan 22, 20; m 43; c 2. CHEMICAL ENGINEERING. *Educ:* Univ Calif, Los Angeles, BS, 47; Mass Inst Technol, ScD(chem eng), 52. *Prof Exp:* Chem engr, Montebello Res Lab, Texaco, Inc, 52-58, res chem engr, 59-61, sr res chem engr, 61-62, process rep, 62-67, sr process rep, 67-79, asst mgr process licensing, 79-80, MGR PROCESS LICENSING, TEXACO DEVELOP CORP, 80- *Mem:* Fel Am Inst Chem; Am Inst Chem Engrs; AAAS. *Res:* Process development and design, particularly combustion and heat transfer; synthesis-gas-generation process; coal gasification. *Mailing Add:* 655 Shae Acres Dr Mamaroneck NY 10543

MARION, JAMES EDSEL, b Cana, Va, May 30, 35; m 57; c 2. FOOD SCIENCE, NUTRITION. *Educ:* Berea Col, BS, 57; Univ Ky, MS, 59; Univ Ga, PhD(nutrit), 62. *Prof Exp:* Res asst poultry nutrit, Univ Ky, 57-59; res asst poultry nutrit, Univ Ga, 59-62, asst food technologist, Ga Exp Sta, 62-67, assoc food scientist & head food sci dept, 67-69; asst dir res, Gold Kist Res Ctr, 69-72, dir res, 72-; chmn, Dept Poultry Sci, Univ Fla, Gainesville, 72-88; DEAN, COL AGR, AUBURN UNIV, 88- *Mem:* AAAS; Am Inst Nutrit; Inst Food Technol; Oil Chem Soc; Poultry Sci Asn. *Res:* Feed and nutrition; plant breeding; product development. *Mailing Add:* Marion Col Agr 107 Comer Hall Auburn Univ Auburn University AL 36849-5401

MARION, JERRY BASKERVILLE, nuclear physics; deceased, see previous edition for last biography

MARION, ROBERT HOWARD, b Paterson, NJ, Dec 10, 45; m 71; c 2. MATERIALS SCIENCE, CERAMICS. *Educ:* Stevens Inst Technol, BEng, 67; Northwestern Univ, Evanston, PhD(mat sci), 72. *Prof Exp:* mem tech staff mat sci, Sandia Labs, Albuquerque, NMex, 72-80; MGR PROCESS TECHNOL, CORP RES LAB, AVX CERAMICS CORP, 80- *Mem:* Am Soc Metals; Am Ceramic Soc. *Res:* Structure-property relationships in ceramics and glasses; mechanical properties, fracture, high temperature mechanical testing, nuclear fuels for pulsed reactors, thermal stress resistance, residual stress measurement. *Mailing Add:* 2605 Terry Rd Jackson MS 39204

MARION, WAYNE RICHARD, b Ithaca, NY, June 28, 47; m 69; c 2. AVIAN ECOLOGY, LAND RECLAMATION. *Educ:* Cornell Univ, BS, 69; Colo State Univ, MS, 70; Tex A&M Univ, PhD(wildlife sci), 74. *Prof Exp:* Instr wildlife mgt, Cornell Univ, 74; asst prof, 75-80, ASSOC PROF WILDLIFE ECOL, UNIV FLA, 80- *Concurrent Pos:* Consult, Appl Biol, Inc, Decatur, Ga, 78-86, Fla Power & Light Co, 79-, Bio-Scan, Inc, LeHigh, Fla, 87-, KBN Eng & Appl Scis, Inc, 90-; vis res scientist, Fla Inst Phosphate Res, Bartow, 83-84. *Mem:* Sigma Xi; Am Ornithologists Union; Wildlife Soc. *Res:* Avian ecology; intensive forest management; phosphate mine reclamation. *Mailing Add:* Dept Wildlife & Range Sci Univ Fla 118 Newins-Ziegler Hall Gainesville FL 32611

MARION, WILLIAM W, b Hillsville, Va, Feb 3, 30; m 54; c 4. FOOD SCIENCE. *Educ:* Berea Col, BS, 53; Purdue Univ, MS, 55, PhD(food technol), 58. *Prof Exp:* Instr poultry husb, Purdue Univ, 55-58; from asst prof to prof animal sci, 58-74, chmn dept poultry sci, 68-71, PROF FOOD TECHNOL & HEAD DEPT, IOWA STATE UNIV, 74- *Mem:* Inst Food Technol; Am Oil Chem Soc; Poultry Sci Asn; Am Inst Nutrit. *Res:* Structure and composition of muscle lipids; post-mortem biochemical changes in muscle. *Mailing Add:* 2009 Northwestern Ames IA 50010

MARIS, HUMPHREY JOHN, b Ipswich, Eng, Apr 25, 39. SOLID STATE PHYSICS. *Educ:* Imp Col, Univ London, BSc, 60, PhD(physics), 63. *Prof Exp:* Fel physics, Case Inst Cleveland, 63-65; from asst prof to assoc prof, 65-76, PROF PHYSICS, BROWN UNIV, 76- *Concurrent Pos:* Vis fel, Univ EAnglia, 72-73, Chalmers Inst, 73, & Nat Ctr Sci Res, Grenoble, France, 73. *Honors & Awards:* Humboldt Award, 89. *Mem:* Fel Am Phys Soc. *Res:* Low temperature physics; ultrasonics; lattice dynamics. *Mailing Add:* Dept Physics Brown Univ Providence RI 02912

MARISCAL, RICHARD NORTH, b Los Angeles, Calif, Oct 4, 35; m 74; c 4. MARINE BIOLOGY, INVERTEBRATE ZOOLOGY. *Educ:* Stanford Univ, AB, 57, MA, 61; Univ Calif, Berkeley, PhD(zool), 66. *Prof Exp:* Asst entom, Univ Calif, Berkeley, 60-61, asst zool, 61-64, lectr, 66; fac asst, Te Vega & Int Indian Ocean Expeds, Hopkins Marine Sta, Stanford Univ, 64-65; NIH fel, Lab Quant Biol, Univ Miami, 67-68; from asst prof to assoc prof, 68-78, assoc chmn grad studies & res, 77-79, PROF BIOL SCI, FLA STATE UNIV, 78- *Mem:* Fel AAAS; Am Soc Zool; Ecol Soc Am; Am Inst Biol Sci; NY Acad Sci. *Res:* Cnidarian nematocyst physiology, biochemistry and morphology; prey-predator interactions and symbioses between cnidarians, fishes, crustaceans and molluscs; invertebrate behavior, ecology, physiology, electron microscopy and x-ray microanalysis; sensory receptors of cnidarians. *Mailing Add:* Dept Biol Sci Fla State Univ Tallahassee FL 32306

MARISI, DAN(IEL) QUIRINUS, b Sask, Nov 12, 40; m 64; c 4. SPORT PSYCHOLOGY, MOTOR LEARNING & CONTROL. *Educ:* Univ Sask, BEd, 65, BA, 66, MSc, 69; Univ Calif, Berkeley, PhD(phys educ), 71. *Prof Exp:* Lectr phys educ, Univ Sask, 67-70; asst prof, 71-73, ASSOC PROF SPORT PSYCHOL, MCGILL UNIV, 73- *Concurrent Pos:* Co-dir, Motor Sport Res Group, McGill Univ, 82-; vchmn, Int Coun Motor Sport Sci, 88-; psychol consult, Nat Alpine Downhill Ski Team, 90- *Mem:* Am Alliance Health, Phys Educ, Recreation & Dance; Can Psychol Asn. *Res:* Psycho-physiological readiness for competition. *Mailing Add:* McGill Univ 475 Pine Ave W Montreal PQ H2W 1S4

MARISKA, JOHN THOMAS, b Fairbanks, Alaska, Feb 25, 50; m 72; c 2. ASTROPHYSICS, ASTRONOMY. *Educ:* Univ Colo, BA, 72; Harvard Univ, AM, 73, PhD(astron), 77. *Prof Exp:* resident res assoc space sci, 77-79, res physicist, 79-81, ASTROPHYSICIST, E O HULBURT CTR SPACE RES, NAVAL RES LAB, 81- *Mem:* AAAS; Am Astron Soc; Am Geophys Union; Int Astron Union. *Res:* Solar and stellar physics; structure of the solar corona; extreme ultraviolet spectroscopy. *Mailing Add:* Code 4175 Naval Res Lab Washington DC 20375-5000

MARK, DAVID FU-CHI, b Hongkong, China, Dec 10, 50; US citizen. GENE REGULATION, PROTEIN STRUCTURE-FUNCTION. *Educ:* Univ Mass, BA, 73; Harvard Univ, PhD(biochem), 77. *Prof Exp:* Postdoctoral fel, Stanford Univ Med Ctr, 77-79; scientist, Cetus Corp, 79-83, sr scientist, 83-89; EXEC DIR, MERCK SHARP & DOHME RES LABS, 89- *Concurrent Pos:* Mgr, Bioactive Peptides Proj, Cetus Corp, 81-82, dir molecular biol, 82-87, dir new therapeut, 87-89; sci adv, China Nat Ctr Biotechnol Develop, 84-87; adv, Biotechnol Inst, Hong Kong Univ Sci & Technol, 91- *Honors & Awards:* Outstanding Young Scientist in Am, Sci Dig Mag, 84. *Mem:* AAAS; Am Soc Microbiol; Am Asn Immunologists. *Res:* Structure-function relationship of human cytokines; regulation of tumor nerosis factor mRNA transcription by bacterial endotoxin. *Mailing Add:* Dept Microbial Chemotherapeutic & Molecular Genetics Merck Sharp & Dohme Res Labs PO Box 2000 Rahway NJ 07065

MARK, EARL LARRY, b Ogden, Utah, Dec 13, 40; m 62; c 4. PHYSICAL CHEMISTRY. *Educ:* Weber State Col, BS, 65; Univ Idaho, PhD(phys chem), 70. *Prof Exp:* Res chemist, Amalgamated Sugar Co, 70-73; dir res, Water Refining Co, 73-74; PROD MGR, BLACK CLAWSON CO, 74- *Mem:* Am Chem Soc; Filtration Soc; Am Soc Testing & Mats. *Res:* Ion exchange; surface adsorption; use of radiotracers in adsorption studies; activated carbon adsorption; liquid-solid separation; liquid filtration. *Mailing Add:* 3111 Whispering Trails Carmel IN 46032-3954

MARK, HANS MICHAEL, b Mannheim, Germany, June 17, 29; nat US; m 51; c 2. PHYSICS, ENGINEERING. *Educ:* Univ Calif, Berkeley, AB, 51; Mass Inst Technol, PhD(physics), 54. *Hon Degrees:* ScD, Fla Inst Technol, 77; DEng, Polytech Inst NY, 82. *Prof Exp:* Asst, Mass Inst Technol, 52-54, res assoc, 54-55; jr res physicist, Univ Calif, 55-56, physicist, Lawrence Radiation Lab, 56-58; asst prof physics, Mass Inst Technol, 58-60; from assoc prof to prof nuclear eng, Univ Calif, Berkeley, 60-69, chmn dept, 64-69, physicist, Lawrence Livermore Nat Lab, 60-69; leader exp physics div, 60-64, dir, Ames Res Ctr, NASA, 69-77; undersecy, Air Force, 77-79, secy, 79-81; dep adminr, NASA, 81-84; CHANCELLOR, UNIV TEX SYST, 84- *Concurrent Pos:* Lectr, dept appl sci, Univ Calif, Davis, 69-73; consult, Inst Defense Anal, US Army, DC; consult prof, Sch Eng, Stanford Univ, 73-84; consult, US Air Force, US Army & Inst Defense Anal; dir, Pennzoil Corp, 84-87; trustee, Mitre Corp, 85-87. *Mem:* Nat Acad Eng; Am Geophys Union; fel Am Phys Soc; Am Nuclear Soc; fel Am Inst Aeronaut & Astronaut. *Res:* Nuclear and atomic physics; nuclear instrumentation; astrophysics. *Mailing Add:* Off Chancellor Univ Tex Syst 601 Colorado St Austin TX 78701

MARK, HAROLD WAYNE, b Chanute, Kans, May 2, 49. PHYSICAL ORGANIC CHEMISTRY, SULFUR CHEMISTRY. *Educ:* Univ Kans, BS, 71; Northwestern Univ, PhD(chem), 75. *Prof Exp:* res chemist, Phillips Petrol Co, 75-80, sr res chem, 80-88; sr res chemist, 88-90, ANAL & ASSURANCE MGR, WESTVACO CORP, 90- *Mem:* AAAS; Am Chem Soc; Sigma Xi; Am Soc Testing & Mat. *Res:* Sulfur chemistry, catalysis, carbonium ion chemistry and nucleophilic substitution reactions; analytical chemistry. *Mailing Add:* Westvaco Corp PO Box 836 DeRidder LA 70634

MARK, HARRY BERST, JR, b Camden, NJ, Feb 28, 34; m 60; c 3. ELECTROCHEMISTRY, ANALYTICAL CHEMISTRY. *Educ:* Univ Va, BA, 56; Duke Univ, PhD(electrochem), 60. *Prof Exp:* Assoc, Univ NC, 60-62; fel, Calif Inst Technol, 62-63; from asst prof to assoc prof chem, Univ Mich, Ann Arbor, 63-70; PROF CHEM, UNIV CINCINNATI, 70- *Concurrent Pos:* Vis prof, Free Univ Brussels, 70; cong legis Counr, Am Chem Soc, 74- *Mem:* AAAS; Am Chem Soc; Electrochem Soc; NY Acad Sci; Am Inst Chem. *Res:* Heterogeneous electron transfer kinetics; electrical double layer phenomena; electroanalytical techniques; neutron activation analysis; kinetic methods for the analysis of closely related mixtures; bioelectrochemistry; environmental analysis methods. *Mailing Add:* Dept Chem Univ Cincinnati Cincinnati OH 45221

MARK, HERBERT, b Jersey City, NJ, June 10, 21; m 45; c 3. MEDICINE, CARDIOLOGY. *Educ:* Columbia Univ, AB, 42; Long Island Col Med, MD, 45; Am Bd Internal Med, dipl, 53. *Prof Exp:* Resident med, Montefiore Hosp, New York, 48-49; resident, Vet Admin Hosp, Bronx, 49-50; pvt pract, 51-64; assoc prof med, NY Med Col, 64-67; asst prof prev med & med, Albert Einstein Col Med, 67-69; clin assoc prof med, NJ Col Med, 69-72, prof med, 72-75; CHIEF, MED SERV, VET ADMIN MED CTR, BRONX, NY, 75-; PROF, MT SINAI SCH MED, 75- *Concurrent Pos:* Fel cardiol, Montefiore Hosp, New York, 50-51, mem staff, assoc attend physician, 61-67, attend physician, 67-75; attend physician, Vet Admin Hosp, Bronx; assoc chief med, chief cardiol & assoc attend physician, Bird S Coler Hosp, 64-67; assoc attend physician, Flower & Metrop Hosps, 64-67; dir ambulatory serv, Montefiore-Morrisania Affiliation, 67-69; chief med, Jersey City Med Ctr, 69-75; adj attend physician, Med Serv, Mt Sinai Hosp, New York, 75- *Mem:* Am Fedn Clin Res; fel Am Col Physicians; fel Am Col Cardiol; Am Heart Asn. *Res:* Vectorcardiography; electrophysiology; congenital heart disease. *Mailing Add:* Bronx VA Hosp 130 W Kingsbridge Rd Bronx NY 10468

MARK, HERMAN FRANCIS, b Vienna, Austria, May 3, 95; nat US; m 22; c 2. POLYMER CHEMISTRY, ENGINEERING. *Educ:* Univ Vienna, PhD, 21, Dr rer nat, 56. *Hon Degrees:* EngD, Univ Leige, 49; PhD, Uppsala Univ, 42, Lowell Technol Inst, 57 & Munich Tech Univ, 60. *Prof Exp:* Instr physics & phys chem, Univ Vienna, 19-21; instr org chem, Univ Berlin, 21-22; from res fel to group leader, Kaiser Wilhelm Inst, Dahlem, 22-26; res chemist, I G Farben-Indust, 27-28, group leader, 28-30, asst res dir, 30-32; prof chem, Univ Vienna, 32-38; adj prof org chem, 40-42, prof, 42-46, dir polymer res inst, 46-70, EMER DEAN, POLYTECH UNIV NEW YORK, 70- *Concurrent Pos:* Assoc prof, Karlsruhe Tech Inst, 27-32; tech consult, US Navy, Qm Corps, US Army; NSF ed, J Polymer Sci, J Appl Polymer Sci; Series on Highpolymer, Rev in Polymer Sci, Resins, Rubbers, Plastics & Natural & Synthetic Fibers; chmn tech comt wood chem, Food & Agr Orgn, UN; chmn comn macromolecules, Int Union Pure & Appl Chem; chmn comt macromolecules, Nat Res Coun; vpres in-chg proj res, Am Comt, Weizmann Inst, Israel & Gov Inst; chmn, Gordon Res Conf Macromolecules & Textiles. *Mem:* Nat Acad Sci; AAAS; Am Chem Soc; fel Am Phys Soc; Soc Rheol. *Res:* Use of x-rays and electrons in the synthesis, characterization, reactions and properties of natural and synthetic macromolecules. *Mailing Add:* 333 Jay St Polytech Univ Brooklyn NY 11201

MARK, J CARSON, b Lindsay, Ont, July 6, 13; US citizen; m 35; c 6. MATHEMATICS, MATHEMATICAL PHYSICS. *Educ:* Univ Western Ont, BA, 35; Univ Toronto, PhD(math), 38. *Prof Exp:* Instr math, Univ Man, 38-43; scientist, Montreal Lab, Nat Res Coun Can, 43-45; scientist, Los Alamos Sci Lab, 45-46, mem staff, theoret physics div, Los Alamos Sci Lab, Univ Calif, 46-73, div leader, 47-73; MEM ADV COMT, REACTOR SAFEGUARDS OF US NUCLEAR REGULATORY COMN, 76- *Concurrent Pos:* Mem, Sci Adv Bd, US Air Force; sci adv, US Deleg, Conf Experts Means of Detection Nuclear Explosions, Geneva, 58. *Mem:* Am Math Soc; Am Phys Soc. *Res:* Finite group theory; transport theory; hydrodynamics; neutron physics. *Mailing Add:* Los Alamos Nat Lab MS 210 PO Box 1663 Los Alamos NM 87544

MARK, JAMES EDWARD, b Wilkes-Barre, Pa, Dec 14, 34; m 64, 90; c 2. POLYMER CHEMISTRY. *Educ:* Wilkes Col, BS, 57; Univ Pa, PhD(phys chem), 62. *Prof Exp:* Res chemist, Rohm & Haas Co, 55-56; res asst, Stanford Univ, 62-64; asst prof chem, Polytech Inst Brooklyn, 64-67; from asst prof to prof chem, Univ Mich, Ann Arbor, 67-77; PROF CHEM, CHMN PHYS CHEM DIV & DIR POLYMER RES CTR, UNIV CINCINNATI, 77- *Concurrent Pos:* Consult var industs, 63-; vis prof, Stanford Univ, 73-74; spec res fel, NIH, 75-76; lectr short course prog, Am Chem Soc, 73- *Mem:* AAAS; Am Chem Soc; Am Phys Soc; NY Acad Sci. *Res:* Statistical properties of chain molecules; elastic properties of polymer networks. *Mailing Add:* Dept of Chem Univ of Cincinnati Cincinnati OH 45221

MARK, JAMES WAI-KEE, b Calcutta, India, Aug 29, 43; US citizen. PLASMA PHYSICS, ELECTRIC ENGINEERING. *Educ:* Univ Calif, Berkeley, BS, 64; Princeton Univ, PhD(astrophys sci), 68. *Prof Exp:* Res assoc plasma physics, Plasma Physics Lab, Princeton Univ, 68-69; C L E Moore instr appl math, Mass Inst Technol, 69-70. from asst prof to assoc prof, 70-79, RES SCI, INERTIAL FUSION ENERGY, LAWRENCE LIVERMORE NAT LAB, 79- *Concurrent Pos:* Vis scientist, Kitt Peak Nat Observ, 75-76,; vis assoc prof, dept phys & astron, Univ Md & vis scientist, Univ Groningen, The Netherlands, 77; res scientist, Mass Inst Technol Res Lab Electron, 77-79. *Mem:* NY Acad Sci; Am Phys Soc; Inst Elec & Electron Eng; Int Astron Union; Soc Indust & Appl Math. *Res:* Inertial-fusion targets; gaussian-quadrature beam illuminization; ion beam fusion; galaxy spiral structure and bending waves; high current particle beams; magnetized plasmas; ballooning modes second stable region; magnetic-reconnection; asymptotics and numerical simulations; rotating stars. *Mailing Add:* Lawrence Livermore Nat Lab Mail Stop L-297 PO Box 808 Livermore CA 94550

MARK, JON WEI, b Toysun, China; Can citizen; m 64; c 2. COMPUTER COMMUNICATIONS, SPREAD SPECTRUM COMMUNICATIONS. *Educ:* Univ Toronto, BASc, 62; McMaster Univ, MEng, 68, PhD(commun), 70. *Prof Exp:* Engr, Can Westinghouse Co, Ltd, 62-67, sr engr, 68-70; from assoc prof to assoc chmn dept, 70-82, chmn dept, 84-90, PROF ELEC ENG, UNIV WATERLOO, ONT, 78- *Concurrent Pos:* Consult, Westinghouse, Can, Inc, 74-, Can Dept Commun, 78-; res scientist, IBM, Thomas J Watson Res Ctr, 76-77; resident consult, Bell Labs, Murray Hill, NJ, 82-83; assoc ed, Inst Elec Electronic Engrs Trans Commun, 83-85, ed, 86-89. *Honors & Awards:* Fel, Inst Elec Electronic Engrs. *Mem:* Inst Elec Electronic Engrs. *Res:* Computer communication; local and metropolitan area networks; multiple access systems; network modeling and analysis; digital signal processing; antijam and spread spectrum communication; integrated services networks; image and speech coding. *Mailing Add:* Dept Elec & Computer Eng Univ Waterloo Waterloo ON N2L 3G1 Can

MARK, LESTER CHARLES, b Boston, Mass, July 16, 18; m 46; c 2. MEDICINE. *Educ:* Univ Toronto, MD, 41; Am Bd Anesthesiol, dipl, 52. *Prof Exp:* Intern, Jewish Mem Hosp, 41-43; asst resident surg, Grace Hosp, New Haven, Conn, 43; resident anesthesiol, Hosp Spec Surg, New York, 47-48; clin instr, Col Med, State Univ NY, 52-53; assoc, 53-54, from asst prof to prof, 54-84, EMER PROF ANESTHESIOL, COL PHYSICIANS & SURGEONS, COLUMBIA UNIV, 84- *Concurrent Pos:* Res fel, Res Serv, NY Univ-Bellevue Med Ctr & Goldwater Mem Hosp, 48-51; Am Heart Asn res fel, 49-51; travel award, Int Cardiol Cong, Paris, 50; Guggenheim fel, 60-61; Macy fac scholar, Switz, 74-75; asst adj anesthesiologist, Jewish Mem Hosp, New York, 47-52; asst clin vis anesthesiologist, Jewish Mem Hosp, New York, 47-52; asst clin vis anesthesiologist, Goldwater Mem Hosp, 48-50; dir anesthesiol, Brunswick Gen Hosp, Amityville, & anesthesiologist, SNassau Communities Hosp, Oceanside, 51-53; dir anesthesiol, Freeport Hosp, assoc vis anesthesiologist, Kings County Hosp, Brooklyn & consulting anesthesiologist, Vet Admin Hosp, Northport, 52-53; from asst attend anesthesiologist to assoc attend anesthesiologist, Presby Hosp, New York, 53-65, attend anesthesiologist, 65-84, collab med & exp pharmacol, 59-67, consult, 84-; Fulbright res prof, Denmark, 60-61; actg consult, WHO Anaesthesia Ctr, Copenhagen, 60-61; consult, Coun Drugs, AMA, 62-84; mem, adv comt respiratory & anesthetic drugs, Food & Drug Admin, 66-70, over-the-counter hypnotics, tranquillizers & sleep-aids rev panel, 72-78; mem pharmacol-toxicol rev comt, Nat Inst Gen Med Sci, 68-70, prog comt, 70-72, chmn, 71-72; mem prof adv bd, Found Thanatology, 68-, exec comt, 74-; China Med Bd vis prof, Sapporo Med Col, Japan, 67; guest scientist, Med Dept & vis attend physician, Med Res Ctr, Brookhaven Nat Lab, 68-71; Fulbright sr scholar, Sch Pharmaceut Sci, Victorian Col Pharm, Melbourne, Australia, 81; vol anesthesiologist, Proj Hope, Grenada, 84-85 & 85-86. *Mem:* Fel AAAS; Am Soc Anesthesiol; Am Soc Pharmacol & Exp Therapeut; sr mem Asn Univ Anesthet; fel Am Col Anesthesiol; Am Acad Acupunture; Am Col Acupunture; Am Pain Soc; Am Soc Clin Hypnosis; Int Soc Hypnosis. *Res:* Barbiturates; drug metabolism and distribution; mechanisms of drug action in man; thanatology; hypnosis; acupuncture. *Mailing Add:* Fernwood Corner PO Box 317 Otis MA 01253

MARK, LLOYD K, b Columbus, Ohio, Feb 19, 25; m 51; c 4. OBSTETRICAL ULTRASOUND, DIAGNOSTIC RADIOLOGY. *Educ:* Ohio State Univ, MD, 50; Am Bd Radiol, dipl, 54; Am Bd Nuclear Med, dipl, 72. *Prof Exp:* Radiologist, Michael Reese Hosp, 55-57 & Mt Sinai Med Ctr, Milwaukee, 57-77; asst clin prof radiol, Med Col Wis, 58-77; PROF RADIOL, SCH MED, TEX TECH, 77-; DIR, DEPT RADIOL, R E THOMASON GEN HOSP, EL PASO, 77- *Mem:* Fel Am Col Angiography; Am Col Radiol; Soc Nuclear Med; Am Inst Ultrasound Med; Am Col Chest Physicians; Am Col Cardiol; Radiol Soc NAm. *Res:* Obstetrical and abdominal angiography; use of radionuclides in liver and biliary disease. *Mailing Add:* Thomason Gen Hosp 4815 Alameda Ave El Paso TX 79915

MARK, MELVIN, b St Paul, Minn, Nov 15, 22; m 51; c 3. THERMAL DESIGN, UNIVERSITY ORGANIZATION & CURRICULUM DESIGN. *Educ:* Univ Minn, BME, 43, MS, 46; Harvard Univ, ScD(mech eng), 50. *Prof Exp:* Instr physics & mech eng, NDak Agr Col, 43-44; instr mech eng, Univ Minn, 45-47; proj engr, Aircraft Gas Turbine Div, Gen Elec Co, 50-52; sr mech engr, Raytheon Co, 52-54, sect head, appl mech group, 54-56; prof mech eng, Lowell Technol Inst, 57-59, dean fac, 59-62; dean col eng, Northeastern Univ, 68-79, prof mech eng, 63-84, provost & sr vpres, 79-84; CONSULT, 84- *Concurrent Pos:* Consult engr to various industs, 56-; pres, Cambridge Develop & Eng Corp, 56-59; vis lectr, Brandeis Univ, 58; vis prof, Univ Mass, 84-86. *Mem:* Fel Am Soc Mech Engrs; Am Soc Eng Educ; Sigma Xi. *Res:* Photoelasticity; residual stresses in welding; viscosity of lubricants; aircraft gas turbine cooling; aerodynamic loads on radar antennas; heat transfer in electronic equipment; thermodynamics; author or co-author of numerous publications. *Mailing Add:* 17 Larch Rd Waban MA 02168

MARK, ROBERT, b New York, NY, July 3, 30; m 55; c 3. EXPERIMENTAL MECHANICS, ARCHITECTURAL HISTORY. *Educ:* City Col NY, BCE, 52; State Univ NY, PE, 58. *Prof Exp:* Stress analyst, Combustion Eng Nuclear Power Div, 52-57, res staff engr, 57-64, PROF CIVIL ENG, PRINCETON UNIV, 64- *Mem:* Sigma Xi; fel Soc Exp Mech. *Res:* Application of modern engineering analysis to historic structures and construction. *Mailing Add:* Sch Archit Princeton Univ Princeton NJ 08544

MARK, ROBERT VINCENT, b Jamaica, NY, Dec 22, 42; c 3. ORGANIC CHEMISTRY. *Educ:* St John's Univ, NY, BS, 64, MS, 66, PhD(org chem), 71. *Prof Exp:* From instr to asst prof gen & org chem, 70-77, spec asst actg pres, 76-77, chmn dept chem, 77-81, assoc prof chem, 77-84, PROF CHEM, AGR & TECH COL, STATE UNIV NY, FARMINGDALE, 84-, DEAN, SCH ARTS & SCI, 81- *Concurrent Pos:* NSF traineeship, 70; res assoc, Long Island Jewish-Hillside Med Ctr, 73-74; consult, Pall Corp, 75-77. *Mem:* Am Chem Soc; Am Asn Higher Educ. *Res:* Preparation and mass spectral characteristics of small ring heterocyclic compounds. *Mailing Add:* Five Browning Dr Greenlawn NY 11740-3103

MARK, ROGER G, b Boston, Mass, June 4, 39; m 66; c 4. BIO-ELECTRICAL ENGINEERING, INTERNAL MEDICINE. *Educ:* Mass Inst Technol, BS, 60, PhD(elec eng), 66; Harvard Med Sch, MD, 65. *Prof Exp:* Intern & resident internal med, Harvard Med Serv-Boston City Hosp, 65-67; med officer, Spec Weapons Defense, US Air Force, 67-69; instr med, Harvard Med Sch, 69-72; from asst prof to assoc prof, 69-88, PROF ELEC ENG, MASS INST TECHNOL, 88-, GROVER HERMAN PROF HEALTH SCI & TECHNOL, 90-; ASST PROF MED, HARVARD MED SCH, 72- *Concurrent Pos:* Co-dir Div Health Sci & Technol, Mass Inst Technol, Harvard, 85- *Res:* Biomedical instrumentation; medical care delivery systems; cardiovascular physiology. *Mailing Add:* Rm E25-519 Mass Inst Technol Cambridge MA 02139

MARK, SHEW-KUEY, b China, Aug 8, 36; Can citizen; m 61; c 2. EXPERIMENTAL NUCLEAR PHYSICS. *Educ:* McGill Univ, BSc, 60, MSc, 62, PhD(nuclear physics), 65. *Prof Exp:* Nat Res Coun Can fel, Univ Man, 65-66; from asst prof to assoc prof, 66-75, dir, Foster Radiation Lab, 71-79, chmn dept, 82-90, PROF PHYSICS, MCGILL UNIV, 75- *Concurrent Pos:* Sr vis scientist, Nat Ctr Sci Res, France, 79-80. *Mem:* Can Asn Physicists. *Res:* Nuclear reactions; spectroscopy; structural studies; high energy heavy ion collision. *Mailing Add:* Dept Physics McGill Univ Montreal PQ H3A 2T8 Can

MARKAKIS, PERICLES, b Cassaba, Turkey, Mar 3, 20; nat US; m 53; c 3. FOOD SCIENCE, NUTRITION. *Educ:* Univ Salonika, Greece, BS, 42 & 49; Univ Mass, MS, 52, PhD(food technol), 56. *Prof Exp:* Instr food sci, Univ Salonika, Greece, 42-50; asst res prof food technol, Univ Mass, 55-56; sr food technologist, DCA Food Industs, Inc, 56-57; res food technologist, Univ Calif, 57-59; from asst prof to prof, 59-90, EMER PROF FOOD SCI, MICH STATE UNIV, 90- *Concurrent Pos:* Tech assignments, Latin Am, Asia, Africa & Europe. *Honors & Awards:* Fulbright Study Award. *Mem:* Am Chem Soc; Inst Food Technologists; NY Acad Sci; Sigma Xi. *Res:* chemistry and technology of foods; irradiation preservation of foods. *Mailing Add:* Dept Food Sci Mich State Univ E Lansing MI 48823

MARKEE, KATHERINE MADIGAN, b Cleveland, Ohio, Feb 24, 31. TECHNICAL MANAGEMENT. *Educ:* Trinity Col, Washington, DC, BA, 53; Teachers Col, Columbia Univ, MA, 62; Case Western Reserve Univ, MSLS, 68. *Prof Exp:* From instr library sci to asst prof library sci & medline analyst, 68-86, ASSOC PROF COMPUT RES LIBRARIAN, PURDUE UNIV LIBRARY, 86- *Honors & Awards:* Sigma Xi Res Award, 86. *Mem:* Am Asn Univ Profs; Am Soc Info Sci; Am Library Asn; Spec Libraries Asn; Med Library Asn. *Res:* To provide information sources that are accessible by computers to users. *Mailing Add:* Purdue Univ Library West Lafayette IN 47907

MARKEES, DIETHER GAUDENZ, b Basel, Switz, Oct 16, 19; nat US; c 1. ORGANIC CHEMISTRY. *Educ:* Univ Basel, Dr phil, 46. *Prof Exp:* Res fel med chem, Univ Va, 47-48; res assoc E R Squibb & Sons, 49-53 & Amherst Col, 53-58; from asst prof to prof, 58-89, EMER PROF CHEM, WELLS COL, 89- *Mem:* Am Chem Soc; Swiss Chem Soc. *Res:* Medicinal chemistry; chemistry of heterocycles; synthetic organic chemistry. *Mailing Add:* Box 163 Main St Aurora NY 13026

MARKELL, EDWARD KINGSMILL, b Brooklyn, NY, Apr 14, 18; m 53; c 2. TROPICAL MEDICINE. *Educ:* Pomona Col, BA, 38; Univ Calif, PhD(zool), 42; Stanford Univ, MD, 51. *Prof Exp:* Asst zool, Univ Calif, 38-41; intern, Stanford Univ Hosps, 50-51; asst prof infectious dis, Sch Med, Univ Calif, Los Angeles, 51-58; mem dept internal med, Kaiser Found Med Ctr, 58-85; CLIN PROF MED & TROP MED, SCH MED, UNIV CALIF, SAN FRANCISCO, 86- *Concurrent Pos:* Markle fel, 52-57; Clin assoc prof prev med, Sch Med, Stanford Univ, 61-70, clin prof, 70-86, emer prof, 86. *Mem:* Royal Soc Trop Med & Hyg; Am Soc Parasitol; Am Soc Trop Med & Hyg. *Res:* Parasitic diseases of man; filariasis. *Mailing Add:* 28 Senior Ave Berkeley CA 94708-2212

MARKELS, MICHAEL, JR, b New York, NY, Feb 4, 26; div; c 4. EXECUTIVE MANAGEMENT. *Educ:* Columbia Univ, BS, 48, MS, 49, DES, 57. *Prof Exp:* Technologist, Wood River Refinery, Shell Oil Co, 49-52; engr heat transfer opers & res assoc, Columbia Univ, 52-57; dir advan technol dept, Atlantic Res Corp, Va, 57-68, asst gen mgr res div, 68-69; PRES, VERSAR INC, 69- *Concurrent Pos:* Dep dir corp res lab, Susquehanna Corp, Va, 68-69. *Mem:* AAAS; Nat Soc Prof Engrs; Am Nuclear Soc; Am Inst Aeronaut & Astronaut; Am Inst Chem Engrs; Am Chem Soc. *Res:* Mass transfer operations; fluid mechanics; small particle dynamics; rheology; research administration; general management. *Mailing Add:* PO Box 1549 Springfield VA 22151

MARKENSCOFF, PAULINE, b Serres, Greece, Apr 30, 52. COMPUTER ARCHITECTURE. *Educ:* Nat Tech Univ, Greece, dipl, 75; Univ Minn, MS, 77, PhD(elec eng), 80. *Prof Exp:* ASST PROF COMPUT ENG, UNIV HOUSTON, 80- *Mem:* Inst Elec & Electronics Engrs; Asn Comput Mach; Greek Chamber Engrs. *Res:* Modeling of computer systems; computer architecture; performance evaluation and distributed processing. *Mailing Add:* Dept Elec Eng Univ Houston 4800 Calhoun Rd Houston TX 77004

MARKER, DAVID, b Atlantic, Iowa, Mar 20, 37; m 66; c 2. THEORETICAL PHYSICS. *Educ:* Grinnell Col, BA, 59; Pa State Univ, MS, 62, PhD(physics), 66. *Prof Exp:* From asst prof to assoc prof, 65-72, assoc dean nat sci, 73-74, PROF PHYSICS, HOPE COL, 72-, PROVOST, 74- *Mem:* Am Phys Soc; Sigma Xi. *Res:* Theoretical high energy physics; calculation of nucleon-nucleon bremsstrahlung cross sections; analytic approximation theory. *Mailing Add:* 6001st St W Mt Vernon IA 52314

MARKER, LEON, b Lancaster, Pa, Jan 6, 22; m 53; c 3. PHYSICAL CHEMISTRY. *Educ:* Temple Univ, AB, 47; Univ Utah, PhD(phys chem), 51. *Prof Exp:* Res chemist, Gen Res Labs, Olin Industs, 52-57, res chemist, Film Res Dept, Olin Mathieson Chem Co, Conn, 57-63; head polymer physics sect, Res & Develop Div, 63-73, sr res assoc, 73-84, MGR POLYMER PHYSICS & ENG MECH, GEN, TIRE & RUBBER CO, 84- *Mem:* AAAS; Am Chem Soc; Soc Rheol; Sigma Xi. *Res:* Polymer physics; rheology; polymer processing; physical chemistry of polymers; chemical kinetics; kinetics of electrode reactions. *Mailing Add:* 3077 W Edgerton Rd Cuyahoga Falls OH 44224

MARKER, THOMAS F(RANKLIN), b Denver, Colo, June 25, 19; m 44; c 2. ELECTRONICS ENGINEERING, GAS & OIL EXPLORATION TECHNOLOGY DEVELOPMENT. *Educ:* Yatesbury Sch Eng, cert elec eng, 42; Malvern Col, Eng, cert elec eng, 43. *Prof Exp:* Mil observer, Am Embassy, London, Eng, 41-43; sr engr, Fed Telecommun Labs, 46-48; staff mem, Univ Calif, Los Alamos Labs, 48-50; mem staff, Sandia Corp, 50-52, mgr electronic develop, 52-56, mgr advan data systs, 56-68, mgr nuclear test dept, 68-70, mgr patent dept, 70-85; CONSULT EXPLOR TECHNOL DEVELOP, 85- *Mem:* Sr mem Inst Elec & Electronics Engrs. *Res:* Radar systems; management of advanced data systems development; patent management; new systems for oil and gas geophysical exploration. *Mailing Add:* 7613 Pickard Ave NE Albuquerque NM 87110-1530

MARKERT, CLAUS O, b Frankfort, Ger, Sept 8, 46; m 71; c 2. CELL CULTURE, CULTURED INTERVERTEBRAL DISCS. *Educ:* Johann Wolfgang Goethe Univ, dipl, 73, Dr phil nat(biochem & cell biol), 76. *Prof Exp:* Res asst physiol chem, Frankfort Univ Med Sch, 79-85; LAB ASST, PHARMACEUT BR, BASF CORP, KNOLL AG, LUDWIGSHAFEN, 85- *Mem:* Am Soc Cell Biol. *Res:* Use of human cell cultures for the screening of therapeutically attractive substances; rationale is either to revert a pathologically dysregulated cellular function or to trigger a latent repair mechanism. *Mailing Add:* Theodor Stern Kai 7, Knoll A G Postfach 210805 Ludwigshafen D-6700 Germany

MARKERT, CLEMENT LAWRENCE, b Las Animas, Colo, Apr 11, 17; m 40; c 3. DEVELOPMENTAL GENETICS, ENZYMOLOGY. *Educ:* Univ Colo, BA, 40; Univ Calif, Los Angeles, MS, 42; Johns Hopkins Univ, PhD(biol), 48. *Prof Exp:* Merck-Nat Res Coun fel, Calif Inst Technol, 48-50; from asst prof to assoc prof zool, Univ Mich, 50-57; prof biol, John Hopkins Univ, 57-65; prof biol, Yale Univ, 65-85, chmn dept, 65-71, dir, Ctr Reproductive Biol, 74-85; DISTINGUISHED UNIV RES PROF ANIMAL SCI & GENETICS, NC STATE UNIV, 86- *Concurrent Pos:* Panelist, NSF, 59-63, Am Cancer Soc, 63-65 & Subcomt Marine Biol, President's Sci Adv Comt, 65-66; trustee, Bermuda Biol Sta Res, 59-83, life trustee, 83-, trustee, Biosci Info Serv, 76-81, chmn bd, 81; mem comts, Nat Acad Sci, 63-81; managing ed, J Exp Zool, 63-85; co-chmn, Develop Biol Cluster, President's Biomed Res Panel, 75; coun mem, Am Cancer Soc, 76-78; mem bd sci adv, LaJolla Cancer Res Found, 77-86, Jane Coffin Fund Med Res, 79-87 & Dept Path, Health Sci Ctr, Univ Colo, 79-89; mem, Comt Animal Models & Genetic Stocks, Nat Res Coun, 79- 83; co-ed, Develop Genetics, 79-; mem, Coun Am Acad Arts & Sci, 81-84, Peromyscus Colony Adv Comt, 86- & Summer Res Confs Adv Comt, Fedn Am Socs Exp Biol, 87-93. *Honors & Awards:* Hirai Prize, 89. *Mem:* Nat Acad Sci; Am Soc Naturalists (vpres, 67); Am Soc Zoologists (pres, 67); Soc Develop Biol (pres, 63-64); Am Inst Biol Sci (pres, 66); Soc Study Reproduction. *Res:* Molecular control of stress protein synthesis during development; parthenogenesis and fertilization; differentiation of the gamete genome. *Mailing Add:* Dept Animal Sci Box 7621 NC State Univ Raleigh NC 27695-7621

MARKESBERY, WILLIAM RAY, b Florence, Ky, Sept 30, 32; m 58; c 3. NEUROLOGY, NEUROPATHOLOGY. *Educ:* Univ Ky, BA, 60, MD, 64. *Prof Exp:* Resident neurol, Col Physicians & Surgeons, Columbia Univ, 65-67, instr, 68-69, asst neurologist, Vanderbilt Clin, Columbia Presby Med Ctr, 68-69; asst prof path & neurol, Sch Med & Dent, Univ Rochester, 69-72; assoc prof, 72-77, PROF NEUROL & PATH, UNIV KY, 77-, DIR, SANDERS-BRAUN RES CTR AGING, 79-, PROF ANAT, 80- *Concurrent Pos:* USPHS-NIH-Nat Inst Neurol Dis & Blindness spec fel, Col Physicians & Surgeons, Columbia Univ, 67-69; USPHS res grant, Univ Rochester Med Ctr, 69-70; assoc neurologist, Strong Mem Hosp, 69-, asst pathologist, 70-; assoc neurologist, Univ Hosp, Univ Ky, 72; USPHS-NIH res grant, Univ Ky, 77-79; mem, Path A Study Sect, Nat Inst Aging, 82-85; mem, Adv Panel Assessment Dis Causing Dementia, Off Technol Assessment, US Cong, 85-86; prin investr, Alzheimers Dis Res Ctr, NIH (Nat Inst Aging), 85-, prog proj grant, biochem & morphol studies Alzheimers Dis. *Mem:* Am Acad Neurol; Am Asn Neuropath; Am Neurol Asn. *Res:* Aging of the central nervous system; Alzheimers disease and dementia; ultrastructure of central nervous system tumors; neuroimmunology and muscle diseases. *Mailing Add:* Dept Neurol & Path Univ Ky Med Ctr Lexington KY 40536

MARKEVICH, DARLENE JULIA, b Elizabeth, NJ, Aug 31, 49. ATOMIC PHYSICS. *Educ:* New York Univ, BA, 71, MS, 73, PhD(physics), 81. *Prof Exp:* Staff scientist, Lawrence Berkeley Lab, 81-83; PHYSICIST, US DEPT ENERGY, 83- *Res:* Cross section for inner-shell x-ray production in atomic collision processes; atomic inner-shell fluorescence yields. *Mailing Add:* ER-542 GTN US Dept Energy Washington DC 20585

MARKEY, SANFORD PHILIP, b Cleveland, Ohio, June 15, 42; m 66; c 2. ORGANIC CHEMISTRY, PHARMACOLOGY. *Educ:* Bowdoin Col, AB, 64; Mass Inst Technol, PhD(chem), 68. *Prof Exp:* Instr pediat, Med Sch, Univ Colo, 69, asst prof, 69-74, asst prof pharmacol, 71-74; res scientist pharmacol, 74-82, CHIEF, SECT ANAL BIOCHEM, NIMH, 82- *Concurrent Pos:* NIH grant mass spectrometry, Med Sch, Univ Colo, 70-74; assoc ed, Org Mass Spectrometry, 72-74. *Mem:* Am Chem Soc; Am Soc Mass Spectrometry. *Res:* Mass spectrometry applied to clinical research; drug-induced Parkinson's disease. *Mailing Add:* Lab Clin Sci Nat Inst Ment Health Bethesda MD 20892

MARKEY, WINSTON ROSCOE, b Buffalo, NY, Sept 20, 29; m 55; c 3. ENGINEERING. *Educ:* Mass Inst Technol, SB, 51, ScD(instrumentation), 56. *Prof Exp:* Instr aeronaut eng, 54-56, mem res staff, Instrumentation Lab, 56-57, from asst prof to assoc prof aeronaut & astronaut, 57-66, PROF AERONAUT & ASTRONAUT, MASS INST TECHNOL, 66-, DIR EXP ASTRON LAB, 61- *Concurrent Pos:* Consult, United Aircraft Corp, 62-; chief scientist, US Dept Air Force, 64-65, mem sci adv bd, 66-69. *Mem:* Assoc fel Am Inst Aeronaut & Astronaut. *Res:* Navigation system design; instrumentation for physical measurements. *Mailing Add:* 11 Edgewood Rd Lexington MA 02173

MARKGRAF, J(OHN) HODGE, b Cincinnati, Ohio, Mar 16, 30; m 57; c 2. ORGANIC CHEMISTRY. *Educ:* Williams Col, BA, 52; Yale Univ, MS, 54, PhD, 57. *Prof Exp:* Fel, Univ Munich, 56-57; chemist, Procter & Gamble Co, 58-59; from asst prof to assoc prof, 59-69, PROF CHEM, WILLIAMS COL, 69- *Concurrent Pos:* Provost, Williams Col, 80-83, vpres alumni rels & develop, 85- *Mem:* Am Chem Soc; Sigma Xi. *Res:* Physical organic studies of heterocyclic systems. *Mailing Add:* Dept of Chem Williams Col Williamstown MA 01267

MARKHAM, CHARLES HENRY, b Pasadena, Calif, Dec 24, 23; m 45, 71; c 6. NEUROLOGY, NEUROPHYSIOLOGY. *Educ:* Stanford Univ, BS, 47, MD, 51; Am Bd Psychiat & Neurol, dipl, 59. *Prof Exp:* Teaching fel neurol, Harvard Med Sch, 54-55; from instr to assoc prof, 56-71, PROF NEUROL, SCH MED, UNIV CALIF, LOS ANGELES, 71- *Concurrent Pos:* Consult, Wadsworth Vet Admin Hosp, Los Angeles, 56- *Mem:* Am Epilepsy Soc; Am Neurol Asn; Am Acad Neurol; Soc Neurosci; Barany Soc; Dyatonia Med Res Found; Am Parkinson's Dis Asn. *Res:* Vestibular, brain-stem and basal ganglia physiology; Parkinson's disease and other movement disorders. *Mailing Add:* Dept Neurol Univ Calif Sch Med Los Angeles CA 90024

MARKHAM, CLAIRE AGNES, b New Haven, Conn, Aug 12, 19. PHOTOCATALYSIS, PHOTOSENSITIZED REACTIONS. *Educ:* St Joseph Col, BA, 40; Cath Univ Am, PhD(chem), 52. *Hon Degrees:* LHD, St Joseph Col, 89. *Prof Exp:* Instr chem & phys, Sacred Heart High Sch, Waterbury, Conn, 45-49; fel chem, Norwegian Inst Technol, Trondheim, 67; sr fac fel, Calvin Lab, Univ Calif, Los Angeles, Berkeley, 67-68; undersecy energy, Off Policy & Mgt, State of Conn, 77-79; dean, grad div, 80-87, MEM FAC CHEM, ST JOSEPH COL, 52-, ASST PRES ACAD AFFAIRS, 87- *Concurrent Pos:* Res grants, NSF, 60-78, US Air Force, 60-65, Am Chem Soc & Petrol Res Fund, 65-76; mem, adv group, Dept Environ Protection, State of Conn, 62-67; sci adv, League Women Voters, Conn, 70-78; trustee, Mt St Joseph Acad, West Hartford, Conn, 72-80; mem, bd dirs, Conn Resource Recovery Authority, 77-79; chmn, adv bd, Conn State Environ Mediation Ctr, 82-88; dir, NSF Teacher Enhancement Prog, 87-89. *Honors & Awards:* Int Travel Award, NSF, 74. *Mem:* AAAS; Am Chem Soc; Sigma Xi. *Res:* Theoretical calculations of rates and mechanisms of reactions on catalytic surfaces; photochemical studies on energy transfer in photochemical reactions in solution; photovoltaic and photocatalytic studies on semiconductor surfaces. *Mailing Add:* St Joseph Col 1678 Asylum Ave W Hartford CT 06117

MARKHAM, ELIZABETH MARY, b New Haven, Conn, Oct 12, 29. MATHEMATICS. *Educ:* St Joseph Col, Conn, BA, 51; Univ Notre Dame, MS, 60, PhD(math), 64. *Prof Exp:* Teacher high sch, Conn, 54-59; instr math, St Joseph Col, Conn, 64-65; teacher, Our Lady of Mercy Acad, 65-66; from asst prof to assoc prof, 66-80, PROF MATH, ST JOSEPH COL, CONN, 80-, CHMN DEPT, 68-, chmn natural sci div, 78-81. *Concurrent Pos:* Dir & instr, NSF in serv inst high sch math teachers, 66-69; instr, Cent Conn State Col, 68; NSF consult, US AID Inst High Sch Math Teachers, Ramjas Col, Delhi Univ, 68; mem, Conn State Adv Comt Math, 69-74; dir & researcher, NSF Cause Grant, 76-79; chmn natural sci div, St Joseph Col, 78-81; NSF fac fel opers res, Sch Orgn & Mgt, Yale Univ, 80-81 & acturial rse dept, Hartford Group. Conn, 81. *Mem:* Math Asn Am; Am Math Soc; Nat Coun Teachers Math. *Res:* Mathematical modeling and applications. *Mailing Add:* Dept Math St Joseph Col West Hartford CT 06117

MARKHAM, GEORGE DOUGLAS, b Washington, DC, Oct 23, 51; m 83; c 1. ENZYMOLOGY, PROTEIN CHEMISTRY. *Educ:* Col William & Mary, BS, 72; Univ Pa, PhD(biochem), 77. *Prof Exp:* Fel, Univ Pa, 77-78; staff fel, NIH, 78-81; assoc mem, 81-86, MEM, INST CANCER RES, 86- *Mem:* Am Chem Soc; Am Soc Biol Chemists; Sigma Xi. *Res:* Enzyme structure and mechanism, particularly metalloenzymes; application of magnetic resonance spectroscopy to biological problems; metabolism of sulfur containing biochemicals. *Mailing Add:* Inst Cancer Res 7701 Burholme Ave Philadelphia PA 19111

MARKHAM, JAMES J, b Oreland, Pa, Aug 23, 28; m 52; c 8. ANALYTICAL CHEMISTRY, OCEANOGRAPHY. *Educ:* Villanova Univ, BS, 50; Univ Minn, PhD(chem), 58. *Prof Exp:* Res chemist, Whitemarsh Res Lab, Pa Salt Mfg Co, 50-51; from asst prof to assoc prof, 56-67, PROF CHEM & ASSOC DEAN SCI, VILLANOVA UNIV, 68- *Concurrent Pos:* USPHS vis res fel, Dept Inorg & Struct Chem, Univ Leeds, 65-66. *Mem:* Franklin Inst; Am Chem Soc; Royal Soc Chem; AAAS. *Res:* Instrumentation; marine chemistry; water supply and pollution control. *Mailing Add:* 2149 Menlo Ave Glenside PA 19038

MARKHAM, JORDAN JEPTHA, b Samokov, Bulgaria, Dec 25, 16; US citizen; m 43; c 2. PHYSICS. *Educ:* Beloit Col, BS, 38; Syracuse Univ, MS, 40; Brown Univ, PhD(physics), 46. *Prof Exp:* Res physicist, Div War Res, Columbia Univ, Conn, 42-45, NY, 45; fel, Clinton Labs, Tenn, 46-47; instr physics, Univ Pa, 47-48; asst prof, Brown Univ, 48-50; physicist, Appl Physics Lab, Johns Hopkins Univ, 50-53; physicist, Zenith Radio Corp, 53-60; sci adv physics res, IIT Res Inst, Ill Inst Technol, 60-62, prof physics, 62-81; RETIRED. *Concurrent Pos:* Vis prof, Univ Frankfurt, Ger, 65 & Univ Reading, Eng, 77. *Mem:* Fel Am Phys Soc. *Res:* Theory of imperfections in ionic crystals; spectroscopy of solids; color centers. *Mailing Add:* Caroline Meadows Villa 128 Chapel Hill NC 27514

MARKHAM, THOMAS LOWELL, b Apex, NC, Jan 2, 39. ALGEBRA. *Educ:* Univ NC, Chapel Hill, BS, 61, MA, 64; Auburn Univ, PhD(math), 67. *Prof Exp:* Asst prof math, Univ NC, Charlotte, 67-68; from asst prof to assoc prof, 68-82, PROF MATH, UNIV SC, 82- *Mem:* Am Math Soc; Math Asn Am; Sigma Xi; Soc Indust & Appl Math. *Res:* Linear algebra. *Mailing Add:* Dept Math Univ SC Columbia SC 29208

MARKHART, ALBERT H, JR, organic chemistry, polymer chemistry; deceased, see previous edition for last biography

MARKHART, ALBERT HENRY, III, b Malden, Mass, Nov 15, 51; m 78. BOTANY, PLANT PHYSIOLOGY. *Educ:* Gettysburg Col, BA, 73; Duke Univ, MA, 76, PhD(botany), 78. *Prof Exp:* Res assoc plant physiol, Duke Univ, 78-80; asst prof, 80-84, ASSOC PROF, DEPT HORT, UNIV MINN, 84- *Concurrent Pos:* Alexander von Humboldt fel, 79. *Mem:* Am Asn Plant Physiol; Am Soc Hort Sci. *Res:* Biochemical and physiological response of plants to a fluctuating environment; primarily the role of membrane and hormone changes as related to photosynthesis, water permeability and ion transport. *Mailing Add:* Dept Hort Univ Minn 1970 Folwell Ave St Paul MN 55108

MARKIEWICZ, ROBERT STEPHEN, b Worcester, Mass, Apr 18, 47; m 70. SOLID STATE PHYSICS. *Educ:* Mass Inst Technol, SB, 69; Univ Calif, Berkeley, PhD(physics), 75. *Prof Exp:* Res assoc physics, Univ Calif, Berkeley, 75-77; res physicist, Gen Elec Res & Develop Ctr, 77-80; ASSOC PROF PHYSICS, NORTHEASTERN UNIV, 80- *Concurrent Pos:* Fel, Int Bus Mach, 75-76. *Mem:* Am Phys Soc; Mat Res Soc. *Res:* Electron-hole droplets; electron-phonon interaction; intercalated graphite; localization in metals; high-Tc superconductivity. *Mailing Add:* Dept Physics Northeastern Univ 360 Huntington Ave Boston MA 02115

MARKIEWITZ, KENNETH HELMUT, b Breslau, Ger, May 18, 27; nat US; m 57; c 3. POLYMER CHEMISTRY. *Educ:* City Col New York, BS, 51; Columbia Univ, MA, 54, PhD(chem), 57. *Prof Exp:* Chemist, Schwarz Labs, Inc, NY, 51-52; sr chemist, Atlas Powder Co, 57-63; res chemist, 63-67, SR RES CHEMIST, ICI AMERICA INC, 68- *Mem:* Am Chem Soc; Sigma Xi. *Res:* Isolation of natural products; poison ivy; synthesis of alkenyl phenols; carbohydrates; amines; conformational analysis and structural determinations; polymer synthesis; material science. *Mailing Add:* 4012 Greenmount Rd Wilmington DE 19810

MARKING, RALPH H, b Holmen, Wis, Jan 24, 35; m 63. INORGANIC CHEMISTRY. *Educ:* Wis State Univ, La Crosse, BS, 57; Univ Minn, PhD(inorg chem), 65. *Prof Exp:* Assoc prof, 63-74, PROF CHEM, UNIV WIS-EAU CLAIRE, 74- *Mem:* Am Chem Soc; Am Asn Physics Teachers. *Res:* Thermochemistry and thermodynamics; molecular structure. *Mailing Add:* Dept of Chem Univ of Wis Eau Claire WI 54701

MARKIW, ROMAN TEODOR, b Tarnopol, Ukraine, June 25, 23; US citizen; m 50. BIOCHEMISTRY, ORGANIC CHEMISTRY. *Educ:* Univ Conn, BA, 54, PhD(biochem), 66; Rensselaer Polytech, MS, 55. *Prof Exp:* Biochemist, Vet Admin Hosp, 57-62; USPHS grants, Yale Univ, 65-68; res chemist, 68-72, chief biochem res lab, Vet Admin Ctr, 72-77; biochemist, Vet Admin Med Ctr, Canandaigua, NY, 78-87; RETIRED. *Mem:* Am Chem Soc; NY Acad Sci. *Res:* Isolation and identification of peptides in biological fluids; chemical reactions of polynucleotides and derivatives; clinical chemistry. *Mailing Add:* Rte Two Box 448 Kearneysville WV 25430

MARKLAND, FRANCIS SWABY, JR, b Philadelphia, Pa, Jan 15, 36; m; c 2. BIOCHEMISTRY. *Educ:* Pa State Univ, BS, 57; Johns Hopkins Univ, PhD(biochem), 64. *Prof Exp:* Asst prof biochem, Sch Med, Univ Calif, Los Angeles, 66-73; assoc prof, 74-83, actg chmn, 86-88, PROF BIOCHEM, SCH MED, UNIV SOUTHERN CALIF, 83-, VCHMN, 88- *Concurrent Pos:* NIH fel, Sch Med, Univ Calif, Los Angeles, 64-66 & career develop award, 68-73. *Mem:* Am Soc Hemat; Am Soc Biol Chem; Am Chem Soc; Sigma Xi; Endocrine Soc; Int Soc Toxinology; Am Asn Cancer Res; Protein Soc. *Res:* Structure of proteins and relation of structure to function in enzymes; biochemistry of blood coagulation; receptor proteins for steroid hormones; snake venom procoagulants and anticoagulants. *Mailing Add:* Cancer Res Lab No 106 Sch Med Univ Southern Calif 1303 N Mission Rd Los Angeles CA 90033

MARKLAND, WILLIAM R, b Brooklyn, NY, Jan 3, 19; m 42; c 3. COSMETIC CHEMISTRY. *Educ:* Middlebury Col, AB, 41. *Prof Exp:* Lab supvr, Hercules Powder Co, 42-45; chief chemist, John H Breck Inc, 45-57; res group leader hair prep, Revlon, Inc, 57-58; res mgr hair & makeup prods, Chesebrough-Pond's Inc, 58-71; CONSULT COSMETICS & TOILETRIES, 71- *Concurrent Pos:* Ed, Norda Briefs. *Mem:* Am Chem Soc; fel Am Inst Chem; Soc Cosmetic Chem. *Res:* Surfactant and shampoo chemistry; physical and chemical behavior of the hair; transparent microemulsions of mineral oil and water; cosmetic colors and pigments. *Mailing Add:* Box 124 Whately MA 01093

MARKLE, DOUGLAS FRANK, b Terre Haute, Ind, Aug 29, 47; m 71, 81; c 1. ICHTHYOLOGY. *Educ:* Cornell Univ, BS, 69; Col William & Mary, MA, 72, PhD(ichthyol), 76. *Prof Exp:* Ichthyologist & supvr, larval fish identification, Huntsman Marine Lab, 77-85; ASSOC PROF ICHTHYOL, ORE STATE UNIV, 85- *Mem:* Am Soc Ichthyol & Herpetol; Soc Syst Zool; Am Soc Naturalists. *Res:* Systematics of deep-sea fishes; deep-sea ecology; larval fish taxonomy and ecology. *Mailing Add:* Dept Fisheries & Wildlife Ore State Univ Corvallis OR 97331-3803

MARKLE, GEORGE MICHAEL, b Riverside, NJ, Dec 18, 39; m 64; c 3. REVIEW & EVALUATION. *Educ:* Cornell Univ, BS, 62; Rutgers Univ, MS, 70; Command & Gen Staff Col, cert mgt, 79. *Prof Exp:* Res assoc res develop, 65-72, asst prof, 72-78, ASSOC PROF RES DEVELOP, RUTGERS UNIV, 78- *Concurrent Pos:* Officer leader & instr, US Army, 62-; chmn & treas, RCC Compton Entom Trust, 77-; adj fac, Command & Gen Staff Col, 86- *Mem:* Entom Soc Am; Sigma Xi; NY Acad Sci; Regulatory Affairs Prof Soc; Asn Off Analytical Chemists; Am Chem Soc. *Res:* Evaluation of residue, toxicology and efficacy; pesticide and food that reflect the development of data for clearance of safe pest control agents and food and environmental safety. *Mailing Add:* 305 Walnut St Middlesex NJ 08846-1718

MARKLE, RONALD A, b Bellefonte, Pa, May 22, 51; m 78; c 1. CARDIOVASCULAR PHYSIOLOGY, ANATOMY. *Educ:* Pa State Univ, PhD(physiol), 76. *Prof Exp:* ASSOC PROF PHYSIOL & ANAT, NORTHERN ARIZ UNIV, 80- *Mem:* Am Physiol Soc; Sigma Xi; Am Heart Asn; Soc Exp Biol & Med. *Res:* Vascular wall biology with primary interest in events for transition from normal toward disease states. *Mailing Add:* Dept Biol Sci Northern Ariz Univ Box 5640 Flagstaff AZ 86011

MARKLEY, FRANCIS LANDIS, b Philadelphia, Pa, July 20, 39; m 65, 78; c 1. ATTITUDE DYNAMICS, ORBIT DYNAMICS. *Educ:* Cornell Univ, BEP, 62; Univ Calif, Berkeley, PhD(high energy physics), 67. *Prof Exp:* Physicist, Lawrence Radiation Lab, Univ Calif, 67; NSF res fel theoret physics, Univ Md, 67-68; asst prof physics, Williams Col, 68-74; mem tech staff, Computer Sci Corp, 74-78; physicist, Naval Res Lab, 78-85; AEROSPACE ENGR, NASA GODDARD SPACE FLIGHT CTR, 85- *Mem:* Am Asn Physics Teachers; Am Inst Aeronaut & Astronaut; Am Astronaut Soc; Soc Indust & Appl Math. *Res:* Spacecraft attitude and orbit dynamics, estimation, and control. *Mailing Add:* 10317 Wilde Lake Terr Columbia MD 21074-3529

MARKLEY, JOHN LUTE, b Denver, Colo, Mar 6, 41; m 75; c 2. PHYSICAL BIOCHEMISTRY, PROTEIN CHEMISTRY. *Educ:* Carleton Col, BA, 63; Harvard Univ, PhD(biophys), 69. *Prof Exp:* From res chemist to sr res chemist, Merck Inst Therapeut Res, 67-69; USPHS sr fel biophys, Chem Biodynamics Lab, Univ Calif, Berkeley, 70-71; from asst prof to prof chem, Purdue Univ, W Lafayette, 72-84; PROF BIOCHEM, UNIV WIS-MADISON, 84-, DIR GRAD BIOPHYSICS PROG & STEENBUCK PROF BIOMOLECULAR STRUCT, 90- *Concurrent Pos:* USPHS res career develop award, Nat Heart & Lung Inst, 75-80; dir, Purdue Biochem, NMR Lab, 77-85; Fogarty sr int fel, 80-81; dir, Nat Magnetic Resonance Lab, Madison, 85- *Mem:* AAAS; Am Chem Soc; Am Soc Biol Chem; Int Soc Magnetic Resonance; Sigma Xi; Soc Magnetic Resonance Med. *Res:* Structure-function relationships in biological macromolecules; applications of nuclear magnetic resonance spectroscopy to the study of local environments of groups; proteinases and their inhibitors; electron transport proteins; protein-nucleic acid interactions; nuclear magnetic resonance of imaging and spectroscopy of living cells and organisms. *Mailing Add:* Dept Biochem Univ Wis 420 Henry Mall Madison WI 53706-1569

MARKLEY, LOWELL DEAN, b Mishawaka, Ind, Aug 27, 42; m 62; c 2. ORGANIC CHEMISTRY. *Educ:* Manchester Col, BA, 64; Purdue Univ, PhD(org chem), 69. *Prof Exp:* Res chemist, 68-72, res specialist, 72-79, RES ASSOC AGR PRODS DEPT, DOW CHEM CO, 79- *Mem:* Am Chem Soc. *Res:* Synthesis of biologically active organic compounds including pharmaceuticals and agricultural products; synthesis of agricultural products. *Mailing Add:* 1307 Wildwood Midland MI 48640

MARKLEY, WILLIAM A, JR, b Sinking Springs, Pa, Aug 24, 25; m 54; c 4. MATHEMATICS. *Educ:* Bucknell Univ, BS, 49; Univ Pittsburgh, MLitt, 56, PhD(math), 68. *Prof Exp:* From asst prof to assoc prof, 56-70, PROF MATH, MT UNION COL, 70- *Mem:* NY Acad Sci. *Res:* Analysis; summability of infinite series. *Mailing Add:* 1435 Robinwood Alliance OH 44601

MARKO, JOHN ROBERT, b Bayonne, NJ, Jan 28, 38; div; c 1. SOLID STATE PHYSICS, OCEANOGRAPHY. *Educ:* Mass Inst Technol, BS, 59; Syracuse Univ, MS, 63, PhD(physics), 67. *Prof Exp:* Instr physics, Univ BC, 67-68, asst prof, 68-72; res assoc physics, Queen's Univ, 72-73; contract scientist, Pac Region, Marine Sci Directoriate, 74-76; DIR REMOTE SENSING, ARCTIC SCI LTD, 77- *Mem:* Am Phys Soc; Can Meteorol & Oceanog Soc. *Res:* Magnetic resonance; spin-lattice relaxation and electron spin resonance in semiconductors; transport properties of heavily doped semiconductors; arctic oceanography; remote sensing; acoustic devices; physical properties, detection and measurement of icebergs and sea ice and their relationships to environmental driving mechanisms and global climate change. *Mailing Add:* Arctic Sci Ltd 1986 Mills Rd Sidney BC V8L 3S1 Can

MARKO, KENNETH ANDREW, b Bayonne, NJ, Aug 29, 46; m 69; c 3. ARTIFICIAL INTELLIGENCE, DIGITAL CONTROL SYSTEM DIAGNOSTICS. *Educ:* Mass Inst Technol, BS, 68; Univ Mich, MS, 69, PhD(physics), 74. *Prof Exp:* Fel, Physics Dept, Univ Mich, 74-76; RES SCIENTIST, PHYSICS DEPT, ENG & RES STAFF, FORD MOTOR CO, 76- *Mem:* Am Phys Soc. *Res:* Laser spectroscopic techniques for application to combustion research; nonlinear Raman spectroscopy; laser induced fluorescence; laser velocimetry; application of air intelligence techniques to control system diagnostics. *Mailing Add:* Vehicle Electronics Dept Rm E1174 Eng & Res Staff Ford Motor Co PO Box 2053 Dearborn MI 48121

MARKOVETZ, ALLEN JOHN, b Aberdeen, SDak, Apr 17, 33; m 75; c 1. MICROBIOLOGY, BIOCHEMISTRY. *Educ:* Univ SDak, BA, 57, MA, 58; PhD(bact), 61. *Prof Exp:* NIH fel microbial metab, 61-62, from instr to assoc prof, 62-73, PROF MICROBIOL, UNIV IOWA, 73- *Concurrent Pos:* Spec res fel, NIH, Dept Biochem, Univ Calif, Davis, 69-70; Career develop award, NIH, 72-76; mem, NIH Study Sect, 78, 80-84. *Mem:* Am Soc Microbiol; Am Chem Soc. *Res:* Microbial physiology and metabolism; microbial hydrocarbon and ketone metabolism; microbial-insect interactions. *Mailing Add:* Dept of Microbiol Univ of Iowa Iowa City IA 52242-1000

MARKOVITZ, ALVIN, b Chicago, Ill, May 30, 29; m 52; c 4. MOLECULAR BIOLOGY. *Educ:* Univ Ill, BS, 50, MS, 52; Univ Wash, PhD, 55. *Prof Exp:* Fel, Nat Heart Inst, 55-57; instr, La Rabida Inst, 57-59, from asst prof to prof microbiol, 59-84, assoc prof, La Rabida Inst, 64-67, PROF BIOCHEM & MOLECULAR BIOL, UNIV CHICAGO, 84- *Mem:* AAAS; Am Soc Biol Chem; Am Soc Microbiol. *Res:* Regulation of cell division, radiation sensitivity and capsular polysaccharide synthesis in bacteria and their relation to proteolytic enzymes; molecular biology of chicken embryo cartilage differentiation. *Mailing Add:* Dept Biochem & Molecular Biol 920 E 58th St Chicago IL 60637

MARKOVITZ, HERSHEL, b McKeesport, Pa, Oct 11, 21; m 49; c 3. POLYMER PHYSICS, RHEOLOGY. *Educ:* Univ Pittsburgh, BS, 42; Columbia Univ, AM, 43, PhD(phys chem), 49. *Prof Exp:* Mathematician, Kellex Corp, 43-45; asst, Columbia Univ, 45-49; fel, 49-51, sr fel, 51-56, sr fel, Fundamental Res Group Synthetic Rubber Properties, Mellon Inst, 56-69, prof, 67-86, EMER PROF MECH & POLYMER SCI, CARNEGIE-MELLON UNIV, 86- *Concurrent Pos:* Lectr, Univ Pittsburgh, 56-58; vis lectr, Johns Hopkins Univ, 58-59; Fulbright lectr, Weizmann Inst, 64-65; asst ed, J Polymer Sci, 65-68, assoc ed, 69-87; mem gov bd, Am Inst Physics, 70-72; adj prof, Univ Pittsburgh, 72-; mem US Nat Comt Theoret & Appl Mech; mem nat bd dirs, Comt Concerned Scientists. *Honors & Awards:* Bingham Medal, Soc Rheol, 67. *Mem:* Am Chem Soc; Am Phys Soc; Soc Rheol (vpres, 67-69, pres, 69-71); Soc Natural Philos (treas, 65-66). *Res:* Physics of polymers; continuum mechanics; rheology. *Mailing Add:* 21 Disraeli St Jerusalem 92222 Israel

MARKOVITZ, MARK, b Rosario, Argentina, June 3, 38; US citizen. ORGANIC CHEMISTRY, POLYMER CHEMISTRY. *Educ:* City Col New York, 58; NY Univ, PhD(thiophene chem), 63. *Prof Exp:* Res fel, Thiophene Chem, NY Univ, 58-62; CHEMIST, MAT & PROCESSES LAB, GEN ELEC CO, 62- *Mem:* Am Chem Soc. *Res:* Electrical insulating materials; organo-metallic polymers; thermosetting resins; epoxy and silicone resins. *Mailing Add:* 2173 Apple Tree Lane Schenectady NY 12309

MARKOWITZ, ABRAHAM SAM, b New York, NY, July 12, 21; m 48; c 3. IMMUNOLOGY. *Educ:* NY Univ, BA, 48; Univ Southern Calif, MS, 50, PhD(bact), 52. *Prof Exp:* Asst prof bact, San Diego State Col, 52-54; from asst prof to assoc prof, 55-68, PROF MICROBIOL, UNIV ILL COL MED, 68-; HEAD EXP IMMUNOL, HEKTOEN INST MED RES, 58-; CHMN, DIV IMMUNOL, COOK COUNTY HOSP, 74- *Mem:* Am Asn Immunol; Am Soc Cell Biol; NY Acad Sci; Transplantation Soc; Int Soc Nephrology. *Res:* Heterophile antigens and antibodies; autogenous hypersensitivity; immunochemistry. *Mailing Add:* Hektoen Inst Med Res 639 S Wood St Chicago IL 60612

MARKOWITZ, ALLAN HENRY, b Jersey City, NJ, Oct 22, 41; m 66. ASTRONOMY, ASTROPHYSICS. *Educ:* Univ Calif, Los Angeles, AB, 63; Ohio State Univ, MSc, 66, PhD(astron), 69. *Prof Exp:* Analyst sci appln, Comput Sci Corp, 69-71; mgr adv systs studies, Aerojet Electrosysts Co, 71-75; eng specialist, Pomona Div, Gen Dynamics Corp, 75-76; PRES ME ENTERPRISES, 76-; INSTR ASTRON, CITRUS COL, 73- *Concurrent Pos:* Instr astron, Calif Polytech Inst, Pomona, 85-88. *Mem:* Astron Soc Pac; Am Astron Soc. *Res:* Spectroscopy of close binary stars; stellar statistics; peculiar A-type stars; early phases stellar evolution; teaching of undergraduate astronomy. *Mailing Add:* 2650 Country Club Dr Glendora CA 91740

MARKOWITZ, DAVID, b Paterson, NJ, Mar 24, 35; m 61; c 3. SOLID STATE PHYSICS, BIOLOGICAL PHYSICS. *Educ:* Mass Inst Technol, BS, 56; Univ Ill, PhD(physics), 63. *Prof Exp:* Res assoc physics, Rutgers Univ, 63-65; asst prof, 65-73, ASSOC PROF PHYSICS, Univ Conn, 73- *Concurrent Pos:* Adj prof, New Eng Inst, 70-; vis res scientist, Univ Sussex, 71-72. *Mem:* Am Phys Soc. *Mailing Add:* Dept of Physics Univ of Conn Storrs CT 06268

MARKOWITZ, HAROLD, b New York, NY, Sept 1, 25; m 53; c 4. IMMUNOCHEMISTRY. *Educ:* City Col New York, BS, 47; Columbia Univ, MA, 52, PhD(biochem), 53; Univ Utah, MD, 58. *Prof Exp:* Res assoc immunochem, Columbia Univ, 52-53; res fel med, Univ Utah, 53-59, intern, 58-59, instr, 59-61; consult, Mayo Clin, 61-85, asst prof microbiol, 62-68, ASSOC PROF MICROBIOL, MAYO GRAD SCH MED, UNIV MINN, 68- *Concurrent Pos:* NIH fel, 59-60; assoc attend physician, Salt Lake County Gen Hosp, Utah, 59-61. *Mem:* Am Asn Immunol; Am Soc Microbiol; Soc Exp Biol & Med; NY Acad Sci; Cent Soc Clin Res. *Res:* Carbohydrate chemistry; trace metal and copper metabolism; antigens of erythrocytes and pathogenic fungi; immunohematology; phytohemagglutinins. *Mailing Add:* Dept Microbiol Mayo Clinic 200 First St Rochester MN 55901

MARKOWITZ, MILTON, b New York, NY, June 6, 18; c 4. PEDIATRICS. *Educ:* Syracuse Univ, AB, 39, MD, 43. *Prof Exp:* Asst pediat, Sch Med, Johns Hopkins Hosp, 48-49, instr, 50-55, asst prof, Sch Med, Univ, 55-62, dir pediat rheumatic clins, Children's Med & Surg Ctr, 61-69; assoc pediatrician-in-chief, Sinai Hosp, Baltimore, 63-69; PROF PEDIAT & HEAD DEPT, SCH MED, UNIV CONN HEALTH CTR, 69- *Concurrent Pos:* Pvt pract, 49-52; dir streptococcal dis res lab, Sinai Hosp, Baltimore, 60-69; assoc prof pediat, Sch Med, Johns Hopkins Univ, 62-69. *Mem:* Fel Am Acad Pediat; Am Pediat Soc. *Res:* Rheumatic heart disease. *Mailing Add:* Dept Pediat Univ Conn Health Ctr 263 Farmington Rd Farmington CT 06032

MARKOWITZ, SAMUEL SOLOMON, b Brooklyn, NY, Oct 31, 31; m 58; c 3. NUCLEAR CHEMISTRY, ENVIRONMENTAL CHEMISTRY. *Educ:* Rensselaer Polytech, BS, 53; Princeton Univ, MA, 55, PhD, 57. *Prof Exp:* Jr res assoc nuclear chem, Brookhaven Nat Lab, 55-57; NSF fel, Univ Birmingham, 57-58; from asst prof to assoc prof chem, Univ Calif, Berkeley, 58-72, mem staff, Lawrence Berkeley Lab, 58-64, sr scientist, 64-80, PROF CHEM, UNIV CALIF, BERKELEY, 72-, FAC SR SCIENTIST, LAWRENCE BERKELEY LAB, 80- *Concurrent Pos:* Imp Chem Industs hon fel, Univ Birmingham & Charlotte Elizabeth Proctor fel, Princeton Univ, 57-68; NSF sr fel, fac sci, Univ Paris, 64-65; vis prof, Weizmann Inst Sci, Israel, 73-74; chmn, Calif Sect, Am Chem Soc, 91. *Mem:* Fel AAAS; Am Chem Soc; Am Phys Soc; Sigma Xi. *Res:* Nuclear reactions for analysis of atmospheric aerosols; chemistry of atmospheric aerosols; nuclear reactions at billion-electron-volt energies; fission and spallation; meson-induced reactions; nuclear activation analysis by He-3-induced reactions; radiochemistry; chemical fate of atoms produced via nuclear transformations. *Mailing Add:* Dept of Chem & Lawrence Berkeley Lab Univ of Calif Berkeley CA 94720

MARKOWITZ, WILLIAM, b Poland, Feb 8, 07; nat US; m 43; c 1. ASTRONOMY. *Educ:* Univ Chicago, BS, 27, MS, 29, PhD(astron), 31. *Prof Exp:* Instr math, Pa State Col, 31-32; astronr, US Naval Observ, 36-66, dir time serv, 53-66; prof physics, Marquette Univ, 66-68, Wehr prof, 68-72; ADJ PROF, NOVA UNIV, FLA, 72- *Concurrent Pos:* Ed, Geophys Surv, 72-79. *Mem:* Int Astron Union; Int Union Geod & Geophys; Am Astron Soc; Am Geophys Union. *Res:* Time and frequency; variations in earth rotation; motion of pole; SI Units. *Mailing Add:* 2800 East Sunrise Blvd Apt 15-B Ft Lauderdale FL 33304

MARKOWSKI, GREGORY RAY, b Milwaukee, Wis, June 23, 47. AEROSOL SCIENCE, INTELLIGENCE & CONSCIOUSNESS. *Educ:* Calif Inst Technol BS, 69; Univ Calif, Berkeley, MS, 72. *Prof Exp:* Sr scientist, Meteorol Res Inc, 73-84; CONSULT SCIENTIST, 84- *Mem:* AAAS; Am Asn Aerosol Res; Am Math Asn; Polanyi Soc; Philos Sci Asn. *Res:* Formation and properties of combustion and atmospheric aerosol; increased basin ventilation due to shear induced turbulence; inversion of ill-conditioned data by non-linear methods; influence of forcing function perturbations on chaotic systems, especially climate; theory of intelligence, consciousness and sensation; cybernetic theory of living systems; application of theory of logical types to resolution of quantum mechanical paradoxes. *Mailing Add:* 2009 N Madison ave Altadena CA 91001

MARKOWSKI, HENRY JOSEPH, b Worcester, Mass, July 1, 29; m 54; c 6. ORGANIC POLYMER SYNTHESIS, ELECTRICAL ENGINEERING. *Educ:* Providence Col, BS, 52. *Prof Exp:* Anal chemist, Nitrogen Div, Allied Chem Corp, 52-56; mgr, Markowski's Bakery, RI, 56-57; develop chemist, Lowe Brothers Paint Co, 57-60 & Hysol Corp, 60-64; develop chemist, Insulation Mat Dept, Gen Elec Corp, 64-68; mgr resin develop, P D George Paint & Varnish Co, 68-70; tech dir, Windecker Res, Tex, 70-71; mgr res synthesis, anal & testing, Carboline Co, 71-86; tech dir, Thermal Sci, Inc, 86-90; SR RES CHEMIST, P D GEORGE PAINT & VARNISH CO, 90- *Concurrent Pos:* Consult, Grigsby Bros Coatings Co. *Mem:* Am Chem Soc. *Res:* Polymer synthesis and research; urethane elastomers; epoxy coatings, organo-metallic polymers; alkyl ortho silicates resins, coatings, foams and adhesives used for electrical insulation; corrosion resistance; fire retardance and high temperature applications. *Mailing Add:* 12 Leaside Ct Manchester MO 63011-4029

MARKS, ALFRED FINLAY, b Yorktown Heights, NY, Sept 13, 32; m 71; c 4. AGRICULTURAL CHEMISTRY, PESTICIDE FORMULATIONS. *Educ:* Iowa State Univ, BS, 55. *Prof Exp:* Res chemist, Am Cyanamid Corp, 55-67; sr res chemist, Esso Res & Eng Corp, 67-70; group leader biochem formulations, SDS Biotech Corp, 70-81, mgr environ sci, 81-86; mgr environ sci, 86-89, DIR ENVIRON SCI, RICERCA, INC, 89- *Mem:* Am Chem Soc; fel Am Inst Chemists; Soc Environ Toxicol & Chem. *Res:* Development of residue and environmental fate data on new and existing products, particularly pesticides and animal health drugs, for support of registrations. *Mailing Add:* RICERCA Inc PO Box 1000 Painesville OH 44077-1000

MARKS, BERNARD HERMAN, b Cleveland, Ohio, Apr 21, 21; m 43; c 2. PHARMACOLOGY, BIOCHEMISTRY. *Educ:* Ohio State Univ, BA, 42, MD, 45, MA, 50. *Prof Exp:* Instr pharmacol & biochem, Ohio State Univ, 48-53, asst prof pharmacol, 54-56, from assoc prof to prof, 57-63, chmn dept, 63-73; prof pharmacol & chmn dept, 74-87, PROF PHARMACOL, WAYNE STATE UNIV, 87- *Mem:* AAAS; Am Soc Pharmacol & Exp Therapeut; Soc Exp Biol Med. *Res:* Cellular, cardiovascular and endocrine pharmacology; digitalis; radio-labeled drugs; autonomic pharmacology; autonomic receptors. *Mailing Add:* Dept Pharmacol Wayne State Univ Sch Med 1357 S Rehaud Grosse Point Woods MI 48236

MARKS, BURTON STEWART, b New York, NY, Oct 23, 24; m 48; c 3. POLYMER CHEMISTRY. *Educ:* Univ Miami, Fla, BS, 48, MS, 50; Polytech Inst Brooklyn, PhD(chem), 55. *Prof Exp:* Sr res chemist, Hooker Chem Co, 55-58; adv scientist, Continental Can Co, Inc, Ill, 58-62; STAFF SCIENTIST, LOCKHEED-PALO ALTO RES LABS, 62- *Concurrent Pos:* Adj prof, Niagara Univ, 57-58 & Roosevelt Univ, 60-62. *Mem:* Am Chem Soc; Royal Soc Chem. *Res:* Organic and polymer synthesis and chemistry; monomers; polymerization; resins; coatings; adhesives; foams; composite structures; ceramics; carbides. *Mailing Add:* 3415 Louis Rd Palo Alto CA 94303

MARKS, CHARLES, medicine, for more information see previous edition

MARKS, CHARLES FRANK, b Codroy, Nfld, Oct 23, 38; m 68; c 2. RESEARCH ADMINISTRATION. *Educ:* Macdonald Col, Montreal, BSc, 59; Univ Guelph, MSA, 63; Univ Calif, PhD(nematol), 67. *Prof Exp:* Res officer, Res Sta, St John's, Nfld, 64-67, res scientist, Vineland Sta, Ont, 67-76, dir, Delhi Sta, Ont, 76-81, dir, Res Sta, Agr Can, Harrow, Ont, 81-90, DIR, RES STA, AGR CAN, LONDON, ONT, 90- *Honors & Awards:* Hoechst Award, Can Soc Hort Sci, 85. *Mem:* Soc Nematol; Agr Inst Can. *Mailing Add:* Res Sta Res Br Agr Can Lon ON N6G 2V4 Can

MARKS, COLIN H, b Cardiff, Wales, Oct 8, 33; US citizen; c 1. MECHANICAL ENGINEERING. *Educ:* Carnegie Inst Technol, BSME, 56, MS, 57; Univ Md, PhD(mech eng), 65. *Prof Exp:* From instr to assoc prof, 59-78, PROF MECH ENG, UNIV MD, COLLEGE PARK, 78- *Concurrent Pos:* Vis scientist, Woods Hole Oceanog Inst, 67; vis scholar, Univ Wash, 78; guest worker, Ctr for Fire Res, Nat Bur of Standards, 87. *Mem:* Am Soc Mech Engrs; Combustion Inst. *Res:* Ground vehicle aerodynamics; fluid damping of structures; fire induced flows. *Mailing Add:* Dept Mech Eng Univ Md College Park MD 20740

MARKS, CRAIG, AUTOMOTIVE TECHNOLOGY. *Educ:* Calif Inst Technol, BS, MS & PhD(mech eng). *Prof Exp:* Res engr, Ford Motor Co, 55-57; sr res engr, Gen Motors Res Labs, 57-58, var eng, mgt & exec pos, 58-83; vpres sci & technol, Automotive Worldwide Sector, TRW, Inc, 84-88; VPRES TECHNOL & PRODUCTIVITY PLANNING, ALLIED-SIGNAL, 88- *Honors & Awards:* Horning Mem Award, Soc Automotive Engrs, 61. *Mem:* Nat Acad Eng; fel Soc Automotive Engrs; Am Soc Mech Engrs. *Mailing Add:* Automotive Sector World HQ Allied-Signal Inc PO Box 5029 Southfield MI 48086-5029

MARKS, DARRELL L, b Mountain Home, Idaho, July 23, 36; m 55; c 4. PHYSICS. *Educ:* Northwest Nazarene Col, AB, 58; Mass Inst Technol, MS, 59; Ore State Univ, PhD(biophys), 66. *Prof Exp:* Mem fac, 59-75, CHMN DIV MATH & NATURAL SCI, NORTHWEST NAZARENE COL, 75- *Concurrent Pos:* Mem curric comn, Idaho State, 75-81. *Mem:* AAAS; Am Asn Physics Teachers. *Res:* Science education; microprocessors used in teaching large lab classes. *Mailing Add:* Dept of Physics Northwest Nazarene Col Nampa ID 83686

MARKS, DAVID HUNTER, b White Plains, NY, Feb 22, 39; div; c 1. CIVIL & ENVIRONMENTAL ENGINEERING. *Educ:* Cornell Univ, BCE, 62, MS, 64; Johns Hopkins Univ, PhD(environ eng), 69. *Prof Exp:* From asst prof to assoc prof, 69-75, PROF CIVIL ENG, MASS INST TECHNOL, 75-, HEAD DEPT CIVIL ENG, 85- *Honors & Awards:* Huber Res Prize, Am Soc Civil Engrs, 77. *Mem:* Am Soc Civil Engrs; Am Geophys Union; Am Water Resources Asn; Water Pollution Control Fedn; Opers Res Soc Am. *Res:* Water resource systems; water quality management. *Mailing Add:* Dept Civil Eng 77 Massachusetts Ave Cambridge MA 02139

MARKS, DAWN BEATTY, b West Reading, Pa, July 16, 37; m 59; c 3. BIOCHEMISTRY. *Educ:* Bucknell Univ, BA, 59; Univ Pa, MS, 63, PhD(biochem), 65. *Prof Exp:* From asst prof to assoc prof, 70-90, PROF BIOCHEM, TEMPLE UNIV SCH MED, 90- *Mem:* Am Soc Biol Chemists. *Res:* Chromatin structure and function; regulation of protein synthesis. *Mailing Add:* Dept of Biochem Temple Univ Sch Med 3420 N Broad Philadelphia PA 19140

MARKS, DENNIS WILLIAM, b Madison, Wis, Nov 5, 44; m 68. RELATIVITY & RADIATION HYDRODYNAMICS. *Educ:* Fordham Univ, BS, 66; Univ Mich, Ann Arbor, PhD(astron), 70. *Prof Exp:* Fel & asst prof, David Dunlap Observ, Univ Toronto, 70-71; from asst prof to assoc prof, 71-78, dir planetarium & observ, 73-80, PROF PHYSICS, VALDOSTA STATE COL, 78-, DEPT HEAD PHYSICS, ASTRON & GEOL, 88- *Concurrent Pos:* Mem coun, Am Asn Univ Prof, 78-81; vis prof, Iowa State Univ, 82-83; secy, Physics, Math & Eng sect, Ga Acad Sci, 84-85, chmn, 85-86, counr, 88-; fac assoc, Advan Comput Methods Ctr, Univ Ga, 85. *Mem:* Int Soc Gen Relativity & Gravitation; Am Asn Univ Prof; Am Astron Soc. *Res:* Internal differential rotation of stars; viscosity of gases and radiation; relativistic stellar structure and evolution; cosmic background radiation; quantum metrology; telescope computerization. *Mailing Add:* Dept Phys Astron & Geol Valdosta State Col 1500 N Patterson Valdosta GA 31698

MARKS, EDWIN POTTER, b Detroit, Mich, Apr 22, 25; m 80; c 3. CELL & TISSUE CULTURE, METABOLIC INHIBITORS. *Educ:* Univ Mich, BS, 46; Univ Kans, MA, 48; Kans State Univ, PhD(entom & bot), 60. *Prof Exp:* Prof biol, Washburn Univ, 50-64; entomologist, Metab & Radiation Res Lab, Agr Res Serv, USDA, 65-87; RETIRED. *Concurrent Pos:* Adj prof, NDak State Univ, 67-; nat tech adv, Agr Res Serv, USDA, 82-86. *Mem:* Tissue Cult Asn; Entom Soc Am. *Res:* Insect cell and organ culture; hormones, metabolic inhibitors and secretory products; molecular biology. *Mailing Add:* PO Box 207 Ridge MD 20680

MARKS, GERALD A, b New York, NY, May 17, 49; m 76; c 1. NEUROSCIENCE, NEUROPHYSIOLOGY. *Educ:* City Col New York, BS, 72; City Univ New York, PhD(physiol psychol), 78. *Prof Exp:* Res asst neurosci, Montefiore Hosp & Med Ctr, 75-77; fac assoc, 77-81, res instr, neurosci, 81-83, ASST PROF PSYCHIAT, UNIV TEX SOUTHWESTERN MED CTR, DALLAS, 83- *Concurrent Pos:* Adj lectr psychol, City Col New York, 74-76; NIH fel neurosci, 79-81; NIH grant prin investr, 81-82, 87-91. *Mem:* AAAS; Asn Psychophysiol Study Sleep; Soc Neurosci. *Res:* Basic mechanisms of the central nervous system with an emphasis on those mechanisms relating to sleep behavior. *Mailing Add:* Dept Psychiat Univ Tex Health Sci Ctr Dallas TX 75235

MARKS, GERALD SAMUEL, b Cape Town, SAfrica, Feb 13, 30; m 55; c 2. ORGANIC CHEMISTRY, PHARMACOLOGY. *Educ:* Univ Cape Town, BSc, 50, MSc, 51; Oxford Univ, DPhil(org chem), 54. *Prof Exp:* Res chemist, SAfrican Inst Med Res, 55-56; res assoc porphyrin biosynthesis, Univ Chicago, 57-59; assoc prof pharmacol, Univ Alta, 62-69; head dept pharmacol, 69-88, PROF DEPT PHARMACOL, QUEEN'S UNIV, 88- *Concurrent Pos:* Nat Res Coun Can fel, 56-57; Brit Empire Cancer Campaign fel, Dept Chem Path, St Mary's Hosp, London, 60-62; vis prof, Sch Pharm, Univ Calif, San Francisco, 84-85. *Honors & Awards:* Upjohn Award Pharmacol Soc Can, 86. *Mem:* Pharmacol Soc Can; Am Soc Pharmacol & Exp Therapeut; Toxicol Soc Can. *Res:* Porphyrin biosynthesis; organic nitrates; mechanism of action and pharmacokinetics. *Mailing Add:* Dept Pharmacol Queen's Univ Kingston ON K7L 3N6 Can

MARKS, HENRY L, b Waynesboro, Va, Sept 6, 35; m 59; c 1. ANIMAL GENETICS. *Educ:* Va Polytech Inst, BS, 58, MS, 60; Univ Md, PhD, 67. *Prof Exp:* Res geneticist, 60-67, RES GENETICIST, SOUTH REGIONAL POULTRY BREEDING PROJ, AGR RES SERV, USDA, 67- *Mem:* Poultry Sci Asn; World Poultry Sci Asn; Sigma Xi. *Res:* Design and test animal breeding systems for increasing production by genetic selection. *Mailing Add:* Room 107 Livestock-Poultry Bldg Univ Ga Athens GA 30601

MARKS, JAY STEWART, b Ottawa, Ill, Apr 10, 37; m 61; c 3. FOOD ENGINEERING, CHEMICAL ENGINEERING. *Educ:* Univ Kans, BS, 59, PhD(chem eng), 65. *Prof Exp:* Chem engr, Mallinckrodt, Inc, 63-66, prod supt, 66-69, asst to mfg dir, 69-70, mgr mkt res, 70-73; plant mgr protein, Ralston Purina, 73-78; ASSOC PROF FOOD ENG, PURDUE UNIV, WEST LAFAYETTE, 78- *Mem:* Am Inst Chem Engrs; Inst Food Technologists; Am Soc Agr Engrs. *Res:* Energy conservation in food processing; solar utilization in food processing; physical properties of food. *Mailing Add:* Dept of Agr Eng Purdue Univ West Lafayette IN 47907

MARKS, L WHIT, b Fairfax, Okla, Nov 2, 26; m 48; c 4. PHYSICS. *Educ:* Cent State Univ, Okla, BS, 49; Univ Okla, MS, 51, PhD(physics), 55. *Prof Exp:* Asst physics, Univ Okla, 49-55; instr, 55, from asst prof to prof, 55-88, EMER ADJ PROF PHYSICS, CENT STATE UNIV, OKLA, 88- *Concurrent Pos:* chmn, physics dept, Cent State Univ, Okla, 58-78. *Mem:* Sigma Xi; Am Asn Physics Teachers; Int Solar Energy Soc; Am Scientific Affil; Int Asn Hydrogen Energy; Am Solar Energy Soc. *Res:* Electrode process energetics, particularly analysis of decay of activation overpotential for intermediate stages cathodic hydrogen processes; solar- hydrogen economy development. *Mailing Add:* Dept Physics Cent State Univ Edmond OK 73060-0177

MARKS, LAURENCE D, b London, Eng, July 4, 54; m. ELECTRON MICROSCOPY, SURFACE SCIENCE. *Educ:* Univ Cambridge, BA, 76, PhD(physics), 80. *Prof Exp:* Asst physics, Univ Cambridge, 80-83 & Ariz State Univ, 83-85; asst prof, 85-90, ASSOC PROF MAT SCI, NORTHWESTERN UNIV, 90- *Honors & Awards:* Burton Medal. *Res:* High resolution electron microscopy; catalysis; surface science; ceramic materials. *Mailing Add:* Dept Mat Sci Northwestern Univ Evanston IL 60208

MARKS, LAWRENCE EDWARD, b New York, NY, Dec 28, 41; m 63; c 2. EXPERIMENTAL PSYCHOLOGY. *Educ:* Hunter Col, AB, 62; Harvard Univ, PhD(psychol), 65. *Prof Exp:* Res fel lectr, Harvard Univ, 65-66; res psychologist, 66-69, from asst prof to assoc prof, 70-84, PROF EPIDEMIOL & PSYCHOL, YALE UNIV, 84- *Concurrent Pos:* Asst fel psychol, John B Pierce Found, 66-69, assoc fel, 69-84, fel, 84-, assoc dir, 85-88, actg dir, 89. *Honors & Awards:* Jacob Javits Neurosci Investr Award. *Mem:* Fel Am Psychol Asn; Soc Neurosci; Optical Soc Am; Acoust Soc Am; fel Am Psychol Soc. *Res:* Sensory processes, especially hearing and touch senses; interrelations among the senses; psychophysical measurement; psychology of language, particularly metaphor. *Mailing Add:* John B Pierce Lab 290 Congress Ave New Haven CT 06519

MARKS, LEON JOSEPH, b Providence, RI, Nov 30, 25; m 56; c 2. INTERNAL MEDICINE, ENDOCRINOLOGY. *Educ:* Brown Univ, AB, 44; Johns Hopkins Univ, MD, 48; Am Bd Internal Med, dipl, 56. *Prof Exp:* Intern med, Jewish Hosp, Brooklyn, NY, 48-49; jr resident, Kings County Hosp, 49-50; sr resident, Montefiore Hosp, 50-51; staff physician & dir steroid res lab, 52-73, chief outpatient serv & ambulatory health care, 52-75, assoc chief staff, Ambulatory Health Care, 75-78, CHIEF STAFF, BOSTON VET ADMIN HOSP, 79-; ASSOC PROF MED, SCH MED, BOSTON UNIV, 77-, ASST DEAN, 79- *Concurrent Pos:* Milton res fel, Harvard Univ, 51-52; res fel pediat, Mass Gen Hosp, 51-52; clin instr, Sch Med, Tufts Univ, 61-66, sr clin instr, 66-68, asst prof, 68-, asst dean, 79- *Mem:* Endocrine Soc; Am Geriat Soc; AMA; fel Am Col Physicians; Am Fedn Clin. *Res:* Res: Metabolism and endocrinology, especially adrenal steroid biochemistry and physiology as applied clinically to medicine, surgery and psychiatry; cancer of the prostate. *Mailing Add:* VA Hospital Boston MA 02130

MARKS, LOUIS SHEPPARD, b New York, NY, Dec 13, 17; m 44; c 4. ENTOMOLOGY, MATHEMATICAL BIOLOGY. *Educ:* City Col New York, BS, 39; Fordham Univ, MS, 51, PhD(entom), 54. *Prof Exp:* Statist analyst, US Dept Com, 40-42; prof chem & sanit sci, Am Acad, 46-51; instr biol, Fordham Univ, 51-54, from asst prof to assoc prof, 54-65; prof & head dept, Pace Col, 65-66; chmn dept, 66-78, prof, 66-87, EMER PROF BIOL, ST JOSEPH'S COL, PA, 87- *Concurrent Pos:* Smith Mundt vis prof, Nat Univ Mex, 56; vis prof, Hunter Col, 57; fel, Harvard Univ, 62-64; fel, NC State Univ, 63 & Williams Col, 65; Lilly fel, Barnes Arboretum, 70-81, Univ Pa, 80-84. *Mem:* Fel Linnean Soc; Soc Syst Zool; fel Royal Entom Soc; Soc Bibliog Natural Hist; Soc Study Evolution; Sigma Xi. *Res:* Human pedigree analysis; systematic entomology; taxonomy; morphology and zoogeography of the Lepidoptera; vertebrate coronary circulation; mathematical and evolutionary biology; zoological bibliography; chordate morphology. *Mailing Add:* St Josephs Univ 5600 City Ave Philadelphia PA 19131

MARKS, MEYER BENJAMIN, b Chicago, Ill, Feb 16, 07; m 32; c 2. PEDIATRIC ALLERGY. *Educ:* Univ Ill, BS, 31, MD, 33. *Prof Exp:* Assoc clin prof & dir pediat allergy, 54-70, CLIN PROF PEDIAT DIV ALLERGY/IMMUNOL, SCH MED, UNIV MIAMI, 70- *Concurrent Pos:* Dir pediat, Mt Sinai Hosp, Miami Beach, 50-59; dir pediat allergy clin, Jackson Mem Hosp, Miami, 55-80; consult pediat allergy, Mt Sinai Med Ctr, Miami Beach, 60- *Honors & Awards:* William Beaumont Prize, Univ Ill Col Med, 35; Silver Award, Am Acad Pediat, 75. *Mem:* Am Acad Allergy; Am Col Allergists; Am Acad Pediat. *Res:* Clinical research in all phases of pediatric allergy; identification of the allergic child; prophylaxis of childhood asthma; bruxism in allergic children; prevalence of allergy in school children in subtropical climates; inhalational medications in treatment of asthmatic children; nebulized cromolyn mist in childhood asthma. *Mailing Add:* 4302 Alton Rd No 640 Miami Beach FL 33140-2849

MARKS, NEVILLE, b Dublin, Ireland, Apr 10, 30; m; c 2. NEUROBIOLOGY. *Educ:* Univ London, MSc, 55, PhD(neurochem), 59. *Prof Exp:* Lectr neurochem, Inst Psychiat, Univ London, 57-59; fel biochem, Northwestern Univ, 59-60; neurochem, Ment Health Res Inst, Univ Mich, 60-61; sr res scientist, 61-68, assoc res scientist, 68-70, PRIN RES SCIENTIST, NY STATE RES INST NEUROCHEM & DRUG ADDICTION, 70-; ASSOC PROF, DEPT PSYCHIAT, NY UNIV, 79- *Concurrent Pos:* Ed, Res Methods Neurochem, Vols I-VI, assoc ed, Neurochem Res, 75; consult, Vet Admin Hosp, East Orange, 71-78; assoc ed, Neurochem Int, 80- *Mem:* Am Acad Neurol; Am Soc Neurochem; Int Soc Neurochem; Biochem Soc UK; Am Soc Biol Chemists. *Res:* Protein breakdown and turnover in brain; purification of catabolic enzymes; myelin turnover in experimental demyelination; formation and breakdown of hormonal peptides; neurochemistry of senile disorders and neuropathies. *Mailing Add:* Nathan Kline Inst Phychiat Res Ctr Neurochem Orangeburg NY 10962

MARKS, PAUL A, b New York, NY, Aug 16, 26; m 53; c 3. INTERNAL MEDICINE, BIOCHEMISTRY. *Educ:* Columbia Univ, AB, 45, MD, 49. *Hon Degrees:* Univ Urbino, Italy, 82; PhD, Hebrew Univ, Jerusalem, 87. *Prof Exp:* Res fel med, Med Col, Cornell Univ, 49; intern med, Presby Hosp, New York, 50, asst resident, 51; fel, Col Physicians & Surgeons, Columbia Univ, 52-53; assoc investr, Nat Inst Arthritis & Metab Dis, 53-55; instr med, Sch Med, George Washington Univ, 54-55; instr, 55-56, assoc, 56-57, from asst prof to assoc prof, Col Physicians & Surgeons, Columbia Univ, 57-67, dir hemat training 61-74, prof med, 67-82, prof human genetics & develop, 69-82, chmn dept, 69-70, dean, fac med & pres in chg med affairs, 70-73, vpres Health Sci & dir Cancer Res Ctr, 73-80, Frode Jensen prof med, 74-80; PRES & CHIEF EXEC OFFICER, MEM SLOAN-KETTERING CANCER CTR & MEM, SLOAN-KETTERING INST CANCER RES, 80-; PROF MED, MED COL, CORNELL UNIV, 82-, PROF, GRAD SCH MED SCI, 83- *Concurrent Pos:* Commonwealth Fund fel, 61-62; vis scientist, Lab Cellular Biochem, Pasteur Inst, 61-62; consult, Vet Admin Hosp, 62-69; mem adv panel develop biol, NSF, 64-67; Swiss-Am Found lectr & award in med res, 65; ed-in-chief, J Clin Invest, Am Soc Clin Invest, 67-71; mem adv panel hemat training grants prog, NIH, 69-73, chmn hemat training grants comn, 71-73; trustee, Roosevelt Hosp & St Luke's Hosp, 70-80; mem div med sci, Nat Res Coun, 72-, chmn exec comt, 73-76; Carl R Moore lectr, Sch Med, Wash Univ, 73, honors prog lectr, Sch Med, NY Univ, 73, Rufus Cole lectr, Rockefeller Univ, 77, Mayre guest prof, biochem, Univ Queensland St Lucia, Australia, 78, distinguished lectr Soc Univ Surgeons & Sigma Xi Soc Downstate Med Ctr State Univ NY, 81, Kazanjian lectr, Am Soc Maxillofacial Surgeons, 81; mem jury, Albert Lasker Awards, 74-76 & 80-81, mem adv comt, XV Int Cong Hemat, Israel, 74 mem, rev comt blood dis & blood resources panel, Nat Res & Demonstration Ctr, Nat Heart & Lung Inst, 74, mem Darthmouth Med Sch Conf on Health Systs, 75, Frontier's in Biol Sci Lectr, Case Western Reserve Univ Sch Med, 75; President's Biomed Res Panel, 75-76, Presidents Cancer Panel, 76-79, mem President's Comn on accident at Three Mile Island, 79, ad hoc adv, 1981 White House Conf Aging, 80-81; Frontiers in Biol Sci lectr, Case Western Reserve Univ Sch Med, 75, Sci Coun ad to bd dirs, Radiation Effects Res Found, Japan, 75-77, mem bd gov, Weizmann Inst Sci, 76-; dir, Charles H Resson Found, Inc, 76-, ed-in-chief, Blood, J Am Soc Hemat, 78-; adj prof Rockefeller Univ, 80-, vis physician, Rockefeller Univ Hosp, 80-, mem hon staff med, NY Hosp, 81-, attending physician Mem Hosp Cancer & Allied Dis, 80-, chmn, Sect Med Genetics, Hemat & Oncol, Nat Acad Sci, 80-; mem Comt Planning Study for ongoing study of costs of environ related health effects, 80, chmn head forum Adv Comt, Nat Acad Sci, 80-81; mem bd sci counrs, Div Cancer Treatment, Nat Cancer Inst, 80-83, tech bd, Milbank Mem Fund, 78-83, adv, Leopold Schepp Found, 80-, mem sci adv comt, Mass Gen Hosp, 81-84; bd sci counselors, Div Cancer Treatment, Nat Cancer Inst, 80-81, adv, The Leopold Schepp Found, 80-; mem, Comt Planning Study for an Ongoing Study Costs of Environ-Related Health Effects Inst Med, Nat Acad Sci, 80, chmn, Acad Forum Adv Comt, 80-81, coun Nat Acad Sci, USA, 84-87, mem, Comt Int Affairs, Coun Nat Acad Sci, 85-87, mem Nat Acad Sci Delegation Biol Warfare, Comt Int Security & Arms Control, 86-; ed bd, Cancer Treatment Revs, 81-, mem guest ed bd, Japanese J Cancer Res, 85-, assoc ed, Molecular Reprod & Develop, 88-; mem bd trustees, Ctr Adv Study Behav Sci, Stanford, Calif, 82-83; chmn, Robert Wood Johnson Prog Adv Comt, 83-, nat vis comt, City Univ NY Med Sch, 86-; mem Sci Adv Bd, City of Hope Nat Med Ctr, Durate, Calif, 87-; vis prof, Col France, 88; Mario A Baldini vis prof med, Med Sch, Harvard Univ, 91. *Honors & Awards:* Nat Medal of Sci, 91; Charles Janeway Prize, 49, Joseph Mather Smith Prize, 59, Stevens Triennial Prize, 60, Bicentennial Medal, 68; Swiss Am Found Award Med Res, 65; Harvey Soc Lectr, 70, Mackenzie Lectr, Cooperstown, NY, 71, Carl R Moore Lectr, Washington Univ Sch Med, 73, Rufus Cole Lectr, Rockefeller Univ, 77; Recognition Acad Accomplishments, Chinese Acad Med Sci, 82, Medal Found for Promotion of Cancer Res, Tokyo, Japan, 84; William H Resnick Lectr in Med, Stamford Hosp, 86, Distinguished Fac Lectr, M D Anderson, Univ Tex, 86, Distinguished Oncologist Award, Hipple Cancer Ctr & Kettering Ctr, 87; Centenary Medal, Inst Pasteur; Maurice C Pincoffs Lectr, Univ Md, Baltimore, 87; Charles Apffel Mem Lectr, New Eng Cancer Asn, 88; Robert Wood Johnson Found Medal, 89. *Mem:* Nat Acad Sci; Nat Inst Med; fel Am Acad Arts & Sci, Am Soc Clin Invest (pres, 71-72); Harvey Soc (treas, 67-70, vpres, 72-73, pres, 73-74); Am Soc Hemat; Soc Am Physicians; Am Asn Cancer Res; Int Soc Develop Biol; Am Col Physicians; Soc Develop Biol; Clin Investigation; Am Soc Biochem & Molecular Biol; Am Soc Hemat; Am Asn Cell Biol; Am Asn Clin Oncol; Am Soc Human Genetics; World Med Asn. *Res:* Cellular development; protein synthesis; human genetics; hematology. *Mailing Add:* Beach Hill Rd Bridgewater CT 06752

MARKS, PETER J, US citizen. ENVIRONMENTAL ENGINEERING. *Educ:* Franklin & Marshall Col, BS, 63; Drexel Univ, MS, 65. *Prof Exp:* Mem staff res lab anal methods develop, Lancaster County Gen Hosp, 63-64; VPRES, WESTON, 65- *Mem:* Am Soc Testing & Mat; Water Pollution Control Fedn. *Res:* Environmental analytical laboratory analysis; source emissions and ambient air sampling; wastewater treatment; biological monitoring methods. *Mailing Add:* Roy F Weston Inc Weston Way Westchester PA 19380

MARKS, RICHARD HENRY LEE, b Richmond, Va, Nov 23, 43; m 66; c 2. BIOCHEMISTRY. *Educ:* Univ Richmond, BS, 65; Ind Univ, Bloomington, PhD(biol chem), 69. *Prof Exp:* USPHS fel, Univ Calif, Santa Barbara, 69-70, univ fel, 71-72; asst prof biochem, Col Med & Dent NJ, 72-76; asst prof, 76-77, ASSOC PROF BIOCHEM, SCH MED, E CAROLINA UNIV, 77- *Mem:* Am Soc Biol Chemists; Am Chem Soc; Sigma Xi; Am Heart Asn. *Res:* Structure-function relationships in proteins, especially metalloproteins and lipoproteins. *Mailing Add:* Dept of Biochem East Carolina Univ Sch of Med Greenville NC 27858-4354

MARKS, RONALD LEE, b Jersey Shore, Pa, May 23, 34; m 71; c 3. INORGANIC CHEMISTRY. *Educ:* Lock Haven State Col, BS, 56; Pa State Univ, MEd, 59, EdD(chem educ), 66. *Prof Exp:* PROF CHEM, INDIANA UNIV, PA, 59- *Mem:* Am Chem Soc. *Res:* Hydroboration of aromatic heterocycles. *Mailing Add:* 6457 E Via Algardi Tucson AZ 85715

MARKS, SANDY COLE, JR, b Wilmington, NC, Nov 16, 37; m 62; c 2. ANATOMY. *Educ:* Washington & Lee Univ, BS, 60; Univ NC, DDS, 64; Johns Hopkins Univ, PhD(anat), 68. *Prof Exp:* Res officer, Dent Res Dept, Naval Med Res Inst, Bethesda, Md, 68-70; asst prof, 70-73, assoc prof 73-77, PROF ANAT, MED SCH, UNIV MASS, 77-, COORDR ANAT DONATIONS 76- *Mem:* AAAS; Int Asn Dent Res; Am Asn Anatomists; Am Soc Bone Mineral Res. *Res:* Bone metabolism; calcium homeostasis; tooth eruption. *Mailing Add:* Dept Cell Biol Univ Mass Med Sch 55 Lake Ave N Worcester MA 01605

MARKS, THOMAS, JR, b Rock Hill, SC, May 17, 51. NUCLEAR PHYSICS. *Educ:* Univ SC, BS, 72, PhD(nuclear phys), 77. *Prof Exp:* Res asst, Los Alamos Meson Physics Facil, 75-77; fel nuclear physics, TRIUMF Lab Vancouver, 77-78; STAFF MEM NUCLEAR PHYSICS INSTRUMENTATION, DOE'S LOS ALAMOS SCI LAB, 78- *Concurrent Pos:* Vis staff mem, ERDA's Los Alamos Sci Lab, 77. *Mailing Add:* Mail Stop E 540 Los Alamos Nat Lab Los Alamos NM 87545

MARKS, TOBIN JAY, b Washington, DC, Nov 25, 44; m 85; c 1. CHEMISTRY, INORGANIC CHEMISTRY. *Educ:* Univ Md, College Park, BS, 66; Mass Inst Technol, PhD(chem), 70. *Prof Exp:* From asst prof to assoc prof chem, 70-78, prof chem, 78-, MORRISON PROF CHEM, 87-, PROF MAT SCI, NORTHWESTERN UNIV, ILL, 88- *Concurrent Pos:* Sloan fel; Dreyfus fel; Guggenheim fel. *Honors & Awards:* Frensius Award in Pure & Appld Chem; Sobral Medal; Doolittle Award in Polymeric Mat; Organometallic Chem Award, Am Chem Soc. *Mem:* Am Chem Soc; Sigma Xi; Mat Res Soc. *Res:* Inorganic and organometallic chemistry; structural chemistry in solution; catalysis; solid state chemistry; polymer chemistry; metal ion biochemistry. *Mailing Add:* Dept Chem Northwestern Univ Evanston IL 60208

MARKSON, RALPH JOSEPH, b Feb 25, 31; US citizen; m 67; c 2. ATMOSPHERIC PHYSICS. *Educ:* Reed Col, BA, 56; Pa State Univ, MA, 67; State Univ NY Albany, PhD(atmospheric sci), 74. *Prof Exp:* Res engr, Convair Astronaut, 56-58; physicist, self employed, 58-65; res assoc atmospheric elec, State Univ NY Albany, 67-74; RES ASSOC ATMOSPHERIC PHYSICS, MASS INST TECHNOL, 74- *Concurrent Pos:* Pres, Airborne Res Assocs Inc; res pilot, high altitude atmospheric res aircraft. *Mem:* Am Geophys Union; Am Meteorol Soc; AAAS. *Res:* Atmospheric electrical global circuit; thundercloud electrification; extra-terrestrial modulation of atmospheric electricity and weather; use of atmospheric space charge as an air tracer; remote thermal detection for soaring; maritime convection and fog; development of thunderstorm and lightning detection instrumentation; development of corona elective field instrumentation and optical lightning detection sensing systems. *Mailing Add:* Dept Earth Planetary Sci Mass Inst Technol 77 Mass Ave Cambridge MA 02139

MARKSTEIN, GEORGE HENRY, b Vienna, Austria, June 22, 11; nat US; m 37; c 1. APPLIED PHYSICS. *Educ:* Vienna Tech Univ, Ing, 35, PhD(appl physics), 37. *Prof Exp:* Res physicist, Allgem Gluhlampenfabrics AG, Austria, 37-38; asst seismologist, Shell Petrol Co, Colombia, 39-40, prod supt, Plastic Molding Plant, 42-43; prod supt, Globe Soc Ltd, 44-46; res physicist, Cornell Aeronaut Lab, Inc, 46-50, head combustion sect, 50-56, prin physicist, 56-71; prin res scientist, 71-81, CONSULT, FACTORY MUTUAL RES CORP, 81- *Honors & Awards:* Silver Medal, Combustion Inst, 76, Bernard Lewis Gold Medal, 86. *Mem:* AAAS; Am Phys Soc; Combustion Inst. *Res:* Combustion; fluid dynamics; reaction kinetics; fire research; radiative energy transfer. *Mailing Add:* Factory Mutual Res Corp 1151 Boston-Providence Turnpike Norwood MA 02062

MARKUNAS, PETER CHARLES, b Chicago, Ill, Nov 5, 11; m 41; c 3. ANALYTICAL CHEMISTRY. *Educ:* Shurtleff Col, BS, 34; Univ Ill, MS, 37, PhD(anal chem), 40. *Prof Exp:* Asst, Univ Ill, 37-40; res chemist, Nat Distillers & Chem Corp, 40-41 & Com Solvents Corp, 41-51; dir anal res, R J Reynolds Industs, Inc, 51-72; RETIRED. *Mem:* AAAS; Am Chem Soc; Sigma Xi. *Res:* Instrumentation methods of analysis; chromatography; development of methods for analysis of nitroparaffins and derivatives; penicillin; bacitracin; hexachlorocyclohexanes; complex cations in microanalysis; titrimetry in nonaqueous solvents; functional group analysis. *Mailing Add:* 2425 Westchester Blvd Springfield IL 62704

MARKUS, GABOR, b Budapest, Hungary, June 8, 22; nat US; m 64; c 3. BIOCHEMISTRY. *Educ:* Univ Budapest, MD, 47; Stanford Univ, PhD, 50. *Prof Exp:* Estab investr, Am Heart Asn, 60-63; assoc res prof, State Univ NY Buffalo, 63-67, chmn dept biochem, Roswell Park Div, 67-71; assoc cancer res scientist, 63-67, prin cancer res scientist, 67-79, ASSOC CHIEF CANCER RES SCIENTIST, ROSWELL PARK CANCER INST, 79-; RES PROF BIOCHEM, STATE UNIV NY BUFFALO, 67- *Concurrent Pos:* Coun on Thrombosis, Am Heart Asn. *Honors & Awards:* Jacob F Schoellkopf Medal, Am Chem Soc, 81. *Mem:* AAAS; Am Soc Biol Chemists; Soc Exp Biol & Med; Am Chem Soc; Am Asn Cancer Res. *Res:* Protein structure and conformations; enzyme regulation; biochemistry of fibrinolysis; proteolytic enzymes in cancer; biochemistry of cancer metastasis. *Mailing Add:* Dept Exp Biol Roswell Park Cancer Inst 666 Elm St Buffalo NY 14263

MARKUS, LAWRENCE, b Hibbing, Minn, Oct 13, 22; m 50; c 2. MATHEMATICS. *Educ:* Univ Chicago, BS, 42, MS(meteorol) & MS(math), 47; Harvard Univ, PhD(math), 51. *Prof Exp:* Instr meteorol, Univ Chicago, 42-44, res meteorologist, Atomic Energy Proj, 44; instr math, Harvard Univ, 51-52 & Yale Univ, 52-55; lectr, Princeton Univ, 55-57; from asst prof to assoc prof, 57-60, assoc head dept math, 61-63, PROF MATH, UNIV MINN, MINNEAPOLIS, 60-, DIR, CONTROL & DYNAMICAL SYSTS CTR, 65-, REGENTS PROF, 80- *Concurrent Pos:* Fulbright fel, Paris, France, 50-51; Guggenheim fel, Univ Lausanne, 63-64; Nuffield prof, Univ Warwick, 68-69, dir control theory ctr, 70-; course dir, Int Ctr Theoret Physics, 74; lectr, Int Math Cong, 74; prin lectr, Iranian Math Soc, 75; vis prof, Japan Soc Prom Sci, 76, China, 85; mem panel, Int Math Cong, 78; Sci Res Coun sr vis fel, Imp Col, Eng, 78; mem sci adv comt, US Educ Sci & Cult Orgn, Univ Strasbourg, France, 80; hon prof, Univ Warwick, 85- *Mem:* Am Math Soc; Math Asn Am; Sigma Xi. *Res:* Ordinary differential equations; control theory; differential geometry; cosmology. *Mailing Add:* Dept of Math Univ of Minn Minneapolis MN 55455

MARKUSZEWSKI, RICHARD, b Pinsk, Poland, July 18, 41; US citizen. COAL ANALYSIS, DESULFURIZATION OF COAL. *Educ:* Loyola Univ, BS, 63; Univ Wis, MS, 66; Iowa State Univ, PhD(anal chem), 76. *Prof Exp:* Assoc ed anal chem, Chem Abstracts Serv, 73-75; instr gen & anal chem, Dept Chem, Iowa State Univ, 76-77; res assoc, Ames Lab, Iowa State Univ, 77-79, assoc chem, 79-81, asst dir, 81-85, dir fossil energy, Ames Lab, 85; DIR, MINING & METALL RES INST, IOWA STATE UNIV, 85- *Concurrent Pos:* Vis scientist, Warsaw Tech Univ, Poland, 77; asst prof, Dept Chem, Iowa State Univ, 78-81, grad lectr, 81; sci adminr, Fossil Energy, Ames Lab, 80-81; vis lectr, Polish Acad Sci, 80; adj prof Dept Geol Athos Sci, 89- *Honors & Awards:* Louis Gordon Mem Award, 80. *Mem:* Am Chem Soc; Am Inst Chem Engrs; Am Soc Testing & Mat; Sigma Xi; Geochem Soc; Soc Mining Metall Explor. *Res:* To characterize coal, especially for sulfur and trace elements; develop chemical and physical methods for the removal of sulfur and ash-forming mineral matter; develop on-line monitors for coal; energy; environmental aspects of coal utilization. *Mailing Add:* 111 Metals Develop Bldg Ames Lab Iowa State Univ Ames IA 50011-3020

MARKWALD, ROGER R, b Benton Harbor, Mich, Aug 21, 43; m; c 3. CELLULAR & DEVELOPMENTAL BIOLOGY. *Educ:* Calif State Univ, BS, 65; Colo State Univ, MS, 68, PhD(anat), 69. *Prof Exp:* Fel, Sch Med, Med Univ SC, 69-70, asst prof, Dept Anat, 70-75, grad fac mem, 71-75, res affil, Cardiol Sect, 73-75, assoc prof, Dept Anat, 75; assoc chmn, 77-81, ASSOC PROF, DEPT ANAT, HEALTH SCI CTR, TEX TECH UNIV, 75-, ACTG CHMN, 81- *Concurrent Pos:* Univ res award, 65, NIH Res Career Develop Award, Med Univ SC, 75 & Tex Tech Univ, 76; prin investr grants, Am Heart Asn, 70-72, SC Heart Asn, 73, NIH, 73-75, 76-79, 76-81, 78-81, 79-84 & 81. *Honors & Awards:* Lyndon Baines Johnson Res Award, Am Heart Asn, 77. *Mem:* Sigma Xi; Am Asn Anatomists; Int Soc Develop Biol; Int Soc Cell Biol. *Res:* Capacity for and mechanisms (structural and biochemical) by which extracellular macromolecules mediate genetic expression in cardiac morphogenesis, neural crest development and limb regeneration; cell biological basis of in situ cell movement; cardiac and neural crest related malformations (teratology); author or coauthor of over 50 publications. *Mailing Add:* Dept Anat Med Col Wis 8701 Watertown Plank Rd Milwaukee WI 53226

MARKWELL, DICK ROBERT, b Muskogee, Okla, Feb 20, 25; m 49; c 4. ANALYTICAL CHEMISTRY. *Educ:* Univ Wichita, BS, 48, MS, 50; Univ Wis, PhD(chem), 56. *Prof Exp:* Res & develop coordr, Off Chief Res & Develop, Hq, US Army, 65-67; assoc prof chem, San Antonio Col, 67-74; chemist, Corpus Christi Dept Health, 75-77; supvr chem sect, San Antonio Metrop Health Dist, 77-87; RETIRED. *Mem:* Am Chem Soc. *Mailing Add:* PO Box 65160 San Antonio TX 78265-5160

MARKWORTH, ALAN JOHN, b Cleveland, Ohio, July 13, 37; m 76; c 3. PHYSICS. *Educ:* Case Inst Technol, BSc, 59; Ohio State Univ, MSc, 61, PhD(physics), 69. *Prof Exp:* prin physicist, 66-80, sr res scientist, 80-84, assoc mgr, 84-88; ASSOC MGR & RES LEADER, ENGINEERED METALS & CERAMICS DEPT BATTELLE MEM INST, 88- *Concurrent Pos:* Mem comput simulation mat sci tech activ, Am Soc Metals Int; adj prof physics, Ohio Univ, Athens, Ohio, 89- *Mem:* Am Asn Physics Teachers; Metall Soc Am Inst Mining Metall & Petrol Engrs; Sigma Xi; Am Soc Metals Int; Am Phys Soc. *Res:* Computer-simulation studies of kinetic processes in solids and liquids; atomistic computer simulation; theory of phase transformation kinetics; materials processing in space; nonlinear dynamics applied to materials science. *Mailing Add:* Engineered Metals & Ceramics Dept Battelle Mem Inst 505 King Ave Columbus OH 43201-2693

MARLAND, GREGG (HINTON), b Oak Park, Ill, Sept 16, 42; m 63; c 3. GEOCHEMISTRY. *Educ:* Va Polytech Inst & State Univ, BS, 64; Univ Minn, PhD(geol), 72. *Prof Exp:* Asst prof geochem, Ind State Univ, Terre Haute, 70-75; scientist, Inst for Energy Anal, 75-87; SCIENTIST, OAK RIDGE NAT LAB, 87- *Res:* Environmental geochemistry, energy options and environmental implications; energy resources; global climate change. *Mailing Add:* Environ Sci Div Oak Ridge Nat Lab PO Box 2008 Oak Ridge TN 37831-6335

MARLATT, ABBY LINDSEY, b Manhattan, Kans, Dec 5, 16. NUTRITION. *Educ:* Kans State Univ, BS, 38; Univ Calif, cert, 40, PhD(animal nutrit), 47. *Prof Exp:* Asst home econ, Univ Calif, 40-45; from assoc prof to prof foods & nutrit, Kans State Univ, 45-56; vis prof home econ, Beirut Col Women, 53-54; dir, Sch Home Econ, 56-63, prof, 63-85, EMER PROF NUTRIT & FOOD SCI, COL HOME ECON, UNIV KY, 85- *Concurrent Pos:* Consult, Ky State Col, 68-71 & Bd Int Food & Agr Develop, AID, 78-85. *Honors & Awards:* Sullivan Medal, Univ Ky, 85. *Mem:* Am Home Econ Asn; Am Dietetic Asn; Soc Nutrit Educ. *Res:* Human nutrition; nutrient interrelationships; nutritional status and dietary surveys; pyridoxine requirements. *Mailing Add:* 256 Tahoma Rd Lexington KY 40503

MARLATT, WILLIAM EDGAR, b Kearney, Nebr, June 5, 31; m 56; c 2. ATMOSPHERIC SCIENCE. *Educ:* Nebr State Col Kearney, BA, 56; Rutgers Univ, MS, 58, PhD(soil physics), 61. *Prof Exp:* Res asst forestry, US Forest Serv, 54-55; res asst meteorol, Rutgers Univ, 56-58, asst prof, 58-61; prof atmospheric sci, 61-69, chmn dept earth resources, 70-74, assoc dean, Grad Sch, 67-68, PROF EARTH RESOURCES, COLO STATE UNIV, 75- *Concurrent Pos:* Consult, Nat Bur Stand, 64-67; mem, Colo Natural Resource Ctr Coun, 66-71; consult, Martin Marietta Co, 67-73; mem, Int Biol Prog Biometeorol Panel, Nat Res Coun-Nat Acad Sci, 68-70; sr scientist, Environ Resources Assocs, Inc, 69-70; consult, Manned Spaceflight Ctr, NASA, 70 & Colspan Environ Systs, Inc, 70-71; Int Biol Prog mem, Nat Adv Comt Aerobiol, 70-; consult, Thorne Ecol Found, 71-, mem bd dirs, 78- chmn educ comt, Colo Environ Res Ctr, 71- *Mem:* AAAS; Am Meteorol Soc; Am Astronaut Soc; Am Geophys Union; Coun Agr Sci & Technol; Agr Res Inst; Sigma Xi. *Res:* Remote sensing of atmosphere and earth surface, environment quality, interaction of climate and environment. *Mailing Add:* 3611 Richmond Dr Ft Collins CO 80526

MARLBOROUGH, JOHN MICHAEL, b Toronto, Ont, Aug 1, 40. ASTRONOMY. *Educ:* Univ Toronto, BSc, 62, MA, 63; Univ Chicago, PhD(astron), 67. *Prof Exp:* Lectr, 67, asst prof, 67-70, assoc prof, 70-76, PROF ASTRON, UNIV WESTERN ONT, 76- *Concurrent Pos:* Vis scientist, Dominion Astrophys Observ, Victoria, BC, 73-74. *Mem:* Am Astron Soc; Can Astron Soc; Royal Astron Soc. *Res:* Stellar interiors and evolution; early-type stars with extended atmospheres; radiative transfer; gas dynamics. *Mailing Add:* Dept of Astron Univ of Western Ont London ON N6A 5B9 Can

MARLER, PETER, b London, Eng, Feb 24, 28; US citizen; m 54; c 3. ZOOLOGY. *Educ:* Univ London, BSc, 48, PhD(bot), 52; Cambridge Univ, PhD(zool), 54. *Hon Degrees:* DSc, State Univ NY, Purchase, 83. *Prof Exp:* Res fel, Jesus Col, Cambridge Univ, 54-56; from asst prof to prof zool, Univ Calif, Berkeley, 57-66; dir, Field Res Ctr Ethol & Ecol, 72-81; prof, Rockefeller Univ, 66-89; PROF, UNIV CALIF, DAVIS, 89- *Concurrent Pos:* Guggenheim fel, 64-65; mem, Study Sect Exp Psychol, NIH, 64-69; sr res zoologist, NY Zool Soc, 66-72; mem & chmn comt, Walker Prize Biol, 67-68; dir, Inst Res Animal Behav, 69-72; mem health primate ctr rev comt, NIH, 74-78; external adv, Res Dept, Cent Inst Deaf, 78; mem external adv comt, Duke Primate Ctr, 78-79, Calif Reg Primate Ctr, 78- & Caribbean Primate Res Ctr, 79-86; mem bd sci counsrs, Nat Inst Child Health & Human Develop, 78 & Smithsonian Coun, 79-85. *Honors & Awards:* Elliott Coues Award, Am Ornithologist Union, 74. *Mem:* Nat Acad Sci; fel AAAS; fel Am Acad Arts & Sci; Animal Behav Soc (pres, 69-70); Am Soc Zoologists; Am Ornithologists Union; Am Psych Asn; Am Philol Soc; Dutch Ornithol Soc. *Res:* Behavior of animals, with special reference to the development of vocalizations in birds and primates; processes of communication in animals; field studies of social behavior. *Mailing Add:* Dept Zool Univ Calif Davis CA 95616-8755

MARLETT, JUDITH ANN, b Toledo, Ohio, June 20, 43. NUTRITION. *Educ:* Miami Univ, BS, 65; Univ Minn, PhD(nutrit), 72. *Prof Exp:* Therapeut dietician, Minneapolis Vet Admin Hosp, 66-67; res fel nutrit, Sch Pub Health, Harvard Univ, 73-74; from asst prof to assoc prof, 75-84, dir, coord undergrad & plan IV prog, 85-89, PROF NUTRIT, UNIV WIS-MADISON, 84- *Concurrent Pos:* Prin investr fed grants, NIH, ad hoc mem study sect; mem Am Coun Sci & Health Bd Sci Adv, 88- *Mem:* Am Soc Clin Nutrit; Am Inst Nutrit; Am Dietetics Asn; Inst Food Technologists; Am Asn Cereal Chemists. *Res:* Role of dietary fiber in human nutrition and in the human gastrointestinal tract; nutrient bioavailability. *Mailing Add:* Dept Nutrit Sci Univ Wis 1415 Linden Dr Madison WI 53706

MARLETTE, RALPH R(OY), b Bowdle, SDak, Oct 16, 20; m 43; c 3. CIVIL ENGINEERING. *Educ:* Univ Nebr, BS, 43, MS, 52; Delft Univ, DHE, 61. *Prof Exp:* Res engr, Black & Veatch, Consults, 47-50; PROF CIVIL ENG, UNIV NEBR, LINCOLN, 50- *Concurrent Pos:* Off engr, Garrison Dam, US Corps Engrs, 46-47; consult, President's Mo Basin Surv Comn, 52 & Sanit Dist 1, Lancaster County, Nebr, 57-60; NSF fac fel, Delft Univ, 60-61. *Mem:* Fel Am Soc Civil Engrs. *Res:* Hydraulic engineering; hydrology of small watersheds and ground water yield. *Mailing Add:* 640 S 52nd St Lincoln NE 68510

MARLEY, GERALD C, b Lovington, NMex, Nov 11, 38; m 59. MATHEMATICS. *Educ:* Eastern NMex Univ, BSc, 59; Tex Tech Col, MSc, 61; Univ Ariz, PhD(math), 67. *Prof Exp:* Res engr, Gen Dynamics-Astronaut, 61; lectr math, Univ Ariz, 67; from asst prof to assoc prof, 67-74, PROF MATH, CALIF STATE UNIV, FULLERTON, 74- *Mem:* Am Math Soc; Math Asn Am; Am Sci Affil; Sigma Xi. *Res:* Subdivisions of Euclidean space by convex bodies. *Mailing Add:* Dept Math Calif State Univ Fullerton CA 92634

MARLEY, STEPHEN J, b Blencoe, Iowa, Mar 5, 30; m 53; c 4. AGRICULTURAL ENGINEERING. *Educ:* Iowa State Univ, BS, 59, MS, 60, PhD(agr eng), 65. *Prof Exp:* From instr to assoc prof, 60-74, PROF AGR ENG, IOWA STATE UNIV, 74- *Concurrent Pos:* Vis lectr & Fulbright grant, Univ Col, Dublin, 70-71. *Mem:* Am Soc Agr Engrs. *Mailing Add:* Dept Agri Engr Iowa State Univ 102 Davidson Ames IA 50010

MARLIAVE, JEFFREY BURTON, b Oroville, Calif, Feb 28, 49; m 72; c 1. ICHTHYOLOGY, MARINE ECOLOGY. *Educ:* Univ Wash, BSc, 70; Univ BC, PhD(zool), 75. *Prof Exp:* Res assoc, 76-77, RESIDENT SCIENTIST, VANCOUVER PUB AQUARIUM, 77- *Honors & Awards:* Edward H Bean Award, Am Asn Zool Parks & Aquariums, 79, 81, 83. *Mem:* Am Asn Zool Parks & Aquariums; Am Fish Soc; West Soc Naturalists. *Res:* Ichthyoplankton; laboratory culture of planktonic larvae; behavioral ecology of marine fish larvae; developmental marine bioassays; reproductive ecology of intertidal fishes; ecological genetics of marine life. *Mailing Add:* Vancouver Pub Aquarium PO Box 3232 Vancouver BC V6B 3X8 Can

MARLIN, JOE ALTON, b Naylor, Mo, July 3, 35; m 60; c 2. APPLIED MATHEMATICS. *Educ:* Southeast Mo State Col, BS, 58; Univ Mo-Columbia, MA, 60; NC State Univ, PhD(math), 65. *Prof Exp:* Mem tech staff, Bell Tel Labs, 60-63; instr, 64-66, asst prof, 66-68, assoc prof, 68-78, PROF MATH, NC STATE UNIV, 78- *Mem:* Am Math Soc; Math Asn Am; Sigma Xi. *Res:* Oscillatory and asymptotic behavior of systems of ordinary differential equations which represent equations of motion for mechanical systems; Hamiltonians systems and foliations. *Mailing Add:* 3432 Leonard St Raleigh NC 27607

MARLIN, ROBERT LEWIS, b Bronx, NY, June 28, 37; m 59; c 2. INFORMATION SCIENCE, RESEARCH ADMINISTRATION. *Educ:* Syracuse Univ, AB, 58, MPA, 62; Rutgers Univ, PhD, 78. *Prof Exp:* Asst psychologist, NY State Dept Ment Hyg, 57-59; asst scientist exp psychol, Sterling-Winthrop Res Inst, 59-60, asst scientist statist, 60-62; asst to dir new prod develop, Winthrop Labs, Sterling Drug, Inc, 62-65; coordr med affairs, Knoll Pharm Co, 65-68; coordr prod develop, Schering Corp, 68-69; clin res assoc, Sandoz Pharm Co, 69-71, sr clin res assoc med, 71-75; biomed consult, 75-83; PRES, CAREFORMS INC, 82-; DIR CLIN RES, THOMPSON MED CO, 83- *Concurrent Pos:* Ed newslett, NJ Acad Sci, 70-74; mem staff, Ctr Prof Advan, 78-83; adj asst prof biomed sci, Mass Col Pharm & Allied Health Sci, 81-84, dept pharmacol, City Col NY, 86- *Mem:* AAAS; Drug Info Asn (secy, 68-69, vpres, 70-71); Biomet Soc; NY Acad Sci; Am Statist Asn. *Res:* Information systems for on-line audited data collection in clinical pharmaceutical research, including protocol design, summary reports and statistical analysis. *Mailing Add:* 8 Biscay Dr Parsippany NJ 07054

MARLISS, ERROL BASIL, b Edmonton, Alta, Jan 6, 41; m 69; c 2. MEDICINE, NUTRITION & FOOD SCIENCE. *Educ:* Univ Alta, MD, 64; FRcP(c), 73. *Prof Exp:* Intern, Univ Alta Hosp, Edmonton, 64-65; resident internal med, Royal Victoria Hosp, Montreal, 65-67; Med Res Coun Can postdoctoral fel, Sch Med, Boston Univ, 67-68, Med Sch, Harvard Univ, 68-70 & Inst Clin Biochem, Univ Geneva, Switz, 70-72; from asst prof to prof med, Univ Toronto, 72-82; PROF MED & DIR, NUTRIT & FOOD SCI CTR, MCGILL UNIV, MONTREAL, 82-, GARFIELD WESTON PROF NUTRIT, 84- *Concurrent Pos:* Med Res Coun Can scholar, 72-77; staff physician, Dept Med, Women's Col Hosp, 72-76 & Toronto Gen Hosp, 72-82; mem, Spec Study Sect, NIH, 75 & 76, Metab Spec Study Sect, 81, Med Sci Adv Bd, Juv Diabetes Found Int, 81-83, Policy Adv Group, NIH Diabetes Control & Complications Trial, 82-88, Sci Adv Coun, NIH, 84-88 & Prog Grants Rev Comt, Med Res Coun Can, 91-; chmn, Metab Spec Study Sect, NIH, 81; sr physician, Dept Med, Royal Victoria Hosp & Montreal Gen Hosp, 82-; dir, Clin Invest Unit, Royal Victoria Hosp, 83-; sci officer, Grants Comt Metab, Med Res Coun Can, 83-86; Alta Heritage Found Med Res vis prof, Muttart Diabetes Ctr, 84 & Nutrit-Metab Res Group, 88; vis prof, Joslin Diabetes Ctr & Brigham & Women's Hosp, 84; guest lectr, Julia MacFarlane Diabetes Ctr, Calgary, Alta, 87 & Brazil Diabetes Asn, 89. *Honors & Awards:* Baxter Lectr, Clin Res Soc Toronto, 81; Steve Broidy Lectr, Cedars-Sinai Med Ctr, 82; Ralph W Cooper Lectr, McMaster Univ, 82; Henry Dolger Lectr, Mt Sinai Med Sch, 82; Peter A J Adam Lectr, Case Western Reserve Univ, 82; Charles H Best Lectr, Toronto Diabetes Asn, 82. *Mem:* Am Diabetes Asn; Am Fedn Clin Res; Am Physiol Soc; Am Soc Clin Invest; Am Soc Clin Nutrit; Am Soc Parenteral & Enteral Nutrit; Can Soc Clin Invest; Can soc Endocrinol & Metab; Can Soc Nutrit Sci; Endocrine Soc. *Res:* Food and nutrition; diabetes. *Mailing Add:* McGill Nutrit & Food Sci Ctr Royal Victoria Hosp 687 Pine Ave W Montreal PQ H3A 1A1 Can

MARLOW, KEITH WINTON, b Madison, Kans, Nov 14, 28; m 51; c 3. NUCLEAR PHYSICS. *Educ:* Kans State Univ, BS, 51; Univ Md, PhD, 66. *Prof Exp:* Physicist, 51-67, head, Reactors Br, 67-70, consult, Radiation Technol Div, 71-80, head, Radiation Survivability & Detection Br, Naval Res Lab, 80-84; SR MEM TECH STAFF, SANDIA NAT LABS, 84- *Concurrent Pos:* Physicist, Inst Nuclear Physics Res, Amsterdam, 67-68. *Mem:* AAAS; Am Phys Soc; Sigma Xi. *Res:* Experimental study of nuclear structure, principally by investigating decay of radioactive nuclides; development of instrumentation for detecting low levels of radioactivity; development of nuclear techniques for treaty verification. *Mailing Add:* Sandia Nat Lab Div 9116 Albuquerque NM 87185-5800

MARLOW, RONALD WILLIAM, b San Diego, Calif, Feb 9, 49; m 84; c 2. PHYSIOLOGICAL ECOLOGY, FUNCTIONAL MORPHOLOGY. *Educ:* Univ Calif, Berkeley, BA, 73, PhD(zool), 79. *Prof Exp:* Fel animal behav, Univ Chicago, 79-80; scholar functional morphol, Univ Mich, 80-81; vis asst prof ecol, Univ Santa Clara, Calif, 81; res assoc, Mus Vert Zool, Univ Calif, Berkeley, 82-88; COODR DIV ENDANGERED SPECIES RES, DESERT BIOL RES CTR, DEPT BIOL SCI, UNIV NEV, 89- *Concurrent Pos:* State herpetologist, Dept WildLife, Nev, 88-89; adj prof, Dept Biol Sci, Univ Nev, 88- *Mem:* AAAS; Am Soc Ichthyologists & Herpetologists; Ecol Soc Am; Sigma Xi; Am Soc Zoologists. *Res:* Ecology and physiological ecology of lower vertebrates; evolution of the family Testudinidae; functional morphology in lower vertebrates; conservation biology of amphibians and reptiles. *Mailing Add:* Dept Biol Scis Univ Nev Las Vegas NV 89154-4004

MARLOW, WILLIAM HENRY, b Beaumont, Tex, Mar 1, 44; m 72; c 1. AEROSOLS, CLUSTERS AND ULTRAFINE PARTICLES. *Educ:* Mass Inst Technol, BS, 66; Univ Tex Austin, PhD(physics), 74. *Prof Exp:* Res assoc environ sci, Univ NC, Chapel Hill, 73-74; asst physicist, Brookhaven Nat Lab, 75-76, assoc physicist, 76-79, physicist, 79-86; res scientist, Tex Eng Exp Sta, 85-86, ASSOC PROF NUCLEAR ENG, TEX A&M UNIV, 86- *Concurrent Pos:* Prin investr, US Dept Energy, Brookhaven Nat Lab, 75-86 & US Dept Energy, Tex A&M Univ, 87-; adj asst prof environ med, Med Sch, NY Univ, 81-; organizer-ed, Aerosol Microphysics I: Particle Interactions & Aerosol Microphysics II: Chem Physics of Microparticles; invited lectr, 13th Int Symp Rarefied Gas Dynamics, Novosibirsk, USSR, 82; plenary lectr, Fall Tech Meeting Eastern Sect, Combustion Inst, Atlantic City, 82; consult, Aerospace Corp, 87-; guest lectr, Geselschaft für Aerosolforshung, Karlsruhe, WGer, 88; mem, Sci Rev Panel Air Chem & Physics, Environ Protection Agency, 88- *Mem:* Am Asn Aerosol Res; Am Phys Soc. *Res:* Theory of collective long-range intermolecular forces in cluster and ultrafine particle formation, evolution, and interactions; ultrafine aerosols in environmental health, materials, and other technologies. *Mailing Add:* Nuclear Eng Dept Tex A&M Univ College Station TX 77843-3133

MARLOW, WILLIAM HENRY, b Waterloo, Iowa, Nov 26, 24; m 48; c 5. MILITARY LOGISTICS, SYSTEMS EFFECTIVENESS. *Educ:* St Ambrose Col, BS, 47; Univ Iowa, MS, 48, PhD(math), 51. *Prof Exp:* Instr math, Univ Iowa, 48-51; res assoc, Logistics Res Proj, 51-56, prin investr, 56-69, chmn dept opers res, 71-77, PROF OPERS RES & DIR INST MGR SCI & ENG, GEORGE WASHINGTON UNIV, 69- *Concurrent Pos:* Assoc res mathematician, Univ Calif, Los Angeles, 54-55. *Mem:* Opers Res Soc Am; Math Asn Am; Soc Indust & Appl Math; Inst Mgt Sci. *Res:* Mathematical methods and numerical procedures in operations research and management science; logistics; systems effectiveness. *Mailing Add:* Dept of Opers Res George Washington Univ Sch of Eng & Appl Sci 2121 Eye St NW Washington DC 20052

MARLOWE, DONALD E(DWARD), b Worcester, Mass, Mar 27, 16; m 39; c 2. MECHANICAL ENGINEERING, PHYSICS. *Educ:* Univ Detroit, BCE, 38; Univ Mich, MSE, 39. *Hon Degrees:* ScD, Merrimack Col, 62; DScEng, Milwaukee Sch Eng, 81, Stevens Inst Technol, 81, Villanova Univ, 81. *Prof Exp:* Asst physics, Univ Detroit, 37-38, instr, 40-41; asst res engr, Univ Mich, 39-41; res engr, US Naval Ord Lab, Md, 41-46, chief eval engr, 46-50, asst tech dir, 50-51, assoc dir, 52-55; dean, Sch Eng & Archit, Cath Univ Am, 55-70, vpres for admin, 70-75; exec dir, Am Soc Eng Educ, 75-81; RETIRED. *Concurrent Pos:* Consult, US Res & Develop Bd, Weapon Systs Eval Group, Nat Res Coun & US Navy Bur Ord; dir & mem exec comt, Eng Joint Coun, 60-65; pres, Nat Coun Eng Exam, 66-67; secy, Educ Affairs Coun, Am Asn Eng Socs, 80-81; pres, United Engrs Trustees, 88-90. *Honors & Awards:* Marlowe Award, Am Soc Engr Educ, 81; Centennial Medal, Am Soc Mech Engrs, 80. *Mem:* Fel AAAS; fel & hon mem Am Soc Mech Engrs (vpres, 59-63 & 67-69, pres, 69-70); fel Am Soc Eng Educ; Nat Soc Prof Engrs; Inst Elec & Electronics Engrs. *Res:* Engineering mechanics; mechanical engineering; research administration; engineering education; educational administration. *Mailing Add:* 15402 Short Ridge Ct Silver Spring MD 20906

MARLOWE, EDWARD, b New York, NY, May 5, 35; m 59; c 2. PHARMACEUTICAL CHEMISTRY. *Educ:* Columbia Univ, BS, 56, MS, 58; Univ Md, PhD(pharm chem), 62. *Prof Exp:* Res assoc pharm res & develop, Merck Sharp & Dohme Res Lab, 62-64; sr scientist, Ortho Pharmaceut Corp, 64-67; dir res & develop, Whitehall Labs Div, Am Home Prod, NJ, 67-72; vpres res & develop, Res & Tech Div, Shering-Plough Inc, 72-81; VPRES & PRES CONSUMER PROD, RES & DEVELOP DIV, WARNER LAMBERT CO, 81- *Concurrent Pos:* Vis prof pharmaceut, Univ Tenn, 73-81. *Mem:* Acad Pharmaceut Sci; Soc Cosmetic Chemists; Am Pharmaceut Asn; Cosmetic Toiletry & Fragrance Asn; Sigma Xi; Asn Non-Prescription Mfr. *Res:* Product development, pharmacology, toxicology and photobiology; analytical development; exploratory product design research; drug delivery systems; oral disease research. *Mailing Add:* 56 Kean Rd Short Hills NJ 07078

MARLOWE, GEORGE ALBERT, JR, b Detroit, Mich, May 25, 25; m 53; c 1. HORTICULTURE. *Educ:* George Washington Univ, BS, 49, MS, 50; Univ Md, PhD(hort), 55. *Prof Exp:* Exten specialist hort, Univ Ky, 56-62; assoc prof hort, Ohio State Univ, 62-65; exten specialist veg crops, Univ Calif, Davis, 65-69; prof hort & horticulturist, Inst Food & Agr Sci, Univ Fla, 69-; AT AGENCY INT DEVELOP. *Concurrent Pos:* Chmn, Dept Veg Crops, Univ Fla, 69-72. *Mem:* Am Soc Hort Sci; Am Phys Soc. *Res:* Precision production technology; crop nutrition. *Mailing Add:* US AID/MASERU Agency Int Develop Washington DC 20523

MARLOWE, JAMES IRVIN, b Southport, NC, Sept 23, 32; m 57; c 4. COASTAL & ENVIRONMENTAL GEOLOGY. *Educ:* Fla State Univ, BS, 57; Univ Ariz, PhD(geol), 61. *Prof Exp:* Asst geologist, US Geol Surv, 57; teaching asst geol, Fla State Univ, 57-58; Rockefeller res fel, Univ Ariz, 58-60; geologist, NJ Zinc Co, 60-62; marine geologist, Geol Surv Can, 62-65; res scientist, Marine Sci Br, Dept Energy, Mines & Resources, Govt Can, 65-70; assoc prof geol & oceanog, Miami-Dade Community Col, Fla, 70-73; sr marine geologist, Dames & Moore, consults, 73-80; CONSULT GEOLOGIST, 80- *Concurrent Pos:* Lectr, dept geol, Dalhousie Univ, 64-70. *Mem:* Geol Soc Am; Am Inst Prof Geologists; Asn Ground Water Scientists & Engrs; Sigma Xi. *Res:* Assessment of geohazards at offshore sites; temporal trends in the morphology of sandy coastlines; marine placer deposits; use of marine erosional and depositional features in age-dating coastal faults. *Mailing Add:* Crows Nest Box 222 Rte 1 Hwy 187 Southport NC 28461

MARLOWE, THOMAS JOHNSON, b Fairview, NC, Sept 15, 17; m 45; c 4. ANIMAL GENETICS, ANIMAL PHYSIOLOGY. *Educ:* NC State Univ, BS, 40, MS, 49; Okla State Univ, PhD(animal genetics & physiol), 54. *Prof Exp:* High sch teacher, 40-42; training specialist & asst supvr, Vet Admin, NC, 46-48; asst county agt & livestock specialist, Va Agr Exten Serv, 49-50; instr animal husb, Miss State Univ, 51-52; res asst, Okla State Univ, 52-54; assoc prof animal sci, Va Polytech Inst & State Univ, 54-64, prof, 64-84; RETIRED. *Honors & Awards:* Animal Indust Award, Am Soc Animal Sci, 83. *Mem:* Am Inst Biol Sci; Am Genetic Asn; hon fel Am Soc Animal Sci (secy-treas, 71-74, pres, 75-76); Am Registry Cert Animal Scientists. *Res:* Beef cattle performance testing; heritability of economic traits; genetics and pathology of hereditary dwarfism cattle; effectiveness of selection in beef cattle; cytogenetics; evaluation of sire and dam breed for crossbreeding. *Mailing Add:* 2004 Shadow Lake Dr NW Blacksburg VA 24060

MARMAR, EARL SHELDON, b St Boniface, Man, July 3, 50; m 81; c 2. PLASMA PHYSICS. *Educ:* Univ Man, BSc, 72; Princeton Univ, MS, 75, PhD(physics), 76. *Prof Exp:* Res asst, Physics Dept, Princeton Univ, 72-73, Plasma Physics Lab, 73-76; sponsored res staff, Francis Bitter Nat Magnet Lab, 76-80, prin res scientist, Plasma Fusion Ctr, 80-88, SR RES SCIENTIST,

PHYSICS DEPT, MASS INST TECHNOL, 88- *Mem:* Fel Am Phys Soc. *Res:* Impurity problems in magnetic confinement fusion devices; plasma wall interactions; impurity radiation and power balance; atomic physics of highly stripped materials; particle diffusion and recycling in tokamaks. *Mailing Add:* NW17-105 175 Albany St Cambridge MA 02139

MARMARELIS, VASILIS Z, b Mytilini, Greece, Nov 16, 49; m 89. NONLINEAR SYSTEMS, DATA ANALYSIS. *Educ:* Nat Tech Univ Athens, Dipl, 72; Calif Inst Technol, MS, 73, PhD(eng sci), 76. *Prof Exp:* Lectr & res fel syst sci, Calif Inst Technol, 76-78; from asst prof to assoc prof, 78-88, PROF BIOMED & ELEC ENG, UNIV SOUTHERN CALIF, 88-, DIR, BIOMED SIMULATIONS RESOURCE, 85- *Concurrent Pos:* Vis assoc, Calif Inst Technol, 78-79; chmn biomed eng, Univ Southern Calif, 90- *Mem:* Inst Elec & Electronics Engrs; Int Fedn Automatic Control; Biomed Eng Soc; Soc Comput Simulation. *Res:* Analysis of signals and systems; spectral and correlation analysis; system identification and modeling; random processes and estimation methods; analysis and modeling of physiological systems with emphasis on nonlinear/nonstationary systems. *Mailing Add:* OHE 500 Univ Southern Calif Los Angeles CA 90089-1451

MARMER, GARY JAMES, b Cincinnati, Ohio, Nov 28, 38; m 60; c 3. ENVIRONMENTAL SCIENCE. *Educ:* Case Inst Technol, BS, 60; Auburn Univ, MS, 62; Ohio State Univ, PhD(physics), 68. *Prof Exp:* Physicist high energy physics, Accelerator Res, 68-72, PHYSICIST ENVIRON STUDIES, ARGONNE NAT LAB, 72- *Mem:* Am Water Resources Asn; Air Pollution Control Asn. *Res:* Assess environmental effects of construction and operation of energy facilities. *Mailing Add:* 3210 Indianwood Lane Joliet IL 60435

MARMER, WILLIAM NELSON, b Philadelphia, Pa, July 19, 43; m 73; c 2. ORGANIC CHEMISTRY, LIPID CHEMISTRY. *Educ:* Univ Pa, AB, 65; Temple Univ, PhD(chem), 71. *Prof Exp:* Nat Res Coun-Agr Res Serv res assoc org chem, 70-72, res scientist, Fats & Proteins Res Found, Inc, 72-75, RES CHEMIST, EASTERN REGIONAL RES CTR, USDA, 75-, RES LEADER, HIDES, LEATHER & WOOL RES UNIT, AGR RES SERV, 87- *Concurrent Pos:* Lectr chem, Pa State Univ. *Mem:* Am Chem Soc; Am Oil Chemists' Soc; Sigma Xi; AAAS; Am Leather Chemists Asn; The Fiber Soc. *Res:* Amine oxides; epoxides; O-acylhydroxylamines; fatty acid derivatives; acylations; fabric treatment; mixed anhydrides; lime soap dispersing agents; lipid extraction, analysis and synthesis; wool chemistry. *Mailing Add:* Eastern Regional Res Ctr USDA 600 E Mermaid Lane Philadelphia PA 19118

MARMET, PAUL, b Levis, Que, May 20, 32; m 59; c 4. ATOMIC PHYSICS, MOLECULAR PHYSICS. *Educ:* Laval Univ, BSc, 56, DSc(physics), 60. *Prof Exp:* Asst molecular physics, Commonwealth Sci & Indust Res Orgn, Australia, 60-61; from asst prof to prof physics, Laval Univ, 61-84; sr res officer, Nat Res Coun Can, 84-91; PROF, UNIV OTTAWA, 91- *Concurrent Pos:* Nat Res Coun Can fel, 60-61, grant, 61-, mem adv comt physics, 70-73; Defence Res Bd Can grants, 63-64 & 66-75; fel, Order Can, 81. *Honors & Awards:* Herzberg Medal, Can Asn Physicists, 71; Pariseau Medal, Can-French Asn Advan Sci,76; Rutherford Prize, 60. *Mem:* Can Asn Physicists; Royal Astron Soc Can; fel Royal Soc Can. *Mailing Add:* Physics Dept Univ Ottawa 34 Glinski Ottawa ON K1N 6N5 Can

MARMOR, MICHAEL F, b Aug 10, 41; m; c 2. OPHTHALMOLOGY, RETINAL DISEASES & RESEARCH. *Educ:* Harvard Univ, MD, 66. *Prof Exp:* Staff assoc, Lab Neuropharmacol, Div Spec Mental Health Res, NIMH, 67-70; asst prof ophthal, Univ Calif Sch Med, San Francisco, 73-74; from asst prof to assoc prof, 74-86, PROF OPHTHAL, STANFORD UNIV, 86-, CHMN DEPT, 88- *Concurrent Pos:* Chief, Ophthal Sect, VA Med Ctr, Palo Alto, 74-84; assoc fac mem, Prog Human Biol, Stanford Univ, 82-; vis prof, Mid-Japan Ophthal Soc, Kyoto, 84, Xian Med Univ, China, 88, Kellogg Eye Inst, Univ Mich, 85, Univ Ill, Chicago, 89; dir, physiol sect, Stanford Basic Sci Course Ophthal, Stanford Univ; mem sci adv bd, Nat Retinitis Pigmentosa Found, Calif Med Asn, Calif Asn Ophthal & Northern Calif Soc to Prevent Blindness; bd counselors, Am Acad Ophthal, 82-85, Comt Fed Systs, Pub Health Comt & Rep to Nat Acad Sci Comt Vision; NIH grant, 74- *Honors & Awards:* Serv Award, Nat Retinitis Pigmentosa Found, 81; Honor Award, Am Acad Ophthal, 84; Alcon Res Award Ophthal, 89. *Mem:* Fel Am Acad Ophthal; Asn Res Vision & Ophthal; Soc Gen Physiologists; Asn Univ Professors Ophthal. *Res:* Physiology of retinal adhesion and subretinal fluid transport; electrophysiology of RPE; modification of ischemic damage; models of RPE disease; clinical studies on retinal/RPE dystrophies and age related macular degeneration; application of visual science to art and history. *Mailing Add:* Dept Opthal Rm A-157 Med Ctr Stanford Univ Stanford CA 94305-5308

MARMOR, ROBERT SAMUEL, b Los Angeles, Calif, Nov 8, 43; m 68; c 3. ORGANIC CHEMISTRY. *Educ:* Univ Calif, Los Angeles, BS, 65; Mass Inst Technol, PhD(org chem), 70. *Prof Exp:* Res nat prod, 70-71, res organometallic chem, Mass Inst Technol, 71-72; sr res chemist, 72-75, supvr, 75-80, MGR ORG CHEM SECT, LORILLARD INC, 80- *Mem:* Am Chem Soc. *Res:* Organic synthesis; flavor chemistry; chemistry of tobacco. *Mailing Add:* Amer Cyanamid PO Box 400 Princeton NJ 08543

MARMOR, SOLOMON, b New York, NY, Feb 25, 26; m 54; c 2. ORGANIC CHEMISTRY. *Educ:* City Col New York, BS, 48; Syracuse Univ, PhD(org chem), 52. *Prof Exp:* Res chemist, Becco Chem Div, FMC Corp, 52-56; asst prof chem, Utica Col, Syracuse Univ, 56-62; assoc prof, NMex Highlands Univ, 62-66, head dept, 64-66; assoc prof, 66-70, coordr interdept progs, 68-70, chmn dept, 68-71, actg dean sch natural sci & math, 73-74, PROF CHEM, CALIF STATE UNIV, DOMINGUEZ HILLS, 70- *Mem:* AAAS; Am Chem Soc. *Res:* Epoxides; hypochlorous acid reactions; hydrogen peroxide oxidations of organic compounds. *Mailing Add:* 409 Ave F Redondo Beach CA 90277

MARMUR, JULIUS, b Byelostok, Poland, Mar 22, 26; Can citizen; m 58; c 2. MOLECULAR BIOLOGY, BIOCHEMISTRY. *Educ:* McGill Univ, BS, 46, MS, 47; Iowa State Col, PhD(bact physiol), 51. *Prof Exp:* Mem staff, NIH, 51-52, Rockefeller Inst, 52-54, Pasteur Inst, Paris, 54-55 & Inst Microbiol, Rutgers Univ, 55-56; res assoc chem, Harvard Univ, 56-60; asst prof biochem, Brandeis Univ, 60-61, assoc prof, 61-63; actg chmn dept biochem, 74-76, PROF BIOCHEM, ALBERT EINSTEIN COL MED, 63-, PROF GENETICS, 74- *Mem:* Am Soc Biol Chemists; Am Soc Microbiol. *Res:* Biological and physical-chemical properties of yeast nucleic acids. *Mailing Add:* Dept Biochem Albert Einstein Col Med 1300 Morris Park Ave Bronx NY 10461

MARNER, WILBUR JOSEPH, b Greentown, Ind, Jan 15, 37; m 57; c 2. MECHANICAL ENGINEERING. *Educ:* Purdue Univ, Lafayette, BS, 62, MS, 65; Univ SC, PhD(mech eng), 69. *Prof Exp:* Foreman qual control, Sarkes Tarzian, Inc, 62-63; asst prof mech eng, SDak Sch Mines & Technol, 69-73; res engr, Heat Transfer Res, Inc, 73-80; mem tech staff, 80-88, TECH GROUP SUPVR, JET PROPULSION LAB, CALIF INST TECHNOL, 88- *Concurrent Pos:* Consult, Hanford Eng Develop Lab, Westinghouse Hanford Co, 76-77; lectr, Calif State Univ, Los Angeles, 80-88; NASA-ASEE Summer fac fel, 70; sr partner & treas, VSM Asn, 83-84; lectr, Calif State Univ, Northridge, 88-; chmn, Heat Transfer Div, Am Soc Mech Engrs, 90-91. *Mem:* Fel Am Soc Mech Engrs; Am Soc Eng Educ; Am Inst Aeronaut & Astronaut; Sigma Xi. *Res:* Single-phase convective heat transfer; extended surfaces heat transfer; augmentation of convective heat transfer; fouling of heat transfer surfaces; instrumentation; multiphase flow. *Mailing Add:* 1654 Oakwood Ave Arcadia CA 91006

MARNETT, LAWRENCE JOSEPH, b Kansas City, Kans, Nov 22, 47; m 71; c 2. BIOCHEMISTRY, CARCINOGENESIS. *Educ:* Rockhurst Col, BS, 69; Duke Univ, PhD(chem), 73. *Prof Exp:* Assoc biochem, Karolinska Inst, 73-74; assoc biochem, Wayne State Univ, 74-75, from asst prof to prof chem, 75-89; PROF BIOCHEM, VANDERBILT UNIV, 89- *Concurrent Pos:* Ed, Chem Res Toxicol, 87- *Mem:* Am Chem Soc; Sigma Xi; Am Soc Biochem & Molecular Biol; Am Assoc Cancer Res; AAAS; Soc Toxicol; Int Soc Xenobiotics; Soc Toxicol; Oxygen Soc. *Res:* Prostaglandin biochemistry; chemical carcinogenesis; free radicals in biology and medicine; enzyme mechanism. *Mailing Add:* Biochem Dept Vanderbilt Univ Nashville TN 37232

MAROIS, ROBERT LEO, b Troy, NY, Apr 27, 35; m 61; c 4. PHARMACOLOGY, EDUCATION. *Educ:* Siena Col, NY, BS, 64; Albany Med Col, PhD(pharmacol), 69. *Prof Exp:* Res assoc, State Univ NY Albany, 68-69; PROF PHARMACOL, ALBANY COL PHARM, 69- *Concurrent Pos:* USPHS fel, State Univ NY Albany, 68-69. *Res:* Pharmacology education. *Mailing Add:* Dept Biol Sci Albany Col Pharm 106 New Scotland Ave Albany NY 12208

MAROM, EMANUEL, b Romania, Nov 16, 34; US citizen; m 59; c 2. OPTICAL DATA & SIGNAL PROCESSING, FIBER & INTEGRATED OPTICS. *Educ:* Technion, Israel Inst Technol, BS, 57, MS, 61; Polytech Inst Brooklyn, PhD(elec eng), 65. *Prof Exp:* Res asst electronics, Polytech Inst Brooklyn, 61-65, asst prof, 65-66; mem tech staff image processing, Bendix Res Labs, Southfield, Mich, 66-70; asst tech dir electro-optics, Eljim Ltd, Israel, 70-72; assoc prof electro-optics, Tel Aviv Univ, Israel, 72-76; mem tech staff optical commun, Hughes Res Lab, Malibu, Calif, 76-78; prof & dean elec eng, Tel Aviv Univ, 78-83; SR STAFF ENGR, OPTICAL DATA PROCESSING, HUGHES RES LAB, 83- *Concurrent Pos:* Prof, Tel Aviv Univ, 72- *Mem:* Fel Optical Soc Am; sr mem Inst Elec & Electronics Engrs; Int Comn Optics (vpres, 78-81). *Res:* Holography, optical data and image processing; non-destructive testing; acoustic holography; diffraction theory; optical communications and integrated optics; optical computing. *Mailing Add:* Prof of Electro Optics Fac of Eng Tel Aviv Univ Tel Aviv 69978 Israel

MARON, MELVIN EARL, b Bloomfield, NJ, Jan 23, 24; m 48; c 2. INFORMATION SCIENCE. *Educ:* Univ Nebr, BS, 45, BA, 47; Univ Calif, Los Angeles, PhD(philos), 51. *Prof Exp:* Instr, Univ Calif, Los Angeles, 51-52; tech engr, Int Bus Mach Corp, 52-55; mem tech staff, Ramo-Wooldridge Corp, 55-59; mem sr res staff, Rand Corp, 59-66; prof librarianship, 66-80, PROF, LIBRARY & INFORMATION STUDIES, UNIV CALIF, BERKELEY, 80- *Mem:* AAAS; Asn Comput Mach; Philos Sci Asn. *Res:* Philosophy; computer sciences; cybernetics; theory and foundations of automatic information searching and data retrieval. *Mailing Add:* Dept Libr Studies Univ Calif Berkeley CA 94720

MARON, MICHAEL BRENT, b Long Beach, Calif, Oct 8, 49; m. PULMONARY HEMODYNAMICS, LUNG FLUID BALANCE. *Educ:* Univ Calif, Santa Barbara, BA, 72, PhD(physiol), 76. *Prof Exp:* Res asst, Inst Environ Stress, Univ Calif, Santa Barbara, 73-76; clin asst prof, Dept Phys Educ, Univ Wis-Milwaukee, 77-79; from asst prof to assoc prof physiol, 79-89, interim chmn dept, 85-86, CHMN PHYSIOL, DEPT PHYSIOL, COL MED, NORTHEASTERN OHIO UNIV, 86-, PROF PHYSIOL, 89- *Concurrent Pos:* Am Heart Asn postdoctoral fel, Dept Physiol, Med Col Wis & Res Serv, Vet Admin Ctr, 77-79; ad hoc instr, Dept Phys Educ, Univ Wis-Milwaukke, 77, clin asst prof, 77-79; mem fac, Physiol Grad Prog, Kent State Univ, 81- *Honors & Awards:* Harwood S Belding Award, Am Physiol Soc, 76. *Mem:* Am Physiol Soc; Microcirculatory Soc; Am Col Sports Med; AAAS; NY Acad Sci; Int Soc Lymphology. *Res:* Physiology. *Mailing Add:* Physiol Dept Col Med Northeastern Ohio Univ 4207 State Rd 44 Rootstown OH 44272

MARONDE, ROBERT FRANCIS, b Calif, Jan 13, 20; m 70; c 4. INTERNAL MEDICINE, CLINICAL PHARMACOLOGY. *Educ:* Univ Southern Calif, BA, 41, MD, 44; Am Bd Internal Med, dipl, 51. *Prof Exp:* Resident med, Los Angeles County Gen Hosp, 46-48; asst prof physiol, 48-49, from asst clin prof to assoc clin prof med, 49-63, assoc prof med & pharmacol, 63-68, PROF MED & PHARMACOL, SCH MED, UNIV SOUTHERN CALIF, 68-, CHIEF CLIN PHARMACOL SECT, 70- *Concurrent Pos:*

Consult, Food & Drug Admin, Dept Health & Human Serv, Presidential Task Force Prescription Drugs & Calif State Dept Health. *Mem:* Am Soc Clin Pharmacol & Therapeut. *Res:* Medical computer applications for drug utilization review. *Mailing Add:* Dept Med Pharmacol & Nutrit Sch Med Univ Southern Calif 2025 Zonal Ave Los Angeles CA 90033

MARONEY, SAMUEL PATTERSON, JR, b Wilmington, Del, Feb 3, 26; m 51, 89; c 3. ZOOLOGY. *Educ:* Wesleyan Univ, BA, 50; Univ Del, MA, 53; Duke Univ, PhD(zool), 57. *Prof Exp:* Asst prof, 56-62, ASSOC PROF BIOL, UNIV VA, 62- *Concurrent Pos:* Assoc dean fac, Univ Va, 75-85. *Mem:* Fel AAAS; Am Soc Zoologists; Sigma Xi. *Res:* Cell physiology. *Mailing Add:* Dept Biol Univ Va Charlottesville VA 22901

MARONI, DONNA F, b Buffalo, NY, Feb 27, 38; m 74. BIOTECHNOLOGY, EVOLUTION OF MITOSIS. *Educ:* Univ Wis-Madison, BS, 60, MS, 65, PhD(zool), 69. *Prof Exp:* Proj asst, Dept Zool, Univ Wis-Madison, 60-63, prof assoc, 68-74; Alexander von Humboldt fel, Inst Genetics, Univ Cologne, 74-75; Hargitt fel, Duke Univ, 75-76, res assoc, 76-83, res assoc prof zool, 83-87; sr prog specialist, 87-88, DIR SCI PROG DIV, NC BIOTECHNOL CTR, 88- *Concurrent Pos:* Grants, NSF, 77-79, NIH, 79-82, 79-83, 81-85, 82-87 & 86-91; reviewer grant proposals, NSF, USDA & City New York. *Mem:* Am Soc Cell Biol; Genetics Soc Am; Sigma Xi. *Res:* Science administration; biotechnology; numerous publications on mitosis. *Mailing Add:* NC Biotechnol Ctr PO Box 13547 Research Triangle Park NC 27709-3547

MARONI, GUSTAVO PRIMO, b Merlo, Arg, Nov 20, 41; m 74. DEVELOPMENTAL GENETICS, BIOCHEMICAL GENETICS. *Educ:* Univ Buenos Aires, Lic, 67; Univ Wis, PhD(zool), 72. *Prof Exp:* Res assoc genetics, Dept Zool, Univ NC, 73-74 & Inst Genetics, Univ Cologne, 74-75; from asst prof to assoc prof, Dept Zool, 75-82, ASSOC PROF, DEPT BIOL, UNIV NC, CHAPEL HILL, 83- *Mem:* Genetics Soc Am; AAAS. *Res:* Regulation of sex-linked gene activity in Drosophila; control of alcohol dehydrogenase and other gene-enzyme systems; genetic and molecular analyses of metal-binding proteins. *Mailing Add:* Dept Biol Univ NC Chapel Hill NC 27599-3280

MARONI, VICTOR AUGUST, b Athol, Mass, Sept 8, 42; m 69; c 2. STRUCTURE CHEMISTRY. *Educ:* Worcester Polytech Inst, BS, 64; Princeton Univ, PhD(chem), 67. *Prof Exp:* SR CHEMIST & GROUP LEADER, ARGONNE NAT LAB, 67- *Honors & Awards:* Indust Res 100 Award, 85. *Res:* Chemical and structure properties of inorganic materials including ligand field theory and spectroscopic studies (Raman, nuclear magnetic resonance, electronic absorption); controlled nuclear fusion. *Mailing Add:* 908 Williamsburg Naperville IL 60540-7123

MAROTTA, CHARLES ANTHONY, b New York, NY, Apr 12, 45; m; c 2. MOLECULAR BIOLOGY, PSYCHIATRY. *Educ:* City Col New York, BS, 65; Duke Univ, MD, 69; Yale Univ, MPhil, 72, PhD(molecular biophys & biochem), 75. *Prof Exp:* NIH fel molecular biophys & internal med, Med Sch, Yale Univ, 69-73, fel molecular biophys & internal med, fel internal med, 71-72, fel med res, Clin Res Training Prog, 71-73, res fel human genetics, 72-73; clin fel psychiat, 73-76, res fel, 75-76, from instr to asst prof psychiat biochem, 76-82, ASSOC PROF NEUROSCI & PSYCHIAT, HARVARD MED SCH, 82- *Concurrent Pos:* Fel med, Yale-New Haven Hosp, 70-73; resident psychiat, Mass Gen Hosp, 73-76, asst neurochem, 76-77, clin assoc, 77-78, asst neurochem psychiat, 78-84, dir neurobiol lab & chief biomed studies psychiat, 85- & psychobiologist; William F Milton Fund award, 75, Ethel B Dupont award, 75-76 & Mellon Fac award, Harvard Med Sch, 76-77; asst neurochem, McLean Hosp, 77, asst biochem, 77-81, assoc biochem & dir neurobiol lab, 81-; res career develop award, Nat Inst Med, 80-85; Mc Arthur Found award, 81-84; Sardoz Gerontol Found Award, 88. *Honors & Awards:* Physician Recognition Award, AMA, 72; McKnight Found Award, 77 & 85; Wood Kalb Found Award, 85. *Mem:* AAAS; Soc Neurosci; Am Soc Neurochem; Am Soc Biol Chemists; Int Soc Neurochem. *Res:* Molecular genetics; molecular psychobiology; neurochemistry; neurobiology; Alzheimer's disease. *Mailing Add:* Neurobiol Lab Mass Gen Hosp Bulfinch Four Boston MA 02114

MAROTTA, CHARLES ROCCO, b Atlantic City, NJ, Sept 20, 25; m 55; c 4. NUCLEAR ENGINEERING, MATHEMATICS. *Educ:* St John's Univ, BS, 50; NY Univ, PhD(physics), 55. *Prof Exp:* Instr physics, St Johns Univ, NY, 50-52; scientist reactor physics, Walter Kidde Nuclear Labs, NY, 52-56; sr nuclear eng reactor eng, Am Machine & Foundry, 56-58; sr reactor physicist reactor physics, Nuclear Develop Asn, 58-64; staff physicist physics, Union Carbide Res Inst, 64-69; supv nuclear engr physics, Burns & Roe Archit Eng, 69-71; sr scientist physics, Nuclear Regulatory Comn, 71-88; CONSULT, SANDIA NAT LABS, 88- *Mem:* Am Nuclear Soc; Am Phys Soc. *Res:* Nuclear safety, reactor physics, radiation shielding, mathematical physics and transport theory. *Mailing Add:* 1504 Columbia Ave Rockville MD 20850

MAROTTA, SABATH FRED, b Chicago, Ill, Aug 26, 29. PHYSIOLOGY. *Educ:* Loyola Univ, BS, 51; Univ Ill-Urbana, MS, 53, PhD(physiol), 57. *Prof Exp:* Res assoc animal sci, 57-58, from instr to assoc prof physiol, Col Med, 58-70, res assoc, Aeromed Lab, Med Ctr, 58-60, asst dir, 60-64, assoc dean, Grad Col, 75, assoc dir, Res Resources Ctr, 75-80, PROF PHYSIOL, UNIV ILL, CHICAGO, 70-, DIR, RES RESOURCES CTR, 80- *Concurrent Pos:* Univ Ill adv, Chiengmai Proj, Thailand, 64-66. *Mem:* Aerospace Med Asn; Am Physiol Soc; Soc Exp Biol & Med. *Res:* Neuroendocrinology; biologic rhythms; role of the adrenal cortex in the adaption to environmental stresses. *Mailing Add:* Res Resources Ctr Univ Ill-Chicago m/c 937 PO Box 6998 Chicago IL 60680

MAROULIS, PETER JAMES, b Norfolk, Va, Mar 27, 51. CHROMATOGRAPHY, MASS SPECTROMETRY. *Educ:* Old Dominion Univ, BS, 73; Drexel Univ, MS, 77, PhD(anal chem), 80. *Prof Exp:* Res assoc anal chem & gen chem, Drexel Univ, 80-; AT AIR PROD-CHEM INC, ALLENTOWN, PA. *Mem:* Sigma Xi; Am Chem Soc; Am Geophys Union; AAAS. *Res:* Design and construct instrumentation to measure background levels of sulfur, nitrogen and hydrocarbon gases in the atmosphere, and to interpret the data from a global atmospheric sciences standpoint. *Mailing Add:* Air Prod-Chem Inc PO Box 538 Allentown PA 18105

MAROUSKY, FRANCIS JOHN, b Shenandoah, Pa, Oct 28, 35; m 59; c 5. HORTICULTURE. *Educ:* Pa State Univ, BS, 57; Univ Md, MS, 64; Va Polytech Inst, PhD(hort), 67. *Prof Exp:* Res asst, Univ Md, 62-64; instr, Va Polytech Inst, 64-67; res horticulturist, Agr Res & Educ Ctr, Sci & Educ Admin-Agr Res, USDA, Univ Fla, 67-79; res horticultures & dir, Europ Mkt Res Lab, Rotterdam, Neth, 80-82; res leader, physiol & handling hort crops unit, Agr Res Serv, USDA, Gainesville, Fla, 82-90; PROF, ENVIRON HORT, 90- *Mem:* fel Am Soc Hort Sci; Int Soc Hort Sci; Tissue Cult Asn. *Res:* Market quality and post harvest aspects and senescence of horticultural crops. *Mailing Add:* Dept Environ Hort Univ Fla Gainesville FL 32611

MAROV, GASPAR J, b Unije, Yugoslavia, Jan 3, 20; US citizen; m 46. SUGAR CHEMISTRY. *Educ:* City Col New York, BS, 42; Columbia Univ, MA, 50. *Prof Exp:* Asst food chem, Columbia Univ, 49-50; anal chemist, Thomas J Lipton, Inc, NJ, 50-51; res chemist, Pepsi-Cola Co, 51-57, chief control chemist, 57-76; mgr qual assurance, 76-81, prin scientist, Pepsi Co Inc, 81-85; CONSULT, 85- *Mem:* Am Chem Soc; Sugar Indust Technologists; Soc Soft Drink Technologists; Am Water Works Asn. *Res:* Determination of solids in sugar solutions, syrups and carbonated beverages; measurement of sugar color. *Mailing Add:* 6412 214th St Bayside NY 11364

MARPLE, DENNIS NEIL, b Storm Lake, Iowa, Oct 31, 45; m 66; c 2. ANIMAL PHYSIOLOGY, ANIMAL SCIENCE. *Educ:* Iowa State Univ, BS, 67, MS, 68; Purdue Univ, PhD(physiol), 71. *Prof Exp:* NIH fel swine physiol, Meat & Animal Sci Dept, Univ Wis, 71-73; from asst prof to assoc prof, 73-83, PROF ANIMAL & DAIRY SCI, AUBURN UNIV, 83- *Mem:* Am Soc Animal Sci; Am Meat Sci Asn; Sigma Xi; Endocrine Soc; Am Physiol Soc. *Res:* Endocrinological interactions and their regulation of growth in meat animals; study of animal physiology and meat quality. *Mailing Add:* Dept Animal & Dairy Sci Auburn Univ Auburn AL 36830

MARPLE, DUDLEY TYNG FISHER, b Portland, Ore, Sept 18, 27. PHYSICS. *Educ:* Kenyon Col, AB, 48; Syracuse Univ, MS, 50, PhD(physics), 54. *Prof Exp:* Physicist, Gen Elec Res & Develop Ctr, 54-87; RETIRED. *Mem:* Optical Soc Am; Am Phys Soc. *Res:* Optical properties of solids; energy transmission in optical fibers. *Mailing Add:* 1135 Earl Ave Schenectady NY 12309

MARPLE, STANLEY, JR, b Camden, NJ, Feb 4, 20; m 44; c 2. CHEMICAL ENGINEERING, PHYSICAL CHEMISTRY. *Educ:* Mass Inst Technol, SB, 41, PhD(phys chem), 43. *Prof Exp:* Sr res chemist, refining processes, Houston Res Lab, Shell Oil Co, 43-47, sr technologist, res admin, Head Office, NY, 47-48, group leader, Houston Res Lab, 48-56, asst chief res chem, 56-57, vis scientist, Royal Dutch-Shell Lab, Amsterdam, 57-58, asst chief res chem, Houston, 58-64, engr, Head Office, 64-74, tech mgr refining & chem processes & head office, Eng Dept, 74-85; CONSULT, CHEM ENG, 85- *Concurrent Pos:* Mem tech comt Fractionation Res Inc, 73-; chmn tech comts, Fraction Res, Inc, 79-84. *Mem:* Am Chem Soc; Am Inst Chem Eng. *Res:* Separation processes; distillation tower equipment performance; lubricating oil manufacture; wax products; vapor pressure and PVT of hydrocarbons; energy recovery in separation processes. *Mailing Add:* 810 Soboda Ct Houston TX 77079-4508

MARPLE, STANLEY LAWRENCE, JR, b Tulsa, Okla, Sept 7, 47; m 74; c 3. DIGITAL SIGNAL PROCESSING. *Educ:* Rice Univ, BA, 69, MEE, 70; Stanford Univ, Eng D(elec eng), 76. *Prof Exp:* Mem tech staff, Argosysts, Inc, 71-77; sr mem tech staff, Advent Systs, Inc, 77-79 & Anal Sci Corp, McLean, Va, 80-82; sr scientist, Schlumberger Well Serv, Houston, 83-85; prin investr, Martin Marietta Aero & Naval Systs, Baltimore, 85-89; CHIEF SCIENTIST, ORINCON CORP, SAN DIEGO, 89- *Concurrent Pos:* Assoc ed, Inst Elec & Electronic Engrs Trans Acoust, Speech & Signal Processing, 83-85; instr, Continuing Educ Inst-Europe, Sweden, 85- *Mem:* Fel Inst Elec & Electronic Engrs; sr mem Inst Elec & Electronic Engrs, Acoust Speech & Signal Processing Soc. *Res:* Digital signal processing algorithms, software, and hardware for temporal, spectral, frequency, and spatial domains in sonar, radar, telecommunications, and acoustic well logging. *Mailing Add:* 9891 Broken Land Pkwy Suite 300 Columbia MD 21046

MARPLE, VIRGIL ALAN, b Wendel, Minn, Aug 16, 39; m 62; c 3. MECHANICAL ENGINEERING. *Educ:* Univ Minn, Minneapolis, BME, 62, PhD(mech eng), 70; Univ Southern Calif, MSME, 65. *Prof Exp:* Engr, Aeronutronic Div, Ford Motor Co, 62-65; sr engr, Fluidyne Eng Corp, 65-67; res asst mech eng, 67-70, res assoc, 70-71, from asst prof to assoc prof, 71-83, PROF MECH ENG, UNIV MINN, MINNEAPOLIS, 84-, DIR, MECH ENG CO-OP PROG, 84- *Mem:* AAAS; Am Soc Mech Engrs; Am Indust Hyg Asn. *Res:* Particle technology and aerosol physics; application to particulate air pollution and industrial problems; development of instruments for characterizing the physical properties of aerosol particles. *Mailing Add:* 7949 County Rd 11 Maple Plain MN 55359

MARQUARDT, CHARLES L(AWRENCE), b Chicago, Ill, Dec 12, 36; m 63; c 3. EXPERIMENTAL SOLID STATE PHYSICS. *Educ:* DePaul Univ, BS, 60, MS, 63; Cath Univ Am, PhD, 72. *Prof Exp:* Gen physicist, 63-65, SOLID STATE RES PHYSICIST, US NAVAL RES LAB, 65- *Mem:* Optical Soc Am; Am Phys Soc. *Res:* Electromagnetic theory; ionic transport phenomena; radiation defects in solids; lunar sample analysis; laser damage in semiconductors; photochromic glasses; infrared laser materials. *Mailing Add:* Code 6551 US Naval Res Lab Washington DC 20375

MARQUARDT, DIANA, b LaCrosse, Wis, Oct 4, 54; m 84. IMMUNOLOGY. *Educ:* Wash Univ, St Louis, MD, 79. *Prof Exp:* Asst prof med, 84-90, ASSOC PROF MED, UNIV CALIF, SAN DIEGO, 90- *Mem:* Am Fedn Clin Res; Am Asn Immunologists; Am Acad Allergy & Immunol. *Mailing Add:* Med Ctr Univ Calif 225 Dickinson St San Diego CA 92103

MARQUARDT, DONALD WESLEY, b New York, NY, Mar 13, 29; m 52; c 1. STATISTICS, QUALITY TECHNOLOGY. *Educ:* Columbia Univ, AB, 50; Univ Del, MA, 56. *Prof Exp:* Res engr & mathematician, Exp Sta, E I du Pont de Nemours & Co Inc, 53-57, res proj engr & sr mathematician, 57-64, consult supvr, 64-72, field mgr, 70-73, consult mgr, 72-89, MGR, QUAL MGT & TECHNOL CTR, E I DU PONT DE NEMOURS & CO INC, 89- *Concurrent Pos:* Assoc ed, Technometrics, 74-80; mem, Comt Qual Assurance, Am Nat Standards Inst, 78-, chmn, 83 & 84, dir, 84-; Am Nat Standards Inst rep to Tech Comt Appln Statist Methods, Int Orgn Standards, 79-, Tech Comt Qual Assurance & Qual Mgt, 80-, chmn subcomt statist qual control, 81-83; mem eval panel appl math for Nat Bur Standards, Nat Res Coun, 81-84; mem, Comt Pres Statist Soc, 85-87, Conf Bd Math Sci, 86; chmn, US Tech Adv Group & Head US Deleg, 89-, Ad Hoc Task Force, ISO 9000 series, Vision 2000, 89-90; oversight nomt, Math Sci in Yr 2000; Math Sci & Math Sci Educ Bd, Nat Res Coun; sr examr, Malcolm Baldrige Nat Qual Award, 88; indust rels comt, Oh State Univ, Dept Statist, 87-; bd dirs, Regist Accreditation Bd, 90- *Honors & Awards:* Youden Prize, Am Soc Qual Control, 74, Shewell Prize, 84, Shewhart Medalist, 87. *Mem:* Fel Am Statist Asn (pres-elect, 85, pres, 86-87); fel Am Soc Qual Control; Soc Indust & Appl Math; Sigma Xi; Asn Comput Mach; fel AAAS; Am Inst Chem Engrs; Int Statist Inst. *Res:* Statistics of nonlinear models; biased estimation; strategy of experimentation; smooth regression; mixture models and experiments; computer algorithms; applications in engineering, physical and biological sciences; time series analysis of unequally spaced data; statistical graphics; statistical process control. *Mailing Add:* 1415 Athens Rd Wilmington DE 19803

MARQUARDT, HANS WILHELM JOE, b Berlin, Ger, Aug 28, 38; m 74. PHARMACOLOGY, CANCER. *Educ:* Univ Cologne, MD, 64. *Prof Exp:* Instr pharmacol, Univ Cologne, 64-68; vis investr, Div Pharmacol, Sloan-Kettering Inst Cancer Res, 68-70; vis investr cancer res, McArdle Lab Cancer Res, Univ Wis-Madison, 70-71; ASSOC, 71-74, ASSOC MEM, SLOAN-KETTERING INST CANDER RES, 74- *Concurrent Pos:* Asst prof pharmacol, Grad Sch Med Sci, Cornell Univ, 71-74, assoc prof, 75-; NIH-USPHS res career develop award, 75. *Mem:* Cell Kinetics Soc; Soc Toxicol; Europ Asn Cancer Res; Am Soc Pharmacol & Exp Therapeut; Am Asn Cancer Res. *Res:* Pharmacology and toxicology of antitumor agents and chemical carcinogens; chemical carcinogenesis and mutagenesis in tissue culture. *Mailing Add:* Dept Toxicol Univ Hamburg Med Sch Grindelallee 117 D-2000 Hamburg 13 Germany

MARQUARDT, ROLAND PAUL, b Tulare, SDak, Sept 2, 13. ANALYTICAL CHEMISTRY, ORGANIC CHEMISTRY. *Educ:* Huron Col, SDak, AB, 35; Univ SDak, AM, 36. *Prof Exp:* Chemist, 39-58, ANAL RES SPECIALIST CHEM, DOW CHEM CO, 58- *Mem:* Fel Am Inst Chemists; Am Chem Soc; Sigma Xi. *Res:* Industrial analytical research; analytical method development, including fundamental methods for unsaturation in organic compounds and for residue analysis of pesticides, especially phenoxy acid herbicides. *Mailing Add:* 1212 Baldwin St Midland MI 48640

MARQUARDT, RONALD RALPH, b Bassano, Alta, May 24, 35; m 59; c 4. BIOCHEMISTRY, AVIAN NUTRITION. *Educ:* Univ Sask, BSA, 58; Univ Alta, MSc, 61; Wash State Univ, PhD(animal sci), 65. *Prof Exp:* Asst dist agriculturist, Alta Dept Agr, 58-59; res asst animal sci, Univ Alta, 59-61; res asst, Wash State Univ, 61-65, res assoc biochem, 65-67; assoc prof, 67-77, PROF ANIMAL SCI, UNIV MAN, 77- *Mem:* Am Chem Soc; Can Biochem Soc; Can Nutrit Soc; Can Soc Animal Sci; Am Inst Nutrit; Am Soc Animal Sci. *Res:* Isolation and characterization of mode of action of antimetabolites in rye grain (water-soluble pentosans) and in fababeans (vicine and convicine); mycotoxicosis (ochratoxin and sterigmatocystin) in stored cereal grains. *Mailing Add:* Dept Animal Sci Univ Man Winnipeg MB R3T 2N2 Can

MARQUARDT, WARREN WILLIAM, b Erhard, Minn; m 55; c 4. AVIAN DESEASE, VIROLOGY-IMMUNOLOGY. *Educ:* Univ Minn, BS, 59, DVM, 61, PhD(vet microbiol & biochem), 70. *Prof Exp:* Res fel vet microbiol, Col Vet Med, Univ Minn, 62-65, instr, 65-69; assoc prof, 69-80, prof, dept vet sci, 80-83, PROF VET MICROBIOL, COL VET MED, UNIV MD, 83- *Mem:* Am Vet Med Asn; Am Soc Microbiol; Am Asn Avian Pathologists; World Vet Poultry Asn; Sigma Xi. *Res:* Avian diseases; antigenic assessment of avian viruses using hybridoma technology; monoclonal antibodies; flock health profiling by enzyme linked immunosorbent assay; rapid diagnostics. *Mailing Add:* 7701 Sondra Ct New Carrollton MD 20784

MARQUARDT, WILLIAM CHARLES, b Ft Wayne, Ind, Oct 9, 24; m 48; c 3. PROTOZOOLOGY, PARASITOLOGY. *Educ:* Northwestern Univ, BS, 48; Univ Ill, MS, 50, PhD(zool), 54. *Prof Exp:* Asst, Col Vet Med, Univ Ill, 52-54; from asst prof to assoc prof parasitol, Mont State Col, 54-61; assoc prof biol, DePaul Univ, 61-62; assoc prof parasitol, Univ Ill, 62-66; PROF ZOOL, COLO STATE UNIV, 66- *Concurrent Pos:* Consult, Thorne Ecol Inst, Colo, 72-76; pres, Rocky Mountain Conf Parasitologist, 73. *Mem:* Am Soc Parasitologists; Soc Protozoologists (asst treas, 67-70, pres, 74); Am Soc Zoologists; Am Soc Trop Med & Hyg; Sigma Xi. *Res:* Transmission and host-parasite relationships in parasitic protozoa and helminths. *Mailing Add:* Dept Biol Colo State Univ Ft Collins CO 80523

MARQUART, JOHN R, b Benton Harbor, Mich, Feb 3, 33; c 1. ANALYTICAL CHEMISTRY, PHYSICAL CHEMISTRY. *Educ:* Univ Ariz, BS, 55; Univ Ill, MS, 61, PhD(phys chem), 63. *Prof Exp:* Chem test officer, Dugway Proving Ground, US Army Chem Corps, Utah, 55-58; res assoc physics, Argonne Nat Lab, 62; chemist, Shell Develop Co, 63-68; from assoc prof to prof chem, Mercer Univ, 68-80; assoc prof, 79-84, PROF, CHEM DEPT, EASTERN ILL UNIV, 84-, COORDR, COOP EDUC CHEM, 82- *Concurrent Pos:* res assoc, NSF, 70, 74 & 80; vis prof & sr res assoc, Univ Ill, Champaign-Urbana, 72 & 74-90; vis prof, Univ Ariz, Tucson, 90-91. *Honors & Awards:* Merck Award, 55. *Mem:* Am Chem Soc; Sigma Xi. *Res:* Mass spectroscopy; thermodynamics; solution behavior; high temperature gas phase kinetics; atomic spectroscopy. *Mailing Add:* Chem Dept Eastern Ill Univ Charleston IL 61920

MARQUART, RONALD GARY, b Winnipeg, Man, Apr 22, 38; div. COMMUNICATIONS SCIENCE, CHEMICAL INFORMATION SYSTEMS. *Educ:* Univ Man, BSc, 59; Purdue Univ, MS, 62, PhD(eng), 64. *Prof Exp:* Systs engr, Can Aviation Electronics, Ltd, Man, 59-61; res asst digital coding, Sch Elec Eng, Purdue Univ, 63-64; sr engr surface div, Westinghouse Defense & Space Ctr, 64-68; PRES, FEIN-MARQUART ASSOCS, INC, 68- *Mem:* Asn Comput Mach; Inst Elec & Electronics Engrs; Sigma Xi. *Res:* Computer software systems analysis and design; data management techniques; interactive methods and systems; real-time data collection and analysis systems; chemical information storage and retrieval systems; pattern recognition applications. *Mailing Add:* 110 Beech View Ct Towson MD 21204

MARQUEZ, ERNEST DOMINGO, biological chemistry, for more information see previous edition

MARQUEZ, JOSEPH A, b New York, NY, Nov 5, 30; m 53; c 4. BIOCHEMISTRY. *Educ:* City Col New York, BS, 57; Fairleigh Dickinson Univ, MAS, 70. *Prof Exp:* Lab asst steroid identification & isolation, Columbia Univ, 54-56; lab asst natural prod isolation, 56-58, from res asst to res assoc antibiotic isolation, 58-67, res scientist, 67-68, sr scientist, 68-75, mgr antibiotic dept, 75-78, ASSOC DIR MICROBIOL, SCHERING CORP, 78- *Mem:* AAAS; Am Chem Soc; Am Soc Microbiol; NY Acad Sci; Am Inst Biol Sci. *Mailing Add:* Schering Corp 60 Orange St Bloomfield NJ 07003

MARQUEZ, VICTOR ESTEBAN, b Caracas, Venezuela, Aug 7, 43; m 66; c 4. NUCLEOSIDES, NUCLEOTIDES. *Educ:* Cent Univ Venezuela, BS, 66; Univ Mich, MS, 68, PhD(chem), 70. *Prof Exp:* Fel chem, Nat Cancer Inst, NIH, 70-71; res dir chem, Labs Cosmos, Caracas, 72-77; VIS SCIENTIST CHEM, NAT CANCER INST, NIH, 77- *Mem:* Am Chem Soc; AAAS; Am Asn Cancer Res. *Res:* Design and synthesis of nucleosides as anticancer agents with a specific site of action such as enzyme inhibitors of cytidine deaminase, or cytidine triphosphate synthase. *Mailing Add:* 20020 Dolittle St Gaithersburg MD 20879-1314

MARQUIS, DAVID ALAN, b Pittsburgh, Pa, Jan 16, 34; m 54; c 3. FOREST ECOLOGY. *Educ:* Pa State Univ, BS, 55; Yale Univ, MF, 63, PhD(forest ecol), 73. *Prof Exp:* Res forester, forest ecol, Laconia, NH, 57-65, staff asst forest mgt, Upper Darby, Pa, 65-69, RES PROJ LEADER FOREST ECOL, NORTHEASTERN FOREST EXP STA, WARREN, PA, 70- *Concurrent Pos:* Vis prof, State Univ NY-Syracuse, 80; adj prof, State Univ NY, Col Environ, Sci & Forestry, 78-; vpres & forestry dir, Plessey Pension Investments, Inc, 87-; coordr, NE Decision Model, Northeastern Forest Exp Sta. *Mem:* Soc Am Foresters; Int Union Forestry Res Orgn (Co-Chmn,82-86, chmn, 86-). *Res:* Conduct research on ecology and management of eastern hardwood forests,including environmental factors affecting tree regeneration, stand development, growth and yield,and response to silvicultural treatments. *Mailing Add:* Forestry Sci Lab PO Box 928 Warren PA 16365

MARQUIS, EDWARD THOMAS, b South Bend, Ind, July 10, 39; m 61; c 6. ORGANIC CHEMISTRY. *Educ:* Ind Univ, AB, 61; Univ Tex, PhD(org chem), 67. *Prof Exp:* Res chemist, 66-68, sr res chemist, Jefferson Chem Co, 68-77, sr proj chemist, 77-84, res assoc, 84-87, SR RES ASSOC, TEXACO CHEM CO, TEXACO INC, 87- *Mem:* Am Chem Soc; Sigma Xi. *Res:* Hydrocarbon oxidations and reductions; aromatic and aliphatic isocyanates and their amine precursors; reactions and synthetic use of phosgene; epoxidation of olefins; oligomerization of olefins; alkyl and alkylene carbonates. *Mailing Add:* 9004 Collinfield Dr Austin TX 78758

MARQUIS, JUDITH KATHLEEN, b Van Buren, Maine, Sept 19, 46. TOXICOLOGY, NEUROCHEMISTRY. *Educ:* Trinity Col, BS, 68; Univ Vt, PhD(physiol & biophys), 73. *Prof Exp:* Res fel, Univ Vt Med Sch, 68-71, res asst, 71-72; res assoc biochem, Tufts Univ, 72-74, res fel pharmacol, 75-77, asst prof, 76-81; ASSOC PROF PHARMACOL, BOSTON UNIV SCH MED, 81-, DIR, LAB ANALYSIS TOXICOL, 86-, ASSOC PROF NEUROL, 86- *Concurrent Pos:* Prin investr, Boston Univ Sch Med, 81-; lectr nutrit, Mass Inst Technol, 81-; consult, Toxikon, Inc, 82, Griffin Inc, 85, Becton Dickinson Immunocytometry Systs, 87-; toxicologist, Mass Pesticide Bd, 84- *Mem:* Am Soc Pharmacol & Ecp Therapeut; Soc Toxicol; Am Women Sci; Am Col Toxicologists; Sigma Xi. *Res:* Biochemistry and toxicology of cholinergic functions in the mammalian central nervous system; toxicology of cholinesterase inhibitors and insecticides; neurochemistry of aging and dementia. *Mailing Add:* Tufts Univ Sch Med Boston MA 02111

MARQUIS, MARILYN A, b Salt Lake City, Utah, Sept 26, 26; m 55; c 2. PHYSICAL CHEMISTRY. *Educ:* Univ Utah, BA, 47, PhD(chem), 51; Golden Gate Univ, Calif, MBA, 79. *Prof Exp:* Prin phys chemist, plasma physics dept & market anal group, advan proj, Aerojet-Gen Corp, Sacramento, Calif, 69-70; sr prog coordr, fusion energy & phys res div, US Dept Energy, San Francisco, dep proj mgr, Baca Geothermal Demonstration Power Plant Proj; PRIN, ALDERWOOD ASSOCS, LAFAYETTE, CALIF & ALBUQUERQUE, NMEX, 85- *Mem:* Sigma Xi. *Mailing Add:* 32 Brookdale Ct Lafayette CA 94549

MARQUIS, NORMAN RONALD, b Laconia, NH, Jan 3, 36; m 57; c 3. PHYSIOLOGY, BIOCHEMISTRY. *Educ:* Univ NH, BA, 59, MS, 60; Univ Mich, PhD(physiol), 65. *Prof Exp:* Res fel, Harvard Med Sch, 65-67, teaching fel, 66-67; group leader biochem, 67-70, prin investr, 70-77, prin res assoc, Mead Johnson Res Ctr, 77-79; ASSOC DIR, BRISTOL MYERS, 79- *Concurrent Pos:* Adj prof, Univ Evansville, 74-80. *Mem:* Am Soc Biol Chemists; Soc Exp Biol & Med; Am Physiol Soc; Am Heart Asn; NY Acad Sci. *Res:* Functions of carnitine in lipid metabolism; interrelationships of thrombosis, fibrinolysis and atherosclerosis; role of prostaglandins and cyclic adenosine monophosphate in platelet aggregation and thrombosis; clinical studies, cardiovascular research. *Mailing Add:* Dir Clin Studies Pharm Div Bristol Myers 2404 Penn St Evansville IN 47721-0001

MARQUIS, ROBERT E, b Sarnia, Ont, Jan 21, 34; US citizen; m 57; c 3. MICROBIAL PHYSIOLOGY. *Educ:* Wayne State Univ, BS, 56; Univ Mich, MS, 58, PhD(bact), 61. *Prof Exp:* NATO fel, Univ Edinburgh, 61-62, NSF fel, 62-63; from sr instr to assoc prof, 63-78, PROF MICROBIOL, SCH MED, UNIV ROCHESTER, 78- *Concurrent Pos:* NIH fel, Scripps Inst Oceanog, Univ Calif, San Diego, 70-71. *Mem:* AAAS; Am Soc Microbiol; Undersea Med Soc; Brit Soc Gen Microbiol; Am Asn Dent Res. *Res:* Bacterial physiology; enzymology of bacterial plasma membranes; basic studies of microbial barophysiology; investigation of the physiology of bacteria in dental plaque. *Mailing Add:* Dept Microbiol/Immunol Univ of Rochester Rochester NY 14642

MARQUIT, ERWIN, b New York, NY, Aug 21, 26. PHYSICS, PHILOSOPHY OF SCIENCE. *Educ:* City Univ New York, BEE, 48; Univ Warsaw, MA, 57, DSc(math & phys sci), 63. *Prof Exp:* Res assoc high energy physics, Univ Mich, 63-65; asst prof physics, Univ Colo, 65-66; assoc prof, 66-81, PROF PHYSICS, UNIV MINN, 81- *Concurrent Pos:* Dir & treas, Marxist Educ Press, Minneapolis, 77-; ed, Nature, Soc & Thought, 87. *Mem:* Am Physics Soc; Philos Sci Asn. *Res:* Conceptual foundations of physics; philosophical and methodological problems of science; problems of dialectical materialism. *Mailing Add:* Univ Minn Minneapolis 116 Church St SE Minneapolis MN 55455-0112

MARQUSEE, JEFFREY ALAN, b New York, NY, Sept 30, 54; m 82. MACROMOLECULAR SYSTEMS, POLYMERS. *Educ:* Cornell Univ, BA, 76; Mass Inst Technol, PhD(chem), 81. *Prof Exp:* Teaching asst chem, Mass Inst Technol, 76-77, res asst, 77-81; teaching affil, Stanford Univ, 81-83; teaching assoc, Univ Calif, San Francisco, 83-84, vis asst prof chem, 84-85; RES CHEMIST, POLYMERS DIV, NAT BUR STANDARDS, 85- *Mem:* Am Phys Soc. *Res:* Statistical mechanics of macromolecular systems; structure and dynamics of synthetic and biological polymers and amphiphillic aggregates; dynamic of first order phase transitions. *Mailing Add:* 4517 W Virginia Ave Bethesda MD 20814-4611

MARR, ALLEN GERALD, b Tulsa, Okla, Apr 24, 29; m 48, 70; c 6. MICROBIOLOGY. *Educ:* Univ Okla, BS, 48, MA, 49; Univ Wis, PhD(bact), 52. *Prof Exp:* Proj assoc bact, Univ Wis, 52; instr, 52-54, from asst prof to assoc prof, 54-63, PROF BACT, UNIV CALIF, DAVIS, 63-, DEAN GRAD STUDIES & RES, 69- *Mem:* Am Soc Microbiol. *Res:* Growth and division of bacteria; microbial physiology. *Mailing Add:* Dept Microbiol Univ Calif Davis CA 95616

MARR, DAVID HENRY, b Tillsonburg, Ont, Nov 14, 38; m 58; c 2. SPECTROSCOPY. *Educ:* Univ Western Ont, BSc, 62, PhD(chem), 66. *Prof Exp:* Res fel, Ill Inst Technol, 66-67; res engr, Fairchild Camera & Instrument Corp, 67-68; res chemist, Hooker Chem Corp, 68-70, sr leader spectros, 70-72; sr res chemist, 72-74, res assoc, 74-78, MGR ANALYSIS RES, STAUFFER CHEM CO, 78- *Concurrent Pos:* Ont Grad Fel. *Honors & Awards:* Atkinson Found Award. *Mem:* Am Chem Soc. *Res:* Analytical chemistry; application of spectroscopic techniques to problem solving. *Mailing Add:* Akzo Chem Inc Dobbs Ferry Res Ctr Dobbs Ferry NY 10522

MARR, JAMES JOSEPH, b Hamilton, Ohio, Oct 21, 38; m 63; c 5. INFECTIOUS DISEASES, INTERNAL MEDICINE. *Educ:* Xavier Univ, Ohio, BS, 59; Johns Hopkins Univ, MD, 64; St Louis Univ, MS, 68; Am Bd Internal Med, dipl, 72, cert infectious dis, 74. *Prof Exp:* Am Cancer Soc fel & instr microbiol, Sch Med, St Louis Univ, 67-69; asst prof internal med, 70-75, asst prof path, 73-75, asst prof microbiol, 71-75, assoc prof internal med & path sch med, Wash Univ, 75-76; prof internal med & microbiol & dir div infectious dis, Sch Med, St Louis Univ, 76-82; prof med & biochem & dir, Div Infectious Dis, Univ Colo, 82-90; SR VPRES, SEARLE DISCOVERY RES, 89- *Concurrent Pos:* Fel trop med, USPHS-La State Univ Int Ctr Med Res & Training, Costa Rica, 72; med dir, Microbiol Labs, Barnes Hosp, St Louis, 73-76; consult, Vet Admin Hosp, St Louis, 72-, St Louis Childrens Hosp, 74-76 & Jewish Hosp, St Louis, 75-76, NIH, 82-87; chmn, Comt Chemotherapy Leishmaniasis, WHO, 82-87. *Mem:* Am Fedn Clin Res; fel Am Col Physicians; Infectious Dis Soc Am; Am Soc Clin Invest; Sigma Xi; Am Soc Biol Chemists; Fel Am Acad Microbiol; Am Asn Physicians. *Res:* Metabolic regulation in microorganisms and its relationship to the pathogenesis of intracellular infections in man. *Mailing Add:* Searle 4901 Searle Pkwy Skokie IL 60077

MARR, JOHN MAURICE, b Jefferson City, Mo, June 15, 20; m 49. MATHEMATICS. *Educ:* Cent Mo State Col, BS, 41; Univ Mo, MA, 48; Univ Tenn, PhD, 53. *Prof Exp:* Instr math, Mo Sch Mines, 46-47; from asst instr to instr, Univ Mo, 47-49; asst, Univ Tenn, 49-53; from asst prof to assoc prof, 53-62, PROF MATH, KANS STATE UNIV, 62- *Mem:* Am Math Soc; Math Asn Am. *Res:* Topology; convexity. *Mailing Add:* Dept of Math Kans State Univ Manhattan KS 66504

MARR, ROBERT B, b Quincy, Mass, Mar 25, 32; m 54; c 3. COMPUTED TOMOGRAPHY, MEDICAL IMAGING. *Educ:* Mass Inst Technol, SB, 53; Harvard Univ, MA, 55, PhD(physics), 59. *Prof Exp:* Res assoc theoret physics, 59-61, assoc physicist, 61-64, physicist, 64-69, chmn, Appl Math Dept, 75-78, SR PHYSICIST, BROOKHAVEN NAT LAB, ASSOC UNIVS, INC, 69- *Mem:* Am Math Soc; Soc Magnetic Resonance Med. *Res:* Applied mathematics; computers; applications in physical and biological sciences; theoretical physics. *Mailing Add:* Dept Appl Sci Brookhaven Nat Lab Assoc Univs Inc Upton NY 11973

MARR, WILLIAM WEI-YI, b Kwangtung, China, Sept 3, 36; US citizen; m 62; c 2. SYSTEM SIMULATIONS, MATHEMATICAL MODELING. *Educ:* Marquette Univ, MS, 63; Univ Wis, PhD(nuclear Eng), 69. *Prof Exp:* Asst engr, Taiwan Sugar Corp, 59-61; sr eng analyst, Allis-Chalmers Mfg Co, 63-67; nuclear engr, Nuclear Reactor Safety, 69-80, ENGR, ENERGY ENVIRON SYSTS, ARGONNE NAT LAB, 80- *Mem:* Am Soc Mech Engrs. *Res:* Development of computerized models for analyzing performance of electric/hybrid vehicles; oxygen-enriched diesel technology to increase efficiency and reduce emissions. *Mailing Add:* Argonne Nat Lab Argonne IL 60439

MARRA, DOROTHEA CATHERINE, b Brooklyn, NY, Jan 23, 22; m 47; c 1. SURFACE CHEMISTRY, COLLOID CHEMISTRY. *Educ:* Brooklyn Col, BA, 43. *Prof Exp:* Anal chemist, Matam Corp, 43-44; res chemist, Foster D Snell Inc, 44-69; vpres, Omar Res, Inc, New York, 69-80; vpres, Aerosol Prod Technol, Inc, 81-85; VPRES, COSTECH, INC, 85- *Mem:* AAAS; Sigma Xi; Soc Cosmetic Chemists; fel Am Inst Chemists. *Res:* Creation and development of new products, specifically in cosmetics, toiletries and pharmaceuticals. *Mailing Add:* 107 Fernwood Rd Summit NJ 07901

MARRA, MICHAEL DOMINICK, biochemistry, bioanalysis; deceased, see previous edition for last biography

MARRACK, PHILIPPA CHARLOTTE, b Ewell, Eng, June 28, 45; Brit & US citizen; m 74; c 2. IMMUNOLOGY. *Educ:* Univ Cambridge, BA, 67, PhD(biol sci), 70. *Prof Exp:* Fel immunol, Univ Calif, San Diego, 71-73; fel immunol, Univ Rochester, 71-74, assoc, 74-75, asst prof immunol, 75-79; assoc prof, 80-85, PROF, DEPT BIOPHYS, BIOCHEM & GENETICS, UNIV COLO HEALTH SCI CTR, 85, PROF, DEPT MICROBIOL & IMMUNOL, 88-; MEM, DEPT MED, NAT JEWISH HOSP & RES CTR, 79- *Concurrent Pos:* Mem, Prog Comt, Am Asn Immunologists, 77-79, Immunobiol Study Sect, 80-84 & Ad Hoc Res Briefing Panel Immunol, Nat Acad Sci, 83; vchmn, Gordon Conf Immunochem & Immunobiol, 81, chmn, 83; assoc ed, J Immunol, 81-84, sect ed, 84-87; adv ed, J Exp Med, 85-, assoc ed, Cell, 87-, rev ed, Sci, 88-90; head, Div Basic Immunol, Nat Jewish Ctr Immunol & Respiratory Med, 88-90. *Honors & Awards:* Feodor Lynen Medal, 90; Wellcome Found Lectr, Royal Soc, 90; William B Coley Award, Cancer Res Inst, 91. *Mem:* Nat Acad Sci; Am Asn Immunol; Brit Soc Immunol; Am Heart Asn. *Res:* Study of functional and maturational heterogeneity of mouse T-cells; mode of antigen recognition and action of helper T-cells. *Mailing Add:* Dept Biochem & Biophys Univ Colo Med Ctr 4200 E Ninth Ave Denver CO 80220

MARRAS, WILLIAM STEVEN, b Evanston, Ill, Dec 1, 52; m 80. ERGONOMICS, HUMAN FACTORS ENGINEERING. *Educ:* Wright State Univ, BS, 76; Wayne State Univ, MSIE, 78, PhD(bioeng), 82. *Prof Exp:* Instr probability & human factors, Wayne State Univ, 78-82; asst prof, 82-87, ASSOC PROF BIOENG & ERGONOMICS, OHIO STATE UNIV, 87- *Concurrent Pos:* Prin investr, Isometric vs Isokinetic Back Performed, NIH, 83-85, Hand Tool Analysis, US Bur Mines, 84-88, Eval of Motion Component in Lifting, Indust Comn Ohio, 86- & Nat Inst Disability & Rehab Res, 88- *Honors & Awards:* Res Award, Col Eng, Ohio State Univ, 87. *Mem:* Human Factors Soc; Am Soc Biomech; Ergonomic Soc; Int Found Indust Ergonomics & Safety Res; Inst Indust Engrs; Int Ergonomics Asn. *Res:* Biomechanical assessment and modelling of the back and spine in response to back motions experienced in occupational activities; biomechanics of hand-wrist disorders due to work. *Mailing Add:* Dept Indust & Systs Ohio State Univ Main Campus Columbus OH 43210

MARRAZZI, MARY ANN, b Ann Arbor, Mich, Dec 22, 45. NEUROPHARMACOLOGY, NEUROCHEMISTRY. *Educ:* Univ Minn, BA, 66; Wash Univ, PhD(pharmacol), 72. *Prof Exp:* NIH fel pharmacol, Sch Med, Wash Univ, 72-74; vis investr neuropharmacol, Inst Psychiat, Univ Mo, 74; asst prof, 74-78, ASSOC PROF PHARMACOL, SCH MED, WAYNE STATE UNIV, 78- *Concurrent Pos:* Assoc, Dept Psychiat, Harper Grace Hosp, 81- *Mem:* Soc Neurosci; Am Soc Pharmacol & Exp Therapeut; Am Neurochemistry Soc; Soc Biol Psychiat. *Res:* Central nervous system regulation of metabolic homeostasis and appetite; anorexia nervosa as an addiction; opioids & glucoregulation neurophysiology; neuropharmacology; relation to possible insulin central nervous system actions; energy metabolism in brain; metabolic encephalopathies; metabolism and role in nervous system; microchemical methodology. *Mailing Add:* 962 Lochmoor Grosse Pointe Woods MI 48236

MARRELLO, VINCENT, b Belsito, Italy, Apr 20, 47; Can citizen; m 72; c 1. SOLID STATE PHYSICS. *Educ:* Univ Toronto, BASc, 70; Calif Inst Technol, MS, 71, PhD(elec eng), 74. *Prof Exp:* Fel appl physics, Calif Inst Technol, 74-75; MEM RES STAFF APPL PHYSICS, IBM RES LAB, 75- *Mem:* Am Phys Soc; Sigma Xi. *Res:* Device physics; material physics; optical and electrical properties of amorphous materials. *Mailing Add:* IBM Res Lab Almaden Res 650 Harry Rd San Jose CA 95120

MARRERO, HECTOR G, b Santurce, PR, May 19, 55. SPECTROSCOPY, INSTRUMENTATION DESIGN. *Educ:* Univ PR, BS, 77; Boston Univ, MA, 81, PhD(biophysics), 86. *Prof Exp:* Res asst, Univ PR, 73-76, teaching fel, physics labs, 74-76; teaching fel, physics labs, 78-85, RES ASST, BOSTON UNIV, 70- *Res:* The properties of photosensitive biological membranes; proteins in aqueous media. *Mailing Add:* 1040 N Pleasant St Amherst MA 01002

MARRERO, THOMAS RAPHAEL, b New York, NY, May 21, 36. CHEMICAL ENGINEERING. *Educ:* Polytech Inst Brooklyn, BS, 58; Villanova Univ, MS, 59; Univ Md, College Park, PhD(chem eng), 70. *Prof Exp:* Engr, Nuclear Div, Martin-Marietta Corp, 59-64 & Res Div, WR Grace & Co, 64-65; res assoc gases, Univ Mo-Columbia, 70-71; res specialist, Res & Develop Div, Babcock & Wilcox Co, 71-73; sr engr, Gen Elec Co, 73-78; ASSOC PROF, UNIV MO-COLUMBIA, 79- *Concurrent Pos:* Vis prof, Texas A&M Univ, 78-79. *Mem:* Am Inst Chem Eng; Am Chem Soc; Sigma Xi. *Res:* Transport properties of gases; corresponding states correlations. *Mailing Add:* Univ Mo Columbia MO 65211

MARRESE, RICHARD JOHN, b New York, NY, Sept 1, 31; m 53; c 4. AGRONOMY, PLANT PHYSIOLOGY. *Educ:* Cornell Univ, BS, 53, MS, 55; Rutgers Univ, PhD(agron), 59. *Prof Exp:* Asst prof agron, Iowa State Univ, 59-60; mem staff tech serv agr, Diamond-Shamrock Corp, 60-64; biol coordr plant protection, Tenneco, 64-66; mgr field develop, Esso Agr Res Labs, 66-70; DIR FIELD DEVELOP PLANT PROTECTION, AM HOECHST CORP, 70- *Mem:* Weed Sci Soc Am; Entomol Soc Am; Am Phytopath Soc. *Res:* Discovery and development of candidate compounds in plant protection, especially herbicides. *Mailing Add:* Hoechst Roussel Agr-Vet Co Rte 202-206 PO Box 2500 Somerville NJ 08876-1258

MARRIAGE, LOWELL DEAN, b New Rockford, NDak, June 28, 23; m 49; c 3. WILDLIFE CONSERVATION. *Educ:* Ore State Univ, BS, 48. *Prof Exp:* Aquatic biologist, Fish Comn Ore, 48-56, water resources analyst, 56-60, asst dir, 60-62, regional fisheries biologist, Soil Conserv Serv, 62-71, REGIONAL BIOLOGIST, SOIL CONSERV SERV, USDA, 71- *Mem:* Am Fisheries Soc; Wildlife Soc; Am Inst Fishery Res Biol; Soil Conserv Soc Am. *Res:* Shellfish management and research; water projects effects on fisheries and wildlife populations; anadromous fisheries biology and management; water quality and wildlife habitat management in land and water development projects. *Mailing Add:* 3948 SE Wake St Portland OR 97222

MARRINER, JOHN P, b Dover-Foxcroft, Maine, Nov 15, 48; m 70; c 2. ELEMENTARY PARTICLE PHYSICS. *Educ:* Univ Calif, PhD(physics), 77. *Prof Exp:* Physicist, Lawrence Berkeley Lab, 77-78; PHYSICIST, FERMI NAT ACCELERATOR LAB, 78 - *Res:* High energy physics and accelerators. *Mailing Add:* Fermilab PO Box 500 Batavia IL 60510

MARRIOTT, HENRY JOSEPH LLEWELLYN, b Hamilton, Bermuda, June 10, 17; nat US; m 51; c 4. CARDIOLOGY. *Educ:* Oxford Univ, BA, 41, MA, 43, BM, BCh, 44. *Prof Exp:* House physician, St Mary's Hosp London, 44, resident med officer, Sir Alexander Fleming's Penicillin Res Unit, 45; resident, King Edward Hosp, Bermuda, 45-46; from asst to asst prof med, Med Sch, Univ Md, 48-53, assoc prof med & head div phys diag, 53-62, head div arthritis, 56-59; dir med educ & cardiol ctr, Tampa Gen Hosp, Fla, 62-65; dir coronary care, St Anthony's Hosp, 68-78; CLIN PROF PEDIAT, COL MED, UNIV FLA, 70-; DIR CLIN RES, ROGERS HEART FOUND, 65- *Concurrent Pos:* Fel med, Johns Hopkins Hosp, 46-47; chief EKG dept, Mercy Hosp, Baltimore, Md, 54-62; consult, Vet Admin Hosp, Bay Pines, Fla, 63-83; clin prof med, Sch Med, Emory Univ, 66- *Mem:* Am Heart Asn; fel Am Col Cardiol; fel Am Col Physicians; Brit Med Asn. *Res:* Electrocardiography and clinical cardiology. *Mailing Add:* Rogers Heart Found St Anthony's Hosp St Petersburg FL 33705

MARRIOTT, LAWRENCE FREDERICK, b Browns, Ill, Dec 18, 13; m 38; c 1. AGRONOMY, SOILS. *Educ:* Univ Ill, BS, 35; Univ Wis, MS, 53, PhD(soils), 55. *Prof Exp:* Asst soil exp fields, Univ Ill, 35-42; self employed, 46-51; asst soils, Univ Wis, 51-55; asst prof, Pa State Univ, 55-59, assoc prof soil technol, 59-77; RETIRED. *Mailing Add:* 626 E Waring Ave State Col PA 16801

MARRIOTT, RICHARD, b London, Eng, Nov 17, 29; US citizen; m 52; c 2. MATHEMATICAL PHYSICS, COMPUTER SCIENCE. *Educ:* Univ London, BSc, 54, PhD(atomic physics), 56. *Prof Exp:* Res asst, Univ Col, Univ London, 56; design engr, Atomic Energy Div, Can Westinghouse, Atomic Energy Can Ltd, 56-59; staff mem res div, Radiation Inc, 59-61; consult, Space Sci Lab, Gen Dynamics/Convair, 61-65; res fel, Inst Pure & Appl Physics, Univ Calif, San Diego, 65-67; sr res fel atomic physics, Royal Holloway Col, Univ London, 67-68; PROF CHEM ENG, WAYNE STATE UNIV, 68- *Concurrent Pos:* Consult, Space Sci Lab, Gen Dynamics/Convair, 61-66, Inst Pure & Appl Physics, Univ Calif, San Diego, 68-69, Atomic Energy Res Estab, Harwell, Eng, 68- & Phys Dynamics, Inc, 77-; sr vis fel, Dept Theoret Physics, Queen's Univ, Belfast, 69-70. *Mem:* Fel Brit Phys Soc; Sigma Xi. *Res:* Computer applications; theoretical studies of atomic structure and collision processes. *Mailing Add:* Dept Chem Eng Col Eng Wayne State Univ Detroit MI 48202

MARROCCO, RICHARD THOMAS, b Rochester, NY, July 27, 43; m 70. SENSORY PHYSIOLOGY, NEUROSCIENCES. *Educ:* Univ Calif, Los Angeles, BA, 65; Ind Univ, PhD(psychol), 71. *Prof Exp:* Res assoc, Ind Univ, 65-68; fel, 68-71, fel neurophysiol, Univ Calif, Berkeley, 72-73; asst prof, 73-79, ASSOC PROF SENSORY PHYSIOL, UNIV ORE, 79- *Mem:* Soc Neurosci; AAAS. *Res:* Physiology of color, form and binocular vision in primates. *Mailing Add:* Dept of Psychol Univ Ore Eugene OR 97403

MARRON, MICHAEL THOMAS, b Jan 31, 43; US citizen; m 66; c 2. BIOPHYSICAL CHEMISTRY, PHYSICAL CHEMISTRY. *Educ:* Univ Portland, BS, 64; Johns Hopkins Univ, MA, 65, PhD(theoret chem), 69. *Prof Exp:* Res assoc, Theoret Chem Inst, Univ Wis, 69-70; from asst prof to prof chem, Univ Wis-Parkside, 70-83, chmn dept, 76-79, dean div sci, 79-83; prog mgr bioelectromagnetics, 83-84, PROG MGR MOLECULAR BIOL, OFF NAVAL RES, 85- *Concurrent Pos:* Mem, Biomed Res Inst, Univ Wis-Parkside, 80-85. *Mem:* AAAS; Am Chem Soc; Bioelectromagnetics Soc; Biophys Soc; Am Soc Biochem & Molecular Biol. *Res:* Biophysical chemistry; biological effects of electromagnetic fields; application of computers to problems in molecular biology. *Mailing Add:* Molecular Biol Prog Off Naval Res 800 N Quincy St Arlington VA 22217-5000

MARRONE, MICHAEL JOSEPH, b Lewistown, Pa, July 19, 37; m 61; c 4. SOLID STATE PHYSICS. *Educ:* Univ Notre Dame, BS, 59; Univ Pittsburgh, MS, 61; Cath Univ Am, PhD(physics), 71. *Prof Exp:* RES PHYSICIST, US NAVAL RES LAB, 61- *Mem:* Am Phys Soc; Sigma Xi; Optical Soc Am. *Res:* Optical properties of solids; optical absorption and emission; radiation effects in solids; magneto-optics. *Mailing Add:* Code 6570 US Naval Res Lab Washington DC 20375

MARRONE, PAMELA GAIL, b Middletown, Conn, Oct 14, 56. MICROBIAL PESTICIDES, INTEGRATED PEST MANAGEMENT. *Educ:* Cornell Univ, BS, 78; NC State Univ, PhD(entom), 83. *Prof Exp:* Sr res biologist insect biol, Monsanto Agr Co, 83-84, group leader, 84-88, sr group leader insect control, 88-90; PRES, SUBSID NOVONORDISK, ENTOTECH, INC, 90- *Concurrent Pos:* Chmn, B T Mgt Working Group, 88-91; pres, Monsanto Chap, Sigma Xi, 89-90; bd dirs, Davis Regional Sci Ctr, 91- & Calif Indust Biotechnol Asn, 91-; dean's adv coun, Agr & Environ Sci, Univ Calif, 91-; mem, Calif Chamber of Com, 91- *Mem:* Entom Soc Am; Am Chem Soc; Soc Invert Path; AAAS; Am Mgt Asn; Indust Biotechnol Asn. *Res:* Biological pesticides; biotechnology to resolve insect control problems. *Mailing Add:* Entotech Inc 1497 Drew Ave Davis CA 95616

MARRONE, PAUL VINCENT, b Niagara Falls, NY, Mar 14, 32; m 55; c 5. GAS DYNAMICS, FLUID PHYSICS. *Educ:* Univ Notre Dame, BS, 54; Princeton Univ, MS, 56; Univ Toronto, PhD(aero physics), 66. *Prof Exp:* Res scientist gas dynamics, 56-63, prin res scientist, fluid physics, Cornell Aeronaut Lab, 66-76; prog mgr fluid physics, Calspan Corp, 76-78, asst dept head, Aerodyn Res Dept, 78-80, dept head, 80-85, VPRES EXP RES DIV, CALSPAN ADVAN TECHNOL CTR, 85- *Concurrent Pos:* adj prof, State Univ NY, Buffalo. *Mem:* Am Inst Aeronaut & Astronaut; Sigma Xi. *Res:* High temperature gas dynamics; molecular radiation; chemical kinetics; thermodynamics; computational modeling; experiment methods in gas dynamics. *Mailing Add:* Calspan Adv Technol Ctr PO Box 400 Buffalo NY 14226

MARRS, BARRY LEE, b Newark, NJ, Sept 23, 42; m 66; c 2. MICROBIOL GENETICS, PHOTOSYNTHESIS. *Educ:* Williams Col, BA, 63; Western Reserve Univ, PhD(biol), 68. *Prof Exp:* NSF fel, Univ Ill, Urbana, 67-69; Am Cancer Soc fel, Stanford Univ, 69-71; res assoc microbiol, Ind Univ, Bloomington, 71-72; from asst prof to assoc prof biochem, Sch Med, St Louis Univ, 72-78, prof, 78-83; sr res assoc, Corp Res Lab, Exxon Res & Eng Co, Clinton, NJ, 83-85; res mgr, 85-90, DIR LIFE SCI, CENT RES & DEVELOP, DUPONT, 90- *Mem:* Am Soc Microbiol; Am Soc Biol Chemists. *Res:* Regulation of membrane formation and genetics of photosynthetic bacteria; structure and function of photosynthetic bacterial membranes. *Mailing Add:* Cent Res & Develop E 173/118 Du Pont de Nemours & Co Wilmington DE 19880-0173

MARRS, ROSCOE EARL, b Schenectady, NY, Oct 21, 46; m 74; c 5. ELECTRON-ION COLLISIONS. *Educ:* Cornell Univ, AB, 68; Univ Wash, MS, 69, PhD(physics), 75. *Prof Exp:* Res fel physics, Calif Inst Technol, 75-76; sr res assoc, Ind Univ Cyclotron Facil, 76-78; res assoc, Triumf, Univ BC, 78-79; res asst prof, Univ Wash, 79-80; MEM STAFF, LAWRENCE LIVERMORE NAT LAB, 80- *Honors & Awards:* IR 100 Award, 87. *Mem:* Am Phys Soc; AAAS. *Res:* Experimental studies of highly charged ions, especially in high temperature plasmas and electron beam ion scources; development of the electron beam ion trap for highly charged ions. *Mailing Add:* Lawrence Livermore Nat Lab PO Box 808 L-296 Livermore CA 94550

MARRUS, RICHARD, b Brooklyn, NY, Sept 14, 32. ATOMIC PHYSICS, NUCLEAR PHYSICS. *Educ:* NY Univ, BS, 54; Univ Calif, Berkeley, MA, 56, PhD(physics), 59. *Prof Exp:* Asst, Univ, 54-56, res physicist, Lawrence Radiation Lab, 56-66, assoc prof, Univ, 66-71, PROF PHYSICS, UNIV CALIF, BERKELEY, 71- *Concurrent Pos:* Guggenheim fel, 70-71. *Mem:* Am Phys Soc. *Res:* Atomic beam magnetic resonance spectroscopy; optical pumping. *Mailing Add:* Dept of Physics Univ of Calif 2120 Oxford St Berkeley CA 94720

MARSAGLIA, GEORGE, mathematics, computer science, for more information see previous edition

MARSCHALL, ALBERT RHOADES, b May 5, 21; m; c 5. ENGINEERING. *Educ:* Rensselaer Polytech Inst, BCE, 48, MCE, 48; US Naval Acad, BS, 44. *Prof Exp:* Vpres, George Hyman Construct Co, Bethesda, 77-79; comr pub bldg, Gen Serv Admin, 79-81; CONSULT, 81- *Mem:* Hon mem Am Inst Arch; hon mem Am Pub Works Asn; Soc Am Mil Engrs. *Res:* Construction; energy management; environmental planning. *Mailing Add:* 807 Vassar Rd Alexandria VA 22314

MARSCHALL, CHARLES W(ALTER), b Rosendale, Wis, Sept 25, 30; m 52; c 5. METALLURGY. *Educ:* Univ Wis, BS, 53, MS, 57; Case Inst Technol, PhD(metall), 60. *Prof Exp:* Res engr, Gen Motors Corp, 53; instr metall, Univ Wis, 55-57; asst, Case Inst Technol, 57-59; sr metallurgist, Battelle Mem Inst, 59-65, assoc chief mech metall, 65-66; asst prof mat, Univ Wis, Milwaukee, 66-67; assoc chief mech metall div, 67-71, assoc chief deformation & fracture res div, 71-74, prin scientist, 74-85, SR RES SCIENTIST, BATTELLE COLUMBUS LABS, 85- *Concurrent Pos:* Battelle Seattle Res Ctr fel, 73-74. *Mem:* Am Inst Mining, Metall & Petrol Engrs; Am Soc Metals; Am Soc Testing & Mat. *Res:* Physical metallurgy of steels; controlled-modulus alloy development; stress-corrosion cracking; distortion and residual stress; mechanical behavior of materials; dimensional stability and micromechanical properties of materials for precision applications; fracture resistance of nuclear reactor components; radiation effects on fracture resistance; dynamic strain aging embrittlement; failure analysis; heavy-armor materials. *Mailing Add:* Battelle Columbus Labs Metals & Ceramics Dept 505 King Ave Columbus OH 43201-2693

MARSCHER, WILLIAM DONNELLY, b Utica, NY, Feb 18, 48; m 72; c 3. FINITE ELEMENT ANALYSIS, ROTOR DYNAMIC ANALYSIS. *Educ:* Cornell Univ, BSME, 70, ME, 72; Rensselaer Polytech Inst, MS, 76. *Prof Exp:* Engr, Bendix Electronics Fuel Injection Div, 70-73; sr engr, Pratt & Whitney Aircraft, United Technologies, 73-78; proj supvr, Creare Inc, 78-82; ENG DIR MECH, DRESSER PUMP DIV, DRESSER INDUSTS, 82- *Concurrent Pos:* NASA fel, Cornell Univ, 72; instr, Lebanon Community Col, 80-81; lectr, Concepts ETI Inc, 81-, Univ Va, 86- & Tex A&M Univ, 87-89; mem bd dirs, Soc Tribologists & Lubrication Engrs. *Honors & Awards:* ASLE Hudson Award. *Mem:* Soc Tribologists & Lubrication Engrs; Am Soc Testing & Mat; Am Soc Mech Engrs; Soc Exp Mech. *Res:* Stress analysis, vibrations and rotor dynamics of pumps and compressors; finite element analysis of large problems; experimental modal analysis of rotating components; rotor dynamic analysis; computer aided design; experimental modal analysis. *Mailing Add:* 10 Grace Way Morristown NJ 07960

MARSDEN, BRIAN GEOFFREY, b Cambridge, Eng, Aug 5, 37; m 64; c 2. CELESTIAL MECHANICS, PLANETARY SCIENCES. *Educ:* Oxford Univ, BA, 59, MA, 63; Yale Univ, PhD(astron), 66. *Prof Exp:* Res asst astron, Yale Univ Observ, 59-65; astronr, Smithsonian Astrophys Observ, 65-86; ASSOC DIR PLANETARY SCI, HARVARD-SMITHSONIAN CTR ASTROPHYS, 87- *Concurrent Pos:* Lectr, Harvard Univ, 66-; dir cent bur astron telegrams, Int Astron Union, 68-; dir minor planet ctr, Int Astron

Union, 78- *Honors & Awards:* Merlin Medal, Brit Astron Asn, 65; Goodacre Medal, Brit Astron Assn, 79. *Mem:* Am Astron Soc; Royal Astron Soc; Sigma Xi; Brit Astron Asn; Int Astron Union. *Res:* Orbits of comets, minor planets and natural satellites; celestial mechanics; astrometry; physics of comets. *Mailing Add:* Smithsonian Astrophys Observ 60 Garden St Cambridge MA 02138

MARSDEN, D(AVID) J(OHN), b Provost, Alta, Sept 23, 33; m 59; c 4. AERONAUTICAL ENGINEERING, AERODYNAMICS. *Educ:* Univ Alta, BSc, 55; Col Aeronaut Eng, dipl, 57; Univ Toronto, PhD(rarefied gas dynamics), 65. *Prof Exp:* Asst res officer aerodyn, Nat Res Coun Can, 57-60; res engr, DeHavilland Aircraft of Can Ltd, 64-65; assoc prof mech eng, 65-75, PROF MECH ENG, UNIV ALTA, 75- *Honors & Awards:* FAI Tissandier Dipl, Sport Aviation & Sailplane Aerodynamics, 86. *Mem:* Can Aeronaut & Space Inst. *Res:* Experimental low speed aerodynamics; aerodynamics of sailplanes; aerofoil theory and development. *Mailing Add:* Dept Mech Eng Univ Alberta Edmonton AB T6G 2M7 Can

MARSDEN, HALSEY M, b July 25, 33; US citizen; m 62; c 2. ZOOLOGY, RESEARCH ADMINISTRATION. *Educ:* Univ Conn, BS, 55; Univ Mo, MA, 57, PhD(zool), 63. *Prof Exp:* NIH fel animal behav reproduction, Jackson Lab, 63-65; res biologist, Primate Ecol Sect, Nat Inst Neurol Dis & Stroke, 65-69 & Behav Systs Sect, Lab Brain Evolution & Behav, NIMH, 69-72; res biologist, 72-77, HEALTH SCI ADMINR, DEVELOP NEUROL BR, NAT INST NEUROL & COMMUNICATIVE DIS & STROKE, 77- *Mem:* AAAS; Int Soc Res Aggression. *Res:* Ecology; animal behavior; mammalian reproduction; behavioral-environmental systems; child and human development; primatology. *Mailing Add:* NIH Federal 816 9000 Rockville Pike Bethesda MD 20814

MARSDEN, JERROLD ELDON, b Ocean Falls, BC, Aug 17, 42; m 65; c 1. MATHEMATICS. *Educ:* Univ Toronto, BSc, 65; Princeton Univ, PhD(math), 68. *Prof Exp:* Instr math, Princeton Univ, 68; lectr, 68-69, asst prof, 69-72, assoc prof, 72-77, PROF MATH, UNIV CALIF, BERKELEY, 77- *Concurrent Pos:* Asst prof, Univ Toronto, 70-71. *Mem:* Am Phys Soc. *Res:* Mathematical physics; global analysis; hydrodynamics; quantum mechanics; nonlinear Hamiltonian systems. *Mailing Add:* Dept Math Evans Hall Univ of Calif 2120 Oxford St Berkeley CA 94720

MARSDEN, RALPH WALTER, geology; deceased, see previous edition for last biography

MARSDEN, SULLIVAN S(AMUEL), JR, b St Louis, Mo, June 3, 22; m 48, 63, 71; c 4. PETROLEUM ENGINEERING, PHYSICAL CHEMISTRY. *Educ:* Stanford Univ, BA, 44, PhD(phys chem), 48. *Prof Exp:* Asst, Off Sci Res & Develop, Stanford Univ, 44; res chemist, Manhattan Proj, Tenn Eastman Co, 45; assoc chemist, Stanford Res Inst, 47-50; asst dir, Nat Chem Lab, India, 50-53; assoc prof petrol & natural gas eng, Pa State Univ, 53-57; assoc prof petrol eng, 57-62, PROF PETROL ENG, STANFORD UNIV, 62- *Concurrent Pos:* Fulbright vis prof, Univ Tokyo, 63-64 & Gubkin Inst, Moscow, USSR, 78. *Mem:* Soc Petrol Engrs. *Res:* Flow of non-Newtonian fluids in tubes and porous media; transportation of oil and gas in the arctic; pipeline engineering; petroleum, natural gas and geothermal reservoir engineering; rheology of non-Newtonian fluids; development of offshore and remote natural gas fields. *Mailing Add:* Mitchell Bldg Rm 351 Stanford Univ Stanford CA 94305-2220

MARSH, ALICE GARRETT, b Berrien Center, Mich, Feb 20, 08; m 27; c 2. NUTRITION, FOODS. *Educ:* Emmanuel Missionary Col, BS, 29; Univ Nebr, MS, 38. *Hon Degrees:* ScD, Andrews Univ, 73. *Prof Exp:* Instr, Hinsdale Acad, Ill, 28-30; dietitian, Hindsdale Sanitarium & Hosp, 30-36; instr foods & nutrit, Union Col, 37-39; asst, Human Nutrit Lab, Univ Nebr, 39-44; instr, Pub Sch, 45-47; instr, Union Col, 47, asst prof foods & nutrit, 47-48, assoc prof home econ, 48-50; assoc prof, 50-59, PROF HOME ECON, ANDREWS UNIV, 59- *Concurrent Pos:* Instr, Univ Nebr, 40-44, mem staff, State & Fed Res Exp Sta. *Mem:* Am Dietetic Asn; Am Home Econ Asn. *Res:* Human nutrition, especially response of blood serum lipids to a controlled diet; animal nutrition; effect of food supplementary proteins upon successive generations of animals; dietary and lifestyle relationships to bone mineral density of human subjects. *Mailing Add:* 8254 N Hillcrest Dr Berrien Springs MI 49103

MARSH, BENJAMIN BRUCE, b Petone, NZ, Nov 15, 26; m 52; c 2. MEAT SCIENCE, MUSCULAR PHYSIOLOGY. *Educ:* Univ NZ, BSc, 46, MSc, 47; Cambridge Univ, PhD(biochem), 51. *Prof Exp:* Chemist, Fats Res Lab, Wellington, NZ, 47; biochemist, Low Temperature Res Sta, Cambridge, Eng, 47-51 & Dominion Lab, Wellington, NZ, 51-57; biochemist & dep dir, Meat Indust Res Inst, Hamilton, NZ, 57-71; PROF MEAT & ANIMAL SCI & DIR MUSCLE BIOL & MEAT SCI LAB, UNIV WIS-MADISON, 71- *Honors & Awards:* Distinguished Meats Res Award, Am Meat Sci Asn, 70, Signal Serv Award, 90; Meat Res Award, Am Soc Animal Sci, 78. *Mem:* Am Meat Sci Asn; Am Soc Animal Sci; Inst Food Technologists. *Res:* Early postmortem muscle metabolism; rigor mortis; meat quality; muscular contraction and relaxation; effects of muscle shortening on meat tenderness. *Mailing Add:* Muscle Biol & Meat Sci Lab Univ Wis Madison WI 53706-1181

MARSH, BERTRAND DUANE, b Bellingham, Wash, June 22, 38; m 60; c 2. CHEMICAL ENGINEERING, RHEOLOGY. *Educ:* Univ Wash, BS, 61, MS, 63, PhD(chem eng), 65. *Prof Exp:* Fel, Univ Wis, 66-67; res engr, 67-69, proj scientist, 69-71, group leader process res, 71-74, assoc dir process res, 74-77, GEN MGR RES & DEVELOP, PARMA TECH CTR, CARBON PRODS DIV, UNION CARBIDE CORP, 77- *Concurrent Pos:* Mem bd dirs, Cuyahoga Community Col Found. *Mem:* Am Inst Chem Engrs; Soc Rheol. *Res:* Rheological characterization of viscoelastic and elasticoviscous materials; flow description of rheologically complex fluids in processing equipment. *Mailing Add:* Union Carbide Corp 39 Old Ridgebury Rd J2 Danbury CT 06817

MARSH, BRUCE BURTON, b Dickinson Center, NY, Aug 8, 34; m 60; c 3. PARTICLE-SOLID INTERACTIONS. *Educ:* State Univ NY Albany, BS, 56; Univ Rochester, PhD(physics), 62. *Prof Exp:* Assoc prof, 62-64, PROF PHYSICS, STATE UNIV NY ALBANY, 64- *Concurrent Pos:* Vis scientist, Univ Aarhus, Denmark, 80-81. *Mem:* Am Phys Soc; Am Asn Physics Teachers. *Res:* Electron channeling; corrosion inhibition by ion implantation. *Mailing Add:* Dept of Physics State Univ of NY 1400 Wash Ave Albany NY 12222

MARSH, BRUCE DAVID, b Munising, Mich, Jan 4, 47; m 70; c 2. GEOLOGY. *Educ:* Mich State Univ, BS, 69; Univ Ariz, MS, 71; Univ Calif, Berkeley, PhD(geol), 74. *Prof Exp:* Geophysicist, Anaconda Co, 69-70, geologist, 70-71; asst prof, 74-78, assoc prof, 78-81, PROF EARTH & PLANETARY SCI, JOHNS HOPKINS UNIV, 81- *Concurrent Pos:* Pres elect, secy, volcanology, geochem & petrol sect, Am Geophys Union; fel panel, Nat Res Coun, 79-83, chmn, 83, adv, comt physics & chem earth mat, 85-88, panel planetary geol, NASA, 80-83, solid earth geophys working group, Earth Syst Sci Comt, 84-85, rev panel NSF, 84, 85-88 & ocean drilling prog panel, 84-86; ed, J Volcanology & Geothermal Res, 79, Geol, Geol Soc Am, 80-83, J Petrol, 86-, J Geophys Res, 87-; vis prof, Calif Inst Technol, 85; chmn, US Nat Comt Int Geol Coord Prog, 87- *Mem:* Am Geophys Union; fel Geol Soc Am; fel Soc Explor Geophysicists; fel Royal Astron Soc. *Res:* Physical geology and the physics and chemistry of the generation ascension and general evolution of magma within the earth, as well as the earth's internal global dynamics. *Mailing Add:* Dept Earth & Planetary Sci Johns Hopkins Univ Baltimore MD 21218

MARSH, CEDRIC, b Wigan, Eng, March 2, 24; Can citizen; m 48; c 3. STRUCTURAL ENGINEERING. *Educ:* Cambridge Univ, BA, 44, MA, 66. *Prof Exp:* Tech asst, Royal Aircraft Estab, UK, 44-46; asst chief designer, Struct & Mech, Develop Eng, Ltd, UK, 46-52; chief design, Aluminum Lab, Ltd, Switz, 52-56; sr designer, Aluminum Co, Can, 56-67; PROF, CONCORDIA UNIV, MONTREAL, 69- *Concurrent Pos:* Consult engr, Cedric Marsh, Montreal, 67-, Cardona, Trol, Marsh, Hidalgo, Colombia, 80- & Polygenie Inc, Montreal, 79-; mem, Order Engrs Que. *Honors & Awards:* Raymond C Reese Prize, Am Soc Civil Engrs, 83. *Mem:* Am Soc Civil Engrs; Can Soc Civil Engrs; Inst Struct Engrs. *Res:* Aluminum as a structural and engineering material; large span aluminum roofs; ultimate limit state analysis of components and structures; elastic instability and post-buckling behavior. *Mailing Add:* 15 Parkside Pl Montreal PQ H3H 1A7 Can

MARSH, DAVID GEORGE, b London, Eng, Mar 29, 40; m; c 3. IMMUNOGENETICS, BIOCHEMISTRY. *Educ:* Univ Birmingham, BSc, 61; Cambridge Univ, PhD(biochem), 64. *Prof Exp:* From asst prof to assoc prof, Sch Med, Johns Hopkins Univ, 69-89, asst prof microbiol, 72-77, vchmn, Immunol Coun, 81-82, PROF MED, SCH MED, JOHNS HOPKINS UNIV, 89- *Concurrent Pos:* USPHS fel, Calif Inst Technol, 66-69; USPHS res grants, Johns Hopkins Univ, 69-,; investr, Howard Hughes Med Inst, 76-81; mem asthma & allergic dis task force, Nat Inst Allergy & Infectious Dis, 78-79, ad hoc grant rev comt, NIH, 77- & res career develop award, 71-76; chmn, Immunogenetics Comt, Am Acad Allergy, Allergen Nomenclature Sub Comt (IUIS); chmn, Int Histocompatibility Workshop, HLA & Allergy, 91. *Mem:* Fel Am Acad Allergy; Am Asn Immunol; NY Acad Sci; Am Soc Human Genetics; Col Int Allergol. *Res:* Immunochemistry and molecular genetics of human immune response to allergens. *Mailing Add:* Johns Hopkins Asthma & Allergy Ctr 301 Bayview Blvd Baltimore MD 21224

MARSH, DAVID PAUL, b Seattle, Wash, Dec 10, 34; m 58; c 3. COMPUTER APPLICATIONS IN PHYSICS. *Educ:* DePauw Univ, BA, 57; Univ Calif, Berkeley, PhD(physics), 62. *Prof Exp:* Asst prof physics, Univ Hawaii, 62-63; asst prof, 63-69, ASSOC PROF PHYSICS, UNIV NEV, RENO, 69- *Concurrent Pos:* Pres, 6502 Exchange Prog, 85- *Mem:* AAAS; Am Asn Physics Teachers. *Res:* Physics educational software. *Mailing Add:* Dept Physics Univ Nev Reno NV 89557

MARSH, DEAN MITCHELL, chemical engineering, organic chemistry, for more information see previous edition

MARSH, DONALD CHARLES BURR, b Jackson, Mich, July 20, 26; m 80. NUMBER THEORY. *Educ:* Univ Ariz, BS, 47, MS, 48; Univ Colo, PhD(math), 54. *Prof Exp:* Instr math, Univ Ariz, 48-50; asst prof, Tex Tech Col, 54-55; from instr to assoc prof, 55-66, PROF MATH, COLO SCH MINES, 66- *Concurrent Pos:* Asst to dir, Nat Number Theory Inst, Univ Colo, 59; ed, Aristocrat Dept, The Cryptogram. *Honors & Awards:* C B Warner Award, 78; Damon Award, 80. *Mem:* Sigma Xi; Am Cryptogram Asn (pres, 68-70). *Res:* Heuristics; cryptanalysis; number theory. *Mailing Add:* 5990 W 34th Ave Denver CO 80212

MARSH, DONALD JAY, b New York, NY, Aug 5, 34; m; c 2. PHYSIOLOGY, BIOMEDICAL ENGINEERING. *Educ:* Univ Calif, Berkeley, AB, 55; Univ Calif, San Francisco, MD, 58. *Prof Exp:* NIH fel, 59-63; from asst prof to assoc prof physiol, Sch Med, NY Univ, 63-71; prof biomed eng, Sch Eng, 71-78, PROF & CHMN DEPT PHYSIOL, SCH MED, UNIV SOUTHERN CALIF, 78- *Concurrent Pos:* NIH spec fel, 70-71. *Mem:* Am Physiol Soc; Am Soc Nephrol; Biophys Soc; Soc Gen Physiologists; Soc Math Biol Asn; Inst Elect Electronics Engrs. *Res:* Renal physiology, mechanism of hypertonic urine formation; regulation of glomerular filtration and proximal tubule reabsorption. *Mailing Add:* Mudd Hall 128 Sch Med Univ of Southern Calif Los Angeles CA 90033

MARSH, FRANK LEWIS, b Aledo, Ill, Oct 18, 99; m 27; c 2. ECOLOGY. *Educ:* Emmanuel Missionary Col, Andrews Univ, AB, 27, BS, 29; Northwestern Univ, MS, 35; Univ Nebr, PhD(bot), 40. *Prof Exp:* Instr sci & math, Hinsdale Acad, Ill, 29-34; asst zool, Northwestern Univ, 34-35; from instr to prof biol, Union Col, Nebr, 35-50; prof & head dept, Emmanuel Missionary Col, Andrews Univ, 50-58; researcher, Geo-Sci Res Inst, 58-64; prof 64-71, EMER PROF BIOL, ANDREWS UNIV, 71- *Res:* Ecological entomology and botany; hyperparasitism; origin of species; variation; hybridization. *Mailing Add:* 8254 N Hillcrest Dr Berrien Springs MI 49103

MARSH, FREDERICK LEON, b Richmond, Va, Dec 20, 35; m 64. ANALYTICAL CHEMISTRY, PHYSICAL CHEMISTRY. *Educ:* Blackburn Univ, AB, 58; Univ Minn, PhD(anal chem), 65. *Prof Exp:* Trainee microchem, Northern Util Res & Develop Br, Agr Res Serv, USDA, 57, microchemist, 58; instr chem, Univ Toledo, 64-65; sr res electrochemist, Gould Labs, 65-76, SUPVR ADVAN DEVELOP, AUTOMOTIVE BATTERY DIV, GOULD INC, 76- *Mem:* AAAS; Am Chem Soc; Electrochem Soc. *Res:* Microanalytical chemistry; electrochemistry; electroanalytical chemistry. *Mailing Add:* 1523 Windemere Dr Minneapolis MN 55421

MARSH, GLENN ANTHONY, physical chemistry, corrosion, for more information see previous edition

MARSH, HOWARD STEPHEN, b New York, NY, Feb 4, 42; m 68; c 2. SYSTEMS ENGINEERING, COMMUNICATIONS ENGINEERING. *Educ:* Rensselaer Polytech Inst, BS, 63; Cornell Univ, PhD(physics), 69. *Prof Exp:* Asst physics, Cornell Univ, 63-65, asst, Lab Atomic & Solid State Physics, 64-69; group leader, 79-81, assoc dept head, Mitre Corp, 81-82; spec asst, Supreme Hq, Allied Powers, Europe, 82-88; DEPT HEAD, MITRE CORP, 88- *Res:* Signal transmission and detection; command, control and communications systems. *Mailing Add:* Mitre Corp 7525 Colshire Dr McLean VA 22102

MARSH, JAMES ALEXANDER, JR, b Wilson, NC, Dec 8, 40. MARINE ECOLOGY. *Educ:* Duke Univ, BS, 63; Univ Ga, PhD(zool), 68. *Prof Exp:* Fel environ sci, Univ NC, Chapel Hill, 68-70; from asst prof to prof marine sci, Marine Lab, Univ Guam, 70-88, dir, 76-79, dean grad studies & res, Univ, 84-88, EMER PROF MARINE SCI, MARINE LAB, UNIV GUAM, 88- *Mem:* Ecol Soc Am; Am Soc Limnol & Oceanog; Am Inst Biol Sci; AAAS; Int Soc Reef Studies. *Res:* Coral reef ecology; primary productivity; nutrient and energy cycling in tropical marine ecosystems. *Mailing Add:* Box 5400 Univ Guam Sta Mangilao GU 96923

MARSH, JAMES LAWRENCE, b San Francisco, Calif, Oct 24, 45; m 70; c 2. BIOCHEMISTRY, GENETICS. *Educ:* Univ Calif, Santa Barbara, BA, 68; Univ Wash, PhD(biochem), 74. *Prof Exp:* Fel develop genetics, Bio Ctr, Univ Basel, Switz, 74-76; res assoc, Dept Biol, Univ Va, 77-80; ASST PROF GENETICS, UNIV CALIF, IRVINE, 80- *Mem:* Am Chem Soc; Genetics Soc Am. *Res:* Recombinant DNA techniques and genetic chromosomal analysis of the control of gene extension during embryonic development; inonoclonal antibodies to probe developmentally important cell surface molecules. *Mailing Add:* Develop Biol Ctr Univ Calif Irvine CA 92717

MARSH, JOHN L, b Washington, DC, Feb 25, 16; m 43, 72; c 6. INFORMATION SCIENCE. *Educ:* Univ Ill, PhD(org chem), 41. *Prof Exp:* Res chemist, Hooker Electrochem Co, NY, 41-43; res chemist, CIBA-GEIGY Pharmaceut Co, 43-57, sr info scientist, 57-82; CONSULT, 82- *Mem:* Am Chem Soc. *Res:* Processing of scientific data and drug information by classical and computer techniques; communication of research knowledge; chemical nomenclature; molecular notations; published and proprietary scientific information. *Mailing Add:* 108 Beekman Rd Summit NJ 07901

MARSH, JULIAN BUNSICK, b New York, NY, Jan 21, 26; m 48; c 1. BIOCHEMISTRY. *Educ:* Univ Pa, MD, 47. *Prof Exp:* Intern, Episcopal Hosp, Philadelphia, 47-48; NIH fel biochem, dept res med, Univ Pa, 48-50, instr res med, 50-51, assoc biochem, Grad Sch Med, 52-53, from asst prof to assoc prof, Sch Med & Grad Sch Med, 53-63, prof, Grad Sch med, 63-65, prof biochem & chmn dept, Sch Dent Med, 65-75; prof physiol & prof biochem & chmn dept, 75-84, PROF BIOCHEM, MED COL PA, 84- *Concurrent Pos:* Guggenheim Mem fel, Nat Inst Med Res, Eng, 60-61; chmn, Gordon Conf Lipid Metab, 82; ed, J Lipid Res, 83-87; chmn, Coun on Arteriosclerosis, Am Heart Asn, 86-88; counr, Soc Exp Biol & Med, 80-82. *Honors & Awards:* Sr Invest Achievement Award, Am Heart Asn, 89. *Mem:* AAAS; Am Soc Biochem & Molecular Biol; Am Physiol Soc; Soc Exp Biol & Med (treas, 90-); Sigma Xi; Am Inst Nutrit. *Res:* Action of insulin and other hormones; carbohydrate, chromoprotein and lipoprotein metabolism; experimental nephrosis. *Mailing Add:* 3300 Henry Ave Med Col Philadelphia PA 19129

MARSH, KENNETH NEIL, b Melbourne, Victoria, Australia, Dec, 27, 39; m 63; c 3. FLUIDS & FLUID MIXTURES. *Educ:* Univ Melbourne, Australia, BSc, 61; Univ New Eng, Australia, MSc, 64, PhD(phys chem), 68. *Prof Exp:* Prof chem, Univ New Eng, Australia, 74-83; assoc dir, 83-85, DIR, THERMODYN RES CTR, TEX A&M UNIV, 85- *Concurrent Pos:* Travel fel, Nuffield Dominion, 69-70; consult, Univ Pertanian, Malaysia, 77; chmn, subcomt physicochem measurements & standards, Int Union Pure & Appl Chem, 80-85, mem, phys chem comt; Tex Eng Exp Sta fel, 84-85; ed, J Chem Eng Data, 91- *Honors & Awards:* Rennie Medal, Royal Australian Chem Inst, 72. *Mem:* Am Chem Soc; fel Royal Australian Chem Inst; Am Inst Chem Eng; Royal Soc Chem Gt Brit; Int Union Pure & Appl Chem. *Res:* Experimental and theoretical aspects of thermodynamic properties of pure fluids and fluid mixtures. *Mailing Add:* Thermodyn Res Ctr Tex A&M Univ Col Station TX 77843-3111

MARSH, LELAND C, b Lyons, NY, Nov 19, 28; m 53; c 4. BOTANY. *Educ:* Syracuse Univ, BS, 51, PhD(bot), 62. *Prof Exp:* Asst prof bot & biol, Marshall Univ, 57-60; assoc prof, State Univ NY Col Buffalo, 60-65; PROF BIOL, STATE UNIV NY COL OSWEGO, 65-, CHMN DEPT BOT & PHYSIOL, 71- *Mem:* Bot Soc Am; Soc Study Evolution. *Res:* Botanical research of Typha species, including ecological, genetic and systematic studies; industrial uses of Typha. *Mailing Add:* Dept of Biol State Univ of NY Oswego NY 13126

MARSH, MAX MARTIN, b Indianapolis, Ind, Feb 25, 23; m; c 4. PHYSICAL CHEMISTRY, ANALYTICAL CHEMISTRY. *Educ:* Ind Univ, BS, 47. *Prof Exp:* Analytical chemist, Eli Lilly & Co, 47-56, head analyst, Res Dept, 56-61, res adv, Res Labs, 66-86; vis res scientist, 87-90, ADJ PROF, CHEM DEPT, IND UNIV, 90- *Concurrent Pos:* Dir phys chem res div, Eli Lilly & Co, 67-69; indust prof chem, Ind Univ, 71-76. *Mem:* Am Chem Soc; Sigma Xi. *Res:* Optical analytical techniques; molecular structure-activity relationships; chemistry of nucleic acids. *Mailing Add:* 2311 E Covenanter Dr Bloomington IN 47401-5401

MARSH, PAUL MALCOLM, b Fresno, Calif, Nov 7, 36; m 65; c 2. ENTOMOLOGY. *Educ:* Univ Calif, Davis, BS, 58, MS, 60, PhD(entom), 64. *Prof Exp:* Lab technician, Univ Calif, Davis, 61-63; RES ENTOMOLOGIST, SYST ENTOM LAB, AGR RES SERV, USDA, 64- *Mem:* Entom Soc Am; Am Entom Soc. *Res:* Systematic entomology; taxonomy and biology of parasitic wasps of the family Braconidae. *Mailing Add:* Syst Entom Lab USDA US Nat Mus Washington DC 20560

MARSH, RICHARD EDWARD, b Jackson, Mich, Mar 6, 22; m 47; c 4. PHYSICAL CHEMISTRY. *Educ:* Calif Inst Technol, BS, 43; Univ Calif, Los Angeles, PhD(phys chem), 50. *Prof Exp:* Fel struct of metals, 50-51, fel struct of proteins, 51-55, sr fel, 55-74, res assoc structure of proteins, 74-77, res assoc, 77-81, SR RES ASSOC CHEMIST, CALIF INST TECHNOL, 81- *Concurrent Pos:* Instr, Univ Calif, Los Angeles, 53. *Mem:* Am Crystallog Asn; Sigma Xi. *Res:* Crystal structure analysis; molecular structure; structure of biologic molecules. *Mailing Add:* Dept Chem Bldg 127-72 Calif Inst Technol 1201 E Calif Pasadena CA 91125

MARSH, RICHARD FLOYD, b Portland, Ore, Mar 3, 39; m 59; c 5. VETERINARY VIROLOGY, VETERINARY PATHOLOGY. *Educ:* Wash State Univ, BS, 61, DVM, 63; Univ Wis-Madison, MS, 66, PhD(vet sci), 68. *Prof Exp:* Res veterinarian, Kellogg Co, Battle Creek, Mich, 63-64; NIH trainee vet sci, Univ Wis-Madison, 64-66, NIH spec fel, 66-68; vet officer, Nat Inst Neurol Dis & Stroke, NIH, USPHS, 68-70; res assoc, 70-79, assoc prof, 79-84, PROF & CHMN VET SCI, UNIV WIS-MADISON, 84- *Concurrent Pos:* Romnes Fac Fel, 80. *Mem:* AAAS; Am Soc Microbiol; Am Soc Virol. *Res:* Development and study of animal models of human disease, especially persistent virus infections of the central nervous system. *Mailing Add:* Dept Vet Sci 337 Vet Sci Bldg Univ Wis 1655 Linden Dr Madison WI 53706

MARSH, RICHARD L, b Cleveland, Ohio, Aug 9, 46; m 79; c 2. MUSCLE PHYSIOLOGY, PHYSIOLOGICAL ECOLOGY. *Educ:* Hiram Col, AB, 68; Univ Mich, Ann Arbor, PhD(zool), 79. *Prof Exp:* Teaching assoc biol, Hiram Col, Ohio, 68-70; teacher, Hawken Sch, Gates Mill, Ohio, 70-72; teaching asst zool, Univ Mich, Ann Arbor, 73-75, res asst, 76-79, lectr, 78, postdoctoral scholar physiol, 79-80; postdoctoral scholar, Univ Calif, Irvine, 80-82, lectr, 80-82; asst prof, 82-88, ASSOC PROF PHYSIOL, NORTHEASTERN UNIV, BOSTON, MASS, 88- *Mem:* AAAS; Am Soc Zoologists; Am Ornithologists Union; Am Physiol Soc; Cooper Ornith Soc; Sigma Xi. *Res:* Structural, physiological and biochemical adaptations of skeletal muscle, particularly an analysis of the in vivo mechanical function of skeletal function during locomotion; temperature regulation and seasonal acclimatization in birds. *Mailing Add:* Dept Biol Northeastern Univ 360 Huntington Ave Boston MA 02115

MARSH, ROBERT CECIL, b Lexington, Ky, Feb 27, 44; m 65; c 2. MOLECULAR BIOLOGY. *Educ:* Western Ky Univ, BS, 65; Vanderbilt Univ, PhD(molecular biol), 71. *Prof Exp:* Res assoc biochem, Soc Molecular Biol Res, Stoeckheim, Ger, 71-75; res assoc, Princeton Univ, 75-76; ASST PROF BIOCHEM, UNIV TEX, DALLAS, 76- *Mem:* Am Chem Soc; Sigma Xi. *Res:* Mechanism of action of elongation factors in protein biosynthesis; DNA replication and gene expression by bacteriophage T4; structure of E coli origin of replication. *Mailing Add:* Dept Biol Univ Tex Dallas Box 688 Richardson TX 75083

MARSH, TERRENCE GEORGE, b Winnipeg, Man, Jan 12, 41; US citizen; m 65; c 2. ENVIRONMENTAL BIOLOGY. *Educ:* Earlham Col, AB, 63; Ore State Univ, MS, 65; Univ Ky, PhD(zool), 69. *Prof Exp:* Instr biol, Asbury Col, 68-69; from asst prof to assoc prof, 69-81, PROF BIOL, N CENT COL, ILL, 82- *Mem:* Nat Speleol Soc. *Res:* Bryozoa; insect ecology. *Mailing Add:* Dept of Biol PO Box 3063 Naperville IL 60566-7063

MARSH, WALTON HOWARD, biochemistry, for more information see previous edition

MARSH, WILLIAM ERNEST, b New Brunswick, NJ, Nov 22, 39; m 62; c 2. MATHEMATICAL LOGIC. *Educ:* Dartmouth Col, AB, 62, MA, 65, PhD(math), 66. *Prof Exp:* Asst prof math & chmn dept, Talledega Col, 66-69; asst prof, 69-74, ASSOC PROF MATH, HAMPSHIRE COL, 74- *Mem:* Asn Symbolic Logic; Am Math Soc. *Res:* Model theory; foundations of mathematics; mathematical linguistics; automata theory. *Mailing Add:* 52 Crystal St Worchester MA 01603

MARSH, WILLIAM LAURENCE, b Cardiff, Wales, Apr 21, 26; US citizen; m 52; c 2. IMMUNOHEMATOLOGY, BIOCHEMISTRY OF CELL MEMBRANES. *Educ:* Columbia Pac Univ, PhD(human genetics), 68; FRCPath, 74. *Prof Exp:* Lab chief, Regional Blood Transfusion Ctr, Brentwood, Eng, 55-69; from assoc investr to sr investr, 69-87, SR VPRES RES & HEAD, LINSLEY F KIMBALL RES INST NY BLOOD CTR, 87- *Concurrent Pos:* Sr lectr, Bromley Col Technol, London, 60-69; chmn Inst Med Lab Sci Hemat Bd, London, 61-69. *Honors & Awards:* Dunsford Mem Award, Am Asn Blood Banks, 75; Race Prize Immunohemat, Inst Med Lab Sci, 76; Emily Cooley Award, Am Asn Blood Banks, 88, Grove-Rasmussen Award, 90. *Mem:* Int Soc Blood Transfusion; Am Asn Blood Banks; fel Inst Biol Eng; Brit Soc Hemat. *Res:* Serology, genetics and biochemistry of the human red cell blood groups;; relationship of blood groups to disease. *Mailing Add:* NY Blood Ctr 310 E 67 St New York NY 10021

MARSHAK, ALAN HOWARD, b Miami Beach, Fla, Mar 21, 38; div; c 2. SEMICONDUCTOR PHYSICS, DEVICE ANALYSIS. *Educ:* Univ Miami, BS, 60; La State Univ, MS, 62; Univ Ariz, PhD(elec eng), 69. *Prof Exp:* Res assoc, Solid State Lab, Dept Elec Eng, Univ Ariz, 65-66; instr, 62-65, from asst prof to assoc prof, 69-78, PROF ELEC ENG, LA STATE UNIV, 78-, CHMN DEPT ELEC & COMPUT ENG, 83- *Concurrent Pos:* Grad trainee, NSF, 67-69, div panelist to rev Presidential Young Investr Awards, Washington, DC, 84, tech reviewer, div mat res & div elec comput & systs eng; area supvr, core electronics prog, La State Univ, 69-82; mem, Int Electron Devices Meeting, Inst Elec & Electronics Engrs, 78; organizer, IEDM Spec Session, Wash, DC, 78; vis prof, dept elec eng, Electron Device Res Ctr, Univ Fla, 79-80; bk rev, Holt, Rinehart & Winston, 79-; coorganizer, Workshop Heavy Doping Effects Silicon Devices, Univ Fla, 81; prin invest 16 res proj & grants; tech referee, Var Jour. *Mem:* Fel Inst Elec & Electronic Engrs; Sigma Xi; Electron Devices Soc. *Res:* Semiconductor device physics and analysis; modeling theory; electrical transport in semiconductors; electron and hole transport in degenerate semiconductors and semiconductors with position-dependent band structure; author of 53 technical papers and two books. *Mailing Add:* Dept Elec & Comput Eng La State Univ Baton Rouge LA 70803-5901

MARSHAK, HARVEY, b Brooklyn, NY, Nov 9, 27; m 58, 78; c 2. NUCLEAR PHYSICS. *Educ:* Univ Buffalo, BA, 50; Univ Conn, MA, 52; Duke Univ, PhD(physics), 55. *Prof Exp:* Res assoc, Duke Univ, 54-55; assoc physicist, Brookhaven Nat Lab, 55-62; PHYSICIST, NAT BUR STANDARDS, 62- *Concurrent Pos:* Vis prof, Katholieke Univ, Leuven, Belg, 80-81. *Mem:* Fel Am Phys Soc. *Res:* Nuclear orientation and spectroscopy; neutron physics; low temperature physics. *Mailing Add:* 8413 Peck Pl Bethesda MD 20817

MARSHAK, MARVIN LLOYD, b Buffalo, NY, Mar 11, 46; m 72; c 2. PHYSICS. *Educ:* Cornell Univ, AB, 67; Univ Mich, MS, 69, PhD(physics), 70. *Prof Exp:* Res assoc physics, Univ Mich, 70; res assoc, 70-74, from asst prof to assoc prof, 74-82, dir grad studies, 83-86, PROF PHYSICS, UNIV MINN, 82-, HEAD SCH PHYSICS & ASTRON, 86- *Concurrent Pos:* NSF grant, 77-79 & 81; prin investr, Dept Energy grant, 82-86. *Mem:* Am Phys Soc. *Res:* Experimental elementary particle physics. *Mailing Add:* Dept Physics Univ Minn 116 Church St S E Minneapolis MN 55455

MARSHAK, ROBERT EUGENE, b New York, NY, Oct 11, 16; m 43; c 2. PARTICLE PHYSICS. *Educ:* Columbia Col, AB, 36; Cornell Univ, PhD(physics), 39. *Prof Exp:* From instr to prof physics, Univ Rochester, 39-50, Harris prof & chmn dept, 50-64, distinguished univ prof, 64-70; pres, City Col NY, 70-79; univ distinguished prof, 79-87, EMER UNIV DISTINGUISHED PROF, VA POLYTECHNIC INST, 87- *Concurrent Pos:* Physicist, Radiation Lab, Mass Inst Technol, 42-43 & Dept Sci & Indust Res Gt Brit, 43-44; dep group leader, Los Alamos Sci Lab, 44-46; mem, Inst Advan Study, 48; Guggenheim fel & prof, Sorbonne, 53-54; Guggenheim fel & guest prof, Ford Found, Europ Orgn Nuclear Res, Switz, 60-61, Yugoslavia, Israel & Japan, 67-68; vis prof, Columbia Univ, Univ Mich, Harvard Univ, Cornell Univ & Tata Inst Fundamental Res, India; Niels Bohr vis prof, Madras Univ, 63, Yalta Int Sch, Carnegie-Mellon Univ & Univ Tex; Nobel lectr, Sweden; Solvay Cong, 67; trustee, Atoms for Peace Awards. Secy, High Energy Physics Comn, Int Union Pure & Appl Physics, 57-63; chmn, Int Conf High Energy Physics, 60; mem, Nat Acad Sci Adv Comt Soviet Union & Eastern Europe, 63-66, head deleg to Poland, 64 & Yugoslavia, 65; mem, US Mission to Soviet Labs, 60; chmn vis physics comn, Brookhaven Nat Lab, 65; vis physics comn, Carnegie Mellon Univ, 66-70; mem coun, Nat Acad Sci, 71-74; mem sci coun, Int Ctr Theoret Physics, Trieste, 67-75; mem, US-Japan Sci Comt, 68-72; mem exec comt, Nat Comn, UNESCO, 71-73; mem coun, Nat Acad Sci, 71-74. *Honors & Awards:* Cressy Morrison Prize, NY Acad Sci, 40. *Mem:* Nat Acad Sci; fel AAAS; Am Acad Arts & Sci; Am Phys Soc (pres-elect, 82, pres, 83); Fedn Am Scientists (chmn, 47-48); Am Philos Soc. *Res:* Energy sources of stars; atomic nuclei; neutron diffusion; elementary particles. *Mailing Add:* 202 Fincastle Dr Blacksburg VA 24060

MARSHAK, ROBERT REUBEN, b New York, NY, Feb 23, 23; m 83; c 3. VETERINARY MEDICINE. *Educ:* Cornell Univ, DVM, 45. *Hon Degrees:* Dr Vet Med, Univ Bern, 68. *Prof Exp:* Pvt pract & clin invest, 45-56; chmn dept clin studies, Univ Pa, 61-73, dir, Bovine Leukemia Res Ctr, 66-75, dean, 73-87, PROF MED, SCH VET MED, UNIV PA, 56- *Concurrent Pos:* Mem comt vet med sci, Nat Acad Sci, 74-76; mem, Adv Coun, James A Baker Inst, Cornell Univ, 77-; mem sci adv bd, Sch Vet Med, Hebrew Univ Jerusalem, 84-; mem, Nat Acad Pract, 87- *Mem:* Inst Med-Nat Acad Sci; Am Asn Cancer Res; fel NY Acad Sci; Am Col Vet Internal Med (pres, 75-76); Am Vet Med Asn; AAAS. *Res:* Bovine leukemia; metabolic diseases of cattle. *Mailing Add:* 280 Melrose Ave Merion PA 19066

MARSHALEK, EUGENE RICHARD, b New York, NY, Jan 17, 36; m 62; c 2. NUCLEAR PHYSICS. *Educ:* Queen's Col, NY, BS, 57; Univ Calif, Berkeley, PhD(nuclear struct), 62. *Prof Exp:* NSF fel nuclear physics, Niels Bohr Inst, Copenhagen, Denmark, 62-63; res assoc physics theory group, Brookhaven Nat Lab, 63-65; asst prof, 65-69, assoc prof, 69-78, PROF PHYSICS, UNIV NOTRE DAME, 78- *Honors & Awards:* Alexander von Humboldt US Sr Scientist Award, 85. *Mem:* Sigma Xi; Am Phys Soc; AAAS. *Res:* Nuclear structure theory; nuclear theory, particularly collective effects in atomic nuclei. *Mailing Add:* Dept of Physics Univ of Notre Dame Notre Dame IN 46556

MARSHALL, ALAN GEORGE, b Bluffton, Ohio, May 26, 44; m 65; c 2. BIOCHEMISTRY. *Educ:* Northwestern Univ, BA, 65; Stanford Univ, PhD(chem), 70. *Prof Exp:* From asst prof to assoc prof chem, Univ BC, Vancouver, 69-80; PROF CHEM & BIOCHEM & DIR, CAMPUS CHEM INSTRUMENT CTR, OHIO STATE UNIV, 80- *Concurrent Pos:* NSF fel, 65-69; Alfred P Sloan Found fel, 76-80; distinguished res fel, Oak Ridge Nat Lab, 90. *Honors & Awards:* Chem Instrumentation Award, Am Chem Soc, 90. *Mem:* Am Chem Soc; fel Am Phys Soc; fel AAAS; Sigma Xi; Am Soc Mass Spectrometry; NY Acad Sci; AM Soc Biol Chemists. *Res:* Mass spectrometry; Fourier transform spectroscopy; analytical chemistry. *Mailing Add:* Dept Chem Ohio State Univ 120 W 18th Ave Columbus OH 43210-1173

MARSHALL, ALBERT WALDRON, mathematics, for more information see previous edition

MARSHALL, CARTER LEE, b New Haven, Conn, Mar 31, 36; c 2. PREVENTIVE MEDICINE. *Educ:* Harvard Univ, BA, 58; Yale Univ, MD, 62, MPH, 64; Am Bd Prev Med, dipl, 70. *Prof Exp:* Proj dir, Conn Dept Health, New Haven, 64-65; asst prof prev med, Sch Med, Univ Kans Med Ctr, Kansas City, 67-69; from assoc prof community med to prof community med & med educ, Mt Sinai Sch Med, 69-77, assoc dean, 73-77; PROF MED & PREV MED, COL MED & DENT NJ, NJ MED SCH, DIR, OFF PRIMARY HEALTH CARE EDUC & DIR AMBULATORY CARE, COL HOSP, 77- *Concurrent Pos:* Fel epidemiol & pub health, Yale Univ, 64-65; consult, New York Health Serv Admin, 70-71; univ dean health affairs, City Univ New York, 72-74. *Mem:* AAAS; fel Am Pub Health Asn; Asn Teachers Prev Med; AMA; Am Col Prev Med. *Res:* Health services delivery. *Mailing Add:* Dept Prev Med NJ Col Med & Dent 100 Bergen St Newark NJ 07103

MARSHALL, CHARLES EDMUND, colloid chemistry, soil science; deceased, see previous edition for last biography

MARSHALL, CHARLES F, b St Louis, Mo, Sept 24, 43; m 66; c 2. AERONAUTICAL & ASTRONAUTICAL ENGINEERING, MECHANICAL ENGINEERING. *Educ:* Univ Mo-Rolla, BS, 72. *Prof Exp:* Technician, US Navy Submarine Serv, Ballistic Missiles, 61-68; lead engr, McDonnell Douglas Tactical Missile Systs, 72-80; lead engr, Strategic Ballistic Launch Vehicles, 80-84, eng mgr, Space Transp Architectures, 84-87, eng mgr, Orbit Transfer Vehicles, 87-89, ENG MGR, SPACE LAUNCH SYST HEAVY VEHICLES, MARTIN MARIETTA CORP, 89- *Concurrent Pos:* Ed, Sci Publ Strategic Missiles, 89-91. *Mem:* Soc Automotive Engrs. *Res:* Development of advanced rocket systems for spacecraft launch and orbit transfer vehicles; establish new technology and concepts for rocket launch operations including mission analysis, testing and launch control. *Mailing Add:* 6874 S Lewis Ct Littleton CO 80127

MARSHALL, CHARLES RICHARD, b Australia, Jan 13, 61; US citizen; m 91. MOLECULAR & MORPHOLOGICAL SYSTEMATICS, QUANTITATIVE PALEONTOLOGY. *Educ:* Australian Nat Univ, BS, Hons, 84; Univ Chicago, MS, 86, PhD(evol biol), 89. *Prof Exp:* Postdoctoral molecular biol, Ind Univ, NIH, 89-91; ASST PROF PALEONT & MOLECULAR EVOLUTION, UNIV CALIF, LOS ANGELES, 91- *Mem:* Soc Syst Zool; Soc Study Evoltuion; Paleont Soc; Paleont Res Inst; Australian Asn Paleont. *Res:* Molecular and morphological systematics; molecular evolution; evolutionary biology and paleontology of lungfish and echinoderms, especially echinoids; methods of phylogenetic reconstruction; mathematical approaches for assessing the completeness of the fossil record. *Mailing Add:* Dept Earth & Space Sci Univ Calif Los Angeles CA 90024-1567

MARSHALL, CHARLES WHEELER, b Syracuse, NY, Oct 20, 06; m 39; c 2. VITAMINS & MINERAL NUTRIENTS. *Educ:* Univ Chicago, BS, 31, MS, 33, PhD(biochem), 49. *Prof Exp:* Chemist, Edwal Labs, 38-39; chief control chemist, Lakeside Labs, 39-43; res chemist, US Off Sci Res & Develop, Chicago, 43-45 & G D Searle & Co, 49-67; SCI WRITER, 67- *Honors & Awards:* Trade Book Award, Am Med Writers Asn, 84. *Mem:* Am Chem Soc; Am Coun Sci & Health; Nat Coun Against Health Fraud. *Res:* Synthesis of steroids related to adrenal cortical hormones and steroids with new pharmacological properties; nutritional supplements, especially vitamins and minerals. *Mailing Add:* 6194 Birch Row Dr E Lansing MI 48823

MARSHALL, CLAIR ADDISON, b Bluffton, Ohio, Jan 20, 11; wid; c 1. CHEMICAL ENGINEERING. *Educ:* Univ Mich, BS, 34. *Prof Exp:* Chem engr, Tenn Eastman Co, 34-39, asst supt, Acetic Acid Dept, 39-43; consult, Clinton Eng Works, 43-44, supt, Acid Div Process Improvement, 44-50, sr chem engr, Div Staff, 50-59, supvr acid maintenance eng, 59-63, tech asst to div supt, Tenn Eastman Co, 63-75; RETIRED. *Mem:* Am Inst Chem Eng. *Res:* Organic acids and anhydrides; equipment and process for making anhydrides by direct oxidation of aldehydes. *Mailing Add:* 3755 Peachtree Rd NE Atlanta GA 30319

MARSHALL, CLIFFORD WALLACE, b New York, NY, Mar 11, 28; m 55; c 2. APPLIED MATHEMATICS, APPLIED STATISTICS. *Educ:* Hofstra Col, BA, 49; Syracuse Univ, MA, 50; Polytech Inst Brooklyn, MS, 55; Columbia Univ, PhD, 61. *Prof Exp:* Instr math, Polytech Inst Brooklyn, 50-57; mem staff, Inst Defense Anal, 58-59; prin dynamics engr, Repub Aviation Corp, 59-60; from instr to assoc prof, 60-68, PROF MATH, POLYTECH INST NEW YORK, 68- *Concurrent Pos:* Consult, Urban Inst, 72-76. *Mem:* Math Asn Am; Sigma Xi. *Res:* Combinatorial theory; finite graph theory; probability; time series analysis and forecasting. *Mailing Add:* Polytech Univ Dept Math Rte 110 Farmingdale NY 11735

MARSHALL, DALE EARNEST, b Pinckney, Mich, Aug 13, 34; m 55; c 3. MECHANICAL HARVESTING OF FRUITS & VEGETABLES. *Educ:* Mich State Univ, BS, 60, MS, 75. *Prof Exp:* Jr proj engr mach mfg, Farmhand Div, Daffin Corp, 61-63; proj engr, Chore-Time Equip, Inc, 63-66; agr engr, Fla, 66-69, AGR ENGR, USDA, MICH, 69- *Concurrent Pos:* Sr prin investr, Binat Agr Res & Develop Fund, US & Israel, 80-84. *Honors & Awards:* Eng Concept of Year, Am Soc Agr Engrs, 78; Except Creative Res Agr Eng, Pickle Packers Int, 80. *Mem:* Am Soc Agr Engrs; Am Soc Hort Sci. *Res:* Mechanical harvesting of fruits and vegetables assisting commodity groups, processors, manufacturers and growers. *Mailing Add:* 5411 Marsh Rd Haslett MI 48840

MARSHALL, DELBERT ALLAN, b Topeka, Kans, July 22, 37; m 64. ANALYTICAL CHEMISTRY. *Educ:* Kans State Teachers Col, BS, 59; Kans State Univ, MS, 65, PhD(anal chem), 68. *Prof Exp:* High sch teacher, Kans, 61-63; instr chem, Mo Valley Col, 63-64; from asst prof to assoc prof, 67-77, PROF CHEM, FT HAYS STATE UNIV, 77- *Mem:* Am Chem Soc; Soc Appl Spectros; Coblentz Soc; Sigma Xi. *Res:* Chemistry of metal chelates; atomic absorption spectroscopy. *Mailing Add:* Dept of Chem Ft Hays State Univ 600 Park St Hays KS 67601-4099

MARSHALL, DONALD D, b Woodland, Calif, Aug 8, 34; m 64; c 2. ANALYTICAL CHEMISTRY, INORGANIC CHEMISTRY. *Prof Exp:* Anal chemist, US Bur Mines, Nev, 58-60; asst prof chem, Southern Ore Col, 65-66; from asst prof to assoc prof, 66-73, PROF CHEM, SONOMA STATE UNIV, 73-, CHMN, 79- *Mem:* Am Chem Soc; Sigma Xi. *Res:* Determination of stability constants of inorganic compounds in aqueous solutions; water and air pollution; computer applications in chemistry. *Mailing Add:* Dept of Chem Sonoma State Univ Rohnert Park CA 94928

MARSHALL, DONALD E, b Boston, Mass, Dec 15, 47; m 72; c 3. COMPLEX FUNCTION THEORY, HARMONIC ANALYSIS. *Educ:* Univ Calif-Los Angeles, BA, 70, MA, 72, PhD(math),76. *Prof Exp:* From asst prof to assoc prof, 76-86, PROF MATH, UNIV WASH, 86- *Concurrent Pos:* Res prof math, Mittag- Leffler Inst, Royal Swed Acad Sci, 76-77 & 82-83; vis asst prof, Univ Paris, 77, Univ Calif, Los Angeles, 81-82; vis assoc prof, Univ Calif, San Diego, 86; vis prof, Yale Univ, 88; consult, Jet Propulsion Lab, 66-69 & 75; technician & mathematician, Aerospace Corp, 71-72; prin investr, NSF, 77- *Mem:* Am Math Soc. *Res:* Complex analysis - primarily function theory on the disk and other plane domains. *Mailing Add:* Math Dept GN-50 Univ Wash Seattle WA 98195

MARSHALL, DONALD IRVING, b Houston, Tex, Jan 22, 24; div; c 3. CHEMICAL ENGINEERING. *Educ:* Sam Houston Col, BS, 44; Univ Tex, MA, 46, PhD(chem), 48. *Prof Exp:* Develop assoc, Plastics Div, Union Carbide Corp, 48-58; sr res engr, Western Elec Co, 58-65, res leader, 65-71, sr staff engr, 71-83,; sr staff engr, AT&T Technologies, 83-87; RETIRED. *Concurrent Pos:* Consult, 87- *Mem:* Am Chem Soc; Soc Rheol; Soc Plastics Engrs. *Res:* Rheology; material characterization; extrusion, molding and calendering processes; communication cable design; plastics recycling. *Mailing Add:* 2784 Pheasant Trail Duluth GA 30136

MARSHALL, DONALD JAMES, b Marlboro, Mass, Apr 14, 33; m 54; c 1. PHYSICS. *Educ:* Mass Inst Technol, BS, 54, PhD(geophys), 59; Calif Inst Technol, MS, 55. *Prof Exp:* Dir res, Nuclide Corp, 58-74, gen mgr & pub rels officer, AGV Div, Nuclide Corp, 74-87; GEN MGR, NEW ENG OFF, MAAS, 87- *Mem:* Am Soc Testing & Mat; NY Acad Sci. *Res:* Mass spectrometry; ion physics; electron beam technology; cathodoluminescence of geological materials. *Mailing Add:* Eight Highland Ct Taunton MA 02780

MARSHALL, FINLEY DEE, b Rochester, NY, Aug 1, 30; m 59; c 3. BIOCHEMISTRY. *Educ:* Bucknell Univ, BS, 52; Univ Mo, MS, 59, PhD(biochem), 61. *Prof Exp:* Chemist, Olin-Mathieson Chem Corp, 52-56; res assoc biochem, Sterling-Winthrop Res Inst, 61-62, Univ Iowa, 62-64; from asst prof to assoc prof, 64-75, PROF BIOCHEM, UNIV SDAK, 75- *Concurrent Pos:* Prin investr, NIH grant, 64-86, Huntington's Chorea Found grant, 73-76; vis scientist, div neurosci, City of Hope, 71; vis prof, Univ Iowa, 78. *Mem:* Am Soc Biol Chemists; Am Soc Neurochem; Soc Neurosci. *Res:* Metabolism of carnosine and related dipeptides, particularly in the central nervous system; study of enzymes involved in their metabolism and the effects of drugs on their levels in tissues. *Mailing Add:* Dept Biochem Univ SDak Vermillion SD 57069

MARSHALL, FRANCIS J, b New York, NY, Sept 5, 23; m 52; c 4. AERODYNAMICS, COMPUTER SCIENCES. *Educ:* City Col New York, BME, 48; Rensselaer Polytech Inst, MS, 50; NY Univ, DrEngSci, 55. *Prof Exp:* Engr, Gen Elec Co, 50-51 & Wright Aeronaut Corp, 51-52; group leader, Lab Appl Sci, Univ Chicago, 55-60; PROF AERONAUT & ASTRONAUT, PURDUE UNIV, 60- *Concurrent Pos:* Res mech engr, US Naval Underseas Warfare Ctr, Calif, 66-67; fac fel, NASA-Langley, 69-70; Fulbright scholar, Turkey, 88-89; vis prof, Inst Technol Mara-Midwest Univs Consortium Int Activ, Malaysia, 89. *Mem:* Assoc fel Am Inst Aeronaut & Astronaut; Am Asn Univ Professors; Am Soc Eng Educ. *Res:* Synthesis of computational aerodynamics with computer aided design of aircraft. *Mailing Add:* Sch Aeronaut & Astronaut Purdue Univ West Lafayette IN 47907

MARSHALL, FRANKLIN NICK, b Chicago, Ill, July 5, 33; m 55. PHARMACOLOGY. *Educ:* Univ Iowa, BS, 57, MS, 59, PhD(pharmacol), 61. *Prof Exp:* From pharmacologist to sr pharmacologist, Pitman-Moore Div, Dow Chem USA, 61-65, proj leader, 65-67, group leader, Dow Human Health Res Labs, 67-68, asst head dept pharmacol, Dow Human Health Res Labs, Ind, 68-72; HEAD DEPT PHARMACOL, MARION MERRELL DOW RES INST, 72-, ASSOC SCIENTIST, 73- *Mem:* AAAS; Am Soc Pharmacol & Exp Therapeut. *Res:* Autonomic neuromuscular and renal pharmacology; pharmacology of antibiotics and anesthetic agents; lipid metabolism in neoplasms; blood coagulation and fibrinolysis. *Mailing Add:* Dept Pharmacol Marion Merrell Dow Res Inst PO Box 68470 Indianapolis IN 46268-0470

MARSHALL, FREDERICK J, b Detroit, Mich, Aug 14, 20; m 46; c 7. ORGANIC CHEMISTRY. *Educ:* Univ Detroit, BS, 41, MS, 43; Iowa State Col, PhD(org chem), 48. *Prof Exp:* Res chemist, Eli Lilly & Co, 48-76, res scientist, 76-83; RETIRED. *Mem:* Am Chem Soc; NY Acad Sci. *Res:* Pharmaceuticals; antibiotic structure; radioactive carbon synthesis. *Mailing Add:* 3120 Shady Grove Ct Indianapolis IN 46222

MARSHALL, FREDERICK JAMES, b Vancouver, BC, Feb 11, 25; m 48; c 4. HISTOLOGY. *Educ:* Univ Ore, DMD, 49; Univ Ill, MS, 59; Am Bd Endodont, dipl. *Prof Exp:* Assoc prof histol & endodont, Fac Dent, Univ Man, 59-65; assoc prof oper dent, Sch Dent, Univ Pittsburgh, 65-67; prof endodont & head dept, Col Dent, Ohio State Univ, 67-72; prof endodont & chmn dept, 72-90, EMER PROF ENDODONT ORE SCH DENT, HEALTH SCI UNIV, 90- *Concurrent Pos:* Consult, Vet Admin Hosp, Portland, Ore; vpres, Am Asn Dent Schs; dir, Am Bd Endodontics, 85-91, pres, 89-90. *Honors & Awards:* Fel, Am Col Dentists; Fel, Int Col Dentists. *Mem:* Am Dent Asn; Am Asn Endodont; Int Asn Dent Res; Sigma Xi. *Res:* Endodontic culturing techniques; root canal medications; computer-assisted instruction for the diagnosis of toothache; electron microscopy of dentin. *Mailing Add:* Ore Health Sci Univ Sch Dent 611 SW Campus Dr Portland OR 97201

MARSHALL, GAILEN D, JR, b Houston, Tex, Sept 9, 50; m 78; c 3. ALLERGY-CLINICAL IMMUNOLOGY, DIAGNOSTIC LABORATORY IMMUNOLOGY. *Educ:* Univ Houston, BS, 72; Tex A&M Univ, MS, 75; Univ Tex, Galveston, PhD(immunol), 79; Univ Tex Med Sch, MD, 84. *Prof Exp:* Instr immunol, George Mason Univ, 78-79; intern internal med, Univ Iowa Hosps, 84-85, resident, 85-86; resident internal med, Baptist Mem Hosp, Memphis, 86-88, chief resident internal med, 88-89; assoc med dir, Res Health, Inc, 89-90; clin asst prof, 90-91, ASST PROF ALLERGY-IMMUNOL, MED SCH, UNIV TEX, HOUSTON, 91- *Concurrent Pos:* Lab dir, Biotherapeut, Inc, 86-88; fel, Div Allergy-Immunol, Univ Tenn, 88-89; dipl, Am Bd Internal Med, 88-, Am Bd Allergy-Immunol, 89- & Am Bd Diag Lab Immunol, 90- *Honors & Awards:* President's Award, Am Acad Allergy & Immunol, 86. *Mem:* Am Acad Allergy & Immunol; Am Col Allergy & Immunol; Am Asn Immunologists; Clin Immunol Soc; Am Thoracic Soc; Am Col Physicians. *Res:* Control of IgE production by various immunotherapeutic modalitics in patients with allergic diseases. *Mailing Add:* Div Allergy-Immunol Med Sch Univ Tex PO Box 20708 Houston TX 77225

MARSHALL, GARLAND ROSS, b San Angelo, Tex, Apr 16, 40; m 59; c 4. BIOCHEMISTRY. *Educ:* Calif Inst Technol, BS, 62; Rockefeller Univ, PhD(biochem), 66. *Prof Exp:* From instr to assoc prof, 66-76, PROF PHYSIOL & BIOPHYS, SCH MED, WASH UNIV, 76-, PROF PHARMACOL, 85- *Concurrent Pos:* Fel, Oxford Univ, 66; res assoc, Comput Systs Lab, 68-; estab investr, Am Heart Asn, 70-75; guest investr, Massey Univ, NZ, 75-76. *Mem:* Biophys Soc; Am Physiol Soc; Am Soc Biol Chemists; Am Chem Soc; Am Soc Pharm Exp Ther. *Res:* Solid phase peptide synthesis; conformation of peptides; computer-aided drug design; endocrinology. *Mailing Add:* Dept of Physiol & Biophys Wash Univ Sch of Med 660 S Euclid Ave St Louis MO 63110

MARSHALL, GRAYSON WILLIAM, JR, b Baltimore, Md, Feb 12, 43; m 70; c 2. BIOMATERIALS, DENTAL RESEARCH. *Educ:* Va Polytech Inst & State Univ, BS, 65; Northwestern Univ, PhD(materials sci), 72; DDS, 86. *Prof Exp:* Res assoc materials sci, Design & Develop Ctr, Northwestern Univ, 72-73; from instr to assoc prof biomat, Dent Sch, Northwestern Univ, Chicago, 73-87; PROF RESTORATIVE DENT, UNIV CALIF, SAN FRANCISCO, 87-, CHIEF BIOMAT SCI, 88- *Concurrent Pos:* Nat Inst Dent Res fel, Dent Sch, Northwestern Univ, 72-73, spec dent res award, 75; vis fel dent prosthetics, Univ Melbourne, 81; chmn, Credentials Comt, Acad Dent Mat, 84-; mem bd dirs, Acad Dent Mat, 85- & Oral Biol & Med Sect, NIH, 88-; guest staff scientist, Lawrence Berkeley Lab, 89-; mem grad group bioeng, Univ Calif, San Francisco, 89- *Honors & Awards:* Res Prize, Am Asn Dent Res, 74. *Mem:* Fel AAAS; Int Asn Dent Res; Electron Micros Soc Am; Am Inst Mining, Metall & Petroleum Eng; Am Col Sport Med; fel Acad Dent Mat; Am Dent Asn. *Res:* Use of metals, polymers and ceramics in dentistry and surgery; scanning electron microscopy of enamel, dentin and bone; corrosion resistance of new alloys, amalgams. *Mailing Add:* Dept Restorative Dent Univ Calif San Francisco CA 94143-0758

MARSHALL, HAROLD GENE, b Evansville, Ind, May 7, 28; m 53; c 2. PLANT BREEDING, PLANT GENETICS. *Educ:* Purdue Univ, BS, 52; Kans State Col, MS, 53; Univ Minn, PhD(plant genetics), 59. *Prof Exp:* Asst agron, Kans State Col, 52-53; asst, Univ Minn, 56-58; res agronomist, Oat Sect, USDA, Pa State Unive, University Park, 59-87, adj prof plant breeding, 74-87; RETIRED. *Mem:* Fel Am Soc Agron. *Res:* Nature of winter hardiness of winter oats and the development of winter-hardy varieties; genetics and breeding of spring oats. *Mailing Add:* Rte 2 Box 230 Bellefonte PA 16823

MARSHALL, HAROLD GEORGE, b Bedford, Ohio, May 17, 29; m 51; c 3. MARINE BIOLOGY. *Educ:* Baldwin-Wallace Col, BS, 51; Western Reserve Univ, MS, 53, PhD(biol), 62. *Prof Exp:* Biologist, Nat Dairy Labs, 51-52; instr, Cleveland City Schs, Ohio, 52-58, chmn sci dept, Bedford, 58-62; instr, Western Reserve Univ, 62-63; from asst prof to assoc prof, 63-69, PROF BIOL & CHMN DEPT, OLD DOMINION UNIV, 69- *Concurrent Pos:* NSF res grants, 64-78; environ consult, 72-85; NASA, NSF & Nat Oceanic & Atmospheric Admin res grants, 65-85. *Mem:* AAAS; Am Soc Limnol & Oceanog; Phycol Soc Am; Int Phycol Soc; NY Acad Sci. *Res:* Spatial distribution and ecology of phytoplankton off the eastern coast of the United States, Chesapeake Bay and Caribbean Sea. *Mailing Add:* Dept Biol Old Dominion Univ 5215 Hampton Blvd Norfolk VA 23508

MARSHALL, HEATHER, b Montreal, Can, Apr 28, 49. RADIOANALYSIS. *Educ:* McGill Univ, BSc, 69, MSc, 72; Queen's Univ, Ont, PhD(anal chem), 76. *Prof Exp:* Instr chem, Loyola Col, Montreal, 70-71; fel nuclear chem, McGill Univ, 76-79; res assoc anal chem, Nat Res Coun Can, 79-80; asst prof, Univ Sask, 80-82; chemist anal chem, Sask Res Coun, 82; res assoc, McGill Univ, 83-87; CHEMIST RADIOACTIV, NAT HEALTH & WELFARE CAN, 88- *Mem:* Chem Inst Can; Can Asn Physicists; Sigma Xi. *Res:* Radioanalytical chemistry, radioanalysis of natural radionuclides and environmental transfer of fallout. *Mailing Add:* Bureau Radiation & Med Device's 775 Brookfield Rd Ottawa ON K1A 1C1 Can

MARSHALL, HENRY PETER, b Altoona, Pa, May 12, 24; m 51; c 3. PHYSICAL ORGANIC CHEMISTRY. *Educ:* Pa State Univ, BS, 47; Univ Calif, Los Angeles, PhD(chem), 52. *Prof Exp:* Fel chem, Fla State Univ, 52-53; res chemist, Celanese Corp, 53-56 & Stanford Res Inst, 56-58; consult scientist mat, Lockheed Missiles & Space Co Inc, 58-88; RETIRED. *Mem:* Am Chem Soc; Am Inst Physics; Sigma Xi. *Res:* Study of chemical structural aging effects of non-metallics; identification and kinetic measurements of chemical processes occurring in non-metallics, principally polymers, in all types of environments. *Mailing Add:* 1082 Yorkshire Dr Los Altos CA 94022-7052

MARSHALL, J HOWARD, III, b San Francisco, Calif, Feb 6, 36; m. PHYSICS, ELECTRONICS. *Educ:* Calif Inst Technol, BS, 57, PhD(high energy physics), 65. *Prof Exp:* Sr res engr, Jet Propulsion Lab, Calif Inst Technol, 62-65; vpres prod, Analog Technol Corp, 65-66, chmn bd & vpres advan planning, 66-71, chmn bd & vpres technol, 71-73, pres, 73; CHMN BD

& PRES, MDH INDUSTS, INC, 73- *Mem:* AAAS; sr mem Inst Elec & Electronics Engrs; Am Inst Aeronaut & Astronaut. *Res:* Electronic and system design of instrumentation for spaceborne and earthbound applications involving nuclear physics, high-energy physics, mass spectroscopy, gas chromatography and infrared and ultraviolet radiation; corporate management. *Mailing Add:* MDH Industs Inc 426 W Duarte Rd Monrovia CA 91016

MARSHALL, JAMES ARTHUR, b Oshkosh, Wis, Aug 7, 35; m 57; c 1. ORGANIC CHEMISTRY. *Educ:* Univ Wis, BS, 57; Univ Mich, PhD(chem), 60. *Prof Exp:* USPHS fel org chem, Stanford Univ, 60-62; from asst prof to prof chem, Northwestern Univ, 62-80; prof chem, 80-83, GUY LIPSCOMB PROF, UNIV SC, 84- *Concurrent Pos:* Sloan Found fel, 66-70; Seidel Wooley lectr, Ga Inst Technol, 70, Am-Swiss Found lectr, 71-72; vis prof, Univ Calif, Los Angeles, 68; mem, Nat Acad Sci Panel for Grad Educ in Brazil, 73-77; mem, Med Chem Study Sect, USPHS, 77-80; exec ed, Synthetic Commun, 72-; consult, Ortho Res Corp & Givaudan; mem adv comn, NSF, 80-83; chmn elect, Org Div, Am Chem Soc, 91. *Honors & Awards:* Guenther Award, Am Chem Soc, 79; Russell Award Res, Univ SC, 85; Charles H Stone Award, Am Chem Soc, 86. *Mem:* Am Chem Soc; Chem Soc Japan; Royal Soc Chem; fel AAAS; Brazilian Acad Sci. *Res:* Synthetic organic chemistry related to natural products; stereochemistry and organic reaction mechanisms; trans-cycloalkenes. *Mailing Add:* Dept Chem Univ SC Columbia SC 29208

MARSHALL, JAMES JOHN, b Edinburgh, Scotland, July 7, 43; m 66; c 2. BIOCHEMISTRY, MEDICAL RESEARCH. *Educ:* Univ Edinburgh, BSc, 65; Heriot-Watt Univ, PhD(appl biochem), 69. *Prof Exp:* Res assoc biochem, Sch Med, Univ Miami, 69-71; fel, Royal Holloway Col, Univ London, 71-72; asst prof biochem, Sch Med, Univ Miami, 73-75, assoc prof & asst prof med, 75-80; dir lab biochem res, Howard Hughes Med Inst, 73-80; mem staff, Miles Lab Inc, 80-; AT INST APPL BIOCHEM. *Concurrent Pos:* Ed-in-chief, J Appl Biochem. *Mem:* Biochem Soc; Am Soc Microbiol; Am Chem Soc; Am Soc Biol Chemists; Am Asn Cereal Chemists. *Res:* Structure and mechanism of action of glycoside hydrolases; structure, function and metabolism of polysaccharides; glycoproteins, especially structure, function and synthesis; naturally occurring enzyme inhibitors. *Mailing Add:* 1635 Victoria Dr Elkhart IN 46514-4161

MARSHALL, JAMES LAWRENCE, b Denton, Tex, May 19, 40; m 63, 81; c 2. ORGANIC CHEMISTRY. *Educ:* Ind Univ, Bloomington, BS, 62; Ohio State Univ, PhD(org chem), 66. *Prof Exp:* NIH fel org chem, Univ Colo, 66-67; asst prof chem, NTex State Univ, 67-71, assoc prof, 71-75, prof chem, 75-; MEM STAFF, CELANESE CHEMICAL CO, 81- *Mem:* Am Chem Soc. *Res:* Proton and carbon magnetic resonance studies of carbon-13 labeled compounds; conformational analysis; small polycyclic compounds; ammonia-metal reductions of aromatic compounds; nuclear magnetic resonance studies of small polycyclic compounds. *Mailing Add:* Dept Chem N Tex State Univ Box 5068 Denton TX 76203

MARSHALL, JAMES TILDEN, JR, b Canadian, Tex, July 30, 45; m 68; c 2. FOOD SCIENCE, DAIRY SCIENCE. *Educ:* Tex Tech Univ, BS, 68, MS, 69; Mich State Univ, PhD(food sci), 74. *Prof Exp:* Res asst dairy sci, Tex Tech Univ, 68-69 & food sci, Mich State Univ, 69-73; asst prof dairy & food sci, Miss State Univ, 74-77 & Kans State Univ, 77-80; DIR RES & QUAL ASSURANCE, FRIGO CHEESE CORP, 80- *Mem:* Am Dairy Sci Asn; Inst Food Technologists; Int Asn Milk, Food & Environ Sanitarians. *Res:* Chemical and physical properties of dairy and food products; formulation and processing of marketable products from dairy and other food by-products. *Mailing Add:* Perry's Ice Cream One Ice Cream Plaza Akron NY 14001

MARSHALL, JEAN MCELROY, b Chambersburg, Pa, Dec 31, 22. PHYSIOLOGY. *Educ:* Wilson Col, AB, 44; Mt Holyoke Col, MA, 46; Univ Rochester, PhD, 51. *Prof Exp:* Instr physiol, Mt Holyoke Col, 46-47; from instr to asst prof, Sch Med, Johns Hopkins Univ, 51-60; asst prof, Harvard Med Sch, 60-66; assoc prof, 66-69, PROF BIOL & MED SCI, 69-, E E BRINTZENHOFF PROF MED SCI, BROWN UNIV, 87- *Concurrent Pos:* Res fel pharmacol, Oxford Univ, 54-55; mem physiol study sect, NIH, 67-71 & eng in biol & med training comt, 71-73; mem physiol testing comt, Nat Bd Med Examrs, 71-76, Neurobiol Comt, 78-82. *Mem:* Am Physiol Soc; Soc Gen Physiol; Am Soc Pharmacol & Exp Therapeut; Soc Reprod Biol. *Res:* Electrical and mechanical properties of smooth muscle. *Mailing Add:* Div Biol & Med Sci Brown Univ Providence RI 02912

MARSHALL, JOHN CLIFFORD, b Whitewater, Wis, Jan 23, 35; m 57; c 2. ANALYTICAL CHEMISTRY. *Educ:* Luther Col, Iowa, BA, 56; State Univ Iowa, MS, 58, PhD(chem), 60. *Prof Exp:* Instr chem, State Univ Iowa, 60; fel, Univ Minn, 60-61; from asst prof to assoc prof, 61-74, PROF CHEM, ST OLAF COL, 74- *Concurrent Pos:* NSF res grant, St Olaf Col, 62-64, Petrol Res Fund grant, 64-67; res assoc, Argonne Nat Lab, 68-69; NSF fac res fel sci, Univ NC, 69-70. *Mem:* Am Chem Soc. *Res:* Flame spectroscopy and computer applications in chemistry. *Mailing Add:* Dept of Chem St Olaf Col Northfield MN 55057

MARSHALL, JOHN FOSTER, b Boston, Mass, Sept 5, 48; m 77. NEUROSCIENCE, NEUROPHARMACOLOGY. *Educ:* Williams Col, Mass, BA, 70; Univ Pa, Philadelphia, PhD(psychobiol), 73. *Prof Exp:* Vis scientist histochem, Karolinska Inst, Stockholm, 74-75; asst prof psychobiol, Univ Pittsburgh, 73-77; from asst prof to assoc prof, 77-85, PROF PSYCHOBIOL, UNIV CALIF, IRVINE, 85- *Concurrent Pos:* Mem, Task Force Obesity & Am Pub, Fogarty Ctr, 77; consult, Neurobiol Panel, Nat Sci Found, 80-82; fel, Found Fund Res Psychiat, 74-75 & Alfred P Sloan Found, 80-82. *Mem:* Sigma Xi; AAAS; Soc Neurosci. *Res:* Neural changes occurring within motor regions of the brain during pathology and senescence; nervous system compensation for cell death and how this compensation is affected in advanced age. *Mailing Add:* Dept Psychobiol Univ Calif Irvine CA 92717

MARSHALL, JOHN ROMNEY, b Los Angeles, Calif, Apr 27, 33; m 56; c 5. OBSTETRICS & GYNECOLOGY. *Educ:* Univ Pa, MD, 58; Am Bd Obstet & Gynec, cert, 67. *Prof Exp:* Intern, Los Angeles Gen Hosp, 58-59; instr pharmacol, Sch Med, Univ Pa, 59-60; resident obstet & gynec, George Washington Univ Hosp, 60-63, asst clin prof, 63-69; PROF OBSTET & GYNEC & VCHMN DEPT, SCH MED, UNIV CALIF, LOS ANGELES, 70-; CHMN DEPT, HARBOR GEN HOSP, 70- *Concurrent Pos:* Resident, DC Gen Hosp, 60-63; sr investr, Nat Cancer Inst, 63-69; consult, Long Beach Naval Hosp, Calif, 71-; mem hon fac staff, Mem Hosp Med Ctr, Long Beach, Calif, 71- *Mem:* Fel Am Col Obstet & Gynec; Soc Study Reprod; Endocrine Soc; Soc Gynec Invest; Am Fertility Soc. *Res:* Clinical pharmacology of reproductive biology. *Mailing Add:* 1000 W Carson Torrance CA 90509

MARSHALL, JOSEPH ANDREW, b New Brunswick, NJ, Mar 25, 37; m 58; c 2. ANIMAL BEHAVIOR. *Educ:* Univ Md, BS, 60, PhD(zool), 66. *Prof Exp:* USPHS fel, 66-68; asst prof, 68-73, ASSOC PROF BIOL, WVA UNIV, 73- *Mem:* Animal Behav Soc; Am Soc Ichthyol & Herpet. *Res:* Acoustical behavior of fishes; ecology of fishes. *Mailing Add:* Dept Biol WVa Univ Morgantown WV 26506-6057

MARSHALL, KEITH, b Leeds, Eng,. POWDER TECHNOLOGY, INSTRUMENTATION OF TABLET-CAPSULE MACHINES. *Educ:* Royal Pharmaceut Soc Gt Brit, chemist, 54, dipl biochem anal, 56; London Univ, BS, 63; Univ Bradford, UK, PhD(pharm), 70. *Prof Exp:* Lectr indust pharm, Univ Bradford, 56-76; res dir pharmaceut, Colorcon Inc, 76-78; dir pharm, Inst Appl Pharmaceut Sci, 78-83; ASSOC DIR PHARMACEUT, SMITHKLINE BEECHAM PHARMACEUT, 83- *Concurrent Pos:* Adj prof, Univ RI, 76-, Philadelphia Col Pharm & Sci, 84- *Mem:* Royal Pharmaceut Soc Gt Brit; Am Asn Pharmaceut Scientists; Fedn Int Pharceutique. *Res:* Characterization of powdered solids, including properties contributing to tableting and to bioavailability; application of instrumented presses to tablet formulation and development, and to tablet production problems. *Mailing Add:* 91 Pine Valley Rd Doylestown PA 18901

MARSHALL, KENNETH D, b Detroit, Mich, Apr 21, 41. MODAL ANALYSIS. *Educ:* Adrian Col, BS, 64; Mich State Univ, MS, 66, MBA, 68. *Prof Exp:* Physicist, Res & Develop Ctr, BF Goodrich, 68-70, sr physicist, 70-74, res assoc, 74-79, sr res assoc, 79-85; MGR, TIRE RES, UNIROYAL GOODRICH TIRE CO, 85- *Mem:* Soc Automotive Engrs; Am Phys Soc; Am Soc Testing & Mat. *Res:* Acoustics and vibrations; musical acoustics of stringed instruments; dynamics of rotating machinery; modal analysis; tire-vehicle vibrations. *Mailing Add:* Uniroyal Goodrich Tire Co 9921 Brecksville Rd Brecksville OH 44141

MARSHALL, KNEALE THOMAS, b Filey, Eng, Feb 13, 36; US citizen; m 64. OPERATIONS RESEARCH. *Educ:* Univ London, BSc, 58; Univ Calif, Berkeley, MS, 64, PhD(opers res), 66. *Prof Exp:* Metallurgist, Beaverlodge Oper, Eldorado Mining & Refining Ltd, 58-60, chief metallurgist, 60-62; mem tech staff, Bell Tel Labs, NJ, 66-68; from asst prof to assoc prof, Naval Postgrad Sch, 68-75, adv dep chief naval opers, 78-80, chmn, dept opers res, 80-83, dean, Info & Policy Sci, 83- 89, PROF OPER RES, NAVAL POSTGRAD SCH, 74-, CHIEF NAVAL OPERS CHAIR EMERGING TECHNOL, 90- *Concurrent Pos:* Consult, Res Proj in Higher Educ, Univ Calif, 68-70; assoc ed, Opers Res, 70-78 & Soc Indust & Appl Math J Appl Math, 71-78; vis prof, London Sch Econ, 78; coun mem, Oper Res Soc Am, 83- *Mem:* Opers Res Soc Am. *Res:* Stochastic models of congested systems; theory of manpower and budget planning. *Mailing Add:* Dept Oper Res Code OR/MT Naval Postgrad Sch Code 0223 Monterey CA 93943-5000

MARSHALL, LESLIE BRUSLETTEN, b Minneapolis, MN, 43; div; c 1. FEMALE REPRODUCTIVE HEALTH, INFANT CARE & FEEDING. *Educ:* Grinnell Col, BA, 65; Univ Washington, PhD(neurobiol), 73; Mt Mercy Col, BSN, 90. *Prof Exp:* ASSOC PROF, COL NURSING, UNIV IOWA, 78- *Concurrent Pos:* Neurosurg staff nurse; res officer, Univ Papua New Guinea & Papua New Guinea Inst Appl Social Econ Res; Woodrow Wilson Hon fel; bd chair, Asn Social Anthropologists, Oceania. *Mem:* numerous. *Res:* author one book on infant care and feeding in the South Pacific and co-author of 2 books; numerous journal articles. *Mailing Add:* Col Nursing Univ Iowa Iowa City IA 52242

MARSHALL, LOUISE HANSON, b Perrysburg, Ohio, Oct 2, 08; wid; c 2. NEUROSCIENCE, MAMMALIAN PHYSIOLOGY. *Educ:* Vassar Col, MA, 32; Univ Chicago, PhD(physiol), 35. *Prof Exp:* Asst physiol, Vassar Col, 30-32, instr, 36-37; asst, Univ Chicago, 34-35; physiologist, NIH, 43-65; prof assoc, Div Med Sci, Nat Acad Sci-Nat Res Coun, 65-75; admin analyst, 75-88, ASSOC DIR NEUROSCIENCE HISTORY PROG, BRAIN RES INST, UNIV CALIF, 80- *Concurrent Pos:* Ed, Neurosci Newsletter, 70-75; managing ed, Exp Neurol, 75-88. *Mem:* Am Physiol Soc; Soc Neurosci (secy-treas, 69-70). *Res:* Circulatory and renal response to plasma expanders; peripheral circulation; neuroscience administration; neuroscience history and archives. *Mailing Add:* Brain Res Inst Univ Calif Los Angeles CA 90024-1761

MARSHALL, MARYAN LORRAINE, b New Haven, Conn, Jan 18, 40. PHYSICAL CHEMISTRY. *Educ:* Conn Col, BA, 60; Yale Univ, PhD(phys chem), 65. *Prof Exp:* Instr chem, Randolph-Macon Woman's Col, 64-66, asst prof, 66-72; assoc prof, 72-75, PROF CHEM, CENT VA COMMUNITY COL, 75- *Mem:* AAAS; Am Chem Soc; Nat Sci Teachers' Asn; Sigma Xi. *Mailing Add:* 5804 Navajo Circle Lynchburg VA 24502-1412

MARSHALL, MAURICE K(EITH), b Monroe, Iowa, May 15, 21; m 47; c 3. MECHANICAL ENGINEERING. *Educ:* Purdue Univ, BS(mech eng) & BS(aeronaut eng), 46; Univ Ky, MS, 56. *Prof Exp:* Engr, Otis Elevator Co, 47-49, sales engr, 49-52; from instr to prof, 53-87, EMER PROF MECH ENG, UNIV KY, 87- *Concurrent Pos:* NSF fel, 60-61. *Mem:* Am Soc Mech Engrs; Am Soc Eng Educ. *Res:* Mechanical engineering, particularly fluid flow, heat transfer and propulsion. *Mailing Add:* 1650 Tazwell Dr Lexington KY 40504

MARSHALL, NELSON, b Yonkers, NY, Dec 16, 14; m 40; c 4. BIOLOGICAL OCEANOGRAPHY, MARINE SCIENCES. *Educ:* Rollins Col, BS, 37; Ohio State Univ, MS, 38; Univ Fla, PhD(biol), 41. *Prof Exp:* From instr to asst prof zool, Univ Conn, 41-45; asst prof & fisheries biologist, Marine Lab, Univ Miami, 45-46; assoc prof, Univ NC, 46-47; prof biol, Col William & Mary, 47-51, dean, 49-51, dir, Va Fisheries Lab, 47-51; assoc dir oceanog inst, Fla State Univ, 52-54; vis investr, Bingham Oceanog Lab, Yale Univ, 54-55; dean col lib arts, Alfred Univ, 55-59; dir int ctr marine resource develop, 72-75, prof oceanog, 59-84, prof marine affairs, 75-84, EMER PROF OCEANOG & MARINE AFFAIRS, GRAD SCH OCEANOG, UNIV RI, 84- *Concurrent Pos:* Hon mem bd trustees, Rollins Col; adj prof, Horn Pt Environ Lab, Univ Md, 85-, mem bd visitors, 87- *Mem:* Fel AAAS; Am Soc Limnol & Oceanog; Ecol Soc Am; hon mem Atlantic Estuarine Res Soc. *Res:* Estuarine and coral reef ecology; higher education for marine resource development in developing countries. *Mailing Add:* PO Box 1056 St Michaels MA 21663

MARSHALL, NORMAN BARRY, b Brooklyn, NY, Oct 3, 26; m 52; c 3. PHYSIOLOGY. *Educ:* Long Island Univ, BS, 49; Clark Univ, MA, 52; Harvard Univ, PhD(med sci), 56. *Prof Exp:* Instr physiol, Med Ctr, Duke Univ, 56-58, assoc, 58-59; res assoc nutrit, 59-61, head, Lipid Metab Sect, 61-67, mgr, Dept Biochem, 67-68, mgr, Hypersensitivity Dis Res, 68-79, group mgr, Therapeut Res, 79-81, RES DIR, UPJOHN C0, 81- *Mem:* Am Physiol Soc; NY Acad Sci; AAAS. *Res:* Regulation of food intake; endocrine control of metabolism. *Mailing Add:* 6738 Trotwood Kalamazoo MI 49001

MARSHALL, NORTON LITTLE, b Washington, DC, Dec 30, 27. BOTANY. *Educ:* Pa State Univ, BS, 49; Univ Md, MS, 52, PhD(bot), 55. *Prof Exp:* Asst pathologist, Trop Res Dept, United Fruit Co, Honduras, 49-50; asst, Univ Md, 50-54, instr bot, 55-56; pathologist, Res Dept, Firestone Plantations Co, Liberia, 56-58; from asst prof to prof bot, Auburn, Univ, 58-90; RETIRED. *Mem:* Bot Soc Am; Am Phytopath Soc. *Res:* Plant pathology; microbiology. *Mailing Add:* Dept Bot Auburn Univ Auburn AL 36849

MARSHALL, RICHARD ALLEN, b Madisonville, Tex, Aug 25, 35; m 59; c 2. POLYMER CHEMISTRY. *Educ:* Rice Univ, BA, 57; Ohio State Univ, PhD(org chem), 62. *Prof Exp:* Res chemist, Baytown Res & Develop Div, Exxon Res & Eng Co, 62-64, sr res chemist, 64-68; mem staff, Chem Div, Vulcan Mat Co, Kans, 68-74; assoc scientist, 83-90, SR RES CHEMIST, GOODYEAR RES, 74-, RES ASSOC, 90- *Mem:* Am Chem Soc. *Res:* Charge transfer complexes; exploratory polymers and polymerization processes; process and exploratory research in chlorinated organics; polyvinyl chloride polymerization; particle size analysis. *Mailing Add:* Goodyear Res 142 Goodyear Blvd Rm 334 Akron OH 44305

MARSHALL, RICHARD BLAIR, b Melrose, Mass, July 25, 28; m 53; c 5. PATHOLOGY. *Educ:* Boston Univ, BA, 49, MD, 55. *Prof Exp:* Intern, Detroit Receiving Hosp, 55-56; resident path, Henry Ford Hosp, Detroit, 56-60; from asst prof to prof path, Univ Tex Med Br Galveston, 64-75; PROF PATH, BOWMAN GRAY SCH MED, 75-, DIR ANAT PATH, 75-; MEM STAFF DEPT PATH, NC BAPTIST HOSP, 75- *Mem:* Am Soc Clin Path; US-Can Acad Path; Sigma Xi. *Res:* Cytochemical and ultrastructural studies of human endocrine pathology. *Mailing Add:* Dept Path Bowman-Gray Sch Med 300 S Hawthorne Rd Winston-Salem NC 27103

MARSHALL, RICHARD OLIVER, biochemistry; deceased, see previous edition for last biography

MARSHALL, ROBERT HERMAN, b Decatur, Ill, June 26, 25. INORGANIC CHEMISTRY. *Educ:* Ill State Normal Univ, 47, MS, 50; Univ Ill, PhD(inorg chem), 54. *Prof Exp:* Res chemist, Ethyl Corp, 54-58; assoc prof chem, La Polytech Inst, 58-60; assoc prof, 60-67, actg chmn dept, 70-71, PROF CHEM, MEMPHIS STATE UNIV, 67- *Res:* Organometallic compounds. *Mailing Add:* Dept Chem Memphis State Univ Memphis TN 38152

MARSHALL, ROBERT P(AUL), b Detroit, Mich, Dec 26, 30; m 57; c 2. METALLURGY. *Educ:* Univ Detroit, BS, 54. *Prof Exp:* Engr, Res Lab, Chrysler Corp, 53-54; proj mgr, Savannah River Lab, E I du Pont de Nemours & Co, Inc, 54-67; proj mgr, Battelle Mem Inst, 67-69, assoc mgr metall & ceramics dept, 69-70, mgr metall, 70-74, mgr fuels & mat dept, 74-76, mgr mat dept, 76-79, DIR PROG, PAC NORTHWEST DIV, BATTELLE MEM INST, 79- *Mem:* Am Inst Metall Engrs; Am Soc Metals. *Res:* Physical metallurgy; solid state diffusion; mechanisms of deformation of metals; irradiation effects; mechanical metallurgy; biomaterials; bioengineering. *Mailing Add:* 306 Saint St Richland WA 99352

MARSHALL, ROBERT T, b Halltown, Mo, July 27, 32; m 53; c 4. MICROBIOLOGY, FOOD SCIENCE. *Educ:* Univ Mo, BS, 54, MS, 58, PhD(food microbiol), 60. *Prof Exp:* From instr to asst prof dairy microbiol, 60-65, assoc prof dairy microbiol & mfrs, 65-70, PROF FOOD SCI & NUTRIT, UNIV MO-COLUMBIA, 70- *Mem:* Int Asn Milk, Food & Environ Sanit (pres); Am Dairy Sci Asn (pres); Inst Food Technol; Am Soc Microbiol; Am Pub Health Asn; Asn Anal Chem. *Res:* Functions of ingredients in frozen foods. *Mailing Add:* Univ Mo 122 Eckles Hall Columbia MO 65211

MARSHALL, ROSEMARIE, b Medford, Ore, Jan 28, 43. BACTERIOLOGY, BIOSTATISTICS. *Educ:* Univ Wash, BS, 64; Iowa State Univ, MS, 66, PhD(bact), 68. *Prof Exp:* NIH fel, Retina Found, Harvard Med Sch, 68-70; head dept bact, Grays Harbor Hosp, 70-71; from asst prof to assoc prof, Ga Southern Col, 71-78; ASST PROF MICROBIOL, CALIF STATE UNIV, LOS ANGELES, 78-; MEM INST ANTHROPODOLOGY & PARASITOL, 74- *Mem:* AAAS; Am Soc Microbiol. *Res:* Clinical bacteriology; turnover of macromolecules in vivo in differentiating systems; systems analysis of differentiating systems. *Mailing Add:* Dept Microbiol Calif State Univ 5151 St University Ave Los Angeles CA 90032

MARSHALL, SALLY J, b Racine, Wis, Jan 8, 49; m 70; c 2. BIOMATERIALS, DENTAL RESEARCH. *Educ:* Northwestern Univ, BS, 70; PhD(mat sci & eng), 75. *Prof Exp:* From instr to prof biomat, Northwestern Univ, 74-87; PROF & VCHAIR RES, RESTORATIVE DENT, UNIV CALIF, SAN FRANCISCO, 87- *Concurrent Pos:* Mem, Oral Biol & Med Study Sect, NIH, 80-84; vis fel, Univ Melbourne, Australia, 81; mem bd dir, Acad Dent Mat, 83-89; bd dirs, Am Asn Dent Res, 90- *Mem:* Am Soc Metals; Am Inst Mining, Metall & Petrol Eng; Int Asn Dent Res; Soc Biomat; Acad Dent Mat (treas, 83-85, vpres, 85-87, pres, 87-89; Am Ceramic Soc; Am Asn Dent Res (vpres, 90-91, pres elect, 91-92). *Res:* X-ray diffraction of biomaterials; degradation of dental materials; materials for use in restorative dentistry; amalgams; biomechanics; characterization of dentin. *Mailing Add:* Restorative Dent Univ Calif PO Box 0758 San Francisco CA 94143-0758

MARSHALL, SAMSON A, b Chicago, Ill, Oct 25, 24. SOLID STATE PHYSICS. *Educ:* Ill Inst Technol, BS, 50; Univ Mich, MS, 51; Cath Univ, PhD(physics), 56. *Prof Exp:* Physicist, Nat Bur Standards, 51-53, Naval Ord Lab, 53-56 & Armour Res Found, 56-65; physicist, Argonne Nat Lab, 65-80; MEM FAC, MICH TECHNOL UNIV, 80- *Mem:* AAAS; Am Phys Soc. *Res:* Microwave and radiofrequency spectroscopy of solids and gases. *Mailing Add:* Dept Physics Fisher Hall Michigan Technol Univ Houghton MI 49931

MARSHALL, SAMUEL WILSON, b Dallas, Tex, Sept 8, 34; m 56; c 3. PHYSICS. *Educ:* Va Mil Inst, BS, 55; Tulane Univ, MS, 63, PhD(physics), 65. *Prof Exp:* Jr res engr, Prod Res Div, Humble Oil & Refining Co, 55-56, 59-60; from asst prof to assoc prof physics, Colo State Univ, 65-70; mem staff, Naval Res Lab, 70-77; div dir, Ocean Sci & Technol Lab, Naval Ocean Res & Develop Activity, 77-83; sci adv, Navy Sci Assistance Prog, 83-84, dir, 84-88; SUPV SCIENTIST, BBN SYSTS & TECHNOL, 88- *Mem:* Am Phys Soc; Acoust Soc Am. *Res:* Mossbauer effect; acoustic propagation and scattering. *Mailing Add:* BBN Systems & Technol 1300 N 17th St Arlington VA 22209

MARSHALL, STANLEY V(ERNON), b Long Lane, Mo, Oct 3, 27; m 50; c 3. ELECTRICAL ENGINEERING, ELECTROMAGNETICS. *Educ:* Ore State Univ, BSEE, 54; Univ Mo-Columbia, MS, 65, PhD(elec eng), 67. *Prof Exp:* Mem tech staff radar systs, Bell Tel Labs, 54-63; res physicist, Naval Ord Labs, 67; ASSOC PROF ELEC ENG, UNIV MO-ROLLA, 67- *Mem:* Inst Elec & Electronics Engrs; Nat Soc Prof Engrs. *Res:* Magnetic field sensors; lightning detection. *Mailing Add:* Dept Elec Eng Univ Mo Box 249 Rolla MO 65401

MARSHALL, THEODORE, b Chicago, Ill, Dec 31, 27; m 54; c 3. PLASMA PHYSICS. *Educ:* Ill Inst Technol, BS, 51; Cath Univ Am, PhD(physics), 62. *Prof Exp:* Scientist, US Naval Ord Lab, 55-62; sr consult scientist, Res & Develop Div, Avco Corp, 62-64; mgr exp physics, Parametrics Inc, Mass, 65; assoc prof elec eng, Univ RI, 65-68; chmn dept, 68-80, PROF PHYSICS, SUFFOLK UNIV, 68- *Concurrent Pos:* vis scientist, Argonne Nat Lab, 79. *Mem:* Am Phys Soc; Sigma Xi; AAAS. *Res:* Propagation of electromagnetic waves in ionized gas; physics of fluids; high temperature properties of gases; electron spin resonance in crystalline solids. *Mailing Add:* Dept Physics Suffolk Univ Beacon Hill 41 Temple St Boston MA 02114

MARSHALL, THOMAS C, b Cleveland, Ohio, Jan 29, 35; m 64; c 2. PHYSICS. *Educ:* Case Inst Technol, BS, 57; Univ Ill, MS, 58, PhD(physics), 60. *Prof Exp:* Asst prof elec eng, Univ Ill, 61-62; from asst prof to assoc prof, 62-70, PROF APPL PHYSICS, COLUMBIA UNIV, 70- *Concurrent Pos:* Mem, Study Group Direceted Energy Weapons, Am Phys Soc, 85-87. *Mem:* Fel Am Phys Soc. *Res:* Plasma and atomic physics; microwave scattering and radiation from plasmas; lasers; shock waves; plasma stability; relativistic beams; toroidal containment experiments; free electron lasers. *Mailing Add:* Plasma Lab Columbia Univ New York NY 10027

MARSHALL, VINCENT DEPAUL, b Washington, DC, Apr 5, 43; m; c 2. MICROBIAL PHYSIOLOGY. *Educ:* Northeastern State Col, BS, 65; Univ Okla, MS, 67, PhD(microbiol), 70. *Prof Exp:* From res asst to res assoc microbiol, Univ Okla, 65-70; res assoc biochem, Univ Ill, 70-73; res scientist fermentation microbiol, Fine Chem Div, 73-74, res head microbial control, Prod Control Div, 75, res scientist cancer res, 75-78, from res scientist to sr res scientist, Infectious Dis Div, 78-87, SR RES SCIENTIST, UPJOHN LABS, UPJOHN CO, 87- *Concurrent Pos:* NIH fel, Univ Okla, 68-70, Univ Ill, 71-73; mem, Mem Comt, 87-89, co-chmn, Educ Comt Soc Indust Microbiol, 89- *Honors & Awards:* Fel, Am Acad Microbiol, 90. *Mem:* Am Soc Microbiol; Am Soc Biochem & Molecular Biol; Sigma Xi; Soc Indust Microbiol; Ind Sect Soc Indust Microbiol (pres elect, 90-91). *Res:* Microbial metabolism and transformation of antibiotics, amino acids and terpenes; fermentation microbiology; phagocyte-antibiotic interactions; natural product screening and discovery; numerous publications; patentee in field. *Mailing Add:* Upjohn Co 301 Henrietta St Kalamazoo MI 49007

MARSHALL, WALTER LINCOLN, b Princeton, NJ, May 6, 25; m 53; c 3. PHYSICAL CHEMISTRY. *Educ:* Princeton Univ, AB, 46; Harvard Univ, PhD(chem), 50. *Prof Exp:* Chemist, Mat & Processes Lab, Gen Elec Co, 50-53, Schenectady supvr appl res & insulation mat, 53-58, mgr chem & elec insulation, 58-67, mgr Mat & Processes Lab, Large Steam Turbine & Generator Div, 67-86; RETIRED. *Mem:* Inst Elec & Electronic Engrs; Am Chem Soc. *Res:* Electrical insulation of large rotating electrical apparatus. *Mailing Add:* Belmont Dr Trappe MD 21673

MARSHALL, WAYNE EDWARD, b Washington, DC, Dec 20, 44; m 68; c 1. BIOLOGICAL CHEMISTRY, FOOD CHEMISTRY. *Educ:* Univ Md, BS, 66; Univ Ill, PhD(biol chem), 71. *Prof Exp:* Res assoc physiol, Univ Ill Med Ctr, 71-77; group leader protein prod, Kraft, Inc, 77-85; RES LEADER, SOUTHERN REGIONAL RES CTR, AGR RES SERV, USDA, 85- *Mem:* Am Chem Soc; Sigma Xi. *Res:* Food proteins; food component chemistry; enzymology; calorimetry of food systems. *Mailing Add:* 1000 Rue Corton Slidell LA 70458

MARSHALL, WILLIAM EMMETT, b Chicago, Ill. IMMUNOLOGY, MOLECULAR BIOLOGY. *Educ:* Univ Ill, Urbana, BS, 57, MS, 59, PhD(biochem), 61. *Prof Exp:* Teaching fels biochem, Uppsala Univ, Sweden & Cambridge Univ, Eng, 62-64; asst prof, Med Sch, Univ Minn, Minneapolis, 64-67; dir tech develop, Gen Foods Corp, White Plains, 67-83; PRES, MICROBIAL GENETICS DIV, PIONEER HI-BRED INT, 83-; PRES, MICROBIAL ENVIRON SERV, INC, 90- *Concurrent Pos:* Supvr, Chem Lab, Vet Admin Hosp, Minneapolis, 64-67; chmn, Nat Agr Res & Exten Users Adv Comn, 82-88; mem, task force on competency in agr, Nat Res Coun, 84-86; mem bd dirs, Agrion, 85-89 & Rodale Inst, 85-; mem new develop biotechnol, Off Technol Assessment, US Cong, 86-90, adv bd, Intellectual Property Matters, Gen Agreements Trade & Tariff, 88-; mem Awards & Recognition Comt, Iowa Acad Sci, 87- *Mem:* AAAS; Soc Indust Microbiol; Am Soc Microbiol. *Res:* Biotechnology of naturally-occurring organisms of benefit to agriculture. *Mailing Add:* 701 Foster Dr Des Moines IA 50312

MARSHALL, WILLIAM HAMPTON, b Montreal, Que, Apr 20, 12; nat US; m 37; c 1. WILDLIFE MANAGEMENT. *Educ:* Univ Calif, BS, 33; Univ Mich, MF, 35, PhD(wildlife mgt), 42. *Prof Exp:* Foreman, US Forest Serv, Calif, 33 & Ark, 34, asst conservationist, Mass, 35; asst, Univ Mich, 34-35; instr wildlife mgt, Utah State Col, 36; jr biologist, US Fish & Wildlife Serv, Idaho, 36-43; area supvr, War Food Admin, 43-44, wage control off, 44-45; from assoc prof to prof econ zool, 45-70, from assoc dir to dir, Lake Itasca Forestry & Biol Sta, 55-70, prof, 70-78, EMER PROF WILDLIFE MGT, UNIV MINN, ST PAUL, 78- *Mem:* Hon mem Wildlife Soc. *Res:* Ecology and management of woodcock and grouse. *Mailing Add:* 1215 Evergreen Ct No 325 Clarkston WA 99403-2800

MARSHALL, WILLIAM JOSEPH, b Pittsburgh, Pa, Apr 10, 29; m 56; c 3. PHYSICAL CHEMISTRY. *Educ:* Univ Pittsburgh, BS, 51; Carnegie Inst Technol, MS, 55, PhD, 56. *Prof Exp:* Res supvr, E I du Pont de Nemours & Co, Inc, 56-68, res mgr, Pigments Dept, 68-69, asst lab dir, 69-74, tech supt, Edge Moor Lab, 74-91; RETIRED. *Mem:* Am Chem Soc. *Res:* Pigment technology; solid state chemistry. *Mailing Add:* 1124 Webster Dr Wilmington DE 19803

MARSHALL, WILLIAM LEITCH, b Columbia, SC, Dec 3, 25; m 49; c 2. PHYSICAL CHEMISTRY. *Educ:* Clemson Univ, BS, 45; Ohio State Univ, PhD(phys org chem), 49. *Prof Exp:* Asst chem, Clemson Univ, 44-45; asst chem, Ohio State Univ, 45-46; Naval res fel, Ohio State Univ, 47-49; from chemist to sr chemist, Oak Ridge Nat Lab, 49-57, group leader, 57-74, sr staff scientist, 75-89; RETIRED. *Concurrent Pos:* Guggenheim fel, Van der Waals Lab, Univ Amsterdam, 56-57; mem, Org Comt, First Int Cong High Temperature Aqueous Electrolytes, Eng, 73 & Int Asn Properties of Steam Working Groups, 75-; vis lectr, Tenn Acad Sci, 75-80. *Honors & Awards:* Charles Holmes Herty Gold Medal Award, Am Chem Soc, 77. *Mem:* AAAS; Geochem Soc; Sigma Xi; Am Chem Soc; NY Acad Sci; Am Geochem Union. *Res:* High temperature high presssure thermodynamics of aqueous systems; geothermal energy solubilities; electrical conductance of aqueous electrolytes to 800 degrees centigrade and 4000 atmospheres; chemistry of aqueous homogeneous reactors; aqueous uranium and thorium salt systems; effect of pressure on elastic constants of quartz; constitution of Grignard-type reagents. *Mailing Add:* 101 Oak Lane Oak Ridge TN 37831-4046

MARSHALL, WILLIAM SMITHSON, b Victoria, BC, April 26, 51; m 86; c 2. EPITHELIAL TRANSPORT, COMPARATIVE ENDOCRINOLOGY. *Educ:* Acadia Univ, BSc, 73; Univ BC, PhD(zool), 77. *Prof Exp:* Natural Sci & Eng Res Coun postgrad scholarships, 73-77, postdoc fel & Killam postdoc fel, 77-79; assoc ophthal, La State Univ Med Ctr, 79-81, instr, 81-82; asst prof, 82-87, ASSOC PROF BIOL, ST FRANCIS XAVIER UNIV, 87- *Concurrent Pos:* Univ res fel, Natural Sci & Eng Res Coun, 82-92. *Mem:* Can Soc Zool; Am Soc Zool; Am Physiol Soc; Can Physiol Soc; AAAS. *Res:* Hormonal control of ion transport processes in renal and reproductive epithelia; microspectro fluorometric studies of intracellular ph regulation in epithelia; marine and freshwater fish physiology. *Mailing Add:* Dept Biol St Francis Xavier Univ PO Box 45 Antigonish NS B2G 1C0 Can

MARSHALL, WINSTON STANLEY, b Nashville, Tenn, Jan 16, 37; m 61; c 3. MEDICINAL CHEMISTRY. *Educ:* Vanderbilt Univ, AB, 59; Wayne State Univ, PhD(org chem), 63. *Prof Exp:* Sr org chemist, 63-69, res scientist med chem, 69-72, res assoc, 72-73, head org chem, 73-77, res assoc, 77-81, res adv, 81-83, RES CONSULT, LILLY RES LABS, ELI LILLY & CO, 83- *Mem:* AAAS; Am Chem Soc; Am Rheumatism Asn; NY Acad Sci; Inflammation Res Asn. *Res:* Drug design in the areas of arthritis, immunology, allergy, and asthma, especially leukotriene synthesis modulating drugs and leukotriene antagonists. *Mailing Add:* 16 Blair Patch Rd Bargersville IN 46106

MARSHEK, KURT M, b Clintonville, Wis, Oct 13, 43; m 71; c 2. MECHANICAL ENGINEERING DESIGN. *Educ:* Univ Wis-Madison, BS, 66, MS, 68; Ohio State Univ, PhD(mech eng), 71. *Prof Exp:* Engr, Falk Corp, 64-65 & Perfex Corp, 66; instr eng, State Univ NY, 67-68; spec tech asst, Western Elec Co, 68; instr mech eng, Ohio State Univ, 69-71; prof, Univ Conn, 71-75; prof, Univ Houston, 75-81; PROF MECH ENG, UNIV TEX, AUSTIN, 81- *Concurrent Pos:* Fel, Ames Res Ctr & Stanford Univ, 72; Lectr, Rogers Corp, 74, Pullman-Kellogg Co, 75 & Brown & Root, 77; fel, Gulf Oil, 85. *Honors & Awards:* Ralph R Teetor Award, Soc Automotive Engrs, 73. *Mem:* Soc Automotive Engrs; Soc Mfg Engrs; Am Soc Metals; Am Gear Mfg Asn; Am Soc Mech Engrs. *Res:* Implementation of finite difference method for determining load distributions in engaged machine elements: chains, sprockets, bearings, timing belts, pulleys, and threaded connectors; tribology and stress analysis. *Mailing Add:* Dept Mech Eng Univ Tex Austin TX 78712

MARSHO, THOMAS V, plant biochemistry; deceased, see previous edition for last biography

MARSI, KENNETH LARUE, b Los Banos, Calif, Dec 13, 28; m 55; c 4. PHYSICAL ORGANIC CHEMISTRY. *Educ:* San Jose State Col, AB, 51; Univ Kans, PhD(org chem), 55. *Prof Exp:* Instr chem, Univ Kans, 54; sr res chemist, Sherwin-Williams Co, 55-57; from asst prof to assoc prof chem, Ft Hays Kans State Col, 57-61; from asst prof to assoc prof, 61-70, PROF CHEM, CALIF STATE UNIV, LONG BEACH, 70-, CHMN DEPT, 75- *Concurrent Pos:* Petrol Res Fund grant, 59-61; NIH spec fel, Rutgers Univ, 67-68; NSF grants, 68-79; Danforth assoc, 77-82. *Mem:* Am Chem Soc; Sigma Xi. *Res:* Synthesis and stereochemistry of reactions of organophosphorus compounds; ring closure reactions of compounds leading to phosphorus and nitrogen heterocycles. *Mailing Add:* Dept of Chem Calif State Univ 1250 Bellflower Blvd Long Beach CA 90840-0001

MARSLAND, DAVID B(OYD), b Ft Meade, Fla, Dec 27, 26; m 51; c 4. CHEMICAL ENGINEERING, ENGINEERING ECONOMY. *Educ:* Cornell Univ, BChE, 51, PhD(chem eng), 58. *Prof Exp:* Jr res assoc, Brookhaven Nat Lab, 55-58; res engr, Eng Res Lab, E I du Pont de Nemours & Co, 58-61; from asst prof chem eng to prof, 61-88, EMER PROF CHEM ENG, NC STATE UNIV, 88- *Concurrent Pos:* Consult, Corning Glass Works, 63-66 & 73, Monsanto, 76 & Res Triangle Inst, 77-85; Ford Found resident in eng pract, Esso Res & Eng Co, NJ, 66-67; staff engr, Cost Anal, US Environ Protection Agency, 73-74. *Mem:* Am Asn Univ Professors. *Mailing Add:* Dept Chem Eng NC State Univ Box 7905 Raleigh NC 27695-7905

MARSLAND, T(HOMAS) ANTHONY, b Dundee, Scotland, Apr 24, 37; m 60; c 3. COMPUTER SCIENCE. *Educ:* Univ Nottingham, BSc, 58; Univ Wash, PhD(elec eng), 67. *Prof Exp:* Engr, Eng Elec Co, Luton, Eng, 58-62; res engr, Boeing Co, 62-65; res asst elec eng, Univ Wash, 65-67, asst prof, 67-68; mem tech staff, Bell Tel Labs, 68-70; assoc prof, 70-80, PROF COMPUT SCI, UNIV ALTA, 80- *Concurrent Pos:* Sr indust fel, 80-81; McCalla res prof, 85-86; prof engr, 85- *Mem:* Asn Comput Mach; Sigma Xi; sr mem Inst Elec & Electronics Engrs. *Res:* Chess playing computer programs; distributed computing applications; control of synchronization overhead in multiprocessor systems; deadlock avoidance in distributed systems. *Mailing Add:* Dept Comput Sci Univ Alta Edmonton AB T6G 2H1 Can

MARSOCCI, VELIO ARTHUR, b Corona, NY, June 7, 28; m 55; c 1. ELECTRICAL ENGINEERING. *Educ:* NY Univ, BEE, 53, MEE, 55, EngScD, 64. *Prof Exp:* Instr elec eng, NY Univ, 54-56; from asst prof to assoc prof, Stevens Inst Technol, 56-65; assoc prof, 65-67, acting chmn dept, 69-71, chmn dept, 71-74, acting assoc dean, Col Eng & Appl Sci, 74-76, PROF ELEC ENG, STATE UNIV NY, STONY BROOK, 67-, CLIN PROF HEALTH SCI, 75- *Concurrent Pos:* Consult indust. *Honors & Awards:* Region 1 Award, Inst Elect & Electronics Engrs, Engr of Year Award, Suffolk City Chap, NYSSPE. *Mem:* AAAS; Inst Elec & Electronics Engrs; Am Soc Eng Educ; Am Phys Soc; Nat Soc Prof Engrs. *Res:* Solid state theory; physical electronics; electronic devices; biomedical electronics. *Mailing Add:* Dept Elec Eng State Univ NY Stony Brook NY 11794-2350

MARSTEN, RICHARD B(ARRY), b New York, NY, Oct 28, 25; m 49; c 2. ELECTRONICS ENGINEERING. *Educ:* Mass Inst Technol, SB & SM, 46; Univ Pa, PhD(eng phys), 51. *Prof Exp:* Asst commun & electronics, Mass Inst Technol, 46-49; instr & asst electronics, Moore Sch Elec Eng, Univ Pa, 49-51; engr, Lab Electronics, 51-52; head res & develop sect, Allen B DuMont Labs, 52-55; chief microwave engr, Polarad Electronics Corp, 55; staff engr, Res & Develop Div, Air Assocs, 55-56; proj leader tactical radar systs, Missile & Surface Radar Div, Radio Corp Am, 56-59, mgr radar systs proj, 59-61, mgr adv radar systs, 61, mgr space commun systs, Astro-Electronics Div, 61-64, mgr spacecraft electronics, 64-66, mgr reliability & sr tech staff, 66-67, chief engr, 67-69; dir commun progs, Off Space Sci & Appln, NASA, 69-75; dean, Sch Eng, City Col, City Univ New York, 75-79; mgr, Space Policy & Applns Prog, Off Technol Assessment, US Cong, 80; exec dir, Bd Telecommun & Comput Appln, Nat Acad Sci/Nat Acad Eng, 81-88; dir, advan technol, Vitro Corp, 88-90; EXEC VPRES & CHIEF OPERATING OFFICER, INT RADIO SATELLITE CORP, 90- *Concurrent Pos:* Mem, Nat Technol Comt Commun, Am Inst Astronaut & Aeronaut, 64-66, chmn, 67-70; prog chmn, Commun Satellite Systs Conf, 66; chmn, Tech Specialty Group Info Syst, Am Inst Astronaut & Aeronaut, 70-71; prog chmn, Electronic & Aerospace Systs Conf, 70, adv bd, 72-75, bd dir, 84-; mem commun policy bd, Inst Elec & Electronic Engrs, 72-76; mem space appl study, Panel Space Broadcasting, Nat Acad Sci, 67, chmn panel, Points-to-Point Commun, Space Appln Study, 67-68; consult, Arthur D Little, Inc, 75-, Western Union Space Commun, Inc, 78- & Secy of Space, Govt India, 78-; chmn, Comt Telecommun Policy, 78-80, mem, 80-82; mem bd gov, Aerospace & Electronic Systs Soc, 82-, vpres, Tech Opers, 85-88; mem, Aerospace Res & Develop Policy Ctr, Eng Res & Develop Policy Ctr, Inst Elec & Electronic Engrs, 85- *Honors & Awards:* White House Citation for Sustained Superior Performance, 72; NASA Except Serv Medal, 74, Group Achievement Award, ATS-6 Proj, 74; Apollo-Soyuz Space Medal, 76. *Mem:* Fel Inst Elec & Electronic Engrs; NY Acad Sci; AAAS; Nat Energy Found. *Res:* Network synthesis; computer logical synthesis; microwave devices and circuits; signal processing; radar systems and equipment; spacecraft systems; space electronics and equipment; space exploration, communications and satellite broadcasting. *Mailing Add:* 9531 E Stanhope Rd Kensington MD 20895

MARSTERS, GERALD F, b Summerville, NS, Dec 18, 32; m 54; c 1. MECHANICAL ENGINEERING, AEROSPACE ENGINEERING. *Educ:* Queen's Univ, Ont, BSc, 62; Cornell Univ, PhD(aerospace eng), 67. *Prof Exp:* From asst prof to assoc prof, 67-76, PROF MECH ENG, QUEEN'S UNIV, ONT, 76- *Concurrent Pos:* Consult, Defence Res Bd Can, 67- *Mem:* Am Inst Aeronaut & Astronaut; Am Soc Eng Educ; Nat Soc Prof Engrs; Can Soc Mech Engrs; Am Soc Mech Engrs. *Res:* General fields of heat transfer and low velocity fluid mechanics; vistol aerodynamics. *Mailing Add:* Nat Res Council Montreal Rd Ottawa ON K1A 0R6 Can

MARSTON, CHARLES H, b Lynbrook, NY, Jan 3, 32; m 56; c 4. POWER PLANT SYSTEMS, COAL COMBUSTION. *Educ:* Stevens Inst Technol, ME, 53; Mass Inst Technol, MS, 59, MechE, 61, ScD, 62. *Prof Exp:* Trainee, Allis-Chalmers Mfg Co, 53-54; res engr, Gen Elec Co, 62-71, sr res engr, Space Sci Lab, Space Div, 72-76, mgr MHD Systs & anal, 76-82; ASSOC PROF MECH ENG, VILLANOVA UNIV, 82- *Concurrent Pos:* Lectr, Grad Sch, Pa State Univ, Radnor Ctr, 65-75. *Mem:* Am Soc Mech Engrs; Am Inst Aeronaut & Astronaut; Sigma Xi; Am Soc Eng Educ. *Res:* Power plant systems analysis; magnetohydrodynamic power generation and coal fired combusters; electric arc radiation at high pressure; magnetohydrodynamic acceleration; hypersonic flow in shock tunnels; power plant systems coal combustion. *Mailing Add:* 301 Greene Rd Berwyn PA 19312

MARSTON, GEORGE ANDREWS, b Montague City, Mass, Oct 5, 08; m 34; c 2. HYDRAULIC ENGINEERING. *Educ:* Worcester Polytech Inst, BS, 30, CE, 40; Univ Iowa, MS, 33. *Hon Degrees:* DEng, Worcester Polytech Inst, 58. *Prof Exp:* Field engr, Turners Falls Power Co, Mass, 30-31; instr math & civil eng, Mass State Col, 33-37; jr engr, Bur Reclamation, US Dept Interior, Colo, 34 & US Geol Surv, Mass, 36-37; asst prof civil eng, 37-43, prof, 46-63, head dept, 46-48, acting dean, sch eng, 46-47, dean, 48-63, EMER PROF CIVIL ENG & EMER DEAN SCH ENG, UNIV MASS, AMHERST, 63- *Concurrent Pos:* Jr engr, US Eng Dept, RI, 38-39, asst engr, 39; asst engr, US Geol Surv, Mass, 41; prof eng & dean sch eng, Western New Eng Col, 63-68, prof mech eng, 68-73. *Mem:* Am Soc Civil Engrs; Am Soc Eng Educ; Am Geophys Union; Sigma Xi. *Res:* Rainfall intensity-frequency relations; New England climatology; runoff-frequency. *Mailing Add:* 323 E Pleasant St Amherst MA 01002

MARSTON, PHILIP LESLIE, b Seattle, Wash, Feb 1, 48; m 76; c 1. OPTICS, NONLINEAR ACOUSTICS. *Educ:* Seattle Pac Col, BS, 70; Stanford Univ, MS, 72 & 74, PhD(physics), 76. *Prof Exp:* Res asst, Stanford Univ, 71-76; fel, Yale Univ, 76-78; ASST PROF PHYSICS, WASH STATE UNIV, 78- *Concurrent Pos:* Teaching asst, Stanford Univ, 74; res fel, Alfred P Sloan Found, 80-; consult, Jet Propulsion Lab, 81. *Mem:* Sigma Xi; Am Phys Soc; Acoustical Soc Am; Am Asn Physics Teachers; Optical Soc Am. *Res:* Scattering of acoustical and optical waves: scattering of light and sound by bubbles, radiation pressure of ultrasound, and cavitation of liquids induced by shock wave reflection; quantum liquids. *Mailing Add:* Dept Physics Wash State Univ Pullman WA 99164

MARSTON, ROBERT QUARLES, b Toano, Va, Feb 12, 23; m 46; c 3. MEDICINE. *Educ:* Va Mil Inst, BS, 43; Med Col Va, MD, 47; Oxford Univ, BSc, 49. *Prof Exp:* House officer med, Johns Hopkins Hosp, 49-50; asst resident, Vanderbilt Univ, 50-51; asst resident, Med Col Va, 53-54, from asst prof to assoc prof, 54-61, asst dean, 59-61; vchancellor & dean sch med, Univ Miss, 61-66; assoc dir, Regional Med Prog, NIH, 66-68, adminstr, Health Serv & Ment Health Admin, 68, dir, NIH, 68-73; scholar-in-residence, Univ Va, Charlottesville, 73-74; pres, 74-84, JOINT PROF FISHERIES & AQUACULTURE, EMER PROF MED & EMER PRES, UNIV FLA, 84- *Concurrent Pos:* Markle scholar, Med Col Va, 54-59; mem staff, Armed Forces Spec Weapons Proj, NIH, 51-53, chmn int fels rev panel, 64-66; asst prof, Univ Minn, 58-59; consult rev comt, Div Hosp & Med Facil, Dept Health, Educ & Welfare, 61-66; distinguished fel, Inst Med, Nat Acad Sci; mem bd vis, Charles R Drew Postgrad Sch; chmn bd vis, Air Univ; mem, Fla Coun 100; mem, Vet Admin Scholars Bd; mem bd dirs, Johnson & Johnson; mem coun, Inst Med, Nat Acad Sci; mem, Inst Med Comt Aging; chmn, Nat Asn State Univs & Land Grant Col Comt on Health Policy; mem, Macy Found Comn on Present Conditon & Future of Acad Psychiat in US; chmn, Sloan Found Cognitive Sci Adv Comt. *Mem:* Inst Med-Nat Acad Sci; fel Am Pub Health Asn; hon mem Nat Med Asn; hon mem Am Hosp Asn; Nat Asn State Univs & Land Grant. *Res:* Infectious diseases; medical administration. *Mailing Add:* Univ Fla 7922 NW 71st St Gainesville FL 32606

MARTEL, HARDY C(ROSS), b Pasadena, Calif, Jan 4, 27; m 54; c 5. ELECTRICAL ENGINEERING. *Educ:* Calif Inst Technol, BS, 49, PhD(elec eng), 56; Mass Inst Technol, MS, 50. *Prof Exp:* Res engr elec eng, Jet Propulsion Lab, 50-51, from instr to asst prof, 53-58, ASSOC PROF ELEC ENG, CALIF INST TECHNOL, 58-, EXEC ASST TO PRES, 69-, SECY BD TRUSTEES, 73- *Concurrent Pos:* Mem tech staff, Bell Tel Labs, Inc, 59-60. *Mem:* Inst Elec & Electronics Engrs; Sigma Xi. *Res:* Communication theory; stochastic processes; decision and information theory. *Mailing Add:* 1864 Midlathian Dr Altadena CA 91001

MARTEL, WILLIAM, b New York, NY, Oct 1, 27; m 55; c 4. RADIOLOGY. *Educ:* NY Univ, BS, 50, MD, 53. *Prof Exp:* PROF RADIOL, UNIV MICH, ANN ARBOR, 66- *Mem:* AMA; Radiol Soc NAm. *Mailing Add:* Dept of Radiol Univ of Mich Ann Arbor MI 48109

MARTELL, ARTHUR EARL, b Natick, Mass, Oct 18, 16; m 44, 65; c 8. CHEMISTRY. *Educ:* Worcester Polytech Inst, BS, 38; NY Univ, PhD(chem), 41. *Hon Degrees:* DSc, Worcester Polytech Inst, 62. *Prof Exp:* Asst, NY Univ, 38-40; instr chem, Worcester Polytech Inst, 41-42; from asst prof to prof, Clark Univ, 42-61, chmn dept, 59-61; prof & chmn dept, Ill Inst Technol, 61-66; head dept chem, 66-80, adv to pres for res, 80-82, DISTINGUISHED PROF CHEM, TEX A&M UNIV, 66- *Concurrent Pos:* Res fel, Univ Calif, 49-50; Guggenheim fel, Univ Zurich, 54-55; NSF sr fel, 59-60; fel, Sch Advan Studies, Mass Inst Technol, 59-60; NIH fel, Univ Calif, Berkeley, 64-65; ed, J Coord Chem, 70-80. *Honors & Awards:* Inorg Chem Award, Am Chem Soc, 80. *Mem:* AAAS; Am Chem Soc; fel Am Acad Arts & Sci; hon mem NY Acad Sci; Am Soc Biol Chemists. *Res:* Synthesis; potentiometry; spectroscopy; physical and chemical properties, stabilities and catalytic effects of metal chelate compounds. *Mailing Add:* Dept Chem Tex A&M Univ College Station TX 77843

MARTELL, EDWARD A, b Spencer, Mass, Feb 23, 18; m 42; c 4. RADIOCHEMISTRY, NUCLEAR GEOCHEMISTRY. *Educ:* US Mil Acad, BS, 42; Univ Chicago, PhD(nuclear chem), 50. *Prof Exp:* Prog dir, Armed Forces Spec Weapons Proj, Washington, DC, 50-54; res assoc, Enrico Fermi Inst Nuclear Studies, Univ Chicago, 54-56; group leader atmospheric radioactivity & fallout, Geophys Res Div, Air Force Cambridge Res Lab, Mass, 56-62; RES SCIENTIST, NAT CTR ATMOSPHERIC RES, 62- *Concurrent Pos:* Secy, Int Comn Atmospheric Chem & Global Pollution, Int Asn Meteorol & Atmospheric Physics, 67-75, pres, 75-79. *Mem:* Fel AAAS; Am Geophys Union; Health Phys Soc; Sigma Xi; Radiation Res Soc; Am Chem Soc; hon mem Int Asn Meteorol & Atmospheric Physics. *Res:* Natural radioactivity; discovery of indium-115 beta activity; radiation and fallout effects of nuclear explosions; nuclear meteorology; upper atmosphere composition with rocket samplers; radioactive aerosols; environmental and biological distribution and health effects of primordial and cosmogenic radionuclides. *Mailing Add:* Nat Ctr for Atmospheric Res PO Box 3000 Boulder CO 80307-3000

MARTELL, MICHAEL JOSEPH, JR, b Minneapolis, Minn, May 20, 32. INDUSTRIAL CHEMISTRY, PHARMACY. *Educ:* Univ Minn, BS, 54, PhD(pharmaceut chem), 58; Fairleigh Dickinson Univ, MBA, 75. *Prof Exp:* NIH fel, Univ Ill, 59-60; res chemist, Lederle Labs Div, 60-70, mgr prod develop, 70-75, DIR, OVERSEAS PROD DEVELOP, MED RES DIV, AM CYANAMID CO, 75- *Mem:* Am Chem Soc. *Res:* Tetracycline antibiotics; alkaloids; biopharmaceutics as pertains to product development. *Mailing Add:* Dept 990 Lederle Labs Am Cyanamid Co Pearl River NY 10965-1299

MARTELLOCK, ARTHUR CARL, b Detroit, Mich, Jan 7, 28; m 49; c 3. POLYMER CHEMISTRY, ORGANIC CHEMISTRY. *Educ:* Wayne State Univ, AB, 51; Rutgers Univ, PhD(org chem), 57. *Prof Exp:* Chemist, Silicone Prod Dept, Gen Elec Co, 56-68, specialist silicone rubber develop, 68-70; tech specialist & proj mgr, Spec Mat Eng Area, 70-80, SR TECH SPECIALIST MAT, XEROX CORP, 70- *Mem:* Fel Am Inst Chemists; Am Chem Soc. *Res:* Materials development; polymer molecular structure and rheology; silicone polymer synthesis; characterization and degradation kinetics. *Mailing Add:* 54 Parkridge Dr Pittsford NY 14534

MARTEN, DAVID FRANKLIN, b Springfield, Ill, Feb 1, 48; m 67; c 2. ORGANIC CHEMISTRY, ORGANOMETALLIC CHEMISTRY. *Educ:* Western Ill Univ, BS, 70; Univ Wis, Madison, PhD (chem), 74. *Prof Exp:* Res assoc organometallic chem, Brandeis Univ, 74-76; asst prof, Univ Iowa, 76-77; Asst Prof org chem, Univ Oklahoma 77-83; ASSOC PROF, CHEM, WESTMONT COL, 83- *Mem:* Am Chem Soc; Am Sci Affil. *Res:* Organic synthesis and reaction mechanisms. *Mailing Add:* Dept Chem Westmont Col Santa Barbara CA 93108

MARTEN, GORDON C, b Wittenberg, Wis, Sept 14, 35; m 61; c 1. AGRONOMY. *Educ:* Univ Wis, BS, 57; Univ Minn, MS, 59, PhD(agron), 61. *Prof Exp:* Res agronomist, Agr Res Serv, USDA, 61-72, supvr res agronomist & res leader, 72-89, ASSOC DIR, BELTSVILLE AREA, AGR RES SERV, USDA, 89- *Concurrent Pos:* From asst prof to assoc prof agron & plant genetics, Univ Minn, St Paul, 62-71, prof, 71-; prog chmn, XIV Int Grassland Cong, 81; mem bd dirs, Coun Agr Sci & Technol, 85-90; mem bd trustees, Agron Sci Found, 85-88. *Honors & Awards:* Thornton Distinguished lectr, Texas Tech Univ; Goddard Mem lectr, Univ Tenn; Merit Award, Outstanding Serv Award, Am Forage & Grassland Coun. *Mem:* Fel Crop Sci Soc Am; fel Am Soc Agron; Am Forage & Grassland Coun; Agron Sci Found. *Res:* Lab techniques for evaluating forage crops; nutritive value of forage crops as influenced by genetics and agronomic practices; cattle and sheep grazing response; effects of ecological factors on forage quality. *Mailing Add:* US Dept Agr ARS BARC-W 402 Nat Agr Library Beltsville MD 20705

MARTEN, JAMES FREDERICK, b Liverpool, Eng, Sept 11, 31; m 54; c 2. BIOCHEMISTRY. *Educ:* Univ Leeds, PhD, 56. *Prof Exp:* Gen mgr, Technicon Controls Inc, NY, 64-67; Co-founder, Damon Biomed Sci Inc, 69-73; co-founder & chmn, Delmed Inc, Mass, 73-88; CHMN & TREAS, HOOD LABS INC, MASS, 84- *Concurrent Pos:* Dir, Armendaris Corp, Mo, 75-84; founder & dir, Appl Immunosci, Inc, Calif, 84-87; vis prof, Northeastern Univ Grad Sch Bus Admin, Mass, 84-88; vchmn & chief financial officer, Medchem Products Inc, Mass, 85-; trustee, Eye Res Inst, Mass, 85-88. *Mem:* Fel Royal Soc Chem. *Res:* Conception and design of automated medical diagnostic instrumentation. *Mailing Add:* 78 Nichols Rd Cohasset MA 02025

MARTENS, ALEXANDER E(UGENE), b Schemnitz, Czech, June 27, 23; m 48; c 2. ELECTRONICS, ENGINEERING MANAGEMENT. *Educ:* Breslau Tech Univ, BSEE, 42; Univ Rochester, MS, 64. *Prof Exp:* Dept head electronics, US Army Spec Serv, 46-52; engr, Motorola Ltd, Can, 53-54; vpres eng, Tele-Tech Electronics Ltd, 55-60; engr, Bausch & Lomb, 60-64, dept head electronics res & develop, 64-68, vpres res & develop, Anal Systs Div, 68-76, vpres image anal, 76-83, vpres res & develop, Sci Instruments Group, 81-83; CONSULT TECHNOL, 83- *Concurrent Pos:* Adj prof, Rochester Inst Technol, 78-83. *Honors & Awards:* IR 100 Award, Indust Res Mag, 66 & 68. *Res:* Research and development of biomedical, analytical and metrological instruments and systems; assessment of technologies. *Mailing Add:* 104 Nettlecreek Rd Fairport NY 14450

MARTENS, CHRISTOPHER SARGENT, b Akron, Ohio, Jan 11, 46; m 68; c 2. MARINE CHEMISTRY. *Educ:* Fla State Univ, BS, 68, MS, 69, PhD(chem oceanog), 72. *Prof Exp:* Res technician Antarctic sediment chem, Fla State Univ, 68-69; guest scientist, Lawrence Livermore Radiation Lab, 71-72; res staff marine chem, Dept Geol & Geophys, Yale Univ, 72-74; from asst prof to assoc prof, 74-83, PROF MARINE SCI & GEOL, MARINE SCI PROG, UNIV NC, CHAPEL HILL, 83- *Concurrent Pos:* Guest investr, Woods Hole Oceanog Inst, 81-82, adj scientist, Dept Chem, 83-; chmn, Gordon Res Conf Chem Oceanog, 87. *Mem:* AAAS; Am Soc Limnol & Oceanog; Am Geophys Union; Geochem Soc. *Res:* Chemical processes in organic-rich marine environments, particularly, microbially-mediated gas production and consumption; nutrient regeneration; organic matter remineralization; organic and amino acid cycling; chemical exchanges between sediments; water and atmosphere; tracer studies utilizing radon-222 and lead-210, in situ flux measurements; kinetic modeling. *Mailing Add:* Marine Sci Prog 045A Univ NC Chapel Hill NC 27514

MARTENS, DAVID CHARLES, b Shawano, Wis, Apr 17, 33; m 57; c 2. SOIL SCIENCE. *Educ:* Univ Wis, BS, 60, MS, 62, PhD(soil sci), 64. *Prof Exp:* Asst prof, 64-68, assoc prof soil sci, 68-81, PROF AGRON, VA POLYTECH INST & STATE UNIV, 81- *Mem:* Am Soc Agron; Soil Sci Soc Am; Soil Conserv Soc; Sigma Xi. *Res:* Micronutrient chemistry of soil; diagnosis of chemical factors of soil responsible for abnormal plant growth; by-product disposal. *Mailing Add:* Dept of Agron Va Polytech Inst & State Univ Blacksburg VA 24061

MARTENS, EDWARD JOHN, b Evergreen Park, Ill, July 31, 38; m 59; c 2. NUCLEAR PHYSICS. *Educ:* Mass Inst Technol, BS, 61, MS, 65, PhD(physics), 67. *Prof Exp:* Instr physics, Northeastern Univ, 67-69; sr scientist, Am Sci & Eng, Inc, 69-71; instr, 71-74, ASST PROF INDUST TECHNOL FITCHBURG STATE COL, 74- *Mem:* Am Phys Soc. *Res:* X-ray astronomy. *Mailing Add:* 35 Lewis St Fitchburg State Col 160 Pearl St Newton MA 02158

MARTENS, HINRICH R, b Luebeck, Ger, Apr 21, 34; US citizen; m 57; c 4. MECHANICAL & ELECTRICAL ENGINEERING. *Educ:* Univ Rochester, BSME, 57, MS, 59; Mich State Univ, PhD(elec & mech eng), 62. *Prof Exp:* Instr mech eng, Mich State Univ, 58-60, instr mech & elec eng, 60-62; from asst prof to assoc prof, 62-70, PROF MECH & ELEC ENG, STATE UNIV NY, BUFFALO, 70- *Concurrent Pos:* Res engr, Cornell Aeronaut Lab, 62-67. *Mem:* Inst Elec & Electronics Engrs; Am Soc Eng Educ; Sigma Xi. *Res:* Systems modeling; computer applications. *Mailing Add:* Dept of Elec Eng SUNY at Buffalo-N Campus 207 Bell Hall Buffalo NY 14214

MARTENS, JAMES HART CURRY, b Brooklyn, NY, Jan 2, 01; m 31; c 2. GEOLOGY, MINERALOGY-PETROLOGY. *Educ:* Cornell Univ, CE, 21, MS, 23, PhD(petrog), 26. *Prof Exp:* Asst prof geol, Cornell Univ, 21-24, instr, 24-27; asst geologist, Fla Geol Surv, 27-29; from asst prof to prof geol, W Va Univ, 29-47; assoc res specialist geol, Rutgers Univ, 47-51, prof, 51-66; instr geol, Hunter Col, 66-71; RETIRED. *Concurrent Pos:* Mineralogist, W Va Geol Surv, 30-47. *Mem:* Fel Geol Soc Am; fel Mineral Soc; Am Asn Petrol Geologist. *Res:* Investigation of mineral composition of sand and sandstone, beaches of Atlantic and gulf coast of US. *Mailing Add:* 1417 Sunken Rd Fredericksburg VA 22401

MARTENS, JOHN WILLIAM, b Desalaberry, Man, July 31, 34; m 59; c 4. PLANT PATHOLOGY. *Educ:* Univ Man, BSc, 62; Univ Wis-Madison, PhD(plant path, mycol), 65. *Prof Exp:* RES SCIENTIST CEREAL RUSTS, RES BR, AGR CAN, 65-, HEAD CEREAL DIS, AGR CAN RES STA, WINNIPEG, 88- *Concurrent Pos:* Head plant path sect, Plant Breeding Sta, Njoro, Kenya, Can Int Develop Agency, 68-69, 71-72; vis res scientist, Dept Sci & Indust Res, Christchurch, NZ, 75-76; adj prof, Univ Man, 77- *Mem:* Can Phytopath Soc (pres, 86-); Am Phytopath Soc; Sigma Xi. *Res:* Physiologic specialization in cereal rusts; host resistance; collection, preservation and utilization of wild avena and triticum species. *Mailing Add:* Can Agr Res Sta 195 Dafoe Rd Winnipeg MB R3T 2M9 Can

MARTENS, LESLIE VERNON, b Peoria Heights, Ill, Oct 15, 38; m 61; c 4. DENTISTRY, PUBLIC HEALTH & EPIDEMIOLOGY. *Educ:* Loyola Univ Chicago, DDS, 63; Univ Minn, Minneapolis, MPH, 69. *Prof Exp:* Pvt pract, Ill, 63; US Army Dent Corp, 63-68; lectr prev dent, Sch Dent, 68-69, asst prof maternal & child health, Sch Pub Health, 69-70, asst prof prev med, Sch Dent, 69-71, assoc prof health ecol & assoc chmn div, Sch Dent, 71-80, dir, Gen Pract Residency, 77-81; PROF & SCH PUB HEALTH & PROF & CHMN, DEPT HEALTH ECOL, UNIV MINN, MINNEAPOLIS, 81-; VPRES, VITA DENT CO, IOWA, 87- *Concurrent Pos:* Consult, Cambridge State Hosp & Sch for Ment Retarded, 68-77; pvt pract, 69-; lectr, Schs Nursing & Pharm, Univ Minn, Minneapolis, 70-83 & Normandale State Jr Col, 71-77; consult, Minneapolis Pub Schs, 72-81 & USPHS, 73-, Anoka State Hosp, 81- & Am Dent Asn, 83-; prin investr, Minn Dent Pract Anal Syst, 81- *Mem:* Am Dent Asn; Int Asn Dent Res; Am Pub Health Asn; Behav Sci in Dent Res; fel Int Col Dent; Am Asn Dent Sch. *Res:* Preventive dentistry; health manpower; infection control in dental operations; veterinary dentistry; dental epidemiology; health education; health behavior. *Mailing Add:* Div Health Ecol Univ Minn Sch Dent 513 Delaware St SE 15-136 Moos Tower MN 55455

MARTENS, MARGARET ELIZABETH, MITOCHONDRIAL METABOLISM. *Educ:* Wayne State Univ, PhD(biochem), 80. *Prof Exp:* ASST PROF, DEPT NEUROL, WAYNE STATE UNIV, 85- *Res:* Molecular mechanism of neuromuscular disease. *Mailing Add:* 24745 Rensselaer Oak Park MI 48237

MARTENS, VERNON EDWARD, b St Louis, Mo, Aug 15, 12; m; c 7. PATHOLOGY. *Educ:* St Louis Univ, BS, 35, MD, 37; Am Bd Path, dipl, 48. *Prof Exp:* Intern, St Louis City Hosp, Mo, 37-38; resident med, US Naval Hosps, Chelsea, Mass, 38-39, pathologist, Norman, Okla, 44-45, asst pathologist, Philadelphia, Pa, 45-47, pathologist, 50-51; dir labs, US Navy Med Sch, 51-58; dir labs, Washington Hosp Ctr, 58-84; DIR LABS, LELAND MEM HOSP, 84- *Concurrent Pos:* Fel path, Hosp Univ Pa, 47-50, instr, 50-51; assoc clin prof, Sch Med, George Washington Univ, 63, 65 & 66. *Mem:* AMA; Am Soc Clin Path; Asn Clin Sci (pres, 57-58); Col Am Path; Int Acad Path. *Res:* Clinical and anatomical pathology. *Mailing Add:* HC6 Box 609 Madison VA 22727

MARTENS, WILLIAM STEPHEN, b Pittsburgh, Pa, June 14, 35. ENVIRONMENTAL CHEMISTRY. *Educ:* Rutgers Univ, BS, 56, PhD(anal & inorg chem), 60. *Prof Exp:* Sr res chemist, Int Minerals & Chem Corp, 60-62 & Agr Div, Allied Chem Corp, 62-69; consult, State of Va Health Dept, 69-70; SECT LEADER ENVIRON ENHANCEMENT, US NAVAL SURFACE WEAPONS CTR, 70- *Mem:* Am Chem Soc. *Res:* Sodium polyphosphate analyses; ion exchange; phosphate rock and wet-process phosphoric acid; inorganic polymers; air pollution detector development; environmental assessment, enhancement and control; air, water and solid waste pollution abatement; incineration technology. *Mailing Add:* US Naval Weapons Ctr Code DG-30 Dahlgren VA 22448

MARTENSEN, TODD MARTIN, b Spokane, Wash, Oct 23, 43. BIOCHEMISTRY. *Educ:* Wash State Univ, BS, 66; Iowa State Univ, PhD(biochem), 72. *Prof Exp:* Postdoctoral fel biochem & pharmacol, Stanford Univ Med Ctr, 72-76; fel biochem, Nat Heart, Lung & Blood Inst, NIH, 76-79, staff & sr staff, 80-87, res scientist, Nat Inst Alcohol Abuse & Alcoholism, 87-90; DIR, BIOCHEM PROG, DIV MOLECULAR BIOSCI & BIOCHEM, NSF, 90- *Concurrent Pos:* Mem, Educ Comt, Am Soc Biochem & Molecular Biol. *Mem:* Am Soc Biochem & Molecular Biol; AAAS; Am Chem Soc; Am Soc Cell Biol. *Res:* Biochemistry of the regulation of cellular signal transduction; post-translational modification of proteins by phosphorylation of serine, threonine, and tyrosine residues; substrate and enzymatic regulation. *Mailing Add:* Div Molecular Biosci NSF Washington DC 20550

MARTENSON, RUSSELL ERIC, PROTEIN CHEMISTRY, MOLECULAR BIOLOGY. *Educ:* Harvard Univ, PhD(biochem), 65. *Prof Exp:* RES CHEMIST, NIMH, 70- *Mailing Add:* Lab Cell Biol NIMH Clin Neurosci Br Bethesda MD 20892

MARTH, ELMER HERMAN, b Jackson, Wis, Sept 11, 27; m 57. FOOD MICROBIOLOGY, DAIRY MICROBIOLOGY. *Educ:* Univ Wis, BS, 50, MS, 52, PhD(bact), 54. *Prof Exp:* Asst bact, Univ Wis, 49-54, proj assoc, 54-55, instr, 55-57; bacteriologist, Kraft, Inc, 57-59, from res bacteriologist to sr res bacteriologist, 59-63, group leader bact, 63-66, assoc mgr microbiol, 66; from assoc prof to prof, 66-90, EMER PROF FOOD SCI & BACT, UNIV WIS-MADISON, 90- *Concurrent Pos:* Ed, J Food Protection, Int Asn Milk, Food & Environ Sanit, 67-87; chmn, Intersoc Coun Stand Methods Exam Dairy Prods, 72-78; WHO travel fel, 75; vis prof, Swiss Fed Inst Technol, 81. *Honors & Awards:* Pfizer Award, 75 & Dairy Res Found Award, 80, Am Dairy Sci Asn; Nordica Award, Am Cultured Dairy Prod Inst, 79; Borden Award, 86; Nicholas Appert Award, 87; Babcock-Hart Award, 89. *Mem:* Am Soc Microbiol; Am Dairy Sci Asn; Inst Food Technol; hon mem Int Asn Milk, Food & Environ Sanit; Coun Biol Ed; AAAS. *Res:* Microbiology of dairy and food products; psychrotrophic bacteria; mycotoxins; dairy starter cultures; fermentations; manufacturing of fermented dairy foods; fate of pathogenic bacteria in foods. *Mailing Add:* Dept Food Sci Univ Wis-Madison Madison WI 53706

MARTI, KURT, b Berne, Switz, Aug 18, 36; m 63; c 3. COSMOCHEMISTRY. *Educ:* Univ Berne, MSc, 63, PhD, 65. *Prof Exp:* Res chemist, 65-67, asst res chemist, 67-68, from asst prof to assoc prof, 69-80, PROF COSMOCHEM, UNIV CALIF, SAN DIEGO, 80- *Concurrent Pos:* NASA grant, Univ Calif, San Diego, 71-; prin investr, Lunar Sample Anal, 72- *Honors & Awards:* Guggenhiem fel, 76. *Mem:* AAAS; Am Geophys Union; Meteoritical Soc; Am Chem Soc. *Res:* Isotopic and nuclear cosmochemistry; origin and history of the solar system; products of extinct elements; origin of elements. *Mailing Add:* Dept Chem Univ Calif San Diego La Jolla CA 92037-0317

MARTIGNOLE, JACQUES, b Carcassonne, France, Oct 11, 39; m 62. GEOLOGY. *Educ:* Univ Toulouse, Lic es Sci, 61, Dr 3rd Cycle, 64, Dr Univ, 68, DSc, 75. *Prof Exp:* Nat Coun Arts Can fel, 64-66, lectr geol, 66-68, from asst prof to assoc prof, 68-79, PROF GEOL, UNIV MONTREAL, 79- *Mem:* Geol Asn Can; Asn Study Deep Zones Earth's Crust. *Res:* Precambrian geology; igneous and metamorphic petrology; structural geology. *Mailing Add:* Universiste' de Montereal 6123 Univ Montreal Montreal PQ H3C 3J7 Can

MARTIGNONI, MAURO EMILIO, b Lugano, Switz, Oct 30, 26; nat US; m 53; c 2. VIROLOGY, INVERTEBRATE PATHOLOGY. *Educ:* Swiss Fed Inst Technol, dipl ing agr, 50, PhD(microbiol, entom), 56. *Prof Exp:* Asst entom, Swiss Fed Inst Technol, 50 & 52, entomologist, Swiss Forest Res Inst, 53-56; from asst insect pathologist to assoc insect pathologist & lectr invert path, Univ Calif, Berkeley, 56-63, assoc prof, 63-65; prin microbiologist, USDA Forest Serv, Ore, 65-68, chief microbiologist, 68-85; prof entomol, Ore State Univ, Corvallis, 65-87; CONSULT, 87- *Concurrent Pos:* Consult entomologist, Food & Agr Orgn, UN, Rome, Italy, 52-53; USPHS grants, 58-64; mem trop med & parasitol study sect, NIH, 64-65; consult med zool dept, US Naval Med Res Unit 3, 66-69; mem, Int Comt Nomenclature Viruses, 66-78; consult, NASA, 66-67; vis scientist insect virol, Agr Res Coun, Littlehampton, Gt Brit, 72-73; mem, Ad-Hoc Panel Experts, Univ Calif, Berkeley, 75-82; proj coordr, Microbiol Working Group, Comn Sci & Tech Coop, US/USSR, 77-83; trustee, Soc Invert Path, 67-70, 73-76. *Honors & Awards:* Kern Award & Silver Medal, Swiss Fed Inst Technol, 57; Superior Serv Group Honor Award, USDA, 77. *Mem:* AAAS; Entom Soc Am; Sigma Xi; Am Soc Microbiol; Soc Invert Path. *Res:* Insect pathology, especially viral diseases of insects; pathologic physiology; insect tissue culture; bioassay; large-scale production and safety evaluation of viral preparations. *Mailing Add:* PO Box 14892 Albuquerque NM 87191

MARTIN, AARON J, b Lancaster, Pa, June 2, 28; m 52; c 1. ANALYTICAL CHEMISTRY. *Educ:* Franklin & Marshall Col, BS, 50; Pa State Col, MS, 52, PhD(anal chem), 53. *Prof Exp:* AEC res asst, Pa State Univ, 51-53; res chemist, E I du Pont de Nemours & Co, Inc, 53-58, res supvr, 58-59; dir res, F&M Sci Corp, 59-65, mgr res & eng, F&M Sci Div, Hewlett Packard Co, 65-69; pres, Marlabs, Inc, 69-89; chmn, Advan Microcomputer Systs, Inc, 80-; SECY, SOUTHRIDGE, INC, 82-; DIR, INTERACTIVE MED COMT, INC, 87- *Concurrent Pos:* chmn trustees, Franklin & Marshall Col. *Mem:* AAAS; Instrument Soc Am; Am Chem Soc. *Res:* Polarographic behavior of organic compounds; analytical instrumentation. *Mailing Add:* 102 Redwood Lane Kennett Square PA 19348

MARTIN, ALBERT EDWIN, b Mifflintown, Pa, Nov 25, 31; m 53; c 2. ANALYTICAL CHEMISTRY, PHARMACEUTICAL CHEMISTRY. *Educ:* Franklin & Marshall Col, BS, 53; Univ Calif, Los Angeles, MS, 56; Univ NC, Chapel Hill, PhD(chem), 59. *Prof Exp:* Sr res chemist, Chas Pfizer & Co, Inc, Conn, 59-62; mgr, A H Robins Co, Inc, 62-74, dir anal res, 74-75, dir & asst vpres Good Mfg Prac, 75-82, vpres res & develop dir, 82-90; CONSULT, 90- *Concurrent Pos:* Spec lectr, Va Commonwealth Univ, 67-70; mem revision comt, US Pharmacopeia XIX; consult, Gov Mgt Study Comn, Va; vchmn, Lab Serv Adv Bd, Commonwealth Va, 78- *Mem:* Am Chem Soc. *Res:* Complex solution analysis; instrumental and electrochemical techniques; analytical chemistry of organic compounds; pharmaceutical dosage formulations. *Mailing Add:* 1407 Cummings Dr Richmond VA 23220

MARTIN, ALEXANDER ROBERT, b Can, Oct 12, 28; m 51; c 3. NEUROPHYSIOLOGY. *Educ:* Univ Man, BSc, 51, MSc, 53; Univ London, PhD(biophys), 55; Yale Univ, MA, 68. *Prof Exp:* Asst biophys, Univ Col, Univ London, 53-55; from instr to assoc prof physiol, Col Med, Univ Utah, 57-66; prof, Yale Univ, 66-70; PROF PHYSIOL & CHMN DEPT, SCH MED, UNIV COLO, DENVER, 70- *Concurrent Pos:* Bronfmann fel neurophysiol, Montreal Neurol Inst, 55-57. *Mem:* Am Physiol Soc; Brit Physiol Soc. *Res:* Synaptic transmission. *Mailing Add:* Dept Physiol Container C240 Univ Colo Sch Med 4200 E 9th Ave Denver CO 80220

MARTIN, ALFRED, b Pittsburgh, Pa, May 1, 19; m 46; c 2. PHYSICAL MEDICINAL CHEMISTRY. *Educ:* Philadelphia Col Pharm, BS, 42; Purdue Univ, MS, 48, PhD, 50. *Prof Exp:* From asst prof to assoc prof pharm, Temple Univ, 50-55; from assoc prof to prof, Sch Pharm, Purdue Univ, 55-66; prof, Sch Pharm, Med Col Va, 66-68; prof phys med chem & dean, Sch Pharm, Temple Univ, 68-72; prof & dir, 73-78, Coulter R Sublett prof, 77-88, EMER COULTER R SUBLETT PROF, DRUG DYNAMICS INST, COL PHARM, UNIV TEX AUSTIN, 88- *Concurrent Pos:* Pfeiffer mem res fel, Ctr Appl Wave Mech, France, 62-63; indust consult, 62- *Honors & Awards:* Ebert Medal, Am Pharmaceut Asn, 66; Sturmer Lect Award, Philadelphia, 67; Kauffman Lect Award, Ohio State Univ, 70; Roland T Lakey Hon Lect Award, Col Pharm, Wayne State Univ, 90. *Mem:* Am Chem Soc; Am Pharmaceut Asn; fel Acad Pharmaceut Sci. *Res:* Application of physical chemistry to pharmaceutical and medicinal sciences. *Mailing Add:* Drug Dynamics Inst Col Pharm Univ Tex Austin TX 78712

MARTIN, ARLENE PATRICIA, b Binghamton, NY, June 30, 26. BIOCHEMISTRY. *Educ:* Cornell Univ, BA, 48, MNutritS, 52; Univ Rochester, PhD(biochem), 57. *Prof Exp:* Fel, Sch Med & Dent, Univ Rochester, 57-58, instr biochem, 58-65; asst prof radiol, Jefferson Med Col, 65-67, asst prof biochem, 67-68; assoc prof, 68-74, PROF PATH & BIOCHEM, SCH MED, UNIV MO-COLUMBIA, 74- *Mem:* Fel AAAS; Am Chem Soc; NY Acad Sci; Am Soc Biol Chem. *Res:* Isolation, characterization and function of enzymes concerned with biological oxidation, especially respiratory enzymes; structure-function relationships of mitochondria; membrane lipids and changes during aging. *Mailing Add:* Dept Path Univ Mo Sch Med Columbia MO 65212

MARTIN, ARNOLD R, b Missoula, Mont, Mar 6, 36; m 59; c 4. MEDICINAL CHEMISTRY. *Educ:* Wash State Univ, BS, 59, MS, 61; Univ Calif, San Francisco, PhD(pharm chem), 64. *Prof Exp:* From actg asst prof to assoc prof pharm, Wash State Univ, 64-74, prof pharm chem, 74-77; PROF MED CHEM, UNIV ARIZ, 77- *Concurrent Pos:* Mem, Am Found Pharmaceut Educ. *Mem:* AAAS; Am Chem Soc; Am Pharmaceut Asn; Acad Pharmaceut Sci; Sigma Xi. *Res:* Medicinal chemistry; phenothiazine tranquilizers; tricyclic antidepressants; stereochemical and conformational studies; aminotetralin as analgesics; adrenergic blocking agents. *Mailing Add:* Dept Pharm Sci Col Pharm Univ Ariz Tucson AZ 85721

MARTIN, ARTHUR FRANCIS, b Elkins, WVa, Feb 5, 18; m 53; c 3. SOFTWARE SYSTEMS. *Educ:* Ursinus Col, AB, 38; Mass Inst Technol, PhD(org chem), 42. *Hon Degrees:* ScD, Ursinus Col, 63. *Prof Exp:* Res chemist, Exp Sta, Hercules, Inc, 41-42, sr chemist, Va, 43, asst leader, Cellulose Prod Group, Del, 44, chief chemist & head lab, Cellulose Plant, Va, 45-49, mgr, Va Cellulose Res Div, Exp Sta, 49-53, spec assignment, Argonne Nat Lab, 54-55, actg mgr, Phys Chem Res Div, Res Ctr, 56, sr res chemist, Phys Chem Res Div, 57-58 & Appl Math Div, 59-63, mgr, 63-67, mgr, Opers Res Div, 67-71, sr financial analyst, 71-76, appl math consult, 77-80; RETIRED. *Mem:* Am Chem Soc; Sigma Xi. *Res:* Cellulose and cellulose products; research administration; applied mathematics; operations research. *Mailing Add:* 116 Meriden Dr Hockessin DE 19707-1702

MARTIN, ARTHUR WESLEY, III, b Palo Alto, Calif, July 5, 35; m 58; c 3. THEORETICAL PHYSICS. *Educ:* Harvard Univ, AB, 57; Stanford Univ, MS, 59, PhD(particle physics), 62. *Prof Exp:* Res assoc physics, Argonne Nat Lab, 62-64; asst prof, Stanford Univ, 64-67; assoc prof, Rutgers Univ, 67-69; ASSOC PROF PHYSICS, UNIV MASS, BOSTON, 69- *Mem:* Am Phys Soc. *Res:* Elementary particle physics; dispersion theory; general relativity. *Mailing Add:* Dept of Physics Univ of Mass Harbor Campus Boston MA 02125

MARTIN, ARTHUR WESLEY, b Nanking, Kiangsu, China, Dec 13, 10; US citizen; m 31; c 2. MOLLUSCAN PHYSIOLOGY, IRON TRANSPORT. *Educ:* Univ Puget Sound, BS, 31; Leland Stanford Jr Univ, PhD(physiol), 36. *Prof Exp:* Teaching asst biol, Stanford Univ, 32-34, instr physiol, 36-37; res asst, C B van Niel Hopkins Marine Sta, 35-36; instr to assoc prof zool, Univ Wash, 37-46, assoc prof physiol, 46-48, prof & chmn zool, 48-63, prof zool, 63-81, EMER PROF ZOOL, UNIV WASH, 81- *Concurrent Pos:* Assoc prof, Dept Physiol, Stanford Univ, 44; vis investr, Univ Hawaii, 50, 52 & 54 & Mem Univ, Nfld, Can, 78; mem coun, AAAS, 54-56, Finance Comt, 61-63, chmn, Sect Comp Physiol, 66-67, pres, 52-53; prog dir, Regulatory Biol, NSF, 58-59; vis prof, Dept Animal Physiol, Univ Sao Paulo, Brazil, 64, zool, Univ NSW, Australia, 69-70 & Comp Pharmacol, Irvine Med Sch, Univ Calif, 75; Underwood fel, Cambridge Univ, Eng, 68. *Mem:* Fel AAAS; Am Physiol Soc; Am Soc Zoologists; Western Soc Naturalists; Sigma Xi. *Res:* Metabolic control; muscle hypertrophy and hyperplasia; cephalopod circulatory, renal and reproductive physiology; spider physiology; natilus growth and renal physiology; transferrins in invertebrates; water balance and mucous formation in terrestrial slugs. *Mailing Add:* 17737 15th NW Seattle WA 98177

MARTIN, BARBARA BURSA, b Oak Park, Ill, Aug 2, 34; m 56; c 6. INORGANIC CHEMISTRY, ANALYTICAL CHEMISTRY. *Educ:* Grinnell Col, AB, 56, Pa State Univ, MSc, 59. *Prof Exp:* ASST PROF CHEM (COURTESY), UNIV S FLA, 75- *Concurrent Pos:* Co-ed, Fla Scientist, 84- *Mem:* Am Chem Soc; Sigma Xi; Aquatic Plant Mgt Soc. *Res:* Chelating tendencies; certain aspects of environmental and marine chemistry. *Mailing Add:* Dept Chem Univ S Fla Tampa FL 33620

MARTIN, BERNARD LOYAL, b Whittier, Calif, Jan 1, 28; m 55; c 3. MATHEMATICS. *Educ:* Cent Wash State Col, BA, 55, MEd, 57; Ore State Univ, MS, 64, PhD(math), 66. *Prof Exp:* Instr high schs, Wash, 55-59, chmn dept math, 56-59; from instr to assoc prof, 59-69, from asst dean to dean arts & sci, 66-72, dean sch natural sci & math, 72-80, PROF MATH & COMPUT SCI, CENT WASH STATE UNIV, 69- *Mem:* Math Asn Am. *Res:* Statistics; computer science. *Mailing Add:* Dept Math Cent Wash State Univ Ellensburg WA 98926

MARTIN, BILLY JOE, b Talpa, Tex, May 24, 33; m 55; c 2. CELL BIOLOGY, CYTOCHEMISTRY. *Educ:* Univ Southern Miss, BS, 62, MS, 63; Rice Univ, PhD(biol), 70. *Prof Exp:* Asst prof biol, William Carey Col, 63-66; fel, Inst Pathobiol, Med Univ SC, 70-71, asst prof path, 71-73, asst prof, Sch Dent, 72-73; from assoc prof to prof biol, 73-82, res coordr, Col Sci & Technol, 79-82, PROF BIOL SCI & CHMN DEPT, UNIV SOUTHERN MISS, 82- *Concurrent Pos:* Mem, grad fac, Med Univ SC, 72-73. *Mem:* AAAS; Sigma Xi; Electron Micros Soc Am; Am Soc Cell Biol. *Res:* Ultrastructure of cells specialized for electrolyte transport; dynamics of cell transport; cytochemistry of cell surface; use of lectins as cytochemical tools; experimental oncology of lower vertebrates (teleosts). *Mailing Add:* Univ Southern Miss Box 5018 Southern Sta Hattiesburg MS 39401

MARTIN, BILLY RAY, b Winston-Salem, NC, Apr 25, 43; m 71; c 2. PHARMACOLOGY. *Educ:* Univ NC, AB, 65, PhD(pharmacol), 74. *Prof Exp:* Jr chemist, Res Triangle Inst, 65-69, res analyst, Univ NC, 69-73; fel pharmacol, Uppsala Univ, Sweden, 75-76, Univ Oxford, 76-77; asst prof, Med Col VA, 76-82, assoc prof pharmacol, 82-87; from asst prof to assoc prof, 76-87, PROF PHARMACOL, MED COL VA, VA COMMONWEALTH UNIV, 87- *Concurrent Pos:* Fel, Swedish Med Res Coun, 75-76, Wellcome Trust Found, 76-77; asst prof, Pharmaceut Mfrs Asn, 78-; mem comt, Problems Drug Dependence. *Mem:* Sigma Xi; Am Soc Pharmacol & Exp Therapeut; Soc Toxicol. *Res:* Pharmacology of drugs of abuse such as the pharmacokinetics of marijuana constituents, nicotine and phencyclidine; agents that alter neurotransmission in the brain. *Mailing Add:* Med Col Va Box 613 Richmond VA 23298

MARTIN, BOSTON FAUST, b Tampa, Fla, June 1, 27; m 88; c 4. NEUROSURGERY. *Educ:* Howard Univ, BS, 49; Univ Fribourg, Switzerland, BMS, 54; Univ Geneva, MD, 58; Am Bd Neurol & Orthop Surgeons, cert, 80. *Prof Exp:* Resident to chief, 62-66, fel neurosurgery, NY Univ Med Ctr, 66-67; insular neurosurgeon, US Govt VI, 67-69; asst prof, Univ PR Sch Med, 69-75, interim chief, neurosurgery, 69-70; chief spinal cord injury serv, Vet Admin Med Ctr, 75-84; flight surgeon, Lt Col USAFR, 84-90; CHIEF NEUROSURG, EAST ORANGE GEN HOSP, NJ, 90- *Concurrent Pos:* Sr clin res fel neurosurgery, NY Univ Med Ctr, 66-67; co- investr brain tumor chemotherapy, Nat Cancer Inst, 66-67; med, pvt practice, 84- *Mem:* Am Col Surg; Am Acad Neurol & Orthop Surgeons; Cong Neurol Surg; Soc Neurosci; Int Col Surg. *Res:* Effects of Urecholine on the external urethral sphincter; use of Dantrium in Detrusor Sphincter Dyssynergia; use of Septra in intermittent catheterization; effects of transcutaneous nerve stimulation on vesico- urethral function; etiology, neuropathophysiology and altered responses of spinal cord injury patients. *Mailing Add:* 81 Northfield Ave West Orange NJ 07052

MARTIN, BRUCE DOUGLAS, b Rochester, NY, Apr 8, 34; m 57, 84; c 2. PHARMACEUTICAL CHEMISTRY. *Educ:* Albany Col Pharm, BS, 55; Univ Ill, MS, 59, PhD(pharmaceut chem), 62. *Prof Exp:* From asst prof to assoc prof, 61-68, dean sch pharm, 71-81, actg vpres acad affairs, 81, PROF PHARMACEUT CHEM, DUQUESNE UNIV, 68-, ASSOC VPRES ACAD AFFAIRS, 82- *Concurrent Pos:* Fulbright lectr, Univ Sci & Technol, Ghana, 68-69. *Mem:* Am Chem Soc; Am Pharmaceut Asn. *Res:* Organic synthesis; sulfonamides. *Mailing Add:* 502 Admin Bldg Duquesne Univ Pittsburgh PA 15282

MARTIN, CARROLL JAMES, RESPIRATORY PHYSIOLOGY. *Educ:* Univ Iowa, MD, 40. *Prof Exp:* EMER PROF, VIRGINIA MASON RES CTR, SEATTLE, WASH, 68- *Res:* Lung tissue properties; interstitial matrix. *Mailing Add:* 985 NE Melanee Ct Bremerton WA 98310

MARTIN, CHARLES EVERETT, b Cape Girardeau, Mo, Nov 21, 44; m 67; c 1. MEMBRANE BIOLOGY. *Educ:* Univ Ill, BS, 66; Fla State Univ, PhD(biol), 72. *Prof Exp:* Fel biochem genetics, Univ Tex, Austin, 72-75, membrane biol, 75-78; assoc prof, 78-84, ASST PROF, GENETICS & CELL BIOL, RUTGERS UNIV, 84- *Concurrent Pos:* Fel, NIH, 72-74. *Mem:* Genetics Soc Am; Am Soc Cell Biol; Sigma Xi; Am Soc Biochem & Molecular Biol. *Res:* Biology and chemistry of cell membranes; lipid-protein interactions; genetic aspects of membrane assembly and function; structure-function of fatty acid Desaturases; regulation of unsaturated fatty acid biosynthesis. *Mailing Add:* Nelson Biol Lab Rutgers Univ PO Box 1059 Piscataway NJ 08855-1057

MARTIN, CHARLES EVERETT, b Moscow Mills, Mo, Nov 7, 29; m 52; c 3. VETERINARY PHYSIOLOGY. *Educ:* Univ Mo, BS & DVM, 58; Purdue Univ, Lafayette, MS, 67. *Prof Exp:* Practitioner, Green Hills Animal Hosp, 58-65; from instr to asst prof vet med & surg, Purdue Univ, Lafayette, 65-67; from asst prof to prof vet med & surg, Univ Mo-Columbia, 67-80, chmn dept, 74-80; TECH SERV VET, UPJOHN CO, MICH, 80- *Concurrent Pos:* Mem, NCent Res Comt, 64 & 68- & Nat Pork Producers Res Coord Comt, 70-74; Mo Pork Producers, Agr Exp Sta & USDA grants, Univ Mo-Columbia, 70-76. *Mem:* Am Vet Med Asn; Am Asn Equine Practitioners; Am Col

Theriogenologists; Soc Theriogenology (pres, 85-86); Am Asn Swine Practitioners. *Res:* Bovine, equine and swine reproduction; physiology, endocrinology and pathology of lactation failure in swine. *Mailing Add:* 1108 Lacosta Ct Columbia MO 65203

MARTIN, CHARLES J, b New Castle, Pa, Dec 5, 21; m 45; c 4. BIOCHEMISTRY. *Educ:* Univ Pittsburgh, BS, 44, PhD(chem), 51. *Prof Exp:* Asst, Western Pa Hosp, 49-51; instr path, Western Reserve Univ, 51-53, sr instr biochem, 53-54; res assoc, Sch Med, Univ Pittsburgh, 54-57, asst res prof, 57-63; res assoc prof enzymol & hypersensitivity, 63-67, asst dean acad affairs, 68-72, asst to the pres acad affairs, Univ Health Sci-Chicago Med Sch, 72-75, PROF BIOCHEM, UNIV HEALTH SCI-CHICAGO MED SCH, 67-, V PRES, 77- *Mem:* AAAS; Am Chem Soc; Am Soc Biol Chemists; Am Calorimetry Conf; NY Acad Sci; Sigma Xi. *Res:* Mechanism of enzyme action; protein modifications, calorimetry of biological systems. *Mailing Add:* Chicago Med Sch Univ Health Sci 3333 Green Bay Road North Chicago IL 60064

MARTIN, CHARLES JOHN, b Sloatsburg, NY, Apr 3, 35; m 59; c 2. APPLIED MATHEMATICS. *Educ:* Union Col, NY, BS, 56; Mich State Univ, MS, 57; Rensselaer Polytech Inst, PhD(math), 61. *Prof Exp:* Instr math, Union Col, NY, 58-59; res asst, Rensselaer Polytech Inst, 59-61; sr staff scientist, Res & Advan Develop Div, Avco Corp, 61-66; from assoc prof to prof math, Mich State Univ, 66-75; prof math & head dept, Western Carolina Univ, 75-; AT EMBRY-RIDDLE AERONAUT UNIV. *Concurrent Pos:* NASA res grant, Mich State Univ, 67-71; lectr, Soc Indust & Appl Math, 73-75; NSF proposal reviewer, 76, 83-87; NSF grants, Western Carolina Univ, 79-80, 80-83, 82-85 & 87-90. *Mem:* Am Soc Mech Eng; Asn Comput Mach; Sigma Xi; Inst Elec & Electronics Engrs. *Res:* Mechanics. *Mailing Add:* Dept Eng & Agr Sci Embry-Riddle Aeronaut Univ Daytona Beach FL 32114

MARTIN, CHARLES R, b Cincinnati, Ohio, Nov 20, 53; m 87. MATERIALS SCIENCE, ELECTROCHEMISTRY. *Educ:* Centre Col Ky, BS, 75; Univ Ariz, PhD(chem), 80. *Prof Exp:* Asst prof chem, 81-86, ASSOC PROF, TEX A & M UNIV, 86- *Concurrent Pos:* Consult, Dow Chem Co, 84-88 & Gen Motors Res Lab, 86-87. *Mem:* Am Chem Soc; Electrochem Soc; Electroanal Soc. *Mailing Add:* Dept Chem Tex A&M Univ College Station TX 77843

MARTIN, CHARLES SAMUEL, b Staunton, Va, May 22, 36; c 2. FLUID MECHANICS, HYDRAULICS. *Educ:* Va Polytech Inst, BS, 58; Ga Inst Technol, MS, 61, PhD(civil eng), 64. *Prof Exp:* Hydraul designer, Newport News Shipbldg & Dry Dock Co, 59-60; res asst fluid mech, 60-63, from asst prof to assoc prof, 63-76, PROF CIVIL ENG, GA INST TECHNOL, 76- *Concurrent Pos:* Ford Found fac resident, Harza Eng Co, Chicago, 66-67; Fulbright travel grant, 70-71; Am Soc Mech Engrs John R Freeman fel, 70-71; guest prof, Univ Karlsruhe, 70-71. *Mem:* Am Soc Civil Engrs; Am Soc Mech Engrs; Inst Asn Hydraul Res; Am Soc Eng Educ; Sigma Xi. *Res:* Pressure and hydraulic transients; two-phase flow; free streamline hydrodynamics; waterhammer. *Mailing Add:* Sch Civil Eng Ga Tech Atlanta GA 30332

MARTIN, CHARLES WAYNE, b Shenandoah, Iowa, June 9, 32; m 55; c 2. ENGINEERING MECHANICS. *Educ:* Iowa State Univ, BS, 54, MS, 59, PhD(theoret & appl mech), 62. *Prof Exp:* Asst theoret & appl mech, Iowa State Univ, 57-58, instr, 58-59, res assoc, eng exp sta, 60-62, asst prof eng sci, univ, 62-63; sr mech engr, Melpar Inc, Westinghouse Air Brake Co, 63-65; assoc prof eng mech, 65-75, PROF ENG MECH, UNIV NEBR, LINCOLN, 75- *Concurrent Pos:* US Air Force Pilot, 55-57; Charles A Lindburgh Fund Grant Recipient, 81. *Mem:* Am Soc Eng Educ; fel Am Soc Civil Engrs; Am Inst Aeronaut & Astronaut. *Res:* Structures; finite element analysis; similitude; explosive and impact loading; automatic structural design; wind energy systems. *Mailing Add:* Dept Eng Mech Univ Nebr Lincoln NE 68588-0347

MARTIN, CHARLES WELLINGTON, JR, b Omaha, Nebr, Apr 28, 33; m 59; c 3. PETROLOGY, GEOLOGY. *Educ:* Dartmouth Col, AB, 54; Univ Wis, MS, 59, PhD(geol), 62. *Prof Exp:* From asst prof to assoc prof, 60-71, assoc acad dean, 81-83, Lilly fac fel, 83-84, PROF GEOL, EARLHAM COL, 71- *Concurrent Pos:* Consult, teaching & learning, 84-86. *Mem:* Geol Soc Am; Nat Asn Geol Teachers. *Res:* Petrology; structural and regional geology of western Connecticut Highlands; geology of Northern Honshu, Japan; history of geology of the western United States, 1836-60. *Mailing Add:* Dept Geol Earlham Col Richmond IN 47374

MARTIN, CHARLES WILLIAM, b Kansas City, Mo, July 16, 43; m 68; c 2. FLUOROCARBON CHEMISTRY, GAS PROCESSING. *Educ:* Univ Pa, BS, 65; Univ Kans, PhD(chem), 73. *Prof Exp:* Res chemist process res, 71-72, sr res chemist amine chem, 72-75, res specialist amine chem, 75-79, GAS PROCESSING SPECIALIST, FLUOROCARBON CHEM, DOW CHEM CO, 79- *Concurrent Pos:* Instr, Saganaw Valley Col, 75. *Mem:* Am Chem Soc. *Res:* Development of new amine products and applications with emphasis on the gas processing industry. *Mailing Add:* Dow Chem USA B1407 Bldg Freeport TX 77541

MARTIN, CHRISTOPHER MICHAEL, b New York, NY, Sept 25, 28; m 54; c 2. MEDICINE. *Educ:* Harvard Univ, AB, 49, MD, 53. *Prof Exp:* Intern med, Boston City Hosp, 53-54, asst resident, 56-57; res fel, Thorndike Mem Lab, Boston City Hosp & Harvard Med Sch, 57-59; res fel, Med Found Metrop Boston, Inc, 58-59; from asst prof to assoc prof, Seton Hall Col Med & Dent, 59-65; prof med & pharmacol, Sch Med, Georgetown Univ, 65-70; sr dir med affairs, Merck, Sharp & Dohme Res Labs, 70-77; PROF MED, JEFFERSON MED COL, 70-; EXEC DIR INFECTIOUS DIS, MERCK, SHARP & DOHME RES LABS, 77- *Concurrent Pos:* Dir, Georgetown Med Div, DC Gen Hosp. *Mem:* Am Soc Pharmacol & Exp Therapeut; Am Soc Clin Pharmacol & Therapeut; Am Asn Immunol; Infectious Dis Soc Am; Am Soc Microbiol. *Res:* Infectious diseases; immunology; virology; chemotherapy; virus synthesis; clinical pharmacology; carcinogens. *Mailing Add:* Infectious Dis Res Div Merck Sharp & Dohme Res Labs West Point PA 19486

MARTIN, CONSTANCE R, b Brooklyn, NY, Dec 31, 23; m 43, 71; c 2. ENDOCRINOLOGY. *Educ:* Long Island Univ, BS, 44; Univ Iowa, PhD(physiol), 51. *Prof Exp:* Res assoc physiol & pharmacol, NY Med Col, 50-51; sr physiologist, Creedmoor Inst Psychobiol Studies, 51-53; from instr physiol & pharmacol to asst prof physiol & pharmacol, NY Med Col, 53-57; from asst prof to assoc prof biol, Long Island Univ, 59-63; asst prof physiol, 63-66, assoc prof biol sci, 66-76, PROF BIOL SCI, HUNTER COL, 76- *Concurrent Pos:* Am Cancer Soc Res grant, 65-68. *Mem:* AAAS; Am Physiol Soc; Endocrine Soc; Soc Study Reproduction; Am Soc Zoologists; NY Acad Sci; Soc Bone & Mineral Res. *Res:* Thymus gland function; reproduction physiology; biological rhythms; electrolyte metabolism. *Mailing Add:* Dept Biol Sci Hunter Col 695 Park Ave New York NY 10021

MARTIN, DANIEL WILLIAM, b Georgetown, Ky, Nov 18, 18; m 41; c 4. PHYSICS. *Educ:* Georgetown Col, AB, 37; Univ Ill, MS, 39, PhD(physics), 41. *Hon Degrees:* ScD, Georgetown Col, 81. *Prof Exp:* Asst instr, Univ Ill, 37-41; acoust develop engr, Radio Corp Am, 41-49; supvr engr, Acoust Res, Baldwin Piano Co, 49-57, res dir, 57-70, res & eng dir, D H Baldwin Co, 70-74, res & patent dir, Baldwin Piano & Organ Co, 74-83; CONSULT, ACOUST, 83- *Concurrent Pos:* Instr, Purdue Univ, 41-46; ed, Audio Trans, 54-56; asst prof, Univ Cincinnati, 65-74; editor-in-chief, J Acoust Soc Am, 85- *Mem:* Fel Acoust Soc Am (pres, 84-85); fel Audio Eng Soc (exec vpres, 63-64, pres, 64-65); fel Inst Elec & Electronics Engrs. *Res:* Acoustics of piano, organ, brass wind instruments, auditoriums; sound powered telephones; aircraft intercommunication; microphones; loudspeaker enclosures; reverberation simulation; analog-to-digital encoders; optoelectronics; audio systems. *Mailing Add:* 7349 Clough Pike Cincinnati OH 45244

MARTIN, DAVID E(DWIN), b Elmhurst, Ill, Sept 11, 29; div; c 2. MECHANICAL ENGINEERING. *Educ:* Univ Ill, BS, 53, MS, 56. *Prof Exp:* Instr theoret & appl mech, Univ Ill, 54-56; res engr, 56-57, sr res engr, 57-66, supvr dynamics & stress, 66-67, asst dept head automotive safety res, vehicle res dept, 67-72, asst dir, 72-74, DIR, AUTOMOTIVE SAFETY ENG, ENVIRON ACTIVITIES STAFF, GEN MOTORS RES LABS, 74- *Mem:* Soc Exp Stress Anal; Am Soc Testing & Mat; Soc Automotive Engrs; Sigma Xi. *Res:* Biomechanics; structural research; fatigue of metals. *Mailing Add:* PO Box 1444 1986 Hood Creek New Bern NC 28560-1444

MARTIN, DAVID EDWARD, b Green Bay, Wis, Oct 1, 39. REPRODUCTIVE PHYSIOLOGY, EXERCISE PHYSIOLOGY. *Educ:* Univ Wis, Madison, BS, 61, MS, 63, PhD(physiol), 70. *Prof Exp:* Ford found trainee, Univ Wis Regional Primate Res Ctr, 66-70; asst prof health sci, Ga State Univ, 70-74; collaborating scientist reproductive biol, Yerkes Primate Res Ctr, Atlanta, 70-74; assoc prof health sci, Ga State Univ, 74-80; collaborating scientist, 74-79, AFFIL SCIENTIST REPRODUCTIVE BIOL, YERKES PRIMATE RES CTR, ATLANTA, 79-; PROF HEALTH SCI, GA STATE UNIV, 80- *Concurrent Pos:* Prin investr, NSF res grant, 71-74, NIH res grant, 75-88, NIMH res grant, 76-85 & Vet Admin res grant, 85-90; vis prof, Sch Med, St Georges Univ, Grenada, WI, 79-87; mem, US Olympic comt res funding, 81-90. *Mem:* Am Physiol Soc; Soc Study Reproduction; fel Am Col Sports Med; Am Soc Primatologists; Int Primatol Soc. *Res:* Male and female great ape reproductive physiology; fertility dysfunction in male spinal cord injured patients; cardiopulmonary performance physiology of elite distance runners. *Mailing Add:* Sch Allied Health Professions Ga State Univ Atlanta GA 30303-3083

MARTIN, DAVID LEE, b St Louis, Mo, May 30, 41; m 66; c 2. BIOCHEMISTRY, NEUROCHEMISTRY. *Educ:* Univ Minn, St Paul, BS, 63; Univ Wis-Madison, MS, 65, PhD(biochem), 68. *Prof Exp:* From asst prof to assoc prof chem, Univ Md, College Park, 68-80; res scientist, 80-83, CHIEF LAB NEUROTOXICOL & NERVOUS SYST DISORDERS, WADSWORTH CTR FOR LABS & RES, NEW YORK STATE DEPT HEALTH, 83- *Concurrent Pos:* Vis scientist, Armed Forces Radiobiol Res Inst, Md, 75-77, 78-80, chemist, 77-78; prof environ health & toxicol & adj prof biol sci, State Univ New York, Albany, 85-, chmn, Dept Environ Health & Toxicol, 89- *Mem:* AAAS; Biochem Soc; Am Soc Neurochem; Soc Neurosci; Am Soc Biochem & Molecular Biol; Int Soc Neurochem. *Res:* Membrane transport of small molecules; regulation of neurotransmitter metabolism; functions of glial cells. *Mailing Add:* Wadsworth Ctr Labs & Res New York State Dept Health PO Box 509 Albany NY 12201-0509

MARTIN, DAVID P, agronomy, for more information see previous edition

MARTIN, DAVID WILLIAM, b Chicago, Ill, Mar 7, 42; m 64; c 2. METEOROLOGY. *Educ:* Univ Wis, BS, 64, MS, 66, PhD(meteorol), 68. *Prof Exp:* Meteorologist, Aerophys Br, Phys Sci Lab, Redstone Arsenal, 68-69; asst scientist, 70-75, assoc scientist, 75-79, SR SCIENTIST, SPACE SCI & ENG CTR, UNIV WIS, 79- *Concurrent Pos:* Satellite meteorologist, Global Atmospheric Res Prog Atlantic Trop Exp, Dakar, Senegal, 74; lectr, dept meteorol, Univ Wis, 76; vis prof, Fed Univ Para, Belém, Brazil, 86; mem, Working Group Data Mgt, Global Precipitation Climat Proj, 86- *Mem:* Am Meteorol Soc; Sigma Xi; Royal Meteorol Soc; Am Geophys Union. *Res:* structure, behavior and context of tropical convective systems; applications of meteorological satellites, including measurements of cloud properties, wind and transports of dust and moisture; satellite estimates of rainfall; monsoon systems; history of meteorological satellites. *Mailing Add:* Space Sci & Eng Ctr Univ of Wis 1225 W Dayton St Madison WI 53706

MARTIN, DAVID WILLIAM, JR, b West Palm Beach, Fla, Jan 15, 41; m 64; c 2. MEDICAL GENETICS. *Educ:* Mass Inst Technol, BS, 60; Duke Univ, MD, 64. *Prof Exp:* Res assoc Gordon Tomkins Lab, Lab Molecular Biol Nat Inst Arthritis Metab Dis, Nat Inst Health, 66-69; instr Dept Med & Dept Biochem, Univ Calif San Francisco, 69-70, asst prof med & chief Med Genetic Servs, Dept Med, 70-75, from assoc prof to prof med & biochem, 75-82; VPRES RES, GENENTECH, INC, 83- *Concurrent Pos:* Recombinant Defense Nuclear Agency Adv Comt, Nat Inst Health, 81-85; mem Comt Pub-Pvt Sector Relations in Vaccine Develop, Inst Med, Nat Acad Sci, 83-85; mem Res Develop Coun, Cystic Fibrosis Found, 83-87; bd overseerers, Duke

Univ Comprehensive Cancer Ctr, 85-88; mem Adv Comt Biotechnol Res & Develop Prog, Univ Calif, 86-, bd UCLA Symposia, 86-; mem Forum Drug Develop & Regulation, Inst Med, Nat Acad Sci, 87- *Mem:* Am Fed Clin Res; Am Soc Biol Chemists; Am Soc Clin Inves; Asn Am Physicians. *Res:* Metabolic and genetic basis of inherited diseases amd molecular and cellular aspects of immune dysfunction. *Mailing Add:* Genentech Inc 460 Point San Bruno Blvd South San Francisco CA 94080

MARTIN, DAVID WILLIS, b Philadelphia, Pa, Sept 19, 27; m 50; c 6. PHYSICS. *Educ:* Univ Mich, BS, 50, MS, 51, PhD(physics), 57. *Prof Exp:* Res assoc, Univ Mich, 53-54; resident student assoc nuclear spectros, Argonne Nat Lab, 54-56, res assoc, 56-57; from asst prof to assoc prof physics, 57-65, PROF PHYSICS, GA INST TECHNOL, 65- *Concurrent Pos:* Consult, Oak Ridge Nat Lab, 65-71. *Mem:* Am Phys Soc. *Res:* Ion-molecule reactions in gases at thermal energies; ionization and charge-transfer cross sections in gases at high energies; nuclear spectroscopy. *Mailing Add:* 1722 Wilmont Dr NE Atlanta GA 30329

MARTIN, DEAN FREDERICK, b Woodburn, Iowa, Apr 6, 33; m 56; c 6. INORGANIC CHEMISTRY. *Educ:* Grinnell Col, AB, 55; Pa State Univ, PhD(chem), 58. *Prof Exp:* NSF fel chem, Univ Col, London, 58-59; from instr to asst prof inorg chem, Univ Ill, 59-64; assoc prof, 64-69, affil prof biol, 74-79, dir, Chem Ctr, 79-88, PROF CHEM, UNIV SFLA, 69-, DIR, INST ENVIRON STUDIES, 88- *Concurrent Pos:* USPHS career develop award, Nat Inst Gen Med Sci, 69-74; vis prof physiol & pharmacol, Duke Univ Med Ctr, 70-71. *Honors & Awards:* F J Zimmerman Award, Am Chem Soc. *Mem:* AAAS; Am Chem Soc; Roy Soc Chem; Aquatic Plant Mgt Soc; NY Acad Sci. *Res:* Coordination chemistry; environmental chemistry. *Mailing Add:* Dept Chem Univ SFla 3402 Fowler Ave Tampa FL 33620

MARTIN, DENNIS JOHN, b Berwyn, Ill, Jan 24, 47. BEHAVIORAL ECOLOGY, ORNITHOLOGY. *Educ:* Ill State Univ, BS, 70; Univ NMex, MS, 71; Utah State Univ, PhD(zool), 76. *Prof Exp:* Asst prof, 75-81, ASSOC PROF BIOL, PAC LUTHERAN UNIV, 81- *Concurrent Pos:* Frank M Chapman Mem grant, Am Mus Nat Hist, 73-75; 77-78; NSF grant, 76-79. *Mem:* Sigma Xi; Am Ornith Union; Cooper Ornith Soc. *Res:* Structure and function of communication systems, primarily in vertebrates as they relate to the ecology of the organism. *Mailing Add:* Dept of Biol Pac Lutheran Univ Tacoma WA 98447

MARTIN, DEWAYNE, b Wausau, Wis, July 17, 35; m 57; c 2. PETROLOGY, PLANETARY GEOLOGY. *Educ:* Univ Wis, BS, 57, MS, 60. *Prof Exp:* Asst prof, 61-65, ASSOC PROF GEOL, MINOT STATE UNIV, 65-, DIR, MINOT STATE UNIV OBSERV, 67- *Concurrent Pos:* Co-dir, NASA Regional Space Sci Prog, 73-74; Head, Dept Earth Sci, 69- *Mem:* Nat Asn Geol Teachers; Nat Sci Teachers Asn; Sigma Xi; Int Amateur-Prof Photoelectric Photom Asn. *Res:* Development of computer-driven interactive laser videodisc programs for teaching; petrography and petrology of meteorites. *Mailing Add:* Dept Earth Sci Minot State Univ Minot ND 58701

MARTIN, DON STANLEY, JR, b Indianapolis, Ind, Feb 19, 19; m 49; c 3. PHYSICAL INORGANIC CHEMISTRY. *Educ:* Purdue Univ, BS, 39; Calif Inst Technol, PhD(chem), 44. *Prof Exp:* Asst, Nat Defense Res Comt, Calif Inst Technol, 41-42; res assoc, Northwestern Univ, 42-44; assoc scientist, Manhattan Dist, 44-46; from asst prof to assoc prof, 46-55, chemist, AEC, 46-66, sect chief, Ames Lab, Energy Res & Develop Admin, 66-76, prof, 55-84, EMER PROF CHEM, IOWA STATE UNIV, 84- *Mem:* Am Chem Soc; Am Phys Soc; Sigma Xi. *Res:* Chemistry of the platinum elements; chemical kinetics; optical absorption spectra of single crystals of coordination compounds; radiochemistry; applications of radioactive materials; inorganic chemistry. *Mailing Add:* Dept of Chem Iowa State Univ Ames IA 50011

MARTIN, DONALD BECKWITH, b Philadelphia, Pa, July 24, 27; m 56; c 4. MEDICINE. *Educ:* Haverford Col, AB, 50; Harvard Univ, MD, 54. *Prof Exp:* From intern to asst resident med, Mass Gen Hosp, 54-56, resident, 58, chief resident, 59; Med Found Boston fel, Nat Heart Inst, 60-61; Fulbright res scholar, Nat Ctr Sci Res, France, 62; from instr to asst prof med, Harvard Med Sch, 63-71; asst in med, Mass Gen Hosp, 63-71; assoc prof med, Harvard Med Sch, 71-80; PROF MED, UNIV PA, 80- *Concurrent Pos:* Res fel, Harvard Med Sch & Peter B Brigham Hosp, 56-58, USPHS fel, 57-58; assoc ed, Metabolism, 68 & Diabetes; Guggenheim fel, Univ Geneva, 74-75; assoc physician, Mass Gen Hosp, 71-73, physician, Diabetes Unit, 73-80. *Mem:* Endocrine Soc; fel Am Col Physicians; Am Diabetes Asn; NY Acad Sci; Royal Col Med. *Res:* Academic medicine; diabetes mellitus; intermediate metabolism; glucose transport in mammalian systems. *Mailing Add:* Dept Med Univ Pa Philadelphia PA 19104

MARTIN, DONALD CROWELL, b Floral Park, NY, June 16, 29; m 51; c 2. CHEMICAL ENGINEERING. *Educ:* Univ SC, BS, 51, MS, 60; NC State Univ, PhD(chem eng), 65. *Prof Exp:* From instr to assoc prof chem eng, 60-77, PROF CHEM ENG & COMPUT SCI & HEAD DEPT COMPUT SCI, NC STATE UNIV, 77- *Mem:* Am Inst Chem Engrs; Instrument Soc Am. *Res:* Digital, analog and hybrid simulation of physical systems, process control applications in particular; adaptive process modeling and control. *Mailing Add:* 820 Carlisle St Raleigh NC 27610

MARTIN, DONALD RAY, b Marion, Ohio, Oct 21, 15; m 39; c 2. INORGANIC CHEMISTRY. *Educ:* Otterbein Col, AB, 37; Western Reserve Univ, MS, 40, PhD(inorg chem), 41. *Prof Exp:* Lectr chem, Cleveland Col, Western Reserve Univ, 40-42, lab mgr, Naval Res Proj, 41-43; from instr to asst prof chem, Univ Ill, 43-51; head, Chem Metall Br, Metall Div, US Naval Res Lab, 51-52; lab mgr, Govt Res, Mathieson Chem Corp, 52-56, mgr chem res, Aviation Div, Olin Mathieson Chem Corp, 56-57, assoc dir fuels res, Energy Div, 57-60; dir res, Libbey-Owens-Ford Glass Co, 60-61; dir tech develop, Harshaw Chem Co, 61-63, dir chem res, 63-67,; Vpres, Res & Develop Div, Kewanee Oil Co, 67-68; chmn dept, 69-75, prof, 69-85, EMER PROF CHEM, UNIV TEX, ARLINGTON, 86- *Concurrent Pos:* Res chemist, E I du Pont de Nemours & Co, Inc, 41; mem chem adv comt, Air Force Off Sci Res, 55-60; trustee, Otterbein Col, 62-72. *Mem:* AAAS; Am Chem Soc; Electrochem Soc; Chem Soc. *Res:* Coordination compounds of boron halides; boron and silicon hydrides; reactions with hydrogen fluoride; fluoroborates; gaseous halides; corrosion; hafnium; inorganic nomenclature; glass; electroplating; color in compounds; gold compounds. *Mailing Add:* 3311 Cambridge Dr Arlington TX 76013

MARTIN, DOUGLAS LEONARD, b London, Eng, Nov 11, 30. METAL PHYSICS, THERMAL PHYSICS. *Educ:* Univ London, BSc, 51, PhD(physics), 54, DSc, 70. *Prof Exp:* Nat Res Coun Can fel, 54-55; sci officer physics, Royal Aircraft Estab, Eng, 55-56; from asst res officer to assoc res officer, Nat Res Coun Can, 57-64, sr res officer physics, 64-86; RETIRED. *Honors & Awards:* Huffman Award, Calorimetry Conf, 85. *Mem:* Calorimetry Conf; Can Asn Physicists. *Res:* Solid state physics; calorimetry; cryogenics. *Mailing Add:* 708-200 Rideau Terr Ottawa ON K1M 0Z3 Can

MARTIN, DUNCAN WILLIS, b Durango, Colo, Mar 26, 31; m 64; c 2. PHYSIOLOGY, BIOPHYSICS. *Educ:* Univ NMex, BS, 55, MS, 56; Univ Ill, Urbana, PhD(physiol), 62. *Prof Exp:* Asst biol, Univ NMex, 55-56; asst physiol, Univ Ill, 56-60, res asst, 60-62, USPHS trainee, 62; fel, Marine Biol Lab, Woods Hole, 62; fel biophys, Harvard Univ, 62-65; from asst prof to assoc prof, 65-75, PROF ZOOL, UNIV ARK, 75- *Concurrent Pos:* Vis assoc prof physiol, Yale Univ Sch Med, 74-75; chmn, Ark Game & Fish Comn, 80-85. *Mem:* Am Soc Zool; Am Physiol Soc. *Res:* Membrane physiology; active transport of ions; anion transport. *Mailing Add:* Dept Biol Sci SCEN 632 Univ of Ark Fayetteville AR 72701

MARTIN, EDGAR J, tropical medicine, pharmacology, for more information see previous edition

MARTIN, EDWARD SHAFFER, b Terre Haute, Ind, Jan 14, 39. INORGANIC CHEMISTRY. *Educ:* DePauw Univ, BA, 60; Northwestern Univ, PhD(chem), 67. *Prof Exp:* Lectr chem, 65-66, asst prof, Ind Univ, South Bend, 66-72; scientist, 72-74, sr scientist, 74-78, staff scientist, 78-81, TECH SPECIALIST, ALCOA LABS, ALUMINUM CO AM, 81- *Mem:* Am Chem Soc; AAAS; Sigma Xi. *Res:* Kinetics and thermodynamics applied to the production of high purity alumina and surface coatings on alumina and aluminum hydroxides; synthesis of aluminum compounds. *Mailing Add:* Alcoa Labs Aluminum Co Am Alcoa Ctr PA 15069

MARTIN, EDWARD WILLIFORD, b Sumter, SC, Nov 29, 29; m 57; c 3. EMBRYOLOGY. *Educ:* Fisk Univ, AB, 50; Ind Univ, MA, 52; Univ Iowa, PhD, 62. *Prof Exp:* Actg head dept biol, Fayetteville State Teachers Col, 52; from asst prof to assoc prof zool, Div Natural Sci, 52-73, prof biol, head dept & chmn, 73-81, DEAN, COL ARTS & SCI, PRAIRIE VIEW A&M UNIV, 81- *Mem:* AAAS; Nat Inst Sci; Am Soc Zoologists; Sigma Xi. *Res:* Synergic action and individual actions of streptomycin and aureomycin on Brucella abortus and Brucella melentensis. *Mailing Add:* Dept Biol A&M Univ Prairie View Box 878 Prairie View TX 77446-0878

MARTIN, EDWIN J, JR, b Kansas City, Mo, Dec 1, 25; m 46; c 3. ELECTRICAL ENGINEERING. *Educ:* Mass Inst Technol, SB, 50; Univ Kans, MS, 56, PhD(elec eng), 64. *Prof Exp:* Res engr, Midwest Res Inst, 50-55; sr res engr, Vendo Co, 55-56; res asst, Univ Kans Res Found, 56-59; sr engr, Midwest Res Inst, 59-65; staff engr, elec eng, Wilcox Elec, Inc, Subsid Northrop Corp, 65-87; independent consult, 87-88; TECH SPECIALIST, MCDONNELL DOUGLAS CORP, 88- *Mem:* Sigma Xi. *Res:* Antennas and antenna arrays; radio aids to navigation. *Mailing Add:* 949 Thunderhead Dr St Louis MO 63138

MARTIN, ELDEN WILLIAM, b Frankfort, Kans, Feb 2, 32; m 55; c 4. ANIMAL PHYSIOLOGY, AVIAN NUTRITION & ECOLOGY. *Educ:* Kans State Univ, BS, 54, MS, 59; Univ Ill, PhD(zool, ecol), 65. *Prof Exp:* Res grant & instr physiol, 63-65, asst prof, 65-69, asst chmn dept, 76-77 & 82-86, ASSOC PROF PHYSIOL, BOWLING GREEN STATE UNIV, 69-, VCHAIR DEPT, 86- *Concurrent Pos:* Frank M Chapman Fund & Marcia Brady Tucker travel awards, 62; Peavey Co res grant, 65-68; var fac res grants, Bowling Green State Univ, 65-; Frank M Chapman Mem Fund grant, 68-69; NSF int travel grant, 70; mem working group granivorous birds, Int Biol Prog; mem Int Ornith Cong; Ohio Biol Survey grant, 74-76; vis assoc prof, Dept Poultry Sci, Univ Wis, 79-80; NSF grant Co-pvt investr, 81. *Mem:* AAAS; Am Ornith Union; Wilson Ornith Soc; Am Soc Zool; Int Union Physiol Sci. *Res:* Physiology and physiological ecology of vertebrate animals; amino acid nutrition, bioenergetics and temperature regulation in birds and other animals; effects of gaseous pollutants on the physiology of birds. *Mailing Add:* Dept Biol Sci Bowling Green State Univ Bowling Green OH 43403-0212

MARTIN, ELMER DALE, b Lancaster Co, Pa, Apr 22, 34; m 83; c 2. FLUID DYNAMICS. *Educ:* Franklin & Marshall Co, BA, 57; Rensselaer Polytech Inst, BAeroE, 57, MAeroE, 58; Stanford Univ, PhD(aeronaut, astronaut), 68. *Prof Exp:* Aeronaut res engr, US Air Force, 58-61, RES SCIENTIST, AMES RES CTR, NASA, 61- *Mem:* Assoc fel Am Inst Aeronaut & Astronaut; Am Phys Soc; Am Math Soc; Soc Indust & Appl Math. *Res:* Viscous and compressible flows; gas-kinetic theory; singular-perturbation techniques; computational and mathematical methods; complex analysis. *Mailing Add:* Ames Res Ctr Moffett Field CA 94035

MARTIN, EUGENE CHRISTOPHER, b Evansville, Ind, Dec 17, 25. ORGANIC POLYMER CHEMISTRY. *Educ:* Evansville Col, BA, 49; De Paul Univ, MS, 51; Univ Ky, PhD(chem), 54. *Prof Exp:* From assoc chemist to chemist, Am Oil Co, 54-60; sr res chemist, Southwest Res Inst, 60-71; res chemist, 71-79, HEAD, POLYMER SCI BR, NAVAL WEAPONS CTR, 79- *Mem:* Am Chem Soc. *Res:* Polymer synthesis and modification of polyethylene and polybutadiene, cellulose derivatives, polyurethanes, polyureas and polypeptides for commercial use and biomedical applications; polymer synthesis and modification for membrane separation processes; gelation of liquids; microencapsulation of liquids and solids; organic synthesis. *Mailing Add:* 100 Diana Way Antioch CA 94509

MARTIN, FLOYD DOUGLAS, ichthyology, aquatic ecology, for more information see previous edition

MARTIN, FRANCIS W, b Minneapolis, Minn, Mar 7, 11; m 44; c 3. PHYSICAL CHEMISTRY. *Educ:* Univ Minn, BChem, 33, PhD(phys chem), 38. *Prof Exp:* Instr inorg chem, Univ Mont, 38-39; res assoc, Battelle Mem Inst, 39-40; res chemist, Corning Glass Works, 40-42; res assoc, Radiation Lab, Mass Inst Technol, 42-46; res assoc, Corning Glass Works, 46-74, sr res assoc, 74-76; CONSULT, 76- *Mem:* AAAS; Sigma Xi; Am Ceramic Soc; Brit Soc Glass Technol. *Res:* Glass; glass ceramics. *Mailing Add:* 101 Hornby Dr Painted Post NY 14870

MARTIN, FRANK BURKE, b Cleveland, Ohio, Mar 21, 37; m 61; c 3. BIOSTATISTICS. *Educ:* St Mary's Col, Minn, BA, 58; Iowa State Univ, MS, 66, PhD(statist), 68. *Prof Exp:* Instr math, St Mary's Col, Minn, 60-63; teaching asst statist, Iowa State Univ, 63-65, res assoc, 65-67; from asst prof to assoc prof, Univ Minn, St Paul, 67-78, Exp Sta statistician, 67-78; CONSULT STATISTICIAN, DEPT STATIST, UNIV MINN. *Concurrent Pos:* expert statist witness. *Honors & Awards:* Snedacor Prize, 66. *Mem:* Biomet Soc; Am Statist Asn. *Res:* Data analysis; forensic applications; survey sampling; experiment design; statistics; clinical trials. *Mailing Add:* Dept Statist Univ Minn Minneapolis MN 55455

MARTIN, FRANK ELBERT, b Warrensburg, Mo, Nov 21, 13. PHYSICS. *Educ:* Univ Mo, AB, 34, PhD(physics), 63; Univ Ill, MS, 56. *Prof Exp:* Instr sr high sch, Mo, 38-42; instr physics, Little Rock Jr Col, 42-43; from instr to asst prof, Cent Mo State Col, 43-58; physicist, Metall Div, US Naval Res Lab, DC, 45-54; asst instr math, Univ Ill, 55-56; instr physics, Univ Mo, 58-59; 60-62; instr physics & eng, Penn Valley Community Col, Kansas City, 79-82; from asst prof to assoc prof, Cent Mo State Col, 62-67, prof, 67-79, EMER PROF PHYSICS, CENT MO STATE UNIV, 79- *Concurrent Pos:* NSF equip grant, Cent Mo State Col; postdoctoral summer res fel, Marshall Space Flight Ctr, 71 & 72. *Mem:* AAAS; Am Phys Soc; Sigma Xi; Inst Elec & Electronics Eng; Am Asn Physics Teachers. *Res:* Elastic constants and internal stresses; electric contact transients; low-carbon steel dilatometry; photoelectric emission; electronic structure of semiconductors; cryogenics; charge carriers in crystalline solids. *Mailing Add:* 123 W South St Warrensburg MO 64093-2323

MARTIN, FRANK GARLAND, b New Orleans, La, Oct 9, 32; m 55; c 3. EXPERIMENTAL STATISTICS. *Educ:* Okla State Univ, BS, 54, MS, 55; NC State Univ, PhD(exp statist), 59. *Prof Exp:* Sr scientist, Bettis Atomic Power Lab, Westinghouse Elec Corp, Pa, 58-62; res statistician, Stamford Res Lab, Am Cyanamid Co, Conn, 62-64; assoc prof, 64-84, PROF STATIST, UNIV FLA, 84- *Concurrent Pos:* Consult, Fla Agr Exp Sta. *Mem:* Am Statist Asn; Biomet Soc. *Mailing Add:* Dept of Statist Univ of Fla Gainesville FL 32611

MARTIN, FRANK GENE, b Clarksville, Tenn, Mar 15, 38; m 59; c 2. PHARMACOLOGY. *Educ:* Univ Tenn, BSPh, 59, MS, 65, PhD(pharmacol), 69. *Prof Exp:* Teaching fel pharmacol, Univ Tenn, 63-68; from asst prof to assoc prof, 68-77, prof, Sch Pharm, Univ Kans, 77-87, assoc dean, 84-87; STAFF PHARMACIST, BAPTIST HOSP, 87- *Mem:* Am Pharmaceut Asn. *Res:* Autonomic pharmacology, especially release and degradation of transmitters. *Mailing Add:* 23 Manor Dr Pensacola FL 32405

MARTIN, FRANKLIN WAYNE, b Salt Lake City, Utah, Apr 14, 28; m 56; c 6. GENETICS. *Educ:* Okla Baptist Univ, BS, 48; Univ Calif, PhD(genetics), 60. *Prof Exp:* Sr lab technician, Univ Calif, 54-60; asst horticulturist, Western Wash Exp Sta, 60-61; mem staff, Fed Exp Sta, Agr Res Serv, 61-71, dir, Mayaguez Inst Trop Agr, PR, 71-79; consult Cent Am, 79-80; breeder, Trop Agr Res Sta, Agr Res Serv, 81-83; RETIRED. *Mem:* Bot Soc Am; Am Soc Hort Sci; Soc Econ Bot; Int Soc Trop Root Crops. *Res:* Genetics, breeding and development of tropical root and tuber crops; development of small scale food production systems for the tropics; introduction and development of little known tropical fruits and vegetables; breeding of sweet potatoes for the tropics; prebreeding of tropical tomatoes. *Mailing Add:* 2305 E Second St Lehigh Acres FL 33936

MARTIN, FREDDIE ANTHONY, b Raceland, La, Nov 17, 45; m 69; c 3. PLANT PHYSIOLOGY, PLANT BREEDING. *Educ:* Nicholls State Col, BS, 66; Cornell Univ, MS, 68, PhD(veg crops), 70. *Prof Exp:* Asst prof, 71-76, assoc prof, 76-80, PROF AGRON, LA STATE UNIV, BATON ROUGE, 80- *Concurrent Pos:* Ed, J Am Soc Sugarcane Technologists; mem, Sugarcane Adv Comt, USDA. *Mem:* Crop Sci Soc Am; Am Soc Sugarcane Technologists; Am Soc Agronomists; Plant Growth Regulation Soc. *Res:* Improving the system of breeding sugarcane and developing improved sugarcane varieties for Louisiana. *Mailing Add:* Dept Agron La State Univ Baton Rouge LA 70803

MARTIN, FREDERICK JOHNSON, physical chemistry, for more information see previous edition

MARTIN, FREDERICK N, b Brooklyn, NY, July 24, 31; m 54; c 2. AUDIOLOGY. *Educ:* Brooklyn Col, BA, 57, MA, 58; City Univ New York, PhD(speech), 68. *Prof Exp:* Speech therapist, Lenox Hill Hosp, 57-58; audiologist, Ark Rehab Serv, 58-60 & Bailey Ear clin, 60-66; lectr, Speech & Hearing Ctr, Brooklyn Col, 66-68, asst prof audiol, 68; from asst prof to assoc prof, 68-74, PROF AUDIOL, UNIV TEX, AUSTIN, 74- *Mem:* Fel Am Speech-Lang-Hearing Asn. *Res:* Clinical audiology and normal audition. *Mailing Add:* Dept Speech Commun Univ Tex Austin TX 78712

MARTIN, FREDERICK WIGHT, b Boston, Mass, Feb 16, 36; m 65; c 2. EXPERIMENTAL PHYSICS. *Educ:* Princeton Univ, AB, 57; Yale Univ, MS, 58, PhD(physics), 64. *Prof Exp:* From physicist to sr physicist, Ion Physics Corp, High Voltage Eng Corp, 63-66; asst prof atomic & solid state physics, Aarhus Univ, Denmark, 66-68; res assoc atomic physics, Univ Ky, 68-69, asst prof, 69-70; asst prof physics & astron, Univ Md, College Park, 70-78; PRES, MICROSCOPE ASSOCS, INC, 78- *Concurrent Pos:* Adj assoc prof physics, Worcester Polytech, Mass, 83-; prin investr, var contracts. *Mem:* Am Phys Soc; Inst Elec & Electronics Engrs. *Res:* Penetration of high energy particles in matter; single atomic collisions involving electron capture or loss or x-ray production by heavy ions; channeling, ion implantation and radiation damage in solids; ion microscopy, and ion optics; achromatic quadrupole lenses and correction of lens aberrations. *Mailing Add:* Microscope Assocs Inc 50 Village Ave Dedham MA 02026

MARTIN, G(UY) WILLIAM, JR, b Abbeville, SC, Mar 12, 46; m; c 1. ANALYTICAL CHEMISTRY, ELECTROCHEMISTRY. *Educ:* Erskine Col, BA, 68; Univ NC, Chapel Hill, PhD(anal chem), 77. *Prof Exp:* GROUP LEADER, BURROUGHS WELLCOME CO, 78- *Mem:* Am Chem Soc. *Res:* Non-aqueous electrochemistry; computer interfacing. *Mailing Add:* Burroughs Wellcome Co 3030 Cornwallis Rd Research Triangle Park NC 27709

MARTIN, GAIL ROBERTA, b New York, NY, Apr 12, 44; m 69; c 1. BIOLOGY, EMBRYOLOGY. *Educ:* Univ Wis, Madison, BA, 64; Univ Calif, Berkeley, PhD(molecular biol), 71. *Prof Exp:* Fel embryol, Univ Col London, Eng, 73-75; fel pediatrics, 75-76, from asst prof to assoc prof, 76-86, PROF ANAT, UNIV CALIF, SAN FRANCISCO, 86- *Concurrent Pos:* Mem, Am Cancer Soc Comt, cell & develop biol, 82-86, personnel comt, 90; Fac Res Award, Am Cancer Soc, 79-84. *Mem:* Soc Develop Biol; Brit Soc Develop Biol. *Res:* Teratocarcinoma cell biology and early mouse embryogenesis. *Mailing Add:* Dept of Anat Univ Calif San Francisco CA 94143

MARTIN, GARY EDWIN, b Wilkensburg, Pa, Oct 14, 49; m 80. HETERONUCLEAR NUCLEAR MAGNETIC RESONANCE SPECTROSCOPY. *Educ:* Univ Pittsburgh, BS, 72; Col Pharm, Univ Ky, PhD(med chem), 76. *Prof Exp:* From asst prof to assoc prof med chem, Col Pharm, Univ Houston, 75-90; AT BURROUGHS WELLCOME CO, 90- *Concurrent Pos:* Adj assoc prof chem, Dept Chem, Univ Tex, Arlington, 81- *Mem:* Int Soc Heterocyclic Chem; Am Chem Soc. *Res:* Synthesis of new heterocyclic ring systems; the correlation of molecular geometry with structural features; spectroscopic parameters and the potential role of these features in the central nervous systems activity of related drugs; two-dimensional nuclear magnetic resonance spectroscopy. *Mailing Add:* Burroughs Wellcome Co 3030 Cornwallis Rd Research Triangle Park NC 27709

MARTIN, GEORGE, US citizen. MECHANICAL & METALLURGICAL ENGINEERING. *Educ:* Univ Birmingham, BSc, 50, PhD, 52. *Prof Exp:* Sr develop engr, John Gardom & Co, Eng, 52-56; tech mgr, Chromizing Co, Calif, 56-57; chief metallurgist, Honolulu Oil Co, 57-61; prog mgr metals sci, Los Angeles Div, NAm Aviation, Inc, 61-70; chief advan fabrication dept, McDonnell-Douglas Astronaut Co, 70-72; PRES, CREATIVE METAL CRAFTS INC, 72- *Concurrent Pos:* Lectr, Dept Eng, Univ Calif, Los Angeles, 58-90, adj prof eng, 72-80; consult, 72-; Nat Endowments Arts res fel, 78. *Mem:* Am Soc Testing & Mat; Brit Inst Mech Engrs. *Res:* Materials and production research and development management; manufacturing engineering; material properties, corrosion and fracture; nondestructive testing; blacksmithing. *Mailing Add:* 1708 Berkeley St Santa Monica CA 90404

MARTIN, GEORGE C(OLEMAN), b Everett, Wash, May 16, 10; m 35; c 2. AERONAUTICS. *Educ:* Univ Wash, Seattle, BS, 31. *Prof Exp:* Engr struct res, Boeing Co, 31-35, eng supvr, 35-39, chief struct design, 39-41, staff engr, 41-45, proj engr, 45-47, chief preliminary design, 47-48, proj engr bomber prod design, 48-52, chief engr, 53-58, vpres & gen mgr, Seattle Div, 58-59, vpres & asst gen mgr, Aerospace Div, 59-61, vpres & gen mgr, Seattle Br, Mil Aircraft Systs Div, 61-63, vpres & prog mgt dir, Airplane Div, 63-64, vpres eng, 64-71, vpres design & develop, 71-72, consult, Boeing Co, 72-85; RETIRED. *Honors & Awards:* Hon Pathfinder Award, NW Museum Flight, 87. *Mem:* Fel Am Inst Aeronaut & Astronaut; Aerospace Indust Asn Am. *Res:* Airplane and missile design, development and production. *Mailing Add:* 425 SE Shoreland Dr Bellevue WA 98004

MARTIN, GEORGE C, b San Francisco, Calif, Sept 15, 33; m 53; c 2. POMOLOGY. *Educ:* Calif State Polytech Col, BS, 55; Purdue Univ, MS, 60, PhD(plant physiol), 62. *Prof Exp:* Res asst plant physiol & hort, Purdue Univ, 58-62; res plant physiologist, Crops Res Div, Agr Res Serv, USDA, Wash, 62-67; assoc pomologist, 67-73, PROF, UNIV CALIF, DAVIS, 73- *Concurrent Pos:* Mem Awards, Am Soc Hort Sci, 77, 81 & 83, chair bd dirs, 90-91. *Honors & Awards:* J H Gourley Award in Pomol, Am Soc Hort Sci, 71, Stark Award, 80, Miller Award, 81. *Mem:* Am Philos Soc; fel Am Soc Hort Sci (pres, 89-90); Am Soc Plant Physiol. *Res:* Chemical thinning; mechanism of fruit set, dormancy and rest; use of chemicals to aid mechanical harvest of fruit; measurement of hormones. *Mailing Add:* Dept of Pomol Univ of Calif Davis CA 95616

MARTIN, GEORGE EDWARD, b Batavia, NY, July 3, 32; m 69. GEOMETRY COMBINATORICS. *Educ:* State Univ NY Albany, AB, 54, MA, 55; Univ Mich, PhD(math), 64. *Prof Exp:* Asst prof math, Univ RI, 64-66; from asst prof to assoc prof, 66-84, PROF MATH, STATE UNIV NY ALBANY, 84- *Mem:* Am Math Soc; Math Asn Am. *Res:* Geometry, specializing in tessellations and the foundations of geometry. *Mailing Add:* Dept Math State Univ NY Albany NY 12222

MARTIN, GEORGE FRANKLIN, JR, b Englewood, NJ, Feb 20, 37; m 60; c 2. NEUROANATOMY. *Educ:* Bob Jones Univ, BS, 60; Univ Ala, MS, 63, PhD(anat), 65. *Prof Exp:* From instr to assoc prof, 65-73, PROF ANAT, COL MED, OHIO STATE UNIV, 73- *Concurrent Pos:* NIH res grants, 65-87, NSF res grants, 80-86; mem, Neurobiol Rev Group, (NEUB2), 82-87, chmn, 85-87. *Mem:* AAAS; Soc Neurosci; Am Asn Anat; Pan-Am Asn Anat; Sigma Xi; Nucleolus (pres, 85-86); Int Soc Develop Neurobiol. *Res:* Determining the various connections, functions development and plasticity of somatic motor and sensory systems. *Mailing Add:* Dept Cell Biol Neurobiol & Anat Col Med Ohio State Univ Columbus OH 43210

MARTIN, GEORGE H(ENRY), b Chicago, Ill, June 8, 17. ENGINEERING MECHANICS. *Educ:* Ill Inst Technol, BS, 41, MS, 44; Northwestern Univ, PhD(mech eng), 55. *Prof Exp:* Jr design engr, Teletype Corp, 41-42; from instr to asst prof, Ill Inst Technol, 43-50; lectr, Northwestern Univ, 50-55; assoc prof, 55-79, EMER ASSOC PROF MECH ENG, MICH STATE UNIV, 79- *Mem:* Fel Am Soc Mech Engrs. *Res:* Kinematics and dynamics of machines. *Mailing Add:* 1320 Westview East Lansing MI 48823

MARTIN, GEORGE MONROE, b New York, NY, June 30, 27; m 52; c 4. EXPERIMENTAL PATHOLOGY. *Educ:* Univ Wash, BS, 49, MD, 53. *Prof Exp:* Intern med, surg & gynec, Montreal Gen Hosp, 53-54; asst resident path, Univ Chicago, 54-55, asst, 55-56, instr, 56-57; from asst prof to assoc prof path, 60-68, dir cytogenetics lab, Hosp, 64-68, actg chmn, dept path, 80-81, PROF PATH, UNIV WASH, 68- *Concurrent Pos:* Consult, Firlands Sanitarium, Seattle, Wash, 57-59 & Northern State Hosp, Sedro Woolley, 59-63; mem path B Study sect, NIH, 66-70, mem adult develop & aging res comn, 73-, chmn aging res rev comt, 75-77; Josiah Macy Jr Found fac scholar, Dunn Sch Path, Oxford, 78-79; chmn, nat res plan aging, Nat Inst Aging, 80-82; mem, geriat & geront adv comt, Vet Admin, 81-; dir, Univ Wash Alzheimer Dis Res Ctr, 85-; attend pathologist, 59-, adj prof genetics, 75-, dir med scientist training prog, Univ Wash, 70-73 & 84-88. *Honors & Awards:* Brookdale Award, 81. *Mem:* Am Asn Path; Am Soc Human Genetics; Tissue Cult Asn (pres, 86-88); Genetics Soc Am; Gerontol Soc Am; fel AAAS. *Res:* Mammalian cell culture; somatic cell and human genetics, cytogenetics; cell senescence; genetic aspects of aging; pathogenesis of Alzheimer disease. *Mailing Add:* Dept Path SM-30 Univ Wash Seattle WA 98195

MARTIN, GEORGE REILLY, b Boston, Mass, Jan 20, 33; m; c 3. GERONTOLOGY. *Educ:* Colgate Univ, BS, 55; Univ Rochester, PhD(pharmacol & radiation biol), 58. *Prof Exp:* Res asst, Atomic Energy Proj, Univ Rochester, 55-58; res assoc, Am Dent Asn, Nat Inst Dent Res, NIH, 59-61, pharmacologist, Lab Biochem, 61-66, chief, Connective Tissue Sect, 67-74, actg chief, Lab Biochem, 72-73 & Lab Develop Biol & Anomalies, 74, chief, Lab Develop Biol & Anomalies, 74-88; SCI DIR, GERONT RES CTR, NAT INST AGING, 88- *Concurrent Pos:* Guest worker, Lab Chem Pharmacol, Nat Heart Inst, NIH, 58-59; mem staff, Weizman Inst Sci, 66-67 & Max-Planck Inst Biochem, Munich, Ger, 75-76; co-chmn, Gordon Res Conf Struct Macromolecules, 74; consult, Med Adv Bd, Juv Diabetes Found, 77; mem, Res Adv Bd, Shriners Hosps, 81 & Planning Comt Working Conf Perspectives Ovarian Cancer Older-Aged Women, 91; chmn, Res Adv Bd, Shriners Hosp Crippled Children, 83-85 & Sci Rev Bd, Ludwig Inst Cancer Res, Brazil; Alexander von Humboldt sr scientist award, 84-85; presidential lectr, Am Thoracic Soc, 87; chair, Conf Molecular Aspects Tumor Metastasis, Fedn Am Socs Exp Biol, 90; lectr med, NIH, 91. *Honors & Awards:* G Burroughs Mider Lectr, NIH, 88; Robert R Kohn Lectr, Case Western Reserve Univ, 89. *Mem:* Am Soc Biol Chemists; Am soc Cell Biol; Biophys Soc; Develop Biol Soc; Geront Soc Am; Int Soc Differentiation. *Res:* Radiation biology. *Mailing Add:* Geront Res Ctr Nat Inst Aging 4940 Eastern Ave Baltimore MD 21224

MARTIN, GEORGE STEVEN, b Oxford, Eng, Sept 19, 43; m 69; c 1. VIROLOGY, TUMOR BIOLOGY. *Educ:* Cambridge Univ, BA, 64, PhD(molecular biol), 68. *Prof Exp:* Fel viral oncol, Dept Molecular Biol, Univ Calif, Berkeley, 68-71; staff mem, Imp Cancer Res Fund, London, 71-75; from asst prof to assoc prof, 75-83, PROF VIROL ONCOL, DEPT ZOOL, UNIV CALIF, BERKELEY, 83- *Mem:* Am Soc Microbiol; AAAS; Am Soc Cell Biol; Am Soc Virol. *Res:* Transformation by RNA tumor viruses; transformation of differ- entiating cells. *Mailing Add:* 401 Barker Hall Univ Calif Berkeley CA 94720

MARTIN, GORDON EUGENE, b San Diego, Calif, Aug 22, 25; m 49; c 4. ENGINEERING MECHANICS, ELECTRICAL ENGINEERING. *Educ:* Univ Calif, Berkeley, BS, 47; Univ Calif, Los Angeles, MS, 51; San Diego State Univ, MA, 61; Univ Tex, PhD(elec eng), 66. *Prof Exp:* Elec scientist, Naval Electronics Lab, San Diego, 47-52, A officer-in-chg, Naval Exp Fac, Eleuthera Island, Bahamas, 52-53, res proj officer, US Navy Underwater Sound Lab, New London, 53- 54, sr res physicist, Naval Ocean Systs Ctr, San Diego, 54-80; head, Acoustics Dept, Systs Explor, Inc, 80-82; owner, Martin Acoustics Software Technol, 82-86; PRES, MARTIN ANALYTICAL SOFTWARE TECHNOL, INC, 86- *Concurrent Pos:* Prin investr, US Navy & Naval Ocean Systs Ctr, 54-80; lectr, spec training proj, San Diego State Univ, 60-62; mem, nat comt piezoelec & transducer mat, US Navy, 60-80; mem, standard coordinating comt definitions, Inst Elec & Electronics Engrs & tech comt, transducers & resonators, 72-; prin scientist, MAST Inc & numerous fed govt res projs, 88- *Mem:* NY Acad Sci; fel Acoust Soc Am; sr mem Inst Elec & Electronics Engrs. *Res:* Computer-aided-engineering mathematical modeling theory and software for piezoelectric transducer materials; elements and arrays; inverse piezoelectric transducer/material parameter theory and experimental methods; dissipation in piezoelectric materials; design of vibrating (sonar) systems; eigensystem signal processing for radio and sonar arrays. *Mailing Add:* 2627 Burgener Blvd San Diego CA 92110-1022

MARTIN, GORDON MATHER, b Brookline, Mass, Mar 2, 15; m 40; c 3. PHYSICAL MEDICINE. *Educ:* Nebr Wesleyan Univ, AB, 36; Univ Nebr, MD, 40; Univ Minn, MS, 44. *Prof Exp:* Asst prof phys med, Sch Med, Univ Kans, 44-47; from asst prof to assoc prof, Mayo Grad Sch Med, 47-73, prof phys med, 73-81, EMER PROF & CONSULT MED & REHAB, MAYO MED SCH, UNIV MINN, 81- *Concurrent Pos:* Consult, Mayo Clin, 47-73, sr consult phys med & rehab, 73-; exec secy, Am Bd Phys Med & Rehab, 81- *Mem:* Am Cong Rehab Med; Am Acad Phys Med & Rehab; AMA; Sigma Xi. *Res:* Clinical research. *Mailing Add:* Mayo Clinc Rochester MN 55905

MARTIN, GREGORY EMMETT, NEUROPHARMACOLOGY, ANTIPSYCHOTIC DRUGS. *Educ:* Purdue Univ, PhD(neurobiol), 74. *Prof Exp:* PRIN SCIENTIST, MCNEIL PHARMACEUT, 84- *Res:* Analgesic drugs. *Mailing Add:* Pharmacol Dept Rorer Central Res 680 Allandal Rd King of Prussia PA 19406

MARTIN, HANS CARL, b Winnipeg, Man, Nov 20, 37; m; c 3. MICROMETEOROLOGY. *Educ:* Univ Man, BSc, 58; Univ Western Ont, MSc, 61, PhD(physics), 66. *Prof Exp:* Res scientist meteorol physics, Commonwealth Sci & Indust Res Orgn, Australia, 66-68; res scientist meteorol physics, Can Meteorol Serv, Dept Environ, 69-77; res mgr, Long-Range Transp Air Pollution Prog, Acid Rain, 77-88, DIR, AIR QUAL RES BR, DEPT ENVIRON, 88- *Concurrent Pos:* Lectr, Mgt Training Prog, Fed Govt; Lectr, various univ. *Res:* Air-sea interactions; energy exchange processes near the surface of the earth over land, water and ice; air pollution transport, transformation and deposition and impact on material and man-made receptors. *Mailing Add:* 4905 Dufferin St Downsview ON M3H 5T4 Can

MARTIN, HAROLD ROLAND, medicine; deceased, see previous edition for last biography

MARTIN, HARRY LEE, b Nashville, Tenn, Mar 10, 56; m 79; c 1. ROBOTIC CONTROLS, HUMAN MACHINE INTERFACE. *Educ:* Univ Tenn, BS, 78; Purdue Univ, MS, 79; Univ Tenn, PhD(eng), 86. *Prof Exp:* Develop engr, Oak Ridge Nat Lab, 80-86; FOUNDER & PRES, TELEROBOTICS INT, INC, 86- *Concurrent Pos:* Lectr mech eng, Univ Tenn, 80-81. *Honors & Awards:* IR-100 Award, 84. *Mem:* Am Nuclear Soc; Am Soc Mech Engrs; Nat Soc Prof Engrs. *Res:* Developer of remote control systems for robotic systems for hazardous environments; recipient of four patents. *Mailing Add:* 7325 Oak Ridge Hwy Knoxville TN 37921

MARTIN, HERBERT LLOYD, b Somerville, Mass, Dec 7, 21; m 51; c 6. NEUROLOGY. *Educ:* Boston Univ, BS, 47, MD, 50. *Prof Exp:* Teaching fel neurol, Montreal Neurol Inst, McGill, 57-58; assoc prof clin neurol, 58-69, PROF NEUROL, UNIV VT, 69-, ASSOC CHMN DEPT, 71- *Mem:* AMA; Am Acad Neurol; Am Epilepsy Soc; Am Asn Res Nerv & Ment Dis; Asn Am Med Cols. *Mailing Add:* Dept Neurol Univ Vt One S Prospect St Burlington VT 05401

MARTIN, HORACE F, b Azores, Jan 11, 31; US citizen; m 54; c 7. CLINICAL CHEMISTRY, CLINICAL PATHOLOGY. *Educ:* Providence Col, BS, 53; Univ RI, MS, 55; Boston Univ, PhD(biochem), 61; Brown Univ, MD, 75; SNortheast Sch Law, JD, 91. *Hon Degrees:* MA, Brown Univ, 67. *Prof Exp:* Sr res chemist, Monsanto chem Co, 57-59, group leader life sci, Monsanto Res Corp, 61-63; assoc prof med sci, 66-78, prof, 78-81, PROF PATH, BROWN UNIV, 81- *Concurrent Pos:* Biochemist, RI Hosp, Providence, 63-78, dir clin chem, 79- *Mem:* Am Soc Clin Pathologists; Am Asn Clin Chemists; AMA; fel Am Bd Clin Chem; fel Am Inst Chem; fel Am Inst Clin Scis. *Res:* Clinical pathology; normal values; instrumental methods of analysis. *Mailing Add:* Dept Med Brown Univ Providence RI 02912

MARTIN, HUGH JACK, JR, b San Diego, Calif, Sept 1, 26; m 50; c 6. HIGH ENERGY PHYSICS. *Educ:* Calif Inst Technol, BS, 51, PhD(physics), 56. *Prof Exp:* Res assoc, 55-57, from asst prof to assoc prof, 57-61, PROF PHYSICS, IND UNIV, BLOOMINGTON, 65- *Mem:* Am Phys Soc; Asn Comput Mach; Am Asn Univ Prof; Sigma Xi. *Res:* Experimental high energy, bubble chamber and spark chamber physics; pattern recognition and computer applications. *Mailing Add:* Dept Physics Col Arts & Sci Ind Univ Swain W 267 Bloomington IN 47401

MARTIN, J(AMES) W(ILLIAM), b Lenox, Iowa. ENGINEERING. *Educ:* Kans State Col, BS, 33 & 38; Iowa State Col, MS, 39. *Prof Exp:* Asst, Kans State Col, 33-35, lab asst, 37-38, from instr to assoc prof eng, 40-46; prof eng, Univ Idaho, 46-81; RETIRED. *Concurrent Pos:* Field engr, John Deere Plow Co, 35-37; instr eng, Iowa State Col, 38-39 & Univ Ill, 39-40. *Mem:* Am Soc Eng Educ; fel Am Soc Agr Engrs. *Res:* Agricultural machinery. *Mailing Add:* Good Samaritan Village 640 N Eisenhower St Moscow ID 83843

MARTIN, JACK, b Tuscaloosa, Ala, Aug 11, 27; m 57; c 4. PSYCHIATRY. *Educ:* Univ Ala, BS, 49; Vanderbilt Univ, MD, 53. *Prof Exp:* Intern, Charity Hosp, New Orleans, La, 53-54; resident physician gen psychiat, Cincinnati Gen Hosp, 54-56, res fel child psychiat, Cincinnati Gen Hosp & Child Guid Home, 56-58; from instr to asst prof, 58-63, CLIN PROF PSYCHIAT, UNIV TEX HEALTH SCI CTR DALLAS, 63- *Concurrent Pos:* Med dir & pres, Shady Brook Schs. *Res:* Child development; clinical child psychiatry. *Mailing Add:* 3636 Dickason St Dallas TX 75219

MARTIN, JACK E, b Bogard, Mo, June 4, 31; m 55; c 3. NUTRITION, BIOCHEMISTRY. *Educ:* Univ Mo, BS, 53, MS, 60; Univ Fla, PhD(nutrit), 63. *Prof Exp:* Prod supvr biochem, Monsanto Co, Mo, 63-67; asst res nutritionist, Ralston Purina Co, 67-69; nutritionist, Ceres Land Co, 69-71; NUTRITIONIST, STERLING NUTRIT SERV, INC, 71- *Honors & Awards:* Distinguished Soc Award, Am Soc Agr Consults, 88. *Mem:* Am Soc Animal Sci; Am Soc Agr Consults (pres, 87). *Res:* Effect of mineral nutrition on cellulose digestion in ruminants. *Mailing Add:* Rte 4 407 Highland Dr Sterling CO 80751

MARTIN, JAMES ARTHUR, b Ft Benning, Ga, Aug 9, 44; m 66; c 3. VEHICLE DESIGN, PROPULSION. *Educ:* WVa Univ, BS, 66; Mass Inst Technol, MS, 67, EEA, 69; George Washington Univ, DSc, 82. *Prof Exp:* Aerospace engr, Langley Res Ctr, NASA, 67-90; ASSOC PROF AEROSPACE ENG, UNIV ALA, 91- *Concurrent Pos:* Mem, liquid propulsion tech comt, Am Inst Aeronaut & Astronaut, 80-83; mem, space propulsion comt, Soc Automotive Engrs, 83-; assoc ed, J Spacecraft & Rockets, 83- *Honors & Awards:* NASA Group Achievement Award, 83. *Mem:* Assoc fel Am Inst Aeronaut & Astronaut. *Res:* Evaluation of system and propulsion concepts for advanced Earth-to-orbit vehicles including trajectory optimization, vehicle mass estimation, geometry, aerodynamics, sizing, propulsion system analysis and cost estimation. *Mailing Add:* Aerospace Eng Univ Ala Tuscaloosa AL 35487-0280

MARTIN, JAMES CULLEN, b Dover, Tenn, Jan 14, 28; m 51; c 5. ORGANIC CHEMISTRY. *Educ:* Vanderbilt Univ, BA, 51, MS, 52; Harvard Univ, PhD(chem), 56. *Prof Exp:* From instr to assoc prof, 56-85, DISTINGUISHED PROF CHEM, VANDERBILT UNIV, NASHVILLE, 85- *Concurrent Pos:* Sloan Found fel, 62-66; Guggenheim Mem Found fel, 65-66; assoc mem, Ctr Advan Study, Univ Ill, Urbana, 71-72; Humboldt Found Sr US Scientist Award, 78-79 & 88. *Honors & Awards:* Buck-Whitney Medal, Am Chem Soc, 79. *Mem:* Am Chem Soc; fel AAAS; fel Japan Soc Prom Sci. *Res:* Mechanisms of organic reactions; free-radical reactions; synthesis of compounds expected to show unusual physical properties or reactivity; compounds of hypervalent non-metals; sulfuranes. *Mailing Add:* Dept Chem Vanderbilt Univ Box 1822 Sta B Nashville TN 37235-0002

MARTIN, JAMES CUTHBERT, b Wilson, NC, May 8, 27; m 66. INDUSTRIAL ORGANIC CHEMISTRY. *Educ:* Univ NC, BS, 47. *Prof Exp:* From res chemist to sr res chemist, 48-67, res assoc, 67-72, sr res assoc, 72-78, RES FEL, TENN EASTMAN CO, 78- *Mem:* Am Chem Soc; AAAS; Sigma Xi. *Res:* Chemistry of ketenes; small ring compounds; new polymer systems; applied organic chemistry; exploratory research in catalysis; new product development. *Mailing Add:* Res Lab Tenn Eastman Co Kingsport TN 37662

MARTIN, JAMES EDWARD, b Rock Springs, Wyo, Oct 8, 49. TERTIARY BIOSTRATIGRAPHY, TAXONOMY. *Educ:* SDak Sch Mines & Technol, BS, 71, MS, 73; Univ Washington, PhD(geol), 79. *Prof Exp:* Res assoc paleont, Mus Geol, SDak Sch Mines, 67-74; asst instr, Dept Geol, Univ Washington, 74-77; cur paleont, Thomas Burke Mus, Univ Washington, 77-79; dir, Black Hills Natural Sci Field Sta, 79-88, PROF, DEPT GEOL & GEOL ENG, SDAK SCH MINES & TECHNOL, 79-, CUR, MUS GEOL, 79- *Concurrent Pos:* Consult, Wash Pub Power Supply Systs, 75, Shannon & Wilson Geotech Consult, 77 & Bur Land Mgt, 77; res geologist, SDak Geol Surv; NSF & Nat Geog Soc grants. *Mem:* Geol Soc Am; Soc Vertebrate Paleont; Sigma Xi. *Res:* Vertebrate paleontology, stratigraphy and biostratigraphy; biostratigraphy of Hemphillian deposits in Oregon and Washington, of marine vertebrates in the Pierre Shale of South Dakota, of Miocene sediments in South Dakota; taxonomic studies of Cretaceous, Miocene, and Pleistocene vertebrates; sedimentation of Pennsylvanian-Permian sediments in Black Hills; stratigraphy and structural geology of the Black Hills uplift. *Mailing Add:* Dept Geol & Geol Eng SDak Sch Mines & Technol 500 E St Joseph Rapid City SD 57701

MARTIN, JAMES ELLIS, b Cleveland, Ohio, Dec 25, 52. POLYMER PHYSICS, SCATTERING MEASUREMENTS. *Educ:* Univ Wash, BS, 76, PhD(phys chem), 81. *Prof Exp:* STAFF SCIENTIST, DEPT SOLID STATE RES, SANDIA NAT LAB, 81- *Mem:* Am Chem Soc. *Res:* Theory of the statistical and dynamical properties of polymers in solution and in the bulk state; photon correlation spectroscopy measurement of the dynamical properties of polymers. *Mailing Add:* 5020 Calle De Tierra NE Albuquerque NM 87111

MARTIN, JAMES FRANKLIN, microbiology, for more information see previous edition

MARTIN, JAMES FRANKLIN, b St Mary's, WVa, Mar 20, 17; m 42; c 2. MEDICINE, RADIOLOGY. *Educ:* Marietta Col, AB, 38; Western Reserve Univ, MD, 42. *Prof Exp:* Teaching fel radiol, Western Reserve Univ Hosp, 47-48, demonstr radiol, Western Reserve Univ, 48, from instr to sr instr, 48-50; from asst prof to prof radiol, 50-75, prof med sonics, asst radiol, 75-86, EMER PROF RADIOL, NEUROL & ULTRASOUND, BOWMAN GRAY SCH MED, 86- *Concurrent Pos:* Physician, NC Baptist Hosp; dir postgrad med sonics, Bowman Gray Sch Med, 75- *Mem:* Radiol Soc NAm; Am Roentgen Ray Soc (pres, 81-82); AMA; fel Am Col Radiol; Am Inst Ultrasound Med. *Res:* Clinical radiology; ultrasound. *Mailing Add:* 26803 Grosvenor Pl Winston-Salem NC 27106

MARTIN, JAMES HAROLD, b Collinwood, Tenn, Oct 12, 31; m 54; c 2. DAIRY MICROBIOLOGY. *Educ:* Univ Tenn, BSc, 57; Ohio State Univ, MSc, 58, PhD(microbiol), 63. *Prof Exp:* Res asst dairy tech, Ohio State Univ, 57-58, res assoc, 60-63, asst prof, 63-65; instr dairy mfg, Miss State Univ, 58-60; from asst prof to assoc prof dairy microbiol, Univ Ga, 65-72; prof dairy sci & head dept, SDak State Univ, 72-78 & Clemson univ, 78-85; prof & chmn food sci & nutrit dept, 85-89, PARKER PROF DAIRY FOODS, OHIO STATE UNIV, 89- *Concurrent Pos:* Ed, Cultured Dairy Prod J. *Honors & Awards:* Nordica Int Res Award, 80. *Mem:* Am Dairy Sci Asn; Inst Food Technol. *Res:* Physiology and metabolism of microorganisms common to milk, especially as related to bacterial spores, starter and spoilage organisms. *Mailing Add:* Dept Food Sci & Technol Ohio State Univ Columbus OH 43210

MARTIN, JAMES HENRY, III, b New Orleans, La, Mar 31, 43; m 66; c 4. VERTEBRATE PHYSIOLOGY. *Educ:* Univ Va, BA, 65; Univ Richmond, MS, 67; Univ Tenn, PhD(zool), 70. *Prof Exp:* Instr anat & physiol, Univ Tenn, Knoxville, 68; NIH fel & instr physiol, Med Col Va, Va Commonwealth Univ, 71-73; asst prof natural sci, 73-76, assoc prof, 76-78, PROF BIOL, J S REYNOLDS COMMUNITY COL, 79- *Concurrent Pos:* Vis lectr, Math & Sci Ctr, Richmond, Va, 72-; vis scientist, Va Acad Sci. *Mem:* AAAS; Am Soc Ichthyologists & Herpetologists; Am Inst Biol Sci; Soc Vert Paleont; Am Soc Zoologists. *Res:* Physiology and biophysics of muscle contraction; transport across epithelial membranes; long term biological effect of nuclear war; nuclear disasters. *Mailing Add:* Dept Biol J S Reynolds Community Col Parham Campus Richmond VA 23261-2040

MARTIN, JAMES MILTON, b Waxahachie, Tex, May 15, 14; m 41; c 3. GEOPHYSICS. *Educ:* Univ Okla, BS, 38; Rensselaer Polytech Inst, MS, 46. *Prof Exp:* Seismic explor, Geophys Serv, Inc, 34-35 & 37-39; seismol & seismic prospecting, Magnolia Petrol Co, 40-41 & 47; tech eval, US Naval Ord Lab, Md 42 & 47-57, chief underwater eval dept, 57-73, proj mgr torpedoes, 73-75, RES, SEAMINES IN WARS & MINOR HOSTILITIES, US NAVAL ORD LAB, MD, 75- *Mem:* Sigma Xi; Retired Officers Asn; Reserve Officers Asn US; Naval Reserve Asn. *Res:* Technical evaluation of naval ordnance; geophysics. *Mailing Add:* 314 Williamsburg Dr Silver Spring MD 20901

MARTIN, JAMES PAXMAN, b Cowley, Wyo, Sept 22, 14; m 37; c 2. SOIL MICROBIOLOGY. *Educ:* Brigham Young Univ, BS, 38; Rutgers Univ, PhD(soil microbiol), 41. *Prof Exp:* Asst soil microbiol, NJ Exp Sta, 38-41; coop agent, Soil Conserv Serv, USDA & NJ Exp Sta, 41-43; asst prof bact & asst soil microbiologist exp sta, Univ Idaho, 43-45; from asst chemist to assoc chemist, 45-57, prof, 61-80, EMER PROF SOIL SCI, UNIV CALIF, RIVERSIDE, 81-, CHEMIST, CITRUS RES CTR, 57- *Concurrent Pos:* Soil Sci Res Award, Soil Sci Soc Am, 79; vis prof, Univ Rio, Janeiro, Brazil, 80. *Mem:* Fel AAAS; Soil Sci Soc Am; fel Am Soc Agron. *Res:* Decomposition and stabilization of 14-C-labelled phenolic substances; model and plant lignins; fungal melanins; model humic polymers in relation to soil humus formation; citrus replant problem; influence of pesticides on soil properties. *Mailing Add:* Dept Soil & Environ Sci Univ Calif Riverside CA 92521

MARTIN, JAMES RICHARD, b Schenectady, NY, July 13, 49; m 82; c 1. PHYSIOLOGICAL PSYCHOLOGY. *Educ:* Allegheny Col, BS, 71; Univ Minn, PhD(physiol psychol), 75;. *Prof Exp:* Asst res psychologist physiol psychol, Univ Calif, Los Angeles, 75-77; res scientist, 77-82, BEHAV SCI INST, SWISS FED INST TECHNOL, HOFFMANN-LA ROCHE-BASEL, SWITZ, 84- *Concurrent Pos:* Lab chief & Group leader, E Merck, Darmstadt, WGermany, 82-84. *Mem:* Soc Neurosci; Collegium Int Neuro-Psychopharmacologicum; Europ Brain & Behav Soc; Europ Neurosci Asn. *Res:* Drug dependence; behavioral pharmacology; neuroscience; behavioral science; animal learning; the development of cognitive enhancing compounds using animal models. *Mailing Add:* Pharmaceut Res Div F Hoffman-La Roche Basel Switzerland

MARTIN, JAMES TILLISON, b Bluefield, WVa, Aug 10, 46. BEHAVIORAL PHYSIOLOGY. *Educ:* WVa Univ, AB, 67; Univ Conn, MS, 71; Univ Munich, PhD(zool), 74. *Prof Exp:* Asst pharmacologist, RMI Inst Pharmacol, State Univ Utrecht, 73-74; assoc animal sci, Univ Minn, St Paul, 74-76; from asst prof to assoc prof biol, Stockton State Col, 76-86; ASSOC PROF PHYSIOL, COL OSTEOP, MED OF THE PAC, 86- *Concurrent Pos:* Translr Ger to English, Fr Nat Mus Natural Hist, 73-75; Nat Inst Child Health & Human Develop fel, 74-76; vis asst prof, MIT, 80-81. *Mem:* Soc Neurosci; AAAS; Int Soc Psychoneuroendocrinol. *Res:* Neurochemical aspects of imprinting; neuroendocrine aspects of brain development and behavior; biological basis of sexual orientation. *Mailing Add:* Col Osteop Med of the Pac Pomona CA 91766

MARTIN, JAY RONALD, b Sanford, Maine, Jan 27, 44; m 67; c 2. CHEMICAL ENGINEERING, POLYMER SCIENCE. *Educ:* Lafayette Col, BS, 66; Princeton Univ, PhD(chem eng), 72. *Prof Exp:* Staff scientist, 71-75, sr scientist, Textile Res Inst, 75-76; asst prof, 76-80, ASSOC PROF, LAFAYETTE COL, 80- *Concurrent Pos:* Res assoc, Textile Res Inst, 77-80. *Mem:* Am Inst Chem Eng. *Res:* Dynamic mechanical properties of polymers; polymer flammability; polymer degradation; thermal analysis. *Mailing Add:* Dept of Chem Eng Lafayette Col Easton PA 18042

MARTIN, JEROME, b Bisbee, Ariz, Jan 12, 02; m 28; c 2. CHEMISTRY. *Educ:* Univ Colo, BS, 24; Univ Calif, PhD(phys chem), 28. *Prof Exp:* Res chemist, Com Solvents Corp, 27-39, res dir, 39-57, sci dir, 57-67; CHEM CONSULT, IMC CHEM GROUP INC, 77- *Concurrent Pos:* Consult, 67-77. *Mem:* AAAS; Am Chem Soc; Math Asn Am. *Res:* Catalysis; organic chemistry; antibiotics. *Mailing Add:* IMC Chem Group Inc PO Box 207 Terre Haute IN 47808

MARTIN, JERRY, JR, b Darwin, Okla, Oct 28, 30; m 53; c 4. ANIMAL NUTRITION, ANIMAL PHYSIOLOGY. *Educ:* Okla State Univ, BS, 57, MS, 59, PhD(animal nutrit & physiol), 61. *Prof Exp:* Instr animal sci, Murray State Agr Col, 61-67; ASSOC PROF BIOL & ANIMAL SCI, PANHANDLE STATE UNIV, 67-, CHMN DIV AGR, 73- *Mem:* Am Inst Biol Sci; Am Soc Animal Sci. *Res:* Grain processing; additives in feedlot rations. *Mailing Add:* Dept Animal Sci Panhandle State Univ PO Box 430 Goodwell OK 73939

MARTIN, JIM FRANK, b Millersburg, Iowa, Nov 24, 44; m 76; c 2. NONDESTRUCTIVE EVALUATION, AUTOMATION. *Educ:* Iowa State Univ, BS, 65; Mass Inst Technol, PhD(physics), 71. *Prof Exp:* Engr, Chevron Res Lab, Standard Oil Calif, 65; res asst, Lab Nuclear Sci, Mass Inst Technol, 71-75; res assoc, Stanford Linear Accelerator Ctr, Stanford Univ, 75-78; MEM TECH STAFF, SCI CTR, ROCKWELL INT, 78- *Mem:* Am Phys Soc; Inst Elec & Electronics Engrs; Am Soc Nondestructive Testing; Sigma Xi. *Res:* Research and development of advanced techniques for ultrasonic nondestructive evaluation: including transducer development, digital data acquisition signal and image processing, and device control (robotics). *Mailing Add:* Sogo Hanzomon Bldg 1-7 Kojimachi Chiyoda-Ku Tokyo 102 Japan

MARTIN, JOEL JEROME, b Jamestown, NDak, Mar 27, 39. SOLID STATE PHYSICS. *Educ:* SDak Sch Mines & Technol, BS, 61, MS, 63; Iowa State Univ, PhD(physics), 67. *Prof Exp:* AEC fel, Ames Lab, Iowa State Univ, 67-69; from asst prof to assoc prof, 69-79, PROF PHYSICS, OKLA STATE UNIV, 79- *Concurrent Pos:* Sabbatical leave, Sandia Nat Labs, Albuquerque, NM, 82. *Mem:* Am Phys Soc; Am Asn Physics Teachers; Am Asn Crystal Growth. *Res:* Radiation damage in insulators; thermal conductivity; crystal growth of optical materials. *Mailing Add:* Dept of Physics Okla State Univ Stillwater OK 74078-0444

MARTIN, JOEL WILLIAM, b Durham, NC, June 10, 55; m 81; c 2. CRUSTACEAN LARVAL DEVELOPMENT. *Educ:* Univ Ky, BS, 78; Univ Southwestern La, MS, 81; Fla State Univ, PhD(biol sci), 86. *Prof Exp:* Res assoc, Fla State Univ, 86-88; ASSOC CUR, LOS ANGELES MUS, 89-

Concurrent Pos: Adj asst prof, Univ Southern Calif, 88- *Honors & Awards:* James R Fisher Award, Sigma Xi, 85. *Mem:* The Crustacean Soc; Am Soc Zoologists; Sigma Xi; Soc Systematic Zool. *Res:* Systematics, morphology and evolution in marine and freshwater decapod crustaceans; larval development and morphology and the significance of larval characters in crustacean evolution; morphology and evolution of branchiopod crustaceans; crustacean ecology. *Mailing Add:* Life Sci Div Invert Zool Sect Los Angeles County Mus Natural Hist 900 Exposition Blvd Los Angeles CA 90007

MARTIN, JOHN A, b Ballston Spa, NY, May 6, 35. REAL-TIME SYSTEMS, COMPUTER COMMUNICATIONS. *Educ:* State Univ NY, BS, 61; Yale Univ, MS, 62. *Prof Exp:* Res asst, Forest Soils Lab, Col Forestry, State Univ NY, 60-62; systs programmer, Univ Tenn Comput Ctr, 63-66; sr systs rep, Burroughs Corp, 66-72; SR COMPUTER SCIENTIST, COMPUTER SCI CORP, 80- *Concurrent Pos:* Consult, Oak Ridge Nat Lab, 62-64; independent computer consult & contractor, 72-80. *Mem:* AAAS; Asn Comput Mach; Inst Elec & Electronic Engrs; Inst Elec & Electronic Engrs Computer Soc. *Mailing Add:* Computer Sci Corp 1100 West St Laurel MD 20707

MARTIN, JOHN B(RUCE), b Auburn, Ala, Feb 2, 22; m 43, 63; c 3. CHEMICAL ENGINEERING, MARKETING RESEARCH. *Educ:* Ala Polytech Inst, BS, 43; Ohio State Univ, MSc, 47, PhD(chem eng), 49. *Prof Exp:* Asst chem eng, Ohio State Univ, 46-49; chem engr, Res & Develop Dept, Procter & Gamble Co, 49-77 & Mkt Res Dept, 77-82; sr assoc & dir, Indumar, Inc, 82-87, sr vpres, 87-88; RETIRED. *Concurrent Pos:* Dir, Am Inst Chem Engrs, 68-70 & chmn, Mkt Div, 85; lectr, Univ Cincinnati, 82-88; adj assoc prof, Auburn Univ, 83-88. *Honors & Awards:* Mkt Hall of Fame, Am Inst Chem Engrs, 88. *Mem:* Am Chem Soc; Am Inst Chem Engrs; Am Soc Eng Educ; Chem Mkt Res Asn. *Res:* Detergents; spray drying; organization development; chemical market research. *Mailing Add:* 644 Doepke Lane Cincinnati OH 45231

MARTIN, JOHN CAMPBELL, b Orangeburg, SC, Dec 6, 26; m 50; c 3. ELECTRICAL & SYSTEMS ENGINEERING. *Educ:* Clemson Univ, BS, 48; Mass Inst Technol, MS, 53; NC State Univ, PhD(elec eng), 62. *Prof Exp:* Instr elec eng, Clemson Univ, 48-50; asst, Mass Inst Technol, 50-52; proj engr, Gen Electronic Lab, Mass, 52; design engr, Spencer-Kennedy Lab, 52-53; asst prof physics, 53-55, from asst prof to assoc prof elec eng, 55-66, PROF ELEC ENG, CLEMSON UNIV, 66- *Concurrent Pos:* Consult, Sangamo Elec Co, Ill, 63; NASA res grant, 65-70. *Mem:* Inst Elec & Electronics Engrs. *Res:* Modern control theory; automatic controls; very high frequency filters; frequency multipliers using varactors; minimum phase shift band-pass filters; social system engineering. *Mailing Add:* 210 Riggs Dr Clemson SC 29631

MARTIN, JOHN DAVID, b Chicago, Ill, Nov 8, 39; m 62; c 2. NUCLEAR PHYSICS, ATMOSPHERIC PHYSICS. *Educ:* Va Mil Inst, BS, 61; Col William & Mary, MA, 63; Univ Fla, PhD(physics), 67. *Prof Exp:* Nuclear res officer physics, McClellan Cent Labs, McClellan Air Force Base, 67-70; physicist, 70-76; VPRES TECH, TELEDYNE ISOTOPES, 70- *Mem:* Am Phys Soc. *Res:* Research in measurement of stable and radioactive trace gases in the atmosphere; development of nuclear counting techniques for measurement of environmental-level fission and activation radioisotopes. *Mailing Add:* 23 Twin Oak Dr Montvale NJ 07645

MARTIN, JOHN HARVEY, b Chambersburg, Pa, Jan 20, 32; m 58; c 3. INTERNAL MEDICINE, RHEUMATOLOGY. *Educ:* Gettysburg Col, BA, 54; Temple Univ, MD, 58; Mayo Grad Sch Med, Univ Minn, MS, 62; Am Bd Internal Med, dipl, 65, cert med, 74. *Prof Exp:* From asst to staff med, Mayo Clin, 64-65; from instr to prof med, Temple Univ, 66-78; PROF MED & DIR DIV GEN MED, THOMAS JEFFERSON UNIV, 78-, INTERIM CHMN, DEPT MED, 81- *Concurrent Pos:* Fel rheumatol, Temple Univ, 65-66; P S Hench scholar, Mayo Grad Sch Med, Univ Minn, 66. *Mem:* Fel Am Col Physicians; Am Rheumatism Asn; Am Asn Clin Res. *Res:* Clinical research. *Mailing Add:* 17 Dartmouth Lane Haverford PA 19041

MARTIN, JOHN HOLLAND, b Old Lyme, Conn, Feb 27, 35; m 69; c 2. OCEANOGRAPHY. *Educ:* Colby Col, BA, 59; Univ RI, MS, 64, PhD(oceanog), 66. *Prof Exp:* Assoc scientist oceanog, PR Nuclear Ctr, Univ PR, Mayaguez, 66-69; sr scientist, Hopkins Marine Sta, Stanford Univ, 69-72; from asst prof to assoc prof oceanog, 72-77, PROF OCEANOG, CALIF STATE UNIV, SAN FRANCISCO, 77-, DIR, MOSS LANDING MARINE LABS, 76- *Concurrent Pos:* Dir, Vertex Prog, 79-88; vchmn, Univ Nat Oceanog Lab Syst. *Mem:* Am Soc Limnol & Oceanog; Am Geophysic Union. *Res:* Trace elements in sea water and plankton. *Mailing Add:* Moss Landing Marine Labs Moss Landing CA 95039

MARTIN, JOHN J(OSEPH), b Detroit, Mich, Oct 19, 22; m 48; c 3. ENGINEERING SCIENCE. *Educ:* Notre Dame, BSME, 43, MSME, 50; Purdue Univ, PhD(mech eng), 51. *Prof Exp:* Instr, Mech Eng Lab, Notre Dame, 46-47, calculus, 47; engr, Clark Equip Co, 47-49; res engr, NAm Aviation, Inc, 51-53; chief eng res, Bendix Prods Div, Bendix Aviation Corp, 53-60; mem tech staff, Inst Defense Anal, 60-63 & 64-66, spec asst to vpres res, 66-67, asst to pres, 67-68, dir, Systs Eval Div, 68-69; tech asst for nat security affairs, Off Sci & Technol, Exec Off of the President, 69-73; asst secy res & develop, US Air Force, 74-79; vpres & gen mgr, Advan Technol Ctr, Bendix Corp, 79-83; assoc adminr, NASA, 83-85; CONSULT, 85- *Concurrent Pos:* Consult, US Army Asst Chief Staff Intel, 59; resident consult, UK Royal Aircraft Estab, 63-64. *Res:* Heat transfer and thermodynamics; high altitude and reentry aerophysics, aerodynamics and aeromechanics; physical oceanography. *Mailing Add:* 7818 Fulbright Ct Bethesda MD 20817-3126

MARTIN, JOHN LEE, b Houston, Tex, Nov 19, 23; m 48; c 4. NUTRITION. *Educ:* Southern Methodist Univ, BS, 49; Univ Ark, MS, 53; Tex A&M Univ, PhD(biochem, nutrit), 56. *Prof Exp:* From instr to asst prof chem, Colo State Univ, 55-59; prof & head dept, Baker Univ, 59-60; assoc prof, Colo State Univ, 60-67; prof chem, Metrop State Col, 67-77; PROF FOOD & NUTRIT, TEX TECH UNIV, 78- *Mem:* Sigma Xi; Am Inst Nutrit. *Res:* Selenium-sulfur interrelationships; selenium enhancement of the immune response. *Mailing Add:* 6340 Sycamore Bluff Coloma MI 49038

MARTIN, JOHN MUNSON, b Eldora, Iowa, Nov 8, 48. AGRONOMY, PLANT BREEDING. *Educ:* Iowa State Univ, BS, 71, MS, 74, PhD(agron), 78. *Prof Exp:* ASST PROF AGRON, MONT STATE UNIV, 78- *Mem:* Am Soc Agron; Crop Sci Soc Am; Sigma Xi. *Mailing Add:* Dept of Plant & Soil Sci Mont State Univ Bozeman MT 59715

MARTIN, JOHN PERRY, JR, b Dunbar, Pa; m 81; c 2. PHYSICAL CHEMISTRY, ANALYTICAL CHEMISTRY. *Educ:* Carnegie Inst Technol, BS, 47, MS, 55, PhD(chem), 62. *Hon Degrees:* MHL, Davis & Elkins Col. *Prof Exp:* Asst chem, Metals Res Lab, Carnegie Inst Technol, 47-50; chemist, Dunbar Corp, 50-52; chief chemist, Duraloy Co, 52-59; from asst prof to assoc prof, 62-75, from actg chmn to chmn dept, 62-68, PROF CHEM, DAVIS & ELKINS COL, 75- *Concurrent Pos:* Consult, Pa Wire Glass Co, 50-52. *Mem:* Am Chem Soc; Hist Sci Soc; Sigma Xi. *Res:* X-ray, ultraviolet visible and infra-red spectroscopy; crystal growth in gels; hydrogen bonding of secondary amines with polar organic compounds and of amine complexes; educationally valuable demonstrations and student performable laboratory investigations, both introductory and advanced. *Mailing Add:* Dept of Chem Davis & Elkins Col Sycamore St Elkins WV 26241

MARTIN, JOHN ROBERT, b Lancaster, Pa, Dec 6, 23; m 45; c 3. ANALYTICAL CHEMISTRY. *Educ:* Goshen Col, AB, 44; Pa State Col, MS, 49, PhD(chem), 50. *Prof Exp:* Res chemist, E I Du Pont de Nemours & Co, Inc, 50-51, supvr, 51-64, div head, 64-75, consult supvr occup health, 77-85; ADJ PROF CHEM, MESSIAH COL, GRANTHAM, PA, 85- *Mem:* Am Chem Soc. *Res:* Application of ion exchange to analytical chemistry; analysis of fluoro compounds; functional group analysis; microanalysis. *Mailing Add:* 2105 Foxfire Dr Mechanicsburg PA 17055

MARTIN, JOHN SAMUEL, b Philadelphia, Pa, Oct 5, 48. PHYSIOLOGY, ANIMAL PHYSIOLOGY. *Educ:* Temple Univ, AB, 65; Woman's Med Col, MS, 68; Thomas Jefferson Univ, PhD, 73. *Prof Exp:* Instr biol, Holy Family Col, 71-72; instr, 72-74, asst prof physiol, 74-80, assoc prof oral biol & physiol, Sch Dent, 80-85, ASSOC PROF PHYSIOL, SCH MED, TEMPLE UNIV, 86- *Concurrent Pos:* Smith Kline & French Labs fel, Dept Physiol & Biophys, Sch Dent, Temple Univ, 72-74. *Mem:* Am Physiol Soc; Am Pharmaceut Asn; Inst for Animal Dis Res. *Res:* Gastrointestinal and autonomic physiology; upper airway physiology; membrane phenomena. *Mailing Add:* Dept Physiol Temple Univ Sch Med Philadelphia PA 19140

MARTIN, JOHN SCOTT, b Toronto, Ont, Sept 1, 34; m 57; c 2. SCIENCE EDUCATION, PHYSICAL CHEMISTRY. *Educ:* Univ Toronto, BA, 56; Columbia Univ, PhD(chem), 62. *Prof Exp:* Fel chem, Nat Res Coun Can, 61-63; asst prof, 63-69, ASSOC PROF CHEM, UNIV ALTA, 69- *Mem:* Chem Inst Can; Sigma Xi; AAAS; Asn Develop Comput Instr Systs. *Res:* Computer assisted instruction in chemistry; hydrogen bonding and ionic solvation processes by nuclear magnetic resonance; structure and spectra of bihalide ions; computer analysis of nuclear magnetic resonance spectra of symmetric molecules. *Mailing Add:* Dept Chem Univ Alta Edmonton AB T6G 2G2

MARTIN, JOHN WALTER, JR, pharmaceutical chemistry; deceased, see previous edition for last biography

MARTIN, JOSEPH B, b Bassano, Alta, Oct 20, 38; m; c 4. EDUCATION ADMINISTRATION. *Educ:* Eastern Mennonite Col, BSc, 59; Univ Alta, MD, 62; Univ Rochester, PhD, 71. *Hon Degrees:* AM, Harvard Univ, 78. *Prof Exp:* Rotating intern, Univ Hosp, Edmonton, Alta, 62-63, jr asst resident internal med, 63-64; jr resident neurol, Univ Hosp, Case Western Reserve Univ, Cleveland, Ohio, 64-65, sr resident, 65-66, fel neuropath, Cleveland Metrop Gen Hosp, 66-67; res fel & instr, Depts Med & Anat, Sch Med & Dent, Univ Rochester, NY, 67-69; asst prof, Dept Med & Pediat Specialties, Sch Med, Univ Conn, Hartford, 69-70; from asst prof to assoc prof, Dept Neurol & Neurosurg & Dept Exp Med, McGill Univ, Montreal, Que, 70-75, prof med, Dept Med, 76-78, prof neurol, Dept Neurol & Neurosurg, 76-78, chmn, 77-78; chief, Neurol Serv, Mass Gen Hosp, Boston, 78-89; DEAN, SCH MED, UNIV CALIF, SAN FRANCISCO, 89-; CONSULT NEUROLOGIST, MCLEAN HOSP, BELMONT, MASS, 82- *Concurrent Pos:* Centennial fel, Med Res Coun Can, 67-70; asst physician, Dept Med, Montreal, Gen Hosp, 70-73, assoc physician, 73-75, sr physician, 75-78; counr neurosci, Can Soc Clin Invest, 75-76; neurologist-in-chief, Montreal Neurol Inst, 76-78; Bullard prof neurol, Harvard Med Sch, 78-84, Julieanne Dorn prof, 84-89; chief, neurol serv, Mass Gen Hosp, 78-89, interim gen dir, 81-82; chmn, Sect Neuroendocrinol, NIH, 78-81, Fogarty scholar, 88-89; trustee, Neurosci Res Found, 88-; mem, Bd Biobehav Sci & Ment Dis, Inst Med, 91-; numerous vis professorships. *Honors & Awards:* Bowditch Lectr, Am Physiol Soc, 78; Bernard Pimstone Mem Lectr, Capetown, SAfrica, 80; Bernard J Alpers Lectr, Jefferson Med Col, 81; Marjorie Guthrie Lectr, NIH, 85; MacCallum Lectr, Univ Toronto, 86; Tinsley Harrison Lectr, Univ Ala, 88; Mandel Lectr, Jeanes Hosp, Pa, 86; Arthur Cherkin Mem Lectr, Univ Calif, 90; Hughlings Jackson Lectr, Montreal Neurol Inst & Hosp, 90. *Mem:* Inst Med-Nat Acad Sci; fel Am Acad Neurol; Royal Col Physicians & Surgeons Can; Am Fedn Clin Res; Am Physiol Soc; Endocrine Soc; Int Soc Neuroendocrinol (vpres, 84-88); Soc Neurosci; Am Neurol Asn (pres-elect, 88); Am Asn Clin Invest. *Res:* Neurological sciences; neuroendocrinology; brain peptides. *Mailing Add:* Univ Calif Third Ave & Parnassus San Francisco CA 94143

MARTIN, JOSEPH J, chemical engineering; deceased, see previous edition for last biography

MARTIN, JOSEPH PATRICK, JR, b Lynn, Mass, May 21, 52; m 82. EVOLUTIONARY BIOCHEMISTRY, FREE RADICAL TOXICITY. *Educ:* Harvard Univ, AB, 73; Duke Univ, PhD(zool), 79. *Prof Exp:* Fel biochem, Med Ctr, Duke Univ, 79-81; ASST PROF BIOL, RICE UNIV, 81- *Concurrent Pos:* Prin investr, NIH grant, 83- *Mem:* Soc Study Evolution; AAAS; Am Soc Microbiol; Am Soc Zoologists; Sigma Xi; Genetics Soc Am. *Res:* Free radical basis of oxygen toxicity; evolution and biochemistry of superoxide dismutases and catalases; biochemical genetics of marine organisms; enzymology. *Mailing Add:* 6749 32nd St N Richland MI 49083

MARTIN, JULIA MAE, b Snow Hill, Md, Nov 9, 24. BIOCHEMISTRY. *Educ:* Tuskegee Inst, BS, 46, MS, 48; Pa State Univ, PhD(biochem), 63. *Prof Exp:* Instr chem, Tuskegee Inst, 48-49; from instr to asst prof, Fla Agr & Mech Univ, 49-59; assoc prof, Tuskegee Inst & res assoc, Carver Res Found, 63-66; actg dean, Grad Sch & A&M Col, 74-76, PROF CHEM, SOUTHERN UNIV, BATON ROUGE, 66-, DEAN COL SCI, 78- *Mem:* Am Chem Soc; AAAS; fel Am Inst Chemists; NY Acad Sci; Nat Inst Sci. *Res:* Biochemical abnormalities of red blood cells of patients with hemolytic disorders. *Mailing Add:* Dept of Chem Southern Univ & A&M Col Box 9608 Baton Rouge LA 70813

MARTIN, JULIO MARIO, b Salta, Arg, Sept 16, 22; m 53; c 3. PHYSIOLOGY. *Educ:* Nat Univ La Plata, MD, 50. *Hon Degrees:* Dr medsurg, Oulu Univ, Finland, 83. *Prof Exp:* Res asst, Inst Biol & Exp Med, Univ Arg, 51-55; assoc prof physiol, Nat Univ La Plata, 56-60; res fel exp path, Wash Univ, 61-63; asst prof path, Univ Toronto, 63-71, Assoc prof physiol & pediat, 71-79; asst scientist, 63-70, assoc scientist, 70-75, SR SCIENTIST, RES INST, HOSP SICK CHILDREN, 75-; PROF PHYSIOL & PEDIAT, UNIV TORONTO, 79- *Concurrent Pos:* Squibb res fel, 51-53. *Mem:* Am Diabetes Asn; Can Physiol Soc; Arg Med Asn; Arg Physiol Soc; Endocrine Soc. *Res:* Experimental diabetes; pancreatic islets transplantation; insulin synthesis and release; neuroendocrine control of insulin secretion; relationship between growth hormone and beta-cells activity. *Mailing Add:* Hosp for Sick Children Res Inst 555 University Ave Toronto ON M5G 1X8 Can

MARTIN, KATHRYN HELEN, b Hartford, Conn, Oct 5, 40. HISTOPHYSIOLOGY. *Educ:* Cath Univ Am, AB, 62; Cornell Univ, PhD(zool), 73. *Prof Exp:* Asst prof, 72-77, ASSOC PROF ZOOL, STATE UNIV NY COL OSWEGO, 77- *Mem:* Sigma Xi; Am Soc Mammalogists; AAAS. *Res:* Histophysiological analysis of reproductive and stress phenomena in small mammals. *Mailing Add:* 210 Piez Hall State Oswego Oswego NY 13126

MARTIN, KENNETH EDWARD, b Cheney, Kans, Apr 5, 44; m 65; c 1. MATHEMATICS. *Educ:* St Benedict's Col, Kans, BA, 64; Ind Univ, Bloomington, MA, 66; Univ Notre Dame, PhD(math), 70. *Prof Exp:* From asst prof to assoc prof math, Gonzaga Univ, 70-77; assoc prof math & head dept, Valdosta State Col, 77-; AT UNIV NFLA. *Mem:* Math Asn Am. *Res:* Group theory and its generalizations; algebraic number theory. *Mailing Add:* Univ NFla PO Box 17074 Jacksonville FL 32245-9981

MARTIN, KUMIKO OIZUMI, b Chiba, Japan, Jan 17, 41; m 73. BIOCHEMISTRY, ENDOCRINOLOGY. *Educ:* Univ Tokyo, BA, 63, PhD(biochem), 70; Ochanomizu Univ, MS, 65. *Prof Exp:* Lectr biochem, Bunka Women's Col, 66-67; fac assoc nutrit, Univ Tokyo, 71-73; fel, 70-71, res assoc, res inst skeletomuscular dis, Hosp Joint Dis, 73-79; asst prof, dept pediat, Cornell Univ Med Col, 82-88; res assoc cell biol, 79-82, SR SCIENTIST, DEPT MED, NY UNIV SCH MED, 88- *Mem:* AAAS; Endocrine Soc; Am Soc Hypertension; NY Acad Sci. *Res:* Metabolism of corticosteroids; enzymology. *Mailing Add:* 142 W 26th St 9B New York NY 10001

MARTIN, L(AURENCE) ROBBIN, b Annapolis, Md, Sept 21, 39. PHYSICAL CHEMISTRY. *Educ:* Pomona Col, BA, 61; Mass Inst Technol, PhD(phys chem), 66. *Prof Exp:* Noyes fel, Calif Inst Technol, 66-68; asst prof chem, Univ Calif, Riverside, 68-73; MEM TECH STAFF, AEROSPACE CORP, 73- *Mem:* AAAS; Am Phys Soc; Sigma Xi; Am Chem Soc; Am Geophys Union. *Res:* Aqueous and gas phase chemical kinetics of interest in atmospheric chemistry; gas-aerosol interactions; diamond films; thermodynamics; molecular beam kinetics. *Mailing Add:* Aerophysics Lab Aerospace Corp Box 92957 Los Angeles CA 90009

MARTIN, LARRY DEAN, b Bartlett, Nebr, Dec 8, 43; m 67; c 2. VERTEBRATE PALEONTOLOGY. *Educ:* Univ Nebr, BS, 66, MS, 69; Univ Kans, PhD(biol), 73. *Prof Exp:* From asst prof to assoc prof, 72-86, PROF SYSTS & ECOL, UNIV KANS, 86-; CUR VERT PALEONT, MUS NATURAL HIST, 72- *Concurrent Pos:* Res affil, Univ Nebr State Mus, 72-; mem, US Nat Working Group Neogene Quaternary Boundry, 74-; adj prof geol, Kans State Univ. *Mem:* Soc Vert Paleont; Am Soc Mammalogists; Am Quaternary Asn; Sigma Xi; Am Ornithologist Union. *Res:* Fossil history of certain birds, rodents and Saber-toothed cats with emphasis on functional morphology; relationship between climatic history and vertebrate extinctions; author of over 170 scientific articles and one book. *Mailing Add:* Dept Systematics-ECO Univ Kans Lawrence KS 66045

MARTIN, LAWRENCE LEO, b Charleston, WVa, July 5, 42; m 73; c 2. CENTRAL NERVOUS SYSTEM DRUGS. *Educ:* Univ Md, BS, 66; Ohio State Univ, PhD(med chem), 71. *Prof Exp:* Sr res chemist, Hoechst Roussel Pharmaceut Inc, 72-79, res assoc, 79-81, sr res assoc, 81-87, group leader, 87-91, RES GROUP MGR, HOECHST ROUSSEL PHARMACEUT INC, 91- *Concurrent Pos:* Exchange scientist, Hoechst AG, Ger, 76-77. *Mem:* Am Chem Soc; Am Pharmaceut Asn; Acad Pharmaceut Sci. *Res:* Design and synthesis of potential central nervous system agents including antidepressants, anxiolytics, antipsychotics, and agents for the treatment of cognitive dysfunctions including Alzheimer's disease. *Mailing Add:* Hoechst Roussel Pharmaceut Inc Rte 202-206N PO Box 2500 Somerville NJ 08876-1258

MARTIN, LEROY BROWN, JR, b Elkin, NC, June 6, 26; m 61; c 3. MATHEMATICS, OPERATIONS RESEARCH. *Educ:* Wake Forest Col, BS, 49; NC State Univ, MS, 52; Harvard Univ, MS, 53, PhD(appl math), 58. *Prof Exp:* Appl sci rep, Int Bus Mach Corp, 55-56, spec rep, Serv Bur Corp, 56-59, asst mgr planning & develop, 59-61; from asst prof to prof comput sci, Dir Comput Ctr & asst provost provost, 68-83, PROF MATH, NC STATE UNIV, 83- *Mem:* Am Asn Comput Mach; Inst Mgt Sci. *Res:* Mathematical optimization; numerical analysis; management science. *Mailing Add:* Dept Math NC State Univ Raleigh NC 27695

MARTIN, LESTER W, b Edwards, Mo, Aug 15, 23; m 49; c 5. PEDIATRIC SURGERY. *Educ:* Univ Mo, BS, 44, BSc, 47; Harvard Med Sch, MD, 49; Am Bd Surg, dipl, 57. *Prof Exp:* From asst prof to assoc prof, 57-72, PROF SURG, COL MED, UNIV CINCINNATI, 72-; DIR PEDIAT SURG, CHILDREN'S HOSP, 57- *Mem:* Affil fel Am Acad Pediat; fel Am Col Surg; AMA; Brit Asn Paediat Surg. *Res:* Various aspects of surgery of infancy and childhood; esophageal anomalies and Hirschsprung's disease. *Mailing Add:* Dept Surg-Pediat (54)NECh Univ Cincinnati Cincinnati OH 45221

MARTIN, LINDA SPENCER, b New Orleans, La, May 12, 46; m 76; c 1. OCCUPATIONAL HEALTH & SAFETY, HEALTH CARE WORKER SAFETY. *Educ:* Univ Ala, BS, 68; George State Univ, MS, 72; Emory Univ, PhD(immunol), 82. *Prof Exp:* Res microbiologist parasitic serol, Ctrs for Dis Control, 68-73, res microbiologist, Immunol Br, Div Hart Factors, 73-88, asst to dep dir, Mgt/Leadership Develop Prog, 88-89, DIR HIV/AIDS ACTIV, NAT INST OCCUP SAFETY & HEALTH, CTRS FOR DIS CONTROL, 89- *Concurrent Pos:* Pres, bd trustees, Ctrs for Dis Control, 78-79, mem, Health & Safety Policy Subcomt Biol Safety, 87-89 & secy, Health & Safety Policy Comt, 89-90; treas, Southeast Immunol Conf, 78-81; mem, Workgroup Chem Germicides, Ctrs for Dis Control/Environ Protection Agency/Food & Drug Admin. *Mem:* Am Asn Immunologists; Am Biol Safety Asn; Int AIDS Soc; Sigma Xi; Am Mgt Asn. *Res:* Occupational safety and health for healthcare and public safety workers; inactivation of HIV; biosafety; cellular immunology. *Mailing Add:* NIOSH HIV-AIDS Activity Ctr Disease Control 1600 Clifton Rd F40 Atlanta GA 30333

MARTIN, LLOYD W, b Ada, Okla, Jan 5, 34; m 57; c 4. PRODUCTION & MANAGEMENT, CULTURE & PHYSIOLOGY. *Educ:* Okla State Univ, BS, 58, MS, 61; Mich State Univ, PhD(hort), 67. *Prof Exp:* Supt res, Eastern Okla Field Sta, Okla State Univ, 59-63; technician, Dept Hort, Mich State Univ, 63-67; exten specialist small fruits, 67-70, SUPT RES & ADMIN, N WILAMETTE EXP STA, ORE STATE UNIV, 70- *Concurrent Pos:* Mem, small fruits res, Tri-State Task Force, 78-, Coun Agri Sci Technol, 82, Int Soc Hort Sci, 85-; actg assoc dir, Ore Agr Exp Sta, 86-87. *Mem:* Am Soc Hort Sci; Am Pomol Soc; Soil Sci Soc Am; Crop Sci Soc Am; Am Soc Agron. *Res:* Develop systems of small fruits production and management that minimize cost and harmful effect to the environment while maximizing production and quality of the fruit. *Mailing Add:* N Willamette Res & Exten Ctr 15210 NE Miley Rd Aurora OR 97002

MARTIN, LOREN GENE, b Danville, Ill, Oct 31, 42; div; c 1. CARDIOVASCULAR PHYSIOLOGY, ENDOCRINE PHYSIOLOGY. *Educ:* Ind Univ, AB, 65, PhD(physiol), 69, PhD, 70. *Prof Exp:* Res assoc environ physiol, Ind Univ, 69-70; asst prof physiol, Sch Med, Temple Univ, 70-73; asst prof & assoc prof, Col Med, Univ Ill, Chicago & Peoria, 73-78; assoc prof, Va Commonwealth Univ, 78-81; assoc prof, 81-85, PROF PHYSIOL & DIR MED EDUC, COL OSTEOP MED, OKLA STATE UNIV, 85- *Concurrent Pos:* Adj prof physiol, Gwyned-Mercy Col, 71-73, Philadelphia Community Col, 72-73, Eureka Col, 74-78; adj prof biol, Bradley Univ, 74-78; adj assoc prof optom, Northeastern Okla State Univ, 81-83. *Mem:* Sigma Xi; Am Physiol Soc; Endocrine Soc; Soc Exp Biol & Med; Int Soc Heart Res. *Res:* Environmental effects upon thyroid and heart metabolism; effects of sex steroids upon coronary flow and myocardial anoxic resistance; effects of aging upon endocrine function. *Mailing Add:* Col Ostep Med Okla State Univ 1111 W 17th St Tulsa OK 74107-1898

MARTIN, MALCOLM ALAN, HUMAN RETROVIRUSES. *Educ:* Yale Univ, MD, 62. *Prof Exp:* CHIEF, MOLECULAR MICROBIOL LAB, NIH, 64- *Res:* Acquired Immune Deficiency Syndrome virus. *Mailing Add:* Molecular Microbiol Lab NIH Bldg 5 Rm B-125 9000 Rockville Pike Bethesda MD 20892

MARTIN, MALCOLM MENCER, b Vienna, Austria, Dec 10, 20; US citizen; m 62; c 3. PEDIATRIC ENDOCRINOLOGY. *Educ:* Univ Durham, MB, BS, 45, MD, 52; Am Bd Internal Med, dipl & cert, 66; FRCP, 72. *Prof Exp:* Resident, Postgrad Sch Med, Univ London, 48-50, first asst, Diabetic & Metab Unit, King's Col Hosp, 50-53, registr, Med Unit, 53-56; physician, Out-Patient Dept, Harriet Lane Home, Johns Hopkins Hosp, 56-57; asst med, Peter Bent Brigham Hosp, Boston, 57-59; from asst prof to assoc prof pediat, 59-67, PROF PEDIAT & MED, SCH MED, GEORGETOWN UNIV, 67- *Concurrent Pos:* Ministry Educ fel, Eng, 48-49; Lund res fel, Brit Diabetes Asn, 50-51; King's Col res grant, 52-55; Leverhulme res fel, Inst Clin Res, Middlesex Hosp Med Sch, 56; NIH spec res fel, 57-59; Lederle fac award, 62-65; consult endocrinol & metab dis, Childrens' Convalescent Hosp, DC, 63; mem acad staff, Childrens' Hosp, DC, 63; mem Worcester Found, 66. *Mem:* AAAS; Endocrine Soc; Am Pediat Soc; Am Soc Human Genetics; Am Diabetes Asn; Lawson Wilkins Pediat Endocrine Soc. *Res:* Endocrinology and metabolism, particularly as related to growth and development and diabetes mellitus. *Mailing Add:* Dept Pediat Georgetown Univ Med Ctr Washington DC 20007

MARTIN, MARGARET EILEEN, b Albright, WVa, Oct 17, 15. BIOCHEMISTRY. *Educ:* WVa Wesleyan Col, BS, 40; Georgetown Univ, MS, 54, PhD(chem), 58. *Prof Exp:* Teacher pub schs, WVa, 35-43; med technician, Emergency Hosp, Wash, DC, 47-49 & St Elizabeth's Hosp, 49-52; chemist, US Food & Drug Admin, 52-54, Phys Biol Lab, Nat Inst Arthritis & Metab Dis, 54, Nat Heart Inst, 54-57 & Food Qual Lab, Agr Res Ctr, USDA, 57-62; chemist, Lab Br, St Elizabeth's Hosp, Dept Health, Educ &

Welfare, Wash, DC, 62-78; RETIRED. *Concurrent Pos:* Mem spec ment health res neurochem sect, Nat Inst Ment Health, 70-78. *Mem:* Am Chem Soc; Am Inst Chemists; Am Soc Microbiol; AAAS; NY Acad Sci. *Res:* Basic biochemistry; mental illness; toxicology; clinical chemistry; side effects of thorazine; author of nine publications on the quality of fresh and frozen food. *Mailing Add:* 4006 Rickover Rd Silver Spring MD 20902

MARTIN, MARGARET PEARL, b Duluth, Minn, Apr 22, 15. FOREST BIOMETRY. *Educ:* Univ Minn, BA, 37, MA, 39, PhD(math), 44. *Prof Exp:* Instr biostatist, Univ Minn, 40-41 & Columbia Univ, 42-45; statist consult, Health Dept, NY, 45; asst prof biostatist, Univ Minn, 45-46; from asst prof to assoc prof prev med, Vanderbilt Univ, 47-58; assoc prof biostatist, Sch Hyg & Health, Johns Hopkins Univ, 59-64; asst prof biomet, Univ Minn, St Paul, 67-68; prin biometrician, NCent Forest Exp Sta, USDA, 68-84; RETIRED. *Mem:* Fel Am Statist Asn; Biomet Soc; Sigma Xi. *Res:* Applied statistics; biological fields. *Mailing Add:* 1366 Selby Ave St Paul MN 55104-6301

MARTIN, MARK WAYNE, b Twin Falls, Idaho, June 17, 30; m 52; c 5. GENETICS. *Educ:* Univ Idaho, BS, 52; Cornell Univ, MS, 54, PhD(plant breeding), 59. *Prof Exp:* Geneticist, Agr Res Serv, USDA, Utah State Univ, 59-67, GENETICIST, RES & EXTEN CTR, AGR RES SERV, USDA, Wash, 67- *Mem:* Potato Asn Am; European Asn Potato Res. *Res:* Breeding vegetables; disease resistance, especially resistance to virus diseases. *Mailing Add:* USDA Res & Exten Ctr Prosser WA 99350

MARTIN, MICHAEL, b Vallejo, Calif, Jan 16, 43. POLLUTION BIOLOGY. *Educ:* Univ Calif, Davis, BS, 65; Sacramento State Col, MA, 67; Univ Southern Calif, PhD(biol), 72. *Prof Exp:* Instr physiol & biol, Glendale Community Col, 71; head instr ichthyol & freshwater ecol, NAm Sch Conserv & Ecol, 71, dean, 71-73; assoc water qual biologist, 73-79, SR WATER QUAL BIOLOGIST, CALIF DEPT FISH & GAME, 80- *Mem:* AAAS; Am Soc Ichthyol & Herpet; Am Fisheries Soc; Am Inst Fish Res Biol; Sigma Xi. *Res:* Vertebrate biology, morphology, systematics and ecology of fishes; marine pollution; heavy metal toxicity. *Mailing Add:* 2201 Garden Rd Monterey CA 93940

MARTIN, MICHAEL MCCULLOCH, b Junction City, Kans, Mar 21, 35; m 65; c 2. BIOCHEMICAL ECOLOGY. *Educ:* Cornell Univ, AB, 55; Univ Ill, PhD(org chem), 58. *Prof Exp:* NSF res fel, Mass Inst Technol, 58-59; from instr to assoc prof chem, 59-70, assoc prof zool, 68-70, prof chem & biol, 70-85, PROF BIOL, UNIV MICH, ANN ARBOR, 85- *Concurrent Pos:* Res chemist, Entom Div, USDA, 64; Sloan Found fel, 66-68; NSF fac fel, 76-77; Fulbright Res Award, 86-87. *Mem:* AAAS; Am Inst Biol Sci; Soc Study Evolution; Int Soc Chem Ecol. *Res:* Biochemical aspects of ecological interations; insect-plant interactions; insect-microbial interactions. *Mailing Add:* Div Biol Sci Univ Mich Ann Arbor MI 48104

MARTIN, MONROE HARNISH, b Lancaster, Pa, Feb 7, 07; m 32; c 1. APPLIED MATHEMATICS. *Educ:* Lebanon Valley Col, BS, 28; Johns Hopkins Univ, PhD(math), 32. *Hon Degrees:* DSc, Lebanon Valley Col, 58. *Prof Exp:* Nat res fel, Harvard Univ, 32-33; instr math, Trinity Col, 33-36; from asst prof to prof, 36-68, from actg head dept to head dept, 42-53, from actg dir to dir, Inst Fluid Dynamics & Appl Math, 53-68, res prof, 68-71, EMER RES PROF MATH, UNIV MD, COLLEGE PARK, 71- *Concurrent Pos:* Mem, US Nat Comt Theoret & Appl Math, 53-56; exec secy, Div Math, Nat Acad Sci-Nat Res Coun, 55-57 & 58-59, chmn comt appl math, 58-59, mem, 60-61; Guggenheim fel, 60; hon lectr, Univ St Andrews, 60; consult, Naval Ord Lab. *Mem:* Am Math Soc; Math Asn Am. *Res:* Matrices; dynamics; ergodic theory; mathematical theory of the flow of a compressible fluid; partial differential equations; the flow of a viscous fluid mathematical biology; chemical transport through cells, relativity theory. *Mailing Add:* RR 2 Box 64 Denton MD 21629

MARTIN, NANCY CAROLINE, b Chicago, Ill, Feb 5, 48. MOLECULAR BIOLOGY, GENETICS. *Educ:* Pitzer Col, AB, 70; Harvard Univ, MS, 73, PhD(biol), 75. *Prof Exp:* Fel biochem, Univ Chicago, 74-77; asst prof biochem, Univ Minn, 77-79; from asst prof to assoc prof biochem, Univ Tex Health Sci Ctr, Dallas, 79-87; PROF BIOCHEM, UNIV LOUISVILLE, 87- *Concurrent Pos:* Fel, Am Cancer Soc, 75-77. *Mem:* Am Soc Cell Biol; Am Soc Biol Chemists. *Res:* Organelle biogenesis, control of eukaryotic gene expression; ribonucleic acid, transfer structure and function. *Mailing Add:* Dept Biochem Univ Louisville Health Sci Ctr Louisville KY 40292

MARTIN, NATHANIEL FRIZZEL GRAFTON, b Wichita Falls, Tex, Oct 10, 28; m 54; c 2. ERGODIC THEORY, REAL VARIABLES. *Educ:* NTex State Col, BS, 49, MS, 50; Iowa State Col, PhD(math), 59. *Prof Exp:* Asst math, NTex State Col, 49-50; instr, Midwestern Univ, 50-52; asst, Iowa State Col, 55-58, instr, 58-59; from instr to asst prof, 59-64, asst chmn dept, 74-75, assoc dean grad sch arts & sci, 75-81, assoc prof math, 64-83, PROF MATH, UNIV VA, 83- *Concurrent Pos:* NSF fac fel & res assoc, Univ Calif, Berkeley, 65-66; vis lectr, Copenhagen Univ, 69-70; consult ed, McGraw-Hill Book Co, 72-79; vis fel, Univ Warwick, 82. *Mem:* Am Math Soc; Math Asn Am; Sigma Xi. *Res:* Analysis; real function theory and measure theory, particularly differentiation of set functions; ergodic theory; entropy; isomorphisms of dynamical systems. *Mailing Add:* Dept Math Math-Astron Bldg Univ Va Charlottesville VA 22903

MARTIN, NED HAROLD, b New Brunswick, NJ, May 18, 45. BIO-ORGANIC CHEMISTRY. *Educ:* Denison Univ, AB, 67; Duke Univ, PhD(org chem), 72. *Prof Exp:* Chemist, Res Triangle Inst, 69-70; from asst prof to assoc prof, 72-80, PROF ORG CHEM, UNIV NC, WILMINGTON, 80- *Concurrent Pos:* Postdoctoral, Univ Geneva, 80-81; consult, LaQue Ctr Corrosion Technol, 83-88; premed sci adv, Univ NC, Wilmington, 84- *Mem:* Am Chem Soc; Nat Asn Adv Health Prof; AAAS; Sigma Xi. *Res:* Singlet oxygen chemistry of enamines; alkaloid biosynthesis; synthesis and spectroscopy of xanthones; quantum chemical calculations; molecular modeling. *Mailing Add:* Dept Chem Univ NC 601 S College Rd Wilmington NC 28403-3297

MARTIN, NORMAN MARSHALL, b Chicago, Ill, Jan 16, 24; m 50; c 3. SYSTEMS OF LOGIC. *Educ:* Univ Chicago, MA, 47; Univ Calif, Los Angeles, PhD(philos), 52. *Prof Exp:* Instr philos, Univ Ill, 50-51 & Univ Calif, Los Angeles, 52-53; res assoc, Willow Run Res Ctr, Mich, 53-55; mem tech staff, Space Tech Labs, Thompson-Ramo-Wooldridge, Inc, 55-59, head logic tech group, 59-61; mem tech staff & bd dirs, Logicon, Inc, 61-65, treas, 62-65; assoc prof, Univ Tex, Austin, 66-68, res scientist, Comput Ctr, 66-71, prof philos & comput sci, 68-90, prof elec & comput eng, 74-90, EMER PROF PHILOS & COMPUT SCI, UNIV TEX, AUSTIN, 90- *Concurrent Pos:* Lectr, Univ Calif, Los Angeles, 57-65; consult, Logicon, Inc, 65-74; assoc chmn dept comput sci, Univ Tex, 75-78. *Mem:* Am Math Soc; Math Asn Am; Asn Symbolic Logic; Asn Comput Mach; Am Philos Asn. *Res:* Systems, organization and logical design of digital computing equipment; missile guidance system engineering; applications of digital equipment; mathematical logic, especially many-valued logic; philosophy of language; switching theory. *Mailing Add:* 4423 Crestway Dr Austin TX 78731

MARTIN, PAUL BAIN, b Nixon, Tex, Nov 24, 46; m 71; c 3. ENTOMOLOGY. *Educ:* Tex A&M Univ, BS, 69, MS, 71; Univ Fla, PhD(entom), 76. *Prof Exp:* Assoc entom, Agr Res & Educ Ctr, Quincy, Univ Fla, 71-73; entomologist & fieldman, Peggie Martin & Assocs, Pearsall, Tex, 75-76; asst prof entom, Univ Ga, 77-81; specialist entom, Inst Interam Cooperacao Para Agr, Off Agr Serv, Campo Grande, Brazil, 81-83; ENTOMOLOGIST, TEX DEPT AGR, 86- *Concurrent Pos:* Consult, cotton, grain sorghum pest mgt; res grant award, Univ Ga, 80. *Mem:* Entom Soc Am; AAAS; Int Org Biol Control Noxious Animals & Plants. *Res:* Spittlebug-pasture management in Brazil; host plant resistance habitat management, biological control (fungus) and integrated pest systems management; dynamics of populations of three noctuids; biological control of cotton pests. *Mailing Add:* 605 Elm St Sequin TX 78155

MARTIN, PAUL CECIL, b Brooklyn, NY, Jan 31, 31; m 57; c 3. STATISTICAL & CONDENSED MATTER THEORY. *Educ:* Harvard Univ, AB, 51, AM, 52, PhD(physics), 54. *Prof Exp:* NSF res fel, Univ Birmingham, UK, 55 & Inst Theoret Physics, Denmark, 56; from asst prof to prof physics, 56-82, chmn dept, 72-75, JOHN HASBROUCK VAN VLECK PROF PURE & APPL PHYSICS, HARVARD UNIV, 82-, DEAN, DIV APPL SCI, 77-, ASSOC DEAN, FAC ARTS & SCI, 81- *Concurrent Pos:* Res fel, Sloan Found, 59-62; vis prof, Ecole Normale Superieure, 63 & 66; ed, J Math Physics, 63-66, Ann Physics, 68-, Transp Theory & Statist Physics, 71- & J Statist Physics, 75-79; Guggenheim Found fel, 65 & 71; vis prof, Univ Paris, 71; consult, Brookhaven Nat Lab; mem adv bd, Nat Inst Theoret Physics, Santa Barbara, 80-82, chmn, 78-80; mem bd dirs, Asn Univ Res Astron, 78-85, bd trustees, Assoc Univ Inc, 81-; chmn, Physics Sect, AAAS, 86; counr, Am Phys Soc, 82-84; Mat Res Adv Coun, NSF, 86- *Mem:* Fel Nat Acad Sci; Am Acad Arts & Sci; Am Phys Soc; AAAS; NY Acad Arts & Sci. *Res:* Quantum theory of fields; physics of solids and fluids; statistical physics. *Mailing Add:* Dept of Physics Harvard Univ Cambridge MA 02138

MARTIN, PAUL JOSEPH, b Hammond, Ind, May 22, 36; m 60; c 4. PHYSIOLOGY, BIOMEDICAL ENGINEERING. *Educ:* Univ Tex, Austin, BS, 61; Drexel Univ, MS, 62; Case Western Res Univ, PhD(biomed eng), 67. *Prof Exp:* Res fel cardiovasc physiol, Latter Day Saints Hosp, Salt Lake City, 62-63; biomed res engr, Technol Inc, Dayton, 63-64; res assoc, 67-73, ASSOC CHIEF, CARDIOVASC PHYSIOL, MT SINAI HOSP, 73- *Concurrent Pos:* Consult, Cleveland Clin Found, 70-72; mem chmn, Res Study Sect, Am Heart Asn, 73; from asst prof to prof, dept physiol & biomed eng, Case Western Reserve Univ, 67-85; teaching consult, Ohio Col Pediat Med, 73-; assoc ed, Am J Physiol, 75-81. *Mem:* Am Heart Asn; Am Physiol Soc; Biomed Eng Soc; Inst Elec & Electronics Engrs; Sigma Xi. *Res:* Cardiovascular physiology; nervous control of the heart; applied pharmacokinetics; computers in medicine. *Mailing Add:* Mt Sinai Hosp University Circle Cleveland OH 44106

MARTIN, PAUL SCHULTZ, b Allentown, Pa, Aug 22, 28; m 50; c 3. ECOLOGY. *Educ:* Cornell Univ, BA, 51; Univ Mich, MA, 53, PhD, 56. *Prof Exp:* Rackham fel, Univ Mich & Yale Univ, 55-56; Nat Res Coun Can res fel, Univ Montreal, 56-57; res assoc palynol, 57-62, assoc prof, 62-68, PROF PALYNOL & CHIEF SCIENTIST, PALEOENVIRON STUDIES, GEOCHRONOL LABS, UNIV ARIZ, 68- *Concurrent Pos:* Guggenheim fel, 65-66. *Mem:* AAAS; Soc Study Evolution; Am Soc Nat; Ecol Soc Am. *Res:* Pleistocene biogeography; pollen stratigraphy; faunal extinction and its causes. *Mailing Add:* Dept of Geosci Univ of Ariz Tucson AZ 85721

MARTIN, PETER GORDON, b Owen Sound, Ont, Sept 19, 47; m 80; c 4. ASTROPHYSICS. *Educ:* Univ Toronto, BSc, 68, MSc, 69; Univ Cambridge, PhD(astron), 72. *Prof Exp:* From asst prof to assoc prof, 72-80, PROF ASTRON, UNIV TORONTO, 80- *Mem:* Int Astron Union; Can Astron Soc; Am Astron Soc. *Res:* Interstellar dust; optical polarization; galactic nuclei; late-type stars; interstellar shocks. *Mailing Add:* CITA Univ Toronto 60 St George St Toronto ON M5S 1A7 Can

MARTIN, PETER WILSON, b Glasgow, Scotland, Jan 7, 38; m 65. NUCLEAR PHYSICS. *Educ:* Glasgow Univ, BSc, 60, PhD(nuclear physics), 64. *Prof Exp:* Asst lectr physics, Univ Glasgow, 64-65; from asst prof to assoc prof, 65-80, PROF PHYSICS, UNIV BC, 81- *Res:* Solid state physics; materials science. *Mailing Add:* Dept of Physics Univ of BC 2075 Wesbrook Pl Vancouver BC V6T 1W5 Can

MARTIN, R RUSSELL, b Decatur, Ga, Mar 3, 36; m 61; c 3. MICROBIOLOGY, IMMUNOLOGY. *Educ:* Yale Univ, AB, 56; Med Col Ga, MD, 60. *Prof Exp:* Intern, Boston City Hosp, 60-61; resident med, Jackson Mem Hosp, 61-62 & Eugene Talmadge Mem Hosp, Med Col Ga, 63-65; NIH fel infectious dis, Med Col Ga, 65-67; from asst prof to assoc prof med, Sch Med, Ind Univ, 67-71; from assoc prof to prof med, microbiol & immunol, Baylor Col Med, 71-83; therapeut area clin dir infectious dis res, 83-86, sr dir, 87-89, VPRES, CLIN RES INFECTIOUS DIS, BRISTOL-MYERS CO, 89-; CLIN PROF MED, UNIV CONN HEALTH CTR, 87-

Concurrent Pos: Mead Johnson fel med, 65-66; head, Infectious Dis Sect, Ben Taub Gen Hosp, Tex Med Ctr, 71-83; Nat Inst Allergy & Infectious Dis res career develop award, 74-78; mem, Allergy & Immunol Res Comt, Nat Inst Allergy & Infectious Dis, NIH, 75-77, Adult Pulmonary Scor Rev Comt, Nat Heart, Lung & Blood Inst, 80-81, 86 & 89; chmn, Sci Assembly Microbiol, Infection & Immunity Am Thoracic Soc, 78-79; adj prof chem eng, Rice Univ, Houston, 78-83; clin prof med, State Univ NY, Upstate Med Ctr, 83-87; adj prof med, microbiol & immunol, Baylor Col Med, 83-; assoc clin prof, Dept Internal Med, Yale Univ, New Haven, 88- *Mem:* Fel Infectious Dis Soc Am; fel Am Col Physicians; AAAS; Am Asn Immunologists; Am Fedn Clin Res; NY Acad Sci; Am Soc Microbiol; Am Soc Virol; Am Thoracic Soc; Am Soc Immunol. *Res:* Infectious diseases. *Mailing Add:* Bristol-Myers Co Five Research Pkwy PO Box 5100 Wallingford CT 06492

MARTIN, RALPH H(ARDING), b Youngstown, Ohio, Dec 22, 23; m 45; c 2. CHEMICAL ENGINEERING. *Educ:* Carnegie Inst Technol, BS, 44, MS, 49. *Prof Exp:* Asst proj engr res & develop, Consol Coal Co, 48-53, supvr, Pittsburgh Coke & Chem Co, 53-55, gen mgr, Pitt-Consol Chem Co, 55-58; vpres, Dixon Chem & Res Corp & Dixon Chem Industs, Inc, 58-59; asst to pres, Standard Packaging Corp, 59-60, vpres res & develop, 60-64; exec vpres, gen mgr & dir, C H Dexter & Sons, Inc, 64-66, vchmn, Dexter Int SA, Belg, 65-66, pres, C H Dexter & Sons Co & vpres & dir, Dexter Corp, 66-76, PRES, C H DEXTER DIV, DEXTER CORP, 76- *Mem:* Am Chem Soc; Am Inst Chem Engrs. *Res:* Pulp and paper; packaging materials; graphic arts; hydrocarbon vapors; coal gasification; low temperature carbonization of coal; tar chemicals and refined tar products; spectroscopic analysis; smokeless rocket propellants. *Mailing Add:* Ocean Reef Club 32 Halfway Rd North Key Largo FL 33037

MARTIN, RICHARD ALAN, b Jacksonville, NC, Oct 16, 44; m 68; c 2. THERMOPHYSICS, FLUID MECHANICS. *Educ:* Univ Mich, BSE, 66; Univ S Calif, MS, 71; Iowa State Univ, PhD(aerospace), 75. *Prof Exp:* Res engr gas dynamics, NASA Flight Res Ctr, 67-71; res assoc, aerospace, Iowa State Univ, 75, turbulence, NASA Ames Res Ctr, 76-78; STAFF MEM THERMOPHYSICS, LOS ALAMOS NAT LAB, 77- *Concurrent Pos:* NRC res assoc, 76-78. *Mem:* Am Soc Mech Eng. *Res:* Fluid mechanics (theoretical and experimental); gas dynamics; thermodynamics; turbulence; statistics and applied mathematics; nuclear fuel cycle safety; thermoacoustic engines. *Mailing Add:* MS J576 PO Box 1663 Los Alamos Nat Lab Los Alamos NM 87545

MARTIN, RICHARD BLAZO, b Winchendon, Mass, July 1, 17; m 41; c 4. CHEMISTRY, SCIENCE ADMINISTRATION. *Educ:* Clark Univ, AB, 39, AM, 40, PhD(chem), 49. *Prof Exp:* From instr to asst prof chem, Clark Univ, 46-53; chemist, Res Br, Oak Ridge Opers, Atomic Eenergy Comn, 53-57, chief, 57-59, dep dir lab & univ div, 59-72, asst br chief, Waste Mgt Br, Res & Tech Support Div, 72-73, phys scientist, Classification & Tech Support Br, Oak Ridge Opers, 73-77; RETIRED. *Concurrent Pos:* Consult, US Dept Energy, 78-82, Los Alamos Tech Assoc Inc, 81-83, Martin Marietta Energy Systs, Inc, 84-87, Analysas, Inc, 89- *Mem:* Am Chem Soc; Am Nuclear Soc; Sigma Xi; AAAS; NY Acad Sci. *Res:* Radiochemical processing; isotopic separations; reactions of aliphatic diazo compounds with alicyclic ketones; synthesis of alicyclic ketones; spectrophotometric analysis. *Mailing Add:* 117 Meadow Rd Oak Ridge TN 37830

MARTIN, RICHARD HADLEY, JR, b Worcester, Mass, May 15, 24; m 46; c 3. COMPUTER SCIENCES,. *Educ:* Worcester Polytech Univ, BS, 45; Princeton Univ, MS, 47. *Prof Exp:* Res chemist, Plastics Div, 47-58, res specialist, Hydrocarbons & Polymers Div, Monsanto Co, 58-82; RETIRED. *Mem:* Am Chem Soc; AAAS; Sigma Xi. *Res:* Polymerization, processing and analytical characterization of high polymers. *Mailing Add:* 57 Brewster St Springfield MA 01119

MARTIN, RICHARD HARVEY, b LaPorte, Ind, Aug 27, 32; m 54; c 2. MEDICINE, PHYSIOLOGY. *Educ:* Johns Hopkins Univ, 50-53; Univ Rochester, MD, 57. *Prof Exp:* Intern, Strong Mem Hosp, 57-58; resident, Univ Wash, 58-59 & 61-62, asst med, Div Cardiol, 62-65; asst prof med, 65-68, assoc prof med & physiol, 68-73, dir coronary care unit, 69-75, PROF MED, SCH MED, UNIV MO-COLUMBIA, 73-, DIR DIV CARDIOL, 70- *Concurrent Pos:* Fel, Coun Clin Cardiol, Am Heart Asn; res fel med, Div Cardiol, Univ Wash, 62-64; res fel physiol & biophys, 64-65. *Mem:* Sigma Xi; fel Am Col Cardiol; Am Fedn Clin Res; fel Am Col Physicians. *Res:* Hemodynamic and clinical observations in acute myocardial infarction; ventricular aneurysm; left ventricular function; effect of atrial systole in man. *Mailing Add:* Med Sci Bldg C7 Univ of Mo Columbia MO 65212

MARTIN, RICHARD HUGO, b Hanover, Pa, Aug 16, 36; m 59; c 2. PHYSICAL CHEMISTRY. *Educ:* Gettysburg Col, AB, 58; Pa State Univ, PhD(chem), 65. *Prof Exp:* Res assoc ion molecule res, Pa State Univ, 66-67; sr res chemist, Chem Div, 67-69, sr res chemist, Ecusta Paper Div, 69-74, res assoc, Fine Paper & Film Group, 74-75, sr res assoc, Ecusta Paper & Film Group, 75-78, res mgr, Olin Corp, 78-85; STAFF RES ASSOC, ECUSTA DIV, P H GLATFELTER CO, 85- *Mem:* Am Chem Soc; Sigma Xi. *Res:* Energetics and kinetics of organic reactions; homogeneous and heterogeneous catalysis; reaction mechanisms and substituent effects; vapor phase synthesis; photochemical synthesis; instrumental measurements of dynamic systems; analysis of cigarette smoke; cigarette papers; gas chromatogrady; mass spectrometry. *Mailing Add:* 860 Country Club Rd Brevard NC 28712

MARTIN, RICHARD LEE, b Garden City, Kans, Oct 13, 50; m 85. ATOMIC & MOLECULAR PHYSICS, SOLID STATE PHYSICS. *Educ:* Kans State Univ, BS, 72; Univ Calif, Berkeley, PhD(chem), 76. *Prof Exp:* Res & teaching asst chem, Univ Calif, Berkeley, 72-76; res assoc, Univ Wash, 76-78; MEM STAFF THEORET DIV, LOS ALAMOS SCI LAB, UNIV CALIF, 78- *Concurrent Pos:* Chaim Weizmann fel, Univ Wash, 77-78. *Mem:* Am Chem Soc; Am Phys Soc; Sigma Xi. *Res:* Electronic structure of molecules; interaction of light with matter, especially photoabsorption and photoionization processes; cluster approaches to low-dimensional materials. *Mailing Add:* Theoret Div Mail Stop B 268 Los Alamos Sci Lab Los Alamos NM 87545

MARTIN, RICHARD MCFADDEN, b Somerville, Tenn, Aug 19, 42; m 64; c 2. PHYSICS. *Educ:* Univ Tenn, Knoxville, SB, 64; Univ Chicago, MS, 66, PhD(physics), 69. *Prof Exp:* Mem tech staff, Bell Tel Labs, 69-71; SCIENTIST, XEROX PALO ALTO RES CTR, 71- *Mem:* Am Phys Soc. *Res:* Solid state physics, mainly lattice dynamics of insulators, interaction of light with insulators and semiconductors. *Mailing Add:* Dept Physics, Univ Ill 1110 W Green St Urbana IL 61761

MARTIN, RICHARD MCKELVY, b Los Angeles, Calif, Jan 26, 36; m 54, 74; c 3. PHYSICAL CHEMISTRY. *Educ:* Univ Calif, Riverside, BA, 59; Univ Wis, PhD(phys chem), 63. *Prof Exp:* NIH fel, Harvard Univ, 63-65; from asst prof to assoc prof, 65-79, PROF CHEM, UNIV CALIF, SANTA BARBARA, 79- *Mem:* AAAS; Am Chem Soc; Am Phys Soc. *Res:* Photochemistry; molecular beams; electronic energy transfer; surface chemistry; reactions of electronically excited atoms and molecules with crossed molecular beams and with surfaces. *Mailing Add:* Dept of Chem Univ of Calif Santa Barbara CA 93106

MARTIN, ROBERT ALLEN, b New York, NY, Feb 19, 44; m 69. VERTEBRATE PALEOBIOLOGY, VERTEBRATE PHYSIOLOGICAL ECOLOGY. *Educ:* Hofstra Univ, BA, 65; Tulane Univ, La, MS, 67; Univ Fla, PhD(zool), 69. *Prof Exp:* Asst prof biol, SDak Sch Mines & Technol, 69-72; from asst prof to assoc prof, 72-81, PROF BIOL, FAIRLEIGH DICKINISON UNIV, 81- *Concurrent Pos:* Sigma Xi grant, SDak Sch Mines & Technol, 70-71; NSF grants, 70-72. *Mem:* Soc Study Evolution; AAAS; Soc Vert Paleont; Am Soc Mammal; Am Quaternary Asn. *Res:* Mammalian evolution, ecology, physiology. *Mailing Add:* Dept Biol Fairleigh Dickinson Univ 285 Madison Ave Madison NJ 07940

MARTIN, ROBERT BRUCE, b Chicago, Ill, Apr 29, 29; m 53. BIOPHYSICAL CHEMISTRY, BIOINORGANIC CHEMISTRY. *Educ:* Northwestern Univ, BS, 50; Univ Rochester, PhD(phys chem), 53. *Prof Exp:* Asst prof chem, Am Univ Beirut, 53-56; res fel, Calif Inst Technol, 56-57 & Harvard Univ. 57-59; from asst prof to assoc prof chem, 59-65, chmn dept, 68-71, PROF CHEM, UNIV VA, 65- *Concurrent Pos:* NIH spec fel. Oxford Univ, 61-62; prog dir molecular biol sect, NSF, 65-66. *Mem:* Fel AAAS; Am Chem Soc. *Res:* Structure, equilibrium and mechanism investigations of systems with biological interest; metal ion interactions with proteins and nucleic acids; nuclear magnetic resonance studies; bioinorganic chemistry of aluminum; bioinorganic chemistry of toxicity. *Mailing Add:* Dept Chem Univ Va Charlottesville VA 22901

MARTIN, ROBERT EUGENE, b Monterey, Tenn, July 19, 30; m 56; c 2. ANIMAL ECOLOGY. *Educ:* Tenn Tech Univ, BS, 52; Univ Tenn, MS, 59, PhD(zool), 63. *Prof Exp:* From asst prof to assoc prof, 63-72, PROF BIOL, TENN TECHNOL UNIV, 72- *Mem:* Ecol Soc Am; Am Fisheries Soc. *Res:* Fish and insect population dynamics; biometrics; fish scale structure and development. *Mailing Add:* PO Box 5063 Tenn Technol Univ Cookeville TN 38505

MARTIN, ROBERT FRANCOIS CHURCHILL, b Ottawa, Ont, Nov 3, 41; m 63; c 3. GEOLOGY. *Educ:* Univ Ottawa, Ont, BSc, 63; Pa State Univ, MS, 66; Stanford Univ, PhD(geol), 69. *Prof Exp:* Res assoc geol, Stanford Univ, 68-70; asst prof, 70-74, ASSOC PROF GEOL, MCGILL UNIV, 74- *Concurrent Pos:* Ed, Can Mineralogist, 78- *Mem:* Mineral Soc Am; Mineral Asn Can; Swiss Soc Mineral & Petrog. *Res:* Igneous and metamorphic mineralogy and petrology. *Mailing Add:* Dept Geol Sci McGill Univ 3450 University St Montreal PQ H3A 2A7 Can

MARTIN, ROBERT FREDERICK, b Weehawken, NJ, Nov 10, 38; m 60; c 3. VERTEBRATE & INVERTEBRATE ECOLOGY. *Educ:* Fairleigh Dickinson Univ, BS, 60; Univ Tex, Austin, MA, 64, PhD(zool), 69. *Prof Exp:* Res sci asst zool, Univ Tex, Austin, 64-65, cur vert, Tex Mem Mus, 69-89, dir, Tex Nat Hist Lab, 80-89; RETIRED. *Concurrent Pos:* Lectr zool, Univ Tex, Austin, 75- *Mem:* Asn Trop Biol; AAAS; Am Soc Ichthyologists & Herpetologists; Am Ornithologists Union. *Res:* Reproductive ecology of reptiles and birds; anuran morphology and evolution; human impact on vertebrate populations; ecology of endangered species; ecology of Yucatan avifauna, chitons and beach nesting wasps. *Mailing Add:* Cur Vert Tex Mem Mus 2400 Trinity Austin TX 78705

MARTIN, ROBERT G, b New York, NY, May 21, 35; m 60; c 2. VIROLOGY, CELL BIOLOGY. *Educ:* Harvard Univ, AB, 56, Med Sch, MD, 60. *Prof Exp:* SCIENTIST, NAT INST HEALTH, 60- *Mem:* Am Soc Biol Chemists; Am Soc Microbiol; Genetics Soc; Am Soc Virologists; AAAS. *Res:* Molecular biology of cellular and viral DNA replication; mechanisms of the control of cellular proliferation. *Mailing Add:* Biochem Nat Inst Arthritis Metab & Digestive Dis NIH Bethesda MD 20892

MARTIN, ROBERT LAWRENCE, b Washington, DC, Nov 18, 33; m 55. MAMMALOGY, ZOOLOGY. *Educ:* Univ Maine, BS, 56; Kans State Univ, MS, 59; Univ Conn, PhD, 71. *Prof Exp:* Teacher high sch, Maine, 56-57; from instr to asst prof anat & biol, State Univ NY Col Plattsburgh, 61-64; collabr, Great Smoky Mountain Nat Park, Nat Park Serv, Tenn, 64-65; assoc prof, 66-71, PROF MAMMAL & BIOL, UNIV MAINE, FARMINGTON, 71- *Concurrent Pos:* Res assoc, Mt Washington Observ, NH, 68-74; ed, Bat Res News, 70-75; res assoc, Univ Conn Paraguayan Exped, 73; mem, Univ Conn Chaco Exped, 74 & 75; mem, Ind Bat Recovery Team, US Fish & Wildlife Serv, 75-; hon consult, Chiropiera Specialist Group, Int Union for the Conserv of Nature and Natural Resources, 76- *Mem:* Fel AAAS; Am Soc Mammal; NY Acad Sci; Mammal Soc Brit Isles; fel Zool Soc London. *Res:* Comparative vertebrate anatomy; bat studies; mammalian natural history. *Mailing Add:* RFD 1 Box 1510 New Sharon ME 04955

MARTIN, ROBERT LEONARD, b Seattle, Wash, July 14, 19; m 46; c 4. PHYSICS. *Educ:* Reed Col, BA, 41; Univ Mich, MS, 47, PhD(physics), 56. *Prof Exp:* Res systs analyst, Willow Run Res Lab, Univ Mich, 50-52; asst prof physics, Reed Col, 56-62; from assoc prof to prof physics, 62-89, chmn dept,

63-89, EMER PROF, LEWIS & CLARK COL, 89- *Mem:* Am Asn Physics Teachers; Am Phys Soc. *Res:* Solid state; logic of physics; optical and electrical properties of ionic solids, especially in silver and alkali halides. *Mailing Add:* Dept Physics Lewis & Clark Col 0615 SW Palatine Hill Rd Portland OR 97219

MARTIN, ROBERT O, b Honolulu, Hawaii, Jan 8, 31; m 56; c 3. BIOCHEMISTRY, ORGANIC CHEMISTRY. *Educ:* Univ San Francisco, BS, 56; Univ Calif, Berkeley, PhD(biochem), 59. *Prof Exp:* USPHS fel chem, Kings Col, Newcastle-on-Tyne, 59-61; chemist, Lawrence Radiation Lab, Univ Calif, 61-65; from asst prof to assoc prof biochem, 65-74, PROF BIOCHEM, UNIV SASK, 74- *Concurrent Pos:* Can Med Res Coun vis scientist, Neurol Inst, London, 72-73. *Mem:* fel Chem Inst Can; Can Fedn Biol Sci. *Res:* Alkaloid structures and biosynthesis; neurochemistry; diabetes and other disorders of carbohydrate metabolism. *Mailing Add:* Dept Biochem Univ Sask Saskatoon SK S7N 0W0 Can

MARTIN, ROBERT PAUL, b Hartford, Conn, Mar 10, 43; m 67; c 3. MATHEMATICAL ANALYSIS. *Educ:* Cent Conn State Col, BS, 65, MS, 68; Univ Md, MA, 72, PhD(math), 73. *Prof Exp:* Teacher math, Washington Jr High Sch, New Britain, Conn, 65-66; instr, Northwestern Community Col, 66-69; instr, Univ Pa, 73-76; PROF MATH, MIDDLEBURY COL, 76- *Mem:* Am Math Soc; Math Asn Am. *Res:* Non-abelian harmonic analysis and representation theory of lie groups. *Mailing Add:* Dept of Math Voter Hall Middlebury Col Middlebury VT 05753

MARTIN, ROGER CHARLES, b Janesville, Wis, June 12, 31; m 64. GEOLOGY. *Educ:* Univ Calif, Los Angeles, AB, 53; Univ Idaho, MS, 57; Victoria Univ Wellington, PhD(geol), 63. *Prof Exp:* Geologist, NZ Geol Surv, 59-60 & 63-64; geologist, Calif Dept Water Resources, 64-68; geologist, Earth Resources Opers, NAm Rockwell Corp, 68-73; geologist, Calif State Lands Div, 73-77, GEOLOGIST, CALIF DIV MINES & GEOL, 77-, SR SCIENTIST VOLCANOLOGY & VOLCANO HAZARDS ASSESSMENT COORDR, 80- *Concurrent Pos:* Geol consult, NZ Forest Prod, Ltd, 61-64. *Mem:* Geol Soc Am; Am Geophys Union; Asn Eng Geologists. *Res:* Application of remote sensing to geoscience problems; discrimination of mineral desposits and lithologic bodies by thermal infrared and other multispectral sensors; geothermal research and exploration also environmental problems; program management of statewide assessment of low-moderate temperature geothermal resources. *Mailing Add:* 4830 Winding Way Sacramento CA 95841

MARTIN, RONALD LAVERN, b Devereaux, Mich, Sept 13, 22; m 49; c 6. NUCLEAR PHYSICS. *Educ:* US Naval Acad, BS, 44; Mich State Univ, MS, 48, Univ Chicago, PhD(physics), 52. *Prof Exp:* Res assoc physics, Univ Chicago, 52-53 & Cornell Univ, 53-56; mem staff, Bell Tel Labs, 56-59; sr scientist, TRG, Inc, 59-62; from assoc dir to dir, Particle Accelerator Div, 62-82, SR SCIENTIST, ARGONNE NAT LAB, 82- *Concurrent Pos:* Pres, Acctek Asn, 84- *Mem:* Fel Am Phys Soc. *Res:* High energy nuclear physics; accelerators; laser development; nonlinear properties of ferrites at microwave frequencies. *Mailing Add:* 901 S Kensington La Grange IL 60525

MARTIN, RONALD LEROY, b Beloit, Wis, Sept 14, 32; m 53; c 3. ANALYTICAL CHEMISTRY. *Educ:* Beloit Col, BS, 53; Univ Wis, MS, 55, PhD, 57. *Prof Exp:* Res chemist, Amoco, 57-75, MGR RES & DEVELOP, ADDITIVES DIV, AMOCO PETROL ADDITIVES CO, 75- *Honors & Awards:* IR 100 Prod, 66. *Mem:* Soc Automotive Engrs. *Res:* New lubricant development; spectrochemical analysis; gas chromatography. *Mailing Add:* Amoco Petrol Additives Co PO Box 3011 Naperville IL 60566

MARTIN, ROY A, b Coffee Co, Tenn, Mar 8, 20; m 42; c 2. ELECTRICAL ENGINEERING. *Educ:* Ga Inst Technol, BS, 42, MS, 51. *Prof Exp:* Builder, Fla, 36-37; jr engr, Ga Power Co, 37-42; engr, Western Union Tel Co, NY, 46; res asst, Eng Exp Sta, Ga Inst Technol, 46-47, res engr, asst prof & proj dir indust res, 47-53, spec res eng, 53-65, prin res engr, 65-68, lectr elec eng, 61-68; pres, Packer Eng Assocs, Inc, Ga, 68-71; pres, Roy A Martin Assocs, Inc, Ga, 71-88; PRES, ROY MARTIN, PC, 88- *Concurrent Pos:* Consult, Nat Res Coun, 48 & US Air Force, 57; pvt consult, 52-; asst secy, Res Inst, Ga Inst Technol, 60-68; pres, Hydrol Loader Acceptance Corp, 67- *Mem:* Nat Fire Protection Asn. *Res:* Electro-mechanical devices; instrumentation; patent technology; research administration; principal failure analysis; fire and industrial accident reconstruction; personal injury investigations; technical investigation; forensic science; materials technology and engineering. *Mailing Add:* Roy Martin PC 1874 Piedmont Ave NE Atlanta GA 30324-4839

MARTIN, ROY JOSEPH, JR, b Lutcher, La, Jan 3, 43; m 67; c 2. NUTRITION, BIOCHEMISTRY. *Educ:* Univ Southwestern La, BS, 64; Univ Fla, MS, 65; Univ Calif, Davis, PhD(nutrit), 70. *Prof Exp:* Asst prof nutrit, Pa State Univ, Univ Park, 70-74, assoc prof animal nutrit, 74-; PROF NUTRIT, DEPT FOODS & NUTRIT, UNIV GA, ATHENS. *Mem:* Am Dairy Sci Asn; Am Soc Animal Sci. *Res:* Metabolic regulation of growth and development; effects of early nutritional experiences. *Mailing Add:* Dept Foods & Nutrit Univ Ga Dawson Hall Athens GA 30602

MARTIN, RUFUS RUSSELL, b Decatur, Ga, Mar 3, 36; m 61; c 3. INTERNAL MEDICINE, INFECTIOUS DISEASES. *Educ:* Yale Univ, AB, 56; Med Col Ga, MD, 60. *Prof Exp:* NIH fel infectious dis, 65-67; from asst prof to assoc prof med, Sch Med, Ind Univ, Indianapolis, 67-71; assoc prof, Baylor Col Med, 71-75, prof med & microbiol, 75-; CLIN DIR, INFECT DIS RES, PHARMACEUT RES & DEVELOP DIV, BRISTOL-MYERS CO, WALLINGFORD, CT. *Concurrent Pos:* Attend physician, Vet Admin Hosp, Houston, 71-; NIH career develop award, 75-79. *Mem:* Fel Am Col Physicians; Am Fedn Clin Res; NY Acad Sci; Infectious Dis Soc Am; Am Thoracic Soc. *Res:* Staphylococcal immunology; histamine; lysosomes; leukocytes and inflammation; pulmonary macrophages and smoking. *Mailing Add:* 27 High Field Lane Madison CT 06443

MARTIN, RUSSELL JAMES, b Beaumont, Tex, May 15, 39; m 64; c 2. EPIDEMIOLOGY. *Educ:* Tex A&M Univ, BS, 61, DVM, 63; Univ Mich, MPH, 66. *Prof Exp:* Epidemiol intel serv officer, Ctr Dis Control, USPHS, 63-65, regional pub health vet, 66-72, chief pub health vet, 72-75, commun dis epidemiologist, 75-85, CHIEF DIV INFECTIOUS DIS, ILL DEPT PUB HEALTH, 85- *Concurrent Pos:* Assoc prof vet pub health, Col Vet Med, Univ Ill, 68-; clin assoc prof prev med, Peoria Sch Med, Univ Ill, 75- *Mem:* Am Pub Health Asn; Conf Pub Health Vets (pres, 76-77); Am Bd Vet Pub Health; Am Vet Med Asn; Am Vet Epidemiol Soc; Nat Asn State Pub Health Veterinarians (pres, 75-76, 90-93). *Res:* Zoonotic diseases that occur naturally in the United States, particularly delineating the epidemiology of this group. *Mailing Add:* 219 Wild Rose Lane Rochester IL 62563

MARTIN, S W, b Mount Vernon, Ohio, July 21, 58; m; c 2. GLASS RESEARCH. *Educ:* Capital Univ, BA, 80; Purdue Univ, PhD(chem), 86. *Prof Exp:* ASST PROF MAT SCI & ENG, IOWA STATE UNIV, 86- *Mem:* Am Ceramic Soc; Mat Res Soc; Am Asn Eng Educr. *Res:* Glass structure and properties. *Mailing Add:* Dept Mats Sci 110 Eng Annex Iowa State Univ Ames IA 50011

MARTIN, SAMUEL CLARK, b McNeal, Ariz, Apr 16, 16; m 44; c 2. RANGE CONSERVATION. *Educ:* Univ Ariz, BS, 42, MS, 47, PhD, 64. *Prof Exp:* Range conservationist, Southwestern Forest & Range Exp Sta, US Forest Serv, 42-49, range conservationist, Cent States Forest Exp Sta, 49-55, prin range scientist, Rocky Mountain Forest & Range Exp Sta, 55-79; PROF RANGE MGT, UNIV ARIZ, 79- *Mem:* Soc Range Mgt; Sigma Xi. *Res:* Grazing management; noxious plant control; range revegetation. *Mailing Add:* 4402 E Sixth St Tucson AZ 85711

MARTIN, SAMUEL PRESTON, III, b East Prairie, Mo, May 2, 16; m 70; c 3. INTERNAL MEDICINE, COMMUNITY HEALTH. *Educ:* Wash Univ, MD, 41. *Hon Degrees:* MA, Univ Pa, 71. *Prof Exp:* Am Col Physicians fel, Rockefeller Inst, 48-49; Markle Found fel, Duke Univ, 50-55; prof internal med & chmn dept, Univ Fla, 56-62, provost health affairs, 62-69; vis prof health econ, Harvard Univ, 69-71; dir, Robert W Johnson Clin Scholars Prog, 74-90; exec dir, Leonard Davis Inst, 74-77, PROF MED COL MED & PROF HEALTH CARE SYSTS, WHARTON SCH, UNIV PA, 71- *Concurrent Pos:* Dir, Fla Regional Med Prog, 67-68; mem comt accreditation, US Off Educ, 68-70; Commonwealth & USPHS fels, Harvard Univ & London Sch Hyg, 69-71; dir, Smith Kline Corp, Philadelphia, 72- *Honors & Awards:* Belg Order of Leopold, 47. *Mem:* Asn Am Physicians; Am Asn Immunol; Am Col Physicians; Am Fedn Clin Res (vpres, 54); Am Pub Health Asn. *Res:* Immunology microbiology; medical economics. *Mailing Add:* Sch Med Univ Pa Philadelphia PA 19104

MARTIN, SCOTT ELMORE, b Wilmington, Del, Sept 17, 46; m 71; c 2. FOOD MICROBIOLOGY. *Educ:* Tarkio Col, BA, 68; Wichita State Univ, MS, 70; Kans State Univ, PhD(microbiol), 73. *Prof Exp:* Res asst gen microbiol, Kans State Univ, 70-73; res assoc, Univ Calif, Irvine, 73-75; res assoc, 75-77, asst prof, 77-81, ASSOC PROF FOOD MICROBIOL, UNIV ILL, URBANA, 81- *Mem:* Am Soc Microbiol; Inst Food Technologists; Sigma Xi. *Res:* Examination of sublethally-stressed bacteria; determination of the effects of oxygen toxicity on microbial enumeration; ribosome assembly mechanisms; analysis of the antimutagenic and anticarcinogenic properties of the element selenium. *Mailing Add:* 580 Bevier Hall Univ Ill 905 S Goodwin Ave Urbana IL 61801

MARTIN, SCOTT MCCLUNG, b Charleston, WVa, Mar 2, 43. SCIENCE COMMUNICATIONS. *Educ:* Marshall Univ, Huntington, WVa, BS, 65; Brigham Young Univ, Provo, Utah, MS, 68; Ohio State Univ, Columbus, PhD(develop biol), 73. *Prof Exp:* Asst prof biol, Wittenberg Univ, Springfield, Ohio, 73-75; asst microbiol, Ohio State Univ, Columbus, 75-76 & Univ Mo, Kansas City, 76-78; cur, Bacillus Genetic Stock Ctr, Ohio State Univ, Columbus, 78-79; ASSOC ED BIOCHEM, CHEM ABSTR SERV, COLUMBUS, OHIO, 80- *Concurrent Pos:* Vis asst prof biol, Univ Mo, Kansas City, 78. *Mem:* AAAS; Am Inst Biol Sci; Am Soc Microbiol; Soc Protozoologists. *Res:* Biochemistry literature, primarily in molecular genetics and microbial physiology. *Mailing Add:* 712 Harley Dr Columbus OH 43202

MARTIN, SEELYE, b Northampton, Mass, Sept 22, 40. OCEANOGRAPHY. *Educ:* Harvard Univ, BA, 62; Johns Hopkins Univ, PhD(mech), 67. *Prof Exp:* Res assoc oceanog, Mass Inst Technol, 67-69; res asst prof, 69-77, res assoc, 77-79, RES PROF OCEANOG, UNIV WASH, 79- *Mem:* Am Geophys Union; AAAS. *Res:* desalination of sea ice, marginal ice zone processes, Bering Sea ice. *Mailing Add:* Dept Oceanog Univ Wash Seattle WA 98195

MARTIN, STANLEY BUEL, b Tulsa, Okla, Oct 21, 27; m 51; c 4. PHYSICAL CHEMISTRY. *Educ:* San Jose State Col, AB, 50. *Prof Exp:* Chemist, US Naval Radiol Defense Lab, 50-61, supvry chemist, 61-66; prin res chemist, URS Corp, 66-69; mgr fire res prog, 69-76, dir, fire res dept, Stanford Res Inst Int, 76-; AT STAN MARTIN & ASSOC. *Mem:* Am Chem Soc; Combustion Inst; Soc Fire Protection Engrs. *Res:* Thermal radiation transport and effects; nuclear weapons effects; fire phenomenology; transient heat conduction in solids; thermal decomposition of organic solids; ignition processes; kinetics of pyrolysis; reactions in unsteady-state systems; combustion and fire protection research. *Mailing Add:* 860 Vista Dr Redwood City CA 94062

MARTIN, STEPHEN FREDERICK, b Albuquerque, NM, Feb 8, 46; m 85. SYNTHETIC ORGANIC CHEMISTRY. *Educ:* Univ NMex, BS, 68; Princeton Univ, MA, 70, PhD(org chem), 72. *Prof Exp:* Alexander von Humboldt Found fel org chem, Univ Munich, 72-73; NIH fel, Mass Inst Technol, 73-74; from asst prof to assoc prof, 74-85, PROF ORG CHEM, UNIV TEXAS, AUSTIN, 85- *Concurrent Pos:* NIH career develop award, 80-85. *Mem:* Am Chem Soc; Sigma Xi. *Res:* Design and development of new synthetic methods; chemistry and total synthesis of natural products, particularly alkaloids and terpenes; heterocyclic chemistry; peptide mimics; phospholipids; enzyme inhibitors. *Mailing Add:* Dept of Chem Univ of Tex Austin TX 78712-1167

MARTIN, STEPHEN GEORGE, b Eagle Grove, Iowa, Sept 20, 41; m 65, 84; c 2. ENVIRONMENTAL MANAGEMENT CONSULTING ADMINISTRATION. *Educ:* Univ Wis-Madison, BS, 64; MS, 67; Ore State Univ, PhD(zool), 70. *Prof Exp:* Asst prof zool, Colo State Univ, 70-73; vpres, Ecol Consult, Inc, 73-80; vpres, Environ Res & Tech Inc, 80-85, regional vpres, 85-88, exec vpres, 88-90; pres, Analytikem, 90; PRES, S G MARTIN & ASSOCS INC, 90- *Concurrent Pos:* Affil fac mem, Colo State Univ, 73-; dir, Univ Nat Bank, 80-, chmn, 86-87; dir, Eng Hydraul, Inc, 87-90, Taywood Eng, 89-90, First Nat Bank, Loveland, 90-, S G Martin & Assocs Inc, 90-, Macleod Pharmaceut Inc, 91- *Honors & Awards:* John T Curtis Award, Univ Wis Syst, 64; A Brazier Howell Award, Cooper Ornith Soc, 70; Blosser Award, Tech Assoc Pulp & Paper Indust, 88. *Mem:* AAAS; Ecol Soc Am; Am Ornith Union; Cooper Ornith Soc; The Wildlife Soc; Wilson Ornith Soc. *Res:* Administration of the technical activities of environmental scientists and engineers in applied ecological and environmental research, nationally and internationally; natural resource damage assessment; analysis of environmental impact; applied ecology; hazardous waste site assessment and remediation technologies. *Mailing Add:* 7121 County Rd 9 Wellington CO 80549

MARTIN, SUSAN SCOTT, b Paducah, Ky, May 18, 38. PLANT CHEMISTRY. *Educ:* Univ Colo, BA, 60; Utah State Univ, MS, 68; Univ Calif, Santa Cruz, PhD(biol), 73. *Prof Exp:* Res chemist, Dept Wildlife Resources, Utah State Univ, 60-68; PLANT PHYSIOLOGIST, AGR RES SERV, USDA, 74- *Concurrent Pos:* Ed, J Sugar Beet Res, 87- *Mem:* Phytochem Soc NAm; Bot Soc Am. *Res:* Biochemical aspects of plant-pathogen interaction, including pathogen-produced toxins and phytoalexins; physiological factors affecting sugarbeet quality and sucrose production; resin and glucosinolate chemistry and chemical ecology. *Mailing Add:* Agr Res Serv USDA 1701 Center Ave Ft Collins CO 80526

MARTIN, TELLIS ALEXANDER, b Hickory, NC, May 20, 19; m 49; c 4. ORGANIC CHEMISTRY. *Educ:* Berea Col, BA, 42; Univ Va, MS, 45, PhD(org chem), 48. *Prof Exp:* Asst chem, Berea Col, 40-42, Univ Va, 42-44 & Off Sci Res & Develop, 44-48; res chemist, Gen Aniline & Film Corp, 48-53; sr chemist, Mead Johnson & Co, 53-63, res assoc, 63-70, sr investr, 70-73, prin investr, 73-76, prin res assoc, 76-80, prin res scientist, 80-87; RETIRED. *Mem:* AAAS; Am Chem Soc. *Res:* Synthesis of benzalacetophenones; aminoketones and amino alcohols of phenyl, quinoline and phenylquinoline series for possible use as antimalarials, cancer agents and antitubercular drugs; phthalocyanines; phenanthridines; biphenylsulfones; carbohydrates; steric hindrance; ring-chain tautomerism; cysteines; sulfa drugs; hypnotics; betalactams and aspartic acids; anti-inflammatory, fibrinolytic and mucolytic agents; cardiovascular agents; antihypertensive-diuretic agents and antiarrhythmic agents. *Mailing Add:* 12 Sand Dollar Dr Isle of Palms SC 29451

MARTIN, TERENCE EDWIN, b Adelaide, Australia, Apr 28, 41; m 63; c 2. MOLECULAR BIOLOGY, CELL BIOLOGY. *Educ:* Univ Adelaide, BSc, 62; Cambridge Univ, PhD(biochem), 66. *Prof Exp:* Univ fel, Univ Chicago, 66-68, Am Cancer Soc fel, Univ Wash, 69-70; from asst prof to assoc prof biol, 71-85, PROF MOLECULAR GENETICS & CELL BIOL, UNIV CHICAGO, 85- *Concurrent Pos:* USPHS res grant, Univ Chicago, 71- *Mem:* Brit Biochem Soc; Am Soc Cell Biol. *Res:* Control of gene expression in eukaryotic cells; nucleic acid synthesis and metabolism; control of protein synthesis; molecular basis of cancer; structure of the cell nucleus and nuclear antigens. *Mailing Add:* Dept Molecular Genet & Cell Biol Univ Chicago 920 E 58th St Chicago IL 60637

MARTIN, TERRY JOE, b Baxter Springs, Kans, Dec 28, 47; m 66; c 3. PLANT PATHOLOGY. *Educ:* Kans State Col, Pittsburg, BS, 70; Kans State Univ, MS, 71; Mich State Univ, PhD(plant path), 74. *Prof Exp:* Asst prof plant path, 74-79, from asst prof to prof, 79-86, PROF, WHEAT BREEDER, FT HAYS EXP STA, KANS STATE UNIV, 86- *Mem:* Am Soc Agron; Crop Sci Soc Am. *Res:* Development of improved hard red winter wheat cultivars for Kansas with emphasis on pest resistance. *Mailing Add:* Hays Exp Sta Hays KS 67601

MARTIN, TERRY ZACHRY, b New York, NY, Aug 7, 46; m 82; c 1. PLANETARY ASTRONOMY. *Educ:* Univ Calif, Berkeley, AB, 67; Univ Hawaii, MS, 69, PhD (astron), 75. *Prof Exp:* Res assoc, Inst Astron, Univ Hawaii, 68-74; res geophysicist planetary astron, Dept Earth & Space Sci, Univ Calif, Los Angeles, 75-79; mem tech staff, Infrared Instruments Sect, 79-83, MEM TECH STAFF, ATMOSPHERIC & COMETARY SCI SECT, JET PROPULSION LAB, CALIF INST TECHNOL, 84- *Mem:* Am Astron Soc. *Res:* Planetary astronomy, especially atmospheric composition and thermal behavior; design of infrared instruments; testing of infrared detector systems; planetary spectroscopy; pressure induced absorption, Jupiter and Saturn; Mars atmosphere thermal behavior and opacity. *Mailing Add:* Jet Propulsion Lab 169-237 4800 Oak Grove Dr Pasadena CA 91109

MARTIN, THOMAS A(DDENBROOK), b Cleveland, Ohio, July 7, 24; m 50; c 2. SOFTWARE ENGINEERING. *Educ:* Cleveland State Univ, BEE, 50; Rensselaer Polytech Inst, MEE, 54, PhD(commun), 62. *Prof Exp:* Engr, electro-motive div, Gen Motors Corp, 50; instr electronics, Rensselaer Polytech Inst, 50-55, res assoc, 53-56, asst prof, 55-59; eng specialist, Goodyear Aerospace Corp, 59-63, sr specialist systs eng, 63-68, asst dept mgr weapon syst anal, 68-70; leader strategic progs systs eng, Missile & Surface Radar Div, RCA Corp, 70-74, tech dir, Tradex Radar Site, 74-77, staff tech adv, Govt Systs Div, 78-80, dir tech planning, Aerospace Defense, 80-88; DIR, ENG DEVELOP, GE AEROSPACE, 88- *Concurrent Pos:* Consult, Empire Res Corp, 58-59; lectr, Univ Akron, 65-69. *Mem:* Sr mem Inst Elec & Electronics Engrs; Sigma Xi. *Res:* Navigation; guidance control; missile systems; weapon systems; radar systems; automatic data processing systems. *Mailing Add:* 179 Conawaga Trail Medford Lakes NJ 08055

MARTIN, THOMAS FABIAN JOHN, b Rahway, NJ, June 10, 46; c 2. MOLECULAR ENDOCRINOLOGY, CELL BIOLOGY. *Educ:* Cornell Univ, AB, 68; Harvard Univ, PhD(biophysics), 74. *Prof Exp:* Fel endocrinol, Pharmacol Dept, Harvard Med Sch, 74-78; asst prof, 78-83, ASSOC PROF ZOOL, UNIV WIS, 83- *Mem:* Endocrine Soc; AAAS. *Res:* Molecular mechanisms by which peptide hormones alter cellular function, including intracellular mediators different from cyclic AMP. *Mailing Add:* Zool Res Bldg Univ Wis 1050 Bascom Mall Madison WI 53706

MARTIN, THOMAS GEORGE, III, b Boston, Mass, Jan 14, 31; m 51; c 2. HEALTH PHYSICS. *Educ:* Northeastern Univ, BS, 58; Am Bd Health Physics, dipl, 65. *Prof Exp:* Head radiol safety dept, Controls for Radiation Inc, 58-61, head labs, 61-62; radiation chemist, Mass Inst Technol, 62-63; dir safety & radiation protection, R D & E Ctr, Natick Lab, US Army, 63-86, chief, Biochem Br & sci & adv technol dir, 86-90; RADIOL SAFETY CONSULT, 90- *Concurrent Pos:* Consult tech points of contact, Interdept Comt on Radiation Preservation of Food, 64- *Mem:* Health Physics Soc; Sigma Xi; Conf Radiol Health; NY Acad Sci. *Res:* Induced radioactivity in food sterilized by ionizing radiation; health physics problems associated with particle accelerators; trace metals in foods; activation analysis. *Mailing Add:* 588 Winter St Framingham MA 01701

MARTIN, THOMAS L(YLE), JR, b Memphis, Tenn, Sept 26, 21; m 43; c 3. ELECTRICAL ENGINEERING. *Educ:* Rensselaer Polytech Inst, BEE, 42, MEE, 48; Stanford Univ, PhD(elec eng), 51. *Hon Degrees:* DEng, Rensselaer Polytech Inst, 67. *Prof Exp:* Instr elec eng, Rensselaer Polytech Inst, 46-48; asst prof, Univ NMex, 48-51, assoc prof, 51-53; prof & head dept, Univ Ariz, 53-58, dean, col eng, 58-63; dean, Col Eng, Univ Fla, 63-66; dean, Inst Technol, Southern Methodist Univ, 66-74; pres, 74-87, EMER PRES, ILL INST TECHNOL, 87- *Concurrent Pos:* Consult, Sandia Corp, 51, Wesix Elec Heater Co, 51-56, Gen Eng Co, 52-53, Army Electronic Proving Ground, 55-56, Hughes Aircraft, 57, US Radium Corp, 56-59, Mellonics Corp, 58-59, Surgeon Gen USAF, 59-60 & Tenn Comn Higher Educ, 68-69; consult & mem, Sci Adv Bd to Commanding Gen, Army Electronics Proving Grounds, 58-62; mem bd dirs numerous corp, 68-; chmn, Comt Minorities in Eng, Nat Res Coun, 77-79; spec adv higher educ, Prime Minister, Repub China, 78-81. *Mem:* Nat Acad Eng; Am Soc Eng Educ; fel Inst Elec & Electronic Engrs; Sigma Xi. *Res:* Application of closed circuit talk-back television to engineering education; author of six books; recipient of three patents in field of unipolar radioactive ionizers. *Mailing Add:* Ill Inst Technol Chicago IL 60616

MARTIN, THOMAS WARING, b Cumberland, Md, July 24, 25; m 47; c 4. PHYSICAL CHEMISTRY. *Educ:* Franklin & Marshall Col, BS, 50; Northwestern Univ, PhD(chem), 54. *Prof Exp:* Res fel photochem, Nat Res Coun Can, 54-55; instr chem, Williams Col, Mass, 55-57; from asst prof to assoc prof, 57-66, chmn dept, 67-73, PROF CHEM, VANDERBILT UNIV, 66- *Mem:* AAAS; Am Chem Soc; The Chem Soc; Royal Inst Chem. *Res:* Photochemistry; electron spin resonance and magneto-chemical effects; biochemical oxidoreduction and model systems; chemical kinetics and catalysis; mass spectrometry. *Mailing Add:* Dept Chem Vanderbilt Univ Box 1506 Sta B Nashville TN 37235

MARTIN, TREVOR IAN, b Southall, Eng, Apr 3, 43; m 68; c 2. ANALYTICAL CHEMISTRY, INDUSTRIAL ORGANIC CHEMISTRY. *Educ:* Royal Inst Chem, grad, 69, ARIC, 74, Cert chem, 76; McMaster Univ, PhD(chem), 74. *Prof Exp:* Analytical chemist, Laporte Titanium Ltd, Eng, 64-68; org analyst, Xerox Res Ctr, 74-79, area mgr, Org Chem & Spec Chem, 79-86, mgr, Technol Strategy & Planning, 86-89, MGR, SYNTHETIC CHEM AREA, XEROX RES CTR, CAN, 89- *Honors & Awards:* Pres Achievement Award, Xerox Corp, 83. *Mem:* Am Chem Soc; Royal Inst Chem. *Res:* Organic synthesis of electrophotographic dyes and pigments; elucidation of molecular structure and analysis of trace levels of organic and elemental impurities using high performance liquid chromatography and plasma emission spectroscopy; on line liquid chromatography, mass spectrometry; microencapsulation technology for composite polymeric materials. *Mailing Add:* Xerox Res Centre of Can Ltd 2660 Speakman Dr Mississauga ON L5K 2L1 Can

MARTIN, TRUMAN GLEN, b Wortham, Tex, May 24, 28; m 50; c 3. ANIMAL GENETICS. *Educ:* Tex A&M Univ, BS, 49; Iowa State Univ, MS, 51, PhD(animal breeding & nutrit), 54. *Prof Exp:* Asst dairy husb, Iowa State Univ, 49-51, asst animal breeding, 53-55; from asst prof to assoc prof, 55-63, PROF ANIMAL BREEDING, PURDUE UNIV, 63- *Concurrent Pos:* Vis scientist, Animal Breeding Res Orgn, Edinburgh, Scotland, 78- 79, Instituto Nacional Investigaciones Agraria, Madrid, Spain, 85-86; consult, US-Aid in Portugal, 82, 83, 85, 86. *Mem:* Am Dairy Sci Asn; Am Soc Animal Sci; Brit Soc Animal Prod; Sigma Xi. *Res:* Crossbreeding and selection of dairy and beef cattle; factors affecting body composition of swine, sheep and cattle. *Mailing Add:* Dept Animal Sci Purdue Univ West Lafayette IN 47907

MARTIN, VIRGINIA LORELLE, b Mount Olive, NC, Nov 29, 39. BIOLOGY, PARASITOLOGY. *Educ:* Wake Forest Col, BS, 61; Emory Univ, MS, 63, PhD(biol), 67. *Prof Exp:* From asst prof to assoc prof, 66-81, PROF BIOL, QUEENS COL, NC, 81- *Concurrent Pos:* Sr lectr, Univ Calabar, Nigeria, 75-77. *Mem:* AAAS; Am Soc Parasitol; Am Inst Biol Sci; Sigma Xi. *Res:* fluorescent antibody immunodiagnosis of parasitic diseases; taxonomy and life cycle of Spirorchiidae. *Mailing Add:* Dept of Biol Queens Col 1900 Selwyn Ave Charlotte NC 28274

MARTIN, WAYNE DUDLEY, b Watertown, Ohio, Nov 22, 20; m 53; c 2. GEOLOGY. *Educ:* Marietta Col, BS, 48; Univ WVa, MS, 50; Univ Cincinnati, PhD(geol), 55. *Prof Exp:* Instr geol, Bowling Green State Univ, 51-52; from instr to prof, 52-86, EMER PROF GEOL, MIAMI UNIV, 86- *Mem:* Geol Soc Am; Soc Econ Paleontologists & Mineralogists; Nat Asn Geol Teachers; Am Asn Petrol Geologists; Int Asn Sedimentologists; Am Inst Prof Geologists. *Res:* Petrology of the Cincinnatian Series limestone; sedimentary facies of the Dunkard Basin; petrology of the Cambrian System, NW Wind River Basin, Wyo. *Mailing Add:* Dept Geol Miami Univ Oxford OH 45056-2530

MARTIN, WAYNE HOLDERNESS, b Manchester, Ohio, Mar 15, 31; m 54; c 7. ENVIRONMENTAL SCIENCES, POLYMER CHEMISTRY. *Educ:* Ohio State Univ, BS, 52, PhD(chem), 58. *Prof Exp:* Instr chem, Ohio State Univ, 57-58; from chemist to sr chemist, Plastics Dept, Washington Works, WVa, E I Du Pond de Nemours & Co, Inc, 58-78, sr res chemist, 78-82, environ control coordr, 82-83, Safety, Environ Control, Occup Health Adminr, Plastic Prod & Resins Dept, Exp Sta, 83-86, CONSULT REGULATOR AFFAIRS, E I DU PONT DE NEMOURS & CO, INC, 86- *Mem:* Am Chem Soc. *Res:* Environmental control; industrial hygiene; air pollution; solid waste. *Mailing Add:* E I du Pont de Nemours & Co Inc Polymers M-5616 Wilmington DE 19898

MARTIN, WILFRED SAMUEL, chemical engineering; deceased, see previous edition for last biography

MARTIN, WILLARD JOHN, b Minneapolis, Minn, May 29, 15; m 42; c 5. PHYSICAL CHEMISTRY. *Educ:* Univ Minn, BS, 37; Cornell Univ, PhD(phys chem), 41. *Prof Exp:* Lab asst, Cornell Univ, 37-41; asst prof in-chg chem dept, Univ Maine, Brunswick Campus, 46-49; prof chem, SDak Sch Mines & Technol, 49-78; RETIRED. *Mem:* Am Chem Soc. *Res:* X-ray diffraction; atomic and molecular structure; general chemistry; computer science. *Mailing Add:* 3902 Ponderosa Trail Rapid City SD 57702-2598

MARTIN, WILLIAM BUTLER, JR, b Winchendon, Mass, Aug 31, 23; m 50; c 3. PHYSICAL ORGANIC CHEMISTRY, SPECTROSCOPY. *Educ:* Clark Univ, AB, 48, AM, 49; Yale Univ, PhD(org chem), 53. *Prof Exp:* Lab instr chem, Clark Univ, 47-49 & Yale Univ, 49-50; fel, Hickrill Res Fedn, NY, 52-53; from asst prof to assoc prof, 53-63, prof, 63-89, EMER PROF CHEM, UNION COL, NY, 89- *Concurrent Pos:* NIH spec res fel, Sch Advan Studies, Mass Inst Technol, 59-61; res assoc, Swiss Fed Inst Technol, 67-68 & Univ Basel, 74-75, 81-82, 86 & 89-90; exec officer, NE Reg Meetings, Am Chem Soc, 91- *Mem:* Sigma Xi; Am Chem Soc; Fedn Am Sci; Swiss Chem Soc. *Res:* Biochemistry; synthetic organic chemistry; electron spin resonance in pi-electron systems; distance of electron jumps from one pi system to another isolated in the same molecule; alkaloids and polyamides in plants at various stages of growth; mechanism and stereochemistry. *Mailing Add:* Dept Chem Union Col Schenectady NY 12308

MARTIN, WILLIAM CLARENCE, b Dayton, Ky, Nov 27, 23; m 47; c 3. PLANT TAXONOMY, FLORISTICS. *Educ:* Purdue Univ, BS, 50; Ind Univ, MA, 56, PhD(bot), 58. *Prof Exp:* From asst prof to assoc prof, 58-71, PROF BIOL, UNIV NMEX, 71- *Res:* Floristics; genetics; plant geography; analysis of taxa; distribution of species; studies of threatened and endangered species. *Mailing Add:* Dept Biol Univ NMex Albuquerque NM 87131

MARTIN, WILLIAM CLYDE, b Cullman, Ala, Nov 27, 29; m 59; c 2. ATOMIC SPECTROSCOPY, ATOMIC PHYSICS. *Educ:* Univ Richmond, BS, 51; Princeton Univ, MA, 53, PhD(physics), 56. *Prof Exp:* Instr physics, Princeton Univ, 55-57; physicist, 57-62, chief spectros sect, 62-79, GROUP LEADER ATOMIC SPECTROS, NAT BUR STANDARDS, 79- *Honors & Awards:* Silver Medal Award, US Dept Com, 68, Gold Medal, 81; William F Meggers Award, Optical Soc Am, 83. *Mem:* Am Phys Soc; Optical Soc Am; Am Astron Soc; Int Astron Union. *Res:* Optical atomic spectroscopy; atomic structure. *Mailing Add:* A167 Physics Bldg Nat Bur Standards Gaithersburg MD 20899

MARTIN, WILLIAM DAVID, b Anaconda, Mont, June 24, 42; m 64; c 4. ANATOMY. *Educ:* Carroll Col, Mont, AB, 64; Creighton Univ, MS, 66; Univ Minn, Minneapolis, PhD(vet anat), 73. *Prof Exp:* Instr vet anat, Univ Minn, 66-73; instr, 73-75, asst prof anat, Univ Ky, 75-80; ASSOC PROF ANAT, W VA SCH OSTEOP MED, 80- *Mem:* Am Asn Anatomists; Sigma Xi; Am Asn Vet Anatomists. *Res:* Muscle histochemistry and ultrastructure. *Mailing Add:* Dept of Anat W Va Sch Osteop Med 400 N Lee St Lewisburg WV 24901

MARTIN, WILLIAM EUGENE, b St Joseph, Mo, Dec 19, 41; m 64; c 2. PHYSICS, PHYSICAL OPTICS. *Educ:* San Diego State Univ, BS, 69, MS, 70; Univ Calif, San Diego, PhD(appl physics), 74. *Prof Exp:* Physicist solid state res, US Naval Electronics Lab Ctr, 70-75; physicist & group leader, laser fusion proj, Lawrence Livermore Lab, 75-; AT BAE DYNAMICS GROUP, BRISTOL, ENG. *Mem:* Optical Soc Am; AAAS. *Res:* Laser amplifier and oscillator physics; optical pulse shaping; physics of II-VI and III-V materials; integrated and fiber optics device research; non linear optics. *Mailing Add:* BAE Dynamics Group FPC 267 PO Box 5 Filton Bristol BS12 7QW England

MARTIN, WILLIAM GILBERT, b Shreveport, La, June 15, 31; m 53; c 3. NUTRITION, BIOCHEMISTRY. *Educ:* La Polytech Inst, BS, 56; NC State Univ, MS, 58, WVa Univ, PhD(biochem), 63. *Prof Exp:* Res asst animal nutrit, NC State Univ, 56-58; res asst agr biochem, 58-60, from instr to assoc prof, 60-74, AGR BIOCHEMIST & PROF AGR BIOCHEM, WVA UNIV, 74- *Mem:* Am Inst Nutrit; Am Chem Soc; Soc Exp Biol & Med; Poultry Sci Asn; Am Soc Animal Sci; Sigma Xi. *Res:* Amino acid and mineral metabolism; sulfur metabolism. *Mailing Add:* Comt Agr Biochem WVa Univ PO Box 6108 Morgantown WV 26506

MARTIN, WILLIAM HAYWOOD, III, b Bath Springs, Tenn, Nov 29, 38; div; c 2. PLANT ECOLOGY, FOREST ECOLOGY. *Educ:* Tenn Polytech Inst, BS, 60; Univ Tenn, Knoxville, MS, 66, PhD(bot), 71. *Prof Exp:* PROF BIOL SCI, EASTERN KY UNIV, 69-, DIR, DIV NATURAL AREAS, 77- *Concurrent Pos:* Inst grants, Eastern Ky Univ, 71, 73, 78 & 79; consult, Nat Natural Landmark Theme Studies, Dept Interior; prin investr, Natural Plant Comt Study, Dept Energy; adj prof, Univ Ky; mem, Ecol Exped Yuntai Mountains, People's Repub China, 87. *Honors & Awards:* Ky Wildlife Conservationist of Year, 77. *Mem:* AAAS; Ecol Soc Am; Sigma Xi. *Res:* Relationships among plant components of forests and soil and geologic, topographic parameters; relationship of plant communities and populations to climatic, soil, geologic, topographic and biotic factors; major areas of interest and expertise in forest, natural and cultivated grassland ecosystems. *Mailing Add:* Div of Natural Areas Eastern Ky Univ Richmond KY 40475-0947

MARTIN, WILLIAM L, JR, b St Louis, Mo, Feb 21, 36; m; c 3. MATHEMATICS, ELECTRONICS. *Educ:* Washington Univ, St Louis, BS, 65. *Prof Exp:* Engr, Gen Cable Corp, 62-66 & Conrow Co, 66; engr, 66-69, dir elec insulation & conductors, 69-78, ENGR MKT & SALES, MAGNET WIRE PRODUCTS, P D GEORGE CO, 78-, ENG SALES & SERV, ELEC INSULATING PROD. *Mem:* Inst Elec & Electronics Engrs. *Res:* Commercial evaluation and application of organic electrical insulating systems for magnet wire. *Mailing Add:* P D George Co 5200 N Second St St Louis MO 63147

MARTIN, WILLIAM MACPHAIL, b Heatherdale, PEI, Aug 16, 19; m 45, 86; c 4. NUCLEAR PHYSICS. *Educ:* Queen's Univ, Ont, BSc, 41; McGill Univ, PhD(physics), 51. *Prof Exp:* Jr res assoc physics, Chalk River Labs, 45-46; from asst prof to assoc prof, Queen's Univ, Ont, 51-55; assoc prof, McGill Univ, 55-63, asst chmn dept, 70-84, prof physics, 63-84; RETIRED. *Mem:* Can Asn Physicists; Am Phys Soc. *Res:* Radioactivity; nuclear reactions and isomerism. *Mailing Add:* 136 Pointe Claire Ave Pointe Claire PQ H9S 4M5 Can

MARTIN, WILLIAM PAXMAN, b American Fork, Utah, July 15, 12; m 37; c 3. SOIL MICROBIOLOGY. *Educ:* Brigham Young Univ, AB, 34; Iowa State Col, MS, 36, PhD(soil bact), 37. *Prof Exp:* Instr soil microbiol, Univ Ariz, 37-40, asst prof, 40-45; forest ecologist, Southwestern Forest & Range Exp Sta, US Forest Serv, 45-48; prof agron & bact, Ohio State Univ, 48-54; prof soils & head dept, 54-82, EMER PROF, COL AGR, UNIV MINN, ST PAUL, 82- *Concurrent Pos:* Asst soil microbiologist exp sta, Univ Ariz, 37-45; asst res chemist, Soil Conserv Serv, USDA, 41-44, soil chemist, 44-45, microbiologist, Regional Salinity Lab, Calif, 45; mem comn zero tolerance & residue regulation pesticides, Nat Acad Sci-Nat Res Coun, 64-65; mem agr libr network comt, Inter-Univ Commun Coun Educ Commun, 68-69; bd mem, Agron Sci Found, 71-74; mem, Nat Soybean Crop Improv Coun. *Mem:* Fel AAAS; hon mem Am Soc Agron (pres, 75); fel Soil Sci Soc Am (pres, 66); fel Soil Conserv Soc Am; Coun Agr Sci Technol. *Res:* Soil fertility; soil and water conservation; soil microbiology. *Mailing Add:* 8438 N Breezewood Pl Tucson AZ 85704-0901

MARTIN, WILLIAM RANDOLPH, b Knoxville, Tenn, Apr 19, 22; m 49; c 2. MICROBIOLOGY. *Educ:* Univ Tenn, BA, 47, MS, 50; Univ Tex, PhD(bact), 55. *Prof Exp:* Res asst biophys, Oak Ridge Nat Lab, 47-48; res asst bact, Univ Tex, 51-52, res scientist, 52-55; assoc bacteriologist, Am Meat Inst Found, 55-57; from instr to assoc prof microbiol, Univ Chicago, 57-80, sr adv biol sci, 76-80; prof basic med sci, 80-88, EMER PROF, SCH MED, MERCER UNIV, MACON, GA, 88- *Concurrent Pos:* USPHS career develop award, 60-; Guggenheim fel, 65-66; vis investr, Inst Microbiol, Gottingen, Ger, 65-66. *Mem:* Am Soc Microbiol; NY Acad Sci; Brit Soc Gen Microbiol. *Res:* Microbial metabolism; filamentous fungi and mechanisms of cellular resistance to anti-tumor drugs. *Mailing Add:* 4609 Oxford Circle Macon GA 31210

MARTIN, WILLIAM ROBERT, b Aberdeen, SDak, Jan 30, 21; m 49; c 3. PHARMACOLOGY. *Educ:* Univ Chicago, BS, 48; Univ Ill, MS & MD, 53. *Prof Exp:* Intern, Cook County Hosp, Chicago, Ill, 53-54; from instr to asst prof pharmacol, Univ Ill, 54-57; neuropharmacologist, Nat Inst Drug Abuse, 57-63, dir, Addiction Res Ctr, 63-77; PROF PHARMACOL & CHMN DEPT, UNIV KY, 77- *Concurrent Pos:* Adj assoc prof, Univ Ky, 62, adj prof, Sch Med, 71. *Honors & Awards:* Nathan B Eddy Mem Award, Nat Acad Sci-Nat Res Coun; Commendation Medal, USPHS, 66,. *Mem:* AAAS; Am Soc Pharmacol & Exp Therapeut; Am Soc Clin Pharmacol & Therapeut; Am Col Neuropsychopharmacol; Soc Neurosci. *Res:* Neuropharmacology; clinical pharmacology; drug addiction. *Mailing Add:* Box 764 Midway KY 40347

MARTIN, WILLIAM ROYALL, JR, b Raleigh, NC, Sept 3, 26; m 52; c 2. ORGANIC CHEMISTRY, BUSINESS ADMINISTRATION. *Educ:* Univ NC, AB, 48, MBA, 64; NC State Univ, BS, 52. *Prof Exp:* Chemist, Am Cyanamid Co, 52-54; plant chemist, Dan River Mills, Inc, 54-56; group leader, Union Carbide Corp, 56-59; res assoc & head appl chem res, NC State Univ, 59-63; tech dir, 63-73, EXEC DIR, AM ASN TEXTILE CHEMISTS & COLORISTS, 74- *Concurrent Pos:* US deleg & secy meeting subcomt color-fastness tests, Int Orgn Standardization, NC, 64, US deleg tech comt textiles, London, 65, 70, 75, 80 & Manchester, 85 & 90, secy meetings subcomt color fastness & color measurement & subcomt dimensional stability, Wurzburg, 68, Newton, Mass, 71, Paris, 74, Ottawa, 77, Copenhagen, 81, Manchester, 84, Bad Soden, 87 & Williamsburg, Va, 89; mem comt textiles, Pan Am Standards Comn, Montevideo, 66; spec lectr & adj assoc prof, NC State Univ, 66- *Mem:* Fel Textile Inst; Am Chem Soc; Fiber Soc; fel Soc Dyers & Colourists; Am Asn Textile Chemists & Colorists. *Res:* Business and technical administration. *Mailing Add:* Am Asn Textile Chemists & Colorists PO Box 12215 Research Triangle Park NC 27709-2215

MARTIN, YVONNE CONNOLLY, b St Paul, Minn, Sept 13, 36; m 63; c 2. COMPUTER ASSISTED DRUG DESIGN. *Educ:* Carleton Col, BA, 58; Northwestern Univ, PhD(chem), 64. *Prof Exp:* Res asst, 58-60, sr pharmacologist, 64-70, assoc res fel, 70-74, res fel, 74-85, SR PROJ LEADER, 84-, SR RES FEL, ABBOTT LABS, 85- *Concurrent Pos:* Mem, Carcinogenesis Prog Sci Review Comt, Nat Cancer Inst, 78-81; mem NAS/NRC rev; vis prof, Sch Pharm, Univ Va, 90; collabr, Tripos Assocs. *Mem:* Fel AAAS; Sigma Xi; Am Crystallog Soc. *Res:* Quantitative structure-activity relationships of drugs; receptor mapping based on 3D structure of ligands; 3D database application in drug design. *Mailing Add:* Dept 47E AP9A Abbott Labs Abbott Park IL 60064

MARTINDALE, WALLACE S, b Philadelphia, Pa, Aug 18, 30. ALGEBRA. *Educ:* Amherst Col, BA, 52, MA, 54; Univ Pa, PhD(math), 58. *Prof Exp:* PROF MATH, UNIV MASS, 64- *Mem:* Am Math Soc; Math Asn Am. *Res:* Research into ring theory. *Mailing Add:* Math Dept Univ Mass Amherst MA 01003

MARTINDALE, WILLIAM EARL, b Nashville, Ark, Sept, 4, 23; m 50; c 3. BIOCHEMISTRY. *Educ:* Henderson State Teachers Col, BA, 47; Univ Ark, MS, 49; Univ Ala, MS, 57, PhD(biochem), 62. *Prof Exp:* Biochemist, Thayer Vet Admin Hosp, Nashville, Tenn, 50-54; asst chief radioisotope serv, Birmingham Vet Admin Hosp, 54-62; asst prof chem, Miss State Univ, 62-64; prof chem & chmn dept, Belmont Col, 64-83, prof chem & physics, 83-; RETIRED. *Concurrent Pos:* Vis prof org chem, Trevecca Col, 69- *Res:* Carbohydrate metabolism in thyroid tissues; synovial permeability in arthritis; chronic vitamin B-6 deficiency in rat; folic acid deficiency in chicks. *Mailing Add:* 3614 Brushhill Rd Nashville TN 37216

MARTIN-DELEON, PATRICIA ANASTASIA, b Port Maria, Jamaica, WI; Can citizen; c 2. HUMAN GENETICS. *Educ:* Univ WI, BSc, 67, MSc, 69; Univ Western Ont, PhD(genetics), 72. *Prof Exp:* Res asst human cytogenetics, Univ WI, 67-69; sessional lectr genetics, McGill Univ, 75-76; asst prof, 76-81, ASSOC PROF HUMAN GENETICS, UNIV DEL, 81- *Concurrent Pos:* Fel, McGill Univ, 72-75; UDRF res grant, Univ Del, 77-78. *Mem:* AAAS; Genetic Soc Can; Am Soc Human Genetics; Sigma Xi. *Res:* The role of the aging sperm in the induction of chromosome anomalies in resulting embryos; regulation of the activity of nucleolar organizer regions. *Mailing Add:* Sch Life & Health Sci Univ Del Newark DE 19716

MARTINEAU, RONALD, b Quebec, Que, Apr 1, 46; m 71; c 1. IMMUNOLOGY, BIOCHEMISTRY. *Educ:* Laval Univ, BSc, 67; Univ Rochester, PhD(microbiol), 73. *Prof Exp:* Res fel immunol, Scripps Clin & Res Found, 72-74; asst prof biochem, 75-80, ASSOC PROF BIOCHEM & IMMUNOL, LAVAL UNIV, 80- *Mem:* Can Biochem Soc. *Mailing Add:* Dept Biochem Sch Med Laval Univ Quebec PQ G1K 7P4 Can

MARTINEC, EMIL LOUIS, b Chicago, Ill, July 28, 27; m 54; c 3. EDUCATIONAL ADMINISTRATION, TECHNICAL MANAGEMENT. *Educ:* Ill Inst Technol, BS, 50; Univ Idaho, MS, 57; Northwestern Univ, MBA, 65. *Hon Degrees:* Dr Engr Mgt, Midwest Col Eng, 73. *Prof Exp:* Design engr brake systs, Am Steel Foundries, 50-51; design engr exp equip, Taylor Forge and Pipe Works, 51-52; proj engr furnaces, Standard Oil Co, Ind, 52-55; asst engr exp loop design, Argonne Nat Lab, 55-62, engr reactor design, 62-73, asst div dir admin, Reactor Anal & Safety Div, 73-79, dir prog admin, Energy & Environ Technol Prog, 79-90; CONSULT, 90- *Concurrent Pos:* Prof, Midwest Col Eng, 69-, chmn, Engr Mgt Dept, 69-, acad dean, 72-75, vpres acad affairs, 79-88 & pres, 88-90; chmn, Mgt Div, Am Soc Mech Engrs, 79, Prof Develop, 80 & vpres gen eng, 90-; ed, Eng Mgt Int, Elsevier Publ Co, 85-87. *Honors & Awards:* Edwin F Church Medal, Am Soc Mech Engrs, 85; Distinguished Service Award, Am Soc Mech Engrs, 83. *Mem:* Fel Am Soc Mech Engrs; Am Soc Qual Control; Am Soc Eng Educ; Am Soc Eng Mgt. *Res:* Management of energy and environmental research and development. *Mailing Add:* 5725 Brookbank Rd Downers Grove IL 60516

MARTINEK, GEORGE WILLIAM, b Chicago, Ill, Apr 23, 32. GENETICS. *Educ:* Concordia Teachers Col, Ill, BS, 53; Los Angeles State Col, MA, 60; Univ Calif, Los Angeles, PhD(bot), 68. *Prof Exp:* Teacher, Trinity Lutheran Sch, 53-58; instr biol, Concordia Teachers Col, Ill, 58-62; from asst prof to assoc prof biol, 67-75, PROF BIOL, CALIF STATE POLYTECH UNIV, POMONA, 75- *Mem:* AAAS; Genetics Soc Am; Sigma Xi. *Res:* Genetics of Chlamydomonas reinhardi; recombination. *Mailing Add:* Dept of Biol Sci Calif State Polytech Univ Pomona CA 91768

MARTINEK, ROBERT GEORGE, b Chicago, Ill, Nov 25, 19; m 52. CLINICAL CHEMISTRY. *Educ:* Univ Ill, BS, 41 & 45, MS, 43; Univ Southern Calif, PharmD, 54; Am Bd Bioanal, cert. *Prof Exp:* Pharmacist & pharmaceut chemist, Bates Labs, Inc, 45-47; chemist res lab, Diversey Corp, Victor Chem Works, 47-50; assoc chemist, AMA, 50-55; sr chemist, Mead Johnson & Co, 55-56; clin chemist, Butterworth Hosp, Grand Rapids, Mich, 56-58, Iowa Methodist Hosp, 58-62 & Chicago Dept Health, 62-65; clin biochemist & Chief Lab Improv Sect, Dept Pub Health, Ill, 65-86, dir, 78; RETIRED. *Concurrent Pos:* Assoc ed, J Am Med Technologists, 64-83; consult, Abel Labs & Thornburg Labs, 65; consult, Lab-Line Instruments, 71-, bd dirs, 73-; lectr, dept prev med & community health, Col Med, Univ Ill Med Ctr; ed consult, Med Electronics; mem clin chem adv bd, Ctr Dis Control, USPHS, Atlanta, Ga, 74-86; mem subcomt temperature measurement, Nat Comt Clin Lab Standards 76-86; scientist dir, US Pub Health Serv Reserve Corps, 52-; assoc ed, Jour Med Technol, 84-, AMT Events, 84-86; consult, Sci Supply Co, 87- *Honors & Awards:* Pub Health Serv Award, Ill Assoc Clin Labs, 70. *Mem:* AAAS; AMA; Am Pharmaceut Asn; Am Inst Econ Res; fel Am Inst Chemists; Nat Geog Soc; Sigma Xi. *Res:* Pharmaceutical and detergent chemistry; vitamin assay; sympathomimetic amines; clinical chemistry methodology. *Mailing Add:* 350 W Schaumburg Rd A-326 Schaumburg IL 60194

MARTINELLI, ERNEST A, b Lucca, Italy, Dec 15, 19; nat US; m 46; c 3. NUCLEAR PHYSICS. *Educ:* Univ Calif, BS, 41, PhD(physics), 50. *Prof Exp:* Mem staff radiation lab, Mass Inst Technol, 42-45; instr physics, Stanford Univ, 50-51; physicist radiation lab, Univ Calif, Berkeley, 51-52, Livermore, 52-56; physicist, Aeronutronics Systs, Inc, 56-57 & Rand Corp, 57-71; HEAD DEPT PHYSICS, R&D ASSOCS, 71- *Mem:* Am Phys Soc. *Res:* Nuclear weapon effects; weapon systems. *Mailing Add:* R&D Assocs PO Box 9695 Marina del Rey CA 90291

MARTINELLI, MARIO, JR, b Covington, Va, May 7, 22; m 53; c 1. METEOROLOGY, FORESTRY. *Educ:* Univ Chicago, BS, 44; Duke Univ, MF, 48; State Univ NY, PhD, 56. *Prof Exp:* Forester, Southern Pine Lumber Co, Tex, 48-49; instr & asst, Purdue Univ, 49-54; res meteorologist, Rocky Mountain Forest & Range Exp Sta, US Forest Serv, 54-85; RETIRED. *Mem:* Am Meteorol Soc; Glaciol Soc. *Res:* Watershed management of alpine areas, especially late-lying snow-beds and avalanche research. *Mailing Add:* 2921 Terry Lake Rd Ft Collins CO 80524

MARTINELLI, RAMON U, b Pittsburgh, Pa, Dec 28, 38; m 61. SOLID STATE ELECTRONICS. *Educ:* Dartmouth Col, AB, 60, MS, 62; Princeton Univ, PhD(lasers), 66. *Prof Exp:* MEM TECH STAFF, RCA LABS, 65- *Mem:* Am Phys Soc; Inst Elec & Electronics Engrs. *Res:* Laser and surface physics; electronic photoemission from solids; physical electronics; secondary emission. *Mailing Add:* RCA Labs David Sarnoff Res Ctr Princeton NJ 08543

MARTINEZ, A JULIO, AMEBIC ENCEPHALITIS. *Educ:* Univ Havana, Cuba, MD, 59. *Prof Exp:* PROF PATH, UNIV PITTSBURGH, 75- *Res:* Alzheimer's disease. *Mailing Add:* Div Neuropath Univ Pittsburgh Pittsburgh PA 15213

MARTINEZ, J RICARDO, b El Salvador. EXOCRINE CELL FUNCTION, EPITHELIAL ION TRANSPORT. *Educ:* Tulane Univ, MD, 60. *Prof Exp:* Prof, internal med physiol pharmacol, sch med, Univ El Salvador, 68-72; from assoc prof to prof child health physiol, Sch Med, Univ Mo, 73-88; DIR BIOMED RES, LOVELACE MED FOUND, 88- *Concurrent Pos:* Vis prof, Univ Copenhagen, Denmark, 71, Univ Dundee, Scotland, 77, Max Planck Inst, Frankfurt, 80; res comt mem, Cystic Fibrosis Found, 80-83, study sect mem, oral biol & med study sect, Nat Inst Health, mem, comt prog eval, NIDR, Nat Inst Health, 76-88; ed bd, Archives Oral Biol, 88- *Mem:* Am Physiol Soc. *Res:* Characterization, physiological regulation and ultracellular control of transmembrane ion; transport in exocrine gland cells; signal transduction mechanisms in salivary and pancreatic cells; alteration in ion transport in cystic fibrosis. *Mailing Add:* Lovelace Med Found 2425 Ridgecrest Blvd SE Albuquerque NM 87108

MARTINEZ, JOE L, JR, b Albuquerque, NMex, Aug 1, 44. NEUROBIOLOGY, PSYCHOPHARMACOLOGY. *Educ:* Univ San Diego, BA, 66; NMex Highlands Univ, MS, 68; Univ Del, PhD(physiol psychol), 71. *Prof Exp:* Vis scientist, NEng Regional Primate Res Ctr, 71-72; from asst prof to assoc prof, Dept Psychol, Calif State Col, 72-77; ASSOC RES PSYCHOBIOLOGIST & LECTR, IRVINE, 77-; PROF, DEPT PSYCHOL, UNIV CALIF, BERKELEY. *Concurrent Pos:* Res fel, Univ Del, 67-68; NIMH fel, 75-77; fel, Neth, 79; managing ed, Behav & Neural Biol, 78-; consult ed, Psych Sci & Mex Studies. *Mem:* Fel Am Psychol Asn; Soc Neurosci. *Res:* Neurobiological basis of learning and memory; psychopharmacology; cross cultural psychology; acculturertion. *Mailing Add:* Dept Psychol Univ Calif Berkeley CA 94720

MARTINEZ, JOHN L(UIS), b New Orleans, La, June 19, 22; m 55. MECHANICAL ENGINEERING. *Educ:* Tulane Univ, BE, 43; La State Univ, MS, 50. *Prof Exp:* From instr to assoc prof mech eng, 46-66, asst dean, Sch Eng, 58-76, PROF MECH ENG, TULANE UNIV, 66-, DEAN ADMIS, 76-, DIR PLANNED GIFTS, 83- *Concurrent Pos:* Ford Found lectr, 58-59. *Mem:* Am Soc Eng Educ; Nat Soc Prof Engrs. *Res:* Design, biomechanics. *Mailing Add:* Off Admin Tulane Univ Sch Eng New Orleans LA 70118

MARTINEZ, JORGE, b Willimstadt, Curacao, Nov 1, 45. MATHEMATICS, ALGEBRA. *Educ:* Univ Fla, BA, 66; Tulane Univ, PhD, 70. *Prof Exp:* PROF MATH, UNIV FLA, 69- *Mem:* Am Math Soc. *Mailing Add:* Math Dept Univ of Fla Gainesville FL 32611

MARTINEZ, JOSE E(DWARDO), b Laredo, Tex, Feb 28, 43; m 64; c 3. MECHANICAL ENGINEERING. *Educ:* Tex A&M Univ, BS, 63, ME, 65, PhD(mech eng), 67. *Prof Exp:* Struct engr, Gen Dynamics Corp, 63-64; from instr to assoc prof, 67-75, PROF CIVIL ENG, TEX A&M UNIV, 75- *Concurrent Pos:* Consult, Trans-Alaska Pipeline Serv Co, 68-70; grad faculty assoc mem, Tex A&M Univ, 68-; consult Alyeska Pipeline Serv Co, 70- & Continental Oil Co, 71- *Mem:* Am Soc Civil Engrs; Am Soc Eng Educ; Am Inst Aeronaut & Astronaut. *Res:* Analysis and design of safer highway structures; computer methods in structural mechanics. *Mailing Add:* 2105 Quil Hollow Bryan TX 77801

MARTINEZ, JOSE E(LEAZAR), b Questa, NMex, Jan 13, 22; m 64; c 4. CIVIL ENGINEERING. *Educ:* Univ NMex, BS, 43; Iowa State Univ, MS, 50. *Prof Exp:* From instr to prof civil eng, 51-86, EMER PROF CIVIL ENG, UNIV NMEX, 86- *Mem:* Am Soc Civil Engrs; Am Soc Eng Educ; Nat Soc Prof Engrs. *Res:* Hydrology; applied hydraulics; highway materials; pavement design. *Mailing Add:* 9715 Admiral Nimitz NE Albuquerque NM 87111

MARTINEZ, LUIS OSVALDO, b Havana, Cuba, Dec 27, 27; US citizen; m 55; c 3. RADIOLOGY. *Educ:* Inst Sec Educ, BS, 47; Univ Havana, MD, 54. *Prof Exp:* From instr to prof, 65-76, clin asst prof, 68-70, PROF RADIOL, SCH MED, UNIV MIAMI, 76- *Concurrent Pos:* Counr, Interam Col Radiol, 70-79; chief, Div Diag Radiol, Mt Sinai Med Ctr, 70-, prog dir, Diag Radiol Residency Prog, 70- & assoc dir radiol, 70-; ed, J InterAm Col Radiol. *Honors & Awards:* Recognition Awards, AMA, 71-74; Gold Medal, Interam Col Radiol, 75. *Mem:* Fel Am Col Radiol; Am Roentgen Ray Soc; Radiol Soc NAm; Am Asn Univ Radiologists; Soc Gastrointestinal Radiologists. *Res:* Clinical evaluation of contrast media for intravenous cholangiography. *Mailing Add:* 4300 Alton Rd Miami Beach FL 33140

MARTINEZ, MARIO GUILLERMO, JR, b Havana, Cuba, Mar 6, 24; US citizen; m 49; c 3. ORAL PATHOLOGY. *Educ:* Univ Havana, DDS, 47; Univ Ala, Birmingham, DMD, 64, MS, 68; Am Bd Oral Path, dipl. *Prof Exp:* From instr to asst prof oral path, Sch Dent, Univ Havana, 49-59, prof, 59-60; from asst prof to assoc prof, 67-75, PROF PATH, MED CTR, UNIV ALA, BIRMINGHAM, 71-, DIR CLIN CANCER TRAINING PROG, SCH DENT, 70-, DIR DIV ORAL PATH, MED CTR, 71-, SR SCIENTIST COMPREHENSIVE CANCER CTR, 75- *Concurrent Pos:* Consult, Vet Admin Hosps, Birmingham & Tuscaloosa, Ala; dep med examr, Jefferson County, Ala. *Mem:* Int Asn Dent Res; fel Am Acad Oral Path; fel Am Col Dent; Am Dent Asn; Am Asn Cancer Educ. *Res:* Ultrastructure of giant cell lesion of the jaws; oral oncology. *Mailing Add:* Lyons Harrison Res Bldg 701 S 19th St Birmingham AL 35233

MARTINEZ, OCTAVIO VINCENT, b Jacksonville, Fla, Nov 14, 47. BACTERIOLOGY, CLINICAL MICROBIOLOGY. *Educ:* Univ Miami, BS, 69, PhD(microbiol), 77. *Prof Exp:* Res asst, 77-78, RES INSTR SURG MICROBIOL & RES ASST PROF SURG, SCH MED, 78-, RES ASSOC PROF DEPTS ORTHOP & MICROBIOL & IMMUNOL, UNIV MIAMI, 87- *Mem:* Am Soc Microbiol; Am Soc Clin Path; corresp mem Venezuelan Col Ortho Surg & Traumatology; Infectious Dis Soc Am. *Res:* Surgical microbiology; antimicrobial chemotherapy. *Mailing Add:* Dept Orthop (R-12) Sch Med Univ Miami PO Box 016960 Miami FL 33101

MARTINEZ, RAFAEL JUAN, b Santurce, PR, Feb 28, 27; m; c 3. BACTERIAL PHYSIOLOGY. *Educ:* Univ Southern Calif, AB, 52, PhD, 56. *Prof Exp:* From asst prof to assoc prof bact, 61-69, chmn dept, 71-73, PROF BACT, UNIV CALIF, LOS ANGELES, 69- *Concurrent Pos:* Fulbright fel, 70-71; NIH res grants bact & mycol, 73-77. *Mem:* Am Soc Microbiol; Brit Soc Gen Microbiol. *Res:* Biochemistry of pathogenesis; host-parasite interactions. *Mailing Add:* Dept Microbiol 5304 Life Sci Univ Calif 405 Hilgard Ave Los Angeles CA 90024

MARTINEZ, RICHARD ISAAC, b Havana, Cuba, Aug 16, 44; US citizen; m 78; c 2. ORGANIC REACTION MECHANISMS, NEUTRAL-NEUTRAL & ION-NEUTRAL. *Educ:* McGill Univ, BSc, 64; Univ Calif, Los Angeles, PhD(chem), 76. *Prof Exp:* Lab asst chem, DuPont of Can, Ltd, 62; teaching asst, McGill Univ, 64-65; teaching & res asst, San Diego State Univ, 65-67; chemist, Shell Chem Co, 67-70; res chemist, Univ Calif, Los Angeles, 71-76; RES CHEMIST, DEPT COM, NAT INST STANDARDS & TECHNOL, 76-, PRIN INVESTR, 84- *Concurrent Pos:* Res assoc, Nat Res Coun-Nat Acad Sci, 76-78. *Honors & Awards:* Bronze Medal Award, US Dept Com, 81; I-R 100 Award, Indust Res, 83. *Mem:* Am Chem Soc; AAAS; Am Soc Mass Spectrom; Am Inst Chem. *Res:* Development, standardization and application of MS/MS Tandem Mass Spectrometry for the study of the kinetics of complex organic reaction systems relevant to oxidation and atmospheric chemistry; kinetics and mechanisms of the oxidation chemistry of olefins and sulfur compounds; free-radical reactions; ozone-olefin and ozone-sulfide reactions; established existence of dioxiranes, of metastable excited sulfur dioxide, and of vicinal hydroxy-substituted alkyl and oxoalkyl nitrates and peroxynitrates; patented methods for producing carbocyclic compounds and for flue-gas desulfurization. *Mailing Add:* A260 Chem Nat Inst Standards & Technol Gaithersburg MD 20899

MARTINEZ, ROBERT MANUEL, b Glendale, Calif, Nov 29, 43; m 83; c 3. GENETICS. *Educ:* Niagara Univ, BS, 65; Univ Calif, Berkeley, PhD(genetics), 72. *Prof Exp:* Asst prof biol, Wilkes Col, 71-72; from asst prof to assoc prof biol, 72-85, CHMN, DEPT BIOL, QUINNIPIAC COL, 84-, PROF BIOL, 85- *Concurrent Pos:* Pres, Quinnipiac Fac Fedn, 72-; consult, D'Youville Col, 89. *Mem:* Soc Lit & Sci; Genetics Soc Am; Soc Study Evolution; History Sci Soc; AAAS; Am Inst Biol Sci; fel Soc Health & Human Values. *Res:* Science and literature; history of science. *Mailing Add:* Dept Biol Box 130 Quinnipiac Col Hamden CT 06518

MARTINEZ-CARRION, MARINO, b Felix, Spain, Dec 2, 36; US citizen; m 57; c 2. BIOCHEMISTRY. *Educ:* Univ Calif, Berkeley, BA, 59, MA, 61, PhD(comp biochem), 64. *Prof Exp:* NIH fel biochem, Rome, 64-65; from asst prof to prof chem, Univ Notre Dame, 65-77; PROF BIOCHEM & CHEM DEPT, MED COL VA, VA COMMONWEALTH UNIV, 77- *Concurrent Pos:* NIH career develop award, 72-77; chmn biophysics & biophys chem B study sect, NIH, 77-82; assoc ed, J Protein Chem, 81-; secy gen, Pan-Am Asn, Biochem Soc, 81-; dir, Inst Biotechnol, Va Ctr Innovative Technol, 84-; mem Panel Cell & Molecular Neurobiol, NSF, 85- *Mem:* Spanish Biochem Soc; fel NY Acad Sci; Am Soc Biol Chemists; Am Chem Soc; Biophys Soc; Am Soc Neurochem; Int Soc Neurochem. *Res:* Mechanisms of enzyme action; active center of pyridoxal dependent enzymes; isoenzymes; nuclear magnetic resonance of enzyme-substrate interaction; acetylcholine neuroreceptors; neurochemistry; membrane research; site specific mutagenesis. *Mailing Add:* 109 Biol Sci Bldg Univ Missouri 5100 Rocknell Rd Kansas City MO 64110-2499

MARTINEZ DE PINILLOS, JOAQUIN VICTOR, b Havana, Cuba, Mar 10, 41; US citizen; m 67; c 2. CHEMICAL PHYSICS, CATALYSIS. *Educ:* Univ Miami, BS, 66; Bowling Green State Univ, MA, 68; Univ Fla, PhD(chem physics), 74. *Prof Exp:* Teacher sci & math, Fla City Shelter for Unaccompanied Cuban Children, 62-63; plant chemist anal develop, Cyclo-Chem Corp, 68-70; chmn sci dept, Christopher Columbus High Sch, 64-70; asst physics & chem, Univ Fla, 70-74; develop engr polymer catalysis, Am Cyanamid Co, 74-75; res chemist, Air Prof & Chem, Inc, 75-80, mgr-anal serv, High Purity Mat Lab, 87-88, mgr, Phys Anal Technol Ctr, 88-90, SR RES ASSOC & MGR, ANALYSIS TECHNOL, ELECTRONICS GROUP, AIR PROD & CHEM, INC, 90- *Concurrent Pos:* NSF grants, 67, 68, 69; interim instr, Univ Fla, 73-74. *Mem:* Am Chem Soc; Am Soc Testing & Mat; Nat Assoc Corrosion Engrs; Inst Environ Sci. *Res:* Electron spin resonance of trapped radicals; electron spectroscopy for chemical analysis and auger; catalytic reactions, electron microscopy and surface phenomena; catalysis and process development; corrosion; ultra-high-purity systems; gaseous and particulate contamination as they affect gas deliveries for the electronics industries. *Mailing Add:* 2267 Woodbarn Rd Macungie PA 18062

MARTINEZ-HERNANDEZ, ANTONIO, b Calahorra, Spain, Apr 20, 44; m 68; c 2. EXPERIMENTAL PATHOLOGY, CARDIOVASCULAR PATHOLOGY. *Educ:* Univ Madrid, MD, 68. *Prof Exp:* Instr path, Med Sch, Univ Colo, 73-74, asst prof, 74-77, assoc prof, 77-78; assoc prof, 78-80, prof path, Hahnemann Med Col, 80-86; PROF PATH, THOMAS JEFFERSON UNIV, 86- *Concurrent Pos:* Mem, NIH Pathobiol Study Sect, 78- *Mem:* Am Asn Path; Int Acad Path; Histochem Soc; Cell Biol Soc. *Res:* Biology and pathology of connective tissues; basement membrane, elastic fibers, collagen. *Mailing Add:* Dept Path Jefferson Med Col Tenth & Walnut Sts Philadelphia PA 19107

MARTINEZ-LOPEZ, JORGE IGNACIO, b Santurce, PR, Oct 5, 26; m 50, 83; c 4. INTERNAL MEDICINE, CARDIOLOGY. *Educ:* La State Univ, MD, 50. *Prof Exp:* Intern, Arecibo Dist Hosp, PR, 50-51; physician, Elizabeth, La, 53-54; resident internal med, Charity Hosp, New Orleans, La, 54-57; from instr to prof med, 57-86, PROF EMER, LA STATE UNIV MED CTR, NEW ORLEANS, 86-, PROF MED, TEX TECH UNIV HEALTH SCI CTR, LUBBOCK/EL PASO, TX, 88- *Concurrent Pos:* Vis physician, 57-64, sr vis physician, 64-86, dir, Cardiol Dept, 60-86, mem consult staff, 75, 86, Charity Hosp, New Orleans; chief, Cardiol Clin, Asst Chief, Cardiol Serv, William Beaumont Army Med Ctr, mem consult staff, 86-88; med staff, R E Thomason Gen Hosp, El Paso, Tex. *Mem:* Fel Am Heart Asn; fel Am Col Chest Physicians; fel Am Col Physicians; fel Am Col Cardiol. *Res:* Clinical cardiology; cardiac catheterization and other special diagnostic procedures; electrocardiography. *Mailing Add:* 7740 Cedar Breaks Lane El Paso TX 79904-3522

MARTINEZ-LOPEZ, NORMAN PETRONIO, b Managua, Nicaragua, Mar 31, 43; c 4. DENTISTRY. *Educ:* Univ Costa Rica, DDS, 64; Marquette Univ, MS, 68, MEd, 75, PhD(curric instr), 77; Am Bd Pedodont, dipl, 79. *Prof Exp:* Asst prof dent, Nat Univ Nicaragua, 68-73; asst prof, Marquette Univ, 73-76; assoc prof, 76-81, PROF DENT, SOUTHERN ILL UNIV, EDWARDSVILLE, 81-, HEAD, SECT PEDIAT DENT, 76- *Mem:* Am Dent Asn; Int Asn Dent Res; Am Acad Pediat Dent; Am Asn Educ Res; Fedn Dentaire Int. *Res:* Caries; educational development in higher education; role of behavioral science in dentistry. *Mailing Add:* 2800 College Bldg 274 Alton IL 62002

MARTINEZ-MALDONADO, MANUEL, b Yauco, PR, Aug 25, 37; m 59; c 4. INTERNAL MEDICINE, NEPHROLOGY. *Educ:* Univ PR, San Juan, BS, 57; Temple Univ, MD, 61. *Prof Exp:* Intern, St Charles Hosp, Toledo, Ohio, 61-62; resident internal med, Vet Admin Hosp & Sch Med, Univ PR, San Juan, 62-65; USPHS fel, Univ Tex Southwestern Med Sch Dallas, 65-67; Lederle Labs int fel, 66-67; instr, Univ Tex Southwestern Med Sch Dallas, 67-68; from asst prof to prof, Baylor Col Med, 68-73; actg chmn dept physiol, Sch Med, Univ PR, 74-79, prof Med & Physiol, 72-90; CHIEF MED SERV, ATLANTA VET ADMIN MED CTR, 90- *Concurrent Pos:* Dir chronic dialysis unit, Parkland Mem Hosp, Dallas, Tex, 67-68; attend physician, Ben Taub Gen Hosp, Houston, 68-73 & Methodist Hosp, Houston, 69-73; assoc chief staff for res, Vet Admin Hosp, San Juan, 73, chief med serv, 74-90; mem nat adv bd, Nat Inst Arthritis, Metabolism, Digestive Dis & Kidneys, 79-82; vis prof med, Harvard Univ, 79-80, Vanderbilt Sch Med, 89-90; mem sci adv bd, Nat Kidney Found, 81-84; mem, Gen Med Study B Sect, NIH, chmn, 84-88; vchmn & prof med, Sch Med, Emory Univ, 90-; mem, US Pharmacopeia Cardiovasc & Renal Drugs Adv Panel, 90-95. *Mem:* Inst Med-Nat Acad Sci; Am Soc Clin Invest; Cent Soc Clin Res; Am Fedn Clin Res; Am Soc Nephrology; Am Physiol Soc; Asn Am Physicians; AAAS. *Res:* Renal physiology; electrolyte metabolism; biochemistry of transport; control of renin secretion. *Mailing Add:* Med Serv Atlanta Vet Admin Med Serv 1670 Clairmont Rd Decatur GA 30033

MARTINI, CATHERINE MARIE, b New York, NY, July 7, 24. PHYSICAL CHEMISTRY. *Educ:* Hunter Col, BA, 46; Univ Pa, MS, 48. *Prof Exp:* Tutor chem, Hunter Col, 47-49; asst, 49-53, res assoc, 53-60, res chemist, 61-66, SR RES CHEMIST, STERLING-WINTHROP RES INST DIV, STERLING DRUG, INC, 66- *Mem:* Am Chem Soc; Coblentz Soc. *Res:* Infrared, ultraviolet and nuclear magnetic resonance spectroscopy of organic molecules. *Mailing Add:* 36 Albin Rd Delmar NY 12054

MARTINI, IRENEO PETER, b Dec 14, 35; Can citizen; m 62; c 2. SEDIMENTOLOGY, GLACIAL GEOLOGY. *Educ:* Univ Florence, DrGeolSci, 61; McMaster Univ, PhD(geol), 66. *Prof Exp:* From geologist to sr geologist, Shell Can Ltd, 66-69; PROF SEDIMENTOLOGY, UNIV GUELPH, 69- *Mem:* Soc Econ Paleontologists & Mineralogists; Int Asn Sedimentol. *Res:* Sedimentary geology; sedimentology of Recent and Pleistocene clastic sediments and ancient sedimentary rocks; fabric of soils and sediments; analysis of hydrocarbon potentials of selected regions; study of cold-climate sediments and peat; study of L Permian cool measures. *Mailing Add:* Dept Land Resource Sci Univ Guelph Guelph ON N1G 2W1 Can

MARTINI, MARIO, b Florence, Italy, Mar 24, 39; m 66; c 3. PHYSICS, RESEARCH MANAGEMENT. *Educ:* Univ Bologna, Italy, Dr(physics), 62. *Prof Exp:* Asst prof physics, Univ Bologna, 62-68; assoc prof, Univ Modena, 69-70; tech dir, Simtec Indust, Montreal, 71-73, NRD Div Electron Assoc Can Ltd, 73-74; TECH DIR, NUCLEAR TECHNOL DIV, EG&G ORTEC, OAK RIDGE, TENN, 75-, MKT MGR, DETECTORS. *Concurrent Pos:* Fel physics, Chalk River Nuclear Labs, Ont, 68-69. *Mem:* Sr mem Inst Elec & Electronics Engrs; Can Asn Physicists. *Res:* Semiconductor physics (transport properties), interaction of ionizing radiation with semiconductors. *Mailing Add:* EE&G Ortec Inc 100 Midland Rd Oak Ridge TN 37830

MARTINI, WILLIAM ROGERS, chemical & mechanical engineering, for more information see previous edition

MARTINO, JOSEPH PAUL, b Warren, Ohio, July 16, 31; m 57; c 3. OPERATIONS RESEARCH, SCIENCE ADMINISTRATION. *Educ:* Miami Univ, AB, 53; Purdue Univ, MS, 55; Ohio State Univ, PhD(math), 61. *Prof Exp:* With US Air Force, 53-75, proj engr, Wright Air Develop Ctr, 55-58, mathematician, Air Force Off Sci Res, 60-62, staff scientist, 63-67, opers analyst, Res & Develop Field Univ, Bangkok, Thailand, 62-63, chief tech anal div, Air Force Off Res Anal, 68-71, staff scientist, Avionics Lab, Wright-Patterson AFB, 72-73, dir eng standardization, Defense Electronics Supply Ctr, 73-75; res scientist, 75-80, SR RES SCIENTIST, RES INST, UNIV DAYTON, 80- *Mem:* Fel AAAS; fel Inst Elec & Electronics Eng; assoc fel Am Inst Aeronaut & Astronaut; Inst Mgt Sci. *Res:* Application of operations research to problems of technological change, with emphasis on technological forecasting. *Mailing Add:* Research Inst, Univ Dayton 300 College Park Dayton OH 45469-0120

MARTINS, DONALD HENRY, b Poplar Bluff, Mo, July 31, 45; m 69; c 1. ASTROPHYSICS, PHYSICS. *Educ:* Univ Mo, Columbia, BS, 67, MS, 69; Univ Fla, PhD(astron), 74. *Prof Exp:* Res assoc astron, Nat Res Coun, Johnson Space Ctr, 74-76; res assoc, Houston Baptist Univ, 76; adj fac mem, Univ Houston, 76-77; asst prof astron & physics, Univ Ga, 77-; AT DEPT PHYS CHEM & ASTRON, UNIV ALASKA. *Mem:* Am Astron Soc; Am Inst Physics; Astron Soc Pac; Int Astron Union. *Res:* Active in photoelectric photometry of variable stars; photographic and photoelectric surface photometry of galaxies and globular clusters, with emphasis on nuclear structure. *Mailing Add:* Dept Phys Chem & Astron Univ Alaska 3221 Providence Dr Anchorage AK 99508

MARTINSEK, ADAM THOMAS, b Jersey City, NJ, Dec 30, 53; m 77; c 2. SEQUENTIAL INFERENCE. *Educ:* Harvard Univ, AB, 76; Univ Chicago, MS, 77; Columbia Univ, PhD(statist), 81. *Prof Exp:* Asst prof, 81-86, ASSOC PROF STATIST, UNIV ILL, 86- *Mem:* Am Statist Asn; Inst Math Statist. *Res:* Sequential inference; adaptive methods; density estimation. *Mailing Add:* Dept Statist Univ Ill 725 S Wright St Champaign IL 61820

MARTINS-GREEN, MANUELA M, b Luso, Mexico, Portugal, Dec 30, 47; US citizen. CELL GROWTH REGULATION & INVASION. *Educ:* Univ Lisbon, BS, 70; Univ Calif, MS, 75, PhD, 87. *Prof Exp:* Postdoctoral fel, NRSA, Univ Calif, Berkeley, 87-91; STAFF RES SCIENTIST, LAWRENCE BERKELEY LAB, UNIV CALIF, 91- *Concurrent Pos:* Fulbright travel grant, 73-75; nat res serv award, NIH, 88-91; postgrad researcher, Univ Calif, Davis, 89-91; adj asst prof cell biol, Rockefeller Univ, 91- *Mem:* Am Soc Cell Biol; Soc Develop Biol; Am Cancer Soc; Electron Micros Soc Am. *Res:* Processes involved in cell growth regulation and invasion, whether during embryonic development, wound healing or pathological conditions such as cancer. *Mailing Add:* 1225 Purdue Dr Davis CA 95616

MARTINSON, CANDACE, b Cleveland, Ohio, Jan 12, 49. ENTOMOLOGY. *Educ:* Ohio State Univ, BS, 71, MS, 74, PhD(entom), 77. *Prof Exp:* Technician, Dept Entom, Ohio State Univ, 72-73, asst cur entom, Insects & Spiders, 77-82; programmer analyst, Dollar Savings Bank, 83-88; PROGRAMMER ANALYST, NATIONWIDE INS, 88- *Concurrent Pos:* Illusr entom, NSF grant, 71-72. *Mem:* Audobon Soc. *Res:* Systematics of leafhoppers. *Mailing Add:* 85 Acton Rd Columbus OH 43214

MARTINSON, CHARLIE ANTON, b Orchard, Colo, Sept 15, 34; m 57; c 4. PLANT PATHOLOGY. *Educ:* Colo State Univ, BS, 57, MS, 59; Ore State Univ, PhD(plant path), 64. *Prof Exp:* Asst prof plant path, Cornell Univ, 63-68; ASSOC PROF, IOWA STATE UNIV, 68- *Concurrent Pos:* Consult, Corn Prod Syst, Inc, 72-79. *Mem:* Fel AAAS; Am Phytopath Soc. *Res:* Root diseases of economically important crops; biological control of plant disease; corn diseases; role of toxins in pathogenesis; aflatoxins; disease resistance breeding; international programs in corn production and plant disease control. *Mailing Add:* Dept Plant Path Iowa State Univ Ames IA 50011

MARTINSON, EDWIN O(SCAR), b Seattle, Wash, Mar 10, 10; m 41; c 2. ELECTRICAL ENGINEERING. *Educ:* Univ Wash, BS, 32. *Prof Exp:* Mech engr, Mason Walsh Atkinson Kier Co, Wash, 34-36 & Tenn Valley Authority, 36-37; chief engr, DeBothezat Ventilating Equip Div, Am Mach & Metals, Inc, Ill, 37-41; pres & gen mgr, C S Johnson Co Div, Koehring Co, 41-48, vpres & chief engr, 48-53, pres & gen mgr, Koehring-Waterous, Ltd Div, 53-57, vpres res, Develop & Mfg Cent Off, 57-68; OWNER, MARTINSON ENG, 68-, INTER-ACTIVE CONTROLS, INC, 68- *Mem:* Soc Automotive Engrs; Nat Soc Prof Engrs. *Res:* Construction machinery; hydrostatic hydraulic devices; machines for concrete and asphalt industries; new type power operated indoor swim pools and propellers. *Mailing Add:* 6615 N River Rd Milwaukee WI 53217

MARTINSON, HAROLD GERHARD, b Hartford, Conn, Sept 9, 43; m 68; c 2. MOLECULAR BIOLOGY. *Educ:* Augsburg Col, BA, 65; Univ Calif, Berkeley, PhD(molecular biol), 71. *Prof Exp:* Fel biol, Univ Lethbridge, 71-73 & biochem, Univ Calif, San Francisco, 73-75; ASST PROF CHEM, UNIV CALIF, LOS ANGELES, 75- *Res:* Chromosome structure and chemistry; control of gene expression in eucaryotes. *Mailing Add:* Dept of Chem Univ of Calif/3010 Young Hall 405 Hilgard Ave Los Angeles CA 90024

MARTINSON, IDA MARIE, b Mentor, Minn, Nov 8, 36; m 62; c 2. HOMECARE, BEREAVEMENT. *Educ:* Univ Minn, BS, 60, MNA, 62; Univ Ill, PhD(physiol), 72. *Prof Exp:* Instr nursing, Thorton Jr Col, 67-69; lectr physiol, Med Sch, Univ Minn, 72-82, asst prof nursing, 72-74, assoc prof & dir res, 75-77, prof & dir res nursing, Sch Nursing, 77-82; chmn, 82-90, PROF, DEPT FAMILY NURSING, UNIV CALIF, SAN FRANCISCO, 82- *Concurrent Pos:* Res consult, Clin Ctr, Bethesda, 77-78; proj dir, HEW, 75-77 & 77-79; prin investr, Nat Cancer Inst, 76-79 & Am Cancer Soc, 78-; mem, Adv Coun, Nat Inst Aging, 80-83; mem, Geriat & Geront Adv Comt, Vet Admin, 81-85. *Mem:* Inst Med-Nat Acad Sci; Am Nurses Asn; Sigma Xi. *Res:* Home care for the child with cancer; psychosocial impact of childhood cancer on child-family; childhood cancer. *Mailing Add:* Dept Family Nursing Univ Calif N411Y San Francisco CA 94143-0606

MARTINSONS, ALEKSANDRS, b Russia, Nov 30, 12; US citizen; m 39; c 2. ELECTROCHEMISTRY. *Educ:* Univ Mich, MS, 55. *Prof Exp:* Sr res chemist, Am Potash & Chem Corp, 57-60 & Chem Div, PPG Industs Inc, 60-77; RETIRED. *Mem:* Electrochem Soc. *Res:* Overvoltage; electrowinning and metal deposition; thin film coatings. *Mailing Add:* 135 Westview Ave Wadsworth OH 44281-1132

MARTIRE, DANIEL EDWARD, b New York, NY, June 3, 37; m 61. PHYSICAL CHEMISTRY, ANALYTICAL CHEMISTRY. *Educ:* Stevens Inst Technol, BE, 59, MS, 60, PhD(chem), 63. *Prof Exp:* Instr chem, Stevens Inst Technol, 62-63; NSF fel, Cambridge Univ, 63-64; from asst prof to assoc prof, 64-74, PROF CHEM, GEORGETOWN UNIV, 74- *Concurrent Pos:* Vis prof, Col France, 72, Univ Col Swansea, UK, 76 & Fed Polytech Sch, Lausanne, Switz, 80. *Mem:* Am Asn Univ Profs; Am Chem Soc. *Res:* Thermodynamics and statistical mechanics of liquid crystals and liquid mixtures; theory of retention and selectivity in gas, liquid and supercritical-fluid chromatography; weak organic complexes. *Mailing Add:* Dept Chem Georgetown Univ Washington DC 20057

MARTLAND, CARL DOUGLAS, b Providence, RI, Sept 22, 46; m 69; c 1. CIVIL ENGINEERING. *Educ:* Mass Inst Technol, SB, 68, SM & CE, 72. *Prof Exp:* Res engr, 72-77, res assoc, 77-81, PRIN RES ASSOC & LECTR, MASS INST TECHNOL 81- *Concurrent Pos:* Sr consult, Multisyst, Inc, 78-83, pres, Transp Res Forum, 85-86; mem, Transportation Res Bd, Comt Railroad Intermodal Terminal Design 86-; dir, Rail Group, Ctr Transportation Studies, Mass Inst Technol, 80- *Mem:* Am Railroad Eng Asn; AAAS. *Res:* Freight transportation, with a special emphasis on the use of the techniques of systems analysis in operations management, especially service reliability and equipment utilization. *Mailing Add:* 56 Fairview St Boston MA 02131

MARTNER, SAMUEL (THEODORE), b Prairie du Chien, Wis, Apr 20, 18; m 42, 85; c 2. GEOPHYSICS. *Educ:* Univ Calif, BA, 40, Calif Inst Technol, MS, 46, PhD(geophys), 49. *Prof Exp:* From prod engr to prod control mgr, Los Angeles Shipbldg & Drydock Co, 41-43; ship supvr & asst to repair gen mgr, Todd Shipyards Corp, 43-45; geologist, Standard Oil Co Calif, 46; geologist, Stanolind Oil & Gas Co Div, Standard Oil Co, Ind, 47-48, seismic interpreter, 48, asst party chief, 48-49, party chief, 49-50, tech group supvr, 51-52, res group supvr, 52-58, res sect supvr, Pan Am Petrol Corp, 58-64, div geophysicist, 65-67, asst chief geophysicist, 67-68, geophys res dir, 68-71, res consult, 71-81; CONSULT, 81- *Concurrent Pos:* Consult, Amoco Prod Co, 65-81. *Mem:* Seismol Soc Am; Soc Explor Geophys; Geol Soc Am; Am Geophys Union; Sigma Xi; Europ Asn Explor Geophys. *Res:* Petroleum; earth sciences. *Mailing Add:* 2520 Woodward Blvd Tulsa OK 74114

MARTO, PAUL JAMES, b Flushing, NY, Aug 15, 38; m 61; c 4. HEAT TRANSFER. *Educ:* Univ Notre Dame, BS, 60; Mass Inst Technol, SM, 62, ScD(nuclear eng), 65. *Prof Exp:* From asst prof to prof, 65-77, chmn mech eng, 78-86, DISTINGUISHED PROF, NAVAL POSTGRAD SCH, 85-, DEAN RES, 90- *Concurrent Pos:* Consult, Lawrence Radiation Lab, SKF Industs, Union Carbide Agr Systs, US Dept Energy, Fr Ferodo Ltd & Beltran Assocs, Inc & Modine Mfg Co; vis prof, US Naval Acad, 84-85 & Univ London, 87-88; NATO sr guest scientist fel, Centre D'etudes Nuclear de Grenoble, France, 88, Assoc Tech Ed, Asme Jour of Heat Transfer, 83-89; assoc tech ed, Am Soc Mech Eng J Heat Transfer, 83-89. *Honors & Awards:* Alexander von Humboldt US Sr. Scientist Award, Ger, 89-90. *Mem:* Fel Am Soc Mech Engrs; Am Soc Eng Educ; Am Soc Naval Engrs; Sigma Xi. *Res:* Heat transfer; boiling, condensation; heat pipe operation. *Mailing Add:* Naval Postgrad Sch Monterey CA 93943

MARTON, JOHN PETER, solid state electronics, for more information see previous edition

MARTON, JOSEPH, b Budapest, Hungary, Mar 5, 19; US citizen; m 49; c 1. WOOD AND FIBER CHEMISTRY PAPER CHEMISTRY. *Educ:* Pazmany Peter Univ, Hungary, BS & MS, 42, PhD(org chem), 43. *Prof Exp:* Asst prof org chem, Budapest Tech Univ, 45-47, lectr, 49-56; chemist, Arzola Chem Co, 47-49; res supvr, Res Inst Indust Org Chem, 49-56; res fel wood chem, Chalmers Univ Technol, Sweden, 56-60; res assoc wood & phys org chem, Charleston Res Lab, Westvaco Corp, SC, 60-66, res assoc, Phys Org & Surface Chem, 66-72, sr res assoc & paper making chemist, Fiber Sci, Laurel Res Ctr, 72-87; CONSULT, J&T ASSOCS, SILVER SPRING, MD, 87- *Concurrent Pos:* Adj prof paper chem, State Univ NY, Syracuse, 83- *Honors & Awards:* Paper & Bd Mfg Div Award & Harris O Ware Prize, 86. *Mem:* AAAS; Am Chem Soc; fel Tech Asn Pulp & Paper Indust; Can Pulp & Paper Asn; fel Am Inst Chem. *Res:* Chemistry and reaction of polymeric compounds; analysis and chemistry of fiber surfaces; colloid and surface chemistry of pulp and papermaking and sizing, wet end chemistry; printability of paper; forest improvement, lignin chemistry, enzyme reactions. *Mailing Add:* J&T Assocs 10705 Meadowhill Rd Silver Spring MD 20901-1529

MARTON, LAURENCE JAY, b Brooklyn, NY, Jan 14, 44; m 67; c 1. LABORATORY MEDICINE, CANCER. *Educ:* Yeshiva Univ, BA, 65; Albert Einstein Col Med, MD, 69. *Prof Exp:* Clin assoc cancer med, Baltimore Cancer Res Ctr, Nat Cancer Inst, 71-73; asst res biochemist, Brain Tumor Res Ctr, Dept Neurosurg, 73-74, asst clin prof, Dept Lab Med & Neurosurg, 74-75, asst prof, 75-78, assoc prof, 78-79, asst dir clin chem, Dept Lab Med, 74-75, dir div clin chem, 75-78, actg chmn, 78-79, PROF, LAB MED & NEUROSURG, UNIV CALIF, SAN FRANCISCO, 79-, CHMN, DEPT LAB MED, 79- *Concurrent Pos:* Nat Cancer Inst res career develop award, 75-80 & res grant, 75-; chmn, Gordon Res Conf on Polyamines, 87. *Mem:* Am Asn Cancer Res; Am Asn Clin Chem; AAAS; Acad Clin Lab Physicians & Scientists; Asn Path (chmn); Am Asn Pathologists; Int Acad Path; US & Can Acad Path. *Res:* Biochemical markers for brain tumors; molecular and cellular biology of the polyamines; polyamine biosynthesis inhibitors and analogs as cancer chemotherapeutic agents; polyamine and nucleic acid interactions. *Mailing Add:* Dept Lab Med Univ Calif San Francisco CA 94143-0134

MARTON, RENATA, b Krakow, Poland, July 27, 10; US citizen; m 38; c 2. CHEMISTRY, PAPER SCIENCE ENGINEERING. *Educ:* Univ Jagello, Poland, MS, 34, PhD, 36. *Prof Exp:* Asst chem, Univ Jagello, Poland, 34-38; res asst cellulose derivatives, Inst Chem, Sorbonne, 38-43; from researcher to tech mgr, Chem Indust, Hungary, 45-50; head dept, Pulp & Paper Res Inst, 50-56; Rockefeller Found fel, Austrian Wood Res Inst, 56-57; from asst prof to prof pulp & paper res, State Univ NY Col Forestry, 57-77; prof & sr res assoc, Empire State Paper Res Inst, 77-91; at dept paper sci eng, State Univ NY, Col Environ Sci & Forestry, 91; RETIRED. *Honors & Awards:* Silver Medal, Pulp & Paper Indust, 77 & 80. *Mem:* Tech Asn Pulp & Paper Indust. *Res:* Fundamentals of pulp and papermaking fibers; morphology and nature of coloring materials in wood; pulping technology. *Mailing Add:* Dept Paper Sci Eng State Univ NY Col Environ Sci & Forestry Syracuse NY 13210

MARTONOSI, ANTHONY, b Szeged, Hungary, Nov 7, 28; US citizen; m 59; c 4. BIOCHEMISTRY. *Educ:* Univ Szeged, MD, 53. *Prof Exp:* Asst prof physiol, Univ Szeged, 54-57; Nat Acad Sci res fel biochem, Mass Gen Hosp, Boston, 57-59; assoc, Retina Found, 59-63, asst dir, 64-65; prof biochem, Sch Med, St Louis Univ, 65-79; PROF BIOCHEM, STATE UNIV NY, UPSTATE MED CTR, 79- *Concurrent Pos:* USPHS grant, 59-; estab investr, Am Heart Asn, 61-66; NSF grant, 63- *Mem:* Am Soc Biol Chem; Biophys Soc. *Res:* Biochemistry of muscle contraction; contractile proteins; structure and function of membranes. *Mailing Add:* Dept Biochem Upstate Med Ctr State Univ NY 155 Elizabeth Blackwell St Syracuse NY 13210

MARTSOLF, J DAVID, b Beaver Falls, Pa, Nov 26, 32; m 55; c 3. AGRICULTURAL METEOROLOGY. *Educ:* Univ Fla, BSA, 54, MSA, 62; Univ Mo-Columbia, PhD(atmospheric sci), 66. *Prof Exp:* Asst county agr agt, Agr Exten Serv, Univ Fla, 58-62, asst prof hort, 62-64; res asst surface energy balance study, Univ Mo-Columbia, 64-66; assoc prof agr climat, Pa State Univ, University Park, 66-77, prof microclimat, 77-79; PROF CLIMAT, INST FOOD & AGR SCI, UNIV FLA, GAINESVILLE, 79- *Honors & Awards:* Caroll R Miller Award, Am Soc Hort Sci; Technol Award, NASA, 82 & Cert Appreciation, 83. *Mem:* Am Meteorol Soc; Am Soc Hort Sci. *Res:* Frost protection of citrus and other tree crops; modification of plant microclimate; general energy balance of vegetative canopies. *Mailing Add:* Inst Food & Agr Sci 2121 Fifield Hall Univ Fla Gainesville FL 32611

MARTT, JACK M, b Ashland, Ky, Nov 9, 22; m 48; c 2. INTERNAL MEDICINE, CARDIOLOGY. *Educ:* Univ Mo, BS, 43; Wash Univ, MD, 46. *Prof Exp:* Asst prof internal med, Col Med, Univ Iowa, 55-56; from asst prof to prof, Sch Med, Univ Mo-Columbia, 56-69, dir cardiopulmonary lab, Med Ctr, 65-69; CARDIOLOGIST, SCOTT & WHITE CLIN, 69-; PROF INTERNAL MED, TEX A&M SCH MED, 80- *Concurrent Pos:* Fel coun clin cardiol, Am Heart Asn, 63. *Mem:* Fel Am Col Physicians; fel Am Col Cardiol. *Res:* Cardiology research primarily in atherosclerosis. *Mailing Add:* 2401 S 31st St Temple TX 76508

MARTUCCI, JOHN A, b Charleroi, Pa, Sept 21, 32; m 81; c 4. NUCLEAR POWER PLANT CHEMISTRY, WATER CHEMISTRY FOR POWER GENERATION. *Educ:* Univ Pittsburgh, BS, 54; Carnegie Inst Technol, BSIM, 65. *Prof Exp:* Chemist, Duquesne Light Co, Shippingport Atomic Power Sta, Pa, 56-57, actg radiochemist, 57-58, radiochemist, 58-60, reactor control chemist, 60-66; res engr, Kreisinger Develop Lab, Combustion Eng Inc, 66-67, sr proj engr, 67-68, mgr chem group, Nuclear Power Lab, 68-72, asst to dir, Nuclear Labs, 72-73; dir, Manor Mining & Minerals Co, 66-72, pres & dir, 73-80; DIR ENERGY RESOURCES, INTER-POWER, PA, 87- *Concurrent Pos:* Secy & treas, Laurel Ridge Coal Inc, 75-80; consult, 80-87; gov, Coal Conf, 80; Coal Workgroup, 88. *Mem:* Fel Am Inst Chem; Am Soc Testing & Mat; Am Nuclear Soc; Nat Asn Corrosion Engrs; United Surface Mine Operators. *Res:* Chemistry and corrosion problems for nuclear power plant technology, specifically stress corrosion, fission product release studies, neutron activation product studies and high temperature behavior of inorganic materials in aqueous solutions for steam generation; radioactive waste preparation and management; coal chemical technology; hydrometallurgy of precious metals. *Mailing Add:* 619 Fallowfield Ave R Charleroi PA 15022-1903

MARTY, ROBERT JOSEPH, b Evanston, Ill, July 6, 31; m 62; c 2. FOREST ECONOMICS. *Educ:* Mich State Univ, BS, 54; Duke Univ, MF, 55; Harvard Univ, MPA, 59; Yale Univ, PhD(forestry), 62. *Prof Exp:* Res forester economics, Forest Serv, USDA, 55-65, chief forest econ br, 65-67; assoc prof forestry, 67-71, PROF FORESTRY, MICH STATE UNIV, 71-, PROF RESOURCE DEVELOP, 77- *Mem:* Soc Am Foresters; Econ Asn. *Res:* Timber production economics; economics of public, natural resource programs and policies. *Mailing Add:* Dept Forestry 126 Natural Resource Mich State Univ East Lansing MI 48824

MARTY, ROGER HENRY, b Sterling, Ohio, Oct 16, 42; m 64; c 2. TOPOLOGY, MATHEMATICS GENERAL. *Educ:* Kent State Univ, BS, 64; Pa State Univ, MS, 66, PhD(math), 69. *Prof Exp:* Asst prof math, 69-73, ASSOC PROF MATH, CLEVELAND STATE UNIV, 73- *Concurrent Pos:* Cleveland State Univ grantee, 77, 80 & 89; Ohio Bd of Regents grantee, 86. *Mem:* Am Math Soc; Math Asn of Am; NY Acad Sci; Nat Coun Teachers of Math; Sigma Xi. *Res:* Set-theoretic topology; general topology; set-theory; math education. *Mailing Add:* Dept of Math Cleveland State Univ Cleveland OH 44115

MARTY, WAYNE GEORGE, b LuVerne, Iowa, Feb 14, 32; m 54; c 3. PARASITOLOGY. *Educ:* Westmar Col, BA, 53; Univ Iowa, MS, 59, PhD(zool), 62. *Prof Exp:* Teacher high sch, Iowa, 55-57; PROF BIOL, WESTMAR COL, 59- *Concurrent Pos:* NIH fel malariology, 69-70; instr & researcher, Silliman Univ, Philippines, 78-79. *Mem:* Am Soc Parasitol; Nat Asn Biol Teachers; Sigma Xi. *Res:* Experimental infections of parasites in abnormal hosts, specifically Trichinella Spiralis in chickens. *Mailing Add:* RR 2 LeMars IA 51031

MARTZ, BILL L, b Anderson, Ind, Jan 17, 22; m 48; c 3. MEDICINE. *Educ:* DePauw Univ, AB, 44; Ind Univ, MD, 45; Am Bd Internal Med, dipl. *Prof Exp:* Dir clin invest, Dow Chem Co, 74-83; asst med serv, 53-55, from asst prof to assoc prof, 55-67, PROF MED, SCH MED, IND UNIV, INDIANAPOLIS, 67- *Concurrent Pos:* Assoc med serv, Indianapolis Gen Hosp, 53-55, mem vis staff, Med Serv, 56-; mem, Coun High Blood Pressure Res, Am Heart Asn; res physician, Marion County Gen Hosp, 51-60, dir, Lilly Lab Clin Res, 60-72; chief med, Kansas City Gen Hosp, 72-74. *Mem:* AMA; Am Col Physicians; Am Fedn Clin Res; Am Col Cardiol; Am Soc Clin Pharmacol & Therapeut. *Res:* Cardiovascular renal disease and clinical pharmacology. *Mailing Add:* 216 W Tilden Rd Brownsburg IN 46112

MARTZ, DOWELL EDWARD, b Livonia, Mo, Sept 29, 23; m 50; c 4. PHYSICS. *Educ:* Union Col, Nebr, BA, 50; Vanderbilt Univ, MS, 53; Colo State Univ, PhD, 68. *Prof Exp:* Physicist, US Naval Ord Lab, 53-61; assoc prof physics, Pac Union Col, 61-62; sr res physicist, Calif Inst Technol, 62-64; assoc prof physics, 64-74, PROF PHYSICS, PAC UNION COL, 74- *Concurrent Pos:* Consult, Calif Inst Technol, 61- & Ames Res Lab, NASA, 64-65. *Mem:* Optical Soc Am; Sigma Xi; Am Inst Physics. *Res:* Infrared optics, detectors, astronomy and space research; application to long wavelength. *Mailing Add:* 546 Shank Ct Grand Junction CO 81503

MARTZ, ERIC, b Columbus, Ind, Apr 30, 40; m 71; c 2. CELLULAR IMMUNOLOGY. *Educ:* Oberlin Col, AB, 63; Johns Hopkins Univ, PhD(biol), 69. *Prof Exp:* Fel, dept biol, Princeton Univ, 69-70; fel, dept pathol, Harvard Med Sch, 70, assoc prof immunol, 77-81; ASSOC PROF IMMUNOL, DEPT MICROBIOL, UNIV MASS, AMHERST, 81- *Concurrent Pos:* Assoc ed, J Immunol, 77-79. *Mem:* Am Asn Immunologists. *Res:* Mechanism of T lymphocyte mediated killing; cell adhesion; role of calcium in lymphocyte function. *Mailing Add:* Dept Microbiol Univ Mass Amherst MA 01003

MARTZ, FREDRIC A, b Columbia City, Ind, May 24, 35; m 59; c 5. ANIMAL NUTRITION, DAIRY SCIENCE. *Educ:* Purdue Univ, BS, 57, MS, 59, PhD(dairy sci), 61. *Prof Exp:* Instr dairy sci, Purdue Univ, 60-61; from asst prof to assoc prof, 61-73, forage livestock res coord agr, 73-78, chmn, dairy sci dept, 78-82, PROF DAIRY SCI, UNIV MO-COLUMBIA, 73-; SUPT, NMO CENTERS, 90- *Concurrent Pos:* NIH grant, 69; vis assoc prof, Cornell Univ, 71-72; USDA grants, 80 & 81; res scientist, USDA, 82-90; Mo Dept Com grant, 87. *Honors & Awards:* Res Award, NSF, 75. *Mem:* Am Inst Nutrit; Am Dairy Sci Asn; Am Soc Animal Sci; Am Forage & Grassland Soc. *Res:* Digestibility of feedstuffs for ruminant; regulation of food intake in ruminant; recycling of fibrous wastes through ruminant feeds; forage utilization; mineral bioavailability. *Mailing Add:* Dept Animal Sci S-142 Animal Res Ctr Univ Mo Columbia MO 65211

MARTZ, HARRY EDWARD, JR, b Albany, NY, Feb 14, 57; m 82. PERTURBED ANGULAR CORRELATION MEASUREMENTS. *Educ:* Siena Col, BS, 79; Fla State Univ, MS, 83, PhD(nuclear chem), 86. *Prof Exp:* NUCLEAR CHEM COMPUT TOMOGRAPHY, NONDESTRUCTIVE EVAL DIV, LAWRENCE LIVERMORE NAT LAB, 86- *Mem:* Am Phys Soc; Am Chem Soc. *Res:* Computed tomography and x-ray radiography for nondestructive evaluation of various materials and parts before and after assembly. *Mailing Add:* Nondestructive Eval Sect Lawrence Livermore Nat Lab L-333 PO Box 808 Livermore CA 94550

MARTZ, HARRY FRANKLIN, JR, b Cumberland, Md, June 16, 42; m 64; c 2. STATISTICS, OPERATIONS RESEARCH. *Educ:* Frostburg State Col, BS, 64; Va Polytech Inst, PhD(statist), 68. *Prof Exp:* From asst prof to assoc prof indust eng & statist, Tex Tech Univ, 67-77; MEM STAFF, ENERGY SYSTS & STATIST, LOS ALAMOS SCI LAB, 77- *Concurrent Pos:* NASA grants, 69-71. *Mem:* AAAS; Inst Math Statist; Am Inst Indust Eng; Am Statist Asn. *Res:* Empirical Bayes decision theory; reliability theory; trajectory estimation and filter theory; stochastic processes. *Mailing Add:* Los Alamos Nat Lab PO Box 1663 Mail Stop K575 Los Alamos NM 87545

MARTZ, LYLE E(RWIN), b Grand Rapids, Mich, Feb 15, 22; m 52; c 2. CHEMICAL ENGINEERING. *Educ:* Univ Mich, BSE, 43. *Prof Exp:* Chem engr, Dow Chem USA, 43-44 & 46-64, proj leader, 64-67, sr process engr, 67-71, res engr, 71-72, res specialist, 72-82; RETIRED. *Mem:* Am Chem Soc; Am Inst Chem Engrs. *Res:* Pilot plant operation in research and development of industrial organic chemicals; design of organic chemical processes. *Mailing Add:* 2008 Airfield Lane Midland MI 48640

MARUCA, ROBERT EUGENE, b Buckhannon, WVa, Nov 25, 41; m 62; c 3. PHYSICAL CHEMISTRY, INORGANIC CHEMISTRY. *Educ:* WVa Wesleyan Col, BS, 63; Cornell Univ, PhD(chem), 66. *Prof Exp:* NIH fel chem, Ind Univ, 66-68; asst prof, Miami Univ, 68-72; assoc prof, 72-80, chmn, Div Natural Sci, 75-86, CHARLES MCCLUNG SWITZER CHMN CHEM, ALDERSON-BROADDUS COL, 75-, PROF CHEM, 80- *Concurrent Pos:* Vis res fac, Univ Ky, 88-89. *Res:* Environmental chemistry; quantum mechanical calculations. *Mailing Add:* Dept Chem Alderson-Broaddus Col Philippi WV 26416

MARULLO, NICASIO PHILIP, b Apr 13, 30; US citizen; m 54; c 2. ORGANIC CHEMISTRY. *Educ:* Queen's Col, NY, BS, 52; Polytech Inst Brooklyn, PhD(chem), 61. *Prof Exp:* NIH fel & res assoc chem, Calif Inst Technol, 60-61; from asst prof to assoc prof, 61-74, PROF CHEM, CLEMSON UNIV, 74- *Mem:* Am Chem Soc. *Res:* Organic reaction mechanisms; rate processes by nuclear magnetic resonance; coordination compounds of alkali metal salts. *Mailing Add:* Dept Chem Clemson Univ Clemson SC 29634

MARUSICH, WILBUR LEWIS, nutrition, toxicology; deceased, see previous edition for last biography

MARUSYK, RAYMOND GEORGE, b Yellowknife, NT, Mar 19, 42; m 66; c 1. VIROLOGY. *Educ:* Univ Alta, BSc, 65, MSc, 67; Karolinska Inst, Sweden, Fil dr(virol), 72. *Prof Exp:* Asst prof biochem, 72-73, asst prof med bact, 73-77, assoc prof med microbiol, 77-83, PROF MED MICROBIOL & INFECT DIS, UNIV ALTA, 83-, ASSOC DIR, PROV LAB PUB HEALTH, 88- *Concurrent Pos:* Vis prof & exchange scientist molecular virol, Nat Inst Health & Med Res, Lille, France, 78-79; Can deleg, Fifth Int Cong Virol, 81, chmn, Seventh Int Cong Virol, 87; chmn, Fourth Int Conf Comp Virol, 82, Fifth Int Conf Comp Virol, 86, Sixth Int Conf Comp Virol, 88. *Mem:* Can Soc Microbiologists; Am Microbiol Soc; NY Acad Sci; Tissue Cult Asn; Can Col Microbiologists (pres, 80-81); Soc Res Admin. *Res:* Structural and functional relationships of viral capsid components; immune response to virus vaccines. *Mailing Add:* Prov Lab Pub Health Fac Med Univ Alta Edmonton AB T6G 2J2 Can

MARUTA, HIROSHI, b Tokyo, Japan, Nov 8, 42; m 87; c 4. ANTI-ONCOGENE RESEARCH, PROTEIN DESIGNING & ENGINEERING. *Educ:* Univ Tokyo, BSc, 67, PhD(biochem), 72. *Prof Exp:* Instr biochem, Univ Tokyo, 72-73; NIH int fel, Dept MCDB, Univ Colo, 73-74; vis assoc, Nat Heart, Lung & Blood Inst, NIH, 74-80; sr scientist, Max Planck Inst Biochem, 80-82, Max Planck Inst Psychiat, 82-84; sr scientist molecular biol, Dept Biol, Yale Univ, 84-86; sr scientist, Dept Biol, Univ Calif, 86-87; SR SCIENTIST MOLECULAR BIOL, LUDWIG INST CANCER RES, 88- *Mem:* Am Soc Cell Biol. *Res:* Mutational analysis of several Ras-related or -associated proteins which potentially counteract the oncogenic action of Ras proteins, in order to study their structure-function relationship; to design and create a series of potent anti-cancer agents which are able to reverse malignant transformation caused by Ras and other oncogenes. *Mailing Add:* Ludwig Inst Cancer Res PO Royal Melbourne Hosp Melbourne 3050 Australia

MARUVADA, P SARMA, b Rajahmundry, AP India, Jan 1, 38; m 63; c 2. ELECTRICAL PERFORMANCE OF HIGH VOLTAGE. *Educ:* Andhra Univ, India, BE, 58; Indian Inst Sci, Bangalore, ME, 59; Univ Toronto, Can, MASc, 66 & PhD(elec eng), 68. *Prof Exp:* Sr teaching fel elec eng, Indian Inst Technol, Kharagpur, 59-61; lectr elec eng, MA Col Technol, Bhopal, 61-64; teaching asst elec eng, Univ Toronto, 64-68; researcher high voltage eng, 69-74, prog mgr high voltage eng, 75-81, group mgr, power transmission dept , 81-83, mgr, 83-87, SR RESEARCHER, INST RES HYDRO, QUE, 87- *Concurrent Pos:* Vis prof, Ecole Polytechnique, Montreal, 71-72 & INRS-Energie, Varennes, 71-77. *Mem:* Inst Elec & Electronics Engrs. *Res:* Theoretical and experimental investigation of corona, radio interference, audible noise, electric and magnetic field effects of high voltage AC and DC transmission systems. *Mailing Add:* 817 De Serigny Boucherville PQ J4B 5C5 Can

MARUYAMA, GEORGE MASAO, b Las Animas, Colo, June 15, 18; m 46; c 2. CHEMISTRY. *Educ:* Western State Col Colo, BA, 41; Univ Wis, MS, 48; Am Bd Bioanal, dipl, 67. *Prof Exp:* Biochemist & asst dir lab, Med Assocs Clin, 48-86; RETIRED. *Mem:* Am Asn Bioanalysts (pres-elect, 65-66, pres, 66-67); Am Chem Soc; Am Asn Clin Chem; fel Am Inst Chemists. *Res:* Clinical chemistry. *Mailing Add:* 1650 Atlantic St Dubuque IA 52001-5805

MARUYAMA, KOSHI, b Sapporo, Hokkaido, Japan, Feb 19, 32; m 61; c 3. SURGICAL & ANATOMICAL PATHOLOGY, VIRAL ONCOLOGY. *Educ:* Univ Hakkaido, MD, 57, PhD(med sci & path), 62. *Prof Exp:* Med intern, Sapporo Munic Gen Hosp, 57-58, assoc med internal med, Dept Internal Med, 61-62; staff mem path, Dept Path, Nat Inst Leprosy Res, 63-65, Nat Cancer Ctr Res Inst, 65-67; asst prof virol, M D Anderson Cancer Ctr, Univ Tex, 67-74, chief, Sect Leukemia-Lumphoma Studies, 68-75, assoc prof virol, Grad Sch Biomed Sci, 74-75; DEPT HEAD PATH, CHIBA CANCER CTR RES INST, 75- *Concurrent Pos:* Leukemia Soc Am scholar, 68; vis prof, Toho Univ, 76-, Tokyo Women's Med Col, 85-88; assoc ed, Japanese J Cancer Clin, 78-, Yearbk Cancer, 78-79; lectr, Chiba Prefectural Col Nursing & Med Technol, 78-, Univ Tottori Sch Med, 87-88. *Mem:* Fel NY Acad Sci; corresp mem Am Asn Cancer Res; Am Soc Microbiol; Am Asn Pathologists; fel Japanese Path Soc; fel Japanese Cancer Asn. *Res:* Transmission of Mycobacterium leprae to foot-pads of laboratory mice; isolation and characterization of a retrovirus from lymphoma patients and development of diagnostic methods for detection of this virus; demonstration of chromosomal and gene rearrangements of human cells infected with this retrovirus; development of a vaccine against bovine leukemia virus infection. *Mailing Add:* Dept Path Chiba Cancer Res Inst 666-2 Nitona-cho Chiba 280 Japan

MARUYAMA, YOSH, b Pasadena, Calif, Apr 30, 30; m 54; c 4. RADIOTHERAPY, RADIOBIOLOGY. *Educ:* Univ Calif, Berkeley, AB, 51; Univ Calif, San Francisco, MD, 55. *Prof Exp:* Intern, San Francisco Hosp, Calif, 55-56; residency, Mass Gen Hosp, Boston, 58-61; James Picker advan acad fel, Stanford Univ, 62-64, traveling fel, Eng, France, Scand, 64; from asst prof to assoc prof radiol, Col Med Sci, Univ Minn, 64-70, admin dir div radiother, 68-70; PROF RADIATION MED, CHMN DEPT & DIR, RADIATION CANCER CTR, COL MED, UNIV KY, 70- *Concurrent Pos:* Consult, Vet Admin Hosp; assoc ed, Appl Radiol; pres local chap, Sigma Xi, 84, Ky Am Cancer Soc, 86. *Honors & Awards:* Nat Award, Am Cancer Soc, 88. *Mem:* AAAS; Radiation Res Soc; Am Asn Cancer Res; Radiol Soc NAm; Am Soc Therapeut Radiol; Sigma Xi; fel Am Col Radiol. *Res:* Radiation medicine, tumor cell biology; mouse leukemia; cell kinetics; neutron therapy; expert on Cf-252 neutron brachytherapy and radiotherapy of human cervix, gynecological, brain and other malignancies; extensive research studies of cells, tissues and tumor effects of neutrons and chemotherapy in biological systems. *Mailing Add:* Dept of Radiation Med Univ of Ky Med Ctr Lexington KY 40506

MARVEL, JOHN THOMAS, b Champaign, Ill, Sept 14, 38; div; c 3. ORGANIC CHEMISTRY, BIOCHEMISTRY. *Educ:* Univ Ill, AB, 59; Mass Inst Technol, PhD(chem), 64. *Prof Exp:* Res assoc agr biochem, Univ Ariz, 64-65, asst agr biochemist, 65-68; sr res chemist, 68-72, sr res group leader, Monsanto Agr Prod Co, 72-75, mgr res, 75-78, assoc dir res, 78-79, gen mgr, res div, 81-85, DIR RES, MONSANTO AGR PROD CO, MONSANTO CO, 81-, GEN MGR, SCI & TECHNOL EUROPE-AFRICA, 85- *Honors & Awards:* Thomas-Hochwalt Sci & Tech Award, Monsanto Co, 84. *Mem:* AAAS; Am Chem Soc; Royal Chem Soc; NY Acad Sci; Weed Sci Soc Am. *Res:* Synthesis of carbohydrates and nucleic acids; nuclear magnetic resonance spectroscopy; mass spectrometry; photochemistry; pesticide metabolism; synthesis and plant growth regulators, plant biochemisty, cell biology. *Mailing Add:* Ethyl Corp 645 High Lake Dr PO Box 14799 Baton Rouge LA 70898

MARVEL, MASON E, b Brewton, Ala, Dec 11, 21; m 45; c 3. HORTICULTURE, PLANT PATHOLOGY. *Educ:* Univ Mass, BS, 50; Va Polytech Inst, MS, 52; WVa Univ, PhD, 70. *Prof Exp:* Instr hort, WVa Univ, 51-56; tech rep, Calif Chem Corp, 56-57; from asst prof to assoc prof, Univ Fla, 57-70, prof veg crops, 70-80; RETIRED. *Concurrent Pos:* Consult, United Brands, 67-70, Hanover Brands Inc, 73-75, AID & World Bank, 78-; Chief party, Contract Team to Nat Agr Ctr, Saigon, SVietnam, 70-72, asst dir tech assistance, Int Progs Ctr Trop Agr Inst, 72-75, team leader, Pulse Prod Prog, Near East Found, Ethiopia, 75-77; veg crops scientist, Int Agr Develop Serv, Bandung, Indonesia, 79; gen mgr, Albaraka, Agr Develop, Riyadh Saudi Arabia, 83-84; horticulturist veg, Coop Ext Serv, Auburn Univ, 84-86, dir, Int Progs, 86- *Mem:* Am Soc Hort Sci; Asn Univ Dirs Int Agr Progs. *Res:* Tropical vegetable crops production and marketing; pest management; minimum tillage; mulching and drip irrigation. *Mailing Add:* 3805 Heritage Pl Opelika AL 36801-7615

MARVELL, ELLIOT NELSON, organic chemistry, for more information see previous edition

MARVIN, HENRY HOWARD, JR, b Lincoln, Nebr, Mar 9, 23; m 44; c 2. PHYSICAL CHEMISTRY. *Educ:* Univ Nebr, BA, 47; Univ Wis, PhD(chem), 50. *Prof Exp:* Res assoc phys chem, Res Lab, Gen Elec Co, 50-53, liaison scientist chem, 53-55, personnel adminr, 56-58, mgr solid state chem, 59-61, mgr eng capacitor dept, 62-64, mgr lighting res lab, Ohio, 64-69, gen mgr, High Intensity Quartz Lamp Dept, 69-75; dir div solar energy, Dept Energy, 75-79; VPRES TECHNOL, BRUNSWICK CORP, 79- *Mem:* Am Chem Soc. *Mailing Add:* 965 Forest Way Glencoe IL 60022-1208

MARVIN, HORACE NEWELL, b Camden, Del, Apr 20, 15; m 40; c 5. ANATOMY, ENDOCRINOLOGY. *Educ:* Morningside Col, BA, 36; Univ Wis, MA, 38, PhD(zool), 41. *Hon Degrees:* DSc, Morningside Col, 69, Univ Ark Med Sci, 88. *Prof Exp:* Asst, Morningside Col, 32-36; asst zool, Univ Wis, 36-41; spec investr, Dept Genetics, Carnegie Inst, 41-42; from instr to asst prof anat, Med Sch, Univ Ark, 42-48, head dept biol res, Univ Tex M D Anderson Hosp Cancer Res, 48-49; assoc prof anat, Col Med, Univ Ark Med Sci Campus, Little Rock, 49-58, prof & head dept, 58-67, assoc dean, 65-77, actg chmn, 83-84, prof anat, 58-85; RETIRED. *Concurrent Pos:* Consult, Radiopath, Los Alamos AEC, 57; vis prof, Univ Lagos, 63; Fulbright fel, 63; Commonwealth fel, 65. *Mem:* Am Asn Anat; Asn Am Med Cols; Sigma Xi. *Res:* Reproductive endocrinology; curriculum. *Mailing Add:* 4015 N Lookout Rd Little Rock AR 72205

MARVIN, PHILIP ROGER, b Troy, NY, May 1, 16; m 42. SOLID STATE SCIENCE. *Educ:* Rensselaer Polytech Inst, BS, 37; Ind Univ, DCS, 51; La Salle Col, LLB, 54. *Prof Exp:* Engr, Gen Elec Co, 37-42; dir chem & metall eng, Bendix Aviation Corp, NY, 43-44; dir res & develop, Milwaukee Gas Specialty Co, 45-52; vpres & dir, Commonwealth Eng Co, Ohio, 52-54 & Am Viscose Corp, 54-56; mgr res & develop div, Am Mgt Asn, 56-64; pres, Clark, Cooper, Field & Wohl, 64-65; dean prof develop, Univ Cincinnati, 65-73, prof Int Mgt, 73-87; UNIV NEW HAVEN, 88- *Concurrent Pos:* Lectr, Bridgeport Eng Inst, 37-43, Jr Col Conn, 40-41 & war training prog, Yale Univ, 41-44; lectr, US Air Force Inst Technol, 53-54; consult, NASA, 66-73 & Am Tel & Tel, 76- *Mem:* AAAS; Am Inst Aeronaut & Astronaut; Am Defense Preparedness Asn; Inst Elec & Electronics Eng; Am Soc Int Law. *Res:* Bio-mechanics; physics of the solid state; metallurgy of electrical steel and beryllium copper; x-ray diffraction; design of electronic controls; thermoelectric phenomena; rectification phenomena. *Mailing Add:* 11 Canbourne Way Madison CT 06443-3446

MARVIN, RICHARD F(REDERICK), b Bozeman, Mont, May 1, 26; m 55; c 2. GEOLOGICAL ENGINEERING, GEOLOGY. *Educ:* Mont Sch Mines, BS, 50, MS, 52. *Prof Exp:* Geologist, Petrol & Geochem Br, US Geol Surv, 52-57, Ground Water Br, 57-61, Isotope Geol Br, 61-87; RETIRED. *Mem:* Geol Soc Am. *Res:* Determination and evaluation of potassium-argon ages pertaining to geologic investigations. *Mailing Add:* 2470 Miller St Lakewood CO 80215-1321

MARVIN, URSULA BAILEY, b Bradford, Vt, Aug 20, 21; m 52. PLANETARY GEOLOGY, HISTORY OF GEOLOGY. *Educ:* Tufts Col, BA, 43; Harvard Univ, MS, 46, PhD, 69. *Prof Exp:* Asst silicate chem, Univ Chicago. 47-50; mineralogist, Union Carbide Ore Co, NY, 53-58; instr mineral, Tufts Univ, 58-61; coordr, Fed Women's Prog, 74-77, GEOLOGIST, SMITHSONIAN ASTROPHYS OBSERV, 61- *Concurrent Pos:* Assoc Harvard Col Observ, 65-; lectr, Tufts Univ, 68-69 & Harvard Univ, 74-; mem, bd trustees, Tufts Univ, 75- & Univs Space Res Assoc, 79-84, chmn, 82-83; mem, Lunar Sample Anal Planning Team, NASA, 76-78; vis prof, Dept Chem, Ariz State Univ, 78; mem, Antarctic Search Meteorites Team, NSF, 78-79 & 81-82; rep, cosmic mineral, Int Mineral Asn, 82-87; secy-gen, Int Comn Hist Geol Sci, 89-; mem Geol & Paleontol of Seymour Island, Antarctica exped, NSF, 85, Adv Comt to Div Polar Progs, NSF, 83-86, Lunar & Planetary Sci Coun, Univ Space Res Asn, 87- *Honors & Awards:* Hist Geol Award, Geol Soc Am, 86. *Mem:* Mineral Soc Am; fel Geol Soc Am; fel Meteoritical Soc (vpres, 73-74, pres, 75-76); Am Geophys Union; Sigma Xi; fel AAAS; Hist Earth Sci Soc (pres, 91). *Res:* Mineralogy and petrology of meteorites and lunar samples; history of geology; geological mapping of Galilean Satellites. *Mailing Add:* Harvard-Smithsonian Ctr Astrophys 60 Garden St Cambridge MA 02138

MARWAH, JOE, b EAfrica, May 27, 52; m 78. PSYCHOPHARMACOLOGY, NEUROBIOLOGY. *Educ:* Univ London, England, BSc, 74; Univ Alta, PhD(pharmacol), 78. *Prof Exp:* Instr pharmacol, Univ Colo Sch Med, 79-80; asst prof psychiat, Yale Univ Sch Med, 80-81; assoc prof pharmacol, Sch Med, Univ Ind, 81-84; PROF PATH, UNIV MED DENT NJ, 85- *Concurrent Pos:* Fel, Univ Colo Health Sci Ctr, 78-79. *Honors & Awards:* A E Bennett Award, Soc Biol Psychiat, 81. *Mem:* Am Physiol Soc; Am Soc Pharmacol & Exp Therapeut; Soc Neurosci; Soc Biol Psychiat; Pharmacol Soc Can. *Res:* Psychopharmacology and neuropharmacology of central nervous system and substances affecting central nervous system. *Mailing Add:* NIH Bldg 31-1B62 Bethesda MD 20892

MARWICK, ALAN DAVID, b Fraserburgh, Scotland, July 17, 46. HYDROGEN IN INTERFACES, RADIATION EFFECTS IN SUPERCONDUCTORS. *Educ:* Univ Sussex, Brighton, UK, BSc, 67, PhD(exp physics), 71. *Prof Exp:* Res fel, Metall Div, Atomic Energy Res

Estab, Harwell, UK, 71-73, sr sci officer, 73-79, prin sci officer, 79-85; RES STAFF MEM, IBM RES, T J WATSON RES LAB, 85- *Concurrent Pos:* Vis scientist, IBM Res Div, T J Watson Res Lab, 81-82. *Mem:* Am Phys Soc; Bohmische Phys Soc; Mat Res Soc. *Res:* Ion beam effects on matter, especially high-Tc superconductors; ion beam analysis, especially hydrogen in materials and their interfaces. *Mailing Add:* IBM Res Div T J Watson Res Lab PO Box 218 Yorktown Heights NY 10598

MARWIL, S(TANLEY) J(ACKSON), b Henderson, Tex, Aug 13, 21; m; c 2. CHEMICAL ENGINEERING. *Educ:* Agr & Mech Col, Tex, BS, 43, MS, 47. *Prof Exp:* Staff asst data correlation, 47-49, group leader adsorption develop, 49-50, staff asst surv & correlation, 50-51, staff engr, 51-55, staff engr oil prod develop, 55-56, group leader crystallization develop, 56-57, group leader exp plastics, 57-60, sect mgr, Chem Processes Sect, 60-66, Explor Chem Sect, 66-69 & Chem Develop Sect, Org Chem Br, Res & Develop Div, 69-75, sect supvr, Chem Develop Sect, Chem Br, 75-80, supvr, chem develop sect, Chem Develop Br, Res & Develop Div, Phillips Petrol Co, 80-85; mgr, 85-91, PRES, MARCO ENG, 91-; DIR, TEX INST ADVAN CHEM TECHNOL, 88- *Concurrent Pos:* Invited speaker, Lindsay Lectr Series, Dept Chem Engr, Tex A&M, 85. *Honors & Awards:* Andre Wilkins Award. *Mem:* Fel Am Inst Chem Engrs. *Res:* High temperature and high pressure thermal and catalytic reactions; high vacuum distillation; high density particle-form polyethylene; hydrocarbon separations, including crystallization, extraction and adsorption; semi-works production of high productivity catalysts, specialty olefins, alkylaromatics and other petrochemicals. *Mailing Add:* 5700 SE Baylor Pl Bartlesville OK 74006

MARWIN, RICHARD MARTIN, b Minneapolis, Minn, Dec 10, 18; m 42; c 1. MEDICAL MICROBIOLOGY, MEDICAL MYCOLOGY. *Educ:* Univ Minn, BA, 41, MS, 43, PhD(bact), 47; Am Bd Med Microbiol, dipl, 62. *Prof Exp:* Teaching asst med bact, Med Sch, Univ Minn, 41-44, instr, 45-48; chmn dept, Univ NDak, 48-62, from assoc prof to prof bact, 48-80, prof microbiol, 80-83, EMER PROF, MED SCH, UNIV NDAK, 83- *Concurrent Pos:* Chief lab bact & blood bank, Univ Minn Hosps, 45-47. *Mem:* Am Soc Microbiol; Mycol Soc Am. *Res:* Culture media modification for growth of pathogenic bacteria; effects of chemicals on pathogenic fungi of man. *Mailing Add:* 1519 Chestnut Grand Forks ND 58202

MARWITZ, JOHN D, b Brownwood, Tex, March 6, 37; m 59; c 5. ATMOSPHERIC SCIENCE. *Educ:* Colo State Univ, BS, 59, MS, 65; McGill Univ, PhD(meteorol), 71. *Prof Exp:* Res asst meteorol, Colo State Univ, 62-65, instr, 65-67; assoc prof, 67-78, PROF ATMOSPHERIC SCI, UNIV WYO, 78- *Mem:* Am Meteorol Soc; Sigma Xi; Royal Meteorol Soc; AAAS. *Res:* Severe thunderstorms; hailstorm modification; the dynamics, kinematics, precipitation processes and modification of orographic winter storms; wind energy and characteristics. *Mailing Add:* Box 3038 Univ Sta Laramie WY 80210

MARX, DONALD HENRY, b Ocean Falls, BC, Oct 3. 36; US citizen; m 57; c 5. PLANT PATHOLOGY, SOIL MICROBIOLOGY. *Educ:* Univ Ga, BSA, 61, MS, 62; NC State Univ, PhD(plant path), 66. *Prof Exp:* PLANT PATHOLOGIST, FORESTRY SCI LAB, SOUTHEASTERN FOREST EXP STA, 62- *Concurrent Pos:* Adv, Int Union Forest Res Orgn, 63. *Honors & Awards:* Arthur Fleming Award, 75; Barrington Moore Award, 77; Ruth Allen Award, 77; Cong Medal of Achievement, Int Cong Plant Protection, 88; Marcus Wallenberg Prize, 90. *Mem:* AAAS; Am Phytopath Soc; Am Inst Biol Scientists. *Res:* Mycorrhizae of conifers and hardwoods; ecology and parasitism of soil-borne organisms; reforestation of adverse sites by use of specific mycorrhizae. *Mailing Add:* Forestry Sci Lab Carlton St Southeastern Forest Exp Sta Athens GA 30602

MARX, EGON, b Cologne, Ger, Apr 4, 37; m 65; c 2. WAVE OPTICS, RELATIVISTIC QUANTUM MECHANICS. *Educ:* Univ Chile, EE, 59; Calif Inst Technol, PhD(physics), 63. *Prof Exp:* Assoc investr physics, Univ Chile, 63-65, independent investr, 65; asst prof, Clarkson Col Technol, 65-67 & Drexel Univ, 67-72; physicist, Harry Diamond Labs, 72-80; PHYSICIST, NAT INST STANDARDS & TECHNOL, 80- *Res:* Theory and computation of electromagnetic wave propagation and scattering; classical and quantized fields, including probability amplitudes in relativistic quantum mechanics; discrete computer simulation of communications systems. *Mailing Add:* Nat Inst Standards & Technol Gaithersburg MD 20899

MARX, GEORGE DONALD, b Antigo, Wis, Apr 30, 36; m 64; c 2. AGRICULTURE, ANIMAL SCIENCE & NUTRITION. *Educ:* Univ Wis-River Falls, BS, 58; SDak State Univ, MS, 60; Univ Minn, Minneapolis, PhD(animal sci), 64. *Prof Exp:* Farm planner, Soil Conserv Serv, USDA, 57-58; asst dairy sci, SDak State Univ, 58-60; asst animal sci, Univ Minn, Minneapolis & St Paul, 60-64; from instr to assoc prof, 64-77, PROF ANIMAL SCI, AGR EXP STA, UNIV MINN, CROOKSTON, 77- *Mem:* Am Dairy Sci Asn; Am Soc Animal Sci. *Res:* Animal management and nutrition. *Mailing Add:* Agr Exp Sta Univ of Minn Crookston MN 56716

MARX, GERTIE F, b Frankfurt am Main, Ger, Feb 13, 12; US Citizen; m 40. ANESTHESIOLOGY. *Educ:* Univ Bern, MD, 37. *Hon Degrees:* Dr Med hc, Johannes Gutenberg-Univ, Mainz, Ger, 86. *Prof Exp:* From asst attend to assoc attend anesthesiologist, Beth Israel Hosp, New York, 43-55; from asst prof to assoc prof, 55-70, PROF ANESTHESIOL, ALBERT EINSTEIN COL MED, 70-; ATTEND ANESTHESIOLOGIST, BRONX MUNIC HOSP CTR, 55- *Concurrent Pos:* Attend anesthesiologist, Bronx Vet Admin Hosp, 66-72, consult, 72-84. *Honors & Awards:* Gold Medal, Obstet Anaesthetist Asn, Eng, 80, 88; Distinguished Serv Award, Am Soc Anesthesiol, 89 & 90; Distinguished Serv Award, Am Soc Regional Anesthesia, 90. *Mem:* Fel Am Soc Anesthesiol; AMA; fel NY Acad Med; NY Acad Sci; assoc fel, Am Col Obstetricians & Gynecologists. *Res:* Obstetric anesthesia. *Mailing Add:* Dept Anesthesiol Albert Einstein Col Med Bronx NY 10461

MARX, HYMEN, b Chicago, Ill, June 27, 25; m 50; c 2. HERPETOLOGY. *Educ:* Roosevelt Univ, BS, 49. *Prof Exp:* Asst cur, Div Amphibians & Reptiles, 50-64, assoc cur, 65-73, CUR, FIELD MUS NATURAL HIST, 73-, HEAD DIV AMPHIBIANS & REPTILES, 70-79 & 84- *Concurrent Pos:* Consult, US Naval Med Res Unit, Egypt, 53; vis scientist, NSF Field Mus, 67-71; lectr, Univ Chicago, 73- *Mem:* Am Soc Ichthyol & Herpet; Soc Study Amphibians & Reptiles. *Res:* Reptiles; systematics; North Africa and Southwestern Asia herpetology; zoogeography of Old World reptiles; phyletic character analysis; phylogeny of vipers; phylogenetic theory. *Mailing Add:* Div Amphibians & Reptiles Field Mus Natural Hist Roosevelt Rd & Lakeshore Dr Chicago IL 60605

MARX, JAMES JOHN, JR, b Paris, Tex, Dec 17, 44; m 73; c 3. IMMUNOPATHOLOGY. *Educ:* St Vincent Col, BA, 66; WVa Univ, MS, 70, PhD(microbiol), 73. *Prof Exp:* RES SCIENTIST IMMUNOL, MARSHFIELD MED FOUND, 73- *Concurrent Pos:* Investr grants, Am Lung Asn, 74 & Nat Heart & Lung Inst, 75; lectr, Sch Nursing St Joseph's Hosp, 74-79; co investr, Nat Inst Occup Safety & Health, 76, Muscular Dystrophy Asn, 79, Elsa Pardee Found, & Marshfield Clin. *Mem:* Am Acad Allergy; Am Asn Immunologists; Am Lung Asn; Am Soc Microbiol; Am Soc Clin Immunol. *Res:* Immunologic mechanisms involved in occupational diseases; drug sensitivity of human tumor cells; clinical immunology. *Mailing Add:* Marshfield Med Found 1000 N Oak Ave Marshfield WI 54449

MARX, JAY NEIL, b New York, NY, Nov 30, 45; m 77; c 2. SYNCHROTRON RADIATION PHYSICS, HIGH ENERGY & ACCELERATOR PHYSICS. *Educ:* Columbia Univ, AB, 66, MS, 69, PhD(physics), 70. *Prof Exp:* Fac mem physics, Yale Univ, 70-77; div fel, 77-81, dep div head, 85-87, SR PHYSICIST, LAWRENCE BERKELEY LAB, 81-, DEP ASSOC DIR, 87- *Concurrent Pos:* Vis scientist, Lawrence Berkeley Lab, 75-76; vis physicist, Brookhaven Nat Lab, 76; consult, US Dept Energy, Off Energy Res, 83; dir, Advan Light Source. *Mem:* Am Phys Soc; Sigma Xi; AAAS. *Res:* Third generation synchroton radiation source being constructed for operation. *Mailing Add:* One Cyclotron Rd M/S 46-161 Berkeley CA 94720

MARX, JOHN NORBERT, b Columbus, Ohio, Oct 31, 37; m; c 2. ORGANIC CHEMISTRY. *Educ:* St Benedict's Col, Kans, BS, 62; Univ Kans, PhD(org chem), 65. *Prof Exp:* Fel org chem, Cambridge Univ, 65-66 & Johns Hopkins Univ, 66-67; asst prof, 67-73, ASSOC PROF ORG CHEM, TEX TECH UNIV, 73- *Mem:* Am Chem Soc; Am Inst Chemists; The Chem Soc. *Res:* Structural determination and synthesis of natural products, especially terpenes and steroids; new synthetic methods; stereochemistry; cyclohexadienone rearrangements; migrations of electronegative groups. *Mailing Add:* Dept Chem Tex Tech Univ Lubbock TX 79409

MARX, JOSEPH VINCENT, b Joplin, Mo, Mar 19, 43; m 67. CLINICAL BIOCHEMISTRY. *Educ:* Johns Hopkins Univ, AB, 65; State Univ NY Upstate Med Ctr, PhD(biochem), 69. *Prof Exp:* From assoc scientist to sr scientist biochem, 69-73, PRIN SCIENTIST APPL SCI, ORTHO DIAGNOSTICS, INC, 74- *Mem:* Am Asn Clin Chemists; Am Chem Soc; Soc Cryobiol; NY Acad Sci. *Res:* Applied research and development including techniques in protein separation and purification, clinical enzymology, blood coagulation and clinical hematology. *Mailing Add:* Sidney Rd Pittstown NJ 08867

MARX, KENNETH ALLAN, b US. BIOPHYSICAL CHEMISTRY. *Educ:* Univ San Diego, BS, 68; Univ Calif, Berkeley, PhD(chem), 73. *Prof Exp:* Res assoc chem, Univ Calif, Berkeley, 73-74; fel epigenetics, Muscular Dystrophy Asn Am, Univ Edinburgh, 74-76 & cell biol, Worcester Found, 76-77; ASST PROF CHEM, DARTMOUTH COL, 77- *Concurrent Pos:* Prin investr, NIH grants. *Mem:* Am Chem Soc; Biophys Soc; Am Soc Cell Biol. *Res:* Structure and function of chromosomes; organization of nucleic acids in bacteriophage and viruses; unusual nuclease-resistancy; conserved DNA sequences in eukaryotic chromosomes. *Mailing Add:* Dept Chem Lowel Univ One University Ave Lowell MA 01854

MARX, KENNETH DONALD, b Amity, Ore, Mar 13, 40; m 63; c 2. PHYSICS, ELECTRICAL ENGINEERING. *Educ:* Ore State Univ, BS, 61; Univ Calif, Davis, MS, 65, PhD(appl sci), 68. *Prof Exp:* Staff mem, Sandia Labs, 61-65 & 68-74; physicist, Lawrence Livermore Lab, 74-79; MEM STAFF, SANDIA NAT LABS, 79- *Concurrent Pos:* Lectr, Univ Calif, Davis/Livermore, 69-79. *Mem:* Am Phys Soc. *Res:* Fluids; combustion science; computational physics. *Mailing Add:* 1231 Regent Pl Livermore CA 94550

MARX, MICHAEL, b Stuttgart, Ger, Nov 10, 33; US citizen; m 63; c 3. ORGANIC CHEMISTRY. *Educ:* Dartmouth Col, BA, 54; Columbia Univ, MA, 64, PhD(chem), 66. *Prof Exp:* Chemist, Lederle Labs Div, Am Cyanamid Co, 54-63; fel, Stanford Univ, 66-67; chemist, 67-69, dept head, Synthetic Org Chem, 69-74, asst dir, Inst Org Chem, 74-80, SR SCIENTIST, BASIC CHEM RES, SYNTEX RES CTR, 80- *Mem:* Am Chem Soc; Royal Soc Chem. *Res:* Synthetic methods; synthesis and transformations of steroids and terpenoid natural products; synthesis of medicinal agents. *Mailing Add:* 10512 W 102nd Terr Overland Park KS 66210

MARX, MICHAEL DAVID, b Durban, SAfrica, July 14, 46; US citizen. EXPERIMENTAL PARTICLE PHYSICS. *Educ:* City Col New York, BS, 67; Mass Inst Technol, PhD(physics), 74. *Prof Exp:* Res assoc physics, Mass Inst Technol, 74-75; asst physicist, Brookhaven Nat Lab, 75-77, assoc physicist, 77-80; ASST PROF, STATE UNIV NY, STONY BROOK, 80- *Mem:* Am Phys Soc. *Res:* Measurement of neutrino elastic scattering off electrons and protons; search for exotic dibaiyon states; hadronic total cross sections at high energies; low energy proton-antiproton resonances. *Mailing Add:* Dept Physics State Univ NY Stony Brook NY 11794

MARX, PAUL CHRISTIAN, b Los Angeles, Calif, Jan 21, 29; m 66; c 1. PHYSICAL CHEMISTRY. *Educ:* Univ Calif, Los Angeles, BS, 51; Northwestern Univ, PhD(chem), 55. *Prof Exp:* Res chemist, Standard Oil Co Calif, 54-57; sr res chemist, Gillette Co, 57-61; mem tech staff high temperature chem, Aerospace Corp, 61-71; consult, 71-73; MGR RES & DEVELOP, FORTIN INDUSTS, 73- *Mem:* Am Chem Soc; Sigma Xi. *Res:* Thermal analysis of polymers; electro forming of thin metallic foils; electrogeochemistry. *Mailing Add:* Fortin Industs 12840 Bradley Ave Sylmar CA 91342

MARX, STEPHEN JOHN, b New York, NY, Nov 23, 42; m 74; c 2. ENDOCRINOLOGY. *Educ:* Yale Univ, BA, 64; Johns Hopkins Univ, MD, 68; Am Bd Internal Med, cert, 74; Am Bd Endocrinol, cert, 76. *Prof Exp:* Med intern, Mass Gen Hosp, 68-69, med resident, 69-70 & 72-73; clin assoc, 70-72, SR INVESTR METAB DIS, NAT INST ARTHRITIS & DIGESTIVE DIS, NIH, 73- *Honors & Awards:* Fuller Albright Award, Am Soc Bone & Mineral Res. *Mem:* Endocrine Soc; Am Fedn Clin Res; AAAS; Am Soc Bone & Mineral Res. *Res:* Mechanism of action of hormones, hormone-receptor interaction, disorders of calcium metabolism, diagnosis and treatment of disorders of parathyroid gland. *Mailing Add:* Metab Dis Br NIH Bldg 10 Rm 9C101 Bethesda MD 20892

MARXHEIMER, RENE B, b Cologne, Ger, Mar 14, 23; US citizen; m 58; c 2. ELECTRICAL ENGINEERING. *Educ:* Univ Lausanne, Ing Elec, 47; Univ Calif, Berkeley, MS, 52. *Prof Exp:* Elec engr, Soc Electroradiol, Paris, France, 47; elec draftsman, Pac Gas & Elec Co, Calif, 48; instructional asst elec eng, Univ Calif, Berkeley, 48-50; elec designer, Bechtel Corp, 50-51; jr res engr, Inst Eng Res, Richmond Field Sta, Univ Calif, 51-52; elec engr, Henry Kaiser Ctr, Kaiser Engrs, 52-53; elec engr & lighting specialist, Div Hwy & San Francisco Bay Toll Crossings, Calif Dept Pub Works, 53-57; illum engr & in chg indust lab, Globe Illum Co, 57-59; from asst prof to assoc prof eng, San Francisco State Col, 59-69, PROF ENGR, SAN FRANCISCO STATE UNIV, 69- *Concurrent Pos:* Asst elec engr, Westinghouse Elec Corp, Calif, 48-50; elec designer, Brown & Caldwell, 50-51; mem, Nat Roadway Lighting Comt, 56-62; NSF stipends, Numerical Anal Inst, 60-61, Appl Probability & Statist Inst, 64; consult, McGraw Hill Co, 61-64; eng consult, Holden-Day Pub Co, Calif, 67-69 & Electrotest Co, 82-87. *Mem:* Sr mem Inst Elec & Electronics Engrs; Am Soc Eng Educ; Nat Soc Prof Engrs; Engrs & Scientists of France. *Mailing Add:* Dept Eng SFSU 1600 Holloway Ave San Francisco CA 94132

MARY, NOURI Y, b Baghdad, Iraq, June 25, 29; m 58; c 2. PHARMACOGNOSY. *Educ:* Univ Baghdad, PhC, 51; Ohio State Univ, MSc, 53, PhD(pharm, pharmacog), 55. *Prof Exp:* Asst prof pharmacog, Col Pharm, Univ Baghdad, 56-60, assoc prof & actg dean, 60-61; vis res scientist, Sch Pharm, Univ Calif, San Francisco, 61-63; fel, Univ Conn, 63-65; assoc prof, 65-72, PROF PHARMACOG, 72-, ASSOC DEAN, ARNOLD & MARIE SCHWARTZ COL PHARM & HEALTH SCI, LONG ISLAND UNIV, 82- *Mem:* Am Pharmaceut Asn; Am Soc Pharmacog; Acad Pharmaceut Sci; Am Asn Col Pharm; Am Asn Clin Chem. *Res:* Chemical and biochemical studies of natural products. *Mailing Add:* 75 DeKalb Ave at Univ Plaza Brooklyn NY 11201

MARYANOFF, BRUCE ELIOT, b Philadelphia, Pa, Feb 26, 47; m 71. SYNTHETIC ORGANIC CHEMISTRY, MEDICINAL CHEMISTRY. *Educ:* Drexel Univ, BS, 69, PhD(org chem), 72. *Prof Exp:* Fel phys org chem, Princeton Univ, 72-74; med chemist, McNeil Pharmaceut, 74-83, sr res fel org chem, 84-87, distinguished res fel, Janssen Res Found, Johnson & Johnson, 87-89, DISTINGUISHED RES FEL & ASST DIR, R W JOHNSON PHARMACEUT RES INST, JOHNSON & JOHNSON, 89- *Honors & Awards:* Philip B Hofmann Res Award, Johnson & Johnson, 78 & 87. *Mem:* Am Chem Soc; Royal Soc Chem; Sigma Xi; NY Acad Sci; Int Union Pure Appl Chem; AAAS. *Res:* Synthesis of biologically active compounds; new synthetic reactions and processes; isoquinoline and indole alkaloids; stereochemistry and asymmetric synthesis; carbohydrate chemistry; central nervous system agents; peptidomimetics; enzyme inhibitors. *Mailing Add:* R W Johnson Pharmaceut Res Inst Spring House PA 19477

MARYANOFF, CYNTHIA ANNE MILEWSKI, b Ringtown, Pa, Nov 27, 49; m 71. INORGANIC CHEMISTRY, MEDICINAL CHEMISTRY. *Educ:* Drexel Univ, BS, 72; Princeton Univ, MA, 74, PhD(org chem), 76. *Prof Exp:* Fel heterocyclic chem, Princeton Univ, 76-77; assoc sr investr, Smith Kline & French Labs, 77-81; group leader chem develop, McNeil Pharmaceut, 81-84, sect head, Chem Process Res, 84-88; HEAD, CHEM PROCESS RES, SPRING HOUSE PA & RARITAN, NJ SITES, RW JOHNSON PHARMACEUT RES INST, 88- *Concurrent Pos:* Mem adv bd, Petrol Res Fund, Am Chem Soc, 86-89; mem adv bd, Med Chem, NIH, 88-92 & med chem study sect, Div grants, 88-92. *Honors & Awards:* Philip B Hofmann Res Scientist Award, Johnson & Johnson, 85. *Mem:* Am Chem Soc; Sigma Xi; AAAS; Am Inst Chem. *Res:* Synthesis of biologically active molecules; bioinorganic and organometallic chemistry; development of synthetic methods; stereochemistry and asymmetric synthesis; investigation of reaction mechanisms. *Mailing Add:* RW Johnson Pharmaceut Res Inst Spring House PA 19477-0776

MARYANSKI, FRED J, b Bayonne, NJ, July 21, 46. COMPUTER SCIENCE, DATABASE MANAGEMENT. *Educ:* Providence Col, BS, 68; Stevens Inst Technol, MS, 71; Univ Conn, PhD(computer sci), 74. *Prof Exp:* PROF COMPUTER SCI, UNIV CONN, 83-, ASSOC PROVOST, 89- *Concurrent Pos:* Mem coun, Asn Comput Mach, 87-89. *Mem:* Asn Comput Mach; Inst Elec & Electronics Engrs. *Mailing Add:* Gully Hall U-86 Univ Conn Storrs CT 06269-2086

MARZELLA, LOUIS, b Italy, May 14, 48; US citizen. CELL PATHOBIOLOGY. *Educ:* Loyola Col, BS, 70; Univ Md, MD, 74; Karolinska Inst, PhD(path), 79. *Prof Exp:* asst dir res prog, 80-83, DIR RES HYPERBARIC MED, MD INST EMERGENCY MED SERV SYST, 86-; ASSOC PROF PATH, SCH MED, UNIV MD, 88- *Concurrent Pos:* Asst prof, Sch Med, Univ MD, 80-87. *Mem:* Am Asn Pathologists; Am Soc Cell Biol; AAAS; Shock Soc; Undersea Med Soc. *Res:* Regulation of the function of lysosomes in cells; pathologic mechanisms that cause the loss of integrated cell functions and lead to cell death. *Mailing Add:* Dept Path Rm 7-23 MSTF Univ Md 10 S Pine St Baltimore MD 21201

MARZETTI, LAWRENCE ARTHUR, b Mt Vernon, Ohio, Apr 17, 17; m 42; c 5. APPLIED STATISTICS. *Educ:* Morehead State Univ, AB, 39. *Prof Exp:* Opers head 1940 census, Pop Div, Bur Census, 40-42, surv statistician, 46-51, asst budget officer, Budget Off, 52-56, chief overseas consult foreign census & statist, Int Statist Progs, 56-70, tech adv statist legis, Subcomt Census & Statist, Post Off & Civil Serv Comt, US House Rep 93rd Cong, 71-73; spec asst, Econ Census Staff, Bur Census, 73-74; statist consult, 75-80; RETIRED. *Concurrent Pos:* Census consult, Tech Coop Admin, Amman, Jordan, 52; consult, US Bur of Census, 78; head, statist adv team, Qatar, 78, researcher for develop of cent statist off, 79, consult, 80. *Honors & Awards:* Meritorious Award, Dept Com, 52. *Mem:* Am Statist Asn. *Res:* Foreign census methodology; mid-decade census; census confidentiality; national vote registration and election practice; international statistical consultation. *Mailing Add:* 2587 Golfers Ridge Rd Annapolis MD 21401-6915

MARZKE, ROBERT FRANKLIN, b Worcester, Mass, Oct 19, 38; m 62; c 3. MAGNETIC RESONANCE, SURFACES. *Educ:* Princeton Univ, BA, 59; Columbia Univ, PhD(physics), 66. *Prof Exp:* Res assoc, Univ NC, 66-70; asst prof, 70-79, ASSOC PROF PHYSICS, ARIZ STATE UNIV, 79- *Concurrent Pos:* Co-prin investr, Nat Sci Found, 76-88 & Petrol Res Fund, 77-79. *Mem:* Am Phys Soc; Calif Catalysis Soc. *Res:* Physical and chemical properties of condensed matter by magnetism and magnetic resonance; intercalation compounds; small metallic particles; catalysis. *Mailing Add:* Dept Physics Ariz State Univ Tempe AZ 85287

MARZLUF, GEORGE A, b Columbus, Ohio, Sept 29, 35; m 60; c 4. GENETICS, BIOCHEMISTRY. *Educ:* Ohio State Univ, BSc, 57, MS, 60; Johns Hopkins Univ, PhD(genetics), 64. *Prof Exp:* NSF fel biochem genetics, Sch Med, Univ Wis, 64-66; asst prof biol, Marquette Univ, 66-69; assoc prof, 70-75, dir, Prog Molecular, Cellular & Develop Biol, 80-83, PROF BIOCHEM, OHIO STATE UNIV, 75-, CHMN, DEPT BIOCHEM, 86- *Concurrent Pos:* Career develop award, NIH, 75-80, & grants biochem genetics, 67-; NSF grants molecular genetics, 78-82; vis scientist, Beatson Inst, Scotland, 76; USDA grant genetic transformation, 85-87. *Mem:* AAAS; Genetics Soc Am; Am Soc Microbiol; Am Asn Biol Chemists. *Res:* Developmental and biochemical genetics; synthesis, allosteric control and turnover of enzymes and permeases in higher organisms; control of differential gene action; molecular genetics. *Mailing Add:* Dept Biochem Ohio State Univ Columbus OH 43210

MARZLUFF, WILLIAM FRANK, JR, b Washington, DC, May 7, 45; m 66; c 2. BIOCHEMISTRY. *Educ:* Harvard Univ, AB, 67; Duke Univ, PhD(biochem), 71. *Prof Exp:* NIH fel biol, Johns Hopkins Univ, 71-74; from asst prof to assoc prof, 74-81, PROF CHEM, FLA STATE UNIV, 82- *Concurrent Pos:* Dir, Prog Molecular Biophysics, Fla State Univ, 85- *Mem:* Am Soc Biochem & Molecular Biol; Am Soc Microbiol. *Res:* Organization and expression of histone genes and genes coding for small nuclear RNAs; RNA synthesis in sea urchin development; RNA transcription and processing. *Mailing Add:* 2014 Travis Ct Tallahassee FL 32303

MARZOLF, GEORGE RICHARD, b Columbus, Ohio, Dec 13, 35; m 58; c 2. LIMNOLOGY. *Educ:* Wittenberg Col, AB, 57; Univ Mich, MS, 61, PhD(zool), 62. *Prof Exp:* From asst prof zool to assoc prof biol, 62-75, assoc dir div biol, 73-75, prof biol, Kans State Univ, 75-88; prof biol, Murray State Univ, 88-91; BIOLOGIST, WATER RESOURCES DIV, US GEOL SURV, 91- *Concurrent Pos:* Vis prof zool, Univs Wis, Okla, Ore & Mich State, 66, 67, 75 & 77; mem, Water Sci Technol Bd, Nat Res Coun/Nat Acad Sci, 85-90; vis res scientist, US Geol Surv, 86. *Mem:* Int Asn Theoret & Appl Limnol; Am Soc Limnol & Oceanog; Ecol Soc Am; NAm Beuthological Soc. *Res:* Reservoir limnology; prairie streams; dissolved organic carbon; zooplaukton ecology. *Mailing Add:* US Geol Surv Marine St Sci Ctr Boulder CO 80303

MARZULLI, FRANCIS NICHOLAS, b New York, NY, Feb 2, 17; m 45; c 2. PHARMACOLOGY, TOXICOLOGY. *Educ:* St Peters Col, BS, 37; Johns Hopkins Univ, MA, 40, PhD(physiol), 41. *Prof Exp:* Aquatic biologist, US Fish & Wildlife Serv, 41-43; toxicologist, Dugway Proving Ground, Utah, 46-47; toxicologist, physiologist & chief field toxicol br, Army Chem Res & Develop Labs, 47-63; chief dermal toxicity br, Pharmacol Div, 63-73, spec assignment, Med Ctr, Univ Calif, San Francisco, 73-75; sr scientist, Food & Drug Admin, 75-80; sr toxicologist, Nat Res Coun-Nat Acad Sci, 80-87; CONSULT TOXICOL, 88- *Concurrent Pos:* Exchange scientist, Chem Defense Exp Estab, Eng, 60-61. *Mem:* Soc Invest Dermat; Am Col Toxicol & Med; Soc Cosmetic Chem; Soc Toxicol; Asn Res Vision & Ophthal. *Res:* Environmental, skin and eye physiology; dermatotoxicology. *Mailing Add:* 8044 Park Overlook Dr Bethesda MD 20817-2724

MARZZACCO, CHARLES JOSEPH, b Philadelphia, Pa, May 1, 42; m 64; c 1. PHYSICAL CHEMISTRY. *Educ:* Temple Univ, AB, 64; Univ Pa, PhD(chem), 68. *Prof Exp:* Grant, Princeton Univ, 68-69, instr chem, 69-70; asst prof, NY Univ, 70-73; asst prof, 73-76, PROF CHEM, RI COL, 76- *Concurrent Pos:* Mary Tucker Thorp prof, 85-86. *Mem:* Am Chem Soc; Int-Am Photochem Soc; Sigma Xi. *Res:* Photophysical and photochemical properties of organic molecules. *Mailing Add:* Dept of Phys Sci RI Col Providence RI 02908

MASAITIS, CESLOVAS, b Kaunas, Lithuania, Mar 2, 12; US citizen; m 40; c 1. APPROXIMATION THEORY, APPLIED MATH. *Educ:* Univ Tenn, PhD(math), 56. *Prof Exp:* Asst astron, Univ Kaunas, Lithuania, 37-40; asst astron & math, Univ Vilnius, Lithuania, 40-44; instr math & physics, Nazareth Col, Ky, 50-52, math, Univ Ky, Lexington, 52-53 & Univ Tenn, Knoxville, 53-56; chief appl math, Anal Br, Ballistic Res Lab, Aberdeen, Md, 56-80; res asst prof consult, Med Sch, Univ Md, 63-76; vis prof appl math, Tech Inst Kaunas, Lithuania, 90; RETIRED. *Mem:* Lithuania Cath Acad Sci; Am Math Asn. *Res:* Data analysis; approximation theory; mathematical modelling. *Mailing Add:* PO Box 442 Thompson CT 06277

MASAMUNE, SATORU, b Fukuoka, Japan, July 24, 28; m 56; c 2. ORGANIC CHEMISTRY. *Educ:* Tohuku Univ, Japan, AB, 52; Univ Calif, PhD(chem), 57. *Prof Exp:* Postdoctoral fel, Univ Wis, 56-59, lectr, 59-61; fel, Mellon Inst, 61-64; from assoc prof to prof chem, Univ Alta, 64-79; PROF CHEM, MASS INST TECHNOL, 78- *Concurrent Pos:* Ed, Organic Syntheses Inc, 70-78, Rev Chem Interim, 76. *Honors & Awards:* Am Chem Soc Award, 78; Centenary lectr, Royal Soc Chem, 80; Hamilton Award, 84; Arthur C Cope Scholar Award, 87. *Mem:* Am Chem Soc; The Chem Soc; fel Royal Soc Can; Japan Chem Soc; Soc German Chemists; fel Am Acad Arts & Sci. *Res:* Organic synthesis of biologically important compounds; chemistry of main group IVA; enzymology of carbon and carbon bond forming reactions. *Mailing Add:* Dept of Chem Mass Inst Technol 77 Mass Ave Cambridge MA 02139

MASANI, PESI RUSTOM, b Bombay, India, Aug 1, 19. MATHEMATICS. *Educ:* Univ Bombay, BSc, 40; Harvard Univ, MA, 42, PhD(math), 46. *Prof Exp:* Teaching fel math, Harvard Univ, 43-45; mem, Inst Advan Study, Princeton, NJ, 46-48; sr res fel, Tata Inst Fundamental Res, Bombay, 48-49; prof math & head dept, Inst Sci, Bombay, 49-59; vis lectr, Brown Univ, 59-60; prof math, Ind Univ, Bloomington, 60-72; prof, 72-73, UNIV PROF MATH, UNIV PITTSBURGH, 73- *Concurrent Pos:* Vis lectr, Harvard Univ & Mass Inst Technol, 57-58; vis prof, Math Res Ctr, Univ Wis, 65-66 & Statist Lab, Cath Univ Am, 66-67; vis researcher, Battelle Seattle Res Ctr, 69-70; vis prof, Fed Polytech Sch, Lausanne, 75; Alexander von Humboldt vis sr scientist, Fed Repub Ger, 79-80. *Mem:* Am Math Soc; Soc Indust & Appl Math; Math Asn Am. *Res:* Noncommutative analysis, specifically the factorization of operator-valued functions; prediction and filter theory of stationary stochastic processes; Hilbert spaces, specifically isometric flows, spectral integrals and vector-valued measures; ordinary linear differential systems. *Mailing Add:* Univ Pittsburgh Pittsburgh PA 15260

MASARACCHIA, RUTHANN A, b St Louis, Mo, Oct 24, 42; m 79; c 1. BIOCHEMISTRY. *Educ:* Univ Mass, PhD(biochem), 69. *Prof Exp:* ASSOC PROF BIOCHEM, UNIV NTEX, 84- *Concurrent Pos:* Med Biochem Study Sect, NIH, 88- *Mem:* Am Asn Parasitol; Sigma Xi; Am Soc Biochem & Molecular Biol; Am Asn Univ Women. *Res:* Protein phosphorylatern and metabolic regulation. *Mailing Add:* Biochem Dept Univ NTex Denton TX 76203

MASAT, ROBERT JAMES, b Greeley, Colo, Sept 6, 28; m 64; c 2. COMPARATIVE PHYSIOLOGY, BIOCHEMISTRY. *Educ:* Univ Portland, BSc, 54; Wash State Univ, MSc, 58; St Louis Univ, PhD(biol), 64. *Prof Exp:* Instr biol, Rockhurst Col, 58-61, head dept, 59-61; res assoc cryobiol, Am Found Biol Res, 62; res assoc biochem, Med Ctr, La State Univ, 64, Nat Heart Inst fel, 64-66; Am Found Biol Res, 66-68; assoc prof biol, 68-73, prof biol, St Ambrose Col, 73-78, chmn div natural & math sci, 75-78; assoc prof biochem & physiol, 78-79, prof, 79-85, EMER PROF CHEM, PALMER COL CHIROPRACTIC, 85-; CLIN CHEM CONSULT, 85- *Mem:* AAAS; Am Soc Zool; Am Chem Soc; Am Physiol Soc. *Res:* Comparative biochemistry and physiology of plasma proteins; physiology of biological cyclic phenomena of hibernation and migration; active transport of substances across living membranes; physiology of hypothermia. *Mailing Add:* 4556 Sheridan St Davenport IA 52806

MASCARENHAS, JOSEPH PETER, b Nairobi, Kenya, Nov 19, 29; m 60; c 2. DEVELOPMENTAL BIOLOGY. *Educ:* Univ Poona, BSc, 52, MSc, 54; Univ Calif, Berkeley, PhD(plant physiol), 62. *Prof Exp:* Res officer, Parry & Co, India, 53-56; instr biol, Amherst Col, 62-63; Res Corp Brown-Hazen Fund grant & instr bot, 63-64, asst prof, 64-67; res assoc biol, Mass Inst Technol, 67-68; assoc prof, 69-74, PROF BIOL, STATE UNIV NY ALBANY, 74- *Concurrent Pos:* NSF grants, Wellesley Col, Mass Inst Technol & State Univ NY Albany, 65-; vis asst prof, Mass Inst Technol, 66-67; res found fel & grant, State Univ NY Albany, 69-74; prog dir develop biol, NSF, Washington, DC, 85-86, 86-87. *Mem:* Fel AAAS; Int Soc Develop Biol; Soc Develop Biol; Bot Soc Am; Am Soc Plant Physiol; Intl Soc Plant Mol Biol. *Res:* Molecular control of plant development. *Mailing Add:* Dept of Biol Sci State Univ NY 1400 Washington Ave Albany NY 12222

MASCHERONI, P LEONARDO, b Tucuman, Arg, July 20, 35; US citizen; m 67; c 2. THEORETICAL PHYSICS. *Educ:* Univ Cuyo, Arg, BS, 62; Univ Calif, Berkeley, PhD(physics), 68. *Prof Exp:* Res assoc & lectr physics, Temple Univ, 68-70, vis asst prof chem & physics, 70-71; res scientist assoc physics, Ctr Statist Mechanics & Thermodynamics, 71-74; res scientist assoc physics, Fusion Res Ctr, Univ Tex, Austin, 74-77; sr res scientist, Sci Appln, Inc, La Jolla, 77-79; STAFF SCIENTIST, LOS ALAMOS NAT LAB, 79- *Mem:* Am Phys Soc; AAAS. *Res:* Current research in plasma physics; laser interaction with matter and laser fusion; turbulent heating of a plasma, anomalous transport; fusion research in Tokomak systems; inertial confinement fusion. *Mailing Add:* 1900 Camino Moro Los Alamos NM 87544

MASCIANTONIO, PHILIP, b Monongahela City, Pa, Mar 14, 29; m 50; c 5. PHYSICAL CHEMISTRY, ORGANIC CHEMISTRY. *Educ:* St Vincent Col, BS, 50; Carnegie Inst Technol, MS, 57, PhD(chem), 60. *Prof Exp:* Asst chemist, Robertshaw-Fulton Div, Nat Roll & Foundry Co, 50, chief chemist, 51; prod supvr dyestuffs, Pittsburgh Chem Co, 51-55; sr res chemist, Appl Res Lab, 55-67, sect supvr, 67-69, div chief chem, 69-74, asst dir environ control, 74-75, dir environ control, 75-80, VPRES ENVIRON AFFAIRS, USX CORP, 80- *Concurrent Pos:* Instr, Carnegie Inst Technol, 62-64; abstractor, Chem Abstr, 63-70. *Mem:* Am Chem Soc; Water Pollution Control Fedn; Air Pollution Control Asn; Am Iron & Steel Inst. *Res:* Chemistry, properties and structure of coal; process research of chemicals polymer properties and development of air and water pollution abatement systems. *Mailing Add:* USX Corp 4000 Tech Ctr Dr Monroeville PA 15146-3057

MASCIOLI, ROCCO LAWRENCE, b Mt Carmel, Pa, May 7, 28; m 54; c 7. ORGANIC CHEMISTRY, POLYMER CHEMISTRY. *Educ:* Bucknell Univ, BS, 52; Univ Pa, MS, 54, PhD(chem), 57. *Prof Exp:* Res chemist, Houdry Process Corp Labs, Air Prod & Chem, Inc, 56-57, proj dir, 67-69, sect head appln res & develop, 69-71, asst dir res & develop, Chem Additives Div, 71-79, dir res & performance, 79-82; DIR RES & PERFORMANCE & TECH ADV, ARCO CHEM CO, 83- *Mem:* Am Chem Soc. *Res:* Nitrogen chemistry; catalysis; urethane polymers. *Mailing Add:* Pennell Manor 205 Kevin Lane Media PA 19063

MASDEN, GLENN W(ILLIAM), b Denver, Colo, Jan 17, 33; m 58; c 2. ENERGY CONVERSION, COMPUTER SCIENCE. *Prof Exp:* Elec engr, Univac Div, Remington Rand Corp, 55-56; instr elec eng, Univ Colo, 56-57; from instr to assoc prof, 57-72, PROF ENG, WALLA WALLA COL, 72- *Mem:* Inst Elec & Electronics Engrs. *Res:* Photovoltaics. *Mailing Add:* Sch of Eng Walla Walla Col 204 S College Ave College Place WA 99324

MASE, DARREL JAY, b Delphos, Kans, Nov 13, 05; m 35; c 2. SPEECH PATHOLOGY. *Educ:* Emporia Teachers Col, BS, 28; Univ Mich, MA, 32; Columbia Univ, PhD(speech path & clin psychol), 41. *Hon Degrees:* DSc, Col Med & Dent NJ, 81. *Prof Exp:* Teacher speech & math, Mankato High Sch, Kans, 28-29; prof speech correction, Calif State Teachers Col, 33-40; prof spec educ, NJ State Teachers Col, 40-46; prof speech & educ, 50-71, coordr, Fla Ctr Clin Serv, 50-58, dean, 58-71 EMER PROF SPEECH & EDUC & EMER DEAN, COL HEALTH RELATED PROFESSIONS, UNIV FLA, GAINESVILLE, 71- *Concurrent Pos:* Consult, J Hillis Miller Health Ctr, Univ Fla, 50-; vis prof, Los Angeles State Col, 53 & Ark Polytech Col, 58; adj prof, Dept Community Health & Family Med, Univ Fla, 76- *Honors & Awards:* Hon Award, Asn Schs Allied Health Professions, 72; W F Faulkes Award, Nat Rehab Asn, 72. *Mem:* Fel Am Asn Mental Deficiency; Am Asn Social Psychiat; fel Am Speech & Hearing Asn; Am Soc Allied Health Professions; Int Soc Rehab Disabled. *Res:* Social and behavioral sciences; rehabilitation; physical disabilities; speech pathology; education of exceptional children; manpower in allied health professions. *Mailing Add:* Dept Community Health & Family Med Univ Fla Med Col Gainesville FL 32610

MASEK, BRUCE JAMES, b Chicago, Ill, July 15, 50; m 72; c 2. BEHAVIORAL MEDICINE, PEDIATRIC HEALTH PSYCHOLOGY. *Educ:* Bradley Univ, BS, 72, MA, 74; Auburn Univ, PhD(clin psychol), 77. *Prof Exp:* Instr psychiat, Johns Hopkins Sch Med, 78-79, asst prof med psychol, 79-80; instr, 80-84, ASST PROF PEDIAT PSYCHOL, HARVARD MED SCH, 84- *Concurrent Pos:* Pre-doctoral clin psychol intern, Mendota Mental Health Inst, Madison, Wis, 76-77; post-doctoral fel, dept pediat, Johns Hopkins Univ Sch Med, 77-78; staff psychologist, dept psychiat, Johns Hopkins Hosp, 78-80 & Children's Hosp, Boston, 80-; vis prof, US Air Force, Wilford Hall Med Ctr, 82, 84 & 90; clin assoc, Mass Gen Hosp, Boston, 86-; dir, behav med, Children's Hosp, child psychol, Mass Gen Hosp, 89- *Mem:* Asn Advan Behav Ther; Soc Behav Med; Psychophysiol Res Soc; Am Psychol Asn; Soc Pediat Psychol. *Res:* Biobehavioral intervention in chronic illness and stress-related disorders in pediatric populations; psychophysiological assessment of pain behavior. *Mailing Add:* Children's Hosp Med Ctr 300 Longwood Ave Boston MA 02115

MASEK, GEORGE EDWARD, b Norfolk, Va, Feb 10, 27; m 55; c 2. PHYSICS. *Educ:* Stanford Univ, PhD(physics), 56. *Prof Exp:* Res assoc physics, Hansen Lab, Stanford Univ, 55-56; instr, Princeton Univ, 56-57; from asst prof to prof, Univ Wash, 57-65; assoc prof, 65-67, PROF PHYSICS, UNIV CALIF, SAN DIEGO, 67- *Concurrent Pos:* Dir, Inst Res Particle Acceleration, Univ Calif, 77-; mem, Sci & Educ Adv Comt, Livermore Res Lab, 79- *Mem:* Am Phys Soc. *Res:* Elementary particle and high energy physics. *Mailing Add:* Dept Physics Univ Calif at San Diego La Jolla CA 92093

MASEL, RICHARD ISAAC, b Philadelphia, Pa, Mar 21, 51. CATALYSIS, ELECTRONIC MATERIALS. *Educ:* Drexel Univ, BS, 72, MS, 73; Univ Calif, Berkeley, PhD(chem eng), 77. *Prof Exp:* From asst prof to assoc prof, 78-87, PROF CHEM ENG, UNIV ILL, 87- *Concurrent Pos:* Exxon fel solid state, Am Chem Soc, 82; NSF presidential young investr award, 84. *Mem:* Am Inst Chem Engrs; Am Vacuum Soc; Electrochem Soc; Am Chem Soc; NAm Catalysis Soc; Mat Res Soc. *Res:* Catalysis; electronic materials production; pollution control; effects of surface structure on rate processes on solid surfaces; unexplored orbital symmetry effects; surface chemistry of carbon incorporation during metal-organic chemical vapor deposition; effect of surface structure on instabilities in catalytic reactions. *Mailing Add:* Chem Eng Dept Univ Ill 1209 W California St Urbana IL 61801

MASELLI, JAMES MICHAEL, b Pottsville, Pa, Mar 29, 35; m 61; c 2. INORGANIC CHEMISTRY. *Educ:* Lafayette Col, AB, 57; Univ Pa, PhD(inorg chem), 61. *Prof Exp:* Univ fel, Harvard Univ, 61-62; sr res chemist, 63-66, res supvr, 66-68, mgr, 68-75, dir, Clarksville, 75-76; vpres res & develop, Emission Control Dept, 76-78, VPRES RES, DAVISON CHEM DIV, W R GRACE & CO, 78- *Mem:* Am Chem Soc. *Res:* Inorganic synthesis and catalysis. *Mailing Add:* 6413 Amherst Ave Columbia MD 21046

MASELLI, JOHN ANTHONY, b New York, NY, Feb 18, 28; m 48; c 2. FOOD CHEMISTRY. *Educ:* City Col New York, BS, 47; Fordham Univ, MS, 49, PhD(chem), 52. *Prof Exp:* Instr chem, Fordham Univ, 48-52; chemist, Fleischmann Labs, Standard Brands, Inc, 52-57, mgt staff asst, 57-59, dept dir, 59-62, dir res, Fleischmann Mfg Div, 62-64, prod develop mgr M&M Candies, NJ, 64-67; group dir food ingredients, Standard Brands, 79-80, vpres corp technol, 80-81, corp officer, 81-82; vpres corp technol, Nabisco Brands, 81-82, vpres Wilton Technol Ctr, 82-84, vpres & gen mgr, biotechnol ventures, 84-85; vpres opers, 67-79, pres, OZ Food Corp, 70-79; founder, Lubin-Maselli Labs, 73-79; vpres corp res & develop & corp officer, R J Nabisco, Inc, 85-87; SR VPRES, PLANTERS LIFE SAVERS CO, 87- *Concurrent Pos:* Mem bd dirs & chmn res comt, Nat Peanut Coun; mem bd dirs, NC Biotechnol Ctr & USA Cultor. *Mem:* Am Soc Bakery Eng; Am Asn Cereal Chem; Am Asn Candy Technol; Inst Food Technol; Am Chem Soc. *Res:* Food, cereal and yeast chemistry; candy technology. *Mailing Add:* Planters Life Savers Co 1100 Reynolds Blvd Winston-Salem NC 27102

MASER, CHRIS, b Bronxville, NY, Oct 13, 38; m 81; c 1. ECOSYSTEM FUNCTIONING, LAND MANAGEMENT. *Educ:* Ore State Univ, BS, 62, MS, 66. *Prof Exp:* Vert zoölogist, Yale Univ Prehist Exped, Nubia, Egypt, 63-64; mammalogist, US Naval Med Res, Cairo, Egypt & Nepal, 66-67; mem staff, US Dept Interior, 67-75, res wildlife biologist, Bur Land Mgt, 75-87; ASST PROF FORESTRY, ORE STATE UNIV, 80- *Concurrent Pos:* Prin investr, Ore Coast Ecol Surv, Puget Sound Mus Nat Hist, Univ Puget Sound, Tacoma, Wash, 70-73; mem, Nat Adv Comt, H J Andrews Exp Forest Exp Ecol Res, 77-79; prin investr, Role of Large Rotting Wood, Oregon Coniferous Forests, 81-87; consult, sustainable forestry, 87- *Mem:* Am Soc Mammalogists; Soc Icthyol & Herpetol; Soc Range Mgt; Sigma Xi. *Res:* Conduct ecological research in forest lands and range lands and apply (through synthesis) such research to management of public lands and cultural landscapes. *Mailing Add:* c/o Dept Forest Sci Ore State Univ Corvallis OR 97331

MASER, MORTON D, b Hagerstown, Md, Nov 24, 34; m 55; c 2. CELL BIOLOGY. *Educ:* Univ Pa, AB, 55; Univ Pittsburgh, PhD(biophys), 62. *Prof Exp:* Res asst, Mellon Inst, 56-58, res assoc, 58-60, jr fel, 60-62, fel, 63; res assoc, Marine Biol Lab, Woods Hole, 62; res assoc, Biol Labs, Harvard Univ, 62-64, lectr, 64-66; dir cell biol lab, Millard Fillmore Hosp, Buffalo, 66-70; asst prof biol, Erie County Community Col, 70; assoc prof continuing educ, Northeastern Univ, 70-73; assoc prof path, Creighton Univ, 73-77; coordr continuing educ, Marine Biol Lab, 77-79, asst dir educ & res serv, 79-82; PRES & DIR, WOODS HOLE EDUC ASSOCS, 82- *Mem:* AAAS; Electron Micros Soc Am; Am Soc Cell Biol; Int Soc Stereology. *Res:* Electron microscopical techniques; computer sciences, software systems. *Mailing Add:* Woods Hole Educ Asn PO Box EM Woods Hole MA 02543

MASERICK, PETER H, b Washington, DC, Feb 8, 33; m 56; c 4. MATHEMATICS. *Educ:* Univ Md, BS, 55, MA, 57, PhD(math), 60. *Prof Exp:* NSF fel math, Univ Wis, 63-64; asst prof, 64-71, ASSOC PROF MATH, PA STATE UNIV, UNIVERSITY PARK, 71- *Res:* Functional analysis; convexity. *Mailing Add:* Dept of Math Pa State Univ University Park PA 16802

MASERJIAN, JOSEPH, b Albany, NY, Feb 10, 29; m 53; c 4. SEMICONDUCTORS, MICROELECTRONICS. *Educ:* Rensselaer Polytech Inst, BS, 52; Univ Southern Calif, MS, 55; Calif Inst Technol, PhD(mat sci), 66. *Prof Exp:* Mem tech staff, Semiconductor Div, Hughes Aircraft Co, 52-60; mem tech staff, 60-64, supvr, 64-86, SR RES SCIENTIST, SEMICONDUCTOR TECHNOL GROUP, JET PROPULSION LAB, CALIF INST TECHNOL, 85- *Concurrent Pos:* Vis prof, Chalmers Inst Technol, 72-73, Jubilee prof, 82; vis scientist, Univ Calif, Santa Barbara, 87-88. *Honors & Awards:* Exceptional Scientific Achievement Award, NASA, 85. *Mem:* Am Phys Soc; Am Vacuum Soc; sr mem Inst Elec & Electronics Engrs; Sigma Xi. *Res:* Physics and chemistry of semiconductor interfaces; thin films; interface physics; reliability of semiconductor devices; III-V quantum well devices; infrared detectors; optical modulators. *Mailing Add:* 24638 Brighton Dr Valencia CA 91355

MASH, ERIC JAY, b Brooklyn, NY, Feb 6, 43; m 69. CHILD & FAMILY PSYCHOPATHOLOGY,. *Educ:* City Univ NY, BBA, 64; Temple Univ, MA, 65; Fla State Univ, PhD(clin psychol), 69. *Prof Exp:* From asst prof to assoc prof, 69-81, PROF PSYCHOL, UNIV CALGARY, 81- *Concurrent Pos:* Vis prof, Ore Health Sci Univ, 76-77, 83-84 & 87-88; assoc ed, Behav Assessment, 81-83. *Mem:* Fel Am Psychol Asn; fel Can Psychol Asn; Soc Res Child Develop; Asn Advan Behav Ther; Soc Pediat Psychol. *Res:* Psychological risk factors in families in hyperactive, conduct disordered and abused children; behavioral assessment of child and family disorders. *Mailing Add:* Dept Psychol Univ Calgary Calgary AB T2N 1N4 Can

MASHALY, MAGDI MOHAMED, b Alexandria, Egypt, Feb 5, 44; US citizen; m 84; c 1. ENDOCRINOLOGY. *Educ:* Cairo Univ, BS, 64, MS, 70; Univ Wis, MS, 73, PhD(physiol), 76. *Prof Exp:* Res assoc, Miss State Univ, 76-78; asst prof, 78-87, ASSOC PROF PHYSIOL, PA STATE UNIV, 87- *Concurrent Pos:* Fulbright scholar, 89-90. *Mem:* Poultry Sci Asn; AAAS; Soc Study Fertility. *Res:* Role of endocrine system in affecting avian reproduction; relationships between endocrine and immune systems in chickens. *Mailing Add:* Dept Poultry Sci Penn State Univ University Park PA 16802

MASHBURN, LOUISE TULL, b Wayne, Pa, Aug 27, 30; m 58. BIOCHEMISTRY, IMMUNOLOGY. *Educ:* Westhampton Col, BA, 52; Duke Univ, PhD(biochem), 61. *Prof Exp:* Fel biochem, Univ Del, 61-63, res assoc, 63-64; res assoc, Res Inst, Hosp Joint Dis, 64-76; assoc prof, Ctr Health Sci, Univ Tenn, 76-84; Karnofsky fel immunol, St Jude Children's Hosp, 84-85; ASSOC PROF IMMUNOL, WVA SCH OSTEOP MED, 85- *Concurrent Pos:* Asst prof, Mt Sinai Sch Med, 71-76; USPHS spec res fel, 65-67, grant, 66-75; Am Cancer Soc grant, 68-74; Leukemia Soc Am scholar, 71-76. *Mem:* Am Chem Soc; Am Soc Biol Chemists; NY Acad Sci; Am Asn Cancer Res. *Res:* Biochemistry of malignant diseases; protein phosphorylation; enzymology in therapy; tissue culture; immune mechanisms of surveillance. *Mailing Add:* Box 206 Lewisburg WV 24901-0206

MASHBURN, THOMPSON ARTHUR, JR, b Morganton, NC, Oct 9, 36; m 58. BIOCHEMISTRY. *Educ:* Univ NC, AB, 56; Duke Univ, PhD(org chem), 61. *Prof Exp:* Res chemist, Plastics Dept, E I du Pont de Nemours & Co, Inc, 60-62, res chemist, Org Chem Dept, 62-64; res assoc biochem, Res Inst Skeletomuscular Dis, Hosp Joint Dis, New York, 64-76; ASSOC PROF BIOCHEM, UNIV TENN CTR HEALTH SCI, MEMPHIS, 76- *Concurrent Pos:* Advan res fel, Am Heart Asn, Inc, 65-69, estab investr, 69-74; res asst prof biochem, Mt Sinai Sch Med, 73-76. *Mem:* AAAS; Am Soc Biol Chemists; Am Chem Soc. *Res:* Structure, chemistry and biochemistry of connective tissue and mucopolysaccharides; structure and chemistry of natural highpolymers; sugar chemistry. *Mailing Add:* Box 206 Lewisburg WV 24901-0206

MASHHOON, BAHRAM, b Tehran, Iran, Sept 9, 47; m 78; c 1. THEORY OF GRAVITATION, THEORY OF RELATIVITY. *Educ:* Univ Calif, Berkeley, AB, 69; Princeton Univ, PhD(physics), 72. *Prof Exp:* Asst prof physics, Arya-Mehr Univ, Tehran, Iran, 72-73; res assoc, Princeton Univ, 73-74; postdoctoral fel, Univ Md, College Park, 74-76; res assoc & instr, Univ Utah, 76-78; lectr & res fel, Calif Inst Technol, 78-80; res fel, Univ Cologne, Fed Repub Ger, 80-85; ASSOC PROF PHYSICS, UNIV MO, COLUMBIA, 85- *Mem:* Am Phys Soc; Sigma Xi. *Res:* Theory of gravitation; nonlocal theory of accelerated observers; theory of measurement and wave phenomena in a gravitational field; hypothesis of locality in the theory of relativity; principle of complementerity of absolute and relative motion as well as gravitational radiation and black holes. *Mailing Add:* Dept Physics & Astron Univ Mo Columbia MO 65211

MASHIMO, PAUL AKIRA, b Osaka, Japan, Oct 25, 26; m 55; c 2. ORAL MICROBIOLOGY. *Educ:* Osaka Dent Univ, Japan, DDS, 48; Kyoto Med Univ, Japan, PhD(microbiol), 55. *Prof Exp:* Instr oral surg, Osaka Dent Univ, 48-50, lectr microbiol, 50-53, from asst prof microbiol to assoc prof pub health, 53-65; prof res assoc oral biol, 66, asst res prof, 66-67, asst prof oral biol, 68-69, ASSOC PROF ORAL BIOL, SCH DENT, STATE UNIV NY BUFFALO, 70-, MEM FAC GRAD SCH, 69- *Concurrent Pos:* Louise C Ball fel, Sch Dent & Oral Surg, Columbia Univ, 56-58; Japan Soc fel, 56-67; fac honor, Osaka Dent Univ, 72-; vis prof, Gifu Col Dent, 78-; hon consult, Sunstar Co, Osaka, 79-, Belmont/Takara Co, NJ, IPD Co, Osaka. *Mem:* Am Soc Microbiol; Int Asn Dent Res; NY Acad Sci; Sigma Xi. *Res:* Oral microbiology; immunology; dentistry. *Mailing Add:* 639 Cottonwood Dr Williamsville NY 14221

MASI, ALFONSE THOMAS, b New York, NY, Oct 29, 30; m 60; c 4. INTERNAL MEDICINE. *Educ:* City Col New York, BS, 51; Columbia Univ, MD, 55; Johns Hopkins Univ, MPH, 61, DrPH(epidemiol), 63. *Prof Exp:* Intern med, Osler Serv, Johns Hopkins Hosp, 55-56; sr asst surgeon, Commun Dis Ctr, USPHS, 56-58; asst resident, Johns Hopkins Hosp, 58-59; assoc resident, Med Ctr, Univ Calif, Los Angeles, 59-60; res fel epidemiol, Sch Hyg & Pub Health, Johns Hopkins Univ, 60-63, asst prof epidemiol, 63-65; instr med, Sch Med, 63-67, assoc prof epidemiol, Sch Hyg & Pub Health, 65-67; prof med & prev med & chief sect rheumatol, Dept Med, Univ Tenn Ctr Health Sci, Memphis, 67-72, dir div connective tissue dis, 72-78; prof med & head dept, Sch Med, 78-85, PROF EPIDEMIOL, SCH PUB HEALTH, UNIV ILL, PEORIA, 79-, PROF MED, SCH MED, UNIV ILL, PEORIA, 85- *Concurrent Pos:* Consult, Radiol Health Res Br, USPHS, 63-67, Nat Inst Arthritis & Metab Dis spec fel, 63-66; res geog epidemiol sect, Vet Admin, 64-66 & 78-80; mem subcomt epidemiol use of hosp data, Nat Comt Health & Vital Statist, 65-69; consult, US Food & Drug Admin, 65-70; sr investr, Arthritis Found, 66-71; Russell L Cecil fel award, 70-71; mem arthritis training grant comt, Nat Inst Arthritis & Metab Dis, 71-73. *Mem:* Am Fedn Clin Res; Am Rheumatism Asn; fel Am Pub Health Asn; fel Am Col Physicians; fel Am Col Epidemiol. *Res:* Application of epidemiologic methods, such as population studies, community-wide hospital surveys and case-control investigations to research in chronic diseases in order to define better their causes and pathogenesis. *Mailing Add:* Dept of Med 1 Illini Dr PO Box 1649 Peoria IL 61656

MASI, JAMES VINCENT, b Norwalk, Conn, Sept 21, 38; m 64; c 6. PHYSICS, MATERIALS SCIENCE. *Educ:* Fairfield Univ, BS, 60; Long Island Univ, MS, 70; Univ Del, PhD, 81. *Prof Exp:* Physicist res & develop, Transitron Electronics Corp, 60-61; teacher-researcher, Boston Col, 61-62; sr physicist, Space Age Mat, 62-65, Servo Corp, 65-66 & Hartman Systs Co, 66-69; sr engr-corp consult, Bunker-Ramo Corp, 69-73; dir res & develop, Innotech Corp, 73-74; consult, Optix Corp, 74-75; vpres, UCE Inc, 75-77; prog develop mgr, Inst Energy Conversion, Univ Del, 77-80; PROF ELEC ENG, WESTERN NEW ENG COL, 80-; DIR RES, SHRINERS HOSP, 89- *Honors & Awards:* AT&T Found Award, 86. *Mem:* Inst Elec & Electronics Engrs; Soc Info Display; Electrochem Soc; Am Soc Metals Int; AAAS; Am Ceramic Soc; Am Soc Mech Engrs; Am Vacuum Soc. *Res:* Solid state electronic devices, materials; electro-optics devices and materials; solar energy; photovoltaic devices and materials; microwave properties of materials; biomechanics. *Mailing Add:* Dept Elec Eng Western New Eng Col 1215 Wilbraham Rd Springfield MA 01119

MASKAL, JOHN, b Garfield, NJ, Sept 27, 18; m 43; c 2. PHYSICAL CHEMISTRY. *Educ:* Syracuse Univ, BS, 40; Mich State Univ, MS, 42. *Prof Exp:* Supvr anal lab, Evansville Ord Plant, 42-44; engr mat testing, Chrysler Corp, 44-45; high sch teacher, Mich, 46-47; mgr plating plant, Ludington Plating Co, 47-50; supvr, Denham Mfg Co, 50-51; supvr, Res Lab, Dow Chem USA, 51-60, dir, Gen Lab, 60-80; RETIRED. *Concurrent Pos:* Group leader, Ludington Prod Labs, 72, mgr environ qual control, 73. *Mem:* Am Chem Soc. *Res:* Analytical and physical chemical research involving inorganic chemical manufacture. *Mailing Add:* 712 Dexter Ludington MI 49431

MASKEN, JAMES FREDERICK, b Frederick, Md, Apr 4, 27; m 59; c 3. PHYSIOLOGY, BIOCHEMISTRY. *Educ:* NY Univ, BA, 53; Colo State Univ, MS, 60, PhD(physiol), 65. *Prof Exp:* Lab asst pharmacol, Wm R Warner Co, 50-52 & Nepera Chem Co, 52-53; res technician, Surg Dept, Sinai Hosp, Baltimore, Md, 54-55; lab technician, Biochem Div, Toni Co, 55; res asst endocrinol, 55-62; from instr to asst prof physiol, 62-69, assoc prof, 69-77, PROF PHYSIOL, COLO STATE UNIV, 77- *Concurrent Pos:* Vis assoc prof, Univ Calif, 70-71. *Mem:* AAAS; Am Physiol Soc; Can Physiol Soc; Soc Study Reproduction. *Res:* Reproductive physiology; neuroendocrinology. *Mailing Add:* Dept of Physiol & Biophys Colo State Univ Ft Collins CO 80523

MASKER, WARREN EDWARD, b Honesdale, Pa, July 8, 43; m 85; c 1. MOLECULAR BIOLOGY. *Educ:* Lehigh Univ, BS, 65; Univ Rochester, PhD(physics), 70. *Prof Exp:* Fel, Univ Rochester, 69-71; fel, Stanford Univ, 71-73; fel, Med Sch, Harvard Univ, 73-74; mem res staff, biol div, Oak Ridge Nat Lab, 75-84; DEPT BIOCHEM, TEMPLE UNIV, 84- *Concurrent Pos:* Am Cancer Soc fel, 70; Helen Hay Whitney Found fel, 71; mem microbiol genetics, NIH Study Sect, 86-88. *Mem:* Am Phys Soc; Am Soc Biochem &

Molecular Biol; Am Soc Microbiol; fel Am Soc Adv Sci. *Res:* Study of DNA replication and the molecular mechanism of DNA repair in bacteria and bacteriophage. *Mailing Add:* Dept Biochem Temple Univ Philadelphia PA 19140

MASKIN, ERIC S, b New York, NY, Dec 12, 50; c 2. APPLIED MECHANISM. *Educ:* Harvard Univ, AB, 72, AM, 74, PhD(math), 76. *Hon Degrees:* Cambridge Univ, AM, 76. *Prof Exp:* Res fel, Jesus Col, Cambridge Univ, 76-77; from asst prof to prof econ, Mass Inst Technol, 77-84; PROF ECON, HARVARD UNIV, 85- *Concurrent Pos:* Am ed, Rev Econ Studies, 77-82; assoc ed, Social Choice & Welfare, 83; ed, Quart J Econ, 84-90; adv ed, J Risk & Uncertainty, 87; assoc ed, Games & Econ Behav, 88; mem coun, Economet Soc, 90- *Mem:* Economet Soc. *Mailing Add:* Dept Economics Harvard Univ Littauer 308 Cambridge MA 02138

MASKIT, BERNARD, b New York, NY, May 27, 35; m; c 3. MATHEMATICS. *Educ:* NY Univ, AB, 57, MS, 62, PhD(math), 64. *Prof Exp:* Mem, Inst Advan Study, 63-65; from asst prof to assoc prof math, Mass Inst Technol, 65-70, Sloan Found fel, 70-71; chmn dept, 74-75, PROF MATH, STATE UNIV NY STONY BROOK, 71- *Mem:* Am Math Soc. *Res:* Riemann surfaces; Kleinian groups and low dimensional topology. *Mailing Add:* Dept Math State Univ NY Stony Brook NY 11794-3651

MASLACH, GEORGE JAMES, b San Francisco, Calif, May 4, 20; m 43; c 3. AERONAUTICAL ENGINEERING. *Educ:* Univ Calif, Berkeley, BS, 42. *Prof Exp:* Staff mem, Radiation Lab, Mass Inst Technol, 42-45 & Gen Precision Labs, NY, 45-49; res engr, 49-52, assoc prof mech design, 52-58, asst dir, Inst Eng Res, 56-58, assoc prof aeronaut eng, 58-59, chmn, Div Aeronaut Sci, 60-63, actg dean, Col Eng, 63, dean, 63-72, provost, Prof Schs & Cols, 72-81, vchancellor, Res & Acad Serv, 81-83, PROF AERONAUT ENG, UNIV CALIF, BERKELEY, 59- *Concurrent Pos:* Consult, Wright Air Develop Ctr, US Air Force, 54-55, Martin Aircraft Co, 58-60, Missile & Space Vehicles Dept, Gen Elec Co, 58-62 & Aeronaut Div, Ford Motor Co, 60-63; mem, Adv Group for Aeronaut Res & Develop, NATO, 60-; mem adv comt, US Naval Postgrad Sch, 64-65, US Naval Acad, 66-; mem tech adv bd, US Dept Commerce, 64- *Mem:* Am Soc Mech Engrs. *Res:* Rarefied gas dynamics and heat transfer; fluid mechanics; low density aerodynamics facilities. *Mailing Add:* 265 Panoramic Way Berkeley CA 94704

MASLAK, PRZEMYSLAW BOLESLAW, b Wroclaw, Poland, May 26, 54; m 78; c 2. PHYSICAL ORGANIC CHEMISTRY. *Educ:* Tech Univ, Wroclaw, Poland, MS, 78; Univ Ky, Lexington, PhD(chem), 82. *Prof Exp:* Res assoc chem, Columbia Univ, NY, 82-85; ASST PROF CHEM, PA STATE UNIV, 85- *Mem:* Am Chem Soc; AAAS. *Res:* Physical organic chemistry; chemistry of radical ions and radicals; electron transfer; molecular electronic devices; synthesis of theoretically interesting molecules; design of molecular catalysts. *Mailing Add:* Chem Dept Pa State Univ 152 Davey Lab University Park PA 16802

MASLAND, RICHARD HARRY, b Philadelphia, Pa, June 12, 42; m 86. NEUROBIOLOGY, PHYSIOLOGY. *Educ:* Harvard Univ, AB, 64; McGill Univ, PhD, 68. *Prof Exp:* Fel neurophysiol, Med Sch, Stanford Univ, 68-71; res assoc, 71-75, from asst prof to assoc prof physiol, 75-89, PROF NEUROSCI, HARVARD MED SCH, 89- *Concurrent Pos:* Prof ophthal, Mass Eye & Ear Infirmary, 90- *Honors & Awards:* NIH res career develop award, 77- *Mem:* Asn Res Vision & Ophthal; Neurosci Soc; Am Physiol Soc. *Res:* Biology of the retina. *Mailing Add:* Mass Gen Hosp Boston MA 02114

MASLEN, STEPHEN HAROLD, b Cleveland, Ohio, Jan 28, 26; m 51; c 3. APPLIED MATHEMATICS. *Educ:* Rensselaer Polytech Inst, BAeroEng, 45, MAeroEng, 47; Brown Univ, PhD(appl math), 52. *Prof Exp:* Aeronaut res scientist, Nat Adv Comt Aeronaut, 47-58, chief plasma physics br, Advan Propulsion Div, Lewis Res Ctr, NASA, 58-60; prin res scientist, Martin Co, 60-67, assoc dir, Martin Marietta Labs, 67-89; RETIRED. *Mem:* Nat Acad Eng; fel Am Inst Aeronaut & Astronaut. *Res:* Fluid dynamics. *Mailing Add:* 1315 Margarette Ave Baltimore MD 21204

MASLOW, DAVID E, b Brooklyn, NY, July 6, 43; m 72; c 3. ONCOLOGY, CELL BIOLOGY. *Educ:* Brooklyn Col, BS, 63; Univ Pa, PhD(zool), 68. *Prof Exp:* From cancer res scientist to cancer res scientist II, Roswell Park Mem Inst, 68-77, cancer res scientist IV, 77-89; asst prof, Dept Biophys, State Univ NY, Buffalo, 78-89, adj asst prof, Dept Oral Biol, Sch Dent Med, 87-89; HEALTH SCIENTIST ADMIN, NAT CANCER INST, 89- *Concurrent Pos:* Icrett fel, UICC, 83. *Mem:* Am Asn Cancer Res; AAAS. *Res:* Cell interactions and behavior in neoplastic and developing systems; invasion and metastasis. *Mailing Add:* Grants Rev Br Nat Cancer Inst Westwood Bldg Rm 820 Bethesda MD 20892

MASNARI, NINO A, b Three Rivers, Mich, Sept 20, 35; m 57; c 3. ELECTRONIC MATERIALS PROCESSING, SUBMICRON SEMICONDUCTOR DEVICES. *Educ:* Univ Mich, BS, 58, MS, 59, PhD(elec eng), 64. *Prof Exp:* Res assoc & lectr elec eng, Univ Mich, 64-67; res engr, Gen Elec Res & Develop Ctr, 67-69; assoc prof elec eng, Univ Mich, 69-77, prof & lab dir, 77-79; dept head, 79-88, PROF ELEC ENG, NC STATE UNIV, 79-, DIR, 88- *Concurrent Pos:* Dir, Electron Physics Lab, Univ Mich, 75-79, NSF Eng Res Ctr Advan Electronic Mat Processing, 88-; assoc ed, Inst Elec & Electronic Engrs Transactions on Electron Devices, 77-81; conf chmn, 1984 Inst Elec & Electronics Engrs Int Electron Devices Meeting, 80-84; chmn, Inst Elec & Electronics Engrs David Sarnoff Award Comt, 85-86. *Mem:* fel Inst Elec & Electronics Engrs; Am Soc Eng Educators; Mat Res Soc. *Res:* Semiconductor materials and technology; advanced electronic materials processing; high-speed devices; novel heterojunction devices; field effect transistors; electron beam devices. *Mailing Add:* NC State Univ Box 7920 Raleigh NC 27695

MASNER, LUBOMIR, b Prague, Czech, Apr 18, 34; Can citizen; m 61; c 2. ENTOMOLOGY. *Educ:* Charles Univ, BSc, 54, MSc, 57; Czech Acad Sci, PhD(entom), 62. *Prof Exp:* Res scientist, Inst Entom, Czech Acad Sci, 57-68; RES SCIENTIST ENTOM, BIOSYST RES INST, AGR CAN, 69- *Concurrent Pos:* Vis scientist, Brit Mus Natural Hist, London, 61, Smithsonian Inst, 64 & Harvard Univ, 66; Nat Res Coun Can fels, 65-66 & 68-69. *Mem:* Entom Soc Can. *Res:* Biosystematics of parasitic wasps of the superfamily Proctotrupoidea (Hymenoptera); taxonomy; evolution; geological history; ecology. *Mailing Add:* 129 Woodfield Dr Nepean ON K2G 0A1 Can

MASNYK, IHOR JAREMA, b Mostyska, Ukraine, Sept 17, 30; m 56; c 2. CHEMISTRY, BIOLOGICAL SCIENCES. *Educ:* Univ Chicago, BA, 53, MS, 58 & PhD(chem), 62. *Prof Exp:* Head, Drug Procurement Distrib Sect, Cancer Chemother Nat Serv Ctr, Nat Cancer Inst, NIH, 62-66; dir chief drug control, Dept Health, 66-68; head, Biol Activ Hormones Sect, Gen Labs & Clin, 68-73, chief, Planning & Anal Br, 73-78, DIR, EXTRAMURAL RES PROG, NIH, 78-, DEP DIR SCI ADMIN, DIV CANCER BIOL, DIAG & CENTERS, NAT CANCER INST, 78- *Concurrent Pos:* Assoc dir, Off Int Affairs, Nat Cancer Inst, 83-88. *Mem:* AAAS. *Res:* Starting with synthesis of steroid markers for breast cancer patients, organized and managed serum bank for tumor markers. *Mailing Add:* NIH Bldg 31 Rm 3A03 Bethesda MD 20892

MASO, HENRY FRANK, b Perth Amboy, NJ, Nov 20, 19; m; c 3. COSMETIC CHEMISTRY. *Educ:* City Col NY, BS, 40. *Prof Exp:* Asst bur biol res, Rutgers Univ, 41; jr chemist, Philadelphia Navy Yd, 41-44; sr res chemist, Johnson & Johnson, 44-57; dir tech serv, Am Cholesterol Prod, Inc, 57-70, vpres, 70-79, sr vpres, AmerChol, CPC Int Inc, 80-86; RETIRED. *Concurrent Pos:* Mem, Praesidium Int Fedn Socs Cosmetic Chemists rep USA Soc, 78, adv, 85-; mem sci adv comt, Cosmetic, Toiletry, & Fragrance Asn, 81-; pub relations secy, Int Fedn Socs Cosmetic Chemists, 87-; consult, H F Maso Assoc, 86-; course dir, Ctr Prof Advan, 86- *Honors & Awards:* Medalist, Soc Cosmetic Chemists, 71. *Mem:* Am Chem Soc; Soc Cosmetic Chemists (pres, 67); Int Fedn Socs Cosmetic Chemists (pres, 83). *Res:* Manufacture and utilization of raw materials for cosmetics, dermatologicals and pharmaceuticals. *Mailing Add:* H F Maso Assocs 1050 George St 12 M New Brunswick NJ 08901

MASON, ALLEN SMITH, b Tulsa, Okla, Dec 9, 32; m 55; c 1. ATMOSPHERIC RADIOACTIVITY, RESEARCH AVIATION. *Educ:* Kans State Univ, BS, 54; Fla Inst Technol, MS, 67; Univ Miami, PhD(marine & atmospheric sci), 74. *Prof Exp:* Chemist inorg chem, FMC Labs, 60; mem tech staff, RCA Labs, 61-63; supt tech eval, Pan Am World Airways, 63-69; res assoc prof marine & atmospheric chem, Univ Miami, 74-80; STAFF MEM, LOS ALAMOS NAT LAB, 81- *Concurrent Pos:* Proj leader, Los Alamos Nat Lab, 86-; adj assoc prof, Univ Miami, 81-; brig gen, USAFR, 86-90. *Mem:* Am Chem Soc; Am Geophys Union. *Res:* Atmospheric tracer studies, lofting of dust by very large explosions. *Mailing Add:* Los Alamos Nat Lab, MSJ514 INC-7 Los Alamos NM 87545

MASON, ARTHUR ALLEN, b St Louis, Mo, May 6, 25; m 64. INFRARED PHYSICS. *Educ:* Univ Okla, BS, 51; Univ Tenn, PhD(physics), 63. *Prof Exp:* From asst prof to assoc prof, 64-74, asst dean, 76-80, assoc dean, 80-86, PROF PHYSICS, SPACE INST & DEPT PHYSICS & ASTRON, UNIV TENN, 74- *Concurrent Pos:* Consult Air Force, Arnold Eng Develop Ctr, 64-; dir ann short course, Infrared Physics & Technol, 70- *Mem:* Am Phys Soc; Optical Soc Am; Am Inst Aeronaut & Astronaut; Sigma Xi. *Res:* Infrared radiation intensity spectroscopy of gases. *Mailing Add:* Rte 7 PO Box 7324-E Angwen Ave Manchester TN 37355

MASON, BERYL TROXELL, b Victoria, BC, Jan 21, 07; US citizen; m 35; c 2. NEUROLOGY, PSYCHIATRY. *Educ:* Univ Wash, BS, 29; Univ Chicago, MS, 32, MD, 36. *Prof Exp:* Resident, Univ Chicago, 36; intern, St Margaret's Hosp, Pittsburgh, 36-37; instr bact & pub health, Col Med, Univ Ill, 37-38; instr phys diag, Univ NC, 42-44; pvt pract, NC, 42-44 & Ill, 45-53; res assoc neurol & neurosurg, Col Med, Univ Ill, 53-58, consult, 58-61; actg chief neurol, Vet Admin Hosp, Topeka, Kans, 61-64; clin dir, Ark Rehab Serv, 64-69; study of Mid-East dis, Istanbul, Turkey, 69-71; CONSULT IN RES, MATROX LABS, 71- *Res:* Central nervous system; epilepsy; brain x-radiation; public health; radiation; neurological disorders. *Mailing Add:* 5059 N Jones Rd Oak Harbor WA 98277

MASON, BRIAN HAROLD, b Port Chalmers, NZ, Apr 18, 17; m 43. GEOCHEMISTRY. *Educ:* Univ NZ, MSc, 38; Univ Stockholm, PhD(mineral), 43. *Prof Exp:* Res officer, NZ Govt, 43-44; sr lectr geol, Univ NZ, 44-47; assoc prof mineral, Ind Univ, 47-53; cur phys geol & mineral, Am Mus Natural Hist, 53-65; RES CUR, DIV METEORITES, US NAT MUS, 65- *Mem:* Fel Mineral Soc Am (pres, 65-66); Geochem Soc (pres, 64-65); Royal Soc NZ; Swedish & Norweg Geol Soc. *Res:* Geochemistry; petrology; regional geology; meteorites. *Mailing Add:* Smithsonian Inst Div of Meteorites Washington DC 20560

MASON, CAROLINE FAITH VIBERT, b Harrogate, Eng, Feb 24, 42; US citizen; m 69; c 2. INORGANIC CHEMISTRY. *Educ:* Univ London, BSc, 64, PhD(chem), 67. *Prof Exp:* Fel, State Univ NY Buffalo, 67-68; chemist, Howmet Corp, Dover, NJ, 69-70; assoc scientist chem, Ortho Res Found, Raritan, NJ, 70-71; biochemist, Los Alamos Med Ctr, 72-74, STAFF MEM CHEM, LOS ALAMOS NAT LAB, 75- *Concurrent Pos:* Consult, Particle Technol Inc, Coulter Electronics, 73-75; vis scholar, Cornell Univ, 90-91. *Mem:* The Chem Soc; Am Chem Soc. *Res:* Thermochemical cycles for the decomposition of water to hydrogen and oxygen and related problems such as catalysis, separation of gases, kinetics and materials problems. *Mailing Add:* 148 Piedra Loop Los Alamos NM 87544

MASON, CHARLES EUGENE, b Brighton, Colo, Aug 28, 43; m 89; c 3. PEST MANAGEMENT, BIOLOGICAL CONTROL. *Educ:* Colo State Univ, BS, 68; Univ Mo, MS, 71; Kans State Univ, PhD(entom), 73. *Prof Exp:* Res asst entom, Univ Mo, 68-71 & Kans State Univ, 71, asst instr, 72-73; asst entomologist, Univ Ariz, 73-75; ASSOC PROF ENTOM, UNIV DEL, 75- *Concurrent Pos:* mem, Legional Res Comt Stalk Boring Lepidoptera, USDA, Secy 85-90, chair, 90- *Mem:* Entom Soc Am; Am Entom Soc (pres, 81-84). *Res:* Biological control, phenology models, pheromones, genetics and ecology of European corn borer. *Mailing Add:* Dept of Entom Univ of Del Newark DE 19717-1303

MASON, CHARLES MORGAN, b Kenora, Ont, July 7, 06; US citizen; m 29; c 3. PHYSICAL CHEMISTRY. *Educ:* Univ Ariz, BS, 28, MS, 29; Yale Univ, PhD(phys chem), 32. *Prof Exp:* Asst, Yale Univ, 29-31; from asst prof to assoc prof, Univ NH, 32-41; phys chemist, US Bur Mines, 41-43; res chemist, Tenn Valley Authority, 43-48; phys chemist, US Bur Mines, 48-50, chief explosives res sect, 50-56, phys res sect, 56-60, proj coordr explosive res ctr, 60-69, supvry res chemist, Pittsburgh Mining & Safety Res Ctr, 69-77; RETIRED. *Concurrent Pos:* Explosives consult, Aluminum Co Am, 72-78. *Honors & Awards:* Garland Scholar, Yale Univ. *Mem:* Am Chem Soc. *Res:* Thermodynamic properties of phosphates, rare earths and barium salts; water adsorption of glue; metallurgy of aluminum and lithium; magneto chemistry; non-metallic minerals; ignition of fire-damp; explosives and explosion phenomena; hazardous chemicals; ammonium nitrate. *Mailing Add:* 2200 West Liberty Ave #204 Pittsburgh PA 15226-1504

MASON, CHARLES PERRY, b Newport, RI, Aug 12, 32; m 58; c 2. BOTANY. *Educ:* Univ RI, BS, 54; Univ Wis, MS, 58; Cornell Univ, PhD(bot), 61. *Prof Exp:* Instr bot, Univ Wis, Milwaukee, 57-58; from asst prof to assoc prof biol, Hamline Univ, 61-67; assoc prof, 67-80, PROF BIOL, GUSTAVUS ADOLPHUS COL, 80- *Mem:* Sigma Xi; Phycol Soc Am; Int Phycol Soc. *Res:* Toxins produced by blue-green algae; unialgal growth of Anabaena and Dictyosphaerium in Lake Itasca; algal protein studies using acrylamide gel electrophoresis; ecology of Cladophora in farm ponds; effect of temperature shock on DNA content of beta-chromosome containing nuclei in maize. *Mailing Add:* Dept Biol Gustavus Adolphus Col St Peter MN 56082

MASON, CHARLES THOMAS, JR, b Joliet, Ill, Mar 26, 18; m 43; c 1. BOTANY. *Educ:* Univ Chicago, BS, 40; Univ Calif, MA, 42, PhD(bot), 49. *Prof Exp:* Instr bot, Univ Wis, 49-53; from asst prof to assoc prof, 53-62, PROF BOT, UNIV ARIZ, 62-, BOTANIST & CUR HERBARIUM, 53- *Mem:* Fel AAAS; Sigma Xi; Am Soc Plant Taxon; Int Asn Plant Taxon; Am Inst Biol Sci. *Res:* Cytotaxonomy of angiosperms, Limnanthaceae and Gentianaceae; flora of Arizona. *Mailing Add:* Herbarium-113 Shantz Bldg Univ Ariz Tucson AZ 85721

MASON, CONRAD JEROME, b Detroit, Mich, Jan 12, 32. MICROMETEOROLOGY, BIOMETEOROLOGY. *Educ:* Univ Mich, BS, 53; Univ Calif, Berkeley, MA, 55, PhD(physics), 60. *Prof Exp:* Assoc res physicist, Radiation Lab, 60-63 & High Altitude Eng Lab, 63-70, lectr, Dept Meteorol & Oceanog, 70-74, RES SCIENTIST & LECTR, DEPT ATMOSPHERIC & OCEANIC SCI, UNIV MICH, ANN ARBOR, 74-, RES SCIENTIST, BIOL STA, 79- *Concurrent Pos:* Pres, Aeromatrix Inc, Ann Arbor, 76-; vis prof dept plant physiol & entomol, Univ RI, 77-78; adj prof dept entomol, Mich State Univ, East Lansing, 79- *Mem:* Am Phys Soc; Am Geophys Union; Am Meteorol Soc; Air Pollution Control Asn; Int Asn Aerobiol. *Res:* Atmospheric science; air pollution; interaction of biological organisms with their physical environment. *Mailing Add:* Comp Syst Consult & Support Serv Univ Mich 611 Church St 2nd Floor Ann Arbor MI 48109-3056

MASON, CURTIS LEONEL, b Daingerfield, Tex, Oct 9, 19; m 42; c 2. PESTICIDES. *Educ:* Tex Agr & Mech Col, BS, 40, MS, 42; Univ Ill, PhD(plant path), 47. *Prof Exp:* Agent bur plant indust, soils & agr eng, USDA, Tex, 39-42; asst, Univ Wis, 42-43; spec asst, Univ Ill, 46-47; assoc pathologist, Tex A&M Univ, 47-48; asst prof plant path, Univ Ark, 48-52; plant pathologist, Niagara Chem Div, Food Mach & Chem Corp, 52-54; asst sales mgr, 54-57, mgr tech serv, 57-59, regional mgr, 59-62; microbiologist, Buckman Labs, Inc, 62-65, area mgr, 65-70; exten plant pathologist, Univ Ark, Little Rock, 71-77, pesticide coordr, Agr Exten Serv, 77-85; southern regional coordr, Nat Agr Pesticide Impact Assessment prog, Ark Agr Exp Sta, USDA, 77-85; RETIRED. *Mem:* Am Phytopath Soc. *Res:* Diseases of cotton, orchard crops and peaches; testing of fungicides; fungicidal action of 8-quinolinol and some of its derivatives; industrial microorganism control. *Mailing Add:* 4712 Hampton Rd Little Rock AR 72114

MASON, D(AVID) M(ALCOLM), b Los Angeles, Calif, Jan 7, 21; m 53. CHEMICAL ENGINEERING, CHEMISTRY. *Educ:* Calif Inst Technol, BS, 43, MS, 47, PhD(chem eng), 49. *Prof Exp:* Chem engr, Standard Oil Co, 43-46; instr chem eng, Calif Inst Technol, 49-51, supvr appl phys chem group, Jet Propulsion Lab, 52-55; assoc prof chem eng, 55-57, chmn dept, 55-72, assoc dean undergrad studies, 73-76, assoc dean eng, 79-82, prof chem eng & chem, 57-86, EMER PROF CHEM ENG & CHEM, STANFORD UNIV, 87- *Concurrent Pos:* NSF fel, Imp Col, Univ London, 64-65; liaison scientist, Office Naval Res, London, 72-73; vis prof, Imperial Col, Univ London, 78-79 & 86. *Honors & Awards:* Founders Award, Am Inst Chem Engrs, 84; David M Mason lectr, 75. *Mem:* Fel Am Inst Chem Engrs; Am Chem Soc; Am Electrochem Soc. *Res:* Investigation of electrocatalysis and transport processes involved in solid oxide electrolytes used in electrochemical reactors for fuel cells; pollution abatement; methane activation. *Mailing Add:* 148 Doud Dr Los Altos CA 94022

MASON, D(ONALD) R(OMAGNE), b Urbana, Ill, Aug 19, 20; m; c 3. CHEMICAL ENGINEERING. *Educ:* Univ Ill, BS, 42; Univ Minn, PhD(chem eng), 49. *Prof Exp:* Chem engr, Plastics Dept, E I du Pont de Nemours & Co, 42-44; asst chem eng, Univ Minn, 46-48; mem tech staff, Bell Tel Labs, Inc, NJ, 49-52, 53-56; assoc prof chem eng, Univ Mich, 56-61, prof, 61-65; sr scientist, Phys Electronics Div, Radiation, Inc, 65-67, Microelectronics Div, 67- 70, sr scientist, Harris Semiconductor Div, Harris-Intertype Corp, 70-77; head, depts environ sci & eng, 77-84, PROF CHEM ENG, FLA INST TECHNOL, 77-, HEAD DEPT, 84- *Concurrent Pos:* Fulbright scholar, Univ Nancy, 52-53; adj asst prof, NY Univ, 53-54; adj instr, Polytech Inst Brooklyn, 54-56; exchange prof, Ecole Normale Superieure, Paris, 63-64; adj prof, Fla Inst Technol, 66-77; indust consult, 77- *Mem:* Am Chem Soc; Electrochem Soc; Inst Elec & Electronics Engrs; Am Inst Chem Engrs; Am Phys Soc. *Res:* Continuous stirred tank reactor systems; semiconducting materials; chemical process dynamics; integrated circuits; processing electrochemistry. *Mailing Add:* 504 S Riverside Dr Indialantic FL 32903-4350

MASON, DAVID DICKENSON, b Adingdon, Va, Jan 22, 17; m 44; c 2. APPLIED STATISTICS. *Educ:* King Col, BA, 36; Va Polytech Inst, MS, 38; NC State Col, PhD(agron), 48. *Prof Exp:* Asst agronomist, Exp Sta, Va Polytech Inst, 38-39 & Miss State Col, 41; asst, NC State Col, 41 & 45-47; asst prof agron, Ohio State Univ, 47-49; biometrician, Bur Plant Indust, USDA, 49-53; prof, 53-81, head dept & head, Inst Statist, 63-81, EMER PROF STATIST, NC STATE UNIV, 81- *Concurrent Pos:* Statist consult, Res Div, United Fruit Co, Mass, 57-; chmn, Southern Regional Ed Bd Comt on Statist, 73-75. *Mem:* Fel Soil Sci Soc Am; fel Am Soc Agron; fel Am Statist Asn; Biomet Soc. *Res:* Applied statistics; soil and plant science. *Mailing Add:* Dept of Statist NC State Univ Box 8203 Raleigh NC 27695

MASON, DAVID LAMONT, b Warren, Pa, Dec 24, 34; m 63; c 1. BOTANY. *Educ:* Edinboro State Col, BS, 63; Univ Wis, MS, 67, PhD(bot), 70. *Prof Exp:* Teaching asst gen bot, Univ Wis, 63-64, teaching assoc, 64-65, res asst mycol, 65-69; from asst prof to assoc prof, 69-80, PROF BIOL, WITTENBURG UNIV, 80- *Mem:* Am Phytopath Soc; Sigma Xi. *Res:* Fungal parasitism; electron microscopy of human tumors, cancers and autoimmune diseases. *Mailing Add:* 2427 Rebecca Dr Springfield OH 45503

MASON, DAVID MCARTHUR, b Sidney, Mont, Jan 6, 12; m 38; c 4. CHEMICAL ENGINEERING, PHYSICAL CHEMISTRY. *Educ:* Mont State Col, BS, 34; Univ NDak, MS, 35. *Prof Exp:* Asst chemist, Socony-Vacuum Oil Co, 37-38; analyst, Standard Oil Develop Co Div, Standard Oil Co, 38-40, anal chemist, 40-52; supvr, Analysis Lab, 52-62, sr chemist, 62-77, SR ADV, INST GAS TECHNOL, 77- *Concurrent Pos:* Mem, Comm Solubility Data, Int Union Pure & Appl Chem. *Honors & Awards:* R H Glenn Award, Am Soc Testing & Mat, 87. *Mem:* Am Chem Soc; Am Soc Testing & Mat. *Res:* Correlation of coal properties; coal petrography; properties of gaseous fuels; coal ash chemistry; phase equilibria. *Mailing Add:* 5434 S Blackstone Ave Chicago IL 60615

MASON, DAVID THOMAS, b Berkeley, Calif, Jan 7, 37. LIMNOLOGY. *Educ:* Reed Col, BA, 58; Univ Calif, Davis, MA, 61, PhD(zool), 66. *Prof Exp:* Lectr zool, Univ Calif, Davis, 65; from asst prof to assoc prof, 66-80, PROF BIOL, FAIRHAVEN COL, WESTERN WASH UNIV, 81- *Concurrent Pos:* Asst prof, Univ Calif, Berkeley, 69-71. *Mem:* Am Soc Limnol & Oceanog; Sigma Xi; Int Soc Limnol; Ecol Soc Am. *Res:* Physical and biological limnology; saline lakes; arctic salt marshes; riparian ecology. *Mailing Add:* Fairhaven Col Western Wash Univ Bellingham WA 98225

MASON, DEAN TOWLE, b Berkeley, Calif, Sept 20, 32; m 57; c 2. CARDIOVASCULAR DISEASES. *Educ:* Duke Univ, BA, 54, MD, 58; Am Bd Internal Med, dipl, 65; Am Bd Cardiovasc Dis, dipl, 66. *Prof Exp:* From intern to asst resident, Osler Med Serv, Johns Hopkins Hosp, 58-61; asst resident med, Med Ctr, Duke Univ, 59-60; clin assoc, Cardiol Br, Nat Heart Inst, 61-63, asst sect chief cardiovasc diag, sr investr & attend physician, 63-68; prof med & physiol & chief sect cardiovasc med, Sch Med, Univ Calif, Davis, 68-82; PHYSICIAN-IN-CHIEF, WESTERN HEART INST, 83-; CHMN, DEPT CARDIOVASC MED, ST MARY'S MED CTR, SAN FRANCISCO, CALIF, 83- *Concurrent Pos:* Consult, Surg Br, Nat Heart Inst & Clin Ctr, NIH, 61-68; from clin asst prof to clin assoc prof med, Sch Med, Georgetown Univ, 65-68; fel, Coun on Circulation, Am Heart Asn, 66-, fel, Coun Clin Cardiol, 67-; consult, US Naval Med Ctr, Bethesda, Md, 67-68, Letterman Army Gen Hosp, San Francisco, Calif, 68-, David Grant Med Ctr, Travis AFB, Calif, NIH, Nat Aeronaut & Space Admin, Vet Admin & NSF; mem, Am Bd Internal Med; mem adv comt, US Pharmacopeia, 70, NIH Lipid Metab, 71 & NASA Life Sci, 73-; ed-in-chief, Am Heart J, 80- *Honors & Awards:* Am Therapeut Soc Award, 65 & 73. *Mem:* Am Soc Clin Invest; Am Physiol Soc; fel Royal Soc Med; Am Soc Pharmacol & Exp Therapeut; Am Col Cardiol (pres, 77-78). *Res:* Adult and pediatric clinical cardiology; cardiac catheterization and diagnosis; cardiovascular medicine, physiology, biochemistry and pharmacology; author or coauthor of over 1000 publications. *Mailing Add:* Western Heart Inst 450 Stanyan St San Francisco CA 94117

MASON, DONALD FRANK, b Chicago, Ill, Mar 17, 26; m; c 4. PHYSICAL CHEMISTRY, APPLIED MATHEMATICS. *Educ:* Univ Ill, BS, 49; Univ Wis, PhD(chem), 53. *Prof Exp:* Asst, Naval Res Lab, Univ Wis, 49-52; res assoc chem eng, Northwestern Univ, 52-55, asst prof, 55-58; assoc chemist, Argonne Nat Lab, 58-62; assoc prof chem, Natural Sci Div, Ill Teachers Col, Chicago-North, 62-68; prof chem, Northeastern Ill Univ, 68-86; RETIRED. *Concurrent Pos:* Consult, Vern Alden Co, 54-55. *Mem:* AAAS; Am Chem Soc; Sigma Xi. *Res:* Heterogeneous reaction kinetics; mass spectrometry; instrumentation. *Mailing Add:* 2300 Colfax St Evanston IL 60201

MASON, DONALD JOSEPH, b Kokomo, Ind, July 24, 31; m 53; c 4. MICROBIOLOGY. *Educ:* Purdue Univ, BS, 53, MS, 55, PhD, 58. *Prof Exp:* Res assoc microbiol, 58-66, sect head anal microbiol, 66-68, MGR FED DRUG ADMIN, UPJOHN CO, 68- *Mem:* Am Soc Microbiol; AAAS; Am Fedn Clin Res; Sigma Xi. *Res:* Microbial cytology and biochemistry; antibiotic production. *Mailing Add:* Upjohn Co 301 Henrietta St Kalamazoo MI 49006

MASON, EARL JAMES, b Marion, Ind, Aug 26, 23; m 46; c 2. PATHOLOGY, MICROBIOLOGY. *Educ:* Ind Univ, BS, 44, AB & MA, 47; Ohio State Univ, PhD(bact), 50; Western Reserve Univ, MD, 54. *Prof Exp:* Damon Runyon Cancer fel, 54-56; fel path, Postgrad Sch Med, Univ Tex, 58-59; asst prof path, Col Med, Baylor Univ, 59-60; asst pathologist, Michael Reese Hosp, Chicago, 60-61; assoc pathologist, Mercy Hosp, Chicago, 61-65, chmn dept biol sci, 62-65; DIR LABS, ST MARY MERCY HOSP, 65-; CLIN PROF PATH, SCH MED, INDIANA UNIV, 76- *Concurrent Pos:* From intern to resident, Case Western Reserve Univ, 54-56; assoc prof, Dept Path, Chicago Med Sch, 65- *Mem:* Am Asn Path & Bact; Am Asn Cancer Res; Am Soc Exp Path; Am Soc Hemat. *Res:* Mechanism of action of viruses on cells; cellular production of antibodies; thrombocytopathic action of viruses. *Mailing Add:* Mercy Hosp 540 Tyler St Gary IN 46402

MASON, EARL SEWELL, b Grand Forks, NDak, June 9, 35; m 58; c 3. LAW, BUSINESS ADMINISTRATION. *Educ:* Univ NDak, BSCE, 57, JurD, 73; Utah State Univ, PhD(civil eng), 65; ND State Univ, MBA, 88. *Prof Exp:* Civil engr, Bur Reclamation, US Dept Interior, 57, hydraul engr, 63; design engr, Aerospace Div, Boeing Airplane Co, 60; res asst meteorol, Colo State Univ, 60-61; asst prof civil eng, Univ Utah, 64-68; assoc prof, 68-75, PROF CIVIL ENG, UNIV NDAK, 75- *Concurrent Pos:* Prin investr, NSF res initiation grant, Univ Utah, 66-68; assoc dir, NDak Water Resources Res Inst, 75-87. *Mem:* Am Soc Civil Engrs; Am Soc Eng Educ; Nat Soc Prof Engrs. *Res:* Large open channel roughness; groundwater and surface water hydrology; meteorology; law. *Mailing Add:* Dept Civil Eng Univ NDak Grand Forks ND 58202

MASON, EDWARD ALLEN, b Atlantic City, NJ, Sept 2, 26; m 52; c 4. CHEMICAL PHYSICS. *Educ:* Va Polytech Inst, BS, 47; Mass Inst Technol, PhD(phys chem), 51. *Hon Degrees:* MA, Brown Univ, 68. *Prof Exp:* Res assoc chem, Mass Inst Technol, 50-52; Nat Res Coun fel, Univ Wis, 52-53; asst prof chem, Pa State Univ, 53-55; from assoc prof to prof molecular physics, Inst Molecular Physics, Univ Md, 55-67, dir, 66-67; PROF CHEM & ENG, BROWN UNIV, 67-, NEWPORT ROGERS PROF CHEM, 83- *Concurrent Pos:* Vis prof, Harvard, Mass Inst Technol, 75, Leiden, 81-82. *Honors & Awards:* Sci Achievement Award, Wash Acad Sci, 62. *Mem:* AAAS; Am Asn Physics Teachers; fel Am Phys Soc. *Res:* Molecular and ionic scattering and transport; equation of state of gases; theory of transport phenomena; membrane transport; intermolecular forces; statistical mechanics. *Mailing Add:* Dept of Chem Brown Univ Providence RI 02912

MASON, EDWARD ARCHIBALD, b Rochester, NY, Aug 9, 24; m 50; c 6. NUCLEAR ENGINEERING. *Educ:* Univ Rochester, BS, 45; Mass Inst Technol, SM, 48, ScD(chem eng), 50. *Prof Exp:* From instr to asst prof chem eng, Mass Inst Technol, 49-53; sr engr, Ionics, Inc, 53-54, dir res, 54-57; from assoc prof to prof nuclear eng, Mass Inst Technol, 57-75, head dept, 71-75; comnr, US Nuclear Regulatory Comn, 75-77; VPRES RES, AMOCO CORP, 77- *Concurrent Pos:* NSF sr fel, 65-66; consult govt agencies & indust co. *Honors & Awards:* Robert E Wilson Award, Am Inst Chem Engrs. *Mem:* Nat Acad Eng; Am Chem Soc; fel Am Acad Arts & Sci; fel Am Nuclear Soc; fel NY Acad Sci; Am Inst Chem Engrs. *Res:* Research management; nuclear fuel and power systems; venture start-ups. *Mailing Add:* Amoco Res Ctr PO Box 400 Naperville IL 60566

MASON, EDWARD EATON, b Boise, Idaho, Oct 16, 20; m 44; c 4. SURGERY. *Educ:* Univ Iowa, BA, 43, MD, 45; Univ Minn, PhD(surg), 53. *Prof Exp:* Intern surg, Univ Minn Hosps, 45-46, fel surg, 48-52; from asst prof to assoc prof, 53-60, PROF SURG, COL MED & UNIV HOSPS, UNIV IOWA, 60-, CHMN GEN SURG, 78-, actg head, 81-82. *Mem:* AAAS; Soc Univ Surgeons; Soc Exp Biol & Med; AMA; Am Col Surgeons. *Res:* Diseases of thyroid, parathyroid and gastrointestinal tract; pneumoperitoneum in giant hernia repair; side-to-side spenorenal shunt; fluid, electrolyte and nutritional balance; gastric bypass for obesity; fatty acid toxicity; vertical banded gastroplasty for obesity. *Mailing Add:* Dept Surg Univ Iowa Hosps Iowa City IA 52242

MASON, ELLIOTT BERNARD, b Detroit, Mich, July 29, 43; m 71; c 3. PHYSIOLOGY. *Educ:* Loyola Univ, Chicago, BS, 65; Wayne State Univ, MS, 69, PhD(biol), 72. *Prof Exp:* Asst prof biol, George Mason Col, Univ Va, 71-73; from asst prof to assoc prof, 73-83, PROF BIOL, STATE UNIV NY COL CORTLAND, 83- *Mem:* AAAS; Sigma Xi. *Res:* Writer in area of human physiology; human anatomy and physiology. *Mailing Add:* Dept Biol Sci State Univ NY Col, PO Box 2000 Cortland NY 13045

MASON, GEORGE ROBERT, b Rochester, NY, June 10, 32; m 56; c 3. SURGERY, PHYSIOLOGY. *Educ:* Oberlin Col, BA, 55; Univ Chicago, MD, 57; Stanford Univ, PhD(physiol), 68. *Prof Exp:* Teaching asst path, Univ Chicago, 54-56, teaching asst physiol, Stanford Univ, 60-62, actg instr surg, 65-66, from instr to assoc prof, 66-71; prof surg & physiol & chmn dept surg, Univ Md, Baltimore City, 71-80; PROF SURG & CHMN DEPT SURG, UNIV CALIF, IRVINE, 80- *Concurrent Pos:* Giannini fel, 66, Markle scholar acad med, Univ Md, 69-74; examnr, Am Bd Surg, 77-80, dir, 80-86, Residency Rev Comt Surg, 81-87; mem bd dirs regional planning, Coun Emergency Med Serv Develop Corp; chief surg, Med Ctr, Univ Calif, Irvine; attending surg, Long Beach Vet Admin Med Ctr, Long Beach Mem Hosp & St Joseph's Hosp. *Mem:* Am Gastroenterol Asn; Am Gastroenterol Asn; Am Col Surgeons; Am Surg Asn; Soc Clin Surg; Am Physiol Soc; Sigma Xi. *Res:* Gastrointestinal physiology, particularly autonomic control of visceral function; thoracic surgery. *Mailing Add:* PO Box 619 Elmhurst IL 60126-0619

MASON, GRANT WILLIAM, b Waialua, Hawaii, Aug 8, 40; m 64; c 5. COSMIC RAY PHYSICS. *Educ:* Brigham Young Univ, BA, 61; Univ Utah, PhD(physics), 69. *Prof Exp:* Res assoc & assoc instr physics, Univ Utah, 68-69; sci co-worker, Physics Inst, Aachen Tech Univ, 69-70; from asst prof to assoc prof, 70-79, PROF PHYSICS, BRIGHAM YOUNG UNIV, 79-, DEAN, COL PHYS & MATH SCI, 85- *Mem:* Am Phys Soc; Am Asn Physics Teachers; Sigma Xi. *Res:* High energy cosmic ray studies; plasma physics. *Mailing Add:* Dept Physics & Astron Brigham Young Univ Provo UT 84602

MASON, GRENVILLE R, b Sask, 34. NUCLEAR & PARTICLE PHYSICS. *Educ:* Univ BC, BASc, 56; McMaster Univ, MEng, 59; Univ Alberta, PhD(physics), 64. *Prof Exp:* Engr, Can Westinghouse Co, Ltd, 56-58; lectr, 62-64, from instr to assoc prof, 64-80, PROF PHYSICS, UNIV VICTORIA BC, 80- *Mem:* Can Asn Physicists; Am Phys Soc. *Res:* Mesonic atoms; experimental particle physics. *Mailing Add:* Dept Physics & Astron Univ Victoria Victoria BC V8W 2Y2 Can

MASON, HAROLD FREDERICK, b Porterville, Calif, Feb 15, 25; m 54; c 3. PHYSICAL CHEMISTRY. *Educ:* Cornell Univ, BChE, 50; Univ Wis, PhD(phys chem), 55. *Prof Exp:* Chem engr, Rohm & Haas Co, Pa, 50-51; res chemist, Chevron Res Co, Standard Oil Co, Calif, 54-59, group supvr, 59-64, sect supvr, 64-67, mgr petrol process develop div, 67-71, mgr petrol process res div, 71-82, mgr synthetic fuels div, 82-85, mgr residuum conversion div, 85-86; RETIRED. *Mem:* Am Chem Soc; Am Inst Chem Eng. *Res:* Chemical reaction kinetics; catalysis; petroleum processing; hydrogenation and hydrocracking; solid state reactions. *Mailing Add:* 553 Monarch Ridge Dr Walnut Creek CA 94596-2920

MASON, HAROLD L, b Compton, Calif, Apr 17, 01. ENDOCRINOLOGY. *Educ:* Univ Chicago, PhD(org chem), 27. *Prof Exp:* Head biochem sect, Mayo Clin, 57-63; prof biochem, Univ Minn Mayo Clin, 63-66; RETIRED. *Mem:* Am Chem Soc; Am Soc Biol Chemists; Endocrine Soc. *Mailing Add:* 211 Second St NW Apt 1010 Rochester MN 55901

MASON, HARRY LOUIS, b Louisville, Ky, May 1, 35; m 57. MECHANICAL ENGINEERING. *Educ:* Univ Ky, BS, 56, MS, 59. *Prof Exp:* Engr, Gen Elec Co, 56-57; instr mech eng, 57-59, ASST PROF MECH ENG, UNIV KY, 59- *Concurrent Pos:* Consult, Stephen Watkins Inc, 65-, Minister Press Co, 79, Allis Chalmers, 80; sr engr, Monsanto, 66-90. *Mem:* Am Soc Eng Educ. *Res:* Machine design and mechanisms; dynamic valve; nuclear blast closure device; 3 United States patents. *Mailing Add:* Dept Mech Eng Univ Ky Lexington KY 40506

MASON, HENRY LEA, mechanical engineering, for more information see previous edition

MASON, HERMAN CHARLES, public health, immunology; deceased, see previous edition for last biography

MASON, J(OHN) PHILIP HANSON, JR, b Richmond, Va, May 8, 30; m 53; c 3. AGRICULTURAL ENGINEERING. *Educ:* Va Polytech Inst, BS, 51, BS, 55, MS, 56; Univ Mo, PhD(agr eng), 62. *Prof Exp:* From asst prof to assoc prof, 55-71, head dept, 69-79, PROF AGR ENG, VA POLYTECH INST & STATE UNIV, 71- *Mem:* Sr mem Am Soc Agr Engrs. *Res:* Farm structures; alternate sources of energy for agriculture; temporary storage of hay. *Mailing Add:* Dept Agr Eng Va Polytech Inst & State Univ Blacksburg VA 24061

MASON, JAMES IAN, b Skipton, Yorkshire, Eng, Dec 1, 44; m 68; c 4. BIOCHEMISTRY. *Educ:* Univ Edinburgh, BSc, 66, PhD(biochem), 70. *Prof Exp:* Res fel, Southwestern Med Sch, Univ Tex, 70-72, asst prof biochem, 72-73; res fel, Univ Edinburgh, 73-77; staff scientist biochem, Worcester Found Exp Biol, 77-80; RES ASST PROF BIOCHEM & OBSTET-GYNEC, UNIV TEX HEALTH SCI CTR, DALLAS, 80- *Concurrent Pos:* Prin investr, Nat Cancer Inst grant, 77- *Mem:* Biochem Soc; Soc Endocrinol; Endocrine Soc; Am Soc Biol Chemists. *Res:* Membrane-bound hydroxylation enzymes; mechanism of tropic hormone action; steroid biosynthesis. *Mailing Add:* Biochem & Obstet-Gynec Depts Univ Tex Health Sci Ctr 5323 Harry Hines Blvd Dallas TX 75235

MASON, JAMES MICHAEL, b Kingsport, Tenn, Mar 19, 43; m 69. HISTOCOMPATIBILITY TESTING, PATERNITY EVALUATION. *Educ:* Memphis State Univ, BS, 66; Univ Tenn, PhD(exp path), 71. *Prof Exp:* From instr to assoc prof path, Univ Tenn, Memphis, 71-89; DIR PARENTAGE EVAL & HISTOCOMPATIBILITY TESTING, ROCHE BIOMED LABS INC, BURLINGTON, 89- *Concurrent Pos:* Consult, Chief Med Examr, State of Tenn, 71-89, paternity eval, Roche Biomed Lab, 86-89; mem comt, Histocompatibility Testing, Am Asn Blood Banks, 87-89, chmn, Transplantation Immunol Comt, 89- *Mem:* AAAS; Am Asn Blood Banks; Int Soc Blood Transfusion; Am Soc Histocompatibility & Immunogenetics. *Res:* Cryopreservation of platelets; histocompatibility typing techniques; DNA typing techniques and their use in parentage evaluation. *Mailing Add:* Dept Paternity Eval Roche Biomed Labs Inc 1447 York Court Burlington NC 27215

MASON, JAMES O, b Salt Lake City, Utah, June 19, 30; m; c 7. VIROLOGY. *Educ:* Univ Utah, BA, 54, MD, 58; Harvard Univ, MPH, 63, PhD(pub health), 67. *Prof Exp:* Intern, John's Hosp, 58-59; resident, Peter Brent Brigham Hosp, Boston, 61-62; USPHS, 59-68 & 69-70; chief infectious dis, Latter-Day Saints Hosp, 68-69; comnr health serv, Latter-Day Church, 70-76; assoc prof & chmn, Div Community Med, Univ Utah Col Med, 78-79; exec dir, Utah Dept Health, 79-83; dir, Ctr Dis Control & adminr, Toxic Substances & Dis Registry, 83-89; HHS ASST SECY HEALTH & HEAD PUB HEALTH SERV, DEPT HEALTH & HUMAN SERV, 89- *Concurrent Pos:* US deleg, WHO, Geneva, Switz. *Mem:* Inst Med-Nat Acad Sci; Am Med Asn; Am Soc Trop Med & Hyg; fel Am Pub Health Asn. *Mailing Add:* 200 Independence Ave SW Rm 7166 Washington DC 20201

MASON, JAMES RUSSELL, b New York, NY, Oct, 31, 54; m 76; c 3. LEARNING & MOTIVATION, PSYCHOPHYSICS. *Educ:* De Pauw Univ, BA, 76; Clark Univ, MA, 78, PhD(psychol), 80. *Prof Exp:* Asst prof, 82-84, ASSOC PROF, MONELL CHEM SENSES CTR, 84-; PROJ LEADER, USDA, 86- *Concurrent Pos:* Adj assoc prof, Dept Biol, Univ Pa, 84- *Mem:* Psychonomic Soc; Asn Chemoreception Sci; Animal Behavior Soc; Am Ornith Union; Wildlife Soc. *Res:* Sensory evaluation of vertebrate species (with emphasis on olfaction, trigeminal chemoreception and taste); transduction, perception and practical applications (wildlife management). *Mailing Add:* 3500 Market St Monell Chem Senses Ctr Philadelphia PA 19104

MASON, JAMES WILLARD, b Hollywood, Calif, Apr 5, 33; m 56; c 2. POLYMER CHEMISTRY, ANALYTICAL CHEMISTRY. *Educ:* Univ Calif, BS, 56, PhD(org chem), 60. *Prof Exp:* Chemist, Papermate Pen Co, 56; res assoc med chem, Merck Sharp & Dohme Res Labs, 60-64; sr scientist, Aeronutronic Div, Philco-Ford Corp, 64-69; scientist, Havens Int, 69; prin scientist & consult, Aeronautronic Div, Ford Aerospace & Communs Corp, 69-77, supvr process control, 77-78; vpres, Am Thermoform Corp, 78-80; pres, J Mason Assocs, Inc, 80-87; pres, Alex Jacobs Sales Co, Inc, 83-86; MGR, ANALYSIS SERV CTR, BAXTER HEALTHCARE CORP, 87- *Mem:* Am Chem Soc; Soc Plastic Engrs. *Res:* Plastics polymer chemistry; adhesives; water and waste treatment; membrane processes; analytical chemistry. *Mailing Add:* Baxter Healthcare Corp 2132 Michelson Dr Irvine CA 92715

MASON, JESSE DAVID, US citizen. MATHEMATICS. *Educ:* Univ Mo, Kansas City, BS, 62; Univ Calif, Riverside, PhD(math), 68. *Prof Exp:* Prod designer, Vendo Co, 58-62; dynamics engr, Gen Dynamics, Pomona, 62-65; res assoc math, Univ Calif, Riverside, 65-67; asst prof, Calif State Univ, San Bernardino, 67-68; asst prof, Univ Ga, 68-71; PROF MATH, UNIV UTAH, 71- *Concurrent Pos:* Adj assoc prof indust eng, Univ Utah, 77- *Mem:* Inst Math Statist; Am Statist Asn. *Res:* Limit theorems in probability theory and stochastic differential equations. *Mailing Add:* Dept Math Unit Utah Salt Lake City UT 84112

MASON, JOEL B, b Syracuse, NY, June 17, 55. NUTRITION. *Educ:* Univ Chicago, MD, 81. *Prof Exp:* ASST PROF MED, TUFTS UNIV SCH MED, 88- *Concurrent Pos:* Investr nutrit, USDA Human Nutrit Res Ctr, Tufts Univ, 89-; dir, Nutrit Support Ser, New Eng Med Ctr, 89- *Mem:* Am Inst Nutrit; Am Soc Clin Nutrit; Am Gastroenterol Soc; Am Fedn Clin Res; Am Soc Parenteral & Enteral Nutrit. *Res:* Clinical implications of folate and B12 metabolism. *Mailing Add:* 711 Washington St Boston MA 02111

MASON, JOHN FREDERICK, b Los Angeles, Calif, Nov 25, 13; m 39; c 4. PETROLEUM GEOLOGY. *Educ:* Univ Southern Calif, AB, 34, AM, 35; Princeton Univ, PhD(geol), 41. *Prof Exp:* Field geologist, Socony-Vacuum Oil Co, Egypt, 37-40; instr earth sci, Univ Pa, 41-42; field geologist, Venezuelan Atlantic Ref Co, Barcelona & Caracas, 42-46; geologist, Foreign Prod Dept, Atlantic Ref Co, Pa, 46-51; asst to gen mgr, Foreign Opers Dept, Union Oil Co, Calif, 52-54; mgr explor, Standard Vacuum Oil Co, India, 54-56, resident mgr, Prod Div, Pakistan, 56-59; staff geologist, Foreign Dept, Continental Oil Co, 59-65, sr explor adv, 65-75; RETIRED. *Concurrent Pos:* Field geologist, Pa Geol & Topog Surv, 41-; consult, 75-; consult, UN Dept Tech Coop Develop, People's Repub China & India, 80-84. *Mem:* Am Asn Petrol Geologists; Am Geol Inst; fel Geol Soc Am. *Mailing Add:* 240 Fisher Place Princeton NJ 08540-6444

MASON, JOHN GROVE, b Louisville, Ky, Dec 4, 29; m 56; c 2. ANALYTICAL CHEMISTRY. *Educ:* Univ Louisville, BS, 50; Ohio State Univ, PhD(chem), 55. *Prof Exp:* Instr chem, Ill Inst Technol, 56-59; assoc prof, 59-66, PROF CHEM, VA POLYTECH INST & STATE UNIV, 66- *Mem:* Am Chem Soc. *Res:* Polarography; electrode processes. *Mailing Add:* Dept of Chem Va Polytech Inst & State Univ Blacksburg VA 24061

MASON, JOHN HUGH, b Batavia, NY, Mar 8, 29; m 53; c 2. POLYMER CHEMISTRY, ABRASIVES TECHNOLOGY. *Educ:* Univ Rochester, BS, 50; Carnegie Inst Technol, PhD(org chem), 55. *Prof Exp:* Res chemist, Union Carbide Plastics Co Div, Union Carbide Corp, 54-61; sr res assoc, Carborundum Co, 62-70, projs mgr, 70-73, sr develop assoc, 73-75, mgr abrasive bond develop, 75-83; pres, Mason Abrasives Inc, 83-88; DIR ENG, CARBORUNDUM ABRASIVES CO, LOGAN, OHIO, 88- *Honors & Awards:* IR-100 Award, 70, 73 & 75. *Mem:* Soc Advan Mat & Process Eng; Am Chem Soc. *Res:* Inorganic fibers and composites; analytical chemistry; resin development and characterization; abrasive development and characterization; granted several patents; author of various publications. *Mailing Add:* 5205 Brookfield Lane Clarence NY 14031

MASON, JOHN L(ATIMER), b Los Angeles, Calif, Nov 8, 23; m 54; c 4. CHEMICAL ENGINEERING. *Educ:* Univ Chicago, BS, 44; Calif Inst Technol, BS, 47, MS, 48, PhD(chem eng), 50. *Prof Exp:* Group supvr preliminary design, AiResearch Mfg Co Div, Garrett Corp, 50-57, sr proj engr heat transfer systs, 57-58, chief preliminary design, 58-60, chief engr, 60-68, dir eng, 68-72, vpres eng, AiResearch Mfg Co Div, Allied Signal, 72-89; RETIRED. *Concurrent Pos:* Mem NASA res adv comt mech power plant systs, 59-60, comt nuclear systs, 60-62 & comt biotech & human res, 63-67; vchmn, Cryogenic Eng Conf Planning Bd, 64-67, chmn, 68; mem cryogenics evaluation panel, Nat Bur Standards, 68- *Mem:* Assoc fel Am Inst Aeronaut & Astronaut; Soc Automotive Engrs. *Res:* Heat transfer; power and propulsion; systems engineering. *Mailing Add:* 1132 Via Mirabel Palos Verdes Estates CA 90274

MASON, JOHN WAYNE, b Chicago, Ill, Feb 9, 24; m 50; c 3. NEUROENDOCRINOLOGY. *Educ:* Ind Univ, AB, 44, MD, 47. *Prof Exp:* Asst physiol, Ind Univ, 43-45; intern surg, NY Hosp-Cornell Med Ctr, 47-48, resident path, 48-50; pathologist, Ft Riley, Kans & Brooke Army Hosp, Tex, 50-53; chief neuroendocrinol dept, Walter Reed Army Inst Res, 53-74, sci adv, Div Neuropsychiat, 74-77; PROF DEPT PSYCHIAT, SCH MED, YALE UNIV, 77- *Mem:* Endocrine Soc; Am Psychosom Soc (pres, 70). *Res:* Stress, psychoendocrine and psychosomatic mechanisms. *Mailing Add:* Dept Psychiat Yale Univ Sch Med New Haven CT 06510

MASON, LARRY GORDON, b Wyandotte, Mich, Jan 19, 37. POPULATION BIOLOGY. *Educ:* Univ Mich, BS, 58, MA, 59, Univ Kans, PhD(entom), 64. *Prof Exp:* Res assoc biol, Stanford Univ, 64-65; asst prof, 66-72, ASSOC PROF BIOL, STATE UNIV NY,ALBANY, 72- *Mem:* Soc Study Evolution; Am Soc Nat. *Res:* Population phenomena, especially quantitative aspects, in natural animal populations. *Mailing Add:* Dept Biol State Univ NY Albany 1400 Washington Ave Albany NY 12222

MASON, MALCOLM, b Darlington, England, Apr 17, 45; m 80. EXPLORATION GEOCHEMISTRY, BIOGEOCHEMISTRY. *Educ:* London Univ, BSc, 66, MSc & DIC, 67, PhD(geochem), 74. *Prof Exp:* Explor geologist, Int Nickel, SAfrica, Inco Ltd, 71-72, explor geochemist, Brazil, 72-75, staff geochemist, Brazil, 75-78, staff geochemist, Colo, 78-81, mgr explor geochem, 81-88; MGR EXPLOR & GEOCHEM, AM COPPER & NICKEL CO INC, 88- *Concurrent Pos:* Geochem explor of base & precious metals in Africa, Brazil, Mex, Can, Eire & USA. *Mem:* Asn Explor Geochemists. *Res:* Geochemical, biogeochemical and geobotanical research in Southwest Africa in an exploration program for stratabound copper deposits. *Mailing Add:* 4860 Robb St #201 Wheat Ridge CO 80033-2106

MASON, MARION, b Toronto, Ont, Nov 29, 33; US citizen. NUTRITION. *Educ:* Miami Univ, BS, 55; Ohio State Univ, MS, 59; Cornell Univ, PhD(nutrit), 69. *Prof Exp:* Instr nutrit, Univ Rochester, 56-58; consult, Vis Nurse Asn, Chicago, 59-63; asst prof, Univ Rochester, 63-66; assoc prof med dietetics, Ohio State Univ, 69-72; RUBY WINSLOW LINN PROF NUTRIT, SIMMONS COL, 73- *Concurrent Pos:* Clin consult dietetics, Peter Bent Brigham Hosp, Boston, 73-77; res assoc, Eastman Dent Ctr, Rochester, NY, 74-75; vis prof, Univ Rochester, 75-77; NIH Extramural Assoc, 85-86. *Honors & Awards:* Mary Schwartz lectr, Greater New York City Diet Asn, 84. *Mem:* Am Dietetic Asn; Sigma Xi; Soc Nutrit Educ; Am Asn Univ Prof. *Res:* Health care compliance; intervention and ethics; clinical dietetic practice; cost/benefit methodology in clinical dietetics. *Mailing Add:* Simmons Col 300 Fenway Boston MA 02115

MASON, MAX GARRETT, b Roanoke, Va, Jan 15, 44; m 67; c 2. SURFACE PHYSICS. *Educ:* Johns Hopkins Univ, BA, 65, PhD(chem), 70. *Prof Exp:* Sr res assoc chem, Univ Southern Calif, 70-72; sr res chemist, 72-78, RES ASSOC, KODAK RES LABS, EASTMAN KODAK CO, 78- *Mem:* Am Vacuum Soc. *Res:* Ultraviolet and x-ray photoemission studies of solid surfaces; chemistry and physics of adsorbed species. *Mailing Add:* Dept Tech Units Bldg 82-C Floor Two Eastman Kodak Rochester NY 14650-2132

MASON, MERLE, b Coldspring, Mo, Aug 9, 20; m 42; c 1. BIOCHEMISTRY. *Educ:* Univ Iowa, BS, 47, PhD(biochem), 50. *Prof Exp:* From instr to prof biochem, 50-82, EMER PROF, UNIV MICH, ANN ARBOR, 83- *Mem:* Am Chem Soc; Am Soc Biol Chemists. *Res:* Amino acid metabolism; steroid metabolism. *Mailing Add:* 17423 102nd Dr Sun City AZ 85373-1614

MASON, MORTON FREEMAN, b Pasadena, Calif, Nov 12, 02; m 29; c 2. BIOCHEMISTRY. *Educ:* Ore State Col, BSc, 25; Duke Univ, PhD(biochem), 34. *Prof Exp:* Asst chem, Exp Sta, Mich State Col, 26-30; asst biochem, Sch Med, Duke Univ, 32-34; from instr to assoc prof, Sch Med, Vanderbilt Univ, 34-44; prof path chem, 44-55, prof forensic med & toxicol, 55-78, EMER PROF FORENSIC MED & TOXICOL, UNIV TEX HEALTH SCI CTR DALLAS, 78- *Concurrent Pos:* Toxicologist, Dallas City-County, 44-74, dir, Criminal Invest Lab, 55-74; chemist, Parkland Mem Hosp, 44-74; sr consult, US Vet Admin, 46-74. *Mem:* Am Soc Biol Chem; Am Chem Soc; Am Acad Forensic Sci (pres, 73-74); Am Asn Clin Chem; Am Indust Hyg Asn; Sigma Xi. *Res:* Analytical toxicology. *Mailing Add:* Dept Path Univ Tex Southwestern Med 5323 Harry Hines Blvd Dallas TX 75235

MASON, NORBERT, b Karlsruhe, Ger, Feb 10, 30; US citizen; m 56; c 2. CHEMICAL ENGINEERING. *Educ:* Univ Minn, BS, 54, MS, 55; Case Western Reserve Univ, PhD, 69. *Prof Exp:* Res engr, B F Goodrich Res Ctr, 55-64; proj engr, Case Western Reserve Univ, 64-66, asst, 66-68, res assoc, 68-70, sr res assoc, 70-72; SR RES ASSOC, WASH UNIV, 72- *Mem:* Am Inst Chem Engrs; Am Chem Soc. *Res:* Microencapsulation; granulation. *Mailing Add:* 645 Langton Dr Clayton MO 63105-2416

MASON, NORMAN RONALD, b Rochester, Minn, Nov 20, 29; m 53; c 2. NEUROSCIENCE, RECEPTOR BINDING. *Educ:* Univ Chicago, AB, 50, BS, 53; Univ Utah, MA, 56, PhD(biochem), 59. *Prof Exp:* From res instr to res asst prof biochem, Endocrinol Lab, Sch Med, Univ Miami, 59-64; SR SCIENTIST, RES LABS, ELI LILLY & CO, 64- *Concurrent Pos:* Investr, Howard Hughes Med Inst, 59-64. *Mem:* AAAS; Endocrine Soc; Am Chem Soc; Soc Neurosci. *Res:* Endocrinology; ovarian function; gonadotropin action; cyclic nucleotides; steroid hormone synthesis and metabolism; neurotransmitter receptor binding; hormone action. *Mailing Add:* Lilly Res Labs Lilly Corp Ctr Indianapolis IN 46285

MASON, PERRY SHIPLEY, JR, b Lubbock, Tex, Oct 2, 38; m 60; c 2. ORGANIC CHEMISTRY. *Educ:* Harding Col, BS, 59; La State Univ, PhD(org chem), 63. *Prof Exp:* Asst prof sci, Okla Christian Col, 63-64; res assoc chem, Grad Inst Technol, Univ Ark, 64-66; asst prof, Ark State Col, 66-71; PROF CHEM & HEAD DEPT, LUBBOCK CHRISTIAN COL, 71- *Concurrent Pos:* NIH fel, 63-66. *Mem:* Am Chem Soc. *Res:* Organometallic chemistry; reaction mechanism; gas chromatography. *Mailing Add:* Dept of Chem Lubbock Christian Col 5601 W 19th St Lubbock TX 79407

MASON, RICHARD CANFIELD, b Indianapolis, Ind, Aug 12, 23; m 44; c 2. PHYSIOLOGY. *Educ:* Ind Univ, AB, 48, PhD(zool), 52. *Prof Exp:* Asst, Ind Univ, 48-49; res assoc, Merck Inst Therapeut Res, 52-56; asst prof physiol, Seton Hall Col Med & Dent, 56-61; asst prof, Col Physicians & Surgeons, Columbia Univ, 61-71, asst dean student affairs, 70-71; assoc prof physiol & assoc dean admis & student affairs, 71-76, ASSOC PROF PHYSIOL & BIOPHYSICS, UNIV MED & DENT NJ-RUTGERS MED SCH, 77- *Concurrent Pos:* Actg dean admis, NJ Sch Osteop Med, 76-77. *Mem:* AAAS; Am Soc Zool; Am Physiol Soc; Harvey Soc; Sigma Xi. *Res:* Renal physiology; medical education. *Mailing Add:* Univ Med & Dent NJ Rutgers Med Sch PO Box 101 Piscataway NJ 08854

MASON, RICHARD RANDOLPH, b St Louis, Mo, Oct 3, 30; m 56; c 4. FORESTRY, ENTOMOLOGY. *Educ:* Univ Mich, BS, 52, MF, 56, PhD(forestry), 66. *Prof Exp:* Forest entomologist, Bowaters Southern Paper Corp, 56-58, res forester, 58-65; res entomolgist, Forestry Sci Lab, 65-82,

PRIN INSECT ECOLOGIST, FORESTRY & RANGE SCI LAB, LAGRANDE, 82- *Honors & Awards:* Super Serv Award, USDA, 89. *Mem:* Soc Am Foresters; Entom Soc Am; AAAS; Entom Soc Can; Ecol Soc Am; Am Arachnological Soc. *Res:* Protection of forests from destructive insect pests, emphasizing their detection, evaluation and population dynamics. *Mailing Add:* 1401 Gekeler Lane LaGrande OR 97850

MASON, ROBERT C, b Anthony, Idaho, July 9, 20; m 46; c 3. PHARMACOLOGY, MEDICINAL CHEMISTRY. *Educ:* Univ Utah, BS, 50; Univ Wis, PhD(pharmaceut chem), 54. *Prof Exp:* Prof med chem, 54-87, EMER PROF, UNIV UTAH, 87- *Mem:* Am Chem Soc; Am Pharmaceut Asn. *Res:* Isolation, characterization and synthesis of natural products and related substances. *Mailing Add:* Col Pharm Univ Utah Salt Lake City UT 84112

MASON, ROBERT EDWARD, b Thunder Bay, Ont, Jan 21, 34; m 57; c 3. STATISTICS, ECOLOGY. *Educ:* Univ Toronto, BSA, 57, MSA, 62; NC State Univ, PhD(statist), 71. *Prof Exp:* Dist biologist, Ont Dept Lands & Forests, 57-59, fish & wildlife supvr, 59-64; from asst statistician to assoc statistician, NC State Univ, 65-71; statistician, 71-73, SR STATISTICIAN, RES TRIANGLE INST, 73- *Mem:* Am Statist Asn; Biomet Soc; Sigma Xi. *Res:* Design and analysis of probability samples; nonlinear variance estimation; statistical ecology. *Mailing Add:* Ctr Res Statist Res Triangle Inst Research Triangle Park NC 27709

MASON, ROBERT PAIGE, b Cambridge, Mass, Sept 19, 31; m 56; c 2. PHOTO-OPTICAL INSTRUMENT DESIGN, TECHNICAL PROBLEM SOLVING. *Educ:* Amherst Col, BA, 54; Mass Inst Technol, BS, 54. *Prof Exp:* Dep br chief, Physics Br, US Govt, Washington, DC, 54-59; vpres res & develop, Photomechanisms Inc, 59-70; tech dir, Remak Ltd, 70-72; sr res engr, Log Etronics, 72-76; vpres new bus technol, Hunter Assoc Lab, 76-90; RETIRED. *Concurrent Pos:* Nat tech comt, Int Comn Illumination; mem, Intersoc Color Coun; bd dirs, Tech Asn Graphic Arts, 89-92. *Mem:* Soc Photog Scientists & Engrs (vpres eng, 74-76); Tech Asn Graphic Arts. *Res:* Technology of photographic imaging systems, their development and implementation; image quality assessment; image color comparison by objective quantitative means. *Mailing Add:* PO Box 47 Bozman MD 21612

MASON, ROBERT THOMAS, b Hartford, Conn, Dec 18, 59. CHEMICAL ECOLOGY OF VERTEBRATE PHEROMONES, REPRODUCTIVE BIOLOGY. *Educ:* Col Holy Cross, Worcester, Mass, BA, 82; Univ Tex, Austin, PhD(zool), 87. *Prof Exp:* Sr staff fel, Lab Biophys Chem, Nat Heart, Lung & Blood Inst, NIH, 87-91; ASST PROF PHYSIOL, DEPT ZOOL, ORE STATE UNIV, 91- *Mem:* AAAS; Int Soc Chem Ecologists; Am Soc Zoologists; Am Soc Icthylogists & Herpetologists; Am Soc Mass Spectros; Asn Chemoreception Sci. *Res:* Natural products chemistry; sex pheromones in reptiles and how pheromones mediate reproductive behavior and physiology of both sexes. *Mailing Add:* Dept Zool Ore State Univ Corvallis OR 97331-2914

MASON, RODNEY JACKSON, b New York, NY, Feb 27, 39; m 69; c 2. PLASMA PHYSICS, COMPUTER SIMULATION. *Educ:* Cornell Univ, BA, 60, PhD, 64. *Prof Exp:* Fulbright grant, Inst Plasma Physics, Garching, WGer, 64-65; asst prof aeronaut & astronaut, Mass Inst Technol, 65-67; mem tech staff, Bell Tel Labs, 67-72; STAFF MEM, LOS ALAMOS NAT LAB, 72- *Concurrent Pos:* Scientist astronaut finalist, 67; consult, Can NSERC, 83-, Jaycor Corp, 87-, Dept Nuclear Eng, Univ Ill, Urbana-Champaign; vis fel, Imp Col Univ London, 84; vis scientist, Lab for Plasma Studies, Cornell Univ, 90-91. *Mem:* Fel Am Phys Soc; sr mem, IEEE. *Res:* Kinetic theory of shock formation and structure; computer simulation of ion-acoustic and magnetosonic collisionless shocks; computational physics; laser-plasma interaction studies; transport in laser fusion pellets; pellet design implosion and thermonuclear burn physics; long time scale-implicit plasma simulation in one and two dimensions; pulsed power physics; plasma opening switch studies. *Mailing Add:* Div X MS-E531 Los Alamos Nat Lab Los Alamos NM 87545

MASON, RONALD GEORGE, b Southampton, Eng, Dec 24, 16; m 46. CRUSTAL STUDIES, TECTONOPHYSICS. *Educ:* Univ London, BSc, 38, MSc, 39, PhD(geophys), 51. *Prof Exp:* Lectr geophys, Imp Col, London, 47-64; asst res geophysicist, Scripps Inst, Univ Calif, 52-62; reader, 64-67, prof geophys, 67-84, SR RES FEL, IMP COL, UNIV LONDON, 84-; RES AFFIL, HAWAII INST GEOPHYS, UNIV HAWAII, 63- *Mem:* AAAS; Seismol Soc Am; Soc Explor Geophys; Am Geophys Union; Europ Asn Explor Geophys; Royal Astron Soc. *Res:* Crustal structure of the earth; earthquake movements at crustal plate boundaries; earthquake mechanisms. *Mailing Add:* Dept Geol Imp Col London SW7 2BP England

MASON, THOMAS JOSEPH, b St Louis, Mo, Aug 8, 42. SCREENING & EARLY DETECTION, BIOCHEMICAL EPIDEMIOLOGY. *Educ:* St Bernard Col, BA, 64; Univ Ga, MS, 68, PhD(statist & comput sci), 73. *Prof Exp:* Aerospace engr, NASA Manned Spacecraft Ctr, 64-65; statistician epidemiol, Ctr Dis Control, 67-69; statistician epidemiol, NIH Nat Cancer Inst, 71-87, chief, pop studies sect, 78-87; DIR EPIDEMIOL RES, FOX CHASE CANCER CTR, 87- *Concurrent Pos:* Adj prof, Sch Med, Georgetown Univ, 76-, Univ SFla, 88- & Thomas Jefferson Univ Sch Med, 88- *Mem:* Sigma Xi; Am Soc Prev Oncol; AAAS. *Res:* Bladder cancer screening with emphasis on occupational cohorts and determinants of individual susceptibility; biochemical and molecular epidemology; geographically determined risks for cancer; computer applications in epidemiology. *Mailing Add:* Fox Chase Cancer Ctr 510 Township Line Rd Cheltenham PA 19012-2009

MASON, THOMAS OLIVER, b Cleveland, Ohio, Oct 14, 52; m 74. MATERIALS SCIENCE, CERAMICS ENGINEERING. *Educ:* Pa State Univ, BS, 74; Mass Inst Technol, PhD(mat sci & eng), 77. *Prof Exp:* Asst prof, 78-83, ASSOC PROF MAT SCI & ENG, NORTHWESTERN UNIV, 83- *Concurrent Pos:* NATO fel, Inst Phys Chem & Electrochem, Tech Univ Hannover, W Ger, 77-78. *Mem:* Am Ceramic Soc; Am Soc Eng Educ; Mat Res Soc; Nat Inst Ceramic Engrs. *Res:* Bulk and point defect thermodynamics in the solid state including phase equilibria; high temperature electrical properties of materials; ceramic processing; Möss Bauer effect. *Mailing Add:* Dept Mat Sci & Eng Technol Inst Northwestern Univ Evanston IL 60208

MASON, TIM ROBERT, b Hereford, Tex, Apr 26, 30; m 53; c 3. ANIMAL NUTRITION, REPRODUCTIVE PHYSIOLOGY. *Educ:* Abilene Christian Col, BS, 53; Tex Tech Col, MS, 55; Tex A&M, PhD(animal nutrit), 63. *Prof Exp:* High sch teacher, Tex, 55-56; instr animal husb, Abilene Christian Col, 56-59, asst prof, 63-64; res asst, Tex A&M, 59-61 & 62-63; dir agr develop, Tex Power & Light Co, 61-62; dir livestock res, Beacon Div, Textron, Inc, 64-65; dir tech serv, 65-66; assoc prof animal husb, 66-68, prof agr, Tarleton State Univ, 68-77; CONSULT, SPAN-TEX CONSULT SERV, 77- *Mem:* Am Dairy Sci Asn; Am Soc Animal Sci. *Res:* Beef cattle; sheep, dairy and swine nutrition research. *Mailing Add:* 1201 Kendall Lane McWeese State Univ 4100 Ryan St Sulpher Springs TX 75482

MASON, V BRADFORD, b Boston, Mass, Nov 19, 42; m 65; c 2. ELECTRICAL ENGINEERING. *Educ:* Univ Mass, BS, 69; Univ Mich, MSE, 70, PhD(elec eng), 72. *Prof Exp:* Sr res geophysicist, Shell Develop Co, 72-77; STAFF SCIENTIST, SRI INT, 77- *Res:* Digital signal processing; radar; sonar. *Mailing Add:* 2564 Greer Rd Palo Alto CA 94303

MASON, W ROY, III, b Charlottesville, Va, Feb 6, 43; m 63; c 3. INORGANIC CHEMISTRY. *Educ:* Emory Univ, BS, 63, MS & PhD(chem), 66. *Prof Exp:* Instr chem, Emory at Oxford, summer 64; res fel, Calif Inst Technol, 66-67; from asst prof to assoc prof, 67-80, PROF CHEM, NORTHERN ILL UNIV, 80- *Concurrent Pos:* Vis res fel, H C Orsted Inst, Univ Copenhagen, Denmark, 74-75. *Mem:* Am Chem Soc; Sigma Xi. *Res:* Heavy metal coordination compounds; electronic structure and reactivity; molecular orbital and ligand field theory; optical electronic spectroscopy. *Mailing Add:* Dept Chem Northern Ill Univ DeKalb IL 60115

MASON, WILLIAM BURKETT, clinical chemistry, for more information see previous edition

MASON, WILLIAM C, US citizen. CIVIL ENGINEERING. *Educ:* Mont State Univ, BS, 63; Univ Wash, MS, 69; Am Acad Environ Engrs, dipl. *Prof Exp:* Mfg engr, Gen Elec Co, 63-67; res asst, Univ Wash, 67-69; environ engr, Ga Environ Protection Div, 69-72; applicatons engr, Gen Environ Equip Inc, 72-73; mem staff, Weston, 73-; AT LINCOLN LAB, MASS INST TECHNOL. *Mem:* Water Pollution Control Fedn; Am Soc Civil Engrs; Am Acad Environ Eng. *Res:* Industrial treatment; pollution abatement; sludge dewatering and disposal; hazardous waste disposal. *Mailing Add:* Lincoln Lab MS V 128 Mass Inst Technol PO Box 73 Lexington MA 02173-9108

MASON, WILLIAM HICKMON, b Bradford, Ark, June 16, 36; m 55; c 3. ZOOLOGY. *Educ:* Ark Polytech Col, BS, 58; Univ Ga, MEd, 64, DEd(sci educ), 66. *Prof Exp:* From asst prof to assoc prof, 66-75, PROF ZOOL & ENTOM, AUBURN UNIV, 75-, COORDR GEN BIOL, 68- *Mem:* Am Inst Biol Sci; AAAS; Entom Soc Am; Ecol Soc Am. *Res:* Ecosystem analysis through the use of radionuclide cycling; improvement of undergraduate teaching through the use of audiotutorial and modular concepts. *Mailing Add:* Col Sci & Math Auburn Univ Exten Cottage Auburn AL 36849

MASON, WILLIAM VAN HORN, b Pittsburgh, Pa, Jan 8, 30; m 65; c 2. MEDICINE. *Educ:* Harvard Univ, AB, 51; Baylor Univ, MD, 61. *Prof Exp:* Jr geophysicist, Humble Oil & Refining Co, 54-56; physician, Hood River Med Group, 62-65; res physician, 65-76, DERMATOLOGIST, LOVELACE FOUND, 76- *Concurrent Pos:* NIH grants, 65- *Res:* Advanced diagnostic instrumentation; physiology of unusual environments; clinical dermatology. *Mailing Add:* 201 Cedar St SE No 104 Albuquerque NM 87106

MASORO, EDWARD JOSEPH, b Oakland, Calif, Dec 28, 24; m 47. PHYSIOLOGY, BIOLOGICAL GERONTOLOGY. *Educ:* Univ Calif, AB, 47, PhD(physiol), 50. *Prof Exp:* Asst physiol, Univ Calif, 47-48; asst prof, Queens Univ, 50-52; from asst prof to assoc prof, Med Sch, Tufts Univ, 52-62; from res assoc prof to res prof, Univ Wash, 62-64; prof & chmn dept, Med Col Pa, 64-73; PROF PHYSIOL & CHMN DEPT, UNIV TEX HEALTH SCI CTR, SAN ANTONIO, 73- *Concurrent Pos:* Mem, Nat Inst Aging, Aging Rev Comt, 81-84, Nat Inst Aging Bd Sci Counselors, 85; ed, Biol Sci, Exp Aging Res; assoc ed, J Gerontol. *Honors & Awards:* Allied Signal Achievement Award in Aging, 89; Robert W Kleemeiler Award, Geront Soc Am, 90. *Mem:* Am Physiol Soc; Am Chem Soc; Can Biochem Soc; Can Physiol Soc; Am Soc Biol Chemists; AAAS; fel Geront Soc Am. *Res:* Retardation of aging processes by nutritional means; adipose tissue structure and function. *Mailing Add:* Dept Physiol Univ Tex Health Sci Ctr San Antonio TX 78284-7756

MASOUREDIS, SERAFEIM PANOGIOTIS, b Detroit, Mich, Nov 14, 22; m 43; c 2. HEMATOLOGY, IMMUNOHEMATOLOGY. *Educ:* Univ Mich, AB, 44, MD, 48; Univ Calif, Berkeley, PhD(med physics), 52. *Prof Exp:* Clin instr med, Univ Calif, San Francisco, 50-52, res assoc med physics, Donner Lab, Berkeley, 54; from asst prof to assoc prof path, Sch Med, Univ Pittsburgh, 55-59, asst dir cent blood bank, 55-59; assoc prof prev med, Sch Med, Univ Calif, San Francisco, 59-62, assoc prof med, 62-66, assoc prof clin path & lab med, 66-67; prof med & microbiol, Sch Med, Marquette Univ, 67-69; PROF PATH & DIR UNIV HOSP BLOOD BANK, SCH MED, UNIV CALIF, SAN DIEGO, 69- *Concurrent Pos:* Res assoc, Cancer Res Inst, 59-67; chief H C Moffitt Blood Bank, 62-67; spec fel, Univ Lausanne, 65-66; exec dir, Milwaukee Blood Ctr, 67-69. *Honors & Awards:* Emily Cooley Mem lectr, 73; Karl Landsteiner Mem Award, 79. *Mem:* Am Asn Cancer Res; Am Asn Immunol; Am Soc Hemat; Am Soc Clin Invest; Brit Soc Immunol; Sigma Xi. *Res:* Blood group antigens; red cell membranes; membrane ultrastructure; immunological reactions involving red cell, hemolytic anemias. *Mailing Add:* 2745 Inverness Ct La Jolla CA 92037

MASOVER, GERALD K, b Chicago, Ill, May 12, 35; m 59; c 3. MICROBIOLOGY, PHARMACY. *Educ:* Univ Ill, BS, 57, MS, 70; Stanford Univ, PhD(med microbiol), 73. *Prof Exp:* Res assoc med microbiol, Med Sch, Stanford Univ, 73-76, res assoc surg, 76-80; Mem staff, Bruce Lyon Mem Res Labs, Children's Hosp & Med Ctr, 80-83; mem staff, Hana Biologics, 83-85; MEM STAFF, GENETECH, 86- *Concurrent Pos:* NSF fel, Stanford Univ, 70-73; consult, Durrum Chem Corp, 75, Bactilabs, Inc, 76 & WHO, 75-; asst, Univ Ill Chicago Circle, 69 & Stanford Univ, 72; researcher, Univ Ill, 69; pharmacist, Calif & Ill, 57- *Mem:* Am Soc Microbiol; AAAS; Int Orgn Mycoplasmologists; Sigma Xi; Fedn Am Scientists; Am Tissue Cult Asn; Parenteral Drug Asn; Int Soc Pharmaceut Eng; Parenteral Drug Asn; Int Soc Pharmaceut Engr. *Res:* Definition of life using smallest free living cell as model; host-parasite interaction; aging and development; cell culture; pharmaceutical MFG QC-microbiology/environmental control. *Mailing Add:* Genentech 499 Point San Bruno Blvd S San Francisco CA 94110

MASRI, MERLE SID, b Jerusalem, Palestine, Sept 12, 27; nat US; m 52; c 4. AGRICULTURAL CHEMISTRY, MAMMALIAN PHYSIOLOGY. *Educ:* Univ Calif, AB, 50, PhD(physiol), 53. *Prof Exp:* Res assoc hemat, Michael Reese Hosp, Chicago, Ill, 54-56; res chemist pharmacol, 56-71, RES CHEMIST FIBER SCI, WESTERN REGIONAL RES LAB, USDA, 71- *Mem:* AAAS; Am Asn Cereal Chem; NY Acad Sci; Am Chem Soc. *Res:* Chemistry metabolism and pharmacology of mycotoxins; toxicology; fiber science, especially wool; protein chemistry; metallic ion interactions with proteins and bio polymers; polymers and enzyme immobilization; pollution abatement. *Mailing Add:* Nine Commodore Dr No 401 Emeryville CA 94608-1633

MASRI, SAMI F(AIZ), b Beirut, Lebanon, Dec 9, 39; m 64; c 3. MECHANICAL ENGINEERING. *Educ:* Univ Tex, BS, 60, MS, 61; Calif Inst Technol, MS, 62, PhD(mech eng), 65. *Prof Exp:* Res fel mech eng, Calif Inst Technol, 65-66; from asst prof to assoc prof eng, 66-76, PROF ENG, UNIV SOUTHERN CALIF, 76- *Mem:* Am Soc Civil Engrs; Am Soc Mech Engrs; Am Inst Aeronaut & Astronaut; Inst Elec & Electronics Engrs. *Res:* Applied mechanics; shock and vibration; structural dynamics. *Mailing Add:* Dept Civil Eng Univ Southern Calif Los Angeles CA 90089-0242

MASRY, SALEM EL, b Aug 2, 38; Can citizen. PHOTOGRAMMETRY. *Educ:* Ain Shams Univ, Cairo, BSc, 60; Univ Col, dipl, 62 & 63, Univ London, PhD(photogram), 66. *Prof Exp:* Fel photogram, 66-67, res assoc, 67-70, asst prof, 70-77, ASSOC PROF PHOTOGRAM, UNIV NB, FREDERICTON, 77- *Mem:* Am Soc Photogram. *Res:* Applications of real-time control to photogrammetric instruments. *Mailing Add:* Dept Survey Eng Univ NB College Hill Box 4400 Fredericton NB E3B 5A3 Can

MASRY, SOUHEIR EL DEFRAWY, biochemical pharmacology, drug metabolism, for more information see previous edition

MASSA, DENNIS JON, b Myrtle Beach, SC, Sept 29, 45; m 66; c 3. PHYSICAL CHEMISTRY, POLYMER PHYSICS. *Educ:* Bradley Univ, BA, 66; Univ Wis-Madison, PhD(phys chem), 70. *Prof Exp:* NSF fel phys biochem, Univ Calif, San Diego, 70-71; sr res chemist, 71-78, res lab head, 85-88, RES ASSOC, RES LABS, EASTMAN KODAK CO, 79- *Concurrent Pos:* Adj fac, Univ Rochester, 83- *Mem:* Am Chem Soc; Am Phys Soc; Sigma Xi; Soc Plastics Engrs. *Res:* Polymer physics; physical chemistry of polymers and biopolymers; molecular motion in the solid state; polymer rheology. *Mailing Add:* 40 Deer Creek Rd Pittsford NY 14534-4146

MASSA, FRANK, technical management, physics; deceased, see previous edition for last biography

MASSA, LOUIS, b Aug 4, 40; US citizen. CHEMICAL PHYSICS. *Educ:* LeMoyne Col, BS, 61; Clarkson Col, MS, 62; Georgetown Univ, PhD(physics), 66. *Prof Exp:* Res fel chem, Brookhaven Nat Lab, 66-69; assoc prof, 69-76, PROF CHEM, HUNTER COL, 76- *Concurrent Pos:* Petrol Res Fund grant, Hunter Col, 70-, City Univ New York Res Found grant, 71- *Mem:* AAAS; Am Phys Soc; Am Chem Soc. *Res:* Theoretical chemical physics; quantum mechanics. *Mailing Add:* Dept Chem Hunter Col New York NY 10021

MASSA, TOBIAS, b Brooklyn, NY, Dec 14, 50; m 74; c 2. RESEARCH ADMINISTRATION, ZOOLOGY. *Educ:* State Univ NY, BA, 72; City Univ New York, PhD(biomed sci), 79; Am Bd Toxicol, dipl, 83. *Prof Exp:* Assoc dir toxicol, Pfizer Pharmaceut, 86-89, assoc dir & group leader toxicol, 89-90; sr scientist toxicol, 78-81, prin scientist toxicol, 81-86, assoc dir regulatory affairs, 90-91, DIR REGULATORY AFFAIRS, SCHERING PLOUGH RES, 91- *Concurrent Pos:* Adj prof, Dept Biol, Upsala Col, 85-86. *Mem:* Soc Toxicol; Am Soc Pharmacol & Exp Therapeut; Regulatory Affairs Professionals Soc; Am Col Toxicol. *Res:* Toxicology of pharmaceutical compounds; regulation of recombinant biological agents. *Mailing Add:* Schering-Plough Res 2000 Galloping Hill Rd Kenilworth NJ 07033

MASSALSKI, T(ADEUSZ) B(RONISLAW), b Warsaw, Poland, June 29, 26; US citizen; m 53; c 3. PHYSICAL METALLURGY. *Educ:* Univ Birmingham, BSc, 52, PhD, 54, DSc, 64. *Hon Degrees:* DSc, Univ Warsaw, 73. *Prof Exp:* Res fel, Inst Study Metals, Univ Chicago, 54-56; lectr phys metall, Univ Birmingham, 56-59; sr res fel & head metal physics, Mellon Inst, Carnegie-Mellon Univ, 59-62, staff fel & mem pres adv comt, 62-72, prof metal physics, Mellon Inst Sci, 66-75. *Concurrent Pos:* Vis prof, Univ Buenos Aires, 62, Calif Inst Technol, 62, Stanford Univ, 63, Univ Calif, 63 & 64, Inst Physics Bariloche, Arg, 66 & 70 & Harvard Univ, 69; Guggenheim fel, Oxford Univ, 65-66; consult, Lawrence Livermore Nat Lab, Univ Calif, 65-, Nat Bur Standards, Washington, DC, 79- & Oak Ridge Nat Lab, 80-; exchange prof, Krakov, Poland, 68; NAVSEA chair prof, Naval Postgrad Sch, Monterey, Calif, 84-85. *Honors & Awards:* Medal, Polish Mining & Metall Acad, 73; Henry Krumb Lectr, Am Inst Mining, Metall & Petrol Engrs, 74,; McDonald Mem Lectr, Can, 76; Hume-Rotheny Award, Am Inst Mining & Petrol Engs, 82. *Mem:* Am Phys Soc; Am Inst Mining, Metall & Petrol Engrs; fel Am Soc Metals; fel Brit Inst Physics; fel Brit Inst Metallurgists; Am Geophys Union; Brit Metals Soc; foreign mem Polish Acad Sci; Polish Inst Arts & Sci Am; foreign mem Ger Acad Sci(Göttinger). *Res:* Theory of alloy phases; transformations and crystallographic relationships in metals and alloys; metal physics. *Mailing Add:* Sci Hall 4311 Schenley Park Pittsburgh PA 15213

MASSARI, V JOHN, b Bari, Italy, June 2, 45; US citizen. NEUROPHARMACOLOGY. *Educ:* Vanderbilt Univ, PhD(pharmacol), 75. *Prof Exp:* ASSOC PROF PHARMACOL, COL MED, HOWARD UNIV, 77- *Mem:* Am Soc Pharmacol & Exp Therapeut; Soc Neurosci; NY Acad Sci. *Mailing Add:* Dept Pharmacol Col Med Howard Univ 520 West St NW Washington DC 20059

MASSARO, DONALD JOHN, b Jamaica, NY, Aug 7, 32; m 57; c 2. MEDICINE. *Educ:* Hofstra Col, BA, 53; Georgetown Univ, MD, 57. *Prof Exp:* Am Thoracic Soc fel, 60-62; from instr to asst prof med, Georgetown Univ, 62-67; assoc prof, Duke Univ, 67-68; assoc prof, George Washington Univ, 68-72, prof med, 72-76; chief chest sect, Vet Admin Hosp, 68-76; PROF MED & PHYSIOL, SCH MED, UNIV MIAMI, 76- *Concurrent Pos:* fel physiol chem, Johns Hopkins Univ, 64-65; med investr, Vet Admin, 70- *Mem:* Am Physiol Soc; Am Soc Clin Invest; Soc Exp Biol & Med; Am Fedn Clin Res; Am Thoracic Soc. *Res:* Pulmonary diseases; lung biochemistry; phagocytosis; pulmonary physiology. *Mailing Add:* Dept Med Georgetown Univ Med Ctr 3900 Resernor Rd NW Preclin Sci Bldg Washington DC 20007

MASSARO, EDWARD JOSEPH, b Passaic, NJ, June 7, 33; m 78; c 4. CELL PHYSIOLOGY. *Educ:* Rutgers Univ, AB, 55; Univ Tex, MA, 58, PhD(biochem), 62. *Prof Exp:* Instr biol, Blinn Col, Tex, 56-57; USPHS fel biochem, Univ Tex, 62-63; USPHS fel physiol chem & biol, Med Sch, Johns Hopkins Univ, 63-65, res assoc biol, 65; res assoc & instr, Yale Univ, 65-68; from asst prof to prof biochem, State Univ NY Buffalo, 68-78, res prof, 78; dir chem carcinogenesis, Mason Res Inst, Mass, 77-78, dir toxicol, 78; prof toxicol, Dept Vet Sci & dir ctr Air Environ Studies, Pa State Univ, 78-83; dir, Inhalation Toxicol Div, 83-85, DIR DOSIMETRY & FLOW CYTOMETRY & SR SCIENTIST, US ENVIRON PROTECTION AGENCY EFFECTS RES LAB, RES TRIANGLE PARK, NC, 85-; PROF, CTR BIOCHEM ENG, DUKE UNIV, DURHAM, NC, 90- *Concurrent Pos:* Adj prof toxicol, Duke Univ, Durham, 86-; Rutgers scholar, Rutgers Univ, 54; fel Rachel Carson Col, Univ NY, Buffalo, NY, 68-79; adj prof toxicol, Univ NC, Chapel Hill, 84- *Honors & Awards:* Sigma Xi Res Award, Univ Tex Med Branch, 62. *Mem:* fel AAAS; Am Soc Biol & Chem Molecular Biol; Am Asn Pathologists; Am Soc Pharmacol & Exp Therapeut; Am Soc Cell Biol; Biophys Soc; Soc Toxicol. *Res:* Cell biology; metals toxicology; biochemistry; cell physiology, biology and toxicology; cellular pathology. *Mailing Add:* Ctr Biochem Eng Teer Bldg Duke Univ Durham NC 27706

MASSE, ARTHUR N, b Columbus, Ohio, May 1, 28; m 55; c 3. CHEMICAL ENGINEERING. *Educ:* Ohio State Univ, BS, 51. *Prof Exp:* Jr engr, Raw Mat Lab, BF Goodrich Co, Ohio, 51 & 53, shift foreman acrylonitrile prod, Ky, 53-56, mem tech dept, 56-61; chem engr, res & develop water pollution control, 61-68, chief pilot plants, US Dept Interior, 68-77; chief prod control, Environ Protection Agency, 77-80, ENVIRON ENGR, TECH EVAL SERV, NAT ENFORCEMENT INVEST CTR, 80- *Concurrent Pos:* Technol transfer specialist, Am Water Works Asn Res Found, 86-91. *Mem:* Am Inst Chem Eng; Water Pollution Control Fedn; Am Water Works Asn; N Am Lake Mgt Soc. *Res:* Research and development work on physical and chemical processes to supplement or replace conventional biological treatment of municipal wastewaters; drinking water treatment. *Mailing Add:* 2444 W Park Lane Litteton CO 80120

MASSEE, TRUMAN WINFIELD, b Joseph, Ore, May 5, 30; m 51; c 3. SOIL FERTILITY. *Educ:* Ore State Univ, BS, 52, AgM, 53; Mont State Univ, PhD, 73. *Prof Exp:* Soil scientist, Northern Mont Br Exp Sta, Agr Res Serv, 55-57, res soil scientist, Tetonia Br Exp Sta, Univ Idaho, 58-64, RES SOIL SCIENTIST, SNAKE RIVER RES CTR, AGR RES SERV, USDA, 65- *Mem:* Am Soc Agron; Soil Conserv Soc Am; Soil Sci Soc Am; Sigma Xi. *Res:* Dryland soil moisture-fertility-plant growth relationships. *Mailing Add:* Rte 1 Jerome ID 83338

MASSEL, GARY ALAN, b Trenton, NJ, May 5, 39; m 59; c 2. TECHNICAL MANAGEMENT, SYSTEMS SCIENCE. *Educ:* NC State Univ, BS, 61, PhD(physics), 67. *Prof Exp:* Asst physics, NC State Univ, 61 & 63; aerospace engr, Langley Res Ctr, NASA, Va, 65-67; res staff mem, Inst Defense Anal, 67-70; dir army planning, Dept Defense, 70-72, dir naval planning, 72-73; assoc adminr planning, Dept Health, Educ & Welfare, 73-75; sr vpres corp develop, Sci Applns Int, 75-81; dir prod develop, Int Paper Co, 81-86; VPRES ENG & DEVELOP, PACKAGING CORP AM, 86- *Concurrent Pos:* Consult, Defense Atomic Support Agency, 68-; chmn, Comsysts Corp, 78-80; dir, JRB Assocs, 78-; res adv comt, Nat Asn Food Processors, 84-86 & Inst Paper Chemists, 87- *Mem:* Am Phys Soc; Opers Res Soc Am; Am Pub Health Asn; Tech Asn Pulp & Paper Indust; Soc Plastic Engrs. *Res:* Computer modeling of many-body systems; high-temperature hydrodynamics; weapon systems analysis; health systems planning; energy systems analysis and planning; solar energy. *Mailing Add:* 834 Sheridan Rd Glencoe IL 60022

MASSELL, PAUL BARRY, b Boston, Mass, June 26, 48. NON-LINEAR DIFFERENTIAL EQUATIONS. *Educ:* Univ Chicago, AB, 70; City Univ New York, PhD(math), 75. *Prof Exp:* Sr programmer, Nat Bur Econ Res, 74-75; res analyst, JWK Int Corp, 75-76; statist programmer, Battelle Mem Inst, 76-78; software engr, Hadron, 79-80; asst prof, US Naval Acad, 81-88; CONSULT, MATH, 89- *Concurrent Pos:* Lectr math, Brooklyn Col, 71-74; lectr, Johns Hopkins Univ, 88- *Mem:* Am Acoust Asn; Math Asn Am; Soc Indust & Appl Math. *Res:* Dynamical systems. *Mailing Add:* 650 Americana Dr No T3 Annapolis MD 21403

MASSELL, WULF F, b Germany, June 2, 43; US citizen; m 68; c 2. EXPLORATION SEISMOLOGY, INFORMATION THEORY. *Educ:* Univ Minn, BS, 66; Ind Univ, MA, 69, PhD(geophys), 72. *Prof Exp:* Lab mgr, US Antarctic Res Prog, 66-67; asst prof, Univ Tex, Austin, 72-77; sr staff geophysicist, Amoco Prod Co, 77-80; dir, res & develop explor serv div, Geosource Inc, 80-83, consult geophysicist, 83-85; CONSULT, 85- *Concurrent Pos:* Vis prof, Fed Univ Bahia, Brazil, 85-86. *Mem:* Soc Explor Geophysicists; Soc Petrol Eng; Europ Asn Explor Geophysicists; Asn Comput Mach. *Res:* Remote sensing, specifically three dimensional seismic reflection data acquisition, processing and interpretation to make decision relative to the world wide exploration for oil and gas; 3D visualization and graphics workstation design. *Mailing Add:* 4002 El James Dr Spring TX 77388

MASSENGALE, M A, b Monticello, Ky, Oct 25, 33; m 59; c 2. AGRONOMY, CROP PHYSIOLOGY. *Educ:* Western Ky Univ, BS, 52; Univ Wis, MS, 54, PhD(agron), 56. *Hon Degrees:* LHD, Nebr Wesleyan Univ, 87. *Prof Exp:* Asst agron, Univ Wis, 52-56; from asst prof & asst agronomist to assoc prof & assoc agronomist, Univ Ariz, 58-65, head dept agron & plant genetics, 66-74, prof & agronomist, 65-76, assoc dean col agr & assoc dir, Ariz Agr Exp Sta & Coop Ext Serv, 74-76; prof agron & vchancellor agr & natural resources, 76-81, chancellor, 81-91, PRES, UNIV NEBR, LINCOLN, 91- *Concurrent Pos:* Assoc ed, Agron J & Crop Sci, 69-72; consult in Brazil, Saudi Arabia, & USSR; mem nat coord comt for cotton res, 74; consult, crop prod, Indonesia, Morocco, 87, educ progs, Repub of China, Australia & New Zealand, 88. *Mem:* Fel AAAS; fel Am Soc Agron; fel Crop Sci Soc Am (pres, 72-73); Am Soc Plant Physiol; Soil Conserv Soc Am. *Res:* Forage crops physiology, production and management; water-use efficiency, photosynthesis, respiration and dry-matter production. *Mailing Add:* Univ Nebr 3835 Holdrege St Lincoln NE 68588-0745

MASSENGILL, RAYMOND, b Bristol, Va, Dec 8, 37; m 59; c 4. SPEECH PATHOLOGY. *Educ:* Univ Tenn, BS, 58, MS, 59; Univ Va, EdD(speech path & audiol), 68. *Prof Exp:* Dir speech path, audiol & speech sci, Palmer Rehab Ctr, Tenn, 60-62; dir speech path & speech sci & dir speech sci lab, Med Ctr, Duke Univ, 64-73, assoc prof med speech path, Div Plastic, Maxillofacial & Oral Surg, 71-77; PROF OTOLARYNGOL, JAMES H QUILLEN COL MED, E TENN STATE UNIV, 77-, ASST DEAN & DIR MED EDUC, SCH MED, 77- *Concurrent Pos:* NIH grant, 67-; United Med Res Found grant, 67-68; Nat Inst Dent Res grant; consult speech path, audiol & speech sci, Univ Tenn, 67- & Nat Inst Dent Res, 68- *Mem:* AAAS; fel Am Speech & Hearing Asn; Asn Am Med Cols. *Res:* Speech physiology as it relates to oral and pharyngeal mechanisms and how this mechanism is altered due to certain plastic surgery procedures. *Mailing Add:* Sch Med Bristol Regional Med Ctr E Tenn State Univ Bristol TN 37620

MASSERMAN, JULES HOMAN, b Chudnov, Poland, Mar 10, 05; nat US; m 43. NEUROPHYSIOLOGY, PSYCHOANALYSIS. *Educ:* Wayne State Univ, MB, 30, MD, 31. *Prof Exp:* Resident neurol, Stanford Univ, 31-32; asst psychiatrist, Johns Hopkins Univ, 32-35; resident psychiat, Univ Chicago, 35-36, from instr to asst prof, 36-46; assoc prof, 46-50, PROF NEUROL & PSYCHIAT, NORTHWESTERN UNIV, CHICAGO, 50-, CO-CHMN DEPT, 64- *Concurrent Pos:* Chief consult, Downey Vet Hosp, 46-; sci dir, Nat Found Psychiat Res, 46-; consult, Great Lakes Naval Hosp, 47- & WHO, 50-; H M Camp lectr, 64; Karen Horney lectr, 65; dir ed, Ill State Psychiat Inst; vis prof psychiat, Univ Louis & Univ Zagreb. *Honors & Awards:* Lasker Award, Am Pub Health Asn, 47; Sigmund Freud Award, Am Asn Psychoanal, 80. *Mem:* Soc Biol Psychiat (pres, 57-58); Int Asn Social Psychiat (pres, 69-); Am Asn Social Psychiat (pres, 78-79); fel Am Psychiat Asn (vpres, 74-75, secy, 75-77, pres, 78-79); Acad Psychoanal (pres, 57-58); World Asn Social Psychiat (pres, 82-). *Res:* Experimental neuroses; physiology of emotion; music; occultisms; dynamics of language; dynamics of phantasy; dynamics of political action; 18 books, 432 articles. *Mailing Add:* 2231 E 67th St Chicago IL 60649

MASSEY, DOUGLAS GORDON, b Clinton, Ont, Oct 14, 26; m 66; c 3. MEDICINE. *Educ:* Univ Toronto, MD, 51; MRCPE, 61; McGill Univ, MSc, 63; FRCP(C), 63; FACP, 68; FRCPE, 79. *Prof Exp:* Consult, Estab Pulmonary Labs, Repatriation Dept, Australia, 55-57; dir pulmonary lab, Hosp St Luke, Montreal, Que, 64-66; from asst prof to assoc prof med, Univ Sherbrooke, 66-73, dir serv pneumology, Univ Hosp, 66-72; PROF MED, SCH MED, UNIV HAWAII, MANOA, 73- *Concurrent Pos:* Sir Edward Beatty fel, McGill Univ, 62-63; grants, Med Res Coun Can, 66-68, Nat Cancer Inst, 66-68, Inst Occup & Environ Health, 67-68 & Minister of Educ, 67-69. *Mem:* Fel Am Col Chest Physicians. *Res:* Medical education, curriculum development, programmed texts; asthma; asbestosis; herbal medicine. *Mailing Add:* Univ Hawaii Sch Med 2230 Liliha St Honolulu HI 96817

MASSEY, EDDIE H, b Canadian, Tex, July 14, 39; m 57; c 2. PHARMACEUTICAL CHEMISTRY. *Educ:* McMurry Col, BA, 61; Vanderbilt Univ, PhD(org chem), 66. *Prof Exp:* sr pharmaceut chemist 66-72, RES SCIENTIST, DEPT PHARMACEUT RES, ELI LILLY & CO, 72- *Mem:* Am Chem Soc. *Res:* chemistry and chemical modification of macrolide antibiotics. *Mailing Add:* 8337 Hi Vu Dr Indianapolis IN 46227-2703

MASSEY, FRANK JONES, JR, b Portsmouth, NH, Nov 22, 19; m 43; c 2. MATHEMATICAL STATISTICS. *Educ:* Univ Calif, AB, 41, MA, 44, PhD(math statist), 47. *Prof Exp:* Asst prof math, Univ Md, 47-48; from asst prof to assoc prof, Univ Ore, 48-56; prof prev med & pub health, 59-80, PROF BIOSTATIST, UNIV CALIF, LOS ANGELES, 59-, PROF BIOMATH, 70- *Concurrent Pos:* Ford fel, 53-54. *Mem:* Am Statist Asn; Am Math Asn; Inst Math Statist; Biomet Soc. *Res:* Non-parametric statistical analysis. *Mailing Add:* 107 Larkin Pl Santa Monica CA 90402

MASSEY, FREDRICK ALAN, b Birmingham, Ala, Dec 20, 38; m 67; c 1. APPLIED MATHEMATICS. *Educ:* Samford Univ, BS, 61; Auburn Univ, 63, PhD(math), 66. *Prof Exp:* Asst prof math, Auburn Univ, 66-67; from asst prof to assoc prof, 67-74, interim chmn dept, 77-78, PROF MATH, GA STATE UNIV, 74-, CHMN DEPT, 78- *Concurrent Pos:* Mem Urban Life Fac, Ga State Univ, 77-81, Pub & Urban Affairs Fac, 81. *Mem:* Am Math Soc; Math Asn Am; Soc Indust & Appl Math; Asn Comput Mach. *Res:* Theoretical physics; economic theory; nonlinear programming. *Mailing Add:* Dept Math & Comput Sci Ga State Univ Atlanta GA 30303

MASSEY, GAIL AUSTIN, b El Paso, Tex, Dec 2, 36; m 60. LASERS. *Educ:* Calif Inst Technol, BS, 59; Stanford Univ, MS, 67, PhD(elec eng), 70. *Prof Exp:* Engr, Raytheon Co, Santa Barbara, 59-63; sr eng specialist, Electro-optics orgn, GTE Sylvania, 63-72; prof appl physics, Ore Grad Ctr, 72-80; PROF ELEC ENG, SAN DIEGO STATE UNIV, 81- *Concurrent Pos:* Consult, various indust & gov labs, 75- *Mem:* Fel Optical Soc Am; Inst Elec & Electronics Engrs; Soc Photo-Optical Instrumentation Engrs. *Res:* Nonlinear optical devices; ultraviolet and wavelength-tunable lasers; electron beams and microscopy. *Mailing Add:* Dept Elec Eng San Diego State Univ San Diego CA 92182

MASSEY, HERBERT FANE, JR, b Kerrville, Tenn, Jan 23, 26; m 51. AGRONOMY, SOIL FERTILITY. *Educ:* Univ Tenn, BS, 49; Univ Wis, MS, 50, PhD(soils), 52. *Prof Exp:* Asst, Univ Wis, 49-52; res agronomist, Int Minerals & Chem Corp, 52-53; from asst prof to assoc prof, 53-60, dir off Int Prog Agr, 75-85, PROF AGRON, UNIV KY, 60-, DIR DIV REGULATORY SERV, 70- *Concurrent Pos:* Vis prof, San Carlos Univ, Guatemala, 59-60, Univ Indonesia, 61-64 & Thailand, 67-70. *Mem:* Soil Sci Soc Am; Am Soc Agron; Int Soc Soil Sci; Sigma Xi. *Res:* Soil chemistry and fertility; micro element studies; tropical agriculture. *Mailing Add:* 300 Garrigus Bldg Univ Ky Lexington KY 40546-0215

MASSEY, JAMES L, b Wauseon, Ohio, Feb 11, 34; m 58; c 4. ELECTRICAL ENGINEERING. *Educ:* Univ Notre Dame, BS, 56; Mass Inst Technol, SM, 60, PhD(elec eng), 62. *Hon Degrees:* DTech, Univ Lund, Sweden, 90. *Prof Exp:* From asst prof to prof, Univ Notre Dame, 62-76, Frieimann prof elec eng, 72-77; prof syst sci, Univ Calif, Los Angeles, 77-80; PROF DIGITAL TECH, SWISS FED INST TECHNOL, 80- *Concurrent Pos:* Vis assoc prof, Mass Inst Technol, 66-67, vis prof, 77-78; guest prof, Tech Univ Denmark, 71-72. *Honors & Awards:* W R G Baker Award, Inst Elec & Electronics Engrs, 87; Shannon lectr, Inst Elec & Electronics Engrs Info Theory Soc, 88. *Mem:* Nat Acad Eng; Int Asn Cryptologic Res; Sigma Xi; fel Inst Elec & Electronic Engrs. *Res:* Information theory; coding theory; cryptography. *Mailing Add:* Swiss Fed Inst Tech Inst Signal & Info Proc Eth Zentrum Zurich 8092 Switzerland

MASSEY, JIMMY R, b Mart, Tex, July 9, 40; m 62. BOTANY, PLANT TAXONOMY. *Educ:* NTex State Univ, BSEd, 62; Tex A&M Univ, MS, 65; Univ Okla, PhD(bot), 71. *Prof Exp:* Instr bot, Tex A&M Univ, 64-65; vis scholar bot & genetics, Okla Col Lib Arts, 70-71; herbarium adminr, 73-83, CUR HERBARIUM, UNIV NC, CHAPEL HILL, 71- ADJ ASSOC PROF, 80-, HERBARIUM DIR, 83- *Mem:* Int Asn Plant Taxon; Am Soc Plant Taxonomists; Sigma Xi; Am Inst Biol Sci. *Res:* Vascular flora of southeastern United States; monograph polygalaceae; pollination-reproductive biology; species biology of threatened and endangered plant species. *Mailing Add:* Cur Herbarium Univ NC Dept Bot Chapel Hill NC 27515

MASSEY, JOE THOMAS, b Raleigh, NC, Apr 22, 17; m 41; c 2. BIOMEDICAL ENGINEERING. *Educ:* NC State Col, BS, 38; Johns Hopkins Univ, PhD(physics), 53. *Prof Exp:* Instr elec eng, Clemson Col, 38-39; instr eng mech, NC State Col, 39-41; asst to dir, 72-74, DIR BIOMED PROGS, JOHNS HOPKINS UNIV, 74-, PRIN STAFF MEM, APPL PHYSICS LAB, 46-, ASSOC PROF BIOMED ENG, SCH MED, 68- *Concurrent Pos:* Lectr, Johns Hopkins Univ, 57. *Mem:* NY Acad Sci; Inst Elec & Electronics Engrs; Sigma Xi. *Res:* Microwave plasma and physics; laser physics. *Mailing Add:* 10111 Parkwood Dr Bethesda MD 20814

MASSEY, JOHN BOYD, b Memphis, Tenn, Feb 22, 50. LIPOPROTEIN METABOLISM, MEMBRANE PHYSICAL CHEMISTRY. *Educ:* Univ Tenn, BA, 72, PhD(biochem), 77. *Prof Exp:* Res assoc, 77-80, INSTR, BAYLOR COL MED, 80- *Mem:* Am Chem Soc. *Res:* Dynamics and thermodynamics of lipid-protein associations in plasma lipoproteins; mechanisms of human plasma lipoprotein lipid transfer; role of plasma lipoproteins in vitamin E metabolism. *Mailing Add:* Methodist Hosp 6565 Fannin MS A601 Houston TX 77030-2707

MASSEY, L(ESTER) G(EORGE), b Madison, Wis, Dec 9, 18; m 42; c 3. CHEMICAL ENGINEERING. *Educ:* Univ Wis, BS, 42, MS, 47, PhD(chem eng), 50. *Prof Exp:* Instr chem eng, Univ Wis, 46-50; chem engr pilot plants, Am Cyanamid Co, 50-51; res chem engr, Universal Oil Prod Co, 51-55, process engr, 55-57, asst head comput dept, 57-59, head, 59-66, mgr, 66-67; assoc dir res, Consol Natural Gas Serv Co, Inc, 67-77, mgr process develop, 77-83; RETIRED. *Concurrent Pos:* CNE rep, tech adv comt, Fuel Gas Assoc, 68-71; Proj adv, Am Gas Asn/US Govt Coal Gasification Prog, 72-76; panelist, Akron Ohio Energy Forum, 73, Am Chem Soc, 74, Nat Bureau Standards, 79 & 80-81; speaker-lectr, Am Soc Metals, 73 & 76, Inst Gas Technol, 77, Univ S Calif, 80; chmn, Gordon Res Conf, Fuel Sci, 78-79. *Mem:* AAAS; Am Chem Soc; Am Inst Chem Engrs; fel Am Inst Chem; Sigma Xi. *Res:* Physical and chemical equilibrium; chemical thermodynamics; mass and heat transfer; heterogeneous catalysis and kinetics; coal conversion technology; author of technical material and holds numerous US patents. *Mailing Add:* 192 S Strawberry Lane Chagrin Falls OH 44022

MASSEY, LINDA KATHLEEN LOCKE, b Oklahoma City, Okla, Aug 27, 45. HUMAN NUTRITION. *Educ:* Univ Okla, BS, 66, PhD(microbiol), 71. *Prof Exp:* NIH fel microbiol, Health Sci Ctr, Univ Okla, 71-72, instr, 72-73; res assoc, Cancer Sect, Okla Med Res Found, 73-74; asst prof biochem, Okla Col Osteopath Med & Surg, 74-78; from asst prof to assoc prof, 78-89, PROF,

FOOD SCI & HUMAN NUTRIT, WASH STATE UNIV, 89- *Concurrent Pos:* adj prof, foods, nutrit & inst mgt, Okla State Univ, 78; sr int fel, NIH, Univ BC, 91. *Mem:* Am Dietetic Asn; Am Soc Bone Mineral Res; Am Inst Nutrit. *Res:* Effects of diet and caffeine on calcium metabolism. *Mailing Add:* Food Sci & Human Nutrit Wash State Univ Spokane WA 99204-0399

MASSEY, MICHAEL JOHN, b Madison, Wis, July 7, 47; m 70, 79; c 2. ENVIRONMENTAL MANAGEMENT, FUEL TECHNOLOGY. *Educ:* Univ Wis-Madison, BS, 70; Carnegie-Mellon Univ, MS, 72, PhD(ChE), 74. *Prof Exp:* Res asst gas chromatography, dept chem eng, Univ Wis-Madison, 66-70; instr chem eng, Carnegie-Mellon Univ, 70-74, asst prof chem eng & pub affairs, 74-78; mgr process eng, Environ Res & Tech, Inc, 78-86; VPRES & GEN MGR, ENSR TECHNOL INC, 86- *Concurrent Pos:* Alt mem, Allegheny County Air Pollution Adv Bd, 73-; consult, US Dept Com, tech adv bd, 70-71, US Environ Protection Agency, Econ & Eval Br, 71, Alan Wood Steel Co, 73-74, Ill Inst Technol, 75-78 & Conoco Coal Develop Co, 75-78. *Mem:* Am Inst Chem Eng; Am Soc Eng Educ. *Res:* Hazardous waste engineering; technology development; fossil-fuels processing, wastewater treatment bioremediation, electrochemistry; toxicology of coal conversion processing. *Mailing Add:* 3123 University Blvd Houston TX 75474

MASSEY, PHILIP LOUIS, b New York, NY, Feb 28, 52. STELLAR ASTRONOMY. *Educ:* Calif Inst Technol, BS & MS, 75; Univ Colo, Boulder, PhD(astrophysics), 80. *Prof Exp:* Teaching asst, Dept Astro-geophysics, Univ Colo, 75-76, res asst, Joint Inst Lab Astrophysics, 77-80; RES ASSOC ASTRON, DOMINION ASTROPHYS OBSERV, NAT RES COUN, 80-; AT KITT PEAK NAT OBSERV. *Mem:* Am Astron Soc; Am Asn Variable Star Observers. *Res:* Observational studies of very hot and massive stars, particularly Wolf-Rayets and O stars in both the Milky Way and the neighboring galaxies of the local group. *Mailing Add:* Nat Optical Astron PO Box 26732 Tucson AZ 85726

MASSEY, ROBERT UNRUH, b Detroit, Mich, Feb 23, 22; m 43; c 2. INTERNAL MEDICINE. *Educ:* Wayne Univ, MD, 46. *Prof Exp:* From intern to resident med, Henry Ford Hosp, 46-50; assoc, Lovelace Clin, 50-68, chmn dept, 58-68; clin assoc, Sch Med & Res, Univ NMex, 62-68; assoc dean, 68-71, dean, 71-85, actg exec dir & actg vpres health affairs, 75-76, head Div Health Care Admin, 85-87, PROF MED, MED SCH, UNIV CONN, 68- *Concurrent Pos:* Consult, West Interstate Comn Higher Educ, 58-60; dir educ, Lovelace Found Med Educ, 60-68; consult, NMex Regional Med Prog, 65-68; mem accreditation comn, Am Asn Med Clins, 66-72; ed, J Hist Med & Allied Sci, 87- *Mem:* AAAS; Sigma Xi; Am Col Physicians; Soc Med Adminrs; Asn Am Med Cols; Am Asn Hist Med. *Res:* Clinical endocrinology and diabetes; medical education; medical care and medical administration; medical history. *Mailing Add:* Sch of Med Univ of Conn Farmington CT 06032

MASSEY, VINCENT, b Berkeley, Australia, Nov 28, 26; m 50; c 3. BIOCHEMISTRY. *Educ:* Univ Sydney, BSc, 47; Cambridge Univ, PhD(biochem), 53. *Prof Exp:* Res officer biochem, Commonwealth Sci & Indust Res Orgn, Australia, 47-50; Imp Chem Industs Res fel, 53-55; mem res staff, Edsel B Ford Inst, Mich, 55-57; from lectr to sr lectr, Univ Sheffield, 57-63; PROF BIOL CHEM, SCH MED, UNIV MICH, ANN ARBOR, 63- *Concurrent Pos:* Guest prof, Univ Konstanz, 75-, Yokohama City Univ, 88. *Honors & Awards:* Fel Royal Soc London. *Mem:* Am Soc Biol Chemists; Am Chem Soc; Biochem Soc. *Res:* Basic enzymology; mechanisms of enzyme reactions, especially of flavoproteins and metalloflavoproteins. *Mailing Add:* Dept Biol Chem Univ Mich Sch Med Ann Arbor MI 48109

MASSEY, WALTER EUGENE, b Hattiesburg, Miss, Apr 5, 38; m 69; c 2. THEORETICAL SOLID STATE PHYSICS. *Educ:* Morehouse Col, BS, 58; Wash Univ, MA & PhD(physics), 66. *Hon Degrees:* DSc, Lake Forest Col, Ill, 81; Williams Col, Williamston, Mass, 81; Elhurst Col, Ill, 82; Atlanta Univ, Ga, 82; Rutgers Univ, New Brunswick, NJ, 84; Morehouse Col, Atlanta, 84; Marquette Univ, Milwaukee, 87; Boston Col, Mass, 87. *Prof Exp:* Instr physics, Morehouse Col, 58-59; from fel to physicist, Argonne Nat Lab, 66-68; asst prof, Univ Ill, Urbana, 69-70; assoc prof, Brown Univ, 70-75, prof physics & dean col, 75-79; vpres res, 83-84, PROF PHYSICS, UNIV CHICAGO, 79-; VPRES RES, ARGONNE NAT LAB, 84-; DIR, NSF, 91- *Concurrent Pos:* Mem, Physics Rev Comt, Nat Acad Sci-Nat Res Coun, 72-75, mem Adv Comt on Eastern Europe & USSR, 73-76; mem, Adv Panel, Div Physics, NSF, 75-77; mem, Nat Sci Bd, 78-84; mem, bd trustees, Wash Univ, 80-81, Brown Univ, 80-; mem, Corp Woods Hole Oceanographic Inst, 80-83, pres, 86-; vis comt, Col Arts & Scis, Nortwestern Univ, 82-85, vis comt dept physics, Mass Inst Technol, 82-; Gov's Comn Sci & Technol, Ill, 82-; bd dirs, Amoco Corp, 82-, First Nat Bank Chicago, 83-; bd trustees, Rand Corp, 83-; chmn bd, Chicago High Tech Asn, 84-, bd dirs, Motorola, Inc, 84-; adv comt sci & math educ, NSF, 85-; vis comt dept physics, Harvard Univ, 85-; bd trustees, Ill Math & Sci Acad, 85-; bd dir, Tribune Co, 87- *Mem:* Am Physical Soc; Am Asn Physics Teachers; fel AAAS (bd dirs, 81-85, pres, 88); NY Acad Sci; Sigma Xi; Am Nuclear Soc. *Res:* Many-body problem; quantum liquids and solids; theory of classical liquids; solid state theory. *Mailing Add:* NSF Rm 520 1800 G St NW Washington DC 20550

MASSEY, WILLIAM S, b Granville, Ill, Aug 23, 20; m 53; c 3. MATHEMATICS. *Educ:* Univ Chicago, BS, 41, MS, 42; Princeton Univ, PhD(math), 48. *Prof Exp:* Off Naval Res fel, Princeton Univ, 48-50; from asst prof to assoc prof math, Brown Univ, 50-54; vis assoc prof, Princeton Univ, 54-55; prof, Brown Univ, 55-60; chmn dept, 68-71, PROF MATH, YALE UNIV, 60- *Concurrent Pos:* Assoc ed, Ind Univ Math J, 75-85. *Mem:* Am Acad Arts & Sci; Am Math Soc; Math Asn Am. *Res:* Algebraic topology. *Mailing Add:* Dept Math Box 2155 Yale Sta Yale Univ New Haven CT 06520

MASSIAH, THOMAS FREDERICK, b Montreal, Que, Aug 26, 26; m 51; c 1. ORGANIC CHEMISTRY. *Educ:* Sir George Williams Univ, BSc, 47; McGill Univ, MSc, 56; Univ Montreal, PhD(org chem), 62. *Prof Exp:* Chief control chemist, Dewey & Almy Chem Co, Que, 47-53; demonstr chem, McGill Univ, 53-56; res chemist, Merck & Co Ltd, Que, 56-59; sr demonstr chem, Univ Montreal, 59-62; chemist, Ayerst, McKenna & Harrison Ltd, 62-66; group leader chem develop, Can Packers Ltd, 66-84; consult, 85; ASSOC PROF CHEM, SENECA COL, ONT, 86-; PRES, INNOCHEM CONSULTS LTD, ONT, 86- *Concurrent Pos:* Lectr, Sir George Williams Univ, 49-64; mem, Drug Qual & Therapeut Comt, Ont Ministry Health, 86-89. *Mem:* Fel Chem Inst Can. *Res:* Organic, biochemical and medicinal chemistry; antibiotics, bile acids, enzymes; pharmaceuticals; steroids. *Mailing Add:* Innochem Consult Ltd 802-65 Huntingdale Blvd Agincourt ON M1W 2P1 Can

MASSIE, BARRY MICHAEL, b St Louis, Mo, May 23, 44; c 2. CARDIOVASCULAR DISEASE, CARDIOVASCULAR FUNCTION. *Educ:* Harvard Univ, BA, 66; Columbia Univ, MD, 70. *Prof Exp:* Harvard Univ fels, 62-66; chief resident med, Bellevue Hosp, NY Univ Med Ctr, 70-74; cardiol fel, 75-78, asst prof, 78-83, ASSOC PROF MED, UNIV CALIF, SAN FRANCISCO, 83-; CHIEF, HYPERTENSION UNIT & DIR, CORONARY CARE UNIT, SAN FRANCISCO VET ADMIN HOSP, 78- *Concurrent Pos:* Vis prof, Univ Oxford, 85-86; assoc staff, Cardiovasc Res Inst, San Francisco, 81-; consult & lectr, various orgn, 78-; fel Coun Clin Cardiol & High Blood Pressure, Am Heart Asn, 78- *Mem:* Am Heart Asn; fel Am Col Cardiol; Soc Magnetic Resonance Med; Am Fed Clin Res; Am Osteop Asn. *Res:* Cardiovascular physiology and pharmacology; cardiac function and metabolism in hypertension, congestive heart failure. *Mailing Add:* Cardiology Div IIIC Vet Affairs Med Ctr 4150 Clement St San Francisco CA 94121

MASSIE, EDWARD, cardiology; deceased, see previous edition for last biography

MASSIE, HAROLD RAYMOND, b Brisbane, Australia, Jan 31, 43; US citizen; m 70; c 4. MOLECULAR BIOLOGY. *Educ:* San Diego State Col, AB, 64; Univ Calif, San Diego, PhD(chem), 67. *Prof Exp:* NIH res fel chem & tutor biochem, Harvard Univ, 67-70; RES SCIENTIST, MASONIC MED RES LAB, 70- *Mem:* AAAS; Am Soc Biol Chemists; Am Aging Asn. *Res:* Aging. *Mailing Add:* Masonic Med Res Lab Aging Prog 2150 Bleecker St Utica NY 13503

MASSIE, N A (BERT), b Inglewood, Calif, Jan 4, 44. ELECTRO-OPTICS, LASERS & SYSTEMS. *Educ:* Univ Calif, Los Angeles, PhD(physics), 74. *Prof Exp:* Proj engr, Rockwell Int, 75-83; dept mgr, Western Res, 83-86; PROJ MGR, LAWRENCE LIVERMORE NAT LAB, 86- *Mem:* Am Physics Soc; Optical Soc Am; Int Soc Optical Eng. *Res:* Segmented mirror; heterodyne interferometer. *Mailing Add:* 231 Market Pl No 352 San Ramon CA 94583

MASSIE, SAMUEL PROCTOR, b North Little Rock, Ark, July 3, 19; m 47; c 3. CHEMISTRY. *Educ:* Agr Mech & Normal Col, Ark, BS, 38; Fisk Univ, MA, 40; Iowa State Univ, PhD(org chem), 46. *Hon Degrees:* LLD, Univ Ark, 70; DSc, Lehigh Univ, 85, LHD, Bowie State Univ, Md, 90. *Prof Exp:* Lab asst chem, Fisk Univ, 39-40; assoc prof math, Agr Mech & Normal Col, Ark, 40-41; res assoc chem, Iowa State Univ, 43-46; instr, Fisk Univ, 46-47; prof & head dept, Langston Univ, 47-53, Fisk Univ, 53-60 & Howard Univ, 62-63; assoc prog dir, NSF, 60-63; pres, NC Col Durham, 63-66; chmn dept, 77-81, PROF CHEM, US NAVAL ACAD, 66- *Concurrent Pos:* Sigma Xi lectr, Swarthmore Col, 57, IIT, 91; bd trustees, Col of Wooster, 66-87; chmn, Md State Bd Comm Col, 68-89, Div Chem Ed, 76, Gov Sci Adv Coun, Md, 79-89; PRF Adv Bd, 81-86; distinguished vis prof, Va State Univ, 83-84, Dillard Univ, 84-85, UMES, 86-88; Smithsonian Exhibit Adv Comm, 89- *Honors & Awards:* Mfg Chem Asn Award, 61; Henry Hill Lect, NOBCChE, 85; White House Initiative Lifetime Achievement Award, 89; Nat Black Col Hall of Fame in Sci, 89. *Mem:* Am Chem Soc; Sigma Xi. *Res:* Med chem-antibact agents, antiradioactive agents, chem warfare antiagents, environmental chem studies, patent, antibacterial agents, heterocyclic chem, especially phenothiazine, organosilicon chem. *Mailing Add:* Dept of Chem US Naval Acad Annapolis MD 21402

MASSIER, PAUL FERDINAND, b Pocatello, Idaho, July 22, 23; m 48, 78; c 2. PROPULSION, DIRECT THERMAL-TO-ELECTRIC ENERGY CONVERSION. *Educ:* Univ Colo, BS, 48; Mass Inst Technol, MS, 49. *Prof Exp:* Design engr, Maytag Co, 49-50; res engr, Boeing Co, 50-55; res engr, 55-58, group supvr, 58-81, exec asst, 81-83, task mgr, 83-88, MEM TECH STAFF JET PROPULSION LAB, CALIF INST TECHNOL, 89- *Concurrent Pos:* Mem, Propulsion Noise Res Coord Group, NASA, 72; tech comt mem, Nuclear Propulsion, Am Inst Aeronaut & Astronaut, 68-70, Aeroacoustics, 73-75 & 79-81, Propellants & Combustion, 88-89. *Honors & Awards:* Apollo Achievement Award, NASA, 69, Basic Noise Res Award, 80. *Mem:* AAAS; Planetary Soc; Sigma Xi; Am Inst Aeronaut & Astronaut (secy, 73-75). *Res:* Thermally ionized gas flows with heat transfer, accelerating and decelerating supersonic flows with shock wave boundary layer interactions and heat transfer, noise generation and emission in supersonic jet flows; metastable states and matter-antimatter anihilation for interstellar rocket propulsion; direct thermal to electric energy conversion, confined vortex flows in supersonic nozzles and diffusers, shock wave boundary layer interactions in supersonic diffusers. *Mailing Add:* 1000 N First Ave Arcadia CA 91006

MASSIK, MICHAEL, b Buffalo, NY, Feb 4, 58. FLAVOR & FRAGRANCE ANALYSIS, BEVERAGE TECHNOLOGY. *Educ:* Univ Ky, BS, 80, MS, 82. *Prof Exp:* Toxicologist, Path & Cytol Labs, 78-82; res asst, Univ Louisville, 82-86; ANALYTICAL SERV MGR, BROWN-FORMAN BEVERAGE CO, 86- *Concurrent Pos:* Consult, Toxicol Consults, 80-84. *Mem:* Am Chem Soc; Asn Off Analytical Chemists; Am Soc Mass Spectrometry; Inst Food Technologists. *Res:* Alcohol beverages, flavors and fragrances; deduce chemical basis of flavor response and sensory characteristics; classification of chemical basis of flavors and beverages using instrumental analysis and sensory science. *Mailing Add:* Brown-Forman Beverage Co PO Box 1080 Louisville KY 40201

MASSINGILL, JOHN LEE, JR, b Lufkin, Tex, Aug 18, 41; m 63; c 2. INDUSTRIAL ORGANIC CHEMISTRY. *Educ:* Tex Christian Univ, BA, 63, MS, 65, PhD(chem), 68. *Prof Exp:* Sr res chemist, Basic Res Dept, 68 & Hydrocarbon Process Res Dept, 70-73, res specialist, Hydrocarbon Process Res Dept, 73-76, res leader, 76-83, DEVELOP ASSOC, RESINS RES DEPT, TEX DIV, DOW CHEM USA, FREEPORT, 84- *Concurrent Pos:* Consult, Ionics Res, Inc, 71-72, Network Mkt. *Honors & Awards:* Brazosport Sect Award for Outstanding Prom Sci, Am Chem Soc, 90. *Mem:* AAAS; Am Chem Soc; Sigma Xi. *Res:* Hydrocarbon utilization; new product research and development; new process research and development. *Mailing Add:* 410 Forest Dr Lake Jackson TX 77566

MASSION, WALTER HERBERT, b Eitorf, Ger, June 4, 23; nat US; m 56; c 3. ANESTHESIOLOGY, PHYSIOLOGY. *Educ:* Univ Cologne, BS, 47; Univ Heidelberg, MD, 51; Univ Bonn, DrMed, 51. *Prof Exp:* Intern med, Med Ctr, Univ Zurich, 51-52; trainee anesthesiol, Anesthesiol Ctr, WHO, Denmark, 52-53; asst prof physiol, Med Sch, Univ Basel, 53-54; asst resident anesthesiol, Med Sch, Univ Rochester, 54-56; from asst prof to assoc prof anesthesiol, Col Med, Univ Okla, 58-67, assoc prof physiol & res surg, 66-71, adj prof cardiorespiratory sci, 75-81, prof physiol, 71-88, PROF ANESTHESIOL, COL MED, UNIV OKLA, 67- *Concurrent Pos:* Fel physiol, Med Sch, Univ Rochester, 54-56; fel, Cardiovasc Res Inst, Sch Med, Univ Calif, 59-60; NIH res career develop award, 61-71; John A Hartford Found res grant, 67-71; guest prof, Tech Univ Munich, Germany, 74-75; mem sci coun, Am Heart Asn; Fulbright scholar, Aachen, WGermany, 84-85; ed, Prog Critical Care Med, 84- *Honors & Awards:* Alexander von Humboldt Prize, 73; Order of Merit, Fed Repub Ger. *Mem:* AAAS; Am Physiol Soc; Am Soc Anesthesiol; Int Anesthesia Res Soc. *Res:* Respiration; circulation; shock; vasoactive polypeptides; extracorporeal circulation. *Mailing Add:* Dept Anesthesiol Univ Okla Hlth Sci Ctr Oklahoma City OK 73190

MASSLER, MAURY, dentistry; deceased, see previous edition for last biography

MASSO, JON DICKINSON, b Sewickley, Pa, Nov 12, 41; m 66; c 4. THIN FILM COATING. *Educ:* Drexel Univ, BS, 64; Colo State Univ, MS, 67, PhD(physics), 70. *Prof Exp:* PRIN PHYSICIST, AM OPTICAL CORP, 80- *Mem:* Am Vacuum Soc; Soc Vacuum Coaters. *Res:* Development of thin film; vacuum deposited coatings for plastic lenses including ophthalmic lenses; coatings include anti-reflective, scratch resistant and IR reflecting. *Mailing Add:* 175 Carpenter Rd Whitinsville MA 01558

MASSOF, ROBERT W, b Pittsburgh, Pa, Jan 2, 48; m 74; c 2. OPHTHALMOLOGY, VISION SCIENCE. *Educ:* Hamline Univ, BA, 70; Ind Univ, PhD(physiol optics), 75. *Prof Exp:* ASSOC PROF OPHTHAL, SCH MED, JOHNS HOPKINS UNIV, 75-, DIR, WILMER VISION RES & REHAB CTR, 76- *Concurrent Pos:* Dir, LKC Technol Inc, 87- *Mem:* Fel Optical Soc Am; fel Am Acad Optom; Asn Res Vision Ophthal. *Res:* Visual sensation & performance in diseases of the visual system; retinitis pigmentosa; age-related macular degeneration. *Mailing Add:* Wilmer B-20 John Hopkins Hosp Baltimore MD 21205

MASSON, D(OUGLAS) BRUCE, b Corvallis, Ore, Dec 19, 32; m 58; c 3. PHYSICAL METALLURGY. *Educ:* Wash State Univ, BS, 54; Univ Chicago, MS, 56, PhD(chem), 58. *Prof Exp:* Asst prof mech eng, Rice Inst, 58-60; assoc prof phys metall, 60-71, PROF PHYS METALL, WASH STATE UNIV, 71-, CHMN DEPT MAT SCI & ENG, 75- *Mem:* Am Soc Metals; Am Inst Mining, Metall & Petrol Engrs. *Res:* Phase equilibria and transformations in metals; solid-state reaction kinetics; x-ray diffraction; crystal chemistry; thermodynamics of alloys. *Mailing Add:* Dept Mat Sci & Eng Wash State Univ Pullman WA 99164-2920

MASSOPUST, LEO CARL, JR, b Milwaukee, Wis, Nov 12, 20; m 43; c 2. ANATOMY. *Educ:* Marquette Univ, BS, 43, MS, 47; Univ Colo, PhD, 53. *Prof Exp:* Asst prof biol, Westminster Col (Mo), 47-48; instr anat, Univ Colo, 48-54; neuroanatomist, NIH, 54-58; sr res physiologist, Southeast La Hosp, 58-60; dir div neurophysiol, Cleveland Psychiat Inst & Hosp, 60-73; ASSOC PROF ANAT, MED SCH, ST LOUIS UNIV, 74- *Mem:* Am Asn Anat; Am Physiol Soc; Soc Exp Biol & Med; Am Acad Neurol; Soc Neurosci. *Res:* Physiology of ergot alkaloid; hypothermia; electrophysiology of vision; psychophysiology of audition; neurophysiology of brain function; hodology of central nervous system pathways. *Mailing Add:* Sch of Med 1402 S Grand Blvd St Louis MO 63104

MASSOUD, HISHAM Z, b Cairo, Egypt, July 31, 49; US citizen; m 83; c 1. BIOENGINEERING & BIOMEDICAL ENGINEERING, OTHER PHYSICS. *Educ:* Cairo Univ, BS, 73, MSc, 75; Stanford Univ, MS, 76, PhD(elec eng), 83. *Prof Exp:* Asst prof, 83-86, ASSOC PROF ELEC ENG, DUKE UNIV, 86-, LAB DIR, SEMICONDUCTOR RES LAB, 90- *Concurrent Pos:* Mem, Tech Prog Planning Comt, Electrochem Soc, 88-, Inst Elec & Electronics Engrs, Int Electron Devices Meeting, 90-92 & Semiconductor Interface Specialists Conf, 91-93. *Mem:* Inst Elec & Electronics Engrs; Am Vacuum Soc; AAAS. *Res:* Solid-state devices and materials; processing technology-characterization and modeling; microstructure applications in medical and biological research. *Mailing Add:* Dept Elec Eng Duke Univ Durham NC 27706

MASSOUD, MONIR FOUAD, b Fayoum, Egypt, June 13, 30; m 57; c 2. ENGINEERING, SOLID MECHANICS. *Educ:* Cairo Univ, BSc, 51, dipl higher study, 57; Rensselaer Polytech Inst, MSc, 61, PhD(mech eng), 63. *Prof Exp:* Design engr, Heliopolis Aircraft Factory, Egypt, 51-59; res asst design, Rensselaer Polytech Inst, 60-63; assoc prof solid mech, Cairo Polytech Inst, 63-67; Norweg AID fel & researcher, Tech Univ Norway, 67-68; assoc prof 68-74, PROF MECH ENG, UNIV SHERBROOKE, 74-, CHMN DEPT, 84- *Honors & Awards:* Robert W Angus Medal, Can Soc Mech Eng, 86. *Mem:* Am Soc Mech Engrs; fel Can Soc Mech Engrs; Am Acad Mech; fel Engr Inst Can. *Res:* Applied dynamics; investigation of dynamic behavior of mechanical engineering elements and systems; methods and probabilistic approaches to improve the reliability and optimize the behavior of a mechanical design. *Mailing Add:* Dept Mech Eng Fac Appl Sci Univ Sherbrooke 2500 University Blvd Sherbrooke PQ J1K 2R1 Can

MASSOVER, WILLIAM H, b Chicago, Ill, 41; m 69. CELL BIOLOGY, STRUCTURAL BIOLOGY. *Educ:* Univ Chicago, AB, 63, MD, 67, PhD(cell biol), 70. *Prof Exp:* NATO fel electron micros, Lab Electron Optics, CNRS, Toulouse, France, 71-72; res assoc dept physics, Ariz State Univ, 72-73; asst prof biol, Brown Univ, 73-79; ASSOC PROF ANAT, NJ MED SCH, UNIV MED & DENT NJ, 79- *Concurrent Pos:* Mem spec study sect, NIH, 75, 77 & 90, external adv comt for HVEM at NY State Dept Health, Albany, 82-85; outside reviewer, NSF, 77-79, 81, 86 & 89; outside reviewer, NIH, 88; assoc ed book reviews, EMSA Bull, 84-; dir, Biol Sci, Electron Micros Soc Am, 84-87; outside reviewer, Agr Res Serv/USDA, 89; vis assoc prof, Dept Cell Biol, NY Univ Med Ctr, 90-91. *Mem:* Am Soc Cell Biol; Biophys Soc; Electron Micros Soc Am; Electrophoresis Soc. *Res:* Biochemistry and immunology of ferritin heterogeneity; biochemical pathology of iron storage in liver cancer; high-resolution ultrastructure of biological macromolecules and supramolecular systems; cancer cell biology. *Mailing Add:* Dept Anat NJ Med Sch Univ Med & Dent NJ Newark NJ 07103-2757

MAST, CECIL B, b Chicago, Ill, Feb 21, 27; m 59; c 2. GEOMETRY, THEORETICAL PHYSICS. *Educ:* De Paul Univ, BS, 50; Univ Notre Dame, PhD(physics), 56. *Prof Exp:* Instr physics, 56-57, from instr to asst prof math, 59-63, ASSOC PROF MATH, UNIV NOTRE DAME, 63- *Concurrent Pos:* Vis lectr, St Andrews, 65-66. *Mem:* Am Phys Soc; Am Math Soc; Math Asn Am. *Res:* Nuclear physics and group representations as used in nuclear models; relativity theory; differential geometry and lie groups; foundations of physics. *Mailing Add:* 311 E Pokagon South Bend IN 46617

MAST, MORRIS G, b Kalona, Iowa, Dec 8, 40; m 64; c 2. POULTRY AND EGG PRODUCTS TECHNOLOGY. *Educ:* Goshen Col, BS, 62; Ohio State Univ, MS, 69, PhD(food sci), 71. *Prof Exp:* Diag parasitologist, Evanston Hosp, Evanston, Ill, 62-65; mgr qual control, V F Weaver, Inc, New Holland, Pa, 65-67; res assoc poultry sci, Ohio State Univ, 67-71; PROF FOOD SCI, PA STATE UNIV, UNIV PARK, 71- *Concurrent Pos:* Scientist, Spelderholt Inst for Poultry Res, The Netherlands, 80-81. *Honors & Awards:* Res Award, Am Egg Bd, 83. *Mem:* Coun Agr Sci & Technol; Poultry Sci Asn; Inst Food Technologists. *Res:* Microbiological, biochemical, and organoleptic changes occurring in poultry, egg, and seafood products during processing and storage. *Mailing Add:* Dept of Food Sci Pa State Univ University Park PA 16802

MAST, P(LESSA) EDWARD, b Bourbon, Ind, Oct 18, 26; m 47; c 7. ENGINEERING ADMINISTRATION. *Educ:* Purdue Univ, BSEE, 48, MSEE, 50; Univ Ill, PhD(elec eng), 58. *Prof Exp:* From assoc prof to prof elec eng, 52-87, assoc head, Elec & Comput Eng dept, 86-87, EMER PROF, UNIV ILL, URBANA, 87- *Concurrent Pos:* Sr scientist, Aeronutronic Div, Philco-Ford Corp, 64-65; distinguished vis prof, US Air Force Acad, Colo, 84-86. *Mem:* Inst Elec & Electronics Engrs. *Res:* Antennas; electromagnetic theory and scattering. *Mailing Add:* 2208 S Cottage Grove Urbana IL 61801

MAST, RICHARD F(REDRICK), b Chicago, Ill, Oct 4, 31; m 57; c 5. PETROLEUM ENGINEERING, GEOLOGY. *Educ:* Univ Ill, BS, 57, MS, 60. *Prof Exp:* Asst petrol engr, Ill State Geol Surv, 59-64; petrol engr, Co Francais des Petrol, 64 & Francore Labs, 64-65; assoc petrol engr, Ill State Geol Surv, 65-73; mem staff, 73-76, chief, Oil & Gas Resources Br, 76-81, regional geologist, Cent Region, 81-84, RES GEOLOGIST, US GEOL SURV, 85- *Concurrent Pos:* Chmn subcomt, Future Gas Requirements & Supply Comt, 65-69; mem eng comt, Interstate Oil Compact Comn, 69-71. *Mem:* Am Inst Mining, Metall & Petrol Engrs; Am Asn Petrol Geologists; Geol Soc Am. *Res:* Properties of rocks; especially continuity, inhomogeneity, grain orientation and permeability; petroleum geochemistry; Tertiary recovery methods; resource analysis. *Mailing Add:* Eight Skyline Dr Denver CO 80215

MAST, ROBERT F, b Springfield, Ill, May 20, 34. ENGINEERING. *Educ:* Univ Ill, Urbana, BA, 57. *Prof Exp:* Design engr, Anderson, Birkeland, Anderson & Mast, Tacoma, Wash, 59-62, partner, 63-65; exec vpres, ABAM Engrs Inc, 66-72, pres, 72-86, CHMN BD, BERGER/ABAM ENGRS INC, FEDERAL WAY, WASH, 86- *Concurrent Pos:* Mem, Planning Comt, Am Concrete Inst, Tech Activ Comt, Comt Responsibility Concrete Construct; mem, Spec Comt Offshore Installations, Am Bur Shipping. *Honors & Awards:* T Y Lin Award, Am Soc Civil Engrs. *Mem:* Nat Acad Eng; Prestressed Concrete Inst; fel Am Concrete Inst; Am Soc Civil Engrs; Reinforced Concrete Res Coun; Soc Naval Architects & Marine Engrs; Marine Technol Soc. *Res:* Author of various publications. *Mailing Add:* BERGER/ABAM Engrs Inc 33301 Ninth Ave S Federal Way WA 98003

MAST, ROY CLARK, b Wheeling, WVa, Nov 28, 24; m 48; c 3. PHYSICAL CHEMISTRY, INORGANIC CHEMISTRY. *Educ:* Univ Cincinnati, BS, 49, MS, 51, PhD(inorg chem), 53. *Prof Exp:* Student asst instr, Univ Cincinnati, 49-51; RES CHEMIST PHYS & INORG CHEM, MIAMI VALLEY LABS, PROCTER & GAMBLE CO, 52- *Mem:* Am Chem Soc; AAAS. *Res:* Surfactant solutions; adsorption of surfactants; emulsion formation; surface chemistry; phase studies; fundamentals of detergency; micellar solubilization; diffusion studies; polymer properties. *Mailing Add:* 3336 Nandale Dr Cincinnati OH 45239

MAST, TERRY S, b Los Angeles, Calif, Jan 2, 43; m 89; c 2. LARGE TELESCOPE DESIGN. *Educ:* Calif Inst Technol, BS, 64; Univ Calif, Berkeley, PhD(physics), 71. *Prof Exp:* Res physicist res, Univ Calif, Berkeley, 71-91. *Res:* Particle astrophysics; optics; telescope design. *Mailing Add:* 77-6469 Alii Dr No 324 Kailua-Kona HI 96740

MASTALERZ, JOHN W, b Mass, Mar 16, 26; m 54; c 3. FLORICULTURE, HORTICULTURE. *Educ:* Univ Mass, BS, 48; Purdue Univ, MS, 50; Cornell Univ, PhD(floricult), 53. *Prof Exp:* Asst prof res floricult, Waltham Field Sta, Univ Mass, 52-56; prof floricult, Pa State Univ, 56-86; RETIRED. *Concurrent Pos:* Ed, Pa Flower Growers, 56-85. *Mem:* AAAS; Am Soc Agron; Crop Sci Soc Am; Am Soc Horticult Sci; Am Hort Soc. *Res:* Post-harvest life of cut flowers; photoperiodic, temperature, soil mixture and fertilization requirements of flower crops; growth regulators; greenhouse environment. *Mailing Add:* 51 Fairview Ave South Chatham MA 02659

MASTASCUSA, EDWARD JOHN, b Pittsburgh, Pa, June 27, 38; m 60; c 6. ELECTRICAL ENGINEERING. *Educ:* Carnegie Inst Technol, BS, 60, MS, 61, PhD(elec eng), 64. *Prof Exp:* Asst prof elec eng, Univ Wyo, 66-68; asst prof, 68-73, ASSOC PROF ELEC ENG, BUCKNELL UNIV, 73-, CHMN DEPT, 77- *Mem:* Soc Comput Simulation; Sigma Xi. *Res:* Digital computation of system and circuit response; engineering creativity. *Mailing Add:* Dept Elec Eng Bucknell Univ Lewisburg PA 17837

MASTELLER, EDWIN C, b Independence, Iowa, Aug 11, 34; m 57; c 2. BIOLOGY, ENTOMOLOGY. *Educ:* Northern Iowa Univ, BS, 58; Univ SDak, MA, 61; Iowa State Univ, PhD(entom), 67. *Prof Exp:* Pub sch instr, Minn, 58-64; from asst prof to assoc prof, 67-84, PROF BIOL, BEHREND COL, PA STATE UNIV, 84- *Concurrent Pos:* Fulbright-Hays sr res fel, Ger, 74-75. *Mem:* Am Inst Biol Sci; Entom Soc Am; NAm Benthol Soc (pres, 84); Sigma Xi. *Res:* Tropical and temperate latitude aquatic insect comparisons; aquatic insect emergence-phenology; Ephemeroptera, Plectoptera, Trichoptera and Diptera. *Mailing Add:* Behrend Campus Pa State Univ Erie PA 16563

MASTEN, MICHAEL K, b Gainesville, Tex, Nov 11, 39; m 64; c 1. ELECTRICAL ENGINEERING. *Educ:* Univ Tex, Austin, BS, 63, MS, 65, PhD(elec eng), 68. *Prof Exp:* Teaching asst, Univ Tex, Austin, 64-68; mem tech staff, 68-79, sr mem tech staff elec eng, 80-88, FEL, TEXAS INSTRUMENTS INC, 89- *Mem:* Inst Elec & Electronic Engrs; fel Inst Elec & Electronic Engrs. *Res:* Control systems; pattern recognition; adaptive systems; line-of-sight stabilization and target tracking. *Mailing Add:* Tex Instruments Inc PO Box 405 MS 3478 2501 S Hwy 121 Lewisville TX 75067

MASTENBROOK, S MARTIN, JR, b Ft Worth, Tex, Mar 24, 46. BIOINSTRUMENTATION, PULMONARY PHYSIOLOGY. *Educ:* Tex Tech Univ, BSEE, 69; Univ Wis, Madison, MS(physiol) & MS(elec eng), 73, PhD(elec eng, bioeng), 77. *Prof Exp:* Res asst elec eng, Univ Wis, Madison, 70-74, fel bioeng, 74-77, fel pulmonary physiol, 77-78; res fel pulmonary physiol, Univ Calif, San Diego, 78-79, asst res physiologist, 79-81; asst prof biomed eng, Boston Univ, 81-; AT AT&T BELL LABS. *Concurrent Pos:* NIH fel, Univ Wis, Madison, 77-78, Am Lung Asn fel, 74-76. *Mem:* Inst Elec & Electronics Engrs; Biomed Eng Soc; Am Thoracic Soc; Am Soc Eng Educ; Am Heart Asn. *Res:* Physiology of pulmonary gas exchange; mathematical modeling of the lung; biomedical instrumentation applied to pulmonary physiology and medicine. *Mailing Add:* AT&T Bell Labs Rm 1a-07 75 Foundation Ave Ward Hill MA 01835

MASTERS, BETTIE SUE SILER, b Lexington, Va, June 13, 37; m 60; c 2. MICROSOMAL METABOLISM EICSANOIDS, STRUCTURE-FUNCTION STUDIES FLAVOPROTEINS. *Educ:* Roanoke Col, BS, 59; Duke Univ, PhD(biochem), 63. *Hon Degrees:* DSc, Roanoke Col, 83. *Prof Exp:* Res assoc biochem, Duke Univ, 65-67, assoc, 67-68; from asst prof to prof biochem, Univ Tex Health Sci Ctr Dallas, 68-82, prof surg & dir biochem burn res, 79-82; prof & chmn biochem, Med Col Wis, Milwaukee, 82-90; PROF CHEM, ROBERT A WELCH FOUND, 90- *Concurrent Pos:* Am Heart Asn estab investr, 68-73, prin investr, Nat Heart & Blood Inst, 70-; ed bd, J Biol Chem, 76-81; mem pharmacol-toxicol prog res rev comt, Nat Inst Gen Med Sci, NIH, 75-79; Am Cancer Soc fel biochem, Duke Univ, 63-65, Am Heart Asn advan res fel, 66-68; vis prof, Japan Soc Promotion Sci, 78, Osaka Univ & Kyushu Univ; mem bd sci couns, Nat Inst Environ Health Sci, 82-86, chmn 84-86; prin investr, Nat Inst Gen Med Sci, 82-; mem, physical biochem study sect, Am Cancer Soc, 88-90, res rev comt, 89- *Mem:* AAAS; Am Soc Biochem & Molecular Biol; Am Chem Soc; Am Soc Pharmacol & Exp Therapeut; Am Soc Cell Biol. *Res:* Microsomal electron transport in various tissues with specific reference to nicotinamide adenine dinucleotide phosphate-cytochrome c (P-450) reductase; cytochrome P450-mediated W-hydroxylation of eicosanoids, including prostaglandins. *Mailing Add:* Dept Biochem Univ Tex Health Sci Ctr 7703 Floyd Curl Dr San Antonio TX 78284-7760

MASTERS, BRUCE ALLEN, b Terre Haute, Ind, Nov 3, 36; m 63. MICROPALEONTOLOGY. *Educ:* Univ Valparaiso, BS, 59; Univ Calif, Berkeley, MA, 62; Univ Ill, Urbana, PhD(geol), 70. *Prof Exp:* Jr geologist, Humble Oil & Ref Co, 62-63, asst geologist, 63-65, assoc geologist, 65; from asst prof to assoc prof geol, Hartwick Col, 69-74; sr res scientist, Amoco Prod Co, 74-77, staff res scientist, 77-82, res assoc, 83-90, SPEC RES ASSOC, AMOCO PROD CO, 90- *Mem:* Am Asn Petrol Geol; Paleont Res Inst; Paleont Soc; Cushman Found; Swiss Geol Soc; Brit Micropaleont Soc. *Res:* Morphology, taxonomy, phylogeny, paleoecology and biostratigraphy of Mesozoic and Cenozoic planktonic foraminifers, benthonic foraminifers, nannoconids, calcispheres and tintinnids. *Mailing Add:* Amoco Prod Co Res Ctr PO Box 3385 Tulsa OK 74102

MASTERS, BURTON JOSEPH, b Casper, Wyo, Sept 8, 29; m 53; c 5. SOLID STATE SCIENCE. *Educ:* Univ Calif, Los Angeles, BS, 50; Ore State Col, PhD(chem), 54. *Prof Exp:* Staff mem, Los Alamos Sci Lab, 54-63; sr chemist, IBM Corp, 63-85; ASST PROF, SAN JOSE STATE UNIV, 85- *Mem:* Am Phys Soc. *Res:* Diffusion and ion implantation in silicon; development of silicon integrated circuits; alpha particle induced soft errors in silicon memories; silicon micromechanics; diffusion and ion implantation of semiconductor devices and circuits. *Mailing Add:* 6138 Del Canto Dr San Jose CA 95119

MASTERS, CHARLES DAY, b Pawhuska, Okla, Aug 4, 29; m 53; c 3. PETROLEUM GEOLOGY, GLOBAL RESOURCES ASSESSMENT. *Educ:* Yale Univ, BS, 51, PhD(geol), 65; Univ Colo, MS, 57. *Prof Exp:* Hydrographic officer, US Navy, 52-54; explor geologist, Pan Am Petrol Corp, 57-68, res geologist, 68-70; chmn div sci & math, WGa Col, 70-73; chief, Off Energy Resources & actg chief, Off Marine Geol, 73-80, RES GEOLOGIST WORLD ENERGY RESOURCES, US GEOL SURV, 80- *Mem:* AAAS; Sigma Xi; Geol Soc Am; Am Asn Petrol Geologists. *Res:* World basin analysis relative to petroleum occurrence. *Mailing Add:* 1028 Walker Rd Great Falls VA 22066

MASTERS, CHRISTOPHER FANSTONE, b Ashridge, Eng, Dec 26, 42; US citizen; m 66; c 2. MATHEMATICS. *Educ:* Doane Col, AB, 64; Fla State Univ, MS, 66; Univ Northern Colo, DA, 74. *Prof Exp:* Instr math, Fla Southern Col, 66-68; from asst prof to assoc prof, 68-83, PROF MATH, DOANE COL, 83- *Mem:* Math Asn Am; Nat Coun Teachers Math. *Mailing Add:* Dept of Math Doane Col Crete NE 68333

MASTERS, EDWIN M, b Everette, Mass, Nov 21, 31; m 64; c 4. ANATOMY. *Educ:* Harvard Univ, AB, 52; Ind Univ, AM, 55; Univ Minn, PhD(anat), 65. *Prof Exp:* Instr anat, Univ Pittsburgh, 58-64; instr, 64-65, asst prof, 65-75, ASSOC PROF ANAT, JEFFERSON MED COL, 75- *Concurrent Pos:* NIH fel, 51-64. *Mem:* AAAS; Am Asn Anat; NY Acad Sci. *Res:* Physiology of fat cells in tissue culture; histogenesis of elastic tissue; survival of homologous grafts in the brain. *Mailing Add:* 1020 Locust Philadelphia PA 19107

MASTERS, FRANK WYNNE, b Pittsburgh, Pa, Nov 1, 20; m; c 3. PLASTIC SURGERY. *Educ:* Hamilton Col, AB, 43; Univ Rochester, MD, 45; Am Bd Plastic Surg, dipl, 55; Am Bd Surg, dipl, 56. *Prof Exp:* Intern, Strong Mem Hosp, Rochester, NY, 45-46; trainee gen surg, 48-51; trainee plastic surg, Med Ctr, Duke Univ, 51-53, assoc, 53-54; chief plastic surg, Charleston Mem Hosp, WVa, 54-58; from asst prof to assoc prof, 58-67, vchmn dept surg, 72-77, assoc dean clin affairs, 73-76, chief sect plastic surg, 72-80, PROF PLASTIC SURG, UNIV KANS MED CTR, KANSAS CITY, 67-, CHMN DEPT SURG, 77- *Concurrent Pos:* Consult, Vet Admin Hosps, Kansas City, Mo & Wadsworth, Kans; mem, Am Bd Plastic Surg, 68-74, co-chmn exam comt, 68-69, chmn, 69-73, rep, Am Bd Med Spec, 71-74, mem exec comt, 72-74, chmn, 73-74. *Mem:* Am Burn Asn; Am Asn Surg Trauma; Am Cleft Palate Asn; Am Soc Plastic & Reconstruct Surg; fel Am Col Surg; Sigma Xi. *Mailing Add:* Dept Surg Univ Kans Med Ctr 39th St & Rainbow Blvd Kansas City KS 66103

MASTERS, JOHN ALAN, b Shenandoah, Iowa, Sept 20, 27; m 51; c 5. PETROLEUM GEOLOGY. *Educ:* Yale Univ, BA, 48; Univ Colo, MS, 51. *Prof Exp:* Dist geologist, AEC, 51-53; chief geologist, Kerr-McGee Oil Industs, Inc, 53-66, mgr Can explor, Kerr-McGee Corp, 66-69, pres, Kerr-McGee Can Ltd, 69-73; PRES, CAN HUNTER EXPLOR, 73- *Honors & Awards:* Mattson Award, Asn Petrol Geol, 57. *Mem:* Geol Soc Am; Am Asn Petrol Geol. *Res:* Stratigraphy; oil exploration by means of subsurface and surface geology. *Mailing Add:* Can Hunter Explor 700-435 Fourth Ave SW Calgary AB T2P 3A8 Can

MASTERS, JOHN EDWARD, b Greeneville, Tenn, June 20, 13; m 38; c 2. ORGANIC CHEMISTRY. *Educ:* Tusculum Col, AB, 36; Univ Tenn, MS, 38. *Prof Exp:* Resin chemist high polymers, Jones-Dabney Co, 39-46, chief chemist resin div, 46-49; res dir, Devoe & Raynolds Co, 49-65; mgr, Trade Sales Labs, Celanese Coatings Co, Jeffersontown, 65-69, sr res assoc, 69-79; RETIRED. *Mem:* Am Chem Soc. *Res:* Exploratory research in the field of high polymers and protective coatings. *Mailing Add:* 9207 Darley Dr Louisville KY 40241

MASTERS, JOHN MICHAEL, b Cincinnati, Ohio, Aug 29, 42; m 63; c 2. GEOLOGY. *Educ:* Univ Cincinnati, BS, 64, MS, 66. *Prof Exp:* Geologist petrol, Chevron Oil Co, 66-71; asst geologist, 71-80, ASSOC GEOLOGIST INDUST MINERALS, ILL STATE GEOL SURV, 80- *Mem:* Soc Mining Engrs; Am Inst Mining, Metall & Petrol Engrs; Sigma Xi. *Res:* Geology of industrial minerals, specializing in sand and gravel resources, beneficiation and utilization. *Mailing Add:* 2005 Easy St Urbana IL 61801

MASTERS, LARRY WILLIAM, b Martinsburg, WVa, Nov 18, 41; m 63; c 2. MATERIALS SCIENCE. *Educ:* Shepherd Col, BS, 63; Am Univ, MS, 68. *Prof Exp:* chemist, Hazleton Labs, Falls Church, Va, 63-64; res chemist, Nat Bur Standards, 64-77; GROUP LEADER, NAT INST STANDARDS & TECHNOL, GAITHERSBURG, MD, 77- *Concurrent Pos:* Chmn, Comt Prediction Serv Life, Int Union Testing & Res Lab Mat & Struct; mem, US Tech Adv Group, Int Standardization Orgn, TC 180 on Solar Energy, 81 -; chmn comt E44 Solar Energy Conversion , Am Soc Testing & Mat; US deleg, Int Union Testing & Res Lab Mat & Struct. *Honors & Awards:* Bronze Medal, Dept Commerce. *Mem:* Am Chem Soc; Am Soc Testing & Mat; Int Union Testing & Res Lab Mat & Struct. *Res:* Service life prediction of polymeric building materials; materials for use in solar energy systems; author or co-author of over 75 publications. *Mailing Add:* Rm B-348 Bldg 226 Nat Inst Standards & Technol Gaithersburg MD 20899

MASTERS, ROBERT WAYNE, b Ft Wayne, Ind, May 25, 14; m 41. ELECTRONICS. *Educ:* Univ Ala, BS, 38; Ohio State Univ, MS, 41; Univ Pa, PhD, 57. *Prof Exp:* Adv develop engr, RCA Victor Div, Radio Corp Am, 41-49; assoc supvr, Antenna Lab, Ohio State Univ, 49-58; staff engr res supvr, Boeing Airplane Co, 58-60; mgr, Antenna Dept, Melpar, Inc, 60-62, mem systs staff to vpres eng, 62-64; staff scientist, DECO Electronics, 65-67; fel engr, DECO Commun Dept, Westinghouse Elec Corp, 67-68; vpres res eng, 68-70, vpres & secy, 70-75, PRES & CHMN, ANTENNA RES ASSOCS, INC, 75- *Concurrent Pos:* Prof, Ohio State Univ, 56-58. *Mem:* Fel Inst Elec & Electronics Engrs; Sigma Xi. *Res:* Antennas, including television transmitting, antenna complexes, radio-frequency transmitting systems; propagation. *Mailing Add:* Antenna Res Assoc 11317-19 Frederick Ave Beltsville MD 20705

MASTERS, WILLIAM HOWELL, b Cleveland, Ohio, Dec 27, 15; m 71; c 2. OBSTETRICS & GYNECOLOGY. *Educ:* Hamilton Col, BS, 38; Univ Rochester, MD, 43; Am Bd Obstet & Gynec, cert, 51; Am Bd Cytol, cert, 70. *Hon Degrees:* ScD, Hamilton Col, 73. *Prof Exp:* Intern path, 44, asst obstet & gynec, 44-47, from instr to assoc prof obstet & gynec, 47-64, dir, div reproductive biol, 60-63, assoc prof clin obstet & gynec, 64-69, PROF CLIN OBSTET & GYNEC, SCH MED, WASH UNIV, 69-, LECTR, HUMAN SEXUALITY IN PSYCHIAT, 81-; CHMN BD, MASTERS & JOHNSON INST, 81- *Concurrent Pos:* Intern, Barnes Hosp, 43 & 45, asst resident, 44, resident, 46-47; intern, St Louis Maternity Hosp, 43, asst resident 44, resident, 45-46; assoc obstetrician & gynecologist, Barnes Hosp, St Louis Maternity Hosp & Wash Univ Clins; assoc gynecologist, St Louis Children's Hosp; asst attending obstetrician & gynecologist, Jewish Hosp; consult gynecologist, St Louis City Infirmary; dir, Reproductive Biol Res Found, 64-73; co-dir, Masters & Johnson Inst, 73-80. *Honors & Awards:* Paul H Hoch Award, Am Psychopath Asn, 71; Sex Info & Educ Coun US Award, 72; Am Asn Sex Educr, Counr & Therapists Award, 78; Biomed Res Award, World Asn Sexology, 79; Edward Henderson Lect Award, Am Geriat Soc, 81. *Mem:* AAAS; Am Fertil Soc; Endocrine Soc; NY Acad Sci; Am Geriat Soc; fel Am Col Obstet & Gynec. *Res:* Infertility and sterility; geriatric endocrinology; sexual inadequacy. *Mailing Add:* Masters & Johnson Inst Wash-Univ Med 660 S Euclid Ave St Louis MO 63108

MASTERSON, KLEBER SANLIN, JR, b San Diego, Calif, Sept 26, 32; m 57; c 2. OPERATIONS ANALYSIS, ADVANCED COMPUTATIONAL TECHNOLOGIES. *Educ:* US Naval Acad, BS, 54; US Naval Postgrad Sch, MS, 60; Univ Calif, PhD(physics), 63. *Prof Exp:* Comput programmer, US Navy Electronics Lab, San Diego, 58-59; dep dir combat data syst, US Navy Electronics Lab & USS Wright, 63-64; head, Sea Cont Forces Group, Syst Analysis Div, Off Chief Naval Opers, 71-74; proj mgr develop & procurement combat syst, Naval Mat & Sea Syst Commands, Antiship Missile Defense Proj, 74-77, exec asst & naval aide to secy navy, Dept Navy, 77-79, asst dep comdr, Naval Sea Systs Command, antiair warfare & anti-ship combat systs, 79-81, chief Studies Anal & Games Agency, Off Joint Chiefs Staff, 81-82; prin & leader anal Computational Technologies Pract, 82-88; VPRES & PARTNER, BOOZ, ALLEN & HAMILTON INC, 88- *Concurrent Pos:* Mem, Bd Control, US Naval Inst, 73-82; dir, Mil Opers Soc, 84-88. *Mem:* Am Phys Soc; Sigma Xi; Mil Opers Res Soc (pres, 88-89). *Res:* Plasma physics; theoretical nuclear physics, many-body problem; compilers; radars; anti-aircraft and anti-missile systems; electronic warfare systems; gun and missile systems command and control systems; systems analysis, strategic and naval, ground, air forces; neural network technology and AI; competitive strategis simulation for industry. *Mailing Add:* 101 Pommander Walk Alexandria VA 22314

MASTERTON, WILLIAM LEWIS, b Conway, NH, July 24, 27; m 53; c 2. GENERAL CHEMISTRY. *Educ:* Univ NH, BS, 49, MS, 50; Univ Ill, PhD(chem), 53. *Prof Exp:* Instr chem, Univ Ill, 53-55; from instr to assoc prof, Univ Conn, 55-66, prof chem, 66-87; RETIRED. *Mem:* Am Chem Soc; Sigma Xi; Am Assoc Univ Prof. *Res:* Thermodynamics of solutions; activity coefficients of electrolytes; solubility of gases in salt solutions; author textbook on general chemistry. *Mailing Add:* One Ridge Rd Storrs CT 06268

MASTIN, CHARLES WAYNE, b Salinas, Calif, Apr 23, 43; m 71; c 2. MATHEMATICAL ANALYSIS. *Educ:* Austin Peay State Col, BS, 64; Miami Univ, MS, 66; Tex Christian Univ, PhD(math), 69. *Prof Exp:* Asst prof math, Miss State Univ, 69-75; vis scientist, Inst Comput Appln Sci & Eng, NASA Langley Res Ctr, 75-76; ASSOC PROF MATH, MISS STATE UNIV, 76- *Mem:* Sigma Xi; Soc Indust & Appl Math; Asn Comput Mach. *Res:* Practical application of transformation methods to the solution of fluid dynamics problems. *Mailing Add:* Dept Math Miss State Univ Drawer MA Mississippi State MS 39762

MASTRANGELO, MICHAEL JOSEPH, b Phoenixville, Pa, Oct 3, 38; m 64; c 3. MEDICINE, IMMUNOLOGY. *Educ:* Villanova Univ, BS, 60; Johns Hopkins Univ, MD, 64. *Prof Exp:* Instr, Sch Med, Thomas Jefferson Univ, 70-73; asst prof med, Temple Univ, 74-77, ASSOC PROF, 77-; RES PHYSICIAN & CHIEF MELANOMA UNIT, FOX CHASE CANCER CTR, 72-; AT MED COL VA, RICHMOND. *Mem:* Am Asn Cancer. *Res:* Am Soc Clin Oncol; Am Fedn Clin Res; AAAS; Am Col Clin Pharmacol. Res: Tumor immunology; tumor biology; cancer therapy. *Mailing Add:* Am Oncol Hosp Central & Shelmire Ave Philadelphia PA 19111

MASTRANGELO, SEBASTIAN VITO ROCCO, b New York, NY, July 1, 25; m 49; c 2. PHYSICAL CHEMISTRY. *Educ:* Queens Col, NY, BS, 47; Pa State Univ, MS, 48, PhD(chem), 51. *Prof Exp:* Chemist, Barrett Div, Allied Chem Corp, 51-52; phys chemist, Dextran Corp, 52-53, dept supvr, 53-56; res chemist, Jackson Lab, E I du Pont de Nemours & Co, Inc, 56-62, res assoc, Exp Sta, 62-64, res supvr, 64-69, res fel, Exp Sta, 69-78, RES FEL, JACKSON LAB, E I DU PONT DE NEMOURS & CO, INC, 78- *Mem:* Am Chem Soc; Am Phys Soc; Am Inst Chem. *Res:* Low temperature purification; third law thermodynamics; adsorption thermodynamics; adsorption thermodynamics at liquid helium temperatures; high temperature adiabatic calorimetry; molecular weight distribution of high polymers; raman and infrared spectroscopy; free radical chemistry; electrochemistry; theory of fine particle nucleation and growth. *Mailing Add:* 15 Yorkridge Trail PO Box 73 Hockessin DE 19707-0073

MASTRO, ANDREA M, b Sewickley, Pa, Sept 8, 44; m 73; c 1. CELL BIOLOGY, CELL PHYSIOLOGY. *Educ:* Carlow Col, BA, 66; Pa State Univ, MS, 68, PhD(biol), 71. *Prof Exp:* Fel oncol, McArdle Lab, Univ Wis, 71-73; res fel cell biol, Imp Cancer Res Fund Lab, London, 74-75; from res assoc to asst prof biochem & biophys, 75-83, assoc prof microbiol, 84-87, PROF MOLECULAR & CELL BIOL, PA STATE UNIV, 87- *Concurrent Pos:* Res grants, NIH-Nat Inst Gen Med Sci, 76-81, NIH-Nat Can Inst, 78-, Off Naval Res, 79-80 & NSF, 83-84; Damon Runyon fel, Damon Runyon Mem Fund Cancer Res, 71-73; vis prof, Ger Cancer Res Ctr, Heidelberg, 85. *Honors & Awards:* Res Career Develop Award, NIH, Nat Cancer Inst, 82-87. *Mem:* AAAS; Am Soc Cell Biol; Tissue Culture Asn; Am Asn Cancer Res; Am Asn Women Sci; Am Asn Immunologists. *Res:* Role of the protein kinases in the control of cell growth in mammalian cells; role of cell-cell interactions in cell growth and differentiation; lymphocyte proliferation; phorbol esters. *Mailing Add:* 431 S Frear Pa State Univ University Park PA 16802

MASTROIANNI, LUIGI, JR, b New Haven, Conn, Nov 8, 25; m 57; c 3. OBSTETRICS & GYNECOLOGY. *Educ:* Yale Univ, AB, 46; Boston Univ, MD, 50; Am Bd Obstet & Gynec, cert 59 & 74, recert, 81. *Hon Degrees:* DSc, Boston Univ, 74. *Prof Exp:* From instr to asst prof obstet & gynec, Sch Med, Yale Univ, 55-61; prof, Univ Calif, Los Angeles, 61-65; chmn dept, 65-87, WILLIAM GOODELL PROF OBSTET & GYNEC, UNIV PA SCH MED, 65-, DIR, DIV HUMAN REPRODUCTION, 87- *Concurrent Pos:* Ed, J Fertil & Steril; res fel infertility & endocrinol, Harvard Med Sch, 54-55; mem, Corp Marine Biol Lab, Woods Hole, Mass; prin investr, Mellon Found & Rockefeller Found Training grants, 87-90, NIH, 87-; assoc ed, Fertil & Steril, 65-70, ed-in-chief, 70-75; bd dirs, Sex Info & Educ Coun US, 68-70, Am Fertil Soc, 69-71, Soc Study Reproduction, 72-74 & Int Soc Develop Biologists, 80-83; chmn, Primate Res Ctr Adv Comt, NIH, 68-69, Comt Contraceptive Develop, Nat Acad Sci & Inst Med, 88-; bd sci counselors, Nat Inst Child Health & Human Develop, 80-83; mem, Basic Res adv comt, Nat Found March Dimes, 80-, coun, Asn Professors Obstet & Gynec, 76-79; pres, IXth World Cong Fertil & Steril, 76-77. *Mem:* Endocrine Soc; Am Fertil Soc (pres, 76-77); Am Gynec Soc; fel Am Col Obstetricians & Gynecologists; fel Am Col Surg; Sigma Xi; Am Gynec & Obstet Soc (vpres, 89); Am Physiol Soc; Soc Gynec Invest; Soc Exp Biol & Med; Soc Study Reproduction. *Res:* Human infertility; reproductive physiology. *Mailing Add:* Hosp Univ Pa 3400 Spruce St Philadelphia PA 19104-4283

MASTROMARINO, ANTHONY JOHN, b Brooklyn, NY, June 13, 40; m 73; c 2. RESEARCH ADMINISTRATION, CANCER. *Educ:* Iona Col, BS, 61; Syracuse Univ, MS, 71; Baylor Col Med, PhD(exp biol, microbiol), 75. *Prof Exp:* Res assoc microbiol, Naylor Dana Inst Dis Prev, Am Health Found, 74-76; asst dir sci opers, Nat Large Bowel Cancer Proj, Univ Tex M D Anderson Cancer Ctr, 76-80, assoc sci dir, 80-84, asst vpres res, 83-90, ASST BIOLOGIST, DEPT MED ONCOL & GASTROENTEROL, UNIV TEX M D ANDERSON CANCER CTR, 78-, ASSOC VPRES RES, 90- *Mem:* Am Soc Microbiol; Asn Gnotobiotics (vpres, 79-80, pres, 80-81); Int Soc Prev Oncol; AAAS; NY Acad Sci; Soc Exp Biol & Med; Am Cancer Soc. *Res:* Bacterial products for risk assessment of colon cancer; colon tumor inhibition; using natural dietary products; interaction of diet sterols and intestinal anaerobes in metabolic epidemiology of colon cancer. *Mailing Add:* Univ Tex M D Anderson Cancer Ctr 1515 Holcombe Blvd MD 101 Houston TX 77030

MASTRONARDI, RICHARD, b New York, NY, Nov 14, 47; wid; c 1. ENERGY RESEARCH. *Educ:* Rensselaer Polytech Inst, BS, 69; Northeastern Univ, MBA, 75. *Prof Exp:* Design engr, Com Airplane Div, Boeing Co, 69-70; flight test engr, Sikorsky Aircraft, div United Technol, 70; proj engr, Atkins & Merrill Inc, 70-75; sr vpres, Am Sci & Eng Inc, 75-87; VPRES, TECOGEN INC, SUBSID THERMO ELECTRON CORP, 87- *Mem:* Am Inst Aeronaut & Astronaut; Am Soc Mech Engrs. *Res:* Development of new technology in the field of alternate fuel vehicles, imaging telescopes for astronomy, energy systems, heat- transfer technology, mechanical structures; non-destructive testing; medical and security applications using state-of-the-art mechanical, electrical and computer systems. *Mailing Add:* 55 South St Medfield MA 02052

MASTROPAOLO, JOSEPH, BIOMECHANICS, MUSCLE PHYSIOLOGY. *Educ:* Univ Iowa, PhD(phys educ), 57. *Prof Exp:* PROF EXERCISE PHYSIOL & BIOMECHANICS, CALIF STATE UNIV, LONG BEACH, 68- *Mailing Add:* Dept of Phys Educ Calif State Univ 1250 Bellflower Blvd Long Beach CA 90840

MASTROTOTARO, JOHN JOSEPH, b Worcester, Mass, July 17, 60. BIOSENSORS, CARDIAC ELECTRODES. *Educ:* Col Holy Cross, BA, 82; Duke Univ, MS, 84, PhD(biomed eng), 89. *Prof Exp:* SR SCIENTIST, MED DEVICES DIV, ELI LILLY & CO, 89- *Concurrent Pos:* Prin investr, NC Biotechnol Ctr Grant, 87-88. *Mem:* Inst Elec & Electronic Engrs. *Res:* Development of an implantable electroenzymatic glucose sensor; develop implantable cardiac recording electrodes. *Mailing Add:* Lilly Corp Ctr Drop Code 0811 Eli Lilly & Co Indianapolis IN 46285

MASUBUCHI, KOICHI, b Otaru, Hokkaido, Japan, Jan 11, 24; m 50. MARINE ENGINEERING. *Educ:* Univ Tokyo, BS, 46, MS, 48, DEng(naval archit), 59. *Prof Exp:* Res engr, Shipbuilding Lab, Japan, 48-50; res engr, Transp Tech Res Inst, 50-53, chief methods fabrication sect, 53-58; vis fel & consult, Battelle Mem Inst, 58-62, res assoc, 62; chief methods fabrication sect, Transp Tech Res Inst, 62-63; chief welding mech sect, Ship Res Inst, 63; res assoc, Battelle Mem Inst, 63-64, fel, 65-66, tech adv technol div, 66-68; assoc prof naval archit, 68-71, PROF OCEAN ENG & MAT SCI, MASS INST TECHNOL, 71- *Honors & Awards:* R D Thomas Mem Award, Am Welding Soc, 77. *Mem:* Am Welding Soc; Am Soc Metals; Japanese Welding Soc; Japanese Soc Naval Archit; Am Soc Mech Engrs. *Res:* Naval architecture; welding fabrication of ships and other structures; strength of welded structures; residual stresses; brittle fracture; welding of high strength steels and other materials. *Mailing Add:* Dept Ocean Eng Mass Inst Technol 77Massachusetts Ave Cambridge MA 02139

MASUELLI, FRANK JOHN, b Masio, Italy, Apr 16, 21; nat US; m 63; c 2. ORGANIC CHEMISTRY. *Educ:* Manhattan Col, BS, 42; Va Polytech Inst, MS, 48, PhD(chem), 53. *Prof Exp:* Instr chem, Va Polytech Inst, 46-53; chemist, 53-55, supvr chem, 55-60, CHIEF PROPELLANTS RES BR, PICATINNY ARSENAL, 60- *Mem:* Am Chem Soc; Tech Asn Pulp & Paper Indust; Am Inst Chemists; Sigma Xi. *Res:* Nitrocellulose chemistry; artillery and rocket propellants. *Mailing Add:* 344 Diamond Spring Rd Box 1271 Denville NJ 07834

MASUI, YOSHIO, b Kyoto, Japan, Oct 6, 31; m 59; c 2. DEVELOPMENTAL BIOLOGY. *Educ:* Kyoto Univ, BSc, 53, MS, 55, PhD(zool), 61. *Prof Exp:* Lectr biol, Konan Univ, Japan, 58-65, asst prof, 65-68; lectr, Yale Univ, 69; assoc prof, 69-78, PROF ZOOL, UNIV TORONTO, 78- *Mem:* Int Soc Develop Biol; Soc Develop Biol; Can Soc Cell Biol; Can Soc Zoologists. *Res:* Developmental biology relating to nucleocytoplasmic interactions in early development and gametogenesis. *Mailing Add:* Dept Zool Univ Toronto 25 Harbord St St George Campus Toronto ON M5S 1A1 Can

MASUR, SANDRA KAZAHN, b New York, NY, Nov 27, 38; m 59, 89; c 2. CELL BIOLOGY, ENDOCRINOLOGY. *Educ:* City Col NY, BA, 60; Columbia Univ, MA, 63, PhD(cell biol), 67. *Prof Exp:* Lectr biol, City Col NY, 60-61, asst prof, 67-68; instr, 68-73, res assoc, 73-75, adj asst prof, 75-79, asst prof, 79-87, ASSOC PROF PHYSIOL & BIOPHYSICS, MT SINAI SCH MED, CITY UNIV NY, 88- *Concurrent Pos:* Res assoc, Columbia Univ, 71-82; prin investr, Mt Desert Island Biol Lab, 80, 81, 83 & 84; vis scientist, Europ Molecular Biol Lab & Nuclear Studies Ctr, 85; vis prof, Univ Iowa, 88. *Honors & Awards:* Res Career Develop Award, NIH, 79; Irma T Hirschl Res Scientist. *Mem:* Am Soc Cell Biol; Am Physiol Soc; AAAS; Am Soc Nephrology; NY Soc Electron Micros. *Res:* Cell biology of signal transduction across the plasma membrane of corneal fibroblasts and urinary epithelia. *Mailing Add:* Box 1218 Dept Physiol & Biophys Mt Sinai Sch Med 100th St & Fifth Ave New York NY 10029-6574

MASUREKAR, PRAKASH SHARATCHANDRA, b Bombay, India, Jan 23, 41; m 68; c 2. INDUSTRIAL MICROBIOLOGY, BIOCHEMICAL ENGINEERING. *Educ:* Univ Bombay, BSc, Hons, 62, BSc, 64, MSc, 66; Mass Inst Technol, SM, 68, PhD(biochem eng), 73. *Prof Exp:* Sr res chemist biochem eng, Eastman Kodak Co, 73-80; sr res bioengr, W R Grace & Co, 80-81; SR RES FEL, MERCK SHARP & DOHME RES LABS, 81- *Mem:* Am Soc Microbiol; Soc Indust Microbiol. *Res:* Microbiology and engineering of industrial fermentations; microbial physiology and genetics; biological conversions and enzyme technology. *Mailing Add:* Merck Sharp & Dohme Res Labs PO Box 2000 Rahway NJ 07065

MASURSKY, HAROLD, b Ft Wayne, Ind, Dec 23, 23; m 52; c 4. ASTROGEOLOGY. *Educ:* Yale Univ, BS, 43, MS, 51; Northern Ariz Univ, DSc, 80. *Prof Exp:* Geologist, 51-67, chief br astrogeol studies, 67-71, chief scientist, 71-76, SR SCIENTIST, CTR ASTROGEOL, US GEOL SURV, 77- *Concurrent Pos:* Team leader & prin investr, TV exp, Mariner Mars, 71; co-investr, Apollo Field Geol Team, Apollo 16 & 17, mem Apollo Orbital Sci Photog Team, Apollo Site Selection Group, leader, Viking Landing Site Staff, dep team leader, Orbiter Visual Imaging System, Viking Mars 75; comt space res rep Inter-Union Comn for Studies of the Moon; pres, Int Astron Union Working Group on Nomenclature; Moon, Mars, Venus del, USA-USSR planetary data exchange; radar team mem & chmn, Surface & Interiors Group, Venus Pioneer, 78; interdisciplinary scientist, Voyager Imaging Team, Galileo & Imaging Spectros Teams, 82; mem radar team, Venus Radar Mapper, 82; mem comt space res, Int Coun Sci Unions. *Mem:* Geol Soc Am; Am Geophys Union; AAAS; Geochem Soc; Meteoritical Soc; Am Astron Soc. *Res:* Geology of Owl Creek Mountains, Wyoming; uranium bearing coal in the Red Desert, Wyoming; structure stratigraphy and volcanic rocks in central Nevada; stratigraphy and structure of the moon; geology of Mars; crustal formation, eolian deposits, volcanic history, channel formation; impact history; geology of Venus and satellites of Jupiter, Saturn and Uranus. *Mailing Add:* US Geol Surv 2255 N Gemini Dr Flagstaff AZ 86001

MASUT, REMO ANTONIO, b Morteros, Cordoba, Arg, July 20, 48; Can citizen; m 74; c 1. III-V COMPOUND SEMICONDUCTORS, METAL ORGANIC CHEMICAL VAPOR DEPOSITION. *Educ:* Licenciate, Univ Cordoba, Arg, 74; Univ Mass Amherst, PhD(physics), 82. *Prof Exp:* Tech asst physics, Univ Cordoba, Arg, 73-76; tech asst & res asst physics, Univ Mass, 77-82; postdoctoral res assoc chem, Amherst Col, 82-84; res assoc officer, Nat Res Coun, Can, 84-87; researcher, 87-90, PROF SOLID STATE PHYSICS, DEPT ENG PHYSICS, ECOLE POLYTECHNIQUE, CAN, 90- *Mem:* Am Phys Soc; Mat Res Soc; Can Asn Physicists. *Res:* Growth and characterization of III-V compound semiconductor materials and advanced structures; development of metal organic chemical vapor deposition technology; study of growth mechanisms; high resolution x-ray crystal diffraction analysis of thin films. *Mailing Add:* Dept Eng Physics Ecole Polytechnique CP 6079 Sta A Montreal PQ H3C 3A7 Can

MASYS, DANIEL R, BIOMEDICAL COMMUNICATIONS. *Prof Exp:* DIR, LISTER HILL NAT CTR BIOMED COMMUN, NAT LIBR MED, NIH, 86- *Mailing Add:* NIH Nat Libr Med Lister Hill Nat Ctr Biomed Commun Bldg 38A Rm 7N707 Bethesda MD 20892

MATA, LEONARDO J, b Dota, Costa Rica, Dec 6, 33; m 56; c 4. PUBLIC HEALTH, VIROLOGY. *Educ:* Univ Costa Rica, BS, 56; Univ PR, dipl, 58; Harvard Univ, MS, 60, DSc(trop pub health), 62. *Prof Exp:* Chief bact & parasitol, San Juan de Dios Hosp, Costa Rica, 56-59; lab instr microbiol, Sch Nursing, Univ Costa Rica, 57-58; chief enteric bact, Inst Nutrit Cent Am & Panama, 59; lab instr parasitol, Harvard Med Sch, 61-62; chief microbiol, Inst Nutrit Cent Am & Panama, Guatemala, 62-75 & prof microbiol, Sch Nutrit, 66-75; DIR INST RES HEALTH & PROF, UNIV COSTA RICA, 75- *Concurrent Pos:* Nat Inst Allergy & Infectious Dis grant, Inst Nutrit Cent Am & Panama, Guatemala, 62-71; mem, Pan Am Health Orgn-WHO-Inst Nutrit Cent Am & Panama internal coun, 65-74; mem, US-Japan coop proj, NIH grant, Guatemala, 68-71; US Armed Forces Res & Develop Command grant, Cent Am, 68-71; vis prof, San Carlos Univ Guatemala, 65-70 & sch med, Univ El Salvador, 70-; mem informal study group, WHO, 71-; dir, Inst Res Health, 75-; vis lectr, Harvard Sch Pub Health; clin prof, Univ Wash. *Honors & Awards:* UNESCO Sci Prize, 80. *Mem:* AAAS; Am Soc Trop Med & Hyg; Latin Am Nutrit Soc; NY Acad Sci; Am Soc Microbiol. *Res:* Tissue culture and virology; enteric microbiology; nutrition and human growth; tropical public health; research on public health interventions. *Mailing Add:* Populahon Sci Sch Public Health Harvard Univ 665 Huntington Ave Boston MA 02115

MATAGA, PETER ANDREW, b Auckland, NZ, Dec 23, 59; NZ & Can citizen; m 86; c 1. FRACTURE MECHANICS, MATERIALS PROCESSING. *Educ:* Univ Auckland, NZ, BSc & BE, 82; Harvard Univ, SM, 83, PhD(eng sci), 86. *Prof Exp:* Asst res engr, Mat Dept, Univ Calif, Santa Barbara, 86-89; ASST PROF AEROSPACE & MECH ENG SCI DEPT, UNIV FLA, 89- *Concurrent Pos:* Consult, Mobil Solar Energy Corp, 85-86; lectr, Mech Eng Dept, Univ Calif, Santa Barbara, 87; NSF presidential young investr award, 90. *Mem:* Am Soc Mech Engrs; Soc Eng Sci; Am Ceramic Soc; Am Acad Mech; Am Soc Eng Educ. *Res:* Fracture mechanics of brittle, ductile and composite materials; interfacial fracture; processing of composite materials; sintering and hot pressing; steady-state growth process. *Mailing Add:* Dept AEMES Univ Fla 231 Aerospace Bldg Gainesville FL 32611

MATALON, MOSHE, b Cairo, Egypt, Feb 25, 49; Israel citizen; m 78; c 2. FLAME PROPAGATION, STABILITY THEORY. *Educ:* Tel Aviv Univ, Israel, BSc, 69, MSc, 72; Cornell Univ, PhD(mech & aerospace), 77. *Prof Exp:* Res assoc & instr theoret & appl mech, Cornell Univ, 77-80; asst prof aerodyn labs, Polytech Inst NY, 78-80; from asst prof to assoc prof, 80-91, PROF ENG SCI & APPL MATH, NORTHWESTERN UNIV, 91- *Concurrent Pos:* Vis prof, Tel Aviv Univ, 85, Technion-Israel, EPFL, Lausanne, Switz, 89, consult, ICASE, NASA, Langley, 88. *Mem:* Am Inst Aeronaut & Astronaut; Am Phys Soc; Soc Indust & Appl Math; Combustion Inst. *Res:* Applied mathematics; asymptotic and singular perturbation; combustion processes; flame propagation and stability; theoretical fluid mechanics. *Mailing Add:* Eng Sci & Appl Math Dept MEAS Northwestern Univ 2145 Sheridan Rd Evanston IL 60208

MATALON, SADIS, b Athens, Greece, Oct 6, 48; US citizen; m 75; c 1. PULMONARY PHYSIOLOGY. *Educ:* Macalester Col, BA, 70; Univ Minn, Minneapolis, MS, 73, PhD(physiol), 75. *Prof Exp:* Syst analyst biomed sci, Univ Minn, Minneapolis, 73-75; assoc physiol & pediat, Children's Mem Hosp & Northwestern Univ, Chicago, 75-76; from res asst to assoc prof physiol, State Univ NY, Buffalo, 76-87, res assoc prof path, 84-87; PROF ANESTHESIOL, PHYSIOL & BIOPHYS, UNIV ALA, BIRMINGHAM, 87- *Concurrent Pos:* Fulbright scholar, 66-70; ad-hoc reviewer, NIH. *Mem:* Am Physiol Soc; Am Thoracic Soc; Sigma Xi. *Res:* Oxygen toxicity; solute transport across the alveolar capillary membrane; ion transport; cellular mechanisms of oxygen injury. *Mailing Add:* Dept Anesthesiol Univ Ala 619 S 19th St Birmingham AL 35233

MATANOSKI, GENEVIEVE M, b Salem, Mass, Aug 26, 30; c 5. PEDIATRICS, EPIDEMIOLOGY. *Educ:* Radcliffe Col, AB, 51; Johns Hopkins Univ, MD, 55, MPH, 62, DrPH, 64. *Prof Exp:* Intern & asst resident pediat, Johns Hopkins Hosp, 55-57, res assoc epidemiol, 57-60, from instr to assoc prof, 60-75, PROF EPIDEMIOL, JOHNS HOPKINS UNIV, 75- *Concurrent Pos:* Assoc prof, Schs Med & Dent, Univ Md; NIH grants, 65-66 & 70- *Mem:* AAAS; NY Acad Sci; Soc Epidemiol Res; Int Epidemiol Asn; Am Pub Health Asn; Soc Occup & Environ Health; Am Epidemiol Soc. *Res:* Cancer risks from occupational and environmental exposures to radiation and other agents; evaluation of health programs; family-based population studies; dental disease, especially oral cancer and the role of immunology in periodontal disease; rheumatic fever and streptococcal infections; infant mortality and congenital malformations. *Mailing Add:* Sch Hyg & Pub Health Johns Hopkins Univ Baltimore MD 21205

MATAR, SAID E, b Alexandria, Egypt, Apr 2, 34; Can citizen; m 73; c 2. APPLICATION OF FINITE ELEMENT TECHNIQUES, HEAT TRANSFER FROM IONIZED GASES. *Educ:* Univ Alexandria, Egypt, 57; Okla State Univ, MSc, 61; Northwestern Univ, PhD(mech eng), 66. *Prof Exp:* Asst prof mech eng, Univ Alexandria, 66-68; postdoctoral fel, Univ Toronto, 68-70; lectr, 70-80, PROF MECH ENG, RYERSON, 85-, CHMN, 89- *Concurrent Pos:* Mem, Acad Requirement Comt, Asn Prof Engrs Ont, 85-, chmn, Prof Pract & Ethics Subcomt, 87- *Mem:* Asn Prof Engrs. *Res:* Gas dynamics; heat transfer from dissociated and ionized gases; application of finite element techniques. *Mailing Add:* Dept Mech Eng Ryerson Polytech Inst 350 Victoria St Toronto ON M5B 2K3 Can

MATARAZZO, JOSEPH DOMINIC, b Italy, Nov 12, 25; US citizen; m 49; c 3. NEUROPSYCHOLOGY, HEALTH PSYCHOLOGY. *Educ:* Brown Univ, BA, 46; Northwestern Univ, MS, 50, PhD(psychol), 52; Am Bd Prof Psychol, dipl clin psychol, 57, dipl clin neuropsychol, 83. *Hon Degrees:* LHD, Pac Grad Sch Psychol, Menlo Park, Calif, 88. *Prof Exp:* Fel, Sch Med, Washington Univ, St Louis, 50-51, instr, 51-53, asst prof med psychol, 53-55; res assoc, dept psychiat, Med Sch, Harvard Univ, 55-57; PROF MED PSYCHOL & CHMN DEPT, SCH MED, ORE HEALTH SCI UNIV, PORTLAND, 57- *Honors & Awards:* Hofheimer Prize, Am Psychiat Asn, 62. *Mem:* Fel Am Psychol Asn; Asn Am Med Cols; AAAS; Int Coun Psychologists (pres, 76-77); Psychonomic Soc; Int Asn Appl Psychol; Acad Behav Med Res (pres, 82-83); Am Psychol Asn (pres, 89-90). *Res:* Smoking, diet and other life style risk factors; intellectual and memory deficits in brain disorders. *Mailing Add:* Dept Med Psychol Sch Med Ore Health Sci Univ 3181 SW Sam Jackson Park Rd Portland OR 97201

MATARÉ, HERBERT FRANZ, b Aachen, W Ger, Sept 22,12, US citizen; m 39; c 3. SOLID STATE PHYSICS, ELECTRONICS. *Educ:* Aachen Tech Univ, MS, 39; Tech Univ Berlin, PhD(electronics), 42; Ecole Normale Superieure Univ, Paris, PhD(solid state), 50. *Prof Exp:* Dir, microwave receiver lab, Telefunken Berlin, W Ger, 39-45; dir, semiconductor lab, Westinghouse, Paris, France, 46-52; pres, Intermetall Inc, Dusseldorf, W Ger, 52-56; head, semiconductor res & develop, Gen Tel & Electronics Corp, NY, 56-59; dir, res Tekade semiconductor dept, Neurenberg, WGer, 59-61; head, quantum physics dept, Bendix Corp Res Labs, 61-65; asst chief engr, McDonnell-Douglas Corp, 64-66; sci adv solid state, Rockwell Int, 66-69; CONSULT ELECTRONICS, INT SOLID STATE ELECTRONICS CONSULT, 71-; DIR, COMPOUND CRYSTALS, LTD, LONDON, 86- *Concurrent Pos:* Consult, Intermetall Corp, 53-55 & US Army Electronics Command, 53-56; conf chmn, int meetings, Electrochem Soc, Chicago, 55, NY 58 & 69; vis prof, Univ Calif, Los Angeles, 68-69 & Calif State Univ, Fullerton, 69-70. *Mem:* AAAS; Am Phys Soc; Electrochem Soc; Emer Mem NY Acad Sci; Life fel Inst Elec & Electronic Engrs; Am Vacuum Soc; Mat Res Soc. *Res:* Compound semiconductors; crystal growth; III-V-ternary compounds; epitaxy; solid state electronics; solar cells; author of over 100 publications and four books; awarded 60 patents. *Mailing Add:* PO Box 49177 Los Angeles CA 90049

MATA-TOLEDO, RAMON ALBERTO, b Caracas, Venezuela, Aug 31, 49; m 69; c 3. SOFTWARE ENGINEERING, DATA BASE DESIGN. *Educ:* Caracas Pedag Inst, BS, 72; Fla Inst Technol, MS & MBA, 78; Kans State Univ, DSc(comput sci), 84. *Prof Exp:* Asst prof math and head, Francisco De Miranda Univ, 74-76; teaching asst comput sci, Kans State Univ, 79-83; syst analyst, Comput Info Systs, 81-83; asst prof comput sci, Fla Inst Technol, 83-88; ASST PROF COMPUT SCI, JAMES MADISON UNIV, HARRISONBURG, VA, 88- *Concurrent Pos:* Consult, comput & info systs, 83- *Mem:* Asn Comput Mach; Inst Elec & Electronics Engrs. *Res:* Development of software metrics that will help to identify programs of good quality; databases and natural language processing. *Mailing Add:* Computer Sci Dept James Madison Univ Harrisonburg VA 22801

MATCHA, ROBERT LOUIS, b Omaha, Nebr, Oct 22, 38; m 80; c 3. THEORETICAL CHEMISTRY. *Educ:* Univ Omaha, BA, 60; Univ Wis-Madison, PhD(theoret chem), 65. *Prof Exp:* Fel, IBM Corp, Calif, 65-66 & Battelle Inst, 66-67; from asst prof to assoc prof, 67-80, PROF THEORET CHEM, UNIV HOUSTON, 80- *Concurrent Pos:* Consult, Battelle Inst, 70-74, Exxon Prod Res, 80-81. *Mem:* AAAS; Am Phys Soc; Am Chem Soc; fel Am Inst Chemists. *Res:* Theoretical study of electromagnetic interactions; effects of nuclear motion on expectation values, molecular Hartree Fock calculations; compton profiles; interactions between molecular ions and surfaces. *Mailing Add:* Dept Chem/4800 Calhoun Rd Univ Houston Cullen Blvd Houston TX 77204

MATCHES, ARTHUR GERALD, b Portland, Ore, Jan 28, 29; m 52; c 3. AGRONOMY. *Educ:* Ore State Univ, BS, 52, MS, 54; Purdue Univ, PhD(crop physiol & ecol), 60. *Prof Exp:* Asst farm crops, Ore State Univ, 52-54; instr agron, Purdue Univ, 56-60; asst prof, Southeastern Substa, NMex State Univ, 60-61; prof agron, Univ Mo-Columbia, 61-81; JESSIE W THORNTON DISTINGUISHED PROF PLANT & SOIL SCI, TEX TECH UNIV, 81- *Concurrent Pos:* Res agronomist, Agr Res Serv, USDA, 61-81. *Honors & Awards:* Medallion Award, Am Forage & Grassland Coun, 82. *Mem:* Am Forage & Grassland Coun (pres, 77); Soc Range Mgt; fel Am Soc Agron; fel Crop Sci Soc Am; Am Soc Animal Sci; AAAS. *Res:* Pasture systems for nearly year long grazing; use of multiple assignment tester animals in grazing trials; pasture research methods; current research is highly orientated towards the development of forage-livestock grazing systems for the semi-arid Southern High Plains; recognized nationally and internationally for research on grazing systems, plant/animal interactions, forage-livestock research methods and pasture management (grasses and legumes); more than 160 scientific publications. *Mailing Add:* Dept Agron Hort & Entom Texas Tech Univ Lubbock TX 79409

MATCHES, JACK RONALD, b Portland, Ore, May 20, 30; m 54; c 2. FOOD SCIENCE, FISH TECHNOLOGY. *Educ:* Ore State Univ, BS, 57, MS, 58; Iowa State Univ, PhD(microbiol), 63. *Prof Exp:* Res assoc microbiol & food sci, Iowa State Univ, 58-63; sr microbiologist, 63-65, asst prof microbiol & food sci, 65-68, assoc prof, 68-75, PROF FISHERIES, COL FISHERIES, UNIV WASH, 75- *Mem:* Inst Food Technologists; Am Soc Microbiol; Int Asn Milk, Food & Environ Sanit. *Res:* Food microbiology; low temperature microbiology; anaerobic microbiology; fish technology and decomposition. *Mailing Add:* Dept Food Sci Univ Wash Seattle WA 98195

MATCHETT, ANDREW JAMES, b Chicago, Ill, Jan 30, 50; m 76; c 3. ALGEBRA, STATISTICS. *Educ:* Univ Chicago, BS, 71; Univ Ill, PhD(math), 76. *Prof Exp:* Teaching fel math, Univ Ill, Urbana-Champaign, 71-73, asst, 73-76; asst prof math, Tex A&M Univ, 76- AT MATH DEPT, UNIV WIS-LA CROSSE. *Mem:* AAAS; Am Math Soc; Math Asn Am. *Res:* Group representation theory, operatory theory, K-theory; representations of the symmetric groups; statistics; signal processing. *Mailing Add:* Dept of Math Univ Wis 1725 State St La Crosse WI 54601

MATCHETT, WILLIAM H, b Pinehurst, NC, Jan 4, 32; m 54; c 2. MICROBIAL PHYSIOLOGY. *Educ:* Univ Ill, BS, 53, MS, 58, PhD(plant physiol), 60. *Prof Exp:* Asst bot, Univ Ill, 57-60; NSF fel microbiol, Sch Med, Yale Univ, 60-61; res fel, Univ Calif, San Diego, 61-63; res scientist biol, Hanford Labs, Gen Elec Co, 63-65; mgr cell biol, Pac Northwest Lab, Battelle Mem Inst, 65-69, coordr life sci, Seattle Res Ctr, 69-70; prof bot & biol sci, Wash State Univ, 70-77, assoc dean, Grad Sch, 71-74, chmn dept bot, 71-76; DEAN GRAD SCH, NMEX STATE UNIV, 77- *Concurrent Pos:* Vis lectr, Wash State Univ, 65, adj assoc prof chem, 67-71. *Mem:* AAAS; Am Soc Microbiol; Am Soc Biol Chemists; Sigma Xi. *Res:* Biochemical genetics of Neurospora crassa; metabolism of tryptophan in Neurospora; enzymology of tryptophan biosynthetic enzymes. *Mailing Add:* Grad Sch NMex State Univ Box 36 Las Cruces NM 88003

MATCOVICH, THOMAS J(AMES), b New York, NY, Jan 15, 29; m 55; c 3. PHYSICS, ELECTRICAL ENGINEERING. *Educ:* Cooper Union, BEE, 50; Univ Pa, MS, 56; Temple Univ, PhD(physics), 61. *Prof Exp:* Mem staff, Gen Elec Co, NY, 50-53 & Philco Corp, Pa, 53-55; eng mgr, Molecular Systs Dept, Univac Div, Sperry Rand Corp, 55-66; assoc prof elec eng, Drexel Univ, 66-71, prof & chmn, electrophys advan study group, 71-; RES, APPL MICROELECTRONICS INC. *Concurrent Pos:* Lectr, St Joseph's Col, Pa, 64-66; partner, C&M Assocs, 67- *Mem:* Am Inst Physics; Inst Elec & Electronics Engrs. *Res:* Microelectronics; computer memory systems; magnetism. *Mailing Add:* 1667 Ludwell Dr Ambler PA 19002

MATEER, FRANK MARION, b Pittsburgh, Pa, June 21, 21; m 44; c 3. MEDICINE. *Educ:* Univ Pittsburgh, BS, 41, MD, 44; Am Bd Internal Med, dipl, 52. *Prof Exp:* From instr to asst prof res med, Univ Pittsburgh, 50-62; dir, 65-70, mem sr staff & chief div med educ & res, 70-73, DIR RENAL UNIT, WESTERN PA HOSP, 73-; CLIN PROF MED, UNIV PITTSBURGH, 60- *Concurrent Pos:* Am Heart Asn estab investr, Univ Pittsburgh, 54-59; clin asst prof med, Univ Pittsburgh, 62-72; sr teaching fel physiol, Univ Pittsburgh, 47-48, res fel med, 48-50, Am Heart Asn res fel, 52-54. *Mem:* AAAS; Am Diabetes Asn; Am Fedn Clin Res; fel Am Col Physicians; Am Heart Asn. *Res:* Renal disease; endocrinology. *Mailing Add:* 4815 Liberty Ave Pittsburgh PA 15224

MATEER, NIALL JOHN, b Gerrards Cross, UK, Nov 17, 50. PALEONTOLOGY. *Educ:* Durham Univ, BSc Hons, 73; Uppsala Univ, FD, 78. *Prof Exp:* Res asst geol, Uppsala Univ, 73-78; high sch instr biol & geog, Enkoping, Sweden, 78-80; DEPT CHMN GEOL, MCMURRY COL, ABILENE, 81- *Concurrent Pos:* Lectr, Benin Univ, Nigeria, 78; res asst, NMex Bur Mines & Mineral Res, 80. *Mem:* Soc Econ Paleontologists & Mineralogists; Geol Soc Am; Soc Vert Paleontologists; Paleont Soc. *Res:* Paleoenvironment analysis and paleoecology of the cretaceous basins of the western interior, concentrating primarily on fossil vertebrates. *Mailing Add:* 1467 N 17th Laramie WY 82070

MATEER, RICHARD AUSTIN, b Ashland, Ky, July 30, 40; m 62; c 2. PHYSICAL ORGANIC CHEMISTRY. *Educ:* Centre Col, BA, 62; Tulane Univ, PhD, 66. *Prof Exp:* ASSOC PROF CHEM, UNIV RICHMOND, 66-, DEAN, 75- *Mem:* Sigma Xi. *Res:* Synthetic photochemistry; organic reaction mechanisms; nuclear magnetic resonance and infrared spectroscopy; organometallics. *Mailing Add:* 8416 Pamela Dr Richmond VA 23229

MATEER, RICHARD S(HELBY), b Fredericktown, Mo, Sept 2, 23; m 52; c 3. METALLURGICAL ENGINEERING. *Educ:* Mo Sch Mines, BS, 44; Carnegie Inst Technol, MS, 47; Univ Pittsburgh, PhD(metall eng), 50. *Prof Exp:* Jr engr, Western Elec Co, 44-45; res fel, Mellon Inst, 50-52; res engr, Kaiser Aluminum & Chem Corp, 52-54; assoc prof metall eng, Univ Pittsburgh, 54-58; head dept mining & metall eng, 58-69, prof metall eng, 58-80, PROF DEPT CIVIL ENG, 80- AT DEPT MINING ENG, UNIV KY. *Concurrent Pos:* Consult, Oak Ridge Nat Lab, 58-; vis prof, Imp Col, Univ London, 65. *Mem:* Am Soc Metals; Am Soc Eng Educ; Nat Soc Prof Engrs; Am Inst Mining, Metall & Petrol Engrs. *Res:* Surface tension of molten metals; powder metallurgy; nuclear and dental amalgam alloys. *Mailing Add:* Dept Mining Eng Univ Ky Lexington KY 40506

MATEESCU, GHEORGHE D, b Pitesti, Romania, Nov 11, 28; US citizen. SURFACE CHEMISTRY, METALLURGICAL SPECTROSCOPIC ANALYSIS. *Educ:* Univ Bucharest, Licantiate, 51; Case Western Reserve Univ, Cleveland, PhD(chem), 71. *Prof Exp:* Asst prof pharmaceut chem, Univ Bucharest, 57-64; sr res fel chem, Romanian Acad Sci, 64-67; res assoc chem, 68-71, DIR CHEM, MAJ ANAL INSTR FACIL, CASE WESTERN RESERVE UNIV, 71-, PROF CHEM, 78- *Concurrent Pos:* Vis prof, Univ Paris, 77 & Univ Marsrille, 82. *Honors & Awards:* G Spacu Award, Romanian Acad Sci, 66. *Mem:* Am Chem Soc; Soc Magnetic Resonance Med. *Res:* Magnetic resonance microimaging and spectroscopy; vibrational, photoelectron and ion spectroscopy in biomedical, agricultural and materials research. *Mailing Add:* Dept Chem Case Western Reserve Univ Millis Sci Ctr Cleveland OH 44106

MATEJA, JOHN FREDERICK, b New Castle, Pa, June 3, 50; m 78; c 2. NUCLEAR PHYSICS, SCIENCE EDUCATION. *Educ:* Univ Notre Dame, BS, 72, PhD(nuclear physics), 76. *Prof Exp:* Res assoc nuclear physics, Fla State Univ, 76-78; from asst prof to prof physics, Tenn Technol Univ, 78-88; STUDENT PROG LEADER, DIV EDUC PROGS, ARGONNE NAT LAB, 86- *Mem:* Am Phys Soc; Sigma Xi. *Res:* Low energy nuclear physics with a current emphasis on heavy ion induced reactions. *Mailing Add:* Div Educ Prog Argonne Nat Lab Argonne IL 60439

MATEKER, EMIL JOSEPH, JR, b St Louis, Mo, Apr 25, 31; m 54; c 3. GEOPHYSICS. *Educ:* St Louis Univ, BS, 56, MS, 59, PhD(geophys), 64. *Prof Exp:* Mgr, geophys res, Western Atlas Int Inc, Houston, Tex, 69-70, vpres, res & develop, 70-74, vpres, Western Geophys Co Am, 74-87, Aero Serv Div, 74-90, VPRES, WESTERN ATLAS INT INC, HOUSTON, TEX, 87-, VPRES TECHNOL, WESTERN GEOPHYS DIV, 90- *Concurrent Pos:* pres, Litton Resources Systs, 77 & Westrex Limited, 74-77; mem bd registration for geologists & geophysicists, State Calif. *Mem:* AAAS; Am Geophys Union; Seismol Soc Am; Soc Explor Geophys; Europ Asn Explor Geophys; Sigma Xi. *Res:* Generation of seismic waves; seismic vibrations; tectonics of stable interior; exploration geophysics; inertial guidance for marine exploration; lithology measurements from seismic reflections; marine seismic energy sources; deep solid earth geophysics. *Mailing Add:* 419 Hickory Post Houston TX 77079

MATELES, RICHARD I, b New York, NY, Sep 11, 35; m 56; c 3. BIOTECHNOLOGY. *Educ:* Mass Inst Technol, BS, 56, MS, 57, ScD, 59. *Prof Exp:* Fel, Microbiol Lab, Delft Univ Technol, 59-60; from instr biochem engr to assoc prof, Dept Nutrit & Food Sci, Mass Inst Technol, 60-70; prof appl microbiol, Inst Microbiol, Hebrew Univ, 68-80; asst dir & dir res, 80-81, vpres res, Stauffer Chem Co, 80-88; sr vpres, IIT Res Inst, 88-90; SR VPRES, CANDIDA CORP, 90- *Concurrent Pos:* Vis lectr, Dept Chem Eng, Univ Calif, Berkeley, 65; consult, Nat Coun Res & Develop, Israel, 65-80; mem, Panel World Food Supply, Presidents' Sci Adv Comt, 66-67; dir, Fermentation Unit, Jerusalem, 68-77; co-ed, Microbiol Series, Marcel Dekker Inc, 78- *Honors & Awards:* Sigma Xi. *Mem:* Am Chem Soc; Am Soc Microbiol; Soc Gen Microbiol; Inst Food Technologists; Am Inst Chem Eng. *Res:* Fermentation kinetics; continuous culture; plant cell culture; single cell protein; applied microbiology; biochemical engineering. *Mailing Add:* Candida Corp 175 W Jackson Chicago IL 60604

MATES, ROBERT EDWARD, b Buffalo, NY, May 19, 35; m 60; c 3. BIOENGINEERING, MECHANICAL ENGINEERING. *Educ:* Univ Rochester, BS, 57; Cornell Univ, MS, 59, PhD(mech eng), 63. *Prof Exp:* Instr, Cornell Univ, 58-61; from asst prof to assoc prof, 62-69, chmn, Dept Mech Eng, 67-70, 79-82, res assoc prof, Dept Med, 72-85, PROF MECH & AEROSPACE ENG, STATE UNIV NY, BUFFALO, 69-, RES PROF, DEPT MED, 86- *Concurrent Pos:* NIH spec res fel, Dept Med, State Univ NY Buffalo, 70-71; consult, Cornell Aeronaut Lab, 62-69, chmn mech eng, 67-70, 79-82; assoc ed, J Biomech Eng, 76-83; NIH fel, State Univ NY, Buffalo, 78-79, dir undergrad studies, mech eng, 86-89, dir, Ctr Biomed Eng, 89- *Mem:* Fel Am Soc Mech Engrs; Am Soc Eng Educ; Am Heart Asn; Am Physiol Soc; Biomed Eng Soc. *Res:* Biomechanics of the cardiovascular and pulmonary systems, particularly coronary blood flow and its regulation. *Mailing Add:* 153 Bidwell Pkwy Buffalo NY 14222

MATESE, JOHN J, b Chicago, Ill, May 1, 38; m 66; c 3. THEORETICAL PHYSICS. *Educ:* DePaul Univ, BS, 60; Univ Notre Dame, PhD(physics), 66. *Prof Exp:* Lectr physics, Univ Notre Dame, 65-66; asst prof, La State Univ, Baton Rouge, 66-74; asst prof, 74-77, ASSOC PROF PHYSICS, UNIV SOUTHWESTERN LA, 77- *Mem:* Am Phys Soc. *Res:* Atomic physics. *Mailing Add:* 107 Harwell St Lafayette LA 70503

MATESICH, MARY ANDREW, b Zanesville, Ohio, May 5, 39. PHYSICAL CHEMISTRY. *Educ:* Ohio Dominican Col, BA, 62; Univ Calif, Berkeley, MS, 63, PhD(chem), 66. *Prof Exp:* Asst prof chem, 65-70, chmn dept, 65-73, acad dean, 73-78, ASSOC PROF CHEM, OHIO DOMINICAN COL, 70-, PRES, 78- *Concurrent Pos:* Petrol Res Fund grant, Ohio Dominican Col, 65-68; NSF grant, Case Western Reserve Univ & Ohio Dominican Col, 69-72. *Res:* Ion transport in membranes; nonaqueous solvents; transport processes in solution; solution thermodynamics. *Mailing Add:* Ohio Dominican Col 1216 Sunbury Rd Columbus OH 43219

MATEY, JAMES REGIS, b McKeesport, Pa, June 30, 51; m 73; c 4. PHYSICS. *Educ:* Carnegie-Mellon Univ, BS, 73; Univ Ill, Urbana, MS, 74, PhD(physics), 78. *Prof Exp:* MEM TECH STAFF RES & DEVELOP, DAVID SARNOFF RES CTR, RCA CORP, 78- *Mem:* Am Phys Soc. *Res:* Development of new hardware and software tools to study material surfaces; development of new microscopier and sensors; image processing. *Mailing Add:* David Sarnoff Res Ctr Princeton NJ 08543-5300

MATHAI, ARAK M, b Palai, India, Apr 28, 35; m 64; c 3. MATHEMATICAL STATISTICS. *Educ:* Univ Kerala, BSc, 57, MSc, 59; Univ Toronto, MA, 62, PhD(math statist), 64. *Prof Exp:* Lectr math, St Thomas Col, Univ Kerala, 59-61; Commonwealth scholar statist, Univ Toronto, 61-64; asst prof, 64-68, assoc prof, 68-78, PROF MATH, MCGILL UNIV, 79- *Concurrent Pos:* Ed, Can J Statist, 74-77. *Honors & Awards:* Gold Medal, Univ Kerala, 59. *Mem:* Exec mem Statist Sci Asn Can (secy, 72-74); fel Inst Math Statist; Statist Soc Can; Int Statist Inst; Nat Acad Sci, India. *Res:* Statistical distributions; multivariate analysis; axiomatic foundations of statistical concepts; special functions and complex analysis; functional equations. *Mailing Add:* Dept Math/Statist McGill Univ 805 Sherbrooke St W Montreal PQ H3A 2K6 Can

MATHENY, ADAM PENCE, JR, b Stanford, Ky, Sept 6, 32; m 67; c 2. DEVELOPMENTAL PSYCHOLOGY, PSYCHOBIOLOGY. *Educ:* Columbia Univ, BS, 58; Vanderbilt Univ, PhD(psychol), 62. *Prof Exp:* Sr engr human factors, Martin Aerospace Systs, 62-63; instr pediat, Med Sch, Johns Hopkins Univ, 63-65; staff fel, Nat Inst Child Health & Human Develop, 65-67; from asst prof to assoc prof, 67-76, PROF PEDIAT, MED SCH, UNIV LOUISVILLE, 76- *Concurrent Pos:* Out patient psychologist, Johns Hopkins Hosp, 63-65; clin instr, Georgetown Univ, 65-67; instr, Univ Louisville Col, 68-; chmn, Ky Task Force Except Children, 71-72; consult, Southeastern Ind Rehab Ctr, 72-75; bd mem, Ky Examr Speech & Hearing, 75-78. *Mem:* AAAS; Am Psychol Asn; Soc Res Child Develop; Int Soc Twin Studies; Sigma Xi; Behav Genetics Asn; Am Psychol Soc; Int Soc Study Behav Develop. *Res:* Cognitive and affective development; children; behavioral genetics; medical counseling; studies of attention. *Mailing Add:* Dept of Pediat Sch of Med Univ Louisville Louisville KY 40292

MATHENY, ELLIS LEROY, JR, b Harrisonburg, Va, Nov 9, 39; div; c 3. ENTOMOLOGY. *Educ:* James Madison Univ, BS, 65; Univ Tenn, MS, 68, PhD(entom), 71. *Prof Exp:* Asst prof biol, Motlow State Col, Tullahoma, Tenn, 71-76; from asst prof to assoc prof, 76-85, PROF ENTOM & DIR, INT TRAINING DIV, UNIV FLA, 85- *Concurrent Pos:* NSF grant, Univ Fla, 77-79; mem, USAID proj, Cameroon, Africa; actg head, Training Ctr, Int Rice Res Inst, Philippines. *Mem:* Entom Soc Am; Sigma Xi; Am Registry Prof Entomologists. *Res:* Turf insects; mole crickets; sod webworms; educational technology; development of self-learning materials. *Mailing Add:* Dept Entom & Nematol Univ Fla 3103 McCarty Hall-IFAS Gainesville FL 32611

MATHENY, JAMES DONALD, b Jackson, Miss, Dec 22, 25; m; c 2. MECHANICAL ENGINEERING. *Educ:* Univ SC BS, 45; Univ Tex, BS, 50, MS, 56, PhD (mech eng), 59. *Prof Exp:* Instr mech eng, Univ Tex, 52-56; asst prof, La Polytech Inst, 56-57; sr res engr, Convair, Tex, 57-58; assoc prof mech eng, Univ Ala, 58-63, prof, eng mech, 63-69; prof mech eng & head dept, Col Eng, Univ Wyo, 70-77; DEAN & PROF ENG, CALIF STATE UNIV, FRESNO, 77- *Mem:* Am Soc Mech Eng; Am Inst Aeronaut & Astronaut; Am Soc Eng Educ; Soc Natural Philos. *Res:* Heat transfer; naval aircraft landing systems; stress analysis of landing field materials; restrained buried pipe. *Mailing Add:* Calif State Univ Sch Eng Fresno CA 93740

MATHENY, JAMES LAFAYETTE, b Vicksburg, Miss, Aug 28, 43; m 65. PHARMACOLOGY. *Educ:* Delta State Col, BS, 67; Univ Miss, PhD(pharmacol), 71. *Prof Exp:* Asst prof pharmacol, Med Col Ga, 71-77; ASSOC PROF BIOL, COL DENT, UNIV KY, 78- *Concurrent Pos:* NIH, Ga Heart Assoc, Fight for Sight Inc grant, 71-72. *Mem:* AAAS; NY Acad Sci; fel Am Col Clin Pharmacol; Am Heart Asn; Am Soc Pharmacol & Exp Therapeut; Sigma Xi. *Res:* Autonomic-cardiovascular pharmacology. *Mailing Add:* Dept of Oral Biol Univ of Ky Col of Dent Lexington KY 40506

MATHER, BRYANT, b Baltimore, Md, Dec 27, 16; m 40. CIVIL ENGINEERING. *Educ:* Johns Hopkins Univ, AB, 36. *Hon Degrees:* DSc, Clarkson Col, 78. *Prof Exp:* Asst cur mineral, Chicago Mus Natural Hist, 39-41; geologist, Cent Concrete Lab, US War Dept, NY, 41-42, concrete res engr, 42-46; chief, Spec Invest Br, Concrete Div, Waterways Exp Sta, 46-65, from asst chief to chief, 65-78, CHIEF, STRUCT LAB, CORPS OF ENGRS, 78- *Concurrent Pos:* Mem comts, Transp Res Bd, Nat Acad Sci-Nat Res Coun & US Comt Large Dams; res assoc, Fla Dept Agr, 68-, Am Mus Natural Hist, Miss Mus Natural Sci & Miss Entom Mus, 80- & Miss Mus Coun, 81-84. *Honors & Awards:* Cavanaugh Award, Am Soc Testing & Mat, 90. *Mem:* Fel AAAS; Am Soc Testing & Mat (pres, 75-76); Meteoritical Soc; Lepidop Soc; Am Concrete Inst (pres, 64). *Res:* Structural geology; composition and properties of concrete and concrete aggregates; methods of testing; butterflies and moths of Mississippi. *Mailing Add:* 213 Mt Salus Rd Clinton MS 39056

MATHER, EDWARD CHANTRY, b Iowa City, Iowa, Apr 7, 37; m 58; c 2. VETERINARY MEDICINE, ACADEMIC ADMINISTRATION. *Educ:* Iowa State Univ, DVM, 60, Univ Mo, MS, 68, PhD(reprod physiol), 70. *Prof Exp:* Pvt vet pract, 60-66; instr vet med, Univ Mo-Columbia, 66-68, res assoc, 68-70, from asst prof to assoc prof, 70-74, dir theriogenology lab, 68-73; assoc prof vet med & head div theriogenology, Univ Minn, 74-78; PROF & CHMN LARGE ANIMAL SURG & MED, MICH STATE UNIV, EAST LANSING, 78- *Concurrent Pos:* Adv prog appl res on fertil regulation, AID, 74-80; external examr, Nigeria, 78; mem coun educ, Am Vet Med Asn, 83-89; Sabbital Leave, Int Atomic Energy Agency, Vienna, 87. *Honors & Awards:* Dipl, Am Col Theriogenology. *Mem:* Soc Theriogenology; Am Asn Vet Med Cols. *Res:* Effect of seminal constituents on endometrial metabolism; endocrinological variations in large mammals as affected by reproductive pathology; cost and benefits of animal health programs. *Mailing Add:* Vet Clinical Ctr/D201 Vet Clin Ctr Mich State Univ East Lansing MI 48824

MATHER, IAN HEYWOOD, b Cheadle, Cheshire, Eng, June 24, 45; m 74; c 2. BIOCHEMISTRY, CELL BIOLOGY. *Educ:* Univ Wales, BSc, 66, PhD(biochem), 71. *Prof Exp:* Res fel biochem, Univ Kent, Canterbury, 70-72; res assoc, Purdue Univ, 73-75; from asst prof to assoc prof, 75-85, PROF BIOCHEM & CELL BIOL, UNIV MD, 85- *Concurrent Pos:* NSF grants, 78-89, USDA grants, 81 & 85-89; Fogarty fel, 89-90. *Mem:* AAAS; Am Dairy Sci Asn; Brit Biochem Soc; Am Soc Cell Biol; NY Acad Sci. *Res:* Biochemistry and physiology of milk secretion; epithelial cell polarity. *Mailing Add:* Dept Animal Sci Univ MD College Park MD 20742

MATHER, JANE H, b Green Bay, Wis, July 16, 22. BIOCHEMISTRY. *Educ:* Univ Wis, BA, 44; Univ Chicago, PhD(biochem), 63. *Prof Exp:* Chem analyst, Western Elec Co, Ill, 44-45; res chemist, Armour Res Labs, Ill, 45-53; electron microscopist, Northwestern, 54-56, res asst, Univ Chicago, 56-62; asst prof biochem, Ill Inst Technol, 62-65; ASSOC PROF CHEM, GA STATE UNIV, 65- *Concurrent Pos:* Consult, Armour Res Labs, 53-57; USPHS res grants, Ill Inst Technol & Ga State Univ, 65-68. *Mem:* AAAS; Am Chem Soc. *Res:* Biochemical intermediary metabolism and enzymology; analytical chemistry. *Mailing Add:* 1016 Willivie Dr Decatur GA 30033-4131

MATHER, JENNIE POWELL, CELL CULTURE, ENDOCRINOLOGY. *Educ:* Univ Calif, San Diego, PhD(biol), 75. *Prof Exp:* Sr scientist, 84-88, STAFF SCIENTIST, GENENTECH, INC, 88- *Mem:* Endocrine Soc; Am Soc Cell Biol; AAAS. *Mailing Add:* Genentech Inc 460 Point San Bruno Blvd San Francisco CA 94080

MATHER, JOHN CROMWELL, b Roanoke, Va, Aug 7, 46; m 80. ASTROPHYSICS, INFRARED SPECTROSCOPY. *Educ:* Swarthmore Col, BA, 68; Univ Calif, Berkeley, PhD(physics), 74. *Prof Exp:* Nat Res Coun fel astrophysics, Goddard Inst Space Studies, 74-76, ASTRONOMER ASTROPHYSICS, GODDARD SPACE FLIGHT CTR, NASA, 76- *Concurrent Pos:* Lectr astron, Columbia Univ, 75-76; prin investr, far infrared absolute spectrophotom & proj scientist, Cosmic Background Explorer Satellite, 76- *Mem:* Am Phys Soc; Am Astron Soc; Sigma Xi. *Res:* Cosmology; the Big Bang. *Mailing Add:* NASA Goddard Space Flight Ctr Code 685 Greenbelt MD 20771

MATHER, JOHN NORMAN, b Los Angeles, Calif, June 9, 42; m 71; c 4. TOPOLOGY, ANALYSIS & FUNCTIONAL ANALYSIS. *Educ:* Harvard Univ, BA, 64; Princeton Univ, PhD(math), 67. *Prof Exp:* Vis prof, Inst Higher Sci Studies, 67-69; from assoc prof to prof, Harvard Univ, 69-75; PROF MATH, PRINCETON UNIV, 75- *Concurrent Pos:* Vis prof math, Princeton Univ, 74-75, Inst Higher Sci Studies, 82-83. *Honors & Awards:* John J Carty Medal, Nat Acad Sci, 78. *Mem:* Am Math Soc. *Res:* Study of dynamical systems especially in dimension 2 with reference to questions of transitivity and ergodicity. *Mailing Add:* Fine Hall Washington Rd Princeton NJ 08544

MATHER, JOHN RUSSELL, b Boston, Mass, Oct 9, 23; m 46; c 3. CLIMATOLOGY. *Educ:* Williams Col, BA, 45; Mass Inst Technol, BS, 47, MS, 48; Johns Hopkins Univ, PhD(climat), 51. *Prof Exp:* Asst prof, Johns Hopkins Univ, 51-53; adj assoc prof, Drexel Inst Technol, 57-60; chmn, 65-89, PROF GEOG, UNIV DEL, 61- *Concurrent Pos:* Res assoc climatologist, Lab Climat, 48-55, prin res scientist, 55-63; pres, C W Thornthwaite Assocs, 63-72; consult, World Meteorol Orgn, Yugoslavia, 57; vis lectr, Univ Chicago, 57-61; State Climatologist Del, 78-; coun mem, Am Geog Soc. *Mem:* Am Meteorol Soc; Am Geog Soc (secy, 83-); Am Geophys Union; Asn Am Geog (pres, 90-92); Am Water Res Asn; fel AAAS. *Res:* Water budget; evapotranspiration; water resources; applied climatology; hydroclimatology; global change. *Mailing Add:* Dept Geog Univ Del Newark DE 19716

MATHER, KATHARINE KNISKERN, b Ithaca, NY, Oct 21, 16; m 40. GEOLOGY. *Educ:* Bryn Mawr Col, AB, 37. *Hon Degrees:* Dsc, Clarkson Col Technol, 78. *Prof Exp:* Geologist, Cent Concrete Lab, NY, 42-44, engr concrete res, 44-46; chief petrog & x-ray br, Waterways Exp Sta, US Army Corps Engrs, 48-76, chief eng sci div, 76-78, geologist, Concrete Lab, 46-78 & spec tech asst, Struct Lab, 78-82; RETIRED. *Concurrent Pos:* Chmn comt basic res cement & concrete, Transp Res Bd, Nat Acad Sci-Nat Res Coun; chmn subcomt on Vol Change Concrete, Am Soc Testing & Mat Comt c-9, Subcomt on Sulfate Resistance Cement, Am Soc Testing Mat Comt c-1; mem, Am Concrete Inst Comts, Hist, Durability, Hydraulic Cement, Rehabilitation. *Honors & Awards:* Thompson Award, Am Soc Testing & Mat, 53; Wason Res Medal, Am Concrete Inst, 53; Arthur R Anderson Award, Am Concrete Inst, 82; Fed Woman's Award, 63. *Mem:* Fel Mineral Soc Am; Am Ceramic Soc; Am Concrete Inst fel, 68-71; Am Inst Mining, Metall & Petrol Eng; Clay Minerals Soc (secy, 64-67, pres, 73); hon mem Am Soc Testing & Mat; hon mem Am Concrete Inst. *Res:* Constitution and microstructure of concrete, its constituents and alteration products; effects of variation in composition and exposure on properties of concrete. *Mailing Add:* 213 Mt Salus Rd Clinton MS 39056

MATHER, KEITH BENSON, b Adelaide, S Australia, Jan 6, 22; m 46; c 2. GEOPHYSICS, NUCLEAR PHYSICS. *Educ:* Univ Adelaide, BSc, 42, MSc, 44. *Hon Degrees:* DSc, Univ Alaska, 68. *Prof Exp:* Demonstr physics, Univ Adelaide, 43-45, lectr, 46; Sci & Indust Endowment Fund stud & asst, Wash Univ, 46-48; Imp Chem Indust fel, Birmingham, 49-50; lectr, Ceylon, 50-51; res officer, Commonwealth Sci & Indust Res Org Australia, 52-54; sr res officer, Australian AEC, 54-56, physicist-in-chg Antarctic Div, 56-58; lectr physics, Melbourne, 58-61; assoc prof geophys, Geophys Inst, Univ Alaska, 61-62, asst dir, 63, prof physics & dir geophys, 63-76, vice chancellor res & advan study, 76-85; EMER DIR GEOPHYS INST & EMER PROF PHYS, UNIV ALASKA, 86- *Concurrent Pos:* Sci corresp, Melbourne AGE, 54-61; Fulbright travel grant, 61-62; mem polar res bd, Nat Acad Sci - Nat Res Coun, 70, vchmn, 72-77; mem bd govs, Arctic Inst NAm, 70, vchmn, 74-75; mem, Rhodes Scholar selection comt, Alaska, 71, state secy, 77-85. *Honors & Awards:* Polar Medal, Commonwealth Australia, 60. *Mem:* Am Geophys Union; Nat Coun Univ Res Adminrs; Inst Physics & Phys Soc London; Australian Inst Physics; Archaeol Inst Am; fel Arctic Inst N Am; fel AAAS. *Res:* Optical spectroscopy; nuclear scattering and reaction studies; cosmic radiation; geomagnetism and aurora; katabatic winds; polar geophysics; theory of road corrugation and relaxation oscillations; university research administration. *Mailing Add:* 2023 Stone Crest Dr Eugene OR 97401-1728

MATHER, ROBERT EUGENE, b Gogoi Mission Station, Portuguese E Africa, Nov 12, 18; US citizen; m 43; c 2. ANIMAL BREEDING, STATISTICAL ANALYSIS. *Educ:* Purdue Univ, BS, 39; Univ Md, MS, 41; Univ Wis, PhD(dairy husb, genetics), 46. *Prof Exp:* Grad asst dairy husb, Univ Md, 39-41 & Univ Wis, 41-45; from asst dairy husbandman to assoc dairy husbandman, Exp Sta, Va Polytech Inst, 45-48; assoc prof dairy husb, Rutgers Univ, New Brunswick, 48-59, from assoc res specialist to res specialist, Dairy Res Ctr, Exp Sta, 48-70, asst to dir, Agr Exp Sta for Statist & Comput Consult, 70-77, prof statistical genetics, 59-79, mgt info serv, Cook Col, Rutgers Univ, New Brunswick, 79-88; RETIRED. *Mem:* Am Dairy Sci Asn; Sigma Xi. *Res:* Dairy cattle genetics; use of statistics and computers in agricultural research. *Mailing Add:* Nine Sterling Ct East Brunswick NJ 08616

MATHER, ROBERT LAURANCE, b Clarksville, Iowa, Oct 1, 21; m 56; c 2. PHYSICS. *Educ:* Iowa State Univ, BS, 42; Columbia Univ, AM, 47; Univ Calif, PhD(physics), 51. *Prof Exp:* Physicist, Naval Ord Lab, 42-44, Radio Corp Am, 44-46 & Nevis Cyclotron Lab, Columbia Univ, 46-47; asst physics, Univ Calif, 47-48, physicist, Lawrence Radiation Lab, 48-51; physicist, Atomic Energy Div, NAm Aviation, Inc, 51-52; physicist, Nuclear Radiation Physics Br, US Naval Radiol Defense Lab, 52-69, eng physicist, Naval Ocean Systs Lab, 69-85, computer sci, 85-87; RETIRED. *Mem:* Am Phys Soc; Inst Elec & Electronic Engrs; Sigma Xi. *Res:* Radar and communications; electronics; accelerators; Cerenkov radiation; nuclear weapon residual radiation; radiation physics. *Mailing Add:* 100 Bay Pl Apt 2110 Oakland CA 94610

MATHER, ROGER FREDERICK, b London, Eng, May 27, 17; US citizen; m 43, 73; c 2. NUCLEAR ENERGY & POWER CONVERSION, ACOUSTICS OF WIND INSTRUMENTS & THE VOICE. *Educ:* Cambridge Univ, Eng, BA, 38; Mass Inst Technol, Cambridge, Mass, MSc, 40. *Hon Degrees:* MA, Cambridge Univ, Eng, 41. *Prof Exp:* Res metallurgist, Inland Steel Co, E Chicago, Ind, 40-42; chief metallurgist, Willys-Overland Motors, Toledo, Ohio, 42-46 & Kaiser-Frazer Corp, Willow Run, Mich, 46-50; proj mgr, US Steel Corp, Pittsburgh, Pa, 50-61; dir res & eng, Mine Safety Appliances, Pittsburgh, Pa, 61-62; res staff, E I du Pont de Nemours, Wilmington, Del, 62-63; chief, Nuclear Power Technol Br, NASA, Cleveland, Ohio, 63-73; ADJ PROF MUSIC, UNIV IOWA, IOWA CITY, 73- *Concurrent Pos:* Lectr, Univ Toledo, Ohio, 42-44; instr, Kirkwood Community Col, Iowa City, Iowa, 81-83; mem numerous prof, admin & govt adv comts. *Res:* Improved methods of playing the flute and other wind instruments and of using the voice for speaking and singing; improved instruments of the flute and recorder families and their maintenance; music and acoustics; author of various publications. *Mailing Add:* 308 Fourth Ave Iowa City IA 52245

MATHERS, ALEXANDER PICKENS, b Matherville, Miss, Sept 17, 09; m 33; c 1. ORGANIC CHEMISTRY. *Educ:* Univ Fla, BS, 31; Tulane Univ, MS, 46; George Washington Univ, PhD(chem), 56. *Prof Exp:* Self employed, 31-38; teacher high sch, Miss, 38-41; chemist, 41-55, from asst chief to chief Alcohol & Tobacco Tax Lab, US Treas Dept, 55-73; CONSULT WINE & DISTILLED SPIRITS INDUSTS, 73- *Concurrent Pos:* Owner & operator, Exp Winery & Vineyard, 74-85. *Mem:* AAAS; Am Inst Chemist (pres elect, 74-75, pres, 75-76); Asn Official Anal Chem (vpres, 65-66, pres, 66-67); Am Chem Soc; NY Acad Sci. *Res:* Analytical chemistry; effect of skins, seed and pulp on the fermentation of juice of vitis Rotundifolia; pilot operation for commercial production of white muscadine juice. *Mailing Add:* Dept Human Serv & Home Econ Ashland Col 401 College Ave Ashland OH 44805

MATHES, KENNETH NATT, b Schenectady, NY, June 30, 13; wid; c 2. ELECTRICAL PROPERTIES POLYMERS. *Educ:* BSEE, Union Col, 35. *Prof Exp:* Engr elec insulation, Gen Eng Lab, Gen Elec Co, 37-40, supvr, 40-48, asst div engr, 48-52, supvr, Mat & Process Lab, Turbine Gen Div, 52-56, insulation systs engr, Gen Eng Lab, 56-65 & Corp Res & Develop, 65-78; CONSULT, 78- *Concurrent Pos:* Consult, Rensselaer Polytech Inst, 46-52, NASA, US Navy & US Air Force, 52-78 & Nat Res Coun, 56-68; tech adv & chief deleg, US Nat Comt, Int Electrotech Comn, 62-84; consult, 78-; mem, short course staff, Univ Wis, Madison, 78- & Univ Calif, Los Angeles, 80- *Honors & Awards:* Res Award, NASA, 68; Arnold Scott Award, Am Soc Testing & Mat, 71; Centennial Award, Inst Elec & Electronics Engrs, 85. *Mem:* Fel Inst Elec & Electronics Engrs; fel Am Soc Testing & Mat; fel AAAS; Am Chem Soc. *Res:* Physical and electrical characteristics of electrical insulation; insulating materials; cryogenics; thermal aging; voltage aging; environmental aspects; surface failure; development of national and international standards for insulating materials. *Mailing Add:* 2052 Baker Ave Schenectady NY 12309-4132

MATHES, MARTIN CHARLES, b Amherst, Ohio, Feb 18, 35; m 57; c 3. PLANT PHYSIOLOGY. *Educ:* Miami Univ, BA, 57; Univ Md, MS, 59, PhD(plant physiol), 61. *Prof Exp:* Asst bot, Univ Md, 57-61; res aide plant physiol, Inst Paper Chem, Lawrence Univ, 61-64; asst prof, Univ Vt, 64-67; assoc prof, 67-74, PROF BIOL COL WILLIAM & MARY, 74- *Mem:* AAAS; Am Soc Plant Physiol. *Res:* Growth regulators; plant tissue cultures. *Mailing Add:* Dept of Biol Col of William & Mary Williamsburg VA 23185

MATHESON, ALASTAIR TAYLOR, b Vancouver, BC, Can, Oct 10, 29; m 59; c 2. BIOCHEMISTRY, CELL BIOLOGY. *Educ:* Univ BC, BA, 51, MSc, 53; Univ Toronto, PhD(biochem), 58. *Prof Exp:* Res fel enzymol, Nat Res Coun Can, 58-59; res assoc biophys, Johns Hopkins Univ, 59-60; from asst res officer to assoc res officer, Nat Res Coun Can, 60-70, sr res officer cell biochem, 70-77; prof & chmn dept biochem & microbiol, 77-85, DEAN SCI, UNIV VICTORIA, 85- *Concurrent Pos:* Adj prof biol, Carleton Univ, Ottawa, 74-77. *Mem:* Am Soc Biol Chem; Can Biochem Soc; fel Royal Soc Can; Can Soc Cell Biol; Can Soc Microbiol. *Res:* The structure and function of archaebacterial ribosomes; molecular evolution and molecular biology. *Mailing Add:* Dept Biochem & Microbiol Univ Victoria Box 1700 Victoria BC V8W 2Y2 Can

MATHESON, ARTHUR RALPH, b Kansas City, Mo, Oct 7, 15; m 41; c 4. INORGANIC CHEMISTRY. *Educ:* Univ Ill, BS, 40, MS, 47, PhD(inorg chem), 48. *Prof Exp:* Anal chemist, Univ Ill, 38-40; control chemist synthetic paints, Cook Paint & Varnish Co, Mo, 40; asst chem, Univ Ill, 46-47, asst, Off Naval Res Contract, 47-48; res chemist, Hanford Works, Gen Elec Co, 48-51; dir tech admin div, Schenectady Opers Off, US AEC, 51-54; mgr contract admin dept & asst to pres, M & C Nuclear, Inc, Mass, 54-59; mgr reprocessing, Sylvania-Corning Nuclear Corp, 59-60; sales mgr nuclear fuels dept, Spencer Chem Co, Mo, 60-61; mgr mat appln, Gen Atomic Div, Gen Dynamics Corp, 61-70, asst mgr uranium mkt, Gulf Gen Atomic Co, 70-72, mgr indust & govt activities, Uranium Supply & Distrib, Gulf Energy & Environ Systs, 72-73; consult, Sci Applns Inc, 73-82; RETIRED. *Concurrent Pos:* Battelle Mem Inst, 74-81, Energy Inc, 75-76, Gen Atomic Co, 76-84, Magnesep Corp, 78-81. *Mem:* fel Am Inst Chemists; Am Chem Soc; Am Nuclear Soc. *Res:* Nuclear fuels; reprocessing; coated particle fuels. *Mailing Add:* 4320 N Lane Del Mar CA 92014

MATHESON, DALE WHITNEY, b Waterville, Maine, Aug 18, 37; m 59; c 4. CHEMICAL MUTAGENESIS & CARCINOGENESIS. *Educ:* Bowdoin Col, AB, 60; Mich State Univ, MS, 64; Univ Pa, PhD(anat), 69. *Prof Exp:* From instr to asst prof anat, Milton S Hershey Med Ctr, Pa State Univ, 69-73; cell biologist, Litton Bionetics, Inc, 73-75, chief mammalian genetics, 75-78, assoc dir genetic toxicol, 76-78; mgr in vitro toxicol, Stauffer Chem Co, 78-87, mgr cell & reproductive toxicol, 88-90; MGR EXP TOXICOL, CIBA-GEIGY CORP, 90- *Concurrent Pos:* Clin assoc, Univ Conn Health Ctr, 78-, adj res assoc prof, Sch Pharm, 80- *Mem:* Environ Mutagen Soc; Genetic Toxicol Asn; AAAS; Tissue Culture Asn. *Res:* Mechanisms of chemically induced mammalian mutagenesis and carcinogenesis. *Mailing Add:* Ciba-Geigy Corp 400 Farmington Ave Farmington CT 06032

MATHESON, DAVID STEWART, b Camrose, Alta, Dec 28, 45; m 68; c 2. PEDIATRIC IMMUNOLOGY. *Educ:* Univ Calgary, BSc Hons, 67, MD, 74; Univ Waterloo, Math, 68; FRCP(Can), 79. *Prof Exp:* Intern pediat, Foothills Hosp, Calgary, 74-75; resident, Hosp Sick Children, Toronto, 75-76, Med Res Coun, res fel immunol, 76-79; from asst prof to assoc prof pediat & internal med, Fac Med, Univ Calgary, 79-87; ASSOC PROF PEDIAT, FAC MED, UNIV BC, 87- *Concurrent Pos:* Alta Heritage Found Med Res scholar, 81-87; vpres med, BC Children's Hosp, 89- *Mem:* Am Asn Immunologists; Am Fedn Clin Res; Can Soc Clin Invest; Can Soc Immunol. *Res:* Cellular immune regulation; immunodeficiency; immune response modification. *Mailing Add:* Dept Pediat Univ BC Children's Hosp Vancouver BC V6H 3V4 Can

MATHESON, DE LOSS H(EALY), b Ont, Feb 3, 08; m 42; c 1. ENGINEERING. *Educ:* Univ Toronto, BASc, 29, MASc, 31. *Prof Exp:* Chemist & bacteriologist, Hamilton Filtration Plant, 33-55; dir, Munic Labs, City Hamilton, 55-72; consult, 72-80; RETIRED. *Mem:* Am Chem Soc; Am Water Works Asn; Air Pollution Control Asn; fel Chem Inst Can. *Res:* Water purification; water and air pollution. *Mailing Add:* 78 Mountain Ave Hamilton ON L8P 4G2 Can

MATHEW, CHEMPOLIL THOMAS, b Kerala, India, Jun 3, 33; m 67; c 2. ORGANIC OXIMES, HYDROXYLAMINE DERIVATIVES. *Educ:* Kerala Univ, BSc, 53; Bombay Univ, MSc, 56; Calcutta Univ, D Phil, 64. *Prof Exp:* Lectr, chem, Kerala Univ, India, 56-60; postdoctoral, John's Hopkins Univ, Baltimore, 64-66, postdoctoral & Harvard Univ, Mass, 66-67; res chem, Allied Chem Corp, 67-80, sr res assoc, 80-85, DIR, R&D, ALLIEDSIGNAL, INC, 86- *Concurrent Pos:* lectr chem, Kerala Univ, 55-60. *Mem:* NY Acad Sci; Am Asn Advan Sci. *Res:* Process and product development activities based on hydroxylamine such as oximes, substituted hydroxylamine. *Mailing Add:* 19 Openaki Rd Randolph NJ 07869

MATHEW, MATHAI, b Mavelikara, India; US citizen; m 67. X-RAY CRYSTALLOGRAPHY, INORGANIC CHEMISTRY. *Educ:* Univ Kerala, India, BS, 53; Univ Agra, India, MS, 56; Univ Western Ont, PhD(chem), 66. *Prof Exp:* Lectr chem, Cath Col, India, 56-58 & 60-62; res asst, Atomic Energy Estab, India, 58-60; fel, Nat Res Coun Can, 65-67, Univ Waterloo, 67-70; res assoc, Univ Fla, 70-74; res assoc chem, 75-77, CHIEF RES SCIENTIST, DIV CRYSTALLOG, AM DENT ASN HEALTH FOUND, RES UNIT, NAT BUR STANDARDS, 77- *Mem:* Am Chem Soc; Am Crystallog Asn; Int Asn Dent Res. *Res:* X-ray crystallographic structural studies of dental materials and of compounds related to constituents of tooth, bone and dental calculus; phosphate minerals; substituted apatites. *Mailing Add:* Am Dent Asn Health Found Nat Inst Studies & Technol Gaithersburg MD 20899

MATHEWES, DAVID A, b Gastonia, NC, Sept 22, 31; m 57; c 4. ORGANIC CHEMISTRY. *Educ:* Davidson Col, BS, 53; Univ Kans, MS, 55; Duke Univ, PhD(chem), 63. *Prof Exp:* Teaching asst chem, Univ Kans, 53-55; instr, Ga Inst Technol, 55-57; instr, Hampden-Sydney Col, 57-58; res asst, Duke Univ, 58-62; asst prof, 62-67, head dept, 67-69, PROF CHEM, WESTERN CAROLINA UNIV, 67- *Concurrent Pos:* Instr, Westminster Schs, 56-57; vis prof, Univ Stirling, 77-78. *Mem:* AAAS; Am Chem Soc; Sigma Xi; Royal Soc Chem. *Res:* Curriculum development; science education; computer-aided instruction. *Mailing Add:* RR 1 Box 530 Canton NC 28716-9749

MATHEWES, ROLF WALTER, b Berleburg, WGer, Nov 11, 46; Can citizen; m 72. PALYNOLOGY, PALEOCLIMATOLOGY. *Educ:* Simon Fraser Univ BC, Can, BSc; Univ BC, PhD(bot). *Prof Exp:* Vis asst prof biogeog, Simon Fraser Univ, 73; Nat Res Coun fel palynology, Sch Bot, Cambridge Univ, Eng, 74; from asst prof to assoc prof, 75-87, PROF BIOL, SIMON FRASER UNIV, 87- *Concurrent Pos:* Environ consult plant ecol, F F Slaney & Co Ltd, Vancouver, 74-75; res fel, Alexander von Humboldt Found, Germany, 82; assoc ed, Can J Bot, 89- *Mem:* Can Asn Palynologists (pres, 86-87); Can Bot Asn; Am Quaternary Asn; Am Asn Stratig Palynologists. *Res:* Palynology and paleoecology of postglacial vegetation; application of pollen analysis to archaeological and paleoclimatic problems. *Mailing Add:* Simon Fraser Univ Burnaby BC V5A 1S6 Can

MATHEWS, A L, b Whittier, NC, Mar 28, 40; m 60; c 3. CHEMISTRY. *Educ:* Western Carolina Univ, BS, 61; Univ Miss, PhD(phys chem), 65. *Prof Exp:* Asst prof phys chem, Western Carolina Univ, 65-69; admin asst biochem, Mich State Univ, 69-76; admin mgr, dept chem, 76-87, DIR, PHYS FACIL, OHIO STATE UNIV, 87- *Mem:* AAAS; Am Chem Soc. *Res:* Research administration; communication and information exchange in solving significant problems; use of computers in designing experiments; thermodynamics of multiple phase systems. *Mailing Add:* 6606 Evening St Worthington OH 43085-3013

MATHEWS, BRUCE EUGENE, b Peru, Ill, June 1, 29; m 58; c 4. ELECTRICAL ENGINEERING. *Educ:* Univ Fla, BEE, 52, MSE, 53, PhD(elec eng), 64. *Prof Exp:* Engr, NAm Aviation, Inc, 55-56; asst prof elec eng, Univ Fla, 56-57, res assoc, 57-64, assoc prof, 64-69; chmn dept elec eng & commun sci, 69-78, PROF ELEC ENG & COMMUN SCI, UNIV CENTRAL FLA, 69-, ASST DEAN ENG, 78- *Concurrent Pos:* Consult, Vitro Corp Am, 57, Ohio State Res Found, 58, Northrop Corp, 59, Lockheed Missile & Space Co, 60 & Scott Aviation Corp, 62-63. *Mem:* Inst Elec & Electronics Engrs; Am Soc Engr Educr. *Res:* Electromagnetic fields; sensory aids for the handicapped; creative problem solving. *Mailing Add:* Col Eng Univ Central Fla Orlando FL 32816

MATHEWS, C A, DESIGN ENGINEERING, PROTECTIVE RELAYS. *Educ:* Los Angeles State Univ, BSEE, 43. *Prof Exp:* DESIGN ENGR, GEN ELEC CO, 43- *Mem:* Fel Inst Elec & Electronics Engrs. *Mailing Add:* 1149 Providence Rd Springfield PA 19064

MATHEWS, CHRISTOPHER KING, b New York, NY, May 5, 37; m 60; c 2. BIOCHEMISTRY. *Educ:* Reed Col, BA, 58; Univ Wash, PhD(biochem), 62. *Prof Exp:* Asst prof biol, Yale Univ, 63-67; from assoc prof to prof biochem, Col Med, Univ Ariz, 67-77; PROF & CHMN DEPT BIOCHEM & BIOPHYS, ORE STATE UNIV, 78- *Concurrent Pos:* USPHS fel biochem, Univ Pa, 62-63; Am Cancer Soc scholar, Univ Calif, San Diego, 73-74; mem & chmn, virol & microbial chem study sect, NIH, 77-81; Eleanor Roosevelt Int Cancer fel, Stockholm, Sweden, 84-85. *Honors & Awards:* Discovery Award, Med Res Found, Ore, 86. *Mem:* AAAS; Am Soc Biochem & Molecular Biol; Am Soc Cell Biol; Am Soc Microbiol; Am Chem Soc; Am Asn Univ Professors; Am Soc Virol. *Res:* Nucleic acid enzymology, particularly deoxyribonucleotide biosynthesis and its regulation; fidelity and control of DNA replication; mechanisms of mutagenesis; genetic and metabolic control in virus-infected bacteria and in cultured mammalian cells; metabolism of coenzymes, nucleotides, and nucleic acids. *Mailing Add:* Dept Biochem & Biophys Ore State Univ 535 Weniger Hall Corvallis OR 97331-6503

MATHEWS, COLLIS WELDON, b Troy, Ala, July 19, 38; m; c 2. PHYSICAL CHEMISTRY, MOLECULAR SPECTROSCOPY. *Educ:* Univ Ala, BS, 60; Vanderbilt Univ, PhD(phys chem), 65. *Prof Exp:* Res assoc ultraviolet spectros, Vanderbilt Univ, 64-65; fel div pure physics, Nat Res Coun Can, 65-67; asst prof, 67-72, ASSOC PROF PHYS CHEM, OHIO STATE UNIV, 72-, VCHMN GRAD STUDIES CHEM, 88- *Mem:* Am Chem Soc; fel Optical Soc Am. *Res:* Investigations of high-resolution visible and ultraviolet molecular spectra for the purposes of obtaining their geometric and electronic structures especially of unstable molecular species. *Mailing Add:* Dept Chem Ohio State Univ 120 W 18th Ave Columbus OH 43210

MATHEWS, DONALD R(ICHARD), b Madera, Calif, Nov 23, 31; m 58; c 3. NUCLEAR ENGINEERING. *Educ:* Univ Calif, Berkeley, BS, 58, MS, 59; Mass Inst Technol, PhD(nuclear eng), 66. *Prof Exp:* Nuclear engr, Aerojet Gen Nucleonics, Calif, 59-61, sr nuclear engr, Univ Idaho, 61-62; staff assoc thermionic reactor design, Gen Atomic Div, Gen Dynamics Corp, 66-67; staff mem, Gulf Gen Atomic Inc, 67-70, mgr, 70-72, mem staff, 72-78, SR STAFF MEM, REACTOR PHYSICS METHODS DEVELOP BR, GULF GEN ATOMIC CO, 78- *Mem:* Am Nuclear Soc. *Res:* Nuclear reactor physics methods development, both analytic and computational aspects; neutron cross section evaluation and processing; gas-cooled thermal and fast reactor design; thermionic reactor design. *Mailing Add:* TSI Wurenlingen CH 5303 Switzerland

MATHEWS, F(RANCIS) SCOTT, b Albany, Ore, Mar 2, 34; m 59; c 3. BIOCHEMISTRY. *Educ:* Univ Calif, BS, 55; Univ Minn, Minneapolis, PhD(phys chem), 59. *Prof Exp:* Corp fel chem, Harvard Univ, 59-61; USPHS res fel biol, Mass Inst Technol, 61-63; spec fel protein crystallog, Lab Molecular Biol, 63-65, assoc, 66-80, PROF CELL BIOL, SCH MED, WASH UNIV, 80- *Honors & Awards:* Fogarty Sr Int Fel, 83. *Mem:* Am Crystallog Asn; Biophys Soc; Am Chem Soc. *Res:* X-ray crystallographic study of biological materials, especially the structure and function of cytochromes and flavoenzymes. *Mailing Add:* Dept Cell Biol & Physiol Wash Univ Sch Med 4566 Scott Ave Box 8101 St Louis MO 63110

MATHEWS, FREDERICK JOHN, b Columbus, Wis, Dec 20, 18; m 52; c 2. ORGANIC CHEMISTRY. *Educ:* Carroll Col, BA, 40; Univ Wis, PhD(org chem), 43. *Prof Exp:* Asst, Univ Wis, 41-43; res chemist, Rohm & Haas Co, Pa, 43-46; asst prof chem, Kent State Univ, 46-47; from asst prof to assoc prof, Beloit Col, 47-59; PRES, LAB CRAFTSMEN INC, 59- *Res:* Benzoquinoline compounds synthesis; silicones; design and manufacture of scientific equipment. *Mailing Add:* 2925 Bartells Dr Beloit WI 53511

MATHEWS, HARRY T, b Atlanta, Ga, Nov 13, 31; m 59; c 3. MATHEMATICS. *Educ:* Ga Inst Technol, BS, 59; Tulane Univ, PhD(math), 64. *Prof Exp:* Asst prof math, Wayne State Univ, 63-65; from asst prof to prof math & head dept, 65-73, PROF MATH, UNIV TENN, KNOXVILLE, 73- *Mem:* Am Math Soc; Math Asn Am. *Res:* Boundary behavior of functions of a complex variable. *Mailing Add:* Univ Tenn Knoxville TN 37916

MATHEWS, HENRY MABBETT, b Thomasville, Ga, May 19, 40; m 62; c 3. MEDICAL PARASITOLOGY. *Educ:* Univ Ga, BS, 62; Emory Univ, MS, 65, PhD(biol), 67. *Prof Exp:* Resident microbiol, Nat Commun Dis Ctr, 67-69; RES MICROBIOLOGIST, CTR DIS CONTROL, 69- *Mem:* Am Soc Parasitologists; Am Soc Trop Med & Hyg; Soc Protozoologists. *Res:* Biochemistry and immunology of protozoa. *Mailing Add:* 1192 Denison Dr Clarkstown GA 30021

MATHEWS, HERBERT LESTER, b Johnstown, Pa, Oct 5, 49; m 73; c 3. TUMOR IMMUNOLOGY, IMMUNE REGULATION. *Educ:* Washington & Jefferson Col, BA, 71; WVa Univ, MS, 74, PhD(med microbiol), 77. *Prof Exp:* Training fel med microbiol, dept microbiol, WVa Univ, 72-77; res fel immunol, dept med, Nat Jewish Hosp, NIH fel, 78-80; ASSOC PROF MICROBIOL & IMMUNOL, DEPT MICROBIOL, STRITCH SCH MED, LOYOLA UNIV, CHICAGO, 80- *Mem:* Am Asn Immunologists; AAAS; Sigma Xi; Am Soc Microbiol. *Res:* Means and mechanisms of immune response regulation to tumor and microbial determinants with emphasis on the role of lymphocytes and cytokines. *Mailing Add:* Dept Microbiol Loyola Univ Med Col 2160 S First Ave Maywood IL 60153

MATHEWS, JEROLD CHASE, b Des Moines, Iowa, Sept 12, 30; m 59; c 2. MATHEMATICS. *Educ:* Iowa State Univ, BS, 55, MS, 57, PhD(math), 59. *Prof Exp:* Asst prof math, Univ Okla, 60-61; mathematician, Mathematica, Inc, NJ, 61-62; from asst prof to assoc prof, 62-65, PROF MATH, IOWA STATE UNIV, 70- *Concurrent Pos:* Fulbright lectr, Ghana, 72-73. *Mem:* Math Asn Am; Am Math Soc; Nat Coun Teachers Math. *Res:* History of almost periodic functions; biography of Harold Bohr. *Mailing Add:* Dept of Math Iowa State Univ Ames IA 50011

MATHEWS, JOHN DAVID, b Kenton, Ohio, Apr 3, 47; m 69; c 2. IONOSPHERIC PHYSICS, RADAR SIGNAL PROCESSING. *Educ:* Case Inst Technol, BS, 69; Case Western Reserve Univ, MS, 72, PhD(elec eng), 72. *Prof Exp:* Res assoc, 72-73, sr res assoc, 73-75, from asst prof to assoc prof, 75-85, PROF ELEC ENG & APPL PHYSICS, CASE WESTERN RESERVE UNIV, 85- *Concurrent Pos:* Vis scientist, Nat Astron & Ionosphere Ctr, Puerto Rico, 72-75; adj prof & asst mem grad fac, Elec Eng Dept & Ionosphere Res Lab, Pa State Univ, 78-83; consult, Nat Astron & Ionosphere Ctr, 78- *Mem:* Am Geophys Union; Sigma Xi; Int Union Radio Scientists; sr mem Inst Elec & Electronics Engrs. *Res:* Experimental and theoretical investigation of the physics and chemistry of the earth's upper atmosphere and ionosphere; radar scattering theory and signal processing; electromagnetic theory. *Mailing Add:* Case Western Reserve Univ Glennan Bldg Rm 715 Cleveland OH 44106

MATHEWS, JOSEPH F(RANKLIN), b Rochester, NY, Nov 28, 33; m 67; c 2. CHEMICAL ENGINEERING. *Educ:* Univ Rochester, BS, 55; Univ Tex, MS, 57, PhD(chem eng), 60. *Prof Exp:* Res engr, Synthetic Rubber Div, Shell Chem Co, 60-66; asst prof chem & chem eng, Univ Sask, 66-68, assoc prof, 68-74, prof chem eng, 74- *Mem:* Am Inst Chem Engrs; Am Chem Soc; Can Soc Chem Engrs. *Res:* Reaction kinetics and catalysis; reactor design; alternative uses for bio-mass. *Mailing Add:* Dept Chem Eng Monash Univ Victoria 3168 Australia

MATHEWS, KENNETH PINE, b Schenectady, NY, Apr 1, 21; m 52, 75; c 3. ALLERGY, INTERNAL MEDICINE. *Educ:* Univ Mich, AB, 41, MD, 43. *Prof Exp:* From asst prof to prof, 61-86, EMER PROF INTERNAL MED, UNIV MICH, 86- *Concurrent Pos:* Consult, Ann Arbor Vet Admin Hosp, 60-86; ed, J Allergy & Clin Immunol, 68-72; mem training grant comt, Nat Inst Allergy & Infectious Dis, 71-73; chmn allergy & immunol res comt, NIH, 73-75; mem, Am Bd Allergy & Immunol, 77-82; res rev comt, 81-87, chmn, Allergy & Immunol, 83-86; adj mem, Scripps Clin Res Found, 86. *Honors & Awards:* Distinguished Serv Award, Am Acad Allergy, 76. *Mem:* Am Acad Allergy (pres, 64-65); Am Col Physicians; Am Asn Immunologists; Am Thoracic Soc; Am Fed Clin Res. *Res:* Basic and clinical research in allergy; mechanisms of urticaria and angioedema; complement. *Mailing Add:* Dept Molecular & Exp Med Scripps Clin Res Found 10666 N Torres Pines Rd La Jolla CA 92037

MATHEWS, LARRY ARTHUR, b Bremerton, Wash, Mar 23, 36; m 71; c 3. ATMOSPHERIC SCIENCES, AEROSOL PHYSICS & CHEMISTRY. *Educ:* Univ Wash, BS, 59; Univ Utah, ME, 69, PhD(chem eng), 70. *Prof Exp:* Asst engr, Ga-Pac Corp, 58; technologist, Shell Oil Co, 59-60; assoc engr chem milling res, Boeing Co, 61; res assoc chem eng, Univ Utah, 64-69; res chem engr atmospheric & aerosol physics, 70-79, RES PHYS CHEMIST AEROSOL CHEM & PHYS, NAVAL WEAPONS CTR, 80- *Concurrent Pos:* Consult dust storms, Great Basin Unified Air Pollution Control Dist & Inyo County Dist Atty Off, 76-78; mem, Owens Dry Lake Task Force Comt, 80-, Joint Tri-Servs Coord Group/Munitions Effectiveness for Smokes &

Obscurants, 83- *Mem:* Sigma Xi. *Res:* Weather modification; rate of solution of ice nuclei in water drops; collection efficiencies of cloud drops; cloud and aerosol physics; air pollution; desert dust storms; fluid flow; earth sciences; both theoretical and experimental work; military smoke and obscurants; solid propellant rocket plumes. *Mailing Add:* Eng Sci Div Naval Weapons Ctr Code 3892 China Lake CA 93555

MATHEWS, M(AX) V(ERNON), b Columbus, Nebr, Nov 13, 26; m 47; c 3. COMPUTER MUSIC. *Educ:* Calif Inst Technol, BS, 50; Mass Inst Technol, ScD, 54. *Prof Exp:* mem tech staff, 55-61, head, 61-62, dir, 62-85, mem tech staff, Acoust & behav res ctr, AT&T Bell Labs, 85-87; PROF OF MUSIC (RES), STANFORD UNIV, 87- *Concurrent Pos:* Sci adv, Inst de Recherche et Coord Acoustique/Musique, Paris, France, 74-80. *Honors & Awards:* David Sarnoff Gold Medal, Inst Elec & Electronics Engrs, 73. *Mem:* Nat Acad Sci; Nat Acad Eng; Inst Elec & Electronics Engrs; Audio Eng Soc; Acoust Soc Am; Am Acad Arts & Sci. *Res:* Speech coding; computer music; digital computer technology. *Mailing Add:* Ctr Comput Res Music & Acoust Music Dept Stanford Univ Stanford CA 94305

MATHEWS, MARTIN BENJAMIN, b Chicago, Ill, May 30, 12. CONNECTIVE TISSUES. *Educ:* Univ Chicago, PhD(chem), 49. *Prof Exp:* prof, 70-83, EMER PROF PEDIAT & BIOCHEM, UNIV CHICAGO, 83- *Mailing Add:* Dept Pediat Univ Chicago 5825 Maryland Ave Box 413 Chicago IL 60637

MATHEWS, NANCY ELLEN, b Kettering, Ohio, Mar 9, 58. WILDLIFE ECOLOGY, BEHAVIORAL ECOLOGY. *Educ:* Pa State Univ, BS, 80; State Univ NY, MS, 82, PhD(forest biol), 89. *Prof Exp:* Res asst, Col Environ Sci & Forestry, 80-82 & 85-89; scientist energy measurements, EG&G Inc, 82-85; res asst ecologist, Ecol Lab, Univ Ga, Savannah River, 89-90; RES BIOLOGIST ECOL & ASST UNIT LEADER, US FISH & WILDLIFE SERV, TEX COOP FISH & WILDLIFE RES UNIT, TEX TECH UNIV, 90- *Concurrent Pos:* Scientist aerial measurements opers, EG&G, Inc, 84-85; mem, Mem Comt, Ecol Soc, 90- *Mem:* AAAS; Am Soc Mammalogists; Am Soc Naturalists; Animal Behav Soc; Ecol Soc; Wildlife Soc. *Res:* Behavioral ecology with emphasis on social organization and predator-prey interactions; conservation biology with emphasis on endangered species and the conservation of genetic diversity. *Mailing Add:* Range & Wildlife Dept Goddard Bldg Rm 9 Tex Tech Univ Lubbock TX 79409

MATHEWS, ROBERT THOMAS, b Indianapolis, Ind, Aug 30, 19; m 52; c 4. ASTRONOMY, HISTORY OF SCIENCE. *Educ:* Wesleyan Univ, BA, 40; Univ Calif, MA, 54. *Prof Exp:* From jr astronr to asst astronr, US Naval Observ, 42-44; observing asst, Lick Observ, Mt Hamilton, 47-48; instr astron, Wesleyan Univ, 48-54; from instr to asst prof astron & math, 54-67, assoc prof, 67-81, prof, 81-84, EMER PROF ASTRON, CARLETON COL, 84- *Mem:* AAAS; Am Astron Soc; Sigma Xi; Am Asn Univ Professors. *Res:* Stellar parallax; photoelectric and spectroscopic study of galactic star clusters; visual double stars; history of science. *Mailing Add:* 405 Nevada St Northfield MN 55057

MATHEWS, W(ARREN) E(DWARD), b Osborne, Kans, Nov 10, 21; m 49, 71; c 3. SYSTEMS ANALYSIS, ELECTRO-OPTICAL SYSTEMS. *Educ:* Ohio Wesleyan Univ, AB, 42; Mass Inst Technol, BS & MS, 44; Calif Inst Technol, PhD(physics), 53. *Prof Exp:* Mem tech staff, Radio Res Dept, Bell Tel Labs Inc, NJ, 46-49; mem tech staff, 50-54, head adv planning staff, 54-57, corp dir planning, 57-60, dir Infrared Labs, 60-66, assoc dir Res & Develop Div, 62-66, mgr Missile Systs Div, 66-70, assoc mgr Systs Divs, 70-71, mgr Equip Eng Divs, 71-74, asst group exec, Electro-Optical & Data Systs Group, 74-75, corp dir, 75-82, staff vpres, Prod Effectiveness, 82-85, staff vpres, Tech Mgt Planning, Hughes Aircraft Co, 85-86. *Concurrent Pos:* Mem nat exec comt, Infrared Info Symp, 60-65; mem bd dir, Santa Barbara Res Ctr, 60-66; gen chmn, Winter Conv Aerospace & Electronic Systs, 71; mem bd gov, Electronic Indust Asn, 76-86. *Mem:* Fel Inst Elec & Electronics Engrs; assoc fel Am Inst Aeronaut & Astronaut. *Res:* Developed coupled transmission line theory of traveling wave amplifiers, closed-form solution for miss distances of homing navigation systems; holder of two patents. *Mailing Add:* 1010 Centinela Ave Santa Monica CA 90403-2341

MATHEWS, WALTER KELLY, b Columbus, Ga, Jan 16, 37; m; c 2. ORGANIC CHEMISTRY, BIOCHEMISTRY. *Educ:* Univ Ga, BS, 60, MS, 61; Univ Louisville, PhD(chem), 67. *Prof Exp:* Chemist, Sinclair Res, Inc, Ill, 61-63; instr gen chem, Wingate Col, 63-64; vis instr org chem, Univ Louisville, 65-66; ASSOC PROF ORG CHEM, GA SOUTHWESTERN COL, 67- *Mem:* Am Chem Soc; Sigma Xi. *Res:* Cationic polymerization mechanisms; selected oxidation processes; mechanisms of counterion binding to colloids and surfactants. *Mailing Add:* 115 Springdale Dr Y841 Ga Southwestern Col Americus GA 31709

MATHEWS, WILLIAM HENRY, b Vancouver, BC, Feb 2, 19; m 48; c 3. GEOLOGY. *Educ:* Univ BC, BASc, 40, MASc, 41; Univ Calif, Berkeley, PhD(geol), 48. *Prof Exp:* Assoc mining engr, BC Dept Mines, Can, 42-49; asst prof geol, Univ Calif, 49-51; from assoc prof to prof, 51-84, head dept, 64-71, EMER PROF GEOL, UNIV BC, 84- *Concurrent Pos:* Nat Res Coun Can sr fel, 63-64; mem, Can Nat Comt for Int Hydrologic Decade, 64-74; mem, Int Comt Marine Geol, 66-; mem, Can Nat Adv Comt Res Geol Sci, 67-69; chmn standing comt solid earth sci, Pac Sci Asn, 67-71; Killam sr fel, 71-72. *Honors & Awards:* Miller Medal, Royal Soc Can, 89. *Mem:* Fel Geol Soc Am; fel Royal Soc Can; hon fel Geol Asn Can; Int Glaciol Soc; Am Qua; Can Qua. *Res:* Geomorphology and glacial geology; glaciology, sedimentology and geological oceanography; sub-glacial vulcanism. *Mailing Add:* Dept Geol Sci Univ BC Vancouver BC V6T 2B4 Can

MATHEWS, WILLIS WOODROW, b Wendling, Ore, May 27, 17; m 42; c 3. EMBRYOLOGY. *Educ:* Ore State Col, BA, 40; Univ Wis, PhD(zool), 45. *Prof Exp:* Asst zool, Univ Wis, 40-44; from instr to asst prof biol, Univ Chattanooga, 44-47; from asst prof to assoc prof, 47-82, from actg chmn dept to chmn dept, 62-65, EMER PROF BIOL, WAYNE STATE UNIV, 82- *Concurrent Pos:* Vis lectr, Oakland Univ, 74-75 & 82. *Mem:* AAAS; Am Soc Zool. *Res:* Experimental embryology of chick; microscopy; growth factors; descriptive embryology. *Mailing Add:* 22443 Bayview Dr St Clair Shores MI 48081

MATHEWSON, CHRISTOPHER COLVILLE, b Plainfield, NJ, Aug 12, 41; c 2. GEOLOGY. *Educ:* Case Inst Technol, BS, 63; Univ Ariz, MS, 65, PhD(geol eng), 71. *Prof Exp:* From asst prof to assoc prof,71-82, PROF GEOL,TEX A&M UNIV,82-,DIR, CTR ENG GEOSCIENCES,82- *Concurrent Pos:* Instr geol eng, Univ Ariz, 71; consult, 71- *Honors & Awards:* Claire P Holdredge Award, Asn Eng Geologists, 81. *Mem:* Asn Eng Geologists (pres, 89); Am Soc Civil Engrs; Geol Soc Am; Am Inst Mining Engrs; Am Geophys Union. *Res:* Engineering geology applied to coal mining, urban development, hazardous geologic processes, and natural resources. *Mailing Add:* Dept Geol Tex A&M Univ College Station TX 77843

MATHEWSON, FRANCIS ALEXANDER LAVENS, b New Westminster, BC, Feb 1, 05; m 36; c 2. INTERNAL MEDICINE. *Educ:* Univ Man, MD, 31, BSc, 33; Am Bd Prev Med, dipl & cert, 54; FRCP, 81. *Prof Exp:* Assoc prof med, Fac Med, Univ Man, 45-; RETIRED. *Concurrent Pos:* Physician, Winnipeg Gen Hosp, mem attend staff, 35, chmn, 51-52; mem, Asn Comt Aviation Med Res, Nat Res Coun Can, 42-44; mem, Panel Aviation Med Res, Defense Res Bd, Can, 50-54; chmn, Royal Can Air Force Med Adv Comt, 54; fel, Coun Clin Cardiol, Am Heart Asn; Col Physicians & Surgeons Man Gordon Bell res fel, 33-34. *Mem:* Fel Am Col Cardiol; Asn Life Ins Med Dirs (pres, 68-69); Can Cardiovasc Soc (pres, 57-58); hon mem Can Soc Aviation Med; Defense Med Asn Can (pres, 54-55); hon mem Can Life Ins Med Officers Asn (pres, 55-56). *Res:* Cardiology; prospective epidemiological study of coronary heart disease. *Mailing Add:* 711 Med Arts Bldg 283 Yale St Winnipeg MB R3C 3J5 Can

MATHEWSON, JAMES H, b Norwalk, Conn, Nov 24, 29; m 58; c 3. BIO-ORGANIC CHEMISTRY, OCEANOGRAPHY. *Educ:* Harvard Univ, AB, 51; Johns Hopkins Univ, MA, 57, PhD(org chem), 59. *Prof Exp:* Res assoc chem, Johns Hopkins Univ, 59-60; guest investr, Rockefeller Inst, 60-61; res fel, Univ Calif, Berkeley, 61-63; asst prof chem, Western Wash State Col, 63-64; from asst prof to assoc prof, 64-72, PROF CHEM, SAN DIEGO STATE UNIV, 72- *Concurrent Pos:* USPHS fel, 60-63; Nat Inst Arthritis & Metab Dis res grant, 65-68, NSF sea grant prog res grant, 69-70; Nat Oceanic & Atmospheric Admin res grant, 70-71; actg dir, Bur Marine Sci, San Diego State Univ, 67-70; vis scientist, Lab Chem Enzyme Shell Res, Sittingbourne, Kent, UK, 74-75; dir, Ctr Marine Studies, 86- *Mem:* AAAS; Am Chem Soc; Sigma Xi. *Res:* Organic and biological chemistry, especially tetrapyrroles; environmental chemistry, especially oceanic; science education, especially general education and interdisciplinary courses; chlorophyll chemistry; marine biochemistry; pollution measurement. *Mailing Add:* Dept of Chem San Diego State Univ San Diego CA 92182

MATHEWSON, WILFRED FAIRBANKS, JR, chemical engineering, physical chemistry, for more information see previous edition

MATHEWS-ROTH, MICHELINE MARY, b Mineola, NY, July 26, 34; m 66; c 1. DERMATOLOGY, MICROBIOLOGY. *Educ:* Col St Elizabeth, BS, 56; NY Univ, MD, 61. *Prof Exp:* Intern path, Boston City Hosp, 62-63; res assoc bact, 65-69, assoc bact & immunol, 69-71, assoc microbiol & molecular genetics, 71-74, prin res assoc, 74-83, ASSOC PROF MED, MED SCH, HARVARD UNIV, 83- *Concurrent Pos:* Grants, Med Found, 66-69, NSF, 68-71, NIH, 70, 75-82 & 85-; mem comt environ health & safety, Med Sch, Harvard Univ, 76-; assisting physician med microbiol, Boston City Hosp, 73-74; jr assoc med, Peter Bent Brigham Hosp, 77-82; mem numerous study sects, Nat Cancer Inst, NIH, 80-85; assoc physician, Brigham & Women's Hosp, 82- *Honors & Awards:* Fel Am Soc Clin Invest; Borden Award Med Res, 61. *Mem:* Am Soc Photobiol (pres, 90-91); Am Soc Microbiol; Am Fedn Clin Res; Am Soc Clin Invest; Sigma Xi. *Res:* Photobiology; porphyrias; carotenoid pigments. *Mailing Add:* Channing Lab 180 Longwood Ave Boston MA 02115-5899

MATHIAS, JOSEPH SIMON, b Bombay, India, Oct 28, 25; US citizen; m 56. METALLURGY. *Educ:* Univ Bombay, BS, 44, AB, 46, MS, 48; Univ Calif, MMetE, 51; Lehigh Univ, PhD(metall), 56. *Prof Exp:* Teaching asst physics, Univ Bombay, 44-48; res asst chem, Lehigh Univ, 51-52, res assoc metall, 52-54; chief metallurgist, Superior Metal Corp, 55-56; sr res engr, Jones & Laughlin Steel Corp, 56; group supvr metall, Foote Mineral Co, 56-59; sect mgr mat & processes, 59-66, dept mgr physics & mat, 66-67, dir res, 67-68, dir res & adv techniques, Univac Div, Sperry Rand Corp, 68-79, DIR MFG & HARDWARE RES, SPERRY-UNIVAC, 79- *Concurrent Pos:* Consult, Superior Metal Co, 52-55 & F J Stokes Mach Co, 57-58; lectr, Lehigh Univ, 55-56. *Mem:* Electrochem Soc; Inst Elec & Electronics Engrs; NY Acad Sci. *Res:* Materials and devices for computer memories and peripherals; manufacturing technologies involving electrodeposition, vacuum deposition and sputtering. *Mailing Add:* 105 Thomas Ave Riverton NJ 08077

MATHIAS, LON JAY, b Dec 29, 48; m 66; c 4. POLYMER SYNTHESIS, POLYMER SPECTROSCOPY. *Educ:* Univ Iowa, BS, 71; Univ Mich, MS, 74, PhD(chem), 76. *Prof Exp:* Res fel, Univ Calif-San Diego, 76-77; asst prof, dept chem, Auburn Univ, 77-81; asst prof, 81-83, assoc prof 83-89, PROF, DEPT POLYMER SCI, UNIV SOUTHERN MISS, 89- *Mem:* Am Chem Soc; Sigma Xi. *Res:* High performance polymers and composites; polymerizations in organized media and at interfaces; supernucleophilic pyridine polymer catalysts and reactions; polymeric phase transfer catalysis; macrocyclopolymerizations and Crown Ether polymers; new liquid crystalline polymers and copolymers; polypeptides and polydepsipeptides. *Mailing Add:* Dept Polymer Sci Univ Southern Miss Southern Sta Box 10076 Hattiesburg MS 39406-0076

MATHIAS, MELVIN MERLE, b Columbia City, Ind, Feb 22, 39; m 63; c 3. NUTRITION. *Educ:* Purdue Univ, BS, 61; Cornell Univ, PhD(nutrit), 67. *Prof Exp:* Asst nutrit, Cornell Univ, 62-66; from asst prof to prof nutrit, Colo State Univ, 68-88; PROF NUTRIT, FLA STATE UNIV, 88- *Concurrent Pos:* Fac partic, AEC prog, Donner Lab, Univ Calif, Berkeley, 71; sabbatical, Dept Biochem Nutrit, Hoffmann-La Roche, 74-75 & Dept Human Nutrit, Unilever Res Lab, Netherlands, 81. *Mem:* AAAS; Am Inst Nutrit; Am Oil Chem Soc; Sigma Xi. *Res:* Effects of diet and vitamin deficiencies on intermediary and prostaglandin metabolism. *Mailing Add:* USDA CSRS Aerospace Bldg Rm 329 Washington DC 20250-3444

MATHIAS, MILDRED ESTHER, b Sappington, Mo, Sept 19, 06; m 30; c 4. BOTANY. *Educ:* Wash Univ, AB, 26, MS, 27, PhD(syst bot), 29. *Prof Exp:* Asst, Mo Bot Garden, 29-30; res assoc, NY Bot Garden, 32-36 & Univ Calif, 37-42; herbarium botanist, 47-51, lectr bot, 51-55, from asst prof to prof, 55-74, dir bot garden, 56-74, EMER PROF BOT, UNIV CALIF, LOS ANGELES, 74- *Concurrent Pos:* Asst specialist, Exp Sta, Univ Calif, Los Angeles, 51-55, asst plant systematist, 55-57, vchmn bot dept, 55-66, assoc plant systematist, 57-62; pres, Orgn Trop Studies, 68-70; Secy, Bd Trustees, Inst Ecol, 75-77; exec dir, Am Asn Bot Gardens & Herb Arboreta; pres, Pac Div, AAAS, 77. *Honors & Awards:* Merit Award, Bot Soc Am, 73; Sci Citation, Am Hort Soc, 74; Liberty Hyde Bailey Medal, Am Hort Soc, 80; Medal Hon, Garden Club Am, 82. *Mem:* Fel AAAS; Bot Soc Am (pres, 84); Western Soc Naturalists (pres, 65); Am Soc Plant Taxon (pres, 64); Soc Study Evolution; Am Soc Naturalists; Am Asn Bot Gardens & Arboreta. *Res:* Classification of plants of western United States; monographic studies of the Umbelliferae, especially of North and South America; subtropical ornamental plants; tropical medicinal plants. *Mailing Add:* Dept of Biol Univ of Calif Los Angeles CA 90024-1606

MATHIAS, ROBERT A(DDISON), b Monte Vista, Colo, Jan 14, 27; m 48; c 2. ELECTRICAL ENGINEERING. *Educ:* Univ Colo, BS, 47; Univ Pittsburgh, MS, 49; Carnegie Inst Technol, PhD(elec eng), 55. *Prof Exp:* Engr, Spec Prod Dept, Westinghouse Elec Corp, 47-49; from instr to asst prof elec eng, Carnegie Inst Technol, 51-61; consult, 59-61, adv engr, 61-67, mgr systs simulation & control, 67-75, MGR SYST SCI, WESTINGHOUSE RES LABS, 75- *Concurrent Pos:* NSF fel, 57-59. *Mem:* Sr mem Inst Elec & Electronics Engrs; Inst Mgt Sci; Am Forestry Asn; Am Asn Artificial Intel. *Res:* Magnetic amplifiers; switching and logic control; computer control of real-time systems; systems engineering; planning techniques for urban and social systems; business and technology analysis; industrial strategy planning; image understanding; artificial intelligence. *Mailing Add:* 216 Thornberry Dr Pittsburgh PA 15235

MATHIASON, DENNIS R, b Fairmont, Minn, Feb 6, 41; m 63; c 2. INORGANIC CHEMISTRY, INSTRUMENTAL ANALYSIS. *Educ:* Mankato State Col, BS, 62; Univ SDak, PhD(chem), 66. *Prof Exp:* PROF CHEM, MOORHEAD STATE UNIV, 66- *Mem:* Am Chem Soc; Sigma Xi. *Res:* Analytical studies. *Mailing Add:* Dept Chem Moorhead State Univ 11th St S Moorhead MN 56500

MATHIES, ALLEN WRAY, JR, b Colorado Springs, Colo, Sept 23, 30; m 56; c 2. PEDIATRICS, INFECTIOUS DISEASES. *Educ:* Colo Col, BA, 52; Columbia Univ, MS, 56, PhD(parasitol), 58; Univ Vt, MD, 61. *Prof Exp:* Res assoc path, Col Med, Univ Vt, 57-61, from intern to resident pediat, Los Angeles Co Gen Hosp, 61-64; res assoc, 63-64, from asst prof to prof, 64-71, assoc dean, Sch Med, 70-74, interim dean, 74-75, DEAN SCH MED, UNIV SOUTHERN CALIF, 75-, PROF PEDIAT, 71- *Concurrent Pos:* Head physician commun dis, Los Angeles Co Gen Hosp, 64-75. *Mem:* Am Soc Parasitol; Am Soc Trop Med & Hyg; Soc Pediat Res; Infectious Dis Soc Am; Royal Soc Trop Med & Hyg. *Res:* Infectious diseases; central nervous system infections; tropical medicine. *Mailing Add:* 314 Arroyo Dr South Pasadena CA 91030

MATHIES, JAMES CROSBY, biochemistry; deceased, see previous edition for last biography

MATHIES, MARGARET JEAN, b Colorado Springs, Colo, June 9, 35. MICROBIOLOGY, IMMUNOLOGY. *Educ:* Colo Col, BA, 57; Case Western Reserve Univ, PhD(microbiol), 63. *Prof Exp:* Asst prof biol, Haverford Col, 62-64; vis asst prof zool, Pomona Col, 64-65; from asst prof to assoc prof biol, 65-74, PROF BIOL, JOINT SCI DEPT, CLAREMONT MEN'S, PITZER & SCRIPPS COLS, 74-, CHMN JOINT SCI DEPT, 77- *Mem:* AAAS; Am Soc Microbiol. *Res:* Antibody formation and physicochemical characterization of antibodies, using bacteriophage antigens; cellular immunology, T and B cell interactions; biochemistry; genetics. *Mailing Add:* Joint Sci Dept Claremont McKenna Col 150 E Tenth St Claremont CA 91711

MATHIESON, ALFRED HERMAN, b Union City, NJ, July 6, 17; m 41; c 5. PHYSICS. *Educ:* Pa State Teachers Col, BS, 38; Columbia Univ, MA, 39. *Prof Exp:* Instr physics, Springfield Col, 39-40; from asst prof to assoc prof physics, Univ Mass, 46-84, asst head dept physics & astron, 65-84, spec asst to dean fac nat sci & math, 79-84, EMER PROF, UNIV MASS, AMHERST, 84- *Concurrent Pos:* NSF grant, 64. *Mailing Add:* 285 Shays S Amherst MA 01002

MATHIESON, ARTHUR C, b Los Angeles, Calif, Dec 26, 37; m 58; c 3. BOTANY. *Educ:* Univ Calif, Los Angeles, BA, 60, MA, 61; Univ BC, PhD, 65. *Prof Exp:* From asst prof to assoc prof bot, 65-74, dir, Jackson Estuarine Lab, 72-83, PROF BOT, UNIV NH, 74- *Mem:* Phycol Soc Am; Int Phycol Soc. *Res:* Morphology; distribution and ecology of marine plants in relation to oceanographic factors; coastal processes; ecology; conservation. *Mailing Add:* Dept of Bot Jackson Estuarine Lab Univ of NH Durham NH 03824

MATHIEU, LEO GILLES, b Nicolet, Que, Jan 7, 32; m 56; c 1. BIOCHEMISTRY, MICROBIOLOGY. *Educ:* Univ Montreal, DVM, 56; Cornell Univ, MSc, 58, PhD(nutrit), 60. *Prof Exp:* Asst prof biochem, Col Vet Med, 60-65, asst prof molecular biol, Fac Med, 65-69, assoc prof microbiol, 69-72, PROF MICROBIOL, FAC MED, UNIV MONTREAL, 72- *Concurrent Pos:* Mem exec coun, Grad Sch, Univ Montreal, 71-73; pres, Comt for Med Res & Grad Studies, Fac Med, Univ Montreal. *Mem:* Can Soc Microbiol; Can Vet Med Asn; Am Soc Microbiol. *Res:* Bacterial pathogenicity. *Mailing Add:* Dept of Microbiol Univ of Montreal Fac of Med Montreal PQ H3C 3J7 Can

MATHIEU, RICHARD D(ETWILER), b Trappe, Pa, June 23, 26; m 50; c 3. COMPUTER LITERACY, RESEARCH ADMINISTRATION. *Educ:* Pa State Univ, BS, 52, MS, 54, PhD(aeronaut eng), 61. *Prof Exp:* Instr aeronaut eng, Pa State Univ, 53-61, asst prof, 61; res engr, Space Sci Lab, Missiles & Space Div, Gen Elec Co, 61-63, specialist, 63-64, supvr engr reentry data analysis, Reentry Systs Dept, 64-65; prof aerospace eng & chmn, US Naval Acad, 65-67, sr prof eng, 67-70; liaison scientist, Off Naval Res London, 70-71; dir res, 71-81, dir res & assoc dean, 81-85, VICE ACAD DEAN, US NAVAL ACAD, 85- *Concurrent Pos:* Consult, HRB-Singer, Inc, 60-61 & 65-; adj prof grad ctr, Pa State Univ, 63-65. *Mem:* Am Soc Eng Educ; Sigma Xi; Nat Coun Univ Res Adminr. *Res:* Computers in education; research and development management; educational technology. *Mailing Add:* US Naval Acad Annapolis MD 21402

MATHIEU, ROGER MAURICE, b Montreal, Que, Aug 4, 24; m 49; c 2. RADIOLOGY, PHYSICS. *Educ:* Univ Montreal, BSc, 46, MSc, 48, PhD(physics), 52. *Prof Exp:* Radiation physicist & biophysicist, Montreal Cancer Inst & X-Ray Dept, Hosp Notre Dame, 49-69; radiation physicist & biophysicist, Dept Radiother & Nuclear Med, Maisonneuve-Rosemont Hosp, 69-89; clin prof radiol, Fac Med, Univ Montreal, 70-89; RETIRED. *Honors & Awards:* Croix de Commandeur, France. *Mem:* Fr-Can Soc Radiol; Can Asn Physicists; Can Asn Radiol. *Res:* Radiological physics; biophysics; radiotherapy; nuclear medicine. *Mailing Add:* 3181 Lyall St Montreal PQ H1M 3H2 Can

MATHIPRAKASAM, BALAKRISHNAN, b Virudhunagar, India, Jan 3, 42; US citizen; m 72; c 1. HEAT TRANSFER, THERMOELECTRICS. *Educ:* Virudhunagar Polytech, India, LME, 61; Univ Mysore, ME, 76; Ill Inst Technol, PhD(mech eng), 80. *Prof Exp:* Lectr mech eng, Virudhunagar Polytech, India, 61-68, workshop supt, 68-74; res asst, Ill Inst Technol, 76-79; assoc energy engr, Midwest Res Inst, 80-81, sr energy engr, 81-84, prin engr, 84-89, SR ADV ENG, MIDWEST RES INST, KANSAS CITY, 89- *Concurrent Pos:* Mem adv bd & treas, Int Thermoelec Soc. *Honors & Awards:* Creative Develop Recognition, NASA, 83. *Mem:* Am Soc Mech Engrs; Am Defense Preparedness Asn; Int Thermoelec Soc. *Res:* Heat transfer and thermodynamic analyses of new concepts and processes; energy related research in space conditioning systems and thermal energy storage devices; thermoelectric cooling and heating applications. *Mailing Add:* Midwest Res Inst 425 Volker Blvd Kansas City MO 64110-2399

MATHIS, BILLY JOHN, b Henryetta, Okla, Sept 12, 32; m 57; c 2. LIMNOLOGY. *Educ:* Okla State Univ, BS, 59, MS, 63, PhD(zool), 65. *Prof Exp:* Sci teacher pub schs, Tex, 59-62; from asst prof to assoc prof, 65-72, PROF BIOL, BRADLEY UNIV, 72-, CHMN DEPT, 70- *Mem:* Am Fisheries Soc; Sigma Xi. *Res:* Stream pollution; primary productivity; distribution of heavy metals in aquatic environments. *Mailing Add:* Dept Biol Bradley Univ Peoria IL 61625

MATHIS, JAMES FORREST, b Dallas, Tex, Sept 28, 25; m 48; c 2. CHEMICAL ENGINEERING. *Educ:* Tex A&M Univ, BS, 46; Univ Wis, Madison, MS, 51, PhD(chem eng), 53. *Prof Exp:* Chemist, Humble Oil & Refining, 46-50, mem staff, Exxon Co, USA, 53-61, mgr labs, 61-63, mgr spec prods, 63-65, vpres petrol res, Exxon Res & Eng, 65-68; sr vpres petrochem opers, Imp Oil Ltd, Toronto, 68-71; vpres chem, Exxon Res & Eng, 71-73, vpres technol, Exxon Chem Co, 73-80, vpres sci & technol, Exxon Corp, 80-84; CHMN, NJ STATE COMN SCI & TECHNOL, 88- *Concurrent Pos:* Dir, treas & pres, Chem Indust Inst Toxicol, 74-84; chmn mgt div, Am Inst Chem Engrs, 78-80, dir, 84-86, exec dir, 87-88; mem, Bd Chem Serv Technol, Nat Res Coun, 86-89; Consult, NSF, 85-86, Arthur D Little, Inc, 85-90, Chemshare, 89-; dir, NL Indust, 85-86, Hanlin Corp, 89-, Laser Recording Syst, 89- *Honors & Awards:* Earle B Barnes Award, Am Chem Soc, 84; Robert Jacks Award, Am Inst Chem; Van Antwerpen Award, Am Inst Chem Engrs. *Mem:* Nat Acad Eng; Am Chem Soc; AAAS; Am Inst Chem Engrs; Indust Res Inst; Soc Chem Indust. *Res:* Petrochemical process and product technology; research management. *Mailing Add:* 96 Colt Rd Summit NJ 07901

MATHIS, JAMES L, b Dayton, Tenn, Jan 30, 25; m 48; c 4. PSYCHIATRY. *Educ:* Citadel, 43-44; Univ Mo, 44-45; St Louis Univ, MD, 49; Am Bd Psychiat & Neurol, dipl, 68. *Prof Exp:* Rotating intern, Fitzsimons Gen Hosp, 49-50; resident, Elk City Community Hosp-Clin, Okla, 50-51; gen practr, Crossett Health Ctr, Ark, 51-52; surg asst, Elk City Community Hosp, 52-55; pvt pract, Dayton, Tenn, 55-60; resident psychiat, Med Ctr, Univ Okla, 60-63; from instr to assoc prof, Med Sch, Rutgers Univ, 68-70; prof psychiat & chmn dept, Med Col Va, 70-76; prof & chmn dept psychiat, 76-90, EMER PROF PSYCHIAT, MED SCH, E CAROLINA UNIV, 90- *Concurrent Pos:* Asst chief psychiat serv, Vet Admin, Oklahoma City, 63-64; consult, Peace Corps, 65-69, Job Corps, 68-70 & NJ Correctional Syst, 68-70; asst examr, Am Bd Psychiat & Neurol, 71- *Mem:* Am Psychiat Asn; Am Psychosom Soc; Am Col Psychiat; Am Asn Prof Psychiat. *Res:* Sexuality in medicine; death and dying; drug abuse; sleep and dreams. *Mailing Add:* Dept Psychiat Med Sch E Carolina Univ Greenville NC 27858-4354

MATHIS, JOHN SAMUEL, b Dallas, Tex, Feb 7, 31; m 54; c 5. ASTROPHYSICS. *Educ:* Mass Inst Technol, BS, 53; Calif Inst Technol, PhD(astron), 56. *Prof Exp:* NSF res fel, Yerkes Observ, Chicago, 56-57; asst prof astron, Mich State Univ, 57-59; from asst prof to assoc prof, 59-68, PROF ASTRON, UNIV WIS-MADISON, 68- *Concurrent Pos:* Sr sci awardee, Alexander von Humboldt Found, Ger, 75-76. *Mem:* Int Astron Union; Am Astron Soc. *Res:* Inter- stellar matter. *Mailing Add:* Dept Astron Univ Wis Madison WI 53706

MATHIS, PHILIP MONROE, b Paducah, Ky, June 2, 42; m 64; c 1. BIOLOGY, SCIENCE EDUCATION. *Educ:* Murray State Univ, BS, 64; Mid Tenn State Univ, MS, 67; George Peabody Col, EdS, 71; Univ Ga, EdD(sci educ), 73. *Prof Exp:* Teacher & coord sci, Illmo-Scott City Sch, Mo, 64-67; from instr to assoc prof, 67-84, PROF BIOL, MID TENN STATE UNIV, 84- *Concurrent Pos:* Consult, Sci Manpower, King Personnel, Inc, Tenn, 69-70; Mid Tenn State Univ grant, proj dir, Instructional Develop Proj Biol, 77-81; ed, Soc Col Sci Teachers Publ; mem, Nat Steering Comt, Soc Col Sci Teachers; EERA, Title II Grant, 85. *Mem:* Nat Asn Biol Teachers; Nat Sci Teachers Asn; Soc Col Sci Teachers; Am Genetic Asn; AAAS. *Res:* Lichens as pollution indicators; instructional development in biology; vertebrate karyology; statistical genetics and biometry. *Mailing Add:* Dept Biol Mid Tenn State Univ Murfreesboro TN 37132

MATHIS, ROBERT FLETCHER, b Wheeling, WVa, Jan 22, 46; m 71; c 1. COMPUTER SCIENCE, COMPUTER SOFTWARE SYSTEMS. *Educ:* Ohio State Univ, BSc, 65, MSc, 66, PhD(math), 69. *Prof Exp:* Asst prof comput & info sci, Ohio State Univ, 69-75, asst dean grad sch, 74-75; ASSOC PROF MATH & COMPUT SCI, OLD DOMINION UNIV, 75- *Concurrent Pos:* Consult, Robert Corp & System Develop Corp, 81- *Mem:* Inst Elec & Electronics Engrs; Asn Comput Mach; Sigma Xi. *Res:* software engineering; computer programming languages; realtime and concurrent programming; computer algorithms; functional and numerical analysis. *Mailing Add:* 9712 Ceralene Dr Fairfax VA 22032

MATHIS, RONALD FLOYD, b Los Angeles, Calif, July 26, 42; m 64; c 4. FIBER OPTICS, OPTICAL SIGNAL PROCESSING. *Educ:* Fullerton State Univ, Calif, BA, 66; Univ Mo, Rolla, MS, 71, PhD(physics), 73. *Prof Exp:* Sr physicist, IRT Corp, 74-76; sr engr, Cubic Corp, 76-78; ENG STAFF SPECIALIST, ELECTRONICS DIV, GEN DYNAMICS, 78- *Mem:* Am Phys Soc; Sigma Xi; Optical Soc Am; Soc Photooptical Instrumentation Engrs; Creation Res Soc. *Res:* Experimental atomic physics research and communication theory applied to signal processing; applying photonics to solve signal processing problems; granted 8 patents. *Mailing Add:* 17632 Rancho de Carole Rd Ramona CA 92065

MATHIS, WAYNE NEILSEN, b Price, Utah, July 10, 45; m 70; c 3. DIPTERA, SYSTEMATICS. *Educ:* Brigham Young Univ, BS, 69; Ore State Univ, PhD(entomol), 76. *Prof Exp:* Assoc cur, 76-81, CHMN, ENTOMOL, SMITHSONIAN INST, 81- *Mem:* Entomol Soc Am; Soc Systs Zool; Am Entomol Soc; Great Basin Naturalist. *Res:* Systematics of shore flies (diptera, ephydridae) and other Drosophiloidea families, with emphasis on the Neotropics and Old World. *Mailing Add:* Dept Entom Smithsonian Inst NHB-169 Washington DC 20560

MATHISEN, OLE ALFRED, b Oslo, Norway, Feb 9, 19; nat US; m 48; c 2. POPULATION STUDIES. *Educ:* Univ Oslo, Cand Mag, 41, Cand Real, 45; Univ Wash, PhD, 55. *Prof Exp:* From assoc prof to prof, Fisheries Res Inst, Univ Wash, 64-82; dean, Sch Fisheries & Sci, 83-88, PROF FISHERIES, UNIV ALASKA, JUNEAU, 89- *Concurrent Pos:* Inter-Univ Comt Travel Grants fel, Moscow, 60-61; Fulbright res fel, Oslo, 65-66, USSR Acad Sci, Moscow, 70, Malaysia, 88-89, Vladivostok, 90; consult, Food & Agr Orgn of UN, 73- *Mem:* Am Fisheries Soc; Biomet Soc; Inst Fishery Res Biol; Am Soc Limnol & Oceanog; Sigma Xi. *Res:* Population dynamics, especially of salmonoids; acoustical stock estimation; nekton in upwelling systems; Antarctic krill. *Mailing Add:* PO Box 210443 Auke Bay AK 99821

MATHISON, GARY W(AYNE), b Islay, Alta, Apr 27, 45; m 72; c 4. RUMINANT. *Educ:* Univ Alta, BSc, 67, PhD(animal biochem), 72. *Prof Exp:* Spec lectr animal sci, Univ Sask, 71-72, res assoc, 72; from asst prof to assoc prof, 73-86, PROF ANIMAL SCI, UNIV ALTA, 86- *Mem:* Am Soc Animal Sci. *Res:* Nutrition and digestive physiology of ruminant animals; factors influencing digestive flow, energetics and forage evaluation. *Mailing Add:* Dept Animal Sci Univ Alta Edmonton AB T6G 2P5 Can

MATHISON, IAN WILLIAM, b Liverpool, Eng, Apr 17, 38; m; c 2. ORGANIC & MEDICINAL CHEMISTRY. *Educ:* Univ London, BPharm, 60, PhD(pharmaceut chem), 63, DSc, 76. *Prof Exp:* Res assoc pharmaceut & med chem, Col Pharm, Univ Tenn, Memphis, 63-65, from asst prof to assoc prof med chem, 65-72, prof med chem, Ctr for Health Sci, 72-76; DEAN & PROF MED CHEM, COL PHARM, FERRIS STATE UNIV, 76- *Concurrent Pos:* Prin investr, Marion Labs grant, 65-74 & Beecham Pharmaceut Res grant, 74-79; sr investr grants, NSF, 68-72 & Molecular Design Inc, 81-83; educ consult, 80- *Mem:* Royal Soc Chem; Am Acad Pharmaceut Sci; Brit Pharmaceut Soc; Am Chem Soc; Am Pharmaceut Asn. *Res:* Design and synthesis of organic compounds with potential pharmacodynamic activity; influence of stereochemistry and physicochemical parameters on pharmacological potency. *Mailing Add:* Col Pharm Ferris State Univ 901 South St Big Rapids MI 49307

MATHRE, DONALD EUGENE, b Frankfort, Kans, Jan 5, 38; m 61; c 2. PLANT PATHOLOGY. *Educ:* Iowa State Univ, BS, 60; Univ Calif, Davis, PhD(plant path), 64. *Prof Exp:* Asst prof plant path, Univ Calif, Davis, 64-67; from asst prof to assoc prof, 67-72, PROF PLANT PATH, MONT STATE UNIV, 72- *Mem:* Am Phytopath Soc; Am Soc Agron. *Res:* Soil-borne diseases of cereals and forages. *Mailing Add:* Dept Plant Path Mont State Univ Bozeman MT 59717

MATHRE, OWEN BERTWELL, b Kendall Co, Ill, Nov 26, 29; m 55; c 3. ANALYTICAL CHEMISTRY. *Educ:* Harvard Univ, AB, 51; Univ Minn, PhD(anal chem), 58. *Prof Exp:* Lab helper, Minn Mining & Mfg Co, Minn, 54; res chemist, Electrochem Dept, E I du Pont de Nemours & Co, Inc, Del, 56-58, NY, 58-63, Tenn, 63-65, Del, 65-72, staff chemist, Indust Chem Dept, 72-77, res assoc, 77-81, SR RES ASSOC, DU PONT CHEM DEPT, E I DU PONT DE NEMOURS & CO, INC, 81- *Mem:* Am Chem Soc; Am Soc Testing & Mat; Am Water Works Asn. *Res:* Electrochemistry; instrumental and colorimetric analysis; gas phase catalysis; environmental pollution monitoring; quality control laboratory modernization; gold extraction chemistry. *Mailing Add:* E I DuPont de Nemours & Co 119 Westgate Dr Wilmington DE 19808

MATHSEN, DON VERDEN, b Warren, Minn, Aug 25, 48; m 70; c 4. TECHNICAL MANAGEMENT, SCIENCE POLICY. *Educ:* Univ NDak, BS, 70, MS, 74. *Prof Exp:* Design engr, FMC-Northern Ord, 70-71; adv prog engr, 3M Co, 73-75; vpres, Energy Conserv Systs, 78-79; res engr, Eng Exp Sta, Univ NDak, 75-78, instr mech eng, 75-78, mgr, Eng Exp Sta, 79-81, prin investr, 80-85, dir, 81-85, MGR, ENERGY RES CTR & ADJ PROF, DEPT MECH ENG, UNIV NDAK, 85- *Mem:* Nat Soc Prof Engrs; Am Soc Mech Engrs; Rehab Eng Soc NAm; Am Solar Energy Soc; Nat Water Well Soc. *Res:* Renewable energy systems; rehabilitation technologies; heat pump systems; solar collectors and thermal storage units for solar heating and cooling systems. *Mailing Add:* 1011 19th Ave S Grand Forks ND 58201

MATHSEN, RONALD M, b Minneapolis, Minn, Oct 6, 38; m 62; c 2. MATHEMATICS. *Educ:* Concordia Col, Moorhead, Minn, BA, 60; Univ Nebr, MA, 62, PhD(math), 65. *Prof Exp:* Asst prof math, Concordia Col, Moorhead, Minn, 65-67; fel, Univ Alta, 67-68, asst prof, 68-69; ASSOC PROF MATH, NDAK STATE UNIV, 69- *Concurrent Pos:* Fulbright lectr, Liberia, 73-74. *Mem:* Math Asn Am; Am Math Soc. *Res:* Boundary value problems for ordinary differential equations; generalized convex functions. *Mailing Add:* Dept Math NDak State Univ Sta Fargo ND 58105-5075

MATHUR, CAROLYN FRANCES, b Philadelphia, Pa, Mar 12, 47; m 68; c 2. BIOCHEMISTRY, MICROBIOLOGY. *Educ:* Millersville State Col, BA, 69; Auburn Univ, PhD(biochem), 73. *Prof Exp:* Asst prof biol, Millersville State Col, 74-75; chem, Pa State Univ, Capitol Campus, 76-77; asst prof, 77-80, ASSOC PROF BIOL, YORK COL PA, 80- *Mem:* Am Soc Microbiol; Environ Mutagen Soc. *Res:* Aflatoxin B-1, mode of action, effects on microorganisms, relation to in vitro ageing of human cells, role of dimethyl sulfoxide as a reversing agent. *Mailing Add:* Dept Biol York Col Pa Country Club Rd York PA 17403

MATHUR, DILIP, b Agra, India, Feb 11, 41; m 68; c 2. FISH BIOLOGY. *Educ:* Univ Delhi, India, BSc, 61, MSc, 64; Cornell Univ, MS, 68; Auburn Univ, PhD(fishery mgt), 72. *Prof Exp:* Sr fishery biologist, 67-69, fisheries sect leader, 72-80, PROJ DIR & CHIEF RES OFFICER, ICHTHYOL ASN INC, 80-; MGR ENVIRON SERV, MUDDY RUN LAB, RADIATION MGT CORP, 81- *Mem:* Am Fisheries Soc; Am Inst Fishery Res Biologist. *Res:* Effects of thermal discharges and pumped storage facilities on fishes and fish food organisms; impact of impingement and entrainment of fishes and fish larvae; ecology of fishes; problems related to the operation of hydroelectric stations, particularly instream flow requirements, fish ladders and water quality. *Mailing Add:* Radiation Mgt Corp 1921 River Rd PO Box 10 Drumore PA 17518-0010

MATHUR, MAYA SWARUP, b Amorha, India, July 1, 39; m 64; c 1. EXPERIMENTAL ATOMIC PHYSICS, EXPERIMENTAL MOLECULAR PHYSICS. *Educ:* Univ Allahabad, BSc, 57, MSc, 60, PhD(physics), 69. *Prof Exp:* Res assoc atomic & molecular spectros, dept physics, Univ Man, 76-80, asst prof, dept elec eng, 81-83; vis prof, Univ K, 85-87; CONSULT, 87- *Concurrent Pos:* Sr res fel, Univ Allahabad & res grant, Sigma Xi, 69; fel, Dept Chem, Lakehead Univ, 71-72; fel, Dept Physics, Univ Man, 72-76; vis prof elec eng, Univ Ky, Lexington, 84-87; res grants, Natural Sci & Eng Res Coun, Can, Man Hydro, USAF, Commonwealth Ky Res Bd & Robotic Inst. *Res:* Dielectric and microwave spectroscopy and zero-field-level crossing atomic spectroscopy for the determination of the excited state lifetimes and oscillator strengths; study of collision cross-sections; far infrared, laser Raman and collision-induced light scattering by molecules; inelastic light scattering from surfaces of semiconductors and implanted surfaces and interfaces. *Mailing Add:* Dept Physics Univ Man Winnipeg MB R3T 2N2 Can

MATHUR, PERSHOTTAM PRASAD, b Delhi, India, Jan 19, 38; US citizen; m 72. PHARMACOLOGY. *Educ:* Univ Delhi, BS, 57; Univ Fla, Gainesville, PhD(pharmacol), 68. *Prof Exp:* Res asst, Univ Fla, 63-68; res assoc, Med Col SC, 68-69 & Univ Ga, 69-70; head clin chemist & sr pharmacologist, Dept Path, St Barnabas Hosp, 72, group leader, 73-75, sect head biochem pharmacol, William H Rorrer, Inc Res Div, 75-76; group mgr cardiovasc & automatic pharmacol & spec proj, 77-80, sr prog coordr, 80-82, SCI COORDR CLIN PHARMACOL, A H ROBINS CO, 82- *Concurrent Pos:* NIH fel cardiol, St Luke's Hosp Ctr, NY, 69-72; adj asst prof pharmacol, Med Col Va, 77- *Mem:* Fel Am Col Clin Pharmacol; Am Soc Pharmacol & Exp Therapeut; Am Fedn Clin Res; Am Heart Asn. *Res:* Clinical research - phase I to phase IV development of investigational drug entities, cardiovascular pharmacology with emphasis on myocardial ischemia; coronary blood flow, anti-arrhythmics, anti-hypertensives and adrenergic agents; biochemical mechanisms of drug action, drug metabolism, microsomal mixed function oxidases, adjuvant disease and anti-inflammatory agents. *Mailing Add:* Res Div A H Robins Co 1211 Sherwood Ave Richmond VA 23220

MATHUR, R(ADHEY) M(OHAN), b Alwar, India, Feb 2, 36; m 65; c 2. ELECTRICAL ENGINEERING, ELECTROMAGNETICS. *Educ:* Univ Rajasthan, BSc, 56; Indian Inst Technol, Kharagpur, BTech, 60; Univ Leeds, PhD(elec eng), 69. *Prof Exp:* Lectr elec eng, Univ Jodhpur, 60-64, Malaviya Regional Eng Col, 64-65 & Indian Inst Technol, New Delhi, 65-66; Nat Res Coun Can fel, Univ Man, 69-70, from asst prof to prof, 70-87, head elec eng, 80-87; PROF & DEAN ENG SCI, UNIV WESTERN ONT, 87- *Concurrent Pos:* Nat Res Coun Can res grant & fel, Univ Man, 71-72. *Honors & Awards:* Indian Inst Eng Prize, 64; Centennial Award, Inst Elec & Electronics Engrs, 84. *Mem:* Int Conference on Large High Voltage Elec Systs; Can Elec Asn; Inst Elec & Electronics Engrs. *Res:* Rotating machines; reluctance and stepper motors, transient and steady state performance and design optimization; power systems modeling; HVDC systems; static compensators; power systems. *Mailing Add:* Fac Eng Sci Univ Western Ont London ON N6A 5B9 Can

MATHUR, SUBBI, INFERTILITY, SPERM ANTIBODIES. *Educ:* Univ Madras, India, PhD(microbiol), 67. *Prof Exp:* ASSOC PROF BASIC & CLIN IMMUNOL & MICROBIOL & ASSOC PROF OBSTET, MED UNIV SC, 81- *Mailing Add:* Dept Basic & Clin Immunol & Microbiol Rm 637A Med Univ SC 171 Ashley Ave Charleston SC 29407

MATHUR, SUKHDEV PRASHAD, b Ajmer, Rajasthan, India, Dec 28, 34; Can citizen; m 60; c 3. SOIL SCIENCE, MICROBIOL BIOCHEMISTRY. *Educ:* Univ Delhi, BSc, 57; Ind Agr Res Inst, Assoc, 59; Univ Sask, PhD(soil microbiol), 66. *Prof Exp:* Dist supvr plant protection, Govt Rajasthan, India, 57; lectr chem, Dayanand Agr Col, Ajmer, India, 59-62; RES SCIENTIST SOIL BIOCHEM, RES BR, AGR CAN, 66- *Concurrent Pos:* Int Dev Composts Peat Fisheries Wastes. *Mem:* Can Soc Soil Sci; Soil Sci Soc Am; Int Soc Soil Sci; Am Soc Microbiol; Int Peat Soc. *Res:* Nature and behaviour of humus and organic soils, peats, mucks; soil pesticides and soil pollutants; mitigating decomposition and subsidence of organic soils and classification of organic deposits and terrains; composting; rock phosphate; fisheries wastes; liquid manures. *Mailing Add:* 75 Foxleigh Cres Kanata ON K2M 1B6 Can

MATHUR, SURESH CHANDRA, b Fatehgarh, India, Mar 23, 30; m 63; c 1. NUCLEAR PHYSICS. *Educ:* Univ Lucknow, BS, 48, MS, 50; Univ Tex, PhD(physics), 65. *Prof Exp:* Asst physicist, Dept Atomic Energy, Govt India, 50-58; sr res scientist, Tex Nuclear Corp, 62-67; PROF PHYSICS, UNIV LOWELL, 67-, DIR COMPUT CTR, 71- *Mem:* Am Phys Soc. *Res:* Nuclear radiation detection techniques and instrumentation; nuclear scattering theory and experiments; nuclear particle accelerators; computer programming. *Mailing Add:* Comput Ctr Univ Lowell Lowell MA 01854

MATHUR, VIRENDRA KUMAR, US citizen. OXYGEN SEPARATION FROM AIR, NOX REMOVAL FROM FLUE GASES. *Educ:* Agra Univ, India, BS, 49; Banaras H Univ, India, BS, 53; Univ Mo, Rolla, MS, 61, PhD(chem eng), 70. *Prof Exp:* Scientist, Fuel Res Inst, India, 53-56; from asst prof to prof chem eng, Banaras H Univ, India, 56-74, chmn dept, 72-74; asst prof, Okla State Univ, Stillwater, 70-72; PROF CHEM ENG, UNIV NH, DURHAM, 74- *Concurrent Pos:* Vis prof, Pittsburgh Energy Technol Ctr, US Dept Energy, 78, 80 & 82 & Solar Energy Res Inst, Golden, Colo, 84 & 88; prin investr, US Dept Energy, USAF & var pvt industs; consult, Aeta Corp, Riley Stoker & Tecogen; mem, Energy Comn, Nat Asn State Univ & Land Grant Cols, Wash, DC, 80-85; chmn, Solar Energy Div, Am Inst Chem Engrs. *Mem:* Am Inst Chem Engrs; Am Chem Soc; Am Soc Eng Educ; Tech Asn Pulp & Paper Indust; Nat Asn State Univ & Land Grant Cols; Air & Waste Mgt Asn. *Res:* Coal liquefaction; circulating fluidized beds; NOx removal from flue gases; oxygen separation from air; fuel and chemical production using solar energy; mass and heat transfer; energy and air pollution control; mass transfer; 1ossil fuels; environmental pollution control. *Mailing Add:* Dept Chem Eng Kingsbury Hall Univ NH Durham NH 03824-3591

MATHUR, VISHNU SAHAI, b Asansol, India, Apr 28, 34; m 61; c 3. THEORIES STRONG & ELECTRO-WEAK INTERACTIONS. *Educ:* Delhi Univ, India, BSc, 53, MSc, 55, PhD(physics), 58. *Prof Exp:* Reader, Ctr Advan Studies Theoret Physics & Astrophysics, Delhi Univ, 63-65; vis sr res assoc, 65-68, SR RES ASSOC, UNIV ROCHESTER, 68-, PROF PHYSICS, 70- *Mem:* Am Phys Soc. *Res:* Theoretical particle physics; symmetry principles and group theory; author or co-author of over 100 publications. *Mailing Add:* 45 Hampshire Dr Rochester NY 14618

MATHYS, PETER, b Zurich, Switz, Nov 12, 50. DIGITAL COMMUNICATIONS, CRYPTOGRAPHY & CODING THEORY. *Educ:* Swiss Fed Inst Technol, Zurich, dipl elec eng, 76, PhD(elec eng), 85. *Prof Exp:* Res assoc appl physics, Swiss Fed Inst Technol, Zurich, 77-82, res assoc elec eng, 82-84; vis res scientist, Mass Inst Technol, 85-86; ASST PROF ELEC ENG, UNIV COLO, BOULDER, 86- *Concurrent Pos:* NSF presidential young investr award, 90. *Honors & Awards:* W R G Baker Prize, Inst Elec & Electronics Engrs, 87. *Mem:* Inst Elec & Electronics Engrs. *Res:* Multi-user information theory and coding; data communication networks; communication theory; coding and modulation for magnetical and optical storage media; analysis of algorithms; cryptography; data security. *Mailing Add:* Dept Elec & Computer Eng Univ Colo Boulder CO 80309-0425

MATICK, RICHARD EDWARD, b Pittsburgh, Pa, Nov 25, 33; m 62; c 1. ELECTRICAL ENGINEERING. *Educ:* Carnegie Inst Technol, BS, 55, MS, 56, PhD(elec eng), 58. *Prof Exp:* Staff engr memory res, 58-66, mem tech staff dir res, 66-70, tech asst dir res, 70-80, SR STAFF ENGR, IBM RES CTR, 80- *Mem:* Inst Elec & Electronics Engrs. *Res:* Direct current corona fields; high speed memory devices and systems; ferroelectric-ferromagnetic materials; thin magnetic films; read only memories. *Mailing Add:* Watson Res Ctr Div IBM Corp PO Box 218 Yorktown Heights NY 10598

MATIENZO, LUIS J, b Lima, Peru, Jan 30, 44; US citizen; m 70; c 3. APPLIED SURFACE ANALYSIS, MATERIALS IN MICROELECTRONIC PACKAGING. *Educ:* San Marcos Univ, Peru, BS, 67; Mich State Univ, MS, 69; Univ Md, PhD(chem), 73. *Prof Exp:* Postdoctoral fel, Dept Chem, Case Western Res Univ, 73-74; res scientist, Exp Sta, E I Du Pont de Nemours & Co, Inc, Wilmington, Del, 74-79; sr scientist, Martin Marietta Labs, Baltimore, Md, 79-85; ADV ENGR, SYSTS TECHNOL DIV, IBM CORP, ENDICOTT, NY, 85- *Concurrent Pos:* Rep mem finishes comt, Aluminum Asn Am, 79-82; vis prof, Grad Sch, Nat Sch Eng, Lima, Peru, 78; lectr, dept mat sci & eng, Cornell Univ, 89- *Mem:* Am Inst Physics; Am Vacuum Soc; Sigma Xi. *Res:* Surface science applied to materials science: metal-metal, polymer-metal and polymer-polymer surface interactions; adhesion science; surface modification reactions; ion beam effects on solids; high energy ion beam analysis; two patents, three book chapters and 85 publications. *Mailing Add:* 1211 Cafferty Hill Rd Endicott NY 13760

MATIJEVIC, EGON, b Otocac, Yugoslavia, Apr 27, 22; nat US; m 47. PHYSICAL CHEMISTRY, COLLOID CHEMISTRY. *Educ:* Univ Zagreb, dipl, 44, PhD(chem), 48, Dr habil, 52. *Hon Degrees:* DSc, Lehigh Univ, 77, Maria Curie-Sklodowska Univ, Lublin, Poland, 90. *Prof Exp:* Instr chem, Fac Pharm & Sci, Univ Zagreb, Yugoslavia, 44-47, sr instr phys chem, Fac Sci, 49-52, teacher colloid chem, 52-54 & docent phys & colloid chem, 55-56; res assoc, Res Dept Inst Cinematog, Zagreb, 48; res assoc, 57-59, assoc prof chem, 60-62, assoc dir, Inst Colloid & Surface Sci, 66-68, dir, 68-81, chmn dept, 81-87, PROF CHEM, CLARKSON UNIV, 62-, DISTINGUISHED PROF, 87- *Concurrent Pos:* Asst ed, Croatica Chemica Acta, 53-57; consult, Coulter Corp, Fla, Dow Chem Co, Mich, Int Fluid Cell, Conn, Montedison, Italy, Shipley Co, Mass, XMX Corp, Burlington, Mass & Japan Synthetic Rubber Co, Tsukuba, Japan; res contracts & grants, NSF, USAF, JSR, Japan, Showa-Denko, Japan & XMX Corp; referee, Nato Advan Study Inst; res fel, Dept Colloid Sci, Cambridge Univ, Eng, 56-57; vis prof, Japan Soc Prom Sci, 73, Univ Melbourne, Australia, 76 & Sci Univ Tokyo, Japan, 79; vis scientist, Univ Leningrad, 77; guest lectr, Fed Polytech Inst, Lausanne, Switz, 84. *Honors & Awards:* Kendall Award, Am Chem Soc, 72, Langmuir lectr, 85; Welch Found lectr, Tex, 83; Thomas-Graham Award, Colloid Soc, Berlin, Germany, 85. *Mem:* Am Chem Soc; Colloid Soc Germany; Croatian Chem Soc; Am Water Works Asn; hon mem Int Asn Colloid & Interface Scientists (pres-elect, 81-83, pres, 84-86); Sigma Xi; Chem Soc Japan; foreign mem Yugoslavia Acad Arts & Sci; hon mem Am Ceramic Soc. *Res:* Precipitation processes; coagulation; metal corrosion; photogalvanic phenomena; complex ionic species; heteropoly compounds; ionized monolayers; light scattering; aerosols; monodispersed inorganic and polymer colloids; particle adhesion. *Mailing Add:* Dept Chem Clarkson Univ Potsdam NY 13699

MATILSKY, TERRY ALLEN, b Brooklyn, NY, Mar 29, 47; m 73. ASTROPHYSICS. *Educ:* Univ Mich, BS, 67; Princeton Univ, AM, 69, PhD(astrophys sci), 71. *Prof Exp:* Sr scientist x-ray astron, 71-73, proj scientist, Uhuru Satellite, Am Sci & Eng, 73; res staff x-ray astron, Mass Inst Technol, 74-76; asst prof, 76-81, ASSOC PROF PHYSICS & ASTRON, RUTGERS UNIV, 81- *Concurrent Pos:* Consult, Smithsonian Inst, 78-; prin investr, NASA grant, 77-82; co-investr, NASA satellite proj, 74- *Mem:* Am Astron Soc; AAAS. *Res:* X-ray astrophysics; stellar atmospheres of hot stars. *Mailing Add:* Dept of Physics Rutgers Univ New Brunswick NJ 08903

MATIN, SHAIKH BADARUL, b Agra, India, Feb 21, 44. PHARMACEUTICAL CHEMISTRY, CLINICAL PHARMACOLOGY. *Educ:* Univ Karachi, BSc, 63; Columbia Univ, MS, 65; Univ Calif, PhD(pharmaceut chem), 70. *Prof Exp:* Fel chem, Univ Calif, San Francisco, 70-74; RES SCIENTIST, SYNTEX RES, 74- *Mem:* Am Chem Soc; Am Pharmaceut Asn; Am Soc Mass Spectros; NY Acad Sci. *Res:* Investigation of the pharmacologic profile and time course of action, interaction and mechanism of action of synthetic drugs and naturally occurring compounds on animals and man. *Mailing Add:* 32820 Regents Blvd Union City CA 94587

MATIS, JAMES HENRY, b Chicago, Ill, Mar 3, 41; m 63; c 4. STATISTICS, MATHEMATICAL STATISTICS. *Educ:* Weber State Col, BS, 65; Brigham Young Univ, MS, 67; Tex A&M Univ, PhD(statist), 70. *Prof Exp:* Math statistician, Intermountain Forest & Range Exp Sta, US Forest Serv, 65-67; res assoc statist, 70, from asst prof to assoc prof, 70-79, PROF STATIST, TEX A&M UNIV, 79- *Concurrent Pos:* NIH res career develop award, 74-79; Indo-Am res fel, 84. *Mem:* Fel Am Statist Asn; Biomet Soc; Soc Math Biol; Sigma Xi; AAAS; Int Statist Inst. *Res:* Applied stochastic processes; compartmental analysis; statistical ecology. *Mailing Add:* Dept Statist Tex A&M Univ College Station TX 77843

MATISOFF, GERALD, b Boston, Mass, Apr 27, 51; m 74; c 2. ZEBRA MUSSEL CONTROL, ENVIRONMENTAL SCIENCES. *Educ:* Mass Inst Technol, SB, 73; Johns Hopkins Univ, MA, 75, PhD(geochem), 78. *Prof Exp:* asst prof Earth Sci, 77-83, ASSOC PROF GEOL SCI, CASE WESTERN RESERVE UNIV, 83- *Concurrent Pos:* Consult, Madison & Madison Int, Ecotech, Inc, 81-82 & Cities Serv Oil & Gas, 85; assoc ed, J Great Lake Res, 81-; vis assoc prof geol sci, Cleveland State Univ, 85; ICAIR-Life Systems, 87-; pres, Geoscience Associates Inc, 87-; Fenkberner, Pettis & Strout, 90; adj assoc prof geol sci, Oberlin Col, 91; NSF IL1 Panel, 89; Tech Comt Cuyahoga Remedial Action Plan, 89. *Mem:* Geochem Soc; Am Soc Limnol & Oceanog; Am Geophys Union; Int Asn Great Lakes Res; Am Asn Petrol Geologists; Sigma Xi; Asn Groundwater Scientists & Engrs. *Res:* Redox reactions in groundwater; hydrogeochemistry; lake-groundwater; early diagenetic reactions and chemical fluxes; bioturbation; chemical weathering; solid-aqueous solution interactions; chemical cycles on a global scale; chemical mass balances; deep burial clastic diagenesis; zebra mussel veliger control. *Mailing Add:* Dept Geol Sci Case Western Reserve Univ Cleveland OH 44106

MATJEKA, EDWARD RAY, b San Antonio, Tex, Jan 3, 43; m 71. ORGANIC CHEMISTRY. *Educ:* St Mary's Univ, Tex, BS, 65; Iowa State Univ, PhD(org chem), 74. *Prof Exp:* Proj officer phys sci, US Army Watervliet Arsenal, 70-72; instr chem, Bowling Green State Univ, 73-74; fel org chem, Univ Mass, Amherst, 74-76; asst prof, 76-80, PROF CHEM, BOISE STATE UNIV, 80- *Mem:* Am Chem Soc; Sigma Xi. *Res:* Chemistry of natural products; organometallic reagents, cyclopentadienes. *Mailing Add:* Dept of Chem Boise State Univ Boise ID 83725

MATKIN, ORIS ARTHUR, b Powell, Wyo, Jan 14, 17; m 42; c 3. HORTICULTURE. *Educ:* Univ Calif, Los Angeles, BA, 40. *Prof Exp:* OWNER, SOIL & PLANT LAB, INC, 46- *Honors & Awards:* Res Award, Calif Asn Nurserymen, 74. *Mem:* Am Soc Hort Sci; Soil Sci Soc Am; Int Plant Propagators Soc. *Res:* Soil, plant, water and pathology analyses. *Mailing Add:* Soil & Plant Lab Inc PO Box 6566 Orange CA 92613-6566

MATKOVICH, VLADO IVAN, b Vrboska, Yugoslavia, Feb 17, 24; nat US; m 51; c 2. ENGINEERING & MATERIALS SCIENCE. *Educ:* Univ Zagreb, dipl, 51; Univ Toronto, PhD, 56. *Prof Exp:* Supvr, Aluminum Labs, Ltd, Can, 55-57; eng scientist, Allis-Chalmers Mfg Co, 57-61; res assoc, Carborundum Co, NY, 61-69, proj mgr, Res Br, 69-75, mgr tech br, 75-78; VPRES ENG, PALL BIOMED PROD CORP, 78- *Mem:* Am Chem Soc. *Res:* Crystal chemistry; synthesis and development of high temperature materials; plastics engineering development; plant design and construction; filtration and filter design. *Mailing Add:* Biomed Res & Develop Pall Corp 77 Crescent Beach Rd Glen Cove NY 11542

MATKOWSKY, BERNARD J, b New York, NY, Aug 19, 39; m 65; c 3. COMBUSTION, STOCHASTIC DIFFERENTIAL EQUATIONS. *Educ:* City Col New York, BS, 60; New York Univ, MEE, 61, MS, 63, PhD(math), 66. *Prof Exp:* Prof math, Rensselaer Polytech Inst, 66-77; PROF APPL MATH & PROF MATH, NORTHWESTERN UNIV, 77-, JOHN EVANS CHAIR APPL MATH, 90- *Concurrent Pos:* Fulbright-Hayes fel, US Govt, 72-73; vis prof, Tel-Aviv Univ, 72-73, 76 & 80, Weizmann Inst Sci, 76 & 80; consult, Argonne Nat Lab, 78- & Sandia Nat Lab & Exxon Res & Eng Corp, 80-, Lawrence Livermore Nat lab, 85-; Guggenheim fel, 82-83; ed, SIAM J Appl Math, 76-78, assoc managing ed, 78-, ed, Wave Motion, 79-, Appl Math Lett, 87-, Europ J Appl Math, 90-, Random & Computational Dynamics, 91- *Mem:* Soc Indust & Appl Math; Am Math Soc; AAAS; Combustion Inst; Am Phys Soc; Am Inst Mech. *Res:* Asymptotic and perturbation methods for ordinary and partial differential equations; nonlinear stability and bifurcation theory; stochastic differential equations; applications to fluid dynamics, elasticity, combustion theory and flame propagation; pattern formation and nonlinear dynamics. *Mailing Add:* Dept Eng Sci & Appl Math Technol Inst Northwestern Univ Evanston IL 60208

MATLACK, ALBERT SHELTON, b Washington, DC, Aug 14, 23; m 53; c 2. POLYMER CHEMISTRY, SYNTHETIC ORGANIC & NATURAL PRODUCTS CHEMISTRY. *Educ:* Univ Va, BS, 44; Univ Minn, PhD(org chem), 50. *Prof Exp:* RES SCIENTIST, HERCULES RES CTR, HERCULES INC, 50- *Mem:* Am Chem Soc; AAAS; Sigma Xi. *Res:* Synthesis and polymerization of monomers; preparation of additives for polymers; Ziegler-Natta catalysis; metathesis polymerization. *Mailing Add:* 3751 Mill Creek Rd Hockessin DE 19707

MATLACK, GEORGE MILLER, b Pittsburgh, Pa, June 14, 21; m 43; c 4. RADIOCHEMISTRY. *Educ:* Grinnell Col, AB, 43; Univ Iowa, MS, 46, PhD(chem), 49. *Prof Exp:* Chemist, Iowa Geol Surv, 43-46; asst chem, Univ Iowa, 46-47, res assoc, 47-49; mem staff, 49-82, assoc group leader, 83-88, SR ADVISOR, LOS ALAMOS NAT LAB, UNIV CALIF, 88- *Mem:* Fel AAAS; fel Am Inst Chem; Am Chem Soc; NY Acad Sci; Am Nuclear Soc. *Res:* Radiochemistry of plutonium and fission products; radiation properties of plutonium-238 fuels and environmental effects. *Mailing Add:* 254 San Juan Los Alamos NM 87544

MATLIN, ALBERT R, b Glen Cove, NY, May 14, 55. ORGANIC PHOTOCHEMISTRY, MECHANISTIC ENZYMOLOGY. *Educ:* Bard Col, AB, 77; Yale Univ, MS & MPhil, 79, PhD(chem), 82. *Prof Exp:* Postdoctoral fel, Rockefeller Univ, 82-84; asst prof, 84-90, ASSOC PROF CHEM, OBERLIN COL, 90- *Concurrent Pos:* Vis prof, Rockefeller Univ, 88-89. *Mem:* Am Chem Soc; AAAS; InterAm Photochem Soc. *Res:* Photochemical cycloaddition reactions; enzymatic structure-function studies. *Mailing Add:* Dept Chem Oberlin Col Oberlin OH 44074

MATLIS, EBEN, b Pittsburgh, Pa, Aug 28, 23; m 42; c 6. MATHEMATICS. *Educ:* Univ Pittsburgh, BS, 48; Univ Chicago, MS, 56, PhD(algebra), 58. *Prof Exp:* From instr to assoc prof, 58-67, PROF MATH, COL ARTS & SCI, NORTHWESTERN UNIV, ILL, 67- *Concurrent Pos:* Mem, Inst Advan Study, 62-63. *Mem:* Am Math Soc. *Res:* Homological algebra; theory of rings and modules. *Mailing Add:* Dept Math Northwestern Univ Evanston IL 60208

MATLOCK, DANIEL BUDD, b Seattle, Wash, Aug 6, 47; m 75; c 2. SOMATIC POLYPLOIDY, MOLECULAR EVOLUTION. *Educ:* Univ Calif, Davis, BS, 69; Ore State Univ, MS, 74, PhD(zool), 78. *Prof Exp:* Instr zool, Ore State Univ, 76; instr biol, Cent Ore Community Col, 77-78; asst prof, Univ Guam, 78-82; assoc prof, Southern Utah State Col, 82-84; ASSOC PROF BIOL, SEATTLE UNIV, 84- *Concurrent Pos:* Proj dir, Instrnl Sci Equip Prog grant, NSF, 79-82; co-prin investr, Contract US Navy, 79-80 & Sea Grant, Univ Hawaii, 80-81; prin investr, Hatch grant, USDA, 80-83; proj dir, Col Sci Inst Prog, NSF, 85-87. *Mem:* Sigma Xi; Am Soc Zoologists; Soc Study Evolution; AAAS. *Res:* Cytophotometric and radiographic studies of gene amplification and the development of somatic polyploidy, including the effect of molting hormone on DNA synthesis in polyploid nuclei; biochemical population genetics of tropical island species. *Mailing Add:* Dept Biol Seattle Univ Seattle WA 98122

MATLOCK, GIBB B, b Tom Green Co, Tex, Nov 27, 31; m 57; c 4. SYSTEMS ENGINEERING, MATHEMATICS. *Educ:* Univ Tex, Austin, BA, 57, MS, 59; Southern Methodist Univ, PhD(math statist), 70. *Prof Exp:* Res scientist math, Military Physics Res Lab, Balcomes Res Ctr, 58-59; electron engr, LTV Corp, 59-62; SR MEM TECH STAFF, SYST ENG, TEX INSTRUMENTS, INC, 62- *Mem:* Inst Elec & Electron Engrs; Am Statist Asn; Am Inst Navig. *Mailing Add:* Tex Instruments Inc PO Box 405 Mailstop 3451 Lewisville TX 75067

MATLOCK, HUDSON, b Floresville, Tex, Dec 9, 19; m 42; c 2. FOUNDATION ENGINEERING. *Educ:* Univ Tex, BS, 47, MS, 50. *Prof Exp:* From instr to prof civil eng, Univ Tex, Austin, 48-78, chmn dept, 72-77; vpres res & develop, Earth Tech Corp, 78-85; CONSULT, 85- *Honors & Awards:* James J R Croes Medal, Am Soc Civil Engrs, 68; Offshore Technol Conf Distinguished Achievement Award, 85. *Mem:* Nat Acad Eng; Am Soc Civil Engrs. *Res:* Soil mechanics and foundations; materials of engineering; experimental stress analysis. *Mailing Add:* HCR5 Box 574-655 Kerrville TX 78028

MATLOCK, JOHN HUDSON, b San Angelo, Tex, Nov 23, 44; m 66; c 2. ELECTRONIC MATERIALS, MANUFACTURING, MANAGEMENT. *Educ:* Univ Tex, BES, 67, MS, 69, PhD(mat sci & eng), 70; Southern Ill Univ, MBA, 76. *Prof Exp:* Sr engr, Monsanto Co, St Peters, Mo, 70-71, res specialist, 71-74, supt tech serv, 74-79; sr staff eng, Mostek Corp, Carrollton, Tex, 79-80, mgr mat tech, 80-83; vpres tech, 83-90, EXEC VPRES, S E H AM, VANCOUVER, WASH, 90- *Concurrent Pos:* Adj asst prof physics, Southern Ill Univ, Edwardsville, Ill, 73-76; adj lectr, Wash State Univ, Pullman, Wash, 85; adj prof mech eng & mem grad fac, Ore State Univ, Corvallis, Ore, 85-90; mem, Eng Col Vis Comt, Univ Wash, 85-, Eng & Archit Col Adv Bd, Wash State Univ, 84-, bd dirs, Wash H Educ Telecommun Syst, 85-90 & Wash Tech Ctr, Peer Rev Comt, 88-91. *Mem:* Am Soc Metals; Metall Soc; Electrochem Soc; Mat Res Soc; Am Soc Testing & Mat. *Res:* Silicon material: silicon crystal growth (float zone and czochralski), silicon wafer processing, silicon defects and their effect on integrated circuit performance; published over 40 papers in scientific journals, trade journals and conference proceedings on these topics. *Mailing Add:* 10916 NE 30th Ave Vancouver WA 98686

MATLOCK, REX LEON, b Plain Dealing, La, Nov 27, 34; m 55; c 3. PHYSICS. *Educ:* Northwestern State Univ, BS, 60; La State Univ, Baton Rouge, MS, 65, PhD(physics), 67. *Prof Exp:* Asst prof, 67-70, assoc prof physics, 70-77, CHMN DEPT PHYSICS, LA STATE UNIV, SHREVEPORT, 69-, PROF, 75- *Concurrent Pos:* Grant, La State Univ, Shreveport, 70-71. *Mem:* Am Phys Soc. *Res:* High energy interactions and cosmic ray physics. *Mailing Add:* Dept Physics La State Univ 8515 Youree Dr Shreveport LA 71115

MATLOW, SHELDON LEO, b Chicago, Ill, Aug 24, 28; m 58; c 3. CHEMICAL PHYSICS. *Educ:* Univ Chicago, PhB, 48, BS, 49, PhD(chem), 53. *Prof Exp:* Asst chem, Univ Chicago, 50-52; res assoc, Brookhaven Nat Lab, 53-54; dir chem res, Jefferson Elec Co, Ill, 55; sr physicist, Hoffman Electronics Corp, 57, unit supvr, 57-59, sect mgr, 59, tech coordr, 60; dir res & develop, Intellux, Inc, Calif, 61; pres, Inst Study Solid State, 62; sr scientist, Korad Corp, 63; mgr develop eng, Clevite Corp, 64; consult, 65-75; pres, Parodox Chems, Inc, 75-79; dir polymer res, Southwall Corp, 79-81; pres, Conductimer Corp, 81-83, chmn, 83-85, dir, 81-85; chmn & chief exec officer, Ben-Nace Ltd, 85-90; CHMN & CHIEF EXEC OFFICER, NACE TECHNOL INC, 85- *Concurrent Pos:* Vis scholar, Stanford Univ, 80-82; adj prof, San Jose State Univ, 89- *Mem:* Am Chem Soc; Am Phys Soc; Electrochem Soc; NY Acad Sci. *Res:* Quantum mechanics; solid state physics; materials science and technology; philosophy of science. *Mailing Add:* 2545 Booksin Ave San Jose CA 95125

MATNEY, THOMAS STULL, b Kansas City, Mo, Sept 21, 28; m 54; c 3. BACTERIOLOGY. *Educ:* Trinity Univ, BS, 48, BA, 49, MA, 51; Univ Tex, PhD(bact), 58. *Prof Exp:* Asst res biochemist, Southwest Res Inst, 50-52; med bacteriologist, Res & Develop Lab, US Dept Army, 52-55; assoc biologist & assoc prof biol, Univ Tex M D Anderson Hosp & Tumor Inst, 62-69; assoc prof, 63-70, assoc dean, 70-78, PROF MED GENETICS, UNIV TEX GRAD SCH BIOMED SCI, HOUSTON, 70-, PROF ENVIRON SCI, UNIV TEX SCH PUB HEALTH , HOUSTON, 83- *Concurrent Pos:* Instr, Trinity Univ (Tex), 50-51. *Honors & Awards:* Res Award, Am Soc Hosp Pharmacists, 83. *Mem:* Am Soc Microbiol; Am Acad Microbiol; Environ Mutagen Soc; Sigma Xi. *Res:* Bacterial genetics; radiobiology; biochemistry. *Mailing Add:* Univ Tex Grad Sch Biomed Sci PO Box 20334 Houston TX 77225

MATOCHA, CHARLES K, b Hondo, Tex, Aug 13, 29; m 53; c 2. SPECTROCHEMISTRY. *Educ:* St Marys Univ, Tex, BS, 49. *Prof Exp:* From anal chemist to group leader, Aluminum Co Am, Tex & Pa, 49-73, SCI ASSOC CHEM, ALCOA TECH CTR, ALUMINUM CO AM, 73- *Mem:* Sigma Xi; Soc Appl Spectros. *Res:* Analytical methods for x-ray fluorescence analysis with emphasis on nonmetallic samples; automation of analytical procedures; development of computer systems for mathematical correlation, data handling and automation; chemical analytic automation. *Mailing Add:* 428 Glenview Dr Lower Burrell PA 15068

MATOLYAK, JOHN, b Johnstown, Pa, June 26, 39; m 63; c 2. MAGNETISM, SEMICONDUCTOR MATERIALS. *Educ:* St Francis Col, Pa, BS, 63; Univ Toledo, MS, 66; WVa Univ, PhD(physics), 75. *Prof Exp:* Instr math & physics, St Francis Col, Pa, 63-64; PROF PHYSICS, IND UNIV, PA, 66- *Concurrent Pos:* Consult, Army Night Vision & Electro-optics Lab, 85, NASA, 81-82. *Mem:* Am Phys Soc; Sigma Xi. *Res:* Magnetic and electric properties of solids, particularly the susceptibility, transport properties and magnetostriction of antiferromagnets, amorphous solids and semiconductors. *Mailing Add:* Dept of Physics Ind Univ of Pa Indiana PA 15705

MATON, PAUL NICHOLAS, b Malta Apr 7, 47; Brit citizen; m; c 1. GASTROINTESTINAL PHYSIOLOGY, GASTROINTESTINAL SMOOTH MUSCLE. *Educ:* London Univ, UK, BSc, 68, MBBS, 71, MSc, 78, MD, 82. *Prof Exp:* Sr registrar med, gastroenterol, Royal Postgrad Med Sch, 79-83; vis scientist digestive dis, 83-85, sr investr, 85-89, chief, Clin Invest Digestive Dis Br, 89-90; ASSOC MED DIR, OKLA FOUND DIGESTIVE RES, 90- *Mem:* Am Gastroenterol Soc; Brit Soc Gastroenterol. *Res:* Gut peptides; receptors on gastrointestinal smooth muscle cells; clinical research of gut endocrine tumors. *Mailing Add:* Okla Found Digestive Res 711 Stanton L Young Blvd Suite 501 Oklahoma City OK 73104

MATOSSIAN, JESSE N, b Los Angeles, Calif, Feb 3, 52. PLASMA PHYSICS, MATERIALS SCIENCE. *Educ:* Univ Southern Calif, BS, 75; Stevens Inst Technol, MS, 76, PhD(physics), 83. *Prof Exp:* MEM TECH STAFF, PLASMA PHYSICS DEPT, HUGHES RES LABS, 83- *Mem:* Am Phys Soc; NY Acad Sci; Am Inst Aeronauts & Astronauts; Sigma Xi; Inst Elec & Electronic Engrs. *Res:* Research, development, and application of low pressure plasmas for ion propulsion, plasma contactor, and thin film growth; ion implantation for surface modification of materials; ion emission from solid electrolytes. *Mailing Add:* Plasma Physics Dept Hughes Res Labs 3011 Malibu Canyon Rd Malibu CA 90265

MATOVCIK, LISA M, b Pittsburgh, Pa, June 18, 55. REGULATION OF PARATHYROID HORMONE SECRETION. *Educ:* Oberlin Col, BA, 76; Univ Pittsburgh, MA, 79; Stanford Univ, PhD(physiol), 85. *Prof Exp:* Postdoctoral assoc cell biol, Yale Univ Sch Med, 85-87, postdoctoral fel, 87-89, assoc res scientist, 89-91, ASST PROF SURG, YALE UNIV SCH MED, 91- *Mem:* Am Soc Cell Biol; AAAS. *Res:* Red cell membranes; intracellular membrane trafficking; second messenger regulation of parathyroid hormone secretion and degradation; immunocytochemical localization of cystic fibrosis transmembrane regulator. *Mailing Add:* Dept Surg West Haven Vet Hosp West Haven CT 06516

MATOVINOVIC, JOSIP, b Licko Cerje Croatia, Yugoslavia, Dec 22, 14; US citizen; m 43. MEDICINE. *Educ:* Univ Zagreb, MD, 39. *Prof Exp:* Resident internal med, State Gen Hosp, Zagreb, 40-45; asst prof, Med Sch, Univ Zagreb, 45-46; clin & res fel, Mass Gen Hosp, Harvard Med Sch, 47-48; chief div endocrinol, Univ Zagreb, 48-56, docent, 51-56; res assoc thyroid clin, Mass Gen Hosp, Harvard Med Sch, 56-58, res assoc diabetes clin, 59; from instr to prof, 59-85, EMER PROF INTERNAL MED, MED SCH, UNIV MICH, ANN ARBOR, 85- *Concurrent Pos:* Mem Yugoslav Comn Prev Endemic Goiter, 53, consult endem goiter Eggenberger Found, Switz, 53; mem Int Diabetes Assoc, 55; ed bd, Croatian Med J Lijecnicki Vijesnik, 53-55; lect iodine, Nutrit Dis US Pub Health Serv, WHO, Panam Health Orgn, Food Agr Orgn conf, Princeton Univ, 58; consult endem goiter, Lebanon & Pakistan, 60; lect iodine nutrit, US Food & Nutrit Bd, Nat Acad Sci, 70; dir, Mich Study Endem Goiter & Iodine Nutriture, CDC, 71; consult, Nuclear Med & Goiter, Radiation Ctr, Tata Mem Hosp, Bombay, India, 73; lectr, Panam Health Orgn tech group endem goiter, Guaruja, Sao Paolo, Brazil, 73; consult iodine monogr Life Sci Res Off, Fed Am Soc Exp Biol, Bethesda, Md, 74-76; lect iodine nutriture workshop, AMA, Scottsdale, Ariz, 79; lect evaluation iodine nutriture, Conf Assessment Nutriture, Am Health Found Div Nutrit Food & Drugs Admin, Western Human Nutrit Res Ctr, US Dept Agr, 87; sr adv int, Coun Control Iodine Deficiency Disorders, 87. *Honors & Awards:* McCollum Int Lectr, Cong Nutrit, 81. *Mem:* Endocrine Soc; Am Thyroid Asn; Am Fed Clin Res; Am Inst Nutrit. *Res:* Iodine deficient goiter in humans and animals; transplantable thyroid tumor of the rat: proliferation and differentiation in vivo and in cell structure; pathogenesis of thyroid carcinoma in man. *Mailing Add:* Univ Mich Hosp 1405 E Ann St W-5511 Box 21 Ann Arbor MI 48109

MATSAKIS, DEMETRIOS NICHOLAS, b St Louis, Mo, June 30, 49; m 77; c 2. INTERFERMETRIC ASTROMETRY, MICROWAVE ENGINEERING. *Educ:* Mass Inst Technol, BS, 71; Univ Calif, Berkeley, MS, 72, PhD(physics), 78. *Prof Exp:* Aeronomy asst, Naval Res Labs, 78-79; ASTRONOMER, RES US NAVAL OBSERV, 80- *Mem:* Am Astron Soc; Int Astron Union; Int Union Radio Sci. *Res:* Radio astronomy on an interferometer to study astronomical objects and variable earth rotation; construction of water vapor radiometers to measure tropospheric wet delay; molecular clouds in the interstellar environment. *Mailing Add:* US Naval Observ Washington DC 20392

MATSCH, CHARLES LEO, b Hastings, Minn. GEOLOGY, GLACIAL GEOLOGY. *Educ:* Univ Maine, BA, 59; Univ Minn, MS, 62; Univ Wis-Madison, PhD(geol), 71. *Prof Exp:* Explor geologist petrol, Stand Oil Co Tex, 61-64; instr geol, Univ Minn, Minneapolis, 64-66, asst prof, 66-70; from asst prof to assoc prof, 70-81, PROF & HEAD GEOLOGY DEPT, UNIV MINN, DULUTH, 81- *Concurrent Pos:* Secy, INQUA Comn, Genesis Glacial Sediments, 73-77. *Mem:* Geol Soc Am; Am Quaternary Asn; Nat Asn Geol Teachers; Sigma Xi. *Res:* Glacial geology of the midcontinent of North America; origin of quaternary continental sediments; environmental geology of glaciated terrains; glacial marine sediments. *Mailing Add:* Dept of Geol Univ of Minn Duluth MN 55812

MATSCH, L(EE) A(LLAN), b Chicago, Ill, Feb 21, 35; m 57; c 3. MECHANICAL ENGINEERING. *Educ:* Univ Ariz, BSME, 57; Univ Pittsburgh, MSME, 61; Ariz State Univ, PhD(eng sci), 67. *Prof Exp:* Assoc engr, Atomic Power Dept, Westinghouse Elec Corp, 57-61; supvr, AiRes Mfg Co, Div Garrett Corp, 61-70; mem res staff, Res Div, Ampex Corp, 70-71, eng mgr, Magnetic Tape Mfg Div, 71-76; sr supvr, Aires Mfg Co Ariz, 76-80; chief mech component design, Garrett Engine Div, Garrett Turbine Eng Co, 80-83, chief advan technol, 84-86, chief engr, 86-90; VPRES ENG, ALLIED SIGNAL AEROSPACE CO, 90- *Concurrent Pos:* Fac assoc, Ariz State Univ, 67-68 & 78- *Mem:* Fel Am Soc Mech Engrs; Am Inst Aeronaut & Aeronaut; Soc Automotive Engrs. *Res:* Fluid mechanics; fluid film lubrication; engineering mechanics. *Mailing Add:* 3005 S Fairway Dr Tempe AZ 85282

MATSCHINER, JOHN THOMAS, b Portland, Ore, Dec 2, 27; m 49; c 5. BIOCHEMISTRY. *Educ:* Univ Portland, BS, 50, MS, 51; St Louis Univ, PhD, 57. *Prof Exp:* Res assoc, Univ Va, 51-52; from instr to assoc prof biochem, Sch Med, St Louis Univ, 58-70; PROF BIOCHEM, SCH MED, UNIV NEBR, 70- *Concurrent Pos:* Jane Coffin Childs Fund res fel, Univ Calif, 57-58. *Mem:* Am Soc Biol Chem; Am Inst Nutrit; Sigma Xi. *Res:* Biochemistry and nutrition of vitamin K. *Mailing Add:* 3554 Davenport St Omaha NE 68131

MATSCHKE, DONALD EDWARD, chemical & environmental engineering, for more information see previous edition

MATSEN, JOHN M(ORRIS), b Neenah, Wis, May 30, 36; m 71. CHEMICAL ENGINEERING, FLUIDIZATION. *Educ:* Princeton Univ, BSE, 57; Columbia Univ, MS, 59, PhD(chem eng), 63. *Prof Exp:* Instr chem eng, Columbia Univ, 60-61; engr, Esso Res & Eng Co, 61-71, ENG ASSOC, EXXON RES & ENG CO, 71- *Concurrent Pos:* Ed adv bd mem, Advan Environ Sci & Eng, 77-; tech comt mem, Particulate Solids Res Inc, 78-; chmn, Int Fluidization Conf, 80; co ed, Fluidization, 80. *Mem:* Am Chem Soc; Am Inst Chem Engrs; Air Pollution Control Asn; Sigma Xi. *Res:* Fluidization; pneumatic transport; gas cleaning; particulate emissions and air pollution; separations processes; fluid catalytic cracking; fluidized bed combustion; synthetic fuels. *Mailing Add:* Exxon Res & Eng Co PO Box 101 Florham Park NJ 07932

MATSEN, JOHN MARTIN, b Salt Lake City, Utah, Feb 7, 33; m 59; c 10. MEDICINE, MICROBIOLOGY. *Educ:* Brigham Young Univ, BA, 58; Univ Calif, Los Angeles, MD, 63. *Prof Exp:* From intern to resident pediat, Univ Calif, Los Angeles, 63-66; from asst prof to prof lab med & path, pediat & microbiol, Univ Minn, Minneapolis, 68-74; assoc dean, 79-81, PROF PATH & PEDIAT & DIR CLIN MICRO LAB, UNIV UTAH, 74-, CHMN PATH, 81- *Concurrent Pos:* Fel pediat infectious dis, Univ Minn, Minneapolis, 66-68. *Honors & Awards:* Becton-Dickenson Award, Am Soc Microbiol, 88. *Mem:* Soc Pediat Res; fel Am Acad Microbiol; fel Col Am Pathologists; fel Am Acad Pediat; fel Am Soc Clin Path. *Res:* Pediatric enteric infections; antibiotic evaluation and evaluation of procedures in diagnostic microbiology. *Mailing Add:* Dept Path Univ Utah Sch Med 50 N Medical Dr Salt Lake City UT 84132

MATSON, DENNIS LUDWIG, b San Diego, Calif, Sept 29, 42. PLANETARY SCIENCES. *Educ:* San Diego State Univ, AB, 64; Calif Inst Technol, PhD(planetary sci), 72. *Prof Exp:* Res assoc planetology, Calif Inst Technol, 72-74, sr scientist planetology, 74-76, mem tech staff, 76-80, res scientist & group supvr, Jet Propulsion Lab, 80-85, sr res scientist & mgr, 85-89, CASSIRRI PROJ SCIENTIST, GEOL & PLANETOLOGY SECT, CALIF INST TECHNOL, 90- *Honors & Awards:* Exceptional Sci Achievement Medal, NASA, 83; Astroid named in honor. *Mem:* Am Geophys Union; Am Astron Soc; AAAS; Int Astron Union; Sigma Xi; Geol Soc Am; Comt Space Res; Astron Soc Pac. *Res:* Composition and morphology of planetary surfaces; astronomical photometry and spectroscopy over the entire spectral range from the vacuum ultra-violet through the visible, long wavelength, infrared; two-dimensional photometry and imaging of solar system bodies; theoretical studies of the interactions between planetary surfaces, atmospheres and magnetospheres; spacecraft instrument and mission development in the international environment. *Mailing Add:* Mail Code 183-501 Jet Propulsion Lab Pasadena CA 91103

MATSON, HOWARD JOHN, b Monmouth, Ill, June 8, 21; m 46; c 4. ORGANIC CHEMISTRY, PETROLEUM CHEMISTRY. *Educ:* Monmouth Col, Ill, BS, 43; Pa State Univ, MS, 47. *Prof Exp:* Res asst petrol refining, Pa State Univ, 43-47, res instr, 47-50; chemist, Sinclair Res Labs, Atlantic Richfield Co, 50-53, group leader, 53-55, sect leader, 55-63, res scientist, 63-69, asst mgr prod qual, Atlantic Richfield Co, 69-70, supvr, 70-74, mgr prod specialities, 74-77, mgr specif, Safety & Regulatory Serv Prod, 77-84; RETIRED. *Mem:* Am Chem Soc; Am Inst Chemists; Am Soc Lubrication Engrs; Am Soc Testing & Mat. *Res:* Petroleum product research and development. *Mailing Add:* 139A Comanche Trail Lake Iroquois Loda IL 60948

MATSON, LESLIE EMMET, JR, b Evanston, Ill, Nov 14, 20; m 46; c 1. ELECTRICAL ENGINEERING, AEROSPACE SYSTEMS. *Educ:* Univ Mich, BSE, 42; Univ Pa, MS, 55, PhD, 61. *Prof Exp:* Engr, RCA Corp, 42-55, leader dynamics group, 55-59, mgr systs anal, Missile & Surface Radar Dept & Systs Eng Eval & Res, 59-63; mgr space systs anal, Aerospace Systs Div, 63-66, mgr info & space systs, 66-71, mgr advan systs develop, 71-75; mem tech staff, Charles Stark Draper Lab Inc, 75-87; SR SCIENTIST, PHOTON RES ASSOC INC, 87- *Mem:* IEEE; Am Inst Aeronaut & Astronaut. *Res:* Analysis and synthesis of aerospace electronics systems; tracking and surveillance radar and electro-optical systems; decision and control systems; optimal navigation and information processing systems. *Mailing Add:* PO Box 1814 Roseburg OR 97470-0428

MATSON, MICHAEL, b Offenbach, WGermany, Apr 14, 52; US citizen. SATELLITE REMOTE SENSING. *Educ:* PanAm Univ, BS, 75; Univ Md, MA, 83. *Prof Exp:* Phys scientist, 75-84, hydrologist, 84-87, CHIEF, INTERACTIVE PROCESSING BR, NAT OCEANIC & ATMOSPHERIC ADMIN, 87- *Res:* Satellite remote sensing; snow cover mapping; volcanic eruption detection; fire detection; vegetation monitoring; computer-based interactive image processing; satellite sensor evaluation. *Mailing Add:* Nat Oceanic & Atmospheric Admin World Weather Bldg Rm 510 Washington DC 20233

MATSON, MICHAEL STEVEN, b Ft Wayne, Ind, June 3, 48; m 73; c 2. INORGANIC CHEMISTRY. *Educ:* Purdue Univ, BS, 70; Ind Univ, PhD(chem), 76. *Prof Exp:* Asst chemist, Ames Lab, 75-77; RES CHEMIST, CATALYTIC CHEM, PHILLIPS PETROL CO, 77- *Mem:* Am Chem Soc. *Res:* Catalytic synthesis of specialty chemicals via oxidation, carbonylation, reduction and/or reductive-amination reactions. *Mailing Add:* 6123 Cornell Dr Bartlesville OK 74006-8932

MATSON, TED P, b Ponca City, Okla, Jan 5, 29; m 51; c 3. APPLIED CHEMISTRY, SURFACE CHEMISTRY. *Educ:* Univ Okla, BS, 49, EdM, 51; Okla State Univ, MS, 67. *Prof Exp:* Teacher & coach pub schs, Okla, 51-56; asst res chemist, Res & Develop Dept, Continental Oil Co, 57-59, assoc res chemist, 59-62, res chemist, 62-64, tech adv to managing dir, Condea Petrochem GmbH, Hamburg, Ger, 64-65; res chemist, Res & Develop Dept, Continental Oil Co, 65-66, prod develop coordr, Conoco Chem, 66-72, res group leader surfactants, Continental Oil Co, 72-84; chief tech officer, Vista Chem Co, 85-89; RETIRED. *Concurrent Pos:* Chmn, Second World Conf Detergents, Montreux, Switz, 86. *Honors & Awards:* Charles Allderdier Award, Chem Specialties Mfrs Asn. *Mem:* Am Oil Chem Soc; Am Soc Testing & Mat; Chem Specialties Mfrs Asn (bd dir). *Res:* New product development; oil field chemicals; study of applications and synthesis of surfactants and research and development of evaluation techniques; quality management. *Mailing Add:* 2305 Wood Thrush Ponca City OK 74604

MATSUDA, FUJIO, b Honolulu, Hawaii, Oct 18, 24; m 49; c 6. RESEARCH ADMINISTRATION, CIVIL ENGINEERING. *Educ:* Rose Polytech Inst, BS, 49; Mass Inst Technol, ScD, 52. *Hon Degrees:* DEng, Rose-Hulman Inst Technol, Ind, 75; Dr, Soka Univ, Tokyo, 84. *Prof Exp:* Hydraulic engr, US Geol Surv, 49; res asst, Mass Inst Technol, 50-52, res engr, 52-54; res asst prof civil eng, Univ Ill, 54-55; from asst prof to prof eng, Univ Hawaii, 55-65, chmn, dept civil eng, 60-63, dir, Eng Exp Sta, 62-63; consult struct engr, Park & Yee, Ltd, 58-60; pres, Shimazu, Matsuda, Shimabukuro & Assoc, 60-63; dir, Dept Transp, State Hawaii, 63-73; vpres bus affairs, 73-74, pres & prof eng, 74-84, EXEC DIR, RES CORP, UNIV HAWAII, 84- *Concurrent Pos:* Pres, Pac Coast Asn Port Authorities, 68-69; chmn, Airport Operators Coun Int, Pac Region, 70-73; vpres, Western Col Asn, 79-80, pres, 80-82. *Honors & Awards:* Eng Mgt Award, Am Soc Civil Engrs, 86. *Mem:* Nat Acad Eng; Sigma Xi. *Res:* Response of structures to dynamic loads. *Mailing Add:* 1110 University Ave Rm 408 Honolulu HI 96826

MATSUDA, KEN, b Napa, Calif, Nov 30, 20; m 46; c 2. ORGANIC CHEMISTRY. *Educ:* Univ Md, BS, 44, PhD(chem), 51. *Prof Exp:* Res chemist, Stamford Labs, Am Cyanamid Co, 51-57, sr res chemist, 57-62, group leader, 61-71, proj mgr, 71-74, mgr chem sect, 74-77, mgr water treating & mining chem res & develop, 77-79, catalyst res, 79-80, dir technol assessment, 80-86; RETIRED. *Mem:* Am Chem Soc; Sigma Xi. *Res:* Catalysts; organic flocculants; mining reagents; fluorinated compounds; polynuclear aromatics; photochromic products; vinyl monomers and polymers; cyanogen derivatives; high temperature reaction; coal liquefaction. *Mailing Add:* 29 Lancer Ln Stamford CT 06905

MATSUDA, KYOKO, b Kirya, Japan, Apr 3, 32. PHYSICS, PLASMA PHYSICS. *Educ:* Univ Tokyo, MS, 60; Univ Calif, PhD(physics), 67. *Prof Exp:* MEM STAFF FUSION THEORY, GEN AUTOMIC, 74- *Mem:* Fel Am Phys Soc. *Mailing Add:* 534 Glencrest Dr Solana Beach CA 92075

MATSUDA, SEIGO, b Tokyo, Japan, Feb 26, 25; m 51; c 1. METALLURGY. *Educ:* Yokohama Nat Univ, BEng, 46; Tohoku Univ, Japan, BEng, 50; Mass Inst Technol, ScD(metall), 61. *Prof Exp:* Instr electrochem, Tech High Sch, Japan, 46-47; mem bd dir, Nambu Lumber Co, 47-51; dept res asst, Govt Dept Educ, Tohoku Univ, Japan, 50-51, dept asst, 51-56; res asst, Mass Inst Technol, 56-60; sr metallurgist, Ilikon Co, 60-61; exec vpres & tech dir, Cambridge Metal Res Inc, 61-64; sr res specialist, Monsanto Res Corp, 64-67, advan technologist, 67-69; sr scientist, Biomed Res Lab, Am Hosp Supply Co, 69-70; CHIEF METALLURGIST, THERMO ELECTRON CORP, 70-, DEPT MGR, 75-, VCHMN, 80- *Concurrent Pos:* Lectr, Ctr Continuing Educ, Northeastern Univ, 69-71. *Mem:* Electrochem Soc; Nat Asn Corrosion Engrs; Am Soc Metals; Japan Inst Metals; Electrochem Soc Japan. *Res:* Electrochemistry; corrosion catalysis; technical consulting. *Mailing Add:* Three Charena Rd Wayland MA 01778

MATSUDA, YOSHIYUKI, b Manchuria, China, Dec 7, 43; Japanese citizen; m 71. PLASMA PHYSICS. *Educ:* Kyoto Univ, BS, 66, MS, 68; Stanford Univ, PhD(elec eng), 74. *Prof Exp:* Res assoc plasma physics, Plasma Physics Lab, Princeton Univ, 74-78; PHYSICIST, LAWRENCE LIVERMORE LAB, 78- *Mem:* Am Phys Soc; Inst Elec & Electronics Engrs; AAAS. *Res:* Theoretical and computational study of plasma physics and controlled thermonuclear fusion. *Mailing Add:* 724 Towbridge Rd Danville CA 94526

MATSUGUMA, HAROLD JOSEPH, b Honolulu, Hawaii, Oct 15, 28; m 63; c 1. INORGANIC CHEMISTRY. *Educ:* Univ Hawaii, BA, 51; Univ Ill, MS, 52, PhD(chem), 55. *Prof Exp:* Res assoc chem, Univ Ill, 52-55, assoc, 55; chemist, Explosives Res Sect, US Dept Army, 55-57, chief, Synthesis Unit, 57-59, chief, Off Reactor Requirements & Explosives Res Sect, 59-63, actg chief, Explosives Lab, 63-66, chief, Chem Br, Energetic Mat Div, Arradcom, 66-85; MEM STAFF, APPL ORD TECHNOL, INC, 87- *Mem:* Am Chem Soc; Royal Soc Chem; Am Defense Preparedness Asn; Sigma Xi. *Res:* Synthesis of hydrazine; hydroxylamine derivatives; chemistries of nitrogen, phosphorus and sulfur compounds; chemistry of explosives; explosives safety; relation of chemical constitution to explosive properties; hazard analyses. *Mailing Add:* 19 Kory Rd Newton NJ 07860

MATSUI, SEI-ICHI, CELL BIOLOGY, GENETICS. *Educ:* Hokkaido Univ, Japan, DSc(biol), 74. *Prof Exp:* CANCER RES SCIENTIST, ROSWELL PARK MEM INST, 76-; PROF CELL PHYSIOL, STATE UNIV NY, BUFFALO, 81- *Mailing Add:* Genetics Dept Roswell Park Mem Inst 666 Elm St Buffalo NY 14263

MATSUMOTO, CHARLES, b San Jose, Calif, Mar 25, 32; m 61; c 1. PHARMACOLOGY, BIOCHEMISTRY. *Educ:* San Jose State Col, BA, 53; Univ Idaho, MS, 55; Univ Wash, PhD(pharmacol), 63. *Prof Exp:* Chemist biol lab, US Fish & Wildlife Serv, 58-70; RES ASSOC, LILLY RES LAB, ELI LILLY & CO, 65- *Concurrent Pos:* Exec ed, Life Sci, 70-73; NIH fel, Lab Chem Pharmacol, Nat Heart Inst, 63-65. *Mem:* AAAS; Drug Info Asn; Am Soc Pharmacol & Exp Therapeut; Am Soc Clin Pharmacol & Therapeut; Sigma Xi. *Res:* Autonomic, cardiovascular and biochemical pharmacology. *Mailing Add:* Dept MC 754 Eli Lilly Res Labs 307 E McCarty St Bldg 31-2 Indianapolis IN 46285

MATSUMOTO, HIROMU, b Honolulu, Hawaii, Mar 28, 20. AGRICULTURAL BIOCHEMISTRY. *Educ:* Univ Hawaii, BS, 44, MS, 45; Purdue Univ, PhD(biochem), 55. *Prof Exp:* Asst chem, Exp Sta, Univ Hawaii, 45-49, jr chemist, 49-51; asst biochem, Purdue Univ, 51-54; from asst biochemist to assoc biochemist, 55-66, BIOCHEMIST, EXP STA, UNIV HAWAII, 66- *Concurrent Pos:* Fel, Japan Soc Advan Sci, 75-76. *Mem:* Am Chem Soc; Soc Toxicol; Am Asn Cancer Res. *Res:* Effect of toxic plant constituents on animal metabolism; mimosine, 3-nitropropanoic acid, methylazoxymethanol; metabolic fate in animals of naturally occurring toxicants; cycasin methylazoxymethanol-glucosiduronic acid; toxicology; chemical carcinogenesis. *Mailing Add:* Dept of Agr-Biochem/HKc 319 Univ Hawaii at Monoa 2500 Campus Rd Honolulu HI 96822

MATSUMOTO, HIROYUKI, b Nagasaki, Japan, May 5, 48; m; c 1. NEUROPHYSIOLOGY. *Educ:* Kyoto Univ, BS, 72, MS, 74, PhD(biophysics), 77. *Prof Exp:* Jr researcher, Dept Chem, Univ Hawaii, 77-79; res assoc, Purdue Univ, 79-80, asst res scientist, dept biol sci, 80-85; asst prof, 85-90, ASSOC PROF, DEPT BIOCHEM & MOLECULAR BIOL, UNIV OKLA HEALTH SCI CTR, 91- *Concurrent Pos:* Asst researcher, Dept Chem, Univ Hawaii, 81. *Mem:* Asn Res Vision & Opthalmol; Am Soc Biol Chemists; Sigma Xi; Am Soc Photochem & Photobiol. *Res:* Study of molecular mechanisms of sensory transduction in bovine and fruit fly visual photoreceptor systems; chromophore-protein interaction in rhodopsin and protein modifications involved in visual transduction. *Mailing Add:* Dept Biochem & Molec Biol Univ Okla Health Sci Ctr PO Box 26901 Oklahoma City OK 73190

MATSUMOTO, KEN, b San Bernadino, Calif, Sept 8, 41; m 67; c 2. ORGANIC CHEMISTRY, MEDICINAL CHEMISTRY. *Educ:* Ariz State Univ, BS, 63; Univ Calif, Berkeley, PhD(org chem), 67. *Prof Exp:* Teaching asst, Univ Calif, Berkeley, 63-64; sr org chemist, 69-75, res scientist, 75-81, SR RES SCIENTIST, LILLY RES LABS, 82- *Mem:* Am Chem Soc; Sigma Xi. *Mailing Add:* Lilly Res Labs Eli Lilly & Co Lilly Corp Ctr Indianapolis IN 46285

MATSUMOTO, LLOYD H, NUCLEIC ACID BIOCHEMISTRY, NUCLEAR STRUCTURE & FUNCTION. *Educ:* St Louis Univ, PhD(develop biol), 72. *Prof Exp:* ASST PROF CELL & MOLECULAR BIOL, RI COL, 82- *Mailing Add:* Biol Dept R I Col 600 Mt Pleasant Ave Providence RI 02908

MATSUMOTO, STEVEN G, neurobiology, for more information see previous edition

MATSUMOTO, YORIMI, b Yuba City, Calif, July 29, 26. PHYSIOLOGY. *Educ:* Whittier Col, AB, 50; Univ Calif, Los Angeles, PhD(zool), 64. *Prof Exp:* Instr biophys, Univ Ill, Urbana, 63-66, from asst prof to assoc prof, 66-69; ASSOC PROF PHYSIOL, EMORY UNIV, 69- *Mem:* AAAS; Biophys Soc; Am Soc Zoologists. *Res:* Mechanical analysis of muscular contraction; heat analysis of muscle-contraction; nerve-heat; birefrigency study of invertebrate muscle. *Mailing Add:* Dept of Physiol Emory Univ Sch of Med 1364 Clifton Rd NE Atlanta GA 30322

MATSUMURA, KENNETH N, b Bangkok, Thailand, May 15, 45; m 77; c 1. MOLECULAR BIOLOGY. *Educ:* Univ Calif, Berkeley, BS, 66; Univ Calif, San Francisco, MD, 70; Am Col Angiol, dipl, 87; Am Col Sexology, dipl, 89. *Prof Exp:* Dir res & develop, Immunity Res Lab, 77-87; DIR & COORDR, BIO-ARTIFICIAL LIVER PROJ, GLOBAL DEVELOP, ALIN ACI, 70-, DIR, DIV SCI & TECHNOL DEVELOP, ALIN FOUND, 87- *Concurrent Pos:* Consult, Univ Calif, Berkeley, 74-75; chief, Urgent Care Serv, W Oakland Health Ctr, 75-79, pres & chief staff, W Oakland Health Group, 83-; ed-in-chief, Base J Sci & Technol, 81-; pub health comnr, Berkeley, 85 & 88. *Mem:* AAAS; Soc Sci Study Sex; fel Am Col Angiol; fel NY Acad Sci. *Res:* Original development of bio-artificial organ engineering concepts; artificial liver; artificial pancreas; transplantation immunology; biology of aging; cancer chemotherapy; emergency medical data systems; fertility enhancement and control; genetic engineering; Five patents. *Mailing Add:* One Alin Plaza 2107 Dwight at Shattuck Berkeley CA 94704-2062

MATSUMURA, PHILIP, b San Jose, Calif, Aug 15, 47. MOLECULAR BIOLOGY. *Educ:* Univ Santa Clara, BS, 69; Univ Rochester, PhD(microbiol), 75. *Prof Exp:* Fel, Univ Calif, San Diego, 75-79; asst prof, 79-85, ASSOC PROF, UNIV ILL, CHICAGO, 85- *Mem:* AAAS; Sigma Xi; Am Soc Microbiol. *Res:* Microbial motility and chemotaxis. *Mailing Add:* Dept Biol Sci Univ Ill Chicago IL 60680

MATSUO, KEIZO, b Osaka, Japan, Apr 23, 42. POLYMER CHEMISTRY. *Educ:* Kyoto Univ, BS, 66, MS, 68; Dartmouth Col, PhD(chem), 72. *Prof Exp:* Res assoc chem, 72-74, res instr chem, 74-77, SR RES ASSOC, DARTMOUTH COL, 78- *Mem:* Am Chem Soc; Japan Chem Soc. *Res:* Synthesis of new polymer; characterization; equilibrium and non-equilibrium study of polymer solution; kinetics; play with rotational state model. *Mailing Add:* Geeringstr 35 Zurich 8049 Switzerland

MATSUO, ROBERT R, b Duncan, BC, Feb 28, 32; m 61; c 2. BIOCHEMISTRY. *Educ:* Univ Man, BSc, 57; Univ Alta, PhD(plant biochem), 62. *Prof Exp:* Chemist I, Grain Res Lab, 57-59, chemist III, Durum Wheat Res, 62-66, RES SCIENTIST, GRAIN RES LAB, CAN DEPT AGR, 66- *Mem:* AAAS; Am Asn Cereal Chemists; Chem Inst Can; Prof Inst Pub Serv Can; Can Inst Food Sci & Technol. *Res:* Cereal chemistry; basic and applied research on durum wheat and durum wheat products. *Mailing Add:* 185 Waterloo Winnipeg MB R3N 0S4 Can

MATSUOKA, SHIRO, b Kobe, Japan, May 1, 30; m 57; c 3. POLYMER PHYSICS, PLASTICS ENGINEERING. *Educ:* Stevens Inst Technol, ME, 55; Princeton Univ, MSE, 57, PhD(mech eng), 59. *Prof Exp:* Asst, Princeton Univ, 55-57; supvr res rheol, 63-71, supvr plastics develop, 71-74, HEAD PLASTICS RES & ENG DEPT, AT&T BELL LABS, 74-, MEM TECH STAFF, 59- *Concurrent Pos:* Vis lectr, Stevens Inst Technol, 62-64, vis prof, 64-71; vis prof, Rutgers Univ, 77- *Honors & Awards:* Int Award, Soc Plastics Engrs, 80. *Mem:* Nat Acad Eng; Am Chem Soc; Soc Rheol; fel Soc Plastics Engrs; fel Am Phys Soc. *Res:* Mechanical, electrical and morphological properties of high polymers; molecular relaxation phenomena. *Mailing Add:* AT&T Bell Labs 600 Mountain Ave Rm 7F-202 Murray Hill NJ 07974

MATSUOKA, TATS, b Seattle, Wash, Aug 24, 29; m 64; c 3. VIROLOGY. *Educ:* Univ Minn, BA, 52; State Col Wash, DVM, 59. *Prof Exp:* Asst bacteriologist, Mont State Col, 52-55, asst bacteriologist & virologist, Vet Res Lab, 61-63; practicing vet, Idaho, 59-60; vet diagnostician, Mont Livestock Sanit Bd, 60-61; res vet, 63-67, sr virologist, 67-74, res scientist, 74-84, RES PROJ MGR, GREENFIELD LABS, ELI LILLY & CO, 84- *Mem:* Am Vet Med Asn; Am Soc Microbio; US Animal Health Asn; Conf Res Workers Animal Dis. *Res:* Animal viruses, particularly viral diseases of bovine; antimicrobial evaluation. *Mailing Add:* Lilly Res Labs Bldg 202 Box 708 Greenfield IN 46140

MATSUSAKA, TERUHISHA, b Kyoto, Japan, Apr 5, 26; US citizen; m 50; c 5. GEOMETRY. *Educ:* Kyoto Univ, MS, 49, PhD(math), 54. *Prof Exp:* From instr to asst prof math, Ochanomizu Univ, Japan, 52-54; res assoc, Univ Chicago, 54-57; from assoc prof to prof, Northwestern Univ, 57-61; prof math, 61-81, Irving Schneider prof, 81-83, BERENSON PROF, BRANDEIS UNIV, 84- *Concurrent Pos:* Guggenheim fel, 59. *Mem:* Am Math Soc; Am Acad Arts & Sci. *Res:* Algebra; algebraic geometry. *Mailing Add:* Dept of Math Brandeis Univ Waltham MA 02254-9110

MATSUSHIMA, JOHN K, b Denver, Colo, Dec 24, 20; m 43; c 2. ANIMAL NUTRITION. *Educ:* Colo State Univ, BS, 43, MS, 45; Univ Minn, PhD, 49. *Prof Exp:* Asst animal husb, Colo State Univ, 43-45; from asst prof to prof, Univ Nebr, 49-61; PROF ANIMAL SCI, COLO STATE UNIV, 61- *Mem:* Am Soc Animal Sci; Soc Range Mgt; Am Dairy Sci Asn; Am Inst Nutrit. *Res:* Beef cattle nutrition, feeding and management. *Mailing Add:* Dept of Animal Sci Colo State Univ Ft Collins CO 80523

MATSUSHIMA, SATOSHI, b Fukui, Japan, May 6, 23; nat US; m 55; c 2. ASTRONOMY, ASTROPHYSICS. *Educ:* Univ Kyoto, MS, 46; Univ Utah, PhD(astrophys), 54; Univ Tokyo, DSc, 66. *Prof Exp:* Asst astron, Univ Kyoto, 46-50; res fel & asst, High Altitude Observ & Harvard Col Observ, 50-54; res assoc physics, Univ Pa & Strawbridge Observ, Haverford Col, 54-55; vis astronr, Astrophys Inst & Meudon Observ, Paris, France, 56-57; Humboldt fel, Inst Theoret Physics, Univ Kiel, 57-58; asst prof physics, Fla State Univ, 58-60; assoc prof astron, Univ Iowa, 60-67; actg head, Physics Dept, Pa State Univ, 81-82, prof astron, 67-89, head dept, 76-89, EMER PROF ASTRON, PA STATE UNIV, 89. *Concurrent Pos:* Travel grants, Int Astron Union, 56 & 57, Ger Astron Soc, 57, US Res Corp, 58 & NSF, 65-82; guest astronr, Utrecht Observ, Neth, 56; sr res fel, Calif Inst Technol, 59-61; vis prof, US-Japan Coop Sci Prog, Univs Tokyo & Kyoto, 65-66, Univs Tokyo & Tohoko, 74; consult, Naval Res Lab, 62; mem, Int Astron Union Comns 12 & 36. *Mem:* Am Astron Soc; fel Royal Astron Soc; Am Geophys Union. *Res:* Theory of stellar atmospheres; solar and planetary physics; spectroscopy and spectrophotometry; space and upper atmosphere physics. *Mailing Add:* 155 W 70th St 8F New York NY 10023

MATSUURA, TAKESHI, b Shizuoka, Japan, Dec 22, 36; Can citizen; m 68; c 1. MEMBRANE SEPARATION PROCESSES. *Educ:* Univ Tokyo, BSc, 61, MSc, 63; Tech Univ Berlin, Dr Ing (chem eng), 65. *Prof Exp:* Staff asst synthetic chem, Univ Tokyo, 66-67; res assoc chem eng, Univ Calif, Davis, 67-69; res fel, 69-71, asst res officer, 71-75, assoc res officer, 75-80, SR RES OFFICER CHEM ENG, NAT RES COUN CAN, 81- *Concurrent Pos:* Adj prof, Dept Chem Eng, Univ Ottawa, 85- *Honors & Awards:* Res Award, Int Desalination & Environ Asn, 83. *Mem:* Can Soc Chem Eng; Chem Soc Japan; Am Chem Soc. *Res:* Reverse osmosis; separation by synthetic membranes; water treatment by reverse osmosis; food processing by reverse osmosis; membrane gas and vapor separation; per vaporation. *Mailing Add:* Inst Environ Chem Nat Res Coun Can Ottawa ON K1A 0R9 Can

MATSUYAMA, GEORGE, b Fresno, Calif, Nov 20, 18; m 45; c 2. ELECTROANALYTICAL CHEMISTRY. *Educ:* Univ Calif, BS, 40; Univ Minn, PhD(phys chem), 48. *Prof Exp:* Asst chem, Fresno State Col, 36-38; asst, Univ Minn, 40-43, instr, 43-48; asst prof, Wesleyan Univ, 48-52; res chemist, Union Oil Co, Calif, 52-55, sr res chemist, 55-57, res assoc, 57-59; sr chemist, Beckman Instruments Inc, 59-64, eng specialist, 64-69, res scientist, 70-79; sr staff scientist, Ionetics Inc, 79-81, dir res, 81-84; RETIRED. *Mem:* AAAS; Am Chem Soc; Electrochem Soc. *Res:* Electrometric and volumetric analysis; electroanalytical instrumentation; ion-selective electrodes; gas sensors; enzyme electroanalytical methods. *Mailing Add:* 548 N Stanford Ave Fullerton CA 92631-3336

MATSUZAKI, MASAJI, ELECTRONEUROPHYSIOLOGY, PSYCHOLOGY. *Educ:* Univ Tokyo, PhD(neurophysiol), 67. *Prof Exp:* Res scientist, NY State Div Substance Abuse Serv Testing & Res Lab, 72-88; RES SCIENTIST, DEPT MED, HEALTH SCI CTR, STATE UNIV NY, 88- *Res:* Neuropsychopharmacology. *Mailing Add:* Dept Med Health Sci Ctr State Univ NY 450 Clarkson Ave Box 19 Brooklyn NY 11203

MATT, JOSEPH, organic chemistry, for more information see previous edition

MATTA, JOSEPH EDWARD, b Philadelphia, Pa, July 29, 48. PHYSICS. *Educ:* St Joseph's Col, Philadelphia, BS, 70; Lehigh Univ, MS, 72, PhD(physics), 74. *Prof Exp:* Fel physics, Lehigh Univ, 74-75; res physicist, Mines Safety Res Ctr, 75-80; MEM STAFF, CHEM RES DEVELOP & ENGR CTR, ABERDEEN PROVING GROUND, 80- *Mem:* Am Phys Soc. *Res:* Investigates the atomization and characterization of viscoelastic fluids; aerosol physics. *Mailing Add:* 1007 Londonderry Dr Bel Air MD 21014

MATTA, MICHAEL STANLEY, b Dayton, Ohio, Feb 22, 40; m 62; c 3. BIOLOGICAL CHEMISTRY, ORGANIC CHEMISTRY. *Educ:* Univ Dayton, BS, 62; Ind Univ, PhD(org chem), 66. *Prof Exp:* Sr res chemist, Mound Lab, Monsanto Res Corp, Ohio, 66-68; res assoc biol chem, Amherst Col, 68-69; asst prof, 69-74, assoc prof, 74-77, PROF CHEM, SOUTHERN ILL UNIV, EDWARDSVILLE, 78- *Mem:* AAAS; Am Chem Soc; Sigma Xi. *Res:* Kinetics and mechanism of enzyme action; transfer reactions of borazines, free radical rearrangement and participation. *Mailing Add:* Dept Chem Southern Ill Univ Edwardsville IL 62025

MATTANO, LEONARD AUGUST, b Tampa, Fla, July 8, 17; m 41; c 5. ORGANIC CHEMISTRY. *Educ:* Univ Wis, BS, 41; Mich State Univ, PhD, 48. *Prof Exp:* Chemist, Allis-Chalmers Co, Wis, 41-42 & Dow Chem Co, Mich, 42-45; res chemist, Standard Oil Co, Ind, 48-56; sr res chemist, Dow Chem Co, 56-69; dir chem res, Bissell, Inc, 69-82; RETIRED. *Concurrent Pos:* Consult, Consumer Prod & Guardsman Prod Inc, 84- *Mem:* Am Chem Soc; Sigma Xi. *Res:* Motor oil additives; chelate resins; surfactants; general syntheses; chemical specialties; aerosols. *Mailing Add:* 2325 Ducoma Dr NW Grand Rapids MI 49504

MATTAR, FARRES PHILLIP, quantum electronics, computational physics; deceased, see previous edition for last biography

MATTAUCH, ROBERT JOSEPH, b Rochester, Pa, May 30, 40; m 62; c 2. ELECTRICAL ENGINEERING. *Educ:* Carnegie Inst Technol, BSEE, 62; NC State Univ, MEEE, 63, PhD(elec eng), 67. *Prof Exp:* From asst prof to prof elec eng, 66-83, Wilson prof, 83-84, dir, Semiconductor Device Lab, 67-84, STANDARD OIL CO PROF & CHMN, UNIV VA, 87- *Honors & Awards:* Centennial Medal, Inst Elec & Electronic Engrs, 84. *Mem:* Fel Inst Elec & Electronic Engrs; Sigma Xi. *Res:* Semiconductor materials and devices; millimeter, submillimeter wave heterodyne receiver structures. *Mailing Add:* 2455 Foxpath Ct West Woods Charlottesville VA 22901

MATTAX, CALVIN COOLIDGE, b Sallisaw, Okla, Feb 4, 25; m 49; c 4. PHYSICAL CHEMISTRY. *Educ:* Univ Tulsa, BChem, 50; La State Univ, MS, 52, PhD(phys chem), 54. *Prof Exp:* Eng supvr, Esso Prod Res Co, 55-75, DIV MGR, EXXON PROD RES CO, 76- *Mem:* Am Chem Soc; Am Inst Mining, Metall & Petrol Eng. *Res:* Reservoir engineering; electrochemical kinetics; properties of polymer solutions; fluid mechanics in porous media. *Mailing Add:* 306 Chapel Bell Houston TX 77024

MATTE, J JACQUES, b Neuville, Que, Oct 29, 55; m 77; c 4. NUTRITION, ANIMAL PHYSIOLOGY. *Educ:* Laval Univ, BSc, 78, MSc, 80, PhD(animal sci), 84. *Prof Exp:* RES SCIENTIST SWINE NUTRIT, RES STA, AGR CAN, 81- *Concurrent Pos:* Fel, Animal & Grassland Res Inst, Shinfield, Reading, UK, 84-85; asst ed, Can J Animal Sci, 90- *Mem:* Can Soc Animal Sci; Am Soc Animal Sci. *Res:* Requirements of folic acid in swine primarily through research activities on metabolic status, digestive absorption and production performance. *Mailing Add:* Res Sta Agr Can PO Box 90 Lennoxville PQ J1M 1Z3 Can

MATTE, JOSEPH, III, b Detroit, Mich, Feb 14, 16; m 56, 79; c 3. NUCLEAR POWER PLANT DESIGN AND OPERATION. *Educ:* Wayne State Univ, BSME, 38. *Prof Exp:* Mech engr, Ternsted Mfg Div, Gen Motors Corp, 40-42 & 47-48, Ford Motor Co, 48, Budd Co, 48-57; mech engr, Atomic Power Develop Assoc Inc, 57-63, unit leader mech group, 63-66, sect head, 66-72; proj qual assurance engr, Detroit Edison Co, 73-75; prog mgr, Elec Power Res Inst, 75-86; RETIRED. *Concurrent Pos:* Consult nuclear eng, 86-; mem subcomt, B & PV Code Inserv Inspection, Am Soc Mech Engrs, 77-91; chmn subgroup, Liquid Metal Reactor Inserv Inspection, 86-91. *Mem:* Am Inst Plant Engrs; Am Nuclear Soc; Am Soc Mech Engrs. *Res:* Fuel handling and control mechanisms for large sodium-cooled fast reactors; physical and chemical fundamentals basic to design of sodium system components; inspection and repair of nuclear reactors; management of nuclear plant design projects; light water reactor fire protection research; fast reactor plant design reviews. *Mailing Add:* W 419 17th Ave Spokane WA 99203

MATTEI, JANET AKYÜZ, b Bodrum, Turkey, Jan 2, 43; m 72. ASTRONOMY. *Educ:* Brandeis Univ, BA, 65; Ege Univ, Turkey, 70, PhD, 82; Univ Va, MS, 72. *Prof Exp:* Teacher physics, astron & phys sci, Am Col Inst, Turkey, 67-69; teaching asst astron, Ege Univ, Turkey, 69-70; asst dir, 72-73, DIR ASTRON, AM ASN VARIABLE STAR OBSERVERS, 73- *Mem:* Int Astron Union; Am Asn Variable Star Observers; Am Astron Soc. *Res:* Visual and photometric studies of variable stars, particularly dwarf novae, T Tauri stars and long period variables. *Mailing Add:* Am Asn Variable Star Observers 25 Birch St Cambridge MA 02138

MATTEN, LAWRENCE CHARLES, b Newark, NJ, Sept 1, 38; m 59; c 4. PALEOBOTANY, PLANT MORPHOLOGY. *Educ:* Rutgers Univ, BA, 59; Cornell Univ, PhD(bot), 65. *Prof Exp:* Instr biol, State Univ NY Col Cortland, 64-65; asst prof, 65-70, assoc prof, 70-77, PROF BOT, SOUTHERN ILL UNIV, CARBONDALE, 77-, CHAIR, 89- *Concurrent Pos:* Head ed, Palaeontographica; vis prof, Univ Col N Wales, 72-73. *Mem:* Int Orgn Paleobot; Linnean Soc London; Bot Soc Am. *Res:* Elucidation of Paleozoic flora, especially Devonian plants from eastern United States and Devonian/Mississippian transition floras; study of early seed plants; anatomy of pyritized plant remains. *Mailing Add:* Dept Plant Biol Southern Ill Univ Carbondale IL 62901

MATTENHEIMER, HERMANN G W, b Berlin, Ger, Mar 29, 21; m 43; c 3. BIOCHEMISTRY. *Educ:* Univ Gottingen, MD, 47. *Prof Exp:* Asst physician, Helmstedt Dist Hosp, Ger, 45-49; res asst, Berlin, 49-51; res asst, Free Univ Berlin, 51-55, privat-docent, 55-59; from asst prof to assoc prof biochem, Univ Ill, 59-71, PROF BIOCHEM, RUSH MED COL, 71-, ASSOC CHAIRPERSON, 81- *Concurrent Pos:* Dir clin chem, Presby-St Luke's Hosp, 59-; res fel, Theodor Kocher Inst, Switz, 51-53; Rusk Orsteel Found fel, Carlsberg Lab, Denmark, 55, WHO fel, 56-57. *Mem:* AAAS; Am Chem Soc; NY Acad Sci; Am Soc Biol Chem; Ger Soc Biol Chem; Sigma Xi. *Res:* Cell metabolism; ultramicrotechniques for enzyme determinations in single cells; clinical chemistry; clotting of casein; renal biochemistry. *Mailing Add:* Presby St Lukes Hosp 1753 W Congress Pkwy Chicago IL 60612

MATTERN, MICHAEL ROSS, b Palmerton, Pa, Oct 29, 47. BIOCHEMISTRY, MOLECULAR BIOLOGY. *Educ:* Muhlenberg Col, BS, 69; Princeton Univ, MA, 72, PhD(biochem), 75. *Prof Exp:* Fel biochem lab radiobiol, Univ Calif, San Francisco, 75-78; SR STAFF FEL BIOCHEM, LAB MOLECULAR CARCINOGENESIS, NAT CANCER INST, NIH, 78- *Mem:* Radiation Res Soc; Biophys Soc. *Res:* Molecular and cell biology of DNA replication and DNA repair in mammalian cells; chromosome structure and function. *Mailing Add:* Nat Cancer Inst NIH Bldg 37 Rm 3C27 Bethesda MD 20014

MATTERN, PAUL JOSEPH, b Winnetoon, Nebr, Jan 26, 22; m 56; c 4. ANALYTICAL CHEMISTRY. *Educ:* Univ Northern Iowa, BA, 47; Univ Wis, MS, 51. *Prof Exp:* Instr biochem & nutrit, 53-59, from asst prof to assoc prof, 59-70, PROF AGRON, UNIV NEBR, LINCOLN, 71- *Mem:* Sigma Xi; Am Asn Cereal Chemists. *Res:* Environmental and genetic effects on the chemical, physical and nutritional properties of wheat constituents. *Mailing Add:* 154 Keim Hall East Campus NE 68583

MATTES, FREDERICK HENRY, b Sheboygan, Wis, Feb 18, 41; m 64; c 2. INSTRUMENTAL ANALYSIS. *Educ:* Carroll Col, BS, 63; Ind Univ, PhD(chem), 68. *Prof Exp:* From instr to asst prof chem, Willamette Univ, 67-76; asst prof, 76-79, ASSOC PROF CHEM, HASTINGS COL, 79- *Mem:* Am Chem Soc. *Res:* Electroanalytical chemistry; chemically modified electrodes; Raman spectroscopy. *Mailing Add:* Dept Chem Hastings Col Hastings NE 68902

MATTES, HANS GEORGE, b Washington, DC, Jan 27, 43; m 68; c 3. ELECTRICAL ENGINEERING, PHYSICS. *Educ:* Calif Inst Technol, BS, 64; Univ Southern Calif, MS, 66, PhD(elec eng), 68. *Prof Exp:* Mem tech staff electronics, Bell Tel Lab, 68-70; vis assoc prof, elec eng, Nat Taiwan Univ, 70-71; MEM TECH STAFF ELECTRONICS, BELL TEL LAB, 72- *Concurrent Pos:* Fel NSF, 64-68. *Res:* Man and machine interaction; telecommunications input-output technology. *Mailing Add:* AT&T Bell Labs One Whippany Rd Whippany NJ 07981

MATTES, WILLIAM BUSTIN, b Hartford, Conn, May 5, 53; m 79; c 3. DNA DAMAGE, TRANSCRIPTION REGULATION. *Educ:* Univ Pa, BA, 74; Univ Mich, PhD(biol chem), 81. *Prof Exp:* Postdoctoral fel, Johns Hopkins Sch Pub Health, 81-83; staff fel, Nat Cancer Inst, Bethesda, Md, 83-86; res toxicologist, Stauffer Chem Co, 86-88; SR TOXICOLOGIST, CIBA-GEIGY CORP, 88- *Concurrent Pos:* Adj assoc prof, Sch Pharm, Univ Conn, 90- *Mem:* Environ Mutagen Soc; AAAS; Soc Toxicol. *Res:* Mechanisms by which small molecules such as nitrogen mustards recognize specific DNA sites and the biological implications of this sequence specificity. *Mailing Add:* Ciba-Geigy Corp 400 Farmington Ave Farmington CT 06032

MATTESON, DONALD STEPHEN, b Kalispell, Mont, Nov 8, 32; m 53, 71; c 2. ORGANOMETALLIC CHEMISTRY. *Educ:* Univ Calif, BS, 54; Univ Ill, PhD(chem), 57. *Prof Exp:* Res chemist, E I du Pont de Nemours & Co, 57-58; from instr to assoc prof, 58-69, PROF CHEM, WASH STATE UNIV, 69- *Concurrent Pos:* Sloan Found fel, 66-68; mem comt examrs, Advan Chem Test, Grad Record Exam, 70-84. *Honors & Awards:* Fel AAAS, 88. *Mem:* Am Chem Soc; AAAS; Sigma Xi. *Res:* Boron-substituted carbanions as synthetic intermediates; directed chiral synthesis with boronic esters as intermediates; organometallic reaction mechanisms; amino and amido boronic acids as enzyme inhibitors; tetrametallomethane chemistry; carboranes. *Mailing Add:* Dept Chem Wash State Univ Pullman WA 99164-4630

MATTESON, MICHAEL JUDE, b Everett, Wash, Dec 25, 36; m 63; c 3. AEROSOLS, MASS TRANSFER. *Educ:* Univ Wash, BS, 58, MS, 60; Clausthal Tech Univ, DrEng, 67. *Prof Exp:* USPHS fel radiation biol & biophys, Univ Rochester, 67-69; from asst prof to assoc prof, 69-74, PROF CHEM ENG, GA INST TECHNOL, 79- *Concurrent Pos:* NSF res grants, Sch Chem Eng, Ga Inst Technol, 70-72 & 77-81; Environ Protection Agency air pollution training grant, 70-76; Fulbright-Hays lectureship, Inst Exp Physics, Univ Vienna, Austria, 74-75; res grant, Dept of Energy, 77-79; res grant, Environ Protection Agency, 82-87. *Mem:* Am Inst Chem Engrs; Am Chem Soc; Fine Particle Soc. *Res:* Aerosol characteristics; particle-gas reactions; particle deposition; atomization; surface science and technology. *Mailing Add:* Sch Chem Eng GA Inst of Technol Atlanta GA 30332

MATTFELD, GEORGE FRANCIS, b Port Jefferson, NY, May 2, 41. WILDLIFE ECOLOGY, PHYSIOLOGICAL ECOLOGY. *Educ:* State Univ NY Col Forestry, BS, 62, Col Environ Sci & Forestry, PhD(zool), 74; Univ Mich, MWM, 64. *Prof Exp:* Instr, State Univ NY Col Forestry, 62; res asst, Univ Mich, 64-65; res asst, State Univ NY Col Environ Sci & Forestry, 65-72, res assoc, 72-77, sr res assoc, 77-78; supvry wildlife biologist, 78-81, ENVIRON MGT SPECIALIST, NY STATE DEPT ENVIRON CONSERV, 81- *Concurrent Pos:* Adj assoc prof, State Univ NY Col Environ Sci & Forestry, 78- *Mem:* Wildlife Soc. *Res:* Wildlife ecology and management; energetics of foraging; deer forest relationships. *Mailing Add:* 24 Crow Ridge Rd Voorheesville NY 12186

MATTHAEI, GEORGE L(AWRENCE), b Tacoma, Wash, Aug 28, 23; m 53; c 2. ELECTRICAL ENGINEERING. *Educ:* Univ Wash, BS, 48; Stanford Univ, MS, 49, EE, 51, PhD(elec eng), 52. *Prof Exp:* Asst, Stanford Univ, 49-51; from instr to asst prof elec eng, Univ Calif, 51-55; mem tech staff, Ramo-Woolridge Corp, 55-58; sr res engr, Stanford Res Inst, 58-60, asst head microwave group, 60-62, mgr electromagnetic tech lab, 62-64; PROF ELEC ENG, UNIV CALIF, SANTA BARBARA, 64- *Honors & Awards:* Microwave Prize, Inst Elec & Electronic Engrs, 61, Centennial Medal, 84. *Mem:* Fel Inst Elec & Electronics Engrs; Sigma Xi. *Res:* Microwave and millimeter-wave device research; acoustic devices for electric signal processing; electric circuit synthesis. *Mailing Add:* Dept Elec & Comput Engr Univ Calif Santa Barbara CA 93106

MATTHEIS, FLOYD E, b Ellendale, NDak, Dec 21, 31; m 55; c 5. SCIENCE EDUCATION. *Educ:* Univ NDak, BS, 52; Univ NC, MEd, 59, EdD, 62. *Prof Exp:* Teacher high sch, Minn, 54-58; from asst prof to assoc prof, 60-66, PROF SCI EDUC & CHMN DEPT, E CAROLINA UNIV, 66- *Concurrent Pos:* Dir NSF In-serv Inst Earth Sci for Elem Sch Teachers, 64-65 & Dist IV Nat Sci Teachers Asn, 72-74. *Mem:* AAAS; Asn Educ Teachers Sci; Nat Educ Asn; Nat Asn Res Sci Teaching; Nat Sci Teachers Asn (dir, 72-74). *Res:* Experimental studies in science teaching. *Mailing Add:* Summer Ventures Sci & Math E Carolina Univ 221 Erwin Bldg Greenville NC 27834

MATTHEISS, LEONARD FRANCIS, b Clifton, NJ, Sept 9, 31; m 60; c 5. ELECTRONIC-STRUCTURE THEORY. *Educ:* Fordham Univ, BS, 53; Mass Inst Technol, PhD(physics), 60. *Prof Exp:* Fel, Cavendish Lab, Cambridge, Eng, 61-62; res assoc, solid state & molecular theory group, Mass Inst Technol, 62-63; MEM TECH STAFF, AT&T BELL LABS, 63- *Mem:* Fel Am Phys Soc. *Res:* Electronic structure calculations for transition metals and transition-metal compounds; particular emphasis on the relationship between their electronic and superconducting properties. *Mailing Add:* AT&T Bell Labs 1D 155 PO Box 261 Murray Hill NJ 07974

MATTHES, MICHAEL TAYLOR, b Bay City, Tex, June 2, 51. ELECTRON MICROSCOPY, CELL BIOLOGY. *Educ:* Trinity Univ, BA, 73, MS, 76; Univ Calif, Los Angeles, PhD(exp path), 85. *Prof Exp:* RES ELECTRON MICROSCOPIST, UNIV CALIF, DAVIS, 85- *Concurrent Pos:* Post doctoral fel, Univ Calif, San Francisco, 87- *Mem:* Am Soc Cell Biol; Assoc Res & Vision in Opthal. *Res:* Vision. *Mailing Add:* Dept Ophthal Univ Calif Sch Med San Francisco CA 94143-0730

MATTHES, RALPH KENNETH, JR, b Conway, SC, July 27, 35; m 56; c 3. AGRICULTURAL & BIOLOGICAL ENGINEERING. *Educ:* NC State Univ, BS, 56, MS, 61, PhD(biol & agr eng), 65. *Prof Exp:* From asst to assoc agr eng, NC State Univ, 59-65; from asst prof to assoc prof, 68-72, PROF AGR & BIOL ENG, MISS STATE UNIV, 72- *Mem:* Am Soc Agr Engrs; Nat Soc Prof Engrs; Am Soc Eng Educ; Soc Am Forestry. *Res:* Drying and storage of seed under tropical condition; forest engineering research with emphasis in tree harvesting systems, site preparation for tree plantings, and fuel use during forestry operations. *Mailing Add:* Dept Agr & Biol Eng Box 5465 Miss State Univ Mississippi State MS 39762-5465

MATTHES, STEVEN ALLEN, b Charlottesville, Va, Oct 5, 50; c 4. ICP SPECTROSCOPY, MICROWAVE DISSOLUTION. *Educ:* Ore State Univ, BS, 74. *Prof Exp:* Grad teaching asst gen chem & spectros, Ore State Univ, 74-77; RES CHEMIST, BUR MINES, US DEPT INTERIOR, 77- *Mem:* AAAS; Am Chem Soc; Soc Appl Spectros; Sigma Xi. *Res:* Specialize in innovative sample dissolution techniques in preparation for ICP or AAS spectroscopy, particularly microwave dissolution methods and in high resolution sequential ICP methods for solving element analysis problems; development of PC based LIMS systems. *Mailing Add:* US Bur Mines 1450 Queen Ave SW Albany OR 97321

MATTHEWS, BENJAMIN F, PLANT MOLECULAR GENETICS. *Educ:* Univ Scranton, BS, 71; Syracuse Univ, PhD(biol), 76. *Prof Exp:* Res assoc, Agron Dept, Univ Ill, 76-79; plant biochemist, Cell Culture & Nitrogen Fixation Lab, 80-85, RES LEADER, PLANT MOLECULAR GENETICS LAB, USDA, 85- *Concurrent Pos:* Adj asst prof biol, Va Commonwealth Univ, 82- *Mem:* Am Soc Plant Physiologists; Int Soc Plant Molecular Biol; Int Asn Plant Tissue Culture; Tissue Culture Assoc; Am Asn Adv Sci. *Res:* Regulation of amino acid biosynthesis during plant development to understand metabolism of the gene, mRNA and protein levels; gene structure and expression and protein structure and regulation of expression are examined; gene transfer. *Mailing Add:* USDA-ARS Plant Molecular Biol Lab Beltsville MD 20705

MATTHEWS, BRIAN WESLEY, b SAustralia, May 25, 38; m 63; c 2. MOLECULAR BIOLOGY, X-RAY CRYSTALLOGRAPHY. *Educ:* Univ Adelaide, BSc, 59, Hons, 60, PhD(physics), 64, DSc, 87. *Prof Exp:* Mem staff, Med Res Coun Lab Molecular Biol, Eng, 63-66; vis assoc molecular biol, NIH, 67-68; assoc prof, Univ Ore, 69-72, dir, Inst Molecular Biol, 80-83, chmn dept physics, 85-86, PROF & RES ASSOC PHYSICS & MOLECULAR BIOL, UNIV ORE, 72-, DIR, INST MOLECULAR BIOL, 89- *Concurrent Pos:* Sloan Res Found fel, 71; Guggenheim Mem Found fel, 77; mem, US Nat Comt Crystallog, 80-85, 88-90; counr, Protein Soc, 90- *Mem:* Nat Acad Sci; Am Crystallog Asn; Am Acad Arts & Sci. *Res:* Protein structure and function; crystallography. *Mailing Add:* Inst Molecular Biol Univ Ore Eugene OR 97403

MATTHEWS, BURTON CLARE, b Kerwood, Ont, Dec 16, 26; m 51; c 2. SOIL CHEMISTRY. *Educ:* Ont Agr Col, BSA, 47; Univ Mo, AM, 48; Cornell Univ, PhD, 52. *Prof Exp:* From asst prof to assoc prof soil classification, Ont Agr Col, Guelph, 48-55, prof soil fertil & chem, 55-62, head dept soil sci, 62-66, acad vpres, 66-70; pres, Univ Waterloo, 70-81, vchancellor, 78-81; chmn, Ont Coun Univ Affairs, 81-83; PRES, UNIV GUELPH, 83- *Concurrent Pos:* Fel natural sci, Nuffield Found, 60-61; dir, Ont Educ Commun Authority, 72-78, Campbell Soup Co Ltd, 79-, Mutual Life Assurance Co, 82- & NEXA Corp, 81- *Mem:* AAAS; Can Soil Sci Soc. *Res:* Soil genesis, classification and fertility. *Mailing Add:* Univ Guelph Guelph ON N1G 2W1 Can

MATTHEWS, CHARLES ROBERT, b Philadelphia, Pa, May 12, 46; m 68; c 3. BIOPHYSICAL CHEMISTRY. *Educ:* Univ Minn, BS, 68; Stanford Univ, MS, 69, PhD(chem), 74. *Prof Exp:* Fel biochem, Stanford Univ, 74-75; ASST PROF CHEM, PA STATE UNIV, UNIVERSITY PARK, 75- *Mem:* Am Chem Soc; Sigma Xi. *Res:* Conformational changes in biological macromolecules; mechanisms of reversible unfolding transitions in proteins; effect of missense mutations on protein folding and stability; chemical trapping of intermediates in protein folding. *Mailing Add:* Dept of Chem 152 Davey Lab Pa State Univ University Park PA 16802

MATTHEWS, CHARLES SEDWICK, b Houston, Tex, Mar 27, 20; m 45; c 2. EARTH SCIENCES. *Educ:* Rice Inst, BS, 41, MS, 43, PhD(phys chem), 44. *Prof Exp:* Engr chem plant design, Shell Develop Co, 44-48, chemist, 48-56, sr res assoc, 56-66, mgr exploitation eng, Shell Oil Co, 66-67, dir prod res, Shell Develop Co, 67-72; mgr eng, Shell Oil Co, 72-73, sr petrol engr consult, 73-89; RETIRED. *Concurrent Pos:* Distinguished lectr, Soc Petrol Engrs, 68 & Nat Acad Eng, 85; consult, 89- *Honors & Awards:* Lester C Uren Award, Soc Petrol Engrs, 75. *Mem:* Nat Acad Eng; Soc Petrol Engrs; Am Petrol Inst. *Res:* New methods for recovery of petroleum; behavior of petroleum reservoirs; geothermal energy; recovery from tar sands and oil shale. *Mailing Add:* 5307 S Braeswood Blvd Houston TX 77096

MATTHEWS, CLIFFORD NORMAN, b Hong Kong, Dec 20, 21; nat US; m 47; c 2. COSMOCHEMISTRY. *Educ:* Univ London, BSc, 50; Yale Univ, PhD(chem), 55. *Prof Exp:* Lab supt, Birkbeck Col, Univ London, 46-48; res chemist, Conn Hard Rubber Co, 50-51, Diamond Alkali Co, 55-59 & Monsanto Co, 59-69; prof chem, Univ Ill, Chicago, 69-91; RETIRED. *Mem:* AAAS; Am Chem Soc; Royal Soc Chem; Am Astron Soc. *Res:* Chemical evolution: origin of molecules in biochemistry, geochemistry and galactochemistry, cosmochemistry and the origin of life. *Mailing Add:* Dept Chem M/C 111 Univ Ill Box 4348 Chicago IL 60680

MATTHEWS, DAVID ALLAN, b Washington, DC, Feb 5, 43; m 67; c 2. BIOPHYSICAL CHEMISTRY. *Educ:* Earlham Col, AB, 65; Univ Ill, PhD(chem), 71. *Prof Exp:* Fel chem, 71-76, assoc res chemist, Univ Calif, San Diego, 76-85; DIR PROTEIN CRYSTALLOG, AGOURON PHARMACEUT, 85- *Concurrent Pos:* Jane Coffin Childs Mem Fund Med Res fel, 72-74; Nat Cancer Inst fel, 74-76. *Res:* X-ray crystallographic studies of molecular structure and enzyme mechanisms; structure based design of protein inhibitors. *Mailing Add:* Agouron Pharmaceut 11025 N Torrey Pines Rd La Jolla CA 92037

MATTHEWS, DAVID LESUEUR, b Ottawa, Ont, May 10, 28; m 56; c 3. SPACE PHYSICS. *Educ:* Queen's Univ, Ont, BSc, 49; Princeton Univ, PhD(physics), 59. *Prof Exp:* Res officer, Nat Res Coun Can, 49-53; instr physics, Princeton Univ, 57-59; lectr, Carleton Univ, 59-60; sci officer, Defense Res Telecommun Estab, 60-66; RES ASSOC PROF, INST PHYS SCI & TECHNOL, UNIV MD, COLLEGE PARK, 66- *Concurrent Pos:* Rocket sect leader, Defense Res Telecommun Estab, 63-65; prin investr, NASA & NSF grants, 66- *Honors & Awards:* Group Achievement Award, NASA, 81. *Mem:* Am Geophys Union; Am Phys Soc; Inst Elec & Electronics Engrs. *Res:* Space and upper atmosphere physics; electron precipitation and wave-particle interactions; mesospheric density; time-series analysis. *Mailing Add:* IPST Univ Md College Park MD 20742-2431

MATTHEWS, DAVID LIVINGSTON, b New York, NY, Mar 13, 22; m 44, 85; c 4. AGRONOMY, PLANT BREEDING. *Educ:* Rutgers Univ, BS, 48, MS, 50. *Prof Exp:* Asst farm crops, Rutgers Univ, 48-50; tech specialist radiation genetics, Brookhaven Nat Lab, 50-52; plant breeder, Eastern State Farmers Exchange, 52-54, mgr corn res, 55-64; mgr seed res, Agway Inc, 65-66, dir farm eval & seed res, 66-68, dir crops res, 68-81, dir, Crops Res & Develop, 81-85; CULT EXCHANGE OFFICER, MATTHEWS ASSOC INC, 86- *Concurrent Pos:* Dir, Farmers Forage Res Coop, 65-68; mem, Plant Variety Protection Bd, 83, 86 & 88. *Honors & Awards:* Exten Indust Award, Am Soc Agron, 80, Career Serv Award, 90. *Mem:* Am Soc Agron; Crop Sci Soc Am; Am Soc Hort Sci; Coun Agr Sci & Technol. *Res:* Mulch and irrigation management of vegetables; dairy and poultry manure management for optimum crop returns; forage nutrient conservation through crop management and use of chemical preservatives; high yield research with corn cereals, alfalfa and potatoes; computer applications to crop input recommendations and field history records; participated in developing a dairy industry proposal for Nigerian development company in Nigeria; growing Holstein bull calves to 1000 pound steers in 12 months and market studies of the meat produced from them; studies of by-product utilization for the brewery industry; determine the probility that the USSR can become self sufficient in food production by the year 2000; market studies of fresh aeduwx corn as a new sweet corn genotype pennfresh ADX. *Mailing Add:* David Matthews Assoc Inc 4141 Bayshore Blvd No 1601 Tampa FL 33611-1802

MATTHEWS, DEMETREOS NESTOR, b Port Chester, NY, June 28, 28; m 53; c 2. ORGANIC CHEMISTRY. *Educ:* Rutgers Univ, BS, 49; Polytech Inst Brooklyn, PhD(chem), 60. *Prof Exp:* Chemist, Res Labs, Air Reduction Co, Inc, 52-54; res sci, Res Ctr, US Rubber Co, 59-67, sr res sci, 67-84, res assoc org chem, Uniroyal Res Ctr, 84-88; ASSOC CHEMIST, MOBIL CHEM CO, PENNINGTON, NJ, 88- *Mem:* Am Chem Soc; Sigma Xi. *Res:* Organic reaction mechanisms; Diels-Alder reactions; solution polymerization; free radical reactions; correlation of mechanism with structure. *Mailing Add:* Four Forest Lane Ewing NJ 08628

MATTHEWS, DENNIS L, b Dilheart, Tex, July 15, 48. PHYSICS. *Educ:* Univ Tex, Austin, MS, 72, PhD(physics), 75. *Prof Exp:* GROUP RES PHYSICS, LAWRENCE LIVERMORE LAB, 72- *Mem:* Fel Am Phys Soc. *Mailing Add:* Lawrence Livermore Lab PO Box 808 Livermore CA 94550

MATTHEWS, DOYLE JENSEN, b Liberty, Idaho, Apr 13, 26; m 46; c 5. ANIMAL BREEDING. *Educ:* Utah State Univ, BS, 50, MS, 51; Kans State Univ, PhD, 59. *Prof Exp:* From instr to assoc prof animal husb, 51-65, asst dean, Col Agr, 65-69, assoc dean, 69-71, PROF ANIMAL, DAIRY & VET SCI, UTAH STATE UNIV, 66-, DEAN, COL AGR, 71-, DIR, AGR EXP STA, 74- *Mem:* Am Soc Animal Sci. *Res:* Improvement of carcass characteristics and productivity of meat animals through application of breeding techniques and methods. *Mailing Add:* Col of Agr Utah State Univ Logan UT 84322-4800

MATTHEWS, DWIGHT EARL, b Greencastle, Ind, Sept 10, 51; m 72; c 2. METABOLISM, MASS SPECTROMETRY. *Educ:* De Pauw Univ, BA, 73; Ind Univ, PhD(chem), 77. *Prof Exp:* From instr to asst prof, 77-86, ASSOC PROF MED & SURGERY, CORNELL UNIV MED COL,86- *Concurrent Pos:* Vpres, Prevo's Inc, Greencastle, Ind, 75-85; consult, 80- *Mem:* Am Fedn Clin Res; Am Soc Clin Nutrit; Am Physiol Soc; Am Inst Nutrit. *Res:* Investigation of control of energy and protein metabolism in humans under pathophysiological states of stress and trauma; development of stable isotope tracer methods for measuring in vivo kinetics of metabolite turnover. *Mailing Add:* Cornell Univ Med Col Rm A 328 1300 York Ave New York NY 10021

MATTHEWS, E(DWARD) K, b Staten Island, NY, Apr 10, 30; m 52; c 3. MECHANICAL ENGINEERING. *Educ:* Mass Inst Technol, SB, 52; Univ Ill, MS, 55, PhD(mech eng), 60. *Prof Exp:* Res engr, Esso Res & Eng Co, 59-65, sect head new uses, 65-66, mgr gen tech serv, 66-70; pres, E P G Comput Serv, Inc, 70-72; PRES, SYSTECH BUS SYSTS, 72- *Concurrent Pos:* At pyromete instrument Cok. *Res:* Heat transfer; building materials; petroleum processes and products. *Mailing Add:* Pyrometer Instrument Co Industrial Pkwy Northvale NJ 07647

MATTHEWS, E(DGAR) W(ESLEY), JR, b Johnstown, Pa, May 19, 25; m 49; c 2. ELECTRICAL ENGINEERING. *Educ:* Rensselaer Polytech Inst, BEE, 46, MEE, 50; Harvard Univ, PhD(appl physics), 54. *Prof Exp:* Jr engr, Westinghouse Res Labs, 46-47; instr elec eng, Rensselaer Polytech Inst, 47-49; proj engr, Sperry Gyroscope Co, 53-54, sr engr, 54-55; from asst prof to assoc prof elec eng, Rensselaer Polytech Inst, 55-58; engr microwave res, Radio Corp Am, 58-60, eng leader, 60-62; dept head, Sperry Microwave Electronics Corp, 62-65; mgr solid state component res & develop, Watkins-Johnson Co, 65-66; sr staff specialist, Sylvania Electronic Systs-W, 66-68; independent consult, 68-70; staff scientist, Philco-Ford Corp, 70-76, PRIN ENGR, FORD AEROSPACE & COMMUN CORP, 76- *Mem:* Sr mem Inst Elec & Electronics Engrs. *Res:* Microwave theory and techniques; microwave devices and antennas. *Mailing Add:* Ford Aerospace & Comm Corp 3825 Fabian Way G-43 Palo Alto CA 94303

MATTHEWS, EDWARD WHITEHOUSE, b Annapolis, Md, Mar 7, 36. ANALYTICAL CHEMISTRY, ORGANIC CHEMISTRY. *Educ:* US Mil Acad, BS, 58; Univ Tex, MS, 71; Rutgers Univ, PhD(environ sci), 78. *Prof Exp:* Res chemist org chem, US Geol Surv, 76-81, res chemist trace metals, 81-86; PROF, OGLETHORPE UNIV, 85- *Concurrent Pos:* Res fel, Rutgers Univ, 70-71. *Mem:* Am Chem Soc; Asn Off Anal Chem; Am Soc Testing & Mat; Sigma Xi. *Res:* Organic chemicals found in association with waters, wastewaters, fish, plant tissues and sedimentary deposits. *Mailing Add:* 535 Eagles Landing Dr Alpharetta GA 30201

MATTHEWS, FLOYD V(ERNON), JR, b Parksley, Va, Aug 4, 27; m 46; c 6. AGRICULTURAL ENGINEERING. *Educ:* Va Polytech Inst, BS, 51; Okla State Univ, MS, 51; Mich State Univ, PhD, 66. *Prof Exp:* Layout draftsman, Int Harvester Co, 51-55; from asst prof to assoc prof agr eng, Univ Md, College Park, 55-68; assoc prof, 68-70, chmn dept, 70-86, PROF AGR ENG, CALIF STATE POLYTECH UNIV, POMONA, 86- *Concurrent Pos:* Agr consult, Ministry Educ, Greece, 72; Ministry Agr Egypt, 82-84. *Mem:* Am Soc Agr Engrs; Am Soc Eng Educ. *Res:* Agricultural and food processing. *Mailing Add:* 890 Butte St Claremont CA 91711

MATTHEWS, FREDERICK WHITE, b Carbonear, Nfld, Nov 27, 15; m 43; c 4. CHEMISTRY, INFORMATION SCIENCE. *Educ:* Mt Allison Univ, BSc, 36; McGill Univ, PhD(phys chem), 41. *Prof Exp:* Res chemist, Can Industs Ltd, Montreal, 41-51, asst mgr prod, 51-56, mgr info serv, 56-69; mgr, Cent Tech Info Unit, Imp Chem Industs, Ltd, Eng, 69-72; prof info sci, Sch Libr Serv, Dalhousie Univ, 72-85; RETIRED. *Concurrent Pos:* Chmn data comn, Int Union Crystallog, 48-72; mem adv bd sci & tech info, Nat Res Coun Can, 72-77; consult, Info Systs, Imp Chem Industs, Eng, 72-76, Indexing Systs, Inst Jamaica, Kingston, 74-76 & Dartmouth Regional Libr, NS, 75; consult paper conserv, 86- *Mem:* Chem Inst Can; Can Asn Info Sci; Am Soc Info Sci. *Res:* Library catalogue systems; x-ray diffraction powder data; systems for data retrieval; systems for storage and retrieval of information on computers. *Mailing Add:* Dalhousie Univ 1168 Studley Ave Halifax NS B3H 3R7 Can

MATTHEWS, GARY JOSEPH, b Denver, Colo, Aug 6, 42; m 64; c 3. ORGANIC CHEMISTRY, TECHNICAL MANAGEMENT. *Educ:* Colo State Univ, BS, 64; Univ Colo, Boulder, PhD(org chem), 68. *Prof Exp:* Syntex res fel, Inst Org Chem, Syntex Res, Palo Alto, 68-69, res chemist, Arapahoe Chems Div, Syntex Corp, 69-72, group leader, 72, mgr res, 72-76, dir res & develop, 76-81, vpres res & develop, Arapahoe Chems Inc, 81-83, dir res & develop, Syntex Chem Inc, 83-84, DIR CHEM TECHNOL, SYNTEX USA INC, 84- *Mem:* Am Chem Soc. *Res:* Process research and development on the production of fine organic chemicals and bulk drug substances. *Mailing Add:* Syntex USA Inc 3401 Hillview Ave Palo Alto CA 94303

MATTHEWS, HARRY ROY, b Faversham, Kent, UK, May 25, 42; m 79. CHROMOSOMAL PROTEINS, PROTEIN KINASES. *Educ:* Univ London, BSc, 63, PhD(biophys), 68. *Prof Exp:* Fel, Univ Hong Kong, 63-64; sr lectr physics & biochem, Portsmouth Polytech, 67-80; assoc prof, 80-85, PROF BIOL CHEM, UNIV CALIF, DAVIS, 85- *Concurrent Pos:* Prin investr, Sci Res Coun, Eng, 73-80, Cancer Soc UK, NATO & Am Cancer Soc, 76- & NIH, 83-87, NSF, 87-91; vis scientist, Rockefeller Univ, 75-78. *Mem:* Am Soc Biochem & Molecular Biol; AAAS. *Res:* Chromosome structure and function emphasizing protein kinases and polyamines in the cell nucleus. *Mailing Add:* Dept Biol Chem Sch Med Univ Calif Davis CA 95616

MATTHEWS, HAZEL BENTON, JR, b Hertford, NC, Feb 8, 40; m 65; c 2. ENVIRONMENTAL HEALTH. *Educ:* NC State Univ, BS, 63, MS, 65; Univ Wis-Madison, PhD(entom), 68. *Prof Exp:* NIH grant, Univ Calif, Berkeley, 68-70; staff fel chem, Nat Inst Environ Health Sci, NIH, 70-71, sr staff fel, 71-74, res chem, 74-79, head chem disposition, 79-89, CHIEF, EXP TOXICOL BR, NAT INST ENVIRON HEALTH SCI, NIH, 89- *Concurrent Pos:* Adj prof, NC State Univ, 79- *Mem:* Soc Toxicol. *Res:* Studies of chemical disposition and mechanisms of chemical toxicity to determine the chemical and biochemical interactions which account for toxicity and/or carcinogenicity following acute or chronic exposure to chemicals. *Mailing Add:* Div of Toxicol & Res Testing Nat Inst Environ Health Sci NIH PO Box 12233 Research Triangle Park NC 27709

MATTHEWS, HEWITT WILLIAM, b Pensacola, Fla, Dec 1, 44; m 69; c 2. PHARMACEUTICAL CHEMISTRY. *Educ:* Clark Col, BS, 66; Mercer Univ, BS, 68; Univ Wis, MS, 71, PhD(pharm, biochem), 73. *Prof Exp:* From asst prof to prof pharm, 73-83, dir res, 75-81, asst dean serv, 81-83, asst provost, 83-85, ASSOC DEAN, SCH PHARM, MERCER UNIV, 85- *Concurrent Pos:* Vis scientist, Ctr Dis Control, 87 & 88. *Mem:* Sigma Xi; AAAS; Am Asn Cols Pharm; Nat Inst Sci. *Res:* Pharmacologically active agents from microbial origin; effect of antibiotics or enzymes from biological systems. *Mailing Add:* Mercer Univ Sch Pharm 345 Boulevard NE Atlanta GA 30312

MATTHEWS, JAMES B, b Ft Benning, Ga, May 15, 33; m 53; c 3. AEROSPACE & MECHANICAL ENGINEERING. *Educ:* Rose Polytech Inst, BS, 54; Mass Inst Technol, MS, 59; Univ Ariz, PhD(aerospace eng), 66. *Prof Exp:* Design engr, Collins Radio Co, 54; instr mech eng, Rose-Hulman Inst Technol, 56-58, asst prof, 59-63, prof & chmn dept, 66-78, dean fac, 70-78; dean eng, Bradley Univ, 81-83; prof mech eng & chmn dept, 78-81, DEAN ENG & APPL SCI, WESTERN MICH UNIV, 83- *Mem:* Am Soc Mech Engrs; Am Soc Eng Educ; Soc Automotive Engrs; Nat Soc Prof Engrs; Asn Energy Engrs. *Res:* Vibrations and structural dynamics. *Mailing Add:* Dept Aerotech TAE Ariz State Univ Tempe AZ 85287-6404

MATTHEWS, JAMES FRANCIS, b Winston-Salem, NC, Sept 14, 35; m 61; c 2. CYTOLOGY, PLANT TAXONOMY. *Educ:* Atlantic Christian Col, BA, 57; Cornell Univ, MS, 60; Emory Univ, PhD(cytol), 62. *Prof Exp:* Asst prof biol, Western Ky State Univ, 62-64; from asst prof to assoc prof, 64-72, PROF BIOL, UNIV NC, CHARLOTTE, 72- *Concurrent Pos:* Grants & contracts officer, 74-75. *Mem:* Int Asn Plant Taxon; Am Asn Plant Taxon. *Res:* Speciation of plants endemic to the granite outcrops of the southeastern Piedmont; endangered and threatened plant species and habitats; floristics of urban areas. *Mailing Add:* Dept Biol Univ NC Charlotte NC 28223

MATTHEWS, JAMES HORACE, b Campbellton, NB, Mar 1, 30; m 54; c 5. NUCLEAR PHYSICS. *Educ:* Mt Allison Univ, BSc & cert eng, 51; Dalhousie Univ, MSc, 54; Univ London, PhD(physics), 57. *Prof Exp:* From asst prof to assoc prof, 57-69, PROF PHYSICS, MT ALLISON UNIV, 69- *Concurrent Pos:* Marjorie Young Bell fel, Univ Sussex, 66-67; vis prof, Univ Toronto, 74-75; vis prof, Univ Strathelyde, 88-89. *Mem:* Can Asn Physicists; Brit Inst Physics. *Res:* Theoretical nuclear physics; nuclear models; solar energy systems; picture enhancement. *Mailing Add:* Dept Physics Mt Allison Univ Sackville NB E0A 3C0 Can

MATTHEWS, JAMES LESTER, b Denton, Tex, July 3, 26; m 50; c 3. MICROSCOPIC ANATOMY. *Educ:* NTex State Col, BS, 48, MS, 49; Univ Ill, PhD, 55. *Prof Exp:* Asst biol, NTex State Col, 47-48; instr biol & chem, Cisco Jr Col, 49-52; asst physiol, Univ Ill, 52-55; from res asst to assoc prof anat & physiol, 55-60, PROF MICROS ANAT & CHMN DEPT HISTOLMICROS ANAT, BAYLOR COL DENT, 60-, ASSOC DEAN, BAYLOR UNIV MED CTR, 74-, EXEC DIR, BAYLOR RES FOUND, 85- *Mem:* Am Physiol Soc; assoc Soc Exp Biol & Med; Am Asn Anat. *Res:* Physiology and fine structure of bone and connective tissues; photobiology. *Mailing Add:* Baylor Res Found 3812 Elm St Dallas TX 75246

MATTHEWS, JOHN BRIAN, oceanography, for more information see previous edition

MATTHEWS, JUNE LORRAINE, b Cambridge, Mass, Aug 1, 39. NUCLEAR PHYSICS, INTERMEDIATE ENERGY PHYSICS. *Educ:* Carleton Col, BA, 60; Mass Inst Technol, SM, 62, PhD(physics), 67. *Prof Exp:* NSF fel physics, Glasgow Univ, 68-71; res assoc, Rutgers Univ, 71-72; from asst prof to assoc prof, 72-82, PROF PHYSICS, MASS INST TECHNOL, 82- *Concurrent Pos:* Mem, NSF Adv Panel Physics, 75-78, Am Phys Soc Div Nuclear Physics Prog Comt, 80-82, Am Phys Panel Pub Affairs, 81-84, Los Alamos Meson Physics Facil Prog Adv Comt, 82-85, prog adv comt, Bates Linear Accelerator Ctr, Mass Inst Technol, 85-88; Benedict distinguished vis prof physics, Carleton Col, 83; vis prof physics, Yale Univ, 84; chmn, Los Alamos Meson Physics Fac Users Group Bd Dir, 87; councillor, Am Phys Soc, 86-; mem, Exec Com Am Phys Soc, 86-88; vis prof physics, Oberlin Col, 88. *Mem:* Fel Am Phys Soc; Sigma Xi; Am Asn Physics Teachers; AAAS. *Res:* Interactions of photons with nuclei; study of nucleon momentum distributions, short-range correlations and meson exchange effects; few-body problems; pion-nucleus interaction mechanisms. *Mailing Add:* 10A Kirkland Rd Cambridge MA 02138

MATTHEWS, KATHLEEN SHIVE, b Austin, Tex, Aug 30, 45; m 67. BIOCHEMISTRY. *Educ:* Univ Tex, Austin, BS, 66; Univ Calif, Berkeley, PhD(biochem), 70. *Prof Exp:* Am Asn Univ Women fel, Sch Med, Stanford Univ, 70-71, Giannini Found fel, 71-72; from asst to prof, 71-89, WEISS PROF BIOCHEM, RICE UNIV, 89-, CHAIR, 87- *Concurrent Pos:* NSF fel. *Mem:* Am Soc Biol Chemists; AAAS; Soc Neurosci; Am Chem Soc. *Res:* Chemistry and molecular biology of proteins; studies on the lactose and tryptophan repressor proteins from Escherichia coli, including chemical modification, spectroscopy and other physical methods; lymphocyte metabolism. *Mailing Add:* Dept Biochem & Cell Biol Rice Univ PO Box 1892 Houston TX 77251

MATTHEWS, LEE DREW, b Platteville, Wis, Mar 10, 43; m 69; c 1. PHYSICS. *Educ:* Wis State Col, BS, 64; Univ Vt, MS, 67, PhD(physics), 69. *Prof Exp:* Asst prof, 69-77, ASSOC PROF PHYSICS, SOUTHERN CONN STATE COL, 77- *Mem:* Am Inst Physics; Am Phys Soc; Am Asn Physics Teachers. *Res:* Surface physics; physics education. *Mailing Add:* Dept of Physics Southern Conn State Col New Haven CT 06515

MATTHEWS, LESLIE SCOTT, b Baltimore, Md, Sept 18, 51; m 81; c 2. SPORTS MEDICINE, ARTHROSCOPIC SURGERY. *Educ:* Johns Hopkins Univ, BA, 73; Baylor Col Med, MD, 76. *Prof Exp:* ASST CHIEF ORTHOP SURG, UNION MEM HOSP, 81-; ASST PROF ORTHOP SURG, JOHNS HOPKINS HOSP, 81- *Mem:* Am Acad Orthop Surg; Arthroscopy Asn NAm; Am Col Surgeons. *Res:* techniques of arthroscopic surgery, primarily arthroscopy of the shoulder joint. *Mailing Add:* Dept Orthop Surg Union Mem Hosp 201 E University Pkwy Baltimore MD 21218

MATTHEWS, MARTIN DAVID, b Elizabeth, NJ, Dec 10, 38; m 64; c 1. SEDIMENTATION, GEOCHEMISTRY. *Educ:* Allegheny Col, BS, 60; WVa Univ, MS, 63; Northwestern Univ, PhD(geol), 73. *Prof Exp:* Asst prof geol, Wash State Univ, 72-74; res geologist, Gulf Oil, 74-76, sr res geologist, 76-77, dir geol sect, 77-79, sr staff geologist, 79-81, sr res assoc, Gulf Res & Develop Co, 81, mission coordr subsurface processes geochem, 81-82, mgr geochem & minerals, 82-83, consult, 83-84; SR SCIENTIST, HOUSTON RES CTR, TEXACO, 84- *Concurrent Pos:* Test site dir, Geosat, Inc, 77-83. *Mem:* Soc Econ Paleonotol & Mineral; Geol Soc Am; Int Asn Sedimentol; Am Asn Petrol Geologists. *Res:* Flocculation of river sediments; sedimentology; clay mineralogy; geostatistics; remote sensing; basin evaluation; inorganic and petroleum geochemistry; surface stratigraphic evolution of rifts. *Mailing Add:* 3007 Deer Creek Sugarland TX 77478

MATTHEWS, MARY EILEEN, b Rochester, NY, May 22, 38. FOOD SCIENCE, NUTRITION. *Educ:* Drexel Univ, BS, 60; Okla State Univ, dipl, 61, MS, 62; Univ Wis-Madison, PhD(food sci), 70. *Prof Exp:* Asst nutritionist, Nat Diet Heart Study, Johns Hopkins Hosp, 63-65; res asst, 65-67, asst prof, 70-74, assoc prof, 74-79, PROF FOOD SCI, UNIV WIS-MADISON, 79- *Concurrent Pos:* USPHS grant, 72-73; Am Dietetic Asn & Dept Health, Educ & Welfare grant, Loma Linda Univ, 73-74; prog planning chair, Symposium, Hosp Patient Feeding Syst, US Army Natick Res & Develop Labs, 81. *Mem:* Am Dietetic Asn; Am Sch Food Serv Asn; Inst Food Technol. *Res:* Quality and safety of food produced and served in foodservice systems; optimal use of management resources in foodservice systems. *Mailing Add:* Dept Food Sci 201 Babcock Hall Univ Wis 1605 Linden Dr Madison WI 53706

MATTHEWS, MURRAY ALBERT, b Houston, Tex, June 16, 43; m 69; c 1. ANATOMY, NEUROPATHOLOGY. *Educ:* Univ St Thomas, BA, 65; Univ Tex Med Br, MA, 67, PhD(anat), 70. *Prof Exp:* PROF ANAT, MED CTR, LA STATE UNIV, NEW ORLEANS, 72- *Concurrent Pos:* NIH trainee, Brain Res Inst, Med Sch, Univ Calif, Los Angeles, 70-72; Schlieder Educ Found res grant, 74; Nat Inst Neurol & Commun Disorders & Stroke grant, Nat Inst Dent res grant. *Mem:* Am Asn Anat; Soc Neurosci; Am Acad Dent Radiol. *Res:* Central nervous system trauma; spinal cord injury; reaction of neurons to mechanical or ischemic injury; reactive changes in nonneuronal, vascular elements; reaction of irigeminal system to injury. *Mailing Add:* Dept Anat La State Univ Med Ctr 1100 Florida Ave New Orleans LA 70119

MATTHEWS, N(EELY) F(ORSYTH) J(ONES), b Clinton, NC, Aug 9, 31. SOLID STATE PHYSICS, ELECTRICAL ENGINEERING. *Educ:* George Washington Univ, BS, 57, MS, 59; Princeton Univ, MA, 62, PhD(solid state physics), 64. *Prof Exp:* Instr elec eng, George Washington Univ, 57-59; from asst prof to assoc prof, 64-76, PROF ELEC ENG, NC STATE UNIV, 76- *Mem:* Inst Elec & Electronics Engrs; Sigma Xi. *Res:* Electronic and optical properties of cadmium sulfide; photoconductivity; recombination and generation of charge; optical absorption; luminescence. *Mailing Add:* 2429 Coley Forest Pl Raleigh NC 27612

MATTHEWS, R(OBERT) B(RUCE), b Red Bank, NJ, Mar 7, 42; m 66; c 3. CERAMICS, NUCLEAR FUELS. *Educ:* Pa State Univ, BS, 64; Univ Denver, MS, 66; Univ Col Swansea, Wales, PhD(mat sci), 70. *Prof Exp:* Mem staff, Atomic Energy of Can Ltd, 70-78; res scientist, Northwest Labs, Battelle Mem Inst, 78-80; res scientist, 80-83, PROJ MGR, LOS ALAMOS NAT LABS, 83- *Mem:* Am Ceramic Soc. *Res:* Fabrication development properties and irradiation performance of ceramics and nuclear fuels, ie, oxides, carbides, nitrides and silicides. *Mailing Add:* 6637 Moly Dr Falls Church VA 22046

MATTHEWS, RICHARD FINIS, b Cullman, Ala, June 1, 29; m 55; c 2. FOOD CHEMISTRY, BIOCHEMISTRY. *Educ:* Univ Fla, BSA, 52; Cornell Univ, MS, 57, PhD(food sci), 60. *Prof Exp:* Assoc technologist, Res Ctr, Gen Foods Corp, 60-63; group leader tea chem, T J Lipton Res Ctr, NJ, 63-65; assoc prof, 65-73, PROF FOOD TECHNOL, UNIV FLA, 73- *Mem:* Am Chem Soc; fel Inst Food Technologists. *Res:* Natural products chemistry; flavor chemistry; processing horticultural crops. *Mailing Add:* Dept Food Sci Univ Fla Gainesville FL 32605

MATTHEWS, RICHARD JOHN, JR, b Scranton, Pa, Apr 11, 27; m 53; c 4. PHARMACOLOGY. *Educ:* Philadelphia Col Pharm, BS, 51; Jefferson Med Col, MS, 53, PhD(pharmacol), 55. *Prof Exp:* Head pharmacol res sect, Upjohn Co, Mich, 56-62; pres, Pharmakon, Inc, Pa, 62-65; dir pharmacol, Union Carbide Corp, 65-69; DIR RES, PHARMAKON LABS, 69- *Mem:* Am Soc Pharmacol & Exp Therapeut. *Res:* Action of drugs on synapse in peripheral and central nervous system, especially neurohumoral agents; neuropharmacology of psychotherapeutic drugs and effects of extracts of blood from schizophrenics on the central nervous system. *Mailing Add:* Res Dept Pharmakon Labs Waverly PA 18471

MATTHEWS, ROBERT A, b Augusta, Ga, June 16, 26. ENGINEERING GEOLOGY, HYDROGEOLOGY. *Educ:* Univ Calif, Berkeley, AB, 53. *Prof Exp:* SR LECTR GEOL, UNIV CALIF, DAVIS, 71- *Mem:* Fel Geol Soc Am; fel Sigma Xi; Asn Eng Geologists. *Res:* Groundwater hydrogeology; environmental geology. *Mailing Add:* Dept Geol Univ Calif Davis CA 95616

MATTHEWS, ROBERT WENDELL, b Detroit, Mich, Feb 17, 42; m 63; c 6. ENTOMOLOGY. *Educ:* Mich State Univ, BS, 63, MS, 65; Harvard Univ, PhD(biol), 69. *Prof Exp:* Asst prof, 69-74, assoc prof, 74-79, PROF ENTOM, UNIV GA, 79- *Concurrent Pos:* NSF res assoc, Commonwealth Sci & Indust Res Orgn, Canberra, Australia, 69-70 & Inst Miguel Lillo, Tucuman, Arg, 72; prin investr, NSF res grant, 75, 78 & 84; vis prof, Neurobiol & Behav, Cornell Univ, 79-80. *Mem:* Am Inst Biol Sci; Entom Soc Am; Animal Behav Soc; Int Union Study Social Insects. *Res:* Behavior; systematics; ecology and evolution of Hymemoptera, especially Braconidae, Sphecidae and Vespidae; social insects. *Mailing Add:* 655 Riverview Rd Athens GA 30606

MATTHEWS, ROBLEY KNIGHT, b Dallas, Tex, Oct 6, 35; m 59; c 4. SEDIMENTOLOGY. *Educ:* Rice Univ, BA, 57, MA, 63, PhD(geol), 65. *Prof Exp:* Petrol geologist, Pan Am Petrol Corp, 57-58 & Am Int Oil Co, Libya, 58-60; geologist, Marine Geophys Serv, 60-63; asst prof, 64-71, chmn dept, 71-77, PROF GEOL, BROWN UNIV, 71- *Mem:* Geol Soc Am; Am Asn Petrol Geologists; Soc Econ Paleont & Mineral; Am Geophys Union. *Res:* Stratigraphic modeling; cenozoic sea level history; dynamics of climate change; physical and chemical aspects of carbonate deposition and diagenesis; finite resources. *Mailing Add:* Dept Geol Sci Brown Univ Brown Sta Providence RI 02912-1846

MATTHEWS, ROWENA GREEN, b Cambridge, Eng, Aug 20, 38; US citizen; m 60; c 2. ENZYMOLOGY, METABOLIC REGULATION. *Educ:* Radcliffe Col, BA, 60; Univ Mich, Ann Arbor, PhD(biophys), 69. *Prof Exp:* Instr biol, Univ SC, 63-64; fel biol chem, 71-74, res investr, 74-75, asst prof 75-81, assoc prof biol chem, 81-86, PROF BIOL CHEM & RES SCIENTIST, BIOPHYSICS, UNIV MICH, ANN ARBOR, 86-, ASSOC CHMN, DEPT BIOL CHEM, 88- *Concurrent Pos:* Res chemist, Vet Admin Hosp, 75-78; Estab investr, Am Heart Asn, 78-83; mem, Phys Biochem Study Sect, NIH, 82-86; prog chmn, Biol Chem Div, Am Chem Soc, 85; mem, exec comt, Biol Chem Div, Am Chem Soc, 86-88; mem comt, Am Soc Biol Chem & Mol Biol, 88; secy, Biol Chem Div, Am Chem Soc, 90-93. *Mem:* Am Chem Soc; Biophys Soc; Sigma Xi; AAAS; Am Soc Biochem & Mol Biol; Am Soc Microbiol. *Res:* Catalytic mechanisms of folate-dependent enzymes; catalytic mechanisms of flavoprotein dehydrogenases; regulation of folate metabolism; biochemical correlates of heat shock in Escherichia coli. *Mailing Add:* Dept of Biol Chem Univ of Mich Ann Arbor MI 48109

MATTHEWS, RUTH H, b Cambridge, Md, Feb 12, 26. HUMAN NUTRITION. *Educ:* Univ Md, BS; Columbia Univ, MA. *Prof Exp:* Nutritionist, Human Nutrit Info Serv, 69-77, supvry nutritionist & head, Plant Prod Sect, 77-87, CHIEF, NUTRIENT DATA RES BR, HUMAN NUTRIT INFO SERV, USDA, HYATTSVILL, MD, 87- *Mem:* Am Inst Nutrit; NY Acad Sci; Inst Food Technologists; Am Oil Chemists Soc; Asn Off Anal Chemists; Am Asn Cereal Chemists; Am Dietetic Asn. *Res:* Nutrition; food technology. *Mailing Add:* Human Nutrit Info Serv USDA Fed Bldg 6505 Belcrest Rd Hyattsville MD 20782

MATTHEWS, SAMUEL ARTHUR, ENDOCRINES OF LOWER VETEBRATES. *Educ:* Harvard Univ, PhD(zool), 28. *Prof Exp:* Prof biol, Williams Col, 37-70; RETIRED. *Res:* Respiratory systems of lower vertebrates. *Mailing Add:* Williams Col 130 Woodcock Rd Williamstown MA 01267

MATTHEWS, STEPHEN M, b New York, NY, Oct 25, 38; m 71; c 3. DECOMPOSITION OF HALOGENATED HYDROCARBONS USING INTENSE BREMSSTRAHLUNG, PRODUCTION OF PLASMA X-RAYS FOR COMMERCIAL PURPOSES. *Educ:* Antioch Col, Yellow Springs, Ohio, BS, 61; NY Univ, MS, 64, PhD(physics), 68. *Prof Exp:* Lectr physics, City Col NY, 63-67; sr staff physicist, Physics Int Corp, 79-82; SR PHYSICIST, LAWRENCE LIVERMORE NAT LAB, 68-79 & 82- *Concurrent Pos:* Prin, Accident Anal Assocs, Oakland, Calif, 83-; sci adv, Photonics Spectra, 86- *Mem:* AAAS; Am Phys Soc. *Res:* Production of radiation from high power electron accelerators for processing food, for breakdown of toxic wastes, and environmental restoration; advanced technology spin-off from weapons research to commercial purposes; design and testing of new detection methods for measuring underground nuclear explosions; accident analysis and reconstruction for legal ends. *Mailing Add:* Lawrence Livermore Nat Lab L-629 Box 808 Livermore CA 94550

MATTHEWS, THOMAS ROBERT, b Deadwood, SDak, Dec 24, 39; m 63; c 3. MICROBIOLOGY, BIOCHEMISTRY. *Educ:* Univ Wyo, BS, 64, MS, 66; Univ Ind, PhD(microbiol), 72. *Prof Exp:* Microbiologist antifungal chemother, Eli Lilly & Co, 66-69; div dir antimicrobial & antiviral chemother, ICN Pharmaceut, 72-77; head antimicrobial antiviral chemother, 77-85, SR HEAD ANTIMICROBIOL RES, SYNTEX RES, 85- *Mem:* Am Soc Microbiol. *Res:* Experimental antimicrobial and antiviral chemotherapy; immunomodulation of infectious disease; mechanism of action of antiviral or antimicrobial agents. *Mailing Add:* Div Syntex USA & Co Inc Syntex Res 3401 Hillview Ave Palo Alto CA 94304

MATTHEWS, VIRGIL EDISON, b LaFayette, Ala, Oct 5, 28; div; c 3. ORGANIC & POLYMER CHEMISTRY. *Educ:* Univ Ill, BS, 51; Univ Chicago, SM, 52, PhD(chem), 55. *Prof Exp:* Teaching asst org chem, Univ Chicago, 51-52; res chemist, Res & Develop Dept, Chem Div, Union Carbide Corp, 54-67, Chem & Plastics Div, 67, prog scientist, 67-75, develop sci, Chem & Plastics Div, Union Carbide Corp, 75-86; PROF, CHMN, CHEM DEPT, STATE COL, WVA, 86- *Concurrent Pos:* Instr, WVa State Col, 55-60, part-time assoc prof & prof, 60-70. *Mem:* Fel AAAS; Am Chem Soc; fel Am Inst Chemists; Sigma Xi. *Res:* Synthesis, structure properties and uses of polymers; free radical chemistry; organic synthesis; elastomers and polyurethanes; polymeric composites; fibers; synthetic hydrogels; polymer-anchored catalysts; organometallic chemistry. *Mailing Add:* 835 Carroll Rd Charleston WV 25314

MATTHEWS, WILLIAM HENRY, III, geology, deceased, see previous edition for last biography

MATTHEWS, WILLIAM JOHN, b Memphis, Tenn, Nov 11, 46; m 68; c 2. ICHTHYOLOGY, FISH ECOLOGY. *Educ:* Ark State Univ, BSE, 68, MS, 73; Univ Okla, PhD(zool), 77. *Prof Exp:* Asst prof biol, Roanoke Col, 77-79; RES ASSOC & ASST PROF ICHTHYOL & FISH ECOL, ZOOL DEPT & BIOL STA, UNIV OKLA, 79- *Concurrent Pos:* Assoc ed, Lower Vertebrates, Southwestern Asn Naturalists, 80-; cur fish, Okla Mus Nat Hist, Univ Okla, 81- *Mem:* Am Soc Ichthyologists & Herpetologists; Ecol Soc Am; Sigma Xi; Southwestern Asn Naturalists; Am Fisheries Soc. *Res:* Ecology and systematics of North American freshwater fishes, with emphasis on adaptation of fishes to harsh environments, fish community structure, resource use in fish communities, distributional ecology of fishes, and predator-prey interactions in reservoir fisheries to include larval and adult fish. *Mailing Add:* Dept Zool Univ Okla Star Rte B Kingston OK 73439

MATTHIAS, JUDSON S, b Schofield Barracks, Hawaii, Oct 6, 31; m 56; c 4. CIVIL ENGINEERING. *Educ:* US Mil Acad, BS, 54; Ore State Univ, MS, 63; Purdue Univ, PhD(civil eng, transp), 67. *Prof Exp:* Instr civil eng, Ore State Univ, 62-64 & Purdue Univ, 64-67; asst prof eng, 67-71, assoc prof, 71-81, PROF ENG, ARIZ STATE UNIV, 81- *Concurrent Pos:* Mem, Hwy Res Bd, Nat Acad Sci-Nat Res Coun, 64- *Mem:* Am Soc Civil Engrs; Inst Transp Engrs; Sigma Xi; Am Rd & Transp Builders Asn. *Res:* Transportation planning; urban transportation; traffic engineering; accident reconstruction. *Mailing Add:* 2032 E Luguna Dr Tempe AZ 85281

MATTHIJSSEN, CHARLES, b Amsterdam, Holland, July 26, 31; nat US; m 57; c 2. MICROBIOLOGY, BIOCHEMISTRY. *Educ:* Upsala Col, BS, 51; Rutgers Univ, MS, 55, PhD(microbiol), 57. *Prof Exp:* Res chemist, P Ballentine & Sons, 57-59; vchmn dept endocrinol, Southwest Found Res & Educ, 59-74, assoc scientist, 65-74; asst dir clin res endocrinol, 74-78, ASSOC DIR RES METABOLIC & INFECTIOUS DIS, HOECHST-ROUSSEL PHARM INC, 78- *Mem:* AAAS; Am Chem Soc; Endocrine Soc; fel Am Inst Chemists; Am Soc Microbiol; Int Dis Soc Am. *Res:* Clinical investigations. *Mailing Add:* Hoechst-Roussel Pharm Inc Rte 202-206 N Somerville NJ 08876

MATTHYS, ERIC FRANCOIS, b Brussels, Belg, Feb 22, 56. HEAT TRANSFER, FLUID MECHANICS. *Educ:* Brussels Univ, Engr, 78; Calif Inst Technol, MS, 80, PhD(mech eng), 85. *Prof Exp:* Res engr, Ecole Polytech, 78-79; res asst fluid mechanics & heat transfer, Calif Inst Technol & Jet Propulsion Lab, 79-85; asst prof, 85-91, ASSOC PROF FLUID MECH & HEAT TRANSFER, UNIV CALIF, SANTA BARBARA, 91- *Concurrent Pos:* Prin investr, var grants, 85-; consult, var co, 85-; NSF presidential young investr, 89; vis prof, Univ Calif, San Diego, 90. *Mem:* Am Soc Mech Engrs; Metall Soc; Sigma Xi; Soc Rheology; Am Soc Metals. *Res:* Heat transfer and fluid mechanics; materials processing; rheology; thermal distribution systems; biofluids; instrumentation development; numerical modeling. *Mailing Add:* Mech Eng Dept Univ Calif Santa Barbara CA 93106

MATTHYSSE, ANN GALE, b Chicago, Ill, Oct 25, 39; div; c 1. MICROBIOLOGY, BOTANY PHYTOPATHOLOGY. *Educ:* Radcliffe Col, AB, 61; Harvard Univ, PhD(biol), 66. *Prof Exp:* Lectr biol, Harvard Univ, 70-71; asst prof microbiol, Sch Med, Ind Univ Indianapolis, 71-75; asst prof bot, 75-77, assoc prof biol, 77-89, PROF BIOL, UNIV NC, 90- *Concurrent Pos:* NIH fel, Calif Inst Technol, 66-69 & Harvard Med Sch, 69-70. *Mem:* AAAS; Am Soc Microbiol; Am Soc Plant Physiol; Am Phytopathol Soc; Plant Molecular Biol Asn. *Res:* Bacterial diseases of plants; molecular plant pathology; molecular biology of plants; bacterial attachment to plants. *Mailing Add:* Dept Biol Univ NC Chapel Hill NC 27599-3280

MATTHYSSE, STEVEN WILLIAM, b New York, NY, Aug 27, 39; m 62; c 1. PSYCHOBIOLOGY, PSYCHIATRY. *Educ:* Yale Univ, BS, 59, BA, 60; Harvard Univ, PhD(clin psychol), 67. *Prof Exp:* Asst prof, Pitzer Col, 66-69; asst prof, 70-78, ASSOC PROF PSYCHOBIOL, HARVARD MED SCH, 78- *Concurrent Pos:* Marks Found fel, Harvard Univ, 70-71; res dir, Schizophrenia Res Prog, Scottish Rite, 72- *Mem:* AAAS; Soc Neurosci; Am Soc Neurochem; Asn Res Nerv & Ment Dis. *Res:* Mathematical neuroanatomy; biological aspects of schizophrenia; theoretic genetics. *Mailing Add:* McLean Hosp 115 Mill St Belmont MA 02178

MATTICE, JACK SHAFER, b Hobart, NY, Aug 25, 41; m 67. AQUATIC ECOLOGY, ENVIRONMENTAL TOXICOLOGY. *Educ:* State Univ NY Stony Brook, BS, 63; Syracuse Univ, PhD(invert zool), 71. *Prof Exp:* fel, Nat Acad Sci-Polish Acad Sci, 70-71; res ecologist aquatic ecol, Oak Ridge Nat Lab, Union Carbide Corp, 72-81; proj mgr, 81-87, SR PROJ MGR, ELEC POWER RES INST, 87- *Concurrent Pos:* Adj asst prof, Tenn Technol Univ, 81. *Mem:* Ecol Soc Am; Soc Environ Toxicol & Chem; Am Fisheries Soc; Sigma Xi; Am Soc Testing & Mat; NAm Benthological Soc; AAAS. *Res:* Population regulation in fishes; impacts of power plants on aquatic biota; acidification effects on biota; aquatic effects of hydro generation. *Mailing Add:* Environ Div Elec Power Res Inst 3412 Hillview Ave PO Box 10412 Palo Alto CA 94303

MATTICE, WAYNE LEE, b Cherokee, Iowa, July 9, 40; m 65; c 1. PHYSICAL BIOCHEMISTRY. *Educ:* Grinnell Col, BA, 63; Duke Univ, PhD(biochem), 68. *Prof Exp:* USPHS fel, Fla State Univ, 68-70; from asst prof to prof, 70-85, BOYD PROF CHEM, LA STATE UNIV, BATON ROUGE, 85- *Honors & Awards:* Creative Polymer Chem Award, 83. *Mem:* AAAS; Am Chem Soc; Biophys Soc. *Res:* Physical chemistry of polymers. *Mailing Add:* 581 Northwood Dr Akron OH 44313-5307

MATTICK, JOSEPH FRANCIS, b Hudson, Pa, Nov 16, 18; m 52; c 1. BIOCHEMISTRY, BACTERIOLOGY. *Educ:* Pa State Univ, BS, 42, PhD(dairy technol), 50. *Prof Exp:* Asst prof dairy technol, Univ Md, 50-52, assoc prof, 53-58; tech consult, Venezuela, 58-60; assoc prof dairy technol, 60-65, PROF DAIRY SCI, UNIV MD, COLLEGE PARK, 65-, CHMN DEPT, 74-, CHMN FOOD SCI PROG, 78- *Concurrent Pos:* Consult, Interam Develop Bank & World Bank. *Mem:* Am Dairy Sci Asn; Inst Food Technol; Sigma Xi. *Res:* Products development; curriculum of food science; food processing waste disposal acid whey utilization. *Mailing Add:* 4621 Harvard Rd College Park MD 20740

MATTICS, LEON EUGENE, b Butte, Mont, Mar 2, 40; m 67; c 2. NUMBER THEORY. *Educ:* Mont State Univ, BS, 63, PhD(math), 67. *Prof Exp:* Instr math, Mont State Univ, 66-67; PROF MATH, UNIV SOUTH ALA, 67- *Mem:* Math Asn Am. *Res:* Problem solving; number theory. *Mailing Add:* Dept Math Univ S Ala 307 University Blvd Mobile AL 36688

MATTINA, CHARLES FREDERICK, b Elizabeth, NJ, Dec 5, 44; m 69; c 3. PHYSICAL CHEMISTRY. *Educ:* Providence Col, BS, 66; Yale Univ, PhD(phys chem), 69. *Prof Exp:* Asst prof chem, Albertus Magnus Col, 69-71; res chemist, Kimberly-Clark Corp, 71-74, head chem sect, Schweitzer Div, 74-84, tech dir specialty prods, 84-88; DIR RES & DEVELOP, GDE ANAL, 88- *Concurrent Pos:* Tech dir, Inst Environ Sci, 87- *Mem:* Am Chem Soc; Sigma Xi; Inst Environ Sci. *Res:* Electrolytic conductance; viscosity of ionic solutions; condenser paper; cleanroom technology; semiconductors; microelectronics. *Mailing Add:* 172 Hubbard St Lenox MA 01240

MATTINA, MARY JANE INCORVIA, b New York, NY, July 31, 44; m 69; c 3. MASS SPECTROMETRY. *Educ:* Barnard Col, New York, BA, 66; Yale Univ, New Haven, MPhil, 69, PhD(chem), 70. *Prof Exp:* Instr org chem, Albertus Magnus Col, New Haven, 70-71 & Simon's Rock Col, Gt Barrington, Mass, 73-77; mem res staff, Dept Chem, Emory Univ, Atlanta, 86-87; ASSOC SCIENTIST, DEPT ANALYTICAL CHEM, CONN AGR EXP STA, 88- *Mem:* Am Soc Mass Spectrometry; Am Chem Soc. *Res:* Application of mass spectrometry for the solution of analytical problems in environmental and natural products areas; use of negative ion mass spectrometry techniques and ion source reactions associated with these techniques. *Mailing Add:* Conn Agr Exp Sta 123 Huntington St New Haven CT 06504

MATTINGLY, GLEN E, b Provo, Ark, Oct 31, 32; m 54; c 2. MATHEMATICS. *Educ:* Sam Houston State Univ, BS, 56, MS, 57; NMex State Univ, PhD(math), 65. *Prof Exp:* From instr to assoc prof, 56-67, PROF MATH & DIR DEPT, SAM HOUSTON STATE UNIV, 67- *Mem:* Am Math Soc; Math Asn Am. *Res:* Topological semi-groups; semi-topological groups; topological modules. *Mailing Add:* Dept Math Sam Houston State Univ Huntsville Box 2206 Huntsville TX 77341

MATTINGLY, RICHARD FRANCIS, obstetrics & gynecology; deceased, see previous edition for last biography

MATTINGLY, STEELE F, b Trinity, Ky, Aug 28, 27; m 49; c 2. ANIMAL HUSBANDRY. *Educ:* Berea Col, BS, 50; Auburn Univ, DVM, 55; Am Col Lab Animal Med, dipl, 64. *Prof Exp:* Assoc teacher high sch, Ky, 50-51; mem staff primate test animals, Allied Labs, Pitman Moore Co Div & Dow Chem Co, 55-57, unit head test animals, 57-62; prod mgr, Lab Supply Co, 62-65; DIR DEPT LAB ANIMAL MED, COL MED, UNIV CINCINNATI, 65- *Concurrent Pos:* Consult, Vet Admin, Ohio, 65- *Mem:* Am Vet Med Asn; Am Asn Lab Animal Sci (nat pres, 82); NY Acad Sci. *Res:* Husbandry of laboratory animals; laboratory animal medicine; germ free life and its relationship to other animal research. *Mailing Add:* Dept Lab Animal Med Univ Cincinnati Cincinnati OH 45267

MATTINGLY, STEPHEN JOSEPH, b Evansville, Ind, Mar 3, 43; m 63; c 4. MICROBIAL PHYSIOLOGY. *Educ:* Univ Tex, Austin, BA, 65; Villanova Univ, MS, 68; Med Col Ga, PhD(microbiol), 72. *Prof Exp:* Res assoc, Sch Med, Temple Univ, 72-74; from asst prof to assoc prof, 74-86, PROF MICROBIOL, UNIV TEX HEALTH SCI CTR, 86- *Concurrent Pos:* Nat Inst Dent Res fel, 74. *Honors & Awards:* Alumnus Award, Med Col Ga, 86. *Mem:* Am Soc Microbiol; Sigma Xi. *Res:* Bacterial physiology; regulation of cell wall and polysaccharide biosynthesis; physiology and genetics of group B streptococci, mucoid Pseudomonas aeruginosa. *Mailing Add:* Dept Microbiol Univ Tex Health Sci Ctr 7703 Floyd Curl Dr San Antonio TX 78284

MATTINSON, JAMES MEIKLE, b Maracaibo, Venezuela, Aug 28, 44; US citizen. GEOCHRONOLOGY, PETROLOGY. *Educ:* Univ Calif, Santa Barbara, BA, 66, PhD(geol), 70. *Prof Exp:* Fel geochronology, Geophys Lab, Carnegie Inst, Washington, 70-73; lectr geol, 73-76, asst prof, 77-81, ASSOC PROF GEOL, UNIV CALIF, SANTA BARBARA, 81- *Concurrent Pos:* fel, Geol Soc Am. *Mem:* Am Geophys Union; Geol Soc Am; AAAS; Sigma Xi. *Res:* Igneous rocks, especially calc-alkaline ingeous complexes and ophiolitic complexes. *Mailing Add:* Dept Geol Sci Univ Calif Santa Barbara CA 93106

MATTIS, ALLEN FRANCIS, b Spooner, Wis, May 3, 47; m 75. PETROLEUM GEOLOGY. *Educ:* Univ Wis-Superior, BS, 69; Univ Minn, Duluth, MS, 72; Rutgers Univ, MPhil, 74, PhD(geol), 75. *Prof Exp:* Geologist, Texaco, Inc, 75-80; GEOLOGIST, AMERADA HESS CORP, 80- *Mem:* Geol Soc Am; Am Asn Petrol Geologists. *Res:* Sedimentation; provenance; regional tectonics. *Mailing Add:* 5314 Wigton Houston TX 77096-5115

MATTIS, DANIEL CHARLES, b Brussels, Belg, Sept 8, 32; nat US; m 58; c 2. SOLID STATE PHYSICS. *Educ:* Mass Inst Technol, BS, 53; Univ Ill, MS, 54, PhD(physics), 57. *Prof Exp:* Asst, Univ Ill, 54-57; asst, Nat Ctr Sci Res, France, 57-58; physicist, Res Ctr, Int Bus Mach Corp, 58-65; from assoc prof to prof physics, Belfer Grad Sch Sci, Yeshiva Univ, 65-78; Thomas Potts Prof physics, Polytech Inst NY, 78-81; PROF PHYSICS, UNIV UTAH, 80- *Concurrent Pos:* Sterling lectr, Yale Univ, 66; adj prof physics, State Univ NY Buffalo & Univ Utah, 78- *Mem:* NY Acad Sci; fel Am Phys Soc. *Res:* Theoretical investigation of electronic properties, especially the theory of electrical conduction, with applications to metals and semiconductors; many-body theory of metal alloys; theory of magnetism. *Mailing Add:* Dept Physics Univ Utah Salt Lake City UT 84112

MATTISON, DONALD ROGER, b Minneapolis, Minn, April 28, 44; m 67; c 2. PUBLIC HEALTH & EPIDEMIOLOGY. *Educ:* Augsburg Col, Minn, BA, 66; Mass Inst Technol, 68; Columbia Univ, Col Physicians & Surgeons, NY, MD, 73. *Prof Exp:* Resident obstet & gynec, Dept Obstet & Gynec, Presby Hosp, NY, 73-75 & 77-78; res assoc, USPHS, Develop Pharmacol Br, Sect Molecular Toxicol, Nat Inst Child Health & Human Develop, NIH & Biochem Pharmacol Sect, Lab Chem Pharmacol, Develop Therapeut Prog, Nat Cancer Inst, 75-77, med officer, 78-83, chief, Sect Reproductive Toxicol, Pregnancy Res Br, 83-84; from assoc prof interdisciplinary toxicol, Dept Pharmacol, Univ Ark Med Sci, 87-90, prof obstet & gynec, Dept Obstet & Gynec, 87-90; DEAN, GRAD SCH PUB HEALTH & PROF ENVIRON & OCCUP HEALTH, HUMAN GENETICS & OBSTET & GYNEC, UNIV PITTSBURGH, 90- *Concurrent Pos:* USPHS reserve officer, Div Reproductive & Develop Toxicol, Nat Ctr Toxicol & Res, 84-87, actg dir, 87-88, reserve med officer, 88-90; mem, Sci Group Methodologies Safety Eval Chem, 83, Reproductive & Develop Toxicol Adv Group, Nat Toxicol Prog, Nat Inst Environ Health Serv, NIH, 84-87, Sci Adv Bd, 86-87, Steering Comt Biomarkers, Bd Environ Studies & Toxicol, Nat Acad Sci, Nat Res Coun, 86-89 & 87, Sci Adv Panel, Semiconductor Indust Asn, 87-, Comt Oral Contraceptives & Breast Cancer, Inst Med, 89-91; chmn, Food Drug Admin Expert Comt Eval Teratogenicity Dioctyl Sulfosuccinates, 84, Biomarkers Reproductive & Develop Toxicol, Bd Environ Studies & Toxicol, Nat Acad Sci, Nat Res Coun, 86-87, Human Toxicol & Risk Assessment Prog, Bd Environ Studies & Toxicol, Nat Res Coun, Nat Acad Sci, 89-; co-chmn, Reproductive & Develop Toxicol Work Group, Task Force III, Nat Inst Environ Health Serv, 84, Comt Risks Children Pesticides Foods, Bd Environ Studies & Toxicol, Nat Res Coun, Nat Acad Sci, 88-, Comt Risk Assessment Methods, 89- *Honors & Awards:* Am Chem Soc Medal, Am Chem Soc, 66; Thomas F Cock Award Excellence Obstet & Gynec, Columbia Univ, 73. *Mem:* Am Pub Health Asn; Soc Risk Analysis; Am Asn Cancer Res; NY Acad Sci; Am Col Toxicol; Am Fertil Soc; Soc Gynec Invest; Soc Toxicol. *Res:* Mechanism and site of action of reproductive and developmental toxicity; methods for quantitative assessment of reproductive and developmental toxicity; numerous publications. *Mailing Add:* Sch Pub Health Rm 111 Parran Hall Univ Pittsburgh 130 DeSoto St Pittsburgh PA 15260-0001

MATTISON, LOUIS EMIL, b Lincoln, Nebr, Oct 3, 27; m 49; c 3. CHEMISTRY. *Educ:* La State Univ, BS, 49; Univ Del, MS, 50, PhD(org chem), 52. *Prof Exp:* Res chemist, Carothers Lab, Exp Sta, E I du Pont de Nemours & Co, 52-54; from assoc prof to prof chem, Davis & Elkins Col, 56-62; PROF CHEM & CHMN DEPT, KING COL, 63- *Concurrent Pos:* Cottrell res grant, 56-60; chmn dept chem, Davis & Elkins Col, 56-62; res assoc, Univ Ariz, 62-63. *Mem:* AAAS; Am Chem Soc; Sigma Xi; NY Acad Sci; Am Inst Chemists. *Res:* Organic synthesis of chelating agents; metal chelates; coordination compounds; photochemistry. *Mailing Add:* 323 Poplar St Bristol TN 37620

MATTMAN, LIDA HOLMES, b Denver, Colo, July 31, 12; m 44; c 2. BACTERIOLOGY. *Educ:* Univ Kans, AB, 33, MA, 34; Yale Univ, PhD(bact), 40. *Prof Exp:* Bacteriologist, Med Dept, Endicott Johnson, NY, 34; asst, Iowa Hosp, 40-42; res bacteriologist, Nat Res Coun, 42-44, comn airborne infection, 45; mycologist, Santa Rosa Hosp, San Antonio, Tex, 46-47; sr bacteriologist, State Health Labs, Mass, 47-49; from asst prof to prof, 49-82, EMER PROF BIOL, WAYNE STATE UNIV, 82- *Concurrent Pos:* Nat Res Coun fel, Univ Pa, 43-44; prof, Oakland Univ, 84; consult, NSF Women Sci. *Mem:* Am Soc Microbiol; NY Acad Sci; Sigma Xi. *Res:* Surface tension depressants in immunological systems; pathogenic anaerobes; L variants and mycoplasmae. *Mailing Add:* 319 Rivard Grosse Pointe MI 48230-1625

MATTOON, JAMES RICHARD, b Loveland, Colo, Dec 9, 30; m 53; c 2. MOLECULAR BIOLOGY. *Educ:* Univ Ill, BS, 53; Univ Wis, MS, 54, PhD(biochem), 57. *Prof Exp:* From instr to asst prof chem, Univ Nebr, 57-62; asst prof physiol chem, Sch Med, Johns Hopkins Univ, 64-70, assoc prof, 70-79; PROF BIOL, UNIV COLO, COLORADO SPRINGS, 79- *Concurrent Pos:* Fel, Sch Med, Johns Hopkins Univ, 62-64. *Mem:* Am Chem Soc; Am Soc Biol Chemists; Am Soc Microbiol; Genetics Soc Am; AAAS. *Res:* Genetics of mitochondria; yeast molecular biology; mitochondrial biogenesis; oxidative phosphorylation; yeast respiration and mitochondria; lysine biosynthesis; biotechnology. *Mailing Add:* Dept Biol Univ Colo Colo Springs CO 80933

MATTOR, JOHN ALAN, b Oxford, Maine, Jan 15, 32; m 58; c 4. SYNTHETIC ORGANIC CHEMISTRY. *Educ:* Bates Col, BS, 58; Lawrence Univ, MS, 60, PhD(chem), 63. *Prof Exp:* Mem staff, 62-75, SR RES ASSOC, S D WARREN CO, SCOTT PAPER, WESTBROOK, 75- *Mem:* Am Chem Soc; AAAS; Soc Photog Scientists & Engrs. *Res:* Photochemistry; organic photoconductivity; dye sensitization. *Mailing Add:* Box 85 Bar Mills ME 04004

MATTOX, DOUGLAS MILTON, ceramics, glass technology, for more information see previous edition

MATTOX, KARL, b Cincinnati, Ohio, Aug 22, 36; m 57; c 3. PHYCOLOGY. *Educ:* Miami Univ, BS, 58, MA, 60; Univ Tex, PhD(bot), 62. *Prof Exp:* Asst prof bot, Univ Toronto, 62-66; from asst prof to assoc prof, 66-75, PROF BOT, MIAMI UNIV, 75-, CHMN DEPT, 77- *Concurrent Pos:* Res assoc, Great Lakes Inst, 62- *Mem:* Bot Soc Am; Phycol Soc Am. *Res:* Morphology; cytology and evolution of algae. *Mailing Add:* Dept of Bot Miami Univ Oxford OH 45056

MATTOX, RICHARD BENJAMIN, b Middletown, Ohio, May 15, 21; m 48. GEOLOGY. *Educ:* Miami Univ, BA, 48, MS, 49; Univ Iowa, PhD(geol), 54. *Prof Exp:* Instr geol, Miami Univ, 49-50; petrol geologist, Magnolia Petrol Co, 50; asst instr geol, Univ Iowa, 50-52; asst prof, Miss State Col, 52-54; assoc prof, 54-57, PROF GEOL, TEX TECH UNIV, 57- *Concurrent Pos:* Head dept geol, Tex Tech Univ, 64-70. *Mem:* AAAS; Nat Asn Geol Teachers; Geol Soc Am; Am Asn Petrol Geologists. *Res:* Eolian geology; stratigraphy; geology of Colorado plateau. *Mailing Add:* Dept Geosci Tex Tech Univ Lubbock TX 79409

MATTSON, DALE EDWARD, b Newberry, Mich, Apr 5, 34; m 57; c 2. BIOMETRICS. *Educ:* Colo Col, BA, 59; Univ Ill, MA, 61, PhD(educ psychol), 63. *Prof Exp:* Asst prof educ measurement, Univ Wash, 63-64; dir educ res, Am Asn Dent Schs, 64-66 & Asn Am Med Cols, 66-69; dir admis & rec, 69-72, PROF BIOMET, SCH PUB HEALTH, UNIV ILL, 72- *Mem:* Am Pub Health Asn; Am Statist Asn. *Res:* Indices of serial correlation with applications to measures of health statistics; epidemiology of sports injuries. *Mailing Add:* Dept Pub Health MC 922 Univ Ill Col Med PO Box 6998 Chicago IL 60680

MATTSON, DONALD EUGENE, b Chatsworth, Calif, May 19, 34; m 59; c 3. VETERINARY VIROLOGY. *Educ:* Univ Calif, Davis, BS, 57, DVM, 59; Wash State Univ, PhD(microbiol), 66. *Prof Exp:* Asst prof, 67-69, ASSOC PROF VET MED, ORE STATE UNIV, 69- *Mem:* Am Vet Med Asn; Am Asn Vet Lab Diagnosticians. *Res:* Physical, chemical and serological properties of viruses; virus diseases of the newborn, especially bovine. *Mailing Add:* Sch Vet Med Ore State Univ Corvallis OR 97331

MATTSON, FRED HUGH, b Spokane, Wash, Dec 16, 18; m; c 5. NUTRITION. *Educ:* Loyola Univ, Calif, BS, 40; Univ Southern Calif, MS, 42, PhD(biochem), 48; Am Soc Clin Nutrit, cert specialist human nutrit, 71. *Prof Exp:* Res chemist, Procter & Gamble Co, 48-78; PROF MED & DIR, LIPID RES CLIN, UNIV CALIF, SAN DIEGO, 79- *Concurrent Pos:* Adj prof, Univ Cincinnati, 70-78; mem coun arteriosclerosis, Am Heart Asn; fel, Am Inst Nutrit, 88. *Honors & Awards:* Am Chem Soc Award, 69. *Mem:* Am Chem Soc; Am Soc Biol Chem; Am Inst Nutrit; Am Heart Asn. *Res:* Digestion and absorption of fat; nutritive value of fat; diet and cardio-vascular disease. *Mailing Add:* Dept Med M013D Univ Calif San Diego La Jolla CA 92093

MATTSON, GUY C, b Bloomfield, NJ, Jan 3, 27; m 50; c 4. ORGANIC CHEMISTRY, POLYMER CHEMISTRY. *Educ:* Union Col, NY, BS, 49; Univ Fla, PhD(chem), 55. *Prof Exp:* Chemist, Warner-Chilcott Labs, NJ, 49-52; instr chem, Univ Fla, 52-55; res chemist, Dow Chem Co, Mich, 55-60, facil mgr, Fla, 60-64, prod engr, Saginaw Bay, 64-65, proj mgr, Tex, 65-66, dept head, Ind, 66-71; chmn dept, 78-87, PROF CHEM, UNIV FLA, 70- *Mem:* Am Chem Soc. *Res:* Organic synthesis; process development; polymer synthesis. *Mailing Add:* Dept of Chem Univ Cent Fla Orlando FL 32816

MATTSON, HAROLD F, JR, b Ann Arbor, Mich, Dec 7, 30; m 66; c 2. COMBINATORICS & FINITE MATHEMATICS. *Educ:* Oberlin Col, AB, 51; Mass Inst Technol, PhD(math), 55. *Prof Exp:* Mathematician, Air Force Cambridge Res Ctr, 55-60; mathematician, Appl Res Lab, Sylvania Elec Prod, Inc, Gen Tel & Electronics Corp, 60-70, Eastern Opers, 70-71; PROF COMPUT & INFO SCI, SYRACUSE UNIV, 71- *Concurrent Pos:* Ed, Review, Soc Indust & Appl Math, 70-79. *Mem:* Am Math Soc; Math Asn Am; Inst Elec & Electronic Engrs. *Res:* Error-correcting codes; combinatorial analysis; concentrating on the covering radius of error-correcting codes. *Mailing Add:* 4-116 Ctr Sci & Technol Syracuse Univ Syracuse NY 13244-4100

MATTSON, JOAN C, b Austin, Tex, July 11, 35; m 62; c 5. HEMATOPATHOLOGY. *Educ:* Northwestern Univ, BA, 58, MD, 62; Am Bd Path, cert anat path & clin path, 69, cert hemat, 84. *Prof Exp:* Clin instr anat path, Ohio State Univ Hosps, 64-66 & clin path, 67-69, instr, Dept Path, 69-70; from asst prof to prof, Dept Path, Mich State Univ, 70-81; prof, Dept Path, Health Sci Ctr, Univ Tex, 81-83; ADJ PROF, SCH MED TECHNOL, MICH STATE UNIV, 84-; CHIEF HEMATOPATH, DEPT CLIN PATH, WILLIAM BEAUMONT HOSP, 84- *Concurrent Pos:* Am Cancer Soc fel, Ohio State Univ Hosps, 66-68, Nat Cancer Inst postdoctoral fel, 68-69, assoc surg path, 69-70; investr, Hwy Accident Res Team, Ohio State Univ & Nat Safety Coun, 69-70; consult path, Good Samaritan Hosp, 70; mem, Coun Thrombosis, Am Heart Asn, 71-; attend staff mem & consult path, St Lawrence Hosp, 72-81; dir, Renal Diag Prog, Dept Path, Mich State Univ, 72-81, assoc pathologist, Clin Ctr Labs & co-dir, Thrombosis & Hemostasis Res Labs, 78-81, dir hemat & coagulation, Clin Ctr Lab, 79-81, assoc dir, 80-81; vis scientist, Cell Biophysics Univ, Med Res Coun, London & Dept Anat, Harvard Univ, 77; NIH postdoctoral fel, Marine Biol Lab, Woods Hole, Mass, 78; med dir, Capital Area Career Ctr, 80-81; assoc pathologist, Hermann Hosp, Univ Tex, 81-83. *Mem:* Col Am Pathologists; Int Acad Path; Am Heart Asn; Am Soc Clin Pathologists; Int Soc Thrombosis & Hemostasis; Soc Hematopath; Am Asn Pathologists. *Res:* Hematology; pathology. *Mailing Add:* Dept Clin Path William Beaumont Hosp 3601 W 13 Mile Rd Royal Oak MI 48072

MATTSON, MARGARET ELLEN, b Philadelphia, Pa, May 13, 47. BEHAVIORAL MEDICINE, EPIDEMIOLOGY. *Educ:* Holy Family Col, BA, 69; Cornell Univ, PhD(neurobiol), 75. *Prof Exp:* Investr environ health, Environ Control, Inc, 76-78; proj officer behav med & clin traits, Nat Heart, Lung & Blood Inst, 78-, AT DIV CANCER PREV & CONTROL, NAT CANCER INST, NIH. *Mem:* Soc Neurosci; Am Psychol Asn; Soc Clin Traits; Soc Behav Med. *Res:* Psychobiology. *Mailing Add:* Nat Cancer Inst Div Cancer Prev & Control NIH Bldg 31 Rm 10A49 Bethesda MD 20892

MATTSON, MARLIN ROY ALBIN, b Bellingham, Wash, Apr 25, 39. PSYCHIATRY. *Educ:* Univ Wash, BA, 61, MD, 65. *Prof Exp:* Intern & resident med, Cornell's Combined Prog Med, Bellevue Hosp & Mem Hosp, New York, 65-67; Capt US Army Med Corps med & obstet, 67-69; residency chief & resident psychiat, Payne Whitney Clin, NY Hosp, 69-73, act med dir, 74; asst prof psychiat, 73-79, ASSOC PROF CLIN PSYCHIAT, MED COL, CORNELL UNIV, 79- *Concurrent Pos:* Ginsberg fel, Group Advan Psychiat, 71-73; mem, NY County Health Serv Rev Orgn, 77-, bd dirs, 83-, Nat Task Force, Seclusion & Restraint, 81-85, Nat Comt Champus Peer Rev Prog, 84-86, Nat Quality Assurance Comt, Am Psychiat Asn, 88-; chmn, Peer Rev Comt, NY State Psychiat Asn, 85; asst med dir, qual assurance, Payne Whitney Psychiat Clin, 73-89, assoc med dir, 89-; asst med dir Quality Assurance, Westchester Div, 80-89, assoc med dir, 89-90, assoc med dir Quality Assurance, Dept Psychiatry, NY Hosp, 90-; secy, NY County District Br Am Psychiatric Asn, 87-; mem, Pub Health Comt, NY Acad Med, 84- *Mem:* Fel Am Psychiat Asn; Am Acad Psychiat & Law; fel NY Acad Med; Am Med Asn. *Res:* Quality assurance, peer review and utilization review; aspects of hospital psychiatry. *Mailing Add:* Dept Clin Psychiat Cornell Univ Med Col 1300 York Ave New York NY 10021

MATTSON, PETER HUMPHREY, b Evanston, Ill, Apr 3, 32; m 54; c 3. GEOLOGY, GEOPHYSICS. *Educ:* Oberlin Col, BA, 53; Princeton Univ, PhD(geol), 57. *Prof Exp:* Geologist, US Geol Surv, 57-64; from asst prof to assoc prof, 64-72, PROF GEOL, QUEENS COL, NY, 73- *Concurrent Pos:* Chmn dept geol & geog, Queens Col, NY, 65-68; consult, Commonwealth PR, 65-69 & Venezuela, 77-80; indust consult, 67- *Mem:* Geol Soc Am; Am Geophys Union; Earthquake Eng Res Inst. *Res:* Igneous petrology; volcanic rocks; structural geology; geology of Puerto Rico and the Caribbean area; island arcs; seismicity & earthquake hazard. *Mailing Add:* Dept Geol Queens Col City Univ NY Flushing NY 11367

MATTSON, RAYMOND HARDING, b Matchwood, Mich, Oct 10, 20; wid; c 4. ORGANIC CHEMISTRY. *Educ:* Univ Mich, BS, 43; Univ Ill, PhD(chem), 51. *Prof Exp:* Res chemist, Rohm and Haas Co, 43-44, Am Cyanamid Co, 50-52 & mkt develop, Jefferson Chem Co, 52-55; sr mkt res analyst, Am Cyanamid Co, 55-59, tech rep, 59-62, mgr sales develop rubber chem, 62-63; mkt res assoc, Glidden Co, SCM Corp, 63-71, mgr group mkt res, Glidden-Durkee Div, 71-76, dir bus develop, Org Chem Div, 76-78, mkt res assoc, Durkee Foods Div, group mkt res, Glidden Durkee Div, 78-81; PRES, POLARIS ASSOC CONSULT ORGN, 82- *Concurrent Pos:* Pres, Liberty Opinion Res, 83- *Mem:* Am Chem Soc. *Res:* Restricted rotation in aryl amines. *Mailing Add:* 7396 Ober Lane Chagrin Falls OH 44022

MATTSON, RICHARD LEWIS, b Greeley, Colo, May 29, 35; m 57; c 2. COMPUTER SCIENCE, ELECTRICAL ENGINEERING. *Educ:* Univ Calif, Berkeley, BS, 57; Mass Inst Technol, MS, 59; Stanford Univ, PhD(elec eng), 62. *Prof Exp:* Res engr, Lockheed Aircraft Corp, 59-62; asst prof elec eng, Stanford Univ, 62-64; RES STAFF MEM, IBM CORP, 65- *Concurrent Pos:* Assoc prof, Stanford Univ, 64-65. *Mem:* AAAS; Asn Comput Mach; Inst Elec & Electronics Engrs. *Res:* Switching theory; computer system design. *Mailing Add:* IBM Corp Monterey & Cottle Rds San Jose CA 95193

MATTSON, ROY HENRY, b Chisholm, Minn, Dec 26, 27; m 48; c 7. ELECTRICAL ENGINEERING. *Educ:* Univ Minn, BEE, 51, MS, 52; Iowa State Univ, PhD(elec eng), 59. *Prof Exp:* Mem tech staff, Bell Tel Labs, Inc, 52-56; from asst prof to assoc prof elec eng, Iowa State Univ, 56-61; assoc prof, Univ Minn, 61-66; prof elec eng & head dept, Uniz Ariz, 66-88; ACAD VPRES, NAT TECHNOL UNIV, 88- *Concurrent Pos:* Electronics consult; mem, Amphitheater Sch Bd; ed, Trans on Educ, Inst Elec & Electronics Engrs. *Honors & Awards:* Eng of the Year, Southern Chap, Am Soc Prof Engrs, 78; Anderson Prize, Univ Ariz Eng Col, 81; Meritorious Serv, Inst Elect & Electronic Engrs, 82. *Mem:* Fel AAAS; Am Soc Eng Educ; fel Inst Elec & Electronics Engrs. *Res:* Solid state and biomedical electronics; engineering education; microelectronics, solar energy, and biomedical instrumentation. *Mailing Add:* Nat Technol Univ 700 Centre Ave Ft Collins CO 80526

MATTUCK, ARTHUR PAUL, b Brooklyn, NY, June 11, 30; div; c 1. GEOMETRY. *Educ:* Swarthmore Col, AB, 51; Princeton Univ, PhD(math), 54. *Prof Exp:* Res fel math, Harvard Univ, 54-55; C L E Moore instr, 55-57, lectr, 57-58, from asst prof to prof, 58-73, CLASS OF 1922 PROF MATH, MASS INST TECHNOL, 73- *Concurrent Pos:* Chmn, math dept, Mass Inst Technol, 84-89. *Mem:* Am Math Soc; Math Asn Am. *Res:* Algebraic geometry. *Mailing Add:* Dept of Math Mass Inst Technol Cambridge MA 02139

MATUKAS, VICTOR JOHN, b Freeport, Tex, Oct 20, 33; m 61; c 3. EXPERIMENTAL PATHOLOGY. *Educ:* Loyola Univ, La, DDS, 56; Univ Rochester, PhD(path), 66; Univ Colo, Denver, MD, 73. *Prof Exp:* Resident oral surg, Charity Hosp, New Orleans, La, 58-61; asst prof path, Loyola Univ, La, 61-62 & Univ Pa, 66-68; prof stomatol & chmn dept, Sch Dent, Univ Colo, Denver, 71-74; med internship, Univ Ala Hosps, 74-75, investr dent res, 74-80, dir advan educ prog oral & maxillofacial surg, 75-80, ASSOC DEAN, SCH DENT, UNIV ALA, BIRMINGHAM, 78-, SR SCIENTIST, INST DENT RES, 80-, PROF DENT & CHMN DEPT ORAL & MAXILLOFACIAL SURG, 81- *Concurrent Pos:* Spec res fel, Nat Inst Dent Res, 68-71. *Mem:* AAAS; Am Dent Asn. *Res:* Synthesis, metabolism and ultrastructure of collagen and protein-polysaccharide; biological mineralization. *Mailing Add:* Inst Dent Res Sch Dent Univ Ala Birmingham AL 35294

MATULA, DAVID WILLIAM, b St Louis, Mo, Nov 6, 37; m 66; c 3. COMPUTER ARITHMETIC, DATA STRUCTURE. *Educ:* Wash Univ, BS, 59; Univ Calif, Berkeley, PhD(eng sci & operation res), 66. *Prof Exp:* From asst prof to assoc prof comput sci, Wash Univ, 66-74; dept chmn, 74-79, PROF COMPUT SCI, SCH ENG, SOUTHERN METHODIST UNIV, 74- *Concurrent Pos:* Prin investr, NSF, 73-88; vis prof, Univ Karlsruhe, Ger, 74; consult, Control Data, 76-80; distinguished vis prof, Naval Postgrad Sch, 78; vis prof, Stanford Univ, 80; vis prof, Aarhus Univ, Denmark, 80-81; assoc ed, J of Classification; assoc ed, ORSA J on Comput. *Mem:* Asn Comput Mach; Soc Indust & Appl Math; Operations Res Soc Am; Inst Elec & Electronic Engrs Comput Soc; Sigma Xi. *Res:* Computer arithmetic has emphasized the nature of computer number systems and the best procedures for computation subject to finite precision limitation; cluster analysis and classification emphasizing graph theoretic approaches and efficient algorithms for identifying clusters in data. *Mailing Add:* Dept Comput Sci Sch Eng Southern Methodist Univ Dallas TX 75275

MATULA, RICHARD A, b Chicago, Ill, Aug 22, 39; m 59; c 4. MECHANICAL ENGINEERING, COMBUSTION. *Educ:* Purdue Univ, BS, 61, MS, 62, PhD(thermodyn), 64. *Prof Exp:* Instr mech eng, Purdue Univ, 63-64; asst prof mech eng, Univ Calif, Santa Barbara, 64-66 & Univ Mich, 66-68; from assoc prof to prof mech eng, Drexel Univ, 68-76; prof mech eng & dean, Col Eng, La State Univ, Baton Rouge, 76-86; PRES, INST PAPER CHEM, APPLETON, WIS, 86- *Mem:* Am Soc Mech Eng; AAAS; Combustion Inst; Soc Automotive Engrs; Am Soc Eng Educ. *Res:* Combustion kinetics and energy conversion. *Mailing Add:* 3143 St Ives Country Club Pkwy Duluth GA 30136

MATULA, RICHARD ALLEN, b Newark, NJ, Jan 12, 38; m 75; c 1. INFORMATION SCIENCE, PHYSICS. *Educ:* Newark Col Eng, BSEE, 60, MSEE, 63; Purdue Univ, MS, 65; PhD(physics), 73. *Prof Exp:* Asst instr elec eng, Newark Col Eng, 60-63; res assoc, Ctr Info & Numerical Data Analysis & Synthesis, 73-75, asst sr researcher, 75-78, assoc sr reseacher, 78-80; INFO SCIENTIST, BELL LABS, 80- *Concurrent Pos:* Subcomt, Book of the Am Inst Physics, 85-90; mat info comt, ASM Int, 84-, chmn, 85, 87; mem, Orbit adv panel, 88-; mem, NAm User Coun, Chem Abstr Serv, 89- *Mem:* Am Soc Info Sci; ASM Int. *Res:* Information retrieval using online searching; development of bibliographies from online searches; development of evaluated data for physical properties of materials. *Mailing Add:* AT&T Bell Labs 600 Mountain Ave 6B-301 Murray Hill NJ 07974-2070

MATULEVICIUS, EDWARD S(TEPHEN), b Montreal, Que, Sept 4, 42; m 70; c 2. CHEMICAL ENGINEERING, AVIATION FUELS. *Educ:* McGill Univ, BEng, 64; Mass Inst Technol, SM, 66, ScD(chem eng), 70. *Prof Exp:* Sr res engr, Exxon Res & Eng Co, 69-76, sect mgr process technol, Air Prods & Chem Div, 76-78, mgr fuel utilization eng, 78-81, sect head thermal fluids, 81-83; ENG ASSOC, EXXON RES & ENG CO, 83- *Concurrent Pos:* CRC panel leader, Filter Sidestream Sensor, 88-; mem, Ad Hoc Panel, Commingled Fuels. *Res:* Fuel utilization, especially coal, fluidized bed combustion; combustion; heat transfer; fluid mechanics and fluidization; aviation fuels and aviation fuel handling. *Mailing Add:* RD Two Coopersburg PA 18036

MATULIC, LJUBOMIR FRANCISCO, b Potosi, Bolivia, May 8, 23; US citizen; m 53; c 3. QUANTUM OPTICS. *Educ:* State Gym, Yugoslavia, BA, 42; Univ Chile, Lic Math & Physics, 49; Ind Univ, Bloomington, MS, 63; Univ Rochester, PhD(physics), 71. *Prof Exp:* Teacher high sch, Bolivia, 49-50; prof math, Collegio Normal Superior, Bolivia, 50-54; prof math & physics, Leguerrier Classical Inst, Montreal, 54-58; lectr math, Royal Mil Col, Que, 58-60; assoc prof, 63-68 & 70-73, PROF PHYSICS, ST JOHN FISHER COL,

73-, CHMN, 85- *Concurrent Pos:* Instr, Univ San Simon, Bolivia, 49-50; vis scientist, Inst Ruder Boskovic, Univ Zagreb, Yugoslavia, 74 & Ctr Invest Optom, Leon, Mex, 81-82. *Mem:* Arg Math Union; Am Asn Physics Teachers; Optical Soc Am; Am Phys Soc. *Res:* Theoretical investigation of distortionless propagation of electromagnetic fields through nonlinear absorbers, especially the phase modulation of this field due to the interaction with resonant atoms and to the bulk host medium. *Mailing Add:* Dept Physics St John Fisher Col Rochester NY 14618

MATULIONIS, DANIEL H, b Lithuania, Oct 2, 38; US citizen; m 60; c 2. ANATOMY, EMBRYOLOGY. *Educ:* Wis State Univ-Whitewater, BEd, 63; Univ Ill, Urbana, MS, 65; Tulane Univ, PhD(anat), 70. *Prof Exp:* Instr biol, Eastern Ky Univ, 65-67; asst prof anat, Col Med, Univ Ky, Lexington, 71-90; RETIRED. *Concurrent Pos:* Gen Res Support grant, Univ Ky, 70-71; Ky Tobacco Res Inst grant, 71-72. *Mem:* Am Asn Anat. *Res:* Ultrastructural analysis of keratin precursors; glycogen synthesis; ultrastructural analysis of cigarette smoke effects on the respiratory system. *Mailing Add:* 840 Tremont Ave Lexington KY 40502

MATULIS, RAYMOND M, b Broadview, Ill, Apr 20, 39; div; c 2. ANALYTICAL CHEMISTRY. *Educ:* Culver-Stockton Col, BA, 61; Univ Mo-Columbia, MA, 63, PhD(chem), 66. *Prof Exp:* Res chemist, Gulf Res & Develop Co, Merrian, Kans, 66-67; mgr chem lab, Hallmark Cards, Inc, Kansas City Mo, 67-71, mgr res, 67-79; dept head tech serv, analytical res & explor res, Brown & Williamson Tobacco Co, Louisville, Ky, 79-82; lab mgr, 83-88, DIR TECH INFO & RES SERV, KRAFT INC, GLENVIEW ILL, 88- *Concurrent Pos:* Air Force Off Sci Res grant; Kettering-Found grant. *Mem:* Am Chem Soc; Asn Anal Chemists; Am Oil Chemists Soc; Inst Food Technologists. *Res:* Development of physical, chemical, microbiological test methods; process analytical technology; laboratory automation; computer technology. *Mailing Add:* Kraft Tech Center 801 Waukegan Rd Glenview IL 60025-4312

MATUMOTO, TOSIMATU, b Tokyo, Japan, Aug 3, 26; m 55; c 2. GEOPHYSICS. *Educ:* Tokyo Univ, MS, 51, PhD(seismol), 60. *Prof Exp:* Res asst geophys, Earthquake Res Inst, Univ Tokyo, 51-61; res assoc, Lamont-Doherty Geol Observ, Columbia Univ, 60-65, sr res assoc, 66-74; mem staff, Marine Sci Inst, Galveston, Tex, 74-81; PROF GEOL SCI, UNIV TEX, AUSTIN, 81-, MEM STAFF, INST GEOPHYS, 81- *Concurrent Pos:* Mem Japan Antarctic Res Expedition, 56-67. *Mem:* Seismol Soc Am; Am Geophys Union. *Res:* Spectral analysis of seismic waves and its relation to magnitude; study of seismicity and microearthquake in Alaska and central and South America. *Mailing Add:* Inst Geophys Univ Tex 8701 Mopac Blvd Austin TX 78759

MATURI, VINCENT FRANCIS, b New York, NY, Oct 23, 16; c 5. BIOCHEMISTRY, SCIENCE EDUCATION. *Educ:* Cooper Union, BS, 39; NY Univ, MS, 43; Polytech Inst Brooklyn, PhD(chem eng), 48; George Washington Univ, MHCA, 72. *Prof Exp:* Asst div head res biochem, Standard Brands Inc, Stamford, Conn, 49-54; asst dir indust appln, Am Cyanamid Co, New York, 54-60; biochemist, Food & Drug Admin, 60-62; life sci specialist, NASA, 62-64; dep chief sci div, Smithsonian Inst, 64-68; exec secy, Ctr Demonstration Grants, HEW, Washington, DC, 68-71, health scientist adminr, Health Care Technol Div, Nat Ctr Health Serv Res, Dept Health & Human Serv, 71-84; PROF CHEM, GEORGE MASON UNIV, 84- *Concurrent Pos:* Consult, Nat Ctr Health Serv Res & Develop, HEW, 68-69; consult, health care indust. *Mem:* Am Hosp Asn; Am Pub Health Asn; Am Asn Clin Chemists; Am Chem Soc; Am Inst Chemists. *Res:* Health care administration; health care technology; medical information systems, medical devices, computerized scientific data handling; drugs; biochemistry; clinical chemistry; science education. *Mailing Add:* 5531 Bouffant Blvd Alexandria VA 22311

MATURO, FRANK J S, JR, b Nashville, Tenn, Apr 28, 29; m 60; c 3. MARINE BIOLOGY. *Educ:* Univ Ky, BS, 51; Duke Univ, MA, 53, PhD(marine ecol), 56. *Prof Exp:* Instr zool, Duke Univ, 55-57; vis asst prof, Univ NC, 57-58; asst prof biol, 58-64, assoc prof zool, 64-72, PROF ZOOL, UNIV FLA, 72-, DIR MARINE LAB, 70- *Concurrent Pos:* Nat Acad Sci-Nat Res Coun sr vis res assoc, Smithsonian Inst Mus Natural Hist, 65-66. *Mem:* Am Soc Zoologists; Estuarine Res Fedn; Sigma Xi; fel AAAS; Int Bryozool Asn (pres, 71-74). *Res:* Seasonal distribution and settling rates of marine invertebrates; zoogeography, ecology, and systematics of marine Bryozoa; larval behavior, metamorphosis, and astogeny of Bryozoa. *Mailing Add:* Dept of Zool Univ of Fla Gainesville FL 32611

MATURO, JOSEPH MARTIN, III, b Bridgeport, Conn, Nov 15, 42; m 66; c 2. BIOCHEMISTRY, PHYSIOLOGY. *Educ:* Fairfield Univ, BS, 64; Boston Col, PhD(biol), 69. *Prof Exp:* From asst prof to assoc prof, 69-77, PROF BIOL, C W POST COL, LONG ISLAND UNIV, 77- *Concurrent Pos:* NIH fel, Sch Med, Johns Hopkins Univ, 76-77; vis prof, Johns Hopkins Hosp, Baltimore, Md, 77 & St Georges Sch Med, Granada, WIndies & Sch Med, Univ Calgary, Alberta, 81; consult, Nat Cancer Cynology Ctr, Melville, NY, 75-77, Howard Hughes Med Inst, Miami, Fla, 78- *Mem:* AAAS; Sigma Xi (pres 77-78 & 81-); NY Acad Sci. *Res:* Mechanism of action of insulin. *Mailing Add:* Dept of Biol Long Island Univ PO Box Greenvale Brookville NY 11548

MATUSZAK, ALICE JEAN BOYER, b Newark, Ohio, June 22, 35; m 55; c 2. PHARMACY. *Educ:* Ohio State Univ, BS, 58, MS, 59; Univ Kans, PhD(pharmaceut chem), 63. *Prof Exp:* Asst prof pharmaceut chem, 63-67, assoc prof med chem, 75-78, PROF MED CHEM, UNIV PAC, 78- *Concurrent Pos:* Chief investr, NIMH grant, 65-66. *Mem:* Am Chem Soc; Am Pharmaceut Asn; AAAS; Acad Pharm Res Sci; Fedn Int Pharm; Am Inst Hist Pharm. *Res:* Synthesis of small heterocyclic compounds and their biochemical and pharmacological effects; use of audiovisual techniques to improve the teaching of medicinal chemistry; drug biotransformation; status of women scientists. *Mailing Add:* Sch Pharm Univ Pac Stockton CA 95211

MATUSZAK, CHARLES A, b Pittsburgh, Pa, Jan 7, 32; m 55; c 2. PHYSICAL ORGANIC CHEMISTRY. *Educ:* Univ Okla, BS, 52, MS, 53; Ohio State Univ, PhD(org chem), 57. *Prof Exp:* Asst org chem, Ohio State Univ, 53-57; res chemist, Owens-Corning Fiberglas Corp, 57-58; fel org chem, Ohio State Univ, 58-59, Univ Wis, 59-60 & Univ Kans, 60-61 & 62-63; asst prof, Washburn Univ, 61-62; from asst prof to assoc prof, 63-77, PROF ORG CHEM, UNIV PAC, 77- *Mem:* Am Chem Soc; Royal Soc Chem; Sigma Xi. *Res:* Mechanisms; Birch reduction; imidazole compounds; biphenylenes. *Mailing Add:* Dept Chem Univ Pac Stockton CA 95211

MATUSZAK, DAVID ROBERT, b Oct 2, 34; US citizen; m 53; c 3. GEOLOGY. *Educ:* Univ Okla, BS, 55, MS, 57; Northwestern Univ, PhD(geol), 61. *Prof Exp:* Lab asst geol, Univ Okla, 56-57; geologist, Kerr-McGee Oil Industs, Inc, 57-58; lab asst geol, Northwestern Univ, 58-60; res engr, Pan Am Petrol Corp, 61-63, sr res scientist, 63-68, res group supvr, 68-71, sr staff geologist, 71-76, geol assoc, 76-78, supvr explor systs, 78-87, DIR EXPLOR SYSTS & GEOL SYSTS, AMOCO PROD CO, 87- *Mem:* Am Asn Petrol Geologists; Soc Prof Well Log Analysts. *Res:* Use of subsurface data and computers in oil exploration. *Mailing Add:* 1503 Misty Bend Dr Katy TX 77450

MATUSZEK, JOHN MICHAEL, JR, b Worcester, Mass, Apr 16, 35; m 57; c 4. RADON ASSESSMENT, WASTE MANAGEMENT. *Educ:* Worcester Polytech Inst, BS, 57; Clark Univ, PhD(nuclear chem), 62. *Prof Exp:* Scientist, Southeastern Radiol Health Lab, USPHS, 62-64; asst mgr measurements div, Isotopes, Inc, 64-67; mgr physics dept, Teledyne Isotopes, 67-71; dir radiol sci lab, 71-87, RES SCIENTIST, NY STATE HEALTH DEPT, 87- *Concurrent Pos:* Adj prof, Rensselaer Polytech Inst, 81- *Mem:* Am Nuclear Soc; Health Physics Soc. *Res:* Radioactive waste management; radon measurement and risk assessment; nuclear spectroscopy; radiological health; radiochemical procedures; radionuclide transport models. *Mailing Add:* NY State Health Dept Empire State Plaza Albany NY 12201-0509

MATUSZKO, ANTHONY JOSEPH, b Hadley, Mass, Jan 31, 26; m 56; c 4. ORGANIC CHEMISTRY, INORGANIC CHEMISTRY. *Educ:* Amherst Col, AB, 46; Univ Mass, MS, 51; McGill Univ, PhD(org chem), 53. *Prof Exp:* Demonstr chem, McGill Univ, 50-52; from instr to assoc prof, Lafayette Col, 52-58; assoc head chem div, Res & Develop Dept, US Naval Propellant Plant, 58-59, head fundamental processes div, 59-62, polymer div, 62, actg dir res dept, 61; chief, Org Chem Prog, Air Force Off Sci Res, 62-71, prog mgr, Chem Sci Directorate, 71-89; CONSULT, 89- *Concurrent Pos:* Hon fel, Univ Wis, 67-68. *Mem:* Fel AAAS; Am Chem Soc; fel Am Inst Chemists; Cosmos Club; Sigma Xi. *Res:* Organometallics; pyridine derivatives; modifications and properties of cellulose nitrates; phosphonitrilic derivatives; high nitrogen compounds; biotechnology. *Mailing Add:* 4210 Elizabeth Lane Annandale VA 22003-3654

MATWIYOFF, NICHOLAS ALEXANDER, b Ann Arbor, Mich, Aug 19, 37; m 62; c 1. BIOPHYSICAL CHEMISTRY. *Educ:* Mich Col Mining & Technol, BS, 59; Univ Ill, MS, 61, PhD(chem), 62. *Prof Exp:* Fel chem, Stanford Univ, 62-63; asst prof chem, Pa State Univ, 63-68; sect leader nuclear magnetic resonance spectros, 68-72, alternate group leader inorg & phys chem, 72-78, dep div leader, Chem Div, 78-84, MGR STABLE ISOTOPES RESOURCE, LOS ALAMOS NAT LAB, 75-; PROF CELL BIOL & RADIOL, SCH MED, UNIV NMEX, 84- *Concurrent Pos:* Chmn, dept cell biol, dir, Ctr for Non-Invasive Diag & dep dir, Cancer Ctr, Sch Med, Univ NMex, 84- *Mem:* Am Soc Biol Chem; Am Chem Soc. *Res:* Study of structure and dynamics of peptides, enzymes, nucleic acids, and cellular systems; nuclear magnetic resonance. *Mailing Add:* Ctr for Non-Invasive Diag 1201 Yale NE Albuquerque NM 87131

MATYAS, E(LMER) LESLIE, b Hamilton, Ont, June 28, 32; m 54; c 2. GEOTECHNICAL ENGINEERING. *Educ:* Univ Toronto, BASc, 54; Univ London, PhD(soil mech), 63, Imp Col, dipl, 63. *Prof Exp:* Soils engr, Ont Hydro, 54-60; asst prof eng, Carleton Univ, 63-65; assoc prof, 65-84, PROF CIVIL ENG, UNIV WATERLOO, 84- *Concurrent Pos:* Assoc consult, Golder Assocs, 78-80. *Honors & Awards:* Hogentogler Award, Am Soc Testing & Mat, 69. *Mem:* Can Geotech Soc; Am Soc Civil Engrs; Asn Prof Engrs Ont. *Res:* Earthquake resistant design of earth dams; static and dynamic properties of cohesionless soils. *Mailing Add:* Dept Civil Eng Univ Waterloo Waterloo ON N2L 3G1 Can

MATYAS, GARY RALPH, b Berwick, Pa, Apr 30, 56; m 82; c 1. MONOCLONAL ANTIBODY PRODUCTION, CELL BIOLOGY. *Educ:* Pa State Univ, BS, 78; Purdue Univ, PhD(biol), 85. *Prof Exp:* Staff fel biochem res, Membrane Biochem Sect, Lab Molecular & Cellular Neurobiol, Nat Inst Neurol Communicated Disorders & Stroke, NIH, Bethesda, Md, 85-88; RES CHEMIST BIOCHEM RES, DEPT MEMBRANE BIOCHEM, WALTER REED ARMY INST RES, WASHINGTON, DC, 86- *Mem:* Am Soc Biochem & Molecular Biol; AAAS. *Res:* Role of glycolipids in cell growth control, cell adhesion and cancer; modulation of phospholipases during cell growth; production of vaccines and inhibitors to snake venom phospholipase A2 and other biological toxins. *Mailing Add:* Dept Membrane Biochem Walter Reed Army Inst Res Rm 1026 Washington DC 20307-5100

MATYAS, MARSHA LAKES, b Bluffton, Ind, June 11, 57; m 82; c 1. SCIENCE EDUCATION, SCIENCE POLICY. *Educ:* Purdue Univ, BS, 79, MS, 82, PhD(biol educ), 85. *Prof Exp:* PROJ DIR, WOMEN IN SCI, AAAS, 85- *Concurrent Pos:* Chair, Comt Role & Status Women in Biol Educ, Nat Asn Biol Teachers, 84-85, head, Sect Role & Status Women in Biol Educ, 85-; mem, Comt Role & Status Women in Educ Res & Develop, Am Educ Res Asn, 85- *Mem:* Am Educ Res Asn; Nat Asn Biol teachers; Nat Asn Res in Sci Teaching; Nat Sci Teachers Asn; Asn Women in Sci; Asn Women in Comput. *Res:* Role and status of women in scientific professions; improvement of pre-college science education for female, minority, and physically-disabled students; methods to encourage the participation of young women in science studies and careers. *Mailing Add:* AAAS 1333 H NW Washington DC 20005

MATYJASZEWSKI, KRZYSZTOF, b Konstantynow, Poland, Apr 8, 50; US citizen; m 72; c 2. SYNTHETIC POLYMER CHEMISTRY. *Educ:* Tech Univ Moscow, BS & MS, 72; Polish Acad Sci, PhD(polymer chem), 76; Tech Univ Lodz, DSc, 85. *Prof Exp:* Res asst polymers, Polish Acad Sci, 72-76, res assoc, 78-84; postdoctoral fel, Univ Fla, 77-78; res assoc, Nat Ctr Sci Res, Paris, 84-85; asst prof, 85-89, ASSOC PROF CHEM, CARNEGIE-MELLON UNIV, 89- *Concurrent Pos:* Invited prof polymers, Univ Paris, 85; vis prof, Univ Freiburg, Ger, 88 & Univ Paris, France, 90; NSF presidential young investr, 89; panelist, Nat Res Coun, 89; chmn, Polymer Curric Develop, Am Chem Soc, 89-; consult, Dow Corning, 89-90, Gen Elec, 91-, Arco, 91- *Mem:* Am Chem Soc. *Res:* Synthetic polymer chemistry; living polymers; cationic polymerization; inorganic polymers; block copolymers; polymer modifications; polymers for special applications; polymers for electronics. *Mailing Add:* Dept Chem Carnegie-Mellon Univ 4400 Fifth Ave Pittsburgh PA 15213

MATZ, ROBERT, b New York, NY, Aug 5, 31; m 55, 84; c 3. INTERNAL MEDICINE, ENDOCRINOLOGY. *Educ:* NY Univ, BA, 52, MD, 56. *Prof Exp:* From intern to resident, Bronx Municipal Hosp Ctr, 56-60; consult, Obstet Serv, Lincoln Hosp, Bronx, 62-63; NIH trainee metab, 63-64, assoc prof, 71-79, PROF MED, ALBERT EINSTEIN COL MED, 79-; DIR MED, MONTEFIORE-NORTH CENT BRONX AFFIL, 77-, CO-DIR, PRIMARY CARE RESIDENCY PROG INTERNAL MED, 80- *Concurrent Pos:* Vis physician, Montefiore-Morrisania Affil, 64-76, assoc dir med, 64-, head endocrinol & metab, 75-76, attend physician, Montefiore Hosp & Med Ctr, 75-76; attend physician, Bronx Munic Hosp Ctr, 71-76, co-dir, Diabetes Clin, 73-76; mem, Endocrine Dis Adv Comt, New York Dept Health, 72; consult, Health, Educ & Welfare Eval Unit, Albert Einstein Col Med, 73-76; ed, Diabetes Dept Cardiovasc Rev & Reports, 84- *Honors & Awards:* Frederick C Holden Prize, NY Univ Med Sch, 56. *Mem:* Am Diabetes Asn; Harvey Soc; Am Fedn Clin Res; Am Col Physicians; Am Heart Asn; Am Kidney Found. *Res:* Clinical research in diabetes mellitus, diabetic coma, and metabolic acidoses; clinical investigation of methods to improve delivery of health care; hyponatremia; hypothermia. *Mailing Add:* 32 Buena Vista Dr Hastings-on-Hudson NY 10706

MATZ, SAMUEL ADAM, b Carmi, Ill, July 1, 24; m 51; c 4. FOOD SCIENCE. *Educ:* Evansville Col, BA, 48; Kans State Col, MS, 50; Univ Calif, PhD(agr chem), 58. *Prof Exp:* Instr, Kans State Col, 50; chief chemist, Harvest Queen Mill & Elevator Co, 50-51; food technologist cereal & gen prod & chief br, Armed Forces Qm Food & Container Inst, 51-59; supvr refrig dough invests, Borden Foods Co, 59-65; vpres res & develop, Robert A Johnston Co, Wis, 65-69; vpres, Ovaltine Food Prod, Ovaltine Prod Inc, Ill 69-71, vpres res & develope & regulatory affairs 71-82; PRES, PAN-TECH INT INC, 82- *Concurrent Pos:* Dir, Avi Publ Co, 73-86. *Honors & Awards:* Rohland Isker Award, QM Food & Container Assoc. *Mem:* Am Chem Soc; Inst Food Technologists; Sigma Xi. *Res:* Food preservation methods and texture; cereal and flavor chemistry; permeability mechanisms; nutrition; regulatory affairs. *Mailing Add:* 1604 Riveroaks Edinburg TX 78539

MATZEN, MAURICE KEITH, b Columbus, Nebr, Sept 11, 47; m; c 2. PHYSICS, PLASMA PHYSICS. *Educ:* Iowa State Univ, PhD(phys chem), 74. *Prof Exp:* Instr physics, Hastings Col, 72; instr phys chem, Iowa State Univ, 73-74; RES PHYSICIST, SANDIA NAT LAB, 74- *Mem:* Fel Am Phys Soc. *Mailing Add:* 915 Toro SE Albuquerque NM 87123

MATZEN, VERNON CHARLES, b Petaluma, Calif. STRUCTURAL DYNAMICS, SYSTEM IDENTIFICATION. *Educ:* Univ Colo, BSCE, 66; Purdue Univ, MSCE, 68; Univ Calif, Berkeley, PhD(struct mech), 76. *Prof Exp:* Res & develop engr, Elec Boat Div, Gen Dynamics, Groton, Conn, 67-70; asst res engr, Univ Calif, Berkeley, 76-77; ASST PROF STRUCT MECH, NC STATE UNIV, 77- *Concurrent Pos:* Consult, Acures Corp, 75; lectr, Univ Calif, Berkeley, 76-77; consult, Brandt Indust, NC, 78-; prin investr, NSF, 80-82. *Mem:* Am Soc Civil Eng; Am Acad Mech; Earthquake Eng Res Inst; Sigma Xi. *Res:* Formulation of mathematical models of structures using measure response to known time-varying excitations; relationship between size of model, number and placement of sensors; uniqueness of parameters. *Mailing Add:* Dept Civil Eng NC State Univ PO Box 7908 Raleigh NC 27695

MATZEN, WALTER T(HEODORE), b Columbus, Nebr, Sept 30, 22; m 43; c 3. ELECTRICAL ENGINEERING. *Educ:* Iowa State Univ, BS, 43; Agr & Mech Col, Tex, MS, 50, PhD(elec eng), 54. *Prof Exp:* Electronic engr, Stromberg Carlson Co, 46-48; assoc prof elec eng, Agr & Mech Col, Tex, 48-57, res assoc, 53-57; mgr advan components, Tex Instruments, Inc, 57-79; MGR PROD ENG, OPTOELECTRONICS DIV, HONEYWELL, 79- *Mem:* Inst Elec & Electronics Engrs. *Res:* Design and technology of new semiconductor components. *Mailing Add:* 209 Thompson Richardson TX 75080

MATZINGER, DALE FREDERICK, b Alleman, Iowa, Apr 14, 29; m 60; c 2. QUANTITATIVE GENETICS. *Educ:* Iowa State Univ, BS, 50, MS, 51, PhD(plant breeding), 56. *Prof Exp:* Asst plant breeding, Iowa State Univ, 53-56; asst statistician, NC State Univ, 56-57, asst prof statist, 57-58, statist & genetics, 58-60, assoc prof genetics, 60-64, PROF GENETICS, NC STATE UNIV, 64- *Concurrent Pos:* Ed, Tobacco Sci, 84-88. *Mem:* Fel Am Soc Agron; Fel Crop Sci Soc Am; Genetics Soc Am; Am Genetics Asn; AAAS. *Res:* Statistical genetic theory and breeding methodology of self-pollinated plants; appled statistics; plant science. *Mailing Add:* Dept Genetics NC State Univ Raleigh NC 27695-7614

MATZKANIN, GEORGE ANDREW, b Chicago, Ill, June 30, 38; m 63; c 2. NONDESTRUCTRIVE EVALUATION, MATERIALS SCIENCE. *Educ:* St Mary's Col, AB, 60; Univ Fla, MS, 62, PhD(physics), 66; Trinity Univ, MBA, 77. *Prof Exp:* Res assoc metals physics, Argonne Nat Lab, 66-68; vis asst prof physics, Univ Ill, Chicago Circle, 68-69; sr res physicist, Instrumentation Div, 69-83, MGR NONDESTRUCTIVE TESTING INFO ANALYSIS CTR, SOUTHWEST RES INST, 83- *Mem:* AAAS; Am Phys Soc; fel Am Soc Nondestructive Testing; Sigma Xi. *Res:* Nondestructive evaluation research; instrumentation research; nuclear magnetic resonance; magnetic and mechanical properties of materials; Barkhausen phenomena; residual stress; evaluation of metal fatigue; moisture measurement; composite materials. *Mailing Add:* Southwest Res Inst PO Drawer 28510 San Antonio TX 78284

MATZNER, EDWIN ARTHUR, b Vienna, Austria, May 14, 28; m 53; c 1. ORGANIC CHEMISTRY. *Educ:* Calif Inst Technol, BS, 51; Yale Univ, PhD(org chem), 58. *Prof Exp:* Res chemist, Monsanto Co, 58-63, sr res group leader, 63-67, mgr res & develop, 66-77, MEM STAFF INORG RES DEPT, MONSANTO CO, 77-, MGR, ENVIRON AFFAIRS, 87- *Mem:* Am Chem Soc; Soc Chem & Indust; Res & Eng Soc Am. *Res:* Synthetic organic chemistry; solvent extraction; chemistry of phosphates; electrochemistry; chemistry of detergents and surfactants; environmental impact of chemicals. *Mailing Add:* Monsanto Co 800 N Lindbergh Blvd St Louis MO 63167

MATZNER, MARKUS, b Biala, Poland, Mar 19, 29; nat US; m 54; c 1. ORGANIC CHEMISTRY. *Educ:* Univ Brussels, MS, 50, PhD, 53. *Prof Exp:* Res chemist, Tirlemont Refinery, Belg, 53-54 & Probel Labs, 54-56; res & control chemist, Belg Petrol Refinery, 56-59; res chemist, Plastics Div, Union Carbide Corp, 59-63, proj scientist, 63-66, res scientist, 66-68, sr res scientist, 68-71, res assoc, Chem & Plastics Res & Develop Dept, 71-82; CONSULT, 83- *Honors & Awards:* Prix, Belgian Acad Sci, 53. *Mem:* Am Chem Soc; Sigma Xi. *Res:* Organic synthesis; mechanisms of reactions; polymer chemistry; scientific translations. *Mailing Add:* 23 Marshall Dr Edison NJ 08817

MATZNER, RICHARD ALFRED, b Ft Worth, Tex, Jan 2, 42; m 67; c 1. PHYSICS. *Educ:* Univ Notre Dame, BS, 63; Univ Md, College Park, PhD(physics), 67. *Prof Exp:* NSF fac assoc physics, 67-69, asst prof, 69-73, res physicist, Ctr Relativity Theory, 73-77, ASSOC PROF PHYSICS, UNIV TEX, AUSTIN, 73- *Concurrent Pos:* Res fel physics, Wesleyan Univ, 69-70. *Mem:* AAAS. *Res:* General relativity; cosmology; gravitational collapse; geometrical optics; canonical formulations; statistical mechanics. *Mailing Add:* Dept of Physics Univ of Tex Austin TX 78712

MAUCHE, CHRISTOPHER W, b Elmira, NY, June 2, 58; m 84. CATACLYSMIC VARIABLES, X-RAY ASTRONOMY. *Educ:* State Univ NY, Stony Brook, BS, 81; MA, 84 & PhD, Harvard Univ, 87. *Prof Exp:* Res asst, Harvard-Smithsonian Ctr Astrophys, 81-87; postdoctoral fel, Los Alamos Nat Lab, 87-90; POSTDOCTORAL RES STAFF MEM, LAWRENCE LIVERMORE NAT LAB, 90- *Mem:* Am Astron Soc; Am Physics Soc. *Res:* Studies of the x-ray halos of celestial x-ray sources; studies of the winds, disks and boundary layers of cataclysmic variables. *Mailing Add:* Lawrence Livermore Nat Lab L-401 PO Box 808 Livermore CA 94550

MAUCK, HENRY PAGE, JR, b Richmond, Va, Feb 3, 26; c 2. CARDIOLOGY. *Educ:* Univ Va, BA, 48, MD, 52; Am Bd Internal Med, dipl, 59. *Prof Exp:* DIR CARDIAC CATHETERIZATION LAB, MED COL VA, 70-, PROF MED & PEDIAT, 72- *Concurrent Pos:* Am Heart Asn fel, 56-57; consult pediat cardiol, Langley Air Force Hosp; ed consult, Am Heart J. *Mem:* AMA; fel Am Col Physicians; fel Am Col Cardiol; Am Fedn Clin Res; fel Am Heart Asn. *Res:* Neural control of the circulation. *Mailing Add:* Dept Med Med Col Va Box 281 Richmond VA 23298

MAUDERLI, WALTER, b Aarau, Switz, Mar 8, 24; nat US; m 50; c 5. NUCLEAR PHYSICS, RADIATION PHYSICS. *Educ:* Swiss Fed Inst Technol, MS, 49, DSc(physics), 56. *Prof Exp:* Physicist & asst radiol, Univ Zurich, 50-56; physicist, asst prof & head isotope labs, Sch Med, Univ Ark, 56-60; from assoc prof to prof radiation physics, 60-87, physicist, 60-87, prof environ eng sci, 74-87, EMER PROF RADIATION PHYSICS, J HILLIS MILLER HEALTH CTR, COL MED, UNIV FLA, 87- *Concurrent Pos:* Lectr, Grad Inst Technol & Med Ctr, Univ Ark; consult, Vet Admin Hosp, Little Rock. *Mem:* Am Asn Physicists in Med; Soc Nuclear Med; AMA. *Res:* Radiation physics; computer applications in radiology; electronic instrumentation in radiation physics. *Mailing Add:* Dept Radiol Box J-385 JHM Health Ctr Univ Fla Gainesville FL 32610

MAUDERLY, JOE L, b Strong City, Kans, Aug 31, 43; m 65; c 2. LUNG CANCER, ENVIRONMENTAL LUNG DISEASE. *Educ:* Kans State Univ, BS, 65, DVM, 67. *Prof Exp:* Physiologist, 69-80, supv, Pathophysiol Group, 80-88, DIR, INHALATION TOXICOL RES INST, 89-; PRES, LOVELACE BIOMED & ENVIRON RES INST, 89- *Concurrent Pos:* Assoc ed, Lab Animal Sci, 78-82; Nat Res Coun, Inst Lab Animal Resources subcomt on animal models for aging carnivores, 79-79; comt on biol markers, Panel on Pulmonary Toxicol, Nat Acad Sci, 86-; chmn, air pollution health studies adv comt, Elec Power Res Inst, 87-; prog comt, Assembly Environ & Occup Health, Am Thoracic Soc, 80-81, 83-85 & 86-88, secy, 88, chmn elect, 89-90; adj res prof med & clin prof pharm, Univ NMex, 88-; assoc ed, Fundamental Appl Toxicol, 89-; chmn adv bd, NMex Summer Teacher Enrichment Prog, 90- *Mem:* Am Vet Med Asn; Am Soc Vet Physiol & Pharmacol; Am Physiol Soc; World Asn Vet Physiol, Pharmacol & Biochem; Am Thoracic Soc; Comp Respiratory Soc; Soc Toxicol. *Res:* Comparative pulmonary pathophysiology; respiratory function measurements in animals; environmental and occupational lung disease; animal models of human lung diseases; aging of lung; environmental careinogenesis. *Mailing Add:* Inhal Toxicol Res Inst PO Box 5890 Albuquerque NM 87185

MAUDLIN, LLOYD Z, b Miles City, Mont, Feb 20, 24; m 46; c 4. PHYSICS. *Educ:* Univ Calif, Los Angeles, AB, 49; Univ Southern Calif, MS, 52. *Prof Exp:* Electronic scientist, US Naval Ord Test Sta, 51-56, supvry electronic scientist & head simulation br, 56-59, supvry physicist & dir simulation & comput ctr, 59-67, head simulation & analysis div, Naval Undersea Warfare Ctr, 67-71, supvry physicist & head comput sci & simulation div, Naval Undersea Res & Develop Ctr, 71-77, supvry physicist & head comput sci & simulation dept, Naval Ocean Systs Ctr, 77-81; vpres res & develop, Integrated Systs Inc, 82-87; EXEC VPRES, JIL SYSTEMS INC, 88- *Mem:*

Inst Elec & Electronics Engrs; NY Acad Sci. *Res:* Anti-submarine warfare, particularly guidance and control of underwater weapons; computer and simulation analysis of anti-submarine warfare weapons systems; analog and digital computing techniques; physics of underwater sound. *Mailing Add:* JIL Systs Inc 5461 Toyon Rd San Diego CA 92115

MAUDSLEY, DAVID V, b Blackburn, Eng, Dec 13, 40. BIOCHEMICAL PHARMACOLOGY. *Educ:* Univ London, BP, 62, PhD(pharmacol), 66. *Prof Exp:* Lectr pharmacol, Sch Pharm, Univ London, 64-66; staff scientist, 66-75, sr scientist, 75-85, ASSOC, WORCESTER FOUND EXP BIOL, SHREWSBURY, 85- *Concurrent Pos:* Assoc res prof biochem, Univ Mass Med Sch, 77-85. *Mem:* Am Soc Pharmacol & Exp Therapeut; NY Acad Sci. *Res:* Regulation of diamine and polyamine metabolism in proliferating systems. *Mailing Add:* HND Assoc Box 1063 Conway NH 03818

MAUE-DICKSON, WILMA, b Joliet, Ill, Apr 15, 43; m 71; c 1. DEVELOPMENTAL ANATOMY. *Educ:* Rockford Col, BA, 64; Northwestern Univ, MA, 68; Univ Pittsburgh, PhD(interdisciplinary anat), 70. *Prof Exp:* Clin asst, instr, teaching & res asst, Northwestern Univ, 66-68; vpres, Bicom Co, Pittsburgh, 68-78; dir res, otolaryngol & maxillofacial surg, Mercy Hosp, 69-74; assoc dir, basic sci res prog, Cleft Palate Ctr, Univ Pittsburgh, 70-75, asst prof anat, 74-75; dir, craniofacial res, Mailman Ctr Child Develop, Sch Med, Univ Miami, 76-79, assoc prof surg & pediat, 76-79; assoc mgr, educ serv, Cordis Corp, Miami, 80-82; adj assoc prof surg, Sch Med, Univ Miami, 80-; MGR, EDUC SERV, CORDIS CORP, 82-; DIR CONTINUING EDUC, TELECTRONICS,. *Concurrent Pos:* Vol, US Peace Corps, Ethiopia, 64-66; instr anat & physiol, Carlow Col, Pittsburgh, 70-72; adj asst prof speech & theatre arts, Univ Pittsburgh, 74-75. *Honors & Awards:* First Place Award Sci Presentation, Am Speech & Hearing Asn, 70, First Place Award Sci Merit, 75; First Place Award Sci Merit, Southern Med Asn, 81. *Mem:* Am Cleft Palate Asn; Am Speech & Hearing Asn; Asn Res Otolaryngol; Am Asn Anatomists; Sigma Xi. *Res:* Human head/neck anatomy, physiology, embryology, dysmorphology and computed axial tomography. *Mailing Add:* Telectronics PO Box 025202 Miami FL 33102-5202

MAUER, ALVIN MARX, b Le Mars, Iowa, Jan 10, 28; m 50; c 4. MEDICINE. *Educ:* Univ Iowa, BA, 50, MD, 53. *Hon Degrees:* DSc, Westmar Col, 82. *Prof Exp:* Intern, Cincinnati Gen Hosp, 53-54; from jr resident to chief resident pediat, Cincinnati Children's Hosp, 54-56; from asst prof to assoc prof pediat, Col Med, Univ Cincinnati, 59-69, prof, 69-73; dir, St Jude Children's Res Hosp, 73-84; PROF MED & PEDIAT, UNIV TENN, MEMPHIS, 73-, CHIEF DIV HEMATOL-ONCOL, 73-, DIR, CANCER CTR, 73- *Concurrent Pos:* Dir div hemat, Children's Hosp Res Found, 59-; attend pediatrician & dir div hemat & hemat clin, Children's Hosp, Cincinnati, 59-; attend pediatrician, Cincinnati Gen Hosp, 59-; attend hematologist, Vet Admin Hosp, 60-; NIH fel, 56-58; res fel hemat, Univ Utah, 56-59; Am Cancer Soc fel, 58-59; NIH res career develop award, 62-; fel, Div Hemat, Children's Hosp Res Found, 63- *Mem:* Am Asn Cancer Res; Am Pediat Soc; Am Soc Clin Invest; Am Soc Hemat (pres); Asn Am Physicians; Sigma Xi; Am Acad Pediat; Am Soc Clin Oncol; Int Soc Hematol (pres, 88-90); N Y Acad Sci; Soc Pediat Res. *Res:* Cancer detection and screening; cancer clinical therapeutic trials; transformation of human hematopoietic stem cells; leukemia and normal cell and regulation. *Mailing Add:* 4583 Laurelwood Dr Memphis TN 38117

MAUER, IRVING, b Montreal, Que, Feb 7, 27; nat US; m 52, 81; c 4. MUTAGENESIS, DEVELOPMENTAL BIOLOGY. *Educ:* McGill Univ, PhD(genetics), 60. *Prof Exp:* Asst cytol, Sci Serv, Can Dept Agr, 48-49; demonstr genetics, McGill Univ, 54-56, asst, 55-56; res assoc animal genetics, Storrs Agr Exp Sta, Univ Conn, 56-57; sr med writer, Squibb Inst Med Res Div, Olin Mathieson Chem Corp, 57-60; psychiat res fel, NY State Dept Ment Health, 61-62, sr res scientist, 62-67, lectr, 62 & 65; head cytogenetics group, Dept Exp Path, Hoffmann-La Roche Inc, 67-77; GENETICIST, HEALTH EFFECTS DIV, OFF PESTICIDE PROGS, ENVIRON PROTECTION AGENCY, 78- *Mem:* AAAS; Asn Govt Toxicologists; Am Col Toxicol; Am Soc Testing & Mat; Soc Risk Assessment; Genetic Toxicol Asn; Environ Mutagen Soc. *Res:* Cytogenetics of man; mutagenicity testing; teratology; experimental cytogenetics; genetic counseling. *Mailing Add:* 7121 Gordons Rd Falls Church VA 22043-3058

MAUER, S MICHAEL, DIALYSIS, TRANSPLANTATION. *Educ:* McGill Univ, Can, MD, 66. *Prof Exp:* PROF PEDIAT, SCH MED, UNIV MINN, 69- *Mem:* Soc Pediat Res; Cent Soc Clin Res; Am Soc Clin Invest; Am Soc Nephrol; Int Soc Nephrol; Am Asn Pathologists. *Res:* Diabetic nephropathy; renal failure in childhood. *Mailing Add:* Dept Pediat Sch Med Univ Minn Box 491 Mayo Minneapolis MN 55455

MAUERSBERGER, KONRAD, b Lengefeld, Ger, Apr 28, 38; m 64; c 2. AERONOMY. *Educ:* Univ Bonn, Dipl, 64, PhD(physics), 68. *Prof Exp:* Res assoc physics, Univ Bonn, 68-69; res assoc, 69-74, from asst prof to assoc prof, 74-82, PROF PHYSICS, UNIV MINN, MINNEAPOLIS, 82- *Mem:* Am Geophys Union; AAAS. *Res:* Composition and dynamics of Earth's upper atmosphere using mass spectrometers carried on balloons, rockets and satellites; solar-atmospheric interactions at altitudes above twenty kilometers. *Mailing Add:* 148 Physics Bldg Univ Minn Minneapolis MN 55455

MAUGER, JOHN WILLIAM, b Scranton, Pa, Aug 10, 42; m 65; c 1. PHARMACEUTICS. *Educ:* Union Col, NY, BS, 65; Univ RI, MS, 68, PhD(pharmaceut sci), 71. *Prof Exp:* Instr pharm, Univ RI, 69-71; asst prof, 71-80, ASSOC PROF PHARM, SCH PHARM, WVA UNIV MED CTR, 80- *Mem:* Am Pharmaceut Asn. *Res:* Nonelectrolyte solubility; aqueous solutions of pharmaceutical solutes; convective diffusion. *Mailing Add:* Dept Pharm W Va Univ Med Ctr Morgantown WV 26506

MAUGER, RICHARD L, b Fairdale, Pa, Sept 20, 36; m 63; c 1. GEOLOGY. *Educ:* Franklin & Marshall Col, BS, 58; Calif Inst Technol, MS, 60; Univ Ariz, PhD(geol), 66. *Prof Exp:* Asst prof geol, Univ Utah, 66-69; from asst prof to assoc prof, 69-76, PROF GEOL, EAST CAROLINA UNIV, 76- *Mem:* AAAS; Geol Soc Am; Am Geophys Union; Sigma Xi. *Res:* Mineral deposits; isotopic dating and stable isotopes. *Mailing Add:* Dept of Geol ECarolina Univ Greenville NC 27834

MAUGHAN, EDWIN KELLY, b Glendale, Calif, Oct 13, 26; div; c 4. GEOLOGY, PALEOGEOGRAPHY & TECTONICS. *Educ:* Utah State Univ, BS, 50. *Prof Exp:* Geologist, Corps Engrs, US Dept Army, 51; geologist, 51-90, VOL, US GEOL SURVEY, 90- *Concurrent Pos:* Tech adv phosphate deposits, Inventario Minero Nacional Colombia, 67-69 & stratig, Struct & Coal Resources Southeast Ky, 69-72. *Mem:* Geol Soc Am; Am Asn Petrol Geologists; Colombian Asn Advan Sci; Soc Econ Paleont & Mineral (vpres, 82-85); Int Asn Sedimentologists; Paleontological Soc. *Res:* Areal geology vicinity of Great Falls, Mont, Thermopolis, Wyo, Middlesboro, Ky; stratigraphy and phosphate resources in Cretaceous of Colombia; stratigraphy, paleogeography, tectonics and mineral resources (petroleum, salt, phosphate) in Permian, Pennsylvanian and Mississippian rocks of northern Rocky Mountains and Great Basin. *Mailing Add:* US Geol Surv (MS 939) Fed Ctr Box 25046 Denver CO 80225

MAUGHAN, GEORGE BURWELL, b Toronto, Ont, May 8, 10; m 67; c 6. OBSTETRICS & GYNECOLOGY. *Educ:* McGill Univ, MD & CM, 34, MSc, 38; FACS, 50, FRCS(C), 52, FRCOG, 57. *Prof Exp:* Demonstr path & bact, McGill Univ, 34-35, demonstr anat, 39-40, from demonstr to asst prof obstet & gynec, 40-56, prof & chmn dept, 56-75, EMER PROF OBSTET & GYNEC, McGILL UNIV, 77- *Concurrent Pos:* Obstetrician & gynecologist-in-chief, Royal Victoria Hosp, 56-75, hon consult obstet & gynec, 75-; consult obstetrician & gynecologist, Montreal Gen, Reddy Mem, Queen Elizabeth, St Mary's & Jewish Gen, Lakeshore Gen, Montreal Chinese. *Mem:* Can Med Asn; fel Am Col Surg; Soc Obstet & Gynaec Can (past pres); fel Am Asn Obstet & Gynec; Can Gynaec Soc (past pres). *Res:* Intensive care in high risk pregnancy. *Mailing Add:* 446 Mt Stephen Westmount PQ H3Y 2X6 Can

MAUGHAN, OWEN EUGENE, b Preston, Idaho, Jan 3, 43; m 62; c 6. FISH BIOLOGY, AQUATIC ECOLOGY. *Educ:* Utah State Univ, BS, 66; Univ Kans, MA, 68; Wash State Univ, PhD(zool), 72. *Prof Exp:* Proj leader, Nat Res Planning Div, River Basin Studies, 71-72, asst leader & assoc prof fisheries, Va Coop Fishery Res Unit, 72-77, unit leader & assoc prof fisheries, Okla Coop Fish & Wildlife Res Unit, 84-87, UNIT LEADER & PROF, ARIZ COOP FISH & WILDLIFE RES UNIT, US FISH WILDLIFE SERV, 87- *Concurrent Pos:* Fisheries sect leader, Va Polytech Inst & State Univ, 76-77. *Mem:* Am Fisheries Soc; Am Soc Ichthyol & Herpetol; Can Soc Zool. *Res:* Fisheries biology and aquatic ecology; particularly the effects of man caused development on aquatic ecosystems. *Mailing Add:* US Fish & Wildlife Serv Univ Ariz Tucson AZ 85721

MAUGHAN, W LOWELL, CARDIOLOGY, BIOMEDICAL ENGINEERING. *Educ:* Univ Wash, MD, 70. *Prof Exp:* ASSOC PROF MED, JOHNS HOPKINS UNIV, 78- *Mailing Add:* Med Dept Cardiol Div 592 Carnegie Bldg Johns Hopkins Hosp 600 N Wolfe St Baltimore MD 21205

MAUK, ARTHUR GRANT, b Geneva, Ill, Sept 26, 47; m 70. BIOINORGANIC CHEMISTRY. *Educ:* Lawrence Univ, BA, 69; Med Col Wis, MD & PhD(biochem), 76. *Prof Exp:* Res fel chem, Calif Inst Technol, 76-79; from asst prof to assoc prof, 79-89, PROF BIOCHEM, UNIV BC, 89- *Mem:* Am Chem Soc; Am Soc Biol Chemists. *Res:* Relationship between heme protein structure and function; electron transfer mechanisms; heme protein-heme protein interaction; application of chemical modification and site-directed mutagenesis. *Mailing Add:* Dept Biochem Univ BC Copp Bldg Vancouver BC V6T 1W5 Can

MAUK, MARCIA ROKUS, b Wisconsin Rapids, Wis, July 11, 47; m 70. BIOCHEMISTRY. *Educ:* Ripon Col, AB, 69; Med Col Wis, Milwaukee, PhD(biochem), 74. *Prof Exp:* Fel biochem, Med Col Wis, 74-76; res fel, Calif Inst Technol, 76-79; RES ASSOC BIOCHEM, UNIV BC, 81- *Res:* Structure-function relationships in hemeproteins; mechanisms of protein-protein interaction; targeting of lipid vesicles. *Mailing Add:* Dept Biochem Univ BC Vancouver BC V6T 1W5 Can

MAUKSCH, INGEBORG G, NURSING. *Educ:* Univ Chicago, PhD, 76. *Prof Exp:* SR PROG CONSULT, ROBERT WOOD JOHNSON, 76-; LECTR & CONSULT NURSING, 82- *Mem:* Am Nurses Asn. *Mailing Add:* Sch Nursing Vanderbilt Univ Nashville TN 37240

MAUL, GEORGE A, b Brooklyn, NY, July 17, 38; m 61; c 2. OCEANOGRAPHY. *Educ:* State Univ NY, BS, 60; Univ Miami, PhD(phys oceanog), 74. *Prof Exp:* Comm officer, US Coast & Geodetic Surv, 60-69; res oceanographer, 69-84, SUPVRY OCEANOGRAPHER, ATLANTIC OCEANOG & METEOROL LAB, NAT OCEANIC & ATMOSPHERIC ADMIN, 84- *Concurrent Pos:* Adj fac, Rosenstiel Sch Marine & Atmospheric Sci, Univ Miami, 77-; fel, Coop Inst Marine & Atmospheric Studies, 78-; chmn, climate task team wider Caribbean region, UNEP, 88-; vchmn, IOC subcomt Caribbean & adj regions, Fla Acad Sci, 89-92. *Mem:* Am Geophys Union; Marine Technol Soc. *Res:* Application of remote sensing to studies of ocean circulation including currents, tides, productivity, climate, and transport of pollutants; sea level and climate change. *Mailing Add:* Atlantic Oceanog & Meteorol Lab Nat Oceanic & Atmospheric Admin 4301 Rickenbacker Causeway Miami FL 33149-1097

MAUL, GERD G, b Hoyerswerd, Germany, Mar 14, 40. NUCLEAR STRUCTURE & FUNCTION. *Educ:* Univ Tex, PhD(zool), 40. *Prof Exp:* Assoc prof, 73-85, PROF CELL BIOL, WISTAR INST, 85-, PROF LAB MED, UNIV PA. *Mailing Add:* Wistar Inst 36th & Spruce Sts Philadelphia PA 19104

MAUL, JAMES JOSEPH, b Buffalo, NY, Nov 3, 38; m 62; c 2. ORGANIC CHEMISTRY. *Educ:* Canisius Col, BS, 60; Wayne State Univ, PhD(org chem), 66. *Prof Exp:* Sr chemist, 68-75, SR RES CHEMIST, HOOKER CHEM CORP, 75- *Mem:* Am Chem Soc. *Res:* Organo-fluorine chemistry; organo-halogen chemistry. *Mailing Add:* 15 Beaver Ln Grand Island NY 14072

MAULBETSCH, JOHN STEWART, b Brooklyn, NY, May 25, 39; m 75; c 2. MECHANICAL ENGINEERING. *Educ:* Mass Inst Technol, SB, 60, SM, 62, PhD(mech eng), 65. *Prof Exp:* Asst prof mech eng, Mass Inst Technol, 65-67; proj engr, Dynatech Res & Develop Co, 67-70, prin engr, 70-75; prog mgr heat, waste & water mgt, 75-84, sr prog mgr, Air Qual Control, 84-87, EXEC SCIENTIST, EXPLORATORY RES, ELEC POWER RES INST, 87- *Concurrent Pos:* Ford fel, 65-67. *Mem:* Am Soc Mech Engrs. *Res:* Two-phase flow and boiling; environmental control. *Mailing Add:* 90 Lloyden Dr Atherton CA 94025

MAULDIN, RICHARD DANIEL, b Longview, Tex, Jan 17, 43; m 85; c 2. MATHEMATICS. *Educ:* Univ Tex, BA, 65, MA, 66, PhD(math), 69. *Prof Exp:* From asst prof to assoc prof math, Univ Fla, 69-77; assoc prof, 77-79, PROF MATH, UNIV NTEX, 79- *Concurrent Pos:* Consult, Los Alamos Nat Lab. *Mem:* Am Math Soc; Math Asn Am; Asn Symbolic Logic; Sigma Xi. *Res:* Descriptive set theory, measure and probability theory; topology; use of the computer as an experimental tool. *Mailing Add:* Dept Math Univ NTex Denton TX 76203-5116

MAULDING, DONALD ROY, b Evansville, Ind, Aug 15, 36; m 58; c 1. SYNTHETIC ORGANIC CHEMISTRY. *Educ:* Evansville Col, AB, 58; Univ Ind, PhD(org chem), 62. *Prof Exp:* Fel, Ohio State Univ, 62-64; PRIN RES CHEMIST, AM CYANAMID CO, 64- *Honors & Awards:* Am Cyanamid Sci Achievement Award, 85. *Mem:* Am Chem Soc. *Res:* Organic mechanisms; photochemistry and chemiluminescence; fluorescence; rubber chemicals and polyurethanes; agricultural chemicals; synthesis; process development; imidazolinone herbicides. *Mailing Add:* Org Chems Div Am Cyanamid Co PO Box 400 Princeton NJ 08540

MAULDING, HAWKINS VALLIANT, JR, b Foreman, Ark, Dec 21, 35; m 59; c 3. PHYSICAL CHEMISTRY, PHYSICAL PHARMACY. *Educ:* Univ Ark, Little Rock, BS, 58; Univ Minn, Minneapolis, PhD(med chem), 64. *Prof Exp:* Asst prof pharm, Univ Houston, 64-66; SR SCIENTIST & GROUP LEADER PHYS CHEM & PHYS PHARM, SANDOZ PHARMACEUT, HANOVER, 66- *Mem:* Am Pharmaceut Asn; Sigma Xi. *Res:* Theoretical and applied kinetics; complexation; reaction mechanisms; stability of solid products; dosage form design. *Mailing Add:* Corey Lane Rte 24 Mendham NJ 07945

MAULDON, JAMES GRENFELL, b London, Eng, Feb 9, 20; m 53; c 4. PURE MATHEMATICS. *Educ:* Oxford Univ, BA & MA, 47. *Hon Degrees:* MA, Amherst Col, 70. *Prof Exp:* Lectr math, Oxford Univ, 47-68; prof math, 68-80, Walker Prof, 80-90, EMER PROF MATH, AMHERST COL, 90- *Concurrent Pos:* Lectr, St John's Col, Oxford Univ, 50-59, fel, Corpus Christi Col, 50-68; vis prof, Univ Calif, Berkeley, 60-61; chmn fac, Oxford Univ, 66-68; consult, IBM Corp, 74-75; vis fel, Australian Nat Univ, Canberra ACT, 83, 87. *Mem:* Am Math Soc; Math Asn Am. *Res:* Mathematics, including probability, algebra, analysis, geometry and combinatorics; computer languages. *Mailing Add:* Dept Math & Computer Sci Amherst Col Amherst MA 01002

MAUMENEE, IRENE HUSSELS, b Bad Pyrmont, W Ger, Apr 30, 40; m 72; c 2. OPHTHALMOLOGY. *Educ:* Univ Göttingen, MD, 64; Am Bd Ophthal, cert ophthal, 76 & Am Bd Med Genetics, cert med genetics, 82. *Prof Exp:* Res asst, Univ Hawaii, 68, vis geneticist, Pop Genetics Lab, 68-69; fel, Dept Med, Johns Hopkins Univ Sch Med, 69-71; ophthal preceptorship, Wilmer Inst, Johns Hopkins Hosp, 69-73; from asst prof to assoc prof, 72-87, PROF OPHTHAL & MED, WILMER OPHTHAL INST, JOHNS HOPKINS HOSP & DEPT MED, DIV MED GENETICS, MOORE CLINIC, JOHNS HOPKINS HOSP, 87-; DIR, JOHNS HOPKINS CTR HEREDITARY EYE DIS, WILMER INST, 79- *Concurrent Pos:* Participant, design & execution of a genetic & ophthal study, Ponape Dist, Micronesia; consult, John F Kennedy Inst Visually & Mentally Handicapped Children, 74-; dir, Low Vision Clin, Wilmer Inst, 77-; invited lectr, Opening Ceremony, Int Cong Ophthal, San Francisco, 82, Ger Ophthal Soc, Heidelberg, 85, Portuguese Ophthal Soc, Lisbon, 86, Brazilian Ophthal Soc, 87, Burroughs Wellcome vis prof, BritRoyal Soc Med, 87-88, French Ophthal Soc, Paris & French Acad Med, 88; adv, Nat Eye Inst Task Forces, 76 & 81; foreign fel training prog, 84-; mem comt, foreign fel, Pan Am Asn Ophthal; pres, Sixth Cong, Int Soc Genetic Eye Dis, Amsterdam, 86 & Seventh Cong, Lisbon, 88. *Honors & Awards:* Lloyd Morgan Lectureship, Hospital for Sick Children, Toronto, 79; Sir Stanley Davidson Mem lectr, Univ Edinburgh, Scotland, 85. *Mem:* Am Soc Human Genetics; Am Acad Ophthal; Asn Res Vision & Opthal; Int Soc Genetic Eye Dis; AMA; Am Ophthal Soc; Pan Am Asn Opthal. *Res:* Nosology and management in ophthalmic and general medical genetics; population genetics; computer application to genetic analysis; molecular genetics; over 120 publications on human genetics and eye diseases. *Mailing Add:* Wilmer Ophthal Inst Johns Hopkins Hosp Maumenee Bldg Rm 321 Baltimore MD 21205

MAUNDER, A BRUCE, b Holdrege, Nebr, May 13, 34; m 78; c 1. GENETICS, PLANT BREEDING. *Educ:* Univ Nebr, BS, 56; Purdue Univ, MS, 58, PhD(genetics), 60. *Prof Exp:* Plant breeder, DeKalb Pfizer Genetics, 59-61, sorghum res dir, 61-78, actg agron res dir, 78-79, agron res dir, DeKalb Agresearch Inc, 79-82, Agron Res Dir, DeKalb Pfizer Genetics, 82- 89, AGRON RES DIR, DEKALB PLANT GENETICS, 90- *Honors & Awards:* Gerald Thomas Award, Am Soc Agron, Genetics & Plant Breeding Award; Indust Agronomist Award, Crop Sci Soc. *Mem:* Fel Am Soc Agron; fel AAAS. *Res:* Inheritance of male sterility; heterosis as regards sorghum; disease and insect resistance; genetic advances; evolution. *Mailing Add:* DeKalb Plant Genetics Route Two Lubbock TX 79415

MAUNE, DAVID FRANCIS, b Washington, Mo, July 12, 39; m 61; c 2. PHOTOGRAMMETRY, RESEARCH ADMINISTRATION. *Educ:* Univ Mo-Rolla, BSc, 61; Ohio State Univ, MSc, 70, PhD(geod, photogram), 73. *Prof Exp:* Mech engr, Union Elec Co, St Louis, 61; co comdr & opers officer, 656 Eng Topog Battalion, US Army, Ger, 63-66, mapping officer, Hq, Vietnam, 66-67, opers officer, 36 Engr Group, Korea, 70-71, officer-in-chg prod, Mapping & Charting Estab, Royal Engrs, Eng, 73-74, staff officer, Directorate Army Res, Washington, DC, 74-76, topog plans officer, 77-78, battalion comndr, 652 Eng Topog Battalion Hawaii, 78-80, chief training develop & eval, Defense Mapping Sch, 81-88, DIR, ENGR TOPOG LABS US DEPT ARMY, FT BELVOIR, VA, 88- *Mem:* Am Soc Photogram; Soc Am Military Engrs. *Res:* Photogrammetric calibration of scanning electron microscopes; analytical photogrammetry; satellite geodesy; operations research systems analysis; military research and development management; mapping, charting and geodesy training. *Mailing Add:* 7131 Lake Cove Dr Alexandria VA 22310

MAUNSELL, CHARLES DUDLEY, physics, for more information see previous edition

MAUPIN, PAMELA, MUSCLE PROTEINS, ELECTRON MICROSCOPY. *Educ:* Univ Tex, Austin, BS, 71. *Prof Exp:* SR RES TECHNICIAN, CELL BIOL & ORGAN HISTOL, SCH MED, JOHNS HOPKINS UNIV, 78- *Mailing Add:* Dept Cell Biol & Anat Med Sch Johns Hopkins Univ 720 N Wolfe St Baltimore MD 21205

MAURER, ALAN H, b Atlantic City, NJ, Jan 17, 47. MEDICINE. *Educ:* Brown Univ, BA, 69; Univ Pa, Philadelphia, MS, 71; Temple Univ Sch Med, MD, 75. *Prof Exp:* Biomed engineer, Gen Elec Co, 69-71; DIR, CARDIOVASCULAR NUCLEAR MED, TEMPLE UNIV HOSP, 81-, SECT NUCLEAR MED, TEMPLE UNIV HOSP, 83- *Mem:* Am Col Physicians; Soc Nuclear Med; Radiol Soc N Am; Am Fed Clin Res; Am Heart Asn. *Res:* Application of nuclear medicine to the field of cerebrovascular disease, cardiology, gastroenterology and bone imaging. *Mailing Add:* 538 Ballytore Rd Wynnewood PA 19096

MAURER, ARTHUR JAMES, b Winfield, Pa, Apr 16, 42; m 66; c 2. POULTRY SCIENCE, FOOD SCIENCE. *Educ:* Pa State Univ, BS, 64; Cornell Univ, MS, 66, PhD(food Sci), 71. *Prof Exp:* Asst county agent, Pa State Univ, 63-64; res asst, Cornell Univ, 64-70; from asst prof to assoc prof, 70-81, PROF POULTRY SCI, UNIV WIS, 81- *Concurrent Pos:* NIH fel, 64-68; consult, Vol Tech Asst, 76; poultry consult, Wis Nicaragua Partners Prog, 77; consult, Am Soybean Assoc; vis prof, Univ Philippines, 84; int consult, Arg, 86, Honduras, 87, El Salvador, 88, Brazil, 88, Costa Rica, 89, Korea, 89. *Mem:* Poultry Sci Asn; Inst Food Technol; Int Asn Milk, Food & Environ Sanitarians; World Poultry Sci Asn. *Res:* Poultry products technology; processing, preservation, product development and marketing of poultry meat and eggs. *Mailing Add:* Dept Poultry Sci Univ Wis Madison WI 53706

MAURER, BRUCE ANTHONY, b Springfield, Mass, Oct 22, 36; c 3. RESEARCH ADMINISTRATION. *Educ:* St Michael's Col, BA, 58; Univ Mass, MS, 60; Univ Ariz, PhD(microbiol), 66. *Prof Exp:* Asst prof microbiol, Miami Univ, 66-68; sr cancer res scientist, Roswell Park Mem Inst, 68-71; asst res prof virol, Roswell Park Div, State Univ NY, Buffalo, 69-71; dir biol, Assoc Biomedic Systs, 71-73; res scientist immunol, Litton-Bionetics, Inc, Md, 73-78; grants assoc, NIH, 78-79, prog dir immunol, Nat Inst Aging, 79-81 & Nat Cancer Inst, 81-83; exec secy, Hemat Study Sect DRG, NIH, 83-86, exec secy, Virol Study Sect, 86-89, referral officer, DRG, 86-89, CHIEF IMMUNOL, VIROL, PATH REV SECT, DRG, NIH, 89- *Mem:* Sigma Xi. *Res:* Biochemistry of polyoma virus infection in vitro; cell virus interaction of Epstein-Barr virus and human lymphoblastoid cell lines; cellular immunology of human and animal neoplasias; lymphocyte alloantigens of rhesus monkey major histocompatibility complex. *Mailing Add:* Virol Study Sect Div Res Grants NIH Westwood Bldg Rm 309 Bethesda MD 20892

MAURER, DONALD LEO, b Chicago, Ill, Sept 3, 34; m 67; c 6. MARINE ECOLOGY, POLLUTION BIOLOGY. *Educ:* Univ Ill, BS, 56; Univ Wash, MS, 58; Univ Chicago, PhD(paleozool), 64. *Prof Exp:* Res assoc marine ecol, Pac Marine Sta, Calif, 64-65; assoc prof biol, Old Dom Col, 65-67; asst prof, 67-73, assoc prof marine biol, Univ Del, 73-81; PROF MARINE BIOL, CALIF STATE UNIV LONG BEACH, 85- *Concurrent Pos:* Consult, Sanitation districts. *Mem:* Soc Limnol & Oceanog; Sigma Xi. *Res:* Ecology of marine invertebrates; paleoecology; description of macroscopic Benthic invertebrate communities; determination of community and specific response to pollutants. *Mailing Add:* Cal State Univ Long Beach 1250 Bell Flower Blvd Long Beach CA 90840

MAURER, E(DWARD) ROBERT, b San Francisco, Calif, Jan 3, 21; m 55; c 2. HISTORY OF SCIENCE, PHYSICAL SCIENCE. *Educ:* Stanford Univ, BS, 48, MS, 50, PhD(phys sci), 64. *Prof Exp:* Instr phys sci, Chico State Col, 52-55, asst prof, 55-56; vis asst prof, Stanford Univ, 56-58; from assoc prof to prof, 58-83, head dept phys sci, 58-67, dean, Sch Prof Studies, 70-74, asst dean acad affairs, 75, EMER PROF PHYS SCI, CALIF STATE UNIV, CHICO, 83- *Mem:* Fel AAAS; Hist Sci Soc; Soc Hist Technol; Sigma Xi; Soc Hist Alchemy & Chem; Am Canal Soc. *Res:* General history of science and technology, the Industrial Revolution and British chemistry during the 18th and 19th centuries; the role of the history of science in general education. *Mailing Add:* Dept of Geol & Phys Sci Calif State Univ Chico CA 95929-0205

MAURER, FRED DRY, b Moscow, Idaho, May 4, 09; m 35; c 2. VETERINARY PATHOLOGY. *Educ:* Univ Idaho, BS, 34; State Col Wash, BS & DVM, 37; Cornell Univ, PhD(path bact), 48; Am Col Vet Path, dipl, 50. *Prof Exp:* Jr veterinarian dis control, Bur Animal Indust, USDA, 37; asst prof, Univ Idaho & asst bacteriologist, Exp Sta, 37-38; instr path & bact, Vet Col, Cornell Univ, 38-41; staff mem, Vet Res Lab, Vet Corps, US Army, Va, 41-43, lab officer, War Dis Control Sta, Can & Africa, 43-46, Res & Grad Sch, Army Med Ctr, 47-51, animal dis res, 51-54, chief, Vet Path Div, Armed

Forces Inst Path, 54-61, dir, Div Med, Army Med Res Lab, Ft Knox, Ky, 61-64; dir, Inst Trop Vet Med, 74-76, distinguished prof path, 64-76, emer dir, Inst Trop Vet Med, Col Vet Med, 76-77, EMER PROF, TEX A&M UNIV, 77- *Concurrent Pos:* Assoc dean, Col Vet Med, Tex A&M Univ, 64-74. *Honors & Awards:* Int Vet Vet Cong Prize, 68. *Mem:* Am Vet Med Asn; US Animal Health Asn; Am Col Vet Path (pres, 64); Conf Res Workers Animal Dis; Am Asn Lab Animal Sci. *Res:* Virology and pathology of infectious diseases of animals. *Mailing Add:* 2408 Morris Ln Bryan TX 77802

MAURER, GERNANT E, b Sayre, Pa, May 5, 49; m 72. HIGH TEMPERATURE ALLOYS, PROCESS METALLURGY. *Educ:* Johns Hopkins Univ, BES, 71; Rensselaer Polytech Inst, PhD(mat sci), 76. *Prof Exp:* Engr, 76-78, group leader, 78-80, mgr res & develop, 80-84, VPRES TECH, SPEC METALS CORP, 84- *Concurrent Pos:* Comt mem, Int Symp on Superalloys, 76-84, gen chmn, 84-88; mem, High Tempreature Alloy Comt, Soc Testing & Mats, 84-86; adv comt, Am Vacuum Soc, 88. *Mem:* Fel Am Soc Metals Int; Am Vacuum Soc; Soc Testing & Mats. *Res:* High temperature nickel base superalloys; physical metallurgy; alloy design; melt processing; mechanical behavior. *Mailing Add:* 23 Sherman Circle Utica NY 13501

MAURER, HANS ANDREAS, b Frankfurt, Ger, May 7, 13; US citizen. PHYSICS. *Educ:* Univ Munich, BSc, 33; Univ Frankfurt, PhD(appl physics), 37. *Prof Exp:* Sr proj engr, Missile Systs Div, Raytheon Co, 57-61, mgr advan weapons systs, 61-63, mgr Advan Syst Ctr, 63-64, chief engr, 64-65; tech dir to asst div mgr, Missile Div, Aerospace Group, Hughes Aircraft Co, 66-76, tech dir, Missile Systs Group, 76-85, vpres tech group, 83-85; RETIRED. *Concurrent Pos:* Self-employed aerospace consult, Teledyne Elec, Newbury Park Ca. *Mem:* AAAS; Ger Oberth Soc; fel Inst Elec & Electronics Engrs. *Mailing Add:* 4447 Conchita Way Tarzana CA 91356

MAURER, JOHN EDWARD, b Matherville, Ill, Apr 3, 23; m 46; c 4. ORGANIC CHEMISTRY. *Educ:* Augustana Col, AB, 47; Univ Iowa, MS, 48, PhD(org chem), 50. *Prof Exp:* Res assoc, Northwestern Univ, 50-52; res chemist, Rock Island Arsenal, 52-53; from asst prof to prof, Univ Wyo, 53-88, asst head dept, 68-77, head dept, 77-80, EMER PROF CHEM, UNIV WYO, 88- *Concurrent Pos:* Coun representing Wyoming sect, Am Chem Soc Nat Off, 88-91. *Mem:* AAAS; Am Chem Soc; Sigma Xi. *Res:* Halogenations; Van Slyke reactions; natural products. *Mailing Add:* Dept Chem Univ Wyo Box 3838 Laramie WY 82071

MAURER, KARL GUSTAV, b Philadelphia, Pa, Aug 25, 29; m 55; c 3. ENGINEERING MECHANICS, MECHANICAL ENGINEERING. *Educ:* Drexel Inst Technol, BSME, 59; Univ Kans, MS, 62, PhD(fluid mech), 66. *Prof Exp:* Design engr, Gen Dynamics/Astronaut Div, 59-60; asst instr eng mech, Univ Kans, 60-65; asst prof civil eng, Rose Polytech Inst, 66-67; assoc prof mech eng & asst dean col eng & archit, 67-72, PROF MECH ENG & CHMN DEPT, NDAK STATE UNIV, 72- *Mem:* Am Soc Mech Engrs; Am Soc Eng Educ. *Res:* Classical fluid mechanics and boundary layer theory. *Mailing Add:* Mech Engr Dept NDak State Univ Fargo ND 58105

MAURER, PAUL HERBERT, b New York, NY, June 29, 23; m 48; c 3. IMMUNOLOGY. *Educ:* City Col New York, BS, 44; Columbia Univ, PhD(immunochem), 50. *Prof Exp:* Res biochemist, Gen Foods Corp, 44 & 46; instr, City Col New York, 46-51; res assoc, Col Physicians & Surgeons, Columbia Univ, 50-51; asst res prof, Sch Med, Univ Pittsburgh, 51-54, assoc prof immunochem, 54-60; prof microbiol, NJ Col Med & Dent, 60-66; PROF BIOCHEM & HEAD DEPT, JEFFERSON MED COL, 66- *Concurrent Pos:* NIH res career award, 62-66. *Mem:* Am Chem Soc; Am Asn Immunol; NY Acad Sci; Brit Biochem Soc; Sigma Xi; Am Biochem Soc. *Res:* Immunochemistry; biochemistry; protein chemistry. *Mailing Add:* Dept Biochem Med Col Thomas Jefferson Univ 1020 Locust St Philadelphia PA 19107

MAURER, RALPH RUDOLF, reproductive physiology, reproductive endocrinology; deceased, see previous edition for last biography

MAURER, RICHARD ALLEN, b Long Beach, Calif, Mar 21, 47; m 68. PHYSIOLOGY. *Educ:* Univ Calif, Irvine, BS, 69; Univ Calif, Davis, PhD(physiol), 73. *Prof Exp:* Fel, Univ Wis, Madison, 73-77; from asst prof to assoc prof, 77-85, PROF PHYSIOL, UNIV IOWA, 85-, CHAIR, MOLECULAR BIOL PHD PROG, 88- *Concurrent Pos:* Mem molecular biol study sect, NIH, 83-85. *Mem:* Endocrine Soc; Am Soc Biol Chemists; Sigma Xi; Soc Study Reproduction; Am Physiol Soc. *Res:* The molecular basis for the endocrine control of prolactin synthesis and secretion; analysis of the rate limiting factors in the processes of prolactin gene transcription; hormonal control of DNA synthesis and cell differentiation in the anterior pituitary. *Mailing Add:* 5530 B5B Physiol Univ Iowa Iowa City IA 52242

MAURER, ROBERT DISTLER, b St Louis, Mo, July 20, 24; m 51; c 3. APPLIED PHYSICS. *Educ:* Univ Ark, BS, 48; Mass Inst Technol, PhD(physics), 51. *Hon Degrees:* LLD, Univ Ark, 80. *Prof Exp:* Mem physics staff, Mass Inst Technol, 51-52; physicist, Corning Glass Works, 52-62, sr res assoc, 62-63, mgr fundamental physics res, 63-70, mgr appl physics res, 70-78, res fel, Corning Inc, 78-89; CONSULT, 89- *Honors & Awards:* George W Morey Award, Am Ceramic Soc, 76; Morris N Liebmann Award, Inst Elec & Electronic Engrs, 78; Indust Appln Physics Prize, Am Inst Physics, 78; Ericsson Int Prize Telecommun, Swed Acad Eng, 79; John Tyndall Award, Inst Elec & Electronic Engrs, Optical Soc Am, 87; Int Prize New Mat, Am Phys Soc, 89. *Mem:* Nat Acad Eng; fel Am Ceramic Soc; fel Elec & Electronic Engrs; Am Phys Soc. *Res:* Physical behavior of glasses; optical communications. *Mailing Add:* Corning Inc Sullivan Park FR5 Corning NY 14830

MAURER, ROBERT EUGENE, b Uhrichsville, Ohio, 25; m 53; c 1. GEOLOGY. *Educ:* Ohio State Univ, BS, 50; Univ Utah, PhD, 70. *Prof Exp:* Geologist, Texaco, Inc, 52-56; instr geol, Westminster Col, Utah, 57-59, asst prof, 59-66; asst prof, 66-70, ASSOC PROF GEOL, STATE UNIV NY COL OSWEGO, 70- *Concurrent Pos:* Chmn sci div, Westminster Col, Utah, 62-66; chmn dept earth sci, State Univ NY Col Oswego, 67-72. *Mem:* AAAS; Geol Soc Am. *Res:* Surface and subsurface geologic mapping. *Mailing Add:* Dept Earth Sci State Univ NY Col Oswego Oswego NY 13126

MAURER, ROBERT JOSEPH, b Rochester, NY, Mar 26, 13; m 40. SOLID STATE PHYSICS, RESEARCH ADMINISTRATION. *Educ:* Univ Rochester, BS, 34, PhD(physics), 39. *Prof Exp:* Res assoc, Mass Inst Technol, 39-42; instr physics, Univ Pa, 42-43; from asst prof to assoc prof, Carnegie Inst Technol, 43-49; from assoc prof to prof, 49-81, dir, Mat Res Lab, 63-78, EMER PROF PHYSICS, UNIV ILL, URBANA, 81- *Concurrent Pos:* Physicist, Metall Lab, Univ Chicago, 44-45; head physics br, Off Naval Res, 48. *Mem:* Am Phys Soc. *Res:* Self diffusion in solids; electrical properties of solids; photoelectric properties of metals; optical properties of solids. *Mailing Add:* 909 Trail Cross Ct Sante Fe NM 87505

MAURICE, D V, AVIAN NUTRITION. *Educ:* Univ Ga, PhD(nutrit), 78. *Prof Exp:* ASSOC PROF NUTRIT, CLEMSON UNIV, 78- *Mailing Add:* 144 Folger St Clemson SC 29631-1304

MAURICE, DAVID MYER, b London, Eng, Apr 3, 22; m 54; c 3. CORNEAL PHYSIOLOGY, PHARMACOKINETICS. *Educ:* Univ Reading, BSc, 41; Univ London, PhD(physiol), 51. *Prof Exp:* Jr sci officer, Telecommun Res Estab, Ministry Aircraft Prod, 41-46; staff mem ophthal res unit, Med Res Coun, 46-63; reader physiol, Inst Ophthal, Univ London, 63-68; SR SCIENTIST, OPHTHAL, MED SCH, STANFORD UNIV, 68-, ADJ PROF SURG, 74-, PROF OPHTHAL RES, 87- *Concurrent Pos:* Prof Surg Res, Standford Univ, 85-87; vis prof, Tokyo Univ Med Sch, 82; vis prof, Hadassah Med Sch, 63-64; Guggenheim fel, Univ Paris, 80. *Honors & Awards:* Friedenwald Medal, Asn Res Vision & Ophthal, 67. *Mem:* AAAS; Am Physiol Soc; Asn Res Vision & Ophthal; Int Soc Eye Res. *Res:* Vegetative physiology of the eye; physiology and biochemistry of cornea; transport mechanisms; pharmacokinetics of the eye. *Mailing Add:* Dept of Ophthal Stanford Univ Med Ctr Stanford CA 94305

MAURIZI, MICHAEL R, ENZYMOLOGY, PROTEIN CHEMISTRY. *Educ:* Univ Ill, PhD(biochem), 78. *Prof Exp:* EXPERT, NAT CANCER INST, 83- *Mailing Add:* Lab of Cell Biol NIH Bldg 37 Rm 1B07 Bethesda MD 20892

MAURMEYER, EVELYN MARY, b New York, NY, Sept 24, 51. COASTAL GEOLOGY. *Educ:* Smith Col, AB, 72; Univ Del, MS, 74, PhD(geol), 78. *Prof Exp:* Asst prof geol, Franklin & Marshall Col, 78-81; PRES, COASTAL & ESTUARINE RES, INC, 81- *Concurrent Pos:* Vis asst prof, Univ Del, 79-81, adj asst prof, Col Marine Studies, 82- *Mem:* Geol Soc Am; Soc Econ Paleontologists & Mineralogists; The Coastal Soc. *Res:* Barrier island dynamics; beach erosion; sediment transport processes. *Mailing Add:* 41 Harborview Rd Lewes DE 19958

MAURO, ALEXANDER, biophysics; deceased, see previous edition for last biography

MAURO, JACK ANTHONY, b Brooklyn, NY, Feb 21, 16; m 37; c 3. OPTICAL PHYSICS, PHYSIOLOGICAL OPTICS. *Educ:* Columbia Univ, BS & cert, 47; Phila Optical Col, OD, 51. *Prof Exp:* Mgr optics, Equitable Optical Co, 34-43; chief instr theoret optics & math, NY Inst Optics, 47-53; dir eng, Saratoga Div, Espey Mfg Co, 50-55; engr, Gen Eng Lab, Gen Elec Co, 55-60, consult optics engr, Ord Dept, Defense Electronics Div, 60-65, consult engr, Missile & Space Vehicle Div, 65-70; dir res, Shuron Continental Div, Textron Inc, 70-74; CONSULT HIGH ENERGY LASER OPTICAL SYSTS, BATTELLE MEM INST, UNIV DAYTON RES INST & US AIR FORCE LASER WEAPONS LAB, 74- *Concurrent Pos:* Consult, NY Inst Optics, 47-55 & Navigational Inst Am, 47-52; mem, Bd Dirs, Columbia Univ, 48-51; consult, US Army & US Navy Ord Off, 50-51; chmn, Man-Mach Symp, US Army-Gen Elec Co, 58; mem, High Energy Laser Weapons Ad Hoc Comt, US Army Missile Command, US Air Force Weapons Lab & Univ Dayton Res Inst. *Mem:* Am Phys Soc; Optical Soc Am; fel Am Acad Optom; Soc Photo-Optical Instrument Eng; NY Acad Sci. *Res:* Optics and electronics; physical and geometrical optics; optical design. *Mailing Add:* 2581 Lema Rd SE Rio Rancho NM 87124

MAURRASSE, FLORENTIN JEAN-MARIE ROBERT, b Port-au-Prince, Haiti, Dec 2, 40; US citizen; m 84; c 2. PALEOECOLOGY & STRATIGRAPHY, SEDIMENTOLOGY & TECTONICS. *Educ:* Col, Port-au-Prince, Haiti, dipl fin et, 64; Fac Sci, Marseille, France, lic es sci, 66, dipl et sup, 67; Columbia Univ, PhD(stratig-sedimentation), 73. *Prof Exp:* Prof natural sci, Petionville Sec Sch, Haiti, 63-64; sedimentologist, Lamont Doherty Geol Observ, Columbia Univ, 68-69 & Deep Sea Drilling Proj, 70-71; from asst prof to assoc prof geol, 73-86, CHMN DEPT, 84-, PROF GEOL, FLA INT UNIV, 86- *Concurrent Pos:* Adv, Ministry Mines & Energy Resources, Haiti, 76-; vis prof, Fac Sci, Port-au-Prince, Haiti, 80. *Mem:* Geol Soc Am; Am Geophys Union; Soc Econ Paleontologists & Mineralogists; Am Asn Petrol Geologists; Sigma Xi; Geol Soc France; Am Asn Petrol Geologists. *Res:* Radiolarian paleoecology and its implications concerning paleo-climates and paleo-oceanography; study of late Mesozoic to late Cenozoic eras with special emphasis on the Paleogene; Caribbean geology and tectonics. *Mailing Add:* Dept Geol Fla Int Univ Tamiami Trail Miami FL 33199

MAURY, LUCIEN GARNETT, b Hoisington, Kans, Aug 14, 23; m 47; c 3. PHYSICAL CHEMISTRY, ORGANIC CHEMISTRY. *Educ:* Ill Inst Technol, BS, 48; Northwestern Univ, PhD(chem), 52. *Prof Exp:* Res chemist, Hercules Inc, Del, 51-56, supvr, 56-58, mgr explosives res & high pressure lab, 58-61, mgr synthetics res, 61-64, mgr cent res, 64-67, proj mgr, 67-68, dir develop fibers & film, 68-71, dir fibers, 71-73; gen mgr, Hercules Int Dept, 73-74; pres, 74-78, VPRES RES & DEVELOP, HERCULES EUROPE, 78- *Mem:* Sigma Xi. *Res:* Homogeneous and heterogeneous catalysis; nitrogen chemistry. *Mailing Add:* 29 Brandywine Falls Wilmington DE 19806

MAUSEL, PAUL WARNER, b Minneapolis, Minn, Jan 2, 36; m 66; c 3. GEOGRAPHIC INFORMATION SYSTEMS. *Educ:* Univ Minn, BA(chem) & BA(geog), 58, MA, 61; Univ NC, PhD(geog), 66. *Prof Exp:* Instr geog, Mankato State Col, 61-62; from asst prof to assoc prof, Eastern Ill Univ, 65-71; from assoc prof to prof geog, 72-88, dir Remote Sensing Lab, 76-88, GEOG INFO SYSTS COORDR, IND STATE UNIV, TERRE HAUTE, 88- *Concurrent Pos:* Res grants, Eastern Ill Univ, 67-68 & 69-70; researcher, Lab Appln Remote Sensing, Purdue Univ, 72-74; res grants, NSF, 76, Environ Protection Agency, 77, US Forest Serv, 77 & Ind State Univ, 77; training grants, NSF, 78; res grants, Northern Ky Area Develop Dist, 80, Able Energy Co, 80, Environ Protection Agency, 87-91, Oak Ridge Nat Lab, 90, Stennis Space Ctr, 90 & Nat Park Serv, 89. *Mem:* Asn Am Geog; Am Soc Photogram & Rem Sensing. *Res:* Remote sensing of the environment using automatically data processed multispectral sensor data, stressing land use and mineral resources; soils geography; videography applications; geographic information systems research/applications. *Mailing Add:* Dept of Geog & Geol Ind State Univ Terre Haute IN 47809

MAUSNER, JACK, b Poland, Jan 12, 32; m 58; c 2. CHEMISTRY. *Educ:* London Univ, BSc, 52, PhD(chem), 55; Royal Inst Chem, FRIC, 57. *Prof Exp:* Sr scientist, Monsanto Chemicals, 55-58 & E I du Pont de Nemours, 58-60; dir, Helena Rubinstein Ltd, 60-70, corp vpres, Helena Rubinstein Inc, 70-80; vpres res & develop, 80-87, SR VPRES, CHANEL INC, 87- *Concurrent Pos:* Pres, Fragrance Res Fund-Fragrance Found, 89. *Mem:* Fel Royal Inst; NY Acad Sci; Acad Dermat. *Res:* Holder of numerous patents in field of skin care, makeup and fragrance. *Mailing Add:* 150 E 69th St New York NY 10021

MAUSNER, LEONARD FRANKLIN, b New York, NY, Mar 6, 47; m 69; c 3. MEDICAL ISOTOPES, RADIOPHARMACEUTICALS. *Educ:* Mass Inst Technol, BS, 68; Princeton Univ, MA, 72, PhD(chem), 75. *Prof Exp:* Instr chem, Princeton Univ, 74-75; fel chem, Los Alamos Sci Lab, 75-77; mem staff, Argonne Nat Lab, 77-81; MEM STAFF, BROOKHAVEN NAT LAB, 81- *Concurrent Pos:* Consult, Princeton Gamma Tech, 73. *Mem:* Am Phys Soc; Am Chem Soc; Soc Nuclear Med. *Res:* Radionuclide production and research for diagnostic nuclear medicine and radiotherapy. *Mailing Add:* Med Dept Bldg 801 Brookhaven Nat Lab Upton NY 11973

MAUSTON, GLENN WARREN, b St Paul, Minn, Oct 22, 35; m 57; c 3. ENTOMOLOGY. *Educ:* Gustavus Adolphus Col, BS, 57; Iowa State Univ, MS, 59; NDak State Univ, PhD(entom), 69. *Prof Exp:* INSTR BIOL, MESABI COMMUNITY COL, 59- *Res:* Taxonomy and biology of the subfamily Crambinae. *Mailing Add:* Dept Biol Mesabi Community Col Nine Ave & Chestnut St Virginia MN 55792

MAUTE, ROBERT EDGAR, b Colorado Springs, Colo, July 1, 47; m 88; c 4. PETROPHYSICS, APPLIED PHYSICS. *Educ:* Lamar Univ, BS, 69; Ohio State Univ, MS, 71, PhD(nuclear physics), 73. *Prof Exp:* Res physicist, Accurray Inc, 73-77; RES ASSOC, MOBIL OIL CORP, 77- *Concurrent Pos:* Assoc ed, Soc Prof Well Log Anal, 91- *Mem:* Soc Petro Engrs; Soc Prof Well Log Analysts. *Res:* Detection and evaluation of hydrocarbon bearing rock formations; interpretation techniques for multiple measurements (electrical, nuclear, acoustic) made downhole in oil wells with logging tools. *Mailing Add:* PO Box 819047 Dallas TX 75381

MAUTE, ROBERT LEWIS, b Springfield, Ohio, July 1, 24; m 46; c 2. ANALYTICAL CHEMISTRY. *Educ:* Colo Col, BS, 49; Univ Houston, MS, 50. *Prof Exp:* Chemist, Phillips Petrol Co, 50-51; res chemist, Monsanto Co, 51-55, asst group leader, Analysis Group, 55-58, group leader, 58-64, mgr analysis sect, Tex, 64-78; mgr analysis technol, Monsanto Res Corp, 78-82; RETIRED. *Concurrent Pos:* Teaching chem, 83-85; consult, 85-86. *Mem:* Am Chem Soc. *Res:* Instrumental and chemical analyses; physical chemistry; trace environmental analysis of hazardous material; bioanalytical characterization. *Mailing Add:* 419 Meadow Ridge Dr Kerrville TX 78028

MAUTNER, HENRY GEORGE, b Prague, Czech, Mar 30, 25; nat US; m 67; c 3. BIOCHEMISTRY, PHARMACOLOGY. *Educ:* Univ Calif, Los Angeles, BS, 46; Univ Southern Calif, MS, 49; Univ Calif, PhD(chem), 55. *Hon Degrees:* MS, Yale Univ, 67. *Prof Exp:* Lab asst, Univ Southern Calif, 47-49; res chemist, Productol Co, 50; sr res technician, Univ Calif, 51-53, asst, 53-55; from instr to assoc prof pharmacol, Sch Med, Yale Univ, 56-67, prof pharmacol & head sect med chem, 67-70; chmn depts biochem & pharmacol, Sch Med, Tufts Univ, 70-84, prof pharmacol, 70-84, prof biochem, 70-90, EMER PROF BIOCHEM & PHARMACOL, SCH MED, TUFTS UNIV, 90- *Concurrent Pos:* Squibb fel pharmacol, Sch Med, Yale Univ, 55-56; mem, Neurobiol Panel, NSF, 77-80, vis prof, Max Planck Inst Biochem, Munich, Ger, 67-68, Biol Ctr, Univ Basel, Switz, 85, Dept Biophys Chem, Univ Bielefeld, Ger, 87; study sect, Med Chem, NIH, 68, 72, chmn, 71-72; ed bd, J Med Chem, 68-70, 76-81; mem Neurobiol Panel, NSF, 77-80; vis fel, Dept Org Chem, Uppsala; vis scholar, Dept Fine Arts, Harvard Univ. *Mem:* Am Chem Soc; Am Asn Cancer; Royal Soc Chem; Am Soc Biol Chemists; Biophys Soc. *Res:* Heterocyclic chemistry; purines; pyrimidines; pteridines; chemistry of selenium compounds; coenzyme analogs; choline acetyltransterase; electrically excitable membranes; antimetabolites; comparative kinetics of reactions of oxygen, sulfur and selenium isologs; molecular basis of nerve conduction. *Mailing Add:* Dept Biochem Sch Med Tufts Univ Boston MA 02111

MAUTZ, CHARLES WILLIAM, b St Elmo, Ill, Apr 27, 17; m 45; c 1. PHYSICS. *Educ:* Univ Ill, BS, 41, MS, 43; Univ Mich, PhD(physics), 49. *Prof Exp:* Mem staff, Radiation Lab, Mass Inst Technol, 44-45 & Los Alamos Sci Lab, 49-60; mem staff, Gulf Gen Atomic Co, 60-76; MEM STAFF, LOS ALAMOS SCI LAB, 76- *Concurrent Pos:* Vis assoc prof, Univ Mich, 61-62. *Mem:* AAAS; Am Phys Soc. *Res:* Gas dynamics; explosives; lasers; shockwaves; detonation. *Mailing Add:* Five Mariposa Ct Los Alamos NM 87544

MAUTZ, WILLIAM WARD, b Eau Claire, Wis, Apr 13, 43; m 65; c 3. WILDLIFE RESEARCH, WILDLIFE ECOLOGY. *Educ:* Wis State Univ-Eau Claire, BS, 65; Mich State Univ, MS, 67, PhD(wildlife ecol & physiol), 69. *Prof Exp:* From asst prof to assoc prof wildlife ecol, Inst Natural & Environ Resources, Univ NH, 69-75; asst unit leader, Colo Coop Wildlife Res Unit, Colo State Univ, 75-76; from assoc prof to prof wildlife ecol, 76-88, coordr, Wildlife Prog, Inst Natural & Environ Resources, 79-88, CHAIR, DEPT NATURAL RESOURCES, UNIV NH, 88- *Concurrent Pos:* Wildlife consult, 76. *Mem:* Wildlife Soc; Ecol Soc Am. *Res:* Ecological energetics; energy flow studies with wildlife species involving energy requirements and efficiency of food energy utilization; development of procedures for the determination of energy utilization and requirements in wildlife species. *Mailing Add:* Dept Natural Resources Univ NH Durham NH 03824

MAUZERALL, DAVID CHARLES, b Sanford, Maine, July 22, 29; m 59; c 2. PHYSICAL CHEMISTRY. *Educ:* St Michael's Col, BS, 51; Univ Chicago, PhD(chem), 54. *Prof Exp:* Res assoc, 54-59, from asst prof to assoc prof, 59-69, PROF BIOPHYS, ROCKEFELLER UNIV, 69- *Concurrent Pos:* Vis assoc prof, Univ Calif, San Diego, 65-68, adj prof, 68-; Guggenheim fel, 66. *Mem:* Biophys Soc; Am Chem Soc; Am Soc Biol Chemists. *Res:* Mechanism of photochemical and photobiological reactions; porphyrin biochemistry; photosynthesis. *Mailing Add:* Rockefeller Univ New York NY 10021

MAUZEY, PETER T, b Poughkeepsie, NY, Nov 16, 30. ELECTRICAL ENGINEERING. *Educ:* Columbia Univ, BSEE, 52, MSEE, 54, EE, 58. *Prof Exp:* Asst elec eng, Columbia Univ, 52-54, instr, 54-59, assoc elec eng, 59-62; mem tech staff, Data Commun Dept, 62-77, mem tech staff, Data Terminals Dept, 77-81, mem tech staff, Data Appl Eng Dept, Bell Labs, 81-83; MEM TECH STAFF, DATA COMMUN DEVELOP LAB, AT&T, 83- *Concurrent Pos:* Dir eng, Electronic Music Ctr, Columbia Univ & Princeton Univ, 59-62. *Honors & Awards:* Inst Elec & Electronic Engrs Region 1 Award, 89. *Mem:* Inst Elec & Electronic Engrs; Acoust Soc Am; Audio Eng Soc. *Res:* Electronic circuitry for high-quality recording and reproduction of speech and music; hardware and software design for office system planning; data commmunications systems engineering; product planning. *Mailing Add:* AT&T 200 Laurel Ave Middletown NJ 07748-4801

MAUZY, MICHAEL P, US citizen. TECHNICAL MANAGEMENT. *Educ:* Va Polytech Inst, BS; Univ Tenn, MS. *Prof Exp:* Dir eng & mfg, Monsanto Co, St Louis, 51-71; mem staff, Kummer Corp, St Louis, Mo, 71-72; dir, Ill Environ Protection Agency, Springfield, Ill, 72-81; VPRES, WESTON, 81- *Concurrent Pos:* Mem, Water Mgt Subcomt, Mgt Adv Group & Ohio River Valley Sanit Comm; Judge, Am Consult Engrs Coun Excellence Award, 81. *Mem:* Asn State & Interstate Water Pollution Control Admin (pres, 77 & 80); Am Inst Chem Engrs; Am Pub Works Asn; Water Pollution Control Asn. *Res:* Planning, design, construction and operation of research, development and manufacturing facilities; program planning and management, strategy, policy development and evaluation, fiscal control, personnel training, employee development and public involvement; environmental and resource management at state and federal level; compliance evaluation and management. *Mailing Add:* Three Hawthorn Pkwy Suite 400 Vernon Hills IL 60061

MAVERICK, ANDREW WILLIAM, b Los Angeles, Calif, Mar 21, 55; m 83; c 1. PHOTOCHEMISTRY, MOLECULAR RECOGNITION. *Educ:* Carleton Col, BA, 75; Calif Inst Technol, PhD(inorg chem), 82. *Prof Exp:* Asst prof chem, Washington Univ, 81-87; asst prof, 87-90, ASSOC PROF CHEM, LA STATE UNIV, 90- *Concurrent Pos:* Peace Corps volunteer, Ghana, 75-77. *Mem:* Am Chem Soc; Am Asn Univ Professors. *Res:* Inorganic chemistry; transition-metal complexes; redox reactions and photochemistry; spectroscopy and electronic structure; enzyme-mimic reactions of metal complexes; chemical vapor deposition of electronic materials. *Mailing Add:* Dept Chem La State Univ 232 Choppin Hall Baton Rouge LA 70803

MAVIS, JAMES OSBERT, b Mansfield, Ohio, Aug 6, 25; m 53; c 2. FOOD TECHNOLOGY. *Educ:* Ohio State Univ, BSc, 50, MSc, 53, PhD(food technol), 55. *Prof Exp:* Asst, Food Technol, Agr Exp Sta, Univ Ohio, 48-50; area rep, Topco Assocs, Inc, Ill, 50-52; chief customer res, Heekin Can Co, Ohio, 56-57; sect chief, Non-Milk Frozen Foods, Pet Milk Co, 58-61, group mgr, Bakery & Hort Prods, 61-65, assoc dir res, 65-68, tech dir, Frozen Foods Div, Pet Inc, 68-69; dir res & develop, Fairmont Foods, Co, Nebr, 69-70, vpres, 70-72; dir res & develop, Interstate Brands Corp, 72-77; vpres tech serv, Banquet Foods Corp, 77-90; vpres qual, Conagra Consumer Frozen Foods, 84-90; RETIRED. *Mem:* Am Hort Soc; Am Soc Qual Control; Inst Food Technol. *Res:* New bakery and horticultural food products development and engineering and packaging research. *Mailing Add:* 159 Emerald Green Ct Creve Coeur MO 63141

MAVIS, RICHARD DAVID, b Fergus Falls, Minn, Aug 7, 43; m 66; c 2. BIOCHEMISTRY. *Educ:* St Olaf Col, BA, 65; Univ Iowa, PhD(biochem), 70. *Prof Exp:* Asst prof biochem, Dent Med Sch, Northwestern Univ, Chicago, 72-75; asst prof, 75-81, ASSOC PROF, DEPT RADIATION BIOL, BIOPHYS & BIOCHEM, SCH MED & DENT, UNIV ROCHESTER, 81- *Concurrent Pos:* USPHS fel, Wash Univ, 69-72; NIH res grant, 73. *Mem:* AAAS; Soc Toxicol. *Res:* Membrane structure and function; phospholipid metabolism; effect of toxic agents on metabolism and peroxidation of lipids of lung and other tissues; antioxidant protection of lung by vitamin E; mechanism of lipid peroxidation. *Mailing Add:* Cunl Syst Div Mitre Corp 7325 Colshire Dr McLean VA 22102

MAVITY, VICTOR T(HOMAS), JR, b Sewell, Chile, Mar 14, 20; US citizen; div; c 1. CHEMICAL ENGINEERING. *Educ:* Purdue Univ, BS, 41. *Prof Exp:* Trainee, Ashland Oil & Refining Co, 41-42; chem engr, Lago Oil & Transport Co, Ltd, 42-45 & Esso Standard Oil Co, 45-55; tech analyst & writer, Ethyl Corp, 58-59; chem engr oil refining, Res Ctr, Pure Oil Co, 60-65; chem engr oil refining, Res Ctr, Union Oil Co Calif, 65-82; RETIRED. *Res:* Process analysis and development; engineering and economics; manufacture of petroleum and petrochemical products; technical writing and editing. *Mailing Add:* PO Box 3277 Fullerton CA 92634

MAVLIGIT, GIORA M, IMMUNOLOGY, BIOLOGICAL RESPONSE MODIFIERS. *Educ:* Hebrew Univ, Jerusalem, Israel, MD, 66. *Prof Exp:* PROF MED & INTERNIST, UNIV TEX SYST CANCER CTR, M D ANDERSON HOSP & TUMOR INST, 79-, CHIEF CLIN IMMUNOL SERV, DEPT CLIN IMMUNOL & BIOL THER, 84- *Res:* Clinical oncology. *Mailing Add:* Dept Clin Immunol & Biol Ther Univ Tex M D Anderson Cancer Ctr 1515 Holcomb Blvd Houston TX 77030

MAVOR, HUNTINGTON, neurology; deceased, see previous edition for last biography

MAVRETIC, ANTON, b Slovenia, Yugoslavia, Dec 11, 34; m 64; c 2. ATMOSPHERIC SCIENCES. *Educ:* Denver Univ, BS, 59, MS, 61; Pa State Univ, PhD(elec eng), 68. *Prof Exp:* Proj engr elec eng, Mass Inst Technol, 68-78; sr res engr, Harvard Univ, 78-80; ASSOC PROF ELEC ENG, BOSTON UNIV, 80- *Concurrent Pos:* Sr lectr, Northeastern Univ, 69-82; consult, Div Appl Sci, Harvard Univ, 81-; co-prin investr, Air Force Geophys Lab, 82- *Mem:* Inst Elec & Electronics Engrs. *Res:* Space instrumentation and sophisticated electronic circuit design; chip design. *Mailing Add:* Col Eng Boston Univ 44 Cummington St Boston MA 02215

MAVRIDES, CHARALAMPOS, biochemistry, for more information see previous edition

MAVRIPLIS, F, b Thessaloniki, Greece, Jan 31, 20; Can citizen; m 55; c 3. COMPUTATIONAL AERODYNAMIC DESIGN, EXPERIMENTAL AERODYNAMICS. *Educ:* Munich Tech Univ, DiplIng, 41. *Prof Exp:* Sr engr, Canadair Ltd, 52-58, group leader thermodyn, 58-61, mem sr staff in chg space systs res & develop, 61-66, tech develop aerothermodyn, 66-76, sr staff specialist, CL-600 aerodyn develop, 76-80, sect chief challenger aerodyn develop, 80-82, mgr dynamics & loads, 82-84, mgr advan aerodyn, 84-87, mgr flight sci & methods develop, Challenger Div, 87-89, MGR ADVAN TECHNOL, CANADAIR LTD, 89- *Concurrent Pos:* Mem & chmn assoc comt aerodyn, Nat Res Coun Can, 73; mem steering comt, Nat Ctr Excellence; mem res & develop comt, Aircraft Industs Asn, Can, 88. *Honors & Awards:* F C Baldwin Medals, Can Aeronaut & Space Inst, 71 & 73. *Mem:* Fel Can Aeronaut & Space Inst; assoc fel Am Inst Aeronaut & Astronaut. *Res:* Transonic transport aircraft aerodynamic configuration design and development; supercritical wing aerodynamics; CFD methods development. *Mailing Add:* 11455 Pasteur St Montreal PQ H3M 2N8 Can

MAVROIDES, JOHN GEORGE, b Ipswich, Mass, Dec 29, 22; m 52; c 2. PHYSICS. *Educ:* Tufts Col, BS, 44; Brown Univ, MS, 51, PhD(physics), 53. *Prof Exp:* Proj engr, US Naval Underwater Sound Lab, 46-49; fel Brown Univ, 50-51; mem staff, Lincoln Lab, Mass Inst Technol, 52-60, group leader, 60-74, sr staff, 74-84; RETIRED. *Mem:* Electrochem Soc; Fel Am Phys Soc. *Res:* Solid state physics, especially galvanometric effects; magneto-optical studies; magneto-piezo-optics; magneto-acoustic effects; cyclotron resonance; electronic bandstructure; Fermi surfaces; infrared; lasers; electrochemistry; energy conversion. *Mailing Add:* 24 Clinton Dr Yarmouthport MA 02675

MAVROYANNIS, CONSTANTINE, b Athens, Greece, Nov 13, 27; Can citizen; m 61; c 2. THEORETICAL SOLID STATE PHYSICS. *Educ:* Athens Tech Univ, BS, 57; McGill Univ, PhD(phys chem), 61; Oxford Univ, DPhil(math), 63. *Prof Exp:* Nat Res Coun Can NATO sci overseas fel, 61-63; Nat Res Coun Can fel, 63-64; RES OFFICER, NAT RES COUN, 64- *Mem:* Am Phys Soc; Can Asn Physicists; fel Chem Inst Can. *Res:* Optical properties and many-body interactions in solids; modern quantum chemistry; spin wave theory; electromagnetic interactions in solids; quantum optics. *Mailing Add:* Stacey Inst Molecular Sci Nat Res Coun Sussex Dr Ottawa ON K1A 0R6 Can

MAWARDI, OSMAN KAMEL, b Cairo, Egypt, Dec 12, 17; nat US; m 50. PLASMA PHYSICS, ACOUSTICS. *Educ:* Fuad I Univ, Egypt, BSc, 40, MSc, 46; Harvard Univ, AM, 47, PhD(acoust), 48. *Prof Exp:* Transmission engr, Egyptian State Tel & Tel Co, 40-41; lectr physics, Fuad I Univ, 41-46; asst prof elec eng, Mass Inst Technol, 51-56, assoc prof mech & elec eng & mem res lab electronics, 56-60; chmn plasma dynamics & nuclear eng & dir, Plasma Res Prog, Case Western Reserve Univ, 66-75, energy coordr, 75-77, dir, Energy Res Off, 77-81, prof eng, 60-89; PRES, COLLAB PLANNERS, 73- *Concurrent Pos:* Consult, Bolt, Beranek & Newman, Inc, 50-54, Nat Prod Corp, 51-52, Res Found, Lowell Tech Inst, 53-54, Philco Corp, 53-64, Gen Ultrasonics Corp, 55-57, Boeing Airplane Co, 55-57, Pratt & Whitney Div, United Aircraft Corp, 58-64, Conesco, 58-61, Los Alamos Sci Lab, 59-61, 66-73 & Amoco Res Lab, 69-71; Guggenheim fel, 54-55; mem, Inst Advan Study, 69-70; vpres, Auctor Assocs Inc, 70-73; mem, Adv Energy Task Force to Gov, Ohio, 73-74. *Honors & Awards:* Biennial Award, Acoust Soc Am, 52; Centennial Medal, Inst Elec & Electronic Engrs, 84. *Mem:* Fel AAAS; fel Acoust Soc Am; fel Am Phys Soc; fel Inst Elec & Electronics Engrs; NY Acad Sci. *Res:* Controlled fusion research; acoustic holography; electric power systems. *Mailing Add:* Collab Planners Inc 1922 E 107th St Cleveland Heights OH 44106

MAWE, RICHARD C, cell physiology, for more information see previous edition

MAWHINNEY, MICHAEL G, b Honolulu, Hawaii, Aug 29, 45; m 69; c 2. PHARMACOLOGY. *Educ:* Grove City Col, BS, 67; WVa Univ, MS, 69, PhD(pharmacol), 70. *Prof Exp:* Asst prof, 71-75, ASSOC PROF PHARMACOL & UROL, MED CTR, WVA UNIV, 75- *Concurrent Pos:* Consult, Albert Gallatin Sch Dist, 73- & J Urol, 75- *Honors & Awards:* Award, Pharmaceut Mfg Asn Found, 74 & 76. *Mem:* Am Soc Pharmacol & Exp Therapeut; Endocrine Soc. *Res:* Hormonal regulation of the epithelial and stromal elements of normal, aged and neoplastic male accessory sex organs. *Mailing Add:* Dept of Pharmacol & Urol Med Ctr WVa Univ Morgantown WV 26506

MAX, CLAIRE ELLEN, b Boston, Mass, Sept 29, 46; m 74; c 1. PLASMA PHYSICS, ASTROPHYSICS. *Educ:* Radcliffe Col, AB, 68; Princeton Univ, PhD(astrophys sci), 72. *Prof Exp:* Res assoc physics, Univ Calif, Berkeley, 72-74; PHYSICIST, LAWRENCE LIVERMORE LAB, UNIV CALIF, 74-; ASSOC DIR & HEAD, LIVERMORE BR, INST GEOPHYS & PLANETARY PHYSICS, 84- *Concurrent Pos:* Mem physics adv comt, NSF, 78-82; consult progs to interest young women entering sci careers; sabbatical leave, Ecole Polytech, Paris, 81; mem comt on fusion Hybrid Reactors, Nat Acad Sci, 86, Int Security & Arms Control, 86-90. *Mem:* Fel Am Phys Soc; Am Astron Soc; AAAS; Int Astron Union; Am Geophys Union. *Res:* Laser-plasma interactions; applications of plasma physics to astronomical problems; research on physics of high-energy astrophysical plasmas; recent contributions include physics of cosmic-ray acceleration; magnetic field line reconnection and collisionless shock waves; adaptive optics. *Mailing Add:* L-413 Lawrence Livermore Lab Livermore CA 94550

MAX, EDWARD E, b Baltimore, MD, June 4, 45. IMMUNOGENETICS. *Educ:* Harvard Col, BA, 67; Univ Pa, MD, 74, PhD(biochem), 76- *Prof Exp:* Sr investr, NIH, 81-89; CHIEF, GENE REGULATIONS SECT, LI, DCB, FOOD & DRUG ADMIN-CBER, 89- *Res:* Regulation of immunoglobulin gene expression; heavy brain isotype switch. *Mailing Add:* HFB-800 Bldg 29A Rm 2B09 Food & Drug Admin-CBER Bethesda MD 20892

MAX, STEPHEN RICHARD, b Providence, RI, Dec 25, 40; m 77; c 2. BIOCHEMISTRY. *Educ:* Univ RI, BS, 62, PhD(biochem), 66. *Prof Exp:* Asst prof, Col Med, Howard Univ, 67-70; assoc prof neurol, 70-81, PROF NEUROL, SCH MED, UNIV MD, BALTIMORE, 81-, PROF BIOCHEM, 81-, PROF PEDIAT & PATH, 90- *Concurrent Pos:* Nat Inst Neurol Dis & Stroke fel, 68-70; Dysautonomia Found & Frank G Bressler Reserve Fund res grants, 71-72; res grants, Muscular Dystrophy Asn, 75-76, NIH, 75-, NASA, 81-86; guestworker neurochem, Nat Inst Neurol Dis & Stroke, 68-70; lectr fac grad sch, NIH, 72-; vis scientist, Univ Basel, Switz; dir, MD/PhD prog, Sch Med, Univ Md, Baltimore, 83-90, assoc vpres, res, 89-, actg vpres, grad studies & res, 90- *Mem:* Am Chem Soc; Soc Neurosci; Am Soc Neurochem; Int Soc Neurochem; Am Acad Neurol; Endocrine Soc; Am Soc Biochem & Molecular Biol; Int Brain Res Orgn; Sigma Xi; Am Clin Soc; AAAS. *Res:* Neurochemistry; muscle metabolism; neuromuscular diseases; steroid hormones and muscle; neural cell differentiation; neuro-toxicology. *Mailing Add:* Univ Md Grad Sch 257 Howard Hall Baltimore MD 21201

MAXAM, ALLAN M, BIOLOGY. *Educ:* Harvard Univ, PhD(molecular biol). *Prof Exp:* ASST PROF BIOL, DANA-FARBER CANCER INST, HARVARD UNIV, 80- *Mailing Add:* Lab Molecular Biol Rm J313 Dana-Farber Cancer Inst 44 Binney St Boston MA 02115

MAXEY, BRIAN WILLIAM, b Michigan City, Ind, Sept 13, 39; c 3. VETERINARY PHARMACEUTICALS, RESEARCH MANAGEMENT. *Educ:* Purdue Univ, West Lafayette, BS, 61; Mich State Univ, PhD(anal chem), 68. *Prof Exp:* Chemist, Dow Chem Co, 61-65; from sr scientist to sr res scientist, Upjohn Co, 68-73, proj leader, 73-75, res head vet therapeut, 75-76, res mgr vet parasitol & therapeut, 76-83, dir, 83-87, exec dir, parasitol res & develop, 87-90, EXEC DIR, THERAPEUT & LICENSING, UPJOHN CO, 90- *Mem:* AAAS; fel Am Chem Soc; Sigma Xi. *Res:* Agricultural science; veterinary therapeutics; research and development of veterinary pharmaceuticals. *Mailing Add:* Upjohn Co 7923-25-5 Kalamazoo MI 49001

MAXEY, E JAMES, b Bloomington, Ill, Jan 21, 35; m 56; c 3. EDUCATIONAL MEASUREMENT, EDUCATIONAL PSYCHOLOGY. *Educ:* Ill State Univ, BS, 57; Univ Iowa, MS, 59, PhD(elec stat & mgt), 67. *Prof Exp:* Dir res math, Measurement Res Ctr, 67-69; dir res serv, Am Col Testing, 69-72, asst vpres & dir res serv, 72-85, dir inst serv & sr res scientist, 85-87, ASST VPRES INSTITUTIONAL SERVICES & SR RES SCIENTIST, 87- *Concurrent Pos:* Adj asst prof, Univ Iowa, 67-72, adj assoc prof, 72-; vis lectr, Univ Yucatan, Mexico, 83-88, IKIP, Yogyakarta, Indonesia, 86. *Mem:* Am Educ Res Assoc; Am Inst Res; Nat Coun Measurement Educ. *Res:* Predictive validity of ACT assessment for use in difficult postsecondary environments and study the accuracy of self-reported achievements of high school youths. *Mailing Add:* 29 Norwood Circle Iowa City IA 52243

MAXFIELD, BRUCE WRIGHT, b Coronation, Alta, July 15, 39. SOLID STATE PHYSICS, NON DESTRUCTIVE TESTING. *Educ:* Univ Alta, BSc, 61; Rutgers Univ, PhD(physics), 64. *Prof Exp:* Res assoc, Cornell Univ, 64-66, actg asst prof, 66-67, asst sr res assoc physics, 71-76; engr, Lawrence Livermore Nat Lab, 76-81, sect leader & head, Nondestructive Testing Facil, 81-; PRES, SIGMA RES INC. *Concurrent Pos:* Sloan Found res fel. *Mem:* Am Phys Soc; Inst Elec & Electronics Engrs. *Res:* Transport properties in pure metals and alloys, ultrasonic studies in metals and nondestructive testing research and development including ultrasonics, optical holography and eddy currents. *Mailing Add:* Innovative Sci Inc 400 Hester St San Leandro CA 94577

MAXFIELD, FREDERICK ROWLAND, b Brooklyn NY, Oct 21, 49; m 71; c 2. ENDOCYTOSIS RECEPTOR MEDIATED, CELLULAR CALCIUM. *Educ:* Cornell Univ, PhD(chem), 76. *Prof Exp:* Assoc prof pharmacol, Sch Med, New York Univ, 80-87; PROF, PATH & PHYSIOL, COLUMBIA UNIV, 87- *Mem:* Am Soc Cell Biol; NY Acad Sci. *Res:* Mechanism of receptor regulation; intra cellular protein traffic; measurement of PH and cytoplasmic calcium by digital image analysis; cell motility. *Mailing Add:* Dept Path Columbia Univ 630 W 168th St New York NY 10032

MAXFIELD, JOHN EDWARD, b Los Angeles, Calif, Mar 17, 27; m 48; c 4. ALGEBRA. *Educ:* Mass Inst Technol, BS, 47; Univ Wis, MS, 49; Univ Ore, PhD(math), 51. *Prof Exp:* Instr math, Univ Ore, 50-51; mathematician, Naval Ord Test Sta, 51-58, head math div, 58-60; prof math & head dept, Univ Fla, 60-67; prof math & head dept, Kans State Univ, 67-81; DEAN, GRAD SCH & UNIV RES, LA TECH UNIV, 81- *Mem:* Am Math Soc; Sigma Xi; Soc Indust & Appl Math; Math Asn Am; Asn for Women Math. *Res:* Number theory; analog and digital computing techniques; numerical analysis. *Mailing Add:* Grad Sch La Tech Univ PO Box 7923 Ruston LA 71272

MAXFIELD, MARGARET WAUGH, b Conn, Feb 23, 26; m 48; c 4. APPLIED STATISTICS. *Educ:* Oberlin Col, BA, 47; Univ Wis, MS, 48; Univ Ore, PhD(algebra), 51. *Prof Exp:* Mathematician, Naval Ord Test Sta, Calif, 49-53 & 55-60, consult, 60-65; pvt res & writing, 65-75; asst prof, dept statist, Kans State Univ, Manhattan, 77-87; assoc prof, 81-88, PROF & CONSULT STATIST, DEPT MATH & STATIST, LA TECH UNIV, 88- *Concurrent Pos:* Instr & lectr, Univ Calif, Los Angeles, 57-60. *Honors & Awards:* Lester R Ford Award, 67. *Mem:* Am Statist Asn; Math Asn Am; Decision Sci Inst. *Res:* Number theory; statistics. *Mailing Add:* Dept Math & Statist La Tech Univ PO Box 3168 Tech Sta Ruston LA 71272

MAXIM, LESLIE DANIEL, b New York, NY, Feb 27, 41; m 62; c 2. OPERATIONS RESEARCH. *Educ:* Manhattan Col, BChE, 61; State Univ NY, MSc, 63; Stevens Inst Technol, MMS, 66; NY Univ, PhD(opers res), 73. *Prof Exp:* Jr chemist, Nat Starch & Chem Corp, Plainfield, 60-61, res chemist, 61-65, proj supvr phys chem res, 65-68; staff consult, Mathmatica Inc, 68-69, asst dir opers res, 69-71, dir, 71-73, vpres, Mathtech Div, 73-76, sr vpres & mem bd dirs, 76-79; PRES & CHMN, EVEREST CONSULT ASSOCS, 80- *Concurrent Pos:* Vis lectr, Grad Dept Mgt Sci, Stevens Inst Technol, 66-70, adj prof, Newark Col Eng, 70-75; adj fac, Polytech Inst NY, Brooklyn, 74-77. *Mem:* AAAS; Am Inst Chem Eng; Opers Res Soc Am; Am Statist Asn; NY Acad Sci; Sigma Xi. *Res:* Physical chemistry of polymers; statistics; statistical systems analysis. *Mailing Add:* 15 N Main St Cranbury NJ 08512

MAXON, MARSHALL STEPHEN, b Syracuse, NY, June 21, 37; m 58; c 2. X-RAY LASERS, Z-PINCHES. *Educ:* Syracuse Univ, BS, 58; Ind Univ, MS, 60, PhD(physics), 64. *Prof Exp:* Physicist, 63-69, group leader, 69-71, SR PHYSICIST, LAWRENCE LIVERMORE NAT LAB, UNIV CALIF, 71- *Concurrent Pos:* Prin investr, Nickel-like Soft X-Ray Lasers Theory, 84-; chmn Gordon Comt Ultra-Short Wave-Length Lasers, 89. *Mem:* AAAS; Am Phys Soc. *Res:* X-ray emission from plasmas; solvable models in quantum field theory; thermonuclear physics; nonlinear plasma waves; magnetohydrodynamics. *Mailing Add:* Lawrence Livermore Nat Lab L-472 Univ Calif PO Box 5508 Livermore CA 94550

MAXON, WILLIAM DENSMORE, b Detroit, Mich, Dec 8, 26; m 50; c 4. BIOCHEMISTRY. *Educ:* Yale Univ, BE, 48; Univ Wis, MS, 51, PhD(biochem), 53. *Prof Exp:* Asst biochem, Univ Wis, 49-53; res scientist antibiotics, Upjohn Co, 53-56, sect head, 56-67, group mgr, Fermentation Res & Develop, 67-82, dir, Fermentation Prod, 82-84, dir, Fermentation Res & Develop, 84-86; RETIRED. *Mem:* AAAS; Am Chem Soc; Am Soc Microbiol. *Res:* Fermentation technology and kinetics; continuous fermentation; aeration-agitation in fermentations. *Mailing Add:* 400 Burrows Rd Kalamazoo MI 49007

MAXSON, CARLTON J, b Cortland, NY, Apr 19, 36; m 57; c 2. MATHEMATICS. *Educ:* State Univ NY Albany, BS, 58; Univ Ill, MA, 61; State Univ NY Buffalo, PhD(math), 67. *Prof Exp:* Math teacher, Hammondsport Cent Sch, 58-61; from asst prof to assoc prof math, State Univ NY, Col Fredonia, 61-69; assoc prof, 69-74, assoc dean, Col Sci, 81-87, PROF MATH, TEX A&M UNIV, 74- *Concurrent Pos:* Fac res awards, State Univ NY Col Fredonia, 67 & 68; Fulbright res fel, 88. *Mem:* Am Math Soc; Math Asn Am; Edinburgh Math Soc. *Res:* Algebra; semigroups; rings; near-rings; applications of algebraic structures to study of discrete structures. *Mailing Add:* Dept Math Tex A&M Univ College Station TX 77843-3368

MAXSON, DONALD ROBERT, b Claremont, NH, Jan 19, 24; m 57; c 2. PHYSICS. *Educ:* Bowdoin Col, BS, 44; Univ Ill, MS, 48, PhD(physics), 54. *Prof Exp:* Radio engr, US Naval Res Lab, DC, 44-47; res assoc physics, Univ Ill, 54-55; instr, Princeton Univ, 55-58, res assoc, 58-59; from asst prof to assoc prof, 59-67, PROF PHYSICS, BROWN UNIV, 67- *Mem:* Am Phys Soc. *Res:* Neutrino recoil experiments for identification of beta decay interaction; experimental studies of nuclear reactions induced by charged particles and fast neutrons; reaction mechanics and nuclear structure. *Mailing Add:* 85 Mathewson Rd Barrington RI 02806

MAXSON, LINDA ELLEN R, b New York, NY, Apr 24, 43; m 64; c 1. EVOLUTIONARY BIOLOGY, GENETICS. *Educ:* San Diego Univ, BS, 64, MA, 66; Univ Calif, Berkeley, PhD(genetics), 73. *Prof Exp:* Instr biol, San Diego State Univ, 66-68; gen sci teacher, San Diego Unified Sch Dist, 68-69; instr biochem, Univ Calif, Berkeley, 74; asst prof zool, 74-76, Univ Ill, Urbana, 74-76, asst prof genetics, 76-79, assoc prof genetics & develop, 79-84, prof ecol, ethology & evolution, 79-88, dir, Campus Hon Prog, 85-88; HEAD, DEPT BIOL, PENN STATE UNIV, 88- *Concurrent Pos:* Res biochemist, Univ Calif, Berkeley, 73-74; prin investr, NSF Grant, 76-92; res assoc, Dept Vert Zool, Smithsonian Inst, 80-95. *Mem:* Fel AAAS; Soc Study Evolution; Soc Syst Zool; Sigma Xi; Am Soc Ichthyol & Herpetol. *Res:* Evolutionary biology; immunological and gene sequencing studies of evolution in amphibian phylogeny; use of molecules as evolutionary clocks and as probes of population structure and evolution. *Mailing Add:* Dept Biol Penn State Univ 208 Mueller Lab University Park PA 16802

MAXSON, ROBERT E, JR, b San Francisco, Calif, June 12, 51. MOLECULAR BIOLOGY. *Educ:* Univ Calif, Berkeley, AB, 73, PhD(zool), 78. *Prof Exp:* FEL, DEPT MED, SCH MED, STANFORD UNIV, 78-; AT DEPT BIOCHEM, UNIV SOUTHERN CALIF. *Mem:* Soc Develop Biol. *Res:* Structure and function of the histone gene family in the sea urchin. *Mailing Add:* Dept Biochem Univ Southern Calif Univ Pk 2025 Zonal Ave Los Angeles CA 90033

MAXSON, STEPHEN C, b Newport, RI, Apr 13, 38; m 87. PSYCHOBIOLOGY, BEHAVIOR GENETICS. *Educ:* Univ Chicago, SB, 60, PhD(biopsychol), 66. *Prof Exp:* Instr biol & res assoc behav genetics, Univ Chicago, 66-69; from asst prof to assoc prof, 69-84, PROF BIOBEHAVIORAL SCI, UNIV CONN, 84-, PROF PSYCHOL, 85- *Concurrent Pos:* Sabatical Dept Biol, Univ SC, 86. *Mem:* Behav Genetics Asn; Soc Neurosci; fel Int Soc Res Aggression; Sigma Xi. *Res:* Inheritance of behavior in mice and genetic mapping of behavioral loci; genotype-environment interactions in neurobehavioral development; pharmacogenetics of behavior, molecular genetics of mammalian brain and behavior; genetics, development, physiology and pharmacology of aggressive behavior; behavior as an assay for potentially mutagenic and teratogenic substances; genetics, development, physiology and pharmacology of audiogenic and spontaneous seizures. *Mailing Add:* Dept Psychol & Biobehav Sci Prog Univ Conn Storrs CT 06268

MAXUM, BERNARD J, b Bremerton, Wash, Nov 4, 31; m 59; c 5. ELECTROMAGNETICS, SPACE SYSTEMS COMPUTER SCIENCE. *Educ:* Univ Wash, BS, 55; Univ Southern Calif, MS, 57; Univ Calif, Berkeley, PhD(cyclotron wave instabilities), 63. *Prof Exp:* CHIEF, RES & DEVELOP, ROCKWELL INT, 73- *Concurrent Pos:* Hughes fel, Univ Southern Calif & Ford Found fel, Univ Calif, Berkeley; teaching, res, mgt & direction of res & develop projs in electronics, advan sensor systs & technol spinoffs for domestic applns for aerospace & com industs, 55- *Mem:* Am Phys Soc; Inst Elec & Electronic Engrs; Nat Energy Resources Orgn; Sigma Xi. *Res:* Electromagnetic, acoustic electronic and optic systems; computer systems. *Mailing Add:* 26552 Montebello Mission Viejo CA 92691

MAXWELL, ARTHUR EUGENE, b Maywood, Calif, Apr 11, 25; m 46, 64, 88; c 7. OCEANOGRAPHY. *Educ:* NMex State Univ, BS, 49; Univ Calif, San Diego, MS, 52, PhD(oceanog), 59. *Prof Exp:* Asst, Scripps Inst, Univ Calif, 49-50, asst oceanog, 50-52, jr res geophysicist, 52-55; head oceanogr, Geophys Br, Off Naval Res, 55-59, head, 59-65; assoc dir, Woods Hole Oceanog Inst, 65-69, dir res, 69-71, provost, 71-82; DIR, INST GEOPHYS, UNIV TEX AUSTIN, 82- *Concurrent Pos:* Mem, Nat Adv Comt Oceans & Atmosphere, 72-75. *Mem:* AAAS; Marine Technol Soc (vpres, 64-65, pres, 81-82); Sigma Xi; fel Am Geophys Union (pres, 76-78). *Res:* Physical oceanography and geophysics, particularly the measurement and interpretation of heat flow through the ocean floor. *Mailing Add:* Inst Geophys Univ Tex Austin TX 78712

MAXWELL, BRYCE, b Glen Cove, NY, July 26, 19; Wid; c 3. POLYMERS. *Educ:* Princeton Univ, BS, 43, MS, 48. *Prof Exp:* Res assoc, Plastics Lab, Princeton Univ, 48-53, from asst prof to assoc prof mech eng, 53-63, assoc prof chem eng, 63-68, asst dean, 63-69, prof, 68-85, EMER PROF CHEM ENG, PRINCETON UNIV, 85- *Concurrent Pos:* Ed, Soc Rheol Jour, 56-58. *Honors & Awards:* Gold Medal & Int Award in Plastics Sci & Eng, Soc Plastics Engrs, 76. *Mem:* Soc Plastics Engrs; Soc Rheol; Am Soc Testing & Mat; Am Soc Eng Educ; Am Soc Mech Engrs. *Res:* Mechanical properties of polymers; fabrication and materials processing; rheology. *Mailing Add:* Polymer Mat Prog Eng Quandrangle Princeton Univ Princeton NJ 08544

MAXWELL, CHARLES HENRY, b Las Palomas, NMex, July 9, 23; div; c 3. GEOLOGY. *Educ:* Univ NMex, BS, 50, MS, 52. *Prof Exp:* Geologist, Shell Oil Co, 51-52; geologist, Br Mineral Deposits, US Geol Surv, 52-56, Br Foreign Geol, Brazil, 56-61, Br Regional Geol, Ky, 61-63, Br Mil Geol, 63-66, Br Radioactive mat, 66-69, Br Cent Mineral Resources, 69-89, EMER GEOLOGIST, US GEOL SURV, 89- *Honors & Awards:* Mineral named in honor, Maxwellite. *Mem:* Geol Soc Am. *Res:* Economic geology; field interpretive and engineering geology. *Mailing Add:* 950 Allison Lakewood CO 80215

MAXWELL, CHARLES NEVILLE, b Tuscaloosa, Ala, Oct 27, 27; m 52; c 4. MATHEMATICS. *Educ:* Univ Chicago, BS, 49, MS, 51; Univ Ill, PhD(math), 55. *Prof Exp:* Instr math, Univ Mich, 55-58; assoc prof, Univ Ala, 58-63; PROF MATH, SOUTHERN ILL UNIV, CARBONDALE, 63- *Mem:* Am Math Soc. *Res:* Topology; topological transformation groups; algebraic topology. *Mailing Add:* Dept of Math Southern Ill Univ Carbondale IL 62901

MAXWELL, DAVID SAMUEL, b Bremerton, Wash, Feb 13, 31; m 57; c 3. ANATOMY. *Educ:* Westminster Col (Mo), AB, 54; Oxford Univ, BA, 57; Univ Calif, Los Angeles, PhD, 60. *Prof Exp:* From instr to assoc prof, 59-68, PROF ANAT, SCH MED, UNIV CALIF, LOS ANGELES, 68-, VCHMN DEPT, 73-, PROF SURG/ANAT, CHARLES DREW POSTGRAD MED SCH, 74- *Mem:* AAAS; Am Asn Anat; Electron Micros Soc Am; Am Soc Cell Biol; Soc Neurosci; Asn Am Med Cols. *Res:* Electron microscopy; histochemistry and cytochemistry of the nervous system and eye. *Mailing Add:* Dept Anat 73-235 Chs Med Sch Univ Calif 405 Hilgard Ave Los Angeles CA 90024

MAXWELL, DONALD A, b Austin, Tex, Apr 23, 38; m 63; c 4. CIVIL ENGINEERING. *Educ:* Univ Tex, Austin, BS, 62, MS, 64; Tex A&M Univ, PhD(civil eng), 68. *Prof Exp:* Highway engr, Fed Highway Admin, 62-68; mem tech staff, Comput Sci Corp, 69-71; sr assoc, Alan M Voorhers & Assocs, 71-77; assoc prof, 77-83, PROF CIVIL ENG, TEX A&M UNIV, 83- *Mem:* Am Soc Civil Engrs. *Res:* Design and implementation of paratransit systems; evaluation of cost effectiveness of engineered solutions; applied statistics and simulation. *Mailing Add:* 2601 Wayside Dr Bryan TX 77801

MAXWELL, DONALD ROBERT, b Paris, France, Mar 30, 29; US citizen; m 56; c 8. PHARMACOLOGY. *Educ:* Cambridge Univ, BA, 52, MA, 56, PhD(pharmacol), 55. *Prof Exp:* Res attache, Pasteur Inst, Paris, 55-56; pharmacologist, May & Baker Ltd, Dagenham, Eng, 56-69, mgr pharmacol res, 69-74; sr vpres, Preclin Res, 77-90, EXEC VPRES, SCI AFFAIRS, WARNER-LAMBERT/PARKE-DAVIS RES, 90- *Mem:* Fel Royal Soc Med; fel Brit Inst Biol; Brit Pharmacol Soc; Brit Physiol Soc; Int Col Neuropsychopharmacol. *Res:* Psychopharmacology and neuropharmacology in relation to development of new drugs; cardiovascular drugs; anti-allergic drugs. *Mailing Add:* Warner-Lambert/Parke-Davis Pharmaceut Res Ann Arbor MI 48105

MAXWELL, DOUGLAS PAUL, b Norfolk, Nebr, Feb 12, 41; m 64; c 2. PLANT PATHOLOGY. *Educ:* Nebr Wesleyan Univ, BA, 63; Cornell Univ, PhD(plant path), 68. *Prof Exp:* From asst prof to assoc prof, 68-77, PROF PLANT PATH, UNIV WIS-MADISON, 77-, CHMN, 80- *Mem:* Am Phytopath Soc. *Res:* Ultrastructure of fungi; function of fungal microbodies; breeding for disease resistance in forages. *Mailing Add:* Dept Plant Path 284-A Russell Lab Univ Wis Madison WI 53706

MAXWELL, DWIGHT THOMAS, b Manhattan, Kans, Aug 25, 37; m 64. MINERALOGY. *Educ:* Univ Kans City, BS, 59; Mont State Univ, PhD(geol), 65. *Prof Exp:* Asst prof geol, Univ Mo-Kansas City, 64-67 & Northeast La State Col, 67-70; assoc prof earth sci, 70-, assoc prof geol, 81-85, PROF GEOL, NORTHWEST MO STATE COL, 85- *Mem:* Clay Minerals Soc; Mineral Soc Am. *Res:* Clay mineralogy. *Mailing Add:* Dept Geol Northwest Mo State Col Maryville MO 64468

MAXWELL, EMANUEL, b Brooklyn, NY, Dec 16, 12; m. PHYSICS. *Educ:* Columbia Univ, BS, 34, EE, 35; Mass Inst Technol, PhD(physics), 48. *Prof Exp:* Patent examr, US Patent Off, 37; geophysicist, Shell Oil Co, Inc, Tex, 37-41; physicist, Nat Bur Standards, 48-53; staff mem, Radiation Lab, 41-45, res assoc physics, 45-48, mem staff, Lincoln Lab, 53-63, vis assoc prof physics, 58-63, sr scientist, 63-84, consult, Francis Bitter Nat Magnet Lab, 84-89, VIS SCIENTIST, MASS INST TECHNOL, 89- *Mem:* Fel Am Phys Soc. *Mailing Add:* 24 Bates St Cambridge MA 02140

MAXWELL, GEORGE RALPH, II, b Morgantown, WVa, Mar 27, 35; m 59, 87; c 3. ECOLOGY, ORNITHOLOGY. *Educ:* WVa Univ, AB, 57, MS, 61; Ohio State Univ, PhD(zool), 65. *Prof Exp:* Asst prof biol, The Citadel, 65-66; dir, Rice Creek Biol Field Sta, 66-79, PROF BIOL, STATE UNIV NY COL OSWEGO, 66- *Concurrent Pos:* NSF instrnl sci equipment prog res grant, 67-69; State Univ NY Res Found grant-in-aid, 68-70; fel, Univ NC, Chapel Hill, 70; vis scientist, Fla Med Entom Lab, 73. *Honors & Awards:* John J Elliot Mem Award, 76, 83. *Mem:* Am Ornithologists Union; Wilson Ornith Soc; Asn Field Ornithologists. *Res:* Growth of stream mayflies; maintenance behavior of herons; breeding biology of the grackle; impact of winter navigation on the birds of the St Lawrence River; heron and mosquito ecology. *Mailing Add:* Dept Biol State Univ NY Oswego NY 13126

MAXWELL, GLENN, b Kent, Ohio, May 20, 31; m 59; c 3. MATHEMATICS. *Educ:* Kent State Univ, BS, 53, MA, 54; Ohio State Univ, PhD(math), 64. *Prof Exp:* Teacher high sch, Ohio, 54-56; instr, 63-64, ASST PROF MATH, KENT STATE UNIV, 64- *Mem:* Math Asn Am; Am Math Soc. *Res:* Mathematical foundations of set theory and logic. *Mailing Add:* Dept of Math Kent State Univ Kent OH 44242

MAXWELL, JAMES DONALD, b Mississippi Co, Ark, June 2, 40; m 63; c 2. COTTON SEED PROCESSING. *Educ:* Miss State Univ, BS, 62; Cornell Univ, MS, 65; NC State Univ, PhD(crop sci), 68. *Prof Exp:* From asst prof to prof agron, Clemson Univ, 68-79; exec vpres agron, Hollendale Agr Serv, 79-87. *Mem:* Am Soc Agron; Crop Sci Soc Am. *Res:* Soybean breeding with emphasis on insect and disease resistance in southern region of United States. *Mailing Add:* 277 Dallas Dr Greenville MS 38701

MAXWELL, JOHN ALFRED, b Hamilton, Ont, Aug 28, 21; m 53. GEOCHEMISTRY, ANALYTICAL CHEMISTRY. *Educ:* McMaster Univ, BSc, 49, MSc, 50; Univ Minn, PhD(geol, mineral & analytical chem), 53. *Prof Exp:* Metall chemist, Burlington Steel Co, 39-45; asst chem, McMaster Univ, 48-50; analyst, Rock Analysis Lab, Univ Minn, 51-53; geologist, Geol Surv Can, 53-57, head analytical chem sect, 57-67, dir, Cent Labs & Tech Serv, 67-84, spec adv to dir gen, 84-86; RETIRED. *Mem:* Fel Royal Soc Can. *Res:* Methods of rock and mineral analysis; compilation of geochemical data; meteorites; lunar samples. *Mailing Add:* 672 Denbury Ave Ottawa ON K2A 2P3 Can

MAXWELL, JOHN CRAWFORD, b Xenia, Ohio, Dec 28, 14; m 39; c 2. GEOLOGY, TECTONICS. *Educ:* DePauw Univ, BA, 36; Univ Minn, MA, 37; Princeton Univ, PhD(geol), 46. *Hon Degrees:* DSc, DePauw Univ, 88. *Prof Exp:* Reflections seismograph comput, Tex Co, 37; subsurface geologist, Sun Oil Co, 37-40; from instr to assoc prof geol, Princeton Univ, 46-55, prof geol eng, 55-70, chmn dept, 55-66, chmn dept geol, 66-70, chmn interdept prog water resources, 64-70; William Stamps Farish prof, 70-84, EMER PROF GEOL SCI, UNIV TEX, AUSTIN, 84- *Concurrent Pos:* Fulbright scholar, Italy, 52-53, NSF fel, 61-62; chmn earth sci div, Nat Res Coun, 70-72; consult, Adv Comt Reactor Safeguards, Nuclear Regulatory Comn, 74-; chmn, US Nat Comt Geodynamics, 79-83. *Mem:* Geol Soc Am (pres, 72-73); Am Asn Petrol Geologists; Am Geophys Union; Am Geol Inst (pres, 71-72); Ital Geol Soc. *Res:* Geology of Caribbean area and Montana-Wyoming; tectonics of Italian Apennines and California coast ranges; high temperature high pressure on limestone, quartz sand, and sandstone; origin of rock cleavage. *Mailing Add:* Dept Geol Sci Univ Tex Austin TX 78713-7909

MAXWELL, JOHN GARY, b Salt Lake City, Utah, Oct 5, 33; m 80; c 6. SURGERY. *Educ:* Univ Utah, BS, 54, MD, 58. *Prof Exp:* From instr to asst prof, 66-73, ASSOC PROF SURG, COL MED, UNIV UTAH, 73- *Concurrent Pos:* Asst chief, Vet Admin Hosp, 66-76. *Mem:* Am Col Surg; Asn Acad Surg. *Res:* Gastrointestinal surgery; transplantation; vascular surgery. *Mailing Add:* Dept Surg Univ Utah Col Med 50 N Medical Dr Salt Lake City UT 84132

MAXWELL, JOYCE BENNETT, b Merced, Calif, June 18, 41; m 62; c 2. GENETICS. *Educ:* Univ Calif, Los Angeles, AB, 63; Calif Inst Technol, PhD(genetics, biochem), 70. *Prof Exp:* Res asst neurohistochem, Camarillo State Hosp, 69; from asst prof to assoc prof, 70-81, PROF BIOL, CALIF STATE UNIV, NORTHRIDGE, 81- *Mem:* AAAS; Sigma Xi. *Res:* Biochemical genetics; synthesis of serine and glycine by Neurospora crassa; multiple electrophoretic forms of tyrosinase in Neurospora crassa; high mutable serine-dependent strain of Neurospora. *Mailing Add:* Dept of Biol Calif State Univ Northridge CA 91330

MAXWELL, KENNETH EUGENE, b Huntington Beach, Calif, Sept 27, 08; wid; c 3. ENTOMOLOGY. *Educ:* Univ Calif, BS, 33; Cornell Univ, PhD(entom), 37. *Prof Exp:* Jr entomologist, Univ Calif, Riverside, 37-39; technologist, Shell Oil Co, 39-42; mgr agr div, Chemurgic Corp, 45-47; consult, Maxwell Labs, 47-49; entomologist, E I du Pont de Nemours & Co, 49-50; mgr, Insecticide Dept, Agriform Co, 50-53; entomologist, Monsanto Chem Co, 53-59; tech dir, Moyer Chem Co, 59-63; from assoc prof to prof entom, 63-74, EMER PROF BIOL, CALIF STATE UNIV, LONG BEACH, 74- *Mem:* AAAS; Entom Soc Am; Am Chem Soc; Sigma Xi. *Res:* Toxicology of pesticides; environmental toxicology. *Mailing Add:* 16751 Greenview Lane Huntington Beach CA 92649

MAXWELL, LEE M(EDILL), b Los Angeles, Calif, July 17, 30; m 52; c 4. ELECTRICAL ENGINEERING. *Educ:* Univ Okla, BS, 56; Univ Idaho, MS, 59; Univ Colo, PhD(elec eng), 63. *Prof Exp:* From instr to asst prof elec eng, Univ Idaho, 57-63; assoc prof elec eng, Colo State Univ, 63-69, prof, 69-; RETIRED. *Mem:* Inst Elec & Electronics Engrs; Am Soc Eng Educ. *Res:* Network and graph theory. *Mailing Add:* 35901 Well County Rd No 31 Eaton CO 80615-8624

MAXWELL, LEO C, b 1941; c 2. PHYSIOLOGY. *Educ:* Univ Mich, PhD(physiol), 71. *Prof Exp:* Res investr, Univ Mich, 74-77, asst res scientist, 77-79; ASSOC PROF PHYSIOL, UNIV TEX HEALTH SCI CTR, 79- *Mem:* Am Heart Asn; Am Physiol Soc. *Mailing Add:* Dept Physiol Univ Tex Health Sci Ctr 7703 Floyd Curl Dr San Antonio TX 78284

MAXWELL, LOUIS R, b Waterloo, Iowa, May 28, 00; m 25, 57, 75; c 2. ELECTRON DIFFRACTION MOLECULES. *Educ:* Cornell Col, Iowa, BA, 23, Univ Minn, PhD(physics), 27. *Prof Exp:* Res fel, Bartol Res Found, Swarthmore, Pa, 27-31; res physicist, Fixed Nitrogen Lab, US Dept Agr, 31-40; res physicist mine counter-measures, Bur Ships, US Navy, 40-47; sr res scientist, Naval Ord Lab, 47-70; RETIRED. *Mem:* Am Phys Soc. *Res:* Electron diffraction of molecules; nuclear magnetic resonance (NMR) detection of cancer cells in vivo. *Mailing Add:* 5204 Moorland Lane Bethesda MD 20814

MAXWELL, RICHARD ELMORE, b Dallas, Tex, Aug 8, 21; m 42; c 2. BIOCHEMISTRY. *Educ:* Southern Methodist Univ, BS, 43; Univ Ill, PhD(biochem), 47. *Prof Exp:* Asst, Magnolia Petrol Co, Tex, 43-44; asst chem, Univ Ill, 44-47; asst prof, Iowa State Univ, 47-51; sr res chemist, Warner-Lambert/Parke-Davis, 51-57, res leader, 57-64, dir, Lab Biochem, 64-70, DIR, ATHEROSCLEROSIS SECT, PHARMACUET RES DIV, PHARMACOL DEPT, WARNER-LAMBERT/PARKE-DAVIS, 70- *Mem:* Am Asn Clin Chem; fel AAAS; Am Soc Biol Chem; Am Chem Soc; fel Am Inst Chem. *Res:* Actions of drugs and antibiotics on biological systems. *Mailing Add:* 11631 S Monticello Knoxville TN 37922

MAXWELL, ROBERT ARTHUR, b Union City, NJ, Oct 6, 27; m 56; c 3. PHARMACOLOGY. *Educ:* Princeton Univ, PhD(biol/endocrine physiol), 54. *Prof Exp:* Assoc pharmacologist, Ciba Pharmaceut Co, 54-60, assoc dir pharmacol, 60-62; assoc prof, Col Med, Univ Vt, 62-65; HEAD PHARMACOL, WELLCOME RES LABS, 66- *Concurrent Pos:* Vis prof, Col Med, Univ Vt, 66-; adj prof pharmacol & exp med, Med Ctr, Duke Univ, 70-; adj prof pharmacol, Sch Med, Univ NC, Chapel Hill, 73- *Mem:* AAAS; Am Soc Pharmacol & Exp Therapeut; Pharmacol Soc Can; NY Acad Sci. *Res:* Cardiovascular and autonomic pharmacology. *Mailing Add:* Wellcome Res Labs 3030 Cornwallis Rd Research Triangle Park NC 27709

MAXWELL, ROBERT L(OUIS), b Lexington, Tenn, Feb 24, 20; m 46; c 2. MECHANICAL ENGINEERING. *Educ:* Univ Tenn, BS, 44; Case Western Reserve Univ, MS, 46. *Prof Exp:* Mech engr, Nat Adv Comt Aeronaut, 44-46; instr mech eng, Case Western Reserve Univ, 46; from instr to assoc prof, 46-70, PROF MECH ENG, UNIV TENN, KNOXVILLE, 70- *Concurrent Pos:* Consult, Redstone Arsenal, 52-54 & Union Carbide Co, 52- *Mem:* Am Soc Mech Engrs; Soc Exp Stress Anal; Am Soc Eng Educ. *Res:* Stress analysis; vibrations. *Mailing Add:* 607 Dougherty Hall Univ Tenn Knoxville TN 37916

MAXWELL, WILLIAM HALL CHRISTIE, b Coleraine, Northern Ireland, Jan 25, 36; US citizen; m 60; c 4. CIVIL ENGINEERING, FLUID MECHANICS. *Educ:* Queen's Univ Belfast, BSc, 56; Queen's Univ, Ont, MSc, 58; Univ Minn, PhD(civil eng), 64. *Prof Exp:* Res asst hydromech, Nat Res Coun Can, 58; res fel, St Anthony Falls Hydraul Lab, Univ Minn, 61-63; from asst prof to assoc prof, 64-82, PROF CIVIL ENG, UNIV ILL, URBANA, 82- *Concurrent Pos:* NSF res initiation grant, 66-67; tech ed, 76-85, ed-in-chief, Water Int, Int Water Resources Asn, 86-; mem, Am Soc Civil Engrs Hydraul Div Res Comt, 80-84, Fluids Comt, 79-85; Proj Adv Comt, Maritime Transp Res Bd, Nat Res Coun, 81-84. *Mem:* Am Soc Civil Engrs; Am Geophys Union; Int Asn Hydraul Res; Int Water Resources Asn. *Res:* Hydraulic and water resources engineering; bubble screens; hydraulic models; jet diffusion; surface tension; low flow mixing in natural channels. *Mailing Add:* Newmark CE Lab-MC 250 205 N Mathews Ave Urbana IL 61801-2397

MAXWELL, WILLIAM L, b Philadelphia, Pa, July 11, 34; m 69; c 4. OPERATIONS RESEARCH, INDUSTRIAL ENGINEERING. *Educ:* Cornell Univ, BME, 57, PhD(opers res), 61. *Prof Exp:* Asst prof indust eng, 61-64, assoc prof indust eng & opers res, 64-69, PROF OPERS RES, CORNELL UNIV, 69- *Mem:* Asn Comput Mach; Opers Res Soc Am; Inst Mgt Sci; fel Inst Indust Engrs. *Res:* Scheduling theory; digital simulation; production control and data processing systems. *Mailing Add:* Dept Opers Res Cornell Univ Ithaca NY 14853

MAXWORTHY, TONY, b London, Eng, May 21, 33; US citizen; m 79; c 2. GEOPHYSICAL FLUID DYNAMICS, UNSTEADY AERODYNAMICS. *Educ:* Univ London, BSc, 54; Princeton Univ, MSE, 55; Harvard Univ, PhD(mech eng), 60. *Prof Exp:* Res asst, Harvard Univ, 55-60; group supvr, Jet Propulsion Lab, 60-67; assoc prof, 67-70, PROF MECH & AERO ENG, UNIV SOUTHERN CALIF, 70-, CHMN DEPT, 79- *Concurrent Pos:* consult, Jet Propulsion Lab, 68- & Brown-Boveri Cie, 75-; sr fel, Nat Ctr Atmospheric Res, 76; assoc prof, Univ Grenoble, 82- *Honors & Awards:* Von Humboldt Sr Award, 80-81; Smith Int Prof Mech Eng, 88; Otto Laporte Award, Am Phys Soc, 90. *Mem:* Nat Acad Eng; Am Geophys Union; Am Meteorol Union; Am Soc Mech Engrs; fel Am Phys Soc. *Res:* Fluid mechanics; dynamics of planetary atmosphere; remote sensing; wave dynamics. *Mailing Add:* Dept Mech Eng OHE 430 Univ Southern Calif Los Angeles CA 90089-1453

MAY, ADOLF D(ARLINGTON), JR, b Little Rock, Ark, Mar 25, 27; m 48; c 4. ENGINEERING. *Educ:* Southern Methodist Univ, BS, 49; Iowa State Univ, MS, 50; Purdue Univ, PhD, 55. *Prof Exp:* Instr civil eng, Iowa State Univ, 49-50; res asst, Purdue Univ, 50-52; assoc prof, Clarkson Col Technol, 52-56 & Mich State Univ, 56-59; mem tech staff traffic control, Thompson-Ramo-Wooldridge, Inc, 59-62; dir, Chicago Expressway Surveillance Proj, 62-65; assoc prof, 65-67, PROF TRANSP ENG, UNIV CALIF, BERKELEY, 67- *Concurrent Pos:* Mem Hwy Res Bd, Nat Acad Sci-Nat Res Coun, 50- *Honors & Awards:* Von Homboldt Award. *Mem:* Nat Acad Eng; Inst Traffic Engrs; Am Soc Eng Educ. *Res:* Electronic control systems; theory of traffic flow; highway geometric design and planning; traffic operations. *Mailing Add:* Dept of Civil Eng Rm 114 McLaughlin Hall Univ of Calif Berkeley CA 94720

MAY, BILL B, RESEARCH ADMINISTRATION. *Prof Exp:* CHIEF EXEC OFFICER & CHMN BD, ARGO SYSTS, INC, 69- *Mem:* Nat Acad Eng. *Mailing Add:* Argo Systs Inc 310 N Mary Ave Sunnyvale CA 94086

MAY, DONALD CURTIS, JR, b Ann Arbor, Mich, May 31, 17; m 42. OPERATIONS RESEARCH. *Educ:* Univ Mich, AB, 38; Princeton Univ, AM, 40, PhD(math), 41. *Prof Exp:* Instr math, Princeton Univ, 39-40; mathematician, Bur Naval Weapons, 41-63, mathematician opers res, Surface Missile Systs Proj, US Dept Navy, 63-75; systs analyst, Shipbuilding Proj, 75-80; NAVAL SYST ANALYST, APPL PHYSICS LAB, JOHNS HOPKINS UNIV, 81- *Res:* Evaluation of Navy weapon systems. *Mailing Add:* 5931 Oakdale Rd McLean VA 22101

MAY, EVERETTE LEE, b Timberville, Va, Aug 1, 14; m 40, 65; c 4. MEDICINAL CHEMISTRY. *Educ:* Bridgewater Col, AB, 35; Univ Va, PhD(org chem), 39. *Prof Exp:* Res chemist, Nat Oil Prods Co, 39-41; from assoc chemist to sr chemist, NIH, 41-53, from scientist to sr scientist, Commissioned Corps, 53-58, scientist dir, 59, chief sect med chem, Nat Insts Arthritis & Metabolic Dis, USPHS, 60-78; adj prof pharmacol, 74-77, PROF PHARMACOL, MED COL VA, 77- *Concurrent Pos:* Mem expert adv panel drugs liable to cause addiction & comt probs drug dependence, 58-78; chem adv panel mem, Walter Reed Army Inst, 65-; mem ad hoc rev comt, Nat Cancer Inst, 78. *Honors & Awards:* E E Smissman Award Med Chem, Am Chem Soc, 79; Nathan B Eddy Mem Award, 81. *Mem:* Am Chem Soc. *Res:* Surface active agents; vitamins of the B complex; antimalarial agents; analgesic drugs; antitubercular compounds; carcinolytic agents; chemical and pharmacological investigations on central nervous system and anti-inflammatory agents. *Mailing Add:* Dept Pharmacol Med Col Va Richmond VA 23298

MAY, EVERETTE LEE, JR, b Bethesda, Md, May 26, 44; m 65; c 2. MATHEMATICS, COMPUTER SCIENCE. *Educ:* Wake Forest Col, BS, 66; Emory Univ, PhD(math), 71. *Prof Exp:* Asst prof math, Kennesaw Jr Col, 72; asst prof, 72-75, ASSOC PROF, SALISBURY STATE COL, 75- *Concurrent Pos:* Comput consult, 79-; vis assoc prof, Wake Forest Univ, 80-81. *Mem:* Am Math Soc; Math Asn Am. *Res:* Search for a viable notion of spectrum for a nonlinear transformation viable in the sense of yielding useful analytical tools; applying mathematics along with programming principles to the construction of database systems and the solving of problems for small businesses. *Mailing Add:* 606 Irene Ave Salisbury MD 21801

MAY, FRANK PIERCE, b Quincy, Fla, Oct 2, 20; m 45; c 3. CHEMICAL ENGINEERING. *Educ:* Univ Fla, BChE, 48, MSE, 58, PhD(chem eng, math), 61. *Prof Exp:* Chem engr, Am Oil Co, Tex, 48-50; chem engr, US Phosphoric Prod Div, Tenn Corp, Fla, 50-53, night supt, 53-54, dept supvr, 54-55; from instr to assoc prof chem eng, 55-69, PROF CHEM ENG, UNIV FLA, 69- *Mem:* AAAS; Am Inst Chem Engrs; Nat Soc Prof Engrs; Am Chem Soc; Sigma Xi. *Res:* Chemical process dynamics and optimization; chemical processing by by-product recovery and pollution abatement in the citrus processing industry. *Mailing Add:* 2536 NW 18th Way Gainesville FL 32605

MAY, GERALD WARE, b Warrenton, Ga, Feb 27, 40; m 62. MECHANICAL ENGINEERING, STRUCTURAL MECHANICS. *Educ:* Ga Inst Technol, BS, 63, MS(eng mech), 65, MS, 67, PhD(mech eng), 69. *Prof Exp:* Engr, E I du Pont de Nemours & Co, Inc, 63; res engr, Burlington Industs, Inc, Greensboro, NC, 69-72; lead stress analyst on nuclear reactor, Westinghouse Elec Corp, 72-76; PROG MGR NUCLEAR SHIPPING CASK DEVELOP, DEPT ENERGY, US GOVT, 76- *Mem:* Am Soc Mech Engrs. *Res:* Noise abatement engineering; thermal and solid mechanics research. *Mailing Add:* 3597 Pebble Beach Dr Martinez GA 30907

MAY, HAROLD E(DWARD), b New York, NY, Oct 18, 20; m 43; c 2. MECHANICAL ENGINEERING. *Educ:* Columbia Univ, AB, 41, BS, 42. *Prof Exp:* Supvr, Eng Field Group, E I du Pont de Nemours & Co, Inc, 42-50, area supt prod, 50-55, tech mgr, 55-58, asst dir prod, 58-62, asst dir res, Electrochem Dept, 62-68, mgr, Electronic Prod Div, 68-71, dir, Chem Prod Div, 71-72, dir, Indust Specialties Div, 72-74, dir corp purchasing, 74-76, gen mgr energy & mat, 76-77, vpres mat & logistics, E I du Pont de Nemours & Co, Inc, 77-85; RETIRED. *Mailing Add:* Kennett Square Kennett Sq PA 19348-2714

MAY, IRA PHILIP, b New York, NY, Mar 26, 54; m 90; c 1. HYDROGEOCHEMISTRY, COMPUTER MODELING. *Educ:* Johns Hopkins Univ, BA, 77. *Prof Exp:* Grad student geochem, Univ Del, 76-78; geochemist environ control, Versar, Inc, 78-81; geologist environ control, JRB Assocs, Inc, 81-82 & UTD Corp, 82-83; geologist pub interest, Clean Water Action Proj, 83-85; GEOLOGIST ENVIRON GEOL, US ARMY TOXIC & HAZARDOUS MAT AGENCY, 85- *Mem:* Nat Water Well Asn; Computer Oriented Geol Soc; Sigma Xi. *Res:* Geology and geochemistry of hazardous and toxic waste sites; fate and transport of pollutants in the aquatic and ground water environments; remote sensing of environmental problems; ground water flow and contaminant transport modeling. *Mailing Add:* 2610 St Paul St Baltimore MD 21218

MAY, IRVING, b New York, NY, Feb 16, 18; m 40; c 1. ANALYTICAL CHEMISTRY, GEOCHEMISTRY. *Educ:* City Col New York, BS, 38; George Washington Univ, MS, 48. *Prof Exp:* Testing technician, Panama Canal, 39-41; analytical chemist, USPHS, NIH, 41-48; analytical chemist, US Geol Surv, 48-71, chief, Br Analysis Labs, 71-76, chief chemist, 71-80; RETIRED. *Mem:* Am Chem Soc; Sigma Xi; AAAS. *Res:* Geochemical analysis, particularly determination of trace elements. *Mailing Add:* 917 Brentwood Lane Silver Spring MD 20902

MAY, J PETER, b New York, NY, Sept 16, 39; m 63; c 2. ALGEBRAIC TOPOLOGY, HOMOTOPY THEORY. *Educ:* Swarthmore Col, BA, 60; Princeton Univ, PhD(math), 64. *Prof Exp:* From instr to asst prof math, Yale Univ, 64-67; assoc prof, 67-70, PROF MATH, UNIV CHICAGO, 70- *Concurrent Pos:* Mem, Inst Advan Study, 66; vis prof, Cambridge Univ, 71-72, 76; chmn, Dept Math, 85-91, dir Math Disciplines Ctr, 88- *Mem:* Am Math Soc; Asn Am Univ Prof. *Res:* Algebraic topology, especially homotopy theory: infinite loop space theory, stable homotopy theory, equivariant homotopy theory, etc. *Mailing Add:* Dept Math Univ Chicago Chicago IL 60637

MAY, JAMES AUBREY, JR, b Houston, Tex, July 15, 42; m 68; c 3. POLYMER CHEMISTRY, ORGANOMETALLIC CHEMISTRY. *Educ:* Tex Christian Univ, BS, 64, PhD(org chem), 68. *Prof Exp:* Sr res chemist, 68-72, res specialist, 72-76, group leader, 76-81, res mgr, 81-84, SR RES MGR, DOW CHEM CO, USA, 84- *Mem:* Am Chem Soc; Sigma Xi; NY Acad Sci. *Res:* Polymer characterization; gel permeation chromatography; polyolefins; Ziegler catalysis; product research; specialty polyolefin copolymers; process research; free radical catalysis kinetics; electrochemistry; epoxy resins, acrylic resins. *Mailing Add:* 112 Lily St Lake Jackson TX 77566

MAY, JAMES DAVID, b Blue Mountain, Miss, Aug 21, 40; m 61; c 3. POULTRY PHYSIOLOGY. *Educ:* Miss State Univ, BS, 61, MS, 63; NC State Univ, PhD(physiol), 70. *Prof Exp:* Res physiologist, 69-78, res physiologist, Poultry Res Lab, Sci & Educ Admin-Fed Res, 78-83, RES PHYSIOLOGIST, SCENT POULTRY RES LAB, AGR RES SERV, USDA, MISS, 83- *Concurrent Pos:* Adj prof, Miss State Univ, 83- *Mem:* Poultry Sci Asn; World Poultry Sci Asn; Am Physiol Soc; Am Soc Zool. *Res:* Poultry environmental physiology; thyroid metabolism; amino acid metabolism. *Mailing Add:* 105 Apache Starkville MS 39759

MAY, JERRY RUSSELL, b Seattle, Wash, Apr 24, 42; div; c 2. BEHAVIORAL MEDICINE. *Educ:* Western WAsh State Univ, BA, 68; Bowling Green State Univ, PhD(clin psychol), 74. *Prof Exp:* PROF PSYCHIAT, SCH MED, UNIV NEV RENO, 74- *Concurrent Pos:* Assoc dean admin, Sch Med, Univ Nev Reno, 75-; team psychologist, US Alpine Ski Team, 77-; chmn, Sport psychol comts, Sports Med Coun, US plympic Comt 84-; consult, Vet Admin Med Ctr, Reno & others, 76- *Mem:* Am Psychol Assoc; Am Assoc Med Col; Am Col Sports Med. *Res:* Sports psychology, high level achievement; health psychology, stress management, communications skills. *Mailing Add:* Dept Psychiat Sch Med Univ Nev Reno NV 89557

MAY, JOAN CHRISTINE, b Buffalo, NY. ANALYTICAL CHEMISTRY, INORGANIC CHEMISTRY. *Educ:* Nazareth Col Rochester, BS, 65; Univ Wis-Madison, MS, 68; Univ Notre Dame, PhD(anal-inorg chem), 71. *Prof Exp:* Sr chemist analytical chem, 73-74, DIR, ANALYTICAL CHEM BR, BUR BIOLOGICS, DEPT HEALTH, EDUC & WELFARE, USPHS & FOOD & DRUG ADMIN, 74- *Concurrent Pos:* Chemist, Roswell Park Mem Inst, 71-72 & Analytical Chem Div, Nat Bur Standards, Washington, DC, 72-73. *Mem:* Am Chem Soc; Asn Off Analytical Chemists; Sigma Xi; Int Asn Biol Standardization. *Res:* Accurate analytical methods for the determination of trace and macro amounts of organic and inorganic additives, preservatives and other impurities or constituent materials found in vaccines and other biological products. *Mailing Add:* 5894 Lake Ave Orchard Park NY 14127

MAY, JOHN ELLIOTT, JR, b Meriden, Conn, June 4, 21; m 45; c 2. PHYSICS. *Educ:* Wesleyan Univ, BA, 43; Tufts Col, MS, 49; Yale Univ, PhD(physics), 53. *Prof Exp:* Electronics physicist, US Naval Res Lab Field Sta, Mass, 46-49; mem tech staff, Bell Tel Labs, Inc, 52-59, supvr explor develop delay devices, 59-62, ultrasonic amplifier & evaporated film transducer develop, 62-65, head ultrasonic device dept, 65-71, head process capability dept, 71-77, supvr eval digital transmission equip, 77-79, PHYSICIST, BELL LABS, 79- *Mem:* Am Phys Soc; Inst Elec & Electronics Engrs; fel Acoust Soc Am. *Res:* Research and development in ultrasonic devices including delay line geometry, ceramic transducers; elastic wave guide effects; piezoelectric materials; ultrasonic amplification; mechanical filters; microphones; thin film circuits; conductors, resistors and capacitors. *Mailing Add:* 77 Herrick Rd Boxford MA 01921

MAY, JOHN WALTER, b London, Eng, June 25, 36; Can citizen; m 64; c 2. SURFACE PHYSICS, SURFACE CHEMISTRY. *Educ:* Univ BC, BA, 57, MS, 60; Oxford Univ, PhD(chem), 63. *Prof Exp:* Res assoc, Dept Appl Physics, Cornell Univ, 64-68; physicist, Bartol Res Found, Franklin Inst, 68-72; sr res chemist, 72-80, RES ASSOC, EASTMAN KODAK RES LABS, 80- *Mem:* Soc Photogr Scientists & Engrs. *Res:* Surface science, electrostatics, triboelectrification; electrophotography; low mobility charge transport; xeroradiography. *Mailing Add:* Eastman Kodak Res Labs Kodak Park Rochester NY 14650

MAY, KENNETH NATHANIEL, b Livingston, La, Dec 24, 30; m 53; c 2. FOOD TECHNOLOGY. *Educ:* La State Univ, BS, 52, MS, 55; Purdue Univ, PhD(food technol), 59. *Hon Degrees:* DAgr, Purdue Univ, 88. *Prof Exp:* Asst poultry sci, La State Univ, 52-54, res assoc, 54-56; asst state poultry supvr, State Livestock Sanit Bd, La, 54; asst poultry husb, Purdue Univ, 56-58; from asst prof to prof, Univ Ga, 58-68; prof, Miss State Univ, 68-70; dir res & qual assurance, Holly Farms Poultry Industs, Inc, 70-73, vpres res & qual assurance, 73-85, pres & chief exec officer, Holly Farms Foods, 85-88, chmn, 88-89; CONSULT, 90- *Concurrent Pos:* Mem salmonella adv comt, Secy Agr,

75-78; adj prof, NC State Univ, Raleigh, 75-; mem, Nat Adv Comt, Meat & Poultry Insp, 79-81; mem, Nat Comt Micrological Criteria for Foods, 90- *Honors & Awards:* Res Award, Inst Am Poultry Industs, 63; Res Award, Ga Egg Comn, 64; Indust Serv Award, Poultry & Egg Inst Am, 71. *Mem:* Am Poultry Sci Asn; Inst Food Technol; World Poultry Sci Asn. *Res:* Meat yields and processing losses of poultry; nutritive value and bacteriology of poultry products; biochemistry of bruised tissue. *Mailing Add:* 203 McElwee St North Wilkesboro NC 28659

MAY, LEOPOLD, b Brooklyn, NY, Nov 26, 23; m 47; c 2. PHYSICAL BIOCHEMISTRY. *Educ:* City Col New York, BChE, 44; Polytech Inst Brooklyn, MS, 48, PhD, 51. *Prof Exp:* Instr, Polytech Inst Brooklyn, 49-50; res chemist, Columbia Univ, 50-54; res assoc, Med Sch, Univ Md, 54-56, instr, 56-59; from asst prof to assoc prof, 59-82, PROF CHEM, CATH UNIV AM, 82- *Concurrent Pos:* Lectr, Brooklyn Col, 53; instr, Johns Hopkins Univ, 54-57; ed-in-chief, Appl Spectros, 61-64; vis assoc prof, Tel-Aviv Univ, 72-73; vis scientist, Soreg Nuclear Physics Ctr, Israel, 72-73; Nat Acad Sci exchange scientist, Inst Chem Physics, Moscow, USSR, 76-77, 78; vis prof chem, Banaras Hindu Univ, India, 78; vis res prof, Armed Forces Radiobiol Res Inst, Bethesda, Md, 78-82, 87-, Hebrew Univ, Jerusalem, Israel, 84-85; Fulbright-Hays Award, Lima, Peru, 80; fac fel, Goddard Space Flight Ctr, Greenbelt, Md, NASA/Am Soc Eng Educ, 74, 75; fac res assoc, Naval Med Res Inst, Bethesda, Md, 83; chemist, Lawrence Livermore Nat Lab, Livermore, Calif, 88. *Mem:* Am Chem Soc; Soc Appl Spectros (pres, 71). *Res:* Mössbauer spectroscopy of biological materials and chemicals. *Mailing Add:* Dept Chem Cath Univ Am Washington DC 20064

MAY, MICHAEL MELVILLE, b Marseilles, France, Dec 23, 25; nat US; m 52; c 4. PHYSICS. *Educ:* Whitman Col, BA, 44; Univ Calif, PhD(physics), 52. *Hon Degrees:* DSc, Whitman Col, Walla Walla, WA, 76. *Prof Exp:* Res physicist, Lawrence Livermore Lab, Univ Calif, 52-57; vpres, E H Plesset Assocs, 57-60; res physicist, 60-61, div leader, 61-62, assoc dir, 62-64, lectr appl sci, 64-65, dir, 65-71, RES PHYSICIST & ASSOC DIR-AT-LG, LAWRENCE LIVERMORE LAB, UNIV CALIF, 72- *Concurrent Pos:* Vis physicist, Princeton Univ, 71-72; sr personal adv to Secy of Defense for Strategic Arms Limitation Talks & mem US deleg, 74-76. *Honors & Awards:* E O Lawrence Mem Award, Atomic Energy Comn, 70. *Mem:* Am Phys Soc. *Res:* Nuclear explosions; heat and radiation; relativity. *Mailing Add:* Lawrence Livermore Lab Livermore CA 94550

MAY, PAUL S, b Brooklyn, NY, July 12, 31; m 59; c 3. MICROBIOLOGY. *Educ:* City Col New York, BS, 51; Syracuse Univ, MS, 52; Phila Col Pharm, DSc(indust microbiol), 55; Columbia Univ, MPH, 70. *Prof Exp:* Instr bact, Phila Col Pharm, 52-53, instr zool, 54-55; sr res microbiologist, S B Penick & Co, 55-58; asst microbiologist, Beth Israel Hosp, NY, 58-62; sr scientist microbiol, Life Sci Lab, Melpar, Inc, 62-64; lectr, Sch Pub Health, Columbia Univ, 64; asst dir bur labs, New York Dept Health, 64-71, dep asst comn bur labs, 71-86, actg asst comn, 86-88, spec projs, 88-89; CONSULT, 89- *Concurrent Pos:* Adj asst prof, Sch Pub Health & admin med, Columbia Univ & Hunter Col, 78-89. *Mem:* Am Soc Microbiol; Am Pub Health Asn; Am Biosafety Asn. *Res:* Antibiotics; fermentations; medical bacteriology, parasitology and mycology; public health microbiology; laboratory and public health administration; epidemiology. *Mailing Add:* 23 Fairview Lane Orangeburg NY 10962

MAY, PHILIP REGINALD ALDRIDGE, psychiatry, for more information see previous edition

MAY, RALPH FORREST, b Idaho, Ohio, Oct 1, 41; m 63, 78; c 2. AGRICULTURAL CHEMISTRY. *Educ:* Wilmington Col, AB, 63; Ind Univ, Bloomington, MA, 66, PhD(org chem), 67. *Prof Exp:* Res chemist, 67-80, SR RES CHEMIST AGR CHEM, AGR PROD DEPT, E I DU PONT DE NEMOURS & CO, INC, 80- *Res:* Formulation and development of insecticides. *Mailing Add:* Agr Prod Dept du Pont Exp Sta E 402/1108 Wilmington DE 19880-0402

MAY, ROBERT CARLYLE, aquaculture, for more information see previous edition

MAY, ROBERT MCCREDIE, b Sydney, Australia, Jan 8, 36; m 62; c 1. MATHEMATICAL BIOLOGY, EPIDEMIOLOGY. *Educ:* Univ Sydney, BSc, 56, PhD(theoret physics), 59. *Hon Degrees:* DSc, City Univ London, 89, Uppsala Univ Sweden, 90. *Prof Exp:* Gordon MacKay lectr appl math, Harvard Univ, 59-61 & 66; sr lectr & reader physics, Univ Sydney, 62-72, prof, 70-73; PROF, DEPT BIOL, PRINCETON UNIV, 75-, CHMN, RES BD, 77- *Concurrent Pos:* Vis prof, Calif Inst Technol, 67, UK Atomic Energy Authority, 71, Inst Advan Study, 72, Kings Col Res Ctr, 72 & Imperial Col Field Sta, 73, 75-85; vis mem, Inst Advan Study, 71-72 & King's Col, Cambridge, 76; mem comt ecosyst anal, Nat Acad Sci, 73-75; assoc ed, Theoret Pop Biol & Math Biosci, 74- & SIAM J Appl Math & Appl Ecol Abstr, 75- *Honors & Awards:* Pawsey Medal, Australian Acad Sci, 67; Weldon Mem Prize, Oxford Univ, 80; MacArthur Award, Ecol Soc Am, 84; Hitchcock Lectr, Univ Calif, Berkeley, 85; Croonian, Royal Soc, 85. *Mem:* Brit Ecol Soc; Am Soc Naturalists; fel Royal Soc; Am Acad Arts & Sci. *Res:* Basic understanding of dynamic behavior of plant and animal populations and of community ecology; applications to resource management and infectious diseases. *Mailing Add:* Dept Zool Univ Oxford, S Park Rd Oxford 0X13PS England

MAY, SHELDON WILLIAM, b Minneapolis, Minn, June 27, 46; m 68; c 4. BIOCHEMISTRY, NEUROCHEMISTRY. *Educ:* Roosevelt Univ, BS, 66; Univ Chicago, PhD(chem), 70. *Prof Exp:* Sr res chemist, Corp Res Lab, Exxon Res & Eng Co, 70-73; from asst prof to assoc prof, 74-80, PROF CHEM & BIOCHEM, GA INST TECHNOL, 80- *Concurrent Pos:* NSF fel, 67-70, NIH, 70; fel, A P Sloan Found, 77-79; vis prof, Ctr Neurochem, Strasbourg, France, 79 & Weizmann Inst Sci, 85-86; Fulbright Int sr res scholar, 85-86. *Mem:* Am Soc Biochem & Molecular Biol; AAAS; Am Chem Soc; Am Soc Neurosci; Soc Microbiol & Biochem Technol. *Res:* Enzyme chemistry; rational drug design; chemical immunology; molecular neurochemistry; immobilized enzymes; enzyme and biochemical technology. *Mailing Add:* Sch Chem & Biochem Ga Inst Technol Atlanta GA 30332

MAY, SHERRY JAN, Can citizen. LEVEL THEORY, QUANTITATIVE BIOSTRATIGRAPHY. *Educ:* Univ Sask, BA, 68, dipl math, 69, MBA, 82; Univ Waterloo, MM, 70, PhD(appl math), 74. *Prof Exp:* Nat Res Coun Can fel, Univ Western Ont & Univ Sask, 74-76; asst prof math, Acadia Univ, 77-78; ASST PROF MATH, MEM UNIV NFLD, 78- *Concurrent Pos:* Res grants, Can Coun, 75 & NSERC, 79-85. *Mem:* Am Math Soc; Soc Exact Philos; Admin Sci Asn Can. *Res:* Probability kinematics; level functions; computer software to analyze off-shore well data. *Mailing Add:* Dept Math & Statist Mem Univ Nfld St John's NF A1C 5S7 Can

MAY, STERLING RANDOLPH, b Muskogee, Okla, Dec 27, 46; m. HUMAN GENETICS. *Educ:* Univ Kans, BA, 68; Univ Mich, MS, 69, PhD(human genetics), 77. *Prof Exp:* Coordr, Skin Bank, St Agnes Med ctr, 77-79, assoc dir, Skin Bank & Burn Res Lab, 80, dir burn res, 81-83; from res asst prof to res assoc prof, Dept Surg, Sch Med, Hahnemann Univ, 79-83; assoc clin prof, Dept Surg, Med Col Ga, 84-87; VPRES SCI, LIFECELL CORP, 87-; ADJ PROF, DEPT SURG, MED SCH, UNIV TEX, HOUSTON, 87- *Concurrent Pos:* Mem, var adv comts, Am Asn Tissue Banks, Am Burn Asn & Soc Cryobiol, 78-91, Burn Care Reimbursement Policy Eval Proj Panel, Burn Found, 84, Nat Med Adv Comt, Nat Tissue Serv, Am Red Cross, 81-82 & Surg & Bioeng Study Sect, NIH, 90; consult res tissue preserv & distrib, Nat Diabetes Res Interchange, Juv Diabetes Found, 81-83; dir, Southeastern Burn Res Inst & res data coordr, Humana, Inc Burn Ctr Excellence Prog, Humana Hosp, 83-87; chmn, Nat Sci Bd, Tissue Banks Int, 84-89, Adv Comt Skin, Nat Tissue Serv, Am Red Cross, 89-92 & Res Subcomt, Med Adv Comt, 91-92; adj prof, Dept Biol Sci, State Univ NY, Binghamton, 87-; prin investr, Nat Inst Gen Med Sci, NIH, USPHS, 89-92 & 90-93, USN Med Res & Develop Command, 89-92; feature ed, J Burn Care & Rehab, 91- *Mem:* AAAS; Sigma Xi; Am Burn Asn; Am Asn Tissue Banks (secy, 91-93); Soc Cryobiol (vpres, 87-89, pres, 89-91); NY Acad Sci; Int Soc Burn Injuries; Am Soc Microbiol; Tissue Cult Asn; Am Soc Cell Biol. *Res:* Freezing, drying and rehydration of mammalian cells; preservation of dermis for in situ skin reconstitution. *Mailing Add:* LifeCell Corp 3606-A Research Forest Dr Spring TX 77381

MAY, WALTER GRANT, b Saskatoon, Sask, Nov 28, 18; US citizen; m 83; c 3. CHEMICAL ENGINEERING, PHYSICAL CHEMISTRY. *Educ:* Univ Sask, BSc, 39, MSc, 42; Mass Inst Technol, ScD(chem eng), 48. *Prof Exp:* Asst chemist, Brit Am Oil Co, 39-40; asst prof chem eng, Univ Sask, 43-46; res assoc, Esso Res & Eng Co, 48-59; engr rocket propellants, Inst Defense Analysis, 59-60; sr res assoc, Esso Res & Eng Co, 60-67; prof mech eng, Stevens Inst Technol, 67-77; sr sci advisor, Exxon Res & Eng Co, 77-83; prof chem eng, Univ Ill, Urbana-Champaign, 83-91; RETIRED. *Concurrent Pos:* Chmn thermochem panel, Joint Army-Navy-Air Force, 59-60; mem panel thermodyn, Interagency Chem Rocket Propulsion Group, 59-66; sr sci adv, Exxon Nuclear, 73-77; prof, Rensselaer Polytech Inst, 74-77; vis prof chem eng, Univ Ill, Urbana-Champaign, 91- *Honors & Awards:* Award Chem Eng Pract, Am Inst Chem Engrs, 89. *Mem:* Nat Acad Eng; Am Inst Chem Engrs; Am Soc Mech Engrs; Combustion Inst; Nat Acad Eng. *Res:* Chemical reaction kinetics; solid rocket propellants; combustion; thermodynamics; chemical reactor engineering; isotope separations. *Mailing Add:* 916 W Clark Champaign IL 61821-3328

MAY, WALTER RUCH, b Senath, Mo, Aug 4, 37; div; c 3. PHYSICAL INORGANIC CHEMISTRY. *Educ:* Memphis State Univ, BS, 59; Vanderbilt Univ, PhD(chem), 62. *Prof Exp:* Instr chem, Vanderbilt Univ, 59-62; res chemist, Monsanto Co, 62-65, sr res chemist, 65-66; res chemist, Tretolite Div, Petrolite Corp, 66-67, res group leader, Corp Lab, 67-73, indust chem group leader, 73-75, mgr indust & water res, 75-76, mgr indust fuels res, 76-77, mgr indust fuel additives sales, For Oper Dept, 77-78; gen mgr, Specialty Fuel Additives Div, Perolin Co, Inc, 78-80; PRES, SFA TECHNOL, INC, CHARLOTTE, NC, 80- *Mem:* Am Chem Soc; Nat Asn Corrosion Engrs; Sigma Xi; Am Soc Mech Engrs. *Res:* High temperature corrosion; hydrocarbon oxidation; thermal analysis; coordination compounds; heavy petroleum fuel additives. *Mailing Add:* SFA Tech Inc 9303-B Monroe Rd Charlotte NC 28226

MAY, WILLIAM G(AMBRILL), b St Louis, Mo, Dec 30, 37. ELECTRICAL ENGINEERING, INTEGRATED CIRCUITS. *Educ:* Mass Inst Technol, SB & SM, 60, PhD(elec eng), 64. *Prof Exp:* Asst elec eng, Mass Inst Technol, 60-62, from instr to asst prof, 62-66; from asst prof to assoc prof, 70-77, PROF ELEC ENG, UNIV COLO, BOULDER, 78- *Concurrent Pos:* Ford fel eng, 64-66. *Mem:* Am Phys Soc. *Res:* Semiconductor devices and device physics; electrical properties of semiconductors; integrated circuit design and fabrication; optical and microwave devices. *Mailing Add:* Dept Elec Eng Univ Colo Box 425 Boulder CO 80309-0425

MAY, WILLIE EUGENE, b Dozier, Ala, Sept 10, 47; m 69; c 2. HIGH PERFORMANCE LIQUID CHROMATOGRAPHY, TRACE ORGANIC ANALYSIS. *Educ:* Knoxville Col, BS, 68; Univ Md, PhD(anal chem), 77. *Prof Exp:* Sr lab analyst, Oak Ridge Gaseous Diffusion Plant, Tenn, 68-71; res chemist, Separation & Purification Sect, Nat Bur Standards, 71-78, group leader, Liquid Chromatography Anal Res div, 78-83, CHIEF, ORG ANALYSIS RES DIV, NAT INST STANDARDS & TECHNOL, 83- *Honors & Awards:* Bronze Medal, Dept Com, 81, Silver Medal, 85; Arthur S Fleming Award, 87. *Mem:* Am Chem Soc; AAAS; Nat Orgn Professional Advan Black Chemists & Chem Engrs. *Res:* Trace organic analytical chemistry; development of liquid chromatographic methods for determination of individual organic species in complex mixtures and physico-chemical properties of organic compounds. *Mailing Add:* Nat Inst Standards & Technol Bldg 222 Rm B158 Gaithersburg MD 20899

MAYA, LEON, b Mexico City, Mex, Mar 23, 38; US citizen; m 60; c 2. INORGANIC CHEMISTRY, PHYSICAL CHEMISTRY. *Educ:* Nat Univ Mex, BS, 60; Univ Southern Calif, PhD(inorg chem), 73. *Prof Exp:* Supvr qual control lab, Monsanto Mexicana SA, 60-62; sr chemist, Israel Mining Industs Res Inst, 62-68; chemist, Rainbow Beauty Supply, 68-69; fel inorg chem, Univ Southern Calif, 73-74; MEM RES STAFF, OAK RIDGE NAT LAB, 74-

Mem: AAAS; Am Chem Soc; Sigma Xi. *Res:* Synthetic inorganic chemistry and physical chemistry; use of physical methods for structural determination; chemistry of main group elements, particularly boron, silicon, phosphorus and fluorine; transition metals elements ruthenium, zirconium and niobium; actinide elements with emphasis on uranium and neptunium; ceramic materials such as nitrides, carbides and borides. *Mailing Add:* Oak Ridge Nat Lab PO Box 2008 Oak Ridge TN 37831-6119

MAYA, WALTER, b New York, NY, Oct 25, 29; m 85; c 4. ORGANIC CHEMISTRY, INORGANIC CHEMISTRY. *Educ:* Univ Calif, Los Angeles, BS, 54, PhD(org chem), 58. *Prof Exp:* Res chemist, E I du Pont de Nemours & Co, 58-59; specialist fluorine chem, Rocketdyne Div, NAm Aviation, Inc, 59-70; lectr, 71-72, from asst prof to assoc prof, 72-81, PROF CHEM, CALIF STATE POLYTECH UNIV, POMONA, 81- *Concurrent Pos:* Pfizer fel, Univ Ill, 58-59. *Mem:* AAAS; Am Chem Soc; Sigma Xi; Fedn Am Scientists. *Res:* Synthesis of fluorine compounds; physical-organic chemistry. *Mailing Add:* Dept Chem Calif State Polytech Univ Pomona CA 91768

MAYADAS, A FRANK, b Ferozepore, India, Dec 7, 39; US citizen; m 62; c 2. SOLID STATE PHYSICS, PHYSICAL METALLURGY. *Educ:* Colo Sch Mines, MetE, 61; Cornell Univ, PhD, 66. *Prof Exp:* Mem res staff, Watson Res Ctr, IBM Corp, 65-71, mgr thin film & metall group, 71-75, mgr memory & storage res, 75-77, mgr tech planning staff, 77-79, mgr storage systs & technol, San Jose, Calif, Res Lab, 79-81, dir, Tech Planning & Controls, 81-; DIR, SAN JOSE RES LAB, SAN JOSE, CALIF; DIR, IBM CORP & VPRES, IBM RES DIV. *Mem:* Am Phys Soc; Inst Elec & Electronics Engrs. *Res:* Electron microscopy; dislocation relaxation in metals; microwave resonance; anisotropy studies in magnetic thin films; electron scattering mechanisms in thin metal films. *Mailing Add:* IBM T J Watson Res Ctr PO Box 218 Yorktown Heights NY 10598

MAYALL, BRIAN HOLDEN, b Nelson, Eng, Nov 14, 32; US citizen; m 55; c 4. ANALYTICAL CYTOLOGY, CYTOMETRY. *Educ:* Cambridge Univ, BA, 54, MA, 58; Univ Western Ont, MD, 61. *Prof Exp:* Res assoc, Wistar Inst, 62-64; from instr radiol to asst prof radiol sci, Med Sch, Univ Pa, 64-71, assoc prof radiol, 71-72; sect leader cytogenetics & cytomorphometry, Biomed Div, Lawrence Livermore Nat Lab, 77-82, PROF, LAB MED & DIR, PROG ANALYSIS CYTOL, UNIV CALIF, SAN FRANCISCO, 82- *Concurrent Pos:* Pa Plan scholar, Wistar Inst & Univ Pa, 63-65; consult, Med Res Coun, UK, 68 & Nat Cancer Inst, 70-; adj assoc prof radiol, Univ Calif, Davis, 74-82; ed, Cytometry, 80- *Mem:* AAAS; Histochem Soc; Am Soc Cytol; Soc Analysis Cytol; Sigma Xi. *Res:* Quantitative cytochemistry; image cytometry; automated cytology; image analysis of cells and chromosomes. *Mailing Add:* Lab Cell Analysis Dept Lab Med Box 0808 Univ Calif MCB 230 San Francisco CA 94143-0808

MAYALL, NICHOLAS ULRICH, b Moline, Ill, May 9, 06; m 34; c 2. ASTRONOMY. *Educ:* Univ Calif, AB, 28, PhD(astron), 34. *Prof Exp:* Asst, Univ Calif, 28-29; asst comput, Mt Wilson Observ, 29-31; observing asst, Lick Observ, 33-35, asst astronr, 35-42; mem staff, Radiation Lab, Mass Inst Technol, 42-43; res assoc, Calif Inst Technol, 43-45; from assoc astronr to astronr, Lick Observ, 45-60; dir, Kitt Peak Nat Observ, 60-71; RETIRED. *Concurrent Pos:* Trustee, Univs Res Asn, 72-78. *Mem:* Nat Acad Sci; Am Philos Soc; Am Astron Soc; Am Acad Arts & Sci; Int Astron Union. *Res:* Nebular spectroscopy; celestial photography; radial velocities of galactic nebulae, globular star clusters; red shifts and internal motions of galaxies. *Mailing Add:* 7206 E Camino Vecino Tucson AZ 85715

MAYBANK, JOHN, b Winnipeg, Man, Jan 23, 30; m 52; c 2. ATMOSPHERIC PHYSICS. *Educ:* Univ Man, BSc, 52; Univ BC, MSc, 54; Univ London, PhD(meteorol), 59. *Prof Exp:* Sci officer, Physics & Meteorol Sect, Defence Res Bd, 54-61; res officer, Physics Div, Sask Res Coun, 61-70; climatologist, Caribbean Meteorol Inst, Barbados, 70-71; head physics div, Sask Res Coun, 72-83, dir, Environ Div, 83-88; SR ADV, CLIMATE SERV PROG, ENVIRON CAN, 89- *Concurrent Pos:* Res assoc, Univ Sask, 62-66, adj prof, 68- *Mem:* Can Asn Physicists; Royal Meteorol Soc; Can Meteorol & Oceanog Soc. *Res:* Cloud physics; ice nucleation phenomena; atmospheric pollution; agrometeorology; pesticide spray drift; agriculture and climate change. *Mailing Add:* Atmospheric Environ Serv 1000-266 Graham Ave Winnipeg MB R3C 3V4 Can

MAYBEE, JOHN STANLEY, b Washington, DC, Mar 23, 28; m 55; c 6. APPLIED MATHEMATICS. *Educ:* Univ Md, BS, 50; Univ Minn, PhD, 56. *Prof Exp:* Mathematician, David Taylor Model Basin, US Dept Navy, 50-52; asst math, Univ Minn, 52-56; from instr to asst prof, Univ Southern Calif, 56-59; asst prof, Univ Ore, 59-61; from asst prof to assoc prof, Purdue Univ, 61-67; PROF MATH, UNIV COLO, BOULDER, 67- *Concurrent Pos:* Mem, Inst Math Sci, NY Univ, 58-59. *Mem:* Math Asn Am; Soc Indust & Appl Math. *Res:* Applied mathematics; matrix theory; combinatorics. *Mailing Add:* Prog Appl Math Univ Colo Boulder CO 80309-0526

MAYBERGER, HAROLD WOODROW, b New York, NY, Aug 28, 19; m 51; c 3. GYNECOLOGY. *Educ:* Univ Ala, BA, 41; Long Island Col Med, MD, 44; Am Bd Legal Med, dipl, 56; Am Bd Obstet & Gynec, dipl, 61. *Prof Exp:* Intern, St John's Episcopal Hosp, 44-45 & 47-48, resident obstet & gynec, 48-51; mem courtesy staff, Community Hosp, 53, clin asst, 53-55, from asst attend obstetrician & gynecologist to assoc attend obstetrician & gynecologist, 55-58, attend obstetrician & gynecologist & asst attend pathologist, 58-64, chief, 64-84, EMER CHIEF, DIV OBSTET & GYNEC, COMMUNITY HOSP, GLEN COVE, NY, 84- *Concurrent Pos:* Res fel neonatal path, Beth El Hosp, 51-53; assoc prof clin obstet & gynec, State Univ NY Stony Brook; consult, St John's Episcopal Hosp, Brooklyn, 65, Winthrop Univ Hosp, Mineola. *Mem:* AAAS; fel Am Col Legal Med; fel Am Col Surg; fel Am Col Obstet & Gynec; Am Cancer Soc. *Res:* Neonatal pathology; forensic obstetrics. *Mailing Add:* Ten Medical Plaza Glen Cove NY 11542

MAYBERRY, JOHN PATTERSON, b New Haven, Conn, July 17, 29; m 54; c 3. OPERATIONS RESEARCH, PROBABILITY. *Educ:* Univ Toronto, BA, 50; Princeton Univ, MA, 54, PhD(math), 55. *Prof Exp:* Asst econ, Princeton Univ, 50-52, asst appl math, Analysis Res Group, 53-55; engr, Defense Electronic Prod Dept, Radio Corp Am, 55-58; opers analyst, Hq Fifth Air Force, Japan, 58-61, opers analyst, Hq, US Air Force, Washington, DC, 61-64, chief res group mil opers res, 64-67; mathematician, Mathematica Inc, 67, dir math res serv, 67-69; mathematician, Lambda Corp, Va, 69-71; chmn dept, 72-75, PROF MATH, BROCK UNIV, 71-; CONSULT, JOHN P MAYBERRY ASSOCS, 71- *Mem:* Asn Comput Mach; Am Math Soc; Math Asn Am; Soc Indust & Appl Math. *Res:* Topology; graph theory; decision theory; systems analysis; game theory. *Mailing Add:* Dept of Math Brock Univ St Catharines ON L2S 3A1 Can

MAYBERRY, LILLIAN FAYE, b Portland, Ore, May 19, 43; m 75. CELL BIOLOGY, PARASITOLOGY. *Educ:* San Jose State Col, BA, 67; Univ Nev, Reno, MS, 70; Colo State Univ, PhD(zool), 73. *Prof Exp:* Res assoc cell biol, Colo State Univ, 73-74 & Univ Colo, Boulder, 74-76; res affil biol sci, 76-, INSTR, NURSING DEPT, 77-, ADJ PROF BIOL, UNIV TEX, EL PASO, 87- *Concurrent Pos:* Protozoologist, Yugoslavian Int Biol Prog, 75-76. *Mem:* Am Soc Parasitologists. *Res:* Physiology and ecology of host-parasite relationships. *Mailing Add:* Dept Biol Univ Tex El Paso TX 79968-0519

MAYBERRY, THOMAS CARLYLE, b Nashville, Tenn, Oct 16, 25; m 53; c 2. POLYMER CHEMISTRY, ORGANIC CHEMISTRY. *Educ:* Vanderbilt Univ, BA, 49, MS, 51; Univ Del, PhD(chem), 62. *Prof Exp:* Chemist, Old Hickory Textile Fibers, 51-53, res chemist, Rayon Res Lab, 53-57, res supvr, Indust Prod Res Lab, 58-69 & Dacron Res Lab, 69-72, sr res chemist, Carothers Res Lab, 72-76, RES ASSOC, CHATTANOOGA RES & DEVELOP SECT, E I DU PONT DE NEMOURS & CO, INC, 76- *Mem:* Am Chem Soc. *Res:* Polymer and fiber chemistry; physics; engineering. *Mailing Add:* 5919 Lake Resort Dr Chattanooga TN 37443-4643

MAYBERRY, WILLIAM EUGENE, b Cookeville, Tenn, Aug 22, 29; m 53; c 2. ENDOCRINOLOGY. *Educ:* Univ Tenn, MD, 53, Univ Minn, MS, 59; Am Bd Internal Med, dipl, 64. *Hon Degrees:* DHL, Jacksonville Univ, 85. *Prof Exp:* First asst & asst to staff internal med, Mayo Clin, 56-59, from instr to assoc prof med, Mayo Grad Sch Med, Univ Minn, 60-74, chmn dept lab med, 71-75, PROF INTERNAL MED & LAB MED, MAYO MED SCH, MAYO CLIN, 74- *Concurrent Pos:* Fel internal med, Mayo Grad Sch Med, Univ Minn, 56-59; Nat Inst Arthritis & Metab Dis trainee & res fel endocrinol, New Eng Ctr Hosp, 59-60; Am Cancer Soc fel, Nat Inst Arthritis & Metab Dis, 62-64; asst, Sch Med, Tufts Univ, 59-60; consult, Mayo Clin, 60-62, mem bd gov, 71-87, vchmn, 73-75, chmn, 76-87; consult, Mayo Clin, 64-, mem bd trustees, Mayo Found, 71-87, vchmn, 75-86; chmn exec comt & pres, Mayo Found, 86-87, chief exec officer, 76-87, chmn Bd of Develop, 88- *Mem:* Inst Med Nat Acad Sci; Am Thyroid Asn; Am Chem Soc; Am Fedn Clin Res; fel Am Col Physicians; Sigma Xi; Am Med Asn; Am Clin & Climat Asn; Am Col Physicians Execs (Bd of Regents, 83, vpres, 85); Am Acad Med Dir. *Res:* Biochemistry and physiology of the thyroid gland; biosynthesis of thyroxine. *Mailing Add:* Sect Develop Mayo Clin Rochester MN 55905

MAYBERRY, WILLIAM ROY, b Grand Junction, Colo, Nov 30, 38; m 67. MICROBIOLOGY, ANALYTICAL BIOCHEMISTRY. *Educ:* Univ Colo, BA, 61; Western State Col Colo, MA, 64; Univ Ga, PhD(microbiol & biochem), 66. *Prof Exp:* Chemist, AEC, Lucius Pitkin, Inc, Colo, 60-61; asst instr chem, Mesa Col, 62; res assoc microbiol, Univ Ga, 66-67; res assoc, 67-68, asst prof, 68-75, assoc prof, Sch Med, Univ SDak, 75-78; assoc prof, 78-83, PROF MICROBIOL, COL MED, E TENN STATE UNIV, 83- *Mem:* AAAS; Am Chem Soc; Am Soc Microbiol. *Res:* Gas chromatographic analysis of biological materials; growth yields and energy relationships in bacteria; membrane structure and function; lipids/cell surface components of bacteria. *Mailing Add:* Dept of Microbiol E Tenn State Univ Col of Med Johnson City TN 37614

MAYBURG, SUMNER, b Boston, Mass, Feb 21, 26. SOLID STATE PHYSICS. *Educ:* Harvard Univ, BS, 46; Univ Chicago, MS, 48, PhD(physics), 50. *Prof Exp:* Asst, Univ Chicago, 49-50; sr scientist radiation damage to solids, Atomic Power Div, Westinghouse Elec Corp, 50-52; sr engr, Res Labs, Sylvania Elec Prod, Inc, 52-55, engr mgr & chief engr, Semiconductor Div, 55-59; sr eng specialist, Gen Tel & Electronics Lab Div, 59-64; dir radiation effects div, Controls for Radiation, 64-67; CO-FOUNDER, TREAS & MEM TECH STAFF, SEMICONDUCTOR PROCESSING CO, INC, 67- *Mem:* Am Phys Soc; Electrochem Soc; sr mem Inst Elec & Electronics Engrs; NY Acad Sci; Am Inst Physics. *Res:* Dielectric constants; photoconductivity in insulators and semiconductors; lattice defects in semiconductors; semiconductor devices; intermetallic semiconductors; semiconductor lasers; radiation effects in semiconductor materials and devices; surface preparation of crystalline materials. *Mailing Add:* Eight Wittier Pl Apt 8D Boston MA 02114

MAYBURY, PAUL CALVIN, b Rio Grande, NJ, July 20, 24; m 49; c 5. PHYSICAL CHEMISTRY, PHARMACOLOGY. *Educ:* Eastern Nazarene Col, BS, 47; Johns Hopkins Univ, PhD(chem), 52. *Prof Exp:* Sr staff chemist missiles, Appl Physics Lab, Johns Hopkins Univ, 51-52, res assoc chem, Univ, 52-54; from asst prof to assoc prof, Eastern Nazarene Col, 54-61, chmn dept, 56-61; assoc prof, 61-63, chmn dept, 62-74, PROF CHEM, UNIV SFLA, 64- *Concurrent Pos:* Res assoc, Tufts Univ, 54, vis prof, 66; vis scholar, Univ Calif, Los Angeles, 73-74; consult, Diamond Prod Co, Tampa; vpres & dir res, Belmac Corp, St Petersburg, 80-89; consult, Valcor Sci, Ltd, St Petersburg, Fla, 87-89; vpres & dir res, Fluid Life Systs, Inc, Laguna Niguel, Calif. *Mem:* AAAS; Am Chem Soc; fel Am Inst Chemists. *Res:* Boron hydride chemistry, including isotopic exchange studies; high energy particle tracks; heterogenous catalysis; reactions of metal borohydrides; methanation; pharmacological activity of natural products; research administration; principal investigator research project; use of drug complexes for controlled or sustained drug delivery systems. *Mailing Add:* Dept Chem Univ SFla Tampa FL 33620

MAYBURY, ROBERT H, b Lehighton, Pa, Jan 29, 23; m 46; c 5. TECHNOLOGY IN DEVELOPMENT. *Educ:* Eastern Nazarene Col, BS, 44; Boston Univ, MA, 48, PhD(phys chem), 52. *Prof Exp:* Assoc prof chem, Univ Redlands, 54-63; sci sector, UNESCO, Paris, 63-73 & 81-83, Nairobi, Kenya, 73-80; consult, technol in develop, World Bank, Washington, DC, 83-88; EXEC DIR, INT ORGN CHEM SCI IN DEVELOP, 88-; PROF TECHNOL, NETH INT INST MGT, MAASTRICH, 91- *Mem:* Fel AAAS; Am Chem Soc; Sigma Xi. *Res:* Choice and management of technology in developing countries. *Mailing Add:* 3705 S George Mason Dr Falls Church VA 22041

MAYCOCK, JOHN NORMAN, b Ripley, Eng, Dec 27, 37; m 62; c 2. SOLID STATE PHYSICS, CHEMICAL PHYSICS. *Educ:* Univ London, BS, 59, PhD(solid state chem), 62. *Prof Exp:* Scientist, Rias Div, Martin Co, 62-67, sr scientist, 67-69, head chem physics dept, 69-74, assoc dir, 71-74; head, Energy Technol Ctr, Martin Marietta Labs & corp dir energy affairs, 74-77, tech dir, Martin Marietta Cement, 77-78, vpres tech serv, 78-85, DIR ENVIRON MGT & FAC & MAINTENANCE, MARTIN MARIETTA AERO & NAVAL SYSTS, 85- *Mem:* Am Phys Soc; Royal Chem Soc. *Res:* Charge transport in alkali halides; physics of explosives and oxidizers; energy conservation as related to industry. *Mailing Add:* 103 Chesapeake Park Plaza Baltimore MD 21220

MAYCOCK, PAUL DEAN, b Sioux City, Iowa, Sept 2, 35; m 59; c 5. SOLID STATE PHYSICS, SCIENCE ADMINISTRATION. *Educ:* Iowa State Univ, BS, 57, MS, 62. *Prof Exp:* Res assist physics, Ames Lab, AEC, 60-62; mem tech staff, Tex Instruments Inc, 62-67, mgr new prod develop, 67-69, mgr bus develop, 69-71, sr bus analyst mat & elec prod group, 71-75; br chief econ anal, Solar Energy, Energy Res & Develop Admin, 75-76, dir solar energy planning, 76-77; dir photovoltaics, Dept Energy, 77-80; PRES, PHOTOVOLTAIC ENERGY SYSTS INC, 80- *Concurrent Pos:* Mem bd dirs, Am Solar Energy Soc & Solar Energy Indust Asn; publ, PV News. *Mem:* Am Phys Soc; Inst Elec & Electronics Engrs. *Res:* Thermal properties of solids; energy economics. *Mailing Add:* PO Box 290 Casanova VA 22017

MAYCOCK, PAUL FREDERICK, b Hamilton, Ont, Aug 13, 30; m 53; c 3. PLANT ECOLOGY. *Educ:* Queen's Univ, Ont, BA, 54; Univ Wis, MSc, 55, PhD(bot), 57. *Prof Exp:* Demonstr bot & zool, Queen's Univ, Ont, 52-54; lectr bot, McGill Univ, 57-58, from asst prof to assoc prof, 58-69; PROF BOT, ERINDALE COL, UNIV TORONTO, 69- *Concurrent Pos:* Mem staff, Polish Acad Sci, Cracow, 64-65; mem grant selection comt pop biol, Nat Res Coun, Can, 73-75; assoc ed ecology, Can J Bot, 80-; vis prof, Polish Acad Sci, Cracow, 82-83, Waseda Univ, Tokyo, 89-90; fel, Japan Soc Prom Sci, Tokyo, 89-90; secy gen, Int Asn Ecol, 86-90. *Mem:* Arctic Inst NAm; Ecol Soc Am; Can Bot Asn; Ottawa Field Nat; Int Asn Ecol. *Res:* Phytosociology; boreal forests of North America and world; vegetation of central Canada; synecology and autecology of forest species; nature reserves and conservation research; deciduous and deciduous-evergreen forests of the world. *Mailing Add:* Ecol Lab Erindale Col Univ Toronto 3359 Mississauga Rd Clarkson ON L5L 1C6 Can

MAYDAN, DAN, b Tel Aviv, Israel, Dec 20, 35; m 60; c 3. APPLIED PHYSICS, ELECTROOPTICS. *Educ:* Israel Inst Technol, BSc, 57, MSc, 62; Univ Edinburgh, PhD(physics), 65. *Prof Exp:* Supvr instrumentation, Soreq Res Estab, Israel AEC, 57-62, group leader devices, 65-67; mem tech staff, Bell Labs, 67-71, supvr optical scanning & modulation, 71-72, supvr new exposure syst group, 72-80; MEM STAFF, APPL MAT INC, 80-, VPRES. *Mem:* Inst Elec & Electronics Engrs. *Res:* X-ray lithography; acoustooptical devices; high resolution laser recording; display devices. *Mailing Add:* Appl Mat Inc 3050 Bowers Ave Santa Clara CA 95054

MAYDEW, RANDALL C, b Lebanon, Kans, Jan 29, 24; c 3. AERODYNAMICS. *Educ:* Univ Colo, BS, 48, MS, 49. *Prof Exp:* Asst aerodyn, Eng Exp Sta, Univ Colo, 48-49; aeronaut res scientist, Ames Aeronaut Lab, Nat Adv Comt Aeronaut, 49-52; mem staff, Sandia Corp, 52-57, supvr, Exp Aerodyn Div, 57-64, mgr, Aerodyn Dept, 64-88, MGR, SPECIAL PUBL DEPT, SANDIA NAT LABS, 88- *Mem:* Assoc fel Am Inst Aeronaut & Astronaut; Sigma Xi; Supersonic Tunnel Asn (pres, 69-70). *Res:* Transonic, supersonic and hypersonic experimental aerodynamics; wind tunnel design and operation; boundary layer phenomena; heat transfer; ballistics; re-entry phenomena; decelerators; darrieus wind turbines. *Mailing Add:* Org 400 Sandia Nat Labs Albuquerque NM 87110

MAYEDA, KAZUTOSHI, b Santa Monica, Calif, June 17, 28; m 49; c 3. GENETICS. *Educ:* Univ Utah, BS, 57, MS, 58, PhD(genetics), 61. *Prof Exp:* From asst prof to assoc prof, 61-73, genetic counr, Dept Gynec & Obstet, Sch Med, 77-85, PROF BIOL, WAYNE STATE UNIV, 73-; DIR, CYTOGENETICS, OAKWOOD HOSP, DEARBORN, MI, 88- *Concurrent Pos:* Res assoc, Nat Inst Genetics Japan, Mishima, Shizuoka-Ken, 70-71. *Honors & Awards:* Int Heritage Hall of Fame, 88. *Mem:* AAAS; Am Soc Human Genetics; Am Genetics Soc; Am Genetic Asn; Sigma Xi. *Res:* Immunogenetics of Drosophila and human; linkages of human genes and DNA analyses; genetics of human serum proteins; biochemical studies of amniotic fluids and cells. *Mailing Add:* Dept Biol Wayne State Univ Detroit MI 48202

MAYEDA, WATARU, b Shizuoka, Japan, June 21, 28; m 57; c 2. ELECTRICAL ENGINEERING. *Educ:* Utah State Univ, BS, 54; Univ Utah, MS, 55; Univ Ill, PhD(elec eng), 58; Tokyo Inst Technol, PhD(eng), 65. *Prof Exp:* Asst prof elec eng, Univ Ill, 58-59; staff engr, IBM Res Ctr, 59-60; assoc prof elec eng, 60-65, PROF ELEC ENG, UNIV ILL, URBANA, 65- *Mem:* Inst Elec & Electronics Engrs; Inst Elec Commun Eng Japan. *Res:* Linear graph theory especially application of linear graphs to engineering problems such as electrical networks, communication nets, switching networks. *Mailing Add:* Dept Elec Eng 4-127 Csl Univ Ill Urbana IL 61803

MAYELL, JASPAL SINGH, b Sialkot, Punjab, Pakistan, Jan 1, 29; US citizen; m 55; c 2. ANALYTICAL CHEMISTRY. *Educ:* Punjab Univ, India, BS, 50, MS, 52; Univ Tex, Austin, PhD(chem), 62. *Prof Exp:* Teaching asst & res fel chem, Univ Tex, Austin, 58-61; res chemist, Am Cyanamid Co, Stamford, Conn, 62-70; sr scientist chem, DuPont (endo) Pharmaceut, Garden City, NY, 71-76; ASST DIR CHEM, PURDUE FREDERICK CO, YONKERS, NY 77- *Mem:* Am Chem Soc; Am Pharmaceut Soc. *Res:* Solving pharmaceutical analytical chemistry problems with thorough understanding of the chemistry involved using both modern instrumental methods of analysis as well as the classical wet chemical analysis. *Mailing Add:* 40 Jay Rd Stamford CT 06905

MAYER, CORNELL HENRY, b Ossian, Iowa, Dec 10, 21; m 46; c 2. ELECTRICAL ENGINEERING, PHYSICS. *Educ:* Univ Iowa, BS, 43; Univ Md, MS, 51. *Prof Exp:* Electronic engr, 43-49, physicist, 49-68, head radio astron, 68-80, CONSULT, NAVAL RES LAB, 80- *Concurrent Pos:* Mem vis comt, Nat Radio Astron Observ, 69-72; mem nat adv comt, Owens Valley Radio Observ, Calif Inst Technol, 70-75; mem, Arecibo Adv Bd, Nat Astron & Ionosphere Ctr, 75-78; vis res assoc, Cal Tech, 63. *Honors & Awards:* Sigma Xi, NRL Pure Sci Award, 61. *Mem:* Int Astron Union; Am Astron Soc; Int Sci Radio Union; Inst Elec & Electronic Engrs. *Res:* Physical studies of space molecule regions, of the planets and satellites and discrete radio sources. *Mailing Add:* 1209 Villamay Blvd Alexandria VA 22307

MAYER, DAVID JONATHAN, b Mt Vernon, NY, July 18, 42; m 72. NEUROPHYSIOLOGY. *Educ:* City Univ New York, BA, 66; Univ Calif, Los Angeles, PhD(psychol), 71. *Prof Exp:* Asst prof, 72-75, assoc prof physiol, 75-78, PROF PHYSIOL, MED COL VA, 78- *Concurrent Pos:* NIH fel, Brain Res Inst, Univ Calif, Los Angeles, 71-72. *Mem:* Soc Neurosci; Am Physiol Soc; Int Asn Study Pain; Am Pain Soc. *Res:* Neurophysiology of pain and pain inhibitory systems; neuropharmacology of narcotic analgesics. *Mailing Add:* Dept of Physiol Med Col of Va Richmond VA 23298

MAYER, DENNIS T, BIOCHEMISTRY. *Educ:* Univ Ill, PhD, 38. *Prof Exp:* EMER PROF BIOCHEM, UNIV MO, 71- *Mailing Add:* 606 Shamrock Ave Apt 305 Lee's Summit MO 64063

MAYER, EUGENE STEPHEN, b Norwalk, Conn, June 5, 38; m 63; c 1. MEDICAL EDUCATION. *Educ:* Tufts Univ, BS, 60; Columbia Univ, MD, 64; Yale Univ, MPH, 71. *Prof Exp:* Physician, USPHS & US Peace Corps, Ankara, Turkey, 65-67 & Washington, DC, 67-68; dep dir & assoc prof family med & internal med, 71-78, DIR & PROF FAMILY MED & INTERNAL MED, AREA HEALTH EDUC CTR PROG, SCH MED, UNIV NC, CHAPEL HILL, 78-, ASSOC DEAN SCH MED, 78- *Concurrent Pos:* Consult, Bur Health Manpower, 74- *Mem:* Asn Am Med Cols; Asn Teachers Prev Med; AMA. *Res:* Distribution of health manpower and the effect of medical education on this distribution. *Mailing Add:* Dept Health Educ Univ NC Chapel Hill NC 27514

MAYER, FOSTER LEE, JR, b Fletcher, Okla, Nov 17, 42; m 83; c 2. ECOLOGY. *Educ:* Southwestern State Col, BS, 65; Utah State Univ, MS, 67, PhD(toxicol), 70. *Prof Exp:* Leader res sect, US Fish & Wildlife Serv, 70-74, chief biologist, Fish-Pesticide Res Lab, 74-81; res scientist, Columbia Nat Fish Res Lab, 81-84; sect chief, 84-85, BR CHIEF, ENVIRON RES LAB, US ENVIRON PROTECTION AGENCY, 85- *Concurrent Pos:* Res assoc sch forestry, fisheries & wildlife, Univ Mo- Columbia, 71-84. *Honors & Awards:* US Fish & Wildlife Serv Spec Achievement Award, 73, 79 & 81; Except Serv Award, Am Soc Testing & Mat, 90. *Mem:* Soc Environ Toxicol & Chem; Am Chem Soc; Am Fisheries Soc; Am Soc Testing & Mat; Soc Toxicol. *Res:* Toxicology of chemical contaminants in aquatic organisms, including biochemical and physiological aspects; formulation of mathematical models appropriate for prediction of contaminant effects in natural aquatic ecosystems. *Mailing Add:* US EPA Environ Res Lab Sabine Island Gulf Breeze FL 32561

MAYER, FRANCIS X(AVIER), b Muskogee, Okla, Mar 20, 30; m 54; c 5. CHEMICAL ENGINEERING, CHEMISTRY. *Educ:* Univ Tulsa, BCh, 52. *Prof Exp:* Asst lab instr chem, Univ Tulsa, 50-52; chem engr res & develop, 52-55, 57-60, sr engr, 60-64, eng assoc, 64-66, sect head eng design & math sect, 66-67, asst dir design & math sect & desulfurization, 67-71, mgr lab serv, 71-73, res coordr, 73-78, ENGR ADV, EXXON RES & DEVELOP LABS, 78- *Mem:* Am Chem Soc; Am Inst Chem Engrs. *Res:* Fluid hydroforming; petrochemicals; instrumentation and automation; fluid iron ore reduction; hydroconversion; hydrotreating; hydrodesulfurization; magnetically stabilized beds; hydroconversion. *Mailing Add:* 5277 Whitehaven Baton Rouge LA 70808

MAYER, FREDERICK JOSEPH, b Lock Haven, Pa, May 24, 40; m 65; c 2. PHYSICS. *Educ:* Pa State Univ, BS, 62; Case Inst Technol, MS, 65; Case Western Reserve Univ, PhD(physics), 68. *Prof Exp:* Sr res assoc plasma physics, Case Western Reserve Univ, 69-71; res scientist laser fusion, 71-75, sr scientist, 75-76, scientist-at-large, 80-84, dir advan res, 84-88, MGR LASER FUSION, KMS FUSION, INC, 76-, DIR FUSION EXP DIV, 78- *Concurrent Pos:* Adj res scientist, Dept Nuclear Eng, Univ Mich, 78-; Consult, appl physics, 88- *Mem:* AAAS; fel Am Phys Soc; Sigma Xi. *Res:* Experimental and theoretical laser plasma physics; thermonuclear fusion physics; plasma diagnostic techniques and instrumentation; similarity hydrodynamics; research and development funding and science policy. *Mailing Add:* 1417 Dickens Dr Ann Arbor MI 48103

MAYER, GARRY FRANKLIN, b New York, NY, Oct 11, 45; m 71; c 2. POLLUTION ECOLOGY. *Educ:* Queens Col, BA, 66; Harvard Univ, MA, 70, PhD(biol), 72. *Prof Exp:* Res asst biol, Queens Col, 65-66; res asst, Woods Hole Oceanog Inst, 67; teaching asst, Harvard Univ, 68-70; asst prof biol & natural sci, Boston Univ, 72-74; res assoc marine sci, Univ SFla, 74-76; ecologist & oceanographer marine ecosyst anal, New York Bight Proj, Nat Oceanic & Armospheric Admin, 76-80, sr ecologist, Off Marine Pollution Assessment, 80-85, assoc dir environ studies, Nat Sea Grant Prog, 85-89;

DIR, NAT OCEANIC & ATMOSPHERIC ADMIN COOP MARINE EDUC & RES PROG UNIV RI, 90- *Concurrent Pos:* NSF fel biol, Harvard Univ, 66-71, res assoc ichthyol, Mus Comparative Zool, 73-75; adj prof, Marine Sci Res Ctr, State Univ NY, Stony Brook, 78- *Mem:* Am Soc Limnol & Oceanog; Estuarine Res Fedn; AAAS; Sigma Xi. *Res:* Effects of pollution on estuarine, coastal ecosystems for applications problems in environmental management; ecology, evolution, and systematics of marine fishes. *Mailing Add:* URI/NOAA CMER Prog Grad Sch Oceanog Univ RI S Ferry Rd Narragansett RI 02882-1197

MAYER, GEORGE, b Gyor, Hungary, Feb 10, 34; US citizen; m 61, 90; c 1. MATERIALS SCIENCE, METALLURGICAL ENGINEERING. *Educ:* Boston Univ, BS, 57; Univ Okla, MMetE, 63; Mass Inst Technol, PhD(metall), 67. *Prof Exp:* Develop engr, Missile Div, Chrysler Corp, Mich, 57-58; sr res metallurgist, Ilikon Corp, Mass, 61-63; res asst metall, Mass Inst Technol, 63-67; sr res metallurgist, New Enterprise Div, Monsanto Co, 66-68; chief, Phys Mech Br, 68-72, assoc dir, 72-74, dir, Mat Sci Div, US Army Res Off, 74-88; CONSULT, STAFF MEM, INST DEFENSE ANALYSES, 88- *Concurrent Pos:* Adj assoc prof, NC State Univ, 69-75, adj prof, 75-; adj assoc prof, Duke Univ, 74-75; mem, Joint US/USSR Comn Electrometall & Mat, 78-80. *Mem:* ASM Int; fel Am Inst Chemists; Mat Res Soc; Sigma Xi. *Res:* Mechanical behavior of materials; materials failure; corrosion and environment-sensitive properties; crystal growth and characterization; nondestructive characterization; materials processing; deformation and fracture of polymers and composites. *Mailing Add:* 1010 N Terrill St Alexandria VA 22304-1938

MAYER, GEORGE EMIL, b Albany, NY, July 8, 41; m 64. BATTERIES, ELECTROCHEMISTRY. *Educ:* Rensselaer Polytech Inst, BS, 63, PhD(phys chem), 68. *Prof Exp:* Res physics chemist coatings, US Army Aberdeen Res & Develop Ctr, 68-70; assoc dir res & develop batteries, C&D Batteries Div, Eltra Corp, 70-74; mgr electrochem syst develop, Amerace Corp, 74-75; tech dir, SGL Batteries Div, SGL Industs, 75-79; tech supvr, St Joe Minerals Corp, 79-82; dir MIBTC, Res Inst, Carnegie Mellon Univ, 83-90; PRES, ELECTROCHEM & BATTERIES, BATTERY TECHNOL CTR INC, 91- *Concurrent Pos:* Legal expert batteries, various plaintiffs & defendants, 82-; consult, Elec & Hybrid Vehicles Div, US Dept Energy, 83-88, Jet Propulsion Lab, NASA-Dept Energy, 85-88, Batteries P-Document Working Group, 88-90; adj prof electrochem, Chem Dept, Carnagie Mellon Univ, 87-89. *Mem:* Electrochem Soc. *Res:* Battery technol, especially lead-acid batteries; battery design, processing, materials, testing, analysis, failure analysis; expert witness on batteries; electric vehicles; electronics instrumentation; chargers; electroanalytical chemistry. *Mailing Add:* Appl Res Ctr Univ Pittsburgh 865 William Pitt Way Pittsburgh PA 15238

MAYER, GERALD DOUGLAS, b Crowley, La, Jan 2, 33; m 79; c 6. MICROBIOLOGY. *Educ:* Southwestern La Univ, BS, 58, MS, 60; Iowa State Univ, PhD(bact), 64. *Prof Exp:* Res assoc virol, Charles Pfizer & Co, 64-66; sect head, Dept Infectious Dis, 66-74, dept head chemotherapeut, Merrell Res Ctr, 74-85, ASSOC GROUP DIR ANTIBIOTICS/ANTI-INFECTIVES MED RES, MERRELL-DOW PHARMACEUTICALS INC, 85- *Mem:* Am Soc Microbiol; Soc Exp Biol & Med; Can Asn Clin Microbiol & Infectious. *Res:* Interferon and interferon inducers; antimicrobial chemotherapy; virology. *Mailing Add:* Dept Infectious Dis 2110 E Galbraith Rd Cincinnati OH 45215

MAYER, HARRIS LOUIS, b New York, NY, Feb 15, 21; m 46; c 3. THEORETICAL PHYSICS. *Educ:* NY Univ, BA, 40; Columbia Univ, MS, 41; Univ Chicago, PhD(physics), 47. *Prof Exp:* With Div War Res, Columbia Univ, 41-46; group leader theoret physics, Los Alamos Sci Lab, 47-56; dept head, Aeronutronic Systs, Inc, 56-58; vpres, E H Plesset Assoc, Inc, 58-64; spec asst to vpres res, Inst Defense Analysis, 64-68; group dir survivability, 68-71, MEM TECHNOL PLANNING STAFF, AEROSPACE CORP, 71- *Concurrent Pos:* Consult, Avco Mfg Co, 55, Los Alamos Sci Lab, 56- & Jet Propulsion Lab, Calif Tech, 83-; mem nuclear panel, Sci Adv Bd, US Air Force, 58-62 & Lawrence Radiation Lab, Livermore, 60; mem weapons effects bd, Defense Atomic Support Agency, 60; dir nuclear technol seminar study, Advan Res Projs Agency, 68; mem space task group, Aerospace Corp, 69; study utilization space transp syst, NASA, 71. *Mem:* Fel Am Phys Soc. *Res:* Atomic physics; strategic support systems and space systems; energy management and reactor waste disposal; historical projections to year 2000; future space applications; large space structures; manned space operations; macro-engineering principles; meta-program engineering; space power systems; tethered space systems; strategic defense initiative concepts. *Mailing Add:* 30923 Cartier Dr Rancho Palos Verdes CA 90274

MAYER, J(OHN) K(ING), b Amite, La, Dec 2, 07; m 39. CIVIL ENGINEERING. *Educ:* Tulane Univ, BE, 30, ME, 37. *Prof Exp:* Lab asst, Tulane Univ, 29-30; jr engr, Standard Oil Co, NJ, 30-31; engr, Smith & Kanzler, Inc, 31; instr exp eng, 32-38, head dept, 38-40, from asst prof to prof, 38-49, prof mech eng, 49-60, prof civil eng, 60-73, EMER PROF CIVIL ENG, TULANE UNIV, 73- *Concurrent Pos:* Consult engr, 44- *Mem:* Am Soc Mech Engrs; Am Soc Eng Educ. *Res:* Soil mechanics and testing; engineering materials; material testing; experimental stress analysis. *Mailing Add:* 2419 Audubon St New Orleans LA 70125

MAYER, JAMES W(ALTER), b Chicago, Ill, Apr 24, 30; m 52; c 5. PHYSICS, ELECTRICAL ENGINEERING. *Educ:* Purdue Univ, BS, 52, PhD(physics), 60. *Hon Degrees:* DSc, State Univ NY, Albany, 88. *Prof Exp:* Mem tech staff, Hughes Res Labs, 59-62, sect head solid state studies, 62-67; assoc prof elec eng, Calif Inst Technol, 67-71, prof, 71-80; BARD PROF MAT SCI, CORNELL UNIV, 80- *Concurrent Pos:* Scuba instr, Calif Inst Technol, 70-80. *Honors & Awards:* Von Hippel Award, Mat Res Soc, 81; Silver Medal, Univ Catania, Italy, 86. *Mem:* Fel Am Phys Soc; fel Inst Elec & Electronics Engrs; Bohmische Phys Soc; Nat Acad Eng. *Res:* Characteristics of semiconductor nuclear particle detectors; ion beam modification and analysis of materials; ion implantation in semiconductors; thin film interdiffusion and reactions. *Mailing Add:* Elec Eng Dept Bard Hall Cornell Univ Ithaca NY 14853

MAYER, JEAN, b Paris, France, Feb 19, 20; nat US; m 42; c 5. EDUCATION, PHYSIOLOGY. *Educ:* Univ Paris, BLitt, 37, MSc, 39 & 40; Yale Univ, PhD(physiol chem), 48; Sorbonne, DSc(physiol), 50. *Hon Degrees:* Various from several US & foreign Col & Univ, 65-81. *Prof Exp:* Demonstr physiol chem, Yale Univ, 46-48; mem nutrit div, Food & Agr Orgn, UN, 48-49; res assoc pharmacol, George Washington Univ, 49; from asst prof to assoc prof, 50-65, prof nutrit, Harvard Univ, 65-76, lectr hist pub health, 68-76; PRES, TUFTS UNIV, 76- *Concurrent Pos:* Consult, Spec Div, UN, 48; tech secy, Int Comt Calorie Requirements, Food & Agr Off & WHO, 50 & 57, tech secy, Comt Protein Requirements, 57; assoc ed, Nutrit Revs, 51-54; consult, Children's Hosp, Boston, 57-, Ghana Govt, 58 & Ivory Coast Govt, 59; Severinghouse lectr, Med Sch, Univ Ga, 58; nutrit ed, Postgrad Med, 59-; Phi Beta Kappa scholar, 68-69; mem, Ctr Pop Studies, 68-, Consumer Adv Coun, US Dept Energy, 79-, Adv Comt, Oceans & Int Environ Sci Affairs, US State Dept, 79-; spec consult to the President, 69-70; chmn, White House Conf Food, Nutrit & Health, 69; mem, President's Consumer Adv Coun, 70-77; chmn nutrit div, White House Conf on Aging, 71; W O Atwater Mem lectr, Agr Res Serv, USDA, 71; mem, Protein Adv Group, UN, 73-75; gen coordr, US Sen Nat Nutrit Policy Study, 74; vchmn, President's Comn World Hunger, 78- *Honors & Awards:* Silver Medal, Int Physiol Cong, 56; Alvarenga Prize, Col Physicians Philadelphia, 68; Conrad A Elvehjem Award, Am Inst Nutrit, 78; Atwater Medal, 71; Bradford Washburn Award, Boston Mus Sci, 75; Sarah L Poiley Mem Award, NY Acad Sci, 75; Bolton L Corson Medal, Franklin Inst, 78; Lemuel Shattuck Award, Mass Pub Health Asn, 80- *Mem:* AAAS; Am Physiol Soc; Am Inst Nutrit; Am Fedn Clin Res; fel Am Acad Arts & Sci; Foreign Mem French Acad Sci. *Res:* Regulation of food and water intake; obesity; general nutrition. *Mailing Add:* Off Pres Tufts Univ Medford MA 02155

MAYER, JEROME F, b Milwaukee, Wis, Jan 20, 47; m 78; c 1. PETROLEUM REFINING, HYDROPROCESSING. *Educ:* Univ Wis, BS, 70; Mass Inst Technol, MS, 71, PhD(chem eng), 74. *Prof Exp:* SR ENG ASSOC, CHEVRON RES CO, CHEVRON CORP, 74- *Mem:* Am Inst Chem Engrs; Am Chem Soc. *Res:* Petroleum refining catalysts and processes, particularly in the area of hydroprocessing. *Mailing Add:* Chevron Res Co PO Box 1627 Richmond CA 94802

MAYER, JOERG WERNER PETER, b Munich, Ger, Aug 4, 29; nat US; m 55, 65; c 5. MATHEMATICS, COMPUTER SCIENCE. *Educ:* Univ Giessen, dipl, 53, Dr rer nat, 54. *Prof Exp:* Lectr math, Univ Malaya, 54-57; from asst prof to assoc prof, Univ NMex, 57-68; chmn dept, George Mason Col, 68-70; dir, Comput Ctr, 74-76, chmn, Dept Math, 70-82, PROF MATH, LEBANON VALLEY COL, 82- *Concurrent Pos:* PC consult, Lebanon Chem Corp, 83-86. *Res:* Philosophy of technology. *Mailing Add:* Dept Math Lebanon Valley Col Annville PA 17003

MAYER, JULIAN RICHARD, b New York, NY, Feb 12, 29; m 75; c 3. ENVIRONMENTAL MANAGEMENT, ENVIRONMENTAL CHEMISTRY. *Educ:* Union Univ, NY, BS, 50; Columbia Univ, MA, 51; Yale Univ, PhD(chem), 54. *Prof Exp:* Res assoc, Sterling-Winthrop Res Inst, Sterling Drug Co, 54-60, assoc mem, 60-61, group leader, 61-62; asst prog dir, NSF, 62-63; asst dir, Atmospheric Sci Res Ctr, State Univ NY, 63-64; staff assoc, NSF, 64-70; dir environ resources ctr, State Univ NY Col Fredonia, 70-78; dean, 78-85, PROF, ENVIRON STUDIES, HUXLEY COL, WESTERN WASH UNIV, BELLINGHAM, 85- *Concurrent Pos:* Spec consult, NSF, 63; consult, Environ Protection Agency, 72-73 & Union Carbide Corp, 75-78; vis scientist, Environ Protection Agency Region X Lab, Manchester, Wash, 85. *Mem:* AAAS; Am Chem Soc; Am Water Resources Asn. *Res:* Science administration; science policy planning; environmental problems; water quality research and management; chemical and physical limnology; water chemistry. *Mailing Add:* Huxley Col Environ Studies Western Wash Univ Bellingham WA 98225

MAYER, KLAUS, b May 21, 24; US citizen; m 50; c 2. INTERNAL MEDICINE, HEMATOLOGY. *Educ:* Queens Col, BS, 45; Univ Zurich & Groningen, MD, 50; Am Bd Internal Med, cert, 64. *Prof Exp:* Intern, Hosp St Raphael, New Haven, Conn, 50-51; staff mem, Dept Med, Brookhaven Nat Lab, 51-52; resident, Mem Hosp Cancer & Allied Dis, 52-55; res assoc cancer anemia, Sloan-Kettering Inst, 58-59, asst, 59-60; from instr to clin assoc prof, 58-80, PROF CLIN MED, MED COL, CORNELL UNIV, 80-; ASSOC, SLOAN-KETTERING INST, 60- *Concurrent Pos:* Spec fel med, Mem Hosp, 55-56; Damon Runyon fel, Sloan-Kettering Inst, 55-58; clin asst med, Mem Hosp Cancer & Allied Dis, 56-60, from asst attend physician to assoc attend physician, 60-72, attend physician, 72-, dir, Blood Bank & Serol Lab, 66-, dir, Hemat Lab, 71-, assoc chmn, Dept Med Clin Labs, 81; attend hematologist, Hosp Spec Surg, 57-, res hematologist, 58-62, dir blood bank, 58-, assoc scientist, 62-63, sr scientist, 63-; physician to outpatients, New York Hosp, 58-68, assoc attend physician, 68-; from asst vis physician to assoc vis physician, James Ewing Hosp, 59-68; asst vis physician, Bellevue Hosp, 62-68; res collabr, Brookhaven Nat Lab, 65-66; pres, Am Asn Blood Bank, 73-74; prin, ad hoc comt to form Am Blood Comn, 74-75, secy-treas, 75- *Mem:* Am Soc Nuclear Med; Am Soc Hematol; Harvey Soc; fel Am Col Physicians; Int Soc Hemat. *Res:* Application of radioisotopic technique to hematology and transfusion therapy; quantitation of reticuloendothelial function. *Mailing Add:* 1275 York Av New York NY 10021

MAYER, LAWRENCE MICHAEL, b Laredo, Tex, July 26, 49; m 73. GEOCHEMISTRY, CHEMICAL OCEANOGRAPHY. *Educ:* Case Western Reserve Univ, BS, 71; Dartmouth Col, AM, 74, PhD(geol), 76. *Prof Exp:* At Grad Sch Oceanog, Univ RI, Narragansett; AT IRA C DARLING CTR, MAINE. *Mem:* Clay Minerals Soc; Am Soc Limnol & Oceanog; Am Geophys Union. *Res:* Biogeochemical cycling of nutrients, organic compounds, and trace metals; mineral-water interactions. *Mailing Add:* Ira C Darling Ctr Walpole ME 04573

MAYER, MARION SIDNEY, b New Orleans, La, July 25, 35. ENTOMOLOGY, BIOCHEMISTRY. *Educ:* La State Univ, BS, 57; Tex A&M Univ, MS, 61, PhD(entom), 63. *Prof Exp:* RES ENTOMOLOGIST, AGR RES SERV, USDA, 63- *Mem:* AAAS. *Res:* Insect attractants, isolation and behavioral characteristics leading to host or mate location; electrophysiological studies to demonstrate details of nervous activity leading to host or mate locations and thresholds. *Mailing Add:* USDA Agr Res Serv PO Box 14565 Gainesville FL 32604

MAYER, MEINHARD EDWIN, b Seletin, USSR, Mar 18, 29; m 54; c 2. MATHEMATICAL PHYSICS. *Educ:* Bucharest Polytech Inst, Dipl Ing, 51; Parhon Univ, PhD, 57. *Prof Exp:* From instr to assoc prof math physics, Parhon Univ, 49-61; sr res worker theoret physics, Joint Inst Nuclear Res, Dubna, USSR, 57-58; vis prof theoret physics, Univ Vienna, 61-62 & Imp Col, London, 62; vis physicist, Europ Orgn Nuclear Res, Switz, 62; vis assoc prof physics, Brandeis Univ, 62-64; assoc prof theoret physics, Ind Univ, 64-66; PROF MATH & PHYSICS, UNIV CALIF, IRVINE, 66- *Concurrent Pos:* Asst prof, Bucharest Polytech Inst, 50-52; sr res worker, Inst Atomic Physics, Acad Rumania, 51-58; vis assoc physicist, Brookhaven Nat Lab, NY, 63-; vis prof, Inst Advan Sci Studies, Bures-sur-Yvette, France, 70-71 & 78, Tel-Aviv Univ, 71, Swiss Fed Inst Technol, 77-78, Hebrew Univ, 84, Univ Rome, Univ Cologne, Col France, Paris & Univ Hamburg, 85. *Mem:* Am Math Soc; fel Am Phys Soc; Int Asn Math Physics. *Res:* Quantum field theory and statistical mechanics; differential-geometric approach to gauge theory; relativistic statistical mechanics, wavelets, scheme programming language. *Mailing Add:* Dept Physics Univ Calif Irvine CA 92717

MAYER, RAMONA ANN, b Algona, Iowa, May 9, 29. CHEMISTRY. *Educ:* State Univ,Iowa, Iowa City, BA, 56. *Prof Exp:* Info specialist, 56-59, res scientist, 59-77, QUAL ASSURANCE UNIT, MGR BIOL CHEM SCI CTR, BATTELLE-COLUMBUS LABS, 77- *Concurrent Pos:* Abstractor, Chem Abstr Serv, 58-79; vchmn subcomt E47.05 qual assurance, Am Soc Testing & Mat, 81-84, chmn, 84- *Mem:* Am Chem Soc; Am Inst Chemists; Am Soc Testing & Mat; Am Soc Qual Contol; Soc Qual Assurance. *Res:* Director of quality assurance on programs dealing with toxicology, biochemistry, immunology, teratology, chemotherapy, animal behavior, ecology; environmental sciences. *Mailing Add:* Battelle-Columbus Labs 505 King Ave Columbus OH 43201-2631

MAYER, RICHARD F, b Olean, NY, June 2, 29; m 59; c 5. NEUROLOGY. *Educ:* St Bonaventure Col, BS, 50; Univ Buffalo, MD, 54. *Prof Exp:* Intern med, Boston City Hosp, 54-55; resident neurol, Mass Gen Hosp, 56-57, resident neuropath, 58; res asst, Inst Neurol, Univ London, 57-58; instr neurol, Harvard Med Sch, 61-65, assoc, 65-66; assoc prof, 66-68, PROF NEUROL, SCH MED, UNIV MD, BALTIMORE, 68-, DIR NEUROMUSCULAR CLIN & EMG LAB, UNIV HOSP, 69- *Concurrent Pos:* Fel neurol, Mayo Found, Univ Minn, 55-56; NIH res fel, Harvard Med Sch, 60-61; NIH res grant, Boston City Hosp, 60-64; Nat Multiple Sclerosis Soc res grant, 66-67. *Mem:* AAAS; Am Neurol Asn; Am Electroencephalog Soc; Am Acad Neurol; Soc Neurosci. *Res:* Clinical neurophysiology; clinical and experimental animal studies of motor dysfunction; nerve and reflex activity in man; myasthenia gravis-neuromuscular transmission and ultra structure. *Mailing Add:* Dept of Neurol Univ of Md Med Sch Baltimore MD 21201

MAYER, RICHARD THOMAS, b Pensacola, Fla, May 11, 45; m 66, 86; c 2. PHARMACOLOGY. *Educ:* Univ Ga, BS, 67, PhD(entom), 70. *Prof Exp:* Fel entom, Univ Ga, 70-71; res entomologist, 71-77, res leader physiol & biochem, Livestock Insects Res Unit, Vet Toxicol Entom Res Lab, Sci & Educ Admin, Agr Res, 77-84, LAB DIR, US HORT RES LAB, AGR RES SERV, USDA, FLA, 84- *Concurrent Pos:* exec ed, Archives Insect Biochem & Physiol, 83-; Kellog fel, 85. *Honors & Awards:* Alexander von Humboldt Award, 81, 84, 87, 90; Am Registry Prof Entomologist Outstanding Award, 81; Outstanding Scientist Award, Agric Res Serv, 83. *Mem:* Sigma Xi; Am Registry Prof Entomologists; Am Chem Soc; Am Entom Soc. *Res:* Insecticide and hormone metabolism by insects; isolation and characterization of insect metabolic systems; enzyme assay development. *Mailing Add:* USDA Agr Res Serv US Hort Res Lab 2120 Camden Rd Orlando FL 32803

MAYER, STANLEY WALLACE, b New York, NY, Mar 29, 16; m 45; c 2. PHYSICAL CHEMISTRY, BATTERY ELECTROCHEMISTRY. *Educ:* City Col New York, BS, 38; Univ Calif, Los Angeles, PhD(chem), 53. *Prof Exp:* Res scientist, NY Water Dept, 39-41 & US War Dept, 41-43; sr sci staff, Columbia Univ, 43-46, Oak Ridge Nat Lab, 46-48, US Naval Radio Lab, 48-53, US Radioisotope Res Univ, 53-56 & NAm Aviation Inc, 56-61; SR SCIENTIST, PHYS CHEM DEPT, CHEM & PHYSICS LAB, AEROSPACE CORP, 61- *Mem:* Am Chem Soc; Am Nuclear Soc; Am Inst Aeronaut & Astronaut; Am Phys Soc; Combustion Inst. *Res:* High temperature reactions and propulsion; nuclear power; properties of propellants; electrochemistry; biomedical physics; research with electronic computers; lasers. *Mailing Add:* 2235 Malcolm Ave Los Angeles CA 90064

MAYER, STEVEN EDWARD, b Frankfurt am Main, Ger, Feb 11, 29; nat US; m 51, 87; c 2. PHARMACOLOGY, BIOCHEMISTRY. *Educ:* Univ Chicago, BA, 47, BS, 49; Univ Ill, MS, 52, PhD(pharmacol), 54. *Prof Exp:* Sr asst scientist, Lab Chem Pharmacol, Nat Heart Inst, 54-56; from asst prof to prof pharmacol, Emory Univ, 57-69; chief pharmacol div, Univ Calif, San Diego, 69-79, prof pharmacol, 69-87; RES PROF PHARMACOL, VANDERBILT UNIV, 88- *Concurrent Pos:* Fel pharmacol, Wash Univ, 56-57; asst ed, J Am Soc Pharmacol & Exp Therapeut, 61-64; vis scholar, Univ Wash, 65; mem pharmacol study sect, NIH, 65-69, mem neurosci res training B comt; ed, Molecular Pharmacol, 71-74; chmn, Gordon Res Conf Heart Muscle, 72; mem res comt, Am Heart Asn, 73-78, vchmn, 77-78; mem, US-USSR Working Group Myocardial Metab, 73-78; A J Carlson Lectr, Univ Chicago, 75; vis prof, Dept Biochem, Univ Wash, 76; fel, Coun Circulation, Am Heart Asn, 79. *Honors & Awards:* John J Abel Award, Am Soc Pharmacol & Exp Therapeut, 63. *Mem:* AAAS; Am Soc Pharmacol & Exp Therapeut (pres, 77-78); Am Physiol Soc; Am Soc Biol Chemists; Fedn Am Scientists. *Res:* Drug and hormone action on metabolic control mechanisms. *Mailing Add:* Dept Pharmacol Vanderbilt Univ Nashville TN 37232

MAYER, THEODORE JACK, b Bridgewater, SDak, Feb 13, 33; m 59; c 4. PETROLEUM CHEMISTRY. *Educ:* Univ SDak, BA, 55; Pa State Univ, MS, 61; Carnegie Inst Technol, PhD(phys chem), 63; Widener Col, MBA, 77. *Prof Exp:* Res asst petrol chem, Petrol Ref Lab, Pa State Univ, 55-59; res chemist, Res & Develop Div, Sun Oil Co, 63-65, group leader anal, 65-75, supvr, Indus Hyg Lab, 75-78, supvr gasoline prod develop, Suntech, Inc, Sun Oil Co, 78-80, asst prod mgr white oils, 80-81, mgr spec oil technol, 81-82, tech serv engr indust lubricants & engine oils, 82-90, TECH SERV MGR, FUELS & LUBRICANTS, SUNTECH INC, SUN CO INC, 90- *Mem:* Am Chem Soc; Am Soc Lubrication Engrs; Soc Automotive Engrs. *Res:* Composition of petroleum; analysis of petroleum fractions by instrumental methods; laboratory automation; industrial hygiene analytical methods; gasoline additives; gasoline product quality; technology of white oils; technology of refrigeration, electrical and insulating oils; technology of industrial and automotive lubricants; lube surveys; lube training. *Mailing Add:* PO Box 1135 Marcus Hook PA 19061-0835

MAYER, THOMAS C, b Pittsburgh, Pa, Nov 30, 31; m 58. DEVELOPMENTAL BIOLOGY. *Educ:* Univ Tenn, AB, 53; Johns Hopkins Univ, MA, 57; La State Univ, PhD(embryol), 62. *Prof Exp:* Asst prof biol, Greensboro Col, 57-60; from asst prof to assoc prof, 62-67, PROF BIOL, RIDER COL, 67- *Concurrent Pos:* NSF res grants, 63-72. *Mem:* AAAS; Soc Develop Biol; Am Soc Zool. *Res:* Embryogenesis of spotting patterns in mice. *Mailing Add:* Dept Biol Rider Col 2083 Lawrenceville Rd Lawrenceville NJ 08648

MAYER, TOM G, b New York, NY, May 31, 43; m 68; c 2. SPINAL REHABILITATION. *Educ:* Harvard Col, BA, 64; Columbia Col Physicians & Surgeons NY, MD; Am Bd Orthop Surgeons, dipl, 78. *Prof Exp:* Intern surg, Harbor Gen Hosp, 69; resident, San Francisco Orthop Prog, Univ Calif, 69-70 & 72-75; Lt Comdr, US Naval Res, 70-72; vol physician, Bali, Indonesia & Chieng Mai & Mae Sariang, Thailand, 75; orthop surgeon, Los Alamos Med Ctr, 77-81; chief, Spinal Pain Prog, Dallas Rehab Inst, 81-83; res coordr, Div Orthop Surg, 81-85, from asst prof to assoc prof clin orthop surg, 81-88, CLIN PROF ORTHOP SURG, UNIV TEX SOUTHWESTERN MED CTR, 88-; MED DIR, PROD REHAB INST DALLAS ERGONOMICS, 83- *Concurrent Pos:* Clin preceptor, Orthop Rehab, Div Orthop Surg, Univ Tex Southwestern Med Ctr, Dallas; consult, disability & impairment eval comt, Calif Indust Accident Comn, 85-87; dist med adv, US Dept Labor, Dallas Region, 86-87; med dir, Founding Neuro-Rehab Unit, Med City Dallas, 86-88; panel chmn chronic low back pain, Workshop Low Back Pain, Nat Inst Disability & Rehab Res, 87; discussant & consensus comt mem, Int Conf Exercises, Fitness & Health, Can Asn Sports Sci, 88; bd mem & med adv, Back Systs Inc, 84-; med consult, Injury Task Force, Union Pac-Mo Pac Railroad, 86-; mem diag & therap comt, NAm Spine Soc, 86-; consult author, Spine Sect, Am Med Asn & spine disability subcomt, Disability Task Force, Am Acad Orthop Surgeons, 87- *Honors & Awards:* Volvo Award, Int Soc Study Lumbar Spine & Volvo Co Goteborg, 85. *Mem:* Fel Am Col Surgeons; fel Am Acad Orthop Surgeons; mem Int Soc Study Lumbar Spine; mem NAm Spine Soc. *Res:* Over 60 publications on rehabilitation of spinal disorders, measurement of lumbar physical capacity and spinal impairment and disability evaluation issues. *Mailing Add:* PRIDE 1450 Empire Central Suite 400 Dallas TX 75247

MAYER, VERNON WILLIAM, JR, b Newark, NJ, Mar 29, 39; c 2. MICROBIOLOGY, GENETICS. *Educ:* Univ Md, BS, 63, MS, 65, PhD(microbiol), 67. *Prof Exp:* Nat Res Coun-Nat Acad Sci res assoc, 67-68, RES MICROBIOLOGIST, FOOD & DRUG ADMIN, 68- *Mem:* AAAS; Am Soc Microbiol; Environ Mutagen Soc; Genetics Soc Am. *Res:* Yeast genetics; chemical mutagenesis. *Mailing Add:* 4311 Powder Mill Rd Beltsville MD 20705

MAYER, VICTOR JAMES, b Mayville, Wia, Mar 25, 33; m 65; c 2. EARTH SCIENCES, SCIENCE EDUCATION. *Educ:* Univ Wis, BS, 56; Univ Colo, MS, 60, PhD(sci educ), 66. *Prof Exp:* Pub sch teacher, Colo, 60-62; asst prof earth sci, State Univ NY Col Oneonta, 65-67; from asst prof to assoc prof, 67-75, PROF GEOL & MINERAL, NATURAL RESOURCES & SCI EDUC, OHIO STATE UNIV, 75- *Concurrent Pos:* Consult, NY State Dept Educ, 66-67, Pedag Inst Caracas, 71, 73 & 82, UNESCO, 75; dir, Nat Sci Teachers Asn, 84-86, Korean Sci Teachers Inst, 86 & 87, Prog Leadership Earth Systs Educ, 90-; chair educ, AAAS, 87-90; pres Sci Educ Coun Ohio, 87-88. *Mem:* fel AAAS; Nat Sci Teachers Asn; Am Educ Res Asn; Nat Asn Res Sci Teaching; Nat Asn Geol Teachers; Nat Marine Educ Asn. *Res:* Research designs; curriculum development and evaluation teacher education. *Mailing Add:* 111 W Dominion Ohio State Univ Columbus OH 43214

MAYER, WALTER GEORG, b Silberbach, Czech, Mar 13, 27; nat US; m 59. PHYSICS. *Educ:* Hope Col, AB, 53; Mich State Univ, MS, 55, PhD(physics), 58. *Prof Exp:* Physicist high temperature res, Siemens Res Lab, Ger, 58-59; res asst prof ultrasonics, Dept Physics & Astron, Mich State Univ, 59-65; from asst prof to assoc prof, 65-72, PROF PHYSICS, GEORGETOWN UNIV, 72- *Concurrent Pos:* Assoc ed, Inst Elec & Electronics Engrs Trans Sonics & Ultrasonics, 72-83, J Acoust Soc Am, 74-, ed, Ultrasonics, 86-; dir, Sch Phys Acoust, Erice, Sicily, 82-; dir, Ultras Res Lab, Georgetown Univ, 87- *Honors & Awards:* Humboldt Prize, Germany, 80; Medaille d'Argent, French Acoust Soc, 88. *Mem:* Fel Brit Inst Acoust; fel Acoust Soc India; fel Acoust Soc Am. *Res:* Ultrasonics, particularly measurements of wave characteristics by optical methods; application of ultrasonics to solid and liquid state; surface and interfacial waves; nonlinear acoustics. *Mailing Add:* Dept Physics Georgetown Univ Washington DC 20057

MAYER, WILLIAM DIXON, b Beaver Falls, Pa, Oct 5, 28; m 85; c 4. MEDICINE. *Educ:* Colagte Univ, AB, 51; Univ Rochester, MD, 57. *Hon Degrees:* DSc, Univ Osteop Med & Health Sci, 88. *Prof Exp:* Intern path, Sch Med, Univ Rochester, 57-58, resident, 58-59, instr, 58-61; from asst prof to prof path, Univ Mo, Columbia, 61-76, from asst dean to assoc dean, Sch Med, 61-67, dean, 67-74, dir, Med Ctr, 67-74, dir, Health Serv Res Ctr, 75-76; asst chief med dir acad affairs, Vet Admin Cent Off, 76-79; pres, Eastern Va Med

Authority, 79-87; CONSULT, 87- *Concurrent Pos:* Boswell fel, Univ Rochester, 59-61; Markle scholar, 62-67; assoc dir div regional med progs, NIH, 66-67. *Mem:* AAAS; Am Soc Exp Path; Col Am Path; Asn Am Med Cols; AMA; hon mem Nat Bd Med Examiners. *Mailing Add:* 401 Col Place #20 on the Pier Norfolk VA 23510

MAYER, WILLIAM JOHN, b Detroit, Mich, Mar 29, 21; m 51; c 6. PHYSICAL CHEMISTRY. *Educ:* Wayne State Univ, BS, 44, PhD(chem), 50. *Prof Exp:* Anal control chemist, Gelatin Prod Corp, 44-45; lab asst phys chem, Wayne State Univ, 48-50; res chemist, Argonne Nat Lab, 50-56; STAFF RES SCIENTIST, RES LABS, GEN MOTORS CORP, 56- *Honors & Awards:* Arch T Colwell Award, Soc Automotive Engrs, 67. *Mem:* Am Chem Soc; Soc Automotive Engrs; Combustion Inst. *Res:* Radiochemistry and isotopes; chemistry of surfaces; combustion chemistry; radiometric methods applied to automotive engines. *Mailing Add:* 21518 Tanglewood St Clair Shores MI 48082

MAYER, WILLIAM JOSEPH, b Springfied, Ohio, Sept 30, 39; m 71. INFORMATION SCIENCE. *Educ:* Xavier Univ, Ohio, BS, 61; Univ Mich, MS, 63, PhD(pharmaceut chem), 65. *Prof Exp:* Patent chemist, Res Labs, Parke Davis & Co, 65-71; res info assoc, Olin Corp, 71-74; mgr, Tech Info Serv, James Ford Bell Tech Ctr, Gen Mills, Inc, 74-81; pres, B-K Assocs, 81-86; DIR, TECH INFO SERV, HENKEL RES CORP, 86- *Concurrent Pos:* Consult, Inst Food Technologists Info Connection Short Course, 81; adj prof, dept mgt sci, Sch Mgt, Univ Minn, 85-86. *Mem:* Inst Food Technologists; Am Chem Soc. *Res:* Synthesis of organic medicinals; patent development; computerized information search services; technology assessment; technical writing and editing; food technology futures; records and file management systems; identification of emerging scientific specialties and technologies. *Mailing Add:* 2535 Sea Biscuit Ct Santa Rosa CA 95401

MAYER, WILLIAM VERNON, zoology; deceased, see previous edition for last biography

MAYERI, EARL MELCHIOR, b Berkeley, Calif, Dec 10, 40; m 68; c 1. NEUROBIOLOGY. *Educ:* Univ Calif, Berkeley, BA, 63, PhD(biophys), 69. *Prof Exp:* ASST PROF PHYSIOL, UNIV CALIF, SAN FRANCISCO, 71- *Concurrent Pos:* USPHS fel neurophysiol, Med Sch, NY Univ, 59-71; fel, Pub Health Res Inst, New York, 69-71. *Mem:* AAAS; Am Physiol Soc; Soc Neurosci. *Res:* Invertebrate neurobiology and behavior. *Mailing Add:* Dept Physiol Univ Calif Box 0444 San Francisco CA 94143

MAYERLE, JAMES JOSEPH, b Grand Rapids, Minn, Dec 16, 45; m 75; c 2. SOLID STATE CHEMISTRY, INORGANIC CHEMISTRY. *Educ:* St John's Univ, Minn, BA, 67; Columbia Univ, PhD(inorg chem), 72. *Prof Exp:* Res assoc inorg chem, Mass Inst Technol, 72-74; MEM RES STAFF CHEM, IBM RES LAB, 74- *Mem:* AAAS; Am Chem Soc; Am Crystallog Asn. *Res:* Organic and inorganic solid state chemistry. *Mailing Add:* Dept 2H9/103-1 IBM Corp Rochester MN 55901-2884

MAYERNIK, JOHN JOSEPH, b Manville, NJ, July 6, 16; m 46; c 1. MICROBIOLOGY. *Educ:* Rutgers Univ, BSc, 39, PhD(soil chem), 44; Univ Vt, MS, 41. *Prof Exp:* Chief microbiol & sterile prod control lab, Merck & Co, Inc, , 46-75, mgr microbiol serv, 75-83; RETIRED. *Concurrent Pos:* Mem, US Pharmacopoeia Adv Panel on Biol Indicators. *Mem:* Am Soc Microbiol; Am Chem Soc; fel Asn Off Analytical Chemists. *Res:* Microbiological assays of antibiotics, vitamins and amino acids; quality control of pharmaceutical products; evaluation of methods of sterilization and product sterility; evaluation of preservatives and disinfectants; bacterial monitoring of electron irradiation. *Mailing Add:* 42 Jefferson Ave New Brunswick NJ 08901

MAYERS, GEORGE LOUIS, b New York, NY, Feb 22, 38; m 66; c 2. BIO-ORGANIC CHEMISTRY, IMMUNOCHEMISTRY. *Educ:* City Col New York, BS, 60, MA, 64; City Univ New York, PhD(org chem), 67. *Prof Exp:* Fel peptide chem, St John's Univ, NY, 67-70; CANCER RES SCIENTIST V, ROSWELL PARK MEM INST, 70-, ACTG HEAD, DEPT MOLECULAR IMMUNOL, 87- *Concurrent Pos:* Res prof, Roswell Park Div, State Univ NY Buffalo, 75-, assoc res prof, dept chem & microbiol, 81-, assoc res prof & chmn dept microbiol-immunol, 82-; res prof, Niagara Univ, 77- *Mem:* Am Chem Soc; Royal Soc Chem; Sigma Xi; Am Asn Immunologists; Am Soc Biol Chemists. *Res:* Structure of the antibody site; auto-idiotype control of the immune response; characterization and function of cell surface receptors; homogeneous antibody responses; tumor immunology; hybridomas and monoclonal antibodies. *Mailing Add:* 492 Walnut St Lockport NY 14094

MAYERS, JEAN, b New York, NY, June 8, 20; m 45; c 2. NONLINEAR STRUCTURAL MECHANICS. *Educ:* Polytech Inst Brooklyn, BAeroEng, 42, MAeroEng, 48. *Prof Exp:* Aeronaut res scientist, Nat Adv Comt Aeronaut, 48-56; prin engr, Sperry Utah Co, 56-57, eng sect head, 57-59, eng dept head, 59-61; vis assoc prof aerospace struct, 61-63, assoc prof, 63-67, vchmn dept aeronaut & astronaut, 66-71, PROF AEROSPACE STRUCT, STANFORD UNIV, 67- *Concurrent Pos:* Sci adv, US Army Res Off, 62-74; assoc ed, Am Inst Aeronaut & Astronaut J, 67-70; vis prof, Israel Inst Technol, 70; Naval Air Systs Command res chair, US Naval Acad, 78-79. *Mem:* Assoc fel Am Inst Aeronaut & Astronaut. *Res:* Theoretical aerospace structures; aerospace systems synthesis and analysis; optimum structural design; design criteria for high-strength, high-stiffness, light-weight aerospace structures; conventional, sandwich and composite structures. *Mailing Add:* 2550 N Park No 909 Stanford Univ Chevy Chase MD 20815

MAYERS, RICHARD RALPH, b West Brownsville, Pa, July 6, 25; m 49; c 2. THERMAL SYSTEMS, INDUSTRIAL FURNACES. *Educ:* Dartmouth Col, AB, 47; Wesleyan Univ, MA, 57; Texas A&M Univ, MS, 68. *Prof Exp:* Instr physics, Hood Col, 47-49; physicist, Nat Bur Standards, 49-50 & Glenn L Martin Co, 55-56; asst prof physics, Colby Col, 56-61, actg chmn dept, 57-59; from assoc prof to prof & chmn dept, Defiance Col, 61-74; res, Surface Combustion Co, 74-84; proj engr, Handy & Harman Corp, 86-89; ENG CONSULT, 84-86 & 89- *Concurrent Pos:* Vis prof, Univ Surrey, 67-68; vis lectr, Tex A&M Univ, 68. *Mem:* AAAS; Am Phys Soc. *Res:* Energy savings, increased productivity and improved product quality for installations using industrial furnaces and related equipment; protective coatings for automotive tubing. *Mailing Add:* 415 Monroe St Delta OH 43515

MAYERSON, HYMEN SAMUEL, physiology; deceased, see previous edition for last biography

MAYES, BILLY WOODS, II, b Port Arthur, Tex, Feb 6, 41; m 60; c 1. PHYSICS. *Educ:* Univ Houston, BS, 63, MS, 65; Mass Inst Technol, PhD(physics), 69. *Prof Exp:* Fel, 68-69, asst prof, 69-73, ASSOC PROF PHYSICS, UNIV HOUSTON, 73- *Mem:* Am Phys Soc. *Res:* Experimental pion nucleus cross sections. *Mailing Add:* Phys Dept Univ Houston Houston TX 77004

MAYES, MCKINLEY, b Oxford, NC, Oct 7, 30; m 59; c 1. AGRONOMY. *Educ:* NC Agr & Tech Col, BS, 53, MS, 56; Rutgers Univ, PhD(agron), 59. *Prof Exp:* Prof agron, Southern Univ, Baton Rouge, 59-67, dean & coordr CRS res progs, 74-76; COORDR SPEC PROGS, SCI & EDUC ADMIN-COOP RES, USDA, 76- *Mem:* Am Soc Agron; Soil Conserv Soc Am; Sigma Xi. *Res:* Plant breeding, especially field corn and sweet corn improvement. *Mailing Add:* J S Morrill Bldg CSRS-USDA Washington DC 20250

MAYES, PAUL E(UGENE), b Frederick, Okla, Dec 21, 28; m 50; c 6. ELECTRICAL ENGINEERING. *Educ:* Univ Okla, BS, 50; Northwestern Univ, MS, 52, PhD(elec eng), 55. *Prof Exp:* Res assoc, Northwestern Univ, 50-54; res asst prof elec eng, 54-58, assoc prof, 58-63, PROF ELEC ENG, UNIV ILL, 63- *Concurrent Pos:* Consult, Gen Motors, Mich, TRW Inc, Calif; mem, Comn B, Int Union Radio Sci. *Mem:* Fel Inst Elec & Electronics Engrs. *Res:* Electromagnetic theory; antennas, particularly extremely broadband and low-profile types. *Mailing Add:* Col Eng Univ Ill 1406 W Green St Urbana IL 61801

MAYES, TERRILL W, b Evansville, Ind, Sept 4, 41; m 59; c 3. POLARIMETRY & ELLIPSOMETRY, FIBER OPTICS. *Educ:* Western Ky Univ, BS, 63; Vanderbilt Univ, MA, 65, PhD(physics), 67. *Prof Exp:* Asst prof, 67-74, ASSOC PROF PHYSICS, UNIV NC, CHARLOTTE, 74- *Mem:* Am Phys Soc; Am Asn Physics Teachers; Am Optical Soc. *Res:* Optics; fiber optics. *Mailing Add:* Dept of Physics Univ NC Charlotte Charlotte NC 28223

MAYEUX, JERRY VINCENT, b Mamou, La, Apr 22, 37; m 81; c 2. MICROBIOLOGY, PLANT GROWTH REGULATORS. *Educ:* La State Univ, Baton Rouge, BS, 60, MS, 61; Ore State Univ, PhD(microbiol), 64. *Prof Exp:* Nat Acad Sci-Nat Res Coun res assoc exobiol, NASA Ames Res Ctr, 65-66; asst prof microbiol, Colo State Univ, 66-70; sr res scientist, Manned Exp & Life Sci Dept, Martin Marietta Corp, 70-72, chief, Life Sci, 72-74; dir, Res & Develop, Ferma Gro Corp, 74-75; lectr microbiol, Buena Vista Col, Storm Lake, Iowa, 75-76; founder, pres & chmn bd, Dawn Corp, 76-80; founder & pres, Burst Agr Tech Inc, 80-84, chmn bd & dir res & develop, 84-87; PRES, PLANT BIOREGULATOR TECHNOLS, INC, 87- *Concurrent Pos:* Asst prof range sci, Colo State Univ, 69-70, affil prof microbiol, 70-76, Col Eng, 71; consult, NASA Life Sci Shuttle Planning Panel, 74-; consult, Martin Marietta Aerospace, 74-75. *Mem:* Soc Indust Microbiol; AAAS; Am Soc Microbiol; Am Chem Soc; Sigma Xi; Plant Growth Regulator Soc. *Res:* Microbial ecology; soil and water pollution; microbial/plant interaction; plant growth regulators; aerospace biology; waste reutilization; biological control of plant disease. *Mailing Add:* 4201 W 99th St Overland Park KS 66207

MAYEWSKI, PAUL ANDREW, b Edinburgh, Scotland, July 5, 46; US citizen; m 85. ICE & SNOW CHEMISTRY & GEOCHEMISTRY, GLACIOLOGY. *Educ:* State Univ NY, Buffalo, BA, 68; Ohio State Univ, PhD(geol), 73. *Prof Exp:* Res assoc geol, Inst Polar Studies, Ohio State Univ, 68-73; fel, Inst Quaternary Studies, Univ Maine, Orono, 73-75; asst prof, 75-85, PROF GEOL, UNIV NH, 85-, DIR GLACIER RES GROUP, 80- *Concurrent Pos:* Mem res team, Climate-Long Range Invest Mapping & Prediction, 73-; panel mem, W Antarctic Ice Sheet Proj, 75-; Snow & Ice Comt, Am Geophys Union; coordr, univ res devel, Univ NH indust sci task force; field leader & prin investr, Antarctica, Greenland, Himalayas, Iceland & Canadian Rockies. *Mem:* Glaciol Soc; Am Geophys Union. *Res:* Glaciology and glaciochemistry (ice coring, surface and snowpits) to retrieve records of climatic change, anthropogenic effects, specific air mass tracking (i e monsoons) in a global network including Antarctica, Greenland, Himalayas, Iceland, Africa, South America, North America, and Sub-Antarctic Islands. *Mailing Add:* Glacier Res Group Inst Study Earth Oceans & Space Univ NH Durham NH 03824-3525

MAYFIELD, DARWIN LYELL, b Somerset, Ky, Feb 22, 20; m 45; c 2. ORGANIC CHEMISTRY. *Educ:* Bowling Green State Univ, AB & BS, 41; Univ Chicago, MS, 44; Univ Wis, PhD(org chem), 50. *Prof Exp:* Res chemist, Nat Defense Res Comt, 42-43, Off Sci Res & Develop, 43-45 & Rubber Res Bd, 45-47; asst, Univ Wis, 47-50; from asst prof to assoc prof chem, Univ Idaho, 50-56; from asst prof to prof chem, Calif State Univ, 56-90, chmn dept, 64-66, dir res, 67-81, EMER PROF CHEM, CALIF STATE UNIV, LONG BEACH, 90- *Concurrent Pos:* Fulbright lectr, Kasetsart Univ, Bangkok, 55-56 & Ain Shams Univ, Cairo, 66-67; NIH res fel, Nat Sci Res Ctr, Gif-sur-Yvette, France, 62-63. *Mem:* AAAS; Am Chem Soc. *Res:* Chemistry of plant hormones responsible for floral initiation; research administration and federal relations; history of chemistry. *Mailing Add:* Dept Chem Calif State Univ Long Beach CA 90840

MAYFIELD, EARLE BYRON, b Oklahoma City, Okla, Jan 31, 23; m 52; c 7. SPACE PHYSICS. *Educ:* Univ Calif, Los Angeles, BA, 50; Univ Utah, MA, 54, PhD(physics), 59. *Prof Exp:* Physicist, Res Dept, US Naval Ord Test Sta, Calif, 50-59; mem tech staff, Phys Res Lab, Space Tech Labs, Inc, Thompson-Ramo-Wooldridge, Inc, 59-60; MEM TECH STAFF, AEROSPACE CORP, 60- *Concurrent Pos:* Asst, Univ Utah, 54-55; adj prof physics & astron, Calif State Univ, Northridge, 77- *Mem:* Am Phys Soc; Am Astron Soc; Int Astron Union. *Res:* Plasma and solar physics. *Mailing Add:* 1427 Bayview Heights Dr Los Osos CA 93402-4409

MAYFIELD, HAROLD FORD, b Minneapolis, Minn, Mar 25, 11; m 36; c 4. ORNITHOLOGY. *Educ:* Shurtleff Col, BS, 33; Univ Ill, MA, 34. *Hon Degrees:* DSc, Occidental Col, 68 & Bowling Green State Univ, 75. *Prof Exp:* Secy, Wilson Ornith Soc, 48-52, vpres, 53-54 & 58-59, pres, 60-61; secy, Am Ornith Union, 53-58, vpres, 64-66, pres, 66-68; vpres, Cooper Ornith Soc, 73, pres, 74-76; ADJ PROF BIOL, UNIV TOLEDO, 82- *Honors & Awards:* Brewster Mem Award for work on birds of Western Hemisphere, Am Ornithologists Union, 61. *Mem:* Fel AAAS; Cooper Ornith Soc; Wilson Ornith Soc; fel Am Ornithologists Union. *Res:* Bird reproduction and mortality, social parasitism and ecology. *Mailing Add:* 1162 Nannette Dr Toledo OH 43614

MAYFIELD, JOHN EMORY, b Thomasville, NC, Aug 30, 37; m 60; c 2. MYCOLOGY. *Educ:* Livingstone Col, BS, 59; State Univ NY Buffalo, MA, 71, PhD(biol), 72. *Prof Exp:* Teacher sci, Southside Sch, 59-63; teacher driver educ, Rowan Co Schs, 63-64; teacher sci, J F Kennedy Jr High Sch, 64-67; res botanist mycol, Agr Res Serv, USDA, 72-73; asst prof biol, Ala State Univ, 73-76; assoc prof biol, Atlanta Univ, 76-85; AT NC CENT UNIV. *Mem:* Mycol Soc Am; Electron Micros Soc Am; Sigma Xi; Pan-Am Biodeterioration Soc. *Res:* Fungal ultrastructure; microbial interactions; airborne fungal spores and allergies; polyphenol oxidase in wood decay fungi; fungal populations in forest ecosystems. *Mailing Add:* Dept of Biol NC Cent Univ Durham NC 27707

MAYFIELD, JOHN ERIC, b Toledo, Ohio, Dec 21, 41; m 68; c 2. DNA SEQUENCE ORGANIZATION, SUB UNIT VACCINES. *Educ:* Col Wooster, Ohio, BA, 63; Univ Pittsburgh, MS, 65, PhD(biophys), 68. *Prof Exp:* Teaching fel biol, Calif Inst Technol, 68-71, instr, 71; asst prof, Carnegie-Mellon Univ, 71-77; asst prof, Iowa State Univ, 77-79, coordr, molecular, cellular & develop biol prog, 83-86, assoc prof, 79-89, PROF ZOOL & GENETICS, IOWA STATE UNIV, 89- *Concurrent Pos:* Vis assoc, Harvard Univ, 77; vis lectr, Trinity Col & Univ Dublin, Ireland, 81. *Mem:* AAAS; Am Soc Cell Biol. *Res:* Functional organization of DNA in the cell nucleus; development of practical applications of recombinant DNA technology. *Mailing Add:* Dept Zool & Genetics Iowa State Univ Ames IA 50011

MAYFIELD, LEWIS G, b Forsyth, Mont, Oct 23, 22; m 47; c 2. CHEMICAL ENGINEERING. *Educ:* Mont State Col, BS, 44, MS, 50. *Prof Exp:* Jr engr, Enzymes, Inc, 46-48; chem engr oil shale & petrol exp, US Bur Mines, 49-51; from asst prof to assoc prof chem eng, Mont State Col, 51-62; asst prog dir eng, NSF, 62-64, prog dir, 64-71, dep dir, Advan Technol Applns Div, 71-75, sr scientist, Resources Div, 75-78, dep dir, Integrated Basic Res Div, 78-81, dep div dir, Chem & Process Eng Div, 81-84, div dir, Cross-Disciplinary Res, 84-88; RETIRED. *Concurrent Pos:* Asst prog dir eng, Nat Sci Found, 60-61. *Mem:* Am Inst Chem Engrs; Am Soc Eng Educ; Sigma Xi. *Res:* Catalysis; hydrogenation of shale-oil; reaction kinetics; mass transfer; enzyme technology; mineral benefication. *Mailing Add:* 3720 N Vermont Arlington VA 22207

MAYFIELD, MELBURN ROSS, b Island, Ky, Aug 24, 21; m 50; c 1. PHYSICS, SCIENCE EDUCATION. *Educ:* Western Ky State Col, AB & BS, 48; Univ Fla, MS, 50. *Prof Exp:* From instr to asst prof physics, Mercer Univ, 50-55, asst prof math & physics, 55-57; assoc prof physics, Austin Peay State Univ, 57-61, chmn dept, 58-70, dir prog teachers, 68, prof physics, 61-87, dir, Ctr for Teachers, 70-87, vpres develop & field serv, Austin Peay State Univ, 74-87; RETIRED. *Concurrent Pos:* Consult under NSF grant, Acad Yr Inst Jr Col Teachers, Univ Fla, 66-67, consult, 67-68; consult, Proj Reachigh, 68, Inst Energy Anal, 79- *Mem:* Fel AAAS; Am Asn Physics Teachers. *Res:* Radioactive fallout measurement and identification; physics education at high school and college level. *Mailing Add:* Dept Physics Austin Peay State Univ Col St Clarksville TN 37044

MAYHALL, JOHN TARKINGTON, b Greencastle, Ind, Apr 7, 37; m 60. DENTAL ANTHROPOLOGY. *Educ:* DePauw Univ, BA, 59; Ind Univ, Indianapolis, DDS, 63; Univ Chicago, AM, 68, PhD, 76. *Prof Exp:* Res assoc, 71-72, res assoc prof anthrop, 71-76, assoc prof, 76-81, asst prof dent anat, 72-76, assoc prof, 76-81, prof anthrop, 81-87, PROF DENT ANAT, FAC DENT, UNIV TORONTO, 81- *Concurrent Pos:* Res fel dent anthrop, Fac Dent, Univ Toronto, 71-73; abstractor, Oral Res Abstr, 71-78. *Mem:* Can Asn Dent Res; fel Soc Study Human Biol; Human Biol Coun; Am Asn Phys Anthrop; Int Asn Dent Res; Fr Soc Anthrop & Dent-Facial Genetics. *Res:* Dental morphology, genetics and craniofacial growth and development of North American Eskimos and Indians; dental anatomy; forensic odontology; osteology. *Mailing Add:* Fac Dent Univ Toronto 124 Edwards St Toronto ON M5G 1G6 Can

MAYHAN, ROBERT J(OSEPH), b Omaha, Nebr, Dec 22, 38; m 73; c 4. ELECTRICAL ENGINEERING. *Educ:* Purdue Univ, BSEE, 60, MSEE, 62, PhD(elec eng), 66. *Prof Exp:* Instr elec eng, Purdue Univ, 62-65; sr staff scientist, Space Systs Div, Avco Corp, 65-67; from asst prof to assoc prof , 67-84, PROF ELEC ENG, OHIO STATE UNIV, 84- *Mem:* Inst Elec & Electronics Engrs. *Res:* Electromagnetic scattering from plasma coated obstacles; scattering from turbulent media; electromagnetic wave interaction with plasmas as a diagnostic tool and for reentry vehicle communications; highway research, especially automatic guidance systems and vehicular communications; resistance welding instrumentation and control. *Mailing Add:* Dept Elec Eng Ohio State Univ 2015 Neil Ave Columbus OH 43210

MAYHEW, ERIC GEORGE, b London, Eng, June 22, 38; m; c 3. CELL BIOLOGY, DRUG CARRIERS. *Educ:* Univ London, BSc, 60, MSc, 63, PhD(zool), 67. *Prof Exp:* Res asst cell biol, Chester Beatty Res Inst, London, Eng, 60-64; cancer res scientist, 64-68, sr cancer res scientist, 68-72, assoc cancer res scientist, 72-79, CANCER RES SCIENTIST V, ROSWELL PARK MEM INST, BUFFALO, NY, 79- *Concurrent Pos:* Vis scientist, Int Inst Cellular & Molec Path, Brussels, 77-78; assoc res prof biophysics, State Univ NY, Buffalo, 79-, adj assoc prof pharmaceut, 89-; ed, Selective Cancer Therapeut, 89- *Mem:* NY Acad Sci; Am Asn Cancer Res. *Res:* Possible differences between normal and cancer cells and possible exploitation in chemotherapy; use of liposomes and other macromolecular structures as drug delivery systems. *Mailing Add:* Dept Exp Path Roswell Park Cancer Inst Buffalo NY 14263

MAYHEW, THOMAS R, b Monongahela, Pa, Feb 11, 35; m 55; c 2. ELECTRICAL ENGINEERING, PHYSICS. *Educ:* Univ Fla, BEE, 56; Univ Pa, MSEE, 62. *Prof Exp:* Engr, Radio Corp Am, 56-68, group leader electronics, 68-69, eng leader, Defense Microelectronics Dept, 69-76, mgr monolithic arrays, 76-79, mgr LSI appln, Solid State Tech Ctr, 79-90; SR MEM ENGR STAFF, GEN ELEC AEROSPACE, 90- *Mem:* Inst Elec & Electronics Engrs. *Res:* Microelectronics; circuits; digital analytical systems. *Mailing Add:* Gen Elec Aerospace GCSD Front & Cooper Sts Camden NJ 08102

MAYHEW, WILBUR WALDO, b Yoder, Colo, Mar 17, 20; m 48; c 3. VERTEBRATE BIOLOGY, DESERT ECOLOGY. *Educ:* Univ Calif, AB, 48, MA, 51, PhD(zool), 53. *Prof Exp:* Assoc zool, Univ Calif, Davis, 48-50 & 51-53; jr res biologist, Atomic Energy Proj, Univ Calif, Los Angeles, 53-54; instr biol, 54-56, from asst prof to assoc prof zool, 56-69, PROF ZOOL, UNIV CALIF, RIVERSIDE, 69- *Concurrent Pos:* Fulbright lectr, UAR, 65-66; Am consult, All-Indian Inst Ecol, Saurashtra Univ, India, 70; mem US deleg, Binational Conf Educ & Res Life Sci, India, 71; mem adv comt, Calif Desert Conserv Area, US Bur Land Mgt, 77-81. *Honors & Awards:* A Starker Leopold Conservation Award. *Mem:* Am Soc Ichthyologists & Herpetologists; Herpetologists' League; Soc Study Amphibians & Reptiles. *Res:* Ecology and physiology of avian and reptilian reproduction; cliff swallow nesting and migration; ecology of deserts. *Mailing Add:* Dept of Biol Univ of Calif Riverside CA 92521

MAYKUT, MADELAINE OLGA, clinical pharmacology, for more information see previous edition

MAYKUTH, D(ANIEL) J(OHN), b Detroit, Mich, Sept 29, 23; m 50; c 4. METALLURGICAL ENGINEERING. *Educ:* Mich Col Mining, BS, 46; Ohio State Univ, MS, 53. *Prof Exp:* Res engr, Nonferrous Phys Metall Div, Battelle Mem Inst, Columbus, 47-54, asst chief, 54-71, dir, Cobalt Info Ctr, 72-75, prin scientist, Materials Develop Sect, 75-80; mgr, Tin Res Inst, Columbus, 80-88; RETIRED. *Concurrent Pos:* Consult, Mat Adv Bd, Nat Acad Sci, 58-59. *Mem:* Am Soc Metals; Am Inst Mining, Metall & Petrol Engrs. *Res:* Nonferrous physical metallurgy of tungsten, rhenium, chromium, molybdenum, titanium, zirconium, columbium, tantalum, cobalt and their alloys. *Mailing Add:* 796 Overlook Dr Columbus OH 43214

MAYLAND, BERTRAND JESSE, b Racine, Wis, Aug 31, 16; m 40; c 3. CHEMICAL ENGINEERING, CHEMISTRY. *Educ:* Univ Wis, BS, 40; Univ Ill, MS, 42, PhD(chem eng), 43. *Prof Exp:* Asst, Eng Exp Sta, Univ Ill, 40-43; sr chem engr, Phillips Petrol Co, 43-51; sr develop engr, Girdler Construct Div, Chemetron Corp, 51-60, dir res & develop, C & I Girdler Inc, 60-71; vpres, 71-73, PRES & CONSULT, CHENOWETH DEVELOP LABS, 73- *Concurrent Pos:* Sanit engr, City Water Dept, Springfield, Ill, 42; spec lectr, Okla Agr & Mech Col, 47-51; consult, 60- *Mem:* Am Chem Soc; Am Inst Chem Engrs. *Res:* Petroleum and chemical process development and design; physical properties; separations; catalytic engineering and catalysts; synthetic fuels and fertilizers; computer applications and process control; catalytic reactions and purification in pollution abatement. *Mailing Add:* PO Box 99254 Louisville KY 40269-0254

MAYLAND, HENRY FREDERICK, b Greybull, Wyo, Dec 31, 35; m 57; c 2. SOIL SCIENCE. *Educ:* Univ Wyo, BS, 60, MS, 61; Univ Ariz, PhD(agr chem & soils), 65. *Prof Exp:* SOIL SCI & AGR RES SERV, USDA, 73- *Concurrent Pos:* Fed collabr, Utah State Univ, 67-; affil prof, Univ Idaho, 68-; vis fel, Plant, Soil & Nutrit Lab, USDA & Cornell Univ, 73- 74. *Mem:* Am Soc Agron; Soil Sci Soc Am; Soc Range Mgt; Sigma Xi; Am Soc Animal Sci; Coun Agr Sci & Technol. *Res:* Soil-water-plant-animal relations on rangelands. *Mailing Add:* Agr Res Serv USDA Kimberly ID 83341

MAYLIE-PFENNINGER, M F, CYTOLOGY, EMBRYOLOGY. *Prof Exp:* PROF ANAT, COL PHYSICIANS & SURGEONS, COLUMBIA UNIV, 79- *Mailing Add:* Dept Obstet-Gynec Health Sci Ctr Univ Colo Box B-197 4200 E Ninth Ave Denver CO 80262

MAYNARD, CARL WESLEY, JR, b Eveleth, Minn, June 18, 13; m 37. ORGANIC CHEMISTRY. *Educ:* Colo Col, AB, 34; Mass Inst Technol, PhD(org chem), 38. *Prof Exp:* Analytical chemist, Dow Chem Co, Mich, 35-36; res chemist, Jackson Lab, E I Du Pont de Nemours & Co, Inc, 38-52, sr supvr intel, 52-77; CONSULT SYNTHETIC DYES, 77- *Concurrent Pos:* Civilian with Manhattan Proj, 42-44. *Mem:* Am Chem Soc; Am Asn Textile Chemists & Colorists. *Res:* Chemistry and toxicology of synthetic dyes; chemical literature. *Mailing Add:* 114 Cambridge Dr Wilmington DE 19803-2606

MAYNARD, CHARLES ALVIN, b Des Moines, Iowa, Mar 25, 51; m 71; c 1. FOREST GENETICS, AGROBACTERIUM TRANSFORMATION. *Educ:* Iowa State Univ, BS, 74, MS, 77, PhD(forest genetics), 80. *Prof Exp:* Extension asst forestry, Forestry Dept, Iowa State Univ, 75-77, res asst forest genetics, 77-80; res assoc, 80-86, ASSOC PROF FOREST GENETICS, COL ENVIRON SCI, STATE UNIV NY, 86- *Concurrent Pos:* Consult forestry mgt, YMCA Camp, Boone, Iowa, 77; fel, Int Agr Ctr, Wageninger, Neth, 78. *Mem:* Soc Am Foresters; Tissue Culture Asn; Int Soc Plant Molecular Biol. *Res:* Agrobacterium transformation of black cherry and other high value hardwood species. *Mailing Add:* Fac Forestry 216 Marshall Hall State Univ NY Syracuse NY 13210-2787

MAYNARD, CHARLES DOUGLAS, b Atlantic City, NJ, Sept 11, 34; m 58; c 3. NUCLEAR MEDICINE. *Educ:* Wake Forest Univ, BS, 55; Bowman Gray Sch Med, MD, 59. *Prof Exp:* Dir nuclear med, NC Baptist Hosp, Winston-Salem, 66-77, chmn, dept radiol, 77; from instr to assoc prof radiol, 66-73, assoc dean admis, 66-71, assoc dean student affairs, 71-75, PROF RADIOL, BOWMAN GRAY SCH MED, WAKE FOREST UNIV, 73- *Concurrent Pos:* Am Cancer Soc fel, 64-66; James Picker Found scholar radiol res, 66-68; consult ed, J Nuclear Med & Technol, 74; consult nuclear med, Am Registry Radiologic Technologists, 74; guest examr, Am Bd Radiol,

75. *Mem:* Soc Nuclear Med (vpres, 76, pres, 78); Asn Univ Radiologists; Radiol Soc NAm; Am Col Nuclear physicians; Am Col Radiol; Sigma Xi. *Res:* Clinical applications of radionuclides in the diagnoses of disease. *Mailing Add:* Dept of Radiol Bowman Gray Sch of Med Winston-Salem NC 27103

MAYNARD, JAMES BARRY, b Greensboro, NC, Nov 8, 46; m 67; c 2. GEOCHEMISTRY. *Educ:* Duke Univ, BS, 68; Harvard Univ, PhD(geol), 73. *Prof Exp:* From asst prof to assoc prof, 72-85, PROF GEOL, UNIV CINCINNATI, 85- *Concurrent Pos:* Vis assoc prof, Mem Univ Nfld, 79; dept head geol, Univ Cincinnati, 85-90; mem, Associate Rev Panel, Nat Res Coun, 88-, Comt Adv US Geol Surv, 89-91, Ohio Gov Geol Adv Coun, 90-; assoc ed, Econ Geol, 89- *Mem:* Soc Econ Geologists. *Res:* Geochemistry and petrology applied to the study of sedimentary ore deposits; use of sedimentary ores as indicators of global change in Earth history. *Mailing Add:* Dept Geol Univ Cincinnati Cincinnati OH 45221-0013

MAYNARD, JERRY ALLEN, b Reedsburg, Wis, Apr 22, 37; m 57; c 2. CELL BIOLOGY. *Educ:* Univ Northern Iowa, BA, 58; Ind Univ, MS, 61; Univ Iowa, PhD(phys educ, anat & kinesiology), 70. *Prof Exp:* Fel exercise physiol, Univ Iowa, 68-69, res asst, cytol, Dept Orthop, 69-71, res assoc, 71-72, from asst prof to assoc prof anat, Depts Orthop, Phys Educ & Anat, 72-81, PROF ANAT, DEPTS ORTHOP & EXERCISE SCI, UNIV IOWA, 81- *Concurrent Pos:* Fel Dept Orthop Surg, Univ Iowa, 68-71. *Mem:* Anat Soc Gr Brit & Ireland; Orthop Res Soc; Am Asn Univ Prof. *Res:* Ultrastructure, cytochemistry and light microscopy of the musculoskeletal system during growth and development and as it is affected by hyperactivity, inactivity and skeletal growth dysplasias. *Mailing Add:* Dept Orthop Surg Univ Iowa Iowa City IA 52242

MAYNARD, JULIAN DECATUR, b Newport News, Va, Nov 18, 45; m 87; c 1. SOLID STATE PHYSICS. *Educ:* Univ Va, BS, 67; Princeton Univ, MA, 69, PhD(physics), 74. *Prof Exp:* Instr physics, Princeton Univ, 73-74; adj asst prof, Univ Calif, Los Angeles, 74-77; from asst prof to prof, 77-91, DISTINGUISHED PROF PHYSICS, PA STATE UNIV, 91- *Mem:* Fel Am Phys Soc; fel Acous Soc Am; Sigma Xi. *Res:* Quantum liquids and solids and critical phenomena with extensive application of acoustic techniques. *Mailing Add:* Dept Physics 104 Davey Lab Pa State Univ University Park PA 16802

MAYNARD, NANCY GRAY, b Middleboro, Mass, Apr 18, 41; m 69; c 1. BIOLOGICAL OCEANOGRAPHY. *Educ:* Mary Washington Col, Univ Va, BS, 63; Univ Miami, MS, 67, PhD(biol oceanog), 74. *Prof Exp:* Res asst marine biol, Rosenstiel Sch Marine & Atmospheric Sci, Univ Miami, 65-69; res assoc paleoclimat, Climate Long-Range Invest, Mapping & Prediction, NSF, Lamont-Doherty Geol Observ, Columbia Univ, 72-75; res fel appl chem, Harvard Univ, 75-76; environ studies coordr, Alaska Outer Continental Shelf Environ Assessment Prog, US Dept Interior, 76-78; oil spill sci coordr Alaska, Nat Oceanic & Atmospheric Admin, 78-80, southeast US, 80-82; policy analyst, Exec Off Pres, Off Sci & Technol Policy, 82-83; staff dir, Bd Ocean Sci & Policy, Nat Acad Sci, 83-85; res assoc, Jet Propulsion Lab & Scripps Inst Oceanog, 85-87. *Concurrent Pos:* Res assoc oil pollution, Bermuda Biol Sta, St George, 72-76 & Am Petrol Inst, 74-76; Dept Com sci & tech fel, 82-83; Nat Res Coun resident res assoc, NASA & Jet Propulsion Lab, 85-86. *Mem:* AAAS; Am Geophys Union; Am Soc Limnol & Oceanog; Am Polar Soc. *Res:* Remote sensing; primary production; polar biology and oceanography; phytoplankton ecology; paleooceanography; oil pollution biology. *Mailing Add:* Goddard Space Flight Ctr Oceans & Ice Br Code 671 NASA Greenbelt MD 20771

MAYNARD, RUSSELL MILTON, pathology, for more information see previous edition

MAYNARD, THEODORE ROBERTS, b Denver, Colo, Dec 4, 38; m 66; c 2. GEOTECHNICAL ENGINEERING, ENGINEERING GEOLOGY. *Educ:* Colo Sch Mines, BS, 62; Univ Ill, MS, 65. *Prof Exp:* CHIEF SOILS ENGR, DEPT PUBLIC WORKS, BUR ENG, CITY CHICAGO, 65- *Concurrent Pos:* Guest lectr, various univs & socs, 66-; NRC, USNC, Int Asn Eng Geologists, 88-91. *Mem:* Asn Eng Geologists (treas, 83-85, vpres, 85-86 & pres, 86- 87); Am Soc Engrs; Nat Soc Prof Engrs; Int Soc Soil Mech & Found Engrs; Int Asn Eng Geologists; Am Public Works Asn. *Res:* Soil-structure interaction; performance of earth retention systems and deep foundation elements; pavement evaluation and design; embankment behavior; incinerator ash residue. *Mailing Add:* 6261 N Oriole Ave Chicago IL 60631

MAYNE, BERGER C, b Towner, Colo, July 10, 20; m 56; c 2. PLANT PHYSIOLOGY. *Educ:* Western State Col Colo, AB, 46; Univ Utah, PhD(physiol), 58. *Prof Exp:* Res assoc, Univ Minn, 58-62; staff scientist, 62-67, INVESTR, CHARLES F KETTERING RES LAB, 67- *Mem:* Am Soc Plant Physiologists; Biophys Soc; Sigma Xi. *Res:* Photosynthesis. *Mailing Add:* 113 Marshall St Yellow Spg OH 45387

MAYNE, WILLIAM HARRY, geophysics, for more information see previous edition

MAYNERT, EVERETT WILLIAM, b Providence, RI, Mar 18, 20. PHARMACOLOGY, CHEMISTRY. *Educ:* Brown Univ, ScB, 41; Univ Ill, PhD(org chem), 45; Johns Hopkins Univ, MD, 57. *Prof Exp:* Res chemist, Interchem Corp, NY, 45-47; res assoc pharmacol, Columbia Univ, 47-51, assoc, 51-52, asst prof, 52; assoc prof pharmacol & exp therapeut, Johns Hopkins Univ, 52-65; PROF PHARMACOL, COL MED, UNIV ILL-CHICAGO, 65- *Concurrent Pos:* Am Cyanamid fel, Johns Hopkins Univ, 52-57; Lederle med fac award, 57-60; vis prof, Japan Soc Adv Sci, 74. *Mem:* Fel AAAS; Am Chem Soc; Harvey Soc; Am Soc Pharmacol; Soc Exp Biomed. *Res:* Neuropharmacology; drug metabolism; toxicology. *Mailing Add:* 339 W Barry Ave Apt 10 Chicago IL 60657

MAYNES, ALBION DONALD, b Buffalo, NY, Jan 21, 29; Can citizen; m 52; c 2. ANALYTICAL CHEMISTRY. *Educ:* Univ Toronto, BA, 52, MA, 53, PhD(inorg & anal chem), 56. *Prof Exp:* Res assoc, Dept Physics, Univ Toronto, 56-58; anal chemist, Div Geol Sci, Calif Inst Technol, 58-65; asst prof, 65-66, ASSOC PROF CHEM, UNIV WATERLOO, 66- *Concurrent Pos:* Eldorado Mining & Refining Co res grant, 56-58. *Mem:* Chem Inst Can; Meteoritical Soc. *Res:* Trace analysis; analysis of silicate rocks and minerals; analysis of meteorites. *Mailing Add:* Dept of Chem Univ of Waterloo Waterloo ON N2L 3G1 Can

MAYNES, GORDON GEORGE, b Freeport, NY, Nov 4, 46; m 69; c 2. ORGANIC CHEMISTRY, POLYMER CHEMISTRY. *Educ:* St Lawrence Univ, BS, 68; Univ Ill-Champaign, PhD(org chem), 72. *Prof Exp:* Teaching asst chem, Univ Ill, 68-72; from res chemist to sr res chemist polymer chem, E I du Pont de Nemours & Co Inc, 72-80, supvr res & develop, 80-85, tech supvr, 85-88, TECH GROUP MGR, E I DU PONT DE NEMOURS & CO INC, 88- *Mem:* Am Chem Soc. *Res:* Physical organic chemistry; free radical mechanisms; small ring compounds; dye chemistry; condensation polymerization; polymer melt spinning. *Mailing Add:* 107 Country Rd Seaford DE 19973

MAYO, BARBARA SHULER, b Tallahassee, Fla; Jan 5, 45; m 72; c 2. MARINE BIOLOGY. *Educ:* Mary Baldwin Col, BA, 67; Univ Miami, MS, 70, PhD(marine sci), 74. *Prof Exp:* Teaching asst oceanog, Univ Miami, 70-74; consult biologist, marine biol, Consult Marine Biologists, 74-76; dir, 76-78, INSTR MARINE ECOL, PROVINCETOWN CTR FOR COASTAL STUDIES, 75-, ASSOC SCIENTIST, 76- *Concurrent Pos:* Mem Cape Cod Nat Seashore Adv Comn, 77- *Mem:* Sigma Xi. *Res:* Coastal marine ecology; systematics of decapod crustaceans. *Mailing Add:* Provincetown Ctr Coastal Dis Box 826 Provincetown MA 02657

MAYO, CHARLES ATKINS, III, b Washington, DC, Mar 13, 43; m 72; c 2. MARINE BIOLOGY, MARINE FISHERIES SCIENCE. *Educ:* Dartmouth Col, BA, 65; Univ Miami, MS, 69, PhD(marine biol), 73. *Prof Exp:* Fish biologist, SE Fisheries Ctr, Nat Marine Fish Serv, 72-74; chief biologist, marine ecol, Barceloneta Proj, PRASA, PR, 74-75; dir, Cetacean Res Prog, 80-86, CHIEF NATURALIST CETACEAN STUDIES, DOLPHIN III, PROVINCETOWN CTR FOR COASTAL STUDIES, 74-, ASSOC SCIENTIST MARINE BIOL, 76-, DIR RES, CTR COASTAL STUDIES, 86- *Res:* Ecology of coastal marine systems and effects of pollutants; pollutants; early life histories of marine fishes and applied rearing studies; field studies of great whales. *Mailing Add:* 570 Commercial Provincetown MA 02657

MAYO, DANA WALKER, b Bethlehem, Pa, July 20, 28; m 62; c 3. ORGANIC CHEMISTRY. *Educ:* Mass Inst Technol, BS, 52; Ind Univ, PhD(chem), 59. *Prof Exp:* Res chemist, Polychem Dept, Exp Sta, E I du Pont de Nemours & Co, Del, 52; asst, Univ Pa, 52-53 & Ind Univ, 53-55 & 56-57; res assoc, Mass Inst Technol, 59-60, NIH fel, 60-62, fel, Sch Advan Study, 60-62; from asst prof to prof, Bowdoin Col, 62-69, Charles Weston Pickard prof chem, 70-91, chmn dept, 69-75, 77-78 & 81-83, CHARLES WESTON PICKARD RES PROF CHEM, BOWDOIN COL, 91- *Concurrent Pos:* Vis lectr, Mass Inst Technol, 62-71; NIH spec fel chem, Univ Md, 67 & 69-70; consult, Perkin-Elmer Corp; vis scientist, Explosives Res & Develop Estab, Waltham-Abbey, UK, 75. *Honors & Awards:* Chem Health & Safety Award, Am Chem Soc Div, 87. *Mem:* Am Chem Soc; Soc Appl Spectros; Coblentz Soc; AAAS. *Res:* Natural products; applications of infrared spectroscopy to organic chemistry; oil pollution; animal and plant chemical communication substances. *Mailing Add:* Dept Chem Bowdoin Col Brunswick ME 04011

MAYO, DE PAUL, b London, Eng, Aug 8, 24; m 49; c 2. ORGANIC CHEMISTRY, PHOTOCHEMISTRY. *Educ:* Univ London, BSc, 44, MSc, 52, PhD(chem), 54; Univ Paris, Dr es Sci, 70. *Prof Exp:* Res fel chem, Univ Col Hosp, London, Eng, 50-52, Birkbeck Col, London, 52-53, asst lectr, 54-55; lectr, Glasgow Univ, 55-57 & Imp Col, Univ London, 57-59; dir photochem unit, 69-72, prof, 59-90, EMER PROF CHEM, UNIV WESTERN ONT, 90- *Honors & Awards:* Centennial Medal, 67; Palladium Medal, Chem Inst Can, 82; E W R Steacie Award, 85. *Mem:* Am Chem Soc; Chem Inst Can; fel Royal Soc Can; fel Royal Soc; Royal Soc Chem. *Mailing Add:* Dept Chem Univ Western Ont London ON N6A 5B8 Can

MAYO, FRANK REA, physical organic chemistry, polymer chemistry; deceased, see previous edition for last biography

MAYO, JOHN S, b Greenville, NC, Feb 26, 30; m 57; c 4. ELECTRICAL ENGINEERING. *Educ:* NC State Univ, BS, 52, MS, 53, PhD(elec eng), 55. *Prof Exp:* Photoelectric analyzer, NC State Univ, 54-55; mem tech staff comput res, AT&T Bell Labs, 55-58, supvr, T1 carrier syst, 58-60, head, High-Speed Pulse Code Modulation Terminal Dept, 60-67, dir, Underwater Systs Lab, 67-71, exec dir, Ocean Systs Div, 71-73, exec dir, Toll Electronic Switching Div, 73-75, vpres elec technol, 75-79, exec vpres netword systs, 79-91, PRES, AT&T BELL LABS, 91- *Concurrent Pos:* Chmn, Int Solid State Circuits Conf, 61-69, study comt global commun, Nat Security Indust Asn, 73-76; mem bd dirs, Nat Eng Consortium, 74-76 & Polytech Univ NY, 84-; mem Asn US Army, 76-; mem Gov's comt sci & technol, Task Force on Commun, 82-83, sch eng & appl sci comt, Univ Calif, Los Angeles, 83-87, comt eng utilization, Am Asn Eng Soc, 83-, high Technol comt, New York City, Partnership Inc, 87-, comt on technol issues impact Int Competitiveness, 87-; mem indust adv bd, Rensselaer Polytech Inst & adv bd, Col Eng, Univ Calif, Berkeley, 82-; mem Int Coop & Competition in Space Adv Panel, Off Technol Assessment, 82-85; mem eng adv coun, NC State Univ, 83-86; mem, bd dirs, Johnson & Johnson, 86-, Audit Comt, 86-88, Comt Sci & Technol, 86-, Pub Policy Adv Comt, 90-; mem bd dirs, Western Digital Corp, 88-, chmn, Audit Comt, 91- *Honors & Awards:* Alexander Graham Bell Award, Inst Elect & Electronic Engrs, 78, Simon Ramo Medal, 88; Nat Medal of Technol, 90. *Mem:* Nat Acad Eng; fel Inst Elec & Electronic Engrs. *Res:* Research and development of communications devices, circuitry and systems. *Mailing Add:* AT&T Bell Labs Murray Hill NJ 07974-2070

MAYO, JOSEPH WILLIAM, b Greenfield, Mass, Sept 22, 41; m 69; c 2. BIOCHEMISTRY, PEDIATRICS. *Educ:* Univ Mass, Amherst, BS, 63; Mich State Univ, PhD(biochem), 68; Case Western Reserve Univ, MD, 76. *Prof Exp:* Asst prof biochem in pediat, Med Sch, Case Western Reserve Univ, 70-76; intern pediat, Cleveland Metr Gen Hosp & Rainbow Babies & Children's Hosp, 76-77; resident pediat, 77-78, chief resident, 78-79, ASST PROF PEDIAT, DEPT CHILD HEALTH, UNIV MO MED CTR, 79- *Concurrent Pos:* Nat Cystic Fibrosis Res Found fel, Sch Med, Case Western Reserve Univ, 68-70. *Res:* Carbohydrate metabolism; glycoproteins; composition and regulation of human exocrine secretions; primary care pediatrics. *Mailing Add:* Dept Child Health Univ Mo M 228 Med Sci Bldg Columbia MO 65212

MAYO, RALPH ELLIOTT, b Greenville, NC, May 9, 40; m 64; c 3. PHYSICAL CHEMISTRY. *Educ:* Emory Univ, BS, 63, PhD(phys chem), 66. *Prof Exp:* Sr res chemist, Perkin Elmer Corp, 66-68; supvr method develop lab, 68-78, mgr anal serv, Linwood, 77-81, MGR, RES & DEVELOP SYSTS & PRODUCTIVITY, ALLENTOWN, AIR PROD & CHEM, INC, 81- *Mem:* AAAS; Catalysis Soc; Am Chem Soc; Soc Appl Spectros. *Res:* Nuclear magnetic resonance spectroscopy; analytical instrumentation design; laboratory automation and process control; digital processing of scientific data; analytical methods development; mass spectrometry; spectrophotometric analyses. *Mailing Add:* Air Prod & Chem Inc 7201 Hamilton Blvd Allentown PA 18195

MAYO, SANTOS, b Buenos Aires, Arg, June 10, 28; m 59; c 3. MICROELECTRONICS. *Educ:* La Plata Univ, PhD(physics), 54. *Prof Exp:* Assoc res nuclear spectros, Arg AEC, 53-55, head synchrocyclotron lab, 55-68; mgr res & develop, Fate, 68-71; res mem, Cyclotron Lab, Arg AEC, 71-72; head planning, Arg Inst Indust Technol, 72-73; SOLID STATE PHYSICIST, NAT BUR STANDARDS, 74- *Concurrent Pos:* Instr, La Plata Univ, 50-60, assoc prof, 60-61; guest physicist, Brookhaven Nat Lab, 57; assoc res, Radiation Lab, Univ Pittsburgh, 58-59; head nuclear physics dept, Arg AEC, 62-63; Arg rep, Latin Am Physics Ctr, Brazil, 63-; consult physicist, Nat Res Coun, Arg, 64-; res fel, UN Develop Orgn, Nat Bur Standards, 73-74. *Mem:* Arg Physics Asn (gen secy, 68-70, pres, 70-72); Am Physics Soc. *Res:* Beta and gamma nuclear spectroscopy; low and medium energy nuclear reactions; laser induced resonance ionization spectroscopy; resonance ionization mass spectroscopy; charged particle spectroscopy; accelerator techniques; microelectronic technology; surface probe analysis; physics of microelectronic devices; x-ray topography; x-ray diffractometry. *Mailing Add:* Nat Inst Standards & Technol Bldg 225 Rm A305 Gaithersburg MD 20899

MAYO, THOMAS TABB, IV, b Radford, Va, June 15, 32; m 57; c 3. TEACHING UNDERGRADUATE MATHEMATICS. *Educ:* Va Mil Inst, BS, 54; Univ Va, MS, 57, PhD(physics), 60. *Prof Exp:* Asst instr physics, Va Mil Inst, 54-55; sr scientist res lab eng sci, Univ Va, 60-61; from asst prof to assoc prof physics & math, Hampden-Sydney Col, 62-67, asst acad dean, 71-73, assoc acad dean, 73-75 & 76- 77, actg acad dean, 75-76, prof physics, 67-88, PROF MATH, HAMPDEN-SYDNEY COL, 88- *Concurrent Pos:* Res assoc quantum theory proj, Univ Fla, 69-70. *Mem:* Am Phys Soc; Am Asn Physics Teachers; Sigma Xi. *Res:* Teaching pure and applied mathematics, classical physical theory and computer science. *Mailing Add:* Dept Math & Comput Sci Hampden-Sydney Col Hampden-Sydney VA 23943

MAYO, Z B, b Lubbock, Tex, Mar 29, 43; m 71. ENTOMOLOGY. *Educ:* Tex Tech Univ, BS, 67; Okla State Univ, MS, 69, PhD(entom), 71. *Prof Exp:* Res assoc entom, Okla State Univ, 71-72; from asst prof to assoc prof, 72-82, PROF ENTOM, UNIV NEBR-LINCOLN, 82- *Mem:* Entom Soc Am; Am Registry Prof Entomologists; Sigma Xi. *Res:* Biology and genetics of aphid biotypes. *Mailing Add:* Dept Entom 202 P I Univ Nebr Lincoln NE 68503

MAYOCK, ROBERT LEE, b Wilkes Barre, Pa, Jan 19, 17; m 49; c 3. PULMONARY DISEASE. *Educ:* Bucknell Univ, BS, 38; Univ Pa, MD, 42. *Prof Exp:* Instr med, Sch Med, Univ Pa, 46-49, from asst prof to prof, 49-89, EMER PROF MED, SCH MED, UNIV PA, 89-; SR CONSULT, UNIV HOSP, 72- *Concurrent Pos:* Chief women's sect, pulmonary diseases, Fitzsimmons Army Hosp, 52-54; chief pulmonary dis sect, Hosp Univ Pa, 55-72, & Philadelphia Gen Hosp, 59-69; mem bd pulmonary dis, Am Bd Internal Med, 65-76; regent, Am Col Physicians, 72-79; mem bd dirs, Am Lung Asn, 83- *Mem:* AAAS; Sigma Xi; fel Am Col Physicians; fel Am Col Chest Physicians; Am Fedn Clin Res. *Res:* Pulmonary disease and the circulation; tuberculosis, sarcoidosis and pharmacokinetics. *Mailing Add:* Hosp Univ Pa 3400 Spruce St Ravdin Bldg Suite 1 Philadelphia PA 19104

MAYOL, ROBERT FRANCIS, b Springfield, Ill, Nov 11, 41; m 62; c 4. BIOCHEMISTRY, IMMUNOCHEMISTRY. *Educ:* Southern Ill Univ, Carbondale, BA, 64; St Louis Univ, PhD(biochem), 68. *Prof Exp:* USPHS fel, Calif Inst Technol, 68-70; sr scientist biochem endocrinol, Mead Johnson Res Ctr, 70-75, sr investr drug metab, 75-78, prin res scientist, 78-84; RES FEL, BRISTOL-MYERS SQUIBB, CONN, 84- *Concurrent Pos:* Mem assoc fac, Sch Med, Ind Univ, Evansville Ctr, 71-75. *Mem:* Sigma Xi; Endocrine Soc; AAAS; Am Soc Pharm & Exp Therapeut. *Res:* Protein chemistry; development of radioimmunoassays; drug metabolism. *Mailing Add:* Five Research Pkwy Wallingford CT 06455

MAYOR, GILBERT HAROLD, b Detroit, Mich, Sept 12, 39; m 68; c 3. ENDOCRINOLOGY, CLINICAL PHARMACOLOGY. *Educ:* Wayne State Univ, BS, 61, MD, 65. *Prof Exp:* Asst prof, Univ Mich, 72-77; prof, Mich State Univ, 77-86; med rels dir, Upjohn Co, 86-88; DIR MED SERV, BOOTS PHARMACEUT, 88- *Concurrent Pos:* Fel, Am Col Physicians, 74-, Am Col Clin Pharmacol, 86-; chmn, Med Rel Comt, Pharmaceut Mfg Asn, 87-; regent, Am Col Clin Pharmacol, 88-93, chmn, Honors & Awards Comt, 90-92. *Honors & Awards:* AMA Physician Recognition, AMA, 71, 78, 87. *Mem:* Am Col Clin Pharmacol; Am Thyroid Asn; Am Col Physicians; Endocrine Soc; Am Soc Nephrology; Int Soc Nephrology. *Res:* Clinical and pharmacology research involving thyroid hormone, erythromycin and chymopapain. *Mailing Add:* Boots Pharmaceut 300 Tri-State Int Ctr Suite 200 Lincolnshire IL 60069

MAYOR, HEATHER DONALD, b Melbourne, Australia, July 6, 30; m 56; c 2. VIROLOGY, MOLECULAR BIOLOGY. *Educ:* Univ Melbourne, BS, 48, MSc, 50, DSc, 70; Univ London, PhD(biophys), 54. *Prof Exp:* Res officer crystal physics, Defense Res Labs, Melbourne, Australia, 50-51; electron microscopist, Nat Inst Med Res, London, 52-55; res assoc virol, Walter & Eliza Hall Inst Med Res, Melbourne, 55-56; res assoc bacteriol & immunol, Harvard Med Sch, 56-59; from asst prof to assoc prof virol, 60-71, assoc prof microbiol, 71-74, PROF MICROBIOL, BAYLOR COL MED, 74- *Concurrent Pos:* Consult, Res Resources Br, NIH, 70-, AEC, 71- & Univ Tex M D Anderson Hosp & Tumor Inst, Houston, 73- *Honors & Awards:* Award, Ctr Interaction, Man, Sci & Cult, 73; Sir Hiram Maxim Award, 90. *Mem:* Am Asn Immunol; Am Soc Microbiol; Am Asn Cancer Res; Am Soc Cell Biol (treas); Sigma Xi (secy-treas). *Res:* Molecular biology of the growth and development of animal viruses with particular emphasis on viruses which cause cancer and on extremely small DNA-containing viruses; recombinant DNA technology in viral systems; structure of biological macromolecules. *Mailing Add:* Dept of Microbiol & Immunol Baylor Col of Med Houston TX 77030

MAYOR, JOHN ROBERTS, b La Harpe, Ill, July 9, 06; m 34; c 2. GEOMETRY. *Educ:* Knox Col, BS, 28; Univ Ill, AM, 29; Univ Wis, PhD(math), 33. *Hon Degrees:* LLD, Knox Col, 59. *Prof Exp:* Instr math, Univ Wis, 29-31, 32-35 & Milwaukee Exten Div, 35; prof & chmn dept, Southern Ill Univ, 35-47; assoc prof math & educ, Univ Wis, 47-51, prof math & educ & chmn dept educ, 51-54, actg dean sch educ, 54-55; dir educ, Am Asn Advan Sci, 55-74; asst provost res, Div Human & Community Resources, Univ Md, College Park, 74-85; CONSULT, 85- *Concurrent Pos:* Dir math proj & prof, Univ Md, 57-67; dir study accreditation in teacher educ, Nat Comn Accrediting, 63-65; mem adv comt, Sch Math Study Group; consult, Knox Col. *Mem:* AAAS; Am Math Soc; Math Asn Am; Nat Asn Res Sci Teaching; Conf Bd Math Sci (secy, 60-71, treas, 61-71). *Res:* Mapping rational varieties; multiple correspondences in space and hyperspace. *Mailing Add:* 3308 Solomons Ct Silver Spring MD 20906

MAYOR, ROWLAND HERBERT, b Eng, Nov 5, 20; US citizen; m 48; c 3. RUBBER CHEMISTRY. *Educ:* Univ NH, BS, 42, MS, 44; Univ Conn, PhD(org chem), 49. *Prof Exp:* Instr org chem, Univ RI, 48-51; res chemist, 51-55, res sect head, 55-63, mgr stereorubber res, 64-75, asst mgr synthetic rubber res, 75-76, MGR ELASTOMER RES, GOODYEAR TIRE & RUBBER CO, 76- *Mem:* Am Chem Soc. *Res:* Molecular rearrangements; synthesis of diamines; condensation polymerization; synthetic rubber. *Mailing Add:* 1716 Arndale Rd Cuyahoga Falls OH 44224

MAYPER, V(ICTOR), JR, b New York, NY, June 12, 28; m 58; c 2. PHYSICS, ELECTRONICS. *Educ:* Mass Inst Technol, BS, 47, PhD(physics), 53. *Prof Exp:* Res asst, Mass Inst Technol, 47-53; res engr, Newmont Explor Ltd, 53-55; mem tech staff, Hughes Aircraft Co, 55-64; pres, Mayper Assocs, 64-67; chief engr, Jacobi Systs Corp, 67-70; pres, Mayper Assocs, 70-72; jr staff engr, Compata Inc, 72-77; SR STAFF ENGR, OPERATING SYSTS INC, 77- *Concurrent Pos:* Sr engr, Holmes & Narver, Inc, 50-51. *Mem:* Am Phys Soc; Inst Elec & Electronics Engrs; Asn Comput Mach. *Res:* Electronic systems; computer applications; automation. *Mailing Add:* 10061 Riverside Dr Ben Lomond CA 95005

MAYR, ANDREAS, b Mitterham, Ger, Feb 26, 49; m 82; c 2. TRANSITION METAL-CARBON MULTIPLE BONDS. *Educ:* Univ Munich, dipl, 75, PhD(chem), 78. *Prof Exp:* Fel chem, Univ Calif, Los Angeles, 79-81; asst prof chem, Princeton Univ, 81-88; CONSULT, 88- *Concurrent Pos:* Assoc prof, State Univ NY, Stony Brook, 88. *Mem:* Ger Chem Soc; Am Chem Soc. *Res:* Organometallic synthesis; reactivity of transition metal carbene and carbyne complexes; development of new types of reactions for applications in organic synthesis and homogeneous catalysis, or as models in heterogeneous catalysis. *Mailing Add:* Dept Chem State Univ NY Stony Brook NY 11794-3400

MAYR, ERNST, b Kempten, Ger, July 5, 04; m 35; c 2. EVOLUTIONARY BIOLOGY, HISTORY OF SCIENCE. *Educ:* Univ Berlin, PhD(zool), 26. *Hon Degrees:* DPhil, Univ Uppsala, 57 & Univ Paris, 75; DSc, Yale Univ, 59, Univ Melbourne, 59, Oxford Univ, 66, Univ Munich, 68, Harvard Univ, 70, Univ Cambridge & Guelph Univ, 82, Univ Vt, 83. *Prof Exp:* Asst cur zool mus, Univ Berlin, 26-32; from assoc cur to cur, Whitney-Rothschild Collection, Am Mus Natural Hist, 32-53; Agassiz prof zool, 53-75, dir mus comp zool, 61-70, EMER PROF ZOOL, HARVARD UNIV, 75- *Concurrent Pos:* Mem expeds, Dutch New Guinea, 28, Mandate Territory, New Guinea, 29 & Solomon Islands, 29-30; Jesup lectr, Columbia Univ, 41; ed, Soc Study Evolution, 47-49; vpres, 11th Int Zool Cong; pres, 13th Int Ornith Cong. *Honors & Awards:* Leidy Medal, 46; Darwin-Wallace Medal, 58; Brewster Medal, 65; Verrill Medal, 66; Daniel Giraud Eliot Medal, 67; Nat Medal of Sci, 70; Gregor Mendel Medal, 80; Darwin Medal, 83; Balzan Prize, 83. *Mem:* Nat Acad Sci; Am Soc Naturalists; Am Soc Zool; Soc Syst Zool (pres, 66); Soc Study Evolution (secy, 46, pres, 50); Am Acad Arts & Sci; Am Philos Soc. *Res:* Ornithology; evolution; systematics; history and philosophy of biology. *Mailing Add:* Mus Comp Zool Harvard Univ Cambridge MA 02138

MAYRON, LEWIS WALTER, b Chicago, Ill, Sept 20, 32; div; c 2. ENVIRONMENTAL HEALTH CLINICAL ECOLOGY, NUTRITION. *Educ:* Roosevelt Univ, BS, 54; Univ Ill, MS, 56, PhD(biol chem), 59. *Prof Exp:* Chemist, Qm Food & Container Inst, 54; asst biochem, Univ Ill, 54-59; res assoc, Univ Southern Calif, 59-61; asst biochemist, Presby-St Lukes Hosp, 61-62, Tardanbek Labs, 62-63 & Abbott Labs, 63; res assoc, Michael Reese Hosp & Med Ctr, 64-66; consult, 66-68; res chemist, Hines Vet Admin Hosp, 68-79; mem staff, Nuclear Med Serv, Wadsworth Med Ctr, 79-83; CONSULT CLIN ECOL, 83- *Concurrent Pos:* Guest investr, Argonne Nat Lab, 73-79. *Honors & Awards:* Laureat, Genia Czerniak Prize Nuclear Med & Radiopharmacol, Ahavot Zion Found of Israel, 74. *Mem:* Soc Exp Biol & Med; Pan Am Allergy Soc; AAAS; Am Col Allergy Immunol; Am Asn Clin Chemists. *Res:* Biochemistry of immune mechanisms; allergy diagnosis and treatment. *Mailing Add:* 1779 Summer Cloud Dr Thousand Oaks CA 91362

MAYROVITZ, HARVEY N, b Philadelphia, Pa, Dec 31, 44. APPLIED & CLINICAL MICROVASCULAR RESEARCH, NONINVASIVE CARDIOVASCULAR PHYSIOLOGY. *Educ:* Drexel Univ, BS, 64, MS, 67; Univ Pa, PhD(microcirculation), 74. *Prof Exp:* Chief, Exp Lab, 77-87, CHIEF MICROVASCULAR & PHYSIOL STUDIES UNIT, MIAMI HEART INST, 87- *Concurrent Pos:* Adj prof biomed eng, Univ Miami & physiol, Sch Med, Temple Univ, 80-; mem, Coun Circulation, Am Heart Asn. *Mem:* Biomed Eng Soc; Am Physiol Soc; Am Heart Asn. *Mailing Add:* Microvascula & Physiol Studies Unit Miami Heart Inst 4701 N Meridian Ave Miami Beach FL 33140

MAYS, CHARLES EDWIN, b Lincoln, Nebr, May 6, 38; m 63; c 2. PHYSIOLOGY, HERPETOLOGY. *Educ:* Univ Nebr, BS, 63, MS, 65; Ariz State Univ, PhD(zool), 68. *Prof Exp:* Asst prof, 68-74, assoc prof, 74-80, PROF ZOOL, DEPAUW UNIV, 80- *Concurrent Pos:* Du Pont & Nat Sci Found grants, 69-70; Ind Acad Sci grants, 69, 71, 74, 77 & 88; Hughes Found grants, 89-90. *Mem:* AAAS; Herpetologists League; Sigma Xi; Am Soc Ichthyologists & Herpetologists; Soc Study Amphibians & Reptiles; Am Soc Zoologists. *Res:* Natural history and physiological studies of the hellbender salamander and map turtle; growth studies on fish; effects of passive cigarette smoke on murine reproduction and development; genetic versus environmental effects of passive smoking. *Mailing Add:* Dept Zool DePauw Univ Greencastle IN 46135

MAYS, CHARLES WILLIAM, radiation risk, chelation therapy; deceased, see previous edition for last biography

MAYS, DAVID LEE, analytical chemistry, for more information see previous edition

MAYS, JOHN RUSHING, b Jasper, Tex, Apr 22, 34; m 65; c 2. CIVIL ENGINEERING. *Educ:* Lamar Univ, BS, 56; Univ Colo, MS, 60, PhD(civil eng), 67. *Prof Exp:* From instr to assoc prof, Lamar Univ, 56-67; assoc prof, 67-75, PROF CIVIL ENG, UNIV COLO, 75- *Concurrent Pos:* Consult, US Bur Reclamation, 75- *Mem:* Am Soc Civil Engrs; Am Soc Eng Educ; Sigma Xi. *Res:* Dynamic response of nonlinear structures to shock loads. *Mailing Add:* 3895 Garland St Wheat Ridge CO 80033

MAYS, LARRY WESLEY, b Pittsfield, Ill, Feb 7, 48; m 83; c 2. CIVIL ENGINEERING, WATER RESOURCES. *Educ:* Univ Mo, Rolla, Mo Sch Mines, BS, 70, MS, 71; Univ Ill, PhD(civil eng), 76. *Prof Exp:* Civil eng, US Army Eng Explosive Excavation Res Lab, 71-73; res asst water resources, 73-76, vis res asst prof, Univ Ill, 76; from asst prof to assoc prof, 76-86, PROF CIVIL ENG, UNIV TEX, 86-, ENG FOUND ENDOWED PROF, 87-, DIR, CTR RES WATER RESOURCES, 88- *Concurrent Pos:* Consult, US Army Construct Eng Res Lab, 76-77, UN, NATO, World Bank, Tex Atty Gen Off & others. *Mem:* Am Soc Civil Eng; Am Geophys Union; Am Water Resources Asn; Sigma Xi; Int Asn Hydraul Res; Int Water Resources Asn. *Res:* Field of applications of systems analysis techniques to water resources problems; in particular the application of operations research, probability and statistics to hydraulic design and hydrologic analysis; models for ground water management, reservoir operation under flooding conditions, water reuse planning, hydraulic structure reliability analysis and flood levee design; reliability based design of water distribution systems. *Mailing Add:* Dept Civil Eng Univ Tex Austin TX 78712

MAYS, ROBERT, JR, b El Paso, Tex, Jan 13, 47; m 68; c 2. NONLINEAR OPTICS, LASER PHYSICS. *Educ:* Tex Tech Univ, BS, 69, MS, 71; Tex Christian Univ, PhD(physics), 79. *Prof Exp:* Systs engr, Tex Instruments, Inc, 71-77; tech staff, Avionics Div, Int Tel & Tel, 77-78; mem tech staff, Tex Instruments, Inc, 79-82; sr eng specialist, Vought Corp-LTV Co, 82-85; sr tech staff, Tex Instruments, 85-91; DIR ENG, ELECTRO-OPTICS SYSTS DIV, VARCO CO, 91- *Mem:* Inst Elec & Electronics Engrs; Am Phys Soc; Sigma Xi. *Res:* Electromagnetic and high energy beam interactions with matter; electro-optics, quantum electronics and solid state physics; infrared physics. *Mailing Add:* 11545 Pagemill Rd Dallas TX 75243

MAYS, ROLLAND LEE, b Buffalo, NY, Feb 21, 20; m 44; c 4. ANALYTICAL CHEMISTRY. *Educ:* Univ Buffalo, BA, 52. *Prof Exp:* Chemist, Bliss & Laughlin Co, Inc, 42-48, supvr & chief chemist, 48-52; chemist, Linde Div, Union Carbide Corp, 52-58, develop supvr, 58-63, develop mgr, 63-71, mgr technol, Molecular Sieve Dept, Mat Systs Div, 71-75, dir technol, Molecular Sieve Dept, 75-78 & 79-81, dir res, Linde Div, 78-81. *Mem:* AAAS; Am Chem Soc; Sigma Xi. *Res:* Sorption on solid sorbents and heterogeneous catalysis, particularly in zeolites; sorption, catalytic and ion exchange products; process development and process design. *Mailing Add:* 6611 Falconbridge Rd Chapel Hill NC 27514

MAZAC, CHARLES JAMES, b Deming, NMex, May 4, 40; m 57; c 3. FLAME RETARDANT CHEMISTRY, DIOXIN & FURAN CHEMISTRY. *Educ:* NMex State Univ, BS, 62; MS, 66, PhD(phys chem), 68. *Prof Exp:* Engr res chem, Rocketdyne, 62-64; grad asst chem, NMex State Univ, 64-68; supvr photochem, Phys Sci Lab, NMSU, 64-65; sr res chemist, Corpus Christi Tech Ctr, PPG Indust, 68-75, sr res supvr chem, 75-79, head gen res dept, 74-80; sect leader, Res Dept, Tech Ctr, Celanese Chem Co, 79-85; MGR ANALYTICAL CHEM, GREAT LAKES CHEM CORP, 85- *Concurrent Pos:* Adj prof math, Del Mar Jr Col, 69-81. *Mem:* Am Chem Soc; Am Soc Qual Control; Assoc Off Analytical Chemists; Analysis Lab Mgr Asn. *Res:* Heterogeneous and homogeneous catalysis; gas phase kinetics of organosilanes; photochemistry; iodine chemistry; synthesis of olefin oxides; carbonylation chemistry at high pressures; thermochemistry; dioxins and dibenzo furans. *Mailing Add:* Great Lakes Chem Corp PO Box 2200 W Lafayette IN 47906-0200

MAZAHERI, MOHAMMAD, b Tehran, Iran, July 1, 27; US citizen. ORAL-FACIAL GROWTH, CLEFT LIP & PALATE. *Educ:* Univ Tehran, MDD, 49; Howard Univ, DDS, 56; Univ Pa, MSc, 55. *Prof Exp:* Instr prosthetics, Univ Pa, 52-53; chief dent serv, 57-72, dir clin serv, 72-81, CHIEF MED DENT STAFF, LANCASTER CLEFT PALATE CLIN, 82- *Concurrent Pos:* Consult, Walter Reed Army Hosp & Bethesda Navl Hosp, 60; lectr, Sch Dent, Univ Pa, 63-, asst prof prosthetic dent, Grad Sch Med, 64, adj assoc prof restorative dent, Sch Dent Med, 79-; prof surgery, Hershey Med Ctr, Pa State Univ. *Honors & Awards:* First Prize Educ, Ministry Educ, Tehran, Iran, 50. *Mem:* Am Col Prosthodontists; Am Acad Maxillofacial Prosthetics; Am Dent Asn; Am Col Dentists; Am Cleft Palate Asn (pres, 66-67). *Res:* Oral facial growth in patients with oral facial anomalies; maxillary and dental growth based on longitudinal data; effect of prostheses on maxillary growth, speech production and velopharyngeal stimulation. *Mailing Add:* Lancaster Cleft Palate Clin 223 N Lime St Lancaster PA 17602

MAZARAKIS, MICHAEL GERASSIMOS, b Volos, Greece, Apr 25, 47; US citizen; m 90. PARTICLE BEAM PHYSICS, INERTIAL FUSION. *Educ:* Univ Athens, BS, 60; Univ Paris, PhD(physics), 65; Princeton Univ & Univ Pa, PhD(physics), 71; Mass Inst Technol, cert, 76. *Prof Exp:* Sr res assoc physics, Rutgers Univ, 71-73; vpres & dir exp prog, Fusion Energy Corp, 74-77; physicist, Argonne Nat Lab, Univ Chicago, 78-81; physicist, High Energy Beam Physics Div, 81-90, PHYSICIST & SR MEM TECH STAFF, HIGH CURRENT LINEAR ACCELERATOR DIV, SANDIA NAT LABS, 90- *Concurrent Pos:* Dir, Fusion Energy Corp, 72-77. *Mem:* Am Phys Soc; Inst Elec & Electronics Engrs; Sigma Xi; NY Acad Sci. *Res:* Nuclear fusion; plasma physics; nuclear physics; nuclear astrophysics; accelerator physics; nuclear reactor physics and engineering; pulsed power technology. *Mailing Add:* Sandia Nat Labs Div 1242 PO Box 5800 Albuquerque NM 87185

MAZE, JACK REISER, b San Jose, Calif, Sept 28, 37; m 61. BOTANY. *Educ:* Humboldt State Col, BA, 60; Univ Wash, MS, 63; Univ Calif, Davis, PhD(bot), 65. *Prof Exp:* Lectr bot, Univ Calif, Davis, 65-66; asst prof, Univ Toronto, 66-68; from asst prof to assoc prof, 68-80, PROF BOT, UNIV BC, 80- *Mem:* AAAS; Am Inst Biol Sci; Soc Study Evolution; Bot Soc Am; Int Asn Plant Taxon. *Res:* Plant evolution and taxonomy; embryology and floret development in grasses; variation and development in conifers; theories relating ontogeng and phylogeny. *Mailing Add:* 2878 W 14th Vancouver BC V6K 2X4 Can

MAZE, ROBERT CRAIG, b Galveston, Ind, May 24, 34; m 54; c 3. CHEMICAL ENGINEERING, SURFACE CHEMISTRY. *Educ:* Purdue Univ, BS, 59; Iowa State Univ, MS, 67, PhD(chem eng), 70. *Prof Exp:* Supvr qual control, Hercules, Inc, 59-60; mat engr, Martin Marietta Corp, 60-65; sect mgr, Motorola, Inc, 70-78; MEM STAFF, HEWLETT PACKARD, INC, 78- *Concurrent Pos:* US Atomic Energy Comn assistantship, Iowa State Univ, 66-70. *Mem:* Am Inst Chem Eng; Am Chem Soc. *Res:* Thermal analysis; liquid crystals. *Mailing Add:* Hewlett Packard Inc 1020 NE Circle Blvd Corvallis OR 97330-4241

MAZE, THOMAS HAROLD, b St Paul, Minn, June 1, 51; m 79; c 2. TRANSPORTATION ENGINEERING. *Educ:* Iowa State Univ, BS, 75; Univ Calif, Berkeley, ME, 77; Mich State Univ, PhD(civil eng), 82. *Prof Exp:* Field engr, Blasi Construct Co, 75; intern, Metropolitan Transit Comn, 76-77; res assoc, Univ Fla, 77-79; asst prof civil eng, Wayne State Univ, 79-; AT DEPT CIVIL ENG & ENVIRON SCI, UNIV OKLA. *Concurrent Pos:* Prin investr, Southeastern Transp Authority, 80-81 & Urban Mass Transp Admin, 80- *Mem:* Inst Transp Engrs; Am Soc Civil Engrs; Regional Sci Asn. *Res:* Transit plan and operations; operations and planning of transit maintenance and storage facilities. *Mailing Add:* 4815 Utah Dr Ames IA 50010

MAZEL, PAUL, b Norfolk, Va, Nov 27, 25; m 55; c 3. PHARMACOLOGY, BIOCHEMISTRY. *Educ:* Med Col Va, BS, 46; Trinity Univ, MS, 55; Vanderbilt Univ, PhD(pharmacol), 60. *Prof Exp:* Res asst biol, Southwest Found Res & Educ, 54-55; res asst pharmacol, Yale Univ, 55-56; res asst, Vanderbilt Univ, 56-60, instr, 60-61; from asst prof to assoc prof, 61-71, PROF PHARMACOL, GEORGE WASHINGTON UNIV, 71-, PROF ANESTHESIOL, 74- *Concurrent Pos:* USPHS fel, 60-61; lectr, US Naval Dent Sch, 61-62; consult, Datatrol Corp & Mediphone, Inc, 62-63 & Wallace Labs, 65-; vis prof, Fed City Col, 72-74 & Catholic Univ Am, 80-82; dir, Nurse Anesthetists Prog, George Washington Univ Med Ctr; vis scholar/worker, Dept Pharmacol, Univ Cambridge, UK, 85. *Mem:* Soc Toxicol; Am Soc Pharmacol & Exp Therapeut; Am Soc Anesthesiologists; Soc Study Xenobiotic Metab. *Res:* Pharmacology of central nervous system acting drugs; physiological disposition of drugs; barbiturate metabolism; adaptive enzyme formation; membrane permeability; microsomal enzymes; blood-brain barrier; immunochemistry. *Mailing Add:* Dept Pharmacol George Washington Univ Washington DC 20037

MAZELIS, MENDEL, b Chicago, Ill, Aug 31, 22; m 69; c 1. PLANT BIOCHEMISTRY. *Educ:* Univ Calif, BS, 43, PhD(plant physiol), 54. *Prof Exp:* Jr res biochemist, Univ Calif, 54-55; res assoc & instr, Univ Chicago, 55-57; assoc chemist, Western Regional Res Lab, USDA, 57-61; lectr, 61-65, asst biochemist, 61-64, assoc biochemist, 64-73, assoc prof food sci & technol, 65-73, PROF FOOD SCI & TECHNOL, UNIV CALIF, DAVIS, 73-, BIOCHEMIST, 73- *Mem:* Am Soc Biol Chemists; Am Soc Plant Physiol; Biochem Soc; Phytochem Soc Eur; Phytochem Soc NAm; Inst Food Technologists. *Res:* Intermediary metabolism of higher plants; enzymology. *Mailing Add:* Dept of Food Sci & Technol Univ of Calif Davis CA 95616

MAZERES, REGINALD MERLE, b Metairie, La, Feb 15, 34; m 57; c 3. ALGEBRA, NUMBER THEORY. *Educ:* Univ Southwestern La, BS, 59; Auburn Univ, MS, 60, PhD(math), 69. *Prof Exp:* Instr math, Auburn Univ, 62-63; from asst prof to assoc prof, 63-71, PROF MATH, TENN TECHNOL UNIV, 71- *Concurrent Pos:* US Navy, 52-56. *Mem:* Math Asn Am. *Res:* Inflations and enlargements of semigroups. *Mailing Add:* Box 5054 Tenn Technol Univ Cookeville TN 38505

MAZESS, RICHARD B, b Philadelphia, Pa, June 10, 39. MEDICAL PHYSICS. *Educ:* Pa State Univ, BA, 61, MA, 63; Univ Wis-Madison, PhD(anthrop), 67. *Prof Exp:* NIH fel, 67-68, asst prof anthrop, 67-69, from asst prof to assoc prof radiol, 69-85, prof med physics, 85, EMER PROF MED PHYSICS, UNIV WIS-MADISON, 85- *Concurrent Pos:* Pres, Lunar Radiation Corp, 80- *Mem:* Am Asn Physicists Med. *Res:* Radiological measurements of skeleton and body composition; growth and aging. *Mailing Add:* 2526 Gregory St Madison WI 53711

MAZIA, DANIEL, b Scranton, Pa, Dec 18, 12; m 38; c 2. CELL BIOLOGY. *Educ:* Univ Pa, AB, 33, PhD(zool), 37. *Hon Degrees:* PhD, Univ Stockholm, Sweden, 76; ScD, Univ Mo, 88; DHL, Georgetown Univ, 89. *Prof Exp:* Instr zool, Univ Pa, 35-36; Nat Res Coun fel, Princeton Univ & Marine Biol Labs, Woods Hole, 37-38; from asst prof to prof, Univ Mo, 38-50; from assoc prof to prof zool, Univ Calif, Berkeley, 51-79; PROF BIOL SCI, STANFORD UNIV, 79. *Concurrent Pos:* Trustee, Marine Biol Lab, Woods Hole, Mass, 50-58, head physiol, 52-56. *Honors & Awards:* Wilson Medal, Am Soc Cell Biol, 81. *Mem:* Nat Acad Sci; Am Soc Cell Biol; Am Acad Arts & Sci. *Res:* Ionic changes in stimulation; ion accumulation and exchange; chemistry of chromosomes; nuclear and cellular physiology; surface chemistry of enzymes; biochemistry of mitosis; centrosomes. *Mailing Add:* Hopkins Marine Sta Stanford Univ Pacific Grove CA 93950

MAZO, JAMES EMERY, b Bernardsville, NJ, Jan 15, 37; m 59; c 2. APPLIED MATHEMATICS. *Educ:* Mass Inst Technol, BS, 58; Syracuse Univ, MS, 60, PhD(physics), 63. *Prof Exp:* Res assoc physics, Ind Univ, 63-64; MEM TECH STAFF APPL MATH, BELL LABS, 64- *Mem:* Am Phys Soc; Inst Elec & Electronics Engrs. *Res:* Communication theory, noise theory and information theory. *Mailing Add:* AT&T Bell Labs Rm 2c-377 Mountain Ave Murray Hill NJ 07974

MAZO, ROBERT MARC, b Brooklyn, NY, Oct 3, 30; m 54; c 3. THEORETICAL CHEMISTRY. *Educ:* Harvard Univ, AB, 52; Yale Univ, MS, 53, PhD(chem), 55. *Prof Exp:* NSF res fel, Univ Amsterdam, 55-56; res assoc, Univ Chicago, 56-58; asst prof chem, Calif Inst Technol, 58-62; assoc prof chem, 62-65, dir inst theoret sci, 64-67, assoc dean grad sch, 67-71, dir instr theoret sci, 84-87, PROF CHEM, UNIV ORE, 65- *Concurrent Pos:* NSF sr fel & vis prof, Free Univ Brussels, 68-69, prog dir, 77-78; vis prof Tech Univ, Aachen & Weizmann Inst, Rehovoth, 81-82. *Mem:* AAAS; Am Phys Soc; Am Chem Soc. *Res:* Statistical mechanics; kinetic theory; irreversible thermodynamics; intermolecular forces. *Mailing Add:* Dept of Chem Univ of Ore Eugene OR 97403-1253

MAZUMDAR, MAINAK, b Calcutta, India, Nov 19, 35; m 60; c 3. OPERATIONS RESEARCH, STATISTICS. *Educ:* Univ Calcutta, BS, 54, MS, 56; Cornell Univ, PhD(appl probability & statist), 66. *Prof Exp:* Res asst & res assoc opers res, Cornell Univ, 62-66; sr mathematician, 66-70, fel mathematician, 70-74, adv mathematician, Westinghouse Res Labs, 74-; AT DEPT INDUST ENG, UNIV PITTSBURGH. *Concurrent Pos:* Lectr, Univ Pittsburgh, 67- *Mem:* Am Statist Asn; Inst Math Statist. *Res:* Mathematical theory of reliability; application of reliability methods to nuclear engineering. *Mailing Add:* Dept Indust Eng Benedum 1048 Univ Pittsburgh Pittsburgh PA 15261

MAZUMDER, BIBHUTI R, b July 1, 24; Indian citizen; m 51; c 3. SURFACE CHEMISTRY, SOLID STATE CHEMISTRY. *Educ:* Univ Calcutta, BS, 44; Univ Dacca, MS, 47; Howard Univ, PhD(phys chem), 58; FRIC. *Prof Exp:* Chemist, Standard Pharmaceut, India, 48-54; res assoc chem, Cornell Univ, 58-59; chemist, Unilever, Eng, 59-60; head phys chem, Lever Bros, India, 60-67; PROF CHEM, MORGAN STATE UNIV, 67- *Mem:* Sr mem Am Chem Soc. *Res:* Research and development of soaps, detergents and cosmetics. *Mailing Add:* Dept Chem Morgan State Univ Coldspring Lane & Hillen Rd Baltimore MD 21239

MAZUMDER, RAJARSHI, b Dacca, Bangladesh. BIOCHEMISTRY. *Educ:* Univ Calcutta, BSc, 51, MSc, 53; Univ Calif, Berkeley, PhD(biochem), 59. *Prof Exp:* Pool officer biochem, All-India Inst Med Sci, New Delhi, 64-65, asst prof, 65-67; asst prof, 67-73, ASSOC PROF BIOCHEM, MED SCH, NY UNIV, 73- *Concurrent Pos:* Fel biochem, Med Sch, NY Univ, 60-63. *Mem:* Am Soc Biol Chem; Harvey Soc. *Res:* Mechanism of protein synthesis. *Mailing Add:* Dept Biochem NY Univ Med Sch 550 First Ave New York NY 10016

MAZUR, ABRAHAM, b New York, NY, Oct 8, 11; m 40; c 2. BIOCHEMISTRY. *Educ:* City Col New York, BS, 32; Columbia Univ, AM, 34, PhD(biochem), 38. *Prof Exp:* Tutor, 36-38, from instr to prof chem, City Col New York, 38-75, chmn dept, 69-72; vpres res, Lindsley F Kimball Res Inst, NY Blood Ctr, 75-83; RETIRED. *Concurrent Pos:* Carnegie Corp fel, Col Physicians & Surgeons, Columbia Univ, 38-39; Guggenheim fel, 49-50; res assoc biochem, Med Col, Cornell Univ, 41-49, asst prof, 49-66. *Mem:* AAAS; fel Soc Exp Biol & Med; Am Chem Soc; Am Soc Biol Chemists; Harvey Soc. *Res:* Acetylation mechanism; fat metabolism hormone; stilbestrol; components of autotrophic organisms; anticholinesterases; chemical factor in shock; ferritin; iron metabolism. *Mailing Add:* 254 E 68th St New York NY 10021

MAZUR, BARRY, b New York, NY, Dec 19, 37; c 1. NUMBER THEORY, ALGEBRAIC GEOMETRY. *Educ:* Princeton Univ, PhD(math), 59. *Prof Exp:* Jr fel, 59-62, from asst prof to prof, 62-82, WILLIAM PETSCHEK PROF MATH, HARVARD UNIV, 82- *Honors & Awards:* Veblen Prize in Geom, Am Math Soc, Cole Prize in Number Theory. *Mem:* Nat Acad Sci; AAAS. *Mailing Add:* Dept Math Harvard Univ One Oxford St Cambridge MA 02138

MAZUR, ERIC, b Amsterdam, Holland, Nov 14, 57; Netherlands; m 84; c 1. LASER SPECTROSCOPY, EXPERIMENTAL PHYSICS. *Educ:* Univ Leiden, Netherlands, MA, 77, PhD(physics), 81. *Prof Exp:* Res assoc, Huygens Laboratorium, Leidun Univ, 78-81; post doctorate res fel, Div Appl Sci, Harvard Univ, 82-84, from asst prof to assoc prof, 84-90, GORDON MCKAY PROF APPL PHYSICS & PROF PHYSICS, HARVARD UNIV, 90- *Concurrent Pos:* Consult, Northeast Res Assocs, Woburn, Mass, 84-90. *Honors & Awards:* CJ Kok Price, Netherlands, 83; Presidential Young Investr Award, NSF, 88. *Mem:* Fel Am Physical Soc; Optical Soc Am. *Mailing Add:* Harvard Univ Pierce Hall 225 Cambridge MA 02138

MAZUR, JACOB, b Lodz, Poland, Dec 17, 21; nat US; m 51; c 2. POLYMER PHYSICS. *Educ:* Hebrew Univ, MSc, 45, PhD(phys chem), 48. *Prof Exp:* Res fel, Calif Inst Technol, 48-50; vis fel, Univ Chicago, 50-51; res scientist, Weizmann Inst Sci, Israel, 51-55; res assoc, Univ Ill, 55-57; res chemist, Dow Chem Co, 57-60; phys chemist, Nat Bur Standards, 60-86; CONSULT, NAT CANCER INST, FREDERICK, MD, 86- *Concurrent Pos:* Rockefeller Found fel, 49-50. *Honors & Awards:* Morrison Award, NY Acad Sci, 59. *Mem:* Fel Am Phys Soc. *Res:* Theoretical physical chemistry; high polymer physics; statistical mechanics, biophysics, and molecular biology. *Mailing Add:* 4913 Morning Glory Ct Rockville MD 20853

MAZUR, PETER, b New York, NY, Mar 3, 28; m 53; c 1. CELL PHYSIOLOGY, CRYOBIOLOGY. *Educ:* Harvard Univ, AB, 49, PhD(biol), 53. *Prof Exp:* Mem staff, Hq, Air Res & Develop Command, US Air Force, 53-57; NSF fel, Princeton Univ, 57-59; BIOLOGIST, OAK RIDGE NAT LAB, 59- *Concurrent Pos:* Mem, Am Inst Biol Sci Adv Comt, Biol & Med Br, Off Naval Res, 63-66 & Environ Biol Br, NASA, 66-; mem adv bd, Am Type Cult Collection, 66-70; vis lectr, Duke Univ, 67; chmn long-range planning off, Oak Ridge Nat Lab, 70, sci dir biophys & cell physiol, 74-75; prof, Univ Tenn-Oak Ridge Grad Sch Biomed Sci, 70-; mem, Harvard Bd Overseers Vis Comt Biol, 70-; mem space sci bd, Nat Acad Sci, 75- *Mem:* Fel AAAS; Soc Gen Physiol; Biophys Soc; Bot Soc Am; Soc Cryobiol (pres, 73-74); Sigma Xi. *Res:* Low temperature biology; freezing and drying; cell water, membranes and permeability. *Mailing Add:* Biol Div Oak Ridge Nat Lab PO Box Y Oak Ridge TN 37830

MAZUR, ROBERT HENRY, b Indianapolis, Ind, June 15, 24; m 54; c 3. PEPTIDE SYNTHESIS, MEDICINAL CHEMISTRY. *Educ:* Mass Inst Technol, BS, 48, PhD(org chem), 51. *Prof Exp:* NIH fel, Swiss Fed Inst Technol, 51-52; res chemist, G D Searle & Co, 52-85 & NutraSweet Co, 86-88; RETIRED. *Concurrent Pos:* Nat Cancer Inst fel, Cambridge, 56-57. *Mem:* Am Chem Soc. *Res:* Molecular rearrangements; reaction mechanisms; alkaloids; steroids; peptides. *Mailing Add:* 523 Maple Ave Wilmette IL 60091

MAZUR, STEPHEN, b Baltimore, Md, Apr 9, 45; m 69. ORGANIC CHEMISTRY. *Educ:* Yale Univ, BS, 67; Univ Calif, Los Angeles, MS, 69, PhD(chem), 71. *Prof Exp:* Assoc chem, Univ Calif, Los Angeles, 69-70; NSF fel & res assoc, Columbia Univ, 71-72, lectr, 72-73; asst prof chem, Univ Chicago, 73-79; MEM STAFF, CENT RES DEPT, E I DU PONT DE NEMOURS & CO, INC, 79- *Mem:* Sigma Xi. *Res:* Physical organic chemistry; polymers; electron transfer processes. *Mailing Add:* 4601 Weldin Rd Wilmington DE 19803

MAZUREK, THADDEUS JOHN, b Tarnogrod, Poland, Aug 11, 42; US citizen; m 74; c 2. MAGNETOHYDRODYNAMICS, IONOSPHERIC PLASMA DISTURBANCES. *Educ:* Fordham Col, BS, 66; Yeshiva Univ, MA, 68, PhD(astrophys), 73. *Prof Exp:* Res assoc astrophys, Yeshiva Univ, 73 & Harvard Col Observ, 73-74; res scientist, Univ Tex, 74-78; guest prof, Nordita-Bohr Inst, 78-79; asst prof physics, State Univ NY, Stony Brook, 79-82; RES PHYSICIST PLASMA PHYSICS, MISSION RES CORP, 82- *Concurrent Pos:* Prin investr, NSF, 76-78 & Dept Energy, 80-82. *Mem:* Am Astron Soc; Int Astron Union; NY Acad Sci; AAAS. *Res:* Theory of plasma and neutral gas flows for disturbed environments within the upper atmosphere; application of computer techniques to evolutionary studies of space plasmas injected in and above the ionosphere. *Mailing Add:* 4580 Nueces Dr Santa Barbara CA 93110

MAZURKIEWICZ, JOSEPH EDWARD, b Brooklyn, NY, Mar 19, 42. CELL BIOLOGY, NEUROBIOLOGY. *Prof Exp:* NIH res fel anat, Yale Univ Sch Med, 73-75, res assoc cytol & cell biol, 75-78; from asst prof to assoc prof, 78-86, PROF ANAT, ALBANY MED COL, UNION UNIV, 81- *Concurrent Pos:* NIH fel Gen Med Sci, 73-75; NSF instrumentation grant, 80-82; prin investr, NIH res grant, 85-86 & 88-, NIH spec study sect, 86 & 87. *Mem:* Am Soc Cell Biol; Electron Micros Soc Am; Histochem Soc; Sigma Xi; Am Asn Anatomists. *Res:* Structure and function of biomembranes especially of ion and water transporting epithelia; immunocytochemistry; electron microscopy; cytochemistry; Pathogenesis of motor neuron degineation diseases using animal model in monise designated as Mnd: motor neuron degeneration monse. *Mailing Add:* Dept Anat Albany Med Col 47 New Scotland Ave Albany NY 12208

MAZURKIEWICZ-KWILECKI, IRENA MARIA, b Poland; nat Can; m 57; c 1. BRAIN HISTAMINE REGULATION IN AGING, ALZHEIMER DISEASE & STRESS. *Educ:* Jagellonian Univ, MPharm, 47; McGill Univ, MSc, 55, PhD(pharmacol), 57. *Prof Exp:* Res asst pharmacol, Sch Med, Jagellonian Univ, 47-48; res asst, McGill Univ, 53-57; res pharmacologist, Food & Drug Labs, Dept Nat Health & Welfare, Ont, 57-59; res pharmacologist, Prov Labs, Ministry of Health, Que, 59-60; from asst prof to prof, 60-89, actg head dept, 64-65, EMER PROF PHARMACOL, FAC MED, DEPT PHARMACOL, UNIV OTTAWA, 89- *Concurrent Pos:* Am Med Life Ins Fund Med Res Found fel, 57; vis prof, Polish Acad Sci, 74-75; founder & pres, Can Histamine Res Group, 83-85; prin investr, OMHF, Can. *Mem:* AAAS; Fr-Can Asn Advan Sci; NY Acad Sci; Am Soc Pharmacol & Exp Therapeut; Pharmacol Soc Can. *Res:* Pharmacology of central nervous system; neuropharmacology; brain histamine regulation in aging, Alzheimer disease and in stress. *Mailing Add:* Dept Pharmacol Fac Med Univ Ottawa 451 Smyth Rd Ottawa ON K1H 8M5 Can

MAZZENO, LAURENCE WILLIAM, b New Orleans, La, Sept 4, 21; m 44; c 4. ORGANIC CHEMISTRY. *Educ:* Loyola Univ, La, BS, 42; Univ Detroit, MS, 44. *Prof Exp:* Asst chem, Univ Detroit, 42-44; chemist, Southern Regional Res Lab, Bur Agr & Indust Chem, USDA, 44-53 & Southern Utilization Res Br, 53-58, head new prod invests, Southern Utilization Res Div, 58-59, head chem modification invests, 59-61, asst to dir indust develop, 61-69, res chemist, Cotton Finishes Lab, Southern Mkt & Nutrit Res Div, 69-74, head Tech & Econ Anal Res, Southern Regional Res Ctr, 74-78; RETIRED. *Concurrent Pos:* Lectr chem, Our Lady Holy Cross Col, 78- *Mem:* Am Chem Soc; Sigma Xi. *Res:* Textile finishing, including chemical modification and resin finishing of cotton; flame and weather resistant finishes for cotton textiles. *Mailing Add:* 944 Beverly Garden Dr Metairie LA 70002

MAZZITELLI, FREDERICK R(OCCO), b Pittston, Pa, Mar 4, 24; wid; c 9. AERONAUTICAL ENGINEERING, OPERATIONS ANALYSIS. *Educ:* Pa State Univ, BS, 47. *Prof Exp:* Proj engr, Convertawings, Inc, 53-55; chief res aerodynamicist, Vertol Aircraft Corp, 55-57, chief design analysis, 57-58; preliminary design engr, Grumman Aerospace Corp, 59-61, chief transp, logistics opers analysis, 61-67, mgr transp systs, 67-76, mgr advan studies & analysis dep div, 76-85, dir opers analysis, Corp Develop, 86-89, dir opers analysis, strategic & mkt planning, Grumman Corp, 89-90; RETIRED. *Concurrent Pos:* Consult transp adv comt, Huntington Town Planning Bd, 67-69 & Aviation Adv Comn, 71. *Mem:* Am Helicopter Soc; assoc fel Am Inst Aeronaut & Astronaut. *Res:* Design of vertical take-off and landing aircraft; transportation systems and logistic functions of commercial and military aircraft systems; business analysis of domestic and international airlines; military and commercial aircraft requirements analysis. *Mailing Add:* RD 1 Box 1106 Lake Ariel PA 18436

MAZZOCCHI, PAUL HENRY, b New York, NY, May 6, 39; m 61. ORGANIC CHEMISTRY. *Educ:* Queens Col, NY, BS, 61; Fordham Univ, PhD(org chem), 66. *Prof Exp:* NIH fel org chem, Cornell Univ, 65-67; from asst prof to assoc prof, 67-77, PROF ORG CHEM, UNIV MD, COLLEGE PARK, 77- *Mem:* Interam Photochem Soc; Am Chem Soc. *Res:* Organic photochemistry; synthetic chemistry. *Mailing Add:* Dept Chem Univ Md College Park MD 20740

MAZZOLENI, ALBERTO, b Milan, Italy, Sept 12, 27; US citizen; c 2. CARDIOLOGY. *Educ:* Univ Milan, MD, 52; Am Bd Internal Med, dipl, 63; Am Bd Cardiovasc Dis, dipl, 68; Am Bd Internal Med, dipl, 77. *Prof Exp:* Intern med, Miriam Hosp, Providence, RI, 57-58; res asst, Lemuel Shattuck Hosp, Boston, 58-59 & Boston City Hosp, 59-60; from asst prof med to assoc prof clin med, Sch Med, Univ Ky, 61-72; asst chief internal med, 64-73, chief cardiol, 67-72, dir, Coronary Care Unit, 73-77, DIR HEART STA, VET ADMIN HOSP, LEXINGTON, 78-; ASSOC PROF MED, SCH MED, UNIV KY, 72- *Concurrent Pos:* Res fel cardiol, Beth Israel Hosp, Boston, 55-57 & 60-61. *Mem:* Fel Am Col Cardiol; fel Am Col Physicians. *Res:* Electrocardiogram diagnosis of cardiac hypertrophy; component heart weights. *Mailing Add:* 3772 Gloucester Dr Lexington KY 40511

MAZZONE, HORACE M, b Franklin, Mass, May 19, 30; m 62; c 1. MICROBIOLOGY. *Educ:* Boston Col, BS, 51, MS, 53; Univ Wis, PhD(biochem), 59. *Prof Exp:* Res assoc biochem, Long Island Biol Asn & dept genetics, Carnegie Inst, 59; fel pharmacol, Harvard Med Sch, 59-61; res assoc path, Children's Hosp Med Ctr & Children's Cancer Res Found, Boston, 61-63; res assoc biol, Mass Inst Technol, 63-65; biochemist, 65-79, MICROBIOLOGIST, FOREST INSECT & DIS LAB, USDA, 80- *Concurrent Pos:* Res assoc physics, Mass Gen Hosp, 64-65; lectr, Yale Univ, 72- *Mem:* AAAS; Am Chem Soc; Am Soc Cell Biol; Tissue Cult Asn. *Res:* Viruses; properties of infectious agents; tissue culture. *Mailing Add:* 199 Center Rd Easton CT 06425

MAZZUCATO, ERNESTO, b Padova, Italy, July 7, 37; m 64; c 3. PLASMA PHYSICS. *Educ:* Univ Padova, Dr phys, 60; Univ Roma, Libero Docente (plasma physics), 70. *Prof Exp:* Sr res physicist plasma physics, Univ Padova, 60-62 & Comt Nat Nuclear Energy, 62-72; RES PHYSICIST PLASMA PHYSICS, PRINCETON UNIV, 72- *Mem:* Am Phys Soc. *Res:* Basic plasma physics; study of confinement, stability and heating of plasmas in toroidal magnetic configurations; controlled thermonuclear fusion research. *Mailing Add:* Princeton Univ PO Box 451 Princeton NJ 08543

MAZZULLO, JAMES MICHAEL, b Brooklyn, NY, Aug 28, 55; m 82. STRATIGRAPHY-SEDIMENTATION, MARINE SCIENCES. *Educ:* Brooklyn Col, BS, 77; Univ SC, MS, 79, PhD(geol), 81. *Prof Exp:* ASST PROF GEOL, TEX A&M UNIV, 81- *Concurrent Pos:* Staff scientist, Ocean Drilling Prog, 84- *Honors & Awards:* Presidential Young Investr Award, NSF, 85. *Mem:* Am Asn Petrol Geologists; Geol Soc Am; Int Asn Sedimentologists; Soc Econ Paleontologists & Mineralogists. *Res:* Sediment provenance and transport history; quartz-grain shapes and surface textures; sedimentology and stratigraphy of Permian Basin, Texas; sandstone diagenesis and reservoir characterization; eolian sedimentation; grain size measurement and analysis. *Mailing Add:* Dept Geol Tex A&M Univ College Station TX 77843

MEACHAM, ROBERT COLEGROVE, b Moultrie, Ga, May 1, 20; m 43; c 4. COMBINATORICS & FINITE MATHEMATICS. *Educ:* Southwestern at Memphis, AB, 42; Brown Univ, ScM, 48, PhD(appl math), 49. *Prof Exp:* Instr math, Carnegie Inst Technol, 49-50, asst prof, 50-54; assoc prof, Univ Fla, 54-60; prof, 60-90, EMER PROF MATH, ECKERD COL, 90- *Concurrent Pos:* Consult, RCA Serv Co, 57-64; NSF fel comput sci, Stanford Univ, 65-66; pres, Fla Sect, Math Asn Am, 67-68; mem panel on res, Sch Math Study Group, 68-72; vis prof, Fla State Univ, 61, Univ Tenn, 62, Clemson Univ, 84. *Mem:* Am Math Soc; Math Asn Am (mem Bd Gov, 74-77); Soc Indust & Appl Math. *Res:* Mechanics; numerical analysis; differential equations. *Mailing Add:* Dept Math Eckerd Col PO Box 12560 St Petersburg FL 33733

MEACHAM, ROGER HENING, JR, b Richmond, Va, Sept 10, 42; m 65; c 5. PHARMACOLOGY, DRUG METABOLISM. *Educ:* Univ Richmond, BS, 65, MS, 67; Med Col Va, PhD(pharmacol), 71. *Prof Exp:* Sr scientist metab chem, Wyeth Labs, 71-77, supvr, Biochem Pharm Unit, Drug Disposition Sect, 77-85; group leader drug metab, 85-86, dept mgr, 86-88, DIR, RORER GROUP, 88- *Concurrent Pos:* mem steering comt, Delaware Valley Drug Metab Discussion Group, 79-85. *Mem:* Sigma Xi; Int Soc Study Xenobiotics; Am Soc Pharmacol & Exp Therapeut; NY Acad Sci; Am Asn Pharmaceut Scientists. *Res:* Drug biotransformation; drug analysis; pharmacokinetics; drug-interaction. *Mailing Add:* Pharmaceut Res & Develop Div Drug Metab Rhone-Poulenc Rorer Cent Res 800 Bus Ctr Dr Horsham PA 19044

MEACHAM, WILLIAM FELAND, b Washington, DC, Dec 12, 13; m 44; c 4. SURGERY. *Educ:* Western Ky State Col, BS, 36; Vanderbilt Univ, MD, 40; Am Bd Surg, dipl, 47; Am Bd Neurol Surg, dipl, 48. *Prof Exp:* Intern surg, Univ Hosp, 40-41, asst surg, Sch Med, 41-43, instr, 43-44, from asst prof to assoc prof clin surg, 47-53, assoc prof neurol surg, 53, CLIN PROF NEUROL SURG & CHMN DEPT, SCH MED, VANDERBILT UNIV, 54-, ASSOC DIR, SURG SERV SECT, 75-, EMER CLIN PROF NEUROSURG, 85- *Concurrent Pos:* Howe fel neurosurg, Sch Med, Vanderbilt Univ, 45-47; asst resident, Vanderbilt Univ Hosp, 41-43, resident surgeon, 43-44, asst vis surgeon, 44-, assoc vis surgeon, Outpatient Serv, 44-, neurosurgeon-in-chief; vol asst, Montreal Neurol Inst, 47-; asst prof, Meharry Med Col, 50; attend neurosurgeon & consult, hosps; chmn, Dept Neurol Surg. *Mem:* Neurosurg Soc Am (pres, 52); Am Asn Neurol Surg; Soc Univ Surg; AMA; Am Col Surg. *Res:* Intracranial tumors and aneurysms; steroatactic surgery. *Mailing Add:* 539 Med Art Bldg Nashville TN 37212

MEACHAM, WILLIAM ROSS, b Ft Worth, Tex, Jan 12, 23; m 50; c 3. VERTEBRATE ZOOLOGY. *Educ:* Agr & Mech Col, Tex, BS, 48; NTex State Col, MS, 50; Univ Tex, PhD(vert zool), 58. *Prof Exp:* Head dept, Univ Tex, 63-77, from asst prof to prof biol, 50-89; RETIRED. *Res:* Animal ecology; evolution; genetics; vertebrate population dynamics. *Mailing Add:* 1905 Lanewood Ft Worth TX 76105

MEAD, ALBERT RAYMOND, b San Jose, Calif, July 17, 15; m 42; c 2. MALACOLOGY. *Educ:* Univ Calif, BS, 38; Cornell Univ, PhD(zool), 42. *Prof Exp:* With Marine Biol Lab, 41-42; instr, US Army Col, Brit WAfrica, 44-45; from instr to prof zool, 46-76, head dept, 56-67, coordr marine sci prog, 67-70, cur invert, 67-71, coordr undergrad prog, Biol Sci, 70-76, assoc dean, Liberal Arts Col, 76-80, prof gen biol, 80-85, EMER PROF ECOL & EVOLUTIONARY BIOL, UNIV ARIZ, 85- *Concurrent Pos:* Res fel, Univ Calif, 46; res assoc, Pac Sci Bd, nat Res Coun, 48 & 49 & NSF, 54; Pac Sci Coun observer, UNESCO Adv Comt Humid Tropics Res, 61; mem, Invert Consults Comt Pac, Pac Sci Bd, Nat Res Coun-Nat Acad Sci, 63-; guest scientist, Royal Mus Cent Africa, Tervuren, Belg, 74-75, 77 & 81; distinguished vis scientist, Univ Mich, 83. *Mem:* Soc Invert Path; AAAS; Am Soc Zool; Am Malacol Union (pres, 63); Am Soc Zool Nomenclature. *Res:* Giant African land snail ecology and control; comparative genital anatomy and physiology of Gastropoda; speciation and taxonomy of Achatinidae; economic malacology; molluscan pathology. *Mailing Add:* 310 Bio Sci W Univ Ariz Tucson AZ 85721

MEAD, CARVER ANDRESS, b Bakersfield, Calif, May 1, 34; m 54; c 3. PHYSICS COMPUTATION. *Educ:* Calif Inst Technol, BA, 56, MS, 57, PhD, 60. *Hon Degrees:* Dr, Univ Lund, 87, Univ Southern Calif, 91. *Prof Exp:* From instr to prof elec eng & computer sci, 57-80, GORDON & BETTY MOORE PROF COMPUTER SCI, CALIF INST TECHNOL, 80- *Concurrent Pos:* Consult, NASA Adv Coun & Comt, 78; Walker-Ames distinguished vis prof, Univ Wash, 90. *Honors & Awards:* T D Callinan Award, Electrochem Soc, 71; Centennial Medal, Inst Elec & Electronic Engrs, 84; Harry Goode Mem Award, Am Fedn Info Processing Socs, 85; Walter B Wriston Lectr in Pub Policy Award, 87. *Mem:* Nat Acad Sci; Nat Acad Eng; fel Am Phys Soc; foreign mem Royal Swed Acad Eng Sci. *Res:* Electron transport in thin films; semiconductor surface barriers. *Mailing Add:* Dept Computer Sci 139-74 Calif Inst Technol Pasadena CA 91125

MEAD, CHESTER ALDEN, b St Louis, Mo, Dec 9, 32; m; c 2. MOLECULAR QUANTUM MECHANICS, SYMMETRY IN CHEMISTRY. *Educ:* Carleton Col, BA, 54; Wash Univ, PhD, 57. *Prof Exp:* Res assoc chem, Brookhaven Nat Lab, 57-58; from asst prof to assoc prof, 58-66, PROF PHYS CHEM, UNIV MINN, MINNEAPOLIS, 66- *Concurrent Pos:* Consult, Brookhaven Nat Lab, 59-63; vis prof, Birkbeck Col, London, 64-65, Free Univ, Berlin, 71-72, Technische Hochschule, Aachen, Ger, 80-81. *Mem:* Am Phys Soc; Am Chem Soc. *Res:* Quantum mechanics; algebraic techniques in theoretical chemistry; molecular quantum mechanics, especially corrections to adiabatic approximation; generalized entropy in irreversible thermodynamics. *Mailing Add:* Dept of Chem Univ of Minn Minneapolis MN 55455

MEAD, DARWIN JAMES, b Dowagiac, Mich, June 27, 10; m 36; c 3. CHEMISTRY. *Educ:* Kalamazoo Col, AB, 32; Brown Univ, ScM, 33, PhD(chem), 36. *Prof Exp:* Instr chem, Colby Col, 36-38; res chemist, Gen Elec Co, 38-46; assoc prof, 46-75, EMER ASSOC PROF PHYSICS, UNIV NOTRE DAME, 75- *Res:* Conductance of electrolytes; properties of high polymers; dielectric properties of polymers. *Mailing Add:* 1101 Cleveland Ave South Bend IN 46628

MEAD, EDWARD JAIRUS, b Cleveland, Ohio, Oct 3, 28; m 59; c 2. INORGANIC CHEMISTRY. *Educ:* Va Mil Inst, BS, 49; Purdue Univ, MS, 52, PhD(inorg chem), 55. *Prof Exp:* Instr chem, Va Mil Inst, 49-50; res chemist, E I du Pont de Nemours & Co, Inc, 54-63, res supvr, 64-65, prod develop mgr, 65-73, dir lab, Dept Pigments, Exp Sta, 74-78, mgr New Bus Develop, 78-80; CONSULT, 91- *Mem:* Am Chem Soc. *Res:* Substituted borohydrides; textile fibers; synthesis and properties of organic and inorganic pigments; superconductivity. *Mailing Add:* Four Leigh Ct Wilmington DE 19808

MEAD, FRANK WALDRETH, b Columbus, Ohio, June 11, 22; wid; c 2. ENTOMOLOGY. *Educ:* Ohio State Univ, BS, 47, MS, 49; NC State Univ, PhD(entom), 68. *Prof Exp:* Asst entom, Ohio State Univ, 48-49; scout Japanese beetle control proj, Bur Entom & Plant Quarantine, USDA, 48, biol aid, Div Forest Insect Invest, 50-53; entomologist, 53-71, TAXONOMIC ENTOMOLOGIST, DIV PLANT INDUST, FLA DEPT AGR & CONSUMER SERV, 71- *Concurrent Pos:* Res asst insect mus, NC State Univ, 58-60; courtesy assoc prof, Dept Entom & Nematol, Univ Fla; adj assoc prof, Dept Entom, Fla A&M Univ. *Honors & Awards:* Cert Appreciation for Serv Rendered in Field of Entom, Fla Entom Soc, 75 & 82. *Mem:* Entom Soc Am; Am Registry Prof Entomologists; Soc Syst Zool; Am Birding Asn; Audubon Soc. *Res:* Fulgoroidea, especially Oliarus (Cixiidae); Auchenorrhynchous Homoptera; Culicidae; Heteroptera. *Mailing Add:* Div Plant Indust PO Box 1269 Fla Dept Agr & Consumer Serv Gainesville FL 32602

MEAD, GILBERT DUNBAR, b Madison, Wis, May 31, 30; m 51, 68; c 4. GEOPHYSICS. *Educ:* Yale Univ, BS, 52, MA, 53; Univ Calif, Berkeley, PhD(physics), 62; Univ Md, Baltimore, JD, 91. *Prof Exp:* Instr sci high sch, Calif, 53-55; res asst, Lawrence Radiation Lab, Univ Calif, 57-62; physicist lab theoret studies, Goodard Space Flight Ctr, NASA, 62-68, physicist lab space physics, 68-73, head, Geophys Br, 73-80, geophysicist, Crustal Dynamics Proj, Goodard Space Flight, Ctr, NASA, 80-87; ATTY PVT PRACT, 91- *Concurrent Pos:* Lectr dept space sci & appl physics, Cath Univ Am, 64-67. *Mem:* Fel Am Phys Soc; Am Geophys Union. *Res:* Experimental high energy physics; space and magnetospheric physics; Jupiter's magnetosphere; geomagnetism; magnetospheric models; plate tectonics; models of plate motion; geodesy. *Mailing Add:* 7724 Hanover Pkwy, #302 Greenbelt MD 20770

MEAD, GILES WILLIS, b New York, NY, Feb 5, 28. ICHTHYOLOGY. *Educ:* Stanford Univ, AB, 49, AM, 52, PhD(biol), 53. *Prof Exp:* Fishery res biologist, US Fish & Wildlife Serv, 49-51, syst zoologist, 51-54, dir ichthyol lab, 56-60; cur fishes, Mus Comp Zool, Harvard Univ, 60-70; dir, Los Angeles County Mus Natural Hist, 70-90; RETIRED. *Concurrent Pos:* Chmn, Calif Natural Areas Coord Coun, 70- *Res:* Systematics, distribution and ecology of oceanic fishes. *Mailing Add:* 3029 Atlas Peak Rd Napa CA 94558

MEAD, JAMES FRANKLYN, b Evanston, Ill, Oct 24, 16; m 42; c 3. BIOCHEMISTRY. *Educ:* Princeton Univ, AB, 38; Calif Inst Technol, PhD(org chem), 42. *Prof Exp:* Asst, Calif Inst Technol, 42; from instr to asst prof, Occidental Col, 45-48; res coordr, Off Naval Res, Calif, 48; head synthetic br, Biochem Dept, Atomic Energy Proj, 48-50, res biochemist & chief biochem div, 50-69, assoc clin prof physiol chem, 51-56, prof, 56-63, prof biol chem & biophys, Med Sch, 63-69, PROF BIOL CHEM, MED SCH & ASSOC DIR LABS NUCLEAR MED & RADIATION BIOL, UNIV CALIF, LOS ANGELES, 69-, PROF PUB HEALTH, 73- *Concurrent Pos:* NIH career res award. *Honors & Awards:* E A Bailey Award, 71; Am Oil Chem Soc Award, 80. *Mem:* Am Chem Soc; Am Soc Biol Chem; Am Oil Chem Soc; Sigma Xi. *Res:* Lipid, brain lipid and fatty acid metabolism; essential fatty acids; lipid and membrane peroxidation. *Mailing Add:* 1210 Las Lomas Pl Pacific Pallisades CA 90272

MEAD, JAMES IRVING, b Tucson, Ariz, Aug 4, 52; m 80; c 1. VERTEBRATE PALEONTOLOGY, PALEOENVIRONMENTAL RECONSTRUCTION. *Educ:* Univ Ariz, BA, 74, MS, 79, PhD(geosci), 83. *Prof Exp:* Res asst, dept geosci, Univ Ariz, 77-79; asst dir, Hot Springs Mammoth Site, SDak, 83-88; vis asst prof, 85-86, ASST PROF GEOL, NORTHERN ARIZ UNIV, 86-, ASSOC DIR, QUATERNARY STUDIES PROG, 87- *Concurrent Pos:* Res assoc, Ctr Study Early Man, Inst Quaternary Studies, Univ Maine, 83-85, ed, 84-; Colo Plateau scholar-in-residence, Mus Northern Ariz, 85-86; curator, Lab Quaternary Paleontol, 89-; res assoc, San Bernardino Co Mus. *Mem:* Nat Speleol Soc; Am Quaternary Asn; Soc Vert Paleont; Am Soc Mammalogists; Soc Study Amphibians & Reptiles; Sigma Xi; Geol Soc Am. *Res:* Reconstruction of Quaternary environments of arid North America; plant-animal relationships; excavation of dry cave deposits and fossil packrat nests; osteology of mammals and reptiles. *Mailing Add:* Dept Geol Northern Ariz Univ Box 6030 Flagstaff AZ 86011

MEAD, JAYLEE MONTAGUE, b Clayton, NC, June 14, 29; m 68. ASTRONOMY. *Educ:* Univ NC, BA, 51; Stanford Univ, MA, 54; Georgetown Univ, PhD(astron), 70. *Prof Exp:* Eng asst math, Knolls Atomic Power Lab, Gen Elec Co, 51-52; teacher, Van Antwerp Sch, NY, 52-53; counsr & instr, Univ NC, 54-56; mathematician opers res off, Johns Hopkins Univ, 57-59; mathematician, Goddard Space Flight Ctr, NASA, 59-68, astrom lab theoret studies, 68-71, astrom lab optical astron, 71-77, asst chief lab astron & solar physics, 77-87, ASSOC CHIEF SPACE DATA & COMPUT DIV, GODDARD, SPACE FLIGHT CTR, NASA, 87- *Mem:* Am Astron Soc; Am Geophys Union; Int Astron Union; Sigma Xi. *Res:* Statistical astronomy; stellar dynamics; planetary atmospheres; planet Mars; star catalogues and computerized astronomy data retrieval systems. *Mailing Add:* Code 930 Goddard Space Flight Ctr Greenbelt MD 20771

MEAD, JERE, PULMONARY MECHANICS, CHEST WALL MECHANICS. *Educ:* Harvard Univ, MD, 46. *Prof Exp:* CECIL K & PHILIP DRINKER PROF ENVIRON PHYSIOL, SCH PUB HEALTH, HARVARD UNIV, 76- *Mailing Add:* PO Box 601 Southwest Harbor ME 04679

MEAD, JUDSON, b Madison, Wis, Sept 16, 17; m 44; c 3. GEOPHYSICS. *Educ:* Mass Inst Technol, BS, 40, PhD(geophys), 49. *Prof Exp:* Proj supvr, Airborne Instruments Lab, 41-45; from asst prof to assoc prof, 49-60, dir geol field sta, 65-70, PROF GEOPHYS, IND UNIV, BLOOMINGTON, 60-, DIR GEOL FIELD STA, 74- *Mem:* Geol Soc Am; Am Geophys Union; Am Inst Mining, Metall & Petrol Eng. *Res:* Structure of the crust; exploration geophysics. *Mailing Add:* Dept Geol Ind Univ Bloomington IN 47401

MEAD, LAWRENCE MYERS, b Plainfield, NJ, May 11, 18; m 42; c 4. STRUCTURED DESIGN, AERONAUTICAL SYSTEM ENGINEERING. *Educ:* Princeton Univ, BSE, 40, CE, 41. *Prof Exp:* Stress analyst, Grumann Aerospace Corp, 41-47, from asst proj engr to proj engr, 47-64, head preliminary design, 64-69, from vpres to sr vpres tech opers, 72-83, VPRES PRELIMINARY DESIGN, GRUMANN AEROSPACE CORP, 69-, CONSULT, 83- *Concurrent Pos:* Comt, Educ & Utilization of the Engr, Nat Res Coun, 83-85; chmn, aircraft & space vehicles, tech comt 20, Int Standards Orgn, Geneva, 83-88; mem, adv coun to aerospace & mech dept, Princeton Univ, 83- *Mem:* Nat Acad Eng; fel Am Inst Aeronaut & Astronaut; Soc Logistic Engrs; Soc Advan Mat & Process Eng. *Res:* Design and management of the design of naval carrier base aircraft; engineering and technology department management. *Mailing Add:* 16 Baycrest Huntington NY 11743

MEAD, MARSHALL WALTER, b Franklin, Ind, Jan 15, 21; m 47; c 3. ANALYTICAL CHEMISTRY. *Educ:* Franklin Col, AB, 42. *Prof Exp:* Analytical chemist, Ala Ord Works, E I du Pont de Nemours & Co, 42-43; chief analytical chemist, Oldbury Electrochem Co, 48-54, prod supt, Miss Works, 54-55; admin asst res & develop, Nat Aniline Div, Allied Chem Corp, 55-62; mgr local sect activ off, Am Chem Soc, 62-69, asst dir mem activ div, 69-71, head, 71-82, head, mem statist & planning, 82-85; RETIRED. *Mem:* Am Chem Soc. *Res:* Analytical chemistry of chlorates; phosphorus compounds; perchlorates; instrumental methods of analysis; research management; membership demography and statistics. *Mailing Add:* 710 Capri Estates CT Arnold MD 21012

MEAD, RICHARD WILSON, b Los Angeles, Calif, Jan 7, 41; m 64; c 2. HYDROMETALLURGY. *Educ:* Univ Denver, BChemE, 63, MS, 66; Univ Ariz, PhD(chem eng), 71. *Prof Exp:* Process engr, Shell Chem Co, 63-65; asst instr chem eng, Univ Ariz, 66-68, instr, 68-71, res assoc, 71-72; res engr, Phelps Dodge Corp, 72-74; asst prof, 74-78, ASSOC PROF CHEM ENG, UNIV NMEX, 78- *Concurrent Pos:* Consult, Los Alamos Nat Lab, 79-85, Extraction Res & Develop, Inc, 80-81, Bur Bus & Econ Res, Univ NMex, 75-85 & NMex Environ Improvement Agency, 75-76, 87. *Mem:* Am Inst Chem Engrs; Am Inst Mining, Metall & Petrol Engrs; Am Soc Eng Educ. *Res:* Development of new processes or units which facilitate the recovery of metals from novel sources or waste streams; leaching studies and metal ion mass transfer; economic process evaluation. *Mailing Add:* Dept Chem & Nuclear Eng Univ NMex Albuquerque NM 87131

MEAD, ROBERT WARREN, b Yonkers, NY, Mar 3, 40; m 61; c 3. ANIMAL PARASITOLOGY, INVERTEBRATE ZOOLOGY. *Educ:* Colo State Univ, BS, 62, MS, 63, PhD(zool), 68. *Prof Exp:* Asst prof biol, Davis & Elkins Col, 65-67; NIH fel, Univ Mass, Amherst, 68-70; from asst prof to assoc prof, Sch Med, Univ Nev, Reno, 70-81, chmn dept, 76-79, chmn fac senate, 88-89, PROF BIOL, SCH MED, UNIV NEV, RENO, 81- *Concurrent Pos:* Vis res scientist, Univ Calif, Irvine, 82; prog officer, Am Soc Parasitol, 82-87 & Am Micros Soc, 90- *Mem:* AAAS; Am Soc Zool; Am Soc Parasitol; Rocky Mountain Conf Parasitologists (pres, 78-79); Sigma Xi; Am Micros Soc; NY Acad Sci. *Res:* Cell and developmental biology of parasitic and free living invertebrates; invertebrate zoology. *Mailing Add:* Dept Biol Univ Nev Reno NV 89557-0015

MEAD, RODNEY A, b Moline, Ill, Apr 28, 38; m 61; c 2. REPRODUCTIVE PHYSIOLOGY, ENDOCRINOLOGY. *Educ:* Univ Calif, Davis, AB, 60, MA, 62; Univ Mont, PhD(zool), 66. *Prof Exp:* USPHS fel steroid biochem, Col Med, Univ Utah, 66-68; from asst prof to assoc prof, 68-76, PROF ZOOL, UNIV IDAHO, 76- *Concurrent Pos:* Mem biol reprod study sect, Inst Child Health & Human Develop, 75-79. *Honors & Awards:* A Brazier Howell Award, Am Soc Mammalogists, 66. *Mem:* AAAS; Soc Study Reproduction; Am Soc Mammal; Am Soc Zool; Sigma Xi. *Res:* Hormonal control of delayed implantation in mustelids. *Mailing Add:* Dept Biol Sci Univ Idaho Moscow ID 83843

MEAD, S WARREN, b New Brunswick, NJ, Jan 26, 23; m 54; c 2. PHYSICS. *Educ:* Univ Calif, BA, 48, PhD(physics), 57. *Prof Exp:* Assoc math, Lawrence Livermore Lab, Univ Calif, 49-50, physicist, 57-87; CONSULT, 88- *Mem:* Am Phys Soc; Sigma Xi. *Res:* Lasers; laser-produced plasmas; environmental sciences; radioactivity. *Mailing Add:* 7793 Fairbrook Ct Pleasanton CA 94566

MEAD, WILLIAM C, b Hazleton, Pa, Dec 6, 46; m 69; c 1. LASER PLASMA INTERACTIONS, HYDRODYNAMICS SIMULATIONS. *Educ:* Syracuse Univ, BS, 68; Princeton Univ, MA, 70, PhD(physics), 74. *Prof Exp:* Physicist, Lawrence Livermore Nat Lab, 73-83; PHYSICIST, LOS ALAMOS NAT LAB, 83- *Mem:* Fel Am Phys Soc. *Res:* Behavior of laser-plasma coupling processes and their effects on inertial confinement fusion targets: plasma instabilities, thermal transport, radiative processes, and hydrodynamics; design, analysis and numerical simulations of experiments; fluid instabilities; applications of neurol networks. *Mailing Add:* Los Alamos Nat Lab PO Box 1663 MS F645 Los Alamos NM 87545

MEAD, WILLIAM J(ASPER), b Columbus, Ohio, Dec 29, 27; m 50, 75, 82; c 3. CHEMICAL ENGINEERING. *Educ:* Ohio State Univ, BChE, 48, ChE, 62; Stevens Inst Technol, MS, 61. *Prof Exp:* Chem engr, Colgate Palmolive Co, 50-51; chem engr, Whitehall Labs, Inc, Am Home Prod Corp, 51-56, asst tech dir, Home Prod Int, Ltd, 56-58, tech dir, 58-65; dir mfg, Alberto-Culver Co, Ill, 65-67, vpres, 66-67; DIR MFG, COMBE, INC, 69-, VPRES, 77- *Concurrent Pos:* Ed-in-chief, Encyclop Chem Process Equip, 64- *Mem:* AAAS; Am Inst Chem Engrs; Am Chem Soc; Soc Cosmetic Chemists; Am Inst Chemists; Am Soc Qual Control. *Res:* Manufacture, formulation, quality control, plant design, packaging, equipment selection and layout for pharmaceuticals; cosmetics, insecticides, waxes, polishes and other household chemical specialty products. *Mailing Add:* Combe Inc 1101 Westchester Ave White Plains NY 10604

MEADE, ALSTON BANCROFT, b Jamaica, WI, June 28, 30; m 57; c 5. ENTOMOLOGY. *Educ:* Fisk Univ, BA, 56; Univ Minn, MS, 59, PhD(entom), 62. *Prof Exp:* From res biologist to sr res biologist, 64-81, RES ASSOC, AGR PROD DEPT, E I DU PONT DE NEMOURS & CO INC, 81- *Mem:* Entom Soc Am; Royal Entom Soc London; Sigma Xi; Nat Inst Sci; Int Soc African Scientists. *Res:* Resistance of plants to insect attack; ecology of Empoasca fabae and Macrosteles fascifrons; insecticidal controls of vegetable pests; development of new insecticides; insect attractants. *Mailing Add:* E I duPont Stine-Haskell Bldg 200 Wilmington DE 19898

MEADE, DALE M, b Lodi, Wis, Aug 7, 39; div; c 2. PLASMA PHYSICS. *Educ:* Univ Wis, BA, 61, MS, 62, PhD(physics), 65. *Prof Exp:* Res assoc physics, Univ Wis, 65-66; res assoc, Princeton Univ, 66-67; from asst prof to prof physics, Univ Wis-Madison, 67-73; res physicist, 73-78, sr res physicist, 78-80, HEAD, EXP DIV, PLASMA PHYS LAB, PRINCETON UNIV, 80-, HEAD TFTR PROJ, 86- *Mem:* Fel Am Phys Soc. *Res:* Experimental studies of equilibrium and stability of plasma confined by magnetic fields with emphasis on applications in controlled thermonuclear fusion. *Mailing Add:* Princeton Plasma Physics Lab PO Box 451 Princeton NJ 08543

MEADE, GRAYSON EICHELBERGER, b Palacios, Tex, Apr 8, 12; m 37; c 4. GEOLOGY. *Educ:* Univ Nebr, AB, 35, MA, 37; Univ Chicago, PhD(geol), 46. *Prof Exp:* From instr to assoc prof geol, Tex Tech Col, 41-46; asst geologist, Bur Econ Geol, Tex, 44-45; geologist, Tex Mem Mus, 46-49; assoc prof, Tex Tech Col, 49-52; geologist, Union Oil Co, Calif, 52-58, chief geologist, Union Oil Co Can, 58-61, staff geologist, 61-72; CONSULT GEOLOGIST, 72- *Concurrent Pos:* Sessional instr, Dept Archaeol, Univ Calgary, 68-72. *Mem:* Fel Geol Soc Am; Soc Vert Paleont; Am Asn Petrol Geol. *Res:* Cenozoic and petroleum geology; vertebrate paleontology; Devonian stratigraphy. *Mailing Add:* Agate Springs Ranch Harrison NE 69346

MEADE, JAMES HORACE, JR, b Vicksburg, Miss, Nov 1, 32; m 58; c 3. BIOMETRICS. *Educ:* Miss State Univ, BS, 54, MS, 59; Univ Fla, PhD(animal genetics), 61. *Prof Exp:* Fel biomath, NC State Col, 61-62, asst statistician, 62-63; from asst prof to assoc prof biostatist, Med Ctr, Univ Ala, 63-65, sr biostatistician, 63-65; assoc prof biomet, 65-69, head biomet div, 69-77, PROF BIOMET, MED SCH, UNIV ARK, LITTLE ROCK, 69-80; PROF DEPT MATH & STATIS, MISS STATE UNIV, 80- *Concurrent Pos:* Consult, Vet Admin, 66-80. *Mem:* Biomet Soc; Am Statist Asn. *Res:* Applications of mathematics and statistics in biological research; teaching statistics to biologists. *Mailing Add:* Dept Math & Stat Miss State Univ, Drawer MA Mississippi State MS 39762

MEADE, JOHN ARTHUR, b Coldwater, Mich, Aug 29, 28; m 49; c 2. AGRONOMY. *Educ:* Univ Md, BS, 54, MS, 55; Iowa State Col, PhD(plant physiol), 58. *Prof Exp:* From asst prof to assoc prof agron, Univ Md, 58-66; EXTEN SPECIALIST, RUTGERS UNIV, NEW BRUNSWICK, 66- *Mem:* Weed Sci Soc Am. *Res:* Herbicides. *Mailing Add:* Dept Crop Sci Cook Col PO Box 231 New Brunswick NJ 08903

MEADE, REGINALD ESON, food science, food technology; deceased, see previous edition for last biography

MEADE, ROBERT HEBER, JR, b Brooklyn, NY, Dec 27, 30; m 56; c 3. GEOLOGY, POTAMOLOGY. *Educ:* Univ Okla, BS, 52; Stanford Univ, MS, 57, PhD(geol), 60. *Prof Exp:* Geologist, Calif Co, 52 & 55-56; GEOLOGIST, US GEOL SURV, 57- *Concurrent Pos:* Assoc ed, J Geophys Res, 74-76; adj prof, State Univ NY, Stony Brook, 75-83. *Mem:* Soc Econ Paleont & Mineral; Int Asn Sedimentol; Am Geophys Union; Am Soc Civil Eng; fel Geol Soc Am. *Res:* Sedimentology; erosion, transport, deposition and compaction of sediments; river morphology; coastal hydrology and oceanography. *Mailing Add:* US Geol Surv MS 413 Denver CO 80225-0046

MEADE, THOMAS GERALD, b Pound, Va, Sept 3, 37. PARASITOLOGY, INVERTEBRATE ZOOLOGY. *Educ:* Whitman Col, BA, 59; Purdue Univ, MS, 62; Ore State Univ, PhD(zool), 65. *Prof Exp:* PROF PARASITOL, SAM HOUSTON STATE UNIV, 75- *Mem:* Am Soc Parasitol. *Res:* Helminth parasites of fishes; immuno-parasitology; larval trematode snail interaction. *Mailing Add:* Dept Biol Sci Sam Houston State Univ Huntsville TX 77341

MEADE, THOMAS LEROY, b Center Junction, Iowa, July 4, 20; m 42; c 2. ANIMAL NUTRITION. *Educ:* Univ Fla, BS, 50, MS, 51, PhD, 53. *Prof Exp:* Asst animal nutritionist, Univ Fla, 53-54; animal nutritionist, Chas Pfizer & Co, 54-55; vpres & dir res, Hayne Prod Inc, 55-63; dir res, J Howard Smith, Inc, 63-68; from assoc prof to prof fisheries & marine technol, 68-77, prof animal sci, 77-81, PROF AQUACULT SCI & PATH, UNIV RI, 81- *Res:* Marine resource utilization, primary efforts in process and product development for industrial fisheries; aquaculture systems development, including nutrition and physiology of salmonoids. *Mailing Add:* Dept Path-Fisheries Univ RI Kingston RI 02881

MEADER, ARTHUR LLOYD, JR, b Clarksville, Tenn, Dec 13, 20; m 43; c 2. PETROLEUM CHEMISTRY. *Educ:* Univ Ky, BS, 41; Univ Wis, MS, 44, PhD(chem), 47. *Prof Exp:* From assoc res chemist to res chemist, Calif Res Corp Div, Standard Oil Co, Calif, 47-62, sr res chemist, Chevron Res Co, 62-67, sr res assoc, 67-83; RETIRED. *Mem:* Am Chem Soc. *Res:* Surface-active chemicals; plastics; fibers; surface coatings; elastomers; asphalt specialties. *Mailing Add:* 2023 Los Angeles Ave Berkeley CA 94707

MEADER, RALPH GIBSON, b Eaton Rapids, Mich, Sept 6, 04; m 28; c 1. ANATOMY. *Educ:* Ohio Wesleyan Univ, AB, 25; Hamilton Col, AM, 27; Yale Univ, PhD(comp anat), 32. *Hon Degrees:* LLD, Philadelphia Col Osteop Med, 56; ScD, Ohio Wesleyan Univ, 58. *Prof Exp:* From instr to asst prof biol, Hamilton Col, 25-28; instr, Wesleyan Univ, 28-29; from instr to assoc prof anat, Sch Med, Yale Univ, 31-48; chief, Cancer Res Grants Br, Nat Cancer Inst, 48-53, chief, Res Grants & Fels Br, 53-60, assoc dir grants & training, 60-65; dep dir res admin & exec secy comt on res, 65-76, CONSULT ON GRANTS CONTRACTS, MASS GEN HOSP, 76-; CONSULT, NAT CANCER INST, 65- *Concurrent Pos:* Biologist, State Biol Surv, NY, 29-30; Blossom fel neuroanat, Yale Univ, 29-31; mem corp, Bermuda Biol Sta, 32-; Rockefeller Found fel neurol, Univ Amsterdam & Ctr Inst for Brain Res, 38-39; asst to dir bd sci advs, Jane Coffin Childs Mem Fund Med Res, 42-43, asst dir, 43-48; exec secy, Nat Adv Cancer Coun, 47-65; mem selection & scheduling comt, Gordon Res Conf, 65-80; incorporator, Spaulding Youth Ctr Inc, 67-, trustee, 67-85, pres, 72-75, emer trustee, 85-; incorporator, Eunice Kennedy Shriver Ctr, Ment Retardation Inc, 69-, trustee, 69-83, pres, 77-83; mem biomed libr rev comt, Nat Libr Med, 74-76, chmn, 75-76. *Mem:* Fel AAAS; Soc Develop & Growth; Sigma Xi; Am Asn Anat; Am Asn Cancer Res. *Res:* Comparative anatomy of nervous system; neuroanatomy of teleosts; electrical characteristics of living organisms; history of medicine; sequence of nerve degeneration. *Mailing Add:* Calef Farm R1 Franklin NH 03235

MEADOR, NEIL FRANKLIN, b Sweet Springs, Mo, Sept 19, 38; m 58; c 2. AGRICULTURAL ENGINEERING. *Educ:* Univ Mo, Columbia, BSc, 61; Va Polytech Inst, MSc, 63; Mich State Univ, PhD(agr eng), 68. *Prof Exp:* Instr agr eng, Va Polytech Inst, 61-62; instr, Mich State Univ, 63-67; assoc prof, 67-72, PROF AGR ENG, UNIV MO-COLUMBIA, 72- *Mem:* Am Soc Agr Engrs; Sigma Xi. *Res:* Structural research concerning farm and light industrial buildings. *Mailing Add:* Dept of Agr Eng 103 T-12 Univ of Mo Columbia MO 65211

MEADOWS, ANNA T, b Cherbourg, France, Apr 30, 31; US citizen; m 77. PEDIATRICS, EPIDEMIOLOGY. *Educ:* Queens Col, BA, 52; NY Univ, MA, 53; Med Col Pa, MD, 69. *Prof Exp:* from instr to assoc prof, 72-85, PROF PEDIAT, SCH MED, UNIV PA, 85- *Concurrent Pos:* Asst physician oncol, Children's Hosp Philadelphia, 74-76, assoc physician med, 76-79, sr physician, Dept Med, 79-; dir epidemiol, etiol & genetics, Children's Cancer Res Ctr, 80- *Mem:* Am Soc Clin Oncol; Int Soc Pediat Oncol; Am Asn Cancer Res; Am Soc Pediat Hematol-Oncol. *Res:* Investigations of pediatric cancer etiology, epidemiology, late effects of cancer therapy, and information exchange, utilizing tumor registries originating from a pediatric cancer center, a geographic region, and an international group. *Mailing Add:* Children's Hosp of Philadelphia 34th & Civic Center Blvd Philadelphia PA 19104

MEADOWS, BRIAN T, b London, Eng, May 20, 40; m 63; c 2. HIGH ENERGY PHYSICS. *Educ:* Oxford Univ, BA, 62, MA, 67, PhD(physics), 67. *Prof Exp:* Sci Res Coun Gr Brit res fel high energy physics, Oxford Univ, 66-67; res assoc, Syracuse Univ, 67-68, vis asst prof, 68-69, asst prof, 69-72; assoc prof, 72-81, PROF HIGH ENERGY PHYSICS, UNIV CINCINNATI, 81- *Concurrent Pos:* Prog dir, Exp Particle Physics, NSF, 91-92. *Mem:* Am Phys Soc; Am Asn Univ Professors. *Res:* Experimental high energy physics. *Mailing Add:* Dept Physics 110 Braun Univ Cincinnati Cincinnati OH 54221

MEADOWS, CHARLES MILTON, b Merryville, La, Nov 8, 12; m 41; c 2. ENTOMOLOGY. *Educ:* La State Norm Col, AB, 36; La State Univ, MS, 38; Ohio State Univ, PhD, 42. *Prof Exp:* Asst entom, Ohio State Univ, 38-42; in-charge cotton insect invest, 42-44; tech rep, Sherwin-Williams Co, 46-50; pres & gen mgr, Southwest Sprayer & Chem Co, 50-80; RETIRED. *Concurrent Pos:* Consult, Tenneco Oil Co, Tex, 64-74; consult, agr chem, 74-80. *Mem:* AAAS; Nat Agr Chem Asn; Entom Soc Am. *Res:* Toxicity of insecticides; development and application of selective weed killers; spray machinery. *Mailing Add:* Southwest Sprayer & Chem Co 2632 Cedar Ridge Waco TX 76708

MEADOWS, GARY GLENN, b American Falls, Idaho, June 6, 45; m 68; c 2. PHARMACY, PHARMACOGNOSY. *Educ:* Idaho State Univ, BS, 68, MS, 72; Univ Wash, PhD(pharmaceut sci), 76. *Prof Exp:* assoc prof, 82-89, PROF PHARMACOGNOSY, COL PHARM, WASH STATE UNIV, 89- *Concurrent Pos:* Chmn, Dept Pharmaceut Sci. *Mem:* Am Inst Nutrit; Am Soc Exp Pharm Ther; Res Soc Alcoholism; NY Acad Sci; Am Asn Cancer Res; AAAS; study sect mem, Nat Inst Alcohol Abuse & Alcoholism; Metastasis Res Soc. *Res:* Nutrition and cancer with emphasis on nutritional modulation of metastasis, tumor heterogeneity and drug resistance; alcohol, immunology, and cancer. *Mailing Add:* Col Pharm Wash State Univ Pullman WA 99164-6510

MEADOWS, GEOFFREY WALSH, b Bury, Eng, Jan 16, 21; m 45; c 3. INDUSTRIAL CHEMISTRY. *Educ:* Univ Manchester, BSc, 42, MSc, 43, PhD(chem), 48. *Prof Exp:* Res chemist, Shell Co, Eng, 43-45; asst lectr, Univ Manchester, 45-49; Nat Res Coun Can fel, 49-51; res chemist, E I du Pont de Nemours & Co, Inc, 51-65, res assoc, 65-66, res supvr, 66-72, res assoc, 72-80; RETIRED. *Mem:* Am Chem Soc; Sigma Xi. *Res:* Ionic catalysed polymerization; physical properties of polymers; reaction kinetics; high temperature synthesis. *Mailing Add:* 139 E Sickles St Kennett Square PA 19348-2920

MEADOWS, GUY ALLEN, b Detroit, Mich, May 5, 50; m 73. PHYSICAL OCEANOGRAPHY. *Educ:* Mich State Univ, BS, 72, MS, 74; Purdue Univ, PhD(marine sci), 77. *Prof Exp:* Prod design engr, Ford Motor Co, 72; res instr, Great Lakes Coastal Res Lab, Purdue Univ, 73-77, res coordr phys oceanog, 74-77; ASST PROF PHYS OCEANOG, UNIV MICH, ANN ARBOR, 77- *Concurrent Pos:* Instr, Purdue Univ, 74-76. *Mem:* Sigma Xi; Am Geophys Union; Int Asn Great Lakes Res. *Res:* Coastal hydrodynamics; nearshore dynamics and thermally driven circulations. *Mailing Add:* Naval Arch & Mar Eng Univ Mich Col Eng North Campus Ann Arbor MI 48109

MEADOWS, HENRY E(MERSON), JR, b Atlanta, Ga, May 27, 31; m. ELECTRICAL ENGINEERING. *Educ:* Ga Inst Technol, BEE, 52, MSEE, 53, PhD(elec eng), 59. *Prof Exp:* Asst elec eng, Ga Inst Technol, 54-55, instr, 55-58; mem tech staff, Bell Tel Labs, Inc, 59-62; from asst prof to assoc prof, 62-70, PROF ELEC ENG, COLUMBIA UNIV, 70- *Concurrent Pos:* Consult, 62-; consult & vpres, Technol Consult, Inc, 63-70; NSF grants, 64-66 & 67-71; prin investr, Off Naval Res contract, 64-67; NASA grant, 67-72; Ford Found eng resident, Western Develop Labs Div, Philco-Ford Corp, 68-69. *Mem:* Inst Elec & Electronics Engrs; Tensor Soc; Simulation Coun. *Res:* Systems engineering, controls, network theory, simulation, applied mathematics, mathematical economics. *Mailing Add:* Seeley W Mudd Bldg Columbia Univ New York NY 10027

MEADOWS, JAMES WALLACE, JR, b Meridian, La, Aug 16, 23; m 50; c 2. NUCLEAR PHYSICS. *Educ:* La Polytech Inst, BS, 44; La State Univ, MS, 48, PhD(chem), 50. *Prof Exp:* Anal chemist, Cities Serv Refining Corp, La, 44-46; asst, La State Univ, 46-50; tech assoc, Harvard Univ, 50-58; assoc chemist, 58-70, CHEMIST, ARGONNE NAT LAB, 70- *Mem:* Am Chem Soc; Am Phys Soc; Am Nuclear Soc. *Res:* Nuclear reactions; neutron diffusion; fission; neutron physics. *Mailing Add:* Eng Physics Div Argonne Nat Lab Bldg 314 9700 S Cass Ave Argonne IL 60439

MEADOWS, MARK ALLAN, b Dallas, Tex, Aug 4, 57; m 83. INVERSE THEORY, ELASTIC-WAVE ANISOTROPY. *Educ:* Univ Va, BA(physics) & BA(math), 79; Univ Calif, Berkeley, MS, 82, PhD(eng geosci), 85. *Prof Exp:* Res asst eng geosci, Univ Calif, Berkeley, 81-83, teaching asst geophysics & math, 83-84; res geophysicist, 82-83, SR RES GEOPHYSICIST, CHEVRON OIL FIELD RES CO, 85- *Mem:* Soc Explor Geophysicists; Am Geophys Union; Soc Indust & Appl Math. *Res:* Imaging and inversion of anisotropic wave fields; acquisition and processing of multicomponent seismic data; elastodynamic forward modeling; inverse scattering theory. *Mailing Add:* Geophysics Div Chevron Oil Field Res Co PO Box 446 La Habra CA 90633

MEADOWS, W ROBERT, b Chicago, Ill, Feb 3, 19. INTERNAL MEDICINE, CARDIOLOGY. *Educ:* Northwestern Univ, BS, 41, MD, 44. *Prof Exp:* Intern, Cook County Hosp, 44, resident med, 47-48, 49-50; ward physician med, Vet Admin Hosp, Livermore, Calif, 53-55, ward physician, Palo Alto, 55-60, asst chief cardiopulmonary lab, Hines, Ill, 60-72, chief cardiac catheterization lab, 72-74, chief graphics sect cardiol, 74-86; RETIRED. *Concurrent Pos:* Fel hemat, Cook County Hosp, Chicago, Ill, 50-51; fel cardiol, New Eng Deaconess Hosp, Boston, Mass, 59-60; asst clin prof, Stritch Sch Med, Loyola Univ Chicago, 61-62, asst prof, 62-65, assoc prof, 65- *Mem:* Fel Am Col Physicians; fel Am Col Cardiol. *Res:* Surgery for coronary artery disease (stable angina). *Mailing Add:* PO Box 543 Elgin IL 60121

MEADS, JON A, SYSTEM SOFTWARE, COMPUTER GRAPHICS. *Prof Exp:* Proj mgr, Teletronix, 71-78; info serv mgr, Intel, 78-81; consult, Jon Meads & Assocs, 81-87; SCI STAFF, BELL NORTHERN RES, 87- *Mem:* Asn Comput Math. *Res:* System and user interface design and development, particularly interactive graphic system. *Mailing Add:* Dept 9Z10 Bell Northern Res Ltd PO Box 3511 Sta C Ottawa ON K1Y 4H7 Can

MEADS, MANSON, b Oakland, Calif, Mar 25, 18; m 45; c 1. MEDICINE. *Educ:* Univ Calif, AB, 39; Temple Univ, MD, 43. *Hon Degrees:* DSc, Temple Univ, 56. *Prof Exp:* Asst med, Thorndike Mem Lab, Harvard Med Sch, 44-46, asst bact & immunol, 46-47; instr med, Bowman Gray Sch Med, 47-50, from asst prof to prof internal med, 51-83, from assoc prof to prof prev med & dir dept, 51-57, assoc dean, 55-58, acad dean, 58-59, from exec dean to dean, 59-71, vpres health affairs, 67-83 & dir med ctr, 74-84; RETIRED. *Concurrent Pos:* Ernst fel, Thorndike Mem Lab, Harvard Med Sch, 46-47; Markle scholar, 48-53; med officer, vis prof & adv, USPHS, Thailand, 53-55. *Mem:* Am Soc Clin Invest; AMA; fel Am Col Physicians. *Res:* Medical school administration. *Mailing Add:* 1855 Meadowbrook Dr Winston-Salem NC 27104

MEADS, PHILIP F, b Oakland, Calif, Dec 4, 07; m 36; c 3. PHYSICAL CHEMISTRY. *Educ:* Univ Calif, Berkeley, BS, 28, PhD(chem), 32. *Prof Exp:* Chief chemist, Calif & Hawaiian Sugar Co, 53-59; tech dir, Calif & Hawaiian Sugar Co, Crockett, 59-72, asst refinery mgr environ affairs, 72-76; RETIRED. *Concurrent Pos:* Partic cane sugar refiners res proj, US Nat Comt Sugar Analysis; mem, Int Comn Uniform Methods Sugar Anal & Indust Adv Comn, Sugar Res Found. *Honors & Awards:* Hon Award, Sugar Indust Technicians, 69. *Mem:* Am Chem Soc; Inst Food Technol; Sugar Indust Technologists (pres, 71-72). *Res:* Refining of cane sugar; sugar analysis and products; sugar refinery wastewater treatment. *Mailing Add:* 33 Linda Ave 2603 Oakland CA 94611

MEADS, PHILIP FRANCIS, JR, b Oakland, Calif, May 19, 37; m 66; c 3. THEORETICAL PHYSICS, COMPUTER SCIENCE. *Educ:* Univ Calif, Berkeley, AB, 58, PhD(physics), 63. *Prof Exp:* Asst physics, Lawrence Radiation Lab, Univ Calif, 59-63; physicist, Midwest Univ Res Asn, Wis, 63-65; physicist, William M Brobeck & Assocs, 65-79; CONSULT, 79- *Concurrent Pos:* Consult, William M Brobeck & Assocs, 60, Argonne Nat Lab, Lawrence Radiation Lab, Univ Calif, 65-, Los Alamos Sci Lab, 70-, Hanford Eng Develop Lab, 75- & Fermi Nat Accelerator Lab, 78- *Mem:* AAAS; Health Physics Soc; Am Phys Soc; Audio Eng Soc. *Res:* Aberrations of quadrupole focusing magnets; accelerator design, particularly injection and extraction studies; optical design of beam transport systems; digital computer systems and subsystems, particularly microprocessors. *Mailing Add:* 7053 Shirley Dr Oakland CA 94611

MEAGHER, DONALD JOSEPH, b Albany, NY, May 2, 49; m 71; c 3. SOLID MODELING, THREE-D MEDICAL IMAGING. *Educ:* Rensselaer Polytech Inst, BS, 71, MEng, 72, MEng, 79, PhD(elec eng), 82. *Prof Exp:* Assoc electronics engr, Cornell Aeronaut Lab, Buffalo, NY, 72-74; mem tech staff, Argo Systs, Palo Alto, Calif, 74-77; mgr, Interaction Comput Graphics Ctr, Rensselaer Polytech Inst, Troy, NY, 78-80, instr comput-aided mfg, 80-81; VPRES TECHNOL & GEN MGR, SOLID MODEL DIV, PHOENIX DATA SYSTS, ALBANY, NY, 82- *Concurrent Pos:* Adj prof, Rensselaer Polytech Inst, 85- *Mem:* Inst Elec & Electronics Engrs; Asn Comput Mach; Sigma Xi. *Res:* Development of efficient algorithms and associated machine architectures for solid modeling (the representation, manipulation, analysis and display for computerized models of 3-D solid objects); applications include medical diagnosis and treatment planning, seismic analysis, computer-aided design and computer-aided manufacturing. *Mailing Add:* 4404 Adragna Ct San Jose CA 95136

MEAGHER, JAMES FRANCIS, b Sydney, NS, Can, Oct 23, 46; m 73; c 2. PHOTOCHEMISTRY & GAS KINETICS, FIELD MEASUREMENT OF TRACE POLLUTANTS. *Educ:* St Francis Xavier, BSc, 67; Cath Univ Am, PhD(phys chem), 71. *Prof Exp:* Postdoctoral phys chem, Univ Wash, 71-74; postdoctoral atmospheric chem, Pa State Univ, 74, res staff, 75-75; group leader atmospheric chem, 76-85, MGR ATMOSPHERIC RES, TENN VALLEY AUTHORITY, 85- *Concurrent Pos:* Interagency Sci Comt, Nat Acid Precipitation Assessment Prog, 87-; Air Adv Comt, Elec Power Res Inst, 89- *Mem:* Am Chem Soc; Am Geophys Union; Sigma Xi. *Res:* Atmospheric chemistry of sulfur and nitrogen oxides; smog chamber, numerical modelling, ground based and aircraft measurements; oxidant formation in the rural atmosphere; effects of air pollutants on crops and forests. *Mailing Add:* Tenn Valley Authority CEB 2A Muscle Shoals AL 35660

MEAGHER, MICHAEL DESMOND, b Nelson, BC, Nov 27, 33; m 65; c 2. FORESTRY. *Educ:* Univ BC, BSF, 57, PhD(forestry), 76; Univ Toronto, MScF, 63. *Prof Exp:* Forester-in-training silvicult, BC Forest Serv, 57-61; lectr dendrol, silvicult & urban forestry, Univ Toronto, 63-67, forestry, Univ BC, 70-71; forester silvicult & genetics, BC Forest Serv, 72-84; GENETICIST, CAN FOREST SERV, 84- *Res:* Genetics research and breeding of western white pine (pinus monticola) for reforestation in British Columbia. *Mailing Add:* 666 Jones Terrace 506 W Burnside Rd Victoria BC V8Z 2L7 Can

MEAGHER, RICHARD BRIAN, b Chicago, Ill, Sept 30, 47; m 68; c 1. MOLECULAR GENETICS, ENZYMOLOGY. *Educ:* Univ Ill, BS, 69; Yale Univ, MPhil, 71, PhD(biol), 73. *Prof Exp:* Am Cancer Soc fel biochem, Univ Calif, Berkeley, 73-74, lectr, 73-74; NIH res fel biochem & microbiol, Univ Calif, San Francisco, 74-76; mem staff microbiol, Univ Ga, 76-; AT UNIV SAN FRANCISCO. *Mem:* Am Chem Soc; Plant Molecular Biol Asn; Am Soc Microbiol. *Res:* Evolution of biochemical pathways and their regulation; molecular cloning and expression of higher plant genes; techniques of genetic engineering applied to the developing of new plant phenotypes. *Mailing Add:* Dept Genetics Univ Ga Athens GA 30602

MEAKIN, JAMES WILLIAM, b Smith Falls, Ont, May 28, 29; m 53; c 2. INTERNAL MEDICINE. *Educ:* Queen's Univ, Ont, MD, CM, 53; Univ Toronto, MA, 57; FRCP(C), 60. *Prof Exp:* Fel med, Harvard Univ & asst med, Peter Bent Brigham Hosp, 57-59; clin teacher, 60-65, assoc, 65-68, ASSOC PROF MED, FAC MED, UNIV TORONTO, 68-; PHYSICIAN, ONT CANCER INST & PRINCESS MARGARET HOSP, 60-; PRES, ONT CANCER TREAT & RES FOUND, TORONTO, 79- *Concurrent Pos:* Am Col Physicians fel, Harvard Univ, 57-58, Life Ins Med Res Fund fel, 58-59. *Honors & Awards:* Starr Medal, 57. *Mem:* Can Med Asn; Am Soc Clin Oncol; Can Soc Clin Invest; Endocrine Soc. *Res:* Effect of steroid hormones on cells, particularly tumour cell growth in animals and man. *Mailing Add:* Ont Cancer Treat & Res Found 7 Overlea Blvd Toronto ON M5S 1A8 Can

MEAKIN, JOHN C, b Brisbane, Australia, Mar 13, 46. MATHEMATICS, ALGEBRA. *Educ:* Univ Queensland, BSc hons, 68; Morash Univ, PhD(math), 69. *Prof Exp:* PROF MATH, UNIV NEBR, 70- *Mem:* Am Math Soc; Math Asn Am; Am Asn Univ Prof; Australian Math Soc. *Res:* Research in semigroups in connection with autonata languages, combinatorial group theory and logic. *Mailing Add:* Dept Math & Stats Univ Nebr Lincoln NE 68588-0434

MEAKIN, JOHN DAVID, b Nottingham, Eng, Feb 11, 34; m 57; c 5. MATERIALS SCIENCE, PHOTOVOLTAICS. *Educ:* Univ Leeds, BSc, 55, PhD(metall), 57. *Prof Exp:* Vis res assoc, Franklin Inst, 58-60; res fel, King's Col, Durham, 60-61, Imp Chem Indust sr res fel, 61-62; sr res scientist, Res Labs, Franklin Inst, 62-65, sr staff scientist, 65-66, prin staff scientist, 66-69, lab mgr, 69-74; PROF MAT SCI, MECH & AEROSPACE ENG & SR SCIENTIST, INST ENERGY CONVERSION, UNIV DEL, 74- *Concurrent Pos:* Lectr, Univ Pa, 66; Vis prof, Univ Del, 67, Murdoch Univ, 83; chmn, Dept Mech Eng, Univ Del, 87- *Mem:* sr mem Inst Elec & Electronics Engrs. *Res:* Structure and properties of materials; thin film solar cells; solar energy utilization. *Mailing Add:* Dept Mech Eng Univ of Del Newark DE 19716

MEAKIN, PAUL, b Burton-on-Trent Staffs, Eng, Mar 29, 44; m 85; c 1. PHYSICAL CHEMISTRY, CONDENSED MATTER PHYSICS. *Educ:* Manchester Univ, BSc, 65; Univ Calif, Santa Barbara, PhD(chem), 69. *Prof Exp:* Chemist, E I du Pont de Nemours & Co Inc, 69-75, groupleader, 75-76, res supvr, 76-81, res leader, 81-90; CONSULT. *Mem:* Mat Res Soc; fel Am Phys Soc. *Res:* Magnetic resonance, nuclear magnetic resonance and electron spin resonance; polymer chemistry and physics; chemical dynamics of transition metal complexes; solid state chemistry and physics, solid electrolytes; atmospheric chemistry; non equilibrium growth and aggregation processes; statistical physics. *Mailing Add:* Cent Res & Develop Dept Exp Sta E I du Pont de Nemours & Co Inc Wilmington DE 19880-0356

MEAL, HARLAN C, b Rush Co, Ind, Jan 31, 25; m 53; c 3. OPERATIONS MANAGEMENT. *Educ:* Harvard Univ, AB, 50, MA, 53, PhD(phys chem), 54. *Prof Exp:* Opers analyst oper res off, Johns Hopkins Univ, 53-57; opers analyst, Dunlap & Assocs, Inc, 57-59; opers analyst, Arthur D Little, Inc, 59-72, head logistics unit, 72-76; sr lectr, Sloan Sch of Mgt, Mass Inst Technol, 76-87; MGT CONSULT, 76- *Mem:* Am Prod & Inventory Control Soc. *Res:* Industrial operations research; management of production; distribution and service operations; industrial logistics; management systems for industrial operations; decision rules, control systems and organizational structures. *Mailing Add:* 12 Ellery Square Cambridge MA 02138

MEAL, JANET HAWKINS, b Plymouth, Mass, Aug 24, 27; m 53; c 3. PRODUCTION PLANNING SYSTEMS, MANAGEMENT CONTROL SYSTEMS. *Educ:* Wellesley Col, AB, 49; Harvard Univ, PhD(chem), 53. *Prof Exp:* Nat Res Coun, Nat Bur Standards res fel, 53-55; opers res analyst, Appl Physics Lab, Johns Hopkins Univ, 56-57; sr res assoc, Arthur D Little Inc, 60-65 & 71-80; prin consult, Digital Equip Corp, 81-87; RETIRED. *Res:* Production systems and their associated control systems; design of production planning and control systems. *Mailing Add:* 12 Ellery Sq Cambridge MA 02138

MEAL, LARIE L, b Cincinnati, Ohio, June 15, 39. PHYSICAL CHEMISTRY, ANALYTICAL CHEMISTRY. *Educ:* Univ Cincinnati, BS, 61, PhD(phys chem), 66. *Prof Exp:* Res chemist, US Indust Chem Co, Nat Distillers & Chem Corp, 66-67; instr chem, Ohio Col Appl Sci, 68-69; from asst prof to assoc prof, 69-90, PROF CHEM TECHNOL, UNIV CINCINNATI, 90- *Concurrent Pos:* Consult & chem analyst, Ins Investrs, Fire Dept, 74- *Mem:* AAAS; Am Chem Soc; NY Acad Sci. *Res:* Second derivative ultraviolet spectrometry; chemical analysis of polymers. *Mailing Add:* 2231 Slane Ave Norwood OH 45212

MEALEY, EDWARD H, b Boston, Mass, July 28, 25; m 51; c 3. BIOCHEMISTRY. *Educ:* Tufts Col, BS, 48; Univ Kans, PhD(biochem), 60. *Prof Exp:* Supvr chem test unit lab blood & blood prod, Div Biologics Standards, NIH, 60-63, chief blood & blood derivatives sect, 64-67; dir qual assurance, Hyland Labs, 67-70; sr vpres & tech dir, Int Clin Lab Sci, 70-80; vpres qual assurance, 81-83, SR VPRES TECH OPERS, RES & DEVELOP QUAL ASSURANCE REG AFFAIRS, ALPHA THERAPEUT CORP, 83- *Mem:* AAAS; Am Pub Health Asn; NY Acad Sci; Am Chem Soc; Sigma Xi. *Res:* Physical and chemical studies on whole blood; plasma and plasma protein solutions. *Mailing Add:* 17646 Fremont St Fountain Valley CA 92708

MEALEY, JOHN, JR, b Providence, RI, Aug 30, 28; m 52; c 3. NEUROSURGERY. *Educ:* Brown Univ, BA, 49; Johns Hopkins Univ, MD, 52; Am Bd Neurol Surg, dipl, 62. *Prof Exp:* Intern surg, Johns Hopkins Hosp, 52-53; clin & res fel neurosurg, Harvard Sch Med, 55-56; from asst resident to resident, Mass Gen Hosp, 56-60; from instr to assoc prof surg, 60-68, PROF NEUROL SURG, SCH MED, IND UNIV, INDIANAPOLIS, 69- *Mem:* AMA; Am Asn Neurol Surg; Cong Neurol Surg; Am Col Surg; Sigma Xi. *Res:* Brain tumors; radioactive methods; chemotherapy; head injuries. *Mailing Add:* Neurol Surg Dept EM 142 Sch Med Ind Univ 1100 W Michigan St Indianapolis IN 46223

MEANS, ANTHONY R, b Bartlesville, Okla, May 7, 41. ENDOCRINOLOGY, CELL BIOLOGY. *Educ:* Okla State Univ, BA, 63, MS, 64; Univ Tex, Austin, PhD(physiol), 67. *Prof Exp:* Res assoc molecular biol, Southwest Found Res & Educ, 68-69; asst prof obstet & gynec, Med Sch, Vanderbilt Univ, 69-72, asst prof physiol & asst dir, Ctr Pop Res, 71-72; assoc prof, 73-75; prof cell biol & assoc dir, Ctr Pop Res, Baylor Col Med, 75-91; PROF & CHAIR PHARMACOL, DUKE UNIV MED CTR, 91- *Concurrent Pos:* Res fel, Australian Res Coun, Russell Grimwade Sch Biochem, Univ Melbourne, 67-68. *Honors & Awards:* Edwin B Astwood Award, Endocrine Soc, 80. *Mem:* Soc Study Reproduction; Am Soc Cell Biol; Endocrine Soc; Am Soc Biol Chem. *Res:* Calmodulin and cyclic nucleotide regulation of cell function. *Mailing Add:* Dept Cell Biol Baylor Col Med Houston TX 77030

MEANS, CRAIG RAY, b Shreveport, La, Aug 16, 22; m 75. PROSTHODONTICS. *Educ:* Southern Univ, BS, 50; Howard Univ, DDS, 54; Ohio State Univ, MSc, 63. *Prof Exp:* Asst prof prosthodont, 61-62 & 64-66, supvr dent technicians, 64-67, assoc prof & chief div removable partial & complete dentures, 66-68, actg chmn dept removable prosthodont, 68-69, assoc prof & chmn dept, 69-70, assoc dean undergrad affairs, 70-81, prof, 70-85, dir continuing dent educ, 82-85, EMER PROF REMOVABLE PROSTHODONT, COL DENT, HOWARD UNIV, 85- *Mem:* Am Dent Asn; Nat Dent Asn; Am Prosthodont Soc. *Res:* Damage to the oral tissues resulting from the use of home reline denture materials; temporomandibular joint function in complete denture patients. *Mailing Add:* 9823 Hedin Dr Silver Spring MD 20903

MEANS, D BRUCE, b Los Angeles, Calif, Mar 9, 41; div; c 2. ECOLOGY, HERPETOLOGY. *Educ:* Fla State Univ, BS, 68, MS, 72, PhD(ecol), 75. *Prof Exp:* Teaching asst, Fla State Univ, 68-70; Gerald Beadel res scholar, 70-74; asst dir, Tall Timbers Res Sta, 74-77, dir, 78-84; PRES, COASTAL PLAINS INST, 84- *Concurrent Pos:* Adj asst prof, Fla State Univ, 76-82, adj assoc prof, 82-89, adj asst prof, Univ Fla, 82-, adj prof, 89-; adj cur herpet, Fla State Mus, 84-; res assoc zool, US Nat Mus Smithsonian, 89- *Honors & Awards:* W Kelly Mosley Environ Award, 82. *Mem:* AAAS; Ecol Soc Am; Soc Study Evolution; Am Soc Naturalists; Am Soc Ichthyologists & Herpetologists; Sigma Xi; Soc Study Amphibians & Reptiles. *Res:* Population biology; ecology of reproduction and life history phenomena; evolution at the species and population level; geographical ecology; fire ecology; tropical ecology. *Mailing Add:* Dept Biol Sci Fla State Univ Tallahassee FL 32306

MEANS, GARY EDWARD, b Wykoff, Minn, Aug 31, 40. BIOCHEMISTRY. *Educ:* San Jose State Col, BS, 64; Univ Calif, Davis, PhD(biochem), 68. *Prof Exp:* USPHS fel, Virus Lab, Univ Calif, Berkeley, 68-70; from instr to asst prof, 71-79, Actg Chmn Dept, 80-86, ASSOC PROF BIOCHEM, OHIO STATE UNIV, 79- *Mem:* Am Chem Soc; Am Soc Biol Chemists. *Res:* Protein chemistry; structure-function relationships of proteins, enzyme mechanisms. *Mailing Add:* Dept of Biochem Ohio State Univ 484 W 12th Ave Columbus OH 43210

MEANS, JEFFREY LYNN, b Hinsdale, Ill, July 16, 52; m 75; c 3. GEOCHEMISTRY, GEOLOGY. *Educ:* Bucknell Univ, BS, 74; Princeton Univ, MA, 76, PhD(geol), 81. *Prof Exp:* Res asst geochem, Oak Ridge Nat Lab, 74 & Princeton Univ, 74-78; DEPT MGR, ENVIRON TECHNOL, COLUMBUS LABS, BATTELLE MEM INST, 78- *Concurrent Pos:* Prin investr, Battelle Columbus Lab, contract, 79-; consult, Oak Ridge Nat Lab, 75-77. *Mem:* Sigma Xi. *Res:* Organic, trace element, isotope and aqueous geochemistry, especially hazardous waste disposal, radioactive waste management, stabilization and solidification; hazardous waste treatment. *Mailing Add:* Battelle Columbus Labs 505 King Ave Columbus OH 43201

MEANS, LYNN L, b Kansas City, Mo, Jan 26, 14; m 36; c 4. METEOROLOGY, PHYSICAL SCIENCE. *Educ:* Univ Chicago, BS, 42, MS, 44. *Prof Exp:* Instr & res assoc, Univ Chicago, 42-45; forecaster, Chicago Forecast Ctr, US Weather Bur, Nat Oceanic & Atmospheric Admin, 45-47, res forecaster, 47-55, leading analyst, Nat Weather Anal Ctr, 55-59, chief pub & agr forecast sect, 59-65, dep to dir user affairs, Environ Sci Serv Admin, 65-68, sr prog analyst, 68-74, consult & gen phys scientist, 75-79; RETIRED. *Concurrent Pos:* Consult air weather serv, US Army Air Force, 44-46. *Mem:* Am Meteorol Soc; Sigma Xi. *Res:* Applied meteorology; forecast improvement; economic benefits of science services. *Mailing Add:* 4901 Stan Haven Rd Temple Hills MD 20748

MEANS, WINTHROP D, b Brooklyn, NY, Feb 7, 33; m 63; c 3. EXPERIMENTAL STRUCTURAL GEOLOGY, DEFORMATION THEORY. *Educ:* Harvard Univ, AB, 55; Univ Calif, Berkeley, PhD(geol), 60. *Prof Exp:* Lectr geol, Univ Otago, NZ, 60-64, sr lectr, 64; res fel, dept geophys, Australian Nat Univ, 64-65; assoc prof, 65-77, PROF GEOL, STATE UNIV NY, ALBANY, 77- *Concurrent Pos:* Panelist Earth Sci Prog, NSF, 83-86, prin investr, grants; chmn dept geol, State Univ NY, Albany. *Mem:* Geol Soc Am. *Res:* Small-scale structure in deformed rocks and analog materials; grain-scale structures in ductile materials; observation of microstructural changes during deformation; geometrical and kinematic theory of deformation. *Mailing Add:* Dept Geol Sci State Univ NY 1400 Washington Ave Albany NY 12222

MEARES, CLAUDE FRANCIS, b Wilmington, NC, Sept 25, 46; m 63; c 3. PHYSICAL CHEMISTRY, BIOLOGICAL CHEMISTRY. *Educ:* Univ NC, BS, 68; Stanford Univ, PhD(phys chem), 72. *Prof Exp:* From asst prof to assoc prof, 72-82, PROF CHEM, UNIV CALIF, DAVIS, 82- *Concurrent Pos:* Nat Cancer Inst grant, 74-; Nat Inst Med Sci grant, 78-; res career develop award, Nat Cancer Inst, NIH, 79-84; mem, metallobiochem study sect, NIH, 82-86; mem, educ affairs comt, Am Soc Biol Chemists, 83-86. *Honors & Awards:* Von Hevesy Prize for Nuclear Med, Von Hevesy Comt & Soc Nuclear Med, 74. *Mem:* Am Chem Soc; Am Soc Biol Chemists; Biophys Soc. *Res:* Metals and bifunctional chelating agents in biology and medicine; structure and mechanism of ribonucleic acid polymerases; photo affinity labelling, fluorescence, energy transfer, radiopharmaceuticals, bleomycin. *Mailing Add:* Dept Chem Univ Calif Davis CA 95616

MEARNS, ALAN JOHN, b Los Angeles, Calif, Oct 4, 43; m; m 74; c 2. FISHERIES, POLLUTION BIOLOGY. *Educ:* Calif State Univ, Long Beach, BS, 65, MA, 68; Univ Wash, PhD(fisheries), 71. *Prof Exp:* Biologist, Allan Hancock Found, Univ Southern Calif, Los Angeles/Arctic Res Lab, Barrow, Alaska, 65-66; res assoc, Fisheries Res Inst, Univ Wash, 68-70; consult physiol, Auke Bay Lab, Nat Marine Fisheries Serv, 70; sr environ scientist pollution biol, Southern Calif Coastal Water Res Proj, 71-73, dir, Biol Div, 73-80; SR ECOLOGIST, OCEAN ASSESSMENTS DIV, NAT OCEANIC & ATMOSPHERIC ADMIN, SEATTLE, 80- *Concurrent Pos:* Environ Protection Agency grant, Corvallis, Ore, 72-75; consult, Calif State Water Resources Control Bd, 75-; Nat Oceanog & Atmospheric Admin grant & NY Bight Mesa Proj grant, 77; water qual comt, Am Fisheries Soc, 77-; Bur of Land Mgt subcontract, 77-78; mem res comt, Water Pollution Control Fedn, 78-; NSF grant, 78-81; vice dir, NW Wash Dist, Am Inst Fishery Res Biol, 81-82; dir, Wash dist, Am Inst Fish Res Biol; vchair, Santa Monica Bay Restoration Prog; adv, Sydney Water Bd, Australia, Environ Protection Agency Broremediation Action comt. *Mem:* Am Inst Fishery Res Biologists; Am Fisheries Soc; Sigma Xi; AAAS. *Res:* National review of trends in pesticides and PCBs in United States coastal zone; planning and coordinating national and west coast marine pollution and monitoring programs; developing alternative strategies for marine waste disposal; conducting research on pollutant flow through marine food webs; effect of oil spills and treatment technology, including bioremediation, on marine life; assessment of longterm recovery from Exxon Valdez oil spill. *Mailing Add:* Ocean Assessments Div Nat Oceanic & Atmospheric Admin 7600 Sand Pt Way NE Seattle WA 98115

MEARS, BRAINERD, JR, b Williamstown, Mass, June 24, 21; m 48; c 4. GEOMORPHOLOGY. *Educ:* Williams Col, AB, 43; Columbia Univ, PhD(geol), 50. *Prof Exp:* Lectr geomorphol, Columbia Univ, 47-49; from asst prof to prof, 49-89, EMER PROF GEOL, UNIV WYO, 89- *Concurrent Pos:* Mem, US Geol Surv; Sang distinguished exchange prof, Univ Wyo. *Mem:* AAAS; Geol Soc Am; Int Union Quaternary Res. *Res:* Pleistocene geology. *Mailing Add:* Dept of Geol Univ of Wyo Laramie WY 82071

MEARS, DANA CHRISTOPHER, b Pittsburgh, Pa, Sept 4, 40; m 64; c 2. CHEMICAL METALLURGY, MEDICINE. *Educ:* Cornell Univ, BA, 62, Cambridge Univ, PhD(metall), 65; Oxford Univ, MD, 69; MRCP, UK, 72. *Prof Exp:* Res asst metall, Cent Res Lab, Broken Hill Proprietary Co, Ltd, 62; res worker, Cambridge Univ, 62-66; res fel physiol, Nuffield Dept Orthop Surg, Oxford Univ, 66-70, house physician, Radcliffe Infirmary, 70; intern surg, Univ Pittsburgh, 70-71; sr house officer rheumatol, Nuffield Orthop Ctr, Oxford Univ, 71-72, registr in metab med, Nuffield Dept Orthop Surg, 72; resident orthop surg, Children's Hosp Pittsburgh, 72-75; ASST PROF ORTHOP SURG, UNIV PITTSBURGH, 75- *Concurrent Pos:* Mem, Brit Standards Comt Surg Implants, 64- *Mem:* Am Soc Metals; Brit Corrosion & Protection Asn; Brit Corrosion Sci Soc. *Res:* Corrosion and passivity; selection of materials for surgical implants; bone physiology, the relationship between mechanical stress on bone and bone cell metabolism. *Mailing Add:* Orthop Surg Univ Pittsburgh Sch Med Pittsburgh PA 15261

MEARS, DAVID ELLIOTT, b Hamilton, Ohio, Feb 26, 39; m 66; c 1. CHEMICAL ENGINEERING, CATALYSIS. *Educ:* Carnegie Inst Technol, BS, 61; Univ Calif, Berkeley, PhD(chem eng), 65. *Prof Exp:* Res asst chem eng, Univ Calif, Berkeley, 61-65; res engr, 65-68, sr res engr, 68-71, technol sales engr, 71-74, res assoc, 74-75, SUPVR PETROCHEM PROCESS RES, RES CTR, UNION OIL CO CALIF, 76- *Mem:* Am Chem Soc; Am Inst Chem Engrs. *Res:* Heterogeneous chemical kinetics; catalysis; transport effects; petrochemical and refining processes. *Mailing Add:* 212 Lido Pl Fullerton CA 92635

MEARS, DAVID R, b Brooklyn, NY, May 1, 36; m 68; c 2. BIOLOGICAL & AGRICULTURAL ENGINEERING. *Educ:* Rutgers Univ, BS, 58, MS, 61, PhD(eng mech), 68. *Prof Exp:* Asst instr agr eng, Rutgers Univ, 58-60; asst prof sci, Cuttington Col, Liberia, 60-64; teaching asst eng, Univ Calif,

Davis, 64-65; res assoc agr eng, 65-69, from asst prof to assoc prof biol & agr eng, 69-78, PROF BIOL & AGR ENG, RUTGERS UNIV, 78-; ASSOC DIR PROG PLANNING & DEVELOP, NJ AGR EXP STA, 81- *Mem:* Am Soc Agr Engrs. *Res:* Mechanical properties of biological materials; mechanization of fruit and vegetable harvesting; engineering agricultural systems for livestock housing, forage production and greenhouses; solar energy for greenhouses; greenhouse engineering; waste heat for greenhouses. *Mailing Add:* Dept Agr Eng Rutgers Univ New Brunswick NJ 08908

MEARS, GERALD JOHN, b Peace River, Alta, Nov 25, 38; m 63; c 3. ANIMAL PHYSIOLOGY, ENDOCRINOLOGY. *Educ:* Univ Alta, BSc, 63, MSc, 66; Univ Calif, Davis, PhD(physiol), 71. *Prof Exp:* Res asst animal physiol, Univ Calif, Davis, 63-71; fel med physiol, Univ Calgary, 71-73, res assoc fetal pharmacol, 73-74, Med Res Coun Can prof asst, 74-78; RES SCIENTIST ANIMAL PHYSIOL, RES BR, AGR CAN, 78- *Concurrent Pos:* Res fel animal physiol, Univ Calif, Davis, 67-68; fel med physiol, Univ Calgary, 71-73. *Mem:* Agr Inst Can; Can Soc Animal Prod; Can Physiol Soc; Can Fedn Biol Socs; Sigma Xi. *Res:* The placental transfer of drugs and hormones between the fetus and mother, development of the fetal endocrine system, the relationship between hormones and growth, and ovine reproduction. *Mailing Add:* Livestock Sci Sect Agr Can Res Sta PO Box 3000 Main Lethbridge AB T1J 4B1 Can

MEARS, JAMES AUSTIN, b Baytown, Tex, May 18, 44; m 66; c 3. BIOCHEMISTRY, SYSTEMATICS. *Educ:* Univ Tex, Austin, BA, 66, PhD(biol), 70. *Prof Exp:* asst cur, 70-80, ASSOC CUR, DEPT BIOL, ACAD NATURAL SCI PHILADELPHIA, 80- *Concurrent Pos:* Adj asst prof, Univ Pa, 70-80, adj assoc prof, 80-; NSF grants, Smithsonian Inst, 71 & Acad Natural Sci, 71-78; res assoc, Morris Arboretum, Philadelphia, 71-; Am Philos Soc grants, 75 & 80; mem bd dirs, Henry Found Bot Res; ed & publ, Chem Plant Taxon Newsletter. *Mem:* Am Soc Plant Taxon; Phytochem Soc NAm; Am Inst Biol Sci; Sigma Xi. *Res:* Studies of the processes of organism and molecule evolution, primarily through analyses of morphological and biochemical characteristics. *Mailing Add:* 49 Constitution Ct Wayne PA 19087-5826

MEARS, WHITNEY HARRIS, chemistry; deceased, see previous edition for last biography

MEASAMER, S(CHUBERT) G(ERNT), b Wilder, Tenn, Aug 7, 13; m 39; c 2. CHEMICAL ENGINEERING. *Educ:* Tenn Polytech Inst, BS, 35; Iowa State Col, MS, 37, PhD(chem eng), 41. *Prof Exp:* Instr chem, Eng Unit Opers Lab, Iowa State Col, 38-41; res chem engr, Plastics Dept, E I Du Pont de Nemours & Co, Inc, NJ, 41-47, group leader, 47-49, res supvr, Film Dept, NY, 49-53, process develop supvr, Ohio, 53-64, process develop supt, E I du Pont de Nemours, Luxembourg SA, 64-70, staff engr expansion, 70-72, staff engr, Circleville Lab, 72-73, sr engr Teflon, 73-78; RETIRED. *Mem:* Am Chem Soc; Am Inst Chem Engrs. *Res:* Extraction of oil from soybeans; product and process development work on plastics; polymer films formation; handling and coating. *Mailing Add:* 1315 Bristol Ct Circleville OH 43113

MEASDAY, DAVID FREDERICK, b London, Eng, July 4, 37; Can citizen; m 65; c 2. PHYSICS AT MESON FACTORIES, PHYSICS AT A KAON FACTORY. *Educ:* Oxford Univ, BA, 59, PhD(nuclear physics), 62. *Prof Exp:* Res fel, Harvard Univ, 62-65; res assoc, CERN, 65-68, staff physicist, 68-70; assoc prof, 70-75, PROF PHYSICS, UNIV BC, 75-, ASSOC DEAN SCI, 90- *Concurrent Pos:* Sr Killam fel, Centre d'Etudes Nucleaires, Saclay, 77-78; sr Killam fel, CERN, 84-85; vis prof, Geneve Univ, 84. *Mem:* Fel Am Phys Soc; Can Asn Physicists; Europ Phys Soc. *Res:* Intermediate energy nuclear physics; low-energy particle physics; physics of exotic atoms; muon catalyzed fusion; chemistry of pionic hydrogen atoms. *Mailing Add:* Physics Dept Univ BC Vancouver BC V6T 2A6 Can

MEASEL, JOHN WILLIAM, b Texarkana, Tex, Sept 12, 40; m 65; c 1. IMMUNOLOGY, MEDICAL MICROBIOLOGY. *Educ:* Henderson State Univ, BA, 63; ETex State Univ, MA, 64; Univ Okla, PhD(microbiol), 70. *Prof Exp:* Fel vet path, Purdue Univ, 70-72; sr res scientist biochem, Armour Pharm Co, 72-75; asst prof microbiol, Kirksville Col Osteop Med, 75-78; assoc prof microbiol, 78-89, BIOTHERAPEUT N TEX,TEX COL OSTEOP MED,87-; CLIN IMMUNOLOGIST, ARLINGTON CANCER CTR, 88- *Mem:* Am Soc Microbiol; Am Rheumatism Asn; Soc Analytical Cytol. *Res:* Immunology, particularly cancer immunology; flow cytometery. *Mailing Add:* Dept Path Scott & White Clin & Hosp 2401 S 31st St Temple TX 76508

MEASURES, RAYMOND MASSEY, b London, Eng, Feb 17, 38; m 62; c 3. LASER PHYSICS, FIBER OPTIC SENSORS SMART STRUCTURES. *Educ:* Univ London, BSc, 60, Imp Col, dipl, PhD(physics), 64. *Prof Exp:* From asst prof to assoc prof physics, 64-77, PROF APPL SCI & ENG, INST AEROSPACE STUDIES, UNIV TORONTO, 77- *Concurrent Pos:* Assoc dir, Inst Aerospace Studies, Univ Toronto. *Mem:* Int Soc Optical Eng. *Res:* Fiber optic based smart structures; fiber optic sensors. *Mailing Add:* Inst Aerospace Studies Univ Toronto 4925 Dufferin St Downsview ON M6K 1Y9 Can

MEATH, WILLIAM JOHN, b Toronto, Ont, Apr 8, 36; m 60; c 2. INTERMOLECULAR FORCES, LASER-MOLECULE INTERACTIONS. *Educ:* Carleton Univ, BSc, 60; Univ Wis, PhD(chem), 65. *Prof Exp:* Proj assoc theoret chem inst, Univ Wis, 65; from asst prof to assoc prof chem, 65-71, PROF CHEM, UNIV WESTERN ONT, 71- *Concurrent Pos:* Vis asst prof, Theoret Chem Inst, Univ Wis, 66; vis prof, Inst Perlla Ricerca Sci E Technol, Univ Di Trento, 79; sci & eng res coun sr vis fel & hon res fel, Dept Chem & Math, Univ Col, Univ London, 83, Royal Soc-Natural Sci & Eng Res Coun Can exchange fel & hon res fel, dept chem & math, Univ Col, Univ London, 88. *Mem:* Am Phys Soc; fel Chem Inst Can. *Res:* Atomic and molecular quantum mechanics; intermolecular forces; stationary state and time dependent atomic and molecular properties; laser-molecule interactions. *Mailing Add:* Dept Chem Univ Western Ont London ON N6A 5B7 Can

MEBUS, CHARLES ALBERT, b Paterson, NJ, Sept 10, 32; m 55; c 3. VETERINARY PATHOLOGY. *Educ:* Cornell Univ, DVM, 56; Kans State Univ, MS, 62, PhD(vet path), 63. *Prof Exp:* Pvt pract, Del, 58-60; assoc prof vet path, Kans State Univ, 63-65; prof vet sci, Univ Nebr, Lincoln, 65-77; res leader path, Plum Island Animal Dis Ctr, 77-88; CHIEF, FOREIGN ANIMAL DIS DIAG LAB, 80- *Mem:* Am Vet Med Asn; Am Col Vet Path. *Res:* Viral animal diseases. *Mailing Add:* USDA APHIS NVSL Foreign Animal Dis Diag Lab PO Box 848 Greenport NY 11944

MECCA, CHRISTYNA EMMA, b Brooklyn, NY, Oct 23, 36. BIOLOGY, BIOCHEMISTRY. *Educ:* George Washington Univ, BS, 60, MS, 63, PhD(biol), 69. *Prof Exp:* Med biol technician, NIH, 58-60, biologist, 60-62, chemist, 62-65; instr gen biol, Montgomery Col, 68-69; res biologist, Bur Radiol Health, USPHS, 69-70; staff biologist, Coastal Plains Ctr Marine Develop Serv, Washington, DC, 70-72; prog analyst, NIH, 72-74; analyst-biologist, Smithsonian Inst Sci Info Exchange, 75-76; lectr biol, Wheeling Col, 78; res assoc dept biochem, Sch Med, WVa Univ, 79-82; admin officer, G M Hanmack, Esquire, 84-88; vis asst prof biol, WVa Univ, 89-90; lectr biol, 89-90, ASST PROF BIOL, WAYNESBURG COL, WAYNESBURG, PA, 90- *Mem:* AAAS. *Res:* Qualitative biochemical characteristics of cellular and organismic systems. *Mailing Add:* Rte 5 Box 548 Morgantown WV 26505

MECCA, STEPHEN JOSEPH, b New York, NY, Jan 15, 43; m 64; c 3. ENERGY SYSTEMS, SOFTWARE ENGINEERING. *Educ:* Providence Col, BS, 64, MS, 66; Rensselaer Polytech Inst, PhD(physics), 69. *Prof Exp:* Assoc prof physics, 69-80, vpres acad affairs, 82-85, PROF ENG PHYSICS SYSTEMS, PROVIDENCE COL, 80-; VPRES, GENLEX, 85- *Mem:* Am Phys Soc; Am Inst Physics; Am Asn Physics Teachers. *Res:* Nuclear physics; especially nuclear spectroscopy and photonuclear physics; systems approach to complex problem solving; systems analysis, systems science and engineering; energy management; solar systems. *Mailing Add:* Dept Physics Providence Col Providence RI 02918

MECH, LUCYAN DAVID, b Auburn, NY, Jan 18, 37; div; c 4. WILDLIFE ECOLOGY. *Educ:* Cornell Univ, BS, 58; Purdue Univ, PhD(vert ecol), 62. *Prof Exp:* NIH fel animal movements & telemetry, Univ Minn, Minneapolis, 63-64, res assoc, 64-66; res assoc biol, Macalester Col, 66-69; WILDLIFE RES BIOLOGIST, US FISH & WILDLIFE SERV, 69- *Concurrent Pos:* Adj prof, Univ Minn, 79- *Honors & Awards:* Spec Achievement Award, US Fish & Wildlife Serv, 70 & 81 & Civil Servant of the Year Award, 73; Terrestrial Wildlife Publ Award, Wildlife Soc, 72. *Mem:* Am Soc Mammal; Ecol Soc Am; Sigma Xi; Wildlife Soc. *Res:* Predator-prey relations; mammal behavior and natural history; animal movements and factors affecting them; ecology, behavior and sociology of wolves; spatial organization of mammals; telemetry and radio-tracking. *Mailing Add:* 1992 Folwell Ave St Paul MN 55108

MECH, WILLIAM PAUL, b La Crosse, Wis, Mar 10, 42; m 64, 83; c 3. MATHEMATICS. *Educ:* Wash State Univ, BA, 64; Univ Ill, MS, 65, PhD(math), 70. *Prof Exp:* From asst prof to assoc prof, 70-78, chmn dept, 75-80, PROF MATH, BOISE STATE UNIV, 78- *Concurrent Pos:* Dir honors prog, Boise State Univ, 70-; pres, Nat Collegiate Honors Coun, 80-81, exec secy/treas, 87- *Mem:* AAAS; Math Asn Am; Am Math Soc. *Res:* Analysis and functional analysis; extension of positive operators; graphs of groups; spectral analysis. *Mailing Add:* Dept Math Boise State Univ Boise ID 83725

MECHAM, JOHN STEPHEN, b Austin, Tex, Feb 29, 28; wid; c 2. HERPETOLOGY, EVOLUTIONARY BIOLOGY. *Educ:* Univ Tex, BA, 50, PhD(zool), 55; Univ Fla, MS, 52. *Prof Exp:* Asst prof zool, Univ Tulsa, 55-56; from asst prof to assoc prof, Auburn Univ, 56-65; assoc prof, 65-69, PROF ZOOL, TEX TECH UNIV, 69- *Concurrent Pos:* NSF res grants, 58-61, 62-65, 67-68 & 69-71. *Mem:* Am Soc Ichthyol & Herpet; Soc Study Evolution; Soc Syst Zool; Soc Study Amphibians & Reptiles. *Res:* Systematics and evolutionary mechanisms of anuran amphibians. *Mailing Add:* Dept Biol Sci Tex Tech Univ Lubbock TX 79409

MECHAM, MERLIN J, b Neola, Utah, Jan 31, 23; m 47; c 2. SPEECH PATHOLOGY, AUDIOLOGY. *Educ:* Brigham Young Univ, BA, 48; Utah State Univ, MS, 49; Ohio State Univ, PhD(speech path, audiol), 54. *Prof Exp:* Instr speech, Utah State Univ, 49-50; instr, Ohio State Univ, 52-54; assoc prof speech path, Brigham Young Univ, 54-61; dir speech path & audiol, 61-76, actg chmn dept speech, 69-70, PROF SPEECH, UNIV UTAH, 61- *Concurrent Pos:* Book abstractor, DSH Abstracts, 60-; consult, Utah State Training Sch, American Fork, 68-; Am Speech & Hearing Asn accrediting site visitor, Am Bd Examr, 70-77; consult ed, J Speech & Hearing Disorders, 74-77. *Mem:* Fel Am Speech & Hearing Asn; Am Asn Mental Retardation. *Res:* Developmental aspects of normal and disordered audiolinguistic skills in children; exploratory model of neurolinguistic dysfunction. *Mailing Add:* 1785 S 400 East Bountiful UT 84010

MECHAM, ROBERT P, b July 13, 48. CONNECTIVE TISSUE BIOLOGY, ELASTIN. *Educ:* Boston Univ, PhD(biochem), 77. *Prof Exp:* Res assoc prof med, 77-86, assoc prof, 86-89, PROF CELL BIOL & MED, JEWISH HOSP, WASH UNIV, 89- *Res:* Cellular differentiation. *Mailing Add:* Respiratory & Clin Care Div Jewish Hosp Wash Univ Med Ctr 216 S Kings Hwy St Louis MO 63110

MECHANIC, DAVID, b New York, NY, Feb 21, 36. MEDICAL SOCIOLOGY & SOCIAL PSYCHOLOGY. *Educ:* City Col NY, BA, 56; Stanford Univ, MA, 57, PhD, 59. *Prof Exp:* Nat Inst Mental Health trainee, Univ NC, 59-60; co-dir grad training mental health, 60-62, co-dir grad training prog med sociol & mental health, Univ Wis, 62-65; pub health serv spec fel, social psychiat res unit, Brit Med Res Coun, Maudsley Hosp & Inst Psychiat, Univ London, 65-66; prof sociol, Univ Wis, 65-73, John Bascom prof med, 73-79, dir, Ctr Med Sociol & Health Serv Res, 72-79; dir, Health & Health Care Serv Res Coord Coun, 79-85; dean fac arts & sci, 80-84, UNIV PROF RUTGERS UNIV, NEW BRUNSWICK, 80-, RENE DUBOIS PROF BEHAV SCI, 84-, DIR INST HEALTH, HEALTH CARE POLICY & AGING RES, 85- *Concurrent Pos:* Ed consult, Sociometry: J Res in Social

Psychol, 63-65; Spec Pub Health Serv fel, 65-66; mem, Medicaid Res Adv Consult Panel, Social & Rehab Serv, HEW, 67-71; mem, Epidemiol Studies Rev Comt, NIMH, 68-72, chmn, 70-72, mem task force on nursing, 86-87, panel nat reporting syst, 89-90, chair, Res Resources Panel, 89-90; mem, Panel Health Servs, Bd Med, Nat Acad Sci, 69-73, Comt on Health Status, 73-75, Comt Health Sci Policy, 76-77, Comt Study Nat Needs for Biomed & Behav Res Personnel, Comn Human Resources, 75-79, Comt Health Serv Res, 77-78, Steering Comton Health Care of Minorities & Handicapped, 79-81, Bd Health Care Serv, 81-87, Panel Prev Disability, 89-90, Panel Nat Health Care Survey, 90-, vchair, Study Comt for Pain, Disability & Chronic Illness Behav, 85-87; mem, Pres Sci Adv Comt Panel Health Serv Res & Develop, 71-72; assoc ed, J Health & Social Behav, 65-69 & 72-75, acad ed, Wiley-Intersci Ser Health, Med & Soc, 73-78; consult, Robert Wood Johnson Found, 72-87, spec mental health consult, Prog Chronically Mentally Ill, 85-87, Comn Med Educ, 90-; mem bd sci counr, Ctr Prev Premature Arteriosclerosis, Rockefeller Univ, 73-76; John Simon Guggenheim Found fel, 77-78; chmn, Pres Comn Mental Health Panel on Problems, Scope & Boundaries, 77-78; bd trustees, Health Res & Educ Trust NJ, 80-82, Asn Advan Mentally Handicapped, 80-81; mem, Nat Adv Comt on Nat Hospice Study, 80-84; chair, Sect Social, Econ & Polit Sci, AAAS, 84-86, Subcomt Mental Health Statist, Nat Comt Vital & Health Statist, 90; mem, expert adv panel mental health, WHO, 84-89; mem, Nat Adv Coun Aging, NIH, 82-87, Treatment Comt on Reduction of Cancer Mortality, Nat Cancer Inst, 84-85; mem nat adv coun, Hogg Found Mental Health, 87-90; mem health adv bd, Gen Acct Off, US Cong, 87-; mem, Nat Comt Vital & Health Statist, Dept Health & Human Serv, 88-92; mem, Comn Behav & Social Sci & Educ, Nat Res Coun, 89-; consult to various govt agencies, non-profit med assns, found, univs & health progs. *Honors & Awards:* Distinguished Med Sociologist Award, Am Sociol Asn, 83; First Carl Taube Award, Mental Health Sect, Am Pub Health Asn, 90. *Mem:* Inst Med-Nat Acad Sci; Nat Acad Sci; fel Acad Behav Med Res; fel Soc Behav Med; Am Sociol Asn; Am Asn Pub Health; fel AAAS. *Res:* Organization of medical and psychiatric care; adaptation to stress; decision-making processing in medicine and psychiatry; illness behavior; comparative medical organization; evaluation research and social policy. *Mailing Add:* Inst Health Health Care Policy & Aging 30 College Ave New Brunswick NJ 08903

MECHANIC, GERALD, b New York, NY, Jan 7, 27; m 52; c 2. ORGANIC CHEMISTRY, BIOCHEMISTRY. *Educ:* City Col New York, BS, 51; NY Univ, MS, 53, PhD(chem), 58. *Prof Exp:* Asst acetylene chem, NY Univ, 52-53, res chemist biochem, 57-58; head biochem res lab, Manhattan State Hosp, 58-59; res assoc, Inst Med Res & Studies, NY, 59-60; res fel orthop surg, Mass Gen Hosp, 60-69, asst biochem, 66-69; res assoc biol chem, Harvard Med Sch, 63-69; assoc prof, 69-72, PROF ORAL BIOL, SCH DENT & PROF BIOCHEM, SCH MED, UNIV NC, CHAPEL HILL, 72- *Mem:* AAAS; Am Chem Soc; NY Acad Sci; Royal Soc Chem; Am Soc Biol Chemists. *Res:* Chemistry of amino acids, peptides and proteins, with special reference to connective tissue. *Mailing Add:* Dent Res Ctr Univ NC CB 7455 Rm 212 Chapel Hill NC 27514

MECHERIKUNNEL, ANN POTTANAT, b Kerala State, India, Dec 28, 34; US citizen; m 61; c 2. AEROSPACE TECHNOLOGY & SPACE OPTICS. *Educ:* Univ Madras, BS, 55; Univ Kerala, MS, 58; George Washington Univ, PhD(chem), 70. *Prof Exp:* Nat Res Coun-NASA res assoc atmosphere, 75-78, aerospace technol physicist atmosphere & ocean, 78-88, REMOTE SENSING STUDIES, GODDARD SPACE FLIGHT CTR, GREENBELT, MD, 88- *Concurrent Pos:* Aerospace technol mgr, Space Sta Prog, Johnson Space Flight Ctr, NASA, Houston, Tex, 85-86. *Honors & Awards:* NASA Achievement Award, Earth Radiation Proj, 85; NASA Spec Achievement Award, 88. *Mem:* Am Inst Aeronaut & Astronaut; Illum Eng Soc; AAAS. *Res:* Remote sensing of the earth-atmosphere-ocean system to understand weather and climate processes; solar radiation data analysis and interpretation; earth observation system (EOS). *Mailing Add:* Code 920-1 Goddard Space Flight Ctr Greenbelt MD 20771

MECHLER, MARK VINCENT, b Fredericksburg, Tex, Feb 5, 25; m 57; c 3. PHYSICS. *Educ:* Univ Tex, BA, 51, MA, 57, PhD(physics), 67. *Prof Exp:* Res scientist, Defense Res Lab, Univ Tex, 51-57; res engr, Collins Radio Co, 57-58; from res scientist to head underwater missile div, Defense Res Lab, Univ Tex, Austin, 58-69; sr scientist res div, Unitech, Inc, 69-84; RES SPECIALIST, 3M, 84- *Mem:* Acoust Soc Am; Inst Elec & Electronics Engrs. *Res:* Underwater sound; electro-acoustic transducers; sound propagation and scattering; electronics. *Mailing Add:* 6703 Haney Dr Austin TX 78723

MECHLIN, GEORGE FRANCIS, JR, b Pittsburgh, Pa, July 23, 23; m 49. PHYSICS. *Educ:* Univ Pittsburgh, BS, 44, MS, 47, PhD(physics), 50. *Prof Exp:* Sr scientist, Bettis Atomic Power Div, Westinghouse Elec Corp, 49-57, dir adv systs eng, Sunnyvale Div, Calif, 57-64, mgr missile launching & handling, 64-68, gen mgr, Underseas Div, Md, 68-71, gen mgr astronuclear & oceanic div, Md, 71-73, vpres, Res & Develop, 73-87; RETIRED. *Concurrent Pos:* Mem, Res Adv Comt, US Coast Guard, 73-75; vchmn, Marine Bd, Nat Res Coun, 75-79; mem naval res adv comt, Lab Adv Bd for Naval Ships, 75-78; dir, Pittsburgh Broadcasting Co, 80-84. *Mem:* Nat Acad Eng; Am Phys Soc. *Mailing Add:* 960 Via Malibu Aptos CA 95003

MECHOLSKY, JOHN JOSEPH, JR, b Philadelphia, Pa, July 24, 44; m 66; c 3. MATERIALS SCIENCE ENGINEERING. *Educ:* Catholic Univ Am, BCE, 66, MCE, 68, PhD(mats sci), 73. *Prof Exp:* Civil engr, Naval Fac Eng Command, Washington, DC, 66-67; structural engr, Naval Ship Res & Develop Ctr, 67-72; ceramic res engr, Naval Res Labs, DC, 72-79; MEM TECH STAFF, SANDIA NAT LABS, 79- *Concurrent Pos:* Res asst mat sci, Cath Univ Am, 68-73. *Mem:* Am Soc Testing & Mat; Am Ceramic Soc; Sigma Xi. *Res:* Quantitative fracture surface analysis of brittle materials; identification of toughening mechanisms in glass ceramics; production of reinforced glasses; understanding environmentally assisted stress corrosion in glasses; author or co-author of over 90 publications. *Mailing Add:* 246 Woodland Dr State College PA 16803

MECHTLY, EUGENE A, b Red Lion, Pa, Feb 14, 31; m 65; c 3. RADIOPHYSICS. *Educ:* Western Md Col, BS, 52; Pa State Univ, MS, 58, PhD(physics), 62. *Prof Exp:* Physicist, Army Missile Labs & Marshall Space Flight Ctr, NASA, 54-65; res assoc elec eng, 65-67, asst prof, 67-69, asst dean col eng, 75-76 & 78-79, ASSOC PROF ELEC ENG, UNIV ILL, URBANA, 69-, EXEC OFFICER, GRAD PROG, DEPT ELECT & COMP ENGR, 81- *Concurrent Pos:* Mem comn G, US Nat Comt, Int Union Radio Sci; chmn metrication comt, Am Soc Eng Educ, 78-80; Fulbright scholar. *Mem:* Am Geophys Union; Am Soc Eng Educ; Am Soc Testing & Mat; Int Union Radio Sci. *Res:* Propagation of radio waves in the ionosphere; physics of the upper atmosphere; metrology. *Mailing Add:* 804 Mumford Dr Urbana IL 61801

MECKLENBURG, ROY ALBERT, b Elmhurst, Ill, Feb 10, 33; m 60; c 3. PLANT PHYSIOLOGY, MICROCLIMATOLOGY. *Educ:* Mich State Univ, BS, 58; Cornell Univ, MS, 61, PhD(agr), 63. *Prof Exp:* Asst ornamental hort, Cornell Univ, 58-63; asst prof landscape hort, Mich State Univ, 63-70, assoc prof hort, 70-76, prof, 76-77; PRES, CHICAGO HORT SOC & DIR, CHIGAGO BOT GARDEN, 77- *Mem:* Am Asn Bot Gardens & Arboreta; Am Soc Hort Sci; Sigma Xi. *Res:* Physiology of low temperature hardiness in higher plants; the effect of plants on urban noise, dust and microclimate. *Mailing Add:* 1036 Byerly Way Orlando FL 32818

MECKLER, ALVIN, b New York, NY, Apr 20, 26; m 47; c 3. THEORETICAL PHYSICS. *Educ:* City Col New York, BS, 47; Mass Inst Technol, PhD(physics), 52. *Prof Exp:* Mem staff solid state physics, Lincoln Lab, Mass Inst Technol, 52-55; chief div phys sci, Nat Security Agency, Md, 55-67; ASSOC PROF PHYSICS, UNIV MD, BALTIMORE COUNTY, 67- *Mem:* Am Phys Soc. *Mailing Add:* Univ Md Baltimore City 5401 Wilkens Ave Catonsville MD 21228

MECKSTROTH, GEORGE R, b Cincinnati, Ohio, Aug 26, 35; m 57; c 2. RADIOLOGICAL PHYSICS. *Educ:* Univ Cincinnati, BS, 58, MS, 60; PhD(radiol physics), 63. *Prof Exp:* ADJ PROF RADIOL, SCH MED, TULANE UNIV, 64- *Concurrent Pos:* Consult, Charity Hosp La, New Orleans, 65-, USPHS Hosp, 65-, Vet Admin Hosp, 65-, West Jefferson Gen Hosp, 68-, Hotel Dieu Hosp, 68-, East Jefferson Gen Hosp, 70- & St Charles Gen Hosp, 73- *Mem:* Am Col Radiol; Am Asn Univ Prof; Am Asn Physicists in Med; Health Physics Soc; Soc Nuclear Med. *Mailing Add:* 5311 Marcia Ave New Orleans LA 70214

MECKSTROTH, WILMA KOENIG, b Auglaize Co, Ohio, Feb 12, 29; div; c 3. PULSE RADIOLYSIS. *Educ:* Ohio State Univ, BS, 49, PhD(phys chem), 68. *Prof Exp:* Instr chem, Urbana Univ, 59-62; res scientist, Anchor Hocking Corp, 67-68; prof chem, Ohio State Univ, Newark, 68-88; RETIRED. *Concurrent Pos:* Vis prof, Univ Del, 83-84. *Mem:* Am Soc Mass Spectrometry; Am Chem Soc. *Res:* Ion cyclotron resonance mass spectrometry of metal carbonyls in the gas phase; ion chemistry of the congeneric Group 6 and 7 metal carbonyls using double resonance and ion-trapping techniques; pulse radiolysis. *Mailing Add:* 944 Brice Reynoldsburg OH 43068

MEDAK, HERMAN, b Vienna, Austria, Apr 26, 14; nat US; m 45; c 4. ORAL PATHOLOGY. *Educ:* Univ Toledo, BS, 43; Northwestern Univ, MS & DDS, 46; Univ Ill, PhD(anat), 59; Am Bd Oral Path, dipl, 64; Univ Vienna, MD, 73. *Prof Exp:* Med technician, Flower Hosp, Toledo, Ohio, 39-43; med technician, Chicago Wesley Mem Hosp, 43-47; res asst histol, 48-51, from instr to prof oral path, 53-67, actg head dept, 64-67, PROF PREV MED & COMMUNITY HEALTH, COL MED, UNIV ILL MED CTR, 67-, CHIEF CLIN ORAL PATH, DEPT ORAL DIAG, COL DENT, 67-, HEAD DEPT ORAL DIAG, 77- *Concurrent Pos:* Dent consult, Ill Res Hosp & Tumor Clin, 48-53. *Mem:* AAAS; Am Dent Asn; Am Soc Clin Path; Am Acad Dent Med; Am Acad Oral Path; Sigma Xi. *Res:* Effect of irradiation on teeth and oral structures; epithelium of the oral mucosa; oral cytology. *Mailing Add:* Kostner Lincolnwood IL 60646

MEDALIA, AVROM IZAK, b Boston, Mass, Feb 3, 23; m 43, 56; c 4. COLLOID CHEMISTRY, RUBBER CHEMISTRY. *Educ:* Harvard Univ, AB, 42; Univ Minn, PhD(anal chem), 48. *Prof Exp:* Asst, Cornell Univ, 42-43; chemist, Brookhaven Nat Lab, 49-52; asst dir polymer res, Boston Univ, 52-55; sr res chemist, Godfrey L Cabot, Inc, 56-58, head fundamental res sect, Cabot Corp, 59-62, assoc dir, Res Carbon Black Div, 63-70, group leader, Res & Develop Div, 70-80, sr scientist, 80-84. *Concurrent Pos:* Consult, 84- *Honors & Awards:* Melvin Mooney Award for Distinguished Technol, Am Chem Soc, 87. *Mem:* Am Chem Soc; Soc Rheology. *Res:* Colloids; polymers; carbon black (properties and applications); rubber (mechanical behavior, reinforcement); poly-soaps; emulsion polymerization. *Mailing Add:* 30 Dorr Rd Newton MA 02158

MEDALIE, JACK HARVEY, b Buhl, Minn, Jan 8, 22; m; c 3. FAMILY MEDICINE. *Educ:* Witwatersrand Univ, Johannesburg, BSc, 41, MD, BCh, 45. *Hon Degrees:* Master of Pub Health, Harvard Sch Pub Health, Boston, 58. *Prof Exp:* Instr, Dept Anat, Witwatersrand, Johannesburg, 42-43; sr lectr, Dept Social Med, Hebrew Univ, Jerusalem, 62-66; assoc prof & chmn, 66-69, prof & chmn, Dept Family Med, Tel- Aviv Univ, Israel, 69-74; prof & chmn, Dept Family Med, 75-87, PROF MED & COMMUNITY HEALTH, CASE WESTERN RESERVE UNIV, 76-, DOROTHY JONES WEATHER HEAD CHAIR OF MED, 78- *Concurrent Pos:* Vis prof, Family Med & Epidemiology, Univ NC, Chapel Hill, 73-74; mem Task Force Health Consequences Bereavement, 82-85; dir, Dept Family Practice, Univ Hosps, Cleveland, 82-; mem, Nat Task Force, Preventive Servs, Washington, DC, 84-; mem, Mem Comt, Inst Med, 84-; chmn, Nat Task Force on the Family Soc, Teachers Family Med, 85- *Honors & Awards:* Magnes Prize, Hebrew Univ, 57-; Curtis Hames Career Res Award, Soc Teachers Family Med, 88, Cert of Excellence, 88; Maurice Saltzman Award, 88. *Mem:* Inst Med-Nat Acad Sci; Soc Teachers Family Med; Am Acad Family Physicians; Soc Behav Med. *Res:* Family medicine; published numerous articles in various publications; heart disease, hypertension, diabetes; promotions of health; geriatrics, family therapy; stress and illness; mentoring. *Mailing Add:* Dept Family Med Case Western Univ 2109 Adelbert Cleveland OH 44106-4901

MEDARIS, L GORDON, JR, b Memphis, Tenn, July 14, 36; div; c 3. PETROLOGY. *Educ:* Stanford Univ, BS, 58; Univ Calif, Los Angeles, PhD(geol), 66. *Prof Exp:* Assoc prof, 71-77, PROF GEOL, UNIV WIS-MADISON, 77- *Mem:* Geol Soc Am; Mineral Soc Am; Geochem Soc; Am Geophys Union. *Res:* Igneous and metamorphic petrology; petrology of alpine peridotites and calc-alkaline plutons in the Klamath Mountains Province, California; precambrian geology of Wisconsin; crustal garnet peridotites in Norway & Czechoslovakia. *Mailing Add:* Dept Geol & Geophys Univ Wis Madison WI 53706

MEDCALF, DARRELL GERALD, b Tillamook, Ore, Feb 10, 37; m 60; c 3. CARBOHYDRATE CHEMISTRY. *Educ:* Lewis & Clark Col, BA, 59; Purdue Univ, MS, 62, PhD(biochem), 64. *Prof Exp:* Asst prof cereal technol, NDak State Univ, 63-67; from assoc prof to prof chem, Univ Puget Sound, 67-78; lab mgr, Gen Foods Corp, 78-80, sr lab mgr, 80-83; dir, Basic Res, Hershey Foods, 83-85, dir, Prod Develop, 85-87; VPRES RES, KRAFT GEN FOODS, 87- *Mem:* Am Chem Soc; Am Asn Cereal Chemists; Sigma Xi. *Res:* Organic chemistry of carbohydrates, particularly polysaccharides; plant biochemistry, cereal chemistry. *Mailing Add:* Kraft Tech Ctr R & D 801 Waukegan Rd Glenview IL 60025

MEDEARIS, DONALD N, JR, b Kansas City, Kans, Aug 22, 27; m 56; c 4. PEDIATRICS, MICROBIOLOGY. *Educ:* Univ Kans, AB, 49; Harvard Med Sch, MD, 53; Am Bd Pediat, cert. *Prof Exp:* Intern internal med, Barnes Hosp, St Louis, Mo, 53-54; resident pediat, Children's Hosp, Cincinnati, Ohio, 54-56; res fel, Harvard Med Sch & Res Div Infectious Dis, Children's Med Ctr, Boston, 56-58; asst prof pediat, Sch Med, Johns Hopkins Univ, 58-63, asst prof microbiol, 59-63, assoc prof pediat & microbiol, 63-65; prof pediat, Sch Med, Univ Pittsburgh, 65-75, chmn dept, 65-69, dean, 69-75; PROF PEDIAT, HARVARD MED SCH & CHIEF, CHILDREN'S SERVS, MASS GEN HOSP, 77- *Concurrent Pos:* Med dir, Children's Hosp Pittsburgh, 65-69. *Mem:* Inst Med-Nat Acad Sci; AAAS; Am Asn Microbiol; Soc Pediat Res; Infectious Dis Soc Am; Am Pediat Soc. *Res:* Cytomegalo virus infections. *Mailing Add:* Children Serv Mass Gen Hosp 32 Fruit St Boston MA 02114

MEDEARIS, KENNETH GORDON, b Peoria, Ill, Aug 5, 30; m 53; c 3. CIVIL ENGINEERING, ENGINEERING MECHANICS. *Educ:* Univ Ill, BS, 52, MS, 53; Stanford Univ, PhD(eng), 62. *Prof Exp:* Stress analyst, Sandia Corp, NMex, 57-58; from asst prof to assoc prof civil eng, Univ NMex, 58-62; assoc prof eng, Ariz State Univ, 62-63; eng consult, 62-66; prof & dir comput ctr, Colo State Univ, 66-69; PRES, KENNETH MEDEARIS & ASSOCS, RES, ENG & COMPUT CONSULTS, 69- *Concurrent Pos:* State of Calif res grants, 61-65; affiliate prof, Colo State Univ, 69-; chmn, Larimer Co Comput Comn, 74- *Mem:* Seismol Soc Am; Am Soc Civil Engrs; Sigma Xi; Int Standards Orgn; Univ Coun Earthquake Eng Res; Aircraft Owners & Pilots Asn. *Res:* Structural dynamics; applied mathematics; computers; vibration analysis. *Mailing Add:* 1901 Seminole Dr Ft Collins CO 80525

MEDEIROS, ROBERT WHIPPEN, organic chemistry; deceased, see previous edition for last biography

MEDICI, PAUL T, b New York, NY, May 10, 19; m 43; c 3. HEMATOLOGY, ENDOCRINOLOGY. *Educ:* St John's Univ, NY, BS, 42 & 48, MS, 51; NY Univ, PhD, 56. *Prof Exp:* Instr biol sci, Col Pharm, 48-52, asst prof bact & pub health, 52-56, from assoc prof to prof, 56-65, dean grad sch arts & sci, 69-85, PROF HEMAT & ENDOCRINOL, GRAD SCH, ST JOHN'S UNIV, NY, 65-, CHMN DEPT BIOL, 65-, ASST VPRES GRAD STUDIES, LIBERAL ARTS & EDUC, DEAN SCH EDUC, 85- *Concurrent Pos:* Lectr, Guggenheim Dent Clin, New York, 51-53; Dean Sch Educ, Dean Grad Arts & Sci, 85- *Mem:* Fel AAAS; fel NY Acad Sci; Am Asn Univ Prof; Asn Military Surgeons US; Am Asn Higher Educ; Sigma Xi. *Res:* Endocrinology of blood. *Mailing Add:* 8914 89th St Woodhaven NY 11421

MEDICK, MATTHEW A, b New York, NY, Jan 20, 27; m 53; c 5. ENGINEERING MECHANICS, APPLIED MATHEMATICS. *Educ:* Univ Ill, BS, 48; NY Univ, MS, 50, PhD(eng mech), 58. *Prof Exp:* Lectr math, Pace Col, 50-51; appl sci rep, Int Bus Mach Corp, 51-53; lectr math, City Col New York, 53-55; res assoc civil eng & eng mech, Columbia Univ, 55-57; sr staff scientist, Res & Advan Develop Div, Avco Corp, 57-62; PROF MECH ENG, MICH STATE UNIV, 62- *Mem:* Am Soc Mech Engrs; Am Math Soc; Math Asn Am; Soc Indust & Appl Math; Soc Eng Sci. *Res:* Wave motion; impact and vibrations in solids; continuum mechanics; asymptotic phenomena; functional analysis; biomechanics. *Mailing Add:* Dept Mech Eng Mich State Univ East Lansing MI 48824

MEDICUS, HEINRICH ADOLF, b Zurich, Switz, Dec 24, 18; m 61. NUCLEAR PHYSICS, HISTORY PHYSICS. *Educ:* Swiss Fed Inst Technol, DrScNat(physics), 49. *Prof Exp:* Res assoc physics, Swiss Fed Inst Technol, 43-50; visitor radiation lab, Univ Calif, 50-51; guest, Mass Inst Technol, 51-52, instr physics, 52-54, vis asst prof, 54-55; assoc prof, 55-72, prof physics, 72-86, EMER PROF, RENSSELAER POLYTECH INST, 87- *Concurrent Pos:* Swiss Nat Scholarship, 50-52; vis scientist, Atomic Energy Res Estab, Harwell, Eng, 67-68 & Swiss Inst Nuclear Res, Villigen, 75-76. *Mem:* Am Phys Soc; Swiss Phys Soc. *Res:* Radioactivity; meson physics; photonuclear reactions; nuclear structure; history of physics. *Mailing Add:* Dept Physics Rensselaer Polytech Inst Troy NY 12180-3590

MEDIN, A(ARON) LOUIS, b Baltimore, Md, Oct 2, 25; m 50; c 4. SIMULATION & TRAINING, CHEMICAL ENGINEERING. *Educ:* Johns Hopkins Univ, BE, 48; Ohio State Univ, PhD(chem eng), 51. *Prof Exp:* Asst, USPHS, Ohio State Univ, 49-50; chem engr, US Atomic Energy Comn, 51-53; res engr, Ford Motor Co, 53-55; chief nuclear chem technol, Alco Prod, Inc, 55-58; res engr nuclear appln, US Steel Corp, 58-62; sr proj scientist, Avco Corp, 63-65; mgr sci applns dept, Int Bus Mach Corp, 65-67, mgr advan med applns, IBM Corp, 67-70, mgr environ & health sci, 70-72; dir environ & life sci, Off Dir Defense Res & Eng, Dept Defense, 72-74; sr govt analyst, IBM Corp, 75-78, mgr develop prog, 79-87; EXEC DIR, INST SIMULATION & TRAINING, UNIV CENT FLA, 87- *Honors & Awards:* Outstanding Contrib Award, IBM Corp, 69. *Mem:* Am Inst Chem Engrs; Am Inst Aeronaut & Astronaut; Nat Security Indust Asn; Am Defense Preparedness Asn. *Res:* Aerospace applications; space experimentations; nuclear power technology; corrosion, materials and water technology, data processing and medical applications; defense systems; command and control systems; training device systems; advanced semi-conductor applications; simulation and training; data processing systems. *Mailing Add:* Inst Simulation & Training Univ Cent Fla Orlando FL 32816-0054

MEDINA, DANIEL, b New York, NY, Mar 6, 41; m 63; c 3. ONCOLOGY. *Educ:* Univ Calif, Berkeley, BA, 63, MA, 66, PhD(zool), 69. *Prof Exp:* PROF CELL BIOL, BAYLOR COL MED, 69- *Concurrent Pos:* USPHS res grant chem carcinogenesis, Baylor Col Med, 71-82; assoc ed, Cancer Res, Breast Cancer Res & Treat. *Mem:* AAAS; Am Asn Cancer Res; Am Soc Cell Biol. *Res:* Tumor biology; chemical carcinogenesis of mouse mammary glands; chemical-virus interactions; biology of preneoplastic lesions. *Mailing Add:* Dept Cell Biol Baylor Col Med One Baylor Plaza Houston TX 77030

MEDINA, JOSE ENRIQUE, b Santurce, PR, May 1, 26; m 48, 75; c 3. DENTISTRY. *Educ:* Univ Md, DDS, 48. *Prof Exp:* From instr to prof oper dent, Baltimore Col Dent Surg, Sch Dent, Univ Md, 48-66, from actg head dept to head dept, 57-66, asst dean & prof clin dent, 64-67; from assoc dean to dean, Col Dent, 67-74, dir health ctr space planning & utilization, 74-76, asst vpres facil planning & oper, 76-86, PROF CLIN DENT, UNIV FLA, 67- *Concurrent Pos:* Spec lectr, Walter Reed Army Med Ctr, Washington, DC, 58-64; consult, Univ Md Hosp, 58-67, USPHS Hosp, 60-67 & US Naval Dent Sch, 63-67 & 87-90, Veterans Hosp, Gainesville, Fla, 67-84, Veterans Hosp, Miami, Fla, 70-74, Fund for Dent Health, Chicago, Ill, 74- 79 & Zoller Mem Dent Clin, Chicago, Ill, 85; hon prof, San Carlos Univ Guatemala, 60; mem Nat Adv Coun, Nat Inst Dent Res, NIH, 72-76. *Honors & Awards:* Distinguished Serv Award, Fla Dent Asn, 78-; George M Hollenback Mem Award, Acad Oper Dent, 85; Distinguished Mem Award, Am Acad Gold Foil Oper, 86; William John Gies Award, Am Col Dent, 90. *Mem:* Fel AAAS; fel Int Col Dentist; fel Am Col Dent; Int Asn Dent Res; hon mem Guatemala Dent Assoc; Am Dent Asn; Am Acad Gold Foil Oper; Acad Bd Oper Dent; Acad Oper Dent; hon mem Am Acad Oral Med; hon fel Acad Gen Dent; fel Royal Soc Health. *Res:* Restorative procedures; new materials for dental use. *Mailing Add:* 5002 NW 18 Pl Gainesville FL 32605

MEDINA, MARJORIE B, b Capiz, Philippines, Dec 29, 45; US citizen. FOOD SCIENCE, NUTRITION. *Educ:* Univ Santo Tomas, Manila, BS, 64; Rutgers Univ, MS, 74, PhD(food sci), 78. *Prof Exp:* asst prof, Univ Vt, 77-79, lectr food sci, 77-80; res chemist, 79-89, LEAD SCIENTIST, EASTERN REGIONAL RES CTR, AGR RES SERV, USDA, 90- *Concurrent Pos:* Sr res tech, Cornell Univ Med Ctr, 68-71; adj prof, Rutgers Univ, 84-88; mem, Int Interest Group, Biorecognition Technol. *Mem:* Inst Food Technologists; Am Chem Soc; Asn Official Analytical Chemists. *Res:* Protein quantitation and protein quality analysis; applications of scanning electron microscopy and x-ray fluorescent analysis food systems; carbohydrate analysis; development of microanalytical methods for veterinary drug residues (anabolic hormones, sulphur drugs and antibiotics); immunochemical, biospecific assays and chromatgraphic techniques; food proteins chemistry. *Mailing Add:* USDA ERRC Philadelphia PA 19118

MEDINA, MIGUEL ANGEL, b Laredo, Tex, July 5, 32; m 63; c 3. PHARMACOLOGY, BIOCHEMISTRY. *Educ:* St Mary's Univ, Tex, BS, 57, MS, 63; Southwestern Med Sch, Dallas, PhD(pharmacol), 68. *Prof Exp:* Jr chemist, Res & Develop Div, Am Oil Co, Tex, 57-59; res biochemist, Sch Aerospace Med, Brooks AFB, 59-64, res pharmacologist, 67-70; PROF PHARMACOL, UNIV TEX HEALTH SCI CTR, SAN ANTONIO, 70-, ASST DEAN, 86- *Concurrent Pos:* Lectr, St Mary's Univ, Tex, 67-70. *Mem:* Am Chem Soc; Am Soc Pharmacol & Exp Therapeut. *Res:* Brain and drug metabolism; histamine. *Mailing Add:* 712 W Marshall St San Antonio TX 78212-4940

MEDINA, MIGUEL ANGEL, JR, b Havana, Cuba, Dec 9, 46; US citizen; m 76; c 1. WATER RESOURCES, ENVIRONMENTAL ENGINEERING. *Educ:* Univ Ala, BSCE, 68, MSCE, 72; Univ Fla, PhD(water resources & environ eng), 76. *Prof Exp:* Asst post engr design & construct, 3rd US Army Hq, Ft McPherson, Ga, 70-71; researcher water pollution, Univ Ala, 71-72; researcher urban stormwater, Univ Fla, 72-76; asst prof, 76-80, PROF CIVIL ENG, DUKE UNIV, 80- *Concurrent Pos:* Prin investr, US Environ Protection Agency, 77-79, consult, 78-; rep, Nat Water Data Exchange, US Geol Surv, 77-; prin investr, NSF, 78-80; consult, Univ Fla Indust & Exp Sta, 68-, Tech Adv Serv for Atty, 78- & Technol Dept & Appln Br, Environ Protection Agency, Athens, Ga; prin investr, Off Water Res & Technol, 80-81 & NC Water Resources Res Inst, 81-82. *Mem:* Am Geophys Union; Am Soc Civil Engrs; Am Water Resources Asn; Asn Environ Eng Prof; Sigma Xi. *Res:* Mathematical modeling and computer simulation of pollutant transport systems, natural and man-made, throughout hydrologic cycle; operational hydrology. *Mailing Add:* Dept Civil Eng Duke Univ Durham NC 27706

MEDITCH, JAMES S, b Indianapolis, Ind, July 30, 34; m 64; c 2. TELECOMMUNICATION NETWORK ARCHITECTURES. *Educ:* Purdue Univ, BSEE, 56, PhD, 61; Mass Inst Technol, SM, 57. *Prof Exp:* Mem tech staff, Aerospace Corp, Calif, 61-65; assoc prof elec eng, Northwestern Univ, 65-67; staff mem, Boeing Sci Res Labs, Wash, 67-70; from assoc prof to prof, Elec Eng, Univ Calif, Irvine, 70-77; chmn, 77-85, assoc dean, Col Eng, 87-90, PROF, DEPT ELEC ENG, UNIV WASH, SEATTLE, 77- *Concurrent Pos:* Consult, indust & govt, 70-; assoc ed, Inst Elec & Electronics Engrs Transactions on Automatic Control, 73-75; mem, bd dir, Orincon Corp, San Diego, Calif, 73-88. *Honors & Awards:* Centennial Medal, Inst Elec & Electronic Engrs, 84. *Mem:* Fel Inst Elec & Electronic Engrs. *Res:* Telecommunication networks; modeling, analysis, design and performance optimization of terrestrial, satellite, ground-radio, local area and wide-area networks; broadband integrated services digital networks; switching and traffic theory; multimedia communications. *Mailing Add:* Dept Elec Eng FT-10 Univ Wash Seattle WA 98195

MEDLER, JOHN THOMAS, b Las Cruces, NMex, May 28, 14; m 64; c 4. ENTOMOLOGY. *Educ:* NMex State Col, BS, 36, MS, 37; Univ Minn, PhD(entom), 40. *Prof Exp:* From asst prof to prof, 46-79, EMER PROF ENTOM, UNIV WIS-MADISON, 79-; HON ASSOC, BERNICE P BISHOP MUS, 79- *Concurrent Pos:* Chmn, dept plant sci, Univ Nigeria, chief party Wis-USAID contract, Fac Agr, 68-75; proj dir, Midwest Univ Consortium Int Activ-USAID-Govt Indonesia Higher Educ contracts, Indonesia, 76-79. *Res:* Taxonomy and phylogeny of Flatidae (Homoptera). *Mailing Add:* Dept Entom Bernice P Bishop Mus PO Box 19000-A Honolulu HI 96817-0916

MEDLEY, SIDNEY S, b Vancouver, BC, July 12, 41; m 85. CONTROLLED FUSION RESEARCH, TOKAMAK DIAGNOSTICS. *Educ:* Univ BC, BSc, 63; MSc, 64, PhD (physics), 68. *Prof Exp:* Fel, Nat Res Coun Can, 68; fel plasma physics, Culham Lab, UkAEA, 68-70; res staff plasma physics, Fusion Res Ctr, Univ Tex, Austin, 71-77; PRIN RES PHYSICIST PLASMA PHYSICS, PRINCETON PLASMA PHYSICS LAB, 77- *Mem:* Can Asn Physicists; Am Phys Soc. *Res:* Design and development of diagnostic instrumentation in controlled thermonuclear fusion research; optical, ion spectrometry and nuclear diagnostic techniques; organization and management of fusion diagnostic projects. *Mailing Add:* Plasma Physics Lab Princeton Univ Princeton NJ 08543

MEDLIN, GENE WOODARD, b Greensboro, NC, Oct 5, 25; m 45; c 5. MATHEMATICS. *Educ:* Wake Forest Col, BS, 48; Univ NC, MA, 50, PhD(math), 53. *Prof Exp:* Assoc prof math, Wake Forest Col, 52-56; mathematician, Oak Ridge Nat Lab, 56-57; NSF grant, Swiss Fed Inst Tech, 57-58; assoc prof, 58-65, PROF MATH, STETSON UNIV, 65-, CHMN DEPT, 58- *Concurrent Pos:* Vis lectr grad sch, Univ Tenn, 56-57. *Mem:* Am Math Soc; Math Asn Am; Asn Comput Mach. *Res:* Matrix theory. *Mailing Add:* Dept Math-Comput Sci Stetson Univ Box 8332 Deland FL 32720

MEDLIN, JULIE ANNE JONES, b Battlecreek, Mich, Apr 30, 36. ENVIRONMENTAL SCIENCES, LICHENS. *Educ:* Univ Mich, BS, 58; Western Mich Univ, MA, 63, PhD(environ sci), 80. *Prof Exp:* Instr ecol & oceanog, Western Mich Univ, 78-80; ASST PROF ECOL & PHYSIOL, NAZARETH COL, 81- *Concurrent Pos:* Consult, Geol Dept, Western Mich Univ, 80- *Mem:* AAAS; Am Bryol Soc; British Lichen Soc. *Res:* Uptake of heavy metals in plants; distribution of heavy metals and ions that affect the uptake in the plant; use of lichens as indicators of air pollution. *Mailing Add:* 1232 Peninsula Dr Trevor City MI 49684

MEDLIN, WILLIAM LOUIS, b Harlingen, Tex, Aug 25, 28; m 58. SOLID STATE PHYSICS. *Educ:* Univ Tex, BS, 51, MS, 54, PhD(physics), 56. *Prof Exp:* Sr res assoc, Mobil Oil Corp, 56-57, res assoc, 67-82, SR RES ASSOC, DALLAS RES LAB, MOBIL RES & DEVELOP CORP, 82- *Res:* Rock mechanics, geophysics wave propagation; luminescence; color centers. *Mailing Add:* Dallas Res Lab Mobil Res & Develop Corp PO Box 819047 Dallas TX 75381

MEDNIEKS, MAIJA, CELL BIOLOGY. *Educ:* DePaul Univ, PhD(biochem), 76. *Prof Exp:* Res scientist med biochem, NIH, 80-88; ASST PROF, UNIV CHICAGO, 88- *Res:* Cyclic nucleotides in protein secretion. *Mailing Add:* Dept Pediat Univ Chicago Wyler Childrens Hosp 5841 S Maryland Ave Chicago IL 60637

MEDOFF, GERALD, b New York, NY, Nov 9, 36; m 60; c 2. MICROBIOLOGY. *Educ:* Columbia Univ, AB, 58; Wash Univ, MD, 62; Am Bd Internal Med, dipl. *Prof Exp:* Fel infectious dis, Mass Gen Hosp, Boston, 65-68, instr med & pediat, Harvard Med Sch, 68-70; from asst prof to assoc prof med, 70-76, asst prof microbiol, 71-76, PROF MED & ASSOC PROF MICROBIOL & IMMUNOL, SCH MED, WASH UNIV, 76-, CHIEF INFECTIOUS DIS DIV, 72- *Res:* Mycology; infectious diseases; medicine. *Mailing Add:* Dept of Med Wash Univ Sch of Med 660 S Euclid Ave St Louis MO 63110

MEDOFF, JUDITH, FUNGAL DIMORPHISM, CYTOSKELETON. *Educ:* Brandeis Univ, PhD(biol), 66. *Prof Exp:* PROF BIOL, ST LOUIS UNIV, 73- *Mailing Add:* Dept Biol St Louis Univ 3507 Laclede St Louis MO 63103

MEDORA, RUSTEM SOHRAB, b Deolali, India, May 4, 34; m 64; c 2. PHARMACOGNOSY, BIOLOGY. *Educ:* Gujarat Univ, India, BPharm, 58, MPharm, 60; Univ RI, PhD(pharmaceut sci), 65. *Prof Exp:* Tutor pharmacog, L M Col Pharm, Gujarat Univ, 58-61; asst, Univ RI, 61-65; asst prof, Idaho State Univ, 65-66; Nat Res Coun Can fel bot, McGill Univ, 66-67; asst prof pharmacog, 67-72, assoc prof, 72-79, PROF PHARM, SCH PHARM, UNIV MONT, 79- *Concurrent Pos:* Smith Kline Found, Mont Heart Asn & Miles Lab, Title I, HEA, res grants. *Mem:* Int Tissue Cult Asn; Sigma Xi; Am Soc Pharmacog; Soc Econ Bot; Tissue Cult Asn. *Res:* Pharmacognosy, tissue culture of plants of medicinal interest and gerontology. *Mailing Add:* Sch Pharm Univ Mont Missoula MT 59812

MEDRUD, RONALD CURTIS, b Tracy, Minn, July 9, 34; m 59; c 2. X-RAY CRYSTALLOGRAPHY. *Educ:* Augustana Col, SDak, BA, 56; Univ Iowa, PhD(phys chem), 63. *Prof Exp:* Nat Acad Sci-Nat Res Coun res assoc, US Naval Ord Lab, 63-64; sr res chemist, Corning Glass Works, 64-77; RES ASSOC, CHEVRON RES & TECHNOL CO, 77- *Concurrent Pos:* mem, Joint Comt Powder Diffraction Standards, Int Ctr Diffraction Data. *Mem:* Am Crystallog Asn; Am Chem Soc. *Res:* X-ray crystallography, x-ray powder diffraction and microscopy for materials evaluation; molecular modeling. *Mailing Add:* Chevron Res & Technol Co 100 Chevron Way Richmond CA 94802

MEDVE, RICHARD J, b California, Pa, Jan 28, 36; m 58; c 5. PLANT ECOLOGY. *Educ:* California Univ, Pa, BS, 57; Kent State Univ, MA, 59; Ohio State Univ, PhD(bot), 68. *Prof Exp:* Counr, Kent State Univ, 57-58; teacher pub schs, 58-66; TEACHER BIOL, SLIPPERY ROCK UNIV, 66- *Concurrent Pos:* Consult, Aquatic Ecol Assocs. *Mem:* Torrey Bot Club; Nat Asn Biol Teachers. *Res:* Mycorrhizae; stripmine revegetation; edible wild plants. *Mailing Add:* Dept Biol Slippery Rock Univ Slippery Rock PA 16057

MEDVED, DAVID BERNARD, electrooptics, for more information see previous edition

MEDWADOWSKI, STEFAN J, b Lodz, Poland; US citizen. SHELL STRUCTURES, SPATIAL STRUCTURES. *Educ:* Univ Calif, Berkeley, PhD(civil eng), 56. *Prof Exp:* PRIN, STEFAN J MEDWADOWSKI CONSULT STRUCT ENGR, 58- *Concurrent Pos:* Adj prof, Dept Civil Eng, Univ Calif, Berkeley, 78-86; chmn, Shell Comt, Am Soc Civil Engrs & Am Concrete Inst, 79-83. *Honors & Awards:* Torroja Medal, Int Asn Shell & Spatial Struct, 90. *Mem:* Fel Am Soc Civil Engrs; fel Am Concrete Inst; Int Asn Shell & Spatial Struct (vpres, 83-91, pres, 91-). *Res:* Applying mathematics and computer techniques to design of structures; aesthetically satisfying buildings. *Mailing Add:* 111 New Montgomery Suite 303 San Francisco CA 94105

MEDWAY, WILLIAM, b Man, Can, Feb 23, 27; m 71. VETERINARY MEDICINE, CLINICAL MEDICINE. *Educ:* Univ Man, BS, 47; Ont Vet Col, DVM, 54; Cornell Univ, PhD(physiol), 58. *Hon Degrees:* MA, Univ Pa, 71. *Prof Exp:* Instr biochem, Ont Agr Col, Univ Toronto, 48-49; asst physiol chem, Cornell Univ, 54-58; assoc med, Univ Pa, 58-60; asst prof physiol & res assoc, Ont Vet Col, 60-62; from asst prof to assoc prof, 62-68, PROF CLIN LAB MED, UNIV PA, 68- *Concurrent Pos:* Ed newsletter, Int Asn Aquatic Animal Med, 69-74 & 77-80. *Mem:* Am Soc Vet Clin Path (pres, 68-69); Int Asn Aquatic Animal Med (secy-treas, 69-74, pres, 74-75); Am Soc Vet Physiol & Pharmacol; Am Physiol Soc; Am Vet Med Asn. *Res:* Clinical chemistry; clinical pathology as applied to aquatic animals; veterinary diagnostics; veterinary physiology. *Mailing Add:* 1540 Peter Cheeseman Rd Blackwood NJ 08012

MEDWICK, THOMAS, b Jersey City, NJ, Oct 15, 29. ANALYTICAL CHEMISTRY. *Educ:* Rutgers Univ, BS, 52, MS, 54; Univ Wis, PhD(pharmaceut chem), 58. *Prof Exp:* Res analyst, Merck & Co, Inc, NJ, 58-60; asst prof pharmaceut chem, Col Pharm, Rutgers Univ, Newark, 60-63, assoc prof, 63-68, PROF PHARMACEUT CHEM, COL PHARM, RUTGERS UNIV, NEW BRUNSWICK, 68-, CHMN DEPT PHARMACEUT CHEM, 79- *Concurrent Pos:* Mem, Nat Formulary Bd, Am Pharmaceut Asn, 65-74 comt Food Chem Codex, Nat Res Coun, Nat Acad Sci, 74-83, US Pharmacopeial Rev Comt, 75-; sci adv, NY Dist, Food & Drug Admin, 70-79. *Honors & Awards:* Harvey W Wiley Award, 79. *Mem:* Am Pharmaceut Asn; Am Chem Soc; fel Am Inst Chemists; Sigma Xi; Am Soc Qual Control; fel Acad Pharmaceut Sci of Am Pharmaceut Asn; fel Am Asn Pharmaceut Scientists. *Res:* Use of NMR in drug analysis; acid-base reactions in nonaqueous solvents; analysis of pharmaceuticals and drugs; improvement of standards for drugs. *Mailing Add:* 238 McAdoo Ave Jersey City NJ 07305

MEDWIN, HERMAN, b Springfield, Mass, Apr 9, 20; m 45. ACOUSTICAL OCEANOGRAPHY. *Educ:* Worcester Polytech Inst, BS, 41; Univ Calif, Los Angeles, MS, 48, PhD(physics), 54. *Prof Exp:* Asst physics, Univ Calif, Los Angeles, 46-53, res assoc, 53-54; consult acoustics, Bolt, Beranek & Newman, Inc, 54-55; from assoc prof to prof, 55-80, EMER PROF PHYSICS, US NAVAL POSTGRAD SCH, 80-; OWNER, OCEAN ACOUST ASSOCS, 80- *Concurrent Pos:* Instr, Los Angeles City Col, 48-54; liaison scientist, Off Naval Res, London, 61-62, ed, Europ Sci Notes, 62; consult, Hudson Labs, Columbia Univ, 64-68; vis prof, Imp Col, Univ London, 65-66 & Nat Defense Acad, Yokosuka, Japan, 81; vchmn acoustics panel, Nat Acad Sci-Nat Res Coun Physics Surv Comt, 70-71; vis scientist, Royal Australian Naval Res Lab, 72-73, Academia Sinica, People's Repub China, 85 & Woods Hole Oceanog Inst, 85; vpres, Chmn Tech Coun, Acoust Soc Am, 87-88, chmn, Acoust Oceanog Group, 89- *Honors & Awards:* Res Award, Naval Postgrad Sch Chap, Sigma Xi, 72. *Mem:* Fel Acoust Soc Am; Am Geophys Union; Inst Noise Control Eng; Catgut Acoust Soc. *Res:* Surface and volume scattering of sound in the sea; effects of high intensity sounds; acoustic diffraction; microbubbles at sea. *Mailing Add:* Dept Physics US Naval Postgrad Sch Monterey CA 93943

MEDZIHRADSKY, FEDOR, b Kikinda, Yugoslavia, Feb 4, 32; m 67; c 2. NEUROCHEMISTRY, MOLECULAR PHARMACOLOGY. *Educ:* Tech Univ Munich, MS, 61, PhD(biochem), 65. *Prof Exp:* Instr biochem, Univ Munich, 65-66; from asst prof to assoc prof biochem, 69-81, assoc prof pharmacol, 75-81, PROF BIOCHEM & PHARMACOL, MED SCH, UNIV MICH, ANN ARBOR, 81- *Concurrent Pos:* NIH fel, Univ Wis, 66-67; Nat Inst Neurol Dis & Blindness trainee, Wash Univ, 67-69; vis assoc prof pharmacol, Stanford Univ Med Ctr, 75-76; Nat Res Serv Award, USPHS, 75-76; vis prof, Univ Calif, San Diego, 83-84. *Mem:* Am Soc Biol Chemists; Am Soc Pharmacol & Exp Therapeut; Am Soc Neurochem; AAAS; Ger Soc Biol Chemists. *Res:* Membrane receptor mechanisms with focus on the opioid receptors; mechanisms of signal transduction; membrane biochemistry. *Mailing Add:* Dept Biol Chem Univ Mich Med Sch, 6440 Med Sci Bldg I Ann Arbor MI 48109

MEDZON, EDWARD LIONEL, b Winnipeg, Man, May 26, 36; m 61; c 3. VIROLOGY. *Educ:* Univ Man, BSc, 57, MSc, 60; McGill Univ, PhD(virol, immunol), 64. *Prof Exp:* Instr microbiol, Univ Mich, 63-65; asst prof, 65-69, ASSOC PROF MICROBIOL, UNIV WESTERN ONT, 69- *Concurrent Pos:* Vis lectr, Eastern Mich Univ, 64-65; vis scientist, NIH, 74, Fla Inst Technol, Med Res Inst, 75; LCDC, Health & Welfare, Can, 87; Inst Armand-Frappier, 86; chmn, Can Soc Micro Meetings Comt, 79-88, vpres, 87-89, pres, 89-90. *Mem:* AAAS; Am Soc Microbiol; Can Soc Microbiol; Can Soc Cell Biol; NY Acad Sci; Am Soc Virol. *Res:* Respiratory virus infection and immunity; Clostridium difficile toxins; computer based learning; applications of cloned viral genes; Clostridium difficile toxins. *Mailing Add:* Dept Microbiol & Immunol Health Sci Ctr Univ Western Ont London ON N6A 5C1 Can

MEE, JACK EVERETT, inorganic chemistry, solid state electronics, for more information see previous edition

MEECH, JOHN ATHOL, b Toronto, Ont, Jan 16, 47; m 72; c 2. SURFACE CHEMISTRY, PROCESS CONTROL. *Educ:* McGill Univ, BEng, 70; Queen's Univ, MSc, 75, PhD(mineral eng), 79. *Prof Exp:* Sr asst metal eng, Roan Consolidated Mines Ltd, Zambia, 70-73; lectr, Queen's Univ, Ont, 74-77, asst prof, 77-82, assoc prof mineral processing, 82-89; ASSOC PROF, UNIV BC, 89- *Mem:* Can Inst Mining & Metall. *Res:* Flotation and surface chemistry; chrysocolla ores; process development and feasibility studies; computer simulation; iron ore agglomeration; fine particle recovery techniques; environmental control of mining effluents; pyrrhotite oxidation; expert systems development. *Mailing Add:* Dept Mining & Mineral Process Eng Univ BC 6350 Stores Rd Vancouver BC V6T 1W5 Can

MEECHAM, WILLIAM CORYELL, b Detroit, Mich, June 17, 28; m 48; c 2. MATHEMATICAL PHYSICS, CLASSICAL PHYSICS. *Educ:* Univ Mich, BS & MS, 48, PhD(physics), 54. *Prof Exp:* Asst physics, Univ Mich, 48-53 & Brown Univ, 53-54; res assoc, Univ Mich, 54-55, assoc res physicist, 55-56, instr, 56-57, asst prof, 57-60, res physicist & head fluid & solid mech lab, 59-60; prof fluid mech, Univ Minn, Minneapolis, 60-66; sr scientist, Lockheed Palo Alto Res Labs, 66-67; head div appl mech, Col Eng, 68-69, PROF FLUID MECH & ACOUST, UNIV CALIF, LOS ANGELES, 67- *Concurrent Pos:* Res assoc, Univ Calif, San Diego, 63; consult, TRW, Inc, 59-65, Rand Corp, Calif, 64-72, Inst Sci & Technol, Univ Mich, 60-67 & Bolt, Beranek & Newman, Inc, 68-74; consult, Res & Develop Assoc, 74-76, Aerospace Corp, 75-, Arete Assoc, 76- & Calif Res Tech, 84- *Mem:* Am Phys Soc; fel Acoust Soc Am; assoc fel Am Inst Aeronaut & Astronaut; acoustics; diffraction theory; stochastic processes; wave propagation problems. *Mailing Add:* Sch Eng & Appl Sci Univ Calif Los Angeles CA 90024

MEECHAN, CHARLES JAMES, b Usk, Wash, Aug 7, 28; m 51; c 6. SOLID STATE PHYSICS. *Educ:* Ore State Col, BS, 51. *Prof Exp:* Res physicist, Atomics Int Div, NAm Aviation, Inc, 51-61, staff physicist, Sci Ctr, 61-63, res adv, Corp Off, 63-67, exec dir res & eng, NAm Rockwell Corp, 67-69, vpres indust systs, 69-71, vpres & dir, Sci Ctr, 71-72, vpres res & eng, 72-78, exec vpres, 78-81, vpres strategic planning, energy systs group, Rockwell Int Corp, 81-84; RETIRED. *Mem:* Am Phys Soc; Sigma Xi. *Res:* Experimental research in study of lattice imperfections; radiaton damage and diffusion phenomena in solids; Mossbauer spectroscopy; music composer and video director. *Mailing Add:* 22647 Ventura Blvd Woodland Hills CA 91364

MEECHAN, ROBERT JOHN, b Newport, Wash, Aug 25, 26; m 53; c 3. PEDIATRICS. *Educ:* Ore State Col, BA, 51; Univ Ore, MS & MD, 53. *Prof Exp:* From instr to assoc prof, 57-68, asst dean Med Sch Admis, 86-87, PROF PEDIAT, MED SCH, UNIV ORE, 68- *Mailing Add:* Dept Pediat Univ Ore Sch Med 3181 SWS Jackson Park Rd Portland OR 97201

MEEDEL, THOMAS HUYCK, b Mt Pleasant, Iowa, May 11, 49. DEVELOPMENTAL GENETICS. *Educ:* Nebr Wesleyan Univ, BS, 71; Univ Pa, PhD(mircobiol), 77. *Prof Exp:* Trainee, Wistar Inst Anat Biol, 77-79, res assoc, 79-81; res assoc, Boston Univ Marine Prog, 81-85; ASST SCIENTIST, MARINE BIOL LAB, 85- *Concurrent Pos:* Vis assoc, Calif Inst Technol, 85. *Mem:* Sigma Xi; Int Soc Develop Biologists; Soc Develop Biol. *Res:* Cell fates during embryonic development; ascidians, with acetylcholinesterase as the primary marker of differentiation; cytoplasmic factors localized in the egg playing a key role in determining cell fate; nature of cytoplasmic factors and their mechanisms. *Mailing Add:* Marine Biol Lab Woods Hole MA 02543

MEEDER, JEANNE ELIZABETH, b Erie, Pa, Mar 5, 50. FOOD PRODUCT DEVELOPMENT, LEGAL & REGULATORY FOOD ISSUES. *Educ:* Houghton Col, BA, 72. *Prof Exp:* Lab asst, Welch Foods, Inc, 72-73, tech asst, 73-74, assoc tech asst, 74-76, sr tech asst, Welch Foods, Inc, 76; res & develop technologist, Stouffer Foods Corp, 76-77, entree team leader, 77-82; sr food technologist, Frozen & Specialty, DelMonte, 82-84, prin food technologist, 84-85, mgr, develop projs, 86, mgr, prod develop, Foodservice, 85-86, mgr, explor res, DelMonte-USA, 86-87, consult, 87-88; dir res & develop/qual assurance, Kibun Prod Int, Inc, 88-90; dir res & develop/food design, 90-91, VPRES FOOD DESIGN & NEW BUS DEVELOP, JMS GROUP, INC, 91- *Concurrent Pos:* Consult, Meeder Assocs, 87-; dir res & develop, HFI Foods, Inc, 90-; tech comt & surimi comt, Nat Fisheries Inst, 88-; refrig foods comt, 89-90, meat & poultry labeling task force, Am Frozen Food Inst, 91- *Mem:* Inst Food Technologists; Am Home Economists Bus; Int Microwave Power Inst; Nat Fisheries Inst; Am Frozen Food Inst; Nat Food Processors Asn. *Res:* Developing extensive variety of food products for retail, foodservice, and industrial markets; produced viable frozen and shelfstable entrees, side dishes, dinners, desserts, beverages, and nutritional/diet products. *Mailing Add:* JMS Group Inc 18368 Redmond-Fall City Rd Redmond WA 98052

MEEGAN, CHARLES ANTHONY, b Buffalo, NY, Sept 24, 44. ASTROPHYSICS. *Educ:* Rensselaer Polytech Inst, BS, 66; Univ Md, PhD(physics), 73. *Prof Exp:* Res assoc astrophys, Rice Univ, 74-75, Univ Ala, Huntsville, 76 & Nat Res Coun, 76-78; SPACE SCIENTIST ASTROPHYS, NASA MARSHALL SPACE FLIGHT CTR, 78- *Mem:* Am Phys Soc; Am Astron Soc; Sigma Xi. *Res:* Cosmic ray astrophysics and medium-energy gamma-ray astronomy. *Mailing Add:* Mail Code ES-62 Marshall Space Flight Ctr Huntsville AL 35812

MEEHAN, EDWARD JOSEPH, b Oakland, Calif, July 21, 12; m 45; c 3. CHEMISTRY. *Educ:* Univ Calif, BS, 33, PhD(phys chem), 36. *Prof Exp:* Instr chem, Univ Calif, 36-39; from instr to prof chem, 39-82, EMER PROF CHEM, UNIV MINN, MINNEAPOLIS, 82- *Concurrent Pos:* With Off Rubber Reserve, 44. *Mem:* Am Chem Soc; Optical Soc Am. *Res:* Absorption spectra of solids; spectrophotometry; physical and chemical properties of high polymers; light scattering; reaction mechanisms. *Mailing Add:* Dept of Chem Univ of Minn Minneapolis MN 55455

MEEHAN, JOHN PATRICK, b San Francisco, Calif, May 22, 23; m 49; c 4. PHYSIOLOGY. *Educ:* Univ Southern Calif, MD, 48. *Prof Exp:* Instr, 47-49, asst prof, 49-51 & 54-55, assoc prof, 55-62, PROF PHYSIOL, SCH MED, UNIV SOUTHERN CALIF, 62-, CHMN DEPT, 66- *Mem:* AAAS; Aerospace Med Asn. *Res:* Aviation physiology; central nervous system control of the vascular system; cardiovascular and respiratory physiology; aerospace medicine. *Mailing Add:* 460 W Walnut Ave Arcadia CA 91006

MEEHAN, THOMAS (DENNIS), b Youngstown, Ohio, 1942; m 78. BIOLOGICAL CHEMISTRY, BIOPHYSICS. *Educ:* Univ Akron, BS, 65; St Louis Univ, PhD(biochem), 73. *Prof Exp:* Researcher chem carcinogenesis, Lab Chem Biodynamics, Univ Calif, 74-76, jr staff biochemist, 76-78; sr res assoc, Mich Molecular Inst, 78-84; DEPT PHARM & PHARMACEUT CHEM, UNIV CALIF SCH PHARM, 85- *Concurrent Pos:* Asst prof biochem, Mich Molecular Inst, 78-84; consult, indust & univ & prin invest, 78- *Mem:* AAAS; Am Chem Soc; Biophys Soc; Am Asn Cancer Res. *Res:* Mutation mechanisms of mutagenesis; alteration of gene expression by chemical carcinogens; physical and chemical interactions between carcinogens and DNA; synthetic charat modified antisense oligodeoxynucleotides. *Mailing Add:* Dept Pharm Box 0446 Univ Calif Sch Pharm San Francisco CA 94143

MEEHAN, WILLIAM ROBERT, b Buffalo, NY, Apr 9, 31; m 63; c 3. FISH BIOLOGY. *Educ:* Univ Buffalo, BA, 52; Univ Ore, MA, 55; Mich State Univ, PhD(fisheries, wildlife), 58. *Prof Exp:* Res biologist, Alaska Dept Fish & Game, 58-66; FISHERY RES BIOLOGIST, FORESTRY SCI LAB, US FOREST SERV, 66- *Mem:* Am Fisheries Soc; Pac Fishery Biologists; Am Inst Fishery Res Biologists. *Res:* Wildlife biology; salmon investigations; aquatic entomology. *Mailing Add:* Forestry Sci Lab US Forest Serv PO Box 20909 Juneau AK 99802-0909

MEEK, DEVON WALTER, b River, Ky, Feb 24, 36; m 65; c 2. SYNTHETIC INORGANIC CHEMISTRY. *Educ:* Berea Col, BA, 58; Univ Ill, MS, 60, PhD(inorg chem), 61. *Prof Exp:* Asst chem, Univ Ill, 58-60, res fel, 60-61; from asst prof to assoc prof inorg chem, 61-69, chmn dept, 77-81, PROF INORG CHEM, OHIO STATE UNIV, 69- *Concurrent Pos:* Vis assoc prof, Northwestern Univ, 67; sr res fel, Univ Sussex, England, 74; John Simon Guggenheim fel, 81-82. *Mem:* Am Chem Soc; Sigma Xi; Royal Soc London. *Res:* Studies of the syntheses, electronic and magnetic properties and structures of transition metal complexes with unusual coordination number; homogeneous catalysis and activition and/or stabilization of small molecules; phosphorus-31 nuclear magnetic resonance studies and x-ray crystallography for structural characterization; organometallic chemistry. *Mailing Add:* Dept Chem Ohio State Univ Columbus OH 43210

MEEK, EDWARD STANLEY, pathology, biology, for more information see previous edition

MEEK, JAMES LATHAM, b San Antonio, Tex, Apr 10, 37; m 56; c 3. MATHEMATICS. *Educ:* Univ Tex, BA, 62, MA, 63, PhD(math), 67. *Prof Exp:* Instr math, San Antonio Col, 63-64; res assoc acoust & math, Defense Res Lab, Univ Tex, Austin, 67; asst prof, 67-74, ASSOC PROF MATH, UNIV ARK, FAYETTEVILLE, 74- *Concurrent Pos:* Consult, Defense Res Lab, Univ Tex, Austin, 67-68. *Mem:* Am Math Soc; Math Asn Am. *Res:* Underwater acoustics; boundary behavior of analytic, harmonic, and subharmonic functions; harmonic analysis. *Mailing Add:* Dept of Math Univ of Ark Fayetteville AR 72701

MEEK, JOHN SAWYERS, b Madison, Wis, Aug 12, 18; m 45; c 2. ORGANIC CHEMISTRY. *Educ:* Univ Wis, BA, 41; Univ Ill, MS, 44, PhD(org chem), 45. *Prof Exp:* Asst inorg chem, Univ Ill, 41-44, Allied Chem & Dye fel, 45; from instr to prof, Univ Colo, Boulder,45-87; RETIRED. *Concurrent Pos:* Asst, Univ Wis, 43. *Mem:* Am Chem Soc. *Res:* Diels-Alder reactions; Bridgehead compounds. *Mailing Add:* 1911 Columbine Ave Boulder CO 80302

MEEK, JOSEPH CHESTER, JR, b Sabetha, Kans, July 16, 31; m 54; c 3. INTERNAL MEDICINE, ENDOCRINOLOGY. *Educ:* Univ Kans, AB, 54, MD, 57. *Prof Exp:* Intern, San Diego County Gen Hosp, Calif, 57-58; resident, Univ Kans Med Ctr, Kansas City, 58-60; res asst space med, US Naval Sch Aviation Med, 60-62; from instr to assoc prof, 64-75, vchancellor academic affairs, 81-85, PROF MED, UNIV KANS MED CTR, KANSAS CITY, 75-, CHMN DEPT INTERNAL MED, WICHITA, 85- *Concurrent Pos:* Am Col Physicians Mead Johnson scholar, Univ Kans, 59-60; fel endocrinol, Scripps Clin & Res Found, La Jolla, Calif, 62-63, trainee, 63-64; attending physician, Vet Admin Hosp, 64- *Mem:* Am Fedn Clin Res; Am Thyroid Asn; Am Diabetes Asn; fel Am Col Physicians; Endocrine Soc; Sigma Xi. *Res:* Metabolism; long acting throid stimulator; insulin A and B chains. *Mailing Add:* Med Sch Univ Kans 1010 N Kansas Wichita KS 67214

MEEK, VIOLET IMHOF, b Geneva, Ill, June 12, 39; wid; c 2. INORGANIC CHEMISTRY. *Educ:* St Olaf Col, BA, 60; Univ Ill, MS, 62, PhD(inorg chem), 64. *Prof Exp:* Instr chem, Mt Holyoke Col, 64-65; from asst prof to prof chem, Ohio Wesleyan Univ, 65-84, chmn dept, 75-79, dean educ serv, 80-84; dir annual progs, Coun Independent Cols, 84-86; ASSOC DIR DEVELOP, OHIO STATE UNIV RES FOUND, 86- *Mem:* AAAS; Am Chem Soc; Nat Coun Univ Res Admin; Am Asn Higher Educ. *Res:* Nature of the creative process as it applies to the sciences; the interaction of the variables in the structure of the academic workplace and their effect on the creativity and scholarly productivity of academic researchers. *Mailing Add:* Ohio State Univ Res Found 1960 Kenny Rd Columbus OH 43210-1063

MEEKER, DAVID LYNN, b Burlington, Iowa, Sept 17, 50; m 72; c 4. PRODUCTION & MARKETING TECHNOLOGY. *Educ:* Iowa State Univ, BS, 72, MS, 73, PhD(animal sci), 85. *Prof Exp:* Supt, Res Sta, Iowa State Univ, 80-85; dir, 85-90, VPRES, RES & EDUC, NAT PORK PRODUCERS COUN, 90- *Concurrent Pos:* Mem, Animal Health Res Adv Bd, USDA, 80-90. *Mem:* Am Soc Animal Sci; Am Meat Sci Asn; Coun Agr Sci & Technol. *Res:* Administration of pork industry research programs. *Mailing Add:* 207 N Main Huxley IA 50124

MEEKER, LOREN DAVID, b Plymouth Falls, Ore, Apr 28, 32. APPLIED MATHEMATICS. *Educ:* Ore State Univ, BS, 59; Stanford Univ, MS, 62, PhD(math), 65; Univ Astor, Eng, MSc, 69. *Prof Exp:* PROF MATH, UNIV NH, 70- *Mem:* Am Math Soc; Soc Indust & Appl Math; Math Asn Am; Biometric Soc; Sigma Xi; Am Asn Univ Prof. *Res:* Modeling and analysis of mammary cancer experiments; theory application of time control systems. *Mailing Add:* Math Dept Univ NH Durham NH 03824

MEEKER, RALPH DENNIS, b Chicago, Ill, Nov 15, 45; m 68; c 2. PHYSICS. *Educ:* Ill Benedictine Col, BS, 67; Iowa State Univ, PhD(physics), 70. *Prof Exp:* From asst prof to assoc prof, 70-79, chmn dept, 72-75, PROF PHYSICS, ILL BENEDICTINE COL, 79-, CHMN DEPT, 78, CHMN SCI DIV, 81- *Concurrent Pos:* Resident assoc, Argonne Nat Lab, consult, 71-76; pres, Exradin, Inc, 78- *Mem:* Am Phys Soc; Am Asn Physics Teachers. *Res:* Medical physics (radiation dosimetry). *Mailing Add:* Dept Physics Ill Benedictine Col 5700 College Rd Lisle IL 60532

MEEKER, THRYGVE RICHARD, b Pottstown, Pa, Mar 9, 29; m 54; c 3. PHYSICAL CHEMISTRY. *Educ:* Ursinus Col, BS, 51; Univ Del, MS, 54, PhD(phys chem), 56. *Prof Exp:* MEM TECH STAFF, BELL LABS, 55- *Mem:* Am Chem Soc; Am Phys Soc; Am Inst Chemists; Inst Elec & Electronics Eng; NY Acad Sci. *Res:* Chemical kinetics; spectroscopy; ultrasonics and elastic properties; piezoelectric, dielectric and ferroelectric phenomena; applied mathematics; quantum mechanics; thermodynamics; wave phenomena. *Mailing Add:* 2956 Lindberg Ave Allentown PA 18103

MEEKER, WILLIAM QUACKENBUSH, JR, b New York, NY, Nov 28, 49; m 75; c 1. STATISTICS. *Educ:* Clarkson Col Technol, BS, 72; Union Col, MS, 73, PhD(admin & eng systs), 75. *Prof Exp:* Res fel statist, Inst Admin & Mgt, Union Col, 73-75; asst prof, 75-78, assoc prof, 78-81, PROF STATIST, IOWA STATE UNIV, 81- *Concurrent Pos:* Statistician, Corp Res & Develop, Gen Elec Co, 73-75; statist consult, Bell Lab, 78-81. *Mem:* Am Soc Qual Control; Am Statist Asn; Inst Math Statist; Biomet Soc; AAAS; Sigma Xi. *Res:* Applied areas of statistics, including life data analysis, time series analysis,, sequential analysis statistical computing and experimental design. *Mailing Add:* Statist Lab Iowa State Univ 326 Sendcor Hall Ames IA 50011

MEEKINS, JOHN FRED, b Boston, Mass, Oct 4, 37; m 61; c 2. RADIO ASTRONOMY, X-RAY ASTRONOMY. *Educ:* Bowdoin Col, BA, 59; Cath Univ Am, PhD(physics), 73. *Prof Exp:* RES PHYSICIST, NAVAL RES LAB, 59- *Mem:* Am Astron Soc; Sigma Xi. *Res:* Astrophysics, especially concerning high temperature astrophysical plasmas. *Mailing Add:* Naval Res Lab Code 4210 Washington DC 20375-5000

MEEKS, BENJAMIN SPENCER, JR, b Florence, SC, Nov 17, 24; m 49; c 2. ORGANIC CHEMISTRY. *Educ:* Univ SC, BS, 44; Cornell Univ, PhD(org chem), 51. *Prof Exp:* Res chemist, US Rubber Co, 51-52 & Tenn Eastman Co, 52-56; assoc prof chem, Mercer Univ, 56-58; Univ Ky Contract Team, Bandung Inst Tech, 58-62; assoc prof, 62-65, chmn dept, 74-81, PROF CHEM, MOORHEAD STATE UNIV, 65-, DIR PROG INDUST CHEM, 81- *Concurrent Pos:* Res grant, Univ Col, London, 69-70; res assoc, Univ Minn, 77-78; Fulbright Lectr, Univ Dhaka, Bangladesh, 84-85. *Mem:* AAAS; Am Chem Soc; Royal Soc Chem; Sigma Xi. *Res:* General organic synthesis; ionic reaction mechanisms. *Mailing Add:* Dept of Chem Moorhead State Univ Moorhead MN 56560

MEEKS, FRANK ROBERT, b Ft Worth, Tex, Dec 5, 28. PHYSICAL CHEMISTRY. *Educ:* Tex Christian Univ, BA, 49; Polytech Inst of NY, PhD, 56. *Prof Exp:* From asst prof to assoc prof, 57-77, PROF PHYS CHEM, UNIV CINCINNATI, 77- *Concurrent Pos:* Res scholar, Univ Montpellier, France, 63-64. *Mem:* Am Chem Soc; Sigma Xi. *Res:* Plasma theory; statistical mechanics; thermodynamics of irreversible processes. *Mailing Add:* Dept of Chem Univ of Cincinnati Cincinnati OH 45221

MEEKS, ROBERT G, b Columbus, Ohio, Jan 24, 42; m 62; c 1. TOXICOLOGY. *Educ:* Otterbein Col, BSc, 72; Ohio State Univ, PhD(pharmacol & toxicol), 77. *Prof Exp:* Med technologist, Childrens Hosp, 62-66; head, Protein Hormone & Radioimmunoassay Res & Develop Sect, G D Searle Reference Lab, 66-72; sr staff fel, Dept Health, Educ & Welfare Nat Inst Health, Nat Cancer Inst, 77-80; SR TOXICOLOGIST, SOUTHERN RES INST, 80-, HEAD TOXICOL, 81- *Concurrent Pos:* Adj asst prof, Dept Pub Health, Sch Med & Sch Community Health, Univ Ala, Birmingham, 80- *Mem:* Am Asn Clin Chemists; AAAS; Am Chem Soc; NY Acad Sci; Biophys Soc. *Res:* In vivo and in vitro toxicity of antitumor drugs, anticonvulsants and retinoids as well as other compounds under development for potential use in man. *Mailing Add:* 204 Hickory Park Ct Antioch TN 37013-4310

MEEKS, WILKISON (WINFIELD), b Pittsburgh, Pa, Apr 4, 15; m 46; c 2. ACOUSTICS. *Educ:* Maryville Col, AB, 37; Northwestern Univ, MS, 39, PhD(physics), 41. *Prof Exp:* Contract employee, US Naval Ord Lab, 41-44; staff physicist, Haskins Labs, Inc, NY, 44-46; assoc prof physics, Western Md Col, 46-47 & Southern Ill Univ, 47-48; asst prof, Western Reserve Univ, 48-55; physicist res ctr, B F Goodrich Co, Ohio, 55-58; assoc prof physics, 58-60, chmn dept, 60-68, PROF PHYSICS, ROSE-HULMAN INST TECHNOL, 60- *Mem:* Am Phys Soc; Acoust Soc Am; Am Asn Physics Teachers; Sigma Xi. *Res:* Magneto-mechanical effects; underwater sound. *Mailing Add:* Rose Hulman Inst Technol Terre Haute IN 47803

MEELHEIM, RICHARD YOUNG, b Cape Charles, Va, Aug 30, 25; m 51; c 3. PHYSICAL CHEMISTRY. *Educ:* Univ Va, BS, 50, PhD(chem), 58. *Prof Exp:* Anal chemist, Monsanto Chem Co, 50-54; res chemist fiber surface res, E I du Pont De Nemours & Co, Inc, 58-62, chemist, Dacron Res Lab, 62-67, sr res chemist, 67-84; RETIRED. *Mem:* Sigma Xi. *Res:* Research and development fibers. *Mailing Add:* PO Box 831 204 Gordon St Beaufort NC 28516

MEEM, J(AMES), LAWRENCE, JR, b Brooklyn, NY, Dec 24, 15; m 40; c 2. NUCLEAR ENGINEERING. *Educ:* Va Mil Inst, BS, 39; Univ Ind, MS, 47, PhD(physics), 49. *Prof Exp:* Instr chem, Va Mil Inst, 39-40; aeronaut res scientist, Nat Adv Comt Aeronaut, Va, 41-43, Ohio, 44-46; reactor physicist, Oak Ridge Nat Lab, 50-55; chief reactor scientist, Alco Prod, Inc, 55-57; prof, 57-81, chmn dept, 57-77, EMER PROF NUCLEAR ENG, UNIV VA, 81- *Concurrent Pos:* Consult, NASA, 60-67, Westinghouse Elec Corp, 63-67 & Army Foreign Sci & Technol Ctr, Charlottesville, Va, 81-90; vis mem staff, Los Alamos Sci Lab, 67-68; vis consult, Sandia Labs, NMex, 77-78. *Mem:* Am Phys Soc; fel Am Nuclear Soc; Am Soc Eng Educ. *Res:* Alternative nuclear fuel cycles; plutonium proliferation risk. *Mailing Add:* 2401 Old Ivy Rd Univ Village No 1201 Charlottesville VA 22901

MEEN, RONALD HUGH, b Can, Nov 25, 25; m 68. SYNTHETIC ORGANIC CHEMISTRY. *Educ:* Univ Toronto, BA, 47, MA, 49, PhD(chem), 53. *Prof Exp:* Fel & res assoc org chem, Iowa State Col, 53-54; res chemist, Tenn Eastman Co, 54-86; RETIRED. *Mem:* Sigma Xi. *Res:* Chemistry; organic chemical development. *Mailing Add:* 2121 Cypress St Kingsport TN 37664

MEENAN, PETER MICHAEL, b New York, NY, Nov 20, 42; m 70; c 2. COMPUTER SCIENCE, SYSTEMS ENGINEERING. *Educ:* Manhattan Col, BS, 64; Univ Ariz, MS, 66; Union Col, NY, MS, 70, PhD(admin, eng systs), 74. *Prof Exp:* Nuclear engr, Knolls Atomic Power Lab, 66-70; syst engr, 70-71, mgr syst anal & simulation tech, 72-77, mgr indust info systs, 78-79, mgr electronic planning & res, 80-83, mgr info syst appln serv, 84-87, MGR, COMPUT GRAPHICS & SYSTS, GEN ELEC RES & DEVELOP CTR, 88- *Concurrent Pos:* Adj prof comput sci, State Univ NY, Albany, 76- *Mem:* Inst Elec & Electronics Engrs; Soc Comput Simulation. *Res:* Computer visualization, interactive graphics; software engineering; office automation; mathematical methods, work stations and supercomputers. *Mailing Add:* 11 Knollwood Dr Schenectady NY 12302

MEENTEMEYER, VERNON GEORGE, b Centralia, Ill, Nov 7, 42; m 66; c 2. CLIMATOLOGY. *Educ:* Southern Ill Univ, BA, 65, MA, 68, PhD(climat), 71. *Prof Exp:* Asst prof phys geog, Southern Ill Univ, 73; asst prof, 73-80, ASSOC PROF PHYS GEOG, UNIV GA, 80- *Mem:* Asn Am Geogrs; AAAS; Ecol Soc Am; Sigma Xi. *Res:* Climatic influences on decomposer food chains; atmospheric hazards; impact on natural ecosystems, and probability mapping; radiant energy flows in forests; terretrial carbon balance; computer mapping of ecosystem dynamics. *Mailing Add:* Dept of Geog Univ of Ga Athens GA 30602

MEERBAUM, SAMUEL, b Brno, Czech, May 1, 19; US citizen; m 46; c 3. BIOENGINEERING, PHYSIOLOGY. *Educ:* Mass Inst Technol, BS & MS, 46; Univ Calif, Los Angeles, PhD(bioeng & physiol), 71. *Prof Exp:* Sect head, Propulsion & Indust Res, M W Kellogg Co, 46-62; asst chief engr, Arde Inc, 62-65; prin scientist, Rocketdyne, NAm Rockwell, 65-69; sr res scientist cardiol, Cedars Sinai Med Ctr, 69-86; CONSULT MED, 86- *Mem:* Am Heart Asn; Am Col Cardiol; Biomed Eng Soc; Am Asn Med Instrumentation; Sigma Xi. *Res:* Biomedical sciences, specifically cardiovascular diagnostic and intervention concepts; author of numerous technical publications and two books. *Mailing Add:* 5741 El Canon Ave Woodland Hills CA 91367

MEERBOTT, WILLIAM KEDDIE, b Jersey City, NJ, Aug 31, 18; m 43; c 2. PETROLEUM CHEMISTRY. *Educ:* St Peters Col, BS, 40; Lehigh Univ, MS, 42. *Prof Exp:* Sr res chemist, Shell Oil Co, 49, group leader, 51-66, staff res chemist, 66-72; staff res chemist, Shell Develop Co, 72-78, sr staff res chemist, 78-81; RETIRED. *Mem:* Am Chem Soc. *Res:* Catalysis in the field of petroleum chemistry and related to hydro processing, desulfurization and catalytic reforming. *Mailing Add:* 3006 Winslow Houston TX 77025

MEEROVITCH, EUGENE, b Vladivostok, Russia, July 11, 19; Can citizien; m 61; c 2. PARASITOLOGY. *Educ:* St John's Univ, China, BSc, 47; McGill Univ, MSc, 53, PhD(parasitol), 57. *Prof Exp:* Res asst parasitol, Hebrew Univ, Israel, 48-53; res asst, 53-57, from asst prof to assoc prof, 57-71, dir, Inst Parasitol, 78-84, PROF PARASITOL, MACDONALD COL, McGILL UNIV, 71- *Concurrent Pos:* US Acad Sci Donner fel, Nat Inst Med Res, Eng, 58-59; consult, WHO, 69-; vis sr scientist, Wellcome Res Labs, Eng, 75-76. *Mem:* AAAS; Am Soc Parasitol; Am Soc Trop Med & Hyg; Royal Soc Trop Med & Hyg; Can Soc Zoologists. *Res:* Amoebiasis, especially immunology, serology and host-parasite relations. *Mailing Add:* Inst Parasitol McGill Univ McDonald Campus Ste Anne de Bellevue PQ H9X 1C0 Can

MEESE, JON MICHAEL, b Indianapolis, Ind, Aug 5, 38; m 63; c 1. SOLID STATE PHYSICS. *Educ:* Univ Cincinnati, BS, 61; Purdue Univ, Lafayette, MS, 64, PhD(physics), 70. *Prof Exp:* Jr physicist, Wabash Magnetics, 60-61; asst instr solid state physics, Purdue Univ, Lafayette, 61-62, assoc instr, 62-65, res asst, 65-70; res physicist, Univ Dayton, 70-76; from asst prof to prof physics, Univ Mo, 76-84, sr res physicist & group leader radiation effects, Res Reactor Facil, 76-84, from assoc prof to prof elec eng, 81-84; res physicist, Amoco Corp, 84-89; CHAIR, ELEC ENG DEPT, UNIV MO, COLUMBIA, 89- *Concurrent Pos:* In-house contractor, Aerospace Res Labs, Wright-Patterson AFB, 70-72; adj prof physics, Wright State Univ, 75- *Mem:* Am Phys Soc. *Res:* Radiation damage and ion implantation in semiconductors; luminescence; electro-optical devices; solar cells; neutron transmutation doping of semiconductors; molecular beam epitaxy; OMVPE; DLTS. *Mailing Add:* 2301 Limerick Lane Columbia MO 65203

MEETER, DUANE ANTHONY, b Hammond, Ind, Apr 20, 37; m 60; c 2. STATISTICS. *Educ:* Univ Mich, Ann Arbor, AB, 60; Univ Wis-Madison, MS, 61 & 62, PhD(statist), 64. *Prof Exp:* From asst prof to assoc prof, 64-76, PROF STATIST, FLA STATE UNIV, 76-, DIR DEPT, 75- *Concurrent Pos:* Vis prof statist, Univ Victoria, BC, 76-77. *Mem:* Am Statist Asn. *Res:* Biostatistics; sequential design of experiments; nonlinear design; computer applications. *Mailing Add:* Dept of Statist Fla State Univ Tallahassee FL 32306

MEETZ, GERALD DAVID, b Aurora, Ill, Aug 22, 37; m 81; c 1. IMMUNOLOGY, CELL BIOLOGY. *Educ:* NCent Col, Ill, BA, 59; Univ Ill, MS, 67, PhD(anat), 69; Calif State-Fullerton, MS, 84. *Prof Exp:* USPHS fel cellular, molecular & develop biol, Univ Colo, 69-70; fel pediat, Univ Minn, Minneapolis, 70-71, fel path, 71-72; asst prof anat, Med Col Va, 72-75; asst prof anat, Sch Dent, Marquette Univ, 73-79; asst prof anat, Southern Calif Col Optom, 79-85; assoc prof, Col Osteop Med, Kansas City, Mo, 85-87; ASSOC PROF, UNIV OSTEOP MED & HEALTH SCI, 87-, DEPT CHMN, 90- *Mem:* Am Soc Cell Biol; Am Asn Anatomists; Am Soc Microbiol; Reticuloendothelial Soc; Sigma Xi. *Res:* Role of histones in development of avian erythrocyte; gene activation in lymphoid cells; immunobiology of rheumatoid arthritis; anti-tumor activity of shark serum. *Mailing Add:* Dept Anatomy 3200 Grand Des Moines IA 50312

MEEUSE, BASTIAAN J D, b Sukabumi, Indonesia, May 9, 16; US citizen; m 42; c 2. BIOCHEMISTRY & POLLINATION BIOLOGY. *Educ:* Univ Leiden, Neth, BSc, 36, Drs, 39; Univ Delft, Neth, DSc, 43. *Prof Exp:* Teacher chem physics biol, Hort Inst, Boskoop, Neth, 39-42; asst technol bot, Univ Delft, Neth, 42-46, chief asst, 46-49, lectr technol bot & biochem, 49-52; from asst prof to prof, 52-86, EMER PROF PLANT PHYSIOL, UNIV WASH, SEATTLE, 86- *Concurrent Pos:* Vis prof microbiol, Univ Delft, Neth, 62-63; sci adv, Oxford Sci Films, Eng, 75-; course dir, Chautauqua-type conf, NSF & AAAS, 83; vis prof bot, Univ Nijmegen, Neth, 85. *Mem:* Am Bot Soc; Am Soc Plant Physiol; Royal Dutch Bot Soc; AAAS; Sigma Xi. *Res:* Carbohydrate biochemistry; enzymology; phycology and plant physiology, especially cyanide-insensitive and thermogenic respiration in arum lilies; floral ecology (pollination and seed dispersal). *Mailing Add:* Bot Dept KB-15 Univ Wash Seattle WA 98195

MEEUWIG, RICHARD O'BANNON, b St Louis, Mo, Dec 8, 27; m 67; c 2. FOREST ECOLOGY. *Educ:* Univ Calif, Berkeley, BS, 51, MS, 60; Utah State Univ, PhD(soil physics), 64. *Prof Exp:* Forester timber mgt, US Forest Serv, 51-55, forester watershed mgt, 56-64, soil scientist, 64-75, RES FORESTER, INTERMOUNTAIN FOREST & RANGE EXP STA, 75- *Concurrent Pos:* Adj prof, Univ Nev, Reno. *Mem:* Soc Range Mgt. *Res:* Ecology and management of pinyon-juniper woodlands. *Mailing Add:* 1200 Monroe Reno NV 89509

MEEZAN, ELIAS, b New York, NY, Mar 5, 42; m 67; c 3. BIOCHEMISTRY, PHARMACOLOGY. *Educ:* City Col New York, BS, 62; Duke Univ, PhD(biochem), 66. *Prof Exp:* Asst prof pharmacol, Duke Univ, 69-70; from asst prof to assoc prof pharmacol, Univ Ariz Med Sch, 70-79; prof & chmn, 79-89, PROF PHARMACOL & DIR, METABOCK DIS RES LAB, UNIV ALA, BIRMINGHAM, 89- *Concurrent Pos:* Helen Hay Whitney fel, 66-69. *Honors & Awards:* Res Career Develop Award, NIH, 77-79. *Mem:* AAAS; Am Soc Pharmacol & Exp Therapeut; NY Acad Sci; Am Soc Biol Chem; Am Diabetes Asn. *Res:* Biochemical pharmacology of diabetes; insulin receptors, action and basement membrane structure; metabolism in isolated renal glomeruli; tubules, brain and retinal microvessels; sulfate transport and metabolism in cystic fibrosis. *Mailing Add:* Dept Pharmacol Univ Ala Birmingham Birmingham AL 35294

MEFFERD, ROY B, JR, b Hico, Tex, Sept 22, 20; m 40; c 2. HUMAN RESOURCES. *Educ:* Tex A&M Univ, BS & MS, 40; Univ Tex, PhD(bact & biochem), 51; Kennedy-Western Univ, MSBA, 91. *Prof Exp:* Soils technologist, Bur Reclamation, 41-42; teacher agr, Erath County Voc Sch, 47-48; res scientist genetics, Univ Tex, Austin, 48-51; dir metab res, Southwest Found Res & Educ, 51-56; dir ment dis res, Interdisciplinary Biochem Inst, Univ Tex, 56-59; dir, Psychiat & Psychosom Res Lab, Vet Admin, 59-80; SR CONSULT PSYCHOMET, BIRKMAN & ASSOCS, INC, 66- *Concurrent Pos:* Prof biol, Trinity Univ, 51-56; prof physiol, Baylor Col Med, 59-; prof psychol, Univ Houston, 70-80; adj prof behav sci, Sch Pub Health, Univ Tex, 75- *Honors & Awards:* Damen Runyon Fel Cancer Res, 52 & 53. *Mem:* Am Psychol Asn; Am Chem Soc; Am Physiol Soc. *Res:* Genes; cancer; environmental physiology; stress; adapatation; mental diseases, especially schizophrenia; measurement and prediction of human behaviors. *Mailing Add:* Baylor Col Med 823 Longview Dr Sugar Land TX 77478

MEFFORD, DAVID ALLEN, b Keokuk, Iowa, Dec 28, 28; m 51; c 3. ANALYTICAL CHEMISTRY. *Educ:* Randolph-Macon Col, BS, 51. *Prof Exp:* Asst biol, Randolph-Macon Col, 50; chemist, 53-73, with qual assurance, 57-62, DIR QUAL ASSURANCE, A H ROBINS CO, INC, 73- *Mem:* AAAS; Am Chem Soc; Am Soc Qual Control; Pharmaceut Mfrs Asn. *Res:* Microelemental analysis; analytical problems related to pharmaceutical products; automated analysis, computer automation. *Mailing Add:* 10200 Hobby Hill Rd Richmond VA 23235

MEGAHAN, WALTER FRANKLIN, b Oceanside, NY, Jan 22, 35; m 56; c 2. EROSION-SEDIMENTATION, WILDLAND HYDROLOGY. *Educ:* NY State Col Environ Sci & Forestry at Syracuse Univ, BS, 57, MS, 60; Colo State Univ, PhD(watershed resources), 67. *Prof Exp:* Regional hydrologist, Intermountain Region, 60-66, RES HYDROLOGIST & RES PROJ LEADER, INTERMOUNTAIN RES STA, US FOREST SERV, 67- *Concurrent Pos:* Vis prof, Forest Ecol Inst, Univ Padova, Italy, 78; chmn, Erosion-Sedimentation Comt, Am Geophys Union, 80-84; mem, Comt Hydrol, Transp Res Bd, Nat Acad Sci, 80-88; chmn, Nat Work Group to Define Soil & Water Res Needs for the Nation, 81; co-chmn, Subj Group Erosion, Floods & Avalanches, Int Union Forest Res Orgn, 81-; co-op scientist, watershed mgt res, Pakistan, 82-89; res fel, East-West Environ & Policy Inst, Honolulu, Hawaii, 83, 85, & 90; assoc ed, Western J Appl Forestry, 85- *Honors & Awards:* Super Serv Award, USDA, 88. *Mem:* Am Geophys Union; Soil Sci Soc Am; Sigma Xi; Int Mountain Soc; Am Inst Hydrol; Int Asn Sci Hydrology. *Res:* Erosion and sedimentation in relation to forest management practices; hydrologic processes in mountainous areas; impacts of forest land use on water quality. *Mailing Add:* Forestry Sci Lab 316 E Myrtle St Boise ID 83702

MEGAHED, SID A, b Nabarouh, Egypt, Sept 27, 41; US citizen; m 69; c 2. BATTERY MATERIALS ENGINEERING, BATTERY CHEMISTRY & ELECTROCHEMISTRY. *Educ:* Alexandria Univ-Egypt, BS, 61; La State Univ-Baton Rouge, MS, 65; Univ Wisc-Madison, PhD(chem), 69. *Hon Degrees:* PD, Univ Wisc-Madison, 74. *Prof Exp:* Teaching asst chem, Univ Wisc-Madison, 69-70; res & develop engr chem eng, ESB Inc, 70-74; develop engr, 74-80, DIR ADVAN TECHNOL CHEM ENG (BATTERIES), RAYOVAC INC, 80- *Concurrent Pos:* Lectr exten, Univ Wisc-Madison, 91-; elec comt, Electrochem Soc, 91-93. *Mem:* Chem Soc; Chem Eng Soc. *Res:* Developing new battery systems for society; developed zinc air system for hearing aid users; diralent silver system for watch users; lithium batteries for new CMOS-SRAM (computers); granted 21 patents. *Mailing Add:* 1413 Mound St Madison WI 53711

MEGARD, ROBERT O, b Garretson, SDak, Dec 4, 33; m 58. LIMNOLOGY, PLANKTON. *Educ:* St Olaf Col, BA, 56; Univ NMex, MS, 58; Ind Univ, PhD(zool), 62. *Prof Exp:* Res fel limnol, Univ Minn, Minneapolis, 61-64, res assoc, 64-67, from asst prof to assoc prof ecol, 67-84, asst prof, Col Biol Sci, St Paul, 71-72, PROF ECOL, COL BIOL SCI, UNIV MINN, MINNEAPOLIS, 85- *Mem:* AAAS; Am Soc Limnol & Oceanog; Ecol Soc Am. *Res:* Limnology ecology of plankton; paleolimnology; biology and ecology of plankton populations. *Mailing Add:* Dept Ecol & Behav Biol Univ Minn 109 Zool Bldg 318 Church St SE Minneapolis MN 55455-0302

MEGARGLE, ROBERT G, b Flushing, NY, Oct 11, 41; m 63; c 1. ANALYTICAL CHEMISTRY. *Educ:* Clarkson Col, BS, 63, PhD(chem), 68. *Prof Exp:* NSF fel chem, Univ Minn, Minneapolis, 66-67; asst prof, Univ Mo-Columbia, 67-72; ASSOC PROF CHEM, CLEVELAND STATE UNIV, 72- *Mem:* Am Chem Soc; Am Soc Testing & Mat. *Res:* Development of instrumentation and laboratory computers for measurement and control of experiments; distributed networks for laboratory experimentation; titration methods of analysis; laboratory robotics. *Mailing Add:* Dept Chem Cleveland State Univ Cleveland OH 44115

MEGAW, WILLIAM JAMES, b Belfast, Northern Ireland, July 8, 24; m 46; c 3. ATMOSPHERIC PHYSICS. *Educ:* Univ Liverpool, BSc, 51, DSc(physics), 73. *Prof Exp:* Sci officer, UK Atomic Energy Authority, 51-53; sr sci officer, Atomic Energy Res Estab, Eng, 53-57, prin sci officer, 57-61, sr prin sci officer, 61-71; dir, Ctr Res Environ Qual, York Univ, 74-84, prof physics, 71-89, chmn dept, 79-89, EMER PROF PHYSICS, YORK UNIV, 89-; PRIN INVESTR, INST SPACE & TERRESTRIAL SCI, 89- *Concurrent Pos:* Mem, Comt Nucleation, Comn Cloud Physics, Int Asn Meteorol & Atmospheric Physics, 65-76; subcomn ions, aerosols & radioactivity, Int Comn Atmospheric Elec, 67-81; consult, Danish AEC, 69-78; ed bd, J Aerosol Sci, 81-89. *Mem:* Fel Brit Inst Physics; Can Asn Physicists; AAAS. *Res:* Physics of particles in the atmosphere; inhibition of condensation on cloud nuclei; fog modification; environmental protection. *Mailing Add:* Dept Physics York Univ 4700 Keele St North York ON M3J 2P3 Can

MEGEL, HERBERT, b Newark, NJ, Nov 10, 26; m 51; c 2. TOXICOLOGY. *Educ:* NY Univ, BS, 48, MS, 50, PhD(exp biol), 54. *Prof Exp:* Res physiologist endocrine physiol, Princeton Labs, 54-59; res physiologist environ physiol, Boeing Co, 59-62; res biochemist, Nat Drug Co, 62-70; sect head immunol, Merrell Res Ctr, Merell Nat Lab, Richardson-Merrell Inc, 70-78, sr toxicologist, Path & Toxicol Dept, Merrell Dow Res Inst, 78-90; RETIRED. *Mem:* AAAS; Soc Toxicol; Am Soc Pharmacol & Exp Therapeut; Soc Exp Biol & Med; Am Asn Immunol. *Res:* Immunopharmacology; immunotoxicology; biochemical pharmacology; endocrine physiology; toxicology. *Mailing Add:* 1250 Forest Ct Cincinnati OH 45215

MEGGERS, WILLIAM F(REDERICK), JR, b Washington, DC, May 12, 24; m 50; c 2. ELECTRONICS ENGINEERING, PHYSICS. *Educ:* Univ Wis, BS, 50, MS, 51. *Prof Exp:* Electronic scientist, Microwave Systs Div, US Naval Ord Lab, Corona, 51-55, head, Talos Missile Countermeasures Br, Countermeasures Div, 55-59, head, Counter-countermeasures Tech Br, 59-61, head, Guide Div, Missile Systs Dept, 61-71; consult, Electronic Systs Dept, Naval Weapons Ctr, 71-75, head of staff, Electronic Warfare Dept, 75-80; RETIRED. *Concurrent Pos:* Antiq horologist, 81-90. *Mem:* Am Watchmakers Inst. *Res:* Electronic systems analysis; missile guidance systems design; microwave and electro-optical components and techniques. *Mailing Add:* 609 W Coral Ave Ridgecrest CA 93555

MEGGISON, DAVID LAURENCE, b Lynn, Mass, Dec 24, 28; m 52; c 4. FOOD TECHNOLOGY. *Educ:* Johns Hopkins Univ, BA, 48; Univ Mass, MS, 50, PhD(food sci), 53. *Prof Exp:* Assoc technologist cent labs, Gen Foods Corp, 52-55, proj leader frozen foods, Birds Eye Labs, 55-59; food prod res lab, Borden Foods Co, NY, 59-62; sect chief, Hunt Foods & Industs, Inc, 62-64, assoc dir res, Prod Develop, Hunt-Wesson Foods, 64-74; vpres res & develop, United Vintners, Inc, 74-75; vpres tech serv, RJR Foods, 75-80; VPRES CORP RES & TECH SERV, DEL MONTE CORP, 80- *Mem:* Inst Food Technol. *Res:* Development of new products and processes in instant foods; frozen and canned foods; wines. *Mailing Add:* Del Monte Corp One Mkt Plaza Box 3575 San Francisco CA 94119

MEGGITT, WILLIAM FREDRIC, b Green Springs, Ohio, Feb 9, 28; m 48; c 1. AGRONOMY. *Educ:* Ohio State Univ, BS, 50, MS, 51; Rutgers Univ, PhD(weed control, farm crops), 54. *Prof Exp:* Res agronomist, Agr Res Serv, USDA, 57-58; asst prof farm crops & weed control, Rutgers Univ, 58-60; assoc prof crop sci & weed control, 60-66, PROF CROP SCI, MICH STATE UNIV, 66- *Mem:* Am Soc Agron; Crop Sci Soc Am; Weed Sci Soc Am. *Res:* Weed control, chemical and cultural means; weed life cycles and competition; soil residues, penetration, translocation, accumulation and sites of action of herbicides. *Mailing Add:* 645 S Maple Apt 1018 Bldg 4 Tempe AZ 85283

MEGHREBLIAN, ROBERT V(ARTAN), b Cairo, Egypt, Sept 6, 22; nat US; m; c 2. RESEARCH & ENGINEERING MANAGEMENT. *Educ:* Rensselaer Polytech Inst, BAeE, 43; Calif Inst Technol, MS, 50, PhD(aeronaut, math), 53. *Prof Exp:* Struct engr, Consol Vultee Corp, 46-47;

civil engr, Fluor Corp, 47; asst proj engr, Guided Missile Develop, Jet Propulsion Lab, Calif Inst Technol, 47-52; sr res engr & lectr, Oak Ridge Sch Reactor Technol, 52-55; chief appl mech group, Aircraft Nuclear Propulsion Proj, Oak Ridge Nat Lab, 55-57, assoc dir gas cooled power reactor proj, 57-58; chief physics sect, Jet Propulsion Lab, Calif Inst Technol, 58-60, chief phys sci div, 60-62, mgr space sci div, 62-68, dep asst lab dir tech div, 68-71; vpres, Cabot Corp, 71-87, dir res & eng, 71-79, gen mgr, Crystals Bus Unit, 85-86, dir corp planning & develop, 86-87; pres, Distrigas Corp, 79-85; RETIRED. *Concurrent Pos:* Guggenheim fel, Calif Inst Technol, 49-51, assoc prof, 60-61. *Mem:* Fel Am Nuclear Soc; assoc fel Am Inst Aeronaut & Astronaut; Sigma Xi. *Res:* High temperature thermodynamics; reactor analysis; author of one book. *Mailing Add:* 440 Woodley Rd Santa Barbara CA 93108

MEGIBBEN, CHARLES KIMBROUGH, b Lexington, Ky, Oct 22, 36; m 57; c 4. MATHEMATICS. *Educ:* Southern Methodist Univ, BS, 59; Auburn Univ, PhD(math), 63. *Prof Exp:* Asst prof math, Tex Tech Col, 63-64; res assoc, Off Naval Res, Univ Wash, 64-65; asst prof, Univ Houston, 65-67; assoc prof, 67-81, PROF MATH, VANDERBILT UNIV, 81- *Concurrent Pos:* NSF res grants, 66-72. *Mem:* Am Math Soc. *Res:* Theory of Abelian groups; rings and modules. *Mailing Add:* Dept Math Vanderbilt Univ Nashville TN 37240

MEGILL, LAWRENCE REXFORD, b Potsdam, Ohio, July 5, 25; m 46; c 2. PHYSICS. *Educ:* Univ Nebr, BSc, 49, MA, 51; Univ Colo, PhD(physics), 59. *Prof Exp:* Mem staff, Los Alamos Sci Lab, Univ Calif, 51-53; physicist inst telecommun sci & aeronomy, Environ Sci Serv Admin, Colo, 55-69; prof physics & elec eng, Utah State Univ, 69-, dir, Ctr Atmospheric & Space Sci, 77-86; DIV DIR, GLOBESAT, 86- *Concurrent Pos:* Prog dir aeronomy, NSF, 74-75; pres, Globesat Inc, 84- *Res:* Atmospheric physics, particularly photometry and spectroscopy. *Mailing Add:* Globesat Div EER Systs 1740 Res Pkwy Logan UT 84321

MEGIRIAN, ROBERT, b New York, NY, June 18, 26; m 57; c 1. PHARMACOLOGY. *Educ:* Colgate Univ, AB, 51; Univ Rochester, MS, 53; Boston Univ, PhD(pharmacol), 57. *Prof Exp:* Res assoc pharmacol, Univ Rochester, 51-53; asst, Boston Univ, 53-56; pharmacologist, US Food & Drug Admin, 57-61; from asst prof to assoc prof, 61-74, PROF PHARMACOL & TOXICOL, ALBANY MED COL, 75-, PROF PHYSIOL, 80-, ASST DEAN, 84- *Concurrent Pos:* Res fel, Boston City Hosp, Mass, 57. *Mem:* Reticuloendothelial Soc; Am Soc Pharmacol & Exp Therapeut; Soc Exp Biol & Med. *Res:* Effect of drugs on the reticuloendothelial system. *Mailing Add:* Asst Dean Albany Med Col Albany NY 12208

MEGNA, JOHN C(OSIMO), chemical & biochemical engineering, for more information see previous edition

MEGO, JOHN L, b Pukwana, SDak, Sept 29, 22; m 53; c 2. BIOCHEMISTRY. *Educ:* Johns Hopkins Univ, PhD(biol), 60. *Prof Exp:* Res assoc plant physiol, Johns Hopkins Univ, 60; res asst neurosurg, Baltimore City Hosps, 60-62, res assoc, 62-65, res chief, 65-67; from assoc prof to prof biol, 67-88, PROF EMER BIOL, UNIV ALA, TUSCALOOSA, 88- *Concurrent Pos:* Co-prin investr, NIH-AEC grants, 64-67; Nat Acad Sci exchange fel, Univ Bratislava, 67; Nat Inst Environ Health Sci grant award, 71-74; Nat Inst Gen Med Sci res grant, 75-78; res fel, Univ de Provence, Marseille, France, Nat Ctr Sci Res, 79. *Mem:* Am Soc Biol Chemists. *Res:* Biochemical properties of lysosomes, membrane receptors and protein turnover; the uptake and degradation of altered serum albumins in lysosomes, including binding to receptors, internalization and conditions required for efficient degradation in lysosomes of the kidney, liver and spleen. *Mailing Add:* Dept Biol Box 1927 Univ Ala Tuscaloosa AL 35487

MEGRAW, ROBERT ARTHUR, b Rochester, Minn, Dec 5, 39; m 69; c 2. FORESTRY, WOOD TECHNOLOGY. *Educ:* Univ Minn, BS, 62, PhD(forest prod eng), 66. *Prof Exp:* Res scientist, Pioneering Dept, Res Div, Weyerhaeuser Co, 66-68, sr scientist, Res & Eng Div, 68-74, mgr, Wood Sci & Morphol Sect, 74-76, mgr, Microstruct & Wood Sci Sect, 76-91, SCI ADV, RES & ENG DIV, WEYERHAEUSER CO, 87- *Honors & Awards:* Wood Award, 67. *Mem:* Forest Prod Res Soc; Soc Wood Sci & Technol; Tech Asn Pulp & Paper Indust. *Res:* Wood and fiber properties, wood quality specialist; tree growth-fiber property relationships; flame spread control in wood and fiber products; x-ray applications to wood and fiber research; scanning electron microscopy; kenaf. *Mailing Add:* Weyerhaeuser Co Weyerhaeuser Technol Ctr Tacoma WA 98477

MEGRAW, ROBERT ELLIS, b Philadelphia, Pa, Feb 10, 30; m 71; c 5. CLINICAL CHEMISTRY. *Educ:* Fla State Univ, BA, 56, MS, 60; Iowa State Univ, PhD(bact), 64; Am Bd Clin Chem, dipl. *Prof Exp:* Fel biochem, Albert Einstein Med Ctr, 64-66; scientist, Warner-Lambert Res Inst, 66-71; res biochemist, Sigma Chem Co, 71-73; mgr unitest chem, Bio-dynamics/bmc, 73-80; mrg res, Ortho Diagnostic Systs Inc, 81-82; TECH DIR, BIO-RAD LABS, 82- *Concurrent Pos:* Lectr chem, Butler Univ, 81. *Mem:* AAAS; Am Chem Soc; Sigma Xi; Soc Exp Biol & Med; Am Asn Clin Chemists. *Res:* Research and development of clinical diagnostic reagents. *Mailing Add:* 18651 Ervin Lane Santa Ana CA 92705

MEGRUE, GEORGE HENRY, b Jamaica, NY, Mar 23, 36; m 58; c 4. GEOCHEMISTRY, RESEARCH ADMINISTRATION. *Educ:* Amherst Col, BA, 57; Columbia Univ, MA, 59, PhD(geol), 62. *Prof Exp:* Res assoc chem, Brookhaven Nat Lab, 62-64, assoc chemist, 64-66; geochemist & cosmochemist, Smithsonian Inst Astrophys Observ & Harvard Col Observ, 66-74; FOUNDER & PRES, MEGRUE MICROANAL SYSTS CO, 74- *Concurrent Pos:* Sci leader, Nat Geog Ethiopian Rift Valley Exped, 69; prin investr, Apollo 12, 14 & 15 Manned Space Flights, NASA, 71-73; co-founder, Foxglove Sch, 75- *Mem:* AAAS; Am Geophys Union; Am Chem Soc; Sigma Xi; NY Acad Sci. *Res:* Thermal history of meteorites; tectonic history of Ethiopian Rift Valley; distribution and origin of helium, neon, and argon isotopes in meteorites and lunar rocks; laser probe mass spectrometry; chemistry of cosmic dust; application of laser microchemical analyses to scientific research and development; spatial 40Ar/39Ar dating; holder of one US patent on the system and technique for gas analysis. *Mailing Add:* Megrue Microanal Systs Co Box 523 New Canaan CT 06840

MEGUERIAN, GARBIS H, b Turkey, Sept 10, 22; nat US; m 51; c 2. PHYSICAL ORGANIC CHEMISTRY. *Educ:* Am Univ, Beirut, BS, 47; Brown Univ, PhD(chem), 50. *Prof Exp:* Fel & res chemist, Harvard Univ, 50-52; chemist, Standard Oil Co Ind, 52-58, group leader, 58-64, RES ASSOC AMOCO OIL CO, 64- *Mem:* Am Chem Soc. *Res:* Reactions of elementary sulfur; oxidation of mercaptans; high temperature oxidation of hydrocarbons; nitrogen oxides control in automotive emissions. *Mailing Add:* 20608 Arcadia Dr PO Box 43 Olympia Fields IL 60461

MEHAFFEY, LEATHEM, III, b Jersey City, NJ, June 7, 41; m 63; c 2. VISION RESEARCH. *Educ:* Columbia Univ, AB, 63; Fordham Univ, MS, 65; Ohio State Univ, PhD(biophys), 71. *Prof Exp:* Res assoc, dept ophthal, Harvard Med Sch, 71-73; asst prof, 79-83, ASSOC PROF BIOL & CHMN DEPT, VASSAR COL, 83- *Concurrent Pos:* Sr vis scientist, dept ophthal, NY Univ Med Sch, 81- *Mem:* Sigma Xi; Asn Res Vision & Ophthal. *Res:* Clinical and pathological physiology of the retina; diagnosis and characterization of inherited retinal degeneration. *Mailing Add:* Box 410 Vassar Col Raymond Ave Poughkeepsie NY 12601

MEHENDALE, HARIHARA MAHADEVA, b Philya, India, Jan 12, 42; m 68; c 2. MEDICAL SCIENCES, ENVIRONMENTAL HEALTH. *Educ:* Karnataka Univ, India, BSc, 63; NC State Univ, MS, 66, PhD(physiol), 69; Am Bd Toxicol, dipl, 81; Acad Toxicol Sci, dipl, 84- *Prof Exp:* Fel entom, Univ Ky, 69-71; NIH vis fel, Analysis & Synthetic Chem Br, Nat Inst Environ Health Sci, 71-72, staff fel, Environ Pharmacol & Toxicol Br, 72-75; from asst prof to assoc prof, 75-80, PROF PHARMACOL & TOXICOL, UNIV MISS MED CTR, 80-, DIR, TRAINING PROG TOXICOL, 82- *Concurrent Pos:* mem, Res Adv & Policy Comt, Miss Heart Asn, 79-82 & Toxicol Study Sect, NIH, 81-84; vis prof, Toxicol Dept, Karolinska Inst, Stockholm, Sweden, 83-84; vis prof, Dept Toxicol, Karolinska Inst, Stockholm, 83-84; adj prof, Geriatric Ctr, Univ Miss Med Ctr, 86-; pres, assoc scientist of Indian Origin in Am, 81-82; UN Develop Prog expert, TOKTEN proj, (Transfer of Knowledge & Technol by Expatriate Nat), India, 85-86; external PhD examr, Sri Venkateswara Univ, 84-, Indian Inst Sci, 86- & Magnalore Univ, 86-; mem bd dir, Am Bd toxicol, 86-90, secy, 87-90; Burroughs Wellcome toxicol scholar, 88-; overseas ed, Indian J Pharmacol, 84- *Honors & Awards:* Toxicol Honors Award, Acad Environ Biol, 88. *Mem:* Soc Toxicol; Am Chem Soc; Int Soc Study Xeno Biotics; Am Soc Pharmacol & Exp Therapeut; Am Thoracic Soc; Sigma Xi; fel Acad Toxicol Sci; Acad Environ Biol (pres, 90-93); Asn Study Liver Dis. *Res:* Use of isolated perfused organs in studies of biochemical mechanisms of chemical toxicity; pharmacokinetics of environmental agents; mechanisms of hepato and pulmonary toxicity; mechanisms of interactive toxicity of combinations of toxic chemicals at nontoxic or ambient levels. *Mailing Add:* Dept Pharm & Toxicol Univ Miss Med Ctr Jackson MS 39216-4505

MEHERIUK, MICHAEL, b Derwent, Alta, June 5, 36; m 63. PLANT BIOCHEMISTRY. *Educ:* Univ Alta, BSc, 57, BEd, 59, PhD(biochem), 65. *Prof Exp:* RES SCIENTIST, CAN DEP AGR, 65- *Mem:* Can Soc Hort Sci; Am Soc Hort Sci. *Res:* Post-harvest physiology; storage disorders and storage conditions for pome fruits, soft fruits and grapes. *Mailing Add:* RR 4 Summerland BC V0H 1Z0 Can

MEHL, JAMES BERNARD, b Minneapolis, Minn, May 5, 39; m 61; c 2. PHYSICAL ACOUSTICS. *Educ:* Univ Minn, BPhys, 61, MS, 64, PhD(physics), 66. *Prof Exp:* Res assoc physics, Univ Ore, 66-68; from asst prof to assoc prof, 68-82, PROF PHYSICS, UNIV DEL, 82-, PHYSICS CHAIRPERSON, 88- *Concurrent Pos:* Physicist, Kernforschunsanlage, Juelich, Ger, 77 & Nat Bur Standards, 79-80. *Mem:* Acoustical Soc Am; Am Asn Physics Teachers; Am Phys Soc; AAAS; Soc Indust & Appl Mech. *Res:* Acoustics; metrology; applied mathematics. *Mailing Add:* Dept of Physics Univ of Del Newark DE 19716

MEHLER, ALAN HASKELL, b St Louis, Mo, May 24, 22; m 43; c 4. BIOCHEMISTRY, SCIENCE EDUCATION. *Educ:* Wash Univ, AB, 42; NY Univ, PhD(biochem), 48. *Prof Exp:* Res assoc, Rheumatic Fever Res Inst, Northwestern Univ, 49; asst prof, Inst Radiobiol & Biophys, Univ Chicago, 49-51; vis scientist, NIH, 51-52, chemist, 52-60, chief enzyme chem sect, Nat Inst Dent Res, 60-65; prof & chmn, dept biochem, Med Col Wis, 65-80, dir, lab enzymol, 80-; PROF, DEPT BIOCHEM, COL MED, HOWARD UNIV. *Concurrent Pos:* Mem, Weizmann Inst, 48; NSF sr fel, Dept Genetics, Sorbonne, 58-59; Guggenheim fel, Inst Molecular Biol, Univ Paris, 72-73. *Mem:* AAAS; Am Chem Soc; Am Soc Biol Chem; Biochem Soc. *Res:* Enzyme chemistry; transfer ribonucleic acid; intermediary metabolism. *Mailing Add:* Dept Biochem Col Med Howard Univ 520 W St NW Washington DC 20059

MEHLER, ERNEST LOUIS, b Amsterdam, Holland, Sept 25, 38; US citizen; m 64; c 2. THEORETICAL STRUCTURAL BIOLOGY. *Educ:* Ill Inst Technol, BS, 60; Johns Hopkins Univ, MA, 64; Iowa State Univ, PhD(theoret chem), 68. *Prof Exp:* Instr theoret chem, Univ Groningen, 71-73; sr res assoc theoret biochem, 74-79, PRIVAT DOZENT, UNIV BASEL, 79- *Concurrent Pos:* Fel, Univ Wash, 68-70. *Mem:* Am Chem Soc; Int Soc Quantum Biol. *Res:* Ab initio methods for large molecules; structure of proteins; molecular aspects of drug reactivity; computerized molecular modelling; computer graphics for molecular modelling. *Mailing Add:* Biocenter Univ of Basel Klingelbergstr 70 Basel 4056 Switzerland

MEHLHAFF, LEON CURTIS, b Lodi, Calif, Apr 28, 40; m 65; c 3. ANALYTICAL CHEMISTRY, POLYMER CHEMISTRY. *Educ:* Univ Calif, Berkeley, BS, 61; Univ Wash, PhD(anal chem), 65, Univ Puget Sound, JD, 89. *Prof Exp:* Chemist plastics dept, E I du Pont de Nemours & Co, 65-68; asst prof, 68-71, PROF ANAL CHEM, UNIV PUGET SOUND, 85-, CHMN DEPT, 90- *Concurrent Pos:* Regional coordr, Wash Shoreline Tech

Adv Bd, 71-73; mem, Hazardous Waste Sect, Wash State Dept Ecol, 76-77. *Mem:* Am Chem Soc. *Res:* Analytical chemistry of polymer systems; instrumental analytical chemistry; electrochemistry; environmental science program; hazardous waste. *Mailing Add:* Dept Chem Univ Puget Sound Tacoma WA 98416

MEHLMAN, MYRON A, b Poland, Dec 21, 34; US citizen; m 60; c 3. ASBESTOS EXPOSURE. *Educ:* City Col New York, BS, 57; Mass Inst Technol, PhD(nutrit & biol sci), 64. *Prof Exp:* Prof biochem, Rutgers Univ, 67-69; prof biochem, Univ Nebr Col Med, 69-73; chief biochem toxicol, US Food & Drug Admin, Washington, DC, 72-73; spec asst toxicol & environ affairs, Off Asst Secy, Dept Health, Educ & Welfare, 73-75; interagency liaison officer, Off Dir NIH, Bethesda, 75-77; dir toxicol, Mobil Corp, 77-89; PROF TOXICOL, R W JOHNSON MED SCH, UNIV MED & DENT NJ, 89- *Concurrent Pos:* Pres, Am Col Toxicol, 78-80, mem coun, NAm Secretaria, 83-92; adj prof med, Mt Sinai Sch Med, 80-; vis prof, Sch Med, NY Univ, 80-; vis prof indust & environ toxicol, Rutgers Col Pharm, 81-; fel, Acad Toxicol Sci, 86; prof environ & commun med, Med Sch, Univ Med & Dent NJ, 89-; pres, Int Soc Exposure Anal, 89-91. *Mem:* Fel Am Col Toxicol; fel Collegium Ramazzine. *Res:* Biochemical and toxicological mechanisms of petroleum products; benzene; alkyl benzene gasoline; pah's; carcinogenials of asbestos exposure; editor of four major primary pier review journals. *Mailing Add:* Seven Bouvant Dr Princeton NJ 08540

MEHNER, JOHN FREDERICK, b Grove City, Pa, July 15, 21. ORNITHOLOGY. *Educ:* Grove City Col, BS, 42; Univ Pittsburgh, MS, 50; Mich State Univ, PhD(zool), 58. *Prof Exp:* Teacher high sch, Pa, 42-59; assoc prof biol, Edinboro State Col, 59-63; assoc prof to prof, 63-86, EMER PROF BIOL, MARY BALDWIN COL, 86- *Concurrent Pos:* Chmn dept, Mary Baldwin Col, 63-86. *Mem:* Am Ornith Union; Wilson Ornith Soc; AAAS; Sigma Xi. *Res:* Ecology and ethology of the evening grosbeak; a checklist of the birds of Augusta County, Virginia; breeding biology of the house finch in Virginia. *Mailing Add:* Dept Biol Mary Baldwin Col Staunton VA 24401-1918

MEHR, CYRUS B, b Tehran, Iran, July 7, 27; US citizen; wid; c 4. MATHEMATICS, ELECTRICAL ENGINEERING. *Educ:* La State Univ, BS, 52; Purdue Univ, MS, 53, PhD(math, elec eng), 64. *Prof Exp:* Develop engr, Elec Prod Co, 53-55; systs engr, Gen Elec Co, 55-57; mem develop lab, Square D Elec Co, 57-59; mem res staff, Int Bus Mach Corp, 63-65; asst prof, 65-67, ASSOC PROF MATH, OHIO UNIV, 67- *Mem:* Math Asn Am. *Res:* Functional analysis; probability. *Mailing Add:* Dept Math Ohio Univ Athens OH 45701

MEHRA, MOOL CHAND, b Lahore, Pakistan, July 31, 36; Can citizen; m 64; c 2. RADIOCHEMISTRY, INORGANIC CHEMISTRY. *Educ:* Univ Rajasthan, BSc, 55, MSc, 57; Laval Univ, DSc(chem), 68. *Prof Exp:* Asst prof chem, Univ Rajasthan, 57-59; sci off radiochem, Atomic Energy Estab, India, 59-65; from asst prof to assoc prof, 68-77, head dept, 69-72, PROF CHEM, UNIV MONCTON, 77- *Concurrent Pos:* Univ Moncton rep, Atlantic Prov Inter-Univ Comt Sci, Chem & Water Resources, 70-72 & Chem Inst Can, 75-79 & 83-87; vis prof, France, 76, Switz, 74, Japan, 82, Italy, 83 & Australia, 86. *Mem:* Fel Chem Inst Can; fel Am Inst Chem; fel Indian Chem Soc. *Res:* Radioanalytical chemistry; analytical chemistry of metals coordination complexes in solution; ion chromatographic separations and determinations. *Mailing Add:* Dept Chem & Biochem Cntr Universitaire De Moncton 98 Chapman Moncton NB E1A 3E9 Can

MEHRA, VINODKUMAR S, b Lahore, Pakistan, Oct 26, 35; m 65; c 2. CHEMICAL ENGINEERING. *Educ:* Univ Bombay, BChemEng, 56; Northwestern Univ, PhD(chem eng), 63. *Prof Exp:* Res engr, 63-71, SR RES ENGR, PLASTIC PROD & RESINS DEPT, E I DU PONT DE NEMOURS & CO, INC, 71- *Res:* Chemical engineering thermodynamics; processing equipment; economic analysis; process development; plastics; plastics processing; polymer synthesis; plastics product development. *Mailing Add:* 2203 Regal Dr Wilmington DE 19810

MEHRABIAN, ROBERT, b Tehran, Iran, July 31, 41; US citizen; m 77. RAPID SOLIDIFICATION, MATERIAL PROCESSING. *Educ:* Mass Inst Technol, BS, 64, ScD(metall), 68. *Prof Exp:* From asst prof to assoc prof metall & mat sci, Mass Inst Technol, 72-75; from assoc prof to prof, Univ Ill, Urbana, 75-79, adj prof metall, 75-79; chief metall, Nat Bur Standards, Dept Com, 80-82, dir, Ctr Mat Sci, 82-83; dean, Col Eng, Univ Calif, Santa Barbara, 83-90; PRES, CARNEGIE-MELLON UNIV, 90- *Concurrent Pos:* Consult, Rheocast Corp, 74-, Army Mat & Mech Res Ctr, 75-, Univ Space Res Admin, 75-, United Technol, 76-, Sandia Labs, 76- & Phrasor Technol, 77-; mem, Nat Mat Adv Bd Amorphous & Metastable Mat, 78. *Honors & Awards:* George Kimball Burgess Mem Lectr, Washington, DC, 80; Henry Marion Howe Medal, Am Soc Metals, 83; Leadership Award, Metall Soc, 91. *Mem:* Nat Acad Eng; Am Inst Metall Engrs; fel Am Soc Metals Int; Am Ceramic Soc; Am Welding Soc; Am Soc Testing & Mat; AAAS. *Res:* Foundry solidification, powder metallurgy, metal matrix composites and refining of alloy scrap; co-inventor of rheocasting and several associated processes; recipient of eight US and over thirty foreign patents; published more than 130 technical papers and edited six books in metallurgy and materials science. *Mailing Add:* Carnegie-Mellon Univ 5000 Forbes Ave Pittsburgh PA 15213

MEHRAN, FARROKH, b Tehran, Iran, June 29, 36; m 63; c 2. PHYSICS. *Educ:* Univ Calif, Berkeley, BS, 59; Harvard Univ, PhD(physics), 64. *Prof Exp:* Fel, Harvard Univ, 64-65; asst prof physics, Sacramento State Col, 65-67; staff physicist, 67-70, MEM RES STAFF, IBM CORP, 70- *Mem:* Am Phys Soc. *Res:* Molecular beams; quantum electronics; electron paramagnetic resonance. *Mailing Add:* Dept of Phys Sci IBM Corp PO Box 218 Yorktown Heights NY 10598

MEHRING, ARNON LEWIS, JR, b Washington, DC, Apr 24, 15; m 36; c 3. POULTRY NUTRITION. *Educ:* Univ Md, BS, 36. *Prof Exp:* Mkt inspector eggs, State Dept Mkt, Exten Serv, Univ Md, 36-39; farm owner, 39-43; farm mgr, 43-45; poultryman in-chg nutrit & breeding invests, Lime Crest Res Lab, 45-63; nutritionist, Limestone Prod Corp, 63-70, qual supvr, 70-79; RETIRED. *Mem:* Poultry Sci Asn; Am Soc Animal Sci; Am Dairy Sci Asn; Am Soc Testing & Mat. *Res:* Mineral nutrition of livestock and poultry. *Mailing Add:* 218 Mountain Rd Flanders NJ 07836

MEHRING, JEFFREY SCOTT, b Cleveland, Ohio, July 6, 42; m 65; c 3. NUTRITION, TOXICOLOGY. *Educ:* Ohio State Univ, BS, 64, MS, 66, PhD(nutrit), 69. *Prof Exp:* Lectr nutrit, Ohio State Univ, 69; lab mgr nutrit, Gaines Nutrit Ctr, Gen Foods Corp, 69-77; dir large animal toxicol, Int Res & Develop Corp, 77-79; consult, Occup Health & Safety & Gen Toxicol, 80-81; VPRES, RES LAB, RES ENTERPRISES, INC, 81- *Concurrent Pos:* Mem subcomt lab animal nutrit, Nat Acad Sci-Nat Res Coun, 74-78. *Mem:* Animal Nutrit Res Coun; Am Soc Animal Sci; Inst Food Technologists. *Res:* Comparative animal nutrition; quantification of nutrient requirements; clinical assessment of nutritional status and applied toxicology; safety evaluation and regulatory compliance. *Mailing Add:* 5767 West R Ave Schoolcraft MI 49087

MEHRINGER, PETER JOSEPH, JR, b Lawrence, Kans, Dec 9, 33; m 54; c 3. PALEOECOLOGY, PALYNOLOGY. *Educ:* Calif State Col, Los Angeles, BA, 59, MA, 62; Univ Ariz, PhD, 68. *Prof Exp:* Instr biol, Glendale Jr Col, 60-61; res assoc geochronology, Univ Ariz, 64-68, asst prof earth sci, 68-69; asst prof anthrop, Univ Utah, 69-71; assoc prof anthrop, 71-78, PROF ANTHROP & GEOL, WASH STATE UNIV, 78- *Honors & Awards:* Roald Fryxell Award, Soc Am Archaeol, 79. *Mem:* AAAS; Ecol Soc Am; Soc Am Archael; Am Quaternary Asn. *Res:* Quaternary biogeography, paleoecolgy, chronology and geology of North America, Egypt and the Sudan. *Mailing Add:* Dept Anthrop Wash State Univ Pullman WA 99164-4910

MEHRKAM, QUENTIN D, b Bethlehem, Pa, June 7, 21; m 67; c 2. ISOTHERMAL HEAT TREATMENT, THERMAL PROCESSING. *Educ:* Lehigh Univ, BA, 43. *Prof Exp:* Prod metallurgist, Thompson Prod, TRW, 43-46; vpres eng & sales, 69-84, METALLURGIST & CHMN, METALL LAB, AJAX ELEC CO, 43-, PRES, 84-; CHIEF EXEC OFFICER, CENT PANEL CO, 85- & INTEGRATED MGT CONTROL SYSTS, 88- *Concurrent Pos:* Lectr metall appln salt bath furnaces, 50-; instr metall, Temple Univ, Philadelphia, 55-58; chmn, Mat Educ Inst, Am Soc Metals, 67-71. *Honors & Awards:* Eisenman Award, Am Soc Metals Int, 77. *Mem:* Fel Am Soc Metals Int. *Res:* Application of salt bath furnaces for heat processing with particular work in isothermal heat treatment; surface cleaning, removal of paints and plastics in molten salt; author of numerous technical publications. *Mailing Add:* Ajax Elec Co 60 Tomlinson Rd Huntingdon Valley PA 19006

MEHRLE, PAUL MARTIN, JR, b Caruthersville, Mo, Dec 13, 45; m 64; c 2. BIOCHEMISTRY, TOXICOLOGY. *Educ:* Southwestern at Memphis, BA, 67; Univ Mo-Columbia, MA, 69, PhD(biochem), 71. *Prof Exp:* Physiologist, Fish-Pesticide Res Lab, Nat Fish Res Ctr, US Dept Interior, 71-81, chief biologist, 81-91; ABC LABS, COLUMBIA, MO, 91- *Concurrent Pos:* Res assoc, Univ Mo-Columbia, 73- *Mem:* Am Chem Soc; Sigma Xi; Am Fish Soc; Soc Toxicol; Soc Environ Toxicol Chem. *Res:* Directing research concerned with evaluating the impacts of chemical contaminants on aquatic resources; conducts and coordinates aquatic toxicology program involving agricultural chemicals, industrial chemicals, and acid rain. *Mailing Add:* 1804 W Broadway Columbia MO 65201

MEHRLICH, FERDINAND PAUL, b Cincinnati, Ohio, 05; m 33; c 3. FOOD SCIENCE. *Educ:* Butler Univ, AB, 27; Univ Wis, PhD(plant physiol), 30. *Prof Exp:* Asst plant physiol, Univ Wis, 28-30; assoc pathologist pineapple res, Pineapple Producers Coop Asn, Hawaii, 30-35; sci adv, Hawaiian Pineapple Co, Ltd, 35-43, dir res, 43-44, asst vpres in-charge res, 44-49; vpres, Res Inst, Int Basic Econ Corp, 50-54, trustee & dir res div, 54-58; sci dir, Qm Food & Container Inst Armed Forces, 58-63, dir, Food Labs, US Army Natick Labs, 63-75; CONSULT & TECH WRITER, 75- *Concurrent Pos:* Consult, Cocoa Res Inst, DC, 56-, Nat Planning Asn, 56- & Arthur D Little, Inc, Mass, 56-; consult to vpres res, Gen Foods Corp, NY, 56-57; mem survs, WAfrica, Belgian Cong, Brazil, Costa Rica, Cuba, Mex, PR, Venezuela & Peru. *Honors & Awards:* Meritorious Civilian Serv award, 64. *Mem:* Fel AAAS; Am Soc Plant Physiol; Inst Food Technol; Nutrit Today; NY Acad Sci. *Res:* Food technology including radiation preservation of food; product and process development; agricultural research and planning; pineapple and coffee production and processing; tropical and subtropical crop production and upgrading. *Mailing Add:* 96 Pilgrim Rd Wellesley MA 02181

MEHROTRA, BAM DEO, b Meerut, India, Sept 4, 33; m 65; c 2. BIOCHEMISTRY. *Educ:* Agra Univ, BSc, 50, MSc, 52; Ind Univ, Bloomington, PhD(biochem), 64. *Prof Exp:* Lectr org chem, J V Col, India, 52-58; fel biochem, Inst Enzyme Res, Univ Wis-Madison, 62-64; asst prof, All-India Inst Med Sci, New Delhi, 65-66; res assoc, Ind Univ, Bloomington, 67-69; PROF CHEM, TOUGALOO COL, 69- *Mem:* Am Chem Soc; Soc Toxicol. *Res:* Chemistry of nucleic acids, their interactions with small ions, and their physical, chemical and biological characteristics; effect of pesticides on mammalian systems. *Mailing Add:* Dept Chem Tougaloo Col Tougaloo MS 39174

MEHROTRA, KISHAN GOPAL, b Kashipur, India, Dec 9, 41; m 71; c 2. MATHEMATICAL STATISTICS. *Educ:* Univ Lucknow, BSc, 60, MSc, 62; Univ Wis-Madison, MS, 69, PhD(statist), 71. *Prof Exp:* Lectr statist, Banaras Hindu Univ, 62-66; asst prof, 71-72, assoc prof, 72-77, PROF STATIST, SYRACUSE UNIV, 77- *Mem:* Inst Math Statist; Am Statist Asn. *Res:* Nonparametric statistics; pattern recognition neurol network. *Mailing Add:* Sch Comput & Info Sci Syracuse Univ Syracuse NY 13224

MEHS, DOREEN MARGARET, b Buffalo, NY, July 27, 44; m 80. ANALYTICAL CHEMISTRY, ENVIRONMENTAL CHEMISTRY. *Educ:* Harpur Col, BA, 66; State Univ NY Binghamton, MA, 72; Univ NMex, PhD(chem), 80. *Prof Exp:* Admin asst & asst instr chem, State Univ NY Binghamton, 66-73; instr, 73-75, asst prof, 75-81, chmn, 81-83, PROF CHEM, FT LEWIS COL, 87-, CHMN, 89- *Concurrent Pos:* Consult, 73- *Mem:* Am Chem Soc; Soc Appl Spectros; Am Inst Chem; Air Pollution Control Asn. *Res:* Spectroscopy; environmental analysis. *Mailing Add:* Dept of Chem Ft Lewis Col Durango CO 81301

MEHTA, ATUL MANSUKHBHAI, b Jamnagar, India, May 4, 49; m 77; c 2. INDUSTRIAL PHARMACY, BIOPHARMACEUTICS. *Educ:* Shivaji Univ, India, BPharm, 72; Univ Md, MS, 75, BS, 76, PhD(pharmaceut), 81. *Prof Exp:* Pharmaceut chemist, Roche Prod Ltd, 72-73; asst instr, Univ Md, 74-75; mem packaging-in-charge staff, US Pharmacoepia, 74-78; group leader, Ayerst Labs, 81-84; mgr res & develop, Nortec Develop Assoc Inc, 84-89; PRES, ELITE LABS INC, 90- *Concurrent Pos:* Staff pharmacist, St Agnes Hosp, 77-81. *Mem:* Am Pharmaceut Asn; Acad Pharmaceut Sci; Controlled Release Soc; Am Asn Pharmaceut Scientists. *Res:* Formulation and process variable effects on the performance of a solid dosage form (pharmaceutical) in vitro and in vivo including stability of such dosage forms; design and development of sustained release dosage forms; scanning electron microscopy. *Mailing Add:* 252 E Crescent Ave Ramsey NJ 07446

MEHTA, AVINASH C, b Rehlu, India, Nov 1, 31; m 70; c 1. ORGANIC CHEMISTRY. *Educ:* Panjab Univ, India, BSc, 52, MSc, 54; Univ Delhi, PhD(org chem), 58. *Prof Exp:* Lectr chem, Deshbandhu Col, New Delhi, India, 57-58 & Univ Delhi, 59-62; res assoc org chem, Univ Mich, 62-64; res scientist, Uniroyal Res Labs, Ont, 64-67; sr org chemist, Arthur D Little, Inc, 68-70; scientist, 70-74, res group leader, 74-77, res assoc, 78-87, SR RES GROUP LEADER, 88-, RES FEL, POLAROID CORP, 88- *Mem:* Am Chem Soc; Royal Soc Chem. *Res:* Organic chemical reaction mechanisms; organic synthesis, polymer chemistry and photographic chemistry; synthesis of novel monomers and polymers and novel heterocycles. *Mailing Add:* 12 Brookside Ave Belmont MA 02178-1014

MEHTA, BIPIN MOHANLAL, b Bombay, India, July 25, 35; m 60. MICROBIAL GENETICS & PHYSIOLOGY. *Educ:* Univ Bombay, BSc, 55, BSc, 57, PhD(microbial genetics & nutrit), 63. *Prof Exp:* Teaching asst microbiol & biochem, Univ Bombay, 60-61; sr res asst microbial genetics, Coun Sci & Indust Res, Govt India, 61-65; teaching asst & res assoc molecular biol, State Univ NY Downstate Med Ctr, 65-66; vis res fel molecular genetics, Sloan-Kettering Inst Cancer Res, 66-69; res assoc microbiol, Univ Ottawa, 69-72; asst prof, Sloan-Kettering Div, Grad Sch Med Sci, Cornell Univ, 75-87; res assoc, 72-74, from assoc mem to asst mem, 74-86, ASST LAB MEM, SLOAN-KETTERING INST CANCER RES, 86- *Concurrent Pos:* Adj prof, Dept Chem, Pace Univ, Pleasantville, NY, 82-85, Dept Chem & Phys Sci, 85-88, Dept Biol, 88-; pres, Sound Unit, Westchester Div, Am Cancer Soc, 88- *Mem:* Am Soc Microbiol; Soc Gen Microbiol; Can Soc Cell Biol; Am Asn Cancer Res; Am Soc Clin Oncol; Soc Nuclear Med. *Res:* Study of the distribution kinetics of cancer chemotherapeutic agents in body fluids and tissues of patients and experimental animals; study of mechanism of resistance to drugs; pharmacokinetics of anticancer agents; genetic recombination in microorganisms; study in tumorogenicity and platelet-derived growth factor; screening of chemical agents for antitumor activity. *Mailing Add:* Mem Sloan-Kettering Cancer Ctr 1275 York Ave New York NY 10021

MEHTA, GURMUKH D, b India, Aug 27, 45; US citizen; m 73; c 3. MISSILE DESIGN & ANALYSIS, TRANSPORT THROUGH MEMBRANES & POROUS MEDIA. *Educ:* Punjab Univ, India, BS, 67; Indian Inst Technol, Kanpur, MS, 69; Brown Univ, PhD(aeronaut eng), 75. *Prof Exp:* Res scientist, Hydronautics, Inc, 74-77; dir tech opers, Inter Technol Corp, 77-81; sr scientist, 81-90, ASST VPRES, SCI APPLNS INT CORP, 90-, DEP DIV MGR, 91- *Concurrent Pos:* Reviewer, J Membrane Sci, 81- *Honors & Awards:* Atlas Award, 85. *Mem:* Am Soc Mech Engrs. *Res:* Systems engineering, design and analysis of cruise missile systems; command, control and communication systems; transport through membranes and porous media; energy research; missile sensors and testing; recipient of three patents. *Mailing Add:* 3331 Monarch Lane Annandale VA 22003

MEHTA, JATINDER S, b Amritsar, India, Oct 4, 39; US citizen; m 82. STATISTICAL ESTIMATION, STATISTICAL TESTING. *Educ:* Panjab Univ, India, BA, 59, MA, 61; Wis Univ, MS, 65, PhD(statist), 68. *Prof Exp:* From asst prof to assoc prof, 68-78, PROF MATH, TEMPLE UNIV, 78- *Mem:* Am Statist Asn. *Res:* Statistical estimation and testing; econometrics; author or coauthor of over 60 publications. *Mailing Add:* Dept Math Temple Univ Broad & Montgomery Sts Philadelphia PA 19122

MEHTA, KISHOR SINGH, b Jodhpur, Rajasthan, India, July 27, 41; US citizen; m 63; c 3. MECHANICAL ENGINEERING, PLASTICS ENGINEERING. *Educ:* Univ Rajasthan, India, BSME, 62; Rochester Inst Technol, MBA, 81. *Prof Exp:* Asst prof mech eng, Univ Jodhpur, India, 62-64; plant mgr, Bk Plastics Pvt Ltd, India, 64-70; tech dir, Jayshree Plastics Pvt Ltd, India, 64-70; mgr, Preprod Ctr, Am Can Co, 70-73; sr engr, Johnson & Johnson, 73-74; sr adv engr, Continental Group, 74-78; mgr polymer processing, Xerox Corp, 78-83; MGR, DESIGN ENG PLASTICS, MOBAY CORP, 83- *Honors & Awards:* Meritorious Serv Award, Soc Plastics Engrs, 85. *Mem:* Fel Soc Plastics Eng (treas, 86-87, secy, 87-88). *Res:* Design for assembly and disassembly; design for manufacturing plastic products and components; failure prevention and failure analysis of plastic application; published papers on design plastic products and components. *Mailing Add:* 408 Pine Villa Dr Gibsonia PA 15044

MEHTA, MAHENDRA, b Jodhpur, India, Sept 28, 52; m 79; c 2. PAPER COATINGS & SPECIALTY PAPERS, PAPER PHYSICS. *Educ:* Jodhpur Univ, India, BSc, 71, MSc, 74, PhD(chem), 78. *Prof Exp:* Res fel, Baylor Univ, Waco, Tex, 79-81 & Gaylord Res Inst, Whippany, NJ, 81-83; specialist II, 84-85, Mead Corp Cent Res, res specialist, 85-87, sr res specialist, 87-89, DIR RES & DEVELOP, MEAD SPECIALTY PAPER DIV, 90- *Concurrent Pos:* Res fels, Coun Sci & Indust Res, India, 75, sr res fel, Univ Grants Comn, New Delhi, 78 & Robert A Welch Found, Baylor Univ, 79-81. *Mem:* Am Chem Soc; Tech Asn Pulp & Paper Indust. *Res:* Polymer synthesis structure-property relationship; polymers as materials to modify paper properties; polymer emulsions and their behavior in coating technologies; paper physics; decorative laminates; friction papers; specialty fillers for coatings and paper manufacture; decorative papers; overlay papers; specialty papers. *Mailing Add:* 109 Mountain Dr Pittsfield MA 01201

MEHTA, N(AVNIT) C(HHAGANLAL), b Bombay, India, Feb 21, 38; m 69; c 2. MECHANICAL ENGINEERING, MATHEMATICS. *Educ:* Univ Baroda, India, BEMech, 60; Mo Sch Mines, MS, 62; Univ Mo-Rolla, PhD(mech), 67. *Prof Exp:* Trainee engr, Ex-Cell-O Pvt Ltd, India, 60; proj engr, Olin Mathieson Chem Corp, 62-64; spec consult new plastic prod develop, K V Industs, Bombay, 64; res engr, Shell Oil Co, 67-68; assoc prof vibrations & eng mech, Tri-State Col, 68-73; mgr vehicle dynamics, Int Harvester, 73-80, chief engr, Vehicle Dynamics & Test, Truck Group Eng, 80-83, mgr Prod Develop & Tech Servs, Int Harvester, 83-90; ENG DIR, TECH SERV, J I CASE CONSTRUCT EQUIP, 90- *Concurrent Pos:* Consult, Magnavox Co, 70 & Hendrickson Tandem Corp, 71- *Honors & Awards:* Ralph R Teetor Award, Soc Automotive Engrs, 71. *Mem:* Soc Automotive Engrs; Am Soc Mech Engrs; Am Soc Eng Educ. *Res:* Nonlinear vibrations of automotive vehicles and other structures. *Mailing Add:* 700 Waters Edge Rd Racine WI 53402

MEHTA, NARIMAN BOMANSHAW, b Bombay, India, Apr 8, 20; nat US; m 54; c 3. MEDICINAL & PHARMACEUTICAL CHEMISTRY. *Educ:* Univ Bombay, BSc, 41, BA, 42; Univ Kans, PhD, 52. *Prof Exp:* Lectr physics, Univ Bombay, 41-46; trainee, J E Seagram & Sons, Inc, Ky, 47-48; fel, Univ Toronto, 53-54; prof chem, Cent State Univ, 54-57; sr res scientist, Wellcome Res Labs, Burroughs Wellcome & Co, Inc, 57-76, prin scientist, 77-86, emer prin scientist, 87-88; RETIRED. *Concurrent Pos:* Consult, Charles F Kettering Found, Ohio, 55-57, Glaxo Inc, 88- *Mem:* Am Chem Soc; Royal Soc Chem; fel Am Inst Chemists. *Res:* Organo-physical and heterocyclic chemistry; reaction mechanisms and kinetics; high vacuum techniques; medicinal chemistry; structure-activity studies in design of drugs of central nervous system; mechanism of reactions; cardiovascular drugs. *Mailing Add:* 4207 Union St Raleigh NC 27609-5558

MEHTA, POVINDAR KUMAR, b Panjab, India. MATERIALS SCIENCE. *Educ:* Delhi Polytech Inst, Chem Eng, 52; NC State Univ, MS, 62; Univ Calif, Berkeley, DE(mat sci & eng), 64. *Prof Exp:* Chem engr, Rohtas Industs, India, 52-61; works mgr, Jaipur Udyog, 66-68; asst prof civil eng, 64-66, assoc prof eng sci, 68-74, PROF ENG SCI, UNIV CALIF, BERKELEY, 74- *Concurrent Pos:* Consult cement chem. *Honors & Awards:* Wagon Medal for Materials Research, Am Concrete Inst, 87. *Mem:* Am Soc Testing & Mat; Am Ceramic Soc; Am Concrete Inst; Int Union Testing & Res Labs Mat & Struct. *Res:* Physical chemistry of cement and concrete; cement technology; expansive cements; high alumina cements and blended cements; utilization of industrial and agricultural wastes for making cementitious materials. *Mailing Add:* Dept of Civil Eng Univ of Calif Berkeley CA 94720

MEHTA, PRAKASH V, b Gujarat, India, Feb 23, 46; US citizen; m 72; c 2. ORGANO-SILICON CHEMISTRY, PROCESS DEVELOPMENT. *Educ:* Sardar Patel Univ, MSc, 69; Polytech Inst Brooklyn, MS, 72; Polytech Inst NY, MS, 74, PhD(polymer chem), 77. *Prof Exp:* Sr res chemist, Kay Fries Inc, Stony Pt, NY, 78-82; group leader, Dynamit Nobel Chems, 82-85, mgr res & develop, 85-86; MGR RES & DEVELOP, SILANE/SILICONE CHEM RES HULS AM INC, 86- *Concurrent Pos:* Adj assoc prof chem, Pace Univ, Pleasantville, NY, 80-87. *Mem:* Am Chem Soc; Asn Off Anal Chemists. *Res:* Organo-silicon chemistry and industrial research; organo-titanates and ziconates; cyanide chemistry; product process development. *Mailing Add:* Huls Am/Petrarch Syst 2741 Bartram Rd Bristol PA 19007

MEHTA, RAJEN, cereal & dairy products, food enzymes, for more information see previous edition

MEHTA, RAJENDRA G, b Dabhoi, India, Aug 31, 47; US citizen; m 76; c 2. CANCER RESEARCH, ENDOCRINOLOGY. *Educ:* Gujarat Univ, Ahmedabad, India, BSc, 66, MSc, 68; Univ Nebr, Lincoln, PhD(zool), 74. *Prof Exp:* Res assoc cancer, Univ Rochester Med Sch, NY, 74-75 & Univ Louisville Sch Med, Ky, 76-77; assoc biochemist, Ill Inst Technol, Chicago, 77-78, res biochemist, 78-79, sr biochemist, 80-89, SCI ADV, CANCER RES, ILL INST TECHNOL, CHICAGO, 89- *Concurrent Pos:* Adj assoc prof med, Sch Med, Univ Ill, Chicago. *Mem:* Am Asn Cancer Res; Endocrine Soc; Int Breast Cancer Res Asn. *Res:* Cancer chemoprevention mechanism of vitamin A; hormone action and breast cancer. *Mailing Add:* Dept Biol Ill Inst Technol Res Inst 10 W 35th St Chicago IL 60616

MEHTA, SUDHIR, b Dhoraji, India, Nov 10, 46; m 75. MATERIALS SCIENCE, EARTH SCIENCES. *Educ:* Indian Inst Technol, BSc, 67, MSc, 69; Lehigh UniY, MS, 71, PhD(geol), 73. *Prof Exp:* Res assoc metall, Lehigh Univ, 74-76, res assoc chem, 76-77, res scientist mat sci, 78-80; RES ENGR, BETHLEHEM STEEL CORP, 80- *Concurrent Pos:* Fel, Pa Sci & Eng Found & NSF grants, 76-77; co-investr, Lunar Sci grant, NASA, 78- *Mem:* Microbeam Anal Soc; Sigma Xi. *Res:* Basic and applied research with the application of modern electron optical techniques, (scanning electron microscope, scanning transmission electron microscope, & transmission electron microscope), to problems in lunar science, catalysis, coal carbonization and ceramic materials. *Mailing Add:* 2937 Jenny Pl Philadelphia PA 19136

MEI, CHIANG C(HUNG), b Wuhan, China, Apr 4, 35; m 65; c 1. FLUID MECHANICS, COASTAL & OCEANOGRAPHIC ENGINEERING. *Educ:* Nat Taiwan Univ, BS, 55; Stanford Univ, MS, 58; Calif Inst Technol, PhD(eng sci), 63. *Prof Exp:* Res fel eng sci, Calif Inst Technol, 63-65; mem tech staff, Nat Eng Sci Co, Calif, 65; from asst prof to assoc prof, 65-76, PROF

CIVIL ENG, MASS INST TECHNOL, 76- *Mem:* Mem Nat Acad Eng; Am Soc Civil Engrs; Sigma Xi. *Res:* Hydrodynamics; coastal and ocean engineering; applied mechanics; porous media mechanics. *Mailing Add:* Dept of Civil Eng Mass Inst of Technol Cambridge MA 02139

MEI, KENNETH K, b Shanghai, China, May 19, 32; m 68; c 2. ELECTROMAGNETIC SCATTERING & RADIATION, CUSTOM DESIGNED AUTENNAS. *Educ:* Univ Wis, BSEE, 59, MS, 60, PhD(elec eng), 62. *Prof Exp:* From asst prof to assoc prof, 62-76, PROF ELEC ENG, UNIV CALIF, BERKELEY, 76- *Concurrent Pos:* Mem US Nat Comt, Comn 6, Int Union Radio Sci, 70-; ed, J Electromagnetics; pres, Electromagnetics Soc. *Mem:* Fel Inst Elec & Electronics Engrs; Electromagnetics Soc. *Res:* Electromagnetic theory, wave radiation and propagation. *Mailing Add:* Dept of Elec Eng & Comput Sci Univ of Calif Berkeley CA 94720

MEIBOHM, EDGAR PAUL HUBERT, b New Orleans, La, Dec 13, 15; m 55; c 4. POLYMER SCIENCE. *Educ:* Guilford Col, BS, 36; Univ NC, MS, 39; Ohio State Univ, PhD(phys chem), 47. *Prof Exp:* Instr chem, Kans State Univ, 41-42; group leader, Nat Defense Res Comt Div Eight, Explosive Res Lab, Pa, 42-45, res sect leader, Los Alamos Sci Lab, 45; res chemist, Cent Res Dept, E I du Pont De Nemours & Co, Inc, Philadelphia, 47-55, sr res chemist, Rayon Res Lab, 55-57, sr res chemist, Dacron Res Lab, 57-62, sr res chemist, Kinston Plant Tech, 62-67, staff chemist, Marshall Lab, 67-81; RETIRED. *Mem:* Am Chem Soc; Am Crystallog Asn. *Res:* Small-angle scattering of x-rays; crystal structures; applications of x-ray diffraction to high polymers; physics and physical chemistry of polymers; polymer characterization. *Mailing Add:* 521 Shadeland Ave Drexel Hill PA 19026

MEIBOOM, SAUL, b Antwerp, Belg, Apr 7, 16; US citizen; m 46; c 2. PHYSICS OF LIQUID CRYSTALS. *Educ:* Univ Delft, PhysEng, 39; Hebrew Univ, Israel, PhD(physics), 55. *Prof Exp:* Instr physics, Hebrew Univ, Israel, 40-48; sr scientist, Weizmann Inst, 48-58; mem tech staff, Bell Labs, 58-86; RETIRED. *Mem:* Am Phys Soc. *Res:* Physical properties of liquid crystals, including surface interactions, flow properties and phase transitions. *Mailing Add:* 25 Holly Glen Lane S Berkeley Heights NJ 07922

MEIBUHR, STUART GENE, b Cleveland, Ohio, Jan 6, 34; m 60. ELECTROCHEMISTRY. *Educ:* Western Reserve Univ, BS, 55, MS, 58, PhD(electrochem), 60. *Prof Exp:* Chemist, Harshaw Chem Co, Ohio, 54-57; technologist appl res lab, US Steel Corp, Pa, 60-63; sr res chemist, Fuel Cell Corp, Mo, 63-64; assoc sr res chemist, Res Lab, Gen Motors Corp, 64-65, sr res chemist, 65-81, staff scientist, 81-87; RETIRED. *Mem:* Am Chem Soc; Sigma Xi; Electrochem Soc; Prof Photog Am; Am Asn Zool Parks & Aquariums. *Res:* Electrodeposition; electrode kinetics; fuel cells; batteries; organic electrolytes; scanning electron microscope x-ray. *Mailing Add:* 1155 Timberview Trail Bloomfield Hills MI 48304

MEIENHOFER, JOHANNES ARNOLD, b Dresden, Ger, Mar 3, 29; c 2. BIOCHEMISTRY, ORGANIC CHEMISTRY. *Educ:* Univ Heidelberg, dipl, 54, PhD(chem), 56. *Prof Exp:* Res assoc, Med Sch, Cornell Univ, 57-59; res assoc, Univ Calif, Berkeley, 59-60; proj chief, Ger Wool Res Inst, Aachen, 61-64; head peptide & protein chem lab, Children's Cancer Res Found, Boston, 65-73; sect chief, 73-80, DIR, CHEM RES DEPT, HOFFMANN-LA ROCHE, INC, NUTLEY, 80- *Concurrent Pos:* Fulbright travel grant, 57-61; res assoc, Farbenfabriken Bayer, A G, Ger, 61-64; assoc, Harvard Med Sch, 69-71, lectr, 71- *Mem:* AAAS; Am Chem Soc; Am Soc Biol Chem; Ger Chem Soc; Ger Soc Biol Chem. *Res:* Peptide and protein chemistry; hormones; antibiotics; antitumor agents. *Mailing Add:* 615 Upper Mountain Ave Upper Montclair NJ 07043-1622

MEIER, ALBERT HENRY, b New Haven, Mo, June 29, 29; m 54; c 3. ZOOLOGY, PHYSIOLOGY. *Educ:* Washington Univ, AB, 56; Univ Mo, MA, 59, PhD(zool), 62. *Prof Exp:* NIH fel, Wash State Univ, 62-64; from asst prof to assoc prof zool, 64-72, PROF ZOOL, LA STATE UNIV, 72- *Concurrent Pos:* NIH res career develop award, 69-74. *Honors & Awards:* Distinguished Res Master, La State Univ, 73. *Mem:* Am Soc Zool; Am Ornith Union; Ecol Soc Am; Soc Exp Biol & Med; Int Soc Chronobiol. *Res:* Comparative endocrinology and physiology of vertebrates; biological rhythms in hormonal control of seasonal and developmental conditions. *Mailing Add:* Dept of Zool La State Univ Baton Rouge LA 70803

MEIER, CHARLES FREDERICK, JR, b Cincinnati, Ohio, Feb 14, 49; m 76; c 2. ELECTROPHYSIOLOGY. *Educ:* Univ Cincinnati, BS, 71; Univ Miami, PhD(pharmacol), 77. *Prof Exp:* Res pharmacologist, Dept Pharmacol, Col Med, Univ Calif, San Francisco, 77-79; ASST PROF PHARMACOL, COL MED, UNIV OKLA, 79- *Concurrent Pos:* Mem, Am Heart Asn. *Mem:* Sigma Xi; Int Soc Heart Res. *Res:* Normal and abnormal cardiac excitation contraction coupling, transmembrane ionic currents and drug effects on these phenomena using isolated adult cardiac cell preparations and voltage clamp techniques. *Mailing Add:* Dept Pharmacol BMSB 753 Health Sci Ctr Univ Okla PO Box 26901 Oklahoma City OK 73190

MEIER, DALE JOSEPH, b The Dalles, Ore, Apr 21, 22; m 48; c 2. POLYMER PHYSICS. *Educ:* Calif Inst Technol, BS, 47, MS, 48; Univ Calif, Los Angeles, PhD(chem), 51. *Prof Exp:* Chemist, Shell Develop Co, 51-55, supvr res, 55-68, exchange scientist, Shell Lab, Amsterdam, 68-69, proj leader, Shell Chem Co, 69-71, supvr res, 71-72; sr res scientist, Midland Macromolecular Inst, 72-78, SR SCIENTIST & PROF POLYMER PHYSICS, MICH MOLECULAR INST, 78- *Concurrent Pos:* Consult, Alza Corp, 73-, Dynapol, 73-79, Lawrence Livermore Lab, 76-, AMP, Inc, 76-86, Dow Chem Co, 79-, Coopervision, 80-90, Nuclear Div, Union Carbide, 81-83, du Pont, 83-, ARCO, 85-, Landec, 89-; adj prof, Case Western Reserve Univ, 78-86 & 90, Cent Mich Univ, 79- & Mich Tech Univ, 86-, & Landec, 89-90; vis prof, Univ Kyoto, 81, 82. *Mem:* Fel Am Phys Soc; Am Chem Soc; Soc Rheology; Mat Res Soc. *Res:* Physics of high polymers and polymer solutions; rheology; statistical mechanics; physics of interfaces. *Mailing Add:* Mich Molecular Inst 1910 St Andrews Dr Midland MI 48640

MEIER, EUGENE PAUL, b Rosenberg, Tex, Oct 3, 42; m 63; c 2. ENVIRONMENTAL CHEMISTRY, ANALYTICAL CHEMISTRY. *Educ:* Tex A&M Univ, BS, 65; Univ Colo, Boulder, PhD(anal chem), 69. *Prof Exp:* Biochemist protein chem br, Med Res Labs, Edgewood Arsenal, 69-70, chemist, Phys Protection Br, Defense Eng & Develop, Defense Systs Div, 71-72; res chemist, Environ Protection Res Br, US Army Med Biomech Res & Develop Lab, 72-78, chief, Analytical Support Br, Las Vegas, 78-79, DIR, QUAL ASSURANCE DIV, ENVIRON MONITORING SYSTS LAB, LAS VEGAS, 79- *Mem:* Am Chem Soc. *Res:* Gas chromatography; trace analysis; metals and organics in water; environmental analysis. *Mailing Add:* 6524 Gumwood Rd Las Vegas NV 89108-4415

MEIER, FRANCE ARNETT, b Lubbock, Tex, Aug 11, 28; m 58; c 3. INDUSTRIAL ENGINEERING, PRODUCTION SYSTEMS. *Educ:* Tex Tech Univ, BS, 51; Univ Houston, MS, 59; Wash Univ, DSc(eng), 66. *Prof Exp:* Indust engr, Ideco of Dresser Industs, 51-55; instr indust eng, Lamar State Col, 55-59, asst prof, 59-61; teaching asst, Wash Univ, 61-64; assoc prof, 65-69, head dept, 67-77, PROF INDUST ENG, UNIV TEX, ARLINGTON, 69- *Concurrent Pos:* Consult gen dynamics, Tex Instruments. *Mem:* Am Inst Indust Engrs; Soc Mfg Engrs; Int Mat Mgt Soc; Am Soc Eng Educ; Sigma Xi. *Res:* Reliability design and testing of the exponential and non-exponential cases; applications of probability and statistics to industrial decisions; productivity improvement for handicapped; computer applications. *Mailing Add:* Dept Indust Eng Box 19017 Univ of Tex Arlington TX 76019

MEIER, GERALD HERBERT, b Pittsburgh, Pa, Nov 22, 42; m 66; c 3. METALLURGY. *Educ:* Carnegie Inst Technol, BS, 64; Ohio State Univ, PhD(metall eng), 68. *Prof Exp:* Res grant, Univ Münster, 68-69; asst prof, 69-77, ASSOC PROF METALL, UNIV PITTSBURGH, 77- *Mem:* Am Soc Metals; Am Inst Metall Engrs; Electrochem Soc. *Res:* Physical metallurgy, particularly oxidation of metals and alloys, strengthening of metals, chemical vapor deposition, thermodynamics of point defects. *Mailing Add:* Dept Metall & Mat Eng Univ Pittsburgh 848 Benedum Pittsburgh PA 15260

MEIER, JAMES ARCHIBALD, b New Salem, NDak, May 6, 36; m 67. POLYMER CHEMISTRY. *Educ:* NDak State Univ, BS, 59, PhD(phys chem), 71. *Prof Exp:* Sr chemist resins, Inmont Corp, 71-74, supvr resin develop, 74-76, tech mgr, Automotive Develop Ctr, 76-81; tech coordr, automotive finishes, 81-84, DIR MKT, ADHESIVES & SEALANTS, PPG INDUSTS, 84- *Mem:* Am Chem Soc; Fedn Socs Coatings Technol. *Res:* Automotive coatings development. *Mailing Add:* Five Residence Knibbeler Av Henri Barbusse St Saulve 59880 France

MEIER, JOSEPH FRANCIS, b Sharon, Pa, Nov 7, 36. POLYMER CHEMISTRY. *Educ:* John Carroll Univ, BS, 58; Univ Akron, MS, 60, PhD(polymer chem), 63. *Prof Exp:* Res chemist, Gen Tire & Rubber Co, 62-63 & 64-66; sr res chemist, Westinghouse Elec Corp, 66-72, mgr elastomers group, 73-78, mgr composites, Plastics & Elastomers, 78-88, ADV SCIENTIST, WESTINGHOUSE ELEC CORP, 88- *Honors & Awards:* Mat Eng Awards, Mat Eng Mag, 67, 68 & 78. *Res:* Development of rigid-brittle phenolic foam for energy absorbing applications; phenolic and melamine resins for high pressure decorative laminates; molded and cast elastomers for missile launch systems, including dynamic and static compressive testing and creep measurements on missile support pads, and both launch tube and missile mounted launch seals; application of resin/reinforcement to structural composites and printed wiring boards (PWB); friction testing of rubber and acoustic emission monitoring during PWB drilling; polymers for signature reduction applications. *Mailing Add:* Westinghouse Elec Corp Beulah Rd Pittsburgh PA 15235

MEIER, MANFRED JOHN, b Milwaukee, Wis, July 17, 29; m 54; c 2. NEUROPSYCHOLOGY. *Educ:* Univ Wis, BA, 52, MS, 53, PhD(psychol), 56. *Prof Exp:* Instr psychol, Univ Wis, 56-57; from asst prof to assoc prof, 57-66, PROF PSYCHOL & DIR NEUROPSYCHOL LAB, UNIV MINN, MINNEAPOLIS, 66- *Concurrent Pos:* Staff psychologist, Clin Psychol, Vet Admin Hosp, Wood, Wis, 56-57; Nat Inst Neurol Dis & Blindness res career develop award, 62-72. *Mem:* Am Psychol Asn; Am Acad Neurol; Am Heart Asn. *Res:* Effects of brain lesions on behavior in man. *Mailing Add:* Dept Psychol & Neurosurg Univ Minn Minneapolis MN 55455

MEIER, MARK FREDERICK, b Iowa City, Iowa, Dec 19, 25; m 55; c 3. GLACIOLOGY. *Educ:* Univ Iowa, BS, 49, MS, 51; Calif Inst Technol, PhD(geol, appl mech), 57. *Prof Exp:* Engr geol, US Bur Reclamation, Wash, 48-49; geologist, US Geol Surv, Alaska, 51; instr geol, Occidental Col, 52-55; Fulbright grant, Innsbruck Univ, 55-56; geologist, US Geol Survey, 56-85; res prof geophys, Univ Wash, 64-85; DIR, INST ARCTIC & ALPINE RES & PROF GEOL SCI, UNIV COLO, 85- *Concurrent Pos:* Mem tech panel, US Nat Comt, Int Geophys Year, 57-59; mem glaciol panel, Comt Polar Res, Nat Acad Sci, 59-68; vis assoc prof, Dartmouth Col, 64; US Nat Comt Int Hydrol Decade, 64-66, chmn working group combined balances & glacial basins, 65-71; pres, Int Comn Snow & Ice, 67-71; dir, World Data Ctr A, Glaciol, 70-76; mem gov bd, Permanent Serv Glacier Fluctuations, Zurich, 65-71; mem, comt geophys data, Geophys Res Bd, Nat Acad Sci, chmn, Glaciology Comt Polar Res Bd, 80-84; mem, Climate Coord Forum, Int Coun Sci Unions, 81-, US Nat Comt Global Change, 84-86 & Div Adv Comts Polar Progs & Earth Sci, NSF, 88-91. *Honors & Awards:* Medal of 150th Anniversary of Discovery of Antarctica, Medal of Inst Geog & Medal of 100th Anniversary of Int Polar Year, Acad Sci, USSR, US Antartic Service Medal; Seligman Crystal, Int Glaciological Soc. *Mem:* Fel AAAS; fel Geol Soc Am; fel Am Geophys Union; fel Arctic Inst NAm; hon mem, Int Glaciol Soc (vpres, 66-69); Int Asn Hydrol Sci (pres, 79-83). *Res:* Sea level rise, seasonal snowcover, glaciers, remote sensing of snow and ice, mountain and arctic hydrology, flow of ice and rock. *Mailing Add:* Inst Arctic & Alpine Res Univ Colo Campus Box 450 Boulder CO 80309-0450

MEIER, MARK STEPHAN, b Tacoma, Wash, Nov 30, 59. PEPTIDE CHEMISTRY. *Educ:* Dartmouth Col, BA, 82; Univ Ore, PhD(chem), 88. *Prof Exp:* Postdoctoral, Univ Tex, Austin, 88-90; ASST PROF, DEPT CHEM, UNIV KY, 90- *Mem:* Am Chem Soc; Am Soc Pharmacog. *Mailing Add:* Dept Chem Univ Ky Lexington KY 40506-0055

MEIER, MICHAEL MCDANIEL, b Chicago, Ill, Oct 14, 40; m 70; c 2. NUCLEAR STRUCTURE. *Educ:* St Procopius Col, BS, 62; Duke Univ, PhD(nuclear physics), 69. *Prof Exp:* Teaching asst physics, Duke Univ, 62-63, res asst nuclear physics, 63-69, res assoc, 69-70; physicist, Nat Bur Standands, 70-80; MEM STAFF, LOS ALAMOS NAT LAB, 80- *Mem:* Am Phys Soc; AAAS; Sigma Xi. *Res:* Gamma ray astronomy; imaging systems. *Mailing Add:* 152 Dos Brazos Los Alamos NM 87544

MEIER, PAUL, b New York, NY, July 24, 24; m 48; c 3. STATISTICS. *Educ:* Oberlin Col, BS, 45; Princeton Univ, MA, 47, PhD(math), 51. *Prof Exp:* Asst prof math, Lehigh Univ, 48-49; res secy, Philadelphia Tuberc & Health Asn, 49-51; res assoc math anal, Forrestal Res Ctr, Princeton Univ, 51-52; biostatist, Sch Hyg & Pub Health, Johns Hopkins Univ, 52-53, from asst prof to assoc prof, 53-57; assoc prof statist, Univ Chicago, 57-62, chmn dept statist, 60-66, 73-74, 83-86, dir, Biomed Comput Facil, 62-69, prof theoret biol, 68-74, actg chmn dept statist, 70-71, PROF STATIST, UNIV CHICAGO, 62-, PROF PHARM & PHYSIOL SCI, 74-, RALPH & MARY OTIS ISHAM PROF, 75-, DISTINGUISHED SERV PROF, 84-, PROF MED, DEPT STATIST & DIV BIOL SCI, 85- *Concurrent Pos:* Mem spec study sect biomath & statist, Nat Inst Gen Med Sci, 65-70, therapeut eval comt, Nat Heart Inst, 67-71 & diet-heart feasibility study rev comt, 68; comt biol effects of atmospheric pollution, Nat Acad Sci-Nat Inst Health spec fel, Sch Hyg & Trop Med, Imp Col, Univ London, 66-67; consult, statist probs to indust & govt. *Honors & Awards:* Statist Sect Recognition Award, Am Pub Health Asn, 86. *Mem:* Fel AAAS; fel Am Statist Asn (vpres, 65-67); fel Inst Math Statist (pres, 85 & 86); Soc Indust & Appl Math; fel Am Thoracic Soc; fel Sigma Xi; fel Am Public Health Asn; fel Am Heart Asn (coun epidemiol, 78); Soc Clin Trials)bd dirs, 79-81, pres, 86-87). *Res:* Estimation from incomplete observations. *Mailing Add:* Dept Statist Univ Chicago 5734 University Ave Chicago IL 60637

MEIER, PETER GUSTAV, b Jaegerndorf, Ger, Aug 18, 37; US citizen; m 66; c 2. AQUATIC ECOLOGY, AQUATIC TOXICOLOGY. *Educ:* Univ Mich, BA, 62, PhD(environ health), 70; Cent Mich Univ, MA, 64. *Prof Exp:* Instr aquatic entom, 68-70, res scientist water qual, 69-70, lectr water pollution, 70-72, from asst prof to assoc prof water qual, 72-84, PROF ENVIRON & IND HEALTH, UNIV MICH, ANN ARBOR, 84- *Concurrent Pos:* Consult, Egyptian Acad Sci, 75-, UNESCO, Brazil, 75-, Pan Am Health Orgn, 76- & Ethyl Corp, Manila Bay, Philippines, 78-; Fulbright Hays fel, Yugoslavia, 79, 82. *Mem:* Am Soc Limnol & Oceanog; Int Asn Theoret & Appl Limnol; Am Entom Soc. *Res:* Water quality monitoring with the use of macroinvertebrates; polychlorinated biphenyl uptake by aquatic insects an excretion; methodology of sampling aquatic communities; aquatic bioassays techniques. *Mailing Add:* 2516 SPH I Univ of Mich Ann Arbor MI 48109

MEIER, ROBERT R, b Pittsburgh, Pa, Nov 21, 40. AERONOMY. *Educ:* Duquesne Univ, BS, 62; Univ Pittsburgh, PhD(physics), 66. *Prof Exp:* Res asst physics, Univ Pittsburgh, 62-66, res assoc, E O Hulburt Ctr Space Res, US Naval Res Lab & Univ Pittsburgh, 66-68, res physicist, 68-83, BRANCH HEAD, US NAVAL RES LAB, 83- *Concurrent Pos:* Assoc ed, J Geophysical Res, Am Geophys Union, 73-75. *Mem:* Am Phys Soc; Am Geophys Union; Sigma Xi. *Res:* Aeronomy, especially airglow, radiative transfer theory, ionospheric physics and model atmospheres; interplanetary medium; comets. *Mailing Add:* Code 4140 Naval Res Lab Washington DC 20375

MEIER, RUDOLF H, b Heiligenstadt, Ger, Feb 27, 18; US citizen; m 52; c 1. OPTICAL PHYSICS. *Educ:* Univ Gottingen, Vordiplom, 46; Univ Jena, dipl physics, 49, Dr rer nat(physics), 51. *Prof Exp:* Assoc res scientist, Zeiss Werke, Ger, 49-53; physicist with Dr J & H Krautkramer, Cologne, 53-54; consult engr, Sperry Prod, Conn, 54-55; sr physicist & group leader, Perkin-Elmer Corp, 55-60; sect supvr, Aeronutronic Div, Philco Corp, Calif, 60-66; br chief, Electro-Optics, 66-73, prin scientist tech staff, Mcdonnell Douglas Astronautics Co, Huntington Beach, 73-; RETIRED. *Mem:* Fel Optical Soc Am; Ger Soc Appl Optics; Inst Elec & Electronics Engr. *Res:* Space optics; radiometry; infrared physics; design of ground test facilities for space optical systems. *Mailing Add:* 11001 Limetree Dr Santa Ana CA 92705

MEIER, WILBUR L(EROY), JR, b Elgin, Tex, Jan 3, 39; m 58; c 3. INDUSTRIAL ENGINEERING, OPERATIONS RESEARCH. *Educ:* Univ Tex, Austin, BS, 62, MS, 64, PhD(opers res), 67. *Prof Exp:* Planning engr, Tex Water Develop Bd, Austin, 62-66; res engr, Univ Tex, Austin, 66-67; from asst prof to prof indust eng, Tex A&M Univ, 67-73, asst head dept, 72-73; prof & chmn dept, Iowa State Univ, Ames, 73-74; prof & head, Sch Indust Eng, Purdue Univ, West Lafayette, 74-81; prof indust eng & dean, Col Eng, Pa State Univ, 81-87; chancellor, Univ Houston Syst, 87-89; DIR, DIV ENG INFRASTRUCT DEVELOP, NSF, 89- *Concurrent Pos:* Off Water Resources res grants, Tex A&M Univ, 67-71, US Post Off Dept contract, 70-71, US Air Force contracts, 78-87; consult, govt & indust, 67- *Mem:* Fel Inst Indust Engrs; Opers Res Soc Am; Am Asn Eng Soc; Am Soc Eng Educ; Nat Soc Prof Engrs; Nat Asn State Univ & Land Grant Cols; fel AAAS. *Res:* Systems engineering; application of operations research techniques in solving public planning problems; development of optimization methods; operations research and optimization; systems engineering; engineering and public policy. *Mailing Add:* NSF 1800 G St NW Washington DC 20550

MEIERAN, EUGENE STUART, b Cleveland, Ohio, Dec 23, 37; m 62; c 2. PHYSICAL METALLURGY, CRYSTALLOGRAPHY. *Educ:* Purdue Univ, BS, 59; Mass Inst Technol, MS, 61, ScD(metall), 63. *Prof Exp:* Sr mem res staff semiconductor technol, Res & Develop Labs, Fairchild Camera & Instrument Corp, Palo Alto, 63-73; mgr qual assurance, 73-84, INTEL FEL, INTEL CORP, 84- *Concurrent Pos:* Res assoc, H H Wills Physics Lab, Bristol Univ, 70-71; lectr, Israel Inst Technol, 71; fel Am Eng Coun, Mass Inst Technol, 83. *Mem:* Electron Micros Soc Am; Am Inst Mining, Metall & Petrol Engrs. *Res:* Effects of crystal defects on properties of semiconductor materials; development of advanced x-ray and electron microscopy techniques for studying semiconductor materials; packaging materials; reliability; semiconductors; process control and statistics; artificial intelligence applications in manufacturing. *Mailing Add:* 5421 E Camello Rd Phoenix AZ 85018

MEIERE, FORREST T, b Atlanta, Ga, Oct 12, 37; m 57; c 2. THEORETICAL PHYSICS. *Educ:* Carnegie Mellon Univ, BS 59; Mass Inst Technol, PhD(physics), 64. *Prof Exp:* Res assoc physics, Mass Inst Technol, 64; asst prof, Purdue Univ, Lafayette, 64-69; assoc prof, 69-72, chmn dept, 69-82, PROF PHYSICS, IND, INDIANAPOLIS, 72-,. *Mem:* Am Phys Soc; Am Asn Physics Teachers. *Res:* Theory of elementary particles; biophysics; imaging. *Mailing Add:* Dept Physics Ind Univ Purdue Univ PO Box 647 Indianapolis IN 46223

MEIERHOEFER, ALAN W, b Humboldt, Tenn, July 24, 44; m 65; c 2. SPECIALTY PAPERS. *Educ:* Southwestern Memphis, BS, 65; Miss State Univ, PhD(chem), 70; Western Carolina Univ, MBA, 78. *Prof Exp:* Polymer res, Am Enka Co, 69-83; res & develop group leader, 84, DIR RES & DIR, C H DEXTER, 83- *Mem:* Am Chem Soc; Fiber Soc. *Res:* Research and development of specialty paper and nonwoven products; wet form technology; utilize special cellulosic and synthetic fibers for filtration, barrier or carrier fabrics. *Mailing Add:* 118 Foxcroft Rd West Hartford CT 06119-1584

MEIGHEN, EDWARD ARTHUR, b Vancouver, BC, Dec 27, 42; m 62; c 2. BIOCHEMISTRY. *Educ:* Univ Alta, BSc, 64; Univ Calif, Berkeley, PhD(biochem), 69. *Prof Exp:* Res fel biochem, Dept Molecular Biol & Virol, Univ Calif, Berkeley, 69; res fel biol, Biol Labs, Harvard Univ, 69-71; asst prof, 71-76, PROF BIOCHEM, MCGILL UNIV, 76- *Mem:* Am Soc Biol Chemists; Can Fedn Biol Socs. *Res:* Enzyme regulation and relationship to subunit structure; mechanisms and control of enzyme induction in bioluminescent bacteria. *Mailing Add:* Dept of Biochem McGill Univ 3655 Drummond St Montreal PQ H3A 2T6 Can

MEIJER, AREND, b Berg en Dal, Neth, Aug 27, 47; US citizen; m 68; c 1. GEOLOGICAL SCIENCES, GEOCHEMISTRY. *Educ:* Univ Calif, Santa Barbara, BA, 69, MA, 71, PhD(geol), 74. *Prof Exp:* Fel geochem, Va Polytech Inst & State Univ, 74-75; fel, Calif Inst Technol, 75-76; asst prof geosci, Univ Ariz, 76-83; PROJ LEADER, LOS ALAMOS NAT LABS, 84- *Concurrent Pos:* Sr staff geochemist, Intra Tech, Austin, Tex, 86. *Mem:* Am Geophys Union; Sigma Xi; Geol Soc Am; Geochem Soc. *Res:* Isotope and trace element geochemistry; origin and evolution of volcanic arc magmas; geochemistry as applied to problems in tectonics; environmental geochemistry. *Mailing Add:* Isotope & Nuclear Chem Los Alamos Nat Lab MSJ514 Los Alamos NM 87545

MEIJER, PAUL HERMAN ERNST, b The Hague, Neth, Nov 14, 21; nat US; m 49; c 5. STATISTICAL MECHANICS, APPLIED MATHEMATICS. *Educ:* Delft Technol Univ, BS, 42; Univ Leiden, PhD(physics), 51. *Prof Exp:* Vis lectr physics, Case Univ, 53-54; res assoc, Duke Univ, 54-55; asst prof, Univ Del, 55-56; assoc prof, 56-60, chmn, 80-83, PROF PHYSICS, CATH UNIV AM, 60- *Concurrent Pos:* Fulbright grant, 53-55 & 78; staff mem, Nat Bur Standards, 58-86; Guggenheim Mem Found grant, Lab Magnetic Resonance, Univ Paris, 64-65; vis prof, Univ Paris, 78 & Univ Nancy, 84 & 87; consult, Naval Ord Lab, Naval Res Lab, Ft Belvoir & Lawrence Radiation Lab. *Mem:* Fel Am Phys Soc; Europ Phys Soc; Neth Phys Soc; Sigma Xi; Int Asn Math Physics; Am Asn Physics Teachers. *Res:* Statistical mechanics; critical phenomena; phase transitions; irreversible thermodynamics; solid state; magnetism; paramagnetic resonance; surface phenomena; superconductivity; mathematical physics; group theory; liquid state; supercooled liquids; scattering theory. *Mailing Add:* Dept Physics Cath Univ Am Washington DC 20064

MEIKE, ANNEMARIE, US citizen. MINERAL PHYSICS, MICROMECHANICS. *Educ:* Cornell Univ, BA, 78; Univ Calif, Berkeley, PhD(geol), 86. *Prof Exp:* Fel, Lawrence Berkeley Lab, 87-88; president's fel, President's Off, Univ Calif, 88-90; PHYSICIST, LAWRENCE LIVERMORE NAT LAB, 90- *Concurrent Pos:* Sr Fulbright fel, Australian Nat Univ, 90-91, vis fel, 91-92. *Mem:* Mat Res Soc; Mineral Soc Am. *Res:* Micro-mechanical and micro-chemomechanical behavior of non- metals, primarily with earth materials applications; mechanisms of stress corrosion cracking; transformational superplasticity in man-made and natural materials; chemical and mechanical degradation of man-made materials over extended periods of time, primarily with application to radioactive waste repositories. *Mailing Add:* Lawrence Livermore Nat Lab L-219 Livermore CA 94550

MEIKLE, MARY B, b Springfield, Mass, Aug 30, 34. AUDITORY PHYSIOLOGY. *Educ:* Vassar Col, AB, 54; Univ Ore Med Sch, MS, 67, PhD(physiol psychol), 69. *Prof Exp:* NIH res fel, Kresge Hearing Res Lab, 69-71, res assoc physiol of the ear, 71-72, asst prof otolaryngol & med psychol, 72-76, ASSOC PROF OTOLARYNGOL & MED PSYCHOL, ORE HEARING RES LAB, ORE HEALTH SCI UNIV, 76- *Concurrent Pos:* Vis lectr, Reed Col, 70. *Mem:* Sigma Xi; Acoustical Soc Am; Asn Res Otolaryngol; AAAS; Soc Neurosci; Ore Acad Otolaryngol. *Res:* Agents that damage the ear; physiology and psychology of the auditory system; tinnitus. *Mailing Add:* Ore Hearing Res Ctr Dept of Otolaryngol Ore Health Sci Univ Portland OR 97201

MEIKLE, RICHARD WILLIAM, organic chemistry, biochemistry; deceased, see previous edition for last biography

MEIKSIN, ZVI H(ANS), b Dessau, Germany, July 9, 26; nat US; m 55; c 4. ELECTRICAL ENGINEERING. *Educ:* Israel Inst Technol, Dipl, 51; Carnegie Inst Technol, MS, 53; Univ Pittsburgh, PhD(elec eng), 59. *Prof Exp:* Design & maintenance engr, Israel, 51-52; design engr, Pa Transformer Co, 53-54; from instr to assoc prof, 54-67, PROF ELEC ENG, UNIV PITTSBURGH, 67- *Concurrent Pos:* Proj engr, Westinghouse Elec Corp, 56-59, sr engr, 63, adv engr, 68; consult, Ohio Med Prod, Bell & Howell Co, RCA, Medrad, IBM, Union Carbide, Essex Int, Bendix & Westinghouse Elec Corp, 59-78 & numerous others US & abroad; mem, various comts, Inst Elec & Electronics Engrs. *Mem:* Fel Inst Elec & Electronics Engrs. *Res:* Biomedical instrumentation; integrated-circuit systems; transducers; transistor circuits; biomedical instrumentation. *Mailing Add:* Dept of Elec Eng Univ of Pittsburgh Pittsburgh PA 15261

MEILING, GERALD STEWART, b Provo, Utah, Sept 12, 36; m 62; c 3. CERAMICS, GLASS SCIENCE & TECHNOLOGY. *Educ:* Univ Utah, BS, 58; Mass Inst Technol, SM, 59, ScD(ceramics), 66. *Prof Exp:* Atomic Energy Comn res grant, Mass Inst Technol, 62-66; res ceramist, Corning Inc, res lab, 66-67, Ion Physics, 67-69, Signetics Res Lab, 69-70, res lab, 71-75, res assoc, 75-77, mgr optical prod develop, 77-82, dir develop, 82-86, dir res, 86-87, VPRES, DIR RES, CORNING INC, 88- *Concurrent Pos:* Mem adv bd, Univ Ill Dept Ceramics, 85-; bd dir, Samsung Corning, Seoul, Korea. *Mem:* Am Ceramic Soc; Inst Elec & Electronics Engrs; Sigma Xi. *Res:* Glass and glass ceramic research; crystal growth; photochromic glass; fine ceramics. *Mailing Add:* Sullivan Sci Park Corning Inc Corning NY 14831

MEILING, RICHARD L, obstetrics & gynecology; deceased, see previous edition for last biography

MEIMAN, JAMES R, b Louisville, Ky, Dec 10, 33. WATERSHED MANAGEMENT. *Educ:* Univ Ky, BS, 55, MS, 59; Colo State Univ, PhD(watershed mgt), 62. *Prof Exp:* Soil conservationist, Soil Conserv Serv, USDA, 55, 57-58 & 59; soil conservationist, Forest Serv, 62; from instr to assoc prof, 62-74, prof watershed mgt, 74-75, dean grad sch, 75-81, DIR INT PROGS, COLO STATE UNIV, 75-, ASSOC VPRES RES, 81- *Concurrent Pos:* Assoc ed, Water Resources Res; hydrologist, Rocky Mountain Forest & Range Exp Sta, US Forest Serv, Ft Collins, Colo, 75-; mem, Snow Res Working Group, NASA, 80- *Mem:* Am Geophys Union; Glaciol Soc; Sigma Xi. *Res:* Water yields and quality of wildland watersheds and impact of land use thereon; snow hydrology. *Mailing Add:* Off VPres Res 202 Admin Colo State Univ Ft Collins CO 80523

MEINCKE, P P M, b Winnipeg, Man, Jan 21, 36; m 58; c 2. PHYSICS. *Educ:* Queen's Univ, Ont, BSc, 59; Univ Toronto, MA, 60, PhD(physics), 63. *Prof Exp:* Asst prof physics, Royal Mil Col, Ont, 62-65; mem tech staff, Bell Tel Labs, 65-67; from asst prof to prof physics, Univ Toronto, 66-77, assoc dean, Erindale Col, 70-72, vprovost, 72-76; pres, 78-85, PROF, UNIV PEI, 78- *Concurrent Pos:* Pres & chmn bd, Inst of Man & Resources, 78-82; chmn, Asn of Atlantic Univs, 79-81; mem, Can Environ Adv Coun, 80-83; chmn, Nat Libr Adv Bd, 82-85; vchmn, Ministers Adv Bd on Can Mil Cols, 84-86; pres, Proteck Inc, 88-; secy-treas, Carratech Inc, 90-, mem, Imapro Bd. *Res:* Low temperature solid state physics; thermal expansion of solids at low temperatures; magnetism; superconductivity; superfluidity; information science; social impact of technology. *Mailing Add:* Univ PEI Charlottetown PE C1A 4P3 Can

MEINDL, JAMES D, b Pittsburgh, Pa, Apr 20, 33; m 61; c 2. SOLID STATE ELECTRONICS. *Educ:* Carnegie Mellon Univ, BS, 55, MS, 56, PhD(elec eng), 58. *Prof Exp:* Engr, New Prod Dept, Westinghouse Elec Corp, 58-59; leader microelectronics develop & circuits area, US Army Electronics Command, 59-62, chief semi-conductor & microelectronics br, 62-65, dir integrated electronics div, 65-67; from assoc prof to prof, Stanford Univ, 67-84, dir, Integrated Circuits Lab, 69-84, John M Fluke prof elec eng, 84-86, dir, Stanford Electronics Labs, 72-76, dir, Ctr Integrated Systs, 81-86, assoc dean res, Sch Eng, 84-86; vpres & provost, inst prof eng & sci, 86, SR VPRES ACAD AFFAIRS & PROVOST, INST PROF ENG & SCI, RENSSELAER POLYTECH INST, 88- *Concurrent Pos:* Lectr, Electronics Eng Dept, Monmouth Col, NJ, 60-67; chmn, Int Solid State Circuits Conf, 69, 28th Ann Conf Eng in Med & Biol, 75, Int Soc Biotelemetry Conf, 82; mem bd, Telesensory Systs, Inc, 71-84; chmn, Solid-State Circuits Coun, Inst Elec & Electronic Engrs, 72; mem, Study Sect on Surg & Bioeng, NIH, 78-82; chmn res bd, Hewlett Packard Corp, 89; mem, Acad Adv Coun, Indust Res Inst, Inc. *Honors & Awards:* A S Fleming Award, 67; J J Ebers Award, Electron Devices Soc, Inst Elec & Electronic Engrs; Beatrice K Winner Award, Inst Elec & Electronic Engrs, 88, Solid-State Circuits Award, 89, Educ Medal, 90. *Mem:* Nat Acad Eng; fel Inst Elec & Electronics Engrs; fel AAAS; Electrochem Soc; Biomed Eng Soc; Sigma Xi; Am Asn Univ Profs; Am Soc Eng Educ. *Res:* Microelectronics; micropower circuits; integrated electronics; medical electronics; ultra large scale integration; author of over 300 technical papers; recipient of six patents. *Mailing Add:* Rensselaer Polytech Inst Troy NY 12180

MEINECKE, EBERHARD A, b Braunschweig, Ger, June 7, 33; m 63; c 1. POLYMER ENGINEERING. *Educ:* Brunswick Tech Univ, DrIng, 59. *Prof Exp:* Res assoc wood prod, Inst Wood Prod, Brunswick Tech Univ, 57-59; asst dir res & develop, Joh Kleinewefers Sohne, Germany, 59-60; NATO res fel forest prod, Wis, 60; fel polymer sci, Polymer Res Inst, Univ Mass, 60-63; from asst prof to assoc prof, 63-72, PROF POLYMER SCI, INST POLYMER SCI, UNIV AKRON, 72-, RES ASSOC, 63- *Res:* Mechanical and optical properties of polymers; rheology; properties of engineering materials. *Mailing Add:* Dept of Mech Eng Univ of Akron Akron OH 44325

MEINEKE, HOWARD ALBERT, b Cincinnati, Ohio, July 2, 21; m 46; c 5. ANATOMY. *Educ:* Maryville Col, BA, 47; Univ Cincinnati, MS, 49, PhD(zool), 53. *Prof Exp:* Asst zool, 47-49, 50-51, instr, 49-50, from instr to prof anat, 51-85, EMER PROF ANAT, COL MED, UNIV CINCINNATI, 85- *Mem:* AAAS; Am Asn Anat. *Mailing Add:* 1339 Delta Ave Cincinnati OH 45208

MEINEL, ADEN BAKER, b Pasadena, Calif, Nov 25, 22; m 44; c 7. ENERGY ECONOMICS, ASTRONOMY. *Educ:* Univ Calif, AB, 47, PhD(astron), 49. *Hon Degrees:* DSc(astron), Univ Ariz, 90. *Prof Exp:* From instr to assoc prof astrophys, Univ Chicago, 50-53; assoc dir, Yerkes & McDonald Observ, 53-56; dir, Kitt Peak Nat Observ, 56-61; prof astron & optical sci, Univ Ariz, 61-84, chmn dept astron, 61-65, dir, Steward Observ, 62-67, dir, Optical Sci Ctr, 67-73; SR SCIENTIST OPTICS, JET PROPULSION LAB, CALTECH, 84- *Concurrent Pos:* Regent, Calif Lutheran Col, 62-71; mem pres comn 9, Int Astron Union, 73-76; consult, Energy Res & Develop Admin, 75-79. *Honors & Awards:* Lomb Medal, Optical Soc Am, 52, Ives Medal, 80; Warner Prize, Am Astron Soc, 54; Goddard Award, Soc Photo-Optical Instrumentation Engrs, 84; Van Biesbroek Award, Astron Soc Pac, 90. *Mem:* Fel Soc Photo-Optical Instrumentation Engrs; Am Acad Arts & Sci; fel Optical Soc Am (pres, 72); Am Astron Soc. *Res:* Energy/gross national product dynamics; solar energy; history of technology; volcanic eruptions; astronomical optics; stellar classification; engineering optics; aurora and airglow physics. *Mailing Add:* Jet Propulsion Lab 4800 Oak Grove Dr Pasadena CA 91109

MEINEL, MARJORIE PETTIT, b Pasadena, Calif, May 13, 22; m 44; c 7. SOLAR ENERGY, ASTRONOMY. *Educ:* Pomona Col, BA, 43; Claremont Col, MA, 44. *Prof Exp:* Ed rocketry, Calif Inst Technol, 44-45; res assoc solar energy, Univ Ariz, 74-84; TECHNICIAN OPTICS, JET PROPULSION LAB, CALIF INST TECHNOL, 84- *Concurrent Pos:* Consult, Off Technol Assessment, US Cong, 74-80; consult, Ariz Solar Energy Res Comn, 75-81; adv coun mem, Am Energy Independence, 78-84. *Mem:* NY Acad Sci; Soc Photo-Optical Instrumentation Engrs. *Res:* Solar energy applications; upper atmospheric phenomena; volcanic eruptions; astronomy, solar and variable stars; astronomical optics. *Mailing Add:* Jet Propulsion Lab 4800 Oak Grove Dr Pasadena CA 91109

MEINERS, HENRY C(ITO), b Pendleton, Ore, Feb 11, 16; m 38; c 3. CHEMICAL ENGINEERING. *Educ:* Ore State Col, BS, 38; Mass Inst Technol, DSc, 42. *Prof Exp:* Asst, Mass Inst Technol, 38-42; proj engr, Union Oil Co Calif, 42, process supvr oleum, 42-44, group leader, La, 44, process supvr, Mfg Dept, Calif, 45, supt cracking, La, 45-49, asst refinery mgr oleum, 49-52, refinery mgr, 52-53, Los Angeles, 53-56, process consult, 56-82; RETIRED. *Mem:* AAAS; Am Chem Soc; Am Inst Chem Engrs; Am Inst Chemists; NY Acad Sci; Inst Mgt Sci. *Res:* Design and operation of petroleum refining equipment. *Mailing Add:* 3909 Via Picaposte Palos Verdes Estates CA 90274-1144

MEINERS, JACK PEARSON, b Walla Walla, Wash, Sept 9, 19; m 45; c 3. PHYTOPATHOLOGY. *Educ:* Wash State Univ, BS, 42, PhD(plant path), 49. *Prof Exp:* Jr pathologist, Forage Div, Bur Plant Indust, USDA, Wash, 46-49, assoc pathologist, Fruit & Veg Div, Idaho, 49-50; from asst prof to assoc prof plant path, Wash State Univ, 50-53; pathologist, Cereal Crops Res Br, Wash, Agr Res Serv, USDA, 53-58, asst chief Md, 58-65, asst dir crops res div, 65-70, leader bean & pea invests, Veg & Ornamentals Res Br, Plant Sci Res Div, 70-72, chmn, Plant Protection Inst, Beltsville Agr Res Ctr, 72-74, chief appl Plant Path Lab, 74-80; RETIRED. *Mem:* Am Phytopath Soc. *Res:* Diseases, breeding and physiology of beans, peas and other edible legumes. *Mailing Add:* 2012 Forest Dale Dr Silver Spring MD 20903

MEINERT, WALTER THEODORE, b Walcott, Iowa, May 18, 22; m 46; c 3. ORGANIC CHEMISTRY. *Educ:* St Ambrose Col, BS, 47; Univ Va, MS, 49. *Prof Exp:* Tech serv rep, Emery Industs, Inc, 49-51, develop rep, 51-53, asst dir develop & tech serv, 53-56, dir, 56-65, gen mgr, Western Opers, Org Chem Div, 65-67, dir int opers, 67-69, vpres int opers, 69-79; pres, DM Int, Inc, 79-90; RETIRED. *Concurrent Pos:* Adv to pres, Unilever-Emery NV, Neth, 63-65. *Mem:* Am Chem Soc; Am Inst Chemists; Commercial Develop Asn. *Res:* Markets and uses for chemicals derived from fat sources; ozone oxidation and polymerization of unsaturated fatty chemicals; fatty acids. *Mailing Add:* 2885 Country Woods Lane Cincinnati OH 45248

MEINHARD, JAMES EDGAR, b Ill, 1919; m 45; c 3. DEVELOPMENT OF SAMPLE INTRODUCTION NEBULIZERS FOR EMISSION, ABSORPTION & MASS SPECTROMETRIES. *Educ:* Univ Wis, BS, 47, PhD(chem), 50. *Prof Exp:* Prob leader analytic separations, Hanford Atomic Prod Oper, Gen Elec Co, 50-56; with Nat Cash Register Co, 56-57 & Hughes Aircraft Co, 57-59; pres, Crystech, Inc, 59-69; pres, J E Meinhard Assocs, 69-80; PRES, J E MEINHARD ASSOCS, INC, 80- *Concurrent Pos:* Mem staff, Astropower Lab, Douglas Aircraft Co, 62-64, NAm Aviation, Inc, 64-71, NAm Rockwell Corp, 71-73 & Dept Physics, Calif State Col, Fullerton, 66-69. *Mem:* Fel AAAS; fel Am Inst Chemists. *Res:* Solid state physics; chemistry; organic semiconductors; development of analytical instrument components; solid state devices; sample introduction devices, techniques, theory. *Mailing Add:* 12472 Ranchwood Rd Santa Ana CA 92705

MEINHARDT, NORMAN ANTHONY, b Davenport, Iowa, Jan 19, 19; m 51; c 5. ORGANIC POLYMER CHEMISTRY. *Educ:* St Ambrose Col, BS, 40; Univ Iowa, PhD(org chem), 49. *Prof Exp:* Asst chemist ferrous analysis, Rock Island Arsenal, 40-44; fel rubber res, Univ Ill, 49-50; res chemist org phosphorous compounds, Lubrizol Corp, 50-53, res supvr develop group, 53-56, fundamental group, 56-61 & lubricant additives sect, 61-65, res supvr ashless dispersants, 65-71, res supvr polymer res, 71-79, proj mgr Japanese motor oil additives, 79-81; RETIRED. *Res:* Highly arylated ethylenes; initiation systems for emulsion polymerization; sulfinic acid reactions; organic phosphorous reactions; additive systems and intermediates for diesel engine lubricants; ashless inhibitor systems. *Mailing Add:* 1884 Dunellon Dr Lyndhurst OH 44124

MEINHOLD, CHARLES BOYD, b Boston, Mass, Nov 1, 34; m 56; c 5. RADIATION PROTECTION. *Educ:* Providence Col, BS, 56; Am Bd Health Physics, cert. *Prof Exp:* AEC fel radiol physics, 56; from jr scientist to sr scientist health physics, 57-71; head, Safety & Environ Protection Div, 72-88, Radiol Sci Div, 88-91, SR SCIENTIST, RADIOL SCI DIV, DEPT NUCLEAR ENERGY, ASSOC UNIV INC, BROOKHAVEN NAT LAB, 91-, PRES, NAT COUN RADIOL PROTECTION & MEASUREMENTS, 91- *Concurrent Pos:* Chmn sci comt oper radiation safety, Nat Coun Radiation Protection & Measurements, 73-90, Basic Radiation Protect Criteria; mem, Int Comn Radiol Protection Main Comn, chmn, Comn 3 Protection in Med, 77-85, & Comn 2 Secondary Standards, 85- *Mem:* Fel Health Physics Soc; Am Nuclear Soc; AAAS; Soc Risk Analysis; Radiation Res Soc; NY Acad Sci; Int Radiation Protect Asn (vpres, 88-92). *Res:* Radiation dosimetry and radiation protection standards development. *Mailing Add:* Dept Nuclear Energy Radiol Sci Div Bldg 703M Brookhaven Nat Lab Upton NY 11973

MEININGER, GERALD A, b Detroit, Mich, Jan 3, 52. CARDIOVASCULAR PHYSIOLOGY, MICROCIRCULATION. *Educ:* Univ Mo, Columbia, PhD(physiol), 81. *Prof Exp:* Res assoc, 81-84, ASST PROF PHYSIOL, TEX A&M UNIV, 84- *Concurrent Pos:* Mem, High Blood Pressure Res Coun, Am Heart Asn. *Mem:* Am Physiol Soc; Microcirculation Soc; Europ Soc Microcirculation; Am Heart Asn. *Mailing Add:* Dept Med Physiol Col Med Tex A&M Univ College Station TX 77843

MEINKE, GERALDINE CHCIUK, b Detroit, Mich, Jan 21, 44; m 69; c 3. IMMUNOLOGY. *Educ:* Madonna Col, BS, 65; Wayne State Univ, PhD(microbiol), 70. *Prof Exp:* Res fel immunol, Dept Immunopath, Scripps Clin & Res Found, 70-74, res assoc immunol, 75-78; vis asst prof, Dept Microbiol, 78-79, ADJ ASST PROF, DEPT MED & MOLECULAR MICROBIOL, COL MED, UNIV ARIZ, 80- *Concurrent Pos:* Leukemia Soc Am spec fel, Scripps Clin & Res Found, 73-75. *Mem:* Am Asn Immunologists. *Res:* Immunological and structural studies of bovine papilloma viral proteins; isolation and characterization of crystallizable fragment receptors for immunoglobulin E from human lymphoblastoid cells. *Mailing Add:* 1501 N Campbell Ave Tucson AZ 85724

MEINKE, WILLIAM JOHN, b Troy, Mich, May 16, 42; m 69; c 3. MICROBIOLOGY, VIROLOGY. *Educ:* Albion Col, BA, 64; Wayne State Univ, MS, 67, PhD(microbiol), 69. *Prof Exp:* Nat Cancer Inst fel, Scripps Clin & Res Found, 69-72, assoc microbiol, 72-77; Assoc prof, 77-82, PROF MICROBIOL & IMMUNOL, COL MED, UNIV ARIZ, 82- *Mem:* AAAS; Am Soc Microbiol; Fed Am Soc Exp Biol; NY Acad Sci. *Res:* Cell regulation in normal and neoplastic cells; mechanisms of virus replication. *Mailing Add:* Dept Microbiol Tucson AZ 85724

MEINKOTH, NORMAN AUGUST, zoology; deceased, see previous edition for last biography

MEINS, FREDERICK, JR, b New York, NY, May 31, 42; m 70; c 2. DEVELOPMENTAL BIOLOGY, TUMOR BIOLOGY. *Educ:* Univ Chicago, BS, 64; Rockefeller Univ, PhD(life sci), 69. *Prof Exp:* Asst prof biol, Princeton Univ, 69-76; assoc prof bot & genetics & develop, Univ Ill, Urbana, 76-81, prof, 81-82; GROUP LEADER, FRIEDRICH MIESCHER INST, BASEL, SWITZ, 80-; PROF PLANT DEVELOP BIOL, UNIV BASEL, SWITZ, 90- *Concurrent Pos:* Ed, Planta, 82- *Mem:* Swiss Soc Plant Physiol; Soc Develop Biol. *Res:* Studies of cell heredity and the stability of the differentiated state using chemical approaches and crown gall tumors as test objects; plant-pathogen interactions. *Mailing Add:* Friedrich Miescher Inst PO Box 2543 CH-4002 Basel Switzerland

MEINSCHEIN, WARREN G, b Slaughters, Ky, Nov 12, 20; m 44; c 3. ORGANIC CHEMISTRY. *Educ:* Univ Mich, BS, 48; Univ Tex, PhD, 51. *Prof Exp:* Res assoc geochem, Field Res Lab, Magnolia Petrol Co Div, Socony Mobil Oil Co, Inc, 51-58; sr chemist, Esso Res & Eng Co, NJ, 58-61, res assoc, 61-66; assoc dean acad progs, Sch Pub & Environ Affairs, 75-80, prof, 66-86, EMER PROF GEOCHEM, IND UNIV, BLOOMINGTON, 86- *Concurrent Pos:* Vchmn, Comt Geochem, Nat Res Coun, Nat Acad Sci, 68-70, chmn, 70-72; mem, Lunar Sample Review Bd, NASA, 70-72; mem, Bd Trustees Univ Space Res Assoc, 84- *Mem:* AAAS; Am Chem Soc; Am Geophys Union; Geol Soc Am; Geochem Soc. *Res:* Geochemistry; origin of petroleum composition of naturally occurring hydrocarbons; paleobiochemistry; evidence for life in Precambrian rocks and meteorites; intramolecular distribution of stable carbon isotopes in organic compounds. *Mailing Add:* 2307 Inverrary Circle Austin TX 78747-1607

MEINTS, CLIFFORD LEROY, b Kansas City, Mo, May 23, 30; m 54; c 4. BACTERIAL METABOLISM. *Educ:* Purdue Univ, BS, 53; Ohio Univ, MS, 54; Univ Okla, PhD(biochem), 57. *Prof Exp:* Asst prof chem, 57-58, dir comput develop, 73-80, Kresge-Carver assoc prof natural sci, 58-64, KRESGE-CARVER PROF NATURAL SCI, SIMPSON COL, 64-, CHMN DIV, 59- *Concurrent Pos:* Nat Cancer Res Inst fel, Sci Res Inst, Ore State Univ, 64-65; vis prof chem, Univ Tex, Austin, 80-81, Univ Mass, Amherst, 87-88. *Mem:* Am Chem Soc. *Res:* Computer applications in chemical education; Wiswesser line notation use in courses. *Mailing Add:* Dept of Chem Simpson Col 701 North C Indianola IA 50125

MEINTS, RUSSEL H, b Clara City, Minn, Apr 13, 39; m 82; c 3. MOLECULAR BIOLOGY, BOTANY. *Educ:* Macalester Col, AB, 60; Kent State Univ, MA, 62, PhD(cell biol), 65. *Prof Exp:* Instr biol, Kent State Univ, 62-63; from asst prof to assoc prof, 65-74, prof zool, Univ Nebr-Lincoln, 74-88, dir Sch Life Sci, 75-82; PROF BOT & PLANT PATH, ORE STATE UNIV, 88- *Concurrent Pos:* Res assoc, Dept Biochem, Univ Chicago & Argonne Cancer Res Hosp, 70-71; NIH fel, 71. *Mem:* AAAS; Am Soc Cell Biologists; Sigma Xi. *Res:* Animal-algae symbiotic associations; algal viruses; gene regulation. *Mailing Add:* Dept Bot & Plant Path Ore State Univ Corvallis OR 97331

MEINTS, VERNON W, b Beatrice, Nebr, Apr 20, 48; m 75; c 3. AGRONOMY. *Educ:* Univ Nebr, Lincoln, BS, 70, MS, 71; UniV Ill, PhD(soil fertil & chem), 75. *Prof Exp:* Asst prof, Mont State Univ, 75-77; exten soils specialist & asst prof soils, Mich State Univ, 77-83; INDEPENDENT CROP CONSULT, AGRI-BUSINESS CONSULTS, INC, 83- *Mem:* Am Soc Agron; Soil Sci Soc Am; Sigma Xi; Nat Alliance Independent Crop Consults. *Res:* Soil and fertilizer nitrogen reactions; fertilizer use efficiency; soil testing and plant analysis. *Mailing Add:* 3547 W Hiawatha Dr Okemos MI 48864

MEINTZER, ROGER BRUCE, b Fargo, NDak, July 5, 27; m 54; c 6. BIOCHEMISTRY. *Educ:* NDak Agr Col, BS, 50, MS, 52; Univ Wis, PhD(biochem), 54. *Prof Exp:* Res assoc biochem, Univ Wis, 54-55; instr med sch, Northwestern Univ, 55-57; assoc prof, NDak State Univ, 57-67; assoc prof, 67-70, PROF CHEM & CHMN DEPT, UNIV LETHBRIDGE, 70-, COORDR CONTINUING EDUC, 74- *Mem:* AAAS; Am Chem Soc; NY Acad Sci; Sigma Xi. *Res:* Mechanism of action of vitamin D and parathyroid hormone; chemistry of citric acid and other organic acids; their assay and metabolism as related to vitamins and endocrine functions. *Mailing Add:* Dept Chem Univ Lethbridge 1503 18th Ave S Lethbridge AB T1K 3M4 Can

MEINWALD, JERROLD, b New York, NY, Jan 16, 27; m 55, 80; c 3. ORGANIC CHEMISTRY, NATURAL PRODUCTS. *Educ:* Univ Chicago, PhB, 47, BS, 48; Harvard Univ, MA, 49, PhD(chem), 52. *Hon Degrees:* PhD, Göteborg Univ, 89. *Prof Exp:* Du Pont fel, Cornell Univ, 52, from instr to prof chem, 52-72; prof, Univ Calif, San Diego, 72-73; prof, 73-80, GOLDWIN SMITH PROF CHEM, CORNELL UNIV, 80- *Concurrent Pos:* Sloan Found fel, 58; Guggenheim fels, 60-61 & 76-77; NIH spec fel, 67-68, Fogarty Scholar in residence, 83-85; vis prof, Rockefeller Univ & Univ Calif, San Diego, 70, fel, Ctr Advan Study Behav Sci, 90-91; mem med chem study sect A, NIH, 64-68; chmn vis comt, Dept Chem, Brookhaven Nat Lab & res dir, Int Ctr Insect Physiol & Ecol, Nairobi, 70-77; consult, Schering Corp, Norwich Eaton Pharmaceut Co & Cambridge Neurosci Res Div, Native Plants, Inc; adv bd mem, Petrol Res Fund, 70-73, Res Corp, 78-83, mem adv coun, Princeton Univ, 79-83 & mem, Chem Adv Comm, NSF, 79-83, AP Sloan Found, 85-91; four-col lectr, Mt Holyoke Col, Smith Col, Amherst Col & Univ Mass, 65; del, 10th Int Biochem Cong, Hamburg, Germany, 76; Camille & Henry Dreyfus distinguished scholar, Mt Holyoke, 80 & Bryn Mawr, 83; A C Cope scholar award, Am Chem Soc, 89. *Honors & Awards:* Louderman Lectr, Wash Univ, 64; A Burger Lectr, Univ Va, 66; Rennebohm Lectr, Univ Wis, Hilldale Lectr, 91; F B Dains Lectr, Univ Kans, 68; F P Venable Lectr, Univ NC, 70; Frontiers Chem Lectr, Case-Western Reserve Univ, 71; Reilly Lectr, Univ Notre Dame & Priestly Lectr, Pa State Univ, 73; Inaugural Dow Lectr, Bucknell Univ, 74; Raymond Lemieux Lectr, Univ Ottawa, 76; E Guenther Award, Am Chem Soc, 83; Russell Marker Lectr, Pa State Univ, 87; Syntex Distinguished Lectr, Univ Colo, 89 & Berliner Lectr, Bryn Mawr Col, 90; Tyler Award for Environ Achievement, 90; G J Esselen Award for Chem Public Serv, 91. *Mem:* Nat Acad Sci; AAAS; fel Am Acad Arts & Sci; Am Chem Soc; fel Japan Soc Promotion of Sci; Int Soc of Chem Ecol (pres, 88); Am Philos Soc. *Res:* Problems of structure, synthesis and reaction mechanism from the field of natural products; synthesis and reactions of highly strained systems; molecular rearrangements; photochemistry; chemical defense mechanisms of arthropods; chemistry of pheromones; chemical ecology. *Mailing Add:* Dept Chem Baker Lab Cornell Univ Ithaca NY 14853-1301

MEINWALD, YVONNE CHU, b Shanghai, China, Feb 24, 29; nat US; div; c 2. ORGANIC CHEMISTRY, BIOCHEMISTRY. *Educ:* Bryn Mawr Col, BA, 52; Cornell Univ, PhD(chem), 55. *Prof Exp:* Res assoc, Cornell Univ, 55-67, lectr chem, 71-76, res assoc, 77-82, SR RES ASSOC CHEM, CORNELL UNIV, 82- *Concurrent Pos:* Lectr chem, Univ Calif, San Diego, 72-73; mem comt scholarly relationships with People's Repub of China, Nat Acad Sci, 74- *Mem:* Am Chem Soc. *Res:* Synthesis, solvolysis and rearrangement reaction of medium-sized rings; synthesis of bisketenes; synthesis and reactions of highly unsaturated cyclobutane derivatives; 1, 2-cycloaddition reactions of tetracyanoethylene; nitrosyl chloride addition reactions; chemistry of arthropod defensive secretions and pheromones; peptide synthesis; protein conformation. *Mailing Add:* 219 Comstock Rd Ithaca NY 14850

MEIROVITCH, L(EONARD), b Maxut, Rumania, Nov 28, 28; nat US; m 60. ENGINEERING. *Educ:* Israel Inst Technol, BS, 53; Univ Calif, Los Angeles, MS, 57, PhD(eng), 60. *Prof Exp:* Struct engr, Water Planning Israel, 53-55, asst head sect, 55-56; asst res engr, Univ Calif, Los Angeles, 56-58, assoc eng, 58-60; staff engr, Int Bus Mach Corp, 60-62; assoc prof eng, Ariz State Univ, 62-67; prof, Univ Cincinnati, 67-71; prof, 71-79, Reynolds Metals prof, 79-83, UNIV DISTINGUISHED PROF ENG, VA POLYTECH INST & STATE UNIV, 83- *Concurrent Pos:* Am Soc Eng Educ-NASA res fel, 64; Nat Acad Sci sr res assoc, Space Mech Div, NASA Langley Res Ctr, 66-67; consult, Goodyear Aerospace Corp, 62-63, Draper Lab, 77- & Intelsat, 80- *Honors & Awards:* Struct, Structural Dynamics & Mat Award, Am Inst Aeronaut & Astronaut, 83, Mech & Control of Flight Award, 87; Japan Soc of Mech Engrs Medal, 89. *Mem:* Fel Am Inst Aeronaut & Astronaut. *Res:* Control of structures; vibrations; classical mechanics; nonlinear analysis; astrodynamics. *Mailing Add:* Dept Eng Sci & Mech Va Polytech Inst & State Univ Blacksburg VA 24061

MEISEL, DAN, b Tel-Aviv, Israel, July 4, 43; m 65; c 2. RADIATION CHEMISTRY, PHOTOCHEMISTRY. *Educ:* Hebrew Univ, Jerusalem, BSc, 67, MSc, 69, PhD(phys chem), 74. *Prof Exp:* Res fel radiation chem, Carnegie-Mellon Univ, 74-76; res fel chem, 76-78, CHEMIST, ARGONNE NAT LAB, 78- *Res:* Photochemistry and radiation chemistry of organic and inorganic systems; fast kinetics in solutions; kinetic effects in colloidal solutions; energy conversion and storage; electron-transfer reactions. *Mailing Add:* Argonne Nat Lab Chem Div 9700 S Cass Ave Argonne IL 60439

MEISEL, DAVID DERING, b Fairmont, WVa, Mar 28, 40; m 62; c 2. ASTRONOMY, ASTROPHYSICS. *Educ:* WVa Univ, BS, 61; Ohio State Univ, MS, 63, PhD(astron), 67. *Prof Exp:* From instr to asst prof astron, Univ Va, 65-70; from asst prof to assoc prof, 70-83, PROF ASTRON, COL ARTS & SCI, STATE UNIV NY, COL GENESEO, 83,- DIR PLANETARIUM & OBSERV, 70- *Concurrent Pos:* Mem US nat comt & mem comn radio astron, Int Union Radio Sci; consult, NASA, 74-; res assoc, Kellogg Observ, Buffalo Mus of Sci, 74-; guest investr, Copernicus Space Telescope, Princeton Univ, 74-82; Nat Res Coun-Nat Acad Sci sr assoc, NASA Goddard Space Flight Ctr, 77-78; assoc Dir, C E K Mees Observ, Univ Rochester, 88- *Mem:* Fel AAAS; Am Astron Soc; Am Meteor Soc; Int Astron Union; fel Royal Astron Soc; Soc Photo Optical Instrumentation Engrs. *Res:* Astrophysical studies of early-type stars; comets and meteors; image processing and electronic imaging of astronomical objects; microcomputer utilization in research and education; astronomical time series analysis. *Mailing Add:* Dept Physics & Astron State Univ NY Geneseo NY 14454

MEISEL, JEROME, b Cleveland, Ohio, Aug 9, 34; m 57, 86; c 2. ELECTRICAL ENGINEERING. *Educ:* Case Inst Technol, BS, 56, PhD(elec eng), 61; Mass Inst Technol, MS, 57. *Prof Exp:* Staff engr, Gilmore Indust Inc, 55-57; from instr to asst prof elec eng, Case Inst Technol, 57-64; mem tech staff, Bell Tel Labs, 65-66; assoc prof, 66-70, actg chmn dept, 85-87, PROF ELEC ENG, WAYNE STATE UNIV, 70-,. *Concurrent Pos:* Consult, Encore Mfg, Inc, 59-, Linde Co, 61, Globe Indust Inc, 61-62, Flex Cable

Corp, 62, Lear Siegler, Inc, 62-63 & M Zucker, Inc, 68- *Mem:* Inst Elec & Electronic Engrs. *Res:* System theory; current instrument transformers; electromechanical energy conversion; real-time computer control systems for electronic telephone switching; planning and operation of large interconnected power systems; power electronics. *Mailing Add:* Dept Elec & Computer Eng Wayne State Univ Detroit MI 48202

MEISEL, SEYMOUR LIONEL, b Albany, NY, Aug 19, 22; m 46; c 3. ORGANIC CHEMISTRY, RESEARCH ADMINISTRATION. *Educ:* Union Col, NY, BS, 44; Univ Ill, MS, 46, PhD(org chem), 47. *Prof Exp:* Res chemist, Socony-Vacuum Oil Co, 47-50; sr res chemist, Mobil Res & Develop Corp, 50-56, asst supvr, 56-58, supvr, 58-61, tech dir, 61-64, mgr appl res & develop div, 64-68, vpres res, 68-87; RETIRED. *Concurrent Pos:* Mem, energy eng bd, Nat Res Coun & bd, Woodrow Wilson Nat Fel Found. *Honors & Awards:* Leo Friend Chem Tech Award, 76. *Mem:* Nat Acad Eng; Am Chem Soc; Am Petrol Inst; Am Inst Chem Engrs; NY Acad Sci. *Res:* Thiophene chemistry; petrochemicals; zeolite catalysis; chemical and petroleum processing; shale oil; combustion; lubrication; additives for fuels and lubricants; petroleum synthetic fuels from coal and natural gas; exploration for and production of oil and gas. *Mailing Add:* 28 Constitution Hill W Princeton NJ 08540

MEISELMAN, HERBERT JOEL, b 1940; c 2. BIORHEOLOGY, HEMORHEOLOGY. *Educ:* Mass Inst Technol, ScD, 65. *Prof Exp:* PROF PHYSIOL & BIOPHYS, SCH MED, UNIV SOUTHERN CALIF, 84- *Mem:* Am Physiol Soc; Biophys Soc; Europ Soc Microcirculation; Int Soc Biorheology; N Am Soc Biorheology. *Res:* Rheology of human blood; cell and membrane properties of RBC and WBC; clinical hemorheology. *Mailing Add:* Dept Physiol & Biophysics Univ Southern Calif 2025 Zonal Ave Los Angeles CA 90033

MEISELMAN, NEWTON, b Mineola, NY, Apr 5, 30; m 61; c 4. BOTANY, BIOLOGY. *Educ:* Syracuse Univ, AB, 51; Hofstra Univ, MS, 52; Rutgers Univ, PhD(bot), 56. *Prof Exp:* Asst, Rutgers Univ, 52-54; res assoc, Brookhaven Nat Lab, 55-56; from asst prof to assoc prof, 56-65, chmn dept, 68-74, PROF BIOL, C W POST COL, LONG ISLAND UNIV, 65- *Concurrent Pos:* Res collabr, Brookhaven Nat Lab, 59-62; researcher, Sect Biol Ultrastruct, Weizmann Inst, Israel, 65-66 & Plant Breeding Inst, Cambridge Univ, Eng, 75. *Mem:* AAAS; Bot Soc Am; Electron Micros Soc Am; Sigma Xi. *Res:* Morphogenesis; radiation morphology and cytology; electron microscopy; plant cytology. *Mailing Add:* 30 Kenswick Lane Huntington Station NY 11746

MEISELS, ALEXANDER, b Berlin, Ger, Feb 18, 26; Can citizen; m 53; c 5. CYTOLOGY. *Educ:* Nat Univ Mex, MD, 51; FRCP(C); Am Bd Path, dipl. *Prof Exp:* Asst dir lab cytol, Nat Cancer Inst, Mex, 56-60; dir lab clin cytol, 60-70, from asst prof to assoc prof, 61-68, PROF PATH, UNIV LAVAL, 68-; DIR DEPT PATH & CYTOL, ST SACREMENT HOSP, QUEBEC, 78- *Concurrent Pos:* Dir, Regional Cytodiag Ctr, St Sacrement Hosp, Quebec, 70- *Honors & Awards:* Maurice Goldblett Cytology Award, 76; Papanicolaou Award, 81. *Mem:* Am Soc Cytol (pres, 91-92); Can Asn Path (pres, 88-89); Int Acad Cytol (secy-treas, 71, 86, pres, 86-89); Can Soc Cytol. *Res:* Clinical cytology, particularly hormone, urinary and vaginal cytology; carcinogenesis, particularly effect of viruses on evolution of cancer of the cervix; condylomata acuminata of genital tract. *Mailing Add:* Dept of Path 1050 Chemin Ste-Foy Quebec PQ G1S 4L8 Can

MEISELS, GERHARD GEORGE, b Vienna, Austria, May 11, 31; nat US; m 58; c 1. PHYSICAL CHEMISTRY, ANALYTICAL CHEMISTRY. *Educ:* Univ Notre Dame, MS, 52, PhD(phys chem), 56. *Prof Exp:* Res assoc, AEC, Univ Notre Dame, 53-56; radiation chemist, Gulf Res & Develop Co, 56-59; chemist, Union Carbide Nuclear Co, 59-63, asst group leader, 63-65; from assoc prof to prof chem, Univ Houston, 65-75, assoc chmn dept, 69-72, chmn dept, 72-75; prof chem & chmn dept, 75-81, interim dean, 81-83, dean arts & sci, Univ Nebr-Lincoln, 83-88; VPRES ACAD AFFAIRS & PROVOST, UNIV S FLA, 88- *Concurrent Pos:* Consult, Tech Div, Union Carbide Corp, 65; chmn gov bd, NSF Regional Instrumentation Ctr Mass Spectrometry, Univ Nebr, 78-88; dir, Coun Chem Res, 81-88; mem, coun Sci Soc Presidents, 85, chmn-elect, 89, chmn, 90; bd dirs, Am Soc Mass Spectrometry, 88-90. *Mem:* AAAS; Am Chem Soc; Am Phys Soc; fel Am Inst Chemists; Am Soc Mass Spectrometry (vpres, 84-86, pres, 86-88). *Res:* Radiation chemistry; mass spectrometry; photochemistry; ion molecule reactions and collision dynamics. *Mailing Add:* Vpres Acad Affairs & Provost Univ S Fla Tampa FL 33620

MEISEN, AXEL, b Hamburg, WGer, Oct 17, 43; m 69; c 2. CHEMICAL ENGINEERING. *Educ:* Univ London, BSc, 65; Calif Inst Technol, MSc, 66; McGill Univ, PhD(chem eng), 70. *Prof Exp:* Lectr chem eng, McGill Univ, 67-68; from asst prof to assoc prof, 74-79, assoc dean appl sci, 76-85, PROF CHEM ENG, UNIV BC, 79-, DEAN APPL SCI, 85- *Concurrent Pos:* Environ engr, Imp Oil Enterprises Ltd, 74-75. *Honors & Awards:* ERCO Award, 86. *Mem:* Fel Can Soc Chem Eng; Asn Prof Engrs. *Res:* natural gas processing; petroleum refining; sulphur. *Mailing Add:* Dept Chem Eng Univ BC Vancouver BC V6T 1W5 Can

MEISENHEIMER, JOHN LONG, b Olney, Ill, June 21, 33; m 56; c 2. CHEMICAL EDUCATION COMPUTER SOFTWARE, METEOROLOGY. *Educ:* Evansville Col, BA, 54; Ind Univ, PhD(org chem), 63. *Prof Exp:* From asst prof to assoc prof, 63-68, PROF CHEM, EASTERN KY UNIV, 68- *Concurrent Pos:* Consult meteorol & chem monitoring, Ky Support Rev, Study Contract Chem Stockpike Disposal Plan, 87. *Mem:* Am Chem Soc. *Res:* Medicinal and heterocyclic chemistry. *Mailing Add:* Dept of Chem Eastern Ky Univ Richmond KY 40475

MEISER, JOHN H, b Cincinnati, Ohio, Nov 21, 38; m 67; c 3. THERODYNAMICS, MATERIAL PROPERTIES & METHANE ACTIVATION. *Educ:* Xavier Univ, BS, 61; Univ Cincinnati, PhD(chem), 66. *Prof Exp:* Asst prof chem, Univ Dayton, 66-69; from asst prof to assoc prof, 69-79, PROF CHEM, BALL STATE UNIV, 79- *Concurrent Pos:* Sr Fulbright Lectr, 85-86. *Mem:* Am Chem Soc; Am Phys Soc; Soc Appl Spectros. *Res:* Diffusion of metals; thermodynamics and material properties; enzyme kinetics; x-ray analysis; methane activation. *Mailing Add:* Dept Chem Ball State Univ Muncie IN 47306

MEISER, K(ENNETH) D(ONALDSON), chemical engineering; deceased, see previous edition for last biography

MEISER, MICHAEL DAVID, b Reading, Pa, Sept 28, 53; c 1. MECHANICAL PROPERTIES. *Educ:* Pa State Univ, BS, 75, MS, 77, PhD(ceramic sci), 79. *Prof Exp:* RES CERAMIST, AIRCO CARBON, 79- *Mem:* Am Ceramic Soc; Am Carbon Soc. *Res:* Environmental effects on the mechanical properties of graphite and the oxidation behavior of graphite. *Mailing Add:* 9198 McBride River Ave Fountain Valley CA 92708

MEISINGER, JOHN JOSEPH, b Aurora, Ill, Jan 3, 45; m 66; c 3. SOIL SCIENCE, BIOMETRICS. *Educ:* Iowa State Univ, BS, 67; Cornell Univ, PhD(soil sci), 75. *Prof Exp:* SOIL SCIENTIST, AGR RES SERV, USDA, 75- *Mem:* Soil Sci Soc Am; Am Soc Agron. *Res:* Nitrogen transformations in the soil-plant system, with emphasis on soil nitrogen aspects such as mineralization of organic matter, nitrogen fixation, and the study of nitrogen turnover processes with stable isotopes. *Mailing Add:* Bldg 007 Rm 245 BARC-West Beltsville Agr Res Ctr Beltsville MD 20705

MEISKE, JAY C, b Hartley, Iowa, June 22, 30; m 56; c 4. ANIMAL HUSBANDRY. *Educ:* Iowa State Col, BS, 52; Okla State Univ, MS, 53; Mich State Univ, PhD, 57. *Prof Exp:* Asst animal husb, Okla State Univ, 52-53 & Mich State Univ, 53-57; from instr to assoc prof, 57-70, PROF ANIMAL SCI, UNIV MINN, ST PAUL, 70- *Honors & Awards:* Animal Mgt Award, Asn Soc Animal Sci, 77- *Mem:* Am Soc Animal Sci; Am Inst Nutrit. *Res:* Ruminant nutrition; rumen microbiology and biochemistry. *Mailing Add:* Dept Animal Sci Univ Minn Inst Agr St Paul MN 55108

MEISLER, ARNOLD IRWIN, ONCOGENES, GROWTH CONTROL. *Educ:* NY Univ, MD, 56. *Prof Exp:* PROF MED ONCOL, UNIV PITTSBURGH, 79- *Mailing Add:* Vet Admin Med Ctr Univ Pittsburgh University Dr 4200 Fifth Ave Pittsburgh PA 15240

MEISLER, HAROLD, b New York, NY, Feb 7, 31; m 54; c 2. HYDROGEOLOGY. *Educ:* City Col New York, BS, 52; Univ Mich, MS, 53. *Prof Exp:* Geophysicist, Carter Oil Co, 55-56; geologist, Ground Water Br, 56-75, chief NJ dist, 75-79, HYDROLOGIST, US GEOL SURV, 79- *Mem:* Fel Geol Soc Am. *Res:* Hydrogeology of carbonate rocks, sandstones, shales and coastal plains; geomorphology of limestone terrain; computer simulation of aquifer systems; saltwater-freshwater relations. *Mailing Add:* 32 Winding Way W Morrisville PA 19067

MEISLER, MIRIAM HOROWITZ, b New York, NY, Mar 28, 43; m 63; c 2. BIOCHEMICAL GENETICS. *Educ:* Queens Col, BA, 64; Ohio State Univ, PhD(biochem), 68. *Prof Exp:* Cancer res scientist, Roswell Park Mem Inst, 71-73; asst prof biochem, Sch Med, State Univ NY Buffalo, 73-77; assoc prof, 77-82, PROF HUMAN GENETICS, SCH MED, UNIV MICH, ANN ARBOR, 83- *Concurrent Pos:* NIH fel biochem, Roswell Park Mem Inst, 69-70; Nat Found March of Dimes res grant, 74-76 & 77-79; NIH proj grants, 74-77, 78-81, 81-86, 85- & mem, Mammalian Genetics Study Sect, 82-86; co-investr, Dept Energy contract, 77-83. *Mem:* Am Soc Biol Chemists; AAAS; Am Soc Human Genetics; NY Acad Sci. *Res:* Regulation of gene expression in mammals; developmental biochemistry; genetic polymophism of proteins and restriction fragments in human populations. *Mailing Add:* Dept Human Genetics 1500 E Medical Center Ann Arbor MI 48109-0618

MEISLICH, HERBERT, b Brooklyn, NY, Mar 26, 20; m 55; c 3. ORGANIC CHEMISTRY. *Educ:* Brooklyn Col, AB, 40; Columbia Univ, AM, 47, PhD(chem), 51. *Prof Exp:* Chemist control & res, Edgewood Arsenal, 41-43; from teacher to assoc prof , City Col NY, 46-68, prof chem, 68-86; RETIRED. *Concurrent Pos:* Chemist, Med Sch, Columbia Univ, 50-52; fel, Med Sch, NY Univ, 52-53; lectr, Brooklyn Col, 52-58; Sloan-Kettering fel, 55-57. *Mem:* Am Chem Soc. *Res:* Mechanisms of organic reactions; heterocyclic chemistry; structure and reactivity of organic compounds. *Mailing Add:* 338 Lacey Dr New Milford NJ 07646

MEISLING, TORBEN (HANS), b Copenhagen, Denmark, Feb 20, 23; nat US; m 52; c 1. COMPUTER SCIENCE, INDUSTRIAL AUTOMATION. *Educ:* Royal Tech Univ Denmark, MS, 48; Univ Calif, Berkeley, PhD(elec eng), 52. *Prof Exp:* Res engr microwave antennas, Royal Tech Univ Denmark, 48; res engr digital comput & lectr, Univ Calif, 49-52, asst prof, 52-54; mem staff, Digital Comput Div, Lincoln Labs, Mass Inst Technol, 54-56; mgr, Systs Eng Lab, Stanford Res Inst, 56-68, exec dir, 68-71, managing dir, SRI-Europe, 71-74, sr indust consult, 74-77; int technol consult, Palo Alto, Calif, 77-87; pres, Torben Meisling Inc, 77-87; RETIRED. *Res:* Design and application of electronic digital computers; traffic analysis; design of microwave antennas. *Mailing Add:* 2110 Barbara Dr Palo Alto CA 94303-3445

MEISNER, GERALD WARREN, b Mt Kisco, NY, Aug 16, 38; m 62; c 2. ELEMENTARY PARTICLE PHYSICS. *Educ:* Hamilton Col, AB, 60; Univ Calif, Berkeley, PhD(physics), 66. *Prof Exp:* Res assoc & guest lectr physics, Univ Mass, 66-70; ASSOC PROF PHYSICS, UNIV NC, GREENSBORO, 70- *Concurrent Pos:* Res Corp grant, 72. *Mem:* Am Phys Soc; Sigma Xi. *Res:* Experimental high-energy physics. *Mailing Add:* Dept of Physics Univ of NC 1000 Spring Garden Greensboro NC 27412

MEISNER, LORRAINE FAXON, b Chicago, Ill, Nov 9, 31. CYTOGENETICS. *Educ:* Univ Chicago, MA, 58, PhD(genetics & zool), 66; Am Bd Med Genetics, dipl. *Prof Exp:* ASSOC PROF PREV MED & HUMAN ONCOL, UNIV WIS, 74-, CHIEF, CYTOGENETICS SECT, STATE LAB HYG, 68- *Mem:* Am Soc Cell Biol; NY Acad Sci; Am Soc Human Genetics; AAAS. *Res:* Investigation of the role of chromosome changes in evolution of malignancy. *Mailing Add:* State Lab Hyg 465 Henry Hall Univ Wis Madison WI 53706

MEISS, ALFRED NELSON, b Philadelphia, Pa, Mar 27, 18; m 42; c 3. BIOLOGY. *Educ:* Rutgers Univ, BSc, 41, MSc, 43; Yale Univ, PhD(plant sci), 50. *Prof Exp:* Asst veg crops, Rutgers Univ, 43-44 & 46; asst bot, Yale Univ, 46-47; asst biochemist, Conn Agr Exp Sta, 49-52; asst res specialist soils, Rutgers Univ, 52-54, assoc prof, 54-57; sci adv, Ted Bates & Co, 57-68; sr assoc, Sidney M Cantor Assocs, Inc, 68-71, vpres, 71-74; CONSULT, FFE INT, 74- *Mem:* AAAS; Am Chem Soc; Sigma Xi. *Res:* Foods and nutrition; conservation; planning and management of natural resources; international development; industrial product development and marketing. *Mailing Add:* 14 N Main St Cranbury NJ 08512

MEISS, JAMES DONALD, b Billings, Mont. DYNAMICAL SYSTEMS. *Educ:* Univ Wash, BS, 75; Univ Calif, Berkeley, PhD(physis), 80. *Prof Exp:* Res fel physics, Univ Tex, 80-82, res scientist, Inst Fusion Studies, 82-89; ASSOC PROF, UNIV COLO, 89- *Concurrent Pos:* Vis prof, Queen Mary Col, Eng, 83; vis scientist, Inst Theoret Physics, Univ Calif, Santa Barbara, 85, Queen Mary Col, Eng, 86 & Math Inst, Univ Warwick, 87. *Mem:* Fel Am Phys Soc. *Res:* Theoretical physics research in dynamic systems; foundations of transport theory, chaotic and quasiperiodic motion; author of fifty publications. *Mailing Add:* Prog Appl Math Univ Colo Boulder CO 80309

MEISS, RICHARD ALAN, b Philadelphia, Pa, Aug 25, 43; m 65; c 1. PHYSIOLOGY. *Educ:* Univ Del, BA, 65; Univ Ill, Urbana, PhD(physiol), 69. *Prof Exp:* Asst prof physiol, asst prof obstet & gynec & asst prof med biophys, 71-77, ASSOC PROF PHYSIOL, ASSOC PROF OBSTET & GYNEC & ASSOC PROF MED BIOPHYS, MED CTR, IND UNIV, INDIANAPOLIS, 77- *Concurrent Pos:* Nat Heart Inst fel, Harvard Med Sch, 69-71. *Mem:* AAAS; Am Physiol Soc. *Res:* Physiology and mechanical properties of muscle, in particular, cardiac muscle and the smooth muscle of the female reproductive system. *Mailing Add:* Dept Obstet & Gynec Sch Med 1001 Walnut St 102 Med Res Facil Indianapolis IN 46223

MEISSINGER, HANS F, b Germany, Nov 25, 18; US citizen; m 49; c 1. SPACE SYSTEM & MISSION DESIGN. *Educ:* Technical Univ, Berlin, Germany, BE, 42; New York Univ, MS, 49. *Prof Exp:* Res engr, aeronaut, German Aeronaut Inst, Berlin, 42-45; proj eng, flight simulation, Reeves Instruments Corp, NY, 47-54; sr staff, missiles & space systs, Hughes Aircraft, 55-62; sr staff, TRW, Redondo Beach, Calif, 62-90; RETIRED. *Concurrent Pos:* Mem, Am Inst Aeronaut & Astronaut Tech Comt, Power & Elec Propulsion, 68-72; mem, NASA Adv Comt, Power & Elec Propulsion, 70-72. *Mem:* Assoc fel Am Inst Aeronaut & Astrounat. *Res:* Spacecraft and space system design; mission design; guidance, navigation and control; scientific instrument accommodation on spacecraft; satellite servicing in space; rendezvous and docking; global geometry and pointing; published over 40 technical papers. *Mailing Add:* 4157 Don Luis Dr Los Angeles CA 90008

MEISSNER, CHARLES ROEBLING, JR, b Joliet, Ill, May 4, 23; m 44; c 5. GEOLOGICAL MAPPING, MINERAL EXPLORATION. *Educ:* Lehigh Univ, BA, 48. *Prof Exp:* Explor geologist, US Geol Surv, Colo, 48-49; field geologist, Gulf Oil Corp, Okla, 49-52, field coordr, 52, well-site geologist, 52-53; Foreign Opers Admin mutual security mission consult & petrol geologist, Chinese Petrol Corp, Taiwan, 53-55; asst area geologist, Stanolind Oil & Gas Co, Tex, 55-56; sr geologist, Standard-Vacuum Oil Co, Sumatra & India, 56-61; geologist adv, US Geol Surv, Pakiston, 61-66, econ geologist, Saudi Arabia, 66-70, proj geologist, Southwest Va & WVa, 70-80, chief, Gulf Coast Lignite Proj, Miss Embayment, US Geol Surv, 80-84, res geologist, sedimentary cover rock, Saudi Arabia, 84-88; RETIRED. *Mem:* Emer mem Am Asn Petrol Geologists. *Res:* Regional and detailed geologic mapping; mineral investigations; stratigraphy; petroleum geology; coal geology; phosphorite geology; stratigraphic and structural relationship of sedimentary rocks in mineral exploration, including energy minerals -- uranium, petroleum, and coal. *Mailing Add:* 11160 Harbor Ct Reston VA 22091-4316

MEISSNER, GERHARD, b Jan 26, 37; m; c 2. CALCIUM CHANNEL, SARCOPLASMIC RETICULM. *Educ:* Tech Univ Berlin, Germany, PhD(chem), 63. *Prof Exp:* PROF BIOCHEM & PHYSIOL, SCH MED, UNIV NC, 77- *Concurrent Pos:* mem, Coun Basic Sci, Am Heart Asn, estab investr. *Mem:* Ger Chem Soc; Biophys Soc; Am Soc Biol Chemists. *Res:* Excitation contraction coupling in muscle. *Mailing Add:* Dept Biochem CB #7260 Univ NC Chapel Hill NC 27599

MEISSNER, HANS WALTER, b Berlin, Ger, Mar 19, 22; nat US; wid; c 3. EXPERIMENTAL PHYSICS. *Educ:* Univ Munich, BS, 46, MS & PhD, 48. *Hon Degrees:* MS, Stevens Inst Technol, 62. *Prof Exp:* Res assoc physics, Low Temperature Inst, Acad Sci, Bavaria, 48-52; res engr, Heat Transfer Lab, Ill Inst Technol, 52-53; asst prof physics, Johns Hopkins Univ, 53-59; assoc prof, 59-62, prof, 62-86, EMER PROF PHYSICS, STEVENS INST TECHNOL, 86- *Mem:* Sigma Xi; Fel Am Phys Soc. *Res:* Low temperature physics; superconductivity; metal physics; time dependent phenomena in superconductivity; plasma etching. *Mailing Add:* 438 Grandview Terr Leonia NJ 07605-1005

MEISSNER, LOREN PHILLIP, b Los Angeles, Calif, Nov 24, 28; m 49; c 3. COMPUTER SCIENCE, APPLIED MATHEMATICS. *Educ:* Univ Calif, Berkeley, BA, 49, MA, 63, PhD(appl math), 65. *Prof Exp:* mathematician, Lawrence Berkeley Lab, Univ Calif, 59-67, lectr comput sci, 68-82; LECTR DEPT COMPUT SCI, UNIV SAN FRANCISCO, 82- *Mem:* Asn Comput Mach. *Res:* Non-numeric applications of computers; programming languages. *Mailing Add:* Dept Comput Sci Univ San Francisco 2130 Fulton St San Francisco CA 94117

MEISTER, ALTON, b New York, NY, June 1, 22; m 43; c 2. BIOCHEMISTRY. *Educ:* Harvard Univ, BS, 42; Cornell Univ, MD, 45. *Prof Exp:* From intern to asst resident, New York Hosp, 45-46; res investr, NIH, 46-55; prof biochem & chmn dept, Sch Med, Tufts Univ, 56-67; PROF BIOCHEM & CHMN DEPT, MED COL, CORNELL UNIV, 67-; BIOCHEMIST-IN-CHIEF, NEW YORK HOSP, 71- *Concurrent Pos:* Mem comt growth, Nat Res Coun, 54; mem biochem study sect, USPHS, 55-60, mem biochem training comt, 60-63; consult, Am Cancer Soc, 58-61 & 71-74; vis prof, Univ Wash, 59 & Univ Calif, Berkeley, 62; mem, US Nat Comt Biochem (Int Union Biochem), 62-65 & 78-, chmn, 79-82; mem sci adv comn, New Eng Enzyme Ctr, 63-, chmn, 63-67; chmn physiol chem study sect, USPHS, 64-68; mem bd sci counr, Nat Cancer Inst, 68-72, chmn, 72; ed, Advan Enzymol; assoc ed, J Biol Chem & Annual Review of Biochem. *Honors & Awards:* Paul-Lewis Award, Am Chem Soc, 54; William C Rose Award, Am Soc Biol Chem, 84; Chem Indust Inst of Toxicol Award, 85. *Mem:* Nat Acad Sci; Inst Med-Nat Acad Sci; Am Chem Soc; fel Am Acad Arts & Sci; Am Soc Biol Chemists (pres, 77-78); Harvey Soc. *Res:* Biochemistry of amino acids and proteins; enzymology, metabolism, glutamine, glutathione. *Mailing Add:* Med Col Dept of Biochem Cornell Univ 1300 York Ave New York NY 10021

MEISTER, CHARLES WILLIAM, b Hackensack, NJ, Oct 5, 40. PLANT PATHOLOGY, PLANT PHYSIOLOGY. *Educ:* Rutgers Univ, New Brunswick, BS, 63; Univ Nebr, Lincoln, MS, 66; Univ Ariz, PhD(plant path), 72. *Prof Exp:* Res asst, Boyce Thompson Inst Plant Res, 65-66; asst prof biol, Catawba Col, 67-69; res asst agr biochem, Univ Ariz, 72; citrus virologist, Peace Corps, USDA, Fiji, 72-74; plant virologist, USDA, Suva, Fiji, 74-76, IR-4 PROJ COORDR, SOUTHERN REGION, USDA, PESTICIDE RES LAB, UNIV FLA, 76- *Concurrent Pos:* Consult citrus virol, UNDP/Food Agr Org Surv Plant Pests & Dis in SPac, 75 & 76. *Mem:* Am Phytopath Soc; Am Inst Biol Sci. *Res:* Plant virology, pesticides; obtaining date required for the expansion of pesticide labels for new uses; established citrus-virus-indexing programs in Fiji and the Cook Islands; pesticides. *Mailing Add:* Inst Food/Agr Sci Univ Fla Gainesville FL 32611

MEISTER, PETER DIETRICH, b Schaffhausen, Switz, May 24, 20; m 51; c 2. ORGANIC CHEMISTRY. *Educ:* Swiss Fed Inst Technol, MSc, 44, PhD(chem), 47. *Prof Exp:* Fel, Swiss Fed Inst Technol, 47-49 & Nat Res Coun Can, 49-50; res chemist, Upjohn Co, 50-55, sect head, 56-63, mgr, 63-66, asst dir, 66-68, dir supportive res, 68-78, dir Pharmaceut Develop Labs, 78-82; RETIRED. *Concurrent Pos:* Mem bd dirs, Microlife Technics, 70-84. *Mem:* Am Chem Soc; Am Pharmaceut Asn. *Res:* Synthesis of steroids; degradation of lycopodium alkaloids; microbiological transformations of steroids; dosage forms. *Mailing Add:* 2154 Spruceway Lane Ann Arbor MI 48103-2341

MEISTER, ROBERT, b Brooklyn, NY, Feb 16, 25; m 48; c 3. ELECTRICAL ENGINEERING. *Educ:* Cath Univ Am, BSEE, 49, MS, 52, PhD(physics), 58. *Prof Exp:* Instr elec eng, Howard Univ, 51-55; PROF ELEC ENG, CATH UNIV AM, 55-, CHMN DEPT, 70- *Mem:* Inst Elec & Electronics Engrs; Am Soc Eng Educ; Sigma Xi. *Res:* The elastic and anelastic properties of solids using ultrasonic techniques; the dynamic properties of liquids using nuclear magnetic resonance techniques; dulectric properties of materials using optical techniques. *Mailing Add:* Dept of Elec Eng Cath Univ of Am Washington DC 20017

MEISTERS, GARY HOSLER, b Ottumwa, Iowa, Feb 17, 32; m 52; c 2. MATHEMATICS. *Educ:* Iowa State Univ, BS, 54, PhD(math), 58. *Prof Exp:* Instr math, Iowa State Univ, 57-58; res instr, Duke Univ, 58-59; from asst prof to assoc prof, Univ Nebr, 59-63; from assoc prof to prof, Univ Colo, Boulder, 63-72; PROF MATH, UNIV NEBR, LINCOLN, 72- *Concurrent Pos:* Res fel, Res Inst Advan Study, Md, 60-62. *Mem:* Am Math Soc; Math Asn Am; Sigma Xi. *Res:* Almost periodic functions; ordinary differential equations; abstract and functional analysis. *Mailing Add:* 7135 S Hampton Rd Lincoln NE 68506

MEISTRICH, MARVIN LAWRENCE, b Brooklyn, NY, Oct 10, 41; m; c 2. RADIATION BIOLOGY, REPRODUCTIVE BIOLOGY. *Educ:* Rensselaer Polytech Inst BS, 62; Cornell Univ, PhD(physics), 67. *Prof Exp:* Mem tech staff, Bell Tel Labs, 67-69; res assoc biophys, Ont Cancer Inst, 69-72; from asst prof to assoc prof, 72-83, PROF BIOPHYSICS, MD ANDERSON CANCER CTR, UNIV TEX, 83- *Mem:* Am Asn Cancer Res; Cell Kinetics Soc; Am Soc Cell Biol; Am Soc Andrology; Radiation Res Soc; Int Soc Analytical Cytol. *Res:* Biophysical methods for cell separation; biochemical mechanisms in cell differentiation; cytotoxic and mutagenic effects of radiation and chemicals on spermatogenic cells; spermatogenesis; nucleoproteins; cell biology; flow cytometry; cell kinetics. *Mailing Add:* Dept Exp Radiother Univ Tex M D Anderson Career Ctr 1515 Holcombe Blvd Houston TX 77030

MEITES, JOSEPH, b Kishinev, Russia, Dec 22, 13; nat US; m 43. PHYSIOLOGY, ENDOCRINOLOGY. *Educ:* Univ Mo, BS, 38, MA, 40, PhD(exp endocrinol), 47. *Prof Exp:* Asst, Exp Sta, Univ Mo, 40-42, 46-47; from asst prof to assoc prof, 47-53, PROF PHYSIOL, MICH STATE UNIV, 53-, MEM STAFF, AGR EXP STA, 47- *Concurrent Pos:* Weizmann fel, Weizmann Inst Sci, Israel, 55-56; mem subcomt, use hormones in domestic animals, Nat Acad Sci-Nat Res Coun, 60-; mem endocrinol study sect, NIH, 66-70, chmn comn endocrinol, Int Union Physiol Sci, 71-; pres, Int Soc Neuroendocrinol, 72-76; chmn comn endocrinol, Int Union Physiol Sci, 72-; mem coun, Int Brain Orgn, 76-; lectr, Kansas State Univ, 77; Pfizer lectr, Montreal Clin Res Inst, 77; assoc ed, Cancer Res, Am Asn Cancer Res; vis prof, Cairo Univ, 80. *Honors & Awards:* Carl G Hartman Award, Am Soc Study Reprod, 79; Geoffrey Harris Mem lectr, Cambridge, Eng, 81; Distinguished Leadership Award, Endocrine Soc, 81. *Mem:* AAAS; Am Asn Cancer Res; Am Physiol Soc; Endocrine Soc; Soc Exp Biol & Med; emer mem Geront Soc Am; Int Soc Neuroendocrinol. *Res:* Neuroendocrinology as related to aging; brain-pituitary relationships; author or co-author of over 500 publications and 6 books. *Mailing Add:* Dept Physiol Mich State Univ 129 Giltner Hall East Lansing MI 48824-1101

MEITES, LOUIS, b Baltimore, Md, Dec 6, 26; m 78; c 3. PHYSICAL CHEMISTRY, ANALYTICAL CHEMISTRY. *Educ:* Middlebury Col, BA, 45; Harvard Univ, MA, 46, PhD(chem), 47. *Prof Exp:* Instr chem, Princeton Univ, 47-48; from instr to asst prof, Yale Univ, 48-55; from assoc prof to prof, Polytech Inst Brooklyn, 55-68; chmn dept, Clarkson Col Technol, 68-81, prof chem, 68-84; PROF CHEM & CHMN DEPT, GEORGE MASON UNIV, 84- *Concurrent Pos:* Founding ed, Critical Rev in Analytical Chem, 69-74; Comn Electroanalytical Chem, IUPAC, 65-81. *Honors & Awards:* Louis Gordon Mem Award, Talanta, 70; Benedetti-Pichler Award, Am Microchem Soc, 83. *Res:* Thermochemical and electrochemical investigations of reaction kinetics and equilibria; controlled-potential electrolysis, coulometry, and polarography; titrimetric theory; chemical applications of non-linear regression; machine decisions. *Mailing Add:* Dept Chem George Mason Univ 4400 University Dr Fairfax VA 22030

MEITES, SAMUEL, b St Joseph, Mo, Jan 1, 21; m 45; c 1. CLINICAL CHEMISTRY. *Educ:* Univ Mo, AB, 42; Ohio State Univ, PhD(biochem), 50; Am Bd Clin Chem, dipl. *Prof Exp:* Biochemist, Vet Admin Hosp, Poplar Bluff, Mo, 50-52 & Toledo Hosp, Ohio, 53; res asst prof, 54-66, assoc prof, 66-72, PROF PEDIAT, OHIO STATE UNIV, COL MED, 72-, PROF PATH, 74- *Concurrent Pos:* Clin chemist, Children's Hosp, Columbus, Ohio, 54- *Honors & Awards:* Bernard J Katchman Award, Am Asn Clin Chemists, 72, Fisher Award, 81, Miles/Ames Award, 90. *Mem:* Fel AAAS; Am Chem Soc; fel Am Asn Clin Chemists (secy, 75-77). *Res:* Amylase isoenzymes; pediatric micro methods; measurement of jaundice; reference values in pediatric clinical chemistry; history of clinical chemistry; reference values in the elderly. *Mailing Add:* Clin Chem Lab 700 Children's Dr Columbus OH 43205

MEITIN, JOSE GARCIA, JR, b Havana, Cuba, Sept 22, 50; US citizen; m 73; c 3. METEOROLOGY. *Educ:* Fla State Univ, BS, 72, MS, 75. *Prof Exp:* Teaching asst meteorol, Fla State Univ, 72-75; res asst, Coastal Upwelling Ecosysts Prog, 73-75; support scientist meteorol, Nat Ctr Atmospheric Res, 76-79; prof res asst, Coop Inst Res Environ Sci, Univ Colo, 79-82; METEOROLOGIST, NAT OCEANIC & ATMOSPHERIC ADMIN, US DEPT COMMERCE, 82- *Mem:* Am Meteorol Soc. *Res:* Mesoscale research; rainfall studies. *Mailing Add:* 8202 Kincross Dr Boulder CO 80301-4229

MEITLER, CAROLYN LOUISE, b Detroit, Mich, Aug 10, 38; m 63; c 2. MATHEMATICS. *Educ:* Mich State Univ, BS, 60; Univ Nebr, MS, 64; Univ Wis-Milwaukee, PhD(math educ), 86. *Prof Exp:* Adj asst prof math, Marquette Univ, 63-89; ASST PROF MATH, COMPUTER SCI & MATH EDUC, UNIV WIS-CONCORDIA, 89- *Mem:* Nat Coun Teachers Math; Math Asn Am; Am Math Soc. *Res:* Writing of manuals to accompany computer programs that are designed to accompany mathematic textbooks; writing of graphing calculator supplements to accompany textbooks. *Mailing Add:* 12800 N Lake Shore Dr Mequon WI 53092

MEITZLER, ALLEN HENRY, b Allentown, Pa, Dec 16, 28; m 53; c 3. PHYSICS. *Educ:* Muhlenberg Col, BS, 51; Lehigh Univ, MS, 53, PhD, 55. *Prof Exp:* Asst physics, Lehigh Univ, 51-54; mem tech staff, Bell Tel Labs, Inc, NJ, 55-72; PRIN RES SCIENTIST, RES STAFF, FORD MOTOR CO, 72- *Concurrent Pos:* Adj prof, Dept Elec Eng, Univ Mich-Dearborn. *Mem:* Am Phys Soc; fel Acoust Soc Am; fel Inst Elec & Electronics Engrs. *Res:* Solid state physics; ultrasonic devices, ferroelectric ceramic and display devices; automotive emission control systems; silicon-based smart sensors. *Mailing Add:* Ford Motor Co Box 2053 Dearborn MI 48121

MEIXLER, LEWIS DONALD, b New York, NY, July 25, 41; m 64; c 2. ELECTRONIC DESIGN, LASER DEVELOPMENT. *Educ:* City Col New York, BE, 64; Rutgers Univ, MSEE, 71. *Prof Exp:* Engr data commun, Western Union Tel Co, 64-67; design engr spacecraft electronics, Astro Electronics Div, RCA Corp, 67-75; SR MEM PROF TECH STAFF ELECTRONICS ENG, PLASMA PHYSICS LAB, PRINCETON UNIV, 75- *Concurrent Pos:* Adj prof, Trenton State Col, 78- *Mem:* Inst Elec & Electronics Engrs. *Res:* X-ray laser development; pico second electronics. *Mailing Add:* 11 Cherry Brook Lane East Windsor NJ 08520

MEIZEL, STANLEY, b New York, NY, May 1, 38; m 68; c 2. BIOCHEMISTRY. *Educ:* Queens Col, NY, BS, 59; Univ Rochester, PhD(biochem), 66. *Prof Exp:* From asst prof to assoc prof, 67-80, PROF HUMAN ANAT, SCH MED, UNIV CALIF, DAVIS, 80- *Concurrent Pos:* NIH fel develop biol, Yale Univ, 65-67; gamete res 84- *Mem:* Soc Study Reproduction; Am Soc Cell Biol. *Res:* Biochemistry of mammalian fertilization, sperm enzymes, sperm capacitation and the acrosome reaction; sperm egg fusion. *Mailing Add:* Dept Cell Biol & Human Anat Univ of Calif Sch of Med Davis CA 95616-8643

MEKJIAN, ARAM ZAREH, b New York, NY, Sept 26, 41; m 69; c 2. NUCLEAR PHYSICS. *Educ:* Calif Inst Technol, BS, 63; Univ Md, College Park, PhD(physics), 68. *Prof Exp:* Fel, Rutgers Univ, 68-69; Alexander von Humboldt fel, Univ Heidelberg, 69-71; asst prof, 71-74, ASSOC PROF PHYSICS, RUTGERS UNIV, 74- *Concurrent Pos:* NSF grant, 71-72. *Mem:* Am Phys Soc. *Res:* Nuclear structure; nuclear reactions; heavy ion physics; intermediate energy physics. *Mailing Add:* Dept Physics Rutgers Univ Piscataway NJ 08854

MEKLER, ALAN, b Toronto, Ont, Can, Oct 28, 47. SET THEORY ALGEBRA. *Educ:* York Univ, BA, 69; Stanford Univ MS, 71, PhD(math), 76. *Prof Exp:* Vis asst prof math, Univ Western Ont, 79-80; from asst to assoc prof math 80-86, PROF MATH, SIMON FRASER UNIV, 86- *Concurrent Pos:* Vis scientist, Universitat Essen, 80; asst prof math, Auburn Univ, 80. *Mem:* Can Math Soc; Am Math Soc. *Mailing Add:* Dept Math & Statist Simon Fraser Univ Burnaby BC V5A 1S6 Can

MEKLER, ARLEN B, b New York, NY, May 4, 32; m 61; c 6. ORGANIC CHEMISTRY. *Educ:* San Jose State Col, BS, 53; Iowa State Univ, MS, 55; Ohio State Univ, PhD(phys org chem), 58; Temple Univ, JD, 72. *Prof Exp:* E I du Pont de Nemours & Co res fel, Ohio State Univ, 58-59; res chemist, Polychem Dept, E I du Pont de Nemours & Co, 59-61, res chemist, Explosives Dept, 61-63; sr res chemist, Arco Chem Res Div, Atlantic Richfield Co, 63-69; RES ASSOC, RCEO INC, 69- *Concurrent Pos:* Pres, Del Law Ctr, 72- *Mem:* AAAS; Am Chem Soc; Sigma Xi; Royal Soc Chem. *Res:* Benzil-ammonia reaction; synthesis of steroid intermediates; steric effects in addition reaction; stereospecific polymerizations; selective oxidation and oxidative coupling reactions; transition metal-olefin complexes; synthesis and polymerization of small ring compounds; patent law; forensic science. *Mailing Add:* PO Box 2285 Wilmington DE 19899

MEL, HOWARD CHARLES, b Oakland, Calif, Jan 14, 26; m 49; c 3. CELL MEMBRANE BIOPHYSICS, HEMATOPOESIS. *Educ:* Univ Calif, Berkeley, BS, 48, PhD(phys chem), 53. *Prof Exp:* Traffic mgr, Calo Pet Food Co, Calif, 48-50; asst chem, Univ Calif, Berkeley, 50-51, chemist radiation lab, 51-53; Fulbright fel, Free Univ Brussels, 53-55; instr chem, Univ Calif, Berkeley, 55, USPHS fel, spec fel & lectr med physics & biophys, 55-59, from asst prof to assoc prof, 60-74, dir, Lawrence Hall Sci, 81-82, PROF BIOPHYS, UNIV CALIF, BERKELEY, 74-, FAC SR SCIENTIST, LAWRENCE BERKELEY LAB, 60- *Concurrent Pos:* Am Inst Physics vis scientist, 63-; dir training grants, Nat Inst Gen Med Sci, 72-82, dir res, INSERM, Univ Paris, 74-75, dir, Univ Calif Educ Abroad Prog, Bordeaux, Pau, Poitiers, France, 86-89; mem coun, Biophys Soc, 69-71. *Mem:* Fel AAAS; Biophys Soc; Am Chem Soc; Am Asn Physics Teachers; fel Sigma Xi; Int Soc Biorheology; Soc Gen Systs Res. *Res:* Cellular and subcellular biophysics; biological separations including electrophoresis; cellular development and differentiation, hematopoiesis; thermodynamics of open and closed systems. *Mailing Add:* MCB-BCP Donner LB Univ Calif Berkeley CA 94720

MELA, DAVID JASON, b Washington, DC, Sept 25, 58; m 84; c 3. SENSORY EVALUATION, EATING BEHAVIOR. *Educ:* Univ Vt, BS, 79; Pa State Univ, PhD(nutrit), 85. *Prof Exp:* Postdoctoral fel, Monell Chem Senses Ctr, 85-87, staff scientist, 87-88, asst mem, 88-89; HEAD SENSORY STUDIES, DEPT CONSUMER SCI, AFRC INST FOOD RES, 90- *Mem:* Am Inst Nutrit; Am Soc Clin Nutrit; Asn Chemoreception Sci; Inst Food Technologists; Nutrit Soc. *Res:* Relationships between sensory evaluation, nutrition and eating behavior; basis of food acceptance and comsumption; sensory and nutritional aspects of fats. *Mailing Add:* Dept Consumer Sci AFRC Inst Food Res Earley Gate Whiteknights Rd Reading RG6 2BZ England

MELACK, JOHN MICHAEL, b Pittsburgh, Pa, May 27, 47; m 77; c 1. AQUATIC ECOLOGY, LIMNOLOGY. *Educ:* Cornell Univ, AB, 69; Duke Univ, PhD(zool), 76. *Prof Exp:* NSF fel biol, Univ Mich, Ann Arbor, 76-77; from asst prof to assoc prof, 77-87, PROF BIOL, UNIV CALIF, SANTA BARBARA, 87- *Mem:* Am Soc Limnol & Oceanog; Ecol Soc Am; Soc Int Limnol. *Res:* Phytoplankton ecology, especially in tropical African and saline lakes; biogeochemistry and nutrient dynamics in Amazon floodplain lakes; alpine lakes; acid precipitation. *Mailing Add:* Dept Biol Sci Univ Calif Santa Barbara CA 93106

MELAMED, MYRON ROY, b Cleveland, Ohio, Aug 9, 27; m 58; c 2. MEDICINE, PATHOLOGY. *Educ:* Western Reserve Univ, BS, 47; Univ Cincinnati, MD, 50. *Prof Exp:* From asst attend pathologist to assoc attend pathologist, Mem Hosp Cancer & Allied Dis, 58-69, chief cytol serv, 73-79; assoc prof path, Med Sch, 73-79, PROF PATH, CORNELL UNIV MED COL, 79-; chmn dept path, 79-89, ATTEND PATHOLOGIST, MEM HOSP CANCER & ALLIED DIS, 69-; PROF BIOL, SLOAN-KETTERING DIV OF CORNELL UNIV GRAD SCH MED COL, 80- *Concurrent Pos:* Mem, Sloan-Kettering Inst Cancer Res, 78-; consult, USPHS Hosp, Staten Island, NY, 61-64; med res consult, Int Bus Mach Corp, 64-68, NY State Dept Health, 65 & Col Dent, NY Univ, 65; asst vis pathologist, James Ewing Hosp, 65; consult, Hosp Spec Surg, 73-90; attend pathologist, NY Hosp, 80- *Honors & Awards:* Papanicolaou Award, Am Soc Cytol, 75; Goldblatt Award, Int Acad Cytol, 86. *Mem:* AAAS; Am Asn Path; Am Soc Cytol; Am Soc Clin Path; Soc Anal Cytol; Int Acad Path. *Res:* Pathology and cytology of cancer. *Mailing Add:* Sloan-Kettering Cancer Ctr 1275 York Ave New York NY 10021

MELAMED, NATHAN T, b Poland, May 1, 23; nat US; m 67; c 2. CHEMISTRY. *Educ:* City Col New York, BS, 43; Polytech Inst Brooklyn, PhD(chem), 49. *Prof Exp:* Res chemist, Manhattan Dist Proj, SAM Labs, Columbia Univ, 43-45; instr chem, Polytech Inst Brooklyn, 48; res physicist solid state physics, Horizons, Inc, 49-50; mgr optical electronics, 72-74, ADV SCIENTIST, WESTINGHOUSE RES LABS, 50- *Concurrent Pos:* Lectr, Univ Pittsburgh, 55-56; prof, Fed Univ Rio de Janeiro, 73 & 74. *Honors & Awards:* IR-100 Award, 65. *Mem:* Am Phys Soc; NY Acad Sci. *Res:* Luminescence of inorganic solids-fundamental theory; laser physics; semiconductors; spectroscopy. *Mailing Add:* Westinghouse Res Labs 1310 Beulah Rd Pittsburgh PA 15235

MELAMED, SIDNEY, b Philadelphia, Pa, Oct 5, 20; m 43; c 2. COATINGS, BIOPOLYMERS. *Educ:* Philadelphia Col Pharm, BSc, 41; Univ Ill, PhD(org chem), 44. *Prof Exp:* Res chemist, Univ Ill, 43-44; res chemist, Univ Md, 44-45, res chemist, Comt Med Res, 45-46; res chemist, US Navy, 46-47; res chemist, 47-58, lab head, 58-71, supvr new prod develop, Fibers Div, 71-73, pioneering life sci, 73-76, sr res assoc polymer technol, 73-81, dept mgr biocides, 81-84, ASST DIR RES & NEW TECHNOL, ROHM & HAAS CO, 84- *Mem:* AAAS; Am Chem Soc. *Res:* Nylon, polyesters, elastic and antistatic fibers; vinyl and condensation polymers; monomers; polymers for paper, textiles, leather and coatings; ion-exchange resins and fibers; insecticides; fiber and fabric technology; biomedical research; agricultural chemicals; biocides; technology transfer. *Mailing Add:* 8270 Thomson Rd Elkins Park Philadelphia PA 19117

MELAN, MELISSA A, b New Orleans, La, Apr 6, 58. CYTOSKELETON, CELL WALL FORMATION. *Educ:* Loyola Univ, BS, 80; Vanderbilt Univ, PhD(gen biol), 85. *Prof Exp:* Teaching asst biol & zool, Vanderbilt Univ, 80-85; postdoctoral res assoc biophys, Pa State Univ, 86-87; postdoctoral res assoc micros, Worcester Found Exp Biol, 87-91; POSTDOCTORAL RES ASSOC MOLECULAR BIOL, WELLESLEY COL, 91- *Mem:* Am Soc Cell Biol; AAAS; Am Soc Plant Physiologists; Asn Women Sci. *Res:* Molecular and biochemical analysis of plasma membrane and microtubule associated proteins in Arabidopsis thaliana; control of cell wall microfibril patterning by the cortical microtubule array. *Mailing Add:* Dept Biol Sci Wellesley Col Wellesley MA 02181

MELANCON, MARK J, b Chicago, Ill, Sept 17, 39; m 66; c 3. BIOINDICATORS, BIOMONITORING. *Educ:* St Mary's Col, BA, 61; Loyola Univ, Chicago, MS, 64, PhD(biochem), 66. *Prof Exp:* Instr biochem, Loyola Univ, 66; fel & res assoc biochem, Univ Wis, Madison, 66-70, asst prof pharm chem, Milwaukee, 70-74, res assoc, Great Lakes Res, 78-87; RES CHEM, US FISH & WILDLIFE SERV, PATUXENT WILDLIFE RES CTR, 87- , GROUP LEADER, RISK ASSESSMENT, 88- *Concurrent Pos:* Prin investr res grants, 73- & co-investr, 75-; adj asst prof pharmacol & toxicol, Med Col Wis, 80-87; assoc scientist, Mt Sinai Med Ctr, Med Sch, Univ Wis, 83-86. *Mem:* Am Chem Soc; Am Soc Exp Pharmacol & Therapeut; Soc Environ Toxicol & Chem; AAAS; Sigma Xi. *Res:* Uptake, distribution and elimination of drugs and pollutants; effects of environmental pollution on fish, birds and wildlife; analysis of biological samples for pollutants and their metabolites and drugs and their metabolites, mainly by high-performance liquid chromatography, biochemical indicators of pollution in birds and wildlife. *Mailing Add:* Patuxent Wildlife Res Ctr Laurel MD 20708

MELANDER, WAYNE RUSSELL, b Watertown, SDak, Dec 2, 43; m 84; c 6. SEPARATION SCIENCE, ANALYTICAL CHEMISTRY & BIOCHEMISTRY. *Educ:* Mich State Univ, BS, 65; Cornell Univ, PhD(phys chem), 70. *Prof Exp:* Asst prof biochem, Univ Wyo, 70-75; res assoc chromatography, Dept Eng & Appl Sci, Yale Univ, 75-77; chemist enzymol, Clinton Corn Processing Co, Iowa, 77; res assoc chromatography, Eng & Appl Sci Div, Yale Univ, 78-81, res assoc & lectr, Chem Eng Dept, 81-85; chemist, Shell Develop Co, Modesto, Calif, 85-86; chemist, E I du Pont de Nemours & Co, 86-90, PRIN CHEMIST, DUPONT MERCK PHARMACEUT CO, 91- *Mem:* AAAS; Am Chem Soc; Sigma Xi. *Res:* Chromatography, specifically theory of solvent effects in reversed-phase liquid chromatography and physical chemistry of separations; development of novel stationary phases and interpretation of novel separations of biochemicals; pharmaceutical analyses. *Mailing Add:* 3 Cannoneer Circle Chadds Ford PA 19317-9426

MELA-RIKER, LEENA MARJA, b Viiala, Finland, Apr 5, 35; m 83. PHYSIOLOGY, BIOCHEMISTRY. *Educ:* Univ Helsinki, BSc, 60; Turku Univ, MD, 64. *Prof Exp:* USPHS fel, Johnson Res Found, Univ Pa, 65-69, res assoc, Harrison Dept Surg Res, 69-70, from asst prof to assoc prof phys biochem, 70-80; prof physiol & surg, Mich State Univ, 80-83; PROF SURG & BIOCHEM, ORE HEALTH SCI UNIV, 83- *Concurrent Pos:* NIH career develop award, 71-76; mem prog proj adv comt, Nat Inst Neurol & Commun Dis & Stroke, 74-78 & 79-83. *Mem:* AAAS; Am Soc Biol Chemists; Am Soc Neurochem; Am Phsiol Soc; Shock Soc (secy, 80-85). *Res:* Mitochondrial metabolism during stroke, shock and sepsis; structural and functional damage of the cell after injury; cellular calcium homeostasis; cation transport across cell and subcellular membranes. *Mailing Add:* Dept Surgery Sch Med Oregon Health Sci Univ 3181 SW Sam Jackson Portland OR 97201

MELBY, EDWARD C, JR, b Burlington, Vt, Aug 10, 29; m 53; c 4. VETERINARY MEDICINE. *Educ:* Cornell Univ, DVM, 54; Am Col Lab Animal Med, dipl, 67. *Prof Exp:* Pvt pract, Vt, 54-62; from instr to prof & dir dept, Lab Animal Med, Sch Med, Johns Hopkins Univ, 62-74; dean & prof med, NY State Col Vet Med, Cornell Univ, 74-87; AT SMITH KLINE ANIMAL HEALTH PROD, PHILADELPHIA, 88- *Concurrent Pos:* Consult, Vet Admin, 64-74; mem, White House Conf on Health, 65; mem coun accreditation, Am Asn Accreditation Lab Animal Care, 66-73; mem, Inst Lab Animal Resources, Nat Res Coun-Nat Acad Sci, 75-; mem adv comt, Div Res Resources, NIH; consult, Nat Inst Child Health & Human Develop, 71-; pres, Am Col Lab Animal Med, 74-75. *Mem:* AAAS; Am Asn Lab Animal Sci; Am Vet Med Asn; NY Acad Sci; Asn Biomed Res (pres, 80-); Sigma Xi. *Res:* Laboratory animal medicine and comparative pathology, especially lymphoproliferative diseases and transplantation. *Mailing Add:* Smith Kline Animal Health Prod One Franklin Plaza PO Box 7929 Philadelphia PA 19101

MELBY, JAMES CHRISTIAN, b Duluth, Minn, Feb 14, 28; m 55; c 2. MEDICINE. *Educ:* Univ Minn, BS, 51, MD, 53. *Prof Exp:* Lectr endocrine biochem, Med Sch, Univ Minn, 58, instr med & dir clin chem, 58-59; asst prof med & biochem, Sch Med, Univ Ark, 59-62; assoc prof, 62-69, PROF MED, SCH MED, BOSTON UNIV, 69-, PROF PHYSIOL, 71-, MEM FAC, DIV MED SCI, GRAD SCH, 71- *Concurrent Pos:* Consult, Merck & Co, 55-65; head sect endocrinol, Evans Mem Hosp, 62- *Mem:* AAAS; Am Soc Clin Invest; Asn Am Physicians; Am Chem Soc; Endocrine Soc; Sigma Xi. *Res:* Metabolism of steroid hormones, physical interaction of steroid hormones and macromolecules; endocrinology; internal medicine. *Mailing Add:* Med Dept Boston Univ Med Ctr 80 E Concord St Boston MA 02118

MELCHER, ANTONY HENRY, Can Citizen; m 53; c 2. HISTOLOGY, CELL BIOLOGY. *Educ:* Univ Witwatersrand, BDS, 49, HDD, 58, MDS, 60; Univ London, PhD(morphol), 64, DSc, 85. *Prof Exp:* Chmn, grad dept dent, Univ Toronto, 73-74, dir, Group Periodont Physiol, Med Res Coun Can, 73-83, assoc dean, Sch Grad Studies, 84-88, PROF DENT, FAC DENT, UNIV TORONTO, 69-, VPROVOST, 88- *Concurrent Pos:* Leverhulme Found res fel, Royal Col Surgeons Eng, 64-67, col res fel morphol, 67-69; assoc ed, J Gerodontics; assoc ed, Int J Dent Implantology, 85-; counr & exec comt, Med Res Coun Can; mem sci adv comt, Alta Heritage Fund Med Res. *Mem:* Int Asn Dent Res (pres, 82-83); Am Soc Cell Biol; hon mem Am Acad Periodont. *Res:* Repair of bone; dental implants; structure and function of periodontium. *Mailing Add:* Off Vpres & Provost Univ Toronto Simcoe Hall Toronto ON M5S 1A1 Can

MELCHER, CHARLES L, b Ft Worth, Tex, May 25, 52; m 75; c 3. SCINTILLATOR DETECTORS, FLUORESCENCE. *Educ:* Rice Univ, BA, 74; Wash Univ, MA, 76, PhD(physics), 80. *Prof Exp:* Postdoctoral, Calif Inst Technol, 80-83; RES SCIENTIST, SCHLUMBERGER-DOLL RES, 83- *Mem:* Am Phys Soc; Inst Elec & Electronic Engrs; Am Asn Crystal Growers; Mat Res Soc; Meteorol Soc. *Res:* Investigation of scintillation, fluorescence, and luminescence of inorganic crystalline materials; interaction of ionizing radiation with solid materials; crystal growth of novel inorganic materials. *Mailing Add:* Schlumberger-Doll Res Old Quarry Rd Ridgefield CT 06877-4108

MELCHER, JAMES RUSSELL, electrical engineering, continuum electromechanics; deceased, see previous edition for last biography

MELCHER, ROBERT LEE, b Marshalltown, Iowa, Jan 27, 40; m 66; c 3. SOLID STATE PHYSICS. *Educ:* Southern Methodist Univ, BS, 62; Wash Univ, MA, 65, PhD(physics), 68. *Prof Exp:* Res assoc physics, Cornell Univ, 68-70; MEM RES STAFF, WATSON RES CTR, IBM CORP, 70- *Mem:* Fel Am Phys Soc; Inst Elec & Electronics Engrs. *Res:* Elastic and magnetoelastic properties of materials; nuclear spin-phonon interactions; elastic properties of materials undergoing phase transitions; polarization echoes in piezoelectric and magnetoelastic materials, echo holography in piezoelectric semiconductors; low temperature physics; photoacoustic and pyroelectric phenomena; semiconductor laser technology. *Mailing Add:* IBM Watson Res Ctr PO Box 218 Yorktown Heights NY 10598

MELCHER, ULRICH KARL, b London, Eng, July 7, 45; US citizen; m 68; c 2. MOLECULAR BIOLOGY. *Educ:* Univ Chicago, BS, 65; Mich State Univ, PhD(biochem), 70. *Prof Exp:* Fel molecular biol, Univ Aarhus, Denmark, 70-71; res scientist immunol, Med Ctr, NY Univ, 72; res asst microbiol, Univ Tex Health Sci Ctr, Dallas, 72-74, asst prof, 74-75; from asst prof to assoc prof, 75-83, PROF BIOCHEM, OKLA STATE UNIV, 83- *Concurrent Pos:* NATO fel, 70-71 & NIH fel, 73-74; Fulbright scholar, 83. *Mem:* Am Soc Plant Physiologists; Am Soc Biochem Molecular Biol; Am Phytopath Soc; Am Soc Microbiol. *Res:* Genetic exchanges in plants; viral evolution; plant virology. *Mailing Add:* Dept of Biochem PS II Okla State Univ Stillwater OK 74078

MELCHING, J STANLEY, b New York, NY, Mar 4, 23; m 41; c 2. PLANT PATHOLOGY. *Educ:* Univ Maine, BS, 54, MS, 56; Cornell Univ, PhD(plant path), 61. *Prof Exp:* Res plant pathologist, Field Crops & Animal Prods Br, USDA, Watseka, Ill, 61-63 & Biol Br, Corps Div, US Army Biol Labs, Ft Detrick, Frederick, Md, 63-70; RES PLANT PATHOLOGIST, CROP RESPONSE FOREIGN DIS RES UNIT, PLANT DIS RES LAB, SCI & EDUC ADMIN-AGR RES, USDA, 70- *Mem:* Am Phytopath Soc; Am Soc Plant Physiologists; Am Inst Biol Sci; Sigma Xi. *Res:* Disease dynamics of the rusts of corn and the soybean rust; development of quantitative techniques for measurement of the infection process, disease spread and special instrumentation for assessment of environmental factors. *Mailing Add:* 1608 Rock Creek Dr Apt Eight Frederick MD 21701

MELCHIOR, DONALD L, b June 19, 45. PHYSICAL PROPERTIES OF LIPIDS, CALORIMETRY. *Educ:* Antioch Col, BS, 67; Yale Univ, MPhil, 69, PhD(molecular biophysics & biochem), 72. *Prof Exp:* Res assoc chem dept, Brown Univ, 72-78, asst res prof chem, 78-80, asst res prof chem & med scis, 80-81; ASSOC PROF BIOCHEM, SCH MED, UNIV MASS, 81- *Concurrent Pos:* Vis researcher, Nat Ctr Sci Res, Gif-sur-Yvette, France, 74. *Res:* Regulation of biological processes by lipids; membrane structure; standard biochemical and biophysical techniques, specialities include differential scanning, calorimetry and stopped-flow spectrophotometry. *Mailing Add:* Biochem Dept Sch Med Univ Mass 55 Lake Ave N Worcester MA 01665

MELCHIOR, JACKLYN BUTLER, b Sacramento, Calif, May 19, 18; m 39; c 2. BIOCHEMISTRY. *Educ:* Univ Calif, BS, 40, PhD(biochem), 46. *Prof Exp:* Res assoc chem, Northwestern Univ, 46-49; from instr to asst prof biochem, Sch Med, Loyola Univ, Ill, 49-57, assoc prof pharmacol, 57-59; asst dean, Chicago Col Osteop Med, 59-77, prof biochem & chmn, Dept Basic Sci, 59-80, dean, 77-80; RETIRED. *Concurrent Pos:* Lederle Med Fac Award, 57. *Mem:* Am Chem Soc; Am Soc Biol Chem. *Res:* Enzyme chemistry; protein conformation. *Mailing Add:* 2601 Col Ave No 212 Berkeley CA 94704

MELCHIOR, ROBERT CHARLES, b Fargo, NDak, July 22, 33; m 58; c 3. PALEOBOTANY, STRATIGRAPHY. *Educ:* Moorhead State Col, BS, 58; Univ Minn, MS, 60, PhD(paleobot), 65. *Prof Exp:* PROF BIOL, BEMIDJI STATE UNIV, 64- *Concurrent Pos:* Res assoc, Sci Mus of Minn, 74- *Mem:* Soc Econ Paleontologists & Mineralogists; Am Bot Soc; Sigma Xi; Am Asn Stratig Palynologists. *Res:* Paleoecology of the Paleocene, Glacial and Pleistocene stratigraphy of northern Minnesota. *Mailing Add:* Div Sci & Math Bemidji State Univ Bemidji MN 56601

MELDAU, R(OBERT) F(REDERICK), b Pasadena, Calif, Feb 6, 29; m 51; c 3. THERMAL OIL RECOVERY, RESERVOIR ENGINEERING. *Educ:* Univ Calif, BS, 50; Calif Inst Technol, MS, 55. *Prof Exp:* Res engr, La Habra Lab, Chevron Res Corp, Standard Oil Co Calif, 55-61, sec recovery engr, 61-64; res eng, Phillips Petrol Co, Okla, 64-69, sect mgr pilot projs, 69-70, chief engr, Venezuela, 70-73, staff dir improved oil recovery, 73-76; mgr enhanced oil recovery, Husky Oil Co, 76-78; ENG CONSULT, 78- *Mem:* Soc Petrol Engrs. *Res:* Enhanced recovery of crude oil by steam flooding, underground combustion and chemical flooding. *Mailing Add:* Dolphin Consults 174 Tally Ho Santa Maria CA 93455

MELDNER, HEINER WALTER, b Koenigsberg/Prussia, Ger, May 21, 39; US citizen; m 76; c 2. PHYSICS. *Educ:* Univ Frankfurt, PhD(theoret physics), 65. *Prof Exp:* NATO fel theoret physics, Lawrence Radiation Lab, Univ Calif, Berkeley, 65-67; pvt univ lectr, Univ Berlin, 67-69; assoc prof, Univ Calif, San Diego, 69-75; PHYSICIST NUCLEAR WEAPONS DESIGN, LAWRENCE LIVERMORE NAT LAB, UNIV CALIF, 76- *Concurrent Pos:* Pres, Cordtran Corp, Moss Beach, Calif, 79-; tech consult. *Mem:* Am Phys Soc. *Res:* Inertial confinement fusion and fission; nuclear theory; computer modeling of complex physical processes. *Mailing Add:* Lawrence Livermore Nat Lab L-18 PO Box 808 Livermore CA 94550

MELDON, JERRY HARRIS, b New York, NY, Oct 23, 47. MASS TRANSFER, MEMBRANE PROCESSES. *Educ:* Cooper Union, BE, 68; Mass Inst Technol, PhD(chem eng), 73. *Prof Exp:* Res fel physiol, Odense Univ, Denmark, 73-77; ASST PROF CHEM ENG, TUFTS UNIV, 77- *Concurrent Pos:* Consult, Prototech Co, Newton Highlands, Mass, 80, US Army Mat & Mech Res Ctr, Watertown, Mass, 82. *Mem:* Inst Chem Engrs;

AAAS; Sigma Xi. *Res:* Analysis of mass transfer and separation processes; diffusion and chemical reaction in absorber and membrane systems; physiological transport processes; electrochemical systems; polymer membrane permeation; blood gas chemistry. *Mailing Add:* Dept Chem Eng Tufts Univ Medford MA 02155

MELDRIM, JOHN WALDO, b Glendale, Calif, Sept 29, 41; m 64; c 3. FISH BIOLOGY, ICHTHYOLOGY. *Educ:* Occidental Col, BA, 63; Univ Wash, PhD(fisheries), 68. *Prof Exp:* Sr res biologist, 68-76, tech dir exp studies, Ichthyol Assocs, Inc, 76-85; sr aquatic biol, 84-86, HEAD, AQUATIC RESOURCES & FISHERIES, HARZA ENG CO, 86- *Mem:* Am Fisheries Soc; Sigma Xi; Water Pollution Control Fedn; Estuarine Res Fedn. *Res:* Tolerance and behavioral responses of estuarine, freshwater and marine fishes and motile macroinvertebrates to temperature and to chemicals; fish passage and hydroelectric projects. *Mailing Add:* 192 Newton Ave Glen Ellyn IL 60137

MELDRUM, ALAN HAYWARD, b Lethbridge, Alta, Can, May 24, 13; nat US; m 47. INDUSTRIAL ENGINEERING. *Educ:* Univ Alta, BS, 38; Univ Okla, BS, 46, MS, 49; Pa State Univ, PhD(petrol eng), 54. *Prof Exp:* Coke oven chemist, Algoma Steel Corp, Can, 38-44; ref chemist, Brit Am Oil Ref Co, Ltd, 44-45; res assoc, Pa State Univ, 47-54; from assoc prof to prof petrol eng, 54-61, prof indust eng, 62-80, EMER PROF INDUST ENG, UNIV NDAK, 80- *Mem:* Nat Soc Prof Engrs. *Res:* Three phase equilibria for carbon dioxide-hydrocarbon mixtures; petroleum consulting; industrial consulting for plant layout, organization and wage administration. *Mailing Add:* 512 Columbia Rd N Grand Forks ND 58203

MELE, FRANK MICHAEL, b Englewood, NJ, Aug 15, 35; m 61; c 2. EDUCATION ADMINISTRATION, SCIENCE EDUCATION. *Educ:* Davis & Elkins Col, BS, 58; Montclair State Col, MA, 64; Fordham Univ, EdD(educ), 78. *Prof Exp:* Prof, 64-80, CHMN DEPT BIOL, JERSEY CITY STATE COL, 80- *Concurrent Pos:* Lectr, Fairleigh Dickinson Univ, 69-70, Englewood Cliffs Col, 70-74 & Sch Nursing, St Francis Hosp, 76-; consult, Nat Sci Found, 78. *Mem:* Nat Sci Teachers Asn; AAAS; Soc Col Sci Teachers. *Res:* Piagetian Theory; biology and physical science. *Mailing Add:* Dept Biol Jersey City State Col 2039 Kennedy Blvd Jersey City NJ 07305

MELECA, C BENJAMIN, b Batavia, NY, Nov 8, 37; m 58; c 4. MEDICAL EDUCATION. *Educ:* State Univ NY Col Brockport, BS, 63; Syracuse Univ, MS, 66, PhD(sci educ), 68. *Prof Exp:* High sch teacher, NY, 63-65; instr biol, Syracuse Univ, 67-68; asst dir, Biol Core Prog, 68-70, from asst prof to assoc prof biochem, 68-74, dir introductory biol prog, 70-74, assoc prof prev med, 74-78, PROF, DEPT FAMILY MED, OHIO STATE UNIV, 78-, DIR DIV RES & EVAL IN MED EDUC, COL MED, 74-, PROF, DEPT INTERNAL MED, 85- *Concurrent Pos:* Pres, Int Cong Individualized Audio Instrn, 70-72; chmn, Audio-Tutorial Conf, 71, chmn conf comt, 71; chmn, Promotion & Tenure Comt, Dept Family Med, 80-83, 89-; mem, Nat Adv Coun, Am Podiat Asn, 79-82; co-prin investr, & proj dir, Cancer Educ Prog, 85- *Mem:* Am Educ Res Asn; Am Chem Soc; Eastern Educ Res Asn; Am Asn Med Cols. *Res:* Medical education; biological sciences education; independent study and computer based education; pre and post MD education; research, evaluation, and development in undergraduate medical education; instructional study of clinical teaching skills and strategies; development; academic and curricular affairs. *Mailing Add:* 076 Health Sci Libr Ohio State Univ Col of Med 376 W Tenth Ave Columbus OH 43210

MELECHEN, NORMAN EDWARD, b New York, NY, Jan 26, 24; m 74; c 3. GENETICS. *Educ:* Columbia Univ, AB, 44; Univ Pa, PhD(zool), 54. *Prof Exp:* Spec investr genetics, Carnegie Inst Wash, 54-56; instr, 56-57, sr instr, 57-58, from asst prof to assoc prof, 58-64, PROF MICROBIOL, SCH MED, ST LOUIS UNIV, 64- *Concurrent Pos:* Commonwealth Fund fel, Stanford Univ, 66-67. *Mem:* Genetics Soc Am; Am Soc Microbiol. *Res:* Genetics of the response to environmental stress; genetics and chemistry of bacteriophage infection and induction. *Mailing Add:* Dept of Microbiol St Louis Univ Sch of Med St Louis MO 63104

MELEHY, MAHMOUD AHMED, b Egypt, June 10, 26; nat US; m 57; c 3. ELECTRICAL ENGINEERING. *Educ:* Cairo Univ, BS, 47; Ohio State Univ, MS, 49; Univ Ill, MS, 50, PhD(elec eng), 52. *Prof Exp:* Asst prof elec eng, Univ Ark, 52-53; lectr, Ain Shams Univ, Cairo, 53-54; assoc prof, Univ Alaska, 54-55; asst prof, Mich State Univ, 55-58; from asst prof to assoc prof, 58-71, PROF ELEC ENG, UNIV CONN, 71- *Concurrent Pos:* Consult, Shockley Transistor, 60 & Elec Boat Div, Gen Dynamics Corp, 64- *Mem:* AAAS; Am Phys Soc; Inst Elec & Electronics Engrs; Sigma Xi. *Res:* Consequences of new thermodynamic theory, especially unified theories for p-n junctions, heterojunctions, Schottky diodes, thermoelectricity, superconductivity and superfluidity. *Mailing Add:* Dept of Elec Eng Univ of Conn Storrs CT 06268

MELENDRES, CARLOS ARCIAGA, b Manila, Philippines, Nov 4, 39; US citizen; m 63; c 3. ELECTROCHEMISTRY, MATERIALS CHEMISTRY. *Educ:* Mapua Inst Technol, Manila, BS, 61; Univ Calif, Berkeley, MS, 66, PhD(phys chem), 68. *Prof Exp:* Sr scientist physics, Philippine Atomic Res Ctr, 68-71; lectr physics, Univ Philippines, 70-71; postdoctoral fel phys & electrochem, Inst Phys & Electrochem, Univ Karlsruhe, Ger, 71-72; sr develop engr electrochem eng, Olin Corp Res Ctr, 72-74; SR SCI STAFF MEM PHYS ELECTROCHEM, UNIV CHICAGO, ARGONNE, 74- *Concurrent Pos:* Tech consult, United Nations Develop Prog, 82-; dir, Advan Study Inst Interfacial Spectroelectrochem, Spain, NATO, 88 & Advan Study Inst Corrosion, Viana do Castelo, Portugal, 89. *Mem:* Electrochem Soc; Soc Electroanal Chem; Sigma Xi (treas, 90-91); Philippie Am Acad Sci & Eng. *Res:* Development and application of in-situ spectroscopic and electrochemical techniques to study the structure of solid-liquid interfaces; structure and transport properties of electrochemically formed surface films on metals; author of over 80 publications. *Mailing Add:* Argonne Nat Lab Argonne IL 60439

MELERA, PETER WILLIAM, b Union City, NJ, Feb 19, 42; m 63; c 2. CELL BIOLOGY. *Educ:* Univ Ga, BS, 65, PhD(bot), 69. *Prof Exp:* Asst prof biochem, Cornell Grad Med Sch, 76-78; res assoc, Walker Lab, Sloan-Kettering Inst Cancer Res, 72-75, assoc, 75-87, lab head, 76-88, assoc mem, 87-88; PROF BIOCHEM & ONCOL, SCH MED, UNIV MD, BALTIMORE, 88-, DIR, MOLECULAR & CELL BIOL GRAD PROG, 89- *Concurrent Pos:* NIH fel, McArdle Lab Cancer Res, Univ Wis-Madison, 69-72; mem exp therapeut study sect, NIH, 87-; mem, NSF Cell Biol Adv Panel, 85-86. *Mem:* Am Chem Soc; Am Soc Cell Biol. *Res:* Molecular biology and genetics of anti-tumor drug resistance; mechanisms of somatic cell gene amplification in mammalian cells; recombinent DNA technology; molecular biology. *Mailing Add:* Dept Biochem Univ Med Sch Med 660 W Redwood St Baltimore MD 21201

MELESE, GILBERT B(ERNARD), b Paris, France, Oct 17, 26; m 54, 86; c 4. NUCLEAR ENGINEERING. *Educ:* Univ Paris, AeroEng, 49; Johns Hopkins Univ, PhD(fluid mech), 54. *Prof Exp:* Asst aeronaut, Johns Hopkins Univ, 50-51, 52-54; group leader, Saclay Nuclear Res Ctr, France, 54-57; from asst prof to assoc prof mech eng, Columbia Univ, 57-60; mem res staff, Gen Atomic Div, Gen Dynamics Corp, 60-67; sr tech adv, Gen Atomic Co, 67-82; sci attache, State Dept, Washington DC, 84-87; LECTR NUCLEAR ENG, UNIV CALIF, BERKELEY, 88- *Concurrent Pos:* Consult, Repub Aviation Corp, 58-60; adj assoc prof, Columbia Univ, 60-61; lectr, exten, Univ Calif, San Diego, 60-62; vis prof, MIT, 83, Naval Postgrad Sch, 83-84 & Univ I11 Urbana, 84. *Mem:* AAAS; fel Am Nuclear Soc; Am Soc Mech Engrs; assoc fel Am Inst Aeronaut & Astronaut; Sigma Xi. *Res:* Heat transfer in nuclear reactors; conduction heat transfer; design of thermal and fast gas-cooled power reactors. *Mailing Add:* 15 Heather Ct Seaside CA 93955

MELFI, LEONARD THEODORE, JR, b Charleston, SC, Sept 8, 37; m 60; c 2. PHYSICS, ATMOSPHERIC SCIENCES. *Educ:* The Citadel, BS, 59; Col William & Mary, MA, 64; Fla State Univ, PhD(meteorol), 71. *Prof Exp:* Aerospace technologist & physicist, Langley Res Ctr, NASA, 59-89; RES PROF, OLD DOMINION UNIV, NORFOLK, VA, 90- *Mem:* Sigma Xi. *Res:* Atmospheric physics as applied to the terrestrial thermosphere; materials science for space application; vacuum instrumentation and technology, particularly mass spectrometry. *Mailing Add:* 108 Carrs Hill Rd Williamsburg VA 23185

MELFORD, SARA STECK, b Greenville, Ohio, Feb 3, 42; m 71. INORGANIC CHEMISTRY, MASS SPECTROMETRY. *Educ:* Bowling Green State Univ, BS, 64; Northwestern Univ, PhD(chem), 68. *Prof Exp:* NSF fel chem, Rice Univ, 68-70; asst prof, 70-81, ASSOC PROF CHEM, DEPAUL UNIV, 81- *Mem:* Am Chem Soc; Am Asn Mass Spectrometry; AAAS. *Res:* High temperature inorganic chemistry, mass spectrometry and solid waste chemistry. *Mailing Add:* Dept of Chem DePaul Univ Chicago IL 60614

MELGARD, RODNEY, b Carrington, NDak, Feb 24, 36; m 60; c 3. RADIOCHEMISTRY, NUCLEAR PHYSICS. *Educ:* Jamestown Col, BS, 57. *Prof Exp:* Actinide element group leader, Tracerlab Inc, 58-61; supvr radioactivity measurement, 61-73, mgr lab opers, 74-79, GEN MGR, ENVIRON DIV, FLE CORP, RICHMOND, 80- *Mem:* Am Chem Soc; Am Inst Chemists; Am Nuclear Soc; Health Physics Soc. *Res:* High sensitivity studies of natural and man-made radiation related to environmental surveillance programs; geothermal radiochemistry; reactor coolant chemistry; nuclear rocket ablation studies; snap devices; alpha and gamma spectroscopy; tranuranic translocation and uptake studies. *Mailing Add:* 8205 Cherry Hills Dr NE Albuquerque NM 87111

MELGES, FREDERICK TOWNE, psychiatry, experimental psychology; deceased, see previous edition for last biography

MELHADO, L(OUISA) LEE, b Bryn Mawr, Pa, July 9, 45; m 71; c 2. PHOTOAFFINITY LABELING, RADIOLABELING. *Educ:* Carnegie-Mellon Univ, BA, 68; Wash Univ, St Louis, MA, 72, PhD(org chem), 76. *Prof Exp:* Vis lectr & res assoc chem & biochem, 76-82, vis asst prof & res assoc chem, 82-84, DIR RADIOISOTOPE LAB, UNIV ILL, URBANA, 84- *Mem:* Am Chem Soc; Am Soc Biol Chemists; Sigma Xi. *Res:* Micellar catalysis; aromatic azides; mechanism of action of plant hormones, especially auxins, using the techniques of photoaffinity labeling, radiolabeling, and fluorescence spectroscopy. *Mailing Add:* 612 W Ohio St Urbana IL 61801

MELHORN, WILTON NEWTON, b Sistersville, WVa, July 8, 20; m 61; c 2. GEOMORPHOLOGY, QUATERNARY GEOLOGY. *Educ:* Mich State Univ, BS, 42, MS, 51; NY Univ, MS, 43; Univ Mich, PhD(geol), 55. *Prof Exp:* Hydrogeologist, Geol Surv, Mich, 46-49; meteorologist, US Weather Bur, 49-50; from asst prof to assoc prof eng geol, 55-67, head dept geosci, 67-71, PROF ENG GEOL, PURDUE UNIV, WEST LAFAYETTE, 67- *Concurrent Pos:* Vis prof, Univ Ill, 60-61; vis prof, Univ Nev, Reno, 71-72, adj prof, Mackay Sch Mines, 73-82. *Mem:* Fel AAAS; fel Geol Soc Am; Am Meteorol Soc; Soc Econ Paleontologists & Mineralogists; Am Asn Petrol Geologists. *Res:* Geomorphology and Pleistocene geology; archaeological geology; remote sensing; geomorphology of arid lands. *Mailing Add:* Dept Earth & Atmospheric Sci Purdue Univ West Lafayette IN 47907

MELI, ALBERTO L G, b Florence, Italy, Sept 19, 21; US citizen; m 55; c 1. PHYSIOLOGY. *Educ:* Univ Pisa, DVM, 47; Univ Milan, PhD(physiol), 58, PhD(pharmacol), 69. *Prof Exp:* Lectr physiol, Univ Pisa, 47-49; group leader pharmacol, C Erba Res Inst, Italy, 50-57; group leader endocrinol, Vister Labs, 58-59; group leader pharmacol, Sterling-Winthrop Res Inst, NY, 60-61; sr res assoc physiol, Warner-Lambert Res Inst, NJ, 61-67; dir pharmacol dept, Vister Labs, Italy, 67-68, dir biol res, 68-69; DIR RES, MANARINI LABS, 69- *Mem:* AAAS; Soc Study Reproduction; Endocrine Soc; NY Acad Sci; Soc Exp Biol & Med; Am Physiol Soc; Am Pharmacol Soc. *Res:* General endocrinology and pharmacology; reproductive physiology; metabolism. *Mailing Add:* Lab Chimico Farmaceutico A Menarini Via Sette Santi 1 Florence 50131 Italy

MELIA, FULVIO, b Gorizia, Italy, Aug 2, 56; m 80; c 3. HIGH-ENERGY ASTROPHYSICS, THEORETICAL ASTROPHYSICS. *Educ:* Melbourne Univ, Australia, BSc, 78, MSc, 80; State Univ NY, Stony Brook, MA, 80; Mass Inst Technol, PhD(physics), 85. *Prof Exp:* Postdoctoral fel, Enrico Fermi Inst, Univ Chicago, 85-87; from asst prof to assoc prof physics, Northwestern Univ, 87-91; ASSOC PROF PHYSICS, UNIV ARIZ, 91- *Concurrent Pos:* NSF presidential young investr, 88; Alfred P Sloan researcher, 89; consult, Space Sta Rev Panel, Space Grant Col Prog & Gamma Ray Astron Panel, NASA, 89 & Extragalactic Astron Prog, NSF, 90; mem, High-Energy Astrophys Working Group, NASA, 91-93. *Mem:* Int Astron Union; Am Astron Soc; Am Phys Soc; Am Asn Univ Professors; NY Acad Sci; Sigma Xi. *Res:* Theoretical modeling of the physical processes associated with the interaction of matter and radiation in superstrong gravitational and magnetic fields and the interpretation of relevant astrophysical data. *Mailing Add:* Dept Physics Univ Ariz Tucson AZ 85721

MELIA, MICHAEL BRENDAN, b Blackpool, Eng, Jan 20, 49; US citizen; m 70; c 2. PALYNOLOGY. *Educ:* Northwestern State Univ, BS, 73; State Univ NY, Oneonta, MA, 75; Mich State Univ, PhD(geol), 80. *Prof Exp:* RES SPECIALIST, EXXON CO, 79- *Mem:* Am Asn Stratigraphic Palynologists. *Res:* Distribution of palynomorphs in aerosols and deep-sea sediments off the coast of Northwest Africa; palynological age determination of rocks drilled by oil companies. *Mailing Add:* Exxon Prod Res Co PO Box 2189 Houston TX 77252-2189

MELICH, MICHAEL EDWARD, b Moab, Utah, Feb 22, 40; m 70. SPACE SYSTEMS ENGINEERING. *Educ:* Stanford Univ, BS, 61; Univ Utah, MS, 63; Rice Univ, MA, 66, PhD(physics-theoret), 67. *Prof Exp:* Opers analyst, Oper Eval Group, Ctr Naval Anal, Univ Rochester, 67-76; dir, Command & Control Res Br, Naval Res Lab, 76-86; chair combat systs eng, 85-90, PROF DEPT PHYSICS, NAVAL POSTGRAD SCH, 90- *Concurrent Pos:* Consult, Naval Ctr Space Technol, 85-, Packard Comn Defense Reorganization, 86, Smithsonian Inst, 86- & USA-Japan Forum, 88- *Mem:* Am Phys Soc; Sigma Xi; AAAS; Am Soc Naval Engrs. *Res:* Proper application of information technology; modelling of complex physical systems. *Mailing Add:* 1224 Meigs Dr Niceville FL 32578

MELICK, WILLIAM F, urology; deceased, see previous edition for last biography

MELICKIAN, GARY EDWARD, b Los Angeles, Calif, Apr 2, 35; m 55; c 2. FORENSIC SERVICES, LITIGATION SUPPORT. *Educ:* Colo Sch Mines, Golden, BS, 59; Univ S Calif, Los Angeles, MS, 67. *Prof Exp:* Geologist, Humble Oil & Refining Co, 59; asst civil engr, Los Angeles Co Flood Control Dist, 59-60; staff geophysist, Dames & Moore, 60-61, proj geologist, 62-64, personnel mgr, 65-66, pub rel mgr, 67-69, dir mining, 70-80 & dir tech serv, 80-84; pres, Consult Networks, 85-90; DIR, INDUST MKT, AM GAS ASN, 90- *Concurrent Pos:* US deleg, Int Geol Cong, 68; rep, Am Geol Inst, 69-71; pres, Am Inst Prof Geologists, Calif Sect, 71, nat dir, 72-73; nat dir, Soc Mining Engrs, 72-74, chmn, Peele Award Comt & Publ & Prog Comt. *Mem:* Fel Geol Soc Am; Am Inst Prof Geologists (secy-treas, 82); Soc Mining Engrs; Asn Eng Geologists; Am Asn Petrol Geologists; Nat Soc Prof Engrs. *Res:* Arid land development; desert geomorphology; natural hazards including earthquakes, subsidence, volcanism, flooding, severe weather conditions, collapsing soils, brush fire control, coastal erosion and cliff retreat, landslides, erosion and mass wasting. *Mailing Add:* 3711 Windom Pl NW Washington DC 20016

MELILLO, DAVID GREGORY, b Newark, NJ, Dec 3, 47; m 71. ORGANIC CHEMISTRY. *Educ:* Rutgers Univ, BA, 69; Mass Inst Technol, PhD(org chem), 73. *Prof Exp:* Sr res fel process res, 72-88, DIR LABELED COMPOUND SYNTHESIS, MERCK SHARP & DOHME RES LABS, 88- *Mem:* Am Chem Soc; Int Isotope Soc. *Res:* Discovery and development of commercially viable syntheses of biologically active organic compounds, including beta-lactam antibiotics, amino acids, steroids and heterocycles; synthesis of bioactive molecules with radioactive tracer elements. *Mailing Add:* 2637 Crest Lane Scotch Plains NJ 07076-1513

MELIN, BRIAN EDWARD, b Berwyn, Ill, Dec 10, 43; m 68; c 2. INSECT PATHOLOGY. *Educ:* Carthage Col, AB, 58; Bowling Green State Univ, AM, 69; Univ Ill, PhD(entom), 78. *Prof Exp:* Entomologist, Velsicol Chem Corp, 78-79; res entomologist, Crop Protection Div, Sandoz, Inc, 79-80; RES ENTOMOLOGIST, AGR CHEM DIV, ABBOTT LABS, 80- *Mem:* Soc Invertebrate Path; Entom Soc Am. *Res:* Microbial insect pathogens and the development of these agents into commercial insecticides and miticides. *Mailing Add:* Dept 91M 14th St Sherian Rd North Chicago IL 60064

MELINE, ROBERT S(VEN), chemical engineering; deceased, see previous edition for last biography

MELIS, ANASTASIOS, b Athens, Greece, Sept 10, 47; US citizen; m 76; c 3. MOLECULAR BIOLOGY. *Educ:* Univ Athens, Greece, BS, 70; Fla State Univ, PhD(molecular biophys), 75. *Prof Exp:* Researcher, Greek Atomic Energy Comn, 75-76, Weizmann Inst Sci, 76-77; Embo fel, Europ Molecular Biol Orgn, 77-79; Carnegie fel, Carnegie Inst, Stanford Univ, 79-81; from asst prof to assoc prof, 81-86, PROF PLANT BIOCHEM, UNIV CALIF, BERKELEY, 86- *Concurrent Pos:* Vis prof, Univ Lund, Sweden, 85, Univ Leeds, Eng, 88 & Plant Indust Div, Commonwealth Sci & Indust Res Orgn, 89. *Mem:* AAAS; Am Soc Photobiol; Am Soc Biochem & Molecular Biol; Am Soc Plant Physiologists. *Res:* Plant responses to the environment; chloroplast development and function in selective mutants and in wild type plants grown under physiological and adverse conditions. *Mailing Add:* Plant Biol Dept Univ Calif 411 Genetics & Plant Biol Bldg Berkeley CA 94720

MELISSINOS, ADRIAN CONSTANTIN, b Thessaloniki, Greece, July 28, 29; m 60; c 2. PARTICLE PHYSICS. *Educ:* Mass Inst Technol, MS, 56, PhD, 58. *Prof Exp:* Asst, Univ Athens, Greece, 54-55 & Mass Inst Technol, 55-58; from instr to assoc prof, 58-67, chmn dept physics & astron, 74-77, PROF PHYSICS, UNIV ROCHESTER, 67- *Concurrent Pos:* Guest physicist, Brookhaven Nat Lab, 63-; vis scientist, Europ Orgn Nuclear Res, 68-69, 77-78 & 89-90. *Mem:* Fel Am Phys Soc; Nat Acad Greece. *Res:* Experimental investigation of the interactions and properties of elementary particles; search for gravitational interactions at very high energies; novel methods for acceleration to very high energies. *Mailing Add:* Dept of Physics & Astron Univ of Rochester Rochester NY 14627

MELIUS, PAUL, b Livingston, Ill, Nov 21, 27; m 53; c 4. BIOCHEMISTRY. *Educ:* Bradley Univ, BS, 50; Univ Chicago, MS, 52; Loyola Univ, Ill, PhD(biochem), 56. *Prof Exp:* Chemist, Nat Aluminate Corp, 52-53; biochemist, Med Sch, Northwestern Univ, 56-57; assoc prof, 57-65, PROF BIOCHEM, AUBURN UNIV, 65- *Concurrent Pos:* NIH spec fel biochem, Univ Ky, 62 & Univ Calif, Los Angeles, 68; vis prof, Univ Athens, Greece & Univ Miami, 76; distinguished grad lectr, Auburn Univ, 83. *Mem:* Am Chem Soc; Brit Biochem Soc; Am Soc Biol Chemists. *Res:* Thermal polymerization of amino acids, pyridoxine analogs, platinum complexes and enzymes and protein chemistry; metabolism of polyaromatic hydrocarbons. *Mailing Add:* Dept Chem Auburn Univ Auburn AL 36849-5312

MELKANOFF, MICHEL ALLAN, b Russia, July 3, 23; nat US; wid; c 1. DATA BASES, COMPUTER-AIDED DESIGN. *Educ:* NY Univ, BS, 43; Univ Calif, Los Angeles, MA, 50, PhD(physics), 55. *Prof Exp:* Aeronaut engr, 43-44 & 46-47; vis asst prof physics, 56-58, from asst res physicist to assoc res physicist, 58-61, assoc prof eng, 62-66, chmn dept, 69-77, PROF ENG, UNIV CALIF, LOS ANGELES, 66-, DIR MFG ENG PROG, 81- *Concurrent Pos:* Vis physicist, Saclay Nuclear Res Ctr, France, 61-62; consult, Thompson Ramo Wooldridge, Inc, 59-60, Opers Res Ctr, France, 62, Mass Inst Technol, 63, Univ Hawaii, 64, Univ Md, 65, Douglas Aircraft Inc, 65, Rand Corp, 65-66, Nat Bank of Mex, 71, US Army Comput Syst Command, 71, IBM, IBM World Trade, Lockheed, Hughes, Nat Acad Eng, Northrop & US Air Force. *Mem:* Asn Comput Mach; Soc Mfg Engrs; Inst Elec & Electronics Engrs. *Res:* Digital computers; computer languages, compilers, design automated design; file organization; management; information systems; robotics; computer-aided manufacturing; simulation; graphics. *Mailing Add:* Dept Comput Sci Univ Calif Sch Eng & Appl Sci Boelter Hall 3532 Los Angeles CA 90024

MELKONIAN, EDWARD, b Alexandria, Egypt, June 29, 20; US citizen; m 54; c 3. PHYSICS. *Educ:* Columbia Univ, AB, 40, AM, 41, PhD(physics), 49. *Prof Exp:* Asst physics, Columbia Univ, 41-42; tech engr, Carbide & Carbon Chem Corp, Tenn, 45-46; res scientist, Atomic Energy Comn Contract, 46-63, assoc prof, 63-68, PROF NUCLEAR SCI & ENG, COLUMBIA UNIV, 68- *Concurrent Pos:* Res scientist, Nat Defense Res Comt, Off Sci Res & Develop & Manhattan Dist, Columbia Univ, 41-44; res assoc, Atomic Energy Res Estab, Harwell, Eng, 58-59. *Mem:* AAAS; fel Am Phys Soc; Am Nuclear Soc. *Res:* Porous membranes; gas flow; mathematical study of diffusion cascade; neutron spectroscopy of gases; neutron resonances and physics; fission and nuclear reactor physics; application of computers to scientific problems; nuclear engineering. *Mailing Add:* 35 Claremont Ave New York NY 10027

MELL, GALEN P, b Modesto, Calif, Sept 20, 34; m 61; c 3. BIOCHEMISTRY. *Educ:* Univ Idaho, BS, 56; Univ Wash, PhD(biochem), 61. *Prof Exp:* Res fel folic acid, Scripps Clin & Res Found, 64-68; from asst prof to assoc prof biochem, 68-80, Am Cancer Soc grant, 70-72, PROF BIOCHEM, UNIV MONT, 80- *Mem:* Am Chem Soc; AAAS. *Res:* Enzymology; clinical biochemistry; role of folic acid coenzymes in intermediary metabolism; proteases from plant tissue cultures. *Mailing Add:* Div Biol Sci Univ Mont Missoula MT 59812

MELLBERG, JAMES RICHARD, b Manitowac, Wis, June 3, 32; m 56; c 3. DENTAL CHEMISTRY, CHEMISTRY. *Educ:* Wis State Col, Oshkosh, BS, 55; Loyola Univ Chicago, MS, 60. *Prof Exp:* Res chemist med, Kendall Co, 58-68, head dent res, 68-75; from res assoc dent to assoc res fel, 75-89, RES FEL, COLGATE PALMOLIVE CO, PISCATAWAY, 89- *Concurrent Pos:* Consult, Great Lakes Naval Dent Res Inst, 72-; adj assoc prof, Univ Ill, 74-75. *Honors & Awards:* H Trendley Dean Award, Int Asn Dent Res, 88. *Mem:* Int Asn Dent Res; Am Chem Soc; Europ Asn Dent Res (ORCA). *Res:* Use of topical and systemic fluoride for dental cavities inhibition; dental calculus and plaque; dental abrasives; demineralization-remineralization. *Mailing Add:* Six Addison Dr Pottersville NJ 07979

MELLBERG, LEONARD EVERT, b Springfield, Mass, Dec 18, 35; m 87; c 2. ACOUSTICS. *Educ:* Univ Mass, BS, 61; Trinity Col, MS, 68. *Prof Exp:* Res physicist, Navy Underwater Sound Lab, Naval Underwater Systs Ctr, 61-68, NATO Saclant ASW Res Ctr, Italy, 68-72, res physicist, 72-91; SR SCIENTIST, MARINE ACOUSTICS INC, 91- *Mem:* Fel Acoust Soc Am; Am Geophys Union; Inst Elec & Electronic Engrs; Am Inst Aeronauts & Astronauts. *Res:* Theoretical and experimental studies of the relationships between underwater acoustic propagation and the environment; author of over 70 scientific publications. *Mailing Add:* 20 Willow Ave Middletown RI 02840

MELLEN, ROBERT HARRISON, b New Haven, Conn, Nov 12, 19; m 42; c 2. PHYSICS. *Educ:* Wesleyan Univ, BA, 41; Univ Conn, MA, 53, PhD, 55. *Prof Exp:* Res physicist, US Naval Res Lab, DC, 41-46, Underwater Sound Lab, Conn, 46-64; marine electronics off, Avco Corp, 64-67; mem staff, Naval Underwater Systs Ctr, New London, 67-81; res physicist, Mar Inc, East Lyme, Conn, 81-83 & BK Dynamics Inc, New London, Conn, 83-84; res physicist, Psi Marine Sci, New London, Conn, 84-87; RES PHYSICIST, KILDARE CORP, NEW LONDON, CONN, 87- *Mem:* Inst Elec & Electronics Eng; Am Phys Soc; Acoust Soc Am. *Res:* Underwater sound. *Mailing Add:* Kildare Corp 95 Trumbull St New London CT 06320

MELLEN, WALTER ROY, b Newark, NJ, Mar 10, 28; m 50; c 5. SCIENCE EDUCATION. *Educ:* Mass Inst Technol, SB, 48; Lowell Technol Inst, MS, 62. *Prof Exp:* Tech writer, Sperry Gyroscope Co, 50-53; instr physics, Adelphi Col, 53-55; tech writer, Sperry Gyroscope Co, 55-56; asst prof, Alfred Univ, 56-59; asst prof elec eng, 59-63, ASSOC PROF PHYSICS, LOWELL TECHNOL INST & UNIV LOWELL, 63- *Mem:* Am Asn Physics Teachers; Am Phys Soc. *Res:* Relativistic electromagnetic wave theory and its applications to standing waves, de Broglie waves, the Compton effect and the fine structure constant; electrostatic demonstrations; oscillations of a gas balloon due to a temperature gradient. *Mailing Add:* Dept of Physics Univ of Lowell Lowell MA 01854

MELLETT, JAMES SILVAN, b New York, NY, July 12, 36; m 61; c 2. GEOPHYSICS, PALEONTOLOGY. *Educ:* Iona Col, BS, 59; Columbia Univ, MA, 64, PhD(geol), 66. *Prof Exp:* Asst prof biol, Iona Col, 63-67; asst prof, 67-71, ASSOC PROF GEOL, NY UNIV, 71- *Mem:* Sigma Xi; AAAS; Soc Vert Paleont. *Res:* Paleobiology of fossil mammals; remote sensing and ground penetrating radar applications in geology; environmental geology. *Mailing Add:* Dept Biol 1009 Main NY Univ New York NY 10003

MELLETTE, RUSSELL RAMSEY, JR, b Orangeburg, SC, Dec 11, 27; m 51; c 5. CHILD PSYCHIATRY. *Educ:* Clemson Col, BS, 46; Med Col SC, MD, 50; Am Bd Psychiat & Neurol, dipl gen psychiat & child psychiat. *Prof Exp:* Intern, Wayne County Gen Hosp, Eloise, Mich, 50-51; resident psychiat, Edgewood Sanatorium, Orangeburg, SC, 51-52, clin dir, 52, med dir, 53; staff psychiatrist, US Army Hosp, Ft Sam Houston, Tex, 53-54; resident med neuropsychiat, Med Ctr Hosps, Charleston, SC, 55-56; from asst resident to resident & jr clin instr, Neuropsychiat Inst & Children's Psychiat Hosp, Sch Med, Univ Mich, Ann Arbor, 56-58, instr neuropsychiat, 58; from asst prof to assoc prof psychiat, 58-82, chief child psychiat sect, 63-73, assoc prof pediat, 67-82, assoc prof behav sci, 73-82, actg chief, Youth Serv Div, Dept Psychiat, 73-74, dir Youth Outpatient Serv & Pediat Liaison & Consult Serv, Dept Psychiat, 74-76, dean Div Continuing Educ, 79-84, COORDR, CONTINUING EDUC, DEPT PSYCHIAT, MED UNIV SC, 77-, PROF PEDIAT & PROF PSYCHIAT & BEHAV MED, 82-, ASSOC DEAN, COL MED, 82- *Concurrent Pos:* Dir, Charleston County Ment Health Clin, SC, 58-60; sr teaching fel & chief children's psychiat unit, Sch Med, Univ NC, Chapel Hill, 60-61; ment health consult child psychiat, SC Dept Ment Health, 61-69; consult child psychiat, Child Develop Clin, Dept Pediat, Med Col SC, 61-66, Sch Proj, NIMH, 61-74; consult psychiat, Vet Admin Hosp, Charleston, SC, 77-78; consult, State Ment Planning Comt, 79-; mem Emotional-Behav Sub-Comt, Gov Primary Health Care Task Force, 83-; mem adv coun Continuum Care Emotionally Disturbed Children, 83-; med dir, Sc Consortium Community Teaching Hosp, 83- *Mem:* Fel Am Geriat Soc; fel Am Orthopsychiat Asn; fel Am Psychiat Asn; fel Am Acad Child Psychiat; Soc Med Schs' Dirs Continuing Educ. *Res:* Inpatient child psychiatry sevices; postgraduate education of nonpsychiatric physicians; voodoo; bromism; iatrogenic illnesses; involutional depressive states; infectious mononucleosis. *Mailing Add:* Dept Psychiat & Behav Sci Med Univ SC 171 Ashley Ave Charleston SC 29403

MELLGREN, RONALD LEE, b Des Moines, Iowa, Nov 6, 46; m 69; c 3. ENZYMOLOGY, PROTEIN CHEMISTY. *Educ:* Drake Univ, BA, 69; Iowa State Univ, PhD(biochem), 76. *Prof Exp:* Teaching assoc biochem, Sch Med, Univ Miami, 76-78; asst prof, 78-84, ASSOC PROF PHARMACOL, MED COL OHIO, 84- *Concurrent Pos:* Mem, Basic Sci Coun, Am Heart Asn, 83- *Honors & Awards:* Estab Investr Award, Am Heart Asn, 84. *Mem:* Am Soc Biol Chemists; Int Soc Heart Res. *Res:* Regulatory aspects of the enzymes involved in intermediary metabolism with emphasis on protein phosphorylation and de-phosphorylation; regulation of non-lysosomal intracellular proteolysis. *Mailing Add:* Dept Pharmacol Med Col Ohio CS No 10008 Toledo OH 43699

MELLIERE, ALVIN L, b Praire du Rocher, Ill, Aug 9, 39; m 61; c 3. NUTRITION. *Educ:* Univ Ill, BS, 61, MS, 64, PhD(nutrit biochem), 65. *Prof Exp:* Asst prof nutrit, Univ Minn, 65-67; res scientist, 67-79, res mgr, 79-83, HEAD APPLICATIONS RES, LILLY RES LABS, ELI LILLY & CO, 79- *Mem:* AAAS; Am Soc Animal Sci. *Res:* Dietary and chemical factors affecting the metabolism of nutrients in animals. *Mailing Add:* 1710 Chapman Dr Greenfield IN 46140

MELLIES, MARGOT J, b St Louis, Mo, Feb 18, 42. LIPID RESEARCH, PHARMACEUTICAL RESEARCH. *Educ:* St Louis Univ Sch Med, MD, 67. *Prof Exp:* Clin instr pediat, St Louis Children's Hosp, 67; fel & instr pediat, Johns Hopkins Univ, 68-70; res instr & clin pediat nephrology fel, State Univ NY Buffalo, 70-71; postgrad fel, Dept Pediat, Cincinnati Children's Hosp, 71-73; jr res assoc, Dept Pediat & Internal Med, Univ Cincinnati, 73-75, from asst prof to assoc prof med, 75-87; DIR CLIN RES, PHARMACEUT RES DIV, BRISTOL-MYERS SQUIBB CO, 90- *Concurrent Pos:* Clin pediat renal fel, Nat Kidney Found & NY State Kidney Found, Buffalo Children's Hosp, 70-71; dir, Clermont County Children's Clin, 73-86, Lipid Clin, Univ Cincinnati, 82-87 & NIH, Nat Heart, Lung & Blood Inst, Lipid Res Clin, 86-87; dir, Gen Clin Res Ctr-Outpatient Ctr, 83-87 & assoc dir, 84-87; consult, Procter & Gamble Co; mem, Fac Comt Human Res, 83-85 & Gen Clin Res Ctr Adv Comt, 84-87. *Mem:* Am Acad Pediat; Am Fedn Clin Res; Soc Pediat Res; Am Inst Nutrit; Am Soc Clin Nutrit; Am Heart Asn. *Res:* Nutritional and pharmacologic control of hyperlypidemia and atherosclerotic risk factors prevention of atherosclerosis. *Mailing Add:* Pharmaceut Res Inst Bristol-Myers Squibb Co PO Box 4000 Princeton NJ 08543

MELLIN, GILBERT WYLIE, b Manorville, Pa, Sept 22, 25; m 55; c 2. MEDICINE, PEDIATRICS. *Educ:* Bethany Col, BS, 45; Johns Hopkins Univ, MD, 49; Am Bd Pediat, dipl, 54. *Prof Exp:* Intern med, Med Ctr, Univ Pittsburgh, 49-50; from jr resident to sr resident pediat, Bellevue Hosp, New York, 50-52; instr, Med Sch, NY Univ, 52-53; clin instr, Med Sch, Georgetown Univ, 53-55; actg chmn dept, 70-71, instr, 55-57, assoc, 57-58, asst prof, 58-67, ASSOC PROF PEDIAT, COL PHYSICIANS & SURGEONS, COLUMBIA UNIV, 67- *Concurrent Pos:* Dir fetal life study, Columbia-Presby Med Ctr; chief resident, Bellevue Hosp, 52-53; jr assoc, Children's Hosp, Washington, DC, 53-55; asst pediatrician, Babies Hosp & Vanderbilt Clin, Presby Hosp, New York, 55-57, asst attend pediatrician, 57-67, assoc attend pediatrician, 67-, actg dir pediat serv, 70-71, dir qual assurance, 77-; mem tech adv comt cleft palate, City New York Dept Health, 61-77; proj consult, Children's Bur, US Dept Health, Educ & Welfare & Govt Pakistan, 67-70; mem bd dirs, NY County Health Serv Rev Orgn, 72-89; sr consult, Helen Hayes Hosp, 74- *Mem:* Am Acad Pediat; Am Pediat Soc; Soc Pediat Res; Harvey Soc. *Res:* Epidemiological approach to fetal life and pregnancy outcome by means of direct prospective observation; establishment of documented magnetic tape data set banks for rapid data tabulation and analysis by electronic computer. *Mailing Add:* Col Physicians & Surgeons Columbia Univ New York NY 10032-3784

MELLIN, THEODORE NELSON, b Paterson, NJ, Dec 24, 37; m 59, 77; c 3. PHYSIOLOGY, BIOCHEMISTRY. *Educ:* Univ Vt, BS, 59; Univ Maine, MS, 61; Purdue Univ, PhD(reproductive physiol), 65. *Prof Exp:* Res asst animal sci, Univ Maine, 59-61; res asst reproductive physiol, Wash State Univ, 61-62; res asst, Purdue Univ, 62-63, res fel, 63-65; SR RES FEL, MERCK INST THERAPEUT RES, 66- *Concurrent Pos:* Nat Cancer Inst fel steroid biochem, Worcester Found Exp Biol, 64-66; trainee electrophysiol methods, Cold Spring Harbor Lab, NY, 79. *Mem:* Soc Study Reproduction; Soc Neurosci; Sigma Xi; Soc Invest Dermatol. *Res:* Reproductive physiology and endocrinology of animals; metabolism of steroid sex hormones; development and application of steroid sex hormone assays to the bovine; rumen microbiology; animal growth promotion; invertebrate neurophysiology; schistosome neuropharmacology; biochemistry of protozoan parasites; polypeptide growth factors and wound healing. *Mailing Add:* Dept Growth Factor Res Merck Inst Therapeut Res PO Box 2000 Rahway NJ 07065

MELLINGER, CLAIR, b Ephrata, Pa, Mar 29, 42; m 66; c 2. PLANT ECOLOGY. *Educ:* Eastern Mennonite Col, BS, 64; Univ NC, Chapel Hill, PhD(bot), 72. *Prof Exp:* Asst instr, 64-65, assoc prof, 70-81, PROF BIOL, EASTERN MENNONITE COL, 81- *Mem:* AAAS; Ecol Soc Am; Am Inst Biol Sci; Bot Soc Am; Am Ornith Union. *Res:* Population dynamics of plant species under environmental stress, bird-plant interactions. *Mailing Add:* Biol Dept Eastern Mennonite Col Harrisonburg VA 22801

MELLINGER, GARY ANDREAS, b De Kalb, Ill, Apr 27, 43; m 67; c 2. CHEMICAL ENGINEERING, POLYMER ENGINEERING. *Educ:* Northwestern Univ, BS, 66; Mass Inst Technol, SM, 68, ScD(chem eng), 71. *Prof Exp:* Mem res staff, Corp Res & Develop, 71-75, mgr polymer processing, 75-77, mgr plastics processing technol, Maj Appln Labs, 77-81, mgr Plastics Lab & Appl Ctr, 81-90, MGR REFRIGERATOR CABINET DESIGN, GEN ELEC CO, 90- *Mem:* Soc Plastics Engrs. *Res:* Determination of plastics processing-property relationships and their application to design and fabrication of plastic components. *Mailing Add:* 10806 Tattenham Lane Louisville KY 40243

MELLINGER, GEORGE T, urology, for more information see previous edition

MELLINGER, MICHAEL VANCE, b Harrisburg, Pa, Dec 21, 45; m 68; c 2. ECOLOGY, ENVIRONMENTAL MANAGEMENT. *Educ:* Bloomsburg Univ, BA, 67; Syracuse Univ, PhD(plant ecol), 72. *Prof Exp:* Ecologist & proj coordr, Sargent & Lundy Engrs, 72-80; proj mgr & ecologist, Weston, 80-85; pres, Mellinger Environ Consult, 85-86; proj develop mgr, Rollins, 86-88; pres, Mellinger Environ Consult, 88-89; ENVIRON MGR, ROHN & HAAS, 89- *Concurrent Pos:* Mem, Ecol Soc Am Bd Prof Cert, 85-88. *Mem:* Ecol Soc Am; Sigma Xi. *Res:* Structure and function of terrestrial communities; analysis of environmental effects of industrial facilities and hazardous waste sites. *Mailing Add:* 122 Baker Dr Exton PA 19341

MELLINKOFF, SHERMAN MUSSOFF, b McKeesport, Pa, Mar 23, 20; m 44; c 2. MEDICINE. *Educ:* Stanford Univ, BA, 41, MD, 44; Am Bd Internal Med, cert, 51; FRCP, 81. *Hon Degrees:* LHD, Wake Forest Univ, 81 & Jewish Inst Religion, 88. *Prof Exp:* Intern med, Stanford Univ Hosp, 44, asst resident, 44-45; asst resident, Osler Serv, Johns Hopkins Hosp, 47-49, resident & instr, 50-51, physician in charge gastroenterol, Outpatient Dept, 51-53; from asst prof to prof med, Sch Med, Univ Calif, Los Angeles, 53-87, dean, 62-86, distinguished prof, 87-88, EMER PROF MED, SCH MED, UNIV CALIF, LOS ANGELES, 90-; DISTINGUISHED PHYSICIAN, VET ADMIN, 90- *Concurrent Pos:* Fel, Hosp Univ Pa, 49-50; instr med, Johns Hopkins Univ, 51-53; attend consult, Wadsworth Gen Hosp, Vet Admin Ctr, 53-; sr attend physician, Harbor Gen Hosp, Torrance, 53-; sci adv panel, Res Prevent Blindness, Inc, 75-78. *Honors & Awards:* Abraham Flexner Award, Asn Am Med Cols, 81. *Mem:* Inst Med-Nat Acad Sci; fel AAAS; Am Gastroenterol Asn; Am Fedn Clin Res; Am Col Physicians; AMA; Am Acad Arts & Sci. *Res:* Gastroenterology. *Mailing Add:* Med Ctr Univ Calif Los Angeles CA 90024

MELLINS, HARRY ZACHARY, b New York, NY, May 23, 21; m 50; c 3. RADIOLOGY. *Educ:* Columbia Univ, AB, 41; Long Island Col Med, MD, 44; Univ Minn, MS, 51; Am Bd Radiol, dipl. *Hon Degrees:* AM, Harvard Univ, 69. *Prof Exp:* From instr to asst prof radiol, Med Sch, Univ Minn, 50-53; clin asst prof, Col Med, Wayne State Univ, 53-56; prof & chmn dept, Col Med, State Univ NY Downstate Med Ctr, 56-69; PROF RADIOL, HARVARD MED SCH, 69- *Concurrent Pos:* AEC-Nat Res Coun fel, Rice Inst, 49-50; nat consult, Surgeon-Gen, US Air Force, 65-79; consult, Vet Admin Hosp, West Roxbury; dir, Div Diag Radiol, Peter Bent Brigham Hosp, Boston, 69-80, dir diag radiol, 80-87, dir educ & trng in radiol, Brigham & Women's Hosp, 87- *Honors & Awards:* Ann Oration Radiol Soc NAm, 83. *Mem:* Am Roentgen Ray Soc (pres, 77-79); Am Col Radiol; Asn Univ Radiol (pres, 69); Radiol Soc NAm; Soc Uroradiol (pres, 75). *Res:* Intestinal obstruction; renal medullary function. *Mailing Add:* 25 Shattuck St Boston MA 02115

MELLINS, ROBERT B, b New York, NY, Mar 6, 28; m 59; c 2. PEDIATRICS, CARDIOPULMONARY PHYSIOLOGY. *Educ:* Columbia Univ, AB, 48; Johns Hopkins Univ, MD, 52; Am Bd Pediat, dipl. *Prof Exp:* Intern pediat, Johns Hopkins Hosp, 52-53; clin instr, Col Med, Univ Ill, 54-55; asst resident pediat, New York Hosp, 55-56; asst resident, Columbia-Presby Med Ctr, 56-57, asst, 57-60, instr, 60-65, assoc, 65-66, from asst prof to assoc prof, 66-75, dir, Pediat Intensive Care Unit, 70-75, PROF PEDIAT, COL PHYSICIANS & SURGEONS, COLUMBIA UNIV, 75-, DIR PEDIAT PULMONARY DIV, 72- *Concurrent Pos:* NY Heart Asn trainee, 61-63, res fel, 63-66; NIH career develop award, 66-71; founder & dir, Chicago Poison Control Ctr, 53-55; mem subcomt poisoning, Am Standards Asn, 54-55; asst pediatrician, Presby Hosp, 57-65, asst attend pediatrician, 65-70, assoc attend pediatrician, 70-75, attend pediatrician, 75; bd mem, Am Lung Asn, 81-86, chmn develop comt, 84-86; founding mem, sect pediatric pulmonology, Am Bd Pediat, 85-; vpres, Am Lung Asn, 87-89. *Honors & Awards:* Stevens Triennial Award, Columbia Univ, 80. *Mem:* Am Fedn Clin Res; Soc Pediat Res; Am Thoracic Soc (pres, 82-83); Am Physiol Soc; NY Acad Sci; Am Pediat Soc. *Res:* Cardiopulmonary research; lung mechanics; effect of maturation and sleep on cardiorespiratory control; respiratory muscle function; transvascular exchange of lung fluid; metabolic activities of the lung; lung growth and development; respiratory muscle function; self regulation in family management of asthma. *Mailing Add:* Col Physicians & Surgeons Columbia Univ 630 W 168th St New York NY 10032

MELLITS, E DAVID, b Philadelphia, Pa, Sept 5, 37; m 58; c 2. BIOSTATISTICS. *Educ:* Johns Hopkins Univ, BES, 59, ScD(biostatist), 65. *Prof Exp:* Systs analyst, Strong Mem Hosp, Rochester, NY, 59-61; asst prof, 65-71, ASSOC PROF BIOSTATIST, SCHS HYG & MED, JOHNS HOPKINS UNIV, 71- *Mem:* Am Statist Asn; Biomet Soc; Am Inst Math Statist; NY Acad Sci; Soc Pediat Res. *Res:* Application of statistics in the biological and medical sciences. *Mailing Add:* Dept Pediat Brady Bldg Rm 308 Johns Hopkins Hosp 600 N Wolfe St Baltimore MD 21205

MELLO, NANCY K, b Brockport, NY, Mar 20, 35. ALCOHOL & DRUG ABUSE, NEUROSCIENCE. *Educ:* Pa State Univ, PhD(psychol), 60. *Prof Exp:* PROF PSYCHOL-NEUROSCI, DEPT PSYCHIAT, HARVARD MED SCH, 74-; CO-DIR, ALCOHOL & DRUG ABUSE RES CTR, MCLEAN HOSP, 74- *Mem:* Fel Soc Behav Med; Soc Neurosci; fel AAAS; fel Am Col Neuropsychopharmacol; fel Am Psychol Asn; fel Am Psychosom Soc. *Mailing Add:* Alcohol Drug Abuse Res Ctr Harvard Med Sch McLean Hosp 115 Mill St Belmont MA 02178

MELLOCH, MICHAEL RAYMOND, b Apr 27, 53. GALLIUM ARSENIDE DEVICE TECHNOLOGY, MOLECULAR BEAM EPITAXY. *Educ:* Purdue Univ, BSEE, 75, MSEE, 76, PhD (elec eng), 81. *Prof Exp:* Design eng, Intel Corp, 76-78; mem tech staff, Tex Instruments Cent Res Labs, 82-84; asst prof, 84-88, ASSOC PROF ELEC ENG, PURDUE UNIV, 88- *Mem:* Am Phys Soc; Inst Elec & Electronic Engrs. *Res:* Molecular beam epitaxy (MBE) to grow materials for investigating novel device concepts and physical phenomena in solids; these devices include GaAs based dynamic memories, ZnSe and GaAs FETS, heterojunction bipolar devices and solar cells. *Mailing Add:* Sch Elec Eng Purdue Univ W Lafayette IN 47907

MELLON, DEFOREST, JR, b Cleveland, Ohio, Dec 18, 34. NEUROPHYSIOLOGY. *Educ:* Yale Univ, BS, 57; Johns Hopkins Univ, PhD(biol), 61. *Prof Exp:* Air Force Off Sci Res-USPHS fel neurophysiol, Stanford Univ, 61-63; asst prof, 63-68, assoc prof, 68-78, PROF BIOL, UNIV VA, 78- *Concurrent Pos:* Guggenheim Mem fel, 66-67. *Mem:* Fel AAAS; Soc Gen Physiol; Soc Neurosci; Am Soc Zoologists. *Res:* Comparative neurophysiology; nerve-muscle interactions in development; crayfish oculomotor organization. *Mailing Add:* Dept Biol Univ Va 270 Gilmer Hall Charlottesville VA 22903

MELLON, EDWARD KNOX, JR, b Rochester, NY, Oct 8, 36; m 66. INORGANIC CHEMISTRY, SCIENCE EDUCATION. *Educ:* Univ Tex, BS, 59, PhD(chem), 63. *Prof Exp:* Instr chem, St Edward's Univ, 62-63; fel, Univ Mich, 63-65, lectr, 65-66; from asst prof to assoc prof, 66-80, PROF CHEM, FLA STATE UNIV, 80- *Honors & Awards:* Nat Catalyst Award, Chem Mfrs Asn. *Mem:* Am Chem Soc; Royal Soc Chem; Sigma Xi; Nat Sci Teachers Asn. *Res:* Chemical education; reactivity network. *Mailing Add:* Dept Chem Fla State Univ Tallahassee FL 32306-3006

MELLON, GEORGE BARRY, b Edmonton, Alta, Aug 5, 31. GEOLOGY. *Educ:* Univ Alta, BSc, 54, MSc, 55; Pa State Univ, PhD(mineral, petrol), 59. *Prof Exp:* Sr res off & head geol div, Res Coun Alta, 58-74; mem staff, Alta Energy & Natural Resources, Dep Minister Energy Resources, 74-86; DEP MINISTER EXEC COUN, ALTA, 86- *Mem:* Geol Asn Can; Can Soc Petrol Geol. *Res:* Sedimentary petrology and applied statistics in geology. *Mailing Add:* 5011 143th St Edmonton AB T6H 4E1 Can

MELLOR, ARTHUR M(CLEOD), b Elmira, NY, Jan 1, 42. MECHANICAL ENGINEERING. *Educ:* Princeton Univ, BSE, 63, MA, 65, PhD(aerospace & mech sci), 68. *Prof Exp:* From asst prof to assoc prof mech eng, Purdue Univ, 67-75, prof, 75-; AT COL ENG, DREXEL UNIV, PHILADELPHIA. *Mem:* Air Pollution Control Asn; Combustion Inst; Soc Automotive Engrs; Sigma Xi. *Res:* Chemical kinetics of combustion-generated air pollution; metal combustion; gas turbine combustor design. *Mailing Add:* Vanderbilt Univ Box 6019 Sta B Nashville TN 37235

MELLOR, DAVID BRIDGWOOD, b Brockton, Mass, June 25, 29; div. POULTRY SCIENCE, FOOD TECHNOLOGY. *Educ:* Pa State Univ, BS, 56; Tex A&M Univ, MS, 57; Purdue Univ, PhD(food tech), 65. *Prof Exp:* Overseas trainee, Int Coop Admin, 57-58, livestock adv, Agency Int Develop, 58-62; poultry adv, IRI Res Inst Inc, 65-67; assoc prof poultry mkt, 67-77, POULTRY MKT SPECIALIST, TEX AGR EXTEN SERV, TEX A&M UNIV, 67- *Concurrent Pos:* Consult, UN Develop Prog, Poultry Res Ctr, Pakistan, 72, Fed Univ Minas Gerias Poultry Mkt Prog, Brazil, 75 & Egg Qual & Control Prog, SAfrica, 78- *Mem:* Poultry Sci Asn; Inst Food Technol. *Res:* Egg quality; broiler yield and evaluation consumer opinion. *Mailing Add:* Dept of Poultry Sci Tex A&M Univ College Station TX 77843

MELLOR, GEORGE LINCOLN, JR, b Yonkers, NY, May 12, 29; m 54; c 2. PLANETARY BOUNDARY LAYERS. *Educ:* Mass Inst Technol, SB, 52, SM, 54, ScD(mech eng), 57. *Prof Exp:* Analytical engr, Pratt & Whitney Aircraft Div, United Aircraft Corp, 52-53; res fel, Gas Turbine Lab, Mass Inst Technol, 53-57; aerodyn specialist, Curtiss-Wright Corp, 57; dir, Geophys Fluid Dynamics Prog, 69-76, PROF FLUID MECH, PRINCETON UNIV, 57- *Concurrent Pos:* Consult, Dynalysis & other industs; Nat Sci Found fel, Cambridge Univ, 62-63; Nat Acad Sci sci exchange fel, Inst Oceanol, Moscow, 70-71; CNOC Res Chair Oceanog, Naval Postgrad Sch, Monterey, Calif, 83. *Honors & Awards:* Robert Knapp Award, Am Soc Mech Engrs, 59. *Mem:* Am Soc Mech Engrs; Am Geophys Union; Am Meteorol Soc. *Res:* Aerodynamics of turbomachinery; boundary layer mechanics; physical oceanography; meteorology. *Mailing Add:* Dept of Mech & Aerospace Eng Princeton Univ Princeton NJ 08540

MELLOR, JOHN, b Guilford, Conn, Apr 28, 33; m 56; c 3. PHYSICAL CHEMISTRY. *Educ:* Univ Conn, BA, 56; Tufts Univ, MS, 58; Mass Inst Technol, PhD(phys chem), 62. *Prof Exp:* Chemist, Gen Elec Co, Conn, 56; res asst phys chem, Tufts Univ, 58; asst chemist, Arthur D Little, Inc, Mass, 59, consult physics, 60-62; res assoc, Brookhaven Nat Lab, 62-64; asst prof, 64-74, ASSOC PROF PHYS CHEM, UNIV BRIDGEPORT, 74-, CHMN DEPT, 77- *Concurrent Pos:* Vis asst physicist, Brookhaven Nat Lab, 65, res collabr, 65-66. *Mem:* Am Crystallog Asn. *Res:* Crystal structure analysis by x-ray and neutron diffraction; molecular motions and lattice dynamics by neutron inelastic scattering. *Mailing Add:* Dept Chem & Geol Univ Bridgeport Bridgeport CT 06601

MELLOR, JOHN WILLIAMS, b Paris, France, Dec 28, 28; US citizen; m 73; c 3. AGRICULTURAL DEVELOPMENT & PRICE POLICY. *Educ:* Cornell Univ, BSc, 50, MSc, 51, PhD(agr econ), 54; Oxford Univ, Eng, dipl, 52. *Prof Exp:* Lectr agr econ, dept agr econ, Cornell Univ, 52-54, asst prof, 54-58, from assoc prof to prof agr econ & Asian studies, 58-77, actg dir, Ctr Int Studies, 61-66, dir, Prog Comp Econ Develop, 73-77; DIR, INT FOOD POLICY RES INST, 77- *Concurrent Pos:* Fulbright fel, Oxford Univ, Eng, 51-52; res fel, Cornell Univ, 53-54; vis prof, Balwant Rajput Col, Agra, India, 59-60, Rockefeller Found, Indian Agr Res Inst, New Delhi, 64-65 & Am Univ, Beirut, Lebanon, 68. *Honors & Awards:* Wihuri Int Prize, 85. *Mem:* Fel Am Agr Econ Asn; fel Am Acad Arts & Sci; Int Agr Econ Asn; Am Econ Asn. *Res:* Analysis of the economics of agricultural development, with emphasis on the relationship between growth in the agricultural sector and other sectors; the role of technological change in these processes. *Mailing Add:* Int Food Policy Res Inst 8th Floor 1776 Massachusetts Ave NW Washington DC 20036

MELLOR, MALCOLM, b Stalybridge, Eng, May 24, 33; m 58; c 2. APPLIED PHYSICS, ENGINEERING. *Educ:* Univ Nottingham, BS, 55; Univ Melbourne, MS, 59, DSc(appl sci), 69; Univ Sheffield, PhD(civil & struct eng), 70. *Prof Exp:* Asst engr, State Rivers & Water Supply Comn, Australia, 56; glaciologist, Australian Nat Antarctic Res Expeds, Dept External Affairs, 56-59; contract engr, Snow, Ice & Permafrost Res Estab, US Dept Army, Thayer Sch Eng, Dartmouth Col, 59-61, res engr, Cold Regions Res Lab, 61-75, RES PHYS SCIENTIST, COLD REGIONS RES & ENG LAB, US DEPT ARMY, 75- *Concurrent Pos:* Assoc consult, Creare Inc, 70-74; mem ed bd, Int Glaciological Soc, 72-; mem comt glaciol, Polar Res Bd, 74-78; secy, Int Comn Snow & Ice, 75-83, vpres, 83-87; ed, Cold Regions Sci & Technol, 78-; consult, 70- *Honors & Awards:* Brit Polar Medal, 59; Antarctic Medal, 67; Spec Award, Nat Res Coun, Nat Acad Sci-Nat Acad Eng, 72. *Res:* Applied mechanics; explosives; physics and mechanics of snow, ice and frozen ground; cold regions engineering; glaciology; ocean engineering; machine design. *Mailing Add:* Cold Regions Res & Eng Lab 72 Lyme Rd Hanover NH 03755

MELLOR, ROBERT SYDNEY, b Casper, Wyo, June 25, 31; m 60; c 3. PLANT PHYSIOLOGY. *Educ:* Colo State Univ, BS, 54, MS, 59, PhD(plant physiol), 62. *Prof Exp:* Asst prof, 62-67, ASSOC PROF BIOL, UNIV ARIZ, 67- *Mem:* Am Soc Plant Physiologists; Sigma Xi. *Res:* Plant-water relations; plant biochemistry. *Mailing Add:* Dept Ecol & Evolutionary Biol Univ of Ariz Tucson AZ 85721

MELLORS, ALAN, b Mansfield, Eng, Feb 26, 40; m 86. MEMBRANES, ENZYMES. *Educ:* Univ Liverpool, BSc, 61, PhD(biochem), 64. *Prof Exp:* Res biochemist, Univ Calif, Davis, 64-67; res scientist, Food Res Inst, Can Dept Agr, 67-68; from asst prof to assoc prof, 68-79, PROF CHEM, UNIV GUELPH, 79 - *Concurrent Pos:* Fulbright scholar, 64-67; Nuffield Found scholar, 75-76. *Mem:* Am Soc Biol Chemists; Brit Biochem Soc; Can Biochem Soc; Can Soc Cell Biologists. *Res:* Phospholipases; trypanosomal enzymes; enzymes of pathogens; drug design; structure-activity relationships; molar volume relationships and physical toxicity; membrane events in cell activation. *Mailing Add:* Dept Chem & Biochem Univ Guelph Guelph ON N1G 2W1 Can

MELLORS, ROBERT CHARLES, b Dayton, Ohio, June 18, 16; m 44; c 4. PATHOLOGY, GENERAL MEDICAL SCIENCES. *Educ:* Western Reserve Univ, AB, 37, MA, 38, PhD(biochem), 40; Johns Hopkins Univ, MD, 44. *Prof Exp:* Instr biochem, Western Reserve Univ, 40-42; asst epidemiol, Sch Hyg & Pub Health & asst poliomyelitis, Res Ctr, Johns Hopkins Univ, 42-44; assoc prof path, Grad Sch Med Sci, Cornell Univ, 53-58, assoc dir res, 58-69, dir res, Hosp Spec Surg, 69-84, pathologist-in-chief & dir labs, 58-84, prof path, 61-90, MEM HON STAFF & EMER SCIENTIST, MED COL, CORNELL UNIV, 84-, EMER PROF PATH, 90- *Concurrent Pos:* Spec fel med, Mem Hosp Ctr, New York, 46-47, Am Cancer Soc sr fel, 47-50, Runyan Fund sr fel, 50-53; ed, Anal Cytol & Anal Path Mem, Nat Found Infantile Paralysis Res, 43; assoc, Sloan-Kettering Inst, 50-53; assoc attend pathologist, Mem Hosp & Ewing Hosp, 53-58; mem res adv comt, USPHS, 62-66; attend pathologist, New York Hosp, 71-84. *Mem:* Am Asn Path; Am Soc Biol Chem; Am Asn Immunol; fel Am Soc Clin Path; fel Royal Col Path. *Res:* Experimental pathology; immunopathology; glomerulonephritis; rheumatoid arthritis; autoimmune diseases; systemic lupus erythematosus; cancer; viruses in cancer. *Mailing Add:* Three Hardscrabble Circle Armonk NY 10504

MELLOW, ERNEST W(ESLEY), b St Louis, Mo, Feb 20, 18; m 48; c 3. CHEMICAL ENGINEERING. *Educ:* Univ Mo, BS, 40, MS, 46; Univ Ill, PhD(chem eng), 50. *Prof Exp:* Asst chem eng, Univ Mo, 40-41; chem engr, E I du Pont de Nemours & Co, Inc, 41-42; asst chem eng, Univ Ill, 46-49; asst prof, Univ Mo, 49-52; proj supvr, Callery Chem Co, 52-53; staff chem engr, Res & Develop Dept, Phillips Petrol Co, 52-56, sect mgr, 56-65, asst to mgr eng & test br, Atomic Energy Div, 65-69; PROF CHEM ENG, W VA INST TECHNOL, 69- *Concurrent Pos:* Vis prof, Okla State Univ, 67-68. *Mem:* Am Chem Soc; Am Soc Eng Educ; fel Am Inst Chem Engrs. *Res:* Environmental studies; technology assessment; relations between technology and society. *Mailing Add:* PO Box 14 Charlton Heights WV 25040-0014

MELMON, KENNETH LLOYD, b San Francisco, Calif, July 20, 34; m 58; c 2. CLINICAL PHARMACOLOGY. *Educ:* Stanford Univ, AB, 56; Univ Calif, MD, 59; Am Bd Internal Med, dipl, 66. *Prof Exp:* Intern internal med, Moffitt Hosp, Univ Calif, San Francisco, 59-60, asst resident, 60-61; clin assoc exp therapeut, Nat Heart Inst, 61-64; chief resident med, King County Hosp, Seattle, Wash, 64-65; chief sect clin pharmacol, asst prof med & pharmacol & assoc mem Cardiovasc Res Inst, Sch Med, Univ Calif, San Francisco, 65-68, chief Div Clin Pharmacol, sr staff Cardiovasc Res Inst, prof med & pharmacol, 68-78; chmn dept, 78-85, BLOOMFIELD PROF MED & PHARMACOL, SCH MED, STANFORD UNIV, 85- *Concurrent Pos:* Mosby scholar, 59; consult, NIH, 65-; Burroughs Wellcome scholar clin pharmacol, 66-71; mem consult comt, Food & Drug Admin, Senate Subcomt on Health, House Ways & Means Comt; mem coun basic scis, high blood pressure, res & circulation, Am Heart Asn; mem bd, Am Bd Internal Med, 68; Guggenheim fel exp studies clin biochem & immunopharmacol, 71; NIH spec fel, 71; Nat Coun, AAAS, 85-88. *Mem:* Inst Med-Nat Acad Sci; Am Fedn Clin Res (pres); Am Soc Pharmacol & Exp Therapeut; fel Am Col Physicians; Am Physiol Soc; Am Soc Clin Investr (pres); W Asn Physician (pres). *Res:* Mechanism of action cyclic amp on cell growth and death; hormone receptors on cells; leukocyte; role of vasoactive amines in immunology; carcinoid syndrome. *Mailing Add:* Dept of Med Stanford Univ Sch of Med Stanford CA 94305

MELNGAILIS, IVARS, b Riga, Latvia, Nov 13, 33; US citizen; m 64; c 2. SOLID STATE PHYSICS, ELECTRICAL ENGINEERING. *Educ:* Carnegie Inst Technol, BS, 56, PhD(elec eng), 61. *Prof Exp:* Instr elec eng, Carnegie Inst Technol, 59-60; staff mem, Appl Physics Group, 61-65, asst group leader, 65-71, group leader, 71-75, ASSOC DIV HEAD, SOLID STATE DIV, LINCOLN LAB, MASS INST TECHNOL, 75- *Mem:* Fel Inst Elec & Electronics Engrs; Am Phys Soc. *Res:* Infrared detectors; physics of semiconductor devices; impact ionization in semiconductors at low temperatures; magnetic field effects on plasmas in semiconductors; semiconductor lasers and detectors; integrated optics. *Mailing Add:* Lincoln Lab Mass Inst Technol Lexington MA 02173

MELNGAILIS, JOHN, b Riga, Latvia, Feb 4, 39; US citizen; m 68; c 3. APPLIED PHYSICS. *Educ:* Carnegie Inst Technol, BS, 60, MS, 62, PhD(solid state theory), 65. *Prof Exp:* Assoc engr, Westinghouse Res Lab, 60-65; staff mem, Lincoln Lab, 67-79, prin res scientist, 79-90, SR RES SCIENTIST, MASS INST TECHNOL, 91- *Concurrent Pos:* NSF fel, Max Planck Inst Metall Res, Stuttgart, Ger, 65; res attache, Ctr Nat Res Sci, Bellevue, France, 66-67. *Mem:* Am Phys Soc; sr mem Inst Elec & Electronics Engrs. *Res:* Surface acoustic wave devices; solid state physics; microstructure fabrication technology; focused ion beam fabrication. *Mailing Add:* Mass Inst Technol Rm 39-427A Cambridge MA 02139

MELNICK, DANIEL, b New York, NY, Mar 14, 42; m; c 1. SCIENCE EDUCATION. *Educ:* Univ Wis-Madison, BA, 63, MA, 64, PhD(polit sci), 70. *Prof Exp:* Asst prof, Dept Govt & Politics, Univ Md, College Park, 69-75; analyst, Cong Res Serv, Libr Cong, Am Nat Govt, 75-80, specialist fed statist policy, 80-90, sect head, Sect Pub Opinion, 83-86; dir, Div Sci Resources Studies, 90, SR ADV RES METHODOLOGIES, EDUC & HUMAN RESOURCES DIRECTORATE, NSF, 91- *Concurrent Pos:* Prof staff mem, House Comt on Post Off & Civil Serv Subcomt on Census & Pop, US Cong, 77, Joint Econ Comt, 87-89; consult, Georgetown Univ Med Sch, Inst Health Policy Analysis & Charles Drew Med Sch/Martin Luther King Hosp; mem, Tech Rev Panel, Nat Assessment of Educ Progress, 87-88. *Mem:* Am Polit Sci Asn; Am Statist Asn; AAAS. *Res:* Statistics. *Mailing Add:* Rm 516 NSF 1800 G St NW Washington DC 20550

MELNICK, EDWARD LAWRENCE, b Ann Arbor, Mich, Dec 12, 38; m 63; c 2. MATHEMATICS, STATISTICS. *Educ:* Lehigh Univ, BA, 60; Va Polytech Inst & State Univ, MS, 63; George Wash Univ, PhD(math statist), 70. *Prof Exp:* Math statistician, US Census Bur, Dept of Com, 63-69; from asst prof to assoc prof, 69-79, PROF STATIST, GRAD SCH BUS ADMIN, NY UNIV, 73-, FAC RES GRANT, 69- *Concurrent Pos:* Lectr, Grad Sch, USDA, 64-69; vis prof, Imperial Col, 75-76; statist consult, Morgan Stanley, IBM, Equitable Life Assurance, Nat Westminister Bank & Deutsche Bank; mem, Univ Sen, Univ Financial Affairs, NY Univ; assoc ed, J Forecasting. *Mem:* Am Statist Asn; fel Royal Statist Soc; Inst Math Statist. *Res:* Time series analysis with emphasis upon signal detection and prediction theory; collection and analysis of data. *Mailing Add:* Statists & Opers Res Stern Sch Bus NY Univ New York NY 10003

MELNICK, JOSEPH LOUIS, b Boston, Mass, Oct 9, 14; m 36; c 1. VIROLOGY. *Educ:* Wesleyan Univ, BA, 36, Yale Univ, PhD(biochem), 39; Am Bd Med Microbiol, dipl. *Hon Degrees:* DSc, Wesleyan Univ, 71. *Prof Exp:* Asst physiol chem, Sch Med, Yale Univ, 37-39, from instr to asst prof prev med, 42-49, assoc prof microbiol, 49-54, prof epidemiol, 54-57; chief virus labs, Div Biologics Standards, NIH, 57-58; prof virol & epidemiol & chmn dept, 58-88, dean grad sci, 68-91, DISTINGUISHED SERV PROF, BAYLOR COL MED, 74- *Concurrent Pos:* Finney-Howell Res Found fel, Sch Med, Yale Univ, 39-41, Nat Res Coun fel prev med & pediat, 41-42; Am-Scand Found fel viruses, Karolinska Inst, Sweden, 49; mem, Viral & Rickettsial Registry Comt & Exec Comt, 57-60; mem panel virol & immunol, Comt on Growth, Nat Res Coun, 52-56 & Comt Viral Hepatitis, 71-75; chmn comt echoviruses, Nat Found, 55-57, chmn comt enteroviruses, 57-60; expert adv panel virus dis, WHO, 57-, dir int reference ctr enteroviruses, 61-74, mem comt polio vaccine, 72-, dir Collaborating Ctr Virus Reference & Res, 74-; mem comt live polio virus vaccine, USPHS, 58-61, mem nat adv cancer coun, 65-69; mem viruses & cancer bd, Nat Cancer Inst, 60-62, mem human cancer virus task force, 62-66, mem etiology prog adv comt, 70-73, mem adv comt, Nat Multiple Sclerosis Soc, 80-83; chmn comt enteroviruses, NIH, 60-63; mem bd virus reference reagents, Nat Inst Allergy & Infectious Dis, 62-65, mem allergy & infectious dis training grant comt, 62-65, mem panel picornaviruses, 63-65; nat lectr, Found for Microbiol, 63-64 & 66-67; trustee, George Washington Carver Res Found, Tuskegee Inst, 64-; mem, Int Comt Nomenclature of Viruses, 66-74; chmn, Picornavirus & Papovavirus Comts, 68-81; mem, Surgeon Gen Comt Hepatitis, Dept Army, 66-73; secy-gen, Int Congs Virol, Helsinki, 68 & Budapest, 71; chmn virol sect & mem exec bd, Int Asn Microbiol Socs, 70-75, mem int comt microbial ecology, 72-; mem coun res & clin invest, Am Cancer Soc, 71-75; chmn task group active immunization against viral hepatitis, NIH, 73-74; co-chmn, Duran-Reynals Int Symp Viral Oncol, Barcelona, Spain, 73; ed, Progress Med Virol, 58-, Monographs in Virol, 60-, Intervirol, 73-85; chmn, int conf viruses in water, Am Pub Health Asn & WHO, Mexico City, 74, adv comt viral hepatitis, Ctr Dis Control, USPHS, 89 & adv comt respiratory & enteric viruses, 89; mem, Int Comt Taxon Viruses, 74-, res briefing panel, prev & control viral dis, NaE Acad Sci, Inst Med, 86, bd dirs, Houston Acad Med/Tex Med Ctr Libr, 67-90, chmn, 87-89; lectr virol, Chinese Acad Med Sci, 78 & Shanghai First Med Col, 79; Latin Am prof microbiol, Buenos Aires, 77 & Santiago, 79. *Honors & Awards:* Professorship named in honor. Endowed Joseph L Melnick Professorship of Virol, Baylor Col Med, 87; Polio Hall of Fame, 58; Mod Med Distinguished Achievement Award, 65; Indust Res-100 Award, 71; Inventor of Year Award, 72; Freedman Found Award Res Virol, NY Acad Sci, 73; Maimonides Award, State Israel, 80. *Mem:* Fel AAAS; Am Epidemiol Soc; Soc Exp Biol & Med; Am Asn Immunol; Am Soc Microbiol; Am Soc Virol. *Res:* Virology, especially infectious diseases and cancer. *Mailing Add:* Div Molecular Virol Baylor Col Med Houston TX 77030

MELNICK, LABEN MORTON, b Pittsburgh, Pa, June 10, 26; m 49, 81; c 4. ANALYTICAL CHEMISTRY. *Educ:* Univ Pittsburgh, BS, 49, MS, 50, PhD(anal chem), 54. *Prof Exp:* Anal chemist, Res Lab, Jones & Laughlin Steel Corp, 50, x-ray diffractionist & spectroscopist, 51-52; technologist, Res Lab, US Steel Corp, 53-56, supv technologist, 56-58, sect supvr anal chem, 58-75, div chief physics & anal chem, 75-83; RETIRED. *Concurrent Pos:* Mem, Nat Acad Sci-Nat Res Coun adv panels to anal chem div, Nat Bur Standards, 69-71 & measures for air qual off, 71-73. *Mem:* Am Chem Soc. *Res:* Analysis of raw materials and metals; determination of gases and second phase inclusions in steel; analysis of water. *Mailing Add:* 2819 Shady Ave Pittsburgh PA 15217-2740

MELNICK, RONALD L, b New York, NY, May 19, 43; m 71; c 2. TOXICOLOGY, BIOCHEMISTRY. *Educ:* Rutgers Univ, BS, 65; Univ Mass, MS, 68, PhD(food sci), 71. *Prof Exp:* Fel cell physiol, Univ Calif, Berkeley, 71-73; asst prof biol sci, Polytech Inst New York, 73-80; CHEMIST, NAT TOXICOL PROG, NAT INST ENVIRON HEALTH SCI, 80-, HEAD, EXP TOXICOL UNIT, 85- *Mem:* AAAS; NY Acad Sci; Am Chem Soc. *Res:* Chemical induced toxicity-carcinogenicity in rodents; mechanism of phthalate ester hepatotoxicity; role of reactive intermediates of oxygen reduction in chemical induced toxicities; biochemical mechanisms of chemical toxicants. *Mailing Add:* Nat Toxicol Prog Nat Inst Environ Health Sci PO Box 12233 Research Triangle Park NC 27709

MELNICK, VIJAYA L, m 63; c 1. INFANT NUTRITION & FEEDING, WOMEN & MINORITIES IN THE BIOMEDICAL SCIENCES. *Educ:* Madras Agr Col, India, BSc; Univ Wis-Madison, MS, 61, PhD(pl physiol), 64. *Prof Exp:* Postdoctoral fel, Exp Path, Univ DC, 64-66, from asst prof to assoc prof biol, 70-77, dir, Ctr Appl Res & Urban Policy, 85-89, PROF BIOL, UNIV DC, 77-, SR RES SCHOLAR, CTR APPL RES & URBAN POLICY, 84- *Concurrent Pos:* Vis scientist, Biol & Med Inst, Lawrence Livermore Lab, 72-73; vis prof, Oak Ridge Grad Sch Biomed, Univ Tenn, 74-78; sr staff assoc, Int Ctr Interdisciplinary Studies in Immunol, Georgetown Univ Med Ctr, 78-85, assoc dir, 85-; spec asst policy & bioethics, Nat Inst Aging, NIH, 80-82. *Mem:* AAAS; Am Soc Cell Biol; Sigma Xi; Asn Women Sci; Am Polit Sci Asn. *Res:* Activities directed towards issues related to the health and well being of women and children; investment in the future generation through appropriate policy, development of knowledge base and coordination of services; special interests are disadvantaged populations; bioethics; biomedical policy. *Mailing Add:* Dept Biol & Ctr Appl Res & Urban Policy 4200 Connecticut Ave NW Washington DC 20008

MELNYKOVYCH, GEORGE, b Halych, Ukraine, Oct 14, 24; nat US; m 49; c 2. CELL BIOLOGY. *Educ:* Univ Minn, MS, 53, PhD, 56. *Prof Exp:* Res chemist, Elgin State Hosp, Ill, 58 & Lederle Labs Div, Am Cyanamid Co, 58-63; asst prof microbiol, 63-70, PROF MICROBIOL, SCH MED, UNIV KANS, 70-; RES CAREER SCIENTIST, VET ADMIN HOSP, 63- *Concurrent Pos:* USPHS fel biochem, Univ Tex, 56 & Univ Calif, 56-58. *Mem:* AAAS; Am Soc Cell Biol; Tissue Cult Asn; Am Soc Biol Chem. *Res:* Tissue culture nutrition and metabolism; steroid hormones; membrane biochemistry. *Mailing Add:* US Vet Admin Hosp 4801 Linwood Blvd Kansas City MO 64128

MELOAN, CLIFTON E, b Bettendorf, Iowa, Aug 4, 31; m 57; c 3. ANALYTICAL CHEMISTRY. *Educ:* Iowa State Univ, BS, 53; Purdue Univ, PhD(anal chem), 59. *Prof Exp:* From asst prof to assoc prof, 59-68, PROF CHEM, KANS STATE UNIV, 68- *Concurrent Pos:* Sci adv, Food & Drug Admin, 66- *Mem:* Am Chem Soc; Sigma Xi. *Res:* Liquid-liquid extractions; metal chelates; infrared; gas chromatography; spectrophotometry. *Mailing Add:* Dept of Chem Kans State Univ Manhattan KS 66502

MELOCHE, HENRY PAUL, b Detroit, Mich, Nov 15, 28; div; c 6. MICROBIOLOGY, PUBLIC HEALTH. *Educ:* Univ Detroit, BS, 51; Mich State Univ, MS, 53, PhD(microbiol), 56. *Prof Exp:* Food chemist, Swift & Co, 56-57; microbiologist, Fermentation Lab, Agr Res Serv, USDA, 57-60; res assoc agr chem, Mich State Univ, 60-64; res assoc, Inst Cancer Res, 64-69, asst mem, 69-76; SR RES SCIENTIST & ASSOC DIR, PAPANICOLAOU CANCER RES INST, 77- *Concurrent Pos:* NIH fel, 62-63; adj assoc prof biochem, Papanicloaou Cancer Res Inst, Univ Miami, 77- *Mem:* AAAS; Am Chem Soc; Am Soc Microbiol; Am Soc Biol Chem. *Res:* Enzyme chemistry active site chemistry, sterochemistry. *Mailing Add:* 6946 SW 111th Ct Miami FL 33101

MELONI, EDWARD GEORGE, b East Boston, Mass, Aug 2, 32; div; c 2. INORGANIC CHEMISTRY. *Educ:* Columbia Univ, AB, 53; Tufts Univ, MS, 55; Rutgers Univ, PhD(chem), 61, Northeastern Univ, MBA, 72. *Prof Exp:* Res chemist, Pennsalt Chem Corp, 60-63; res chemist, Esso Res & Eng Co, 63-65; chief chemist, Alfa Inorg Inc, 65-71; tech dir, Ventron Corp, 71-74; technol specialist, Combustion Eng Co, 81-83; DIR MKT, STREM CHEM INC, 83- *Mem:* Am Chem Soc. *Res:* Synthetic inorganic chemistry; inorganic polymers. *Mailing Add:* Lot No Three Zealand Village Seabrook NH 03874

MELOON, DANIEL THOMAS, JR, b Buffalo, NY, Aug 13, 35; m 59; c 5. ANALYTICAL CHEMISTRY. *Educ:* Univ Buffalo, BA, 57, MA, 60, PhD(inorg chem), 63. *Prof Exp:* Sr chemist, Carborundum Co, Niagara Falls, 63-66; from asst prof to assoc prof, 66-80, PROF CHEM, STATE UNIV NY COL BUFFALO, 80- *Concurrent Pos:* Prof, Daemen Col, 81- *Mem:* Am Chem Soc; Creation Res Soc. *Res:* High temperature oxidation catalysts; water pollution and inorganic complex ions. *Mailing Add:* 186 Calvert Blvd Tonawanda NY 14150

MELOON, DAVID RAND, b Buffalo, NY, July 20, 48; m 75. INORGANIC CHEMISTRY, PHYSICAL CHEMISTRY. *Educ:* State Univ NY Buffalo, BA, 70, PhD(chem), 75. *Prof Exp:* Res fel chem, State Univ NY Buffalo, 74-75, lectr, 75-76; res chemist, 76-79, MGR RES & DEVELOP, GEN ABRASIVE DIV, DRESSER INDUSTS, 79- *Mem:* Am Chem Soc; Sigma Xi; Am Inst Chemists. *Res:* Abrasive and refractory materials; kinetics and mechanisms of inorganic and bio-inorganic reactions. *Mailing Add:* 5350 Shawnee Rd Sanborn NY 14132

MELOSH, HENRY JAY, IV, b Paterson, NJ, June 23, 47; m 69; c 3. PLANETARY SCIENCES, GEOPHYSICS. *Educ:* Princeton Univ, AB, 69; Calif Inst Technol, MS, 71, PhD(physics), 73. *Prof Exp:* Res assoc physics, Univ Chicago, 73-74; instr geophys & planetary sci, Calif Inst Technol, 74-76, asst prof planetary sci, 76-78, assoc prof, 78-79; assoc prof geophys, State Univ NY Stony Brook, 79-83; PROF, PLANETARY SCI, LUNAR-PLANETARY LAB, UNIV ARIZ, TUCSON, 83- *Concurrent Pos:* Geophysicist, US Geol Surv, 77-79; consult geosci, Los Alamos Nat Lab, 80-; ed, Reviews Geophys, 89- *Honors & Awards:* Fel, Meteorol Soc, 89; Fel, Geol Soc Am, 89. *Mem:* Am Geophys Union; Sigma Xi. *Res:* Tectonics of planetary lithospheres; crater mechanics; dynamics of earth's lithosphere, especially spreading centers, subduction zones, anelastic response of the earth, rheology; physics of earthquakes and large landslides; origin of the moon. *Mailing Add:* Lunar-Planetary Lab Univ Ariz Tucson AZ 85721

MELOUK, HASSAN A, b Alexandria, Egypt, May 19, 41; US citizen; m 64; c 2. DISEASE RESISTANCE, EPIDEMIOLOGY. *Educ:* Univ Alexandria, BS, 62; Ore State Univ, MS, 67, PhD(plant path), 69. *Prof Exp:* Teaching fel plant path, Wash State Univ, 69-70; res assoc, Ore State Univ, 70-76; PLANT PATHOLOGIST & PROF PLANT PATH, USDA-AGR RES SERV, OKLA STATE UNIV, 76- *Concurrent Pos:* Assoc ed peanut sci, 82-88, pres elect, 88, pres, Am Peanut Res & Educ Soc, 89- *Mem:* Am Phytopath Soc; Am Peanut Res & Educ Soc. *Res:* Peanut diseases with emphasis on disease resistance and epidemiology. *Mailing Add:* 4802 Country Club Ct Stillwater OK 74074

MELOY, CARL RIDGE, b Detroit, Mich, Sept 21, 12; m 32, 65; c 6. ORGANIC CHEMISTRY. *Educ:* Univ Mich, BS, 32, MS, 34; Mich State Univ, PhD(org chem), 42. *Prof Exp:* Instr chem, Highland Park Jr Col, Mich, 34-42; res chemist, Stamford Res Lab, Air Reduction Co, Conn, 42-43; asst prof chem & physics, Baldwin-Wallace Col, 43-45; asst prof phys sci, Mich State Univ, 45-47; from asst prof to prof phys sci & head dept, Univ Ill, 47-64; chmn div sci, Grand Valley State Col, 64-69, prof chem, 64-78; PROF PHYS SCI, URBANA UNIV, 78- *Concurrent Pos:* Chemist, Process Chem Co, Mich, 39-40 & McGean Chem Co, Ohio, 45; res consult, Culligan, Inc, 51-53; Smith-Mundt exchange prof, Kabul Univ, Afghanistan, 61-62; fel, Imp Col, Univ London, 62; vis prof, Univ Kent, 71. *Mem:* AAAS; emer mem Am Chem Soc; Royal Soc Chem; Sigma Xi. *Res:* Organic synthesis; aromatic and heterocyclic compounds. *Mailing Add:* Urbana Univ 635 Boyce St Urbana OH 43078

MELOY, THOMAS PHILLIPS, b New York, NY, Sept 14, 25; m 57; c 1. PHYSICS, METALLURGY. *Educ:* Harvard Univ, AB, 49; Mass Inst Technol, BS, 51, PhD(metall), 60. *Prof Exp:* Metall engr, Gen Elec Co, 51-52, 53-54, instr chem eng, 52-53, dept specialist jet eng, 54-57; from instr to asst prof metall, Mass Inst Technol, 57-61, vis scientist, Lincoln Lab, 61; sr staff scientist, Allis-Chalmers Mfg Co, Wis, 61-67; staff scientist, Westinghouse Airbrake Co & mgr environ & appl sci ctr, Melpar Co, 67-70; vpres res & develop, 70-,; CLAUDE WORTHINGTON BENEDUM PROF MINERAL PROCESSING, W VA UNIV, 67- *Concurrent Pos:* Consult, Mat Tech Co, 58-62, US Coast & Geod Surv, 65-, Off Sci & Technol, 65- & ITT Res Inst, 68-; Ford Found fel, 60-61; instr, Boston Univ, 61-62 & Univ Wis-Milwaukee, 63-64; ed, J Marine Technol, 64-71; dir, Eng Div, Nat Sci Found, 64-66; chmn interagency comt oceanog, Nat Securities Industs Asn Continental Shelf Study, 65-66; dir indust extractive & process res div, Environ Protection Agency, 66-67; chmn, Eng Found Conf Particulate Systs, 66, 69, Rapid Excavation, 70, Coal Mine Safety & Survival, 71; Nat Sci Found travel fel, Poland & Yugoslavia, 71. *Mem:* AAAS; Am Inst Aeronaut & Astronaut; Am Inst Mining, Metall & Petrol Engrs; fel Marine Technol Soc; Am Inst Chem Engrs. *Res:* Behavior of particulate systems involving comminution and accretion; origin of solar system; aerosols, modeling of industrial process, air, water, and oil pollution; oil tagging; particle morphology; separation of particulates; coal cleaning; mineral processing; relating particle shape to behavior in industrial processes. *Mailing Add:* 413 Jefferson St Morgantown WV 26505

MELROSE, JAMES C, b Spokane, Wash, Mar 27, 22; m 52; c 1. COLLOID & SURFACE CHEMISTRY. *Educ:* Harvard Univ, SB, 43; Stanford Univ, PhD(chem), 58. *Prof Exp:* Jr res chemist, Shell Oil Co, 43-46; from res chemist to sr res chemist, Mobil Oil Corp, 47-51 & 54-58, from res assoc to sr res assoc, 58-67; res scientist, Mobil Res & Develop Corp, 67-85. *Concurrent Pos:* Vis lectr, Univ Tex, 64; vis prof, Univ Tex, 74; consult prof, Dept Petrol Eng, Stanford Univ, 88- *Mem:* Fel AAAS; Am Chem Soc; Soc Petrol Engrs; Hist Sci Soc. *Res:* Thermodynamics of interfaces; capillary phenomena in porous media; properties of petroleum reservoir rocks and fluids; processes for enhanced oil recovery; history of colloid and surface chemistry. *Mailing Add:* 7135 Aberdeen Ave Dallas TX 75230

MELROSE, RICHARD B, b Sydney, Australia, April 8, 49; m 70. PARTIAL DIFFERENTIAL EQUATIONS, SPECTRAL THEORY. *Educ:* Univ Toronto, BSc, 69; Australian Nat Univ, BSc, 70; Univ Cambridge, PhD(math), 74. *Prof Exp:* Res asst math, Serv Res Coun, UK, 74-75; res fel, St Johns Col, Eng, 74-76; asst prof, 76-79, PROF MATH, MASS INST TECHNOL, 79- *Honors & Awards:* Bocher Prize, Am Math Soc, 84. *Mem:* Am Math Soc; Am Acad Arts & Sci; Australian Math Soc. *Res:* Differential geometric methods in differential equations and applications to scattering and spectral theory. *Mailing Add:* Mass Inst Technol Rm 2-243 Cambridge MA 02139

MELSA, JAMES LOUIS, b Omaha, Nebr, July 6, 38; m 60; c 6. CONTROL ENGINEERING, DIGITAL SIGNAL PROCESSING. *Educ:* Iowa State Univ, BSEE, 60; Ariz Univ, MSEE, 62, PhD(elec eng), 65. *Prof Exp:* Assoc mem tech staff, Radio Corp Am, 60-61; instr elec eng, Ariz Univ, 61-65, asst prof, 65-67; from assoc prof to prof info & control sci, South Methodist Univ, 67-73; PROF & CHMN ELEC ENG DEPT, UNIV NOTRE DAME, 73- *Concurrent Pos:* Consult, Los Alamos Nat Lab, 65- & Tellabs, Inc, 80- *Mem:* Fel Inst Elec & Electronics Engrs. *Res:* Applications of optimal control and estimation theory to the problem of systems control. *Mailing Add:* Dept of Elec Eng Univ Notre Dame Notre Dame IN 46556

MELSHEIMER, FRANK MURPHY, b Alhambra, Calif, Dec 9, 40; m; c 1. MECHANICAL ENGINEERING, ASTRONOMICAL INSTRUMENTATION. *Educ:* Univ Calif, Berkeley, BS, 63, MSME, 68, DEng, 72. *Prof Exp:* Electrometer engr, Appl Physics Corp, 62-64; instrumentation engr aircraft flight test, Douglas Aircraft Co, 65-67; chief engr astron, Lick Observ, Univ Calif, 72-76; proj prof eng, McDonald Observ, 76; asst prof eng design, Univ Colo, 76-78; ENGR & CONSULT ASTRON, DFM ENG, 79- *Concurrent Pos:* Design engr, T Melsheimer Co, 76-77; instrument designer, Monterey Inst Res Astron, 77-79. *Mem:* Am Astron Soc. *Res:* Digital control; micro computer applications. *Mailing Add:* 8137 Alfalfa Ct Longmont CO 80503-8518

MELSHEIMER, STEPHEN SAMUEL, b Baton Rouge, La, Feb 23, 43; m 65; c 2. CHEMICAL ENGINEERING. *Educ:* La State Univ, Baton Rouge, BS, 65; Tulane Univ, PhD(chem eng), 69. *Prof Exp:* Asst prof, 69-77, assoc prof, 77-81, PROF CHEM ENG, CLEMSON UNIV, 81- *Mem:* Am Inst Chem Engrs. *Res:* Process dynamics and control; applied mathematics. *Mailing Add:* Dept Chem Eng Clemson Univ Clemson SC 29634

MELSON, GORDON ANTHONY, b Sheffield, Eng, July 6, 37; m 62; c 2. INORGANIC CHEMISTRY. *Educ:* Univ Sheffield, BSc, 59, PhD(chem), 62. *Prof Exp:* Res assoc chem, Ohio State Univ, 62-64; lectr, Univ Strathclyde, 64-69; asst prof, Mich State Univ, 69-75; assoc prof, 75-80, PROF CHEM, VA COMMONWEALTH UNIV, 80- *Concurrent Pos:* Chmn chem, Va Commonwealth Univ, 83- *Mem:* Am Chem Soc. *Res:* Catalysis of coal liquefaction; mechanisms of macrocyclic ligand formation. *Mailing Add:* Dept Chem VCU Box 2006 Va Commonwealth Univ Richmond VA 23284-2006

MELTER, ROBERT ALAN, b New York, NY, Mar 20, 35; m 65; c 1. BOOLEAN ALGEBRA, COMBINATURIES. *Educ:* Cornell Univ, AB, 56; Univ Mo, AM, 60, PhD(math), 62. *Prof Exp:* Instr math, Univ Mo, 58-62; asst prof, Univ RI, 62-64 & Univ Mass, 64-67; assoc prof, Univ SC, 67-71; assoc prof, 71-80, PROF MATH, SOUTHAMPTON COL, LONG ISLAND UNIV, 80- *Concurrent Pos:* Vis asst prof, Amherst Col, 67; Fulbright lectr, Univ Niamey, Niger, 74-75; ed consult, Math Rev, 78-80, assoc ed, 80-81. *Mem:* Am Math Soc; Sigma Xi. *Res:* Algebra; abstract distance spaces and valuations; graph theory; binary images. *Mailing Add:* Dept Math Southampton Col Long Island Univ Southampton NY 11968

MELTON, ARTHUR RICHARD, b Ysleta, Tex, Apr 28, 43; m 65; c 2. PUBLIC HEALTH & EPIDEMIOLOGY. *Educ:* Univ Utah, BS, 69; Univ NC, MPH, 74, DrPH, 76. *Prof Exp:* Microbiologist, Div Health, Bur Labs, Utah Dept Social Serv, 70-73; DIR, LAB PROG, SDAK DEPT HEALTH, 76- *Mem:* Am Public Health Asn; Am Soc Microbiol. *Mailing Add:* 6835 S Heather Way West Jordan UT 84084

MELTON, BILLY ALEXANDER, JR, b Wheeler, Tex, Aug 21, 32; m 51; c 2. PLANT BREEDING, GENETICS. *Educ:* NMex Col Agr & Mech Arts, BS, 54; Univ Ill, MS, 56, PhD, 58. *Prof Exp:* Asst plant breeding, Univ Ill, 54-58; FROM ASST PROF TO PROF AGRON, N MEX STATE UNIV, 58- *Honors & Awards:* Distinguished Res Award, NMex State Univ & N K & Co, 67. *Mem:* fel Am Soc Agron; Am Grassland Coun; fel Crop Sci Soc Am; Nat Alfalfa Importers Conf (vpres, 78-80, pres, 80-82); Western Soc Crop Sci (secy, 67, vpres, 70, pres, 71); Western Alfalfa Importers Conf (secy, 73, vpres, 74, pres, 75). *Res:* Genetics of sorghum and forage crops, primarily alfalfa; alfalfa breeding, production, hay packaging and hay quality. *Mailing Add:* Dept Agron NMex State Univ Las Cruces NM 88003

MELTON, CARLTON E, JR, b Allen, Tex, June 1, 24; m 57; c 3. ELECTROPHYSIOLOGY, HUMAN STRESS. *Educ:* NTex State Col, BS, 48; Univ Ill, MS, 50, PhD(physiol), 53. *Prof Exp:* Elec eng asst, Univ Ill, 49-50, asst physiol, 50-53; instr, Med Sch, Western Reserve Univ, 53-55; asst prof, Med Sch, Univ Tex, 55-61; chief electrophysiol, Civil Aeromed Res Inst,

65-84; PROF BIOL, OKLA CITY UNIV, 85- *Concurrent Pos:* Consult human factors, 84- *Mem:* Am Asn Anatomists; Am Physiol Soc; Soc Exp Biol Med; Aerospace Med Asn. *Res:* Hormonal effects on contractile and electrical activity of smooth muscle of the uterus; human stress, fatigue and work tolerance, shift work; ozone toxicity; author of 55 technical publications. *Mailing Add:* 2710 Walnut Rd Norman OK 73072

MELTON, CHARLES ESTEL, b Fancy Gap, Va, May 18, 24; m 46; c 3. CHEMICAL PHYSICS. *Educ:* Emory & Henry Col, BA, 52; Vanderbilt Univ, MS, 54; Univ Notre Dame, PhD(phys chem), 64. *Hon Degrees:* DSc, Emory & Henry Col, 67. *Prof Exp:* Physicist, Oak Ridge Nat Lab, 54-67; head dept, 72-77, PROF CHEM, UNIV GA, 67- *Honors & Awards:* DeFriece Award Physics, Emory & Henry Col, 59. *Res:* Chemical kinetics; mass spectrometry; atmospheric chemistry and physics; catalysis; theoretical chemistry and geochemistry. *Mailing Add:* Rte 2 Box 34 Hull GA 30646

MELTON, JAMES RAY, b Paris, Tex, Aug 24, 40; m 62; c 4. ANALYTICAL CHEMISTRY, SOIL CHEMISTRY. *Educ:* Tex Tech Col, BS, 62; Mich State Univ, MS, 64, PhD(soil sci), 68. *Prof Exp:* PROF AGR ANALYSIS SERV, OFF TEX STATE CHEMIST, TEX A&M UNIV, 68- *Mem:* Sigma Xi; Asn Off Analytical Chemists; Am Soc Agron; Soil Sci Soc Am. *Res:* Development of methods for quantitation of nitrogen in feeds and fertilizers; comparison of methods for analyzing boron in feeds and fertilizers; development of methods for determination of metals by atomic absorption and atomic emission. *Mailing Add:* 1305 Haines College Station TX 77840

MELTON, LEE JOSEPH, III, b Pensacola, Fla, May 17, 44; c 4. EPIDEMIOLOGY, MEDICINE. *Educ:* La State Univ, BS, 65, MD, 69; Univ Mich, MPH, 71. *Prof Exp:* Epidemiologist, US Navy, 67-77; from asst prof to assoc prof, 78-86, PROF, MAYO MED SCH, 86- *Concurrent Pos:* Asst prof epidemiol, Uniformed Serv Univ Health Sci, 76-77; expert comt assessment diabetes data, Nat Diabetes Data Group; consult epidemiol, Mayo Clin, 77-; mem geriat & geront rev comt,Study Sect NIH, 86- *Honors & Awards:* Gaylor Award, Asn Mil Surgeons US, 76. *Mem:* Am Col Prev Med; Soc Epidemiol Res; Am Col Epidemiol; Int Epidemiol Asn; Sigma Xi; Am Epidemiol Soc. *Res:* Epidemiology of diabetes mellitus, osteoporosis and gastrointestinal disease. *Mailing Add:* 925 Sixth Ave SW Rochester MN 55902

MELTON, LYNN AYRES, b Huntsville, Tex, Aug 7, 44; m 67; c 2. PHYSICAL CHEMISTRY, SCIENCE EDUCATION. *Educ:* Calif Inst Technol, BS, 66; Harvard Univ, MA & PhD(chem), 72. *Prof Exp:* From asst prof to assoc prof chem, 71-84, head prog, 76-79, head sci educ, 79-81, PROF CHEM, UNIV TEX, DALLAS, 84- *Concurrent Pos:* Consult, United Technologies Res Ctr, East Hartford, Conn, 81-82. *Mem:* Am Chem Soc; Am Phys Soc; AAAS; Combustion Inst. *Res:* Development of fluorescence based diagnostics; development of analytical instrumentation. *Mailing Add:* Univ Tex MS B2 6 Richardson TX 75083

MELTON, MARILYN ANDERS, b Norristown, Pa, Oct 19, 44; c 2. ORGANIC SYNTHETIC CHEMISTRY. *Educ:* Pa State Univ, BS, 66; Univ Tex, Austin, PhD(org chem), 84. *Prof Exp:* Instr org chem, Southwestern Univ, Georgetown, Tex, 79-80; mgr chem, Kallestad Labs, Inc, Austin, Tex, 84-86; SR STAFF SCIENTIST, RADIAN CORP, AUSTIN, TEX, 86- *Mem:* Am Chem Soc. *Res:* Organic synthetic and organometallic chemistry; immunology. *Mailing Add:* Radian Corp PO Box 201088 Austin TX 78720-1088

MELTON, REX EUGENE, b Ozark, Mo, Dec 4, 21; m 42; c 4. FORESTRY. *Educ:* Univ Mo, BS, 46; Univ Mich, BS & MF, 47. *Prof Exp:* From asst prof to assoc prof, Sch Forest Resouces, Pa State Univ 47-76, dir exp forest, 58-78, asst to dir, 74-88, prof forestry 76-88; RETIRED. *Mem:* Fel Soc Am Foresters. *Res:* Forest entomology and management; silviculture. *Mailing Add:* Sch Forest Resources 107 Ferguson Bldg Pa State Univ University Park PA 16802

MELTON, THOMAS MASON, b Rockville, Va, Jan 8, 27; m 49, 57; c 2. ORGANIC CHEMISTRY, INDUSTRIAL HYGIENE. *Educ:* Col William & Mary, BS, 48; Am Acad Indust Hyg, dipl. *Prof Exp:* Res chemist, Sterling-Winthrop Res Inst, 48-51; sr chemist, Va-Carolina Chem Corp, 51-65; group leader agr chem, Mobil Chem Co, 65-71; chem specialist & consult chem, indust hyg & safety, Travelers Corp, 71-89; RETIRED. *Mem:* Am Chem Soc; Am Indust Hygiene Asn; Am Acad Indust Hygiene. *Res:* Synthesis of insecticides; organic phosphorus and organic medicinal chemicals. *Mailing Add:* 2066 Manakintown Ferry Rd Midlothian VA 23113-9723

MELTON, WILLIAM GROVER, JR, b Oakland, Calif, Jan 1, 23; m 56; c 5. VERTEBRATE PALEONTOLOGY. *Educ:* Univ Mont, BA, 53; Univ Mich, MS, 69. *Prof Exp:* Geologist, US Geol Surv, 53-56; preparator vert paleont, Univ Mich, 57-66; asst, 66-69, lectr, 69-75, adj assoc prof, 75-86, CUR GEOL, UNIV MONT, 66-, EMER PROF, 86- *Mem:* Soc Vert Paleont. *Res:* Paleozoic ray finned fish; conodonts and conodont bearing animals; soft anatomy and classification. *Mailing Add:* Dept Geol Univ Mont Missoula MT 59812

MELTZ, MARTIN LOWELL, b New York, NY, Dec 22, 42; m 70; c 1. EXPERIMENTAL ONCOLOGY, GENETIC TOXICOLOGY. *Educ:* State Univ NY Stony Brook, BS, 63; Univ Rochester, PhD(biophys), 70. *Prof Exp:* AEC fel, Lab Radiobiol, Univ Calif, San Francisco, 69-70, res biophysicist, 70-71; assoc scientist, Southwest Found Res & Educ, 71-79; asst prof, 79-83, ASSOC PROF RADIOL, UNIV TEX HEALTH SCI CTR, 83- *Mem:* AAAS; Biophys Soc; Radiation Res Soc; Environ Mutagen Soc; Am Soc Cell Biol; Bioelectromagnetics Soc. *Res:* Mammalian cell mutagenicity; DNA damage and repair; genetic effects of microwave radiation; combined modality chemotherapy-radiotherapy studies in mammalian cells; multicellular spheroid growth. *Mailing Add:* Dept Radiol Univ Tex Health Sci Ctr 7703 Floyd Curl Dr San Antonio TX 78284

MELTZER, ALAN SIDNEY, b New York, NY, Apr 26, 32; m 57. ASTRONOMY. *Educ:* Syracuse Univ, BS, 53; Princeton Univ, PhD(astron), 56. *Prof Exp:* Res assoc observ, Harvard Col, 56; physicist, Smithsonian Astrophys Observ, 56-57; asst prof, 57-74, assoc prof, 74-78, PROF ASTRON, RENSSELAER POLYTECH INST, 78-, COORDR, LEARNING CTR, 80- *Mem:* Fel AAAS; Am Astron Soc. *Res:* Solar and stellar spectroscopy; interstellar extinction of polarization; science education; developmental education; administration. *Mailing Add:* Dept Astron Rensselaer Polytech Inst Troy NY 12180

MELTZER, ARNOLD CHARLES, b New York, NY, May 9, 36; m 61; c 2. COMPUTER SCIENCE, ELECTRICAL ENGINEERING. *Educ:* George Wash Univ, BS, 58, MSE, 61. *Hon Degrees:* DSc, George Wash, 67. *Prof Exp:* Student trainee elec rates, Fed Power Comn, 56-58, rate engr, 58-60; from instr to assoc prof, 60-74, actg chmn dept elec eng & comput sci, 69-70, chmn, 74-78, PROF COMPUT SCI, GEORGE WASHINGTON UNIV, 74- *Concurrent Pos:* Consult, Fed Power Comn, 61-63 & US govt, 75-; NASA grant, 69-70; Am Soc Eng Educ-Ford Found fac resident, IBM Corp, NY, 70-71. *Mem:* Inst Elec & Electronics Engrs; Asn Comput Mach. *Res:* Synthesis of active filters using distributed parameter networks and a minimum number of active devices; computer architecture; multiprocessors; distributed processing. *Mailing Add:* Sch Eng & Appl Sci George Washington Univ 725 23rd St NW Washington DC 20052

MELTZER, HERBERT LEWIS, b New York, NY, Apr 23, 21; m 49; c 2. BIOCHEMISTRY. *Educ:* Long Island Univ, BS, 42; Columbia Univ, PhD, 50. *Prof Exp:* Res assoc, 51-56, ASST PROF BIOCHEM, COL PHYSICIANS & SURGEONS, COLUMBIA UNIV, 56-; RES SCIENTIST, NY STATE PSYCHIAT INST, 77-; AT CASE WESTERN RESERVE UNIV. *Concurrent Pos:* Assoc res scientist, NY State Psychiat Inst, 52-77. *Mem:* Fel Am Col Clin Pharmacol; Am Soc Biol Chem. *Res:* Neurochemical and behavioral effects of rubidium; membrane transport systems; lithium kinetics in manic-depressive illness; calcium transport in psychiatric illness. *Mailing Add:* NY State Psychiat Inst 722 W 168 St New York NY 10032

MELTZER, HERBERT YALE, b Brooklyn, NY, July 29, 37; m 60; c 2. PSYCHIATRY. *Educ:* Cornell Univ, BA, 58; Harvard Univ, MA, 59; Yale Univ, MD, 63. *Prof Exp:* Res assoc pharmacol & anat, Sch Med, Yale Univ, 59-63; clin assoc, Lab Clin Sci, NIH, 66-68, instr grad training prog, 67-68; from asst prof to assoc prof, 68-74, PROF PSYCHIAT, UNIV CHICAGO, 74-; RES ASSOC, ILL STATE PSYCHIAT INST, 68-, DIR, LAB BIOL PSYCHIAT, 75- *Concurrent Pos:* Teaching fel psychiat, Harvard Med Sch, 64-66; consult, Peter Bent Brigham Hosp & Mass Ment Health Ctr, 65-66 & VISTA, 68-; assoc ed, Schizophrenia Bull. *Mem:* Am Col Neuropsychopharmacol. *Res:* Biological study of mental illness; neuroendocrinology; biochemical and behavioral pharmacology of psychotomimetic drugs. muscle physiology and ultrastructure. *Mailing Add:* Univ Hosp of Cleveland Case West Res Univ 2040 Abington Rd Cleveland OH 44106

MELTZER, MARTIN ISAAC, Zimbabwean citizen. AGRICULTURAL ECONOMICS. *Educ:* Univ Zimbabwe, BSc(agric) hons, 82; Cornell Univ, MS, 87, PhD(agr econ), 90. *Prof Exp:* Proj officer, USAID, Zimbabwe, 83; grad asst, Cornell Univ, 84-89; ASST SCIENTIST, UNIV FLA, 90- *Mem:* Am Agr Econ Asn. *Res:* Potential economic and ecological impact of biotechnology to improve livestock production; economics of transferring biotechnologies to control ticks and tick-born diseases to developing countries, particularly Zimbabwe and Thailand. *Mailing Add:* Dept Infectious Dis Bldg 471 Mowry Rd Gainesville FL 32611-0633

MELTZER, MONTE SEAN, b Jersey City, NJ, Sept 15, 43. MACROPHAGE LYMPHOKINE INTERACTIONS. *Educ:* Georgetown Univ, MD, 69. *Prof Exp:* DIR, PROG NON-SPECIFIC COMMUNITY, DEPT IMMUNOL, WALTER REED ARMY INST RES, 83- *Mem:* Am Acad Dermat; Am Asn Immunol; Reticuloendothelial Soc. *Mailing Add:* Dept Immunol Walter Reed Army Inst Res 9620 Medical Center Dr Suite 200 Rockville MD 20850

MELTZER, RICHARD S, b New York, NY, Aug 1, 42; m 69; c 2. OPTICAL PROPERTIES OF SOLIDS, PHONON PHYSICS. *Educ:* Cornell Univ, AB, 63; Univ Chicago, PhD(chem), 68. *Prof Exp:* Res assoc physics, Johns Hopkins Univ, 68-70; from asst prof to assoc prof, 70-82, PROF PHYSICS, UNIV GA, 82- *Concurrent Pos:* Vis scientist, IBM Corp, 84; ed, J Luminescence, Elsevier, 90- *Mem:* Fel Am Phys Soc. *Res:* Optical properties of solids; laser spectroscopy; interionic interactions; energy transfer; dynamics of the excited state; phonon dynamics. *Mailing Add:* Dept Physics & Astron Univ Ga Athens GA 30602

MELTZER, RICHARD STUART, b New York, NY, Sept 6, 48; m 71; c 2. INTERNAL MEDICINE, CARDIOLOGY. *Educ:* Harvard Univ, BA, 70, MD, 74; Erasmus Univ, Rotterdam, Neth, PhD(cardiol), 82. *Prof Exp:* Fel cardiol, Stanford Univ, 77-79; asst prof, Erasmus Univ, Rotterdam, 79-82; from asst prof to assoc prof, Mt Sinai Med Sch, New York, 83-86; PROF CARDIOL, UNIV ROCHESTER, NY, 86- *Concurrent Pos:* Vis assoc prof cardiol, Tel Aviv Univ, Israel, 82; estab fel, NY Heart Asn, 85; fel, Coun Clin Cardiol, Am Heart Asn. *Honors & Awards:* Clin-Scientist Award, Am Heart Asn, 80. *Mem:* Fel Am Col Physicians; fel Am Col Cardiol; Am Fedn Clin Res; Am Inst Ultrasound in Med; Am Soc Echocardiography; Neth Soc Cardiol; Am Heart Asn; fel Europe Soc Cardiol. *Res:* Uses of ultrasound to diagnose cardiovascular disease, especially in the field of contrast and doppler echocardiography and ultrasound tissue characterization. *Mailing Add:* Cardiol Box 679 Univ Rochester Med Ctr Rochester NY 14642

MELVEGER, ALVIN JOSEPH, b New York, NY, July 9, 37; m 61; c 1. CHEMISTRY, ANALYTICAL CHEMISTRY. *Educ:* Brooklyn Col, BS, 59; Northeastern Univ, MS, 64; Univ Md, College Park, PhD(chem), 68. *Prof Exp:* Scientist mat, Avco Corp, Mass, 60-64; grant, Ctr Mat Res, Univ Md, Col Park, 68-69; res chemist, Allied Chem Corp, 69-72; sect mgr instrumental anal, 72-78, MGR, ANAL CHEM DEPT, ETHICON, INC, 78-, MGR, JOHNSON & JOHNSON POLYMER CHARACTERIZATION CTR, 84- *Mem:* AAAS; Am Chem Soc; Am Phys Soc; Soc Appl Spectros; Am Soc Testing & Mat; Sigma Xi; N Am Thermal Anal Soc. *Res:* Molecular, analytical and high pressure spectroscopy; laser-Raman and infrared spectroscopy of polymers and inorganic materials; structure-property relationships of polymers; polymer characterization; pharmaceutical analysis. *Mailing Add:* Ethicon Inc Somerville NJ 08876

MELVILLE, DONALD BURTON, b Netherton, Eng, Jan 30, 14; nat US; m 40; c 2. BIOCHEMISTRY. *Educ:* Univ Ill, BS, 36, MS, 37, PhD(biochem), 39. *Prof Exp:* Asst biochem, Univ Ill, 36-38; res assoc, Med Col, Cornell Univ, 39-46, from asst prof to assoc prof, 46-60; chmn dept, Biochem, Col Med, Univ Vt, 60-76, prof, 60-79. *Mem:* Am Chem Soc; Am Soc Biol Chem; NY Acad Sci. *Res:* Determination of structure and study of biological effects of biotin; chemistry of penicillin; syntheses with radio isotopes; biochemistry of ergothioneine. *Mailing Add:* 1 Fern St Burlington VT 05401

MELVILLE, JOEL GEORGE, b Meadville, Pa, Feb 5, 43; m 67; c 2. GROUNDWATER HYDRAULICS. *Educ:* Univ Tex, MS, 67; Pa State Univ, BS, 65, PhD(eng mech), 72. *Prof Exp:* Adj asst prof hydraulics, Univ Iowa, 72-76; asst prof, Univ Fla, 76-79; ASSOC PROF HYDRAULICS, AUBURN UNIV, 79- *Mem:* Am Soc Civil Engrs; Am Water Well Asn; Am Geophys Union. *Res:* Fluid mechanical aspects of flow in the small intestine; hydraulic modeling; aquifer thermal energy storage; groundwater transport of contaminants. *Mailing Add:* Dept Civil Eng Auburn Univ Auburn AL 36849

MELVILLE, MARJORIE HARRIS, b Baltimore, Md, Aug 30, 27; m 48; c 3. ORGANIC CHEMISTRY. *Educ:* Agnes Scott Col, BA, 47; Johns Hopkins Univ, MA, 49, PhD(chem), 53. *Prof Exp:* From asst prof to prof, 64-89, EMER PROF CHEM, SAN ANTONIO COL, 89- *Mem:* Am Chem Soc; Sigma Xi. *Mailing Add:* PO Box 33985 San Antonio TX 78265-3985

MELVILLE, ROBERT S, b Worcester, Mass, Nov 20, 13. CLINICAL CHEMISTRY. *Educ:* Clark Univ, BS, 37; State Univ Iowa, PhD(biochem), 50. *Prof Exp:* Dir standards & standardization, 80-82, CONSULT, CLIN LAB SCI, FOOD & DRUG ADMIN, 81- *Mem:* Fel Am Asn Clin Chemists; fel Am Inst Chemists; sr mem Instrument Soc Am. *Mailing Add:* 1112 Kenilworth Ave PO Box 56 Garrett Park MD 20896

MELVIN, CRUSE DOUGLAS, b Woodville, Tex, June 5, 42; m 64; c 2. PHYSICS, MATHEMATICS. *Educ:* Stephen F Austin State Univ, BS, 64, MS, 65; Tulane Univ, PhD(physics), 71. *Prof Exp:* Instr physics, Stephen F Austin State Univ, 65-66; asst prof, Nicholls State Univ, 66-70; sr res scientist phys anal, Sci Res Lab, Ford Motor Co, 71-72, supvr eng res staff, 72-77; assoc prof, 77-80, PROF PHYSICS, DELTA STATE UNIV, 80- *Mem:* Am Asn Physics Teachers. *Res:* Scanning and transmission electron microscopy; surface analysis, x-ray analysis and neutron analysis methods; computer science. *Mailing Add:* 105 George Silsbee TX 77656

MELVIN, DONALD WALTER, b Manchester, NH, Sept 17, 29; m 59, 82; c 2. ELECTRICAL ENGINEERING. *Educ:* Univ NH, BS, 55; Yale Univ, MEng 57; Syracuse Univ, PhD(elec eng), 71. *Prof Exp:* From instr to asst prof, 57-65, asst dean, 77-78, ASSOC PROF ELEC ENG, COL ENG & PHYS SCI, UNIV NH, 65-, ASSOC DEAN, 78- *Concurrent Pos:* Instr, Syracuse Univ, 61-63. *Mem:* Inst Elec & Electronics Engrs; Sigma Xi. *Res:* Ocean engineering; energy systems. *Mailing Add:* Dean's Off Kingsbury Hall Univ NH Durham NH 03824

MELVIN, DOROTHY MAE, b Fayetteville, NC. MEDICAL PARASITOLOGY. *Educ:* Univ NC, Greensboro, AB, 42; Univ NC, Chapel Hill, MS, 45; Rice Univ, PhD(parasitol), 51. *Prof Exp:* Training officer & med parasitologist, Ctr Dis Control, 45-49, 51-62, chief parasitol training unit, 62-74, chief parasitol training br, 74-85; RETIRED. *Concurrent Pos:* La State Univ fel trop med, PR & Haiti, 58; asst prof, Sch Med, Emory Univ, 67- *Mem:* Am Soc Trop Med & Hyg; Am Soc Parasitol. *Res:* Methodology of technical training for health care personnel. *Mailing Add:* 2418 Kingscliff Dr NE Atlanta GA 30345

MELVIN, E(UGENE) A(VERY), b Baltimore, Md, Oct 11, 19; m 46; c 3. ELECTRICAL ENGINEERING. *Educ:* Johns Hopkins Univ, BE, 41; State Univ NY Buffalo, MS, 50. *Prof Exp:* Elec designer, Glenn L Martin, 41; instr electronics, State Univ NY Buffalo, 47-50; proj engr, Frederick Res Corp, 50-58, tech dir, 58-59; eng mgr, Martin Co, Fla, 59-67; dir, Tech Support Div, 67-75, staff asst resources, 75-, dir, Naval Air Testing Ctr; CONSULT. *Concurrent Pos:* Instr, Millard Fillmore Col, 48; res engr, Fredric Flader, Inc, 48-50. *Mem:* Am Soc Eng Educ; Am Ord Asn; Inst Elec & Electronics Engrs; Am Inst Aeronaut & Astronaut. *Res:* Electrical analogs; servomechanisms; ground-support equipment and checkout; instrumentation. *Mailing Add:* Wood-Ivey Systs Corp PO Box 4609 Winter Park FL 32793

MELVIN, JOHN L, b Columbus, Ohio, May 26, 35; m 57; c 4. PHYSICAL MEDICINE & REHABILITATION, ELECTROMYOGRAPHY. *Educ:* Ohio State Univ, BSc, 55, MD, 60, MMSc, 66; Am Bd Phys Med & Rehab, Cert, 68. *Prof Exp:* From asst prof to assoc prof phys med, Col Med, Ohio State Univ, 66-73; PROF PHYS MED & REHAB & CHMN DEPT, MED COL WIS, 73- *Concurrent Pos:* Mem adv comt, Joint Comt Stroke Facil, 69-77; med dir, Curative Rehab Ctr Milwaukee, 73-; sr attend staff, Milwaukee County Med Complex, 73-; dir phys med, Sacred Heart Rehab Hosp, 74-83; mem consult staff, Milwaukee Children's Hosp, 73-, St Luke's Hosp, 73- & West Allis Mem Hosp, 74-, Good Samaritan Med Ctr, 77-78; staff physician, Zablocki Vet Admin Ctr, 76-; consult, Brown Univ, 78, Nat Cancer Inst, 78-79 & Univ Calif, San Francisco, 80; Switzer fel, Nat Rehab Asn, 78; Ohio State Univ Mershon Ctr, 80-82, Emory Univ Sch Med, 82, Sister Kenny Inst, 85, UN High Comn Refugees, 86; mem, Adminr Adv Comn Rehab, Vet Admin, 81-88. *Honors & Awards:* Sci Exhib-Gold Medal Award, Am Cong Rehab Med, 71; Walter J Zeiter Award, Am Acad Phys Med & Rehab, 87; Gold Key Award, Am Cong Rehab Med, 88. *Mem:* Am Acad Phys Med & Rehab; Am Cong Rehab Med (pres, 88-89); Am Asn Electromyog & Electrodiag (pres, 79); Asn Acad Physiatrist (pres, 85-87); Int Fedn Phys Med & Rehab; Nat Asn Rehab Facil (pres, 81-83); Coun Med Specialty Socs (pres, 89-90). *Res:* Testing models for delivery of rehabilitation services; conceptualizing, classifying, and describing disability states; identifying electrophysiologic responses associated with diseases of the peripheral nervous system. *Mailing Add:* Dept Phys Med & Rehab Med Col Wis 1000 N 92nd St Milwaukee WI 53226

MELVIN, JONATHAN DAVID, b New York, NY, Apr 7, 47; c 2. ATOMIC PHYSICS, GEOPHYSICS. *Educ:* Yale Univ, BA & MA, 68; Calif Inst Technol, PhD(physics), 74. *Prof Exp:* MPS, ENG & PHYSICS, CALIF INST TECHNOL, 75- *Concurrent Pos:* Partner, Interfield Res Assoc, Santa Monica, Calif, 83-; var teaching, consult, tech & vis positions, univ & indust, US, Mex, Peoples Repub China. *Mem:* AAAS; Am Geophys Union; Am Phys Soc. *Res:* Atomic physics, nuclear accelerators, geochemical techniques for earthquake prediction; high speed computer networks; computer graphics; data acquisition. *Mailing Add:* Calif Inst Technol 104-44 Pasadena CA 91125

MELVIN, LAWRENCE SHERMAN, JR, b Chicago, Ill, Feb 1, 47; m 66; c 2. ORGANIC CHEMISTRY, MEDICINAL CHEMISTRY. *Educ:* Univ Ill, BS, 69; Univ Wis, PhD(org chem), 73. *Prof Exp:* Fel chem, Harvard Univ, 73-75; ASST DIR, PFIZER INC, 88- CENT RES, 75- *Mem:* Am Chem Soc. *Res:* Synthetic organic chemistry; medicinal chemistry; natural product chemistry. *Mailing Add:* 7 Seabury Ave Ledyard CT 06339

MELVIN, MAEL AVRAMY, b Palestine, Mar 27, 13; US citizen; m 46; c 2. PHYSICS, THEORETICAL PHYSICS. *Educ:* Univ Chicago, BS, 33, MS, 35, PhD(physics), 38. *Prof Exp:* Metallurgist, Carnegie Ill Steel Corp, 37-38; assoc metall, Columbia Univ, 38-40, instr, 40-42, instr physics, 42-46, asst prof metal physics, 46-47, assoc prof, 47-48; vis prof, Univ Ore, 51; Guggenheim fel, Princeton Univ, 51-52; prof physics, Fla State Univ, 52-66; prof physics, Temple Univ, 66-80; RETIRED. *Concurrent Pos:* consult, Chem Warfare Serv, US Dept Army, 41, Nat Defense Res Comt, 42, Hazeltine Serv Corp, NY, 43, Heat Transfer Res Lab, 43 & Off Sci Res & Develop, 43-45; Guggenheim fel & vis prof, Univ Upsala, 57-58; Int Atomic Energy Agency exchange prog vis prof, Inst Physics, Bariloche, Arg, 59-60; Nat Res Coun sr resident res assoc, Jet Propulsion Lab, Calif Inst Technol, 71-72; consult, Chem Warfare Serv, US Dept Army, 41, Nat Defense Res Comt, 42, Hazeltine Serv Corp, NY, 43, Heat Transfer Res Lab, 43 & Off Sci Res & Develop, 43-45; vis fel, Dept Physics, Princeton Univ, 76; vis prof, Tata Inst Fundamental Res, Bombay, 81-82, 82-83; Fulbright fel, 81-82. *Mem:* Fel Am Phys Soc. *Res:* Symmetry methods in physics; elementary particles and fields; relativity; astrophysics; cosmology; homogeneous and anisotropic cosmologies. *Mailing Add:* 1300 Orchid Dr Santa Barbara CA 93111

MELVIN, STEWART W, b Bloomfield, Iowa, Nov 24, 41; m 71; c 2. SOIL & WATER CONSERVATION, WASTE MANAGEMENT ENGINEERING. *Educ:* Iowa State Univ, BS, 64, MS, 67, PhD(agr eng), 70. *Prof Exp:* Field engr, Soil Conserv Serv, USDA, 64; asst prof agr eng, Colo State Univ, 70; asst prof, 70-74, assoc prof, 74-78, PROF AGR ENG, IOWA STATE UNIV, 78- *Concurrent Pos:* private consult, Soil Water & Waste Mgt; vis prof, Silsoe Col, UK, 85-86. *Mem:* Am Soc Agr Engrs; Soil & Water Conserv Soc Am; Sigma Xi. *Res:* Soil and water conservation; agricultural water quality; rural water and waste systems; subsurface irrigation; soil compaction effects of agricultural equipment. *Mailing Add:* Dept Agr Eng Iowa State Univ Ames IA 50011

MELVOLD, ROGER WAYNE, b Henning, Minn, Mar 21, 46. GENETICS. *Educ:* Moorhead State Col, BS, 68; Univ Kans, PhD(genetics), 73. *Prof Exp:* Prin res assoc radiation biol (genetics), Med Sch, Harvard Univ, 72-79; asst prof, 79-83, ASSOC PROF MED & MICROBIOL IMMUNOL, NORTHWESTERN UNIV, 83- *Mem:* Am Asn Immunologists; Transplantation Soc; Am Asn Lab Animal Sci. *Res:* Identifying and utilizing mutations of genes affecting tissue transplantation to explore the fine level genetic influences on the mammalian immune system; special emphasis on mouse system; immunogenetics of CNS demyelimating disease. *Mailing Add:* Dept Microbiol & Immunol Med Sch Northwestern Univ 303 E Chicago Ave Chicago IL 60611

MELZACK, RONALD, b Montreal, Que, July 19, 29; m 60; c 2. PSYCHOLOGY, PAIN. *Educ:* McGill Univ, BSc, 50, MSc, 51, PhD(psychol), 54. *Prof Exp:* Res fel physiol, Med Sch, Univ Ore, 54-57 & Univ Pisa, Italy, 58-59; lectr psychol, Univ Col London, Eng, 57-58; assoc prof, Mass Inst Technol, 59-63; PROF PSYCHOL, MCGILL UNIV, 63-, MED SCIENTIST PSYCHIAT, 70- *Concurrent Pos:* Consult scientist, dept neurosurg & dir, pain res, Pain Clin, Montreal Gen Hosp, 70- *Honors & Awards:* Stratton Award, Am Psychopath Asn, 63; Can Anesthesiol Soc Award, 74; Aid Int Med Found Award, 76; Molson Prize, Can Coun, 85; C S Miller Award, Am Acad Oral Med, 86; Distinguished Contrib Sci Award, Can Psychol Asn, 86; E P Taylor Chair Psychol, 86. *Mem:* Fel Royal Soc Can; Am Psychol Asn; Can Psychol Asn; Int Asn Study Pain (pres, 84-87). *Res:* Pain and analgesia; proposal of the gate control theory of pain; development of the McGill Pain Questionnaire. *Mailing Add:* Dept Psychol McGill Univ 1205 Dr Penfield Ave Montreal PQ H3A 1P1 Can

MEMEGER, WESLEY, JR, b Riverdale, Fla, Sept 21, 39; m 63; c 2. ORGANIC CHEMISTRY, ANALYTICAL CHEMISTRY. *Educ:* Clark Col, BS, 61; Adelphi Univ, PhD(org chem), 66. *Prof Exp:* Res chemist, 65-71, sr res chemist, 71-80, res assoc, 80-85, SR RES ASSOC, E I DU PONT DE NEMOURS & CO INC, 85- *Mem:* Am Chem Soc; Sigma Xi; NY Acad Sci. *Res:* Field effects in nucleophilic substitution reactions; reaction kinetics; synthesis and characterization of addition and condensation polymers; high temperature fibers; basic research aimed at new high strength high modulus polymers and fibers; basic and applied research on high temperature insulating cellular materials; liquid crystalline polymers. *Mailing Add:* 711 Coverly Rd Wilmington DE 19802-1909

MEMORY, JASPER DURHAM, b Raleigh, NC, Dec 10, 36; m 61; c 2. PHYSICS. *Educ:* Wake Forest Col, BS, 56; Univ NC, PhD(physics), 60. *Prof Exp:* From asst prof to assoc prof physics, Univ SC, 60-64; assoc prof, 64-67, vprovost & grad dean, 82-86, PROF PHYSICS, NC STATE UNIV, 67-, VPRES RES, UNC SYSTS, 86- *Mem:* Am Phys Soc; Am Asn Physics Teachers. *Res:* Nuclear magnetic resonance; molecular biophysics. *Mailing Add:* Univ NC Gen Admin PO Box 2688 Univ NC Gen Admin Chapel Hill NC 27515-2688

MENA, ROBERTO ABRAHAM, b Merida, Mex, Mar 12, 46; m 69. ALGEBRA. *Educ:* Univ Houston, BS, 68, MS, 71, PhD(math), 73. *Prof Exp:* From asst prof to assoc prof math, Univ Wyo, 73-; AT CALIF STATE UNIV. *Concurrent Pos:* Nat Acad Sci fel, Calif Inst Technol, 80-81. *Res:* Matrix theory and combinatorics. *Mailing Add:* Calif State Univ Long Beach CA 90804

MENAKER, LEWIS, b New York, NY, Apr 15, 42. PREVENTIVE DENTISTRY. *Educ:* Tufts Univ, DMD, 65; Mass Inst Technol, ScD(nutrit biochem), 71. *Prof Exp:* Chmn, dept oral biol, 80-88 & dept prev & community dent, 83-88, SR SCIENTIST, INST Dent, RES, 77-, ASSOC DEAN, SCH DENT, UNIV ALA 81- *Concurrent Pos:* Sr scholar, Ctr Aging, Univ Ala, Birmingham, 84-; vis prof, Nihon Univ, Japan; consult, Nat Inst Dent Res, NIH, Vet Admin; sr scholar, Ctr Health Risk Assessment & Dis Prev. *Mem:* Fel Royal Soc Prom Health; Int Asn Dent Health; Am Asn Dent Res; Sigma Xi; fel Am Col Dentists. *Res:* Clinical studies on agents for preventing oral diseases. *Mailing Add:* Sch Dent Univ Ala Birmingham AL 35294

MENAKER, MICHAEL, b Vienna, Austria, May 19, 34; US citizen; m 55; c 2. COMPARATIVE PHYSIOLOGY, BIOLOGICAL TIMING. *Educ:* Swarthmore Col, BA, 55; Princeton Univ, MA, 58, PhD(biol), 60. *Prof Exp:* NSF fel, Harvard Univ, 59-61, NIH fel, 61-62; from asst prof to prof zool, Univ Tex, Austin, 62-79; prof biol & dir, Inst Neurosci, Univ Ore, 79-86; PROF BIOL & CHMN, BIOL DEPT, UNIV VA, 87-, DIR, HOWARD HUGHES UNDERGRAD RES PROG BIOL SCI, 89- *Concurrent Pos:* Guggenheim mem found fel, 71; Benjamin Meaker vis prof, Univ Bristol, Eng, 86; sr investr, NSF Sci & Technol Ctr Biol Timing, Univ Va, 91- *Mem:* Am Physiol Soc; Am Soc Photobiol; Soc Neurosci; fel AAAS. *Res:* Circadian rhythms; photoperiodic time measurement; non-visual photoreception; pineal and retinal physiology; control of reproduction. *Mailing Add:* Dept Biol Univ Va Charlottesville VA 22903

MENAPACE, LAWRENCE WILLIAM, b Brooklyn, NY, Apr 13, 37; m 60; c 2. ORGANIC CHEMISTRY. *Educ:* St Peter's Col, BS, 60; Univ NH, PhD(org chem), 64. *Prof Exp:* Chemist, Texaco Exp, Inc, 63-65, sr chemist, Texaco, Inc, 65-68; asst prof, 68-71, ASSOC PROF CHEM, MARIST COL, 71- *Concurrent Pos:* Mem, Environ Adv Comt, 73-77; vpres, R-2 Environ Consults, 73-77. *Mem:* Am Chem Soc; Inst Soc, Ethics & Life Sci. *Res:* Mechanism and scope of organotin hydride reductions; fundamentals of chemical vapor plating; synthesis of petroleum based chemicals. *Mailing Add:* Dept of Chem Marist Col Poughkeepsie NY 12601

MENARD, ALBERT ROBERT, III, b Boston, Mass, July 17, 43; m 70; c 2. LOW TEMPERATURE PHYSICS. *Educ:* Amherst Col, BA, 65; Univ Minn, Minneapolis, MS, 69; Univ Fla, PhD(physics), 74. *Prof Exp:* Asst prof physics, WVa State Col, 74-75, Bloomsburg State Col, 75-76 & Washington Col, 76-77; asst prof physics & geol, Washington & Jefferson Col, 77-80; asst prof physics, 80-87, ASSOC PROF PHYSICS, SAGINAW VALLEY STATE UNIV, 88- *Concurrent Pos:* Res grant, US Air Force, 81-82, consult, 83-85. *Mem:* Am Phys Soc; Am Asn Physics Teachers. *Res:* Properties of superfluid helium at ultralow temperatures; transient heat transfer at low temperatures; physics education. *Mailing Add:* Dept Physics Saginaw Valley State Univ Univ Ctr MI 48710

MENASHE, VICTOR D, b Portland, Ore, July 13, 29; m 52; c 2. PEDIATRICS, CARDIOLOGY. *Educ:* Univ Ore, BS, 51, MD, 53. *Prof Exp:* Intern gen med, Univ Hosps & Clins, 53-54, resident pediat, 54-56, from instr to assoc prof, Med Sch, 58-71, asst dean, 72-77, dir crippled children's div, Health Sci Ctr, 72-84, PROF PEDIAT, MED SCH, UNIV ORE, 71- *Concurrent Pos:* Pediat consult, Shriners Hosp Crippled Children, 59- *Mem:* Am Acad Pediat; Am Heart Asn. *Res:* Epidemiology of congenital heart disease. *Mailing Add:* Dept Pediat Ore Health Sci Univ 3181 SW Sam Jackson Park Rd Portland OR 97201

MENASHI, JAMEEL, b Teheran, Iran, Apr 1, 38; m 64; c 1. FINE PARTICLE RESEARCH, SYNTHESIS OF INORGANIC OXIDES. *Educ:* Univ London, BS, 60, PhD(phys chem), 63. *Prof Exp:* Fel chem, 63-64; group leader, Harshaw Chem Co, 64-68; mem tech staff, 68-85, GROUP LEADER & SR MEM TECH STAFF, CABOT CORP, 85- *Mem:* Am Chem Soc. *Res:* Kinetics of electron exchange reactions; inorganic and organic pigment systems; thermodynamic constants of complexes; synthesis and kinetics of catalytic systems; hydrometallurgy; synthesis of oxides for ceramic dielectrics; properties of fine particles. *Mailing Add:* 68 Gleason Rd Lexington MA 02173

MENCO, BERNARD, b Arnhem, Neth, Jan 9, 46. CHEMICAL SENSES, CELL BIOLOGY. *Educ:* Univ Wageningen, Neth, BS, 68, MS, 72, PhD(cell biol), 77. *Prof Exp:* Res officer biochem, Univ Warwick, Coventry, UK, 72-75; sr res assoc olfactory res, State Univ Utrecht, Neth, 75-81; res assoc olfactory res, Klinikum, Univ Essen, Ger, 82; RES ASSOC PROF OLFACTORY RES, NORTHWESTERN UNIV, EVANSTON, ILL, 82- *Concurrent Pos:* Ed, J Electron Micros Res & Tech, 89- *Honors & Awards:* Takasago Award for Res in Olfaction, 90. *Mem:* Europ Neurosci Asn; Am Soc Cell Biol; Soc Neurosci; Am Asn Chemoreception Res; Europ Chemoreception Orgn; Electron Microscope Soc Am. *Res:* Chemical senses, smell and taste; try to localize the sensory apparatus using advanced electron microscope techniques in combination with (immuno) cytochemistry. *Mailing Add:* Neurobiol & Physiol Dept Northwestern Univ O T Hogan Hall Evanston IL 60208-3520

MENCZEL, JEHUDA H, b Vienna, Austria, Jan 29, 36; US citizen; m 71. ENVIRONMENTAL CHEMISTRY. *Educ:* Univ Minn, Minneapolis, BA, 61; Rutgers Univ, PhD(phys chem), 67. *Prof Exp:* NSF res asst, Rutgers Univ, 65-66; staff scientist, Aerospace Res Ctr, Singer-Gen Precision, Inc, 66-69 & Res Labs, Olivetti Corp, Am, 69-71; chemist, 71-73, SECT CHIEF AIR FACIL BR, US ENVIRON PROTECTION AGENCY, 73- *Mem:* Am Chem Soc; Air Pollution Control Asn. *Res:* Photochemistry and gas phase kinetics; photoconductivity; thermochromic phenomena; electrochemical processes in conjunction with non-impact printing; waste water treatment; air pollution. *Mailing Add:* Ten Rock Spring Ave West Orange NJ 07052

MENDALL, HOWARD LEWIS, b Augusta, Maine, Nov 21, 09; m 33. WILDLIFE BIOLOGY, ORNITHOLOGY. *Educ:* Univ Maine, BA, 31, MA, 34. *Prof Exp:* Asst zool, Univ Maine, 34-36; wildlife technician, US Resettlement Admin, 36-37; asst leader, Maine Coop Wildlife Res Unit, US Bur Sport Fisheries & Wildlife & prof wildlife resources, 37-42, leader, 42-77, EMER PROF, UNIV MAINE, ORONO, 77- *Mem:* Hon mem Wildlife Soc; fel Am Ornithologists Union. *Res:* Field ornithology and general wildlife ecology, especially food habits; habitat influences and breeding biology; fish-eating birds, woodcock and waterfowl. *Mailing Add:* 97 Eastern Ave Box 133 Brewer ME 04412

MENDE, THOMAS JULIUS, b Budapest, Hungary, Oct 3, 22; m 49; c 2. BIOCHEMISTRY. *Educ:* Univ of Sciences, Budapest, PhD(org chem), 48. *Prof Exp:* Res assoc chem embryol, NY Univ, 49-50; res asst prof biochem, 54-58, asst prof, 58-60, assoc prof, 60-74, PROF BIOCHEM, SCH MED, UNIV MIAMI, 74- *Concurrent Pos:* Res fel enzymol, Med Sch, Univ Budapest, 48; res fel, Lobund Inst, Univ Notre Dame, 50-54. *Mem:* Am Chem Soc; Am Soc Pharmacol & Exp Therapeut; Int Soc Haemostasis & Thrombosis; Sigma Xi; fel Gerontological Soc. *Res:* Natural products chemistry; blood coagulation; thrombosis. *Mailing Add:* 7035 SW Florida Ave Miami FL 33143

MENDEL, ARTHUR, b Ger, Dec 14, 31; nat US; m 56; c 1. ORGANIC CHEMISTRY. *Educ:* Univ Ill, BS, 54; Univ Mo, MA, 56, PhD, 58. *Prof Exp:* Res chemist, Petrol chem, Inc, 58-60; res chemist, Minn Mining & Mfg Co, St Paul, 60-76, sr environ specialist environ chem, 76-81; PATENT LIAISON, 3-M CO, 81- *Concurrent Pos:* patent agt, 85- *Mem:* Am Chem Soc. *Res:* Organic synthesis; medicinals; organometallic and analytical organic chemistry; chromatography; electrophoresis. *Mailing Add:* 4525 Oak Leaf Dr White Bear Lake MN 55127

MENDEL, FRANK C, b Cheyenne, Wyo, Dec 2, 46; m 68; c 2. ANATOMY. *Educ:* Calif State Univ, San Diego, BA, 69; Univ Calif, Davis, MA, 71, PhD(anthrop), 76. *Prof Exp:* ASST PROF ANAT, STATE UNIV NY BUFFALO, 76- *Concurrent Pos:* Adj asst prof, State Univ NY Buffalo, 77- *Mem:* Am Asn Phys Anthropologists; Am Asn Anatomists; Sigma Xi. *Res:* Adaptive advantages of suspensory behavior; form and function in masticatory apparatus of primitive mammals. *Mailing Add:* Anat Sci 317 Farber Hall 3435 Main St Buffalo NY 14214

MENDEL, GERALD ALAN, b New York, NY, May 9, 29; m 69; c 4. INTERNAL MEDICINE, HEMATOLOGY. *Educ:* Col William & Mary, BS, 50; Washington Univ, MD, 54; Am Bd Internal Med, dipl, 62; Am Bd Hematol, dipl, 80. *Prof Exp:* Intern, Univ Chicago, 54-55, resident med, 57-60, from instr to asst prof, 60-66; ASST PROF MED, SCH MED, NORTHWESTERN UNIV, EVANSTON, 66- *Concurrent Pos:* Schweppe Found grant, 62-65. *Honors & Awards:* Joseph A Capps Prize, 61. *Mem:* Am Fedn Clin Res; Am Soc Hemat. *Res:* Iron metabolism. *Mailing Add:* 636 Church St Evanston IL 60201

MENDEL, JERRY M, b New York, NY, May 14, 38; m 60; c 2. SIGNAL PROCESSING, EXPLORATION GEOPHYSICS. *Educ:* Polytech Inst Brooklyn, BME, 59, MEE, 60, PhD(elec eng), 63. *Prof Exp:* Instr elec eng, Polytech Inst Brooklyn, 60-63; res specialist, McDonnell Douglas Astronaut Co, Huntington Beach, 63-66, sr engr, 66-74; res assoc prof, 74-79, res prof, 79-80, dir, Geosignal Processing Prog, 80-83, PROF ELEC ENG, UNIV SOUTHERN CALIF, 80-, CHMN, ELEC ENG SYSTEMS DEPT, 84- *Concurrent Pos:* Lectr, Univ Calif, Los Angeles, 64-71; ed, Inst Elec & Electronics Engrs on Automatic Control, 72-74. *Honors & Awards:* Centennial Award, Inst Elec & Electronic Engrs, 84. *Mem:* Fel Inst Elec & Electronics Engrs; distinguished mem Control Systs Soc (pres, 86); Soc Explor Geophysicists; Europ Asn Signal Processing. *Res:* Estimation and identification theories as applied to problems in reflection seismology; higher-order spectral estimation; neural networks; intelligent signal processing. *Mailing Add:* Dept Elec Eng Systs Univ Southern Calif Los Angeles CA 90089-0781

MENDEL, JOHN RICHARD, b St Paul, Minn, Nov 24, 36; m 68. COLLOID CHEMISTRY, POLYMER CHEMISTRY. *Educ:* Univ Wash, BS, 58; Boston Col, MS, 69. *Prof Exp:* Chemist adhesives, Am Marietta Co, 59-61; chemist dispersions, Hercules Inc, 68-70; sr res chemist photo develop, 70-85, UNIT DIR DISPERSION TECHNOL, EASTMAN KODAK CO, 86- *Concurrent Pos:* Sr Tech Assoc, Eastman Kodak co, 83. *Mem:* Am Chem Soc; Soc Photog Scientists & Engrs. *Res:* Polymer colloids; synthesis, characterization and determination of all physical and chemical properties; physical and chemical properties of silver halide chemistry; precipitation, sensitization and sensitometric response of light sensitive materials. *Mailing Add:* 158 Lake Lea Rd Rochester NY 14617

MENDEL, JULIUS LOUIS, b Amarillo, Tex, Jan 13, 25. CLINICAL CHEMISTRY. *Educ:* Univ Tex, BA, 46; Univ Southern Calif, MS, 49, PhD(biochem), 50. *Prof Exp:* Biochemist, US Vet Admin, 51-63; res chemist, US Vet Admin, 51-63 & US Food & Drug Admin, Washington, DC, 63-66; asst to dir path, Cent Off, Vet Admin, 66-81; RETIRED. *Mem:* Am Asn Clin Chemists. *Res:* data processing in clinical laboratories. *Mailing Add:* 16202 Fallkirk Dr Dallas TX 75248

MENDEL, MAURICE I, b Colorado Springs, Colo, Oct 6, 42; m 90; c 3. AUDIOLOGY, SPEECH & HEARING SCIENCES. *Educ:* Univ Colo, Boulder, BA, 65; Washington Univ, MS, 67; Univ Wis-Madison, PhD(audiol), 70. *Prof Exp:* Asst prof audiol, Univ Iowa Hosp, 70-74, assoc res scientist, 75-76; assoc prof, Univ Calif, Santa Barbara, 76-84, prof audiol, 84-88; CHAIR, DEPT AUDIOL & SPEECH PATH, MEMPHIS STATE UNIV, 88- *Concurrent Pos:* Prog dir, Speech & Hearing Sci, Univ Calif, Santa Barbara, 80-82. *Mem:* AAAS; fel Am Speech, Lang & Hearing Asn; Int Elec Response Audiol Study Group; Int Soc Audiol; Sigma Xi; fel Soc Ear Nose & Throat Advance in Children. *Res:* Middle components of the auditory evoked potentials, and their subsequent clinical application to hearing testing. *Mailing Add:* Memphis Speech & Hearing Ctr 807 Jefferson Memphis TN 38105

MENDEL, VERNE EDWARD, b Lewistown, Mont, Apr 28, 23; m 46, 73, 85; c 6. PHYSIOLOGY. *Educ:* Univ Idaho, BSc, 55, MSc, 58; Univ Calif, PhD(animal physiol), 60. *Prof Exp:* Asst prof animal physiol, Univ Alta, 60-63; asst physiologist & lectr, 63-71, assoc prof, 71-75, chmn dept, 73-78, PROF ANIMAL PHYSIOL, UNIV CALIF, DAVIS, 75-, CHMN DEPT, 82- *Mem:* Am Physiol Soc; Soc Neurosci; Am Asn Univ Prof; Sigma Xi. *Res:* Chemical and physiological basis of food intake control. *Mailing Add:* Dept Animal Physiol Univ Calif Davis CA 95616

MENDEL, WERNER MAX, psychiatry; deceased, see previous edition for last biography

MENDELHALL, VON THATCHER, b Soda Springs, Idaho, Nov 1, 37; m 56; c 4. FOOD SCIENCE. *Educ:* Utah State Univ, BS, 62, MS, 67; Ore State Univ, PhD(food sci), 70. *Prof Exp:* Asst prof food sci, Univ Fla, 70-72; asst prof, 72-77, ASSOC PROF FOOD SCI, UTAH STATE UNIV, 77- *Mem:* Inst Food Technologists. *Res:* Protein degradation in shellfish; formaldehyde production in frozen fish tissue; lipid oxidation in Florida mullet and turkey products. *Mailing Add:* Dept Nutrit/Food Sci Utah State Univ Logan UT 84322

MENDELL, JAY STANLEY, b New York, NY, Mar 13, 36; m 61; c 2. TECHNOLOGICAL INNOVATION, NEW VENTURES. *Educ:* Rensselaer Polytech Inst, BS, 56, PhD(physics), 64; Vanderbilt Univ, MA, 58. *Prof Exp:* Health physicist, Oak Ridge Nat Lab, 57-58; asst physics, Rensselaer Polytech Inst, 58-60, asst elec eng, 60-63; asst proj engr, Pratt & Whitney Aircraft Div, United Aircraft Corp, 63-68, sr staff analyst, Advan Planning, 68-73; assoc prof, Sch Technol, Fla Int Univ, 73-76; PROF, COL URBAN & PUB AFFAIRS, FLA ATLANTIC UNIV, 76- *Concurrent Pos:* Consult, Jay S Mendell & Assoc, 73-83, innovation ed, The Futurist, World Future Soc, 69-; contrib ed, Planning Digest, 71-74 & Planning Rev, 74-; mem adv bd, Technol Forecasting & Social Change, 71-76; contrib ed, Brain & Strategy, 79-83; consult, Silicon Beach Consultant, Inc, 84-86; mem adv bd, Futures Res Quart; ed, New Directions in Pub Admin Res, 85- *Mem:* World Future Soc; Inst Elec & Electronics Engrs; fel Am Asn Social Psychiat. *Res:* Creativity; corporate planning; technological innovation; futures research. *Mailing Add:* 11295 NW 38th St Coral Springs FL 33065

MENDELL, LORNE MICHAEL, b Montreal, Que, Nov 6, 41; m 67; c 2. NEUROPHYSIOLOGY. *Educ:* McGill Univ, BSc, 61; Mass Inst Technol, PhD(neurophysiol), 65. *Prof Exp:* From asst prof to assoc prof physiol, Med Ctr, Duke Univ, 68-80; PROF DEPT NEUROBIOL & BEHAV, SUNY, STONY BROOK, NY 80-, CHMN, 88- *Concurrent Pos:* USPHS fel, Harvard Med Sch, 65-68; USPHS grant, Med Ctr, Duke Univ, 69-; NIH career develop award, 71-76; mem neurobiol adv panel, NSF, 75-78; mem staff, dept anat, Univ Col London, 76-77; Macy Found Fac Scholar, 76-77; neurobiol study sect, NIH, 79-81; ed in chief, J Neurophysiol, 83-89; counr, Soc Neurosci, 86-90; Neurosci Javits Award, 88-95. *Mem:* AAAS; Am Physiol Soc; Soc Neurosci. *Res:* Neurobiology. *Mailing Add:* Dept Neurobiol & Behav SUNY Stony Brook NY 11794

MENDELL, NANCY ROLE, b Boston, Mass, June 12, 44; m 67; c 2. BIOSTATISTICS, GENETICS. *Educ:* Smith Col, BA, 66; Harvard Univ, MSc, 68; Univ NC, PhD(biostat), 72. *Prof Exp:* Fel immunol, Duke Univ, 72-73, instr, 73-74, assoc immunol & community health sci, 74-77, lectr, Sch Nursing, 75-80, med res asst prof, 77-80, med res asst prof community & prev med, 80-81; ASST PROF APPL MATH, STATE UNIV NY STONY BROOK, 81- *Concurrent Pos:* Mem transplantation & immunol adv comt, NIH, 75-79; hon res fel, London Hosp Med Col, 76-77. *Mem:* Am Soc Human Genetics; Royal Stat Soc; Sigma Xi; Biometrics Soc. *Res:* Statistical genetics; quantitative immunology. *Mailing Add:* Dept Appl Math & Statist State Univ NY Stony Brook NY 11794

MENDELL, ROSALIND B, b New York, NY, Oct 20, 20; m 41; c 2. COSMIC RAY PHYSICS. *Educ:* Hunter Col, BA, 40; Cornell Univ, MS, 42; NY Univ, PhD(physics), 63. *Prof Exp:* Instr physics, NY Univ, 42-43; physicist, US Bur Standards, 44-46; assoc res scientist, 63-74, adj assoc prof, 72-75, RES SCIENTIST, NY UNIV, 74-, assoc res prof, 75-80, ADJ ASSOC PROF, NY UNIV, 80- *Concurrent Pos:* NSF vis prof, Lehman Col, 84-85; prof, Manhatty Anville Col, 87- *Honors & Awards:* Medallist, Societe D'Encouragement Au Progres. *Mem:* Am Phys Soc; Am Geophys Union. *Res:* Study of neutrons in cosmic radiation; cosmic ray modulation; solar neutrinos. *Mailing Add:* 89 Joyce Rd Hartsdale NY 10530

MENDELSOHN, LAWRENCE BARRY, b Brooklyn, NY, Apr 19, 34; m 58; c 3. PHYSICS. *Educ:* Brooklyn Col, BS, 55; Columbia Univ, MA, 59; NY Univ, PhD(physics), 65. *Prof Exp:* Physicist, Combustion Eng, Conn, 56-57, Walter Kidde Nuclear Labs, 58-59 & Tech Res Group, Inc, 59-62; instr physics, Cooper Union, 62-65; from asst prof to assoc prof, Polytech Inst Brooklyn, 65-73; prof physics, New Sch Lib Arts, Brooklyn Col, 73-80; adj res prof physics, Polytech Inst NY, 73-79; PROF PHYSICS, BROOKLYN COL, 80- *Concurrent Pos:* Consult, Sandia Corp, 69-75. *Mem:* Am Phys Soc. *Res:* Many body problem, particularly calculation of correlation effects in atoms and molecules; industrial experience comprises; nuclear reactor and shielding calculations; x-ray scattering cross sections. *Mailing Add:* Physics Dept Brooklyn Col Bedford Ave & Ave H Brooklyn NY 11210

MENDELSOHN, MARSHALL H, b Chicago, Ill, Apr 5, 46; m 82; c 3. INORGANIC CHEMISTRY. *Educ:* Ill Inst Technol, BS, 67; Univ Calif, Berkeley, PhD(chem), 72. *Prof Exp:* Res asst metal hydrides, Univ Conn, 72; chemist phosphors, US Radium Corp, 73-74 & Zipcor, Inc, 75; asst chemist, 75-80, chemist metal hydrides, 80-86, CHEMIST FLUE-GAS CONTROL SYSTS, ARGONNE NAT LAB, 87- *Mem:* Am Chem Soc; Sigma Xi. *Res:* Research and development on aqueous scrubber systems for simultaneous removal of sulfur dioxide and nitrogen oxides from flue-gas streams. *Mailing Add:* Argonne Nat Lab 9700 S Cass Argonne IL 60439

MENDELSOHN, MORRIS A, b Pittsburgh, Pa, Nov 13, 28; m 51; c 8. CHEMISTRY, CHEMICAL ENGINEERING. *Educ:* Univ Pittsburgh, BS, 48, MS, 54, PhD(chem), 60. *Prof Exp:* Engr, US Bur Mines, 48-52 & Shell Chem Corp, 52-53; res engr, 53-61, sr engr, 61-64, fel scientist, 64-70, ADV SCIENTIST, RES & DEVELOP CTR, WESTINGHOUSE ELEC CORP, 70- *Honors & Awards:* Soc Plastics Engrs Awards, 68 & 70. *Mem:* Am Inst Chem Engrs; Am Chem Soc; Soc Plastics Engrs. *Res:* composites of inorganic glasses and organic polymers; autooxidation of metal chelates; kinetics of hydrolytic and oxidative degradation of polymers; electrical insulating materials; solar collector sealants; shock isolation systems; vibration damping materials; relationships between mechanical properties and chemical composition of polymers; development of polymeric materials having special physical and/or chemical properties. *Mailing Add:* Res & Develop Ctr Westinghouse Elec Corp 1310 Beulah Rd Churchill Boro Pittsburgh PA 15235

MENDELSOHN, MORTIMER LESTER, b New York, NY, Dec 1, 25; m 48; c 3. BIOPHYSICS, CANCER. *Educ:* Harvard Univ, MD, 48; Cambridge Univ, PhD, 58. *Prof Exp:* Intern med, Mass Gen Hosp, 48-49; resident, Mem Ctr, NY, 49-52; from asst prof to prof radiol, Sch Med, Univ Pa, 57-72; dir biomed div, 72-76, ASSOC DIR BIOMED & ENVIRON RES, LAWRENCE LIVERMORE LAB, UNIV CALIF, 76- *Concurrent Pos:* Res fel, Sloan-Kettering Inst Cancer Res, 52-53; Am Cancer Soc Brit-Am exchange fel, 55-57; mem, comput res study sect, NIH, 67-70, chmn, 70-71. *Mem:* Radiation Res Soc; Histochem Soc; Soc Anal Cytol (pres, 78); Am Asn Cancer Res; Environ Mutagen Soc (pres-elect, 78, pres, 79). *Res:* Experimental and clinical cancer research; radiation effects; cell division; biophysical cytology; flow cytometry; computer analysis of cell images; environmental mutagenesis. *Mailing Add:* Lawrence Livermore Lab Bio-Med & Environ Res Univ Calif PO Box 808 Prog Bio-Med Div Livermore CA 94551

MENDELSOHN, NATHAN SAUL, b Brooklyn, NY, Apr 14, 17; m 40; c 2. MATHEMATICS. *Educ:* Univ Toronto, BA, 39, MA, 40, PhD, 42. *Prof Exp:* Supvr munitions gauge lab, Nat Res Coun Can, 42; scientist, Proof & Develop Estab, Que, 42-45; lectr math, Queen's Univ, Ont, 45-47; PROF, UNIV MAN, 47-, HEAD DEPT, 63-, DISTINGUISHED PROF, 81- *Honors & Awards:* Henry Marshall Tory Gold Medal, Royal Soc Can. *Mem:* Soc Indust & Appl Math; Am Math Soc; Math Asn Am; fel Royal Soc Can; Can Math Cong (pres). *Res:* Abstract algebra and geometry; combinatory statistics; a group-theoretic characterization of the general projective collineation group; ballistics; theory of error in computing machines; graph and matroid theory and geometry. *Mailing Add:* Dept of Math Univ of Man Winnipeg MB R3T 2N2 Can

MENDELSOHN, RICHARD, b Montreal, Can, July 1, 46; m 71; c 1. MOLECULAR VIBRATIONAL SPECTROSCOPY, LIPID-PROTEIN INTERACTION. *Educ:* McGill Univ, BSc, 67; Mass Inst Technol, PhD(chem), 72. *Prof Exp:* Res fel biophys, Kings Col, London Univ, 72-73; res assoc chem, Nat Res Coun Can, 73-76; PROF CHEM, RUTGERS UNIV, NEWARK, NJ, 76- *Concurrent Pos:* Mem biophys chem study sect, NIH, 83-86. *Mem:* Am Chem Soc; Biophys Soc; Sigma Xi. *Res:* Organization of biological membranes; infrared spectroscopy. *Mailing Add:* Dept Chem Rutgers Univ 73 Warren St Newark NJ 07102

MENDELSON, BERT, mathematics; deceased, see previous edition for last biography

MENDELSON, CARL VICTOR, b Altadena, Calif, Jan 11, 53; m 81. MICROPALEONTOLOGY OF CYANOBACTERIA, PALYNOLOGY. *Educ:* Univ Calif-Los Angeles, AB, 76, PhD(geol), 81. *Prof Exp:* asst prof geol, 81-87, ASSOC PROF GEOL, BELOIT COL, 87- *Mem:* Paleont Soc; Geol Soc Am; Palaeont Asn London; Am Asn Stratig Palynologists; Nat Asn Geol Teachers; Sigma Xi. *Res:* Origin and early evolution of life; evolution of organic-walled microfossils, including acritarchs; geologic times: Archean, Proterozoic, early Paleozoic. *Mailing Add:* Dept Geol Beloit Col 700 College St Beloit WI 53511-5595

MENDELSON, CAROLE RUTH, MOLECULAR ENDOCRINOLOGY, DEVELOPMENTAL BIOLOGY. *Educ:* Rutgers Univ, PhD(zool), 70. *Prof Exp:* ASST PROF BIOCHEM & ENDOCRINOL, HEALTH SCI CTR, UNIV TEX, DALLAS, 76- *Mailing Add:* 2107 Cross Creek Ct Arlington TX 76017

MENDELSON, ELLIOTT, b New York, NY, May 24, 31; m 59; c 3. MATHEMATICAL LOGIC. *Educ:* Columbia Univ, AB, 52; Cornell Univ, MA, 54, PhD(math), 55. *Prof Exp:* Instr math, Univ Chicago, 55-56; jr fel, Harvard Univ, 56-58; J F Ritt instr, Columbia Univ, 58-61; assoc prof, 61-64, PROF MATH, QUEENS COL, NY, 64- *Concurrent Pos:* Consult ed, Notre Dame J Formal Logic. *Mem:* Am Math Soc; Math Asn Am; Asn Symbolic Logic. *Res:* Axiomatic set theory. *Mailing Add:* Dept of Math Queens Col Flushing NY 11367

MENDELSON, EMANUEL SHARE, b Philadelphia, Pa, July 31, 09; m 36, 84; c 2. PHYSIOLOGY, ACOUSTICS. *Educ:* Univ Pa, BS, 31. *Prof Exp:* Res scientist, US Navy Med Res, 44-69; RETIRED. *Mem:* Am Physiol Soc; Acoust Soc Am; Sigma Xi. *Res:* Physiology of body heat regulation; emergency seat ejection from aircraft; high altitude breathing equipment;

aircraft engine noise hazards; middle ear muscle reflex registration in man; predictability of reduced reflex reactivity of middle ear muscles to progressively louder stimuli; design of stethoscopes for use in noisy environments; aviation physiology. *Mailing Add:* PO Box 103 Harleysville PA 19438

MENDELSON, JACK H, b Baltimore, Md, Aug 30, 29; m 52; c 3. MEDICINE, PSYCHIATRY. *Educ:* Univ Md, MD, 55. *Prof Exp:* Intern med serv, Boston City Hosp, 55-56; asst, 59-71, PROF PSYCHIAT, HARVARD MED SCH, 71-; DIR, ALCOHOL & DRUG ABUSE RES CTR, McLEAN HOSP, 73- *Concurrent Pos:* Teaching fel psychiat, Harvard Med Sch, 56-59; res fel psychiat, Boston City & Mass Gen Hosps, 56-59; consult, Washingtonian Hosp, Boston, 58-59; consult, psychiat res labs, Sch Med, Univ Md, 59-; asst, Mass Gen Hosp, 59-; dir dept psychiat, Boston City Hosp, 71-73. *Mem:* Am Psychiat Asn; Asn Res Nerv & Ment Dis; Endocrine Soc; Am Soc Pharmacol & Exp Therapeut; Sigma Xi. *Res:* Psychiatric research, especially in alcohol and drug abuse. *Mailing Add:* Alcohol & Drug Abuse Res Ctr McLean Hosp Belmont MA 02178

MENDELSON, KENNETH SAMUEL, b Chicago, Ill, Aug 24, 33; m 61; c 3. PHYSICS. *Educ:* Ill Inst Technol, BS, 55; Purdue Univ, MS, 57, PhD(physics), 63. *Prof Exp:* Assoc physicist, IIT Res Inst, 62-65; from asst prof to assoc prof, 65-81, PROF PHYSICS, MARQUETTE UNIV, 81- *Concurrent Pos:* Sr sci fel, NATO, 74. *Mem:* Am Phys Soc; Am Asn Physics Teachers. *Res:* Theory of inhomogeneous media; diffusion in porous media. *Mailing Add:* Physics Dept Marquette Univ Milwaukee WI 53233

MENDELSON, MARTIN, b New York, NY, Apr 16, 37; m 58; c 2. MEDICAL INFORMATICS. *Educ:* Cornell Univ, AB, 58; Calif Inst Technol, PhD(biol), 62; State Univ NY Stony Brook, MD, 76. *Prof Exp:* Res assoc physiol, Col Physicians & Surgeons, Columbia Univ, 61-63; from instr to assoc prof, Sch Med, NY Univ, 63-71; assoc prof physiol, Health Sci Ctr, State Univ NY, Stony Brook, 71-76, mem adj staff, Dept Med, Div Neurol, Nassau County Med Ctr, 76-77; resident family pract, Emanuel Hosp, Portland, 77-81; asst prof family pract, Sch Med, Univ Calif, Davis, Sacramento, 81-85; clin assoc prof family med, Univ Wash Med Sch, Tacoma, 85-89; CONSULT, MED INFO SYSTS, TACOMA, 89- *Concurrent Pos:* Mem, Corp Marine Biol Lab, Woods Hole, Mass; Grass Found fel neurophysiol, 62; physician adv, Dept Social & Health Serv, Olympia, Wash, 89-90; consult, Med Informatics Lab, USPHS, Seattle, 89-, Div Mental Health, Olympia, Wash, 91- *Mem:* Am Acad Family Physicians; Am Med Informatics Asn. *Res:* Sensory mechanisms in Crustacea and mammals; neuromuscular transmission in Crustacea; central nervous mechanisms of rhythmicity and integration in Crustacea; role of chaos in feeding rhythms in mollusc CNS; effects of computerized feedback on medical record quality; can on-line quality assurance be built into automated medical/patient record systems; utilization of clinical experience in design of automated decision-support systems. *Mailing Add:* 812 N K St No 209 Tacoma WA 98403

MENDELSON, MYER, b Lithuania, Dec 5, 20; nat US; m 56; c 1. PSYCHIATRY. *Educ:* Dalhousie Univ, BA, 45, BSc, 46, MD, CM, 50. *Prof Exp:* Instr psychiat, Johns Hopkins Univ, 54-56; asst prof, Dalhousie Univ, 56-58; from asst prof to assoc prof, 58-71, PROF CLIN PSYCHIAT, SCH MED, UNIV PA, 71- *Mem:* Am Psychiat Asn; Can Psychiat Asn. *Res:* Theoretical models in psychoanalysis; depression; manic-depressive illness; psychopharmacology; obesity. *Mailing Add:* 1220 Wyngate Rd Wynnewood PA 19096

MENDELSON, NEIL HARLAND, b New York, NY, Nov 15, 37; m 59; c 2. GENETICS, CELL BIOLOGY. *Educ:* Cornell Univ, BS, 59; Ind Univ, PhD(genetics, bact), 64. *Prof Exp:* Asst prof biol sci, Univ Md, Baltimore County, 66-69; assoc prof, 69-74, prof microbiol & med technol, 74-78, head dept, 79-83, prof cellular & develop biol, Univ Ariz, 78-83, PROF MOLECULAR & CELLULAR BIOL, 83- *Concurrent Pos:* NSF fel, Med Res Coun Microbial Genetics Res Unit, Hammersmith Hosp, London, Eng, 65-66; vis scientist, Unite de physiol cellulaire, dept de biochimie et genetique microbienne, Inst Pasteur, Paris, 76-77; Nat Inst Gen Med Sci res career develop award, 73-77; prin investr res grant, Div Biol & Med Sci, NSF, 67-69, 69-71 & 82-84 & Nat Inst Gen Med Sci, 71-82, 84-87, 90-93. *Mem:* Fel AAAS; Am Soc Microbiol; Genetics Soc Am; Sigma Xi; fel Am Acad Microbiol. *Res:* Molecular and microbial genetics; genetic control of DNA replication, growth and cell division in Bacillus subtilis; helical growth of Bacillus subtilis, the theory of helical clocks and DNA segregation in bacteria; biomechanics of bacteria; bacterial macrofibers. *Mailing Add:* Dept Molecular & Cellular Biol Univ of Ariz Tucson AZ 85721

MENDELSON, ROBERT ALEXANDER, JR, b Los Angeles, Calif, Jan 24, 41; m 83; c 2. NUCLEAR PHYSICS, MOLECULAR BIOLOGY. *Educ:* Occidental Col, AB, 62; Univ Iowa, MS, 64, PhD(physics), 68. *Prof Exp:* Res assoc physics, Lawrence Berkeley Lab, Univ Calif, 68-71; res assoc biophys, 71-79, assoc prof, 79-84, PROF BIOPHYS, UNIV CALIF, SAN FRANCISCO, 84- *Mem:* Am Phys Socy; Biophys Soc. *Res:* Structural biophysics of contractile proteins; neutron diffraction and scattering studies. *Mailing Add:* 326 Santiago St San Francisco CA 94116

MENDELSON, ROBERT ALLEN, b Cleveland, Ohio, Dec 17, 30; m 71; c 2. POLYMER SCIENCE, RHEOLOGY. *Educ:* Case Western Reserve Univ, BS, 52, PhD(chem), 56. *Prof Exp:* Res chemist, Monsanto Chem Co, 56-61, res specialist, 61-69, sci fel, 69-89, SR SCI FEL, MONSANTO CO, 89- *Honors & Awards:* Arthur K Doolittle Award, Am Chem Soc, 82. *Mem:* AAAS; Am Chem Soc; Soc Rheol(secy, 74-78, vpres, 88-90, 90-92); Soc Plastics Engrs. *Res:* Polymer rheology and processing; polymer mechanical properties; polymer blends; polymer molecular weight and structure characterization; polymer solution properties. *Mailing Add:* MCC Technol PASC Monsanto Chem Co 730 Worcester St Springfield MA 01151

MENDELSON, WILFORD LEE, b Baltimore, Md, July 8, 37. ORGANIC CHEMISTRY. *Educ:* Johns Hopkins Univ, AB, 58, MA, 60, PhD(chem), 63. *Prof Exp:* Sr chemist, 63-75, SR INVESTR, SMITH KLINE & FRENCH LABS, 75- *Mem:* Am Chem Soc; Royal Soc Chem; Sigma Xi. *Res:* process chemistry. *Mailing Add:* Smith Kline French Labs 1500 Spring Garden St Philadelphia PA 19101

MENDELSSOHN, ROY, b Philadelphia, Pa, Dec 13, 49. OPERATIONS RESEARCH. *Educ:* Harvard Univ, BA, 71; Yale Univ, MFS, 73, MPhil, 75, PhD(environ studies), 76. *Prof Exp:* opers res analyst, Honolulu Lab, 76-81, OPERS RES ANALYST, PAC ENVIRON GROUP, SOUTHWEST FISHERIES CTR, NAT MARINE FISHERIES SERV, MONTEREY, CALIF, 81- *Mem:* Opers Res Soc Am; Inst Mgt Sci; Am Statist Asn; Biomet Soc; Soc Indust & Appl Math. *Res:* Dynamic programming and the optimal management of renewable resources; multi-objective decision making; statistical analysis of population dynamics. *Mailing Add:* PO Box 831 Pac Environ Group Monterey CA 93942

MENDENHALL, CHARLES L, b Chicago, Ill, Aug 20, 31. HEPATOLOGY. *Educ:* Univ Nebr Col Med, MD, 56; Univ Okla, PhD(biochem), 68. *Prof Exp:* CHIEF, DIGESTIVE DIS, VET ADMIN MED CTR, 73-; PROF MED, COL MED, UNIV CINCINNATI, 80- *Mailing Add:* Vet Admin Med Ctr 3200 Vine St Cincinnati OH 45220

MENDENHALL, GEORGE DAVID, b Iowa City, Iowa, Feb 12, 45; m 73; c 2. PHYSICAL ORGANIC CHEMISTRY, POLYMER CHEMISTRY. *Educ:* Univ Mich, BS, 66; Harvard Univ, PhD(chem), 71. *Prof Exp:* Fel, Nat Res Coun Can, 71-73; fel, Stanford Res Inst, 73-74; mem staff chem, Columbus Labs, Battelle Mem Inst, 74-80; assoc prof, 80-82, PROF, MICH TECHNOL UNIV, 83- *Mem:* Am Chem Soc; Int Photochem Soc. *Res:* Reactions of ozone, singlet molecular oxygen; electron spin resonance studies of organic radicals; kinetics of smog-producing reactions; composition of atmospheric aerosols; autoexidation processes studied by chemiluminescence. *Mailing Add:* Chem Dept Mich Technol Univ Haughton MI 49931

MENDENHALL, ROBERT VERNON, b Geneva, Ind, Dec 27, 20; m 44; c 3. ANALYSIS & FUNCTIONAL ANALYSIS. *Educ:* Ohio State Univ, BA, 47, MA, 49, PhD(math), 52. *Prof Exp:* Instr math, Ohio State Univ, 47-53; mathematician, NAm Aviation, Inc, 53-55 & Vitro Labs, Inc, 55; asst prof math, Univ Miami, 55-62; assoc prof, 62-66, prof, 66-89, EMER PROF MATH, OHIO WESLEYAN UNIV, 89- *Concurrent Pos:* Consult, NSF Math Insts, India, 65, 66 & 70. *Mem:* Am Math Soc; Math Asn Am. *Res:* Measure theory; integration. *Mailing Add:* 129 Oak Hill Ave Delaware OH 43015

MENDENHALL, WILLIAM, III, b Pa, Apr 20, 25; m 49; c 2. STATISTICS. *Educ:* Bucknell Univ, BS, 45, MS, 50; NC State Col, PhD(statist), 57. *Prof Exp:* Asst prof statist, NC State Col, 58-59; assoc prof math, Bucknell Univ, 59-63; PROF STATIST, UNIV FLA, 63- *Concurrent Pos:* Assoc statistician, London Sch Econ, 57-58; consult, Westinghouse Elec Co, Pa, 59-60, Armstrong Cork Co, Pa, 60-, Burroughs Corp, Mich, 60-61, Lewis Res Ctr, NASA, Ohio, 60-61, Merck & Co, Pa, 61-62 & WVa Pulp & Paper Co, 65. *Mem:* Am Statist Asn; Inst Math Statist; Royal Statist Soc. *Res:* Design of experiments; distribution theory. *Mailing Add:* Dept Radiation Ther Univ Fla Med Col J Hills Miller Health Ctr Gainesville FL 32610

MENDES, ROBERT W, b Fall River, Mass, Apr 6, 38; m 60; c 1. INDUSTRIAL PHARMACY. *Educ:* New Eng Col Pharm, BS, 60; Univ NC, MS, 64, PhD(pharm), 66. *Prof Exp:* Dir, Pfeiffer Labs, 79-87; From asst prof to assoc prof pharm, 65-78, PROF INDUST PHARM, MASS COL PHARM, 79- *Concurrent Pos:* Consult & grantee, several material suppliers and mfg concerns, 71-; mem, Eastern Regional Indust Pharmaceut Technol Sect Planning Comt, 80-84, Tableting Specifications Comt, Acad Pharmaceut Sci, 79-81. *Mem:* Am Asn Pharmaceut Scientists; Am Asn Col Pharm; Soc Cosmetic Chemists. *Res:* Development and evaluation of pharmaceutical dosage forms and excipient materials; effect of formulation and process variables on bioavailability and bioequivalence of drug products. *Mailing Add:* Dept Pharm Mass Col Pharm 179 Longwood Ave Boston MA 02115-5896

MENDEZ, EMILIO EUGENIO, b Lerida, Spain, Jan 22, 49. SEMICONDUCTOR PHYSICS, LOW-DIMENSIONALITY MATERIALS. *Educ:* Complutense Univ Madrid, BS, 71; Mass Inst Tech, PhD(physics), 79. *Prof Exp:* Res staff mem, 79-84, mgr quantum structures group, 84-85, res staff mem, 85-88, MEM TECH STAFF, RES DIV, INT BUS MACH, 88- *Concurrent Pos:* Adv comt mem, Nat Ctr Mat Sci, Spain, 87- *Mem:* Fel Am Phys Soc. *Res:* Electronic properties of semiconductor heterostructures, studied by transport, optical and magnetic experiments; novel phenomena for optoelctronic applications. *Mailing Add:* IBM T J Watson Res Ctr PO Box 218 Yorktown Heights NY 10598

MENDEZ, EUSTORGIO, b Panama City, Panama, Mar 1, 27; m 60; c 2. ZOOLOGY, MEDICAL ENTOMOLOGY. *Educ:* Univ Panama, BS, 50; Univ Calif, Berkeley, MS, 54; Mich State Univ, PhD(zool), 76. *Prof Exp:* Entomologist, Ministry of Health, Panama, 55-57; ZOOLOGIST, GORGAS MEM LAB, 57-; PROF ZOOL, UNIV PANAMA, 65- *Concurrent Pos:* Guggenheim Mem Found fel, 61-62; mem, Panama Nat Comn Wildlife Conserv, 66-; consult, Int Union Conserv Nature & Natural Resources, 72-; fel, Orgn Am States, 74-75. *Mem:* Sigma Xi. *Res:* Mammalogy; entomology; wildlife conservation; applied ecology. *Mailing Add:* Gorgas Mem Lab APO Miami FL 34002-0012

MENDEZ, VICTOR MANUEL, b San Antonio, Tex, May 14, 44; c 3. ORGANIC CHEMISTRY, ANALYTICAL CHEMISTRY. *Educ:* St Mary's Univ, Tex, BS, 66; Univ Tex, Austin, MA, 77. *Prof Exp:* Asst res chemist, Southwest Res Inst, 66-67; chem officer, US Army Chem Corps, 68-71; res chemist, Southwest Res Inst, 71-73; RES CHEMIST, SOUTHWEST

FOUND RES & EDUC, 73- *Mem:* Am Chem Soc. *Res:* Production and monitoring of pollutant atmospheres used in toxicological studies; automation of laboratory experiments through the use of microcomputers. *Mailing Add:* Southwest Found for Res & Educ 8848 W Commerce St PO Box 28147 San Antonio TX 78284

MENDICINO, JOSEPH FRANK, b Cleveland, Ohio, Nov 22, 30; m 52; c 4. BIOCHEMISTRY. *Educ:* Case Western Reserve Univ, BS, 53; PhD, 57. *Prof Exp:* NSF res fel, Inst Biochem Invest, Argentina, 59-60 & Case Western Reserve Univ, 60-62; asst prof agr biochem, Ohio State Univ, 62-68; ASSOC PROF BIOCHEM, UNIV GA, 68- *Mem:* Am Chem Soc; Am Soc Biol Chem. *Res:* Enzymology. *Mailing Add:* Dept Biochem Univ Ga Boyd Grad Study Res Ctr Athens GA 30602

MENDILLO, MICHAEL, b Providence, RI, Aug 22, 44; c 2. SPACE PHYSICS, ASTRONOMY. *Educ:* Providence Col, BS, 66; Boston Univ, MA, 68, PhD(physics & astron), 71. *Prof Exp:* Asst prof astron, Boston Univ, 71-72; Nat Acad Sci-Nat Res Coun resident res assoc space sci, Air Force Cambridge Res Lab, Bedford, Mass, 72-74; from asst prof to prof astron, 74-85, assoc dean grad sch, 78-87, PROF ASTRON, BOSTON UNIV, 85- *Mem:* AAAS; Am Geophys Union; Am Astron Soc; Int Union Radio Sci; Sigma Xi. *Res:* Solar-terrestrial relations; artificial modification of ionosphere, history of astronomy and geophysics; antiquarian astronomical maps and charts, 1500-1900. *Mailing Add:* 110 Pleasant St Lexington MA 02173

MENDIS, DEVAMITTA ASOKA, b Colombo, Sri Lanka, Feb 13, 36; US citizen; m 75. COMETARY PHYSICS, MAGNETOSPHERIC PHYSICS. *Educ:* Univ Ceylon, Colombo, BSc, 60; Univ Manchester, Eng, PhD(astrophysics), 67, DSc, 78. *Prof Exp:* Lectr math, Univ Ceylon, Colombo, 60-64 & 67-69; asst res physicist, 70-75, assoc res physicist, 75-78, res physicist, 78-86, PROF SPACE SCI, UNIV CALIF, SAN DIEGO, 86- *Concurrent Pos:* Assoc ed, J Geophys Res, 86-; prin investr, NASA & NSF grants. *Honors & Awards:* Corresp mem Int Acad Astronaut. *Mem:* Am Geophys Union; Am Astron Soc. *Res:* Cometary physics; magnetospheric physics; physics of dusty plasmas. *Mailing Add:* Dept Elec & Comput Eng Univ Calif San Diego La Jolla CA 92093-0114

MENDIS, EUSTACE FRANCIS, b Colombo, Ceylon, June 22, 37; m 71. SOLID STATE PHYSICS. *Educ:* Univ Ceylon, BSc, 58; Univ Wis, PhD(physics), 68. *Prof Exp:* Fel physics, Univ NB, 69-70; lectr, Univ of Toronto, 70-75; head, Physics Dept, Ont Sci Ctr, 75-78, chief scientist, 78-84; dir, Technol Ctr Silicon Valley, 85-87; PRES, EUSTACE MENDIS COMPUTER CONSULT, INC, 87- *Concurrent Pos:* Partner, Proj Consortium, 89- *Mem:* Am Phys Soc; Can Asn Physicists; Ceylon Asn Advan Sci; Sigma Xi. *Res:* Nuclear magnetic resonance in ferromagnets. *Mailing Add:* 489 King St W Suite 200 Toronto ON M5V 1L3 Can

MENDLER, OLIVER J(OHN), b Bacsalmas, Hungary, June 21, 27; US citizen; m 57; c 3. MECHANICAL ENGINEERING. *Educ:* Mass Inst Technol, SB, 55; Univ Pittsburgh, MS, 59, PhD(mech eng), 63. *Prof Exp:* Engr, Bettis Atomic Power Lab, West Mifflin, 55-72, prin engr, Water Reactor Div, Westinghouse Elec Corp, Pittsburgh, 72-90; RETIRED. *Res:* Heat transfer and fluid flow; thermal design of nuclear reactors; high pressure-high temperature tests in two-phase flow; steam generator accident transient tests including loss of feedwater, tube rupture and steam line break; steam generator performance under steady state and accident transient conditions. *Mailing Add:* 134 Carnold Dr Munhall PA 15120

MENDLOWITZ, HAROLD, b New York, NY, Aug 23, 27; m 50; c 3. THEORETICAL PHYSICS, OPTICS. *Educ:* City Col New York, BS, 47; Columbia Univ, AM, 48; Univ Mich, PhD(physics), 54. *Prof Exp:* Asst prin & chmn dept, Beth Yehudah Schs, Mich, 48-49; asst exp physics, Columbia Univ, 49-50; physicist, Nat Bur Standards, 51-52; instr aeronaut eng, Univ Mich, 52-53, asst theoret physics, 53-54; theoret physicist, Nat Bur Standards, 54-65; PROF PHYSICS, HOWARD UNIV, 65- *Concurrent Pos:* Sr res fel, Hebrew Univ, Israel, 61-62; consult, Nat Bur Standards; vis prof, Hebrew Univ, Israel, 71-72; mem staff, Nat Bur Standards, 78-79. *Mem:* Fel Am Phys Soc; Sigma Xi; AOJS. *Res:* Photomeson production; electron physics, scattering, interference and polarization; Dirac theory; optical properties of solids; magnetic moment of electron; characteristic electron energy losses in solids; atomic spectroscopy; transition probabilities; radiation theory, cerenkov radiation, channeling radiation. *Mailing Add:* Dept Physics & Astron Howard Univ Washington DC 20059

MENDLOWITZ, MILTON, b New York, NY, Dec 30, 06; m 40; c 3. INTERNAL MEDICINE, CARDIOLOGY. *Educ:* City Col New York, AB, 27; Univ Mich, MD, 32. *Prof Exp:* Clin asst, 39-40, sr asst, 40-42, adj physician, 46-53, assoc attend physician, 53-59, Joe Lowe & Louis Price prof, Med Sch, 72-77, emer prof med, 77-89, Dept Physiol, 77-80, DEPT MOLECULAR BIOL, INST MEMBRANE & POLYPEPTIDE RES, MT SINAI MED SCH, 89- *Concurrent Pos:* Blumenthal fel, Mt Sinai Hosp, 36; Libman fel, Michael Reese Hosp, Chicago, Ill, 37 & Univ Col Hosp, London, 38; Dazian fel, Mt Sinai Hosp, 42; res fel, Goldwater Mem Hosp, 51 -; pvt pract, 39-42 & 46-83; sr physician, NY Regional Off, US Vet Admin, 46-52; asst clin prof med, Columbia Univ, 56-61, assoc clin prof, 66 -; clin prof, Mt Sinai Med Sch, 66-72, consult, 75 -; attending physician, Mt Sinai Hosp, 59-72. *Honors & Awards:* Jacoby Medal, Mt Sinai Hosp, 65. *Mem:* Fel AAAS; Am Soc Clin Invest; Am Physiol Soc; Soc Exp Biol & Med; AMA; fel Col Chest Physicians; fel Am Col Physicians; fel Am Col Cardiologists. *Res:* Physiology and pathological physiology of digital circulation; mechanism of heart failure; physiological effects of coronary occlusion and pulmonary embolism; mechanism and treatment of hypertension; polypeptide and membrane research; uterine and vascular physiology cloning of phospholipase C genes; coauthor of 380 scientific papers or book chapters and 5 books. *Mailing Add:* 1200 Fifth Ave New York NY 10029

MENDOZA, CELSO ENRIQUEZ, b Bocaue, Bulacan, Philippines, Mar 28, 33; m 68; c 3. BIOCHEMISTRY, ENTOMOLOGY. *Educ:* Univ Philippines, BS, 59; Iowa State Univ, MS, 61, PhD(entom), 64. *Prof Exp:* Res assoc med entom, Cornell Univ, 64-65; Nat Res Coun Can fel pesticide residue anal, Health Protection Br, Nat Health & Welfare Dept, Can, 65-67, res scientist I, 67-69, res scientist II, 69-79, biologist, regulatory toxicol, 79-81; DEFENSE SCIENTIST & TOXICOLOGIST, DEPT NAT DEFENSE, CAN, 81- *Concurrent Pos:* Assoc referee esterase methods, Asn Off Analytical Chem, 70-; NATO spec study travel grant, Alta, Can, 71; vis scientist, Biochem Dept, Arrhenius Lab, Stockholm Univ, 73-74; NATO ecotoxicol travel grant, Univ Surrey, Guilford, Eng, 77; mem, Pesticide Ref Standards Comt, Entom Soc Am, 77-81, chmn, 78-81; Task Force Pesticides Toxicol Rev, Health Protection Br, Nat Health & Welfare Can, 79-81; adv, Animals Res Comt, Soc Toxicol, 83-84. *Mem:* AAAS; Am Chem Soc; Entom Soc Am; Philippine Entom Soc; NY Acad Sci; Soc Toxicol Can; Int Soc Study Xenobiotics (pres, 80-); Soc Environ Toxicol & Chem; Soc Toxicol. *Res:* Development of analytical method for pesticide residues in foods; chromatographic-enzyme inhibition techniques for insecticides; toxicological and biochemical determination of pesticide effects on animals particularly neonates; evaluation of toxicological data in support of submission for pesticide registration; general, acute, subacute, dermal and ocular toxicology; teratology, therapeutic and preventive agents efficacy and toxicology; venom toxicology and analysis; consulting in general and environmental toxicology. *Mailing Add:* 3913 Waterton Crescent Abbotsford BC V2S 6B7 Can

MENDUKE, HYMAN, b Warsaw, Poland, Aug 20, 21; US citizen; m 87; c 1. BIOSTATISTICS. *Educ:* Univ Pa, BA, 43, MA, 48, PhD(econ statist), 52. *Prof Exp:* Instr soc & econ statist, Univ Pa, 47-53; asst prof biostatist, 53-58, assoc prof, 58-63, prof community health & prev med, 63-79, dir sponsored progs, 63-83, PROF PHARMACOL, JEFFERSON MED COL, 78- *Concurrent Pos:* Adj prof statist eval clin data, Philadelphia Col Pharm, 75 - *Mem:* Am Statist Asn; Biomet Soc; Sigma Xi. *Res:* Applied statistics in the design of surveys, clinical trials and laboratory experiments and in the analysis, interpretation and presentation of results. *Mailing Add:* Jefferson Med Col 1020 Locust St Philadelphia PA 19107

MENEELY, GEORGE RODNEY, physiology, biophysics; deceased, see previous edition for last biography

MENEES, JAMES H, b Checotak, Okla, Nov 24, 29; m 53; c 1. HISTOLOGY, EMBRYOLOGY. *Educ:* San Jose State Col, AB, 53; Cornell Univ, MS, 57, PhD(entom), 59. *Prof Exp:* Asst, Cornell Univ, 56-59; PROF ENTOM, LONG BEACH STATE COL, 59- *Concurrent Pos:* Res grants, USDA, 56-59 & Univ Southern Calif, 79-81. *Mem:* Entom Soc Am; Am Soc Parasitol. *Res:* Insect anatomy; arthropod histology, embryology and physiology. *Mailing Add:* Dept Biol Sci Calif State Col 1250 Bellflower Blvd Long Beach CA 90840

MENEFEE, EMORY, b Wichita Falls, Tex, June 30, 29; m 53; c 3. PHYSICAL CHEMISTRY, POLYMER CHEMISTRY. *Educ:* Tex Tech Col, BS, 50; Mass Inst Technol, PhD(phys chem), 56. *Prof Exp:* Chem engr, Amarillo Helium Plant, US Bur Mines, Tex, 50-52; chemist, Plastics Dept, E I du Pont de Nemours & Co, Del, 56-60; chemist, wool lab, Western Regional Res Lab, US Dept Agr, 60-82. *Concurrent Pos:* Consult hair chem, 82- *Mem:* Soc Rheol. *Res:* Viscoelasticity of molten polymers; rheology and physical chemistry of wool fibers and human hair; crosslinking; hair growth measurement. *Mailing Add:* 5313 Rosalind Ave Richmond CA 94805

MENEFEE, MAX GENE, b Perry, Mo, Mar 30, 25; m 51; c 3. FAMILY MEDICINE. *Educ:* Wash Univ, PhD(anat), 56; State Univ NY, MD, 61. *Prof Exp:* Spec lectr surg, Col Med, State Univ NY Upstate Med Ctr, 57-61; asst prof anat, 61-66, assoc prof path & anat, 66-81, PRECEPTOR FAMILY MED, COL MED, UNIV CINCINNATI, 81- *Concurrent Pos:* USPHS fel, 56-57; consult, US Vet Admin Hosp, 71-81; spec residency family med, Univ Hosp, Univ Cincinnati, 80-81. *Mem:* Am Asn Anat; Sigma Xi. *Res:* Morphological aspects of vascular transport and disease; arthritis and rheumatism. *Mailing Add:* Dept Path Univ Cincinnati Med Ctr Cincinnati OH 45229

MENEFEE, ROBERT WILLIAM, b Akron, Ohio, Aug 8, 29; m 54; c 3. SCIENCE EDUCATION. *Educ:* Univ Akron, BS, 52; Kent State Univ, ME, 57; Ohio State Univ, PhD(sci ed), 65. *Prof Exp:* Teacher pub schs, Ohio, 54-62; instr unified sci, Ohio State Univ Sch, 63-65, asst prof zool, Ohio State Univ, 65-67, core prog dir biol, 67-68, asst dean & core dir biol sci, 68-69; assoc prof of sci teaching, Univ Md, College Park, 69-71; chmn div math sci, 71-78, dean, Inst Natural Sci, 78-79 & 81-84, DEAN SCI & MATH, MONTGOMERY COL, GERMANTOWN, 84- *Concurrent Pos:* Prog dir NSF, Washington, DC, 79-81. *Honors & Awards:* Int Cong Individual Instr Award, 80; Outstanding Serv, NSF, 81. *Mem:* Fel AAAS; Soc Col Sci Teachers (pres, 83); Nat Sci Teachers Asn. *Res:* Televised biology instruction; individualized instruction in biology; science curriculum development; Nat Science Foundation-science education; high technology curriculum development; two plus two program development. *Mailing Add:* Sci-Math Cluster Montgomery Col Germantown MD 20874

MENEGHETTI, DAVID, b Chicago, Ill, May 8, 23; m 50; c 2. NUCLEAR PHYSICS. *Educ:* Univ Chicago, BS, 44; Ill Inst Technol, PhD(physics), 54. *Prof Exp:* Asst physicist, Argonne Nat Lab, 46-49 & 52-54; lab asst nuclear physics, Med Sch, Univ Ill, 51-52; assoc physicist, Armour Res Found, Ill Inst Technol, 54-55; SR PHYSICIST, ARGONNE NAT LAB, 55- *Concurrent Pos:* Consult, Centro di Calcolo & lectr, Univ Bologna, 62-63. *Mem:* Am Phys Soc; fel Am Nuclear Soc. *Res:* Neutron physics; neutron diffraction; magnetics and antiferromagnetics; reactor physics. *Mailing Add:* Argonne Nat Lab 9700 Cass Ave Argonne IL 60439-4842

MENENDEZ, MANUEL GASPAR, b New York, NY, June 15, 35; m 58; c 3. ATOMIC PHYSICS, MOLECULAR PHYSICS. *Educ:* Univ Fla, BChE, 58, PhD(chem physics), 63. *Prof Exp:* Fel chem physics, Oak Ridge Nat Lab, 63-65; atomic physicist, Nat Bur Standards, 65-66; staff scientist, Martin Marietta Corp, 66-69; assoc prof, 69-81, PROF PHYSICS, 81-, HEAD, DEPT PHYSICS & ASTRON, UNIV GA, 85- *Concurrent Pos:* Consult, fel, Joint Inst Lab Astrophy, Univ Colo, 79-80, Martin Marietta Corp, 70-72; guest scientist, Hahn-Meitner-Institut, Berlin, 86, FOM Lab Atomic & Molec Physics, Amsterdam, 87-; vis prof, Univ Paris-Orsay, 87, guest prof, Albert-Ludwigs Universitat, Freiburg, Ger, 88; guest prof, Nat Ctr for Sci Res, Univ Paris France, 90. *Mem:* Fel Am Phys Soc; NY Acad Sci. *Res:* Ionization mechanisms at intermediate and low energies; molecular aspects of ion-atom collisions; electron dynamics of atomic collisions. *Mailing Add:* Dept Physics & Astron Univ Ga Athens GA 30602

MENES, MEIR, b Berlin, Ger, Oct 3, 25; nat US. PHYSICS. *Educ:* Cooper Union, BE, 48; NY Univ, PhD(physics), 52. *Prof Exp:* Physicist, Res Labs, Westinghouse Elec Corp, 52-60; ASSOC PROF PHYSICS, POLYTECH INST NY, 60- *Mem:* Am Phys Soc. *Res:* Electrical discharges through gases; nuclear magnetic resonance; acoustic studies of solids. *Mailing Add:* Dept Physics Polytech Inst NY 333 Jay St Brooklyn NY 11201

MEÑEZ, ERNANI GUINGONA, b Manila, Philippines, Aug 15, 31; m 61; c 4. TROPICAL & SUBTROPICAL MARINE PHYCOLOGY, SEAGRASSES. *Educ:* Univ Philippines, BS, 54; Univ Hawaii, MS, 62; Univ NH, PhD, 80. *Prof Exp:* Res asst bot, Univ Philippines, 53-54; instr bot & zool, Southeastern Col, Philippines, 54-58; asst botanist, Univ Hawaii, 58-61 & Univ BC, 62-64; SUPVR ALGAE, SMITHSONIAN OCEANOG SORTING CTR, SMITHSONIAN INST, 64-, DIR, 88- *Concurrent Pos:* Dir, Mediterranean Marine Sorting Ctr, Tunisia, 73-75; sr adv, Philippines Sci & Technol Adv Coun, Washington, DC. *Mem:* Phycol Soc Philippines; Int Phycol Soc. *Res:* Taxonomy and ecology of tropical and subtropical benthic marine algae and seagrasses; marine floristics of Tunisia. *Mailing Add:* Smithsonian Oceanog Sorting Ctr Smithsonian Inst Washington DC 20560

MENEZES, JOSE PIEDADE CAETANO AGNELO, b Curtorim, India, July 25, 39; Can citizen; m 70; c 3. IMMUNOVIROLOGY, TUMOR BIOLOGY. *Educ:* Nat Col Goa, India, dipl, 58; Univ Perugia, DVM, 63; Pasteur Inst, Paris, dipl bact, 65; Univ Montreal, MS, 67; Univ Ottawa, PhD(microbiol), 71. *Prof Exp:* asst prof, 73-77; assoc prof microbiol & immunology & head lab immunovirol, 77-81, PROF & HEAD LAB IMMUNOVIROL, PEDIAT RES CTR, FAC MED, UNIV MONTREAL, 81- *Concurrent Pos:* Sr res scholar, Med Res Coun Quebec, 78-85, Med Res Coun Can fel, 67-73, 78-85, Univ Ottawa, 68-71, & Dept Tumor Biol, Karolinska Inst, Sweden, 71-73; Med Res Coun Can scholar, Univ Montreal, 73-78; govt fel State Inst Animal Virus Res, Denmark, 64; French govt fel Pasteur Inst, Paris 64-65; mem, grants comt Microbiol & Infectious Dis, Med Res Coun Can, 80-82, rev comt Res Ctr Prog, Ministry Educ Que, 84, int comt Astra Res Award Basic Res in Sexually Transmitted Dis, grants panel 85-87, Nat Health & Res Develop Prog, Dept Health & Welfare, Can, 86-88; mem, Med Oncol Peer Rev Comt, Can Cancer Res Soc, 86-87, chmn, 87-; mem, adv comt Med Res & Grad Studies, Univ Montreal, 80-84, joint res comt, Pediat Res Ctr, 84-, comt Grad Studies & Res, Dept Microbiol & Immunol, 87-, adv comt on res, 88-; mem governing bd, Int Asn Res Epstein-Barr Virus & Assoc Dis, 88-; reviewer, grant appln numerous Can organ. *Honors & Awards:* Astra Can Res Award in Herpes, 84. *Mem:* Am Soc Microbiol; Int Asn Comp Res Leukemia & Related Dis; Am Asn Cancer Res; NY Acad Sci; Am Soc Virol; Int Asn Eppstein-Barr Virus & Related Dis. *Res:* Virology; cellular immunology; tumor biology; cell culture; cell biology; immunopharmacology. *Mailing Add:* Lab Immunovirol Pediat Res Ctr Ste-Justine Hosp Montreal PQ H3T 1C5 Can

MENG, H C, LIPID METABOLISM DISEASES, GASTROINTESTINAL HORMONES. *Educ:* Univ Toronto, MD, 41; Northwestern Univ, PhD(physiol), 47. *Prof Exp:* PROF MOLECULAR PHYSIOL & BIOPHYSICS & SURG, SCH MED, VANDERBILT UNIV, 66- *Mailing Add:* Dept Physiol Vanderbilt Univ Sch Med Nashville TN 37232

MENG, HEINZ KARL, b Baden, Ger, Feb 25, 24; nat US; m 53; c 2. BIOLOGY. *Educ:* Cornell Univ, BS, 47, PhD(ornith), 51. *Prof Exp:* From asst prof to assoc prof, 51-61, PROF BIOL, STATE UNIV NY COL NEW PALTZ, 61- *Mem:* Assoc mem Wildlife Soc; assoc mem Am Ornith Union. *Res:* Ornithology; entomology; falconry; vertebrate zoology. *Mailing Add:* Dept Biol State Univ New York Col New Paltz NY 12561

MENG, KARL H(ALL), b Rochester, NY, Nov 7, 11; m 43; c 3. CHEMICAL ENGINEERING. *Educ:* Univ Rochester, BS, 34, MS, 36, PhD(org chem), 39. *Prof Exp:* Asst chem, Univ Rochester, 34-38; instr, Nazareth Col, 39-41; chemist, Distillation Prod, Inc, Eastman Kodak Co, 41-42, chem engr & chem plant supvr, Distillation Prod Industs Div, 42-65, staff asst, Photomat Div, 65-69, staff asst, Res Labs, Admin Div, 70-76; RETIRED. *Mem:* Am Chem Soc; Sigma Xi; Am Inst Chem Engrs. *Res:* Sources and methods for making fish liver oils; stability of vitamin concentrates; adsorption processes; manufacturing methods for Vitamin A and E and derivatives; production of synthetic vitamin A and fine chemicals; process and product development; finance. *Mailing Add:* 40 Hollyvale Rochester NY 14618

MENG, SHIEN-YI, b Kirin, China, Oct 19, 29; m 70; c 2. ANTENNAE, MICROWAVES. *Educ:* Cheng Kung Univ, Taiwan, BS, 53; Okla State Univ, MS, 58; Ohio State Univ, PhD(elec eng), 68. *Prof Exp:* Jr engr, Taiwan Prov Govt, China, 54-56; designer, Ramseyer & Miller, Inc, NY, 58-59; res assoc radio astron, Radio Observ, Ohio State Univ, 66-68; asst prof, 68-74, assoc prof, 74-80, PROF ELEC ENG, CALIF POLYTECH STATE UNIV, SAN LUIS OBISPO, 80- *Concurrent Pos:* Res engr, SRI Int, Menlo Park, Calif, 79; engr, Lockheed Missile & Space Co, 80; consult, TRW Electronic Syst Group, 84-85; consult, Lawrence Livermore Nat Lab, 86-87. *Mem:* Inst Elec & Electronics Engrs. *Res:* Electromagnetic theory; antennas; radio wave propagation; microwave engineering. *Mailing Add:* Dept Elec Eng Calif Polytech State Univ San Luis Obispo CA 93407

MENGALI, UMBERTO, b Brindisi, Italy, Sept 25, 36; m 61; c 4. COMMUNICATION ENGINEERING. *Educ:* Univ Pisa, Laurea, 61. *Prof Exp:* From asst prof to assoc prof elec eng, 63-75, FULL PROF TELECOMMUN, DEPT INFO ENG, UNIV PISA, 75- *Concurrent Pos:* Consult, Telettra, 80-; ed, Inst Elec & Electronic Engrs Trans Commun, 81- *Mem:* Fel Inst Elec & Electronic Engrs. *Res:* Communication theory and digital signal processing as applied to communications; modulation techniques and synchronization algorithms for digital transmissions. *Mailing Add:* Dept Info Eng Univ Pisa Pisa 56100 Italy

MENGE, ALAN C, b Marengo, Ill, Apr 8, 34; m 57; c 4. REPRODUCTIVE PHYSIOLOGY. *Educ:* Univ Ill, BSc, 56; Univ Wis, MSc, 58, PhD(endocrinol), 61. *Prof Exp:* Asst prof animal sci, Rutgers Univ, 61-65, assoc prof, 65-67; ASSOC PROF REPROD BIOL, MED CTR, UNIV MICH, ANN ARBOR, 67- *Concurrent Pos:* Mem, Int Coord Comt Immunol of Reprod USA, 58; vis scientist, Uppsala Univ, Sweden, 79-80. *Mem:* Soc Study Reproduction; Int Soc Immunol Reproduction (vpres, 78-82); Brit Soc Study Fertil; Sigma Xi; Soc Exp Biol & Med. *Res:* Problems related to endocrine and immunologic causes of infertility. *Mailing Add:* 2101 Carhart St Ann Arbor MI 48104

MENGE, BRUCE ALLAN, b Minneapolis, Minn, Oct 5, 43; m 71; c 2. ECOLOGY. *Educ:* Univ Minn, Minneapolis, BA, 65; Univ Wash, PhD(ecol), 70. *Prof Exp:* Ford Found fel, Univ Calif, Santa Barbara, 70-71; asst prof biol, Univ Mass, Boston, 71-76; from asst prof to assoc prof, 76-85, PROF ZOOL, ORE STATE UNIV, 85- *Concurrent Pos:* Vis prof, Univ Guam, 80, Kristineberg Marine Sta, Sweden, 84, Univ Quebec-Rimouski, 86, Univ Umen, Sweden, 89, Univ Lund, Sweden, 89; chairperson, Ecol Div, Am Soc Zool, 87-88. *Honors & Awards:* Mercer Award, Ecol Soc Am, 79. *Mem:* Ecol Soc Am; Am Soc Nat; Am Soc Zool; Soc Study Evolution; AAAS; Am Soc Limnol & Oceanog. *Res:* Population and community ecology in the marine environment; effect of biological interactions; life history strategies; tropical marine ecology. *Mailing Add:* Dept of Zool Ore State Univ Corvallis OR 97331

MENGE, JOHN ARTHUR, b Minneapolis, Minn, Feb 24, 45; m 68; c 2. PLANT PATHOLOGY, MYCOLOGY. *Educ:* Univ Minn, BS, 67, MS, 69; NC State Univ PhD(plant path), 75. *Prof Exp:* ASST PROF PLANT PATH, UNIV CALIF, RIVERSIDE, 74- *Mem:* Am Phytopath Soc; Mycol Soc Am. *Res:* Fungus diseases of citrus; the mycorrhizal association of plants. *Mailing Add:* Dept Plant Path Univ Calif 900 University Ave Riverside CA 92521

MENGEBIER, WILLIAM LOUIS, b New York, NY, Dec 2, 21; m; c 3. PHYSIOLOGY. *Educ:* The Citadel, BS, 43; Oberlin Col, MA, 49; Univ Tenn, PhD(zool), 53. *Prof Exp:* Instr chem & biol, The Citadel, 46-49, asst prof, 49-54; prof biol, Madison Col, Va, 54-67; PROF BIOL, BRIDGEWATER COL, 67- *Res:* Effects of anoxia on cellular respiration; relative survival times of hibernators and mammals to anoxia; cellular physiology; effect of vertebrate hormones on invertebrates. *Mailing Add:* Dept Biol Bridgewater Col PO Box 147 Bridgewater VA 22812

MENGEL, DAVID BRUCE, b East Chicago, Ind, May 1, 48; m 68; c 2. SOIL FERTILITY, CROP PRODUCTION. *Educ:* Purdue Univ, BS, 70, MS, 72; NC State Univ, PhD(soils), 75. *Prof Exp:* Asst prof soil fertility & plant nutrit, Rice Exp Sta, La State Univ, 75-79; from asst prof to assoc prof, 79-86, PROF SOIL FERTILITY & PLANT NUTRIT, DEPT AGRON, PURDUE UNIV, 86- *Mem:* Soil Sci Soc Am; Am Soc Agron. *Res:* Nitrogen fertilization of cereal crops (corn, wheat), particularly as related to reduced tillage systems; soil testing; cropping systems and irrigation. *Mailing Add:* Dept Agron Purdue Univ West Lafayette IN 47907

MENGEL, J T, b Ballston Lake, NY, 18. DATA SYSTEMS. *Educ:* Union Col, BS, 39. *Prof Exp:* Dir tracking & data systs, Goddard Space Ctr, NASA, 58-73; RETIRED. *Mem:* Fel Inst Elec & Electronics Engrs. *Mailing Add:* 2112 Creekwood Ct Ft Collins CO 80525

MENGEL, ROBERT MORROW, zoology; deceased, see previous edition for last biography

MENGELING, WILLIAM LLOYD, b Elgin, Ill, Apr 1, 33; m 58; c 2. VETERINARY VIROLOGY. *Educ:* Kans State Univ, BS, 58, DVM, 60; Iowa State Univ, MS, 66, PhD(microbiol, biochem), 69; Am Col Vet Microbiol; Dipl. *Prof Exp:* Vet, St Francis Animal Hosp, Albuquerque, NMex, 60-61; res vet, 61-76, CHIEF VIROL RES LAB, NAT ANIMAL DIS CTR, 76- *Concurrent Pos:* Mem bd gov, Am Col Vet Microbiol, 76-79, chmn, 77-78; co ed, Dis of Swine, Iowa State Univ Press, 81; bd dirs, Conf Res Workers in Animal Diseases, 81-85, vpres, 86, pres, 87; collab prof, Iowa State Univ. *Honors & Awards:* George Fleming Prize, British Vet J, 78; Vet Med Res Award, Am Vet Med Asn, 89. *Mem:* Am Vet Med Asn; Animal Health Asn; Am Col Vet Microbiol; Am Soc Virol. *Res:* Respiratory and reproductive diseases of swine; virology. *Mailing Add:* Dept Vet Microbiol Iowa State Univ Ames IA 50011

MENGENHAUSER, JAMES VERNON, b Armour, SDak, Oct 12, 33; m 60; c 1. PHYSICAL CHEMISTRY, PETROLEUM CHEMISTRY. *Educ:* Univ SDak, BA, 54; NMex State Univ, MS, 64; Univ Colo, PhD(chem), 69. *Prof Exp:* Chemist, Ames Lab, AEC, 54-55; chemist, El Paso Natural Gas Co, 59-61; chemist, 68-89, SUPVRY CHEMIST, US ARMY BELVOIR RES, DEVELOP & ENG CTR, FT BELVOIR, 89- *Mem:* Am Chem Soc; Sigma Xi. *Res:* Nonflammable hydraulic fluids; nuclear magnetic resonance; solid film lubricants; hydrocarbon thermal and oxidation stability; petroleum fuel additives. *Mailing Add:* 8905 Camfield Dr Alexandria VA 22308

MENGER, EVA L, b South Bend, Ind, Feb 18, 43; m 64, 77; c 2. CHEMICAL PHYSICS. *Educ:* Carleton Col, BA, 64; Harvard Univ, MA, 65, PhD(chem), 68. *Prof Exp:* Asst ed, Accts Chem Res, 69-74; asst prof chem, Univ Calif, Santa Cruz, 74-78; sr res chemist, Allied Chem Corp, 78-80, assoc dir forward res, 80-87; VPROVOST & PROF CHEM, UNIV VA, 87- *Mem:* Am Chem

Soc; Sigma Xi. *Res:* Special interest focuses on nonsteady state effects in diffusion controlled processes and excited state reactions; heterogeneous catalysis; industrial chemicals and processes. *Mailing Add:* Univ Va Brooker House Charlottesville VA 22906

MENGER, FRED M, b South Bend, Ind, Dec 13, 37; m 62. ORGANIC CHEMISTRY. *Educ:* Johns Hopkins Univ, AB, 58; Univ Wis, PhD(chem), 63. *Prof Exp:* NIH fel, Northwestern Univ, 64-65; from asst prof to assoc prof, 65-72, PROF CHEM, EMORY UNIV, 72- *Honors & Awards:* NIH career develop award, 70-75. *Mem:* Am Chem Soc. *Res:* Bioorganic and physical organic chemistry; interfaces and colloidal systems; reaction mechanisms; enzyme models. *Mailing Add:* Dept Chem Emory Univ Atlanta GA 30322

MENGOLI, HENRY FRANCIS, b Plymouth, Mass, June 8, 28; m 54; c 2. IMMUNOLOGY. *Educ:* Boston Univ, AB, 50; Cath Univ, MS, 53, PhD(biol), 57. *Prof Exp:* Bacteriologist, Clin Ctr, NIH, 57-59, res biologist, Peripheral Blood Proj, Collab Area, Diag Res Br, Nat Cancer Inst, 59-63; instr, Cancer Res Lab, 63-65, asst prof path & microbiol, 65-73, asst prof microbiol, 73-75, ASSOC PROF MICROBIOL, IMMUNOL RES LAB, MED CTR, WVA UNIV, 75- *Mem:* NY Acad Sci; Am Chem Soc; Am Asn Microbiol; Sigma Xi. *Res:* Non-immunological mechanisms of host tissue destruction; host-microbial interactions in oral disease; cationic proteins of host tissues. *Mailing Add:* Dept of Microbiol WVa Univ Med Ctr Morgantown WV 26506

MENGUY, RENE, b Prague, Czech, Feb 4, 26; nat US; m; c 2. SURGERY. *Educ:* Univ Paris, MD, 51, Univ Minn, PhD, 57. *Prof Exp:* Fulbright grant, Am Hosp Chicago, Ill, 51-52, fel, Mayo Clin, 52-57; from instr to asst prof surg, Med Ctr, Univ Okla, 57-58; from assoc prof to prof, Med Ctr, Univ Ky, 61-65, assoc prof physiol, 64-65; prof surg & chmn dept, Univ Chicago, 65-71; PROF SURG, SCH MED & DENT, UNIV ROCHESTER, 71-; EMER SURGEON-IN-CHIEF, GENESEE HOSP, 77- *Concurrent Pos:* Markle scholar, 58; asst chief surgeon Vet Admin Hosp, Okla, 59-61; surgeon-in-chief, Genesee Hosp, 71-77. *Mem:* Am Col Surgeons; Am Surg Asn; Soc Univ Surgeons; Fr Acad Surg; Fr Cong Surg. *Res:* Surgery; gastroenterological surgery; experimental biology. *Mailing Add:* Genesee Hosp 224 Alexander St Rochester NY 14607

MENHINICK, EDWARD FULTON, b Cambridge, Mass, May 18, 35; m 61; c 3. ECOLOGY, PHYSIOLOGY. *Educ:* Emory Univ, BA, 57; Cornell Univ, MS, 60; Univ Ga, PhD(zool), 63. *Prof Exp:* Fel, Health Physics Div, Oak Ridge Nat Lab, 63-65; from asst prof to assoc prof, 65-77, PROF BIOL, UNIV NC, CHARLOTTE, 77- *Mem:* Ecol Soc Am; Entom Soc Am; Sigma Xi. *Res:* Ichthyology; water pollution; radiation ecology; environmental physiology; statistical analysis of density, diversity and energy flow. *Mailing Add:* Dept Biol Univ NC Charlotte NC 28223

MENIUS, ARTHUR CLAYTON, JR, b Salisbury, NC, Apr 30, 16; m 46; c 1. GENERAL PHYSICS. *Educ:* Catawba Col, AB, 37; Univ NC, PhD(physics), 42. *Hon Degrees:* DSc, Catawba Col, 69. *Prof Exp:* Instr & asst prof physics, Clemson Col, 42-44; physicist, John Hopkins Appl Physics Lab, 44-46; prof physics, Clemson Col, 46-49, NC State, 49-81; emer dean, Sch Phys & Math Sci, NC State, 60-81; RETIRED. *Concurrent Pos:* Sr physicist & head battery group proximity fuse proj, 45-46; consult nuclear aircraft propulsion, Sverdrup & Parcel, 50-51, reactor design, Monsanto Chem, 51-53, missiles develop & instrumentation, lasers, US Army Rocket & Guided Missile Agency, 56-74, US Atomic Energy Comn, 57-59; reactor design, Babcock & Wilcox & Am Mach & Foundry, 54-56, & other various subj from diff Inst & corp, 59-75; dir, Summer Inst Basic Nuclear Eng, Am Soc Elec Eng, Atomic Energy Comn, 58-60; head, dept physics, NC State 56-60, asst dean eng, 55-56, grad admr, Sommer Inst Nuclear Energy Indust, 55-57; mem exec comt, NC Sci & Technol Bd, Tech Adv Coun, Carolina Capital Corp, adv comt, Sci & Technol Appl Prog, Res Triangle Park, 77-81, bd, Oak Ridge Assoc Univ, 69-75; Raleigh United Fund; chmn, Gov Sci Adv Comt NC, Atomic Energy Adv Bd State NC, subcomt radiation standards, thermonuclear processes session, Nuclear Cong, Cleveland, 59, educ comt Carolina Sect, Am Inst Aeronauts & Astronauts, physics liaison comt, Southeastern Sect Am Soc Elec Eng; vchmn, Oak Ridge Assoc Univ Coun, 68-69; interim adv comt, Off Indust, Tourist & Community Resources, NC State Dept Natural & Econ Resources, 72. *Mem:* Fel Am Phys Soc; Am Inst Aeronauts & Astronauts; Am Nuclear Soc; Sigma Xi; Am Inst Physics. *Res:* Physical mathematics; Over five publications on physical science and technology. *Mailing Add:* 541 Hertford St Raleigh NC 27609

MENKART, JOHN, b Prague, Czech, Aug 20, 22; m; c 4. COSMETIC CHEMISTRY, RESEARCH ADMINISTRATION. *Educ:* Univ Leeds, BSc, 44, PhD(textile chem), 46. *Prof Exp:* Res chemist, Denham & Hargrave, Ltd, 46-48; sci liaison officer, Int Wool Secretariat, Eng, 48-50; res chemist, Patons Baldwins, Ltd, 50-53; asst dir res, Textile Res Inst, Princeton Univ, 54-58; group leader, Harris Res Labs Inc, Gillette Co, 58-65, from asst dir to assoc dir, 65-67, vpres, Gillette Res Inst, 67-68, pres, 68-71; vpres, Clairol Inc, 71-77, sr vpres technol, 77-87; PRES, SHAGBARK RES INC, 87- *Honors & Awards:* CIBS Award, Cosmetics, Toiletries & Fragrances Asn, 79; Maison de Navarre Medal, Soc Cosmetic Chem, 84. *Mem:* Soc Cosmetic Chem; Am Chem Soc; Fiber Soc. *Res:* Chemistry and physical properties of fibers; formulation and properties of topical products. *Mailing Add:* Shagbark Res Inc Barn Hill Rd Greenwich CT 06831-2802

MENKE, ANDREW G, b Lithuania, Aug 24, 44; US citizen; m 71; c 1. THIN FILM TECHNOLOGY, LOW EMISSIVITY COATINGS. *Educ:* Wayne State Univ, BS, 67; Univ Toledo, MS, 73. *Prof Exp:* Asst chem, Wayne State Univ, 63-67; sr res chemist, Libbey-Owens-Ford Co, Ohio, 68-75; dir res & develop, Anglass Indust Inc, 75-77; sr prod & process engr, Donnelly Mirrors, 77-78; mgr glass technol, 78-80, VPRES MFG & GLASS TECHNOL, ARDCO INC, 80- *Mem:* Int Solar Energy Soc; Am Chem Soc; Am Vacuum Soc; Int Soc Optical Eng. *Res:* Selective coating for solar collectors; low emissivity coatings for different applications; work on organotin compounds used to produce conductive tin oxide by pyrolysis of same; properties of these compounds and other related organometallic compounds; transparent conductive coatings on glass; glass bending and tempering. *Mailing Add:* Ardco Inc 12400 S Laramie Ave Chicago IL 60658

MENKE, DAVID HUGH, b St Louis, Mo, Apr 26, 51; m 73; c 5. GEOPHYSICS, ASTROPHYSICS. *Educ:* Univ Calif, Los Angeles, AB, 72, MS, 75, PhD(educ), 80. *Prof Exp:* Prof astron, Santa Monica Col, 73-77, & Pierce Col & Calif State Univ, 76-80; prof astron & dir Ashcroft Observ, Southern Utah State Col, 80-81; PROF ASTRON & EXEC DIR COPERNICAN SPACE SCI CTR, CENT CONN STATE UNIV, 81- *Concurrent Pos:* Lectr astron, Griffith Observ, 70-76; ed, Copernican Observer, 82; astronr & consult, Sitmar Cruise Lines, 85; sci corresp, WPOP Radio Sta, 85. *Mem:* Int Planetarium Soc. *Res:* Arches astronomy; cosmology; planetarium science. *Mailing Add:* Copernican Observ Cent Conn State Univ 1615 N Stanley St New Britain CT 06050

MENKES, HAROLD A, pulmonary medicine; deceased, see previous edition for last biography

MENKES, JOHN H, b Vienna, Austria, Dec 20, 28; US citizen; m 57; c 3. PEDIATRICS, NEUROLOGY. *Educ:* Univ Southern Calif, AB, 47, MS, 51; Johns Hopkins Univ, MD, 52. *Prof Exp:* Intern & asst resident pediat, Boston Children's Hosp, Mass, 52-54; asst prof neurol med & assoc prof pediat, Johns Hopkins Univ, 60-66; prof, 66-70, clin prof, 70-89, EMER PROF, NEUROL & PEDIAT, UNIV CALIF, LOS ANGELES, 89- *Concurrent Pos:* Fel pediat neurol, NY, 57-60; Joseph P Kennedy, Jr scholar ment retardation, 60-66; distinguished lectr, Louisville Pediat Soc, 85. *Honors & Awards:* Hower Award, Child Neurol Soc, 80. *Mem:* AAAS; fel Am Acad Neurol; fel Am Neurol Asn; Am Pediat Soc; Soc Pediat Res. *Res:* Metabolic disorders of the nervous system; child neurology. *Mailing Add:* 405 S Beverly Dr Suite 300 Beverly Hills CA 90212

MENKES, SHERWOOD BRADFORD, b Bradford, Pa, Mar 14, 21; m 50; c 4. MACHINE DESIGN. *Educ:* Columbia Col, BA, 41; Columbia Univ, BS, 42, MS, 54. *Prof Exp:* Engr, Western Elec Co, 42-49; PROF MECH ENG, CITY COL NEW YORK, 49- *Concurrent Pos:* Consult, 50-; forensic engr, 72-; ed, Marcel Dekker, Inc, 78- *Mem:* Am Soc Mech Eng; Am Soc Eng Educ. *Res:* Structural dynamics, especially in those situations in which the loads exceed the elastic limit for the structure, to the point of prompt failure. *Mailing Add:* Dept Mech Engr CCNY Convent Ave & 138th St New York NY 10031

MENLOVE, HOWARD OLSEN, b Mayfield, Utah, Oct 23, 36; m 58; c 4. NUCLEAR ENGINEERING, PHYSICS. *Educ:* Univ Calif, Berkeley, BS, 59; Univ Mich, Ann Arbor, MS, 61; Stanford Univ, PhD(nuclear eng), 66. *Prof Exp:* Physicist, Lockheed Palo Alto Res Lab, 61-66; scientist, Nuclear Res Ctr, Karlsruhe, Ger, 66-67; group leader nuclear safeguards, 67-80, proj mgr int safeguards, 80-85, LOS ALAMOS FEL, LOS ALAMOS NAT LAB, 85- *Concurrent Pos:* Consult, Int Atomic Energy Agency, 74-77. *Honors & Awards:* Radiation Indust Award, Am Nuclear Soc, 85. *Mem:* Am Nuclear Soc; Am Phys Soc; Inst Nuclear Mat Mgt. *Res:* Neutron cross-section measurements; fission physics; nondestructive assay instrumentation research and development; international safeguards research; nuclear cold fusion. *Mailing Add:* Los Alamos Nat Lab PO Box 1663 Los Alamos NM 87545

MENN, JULIUS JOEL, b Free City of Danzig, Feb 20, 29; US citizen. PESTICIDE TOXICOLOGY, PESTICIDE METABOLISM. *Educ:* Univ Calif, BSc, 53, MSc, 54, PhD(toxicol), 58. *Prof Exp:* Res asst toxicol, Univ Calif, Berkeley, 53-57; sect head, insecticide res, Stauffer Chem Co, 57-67, sr sect mgr, pesticide biochem, 67-74, mgr biochem, dept life sci, 74-79; dir agr res, agr chem, Zoecon Corp, 79-84, vpres lic & technol, 84-85; prof toxicol, San Jose State Univ, 79-84; nat prog leader, Crop Protection, USDA-AGR Res Serv, 85-88, ASSOC DEP, AREA DIR-USDA ARS, BELTSVILLE, MD, 88- *Concurrent Pos:* NSF lectr, Univ Calif, Davis, 65; invited speaker spec conf fate of pesticides in environ, Nat Acad Sci-Nat Res Coun, 71; mem team, Fate of Pesticides in Environ, US-USSR, 74-85; insecticide resistance comt, Entom Soc Am, 78-85; team leader, US-USSR Biol Control Team, USDA, 85-; adj prof entomol, Univ Md, 87-; chmn, Gordon Res Conf Agr Sci, 88-89. *Honors & Awards:* Burdick & Jackson Award, Am Chem Soc, 78. *Mem:* AAAS; fel Am Chem Soc; Entom Soc Am; NY Acad Sci; Soc Toxicol; Am Soc Pharmaceut Exp Therapeut; Int Soc Study Xenobiotics. *Res:* Metabolic fate and mode of action of synthetic organic pesticides; discovery of insect growth regulators, neurohormones and biorational insect control agents; integrated pest management. *Mailing Add:* USDA ARS BARC-West Bldg 003 Beltsville MD 20705

MENNE, THOMAS JOSEPH, b St Louis, Mo, May 13, 34; m 56; c 1. THEORETICAL PHYSICS. *Educ:* St Louis Univ, BS, 56; Univ Calif, Los Angeles, MS, 58, PhD(physics), 63. *Prof Exp:* Mem tech staff, Hughes Aircraft Co, 56-59; lectr physics, Loyola Univ Los Angeles, 60-62; res scientist, 62-65, assoc scientist, 65-69, scientist, 69-70, mgr res, 70-81, fusion physics, 81-84, ELECTROMAGNETICS, McDONNELL DOUGLAS CORP, 84- *Concurrent Pos:* Lectr, Washington Univ, 65-67; Hughes Master's fel, 56. *Mem:* Am Phys Soc; Sigma Xi; Inst Elect & Electronic Engrs. *Res:* Theoretical research on laser dynamics, electron spin resonance spectroscopy, crystal field theory and paramagnetic ion-lattice phonon interactions; magnetic and inertial confinement fusion; particle beams, cosmic rays and magnetic resonance characterization of polymers; research direction in chemical and molecular laser development; management of electromagnetic scattering technology. *Mailing Add:* 13716 Mason Green Ct Manchester MO 63011

MENNEAR, JOHN HARTLEY, b Flint, Mich, Apr 25, 35; m 56; c 4. HEALTH HAZARD ASSESSMENT. *Educ:* Ferris Inst, BS, 57; Purdue Univ, MS, 60, PhD(pharmacol), 62. *Prof Exp:* Pharmacologist, Hazleton Labs, Inc, 62-63; Pitman-Moore Div, Dow Chem Co, 63-66; from asst prof to assoc prof, Purdue Univ, 66-72, prof toxicol, 72-76; toxicologist, Chem Indust Inst Toxicol, 76-78; dir toxicol, Baxter Travenol, 78-80; expert toxicol, Nat Inst Environ Health Sci, 80-86; PROF & CHMN PHARMACEUT SCI, SCH PHARM, CAMPBELL UNIV, 87- *Mem:* Am Soc Pharmacol & Exp Therapeut; Soc Toxicol. *Res:* Toxicology; applied pharmacology. *Mailing Add:* Biomed Res Assocs Corp 1381 Kildaire Farm Rd Suite 153 Cary NC 27511

MENNEGA, AALDERT, b Assen, Netherlands, July 3, 30; US citizen; m 58; c 5. ANATOMY, PHYSIOLOGY. *Educ:* Calvin Col, AB, 57; Mich State Univ, MA, 60, PhD(anat, zool), 64. *Prof Exp:* Lab technologist, Grand Rapids Osteop Hosp, Mich, 58-59; med technologist, E W Sparrow Hosp, Lansing, 59-64; from asst to assoc prof, 64-75, PROF BIOL, DORDT COL, 75-, CHMN DEPT, 70- *Mem:* Creation Res Soc. *Res:* Respiratory system, especially of birds; histology and gross anatomy of birds, mammals and snakes; embryology. *Mailing Add:* Dept Biol Dordt Col Sioux Center IA 51250

MENNINGA, CLARENCE, b Otley, Iowa, Apr 6, 28; m 49; c 7. NUCLEAR CHEMISTRY, HISTORY & PHILOSOPHY OF SCIENCE. *Educ:* Calvin Col, BA, 49; Western Mich Univ, MA, 59; Purdue Univ, PhD(chem), 66. *Prof Exp:* Chemist, Maytag Co, 50-56; teacher, Grand Rapids Christian High Sch, 56-61; chemist, Lawrence Radiation Lab, 66-67; prof, 67-90, EMER PROF GEOL, CALVIN COL, 90- *Mem:* AAAS; Nat Sci Teachers Asn. *Res:* Composition of meteorites; history and philosophy of science. *Mailing Add:* Dept Geol Calvin Col Grand Rapids MI 49546

MENNINGER, JOHN ROBERT, b Columbus, Ohio, July 29, 35; m 60; c 2. MOLECULAR BIOLOGY, GERONTOLOGY. *Educ:* Harvard Univ, AB, 57, PhD(biochem), 64. *Prof Exp:* Whitney Found vis res fel molecular genetics, Med Res Coun Lab Molecular Biol, Cambridge Univ, 63-66; asst prof biol, Univ Ore, 66-72, res assoc molecular biol, Inst Molecular Biol, 66-72; assoc prof, 72-78, PROF BIOL, UNIV IOWA, 78-, CHMN DEPT, 85- *Concurrent Pos:* Prog dir, Cellular & Molecular Biol Training Grant, Univ Iowa Grad Col, 75-; vis res scientist, Fogarty Int fel, Nat Inst Med Res, Mill Hill, London, 79-80. *Mem:* Am Soc Biol Chemists. *Res:* Mechanism and control of information transfer in biological systems; protein biosynthesis; peptidyl-tRNA metabolism; mechanisms of cellular aging; accuracy of protein synthesis. *Mailing Add:* Dept Biol Univ Iowa Iowa City IA 52242

MENNINGER, KARL AUGUSTUS, b Topeka, Kans, July 22, 93; m 16, 41; c 4. PSYCHIATRY. *Educ:* Univ Wis, AB, 14, MS, 15; Harvard Univ, MD, 17. *Hon Degrees:* DSc, Washburn Univ, 49, Univ Wis, 65; LHD, Park Col, 55, St Benedict's Col, Kans, 63; LLD, Jefferson Med Col, 56, Parsons Col, 60, Kans State Univ, 62, Baker Univ, 65. *Prof Exp:* Intern, Kansas City Gen Hosp, Mo, 17-18; asst neuropath, Harvard Med Sch, 18-20; prof ment hyg, criminol & abnormal psychol, Washburn Col, 23-40; dean, Menninger Sch Psychiat, 46-70, dir educ, Menninger Found, 46-70, chmn bd trustees, 54-70, mem educ comt, 67-70, chief staff, Menninger Clin, 25-46, 52-70; prof med, Sch Med, Univ Kans, 70-76, univ prof at large, 76-; AT DEPT PSYCHIAT, UNIV HEALTH SCI-CHICAGO MED. *Concurrent Pos:* Asst, Med Col, Tufts Univ, 18-19; from asst to instr, Boston Psychopathic Hosp, 19; ed-in-chief, Bull, Menninger Clin, 20-; col asst, Topeka State Hosp, Kans, 20, chief consult, 48-; adv, Surg Gen, US Army, 45; consult, Fed Bur Prisons, Off Voc Rehab, Dept Health, Educ & Welfare, 48-; consult, Vet Admin Hosp, Topeka, 48-; mgr, Winter Vet Admin Hosp, 45-48, chmn, Dean's Comt & sr consult, 48-55; consult, Forbes AFB Hosp, 58- & Stone-Brandel Ctr, Chicago; dir, Topeka Inst Psychoanal, 60-; mem, Adv Comt, Int Surv Correctional Res & Pract, Calif, 60-; prof-at-large, Univ Kans; neuropsychiatrist, Stormont-Vail Hosp, Topeka; vis prof, Med Sch, Univ Cincinnati, trustee, Albert Deutsch Mem Found, 61; trustee, Aspen Inst Humanistic Studies, 61-64; consult, Inst Mgt Bd Social Welfare, State Kans; consult, Res Inst, Boston Univ, Europ Educ Ctr, Asn Migros Schs, Zurich; mem, Nat Cong Am Indian; mem, Kans Bd, John F Kennedy Mem Libr, Bd Overseers, Lemberg Ctr Study Violence, Brandeis Univ & Spec Comt Psychiat, Off Sci Res & Develop. *Mem:* Fel AMA; fel Am Psychiat Asn; fel Am Psychoanal Asn (pres, 41-43); fel Am Col Physicians; Am Orthopsychiat Asn (secy, 26, pres, 27); hon fel Am Asn Suicidology. *Res:* Influenza and mental diseases; psychological factors in somatic disease; suicide; hypertension; industrial and military psychiatry; psychiatric education; criminology; penology; religion. *Mailing Add:* Menninger Found Box 829 Topeka KS 66601

MENNINGER, WILLIAM WALTER, b Topeka, Kans, Oct 23, 31; m 53; c 6. FORENSIC, PSYCHOANALYSIS. *Educ:* Stanford Univ, AB, 53; Cornell Univ, MD, 57; Am Bd Psychiat & Neurol, dipl, Am Bd Forensic Psychiatry, dipl. *Hon Degrees:* LHD, Middlebury Col, Vt, 82; Ottawa Univ, 86; DSc, Washburn Univ, 82. *Prof Exp:* Intern, Harvard Med Serv & Boston City Hosp, 57-58; resident psychiat, Menninger Sch Psychiat, 58-61; comdg med officer, USPHS, 61; chief med officer & psychiatrist, Fed Reformatory, El Reno, Okla, 61-63; assoc psychiatrist, Peace Corps, 63-64; coordr develop, 67-69, dean, Karl Menninger Sch Psychiat & Ment Health Sci, 84-90, STAFF PSYCHIATRIST, MENNINGER FOUND, 65-, CHIEF STAFF, MENNINGER CLIN, 84- *Concurrent Pos:* Mem staff, USPHS, 59-64; mem nat adv health coun, HEW, 67-71; mem, Nat Comn Causes & Prevention Violence, 68-69; clin supvr, Topeka State Hosp, 69-70, sect dir, 70-72, asst supt, clin dir & dir residency training, 72-81; mem, Fed Prison Facil Planning Coun, 70-73; ed, Psychiat Digest, 71-74; mem adv bd, Nat Inst Corrections, 74-88, chmn, 80-82; adj prof, Washburn Univ & San Francisco Theol Sem; lectr, Menninger Sch Psychiat, Topeka Inst Psychoanalysis; consult, US Bur Prisons & US Secret Serv; clin prof, Kans State Univ Sch Med. *Mem:* Inst of Med of Nat Acad Sci; AMA; Am Acad Psychiat & Law; AAAS; fel Am Col Physicians; fel Am Col Psychiat; Am Psychoanal Asn; fel Am Psychiat Asn. *Mailing Add:* Menninger Found Box 829 Topeka KS 66601

MENNITT, PHILIP GARY, b Battle Creek, Mich, Mar 29, 37; m 61; c 4. PHYSICAL CHEMISTRY. *Educ:* Providence Col, BS, 58; Mass Inst Technol, PhD(phys chem), 62. *Prof Exp:* From instr to asst prof, 64-73, ASSOC PROF CHEM, BROOKLYN COL, 73- *Mem:* Am Chem Soc. *Res:* Nuclear magnetic resonance spectroscopy. *Mailing Add:* Dept Chem Brooklyn Col Brooklyn NY 11210

MENON, MANCHERY PRABHAKARA, b India, Aug 31, 28; m 55; c 2. NUCLEAR CHEMISTRY, RADIOCHEMISTRY. *Educ:* Univ Madras, BSc, 49; Agra Univ, MSc, 53; Univ Ark, PhD(nuclear chem, radiochem), 63. *Prof Exp:* Teacher sci, St Mary's CGH Sch, Kerala, India, 49-50; HS Vadayar, Kerala, 50-51; lectr chem, NSS Col, 53-55; lectr & head dept, Moulmein Col, Rangoon, 55-59; res asst, Univ Ark, 59-63; res assoc nuclear sci, Mass Inst Technol, 63-64; asst prof chem & asst res chemist, Activation Anal Lab, Tex A&M Univ, 64-67; assoc prof, 67-71, PROF CHEM, SAVANNAH STATE COL, 71- *Concurrent Pos:* Res grants, NSF, NASA, Am Chem Soc-PRF, Res Corp, Environ Protection Agency, NIH, DOE. *Mem:* Am Chem Soc; Am Chem Inst. *Res:* Nuclear fission; natural radioactivity; fall-out from nuclear detonation; radio reagent methods of analysis for five elements; trace element release from sediments; lactate dehydrogenase iroenzymes in clinical chemistry; solution chemistry of Ln F3; author or coauthor of forty-five publications. *Mailing Add:* PO Box 20426 Savannah GA 31404

MENON, PREMACHANDRAN R(AMA), b Tapah, Malaysia, June 29, 31; US citizen; m 59; c 1. ELECTRICAL ENGINEERING, COMPUTER SCIENCE. *Educ:* Banaras Hindu Univ, BSc, 54; Univ Wash, PhD(elec eng), 62. *Prof Exp:* asst prof elec eng, Univ Wash, 62-63; mem tech staff, Bell Tel Labs, 63-86; PROF ELEC & COMPUT ENG, UNIV MASS, 86- *Mem:* Fel Inst Elec & Electronic Engrs. *Res:* Switching theory; diagnosis of digital circuits; simulation. *Mailing Add:* Dept Elect & Comp Eng Univ Mass Amherst MA 01003

MENON, THUPPALAY K, b Kerala, June 13, 28; m 62; c 2. RADIO ASTRONOMY, COSMOLOGY. *Educ:* Annamalai Univ, India, BS, 47; Indian Inst Sci, DIISc, 50; Harvard Univ, MS, 53, PhD(astron), 56. *Prof Exp:* Lectr astron, Harvard Univ, 56-58; asst prof astron, Univ Pa, 58-60; assoc scientist astron, Nat Radioastron Observ, 60-63, scientist, 63-68; prof astron, Univ Hawaii, 68-71, Tata Inst, Bombay, 71-79; PROF & HEAD ASTRON, UNIV BC, 79- *Mem:* Am Astron Soc; Royal Astron Soc; Indian Acad Sci; Int Astron Union; Astron Soc Pac. *Res:* Dynamics of interstellar medium; structure and evolution of extragalatic radio sources; cosmology. *Mailing Add:* Dept Geophys & Astron Univ BC 2219 Main Mall Vancouver BC V6T 1Z4 Can

MENQ, CHIA-HSIANG, b Chia-Yi, Taiwan, July 29, 56. SYSTEM DYNAMICS & CONTROLS, ROBOTICS & AUTOMATION. *Educ:* Nat Tsing-Hua Univ, Taiwan, BS, 78; Carnegie-Mellon Univ, MS, 82, PhD(mech eng), 85. *Prof Exp:* Asst prof, 85-90, ASSOC PROF MECH ENG, OHIO STATE UNIV, 90- *Concurrent Pos:* NSF presidential young investr, 89. *Mem:* Am Soc Mech Engrs; Soc Mfg Engrs; Inst Elec & Electronics Engrs. *Res:* Robot calibration; control of robotic manipulators and mechanical systems; automated dimensional inspection using coordinate measuring machines; concurrent process planning for machining and inspection in manufacturing. *Mailing Add:* Dept Mech Eng Ohio State Univ Columbus OH 43210-1107

MENSAH, PATRICIA LUCAS, b Washington, DC, Feb 7, 48. NEUROANATOMY, NEUROSCIENCES. *Educ:* Howard Univ, BS, 70; Univ Calif, Irvine, PhD(biol sci), 74. *Prof Exp:* Psychologist neuropsychol, NIMH, 70; fel psychiat, State Univ NY Stony Brook, 74-75; instr neuroanat, Univ Calif, Irvine, 75, NIMH fel psychobiol, 75-76; ASST PROF ANAT, UNIV SOUTHERN CALIF, 76- *Concurrent Pos:* mem, Minority Women in Sci Network, AAAS. *Mem:* Am Asn Anatomists; Soc Neurosci. *Res:* Detailed neuroanatomical analyses of projection patterns within the caudate nuclei of the mammalian striatum; dendritic specialization: a look at the specific synaptic contacts of individual dendrites of the same neuron; neuronal cell death during development. *Mailing Add:* Dept Anats Ave Cleveland Chiropractic Col 590 N Vt Ave Los Angeles CA 90004

MENSER, HARRY ALVIN, JR, b Pittsburgh, Pa, Dec 30, 30; m 60; c 2. PLANT PHYSIOLOGY. *Educ:* Univ Delaware, BS, 54; Univ Md, MS, 59, PhD(agron, bot), 62. *Prof Exp:* Asst county agent, Agr Exten Serv, Univ Md, 55-57, asst agron, Univ Md, 57-59; res technician, Tobacco Lab, Plant Genetics & Germplasm Inst, Sci & Educ Admin-Agr Res, US Dept Agr, 60-63, res plant physiologist, 63-80; assoc prof hort, Univ Idaho, 80-88; RETIRED. *Mem:* Sigma Xi; Am Soc Hort Sci; Am Soc Agron; Crop Sci Soc Am. *Res:* Analysis of heavy metals as environmental hazards; land recycling of municipal wastes; revegetation of lands disturbed by surface mining; fruit and vegetables; nursery and ornamentals; christmas trees; grain and forages; biomass for fuels; cooperative extension service. *Mailing Add:* 6370 Kanisfu Shores Circle Sand Point ID 83864

MENSING, RICHARD WALTER, b Hackensack, NJ, Sept 20, 36; m 57; c 3. STATISTICS. *Educ:* Valparaiso Univ, BS, 60; Iowa State Univ, MS, 65, PhD(statist), 68. *Prof Exp:* Engr, Gen Elec Co, 60-62; from instr to assoc prof statist, Iowa State Univ, 66-77; STATISTICIAN, LAWRENCE LIVERMORE LAB, 77- *Mem:* Inst Math Statist; Am Statist Asn; Soc Risk Analysis. *Res:* Engineering and industrial statistics; applied probability. *Mailing Add:* Lawrence Livermore Lab PO Box 808 Livermore CA 94550

MENTE, GLEN ALLEN, b Wheatland, Iowa, Feb 25, 38; m 59; c 2. ANIMAL NUTRITION. *Educ:* Iowa State Univ, BS, 61, MS, 63. *Prof Exp:* Nutritionist, Farmers Coop Soc, 63-64; nutritionist, Kent Feeds Inc, 64-67, mgr, 67-74, vpres, 74-81, BD DIRS & SR VPRES NUTRIT & PROD DEVELOP, KENT FEEDS INC, 81- *Mem:* Am Soc Animal Sci; Agr Res Inst; Nutrit Coun Am Feed Mgrs; Nutrit Feed Ingredients Asn. *Res:* Continuous evaluation of nutrient requirements for swine, beef, dairy, poultry, turkeys, lambs, horses, fish, pets and miscellaneous animals. *Mailing Add:* RR 1 Box 320 Muscatine IA 52761

MENTON, DAVID NORMAN, b Mankato, Minn, July 1, 38; m 61; c 2. ANATOMY, HISTOLOGY. *Educ:* Mankato State Univ, BS, 59; Brown Univ, PhD(biol), 66. *Prof Exp:* Instr anat, 66-70, asst prof anat & path, 70-76, ASSOC PROF ANAT & PATH, SCH MED, WASH UNIV, 76- *Honors & Awards:* Silver Award, Am Acad Derm, 71. *Mem:* AAAS; Am Asn Anat; Soc Invest Dermat. *Res:* Fine structure of skin; barrier function of the stratum corneum; effects of essential fatty acid deficiency of skin. *Mailing Add:* Dept Anat & Neurobiol Wash Univ Sch Med Box 8108 660 Euclid Ave St Louis MO 63110

MENTON, ROBERT THOMAS, b Jamaica, NY, Dec 12, 42; m 67; c 1. ACOUSTICS. *Educ:* Cooper Union, BEng, 64; Univ Conn, MS, 67; Catholic Univ Am, PhD(eng acoustics), 73. *Prof Exp:* Mech engr, Naval Underwater Systems Ctr, 64-81; mech engr, 81-83, RES PHYSICIST, NAVAL RES LAB, 83- *Concurrent Pos:* Lectr, acoustic radiation & scattering, Catholic Univ Am. *Mem:* Acoust Soc Am; Sigma Xi. *Res:* Deterministic, random and turbulence excited responses of fluid loaded plates and shells; noise in sonar systems; analytic and numerical radiation and scattering by elastic and rigid bodies. *Mailing Add:* Naval Res Lab Code 5120 4555 Overlook Ave SW Washington DC 20375-5000

MENTONE, PAT FRANCIS, b Chicago, Ill, May 25, 42; m 70; c 3. INORGANIC CHEMISTRY, ELECTROCHEMISTRY. *Educ:* Col St Thomas, BS, 64; Univ Minn, PhD(inorg chem), 69. *Prof Exp:* Sr chemist, Buckbee Mears Co, 69-73, plating prod mgr, 73-74, eng supvr, 74-75, mgr plating opers, 75-80, managing dir, Singapore, 81-84; founder, Mentone Component Technol Corp, 85-89; EXEC VPRES, PMX INDUSTS, 90- *Concurrent Pos:* Consult to Asian Electronic Component Mfrs. *Mem:* Am Chem Soc; Am Electroplaters Soc. *Res:* New product development and manufacturing engineering for microelectronic components. *Mailing Add:* 1916 Oak Knolls Ct SE Cedar Rapids IA 52403

MENTZER, JOHN R(AYMOND), b Arch Spring, Pa, June 16, 16; m; c 2. PHYSICS. *Educ:* Pa State Univ, BS, 42, MS, 48; Ohio State Univ, PhD(physics), 52. *Prof Exp:* Res assoc antennas, Ohio State Univ, 48-52; mem staff propagation theory, Lincoln Lab, Mass Inst Technol, 52-54; from assoc prof to prof elec eng, Pa State Univ, 54-57, prof eng sci, 57-88, head dept, 74-88; RETIRED. *Concurrent Pos:* Consult, Sandia Corp, 59-, Bell Tel Labs, 65-88 & Locus, Inc, 87-91. *Mem:* AAAS; sr mem Inst Elec & Electronics Engrs; Soc Indust & Appl Math; Am Soc Eng Educ. *Res:* Scattering and diffraction of electromagnetic waves; magnetoplasma dynamics; ionospheric and tropospheric propagation; high energy particle dynamics and interactions with electromagnetic fields. *Mailing Add:* 557 Clarence Ave State College PA 16803

MENYHERT, WILLIAM ROBERT, b Budapest, Hungary, Jan 7, 35; US citizen; m 64; c 3. POLYMER CHEMISTRY, CONDUCTIVE POLYMERS. *Educ:* Med Univ Budapest, dipl, 56, ScD, 64; Georgetown Univ, dipl, 57. *Prof Exp:* Chemist, Ohio Valley Gen Hosp, Wheeling WVa, 57-58; biochemist, Oscar B Hunter Mem Lab, Washington, DC, 58-60; asst, Boston Univ, 60 & Harvard Univ, 61; res scientist, Lahey Found, Boston, 62 & Bionics Systs Res, Alexandria, Va, 62-63; asst prof chem, Univ Md, College Park, 63-64; asst prof pharmacol chem, Howard Univ, Washington, DC, 64-65; pres, Menyhert Labs, Inc & Chem-Pox Co, Washington, DC, 65-73; vpres res & develop, Superior Polymers Co, Inc, Easton, Md, 73-76; dir, Environ Sci Cons, Nags Head, NC, 76-79; vpres & dir, Inovations, Inc, St Michaels, Md, 80-82; dir polymer div, Innova, Inc, Clearwater, Fla, 82-83; HEAD DEPT RES, SECURITY TAG SYSTS, INC, ST PETERSBURG, FLA, 84- *Concurrent Pos:* Consult, US Armed Forces, 66-; mem, Pres Coun Consumer Affairs, 68-70 & Pres Coun Environ, 69-72; courtesy prof chem, Univ SFla, Tampa, 83- *Honors & Awards:* Pres Citation, White House, 70; Pres Medal Merit, White House, 87. *Mem:* Fel Am Inst Chemists; Am Chem Soc; Royal Soc Chem; NY Acad Sci. *Res:* Conductive polymers; polymer growth using unique oxi-redox systems; conductive states in semi-crystalline matrixes; new polymers via metal-organo ligands; polymer sensors; polymerized amorphous alloys. *Mailing Add:* American Hi-Tech 4695 Ulmerton Rd Suite 420 Clearwater FL 34622

MENZ, LEO JOSEPH, b Erie, Pa, Mar 7, 27; m 54; c 6. BIOPHYSICS, CRYOBIOLOGY. *Educ:* Gannon Col, BS, 49; St Louis Univ, MS, 52, PhD(biophys), 57. *Prof Exp:* Asst prof biol & dir dept, Gannon Col, 54-57; res assoc, Am Found Biol Res, 57-68; ASSOC PROF SURG, ST LOUIS UNIV, 68- *Concurrent Pos:* Consult, US Vet-Cochran Hosp, St Louis; adj assoc prof biol, St Louis Univ, 80- *Mem:* Sigma Xi; Soc Cryobiol; Electron Micros Soc Am; Biophys Soc; Soc Cell Biol. *Res:* Effects of low temperature on biological material; freeze-drying; correlation of ultrastructural changes and function of frozen-thawed tissues, such as mammalian nerve, heart and blood cells; toxicity of cryoprotectants; hypothermic perservation of mammalian heart and kidney; ultrastructure; nature of freezing injury; scanning EM study of microvascular corrosion casts of kidneys from rabbits subjected to hypovolemic shock; development and implementation of mineature ultrasonic ceramic dimension gauges used in cardiovascular studies. *Mailing Add:* Dept Surg 1402 S Grand Blvd St Louis Univ St Louis MO 63104

MENZEL, BRUCE WILLARD, b Waukesha, Wis, Aug 23, 42; m 69; c 2. ICHTHYOLOGY, FISHERIES. *Educ:* Univ Wis, BS, 64; Marquette Univ, MS, 66; Cornell Univ, PhD(vert zool), 70. *Prof Exp:* From asst prof to assoc prof, 70-80, PROF ANIMAL ECOL, IOWA STATE UNIV, 80-, CHMN DEPT, 85- *Concurrent Pos:* Ed, Iowa State J Res, 87- *Mem:* Am Soc Ichthyol & Herpet; Am Fisheries Soc; Ecol Soc Am. *Res:* Systematics; ecology and behavior of North American freshwater fishes. *Mailing Add:* Dept Animal Ecol 124 Sci Li Iowa State Univ Ames IA 50011

MENZEL, DANIEL B, b Cincinnati, Ohio, Sept 27, 34; m 56; c 1. TOXICOLOGY, PHARMACOLOGY. *Educ:* Univ Calif, Berkeley, BS, 56, PhD(biol chem), 61. *Prof Exp:* Asst specialist nutrit, Univ Calif, Berkeley, 62, 65, asst prof food sci & asst biochemist, Inst Marine Sci, 62-67, asst prof food sci, Univ & asst biochemist, Agr Exp Sta, 65-67; res assoc biol, Pac Northwest Labs, Battelle Mem Inst, Wash, 67-68, mgr nutrit & food technol sect, 68-69; dir clin res, Ross Labs Div, Abbott Labs, Ohio, 69-71; assoc prof, Duke Univ, 71-78, head, Div Pharmacol, 75-77, coordr environ health affairs, 75-89, PROF PHARMACOL & EXP MED, DUKE UNIV, 78- & PROF & CHMN COMMUNITY & ENVIRON MED, 89- *Concurrent Pos:* Biochemist, US Bur Com Fisheries, 62-65; consult, Life Sci Div, Ames Res Ctr, NASA, 63-71; consult, Environ Protection Agency, 71-, mem, Sci Adv Bd, 78-; mem, Toxicol Study Sect, NIH, 75-80; mem, Acad Coun, Duke Univ, 77-78; ed, Toxicol Letts, 77- & Res Methods in Toxicol, 78-; mem, Nat Libr Med Adv Bd, 78-; Foggarty sr fel, 80; mem, Bd Toxicol & Environ Health Hazards & chmn, Comt Drinking Water & Health, Nat Acad Sci, 81; Alexander von Humboldt prize, 80. *Mem:* AAAS; Am Chem Soc; Am Oil Chem Soc; Entom Soc Am; NY Acad Sci. *Res:* Mechanisms of aging; lipid oxidation and vitamin E; biochemistry of fat soluble vitamins; environmental effects on the lung; chronic lung disease. *Mailing Add:* Whitby Bldg Rm 100 Univ Calif Irvine CA 92717

MENZEL, DAVID WASHINGTON, b India, Feb 22, 28; m 52. OCEANOGRAPHY. *Educ:* Elmhurst Col, BS, 49; Univ Ill, MS, 52; Univ Mich, PhD(fisheries), 58. *Prof Exp:* Res biologist, Bermuda Biol Sta, 57-63; assoc scientist, Woods Hole Oceanog Inst, 63-70; DIR, SKIDAWAY INST OCEANOG, 70- *Honors & Awards:* Distinguished Assoc Award, Dept Energy. *Mem:* AAAS; Am Soc Limnol & Oceanog. *Res:* Ecology and physiology of marine plankton; marine chemistry and biology. *Mailing Add:* Skidaway Inst of Oceanog PO Box 13687 Savannah GA 31416

MENZEL, ERHARD ROLAND, US citizen. MOLECULAR SPECTROSCOPY. *Educ:* Wash State Univ, BS, 67, PhD(physics), 70. *Prof Exp:* Fels, Simon Frazer Univ, 70-72, Purdue Univ, 72-73 & Univ Ky, 73-74; mem sci staff physics, Xerox Res Centre Can Ltd, 75-79; ASST PROF PHYSICS, TEX TECH UNIV, 79- *Mem:* Am Chem Soc; Am Phys Soc. *Res:* Luminescence spectroscopy of pigments and dyes; electron spin resonance of free radicals and transition metal complexes; fingerprint detection with use of lasers. *Mailing Add:* Physics Dept Tex Tech Univ Lubbock TX 79409

MENZEL, JOERG H, b Kassel, Ger, July 27, 39; US citizen; m; c 4. NUCLEAR SCIENCE. *Educ:* Rensselaer Polytech Inst, BME, 62, MS, 64, PhD(nuclear sci), 68. *Prof Exp:* Staff mem res & develop safeguards, Los Alamos Sci Lab, 68-73; first officer int safeguards, Int Atomic Energy Agency, 73-75; staff mem res & develop safeguards, Los Alamos Sci Lab, 75-76; phys sci officer, Arms Control & Disarmament Agency, 76-78, chief nuclear safeguards staff, 78-81, chief nuclear safeguards & technol div, 81-89; STAFF ON SITE INSPECTION AGENCY, 89- *Concurrent Pos:* Sr adv, Off of the Ambassador-at-Large, Non-Proliferation Policy Dept, 85-87. *Mem:* Inst Nuclear Mat Mgt. *Res:* Non-destructive analysis of nuclear materials; nuclear technology; international safeguards; peaceful uses of nuclear energy; nuclear non-proliferation policy. *Mailing Add:* On Site Inspection Agency Washington DC 20041-0498

MENZEL, ROBERT WINSTON, marine biology; deceased, see previous edition for last biography

MENZEL, RONALD GEORGE, b Independence, Iowa, Jan 23, 24; m 52; c 2. SOIL CHEMISTRY. *Educ:* Iowa State Col, BS, 47; Univ Wis, PhD(soil chem), 50. *Prof Exp:* Soil scientist, Agr Res Serv, US Dept Agr, 50-69, dir, 69-80, res leader, Water Qual Mgt Lab, 77-85; RETIRED. *Concurrent Pos:* Ed, J Environ Qual, 77-83. *Mem:* AAAS; Soil Sci Soc Am; fel Am Soc Agron; Am Chem Soc. *Res:* Reactions of copper and zinc in soils and availability to plants; uptake of nuclear fission products by plants; relation of agricultural chemicals and fertilizers to water pollution. *Mailing Add:* Water Qual & Watershed Res US Dept Agr PO Box 1430 Durant OK 74702

MENZEL, WOLFGANG PAUL, b Heidenheim, Ger, Oct 5, 45; US citizen. THEORETICAL SOLID STATE PHYSICS. *Educ:* Univ Md, BS, 67; Univ Wis, MS, 68, PhD(physics), 74. *Prof Exp:* Res asst solid state physics, Univ, 72-74, proj asst, 74-75, proj assoc radiometry, 75-77, ASST SCIENTIST, SPACE SCI & ENG CTR, UNIV WIS-MADISON, 77- *Mem:* Sigma Xi. *Res:* Investigation of optical properties of solids using the method of linear combinations of atomic orbitals; real time analysis of the atmosphere with infrared radiometric satellite probing. *Mailing Add:* 109 N Prospect Ave Madison WI 53705

MENZER, FRED J, b Memphis, Tenn, Feb 8, 33. GEOLOGICAL MANAGEMENT. *Educ:* Cole Col, BS, 59; Univ Wash, MS, 60, PhD(geol), 64. *Prof Exp:* Prof, geol, Western State Col, Colo, 70-81; CHIEF GEOLOGIST, FMC GOLD CO, DENVER, 81- *Mem:* AAAS; Geol Soc Am; Soc Econ Geologists; Mine Engr Geol Soc. *Mailing Add:* 4250 S Olive No 113 Denver CO 80237

MENZER, ROBERT EVERETT, b Wash, DC, Dec 21, 38; m 62; c 3. INSECT TOXICOLOGY. *Educ:* Univ Pa, BS, 60; Univ Md, MS, 62; Univ Wis, PhD(entom, biochem), 64. *Prof Exp:* Res asst entom, Univ Md, 61-62, instr, 62; instr, Univ Wis, 64; from asst prof to assoc prof, Univ Md, College Park, 64-73, assoc dean grad studies, 74-76, actg dean grad studies, 76-80, prof entom, 73-89; DIR, ENVIRON RES LAB, US ENVIRON PROTECTION AGENCY, GULF BREEZE, 89- *Concurrent Pos:* Mem toxicol study sect, NIH, 71-75, chmn, 73-75; chmn Fed Insecticide, Fungicide & Rodenticide Act, Sci Adv Panel, Environ Protection Agency, 82-84; chmn, Hazardous Substances Databank Sci Rev Panel, Nat Libr Med, 74- *Mem:* AAAS; Am Chem Soc; Soc Toxicol; Soc Environ Toxicol & Chem. *Res:* Pesticide chemistry and toxicology; metabolism of organophosphorus insecticides; insect biochemistry. *Mailing Add:* Environ Res Lab US Environ Protection Agency Sabine Island Gulf Breeze FL 32561-5299

MENZIE, DONALD E, b DuBois, Pa, Apr 4, 22; m 46; c 4. ENGINEERING. *Educ:* Pa State Univ, BS, 42, MS, 48, PhD, 62. *Prof Exp:* Jr marine engr, Philadelphia Navy Yard, 43-46; asst petrol eng, Pa State Univ, 46-48, instr, 48-51; from asst prof to prof, 51-72, dir, Sch Petrol Eng, 63-72, dir petrol eng educ, 63-79, PROF PETROL & GEOL ENG, UNIV OKLA, 72-, ASSOC EXEC DIR, ENERGY RESOURCES CTR, 79- *Concurrent Pos:* Consult, 51-; Halliburton distinguished lectr award, Col Eng, Univ Okla, 81-84. *Mem:* Am Inst Mining, Metall & Petrol Engrs; Sigma Xi. *Res:* Petroleum, reservoir and geological engineering; secondary recovery. *Mailing Add:* 1503 Melrose Dr Norman OK 73069

MENZIE, ELMER LYLE, agricultural policy & marketing, agricultural development, for more information see previous edition

MENZIES, CARL STEPHEN, b Menard, Tex, Mar 6, 32; m 52; c 2. ANIMAL HUSBANDRY. *Educ:* Tex Tech Col, BS, 54; Kans State Univ, MS, 56; Univ Ky, PhD(animal nutrit), 65. *Prof Exp:* Asst county agent, Tex Exten Serv, 54; res asst animal husb, Kans State Univ, 54-55, from instr to assoc prof, 55-68; prof animal sci & head dept, SDak State Univ, 68-72; PROF ANIMAL SCI & RESIDENT DIR RES, TAMU AGR RES & EXTEN CTR, TEX A&M UNIV, 72- *Honors & Awards:* Silver Ram Award, Am Sheep Producers Coun. *Mem:* Am Soc Animal Sci; Sigma Xi. *Res:* Nutrition and breeding of sheep. *Mailing Add:* 7887 N Hwy 87 San Angelo TX 76901

MENZIES, JOHN, b Alyth, Scotland, June 29, 49; Can citizen; m 72; c 3. SUBGLACIAL PROCESSES, DRUMLIN FORMATION. *Educ:* Univ Aberdeen, UK, BSc, 71; Univ Edinburgh, UK, PhD(geog & geomorphol), 76. *Prof Exp:* Higher sci officer, Soil Surv Dept, Macaulay Inst Soil Res, Aberdeen, UK, 75-77; assoc prof geog, Brock Univ, 77-84, chmn dept, 82-84, assoc prof, 84-90, PROF GEOG & GEOL SCI, BROCK UNIV, 90- *Mem:* Fel Geol Asn Can; Int Glaciol Soc; Quarternary Asn Can; Quarternary Res Asn UK; Soc Econ Paleont & Mineral. *Res:* Glacial geology; geomorphology; glacial sediments and processes of glacial deposition and erosion; drumlin research. *Mailing Add:* Dept Geog & Geol Brock Univ St Catherines ON L2S 3A1 Can

MENZIES, ROBERT ALLEN, b San Francisco, Calif, Nov 13, 35; m 58; c 3. BIOCHEMISTRY. *Educ:* Univ Fla, BS, 60, MS, 62; Cornell Univ, PhD(phys biol), 66. *Prof Exp:* Res chemist, Aging Res Lab, Vet Admin Hosp, Baltimore, Md, 65-67; asst prof biochem, La State Univ Med Ctr, New Orleans, 67-73; asst prof, 73-74, assoc prof biochem & ocean sci, 77-79, assoc prof biochem, Life Sci Ctr, Nova Univ, 75-, prof biochem & oceanog, 79-; PROF DEPT BIOCHEM, SCH MED, ST GEORGE'S UNIV, GRENADA. *Mem:* Am Soc Biol Chem; Am Chem Soc; Biophys Soc; NY Acad Sci; Am Soc Cell Biol. *Res:* Biochemical systematics with applications to evolution and fisheries; metabolism of organic pollutants by marine aguatic organisms; relationship of oceanic currents with gene flow and larval dispersal and recruitment. *Mailing Add:* 3515 E Fletcher Ave Tampa FL 33619

MENZIES, ROBERT THOMAS, b San Francisco, Calif, Mar 31, 43; m 66; c 1. LASER SPECTROSCOPY, ATMOSPHERIC SCIENCES. *Educ:* Mass Inst Technol, BS, 65; Calif Inst Technol, MS, 67, PhD(physics), 70. *Prof Exp:* Scientist laser physics, 70-75, mem tech staff atmospheric studies, 75-88, SR RES SCIENTIST, JET PROPULSION LAB, CALIF INST TECHNOL, 88- *Concurrent Pos:* Vis assoc, Calif Inst Technol, 72-77; vis scientist, Chalmers Inst Technol, Goteborg, Sweden, 74. *Mem:* AAAS; Am Geophys Union; fel Optical Soc Am; Inst Elec & Electronic Engrs Lasers & Electro Optics Soc; Am Meteorol Soc. *Res:* Atmospheric measurements with laser and heterodyne detection techniques; lidar development. *Mailing Add:* Jet Propulsion Lab 4800 Oak Grove Dr MS 169-214 Pasadena CA 91109

MENZIN, MARGARET SCHOENBERG, b New York, NY, Nov 17, 42; m 68; c 2. MATHEMATICS. *Educ:* Swarthmore Col, BA, 63; Brandeis Univ, MA, 67, PhD(math), 70. *Prof Exp:* From instr to asst prof, 69-73, chmn dept, 71-77, assoc prof, 73-80, PROF MATH, SIMMONS COL, 80-, CO-DIR, PROG COMPUT SCI, 80- *Concurrent Pos:* Consult & dir, Design Technol Corp, 69- *Mem:* Am Math Soc; Math Asn Am; Asn Women in Math; Women in Sci & Eng; Sigma Xi. *Res:* Ring theory; linear programming; mathematical models; computer sciences. *Mailing Add:* 26 Mason St Lexington MA 02173

MEOLA, ROBERT RALPH, b Newark, NJ, Jan 27, 27; m 48; c 2. ELECTRICAL ENGINEERING. *Educ:* Newark Col Eng, BS, 46; Stevens Inst Technol, MS, 49. *Prof Exp:* Jr engr, Nat Union Radio Corp, 46-47; instr elec eng, Newark Col Eng, 47-51; engr, Western Elec Co, 51-53; from instr to assoc prof, 53-69, PROF ELEC ENG, NEWARK COL ENG, NJ INST TECHNOL, 69-; RETIRED. *Concurrent Pos:* Consult, Int Tel & Tel Corp, NJ, 58-60, Hewlett Packard Co, 61-64, 66, NJ Bell Tel Co, 65 & Calculagraph Co, 69-72. *Mem:* Am Soc Eng Educ; sr mem Inst Elec & Electronics Engrs; Nat Soc Prof Engrs. *Res:* Analysis and design of transistor circuits. *Mailing Add:* 1810 Aberideen Rd Lakewood NJ 07103

MEOLA, ROGER WALKER, b Cleveland, Ohio, Aug 25, 34; m 56. INSECT PHYSIOLOGY, MEDICAL-VETERINARY ENTOMOLOGY. *Educ:* Ohio State Univ, BS, 56, MS, 58, Ph(entom), 63. *Prof Exp:* Entomologist, Fla State Bd Health, 64-69; res assoc, Univ Ga, 70-72; res assoc, 73-74, from lectr to assoc prof, 75-87, PROF ENTOM, TEX A&M UNIV, 88- *Mem:* AAAS; Am Mosquito Control Asn; Am Soc Zoologists; Entom Soc Am; Sigma Xi. *Res:* Endocrine regulation of mosquito reproduction and behavior; physiology of diapause and insect microsurgical techniques. *Mailing Add:* Dept Entom Tex A&M Univ College Station TX 77840

MEOLA, SHIRLEE MAY, b Canton, Ohio, Dec 7, 35; m 56. NEUROENDOCRINOLOGY, IMMUNOCYTOCHEMISTRY. *Educ:* Ohio State Univ, BSc, 58, MSc, 62, PhD(zool & entom), 70. *Prof Exp:* Res assoc entom, Entom Res Ctr, Vero Beach, Fla, 64-69 & Univ Ga, 69-72; RES ENTOMOLOGIST, TOXICOL & ENTOM RES LAB, SCI & EDUC ADMIN-AGR RES, USDA, 72- *Concurrent Pos:* Vis prof & grad adv, Dept Entom, Tex A&M, College Station. *Mem:* Am Soc Zoologist; Am Soc Cell Biologists; Entom Soc Am; Am Soc Electron Microscopists; Soc Neurosci. *Res:* Histological and ultrastructural studies of the morphology of Diptera, especially the neuroendocrine and reproductive systems and growth regulators in insects. *Mailing Add:* USDA ARS FAPRL RR 5 Box 810 College Station TX 77845

MERBS, CHARLES FRANCIS, b Neenah, Wis, Sept 3, 36; m 62; c 2. PHYSICAL ANTHROPOLOGY. *Educ:* Univ Wis-Madison, BS, 58, MS, 63, PhD(anthrop, genetics), 69. *Prof Exp:* From instr to assoc prof anthrop, Univ Chicago, 63-73; chmn dept, 73-79, assoc prof, 73-74, PROF ANTHROP, ARIZ STATE UNIV, 74- *Concurrent Pos:* Res assoc, San Diego Mus Man, 81- *Mem:* Fel Arctic Inst NAm; Soc Am Archaeol; Am Asn Phys Anthropologists; Paleopath Asn; Am Acad Forensic Sci; Canadian Asn Phys Anthropologists. *Res:* Physical anthropology; human osteology; forensic anthropology; disease ecology; paleopathology; vertebral pathology; trauma; Arctic, southwestern United States and northeastern Africa. *Mailing Add:* Dept Anthrop Ariz State Univ Tempe AZ 85287-2402

MERCADO, TERESA I, b Ponce, PR, Oct 15, 21. HISTOCHEMISTRY, CYTOCHEMICAL PATHOLOGY. *Educ:* Col Mt St Vincent, BS, 43; Catholic Univ Am, MS, 47, PhD(biol physiol), 50. *Prof Exp:* Instr biol, Dunbarton Col, DC, 47-48; RES PHYSIOLOGIST, LAB PARASITIC DIS, NAT INST ALLERGY & INFECTIOUS DIS, NAT INST HEALTH, 49- *Concurrent Pos:* Consult, J Nat Malaria Soc, 51-52; lectr, Biol Dept, Clarke Col, 66; mem, Task Force & Int Cong Parasitiol, Agency Int Develop, 70; speaker, Soc Advan Chicanos & Native Am Sci, 80. *Mem:* Sigma Xi; AAAS; Am Soc Parasitologists; Am Soc Cell Biol; Am Soc Trop Med & Hyg. *Res:* Pathologic physiology, histochemistry, biochemistry and cytochemistry of parasitic diseases, primarily experimental malaria and typanosomiases with emphasis on Typanosoma cruzi; interactions between T:cruzi and Pseudomonas fluorescens; one US patent. *Mailing Add:* Lab Parasitic Dis Twinbrook-II Rm 27 NIH 12441 Parklawn Dr Rockville MD 20852

MERCADO-JIMENEZ, TEODORO, b Arecibo, PR, Jan 4, 35; m 58; c 5. ELECTRICAL ENGINEERING, CONTROL SYSTEMS. *Educ:* Univ PR, Mayaguez, BSEE, 57; Univ Mich, Ann Arbor, MS, 61; Tex A&M Univ, PhD(elec eng), 65. *Prof Exp:* Instr elec eng, Univ PR, Mayaguez, 57-60; instr, Tex A&M Univ, 62-64; from asst prof to assoc prof, 64-80, PROF ELEC ENG, UNIV PR, MAYAGUEZ, 80- *Concurrent Pos:* Am Soc Eng Educ summer fac fel, Goddard Space Flight Ctr, NASA, 65 & 66; vis prof, Univ Oriente, Venezuela, 70-71. *Mem:* Inst Elec & Electronics Engrs; PR Col Eng, Archit & Surveyors; PR Soc Elec Engrs. *Res:* Circuit theory; instrumentation; applied mathematics; education. *Mailing Add:* Dept Elec Eng Univ PR Mayaguez PR 00708

MERCER, ALEXANDER MCDOWELL, b Belfast, N Ireland, Mar 2, 31; m 57; c 3. ORTHOGONAL POLYNOMIALS, APPROXIMATION THEORY. *Educ:* Queen's Univ, Belfast, BSc(mech eng), 52, BSc(math), 54, PhD(math), 57. *Prof Exp:* Asst lectr math, Queen's Univ, Belfast, 54-57, lectr, 59-62 & 64-68; lectr, Queens Col, Dundee, 57-59 & Univ WI, Jamaica, 62-64; sr lectr, Univ WI, Barbados, 68-69; assoc prf, 69-72, PROF MATH, UNIV GUELPH, CAN, 72- *Concurrent Pos:* Vis prof, Univ WI, Barbados, 89-90. *Mem:* Math Asn Am. *Res:* Real analysis such as, biorthogonal expansions associated with non-self-adjoint ordinary differential equations, special functions, approximation theory, distribution of the zeros of orthogonal polynomials and numerical solutions of problems in fluid dynamics and heat transfer. *Mailing Add:* Dept Math & Statist Univ Guelph Guelph ON N1G 2W1 Can

MERCER, EDWARD EVERETT, b Buffalo, NY, Mar 5, 34; m 57; c 6. PHYSICAL INORGANIC CHEMISTRY. *Educ:* Canisius Col, 55; Purdue Univ, PhD(phys chem), 60. *Prof Exp:* Res assoc chem, Lawrence Radiation Lab, Univ Calif, 60-61; from asst prof to assoc prof, 61-73, asst head dept, 73-74, PROF CHEM, UNIV SC, 73-, ASST DEAN SCI & MATH, 79- *Mem:* Am Chem Soc; Sigma Xi. *Res:* Thermodynamics and kinetics of metal complexes in solution; chemistry of ruthenium. *Mailing Add:* Dept Chem Univ SC Columbia SC 29208

MERCER, EDWARD KING, b Santa Barbara, Calif, July 1, 31; m 63; c 2. PLANT NEMATOLOGY, FRESH WATER BIOLOGY. *Educ:* Univ Calif, Santa Barbara, BA, 58, MA, 62; Auburn Univ, PhD(plant nematol), 68. *Prof Exp:* Res asst plant nematol, Univ Calif, Riverside, 63-65; res asst, Auburn Univ, 65-68; assoc prof, 68-76, PROF BIOL SCI, CALIF STATE POLYTECH UNIV, POMONA, 76- *Mem:* AAAS; Am Inst Biol Sci; Am Soc Limnol & Oceanog; Soc Nematol; Asn Meiobenthologists. *Res:* Energy requirements and feeding behavior of free-living fresh water nematodes; nematodes in relation to water quality. *Mailing Add:* Dept Biol Sci Calif State Polytech Univ 3801 W Temple Ave Pomona CA 91768

MERCER, HENRY DWIGHT, b Blakely, Ga, Feb 20, 39; m 60; c 2. VETERINARY PHARMACOLOGY, VETERINARY TOXICOLOGY. *Educ:* Univ Ga, BS, 60, DVM, 63; Univ Fla, MS, 66; Am Bd Vet Toxicol, dipl, 73, Ohio State Univ, PhD, 76. *Prof Exp:* Practitioner vet med, Houston Animal Clin, Blakely, 64-65; NIH fel, Univ Fla, 65-66; br chief vet med, Div New Animal Drugs, Food & Drug Admin, 66-68, actg dir, Div Vet Res, 68-72 & 74-77, dept dir, 72-74; PROF, MISS STATE UNIV, 77-, DEAN, COL VET MED, 87- *Concurrent Pos:* Rating bd mem, Civil Serv Comn, 70-; mem res comt, Food & Drug Admin Task Force, 71 & Nat Mastitis Coun, 73-; fel, Ohio State Univ, 75-76; dir, Animal Health Ctr, Miss State Univ, 78-81, dir, Acad Prog, 79-80, admin officer, Col Vet Med, 79-82. *Mem:* Am Vet Med Asn; Sigma Xi; Am Soc Vet Physiol & Pharmacol; Am Col Vet Toxicol; Am Acad Pharmacol & Therapeut. *Res:* Clinical pharmacology in domestic animals, specifically, the metabolism kinetics and pharmacokinetics of veterinary drugs in food producing animals. *Mailing Add:* PO Drawer V Col Vet Med Mississippi State MS 39762

MERCER, JAMES WAYNE, b Panama City, Fla, Dec 23, 47; m 69. HYDROGEOLOGY. *Educ:* Fla State Univ, BS, 69; Univ Ill, MS, 72, PhD(geol), 73. *Prof Exp:* hydrologist, US Geol Surv, 72-79; PRES, GEOTRANS, INC, 79- *Concurrent Pos:* Asst prof, George Washington Univ, 80- *Mem:* Am Geophys Union; Soc Petrol Engrs; Am Soc Civil Engrs; Geological Soc Am; Am Water Resources Asn. *Res:* Development of theoretical and numerical models for simulating hydrogeologic processes, with emphasis on geothermal systems, hazardous waste, radioactive waste, and salt water intrusion. *Mailing Add:* Dept Geol George Washington Univ 2121 Eye St NW Washington DC 20052

MERCER, KERMIT R, b Brockport, NY, June 1, 33; m 53; c 2. MICROWAVE INSTRUMENTATION, RADIO ENGINEERING. *Educ:* State Univ Col, Brockport, BS, 71. *Prof Exp:* Res asst electronic lab, Delco Div, Gen Motors, Rochester, NY, 51-53; teacher, Radio/Electronics Sch, USAF, Scott AFB, 53-57; res assoc, Physics Dept, Gen Dynamics/Electronics, Rochester, NY, 57-68; res assoc, Biochem Dept, Eastman Dental Ctr, Rochester, NY, 68-69; sci teacher, Wemoco Sch, Monroe County, NY, 71-72; ASSOC BIOPHYSICS, DEPT BIOPHYSICS, UNIV ROCHESTER, 72- *Mem:* Am Asn Physics Teachers; Inst Elec & Electronics Engrs; Nat Asn Radio & Telecommun Engrs. *Res:* Characterization of free radical damage to DNA constituent molecules utilizing electron spin resonance spectroscopy. *Mailing Add:* Dept Biophysics Univ Rochester Rochester NY 14642

MERCER, LEONARD PRESTON, II, b Fort Worth, Tex, Jan 16, 41; m 63; c 3. NUTRITIONAL BIOCHEMISTRY. *Educ:* Univ Tex, Austin, BS, 68; La State Univ, PhD(biochem), 71. *Prof Exp:* NIH fel, Med Sch, Univ Ala, Birmingham, 71-73; instr, Univ S Ala, 73-74, asst prof biochem, 74-77; from asst prof to prof biochem, Sch Med & Dent, Oral Roberts Univ, 77-90, chmn dept, 79-90; PROF & CHMN, DEPT NUTR & FOOD SCI, UNIV KY, 90- *Mem:* Am Inst Nutrit; Am Chem Soc; Soc Math Biol; Am Soc Biochem & Molecular Biol; fel Am Col Nutrit. *Res:* Mathematical analysis of biochemical responses to nutritional stimuli; prediction of nutritional responses of proteins and amino acids and comparison of biological efficacy of alternate nutrient sources; neuroregulation of appetite. *Mailing Add:* Dept Nutritional Food Sci Univ Ky 212 Funkhouse Bldg Lexington KY 40506-0050

MERCER, MALCOLM CLARENCE, b St John's, Nfld, June 20, 44. FISHERIES MANAGEMENT. *Educ:* Mem Univ Nfld, BSc, 65, MSc, 68. *Prof Exp:* Scientist, Fisheries Res Bd Can, 65-69, res biologist, 69-74, sect head shellfish, 74-75, prog head marine fisheries mgt pelagic & shellfish, Nfld Biol Sta, 75-76, sr policy prog adv, Marine Mammals, Dept Environ Fish & Marine Serv, 76-78, assoc dir, Fisheries Res Br, Resource Serv Directorate, Fisheries Environ Fish & Oceans Serv, Ottawa, 78-79, dir, 79-81, regional sci dir, Dept Fisheries & Oceans, Nfld Region, 81-90, DIR, SPEC ASSIGNMENT, OFF ADMIN SCI, DEPT FISHERIES & OCEANS, NFLD REGION, 90- *Concurrent Pos:* Can Whaling Comnr, 78-81. *Mem:* Sigma Xi; Int Whaling Comn (vchmn, 78-81). *Res:* Research administration; resource management. *Mailing Add:* Dept Fisheries & Oceans PO Box 5667 St John's NF A1C 5X1 Can

MERCER, PAUL FREDERICK, b Guelph, Ont, Apr 21, 36; m 62; c 1. PHYSIOLOGY. *Educ:* Univ Toronto, DVM, 59; Cornell Univ, PhD(phys biol), 64. *Prof Exp:* Med Res Coun Can fel biol chem, Copenhagen Univ, 63-64; asst prof physiol, Univ Alta, 64-67; from asst prof to assoc prof, 67-80, PROF PHYSIOL, UNIV WESTERN ONT, 80- *Mem:* Soc Nephrology; Can Physiol Soc; Am Physiol Soc. *Res:* Renal physiology. *Mailing Add:* Dept of Physiol Univ of Western Ont London ON N6A 5C1 Can

MERCER, ROBERT ALLEN, b Providence, RI, Aug 2, 42; m 66; c 2. SYSTEMS ENGINEERING. *Educ:* Carnegie-Mellon Univ, BS, 64; Johns Hopkins Univ, PhD(physics), 69. *Prof Exp:* Res assoc physics, Johns Hopkins Univ, 69-70; asst prof, Ind Univ, Bloomington, 70-73; mem tech staff, Bell Labs, 73-75, tech supvr, 75-79; div mgr, AT&T Co, 79-81; DIR, BELL LABS, 81- *Mem:* Inst Elec & Electronics Engrs. *Res:* Telecommunications systems engineering. *Mailing Add:* 5201 Holmes Pl Boulder CO 80303

MERCER, ROBERT LEROY, b San Jose, Calif, July 11, 46; m 67; c 3. COMPUTER SCIENCE. *Educ:* Univ NMex, BSc, 68; Univ Ill, Urbana, MS, 70, PhD(comput sci), 72. *Prof Exp:* RES ASSOC AUTOMATIC CONTINUOUS SPEECH RECOGNITION, INT BUS MACH CORP, 72-, MGR, REAL TIME SPEECH RECOGNITION. *Mem:* Inst Elec & Electronics Engrs. *Res:* Theoretical nuclear physics at intermediate energies; automatic recognition of continuous speech. *Mailing Add:* Thomas J Watson Res Ctr IBM Res Ctr PO Box 218 Yorktown Heights NY 10598

MERCER, ROBERT WILLIAM, MOLECULAR BIOLOGY, MEMBRANE TRANSPORT. *Educ:* Syracuse Univ, PhD(cell physiol), 80. *Prof Exp:* Res assoc, Dept Physiol, Sch Med, Yale Univ, 80-87; ASST PROF CELL BIOL & PHYSIOL, SCH MED, WASH UNIV, 87- *Mailing Add:* Dept Cell Biol & Physiol Sch Med Wash Univ PO Box 8228 660 S Euclid Ave St Louis MO 63110

MERCER, SAMUEL, JR, b Philadelphia, Pa, Sept 4, 20; m 44; c 8. MECHANICAL ENGINEERING. *Educ:* Drexel Inst Technol, BS, 43; Mich State Col, MS, 50; Purdue Univ, PhD(mech eng), 56. *Prof Exp:* Draftsman, Mech Design Sect, Glenn L Martin Co, 46-47; from instr to asst prof mech eng, Mich State Col, 47-53, from asst prof to assoc prof appl mech, 53-59; prof mech eng & head dept, 59-66, assoc dean col eng, 66-70, dir continuing educ, 71-72, dean continuing & coop educ, 72-77, DEAN CONTINUING EDUC, DREXEL UNIV, 77- *Mem:* Am Soc Mech Engrs; Am Soc Eng Educ. *Res:* Kinematics; dynamics; vibration; machine design; dynamics of cam actuated systems and automobile skid testing. *Mailing Add:* Off of Continuing Prof Educ Drexel Univ 32nd & Chestnut Sts Philadelphia PA 19104

MERCER, SHERWOOD ROCKE, b Manchester, Conn, June 27, 07; m 33; c 3. HISTORY OF MEDICINE, HISTORY OF SCIENCE. *Educ:* Wesleyan Univ, AB, 29, AM, 30. *Hon Degrees:* LLD, Philadelphia Col Textiles & Sci, 57. *Prof Exp:* Instr hist & eng, Pub Schs, Conn, 30-42; assoc, Harvard Univ, 42-44; chmn div appl sci, Elmira Col, 44-45; consult higher educ, Conn Dept Pub Instr, 45-46; dean fac, Muhlenberg Col, 46-54; prof hist med & osteop, 54-76, dean, 54-69, vpres educ affairs, 67-76, EMER PROF HIST MED & OSTEOP, PHILADELPHIA COL OSTEOP MED, 76-, ARCHIVIST, 88- *Concurrent Pos:* Consult, Columbia Univ, 54-75. *Mem:* Am Osteop Asn; Am Asn Cols Osteop Med (secy-treas, 71-). *Res:* Nature, structure and teaching for a liberal education, particularly in a society heavily influenced by pure and applied science. *Mailing Add:* 13 Thompson Dr Havertown PA 19083

MERCER, THOMAS T, b Victoria, BC, Dec 30, 20; m 42; c 3. INDUSTRIAL HYGIENE, HEALTH PHYSICS. *Educ:* San Jose State Col, AB, 49; Univ Rochester, PhD(indust hyg), 57. *Prof Exp:* Health physicist & instr, Univ Wash, 53-55; res assoc aerosol physics, Atomic Energy Proj, Univ Rochester, 55-57, chief aerosol physics sect, 57-59; nuclear physicist, US Naval Radiol Defense Lab, San Francisco, 59-61; head dept aerosol physics, Lovelace Found Med Educ & Res, NMex, 61-65; assoc prof, 65-70, PROF RADIATION BIOL & BIOPHYS, UNIV ROCHESTER, 70- *Mem:* AAAS; Am Indust Hyg Asn; Health Phys Soc. *Res:* Production and characterization of airborne particulates; interaction of radioactive vapors with particles. *Mailing Add:* 3837 Glen Oaks Manor Dr Sarasota FL 34232

MERCER, WALTER RONALD, b Ft Wayne, Ind, Mar 9, 41; m 64; c 2. QUALITY MANAGEMENT, PRODUCT DESIGN. *Educ:* Ind Inst Technol, BS, 64; St Francis Col, MSBA, 74. *Prof Exp:* Design engr, Int Harvester Co, 64-67, proj engr, 67-72, prod engr, 72-80, legis mgr, 80-81, staff engr, 81-83; mgr prod eng, Auburn Gear, Inc, 83-84, dir, tech opers, 84-88; qual mgr, 88-89, MGR ENG, WHEELTEK DIV, AMCAST INDUST, 89- *Mem:* Soc Automotive Engrs; Am Soc Qual Control. *Res:* Gear and mechanical transmission technologies; manufacturing engineering and facilities management and low pressure perminate mold casting technologies; manufacturing engineering. *Mailing Add:* RR 4 Box 201 Fremont IN 46737

MERCEREAU, JAMES EDGAR, b Sharon, Pa, Apr 3, 30; m 50; c 3. PHYSICS. *Educ:* Pomona Col, BA, 53; Univ Ill, MS, 54; Calif Inst Technol, PhD, 59. *Hon Degrees:* DSc, Pomona Col, 68. *Prof Exp:* Asst, Univ Ill, 53-54; physicist, Hughes Res Labs, 54-59; asst prof, Calif Inst Technol, 59-62; prin scientist, Sci Labs, Ford Motor Co, Calif, 62-65, mgr cryogenics, 65-69; prof physics, 69-74, prof physics & appl physics, Calif Inst Technol, 74-86; PROF RESIDENCE, UNIV CALIF, IRVINE, 86- *Concurrent Pos:* Consult, Hughes Aircraft Co, 59-60 & Aerospace Corp, 60-62; vis assoc, Calif Inst Technol, 64-65, res assoc, 65-; prof, Univ Calif, Irvine, 65-69. *Mem:* Fel Am Phys Soc. *Res:* Cryogenics; ferromagnetism; quantum electronics; microwaves; nuclear accelerators. *Mailing Add:* 24652 El Camino Capistrano Dana Point CA 92629

MERCER-SMITH, JAMES A, b Bremarton, Wash, Jan 22, 53; m 75; c 2. THERMONUCLEAR PROCESSES, STAR FORMATION. *Educ:* Yale Univ, MPhil, 79, PhD(astrophysics), 80. *Prof Exp:* Post-doctoral fel, Harvard Col Observ, 80-83; STAFF MEM DESIGN PHYSICIST, LOS ALAMOS NAT LAB, 83- *Mem:* Am Phys Soc; Am Astron Soc; Sigma Xi. *Mailing Add:* X-2 Mail Stop B220 Los Alamos Nat Lab Los Alamos NM 87545

MERCHANT, BRUCE, b Elgin, Nebr, Mar 26, 35; m 61; c 3. ANTIBODY SPECIFICITY. *Educ:* Univ Chicago, MD, 63, PhD(path), 64. *Prof Exp:* Staff assoc, Lab Immunol, NIH, 64-66, career scientist, 66-73; Dir immunohematology, Off Biologics, Food & Drug Admin, 73-84; MED DIR, HYBRITECH INC, 84- *Mem:* NY Acad Sci; Am Asn Immunologists; Am Med Asn; Am Soc Zoologists; Sigma Xi. *Res:* Antibody formation; immunogenetics; regulation of the immune response; kinetics of immunoglobulin synthesis and secretion; immune response to chemical Haptens; transplantation; graft versus host reactons; transition and modulation of immunocyte function; ontogeny of the immune response; natural antibody production, microbial and environmental antigens; immune responsiveness in anexic and athymic animals; autoimmune diseases; transfer factor; anti-lymphocyte globulin; liposome immunogens; hybridoma applications. *Mailing Add:* Dept Therapeut Med Affairs Hybritech PO Box 269 San Diego CA 92196

MERCHANT, DONALD JOSEPH, b Biltmore, NC, Sept 7, 21; m 43; c 3. MICROBIOLOGY. *Educ:* Berea Col, AB, 42; Univ Mich, MS, 47, PhD(bact), 50. *Prof Exp:* From instr to prof bact, Univ Mich, 48-69; dir, W Alton Jones Cell Sci Ctr, Tissue Cult Asn, 69-73; prof & chmn, Microbiol & Immunol, 73-86, dir, Tidewater Region Cancer Net, 77-86, EMER PROF, MICROBIOL & IMMUNOL, EASTERN VA MED SCH, NORFOLK, 86- *Concurrent Pos:* Mem working cadre, Nat Prostatic Cancer Proj, Nat Cancer Inst, 72-79 & 84-86. *Mem:* Am Acad Microbiol; Am Soc Microbiol; Asn Community Cancer Ctrs; Tissue Cult Asn (vpres, 60-64, pres, 64-66); Am Soc Cell Biol; Sigma Xi; NY Acad Sci. *Res:* Tissue culture techniques; cell growth and metabolism; cancer cell biology; pathogenesis of infectious disease; development and characterization of in vitro models for cancer research; characterization of differentiation markers for the study of prostate. *Mailing Add:* 2433 Spindrift Rd Virginia Beach VA 23451

MERCHANT, HENRY CLIFTON, b Washington, DC, Aug 7, 42; m 65; c 4. ECOLOGY, ZOOLOGY. *Educ:* Univ Md, College Park, BS, 64, MS, 66; Rutgers Univ, New Brunswick, PhD(zool), 70. *Prof Exp:* Instr zool, Rutgers Univ, 70; asst prof, 70-75, ASSOC PROF BIOL, GEORGE WASHINGTON UNIV, 75- *Mem:* AAAS; Ecol Soc Am; Am Inst Biol Sci. *Res:* Bioenergetics of species, populations and communities. *Mailing Add:* Dept of Biol Sci George Washington Univ Washington DC 20052

MERCHANT, HOWARD CARL, b Mt Vernon, Wash, Jan 9, 35; m 60; c 2. MECHANICAL ENGINEERING, VIBRATION ENGINEERING. *Educ:* Univ Wash, BS, 56; Mass Inst Technol, SM, 57; Calif Inst Technol, PhD(mech eng), 61. *Prof Exp:* Asst prof, Univ Wash, 61-63; tech staff mem, Livermore Lab, Sandia Corp, 63-65, analysis tech group leader, 65; analysis engr, Physics Int Co, 65-66, head vulnerability dept, 66-67; from assoc prof to prof, Univ Wash, 67-82, affil prof mech eng, 82-89; PRES & PRIN ENGR, MERENCO, INC, 82- *Concurrent Pos:* Consult, Marine Systs Div, Honeywell, Alliant Techsysts, 67-, Westinghouse Hanford Co, 72-90, Boeing Aerospace & Boeing Com Airplane Co, 79-; adj prof geophysics, Univ Wash, 74-82; UNC Nuclear Industs, Inc, 84-88; Los Alamos Nat Lab, 85-; nat chmn, Comn Shock & Vibrations, Am Soc Mech Engrs, 85-86, Comn Vibrations & Sound, 90-; Procter & Gamble, 87- *Honors & Awards:* Charles Bassett II Award, ISA. *Mem:* Fel Am Soc Mech Engrs; Acoust Soc Am; Soc Naval Architects & Marine Engrs; Soc Automotive Engrs; Earthquake Eng Res Inst. *Res:* Applied mechanics; vibration of equipment and structures; shock and vibration instrumentation and specifications; acoustics. *Mailing Add:* Merenco inc 1426 112th Ave NE Bellevue WA 98004

MERCHANT, MYLON EUGENE, b Springfield, Mass, May 6, 13; m 37; c 3. COMPUTER-INTEGRATED MANUFACTURING. *Educ:* Univ Vt, BS, 36; Univ Cincinnati, DSc(physics), 41. *Hon Degrees:* DSc, Univ Vt, 73; Univ Salford, Eng, 80. *Prof Exp:* Res physicist, 40-48, sr res physicist, 48-51, asst dir res, 51-57, dir phys res, 57-63, dir sci res, 63-69, dir res planning, 69-81, prin scientist mfg res, Cincinnati Milacron Inc, 81-83; dir advan mfg res, Metcut Res Assoc, Inc, 83-90; SR CONSULT, INST ADVAN MFG SCI, 90- *Concurrent Pos:* Adj prof mech eng, Univ Cincinnati, 64-69; vis prof, Univ Salford, Eng, 73-; Nat Res Coun Mat Adv Bd, 62-66, Mfg Studies Bd, 80-85;

Regent lectr, Univ Calif, Los Angeles, 76-77; vis prof, Univ Wisconsin-Duluth, 90. *Honors & Awards:* Richards Mem Award, Am Soc Mech Engrs, 59; Res Medal, Soc Mfg Engrs, 68; Tribology Gold Medal, Inst Mech Engrs, UK, 80; George Schlesinger Prize, City Berlin, 80; Otto Benedikt Prize, Comput & Automation Inst, Hungary, 81; M Eugene Merchant Mfg Medal, Am Soc Mech Engrs & Soc Mfg Engrs, 86. *Mem:* Nat Acad Eng; Soc Mfg Engrs (pres, 76-77); Am Soc Mech Engrs (vpres, 73-75); Int Inst Prod Eng Res (pres, 68-69); Am Soc Metals Int; Belg Soc Mech Engrs. *Res:* Physics of manufacturing processes, friction, lubrication and wear; systems approach to manufacturing; manufacturing systems; computer-integrated manufacturing; computer automated factory; future of manufacturing. *Mailing Add:* Inst Advan Mfg Sci 1111 Edison Dr Cincinnati OH 45216

MERCHANT, PHILIP, JR, b Bay City, Tex, Jan 13, 43; m 64; c 3. CHEMISTRY. *Educ:* Tex Southern Univ, BS, 64, MS, 65; Univ Houston, PhD(chem), 71. *Prof Exp:* Chemist, Dow Chem Co, 66-68 & Petro-Chem Corp, 68-70; from res chemist to res group leader, 71-88, govt coordr, 88-90, PROG OFFICER, EXXON EDUC FOUND, 90- *Mem:* Nat Orgn Black Chemist & Chem Engrs. *Mailing Add:* 1741 Snowmass Plano TX 75025

MERCHANT, ROLAND SAMUEL, SR, b New York, NY, Apr 18, 29; m 70; c 3. HOSPITAL ADMINISTRATION, BIOSTATISTICS. *Educ:* NY Univ, BA, 57, MA, 60; Columbia Univ, MS, 63, MSHA, 74. *Prof Exp:* Asst statistician, New York City Dept Health, 57-60, statistician, 60-63; statistician, NY Tuberc & Health Asn, 63-65; biostatistician, Inst Surg Studies, Montefiore Hosp & Med Ctr, 65-72; admin resident, Roosevelt Hosp, 73-74; dir health & hosp mgt, NY City Dept Health, 74-76; asst adminer & adminr, West Adams Community Hosp, 76; spec asst to assoc vpres med affairs, Med Ctr, Stanford Univ, 77-82; dir mgt & strategic planning, Stanford Univ Hosp, 82-90; VPRES STRATEGIC PLANNING, CEDARS-SINAI MED CTR, 90- *Mem:* AAAS; fel Am Pub Health Asn; Am Statist Asn; Biomet Soc; Inst Math Statist; fel Am Col Healthcare Execs. *Res:* Application of biostatistical techniques to administrative methodology in health care delivery systems. *Mailing Add:* 27335 Park Vista Dr Agoura Hills CA 91301

MERCHANT, SABEEHA, b Bombay, India, Aug 31, 59; US citizen. CELL BIOLOGY. *Educ:* Univ Wis-Madison, BS, 79, PhD(biochem), 83. *Prof Exp:* Postdoctoral fel, Harvard Univ, 84-87; ASST PROF BIOCHEM, UNIV CALIF, LOS ANGELES, 87- *Mem:* Am Soc Biochem & Molecular Biol; Am Chem Soc; Am Soc Plant Physiologists; Int Soc Plant Molecular Biol; AAAS. *Res:* Biosynthesis of metalloproteins of the photosynthetic electron transfer chain with emphasis on Cu-regulation of transcription and post-translational assembly of heme & Cu proteins. *Mailing Add:* Dept Chem & Biochem Univ Calif 405 Hilgard Ave Los Angeles CA 90024-1569

MERCIER, PHILIP LAURENT, physical chemistry, for more information see previous edition

MERCKX, KENNETH R(ING), b Chicago, Ill, July 2, 26; m 54; c 3. ENGINEERING MECHANICS. *Educ:* Northwestern Univ, BS, 50; Stanford Univ, PhD(eng mech), 53. *Prof Exp:* Res assoc, Battelle-Northwest, 65-70; sr res assoc, Westinghouse Hanford Co, 70-72; STAFF CONSULT, ADVAN NUCLEAR FUELS CORP, 72- *Concurrent Pos:* Affiliate assoc prof, Joint Ctr Grad Study, 57- *Mem:* Fel Am Soc Mech Engrs; Am Nuclear Soc. *Res:* Material behavior; stress analysis; reactor fuel design. *Mailing Add:* Advanced Nuclear Fuels Corp 2101 Horn Rapids Rd Richland WA 99352

MERDINGER, CHARLES J(OHN), b Chicago, Ill, Apr 20, 18; m 44; c 4. CIVIL ENGINEERING, EDUCATION ADMINISTRATION. *Educ:* US Naval Acad, BS, 41; Rensselaer Polytech Inst, BCE, 45, MCE, 46; Oxford Univ, PhD(phys sci), 49. *Hon Degrees:* DrHLett, Sierra Nevada Col, 87. *Prof Exp:* Officer-in-charge construct, US Navy, Panama, CZ, 46-47, design coordr, Bur Yards & Docks, Washington, DC, 49-51, asst pub works officer, Naval Shipyard, Wash, 51-53, pub works officer, Adak, Alaska, 53-54, Naval Air Sta Miramar, 54-56, commanding officer & dir, Naval Civil Eng Lab, 56-59, pub works officer, Fleet Activities, Yokosuka, Japan, 59-62, head dept Eng, hist & govt, US Naval Acad, 62-65, asst comdr opers & maintenance, Naval Facil Eng Command, 65-67, pub works officer, Da Nang, Vietnam, 67-68, commanding officer, Western Div, Naval Facil Eng Command, 68-70; pres, Washington Col, 70-73; vpres, Aspen Inst Humanistic Studies, Colo, 73-74; dep dir, Scripps Inst Oceanog, Calif, 74-80; dir, Avco Corp, 78-85; RETIRED. *Concurrent Pos:* Mem, Southern Regional Educ Bd, 71-73; mem, Nat Comn Hist & Heritage of Am Civil Eng; alumni trustee, US Naval Acad, 71-74; mem coun, Rensselaer Polytech Inst, 72; trustee, Found Ocean Res, 76-80; chmn bd trustees, Sierra Nev Col, 80-87, emer, 87-; trustee, Desert Res Inst Found, Nev; Nev secy, Rhodes Scholar, 82-89; comnr, NW Asn Comn on Cols, 87- *Honors & Awards:* Nat Hist & Heritage Award, Am Soc Civil Engrs, 72. *Mem:* Fel Am Soc Civil Engrs; Nat Soc Prof Engrs; fel Soc Am Mil Engrs; fel Explorer's Club; Sigma Xi. *Mailing Add:* 726 Tyner Way PO Box 7249 Incline Village NV 89450

MERDINGER, EMANUEL, b Austria, Mar 29, 06; nat US; m 53. BIOCHEMISTRY. *Educ:* Prague German Univ, Master Pharmacol, 31; Univ Ferrara, Or Pham, 34, Dr Chem, 35, Dr Natural Sci, 39. *Prof Exp:* Prof sch eng, Univ Ferrara, 36-38, 45-47; from asst prof to prof chem, Roosevelt Univ, 47-72; BIOCHEM RESEARCHER, USDA & DISTINGUISHED PROF, DEPT ENTOM, UNIV FLA, GAINESVILLE, 77- *Concurrent Pos:* Mem res dept dermat, Univ Chicago; abstractor, Chem Abstr, 49-62; Abbott Labs annual res grants, Roosevelt Univ, 47- 72, Ill State Acad Sci grants, 68-71; pres, Nat Acad Sci, exchange scientist Romanian Acad Sci, 71-72, 75 & 80, Bulgarian Acad Sci, 74-75 & Germany, 82; res assoc, Loyola Univ Stritch Sch Med, 72-, distinguished lectr, 74-76. *Mem:* Am Chem Soc; Am Soc Microbiol; hon mem Union Socs Med Sci Romania; hon mem Balkan Med Union. *Res:* Microbiological biochemistry; lipid and carbohydrate metabolism and enzymology of yeasts and fungi; fungal pigments and some of their chemotherapeutic properties; relation between Pullularia pullulans, a fungus, and arthritis. *Mailing Add:* 4908 NW 16th Pl Gainsville FL 32605

MEREDITH, CAROL N, b Santiago, Chile, Feb 2, 48. METABOLISM. *Educ:* Mass Inst Technol, PhD(nutrit, biochem & metab), 82. *Prof Exp:* SCIENTIST, HUMAN NUTRIT RES CTR AGING, USDA, 82- *Mailing Add:* Div Clin Nutrit Univ Calif Sch Med TB 156 Davis CA 95616

MEREDITH, DALE DEAN, b Centralia, Ill, Mar 24, 40; m 65; c 2. CIVIL ENGINEERING, WATER RESOURCES. *Educ:* Univ Ill, Urbana, BS, 63, MS, 64, PhD(civil eng), 68. *Prof Exp:* Asst prof civil eng, Univ Ill, Urbana-Champaign, 68-73; assoc prof 73-79, PROF CIVIL ENG, STATE UNIV NY, BUFFALO, 79-, CHMN DEPT, 87- *Concurrent Pos:* Prin investr & US Dept Interior res grants, Univ Ill, Urbana-Champaign, 68-73 & State Univ NY, Buffalo, 73-80; prin investr & US Dept Com res grant, State Univ NY, Buffalo, 76-90; ed, J Water Resources Planning & Mgt, Am Soc Civil Engrs, 82-84 & Water Resources Bull, Am Water Resources Asn, 90- *Mem:* Am Soc Civil Engrs; Am Geophys Union; Am Water Resources Asn; Water Pollution Control Fedn; Inst Mgt Sci. *Res:* Water resources system design; planning, operation management and optimization; hydrology. *Mailing Add:* Dept Civil Eng State Univ NY Buffalo NY 14260

MEREDITH, DAVID BRUCE, b Dover, Ohio, July 11, 50; m 78; c 2. BUILDING ENERGY USAGE-DESIGN, SOLAR THERMAL APPLICATIONS. *Educ:* Ohio State Univ, BS, 72; Colo State Univ, MS, 78; Am Consult Engrs Coun, cert, 84; Emergency Mgt Inst, cert, 88. *Prof Exp:* Pollution control engr, Procter & Gamble, 72-74, thermal engr, 74-76; grad res asst, solar thermal applications, Colo State Univ, 77-79; ASSOC PROF BLDG ENERGY SYSTS, PENN STATE UNIV, FAYETTE CAMPUS, 79- *Concurrent Pos:* Instr solar applications, Kent State Univ, Tuscarawas Campus, 78; div chair & prog chair, Am Soc Eng Educ, 85-88; chair, Educ Div, Am Solar Energy Soc, 88-90; mem, Accreditation Comt, Am Soc Heating Refrig & Air Conditioning Engrs, 89-91; alt deleg, Tech Accrediting Comn, Accreditation Bd Eng Sci & Technol, 91- *Mem:* Am Solar Energy Soc; Am Soc Eng Educ (secy-treas, 85-88); Am Soc Heating Refrig & Air Conditioning Engrs. *Res:* Education methods in the area of building energy use; analysis of the professional pipeline to the industry. *Mailing Add:* Pa State Univ PO Box 519 Uniontown PA 15401

MEREDITH, FARRIS RAY, b Denver, Colo, Mar 15, 29; m 50; c 3. BOTANY, SOILS. *Educ:* Colo State Univ, BS, 51; NMex Highlands Univ, MS, 58; Wash State Univ, PhD(bot, plant ecol), 65. *Prof Exp:* Asst prof bot, Humboldt State Col, 63-65; asst prof, NMex Highlands Univ, 65-66; from asst prof to assoc prof, 66-74, PROF BOT, HUMBOLDT STATE UNIV, 74- *Mem:* AAAS; Ecol Soc Am; Torrey Bot Club. *Res:* Plant autecology and synecology; autecology and physiology of coniferous trees. *Mailing Add:* Dept Biol Humboldt State Univ Arcata CA 95521

MEREDITH, HOWARD VOAS, child growth, morphology, for more information see previous edition

MEREDITH, JESSE HEDGEPETH, b Fancy Gap, Va, Mar 19, 23; m; c 3. MEDICINE. *Educ:* Elon Col, BA, 43; Western Reserve Univ, MD, 51; Am Bd Surg, dipl, 58; Am Bd Thoracic Surg, dipl, 59. *Prof Exp:* Intern med, Bellevue Hosp, New York, 51-52; asst surgeon, NC Baptist Hosp, Winston-Salem, 52-56, resident gen & thoracic surg, 56-57, cardiovasc surg, 57-58; asst surgeon, 52-58, from instr to assoc prof, 58-70, PROF SURG, BOWMAN GRAY SCH MED, 70-, DIR SURG RES, 59- *Concurrent Pos:* NIH res fel, 56-57, spec res fel, 59-62; res fel, Bowman Gray Sch Med, 58. *Mem:* Fel Am Col Surgeons; Am Asn Thoracic Surgeons; AMA; Am Soc Artificial Internal Organs; Sigma Xi. *Res:* Cardiovascular surgery and physiology; cancer chemotherapy; biomedical engineering; cadaver blood in transfusions; kidney transplantation. *Mailing Add:* Dept Surg Bowman Gray Sch Med Winston-Salem NC 27103

MEREDITH, LESLIE HUGH, b Birmingham, Eng, Oct 23, 27; m 48; c 3. SPACE PHYSICS. *Educ:* Univ Iowa, BA, 50, MS, 52, PhD(physics), 54. *Prof Exp:* Res assoc, Univ Iowa, 51-53; asst, Proj Matterhorn, Princeton Univ, 53-54; sect head, Naval Res Lab, 54-58, br head, 58; br head, Goddard Space Flight Ctr, 58-59, div chief space sci, 59-70, dep dir space & earth sci directorate, 70-72, asst dir, 72-79, dir applns, 79-84, assoc dir, 84-87; GROUP DIR, AM GEOPHYS UNION, 87- *Mem:* AAAS; Am Phys Soc; Am Geophys Union (gen secy, 80-84); Royal Astrophys Soc. *Mailing Add:* 1241 Cresthaven Dr Silver Spring MD 20903

MEREDITH, ORSELL MONTGOMERY, b Jamestown, NY, Oct 19, 23; m 49; c 1. RADIOLOGICAL HEALTH, RESEARCH ADMINISTRATION. *Educ:* Univ Chicago, BS, 48; Univ Southern Calif, MS, 51, PhD(pharmacol, toxicol), 53; Am Univ, Washington, DC, MS, 74. *Prof Exp:* Asst pharmacol & toxicol, Sch Med, Univ Southern Calif, 49-52; pharmacologist, Carlborg Labs, Calif, 52-53; chief nuclear physiol sect & asst res pharmacologist, Lab Nuclear Med & Radiation Biol, Med Ctr, Univ Calif, Los Angeles, 53-62; res scientist, Lockheed Missiles & Space Co, Calif, 62-66; tech mgr, US Naval Radiological Defense Lab, 66-69; opers res analyst, Nuclear Prog Off, Adv Planning & Analysis Staff, Naval Ord Lab, 69-75; exec secy, Spec Progs Br, Div Res Grants, NIH, 74-78; EXEC SECY, NAT CANCER INST, 78- *Concurrent Pos:* Consult, Nuclear Div, Am Electronics Inc, Calif. *Mem:* AAAS; Soc Nuclear Med; Radiation Res Soc; NY Acad Sci. *Res:* Catecholamine action on intestinal smooth muscle; anticholinesterase action of organic phosphate insecticides; inhalation toxicity of radioactive fallout debris; radioisotope clinical diagnosis; bioastronautics; mammalian radiation biology; operations research. *Mailing Add:* Exec Secy Nat Cancer Inst NCI NIH 5333 Westbard Ave Bethesda MD 20892

MEREDITH, ROBERT E(UGENE), b Santa Barbara, Calif, Feb 25, 28; m 53; c 4. CHEMISTRY, CHEMICAL ENGINEERING. *Educ:* Univ Calif, BS, 56, PhD(chem eng), 59. *Prof Exp:* From asst prof to assoc prof, 59-86, EMER PROF CHEM ENG, ORE STATE UNIV, 86- *Concurrent Pos:* Nat Acad Sci sr res assoc, Jet Propulsion Lab, 66-67; vis prof, Univ Calif, Berkeley, 71; consult, Comt Critical Mat Technol, Nat Res Coun, 73-74, Electric Power Res Inst, 75-78 & US Dept Energy, 81-82, 85-87. *Mem:* Electrochem Soc;

Am Inst Chem Engrs; Nat Asn Corrosion Engrs. *Res:* Electrochemical processes; fuel cells; mass transfer; thermodynamics and kinetics; chemical engineering plant design; corrosion. *Mailing Add:* Dept Chem Eng Ore State Univ Corvallis OR 97331-2702

MEREDITH, RUBY FRANCES, b Sedalia, Mo, Feb 6, 48. RADIATION, CANCER. *Educ:* Univ Mo, BA, 69; Ind Univ, AM, 71, PhD(genetics), 74. *Prof Exp:* Asst prof biol, Baylor Univ, 74-75; fel, 75-76, res assoc viral oncogenesis, Cancer Res Unit, 76-77, assoc head exp hemat sect, Allegheny Gen Hosp, 77-82; at dept radiol, Med Col Va, 83-87; ASST PROF RADIATION ONCOL, UNIV ALA HOSP, 87- *Honors & Awards:* Harold C Bold Award, Phycol Soc Am, 74; Outstanding Contrib Award, Health Res Serv Found, 78; Landacre Soc Res Award, 82. *Mem:* Genetics Soc Am; Int Soc Exp Hemat; spec fel Leukemia Soc Am; Radiation Res Soc; Am Soc Therapeut Radiol & Oncol. *Res:* Pathology, genetics and treatment of murine viral leukemogenesis including chemotherapy, radiotherapy, bone marrow transplantation, immunotherapy and combinations of these; clinical and experimental; hyperthermia; radio labeled antibody therapy. *Mailing Add:* 1537 Camden Ave Birmingham AL 35226-3208

MEREDITH, STEPHEN CHARLES, BIOLOGY. *Educ:* Washington Univ, MD, 74; Univ Chicago, PhD(biochem), 82. *Prof Exp:* ASSOC PROF PATH BIOCHEM, DEPT PATH, UNIV CHICAGO, 79- *Res:* Surface phenomena in biology; lipid-protein interactions and lipoproteins; mineral-protein interactions. *Mailing Add:* Dept Path Univ Chicago 5841 S Maryland Ave Chicago IL 60637

MEREDITH, WILLIAM EDWARD, b Dennison, Ohio, Nov 30, 32; m 57; c 3. MICROBIOLOGY. *Educ:* Ohio Univ, BSc, 59; Ohio State Univ, MSc, 61, PhD(microbiol), 64. *Prof Exp:* Asst microbiol, Ohio State Univ, 59-62, Ohio State Univ Res Found, 62-64; microbiologist, Hess & Clark Div, Richardson-Merrell Inc, 64-68; PROF BIOL, ASHLAND COL, 68-, CHMN DEPT, 77- *Concurrent Pos:* Researcher, Ohio Agr Res & Develop Ctr, 69, 70, 73. *Mem:* AAAS; Am Soc Microbiol. *Res:* General microbiology; immunology; microbial physiology. *Mailing Add:* Dept Biol Ashland Col College Ave Ashland OH 44805

MEREDITH, WILLIAM G, b Fairmont, WVa, May 16, 33; m 55; c 3. ECOLOGY. *Educ:* Fairmont State Col, AB, 55; WVa Univ, MS, 57; Univ Md, PhD(ecol), 67. *Prof Exp:* From instr to assoc prof, 57-71, chmn dept sci & math, 68-75, assoc dean, 81-82, dean, 82-87 PROF BIOL, MT ST MARY'S COL, MD, 71- *Concurrent Pos:* dean, Mt St Mary's Col, 82-87. *Mem:* AAAS; Am Inst Biol Sci; Sigma Xi. *Res:* Comparative ecology and physiology of crayfishes; food habits of freshwater fishes; distribution of crayfishes; ecology of gypsy moth controls. *Mailing Add:* Dept of Sci Mt St Mary's Col Emmitsburg MD 21727

MERESZ, OTTO, b Rima-Sobota, Czech, Jan 16, 32; m 55; c 1. ANALYTICAL CHEMISTRY, ORGANIC CHEMISTRY. *Educ:* Budapest Tech Univ, Dipl org chem, 56; Univ London, PhD(org chem), 65. *Prof Exp:* Tech officer, Imp Chem Indust Ltd, 57-58; res chemist, Res Inst, May & Baker Ltd, 58-61, sect head synthetic perfumes, 61-65, dept head, 66-67; asst prof chem, Univ Toronto, 67-73; dir res, Kemada Res Corp, 73-74; mgr, Org Chem Sect, 74-84, SR SCI ADV, ONT MINISTRY OF THE ENVIRON, 84- *Concurrent Pos:* Fel, Univ Toronto, 65-66; consult, Addiction Res Found, Ont, 74- *Mem:* The Chem Soc; Am Chem Soc; Chem Inst Can. *Res:* Environmental chemistry; correlation between chemical structure and odor; synthetic and structural organic chemistry; trace-organic analysis. *Mailing Add:* PO Box 213 Resources Rd Rexdale ON M9W 5L1 Can

MEREU, ROBERT FRANK, b Alta, Nov 1, 30; m 61; c 3. GEOPHYSICS, SEISMOLOGY. *Educ:* Univ Western Ont, BSc, 52, PhD(physics), 62; Univ Toronto, MA, 53. *Prof Exp:* From asst prof to assoc prof, 63-74, distinguished res prof, 88-89, PROF GEOPHYS, UNIV WESTERN ONT, 74- *Mem:* Seismol Soc Am; Am Geophys Union; Can Geophys Union; Geol Asn Can; Soc Explor Geophysicists. *Mailing Add:* Dept of Geophys Univ of Western Ont London ON N6A 5B7 Can

MERGEN, FRANCOIS, forest genetics; deceased, see previous edition for last biography

MERGENHAGEN, STEPHAN EDWARD, b Buffalo, NY, Apr 12, 30; m 55; c 3. IMMUNOLOGY, MICROBIOLOGY. *Educ:* Allegheny Col, BS, 52; Univ Buffalo, MA, 54; Univ Rochester, PhD(bact), 57. *Prof Exp:* Res microbiologist, 58-65, chief immunol sect, 65-69, CHIEF LAB MICROBIOL & IMMUNOL, NAT INST DENT RES, 69- *Concurrent Pos:* Fel, Univ Rochester, 57-58; vis prof, Univ Heidelberg, WGer, 81-82. *Honors & Awards:* Basic Res Oral Sci, Int Asn Dent Res, 66; US Sr Scientist Award, Alexander Von Humboldt Found, 81; Res in Periodont Dis Award, Int Asn Dent Res, 82. *Mem:* Am Soc Microbiol; Soc Exp Biol & Med; Infectious Dis Soc Am; Int Endotoxin Soc; Am Asn Immunol; Int Asn Dent Res. *Res:* Host-parasite interactons in oral and systemic disease; endotoxic lipopolysaccharides; inflammatory mediators from lymphocytes and macrophages. *Mailing Add:* Lab Microbiol & Immunol Rm 332 Bldg 30 Nat Inst Dent Res Bethesda MD 20892

MERGENS, WILLIAM JOSEPH, b Queens, NY, July 26, 42; m 65; c 3. ANALYTICAL CHEMISTRY. *Educ:* St Johns Univ, BS, 64; Seton Hall Univ, MS, 70, PhD(chem), 76. *Prof Exp:* RES LEADER, VITAMINS RES & DEVELOP, HOFFMANN-LA ROCHE INC, 64- *Mem:* Am Chem Soc; Sigma Xi; Inst Food Technologists; Am Asn Pharmaceut scientist; AAAS. *Res:* Formulation and analysis; product development chemistry; chemical carcinogenesis; pharmaceutical, food and agricultural product vitamin fortification. *Mailing Add:* 20 Coolidge Ave West Caldwell NJ 07006

MERGENTIME, MAX, b Brooklyn, NY, Apr 2, 14; m 50; c 4. FOOD CHEMISTRY. *Educ:* Cornell Univ, BS, 35, MS, 36; Ore State Col, PhD(food tech), 41. *Prof Exp:* Processed foods inspector, Prod & Mkt Admin, USDA, 41-45; chief chemist, Sunshine Packing Corp, Pa, 45-50; head juice dept, Frigid Food Prod, Inc, 50-88; RETIRED. *Mem:* Inst Food Technol. *Res:* Low temperature studies rate; reaction proteolytic enzyme of peas. *Mailing Add:* 2685 Lahser Rd Bloomfield Hills MI 48103

MERGLER, H(ARRY) W(INSTON), b Chillicothe, Ohio, June 1, 24; m 48; c 3. ELECTRICAL ENGINEERING. *Educ:* Case Inst Technol, BS, 48, MS, 58, PhD(eng), 56. *Prof Exp:* Aeronaut res scientist, Nat Adv Comt Aeronaut, 48-56; from asst prof to assoc prof control eng, 56-61, prof eng, 62-73, LEONARD CASE PROF ELEC ENG, CASE WESTERN RESERVE UNIV, 73-, DIR, DIGITAL SYSTS LAB, 67- *Concurrent Pos:* Consult to various indust concerns; consult ed, Control Eng Mag; pres, Digital/Gen Corp, 68-71. *Honors & Awards:* Lamme Medal, Inst Elec & Electronics Engrs, 78, Centennial Medal, 84. *Mem:* Nat Acad Eng; Indust Electronics & Control Instrumentation (pres, 76-78); fel Inst Elec & Electronic Engrs (vpres, 89); Soc Naval Archit & Marine Engrs. *Res:* Application of digital computer techniques to problems of digital guidance and control systems. *Mailing Add:* Dept Elec Eng Case Western Reserve Univ Cleveland OH 44106

MERIAM, JAMES LATHROP, b Columbia, Mo, Mar 25, 17; m 40; c 2. STATICS, DYNAMICS. *Educ:* Yale Univ, BE, 39, MEng, 41, PhD(mech eng), 42. *Prof Exp:* From instr to prof mech eng, Univ Calif, Berkeley, 42-63; dean & prof, Sch Eng, Duke Univ, 63-72; prof mech eng, Calif Polytech State Univ, 72-80; vis prof, Univ Calif, Santa Barbara, 80-90; RETIRED. *Concurrent Pos:* Chmn, Grad Studies Div, Am Soc Eng Educ, 59-60, Mech Div, 74-75. *Mem:* Fel Am Soc Mech Engrs; affil Am Soc Testing & Mat; hon mem Am Soc Eng Educ. *Res:* Photoelasticity; thermal stresses; dynamics; author of textbooks in engineering mechanics and various other publications. *Mailing Add:* 4312 Marina Dr Santa Barbara CA 93110

MERICLE, MORRIS H, b Toledo, Iowa, Mar 26, 25; m 59. ELECTRICAL ENGINEERING. *Educ:* Iowa State Univ, BS, 47, MS, 56, PhD(elec eng), 63. *Prof Exp:* Elec engr, Repub Steel Corp, 47-51; from instr to asst prof elec eng, Iowa State Univ, 53-58; sr engr, Autonetics Div, NAm Aviation Inc, 58-60; asst prof, 60-63, ASSOC PROF ELEC ENG, IOWA STATE UNIV, 63- *Concurrent Pos:* Consult, Iowa Methodist Hosp, Des Moines, 62- *Mem:* Inst Elec & Electronics Engrs. *Res:* Pattern recognition of biological signals. *Mailing Add:* Dept of Elec Eng 120 Coover Hall Iowa State Univ Ames IA 50011

MERICLE, R BRUCE, b Omaha, Nebr, June 4, 38; m 63; c 3. MATHEMATICS. *Educ:* Iowa State Univ, BS, 60; Univ Md, College Park, MS, 64; Wash State Univ, PhD(math), 70. *Prof Exp:* Instr math, Univ Maine, 64-66 & Wash State Univ, 66-70; asst prof, Mankato State Col, 70-74; dir acad comput serv, Mich Technol Univ, 74-77; assoc prof, 77-79, PROF, MANKATO STATE UNIV, 80- *Mem:* Am Math Soc; Math Asn Am. *Res:* Measure theory; measures in topological spaces. *Mailing Add:* Dept Math Mankato State Univ Mankato MN 56001

MERICOLA, FRANCIS CARL, inorganic chemistry; deceased, see previous edition for last biography

MERIFIELD, PAUL M, b Santa Monica, Calif, Mar 17, 32; m 68; c 2. GEOLOGY. *Educ:* Univ Calif, Los Angeles, AB, 54, MA, 58; Univ Colo, PhD(geol), 63. *Prof Exp:* Res scientist, Lockheed-Calif Co, 62-64; dir geosci, Earth Sci Res Corp, 64-73. *Concurrent Pos:* Adj prof, Univ Calif, Los Angeles; partner, Lamar-Merifield, 64-89. *Mem:* Geol Soc Am; Asn Eng Geol. *Res:* Interpretation of satellite photography; age and origin of the earth-moon system; remote sensing; engineering and environmental geology. *Mailing Add:* 3411 Wade St Los Angeles CA 90066

MERIGAN, THOMAS CHARLES, JR, b San Francisco, Calif, Jan 18, 34; m 59; c 1. INFECTIOUS DISEASES, VIROLOGY. *Educ:* Univ Calif, Berkeley, BA, 55; Univ Calif, San Francisco, MD, 58; Am Bd Internal Med, dipl, 65. *Prof Exp:* Intern med, Boston City Hosp, Mass, 58-59, asst resident, 59-60; clin assoc, Nat Heart Inst, 60-62; assoc, Nat Insts Arthritis & Metab Dis, 62-63; from asst prof to assoc prof med, 63-72, dir diag microbiol lab, 66-72, prof med, 72-80, GEORGE E & LUCY BECKER PROF MED, SCH MED, STANFORD UNIV, 80-, CHIEF DIV INFECTIOUS DIS & HOSP EPIDEMIOLOGIST, 66-, DIR DIAG VIROL LAB, 69- *Concurrent Pos:* Mem microbiol training grant comt, Nat Inst Gen Med Sci, 69-73, mem virol study sect, Div Res Grants, NIH, 74-78. *Honors & Awards:* Borden Award for Outstanding Res, 73. *Mem:* Inst Med-Nat Acad Sci; Am Asn Immunol; Am Fedn Clin Res; Am Soc Clin Invest; Am Soc Microbiol. *Res:* Host responses to viral infections and antiviral agents. *Mailing Add:* Div of Infectious Dis Stanford Univ Sch of Med Stanford CA 94305-5107

MERILAN, CHARLES PRESTON, b Lesterville, Mo, Jan 14, 26; m 49; c 2. DAIRY HUSBANDRY. *Educ:* Univ Mo, BS, 48, AM, 49, PhD(dairy husb), 52. *Prof Exp:* Instr dairy husb, 50-52, bact & prev med, 52-53, from asst prof to assoc prof, 53-59, chmn dept, 61-62, assoc dir agr exp sta, 62-63, PROF DAIRY HUSB, UNIV MO-COLUMBIA, 59- *Mem:* AAAS; Am Chem Soc; Am Soc Animal Sci; Am Dairy Sci Asn; Soc Cryobiol; Sigma Xi. *Res:* Cellular physiology; biophysics; reproductive physiology. *Mailing Add:* Dairy Sci S-141 Animal Sci Ctr, Univ of Mo Columbia MO 65211

MERILO, MATI, b Tallinn, Estonia, Jan 23, 44; Can citizen; m 71; c 3. MULTI PHASE FLOW, HEAT TRANSFER. *Educ:* McGill Univ, BEng, 66; Case Inst Technol, MS, 68; Case Western Reserve Univ, PhD(mech eng), 72. *Prof Exp:* Engr, D Q Kern Assoc, 68-71; res engr, Chalk River Nuclear Lab, Atomic Energy Can, Ltd, 71-77; PROJ MGR, ELEC POWER RES INST, 77- *Mem:* Am Soc Mech Engrs; Am Nuclear Soc. *Res:* Nuclear reactor safety; transient two-phase flow and heat transfer; aerosol transport, deposition, scrubbing and filtration. *Mailing Add:* Elec Power Res Inst 3412 Hillview Ave Palo Alto CA 94303

MERIN, ROBERT GILLESPIE, b Glens Falls, NY, June 16, 33; m 58; c 3. ANESTHESIOLOGY, PHARMACOLOGY. *Educ:* Swarthmore Col, BA, 54; Cornell Univ, MD, 58. *Prof Exp:* Instr anesthesiol, Albany Med Col, 63-66, res assoc pharmacol, 65-66; from asst prof to prof anesthesiol, Sch Med, Univ Rochester, 66-81, assoc prof pharmacol, 72-81; PROF ANESTHESIOL, UNIV TEX HEALTH SCI CTR, HOUSTON, 81- *Concurrent Pos:* NIH career develop award, Sch Med, Univ Rochester, 72-77; consult, Vet Admin Hosp, Albany, 63-66. *Mem:* Am Soc Anesthesiol; Int Anesthesia Res Soc; Am Soc Pharmacol & Exp Therapeut. *Res:* Effect of anesthesia on the cardiovascular system and metabolism. *Mailing Add:* Dept Anesthesiol 6431 Fannin 5020 Houston TX 77030

MERINEY, STEPHEN D, b Durham, NC, July 27, 60. NEUROSCIENCE, DEVELOPMENTAL BIOLOGY. *Educ:* Univ NH, BA, 82; Univ Conn, PhD(neurosci), 86. *Prof Exp:* RES ASST, UNIV CONN, 85- *Mem:* Soc Neurosci; AAAS. *Res:* Experimental analysis of the competition for survival among developing motoneurons; development of synaptic transmission of ganglionic and neuromuscular synapses. *Mailing Add:* Dept Physiol & Neurobiol Univ Conn 75 N Eagleville Rd Storrs CT 06268

MERIWETHER, JOHN R, b Beaumont, Tex, May 22, 37; m 56; c 4. NUCLEAR PHYSICS. *Educ:* Univ Southwestern La, BS, 58, MS, 59; Fla State Univ, PhD(nuclear physics), 62. *Prof Exp:* Assoc, Lawrence Radiation Lab, 62-65, staff physicist, 65-66; asst prof nuclear physics & comput sci, 66-71, assoc prof, 71-75, chmn dept, 71-81, PROF PHYSICS, ACADIANA RES LAB, UNIV SOUTHWESTERN LA, 75- *Concurrent Pos:* Ed, Proc Third Geopressured-Geothermal Energy Conf, Lafayette, 78. *Mem:* Am Phys Soc; Sigma Xi. *Res:* Environmental radioactivity gamma-ray spectroscopy; applied atomic physics-proton induced x-ray emission as an analytical method. *Mailing Add:* Rte 3 Box 258 Arnaudville LA 70512

MERIWETHER, JOHN WILLIAMS, JR, b Louisville, Ky, Apr 14, 42; m 73; c 1. AERONOMY. *Educ:* Mass Inst Technol, SB, 64; Univ Md, PhD(physics), 70. *Prof Exp:* Nat Acad Sci res assoc, Goddard Space Flight Ctr, 69-71; res assoc atmospheric physics, Univ Mich, 71-73; staff physicist atmospheric physics, PhotoMetrics, Inc, 73-74; res assoc ionospheric physics, Arecibo Observ, Cornell Univ, 75-79; assoc res scientist, 79-84, RES SCIENTIST SPACE PHYSICS, UNIV MICH, 85- *Mem:* Am Inst Physics; Am Geophys Union. *Res:* Aeronomy of the earth's atmosphere by means of high spectrol resolution observations of airglow and auroral emissions from remote ground-based statious as supplemented with incoherent scatter radar observations of the ionosphere. *Mailing Add:* Hanscom AFB Beford MA 01731

MERIWETHER, LEWIS SMITH, b Washington, DC, May 23, 30; m 53; c 3. PHYSICAL CHEMISTRY, ORGANIC CHEMISTRY. *Educ:* Harvard Univ, AB, 52; Univ Chicago, PhD(chem), 56. *Prof Exp:* Res chemist, 55-59, SR RES CHEMIST, AM CYANAMID CO, 59-, GROUP LEADER, 60- *Concurrent Pos:* Cyanamid Sr Award, 64-65. *Mem:* NY Acad Sci; Am Chem Soc. *Res:* Homogeneous catalysis; transition metal complexes; polymerization; photochemistry; enzyme model systems; membranes; surgical adhesives; artificial kidney systems; biomaterials; biocides; cosmetics; pharmaceuticals. *Mailing Add:* Chem Res Div Am Cyanamid Co 98 Twin Oak Lane Wilton CT 06897-2738

MERKEL, FREDERICK KARL, b Athens, Greece, May 27, 36; US citizen; m 76; c 2. VASCULAR SURGERY. *Educ:* Univ Cincinnati, BS, 57; Johns Hopkins Univ, MD, 61; FACS, 72. *Prof Exp:* Instr surg, Med Ctr, Univ Colo, 69-70; asst prof & coordr transplantation, Med Sch, Northwestern Univ, 70-72; assoc prof, immunol & med & dir transplantation, 72-83, PRES & MED DIR, CHICAGO REGIONAL ORGAN & TISSUE BANK, RUSH-PRESBY-ST LUKE'S MED CTR, 83- *Concurrent Pos:* Chief gastrointestinal & transplant surg, Denver Gen Hosp, 69-70; dir transplantation, Children's Mem Hosp, Chicago, 70-72; consult surg, St Therese Hosp, Waukegan, Ill, 72-, Martha Wash Hosp, Chicago, 73-,Mt Sinai Hosp Med Ctr, Chicago, 74- & Ingalls Mem Hosp, Harvey, 78-; assoc attend surgeon & assoc attend physician, Rush-Presby-St Luke's Med Ctr, 73- *Honors & Awards:* Peter F Salisbury Award, Am Soc Artificial Organs, 78. *Mem:* Am Soc Transplant Surgeons (pres, 78-79); Transplantation Soc; Brit Transplantation Soc; Am Col Surgeons; Europ Soc Surg Res. *Res:* Organ transplantation, especially pancreas and kidney; organ preservation; transplant immunology. *Mailing Add:* Suite 374 1725 W Harrison St Chicago IL 60629

MERKEL, GEORGE, b San Francisco, Calif, Sept 2, 29; m 59; c 1. NUCLEAR RADIATION EFFECTS, ANTENNA THEORY. *Educ:* Callf Inst Technol, BS, 51; Univ Calif, Berkeley, PhD(nuclear physics), 63; George Washington Univ, MS, 75I. *Prof Exp:* Res asst physicist, Lawrence Radiation Lab, 53-62; res assoc, Univ Rochester, 63-64; physicist, Gen Atomic, San Diego, 64-67; physicist, Off Naval Res, 67-68; PHYSICIST DEPT DEFENSE, HARRY DIAMOND LABS, US ARMY LAB COMMAND, 68- *Concurrent Pos:* Reviewer, Nuclear & Plasma Soc, Inst Elec & Electronics Engrs, 80-81. *Mem:* Am Phys Soc; Inst Elec & Electronics Engrs; Sigma Xi; AAAS; Am Geophys Union. *Res:* Nuclear reactions; charged particle transport; nuclear structure program; theoretical and experimental investigations of antenna response in ionized media; intense relativistic electron beam propagation; relativistic electron beam physics. *Mailing Add:* US Army Lab Command Harry Diamond Lab 2800 Powder Mill Rd Adelphi MD 20783

MERKEL, JOSEPH ROBERT, b Alburtis, Pa, Dec 21, 24; m 48; c 1. MICROBIAL BIOCHEMISTRY, MARINE MICROBIOLOGY. *Educ:* Moravian Col, BS, 48; Purdue Univ, MS, 50; Univ Md, PhD(bact), 52. *Prof Exp:* Waksman-Merck fel, Rutgers Univ, 52-53, res assoc, 53-54, res investr, Inst Microbiol, 54-55; dir, Ft Johnson Marine Biol Lab, Col Charleston, 55-62; assoc prof biochem, Ctr Marine & Environ Studies, Lehigh Univ, 62-65, prof dept chem & marine microbiologist, 65-88; RETIRED. *Concurrent Pos:* Researcher, Lehigh Univ, 88- *Honors & Awards:* Wakman Merch Award, 52; Labor Found Award, 53. *Mem:* Am Soc Biochem & Molecular Biol; Am Chem Soc; Am Soc Microbiol. *Res:* Proteolytic enzymes of marine bacteria; collagenases; wound healing. *Mailing Add:* Seeley G Mudd Bldg Dept Chem Lehigh Univ Bethlehem PA 18015

MERKEL, PAUL BARRETT, b Rochester, NY, May 14, 45. PHOTOCHEMISTRY, PHOTOGRAPHIC CHEMISTRY. *Educ:* St John Fisher Col, BS, 67; Univ Notre Dame, PhD(chem), 70. *Prof Exp:* Res assoc chem, Univ Calif, Riverside, 70-71; SR RES CHEMIST, EASTMAN KODAK CO, 71- *Mem:* Am Chem Soc; Sigma Xi. *Res:* In photochemistry and photophysics, interests include laser and flash photolysis, photo-oxidation reactions, electronic excitation and luminescence; research in photographic chemistry involves kinetics and mechanisms, thermoanalytical methods and dye imaging. *Mailing Add:* 525 Westfield St Rochester NY 14619

MERKEL, ROBERT ANTHONY, b Marshfield, Wis, Feb 7, 26; m 54; c 4. MEAT SCIENCE. *Educ:* Univ Wis-Madison, BS, 51, MS, 53, PhD(meat sci, biochem), 57. *Prof Exp:* From asst prof to assoc prof meat sci, Kans State Univ, 57-62; assoc prof, 62-67, prof meat sci, 67-77, PROF ANIMAL HUSB & FOOD SCI & HUMAN NUTRIT, MICH STATE UNIV, 77- *Concurrent Pos:* Sect ed meat sci & muscle biol, J Animal Sci, 74-76. *Mem:* AAAS; Am Meat Sci Asn; Am Soc Animal Sci; Inst Food Technologists; Sigma Xi. *Res:* Differentiation, histogenesis and growth of muscle and adipose tissues; biosynthesis of muscle proteins and subcellular and molecular study of meat tenderness. *Mailing Add:* Dept Animal Husb Mich State Univ Rm 100 Meat Lab East Lansing MI 48823

MERKEL, TIMOTHY FRANKLIN, b Jersey Shore, Pa, June 24, 42; m 66; c 2. ORGANIC POLYMER CHEMISTRY. *Educ:* Lycoming Col, AB, 64; Pa State Univ, MS, 66; Univ Mich, PhD(org chem), 73. *Prof Exp:* Res chemist, Whitmoyer Labs Inc, Rohm & Haas Co, 66-68; mgr prod develop, Sartomer Co, 73-78; mgr long range res & develop, 78-79, MGR FRICTION MATS RES, ABEX CORP, 79- *Mem:* Am Chem Soc; AAAS; Sigma Xi. *Res:* Specialty monomers, cyclic azo compounds; friction materials. *Mailing Add:* Friction Product Div ABEX Corp PO Box 3250 Winchester VA 22601-2450

MERKELO, HENRI, b Borky, Ukraine, June 12, 39; US citizen. QUANTUM ELECTRONICS, ULTRAHIGH SPEED ELECTRONICS. *Educ:* Col Moderne, France, CAP, 60; Univ Ill, PhD(elec eng, physics), 66. *Prof Exp:* McDonald-Douglas Co, 62; assoc prof elec & comput eng, 70-78, DIR, QUANTUM ELECTRONIC RES LAB, UNIV ILL, URBANA, 78- *Concurrent Pos:* NSF grants, Univ Ill, 67-, Cottrell Found grant, 68-70; consult, 80-; Ford Found fel & Indust grants, 82-; dir, Picosecond Digital Electronics, Univ Calif, Santa Barbara, 82-89; consult ultrahigh speed digital electronics & microelectronic packaging; IBM fel. *Mem:* Am Inst Physics; Inst Elec & Electronics Engrs. *Res:* Lasers; picosecond and femtosecond optical electronics; luminescence; photosynthesis; ultrahigh speed electronic devices; microelectronic packaging; modeling and simulation of ultrahigh speed digital signals; software systems. *Mailing Add:* Dept Elec Eng & Computer Eng Univ Ill Urbana IL 61801

MERKEN, HENRY, b Peabody, Mass, Oct 14, 29; m 53; c 4. POLYMER CHEMISTRY. *Educ:* Northeastern Univ, BS, 53. *Prof Exp:* Asst engr, Res & Develop Dept, Am Polymer Corp, Mass, 49-53, develop engr, 53; develop engr, Polyco Dept, Borden Co, 55-56; develop engr, Polyvinyl Chem, Inc, 56-64, dir mfg, 64-68, asst to pres, 68-71, vpres int, 71-72; DIR INT OPER, BEATRICE CHEM, 72-; EXEC VPRES, SANNCOR INDUSTS LEOMINSTER, MASS, 83- *Concurrent Pos:* Instr, Lowell Tech Inst, 59-62. *Mem:* Am Chem Soc; Am Inst Chem Eng. *Res:* Organic chemistry; emulsion polymers; plasticizers. *Mailing Add:* 45 Weatherly Dr Salem MA 01970

MERKEN, MELVIN, b Peabody, Mass, Jan 19, 27; m 56; c 3. CHEMISTRY, SCIENCE EDUCATION. *Educ:* Tufts Univ, BS, 50, AM, 51; Boston Univ, EdD(sci ed), 67. *Prof Exp:* Teacher high schs, Conn, 51-58; assoc prof, 58-67, chmn dept, 58-75, PROF CHEM, WORCESTER STATE COL, 67- *Mem:* Fel AAAS; fel Am Inst Chem; Am Chem Soc; Am Asn Physics Teachers. *Res:* Teaching science to non-scientists in general education program at college level; promotion of scientific literacy and understanding; environmental chemistry. *Mailing Add:* Dept Natural & Earth Sci Worcester State Col Worcester MA 01602-2597

MERKER, MILTON, b New York, NY, Sept 15, 41; m 63; c 2. NUCLEAR PHYSICS, COSMIC RAY PHYSICS. *Educ:* City Col New York, BS, 63; NY Univ, MS, 65, PhD(physics), 70. *Prof Exp:* Res asst physics, NY Univ, 65-70; res assoc, Univ Pa, 69-71, res asst prof astrophys, 71-73, asst prof & chmn dept, 73-77; staff scientist, Sci Applns, Inc, 77-85; SR STAFF SCIENTIST, S-CUBED, 85- *Mem:* Sigma Xi; Am Phys Soc; Am Geophys Union. *Res:* Radiation transport and effects of cosmic rays; astrophysical spallation and heavy-ion reactions; quasars; nuclear cascade; high-energy shielding and radiologic dosimetry; atmospheric neutrons. *Mailing Add:* S-CUBED 3398 Carmel Mountain Rd San Diego CA 92121

MERKER, PHILIP CHARLES, b New York, NY, July 23, 22; m 52; c 2. PHARMACOLOGY, TOXICOLOGY. *Educ:* Brooklyn Col, BA, 46; Long Island Univ, BS, 51; Purdue Univ, MS, 53, PhD, 55. *Prof Exp:* Lab asst physiol, Brooklyn Col, 46-47; teaching asst mat med, Long Island Univ, 48-51; asst pharm, Purdue Univ, 51-53; asst, Sloan-Kettering Inst Cancer Res, 56-62, head sect, 58-62, assoc mem, 62; prof pharmaceut, Col Pharm, Univ Tenn, 62-64; chmn dept pharmacol & animal sci, Col Pharmaceut Sci, Columbia Univ, 65, prof pharmacol & chmn div biol sci & pharmacol, 65-72; pharmacologist, Arthur D Little, Inc, 72-77; assoc dir, 77-80, DIR, PHARMACOL & TOXICOL, VICKS RES DIV, 80- *Concurrent Pos:* Res fel, Sloan-Kettering Inst Cancer Res, 54-56; asst prof, Sloan-Kettering Div, Cornell Univ, 58-62. *Mem:* AAAS; Am Asn Cancer Res; Am Soc Exp Path; Am Soc Pharmacol & Exp Therapeut; Soc Toxicol. *Res:* Experimental cancer chemotherapy; chemotherapy. *Mailing Add:* Vicks Res Ctr One Far Mill Crossing Shelton CT 06484

MERKER, STEPHEN LOUIS, b Cleveland, Ohio, Dec 4, 41; m 41; c 3. LIVING SYSTEMS THEORY, GENERAL SYSTEMS THEORY. *Educ:* Univ Louisville, BS, 69, MBA, 73 & PhD(systs sci), 80. *Prof Exp:* Res physicist, Brown Williamson Tobacco Corp, 69-78; res assoc, 80-87, CHAIR,

SYST SCI INST, UNIV LOUISVILLE, 87- *Concurrent Pos:* Consult, Mercer-Meidinger-Hansen, 81-; exec bd dirs, 85-88, secy to the council, Int Soc Gen Systs Res, 88- *Mem:* Int Soc Gen Systs Res. *Res:* Application of living systems theory process analysis in study of organizations; design and use of survey instruments to measure attitudes and perception. *Mailing Add:* Systs Sci Inst Univ Louisville Louisville KY 40292

MERKES, EDWARD PETER, b Chicago, Ill, Apr 14, 29; m 56. MATHEMATICS. *Educ:* DePaul Univ, BS, 50; Northwestern Univ, PhD, 58. *Prof Exp:* Lectr math, De Paul Univ, 50-54, instr, 56-58, asst prof, 58-59; asst prof, Marquette Univ, 59-62, assoc prof, 62-63; assoc prof, 63-69, head dept, 70-77, PROF MATH, UNIV CINCINNATI, 69- *Concurrent Pos:* Vis assoc prof, Math Res Ctr, Univ Wis, 62-63. *Mem:* Am Math Soc; Math Asn Am. *Res:* Complex variable and continued fractions. *Mailing Add:* Math Dept/025 Univ Cincinnati Cincinnati OH 45221

MERKLE, F HENRY, b Newark, NJ, Aug 31, 31; m; c 3. PHARMACEUTICAL CHEMISTRY. *Educ:* Rutgers Univ, BS, 54, MS, 61, PhD(pharmaceut sci), 64. *Prof Exp:* Instr pharm, Rutgers Univ, 58-62, lectr, 62-63; res scientist, Res Ctr, FMC Corp, NJ, 64-65; sr res scientist, 65-70, dept head pharmaceut prod develop, 70-75, MGR PHARM PROD DEVELOP, BRISTOL MYERS PROD, 75- *Mem:* Am Pharmaceut Asn; Am Chem Soc. *Res:* Pharmaceutical analysis, products, and development. *Mailing Add:* 2217 Shawnee Path Scotch Plains NJ 07090

MERKLE, OWEN GEORGE, plant breeding, genetics, for more information see previous edition

MERKLE, ROBERTA K, b Wilmington, Del, June 11, 55. GLYCOPROTEINS. *Educ:* Va Polytech Inst, PhD(microbiol), 82. *Prof Exp:* Fel, cell biol, Health Sci Ctr, Univ Tex, 82-83; FEL, CELL BIOL, GRAD STUDIES RES CTR, ATHENS, GA, 84- *Mem:* Am Soc Cell Biol. *Mailing Add:* Complex Carbohydrates Res Ctr Univ Ga CCRC 220 Riverbend Rd Athens GA 30602

MERKLEY, DAVID FREDERICK, b Pipestone, Minn, Apr 23, 45; m 70; c 2. VETERINARY SURGERY. *Educ:* Univ SDak, BA, 67; Iowa State Univ, DVM, 71; Mich State Univ, MS, 75; Am Col Vet Surg, dipl, 77. *Prof Exp:* Asst prof vet surg, Mich State Univ, 74-79; assoc prof, 79-85, PROF VET SURG, COL VET MED, IOWA STATE UNIV, AMES, 85- *Mem:* Am Animal Hosp Asn; Am Vet Med Asn; Am Col Vet Surg. *Res:* Total urinary diversion in dogs. *Mailing Add:* Dept Vet Clin Sci Iowa State Univ 1492 Vet Med Ames IA 50011

MERKLEY, WAYNE BINGHAM, b Murray, Utah, Apr 1, 41; m 59; c 4. ECOLOGY OF REGULATED STREAMS, AQUATIC ECOLOGY. *Educ:* Univ Utah, BS, 63, MA, 66, PhD(limnol), 69. *Prof Exp:* Instr biol, Univ Utah, 68; from asst prof to assoc prof, 69-80, PROF BIOL, DRAKE UNIV, 80- *Mem:* Am Inst Biol Sci; Am Soc Limnol & Oceanog; Water Pollution Control Fedn; NAm Benthol Soc. *Res:* Ecological impact of impoundments and urban areas on aquatic environments in large prairie rivers. *Mailing Add:* Dept of Biol Drake Univ 25th St & Univ Ave Des Moines IA 50311

MERLIE, JOHN PAUL, b Vineland, NJ, Dec 4, 45; m 73; c 3. MOLECULAR NEUROBIOLOGY. *Educ:* Villanova Univ, BS, 67; Univ Penn, PhD(microbiol), 73. *Prof Exp:* Asst molecular biol, Pasteur Inst, Paris, 73, neurosci, 74-76 & Salk Inst, 76-78; asst prof biol, Univ Pittsburgh, 78-83; assoc prof pharmacol, 84-87, PROF PHARMACOL, WASH UNIV, 87- *Concurrent Pos:* Fel neurosci, Alfred P Sloan Foun, 76. *Honors & Awards:* Javits Neurosci Investr Award, 89. *Mem:* Am Soc Cell Biol; Soc Neurosci; Am Soc Pharmacol & Exp Therapeut; Am Soc Biol Chem & Molecular Biol; Soc Gen Physiologists; Am Soc Microbiol. *Res:* Molecular mechanisms in synapse formation; regulation of synthesis of synaptic proteins; acetylchoine receptor-channels. *Mailing Add:* Dept Molecular Biol & Pharmacol Wash Univ Med Sch Box 8103 St Louis MO 63110

MERLIN, ROBERTO DANIEL, b Buenos Aires, Argentina, Aug 12, 50. INELASTIC LIGHT SCATTERING, OPTICAL PROPERTIES. *Educ:* Univ Buenos Aires, Licenciatura, 73; Univ Stuttgart, WGer, Dr rer nat, 78. *Prof Exp:* Res assoc, Univ Ill, Urbana, 78-80; from asst prof to assoc prof, 80-89, PROF PHYSICS, UNIV MICH, ANN ARBOR, 89- *Concurrent Pos:* Vis prof, Max-Planck-Inst FKF, Stuttgart, 87. *Honors & Awards:* Von Humboldt Fel, 86. *Mem:* Am Phys Soc. *Res:* Semiconductor heterostructures and superlattices. *Mailing Add:* Dept Physics Univ Mich Ann Arbor MI 48109-1120

MERLINI, GIAMPAOLO, b Quinzano D'Oglio, Italy, Sept 26, 51; m 85; c 2. HEMATOLOGY, CLINICAL CHEMISTRY. *Educ:* Univ Pavia, MD, 76, dipl clin chem, 79, dipl hemat, 82, dipl internal med, 87. *Prof Exp:* Asst prof clin chem, 79-84, ASST PROF INTERNAL MED, UNIV PAVIA, 84- *Concurrent Pos:* Mem, Protein Comn, Ital Soc Clin Biochem, 85-; prin investr, Nat Inst Res, 90- *Mem:* Am Soc Immunol. *Res:* Plasma cell dyscrasias; pathophysiology and treatment. *Mailing Add:* Inst Clin Med II Sci Inst Policlin S Matteo Pavia I-27100 Italy

MERLINO, GLENN T, b New York, NY, Aug 25, 53; m 75. MOLECULAR GENETICS, CANCER RESEARCH. *Educ:* Adelphi Univ, NY, BA, 75; Univ Mich, Ann Arbor, PhD(biol sci), 80. *Prof Exp:* Grad fel biol sci, Univ Mich, 75-80; FEL, NAT CANCER INST, NIH, 80-; ADJ ASST PROF, GEORGE WASH UNIV, 85- *Concurrent Pos:* Fel, Cystic Fibrosis Found, 80 & Arthritis Found, 81- *Mem:* AAAS; Am Soc Zoologists; NY Acad Sci. *Res:* Regulations of enkaryotic genes during cellular differentiation; examination of the structure and regulation of cellular protoarcogenes, including epidermal growth factor receptor. *Mailing Add:* 8635 McHenry St Vienna VA 22180

MERMAGEN, WILLIAM HENRY, b New York, NY, May 21, 35; m 60; c 4. BALLISTICS, AERONAUTICS. *Educ:* Fordham Univ, BS, 57; Univ Del, MS, 66. *Prof Exp:* Res physicist aeroballistics, 57-67, phys sci adminr, 67-90, CHIEF, SYSTS ENG & CONCEPT ANALYSIS DIV, US ARMY BALLISTIC RES LAB, 91- *Concurrent Pos:* Fel, US Army Ballistic Res Lab, 75. *Mem:* Assoc fel Am Inst Aeronaut & Astronaut. *Res:* Aeroballistics including missile dynamics, flight mechanics, heat transfer, fluid flow, measurement systems and liquid-filled projectiles. *Mailing Add:* Webster Village Havre de Grace MD 21078

MERMEL, THADDEUS WALTER, b Chicago, Ill, Sept 12, 07; m 30; c 3. ELECTRICAL ENGINEERING, CIVIL ENGINEERING. *Educ:* Univ Ill, BS, 30. *Prof Exp:* Engr, Bur Reclamation, US Dept Interior, 33-73, asst to comnr res, 64-71, asst to comnr sci affairs, 71-73; CONSULT CIVIL ENG, PROCUREMENT SPECIALTY, CONTRACT ADMIN ARBITRATION, 73- *Concurrent Pos:* Mem fed construct coun, Nat Acad Sci, 55-, mem comt on construct mgt, Hwy Res Bd & alt mem adv bd, Off Critical Tables; chmn comt world register of dams, Int Comn Large Dams, 60-; consult, Overseas Adv Assocs, 73-76 & World Bank, 73-91; mem, World Bank Mission to Pakistan, India, Nepal, Indonesia, Hungary, Syria; mem, Am Arbit Asn. *Honors & Awards:* Gold Medal, US Dept Interior. *Mem:* AAAS; Am Soc Civil Engrs; Inst Elec & Electronics Engrs; Int Comt on Large Dams; Int Soc Rock Mech. *Res:* Dams; hydroelectric plant equipment; generators; turbines; underground high-voltage transmission; water resource development. *Mailing Add:* 4540 43rd St NW Washington DC 20016-4547

MERMELSTEIN, ROBERT, b Mukacevo, Czech; Can citizen. POLYMER CHEMISTRY. *Educ:* Sir George Williams Univ, BSc, 57; Univ Alta, PhD(phys org chem), 64. *Prof Exp:* Fel, Brandeis Univ, 64-65 & Childrens' Cancer Res Found, Boston, 65-66; SCIENTIST POLYMER CHEM, XEROX CORP, 66- *Mem:* AAAS; Am Chem Soc; Environ Mutagen Soc. *Res:* Synthesis, characterization of vinyl and condensation polymers; biopolymers; structure-activity relationships; rheological behavior; kinetics and mechanism of organic reactions; microbiology of nitroarnenes. *Mailing Add:* 345 Pelham Rd Rochester NY 14610-3352

MERMIN, N DAVID, b New Haven, Conn, Mar 30, 35; m 57; c 2. PHYSICS. *Educ:* Harvard Univ, AB, 56, AM, 57, PhD(physics), 61. *Prof Exp:* NSF fel physics, Univ Birmingham, 61-63; res assoc, Univ Calif, San Diego, 63-64; from asst prof to assoc prof, 64-72, PROF PHYSICS, CORNELL UNIV, 72-, DIR, LAB ATOMIC & SOLID STATE PHYS, 84- *Concurrent Pos:* Alfred P Sloan Found fel, 66-70; John Simon Guggenheim Found fel, 70-71. *Mem:* Nat Acad Sci; fel Am Phys Soc. *Res:* Theoretical solid state and statistical physics. *Mailing Add:* Dept Physics Cornell Univ Ithaca NY 14853-2501

MERNER, RICHARD RAYMOND, b Chicago, Ill, Sept 23, 18; m 51; c 2. INDUSTRIAL ORGANIC CHEMISTRY, SCIENCE ADMINISTRATION. *Educ:* Univ Ill, BS, 39; Northwestern Univ, PhD(chem), 49. *Prof Exp:* Asst chem, Univ Mo, 39-40; asst chem electrochem dept, Res & Develop, E I du Pont de Nemours & Co, Inc, 40-44, org chem dept, Res, 49-53, tech supvr process develop, 53-67, supvr tech employ & personnel develop, 67-76, mgr distrib regulatory compliance, 76-78; CONSULT & PRES, MERNER ASSOCS, 77- *Concurrent Pos:* Prof & mgt lectr, Col Bus & Econs, Univ Del, 70-87. *Mem:* AAAS; Am Chem Soc; Sigma Xi. *Res:* Development research; intermediates; dyes and pigments; fluorocarbons; management science; behavior science. *Mailing Add:* RD 2 Box 326 Sullivan Rd Avondale PA 19311

MEROLA, A JOHN, b Freehold, NJ, July 21, 31; m 58; c 3. BIOCHEMISTRY, MICROBIOLOGY. *Educ:* Univ Tex, BA, 53; Rutgers Univ, MS, 59, PhD(bact), 61. *Prof Exp:* Fel biochem, Enzyme Inst, Univ Wis, 61-63; res biologist, Sterling-Winthrop Res Inst, 63-65; from asst prof to assoc prof, 66-73, PROF PHYSIOL CHEM, OHIO STATE UNIV, 73- *Res:* Energy conservation; drug and cholesterol metabolism; electron transport. *Mailing Add:* Dept Physiol Chem Ohio State Univ Col Med 333 W 10th Ave Columbus OH 43210

MEROLA, JOSEPH SALVATORE, b Pittsburgh, Pa, Sept 27, 52; m 77; c 4. HOMOGENEOUS CATALYSIS. *Educ:* Carnegie-Mellon Univ, BS, 74; Mass Inst Technol, PhD(inorg chem), 78. *Prof Exp:* Sr chemist, Corp Res Labs, Exxon Res & Eng Co, 78-87; asst prof, 87-91, ASSOC PROF CHEM, VIRGINIA TECH, 91- *Mem:* Am Chem Soc; Int Union Pure & Appl Chem. *Res:* Synthesis and study of organometallic transition metal compounds capable of directing and influencing the reactivity of organic molecules, especially saturated hydrocarbons. *Mailing Add:* Dept Chem Va Tech Blacksburg VA 24061

MERONEY, ROBERT N, b Chicago, Ill, Oct 4, 37; m 65; c 2. MECHANICAL ENGINEERING, FLUID MECHANICS. *Educ:* Univ Tenn, BS, 60; Univ Calif, Berkeley, MS, 64, PhD(mech eng), 65. *Prof Exp:* Engr, US Naval Ord Lab, 60-65; from asst prof to assoc prof civil eng, 65-75, PROF CIVIL ENG, COLO STATE UNIV, FT COLLINS, 75- *Concurrent Pos:* Clean Air Act fel, 72-73; Fulbright Hays fel, 77-78; Erskine lectr, Univ Canterbury, 77-78; Alexander von Humboldt Award, 80-81. *Mem:* Am Soc Mech Engrs; Am Inst Aeronaut & Astronaut; Am Soc Eng Educ; Am Meteorol Soc; Am Soc Civil Engrs; Sigma Xi. *Res:* Heat transfer and transpiration of turbulent boundary layers; meteorological fluid mechanics; environmental simulation; air pollution. *Mailing Add:* Eng Res Foothills Campus Colo State Univ Ft Collins CO 80523

MERONEY, WILLIAM HYDE, III, b Murphy, NC, Dec 27, 17; m 52. INTERNAL MEDICINE. *Educ:* Univ NC, BS, 43; NY Univ, MD, 45; Am Bd Internal Med, dipl. *Prof Exp:* Instr pharmacol, Sch Med, Univ NC, 43; Med Corps, US Army, 46-75; instr internal med, Yale Univ, 50-51, lectr, 51-52, chief renal insufficiency ctr, Korea, 53, from res clinician to chief dept metab, Walter Reed Army Inst Res, 53-57, dep dir inst, 61-64, dir trop res med lab, San Juan, PR, 57-61, chief res div, Med Res & Develop Command, Washington, DC, 64-65, dep dir personnel & training directorate, Off Surgeon

Gen, 66-68, dir & commandant, Walter Reed Army Inst Res, 68-71, commanding gen, Walter Reed Gen Hosp, 71-72, commanding gen, Madigan Army Med Ctr, Med Corps, US Army, 72-75; consult med, 75-77, DIR PROF SERV, DEPT HEALTH, PROVIDENCE, RI, 77- *Concurrent Pos:* Asst clin prof, Sch Med, Georgetown Univ, 57; clin assoc prof, Univ PR, 57-61; consult, Bayamon Dist Hosp, PR & Surgeon Gen, US Army. *Mem:* AAAS; Endocrine Soc; Soc Exp Biol & Med; AMA; fel Am Col Physicians. *Res:* Metabolic processes. *Mailing Add:* Dept Health Rhode Island Providence RI 02908

MERRELL, DAVID JOHN, b Bound Brook, NJ, Aug 20, 19; m 45; c 4. GENETICS. *Educ:* Rutgers Univ, BS, 41; Harvard Univ, MA, 47, PhD(zool), 48. *Prof Exp:* From instr to assoc prof, 48-64, PROF GENETICS & ECOL, UNIV MINN, MINNEAPOLIS, 64- *Mem:* Soc Study Evolution; Genetics Soc Am; Am Genetics Asn; Am Soc Nat; AAAS; Behav Genetics Asn. *Res:* Ecological and behavioral genetics. *Mailing Add:* 1511 Chelmsford St St Paul MN 55108

MERRIAM, CHARLES WOLCOTT, III, b Birmingham, Ala, Mar 31, 31; m 54; c 2. ELECTRICAL ENGINEERING. *Educ:* Brown Univ, ScB, 53; Mass Inst Technol, MS, 55, ScD(elec eng), 58. *Prof Exp:* Asst prof elec eng, Mass Inst Technol, 58-64; prof, Cornell Univ, 64-71; PROF ELEC ENG & CHMN DEPT, UNIV ROCHESTER, RIVER CAMPUS, 71- *Concurrent Pos:* Consult, Air Res & Develop Command, US Air Force, 58-59; elec engr, Res Lab, Gen Elec Co, 59-64; adj prof, Rensselaer Polytech Inst, 60- *Mem:* Assoc Inst Elec & Electronics Engrs. *Res:* Optimization theory; computations; feedback control. *Mailing Add:* Dept of Elec Eng Univ of Rochester Wilson Blvd Rochester NY 14627

MERRIAM, DANIEL FRANCIS, b Omaha, Nebr, Feb 9, 27; m 46; c 5. GEOLOGY. *Educ:* Univ Kans, BS, 49, MS, 53, PhD, 61; Univ Leicester, MSc, 69, DSc, 75. *Prof Exp:* Geologist, Union Oil Co, Calif, 49-51; asst instr geol, Univ Kans, 51-53, instr, 54; geologist, Kans Geol Surv, 53-58, div head basic geol, 58-63, chief geol res, 63-71; Jessie Page Heroy prof geol, Syracuse Univ, 71-81, chmn dept, 71-80; ENDOWMENT ASN DISTINGUISHED PROF NATURAL SCI & CHMN DEPT, WICHITA STATE UNIV, 81- *Concurrent Pos:* Res assoc, Univ Kans, 63-71; vis res scientist, Stanford Univ, 63; Fulbright-Hays sr res fel, UK, 64-65; dir, Am Geol Inst Int Field Inst, Japan, 67; vis prof geol, Wichita State Univ, 68-70; ed-in-chief, J Math Geol, 68-76; vis geol scientist, Am Geol Inst, 69; partic, Proj Compute, Dartmouth Col, 74; consult, Nat Gas Surv, Fed Power Comn, 72-75 & 78 & chmn, supply-tech adv comt, 75-77; mem ad hoc panel, info mgt comt on remote sensing progs for earth resource surv, Nat Acad Sci-Nat Res Coun, 72-73 & chmn, US Nat Comt for Int Geol Correlation Prog, 76-79; mem, US Nat Comn for UNESCO, 79- *Honors & Awards:* Erasmus Haworth Grad Award Geol, Univ Kans, 55; Esso Distinguished Lectr, Univ Sydney, 79; William Christian Krumbein Medal, Int Asn Math Geol, 81. *Mem:* Fel AAAS; fel Geol Soc London; Int Asn Math Geol (pres, 76-80); fel Geol Soc Am; Am Asn Petrol Geol; Soc Econ Paleontologists & Mineralogists; Nat Asn Geol Teachers; Sigma Xi. *Res:* Carboniferous and Mesozoic stratigraphy; geologic history of the Midcontinent; cyclic sedimentation; petroleum geology; computers and computer applications in the earth sciences; quantitative stratigraphic analysis. *Mailing Add:* Dept Geol Wichita State Univ Wichita KS 67208

MERRIAM, ESTHER VIRGINIA, b Pittsburgh, Pa, Apr 9, 40; m 63; c 2. GENETICS, BIOCHEMISTRY. *Educ:* Elizabethtown Col, BS, 62; Univ Wash, PhD(biochem), 66. *Prof Exp:* USPHS fel biol div, Oak Ridge Nat Lab, 66-67; fel, Calif Inst Technol, 67-69; actg asst prof molecular biol in bact, Univ Calif, Los Angeles, 69-71; asst prof, Loyola Univ Los Angeles, 71-74, assoc prof, 74-82, PROF BIOL, LOYOLA MARYMOUNT UNIV, 82- *Mem:* Am Soc Microbiol; Genetics Soc Am. *Res:* Nucleic acid interactions. *Mailing Add:* Dept Biol Loyola Marymount Univ Los Angeles CA 90045

MERRIAM, GEORGE RENNELL, JR, b Harrisburg, Pa, May 22, 13; m 36; c 4. MEDICINE. *Educ:* Brown Univ, AB, 34; Columbia Univ, MD, 41; Am Bd Ophthal, dipl, 49. *Prof Exp:* from instr ophthal to prof, 49-68; cons opthalmologist, Harlem Hosp, 69-88; RETIRED. *Concurrent Pos:* from asst to attend ophthalmologist, Presby Hosp, NY, 49-; from asst opthalmologist to ophthalmologist, Mem Hosp, NY, 49-69; assoc ophthalmologist, Francis Delafield Hosp, 51-70. *Mem:* Am Ophthal Soc; Am Radium Soc; Asn Res Ophthal; fel Am Col Surgeons; AMA; Sigma Xi; Am Acad Ophthal. *Res:* Ophthalmic radiotherapy; cataracts; relative biological effectiveness of various qualities of radiation. *Mailing Add:* 14 Brook Rd Tenafly NJ 07670

MERRIAM, HOWARD GRAY, b Smithville, Ont, July 8, 32; m 56; c 2. ECOLOGY. *Educ:* Univ Toronto, BSA, 56; Cornell Univ, PhD, 60. *Prof Exp:* Asst gen zool, Cornell Univ, 56-58, asst animal ecol, 58-60; from asst prof to assoc prof animal ecol, Univ Tex, 60-68; ASSOC PROF BIOL, CARLETON UNIV, 68-, CHMN DEPT, 79- *Concurrent Pos:* Consult, Can Ministry State Urban Affairs, 73-74, Can Wildlife Serv, 74-75 & Parks Can, 75- *Mem:* Can Soc Zoologists; Can Soc Environ Biologists; AAAS; Ecol Soc Am; Am Soc Mammal. *Res:* Population ecology and quantitative autecology; ecology of land isopods and marmots; ecology of decomposer ecosystems; heterogeneity in natural systems. *Mailing Add:* Dept Biol Carleton Univ Colonel Bay Dr Ottawa ON K1S 5B6 Can

MERRIAM, JOHN L(AFAYETTE), b Corona, Calif, Nov 27, 11; m 38; c 2. CIVIL ENGINEERING, AGRICULTURAL ENGINEERING. *Educ:* Calif Inst Technol, BSCE, 38. *Prof Exp:* Area engr, Soil Conserv Serv, USDA, 39-56; irrig engr, Ministry Agr, Saudi Arabia, 56-58; prof, 58-78, EMER PROF AGR ENG, CALIF POLYTECH STATE UNIV, 78- *Concurrent Pos:* Consult, Irrig & Drainage Proj, Siwa Oasis, Egypt, Ralph M Parsons Co, 62, Inst Fomento Nac, Managua, Nicaragua, 66, UN Food & Agr Orgn, Saudi Arabia, 69, US Overseas Mission, Thailand, 69-70, Peace Corps, 71, USAID, Tunisia, 69, 75, India, 88-91, Egypt, 90, World Bank, Sri Lanka, 78-85, Pakistan, 89. *Honors & Awards:* Royce J Tipton Award, Am Soc Civil Engrs, 79. *Mem:* Fel Am Soc Civil Engrs; sr mem Am Soc Agr Engrs; Am Geophys Union; US Comt Irrig & Drainage. *Res:* On-farm irrigation efficiency; small project farmer controlled flexible water supply systems; irrigation and drainage. *Mailing Add:* 235 Chaplin Lane San Luis Obispo CA 93405

MERRIAM, JOHN ROGER, b Kenosha, Wis, Jan 6, 40; m 63; c 2. GENETICS. *Educ:* Univ Wis, BS, 62; Univ Wash, MS, 63, PhD(genetics), 66. *Prof Exp:* USPHS fels biol, Oak Ridge Nat Lab, 66-67 & Calif Inst Technol, 67-69; from asst prof to assoc aprof, 69-87, PROF GENETICS, UNIV CALIF, LOS ANGELES, 87- *Concurrent Pos:* Vis fel, Res Sch Biol Sci, Australian Nat Univ, 75-76, genetics & develop, Cornell Univ, 80-81, Cambridge Univ & Churchill Col, 87- 88. *Mem:* AAAS; Genetics Soc Am. *Res:* Developmental genetics of Drosophila; gene regulation; chromosome mechanics; somatic crossing over and mosaic analysis of development. *Mailing Add:* Dept of Biol 2203 Life Sci Bldg Univ of Calif 405 Hilgard Ave Los Angeles CA 90024-1606

MERRIAM, LAWRENCE CAMPBELL, JR, b Portland, Ore, Aug 31, 23; m 47; c 5. FOREST MANAGEMENT, RECREATION RESOURCE MANAGEMENT. *Educ:* Univ Calif, BS, 48; Ore State Univ, MF, 58, PhD(forest mgt), 63. *Prof Exp:* Log scaler-compassman, Shasta Forests Co, Calif, 48; forestry aide, Ore Bur Land Mgt, 49; retail sales millworker, Willamette Nat Lumber Co, 49-50; log pond foreman bookkeeper, M&M Woodworking Co, 50; state parks historian, planner & forester state parks div, Ore State Hwy Dept, 51-59; from asst prof to assoc prof forestry, Univ Mont, 59-66; prof, 66-86, EMER PROF FORESTRY, UNIV MINN, ST PAUL, 86-; COURTESY PROF, ORE STATE UNIV, 86- *Concurrent Pos:* Consult, Bur Land Mgt, DC, 65-66 & UN Food & Agr Orgn, Paraguay, 69; consult, Victoria Forests Comn, Australia, 74-75; visiting prof, Sch Forestry, Ore State Univ, 81; mem, Nat Areas Comt, Soc Am Foresters, 84-86. *Mem:* Soc Am Foresters. *Res:* Park wilderness management and policy; state park history, policy. *Mailing Add:* 3930 NW Elizabeth Pl Corvallis OR 97330

MERRIAM, MARSHAL F(REDRIC), b Ossining, NY, Apr 1, 32; m 53; c 5. ENERGY CONVERSION, MATERIALS SCIENCE. *Educ:* Mass Inst Technol, SB, 53; Carnegie Inst Technol, MS, 58, PhD(physics), 61. *Prof Exp:* Staff scientist, Gen Atomic Div, Gen Dynamics Corp, 60-61; res assoc physics, Univ Calif, San Diego, 61-62; asst prof, 62-66, ASSOC PROF ENG SCI, UNIV CALIF, BERKELEY, 66- *Concurrent Pos:* Staff consult, Gulf Gen Atomic, 61-71, IBM Corp, NY, 63 & United Aircraft Res Labs, Conn, 65; vis prof, Indian Inst Technol, Kanpur, 67-69 & Sch Eng, Univ Sao Paulo, 71; mem res staff, Technol & Develop Inst, East-West Ctr, Hawaii, 71-72; mem vis research staff, Niels Bohr Inst, 77-78. *Mem:* Am Phys Soc; Solar Energy Energy Soc; Am Wind Energy Asn. *Res:* Solar energy devices; electric and magnetic properties of materials; wind energy. *Mailing Add:* Dept of Mat Sci & Eng Univ of Calif Hearst Mining Bldg Berkeley CA 94720

MERRIAM, ROBERT ARNOLD, b Keokuk, Iowa, Apr 30, 27; m 53; c 3. FOREST HYDROLOGY, BIOMASS ENERGY. *Educ:* Iowa State Univ, BS, 51; Univ Calif, Berkeley, MS, 57. *Prof Exp:* Range conservationist, Calif Forest & Range Exp Sta, 53-55, res forester, Pac Southwest Forest & Range Exp Sta, 55-60, Intermountain Forest & Range Exp Sta, 60-63 & Pac Southwest Forest & Range Exp Sta, 63-73; asst mgr Hawaii Water Resources Regional Study, State Hawaii, 73-76; resources mgr forester, Hawaii Div Forestry & Wildlife, 76-89; FORESTRY CONSULT, 89- *Concurrent Pos:* Crown Zellerbach fel. *Mem:* Soc Am Foresters; Am Geophys Union; Sigma Xi; Int Soc Tropical Foresters. *Res:* Watershed management; soil moisture measurement techniques, including neutron probe; interception and fog drip; river basin surveys; biomass fuels; short rotation plantation silviculture. *Mailing Add:* 616 Pamaele St Kailua HI 96734

MERRIAM, ROBERT WILLIAM, b Waverly, Iowa, Nov 21, 23; m 50; c 2. DEVELOPMENTAL BIOLOGY. *Educ:* Univ Iowa, AB, 47; Ore State Univ, MS, 49; Univ Wis, PhD(zool), 53. *Prof Exp:* Asst zool, Ore State Univ, 47-50 & Univ Wis, 50-53; from instr to asst prof, Univ Pa, 53-59; ASSOC PROF BIOL, STATE UNIV NY STONY BROOK, 61- *Concurrent Pos:* Vis fel zool, Columbia Univ, 60-61; NIH spec fel, Oxford Univ, 67-68; vis scientist, Hubrecht Laboratorium, Netherlands, 82. *Honors & Awards:* Lalor Found Awards, 55 & 57. *Mem:* AAAS; Soc Develop Biol; Am Soc Cell Biol. *Res:* Cellular biology; control mechanisms in oogenesis; contractile systems in egg cells. *Mailing Add:* Dept Neurobiol State Univ NY Stony Brook NY 11794

MERRICK, ARTHUR WEST, b Great Falls, Mont, Dec 22, 17; m 45; c 5. PHYSIOLOGY. *Educ:* Univ Mont, AB & BS, 50; Univ Mo, MA, 52, PhD(physiol), 54. *Prof Exp:* Asst physiol, Univ Mo, 51-52, asst instr, 53-54; instr, Univ Kans, 54-55; from asst prof to assoc prof, Med Ctr, Univ Mo-Columbia, 55-68; prof, Ill State Univ, 68-72; HEALTH SCIENTIST ADMINR, NAT HEART, LUNG & BLOOD INST, 72-, CHIEF PROG, PROJ REVIEW SECT. *Concurrent Pos:* Wyeth Drug Corp fel, 61-62; Nat Heart Inst grants; exec secy, Rev Br, Div Extramural Affairs, Nat Heart, Lung & Blood Inst, 73- *Mem:* Fel AAAS; Am Physiol Soc; NY Acad Sci. *Res:* Carbohydrate metabolism of cardiac and nervous tissue; intrinsic nervous system of mammalian heart. *Mailing Add:* 513 Woodridge Dr Columbia MO 65201

MERRICK, JOSEPH M, b Welland, Ont, Mar 20, 30; m 55; c 3. BIOCHEMISTRY, MICROBIOLOGY. *Educ:* Mich State Univ, BS, 51, MS, 53; Univ Mich, PhD(biochem), 58. *Prof Exp:* Arthritis & Rheumatism Found fel, Univ Calif, 59-61; assoc biochem, State Univ NY Buffalo, 61-62, asst prof, 62-65; assoc prof bact & bot, Syracuse Univ, 65-70; PROF MICROBIOL, STATE UNIV NY BUFFALO, 70- *Mem:* Am Soc Microbiol; Am Chem Soc; AAAS; Am Soc Biol Chem. *Res:* Microbiol pathogenesis; mechanism of invasion of epithelial cells by enteric pathogenic bacteria. *Mailing Add:* Dept Microbiol State Univ NY Buffalo NY 14214

MERRIELL, DAVID MCCRAY, b Minneapolis, Minn, Oct 25, 19; m 51; c 2. MATHEMATICS. *Educ:* Yale Univ, BA, 41; Univ Chicago, MS, 47, PhD(math), 51. *Prof Exp:* Instr math, Univ Chicago, 49-51; asst prof, Robert Col, Turkey, 51-54, assoc prof & head dept, 54-57; from asst prof to assoc prof, Univ Calif, Santa Barbara, 57-68; chmn dept math, 62-63, 71-74 & 77-82, prof, 68-85, EMER PROF MATH, VASSAR COL, 85- *Mem:* Am Math Soc; Math Asn Am. *Mailing Add:* Rte 199 Box 17 Penobscot NY 04476

MERRIFIELD, D BRUCE, b Chicago, Ill, June 13, 21; m 49; c 3. PHYSICAL ORGANIC CHEMISTRY. *Educ:* Princeton Univ, BS, 42; Univ Chicago, MS, 48, PhD(phys org chem), 50. *Prof Exp:* Res chemist, Monsanto Co, 50-56; group leader res, Tex-US Chem Co, 56-60, mgr polymer res, 60-63; dir res, Petrolite Corp, 63-68; dir res, Res Ctr, Hooker Chem Corp, 68-70, vpres res & develop, 70-77; vpres technol, Continental Group, Inc, 77-82; asst secy, US Dept Com Technol, 82-89; chief exec officer, Greater Minn Corp, 89-90; PROF MGT, WHARTON BUS SCH, UNIV PA, 89-; CHIEF EXEC OFFICER, PINNACE RES INST DEVELOP CO, 90- *Concurrent Pos:* Consult, Am Electronics Asn. *Mem:* Fel AAAS; fel Am Inst Chemists; Res Dirs Asn; Am Chem Soc. *Res:* Mechanisms of free radical reactions; oxidation mechanisms; polymer and surface chemistry; electrochemistry and electronics. *Mailing Add:* 1316 New Ha Apt 701 Washington DC 20036

MERRIFIELD, PAUL ELLIOTT, b Springvale, Maine, Dec 31, 22; m 44; c 4. COLLOID CHEMISTRY. *Educ:* Colby Col, AB, 47; Rice Inst, AM, 49, PhD(chem), 51. *Prof Exp:* Res chemist, Armstrong Cork Co, 51-59, plant chief chemist, 60-67, mgr felt mfg, 67-81; RETIRED. *Mem:* Am Chem Soc. *Res:* Colloidal properties of fibers; fiber products development. *Mailing Add:* PO Box 184 Springvale ME 04083

MERRIFIELD, RICHARD EBERT, b Seattle, Wash, Feb 18, 29; m 56; c 2. CHEMICAL PHYSICS. *Educ:* Mass Inst Technol, PhD(phys chem), 53. *Prof Exp:* Res chemist, 53-59, RES SUPVR CENT RES DEPT, E I DU PONT DE NEMOURS & CO, INC, 59- *Mem:* Am Phys Soc. *Res:* Molecular spectra and structure; solid state theory; exciton physics; physics of molecular crystals. *Mailing Add:* 2633 Longwood Dr Wilmington DE 19810

MERRIFIELD, ROBERT BRUCE, b Ft Worth, Tex, July 15, 21; m 49; c 6. BIOCHEMISTRY, PEPTIDE CHEMISTRY. *Educ:* Univ Calif, Los Angeles, BA, 43, PhD(chem), 49. *Hon Degrees:* DSc, Univ Colo, 69 & Yale Univ, 71; PhD, Uppsala Univ, 70, Newark Col Eng & Med Col Ohio, 72, Colgate Univ, 77, Boston Col, 84, Univ Barcelona, Spain, 86, Adelphi Univ, 87, Montpellier Univ, France, 88. *Prof Exp:* Chemist, Philip R Park Res Found, 43-44; asst chem, Med Sch, Univ Calif, Los Angeles, 48-49; asst biochem, 49-53, assoc, 53-57, from asst prof to prof, 57-84, JOHN D ROCKEFELLER JR PROF, ROCKEFELLER UNIV, 84- *Concurrent Pos:* Nobel guest prof, Uppsala Univ, 68; assoc ed, Int J Peptide & Protein Res, 69- *Honors & Awards:* Nobel Prize in Chem, 84; Lasker Award Basic Med Res, 69; Gairdner Award, 70; Intra-Sci Award, 70; Award for Creative Work in Synthetic Org Chem, Am Chem Soc, 72, Nichols Medal, 73; Alan E Pierce Award, 79. *Mem:* Nat Acad Sci. *Res:* Development of solid phase peptide synthesis, first synthesis of an enzyme; relation of structure to function in synthetic, biologically active peptides and proteins. *Mailing Add:* Rockefeller Univ 1230 York Ave New York NY 10021-6399

MERRIFIELD, ROBERT G, b Carthage, Mo, July 26, 30; m 52; c 4. SILVICULTURE. *Educ:* Ark Agr & Mech Col, BS, 53; La State Univ, MF, 58; Duke Univ, DF(silvicult), 62. *Prof Exp:* From asst prof to assoc prof forestry, La State Univ, 58-67; PROF FORESTRY, TEX A&M UNIV, 67-, HEAD DEPT FOREST SCI, 69-, ASSOC DIR, TEX AGR EXP STA. *Mem:* Soc Am Foresters; Sigma Xi. *Res:* Artificial regeneration and plantation management of southern pines; intensive culture of pulping hardwood species. *Mailing Add:* Tex Agr Exp Sta Tex A&M Univ College Station TX 77843

MERRIGAN, JOSEPH A, b Maryville, Mo, Feb 8, 40; m 62; c 2. PHOTOGRAPHIC CHEMISTRY. *Educ:* Northwest Mo State Col, BS, 62; Univ Nebr, MS, 65, PhD(phys chem), 66; Mass Inst Technol, SM, 74. *Prof Exp:* Sr res chemist, Eastman Kodak Co, 67-69, res assoc & head silver halide chem lab, 69-74, head radiography lab, Kodak Res Labs, 74-75, head spec processes lab, 75-79, DIR MAT COATING & ENG DIV, KODAK RES LABS, 79- *Res:* Hot atom reactions of neutron irradiated bromine with organic molecules; positronium interactions in solid systems; mechanisms of photographic recording; radiographic recording; photovoltaic cells. *Mailing Add:* Eastman Kodak Res Labs 1669 Lake Ave Rochester NY 14650

MERRIL, CARL R, b Brooklyn, NY, Dec 6, 36; m 61; c 2. MOLECULAR BIOLOGY, MEDICINE. *Educ:* Col William & Mary, BS, 58; Georgetown Univ, MD, 62. *Prof Exp:* Intern med, USPHS Hosp, Boston, Mass, 62-63; res assoc molecular biol, Lab Neurochem Sect, Phys Chem, Nat Inst Mental Health, 63-65, mem staff, 65-69, sr staff scientist, Lab Gen & Comp Biochem, 69-84, chief, sect biochem genetics, CNG, 84-88, chief, Lab Preclin Pharmacol, 88-89, CHIEF, LAB BIOCHEM GENETICS, DIRP, NAT INST MENTAL HEALTH, 89- *Concurrent Pos:* Adj prof genetics, Grad Sch & adj prof biochem, Sch Med, George Wash Univ; ed, Appl & Theoret Electrophoresis. *Honors & Awards:* Outstanding Serv Medal, Pub Health Serv; Surgeon General's Exemplary Serv Medal. *Mem:* AAAS; Biophys Soc; NY Acad Sci; Electrophoresis Soc (pres, 87-); Soc Neurosci; Protein Soc. *Res:* Detection and characterization of cerebrospinal fluid proteins; bacteriophage interactions with eukaryotic systems, quantitative two dimensional electrophoresis of proteins, silver stains for proteins and DNA; galactose metabolism in prokaryotes; galactosemia; inborn metabolic diseases; gene transfer; primary structure of biopolymers; maternally inherited diseases and the mitochondrial genome; genomic linkage studies. *Mailing Add:* Chief Lab Biochem Genetics DIRP Nat Inst Mental Health Bldg 10 Rm 4N224 Bethesda MD 20892

MERRILL, DOROTHY, b Abington, Mass, Jan 1, 27. PHYSIOLOGY. *Educ:* Mass State Col Bridgewater, BS, 47; Univ Mich, AM, 59, PhD(zool), 64. *Prof Exp:* Teacher high sch, Mass, 47-60; from instr to asst prof zool, Smith Col, 64-70; assoc prof biol, Western Col, 70-74; assoc prof, 74-79, prof biol, 79-82, PROF HEALTH SCI, GRAND VALLEY STATE UNIV, 82- *Mem:* Am Soc Zool; Animal Behav Soc; Sigma Xi. *Res:* Neural mechanisms in insect behavior; biology of caddis larvae. *Mailing Add:* Sch Health Sci Grand Valley Univ Allendale MI 49401-9403

MERRILL, E(DWARD) W(ILSON), b New Bedford, Mass, Aug 31, 23; m 48. ENGINEERING. *Educ:* Harvard Univ, AB, 44; Mass Inst Technol, DSc, 47. *Prof Exp:* Res engr, Dewey & Almy Chem Co, 47-50; from asst prof to assoc prof chem eng, 50-64, PROF CHEM ENG, MASS INST TECHNOL, 64- *Concurrent Pos:* Consult, Mass Gen Hosp, Boston, 64- & Dept Surg, Beth Israel Hosp, Boston, 69-; consult to dir, Nat Inst Arthritis & Metab Diseases, 67-; vis res assoc surg, Children's Hosp Med Ctr, Boston, 70- *Mem:* Am Chem Soc; Soc Rheol; fel Am Acad Arts & Sci; Am Inst Chem Engrs. *Res:* Polymer chemistry; rheology; biomedical engineering. *Mailing Add:* Dept Chem Eng Rm 66-568 Mass Inst Technol Rm 66-568 177 Mass Ave Cambridge MA 02139-4301

MERRILL, GARY FRANK, Dec 28, 47; m 66; c 9. CORONARY CIRCULATION, CORONARY PHARMACOLOGY. *Educ:* Weber State Col, BS, 71; Mich State Univ, PhD(physiol), 75. *Prof Exp:* Fel, La State Univ Med Ctr, 75-76; asst prof physiol, 76-82, ASSOC PROF PHYSIOL, RUTGERS UNIV, 82- *Concurrent Pos:* Postdoc fel, La State Univ Med Ctr, 75-76; vis prof physiol, Southwestern Med Sch, Tex Col Osteopath Med, 84-85. *Mem:* Am Physiol Soc; Am Heart Asn; Am Soc Pharmacol Exp Ther; fel Am Col Nutrit. *Res:* Physiological and pharmacological regulation of the coronary vascular bed; interventions such as hypoxia and ischemia are employed; recent attention focused on adenosine deaminase as a metabolic probe. *Mailing Add:* Dept Biol Sci Animal Sci Rutgers Univ New Brunswick NJ 08903

MERRILL, GARY LANE SMITH, b Dexter, Iowa, Dec 3, 39; m 61; c 2. BOTANY. *Educ:* Univ Iowa, BA, 62, MS, 64; Columbia Univ, PhD(bot), 69. *Prof Exp:* Adj asst prof biol sci, Herbert H Lehman Col, 71-76; assoc cur, NY Bot Garden, 69-78; assoc prof bot, Drew Univ, 78-86, chmn, Bot Dept, 80-86; dir bot, Nat Botanic Garden, Santo Domingo, 79-80; ASSOC COORDR, KONZA PRAIRIE RES NATURAL AREA, KANS STATE UNIV, 87- *Mem:* Am Bryol & Lichenol Soc; Int Asn Bryologists. *Res:* Taxonomy and geography of bryophytes; Polytrichaceae; Sphagnaceae. *Mailing Add:* Div Biol Kans State Univ Manhattan KS 66506-4901

MERRILL, GLEN KENTON, b Columbus, Ohio, Aug 28, 35; m 64. GEOLOGY, PALEONTOLOGY. *Educ:* Ohio Univ, BS, 57; Univ Tex, Austin, MA, 64; La State Univ, PhD(geol), 68. *Prof Exp:* Instr geol, Northwestern La State Col, 64; asst prof, Monmouth Col, 68-71; asst prof, Univ Tex, Arlington, 71-74; from asst prof to assoc geol, Col Charleston, 74-82; assoc prof, 82-87, PROF GEOL, UNIV HOUSTON, 87- *Concurrent Pos:* Consult, 77- *Mem:* Geol Soc Am; Paleont Soc; Soc Econ Paleontologists & Mineralogists; Nat Speleol Soc; Paleont Res Inst. *Res:* Biostratigraphy, paleoecology and systematics of late Paleozoic conodonts; carbonate petrography. *Mailing Add:* Dept Nat Sci Univ Houston One Main St Houston TX 77002

MERRILL, HOWARD EMERSON, b Laconia, NH, Aug 6, 30; m 53; c 4. PETROLEUM CHEMISTRY. *Educ:* Stetson Univ, BS, 52; Univ Pittsburgh, PhD(chem), 57. *Prof Exp:* Sr res chemist, Exxon Res Labs, Exxon Co, USA, 57-75, sr staff chemist, Exxon Chem Americas, 75-86; RETIRED. *Res:* Catalysis. *Mailing Add:* 5045 Sequoia Dr Baton Rouge LA 70814

MERRILL, JAMES ALLEN, b Cedar City, Utah, Oct 27, 25; m 49; c 4. OBSTETRICS & GYNECOLOGY, PATHOLOGY. *Educ:* Univ Calif, Berkeley, AB, 45, Univ Calif, San Francisco, MD, 48. *Prof Exp:* Fel path, Harvard Med Sch, 50-51; fel, Cancer Res Inst, Univ Calif, San Francisco, 58-61; from instr to asst prof obstet & gynec, Sch Med, Univ Calif, San Francisco, 57-61, asst clin prof path, 59-61, res asst, Cancer Res Inst, 58-61; PROF GYNEC & OBSTET & HEAD DEPT, SCH MED, UNIV OKLA, 61-, CONSULT PROF PATH, 61-, PROF CYTOTECHNOL, COL HEALTH REL PROFESSIONS, 70- *Concurrent Pos:* Markle scholar med sci, 57-62; consult, Vet Admin Hosp, Oklahoma City, 61-, US Army Hosp, Ft Sill, Okla & Tinker AFB Hosp, Midwest City, 61-; nat consult, Air Force Hosp, Lackland AFB, San Antonio, Tex, 63. *Honors & Awards:* Aesculapian Award, Univ Okla Student Body, 63 & 69; Regents Award Superior Teaching, Univ Okla Bd Regents, 69. *Mem:* Am Asn Obstetricians & Gynecologists; Soc Gynec Invest; Asn Profs Gynec & Obstet (pres, 66-67); Int Soc Advan Humanistic Studies Gynec (pres, 72-73); Am Gynec Soc (treas, 70-75). *Res:* Gynecologic oncology. *Mailing Add:* 4225 Roosevelt Way NE Suite 305 Seattle WA 98105

MERRILL, JERALD CARL, b Las Vegas, Nev, Aug 12, 40; m 63; c 2. PHYSICAL CHEMISTRY. *Educ:* Univ Nev, Reno, BS, 62, PhD(phys chem), 71. *Prof Exp:* US AEC fel, Univ Calif, Davis, 71-72; presidential res intern, Brookhaven Nat Lab, 72-73; lectr chem, Univ Utah, 73-77; asst prof chem, Cent Mo State Univ, 77-78; develop chemist, Amersham Corp, 78-83; GELMAN SCI, INC, ANN ARBOR, 83- *Concurrent Pos:* Lectr, Univ Calif, Davis, 72. *Mem:* Am Chem Soc; Sigma Xi; AAAS. *Res:* Radiochemistry; microemulsions; liquid scintillation counting. *Mailing Add:* Gelman Sci Inc 600 S Wagner Rd Ann Arbor MI 48106

MERRILL, JOHN ELLSWORTH, b Parsonsfield, Maine, May 10, 02; wid; c 2. ASTRONOMY. *Educ:* Univ Boston, AB, 23; Case Inst Technol, MS, 27; Princeton Univ, AM, 29, PhD(astron), 31. *Prof Exp:* Instr math, Case Inst Technol, 24-28, Cleveland Col, 25-28 & Princeton Univ, 29-30; asst prof astron, Univ Ill, 31-32; cur, Buffalo Mus Sci, 32-36; asst, Princeton Univ, 36-37; instr, Hunter Col, 37-38, from asst prof to assoc prof, 38-50; from assoc prof to prof, Ohio Wesleyan Univ & Ohio State Univ, 50-59; sr staff engr, Franklin Inst, 59-61, prin scientist astron, 61-63; prof astron & math & dir, Morrison Observ, 64-67, Dearing prof astron, 67-69, EMER PROF ASTRON, CENT METHODIST COL, 69- *Concurrent Pos:* Am Philos Soc grants, 41-42; dir pilot training prog, Princeton Univ, 42-43, vis asst prof, 43-45, res assoc, 47-63; adj prof, Univ Pa, 59-63; pres comn 42, Int Astron Union, 61-67; vis prof astron, Univ Fla, 69-80, adj prof, 80- *Mem:* AAAS; Am Astron Soc; Int Astronomical Union. *Res:* Photometry of eclipsing variables; solutions for orbits of eclipsing binaries; effects of eccentricity of orbit; tables for facilitating determinations of eccentricity. *Mailing Add:* Astronomy Dept Univ Fla Gainesville FL 32611

MERRILL, JOHN JAY, b Nampa, Idaho, Jan 24, 33; m 53; c 6. PHYSICS. *Educ:* Calif Inst Technol, BS, 55, MS, 56, PhD(physics), 60. *Prof Exp:* Instr physics, Harvey Mudd Col, 59-60; med physicist, Dee Mem Hosp, Ogden, Utah, 60-62; assoc prof physics, Utah State Univ, 62-69; pres, Tronac, Inc, 69-71; PROF PHYSICS, BRIGHAM YOUNG UNIV, 70- *Mem:* Am Phys Soc; Am Asn Physics Teachers. *Res:* Instructional design. *Mailing Add:* Dept Physics-Astron Esc 170 Brigham Young Univ Provo UT 84602

MERRILL, JOHN PUTNAM, medicine; deceased, see previous edition for last biography

MERRILL, JOHN RAYMOND, b Englewood, NJ, Aug 25, 39; m 60; c 4. LOW TEMPERATURE PHYSICS, EDUCATIONAL DESIGN. *Educ:* Swarthmore Col, AB, 61; Cornell Univ, PhD(solid state physics), 66. *Prof Exp:* Instr & res assoc physics, Cornell Univ, 66-67; asst prof, Dartmouth Col, 67-73; assoc prof physics & dir ctr educ design, Fla State Univ, 73-76; instr, NATO Postgrad Inst, Belgium, 76; vpres, dean & prof physics, Hendrix Col, 76-83; vpres, CCX Network, Inc, 83-88; VPRES, MSP, INC, 88- *Concurrent Pos:* Res Corp res grant, 68; AEC res grant, 70-73. *Mem:* AAAS; Am Phys Soc; Am Asn Physics Teachers; Sigma Xi; NY Acad Sci. *Res:* Computers in teaching; low temperature solid state physics; instructional design and development; science education. *Mailing Add:* 108 Saybrook Harbour Bradfordwoods PA 15015

MERRILL, JOHN T, b Oakland, Calif, May 29, 46; m 76; c 2. ATMOSPHERIC CHEMISTRY & PHYSICS, FLUID DYNAMICS. *Educ:* Univ Calif, Berkeley, AB, 68; Univ Ill, MS, 70; Univ Colo, PhD(atmospheric sci), 76. *Prof Exp:* Res assoc air sea interaction, Rosenstiel Sch Marine & Atmospheric Sci, Univ Miami, 76-77, asst prof meteorol & phys oceanog, 77-82; assoc marine scientist, 81-87, ASSOC RES PROF, GRAD SCH OCEANOG, UNIV RI, 87- *Mem:* Am Meteorol Soc; Am Geophys Union. *Res:* Atmospheric transport; geophysical boundary layers. *Mailing Add:* Ctr Atmospheric Chem Studies Grad Sch Oceanog Univ RI Narragansett RI 02882

MERRILL, JOSEPH MELTON, b Andalusia, Ala, Dec 8, 23. MEDICAL CARE. *Educ:* Harvard Med Sch, MD, 48; Am Bd Internal Med, dipl, 56. *Prof Exp:* Intern, Louisville Gen Hosp, Mo, 48-49; intern, Vanderbilt Univ Hosp, 49-50, asst resident med, 50-51; instr med & attend physician, Med Col, Univ Ala, 53-54; asst resident med res, Vet Admin Hosp, Nashville, Tenn, 54-55; res assoc, Postgrad Med Sch, Univ London, 55-56; chief clin physiol, Vet Admin Hosp, 56-64; chief Gen Clin Res Ctrs Br, Div Res Facil, NIH, 64-67; PROF MED, BAYLOR COL MED, 67-, COMM MED, 77- *Concurrent Pos:* Clin investr, Vet Admin Hosp, Nashville, Tenn, 56-59, asst chief radioisotope serv, 57-64, asst dir, Prof Servs for Res, 60-64; instr, Med Sch, Vanderbilt Univ, 60-; Wellcome assoc, Royal Soc Med, 55-56, Univ London, 76-77; vis prof, Harvard Univ, 84-85. *Mem:* Am Heart Asn; Am Fedn Clin Res. *Res:* Application of human factors (ergonomics) to medicine; social psychology of medical practice. *Mailing Add:* 1 Baylor Plaza Houston TX 77030

MERRILL, LELAND (GILBERT), JR, b Danville, Ill, Oct 4, 20; m 49; c 2. ENVIRONMENTAL SCIENCE. *Educ:* Mich State Col, BS, 42; Rutgers Univ, MS, 48, PhD, 49. *Prof Exp:* Asst, Rutgers Univ, 46-49; asst prof entom, Mich State Col, 49-53; exten specialist, Rutgers Univ, New Brunswick, 53-59, res specialist, 59-61, dean Col Agr & Environ Sci, 61-71, dir Inst Environ Studies, 71-76, prof natural resource policy studies, Ctr Coastal & Environ Studies, 76-82; EXEC SECY, NJ ACAD SCI, 85- *Mem:* AAAS; Entom Soc Am; Am Assoc Adv Sci; Am Inst Biol Sci. *Res:* Applied environmental studies; natural resource inventory applications to land use management. *Mailing Add:* NJ Acad Sci Rutgers Univ Box B Beck Hall Piscataway NJ 08854

MERRILL, ROBERT CLIFFORD, JR, b Brooklyn, NY, Apr 23, 58; m 83; c 1. FLUID PHASE THERMODYNAMICS, RESERVOIR ENGINEERING. *Educ:* Univ Va, BS, 80; Univ Notre Dame, MS, 81, PhD(chem eng), 83. *Prof Exp:* Sr res eng, Sohio Petrol Co, 83-86, RESERVOIR ENG IV, BRITISH PETROL CO, 86- *Mem:* Am Inst Chem Engrs; Soc Petrol Engrs-Am Inst Mining, Metall & Petrol Engrs; Am Chem Soc. *Res:* Fluid phase thermodynamics; multiphase phenomena in cryogenic fluids; miscible flooding and enhanced oil recovery. *Mailing Add:* PO Box 201644 Anchorage AK 99520-1644

MERRILL, ROBERT KIMBALL, b Lima, Peru, Oct 11, 45; US citizen; m 72; c 3. EXPLORATION GEOLOGY. *Educ:* Colby Col, AB; Ariz State Univ, MS, PhD. *Prof Exp:* Stratigr, Am Stratig Co, 69-70; staff geologist, Cities Serv Co, 74-78, tech asst vpres, Western Area, 78-81, explor geologist, Western Region, 81-86; explor geologist, Midland Dist, OXY USA, Inc, 86-89; GEOL ADV, ROCKY MOUNTAIN DIST, UNOCAL, 89- *Concurrent Pos:* Regist chmn, annual mgt, Am Inst Prof Geologists; pres, Okla Sect, Am Inst Prof Geologists, 91. *Mem:* Sigma Xi; Geol Soc Am; Am Asn Petrol Geologists; Am Inst Prof Geologists. *Res:* Analysis of the tectonic development within the structural and stratigraphic framework of established hydrocarbon provinces and frontier areas to find oil and gas. *Mailing Add:* 2305 Berryhill Cir Edmond OK 73034

MERRILL, ROBERT P, b Salt Lake City, Utah, Nov 17, 34; m 58; c 6. CHEMICAL ENGINEERING. *Educ:* Cornell Univ, BChE, 60; Mass Inst Technol, ScD(chem eng), 64. *Prof Exp:* Consult, Raytheon Mfg Co, 61-63 & Abcor Inc, 63-64; instr chem eng, Mass Inst Technol, 64; from asst prof to assoc prof, Univ Calif, Berkeley, 64-73; prof, 73-76, HF JOHNSON PROF INDUST CHEM, CORNELL UNIV, 76- *Concurrent Pos:* Consult, Lockheed Missiles & Space Co, 65-72, Stauffer Chem Soc, 65-67, Universal Oil Prods, 66-, Gulf Gen Atomic, 70-73 & Du Pont, 80- *Mem:* AAAS; Am Vacuum Soc; Catalysis Soc; Am Chem Soc; Am Inst Chem Engrs. *Res:* Chemical kinetics; heterogeneous catalysis; physics and chemistry of solid surfaces; atomic and molecular beam scattering from solid surfaces; photoelectron spectroscopy; laser induced surface reactions. *Mailing Add:* Dept Chem Eng Cornell Univ Ithaca NY 14853

MERRILL, RONALD EUGENE, b Salem, Ore, Aug 7, 47; m 74; c 2. SYNTHETIC ORGANIC CHEMISTRY. *Educ:* Mass Inst Technol, BS, 68; Univ Ore, PhD(chem), 73. *Prof Exp:* Fel chem, Syracuse Univ, 73-74; vis asst prof chem, Rochester Inst Technol, 74-77; sr chemist, Lifesysts Co, 77-79; pres, Reaction Design Corp, Hillside, NJ, 79-82; dir res, Balenco Enterprises, Inc, Compton, CA, 83-88; DIR, RES & DEVELOP, CYCLO PROD INC, LOS ANGELES, CA, 90- *Concurrent Pos:* Exec comt enterprise forum, Mass Inst Technol. *Mem:* Am Chem Soc; Am Soc Quality Control. *Res:* Design, development, and applications of chiral catalysts; new synthetic reactions; organometallic chemistry and homogeneous catalysis. *Mailing Add:* 21118 Doble Ave Torrance CA 90502

MERRILL, RONALD THOMAS, b Detroit, Mich, Feb 5, 38; m 61; c 2. GEOPHYSICS. *Educ:* Univ Mich, BS, 59, MS, 61; Univ Calif, Berkeley, PhD(geophys), 67. *Prof Exp:* From asst prof to assoc prof, 67-77, PROF GEOPHYS, UNIV WASH, 77- *Concurrent Pos:* Hon vis fel, Res Sch Earth Sci, Australia Nat Univ, 75, 77 & 79. *Mem:* AAAS; fel Am Geophys Union; Soc Terrestrial Magnetism & Elec Japan. *Res:* Geomagnetism, especially paleomagnetism and rock magnetism. *Mailing Add:* Geophys Prog Univ Wash Seattle WA 98195

MERRILL, SAMUEL, III, b New Orleans, La, Oct 27, 39; m 69; c 2. MATHEMATICAL POLITICAL SCIENCE. *Educ:* Tulane Univ, BA, 61; Yale Univ, MA, 63, PhD(math), 65; Pa State Univ, MS, 80. *Prof Exp:* Instr math, Univ Rochester, 65-67, asst prof, 67-73; assoc prof, 73-81, PROF MATH/COMPUT SCI, WILKES UNIV, 81- *Concurrent Pos:* Vis prof, Biostatist, Yale Univ, 86-87. *Mem:* Am Math Soc; Am Stat Assoc; Public Choice Soc; Math Asn Am. *Res:* Mathematical applications to political science, especially voting power and voting systems, statistical applications, especially biostatistics; functional analysis, especially Banach spaces of analytic functions. *Mailing Add:* Dept Math/Comput Sci Wilkes Univ Wilkes-Barre PA 18766

MERRILL, STEPHEN DAY, b Canandaigua, NY, Mar 27, 39; m 61; c 3. SOIL SCIENCE, SOIL PHYSICS. *Educ:* Dartmouth Col, BA, 61, MA, 63; Univ Calif, Riverside, PhD(soil sci), 76. *Prof Exp:* Sci teacher chem & biol, Drew Sch, San Francisco, Calif, 64-65; chemist soil-plant water rels, USDA, Agr Res Serv, US Salinity Lab, 66-70, physicist, 70-77; RES SOIL SCIENTIST SOIL PHYSICS, NORTHERN GREAT PLAINS RES LAB, AGR RES SERV, USDA, 77- *Concurrent Pos:* Adj prof soils, NDak State Univ, 79-; assoc ed soils, Agron J, 88-90. *Mem:* Am Soc Agron; Soil Sci Soc Am; Soil & Water Conservation Soc. *Res:* Crop root growth and function; dryland soil management and soil physical conditions; soil physics related to soil conservation. *Mailing Add:* Northern Great Plains Res Lab PO Box 459 Mandan ND 58554

MERRILL, WALTER HILSON, b Montgomery, Ala, Oct 27, 47; m 72; c 4. CARDIOTHORACIC SURGERY. *Educ:* Univ S, Tenn, BA, 70; John Hopkins Univ Sch Med, MD, 74. *Prof Exp:* Asst prof, 83-87, ASSOC PROF SURG, VANDERBILT UNIV SCH MED, 87- *Concurrent Pos:* Attending surgeon, Cardiothoracic Surg Serv, Nashville Vet Admin Med Ctr, 83- *Mem:* Soc Univ Surgeons; Am Asn Thoracic Surg; Soc Thoracic Surg; fel Am Col Surgeons; fel Am Col Cardiol. *Res:* Heart and heart-lung transplantation, adult and congenital heart surgery, myocardial protection, reperfusion injury and its amelioration with oxygen free radical scavengers. *Mailing Add:* Dept Thoracic Surg Vanderbilt Clin Rm 2973 Nashville TN 37232

MERRILL, WARNER JAY, JR, b Springfield, Ill, Jan 27, 23; m 45; c 2. STATISTICS. *Educ:* Univ Del, BA, 47; Ohio State Univ, PhD(statist), 56. *Prof Exp:* Asst psychol statist, Ohio State Univ, 49-51; staff statistician, Am Power Jet Co, 51-53; assoc math, IBM, Endicott, NY, 53-56; statistician, Gen Elec Co, NY, 56-61; prof mgt, Rensselaer Polytech Inst, 61-64; mgt scientist, Dunlap & Assocs, 64-65; mem tech staff, Hughes Aircraft Co, 65-66; sr assoc, Planning Res Corp, 66-68; opers res analyst, Nat Inst Justice, Dept Justice, 68-88; RETIRED. *Concurrent Pos:* Prof lectr, George Washington Univ, 68-70; prof lectr, Am Univ, 69-70, adj prof, 70-72. *Mem:* Am Statist Asn. *Res:* Applications of statistics to crime research; systems analysis, including human factors; statistical applications of computers; non-parametric statistics; operations research. *Mailing Add:* 9904 Inglemere Dr Bethesda MD 20817-1547

MERRILL, WILLIAM, b Haverhill, NH, Sept 5, 33; m 61; c 2. FOREST PATHOLOGY, FOREST PRODUCTS. *Educ:* Univ NH, BS, 58; Univ Minn, MS, 61, PhD(plant path), 63. *Prof Exp:* Instr plant path, Univ Minn, 61-64; res staff pathologist & fel, Yale Univ, 64-65; from asst prof to assoc prof, 65-75, PROF PLANT PATH, PA STATE UNIV, UNIVERSITY PARK, 75- *Mem:* Am Phytopath Soc; Sigma Xi. *Res:* Etiology, epidemiology and control of forest tree pathogens; biodeterioration of wood. *Mailing Add:* 210 Buckhout Lab Pa State Univ University Park PA 16802

MERRILL, WILLIAM GEORGE, b Wilmington, Del, Oct 19, 31; m 58; c 3. ANIMAL NUTRITION. *Educ:* Univ Md, BS, 53; Univ Wis, MS, 54; Cornell Univ, PhD(animal nutrit), 59. *Prof Exp:* Asst prof, Exten Div, 59-64, assoc prof animal husb, Dairy Cattle Div, 64-79, PROF ANIMAL SCI & DAIRY MGT, CORNELL UNIV, 79- *Mem:* Am Dairy Sci Asn; Nat Mastitis Coun. *Res:* Dairy husbandry; dairy cattle management systems; milking systems; feeding systems; manure handling. *Mailing Add:* Dept Animal Sci 272 Morrison Hall Ithaca NY 14853

MERRILL, WILLIAM MEREDITH, b Detroit, Mich, Dec 1, 18; m 43; c 3. GEOLOGY. *Educ:* Mich State Univ, BS, 46; Ohio State Univ, MA, 48, PhD(geol), 50. *Prof Exp:* Geologist, Ohio Div Geol Surv, 46-50; from instr to assoc prof geol, Univ Ill, 50-58; prof & chmn dept, Syracuse Univ, 58-63; chmn dept, 63-72, PROF GEOL, UNIV KANS, 63- *Concurrent Pos:* Asst, Ohio State Univ, 47-48, res assoc, Res Found, 55-57; consult, Ohio Div Geol Surv, 50-60; geologist, Nfld Dept Mines, 54 & Res Coun Alta, 58-62; vis scientist, Am Geol Inst, 62 & 64; team capt, Geo-Study, 62-63; chmn panel earth sci teacher prep, Coun Educ Geol Sci, 64-67, mem steering comt, Earth

Sci Curriculum Proj, 65-, mem writers conf, 66. *Mem:* Fel Geol Soc Am; Soc Econ Paleont & Mineral; Am Asn Petrol Geol; Int Asn Sedimentol; Int Asn Math Geol. *Res:* Mesozoic stratigraphy of western United States and Canada; computer simulation in geology. *Mailing Add:* Dept Geol Univ Kans 325 Lin Lawrence KS 66044

MERRIN, SEYMOUR, b Brooklyn, NY, Aug 13, 31; m 63; c 2. PHYSICAL CHEMISTRY, GEOCHEMISTRY. *Educ:* Tufts Univ, BS, 52; Univ Ariz, MS, 54; Pa State Univ, PhD(geochem), 62. *Prof Exp:* Geologist, US Geol Surv, 56-58; asst geochem, Pa State Univ, 59-62; sr assoc chemist, IBM Corp, 62-64; package develop dept mgr, Sperry Semiconductor Div, Sperry Rand Corp, 65-68; independent consult prod & prod develop electronics, 68-69; vpres technol, Innotech Corp, 69-74; div mgr, Exxon Enterprises, 74-78; pres, Computerworks Inc, 79-85; pres, Merrin Resources Inc, 85-88; PRES, MERRIN INFO SERV INC, 88- *Mem:* AAAS; fel Am Inst Chemists; fel Geol Soc Am. *Res:* Glass; experimental petrology; microelectronic processing; phase equilibrium. *Mailing Add:* 143 Buckthorn Way Menlo Park CA 94025

MERRINER, JOHN VENNOR, b Winchester, Va, Sept 13, 41. FISH BIOLOGY, MARINE BIOLOGY. *Educ:* Rutgers Univ, AB, 64; NC State Univ, MS, 67, PhD(zool), 73. *Prof Exp:* Assoc marine scientist, Va Inst Marine Sci, 70-74, actg head dept, 74-75, sr marine scientist & head dept ichthyol, 75-82; CHIEF, DIV FISHERIES, NAT MARINE FISHERIES SERV, BEAUFORT LAB, 82- *Concurrent Pos:* Asst prof marine sci, Col William & Mary & Univ Va, 75-78, assoc prof, 78-82, assoc fac, 82- *Mem:* Am Fisheries Soc; Am Soc Ichthyologists & Herpetologists. *Res:* Ecology and life history of estuarine and marine fishes; fishery management. *Mailing Add:* NMFS SEFC Beaufort NC 28516

MERRIS, RUSSELL LLOYD, b Calif, 1943. ALGEBRA. *Educ:* Harvey Mudd Col, BS, 64; Univ Calif, Santa Barbara, MA, 67, PhD(math), 69. *Prof Exp:* Nat Acad Sci-Nat Res Coun assoc, Nat Bur Stand, 69-71; from asst prof to assoc prof, 71-78, PROF MATH, CALIF STATE UNIV, HAYWARD, 78- *Concurrent Pos:* Fulbright lectr, Pakistan, 73; Exchange Scientist, Nat Acad Sci, Czech, 79; mem Bd Educ, Hayward Unified Sch Dist, 87- *Mem:* Math Asn Am; Soc Indust Appl Math. *Res:* Multilinear algebra; group representation theory; combinatorics. *Mailing Add:* Dept Math Comput Sci Calif State Univ Hayward CA 94542

MERRITT, ALFRED M, II, b Boston, Mass, Apr 10, 37; m 63; c 2. GASTROENTEROLOGY, VETERINARY MEDICINE. *Educ:* Bowdoin Col, AB, 59; Cornell Univ, DVM, 63, Univ Pa, MS, 69. *Prof Exp:* Asst instr vet med, Sch Vet Med, Univ Calif, 63-64; instr vet med, Sch Vet Med, Univ Pa, 64-66, fel, Grad Sch Arts & Sci, 66-69, from asst prof to assoc prof, 72-78; PROF MED, COL VET MED, UNIV FLA, 78- *Concurrent Pos:* Ed comp gastroenterol, Am J Digestive Dis, 72-77; mem res adv bd, Morris Animal Found, 79-82; assoc ed, Vet Gastroenterol; vis scientist, Dept Physiol, Nat Vet Sch, Toulouse, France, 84-85; co-ed, Eq Med & Surg, 4th ed; USDA spec grants reviewer, 86-87. *Mem:* Am Vet Med Asn; Comp Gastroenterol Soc (pres, 75 & 83); NY Acad Sci; Am Gastroenterol Asn; Sigma Xi. *Res:* Neurohumoral control of gastric and pancreatic secretion in swine and horses; gastrointestinal motility in animals. *Mailing Add:* Col Vet Med Box J-136 JHMHC Gainesville FL 32610

MERRITT, CHARLES, JR, b Lynn, Mass, Mar 15, 19; m 42; c 3. ANALYTICAL CHEMISTRY. *Educ:* Dartmouth Col, AB, 41; Univ Vt, MS, 48; Mass Inst Technol, PhD(anal chem), 53. *Prof Exp:* Finish engr, W Lynn Works Lab, Gen Elec Co, 41-46; instr chem, Univ Vt, 46-49; asst, Mass Inst Technol, 49-53; res analalytical chemist, Nat Bur Stand, 53; asst prof analytical chem, Polytech Inst Brooklyn, 53-56; supvry analytical chemist, Natick Labs, US Army, 56-57, head, analytical chem lab, 57-83; RETIRED. *Concurrent Pos:* Mem staff grad sch arts & sci, Northeastern Univ, 56-83; vis lectr, Mass Inst Technol, 65-; adj prof food sci, Univ Mass, 75-; mem subcomt stand ref mat, Nat Acad Sci-Nat Res Coun. *Mem:* AAAS; Inst Food Technol; Am Soc Mass Spectrometry; Am Chem Soc; Soc Appl Spectros. *Res:* Electrodeposition; spectrophotometry; gas chromatography; mass spectrometry; irradiation techniques; determination of the trace components of foodstuffs; pollution and environmental studies. *Mailing Add:* PO Box 478 Glen NH 03838-0478

MERRITT, CLAIR, b Quakertown, Pa, Jan 27, 22; m 43; c 3. SILVICULTURE. *Educ:* Univ Mich, BSF, 43, MF, 48, PhD(forestry), 59. *Prof Exp:* Asst dist forester, Md, 47; self-employed, 48; from instr to asst prof forestry, State Univ NY Col Forestry, Syracuse Univ, 49-56; from assoc prof to prof forestry, Purdue Univ, West Lafayette, 56-87. *Concurrent Pos:* Vpres, Foresters, Inc. *Mem:* Soc Am Foresters. *Res:* Silvics and silviculture, light relationships in forest openings; establishment of hardwood regeneration. *Mailing Add:* 104 Black Hawk Lane West Lafayette IN 47906

MERRITT, DORIS HONIG, b New York, NY, July 16, 23; m 53; c 2. RESEARCH ADMINISTRATION, PEDIATRICS. *Educ:* Hunter Col, BA, 44; George Washington Univ, MD, 52. *Prof Exp:* Exec secy cardiovasc & gen med dis, NIH, 57-60; from asst to assoc prof pediat, Ind Univ, Indianapolis, 61-73, from asst to assoc dean grants admin, 62-68, from asst to assoc dean sponsored prog, 65-70, dean sponsored progs, 70-78, prof pediat, 73-78; spec asst to dir, NIH, 78-87; actg dir, Nat Ctr, Nursing Res, 86-87; ASSOC DEAN & PROF, SCH MED, IND UNIV, 88-, SPEC ASST TO PRES, 88- *Concurrent Pos:* Nat Heart Inst fel cardiovasc dis, Duke Univ, 56-57; Nat Heart & Lung Inst grant & interim dir, Indianapolis Sickle Cell Ctr, Sch Med, Ind Univ-Purdue Univ, 73-76, Lilly endowment urban educ, 74-76; consult div res grants, Nat Heart & Lung Inst, 63-; mem regional adv group, Ind Regional Med Prog, 69-78; mem & chmn biomed libr rev comt, Nat Libr Med, 70-73; chmn, Consortium for Urban Educ, Indianapolis, 71-75; mem bd regents, Nat Libr Med, 74-78; mem bd dirs, Community Serv Coun, 90-91. *Mem:* AAAS; Am Acad Pediat; Sigma Xi. *Res:* Improvement of research administration. *Mailing Add:* FH 302 Ind Univ Sch Med Indianapolis IN 46202-5114

MERRITT, HENRY NEYRON, b Darlington, SC, Nov 9, 19; m 54; c 2. HYPNOTHERAPY, HYPNOANALYSIS. *Educ:* Clemson Univ, BS, 41; Kansas City Univ Physicians & Surgeons, MD, 44; Univ SC, Med, 53; Philathea Col, PhD (psychol), 69. *Prof Exp:* Assoc prof biol, Frostburg Univ, 64-67; assoc prof anatomy, Va Polytech Univ, 67-68; prof & chmn Health, Univ Wisc, Lacross, 68-72; dir Navy Drug Prog, Nat Acad Sci, Jacksonville, Fla, 72-76; DIR PSYCHOTHERAPY CTR, 76- *Concurrent Pos:* Fel Am Bd Med Psychotherapist; consult, 44-64. *Mem:* Fel AAAS; Fel Am Orthopsychiat asn; Soc Med Hypnoanalysts; Fel Nat Soc Classical Hypnotherapists. *Res:* Homeopathic medicine, proving how effective it is in acute disease. *Mailing Add:* 6037 Longchamp Dr Jacksonville FL 32244

MERRITT, J(OSHUA) L(EVERING), JR, b Dundalk, Md, July 28, 31; m 54; c 3. STRUCTURAL & EARTHQUAKE ENGINEERING, ROCK MECHANICS. *Educ:* Lehigh Univ, BS, 52; Univ Ill, MS, 55, PhD(eng), 58. *Prof Exp:* Asst civil eng, Univ Ill, Urbana 52-54, res assoc, 54-58, from asst prof to prof, 58-68, vis prof, 68-69; mgr hard rock silo develop prog, TRW Systs Inc, 68-70, facilities eng, 70-71, asst prog mgr minuteman prog, 68-71; PRES, MERRITT CASES, INC, 71-; DIR, BDM INT, INC, 86- *Concurrent Pos:* Consult, 58-; part-time prof, Univ Ill, Urbana, 68-69; mem-at-large, US Nat Comt Rock Mech, Nat Res Coun, 88- *Mem:* AAAS; Am Soc Civil Engrs; Am Concrete Inst; Am Soc Testing & Mat; Earthquake Eng Res Inst; Int Soc Soil Mech & Found Engrs; Concrete Soc; Am Underground Space Asn; Seismol Soc Am; Int Soc Rock Mechanics. *Res:* Structural analysis and design; strength and behavior of materials and structural systems; behavior of structures and structural systems subjected to transient loads; experimental design and interpretation; tunnel and shaft design and support; shock isolation; earthquake engineering; rock mechanics; geotechnical engineering. *Mailing Add:* BDM Int Inc 657 E Palm Ave Redlands CA 92374-6274

MERRITT, JACK, b Sacramento, Calif, May 2, 18; m 42; c 1. PHYSICS. *Educ:* Pomona Col, AB, 39; Univ Calif, PhD(physics), 53. *Prof Exp:* Admin analyst, US Bur Budget, 46-47; admin officer res div, AEC, 47-49; physicist radiation lab, Univ Calif, 53-54; instrumentation, Shell Develop Co Div, Shell Oil Co, 55-57, Spectros, 57-66; PROF PHYSICS, CLAREMONT MEN'S COL, 66- *Mem:* Am Phys Soc. *Res:* High energy nuclear physics; instrumentation; meson production; spectroscopy; x-rays. *Mailing Add:* 11th & Dartmouth Ave Claremont CA 91711

MERRITT, JAMES FRANCIS, b Wake Co, NC, July 21, 44; m 69; c 3. GENETICS, BOTANY. *Educ:* ECarolina Univ, BS, 66, MS, 68; NC State Univ, PhD(genetics), 73. *Prof Exp:* Asst prof, 73-79, actg chmn dept, 78-79, assoc prof biol & chmn dept, 79-89, DIR, CTR MARINE SCI RES, UNIV NC, WILMINGTON, 89- *Concurrent Pos:* Actg dir, Inst Marine Biomed Res, 86-89. *Mem:* Am Genetic Asn. *Res:* Cytogenetics of Vaccinium species. *Mailing Add:* Ctr Marine Sci Res Univ of NC at Wilmington 601 S Col Rd Wilmington NC 28403

MERRITT, KATHARINE, b Bridgeport, Conn, Apr 11, 38; m 70. IMMUNOLOGY, MICROBIOLOGY. *Educ:* Vassar Col, AB, 60; Univ Mich, MS, 62, PhD(microbiol), 64. *Prof Exp:* Res assoc path & microbiol, Dartmouth Med Sch, 64-66, instr, 66-68, asst prof microbiol, 68-80; ASSOC PROF MICROBIOL, UNIV CALIF, DAVIS, 80- *Mem:* Am Soc Microbiol; Can Soc Immunol; Am Asn Immunol Res. *Res:* Pathogenesis of infectious diseases, host defense mechanisms; host response to orthopaedic implants. *Mailing Add:* Dept Biomed Eng Case Western Reserve Univ 500 Wickenden Bldg Cleveland OH 44106

MERRITT, LAVERE BARRUS, b Afton, Wyo, Mar 11, 36; m 56; c 4. WATER QUALITY, COMPUTER ANALYSIS. *Educ:* Univ Utah, BS, 63, MS, 66; Univ Wash, PhD(civil eng), 70. *Prof Exp:* Civil engr, US Forest Serv, 63-64; Fed Water Qual Admin trainee, Univ Wash, 67-70; instr civil eng, 64-67, PROF CIVIL ENG, BRIGHAM YOUNG, 70-, CHMN, CIVIL ENG, 86- *Concurrent Pos:* Prin investr lake & stream water qual studies, 75-; sanit eng consult. *Mem:* Am Soc Civil Engrs; Am Water Works Asn; Am Acad Environ Engrs; Am Soc Eng Educ; Water Pollution Control Fedn; NAm Lake Mgt Asn; Sigma Xi. *Res:* Water quality research and multidisciplinary investigations; computer applications in sanitary engineering; water quality modeling; sewer design; solid waste management. *Mailing Add:* Brigham Univ 368 Clyde Bldg Provo UT 84602

MERRITT, LYNNE LIONEL, JR, b Alba, Pa, Sept 10, 15; m 37; c 4. ANALYTICAL CHEMISTRY, EDUCATION ADMINISTRATION. *Educ:* Wayne State Univ, BS, 36, MS, 37; Univ Mich, PhD(anal chem), 40. *Hon Degrees:* DSc, Ind Univ, 88. *Prof Exp:* Instr chem, Wayne State Univ, 36-37, 39-42; from asst prof to prof chem, Ind Univ, Blommington, 42-82, assoc dean col arts & sci, 59-62, dir bur instnl res, 60-65, assoc dean faculties, 62-64, actg dean, 63-64, vpres & dean res & advan studies, 65-75, dean res coord & develop, 75-80, spec asst to pres, 75-82, actg dean acad affairs, 82, 85, EMER PROF CHEM, IND UNIV, BLOOMINGTON, 82- *Concurrent Pos:* Vis prof, Calif Inst Technol, 49-50; Guggenheim fel & res assoc, 55-56; pres & dir, Ind Instrument & Chem Corp, 59-; Fulbright fel, Nat Ctr Sci Res, France, 63; actg chmn, Dept Data Processing & Info Systs, Ind Univ, Northwest, 88-91. *Mem:* AAAS; Am Chem Soc; Am Crystallog Asn. *Res:* Organic reagents; instrumental methods of analysis; x-ray diffraction crystal structure determinations. *Mailing Add:* Dept Chem Ind Univ Bloomington IN 47405

MERRITT, MARGARET VIRGINIA, b Springfield, Ohio, June 30, 42. ANALYTICAL CHEMISTRY. *Educ:* Col Wooster, BA, 64; Cornell Univ, PhD(anal chem), 68. *Prof Exp:* Fel electrochem, Univ Calif, Riverside, 68-69; fel, Radiation Res Lab, Mellon Inst, 69-70; asst prof analytical chem, Franklin & Marshall Col, 70-72; res chemist, Upjohn Co, 72-80, RES HEAD, 80-; AT CHEM DEPT, WELLESLEY COL. *Mem:* Am Chem Soc; Sigma Xi. *Res:* High pressure liquid chromatography; trace organic analysis; electron spin resonance; spin labeling; spin trapping; lipid analysis; biomembrane characterization. *Mailing Add:* Chem Dept Wellesley Col Wellesley MA 02181-8203

MERRITT, MELVIN LEROY, b Juneau, Alaska, Nov 12, 21; m 49; c 4. PHYSICS. *Educ:* Calif Inst Technol, BS, 43, PhD(physics), 50. *Prof Exp:* Test engr, Gen Elec Co, 43-46; mem tech staff, Sandia Nat Labs, 50-56, div supvr, 56-85, mgmt staff, 85-89; mem, Advan Sci Inc, 89; MEM, RAY F WESTON INC, 90- *Mem:* AAAS; Am Phys Soc; Am Asn Physics Teachers; Arctic Inst NAm; Health Physics Soc. *Res:* Cosmic rays; shock waves; effects of nuclear weapons; environmental studies and assessments. *Mailing Add:* 1016 Montclaire Dr NE Albuquerque NM 87110

MERRITT, PAUL EUGENE, b Watertown, NY, Oct 23, 20; m 46; c 2. ANALYTICAL CHEMISTRY. *Educ:* NY State Col Teachers, Albany, BA, 42, MA, 47; Rensselaer Polytech Inst, PhD, 54. *Prof Exp:* Line foreman, Gen Chem Defense Corp, 42-43; teacher pub sch, 47-48; asst gen chem, Rensselaer Polytech Inst, 48-51, instr analytical chem, 53-54; from asst prof to assoc prof, St Lawrence Univ, 54-63; chmn dept chem, 70-73, PROF ANALYTICAL CHEM, STATE UNIV NY COL POTSDAM, 63- *Mem:* Fel Am Inst Chemists; Am Chem Soc; Sigma Xi; NY Acad Sci. *Res:* Physico-chemical methods of analysis; infrared analysis of inorganic complexes, minerals and ores; structure studies of complex-inorganic salts; trinitrotoluene. *Mailing Add:* RR1-Box 163A W River Rd Norwood NY 13668-9736

MERRITT, R(OBERT) W(ALTER), b St Joseph, Mich, Feb 2, 13; m 35; c 2. MANUFACTURING CHEMISTRY. *Educ:* Univ Mich, BS, 34, MS, 35, PhD(chem eng), 39. *Prof Exp:* From trainee to dir adhesive prod, Nat Starch Prod Inc, 37-51, asst vpres, 51-56, vpres, 56-63, dir, Nat Starch & Chem Corp, 58-77, exec vpres, 63-77; RETIRED. *Mem:* Sigma Xi. *Res:* Synthetic resin polymers; wood and starch chemistry; manufacturing of adhesive; chemical decomposition of wood by super heated steam at temperature below 270 degrees centigrade. *Mailing Add:* Wyndham Apts No 355 2300 Portage St Kalamazoo MI 49001

MERRITT, RICHARD HOWARD, b Jersey City, NJ, Mar 28, 33; m 55; c 4. HORTICULTURE, ACADEMIC ADMINISTRATION. *Educ:* Rutgers Univ, BSc, 55, MSc, 56, PhD(hort, plant physiol), 61. *Prof Exp:* Lectr pomol, Rutgers Univ, 58-61, from asst prof ornamental hort to assoc prof hort, 61-70, dir res instr & assoc dean, Col Agr & Environ Sci, 62-73, dean instr, Cook Col, 74-81, PROF HORT, RUTGERS UNIV, 70- *Mem:* Am Soc Hort Sci; AAAS; Sigma Xi. *Res:* Educational research; photosynthetic officiency. *Mailing Add:* Horticult Dept Rutgers Univ New Brunswick NJ 08903

MERRITT, RICHARD WILLIAM, b San Francisco, Calif, July 26, 45; m 67; c 2. ENTOMOLOGY. *Educ:* Calif State Univ, San Jose, BA, 68; Wash State Univ, MS, 70; Univ Calif, Berkeley, PhD(entom), 74. *Prof Exp:* From asst prof to assoc prof, 74-84, PROF ENTOM, MICH STATE UNIV, 84- *Concurrent Pos:* Assoc ed, Am Midland Naturalist, 76-80 & Freshwater Invert Biol, 81-85; Fulbright scholar award, 86. *Mem:* Entom Soc Am; NAm Benthological Soc (pres, 84-85); Sigma Xi; Freshwater Biol Asn; Am Inst Biol Sci. *Res:* Aquatic entomology and ecology; filter-feeding ecology and population dynamics of aquatic insects; aquatic biting flies; degradation and processing of organic material by insects; floodplain ecology; biosystematics of Diptera; veterinary entomology. *Mailing Add:* Dept Entom Mich State Univ East Lansing MI 48824-1115

MERRITT, ROBERT BUELL, b Topeka, Kans, Nov 20, 42; m 65; c 1. POPULATION GENETICS. *Educ:* Univ Kans, BA, 64, PhD(zool), 70. *Prof Exp:* Trainee genetics, Univ Rochester, 70-72; from asst prof to assoc prof, 72-84, PROF BIOL, SMITH COL, 84- *Mem:* AAAS; Soc Study Evolution; Genetics Soc Am. *Res:* Genetic structure in natural plants and animal populations. *Mailing Add:* Dept Biol Smith Col Northampton MA 01063

MERRITT, ROBERT EDWARD, b Coudersport, Pa, Aug 1, 30; m 73; c 2. LEATHER CHEMISTRY, CELLULOSE & PAPER CHEMISTRY. *Educ:* Hanover Col, AB, 51; Univ Cincinnati, MS, 53. *Prof Exp:* Dir res, Barrentan Testing & Res Corp, Pa, 53-59; asst ed, Chem Abstracts Serv, 59-63, assoc ed, 63-68, group leader, 68-71, sr assoc indexer, 71-72, SR ASSOC ED, CHEM ABSTRACTS SERV, 72- *Concurrent Pos:* Ed, The Chem Record, Am Chem Soc, 69-86; assoc ed, J Am Leather Chemists Asn, 78-; counr, Am Chem Soc, 87-, Coun Comt Constitution & Bylaws, 88- *Mem:* Am Chem Soc; Am Leather Chemists Asn. *Res:* Leather chemistry, especially synthetic tanning materials and lignosulfonates; polymer chemistry; cellulose and paper chemistry. *Mailing Add:* Chem Abstracts Serv Dept 60 PO Box 3012 Columbus OH 43210

MERRITT, THOMAS PARKER, industrial planning, for more information see previous edition

MERRITT, WILLIAM D, BIOCHEMISTRY, IMMUNOLOGY. *Educ:* Purdue Univ, PhD(biol), 75. *Prof Exp:* ASST RES PROF, GEORGE WASH UNIV MED CTR, 81- *Res:* Ganglioside modulation of immunity-tumor associated gangliosides. *Mailing Add:* Dept Biochem Med Ctr George Washington Univ 111 Michigan Ave NW Washington DC 20010

MERROW-HOPP, SUSAN B, b Feb 22, 17; m 79. CLINICAL NUTRITION, GERIATRIC NUTRITION. *Educ:* Boston Univ, MEd, 44. *Prof Exp:* asst dir, Vt Agr Exp Sta, Univ Vt, 80-86; RETIRED. *Mem:* Sigma Xi; AAAS; Am Dietetic Asn. *Res:* Human nutrition. *Mailing Add:* RR 1 Box 1708 Charlotte VT 05445

MERRYMAN, CARMEN F, IMMUNOLOGY. *Educ:* Univ Havana, MD, 54. *Prof Exp:* ASSOC PROF BIOCHEM, MED COL, THOMAS JEFFERSON UNIV, 68- *Res:* Retinal S-antigen; structure-function relationships. *Mailing Add:* Dept Biochem Med Col Alumni Hall Thomas Jefferson Univ Philadelphia PA 19107

MERSEREAU, RUSSELL MANNING, b Cambridge, Mass, Aug 29, 46; m 68; c 2. DIGITAL SIGNAL PROCESSING, IMAGE PROCESSING. *Educ:* Mass Inst Technol, SB & SM, 69, ScD, 73. *Prof Exp:* Res assoc elec eng, Mass Inst Technol, 73-75; from asst prof to prof, 75-87, REGENTS PROF ELEC ENG, GA INST TECHNOL, 87- *Concurrent Pos:* Assoc ed, Tranactions Acoust, Speech & Signal Processing, Inst Elec & Electronics Engrs, 76-80; vpres & treas, Atlanta Signal Processors, Inc, 81- *Honors & Awards:* Browder J Thompson Award, Inst Elec & Electronic Engrs, 76; Soc Award, Inst Elec & Electronic Engrs, Signal Processing Soc, 90. *Mem:* Fel Inst Elec & Electronic Engrs. *Res:* Development of efficient algorithms for digital processing of multi-dimensional data including sampled photographic images, seismic and sonar arrays, and computer aided tomography. *Mailing Add:* Sch Elec Eng Ga Inst Technol Atlanta GA 30332

MERSKEY, HAROLD, b Sunderland, Eng, Feb 11, 29; UK & Can citizen; m 65; c 3. PSYCHIATRY. *Educ:* Oxford Univ, BA, 50, BM, BCh, 53, DM, 65; FRCP(London); FRCP(C); FRCPsychiat. *Prof Exp:* Lectr, Univ Sheffield, 61-64; consult, Saxondale Hosp, Nottingham, 64-67, Nat Hosps, Nervous Dis, Queen Square, London, 67-76; assoc prof, 76-77, PROF PSYCHIAT, UNIV WESTERN ONT, 77-; DIR EDUC & RES, LONDON PSYCHIAT HOSP, 76- *Concurrent Pos:* Chmn, Comt Taxon, Int Asn Study Pain, 75-, Med & Sci Comt, Soviet Jewry, 71-76, Sci & Adv Comt, Geront Res Coun Ontario, 83-87; mem, Ont Mental Health Found Res Adv Bd, 78-82, Behav Sci Grants Comt, Med Res Coun, Can, 82-83. *Mem:* Am Psychiat Asn; Can Psychiat Asn; Can Med Asn; EEG Soc; Can Pain Soc (pres, 88-91). *Res:* Psychiatric aspects of pain; organic aspects of psychiatry and psychopharmacology; hysteria; social psychiatry, especially abuses of psychiatry. *Mailing Add:* London Psychiat Hosp 850 Highbury Ave London ON N6A 4H1 Can

MERSMANN, HARRY JOHN, b St Louis, Mo, Nov 13, 36; m 59; c 2. LIPID METABOLISM. *Educ:* St Louis Univ, BS, 58, PhD(biol), 63. *Prof Exp:* Res assoc biochem, Auburn Univ, 63-65, Univ Calif, San Francisco, 65-66 & State Univ NY Buffalo, 66-68; asst prof life sci, Ind State Univ, 68-69; biochemist, Shell Develop Co, 69-79; res chemist, Meat Animal Res Ctr, USDA, Clay Center, Nebr, 80-89; RES CHEMIST, CHILDREN'S NUTRIT RES CTR, AGR RES SERV, USDA, HOUSTON, TEX, 89- *Mem:* Am Soc Biol Chemists; Am Inst Nutrit; Soc Exp Biol Med; Am Soc Animal Sci. *Res:* Animal growth and development; lipid metabolism; neonatal biology. *Mailing Add:* USDA-Agr Res Serv Children's Nutrit Res Ctr 1100 Bates St Houston TX 77030

MERSTEN, GERALD STUART, b Brooklyn, NY, Sept 28, 42; m 66; c 2. ELECTRICAL & ELECTRONICS ENGINEERING, COMPUTER ENGINEERING & SCIENCE. *Educ:* City Col NY, BE, 65, ME, 69; City Univ NY PhD(elec/comput eng), 78. *Prof Exp:* Eng sect supvr comput eng, Guid Systs Div, Bendix Corp, 65-78, prin eng, Flight Systs Div, 80-82; eng mgr & basic res & develop & bus area mgr, Adv Tech Syst, Austin Corp, 78-79; eng mgr signal process design & dir eng, Lockheed Elec Co Inc, 82-87; pres Aydin Comput Systs & corp vpres, Aydin Corp, 87-88; VPRES ENG & GEN MGR, ELECTRONICS & COMPUTER CONSULTS, GMA INC, 88- *Concurrent Pos:* Adj lectr elec eng, City Col New York, City Univ NY, 74-75, adj asst prof, 79-81, adj assoc prof, 81-86. *Mem:* Sigma Xi; Inst Elec & Electronic Engrs; Sigma Xi; Nat Mgt Asn. *Res:* Real Time computer systems; high performance general purpose processors; high performance parallel processors (parallel processing); multiprocessing systems; fault tolerant computing; computer graphic and display systems; hardware systems; information and database management systems. *Mailing Add:* 37 Polhemus Terr Whippany NJ 07981

MERTE, HERMAN, JR, b Detroit, Mich, Apr 3, 29; m 52; c 5. MECHANICAL ENGINEERING. *Educ:* Univ Mich, BS, 50 & 51, MS, 56, PhD(mech eng), 60. *Prof Exp:* Instr & lectr mech eng, 59, from asst prof to assoc prof, 60-67, PROF MECH ENG, UNIV MICH, ANN ARBOR, 67- *Concurrent Pos:* NSF sr fel, Munich, Ger, 67-68; vis prof, Inst Reactor & Apparatus Construction, Tech Univ Munich, 74-75. *Mem:* Am Soc Mech Engrs; Am Soc Eng Educ. *Res:* Heat transfer and thermodynamics; phase change dynamics under high and low gravity conditions. *Mailing Add:* Dept Mech Eng Univ Mich Ann Arbor MI 48109-2125

MERTEL, HOLLY EDGAR, b Springfield, Mo, Sept 26, 20; c 1. SYNTHETIC ORGANIC CHEMISTRY, NATURAL PRODUCTS CHEMISTRY. *Educ:* Drury Col, BS, 41; Univ Nev, MS, 43; Univ Southern Calif, PhD(org chem), 50. *Prof Exp:* Res assoc, Columbia Univ, 50-51; RES CHEMIST, MERCK SHARP & DOHME RES LABS DIV, MERCK & CO, INC, 51-, SR RES FEL, 70- *Mem:* AAAS; Am Chem Soc; Royal Soc Chem; Health Physics Soc. *Res:* Radiochemical preparations; cortical steroid hormones. *Mailing Add:* 721 Harding St Westfield NJ 07090-1329

MERTEN, ALAN GILBERT, b Milwaukee, Wis, Dec 27, 41; m 67; c 2. COMPUTER SCIENCES, INFORMATION SYSTEMS. *Educ:* Univ Wis-Madison, BS, 63, PhD(comput sci), 70; Stanford Univ, MS, 65. *Prof Exp:* Asst prof indust & opers eng, 70-74, assoc prof, 74-81, PROF COMPUT & INFO SYST, BUS ADMIN, UNIV MICH, ANN ARBOR, 81-, ASSOC DEAN BUS ADMIN, 82- *Concurrent Pos:* Consult, Timken Co, US Navy & Tex Instruments; chmn, Adv Comt, Wang Inst Grad Studies. *Mem:* Asn Comput Mach; Soc Mgt Info Syst. *Res:* Management and internal control systems; effective use of computer-based information systems; education of managers on implementation and management of computer systems. *Mailing Add:* 303 Malott Hall Ithaca NY 14853

MERTEN, HELMUT L, b Vienna, Austria, Apr 2, 22; US citizen; m 34; c 1. CONDENSATION REACTIONS, CONTINUOUS CHLORINATION REACTIONS. *Educ:* Univ Vienna, PhD(chem), 51. *Prof Exp:* Sci asst selected topics-org reactions, First Chem Inst, Univ Vienna, 48-51, asst, 51-52; postdoctoral fel, Univ Iowa State, 52-54, Univ Toronto, 54-55; sr chemist, Org Chem Dept, Monsanto, St Louis, 55-58, res group leader, 58-72, sr res group leader, Akron, Ohio, 72-87; TECH ADV & CONSULT, HERZOG & HART ENG CO, BOSTON, 87- *Mem:* Soc Catalysis Org Reactions. *Res:* New product and process development in: dielectrics, oil additives, food additives and acidulants, rubber chemicals, antioxidants; new plant design (process) and start-up in: India, Argentina, Brazil, Belgium and United Kingdom; ten US patents and about 100 foreign patents. *Mailing Add:* 1660 Mayflower Lane Hudson OH 44236-3934

MERTEN, ULRICH, b Houston, Tex, Feb 27, 30; m 53; c 2. PHYSICAL CHEMISTRY. *Educ:* Calif Inst Technol, BS, 51; Wash Univ, PhD(chem), 55. *Prof Exp:* Mem staff chem, Knolls Atomic Power Lab, Gen Elec Co, 55-56; mem staff chem, Gen Atomic Div, Gen Dynamics Corp, 56-67, Gulf Gen Atomic, 67-71; vpres chem & minerals, Gulf Res & Develop Co, 71-83; DIR PROG DEVELOP, SIBIA, 83- *Mem:* AAAS; Com Develop Asn. *Res:* High temperature chemistry; membrane phenomena; biotechnology. *Mailing Add:* 4422 Leon St San Diego CA 92107

MERTENS, DAVID ROY, b Jefferson City, Mo, Sept 11, 47; m 72; c 2. RUMINANT NUTRITION, DAIRY SCIENCE. *Educ:* Univ Mo-Columbia, BS, 69, MS, 70; Cornell Univ, PhD(nutrit), 73. *Prof Exp:* Asst prof animal sci, Iowa State Univ, 73-75; assoc prof dairy sci, Univ Ga, 75-84; USDA/ARS US DAIRY FORAGE RES CTR, MADISON, 84- *Mem:* Sigma Xi; Am Dairy Sci Asn; Am Soc Microbiol; Am Soc Animal Sci; AAAS. *Res:* Mathematical and chemical study of ruminal metabolism of forages and fibrous carbohydrates; development of optimal nutrition and management of dairy animals. *Mailing Add:* US Dairy Forage Res Ctr 1925 Linden Dr W Madison WI 53706

MERTENS, FREDERICK PAUL, b Danbury, Conn, June 10, 35; m 62; c 1. PHYSICAL CHEMISTRY, CORROSION. *Educ:* Worcester Polytech Inst, BS, 57, PhD(chem), 65. *Prof Exp:* Res chemist, Columbia-Southern Chem Corp, 57-60; res asst, Worcester Polytech Inst, 60-64; chemist, Texaco Inc, 64-65, sr chemist, 65-69, res chemist, 69-72, proj chemist, 72-80, sr proj chemist, 80-87, TECHNOLOGIST, TEXACO INC, PORT ARTHUR, 87- *Mem:* Am Chem Soc; Nat Asn Corrosion Engrs; Sigma Xi. *Res:* Infrared reflectance study of metallic corrosion and adsorbed molecules; failure analysis; basic and applied catalysis research; evaluation and recommendation of corrosion inhibitors, antifoulants, coatings and materials of construction in petroleum industry; electron spectroscopic study of catalysts. *Mailing Add:* 3216 Lawrence Ave Nederland TX 77627

MERTENS, LAWRENCE E(DWIN), b New York, NY, Mar 6, 29; m 75; c 2. SYSTEMS ENGINEERING, OPTICAL OCEANOGRAPHY RADAR. *Educ:* Columbia Univ, BS, 51, MS, 52, DEngSci, 55. *Prof Exp:* Engr, Bendix Aviation Corp, 52; systs engr, Radio Corp Am, 52-57, staff engr, 57-59, mgr digital commun eng, 59-62, chief scientist, RCA Corp, 62-76, mgr aerostat systs, 76-81, mgr tech anal, RCA Serv Co, 81-86; chief scientist, GE Govt Serv, 86-89; VPRES, SUNTECH, 89- *Concurrent Pos:* Adj prof, Fla Inst Technol, 65- *Mem:* Inst Elec & Electronics Engrs; NY Acad Sci; Marine Technol Soc; Soc Photo-Optical Instrumentation Engrs. *Res:* Range instrumentation; systems engineering; information theory; data processing; oceanography; radar. *Mailing Add:* 690 Pebble Beach Ave NE Palm Bay FL 32905

MERTENS, THOMAS ROBERT, b Ft Wayne, Ind, May 22, 30; m 53; c 2. GENETICS, SCIENCE EDUCATION. *Educ:* Ball State Univ, BS, 52; Purdue Univ, MS, 54, PhD(genetics), 56. *Prof Exp:* Res assoc genetics, Univ Wis, 56-57; from asst prof sci to prof biol, 57-88, CO-DIR, HUMAN GENETICS & BIOETHICS EDUC LAB, BALL STATE UNIV, 77-, DISTINGUISHED PROF BIOL EDUC, 88- *Concurrent Pos:* NSF fac fel, Stanford Univ, 63-64. *Honors & Awards:* Ohaus Award, 86; Distinguished Serv to Sci Educ, Nat Sci Teachers Asn, 87. *Mem:* Fel AAAS; hon mem Nat Asn Biol Teachers (pres, 85); Genetics Soc Am; Am Genetic Asn; fel Ind Acad Sci. *Res:* Plant genetics and taxonomy; cytotaxonomy of genus Polygonum in North America; cytogenetics of Rhoeo; programmed instruction in biology education; how to best teach genetics and human genetics; designing and testing laboratory investigations; needs assessments in human genetics education. *Mailing Add:* Dept of Biol Ball State Univ Muncie IN 47306

MERTES, DAVID H, b Pittsburg, Calif, Nov 22, 29; m 62. CHEMICAL EMBRYOLOGY. *Educ:* San Francisco State Col, BA, 52; Univ Calif, Berkeley, MA, 59, PhD(zool), 66. *Prof Exp:* Mem fac zool, San Joaquin Delta Col, 59-62; Am Cancer Soc Dernham res fel, Univ Calif, Berkeley, 66-68; chmn div sci, 68-69, acad dean, 69-71, pres, Col San Mateo, 71-78; supt pres, Santa Barbara City Col, 78-81; CHANCELLOR, LOS RIOS COMMUNITY COL DIST, 81- *Mem:* AAAS; Am Soc Zool; Am Inst Biol Sci; Soc Develop Biol. *Res:* Biochemical embryology; genetic read-off and protein synthesis during the early embryogenesis of invertebrate embryos, differentiation. *Mailing Add:* State Chancellor Off 1107 Ninth St Sacramento CA 95814

MERTES, FRANK PETER, JR, b Chicago, Ill, Apr 2, 35; m 59; c 1. NUCLEAR ENGINEERING, CHEMICAL ENGINEERING. *Educ:* Northwestern Univ, BS, 58; Iowa State Univ, MS, 59, PhD(nuclear eng), 62. *Prof Exp:* Nuclear engr, Analysis Serv Inc, 62-68; MGR SYST ANALYSIS OFF, TRW ENERGY SYST PLANNING, 68- *Mem:* Am Nuclear Soc; Am Inst Aeronaut & Astronaut. *Res:* Analysis of and computer simulation of competing energy system alternatives; determination of economic and technical feasibility of nuclear power plants, synthetic fuel facilities and solar energy installations. *Mailing Add:* 3223 Foxvale Dr Oakton VA 22124-2239

MERTES, KRISTIN BOWMAN, b Philadelphia, Pa, June 4, 46; m 76. INORGANIC CHEMISTRY. *Educ:* Temple Univ, BA, 68, PhD(chem), 74. *Prof Exp:* Asst prof, 75-81, assoc prof, 81-87, PROF CHEM, UNIV KANS, 87- *Mem:* Am Chem Soc; Am Crystallographer's Asn. *Res:* Structure reactivity aspects of transition metal complexes particularly with macrocyclic ligand systems. *Mailing Add:* Dept Chem Univ Kans Lawrence KS 66045

MERTES, MATHIAS PETER, organic chemistry; deceased, see previous edition for last biography

MERTINS, JAMES WALTER, b Milwaukee, Wis, Feb 18, 43; m 79. ENTOMOLOGICAL PARASITOLOGY. *Educ:* Univ Wis-Milwaukee, BS, 65; Univ Wis-Madison, MS, 67, PhD(entom), 71. *Prof Exp:* Res assoc entom, Univ Wis-Madison, 71-77; asst prof, Iowa State Univ, 77-84; pvt consult, 84-89; ENTOMOLOGIST, USDA, 89- *Concurrent Pos:* NSF fel, Univ Wis-Madison, 69-70. *Mem:* Int Orgn Biol Control; Entom Soc Am; Entom Soc Can. *Res:* Biological insect pest supression; insect pathology; parasitoid-complex interactions; parasitoid life histories and biologies; identification of parasitic arthropods. *Mailing Add:* USDA Nat Vet Serv Lab Pl PO Box 844 Ames IA 50010

MERTON, ROBERT K, b Philadelphia, Pa, Jul 4, 10; m 34; c 3. MEDICAL EDUCATION. *Educ:* Temple Univ, AB, 31; Harvard Univ, MA, 32, PhD(sociol & hist sci), 36. *Hon Degrees:* Numerous from US & foreign Univs. *Prof Exp:* Tutor & instr sociol, Harvard Univ, 36-39; prof & chmn sociol, Tulane Univ, 39-41; from asst prof to prof sociol, Columbia Univ, 41-62, Giddings prof, 63-74, univ prof sociol, 74-79, spec serv prof, 79-84, EMER UNIV PROF, COLUMBIA UNIV, 79- *Concurrent Pos:* Bicentennial lectr, Princeton Univ, 46; Haynes Found lectureship, Claremont Grad Sch, 60; lectureship, NIH, 64; consult, Inst Sci Info, 68-; comt on sci & public policy, Nat Acad Sci, 69-73; comn for joint US-USSR acad study fundamental sci policy, Nat Acad Sci, 74-77; ed adv comn, prog sci, Sloan Found, numerous consult & lectr positions, 46-; adj fac, Rockfeller Univ, 79-; found scholar, Russell Sage Found, 79-; MacArthur Prize fel. *Honors & Awards:* Bacon lectr, Wayne State Univ, 61, Am Philos Soc & Univ Pa, 61; Daniel Coit Gilman lectr, Johns Hopkins Univ, 62; Parsons Prize, Am Acad Arts & Sci, 79; Career of Distinguished Scholarship Award, Am Sociol Asn, 80; Award for Outstanding Support of Biomed Sci, Memorial Sloan- Kettering Cancer Ctr, 81; John Desmond Bernal Award, Soc Social Studies of Sci, 82. *Mem:* Nat Acad Sci; Am Philos Soc; Am Sociol Asn (pres, 57-58); Sociol Res Asn (pres, 68); Am Acad Arts & Sci; Royal Swedish Acad Sci; Soc Social Studies Sci (pres, 75-76). *Res:* Sociology; history; philosophy of science; theoretical sociology. *Mailing Add:* 450 Riverside Dr New York NY 10027

MERTS, ATHEL LAVELLE, b Paragould, Ark, Oct 14, 25; m 47; c 1. ATOMIC PHYSICS, ASTROPHYSICS. *Educ:* Univ Mo-Rolla, BS, 50, MS, 51; Univ Kans, PhD(physics), 57. *Prof Exp:* Instr physics, Univ Tulsa, 51-52; staff mem continuum physics, Los Alamos Sci Lab, 57-64; sr scientist, Gulf-Gen Atomics, staff mem, 65-76, assoc group leader, 76-80, DEP GROUP LEADER ATOMIC PHYSICS OPACITIES, LOS ALAMOS NAT LAB, 80- *Mem:* Am Phys Soc; Am Astrophys Soc; Sigma Xi. *Res:* Study of the effects of collisional excitation and dielectronic recombination processes on the radiative power loss from optically thin plasmas. *Mailing Add:* 125 Aztec Ave Los Alamos NM 87544

MERTZ, DAN, b Allen Co, Ohio, Sept 19, 28; m 56; c 1. PLANT PHYSIOLOGY. *Educ:* Ohio Univ, BA, 54; Univ Tex, PhD(plant physiol), 60. *Prof Exp:* Asst bot & biol, Univ Tex, 54-57, res scientist, 57-60; res assoc bot, 60-61, assoc prof, 61-71, PROF BIOL SCI, UNIV MO-COLUMBIA, 71- *Concurrent Pos:* Visitor, Univ Glasgow, 67-68. *Mem:* Am Soc Plant Physiol; Bot Soc Am; Scand Soc Plant Physiol. *Res:* Physiological and biochemical changes associated with growth and development. *Mailing Add:* Div of Biol Sci 105 Tucker Hall Univ of Mo Columbia MO 65211

MERTZ, DAVID B, b Sandusky, Ohio, July 10, 34. ANIMAL ECOLOGY. *Educ:* Univ Chicago, BS, 60, PhD(zool), 65. *Prof Exp:* NSF vis scholar statist, Univ Calif, Berkeley, 65-66; asst prof biol, Univ Calif, Santa Barbara, 66-69; assoc prof, 69-74, PROF BIOL SCI, UNIV ILL, CHICAGO, 74- *Mem:* Fel AAAS; Am Soc Naturalists; Ecol Soc Am; Soc Study Evolution; Am Soc Zool; Brit Ecol Soc; Soc Pop Ecol. *Res:* Population ecology and ecological genetics of flour beetles. *Mailing Add:* Dept Biol Sci M/C 066 Univ Ill at Chicago Chicago IL 60680

MERTZ, EDWIN THEODORE, b Missoula, Mont, Dec 6, 09; m 36; c 2. BIOCHEMISTRY. *Educ:* Univ Mont, AB, 31; Univ Ill, MS, 33, PhD(biochem), 35. *Hon Degrees:* Dr Agr, Purdue Univ, 77; DSc, Univ Mont, 79. *Prof Exp:* Res chemist, Armour & Co, 35-37; instr biochem, Univ Ill, 37-38; res assoc path, Univ Iowa, 38-40; instr agr chem, Univ Mo, 40-43; res chemist, Exp Sta, Hercules Powder Co, Del, 43-46; from asst prof agr chem to prof biochem, 46-76, EMER PROF BIOCHEM, PURDUE UNIV, WEST LAFAYETTE, 76- *Concurrent Pos:* Consult, Ind State Hosps, 57-75 & US-Japan Malnutrit Panel, 70-73; vis prof, Notre Dame Univ, 76-77; consult agron, Purdue Univ, 76- *Honors & Awards:* McCoy Award, 67; John Scott Award, 67; Hoblitzelle Award, 68; Cong Medal, Fed Land Banks, 68; Spencer Award, Am Chem Soc, 70; Osborne-Mendel Award, Am Inst Nutrit, 72; Edward W Browning Award, Am Soc Agron, 74. *Mem:* Nat Acad Sci; Am Chem Soc; Am Soc Biol Chem; Am Inst Nutrit; Am Asn Cereal Chem. *Res:* Amino acid requirements of humans and animals; purification of plasminogens; biochemistry of mental retardation; opaque-2 and floury-2 high lysine maize and high lysine sorghum. *Mailing Add:* 143 Tamiami Trail St West Lafayette IN 47906-1254

MERTZ, JANET ELAINE, b Bronx, NY, Aug 9, 49; m 80; c 2. MOLECULAR VIROLOGY, MRNA BIOGENESIS. *Educ:* Mass Inst Technol, BS(biol) & BS(elec eng), 70; Stanford Univ, PhD(biochem), 75. *Prof Exp:* Jane Coffin Childs Mem Fund fel, Med Res Coun Lab Molecular Biol, Cambridge, Eng, 75-76; ASSOC PROF ONCOL, McARDLE LAB CANCER RES, UNIV WIS-MADISON, 76- *Mem:* AAAS; Am Soc Microbiol; Am Soc Virol; Am Asn Cancer Res; Am Soc Biochem & Molecular Biol; Fedn Am Scientists. *Res:* Molecular biology of tumor viruses; regulation of transcription; RNA processing, and translation in eucaryotes. *Mailing Add:* McArdle Lab Cancer Res Univ Wis Madison WI 53706

MERTZ, ROBERT LEROY, b Milwaukee, Wis, Jan 4, 34; m 59; c 3. ELECTRICAL ENGINEERING. *Educ:* Marquette Univ, BEE, 56; Mass Inst Technol, SM, 57; Univ Wis, PhD(elec eng), 60. *Prof Exp:* Asst prof elec eng, Marquette Univ, 60-64; staff engr, AC Electronics Div, Gen Motors Corp, 64-68; mgr, radiation effects eng, Gulf Gen Atomic, Inc, 68-70, dep mgr, Gulf Radiation Technol Div, Gulf Energy & Environ Systs Co, San Diego, 70-74; PRES, IRT CORP, 74- *Concurrent Pos:* Consult, AC Electronics Div, Gen Motors Corp, 62-64. *Mem:* Inst Elec & Electronics Engrs. *Res:* Theory of systems and automatic control; effects of radiation on electronic systems. *Mailing Add:* Univ San Diego Alcala Park S-194 San Diego CA 92110

MERTZ, WALTER, b Mainz, Germany, May 4, 23; m 53. NUTRITION, BIOCHEMISTRY. *Educ:* Univ Mainz, MD, 51. *Prof Exp:* Intern surg, County Hosp, Germany, 52-53; asst internal med, Univ Hosp, Univ Frankfurt, 53; res fel nutrit, NIH, 53-56, vis scientist, Exp Liver Dis Sect, 56-61; res biochemist, Walter Reed Army Inst Res, Washington, DC, 61-64, chief dept biol chem, 64-69; chief vitamin & mineral nutrit lab, Human Nutrit Res Div, 69-72, chmn nutrit inst, 72-80, DIR, HUMAN NUTRIT RES CTR, AGR RES SERV, USDA, 80- *Mem:* Am Soc Biol Chem; Am Inst Nutrit. *Res:* Biochemistry and nutrition of trace elements. *Mailing Add:* Human Nutrition Res Ctr Agr Res Serv USDA Beltsville MD 20705

MERTZ, WILLIAM J, b Campbellsport, Wis, Nov 21, 45; m 74; c 2. RESEARCH ADMINISTRATION. *Educ:* Univ Wis, BS, 68; North Inst Ill, MBA, 76. *Prof Exp:* Qual control supvr, Mil Print, Inc, 67-68; PROJ SCIENTIST, JEFFERSON SMURFIT CORP, 70- *Res:* Inks coatings. *Mailing Add:* Jefferson Smurfit Corp Corp Res Ctr 450 E North Ave Carol Stream IL 60488

MERUELO, DANIEL, b Cienfuegos, Cuba, Mar 5, 47; US citizen; m 71; c 2. VIROLOGY, LEUKEMIA. *Educ:* Columbia Univ, BS, 69; Johns Hopkins Univ, PhD(biochem immunol), 74. *Prof Exp:* Asst prof, 82-84, ASSOC PROF, NY UNIV MED CTR, 84- *Concurrent Pos:* Scholar, Leukemia Soc Am, 78-83; Irma Hirschl career scientist award, 78-82; estab investr, Am Heart Asn, 83-88; prin investr grants, NIH, 84-88, reviewer, 84- *Mem:* Am Soc Pathologists; Am Soc Microbiol; Am Heart Asn; NY Acad Sci; Am Asn Immunologists; AAAS. *Res:* Role of viruses in the etiology of leukemia; host genes regulation susceptibility to leukemia; role of immune surveillance; active chromosomal integration sites for oncogenic viruses. *Mailing Add:* Dept Path NY Univ Med Ctr 560 First Ave New York NY 10016

MERVA, GEORGE E, b Guernsey Co, Ohio, Aug 20, 32; m 59; c 5. AGRICULTURAL ENGINEERING, HYDROLOGY. *Educ:* Ohio State Univ, BAE, 60, PhD(hydrol), 67. *Prof Exp:* Asst prof agron, Ohio Agr Res & Develop Ctr, 60-63; res asst agr eng, Ohio State Univ, 63-67; from asst to prof assoc prof, 67-74, PROF AGR ENG, MICH STATE UNIV, 74- *Mem:* Am Soc Agr Eng; Am Soc Eng Educ; Am Inst Hydrol. *Res:* Plant-soil-water relationships and agricultural water management; specializing in drainage and subirrigation. *Mailing Add:* Dept Agr Eng Mich State Univ East Lansing MI 48823

MERYMAN, CHARLES DALE, b Centralia, Ill, Sept 14, 51. AQUATIC ANIMAL PATHOLOGY & FORENSICS. *Educ:* Univ Ill, BS, 73, MS, 74; Univ Metaphysics, PhD(animal behav), 78. *Prof Exp:* Instr, Univ Ill, 73-74; FISH DOCTOR, MERYMAN AQUATIC RESOURCE CTR, RIVERVIEW, FLA, 80- *Concurrent Pos:* Consult, Meryman Environ Enterprises. *Mem:* Int Oceanog Soc; Am Fisheries Soc; Europ Soc Fish Pathologists; Goldfish Soc Am; Int Wildlife Fedn; N Am Native Fish Asn. *Res:* Aquatic animal disease identification and control; rehabilitation and breeding of endangered species. *Mailing Add:* 10408 Bloomingdale Ave Riverview FL 33569

MERYMAN, HAROLD THAYER, b Washington, DC, Feb 5, 21; m 47; c 4. CRYOBIOLOGY, MEDICAL RESEARCH. *Educ:* Long Island Col Med, MD, 46. *Prof Exp:* Intern, US Naval Hosp, Md, 46-47; physiologist, Naval Med Res Inst, 47-54; Am Cancer Soc fel, Yale Univ, 54-56, res fel, Sch Med, 55-57; physiologist, Naval Med Res Inst, Md, 57-68; assoc res dir blood prog, 68-85, HEAD, TRANSPLANTATION LAB, AM NAT RED CROSS, 85- *Honors & Awards:* Grove-Rasmussen Award, Am Asn Blood Banks, 84; Tiffany Award, Am Red Cross, 84; Kamerlngh Ohnes Gold Medal, Neth Asn Refrig, 89. *Mem:* Am Physiol Soc; Biophys Soc; Am Soc Cell Biol; Electron Micros Soc Am; Cryobiol Soc (pres, 81-82); Am Asn Tissue Banks (pres, 81-82). *Res:* Mechanism of freezing and drying injury in biological media; physiology of cold injury; preservation of cells and tissues by freezing; blood and tissue banking; transplantation immunology. *Mailing Add:* Am Red Cross Jerome H Holland Lab 15601 Crabbs Branch Way Rockville MD 20855

MERZ, JAMES L, b Jersey City, NJ, Apr 14, 36; m 62; c 4. SOLID STATE PHYSICS. *Educ:* Univ Notre Dame, BS, 59; Harvard Univ, MA, 61, PhD(appl physics), 67. *Prof Exp:* Mem staff physics, Bell Labs, 66-78; chmn elec eng, 82-84, PROF ELEC ENG, COL ENG, UNIV CALIF, SANTA BARBARA, 78-, ASSOC DEAN RES & DEVELOP, 84- *Concurrent Pos:* Vis lectr, Harvard Univ, 72. *Mem:* Am Phys Soc; sr mem Inst Elec & Electronics Engrs; Electrochem Soc; Mat Res Soc; Sigma Xi. *Res:* Study of semiconducting compounds for optical communications and optoelectronic devices. *Mailing Add:* 756 El Rodeo Rd Santa Barbara CA 93110-1314

MERZ, KENNETH M(ALCOLM), JR, b Philadelphia, Pa, July 28, 22; m 52; c 4. CHEMISTRY, CERAMICS. *Educ:* Susquehanna Univ, BA, 49; Bucknell Univ, MA, 50; Rutgers Univ, PhD(ceramics), 57. *Prof Exp:* Res chemist, Nat Lead Co, 50-53; sr res engr, Carborundum Co, 57-59; res ceramic engr, Cornell Aeronaut Lab, Inc, 59-60; assoc dir res & develop, IRC Inc, Pa, 60-69; MGR RES, PHILADELPHIA LAB, TRW INC, 69- *Mem:* Am Ceramic Soc; Am Chem Soc. *Res:* Electronic components; preparation and properties of resistive materials; electronic ceramics; thin films; thermal expansion; optical, x-ray and electron microscopy; inorganic pigments; electrolytic preparation of titanium; thick films; fiber optic devices; sensors. *Mailing Add:* 1810 Brannocks Neck Rd Cambridge MD 21613

MERZ, PAUL LOUIS, b New Haven, Conn, June 1, 18; m 50; c 4. RUBBER CHEMISTRY, POLYMER CHEMISTRY. *Educ:* Union Col, NY, BS, 40; Yale Univ, PhD(org chem), 51. *Prof Exp:* Res chemist, Beech-Nut Packing Co, 40-43, head, Polymer Lab, 45-47, consult, 47-51; sr res chemist, Naugatuck Chem Div, US Rubber Co, 51-56, group leader, 56-59, sr res specialist, 59-61, proj leader high temperature elastomers, 61-62; plastics chemist, Lawrence Radiation Lab, 62-64; staff scientist, 64-79, sr eng specialist, Gen Dynamics & Convair, 79-82; RETIRED. *Mem:* AAAS; Am Chem Soc; Soc Advan Mat & Process Eng; Sigma Xi. *Res:* Vinyl polymerization; antioxidant and antiozonant research; rubber reclaiming; high temperature and cryogenic seals for aerospace applications; polymeric systems for radiation shielding; advanced structural and thermoformable composites; advanced aerospace adhesives and sealants; materials science. *Mailing Add:* PO Box 33 Ophir OR 97464-0033

MERZ, RICHARD A, b New York, NY, June 21, 48; m 75; c 2. EXPERIMENTAL FLUID MECHANICS, PERFORMANCE TESTING. *Educ:* Rutgers Univ, BS, 70, MS, 72, PhD(mech eng), 75. *Prof Exp:* Asst prof mech eng, Air Force Inst Technol, Wright Patterson AFB, Ohio, 75-79; sr res engr, Gen Motors Res Lab, Gen Motors Corp, Warren, Mich, 79-81; asst prof, 81-87, ASSOC PROF, MECH ENG DEPT, LAFAYETTE COL, EASTON, PA, 87-, DEPT HEAD, 89- *Concurrent Pos:* Fac mem, Mech Eng Dept, Lafayette Col, Easton, Pa, 81. *Mem:* Am Soc Mech Engrs; Am Inst Aeronaut & Astronaut; Am Soc Eng Educ; Sigma Xi. *Res:* Experimental and computational fluid mechanics; wind tunnel testing; performance testing of fluid mechanic systems; instrumentation; computerized data acquisition. *Mailing Add:* Mech Eng Dept Lafayette Col Easton PA 18042

MERZ, TIMOTHY, b Philadelphia, Pa, Jan 11, 27; c 2. CYTOGENETICS, RADIOBIOLOGY. *Educ:* Johns Hopkins Univ, AB, 51, PhD, 58. *Prof Exp:* NIH fel, Johns Hopkins Univ, 58-60, res assoc, 60-61, from asst prof to assoc prof cytogenetics, 64-75; PROF RADIOL & CHMN DIV RADIATION BIOL, MED COL VA, VA COMMONWEALTH UNIV, 75- *Mem:* AAAS; Radiation Res Soc; Am Soc Cell Biol; Genetics Soc Am; Am Soc Human Genetics. *Res:* Chromosome structure and behavior. *Mailing Add:* MCV Sta Med Col Va Commonwealth Univ Box 40 Richmond VA 23298

MERZ, WALTER JOHN, b Cheadle Hulme, Eng, Oct 10, 20; US citizen; m 49; c 2. SOLID STATE PHYSICS. *Educ:* Swiss Fed Inst Technol, MS, 44, PhD(physics), 48. *Prof Exp:* Res asst solid state physics, Mass Inst Technol, 48-51; vis prof, Pa State Univ, 51; mem tech staff, Bell Tel Labs, Inc, 51-56; mem tech staff, RCA Labs, Inc, NJ, 56-57, mgr res, 57-68, dir res, labs, RCA Ltd, Switz, 68-85; RETIRED. *Mem:* Am Phys Soc. *Res:* Ferroelectrics; dielectrics; semiconductors; photoconductors. *Mailing Add:* 61 Kirchbodenstr 8800 Thalwil Switzerland

MERZ, WILLIAM GEORGE, b Orange, NJ, Dec 20, 41. MEDICAL MYCOLOGY, IMMUNOLOGY. *Educ:* Drew Univ, BA, 63; WVa Univ, MS, 65, PhD(microbiol), 68. *Prof Exp:* NIH fel microbiol, Columbia-Presby Med Ctr, 68-70; instr dermat, Col Physicians & Surgeons, Columbia Univ, 70-73; instr dermat, 73-75, asst prof lab med, 73-80, asst prof dermat & epidemiol, 75-80, ASSOC PROF LAB MED, DERMAT & EPIDEMIOL, JOHNS HOPKINS UNIV, 80-; CLIN MICROBIOLOGIST, DEPT LAB MED, JOHNS HOPKINS HOSP, 73- *Concurrent Pos:* Brown-Hazen grant dermat, Columbia Univ, 70-73. *Honors & Awards:* Bot Award, Ciba Pharmaceut Co, 63. *Mem:* Am Soc Microbiol; Mycol Soc Am; Med Mycol Soc of the Americas; Int Soc Human & Animal Mycoses. *Res:* Rapid techniques for the identification of fungi; immune responses to mycotic infections. *Mailing Add:* Dept Lab Med (Path) Johns Hopkins Hosp 720 Rutland Ave Baltimore MD 21205

MERZBACHER, CLAUDE F, b Philadelphia, Pa, Oct 29, 17; m 45; c 2. NATURAL HISTORY, CHEMICAL ENGINEERING. *Educ:* Univ Pa, BS, 39; Claremont Grad Sch, MA, 50; Univ Poitiers, cert, 51; Univ Calif, Los Angeles, EdD, 61. *Prof Exp:* Teacher math high sch, Fla, 45-46; instr math, physics & chem, Oceanside High Sch & Jr Col, 46-47; instr chem, 47-50, from asst prof to assoc prof phys sci, 50-65, chmn dept, 64-69, PROF NATURAL SCI, SAN DIEGO STATE UNIV, 65-, PLANETARIUM LECTR, 53-, COUNR, 54- *Concurrent Pos:* Dir, NSF Coop Col-Sch Sci Prog, 67-69; pvt pract psychother; consult mgt & leadership creativity. *Mem:* AAAS; Am Psychol Asn; Am Soc Clin Hypnosis; Am Chem Soc; fel Am Inst Chemists. *Res:* Affective interference with mathematics performance; design of integrated courses in physical science; statistical methods; noncognitive processes; psychology; unique therapeutic modalities. *Mailing Add:* 7914 La Mesa Blvd No 19 La Mesa CA 92041-5058

MERZBACHER, EUGEN, b Berlin, Ger, Apr 9, 21; nat US; m 52; c 4. THEORETICAL PHYSICS. *Educ:* Istanbul Univ, Licentiate, 43; Harvard Univ, AM, 48, PhD(physics), 50. *Prof Exp:* Mem, Inst Advan Study, 50-51; vis asst prof physics, Duke Univ, 51-52; from asst prof to prof, 52-69, chmn dept, 77-82, KENAN PROF PHYSICS, UNIV NC, CHAPEL HILL, 69- *Concurrent Pos:* NSF fac fel, Inst Theoret Physics, Copenhagen, 59-60, vis prof, Univ Washington, Seattle, 67-68; Sr US Humboldt Award, Univ Frankfurt, Ger, 77. *Honors & Awards:* Thomas Jefferson Award, Univ NC, 72. *Mem:* Fel AAAS; fel Am Phys Soc (vpres, 88, pres elect, 89, pres, 90); Am Asn Physics Teachers. *Res:* Quantum mechanics; atomic and nuclear theory. *Mailing Add:* Dept Physics & Astron CB No 3255 Univ NC Chapel Hill NC 27599-3255

MERZENICH, MICHAEL MATTHIAS, b Lebanon, Ore, May 15, 42; m 66; c 2. NEUROPHYSIOLOGY, NEUROANATOMY. *Educ:* Univ Portland, BS, 64; Johns Hopkins Univ, PhD(physiol), 68. *Prof Exp:* NIH fel, Univ Wis, 68-71; asst prof, 72-75, assoc prof, 75-80, PROF PHYSIOL & OTOLARYNGOL, UNIV CALIF, SAN FRANCISCO, 80-; DIR, COLEMAN MEM LAB, 71- *Concurrent Pos:* Consult, NIH, 74. *Mem:* AAAS; Soc Res Otolaryngol; Acoust Soc Am; Soc Neurosci. *Res:* Auditory neurophysiology; aids for the profoundly deaf; sensation coding; anatomy and physiology of the central auditory nervous system. *Mailing Add:* Dept Physiol & Otolaryngol HSE 863 Univ Calif 300 Parnassus Box 0732 San Francisco CA 94143

MES, HANS, b Eindhoven, Neth, July 1, 44; Can citizen; m 66; c 2. HIGH ENERGY PHYSICS, INTERMEDIATE ENERGY PHYSICS. *Educ:* Univ Ottawa, BSc, 65, PhD(physics), 68. *Prof Exp:* Fel physics, Univ Geneva, 68-69; fel, Carleton Univ, 69-70, res assoc, 70-72; res officer physics, 72-91, DIR, CTR RES PARTICLE PHYSICS, NAT RES COUN CAN, 90- *Mem:* Am Phys Soc; Inst Elec & Electronics Engrs. *Res:* Elementary particle physics. *Mailing Add:* Ctr Res Particle Physics Herzberg Labs Carleton Univ Ottawa ON K1S 5B6 Can

MESA-TEJADA, RICARDO, b Nov 14, 1942; m 84; c 2. IMMUNOPATHOLOGY, IMMUNOCHEMISTRY. *Educ:* Univ Madrid, Spain, MD, 70. *Prof Exp:* Assoc clin prof path, Col Physicians & Surgeons, Columbia Univ, 83-87; DIR RES & IMMUNOCYTOCHEM, METPATH, INC, 87- *Mem:* Histochem Soc; Am Asn Path; AAAS; Int Acad Path. *Res:* Breast carcinoma antigens. *Mailing Add:* Met Path Inc One Malcolm Ave Teterboro NJ 07608-1070

MESCHAN, ISADORE, b Cleveland, Ohio, May 30, 14; m 43; c 4. RADIOLOGY. *Educ:* Case Western Reserve Univ, BA, 35, MA, 37, MD, 39; Am Bd Radiol, dipl, 57. *Prof Exp:* Intern, Cleveland City Hosp, 39-40; resident, Univ Hosps, Case Western Reserve Univ, 40-42, instr radiol, 46-47; prof & head dept, Sch Med, Univ Ark, 47-55; prof radiol & dir, Bowman Gray Sch Med, Wake Forest Univ, 55-77; RETIRED. *Concurrent Pos:* Consult, Walter Reed Army Hosp; chmn comt radiol, Nat Acad Sci-Nat Res Coun, 74. *Honors & Awards:* Gold Medal, Am Col Radiol, 83. *Mem:* Radiol Soc NAm; Radiation Res Soc; Am Roentgen Ray Soc; Soc Nuclear Med; fel Am Col Radiol; Sigma Xi. *Res:* Radioisotopes and nuclear medicine; radiation biology; diagnostic and therapeutic clinical radiology; author of 25 volumes in radiographic Anatomy, diagnostic radiology and approximately 100 papers and abstracts. *Mailing Add:* Bowman Gray Sch Med Wake Forest Univ Winston-Salem NC 27103

MESCHER, ANTHONY LOUIS, b Celina, Ohio, Mar 5, 49; m 75; c 2. LIMB DEVELOPMENT, TISSUE REPAIR. *Educ:* St Joseph's Col, BS, 71; Ohio State Univ, MS, 73, PhD(develop biol), 75. *Prof Exp:* Postdoctoral res assoc cell biol, Salk Inst Biol Studies, 75-77; asst prof embryol, George Washington Univ, 77-81; ASSOC PROF ANAT, IND UNIV, 82- *Honors & Awards:* Singer Medal for Res in Limb Develop & Regeneration, 86. *Mem:* Soc Develop Biol; Am Asn Anatomists; Am Soc Zoologists; AAAS. *Res:* Mechanisms and factors which tissues interact to promote cell proliferation and growth during tissue repair and regeneration. *Mailing Add:* Anat Sect Med Sci Prog Sch Med Ind Univ Myers Hall Bloomington IN 47405

MESCHER, MATTHEW F, MEMBRANE BIOCHEMISTRY, T-CELL RECOGNITION. *Educ:* Harvard Univ, PhD(biol chem), 76. *Prof Exp:* CHIEF, DIV MEMBRANE BIOL, MED BIOL INST, 85- *Mailing Add:* Med Biol Inst 11077 N Torrey Pines Rd La Jolla CA 92037

MESCHI, DAVID JOHN, b East Chicago, Ind, May 1, 24. HIGH TEMPERATURE CHEMISTRY. *Educ:* Univ Chicago, BA, 49, MS, 52; Univ Calif, PhD(chem), 56. *Prof Exp:* Asst res chemist, Inst Eng Res, Univ Calif, 56-59; resident res assoc, Argonne Nat Lab, 59-60; assoc res chemist, 60-65, CHEMIST, INORG MAT RES DIV, LAWRENCE BERKELEY LAB, UNIV CALIF, 65- *Mem:* Am Chem Soc; Am Phys Soc; Am Ceramic Soc; Sigma Xi. *Res:* High temperature physics. *Mailing Add:* Mail Stop 2-100 Lawrence Berkeley Lab Berkeley CA 94720

MESCHIA, GIACOMO, b Milan, Italy, Feb 7, 26; nat US; m 61; c 3. PHYSIOLOGY. *Educ:* Univ Milan, MD, 50. *Prof Exp:* Asst prof physiol, Univ Milan, 51-53; Toscanini res fel, Sch Med, Yale Univ, 53-55; asst prof physiol, Univ Milan, 55-56; res fel, Josiah Macy Found, Sch Med, Yale Univ, 56-58, res asst, 58-59, asst prof, 59-65; assoc prof, 65-69, PROF PHYSIOL, MED CTR, UNIV COLO, DENVER, 69- *Honors & Awards:* Apgar Award, 84. *Mem:* AAAS; Am Physiol Soc; hon fel Am Gynecol Soc. *Res:* Fetal physiology. *Mailing Add:* Dept Physiol Univ Colo Med Ctr Denver CO 80220

MESCHINO, JOSEPH ALBERT, b Cranston, RI, Aug 23, 32; m 54; c 3. BIOTECHNOLOGY, TECHNOLOGY LICENSING. *Educ:* Brown Univ, ScB, 54; Rice Univ, PhD(org chem), 58. *Prof Exp:* NIH fel org synthesis, Mass Inst Technol, 58-59; sr scientist, Johnson & Johnson, Europe, 59-61, group leader chem develop, 61-67, dir chem res, McNeil Labs, 67-80, dir new prod develop, 80-82; dir life sci, Univ Patents, Inc, 82-87; pres, Pharmacepts Assoc, 87-91; BR CHIEF, DIV AIDS, NAT INST ALLERGY & INFECTIOUS DIS, NIH, 91- *Concurrent Pos:* Indust liaison, Univ Va, 88-90. *Mem:* Am Chem Soc; AAAS. *Res:* Biotechnology licensing; development of anti-AIDS drugs. *Mailing Add:* 55 Patrick Dr Fairfield CT 06430

MESECAR, RODERICK SMIT, b Hot Springs, SDak, May 24, 33; m 52; c 4. PHYSICAL OCEANOGRAPHY, ELECTRICAL ENGINEERING. *Educ:* Ore State Univ, BS, 56, MS, 58, EE, 64, PhD(oceanog), 67. *Prof Exp:* Design engr res lab, Raytheon Co, 58-61; asst prof comput res, 61-64, asst prof oceanog res, 64-74, asst chmn dept oceanog, 71-73, ASSOC PROF OCEANOG RES, ORE STATE UNIV, 74- *Concurrent Pos:* Head, Tech Planning & Develop Group, 68- *Mem:* Inst Elec & Electronics Engrs; Marine Technol Soc. *Res:* Application of electronic circuit designs and instrumentation to computer development and oceanographic research. *Mailing Add:* Colo Oceanog Ore State Univ Corvallis OR 97331

MESELSON, MATTHEW STANLEY, b Denver, Colo, May 24, 30; m 87; c 2. MOLECULAR BIOLOGY. *Educ:* Univ Chicago, PhB, 51; Calif Inst Technol, PhD, 57. *Hon Degrees:* DSc, Oakland Univ, 66, Columbia Univ, 71, Univ Chicago, 75, Yale Univ, 87 & Princeton Univ, 88. *Prof Exp:* Asst prof chem, Calif Inst Technol, 58-60; from assoc prof to prof biol, 60-76, CABOT PROF NATURAL SCI, HARVARD UNIV, 76- *Concurrent Pos:* Consult, US Arms Control & Disarmament Agency, 63-73, MacArthur fel, 84, mem, Coun US Nat Acad Sci, 84-87, mem, Coun Smithsonian Inst, 85-; Me Arthur fel, 84. *Honors & Awards:* Nat Acad Sci Prize Molecular Biol, 63; Eli Lilly Award Microbiol & Immunol, 64; Lehman Award, NY Acad Sci, 75; Leo Szilard Award, Am Physical Soc, 78; Presidential Award, NY Acad Sci, 83; Sci Freedom & Responsibility Award, AAAS, 90. *Mem:* Nat Acad Sci; Inst Med-Nat Acad Sci; Coun Foreign Relations; Academia Santa Chiara (Genoa); fel Am Acad Arts & Sci; Am Philos Soc; Royal Soc London. *Res:* Molecular biology of nucleic acids; mechanisms of DNA recombination and repair; gene control and evolution. *Mailing Add:* Dept Biol Biochem & Molecular Biol Harvard Univ 7 Divinity Ave Cambridge MA 02138

MESERVE, BRUCE ELWYN, b Portland, Maine, Feb 2, 17; m 61; c 3. MATHEMATICS. *Educ:* Bates Col, AB, 38; Duke Univ, AM, 41, PhD(math), 47. *Prof Exp:* Teacher, Moses Brown Sch, RI, 38-41; asst math, Duke Univ, 41-42, 45-46; instr, Univ Ill, 46-47, asst prof, 48-54; from assoc prof to prof, Montclair State Col, 54-64, chmn dept, 57-63; prof, 64-82, EMER PROF MATH, UNIV VT, 82- *Mem:* Fel AAAS; Am Math Soc; Math Asn Am; Nat Coun Teachers Math. *Res:* Geometry; mathematical training of prospective teachers; historical evolution of mathematical sciences. *Mailing Add:* 521 S Paseo del Cobre Green Valley AZ 85614-2321

MESERVE, LEE ARTHUR, b Saco, Maine, Mar 9, 44; m 68. ENDOCRINOLOGY, PHYSIOLOGY. *Educ:* Univ Maine, BS, 66; Rutgers Univ, PhD(zool), 72. *Prof Exp:* Vis asst prof biol, Vassar Col, 72-73; asst prof biol sci, 73-81, adj asst prof geront, health & community serv, 77-81, ASSOC PROF BIOL SCI & ADJ ASSOC PROF GERONT, HEALTH & COMMUNITY SERV, BOWLING GREEN STATE UNIV, 81- *Concurrent Pos:* adj assoc prof allied health, Med Col Ohio, 81- *Mem:* AAAS; Am Aging Asn; Am Soc Zoologists; Geront Soc Am; Endocrine Soc. *Res:* Development of endocrine control; endocrine interactions; aging and endocrinology. *Mailing Add:* Dept Biol Sci Bowling Green State Univ Bowling Green OH 43403

MESERVE, PETER LAMBERT, b Buffalo, NY, Sept 22, 45; m 69; c 2. ECOLOGY. *Educ:* Univ Calif, Davis, BA, 67; Univ Nebr-Lincoln, MS, 69; Univ Calif, Irvine, PhD(biol), 72. *Prof Exp:* Asst prof ecol, Cath Univ, Santiago, 73-75; asst prof zool, Univ Idaho, 75-76; PROF BIOL, NORTHERN ILL UNIV, 76- *Concurrent Pos:* Fulbright res fel, 83-84. *Honors & Awards:* Res Fel, Orgn Am States, 84, 89. *Mem:* Ecol Soc Am; Am Soc Mammalogists; Am Soc Naturalists; Brit Ecol Soc. *Res:* Population and community ecology of vertebrates; biogeography; behavioral ecology. *Mailing Add:* Dept Biol Sci Northern Ill Univ De Kalb IL 60115-2861

MESERVEY, ROBERT H, b Hanover, NH, Apr 1, 21; m 53; c 2. SOLID STATE PHYSICS. *Educ:* Dartmouth Col, BA, 43; Yale Univ, PhD(physics), 61. *Prof Exp:* Physicist, US Army Eng Res & Develop Lab, 51-55; consult, Perkin Elmer Corp, 55-60; physicist, Lincoln Lab, 61-63, SR SCIENTIST, FRANCIS BITTER NAT MAGNET LAB, MASS INST TECHNOL, 63- *Mem:* Fel Am Phys Soc. *Res:* Superconductivity; magnetism; low temperature physics; fluid mechanics; optics. *Mailing Add:* MIT Magnet Lab 170 Albany St Cambridge MA 02139

MESETH, EARL HERBERT, b Chicago, Ill, Nov 29, 38; m 81; c 4. ZOOLOGY. *Educ:* Ill Col, BS, 61; Wash Univ, MA, 62; Southern Ill Univ, Carbondale, PhD(zool), 68. *Prof Exp:* From asst prof to assoc prof, 68-78, PROF BIOL, ELMHURST COL, 78- *Mem:* Sigma Xi; Am Ornithologists Union; Cooper Ornith Soc; Soc Study Evolution. *Res:* Behavior, ecology and biology of albatrosses, Diomedea immutabilis; relationship of courtship rituals to nest site selection and pair bonding. *Mailing Add:* Dept Biol Elmhurst Col 190 Prospect Elmhurst IL 60126

MESHII, MASAHIRO, b Hyogo, Japan, Oct 6, 31; m 59; c 2. MATERIALS SCIENCE. *Educ:* Osaka Univ, BEng, 54, MS, 56; Northwestern Univ, PhD(mat sci), 59. *Prof Exp:* Res assoc & lectr mat sci, 59-60, from asst prof to prof mat sci, 60-88, JOHNS EVANS PROF MAT SCI, NORTHWESTERN UNIV, EVANSTON, 88- *Concurrent Pos:* Fulbright Grant, 56; vis scientist, Nat Res Inst Metals, 70-71; guest prof, Osaka Univ, 85. *Honors & Awards:* Howe Medal, Am Soc Metals, 67; Achievement Award, Japan Inst Metals. *Mem:* fel Am Soc Metals; Am Inst Mining, Metall & Petrol Engrs; Am Phys Soc; Electron Micros Soc Am; Japan Inst Metals; fel Japan Soc Promotion Sci. *Res:* Mechanical properties of metals and composites; electron microscopy. *Mailing Add:* Dept of Mat Sci & Eng Northwestern Univ Technol Inst Evanston IL 60208

MESHKOV, SYDNEY, b Philadelphia, Pa, June 5, 27; m 56; c 3. ELEMENTARY PARTICLE PHYSICS. *Educ:* Univ Pa, AB, 47, PhD(physics), 54; Univ Ill, MS, 49. *Prof Exp:* Asst physics, Univ Ill, 47-49; asst instr, Univ Pa, 49-54; asst prof, Univ Del, 54-55; lectr, Univ Pa, 55-56; asst prof, Univ Pittsburgh, 56-62; physicist, Sr Exec Serv, Nat Bur Standards, 62-90; RETIRED. *Concurrent Pos:* Instr, LaSalle Col, 51-52; res assoc, Princeton Univ, 60; res assoc, Weizmann Inst, 61-62; vis assoc, Calif Inst Technol, 73 & 87-88; secy, Aspen Ctr Physics, 75-; vis prof theoret physics, Calif Inst Technol, 77-78 & 81-82; vis prof physics, Univ Calif, Irvine, 81-82, Los Angeles, 82-83 & 88-89; guest scientist, Superconducting Super Collider Lab, 90-91. *Mem:* Fel Am Phys Soc. *Res:* Elementary particle theory. *Mailing Add:* SSC Lab MS 2001 2550 Beckleymeade Ave Dallas TX 75237

MESHRI, DAYALDAS TANUMAL, b Kaloi, WPakistan, Mar 11, 36; m 66; c 2. INORGANIC CHEMISTRY, PHYSICAL CHEMISTRY. *Educ:* Gujarat Univ, India, BSc, 58, MSc, 62; Univ Idaho, PhD(inorg & phys chem), 68. *Prof Exp:* Demonstr chem, St Xavier's Col, India, 58; demonstr, Gujarat Col, 58-61, asst lectr, 61-62; postdoc, assoc, Cornell Univ, 67-69; res chemist, Ozark-Mahoning Co, 69-70, head fluorine res dept, 70-72, dir fluorine & inorg res, Spec Chem Div, 73-87; PRES, ADVAN RES CHEMICALS INC, 87- *Mem:* Fel Am Inst Chemists; Am Chem Soc; AAAS; Electrochem Soc; Sigma Xi. *Res:* Neutron activation analysis; coordination and fluorine chemistry; nitrogen-fluorine, oxygen fluorine chemistry; electrophilic substitution; hydrogen fluoride chemistry. *Mailing Add:* Advan Res Chemicals Inc 1085 Ft Gibson Rd Catoosa OK 74015

MESIROV, JILL PORTNER, b Philadelphia, Pa, May 12, 50. MATHEMATICS. *Educ:* Univ Pa, AB, 70; Brandeis Univ, MA, 71, PhD(math), 74. *Prof Exp:* lectr math, Univ Calif, Berkeley, 74-76; res math, IDA Commun Res Div, 76; at Am Math Soc, Providence, RI, 82-85; DIR MATH SCI RES, THINKING MACH CORP, 85- *Concurrent Pos:* Vis lectr, Princeton Univ, 80 - 81; exec dir, Int Cong Mathematicians, 83-87. *Mem:* Soc Indust & Appl Math; Am Math Soc; Asn Women Math (pres, 88-91); Asn Comput Math; AAAS. *Res:* Parallel computing; applied mathematics. *Mailing Add:* 54 Hurd Rd Belmont MA 02178

MESKIN, LAWRENCE HENRY, b Detroit, Mich, July 21, 35; m 59; c 2. DENTAL EPIDEMIOLOGY. *Educ:* Univ Detroit, DDS, 61; Univ Minn, Minneapolis, MSD, 63, MPH, 64, PhD(epidemiol), 66. *Prof Exp:* Instr oral path, Sch Dent, Univ Minn, Minneapolis, 63-66, assoc prof prev dent & chmn div, 66-68, chmn, Div Health Ecol, 68-81, Hill res prof, Delivery Dent Health Serv, 70-81, lectr pediat, Sch Med, 63-81; DEAN, SCH DENT, UNIV COLO, 81- *Concurrent Pos:* USPHS fel epidemiol, 63-; consult, Cleft Palate Clin, Univ Ill, 64-; partic, Inst Advan Educ Dent Res, 64; WHO traveling fel, 68. *Mem:* Am Dent Asn; Am Acad Oral Path; Int Asn Dent Res; Am Pub Health Asn; Am Asn Dent Sch. *Res:* Preventive dentistry; dental public health; craniofacial malformations; health care delivery. *Mailing Add:* Sch Dent Univ Colo 4200 E 9th Ave Denver CO 80262

MESLER, R(USSELL) B(ERNARD), b Kansas City, Mo, Aug 24, 27; m 51; c 4. CHEMICAL ENGINEERING. *Educ:* Univ Kans, BS, 49; Univ Mich, MS, 53, PhD(chem eng), 55. *Prof Exp:* Process engr, Colgate-Palmolive Co, 49-51; asst Ford nuclear reactor, Univ Mich, 53-55, asst prof nuclear eng & proj engr, Ford nuclear reactor, 55-57; from assoc prof to prof, 57-70, WARREN S BELLOWS DISTINGUISHED PROF CHEM ENG, UNIV KANS, 70- *Concurrent Pos:* Resident res engr, Argonne Nat Lab, 58; consult, Spencer Chem Co, Kerr McGee, Westinghouse Elec, Farmland Industs, Berkeley Nuclear Labs & Cent Elec Generating Bd, England, 75-76, 84-85; fac res participant, Savannah River Lab, E I du Pont de Nemours, Co, Inc, 81. *Honors & Awards:* Robert T Knapp Award, Am Soc Mech Engrs, 67. *Mem:* Am Chem Soc; fel Am Inst Chem Engrs; Am Soc Eng Educ; Am Nuclear Soc; Sigma Xi. *Res:* Heat transfer, especially nucleate boiling; nuclear technology; cavitation; high speed photography. *Mailing Add:* Dept of Chem & Petrol Eng Univ of Kans Lawrence KS 66044

MESLOW, E CHARLES, b Waukegan, Ill, Aug 25, 37; m 59; c 3. ECOLOGY, WILDLIFE RESEARCH. *Educ:* Univ Minn, BS, 59, MS, 66; Univ Wis, PhD(wildlife ecol), 70. *Prof Exp:* Asst prof zool & vet sci, NDak State Univ, 68-71; asst leader, 71-75, LEADER ORE COOP WILDLIFE RES UNIT, ORE STATE UNIV, 75-, PROF WILDLIFE ECOL, 82- *Concurrent Pos:* rep Mt St Helens Sci Adv Bd; mem, various spotted owl conserv & recovery teams. *Mem:* Ecol Soc Am; Wildlife Soc (pres, 84-86); Am Soc Mammal; Am Ornith Union. *Res:* Population dynamics; predation; wildlife ecology; spotted owl. *Mailing Add:* Coop Wildlife Res Unit Ore State Univ Corvallis OR 97331

MESNER, MAX H(UTCHINSON), b Meadville, Mo, Apr 16, 12; m 37; c 1. ELECTRONICS ENGINEERING. *Educ:* Univ Mo, BS, 40. *Prof Exp:* Design engr, RCA Corp, 40-41, res engr, labs, 41-58, proj engr, 58-60, eng group leader, 60-61, eng mgr, astro-electronics div, Princeton, 61-81; RETIRED. *Concurrent Pos:* Dir, Welfare, Cranbury, NJ, 86- *Mem:* Fel Inst Elec & Electronics Engrs; Soc Photo-Optical Instrument Eng; Sigma Xi; assoc fel Am Inst Aeronaut & Astronaut. *Res:* Television camera systems for space vehicles; fourteen US patents. *Mailing Add:* Nine Wynnewood Dr Cranbury NJ 08512

MESNIKOFF, ALVIN MURRAY, b Asbury Park, NJ, Dec 25, 25; m 52; c 4. PSYCHIATRY, PSYCHOANALYSIS. *Educ:* Rutgers Univ, BA, 48; Univ Chicago, MD, 54. *Prof Exp:* Asst chief male serv, NY State Psychiat Inst, 58-60, chief female psychiat serv, 60-65, dir, Wash Heights Community Serv, 65-68; dir, South Beach Psychiat Ctr, 68-75, NY CITY REGIONAL DIR, NY STATE OFF MENT HEALTH, 75-; PROF PSYCHIAT, STATE UNIV NY DOWNSTATE MED CTR, 68- *Concurrent Pos:* Assoc clin prof psychiat, Columbia Univ, 60-68; assoc attend psychiatrist, Columbia Presby Hosp, 68-69; lectr psychiatry, Columbia Univ, 68; attend psychiatrist, Kings County Hosp, 69- & St Vincent's Med Ctr, 70- *Mem:* AAAS; fel Am Psychiat Asn; NY Acad Sci. *Res:* Community psychiatry; design of programs; evaluation of mental health services. *Mailing Add:* Dept Psychiat 360 Central Park W New York NY 10025

MESROBIAN, ROBERT BENJAMIN, b New York, NY, July 31, 24; m 50; c 5. CHEMISTRY. *Educ:* Princeton Univ, BA, 44, MS, 45, PhD(phys chem), 47. *Prof Exp:* Res assoc & proj adminr, Polytech Inst Brooklyn, 47-49, from asst prof to assoc prof polymer chem, 49-54, prof & assoc dir, Polymer Res Inst, 55-57; assoc dir res high polymer chem, Cent Res & Eng Div, 57-58, gen mgr, Gen Packaging Res & Develop Div, 59-64, gen mgr, Cent Res & Eng Div, 64-67, gen mgr res & eng, Continental Can Co, Inc, 67-69, vpres, res & eng, Continental Packaging Co Inc, 69-84; VPRES TECHNOL, VIATECH INC, 84- *Concurrent Pos:* Co-holder, Chaire Franqui lectr, Univ Liege, 47-48; US Educ Found vis prof, State Univ Groningen, 50; consult, Nuclear Eng Div, Brookhaven Nat Lab, 51-; US State Dept adv, Atoms for Peace Conf, Geneva, 58. *Mem:* Am Chem Soc; Soc Plastics Eng; Am Inst Chem; Sigma Xi. *Res:* Synthesis and properties of polymers; oxidation of hydrocarbons; organic peroxides; effects of ionizing radiation on polymers; application of polymers for coatings, adhesives, inks and packaging. *Mailing Add:* 188 Long Lots Rd Westport CT 06880

MESSAL, EDWARD EMIL, b Chicago, Ill, Nov 24, 37; m 67; c 3. MECHANICAL ENGINEERING. *Educ:* Ill Inst Technol, BS, 59, MS, 63, PhD(mech eng), 70. *Prof Exp:* Instr mech eng, Ill Inst Technol, 60-66; from instr to asst prof eng sci, Roosevelt Univ, 66-70; from asst prof to assoc prof, 70-84, PROF MECH ENG TECHNOL & DIR START REHAB ENG CLINIC, IND UNIV-PURDUE UNIV, FT WAYNE, 84- *Concurrent Pos:* Tech writer, Central Soya Co, Inc, 77-78. *Mem:* Assoc Am Soc Mech Engrs; Am Soc Eng Educ. *Res:* Machine design; medical apparatus; hydraulics and pneumatics; stress analysis; rotor dynamics; mathematics. *Mailing Add:* 642 Lyell Ct Ft Wayne IN 46825

MESSENGER, GEORGE CLEMENT, b Bellows Falls, Vt, July 20, 30; m 54; c 3. PHYSICS, ELECTRICAL ENGINEERING. *Educ:* Worcester Polytech Inst, BS, 51; Univ Pa, MS, 57; Calif Coast Univ, PhD, 86. *Prof Exp:* Res scientist solid state physics, Philco Corp, 51-59; eng mgr semiconductor div, Hughes Aircraft Co, 59-61; div mgr transistor div, Transitron Corp, 61-62; staff scientist, Northrop Corp, 63-68; CONSULT ENGR SOLID STATE PHYSICS, 68- *Concurrent Pos:* Lectr, Univ Calif, Los Angeles, 69-74; consult, Defense Nuclear Agency, 70. *Honors & Awards:* Alan Berman Award, Naval Res Lab, 82. *Mem:* Fel Inst Elec & Electronics Engrs; Res Soc Am. *Res:* Radiation effects on electronic components and systems; research and development of solid state electronic components; development of quality control and hardness assurance programs for systems; author of a book. *Mailing Add:* 3111 Bel Air Dr 7F Las Vegas NV 89109

MESSENGER, JOSEPH UMLAH, b Medicine Hat, Alta, Aug 5, 13; US citizen; m 41; c 3. INORGANIC CHEMISTRY, PHYSICAL CHEMISTRY. *Educ:* Univ Calif, Berkeley, AB, 35, BS, 39; Univ Southern Calif, MS, 42. *Prof Exp:* Res chemist, Nat Defense Res Coun, Univ Southern Calif, 41-42 & Univ Chicago, 42-43; from asst res chemist to res chemist, Field Res Lab, Socony Mobil Oil Co, Inc, 43-46, from sr chemist to sr res chemist, 46-50, sr res technologist, 50-54, drilling mud engr, Mobil Oil Can, Ltd, Alta, 54-58, sr staff engr, 58-60, chem eng sect chief, 60-62, eng specialist, Field Res Lab, Mobil Oil Corp, Tex, 62-67, eng assoc, Field Res Lab, Mobil Res & Develop Corp, 67-78; RETIRED. *Concurrent Pos:* Drilling eng consult, 78- *Mem:* Soc Petrol Eng; Am Inst Mining, Metall & Petrol Eng; Sigma Xi; fel Am Inst Chemists. *Res:* Chemical and petroleum engineering; chemical well stimulation, drilling muds, cements, lost circulation, water injection and corrosion; contact catalysis; fluorine, boron and uranium chemistry; drilling engineering. *Mailing Add:* 2906 Gladiolus Lane Dallas TX 75233

MESSENGER, ROGER ALAN, b St Paul, Minn, Aug 26, 43; m 66; c 1. ELECTRICAL ENGINEERING. *Educ:* Univ Minn, BS, 65, MSEE, 66, PhD(elec eng), 69. *Prof Exp:* US Air Force Off Sci Res fel, 69-71, asst prof elec eng, 71-75, ASSOC PROF ELEC ENG, FLA ATLANTIC UNIV, 75- *Mem:* Inst Elec & Electronics Engrs. *Res:* Energy conservation technology. *Mailing Add:* Dept of Elec Eng Fla Atlantic Univ Boca Raton FL 33431

MESSER, CHARLES EDWARD, b Baltimore, Md, Aug 16, 15; m 58. PHYSICAL CHEMISTRY. *Educ:* Johns Hopkins Univ, AB, 36, PhD(phys chem), 40. *Prof Exp:* Res chemist, Biochem Res Found, Del, 41 & Nat Defense Res Comt, US Bur Mines, Pa, 41-42; instr chem, Clarkson Tech, 42-44 & Dartmouth Col, 44-46; from instr to asst prof, 46-53, assoc prof, 53-80, emer prof chem, Tufts Univ, 80; RETIRED. *Mem:* Am Chem Soc; AAAS; Sigma Xi. *Res:* Calorimetry; phase studies. *Mailing Add:* Ten Longwood Dr Apt 408 Westwood MA 02090

MESSER, LOUISE BREARLEY, pediatric dentistry, preventive dentistry, for more information see previous edition

MESSER, WAYNE RONALD, b Cedar Rapids, Iowa, Nov 7, 42; m 64; c 2. ORGANIC CHEMISTRY, PHOTOCHEMISTRY. *Educ:* Iowa State Univ, BS, 64; Univ Ill, PhD(chem), 68. *Prof Exp:* Res chemist, Cent Res Div, Hercules, Inc, 68-77; SR RES CHEMIST, CHEMICAL SCI DIV, HERCULES, INC, 77- *Mem:* Am Chem Soc; Inter-Am Photochem Soc; Sigma Xi. *Res:* Nitrogen heterocycles; cycloadditions; concerted reactions; physical organic chemistry; photopolymerization. *Mailing Add:* Seven Queen Lane RD 2 Landenberg PA 19350

MESSER, WILLIAM SHERWOOD, JR, b Greenfield, Mass, July 28, 58; m 90. NEUROCHEMISTRY, MEDICINAL CHEMISTRY. *Educ:* Springfield Col, BS, 79; Univ Rochester, MS, 82, PhD(neurosci), 85. *Prof Exp:* Asst prof med chem, 85-90, ASSOC PROF MED & BIOL CHEM, UNIV TOLEDO, 90- *Concurrent Pos:* Ad hoc reviewer, Vet Admin Neurobiol, Merit Rev Bd, 91. *Mem:* Soc Neurosci; Am Asn Col Pharm; NY Acad Sci; AAAS; Am Chem Soc. *Res:* Interaction of drugs with neurotransmitter receptors to the development of novel therapies for cognitive disorders. *Mailing Add:* Univ Toledo 2801 W Bancroft St Toledo OH 43606

MESSERLE, LOUIS, b Jersey City, NJ, Apr 6, 53; m; c 1. ORGANOMETALLIC CHEMISTRY. *Educ:* Brown Univ, ScB, 75; Mass Inst Technol, PhD(inorg chem), 79. *Prof Exp:* Fel, Univ Mich, 79-81, Mich Soc fels scholar & asst prof inorg chem, dept chem, 81-84; ASSOC PROF INORG CHEM, DEPT CHEM, UNIV IOWA, 84- *Concurrent Pos:* Lectr, dept chem, Univ Mich, 81-84; Old Gold fel, Univ Iowa, 85 - 86. *Mem:* Am Chem Soc; Sigma Xi. *Res:* Synthetic and mechanistic organotransition metal chemistry; homogeneous catalysis; multinuclear magnetic resonance spectroscopy of transition metal complexes, particularly carbon 13 nuclear magnetic resonance; synthesis and reactivity of metal-metal bonded dinuclear complexes of the early transition metals. *Mailing Add:* Dept Chem Univ Iowa Iowa City IA 52242

MESSERLI, FRANZ HANNES, b Bern, Switz, Aug 1, 42; m 70; c 3. HYPERTENSION. *Educ:* Univ Bern, Switz, BS, 61, MD, 70. *Prof Exp:* DIR, CLIN HYPERTENSION LAB, OCHSNER MED INST, 76-; PROF MED, TULANE MED SCH, 77- *Concurrent Pos:* Ed, J Human Hypertension, 87-; ed, Geriat Cardiovasc Med, 87-; mem, FDA Cardiorenal Adv Comt. *Honors & Awards:* Prize, Swiss Soc Int Med, 84. *Mem:* Fel Am Col Cardiol; fel Am Col Physicians; Am Soc Hypertension; Am Fedn Clin Res; hon mem Southern African Hypertension Soc; hon mem Peruvian Soc Cardiol; corresp mem Ecuadorian Col Physicians & Surgeons. *Res:* Research in the effects of arterial hypertension, obesity, and age on target organs such as the heart, the kidney, and peripheral vascular beds, prevention or reversal of such target disease by specific cardiovascular therapy. *Mailing Add:* Ochsner Clin 1514 Jefferson Highway New Orleans LA 70121

MESSERSCHMITT, DAVID G, ELECTRICAL ENGINEERING. *Educ:* Univ Colo, BS, 67; Univ Mich, MS, 71. *Prof Exp:* From asst prof to assoc prof, 77-81, PROF, DEPT ELEC ENG & COMPUTER SCI, UNIV CALIF, BERKELEY, 81- *Concurrent Pos:* Mem tech staff, Bell Labs, 68-74, supvr, 74-77; dir indust liaison prog, 81-83; bd dirs, Teknekron Infoswitch, 83-86, Coastcom Inc, 88-90; co-founder, Teknekron Commun Syst, Berkeley, 83; vis prof, Nippon Telephon & Telegraph Co, 87; vchmn, Computer Resources, 90-; consult, TRW Vidar, Intel, IBM Almaden Res Lab, Hughes Aircraft Ground Syst Group, Mass Inst Technol Lincoln Labs, GTE Serv Corp, Am

Satellite, Honeywell Corp, Intelsat, Teknekron Infoswitch, Teknekron Commun Syst, Tex Instruments, Silicon Systs, Hitachi, RCA, Databit/ Siemens, Avantek, IBM Develop Lab, Boeing Telecommun, Contel Adv Technol Ctr, Sharp Electronics, Yamaha. *Mem:* Nat Acad Eng; fel Inst Elec & Electronics Engrs; Sigma Xi. *Mailing Add:* Dept Elec Eng & Computer Sci Univ Calif 517 Cory Hall Berkeley CA 94720

MESSERSMITH, DONALD HOWARD, b Toledo, Ohio, Dec 17, 28; m 57; c 4. ENTOMOLOGY, ORNITHOLOGY. *Educ:* Univ Toledo, BEd, 51; Univ Mich, MS, 53; Va Polytech Inst, PhD(entom), 62. *Prof Exp:* Prof biol, Radford Col, 57-64; prof entom, 64-, SCI TEACHING CTR, COL EDUC, UNIV MD, COLLEGE PARK. *Mem:* Entom Soc Am. *Res:* Biology and taxonomy of Culicoides, Forcipomyia and Simuliidae. *Mailing Add:* Sci Teaching Ctr Col Educ Univ MD College Park MD 20742

MESSERSMITH, JAMES DAVID, b Paintsville, Ky, Sept 14, 31; m 60; c 4. FISHERIES MANAGEMENT. *Educ:* Ore State Univ, BS, 53, MS, 58. *Prof Exp:* Res fel, Ore Coop Wildlife Res, 56-58; fishery biologist, 58-60, marine biologist, 61-63, assoc marine biologist, 63-69, sr marine biologist, 69-72, coordr state-fed fisheries mgt progs, 72-75, conserv prog officer & legis coordr, 75-81, ASST DEP DIR OPER, CALIF DEPT FISH & GAME, 81- *Concurrent Pos:* Proj mgr, Dungeness Crab Mgt Proj, Pac Fishery Biol. *Mem:* Am Fisheries Soc; Am Inst Fishery Res Biol. *Res:* Legislation and management with reference to marine fauna of the northeastern Pacific Ocean with emphasis on fish, mollusks and crustaceans of sport and commercial importance. *Mailing Add:* 4445 Northampton Dr Carmichael CA 95608

MESSERSMITH, ROBERT E, b Trenton, NJ, Mar 15, 30; m 57; c 3. VETERINARY MEDICINE. *Educ:* Cornell Univ, DVM, 54. *Prof Exp:* Vet, pvt pract, 54-61; vet, Agr Div, Am Cyanamid Co, NJ, 61-63, mgr swine prog, 63-68; clin vet, Animal Health Res Dept, 68-74, PROF SERV VET, DEPT AGR & ANIMAL HEALTH, CHEM DIV, HOFFMANN-LA ROCHE, INC, 74- *Mem:* Am Vet Med Asn. *Res:* Cause of problems in animal production and development of practical methods of control. *Mailing Add:* Chem Div Hoffmann-La Roche Inc 1521 Angelina Bend Denton TX 76205

MESSICK, ROGER E, b Chicago, Ill, Jan 20, 29; m 50; c 2. APPLIED MATHEMATICS, ENGINEERING. *Educ:* Univ Ill, BS, 51, MS, 52; Calif Inst Technol, PhD(eng sci), 62. *Prof Exp:* Engr, Arnold Eng Develop Ctr, ARO, Inc, 52-53; instr eng math, Calif Inst Technol, 63; asst prof math, Case Western Reserve Univ, 63-67; ASSOC PROF ENG SCI & MATH, UNIV CINCINNATI, 67- *Concurrent Pos:* Consult, Aerojet-Gen Corp, Calif, 57-62; Fulbright grad fel & univ grant, Univ Sydney, 62-63. *Mem:* Am Math Soc; Math Asn Am; Am Asn Physics Teachers; Soc Indust & Appl Math; Am Acad Mech; Sigma Xi. *Res:* Partial differential equations; singular perturbations; asymptotic and numerical approximations; continuum mechanics; elastic shell theory; free surface gravity waves; boundary layer and edge effects. *Mailing Add:* Dept Elec & Comput Eng ML30 Univ of Cincinnati Cincinnati OH 45221

MESSIER, BERNARD, b Montreal, Que, May 4, 26; m 53; c 4. EXPERIMENTAL PATHOLOGY. *Educ:* Univ Montreal, BS, 49; McGill Univ, MSc, 56, PhD(anat), 60. *Prof Exp:* Asst path, Col Physicians & Surgeons, Columbia Univ, 61-62; asst prof med, 64-67, from asst prof to assoc prof anat, 67-75, PROF ANAT, UNIV MONTREAL, 75- *Mem:* Am Asn Anatomists; Can Asn Anatomists. *Res:* Radioautography; cell renewal in normal tissues. *Mailing Add:* Dept of Anat Univ of Montreal Montreal PQ H3C 3J7 Can

MESSIER, DONALD ROYAL, b Springfield, Mass, Oct 3, 32; m 83. CERAMICS ENGINEERING. *Educ:* Alfred Univ, BS, 59; Univ Calif, Berkeley, MS, 61, PhD(eng), 64. *Prof Exp:* Asst ceramic engr, Argonne Nat Lab, 64-68; RES CERAMIC ENGR, ARMY MAT TECHNOL LAB, 68- *Concurrent Pos:* Fel, Am Ceramic Soc. *Mem:* Am Ceramic Soc; Sigma Xi. *Res:* High temperature behavior of inorganic materials; ceramic fabrication processes; nitrogen glass-ceramics. *Mailing Add:* Army Mat Technol Lab SLCMT-EMC Watertown MA 02172-0001

MESSIER, RUSSELL, b Nashua, NH, July 30, 44; m 69; c 2. THIN FILMS, RADIO FREQUENCY-SPUTTERING. *Educ:* Northeastern Univ, BS, 67; Pa State Univ, PhD(solid state sci), 73. *Prof Exp:* Res assoc, Pa State Univ, 73-74; advan res eng, GTE Sylvania, 74-76; from res assoc to sr res assoc, Pa State Univ, 76-81, assoc prof mat res, 81-85, assoc prof eng sci mech, 85-89, PROF ENG SCI MECH, PA STATE UNIV, 89- *Mem:* Inst Elec & Electronics Engrs; Am Vacuum Soc; Mat Res Soc. *Res:* Thin film preparation and characterization using the radio frequency-sputtering technique; characterization of the sputtering process; noncrystalline solid formation; thin films for solar energy conversion. *Mailing Add:* 230D Hammond Bldg Pa State Univ University Park PA 16802

MESSIHA, FATHY S, b Cairo, Egypt, Oct 2, 36. TOXICOLOGY. *Educ:* Univ Bern, Switz, PhD(physio-biochem), 65. *Prof Exp:* Prof path & dir, Div Toxicol, Tex Tech Univ, 80-87; PROF PHARMACOL, UNIV NDAK SCH MED, 87- *Mem:* Am Acad Clin Toxicol; Acad Pharmaceut Sci; Am Soc Pharmacol & Exp Therapeut. *Res:* Neuropharmacology; neurotoxicology; drug metabolism. *Mailing Add:* Dept Pharmacol Univ NDak Sch Med 501 N Columbia Rd Grand Forks ND 58203

MESSINA, CARLA GRETCHEN, b Ames, Iowa, July 22, 37; m 62; c 2. COMPUTER SCIENCE, PHYSICS. *Educ:* Univ Md, BS, 59; George Washington Univ, MS, 62. *Prof Exp:* Physicist, 55-85, SYST ANALYST, NAT BUR STANDARDS & AM CERAMIC SOC, 85- *Honors & Awards:* Silver Metal US Dept Com. *Res:* Data processing; data transformation; scientific text typesetting; data base design. *Mailing Add:* 9800 Marquette Dr Bethesda MD 20817

MESSINA, EDWARD JOSEPH, b Brooklyn, NY, May 28, 37; m 60; c 1. CARDIOVASCULAR PHYSIOLOGY. *Educ:* St John's Univ, BSc, 60; NY Med Col, PhD(physiol), 72. *Prof Exp:* NIH fel, 72-73; from instr to assoc prof, 73-81, PROF PHYSIOL, NY MED COL, 82- *Concurrent Pos:* Study sect, Cardiovascular & Renal, NIH, 84-87. *Honors & Awards:* Herman Tarnower Award, Am Heart Asn, 88. *Mem:* AAAS; Microcirc Soc; NY Acad Sci; Am Physiol Soc. *Res:* Understanding the interrelationships between those local factors which contribute to the regional regulation of blood flow; systemic regulations of blood pressure. *Mailing Add:* Dept Physiol NY Med Col Valhalla NY 10595

MESSINA, FRANK JAMES, b Lawrence, Mass, Jan 24, 55; m 84. INSECT ECOLOGY, INSECT BEHAVIOR. *Educ:* Clark Univ, BA, 76; Cornell Univ, PhD(ecol), 82. *Prof Exp:* Teaching asst intro biol, Cornell Univ, 76-78, res asst, 78-81; fel assoc, Boyce Thomposon Inst Plant Res, 82-85, res assoc, 85-86; ASST PROF, DEPT BIOL, UTAH STATE UNIV, 86- *Mem:* Ecol Soc Am; Soc Study Evolution; Entom Soc Am. *Res:* Feeding and oviposition behavior of phytophagous insects, with reference to biological control; evolution of insect feeding habits and mating systems; insect dispersal. *Mailing Add:* Dept Biol Utah State Univ Logan UT 84322-5305

MESSINEO, LUIGI, b Bronte, Italy, May 25, 26; US citizen; m 68; c 4. BIOCHEMISTRY, BIOPHYSICS. *Educ:* Univ Palermo, Lic clas, 46, PhD(natural sci), 53; Inst Philos, Messina, Italy, 49. *Prof Exp:* Vis investr biochem physiol, Univ Calif, Berkeley, 58-59; vis investr biophys, Univ Pittsburgh, 59-61; res chemist, Vet Admin Hosp, Buffalo, NY, 62-67; dir biochem res lab, Vet Admin Ctr, 67-70; PROF BIOL & CHEM, CLEVELAND STATE UNIV, 70- *Concurrent Pos:* Vis scientist, LaStazione Zoologica, Italy, 53; vis investr, Univ Pittsburgh, 57; Damon Runyon Mem Found res fel, 58-61; Health Res & Serv Found grant, 60-61, Leukemia Found, 62, Health Res Found Western NY, 63, Nat Cancer Inst, 64-67 & Am Heart Asn, 69-70; res assoc, Nat Cancer Inst, 61-62; res asst prof, State Univ NY Buffalo, 62-67 & Xavier Univ, Ohio, 67- *Honors & Awards:* Knight of the Order of Merit, Italian Repub. *Res:* Physiochemical and immunological properties of deoxyribonucleoproteins from normal and abnormal sources; aging. *Mailing Add:* Dept Biol Cleveland State Univ Euclid Ave & E 24th St Cleveland OH 44115

MESSING, FREDRIC, b Brooklyn, NY, June 8, 48; m 70; c 1. PHYSICS. *Educ:* Carnegie-Mellon Univ, BS, 70; Univ Pa, PhD(physics), 75. *Prof Exp:* Res assoc, Lab Nuclear Studies, Cornell Univ, 74-77; asst prof, 77-81, ASSOC PROF PHYSICS, DEPT PHYSICS, CARNEGIE-MELLON UNIV, 81- *Concurrent Pos:* Consult, Deutsches Elektronen Synchrotron, 80. *Mem:* Am Phys Soc. *Res:* Lepton beams; weak interactions; quark models; resonance production and decay. *Mailing Add:* 11545 Daffodil Lane Silver Spring MD 20902

MESSING, JOACHIM W, b Duisburg, WGer, Sept 10, 46; m 75; c 1. MOLECULAR BIOLOGY. *Educ:* Apothekerkammer, BS, 68; Free Univ, MS, 71; LM Univ, Dr Rer Nat, 75. *Prof Exp:* Res fel, Max Planck Inst Biochem, Munich, 75-78; res assoc bact, Univ Calif, Davis, 78-80; from asst prof to prof biochem, Univ Minn, 80-85; dir res, 85-88, UNIV PROF MOLECULAR BIOL, WAKSMAN INST, RUTGERS UNIV, 85-, DIR, 88- *Concurrent Pos:* Vis res assoc biochem & biophys, Univ Calif, San Francisco, 78; instr, Plant Molecular Biol Course, Cold Spring Harbor, NY, 84-88; assoc ed, J Biotechnol, 85-; ed, Gene, 88-; actg chair, Molecular Biol & Biochem Dept, Rutgers Univ, 89-90. *Mem:* AAAS; Am Soc Biol Chem; Am Soc Microbiol; Int Soc Plant Molecular Biol. *Res:* Regulation of gene expression in higher plants, M13 clining, sequencing, gene synthesis. *Mailing Add:* Waksman Inst Rutgers Univ PO Box 759 Piscataway NJ 08855

MESSING, KAREN, b Springfield, Mass, Feb 2, 43. HUMAN GENETICS. *Educ:* Harvard Univ, BA, 63; McGill Univ, MSc, 70, PhD(biol), 75. *Prof Exp:* Res asst biochem, Jewish Gen Hosp, Montreal, 70-71; NIH fel genetics, Boyce Thompson Inst Plant Res, 75-76; PROF, UNIV QUE, MONTREAL, 76- *Concurrent Pos:* Invited researcher, Inst Cancer Montreal, 83-; mem bd dirs, Quebec Sci & Technol Mus, 84-86; Quebec Coun Social Affairs, 84-90; co-dir, Res Group in Work Biol. *Honors & Awards:* Muriel Duckworth Award. *Mem:* AAAS; Genetics Soc Am; Genetics Soc Can; Environ Mutagenesis Soc. *Res:* Molecular study of human spontaneous and induced mutants; methods for detection of genotoxic effects in the work place. *Mailing Add:* Dept Biol Univ Quebec CP 8888 Montreal PQ H3C 3P8 Can

MESSING, RALPH ALLAN, enzymology; deceased, see previous edition for last biography

MESSING, RITA BAILEY, b Brooklyn, NY, July 7, 45; m 65; c 2. NEUROTOXICOLOGY, HEALTH RISK ASSESSMENT. *Educ:* Brooklyn Col, BA, 66; Princeton Univ, PhD(psychol), 70. *Prof Exp:* Asst prof psychol, Rutgers Univ, 69-72; res assoc, Mass Inst Technol, 73-74, Med Found fel neuropharmacol, 74-75; assoc researcher psychobiol & neuropharmacol, Univ Calif, Irvine, 76-81; res assoc, 81-83, asst prof, 83-88, ASSOC PROF, DEPT PHARMACOL, UNIV MINN, 88-; ENVIRON TOXICOLOGIST, MINN DEPT HEALTH, 90- *Concurrent Pos:* Co-prin investr, USPHS grants, Univ Calif, Irvine, 76-81; prin investr, Univ Minn, 83-89. *Mem:* Am Soc Pharmacol & Exp Therapeut; Soc Neurosci; AAAS. *Res:* Environmental mechanisms of neurotoxicity; health risk assessment; environmental review. *Mailing Add:* Div Environ Health Minn Dept Health 925 SE Delaware St Minneapolis MN 55459-0040

MESSING, SHELDON HAROLD, b Apr 6, 47; US citizen; m 67; c 2. ORGANIC CHEMISTRY, POLYMER CHEMISTRY. *Educ:* Brooklyn Col, BS, 67; Polytech Inst Brooklyn, PhD(org chem), 72. *Prof Exp:* res leader, 81-84, SR RES CHEMIST, DOW CHEM CO, 72-, GROUP LEADER, 84- *Mem:* Am Chem Soc; Sigma Xi. *Res:* Physical organic chemistry applied to the synthesis and development of processes for compounds applicable in the agricultural field; pharmaceutical compounds and ion-exchange resins. *Mailing Add:* 2801 Whitewood Dr Midland MI 48640

MESSING, SIMON D, b Frankfurt-am-Main, Ger, July 13, 22; US citizen; m 67; c 1. CULTURAL ANTHROPOLOGY, MEDICAL ANTHROPOLOGY. *Educ:* City Col New York, BSS, 49; Univ Pa, PhD(anthrop), 57. *Prof Exp:* Interdisciplinary res, Behav Inst, Univ Pa, 52-53; asst prof soc sci, Paine Col, 56-58; assoc prof anthrop, Hiram Col, 58-60; assoc prof, Univ SFla, 60-64; researcher & field consult, US AID-Ethiopia, 61-67; prof, 68-89, EMER PROF ANTHROP, SOUTHERN CONN STATE UNIV, 90- *Mem:* AAAS; Am Anthrop Asn; fel Soc Appl Anthrop; fel Am Pub Health Asn; fel Am Pub Health Asn. *Res:* Applied anthropology of Africa, especially in public health attitudes and practices; author of 3 books, many articles and reviews. *Mailing Add:* 58 Shepards Knolls Dr Hamden CT 06514

MESSINGER, HENRY PETER, b Vienna, Austria, July 1, 21; US citizen; m 56; c 2. ELECTRICAL ENGINEERING. *Educ:* Okla State Univ, BS, 43; Ill Inst Technol, MS, 46; Univ Ill, PhD(elec eng), 51. *Prof Exp:* Eng physicist, Capehart Farnsworth Corp, Int Tel & Tel Corp, 50-52; elec engr, Inst Air Weapons Res, Univ Chicago, 52-56; asst prof, 56-63, ASSOC PROF ELEC ENG, ILL INST TECHNOL, 63- *Concurrent Pos:* Mem staff, ITT Kellogg, 58-63, Argonne Nat Lab, 63-64, Amphenol Corp, 66-67, Amtron Inc, 68-70 & Teletype Co, 70-71. *Mem:* Sigma Xi. *Res:* Computer logic and fault diagnostics; operations research in connection with evaluations of air weapons effectiveness; detailed study of magnetic fields in electrical machines; power electronics; microwave engineering. *Mailing Add:* Dept Elec Eng Ill Inst Technol 3300 S Federal St Chicago IL 60616

MESSINGER, RICHARD C, b Apr 3, 30; c 3. ENGINEERING ADMINISTRATION. *Educ:* Univ Cincinnati, BS, 53; Xavier Univ, MBA, 63. *Prof Exp:* Dir, Patent & Com Develop Dept, Cincinnati Milacron Chem, 64-69, Corp Res & Develop, 69-72, vpres res & develop, 72-86, vpres & chief tech officer, 86-90; RETIRED. *Mem:* Nat Acad Eng; Nat Mach Tool Builders Asn; Nat Asn Mfrs. *Mailing Add:* 7360 Algonquin Dr Cincinnati OH 45253

MESSMER, DENNIS A, b Wessington Springs, SDak, Dec 22, 37; m 65; c 1. MICROBIOLOGY, BIOCHEMISTRY. *Educ:* SDak State Univ, BS, 63, MS, 64; Kans State Univ, PhD(bact), 68. *Prof Exp:* From asst prof to assoc prof, 68-85, PROF MICROBIOL, SOUTHWESTERN STATE UNIV OKLA, 85- *Concurrent Pos:* Danforth fel, 71. *Mem:* AAAS; Am Soc Microbiol; Sigma Xi. *Res:* Metabolic interrelationships among bacteria in regard to substrate utilization. *Mailing Add:* S West Okla St Univ 100 Campus Drive Weatherford OK 73096

MESSMER, RICHARD PAUL, b Pittsburgh, Pa, Nov 24, 41; m 67. CHEMICAL PHYSICS. *Educ:* Carnegie Inst Technol, BS, 63; Univ Alta, PhD(theoret chem), 67. *Prof Exp:* Res assoc theoret chem, Mass Inst Technol, 67-68; lectr chem, Univ Alta, 68-69; STAFF MEM, RES & DEVELOP CTR, GEN ELEC CO, 69- *Concurrent Pos:* Vis scientist, Dept Mat Sci & Eng, Mass Inst Technol, 73-83; adj prof, Dept Physics, Univ Pa, 80-; vis prof, Dept Chem & Appl Physics, Calif Inst Technol, 85. *Mem:* AAAS; Am Chem Soc; fel Am Phys Soc. *Res:* Quantum theory of solid state, especially chemically related problems; theoretical studies of surfaces, semiconductors, metals; computer simulations of materials. *Mailing Add:* Res & Develop Ctr Gen Elec Co PO Box 8 Schenectady NY 12301

MESTECKY, JIRI, b Prague, Czech, June 3, 41; m; c 2. IMMUNOLOGY, IMMUNOCHEMISTRY. *Educ:* Charles Univ, Prague, MD, 64. *Hon Degrees:* Doctor Odontologiae honoris cause, Royal Dental Col, Aarhus, Denmark, 83; Medicinae honoris causae doctor, Univ Göteborg, Sweden, 90. *Prof Exp:* Asst, Inst Microbiol & Immunol, Fac Med, Charles Univ, Prague, 61-63; sr res asst immunol, Inst Microbiol, Czech Acad Sci, 63-65; instr, Inst Microbiol, Fac Med, Charles Univ, Prague, 65-66; vis res assoc, 67-68, from instr to asst prof, 68-72, assoc prof, 72-76, PROF MICROBIOL, MED & ORAL BIOL, UNIV ALA, BIRMINGHAM, 76-, SCIENTIST, INST DENT RES, 72-, CANCER RES INST, 73-, ARTHRITIS RES CTR, 79- & CYSTIC FIBROSIS RES CTR, 81- *Concurrent Pos:* Vis fel microbiol, Univ Ala, Birmingham, 67; WHO travel stipend; vis scientist, Rockefeller Univ, NY, 76; vis prof path, Nihon Univ, Tokyo, Japan, 83. *Honors & Awards:* Am Soc Clin Investrs, 82; Asn Am Physicians, 89. *Mem:* Am Asn Immunol; Am Soc Biochem Molecular Biol; Am Asn Pathologists; Am Soc Microbiol; NY Acad Sci; Sigma Xi; Soc Mucosal Immunol (pres, 87-88); Clin Immunol Soc; Am Fed Clin Res. *Res:* Protein chemistry; secretory antibodies; innate immune factors. *Mailing Add:* Dept Microbiol Univ Ala 1919 S 7th Ave Birmingham AL 35294

MESZLER, RICHARD M, b Peekskill, NY, Aug 30, 42; div; c 2. NEUROSCIENCE, ELECTRON MICROSCOPY. *Educ:* NY Univ, BA, 64; Univ Louisville, PhD(anat), 69. *Prof Exp:* NIH res fel anat, Albert Einstein Col Med, 69-71; instr, 71-72, asst prof, 72-76, ASSOC PROF ANAT, SCH DENT, UNIV MD, 76- *Honors & Awards:* Stunkard Prize, NY Univ; Drew Award, Univ Md. *Mem:* Am Soc Cell Biol; Am Asn Anatomists; Soc Neurosci; Am Asn Dental Sch. *Res:* Sensory integration in tectum of pit vipers; synaptology of trigeminal system; thermoreception; reptilian CNS; cell biology. *Mailing Add:* Dept of Anat Univ of Md Dent Sch 666 W Baltimore St Baltimore MD 21201

MESZOELY, CHARLES ALADAR MARIA, b Szekesfehervar, Hungary, Apr 24, 33; US citizen; m 61; c 2. PALEONTOLOGY, PARASITOLOGY. *Educ:* Northeastern Univ, BS, 61; Boston Univ, MA, 63, PhD(biol), 67. *Prof Exp:* Instr biol, Northeastern Univ, 66-68; res assoc biophys, Armed Forces Inst Path, 68-70; PROF BIOL, NORTHEASTERN UNIV, 70- *Concurrent Pos:* Vis lectr dept life sci, Trent Polytech, Nottingham, Eng, 77 & 84. *Mem:* Soc Vert Paleont; Am Soc Ichthyologists & Herpetologists. *Res:* Paleontology, especially evolution and systematics of fossil and recent anguid lizards, and other Cenozoic lower vertebrates; parasitology, especially ultrastructure and ultrastructural changes in the malarial parasite. *Mailing Add:* Dept Biol Northeastern Univ Boston MA 02115

METANOMSKI, WLADYSLAW VAL, b Vienna, Austria, Oct 3, 23; US citizen; m 66; c 1. POLYMER CHEMISTRY, CHEMICAL INFORMATION SCIENCE. *Educ:* Univ London, BSc, 52; Univ Toronto, MASc, 60, PhD(chem eng), 64. *Prof Exp:* Chemist, Analysis & Res Lab, Dearborn Chem Co Ltd, Ont, 52-56, chem engr, Tech Field Serv, 56-58; demonstr chem eng, Univ Toronto, 58-64; asst ed, Am Chem Soc, 64-66, group leader, 66-71, asst to ed, 71-72, mgr develop, 72-88, SR ED ADV, CHEM ABSTR SERV, AM CHEM SOC, 88- *Concurrent Pos:* Chmn, Div Chem Info, Am Chem Soc, 87. *Mem:* Am Chem Soc; Am Soc Info Sci. *Res:* High polymers; electron exchangers; chemical information science; indexing chemical literature; chemical compound nomenclature; vocabulary control; development of computer-based information processing system. *Mailing Add:* 1670 Ardwick Rd Columbus OH 43220

METCALF, ARTIE LOU, b Dexter, Kans, July 5, 29. ZOOLOGY. *Educ:* Kans State Col, BS, 56; Univ Kans, MA, 57, PhD(zool), 64. *Prof Exp:* From instr to assoc prof, 62-68, PROF ZOOL, UNIV TEX, EL PASO, 69- *Mem:* AAAS; Am Malacol Union; Am Quaternary Asn; Conchol Soc Gt Brit & Ireland. *Res:* Systematics and paleoecology of terrestrial mollusks. *Mailing Add:* Dept Biol Sci Univ Tex El Paso TX 79968-0519

METCALF, BRIAN WALTER, b Perth, Western Australia, July 13, 45; m 70. BIO-ORGANIC CHEMISTRY. *Educ:* Univ Western Australia, BSc Hons, 66, PhD(org chem), 70. *Prof Exp:* Fel org chem, Univ Col, Univ London, 70-71, Stanford Univ, 71-73; res scientist, Res Ctr, Merrell Int, 73-78; res scientist org chem, Merrell Res Ctr, Dow Chem Co, 78-79, head, Org Chem, Merrell Dow Pharmaceut, 79-83; dir med chem dept, 83-86, vpres chem res, 87-89, VPRES CHEM & THEOL RES, SK&F LABS, 89- *Mem:* Am Chem Soc. *Res:* Design, synthesis and biochemistry of enzyme inhibitors. *Mailing Add:* Med Chem Dept SK&F Labs 709 Swedeland Rd Swedeland PA 19479

METCALF, DAVID HALSTEAD, b Charleston, SC, Aug 24, 57. MOLECULAR RECOGNITION, INSTRUMENT DESIGN. *Educ:* Col Charleston, BS, 79; Duke Univ, PhD(chem), 85. *Prof Exp:* Postdoctoral fel, 85-86, fac lectr phys chem, 86-91, RES SCIENTIST, DEPT CHEM, UNIV VA, CHARLOTTESVILLE, 90- *Mem:* Am Chem Soc; Sigma Xi. *Res:* Use of chiroptical spectroscopy techniques to investigate the structure and dynamics of chiral metal complexes in solution and in the solid state. *Mailing Add:* Dept Chem Univ Va Charlottesville VA 22901

METCALF, FREDERIC THOMAS, b Oak Park, Ill, Dec 28, 35; m 57; c 2. APPLIED MATHEMATICS. *Educ:* Lake Forest Col, BA, 57; Univ Md, MA, 59, PhD(appl math), 61. *Prof Exp:* Asst engr, Electronics Div, Westinghouse Elec Co, 57-58; mathematician, Phys Chem Div, US Naval Ord Lab, 60-62, res mathematician, Math Dept, 62-63; asst res prof appl math, Inst Fluid Dynamics & Appl Math, Univ Md, 63-66; assoc prof, 66-69, chmn dept, 68-70, PROF MATH, UNIV CALIF, RIVERSIDE, 69- *Concurrent Pos:* Consult, US Naval Ord Lab, 64-66. *Mem:* AAAS; Am Math Soc; Math Asn Am. *Res:* Dynamical systems with two degrees of freedom; finite difference schemes for partial differential equations; inequalities; second order ordinary differential equations; fluid flow about bodies of revolution. *Mailing Add:* Dept Math Univ Calif 900 University Ave Riverside CA 92502

METCALF, HAROLD, b Boston, Mass, June 11, 40; m 63; c 3. PHYSICS, LASER SPECTROSCOPY. *Educ:* Mass Inst Technol, ScB, 62; Brown Univ, PhD(physics), 68. *Prof Exp:* Res assoc physics, Brown Univ, 67-68; res assoc, 68-70, asst prof, 70-74, assoc prof, 74-82, PROF PHYSICS, STATE UNIV NY STONY BROOK, 82- *Concurrent Pos:* Vis assoc prof, Mass Inst Technol, 77-78; consult, Nat Bureau Standards, 81-86; Vis prof, Ecole Normale Superieure, 86-87. *Mem:* fel Am Phys Soc; Optical Soc Am. *Res:* Experimental atomic physics; precision measurements; experimental quantum electrodynamics; level crossing spectroscopy; laser spectroscopy; astrophysics; problems of human visual perception; simple atoms and molecules; quantum beat spectroscopy; Stark and Zeeman spectroscopy; laser cooling and trapping of neutral atoms. *Mailing Add:* Dept of Physics State Univ of NY Stony Brook NY 11790

METCALF, ISAAC STEVENS HALSTEAD, b Cleveland, Ohio, Aug 17, 12; m 41; c 2. GROSS ANATOMY, COMPARATIVE ANATOMY. *Educ:* Oberlin Col, BA, 34; Columbia Univ, MA, 36; Case Western Reserve Univ, PhD(biol), 40. *Prof Exp:* From asst prof to assoc prof biol & chem, The Citadel, 37-57, prof biol, 57-66; prof, 66-80, EMER PROF ANAT, MED UNIV SC, 80- *Mem:* Nat Audubon Soc; Sigma Xi. *Res:* Fresh water biology. *Mailing Add:* Dept Anat Med Univ SC 171 Ashley Ave Charleston SC 29425

METCALF, ROBERT HARKER, b Chicago, Ill, Aug 29, 43; m 68. MICROBIOLOGY. *Educ:* Earlham Col, AB, 65; Univ Wis-Madison, MS, 68, PhD(bact), 70. *Prof Exp:* Asst prof, 70-75, assoc prof, 75-81, PROF BIOL SCI, CALIF STATE UNIV, SACRAMENTO, 81- *Mem:* Am Soc Microbiol; AAAS; Soc Appl Bacteriology; Sigma Xi. *Res:* Food and water microbiology; solar energy applications for serious cooking and water disinfection. *Mailing Add:* Dept Biol Sci Calif State Univ Sacramento CA 95819

METCALF, ROBERT LEE, b Columbus, Ohio, Nov 13, 16; m 40; c 3. TOXICOLOGY, CHEMICAL ECOLOGY. *Educ:* Univ Ill, BA, 39, MA, 40; Cornell Univ, PhD(entom), 43. *Prof Exp:* From asst entomologist to assoc entomologist, Tenn Valley Authority, Ala, 43-46; from asst entomologist to assoc entomologist, Citrus Exp Sta, Univ Calif, Riverside, 46-53, prof entom & entomologist, 53-68; head, Dept Zool, 69-72, PROF ENTOM, UNIV ILL, URBANA-CHAMPAIGN, 68-, DISTINGUISHED PROF BIOL & PROF VET PHARMACOL, 71-, PROF, CTR ADVAN STUDY, 81- *Concurrent Pos:* Vchancellor, Univ Calif, Riverside, 62-67; consult, WHO, USAID, USDA & Tenn Valley Authority; mem, President's Sci Adv Comt & var comts, Nat Acad Sci, 66-81. *Honors & Awards:* Charles T Spencer Award, Am Chem Soc, 66, Int Award Pesticide Chem, 72; Ciba Geigy Award, Entom Soc Am, 77; York Distinguished Lectr, Univ Fla, 87; Robert A Van den Bosch Mem Lectr, Univ Calif, Berkeley, 87. *Mem:* Nat Acad Sci; fel Am Acad Arts & Sci; Am Chem Soc; fel Entom Soc Am (pres, 58); fel AAAS; Am Mosquito

Control Asn; Sigma Xi. *Res:* Insect physiology and toxicology; mosquito control; author of 407 technical publications; awarded 8 US patents and one Canadian patent. *Mailing Add:* Dept Entom Univ Ill 320 Morrill Hall Urbana IL 61801

METCALF, WILLIAM, b Norwood, Mass, Dec 31, 07; m 50. SURGERY. *Educ:* Mass Inst Technol, BSc & MSc, 31; Johns Hopkins Univ, MD, 37. *Prof Exp:* Res fel surg, Johns Hopkins Univ, 38-39; Cushing fel, Sch Med, Yale Univ, 39-40, asst surg, 41-43; asst chief, Vet Admin Hosp, Hines, Maywood, Ill, 47-48; teaching fel, St Vincents Hosp, New York, 50-52; instr, Sch Med, NY Univ, 52-55; from asst prof to assoc prof, 54-62, prof, 62-76, EMER PROF SURG, ALBERT EINSTEIN COL MED, YESHIVA UNIV, 76- *Mem:* AAAS; Am Soc Surg Hand; AMA; Am Col Surg. *Res:* General and hand surgery; surgical metabolism and shock; plasma and plasma expanders; nitrogen metabolism; anabolic steroids; mathematical (exponential) models for cancer survival. *Mailing Add:* Box Seven K North 60 Sutton Place South New York NY 10022

METCALF, WILLIAM KENNETH, b Whitley Bay, Eng, Apr 30, 21; m 44; c 7. HUMAN ANATOMY. *Educ:* Univ Durham, MB & BS, 43; Bristol Univ, MD, 60. *Prof Exp:* From lectr to sr lectr anat, Bristol Univ, 48-64, reader, 64-68; prof, Univ Iowa, 68-73; PROF ANAT & CHMN DEPT, UNIV NEBR MED CTR, 73- *Concurrent Pos:* USPHS fel, 69-73. *Mem:* Am Asn Anat; Anat Soc Gt Brit & Ireland; Brit Soc Hemat; Brit Physiol Soc. *Res:* Hematology; cell kinetics; physical properties of cells; cellular immunology; education, especially use of computers in education. *Mailing Add:* 265 Skyline Dr Elkhorn NE 68022

METCALFE, DARREL SEYMOUR, b Arkansaw, Wis, Aug 28, 13; m 42; c 2. AGRONOMY. *Educ:* Univ Wis, BS, 41; Kans State Univ, MS, 42; Iowa State Univ, PhD(plant physiol, crop breeding), 50. *Prof Exp:* Asst, Kans State Univ, 40-42; prof agron, Iowa State Univ, 46-56, asst dir student affairs, 56-58; dean, Col Agr, Univ Ariz, 78-80, assoc dean & dir, resident instr, Col Agr, 58-82, asst dir & agr exp sta, 58-83; RETIRED. *Concurrent Pos:* Consult, AID, Brazil, 62-73 & Orgn Econ Coop & Develop, Europe, 63-65; mem comt agr educ & nat res, Nat Acad Sci-Nat Res Coun, 66-70; consult, Nat Acad Sci, Egypt, 80-83, Consortium Int Develop, NY emen & Inst Int Educ, 80-82, Somalia & Kenya, 81- & Oman Univ, 82-85. *Mem:* Fel Am Soc Agron. *Res:* Seed production of forage grasses and legumes. *Mailing Add:* Col Agr Univ Ariz Tucson AZ 85721

METCALFE, DEAN DARREL, b Medford, Ore, June 27, 44; m 67; c 3. MAST CELL BIOLOGY. *Educ:* Northern Ariz Univ, BS, 66; Univ Mich, MS, 68; Univ Tenn, MD, 72; Am Bd Internal Med, dipl, 75; Am Bd Allergy & Immunol, dipl, 77; Am Bd Internal Med, Rheumatol, dipl, 79. *Prof Exp:* Resident med, Univ Mich, 72-74; clin assoc allergy-immunol, NIH, 74-77; fel rheumatol, Robert B Brigham Hosp, 77-79; clin investr allergy-immunol, 79-85, HEAD MAST CELL PHYSIOL, LAB CLIN INVEST, NAT INST ALLERGY & INFECTIOUS DIS, NIH, 85- *Concurrent Pos:* Consult, Allergy-Clin Immunol Serv, Walter Reed Army Med Ctr, 84-; mem, Ad Hoc Adv Comt Hypersensitivity Food Constituents, Food & Drug Admin, 85-86; chmn, bd sci adv, Allergy & Immunol Inst, Int Life Sci Inst, 90-; dir, Am Bd Allergy & Immunol, 90-; chmn int coun, Am Acad Allergy & Immunol, 90- *Honors & Awards:* Commendation Medal Outstanding Serv, USPHS, 85, Medal, 91. *Mem:* Am Fedn Clin Res; Am Soc Clin Invest; Asn Am Physicians; Am Acad Allergy & Immunol; Am Rheumatism Asn; Am Asn Immunologists. *Res:* Biology of mast cells, including growth and differentiation, homing and contribution to inflammation; clinical specialist in allergy-immunology, especially of mastocytosis and adverse reactions to foods and additives. *Mailing Add:* Lab Clin Invest Nat Inst Allergy & Infectious Dis NIH Bldg 10 Rm 11C210 Bethesda MD 20892

METCALFE, JAMES, b New Bedford, Mass, Aug 16, 22; m 44; c 4. MEDICINE. *Educ:* Brown Univ, AB, 44; Harvard Univ, MD, 46; Am Bd Internal Med, dipl, 53. *Prof Exp:* Med house officer, Peter Bent Brigham Hosp, Boston, 46-47, asst, 50-51, sr asst, 51-52; ward med officer, US Naval Hosp, Newport, RI, 47-49; instr, Harvard Med Sch, 53-55, assoc med, 55-59, tutor, 57-58, asst prof, 59-61; assoc prof, 61-64, PROF MED, MED SCH, UNIV ORE, 64- *Concurrent Pos:* Res fel, Peter Bent Brigham Hosp, 50-51; res fel physiol, Harvard Med Sch, 49-50, 52-53; fel, Boston Lying-in Hosp, 52-53; assoc physician, Boston Lying-in Hosp, 52-59, vis physician, 59-61; Am Heart Asn estab investr, 53-59; jr assoc, Peter Bent Brigham Hosp, 53-56, assoc, 56-58, sr assoc, 58-61; chmn cardiovasc res, Ore Heart Asn, 61- *Mem:* Am Physiol Soc; Am Fedn Clin Res; Am Clin & Climat Asn; Am Soc Clin Invest. *Res:* Modifications of maternal physiology during pregnancy and their effects on the course of disease. *Mailing Add:* Dept Med Univ Ore Sch Med 3181 SW Sam Jackson Park Rd Portland OR 97201

METCALFE, JOSEPH EDWARD, III, b Fallowfield Twp, Pa, May 27, 38; m 59; c 3. PHYSICAL CHEMISTRY, FUEL TECHNOLOGY. *Educ:* Pa State Univ, BS, 60, MS, 62, PhD(fuel technol), 65. *Prof Exp:* Res asst fuel technol, Pa State Univ, 60-62 & 63-65; sr res chemist, 65-68, res supvr electrokinetics, 68-71 & gasoline phys res, 71-74, supvr technol assessment, 74-76, supvr alt energy sources, 76-83, TECH DIR ALT ENERGY & COAL RES, STAND OIL CO OHIO, 83-; DIR FUELS TECHNOL, BP AMERICA. *Mem:* Am Chem Soc. *Res:* Fused salt batteries; carbon technology; molecular sieves; adsorption; physical properties of gasoline; coal technology; catalysts; coal technology energy research; alcohol motor fuels. *Mailing Add:* BP Am Res & Develop 4440 Warrensville Center Rd Cleveland OH 44128-2837

METCALFE, LINCOLN DOUGLAS, b Melstone, Mont, Feb 11, 21; m 48; c 1. ANALYTICAL CHEMISTRY. *Educ:* Univ Calif, Los Angeles, BS, 43; Univ Chicago, BS, 47. *Prof Exp:* Head analysis sect, Res Div, Armour & Co, 47-60, head analytical res sect, Res Labs, Armour Indust Chem Co, 60-66, asst res dir analytical & phys chem & instrumental res, Res Lab Akzo Chem Div, 66-91; RETIRED. *Concurrent Pos:* Mem lipid analysis comt, Nat Heart Inst, 58-59; mem adv bd, J Chromatographic Sci & Handbook Chromatography; mem bd dirs, Sugar Processing Res, Inc; gov bd, Am Oil Chem Soc, 81-82. *Honors & Awards:* Bond Award, Am Oil Chem Soc, 64, Bailey Medal, 90; Meade Award, Sugar Indust Tech, 86. *Mem:* Am Chem Soc (chmn elect, div chem info, 86); Am Soc Testing & Mat; Am Oil Chem Soc; Sugar Indust Technologists. *Res:* Nonaqueous titrations; gas and high pressure liquid chromatography; infrared and ultraviolet spectrophotometry as applied to lipid and protein chemistry, especially fatty acid derivatives; membrane technology; enzymology. *Mailing Add:* 708 Tenth Ave LaGrange IL 60525

METCOFF, JACK, b Chicago, Ill, Feb 2, 17; m 43; c 2. PHYSIOLOGY, NUTRITION. *Educ:* Northwestern Univ, BS, 38, BM, 42, MD, 43, MS, 44; Harvard Univ, MPH, 44. *Prof Exp:* From asst to assoc pediat, Harvard Med Sch, 48-53, asst prof, 53-56; chmn dept, Michael Reese Hosp, 56-70; PROF PEDIAT BIOCHEM & MOLECULAR BIOL, UNIV OKLA HEALTH SCI CTR, 70-, GEORGE LYNN CROSS RES PROF PEDIAT BIOCHEM & MOLECULAR BIOL, 78- *Concurrent Pos:* Prof, Med Sch, Northwestern Univ, 56-63; prof & chmn dept, Chicago Med Sch, 63-68; Alexander von Humboldt found fel; vis prof, Inst Biochem & Nutrit, Univ Hohenheim, WGer, 88. *Mem:* Fel Am Acad Pediat; Am Physiol Soc; Am Soc Nephrology; Am Soc Clin Nutrit; Am Inst Nutrit; Sigma Xi; Am Soc Pediat Nephrology (pres, 69-70); Am Col Nutrit; Am Acad Cert Med Nutritionists. *Res:* Cell metabolism; electrolyte and renal physiology; relations between intracellular ions and intercellular metabolites during prematurity; normal growth and development of the human infant; severe chronic infantile malnutrition; Kwashiorkor; fetal malnutrition; metabolism of isolated kidneys; leukocyte metabolism; nutrition in pregnancy; renal disease in children; metabolism in uremia. *Mailing Add:* OCMH-Pediat PO Box 26307 Oklahoma City OK 73126

METEER, JAMES WILLIAM, b Columbus, Ohio, Apr 7, 21; m 44; c 4. FOREST MANAGEMENT PLANNING, TECHNICAL ANALYSIS WITH COMPUTER METHODS. *Educ:* Univ Mich, BSF, 44, MF, 47. *Prof Exp:* Asst prof forestry, Agr Exp Sta, Ohio State Univ, 47-54, consult forester, 54-65; from asst prof to prof, 65-84, EMER PROF FORESTRY, SCH FORESTRY, MICH TECHNOL UNIV, 84- *Concurrent Pos:* Chmn comput adv coun, Mich Technol Univ; consult, 84- *Mem:* Fel Soc Am Foresters; Forest Prod Res Soc. *Res:* Forest management and growth investigations; continuous forest inventory control with computer processing of data; computer applications development; forest economic analysis; computers in professional education. *Mailing Add:* 913 Meador St L'Anse MI 49946

METER, DONALD M(ERVYN), b Omaha, Nebr, Aug 10, 31; m 61; c 2. CHEMICAL ENGINEERING. *Educ:* Univ Minn, BS, 53; Princeton Univ, MSE, 55; Univ Wis, PhD(chem eng), 64. *Prof Exp:* Res engr, Edison Labs, Thomas A Edison, Inc, 54-56; instr chem eng, Univ Wis, 57-58; sr res scientist, Squibb Inst Med Res, Olin Mathieson Chem Corp, 62-68, res fel, E R Squibb & Sons, 69-73; res engr, Armour Pharm Co, 73-77; MGR CHEM PROCESS DEVELOP, G D SEARLE & CO, 78- *Mem:* Am Inst Chem Engrs; Am Chem Soc. *Res:* Turbulent flow and viscoelastic properties of dilute non-Newtonian polymer solutions. *Mailing Add:* G D Searle Co 4901 Searle Pkwy Skokie IL 60077-1099

METH, IRVING MARVIN, b Brooklyn, NY, July 27, 29; m 55; c 4. ELECTRONICS, INSTRUMENTATION. *Educ:* City Col New York, BEE, 51; Polytech Inst Brooklyn, MEE, 56. *Prof Exp:* Engr, RCA Labs, 51-57; PROF ELEC ENG, CITY COL NY, 58- *Concurrent Pos:* Vis engr, Brookhaven Nat Lab, 58-; consult, USAEC, 60-65 & Port of NY Authority, 70-75; vis prof, Polytech Inst Brooklyn, 65-66; vis lectr, Bell Tel Labs, 69-70. *Mem:* Sr mem Inst Elec & Electronics Engrs. *Res:* Digital communications, control systems, electronics and electronic systems; bioengineering instrumentation and telemetering of information; electrical safety; applied cryogenics; super conducting power cables; microelectronics. *Mailing Add:* Dept of Elec Eng City Col of NY Convent Ave & 138th St New York NY 10031

METHERELL, ALEXANDER FRANZ, b Canton, China, Aug 21, 39; US citizen; m 64; c 3. RADIOLOGY, ULTRASONIC ENGINEERING. *Educ:* Kingston Polytech Inst, Dip Tech, 61; Bristol Univ, PhD(eng), 64; Univ Miami, MD, 76; Am Bd Radiol, dipl, 81. *Prof Exp:* Res assoc aeronaut, Univ Minn, 64-65; sr engr scientist, Douglas Aircraft Co, 65-66; res scientist acoust holography, Douglas Advan Res Labs, McDonnell Douglas Corp, 66-70, dir med imaging, 70-74; assoc prof radiol, Univ Calif, Irvine, 76-79; radiologist, South Bay Hosp, Redondo Beach, Calif, 79-84; chmn radiol, Costa Mesa Med Ctr Hosp, Calif, 84-86 & MRI CTR, SANTA ANA, CALIF, 86- *Concurrent Pos:* Assoc prof eng, Univ Calif, Los Angeles, 69-70; clin assoc prof radiol, Univ Calif, Irvine, 71-76, adj assoc prof eng, 73-76; assoc ed, Trans Sonics & Ultrasonics, 71-; mem cardiol adv comt, Nat Heart, Lung & Blood Inst, 78-81. *Mem:* Am Inst Ultrasound Med; Alliance Eng Med & Biol; Inst Elec & Electronics Engrs; Am Col Med Imaging. *Res:* Ultrasound imaging; medical imaging. *Mailing Add:* MRI Ctr 1930 Old Tustin Ave Santa Ana CA 92701

METIU, HORIA I, b Clug, Rumania, Mar 7, 40; m 72; c 3. CHEMICAL ENGINEERING, THEORETICAL CHEMISTRY. *Educ:* Politech Inst, Bucharest, BA, 61, Mass Inst Technol, PhD(phys chem), 74. *Prof Exp:* PROF CHEM PHYSICS, UNIV CALIF, SANTA BARBARA, 76- *Honors & Awards:* Solid State Chem Award, Am Chem Soc, 79. *Mem:* Am Chem Soc; fel Am Phys Soc. *Res:* Theoretical chemical physics; dynamics of molecular interaction with solid surface; theory of reaction rate; spectroscopy. *Mailing Add:* Dept Chem Univ Calif Santa Barbara CA 93106

METRIONE, ROBERT M, b Teaneck, NJ, Aug 22, 33; m 57; c 3. BIOCHEMISTRY. *Educ:* Bowling Green State Univ, BS, 55; Univ Nebr, MS, 60, PhD(biochem), 63. *Prof Exp:* Res assoc biochem, Yale Univ, 63-65, asst prof, 65-67; from asst prof to assoc prof, 67-83, PROF BIOCHEM, THOMAS JEFFERSON UNIV, 83- *Mem:* AAAS; Am Chem Soc; Am Soc Biochem & Molecular Biol; The Protein Soc. *Res:* Structure-function relationships of proteolytic enzymes; cathepsins; proteolytic enzymes of the rat visceral yolk sac during mid and late gestation; protein sequencing. *Mailing Add:* Dept Biochem Thomas Jefferson Univ Philadelphia PA 19107

METROPOLIS, NICHOLAS CONSTANTINE, b Chicago, Ill, June 11, 15; m 55; c 3. APPLIED MATHEMATICS, THEORETICAL PHYSICS. *Educ:* Univ Chicago, BS, 36, PhD(physics), 41. *Prof Exp:* Res assoc, Univ Chicago, 41, res assoc, Metall Lab & instr physics, 42; res assoc, Columbia Univ, 42; res assoc & group leader, Los Alamos Sci Lab, 43-46, consult, 46-48, mem staff & group leader 48-57; prof physics, Univ Chicago & Enrico Fermi Inst Nuclear Studies, 57-65, dir inst comput res, 58-65; mem staff, 65-80, sr fel, 81-85, EMER SR FEL, LOS ALAMOS NAT LAB, 85- *Concurrent Pos:* Asst prof, Univ Chicago & Inst Nuclear Studies, 46-48; consult, Argonne Nat Lab, Brookhaven Nat Lab & Lawrence Radiation Lab, Univ Calif; mem adv panel univ comput facilities, NSF, 59- & adv comt comput activ & adv comt res, 73-75; chmn, Comput Adv Group, Atomic Energy Comn, 59-62; mem, UN Tech Mission to India, 61; vis prof, Univ Colo, 64; mem adv comt res, NSF, 73-75; deleg, US-USSR Sci & Technol Exchange, Moscow, 76. *Honors & Awards:* Comput Pioneer Award, Inst Elec & Electronics Engrs, 84; Gillies Mem lectr, Univ Ill, Urbana, 85. *Mem:* Fel Am Phys Soc; Am Math Soc; Soc Indust & Appl Math; Am Acad Arts & Sci. *Res:* Theoretical nuclear physics; electronic computing; logical design of general purpose computers; pure and applied mathematical analysis of inherent error propagation; studies of non-linear differential equations; theoretical investigations of nuclear cascades. *Mailing Add:* Los Alamos Nat Lab MS 210 Los Alamos NM 87545

METRY, AMIR ALFI, b Minia, Egypt, May 4, 42; US citizen; m 66; c 2. ENVIRONMENTAL ENGINEERING. *Educ:* Cairo Univ, BS, 63, MS, 67; Drexel Univ, MS, 69, PhD(environ eng), 73; Am Acad Environ Engrs, dipl, 73. *Prof Exp:* Proj mgr sanit eng, Naim Mahfouz Eng, Cairo, Egypt, 63-68; res assoc environ eng, Drexel Univ, 68-73; dir, MTL Industs, Md, 72-73; vpres waste mgt, Roy F Weston Inc, Pa, 73-78; vpres res & tech serv, IU Conversion Systs Inc, 78-81; VPRES RESIDUALS MGT, ROY F WESTON, INC, 81- *Concurrent Pos:* Adj instr, Cairo Univ, 64-67; adj prof, Drexel Univ, 74-77; adj assoc prof, Temple Univ, 76- *Mem:* Water Pollution Control Fedn; Am Soc Civil Engrs; Inst Environ Sci; Nat Soc Prof Engrs. *Res:* Investigation of the fate of contaminants in the environment; simulation contaminant migration; evaluation of waste treatment technologies; design of pollution control systems; detoxification and management of hazardous wastes; management of radioactive waste. *Mailing Add:* ERM Inc 433 Greenhill Lane Berwyn PA 19312

METS, LAURENS JAN, b Santa Barbara, Calif, Sept 14, 46; m 71; c 3. MOLECULAR BIOLOGY, GENETICS. *Educ:* Pomona Col, BA, 68; Harvard Univ, MA, 70, PhD(biochem & molecular biol), 73. *Prof Exp:* Staff fel phys biol, NIH, 72-75; asst prof biol, Case Western Reserve Univ, 75-80; vis fel, Carnegie Inst, Baltimore, 80-81; PROF MOLECULAR GENETICS & CELL BIOL, UNIV CHICAGO, 81- *Mem:* AAAS; Am Soc Cell Biol; Am Soc Plant Physiologists; Genetics Soc Am; Int Soc Plant Molecular Biol. *Res:* Genetics and biogenesis of chloroplast ribosomes; chloroplast genetics; rhythm biology; photosynthesis; herbicide resistance; molecular mechanisms of photosynthetic light harvesting and trapping; biogenesis of the photosynthetic apparatus, including the photosystems and ribulose bisphosphate carboxylase; genetic analysis of herbicide interaction with the photosystem; chloroplast genetics and molecular biology; mechanisms of cellular differentiation in leaves. *Mailing Add:* Univ Chicago 1103 E 57th St Chicago IL 60637

METSGER, ROBERT WILLIAM, b New York, NY, Apr 27, 20; m 47, 63; c 3. ENVIRONMENTAL HYDROGEOLOGY. *Educ:* Columbia Univ, AB, 48. *Prof Exp:* Geologist, NJ Zinc Co, Inc, 49-55, resident geologist, 55-63, regional geologist, 63-81, chief geologist, 81-88; RETIRED. *Concurrent Pos:* Assoc, Ogdensburg Seismic Observ, 58-81; mem adv bd, NJ State Geol Surv, 82- *Mem:* sr Fel Geol Soc Am; sr fel Soc Econ Geologists; Am Inst Prof Geologists. *Res:* Petrology and structural geology of ore deposits associated with Proterozoic and Paleozoic rocks; earth strain induced by tidal and teleseismic forces; Telluric currents; rock strain and micro-seismic effects associated with deep underground mining; hydrogeology of a karst terrane. *Mailing Add:* 69 Hunters Lane Sparta NJ 07871

METTE, HERBERT L, b Kassel, Ger, Oct 26, 25; US citizen; m 58; c 1. SOLID STATE PHYSICS. *Educ:* Univ Goettingen, dipl physics, 52. *Prof Exp:* Res asst solid state physics, Univ Goettingen, 52-53; scientist, Ger Nat Bur Standards, 53-56; physicist, Res & Develop Lab, 56-58, leader device physics sect, Solid State Devices Div, 58-65, chief integrated devices tech br, Integrated Electronics Div, 65-73, leader advan IC technol team, Microelectronics Div, 73-85, SR MEM, TECH PLANS & PROGS OFF, US ARMY ELECTRONICS TECHNOL & DEVICES LAB, FT MONMOUTH, 85- *Concurrent Pos:* Dep Army mem, Adv Group Electron Devices, Dept of Defense, 76-; tech prog chmn, Govt Microcircuits Applns Conf, 85. *Mem:* Fel Am Phys Soc; sr mem Inst Elec & Electronics Engrs. *Res:* Organic semiconductors; photomagneto and magnetothermal effects in semiconductors; integrated electronics; materials, devices, processes research and automated testing. *Mailing Add:* 650 Valley Rd Brielle NJ 08730

METTEE, HOWARD DAWSON, b Boston, Mass, Aug 6, 39; m 63; c 3. SPECTROCHEMISTRY. *Educ:* Middlebury Col, BA, 61; Univ Calgary, PhD(phys chem), 64. *Prof Exp:* Fel spectros & photochem, Nat Res Coun Can, 64-66 & Univ Tex, Austin, 66-68; from asst prof to assoc prof, 68-81, PROF CHEM, YOUNGSTOWN STATE UNIV, 81- *Concurrent Pos:* Pres, M&M Consults, Inc, 83-; sabbatical leave, Univ Calif, Berkeley, 79-80; trustee bd, Technol Develop Corp, 90- *Mem:* Royal Soc Chem; Am Chem Soc; Sigma Xi. *Res:* Energy transfer and relaxation; primary photochemical events and gas phase kinetics; chemical applications of spectroscopy; thermodynamics of gas phase complexes; photoelectrochemistry of solar energy; analysis and treatment of industrial chemical processes. *Mailing Add:* Dept Chem Youngstown State Univ Youngstown OH 44555

METTEE, MAURICE FERDINAND, b Mobile, Ala, Apr 28, 43; m 68; c 2. AQUATIC BIOLOGY, ECOLOGY. *Educ:* Spring Hill Col, BS, 65; Univ Ala, MA, 67, MS, 70, PhD(biol), 74. *Prof Exp:* Res asst biol, Univ Ala, 74-75; environ biologist, 75-77, DIR, BIOL RESOURCES DIV, GEOL SURV ALA, 77- *Concurrent Pos:* US Forest Serv grant ichthyol, Univ Ala, 74; aquatic biologist, US Air Force Acad, 74; res grant, US Fish & Wildlife Serv, 75-78; consult, Okaloosa Darter, Fla, 75-78; mem, Southeastern Fishes Coun. *Mem:* Am Ichthyol & Herpet; Am Fisheries Soc; Sigma Xi. *Res:* Studies on the systematics, ecology, reproductive behavior, embryology and development of freshwater and marine fishes; endangered and threatened vertebrate life in the southeastern United States; studies on impacts of coal, oil and gas resources on aquatic ecosystems. *Mailing Add:* Ala Geol Surv Biol Resources Div PO Box O Tuscaloosa AL 35486

METTER, DEAN EDWARD, b Champaign, Ill, Aug 1, 32; m 54; c 3. HERPETOLOGY. *Educ:* Eastern Ill Univ, BS, 57; Wash State Univ, MS, 60; Univ Idaho, PhD(zool), 63. *Prof Exp:* Instr zool, Univ Idaho, 63-64; asst prof, 64-69, ASSOC PROF ZOOL, UNIV MO-COLUMBIA, 69- *Mem:* Am Soc Ichthyologists & Herpetologists. *Res:* Distribution and differentiation of amphibian populations. *Mailing Add:* Dept Biol Sci Univ Mo 105 Tucker Hall Columbia MO 65211

METTER, GERALD EDWARD, biostatistics, oncology; deceased, see previous edition for last biography

METTING, PATRICIA J, b Toledo, Ohio, Jan 21, 54; m; c 2. CARDIOVASCULAR PHYSIOLOGY. *Educ:* Univ Toledo, BS, 75; Med Col Ohio, PhD, 80. *Prof Exp:* Asst prof physiol, 79-88, ASSOC PROF PHYSIOL & BIOPHYS, MED COL OHIO, 89- *Concurrent Pos:* Ed Reviewer, Am J Physiol, Hypertension & Res Study Sect, Am Heart Asn; mem, Coun High Blood Pressure Res, Am Heart Asn. *Mem:* Am Physiol Soc; Am Heart Asn; Int Soc Heart Res. *Res:* Adenosine metabolism; purification and regulation of cytosolic 5-nucleotidase; regulation of blood pressure and peripheral blood flow; hypertension; exercise. *Mailing Add:* Dept Physiol & Biophys Med Col Ohio PO Box 10008 Toledo OH 43699-0008

METTLER, JOHN D(ANIEL), JR, electrolytic chemistry, metallurgy, for more information see previous edition

METTLER, RUBEN FREDERICK, b Shafter, Calif, Feb 23, 24; m 55; c 2. ELECTRICAL & AERONAUTICAL ENGINEERING. *Educ:* Calif Inst Technol, BS, 43, MS, 47, PhD(elec & aeronaut eng), 49. *Prof Exp:* From engr to assoc dir radar div, Hughes Aircraft Co, 49-54; spec consult, US Dept Defense, 54-55; prog dir, Thor Prog, TRW Inc, 55-57, Minuteman Prog, 57-58, exec vpres, Space Technol Labs & prog dir, Thor, Atlas, Titan & Minuteman, 58-62, pres, Space Technol Labs, 62-63, pres, Systs Group, 63-65, bd dirs, 65-69, pres & chief oper officer, 69-77, chmn bd & chief exec officer, 77-88; RETIRED. *Concurrent Pos:* Aide to Asst Secy Defense Res & Develop, 54-55; mem, Defense Indust Adv Coun, 62-68; chmn, President's Sci Policy Task Force, 69-70; mem, President's Blue Ribbon Defense Panel, 69-70, Comn Productivity, 81 & Comn on Exec Interchange, 81; vchmn, US-Japan Bus Coun, 85-88, Coun on Competitiveness, 86-88; chmn, Mfg Forum, Nat Acad Eng-Nat Acad Sci, 90- *Honors & Awards:* Distinguished Pub Serv Medal, Dept of Defense; Nat Medal of Honor, Electronic Industs Asn, 90. *Mem:* Nat Acad Eng; fel Inst Elec & Electronic Engrs; fel Am Inst Aeronaut & Astronaut; Am Acad Arts & Sci. *Res:* Systems engineering; systems integration and test. *Mailing Add:* TRW Inc One Space Park E2/1100 Redondo Beach CA 90278

METTRICK, DAVID FRANCIS, b London, Eng, 1932; Can citizen. PARASITOLOGY, PATHOLOGICAL PHYSIOLOGY. *Educ:* Univ Wales, BSc, 54; Univ London, PhD(parasitol), 57, DSc(parasitol), 73. *Prof Exp:* Lectr zool, Univ Rhodesia & Nyasaland, 58-61; sr lectr zool, Univ WI, 62-67; assoc prof, 67-71, assoc chmn dept, 73-75, chmn dept, 75-84, PROF ZOOL, UNIV TORONTO, 71- *Concurrent Pos:* Prof parasitol, Fac Med, Univ Toronto, 71-; pres, Biol Coun Can, 74-79; Can rep to coun, World Fedn Parasitologists, 74-78; chmn, Can Coun Animal Care, 75-77; chmn, Animal Biol Grants Comt, Nat Res Coun Can, 75-76; chmn, Fifth Int Cong Parasitol, 82; ed, J Parasitol, 84-88. *Honors & Awards:* Queen Elizabeth II Silver Jubilee Medal, 77; Sigma Xi, 88. *Mem:* Am Soc Parasitologists; Can Soc Zoologists; Brit Soc Parasitol; fel Royal Soc Can; Can Pub Health Asn. *Res:* Ecology and physiology of intestinal parasites; pathophysiology; membrane transport; metabolism of intestinal parasites; symbiology. *Mailing Add:* 351 Park Lawn Rd Toronto ON M8Y 3K6 Can

METUZALS, JANIS, b Moscow, Russia, June 17, 21; German citizen; m 50; c 4. THEORY OF HELICITY & ASYMMETRY. *Educ:* Univ Latvia, Bachelor, 43; Univ Hamburg, Dr rer nat, 49. *Prof Exp:* Asst cytol & histol, dept zool, Univ Groningen, Neth, 50-54, chief asst, 54-56, sci officer, 56-61; from assoc prof to prof, dept histol, 61-68, dir electron-micros unit, dept anat, 68-79, prof neuroanat, 79-86, RES PROF PATH, DEPT PATH, FAC MED, UNIV OTTAWA, 86-; ASSOC MEM, MCGILL CENTRE FOR STUDIES IN AGE & AGING, MONTREAL GEN HOSP, QUEBEC, CAN, 87- *Concurrent Pos:* Prin investr, Marine Biol Lab, Woods Hole, Mass, 64-; vis prof, dept neurobiol, Univ Goteborg, Sweden, 69-70, dept biochem, Max Planck Inst Biophys & Chem, Gottingen, Germany, , 77-78. *Honors & Awards:* Jungius Award, Univ Hamburg, 50; Olszewski Mem Lectr, Can Soc Neuropathologists, 84. *Mem:* Electron Micros Soc Am; Am Soc Cell Biol; Soc Neurosci; Am Soc Neurochem; Can Asn Anatomists; Am Asn Anatomists. *Res:* Neurofilamentous network in neurons of squid giant axons and in brains of Alzheimer patients, using stereo and immuno electron microscopy; results integrated in general theory of helicity and asymmetry in the organization of the nervous system under normal and pathological conditions. *Mailing Add:* Fac Med Univ Ottawa 451 Smyth Rd Ottawa ON K1H 8M5 Can

METZ, CHARLES EDGAR, b Bayshore, NY, Sept 11, 42; div; c 2. MEDICAL PHYSICS, DIAGNOSTIC PERFORMANCE ANALYSIS. *Educ:* Bowdoin Col, BA, 64; Univ Pa, MS, 66, PhD(radiol physics), 69. *Prof Exp:* From instr to assoc prof, 69-80, dir grad prog med physics, 79-85, PROF RADIOL, UNIV CHICAGO, 80- *Concurrent Pos:* Consult, Int Atomic Energy Agency, 76-78 & Imaging Resources Comt, Col Am Pathologists, 86-; mem, Diag Res Adv Group, Nat Cancer Inst, 80-81; mem, sci comt efficacy studies, Nat Coun Radiation Protection & Measurements, 82-; assoc ed, Radiology, 85-; mem, Comt Performance Assessment Digital Representation Images, Int Comn Radiation Units & Measurements; guest prof, Col Biomed Technol, Osaka Univ, 89. *Mem:* Soc Med Decision Making; Radiol Soc NAm; Am Asn Physicists Med. *Res:* Evaluation of diagnostic performance in terms of signal detection theory and decision analysis; theoretical analysis of medical imaging systems; medical image enhancement by computer. *Mailing Add:* Dept Radiol Univ Chicago Box 429 Chicago IL 60637

METZ, CLYDE, b Gary, Ind, May 3, 40; m 61; c 2. PHYSICAL CHEMISTRY. *Educ:* Rose-Hulman Inst Technol, BS, 62; Ind Univ, PhD(phys chem), 66. *Prof Exp:* Asst, Ind Univ, 62-66; from asst prof to assoc prof chem, Ind Univ-Purdue Univ, Indianapolis, 66-82; DEPT CHEM, COL CHARLESTON, 82- *Mem:* Am Chem Soc; Electrochem Soc. *Res:* Fused salt electrochemistry, phase equilibria, thermodynamics and x-ray crystallography; physical chemistry, general chemistry. *Mailing Add:* Dept Chem Col Charleston Charleston SC 29424

METZ, DAVID A, b Cleveland, Ohio, Sept 10, 33; m 60; c 3. AUDIOLOGY, SPEECH PATHOLOGY. *Educ:* Western Reserve Univ, BA, 60, MA, 65, PhD(aural harmonics), 67. *Prof Exp:* Res asst speech path, Western Reserve Univ, 64-67; asst prof, 67-71, ASSOC PROF AUDIOL, CLEVELAND STATE UNIV, 77- DIR SPEECH PATH & AUDIOL PROG, 67-, CHMN DEPT SPEECH & HEARING, 74- *Mem:* Am Speech & Hearing Asn; Acoust Soc Am. *Res:* Aural harmonics; temporary threshold shift; speech discrimination. *Mailing Add:* Speech & Hearing Clin Cleveland State Univ Euclid Ave & E 24th St Cleveland OH 44115

METZ, DONALD C(HARLES), b Kidder, Mo, Dec 22, 08; m 35; c 3. ELECTRICAL ENGINEERING, INDUSTRIAL ENGINEERING. *Educ:* Purdue Univ, BSEE, 30, MSIE, 49. *Hon Degrees:* Dr Sci, Capital Inst Technol, 74. *Prof Exp:* Engr, Frigidaire Div, Gen Motors Corp, 30-32 & 33-36; statist ed, Bus News Pub Co, 32-33; sales engr, Hughes Heating & Air Conditioning Co, 36-40; supvr elec progs, Purdue Univ, 46-47, supvr instr, 47-50, actg head tech inst, 50-51; dir tech inst, Univ Dayton, 51-63; prin, Tech Col, Ibadan, Nigeria, 63-65; asst dean sch appl arts & sci, Western Mich Univ, 65-67; assoc dean, 67-73, EMER DEAN ENG TECHNOL, SOUTHWEST MINN STATE UNIV, 74- *Concurrent Pos:* Consult, US Air Force Purchasing Surv Comt, 51-52; in-chg Nat Surv Eng Technol Enrollments & Grad, 53-63; mem working group on supporting tech personnel, President's Comt Scientists & Engrs, 56-57; chmn region VII, Nat Surv Tech Inst Ed, 57; comnr eng manpower comn, Engrs Joint Coun, 60-69; tech cols rep, Nigerian Tech & Com Exam Comt, 64-65. *Honors & Awards:* James H McGraw Award for Eng Technol Educ, 77. *Mem:* Am Soc Eng Educ; Nat Soc Prof Engrs; Newcomen Soc NAm. *Res:* Technical manpower as relates to supply and utilization of engineering technicians. *Mailing Add:* 908 S Longwood Loop Mesa AZ 85208

METZ, DONALD J, b Brooklyn, NY, May 18, 24; m 47. RADIATION CHEMISTRY. *Educ:* St Francis Col, NY, BS, 47; Polytech Inst Brooklyn, MS, 49, PhD(phys chem), 55. *Hon Degrees:* DSc, St Francis Col, NY, 86. *Prof Exp:* From assoc chemist to sr chemist, 54-74, head div chem sci, 74-85, HEAD OFF EDUC PROG, BROOKHAVEN NAT LAB, 85- *Concurrent Pos:* Instr chem & physics, St Francis Col, NY, 47-54, from asst prof to assoc prof, 56-63, prof, 63-76. *Mem:* AAAS; fel Am Inst Chem; Am Chem Soc; Sigma Xi. *Res:* Radiation polymerization and radiation chemistry of organic compounds; mechanisms of vinyl polymerizations; biomaterials. *Mailing Add:* 147 Southern Blvd East Patchogue NY 11772

METZ, EDWARD, engineering, materials science, for more information see previous edition

METZ, FLORENCE IRENE, b Willard, Ohio, Sept 1, 29. MATERIAL SCIENCE ENGINEERING, PHYSICAL CHEMISTRY. *Educ:* Case Western Reserve Univ, AB, 51, MS, 56; Iowa State Univ, PhD(phys chem), 60. *Prof Exp:* Res chemist, Lewis Lab, Nat Adv Comt Aeronaut, Ohio, 51-55; instr & res fel, Iowa State Univ, 56-60; sr chemist, 60-63, Midwest Res Inst, 60-63, dir, Germanium Info Ctr, 62-67, prin chemist, 63-67, sr adv chem, 67-68, head, Phys & Anal Chem Sect, 68-72, asst dir phys sci, 72-76, dir chem sci, 76-78, vpres chem & biosci, 78-84; mgr coated prods, 84-87, GEN MGR NEW VENTURES, INLAND STEEL INDUST, 87- *Concurrent Pos:* Lectr, Univ Mo, Kansas City, 61-65. *Mem:* AAAS; Am Chem Soc; Sigma Xi. *Res:* Chemistry of materials; evaluation of effects of radiation on materials; vacuum evaporation of thin metallic films on inorganic oxidizers; analytical chemistry; behavioral sciences; advanced engineering materials. *Mailing Add:* Res Labs-9000 Inland Steel Indust 3001 E Columbus East Chicago IL 46312

METZ, FRED L, b McComb, Ohio, Apr 23, 35; m 60; c 5. BIO-ORGANIC CHEMISTRY, BIOCHEMISTRY. *Educ:* Bowling Green State Univ, BA, 57; Ind Univ, PhD(org chem), 62. *Prof Exp:* Summer res chemist, Monsanto Chem Co, 57 & 59; sr res chemist, T R Evans Res Ctr, Diamond Shamrock Corp, 62-69, res assoc, 69-80; SR CHEMIST, OFF TOXIC SUBSTANCES, US ENVIRON PROTECTION AGENCY, 80- *Honors & Awards:* Bronze Medalist, Environ Protection Agency, 84, 87, 90. *Mem:* Inst Food Technol; AAAS; Am Chem Soc; Am Dairy Sci Asn; Sigma Xi. *Res:* Organic synthesis; amino acids; organic fluorine chemicals; leather chemicals; arsenic chemicals; paper chemicals; immobilized enzymes; food and dairy technology; polymers; environmental sciences. *Mailing Add:* US EPA TS-779 401 M St SW Washington DC 20460

METZ, JOHN THOMAS, b Springfield, Ill, Jan 21, 47. ELECTROPHYSIOLOGY, BIOPSYCHOLOGY. *Educ:* St Louis Univ, BA, 70; Univ Chicago, PhD(biopsychol), 78. *Prof Exp:* Res assoc, 77-79, ASST PROF, DEPT PSYCHIAT, UNIV CHICAGO, 79- *Concurrent Pos:* Instr, Dept Psychol, St Xavier Col, 75, Northeastern Ill Univ, 76 & 77; res scientist, Ill Dept Mental Health, 80- *Mem:* Asn Psychophysiol Study Sleep; Soc Neurosci. *Res:* Pharmacological studies of muscle reflexes, event-related potentials, sleep physiology and their relationships to psychiatric illness. *Mailing Add:* Pet Ctr Univ Chicago Hosps 5841 S Maryland Box 433 Chicago IL 60637

METZ, PETER ROBERT, b Seattle, Wash, Nov 22, 34; m 59; c 3. ELECTRICAL ENGINEERING, COMMUNICATIONS. *Educ:* Univ Wash, BS, 56, PhD(elec eng), 65; Mass Inst Technol, SM, 58. *Prof Exp:* Engr, Boeing Co, 58-61; asst prof elec eng, Univ Wash, 65-72; coordr post-grad progs, Polytech Sch, Fed Univ Paraiba, 72-76; ENGR, BOEING CO, SEATTLE, WASH, 76- *Mem:* Inst Elec & Electronics Engrs. *Res:* Selection of Huffman sequences with desired ambiguity function magnitudes; optical data processing applied to bioengineering. *Mailing Add:* Boeing Co PO Box 3707 Seattle WA 98124

METZ, ROBERT, b New York, NY, June 2, 38; m 61; c 2. STRATIGRAPHY. *Educ:* City Col New York, BS, 61; Univ Ariz, MS, 63; Rensselaer Polytech Inst, PhD(stratig), 67. *Prof Exp:* Asst prof geol, State Univ NY Col Potsdam, 66-67; from asst prof to assoc prof, Newark State Col, 67-80; PROF GEOL, KEAN COL, NJ, 80- *Concurrent Pos:* Sigma Xi grant-in-aid res; NY State grad fel award. *Mem:* Geol Soc Am; Soc Econ Paleontologists & Mineralogists; Nat Asn Geol Teachers; Sigma Xi; Paleont Soc. *Res:* neoichnology and paleoichnology. *Mailing Add:* Dept Geol & Meteorol Kean Col Union NJ 07083

METZ, ROBERT JOHN SAMUEL, b Johannesburg, SAfrica, Jan 23, 29; m 53; c 4. INTERNAL MEDICINE, ENDOCRINOLOGY. *Educ:* Univ Witwatersrand, MB, BCh, 51; Northwestern Univ, MS, 59; Univ Toronto, PhD(physiol), 62. *Prof Exp:* Res fel med & physiol, Northwestern Univ, 57-59; res fel, Banting & Best Dept Med Res, Univ Toronto, 59-61; assoc med, Northwestern Univ, Chicago, 62-64, asst prof, 64-69; chief metab res, 69-71. *Concurrent Pos:* Chief, Med Div, Northwestern Univ, Chicago, 62-65; chief, Diabetes & Metab Serv, Cook County Hosp, Chicago, 62-65; attend physician, Passavant Mem Hosp, Chicago, 65-69; clin asst prof med, Univ Wash, 69-77, clin assoc prof, 77-; attend physician, Virginia Mason Hosp, 69- *Mem:* Am Diabetes Asn; Am Fedn Clin Res; fel Am Col Physicians; Endocrine Soc. *Res:* Diabetes and allied diseases. *Mailing Add:* Va Mason Clin 1100 Ninth Ave Seattle WA 98111

METZ, ROGER N, theoretical physics; deceased, see previous edition for last biography

METZE, GEORGE M, b Newport, RI, June 27, 50; m 82; c 2. HETEROJUNCTION DEVICES, MOLECULAR BEAM EPITAXY. *Educ:* Univ Calif, BS, 72; Univ Ill, MS, 74; Cornell Univ, PhD(elec eng), 81. *Prof Exp:* Staff scientist laser spectros, Lawrence Livermore Lab, 75-78; fel grad student, Cornell Univ, 78-81; staff scientist, heterojunction devices, Mass Inst Technol Lincoln Labs, 81-87; MGR, COMSAT LABS, 87- *Mem:* Sr mem, Inst Elec & Electronic Engrs; Am Phys Soc; Am Vacuum Soc. *Res:* Development of advanced microwave and millimeter-wave components and devices. *Mailing Add:* Comsat Labs 22300 Comsat Dr Clarksburg MD 20871

METZE, GERNOT, b Mahrisch Schonberg, Czech, Nov 1, 30; m 62. ELECTRICAL ENGINEERING. *Educ:* Iowa State Univ, BS, 53; Univ Ill, MS, 55, PhD(digital comput), 58. *Prof Exp:* Res asst, Elec Eng Dept & Digital Comput Lab, 53-57, res assoc, Digital Comput Lab, 58-59, res asst prof, 59-64, res asst prof elec eng, Coord Sci Lab, 65-66, res assoc prof, 66-70, RES PROF ELEC ENG, COORD SCI LAB, UNIV ILL, URBANA, 70- *Concurrent Pos:* Consult, ITT Kellogg, Ill, 65, Scully Int, 65-66, Emerson Elec, 66 & Tex Instruments Inc, 67- *Mem:* Inst Elec & Electronics Engrs. *Res:* Digital systems design; fault-tolerant computing; sequential machine theory; computer arithmetic. *Mailing Add:* Dept Elect Eng Univ Ill Urbana IL 61801

METZENBERG, ROBERT LEE, b Chicago, Ill, June 11, 30; m 54; c 2. BIOCHEMISTRY. *Educ:* Pomona Col, AB, 51; Calif Inst Technol, PhD(biochem), 56. *Prof Exp:* From instr to assoc prof, 55-68, PROF PHYSIOL CHEM, UNIV WIS-MADISON, 68- *Concurrent Pos:* Am Cancer Soc fel, 55-58; Markle investr, 58-; res fel, Univ Zurich, 59-60; USPHS career development awardee, 63-73; genetics study sect, NIH, 69-73; assoc ed, Genetics, 75-90; Am Soc Microbiol Ann Lect, 91. *Mem:* AAAS; Am Soc Biol Chemists; Am Chem Soc; Genetics Soc Am (pres, 90); Am Soc Microbiol; Sigma Xi. *Res:* Mechanism of action of urea cycle enzymes; genetic control of metabolism in Neurospora; control of metabolism in eucaryotes; Genome structure. *Mailing Add:* Univ Wis 687 Med Sci Bldg Madison WI 53706

METZGAR, DON P, b Hastings, Nebr, June 7, 29; m 50; c 3. VIROLOGY, IMMUNOLOGY. *Educ:* Hastings Col, BA, 56; Purdue Univ, PhD(microbiol), 61. *Prof Exp:* NSF fel cell physiol, Purdue Univ, 61-62; res fel virol, Merck Inst Therapeut Res, 62-67; sr res virologist, Nat Drug Co, 67-71; sect head cell biol res, Merrell Nat Labs, 71-77; VPRES OPERS, SR VPRES TECH, CONNAUGHT LABS INC, 77- *Mem:* AAAS; NY Acad Sci. *Res:* Tissue culture; radiochemistry; electron microscopy; vaccine development and relationship to antigenic potentiation. *Mailing Add:* Vpres Tech Connaught Labs Inc 1755 Steeles Ave W Willow Dale ON M2R 3T4 Can

METZGAR, LEE HOLLIS, b Olean, NY, Jan 10, 41; m 61; c 2. POPULATION DYNAMICS, BEHAVIORAL ECOLOGY. *Educ:* State Univ NY Col Fredonia, AB, 62; Univ Mich, MS, 64, MA, 66, PhD(zool), 68. *Prof Exp:* Teacher jr high sch, NY, 62-63; from asst prof to prof, 68-84, dir wildlife biol, 84-90, PROF ZOOL, UNIV MONT, 90- *Mem:* Ecol Soc Am; Wildlife Soc; Sigma Xi. *Res:* Stability properties of mammalian population dynamics; behavioral regulation of numbers; stability and harvest strategies in game populations. *Mailing Add:* 400 N Ave E Missoula MT 59801

METZGAR, RICHARD STANLEY, b Erie, Pa, Feb 2, 30; m 52; c 2. IMMUNOLOGY. *Educ:* Univ Fla, BS, 51; Univ Buffalo, MA, 57, PhD(immunol), 59. *Prof Exp:* Sr cancer res scientist, Roswell Park Mem Inst, 59-62; assoc prof, 62-72, PROF IMMUNOL, SCH MED, DUKE UNIV, 72- *Concurrent Pos:* Mem staff, Yerkes Primate Res Ctr, Emory Univ. *Mem:* AAAS; Am Asn Cancer Res; Am Soc Cell Biol. *Res:* Cancer diagnosis; cancer therapy; tumor biology. *Mailing Add:* Box 3839 Dept Microbiol & Immunol Duke Univ Sch of Med Durham NC 27710

METZGER, A(RTHUR) J(OSEPH), ceramic engineering; deceased, see previous edition for last biography

METZGER, ALBERT E, b New York, NY, Sept 10, 28; m 58; c 2. PLANETARY SCIENCE, RADIATION PHYSICS. *Educ:* Cornell Univ, AB, 49; Columbia Univ, MA, 51, PhD(nuclear chem), 58. *Prof Exp:* Chemist, Sylvania Elec Corp, 51-53; asst, Columbia Univ, 53-54; from scientist to sr scientist, 59-61, res group supvr, Space Sci Div, Jet Propulsion Lab, 61-82, MEM TECH STAFF, CALIF INST TECHNOL, 82- *Honors & Awards:* Except Sci Achievemnet Medal, NASA. *Mem:* Fel AAAS; Am Astrom Soc; Am Phys Soc; Am Geophys Union. *Res:* Geochemistry; gamma ray astronomy; planetary composition; gamma ray spectroscopy; space science instrumentation. *Mailing Add:* 380 Olive Tree Lane Sierra Madre CA 91024

METZGER, BOYD ERNEST, b Hills, Minn, June 13, 34; m 59; c 3. MEDICINE, ENDOCRINOLOGY. *Educ:* State Univ Iowa, MD, 59. *Prof Exp:* Intern internal med, Michael Reese Hosp & Med Ctr, 59-60, resident, 60-63; surgeon, USPHS, 63-65; from instr to assoc prof med, 67-77, PROF MED, MED SCH, NORTHWESTERN UNIV, 77- *Concurrent Pos:* Fel biochem, Wash Univ, 65-67. *Mem:* Am Fedn Clin Res; Am Diabetes Asn; Endocrine Soc; Sigma Xi. *Res:* Intermediary metabolism; metabolism and nutrition in pregnancy; metabolic and hormonal disturbances in diabetes. *Mailing Add:* Ctr for Endocrinol Metab & Nutrit 303 E Chicago Ave Chicago IL 60611

METZGER, CHARLES O, b Nuremberg, Ger, Oct 22, 23; nat US; m 58; c 4. CHEMICAL ENGINEERING. *Educ:* City Col New York, BS, 45; Columbia Univ, MS, 48. *Prof Exp:* Develop engr semi works, Rohm & Haas Co, Pa, 48-50, group leader propellant res, Ala, 50-53, process develop, Pa, 53-58 sect head chem eng, 58-62, asst gen mgr, Rhee Industs Div, RI, 62-66, prod supt, Fayetteville Fibers Plant, 66-71, prod mgr, Fibers Div, 71-75, bus mgr, Carpet Nylon, 75-78, BUS MGR, CARODEL SUBSIDIARIES, ROHM & HAAS CO, PHILADELPHIA, 78- *Mem:* Am Chem Soc; Am Inst Chem Engrs. *Res:* Plastics; fibers. *Mailing Add:* 3 Wexford Ct Cherry Hill NJ 08033

METZGER, DANIEL SCHAFFER, b Greenville, Mich, Sept 3, 36; m 77; c 2. RESEARCH & DEVELOPMENT MANAGEMENT. *Educ:* Kalamazoo Col, BA, 58; Ohio State Univ, MSc, 62, PhD(physics), 65. *Prof Exp:* Vis asst prof physics, Ohio State Univ, 65-66; staff mem exp physics, 66-74, alt group leader, 74-76, group leader, 76-79, assoc div leader, physics div, 79-81, dep div leader, physics, 81-83, prog mgr verification technol, 83-86, DIV LEADER, MECH & ELECTRONIC ENG, LOS ALAMOS NAT LAB, 86- *Concurrent Pos:* Mem eval panel electronics & elec eng, Nat Res Coun-Nat Acad Sci, 80-81. *Mem:* Am Phys Soc; Inst Elec & Electronics Engrs. *Res:* X-ray spectroscopy; nuclear magnetic resonance in solids; spectral and fast transient measurements of radiation associated with nuclear weapons testing; airborne measurements of infrared radiation in aurorae; fiber optics applications in fast analog systems; research and development management. *Mailing Add:* Los Alamos Nat Lab Box 1663 Los Alamos NM 87545

METZGER, DARRYL E, b Salinas, Calif, July 11, 37; m 56; c 4. MECHANICAL ENGINEERING. *Educ:* Stanford Univ, BS, 59, MS, 60, MechE, 62, PhD(mech eng), 63. *Prof Exp:* Instr mech eng, Stanford Univ, 60-61; from asst prof to assoc prof mech eng, 63-70, head dept, 74-88, PROF MECH & AEROSPACE ENG, ARIZ STATE UNIV, 70- *Concurrent Pos:* NSF fel, 61, Ford Found fel, 62-63 & NASA, Am Soc Eng Educ, 64-65; consult, numerous co, US & foreign, 64-; mem heat transfer/Gas Turbine Comt, Am Soc Mech Engrs, 73-, chair, 82-84, honors & awards comt, 88-chair, 91-92; dir thermosci res, US Naval Postgrad Sch, Monterey, Calif, 80-88, Off Naval Technol Chair prof, 89; US deleg, US/China Binat Workshop Heat Transfer, Beijing, Xian, Shanghai, 83-; mem US/China Prog Develop Meeting, NSF, Hawaii, 83; mem, US sci comt Int Heat Transfer Conf, Off Naval Technol, 86; mem, NASA Space Shuttle Main Engine Rev Team, 86-87, Space Eng Prog External Task Team, 87; vis prof, Univ Karlsruhe, Fed Repub Ger, 88-91. *Honors & Awards:* Alexander von Humboldt Sr Res Scientist Award, Fed Repub Ger, 85, 86 & 87; Achievement Award Am Soc Mech Engrs, Japan Soc Mech Engrs, 85. *Mem:* Fel Am Soc Mech Engrs; assoc fel Am Inst Aeronaut & Astronaut; Sigma Xi. *Res:* Heat transfer; fluid mechanics and thermodynamics, particularly convective heat transfer; gas turbine engine heat transfer. *Mailing Add:* 8601 N 49th St Scottsdale AZ 85253

METZGER, DENNIS W, b Suffern, NY, Sept 14, 51. CELLULAR IMMUNOLOGY, IMMUNE REGULATION. *Educ:* Univ Ill, PhD(microbiol), 78. *Prof Exp:* Asst mem, 80-87, ASSOC MEM, DEPT IMMUNOL, ST JUDE CHILDREN'S RES HOSP, 87-; ASST PROF, DEPT MICRO-BIOL & IMMUNOL, UNIV TENN, MEMPHIS, 82- *Mem:* Am Asn Immunol; Clin Immunol Soc; AAAS; Sigma Xi. *Res:* Homeostatic regulation of immune B cell activity and production of monoclonal antibodies. *Mailing Add:* Dept Immunol St Jude Children's Res Hosp PO Box 318 332 N Lauderdale Memphis TN 38101

METZGER, ERNEST HUGH, b Nuremberg, Germany, Oct 22, 23; US citizen; m 56; c 3. GRAVITY INSTRUMENTATION & THEORY, INERTIAL NAVIGATION & INSTRUMENTS. *Educ:* City Col New York, BS, 49; Harvard Univ, MS, 50. *Prof Exp:* Control engr, Bell Aircraft Corp, 50-54, tech dir inertial navig, Bell Aerosysts, 54-60, chief engr, Inertial Inst, 60-70, chief engr, gravity sensors, 70-83, dir gravity sensor systs, 83-86, EXEC DIR ENG, BELL AEROSPACE-TEXTRON, 86- *Concurrent Pos:* Mem, accelerometer criteria comt, NASA, 67-69, mem, Geodesy Comt of Nat Res Coun, Nat Acad of Sci, 87-88; consult, panel future navig systs, Nat Acad Sci, 81- *Honors & Awards:* Thurlow Award, Inst Navig, 84. *Mem:* Inst Elec & Electronics Engrs; Am Inst Aeronaut & Astronaut; AAAS; Inst Navig; NY Acad Sci; Explorers Club. *Res:* Rotating accelerometer gravity gradiometer. *Mailing Add:* 90 High Park Blvd Eggertsville NY 14226

METZGER, GERSHON, b New York, NY, June 25, 35; m. RESEARCH PLANNING, POLICY FORMULATION. *Educ:* Yeshiva Univ, BA, 55; Columbia Univ, MA, 56, PhD, 59. *Prof Exp:* Asst, Columbia Univ, 55-59, Sloan res assoc, 59-60; res chemist, Esso Res & Eng Co, NJ, 60-64; sr chemist, Chem & Phosphates Co, Ltd, Israel, 64-65; assessor of patents, Nat Coun Res & Develop, 65-68, head patent exploitation div, 68-69, dir phys sci div, 69-74, dep div planning, 74-76, dep dir res funds, Ministry of Sci & Develop, 76-81, dep dir Nat Coun Res & Develop, 81-86, DEP DIR, MINISTRY SCI & TECH, 86- *Concurrent Pos:* Instr, Yeshiva Univ, 60-62; assoc prof chem, Jerusalem Col Technol, 73-77. *Mem:* AAAS; Am Chem Soc; Sigma Xi. *Res:* Free radical reaction; fuel cells; organo thiophosphates. *Mailing Add:* Ministry Sci & Tech Bldg 3 PO Box 18195 Kiryat Hamemshala Mizrach Jerusalem 91181 Israel

METZGER, H PETER, b New York, Ny, Feb 22, 31; m 56, 90; c 4. BIOCHEMISTRY, SCIENCE WRITING. *Educ:* Brandeis Univ, BA, 53; Columbia Univ, PhD, 65. *Prof Exp:* Res scientist, NY State Psychiat Inst, 65-66; sr res scientist, NY State Inst Neurochem & Drug Addiction, 66; res assoc biochem, Univ Colo, 66-68; staff scientist, Ball Bros Res Corp, 68-69, mgr adv progs, Environ Instrumentation Dept, 69-70; dir & consult, Colspan Environ Systs, Inc, 70-73; adminr environ affairs, Pub Serv Co, Colo, 74-87; RETIRED. *Concurrent Pos:* Prin investr, USPHS res grant, 67-68; sci ed, Rocky Mt News, Denver, 74-77. *Mem:* AAAS; Am Chem Soc. *Res:* Mechanisms of enzyme action; neurochemistry; biochemical basis of memory; protein hormone production. *Mailing Add:* 2595 Stanford Ave Boulder CO 80303-5332

METZGER, HENRY, b Mainz, Ger, Mar 23, 32; US citizen; m 57; c 3. PROTEIN BIOCHEMISTRY, MEMBRANE RECEPTORS. *Educ:* Univ Rochester, AB, 53; Columbia Univ, MD, 57. *Prof Exp:* Intern internal med, Presby Hosp, NY, 57-58, asst resident, 58-59; res assoc, NIH, 59-61; Helen Hay Whitney Found fel, 61-63; CHIEF BR, 84-, SCI DIR, NAT INST ARTHRITIS, MUSCULOSKELETAL & SKIN DIS, 87- *Concurrent Pos:* Sr investr, Arthritis & Rheumatism Br, Nat Inst Arthritis, Musculoskeletal & Skin Dis, 63-, chief sect chem immunol, 73-; assoc ed, Ann Rev Immunol; vpres, Int Union Immunol Socs, 89-; Am Asn Immunologusts, 90- *Honors & Awards:* Harvey lectr, 84-85. *Mem:* Am Asn Immunol(vpres, 90-); Am Soc Biol Chemists. *Res:* Immunochemistry; protein chemistry; structure of immunoglobulins; membrane receptors. *Mailing Add:* NIAMS/NIH Bldg 10 Room 9N258 Bethesda MD 20892

METZGER, JAMES DAVID, b Columbia, Pa, Aug 23, 52; m 74; c 1. PHYTOHORMONES, SEED DORMANCY. *Educ:* Millersville State Col, BA, 74; Mich State Univ, PhD(bot), 80. *Prof Exp:* STAFF SCIENTIST, AGR RES SERV, US DEPT AGR, 80- *Mem:* Am Soc Plant Physiologists; AAAS; Weed Sci Soc Am. *Res:* Physiological and biochemical basis of seed dormancy and the role of phytohormones in plant growth and development. *Mailing Add:* 2019 Fifth Ave E West Fargo ND 58078

METZGER, JAMES DOUGLAS, b Allentown, Pa, Feb 10, 42; m 64. ORGANIC CHEMISTRY, AGRICULTURAL CHEMISTRY. *Educ:* Univ Nev, Reno, BS, 64, PhD(org chem), 69. *Prof Exp:* Res chemist, 68-78, sr res chemist, 78-88, RES ASSOC, E I DU PONT DE NEMOURS & CO, INC, 88- *Mem:* Am Chem Soc. *Res:* Organic synthesis; agricultural chemicals formulation. *Mailing Add:* 604 Baldwin Lane Carrcroft Crest Wilmington DE 19803

METZGER, MARVIN, corrosion, mechanical behavior; deceased, see previous edition for last biography

METZGER, ROBERT MELVILLE, b Yokohama, Japan, May 7, 40; US citizen; m 70; c 3. CHEMISTRY, CHEMICAL PHYSICS. *Educ:* Univ Calif, Los Angeles, BS, 62; Calif Inst Technol, PhD(chem), 68. *Prof Exp:* Jr res asst chem, Atomics Int Div, NAm Aviation Corp, Calif, 61; res assoc chem, Stanford Univ, 68-71, lectr Italian, 69-71, res assoc chem eng, 71, asst prof, 71-77; from assoc prof to prof chem, Univ Miss, 77-86, Margaret McLean Coulter prof chem, 84-86; PROF CHEM, UNIV ALA, 86- *Concurrent Pos:* Mem, Mat Sci Prog, Univ Ala Syst. *Mem:* Am Chem Soc. *Res:* Solid state chemistry; organic crystals and conductors; crystallography; quantum mechanics; Madelung energy calculations; combustion colorimetry; computers in chemistry; x-ray radial distribution function studies of platinum catalysts; polarizabilities of organic molecules and ions; unimolecular organic rectifiers; high temperature inorganic superconductors. *Mailing Add:* Dept Chem Univ Ala Tuscaloosa AL 35487-0336

METZGER, ROBERT P, b San Jose, Calif, Jan 28, 40; m 68; c 2. BIOCHEMISTRY. *Educ:* Univ Calif, Los Angeles, BS, 61; San Diego State Univ, MS, 63, PhD(chem), 67; Univ Calif, San Diego, PhD(chem), 67. *Prof Exp:* Lectr chem, San Diego State Col, 63-68, from asst prof to assoc prof phys sci, 68-77, PROF NATURAL SCI, SAN DIEGO STATE UNIV, 77-, CHMN DEPT, 87- *Concurrent Pos:* vis res prof, Univ Genova, Italy, 79-; vis sr res scientist, Johns Hopkins Univ, 80; vis scholar, Univ Calif, San Diego, 86-87. *Mem:* Fel AAAS; Am Chem Soc; Sigma Xi; Am Diabetes Asn; NY Acad Sci. *Res:* Enzymology; carbohydrate metabolism; diabetes mellitus. *Mailing Add:* Dept Natural Sci San Diego State Univ San Diego CA 92182-0324

METZGER, SIDNEY, b New York, NY, Feb 1, 17; m 44; c 3. COMMUNICATIONS ENGINEERING. *Educ:* NY Univ, BS, 37; Polytech Inst Brooklyn, MEE, 50. *Prof Exp:* Engr, US Signal Corps Lab, NJ, 39-45; head radio relay div, Fed Telecommun Labs, Int Tel & Tel Corp, 45-54; mgr commun eng, Astro Electronic Prod Div, Radio Corp Am, 54-63; mgr, Eng Div, Commun Satellite Corp, 63-67, chief engr, 68-72, asst vpres, 68-80, chief scientist, 72-82, vpres, 80-82; consult engr, 82-89; RETIRED. *Honors & Awards:* Aerospace Award, Aerospace & Electronics Systs Soc, 75; Int Commun Award, Inst Elec & Electronics Engrs, 76, Koji Kobayashi Computers & Commun Award, 85; Aerospace Commun Award, Am Inst Aeronaut & Astronaut, 84. *Mem:* Nat Acad Eng; fel Inst Elec & Electronics Engrs; fel Am Inst Aeronaut & Astronaut; Sigma Xi. *Res:* Communications satellites; radiation hazards. *Mailing Add:* Apt 1522 10500 Rockville Pike Rockville MD 20852

METZGER, SIDNEY HENRY, JR, b Atlanta, Ga, Mar 29, 29; m 52; c 3. APPLIED CHEMISTRY. *Educ:* Univ Ala, BS, 51; Texas A&M Univ, MS, 56; Univ Ill, PhD(org chem), 62. *Prof Exp:* Chemist, Monsanto Co, 51-54; res chemist, Jefferson Chem Co, 60-62; sr chemist, 62-67, group leader, 67-71, DIR, MOBAY CHEM CO, 71-, MGR ELASTOMERIC A/D, MOBAY CHEM CORP, 73- *Mem:* Am Chem Soc; Soc Automotive Engrs; Am Inst Chemists. *Res:* Organophosphorus chemistry; reactions of epoxides; synthesis and reactions of carbodiimides and isocyanates. *Mailing Add:* Mobay Chem Co Penn-Lincoln Pkwy W Pittsburgh PA 15205-9910

METZGER, THOMAS ANDREW, b Paterson, NJ, July 14, 44; m 70; c 5. PURE MATHEMATICS. *Educ:* Seton Hall Univ, BS, 65; Creighton Univ, MS, 69; Purdue Univ, West Lafayette, PhD(math), 71. *Prof Exp:* Asst prof math, Tex A&M Univ, 71-73; asst prof, 73-79, ASSOC PROF MATH, UNIV PITTSBURGH, 80- *Concurrent Pos:* Vis asst prof, Univ Tenn, 79. *Mem:* Am Math Soc; Math Asn Am. *Res:* Automorphic forms, applications to Riemann surfaces; weighted areal approximation in the complex plane; function theory. *Mailing Add:* Dept Math Univ Pittsburgh Pittsburgh PA 15260

METZGER, WILLIAM JOHN, b Freeport, Ill, Nov 7, 35; m 57, 79; c 2. GEOLOGY. *Educ:* Beloit Col, BS, 57; Univ Ill, MS, 59, PhD(geol), 61. *Prof Exp:* Asst geol, Univ Ill, 57-61; from asst prof to assoc prof geol, State Univ NY Col Fredonia, 61-71, actg chmn dept, 63-65, prof geol, 71-81; SR STAFF GEOLOGIST, DENVER WEST EXPLOR, CONOCO, INC, GOLDEN, COLO, 81- *Mem:* AAAS; fel Geol Soc Am; Am Asn Petrol Geol; Soc Econ Paleont & Mineral; Clay Minerals Soc; Sigma Xi. *Res:* Stratigraphy and sedimentation; petroleum geology. *Mailing Add:* Conoco Inc 851 Werner Ct Casper WY 82601

METZLER, CARL MAUST, b Masontown, Pa, Dec 13, 31; m 53; c 3. PHARMACOKINETICS, MODELING. *Educ:* Goshen Col, BS, 55; NC State Univ, PhD(biomath), 65. *Prof Exp:* Asst prof math, Goshen Col, 60-62; res scientist, 65-70; head res, 70-78, sr statistician, 78-85, sr scientist, 85-88, DISTINGUISHED SCIENTIST, BIOSTATIST, 88- *Mem:* AAAS; Biomet Soc; fel Am Statist Asn; Am Soc Clin Pharmacol & Therapeut; fel Am Asn Pharmaceut Scientists. *Res:* Application of mathematical and statistical methods to chemical biological and medical research; development of mathematical, statistical and computer methodology. *Mailing Add:* 9164-32-1 Upjohn Co Kalamazoo MI 49001

METZLER, CHARLES VIRGIL, b Louisville, Ky, Jan 11, 29; m 52; c 6. CHEMICAL ENGINEERING, ENVIRONMENTAL ENGINEERING. *Educ:* Univ Louisville, BChE, 52, MChE, 53, PhD(chem eng), 60. *Prof Exp:* Instr chem eng, Univ Louisville, 53-60; sr res engr, Rocketdyne Div, NAm Aviation, Inc, 58-60, res specialist, 59-60; from asst prof to assoc prof eng, 60-66, chmn dept eng, 65-67, chmn dept urban studies & assoc dean eng, 70-73, PROF ENG, CALIF STATE UNIV, NORTHRIDGE, 66-, PROF URBAN STUDIES, 73- *Concurrent Pos:* Consult, Atomics Int Div, NAm Aviation, Inc, 60, consult & res specialist, Rocketdyne Div, 60-64; consult, Naval Missile Ctr, 64; consult adv tech staff, Marquardt Corp, 65-66; educ consult, Pac Missile Test Ctr, 73- & Calif State Univs & Cols, 76- *Mem:* Am Soc Eng Educ; Am Inst Chem Engrs. *Res:* High speed gas flow and heat transfer; supersonic combustion; water resources; professional education. *Mailing Add:* Sch of Eng Calif State Univ 1811 Nordhoff St Northridge CA 91330

METZLER, DAVID EVERETT, b Palo Alto, Calif, Aug 12, 24; m 48; c 5. BIOCHEMISTRY. *Educ:* Calif Inst Technol, BS, 48; Univ Wis, MS, 50, PhD(biochem), 52. *Prof Exp:* Res scientist, Univ Tex, 51-53; from asst prof to assoc prof, 53-61, PROF BIOCHEM, IOWA STATE UNIV, 61- *Mem:* Am Chem Soc; Am Soc Photobiol; Am Soc Biol Chem. *Res:* Mechanisms of coenzyme action; electronic absorption spectra of vitamins, coenzymes and proteins. *Mailing Add:* Dept Biochem & Biophys Iowa State Univ A3246 Gilman Ames IA 50011-0061

METZLER, DWIGHT F, b Carbondale, Kans, Mar 25, 16; m 41; c 4. WATER SOURCES, PUBLIC HEALTH ADMINISTRATION. *Educ:* Univ Kans, BS, 40, CE, 47; Harvard Univ, SM, 48; Am Acad Environ Engrs, dipl. *Prof Exp:* Asst engr, Kans Bd Health, 40-41, sanit engr, 46-48; chief engr, Kans State Bd Health, 48-62; from assoc prof to prof civil eng, Univ Kans, 48-66; dep comnr, NY State Dept Health, 66-70 & NY State dept Environ Conserv, 70-74; secy, Kans Dep Health & Environ, 74-79, dir, Water Supply Develop, 79-84; RETIRED. *Concurrent Pos:* Comn officer, USPHS, 42-46, consult, 57-66; housing consult, Chicago-Cook County Health Surv, 46; adv, Govt India, 60, USSR, 62; adv, WHO, 64-84, chmn expert panel solid waste, 71; mem, Water Pollution Bd Int Joint Comn, 67-74, Assembly Eng Nat Res Coun, 77-80; assoc ed, J Int Water Pollution Res, 68-73. *Honors & Awards:* Fuller Award, Am Water Works Asn, 54, Purification Div Award, 58; Bedell Award, Water Pollution Control Fedn, 63; Centennial Award, Am Pub Health Asn, 72, Sedgwick Medal, 81. *Mem:* Nat Acad Eng; hon fel Royal Soc Health; Sigma Xi; Am Pub Health Asn (pres, 64-65); Am Soc Civil Engrs; hon mem Am Waterworks Asn. *Res:* Studies of histoplasmosis in humans and nitrate cyanosis and occurrance of nitrates in water; author or coauthor of over 85 publications. *Mailing Add:* 900 SW 31st St No 325 Topeka KS 66611-2196

METZLER, RICHARD CLYDE, b Cleveland, Ohio, Oct 19, 37; m 60; c 2. PURE MATHEMATICS. *Educ:* Univ Mich, BS, 59; Wayne State Univ, MA, 62, PhD(math), 66. *Prof Exp:* Asst prof, 65-71, ASSOC PROF MATH, UNIV NMEX, 71- *Concurrent Pos:* Assoc dean, Col Arts & Sci, Univ NMex, 80-88, assoc chmn, Dept Math & Statist, 88-90. *Mem:* Math Asn Am; Am Math Soc. *Res:* Ordered topological vector spaces; abstract integration theory. *Mailing Add:* Dept Math & Statist Univ of NMex Albuquerque NM 87131

METZNER, ARTHUR B(ERTHOLD), b Sask, Can, Apr 13, 27; nat US; m 48; c 3. CHEMICAL ENGINEERING. *Educ:* Univ Alta, BSc, 48; Mass Inst Technol, ScD(chem eng), 51. *Hon Degrees:* DASc, Katholieke Univ, Belgium, 75. *Prof Exp:* Res engr, Defense Res Bd, Can, 48; instr chem eng, Mass Inst Technol, 50-51; from asst prof to prof, 53-62, chmn dept, 70-78, H FLETCHER BROWN PROF CHEM ENG, UNIV DEL, 62- *Concurrent Pos:* Res chem engr, Colgate-Palmolive Co, 51-53; instr chem eng, Polytech Inst, Brooklyn, 51-53. *Honors & Awards:* Am Chem Soc Sect Award, 58; Colburn Award, Am Inst Chem Engrs, 58, Walker Award, 70, Lewis Award, 77, Founders' Award, 90; Bingham Medal, Soc Rheology, 77. *Mem:* Nat Acad Eng; Soc Rheology; Am Inst Chem Engrs; Mat Res Soc. *Res:* Transport processes; rheology; fluid mechanics. *Mailing Add:* Dept Chem Eng Univ Del Newark DE 19716

METZNER, ERNEST KURT, b Phoenix, Ariz, Apr 17, 49. ORGANIC CHEMISTRY. *Educ:* Ariz State Univ, BS, 71; Univ Calif, Berkeley, PhD(chem), 74. *Prof Exp:* Fel chem, Stanford Res Inst, 74-75; RES SCIENTIST CALBIOCHEM-BEHRING CORP, AM HOECHST CORP, 75- *Mem:* Am Chem Soc. *Res:* Synthesis of biologically active compounds. *Mailing Add:* 137 Durango Dr Del Mar CA 92014-3421

METZNER, JEROME, b New York, NY, Apr 14, 11; m 32; c 3. BOTANY, CYTOLOGY. *Educ:* City Col New York, BA, 32; Columbia Univ, MA, 33, PhD(bot), 44. *Prof Exp:* Teacher high schs, NY, 44-49; asst prof educ, City Col New York, 49-50; chmn dept biol & gen sci, Jamaica High Sch, 50-53, dept biol & introd sci, High Sch Sci, 53-60 & dept biol, Francis Lewis High Sch, 60-67; prof, 67-81, EMER PROF BIOL, JOHN JAY COL, CITY UNIV N YORK, 81- *Concurrent Pos:* Lectr, Hunter Col, 46-48 & City Col New York, 47-67; admin officer, Education Mission, US Dept Army, Korea, 48; educ dir, Nature Ctrs Young Am, Inc, 59-; mem gifted student comt biol sci curric study, Am Inst Biol Sci. *Mem:* AAAS; Am Micros Soc. *Res:* Protozoology; phycology; cytology. *Mailing Add:* 16 Pinehurst St Lido Beach NY 11561

METZNER, JOHN J(ACOB), b Queens, NY, June 23, 32. ELECTRICAL ENGINEERING. *Educ:* NY Univ, BEE, 53, MEE, 54, ScD(elec eng), 58. *Prof Exp:* Sr res scientist, Electrosci Lab, Sch Eng & Sci, NY Univ, 58-67, asst prof, 65-66, assoc prof elec eng, 67-73; assoc prof, Polytech Inst NY, 73-74; assoc prof, Wayne State Univ, 74-80, prof, 80; prof eng, Oakland Univ, 81-82. *Mem:* Inst Elec & Electronics Engrs. *Res:* Communication theory; techniques for reliable data communication involving error correcting codes and retransmission strategies; efficient utilization of data communication networks; fault tolerant computing. *Mailing Add:* 228 Wooded Way State College PA 16803

MEULY, WALTER C, organic chemistry; deceased, see previous edition for last biography

MEUNIER, MICHEL, b Montreal, Can, Mar 14, 56; m 78; c 1. SEMICONDUCTOR PHYSICS, PROCESSING FOR MICROELECTRONICS. *Educ:* Ecole Polytechnique, Montreal, BScA, 78, MScA, 80; Mass Inst Technol, Boston, PhD(mat sci), 84. *Prof Exp:* Scientist insulation, Inst Res Hydro-Quebec, 84-85; asst prof, 85-89, ASSOC PROF ENG PHYSICS, ECOLE POLYTECHNIQUE, MONTREAL, 89- *Mem:* Am Phys Soc; Am Vacuum Soc; Inst Elec & Electronic Engrs; Can Asn Physicists. *Res:* Semiconductor physics; crystalline and amorphous semiconductors; quantum wells and superlattices; electronic properties; device physics; processing in microelectronics; thin films; laser processing; plasma processing; modeling. *Mailing Add:* Dept Eng Physics Ecole Polytechnique PO Box 6079 Sta A Montreal PQ H3C 3A7 Can

MEUSSNER, R(USSELL) A(LLEN), b Pittsburgh, Pa, July 23, 20. PHYSICAL METALLURGY. *Educ:* Carnegie Inst Technol, PhD(phys metall), 52. *Prof Exp:* Res assoc chem, Forrestal Res Ctr, Princeton, 52-55; METALLURGIST, US NAVAL RES LAB, 55- *Mem:* Sigma Xi; Am Inst Mining Metall & Petrol Engr; Am Soc Metals. *Res:* Gas-metal reactions; superconducting materials. *Mailing Add:* US Naval Res Lab-Code 6320 Washington DC 20375

MEUTEN, DONALD JOHN, b Waterbury, Conn, Oct 4, 48; m 79. PATHOLOGY, ENDOCRINOLOGY. *Educ:* Univ Conn, BS, 70; Cornell Univ, DVM, 74; Ohio State Univ, PhD(vet pathol), 81. *Prof Exp:* Intern vet med, Univ Guelph, 75-76; resident vet pathol, Cornell Univ, 76-77; clin instr vet clin pathol, Ohio State Univ, 77-78, res fel, 78-81; ASST PROF VET PATHOL, TEX A&M UNIV, 81- *Mem:* Am Col Vet Pathologists; Am Vet Med Asn. *Res:* Pathogenesis of hypercalcemia associated with neoplasms in dogs. *Mailing Add:* Rte 4 Box 490 College Station TX 77840

MEUX, JOHN WESLEY, b Little Rock, Ark, Apr 25, 28; m 53; c 2. MATHEMATICS. *Educ:* Henderson State Teachers Col, BS, 53; Univ Ark, MS, 57; Univ Fla, PhD(math), 60. *Prof Exp:* Instr math, Univ Fla, 59-60; asst prof, Kans State Univ, 60-64; from asst prof to assoc prof, 64-77, chmn dept, 64-68, PROF MATH, MIDWEST UNIV, 77- DEAN SCH SCI & MATH, 68- *Mem:* Math Asn Am. *Res:* Orthogonal functions; numerical analysis. *Mailing Add:* Dept Math Midwestern State Univ 3400 Taft Blvd Wichita Falls TX 76308

MEWBORN, ANCEL CLYDE, b Greene Co, NC, Sept 22, 32; m 54; c 2. ALGEBRA. *Educ:* Univ NC, AB, 54, MA, 57, PhD(math), 59. *Prof Exp:* Instr math, Yale Univ, 59-61; from asst prof to assoc prof, 61-70, PROF MATH, UNIV NC, CHAPEL HILL, 70- *Mem:* Am Math Soc; Math Asn Am. *Res:* Structure of non-commutative rings; separable extensions of non-commutative rings. *Mailing Add:* Dept of Math Univ of NC Chapel Hill NC 27514

MEWHINNEY, JAMES ALBERT, b Dayton, Ohio, July 19, 39; m 61; c 3. RADIATION DOSIMETRY, BIOKINETIC MODELING. *Educ:* Wabash Col, BA, 61; Purdue Univ, PhD (bionucleonics), 71. *Prof Exp:* Health physicist, Lovelace Inhalation Toxicol Res Inst, 65-67, radiobiologist, 71-89; HEALTH PHYSICIST, DEPT ENERGY, WASTE ISOLATION PILOT PLANT, 89- *Concurrent Pos:* Educ leave, Purdue Univ, 67-71. *Mem:* Health Phys Soc; Sigma Xi. *Res:* Fate of inhaled radionuclides; radiation dosimetry and biokinetic modeling of internally deposited actinide elements; radiation carcinogenesis; extrapolation of animal data to man. *Mailing Add:* Dept Energy Waste Isolation Pilot Plant PO Box 3090 Carlsbad NM 88220

MEWISSEN, DIEUDONNE JEAN, b Ans, Belg, Oct 25, 24; m 53; c 3. RADIOLOGY, RADIOBIOLOGY. *Educ:* Univ Liege, MD, 50, Agrege, 61; Am Bd Radiol, cert therapeut radiol, 71. *Prof Exp:* HEAD RADIOBIOL LAB, UNIV BRUSSELS, 61-, PROF RADIOBIOL, UNIV, 69- *Concurrent Pos:* mem, Belg Adv Coun Cancer, 63- & Hosps, 64- *Honors & Awards:* Prize, Cong Radiol & Electrol Latin Cult, Lisbon, 57; Dag Hammarskjoeld Int Award, Brussels, 80. *Mem:* AAAS; Radiation Res Soc; Royal Soc Med; Belg Cancer Soc (secy-gen, 71); Am Asn Cancer Res. *Res:* Radiation carcinogenesis; long term effects of radiation; toxicity and carcinogenicity of tritium and tritiated compounds. *Mailing Add:* 950 E 59th St Chicago IL 60637

MEXAL, JOHN GREGORY, b San Antonio, Tex, Oct 14, 46; m 68; c 2. PLANT PHYSIOLOGY, FOREST ECOLOGY. *Educ:* Univ NMex, BS, 69, MS, 71; Colo State Univ, PhD(plant physiol), 74. *Prof Exp:* SCIENTIST, SOUTHERN FORESTRY RES CTR, WEYERHAEUSER CO, 74-; HEAD HORT DEPT, NMEX STATE UNIV. *Mem:* Sigma Xi. *Res:* Tree physiology, soil-plant-water relations; mycorrhizal associations; seed germination; revegetation. *Mailing Add:* NMex State Univ Box 3530 Las Cruces NM 88003-0017

MEYBOOM, PETER, b Barneveld, Netherlands, Apr 26, 34; Can citizen; div; c 4. HYDROGEOLOGY. *Educ:* Univ Utrecht, BSc, 56, MSc, 58, PhD(hydrogeol), 60. *Prof Exp:* Res officer, Alta Res Coun, Can, 58-60; res scientist, Geol Surv Can, Can Fed Dept Energy, Mines & Resources, 60-66, sect head hydrogeol, Inland Waters Br, 66-67, head groundwater subdiv, 67-69, sci adv, Can Fed Dept Finance, 70-71; dir sci policy, Can Fed Dept Environ, 71-73, dir gen sci centre, Can Fed Dept Supply & Serv, 73-75; asst secy, Can Ministry State Sci & Technol, 75-77; dep secy admin policy, Treas Bd Can, 77-84; pub serv adv to Dep Prime Minister Can, Dept Fisheries & Oceans, Can, 84-86, dep minister, 86-91; VPRES ENVIRON AFFAIRS, HILL & KNOWLTON LTD, CAN, 91- *Concurrent Pos:* Distinguished lectr, Can Inst Mining & Metall, 62-63. *Honors & Awards:* Can Centennial Medal, 67. *Res:* Hydrology; science policy. *Mailing Add:* 115 Brighton Ave Ottawa ON K1S 0T3 Can

MEYBURG, ARNIM HANS, b Bremerhaven, Ger, Aug 25, 39; US citizen; m 67; c 1. TRANSPORTATION ENGINEERING & PLANNING. *Educ:* Northwestern Univ, MS, 68, PhD(civil eng), 71. *Prof Exp:* Asst prof transp eng, Cornell Univ, 69-75; assoc prof, 75-78, actg chmn, dept environ eng, chmn, 80-85, PROF TRANSP ENG, CORNELL UNIV, 78- *Concurrent Pos:* Res assoc, Transp Ctr, Northwestern Univ, 68-69; consult, NY State Dept Ed, 74-75, Calif Dept Transp, 75-76, Social Sci Res Inst, Munich, 78-81, Australian road res bd, 76, Univ Sao Paulo, Brazil, 84; vis prof, transp eng, Univ Calif, Irvine, 75, Tech Univ Munich, Ger, 76, 78-79, Univ Sao Paulo, Brazil, 84 & Tech Univ Brunswick, Ger, 85-86; res fel, Alexander von Humboldt Found, 78-79,; partner, Cayuga Anal Serv, 82-; mem, UN Develop Prog, Sao Paulo, Brazil, 84; Ger ministry of transport, 85-87; prin investr res projs, Nat Sci Found, US Dept Transp; Nat Coop Res Hwy Res Prog, NY State Dept Transp. *Honors & Awards:* Humboldt Sr Scientist Award, 84; Fulbright Sr Lectr Award, Brazil, 84. *Mem:* Am Soc Civil Engrs; Oper Res Soc; Sigma Xi; Transp Res Bd; Transp Res Forum. *Res:* Disaggregate travel demand modeling; urban transporation planning; microcomputer-based analysis and modeling of transit operations; transportation-communications interactions; urban goods movement; technology assessment; three books co-authored, two books co-edited. *Mailing Add:* Cornell Univ 315 Hollister Hall Ithaca NY 14853-3501

MEYDRECH, EDWARD FRANK, b Oak Park, Ill, July 21, 43; m 65; c 2. BIOSTATISTICS, EPIDEMIOLOGY. *Educ:* Univ Fla, BS, 65, MS, 67; Univ NC, PhD(biostat), 72. *Prof Exp:* Health serv officer, USPHS, 67-69; asst prof, Va Commonwealth Univ, 72-76; from asst prof to assoc prof, 77-86, PROF BIOSTATIST, UNIV MISS MED CTR, 86- *Mem:* Am Statist Asn; Biomet Soc; Sigma Xi. *Res:* Response surface methodology; epidemiologic studies; nonparametric statistics. *Mailing Add:* Dept Prev Med 2500 N State St Jackson MS 39216

MEYER, ALBERT RONALD, b New York, NY, Nov 5, 41; m 81; c 2. COMPUTER SCIENCE, MATHEMATICAL LOGIC. *Educ:* Harvard Univ, AB, 63, MA, 65, PhD(appl math), 72. *Prof Exp:* Asst prof comput sci, Carnegie-Mellon Univ, 67-69; asst prof elec eng & comput sci, 69-72, assoc dir, Lab Comput Sci, 78-81, PROF COMPUT SCI, MASS INST TECHNOL, 75- *Concurrent Pos:* Prin investr, NSF res grant, 72-; vis prof, Harvard Univ, 78-79. *Mem:* Asn Comput Mach; Math Asn Am; Soc Indust & Appl Math; Am Acad Arts & Sci; Inst Elec & Electronics Engrs. *Res:* Computational complexity; algorithms; logic of programming; automata theory. *Mailing Add:* NE43-315 MIT Lab for Computer Sci 545 Technology Square Cambridge MA 02139

MEYER, ALBERT WILLIAM, b Schenectady, NY, Nov 29, 06; m 31; c 4. PHYSICAL CHEMISTRY. *Educ:* Univ Chicago, BS, 27, PhD(phys chem), 30. *Prof Exp:* Res chemist, E I du Pont de Nemours & Co, 30-31; group leader, A O Smith Corp, Wis, 31-34; dept head, US Rubber Co, 34-54; dir explor res, Diamond Alkali Co, 54-57; head atomic & radiation res, US Rubber Co, 57-59, dir tech personnel & univ rel, 59-63; asst dir res, Stevens Inst Technol, 63-67; exec secy, Plastics Inst Am, 67-71; mgr, Chem Div, Ad-Tech Personnel, 77-80; mem staff, Kelley-Pepper Assoc, 80-83; RETIRED. *Concurrent Pos:* Assoc prof eve div, NY Univ, 46-48; mem Gordon Res Conf, chmn, 51; chmn subcomt continuing educ, Coun Comt Chem Educ, Am Chem Soc, 69-; consult, 72-80; mem coun, Eng & Sci Soc Execs. *Mem:* Soc Plastics Eng; Am Chem Soc; Sigma Xi. *Res:* Liquid ammonia as a solvent; contact catalysis; polymerization, emulsion and oil phase; rubber technology; agricultural chemicals; radiation chemistry; nuclear processes; polymer science and technology. *Mailing Add:* 138 Alexander Ave Upper Montclair NJ 07043

MEYER, ALVIN F, JR, b Shreveport, La, Sept 3, 20; m 42; c 2. ENVIRONMENTAL HEALTH, OCCUPATIONAL HEALTH. *Educ:* Va Mil Inst, BS, 41; Am Acad Environ Eng, dipl, 56; Indust Col Armed Forces, dipl, 62. *Prof Exp:* Chief environ health eng, hq, Air Mat Command, US Air Force, 49-61, hq, Strategic Air Command, 61-65, chief bioenviron eng, Off Surgeon Gen, 62-65, chief biomed sci corps, 65-69; spec asst legis, Off Adminr, Consumer Protection & Environ Health Serv, Dept Health, Educ & Welfare, 69-75; dir off noise abatement & control, Environ Protection Agency, 71-75; PRES, A F MEYER & ASSOC, INC, 75- *Concurrent Pos:* Vis fac mem, Va Mil Inst, 55-; asst prof, Creighton Univ, 55-61; mem nat adv coun environ health, 63-69; chmn environ pollution control comt, Dept Defense, 64-69. *Mem:* Aerospace Med Asn; Am Indust Hyg Asn; Am Soc Civil Eng; Am Pub Health Asn. *Res:* Environmental pollution control from toxic aerospace propellants; engineering control of environmental stresses on man; bioacoustics and noise control; occupational health and safety in fossil energy and industrial processes. *Mailing Add:* 1600 Longfellow St McLean VA 22101

MEYER, ANDREW U, b Berlin, Ger, Apr 21, 27; US citizen; m 64; c 2. ELECTRICAL ENGINEERING, CONTROL SYSTEMS. *Educ:* Northwestern Univ, MS, 58, PhD(elec eng), 61. *Prof Exp:* Develop engr, Assoc Res, Inc, Ill, 50-53; proj engr, Sun Elec Corp, 53-55; assoc elec engr, Armour Res Found, Ill Inst Technol, 56-57; mem tech staff, Bell Tel Labs, NJ, 61-65; assoc prof, 65-68, PROF ELEC ENG, NEWARK COL ENG, NJ INST TECHNOL, NEWARK, 68- *Concurrent Pos:* Vis prof, Mid East Tech Univ, Ankara, 69-70. *Mem:* AAAS; Inst Elec & Electronic Engrs; Soc Indust & Appl Math; Am Soc Eng Educ; Sigma Xi; Int Asn Math & Comput Simulation; Am Asn Univ Professors. *Res:* Automatic control systems; large scale systems; stability theory; biomedical control systems; computer-aided medical diagnosis; modeling of physiological systems; biomedical system analysis and modeling. *Mailing Add:* 746 Ridgewood Rd Millburn NJ 07041

MEYER, AXEL, b Copenhagen, Denmark, Mar 3, 26; nat US; m 50; c 3. METAL PHYSICS. *Educ:* City Univ New York, BS, 48 & 50; Ga Inst Technol, MS, 52; Ill Inst Technol, PhD(physics), 56. *Prof Exp:* Res fel nuclear physics, IIT Res Inst Ill Technol, 52-54; from asst prof to assoc prof physics, Univ Fla, 55-59; solid state physicist, Neutron Physics Div, Oak Ridge Nat Lab, 59-62, Solid State Div, 62-67; assoc prof, 67-69, PROF PHYSICS, NORTHERN ILL UNIV, 69- *Concurrent Pos:* Res Corp grant, 58; vis scientist, Univ Sheffield, Eng, 68, 69, Univ E Anglia, Eng, 71, 73, 78, 80, 84, 85, 86 & 88, Queens Univ Kingston, Ont, Can, 75 & Univ Valladolid, Spain, 90. *Mem:* Am Phys Soc; fel Brit Inst Physics; Fedn Am Scientists; Sigma Xi. *Res:* Theoretical solid state and metal physics, especially electronic structure; structure factors of liquid metals; transport properties and thermodynamics of metals and alloys in both liquid and solid state; long range and core interatomic forces from observed structure factors of liquid metals; 41 research publications. *Mailing Add:* Dept of Physics Northern Ill Univ DeKalb IL 60115

MEYER, BERNARD HENRY, b Cincinnati, Ohio, June 3, 41; m 70; c 1. CHEMICAL ENGINEERING, POLYMER SCIENCE. *Educ:* Univ Cincinnati, BSChE, 64; Univ Akron, MS & PhD(polymer sci), 71. *Prof Exp:* Chem engr, Emery Industs, 64; SCIENTIST, PLASTICS GROUP, KOPPERS CO, 71- *Mem:* Am Chem Soc. *Res:* Impact resistant plastic resins; polymerization kinetics; reverse osmosis; composite resin morphologies and properties. *Mailing Add:* 31 Newton Woods Rd Newtown Square PA 19073-2398

MEYER, BERNARD SANDLER, botany; deceased, see previous edition for last biography

MEYER, BETTY MICHELSON, b Jersey City, NJ. BIOSTATISTICS. *Educ:* Brown Univ, AB, 45, ScM, 46; Univ Mich, MPH, 66, PhD(biostatist), 69. *Prof Exp:* From asst prof to assoc prof biostatist, Sch Pub Health, Univ Mich, 69-76, assoc dir, Ctr Res Dis of Heart, 72-76; dir epidemiol div, Inst Aerobics Res, 76-77, dir res, 77-79; PROF, COL NURSING, TEX WOMAN'S UNIV, 81- *Concurrent Pos:* Mem, Clin Applns & Prev Adv Comt, Div Heart & Vascular Dis, Nat Heart & Lung Inst, 75-76; fel, Epidemiol Coun, Am Heart Asn. *Mem:* Am Statist Asn; Am Women Sci; Biomet Soc; Soc Epidemiol Res; Sigma Xi. *Res:* Epidemiological studies related to heart disease, nutrition and occupational health; longitudinal studies. *Mailing Add:* Col Nursing Tex Woman's Univ 1810 Inwood Rd Dallas TX 75235

MEYER, BRAD ANTHONY, b Amery, Wis, Oct 6, 52; m 76; c 3. FLUID FLOW IN POROUS MEDIA. *Educ:* Univ Wis-Madison, BSc, 76, MSc, 77, PhD(mech eng), 80. *Prof Exp:* MEM TECH STAFF FLUID FLOW HEAT TRANSFER, SANDIA NAT LABS, LIVERMORE, CALIF, 80- *Mem:* Sigma Xi. *Res:* Fluid flow and heat transfer; fluid flow in consolidated porous media and heat transfer in enclosures. *Mailing Add:* 2638 Covey Way Livermore CA 94550-6303

MEYER, BURNETT CHANDLER, b Denver, Colo, Mar 24, 21. MATHEMATICS. *Educ:* Pomona Col, BA, 43; Brown Univ, ScM, 45; Stanford Univ, PhD(math), 49. *Prof Exp:* Asst, Stanford Univ, 46-49; from asst prof to assoc prof math, Univ Ariz, 49-57; from asst prof to prof, 57-90, EMER PROF MATH, UNIV COLO, BOULDER, 90- *Mem:* Am Math Soc; Math Asn Am; Sigma Xi. *Res:* Complex analysis; real analysis; potential theory and history of mathematics. *Mailing Add:* Dept Math Univ Colo Boulder CO 80309-0426

MEYER, CARL BEAT, b Zurich, Switz, May 5, 34; m 61; c 1. PHYSICAL INORGANIC CHEMISTRY. *Educ:* Univ Zurich, PhD(inorg chem), 60; Calif Western Sch Law, JD, 88. *Prof Exp:* Res chemist & fel Lawrence Radiation Lab, Univ Calif, 61-64; from asst prof to prof chem, Univ Wash, 64-86; CONSULT, 86- *Concurrent Pos:* Consult, Lawrence Berkeley Lab, Univ Calif, Berkeley, 64-; NSF grants, 64; consult, US Consumer Prod Safety Comn, 79-, Environ Protection Agency, 80-, US Dept Housing & Urban Develop, 81-; investr, Lawrence Berkeley Lab, Univ Calif, 79-; dir indust res,

Sulphur Inst, Washington, DC, 65-69. *Honors & Awards:* Gold medal, Ger Cellular Plastics Soc, 80. *Mem:* AAAS; Am Phys Soc; Am Chem Soc; Air Pollution Control Asn; Am Bar Asn. *Res:* Inorganic physical chemistry of sulphur-containing compounds; human toxic exposure calculations and modeling; products standards; formaldehyde resins; air chemistry and analysis; toxic wast chemistry and standards. *Mailing Add:* 2701 Second Ave-10 No 302 San Diego CA 92103-6240

MEYER, CARL DEAN, JR, b Greeley, Colo, Nov 22, 42; c 2. MATHEMATICS, COMPUTER SCIENCE. *Educ:* Univ NC, BA, 64; Colo State Univ, MS, 66, PhD(math), 68. *Prof Exp:* PROF MATH, NC STATE UNIV, 68- *Mem:* Am Math Soc; Math Asn Am; Soc Indust & Appl Math. *Res:* Matrix theory; applied linear algebra; numerical analysis. *Mailing Add:* Dept of Math NC State Univ Raleigh NC 27695-8205

MEYER, CHARLES FRANKLIN, mathematics, statistics, for more information see previous edition

MEYER, CHRISTIAN, b Magdeburg, Ger, Mar 26, 43; US citizen; m 66; c 2. STRUCTURAL ENGINEERING & MECHANICS. *Educ:* Univ Calif, Berkeley, MS, 66, PhD(civil eng), 70. *Prof Exp:* Sr systs engr, Albert C Martin & Assocs, 71-73; structure engr & consult, Stone & Webster Eng Corp, 73-78; ASSOC PROF CIVIL ENG, COLUMBIA UNIV, 78- *Mem:* Am Soc Civil Engrs; Am Concrete Inst; Sigma Xi; Int Asn Bridge & Structural Engrs. *Res:* Computer analysis of structures; earthquake engineering; structural design. *Mailing Add:* Dept Civil Eng Columbia Univ New York NY 10027

MEYER, DAVID BERNARD, b Rochester, NY, Jan 20, 23; m 52; c 3. ANATOMY. *Educ:* Wayne State Univ, BA, 48, PhD, 57; Univ Mich, MS, 50. *Prof Exp:* From instr to asst prof biol, Wayne State Univ, 51-57; NIH fel, 58-59; from instr to prof, 60-88, EMER PROF ANAT, SCH MED, WAYNE STATE UNIV, 88- *Concurrent Pos:* Vis prof, Graz Univ, 62-63; vis res anatomist, Carnegie Lab Embryol, Davis, Calif, 75-76; USPHS fel, Wayne State Univ, 57-58. *Mem:* AAAS; Asn Res Vision & Ophthal; Am Asn Anat; Histochem Soc; Pan-Am Asn Anat. *Res:* Ocular development and histochemistry; prenatal ossification of the human skeleton; embryological histochemistry; origin, ultrastructure and chemistry of avian visual cells. *Mailing Add:* Dept Anat & Cell Biol Wayne State Univ Sch Med Detroit MI 48201

MEYER, DAVID LACHLAN, b East Orange, NJ, Dec 26, 43; m 70. INVERTEBRATE PALEONTOLOGY. *Educ:* Univ Mich, BS, 66; Yale Univ, Mphil, 69, PhD(geol), 71. *Prof Exp:* Fel, Smithsonian Inst, 70- 71, biologist, Smithsonian Trop Res Inst, 71-75; from asst prof to assoc prof, 75-85, PROF GEOL, UNIV CINCINNATI, 85- *Mem:* Paleont Soc; Soc Econ Paleontologists & Mineralogists; Geol Soc Am; Sigma Xi; AAAS; Am Soc Zoologists. *Res:* Ecology and functional morphology of recent and ancient crinoids (Echinodermata); coral reef ecology; taphonomy. *Mailing Add:* Dept of Geol Univ of Cincinnati Cincinnati OH 45221

MEYER, DELBERT HENRY, b Maynard, Iowa, Aug 28, 26; m 49; c 5. ORGANIC CHEMISTRY. *Educ:* Wartburg Col, BA, 49; Univ Iowa, PhD(chem), 53. *Prof Exp:* Chemist, Standard Oil Co, Ind, 53-61; chemist, Amoco Chem Corp, 61-67, res supvr, Res & Develop Dept, 67-77, DIR RES, AMOCO CHEM CORP, 77- *Mem:* Am Chem Soc. *Res:* Esterification and oxidation reactions; polymerization; synthetic fibers; flame retardant polymers. *Mailing Add:* 1524 Clyde Dr Naperville IL 60565-1308

MEYER, DIANE HUTCHINS, b Springfield, Mass, Feb 3, 37; m 67; c 4. CELL BIOLOGY. *Educ:* Russell Sage Col, BA, 58; Univ Vt, PhD(zool), 72. *Prof Exp:* Res asst pharmacol res, Sterling-Winthrop Res Inst, 58-59; res technician, Burroughs Wellcome Co, 60-62; res technician med res, Med Col, Univ Vt, 64-68; res officer, Med Col, Univ Western Australia, 73; RES ASSOC BASIC RES, MED COL, UNIV VT, 74- *Res:* Investigation of the role of ribonucleases, esterases and proteases in the turnover of RNA and protein in normal, developing, denervated and dystrophic muscle and in dystrophic human muscle; in vivo and in vitro studies concerning the role of RNase II in cellular defense and its relationship to interferon. *Mailing Add:* Dept Biochem Given Bldg Med Col Univ Vt 185 S Prospect St Burlington VT 05405

MEYER, DONALD IRWIN, b St Louis, Mo, Feb 13, 26; m 50; c 2. PHYSICS. *Educ:* Mo Sch Mines, BS, 46; Univ Wash, PhD(physics), 53. *Prof Exp:* Mem staff, Los Alamos Sci Lab, 46-48; asst prof physics, Univ Okla, 52-54; with Brookhaven Nat Lab, 54-56; ASST PROF, 56,- PROF PHYSICS, UNIV MICH, ANN ARBOR, 56- *Mem:* Am Phys Soc. *Res:* High energy nuclear physics. *Mailing Add:* Dept of Physics Univ of Mich Ann Arbor MI 48109

MEYER, DWAIN WILBER, b Fremont, Nebr, Jan 11, 44; m 66; c 2. AGRONOMY. *Educ:* Univ Nebr, Lincoln, BS, 66; Iowa State Univ, PhD(crop prod & physiol), 70. *Prof Exp:* from asst prof to assoc prof, 70-82, PROF AGRON, NDAK STATE UNIV, 82- *Mem:* Am Soc Agron; Crop Sci Soc Am; Am Forage & Grassland Coun. *Res:* Forage management, production and physiology; irrigated forage problems and systems; techniques of sodseeding in semi-arid grasslands; nutritive value of small grain straws; winter injury in irrigated alfalfa; hay preservation methods; high moisture hay preservatives; legume effects in cash crop production. *Mailing Add:* Crop & Weed Sci Dept NDak State Univ Fargo ND 58105

MEYER, EDGAR F, b El Campo, Tex, July 19, 35; m 65; c 3. STRUCTURAL CHEMISTRY. *Educ:* NTex State Univ, BS, 59; Univ Tex, PhD(chem), 63. *Prof Exp:* Fel lab org chem, Swiss Fed Inst Technol, 63-65; fel dept biol, Mass Inst Technol, 65-67; from asst prof to assoc prof, 65-86, PROF BIOPHYS, TEX A&M UNIV, 86- *Concurrent Pos:* Res collabr chem, Brookhaven Nat Lab, 68-77; acad guest, Swiss Fed Inst Technol, 75-76; vis scientist, Max-Planck-Inst Biochem, 79- *Mem:* AAAS; Sigma Xi; Am Crystallog Asn; Am Chem Soc. *Res:* Crystallographic structure determinations with the assistance of digital computers; x-ray crystallography of biologically related substances; chemical information processing and the application of computational methods to chemical problems; modelling the interaction of small molecules with macromolecular receptors; protein crystallography of receptor & ligand complexes. *Mailing Add:* Dept Biochem & Biophys Tex A&M Univ College Station TX 77843-2128

MEYER, EDMOND GERALD, b Albuquerque, NMex, Nov 2, 19; m 41; c 3. PHYSICAL CHEMISTRY, ENERGY TECHNOLOGY. *Educ:* Carnegie Inst Technol, BS, 40, MS, 42; Univ NMex, PhD(chem), 50. *Prof Exp:* Jr chemist, Harbison Walker Refractories Co, 40-41; instr, Carnegie Inst Technol, 41-42; asst phys chem, US Bur Mines, 42-44; chemist, US Naval Res Lab, 44-46; res div, NMex Inst Mining & Technol, 46-48; head dept sci, Univ Albuquerque, 50-52; prof chem, NMex Highlands, 52-63, head dept, 52-58, dir inst sci res, 58-63, grad dean, 60-63; dean, Col Arts & Sci, 63-75, prof chem, 63-80, vpres res, 75-80, prof energy & natural resources, 81-87, EMER PROF & DEAN, UNIV WYO, 87- *Concurrent Pos:* Grants, Res Corp, AEC, NSF, USPHS, US Dept Energy, US Dept Interior & Am Heart Asn; Fulbright prof, Chile, 59; state sci adv, Wyo, 72-; consult, Los Alamos Sci Lab, NSF & US Dept Health, Social Serv Gen Acct Off; dir, Nat Gov Coun Sci & Technol, 74-, Am Nat Bank, Laramie, 76-, chmn, 81-; chmn, Consortium Univ Res Energy, 78-80; exec consult, Diamond Shamrock, 80; pres, Coal Technol Corp, 81- *Mem:* Fel AAAS; Am Chem Soc; Biophys Soc; fel Am Inst Chem. *Res:* Energy systems; energy policy; thermodynamics; science education; developing a patented process for converting coal into charcoal and liquids by slash hydrophysics which can be mixed to form a uniform; high heat valve pollutant-free fuel that is pipeline transportable. *Mailing Add:* PO Box 3825 Laramie WY 82071-3825

MEYER, EDWARD DELL, b Buffalo, NY, Mar 27, 41; m 64; c 4. MICROBIOLOGY, BIOCHEMISTRY. *Educ:* Univ Ariz, BS, 63, MS, 72, PhD(microbiol), 75. *Prof Exp:* Res asst geosci/chem eng, Univ Ariz, 73-75; asst prof, Madonna Col, 75-79, dir Allied Health, 77-79, chmn, Div Nat Sci, 78-79, asst to pres col, 79-81, dean admin serv, 81-88, assoc prof biol, 79-88; PRES, ST MARY'S COL, 88- *Mem:* Sigma Xi; Am Soc Microbiol. *Res:* Basic biology of psychophilic pseudomonads and psychrophilic yeasts; temperature related adaptations. *Mailing Add:* St Mary's Col Orchard Lake MI 48324

MEYER, EDWIN F, b Chicago, Ill, July 30, 37; m 59; c 6. PHYSICAL CHEMISTRY. *Educ:* DePaul Univ, BS, 59; Northwestern Univ, PhD(phys chem), 62. *Prof Exp:* NATO fel, Queen's Univ, Belfast, 62-63; res assoc phys adsorption, Naval Res Lab, 65-67; from asst prof to assoc prof, 67-78, PROF CHEM, DEPAUL UNIV, 78- *Mem:* Am Chem Soc. *Res:* Vapor pressure measurement; intermolecular interactions; thermodynamics; gas chromatography. *Mailing Add:* Dept of Chem De Paul Univ 2323 N Sem Ave Chicago IL 60614

MEYER, FRANK HENRY, b Brooklyn, NY, July 11, 15; m 46; c 2. SOLID STATE PHYSICS, PHILOSOPHY OF SCIENCE. *Educ:* City Col New York, BS, 36; Polytech Inst Brooklyn, MS, 51; Univ Minn, MA, 68. *Prof Exp:* X-ray crystallogr, Textile Res Inst, NJ, 51-53; res physicist, Continental Oil Co, 54-60; res engr, Kaiser Aluminum & Chem Corp, 60-63; sr develop engr, Univac Div, Sperry Rand Corp, 63-65; teacher pub sch, Wis, 65-66; from asst prof to prof, 66-81, EMER PROF PHYSICS & PHILOS, UNIV WIS-SUPERIOR, 81- *Concurrent Pos:* Ed, Reciprocity, 71-; dir, New Sci Advocates, Inc, pres, 80-81; pres, Int Soc Unified Sci. *Mem:* Am Phys Soc; Am Crystallog Asn; Am Asn Physics Teachers; Fedn Am Sci; Int Soc Unified Sci(pres, 81-). *Res:* Solid state defect physics; solid surface chemistry; philosophy and history of science and education; solid and liquid cohesion theory; three-dimensional time; space time progression; reciprocal system of physics and metaphysics; biophysics. *Mailing Add:* 1103 15th Ave SE Minneapolis MN 55414-2407

MEYER, FRANZ, b Berlin, Ger, July 3, 23; nat US; m 60. BIOCHEMISTRY. *Educ:* Univ Heidelberg, MD, 53. *Prof Exp:* Res assoc, Univ Chicago, 54-60; res fel, Harvard Univ, 60-63; assoc prof, 63-74, PROF MICROBIOL, STATE UNIV NY UPSTATE MED CTR, 74- *Concurrent Pos:* NIH spec fel, 60-62. *Res:* Lipid metabolism in microorganisms and lower invertebrates. *Mailing Add:* 11919 Manorgate Houston TX 77031

MEYER, FRANZ O, b Ruddstadt, Ger, Apr 10, 45; nat US; div; c 2. REEF ECOLOGY, MARINE SCIENCES. *Educ:* State Univ NY, BS, 73; Univ Mich, MS, 75, PhD(geol), 79. *Prof Exp:* Res & teaching fel, Univ Mich, 73-78; explor geologist, Shell Oil Co, 79-84, explor training staff, 84-91; CHIEF EXEC OFFICER, MESU SEA MEDIA, 90- *Mem:* Int Asn Sedimentologists; Soc Econ Paleontologists & Mineralogists; Am Asn Petrol Geologists; CEDAM. *Res:* Sedimentology of carbonate platforms; carbonate platform stratigraphy; ecology of living reef systems; carbonate petrography and diagenesis. *Mailing Add:* PO Box 710925 Houston TX 77271-0925

MEYER, FRED PAUL, b Holstein, Iowa, Aug 15, 31. PARASITOLOGY, FISH DISEASES. *Educ:* Univ Northern Iowa, BA, 53; Iowa State Univ, MS, 57, PhD(parasitol), 60. *Prof Exp:* Teacher high sch, Iowa, 53-56; asst, Iowa State Univ, 56-59; parasitologist & asst dir, Fish Farming Exp Sta, US Fish & Wildlife Serv, 60-73, dir fish control lab, 73-78, dir, Nat Fish Res Lab, 78-90; RETIRED. *Concurrent Pos:* Instr, Miss State, 68-86 & Univ Wis-LaCrosse, 73-90; mem fac, Nat Fisheries Acad, 78-86. *Honors & Awards:* S F Sniezko Award, Am Fisheries Soc, 84; Distinguished Serv Award, US Dept Interior, 90. *Mem:* Am Soc Parasitol; Am Fisheries Soc; Am Inst Fish Res Biol; Sigma Xi. *Res:* Fish parasitology and pathology with reference to diseases of fish; toxicology; chemical and drug registration, environmental stress and fish health; 85 publications. *Mailing Add:* 518 N First St La Crescent MN 55947

MEYER, FRED WOLFGANG, b Heidenheim, W Ger, May 30, 47; US citizen; m 77; c 2. ATOMIC & MOLECULAR PHYSICS. *Educ:* Lawrence Univ, BA, 70; Univ Wis-Madison, MA, 73, PhD(physics), 76. *Prof Exp:* PHYSICIST, OAK RIDGE NAT LAB, 76- *Mem:* Am Phys Soc. *Res:* Electron-transfer between singly and multiply charged ions and atomic hydrogen and alkali metal atoms; crossed beam study of electron loss by hydrogen in collisions with multicharged ions; Rydberg atoms; multi-charged ion source development. *Mailing Add:* PO Box 2008 Bldg 6003 Oak Ridge Nat Lab Oak Ridge TN 37831

MEYER, FREDERICK GUSTAV, b Olympia, Wash, Dec 7, 17; m 46. BOTANY. *Educ:* Wash State Univ, BSc, 39, MSc, 41; Wash Univ, PhD(bot), 49. *Prof Exp:* Lab asst bot, Wash State Univ, 39-51; dendrologist, Mo Bot Garden, 51-56; botanist, New Crops Res Br, Agr Res Serv, USDA, 57-63, RES BOTANIST CHG HERBARIUM, US NAT ARBORETUM, USDA, 63- *Concurrent Pos:* Mo Bot Garden grant, Univ Col & Univ London, 49-51; NSF fel, 55- *Honors & Awards:* Gold Seal, Nat Coun State Garden Clubs; Frank N Meyer Medal, Am Genetic Asn. *Mem:* Linnaean Soc London; Am Soc Plant Taxon; Am Hort Soc; Soc Study Evolution; Int Soc Plant Taxon. *Res:* Taxonomic botany of the flowering plants; studies in Coffea; evolution and taxonomy of cultivated plants; ethnobotany of archeological sites, especially Pompeii, Herculaneum and villas destroyed by Vesuvius; Valeriana; cultivated plants of southeastern United States; taxonomy of cultivated plants. *Mailing Add:* US Nat Arboretum Washington DC 20002

MEYER, FREDERICK RICHARD, b Brooklyn, NY, May 26, 38; m 62; c 2. MAMMALIAN PHYSIOLOGY. *Educ:* Valparaiso Univ, BS, 60; Ind Univ, MA, 62, PhD(physiol), 66. *Prof Exp:* Asst prof biol, Wilson Col, 65-67; PROF BIOL, VALPARAISO UNIV, 67- *Concurrent Pos:* Dir, Valparasiso Overseas Study Ctr, Reutlinger, WGer; Fel NIH Grad Res. *Mem:* Am Physiol Soc; Sigma Xi; AAAS. *Res:* Temperature regulation in the laboratory rat and man. *Mailing Add:* Niels Sci Ctr Rm 243 Valparaiso Univ Valparaiso IN 46383

MEYER, GEORGE G, b Frankfurt, Ger, Nov 13, 31; nat US; m 53; c 3. PSYCHIATRY, SOCIAL PSYCHIATRY. *Educ:* Johns Hopkins Univ, BA, 51; Univ Chicago, MD, 55; Am Bd Psychiat & Neurol, cert psychiat, 64. *Prof Exp:* Resident, 58-61, chief resident psychiat, 60-61, from instr to assoc prof, Univ Chicago, 61-69; from assoc prof to prfo psychiat, 69-82, CLIN PROF, UNIV TEX HEALTH SCI CTR, SAN ANTONIO, 82- *Concurrent Pos:* Nat Inst Ment Health career teacher grant, 61-63; assoc chief, Psychiat Inpatient Serv, Univ Chicago, 61-65, chief, 66-69; consult, Indian Health Serv, USPHS, 68- & NIH, 70- dir, Northwest San Antonio Ment Health Ctr, 69-74; vis lectr, Univ Edinburgh, 66; mem consult staff, Santa Rosa Med Ctr, San Antonio, 71-; mem bd, Econ Develop Corp, Mex-Am Unity Coun, 71-82; mem exec bd, Crisis Ctr, San Antonio, 71-74; psychiat consult, Ecumenical Ctr Bexar County, 72-86; vis prof, Univ Man, 77. *Mem:* Fel Am Psychiat Asn; fel Am Orthopsychiat Asn; Am Group Psychother Asn; Am Asn Med Cols; World Psychiat Asn; Am Col Psychiatrist. *Mailing Add:* 7950 Floyd Curl Dr No 601 San Antonio TX 78229

MEYER, GEORGE WILBUR, b Cleveland, Ohio, Apr 30, 41; m 67; c 2. GASTROENTEROLOGY. *Educ:* Mass Inst Technol, BS, 62; Tulane Univ, MD, 66. *Prof Exp:* Flight surgeon, 432 USAF Hosp, 67-68 & 27 Fighter Interceptor Squadron, 68-69; resident internal med, Pac Med Ctr, San Francisco, 69-72; staff internist, Tachikawa USAF Med Ctr, 72-74; fel gastroenterol, David Grant USAF Med Ctr, 74-76; chief gastroenterol & asst chmn internal med, Keesler USAF Med Ctr, 76-78; asst prof med, Uniformed Serv Univ, 78-80; chief med & prog dir internal med, Wright Patterson USAF Med Ctr, 80-82; chief med & prog dir internal med residency, Wilford Hall USAF Med Ctr, 82-86; chief, hosp serv, US Air Force Acad Hosp, Colorado Springs, 86-88; HOSP COMDR, 1ST MED GROUP, LANGLEY, AFB, VA. *Mem:* Am Gastrointestinal Asn; Am Soc Gastrointestinal Endoscopy; Am Col Physicians; Am Fedn Clin Res; AAAS; Am Med Asn. *Res:* Esophageal motility and motor disorders. *Mailing Add:* 316AD Clin Ramstein Ger APO New York NY 09094-5300

MEYER, GERARD G L, b La Tronche, France, May 13, 41; m 67; c 2. ELECTRICAL ENGINEERING. *Educ:* Univ Calif, Berkeley, MS, 67, PhD(elec eng), 70. *Prof Exp:* Res assoc elec eng, Univ Southern Calif, 70-71, asst prof, 71-72; asst prof, 72-76, assoc prof, 76-81, PROF ELEC ENG, JOHNS HOPKINS UNIV, 81- *Mem:* Inst Elec & Electronics Engrs; Asn Comput Mach; Math Programming Soc; Operations Res Soc. *Mailing Add:* Elec Eng & Comput Sci Dept Johns Hopkins Univ Baltimore MD 21218

MEYER, GLENN ARTHUR, b Baraboo, Wis, Mar 8, 34; m 61; c 3. NEUROSURGERY. *Educ:* Univ Wis, Madison, BS, 57, MD, 60. *Prof Exp:* Surg extern, Univ Wis Hosps, 59-60, res asst neurophysiol, Med Sch, Univ, 63-64, instr neurosurg & staff physician, Univ Hosps, 66; neurosurg consult, St Elizabeth's Hosp, 67-68; assoc prof, Univ Tex Med Br Galveston, 69-72; assoc prof, 72-83, PROF NEUROSURG, MED COL WIS, 83- *Concurrent Pos:* Staff physician, Mendota State Psychiat Hosp, 62-65; med adv, Social Security Admin, 70-72; mem, radiation ther oncol group & pediat oncol group, Am Col Radiol, 85. *Mem:* AAAS; Cong Neurol Surg; Soc Neurosci; Am Col Surg; Am Asn Neurol Surg; Cong Neurosurg Soc (secy, 80, pres, 82). *Res:* Spinal autonomic mechanisms in spinal shock; reconstruction of craniofacial anomalies; pain control with electrical stimulation; chronic monitoring of intracranial pressure with a fully implantable device; intravertricular hemorrhage in premature infants; brain irrigation solutions; neurooncology; surgical TY of epilepsy. *Mailing Add:* 8700 W Wisconsin Ave Milwaukee WI 53226

MEYER, GREGORY CARL, b Willmar, Minn, Feb 10, 18; m 42, 68; c 4. ORGANIC CHEMISTRY. *Educ:* Southwestern Univ, AB, 38; Univ Nebr, MA, 40, PhD(chem), 43. *Prof Exp:* Res chemist, 42-57, tech ed, 57-63, patent chemist, 63-78, LIT CHEMIST, E I DU PONT DE NEMOURS & CO, INC, 78- *Mem:* Am Chem Soc. *Res:* Additives for petroleum products. *Mailing Add:* 3215 Bonnie Hills Dr Los Angeles CA 90068-1322

MEYER, GUNTER HUBERT, b Stettin, Ger, Aug 19, 39; US citizen; m 66; c 3. NUMERICAL ANALYSIS, APPLIED MATHEMATICS. *Educ:* Univ Utah, BA, 61; Univ Md, MA, 63, PhD(math), 67. *Prof Exp:* PROF MATH, GA INST TECHNOL, 71- *Concurrent Pos:* Res mathematician, Mobil Res & Develop Corp, 67-71; consult, UNESCO-Univ Simon Bolivar, Caracas, Venezuela, 74; sr vis fel, Brunel Univ, Uxbridge, Eng, 75 & 78-79; vis scientist, McDonnell Douglas Res Lab, 84; vis fel, CMA, Canberra, ON, 90. *Mem:* Soc Indust & Appl Math. *Res:* Numerical solution of boundary value problems, especially of free boundary problems. *Mailing Add:* Sch of Math Ga Inst of Technol 225 North Ave Atlanta GA 30332

MEYER, HANS-OTTO, b Basel, Switz, Oct 29, 43; m 66; c 2. NUCLEAR PHYSICS. *Educ:* Univ Basel, PhD(exp nuclear physics), 70, Privatdozent, 77. *Prof Exp:* Res assoc nuclear physics, Univ Basel, 66-71, Univ Wis-Madison, 71-73, Los Alamos Sci Lab, NMex, 73-74; asst, Univ Wash, 74, Univ Basel, 74-78; assoc prof, 78-83, PROF NUCLEAR PHYSICS, IND UNIV, BLOOMINGTON, 83- *Mem:* Swiss Phys Soc; Am Phys Soc. *Res:* Reactions and scattering with polarized particles; nuclear structure; medium energy physics; electron cooling, nuclear physics with storage rings. *Mailing Add:* Dept of Physics Ind Univ Bloomington IN 47405

MEYER, HAROLD DAVID, b Indianapolis, Ind, Oct 17, 39; m 68; c 2. NUMERICAL ANALYSIS. *Educ:* Mass Inst Technol, SB & SM, 62; Univ Chicago, MS, 68, PhD(math), 69. *Prof Exp:* Sci & math analyst, Foreign Sci & Technol Ctr, US Army, 69-70, br chief & opers res analyst, Mil Assistance Command, Vietnam, 70-71; asst prof, 71-76, ASSOC PROF MATH, TEX TECH UNIV, 76- *Mem:* Am Math Soc. *Res:* Numerical solution of partial differential equations, finite element approaches, ill-posed problems and representations of solutions to be used numerically in conjunction with such problems. *Mailing Add:* Dept Math Tex Tech Univ PO Box 4319 Lubbock TX 79409

MEYER, HARRY MARTIN, JR, b Palestine, Tex, Nov 25, 28; m 49; c 3. VIROLOGY, PEDIATRICS. *Educ:* Hendrix Col, BS, 49; Univ Ark, MD, 53; Am Bd Pediat, cert, 60. *Prof Exp:* Chief diag sect, Dept Virus & Rickettsial Dis, Walter Reed Army Inst Res, DC, 54-57; asst resident pediat, NC Mem Hosp, 57-59; chief, Gen Virol Sect Div Biologics Standards, NIH, US Pub Health Serv, 59-64 & Chief, Lab Virol Immunol, 64-72; dir Bur Biologics, Food & Drug Admin, Bethesda, 72-82, dir, Ctr Drugs & Biologics, Rockvillle, 82-86; PRES, MED RES DIV, AM CYANAMID CO, PEARL RIVER, NY, 86- *Concurrent Pos:* Expert comt biol standards, WHO, expanded prog immunization comt; sci adv panel, Asn Retarded citizens US; res develop steering comt, Pharmaceut Mfrs Asn, chmn comn med treat drug dependence & abuse. *Honors & Awards:* Chevalier de l'Ordre Nat, Repub of Upper Volta, Africa, 63; Lett of Commendation from Pres, 66; Meritorious Serv Medal, Dept Health, Educ & Welfare, 66, Distinguished Serv Medal, 69 & 82; Mead Johnson Award for Pediat Res, 67; Max Weinstein Award for Med Res, United Cerebral Palsy Asns, 69; Int Award for Distinguished Sci Res, Joseph P Kennedy, Jr Found, 71; Gold Plate Award, Am Acad Achievement, 70; Distinguished Res Award, Asn Retarded Citizens US, 86. *Mem:* Am Pediat Soc; Am Acad Pediat; Am Epidemiol Soc; Soc Pediat Res; AAAS; AMA; Sigma Xi. *Res:* Pharmaceutical research; virus and rickettsial diseases; biologics; infectious diseases; international health; virus vaccine; three patents. *Mailing Add:* 104 Fisher Rd Mahwah NJ 07430

MEYER, HARUKO, b Tokyo, Japan, Jan 17, 29; US citizen; m 60. BIOCHEMISTRY, MICROBIOLOGY. *Educ:* Toho Women's Col Sci, Japan, BS, 49; Tokyo Col Sci, MS, 51; State Univ NY Upstate Med Ctr, PhD(microbiol), 66. *Prof Exp:* RES ASSOC MICROBIOL, STATE UNIV NY UPSTATE MED CTR, 67- *Mem:* Am Soc Microbiol. *Res:* Lipid metabolism of parasitic organisms. *Mailing Add:* Dept Microbiol Upstate Med Ctr State Univ NY 155 Elizabeth Blackwell St Syracuse NY 13210

MEYER, HARVEY JOHN, b St Paul, Minn, July 16, 35; m 62; c 2. GEOLOGY, COMPUTER SCIENCE. *Educ:* Univ Minn, BA, 57; Calif Inst Technol, MS, 59; Pa State Univ, PhD(sedimentary petrol), 64. *Prof Exp:* Res fel, Antarctica Proj, Univ Minn, 62-63; sr res scientist statist geol, Pan Am Petrol Corp, 63-66; mgr data processing, Can Stratig Serv Ltd, 66-71; lectr geol, Univ Calif, Davis, 71-72; GEOL ASSOC, AMOCO PROD CO, 72- *Mem:* Geol Soc Am; Am Asn Petrol Geol; Int Asn Math Geol. *Res:* Digital computer and statistical methods applied to geological problems; hydrocarbon exploration; sedimentary petrology of clastic rocks and Antartica geology. *Mailing Add:* Amoco Prod Co 1670 Broadway Denver CO 80202

MEYER, HEINZ FRIEDRICH, b Suedmoslesfehn, Ger, Jan 4, 32; div; c 4. CHEMISTRY. *Educ:* Univ Frankfurt, Dr phil nat(org chem), 59. *Prof Exp:* Res asst, Ohio State Univ, 60-61 & C H Boehringer, Ger, 61-63; CHEMIST, UPJOHN CO, 63- *Mem:* Am Chem Soc; Soc Ger Chem. *Res:* Alkaloids; antibiotics; steroids. *Mailing Add:* Upjohn Co Unit 1400 Kalamazoo MI 49001-0199

MEYER, HENRY OOSTENWALD ALBERTIJN, b Warwickshire, Eng, Jan 18, 37; m 59; c 5. MINERALOGY, GEOCHEMISTRY. *Educ:* Univ London, BSc, 59, PhD(geol), 62. *Prof Exp:* Res asst mineral, Univ Col, Univ London, 61-66; sr fel mineral & exp petrol, Geophys Lab, Carnegie Inst, Washington, DC, 66-69; sr res assoc, Nat Res Coun, Goddard Space Flight Ctr, NASA, 69-71; assoc prof, 71-74, PROF PETROL, PURDUE UNIV, WEST LAFAYETTE, 74- *Concurrent Pos:* NSF & NASA grants, Purdue Univ, 71-; consult, Harry Winston Inc, Am Metal Climax & Cominco; secy, Comn Gem Mats, Inst Mineral Asn; coun mem, Int Mineral Soc; pres, Am Geol Inst, 88-89. *Mem:* Mineral Soc Gt Brit & Ireland; Mineral Soc Am (secy, 82-87); Am Geophys Union. *Res:* Mineralogy and petrology of ultrabasic igneous rocks, including lunar samples; high pressure phase equilibria studies pertinent to ultrabasic igneous rocks; origin of diamond and kimberlite rock; exploration for diamond; electron microprobe and x-ray diffraction studies. *Mailing Add:* Dept Earth & Atmos Sci Purdue Univ West Lafayette IN 47907

MEYER, HERIBERT, b Eisenstein, Ger, Nov 14, 13; m 45; c 2. FLUID MECHANICS. *Educ:* Munich Tech, Dipl Ing, 42. *Prof Exp:* Asst prof hydraul, Munich Tech, 41-43; design engr, J M Voith Ltd, Ger, 43-45, group leader heat recovery, 45-58; sr scientist, Mead Corp, Ohio, 58-63; res assoc hydrodyn of papermaking, Inst Paper Chem, 63-69, sr res assoc fluid mech,

63-72, prof chem eng & sr res assoc, 69-78; CONSULT ENG SCI, 78- *Mem:* AAAS; Am Asn Univ Professors. *Res:* Inhibited deflection of wall jets; aeromechanics; ordinary and partial differential equations; stress-strain behavior of random assemblages of deformable rods; flow through assemblages of elliptical cylinders; low and high re-number retention dynamics of small particles in fibrous porous media; ideal free jets from nozzels formed by three plates in non-symmetrical configurations; flow between two non-linear boundries moving paralled in opposite directions. *Mailing Add:* 1230 E Pershing St Appleton WI 54911

MEYER, HERMANN, b Frauenfeld, Switz, Mar 29, 27; nat US; m 52; c 3. VETERINARY ANATOMY. *Educ:* Univ Zurich, DVM, 50, Dr med vet(anat), 52; Cornell Univ, PhD(anat), 57. *Prof Exp:* Asst vet anat, Univ Zurich, 51-52; instr, Cornell Univ, 53-56, actg asst prof, 56-57; from asst prof to prof anat, Colo State Univ, 57-72; head vet anat dept, Univ Zurich, 72-73; pub rels dir, Poudre Valley Hosp, 73-75; prof vet anat & physiol, Univ Mo-Columbia, 75-79; PROF VET ANAT, OHIO STATE UNIV, 79- *Concurrent Pos:* Guest auditor, Univ Basel, 51-52; guest lectr, Univ Zurich, 63; consult, Nat Defense Educ Act Title IV grad fel prog, 66-68; vis prof, Cornell Univ, 67-68. *Mem:* Am Vet Med Asn; Am Asn Vet Anat (pres, 80-81); Am Asn Anat; World Asn Vet Anat (treas, 83-); Ger Anat Soc. *Res:* Veterinary anatomy; functional neuromorphology; history of morphology. *Mailing Add:* 3505 Shore Rd Ft Collins CO 80524-1656

MEYER, HORST, b Berlin, Ger, Mar 1, 26; nat Swiss; m 53; c 2. PHASE TRANSITOR, CRITICAL PHYSICS. *Educ:* Univ Geneva, BS, 49; Univ Zurich, PhD(physics), 53. *Prof Exp:* Fel, Swiss Asn Res in Physics & Math for studies in Oxford, 53-55; Nuffield fel Clarendon Lab, Oxford Univ, 55-57; lectr & res assoc, Dept Engr & Appl Physics, Harvard, 57-59; from asst prof to prof, 59-84, FRITZ LONDON PROF PHYSICS, Duke Univ, 84- *Concurrent Pos:* A P Sloan Found fel, 60-64; vis prof, Technische Hochschule, Ger, 65, Tokyo Univ, 80, 81 & 83; travelling fel, Japanese Soc for Promotion Sci, 71, vis scientist, 79; guest scientist, Inst Laue-Langevin, France, 74, 75; Yamada Found fel, Japan, 86; guest scientist, USSR Acad Sci, 88; mem, Nat Res Coun, 89-; chmn, Gordon Conf on solid hydrogeus, 90. *Honors & Awards:* Jesse Beams Prize, Am Phys Soc, 82. *Mem:* Fel Am Phys Soc; Swiss Soc Sci. *Res:* Experimental research on the properties of liquid and solid helium, solid hydrogen and deuterium, magnetic insulators, critical phenomena; author of 203 publications in scientific journals. *Mailing Add:* Dept Physics Duke Univ Durham NC 27706

MEYER, IRVING, b Springfield, Mass, Mar 19, 20; m 53; c 3. ORAL SURGERY, ORAL PATHOLOGY. *Educ:* Univ Mass, BS, 41; Tufts Univ, DMD, 44; Univ Pa, MSc, 50, DSc, 58; Am Bd Oral & Maxillofacial Surg, dipl, 54. *Prof Exp:* Resident oral surg, Metrop Hosp, New York, 48-49; resident, Philadelphia Gen Hosp, Pa, 49-50; dir dept oral maxillofacial surg, Bay State Med Ctr, 74-89; instr oral surg & path, Sch Dent Med, Tufts Univ, 50-60, assoc res prof, 60-65, assoc prof oral path, 65-66, res prof, 66-75, prof oral path, 75-90; RETIRED. *Concurrent Pos:* Instr grad sch med, Univ Pa, 49-51; chief oral surgeon, Wesson Mem Hosp, Springfield, Mass, 51-, secy-treas med staff, 72-74; oral surgeon, Mercy Hosp, 55-; oral surgeon, Bay State Med Ctr, 56-89, pathologist, 57-; ed, Ann Conf Oral Cancer, 60 & 63; lectr, Grad Sch, Boston Univ, 62-70 & Harvard Univ, 67-89; assoc ed & sect ed oral path, J Oral Surg, 65-84; consult ed Oral Surg, Oral Med & Oral Path; pres, New Eng Soc Oral Surg, 63-64; mem, Adv Comt & examr, Am Bd Oral & Maxillofacial Surg, 70-, mem, Bd Dirs, 73-80; oral surgeon, Cancer Div, Western Mass Hosp; consult to var hosps, Mass; chief maxillofacial surg, Baystate Med Ctr, 76-89; physician pvt pract. *Honors & Awards:* William J Gies Mem Award, Am Asn Oral & Maxillofacial Surgeons, 81. *Mem:* Am Bd Oral & Maxillofacial Surg (vpres, 78-79, pres 79-80); fel Am Acad Oral Path; NY Acad Sci; fel Am Asn Oral & Maxillofacial Surg. *Res:* Cancer of oral cavity and adnexia, especially in clinical aspects, etiology, therapy and pathology. *Mailing Add:* 567 Laurel St Longmeadow MA 01106-1916

MEYER, JAMES HENRY, b Lewiston, Idaho, Apr 13, 22; m 80; c 5. NUTRITION. *Educ:* Univ Idaho, BS, 47; Univ Wis, MS, 49, PhD(nutrit), 51. *Prof Exp:* Asst nutrit, Univ Wis, 47-51; from asst prof to prof animal husb & chmn dept, Univ Calif, Davis, 50-63, dean col agr, 63-69, chancellor, 69-87, EMER CHANCELLOR, UNIV CALIF, DAVIS, 87- *Concurrent Pos:* Mem comt animal nutrit, Nat Acad Sci-Nat Res Coun, 65-67 & comn undergrad educ in biol, 82-88, accrediting comn sr cols, univs & schs, 88-90. *Mem:* AAAS; Am Soc Animal Sci; Am Inst Nutrit; Sigma Xi. *Res:* Pasture and fiber nutrition; nutrient requirements; undernutrition; education and research administration. *Mailing Add:* Emer Chancellor's Off Univ Calif Davis CA 95616

MEYER, JAMES HENRY, b St Marys, Pa, July 20, 28; m 60; c 3. METEOROLOGY. *Educ:* Pa State Univ, BS, 53, MS, 55. *Prof Exp:* Meteorologist, Res & Develop Ctr, Intel & Reconnaissance Lab, Griffiss Air Force Base, NY, 54-55; atmospheric physicist, Lincoln Lab, Mass Inst Technol, 55-63; proj meteorologist, Tech Oper Inc, Mass, 63-64; proj mgr, Electromagnetic Res Corp, Md, 64-67; SR STAFF METEOROLOGIST, APPL PHYSICS LAB, JOHNS HOPKINS UNIV, 67- *Concurrent Pos:* Consult meteorologist, 73-; pres, Meteorol Appln, 78- *Mem:* Am Meteorol Soc; Am Geophys Union; Am Soc Testing & Mat; Air Pollution Control Asn. *Res:* Radar and radio meteorology; meteorological instrumentation; atmospheric and cloud physics; power plant siting and atmospheric pollution meteorology; aerosol and fugitive dust; forensic meteorology. *Mailing Add:* 12926 Allerton Lane Silver Spring MD 20904

MEYER, JAMES MELVIN, b West Palm Beach, Fla, Jan 18, 43; m 69; c 2. POLYMER CHEMISTRY. *Educ:* Ind Univ, BS, 64; Northwestern Univ, PhD(inorg chem), 68. *Prof Exp:* Asst prof chem, Univ Ill, 67-69; res chemist, 69-74, div head chem, 74-77, mfg mgr, 78, DIV SUPT, PROCESS, BEAUMONT WORKS, E I DU PONT DE NEMOURS & CO, INC, 78- *Res:* Development of uses for elastomeric polymers. *Mailing Add:* E I du Pont de Nemours & Co Inc, Beaumont Works Div PO Box 3269 Beaumont TX 77704

MEYER, JAMES WAGNER, b Rhineland, Mo, May 22, 20; m 49; c 4. ENERGY CONVERSION & CONSERVATION. *Educ:* Univ Wis, PhB, 48, PhD(physics), 56; Dartmouth Col, MA, 50. *Prof Exp:* Asst physics, Dartmouth Col, 48-49, Univ Calif, 49-50 & Univ Wis, 50-52; mem staff, Lincoln Lab, Mass Inst Technol, 52-57, group leader, 57-59, assoc head radar div, 59-62, assoc head solid state div, 62-63, head radio physics div, 63-65; sr scientist, Educ Serv, Inc, 65-67; mem tech staff, Lincoln Lab, Mass Inst Technol, 67-70, prog mgr, Ctr Space Res, 70-73, prog dir, Energy Lab, Mass Inst Technol, 73-76; asst dir, Plasma Fusion Ctr, 76-80; CONSULT, 80- *Concurrent Pos:* Sr scientist, Aeta Corp, Portsmouth, NH. *Mem:* AAAS; Am Phys Soc; Am Asn Physics Teachers. *Res:* Microwave spectroscopy of solid state; cryogenics, electronics; radio physics; education; energy conservation; alternative energy sources; fusion energy; solar energy; field evaluation of wood pellet gasifier for domestic use; analysis of impact of wind farms on television and radio communications; non intrusive blood pressure sensors for pilots and astronauts. *Mailing Add:* 153 Middleton Rd New Durham NH 03855

MEYER, JEAN-PIERRE, b Lyon, France, Aug 5, 29; US citizen; m 58; c 6. MATHEMATICS, TOPOLOGY. *Educ:* Cornell Univ, BA, 50, MA, 51, PhD(math), 54. *Prof Exp:* Asst prof math, Syracuse Univ, 56; res assoc, Brown Univ, 56-57; from vis asst prof to assoc prof, 57-79, chmn dept math, 85-90, PROF MATH, JOHNS HOPKINS UNIV, 79-; VDIR, JAPAN-US MATH INST, 90- *Mem:* Am Math Soc. *Res:* Algebraic topology. *Mailing Add:* Dept Math Johns Hopkins Univ Baltimore MD 21218

MEYER, JOHN AUSTIN, b St Mary's, Pa, Sept 18, 19; m 55. NUCLEAR CHEMISTRY, RADIATION CHEMISTRY. *Educ:* Pa State Univ, BS, 49, MS, 50; State Univ NY Col Environ Sci & Forestry, PhD(org chem), 58. *Prof Exp:* Asst chemist, Analysis Lab, Speer Carbon Co, Pa, 45-47; chemist, Gulf Res & Develop Co, 50-52; asst head analysis dept, Verona Res Ctr, Koppers Co, 52-54; prof, 58-87, dir, Analysis & Tech Serv, 72-81, EMER PROF NUCLEAR & RADIATION CHEM & RADIOL SAFETY OFF STATE UNIV NY COL ENVIRON SCI & FORESTRY, 87- *Concurrent Pos:* Fel, Oak Ridge Inst Nuclear Studies, 57 & 66; consult various US & foreign industs; vis scientist, Pa State Univ, 80; chmn, Niagara-Finger Lakes Sect, Am Nuclear Soc, 81-82. *Honors & Awards:* Borden Chem Award, Forest Prod Res Soc, 79. *Mem:* AAAS; Am Chem Soc; Am Nuclear Soc; Forest Prod Res Soc; NY Acad Sci. *Res:* Development of wood-polymer materials by the heat-catalyst method; neutron activation analysis; trace analytical methods; radiation chemistry. *Mailing Add:* Dept Chem State Univ NY Col Environ Sci & Forestry Syracuse NY 13210

MEYER, JOHN RICHARD, b St Louis, Mo, Feb 25, 48; m 74; c 3. ENTOMOLOGY, PEST MANAGEMENT. *Educ:* Univ Ill, Urbana, BS, 69; Cornell Univ, MS, 73, PhD(entom), 74. *Prof Exp:* Fel biol, Cornell Univ, 70-74; teacher gen sci, Ithaca Pub Schs, 74-75; asst prof biol, Indiana Univ Pa, 75-76; from asst prof to assoc prof, 76-88, PROF ENTOM, NC STATE UNIV, 88- *Concurrent Pos:* In-serv teacher training, sci educ. *Mem:* AAAS; Entom Soc Am; Nat Asn Cols & Teachers Agr. *Res:* Host plant recognition by insects of economic importance; development of integrated pest management programs on fruit crops in the Southeast. *Mailing Add:* NC State Univ Box 7626 Raleigh NC 27695-7626

MEYER, JOHN SIGMUND, b Princeton, Ill, May 12, 37; m 63; c 3. NON-PARAMETRIC STATISTICS. *Educ:* Wartburg Col, BA, 59; Northwestern Univ, MS, 61; Iowa State Univ, PhD(statist), 73. *Prof Exp:* Instr math, Wartburg Col, 61-67; from asst prof to prof math, Cornell Col, 67-83, chmn dept math, 75-78; assoc prof math, Albion Col, 83-88; HEAD DEPT MATH, MUHLENBERG COL, 88- *Concurrent Pos:* Lectr micro-computers, Am Inst Prof Educ, 77-82; vis assoc prof comput sci, Iowa State Univ, 78-79. *Mem:* Math Asn Am; Am Math Soc. *Res:* Confidence intervals for quantiles of finite populations. *Mailing Add:* Dept Math Muhlenberg Col Allentown PA 18104-5586

MEYER, JOHN STIRLING, b London, Eng, Feb 24, 24; nat US; m 47; c 5. NEUROLOGY. *Educ:* McGill Univ, MD, CM, 48, MSc, 49; Am Bd Psychiat & Neurol, dipl. *Prof Exp:* Demonstr histol, McGill Univ, 45-46, demonstr clin micros, 46-47, fel, 48-49; asst med, Sch Med, Yale Univ, 49-50; demonstr neuropath & teaching fel neurol, Harvard Med Sch, 50-52; USPHS sr res fel, 52-54; instr med, Harvard Med Sch, 54-56; assoc vis physician neurol, Boston City Hosp, 56-57; consult & lectr neurol, US Naval Hosp, Chelsea, Mass, 57; prof neurol & chmn dept, Sch Med, Wayne State Univ, 57-69; chmn dept, 69-76, PROF NEUROL & DIR STROKE LAB, BAYLOR COL MED, 76- *Concurrent Pos:* Asst, Montreal Neurol Inst, McGill Univ, 45-46; from jr intern to sr intern, New Haven Hosp, Conn, 49-50, vis neurologist & supvr EEG lab, 54-56; head dept neurol, Detroit Gen Hosp, 57-69; consult, Grace, Children's Sinai, Detroit Mem & Dearborn Vet Hosps, 57-69, AMA Drug Evaluations, 85; chief neurol dept, Harper Hosp, Detroit, 63-69; chief neurol serv, Methodist & Ben Taub Gen Hosps, Houston, 69-76; consult, Vet Admin & Hermann Hosps, 69-; mem, President's Comn Heart Dis, Cancer & Stroke, 64-65, Nat Adv Coun Neurol Dis & Blindness, 65-69 & Subcomt Cerebrovasc Dis, Nat Heart Inst-Nat Inst Neurol Dis & Blindness Joint Coun, 65-71; sci reviewer, NSF, 78-; adj prof, neuropsychology, Univ Houston, 83-; neurobehavioral merit rev adv comt, Cent Off Vet Admin, Washington DC, 85- *Honors & Awards:* Harold G Wolff Award, 77-79; Mihava Award, 87. *Mem:* Am Neurol Asn; NY Acad Sci; Am Heart Asn; AMA. *Res:* Cerebral blood flow and metabolism studies in migraine and stroke patients; cardiovascular disorders including stroke, migraine, dementia and aging; clinical prevention, diagnosing and treatment and changes in cerebral blood flow and metabolism that accompany stroke. *Mailing Add:* Dept Neurol Baylor Col Med Houston TX 77030

MEYER, JUDY LYNN, b Milwaukee, Wis, May 22, 46; m 73; c 2. STREAM ECOLOGY, ECOSYSTEM ECOLOGY. *Educ:* Univ Mich, BS, 68; Univ Hawaii, MS, 71; Cornell Univ, PhD(ecol & evolution biol), 78. *Prof Exp:* Res assoc, Oceanog Dept, Univ Hawaii, 70-72; from asst prof to assoc prof, 77-89, PROF ZOOL, ZOOL DEPT, UNIV GA, 89- *Mem:* Am Soc Limnol &

Oceanog; Ecol Soc Am; North Am Benthol Soc; AAAS; Int Soc Theoret & Appl Limnol. *Res:* Nutrient dynamics in stream ecosystems with emphasis on dissolved organic carbon and phosphorus; ecosystem analysis of a blackwater river; effects of watershed disturbance on aquatic ecosystems. *Mailing Add:* Zool Dept & Inst Ecol Univ Ga Athens GA 30602

MEYER, KARL, biochemistry; deceased, see previous edition for last biography

MEYER, LAWRENCE DONALD, b Concordia, Mo, Apr 14, 33; m 54; c 3. AGRICULTURAL ENGINEERING, SOILS. *Educ:* Univ Mo, BS, 54, MS, 55; Purdue Univ, PhD(agr eng), 64. *Prof Exp:* Asst soil & water eng, Univ Mo, 54-55; agr engr, USDA, Purdue Univ, 55-73; AGR ENGR/RES LEADER, SEDIMENTATION LAB, USDA, 73- *Concurrent Pos:* From asst prof to assoc prof agr eng, Purdue Univ, 65-73; adj prof agr & biol eng, Miss State Univ, 75- *Honors & Awards:* Hancor Soil & Water Eng Award, Am Soc Agr Engrs, 85. *Mem:* Fel Am Soc Agr Engrs; fel Soil Conserv Soc Am; Soil Sci Soc Am; World Asn Soil & Water Conserv. *Res:* Soil and water conservation engineering; mechanics of the soil erosion process; rainfall simulation; erosion research techniques; physical properties of eroded sediment. *Mailing Add:* USDA Sedimentation Lab PO Box 1157 Oxford MS 38655

MEYER, LEO FRANCIS, b Pittsburgh, Pa, July 19, 29; m 54; c 4. ORGANIC POLYMER CHEMISTRY. *Educ:* Duquesne Univ, BS, 56; Univ Richmond, MS, 66. *Prof Exp:* Res chemist, Gulf Res & Develop Ctr, 56-61; assoc chemist, 61-63, res chemist, 63-69, mgr res ctr, 69-80, DIR RES CTR, PHILIP MORRIS INC, 80- *Mem:* Am Chem Soc. *Res:* Polymers; catalysis; aromatic alkylations; aerosol filtration; gas adsorbants; tobacco processing; cigarette making; plastics extrusion; paper coating. *Mailing Add:* Philip Morris OCR&D PO Box 26603 Richmond VA 23261

MEYER, LEO MARTIN, hematology; deceased, see previous edition for last biography

MEYER, LEON HERBERT, b Navasota, Tex, Sept 4, 26; m 58; c 2. PHYSICAL CHEMISTRY, CHEMICAL ENGINEERING. *Educ:* Ga Inst Technol, BChE, 49, MS, 51; Univ Ill, PhD(chem), 53. *Prof Exp:* Asst, Univ Ill, 52-53; res chemist & engr, Atomic Energy Div, Explosives Dept, 53-64, res mgr, Separations Chem Div, Savannah River Lab, 64-67, dir separations Chem & Eng Sect, 67-69, asst dir, Savannah River Lab, 69-76, prog mgr, AED Tech Div, 76-79, mgr, Special Programs, Savannah River Plant, E I dU Pont De Nemours & Co, Inc, 79-85, PRES, LHM CORP, 86- *Mem:* Am Chem Soc; Am Inst Chem Eng. *Res:* Vapor-liquid equilibrium; nuclear magnetic resonance; fused salt electrolysis; cryogenics; radiochemical separations; nuclear fuel cycle; quality assurance; environmental control; energy conservation; nuclear safety. *Mailing Add:* 2219 Dartmouth Rd Augusta GA 30904-3429

MEYER, LHARY, b New York, NY, 1947. PERCEPTUAL PSYCHOLOGY, STEREO PERCEPTION. *Prof Exp:* Engr, Fantasy Films, 76-78; mgr, Indust Light & Magic, Lucas Film, 78-80; owner, Albedo Eng, 80-82; VPRES & FOUNDER, STEREOGRAPHICS CORP, 82- *Concurrent Pos:* Mem adv comt, Mus Indust & Sci, Chicago, 90-91. *Mem:* Inst Elec & Electronic Engrs; Soc Motion Picture & TV Engrs. *Res:* Stereoscopic displays and systems for science, medicine, government and commercial applications; TV and motion picture engineering. *Mailing Add:* PO Box 1839 Ross CA 94957

MEYER, MARTIN MARINUS, JR, b Wichita, Kans, Dec 24, 36; div; c 2. ORNAMENTAL HORTICULTURE, PLANT PHYSIOLOGY. *Educ:* Kans State Univ, BS, 58; Cornell Univ, MS, 61, PhD(hort), 65. *Prof Exp:* Asst nursery mgr, M Meyer & Son Nursery, Kans, 58-59; asst ornamental hort, Cornell Univ, 59-64; asst prof, 65-71, assoc prof, 71-88, PROF HORT, UNIV ILL, URBANA, 88- *Concurrent Pos:* Fulbright fel, Neth, 64-65. *Mem:* Am Soc Hort Sci; Tissue Cult Asn; Sigma Xi; Am Asn Bot Gardens & Arboreta; Int Plant Propagators Soc. *Res:* Physiology of propagation; growth and development of woody plants; tissue culture propagation. *Mailing Add:* 1207 W Gregory Dept Hort Univ Ill Urbana IL 61801-3838

MEYER, MARVIN CHRIS, b Detroit, Mich, Sept 19, 41; m 66; c 2. BIOPHARMACEUTICS, PHARMACOKINETICS. *Educ:* Wayne State Univ, BS, 63, MS, 65; State Univ NY Buffalo, PhD(pharmaceut), 69. *Prof Exp:* Teaching asst pharmaceut, Wayne State Univ, 64-65 & State Univ NY Buffalo, 66-68; asst prof, 69-75, asst dean grad & res prog, 81-84, PROF PHARMACEUT, UNIV TENN CTR HEALTH SCI, & DIR DIV DRUG METAB & BIOPHARMACEUT, 76-, ASSOC DEAN GRAD & RES PROG, 84- *Concurrent Pos:* Expert, US Food & Drug Admin, 73-76; consult, Pfizer Pharmaceut, 81-, Glaxo, Inc, 80-, Sidmak Labs & Vitarine Pharmaceut, 87- *Honors & Awards:* Mead Johnson Undergrad Res Award, Am Asn Cols Pharm, 70. *Mem:* Am Pharmaceut Asn; fel Acad Pharmaceut Sci; Am Asn Cols Pharm; Am Asn Pharmaceut Sci. *Res:* Study and quantitation of the time course of drugs in humans and animals, especially studies of drug absorption, metabolism, distribution and elimination. *Mailing Add:* Dept Pharmaceut Col Pharm Univ Tenn Memphis TN 38163

MEYER, MARVIN CLINTON, b Jackson, Mo, Dec 20, 07; m 46; c 3. ANIMAL PARASITOLOGY. *Educ:* SE Mo State Univ, BS, 32; Ohio State Univ, AM, 36; Univ Ill, PhD(parasitol), 39. *Prof Exp:* Asst biol, SE Mo State Univ, 30-32; prin high sch, Mo, 32-36; asst zool, Univ Ill, 36-39; instr, Univ Ky, 39-41; adj prof & actg head dept, Douglass Col, Rutgers Univ, 41-42, adj prof, 46; from instr to prof, 46-73, EMER PROF ZOOL, UNIV MAINE, ORONO, 73- *Concurrent Pos:* Fulbright res scholar, NZ, 55-56; mem adv comt, Smithsonian Oceanog Sorting Ctr, 65-; sr vis res assoc, Smithsonian Inst, US Nat Mus, 67-68; mem nat screening comt, Inst Int Educ, 67- *Mem:* Am Micros Soc; Am Soc Parasitol; Am Soc Zool; Soc Syst Zool. *Res:* Parasitology; morphology and taxonomy of Hirudinea; parasites of fish and wildlife; invertebrate morphology; faunistic zoology. *Mailing Add:* Dept of Zool Univ of Maine Orono ME 04469-0146

MEYER, MAURICE WESLEY, b Long Prairie, Minn, Feb 13, 25; m 46; c 2. PHYSIOLOGY, DENTISTRY. *Educ:* Univ Minn, BS, 53, DDS, 57, MS, 59, PhD(physiol), 60. *Prof Exp:* Res fel, 57-60, from instr to asst prof, 60-64, lectr, 61-73, assoc prof, 64-78, PROF DENT PHYSIOL & NEUROL, UNIV MINN, MINNEAPOLIS, 78- *Concurrent Pos:* Res career develop award, 63-69. *Mem:* AAAS; Am Physiol Soc; Sigma Xi; Soc Exp Biol & Med; Int Asn Dent Res. *Res:* Circulation; blood flow in teeth and supporting structures; cerebral blood flow. *Mailing Add:* Dept of Physiol/6-255 Millard Hall Univ Minn 435 Delaware St SE Minneapolis MN 55455

MEYER, NORMAN JAMES, b Wolsey, SDak, Feb 17, 26; m 52; c 2. PHYSICAL CHEMISTRY. *Educ:* Univ SDak, BA, 49; Univ Kans, PhD(chem), 56. *Prof Exp:* Res engr, Continental Oil Co, 54-56; res chemist, Monsanto Chem Co, 56-58; res assoc inorg & nuclear chem, Mass Inst Technol, 58-59; from asst prof to assoc prof, 59-74, PROF CHEM, BOWLING GREEN STATE UNIV, 74- *Concurrent Pos:* Vis prof, Middle East Tech, Ankara, 65-66. *Res:* Hydrolysis of metal ions; ion exchange; thermodynamics of electrolytic solutions. *Mailing Add:* Dept Chem Bowling Green State Univ Bowling Green OH 43403

MEYER, NORMAN JOSEPH, b Wilkes-Barre, Pa, Aug 5, 30. ACOUSTICS. *Educ:* Pa State Univ, BS, 51, MS, 53; Univ Calif, Los Angeles, PhD(physics), 59. *Prof Exp:* Sr engr, HRB-Singer, 51-52; res engr, Lockheed Aircraft Corp, 55-56; res physicist, Ford Aerospace & Com, Aeronutronic Div, 59-61; staff physicist, Marshall Labs, 61-62; sr scientist, West Div, Ling-Temco-Vought Res Ctr, 62-67, dir, 68-70; pres & prin scientist, OAS-Western, 70-72; DIR, WYLE RES-WYLE LABS, 73-; CLIN PROF COMMUNITY & ENVIRON MED, UNIV CALIF COL MED, IRVINE, 78- *Concurrent Pos:* Mem comt hearing & bioacoust, Nat Acad Sci-Nat Res Coun, 68- *Mem:* Acoust Soc Am. *Res:* Experimental acoustics, primarily gases, transducers and instrumentation; noise control technology. *Mailing Add:* 215 N Peck Ave Manhattan Beach CA 90245

MEYER, ORVILLE R, b Cornelius, Ore, Mar 4, 26; m 52; c 3. NUCLEAR ENGINEERING, ELECTRICAL ENGINEERING. *Educ:* Univ Wash, BS, 46; Calif Inst Technol, MS, 49. *Prof Exp:* Engr, Bettis Atomic Power Lab, Westinghouse Elec Corp, 50-53, supvry engr, 53-57, chief test engr, 57-60, mgr reactor control, 60-66; sr staff scientist, Idaho Nuclear Corp, 66-70; mgr protection & control systs, Loft Proj, 70-79, mgr advan control tech off, 80-82, SR ENG SPECIALIST, HUMAN FACTORS RES, EG&G-IDAHO, 83- *Mem:* Am Nuclear Soc; Inst Elec & Electronics Engrs; Sigma Xi. *Res:* Control; instrumentation; protection and safety of nuclear reactors for electric power generation or propulsion; man-machine interface; procedures and training. *Mailing Add:* 6707 S Fifth E Idaho Falls ID 83404

MEYER, PAUL, b Bern, Switz, July 6, 25; US citizen; c 1. PHOTONUCLEAR PHYSICS, NUCLEAR MEDICINE. *Educ:* Univ Wash, BS, 58, MS, 60. *Prof Exp:* Physicist reactor physics, Vallecitos Nuclear Ctr, Gen Elec, 60-66; PHYSICIST NUCLEAR PHYSICS & NUCLEAR MED, LAWRENCE LIVERMORE LAB, 66- *Mem:* Am Phys Soc. *Res:* Photonuclear reactions; nuclear medical diagnostics. *Mailing Add:* Lawrence Livermore Lab 1154 Madison Ave Livermore CA 94550

MEYER, PAUL A, b Philadelphia, Pa, Jan 22, 47; m 77; c 5. ULTRASONIC TRANSDUCERS, PIEZOELECTRIC MATERIALS. *Educ:* Drexel Univ, BSME, 70, MS, 72, PhD(appl mech), 75. *Prof Exp:* Res engr, Ultrasonics Int Inc, 75-76; mgr eng, K B Aerotech, Div SmithKline, 76-82; dir eng, Krautkramer Branson, Smithkline, 82-84, tech dir, Emerson Elec, 84-86, DIR SENSOR RES, KRAUTKAMER BRANSON, EMERSON ELEC, 86- *Mem:* Am Soc Nondestructive Testing; Instrument Soc Am. *Res:* The design and operation of ultrasonic transducers in nondestructive testing and medical applications; published several papers on ultrasonic theory transducer design, signal processing and testing applications. *Mailing Add:* RD 1 Box 255 AA McVeytown PA 17051

MEYER, PAUL RICHARD, b New York, NY, Feb 2, 30; div; c 4. PURE MATHEMATICS, TOPOLOGY. *Educ:* Dartmouth Col, AB, 51, MS, 52; Columbia Univ, MA, 60, PhD(math), 64. *Prof Exp:* Engr, Eastman Kodak Co, NY, 54-55; from lectr to instr math, Columbia Univ, 56-61; asst prof, St John's Univ, 62-64 & Hunter Col, 64-67; assoc prof, 68-71, chmn dept, 75-76 & 77-80, PROF MATH & COMPUTER SCI, LEHMAN COL, 72- *Concurrent Pos:* NSF res grant, 66-68; vis assoc prof, Univ Tex, 68-69; sr vis fel, Westfield Col, Univ London, 71-72; chmn math sect, NY Acad Sci, 76-78; vis res grant, Univ Padua, Italy, 81; vis res grant, Indian Inst Technol, Kanpur, India, 89. *Mem:* Am Math Soc; Math Asn Am; fel NY Acad Sci. *Res:* Spaces of real-valued functions; general topology; digital image processing. *Mailing Add:* Dept Math & Computer Sci Herbert H Lehman Col Bronx NY 10468-1589

MEYER, PETER, b Berlin, Ger, Jan 6, 20; nat US; m 46; c 2. COSMIC RAY PHYSICS. *Educ:* Tech Univ, Berlin, dipl, 42; Univ Gottingen, PhD(physics), 48. *Prof Exp:* Mem staff physics, Univ Gottingen, 46-49; fel, Cambridge Univ, 49-50; mem res staff, Max Planck Inst Physics, Ger, 50-52; res assoc, Inst Nuclear Studies, Univ Chicago, 53-56, from asst prof to prof physics, 56-90, dir, Enrico Fermi Inst, 78-83, chmn, Dept Physics, 86-89, EMER PROF, ENRICO FERMI INST & DEPT PHYSICS, UNIV CHICAGO, 90- *Concurrent Pos:* Consult, NASA; mem Cosmic Ray Comn, Int Union Pure & Appl Physics, 66-72 & Space Sci Bd, Nat Acad Sci, 75-78; chmn, Cosmic Physics Div, Am Phys Soc, 72-73; foreign mem, Max Planck Inst Physics & Astrophys, 73-; Alexander von Humboldt sr US scientist, 84; mem, Space Physics Adv Subcomt, 89- *Mem:* Nat Acad Sci; fel AAAS; Am Geophys Union; Am Astron Soc; fel Am Phys Soc; Sigma Xi. *Res:* Origin of cosmic radiation; astrophysics. *Mailing Add:* Enrico Fermi Inst Univ Chicago 933 E 56th St Chicago IL 60637

MEYER, R PETER, b Buffalo, NY, Dec 7, 43. NEUROANATOMY, GROSS ANATOMY. *Educ:* Heidelberg Col, BS, 70; Temple Univ, PhD(anat), 77. *Prof Exp:* Sr instr, 77-85, ASST PROF ANAT, HAHNEMANN UNIV, 85- *Concurrent Pos:* Assoc ed, Anatom Rec, 85- *Mem:* NY Acad Sci; Sigma Xi;

Cajal Club; Am Asn Anatomists. *Res:* Neural connections of mammalian forebrain; cytology and connections of autonomic nervous system; experimental neuropathology utilizing various light microscopic and electron microscopic methods. *Mailing Add:* Dept Anat MS 408 Hahnemann Univ Broad & Vine Sts Philadelphia PA 19102

MEYER, RALPH A, JR, b Washington, DC, July 3, 43; m 69; c 1. PHYSIOLOGY, ENDOCRINOLOGY. *Educ:* Univ Md, College Park, BS, 65, PhD(zool), 69. *Prof Exp:* Nat Inst Dent Res fel biol, Rice Univ, 69-71; asst prof physiol & pharmacol, Col Dent, NY Univ, 71-73; from asst prof to prof physiol Dent Marquette, 73-90, head div, 73-85; DIR ORTHOP RES BIOL, CAROLINAS MED CTR, 90- *Mem:* AAAS; Am Physiol Soc; Am Soc Zool; Endocrine Soc; Am Soc Bone & Mineral Res. *Res:* Actions of hormones in promoting mineral and bone homeostasis and metabolic bone disease. *Mailing Add:* Orthop Res Lab Carolinas Med Ctr PO Box 32861 Charlotte NC 28232-2861

MEYER, RALPH O, b Covington, Ky, May 28, 38; m 59; c 2. SOLID STATE PHYSICS, NUCLEAR REACTOR ENGINEERING. *Educ:* Univ Ky, BS, 60; Univ NC, PhD(physics), 66. *Prof Exp:* Res assoc physics, Univ Ariz, 65-68; asst metallurgist, Mat Sci Div, Argonne Nat Lab, 68-73; reactor engr, 73-76, SECT LEADER, US NUCLEAR REGULATORY COMN, 76- *Mem:* Am Nuclear Soc. *Res:* Diffusion in solids; reactor fuel analysis; analysis of severe accidents in nuclear power plants. *Mailing Add:* US Nuclear Regulatory Comn Washington DC 20555

MEYER, RALPH ROGER, b Milwaukee, Wis, Feb 18, 40; m; c 4. CELL BIOLOGY, BIOCHEMISTRY. *Educ:* Univ Wis-Milwaukee, BS, 61, Madison, MS, 63 & PhD(zool), 66. *Prof Exp:* Res assoc Yale Univ, 66-67; NIH fel, State Univ NY Stony Brook, 67-69; asst prof, 69-75, assoc prof, 75-79, PROF BIOL SCI, UNIV CINCINNATI, 79- *Concurrent Pos:* NSF grant, Univ Cincinnati, 69-71 & Am Cancer Soc grants, 69-73 & 79-88; NIH res grant, 75-86; vis assoc prof biochem, Stanford Univ, 77-78; vis prof molecular biol, Univ Paris, 83. *Mem:* Am Soc Biochem & Molecular Biol; AAAS; Am Soc Cell Biol; Am Soc Microbiol; Sigma Xi; Protein Soc. *Res:* regulation and mechanism of DNA replication and repair in normal and neoplastic tissues; role of single-stranded DNA-binding protein in E coli and mammalian cells; biochemistry and molecular biology. *Mailing Add:* Dept of Biol Sci Univ of Cincinnati Cincinnati OH 45221-0006

MEYER, RICH BAKKE, JR, b Houston, Tex, Nov 6, 43; m 69. MEDICINAL CHEMISTRY, ORGANIC CHEMISTRY. *Educ:* Rice Univ, BA, 65; Univ Calif, Santa Barbara, PhD(org chem), 68. *Prof Exp:* Res scientist, ICN Pharmaceut, Inc, 70-73, head dept bio-org chem, 73-75; asst prof pharmaceut chem, Univ Calif, San Francisco, 75-80; ASSOC PROF MED CHEM, COL PHARM, WASH STATE UNIV, PULLMAN, 80- *Mem:* Am Chem Soc. *Res:* Design and synthesis of cancer chemotherapeutic agents, enzyme inhibitors and analogs of cyclic adenosine monophosphate; enzyme mechanisms. *Mailing Add:* Microprobe Corp 1725 220th St SE Bothell WA 98021-7499

MEYER, RICHARD ADLIN, b Norwood, Mass, Dec 12, 33; m 56; c 3. NUCLEAR CHEMISTRY & PHYSICS. *Educ:* Northeastern Univ, BS, 56, MS, 58; Univ Ill, Urbana, PhD, 63. *Prof Exp:* Asst chemist, Bird & Son, Inc, Mass, 52-56; res asst chem, Northeastern Univ, 56-58; nuclear physics, Univ Ill, Champaign-Urbana, 59-63, NATO fel nuclear chem, Danish AEC Res Estab, Roskilde, Denmark, 65-66; res scientist nuclear struct, Lawrence Livermore Lab, Univ Calif, 66-75, group leader & prog leader, 75-89; PROG MGR, DEPT ENERGY, US GOVT, 89- *Concurrent Pos:* Lectr, Northeastern Univ, 60. *Mem:* Am Chem Soc; Am Phys Soc; Sigma Xi. *Res:* Nuclear structure; rapid automated chemical separations and spectroscopy measurements; nuclear radiation standards; radiation preservation of foods; electron dosimetry; high energy photonuclear reactions in complex nuclei. *Mailing Add:* Div Nuclear Physics ER-23 Dept Energy Washington DC 20585

MEYER, RICHARD ARTHUR, b Springfield, Mass, June 26, 46; m 67; c 2. NEUROPHYSIOLOGY. *Educ:* Valparaiso Univ, Ind, BS, 68; Johns Hopkins Univ, MS, 72. *Prof Exp:* Assoc staff engr, Appl Physics Lab, 68-74, sr staff engr, 74-83, instr neurosurg, Sch Med, 80-83, asst prof neurosurg & biomed eng, 83-87, PRIN STAFF BIOMED ENGR, APPL PHYSICS LAB, 83-, ASSOC PROF NEUROSURG & BIOMED ENG, SCH MED, JOHNS HOPKINS UNIV, 87-,. *Honors & Awards:* Jacob Javits Neurosci Investr. *Mem:* Inst Elec & Electronics Engrs; Soc Neurosci; Int Asn Study Pain; Sigma Xi. *Res:* Neurophysiological and psychophysical mechanisms of pain sensation; applications of technology to problems in the neurosciences. *Mailing Add:* Johns Hopkins Appl Physics Lab Johns Hopkins Rd Laurel MD 20707

MEYER, RICHARD CHARLES, b Cleveland, Ohio, May 2, 30; m 63; c 2. VETERINARY MICROBIOLOGY, INFECTIOUS DISEASES. *Educ:* Baldwin-Wallace Col, BSc, 52; Ohio State Univ, MSc, 57, PhD(cellulose digestion), 61. *Prof Exp:* Asst microbiol, Ohio State Univ, 56-61, res assoc virol & germ free res, Ohio State Univ Res Found, 61-62; microbiologist, Virol Res Resources Br, Nat Cancer Inst, 62-64; asst prof swine dis & germ free res, Col Vet Med, Univ Ill, Urbana-Champaign, 65-68, assoc prof, 68-73, PROF VET PATH & HYG, COL VET MED & MICROBIOL, SCH LIFE SCI, 73- *Concurrent Pos:* Mem spec review comt, Extramural Activ, Cancer Ther Eval Br, Nat Cancer Inst, 66. *Mem:* AAAS; Am Inst Biol Sci; Am Soc Microbiol; Soc Cryobiol; fel Am Acad Microbiol. *Res:* Porcine and bovine viruses; tissue culture; enteric and respiratory tract infections; diseases of baby pigs; development of germ free techniques to study mixed synergistic infections and host-parasite relationships. *Mailing Add:* Vet Path Dept Univ Ill 2828 VMBSB Urbana IL 61803

MEYER, RICHARD DAVID, b Allentown, Pa, Apr 26, 43. INFECTIOUS DISEASES. *Educ:* Univ Pittsburgh, BS, 63, MD, 67. *Prof Exp:* Intern med, Bellevue Hosp, NY Univ, 67-68, jr asst resident, 68-69; sr asst resident med, Albert Einstein Col Med-Bronx Munic Hosp, 69-70; clin res trainee infectious dis, Mem Hosp, Cornell Univ Med Col, 70-72; Lt Comdr microbiol, US Naval Med Res Inst, 72-74; asst chief, Infectious Dis Sect, Wadsworth Vet Admin Hosp, Los Angeles, 74-82; from asst prof to assoc prof, 74-84, PROF MED, SCH MED, UNIV CALIF, LOS ANGELES; DIR, DIV OF INFECTIOUS DIS, CEDER-SINAI MED CTR, 82- *Mem:* Am Soc Microbiol; fel Infectious Dis Soc Am; Am Fedn Clin Res; fel Am Col Physicians; Am Thoracic Soc. *Res:* Clinical evaluation of antimicrobials; Legionnaires' disease; fungal infections; acquired immune deficiency syndrome. *Mailing Add:* Div Infectious Dis Cedars Sinai Med Ctr Los Angeles CA 90048

MEYER, RICHARD ERNST, b 19. MATHEMATICS, GEOPHYSICS. *Educ:* Swiss Fed Inst Technol, Dipl Mech Eng, 42, Dr Sc Techn, 46; Brown Univ, MA, 62. *Prof Exp:* Jr sci officer math, Brit Ministry Aircraft Prod, 45-46; asst lectr, Univ Manchester, 46-47, Imp Chem Industs res fel, 47-52; sr lectr aeronaut, Univ Sydney, 53-56, reader, 56-57; assoc prof appl math, Brown Univ, 57-59, prof, 59-64; PROF MATH, UNIV WIS-MADISON, 64- *Concurrent Pos:* Vis mem, Courant Inst Math Sci, NY Univ, 63-64; consult, Aeronaut Res Labs, Australian Dept Supply, 55-57 & Rand Corp, 61-62; mem, NSF Postdoctoral Panel, Nat Res Coun, 68-69, chmn panel for math, 70; sr fel, Fluid Mech Res Inst, Univ Essex, 71-72. *Mem:* Fel Australian Acad Sci; Am Geophys Union; Soc Indust & Appl Math. *Res:* Asymptotic analysis; partial differential equations; plasma physics; water waves; meteorology; gas dynamics. *Mailing Add:* Dept of Math Univ of Wis Madison WI 53706

MEYER, RICHARD FASTABEND, b Covington, Ky, Sept 13, 21. ECONOMIC GEOLOGY. *Educ:* Dartmouth Col, AB, 47; Harvard Univ, MA, 50; Univ Kans, PhD(geol), 68. *Prof Exp:* Sr geologist, Humble Oil & Ref Co, 51-61; geologist, US Geol Surv, 64-66; petrol specialist, Off Oil & Gas, US Dept Interior, 66-72; chief, Off Resource Analysis, 78-81, GEOLOGIST, US GEOL SURV, 72-, RES GEOLOGIST, 81- *Mem:* AAAS; Am Asn Petrol Geol; Soc Petrol Eng. *Res:* Petroleum origin and occurrence; stratigraphy; origin and world distribution of heavy crude oil, natural bitumen and oil shale, methods of their assessment and economic potential. *Mailing Add:* Warren Radiation Ther Ctr 16 Mountain Blvd Warren NJ 07059

MEYER, RICHARD LEE, b Independence, Mo, June 5, 31; m 59; c 1. PHYCOLOGY. *Educ:* Mo Valley Col, BS, 54; Univ Minn, PhD(bot), 65. *Prof Exp:* Instr bot & biol, Univ Minn, 58-59; staff scientist, NSF-Int Indian Ocean Exped, 61-62; asst prof phycol & biol, Calif State Univ Chico, 65-68; PROF BOT & BACT, UNIV ARK, FAYETTEVILLE, 68-; CHMN BD DIRS, CEN EMERGENCY MED SERV, INC, 82- *Concurrent Pos:* Vis prof, Univ Minn, 66-75; dir, Eagle Lake Biol, Calilf, 66-68; Water Resources Res Off res grant, 68-; Nat Park Serv, 71-, Corps of Engrs, 72-; consult, Reserve Mining Co, Silver Bay, Minn, 72-76, Minn Pollution Control Agency, 74-76, Limnetics, Inc, 75 & Tenn Valley Authority; endowment fund mgr & bd trustees, Phycol Soc Am, 84-; consult, Beaver Lake Water Dist, 83-; consult, H-D, Inc, 87- *Mem:* AAAS; Bot Soc Am; Int Phycol Soc; Phycol Soc Am; Sigma Xi; Am Micros Soc; Brit Phycol Soc. *Res:* Morphology, cytology, life-history and systematics of the algal class Chrysophyceae and other flagellated algae; algal ecology; desmid morphogenesis. *Mailing Add:* Dept of Bot & Microbiol Univ of Ark Fayetteville AR 72701

MEYER, RICHARD LEE, b Red Wing, Minn, Aug 23, 37; m 61; c 2. AGRICULTURAL ECONOMICS. *Educ:* Univ Minn, BS, 59; Cornell Univ, MS, 67, PhD(agr econ), 70. *Prof Exp:* Peace Corps vol, Chile, 62-64, vol liaison officer, Peace Corps, Washington, DC, 64-65; chief party agr econ res, 70-72, res adv, 72-73, asst prof, 73-74, PROF AGR ECON, OHIO STATE UNIV, 74-, DIR INT PROGS, 81- *Mem:* Am Agr Econ Asn; Am Econ Asn. *Res:* Agricultural development in developing countries; allocation and productivity of agricultural credit; part-time farming and off-farm income. *Mailing Add:* Dept Agri Econ Ohio State Univ 103 Agr Admin Bldg Columbus OH 43210

MEYER, RITA A, GASTRIC MUCOSAL BARRIER, BASAL CELL CARCINOME IN VITRO. *Educ:* Univ Chicago, PhD(anat), 81. *Prof Exp:* RES SCIENTIST, ST PAUL-RAMSEY MED CTR, 81- *Mailing Add:* Dept Path St Paul-Ramsey Med Ctr 1445 Gortner Ave St Paul MN 55108

MEYER, ROBERT BRUCE, b St Louis, Mo, Oct 13, 43; m 66, 80; c 1. SOLID STATE PHYSICS, MATERIALS SCIENCE. *Educ:* Harvard Univ, BS, 65, PhD(appl physics), 70. *Prof Exp:* Res fel & lectr, 70-71,Harvard Univ, from asst prof to assoc prof appl physics, 71-78; assoc prof, 78-85, PROF PHYSICS, BRANDEIS UNIV, 85- *Concurrent Pos:* Sloan Found fel, Harvard Univ, 71-73; NORDITA fel and vis prof, Chalmers Univ, Gotebors, Sweden, 77. *Honors & Awards:* Joliot Curie Professorship & Medal, City of Paris, France, 78. *Mem:* Fel Am Phys Soc. *Res:* Liquid crystals; polymers; condensed molecular systems; colloidal systems. *Mailing Add:* Physics Dept Brandeis Univ Waltham MA 02254

MEYER, ROBERT EARL, b Chicago, Ill, May 18, 32; m 62. PLANT PHYSIOLOGY. *Educ:* Purdue Univ, BS, 54, MS, 56; Univ Wis, PhD(agron), 61. *Prof Exp:* PLANT PHYSIOLOGIST, AGR RES SERV, USDA, 61- *Mem:* Weed Sci Soc Am; Soc Range Mgt; Sigma Xi. *Res:* Development of more efficient methods of brush control on rangeland of the Southwest using good conservation and management practices; evaluation of chemicals; developmental anatomy of woody plants. *Mailing Add:* USDA Dept of Range Sci Tex A&M Univ College Station TX 77843

MEYER, ROBERT F, b Switz, Mar 7, 25; nat US; m 52; c 4. PHARMACEUTICAL CHEMISTRY. *Educ:* Swiss Fed Inst Technol, PhD(chem), 50. *Prof Exp:* With pharmaceut chem, Univ Kans, 50-51; res chemist, 52-58, SR RES CHEMIST, PARKE, DAVIS & CO, 58- *Mem:* Am Chem Soc. *Res:* Chemistry of chloromycetin and analogs; hydroxyl aminderivatives; heterocycles; cardiovascular drugs; diuretics; antianginals; angiotensin converting enzyme inhibitors; hypolipidemics. *Mailing Add:* 5870 Warren Rd Ann Arbor MI 48105

MEYER, ROBERT JAY, b Ithaca, NY, Oct 24, 49; m 71; c 2. GEOPHYSICS. *Educ:* Cornell Univ, BA, 71; Univ Ill, Urbana, MS, 73, PhD(physics), 77. *Prof Exp:* Assoc scientist, Xerox Corp, 77-80; res geophysicist, Gulf Res & Develop Co, 80-84, Shell Develop Co, 84-86; MEM RES STAFF, XEROX CORP, 88- *Mem:* Am Phys Soc; Sigma Xi; Mat Res Soc. *Res:* Adhesion and friction of polymers. *Mailing Add:* 1368 Whalen Rd Penfield NY 14526-1729

MEYER, ROBERT PAUL, b Milwaukee, Wis, Dec 14, 24; m 51; c 5. SEISMOLOGY, GEOPHYSICAL INSTRUMENTATION. *Educ:* Univ Wis-Madison, BS, 48, MS, 50, PhD(geol & geophysics), 57. *Prof Exp:* from asst prof to assoc prof, 59-68, PROF GEOPHYSICS, UNIV WIS-MADISON, 68- *Concurrent Pos:* Exec comt mem, Consortium Continental Reflection Profiling, 71-; mem, Comt Seismol, Nat Res Coun, Nat Acad Sci, 75-78, Comt Seismol, Panel Seismol Studies Continental Lithosphere, 80-83 & Standing Comt, Prog Array Seismic Studies Continental Lithosphere, Inc Res Insts Seismology, 84-85, co-chmn, 84-85, chmn, Comt Instrumentation Develop, 86-; chmn, Oceanog & Limnol Grad Prog, Univ Wis, 75-78, dir, Seismol Lab & Geophysics Comput Fac, 59- *Mem:* Soc Explor Geophysicists; Seismol Soc Am; Am Geophys Union. *Res:* Plate tectonics, seismic properties of convergent margins; crust-mantle parameters from reflection and refraction studies; digital field instrumentation; chirped seismic sources; parameterization of earthquake source properties using digital data. *Mailing Add:* Dept Geol & Geophysics Univ Wis-Madison 1215 W Dayton St Madison WI 53706

MEYER, ROBERT WALTER, b Seattle, Wash, Jan 18, 39; m 63; c 4. EXPERT SYSTEMS FOR WOOD USE, PROPERTIES & USES OF WOOD. *Educ:* Univ Wash, BS, 62, MF, 64; State Univ NY, PhD(wood sci), 67. *Prof Exp:* Res scientist, Western Front Prod Lab, Can Forestry Serv, 67-75; asst prof wood sci, Wash State Univ, 75-79; PROF WOOD SCI & TECHNOL, COL ENVIRON SCI & FORESTRY, STATE UNIV NY, 79- *Concurrent Pos:* Ed, Soc Wood Sci & Technol, 72-79. *Honors & Awards:* Wood Award, Forest Prod Res Soc, 67. *Mem:* Soc Wood Sci & Technol; Forest Prod Res Soc; Int Asn Wood Anatomists. *Res:* Structure and properties of wood; wood use; physical and mechanical properties of juvenile versus adult wood; tropical woods, expert systems. *Mailing Add:* Wood Prod Eng State Univ NY Environ Sci Forestry Syracuse NY 13210-2786

MEYER, ROGER J, b Olympia, Wash, May 14, 28; m 59; c 6. PEDIATRICS. *Educ:* Univ Wash, BS, 51; Wash Univ, MD, 55; Harvard Univ, MPH, 59. *Prof Exp:* Instr pediat, Med Sch, Harvard Univ, 59-62; asst prof, Col Med, Univ Vt, 62-65; assoc prof, Sch Med, Univ Va, 65-68; assoc prof, Sch Med, Northwestern Univ, Evanston, 68-74; assoc prof pub health, Sch Pub Health, Univ Ill, & asst dean continuing educ, 74-76; CLIN PROF HEALTH CARE, SCH PUB HEALTH, PEDIAT SCH MED, UNIV WASH, 76- *Concurrent Pos:* Dir, Infant Welfare Soc, Chicago, 68-70; regional med coordr, Social & Rehab Serv, Dept Health, Educ & Welfare, Chicago, 70-74; mem exec bd, Nat Comt Prev Child Abuse; mem child safety comt, Nat Safety Coun; lectr, Northwestern Univ & Chicago Med Sch; health adv, Mayor's Comt Senior Citizens, Chicago, admin dir, Div Health, Tacoma Pub Schs, 76-80; sr proj dir, Dept Pediat, Madigan Army Med Ctr, Tacoma, Wash, 80-82; pediatrician, Rainier Sch, Buckley, Wash, 82-90; chief, pub health/pac rm Calif bd, US Army Reserves - Marine Corps, 85; physician & consult, Dept Social & Health Servs, State Wash, Region X Health & Human Servs, 90- *Mem:* Am Cong Rehab Med; Am Acad Pediat; Am Pub Health Asn; Am Acad Pediat; Child & Family Health Care Found (pres); Nat Asn Learning Disability. *Res:* Epidemiology and control of childhood injury; diagnosis and management of child and family disorders; community health systems; rehabilitation programs; wellness and health promotion. *Mailing Add:* 709 N Yakima Ave Tacoma WA 98403-2419

MEYER, RONALD ANTHONY, MUSCLE ENERGETICS, EXERCISE PHYSIOLOGY. *Educ:* State Univ NY, Syracuse, PhD(physiol), 80. *Prof Exp:* ASST PROF PHYSIOL, MICH STATE UNIV, 86- *Mailing Add:* Dept Physiol Mich State Univ Giltner Hall East Lansing MI 48824

MEYER, RONALD HARMON, b Walsh, Ill, Dec 30, 29; m 51; c 4. ECONOMIC ENTOMOLOGY. *Educ:* Univ Ill, BS, 51, MS, 56, PhD(econ entom), 63. *Prof Exp:* Asst entomologist, 56-65, assoc entomologist, 65-76, ENTOMOLOGIST, ILL STATE NATURAL HIST SURV, 76- *Mem:* Entom Soc Am. *Res:* Integrated control of insects and mites on fruit crops. *Mailing Add:* Dept Anesthesia Northwestern Univ Med Sch 303 E Chicago Ave Chicago IL 60611

MEYER, RONALD WARREN, plant pathology; deceased, see previous edition for last biography

MEYER, STEPHEN FREDERICK, b Berkeley, Calif, Aug 9, 47. SOLID STATE PHYSICS, THIN FILM TECHNOLOGY. *Educ:* Whitman Col, BA, 69; Stanford Univ, MS, 70, PhD(appl physics), 74. *Prof Exp:* Res assoc physics, Univ Ill, Urbana-Champaign, 73-76; mem staff, Lawrence Livermore Nat Lab, Univ Calif, 76-80; DIR THIN FILM RES, SOUTHWALL TECHNOL, PALO ALTO, CALIF, 80- *Mem:* Am Phys Soc; Am Vacuum Soc. *Res:* Properties of transition metal compounds; applications of thin film technology to materials. *Mailing Add:* Southwall Technol 1029 Corporation Way Palo Alto CA 94303

MEYER, STUART LLOYD, b New York, NY, May 28, 37; c 3. PHYSICS, EDUCATION. *Educ:* Columbia Univ, AB, 57; Princeton Univ, PhD(physics), 62. *Prof Exp:* Res physicist, Nevis Cyclotron Labs, Columbia Univ, 61-63; asst prof physics, Rutgers Univ, 63-67, Rutgers fac fel, Rutherford High Energy Lab, Eng, 66-67; assoc prof physics, 67-77, assoc chmn dept, 68-70, assoc prof decision sci, Grad Sch Mgt, 75-76, assoc prof finance, 76-77, PROF POLICY & ENVIRON & TRANSP, J L KELLOGG GRAD SCH MGT, NORTHWESTERN UNIV, 77- *Concurrent Pos:* Consult, Nat Accelerator Lab, 70-71, NSF, 74-75, Lib Cong, 76, Nat Oceanic & Atmospheric Admin, 76-77, Pace Inst, 78-80, Northwest Reg Educ Lab, Walter Reed Army Med Ctr, Alaska Dept Educ, 78-, Res Med Ctr, 79-80, Tool & Die Inst, Ill Mfrs Asn & Polyurethane Mfrs Asn; chmn gen fac comt, Northwestern Univ, 71-72; vis staff mem, Los Alamos Sci Lab, 72-80; prog dir intermediate energy physics, NSF, 74-75; bd dirs, Speedfam Corp, Colo Video Inc, Televideo Consults, Inc; fel, Ctr Teaching Professions, 74-75. *Mem:* AAAS; Am Phys Soc; NY Acad Sci; Am Asn Physicists Med; Am Mgt Asn. *Res:* High energy interactions; muon physics; weak interactions; counter and spark-chamber techniques; data analysis; probability and statistics; philosophy of science; radiation shielding; neutrino physics; telecommunications; teleconferencing; medical electronics; telemedicine; information science; electronics. *Mailing Add:* Leverone Hall 6-200 Northwestern Univ Evanston IL 60208

MEYER, THOMAS, b Magdeburg, Ger, Mar 23, 46; m 77; c 3. PHYSICS. *Educ:* Univ Calif, Los Angeles, BS, 68, MS, 71, PhD(physics), 76. *Prof Exp:* Adj asst prof physics, Univ Calif, Los Angeles, 76-77; res assoc, Univ Wis-Madison, 77-81; asst prof physics, Tex A&M Univ, 81-88; ASSOC PROF, CALIF STATE UNIV, BAKERSFIELD, 88- *Mem:* Am Phys Soc; Am Asn Physics Teachers. *Res:* High energy proton-antiproton and electron-position collisions and search for magnetic monopoles in cosmic rays. *Mailing Add:* Physics Dept Calif State Univ 9001 Stockdale Hwy Bakersfield CA 93311-1099

MEYER, THOMAS J, b Dennison, Ohio, Dec 3, 41; m 63; c 2. INORGANIC CHEMISTRY. *Educ:* Ohio Univ, BS, 63; Stanford Univ, PhD(chem), 66. *Prof Exp:* NATO fel, Univ Col, Univ London, 66-67; from asst prof to prof, Univ NC, Chapel Hill, 68-82, M A Smith prof, 82-87, chmn dept chem, 85-90, KENAN PROF, UNIV NC, CHAPEL HILL, 87- *Concurrent Pos:* Alfred P Sloan Found fel, 75, Guggenheim fel, 83 & Erskine fel, Univ Canterbury, NZ, 85. *Honors & Awards:* Stone Award, Piedmont Sect, Am Chem Soc, 82, Inorg Chem Award, 90. *Mem:* Am Chem Soc; Am Soc Univ Professors; fel AAAS. *Res:* Kinetics and mechanisms of inorganic and organo-metallic reactions; photochemistry; electrochemistry; catalysis polymers; solar energy conversion; artificial photosynthesis. *Mailing Add:* Dept of Chem Univ of NC Chapel Hill NC 27599-3290

MEYER, VERNON M(ILO), b New Prague, Minn, Dec 20, 24; m 51; c 4. AGRICULTURAL ENGINEERING. *Educ:* Univ Minn, BA, 51, MS, 55, PhD(agr eng), 78. *Prof Exp:* Engr trainee, Fed Land Bank, Minn, 51-52; agr engr, Minn Farm Bur Serv Co, 52-53; asst agr eng, Univ Minn, 53-54, instr, 54-56, res assoc, 56-58; from asst prof to assoc prof, 58-76, PROF AGR ENG, IOWA STATE UNIV, 76-, EXTEN AGR ENGR, 58- *Mem:* Am Soc Agr Engrs; Water Pollution Control Fedn. *Res:* Livestock and poultry environment control and modification; farmstead facility design and energy conservation. *Mailing Add:* 200 Davidson Hall Iowa State Univ Ames IA 50011

MEYER, VICTOR BERNARD, b New York, NY, Dec 17, 20; m 47; c 2. ORGANIC POLYMER CHEMISTRY. *Educ:* City Col New York, BS, 42; Columbia Univ, AM, 49, PhD(org chem), 53. *Prof Exp:* Chemist, Montrose Chem Co, 42 & 46-47; asst, Columbia Univ, 48-50; group leader textile chem, United Merchants Labs, Inc, 53-57; res chemist, W R Grace & Co, 57; sr chemist, Air Reduction Co, Inc, 58-60; chemist & dir, Viburnum Assocs, 60-74, pres & tech dir, Viburnum Resins, Inc, 62-74; res assoc, J P Stevens & Co, Inc, 74-77; group leader, Weyerhaeuser Co, 77-78; mgr latex polymers, Cellomer Corp, 78-80; CONSULT, 80- *Concurrent Pos:* Consult chemist, 62-74. *Mem:* Am Chem Soc; Fedn Soc Coatings Technol; Am Asn Textile Chemists & Colorists. *Res:* Synthetic latex and resin development; emulsion polymers in paints, textiles, nonwoven fabrics and paper; aqueous coatings and adhesives; paint testing and evaluation. *Mailing Add:* 83 Briarwood Dr E Berkeley Heights NJ 07922

MEYER, VINCENT D, b McKees Rocks, Pa, Nov 7, 32; m 62; c 3. PHYSICAL CHEMISTRY, CHEMICAL PHYSICS. *Educ:* Duquesne Univ, BS, 54; Ohio State Univ, PhD(phys chem), 62. *Prof Exp:* Fel electron-impact spectros, Mellon Inst, 62-65; mem tech staff, Gen Tel & Electronics Labs Inc, NY, 65-72 & GTE Sylvania, Inc, Pa, 72-74, sr eng specialist, GTE Sylvania Lighting Ctr, 74-81; eng mgr, 81-90, DIR, TECH SERV, GTE PROD CORP, DANVERS, MA, 90- *Mem:* Am Chem Soc; Am Phys Soc. *Res:* Electron scattering; molecular structure; energy transfer; high vacuum technique; quantum chemistry; cathodoluminescence; electro-optic phenomena; spectroscopy; mass spectrometry; auger electron spectroscopy; x-ray photoelectron spectroscopy. *Mailing Add:* GTE Sylvania 100 Endicott St Danvers MA 01923

MEYER, W(ILLIAM) KEITH, b Sioux Falls, SDak, Dec 21, 29; m 52; c 2. PHYSICAL CHEMSITRY. *Educ:* Morningside Col, BS, 51; Univ Ky, MS, 55; Mich State Univ, PhD(phys chem), 58. *Prof Exp:* Prof chemist, 58-67, sect supvr, 67-68, RES ASSOC, GULF SCI & TECHNOL CO, 68- *Concurrent Pos:* Union Carbide fel, Mich State Univ, 57-58. *Mem:* Am Chem Soc; Soc Petrol Eng; Am Soc Testing Mat; AAAS. *Res:* Secondary oil recovery; drilling fluids; marine transportation; tanker safety. *Mailing Add:* 1331 Balmore Houston TX 77069

MEYER, WALTER, b Chicago, Ill, Jan 19, 32; m 53; c 5. NUCLEAR & CHEMICAL ENGINEERING. *Educ:* Syracuse Univ, BChE, 56, MChE, 57; Ore State Univ, PhD(chem eng), 64. *Prof Exp:* Res asst chem eng, Syracuse Univ, 57; prin chem engr, Battelle Mem Inst, 57-58; instr chem eng, Ore State Univ, 58-64, asst prof, 64; NSF fac fel, Mass Inst Technol & Ore State Univ, 62-63; from asst prof to prof nuclear eng, Kans State Univ, 64-72; prof nuclear eng & chmn, Univ Mo, 72-82, codir, Energy Systs & Resources Prog, 74-82, codir, Energy & Policy Ctr, 81-82; NIAGARA MOHAWK ENERGY PROF, SYRACUSE UNIV, 82-, PROF PUB ADMIN, MAXWELL SCH, 85-, EXEC DIR, MFRS ASSISTANCE CTR, SYRACUSE UNIV, 90- *Concurrent Pos:* Summer res engr, Hanford Atomic Labs, Wash, 59 & 60; consult, GA Pac Co, 64, Kerr McGee Co, 66-67, Gen Physics Corp, 67, Boeing Co, 70, Northeast Utilities Co, Univ Affairs Div, Argonne Nat Lab, 70-71, Wis Elec Power Co, Fed Trade Comn, 76-80, EG&G Idaho Inc, 78, NY Power Authority Nuclear Emergency Educ & Response Team, 84- &

regents degree prog nuclear technol, State Univ NY, 85-; mem, Governor Kans Nuclear Energy Coun, 70; Dir, Dept Energy Summer Inst, 73-78, NSF Summer Inst, 78 & Inst Energy Res, Syracuse Univ, 82-; Exec dir, NY Sci & Technol Found Indust Innovation, Syracuse Univ, 88-90, tech dir, Elec Power Res Inst Knowledge-Based Technol Appln Ctr, 89- *Mem:* Am Inst Chem Engrs; Am Chem Soc; fel Am Nuclear Soc; Am Soc Eng Educ. *Mailing Add:* Inst Energy Res Syracuse Univ 329 Link Hall Syracuse NY 13244-1240

MEYER, WALTER EDWARD, b Hackensack, NJ, Sept 15, 29; m 48; c 2. PHARMACEUTICAL CHEMISTRY. *Educ:* Rutgers Univ, BS, 51; NY Univ, MS & PhD(org chem), 64. *Prof Exp:* Biochemist, 51-53, RES CHEMIST, LEDERLE LABS DIV, AM CYANAMID CO, 53- *Mem:* Am Chem Soc. *Res:* Structure determination on compounds having pharmaceutical interest; synthesis of compounds related to physiologically active materials. *Mailing Add:* 40 Van Arden Ave Suffern NY 10901-6325

MEYER, WALTER H, b Cincinnati, Ohio, Aug 19, 22; m 44; c 3. NUTRITION. *Educ:* Mich State Univ, BS, 48. *Prof Exp:* Proj engr, Procter & Gamble Co, 48-51, supvr qual control, Procter & Gamble Defense Corp, 51-54, sect head toilet goods prod, Procter & Gamble Co, 54-57 & food prod, 57-66, mgr prof & regulatory rels, 66-68, ASSOC DIR FOOD PROD, PROD DEVELOP DEPT, PROCTER & GAMBLE CO, 68- *Concurrent Pos:* Am Med Asn Food Indust Liaison Adv Panel. *Res:* Food safety and regulations. *Mailing Add:* Procter & Gamble Co 6071 Center Hill Rd Cincinnati OH 45224

MEYER, WALTER JOSEPH, b New York, NY, Jan 12, 43; m 67; c 1. MATHEMATICS. *Educ:* Queen's Col, BA, 64; Univ Wis, MS, 66, PhD(math), 70. *Prof Exp:* Asst prof, 69-74, ASSOC PROF MATH, ADELPHI UNIV, 74- *Res:* Graph theory; convex sets. *Mailing Add:* Dept Math Adelphi Univ 100 Brook St Garden City NY 11530

MEYER, WALTER LESLIE, b Toledo, Ohio, Feb 28, 31; m 54; c 3. ORGANIC CHEMISTRY. *Educ:* Univ Mich, BS, 53, MS, 55, PhD(chem), 57. *Prof Exp:* Instr & res assoc chem, Univ Mich, 57; NSF fel, Univ Wis, 57-58; from instr to asst prof chem, Univ Ind, 58-65; assoc prof, 65-68, chmn dept, 67-73, PROF CHEM, UNIV ARK, FAYETTEVILLE, 68- *Mem:* Am Chem Soc; Royal Soc Chem; AAAS. *Res:* Chemistry of natural products; stereochemistry; nuclear magnetic resonance; organic synthesis. *Mailing Add:* Dept of Chem Univ of Ark Fayetteville AR 72701

MEYER, WILLIAM ELLIS, b Bonne Terre, Mo, July 22, 36; m 62; c 3. ANIMAL SCIENCE. *Educ:* Univ Mo, BS, 60, MS, 62, PhD(agr), 65. *Prof Exp:* chmn dept, 70-90, PROF AGR, SOUTHEAST MO STATE UNIV, 65-,. *Mem:* Am Soc Animal Sci. *Res:* Meat technology. *Mailing Add:* Dept Agr Southeast Mo State Univ Cape Girardeau MO 63701

MEYER, WILLIAM LAROS, b Keyser, WVa, May 27, 36; m 67; c 5. ENZYMOLOGY, BIOCHEMICAL REGULATION. *Educ:* Yale Univ, BS, 56; Univ Wash, PhD(biochem), 62. *Prof Exp:* From instr to assoc prof, 62-81, PROF BIOCHEM, UNIV VT, 81-, CHMN CELL BIOL PROG, 81- *Concurrent Pos:* Res group, World Fedn Neurol, 72-; vis assoc prof & NIH spec res fel, Univ Western Australia, 73. *Mem:* AAAS; Am Chem Soc; Am Soc Biol Chemists; NY Acad Sci. *Res:* Physiological control of enzyme activity and turnover in muscle, cartilage and other tissues; proteases and ribonucleases and their relationship to neuromuscular diseases and interferon action; enzyme mechanisms; fructose metabolism; calcium metabolism; insect biochemistry. *Mailing Add:* Dept of Biochem Univ of Vt Col of Med 185 S Prospect St Burlington VT 05405

MEYER, WOLFGANG E(BERHARD), b Berlin, Ger, Aug 2, 10; nat US; m 46; c 1. AUTOMOTIVE ENGINEERING, TRAFFIC SAFETY. *Educ:* Univ Stuttgart, BSME, 33; Univ Hanover, dipl, 35. *Prof Exp:* Res engr, Daimler-Benz Co, Ger, 35-37; asst, Pa State Col, 37-38; eng lab supvr, Am Bosch Corp, Mass, 38-40; res engr, Yale Univ, 41-43; proj engr, Fuel Injection Div, Bulova Watch Co, NY, 44-45; chief engr, Res Eng Corp, Conn, 45-47; assoc prof, 47-51, prof eng res, 51-57, prof mech eng, 57-76, EMER PROF MECH ENG, PA STATE UNIV, 76 - *Concurrent Pos:* Tech consult, indust & govt agencies, 50 -; dir automotive res prog, Transp Inst, Pa State Univ, 68-76; chmn coun tech adv, Pa Air Pollution Comn, 62-63; mem adv group diesel smoke & odor, Calif Motor Vehicle Control Bd, 63-66; chmn comt surface properties-vehicle interaction, Transp Res Bd, 70-76; chmn comt travelled surfaces, Am Soc Testing & Mat, 74-78. *Honors & Awards:* Award of Merit, Am Soc Testing & Mat, 78; Shelburne Award, 83. *Mem:* Fel Soc Automotive Engrs; hon mem Am Soc Testing & Mat. *Res:* Engine combustion; fuel systems; emissions; tire and road friction. *Mailing Add:* WEM Res Bldg B University Park PA 16802

MEYERAND, RUSSELL GILBERT, JR, b Kirkwood, Mo, Dec 2, 33; m 56; c 1. PLASMA PHYSICS. *Educ:* Mass Inst Technol, SB, 55, SM, 56, ScD(plasma physics), 59. *Prof Exp:* Mem staff, res lab electronics, Mass Inst Technol, 56-57; prin scientist, United Aircraft Res Labs, United Technol Res Ctr, 58-64, chief res scientist, 64-67, vpres technol, 79-87, vpres chief scientist, 87-88; CONSULT, 88- *Concurrent Pos:* Consult, Atomic Power Equip Dept, Gen Elec Co, 55-56 & Army Sci Adv Panel, 70-74; mem adv comt, Sch Arts & Sci, Univ Hartford, 65; mem ang adv comt, 65-, vis comt & sponsored res, Dept Physics, 77-, bd trustees, Hartford Grad Ctr, Rensselaer Polytech Inst; adv comt on corp assocs, Am Inst Physics, 71-73; NASA Space Prog Adv Coun, 74-81; chmn eval panel, Quantum Elec Div, Inst Basic Standards, Nat Bur Standards, 70-72; mem ad hoc laser adv panel, NASA Res & Tech Coun, 71-81; panel on productivity enhancement, Off Sci & Technol, Exec Off of the President, 71- *Honors & Awards:* Eli Whitney Award. *Mem:* Nat Acad Eng; Am Phys Soc; fel Am Inst Aeronaut & Astronaut; sr mem Inst Elec & Electronics Engrs; Sigma Xi. *Res:* Plasma and laser physics; electronics. *Mailing Add:* 64 Littel Acres Rd Glastonbury CT 06033

MEYER-ARENDT, JURGEN RICHARD, b Berlin, Ger, Oct 4, 21; US citizen; m 49; c 3. OPTICS. *Educ:* Univ Würzburg, MD, 45; Univ Hamburg, PhD(biophys), 52. *Prof Exp:* Prof physics, Utah State Univ, 60-63 & Univ Colo, 63-66; PROF PHYSICS & OPTICS, PAC UNIV, 66- *Concurrent Pos:* Consult. *Mem:* Am Phys Soc; fel Optical Soc Am. *Res:* Geometrical optics; relativistic optics. *Mailing Add:* Dept of Physics Pacific Univ Forest Grove OR 97116-0364

MEYERHOF, WALTER ERNST, b Kiel, Ger, Apr 29, 22; nat US; m 47; c 2. EXPERIMENTAL PHYSICS, ATOMIC & MOLECULAR PHYSICS. *Educ:* Univ Pa, MA, 44, PhD(physics), 46. *Prof Exp:* Asst instr physics, Univ Pa, 43, res physicist, 44-46; asst prof physics, Univ Ill, 46-49; from asst prof to assoc prof, 49-59, chmn dept, 70-77, PROF PHYSICS, STANFORD UNIV, 59- *Concurrent Pos:* Sloan Found sr res fel, 55-59; Lilly Found tenured fac develop award, 77-78; Alexander von Humboldt US Sr Scientist Award, 80-81. *Mem:* Fel Am Phys Soc; fel AAAS. *Res:* Atomic collisions. *Mailing Add:* Dept Physics Stanford Univ Stanford CA 94305-4060

MEYERHOFER, DIETRICH, b Zurich, Switz, Sept 19, 31; nat US; m 54; c 2. MICROLITHOGRAPHY, OPTOELECTRONICS. *Educ:* Cornell Univ, BEng Phys, 54; Mass Inst Technol, PhD(physics), 58. *Prof Exp:* Asst solid state physics, Mass Inst Technol, 54-56; sr mem tech staff, res labs, RCA Corp, 58-87; SR MEM TECH STAFF, DAVID SARNOFF RES CTR, 87- *Mem:* Inst Elec & Electronics Eng; Am Phys Soc; Sigma Xi. *Res:* Galvanomagnetic and optical measurements in semiconductors and insulators; molecular lasers; holography; physical properties of photosensitive polymers; electrical and optical properties of liquid crystals; micolithography. *Mailing Add:* David Sarnoff Res Ctr Princeton NJ 08543-5300

MEYERHOFF, ARTHUR AUGUSTUS, b Northampton, Mass, Sept 9, 28; m 51; c 3. GEOLOGY. *Educ:* Yale Univ, BA, 47; Stanford Univ, MS, 50, PhD(geol), 52. *Prof Exp:* Geologist, US Geol Surv, 48-52; geologist, Calif Explor Co, Standard Oil Co Calif, 52-56, sr geologist, Cuba Calif Oil Co, 56-59, geophysicist, Chevron Oil Co, 59-60, res geologist, Calif Co, 60-65; pub mgr, Am Asn Petrol Geol, 65-75; prof geol, Okla State Univ, 75-77; PARTNER, MEYERHOFF & COX INC, 75- *Concurrent Pos:* Dir, Tulsa Sci Found, 66-72; pres northeast div, Frontiers Sci Found Okla, Inc, 71-72, mem exec comt, 71-72; vis prof, Univ Calgary, 78; vis scholar, Cambridge Univ, 78, 79. *Honors & Awards:* George C Mattson Award, Am Asn Petrol Geologists. *Mem:* Fel AAAS; fel Geol Soc Am; Soc Sedimentary Geol; Am Asn Petrol Geologists; Asn Earth Sci Ed (pres, 69-70); Am Geophys Union; Sigma Xi; Soc Econ Geologists; Paleontological Soc; Asn Geoscientists Int Develop; Europ Union Geosci. *Res:* Structural geology; geotectonics; stratigraphy; carbonate rock; paleobotany; plate tectonics, continental drift; Caribbean geology; petroleum resources worldwide, with specialty in USSR and People's Repub China. *Mailing Add:* PO Box 4602 Tulsa OK 74159

MEYERHOFF, MARK ELLIOT, b New York, NY, Apr 10, 53. ELECTROANALYTICAL CHEMISTRY, CLINICAL CHEMISTRY. *Educ:* Herbert H Leman Col, BA, 74; State Univ NY, Buffalo, PhD(anal chem), 79. *Prof Exp:* Teaching asst gen chem, State Univ NY, Buffalo, 74-75, res asst analytical chem, 75-79; ASST PROF CHEM & ANALYTICAL CHEM, UNIV MICH, 79- *Concurrent Pos:* Fel analytical chem, Univ Del, 79. *Mem:* Am Chem Soc; AAAS; Am Asn Clin Chemists. *Res:* Development of new ion-selective electrodes, particularly gas sensors and their application as detectors in novel biochemical assay arrangements, including enzyme-labelled competitive binding techniques. *Mailing Add:* Dept Chem Univ Mich Ann Arbor MI 48109

MEYEROTT, ROLAND EDWARD, b Baldwin, Ill, Nov 20, 16; m 44; c 6. ASTROPHYSICS, ATOMIC PHYSICS. *Educ:* Univ Nebr, BA, 38, MA, 40; Yale Univ, PhD(physics), 43. *Prof Exp:* Asst physics, Univ Nebr, 38-40; asst, Yale Univ, 40-41, phys asst, Med Sch, 41, instr physics, 42-47, res assoc, 47-49; sr physicist, Argonne Nat Lab, 49-53 & Rand Corp, 53-56; mgr physics, Missiles & Space Div, Lockheed Aircraft Corp, 56-62, mgr phys sci lab, Lockheed Missiles & Space Co, 62-66, dir sci, res & develop div, 66-69, asst to vpres res & develop div, 69-71, dir sci, res & develop div, 71-73; SCI CONSULT, 73- *Mem:* Fel Am Phys Soc. *Res:* Theoretical molecular physics; spectroscopy; atomic wave function calculations; opacity and equation of state of matter at higher temperatures; effects of minor species on atmosphere. *Mailing Add:* 27100 Elena Rd Los Altos Hills CA 94022

MEYEROWITZ, ELLIOT MARTIN, b Washington, DC, May 22, 51; m 84; c 2. DEVELOPMENTAL BIOLOGY OF DROSOPHILA & ARABIDOPSIS. *Educ:* Columbia Univ, AB, 73; Yale Univ, MPh, 75, PhD(biol), 77. *Prof Exp:* Res fel biochem, Sch Med, Stanford Univ, 77-79; from asst prof to assoc prof, 80-89, PROF BIOL, CALIF INST TECHNOL, 89- *Concurrent Pos:* Alfred P Sloan Found fel, 81; mem bd dirs, Int Soc Plant Molecular Biol, 89- *Mem:* Genetics Soc Am; fel AAAS; Int Soc Plant Molecular Biol; Bot Soc Am. *Res:* Using classical and molecular genetic methods in the study of development in the fly Drosophila and the plant Arabidopsis. *Mailing Add:* Div Biol California Inst Technol Pasadena CA 91125

MEYERS, ALBERT IRVING, b New York, NY, Nov 22, 32; m 57; c 3. ORGANIC CHEMISTRY. *Educ:* NY Univ, AB, 54, PhD(chem), 57. *Prof Exp:* Res chemist, Cities Serv Res & Develop Co, 57-58; from asst prof to prof chem, La State Univ, 58-69, Boyd prof, 69-70; prof, Wayne State Univ, 70-72; PROF CHEM, COLO STATE UNIV, 72- UNIV DISTINGUISHED PROF, 86- *Concurrent Pos:* Res grants, NIH, 58-96, Res Corp, 58-59, New Orleans Cancer Soc, 59-60, Eli Lilly, 65-66, US Army, 66-69, G D Searle, 75-81, Hoffmann-La Roche, 70-75 & Petrol Res Fund, 69-75; vis res fel, Harvard Univ, 65-66; res grants, NSF, 69-94, US Army, 75-82, Petrol Res Fund, 75-78, G D Searle, 75-78, Hoffmann-La Roche, 78, Bristol-Myers Squibb, 83-91, Merck, 88-91; adj Boyd prof, La State Univ, 70-73; consult, G D Searle, 72-82, Midwest Res Inst, 75-79 & Bristol-Myers, 84- & Bristol-Myers Squibb Syntex, 90-; exec comt orgn div, Am Chem Soc, 75-77, chmn, 80-; chmn, Gordon Conf Heterocycles, 73 & Sterochem, 82; sr scientist award,

Alexander von Humboldt Found, W Ger 85-86; Cope scholar, Am Chem Soc, 87. *Honors & Awards:* Distinguished Fac Award, La State Univ, 64; A G Clark Res Award, Colo State Univ, 80; Sigma Xi Award, 81; Silver Medal Centenary Award, Royal Soc Chem, Eng83; Am Chem Soc Synthesis Award, 85. *Mem:* Am Chem Soc; fel Japan Chem Soc; Chem Soc; Sigma Xi; Int Soc Heterocyclic Chem. *Res:* Synthetic organic chemistry; chemistry of heterocyclis compounds; asymmetric syntheses, total synthesis of natural products. *Mailing Add:* Dept Chem Colo State Univ Ft Collins CO 80523

MEYERS, BERNARD LEONARD, b New York, NY, Apr 16, 37; m 58; c 2. STRUCTURAL MECHANICS, MATERIALS SCIENCE. *Educ:* Polytech Inst Brooklyn, BSCE, 58; Univ Mo, MS, 60; Cornell Univ, PhD(struct eng), 67. *Prof Exp:* Asst prof civil eng, Univ Mo, 60-64; prof, Univ Iowa, 67-73; proj mgr, 73-86, vpres & opers mgr, 86-87, VPRES & MGR ENG, BECHTEL CORP, SAN FRANCISCO, CA, 88- *Concurrent Pos:* Consult, Taylor Woodrow, Ltd, London, 71. *Mem:* Am Concrete Inst; Am Soc Civil Engrs; Am Soc Eng Educ. *Res:* Material behavior in civil engineering materials; behavior of structural concrete. *Mailing Add:* Bechtel Corp 50 Beale St PO Box 193965 San Francisco CA 94119-3965

MEYERS, CAL YALE, b Utica, NY, Nov 14, 27. ORGANIC SULFUR CHEMISTRY, REACTION MECHANISMS. *Educ:* Cornell Univ, AB, 48; Univ Ill, PhD(org chem), 51. *Prof Exp:* Res fel, Princeton Univ, 51-53; res chemist, Union Carbide Plastics Co, 53-60; vis res prof, Univ Bologna, 60-63; sr res assoc, Univ Southern Calif, 63; vis scholar, Univ Calif, Los Angeles, 63-64; assoc prof, 64-68, prof chem, 68-86, DISTINGUISHED PROF, SOUTHERN ILL UNIV, CARBONDALE, 86- *Concurrent Pos:* Consult, Heliodyne Corp, 64 & Scripps Clin & Res Found, Calif, 64-; vis res lectr, Kyoto Univ, NSF res grant, 73; vis lectr, Polish Acad Sci, 78 & 86; exchange scientist nominee, Nat Acad Sci-Acad Sci, Germany Dem Repub, 82; vis res lectr, Uppsala Univ, 82; lectr, Nat Coun Sci Res, France, 82; assoc prof, Univ Marseille, 86; assoc prof, French Nat Comt, 86. *Honors & Awards:* Res Award, Union Carbide Corp, 57; Int Travel Award, NSF, 61, 70 & 71; Res Award, Am Chem Soc-Petrol Res Fund, 62; Res Award, Intra-Sci Res Found, 64; Swedish Natural Sci Res Coun Award, 82; French Nat Comt Selection as Prof Associé, 86. *Mem:* Am Chem Soc; Italian Chem Soc; Sigma Xi. *Res:* Organosulfur bonding; electron transfer reactions of anions with perhaloalkanes; isomerizations and eliminations of allylic ethers, sulfides, sulfoxides and sulfones; comparison of nucleophilic and electron-transfer reactivities of anions; aromatic-substitution reactions via electron transfers; radical/anion-radical pair mechanisms; carboxylic acid and non-steroidal estrogens; mechanisms of estrogenic activity; coal-sulfur chemistry. *Mailing Add:* Dept Chem & Biochem Southern Ill Univ Carbondale IL 62901-4409

MEYERS, CAROLYN WINSTEAD, b Hampton, Va, May 11,46; m 68; c 3. CAST ALLOYS & METAL MATRIX COMPOSITES. *Educ:* Howard Univ, BSME, 68; Ga Inst Technol, MSME, 79, PhD(metall), 84. *Prof Exp:* Systs engr, Info Serv Div, Gen Elec C. 68-69; instr graphics & eng, Atlanta Univ & Ga Tech, 72-77; from instr to asst prof mat & deformation, 79-90, ASSOC PROF MECH ENG, GA INST TECHNOL, 90- *Concurrent Pos:* Key prof, Foundry Educ Found, 84-91; NSF presidential young investr, 88; mem, Adv Bd Eng Dir, NSF, 89-, Mat Res Adv Comt, 90- & Unit Mfg Process Comt, Mfg Studies Bd, Nat Res Coun. *Honors & Awards:* Ralph A Teetor Award, Soc Automotive Engrs, 86. *Mem:* Sigma Xi; Am Soc Metals; Am Foundrymen's Soc; Am Soc Mech Engrs; Soc Women Engrs; Soc Automotive Engrs; Mat Soc; Am Soc Eng Educ. *Res:* Solidification processing of metals and metal matrix composites; mechanisms of fracture in metals and alloys; heat treatment kinetics; micromechanisms of wear in tribo-materials. *Mailing Add:* Sch Mech Eng Ga Inst Technol Atlanta GA 30332

MEYERS, DONALD BATES, b Cedar Rapids, Iowa, Jan 30, 22; m 50; c 4. PHARMACOLOGY. *Educ:* Univ Iowa, BS, 44, MS, 48, PhD(pharmaceut chem, pharmacol), 49. *Prof Exp:* Asst pharmaceut chem, Univ Iowa, 47-49; from asst prof to prof pharmacol, Butler Univ, 49-62, Baxter distinguished prof, 58; prof, Univ Tex, 62-63; sr pharmacologist, Eli Lilly & Co, 63-68, res scientist, 68-72, res assoc, Toxicol Div, 72-85, sr res scientist, 85-87; CONSULT, 87- *Concurrent Pos:* Lectr, Sch Nursing, Ind Methodist Hosp, 52-60; lectr, Butler Univ, 65- *Mem:* AAAS; Acad Pharmaceut Sci; Soc Toxicol. *Res:* Toxicology; neuropharmacology; drug metabolism. *Mailing Add:* 638 N State Greenfield IN 46140

MEYERS, EARL LAWRENCE, b Victor, Iowa, Nov 1, 07; m 41; c 2. PHYSICAL CHEMISTRY. *Educ:* Coe Col, BS, 30; Univ Ill, PhD(chem), 34. *Prof Exp:* Res fel, 34; chem consult, 37-38; inspector, US Food & Drug Admin, 39-41, resident inspector, 46-51, new drug off, new drug br, 52-58, chief chemist div new drugs, 58-63, chief controls eval br, 63-66, dir div oncol & radiopharmaceut, 67-74; RETIRED. *Mem:* Fel AAAS; Am Chem Soc; NY Acad Sci; fel Am Inst Chem; Soc Nuclear Med; Acad Pharmaceut Sci. *Res:* X-rays; spectroscopy; rare earths; radiopharmaceuticals; stability of drugs; quality control of drugs. *Mailing Add:* 5225 S Seventh Rd Arlington VA 22204

MEYERS, EDWARD, b New York, NY, Aug 17, 27; m 62; c 1. MICROBIOLOGY. *Educ:* City Col New York, BS, 49; Univ Ky, MS, 51; PhD(bact), Univ Wis, 58. *Prof Exp:* Microbiologist, Nepera Chem Co, NY, 52; med bacteriologist, Ft Detrick, Md, 52-54; RES GROUP LEADER, SQUIBB INST MED RES, 58- *Mem:* Am Soc Microbiol; Sigma Xi; fel Am Acad Microbiol; Japanese Antibiotics Res Asn. *Res:* Area of microbial biochemistry dealing with fermentation, isolation and characterization of antibiotics and other microbial products; full or partial biosynthesis of new compounds; fermentation and analytical techniques. *Mailing Add:* Squibb Inst Med Res Box 4000 Lawrenceville Princeton NJ 08540

MEYERS, FREDERICK H, b Ft Wayne, Ind, June 16, 18; m 47; c 3. PHARMACOLOGY. *Educ:* Univ Calif, MD, 49. *Prof Exp:* Intern, Univ Calif Hosp, 49-50; from instr to asst prof pharmacol, Univ Tenn, 50-53; from asst prof to assoc prof, 53-64, PROF PHARMACOL, SCH MED, UNIV CALIF, SAN FRANCISCO, 64- *Mem:* Sigma Xi; Am Soc Pharmacol & Therapeut Exp; Am Soc Clin Pharm & Therapeut. *Res:* Cardiovascular and autonomic physiology and pharmacology; problems of drug abuse; toxicity of therapeutic agents. *Mailing Add:* 84 Huntington Dr San Francisco CA 94132-1114

MEYERS, GENE HOWARD, b Chicago, Ill, Dec 6, 42; m 71; c 2. PHYSICAL CHEMISTRY, COMPUTER SCIENCES. *Educ:* Univ Ill, Urbana, BS, 64; Univ Calif, Berkeley, PhD(phys chem), 69. *Prof Exp:* Sr systs analyst, 69-85, MGR, COMPUT DEPT, KAISER ALUMINUM & CHEM CORP, PLEASANTON, 85- *Mem:* Asn Comput Mach. *Res:* Laboratory automation; computer operating systems; database; microwave spectroscopy. *Mailing Add:* Kaiser Aluminum & Chem Corp 6177 Sunol Blvd Pleasanton CA 94566

MEYERS, HERBERT, b New York, NY, Nov 15, 31; m 67; c 4. GEOPHYSICS. *Educ:* City Col New York, BS, 58. *Prof Exp:* Geophysicist geomagnetism, Coast & Geod Surv, 58-66; DIV CHIEF GEOPHYS, NAT OCEANIC & ATMOSPHERIC ADMIN, 66- *Concurrent Pos:* Dir, World Data Ctr-A Solid Earth Geophys, 78- *Mem:* Am Geophys Union; Soc Explor Geophysicists; Sigma Xi. *Res:* Geophysical data management, including seismology, geomagnetism and gravity. *Mailing Add:* Nat Geophys Data Ctr, NOAA Mail Code E/GC1 325 Broadway Boulder CO 80303

MEYERS, JAMES HARLAN, b Fountain Springs, Pa, Sept 5, 45; m 66; c 2. GEOLOGY. *Educ:* Franklin & Marshall Col, AB, 67; Ind Univ, Bloomington, MA, 69, PhD(geol), 71. *Prof Exp:* Lectr geol, Ind Univ, Bloomington, 71; asst prof geol, Muskingum Col, 71-80; MEM FAC, DEPT GEOL & EARTH SCI, WINONA STATE UNIV, 80- *Mem:* Geol Soc Am; Am Asn Petrol Geologists; Soc Econ Paleontologists & Mineralogists; Clay Minerals Soc. *Res:* Sedimentary petrology, paleoenvironments of sedimentary rocks; clay mineralogy, relating to provenance and environment of deposition of sedimentary rocks; sedimentology and recent analogs of ancient depositional environments. *Mailing Add:* Dept Geol & Earth Sci Winona State Univ Winona MN 55987

MEYERS, KENNETH PURCELL, b Jamaica, NY; m 57; c 2. CELL BIOLOGY, ENDOCRINOLOGY. *Educ:* NY Univ, AB, 53; Rutgers Univ, New Brunswick, MS, 65, PhD(endocrinol), 67. *Prof Exp:* Sr scientist, Worcester Found Exp Biol, Mass, 67-69; sr pharmacologist, Dept Pharmacol, Hoffmann-La Roche, 69-74, sr scientist, Dept Cell Biol, 74-82 & Dept Allergy & Inflammation, 82-88; CONSULT, 88- *Mem:* Brit Soc Endocrinol; Soc Study Reprod; Sigma Xi. *Res:* Physiology and pharmacology of inflammatory responses. *Mailing Add:* Res Div Hoffmann-La Roche Inc Nutley NJ 07110

MEYERS, LEROY FREDERICK, b New York, NY, June 30, 27. LINEAR ALGEBRA, ANALYSIS & FUNCTIONAL ANALYSIS. *Educ:* Queens Col, NY, BS, 48; Syracuse Univ, MA, 50, PhD(math), 53. *Prof Exp:* Actg asst prof math, Univ Va, 53-54; from instr to asst prof, 54-62, ASSOC PROF MATH, OHIO STATE UNIV, 62- *Concurrent Pos:* Partic, numerical anal training prog, NSF, Nat Bur Standards, 59; Mathematician, Nat Bur Standards, 59-63. *Mem:* Am Math Soc; Math Asn Am. *Mailing Add:* Dept Math Ohio State Univ 231 W 18th Ave Columbus OH 43210-1174

MEYERS, M DOUGLAS, b Mt Sterling, Ill, Jan 30, 33; m 61; c 2. INORGANIC CHEMISTRY, PHYSICAL CHEMISTRY. *Educ:* Univ Ill, BS, 55; Mass Inst Technol, PhD, 59. *Prof Exp:* From res chemist to sr res chemist, Am Cyanamid Co, 59-71; group leader, Kennecott Copper Co, 71-75; tech dir, Kocide Chem Corp, 75-78; develop chemist, Marathon-Morco Co, 78-85; sr res chemist, Penreco, 85-89, STAFF RES CHEMIST, PENNZOIL PROD CO, 89- *Mem:* Am Chem Soc; Soc Tribologist Lubrication Engs. *Res:* Transition metal complexes; inorganic bisulfites; fungicides; aquatic chemicals; sulfonation; compressor lubricants; white mineral oil refining; soluble oils; refrigeration oils. *Mailing Add:* Pennzoil Prod Co PO Box 7569 The Woodlands TX 77387

MEYERS, MARC ANDRE, b Belo Horizonte-Minas Geras, Brazil, Aug 10, 46; m 72; c 2. MECHANICAL & METALLURGICAL ENGINEERING, SHOCK WAVES IN MATERIALS. *Educ:* Fed Univ Minas Geras, BSc, 69; Univ Denver, MSc, 72, PhD (phys metall), 74. *Prof Exp:* Res scientist mat sci, Mil Inst Eng, 74-76; asst prof mat sci, SDak Sch Mines & Technol, 77-79; assoc prof mat sci, NMex Inst Mining & Technol, 79-88; PROF MAT SCI, UNIV CALIF, SAN DIEGO, 88- *Concurrent Pos:* Assoc dir, Ctr Explosives Technol Res, NMex Tech, 83-88; adv to dir, Mat Sci Div, Army Res Off, Res Triangle Park, 85-87; vis scientist, Nat Chem Lab Indust, Tsukuba, Japan, 83; ed, Mat Sci & Eng J, 75-78; mem bd reviewers, Metall Transactions, 75- *Mem:* Am Soc Metals; Am Inst Metall, Mining & Petrol Engrs; Mat Res Soc. *Res:* Materials science with emphasis on high strain rate response of materials and martensitic transformation; written on mechanical behavior of materials. *Mailing Add:* Dept Appl Mech Eng Sci Univ Calif San Diego R-011 La Jolla CA 92093-0411

MEYERS, MARIAN BENNETT, b New York City, NY, Feb 22, 38; m 63; c 2. GENETICS, DRUG RESISTANCE. *Educ:* Barnard Col, AB, 59; Univ Md, MS, 62; Cornell Univ, PhD(biochem), 77. *Prof Exp:* Res chemist, Merck & Co, 62-63; lectr org chem, Barnard Col, 63-64 & Queens Col, 65-72; fel genetics, Albert Einstein Col Med, 77-79; fel, 79-81, RES ASSOC, SLOAN-KETTERING INST, 81- *Mem:* Am Chem Soc; Am Soc Cell Biol; NY Acad Sci; Am Women Sci; AAAS. *Res:* Mechanisms of multidrug resistance; role of gene amplification in the development of multidrug resistance; gene amplification in human neuroblastoma cells. *Mailing Add:* One Century Trail Harrison NY 10528-1701

MEYERS, MARTIN BERNARD, b Newark, NJ, Sept 12, 33; m 62. ORGANIC CHEMISTRY. *Educ:* Polytech Inst Brooklyn, BS, 54; Yale Univ, MS, 56, PhD(chem), 58. *Prof Exp:* NIH fel, Queen's Univ Belfast, 58-59; proj leader steroid res, Gen Mills, Inc, 59-61; Imp Chem Industs fel, Glasgow Univ, 61-64; sr lectr org chem, Col Technol, Belfast, 61-71; sr lectr org chem, Ulster Polytech, 71-84; LECTR ORG CHEM, UNIV ULSTER, 84- *Mem:* Fel Royal Soc Chem. *Res:* Chemistry of color photography; anal of ornamental pollutants. *Mailing Add:* Dept Appl Phys Sci Univ Ulster Jordanstown Newton Abbey Northern Ireland

MEYERS, NORMAN GEORGE, b Buffalo, NY, June 29, 30; m 58; c 4. MATHEMATICS. *Educ:* Univ Buffalo, BA, 52; Ind Univ, MA, 54, PhD(math), 57. *Prof Exp:* From instr to assoc prof, 57-68, PROF MATH, INST TECHNOL, UNIV MINN, MINNEAPOLIS, 68- *Mem:* Am Math Soc. *Res:* Partial differential equations; calculus of variations. *Mailing Add:* Dept Math Univ Minn Vincent Hall Minneapolis MN 55455

MEYERS, PAUL, microbiology, virology, for more information see previous edition

MEYERS, PHILIP ALAN, b Hackensack, NJ, Mar 3, 41; m 65; c 3. ORGANIC GEOCHEMISTRY, PALEOCEANOGRAPHY. *Educ:* Carnegie-Mellon Univ, BS, 64; Univ RI, PhD(oceanog), 72. *Prof Exp:* Chemist, Inmont Corp, 67-68; from asst prof to assoc prof, 72-82, PROF OCEANOG, UNIV MICH, ANN ARBOR, 82- *Concurrent Pos:* Vis scientist, Ind Univ, 79-80; assoc ed, Geophys Res Letters, 79-81, Environ Geol, 80- & Org Geochem, 82-87; actg dir, Great Lakes & Marine Waters Ctr, Univ Mich, 82; mem, Adv Panel on Org Geochem, Joint Oceanog Inst Deep Earth Sampling, 80-83, Adv Panel on Sediments & Ocean Hist, 84-88; vis prof, Kyoto Univ, 90. *Mem:* AAAS; Am Soc Limnol Oceanog; Geochem Soc; Europ Asn Org Geochemists; fel Geol Soc Am; Am Asn Petrol Geol. *Res:* Organic geochemistry of water and sediments; distribution of fatty acids, alcohols and hydrocarbons in natural waters, sediments and organisms; paleoceanography; paleolimnology. *Mailing Add:* 1006 CC Little Bldg Univ of Mich Ann Arbor MI 48109-1063

MEYERS, PHILIP HENRY, b Chicago, Ill, Feb 24, 33; m; c 3. RADIOLOGY. *Educ:* Univ Minn, BA, 52, BS, 53, MD, 55; Am Bd Radiol, dipl. *Prof Exp:* Intern, Kings County Hosp, Brooklyn, 55-56; resident radiol, Bellevue Hosp, NY, 56-57; asst prof, 62-64, assoc prof, 64-77, CLIN ASSOC PROF RADIOL, SCH MED, TULANE UNIV, 77- *Concurrent Pos:* Fel diag radiol, NY Hosp, 59-60; Nat Cancer Inst fel radiation ther & radioisotopes, NY Univ-Bellevue Med Ctr, 60-61; vis radiologist, Charity Hosp, New Orleans, La, 62-; consult radiologist, St Barnabas Hosp Chronic Dis, NY, 62-; consult assoc scientist, Biomed Comput Ctr, Tulane Univ, 64-65. *Mem:* Am Col Radiol; Soc Nuclear Med; Radiol Soc NAm. *Mailing Add:* 3600 Prytania Suite 27 New Orleans LA 70115

MEYERS, ROBERT ALLEN, b Los Angeles, Calif, May 15, 36; m 76; c 4. COAL TECHNOLOGY, CHEMICALS PRODUCTION. *Educ:* San Diego State Col, BA, 59; Univ Calif, Los Angeles, PhD(chem), 63. *Prof Exp:* Fel, Calif Inst Technol, 63-64; sr res chemist, Bell & Howell Res Ctr, 64-66; head, Org Chem Sect, Chem & Chem Eng Dept, 66-73, MGR ENERGY & NATURAL RESOURCES, TRW, INC, 73- *Concurrent Pos:* Mem, US-USSR working group on air pollution control, 74-; Postdoctoral fel, Calif Inst Technol, 64. *Mem:* Am Chem Soc; Am Inst Chem Engrs. *Res:* Organic synthesis chemistry; hydrometallury of sulfur; organic synthesis; aromatic nucleophilic substitution; oxidative mechanisms; polymer synthesis; desulfurization of fossil fuels through chemical reaction; design and construction of pilot test units for fuel processing; author of 6 publications. *Mailing Add:* TRW Inc One Space Park Bldg 01-2270 Redondo Beach CA 90278-1001

MEYERS, SAMUEL PHILIP, b Asbury Park, NJ, Feb 21, 25; m 52; c 3. MARINE MICROBIOLOGY. *Educ:* Univ Fla, BS, 50; Univ Miami, MS, 52; Columbia Univ, PhD(bot), 57. *Prof Exp:* Res aide marine microbiol, Marine Lab, Univ Miami, 52-54, asst prof, 57-61, assoc prof, Inst Marine Sci, 61-68; PROF FOOD SCI & TECHNOL, LA STATE UNIV, BATON ROUGE, 68- *Mem:* AAAS; Mycol Soc; Am Soc Indust Microbiol; Am Soc Microbiol; Brit Soc Gen Microbiol. *Res:* Biology of marine fungi; microbial ecology; bionomics of marine yeasts; ecology of marine nematodes. *Mailing Add:* Dept Food Sci & Technol La State Univ 105 Food Sci Bldg Baton Rouge LA 70803-7505

MEYERS, VERNON J, b Pierre, SDak, Feb 6, 33; m 55; c 2. STRUCTURAL ENGINEERING. *Educ:* SDak Sch Mines & Technol, BS, 55; Purdue Univ, MS, 60, PhD(struct eng), 62. *Prof Exp:* Eng trainee, Gary Sheet & Tin Mill, US Steel Corp, 55-57, engr, 57-59; asst prof civil eng, 62-67, ASSOC PROF CIVIL ENG, PURDUE UNIV, WEST LAFAYETTE, 67- *Mem:* Am Soc Civil Engrs; Am Soc Eng Educ; Am Soc Testing & Mat; Sigma Xi. *Res:* Numerical computer analysis of structures; new methods of forming shell structures; use of new materials in construction. *Mailing Add:* Civil Eng Bldg Purdue Univ Lafayette IN 47907

MEYERS, WAYNE MARVIN, b Aitch, Pa, Aug 28, 24; m 53; c 4. MEDICAL MICROBIOLOGY, PATHOLOGY. *Educ:* Juniata Col, BS, 47; Univ Wis, MS, 53, PhD(microbiol), 55; Baylor Univ, MD, 59. *Hon Degrees:* DSc, Juniata Col, 86. *Prof Exp:* Res assoc, Univ Wis, 51-54; from asst to instr microbiol, Baylor Col Med, 54-59; intern med, Conemaugh Valley Mem Hosp, Johnstown, Pa, 59-60; staff physician, Berrien Gen Hosp, Berrien Ctr, Mich, 60-61; dir, Nyankanda Leprosarium, Burundi, Africa, 61-62; staff physician, Oicha Leprosarium, Beni, Repub Zaire, 62-64; med dir, Kivuvu Leprosarium at Inst Med Evangelique, Kimpese, Repub Zaire, 65-73; prof path, Univ Hawaii Sch Med, Honolulu, 73-75; chief div microbiol & registr leprosy, 75-89, mem AFIP res comt, 78-87, CHIEF, MYCOBACT & REGISTR LEPROSY, ARMED FORCES INST PATH, 89- *Concurrent Pos:* Allergy Found Am fel, 57 & 58; NIH/Leonard Wood Mem res fel path leprosy, Washington, DC, 68-69; consult, Am Leprosy Missions, Zaire, 76, German Leprosy Relief Asn, Hamburg, 80- & educ comt, Nat Bur Leprosy, Repub Zaire, 85-; mem, US Leprosy Panel, US-Japan Coop Med Sci Prog, 77-83; mem bd dirs, Int J Leprosy, 78-, Am Leprosy Missions, Inc, 79-88, chmn, 85-88 & Damien-Dutton Soc Leprosy Aid, Inc, 83-; mem, ad hoc study group, Nat Inst Allergy & Infectious Dis, NIH, 79-82 & 85; res affil, Delta Regional Primate Res Ctr, Tulane Univ, 81-; mem sci adv bd, Leonard Wood Mem, (Am Leprosy Found), 81-85, consult sci dir, 85-; mem, Res Adv Comt, Nat Hansen's Dis Ctr, La, 83-, chmn, 85-; mem, eval team sci res, Schieffelin Leprosy Res & Training Ctr, Karigiri, India, 85; wkshop on Multidrug Ther of Leprosy, Wurzburg, Ger, 86; mem, Planning Wkshop on Nat Prog for Training Leprosy Workers, Nyankunde, Zaire, 87; mem, Corp & Bd Dirs, Gorgas Mem Inst Tropical & Preventative Med, Inc, Washington, DC, 87-91; mem, Biotechnol Immunol Panel, FY 88 Coop Develop Res Prog Rev, US Agency Int Develop, Off Sci Adv, Washington, DC, 87; mem, Wkshop, Armauer Hansen Inst, Wurzburg, Ger, 87; mem, Gen Assembly, Int Fedn Antileprosy Asns, Paris, France, 88; secy-treas, Binford-Dammin Soc Infectious Dis Pathologists, 88-91; med consult,Am Leprosy Missions, Inc, 88-; rep, WHO Interregional Conf Leprosy Control in Africa, Brazzaville, Rep Congo, Int Leprosy Asn, 89; mem, abstract rev bd, Int Acad Path-US & Can Acad Path, Inc, 90-; mem exec comt, Gorgas Mem Inst Trop & Prev Med, Inc, Washington, DC & Panama, 90-91; mem standing comt, Int Fedn Antileprosy Asns, 90-; actg pres, VI Europ Leprosy Res Conf, Santa Margherita Ligure, Genoa, Italy, 90; mem exec bd, WHO, Geneva, 91. *Honors & Awards:* Except Performance Cert & Award, Armed Forces Inst Path, Dept Army, 85 & 88; Damien-Dutton Award, Damien-Dutton Soc Leprosy Aid, Inc, 90. *Mem:* Sigma Xi; Int Acad Path; Int Leprosy Asn; Int Soc Trop Dermat; Am Soc Trop Med & Hyg; Am Sci Affil; Am Soc Microbiol; NY Acad Sci; AAAS. *Res:* Leprosy; filariasis; Mycobacterium ulcerans infections; tropical and parasitic diseases. *Mailing Add:* Div Microbiol Armed Forces Inst Path Washington DC 20306-6000

MEYERS, WILLIAM C, b Vicksburg, Miss, Apr 3, 31; m 54; c 3. PALYNOLOGY, GEOLOGY. *Educ:* Southern Ill Univ, BA, 56; Univ Tulsa, MS, 63, PhD(earth sci), 77. *Prof Exp:* Res geologist, Sinclair Res Labs, 56-68; SR RES SCIENTIST, EXPLOR & PROD RES, CITIES SERV OIL CO, 70- *Mem:* Am Asn Stratig Palynologists; Int Comt Coal Petrol; Comt Paleozoic Palynology. *Res:* Stratigraphic palynology and sedimentary environmental analysis; thermal alteration of kerogen and its organic geochemical products. *Mailing Add:* 11470 E Sixth St Tulsa OK 74128

MEYERS-ELLIOTT, ROBERTA HART, b New York, NY, July 5, 37; m 78; c 2. IMMUNOLOGY. *Educ:* San Diego State Col, BS, 59; Univ Calif, Los Angeles, MS, 62, PhD(med microbiol, immunol), 64. *Prof Exp:* Immunologist, Sch Med, Univ Calif, Los Angeles, 60-62, immunochemist, 62-64; immunochemist, Calif Inst Technol, 64-67; asst prof ophthal, 73-75, assoc prof ophthal, 75-81, PROF OPHTHAL & JULES STEIN EYE INST, SCH MED, UNIV CALIF, LOS ANGELES, 81- *Concurrent Pos:* Nat Inst Allergy & Infectious Dis fel, 64-67; Nat Eye Inst res career develop award, 75- *Mem:* NY Acad Sci; Am Asn Immunol; Asn Res Vision & Ophthal; Reticuloendothelial Soc; Am Soc Microbiol. *Res:* Viral immunology and immunopathology; ocular immunology and inflammation; cellular immunity; autoimmune phenomena and ocular disorders. *Mailing Add:* 5425 S Stanford Dr Nashville TN 37215

MEYERSON, ARTHUR LEE, b East Orange, NJ, June 30, 38; m 61; c 1. MARINE GEOLOGY, MARINE GEOCHEMISTRY. *Educ:* Univ Pa, BA, 59; Lehigh Univ, MS, 61, PhD(geol), 71. *Prof Exp:* Instr geol, Upsala Col, 61-62; from instr to assoc prof, 62-75, chmn dept, 73-79, 82-90, PROF EARTH & PLANETARY ENVIRON, KEAN COL NJ, 75- *Mem:* Atlantic Estuarine Res Soc; AAAS; Geol Soc Am; Soc Econ Paleontologists & Mineralogists; Sigma Xi. *Res:* Holocene stratigraphy; estuarine geochemistry. *Mailing Add:* Dept Geol & Meteorol Kean Col of NJ Union NJ 07083

MEYERSON, BERNARD STEELE, b NY, June 2, 54. SOLID STATE PHYSICS. *Educ:* City Univ New York, PhD(physics), 81. *Prof Exp:* Fel, 80-81, RES STAFF MEM, IBM CORP, 81- *Concurrent Pos:* Adj lectr, physics, City Col New York, 76-80. *Mem:* Am Phys Soc. *Res:* Preparation and analysis of novel semiconducting materials, with emphasis on correlating preparation techniques with resulting transport phenomenon in such films. *Mailing Add:* 235 Calif Rd Yorktown Heights NY 10598

MEYERSON, MARK DANIEL, b Alexandria, Va, Feb 14, 49; m 69; c 2. TOPOLOGY. *Educ:* Univ Md, BS, 71; Stanford Univ, MS, 73, PhD(math), 75. *Prof Exp:* Vis lectr math, Univ Ill, Urbana, 75-78; PROF MATH, US NAVAL ACAD, 78- *Mem:* Am Math Soc; Math Asn Am. *Res:* Geometric topology. *Mailing Add:* Dept of Math US Naval Acad Annapolis MD 21402

MEYERSON, SEYMOUR, b Chicago, Ill, Dec 4, 16; m 43; c 2. CHEMISTRY, MASS SPECTROMETRY. *Educ:* Univ Chicago, SB, 38. *Prof Exp:* Chemist, Res Dept, Standard Oil Co, Ind, 46-61, chemist, Res & Develop Dept, Am Oil Co, 61-62, res assoc, Standard Oil Co, Ind, 62-72, sr res assoc, 72-80, res consult, 80-84; RETIRED. *Mem:* emer mem Am Chem Soc; emer mem Am Soc Mass Spectrometry. *Res:* Mass spectrometry of organic compounds and applications thereof to study of molecular structures and reaction mechanisms. *Mailing Add:* 650 N Tippecanoe St Gary IN 46403

MEYLAN, MAURICE ANDRE, b Cortland, NY, Mar 16, 42; m 83; c 2. GULF COAST GEOLOGY, MARINE MINERAL DEPOSITS. *Educ:* State Univ NY Buffalo, BA, 64; Fla State Univ, MS, 68; Univ Hawaii, PhD(oceanog), 78. *Prof Exp:* Geologist, Offshore Explor, Shell Oil Co, 67-72; lectr, Dept Geol Sci, Univ Wis, Milwaukee, 78-79; asst prof, 80-84, chmn dept, 81-84, ASSOC PROF GEOL, UNIV SOUTHERN MISS, 84- *Concurrent Pos:* Consult, Mining Ventures Div, Shell Oil Co, 76-77; co-investr, Dept Planning & Econ Develop, State Hawaii, 77-78; mem bd dir, Miss Mineral Resources Inst, 81-84. *Mem:* AAAS; Am Asn Petrol Geologists; Am Geophys Union; Geol Soc Am; Soc Econ Paleont Mineral; Sigma Xi. *Res:* Manganese nodule/crust substrate relations in the Southwestern Pacific; hydrocarbon distribution along the Heidelberg-Sand Hill Graben System of Mississippi & Lower Tuscaloosa diagenesis, Mississippi. *Mailing Add:* Dept Geol Southern Sta Box 9247 Hattiesburg MS 39406-9247

MEYN, RAYMOND EVERETT, JR, b Mobile, Ala, Aug 29, 42; m 68; c 2. BIOPHYSICS, RADIATION BIOLOGY. *Educ:* Univ Kans, BS, 65, MS, 67, PhD(radiation biophys), 69. *Prof Exp:* USPHS fel, 69-70, asst physicist, 71-75, assoc physicist & assoc prof biophys, 75-82, PHYSICIST & PROF BIOPHYS, UNIV TEX, M D ANDERSON CANCER CTR, HOUSTON, 82- *Concurrent Pos:* Assoc mem, Univ Tex Grad Sch Biomed Sci, Houston,

71-74, mem, Grad Fac, 74- *Mem:* Radiation Res Soc; Am Asn Cancer Res. *Res:* Replication and repair of DNA in mammalian cells; regulation of growth and division; programmed cell death. *Mailing Add:* Dept Exp Radiother Univ Tex M D Anderson Cancer Ctr Houston TX 77030

MEYRICK, BARBARA O, m; c 1. CELL BIOLOGY, CELL-TO-CELL INTERACTION. *Educ:* Univ London, PhD(exp path), 76. *Prof Exp:* PROF PATH & MED, SCH MED, VANDERBILT UNIV, 85- *Mem:* Am Thoracic Soc; Am Soc Cell Biol; Am Path Soc; Am Heart Asn; AAAS. *Res:* Lung disease. *Mailing Add:* Ctr Lung Res Rm B-1308 Sch Med Vanderbilt Univ Nashville TN 37232

MEYROWITZ, ALAN LESTER, b Brooklyn, NY, Feb 4, 45; m 71; c 2. ARTIFICIAL INTELLIGENCE. *Educ:* Univ Md, BS, 67, MS, 71; George Washington Univ, DSc(comput sci), 80. *Prof Exp:* Systs analyst, Exec off of the Pres of the US, 71-73; comput scientist, Fed Communications Comm, 73-80; COMPUT SCIENTIST, OFF NAVAL RES, 80- *Concurrent Pos:* Vis asst prof, Univ Md, 83-84. *Mem:* Am Asn Artificial Intelligence. *Res:* Artificial intelligence and robotics. *Mailing Add:* 6707 Wooden Spoke Rd Burke VA 22015-4185

MEYSTEL, ALEXANDER MICHAEL, b Leningrad, USSR, Feb 25, 35; US citizen; m; c 1. MULTIRESOLUTIONAL KNOWLEDGE REPRESENTATION, MULTIRESOLUTIONAL PLANNING & CONTROL. *Educ:* Polytech Inst, Odessa, USSR, MSEE, 57; Enims, Moscow, USSR, PhdEE(control theory), 65. *Prof Exp:* Sr scientist, Enims, Moscow, USSR, 64-65, dir lab, Yerevan, Armenia, 65-67, sr staff scientist, Moscow, USSR, 67-71; mgr dept, Informelectro, Moscow, USSR, 71-77; sr staff scientist, Gould, Inc, Rolling Meadows, Ill, 78-79; res & develop dir, Hyperloop, Inc, Chicago, Ill, 79-80; assoc prof, Univ Fla, Gainesville, 80-84; PROF, DREXEL UNIV, PHILADELPHIA, PA, 84- *Concurrent Pos:* Proj leader, SKB-3, Odessa, USSR, 57-63. *Mem:* Inst Elec & Electronics Engrs. *Res:* Intelligent computer architectures; cognitive systems; autonomous mobile robots; theory of multiresolutional decision making. *Mailing Add:* ECE Drexel Univ Philadelphia PA 19104

MEZEI, CATHERINE, b Budapest, Hungary, July 27, 31; Can citizen; m 54; c 1. BIOCHEMISTRY, PHARMACY. *Educ:* Univ Budapest, BS, 54; Univ BC, MS, 60, PhD(biochem), 64. *Prof Exp:* Fel, Ore State Univ, 64-67; from asst prof to assoc prof, 67-85, PROF BIOCHEM, DALHOUSIE UNIV, 85- *Concurrent Pos:* Med Res Coun Can res scholar, Dalhousie Univ, 67-73. *Mem:* Can Biochem Soc; Int Soc Neurochem; Am Soc Neurochem. *Res:* Biochemistry of nerve development. *Mailing Add:* Dept Biochem Dalhousie Univ Halifax NS B3H 4H7 Can

MEZEI, MICHAEL, b Mezokovesd, Hungary, Oct 7, 27; Can citizen; m 54; c 1. PHARMACEUTICS. *Educ:* Med Univ Budapest, Dipl pharm, 54; Ore State Univ, PhD(pharm), 67. *Prof Exp:* Instr pharm, Univ BC, 57-64 & Ore State Univ, 65-67; asst prof, 67-71, assoc prof, 71-79, PROF PHARM, DALHOUSIE UNIV, 79- *Mem:* Am Assoc Pharmaceut Sci; Acad Pharmaceut Sci; Can Pharmaceut Asn; NY Acad Sci. *Res:* Liposomes; formulation of dermatological preparations; biopharmaceutics; new drug delivery systems. *Mailing Add:* Dept Pharm Dalhousie Univ Col Pharm Halifax NS B3H 3J5 Can

MEZEI, MIHALY, b Budapest, Hungary, June 17, 44; US citizen; m 70. PHYSICAL CHEMISTRY. *Educ:* Eotvos Lorand Univ, Budapest, dipl chem, 67, PhD(chem), 72. *Prof Exp:* Res chemist, Hungarian Chem Ind Comput Ctr, 67-72; systs programmer, Young & Rubicam Int, Inc, 73-74; assoc res scientist, NY Univ, 74-76; adj assoc prof, 77-78 & 88-89, sr res assoc statist mech, 76-87, MGR BIOMOLECULAR COMPUT & GRAPHICS, HUNTER COL, 86- *Concurrent Pos:* Adj assoc prof, Manhattan Community Col, 78-87 & Seton Hall Univ, 80-81; mem grad fac, Grad Ctr, City Univ New York. *Mem:* Sigma Xi; Am Chem Soc; NY Acad Sci. *Res:* Statistical thermodynamics of molecular liquids; developments in Monte Carlo simulation methodology; computer simulation of aqueous solutions; free-energy calculation by Monte Carlo methods. *Mailing Add:* Dept Chem 695 Park Ave New York NY 10021

MEZEY, EUGENE JULIUS, b Cleveland, Ohio, Apr 9, 26; m 56; c 2. INORGANIC CHEMISTRY. *Educ:* Ohio Univ, BS, 50; Ohio State Univ, MS, 54, PhD(chem), 57. *Prof Exp:* Asst, Res Found, Ohio State Univ, 52-54, assoc, 54-55, res fel, 55-57; sr res chemist, Pittsburgh Plate Glass Co, 57-60, supvr explor inorganic group, 60-63; SR RES CHEMIST, COLUMBUS DIV, BATTELLE MEM INST, 63- *Concurrent Pos:* Instr, Univ Akron, 59-63; lectr, Ohio State Univ, 64-65, assoc, 65-66. *Mem:* Fel AAAS; Am Chem Soc; Am Ceramic Soc; Am Inst Chem; Int Microwave Power Inst; Sigma Xi. *Res:* Effluent control; minerals and metallurgical processes; fuels and fuel contaminants; technology assessment; hazardous materials disposal; process metallurgy of manganese nodules; plasmas and excited states; chemical process development; carbides; oxides; high energy processes; reactions induced with microwave energy and the use of microwaves in chemical processing. *Mailing Add:* 1516 Cardiff Rd Columbus OH 43221-3910

MEZEY, KALMAN C, b Nagyvarad, Hungary, Sept 18, 09; US citizen; m 35; c 3. MEDICINE, PHARMACOLOGY. *Educ:* Univ Basel, MD, 33. *Hon Degrees:* Dr, Univ Javeriana Bogota, 75. *Prof Exp:* Asst internal med, Univ Hosp, Univ Basel, 33-36; asst, Univ Clin, Univ Vienna, 36-37; prof pharmacol, Pontifical Univ Javeriana, Colombia, 42-58; VPRES MED SCI, MERCK SHARP & DOHME INT, 59- *Concurrent Pos:* Vis prof, Med Sch, Univ Vienna, 72; clin prof med, NJ Med Sch, 74-; attend physician, Vet Admin Med Ctr, East Orange, NJ, 74-; Biol Sci Bogota Award, 52-57. *Honors & Awards:* Cross of Boyaca Award, 57; Kalman C Mezey Chair of Pharmacol, Univ Javeriana Bogota, 75. *Mem:* Am Soc Pharmacol & Exp Therapeut; fel Am Col Clin Pharmacol & Chemother; Am Soc Trop Med & Hyg; AMA. *Res:* Pharmacology of drugs acting on the cardiovascular system; medicinal plants; arrow poisons; clinical pharmacology. *Mailing Add:* Dept Med UMDNJ 33 Hillside Ave Short Hills NJ 07078

MEZEY, PAUL G, b Nagyvarad, Hungary, Apr 28, 43; Can citizen. COMPUTATIONAL & MATHEMATICAL CHEMISTRY. *Educ:* Univ Budapest, MSc, 67, PhD(chem), 70, MSc, 72. *Hon Degrees:* DSc, Univ Sask, 85. *Prof Exp:* Res assoc chem, Hungarian Acad Sci, 67-73; res scientist & lectr, Univ Toronto, 73-77; from asst prof to assoc prof, 77-81, PROF CHEM, UNIV SASK, 81- *Concurrent Pos:* Lectr chem, Univ Budapest, 70-73; mem, sci bd, Int Soc Math Chem. *Mem:* Int-Am Photochem Soc; World Asn Theoret Org Chem; Int Soc Quantum Biol. *Res:* Reaction topology; quantum chemistry of molecular conformational changes and reactions; computer-based quantum chemical molecular design and synthesis planning; convergence properties of approximate molecular wave functions; quantum biochemistry. *Mailing Add:* Dept Chem Univ Sask Saskatoon SK S7N 0W0 Can

MEZGER, FRITZ WALTER WILLIAM, b Bryn Mawr, Oct 19, 28; m; c 4. AERONAUTICAL & ASTRONAUTICAL ENGINEERING, SOFTWARE SYSTEMS. *Educ:* Harvard Col, AB, 48; Univ Cincinnati, PhD(physics), 57. *Prof Exp:* Reactor physicist nuclear energy propulsion aircraft proj, Fairchild Co, 48-51; leader reactor physics group, Aircraft Nuclear Propulsion Dept, Space Div, Gen Elec, Co, 51-54, supvr Nuclear Anal Unit, 54-56, mgr Appl Math Sect, 56-59, Controls & Instrumentation Develop Sect, 59-60, Physics & Math Sect, 60-61, consult physicist, Space Sci Lab & proj scientist, Advan Space Proj Dept, 61-64, mgr space power & propulsion res, Gen Elec Co, 64-68, advan studies, 68-86, mgr resource planning & mgt, Advan Studies & mgr technol develop, Aerospace Group, 87-90; RETIRED. *Mem:* Am Phys Soc; Am Nuclear Soc; Asn Comput Mach; Inst Elec & Electronic Eng; Am Inst Aeronaut & Astronaut. *Res:* Plasma propulsion; magneto-hydrodynamic power generation; laser applications; reactor physics; nuclear engineering; electrodynamics; control engineering; digital computer applications. *Mailing Add:* 219 Hermitage Dr Radnor PA 19087

MEZGER-FREED, LISELOTTE, b Berlin, Germany, Feb 6, 26; US citizen; m 51; c 4. GENETICS, EMBRYOLOGY. *Educ:* Bryn Mawr Col, AB, 46; Wash Univ, MA, 48; Columbia Univ, PhD(zool), 52. *Prof Exp:* Instr biol, Brooklyn Col, 52-53; USPHS fel embryol, Inst Cancer Res, 54-57; instr, Temple Univ, 59-60; res assoc, Bryn Mawr Col, 61-66; res assoc, Inst Cancer Res, 66-70, asst mem, 70-77; prog dir cell biol, NSF, 78-79; res prof, Med Col Pa, 86-87; RETIRED. *Mem:* Am Soc Cell Biol; Am Asn Cancer Res; Genetics Soc Am; AAAS; Soc Develop Biol. *Res:* Gene expression in vertebrate haploid cell lines. *Mailing Add:* 710 Davidson Rd Philadelphia PA 19118

MEZICK, JAMES ANDREW, b Scranton, Pa, July 4, 39; m 70; c 3. DERMATOLOGY, SKIN BIOLOGY. *Educ:* Univ Scranton, BS, 61; Ohio State Univ, PhD(biochem), 69. *Prof Exp:* Fel biochem, St Louis Univ, 69-71; sr res scientist, Pharmaceut Res, Johnson & Johnson, 71-77, asst mgr, New Prod Develop, 77-78, Dermat Biol, 78, Drug Disposition & Metab, 78-79; group leader skin biol, Ortho Pharmaceut Corp, 79-80, group leader dermat, 80-82, res fel dermat pharmacol, 82-88; RES MGR, EXP THERAPEUT, R W JOHNSON PHARMACEUT RES INST, 88- *Honors & Awards:* Johnson Medal, 80. *Mem:* Soc Investigative Dermat; Am Acad Dermat; Am Soc Photobiol; AAAS. *Res:* Design and development of model systems to evaluate compounds for topical and systemic dermatological activity for anti-acne, anti-inflammatory, photodamage applications; retinoid drug discovery. *Mailing Add:* 43 Valley Forge Dr New Brunswick NJ 08816

MEZQUITA, CRISTOBAL, b Montán Castellón, Spain, Nov 30, 44; m 69; c 4. MOLECULAR BIOLOGY OF SPERMATOGENESIS, CHROMATIN STRUCTURE & GENE EXPRESSION. *Educ:* Univ Barcelona, MD, 67, PhD(molecular biol), 74. *Prof Exp:* Post doctoral cell biol, Baylor Col Med, Houston, Tex, 74-75, vis res assoc cell biol, 75-78; assoc prof, 78-81, PROF PHYSIOL, FAC MED, UNIV BARCELONA, SPAIN, 81-, DEAN, 91- *Mem:* Am Soc Cell Biol. *Res:* Nuclear proteins; chromatin structure and the control of gene expression; molecular biology of spermatogenesis; functions of ubiquitin and ubiquitin conjugates during cell differentiation. *Mailing Add:* Dept Physiol Fac Med Casanova 143 Barcelona 08036 Spain

MEZZINO, MICHAEL JOSEPH, JR, b Galveston, Tex, Sept 5, 40; m 65; c 3. MATHEMATICS. *Educ:* Austin Col, BA, 62; Kans State Col Pittsburg, MA, 63; Univ Tex, Austin, PhD(math), 70. *Prof Exp:* Res mathematician, Tracor Inc, Tex, 65-66; instr math, Univ Tex, Austin, 70; asst prof, Southwestern Univ Tex, 70-74; ASSOC PROF MATH, UNIV HOUSTON, CLEAR LAKE CITY, 74- *Concurrent Pos:* Consult, Tracor Inc, Tex, 66-, J & J Marine Diving Co, 74 & Rockwell Int, 80-; chmn dept math, Univ Houston, Clear Lake City. *Mem:* AAAS; Am Math Soc. *Res:* Mathematical modeling; computer graphics. *Mailing Add:* Chmn Dept Math Univ Houston Clear Lake City 2700 Bay Area Blvd Houston TX 77058

MGRDECHIAN, RAFFEE, b New York, NY, Aug 4, 27; m 56; c 3. PULSE TRANSFORMER DESIGN & MAGNETICS SYSTEMS INTEGRATION IN RADAR, HIGH VOLTAGE STRESS ANALYSIS & DESIGN. *Educ:* City Col NY, BEE, 50. *Prof Exp:* Design & test engr, Lewyt Corp, 50-54; chief engr, Carol Electronics, 54-56, Keystone Prod Corp, 56-58; gen mgr, Intecoil/Westbury Electronics, 58-61; vpres & opers mgr, Deflectronics Inc, 61-63; VPRES RES & DEVELOP, AXEL ELECTRONICS INC, 63- *Res:* Design of magnetics and high voltage, primarily radar transmitters for military; developed three channel magnetic switching modulator with core saturating technique to reduce fall time and help eliminate false echo's; author of three publications. *Mailing Add:* 2018 Ladenburg Dr Westbury NY 11590-5920

MI, MING-PI, b Shanghai, China, May 24, 33; m 65. GENETICS. *Educ:* Taiwan Univ, BS, 54; Univ Wis, MS, 59, PhD(genetics), 63. *Prof Exp:* NSF fel genetics, 63-64, asst geneticist, 64-65, from asst prof to assoc prof, 65-71, PROF GENETICS, UNIV HAWAII, 71- *Concurrent Pos:* NIH res grant, 64-67 & 81-83. *Mem:* Am Soc Human Genetics; Biometric Soc; Genetics Soc Am; Am Genetic Asn. *Res:* Statistical and population genetics. *Mailing Add:* Dept Genetics Univ Hawaii Manoa 2500 Campus Rd Honolulu HI 96822

MIALE, JOSEPH NICOLAS, b Johnston, RI, May 9, 19; m 55; c 3. PETROLEUM CHEMISTRY. *Educ:* Providence Col, BS, 40; Tex A&M Univ, MS, 47. *Prof Exp:* Instr chem, Tex A&M Univ, 47; res chemist, Res Dept, Mobil Res & Develop Corp, 47-62, sr res chemist, Cent Res Div, 62-84; RETIRED. *Mem:* Am Chem Soc; Int Cong Catalysis. *Res:* Exploratory research on processes and catalysts for hydrocarbon conversions; issued 95 patents. *Mailing Add:* 25 Merritt Dr Lawrenceville NJ 08648-3131

MIALL, ANDREW D, b Brighton, Eng, Mar 28, 44; Can citizen; m 69; c 2. CLASTIC SEDIMENTOLOGY, BASIN ANALYSIS. *Educ:* Univ London, BSc, 65; Univ Ottawa, PhD(geol), 69. *Prof Exp:* Consult geol mapping, J C Sproule & Assocs, Calgary, 69-71; subsurface geologist, petrol explor, Shell Can Ltd, 71-72; res scientist geol mapping, Geol Surv Can, Calgary, 72-79; assoc prof, 79-84, PROF GEOL, UNIV TORONTO, 84-, ACTG CHMN, DEPT, 86- *Concurrent Pos:* Vis prof, Univ Alta, 77; ed, Geosci Can, 82- *Mem:* Fel Geol Asn Can; Int Asn Sedimentologists; Am Asn Petrol Geologists; Soc Econ Paleontologists & Mineralogists. *Res:* Facies models for fluvial-deltaic sediments; tectonics and sedimentation; geology of Canadian Arctic Islands; methods of sedimentary basin analysis. *Mailing Add:* Dept Geol Univ Toronto Toronto ON M5S 1A1 Can

MIATECH, GERALD JAMES, b Stambaugh, Mich, Dec 31, 22; m 45; c 3. GEOPHYSICS, SPACE SCIENCES. *Educ:* Mich Technol Univ, BS, 49; St Louis Univ, MS, 56; Univ Wis, PhD(geophys), 61. *Prof Exp:* Geophysicist, M A Hanna Co, Mich, 49-50; geophysicist & aerospace engr, US Air Force, 50-65; res scientist, Ames Res Ctr, NASA, Moffett Field, Calif, 66-72; geophysicist & aerospace engr, ESL, Inc, Sunnyvale, Calif, 66-72, sr mem tech staff & eng specialist, 74-82; SR ENG SPECIALIST, FORD AEROSPACE & COMMUN CORP, PALO ALTO, CALIF, 82- *Mem:* AAAS; Am Geophys Union; Soc Explor Geophysicists; Sigma Xi; Am Inst Aeronaut & Astronaut; Am Fedn Astrologers. *Res:* Space systems; geophysical investigations for earth resources (vector aeromagnetometry). *Mailing Add:* PO Box 87 Round Mountain CA 96084

MICALE, FORTUNATO JOSEPH, b Niagara Falls, NY, Aug 11, 32; m; c 3. PHYSICAL CHEMISTRY, COLLOID CHEMISTRY. *Educ:* St Bonaventure Univ, BA, 56; Niagara Univ, BS, 59; Purdue Univ, MS, 61; Lehigh Univ, PhD(phys chem), 65. *Prof Exp:* Res asst prof, 66-70, assoc prof, 70-83, PROF PHYS CHEM, LEHIGH UNIV, 83- *Mem:* Am Chem Soc. *Res:* Colloid and surface properties of inorganic oxides, polmer latexes and carbon; dispersion stability; solution adsorption; printing ink research. *Mailing Add:* Ctr Surface & Coatings Res Sinclair Lab Lehigh Univ Bethlehem PA 18015

MICELI, ANGELO SYLVESTRO, b New York, NY, Dec 24, 13; m 38; c 6. PHYSICAL CHEMISTRY. *Educ:* Wayne Univ, BS, 34, MS, 36; Univ Mich, PhD(chem), 42. *Prof Exp:* From asst to instr chem, Wayne Univ, 34-42; sr scientist, US Rubber Co, 42-48, head chem res & develop, Uniroyal Int, 48-51, from asst mgr res & develop to mgr res & develop, 51-60, sect mgr tires, 60-62, dept mgr, 62-65, dir develop int div, 65-67, dir res & develop, 67-75, dir res & develop, 75-78, MGT CONSULT, UNIROYAL TIRE CO, 78- *Res:* Adhesion of rubber to metal; general industrial adhesives; electrodeposition of alloys, especially brass; resinoid and rubber-bonded grinding wheels; development of gum plastics; kinetics of isotopic exchange reaction; organizational planning; tire technology and product development. *Mailing Add:* 41 Deer Ridge Rd Norris TN 37828

MICELI, JOSEPH N, b New York, NY, Jan 13, 45. ENVIRONMENTAL TOXICOLOGY, THERAPEUTIC AGENTS. *Educ:* Univ Detroit, PhD(biochem), 71. *Prof Exp:* assoc prof pharmacol & pediat, Wayne State Univ, 79-; MORR HOUSE SCH MED, ATLANTA, GA. *Mem:* Soc Pediat Res; Am Asn Clin Chem; Am Soc Pharmacol & Exp Therapeut. *Mailing Add:* Morr House Sch Med 720 Westview Dr SW Atlanta GA 30310

MICELI, MICHAEL VINCENT, b Chicago, Ill, Aug 23, 51; m 74; c 2. OPHTHALMOLOGY, NUCLEAR MAGNETIC. *Educ:* Univ Ill, Champagne, BS, 73; Univ Ill Med Ctr, PhD(biochem), 79. *Prof Exp:* Food technologist, Quaker Oats Co, 73; res asst, Northwestern Univ Med Ctr, 74-75; teaching asst biochem, Univ Ill Med Ctr, 75-79, res assoc, 79-80; fel, Johns Hopkins Univ, 80-83, instr ophthal, 84-86, asst prof ophthal, 86-90; SR STAFF SCIENTIST, TOURO INFIRMARY, LA STATE UNIV, 90- *Concurrent Pos:* Vis scientist, USA-USSR Exchange, Nat Heart Lung & Blood Inst Prog Area 3, Myocardial Bioenergetics, 83; prin investr, metab retinal pigment epithelium, NIH, 87-; adj assoc prof, Dept Ophthal, Tulane Univ. *Mem:* Asn Res Vision & Opthal; Biophys Soc; Sigma Xi; Soc Magnetic Res Med. *Res:* Ocular metabolism and biochemistry studies by cell biology and nuclear spectroscopic techniques; projects include retinal pigment epithelium metabolism, free radical damage and aging of the retina. *Mailing Add:* Touro Infirmary 1401 Foucher St New Orleans LA 70115

MICETICH, RONALD GEORGE, b Madras, India, May 28, 31; Can citizen; m 58; c 4. ORGANIC CHEMISTRY, MEDICINAL CHEMISTRY. *Educ:* Loyola Col, Madras, India, BSc, 52; Univ Madras, MA, 55; Univ Sask, PhD(org chem), 62. *Prof Exp:* Nat Res Coun Can fel, Prairie Regional Lab, Sask, Can, 62-63; res chemist, R&L Molecular Res Ltd, Alta, 63-69; from asst res dir to assoc res dir org & med chem, Raylo Chem Ltd, 69-71, res mgr, 71-75, actg res dir org & med chem, 75-78, res dir pharm chem, 78-81; PROF MED CHEM, UNIV ALTA, 81- *Mem:* Royal Soc Chem; Chem Inst Can; Int Soc Heterocyclic Chem; Am Chem Soc. *Res:* Heterocyclic chemistry; synthetic organic chemistry; organometallic chemistry; chemical modification of B-lactam antibiotics; anti-inflammatory agents; analgesics; immunoregulants; central nervous system active compounds; affinity drugs; antiviral agents; lectius; sulfur chemistry; quindone antibiotics. *Mailing Add:* Synphar Labs Inc 4290 91A St Edmonton AB T6E 5V2 Can

MICH, THOMAS FREDERICK, b Milwaukee, Wis, May 26, 39; m 66. ORGANIC CHEMISTRY. *Educ:* Marquette Univ, BS, 61; Northwestern Univ, MS, 64; State Univ NY Buffalo, PhD(chem), 68. *Prof Exp:* Res assoc, Dartmouth Col, 67-68; sr res chemist, Monsanto Co, 68-69; res chemist, chem dept, 69-73, sr scientist, 73-79, res assoc, 79-81, sr res assoc, 81-83, sect dir antiinfectives, 83-85, sect dir, chem develop, 85-88, sr dir, 88-89, VPRES CHEM DEVELOP, PHARMACEUT RES DIV, WARNER LAMBERT/ PARKE DAVIS & CO, 89- *Mem:* Am Chem Soc; Royal Soc Chem. *Res:* Synthetic organic and medicinal chemistry; antibacterials. *Mailing Add:* Chem Develop Ann Arbor Res Labs Parke Davis & Co 2800 Plymouth Rd Ann Arbor MI 48106

MICHA, DAVID ALLAN, b Villa Mercedes, S Luis, Arg; US citizen; m 65; c 2. CHEMICAL PHYSICS. *Educ:* Nat Univ Cuyo, Arg, Lic Physics, 62; Univ Uppsala, Sweden, Fil Lic, 65, Fil Doctor, 66. *Prof Exp:* Res assoc, Theoret Chem Inst, Univ Wis-Madison, 66-67; asst res physicist, Inst Pure & Appl Phys Sci, Univ Calif, San Diego, 67-69; assoc prof, 69-74, PROF CHEM & PHYSICS, UNIV FLA, 74-, DIR, CTR CHEM PHYSICS, 82- *Concurrent Pos:* Res grants, Am Chem Soc-Petrol Res Fund, 69-72, NSF, 71-73, & 76-92, Nat Res Coun, 71 & 75, Alfred P Sloan fel, 71-73, NATO, 76-78, Nat Resource Comput in Chem, 78 & NASA, 81, Nat Bur Standards & Joint Inst Lab fel, 83; vis prof, Harvard Univ, 72 & 90, Univ Calif, San Diego, 73, Max Planck Inst fur Stromungsforsch, Gottingen, 76, Uppsala Univ, 77 & Imp Col, Univ London, 77, Univ Calif, Santa Barbara, 82, Weizmann Inst, 83, Joint Inst Lab Astrophys, Univ Colo, 83, Oak Ridge Nat Lab, 84-86 & Univ Buenos Aires, 87-88; vis Lamberg Prof, Univ Gothenburg, 70; NSF Adv Comt Advan Sci Computing, 90-92. *Honors & Awards:* Sigma Xi Award, 85; US Sr Scientist Award, A Von Humboldt Found, 76. *Mem:* Fel Am Phys Soc; Am Chem Soc; Sigma Xi. *Res:* Molecular dynamics; electronic structure of matter; intermolecular forces; computational methods in theoretical chemistry. *Mailing Add:* Williamson Hall Univ Fla Gainesville FL 32611

MICHAEL, ALFRED FREDERICK, JR, b Philadelphia, Pa, Aug 10, 28; m 52; c 3. PEDIATRICS, NEPHROLOGY. *Educ:* Temple Univ, MD, 53. *Prof Exp:* Intern, Philadelphia Gen Hosp, 53-54; resident pediat, St Christopher's Hosp Children, Sch Med, Temple Univ, 54-55; from jr resident to sr resident, Children's Hosp & Col Med, Univ Cincinnati, 57-59, chief resident & instr, 59-60; USPHS fel, 60-63, Am Heart Asn estab investr, 63-68, assoc prof, 65-68, PROF PEDIAT, LAB MED & PATH, MED SCH, UNIV MINN, MINNEAPOLIS, 68- , REGENTS PROF & HEAD DEPT PEDIAT, 86- *Concurrent Pos:* Vis investr & Guggenheim fel, Copenhagen, Denmark, 66-67; mem sci adv bd, Nat Kidney Found, 78-82; mem & pres, subspecialty bd pediat nephrol, Am Bd Pediat, 73-79; cert diag lab immunol, 86; Coun Am Soc Nephrol, 88-; bd dirs, Nat Asn Children's Hosps & Related Insts. *Mem:* Soc Pediat Res; Am Asn Path; Am Asn Immunol; Am Soc Nephrology; Am Soc Pediat Nephrology; Asn Am Physicians; Am Soc Clin Invest; Am Fedn Clin Res; Am Pediat Soc. *Res:* Renal disease; immunopathology and mechanisms of kidney disease; basement membranes; diagnostic laboratory immunology. *Mailing Add:* Dept Pediat Univ Minn Hosps Minneapolis MN 55455

MICHAEL, ARTHUR B, b Chilton, Wis, Nov 24, 23. PHYSICAL METALLURGY. *Educ:* Univ Wis, BS, 44; Univ Minn, MS, 47; Mass Inst Technol, ScD, 52. *Prof Exp:* Res engr, Allis Chalmers Mfg Co, 52-55; sr metallurgist, Fansteel Metall Corp, Ill, 55-57, asst dir res, 57-59, dir res, 59-63; mgr metals res, Glidden Co, SCM Corp, 63-68; PROF MECH & MAT ENG & CHMN DEPT MECH ENG, MILWAUKEE SCH ENG, 68- *Res:* Refractory metals; powder metallurgy; electronic materials; materials science. *Mailing Add:* Dept Eng Milwaukee Sch Eng 1025 N Milwaukee Milwaukee WI 53201

MICHAEL, CHARLES REID, b Bucyrus, Ohio, June 30, 39; m 64; c 2. NEUROPHYSIOLOGY. *Educ:* Harvard Col, BA, 61; Harvard Univ, PhD(biol), 65. *Prof Exp:* Fel biophys, Johns Hopkins Univ, 65-68; asst prof, 68-71, PROF PHYSIOL, SCH MED, YALE UNIV, 71- *Mem:* AAAS; Soc Neurosci; Asn Res Vision & Ophthal. *Res:* Vision physiology of the mammalian central nervous system. *Mailing Add:* Dept Physiol Yale Univ Sch Med New Haven CT 06510

MICHAEL, EDWIN DARYL, b Mannington, WVa, Jan 22, 38; m 60; c 2. WILDLIFE MANAGEMENT. *Educ:* Marietta Col, BS, 59; Tex A&M Univ, MS, 63, PhD(wildlife ecol), 66. *Prof Exp:* From asst prof to assoc prof biol, Stephen F Austin State Col, 64-70; assoc prof, 70-74, PROF WILDLIFE MGT, WVA UNIV, 74-, ASSOC WILDLIFE BIOLOGIST, 77- *Mem:* Am Soc Mammal; Wildlife Soc; Copper Ornith Soc. *Res:* Ecology and management of forest wildlife. *Mailing Add:* Div of Forestry WVa Univ Morgantown WV 26506

MICHAEL, ERNEST ARTHUR, b Zurich, Switz, Aug 26, 25; nat US; m 56, 66; c 5. TOPOLOGY. *Educ:* Cornell Univ, BA, 47; Harvard Univ, MA, 48; Univ Chicago, PhD(math), 51. *Prof Exp:* Fel, AEC, Inst Advan Study, 51-52 & Univ Chicago, 52-53; from asst prof to assoc prof, 53-60, PROF MATH, UNIV WASH, 60- *Concurrent Pos:* Mem, Inst Advan Study, 56-57, 60-61 & 68; Guggenheim Found Grant, 60-61; ed, Proc, Am Math Soc, 68-71 & Gen Topology & Applns, 72-; vis, Math Res Inst, Swiss Fed Inst Technol, 73-74; Alexander von Humboldt Found sr Am sci grant, Univ Stuttgart, 78-79; vis prof, Univ Munich, 87-88. *Mem:* Am Math Soc; Math Asn Am. *Res:* General topology. *Mailing Add:* Dept Math Univ Wash Seattle WA 98195

MICHAEL, ERNEST DENZIL, JR, b Lewiston, Maine, Jan 18, 22; m 45; c 3. PHYSIOLOGY, ERGONOMICS. *Educ:* Purdue Univ, BPE, 47; Univ Ill, MS, 49, PhD, 52. *Prof Exp:* Dir athletics high sch, 47-48; instr phys ed, Univ Ill, 49-50 & 51-52; asst, USPHS, 50-51; from instr to assoc prof phys educ & physiol, 52-67, chmn dept ergonomics & phys educ, 73-76, PROF PHYS EDUC, UNIV CALIF, SANTA BARBARA, 67-, RES ASSOC, INST ENVIRON STRESS, 64- *Concurrent Pos:* Am Physiol Soc res fel, Lankenau Hosp, Philadelphia, Pa, 59; res fel, Valley Forge Heart Hosp, 59-60; consult, Water Safety Prog, YMCA, 59-; res prof, Inst Physiol, Univ Glasgow, Scotland, 67-68; Am Acad Phys Educ fel, res prof, Inst Sport, Warsaw, Poland, 78. *Mem:* Am Physiol Soc; Sigma Xi (pres, 78-79); fel Am Col Sports Med; Am Asn Health, Phys Educ & Recreation; NY Acad Sci. *Res:* Effects of training on the cardiovascular system; perception of levels of exertion; body composition and performance. *Mailing Add:* 323 Pebble Hill Ten Santa Barbara CA 93106

MICHAEL, HAROLD LOUIS, b Columbus, Ind, July 24, 20; wid; c 5. TRANSPORTATION ENGINEERING. *Educ:* Purdue Univ, BS, 50, MS, 51. *Prof Exp:* Asst eng transp, 50-51, from instr to assoc prof, 52-62, from asst dir to assoc dir, Joint Hwy Res Proj, 54-77, head, Transp & Urban Eng, 66-78, PROF CIVIL ENG, PURDUE UNIV, W LAFAYETTE, 62-, DIR JOINT HWY RES PROJ, 77-, HEAD SCH CIVIL ENG, 78- *Concurrent Pos:* Mem comt vehicle characteristics & comt origin-destination surv, Hwy Res Bd, Nat Acad Sci-Nat Res Coun, 54-63, chmn comt characteristics traffic flow, 57-63, chmn dept traffic & opers, 64-69, chmn group 3 coun, Oper & Maintenance Transp Facil, 70-76; chmn, W Lafayette Traffic Comn, 56-; consult, Ind Bd Regist Prof Engrs, 56-76, Indianapolis Motor Speedway, 58-60 & USPHS, 65-67; vpres, Ind Hwy for Survival, Inc, 60-64, pres, 65-88; chmn, Lafayette Hwy Tech Comt, 65- & Nat Comt Uniform Traffic Control Devices, 65-74; mem, Nat Comt Uniform Laws & Ord, chmn, 90. *Honors & Awards:* Roy W Crum Award, 78; Theodore M Matson Award, 79; James Laurie Prize, Am Soc Civil Engrs, 81; George S Bartlett Award, 82; Burton W Marsh Award, Inst Transp Engrs, 84. *Mem:* Nat Soc Prof Engrs; Nat Acad Eng; Am Soc Civil Engrs; Inst Traffic Eng (pres, 75); Am Rwy Eng Asn; Am Soc Eng Educ; hon mem, Inst Transp Engrs. *Res:* Traffic engineering; transportation planning and economics; urban transportation planning, traffic safety and urban planning. *Mailing Add:* Civil Eng Bldg West Lafayette West Lafayette IN 47907

MICHAEL, IRVING, b Pittsburgh, Pa, June 28, 29. SPACE PHYSICS, NUCLEAR PHYSICS. *Educ:* George Washington Univ, BSc, 50; Univ Wis, MSc, 51, PhD(physics), 58. *Prof Exp:* Res asst phys chem, Geophys Lab, Carnegie Inst Washington, 50; res asst nuclear physics, 51-58, res assoc, Univ Wis, 58-60; res fel, Univ Notre Dame, 60-62; sr scientist, Northrop Space Labs, Calif, 62-63; res physicist space physics br, Air Force Weapons Lab, NMex, 64; RES PHYSICIST SPACE PHYSICS LAB, AIR FORCE GEOPHYSICS LAB, 64- *Concurrent Pos:* Consult, Radiation Dynamics Corp, NY, 60. *Mem:* Am Phys Soc; Am Geophys Union; Am Vacuum Soc; Int Orgn Vacuum Sci & Technol; Am Soc Mass Spectrometry; Am Meteorol Soc. *Res:* Analysis of satellite plasmas and fields measurements in near-Earth space; low-energy nuclear physics; electrostatic accelerator development; high-voltage breakdown in vacuum; techniques and measurements in extreme-high vacuum; measurements of electric fields in space. *Mailing Add:* 19 Gould Rd Bedford MA 01730

MICHAEL, JACOB GABRIEL, b Rimavska Sobota, Czech, July 2, 31; US citizen; m 58; c 4. IMMUNOLOGY. *Educ:* Hebrew Univ, Israel, BA, 55, MSc, 56; Rutgers Univ, PhD(microbiol), 59. *Prof Exp:* Res assoc & vis scientist, Nat Cancer Inst, 59-61; res assoc, Harvard Med Sch, 61-66; assoc prof, 66-73, PROF MICROBIOL & IMMUNOL, MED CTR, UNIV CINCINNATI, 73- *Concurrent Pos:* Am Soc Microbiol pres fel, 62; USPHS career develop award, 65-73; vis prof, Karolinska Inst, 71, Inst Pasteur, 79 & Univ London, 84. *Mem:* AAAS; Am Soc Microbiol; Am Asn Immunol; NY Acad Sci; Am Acad Microbiol. *Res:* Microbial immunity; regulation of immune response; immediate hypersensitivity. *Mailing Add:* Dept Microbiol Univ Cincinnati Med Ctr Cincinnati OH 45267

MICHAEL, JAMES RICHARD, organic chemistry, for more information see previous edition

MICHAEL, JOE VICTOR, b South Whitley, Ind, Oct 2, 35; m 86; c 4. PHYSICAL CHEMISTRY, CHEMICAL DYNAMICS. *Educ:* Wabash Col, BA, 57; Univ Rochester, PhD(chem), 63. *Prof Exp:* Res assoc, Harvard Univ, 62-64 & Brookhaven Nat Lab, 64-65; asst prof, Carnegie-Mellon Univ, 65-70, assoc prof chem, 70-75; Nat Acad Sci/Nat Res Coun Sr Resident Res Assoc, Goddard Space Flight Ctr, NASA, 75-77; chemist, Brookhaven Nat Lab, 81-87; CHEMIST, ARGONNE NAT LAB, 87- *Concurrent Pos:* Vis prof, Cath Univ Am, 77-81. *Mem:* NY Acad Sci; Am Chem Soc; Sigma Xi; Combustion Inst; AAAS; Int Am Photochemical Soc; Am Geophys Union. *Res:* Photochemistry; chemical kinetics; time-of-flight mass spectroscopy; shock tubes; flow reactors; resonance photometry and fluorescence. *Mailing Add:* Bldg 200 D-183 Argonne Nat Lab Argonne IL 60439

MICHAEL, JOEL ALLEN, b Chicago, Ill, Mar 8, 40; m 65; c 2. COMPUTER-BASED EDUCATION. *Educ:* Calif Inst Technol, BS, 61; McGill Univ, MSc, 64; Mass Inst Technol, PhD(physiol), 65. *Prof Exp:* Carnegie fel neurophysiol, Nat Phys Lab, Teddington, Eng, 65-66; res fel psychiat, Mass Gen Hosp & Harvard Med Sch, Boston, 66-67; asst prof bioeng, Univ Ill, Chicago, 67-70, asst prof physiol, Col Med, 68-70; assoc prof biomed eng & neurol sci, 70-74, actg chmn dept, 74-76, assoc prof, 74-91, PROF PHYSIOL, RUSH MED COL, 91- *Concurrent Pos:* Asst attend bioengr, Presby-St Luke's Hosp, Chicago, 67-70. *Mem:* Am Physiol Soc; Inst Elec & Electronics Eng; Am Educ Res Asn; Asn Develop Comput-based Instr; Cognitive Sci Soc. *Res:* Problem-solving; techniques of tutoring; intelligent tutoring systems. *Mailing Add:* Dept Physiol Rush Med Col Rush-Presby St Luke's Med Ctr 1753 W Congress Chicago IL 60612

MICHAEL, LESLIE WILLIAM, b San Francisco, Calif, Jan 26, 33; m 54. CHEMISTRY. *Educ:* Univ Calif, Berkeley, BA, 58; Fresno State Col, MS, 62; Univ Cincinnati, PhD(chem), 69. *Prof Exp:* Technician, Ortho Div, Chevron Chem Co, 58-60; AMA Educ Res Fund grant environ health, 68-69, res assoc chem, Col Med, Univ Cincinnati, 69-70, from asst prof to assoc prof environ health, 70-77; supvr, 77-80, area dir, 80-, SR INDUST HYGIENIST, OCCUP SAFETY & HEALTH ADMIN, DEPT LABOR. *Mem:* Am Chem Soc; Am Indust Hyg Asn. *Res:* Origin and chemistry of environmental materials with biological interactions; qualitative and quantitative chemical analysis; industrial hygiene; structure and physical chemical constants particularly with regard to essential and toxic metal compounds. *Mailing Add:* Occup Safety & Health Admin 3rd Floor Wing C US Dept Labor 395 Oyster Point Blvd South San Francisco CA 94080

MICHAEL, LLOYD HAL, b Susquehanna, Pa, Nov 14, 42; m 64; c 1. CARDIOVASCULAR PHYSIOLOGY, BIOCHEMISTRY. *Educ:* Moravian Col, BS, 64; Kent State Univ, MS, 66; Univ Ottawa, PhD(med physiol), 73. *Prof Exp:* Instr biol, St Lawrence Univ, 66-67, asst prof, 67-69; sr lectr physiol, Fac of Med, Univ Dar es Salaam, 73-75; instr med, 77-78, ASST PROF MED, SECT CARDIOVASC SCI, BAYLOR COL MED, 78- *Concurrent Pos:* Fel clin immunol, Col Physicians & Surgeons, Columbia Univ, 75; fel myocardial biol, Baylor Col Med, 76-77. *Mem:* Sigma Xi; Int Soc Heart Res; Am Heart Asn. *Res:* Investigation of processes causing faster and shorter duration of heart atrial contraction compared to ventricular contraction; studies to characterize mechanism of sodium-potassium-adenosine triphosphate function and specifically its unique interaction with glycoside. *Mailing Add:* Sect Cardiovasc Sci-Med Baylor Col of Med Houston TX 77030

MICHAEL, MAX, JR, b Athens, Ga, Feb 14, 16; m 44; c 5. INTERNAL MEDICINE. *Educ:* Univ Ga, BS, 35; Harvard Univ, MD, 39. *Prof Exp:* Asst med, Sch Med, Johns Hopkins Univ, 41-42; asst, Emory Univ, 45-47, assoc, 47-50, from asst prof to assoc prof, 52-54; prof, Col Med, State Univ NY, 54-58; exec dir, Jacksonville Hosps, Univ Fla, 58-67, prof med, Col Med, 58-77, asst dean educ prog, 67-77, asst dean Jacksonville prog, 77-90; RETIRED. *Concurrent Pos:* Chief med serv, Vet Admin Hosp, Atlanta, 47-54; dir med serv, Maimonides Hosp, Brooklyn, 54-58; mem bd regents, Nat Libr Med, 68-72. *Mem:* Inst Med-Nat Acad Sci; Am Tuberc Soc; Am Soc Clin Invest; Am Clin & Climat Asn; assoc Am Col Physicians. *Res:* Infectious diseases. *Mailing Add:* 3649 Montclair Dr Jacksonville FL 32217

MICHAEL, NORMAN, b New York, NY, Dec 12, 31; m 63; c 1. INORGANIC CHEMISTRY, MECHANIC ENGINEERING. *Educ:* Columbia Univ, AB, 55. *Prof Exp:* Radiol chemist, US Navy Mat Lab, 54-56; nuclear chemist, Alco Prod Inc, 56-57; scientist, Atomic Power Div, Westinghouse Elec Corp, 57-62; sr res chemist, Astropower Lab, Douglas Aircraft Co, 62-63; chemist, Vallecitos Atomic Lab, Gen Elec Co, 63-66; SR ENGR, STC, WESTINGHOUSE ELEC CORP, 66- *Mem:* Am Chem Soc. *Res:* Oxidation, corrosion and radioactive contamination of metals and alloys; alternate energy systems; fuel cell electrolytes and battery separators; inorganic ion exchangers for water purification. *Mailing Add:* 1080 Evergreen Dr Pittsburgh PA 15235

MICHAEL, PAUL ANDREW, b New York, NY, July 6, 28; m 53; c 2. PHYSICS. *Educ:* NY Univ, AB, 49, PhD, 59; Univ Chicago, BS, 53, MS, 55. *Prof Exp:* Jr test engr, Curtiss-Wright Corp, 53, physicist, Res Div, 55-56; physicist, 58-72, leader meteorol group, 72-75, HEAD ATMOSPHERIC SCI DIV, BROOKHAVEN NAT LAB, 75- *Concurrent Pos:* Instr, NY Univ, 56-58, adj asst prof, 58-59. *Mem:* Am Phys Soc; Am Nuclear Soc; Am Meteorol Soc. *Res:* Reactor and neutron physics; fluid dynamics; atmospheric diffusion. *Mailing Add:* Dept Energy & Environ Brookhaven Nat Lab Upton NY 11973

MICHAEL, RICHARD PHILLIP, b London, Eng, June 9, 24; US citizen; m 58; c 4. PSYCHIATRY, NEUROENDOCRINOLOGY. *Educ:* Univ London, MD, 51, PhD(neuroendocrinol), 60, DSc, 71; FRCPsych, 74. *Prof Exp:* Dir, Primate Behav Res Labs, 67-72; DIR, BIOL PSYCHIAT RES LABS, GA MENTAL HEALTH INST, 72-, PROF ANAT & PSYCHIAT, SCH MED, EMORY UNIV, 71- *Concurrent Pos:* Vis scientist, Clin Neuropharmacol Res Ctr, St Elizabeth's Hosp, Washington, DC, 59-, vis prof, Regional Primate Res Ctr, Univ Wis-Madison, 62-, consult physician psychiat, Bethlehem Royal & Maudsley Hosps, 63-72; reader, Inst Psychiat, Univ London, 68-72; rapporteur, WHO Sci Group Neuroendocrinol & Human Reproduction, Geneva, 64; consult, Brain & Behav Monograph, Sci Policy Studies, Orgn for Econ Coop & Develop, Paris, 72-; mem comn neuroendocrinol, Int Union Physiol Sci, 75-81. *Honors & Awards:* Distinction Award, Ministry Health, UK, 66; Manfred Sakel Award, Soc Biol Psychiat, 72. *Mem:* Primate Soc Gt Brit (pres, 70-73); Int Soc Psychoneuroendocrinol (pres, 75-78); Int Psychoanal Asn; Endocrine Soc; Am Psychoanal Asn. *Res:* Effects of hormones on the brain; biological aspects of psychiatry; ethological studies on the basis of motivation in higher primates; neuroendocrine and behavioral interrelationships. *Mailing Add:* Dept Psychiat Emory Univ Sch Med Atlanta GA 30322

MICHAEL, SANDRA DALE, b Sacramento, Calif, Jan 23, 45. BIOLOGICAL SCIENCES, GENETICS. *Educ:* Calif State Col, Sonoma, BA, 67; Univ Calif, Davis, PhD(genetics), 70. *Prof Exp:* NIH Health Sci Advan Award fel endocrinol, Univ Calif, Davis, 70-73, asst res geneticist, 73-74; from asst prof to assoc prof, 74-88, PROF BIOL SCI, UNIV CTR, STATE UNIV NY, BINGHAMTON, 88- *Concurrent Pos:* Prin investr, Pub Health Serv, Nat Cancer Inst, NIH & NSF grants, 76- *Mem:* Endocrine Soc; Soc Study Reproduction; Soc Study Fertil; Soc Exp Biol & Med; NY Acad Sci; Sigma Xi. *Res:* Role of thymus gland in maturation, aging and carcinogenesis of the female reproductive system; reproduction. *Mailing Add:* Dept Biol Sci State Univ NY PO Box 6000 Binghamton NY 13902-6000

MICHAEL, THOMAS HUGH GLYNN, b Toronto, Ont, May 20, 18; m 42; c 3. CHEMISTRY, SCIENCE ADMINISTRATION. *Educ:* Univ Toronto, BA, 40. *Prof Exp:* Chemist, Ont Res Found, 40-41 & Protective Coatings Lab, Nat Res Coun Can, 41-46; chief chemist, Woburn Chems, Ltd, 46-53; dir res, Howards & Sons Can, Ltd, 53-58; exec dir & secy, Chem Inst Can, 58-85; RETIRED. *Concurrent Pos:* Treas, Youth Sci Found, 61-71; consult, sci soc & conf orgn & mgt. *Honors & Awards:* Can Centennial Medal, 67; Can Silver Jubilee Medal, 78. *Mem:* Fel Royal Soc Chem; Am Chem Soc; fel Chem Inst Can (treas, 53-56); Coun Eng & Sci Soc Exec (pres, 69-70); Can Soc Asn Execs (pres, 71-72). *Res:* Chemistry of synthetic resins, particularly alkyds and plasticizers; chemistry of protective coatings; management of scientific societies and publications. *Mailing Add:* 702-370 Dominion Ave Ottawa ON K2A 3X4 Can

MICHAEL, WILLIAM ALEXANDER, mathematics, for more information see previous edition

MICHAEL, WILLIAM HERBERT, JR, b Richmond, Va, Dec 10, 26; m 52; c 2. SPACE SCIENCES. *Educ:* Princeton Univ, BS, 48, MS, 64, PhD(aerospace sci), 67; Univ Va, MS, 51; Col William & Mary, MA, 62. *Prof Exp:* Res scientist, Nat Adv Comt Aeronaut-NASA, Langley Res Ctr, 48-58, head trajectory analysis group, 58-60, head mission analysis sect, 60-69, head lunar & planetary sci br, 69-70, chief, environ & space sci div, 70-76, sci/eng consult, 76-80; assoc dir, Va Assoc Res Campus, Col William & Mary, Newport News, Va, 80-83; exec dir, eng educ study, 83-85, DIR, SPACE APPLNS BD, NAT RES COUN, 85- *Concurrent Pos:* Prin investr, Lunar Orbiter Selenodesy Exp, Langley Res Ctr, NASA, 65-69, team leader, Viking Mars Missions Radio Sci Team, 69-80; mem work groups figure & motion of moon & laser tracking & appl, Comn 17, Int Astron Union & tracking & dynamics of satellites, Comt on Space Res; adj prof physics, Col William & Mary, 80- *Honors & Awards:* NASA Spec Serv Award, 67, Lunar Orbiter Proj Achievement Award, 68, Apollo Prog Spec Achievement Award, 69; NASA Medal for Except Sci Achievement, 77; I B Laskowitz Aerospace Sci Award, NY Acad Sci, 87. *Mem:* Am Inst Aeronaut & Astronaut; Am Geophys Union; Int Astron Union. *Res:* lunar and planetary exploration; space flight experiments in gravitational fields, geophysics, atmospheric properties and radio science; research administration. *Mailing Add:* 29 Haughton Lane Newport News VA 23606

MICHAEL, WILLIAM R, b Peoria, Ill, May 23, 30; m 53; c 4. TOXICOLOGY, PHARMACOLOGY. *Educ:* Univ Ill, BS, 52; Bradley Univ, MS, 57; St Louis Univ, PhD(biochem), 61. *Prof Exp:* Chemist, Northern Utilization Res Lab, USDA, 54-56; res biochemist, Miami Valley Lab, Procter & Gamble Co, 61- 74, sect head prod develop, 74-90; RETIRED. *Mem:* AAAS; Am Chem Soc; Soc Toxicol; Sigma Xi. *Res:* Drug metabolism; calcium and phosphate metabolism; pharmacokinetics; toxicology; product development. *Mailing Add:* 7432 Pinebrook Dr Cincinnati OH 45224

MICHAELI, DOV, b Tel Aviv, Israel, May 28, 35; US citizen; m 62; c 2. IMMUNOCHEMISTRY, BIOCHEMISTRY. *Educ:* Hebrew Univ, Israel, BS, 60; Univ Calif, Berkeley, PhD(toxicol), 62, Univ Calif, San Francisco, MD, 76. *Prof Exp:* Asst res scientist, Lab Med Entom, Kaiser Found Res Inst, 62-67, assoc res scientist, 67-71; ASSOC PROF BIOCHEM & SURG, SCH MED, UNIV CALIF, SAN FRANCISCO, 71- *Concurrent Pos:* NIH res career develop award. *Mem:* Am Asn Immunol; NY Acad Sci; Am Assoc Clin Immunol. *Res:* Hematology; interaction of platelets with macramolecules; immunochemistry of collagen biochemistry and biology of wound healing from various sources; biochemistry and immunology of connective tissue. *Mailing Add:* Dept Biochem Univ Calif Med Ctr Box 0448 San Francisco CA 94143

MICHAELIS, ARTHUR FREDERICK, b Bronx, NY, July 24, 41; m 64; c 2. DRUG DELIVERY, PHARMACEUTICS. *Educ:* Bucknell Univ, BS, 63; Univ Wis, MS, 65, PhD(pharm), 67; Fairleigh Dickinson Univ, MBA, 76. *Prof Exp:* Sr chemist, Hoffmann-La Roche, Inc, Nutley, 67-70; pres, appl technol div, KV Pharmaceut Corp, St Louis, 79-80; dir res & develop, McNeil Consumer Prods Co, 80-81; vpres res & develop, Menley & James Labs, Ltd, 81-85; founder & pres, Controlled Therapeutics, 85-91; founder & chmn bd, Controlled Therapeutics, Scotland, Ltd, 85-91, Polysysts Healthcare, Ltd, 86-91; dir & pres, Advan Med Inc, AMEX:AMA, 89-91; FOUNDER, CHMN BD & CHIEF EXEC OFFICER, THERAPEUTIC SYSTS, 91- *Concurrent Pos:* Lectr, Sch Pharm, Univ Md, 69. *Mem:* AAAS; Am Pharmaceut Asn; Acad Pharmaceut Sci; fel Am Inst chem; NY Acad Sci; Controlled Release Soc; Am Soc Pharmaceut Engrs. *Res:* Physical chemistry of the ion pair extraction of pharmaceutical amines; physical pharmacy of drugs used in the prophylaxis and treatment of nerve gas casualties; new methods of optimizing drug delivery; high speed liquid chromatography. *Mailing Add:* Third St Anthony Lane Chester Springs PA 19425

MICHAELIS, CARL I, b Paxico, Kans, May 11, 18. ORGANIC CHEMISTRY. *Educ:* Univ Kans, AB, 45, AM, 47; Univ Fla, PhD(org chem), 53. *Prof Exp:* Asst chem, Univ Kans, 45-47; instr, Fla State Univ, 47-50; from instr to asst prof, Univ Fla, 50-54; assoc prof, 54-62, PROF CHEM, UNIV DAYTON, 62-, PREMED ADV, 60- *Mem:* Sigma Xi; Am Chem Soc. *Res:* Quaternary ammonium compounds and their derivatives; epoxides; amines. *Mailing Add:* Dept of Chem Univ of Dayton 300 Col Park Ave Dayton OH 45469-2357

MICHAELIS, ELIAS K, b Wad-Medani, Sudan, Oct 3, 44; m 67; c 1. NEUROCHEMISTRY. *Educ:* Fairleigh Dickinson Univ, BS, 66; St Louis Univ Med Sch, MD, 69; Univ Ky, PhD(physiol & biophys), 73. *Prof Exp:* Spec fel res, Dept Physiol & Biophys, Univ Ky, 72-73; from asst prof to prof, Dept Human Develop & Dept Biochem, 82-87, CHAIR PHARMACOL & TOXICOL, UNIV KANS, 88- *Concurrent Pos:* Dir, Ctr Biomed Res & Higouchi Biosci Res, 88- *Mem:* Am Soc Neurochem; Soc Neurosci; NY Acad Sci; AAAS; Int Soc Biomed Res Alcoholism; Am Soc Biochem & Molecular Biol. *Res:* Characterization of L-glutamate receptors in neuronal membranes; membrane protein isolation and chemical analysis; characterization of membrane transport systems for amino acids, sodium, potassium, and calcium; neuronal membrane biophysics; molecular neurobiology. *Mailing Add:* Dept Pharmacol & Toxicol 5064 Malott Univ Kans Lawrence KS 66045-2505

MICHAELIS, MARY LOUISE, b Denver, Colo, March 30, 43; m 67; c 1. MEMBRANE BIOCHEMISTRY, CALCIUM REGULATION. *Educ:* Webster Col, St Louis, BA, 66; Univ Kans, Lawrence, MA, 68, PhD(neurosci), 78. *Prof Exp:* Asst prof psychol, Webster Col, St Louis, 67-68; lab technician physiol & biophysics, Univ Ky, 70-72; res scientist & prin investr, 78-82, res asst prof biochem, 82-85, ASSOC PROF, DEPT PHARMACOL & TOXICOL, CTR BIOMED RES, UNIV KANS, 86- *Concurrent Pos:* Courtesy asst prof, Dept Human Develop, Univ Kans, 78-; consult, NIH, NSF, Alzheimer's Asn & Am Heart Asn. *Mem:* Soc Neurosci; Int Soc Biomed Res Alcoholism; Int Brain Res Orgn; Asn Women Sci. *Res:* Brain membrane proteins which transport and regulate calcium ions; protein isolation, molecular biological characterization, and the role of calcium transporting proteins in neuronal activity. *Mailing Add:* Ctr Biomed Res Univ Kans 2099 Constant Ave Lawrence KS 66047

MICHAELIS, MICHAEL, b Berlin, June 8, 19; US citizen; m 54; c 2. SCIENCE POLICY, TECHNICAL MANAGEMENT. *Educ:* Univ London, UK, BSc, 40. *Prof Exp:* Group leader, Gen Elec Co Ltd Res Labs, UK, 35-49; dir, Physics Div, Radiochem Ctr, UKAEA, 49-51; sr consult, Arthur D Little, Inc, Cambridge, Mass, 51-61, mgr, Washington, DC, 63-80; exec dir, White House Panel Civilian Technol, 61-63; sr prof, White House Off Sci & Technol, 61-63; PRES, PARTNERS IN ENTERPRISE, INC, 81- *Concurrent Pos:* Exec dir, Res Mgt Adv Panel, US House Rep, CTTE on S&T, 63-68; Exec dir, sr scientist & engr, AAAS, 89-90, chmn, Eng Sect. *Mem:* Sr mem Inst Elec & Electronics Engrs; Inst Elec Engrs; Inst Physics; AAAS. *Res:* Public and private strategies for management of technology; government; industry; academia collaboration. *Mailing Add:* 6812 Meadow Lane Chevy Chase MD 20815

MICHAELIS, OTHO ERNEST, IV, b Lancaster, Pa, Sept 10, 36. CARBOHYDRATE NUTRITION. *Educ:* Univ Md, PhD(nutrit), 73. *Prof Exp:* RES NUTRITIONIST, CARBOHYDRATE NUTRIT LAB, AGR RES SERV, USDA, 73- *Concurrent Pos:* Adj asst prof nutrit, Univ Md, 79- *Mem:* Am Inst Nutrit; N Am Asn Study Obesity. *Res:* Carbohydrate sensitivity. *Mailing Add:* Carbohydrate Nutrit Lab BARC-E USDA Agr Res Serv Rm 317 Bldg 307 Beltsville MD 20705

MICHAELIS, PAUL CHARLES, b Bronx, NY, June 18, 35; m 58; c 1. SOLID STATE PHYSICS, ELECTRICAL ENGINEERING. *Educ:* Newark Col Eng, BS, 64, MS, 67. *Prof Exp:* Technician electronics mech, 59-63, assoc mem tech staff, 63-67, mem tech staff magnetics res & develop, 67-82, TECH SUPVR, ADVAN TECHNOL, BELL LABS, 82- *Honors & Awards:* Morris N Liebmann Award, Inst Elec & Electronics Engrs, 75. *Mem:* Inst Elec & Electronics Engrs; Am Phys Soc; AAAS. *Res:* Magnetic memory devices; magnetoresistance phenomena; optical system interconnection; 12 patents in areas of mechanics, magnetics, packaging, circuits, and optics; currently engineering management in the areas of undersea communications hardware development, fiber optics, underwater acoustic sensors, cables and termination development, shipboard cable repair and system assembly. *Mailing Add:* 103 High Tor Dr Watchung NJ 07060-5408

MICHAELS, ADLAI ELDON, b Alma, Wis, Nov 22, 13; m 40; c 2. PHYSICAL CHEMISTRY. *Educ:* Univ Wis, BS, 35; Ohio State Univ, PhD, 40. *Hon Degrees:* DSc, Wash & Jefferson Col, 84. *Prof Exp:* Asst, Ohio State Univ, 35-39; instr chem, Univ Tenn, 40-43; res chemist, Esso Res & Eng Co, 43-59; from asst prof to prof chem, Wash & Jefferson Col, 67-83, secy fac, 65-83; RETIRED. *Honors & Awards:* Donahue Award, 82. *Mem:* Am Chem Soc. *Res:* Corrosion; electrochemistry; motor fuels; lubricants; fuel and lubricant additives; air pollution. *Mailing Add:* 73 Crest Vue Rd Washington PA 15301

MICHAELS, ALAN SHERMAN, b Boston, Mass, Oct 29, 22; m 51; c 2. CHEMICAL ENGINEERING. *Educ:* Mass Inst Technol, SB, 46, MS, 47, ScD(chem eng), 48. *Prof Exp:* From asst prof to prof chem eng, Mass Inst Technol, 48-66, assoc dir, Soil Stabilization Lab, 50-61; pres, Amicon Corp, Mass, 62-70 & Pharmetrics Inc, 70-72; pres, Alza Res Corp & sr vpres & tech dir, Alza Corp, 72-77; adj prof chem eng & med, Stanford Univ, 77-82; distinguished univ prof, 86-90, EMER DISTINGUISHED UNIV PROF CHEM ENG, NC STATE UNIV, 90- *Concurrent Pos:* Indust consult, 48-; asst tech dir, Seco Venture, 50-51; vis prof chem eng, Univ Col, Univ London, 59, Univ Calif, Berkeley, 77; consult, President's Adv Sci Comn, 61-62; partic conf hemodialysis, NIH, 64-; vis prof, Tech Univ Berlin, 65; consult, Off Saline Water, US Dept Interior, 65-; adj prof chem eng, Mass Inst Technol & Lehigh Univ, 82-84; sci adv, Collagen Corp, Liposome Technol Inc, Bioprod Group of FMC Corp & Moleculon Biotech; vis scholar & consult prof chem eng, Stanford Univ, 70-76; mem, Res Briefing Panel on Chem & Process Eng for Biotechnol, Nat Acad Eng, Comt to Survey Chem Eng & Comt for Bioprocessing Energy-Efficient Production of Chem. *Honors & Awards:* McGraw Hill Outstanding Personal Achievement Award in Chem Eng, 74; Food, Pharmaceut & Bioeng Award, Am Inst Chem Engrs, 77, Mat Eng & Sci Award, 82; Separation Sci & Technol Award, Am Chem Soc, 85. *Mem:* Nat Acad Eng; AAAS; Am Chem Soc; Am Soc Eng Educ; fel Am Inst Chem Engrs; Am Inst Chem; Sigma Xi; fel NY Acad Sci. *Res:* Surface, colloid and polymer chemistry; wetting and adhesion; factors influencing the transmission of gases and vapors through polymers; extracorporeal artificial organs, bioengineering and biomedicine; author or co-author of more than 130 technical papers. *Mailing Add:* 210 Allandale Rd Apt 3A Chestnut Hill MA 02167-3284

MICHAELS, ALLAN, b Aug 22, 44. CELL & MOLECULAR BIOLOGY. *Educ:* Univ Calif, Santa Barbara, PhD(cell biol), 72. *Prof Exp:* ASSOC PROF CELL & MOLECULAR BIOL, BEN GURION UNIV, 77- *Concurrent Pos:* Asst assoc prof, Univ SFla. *Res:* Structure and function of subcellular constituents; membrane biology; gene expression; biotechnology. *Mailing Add:* Dept Life Sci Ben Gurion U Negev Beersheba Israel

MICHAELS, DAVID D, b Cologne, Ger, July 16, 25; nat US; m 53; c 4. OPHTHALMOLOGY. *Educ:* Northern Ill Col Optom, OD, 47; Ill Inst Technol, BS, 51; Chicago Col Optom, MS, 52, DOS, 53; Roosevelt Univ, BS, 56; Univ Ill, MD, 57; Am Bd Ophthal, dipl, 64. *Prof Exp:* Resident surgeon, Dept Ophthal, Cook County Hosp, Ill, 58-60; from instr to assoc prof, 60-80, PROF SURG, UNIV CALIF, LOS ANGELES, 84-; CHMN DEPT OPHTHAL, SAN PEDRO COMMUNITY HOSP, 60- *Concurrent Pos:* Lectr, Loyola Univ, Ill, 54-55; mem attend staff, Los Angeles County Hosp; fac, Univ Calif, Los Angeles Med Sch. *Honors & Awards:* Am Acad Opthol, 84. *Mem:* Fel Optical Soc Am; fel Am Phys Soc; fel Am Acad Optom; fel Am Acad Ophthal & Otolarnygol; fel Int Col Surgeons. *Res:* Clinical diagnosis; physiologic optics. *Mailing Add:* 1441 W Seventh St San Pedro CA 90732

MICHAELS, JOHN EDWARD, b Boston, Mass, Feb 2, 39; m 65; c 2. CELL BIOLOGY. *Educ:* Harvard Univ, AB, 64; Boston Univ, PhD(biol), 70. *Prof Exp:* Fel anat, McGill Univ, 71-74; asst prof, 74-81, ASSOC PROF ANAT, COL MED, UNIV CINCINNATI, 81- *Mem:* Am Asn Anatomists; Am Soc

Cell Biol; AAAS; Electron Micros Soc Am. *Res:* Formation of cell coat glycoproteins and their transport from the Golgi apparatus to the cell surface studied by electron microscopy and related techniques. *Mailing Add:* Dept Anat Col Med Univ Cincinnati Cincinnati OH 45267-0521

MICHAELSEN, TERJE E, b Stavanger, Norway, Aug 20, 42; M 65; c 2. MOLECULAR IMMUNOLOGY. *Educ:* Univ Oslo, PhD(immunol), 75. *Prof Exp:* Res fel, Inst Immunol & Rheumatology, Oslo, 69-75; instr, NY Med Ctr, 75-76; re fel, Inst Immunol & Rheumatol, 76-77; chemist, 77-81, CHIEF, NAT INST PUB HEALTH, 81- *Concurrent Pos:* Chief lab, Nat Inst Pub Health, 81- *Mem:* Am Asn Immunologists. *Res:* Structural and biological studies of human Ig6 subclasses; amyloid fiber proteins, distribution and activity of FCR on lymphocytes; idiotypic studies of human antibodies and lectin isolation-characterization. *Mailing Add:* Nat Inst Pub Health Geitmyrsvelen 75 0462 Oslo 4 Norway

MICHAELSON, I ARTHUR, b New York, NY, Mar 15, 25; m 58; c 3. PHARMACOLOGY. *Educ:* NY Univ, BA, 50; George Washington Univ, PhD(pharmacol), 59. *Prof Exp:* Fel, Lab Chem Pharmacol, Nat Heart Inst, 59-61; USPHS fel, Agr Res Coun Inst Animal Physiol, Cambridge Univ, 61-63; fel, Lab Chem Pharmacol, Nat Heart Inst, 63-65; asst prof, 65-67, assoc prof pharmacol, 67-77, PROF ENVIRON HEALTH, COL MED, UNIV CINCINNATI, 77- *Concurrent Pos:* USPHS res career develop award, 67-72; vis scientist, Toxicol Unit, Med Res Coun, Carshalton, Surrey, Eng, 71-72. *Mem:* Am Soc Pharmacol & Exp Therapeut. *Res:* Intermediary metabolism of drugs and biochemical pharmacology; subcellular localization of biogenic amines and the effect of drugs on synthesis, storage and release; toxicology; effect of metals on brain development and neurochemistry. *Mailing Add:* Dept Environ Health Univ Cincinnati Col Med 3223 Eden Ave ML56 Cincinnati OH 45267

MICHAELSON, JERRY DEAN, b Monterey, Calif, Nov 26, 43; m 66; c 2. COHERENT OPTICS, MICROWAVE INTEGRATED CIRCUITS. *Educ:* Univ Southern Calif, BSEE, 66, MSEE, 71, PhD(elec eng), 79. *Prof Exp:* Mem assoc staff, 64-66, mem tech staff, 66-82, SECT MGR, THE AEROSPACE CORP, 82- *Mem:* Soc Photo Optical Instrumentation Engrs. *Res:* Nonlinear optical processing; high speed gas digital circuits; low noise microwave frequency synthesizers. *Mailing Add:* Aerospace Corp PO Box 92957 Los Angeles CA 90009

MICHAELSON, MERLE EDWARD, b Hudson, Wis, Feb 14, 21; m 44, 85; c 3. PLANT PATHOLOGY. *Educ:* Wis State Col, River Falls, BS, 43; Colo State Univ, MS, 48; Univ Minn, PhD(plant path, bot), 53. *Prof Exp:* Asst prof bot, Colo State Univ, 46-49; res asst plant path, Univ Minn, St Paul, 49-52; asst prof bot, Univ Mo, 52-54; plant pathologist crops res div, Agr Res Serv, SDak Agr Exp Sta, USDA, 54-59; prof biol, St Cloud State Col, 59-67, actg dean grad sch, 66-67; PROF BIOL, UNIV WIS-RIVER FALLS, 67-, ASST DEAN COL ARTS & SCI, 75- *Concurrent Pos:* Chmn dept biol, Univ Wis-River Falls, 67-72. *Mem:* AAAS; Am Phytopath Soc; Mycol Soc Am; Nat Asn Biol Teachers. *Res:* Mycology; diseases of corn and flax. *Mailing Add:* 117 N Cudd Ave River Falls WI 54022

MICHAELSON, S(TANLEY) D(AY), b New York, NY, Sept 4, 13; m 39; c 2. MINING & METALLURGICAL ENGINEERING. *Educ:* Lehigh Univ, BS, 34. *Hon Degrees:* EM, Univ Mont, 60. *Prof Exp:* Engr in chg mining lab, Allis-Chalmers Mfg Co, Wis, 35-37, metall engr mining dept, 37-39, field engr, 39-41, dir basic industs res lab, 46-47; from spec engr to chief engr raw mat, US Steel Corp, Ala, 47-54; chief engr western mining divs, Kennecott Copper Corp, 54-68, chief engr, Metal Mining Div & dir, Eng Ctr, 68-75; CONSULT MINING & METALL ENGR, 75- *Concurrent Pos:* Mem sci comt, Selective Serv Syst, 57-70, adv coun, Col Mines & Mineral Indust, Univ Utah, 58-73, adv panel mining, Nat Acad Sci-Nat Res Coun, 67; adj prof mining eng, Univ Utah, 75-; chmn panel explor, solution mining & underground mining, Comn on Surface Mining & Reclamation, Nat Res Coun, 78-80. *Honors & Awards:* Richards Award, Am Inst Mining, Metall & Petrol Engrs, 62. *Mem:* Col Mining & Metall Soc Am; hon mem Am Inst Mining, Metall & Petrol Engrs (vpres, 60); distinguished mem Soc Mining Engrs (pres, 59); Sigma Xi. *Res:* Open pit and underground mining engineering; mineral beneficiation; pyrometallurgy; mines plant design and mine design; economic evaluations for mineral projects and mines. *Mailing Add:* 1446 Circle Way Salt Lake City UT 84103

MICHAELSON, SOLOMON M, b New York, NY, Apr 23, 22; m 50; c 2. RADIATION BIOLOGY, PHYSIOLOGY. *Educ:* City Col New York, BS, 42; Middlesex Univ, DVM, 46; Am Col Lab Animal Med, dipl. *Prof Exp:* Instr immunol, Univ Ark, 47-48; sr pharmacologist, Eaton Labs, Norwich Pharmaceut Co, 48-53; chief radiation physiol & ther, Atomic Energy Proj, 53-58, from asst prof to assoc prof radiation biol, 58-72, ASSOC PROF MED, SCH MED & DENT, UNIV ROCHESTER, 67-, ASSOC PROF LAB ANIMAL MED, 68-, PROF RADIATION BIOL & BIOPHYSICS, 72- *Concurrent Pos:* Consult, UNRRA, 46-47, Armed Forces Radiobiol Res Inst, 63-70, Walter Reed Army Inst Res, 65-70, Vet Admin, 66-, Nat Acad Sci-Nat Res Coun, 72-, Elec Power Res Inst, 75-84 & Nat Coun Radiation Protection & Measurement, 79-; vis lectr, Am Inst Biol Sci, Environ Protect Agency, 79-, NIH, 82-, Off Sci & Technol Policy, 85-; assoc ed, J Microwave Power, 74-; ed, Radiation & Environ Biophysics; ed, Radiation Res; lect series dir, NATO /AGARD, 75. *Honors & Awards:* Group Achievement Award, NASA. *Mem:* Fel Am Acad & Comp Toxicol; Am Physiol Soc; Radiation Res Soc; Health Physics Soc; Sigma Xi; fel Inst Elec & Electronic Engrs; Int Comn Occup Health; Am Col Lab Animal Med. *Res:* Mechanisms of injury and recovery, especially neuroendocrine physiology and carcinagenesis from electromagnetic radiations; electric and magnetic fields. *Mailing Add:* Dept Biophys Univ Rochester Sch Med & Dent Rochester NY 14642

MICHAL, EDWIN KEITH, b Independence, Kans, Sept 17, 32; m 56; c 4. NEUROPHYSIOLOGY. *Educ:* Kans Wesleyan Univ, BA, 54; Univ Ill, MS, 62, PhD(physiol), 65. *Prof Exp:* Asst prof, 65-72, ASSOC PROF PHYSIOL, OHIO STATE UNIV, 72- *Mem:* Am Physiol Soc; Soc Neurosci. *Res:* Neural mechanisms in behavior; neuroendocrinology; neural control of respiration. *Mailing Add:* Dept Physiol Ohio State Univ Col Med 333 W Tenth Ave Columbus OH 43210

MICHAL, EUGENE J(OSEPH), b Reno, Nev, Oct 23, 22; m 51; c 4. MINING ENGINEERING. *Educ:* Univ Nev, BS, 43; Mass Inst Technol, MS, 47, ScD, 51. *Prof Exp:* Physicist, Mare Island Navy Yard, 43-44; instr metall, Mass Inst Technol, 49-51; res metallurgist, AEC, Watertown Arsenal, 48-49; supvr metall res, Nat Lead Co, 51-68; asst to vpres, INCO Inc, NY, 68-70; dir res & develop, Climax Molybdenum Co, 70-76, vpres, 76-77; pres, Amax Extractive Res & Develop Inc, 77-86; PRES, TOIYABE EXPLOR INC, 86- *Mem:* Am Inst Mining, Metall & Petrol Engrs; Soc Mining Engrs; fel Brit Inst Mining & Metall; Am Inst Chem Engrs; Am Mgt Asn. *Res:* Process metallurgy; hydrometallurgy; electric steelmaking. *Mailing Add:* Toiyabe Explor Inc 14050 Foothill Rd Golden CO 80401

MICHALAK, JOSEPH T(HOMAS), b Philadelphia, Pa, July 1, 32; m 56; c 3. METALLURGY AND PHYSICAL. *Educ:* Drexel Inst, BS, 55; Carnegie Inst Technol, MS, 58, PhD(metall eng), 60. *Prof Exp:* From scientist to sr scientist, US Steel Corp 59-76; RES CONSULT SHEET PROD RES, TECH CTR, US STEEL CORP, 76- *Mem:* Metall Soc. *Res:* Physical and mechanical metallurgy of low-carbon steel sheet and tinplate; strain aging; formability surface topography; cold-rolled, electrogalvinized, hot-dip galvanized. *Mailing Add:* USS Tech Ctr MS-19 4000 Tech Ctr Dr Monroeville PA 15146-3048

MICHALAK, THOMAS IRENEUSZ, b Warsaw, Poland; Can citizen. HEPATITIS, VIRUS-CELL INTERACTION. *Educ:* Warsaw Med Acad, MD, 73; Nat Inst Hyg, Warsaw, PhD(med sci), 76. *Prof Exp:* Res asst immunopath, Nat Inst Hyg, Warsaw, 70-73, res fel, 73-76, asst prof, 76-78, head & adj asst prof electromicros, 78-85; vis asst prof path, Mem Univ Nfld, 82-83, vis scientist immunohistochem, 83-85, asst prof cell sci, 85-90, ASSOC PROF CELL SCI, LIVER RES LAB, MEM UNIV NFLD, ST JOHN'S, CAN, 90- *Concurrent Pos:* Intern internal med, pediat, surg & gynec,80-82. *Mem:* Am Soc Microbiol; Can Asn Gastroenterol; Am Asn Study Liver Dis; Can Asn Study Liver. *Res:* Infectious liver diseases; molecular biology and immunopathogenesis of virus-induced hepatocellular injury; pathological and ultrastructural aspects of human liver diseases; animal models for human viral infections; treatment of hepatitis and other human viral infections; virus cell interaction. *Mailing Add:* Liver Res Lab Fac Med Health Sci Ctr Mem Univ Nfld St John's NF A1B 3V6 Can

MICHALEK, JOEL EDMUND, b Detroit, Mich, Aug 30, 44; m; c 2. STATISTICS. *Educ:* Wayne State Univ, BS, 66, MA, 68, PhD(math statist), 73. *Prof Exp:* Instr math, RI Col, 70-71; asst prof statist, Syracuse Univ, 73-76; MATH STATISTICIAN, US AIR FORCE SCH AEROSPACE MED, 76- *Concurrent Pos:* Consult, dept med, Health Sci Ctr, Univ Tex. *Mem:* Am Statist Asn; Biometric Soc. *Res:* Survival analysis; clinical trials. *Mailing Add:* USAFSAM/EKB Brooks AFB TX 78235-5000

MICHALEK, SUZANNE M, b Chicago, Ill, July 19, 44. MICROBIOLOGY. *Educ:* Ill State Univ, BS, 67, MS, 68; Univ Ala, Birmingham, PhD(microbiol), 76. *Prof Exp:* Res asst microbiol & immunol, Nat Inst Dent Res, NIH, Bethesda, Md, 72-76, postdoctoral fel microbiol & immunol, 77-79; res assoc microbiol, 76-77, from asst prof to assoc prof, 79-85, PROF MICROBIOL, UNIV ALA BIRMINGHAM, 85-, PROF ORAL BIOL, 88- *Concurrent Pos:* Investr, Inst Dent Res, Univ Ala Birmingham, 80-85, scientist, 85- & sr investr, Res Ctr Oral Biol, 88- *Mem:* Am Soc Microbiol; Am Asn Immunologists; Int Asn Dent Res; Am Asn Dent Res; Soc Exp Biol & Med. *Mailing Add:* Microbiol Dept Univ Ala Univ Sta Birmingham AL 35294

MICHALIK, EDMUND RICHARD, b Munhall, Pa, Aug 5, 15; m 46; c 1. APPLIED MATHEMATICS, STATISTICS. *Educ:* Univ Pittsburgh, BA, 37, MA, 40. *Prof Exp:* Instr math, Univ Pittsburgh, 40-42, asst prof math statist, 46-52; sr analyst rev planning, US Dept Army, 52-53; consult appl math & statist, Atlantic Res Co, 53-54; head dept appl math, Mellon Inst, 54-57; sr staff engr, Glass Res Ctr, PPG Indust, Inc, 57-80; RETIRED. *Concurrent Pos:* Consult, flat glass processing & glass strength, 81-90. *Mem:* Sigma Xi; AAAS. *Res:* Use of electronic digital machines in mathematic, statistics, industrial problems; glass strength and glass strengthening. *Mailing Add:* 3711 Spring St West Mifflin PA 15122

MICHALOPOULOS, GEORGE, CELL GROWTH REGULATION, ONCOLOGY. *Educ:* Univ Wis, PhD(liver cell cultures), 77. *Prof Exp:* ASSOC PROF PATH, DUKE UNIV MED CTR, 77- *Res:* Liver pathology. *Mailing Add:* Dept Path Duke Univ Med Ctr PO Box 3432 Durham NC 27710

MICHALOWICZ, JOSEPH C(ASIMIR), b Washington, DC, Mar 4, 16; m 40; c 5. ELECTRICAL ENGINEERING. *Educ:* Cath Univ Am, BEE, 40, MEE, 51. *Prof Exp:* Asst engr, Rural Electrification Admin, USDA, 40-42; from instr to assoc prof elec eng, Cath Univ Am, 42-44, 46-80, head dept, 52-59, asst dean, Sch Eng & Archit, 69-71, dean admis, financial aid & records & registrar, 73-78; RETIRED. *Concurrent Pos:* Cottrell grant, Res Corp, NY, 50-52. *Mem:* Fel Inst Elec & Electronics Engrs; Sigma Xi. *Res:* Electronic instrumentation; electronic automotive fuelmeter. *Mailing Add:* 7509 Wyndale Rd Chevy Chase MD 20815

MICHALOWICZ, JOSEPH VICTOR, b Oct 23, 41; m 63; c 2. MATHEMATICS, SYSTEMS ANALYSIS. *Educ:* Catholic Univ Am, BA, 63, PhD(math), 67. *Prof Exp:* Mathematician, Res Analysis Corp, 66-67; asst prof math, Catholic Univ Am, 67-73; mathematician, Harry Diamond Labs, 73-84; RES PHYSICIST, NAVAL RES LAB, 85- *Concurrent Pos:* Lectr elec eng, Cath Univ Am, 66-67; NSF fel category theory, Bowdoin Col, 69; consult, Res Analysis Corp, 67-70 & Harry Diamond Labs, 70-73. *Mem:* Am Math Soc; Mil Opers Res Soc; Math Asn Am. *Res:* Spaced-based infrared sensors, systems analysis and cost effectiveness studies; research in category theory. *Mailing Add:* 5855 Glen Forest Dr Falls Church VA 22041

MICHALOWSKI, JOSEPH THOMAS, b Newburgh, NY, Dec 17, 43. GEOCHEMISTRY. *Educ:* Marist Col, BA, 68; Univ Calif, Santa Barbara, PhD(chem), 72. *Prof Exp:* Res assoc chem, Univ Calif, Santa Barbara, 72-73, lectr, 73; res assoc biochem, St Louis Univ, 73-74; res scientist geochem, Phillips Petrol Co, 74-84; RES MGR, ENVIRON CHEM, GROUND WATER RES, REMEDIATION & CHARACTERIZATION OF HAZARDOUS WATER SITES, NORTHROP SERV, INC, 86- *Concurrent Pos:* Consult, Geochemistry & Environ Chem, JTM Enterprises, 84-86. *Mem:* Am Geophys Union; Am Chem Soc; AAAS; Geol Soc Am; Sigma Xi. *Res:* Electrochemistry and mechanics of migration of multi-phase fluids through porous rocks; problems of petroleum migration. *Mailing Add:* 7129 Cottington Lane San Diego CA 92139

MICHALOWSKI, RADOSLAW LUCAS, b Poznan, Poland, Dec 9, 51; m 83; c 1. SOIL MECHANICS, GEOTECHNICAL ENGINEERING. *Educ:* Tech Univ Poznan, MSc, 74, PhD(civil eng), 80. *Prof Exp:* Fulbright Fel geotech eng, Dept Civil & Mineral Eng, Univ Minn, 81-84, res assoc, underground Space Ctr, 86-90; ASST PROF, DEPT CIVIL ENG, JOHNS HOPKINS UNIV, 90- *Mem:* Am Soc Civil Engrs; Am Acad Mech; Polish Soc Theoret & Appl Mech; Int Soc Mech Found Eng. *Res:* Mechanics of granular media, constitutive relations; geotechnical, agricultural and chemical engineering applications (e.g. gravity flow in containers, stability of soil mass and bearing capacity problems reinforces soil modeling; soil plasticity and limit analysis; three dimensional, incipient, advanced and post-failure deformation processes. *Mailing Add:* Dept Civil Eng Johns Hopkins Univ Baltimore MD 21218

MICHALSKE, TERRY ARTHUR, b Dunkirk, NY, Jan 19, 53. FRACTURE SURFACE ANALYSIS, FRACTURE MECHANICS. *Educ:* Alfred Univ, BS, 75, PhD(ceramic sci), 79. *Prof Exp:* Fel, Nat Bur Standards, 79-81; MEM TECH STAFF, SANDIA NAT LABS, 81- *Mem:* Am Ceramic Soc. *Res:* Fracture properties of brittle materials; stress corrosion effects in brittle materials; effect of ceramic microstructure on fracture properties. *Mailing Add:* PO Box 1042 Cedar Crest NM 87008

MICHALSKI, CHESTER JAMES, b Detroit, Mich, June 7, 42; m; c 2. MOLECULAR BIOLOGY. *Educ:* Mich State Univ, BS, 65, MS, 67; Univ NC, PhD(biochem), 71. *Prof Exp:* Fel biochem, St Judes Children's Res Hosp, Memphis, 71-72; fel, 72-75, asst prof, 75-79, assoc prof molecular biol & dir multidiscipline labs, 79-88, asst dean, Sch Grad Studies, 84-86, ASST DEAN, RES & GRAD STUDIES MED, MEM UNIV NFLD, 87-, PROF MOLECULAR BIOL, 88- *Mem:* Am Soc Microbiol; Can Soc Microbiol; Asn Multidiscipline Educ Health Sci; Can Biochem Soc; NY Acad Sci. *Res:* Structure and function of the macromolecular components involved in the protein synthesizing system of E coli cells and the molecular events involved in hormone regulation of the fungi. *Mailing Add:* Fac Med Mem Univ Nfld St John's NF A1B 3V6 Can

MICHALSKI, JOSEPH POTTER, b Montgomery, Ala, Jan 27, 45. RHEUMATOLOGY. *Educ:* Dartmouth Col, AB, 66; Univ Md, MD, 70; Am Bd Internal Med, dipl, 76. *Prof Exp:* Clin instr med, Univ Calif, San Fancisco, 77-78, asst prof, Irvine, 78-82; assoc prof, 82-86, PROF MED, SCH MED, LOUISIANA STATE UNIV, 86-, DIR RES, SECT RHEUMATOLOGY & REHAB MED, SCH MED, 82- *Concurrent Pos:* Chief rheumatology sect, Long Beach Vet Admin Med Ctr, Calif, 78-82. *Mem:* Am Asn Immunologists; Am Rheumatism Asn; AAAS; Am Fedn Clin Res; Am Soc Zool; Fedn Am Soc Exp Biol. *Res:* Mechanism of association of HLA-88 with autoimmunity; lung disease; and immunodeficiency in autoimmunity; HLA and rheumatic disease. *Mailing Add:* La State Univ Med Ctr 1542 Tulane Ave New Orleans LA 70112

MICHALSKI, RAYMOND J, environmental chemistry, for more information see previous edition

MICHALSKI, RYSZARD SPENCER, b Kalusz, Poland, May 7, 37. COMPUTER SCIENCES, INTELLIGENT SYSTEMS. *Educ:* Warsaw Tech Univ, BS, 59; Leningrad Polytech Inst, MS, 61; Silesia Tech Univ, Poland, PhD(comput sci), 69. *Prof Exp:* Logical designer comput sci, Inst Math Mach, Polish Acad Sci, Warsaw, 61-62, res scientist, Inst Automatic Control, 62-70; from vis asst prof to assoc prof, 70-82, PROF COMPUT SCI, UNIV ILL, URBANA, 82- *Concurrent Pos:* Lectr comput sci & electronics, State Tech Col, Warsaw, Poland, 64-68; Fulbright fel, US State Dept, 70; res awards, NSF, 75, 77 & 79, Off Naval Res, 80- & Defense Advan Res Proj Agency, 85-; sr fel, British Sci Res Coun, 77. *Mem:* Asn Comput Mach; Pattern Recognition Soc; Sigma Xi; Polish Inst Arts & Sci in Am. *Res:* Machine learning and plausible reasoning; artificial intelligence and cognitive science; applications to agriculture and medicine; expert systems, multiple-valued logic, intelligent systems. *Mailing Add:* Ctr Artificial Intel Rm 301 George Mason Univ 4400 Univ Dr Fairfax VA 22030-4444

MICHALSKY, JOSEPH JAN, JR, b Dayton, Tex, Oct 8, 47; m 74; c 1. ATMOSPHERIC SCIENCE. *Educ:* Lamar Univ, BS, 69; Univ Ky, MS, 71, PhD(physics), 74. *Prof Exp:* Fel astron, Battelle Inst, Ger, 74-76; res scientist, 76-80, SR RES SCIENTIST ATMOSPHERIC SCI, NORTHWEST LABS, BATTELLE MEM INST, 80- *Mem:* Am Geophys Union; Int Solar Energy Soc. *Res:* Spatial and spectral measurements of direct and diffuse insolation and their interpretation. *Mailing Add:* 4005 Georgetown Sq Schenectady NY 12303

MICHALSON, EDGAR LLOYD, b Salem, Ore, Dec 13, 29; m 58; c 3. RESEARCH ADMINISTRATION, INTERDISCIPLINARY RESEARCH. *Educ:* Ore State Univ, BS, 56; Pa State Univ, MS, 59, PhD(agr econ), 63. *Prof Exp:* Agr res economist, farm production, Econ Res Serv, USDA, 63-69; assoc prof, 69-74, PROF AGR ECON, UNIV IDAHO, 74- *Mem:* Am Agr Econ Asn; Am Econ Asn; Am Water Resources Asn; Am Asn Univ Prof. *Res:* Economics of erosion control; agricultural business management research; natural resource economic problems. *Mailing Add:* Dept Agr Econ Univ Idaho Moscow ID 83843

MICHAUD, GEORGES JOSEPH, b Quebec, Que, Apr 30, 40; m 66; c 2. ASTROPHYSICS. *Educ:* Univ Laval, BA, 61, BSc, 65; Calif Inst Technol, PhD(astron), 69. *Prof Exp:* Asst prof, 69-73, assoc prof, 73-79, PROF PHYSICS, UNIV MONTREAL, 79- *Concurrent Pos:* Killam fel, 87-89. *Honors & Awards:* Steacie Prize, 80; Janssen Medal, 82. *Mem:* Am Astron Soc; Can Astron Soc; Int Astron Union. *Res:* The chemical abundance of the elements; nucleosynthesis and diffusion in stellar envelopes and atmospheres. *Mailing Add:* Dept of Physics Univ of Montreal Box 6128 Montreal PQ H3C 3J7 Can

MICHAUD, HOWARD H, b Berne, Ind, Oct 12, 02; m 28; c 1. ENVIRONMENTAL MANAGEMENT. *Educ:* Bluffton Col, AB, 25; Ind Univ, MA, 30. *Prof Exp:* Teacher pub schs, Ind, 25-45; prof, 45-71, EMER PROF CONSERV, PURDUE UNIV, 71- *Concurrent Pos:* chief naturalist, Ind State Parke, 34-44; del, Int Union Conserv Nature & Natural Resources, 48 & 66. *Honors & Awards:* Osborn Wildlife Conserv Award, 59. *Mem:* Conserv Educ Asn (pres, 56-57); Am Asn Biol Teachers (pres, 48); hon mem Soil Conserv Soc Am; Wildlife Soc; Am Nature Study Soc. *Res:* Science and conservation education. *Mailing Add:* 301 E Stadium Ave West Lafayette IN 47906

MICHAUD, RONALD NORMAND, b Madawaska, Maine, July 7, 37; m 58; c 2. MICROBIOLOGICAL CONTROL. *Educ:* Univ Maine, BS, 63; Cornell Univ, MS, 66, PhD(microbiol), 68. *Prof Exp:* Assoc res microbiologist, Sterling Drug Inc, 68-71, group leader, 71-74, res microbiologist, 74-75, head, Bact Sect, 75-80, head, Microbiol Control Sect, Sterling-Winthrop Res Inst Div, 80-89, HEAD, MICROBIOL CONTROL SECT, STERLING RES GROUP, STERLING DRUG INC, 89- *Mem:* Am Soc Microbiol. *Res:* Microbiological control aspects of the pharmaceutical product development of dosage types. *Mailing Add:* 81 Columbia Turnpike Rensselaer NY 12144

MICHAUD, TED C, b Ft Wayne, Ind, Oct 5, 29; m 55; c 3. ZOOLOGY. *Educ:* Purdue Univ, BS, 51; Univ Mich, MS, 54; Univ Tex, PhD(zool), 59. *Prof Exp:* From asst prof to assoc prof, 59-70, PROF BIOL, CARROLL COL, WIS, 70- *Concurrent Pos:* Mem, Demog Inst, Cornell Univ, 71; mem steering comt, Cent States Col Asn Environ Studies Comt; second vpres, Asn Midwestern Col Biol Teachers, 80-81. *Mem:* AAAS; Soc Study Evolution; Am Soc Ichthyol & Herpet. *Res:* Amphibian behavior and evolution. *Mailing Add:* Dept of Biol Carroll Col 100 N East Ave Waukesha WI 53186

MICHEJDA, CHRISTOPHER JAN, b Kielce, Poland, Dec 19, 37; US citizen; m 64; c 1. PHYSICAL ORGANIC CHEMISTRY, BIO-ORGANIC CHEMISTRY. *Educ:* Univ Ill, BS, 59; Univ Rochester, PhD(org chem), 64. *Prof Exp:* NSF fel, Harvard Univ, 63-64; from asst prof to prof org chem, Univ Nebr, Lincoln, 75-78; HEAD CHEM CARCINOGENS, FREDERICK CANCER RES CTR, 78-; ADJ PROF CHEM, UNIV MD, COL PARK,78- *Concurrent Pos:* NIH spec fel, Swiss Fed Inst Technol, 72-73; assoc prog dir for chem dynamics, NSF, 75-77. *Mem:* Am Chem Soc. *Res:* Free radical chemistry; chemical carcinogenesis. *Mailing Add:* Frederick Cancer Res Ctr PO Box B Bldg 538 Frederick MD 21701-1013

MICHEJDA, OSKAR, b Trzyniec, Czech, May 19, 22; m 48; c 1. CIVIL ENGINEERING. *Educ:* Wroclaw Polytech Univ, MScEng, 50, DSc(exp stress anal), 61. *Prof Exp:* Asst prof appl mech & strength of mat, Czestochowa Polytech Univ, 54-65; sr lectr, Univ Khartoum, 65-66; assoc prof, Cooper Union Inst Technol, 66-68; prof civil eng, Ind Inst Technol, 68-74; eng specialist, Burns & Roe, 74-86; INDEPENDENT ENG CONSULT, 86- *Mem:* Fel Am Soc Civil Eng. *Res:* Theory of structures; experimental stress analysis. *Mailing Add:* 4087 Woodview Dr Sarasota FL 34232

MICHEL, ANTHONY NIKOLAUS, b Rekasch, Romania, Nov 17, 35; US citizen; m 57; c 5. ELECTRICAL ENGINEERING, MATHEMATICS. *Educ:* Marquette Univ, BSEE, 58, MS, 64, PhD(elec eng), 68; Tech Univ Graz, Austria, DSc(math), 73. *Prof Exp:* Sr res engr, AC Electronics Div, Gen Motors Corp, Wis, 58-65; from asst prof to prof elec eng, Iowa State Univ, 68-84, Off Naval Res grant, Eng Res Inst, 68-72; chmn, Dept Elec Eng, 84-88, FRANK M FREIMANN PROF ENG, UNIV NOTRE DAME, 84-, MCCLOSKEY DEAN ENG, 88- *Concurrent Pos:* NSF grants, Eng Res Inst, 72-84; assoc ed, Inst Elec & Electronic Engrs Trans on Circuits & Systs, 77-79, ed, 81-83. *Honors & Awards:* Centennial Medal, Inst Elec & Electronic Engrs, 84. *Mem:* Fel Inst Elec & Electronic Engrs; Inst Elec & Electronic Engrs Circuits & Systs Soc (pres, 89). *Res:* Automatic control theory; differential equations; large scale systems; artificial neural networks. *Mailing Add:* 17001 Stonegate Ct Granger IN 46530-9783

MICHEL, BURLYN EVERETT, b Ladoga, Ind, Mar 7, 23; m 46; c 3. PLANT PHYSIOLOGY. *Educ:* Univ Chicago, SB, 48, PhD(bot), 50. *Prof Exp:* Asst prof bot, Univ Iowa, 51-58; assoc prof, 58-64, PROF BOT, UNIV GA, 64- *Concurrent Pos:* Plant physiologist, Agr Res Serv, USDA, 59-62. *Mem:* Am Inst Biol Sci; Bot Soc Am; Am Soc Plant Physiol. *Res:* Plant-water relations and mineral absorption. *Mailing Add:* 344 Beechwood Dr Athens GA 30606

MICHEL, DAVID JOHN, b Denison, Tex, July 24, 42; m 66; c 2. METALLURGY, CRYSTALLOGRAPHY. *Educ:* Univ Mo-Rolla, BSMetE, 64; Pa State Univ, MSMet, 66, PhD(metall), 68. *Prof Exp:* Asst metallurgist, Homer Res Lab, Bethlehem Steel Co, 64; res asst metall, Pa State Univ, 64-68, fel, 68-69; metallurgist, Div Res, US AEC, 69-71; res asst prof mat sci, Univ Cincinnati, 71-72; res metallurgist, 72-73, HEAD HIGH TEMPERATURE METALS SECT, US NAVAL RES LAB, 73- *Concurrent Pos:* Prof lectr, George Washington Univ, 79-; vis scientist, Carnegie-Mellon Univ, 84-85. *Honors & Awards:* Nat Capital Award for Prof Achievement Eng, Washington, DC Coun Eng & Architectural Soc, 78; Res Publ Award, Naval Res Lab, 86. *Mem:* Fel Am Soc Metals; Am Inst Mining, Metall & Petrol Engrs; Am Crystallog Asn; Int Metallog Soc. *Res:* Radiation effects in materials; mechanical behavior of materials; intermetallic compounds; x-ray crystallography; alloy theory; phase equilibria; phase transformations in solids; electron microscopy; semiconductor alloys. *Mailing Add:* Phys Metall Br US Naval Res Lab Code 6326 Washington DC 20375-5000

MICHEL, F CURTIS, b La Crosse, Wis, June 5, 34; m 58; c 2. ASTROPHYSICS, SPACE PHYSICS. *Educ:* Calif Inst Technol, BS, 55, PhD(physics), 62. *Prof Exp:* Res fel astrophys, Calif Inst Technol, 62-63; from asst prof to assoc prof space sci, 63-70, prof physics, space physics & astron, 70-74, chmn dept space physics & astron, 74-79, BUCHANAN PROF ASTROPHYS, RICE UNIV, 74- *Concurrent Pos:* Scientist-astronaut, NASA, 65-69; mem lunar atmosphere working group, planetary atmospheres subcomt, Space Sci Steering Comt, 66-67; mem sch natural sci, Inst Advan Study, 71-72; trustee, Univs Space Res Asn, 75-; Guggenheim fel, Paris France, 79-80; Humboldt sr US scientist, Heidelberg, Germany, 83-84 & 84-85. *Mem:* AAAS; Am Phys Soc; Am Astron Soc; Am Geophys Union. *Res:* Gravitational collapse; particle acceleration; pulsars; magnetospheric tail structure; solar wind interaction with moon and planets; elementary particles; weak magnetism; nuclear parity violation; symmetries; gravitationally induced electric fields. *Mailing Add:* Dept Space Physics & Astron Rice Univ Houston TX 77251-1892

MICHEL, GERD WILHELM, b Darmstadt, Ger, July 4, 30; m 59. NATURAL PRODUCTS CHEMISTRY, CHEMICAL PROCESS DEVELOPMENT. *Educ:* Darmstadt Tech Univ, Dipl chem, 56, Dr rer nat, 59. *Prof Exp:* Res assoc chem, Urbana, Ill, 59-62; res chemist, Res Div, Photo Prod Dept, E I du Pont de Nemours & Co, 62-64, tech serv specialist, 64-65; asst to dir sales indust chem, E Merck A G, Darmstadt, WGer, 66-67; res fel, Squibb Inst Med Res, 67, sr res fel, 67-82, dir, E R Squibb & Sons, New Brunswick, NJ, 82-85. *Mem:* Am Chem Soc; Soc Ger Chem; NY Acad Sci. *Res:* Chemistry of Mannich-bases, nitrones, alkaloids, steroids; antibiotics; photo polymerization; chemical process development; chemical process technology and engineering. *Mailing Add:* Five Oak Pl Province Hill Princeton NJ 08540-4747

MICHEL, HARDING B, b Louisville, Ky, Aug 17, 24; m 48, 70; c 1. MARINE ZOOLOGY. *Educ:* Duke Univ, AB, 46; Univ Miami, MS, 49; Univ Mich, PhD, 57. *Prof Exp:* Asst zool, 46-48, asst instr, 48-50, instr, Marine Lab, 54-57, asst prof, 57-67, assoc prof, Inst Marine Sci, 67-70, PROF BIOL OCEANOG, ROSENSTIEL SCH MARINE & ATMOSPHERIC SCI, UNIV MIAMI, 70 - *Concurrent Pos:* Asst ed, Bull Marine Sci of the Gulf & Caribbean, 52-53. *Mem:* AAAS; Soc Syst Zool; Soc Study Evolution; Marine Biol Asn UK; Sigma Xi. *Res:* Invertebrate embryology; marine zooplankton; distribution of oceanic zooplankton in Caribbean Sea; ecology of estuaries in South Vietnam; zooplankton of the Arabian Gulf. *Mailing Add:* Rosenstiel Sch Marine & Atms Sci 4600 Rickenbacker Causeway Miami FL 33149

MICHEL, HARTMUT, b Ludwigsburg, WGer, July 18, 48. BIOPHYSICS. *Prof Exp:* DIR, MAX-PLANCK INST BIOPHYS, FRANKFURT, GER, 87- *Honors & Awards:* Nobel Prize in Chem, 88; Biophys Prize, Am Phys Soc, 86. *Mailing Add:* Max Planck Inst fur Biophysik Heinrich-strasse 7 6000 Frankfurt 71 Germany

MICHEL, KARL HEINZ, b Marklissa, Ger, Nov 9, 29; m 59. BIO-ORGANIC CHEMISTRY. *Educ:* Weihenstephan Univ, Ger, BS, 56, dipl, 64; Landau Univ, BS, 58. *Prof Exp:* Chem engr, Cent Lab, Swedish Pharmaceut Soc, 59-60; res asst org chem, Royal Inst Pharm, Stockholm, 60-65; res assoc org chem, Iowa State Univ, Ames, 65-66; res assoc, Royal Inst Pharm, Stockholm, 66-67; proj leader, Fleischmann Lab, Stamford, Conn, 67-69; sr biochemist, Eli Lilly & Co, Indianapolis, Ind, 69-75, res scientist, 75-90; RETIRED. *Res:* Isolation, characterization, structure determination and biological evaluation of new antibiotics and other biologically active compounds; designer of chromatography instrumentation, patentee and consultant in fields. *Mailing Add:* 225 E North St Indianapolis IN 46206

MICHEL, KENNETH EARL, b Chicago, Ill, May 22, 30; m 58; c 2. CYTOGENETICS. *Educ:* Northern Ill Univ, BS, 51, MS, 52; Univ Minn, PhD(genetics), 66. *Prof Exp:* Instr sci, Gavin Sch, 54-57; instr biol, Waldorf Col, 57-62; assoc prof, 66-68, prof genetics, 68-77, PROF BIOL, SLIPPERY ROCK STATE COL, 77- *Concurrent Pos:* Chmn dept genetics, Slippery Rock State Col, 66-74; mem, Maize Genetics Coop. *Mem:* Genetics Soc Am; Am Genetics Asn. *Res:* Interrelated behavior of non-homologous chromosomes in maize; chromosome pairing and disjunction. *Mailing Add:* Dept of Biol Slippery Rock State Col Slippery Rock PA 16057

MICHEL, LESTER ALLEN, b Mexico, Ind, Mar 5, 19; m 42; c 5. CHEMISTRY. *Educ:* Taylor Univ, AB, 41; Purdue Univ, MS, 44; Univ Colo, PhD(phys chem), 47. *Prof Exp:* Asst chem, Purdue Univ, 41-44; tech adv, Manhattan Proj, Linde Air Prod Co, NY, 44-45; asst chem, Univ Colo, 45-46; from instr to prof, 47-70, chmn dept, 59-70, VERNER Z REED PROF EMER PROF CHEM, COLO COL, 84- *Concurrent Pos:* Res Corp grant, 48. *Mem:* AAAS; Am Chem Soc. *Res:* Calorimetry; crystal growth; vapor pressures; isothermal flow calorimeter for vapor phase reactions; surface chemistry. *Mailing Add:* Dept of Chem Colo Col Colorado Springs CO 80903

MICHEL, RICHARD EDWIN, b Saginaw, Mich, Oct 31, 28; m 51; c 3. SOLID STATE PHYSICS. *Educ:* Mich State Univ, BS, 50, MS, 53, PhD(physics), 56. *Prof Exp:* Mem tech staff, RCA Labs, 56-62; sr res physicist, Gen Motors Res Labs, 62-73; instr, 71-73, DEAN, LAWRENCE INST TECHNOL, 73- *Res:* Magnetic resonance; magnetic materials; polymers; semiconductors. *Mailing Add:* Lawrence Technol Univ 21000 W Ten Mile Rd Southfield MI 48075

MICHEL, ROBERT GEORGE, b Sheffield, Eng, Jan 21, 49; m 72; c 2. ATOMIC SPECTROMETRY, ANALYTICAL INSTRUMENTATION. *Educ:* Sheffield Polytech, BSc, 71, PhD(anal chem), 74. *Prof Exp:* Fel, Univ Fla, 74-76, Univ Strathclyde, 76-78; asst prof, 79-85, ASSOC PROF ANALYTICAL CHEM, UNIV CONN, 85- *Concurrent Pos:* Mem staff, Inst Mat Sci, Univ Conn, 79-, asst prof, Sch Allied Health, 81-82; Sr Fulbright-Hays Award for Travel, 74; Res Career & Develop Award, NIH, 84- *Mem:* Royal Soc Chem; Am Chem Soc; Soc Appl Spectroscopy; Sigma Xi. *Res:* Atomic and molecular emission; absorption and fluorescence in plasmas; flames and electrothermal atomizers; development of instrumentation and of sensitive, selective and accurate methods for trace analysis of components in materials based on the above spectroscopic techniques. *Mailing Add:* Dept Chem U 60 Univ Conn 215 Glenbrook Rd Storrs CT 06268

MICHELAKIS, ANDREW M, b Greece, Aug 12, 27; US citizen; m 64; c 2. MEDICINE. *Educ:* Athens Col Agr, BS, 52; Univ Kans, MS, 56; Ohio State Univ, PhD(chem), 59; Western Reserve Univ, MD, 64. *Prof Exp:* Asst chem, Univ Kans, 54-55; asst, Ohio State Univ, 55-59; intern, Mt Sinai Hosp, Cleveland, 64-65; resident, Vet Admin Hosp, 65-66; from instr to assoc prof med, Sch Med, Vanderbilt Univ, 66-74, from asst prof to assoc prof pharmacol, 68-74; dir clin pharmacol, 74-77, PROF MED & PHARMACOL, MICH STATE UNIV, 74- *Concurrent Pos:* Fel endocrinol, Vanderbilt Univ, 66-68. *Mem:* Am Fedn Clin Res; Endocrine Soc; Soc Exp Biol & Med; Am Soc Pharmacol & Exp Therapeut. *Res:* Endocrinology; hypertension and cardiovascular diseases; clinical pharmacology. *Mailing Add:* Dept Pharmacol Mich State Univ East Lansing MI 48824

MICHELBACHER, ABRAHAM E, b Riverside, Calif, Apr 12, 99; m 29; c 1. ENTOMOLOGY. *Educ:* Univ Calif, Berkeley, BS, 27, MS, 31, PhD(entom), 35. *Prof Exp:* Lab asst, dept plant physiol, Citrus Exp Sta, Riverside, Calif, 20-24; asst agr, Am (Crystal) Beet Sugar Co, Oxnard, Calif, 27-30; lab asst, 31-35, jr entomologist, 35-43, from asst prof & asst entomologist to prof & entomologist, EMER PROF & EMER ENTOMOLOGIST, DEPT ENTOM, UNIV CALIF, BERKELEY, 61- *Mem:* Fel AAAS; Sigma Xi. *Res:* Taxonomy of symphyla; biology and control of pests of field, fruit, forage, truck and nut crops; supervised control of insects leading to integrated pest management; pollination of cucurbita; penetration of package materials by insects and sanitation on ships. *Mailing Add:* 1701 Thousand Oaks Blvd Berkeley CA 94707

MICHELI, LYLE JOSEPH, b LaSalle, Ill, Aug 9, 40; m 64; c 2. SURGERY, NEUROSCIENCE. *Educ:* Harvard Univ, BA, 62, MD, 66. *Prof Exp:* Asst prof orthop surg, George Wash Univ, 72-74; instr, 74-84, ASST PROF ORTHOP SURG, HARVARD MED SCH, 84- *Concurrent Pos:* Dir, Div Sports Med, Harvard Med Sch, 75-88; assoc ed, Med & Sci Sports & Exercise, 84-88. *Mem:* Am Acad Pediat; Am Acad Orthop Surg; Scoliosis Res Soc; Am Col Sports Med. *Res:* Spinal disorders; sports injuries in children; adaptation to strength training. *Mailing Add:* 300 Longwood Ave Boston MA 02115

MICHELI, ROBERT ANGELO, b San Francisco, Calif, Dec 31, 22; m 77. ORGANIC CHEMISTRY. *Educ:* Univ Calif, BS, 51; Duke Univ, PhD(org chem), 55. *Prof Exp:* Asst, Duke Univ, 51-53; Nat Cancer Inst res fel, Harvard Univ, 54-56; res chemist, Dow Chem Co, 56-58, Western Regional Res Lab, USDA, Calif, 58-62 & Univ Basel, 62-64; res chemist, Hoffmann-LaRoche, Inc, 64-75, res fel, 75-83, sr res fel, 83-85; RETIRED. *Mem:* Am Chem Soc. *Res:* Steroid and prostaglanding chemistry. *Mailing Add:* 1016 Second St Eureka CA 95501

MICHELI, ROGER PAUL, b Denver, Colo, July 26, 49; m 69; c 4. REGULATORY AFFAIRS, PHARMACEUTICAL RESEARCH. *Educ:* Univ Denver, BS, 71; Univ Colo, MS, 73, PhD(phys org chem), 75. *Prof Exp:* Res assoc, Univ Colo, 71-73, sr res assoc, 73-75; postdoctoral fel, Ohio State Univ, 75-76; chief chemist, Acton Analysis Lab, 76-78; sr analytical res chemist, Syntex Chem Inc, 78-80, prin analytical res chemist, 80-83, analytical res & develop mgr, 83-91; MGR, ANALYTICAL RES & REGULATORY SERV, SYNTEX GROUP TECHNOL CTR, 91- *Concurrent Pos:* Expert witness & consult, 76-78. *Mem:* Am Chem Soc; Am Asn Pharmaceut Scientists; AAAS; Pharmaceut Mfrs Asn. *Res:* Bulk pharmaceutical process development; pharmaceutical analysis and characterization; DMF/NDA development. *Mailing Add:* 2075 N 55th St Boulder CO 80301

MICHELMAN, JOHN S, b Portsmouth, Ohio, Apr 19, 38; m 64; c 3. ORGANIC & EMULSION CHEMISTRY. *Educ:* Univ Cincinnati, BS, 60; Harvard Univ, MA, 62, PhD(chem), 65. *Prof Exp:* Asst prof chem, Univ Cincinnati, 66-67; VPRES CHEM, MICHELMAN CHEM, INC, 65- *Mem:* Am Chem Soc. *Res:* Water based coatings; emulsion technology. *Mailing Add:* Michelman Inc 9089 Shell Rd Cincinnati OH 45236

MICHELS, CORINNE ANTHONY, b New York, NY, Jan 2, 43; m 64; c 2. GENE EXPRESSION, YEAST GENETICS. *Educ:* Queens Col, BS, 63; Columbia Univ, MS, 65, PhD(genetics), 69. *Prof Exp:* Res assoc, Columbia Univ, 69-70; res fel, Albert Einstein Col Med, 70-72; asst prof, 72-79, ASSOC PROF BIOL, QUEENS COL, 79- *Mem:* Genetics Soc Am; Am Soc Microbiol; AAAS; Sigma Xi. *Res:* Regulation of gene expression, specifically, in glucose repression in yeast using genetics and gene splicing to study this phenonemon. *Mailing Add:* Dept Biol Queens Col Flushing NY 11367

MICHELS, DONALD JOSEPH, b Brooklyn, NY, Apr 17, 32; m 61; c 6. SOLAR-TERRESTRIAL RELATIONSHIP, SPACE OPTICAL INSTRUMENTATION. *Educ:* St Peter's Col NJ, BS, 54; Fordham Univ, MS, 56; Cath Univ Am, PhD(physics), 70. *Prof Exp:* SUPVRY RES PHYSICIST, NAVAL RES LAB, 61- *Honors & Awards:* Sigma Xi Award for Pure Sci, Naval Res Lab, 83. *Mem:* Optical Soc Am; Am Geophys Union; Am Astron Soc. *Res:* Solar radiation and solar-terrestrial relationships; spacecraft optical instrumentation; electro-optical imaging detectors. *Mailing Add:* E O Hulburt Ctr Space Res Code 4166 Naval Res Lab Washington DC 20375-5000

MICHELS, H(ORACE) HARVEY, b Philadelphia, Pa, Dec 9, 32; m 58, 75; c 2. CHEMICAL PHYSICS. *Educ:* Drexel Inst Technol, BSChE, 55; Univ Del, MChE, 57, PhD, 60. *Prof Exp:* Res engr, G & W H Corson Co, Inc, Pa, 51-53; sr anal engr, 59-62, sr res scientist, 62-68, SR THEORET PHYSICIST, UNITED TECHNOLOGIES CORP, 68- *Concurrent Pos:* Adj asst prof, Rensselaer Polytech, Hartford Grad Ctr, 60-65, adj assoc prof, 65-69, adj prof, 69-72; Nat Bur Stand vis fel, Joint Inst for Lab Astrophys, Univ Colo, 70; vis scholar, Quantum Inst, Univ Calif, Santa Barbara, 71; adj prof, Univ

Hartford, 75- *Mem:* Am Chem Soc; fel Am Phys Soc; Sigma Xi. *Res:* Quantum mechanics of the electronic structure of atoms and molecules; thermochemistry and kinetics of reacting gaseous systems at high temperatures; atomic recombination and transport processes. *Mailing Add:* Physics Dept United Technologies Res Ctr 400 Main St East Hartford CT 06108

MICHELS, LESTER DAVID, b Chicago, Ill, Feb 5, 48; m 69. RENAL PHYSIOLOGY. *Educ:* Univ Minn, BS ChE, 70, PhD(physiol), 75. *Prof Exp:* Process res engr, Dow Chem Co, 70-71; res specialist renal transplant, Dept Surg, Univ Minn, 75-76; LECTR, DEPT PHYSIOL, UNIV MINN & RES PHYSIOLOGIST, MINNEAPOLIS MED RES FOUND, 76- *Mem:* Am Physiol Soc; Am Soc Nephrol; Int Soc Nephrol. *Res:* Glomerular filtration dynamics and permeability in normal and disease states. *Mailing Add:* Sandoz Nutrit 1541 Vernon Ave S Box 370 Minneapolis MN 55440

MICHELS, LLOYD R, b San Francisco, Calif, Aug 2, 16; m 43; c 2. CHEMICAL & NUCLEAR ENGINEERING. *Educ:* Univ Calif, BS, 38; Univ Ill, MS, 40, PhD(chem eng), 41. *Prof Exp:* Asst chem, Univ Ill, 39-41; res chem engr & metallurgist, Permanente Metals Corp, Calif, 41-42; asst chem engr, US Bur Mines, Wash, 42-43, assoc chem engr, Utah, 43-45; develop engr, Titanium Div, Nat Lead Co, 45-51; lead engr, Calif Res & Develop Co, 51-53; tech expert, Magnesium Prod Dept, Dow Chem Co, 53-54; sr engr, Gen Elec Co, 54-63, mgr separations process design eng, 63-66; mgr eng & res, Isochem, Inc, 66-67; prin design engr, Atomic Prod Equip Dept, Gen Elec Co, 67-76; mgr spec projs, Gen Elec Uranium Mgt Corp, 76-80; consult, Nuclear & Chem Processes & Systs, 80-88; pres, Lloyd R Michels, PE, Inc, Consults-Chem & Nuclear Eng, 82-88; COUNR TO SMALL BUS, SR CORPS RETIRED EXECS, 88- *Mem:* Am Chem Soc; Am Inst Chem Engrs; Am Nuclear Soc. *Res:* Magnesium; titanium metal, pigments and tetrachloride; chemical separations related to atomic power reactors; process design and economic evaluation. *Mailing Add:* 1713 Husted Ave San Jose CA 95124-1927

MICHELS, ROBERT, b Chicago, Ill, Jan 21, 36; c 2. PSYCHIATRY, PSYCHOANALYSIS. *Educ:* Univ Chicago, BA, 53; Northwestern Univ, MD, 58; Am Bd Psychiat & Neurol, dipl, 64; Columbia Univ, cert psychoanal med, 67. *Prof Exp:* Res assoc, Lab Clin Sci, NIMH, 62-64; from instr to assoc prof psychiat, Col Physicians & Surgeons, Columbia Univ, 64-74; PROF PSYCHIAT & PSYCHIATRIST IN CHIEF, NY HOSP-CORNELL MED CTR, 74- *Concurrent Pos:* Spec lectr & instr psychiat, Columbia Univ, 60-74, attend psychiatrist, Student Health Serv, 66-74, mem fac & supv & training analyst, Psychoanal Ctr Training & Res, 67-; NIMH career teacher, 64-66; from asst to assoc attend psychiatrist, Vanderbilt Clin & Presby Hosp, 64-74; from asst to attend psychiatrist, St Lukes Hosp Ctr, NY, 66-; from asst examr to dir, Am Bd Psychiat & Neurol, 67-82; secy, Inst Soc, Ethics & Life Sci, 72-77. *Mem:* Fel Am Psychiat Asn; Royal Medico-Psychol Asn; Asn Res Nerv & Ment Dis; Am Psychoanal Asn; Group Advan Psychiat. *Res:* Psychiatric education. *Mailing Add:* Dept Psychiat Cornell Univ Med Col 525 E 68th St New York NY 10021

MICHELSEN, ARVE, b Hamar, Norway, Oct 25, 23; US citizen; m 51; c 2. RADIOBIOLOGY, ELECTRONICS. *Educ:* Univ Ariz, BS, 50; Stanford Univ, MS, 51; Johns Hopkins Univ, PhD(biomed eng), 70. *Prof Exp:* Res engr, Aerojet Gen Corp, 56-58; proj engr, 58-68, DIV & DEPT STAFF ENGR, APPL PHYSICS LAB, JOHNS HOPKINS UNIV, 68-, MEM PRIN PROF STAFF, 78- *Concurrent Pos:* Vis prof, Johns Hopkins Univ, 88- *Mem:* Radiation Res Soc; Am Defense Preparedness Asn. *Res:* Electromagnetic radiating and receiving systems; ordnance devices; flash x-ray systems and radiography; biomedical engineering; radiobiology; electronic systems engineering and integration; air defense systems requirements and performance analysis. *Mailing Add:* Appl Physics Lab Johns Hopkins Univ Johns Hopkins Rd Laurel MD 20707-6090

MICHELSON, EDWARD HARLAN, b St Louis, Mo, June 6, 26; c 6. MALACOLOGY, PUBLIC HEALTH. *Educ:* Univ Fla, BS, 49, MS, 51; Harvard Univ, PhD(biol), 56. *Prof Exp:* Instr biol, Cambridge Jr Col, 51-53; asst, 53-55, res assoc, 55-57, from instr to asst prof, 57-69, assoc mollusks, Mus Comp Zool, 57-77, ASSOC PROF TROP PUB HEALTH, SCH PUB HEALTH, HARVARD UNIV, 69-; AT DEPT PREV MED, UNIFORMED SERVS UNIV HEALTH SCI. *Concurrent Pos:* La State Univ-China Med Bd fel, 59; advisor Schistosomiasis, Pan Am Health Orgn, Orgn Am States, WHO, 70. *Mem:* Am Soc Trop Med & Hyg; Am Soc Parasitol; Am Malacol Union; NY Acad Sci; Netherlands Malacol Soc. *Res:* Ecology of the terrestrial mollusca of Florida; taxonomy of West Indian land and fresh water mollusca; biological control of the intermediate snails host of Schistosomiasis. *Mailing Add:* Dept Prev Med Uniformed Servs Univ Health Sci 4301 Jones Bridge Rd Bethesda MD 20814

MICHELSON, ERIC L, b Philadelphia, Pa, Sept 18, 47; m 77; c 2. CARDIAC ELECTROPHYSIOLOGY, CARDIOVASCULAR PHARMACOLOGY. *Educ:* Univ Pa, BA, 69, MS, 69; Columbia Univ Col Physicians & Surgeons, MD, 73. *Prof Exp:* Intern, Hosp Univ Pa, 73-74, resident, 74-76, fel, 76-78, res fel, 78-79; from asst prof to prof med, Jefferson Med Col, Thomas Jefferson Univ, 79-89; assoc investr, Lankenau Med Res Ctr, 79-87, sr scientist, 87-88; PROF MED, HAHNEMANN UNIV, 89-, DIR RES, LIKOFF CARDIOVASC INST, 88-, DIR, DIV CARDIOL, 89- *Concurrent Pos:* Adj asst prof physiol, Univ Pa Sch Vet Med, 79-82, adj assoc prof, 82-88, adj prof, 88-; chief, Clin Res Unit, Lankenau Med Res Ctr, 79-82, chief, Clin Res, 82-88; prin investr, Clin Investigatorship Award, Nat Heart, Lung & Blood Inst, NIH, 80-88 & grant-in-aid awards, Am Heart Asn; course co-dir & fac, Am Col Cardiol, 79-; consult, Nat Heart, Lung & Blood Inst, NIH, 83-; mem, Extramural Prog Comt, Am Col Cardiol, 82-88. *Mem:* Am Fedn Clin Res; Am Physiol Soc; fel Am Col Physicians; fel Am Col Cardiol (asst treas, 89-); Int Soc Heart Res. *Res:* Studies to determine the electrophysiologic mechanisms responsible for the lethal disorders of cardiac rhythm afflicting patients with chronic ischemic heart disease; electropharmacology of potential new antiarrhythmic drugs; physiology of antihypertensive therapy. *Mailing Add:* Hahnemann Univ MS 470 Broad & Vine Sts Philadelphia PA 19102-1192

MICHELSON, IRVING, b NJ, Jan 4, 22; m 54; c 7. FLUID DYNAMICS, PHYSICAL OCEANOGRAPHY. *Educ:* Ga Inst Technol, BS, 43; Calif Inst Technol, MS, 47, PhD(aeronaut, math), 51. *Prof Exp:* Lectr, Univ Calif, Los Angeles, 51-54; res engr, Odin Assocs, 54-57; prof aeronaut eng & head dept, Pa State Univ, 57-60; PROF AEROSPACE ENG, ILL INST TECHNOL, 60- *Concurrent Pos:* Consult, US Naval Ord Test Sta, 50-, Rand Corp, 51-52, IIT Res Inst, 60-, US Air Force, Argonne Nat Lab, US Naval Observ, C-E-I-R, Inc & Smithsonian Astrophys Observ; vis prof, Univ Nancy, 61-; mem, Adv Comt Pan-Am Policy, 70- *Mem:* Am Astron Soc; Royal Soc New SWales. *Res:* Astrodynamics; orbital and celestial mechanics; aero-dynamics; tides. *Mailing Add:* Dept Mech & Aerospace Eng Ill Inst Technol 7061 N Kedzie Ave No 202 Chicago IL 60645

MICHELSON, LARRY, b Philadelphia, Pa, Nov 6, 52. ANXIETY DISORDERS, ANTISOCIAL BEHAVIOR. *Educ:* Temple Univ, BA, 74; Nova Univ, MS, 75, PhD(clin psychol), 78. *Prof Exp:* Postdoctoral fel, 78-79, asst prof, 79-86, ASSOC PROF, PSYCHIAT & PSYCHOL, UNIV PITTSBURGH, 86-; DEPT PSYCHIAT, SCHL MED, WESTERN PSYCHIAT INSTIT & CLINIC, 86- *Concurrent Pos:* Prin investr, prevention antisocial behav in children, 84-; Spec adv, Carnegie Coun Adolescent Develop violence prevention & Comt Instit Behav & Health Prevention of Drug Abuse, 87- *Mem:* Am Psychol Asn; Soc Psychother Res; Asn Adv Behav Ther. *Res:* Adult anxiety disorders; prevention of antisocial behavior in children. *Mailing Add:* 548 Moore Bldg Pa State Univ University Park PA 16802

MICHELSON, LESLIE PAUL, b New York, NY, June 23, 43; m 70; c 3. HARDWARE SYSTEMS. *Educ:* Adelphi Univ, BA, 66, MS, 68, PhD(exp physics), 75. *Prof Exp:* Lectr physics, Adelphi Univ, 68-74; res collabr, Brookhaven Nat Lab, 70-74, Dept Nuclear Med, Mt Sinai Hosp, 74-75; MGR, BIOL MED ENG LAB COMPUT SERV, UNIV MED & DENT NJ, 75- *Mem:* Am Phys Soc. *Res:* Application of digital computer technology to data acquisition; control and analysis problems in a medical research environment. *Mailing Add:* Univ Med & Dent NJ 185 S Orange Ave Newark NJ 07103

MICHELSON, LOUIS, b Lynn, Mass, Mar 24, 19; m 41; c 1. PHYSICS. *Educ:* Mass Inst Technol, BS, 40. *Prof Exp:* Physicist, Corning Glass Works, 40-41; electronic engr, Sanborn Instruments Co, 45-46; gen mgr, Allied Cement & Chem Co, 46-47; tech dir electromagnetics, US Army Ord Submarine Mine Lab, 47-50; chief, Mine Div, US Naval Ord Lab, 50-51; tech dir torpedo hydrodynamics & acoust, US Naval Underwater Ord Sta, 51-55; mgr rocket engines, Flight Propulsion Lab, Gen Elec Co, 55-60, space environ simulator fac, Missile & Space Vehicle Dept, 60-61, Nimbus Proj, Spacecraft Dept, 61-64, NASA progs, 64-65 & adv requirements, 65-66; pres, Spacerays, 66-67; pres, 67-84, CHMN, LION PRECISION CORP, 84- *Concurrent Pos:* Mem acoust & ord panels, Res & Develop Bd, 47-50 & planning coun & torpedo planning adv comt, Bur Ord, 51-55. *Mem:* Am Mgt Asn; Am Ord Asn; Am Inst Aeronaut & Astronaut; Inst Elec & Electronics Eng. *Res:* Underwater sound and electric phenomena; electromagnetic fields; electronic control systems; rocket propulsion; high vacuum techniques. *Mailing Add:* 25 Beechcroft Rd Newton MA 02158

MICHENER, CHARLES DUNCAN, b Pasadena, Calif, Sept 22, 18; m 40; c 4. ENTOMOLOGY. *Educ:* Univ Calif, BS, 39, PhD(entom), 41. *Prof Exp:* Tech asst entom, Univ Calif, 39-42; from asst cur to assoc cur Lepidoptera & Hymenoptera, Am Mus Natural Hist, 42-48; dir Snow Entom Mus, 74-83; from assoc prof to prof, Univ Kans, 48-59, chmn dept entom, 49-61 & 72-75, actg chmn, Dept Systs & Ecol, 68-69, Elizabeth M Watkins prof, 59-89, Watkins prof systs & ecol, 69-89, EMER PROF ENTOM, UNIV KANS, 89- *Concurrent Pos:* State entomologist, Southern Div, Kans, 49-61; Am ed, Insectes Sociaux, 54-55 & 62-; Guggenheim fel & res prof, Univ Parana, 55-56; pres, Am sect, Int Union Study Soc Insects, 57-60, vpres, Western Hemisphere Sect, 79-80, pres, 77-82; Fulbright scholar, Univ Queensland, 58-59; ed, Evolution, Soc Study Evolution, 62-64; Guggenheim fel, Africa, 66-67; assoc ed, Annual Review Ecology & Systematics, 70-90; res assoc, Am Mus Natural Hist, 49- *Honors & Awards:* A Cressey Morrison Prize, NY Acad Sci, 43; Founders' Award, Entom Soc Am, 81. *Mem:* Nat Acad Sci; Entom Soc Am; Soc Study Evolution (pres, 67); Soc Syst Zool (pres, 68); hon fel Am Entom Soc; Am Acad Art & Sci; Acad Sci Brazil; Royal Entom Soc London; Linnean Soc London; Am Soc Naturalists (pres, 78). *Res:* Biology and taxonomy of bees; behavior of social insects; principles of systematics; bee systematics and behavior. *Mailing Add:* Entom Mus Snow Hall Univ Kans Lawrence KS 66045

MICHENER, CHARLES EDWARD, b Red Deer, Alta, Can, Jan 4, 07; m 36; c 3. GEOLOGY. *Educ:* Univ Toronto, BA, 31; Cornell Univ, MS, 32; Univ Toronto, PhD, 40. *Prof Exp:* Explor geologist, 32-35; geologist, Int Nickel Co, Ltd, 35-39, res geologist, 39-45, vpres,45-69, chief explorer, 55-69; consult geologist, C E Michener & Assoc, Ltd, 69-70; CONSULT GEOLOGIST, DERRY, MICHENER, BOOTH & WOHL, 70- *Honors & Awards:* Blaylock Medal, Can Inst Mining & Metall, 85. *Mem:* Soc Econ Geologists; Am Inst Mining, Metall & Petrol Eng; Can Inst Mining & Metall. *Res:* Examination of Sudbury type ores resulted in discovery of new palladium and platinum mineral ores named Michenerite for the author; research on geophysical airborne techniques resulted in discovery of several important copper-nickel deposits now commonly mined at Thompson Manitoba Canada and Indonesia. *Mailing Add:* 31 Rosedale Rd Apt 306 Toronto ON M4W 2P2 Can

MICHENER, H(AROLD) DAVID, b Pasadena, Calif, Dec 21, 12; m 39; c 4. MICROBIOLOGY. *Educ:* Calif Inst Technol, BS, 34, PhD(plant physiol), 37. *Prof Exp:* Asst, Scripps Inst, Calif, 37-38; jr pomologist, Exp Sta, Univ Hawaii, 38-40; asst, Calif Inst Technol, 40-42; from jr chemist to prin chemist, 42-80, EMER PRIN CHEMIST, WESTERN REGIONAL RES CTR, USDA, ALBANY, CALIF, 80- *Mem:* AAAS; Inst Food Technol; Am Soc Microbiol. *Res:* Heat resistance of bacterial spores; heat resistant fungi; food poisoning and spoilage of microbial origin; microbiological standards for foods; psychrophils; growth and survival of microorganisms at low temperatures; microbiology of frozen and chilled foods. *Mailing Add:* 2616 Etna St Berkeley CA 94704

MICHENER, JOHN WILLIAM, b Wilkinsburg, Pa, May 14, 24; m 53; c 2. PHYSICS. *Educ:* Carnegie Inst Technol, BS, 46, MS & PhD(physics), 53. *Prof Exp:* Asst physics, Carnegie Inst Technol, 42-50; physicist, Owens-Corning Fiberglas Corp, 51-59; head, Dept Physics, 59-73, mgr, Textile Testing Dept, 75-80, MGR, RES SERV DIV, MILLIKEN RES CORP, 80- *Mem:* Am Phys Soc; AAAS; Am Soc Testing & Mat. *Res:* Structure and properties of textile fibers; physics of textiles; static electricity in textile materials; flammability of textile materials; combustion toxicity of textile materials. *Mailing Add:* Milliken Res Corp Res Serv Div M415 Spartanburg SC 29304

MICHENFELDER, JOHN D, b St Louis, Mo, Apr 13, 31; m. ANESTHESIOLOGY. *Educ:* St Louis Univ, MD, 55; Am Bd Anesthesiol, cert, 63; FFARCSI, 82; FFARCSE, 88. *Prof Exp:* Intern, Presby Hosp, Chicago, Ill, 55-56, resident, 56; resident anesthesiol, Mayo Found, 58-61, from instr to assoc prof, Med Sch, 63-71, head, Neurosurg Anesthesia, 68-75, head, St Mary's Sect Anesthesia, 71-75, chmn, Div Anesthesia Res, 87-90, PROF ANESTHESIA, MAYO MED SCH, 75- *Concurrent Pos:* Mem, Stroke Coun, Am Heart Asn; councilman, Asn Univ Anesthetists, 75-78; chmn, subcomt anesthetic action, Am Soc Anesthesiologists, 77-78; mem, Neurol B Study Sect, NIH, 78-80. *Honors & Awards:* Excellence in Res Award, Am Soc Anesthesiologists, 90, Distinguished Serv Award, 90. *Mem:* Inst Med-Nat Acad Sci; Am Soc Anesthesiologists; Int Anesthesia Res Soc; Sigma Xi; Am Asn Neurol Surgeons; Am Heart Asn; Soc Cerebral Blood Flow & Metab; Asn Univ Anesthetists. *Res:* Cerebral metabolism and blood flow; cerebral effects of anesthesia-related interventions. *Mailing Add:* Dept Anesthesiol Mayo Clin 200 SW First St Rochester MN 55905

MICHIE, DAVID DOSS, b Aniston, Ala, Feb 22, 36; m 66; c 2. CLINICAL PHARMACOLOGY, CARDIOVASCULAR PHYSIOLOGY. *Educ:* Trinity Univ, Tex, BS, 58, MSc, 59; Univ Tex, PhD(physiol), 66. *Prof Exp:* Sr res physiologist, Technol Inc, 65-67; asst prof physiol, Med Sch, Creighton Univ, 67-70; asst prof surg, Sch Med, Univ Miami, 70-73; prof physiol & bioeng & chmn dept, Eastern Va Med Sch, 73-78; PRES, CLIN PHYSIOL ASSOCS, 78- *Concurrent Pos:* US Army Inst Surg Res, 60-65. *Mem:* Fel Am Col Cardiol; Am Physiol Soc; Am Soc Clin Pharmacol & Therapeut; fel Am Col Clin Pharmacol. *Res:* Clinincal trials of pharmaceuticals; non-invasive vascular diagnostics; cardiodynamics. *Mailing Add:* Clin Physiol Assoc 3594 Broadway, Ste C Ft Myers FL 33901

MICHIE, JARVIS D, US citizen. CIVIL ENGINEERING. *Educ:* Univ Tex, BS, 55; La State Univ, MS, 61. *Prof Exp:* Res asst, Hydraul Group, Dept Civil Eng, Univ Tex, 54-55; design engr, Struct Design Group, Ethyl Corp, 57-61; design engr, Ezra Meir & Assocs, 62; assoc res engr, Dept Struct Res, 62-63, sr res engr, 63-69, group leader, 69-71, sect mgr, 71-76, dir Struct Systs & Fire Technol Dept, 76-80, DIR STRUCT ENG DEPT, SOUTHWEST RES INST, 80- *Concurrent Pos:* Chmn, Comt Roadside Safety Appurtenances, Transp Res Bd. *Mem:* Am Soc Civil Engrs; Nat Soc Prof Engrs. *Mailing Add:* Dept Struct Eng Southwest Res Inst PO Drawer 28510 San Antonio TX 78284

MICHIELLI, DONALD WARREN, b Queens Village, NY, Aug 15, 34; m 89. EXERCISE PHYSIOLOGY. *Educ:* Springfield Col, BS, 57; Ohio State Univ, MA, 61, PhD(exercise physiol), 65. *Prof Exp:* Res asst, Dept Aviation Med, Ohio State Univ, 62-63; from asst prof to assoc prof physiol, Long Island Univ, 66- 71; adj assoc prof res methods, Hunter Col, City Univ New York, 73-75; adj assoc prof physiol exercise, NY Univ, 76-79; PROF, DEPT PHYS EDUC & DIR, LAB WORK PHYSIOL, BROOKLYN COL, CITY UNIV NEW YORK, 84- *Concurrent Pos:* Prin investr, res grants, City Univ New York, Blood Lipids & Exercise, 73-74, Effects of Exercise Training on Cardiac Contractility, 75-76 & Spectral Anal of Acoust Characteristics of Human Muscle, 86-91. *Honors & Awards:* Dedicated Serv Award for 25 Years of Serv to Mission, Am Col Sports Med. *Mem:* Sigma Xi; fel Am Col Sports Med; AAAS; Am Heart Asn; NY Acad Sci. *Res:* Focusing on a new technique which can quantify the contractive state of muscle tissue through acoustic myography by using a fast fourier to transform a spectral analysis of the frequency and intensity of muscle sounds is created-after further study, it may develop into non-invasive technique for muscle fiber typing and diagnosing muscle disease. *Mailing Add:* Brooklyn Col Bedford & Ave H Brooklyn NY 11210

MICHIELS, LEO PAUL, b Detroit, Mich, Jan 24, 28. PHOTOCHEMISTRY OF CYANOFERRATES, ELECTRONIC SPECTRUM OF CYANOFERRATES. *Educ:* Manhattan Col, AB, 53, MA, 59; Univ Detroit, MS, 64, PhD(chem), 68. *Prof Exp:* Sci dept head chem & physics, St Joseph High Sch, Detroit, 59-64; sci dept head chem, physics & math, De La Salle Col, Detroit, 64-74; adj asst prof, 74-75, asst prof, 75-83, ASSOC PROF CHEM, MANHATTAN COL, RIVERDALE, NY, 83-, DEPT CHMN CHEM, 88- *Mem:* Am Chem Soc; Sigma Xi. *Res:* Synthesis and characterization of cyanoferrate dimers. *Mailing Add:* Dept Chem Manhattan Col 4513 Manhattan Col Pkwy Riverdale NY 10471

MICHL, JOSEF, b Prague, Czech, Mar 12, 39; nat US; m 69; c 2. CHEMISTRY. *Educ:* Charles Univ, Prague, MS, 61; Czech Acad Sci, PhD(chem), 65. *Hon Degrees:* Dr, Georgetown Univ, 90. *Prof Exp:* Fel, Univ Houston, 65-66 & Univ Tex, Austin, 66-67; res chemist, Inst Phys Chem, Czech Acad Sci, 67-68; asst prof, Aarhus Univ, 68-69; fel chem, Univ Utah, 69-70, res assoc prof, 70-71, from assoc prof to prof, 71-86, chmn dept, 79-84; M K Collie-Welch Regents chair chem, Univ Tex, Austin, 86-91; PROF CHEM, UNIV COLO, BOULDER, 91- *Concurrent Pos:* A P Sloan Found fel, 71-75; Alexander von Humboldt sr US scientist award, 80; J S Guggenheim fel, 84-85; ed, Chem Revs, 84-; assoc ed, Theoret Chem Acta, 85- *Honors & Awards:* Robert A Welch Found Lectr, 83; Morris S Kharasch Lectr, Univ Chicago, 84; Landsowne Lectr, Univ Victoria, 85; Herbert C Brown Lectr, Purdue Univ, 89. *Mem:* Nat Acad Sci; Interam Photochem Soc; Royal Soc Chem; Europ Photochem Asn; Am Chem Soc. *Res:* Physical organic chemistry; electronic spectroscopy of organic molecules; low temperature chemistry, especially preparation of new species and photochemical mechanisms; organosilicon chemistry; gas phase ion and cluster chemistry; frozen gas sputtering; author of numerous technical publications. *Mailing Add:* Dept Chem Univ Colo Boulder CO 80309-0215

MICHLMAYR, MANFRED, b Thorn, Poland, Aug 14, 43; Austrian citizen; m 68; c 2. INORGANIC CHEMISTRY, CATALYSIS. *Educ:* Vienna Tech Univ, BSc, 63, MSc, 66, PhD(inorg chem), 67; Univ Calif, MBA, 75. *Prof Exp:* Asst prof inorg chem, Vienna Tech Univ, 66-67; res assoc electrochem, Czech Acad Sci, 67; NSF fel electrochem, Univ Calif, Riverside, 67-68; res chemist, Shell Develop Co, 68-72; sr res chemist, Chevron Res Co, 72-80, sr res assoc, 80-86; STRATEGIC PLANNING CONSULT, 86- *Concurrent Pos:* Session chmn, Gordon Res Conf, 75. *Honors & Awards:* Karoline Krafft Medal, Austrian Govt, 68. *Mem:* Am Chem Soc; Catalysis Soc; Electrochem Soc. *Res:* Catalysis in petroleum and synthetic fuel processing, including search for novel catalysts and new processes; mechanistic studies of heterogeneous catalytic systems; extractive metallurgy; business, mergers and acquisitions. *Mailing Add:* 225 Bush St San Francisco CA 94104

MICKAL, ABE, b Talia, Lebanon, June 15, 13; US citizen; m 42; c 4. OBSTETRICS & GYNECOLOGY. *Educ:* La State Univ, BS, 36, MD, 40; Am Bd Obstet & Gynec, dipl, 51. *Prof Exp:* Instr anat, 45-46, from clin instr to clin assoc prof, 49-59, chmn dept, 59-80, PROF OBSTET & GYNEC, UNIV NEW ORLEANS MED CTR, 59-, EMER CHMN DEPT, 80- *Mem:* Fel Am Col Surgeons; AMA; Am Col Obstetricians & Gynecologists; Asn Univ Profs (pres, 76-77); Soc Gynec Oncol. *Mailing Add:* Dept Obstet & Gynec St Jude Med Ctr New Orleans LA 70112

MICKEL, HUBERT SHELDON, b Bridgeton, NJ, Aug 27, 37; m 79; c 5. NEUROLOGY, EMERGENCY MEDICINE. *Educ:* Eastern Nazarene Col, BS, 58; Harvard Med Sch, MD, 62; Am Bd Neurol & Psychiat, dipl, 71; Emergency Med, dipl, 81. *Prof Exp:* Intern, Mary Fletcher Hosp, Burlington, Vt, 62-63; resident internal med, Royal Victoria Hosp, Montreal, Que, 63-64; resident neurol, Boston City Hosp, 64-67; consult neurol, Travis State Sch, Austin, Tex, 68-70; instr, Harvard Med Sch, 70-71, asst prof, 71-77, asst clin prof neurol, 77-83; assoc neurol, Children's Hosp Med Ctr, 76-83; dir emergency med, Carney Hosp, 81-82; spec expert, Lab Exp Neuropath, Nat Inst Neurol Commun Dis & Stroke, 87-89, spec vol, Lab Biochem, Nat Heart Lung & Blood Inst, 90-91, GUEST RESEARCHER, LAB EXP NEUROPATH, NAT INST DIS & STROKE, NIH, BETHESDA, 91- *Concurrent Pos:* Res fel neurol, Harvard Med Sch, 64-67, NIH spec fel chem, Harvard Univ, 67-68; consult, Boston State Hosp, 70-71; asst neurol, Children's Hosp Med Ctr, 70-76; instr, Sch Med, Boston Univ, 70-; pre-med adv, Leverett House, Harvard Col, 71-79; asst neurol, Beth Israel Hosp, Boston, 71-79; mem consult staff, Emerson Hosp, Concord, 71-75; dir med & res, Wrentham State Sch, Mass, 73-76, dir dept neurol, Wrentham State Sch Div, Children's Hosp Med Ctr, 74-76; hon res assoc, Dept Chem, Harvard Univ, 76-78; affil, Leverett House Harvard Col, 79-87; fel shock-trauma, Wash Hosp Ctr, Wash, DC, 81; emergency physician, Suburban Hosp, 83-91, Barnes Hosp, St Louis, Mo, 89-91, Georgetown Univ Hosp, 90-91; consult emergency med, US Army Surgeon Gen, US Army Reserve, Med Corps, Individual obilization Augmentee Prog, 84-87; clin assoc prof neurol, Uniformed Serv, Univ Health Sci, 85-; vis scientist, Dept Neuropath, Armed Forces Inst Path, Washington, DC, 87; chmn res comt, Soc Acad Emergency Med, 91-93. *Mem:* AAAS; Am Acad Neurol; NY Acad Sci; Am Chem Soc; Am Oil Chemists Soc; fel Am Col Emergency Physicians; Univ Asn Emergency Med; Soc Acad Emergency Med. *Res:* Oxidative stress and demyelination; cerebral ischemia, reperfusion and resuscitation; neurological emergencies; relationship of music to the brain. *Mailing Add:* PO Box 41046 Bethesda MD 20824-1046

MICKEL, JOHN THOMAS, b Cleveland, Ohio, Sept 9, 34; m 59; c 4. PLANT TAXONOMY, PLANT MORPHOLOGY. *Educ:* Oberlin Col, BA, 56; Univ Mich, MA, 58, PhD(fern taxon), 61. *Prof Exp:* From asst prof to assoc prof bot, Iowa State Univ, 61-69; CUR FERNS, NY BOT GARDEN, 69- *Concurrent Pos:* Sigma Xi res grant, 62-63; Iowa State Alumni Asn res grant, 62-63; NSF grant, 63-66 & 69-86; Nat Acad Sci-Nat Res Coun sr vis res assoc, Smithsonian Inst, 67-68; adj prof, City Univ New York, 69-; ed, Fiddlehead Forum, Am Fern Soc, 74- & Brittonia, 76-78. *Mem:* Am Fern Soc (vpres, 70-71, pres, 72-73); Bot Soc Am; Am Soc Plant Taxon; Int Asn Plant Taxon; Brit Pteridological Soc; hon mem, Indian Fern Society. *Res:* Monographic studies in the fern genera Anemia and Elaphoglossum; taxonomic work on the ferns of Mexico, Hispaniola, Trinidad; phylogeny of the ferns. *Mailing Add:* NY Bot Garden Bronx NY 10458

MICKELBERRY, WILLIAM CHARLES, b Seattle, Wash, May 26, 33; m 58; c 4. FOOD SCIENCE. *Educ:* Wash State Univ, BS, 55; Purdue Univ, MS, 60, PhD(food sci), 63. *Prof Exp:* From asst prof to assoc prof food sci & biochem, Clemson Univ, 62-68; mgr prod develop, Western Farmers Asn, 68-73; prod develop mgr, 73-78, res & develop mgr, 78-87, MGR TECH SERV, ORE FREEZE DRY, INC, 87- *Concurrent Pos:* Mem, Res & Develop Assocs, Food & Container Inst, 63-68; mem res coun, Poultry & Egg Inst Am, 68-74. *Mem:* Inst Food Technol. *Res:* Influence of dietary and environmental factors upon the food quality attributes of poultry meats; poultry meat tenderness; freeze dried foods, processes of freeze dried compressed foods research and development. *Mailing Add:* Ore Freeze Dry Inc PO Box 1048 Albany OR 97321

MICKELSEN, JOHN RAYMOND, b Portland, Ore, June 1, 28; m 50; c 4. PHYSICAL CHEMISTRY. *Educ:* Linfield Col, BA, 50; Ore State Col, MA, 53, PhD(phys chem), 56. *Prof Exp:* Instr chem, Ore State Col, 54-55; from instr to asst prof, 55-61, ASSOC PROF CHEM, PORTLAND STATE UNIV, 61- *Concurrent Pos:* Rask-Orsted fel, Copenhagen Univ, 66-67. *Mem:* Am Chem Soc; Am Electroplaters Soc. *Res:* Ionic equilibria in nonaqueous solvents; equilbria of complex ions. *Mailing Add:* Rte 1 PO Box 18 Banks OR 97106

MICKELSEN, OLAF, b Perth Amboy, NJ, July 29, 12; m 39, 53; c 2. NUTRITION. *Educ:* Rutgers Univ, BS, 35; Univ Wis, MS, 37, PhD(biochem & orgchem), 39. *Prof Exp:* Chemist, Univ Hosps, Minn, 39-41; assoc scientist, lab physiol & hyg, Univ Minn, 42-44, from asst prof to assoc prof, 44-48; nutritionist, USPHS, 48-51; scientist, Nat Inst Arthritis & Metab Dis, NIH, 51-62; prof nutrit, Mich State Univ, 62-79; distinguished vis prof, Univ Del,

79-81; EMER PROF, MICH STATE UNIV, 81- *Concurrent Pos:* Consult, US Secy War, 42-43; assoc ed, Nutrit Rev, 55-70; mem, nutrit & metab study sect, NIH, 55-61, animal resources panel, 63-67; mem, White House Conf food, nutrit & health, 69; vis prof, Inst Food Technol & Nutrit Sci, Tehran, Iran, 77-79. *Honors & Awards:* Emmett J Culligan Award, World Water Soc, 72; Fel, Am Inst Nutrit, 83. *Mem:* Am Inst Nutrit (secy, 63-66, pres, 73-74); Am Bd Nutrit; Am Chem Soc; Am Soc Biol Chemists; Brit Nutrit Soc; Soc Exp Biol & Med. *Res:* Human nutrition with special attention to prevention of disease; control of obesity; biochemical and physiological effects of starvation; influence of lacto-ovo-vegetarian diet on maintenance of bone density and prevention of osteoporosis. *Mailing Add:* Belton Bridge Rd Rte 1 Lula GA 30554

MICKELSON, JOHN CHESTER, b Winter, Wis, Nov 16, 20; m 47; c 4. GEOLOGY. *Educ:* Augustana Col, AB, 41; Univ Iowa, MS, 48, PhD(geol), 49. *Prof Exp:* Asst geol, Univ Iowa, 47-49; asst prof, Wash State Univ, 49-54; staff geologist, Sohio Petrol Co, 54-60; sr geologist, DX Sunray Oil Co, 61; assoc prof, 61-66, PROF GEOL & GEOL ENG, SDAK SCH MINES & TECHNOL, 66-, CHMN DEPT, 68-, HEAD GEOL & GEOL ENG, 77- *Mem:* Geol Soc Am; Am Asn Petrol Geol. *Res:* Cretaceous stratigraphy and sedimentation of the Rocky Mountains; geomorphology and Pleistocene geology of Iowa and eastern Washington, particularly loesses. *Mailing Add:* 133 E St Charles Rapid City SD 57701

MICKELSON, JOHN CLAIR, b Canton, SDak, Aug 4, 29; m 52; c 2. MICROBIOLOGY. *Educ:* SDak State Col, BS, 51, MS, 57; Iowa State Univ, PhD(dairy bact), 60. *Prof Exp:* From asst prof to assoc prof, 60-71, PROF MICROBIOL, MISS STATE UNIV, 71- *Mem:* AAAS; Am Soc Microbiol; Am Inst Biol Sci. *Res:* Dairy microbiology; microbial lipases active on butter oil; electrolytic decomposition of human wastes; electrolytic demineralization of algae. *Mailing Add:* 171 First St S Biloxi MS 39530

MICKELSON, MICHAEL EUGENE, b Columbus, Ohio, May 3, 40; m 66; c 2. MOLECULAR SPECTROSCOPY, PLANETARY ATMOSPHERES. *Educ:* Ohio State Univ, BSc, 62, PhD(physics), 69. *Prof Exp:* PROF PHYSICS, DENISON UNIV, 69- *Concurrent Pos:* Physicist, Electromagnetic Metrol Div, Laser & Infrared Standards Lab, Aerospace Guidance & Metrol Ctr, 80-81 & 81-82. *Mem:* Optical Soc Am; Am Astron Soc; Am Asn Physics Teachers; Sigma Xi. *Res:* Spectroscopic studies under high resolution of both laboratory and telescopic spectra of molecules of astrophysical interest. *Mailing Add:* Dept Physics & Astron Denison Univ Granville OH 43023

MICKELSON, MILO NORVAL, b Iowa Co, Wis, Feb 27, 11; m 41; c 4. PHYSIOLOGICAL BACTERIOLOGY. *Educ:* Univ Wis, BS, 35; Iowa State Univ, PhD(physiol bact), 39. *Prof Exp:* Res bacteriologist, Com Solvents Corp, Ind, 39-40; instr bact, Univ Mich, 40-45; sr res bacteriologist, Midwest Res Inst, 45-61; mem staff, Nat Animal Dis Lab, 61-82; RETIRED. *Concurrent Pos:* Assoc, Med Ctr, Kans, 52-61. *Mem:* AAAS; Am Chem Soc; Am Soc Microbiol; Am Acad Microbiol. *Res:* Industrial fermentations; growth requirements of microorganisms; intermediary metabolism of microorganisms. *Mailing Add:* 1803 Meadow Lane Ames IA 50010

MICKELSON, RICHARD W, b Detroit, Mich, Nov 14, 30. CHEMICAL ENGINEERING. *Educ:* Wayne State Univ, BS, 53, MS, 62, PhD(chem eng), 64. *Prof Exp:* Process engr, Naugatuck Chem Div, US Rubber Co, 53-56, process control supvr, 56-58; asst prof chem eng, 62-66, ASSOC PROF CHEM ENG, WAYNE STATE UNIV, 66- *Mem:* Am Inst Chem Engrs; Am Chem Soc; N Am Thermal Anal Soc. *Res:* Kinetics of thermal decomposition; combustion of chars. *Mailing Add:* Dept Chem Eng Wayne State Univ 5950 Cass Ave Detroit MI 48202

MICKELSON, ROME H, b Twin Valley, Minn, Feb 16, 31; m 69, 84. AGRICULTURAL ENGINEERING. *Educ:* NDak State Univ, BS, 55. *Prof Exp:* Asst engr, NDak State Univ, 55; agr engr, Northern Great Plains Res Ctr, Soil & Water Conserv Res Div, Agr Res Serv, USDA, 55-61, agr engr, Southwestern Great Plains Res Ctr, Tex, 61, agr engr, Cent Great Plains Res Sta, Colo, 61-66, actg supt, 66-72, location leader, 72-79, agr engr, 79-87; RETIRED. *Mem:* Am Soc Agr Engrs; Am Geophys Union; Am Soc Agron; fel Soil Conserv Soc Am; Soil Sci Soc Am; Am Water Resources Asn; AAAS. *Res:* Land forming practices and runoff management for moisture conservation on dryland areas; ground water drainage and salinity investigations; water harvest techniques; sprinkler irrigation scheduling and management; deep and reduced tillage for irrigated and dryland systems. *Mailing Add:* 54 Cheyenne Pl Walsenburg CO 81089

MICKENS, RONALD ELBERT, b Petersburg, Va, Feb 7, 43; m 77; c 2. APPLIED MATHEMATICS. *Educ:* Fisk Univ, BA, 64; Vanderbilt Univ, PhD(physics), 68. *Prof Exp:* Lab instr physics, Fisk Univ, 61-64, lectr, 66-67, lectr math, 67-68; NSF res fel physics, Mass Inst Technol, 68-70; from asst prof to assoc prof physics, Fisk Univ, 70-81; PROF PHYSICS & CALLAWAY PROF, CLARK ATLANTA UNIV, 82- *Concurrent Pos:* Vis prof, Howard Univ, 70-71, Mass Inst Technol, 73-74, Atlanta Univ, 78-79, Vanderbilt Univ, 80-81 & Joint Inst for Lab Astrophys, 81-82; res grants, NSF, 71-73 & NASA, 75-, Dept Energy, 83-85, Army Res Off, 86-90. *Mem:* AAAS; Am Phys Soc; Sigma Xi; Soc Indust & Appl Math; European Phys Soc. *Res:* Nonlinear difference and different equations; asymptotic analysis; history and sociology of science; numerical analysis; mathematical biology. *Mailing Add:* 2853 Chaucer Dr SW Atlanta GA 30311

MICKEY, DONALD LEE, b Fairfield, Iowa, Mar 28, 43; m 62; c 2. ASTROPHYSICS. *Educ:* Harvard Univ, AB, 64; Princeton Univ, PhD(astrophys sci), 68. *Prof Exp:* Res assoc astrophys, Princeton Univ, 68-69; res fel physics, Calif Inst Technol, 69-70; asst astronr, 70-77, ASSOC ASTRONR ASTROPHYS, INST ASTRON, UNIV HAWAII, 77- *Res:* Solar physics; spectroscopy. *Mailing Add:* 3355 Paty Dr Honolulu HI 96822-1442

MICKEY, GEORGE HENRY, b Claude, Tex, Jan 26, 10; m 32; c 2. CYTOGENETICS. *Educ:* Baylor Univ, AB, 31; Univ Okla, MS, 34; Univ Tex, PhD(genetics), 38. *Prof Exp:* Asst zool, Univ Okla, 32-34; asst genetics, Univ Tex, 34-35, instr zool, 35-38; from instr to assoc prof, La State Univ, 38-48; assoc prof, Northwestern Univ, Ill, 49-56; prof & chmn dept, La State Univ, 56-59, dean grad sch, 59-60; cytogeneticist, 60-66, prof biol, 66-69, assoc dean grad sch, 69-70, actg dean, 70-71, dean Grad Sch, New Eng Inst, 71-75; CLIN ASSOC CYTOGENETICS, DUKE UNIV MED CTR, 75- *Concurrent Pos:* Guggenheim fel, 48; res fel, Calif Inst Technol & Univ Tex, 48; prin biologist, Oak Ridge Nat Lab, 53; vis prof, Univ Bridgeport, 71. *Mem:* AAAS; Am Soc Nat; Genetics Soc Am; Soc Study Evolution; Am Soc Zool. *Res:* Genetics and cytology of Drosophila; cytology of Romalea; radiation genetics; mutation studies; cytogenetic effects of radio frequency waves; human cytogenetics; tissue culture. *Mailing Add:* Methodist Retirement Home 2616 Erwin Rd Rm 119-1 Durham NC 27705

MICKEY, MAX RAY, JR, b Pagosa Springs, Colo, Mar 24, 23; m 48; c 2. STATISTICS. *Educ:* Va Polytech Inst, BS, 47; Iowa State Col, PhD(statist), 52. *Prof Exp:* Asst prof statist, Iowa State Col, 52-55; assoc mathematician, Rand Corp, 55-58; statistician, Gen Analysis Corp, 58-60 & CEIR, Inc, 60-63; RES STATISTICIAN, DEPT BIOMATH, UNIV CALIF, LOS ANGELES, 63- *Mem:* Economet Soc; Inst Math Statist; fel Am Statist Asn; Int Statist Inst. *Res:* Application of statistical concepts and methods to applied problems of research, particularly in medicine. *Mailing Add:* 340 16th St Santa Monica CA 90402

MICKLE, ANN MARIE, b Columbus, Ohio, Sept 12, 45. PLANT PHYSIOLOGY, AQUATIC BIOLOGY. *Educ:* Ohio State Univ, BSc, 67; Univ Wis, PhD(bot), 75. *Prof Exp:* Res assoc bot, Kellogg Biol Sta, Mich State Univ, 76-77; asst prof, 77-81, ASSOC PROF BIOL, LASALLE UNIV, 81- *Mem:* AAAS; Am Soc Plant Physiologists; Am Soc Limnol & Oceanog; Bot Soc Am; Am Inst Biol Sci; Sigma Xi. *Res:* Aquatic macrophyte physiology; inorganic nutrient uptake rates, use in evaluating environmental nutrient availability and modification of lake inlet waters by littoral flora. *Mailing Add:* Dept Biol LaSalle Univ Philadelphia PA 19141

MICKLE, MARLIN HOMER, b Windber, Pa, July 5, 36. ELECTRICAL ENGINEERING. *Educ:* Univ Pittsburgh, BS, 61, MS, 63, PhD(elec eng), 67. *Prof Exp:* Jr engr, IBM Corp, 62; from asst prof to assoc prof elec eng, 67-75, dir, Comput Eng Prog, 81-84, PROF ELEC ENG, UNIV PITTSBURGH, 75- *Concurrent Pos:* Prog dir, Syst Theory & Appln Prog, NSF, Washington, DC, 74-75; consult, Westinghouse, Contraves Goerz Corp, Pittsburgh, Anal Sci Corp, Reading, Mass, Texas Instruments Inc, Dallas, Tex; vpres, Power Resources, Inc & pres, Mickle Comput Technologies, Inc. *Mem:* Inst Elec & Electronics Engrs; Am Platform Asn. *Res:* Computer systems; socio-economic systems; optimization; electric power systems; design of high performance computing and control systems; microprocessors; associated systems, for example vision and speech recognition; signal processing; development systems. *Mailing Add:* 348 Benedum Eng Hall Univ Pittsburgh Pittsburgh PA 15261

MICKLES, JAMES, b Rochester, NY, May 17, 23; m 46; c 2. MEDICINAL CHEMISTRY. *Educ:* Brigham Young Univ, BS, 44; Purdue Univ, MS, 49. *Hon Degrees:* ScD, Mass Col Pharm, 75. *Prof Exp:* Chemist Qm Corps Proj, Columbia Univ, 45-47; from asst prof to assoc prof, 50-67, dean students & dir admis, 74-77, PROF CHEM, MASS COL PHARM, 67-, VPRES OPER, 77-, DEAN ADMIN, 81- *Mem:* Am Asn Cols Pharm. *Res:* Chelates of pharmacologically active compounds; synthesis of anti-radiation compounds. *Mailing Add:* Mass Col Pharm 179 Longwood Ave Springfield MA 02119

MICKLEY, HAROLD S(OMERS), b Seneca Falls, NY, Oct 14, 18; m 41; c 2. CHEMICAL ENGINEERING. *Educ:* Calif Inst Technol, BS, 40, MS, 41; Mass Inst Technol, ScD(chem eng), 46. *Prof Exp:* Chem engr, Union Oil Co, Calif, 41-42; proj engr, Mass Inst Technol, 43-45, from asst prof to prof chem eng, 46-61, Ford prof eng, 61-70, dir ctr advan eng, 63-70; dir, Stauffer Chem Co, 67-83, vpres technol, 71, exec vpres, 72-81, vchmn, 81-83; RETIRED. *Concurrent Pos:* Chem engr, Artisan Metals Co, Mass, 42-44, Godfrey L Cabot Corp, 44-45, Am Aviation Co, Calif, 46 & Ranger Aircraft Engines Co, NY, 46; consult, Fairchild Engine & Airplane Corp, 46-57 & E I du Pont de Nemours & Co, Inc, 57-67; Naval Ordnance Develop Award, 46. *Mem:* Nat Acad Eng; Am Chem Soc; fel Am Inst Chem Engrs; Am Acad Arts & Sci; fel AAAS. *Res:* Momentum, heat and mass transfer; heterogeneous catalysts; automatic process control; viscoelastic behavior of high polymers; applied mathematics; transport processes in fluids; kinetics and catalysis; industrial chemistry. *Mailing Add:* 11 Pequot Trail Westport CT 06880

MICKLICH, JOHN R, mathematics, for more information see previous edition

MICKO, MICHAEL M, b Trebisov, Czech, Nov 9, 35; Can citizen; m 61; c 2. WOOD SCIENCE, ENGINEERING. *Educ:* Slovak Tech Univ, Bratislava, BEng, 59, PhD(polymer chem), 66; Univ BC, PhD(wood sci & technol), 73. *Prof Exp:* Asst engr polymers, Slovak Acad Sci, 59-61; asst prof fiber technol, Slovak Tech Univ, Bratislava, 66-69; res assoc bioresource eng, Univ BC, 73-77; from asst prof to assoc prof, 77-83, PROF WOOD SCI, UNIV ALTA, 84- *Concurrent Pos:* NATO travel fel wood sci-biomass utilization, 83; NSERC int fel, Slovak Acad Sci, Bratislava, Czech, 84; Norad professorship, Sokoine Univ Agr, Morogoro, Tanzania, 85. *Mem:* Soc Prof Engrs; Forest Prod Res Soc; Soc Wood Sci & Technol; Tech Asn Pulp & Paper Indust; Can Pulp & Paper Asn. *Res:* Chemical and engineering aspects of wood products; wood quality tree improvement; wood energy; energy from forest chemical composition of wood; engineering properties of wood products. *Mailing Add:* Dept Agr Eng Univ of Alta Edmonton AB T6G 2H1 Can

MICKS, DON WILFRED, preventive medicine, community health, for more information see previous edition

MICOZZI, MARC S, b Norfolk, Va, Oct 27, 53; m 83; c 1. BIOMEDICAL ANTHROPOLOGY, NUTRITIONAL EPIDEMIOLOGY. *Educ:* Pomona Col, BA, 74, Univ Penn, MD, 79, PhD (anthropol), 86. *Prof Exp:* Sr invest epidemiol, Nat Cancer Inst, 84-86; ASSOC DIR, ARMED FORCES, INST PATHOL, WASH, DC, 86-; DIR, NAT MUS HEALTH & MED, WASH, DC, 86- *Concurrent Pos:* Adj prof, Uniformed Serv Univ Health Sci, 87-; mem, NATO Adv Study Inst Growth, 82; delegate, White House Conf Children & Youth, 71, Anthropol Standardization Conf, 85; co-investr, Nat Cancer Inst Diet & Cancer Studies, 86- *Mem:* Am Acad Forensic Sci; Am Anthrop Asn; Am Pub Health Asn; NY Acad Sci; Am Asn Physical Anthrop. *Res:* Biomedical anthropology; research on the relations of environment to human health and disease. *Mailing Add:* PO Box 8217 Silver Springs MD 20907

MICZEK, KLAUS A, b Burghausen, Ger, Sept 28, 44; m 70; c 1. PSYCHOPHARMACOLOGY, ETHOLOGY. *Educ:* Paedagogische Hochsch, Berlin, teaching cert, 66; Univ Chicago, PhD(biopsychol), 72. *Prof Exp:* Asst prof psychol, Carnegie-Mellon Univ, 72-76, assoc prof, 76-79; assoc prof, 79-83, PROF PSYCHOL, TUFTS UNIV, 83- *Concurrent Pos:* Prin investr, Nat Inst Drug Abuse & Pittsburgh Found res grants, 74-; res grant, Nat Inst Alcohol Abuse & Alcoholism, 79-; consult, Duphar BV, Weesp, Neth; mem, panel on violence, Nat Acad Sci; Boerhaave prof, Univ Leiden Med Sch, Neth. *Mem:* AAAS; Soc Neurosci; Am Psychol Asn; Behav Pharmacol Soc; Int Soc Res Aggression; Am Primatology Soc; Int Primatology Soc; NY Acad Sci. *Res:* Drugs, primate behavior and aggression. *Mailing Add:* Dept Psychol Tufts Univ Medford MA 02155

MIDDAUGH, RICHARD LOWE, b Salamanca, NY, Oct 2, 38. INORGANIC ELECTROCHEMISTRY. *Educ:* Harvard Univ, AB, 60; Univ Ill, MS, 62, PhD(chem), 65. *Prof Exp:* From asst prof to assoc prof chem, Univ Kans, 64-75; vis assoc prof chem, Northeastern Univ, 76-77, staff scientist, Inst Chem Analysis, Appl & Forensic Sci, 77-78; sr electrochemist, Union Carbide Corp, 78-81; PROJ MGR, EVEREADY BATTERY CO INC, 81- *Mem:* Am Chem Soc; Electrochem Soc. *Res:* Chemistry of battery systems. *Mailing Add:* 224 Cornwall Dr Cleveland OH 44116

MIDDELKAMP, JOHN NEAL, b Kansas City, Mo, Sept 29, 25; m 49, 74; c 4. PEDIATRICS. *Educ:* Univ Mo, BS, 46; Wash Univ, MD, 48. *Prof Exp:* Med intern, D C Gen Hosp, 48-49; asst resident pediat, St Louis Children's Hosp, 49-50 & 52-53, co-chief resident, 53; from instr to assoc prof, 54-70, PROF PEDIAT, SCH MED, WASH UNIV, 70- *Concurrent Pos:* Consult, Barnes & Allied Hosp, 54-, Crippled Children's Servs, Univ Ill, 55- & Univ Mo, 66-; fel internal med, Wash Univ, 60-61, USPHS fel anat, 61-62. *Mem:* Sigma Xi; Am Acad Pediat; Am Soc Microbiol; Am Pediat Soc; Infectious Dis Soc Am; Ambulatory Pediat Assoc. *Res:* Infectious diseases; ambulatory pediatrics. *Mailing Add:* Dept of Pediat 400 S Kings Hwy St Louis MO 63110-1014

MIDDENDORF, DONALD FLOYD, poultry nutrition, biochemistry; deceased, see previous edition for last biography

MIDDENDORF, WILLIAM H, b Cincinnati, Ohio, Mar 23, 21; m 46; c 5. PRODUCT LIABILITY, ELECTRICAL INSULATION. *Educ:* Univ Va, BEE, 46; Univ Cincinnati, MS, 48; Ohio State Uinv, PhD(elec eng), 60. *Prof Exp:* PROF ELEC ENG, UNIV CINCINNATI, 48- *Concurrent Pos:* Dir eng & res, Wadsworth Elec Mfg Co, Inc, 66-86; dir, Nat Elec Mfr Assoc, Univ Cincinnati, Elec Insulation Lab, 77-90; consult, Cincinnati Develop & Mfg Co, 60-66, Allis Chalmers Mfg co, 56-58. *Mem:* Fel Inst Elec & Electronic Engrs; Am Soc Eng Educ. *Res:* Accelerated life tests of electrical insulation to improve consistency of test data; circuit analysis and product design; tracking of insulation under moist conditions with emphasis on use of copper electrodes versus platinum; development of successful procedure for consistent data of insulation arc resistance. *Mailing Add:* Univ Cincinnati Loc 30 Cincinnati OH 45221

MIDDLEBROOK, JOHN LESLIE, b Salem, Ore, Dec 16, 46; m 66; c 3. BIOCHEMISTRY, PHARMACOLOGY. *Educ:* Pac Univ, BS, 68; Duke Univ, PhD(chem), 72. *Prof Exp:* Fel pharmacol, Med Sch, Stanford Univ, 72-74, RES SCIENTIST BIOCHEM & PHARMACOL, US ARMY MED RES INST INFECTIOUS DIS, 75- *Concurrent Pos:* NIH fel, Med Sch, Stanford Univ, 73-74; Arthritis Found fel, 74-75; instr pharmacol, Hood Col, 76- *Mem:* Am Chem Soc; Sigma Xi; Am Soc Biol Chemists. *Res:* Steroid, hormone and toxin receptors; cell biology; protein transport; protein toxins and mechanisms of action. *Mailing Add:* US Army Med Res Path Div Ft Detrick Frederick MD 21701

MIDDLEBROOK, R(OBERT) D(AVID), b Eng, May 16, 29; nat US. ELECTRONICS. *Educ:* Cambridge Univ, BA, 52, MA, 56; Stanford Univ, MS, 53, PhD(elec eng), 55. *Prof Exp:* Sr tech instr electronics, Royal Air Force, Eng, 48-49; asst, Stanford Univ, 53-55; from asst prof to assoc prof, 55-65, PROF ELECTRONICS, CALIF INST TECHNOL, 65- *Concurrent Pos:* Consult, 59- *Honors & Awards:* Award, Nat Prof Group on Indust Engrs, 58; Wm E Newell Power Electronics Award, Inst Elec & Electronics Engrs, 82. *Mem:* Fel Inst Elec & Electronics Engrs. *Res:* New solid-state devices, their development, representation and application; electronics education; power conversion and control. *Mailing Add:* Dept Elec Eng 116-81 Calif Inst Technol Pasadena CA 91125

MIDDLEBROOKS, EDDIE JOE, b Crawford County, Ga, Oct 16, 32; m 58; c 1. WASTEWATER TREATMENT, PHOTOCHEMICAL PROCESSES. *Educ:* Univ Fla, Gainesville, BCE, 56, MSE, 60; Miss State Univ, PhD(environ eng), 66. *Prof Exp:* Asst & assoc prof environ eng, Miss State Univ, 62-67; asst dir, Sanit Eng Res Lab, Univ Calif, Berkeley, 68-70; prof environ eng, Utah State Univ, Logan, 70-82, dean admin, Col Eng, 74-82; Newman chair environ eng, Clemson Univ, SC, 82-83; provost & vpres acad affairs, Tenn Technol Univ, 83-88; provost & vpres acad affairs, 88-90, actg pres, 90, TRUSTEES PROF ENVIRON ENG, UNIV TULSA, OKLA, 90- *Concurrent Pos:* Spec postdoctoral fel, Fed Water Pollution Control Admin, 67; consult, numerous consult firms & industs, 68-; diplomate, Am Acad Environ Engrs, 70-; delegate, US Nat Comn, Int Asn Water Pollution Res & Control, 77-80; dir, Water Pollution Control Fedn, 78-80 & 91-93; mem, Nat Drinking Water Adv Coun, US Environ Protection Agency, 81-83; univ prof & prof civil eng, Tenn Technol Univ, 83-88; prof chem eng, Univ Tulsa, 88-89; mem, Accreditation Bd Eng & Technol, AAAS. *Honors & Awards:* Harrison Prescott Eddy Medal, Water Pollution Control Fedn, 69. *Mem:* Asn Environ Eng Professors (pres, 74); Int Asn Water Pollution Res & Control; fel Am Soc Civil Engrs; Water Pollution Control Fedn. *Res:* Development of low-cost wastewater treatment processes applicable to small communities, industries, and developing nations; use of photosensitive dyes as a means of disinfecting wastewaters and removing hazardous materials. *Mailing Add:* 1115 E 20th St Tulsa OK 74120

MIDDLEDITCH, BRIAN STANLEY, b Bury St Edmunds, Eng, July 15, 45; m 70; c 1. BIOCHEMICAL ECOLOGY, ANALYTICAL BIOCHEMISTRY. *Educ:* Univ London, BSc, 66; Univ Essex, MSc, 67; Glasgow Univ, PhD(chem), 71. *Prof Exp:* Res asst chem, Glasgow Univ, 67-71; vis asst prof lipid res, Baylor Col Med, 71-73, res instr, 74-75; asst prof biophys sci, 75-80, assoc prof, 80-89, PROF BIOCHEM & BIOPHYS SCI, UNIV HOUSTON, 89- *Concurrent Pos:* Hon prof, Eurotech Res Univ. *Mem:* Am Chem Soc; Am Soc Mass Spectrom. *Res:* Biochemical oncology; Mass spectrometry; gas chromatography; natural products chemistry; environmental effects of offshore oil production. *Mailing Add:* Dept Biochem & Biophys Sci Univ Houston Houston TX 77204-5500

MIDDLEHURST, BARBARA MARY, b Penarth, Wales, Sept 10, 15. ASTRONOMY. *Educ:* Cambridge Univ, BA, 36, MA, 47. *Prof Exp:* Observer astron, Univ Observ, St Andrews Univ, 51-54, lectr, Univ, 54-59; res assoc, Yerkes Observ, Chicago, 59-60; res assoc, Lunar & Planetary Lab, Univ Ariz, 60-68; astron ed, Encycl Britannica, 68-72; vis scientist, Lunar Sci Inst, Nassau Bay 73-74. *Concurrent Pos:* Goethe Link fel, Ind Univ, 53, Fulbright travel grant & res assoc, 53-54; Carnegie Trust Scottish Univs res grant, 54; prin investr, Off Naval Res Proj Grant, 63-; mem comt 16, Int Astron Union, 64-; NSF proj grant, 66-68; consult, Lockheed Electronics, 69; consult, Chicago Sch Dist 97, 69-70. *Mem:* Am Geophys Union; Am Astron Soc; Royal Astron Soc. *Res:* Stellar, lunar and planetary research; discovery of the Middlehurst effect, that is, tidally related periodicity in reported shortlived lunar phenomena similar to periodicity in seismic signals recorded from instruments on the moon. *Mailing Add:* 16567 El Camino Real Houston TX 77062

MIDDLETON, ALEX LEWIS AITKEN, b Banchory, Scotland, May 20, 38; Can citizen; m 62; c 3. ZOOLOGY, ECOLOGY. *Educ:* Univ Western Ont, BSc, 61, MSc, 62; Monash Univ, Australia, PhD(zool), 66. *Prof Exp:* Asst prof, 66-70, ASSOC PROF ZOOL, UNIV GUELPH, 70- *Concurrent Pos:* Co-ed, Encyclopedia of Birds, 85- *Mem:* Am Ornith Union; Can Soc Zoologists; Royal Australasian Ornithologists Union; Cooper Ornith Soc. *Res:* Ecology of birds, particularly mating strategies and host/parasite relationships. *Mailing Add:* Dept of Zool Univ of Guelph Guelph ON N1G 2W1 Can

MIDDLETON, ARTHUR EVERTS, b Erie, Pa, June 10, 19; m 41, 72; c 2. SOLID STATE PHYSICS, ELECTRONICS. *Educ:* Westminster Col, BS, 40; Purdue Univ, MS, 42, PhD(physics), 44. *Hon Degrees:* DSc, Westminster Col, 82. *Prof Exp:* Asst physics, Purdue Univ, 40-43, instr, 43-45; res engr, Fed Tel & Radio Co, NJ, 45; res engr, Battelle Mem Inst, 45-47, asst supvr res, 47-51, supvr, 51-53; dir, physics & phys chem labs, P R Mallory & Co, Inc, 53-57; tech counr & group leader, Large Lamp Eng Dept, Gen Elec Co, 58-59; mgr & dir, Solid State Div, Harshaw Chem Co, 59-62; chief scientist, Ohio Semiconductors Div, Tecumseh Prod, Inc, 62-64; vpres & dir, 64-75, EXEC VPRES, SECY & DIR, OHIO SEMITRONICS INC, 75-; EMER PROF ELEC ENG, OHIO STATE UNIV, 84- *Concurrent Pos:* Consult, Adv Group Electronic Parts, 55-57; mem adv panel dielectrics, Nat Adv Bd, 56; dir, N Pittsburgh Syst, Inc, 56-, mem exec comt, 85-; mem, adv panel passive components, Wright Air Develop Ctr, 57; dir, Ohio Semiconductors, Inc, 58-60; lectr, Univ Mich, 65; prof, Ohio State Univ, 65-84. *Mem:* Electrochem Soc; Am Phys Soc; Inst Elec & Electronics Engrs. *Res:* Nuclear reactions in photographic emulsions; galvanomagnetic properties of semiconductors; thermoelectric materials and devices; Hall effect and electroluminescent devices; integrated circuit technology; solid state radiation detection; solar energy convertors; electrophotographic plates; new semiconductors and other electronic components. *Mailing Add:* 1205 Chesapeake Ave Columbus OH 43212

MIDDLETON, BETH ANN, b Madison, Wis, Aug 18, 55. WETLAND ECOLOGY, HERBIVORY. *Educ:* Univ Wis-Madison, BS, 78; Univ Minn, Duluth, MS, 83; Iowa State Univ, PhD(bot), 89. *Prof Exp:* Instr biol & chem, Northland Col, Ashland, Wis, 79-82; postdoctoral, Res Unit Landscape Ecol, Iowa State Univ, 89-90, asst prof wetlands & plant ecol, Dept Bot, 89-90; ASST PROF ECOL & WETLAND ECOL, DEPT PLANT BIOL, SOUTHERN ILL UNIV, 90- *Concurrent Pos:* Statist consult, Computer Ctr, Iowa State Univ, 87; vis prof, G B Pant Univ, Pantnagar, India, 91; Fulbright, Coun Int Exchange Scholars, 90-91; prin investr, Off Res Develop, Southern Ill Univ, 91- *Mem:* Soc Wetland Scientists; Ecol Soc Am; Asn Trop Biologists. *Res:* Effect of goose herbivory on vegetation dynamics in a monsoonal wetland in north-central India; restoration of cypress wetlands along the Cache River in southern Illinois; landscape and tropical ecology. *Mailing Add:* 411 Life Sci II Dept Plant Biol Southern Ill Univ Carbondale IL 62901-6509

MIDDLETON, CHARLES CHEAVENS, b Pilot Point, Tex, Apr 12, 30; m; c 2. EXPERIMENTAL PATHOLOGY. *Educ:* Univ Mo-Columbia, BS & DVM, 58; Mich State Univ, MS, 61; Am Col Lab Animal Med, dipl, 66. *Prof Exp:* Instr vet surg, Univ Pa, 58-59; instr physiol & pharmacol, Mich State Univ, 60-62; fel cardiovasc res, Bowman Gray Sch Med, 62-63, from instr to asst prof lab animal med, 63-66; from asst prof community health & med practices to assoc prof, 66-77, assoc prof vet path, 66-75, PROF MED PATH,

UNIV MO-COLUMBIA, 77-, PROF VET PATH, 75-, DIR SINCLAIR RES FARM, 66- *Concurrent Pos:* Co-investr, Dept Health, Educ & Welfare grants, 67-70 & 68-72; post doctoral thesis adv, Sch Med & Sch Vet Med, Univ Mo-Columbia, 67-; mem coun arterosclerosis Am Heart Asn. *Mem:* Am Inst Biol Sci; Int Primatol Soc; Am Col Lab Animal Med; Am Vet Med Asn; NY Acad Sci; Sigma Xi. *Res:* Atherosclerosis; pathology of laboratory animals. *Mailing Add:* Dept Lab Animal Res SUNY Stony Brook Stony Brook NY 11794

MIDDLETON, DAVID, b New York, NY, Apr 19, 20; m 45, 71; c 4. PHYSICS, STATISTICAL COMMUNICATION THEORY. *Educ:* Harvard Univ, ABES, 42, AM, 45, PhD(physics), 47. *Prof Exp:* Res assoc, Off Sci Res & Develop Proj, Harvard Univ, 42-45, productorial fel, Nat Sci Found, 46 & 47, res fel electronics, 47-49, asst prof appl physics, 49-54; CONSULT PHYSICIST, 54- *Concurrent Pos:* Consult, Govt & Industs, 49-; adj prof, Columbia Univ, 60-61, Rensselaer Polytech Inst, 61-70 & Univ RI, 66- & Rice Univ, 80-87; mem, Naval Res Adv Comt, 70-77, Navy Lab Adv Bd Undersea Warfare, 70-77, Navy Lab Adv Bd Res, 71-77 & US Study Group 1A, Int Radio Consult Comt, 71-; contractor, Off Naval Res, Dept Defense, Inst Telecommun Sci, Off Telecommun & Nat Telecommun & Info Admin, 78-, Dept Com, NASA & Nat Oceanic & Atmospheric Admin; consult, Off Telecommun Policy-Exec Off of Pres, 74-78, Inst Defense Anal, RAND, Raytheon, Sylvania, Gen Elec; US Dept Com, Inst Telecommunications Ser, 78. *Honors & Awards:* Nat Electronic Conf Award, 56; First Prize, Third Int Symp Electromagnetic Compatibility, 79. *Mem:* Fel AAAS; fel Am Phys Soc; Am Math Soc; fel Inst Elec & Electronic Eng; fel Acoust Soc Am; Sigma Xi; fel NY Acad Sci. *Res:* Communication theory in radar, radio, underwater sound, seismology, optics, mechanics, electronics, space sciences; applied mathematics; scattering theory; wave surface oceanography; man-made and natural EM environments; electromagnetic compatability. *Mailing Add:* 127 E 91st St New York NY 10128

MIDDLETON, ELLIOTT, JR, b Glen Ridge, NJ, Dec 15, 25; m 48; c 4. INTERNAL MEDICINE. *Educ:* Princeton Univ, AB, 47; Columbia Univ, MD, 50; Am Bd Internal Med, dipl, 58; Am Bd Allergy, dipl, 62. *Prof Exp:* Intern, Presby Hosp, New York, 50-51, asst resident, 51-53; asst med, Col Physicians & Surgeons, Columbia Univ, 56-57, instr, 57-60, assoc, 60-69; dir, Clin Serv & Res, Children's Asthma Res Inst & Hosp, Nat Asthma Ctr, 69-77; PROF MED & PEDIAT & DIR ALLERGY DIV, STATE UNIV NY, 77- *Concurrent Pos:* Nat Heart Inst clin fel, Presby Hosp, New York, 51-53; clin fel, NIH, 53-54; clin & res fel, Inst Allergy, Roosevelt Hosp, 55; asst, Immunochem Lab, Col Physicians & Surgeons, Columbia Univ, 52; physician pvt pract, 56-69; asst attend physician, Mountainside Hosp, Montclair, NJ, 58-62, assoc attend, 62-; asst physician, Presby Hosp, NY, 60; ed, J Allergy & Clin Immunoi, 83-88. *Honors & Awards:* Distinguished Serv Award, Am Acad Allergy, 91. *Mem:* AAAS; Am Asn Immunologists; Am Acad Allergy (pres, 72-73); Harvey Soc. *Res:* Allergy; immunology; biochemical mechanisms of human allergic reactions by in vitro techniques; chemical mediators of allergic reactions; immunopharmacology. *Mailing Add:* Med-Buffalo Gen SUNY Health Sci Ctr 3435 Main St Buffalo NY 14214

MIDDLETON, FOSTER H(UGH), b Detroit, Mich, Dec 4, 22; m 48; c 4. ELECTRICAL ENGINEERING. *Educ:* Univ Mich, BS, 47; Johns Hopkins Univ, DrEng, 59. *Prof Exp:* Test engr, Ford Motor Co, Mich, 48-50; group leader instrumentation, Sperry Gyroscope Co, 50-52; res staff asst acoust, Johns Hopkins Univ, 52-54, instr elec eng, 54-59; from assoc prof to prof, 59-66, PROF OCEAN ENG & CHMN DEPT, UNIV RI, 66- *Concurrent Pos:* Vis prof, Glasgow Univ, 71-72; panel mem, Marine Bd Panels, Nat Acad Eng, 72- *Res:* Underwater acoustics, ocean instrumentation. *Mailing Add:* Dept of Ocean Eng Univ of RI Kingston RI 02881

MIDDLETON, GERARD VINER, b Capetown, SAfrica, May 13, 31; m 59; c 3. GEOLOGY. *Educ:* Imp Col, Univ London, BSc, 52, dipl & PhD(geol), 54. *Prof Exp:* Geologist, Standard Oil Co Calif, 54-55; lectr, 55-61, assoc prof, 61-67, chmn dept, 59-62 & 78-84, PROF GEOL, McMASTER UNIV, 67- *Concurrent Pos:* Consult, Shell Oil Co, 56-57 & 59. *Honors & Awards:* Logan Medal, Geol Asn Am, 80. *Mem:* Hon mem Soc Econ Paleontologists & Mineralogists; Am Asn Petrol Geol; Geol Asn Can (pres, 87-88); fel Royal Soc Can; hon mem Int Asn Sedimentol (vpres, 78-82). *Res:* Sedimentary petrography; sedimentology. *Mailing Add:* Dept Geol McMaster Univ Hamilton ON L8S 4L8 Can

MIDDLETON, HENRY MOORE, III, b Winston-Salem, NC, Feb 16, 43; m 66; c 3. GASTROENTEROLOGY. *Educ:* Univ NC, Chapel Hill, AB, 65, MD, 69. *Prof Exp:* Intern internal med, Vanderbilt Hosp, Nashville, 69-70, resident, 70-72, fel gastroenterol, 72-74; from asst prof to assoc prof, 74-87, PROF MED, MED COL GA, 87-; ASSOC CHIEF STAFF, VET ADMIN MED CTR, AUGUSTA, 88- *Mem:* Am Asn Study Liver Dis; Am Soc Gastrointestinal Endoscopy; Am Inst Nutrit; Am Soc Clin Nutrit; Am Fedn Clin Res. *Res:* Mechanisms of intestinal absorption of water-soluble vitamins in the rat; whole-organ in vitro and in-vivo models and the brush-border membrane vesicle models. *Mailing Add:* Vet Admin Med Ctr 151 Downtown Div Rm 5B-144 Augusta GA 30910

MIDDLETON, JOHN T(YLOR), b Chicago, Ill, Sept 15, 12; c 4. AIR POLLUTION. *Educ:* Univ Calif, BS, 35; Univ Mo, PhD(bot, plant path), 40. *Prof Exp:* Asst botanist, Univ Mo, 36-39; jr plant pathologist, Exp Sta, Univ Calif, 39-43, asst plant pathologist, 43-48, assoc plant pathologist, 48-54, plant pathologist, 54-57, prof path & chmn dept, Los Angeles & Riverside, 57-63, dir, Statewide Air Pollution Res Ctr, Riverside, 62-67; dir, Nat Ctr Air Pollution Control, Dept Health Educ & Welfare, 67-68, comnr, Nat Air Pollution Control Admin, 68-71, dept asst adminr, US Environ Protection Agency, 71-73; prof mgr & emer prof, Univ Calif, 67-89; environ mgr, WHO, 77-89; CONSULT, US SENATE COMT ENVIRON & PUB WORKS, 73-; SR SCIENTIST, ENVIRON STRATEGOMES CORP, 83- *Concurrent Pos:* Consult, govt & var indust orgns, 49-; mem & past chmn, Calif Motor Vehicle Pollution Control Bd, 60-66; mem, Nat Adv Comn Air Pollution to US Surgeon Gen, 63-66; adv, WHO, 63-, dir, Int Reference Ctr on Air Pollution Control, 71-73, mem, Expert Panel on Air Pollution, 72-; mem exec comt, Hwy Res Bd, Nat Acad Sci-Nat Res Coun, 70-73; adv, Environ Pollution & Hazards, Royal Thai Gov, 79- *Honors & Awards:* Richard Beatty Mellon, Air Pollution Control Asn, 72; Environ Qual, Nat Acad Sci, 76. *Mem:* Air Pollution Control Asn. *Res:* Air pollution control. *Mailing Add:* 2811 Albemarle St NW Washington DC 20008-1037

MIDDLETON, PAULETTE BAUER, b Beeville, Tex, Dec 8, 46; div; c 2. ATMOSPHERIC CHEMISTRY, CHEMICAL PHYSICS. *Educ:* Univ Tex, Austin, BA, 68, MA, 71, PhD(chem), 73. *Prof Exp:* Instr chem, Univ Tex, 73-74, res assoc aerosol physics, dept chem eng, 73-75; res fel, Nat Ctr Atmospheric Res, 75-76; vis scientist, 77-79, SPC staff scientist, 79-87, RES ASSOC, ATMOSPHERIC SCI RES CTR, STATE UNIV NY, ALBANY, 76-77, 87- *Concurrent Pos:* Mem, Nat Acad Sci Climatol Effects Resource Group, Risk/Impact Panel Comt Nuclear & Alt Energy Systs, 76-77, Nat Acad Sci Harpin Nat Parks & Wilderness Areas, 90-91; prin investr, Acid Precipitation Exp, 78-79; mem, Physics & Chem Rev Panel, Environ Protection Agency, 80-, develop projs, Acid Deposition & Air Qual Policy, impacts & modeling, 80-; proj dir, Urban Visual Air Qual Study, 81-84, model Eval & Chem Database Develop, 85-, Denver Air Qual Modeling Study, 90-; mem, Int Global Atmospheric Chem Global Emissions Comt. *Mem:* AAAS; Air Pollution Control Asn. *Res:* Gas and aerosol transport and transformation in the troposphere and stratosphere; assessing impacts of visibility degradation, acid rain and climate change. *Mailing Add:* Nat Ctr Atmospheric Res Boulder CO 80307

MIDDLETON, RICHARD B, b Rockford, Ill, Nov 24, 36; div; c 2. MICROBIAL GENETICS, MEDICAL LAW. *Educ:* Harvard Univ, AB, 58, Am, 60, PhD(biol), 63. *Prof Exp:* Res fel bact genetics, Brookhaven Nat Lab, 62-64; asst prof biol, Am Univ Beirut, 64-65; asst prof genetics, McGill Univ, 65-71; assoc prof, 71-75, prof microbiol & genetics, Mem Univ Nfld, 75-77; assoc dean basic sci, 77-84, PROF MICROBIOL, UNIV MED & DENT, NJ, 77- *Concurrent Pos:* Res grants, Rockefeller Found, 64-65, Med Rds Coun Can, 66-68 & 72-76, Res Corp, 66-68, Nat Res Coun Can, 66-77, Int Cell Res Orgn, Int Lab Genetics & Biophys, Naples, 67, Food & Drug Directorate, Dept Nat Health & Welfare Can, 69-71, World Health Orgn, Ctr Immunol, State Univ NY Buffalo, 71, Tissue Cult Asn, Jones Cell Sci Ctr, Lake Placid, NY, 72, Europ Molecular Biol Orgn, Biozentrum, Univ Basel, Switz, 72, March Dimes, Nat Found, Jackson Lab, Bar Harbor, Maine, 74, Dept Secy State Can, 75, dept Nat Health & Welfare Can, 75-78, NIH, 78-81, Sandoz Inc, 85-87, Off Naval Res, 88-92; vis prof genetics, State Univ NY, Plattsburg, 74, St Georges Univ, Grenada, 78; prof microbiol, Rutgers Univ, 78- *Mem:* AAAS; Am Soc Microbiol; Can Soc Cell Biol; Genetics Soc Am; NY Acad Sci; Sigma Xi. *Res:* Genetic homology of Salmonella typhimurium and Escherichia Coli; fertility of intergeneric crosses of enteric bacteria; legal reform for donation of human tissues for scientific uses. *Mailing Add:* Univ Med & Dent NJ 401 S Central Plaza Stratford NJ 08084

MIDDLETON, ROY, b Atherton, Eng, Oct 3, 27; m 50; c 2. NUCLEAR PHYSICS. *Educ:* Univ London, BSc, 48; Univ Liverpool, PhD(nuclear physics), 51. *Hon Degrees:* MA, Univ Pa, 71. *Prof Exp:* Fel nuclear physics, Univ Liverpool, 51-54, teaching asst, 54-55; prin sci officer, Atomic Weapons Res Estab, Aldermaston, Eng, 55-64; vis prof, 64-65, PROF NUCLEAR PHYSICS, UNIV PA, 65- *Concurrent Pos:* Co-prin investr, NSF tandem accelerator grant; mem physics div rev comt, Oak Ridge Nat Labs, 70-74 & Argonne Nat Lab, 75- *Honors & Awards:* Tom W Bonner Prize, Am Phys Soc, 79. *Mem:* Brit Inst Physics; fel Am Phys Soc. *Res:* Nuclear experimental research; negative ion source development; accelerator development. *Mailing Add:* Dept Physics Univ Penn Tandem Accel Lab 209 E 33rd Philadelphia PA 19104

MIDDLETON, WILLIAM JOSEPH, b Amarillo, Tex, Apr 9, 27; m 48; c 2. ORGANOFLUORINE CHEMISTRY. *Educ:* NTex State Col, BS, 48, MS, 49; Univ Ill, PhD(chem), 52. *Prof Exp:* Res chemist, E I du Pont de Nemours & Co, Inc, 52-84; PROF, URSINUS COL, 85- *Honors & Awards:* Am Chem Soc Award. *Mem:* Am Chem Soc; Sigma Xi. *Res:* Cyanocarbon, organic fluorine, heterocyclic and medicinal chemistry. *Mailing Add:* 95 Ridge Rd Chadds Ford PA 19317

MIDGLEY, A REES, JR, b Burlington, Vt, Nov 9, 33; m 55; c 3. ENDOCRINOLOGY, ANALYTICAL CHEMISTRY. *Educ:* Univ Vt, BS, 55, MD, 58. *Prof Exp:* Sarah Mellon Scaife fel path, Univ Pittsburgh, 58-61; Sarah Mellon Scaife fel, Univ Mich, Ann Arbor, 61-62, res assoc, 62-63, from instr to assoc prof, 63-70, dir, Ctr Human Growth & Develop, 80-82, dir, Consortium Res Develop & Reproductive Biol, 83-88, PROF PATH, UNIV MICH, ANN ARBOR, 70-, DIR, REPRODUCTION SCI PROG, 88- *Concurrent Pos:* Nat Inst Child Health & Human Develop career develop award, 66-76. *Honors & Awards:* Parke Davis Award, Am Soc Exp Path, 70; Ayerst Award, Endocrine Soc, 77; Smith Kline Biol Sci Labs Award, Clin Ligand Assay Soc, 85. *Mem:* Am Asn Path; Soc Study Reproduction; Endocrine Soc; Am Soc Cell Biol; Am Physiol Soc. *Res:* Reproductive endocrinology; immunoendocrinology; development biology. *Mailing Add:* 300 N Ingalls Rm 1125 Univ Mich Ann Arbor MI 48109-0404

MIDGLEY, JAMES EARDLEY, b Kansas City, Mo, Sept 18, 34; m 61; c 6. SYSTEMS THEORY. *Educ:* Univ Mich, BS(eng phys), BS(eng math) & BS (eng mech), 56; Calif Inst Technol, PhD(physics), 63. *Prof Exp:* Res assoc magnetosphere, Univ Tex, Dallas, 63-64, asst prof gravity waves, 64-67, ASSOC PROF PHYSICS, UNIV TEX, DALLAS, 67- *Concurrent Pos:* NSF fel, Calif Inst Technol, 58-61. *Mem:* Sigma Xi. *Res:* Computer operating systems; data structures; computer graphics. *Mailing Add:* Univ Tex Dallas PO Box 830688 Richardson TX 75083

MIDLAND, MICHAEL MARK, b Ft Dodge, Iowa, Jan 1, 46; m 72. ORGANIC CHEMISTRY, ORGANOMETALLIC CHEMISTRY. *Educ:* Iowa State Univ, BS, 68; Purdue Univ, PhD(org chem), 72. *Prof Exp:* Assoc chem, Purdue Univ, 72-75; lectr, 75-76, asst prof, 76-80, ASSOC PROF CHEM, UNIV CALIF, RIVERSIDE, 80- *Concurrent Pos:* Alfred P Sloan fel, 78-82. *Mem:* Am Chem Soc; AAAS; Sigma Xi. *Res:* New chemistry of organoboranes and organolithiums; new synthetic reactions; investigation of reaction mechanisms; asymmetric synthesis. *Mailing Add:* Dept Chem Univ Calif Riverside CA 92521

MIDLARSKY, ELIZABETH, b NY; m 61; c 3. PSYCHOLOGY. *Educ:* Brooklyn Col, City Univ NY, BA, 61; Northwestern Univ, Evanston, MA, 66 & PhD(psychol), 68. *Prof Exp:* Lectr pschol, Northwestern Univ, 65-67; asst prof, Univ Denver, 68-73; dir res, Malcolm X Ment Health Ctr, 74-77; assoc prof human serv, Metrop State Col, 75-77; from assoc prof to prof psychol, Univ Detroit, 77-90, chairperson dept, 78-81, dir geront, Ctr Study Develop & Aging, 81-90; PROF PSYCHOL, COLUMBIA UNIV TEACHERS COL, 90- *Concurrent Pos:* Mem, Initial review group, NIMH, 76-82, prin investr NIH res grants, 82-85 & 87-90; mem, Human Develop Aging Study Sect, NIH, 86-90, eval panel, Nat Heart, Lung & Blood Inst, 88, ad hoc reviewer, 85-; chairperson, Univ Detroit, 78-81, dir Geront Ctr Study Develop & Aging, 81-90; AARP Andrus Found Res Grant, 82-83 & 87-88; post doctoral training fac, Elderly Care Res Ctr, Wayne State Univ, 83-85; vis prof geront, Ctr Geront Studies, Univ Fla, Gainesville, 88-89. *Mem:* Am Psychol Asn; Am Orthopsychiat Asn; Geront Soc Am; Am Psychol Soc; Sigma Xi. *Mailing Add:* Dept Clin Psychol Columbia Univ Teachers Col Box 148 New York NY 10027

MIDLER, MICHAEL, JR, b New York, NY, Aug 15, 36; m 60; c 3. CHEMICAL & BIOCHEMICAL ENGINEERING. *Educ:* Cornell Univ, BChE, 59, PhD(biochem eng), 64. *Prof Exp:* Res assoc, 62-72, res fel, 72-78, sr res fel, 78-90, SR INVESTR, MERCK SHARP & DOHME RES LABS, 90- *Mem:* Am Inst Chem Engrs; Am Chem Soc. *Res:* Crystallization; resolution of stereoisomers; ultrasonic processing; mixing; liquid-liquid extraction; chemical reaction engineering. *Mailing Add:* Chem Eng Res & Develop PO Box 2000 Rahway NJ 07065

MIDLIGE, FREDERICK HORSTMANN, JR, b Hoboken, NJ, June 13, 35; m 61; c 2. VIROLOGY, MICROBIOLOGY. *Educ:* Muhlenberg Col, BS, 57; Lehigh Univ, MS, 59, PhD(biol), 68. *Prof Exp:* From instr to asst prof, 63-72, assoc prof, 72-76, PROF BIOL, FAIRLEIGH DICKINSON UNIV, 76- *Mem:* AAAS; Am Soc Microbiol; Sigma Xi. *Res:* Virus diseases of fish; morphology and maturation of lymphocystis virus. *Mailing Add:* Dept of Biol Fairleigh Dickinson Univ 285 Madison Ave Madison NJ 07940

MIDURA, THADDEUS, b Chicopee, Mass, Dec 2, 31; m 65; c 2. FOOD TECHNOLOGY. *Educ:* Univ Mass, BS, 57, MS, 59; Univ Mich, MPH, 61, PhD(environ health), 64. *Prof Exp:* Sanitarian, Food & Milk Lab, Springfield Health Dept, Mass, 56; sanitarian, Environ Health, Philadelphia Health Dept, Pa, 57; instr food technol, Univ Mass, 58-60; fel, 64-66, res microbiologist, Div Labs, 66-80, CHIEF, MICROBIOL DIS LAB, CALIF STATE DEPT HEALTH SERV, 80- *Concurrent Pos:* Lectr pub health microbiol, Univ Calif, Berkeley, 75-; actg chief, Div Labs, Calif Dept Health Serv, 84-85. *Mem:* Am Soc Microbiol; Am Pub Health Asn; Inst Food Technol. *Res:* Public health microbiology; anaerobic bacteriology and microorganisms significant in food-borne diseases; laboratory aspects of environmental associated disease outbreaks. *Mailing Add:* 2151 Berkeley Way Berkeley CA 94704

MIECH, RALPH PATRICK, b South Milwaukee, Wis, Aug 17, 33; m 57; c 5. BIOCHEMISTRY, PHARMACOLOGY. *Educ:* Marquette Univ, BS, 55, MD, 59; Univ Wis, PhD(pharmacol), 63. *Prof Exp:* Intern med, St Mary's Hosp, Duluth, Minn, 59-60; Nat Cancer Inst fel, Univ Wis, 61-63; asst prof, 63-69, ASSOC PROF MED SCI, BROWN UNIV, 69- *Concurrent Pos:* Nat Inst Neurol Dis & Blindness grant, 69-72; NSF fel, 69-72. *Mem:* Am Col Emergency Physicians. *Res:* Enzymes of nucleotide synthesis; theophylline metabolism; nucleotide metabolism in parasitic organisms; nucleotide metabolism in the brain; purine transport via the blood; inhibitor of bronchial camp phosphodiesterase. *Mailing Add:* Div Biomed Sci Brown Univ Providence RI 02912

MIECH, RONALD JOSEPH, b Milwaukee, Wis, Feb 23, 35; m 60; c 2. MATHEMATICS. *Educ:* Univ Ill, Urbana, BS, 59, PhD(math), 63. *Prof Exp:* Res fel math, Nat Bur Standards, Washington, DC, 63-64; from asst prof to assoc prof, 64-74, PROF MATH, UNIV CALIF, LOS ANGELES, 74- *Res:* Number theory; group theory. *Mailing Add:* Dept Math Univ Calif 405 Hilgard Ave Los Angeles CA 90024

MIED, RICHARD PAUL, b Baltimore, Md, Dec 5, 46; m 70; c 2. PHYSICAL OCEANOGRAPHY, FLUID MECHANICS. *Educ:* Johns Hopkins Univ, BES, 68, PhD(fluid mech), 72. *Prof Exp:* Pres intern oceanog, 72-73, RES SCIENTIST OCEANOG, NAVAL RES LAB, 73- *Mem:* Am Geophys Union; Res Soc NAm; Sigma Xi; Am Meteorol Soc. *Res:* Modeling and simulation of ocean features with horizontal extents of tens to hundreds of kilometers, and time scales of days to months; modeling and analysis on processes causing spatial variation of fine- and microstructure. *Mailing Add:* Naval Res Lab Code 4220 Overlook Ave Washington DC 20375-5000

MIEHLE, WILLIAM, b Ulm, Ger, Mar 31, 15; nat US; m 48; c 2. MATHEMATICS. *Educ:* Mass Inst Technol, SB, 38; Univ Pa, MS, 59. *Prof Exp:* Engr, Radio Corp Am, 40-46; res engr, Philco Corp, 46-49; designer & comput analyst, Res Ctr, Burroughs Corp, 49-56; mem staff, Inst Coop Res, Pa, 56-58; asst prof math, Pa Mil Col, 58-60; asst prof math, Villanova Univ, 60-80; RETIRED. *Concurrent Pos:* Consult, Auerbach Electronics Corp, 60 & Appl Psychol Serv, 61-81. *Res:* Operations research; information retrieval; numerical analysis. *Mailing Add:* 1240 Steel Rd Havertown PA 19083

MIEKKA, RICHARD G(EORGE), b Pontiac, Mich, Oct 18, 33; m 58; c 3. CHEMICAL ENGINEERING. *Educ:* Univ Mich, BS(chem) & BS(chem eng), 56; Mass Inst Technol, SM, 58, ScD(chem eng), 61. *Prof Exp:* Chem engr, WVa Pulp & Paper Co, 61-62; mgr chem res, Amicon Corp, Mass, 62-63, electrochem, Deco Div, 63-64; res chemist, 64-69, res sect head, 69-73, mgr chem res div, 73-81, GROUP MGR TECH DEVELOP, DENNISON MFG CO, FRAMINGHAM, 81- *Mem:* Soc Photog Scientists & Engrs; Am Chem Soc; Am Vacuum Soc; NY Acad Sci; Sigma Xi. *Res:* Nonwoven fabrics; polyelectrolyte structures; fuel cells and related electrochemical devices; electrostatic copying systems; pressure sensitive adhesives, tags, labels, metallized paper and inks. *Mailing Add:* 199 Goodman's Hill Rd Sudbury MA 01776

MIEL, GEORGE J, b Paris, France, Sept 7, 43; US citizen. COMPUTATIONAL MATHEMATICS. *Educ:* Univ Ill, BS, 64, MS, 66; Univ Wyo, PhD(computational math), 76. *Prof Exp:* Mem staff, Aerospace Corp, 85-89; PROF COMPUTATIONAL MATH, UNIV NEV, 78-; RESEARCHER COMPUTER SCI, HUGHES RES LABS, 89- *Concurrent Pos:* Chair, Computational Sci Comt, Aerospace Industs Asn, 89- *Honors & Awards:* Chauvenet Prize, Math Asn Am, 86. *Mem:* Asn Comput Mach; Soc Indust & Appl Math. *Mailing Add:* Box 64 Malibu CA 90265

MIELCZAREK, EUGENIE V, b New York, NY, Apr 22, 31; c 2. SOLID STATE PHYSICS, BIOLOGICAL PHYSICS. *Educ:* Queens Col NY, BS, 53; Catholic Univ, MS, 57, PhD(physics), 63. *Prof Exp:* Physicist, Nat Bur Standards, 53-57; from res asst to res assoc physics, Catholic Univ, 57-62, asst res prof, 62-65; PROF PHYSICS, GEORGE MASON UNIV, 65- *Concurrent Pos:* Vis scientist, NIH, 77-78; mem vis scientist prog, Am Inst Physics, 64-; vis prof, Hebrew Univ Jerusalem, 81. *Mem:* Am Phys Soc; Biophysical Soc; Sigma Xi; Am Asn Physics Teachers; Asn Women Sci. *Res:* Solid state low temperature physics; semiconductors; Mossbauer spectroscopy of metal and biological compounds; biophysics; Fermi surfaces of metals. *Mailing Add:* Dept Physics George Mason Univ Fairfax VA 22030

MIELE, ANGELO, b Formia, Italy, Aug 21, 22; m; m 70. AEROSPACE ENGINEERING. *Educ:* Univ Rome, DrCE, 44, DrAeE, 46. *Prof Exp:* Aerodyn engr, Inst Aircraft Technol, Argentina, 47-50; prof mech, Sch Mil Aviation, 50-52; asst prof aeronaut eng, Polytech Inst Brooklyn, 52-55; prof, Purdue Univ, 55-59; dir astrodyn & flight mech, Sci Res Labs, Boeing Co, 59-64; PROF AEROSPACE & MATH SCI, RICE UNIV, 64-, FOYT FAMILY PROF ENG, 88- *Concurrent Pos:* Consult, Allison Div, Gen Motors Corp, Guided Missiles Div, Douglas Aircraft Co, US Aviation Underwriters, Boeing Commercial Airplane Co; ed-in-chief, J Optimization Theory & Appln, 67-; assoc ed, J Astronaut Sci, 65-, Appl Math & Comput, 75- & Optimal Control Appln & Methods, 79-; ed, Math Concepts & Methods Sci & Eng, 74- *Honors & Awards:* Knight Commander, Order of Merit, Italian Repub, 72; Levy Medal, Franklin Inst, 74; Brouwer Award, Am Astronaut Soc; Pendray Award & Mechanics & Control Flight Award, Am Inst Aeronaut & Astronaut, 82; Shuck Award, Am Automatic Control Coun, 88. *Mem:* Fel Am Astronaut Soc; fel Am Inst Aeronaut & Astronaut; Int Acad Astronaut; Italian Aerotech Asn; corresp mem Acad Sci Turin. *Res:* Dynamics of extraterrestrial, interplanetary and terrestrial flight; flight performance; calculus of variations; astronautical engineering; high speed aerodynamics; computing methods; numerical optimization techniques; numerical analysis; optimal control. *Mailing Add:* Aero-Aeronaut Group Rice Univ 230 Ryon Bldg PO Box 1892 Houston TX 77251

MIELENZ, JONATHAN RICHARD, b Denver, Colo, Mar 25, 48; m 74; c 2. MOLECULAR BIOLOGY, BIOCHEMISTRY. *Educ:* Wittenberg Univ, AB, 70; Univ Ill, MS, 73, PhD(microbiol), 76. *Prof Exp:* NSF fel agron, Plant Growth Lab, Univ Calif, Davis, 76-78; res scientist microbial genetics, 79-80, SR RESEARCHER AND TEAM LEADER, GENETIC ENG, MOFFETT TECH CTR, CORN PROD CORP INT, 80- *Mem:* Sigma Xi; Am Soc Microbiol. *Res:* Improvement of microbial strains used in corn processing through the use of both conventional genetics and recombinant DNA technology. *Mailing Add:* Henkel Res Corp 2330 Circadian Way Santa Rosa CA 95407

MIELENZ, KLAUS DIETER, b Berlin, Ger, May 8, 29; c 2. OPTICS, SPECTROSCOPY. *Educ:* Univ Berlin, BSc, 49; Free Univ Berlin, MSc, 52, PhD(physics), 55. *Prof Exp:* Asst physics, Free Univ Berlin, 49-52; physicist, R Fuess Optical Co, Ger, 52-58 & Nat Bur Standards, 58-60; tech mgr, R Fuess Optical Co, 60-63; proj leader optical masers, Nat Bur Standards, 63-71, proj leader spectrophotom & luminescence spectrometry, 72-79; sci & technol fel & sr policy specialist, US Dept Com, 79-80; CHIEF, RADIOMETRIC PHYSICS DIV, NAT BUR STANDARDS, 81- *Concurrent Pos:* Guest res worker, Inst Appl Spectros, Dortmund, Ger, 57; adj prof, George Washington Univ, 68-72; chmn comt, Int Comn Illum, 73-79, vpres & div dir, 81- *Honors & Awards:* Silver Medal, US Dept Commerce, 66. *Mem:* Fel Optical Soc Am; German Soc Appl Optics. *Res:* Physical optics; spectrochemistry; spectroscopic instruments; spectrophotometry, luminescence spectrometry; radiometry; metrology; thin films; optical masers; vacuum techniques. *Mailing Add:* Radiomet Physics Div Nat Bur Standards Nat Measurement Lab Ctr Radiation Res Gaithersburg MD 20899

MIELKE, EUGENE ALBERT, b Visalia, Calif, Mar 1, 46; m 68; c 3. POMOLOGY, PLANT PHYSIOLOGY. *Educ:* Calif State Polytech Col, San Luis Obispo, BS, 69; Mich State Univ, MS, 70, PhD(hort, pomol), 74. *Prof Exp:* Lectr crop prod, Calif State Polytech Col, San Luis Obispo, 68-69; asst prof hort & pomol, Mich State Univ, 74-75; from asst prof to assoc prof pomol, viticult & enol, Univ Ariz, 75-84; SUPT & PROF HORT, MID-COLUMBIA AGR RES & EXTEN CTR, ORE STATE UNIV, HOOD RIVER, 84- *Concurrent Pos:* regional ed, Pecan Quart, 75-84; Ore rep, Western Region Coord Comt, Flowering & Fruit Set, 85- *Mem:* Am Soc Hort Sci; Am Soc Plant Physiologists; Japanese Soc Plant Physiologists; Am Soc Enologists; Scand Soc Plant Physiol. *Res:* Rootilocks, production systems, flowering, fruit set, and fruit growth and development of pears and apples; extraction, purification, identification and measurement of plant growth substances; identification of physiological factors limiting yield. *Mailing Add:* Mid-Columbia Agr Res & Exten Ctr Ore Univ 3005 Exp Sta Dr Hood River OR 97031

MIELKE, JAMES EDWARD, b Toledo, Ohio, Oct 6, 40; m 66; c 2. GEOCHEMISTRY. *Educ:* Mass Inst Technol, BS, 62; Univ Ariz, MS, 65; George Washington Univ, PhD(geochem), 74. *Prof Exp:* Geologist, Universal Eng Corp, 63-64; geologist, Radiation Biol Lab, Smithsonian Inst, 64-73; SCI POLICY ANALYST, CONG RES SERV, LIBR CONG, 73- *Mem:* AAAS; Am Geophys Union; Marine Tech Soc. *Res:* Science policy in the earth and marine sciences with regard to matters of current and future interest to Congress. *Mailing Add:* 2803 Washington Ave Chevy Chase MD 20815

MIELKE, MARVIN V, b Marshfield, Wis, May 2, 39; m 69. TOPOLOGY. *Educ:* Univ Wis, BS, 60, MS, 61; Ind Univ, PhD(math), 65. *Prof Exp:* Teaching assoc math, Ind Univ, 61-65; mem, Inst Advan Studies, 65-66; res assoc, 66-77, RES PROF MATH, UNIV MIAMI, 77- *Mem:* Am Math Soc; Math Asn Am. *Res:* Differential and algebraic topology. *Mailing Add:* Dept Math Univ Miami Univ Sta Coral Gables FL 33124

MIELKE, PAUL THEODORE, b Racine, Wis, Sept 28, 20; m 46; c 3. MATHEMATICS. *Educ:* Wabash Col, AB, 42; Brown Univ, ScM, 46; Purdue Univ, PhD(math), 51. *Prof Exp:* Instr math, Brown Univ, 43-44; from instr to asst prof, Wabash Col, 46-51; sr group engr digital comput, Dynamics Staff, Boeing Airplane Co, 52-57; assoc prof math, Wabash Col, 57-63, prof & chmn dept, 63-69; assoc dir, Comt Undergrad Prog Math, 69-70, exec dir, 70-71; chmn dept, 71-78, prof, 71-85, EMER PROF MATH, WABASH COL, 85- *Concurrent Pos:* Mem bd gov, Math Asn Am, 72-75; assoc ed math educ, Am Math Monthly, 74-78. *Mem:* Math Asn Am; Am Math Soc. *Res:* Digital computing; linear algebra. *Mailing Add:* Dept Math Wabash Col 308 E Jefferson St Crawfordsville IN 47933

MIELKE, PAUL W, JR, b St Paul, Minn, Feb 18, 31; m 60; c 3. STATISTICS, METEOROLOGY. *Educ:* Univ Minn, BA, 53, PhD(biostatist), 63; Univ Ariz, MA, 58. *Prof Exp:* Asst math, Univ Ariz, 57-58; asst biostatist, Univ Minn, 58-62, lectr, 62-63; from asst prof to assoc prof, 63-73, PROF STATIST, COLO STATE UNIV, 73- *Mem:* Fel Am Statist Asn; Biomet Soc; Am Meteorol Soc. *Res:* Permutation inference procedures; nonparametric techniques; parametric approximations; simulation investigations; quantal assay methods; weather modification studies. *Mailing Add:* Dept of Statist Colo State Univ Ft Collins CO 80523

MIELKE, ROBERT L, US citizen. METROLOGY ENGINEERING, CLEANROOM TECHNOLOGY. *Educ:* Miami Univ, Oxford, BS, 79. *Prof Exp:* Engr phys metrol, Monsanto Res Corp, 79-88; mgr metrol eng, 89-90, MGR PHYS METROL, EG&G MOUND APPL TECHNOLOGIES INC, 91- *Concurrent Pos:* Consult, Monsanto Co, 81-87; standards & practices coordr, Inst Environ Sci, 85-87, chmn, 88- *Honors & Awards:* James R Milden Award, Inst Environ Sci, 91. *Mem:* Inst Environ Sci. *Res:* Cleanroom testing and monitoring, HEPA/ULPA filter testing and other cleanroom related activities; physical metrology; thermometry. *Mailing Add:* EG&G Mound Appl Technologies Inc PO Box 3000 Miamisburg OH 45343-3000

MIENTKA, WALTER EUGENE, b Amherst, Mass, Oct 1, 25; m 54; c 4. MATHEMATICS. *Educ:* Univ Mass, BS, 48; Columbia Univ, MA, 49; Univ Colo, PhD, 55. *Prof Exp:* Instr math, Univ Mass, 49-52; instr & asst, Univ Colo, 52-55; instr, Univ Mass, 55-56; asst prof, Univ Nev, 56-57; from asst prof to assoc prof, 57-70, vchmn dept, 70-75, PROF MATH, UNIV NEBR, LINCOLN, 70-; EXEC DIR, MATH ASN AM, 76- *Concurrent Pos:* Fac fel, Univ Nebr, Lincoln, 60 & 64-65; res scholar, Univ Calif, Berkeley, 64-65. *Mem:* Am Math Soc; Math Asn Am; Indian Math Soc; Sigma Xi. *Res:* Theory of numbers. *Mailing Add:* Rte 5 Box 149 Lincoln NE 68521

MIER, MILLARD GEORGE, b Glendale, Calif, Nov 26, 35; m 60; c 2. SOLID STATE PHYSICS. *Educ:* Occidental Col, AB, 57; Bryn Mawr Col, PhD(physics, math), 67. *Prof Exp:* Instr physics, Mt Holyoke Col, 62-64; res physicist, Owens-Corning Fiberglas Tech Ctr, 64-68; RES PHYSICIST, AIR FORCE WRIGHT AERONAUT LAB, WRIGHT-PATTERSON AFB, 68- *Concurrent Pos:* Adj prof elec eng, Air Force Inst Technol, 68- *Mem:* Am Asn Physics Teachers; Am Phys Soc; Am Asn Univ Prof; Inst Elec & Electronic Engrs; Electrochem Soc; Sigma Xi. *Res:* Optical properties of solids and thin films; thin film electronic properties, especially compound semiconductor, metallic and insulating films; computer memory technology, especially magnetic bubble, electron beam and ion-implant memories. *Mailing Add:* 132 N Walnut St Yellow Springs OH 45387

MIERNYK, JAN ANDREW, b Boulder, Colo, Oct 4, 47; div; c 1. PLANT CELL BIOLOGY. *Educ:* WVa Univ, BA, 74, MS, 77; Ariz State Univ, PhD(bot), 80. *Prof Exp:* Postdoctoral assoc biochem, Queen's Univ, Kingston, Ont, 80-83; postdoctoral fel, Univ Mo, Columbia, 83-84; RES CHEMIST, NAT CTR AGR UTILIZATION RES, AGR RES SERV, USDA, 84- *Concurrent Pos:* Res asst prof, Dept Biochem, Univ Mo, Columbia, 84-; Orgn Econ Coop & Develop fel biochem, Dept Biol, Univ Newcastle-Upon-Tyne, 89. *Mem:* Am Soc Plant Physiologists; Am Soc Cell Biol; Am Soc Biochem & Molecular Biol; AAAS; Int Soc Plant Molecular Biol. *Res:* Plant cell biology; synthesis and targeting of secretory proteins; modification of higher plants through genetic engineering. *Mailing Add:* USDA Agr Res Serv 1815 N University St Peoria IL 61604

MIES, FREDERICK HENRY, b New York, NY, Oct 3, 32; m 53; c 3. QUANTUM CHEMISTRY. *Educ:* City Col New York, BS, 56; Brown Univ, PhD(phys chem), 61. *Prof Exp:* Nat Bur Standards-Nat Res Coun fel phys chem, 61-62, PHYS CHEMIST, NAT BUR STANDARDS, 62- *Mem:* Am Phys Soc. *Res:* Scattering theory; chemical kinetics and energy transfer; pressure broadening and continuum spectroscopy; autoionization and predissociation. *Mailing Add:* B268 Phys Bldg Nat Bur Standards Washington DC 20899

MIESCH, ALFRED THOMAS, b Hammond, Ind, May 10, 27; m 50; c 2. MATHEMATICAL STATISTICS. *Educ:* St Joseph's Col Ind, BS, 50; Ind Univ, MA, 54; Northwestern Univ, PhD, 61. *Prof Exp:* Geol technician, NMex Bur Mines & Mineral Resources, 51-52; asst, Ind Univ, 52-53; res geologist, Regional Geochem Br, US Geol Surv, 53-86, CONSULT, 86- *Concurrent Pos:* Asst, Northwestern Univ, 56-57. *Mem:* Asn Explor Geochemists; Int Asn Math Geol. *Res:* Distribution of minor elements in rocks and ores; Colorado Plateau uranium deposits; statistical methods in geologic and geochemical research; geochemical prospecting; environmental geochemistry. *Mailing Add:* PO Box 1103 Grand Junction CO 81502

MIESCHER, GUIDO, b Zurich, Switz, Dec 13, 21; nat US; m 54; c 1. MICROBIOLOGY. *Educ:* Swiss Fed Inst Technol, dipl, 47, PhD(microbiol, plant path), 49. *Prof Exp:* Res microbiologist, Imcera Corp, 49-85; RETIRED. *Mem:* Am Chem Soc. *Res:* Nutrition and metabolism of plants and microorganisms; development of industrial fermentations. *Mailing Add:* 7 Elks Dr Terre Haute IN 47802

MIESEL, JOHN LOUIS, b Erie, Pa, Nov 26, 41; m 64; c 3. ORGANIC CHEMISTRY. *Educ:* Univ Notre Dame, BS, 62; Univ Ill, PhD(org chem), 66. *Prof Exp:* Sr org chemist, 66-74, RES SCIENTIST, ELI LILLY & CO, 74- *Mem:* Am Chem Soc. *Res:* Synthesis of heterocyclic compounds; structure-activity relationships in insecticides; photochemistry as a synthetic tool. *Mailing Add:* 8744 N Pennsylvania Ave Indianapolis IN 46240

MIESSLER, GARY LEE, b Independence, Kans, Jan 5, 49; m 88; c 2. PHOTOCHEMISTRY. *Educ:* Univ Tulsa, BChem, 70; Univ Minn, PhD(inorg chem), 78. *Prof Exp:* Instr, 78-79, asst prof, 79-84, ASSOC PROF CHEM, ST OLAF COL, 84- *Concurrent Pos:* Prin investr, Res Corp, 81-83 & Petrol Res Grant, Am Chem Soc, 81-83; teacher & scholar, Camille & Henry Dreyfus Found, 81-86. *Mem:* Am Chem Soc. *Res:* Synthesis and photochemistry of transition metal complexes with sulfur ligands; organometallic chemistry and homogeneous catalysis. *Mailing Add:* Dept Chem St Olaf Col Northfield MN 55057

MIETLOWSKI, WILLIAM LEONARD, b Buffalo, NY, Sept 25, 47; m 84. BIOSTATISTICS. *Educ:* Canisius Col, BS, 69; Univ Rochester, MA, 71, PhD(statist), 74. *Prof Exp:* Teaching asst statist, Univ Rochester, 69-73, tech assoc biostatist, Heart Res Follow-up Study, 73-74; res asst prof statist sci, Statist Lab, State Univ NY Buffalo, 74-77; statistician, Sidney Farber Cancer Inst, 78-80; ASSOC DIR STATIST, SANDOZ RES INST, 79- *Concurrent Pos:* Coord statistician, Lung Cancer Group, Vet Admin, 74-79; asst prof biostatist, Harvard Univ Sch Pub Health, 78-79. *Mem:* Am Statist Asn; Biomet Soc; Inst Math Statist; Soc Clin Trials. *Res:* Application of biometric methods to lung cancer clinical trials; multivariate descriptive statistics; discriminant analysis with mixed data; comparison of correlated covariance matrices; application of biometric methods to neurological clinical trials. *Mailing Add:* 19 Ertman Dr Whippany NJ 07981

MIEURE, JAMES PHILIP, b McLeansboro, Ill, July 5, 41; m 66; c 2. CHEMISTRY. *Educ:* Kenyon Col, AB, 63; Purdue Univ, MS, 66; Tex A&M Univ, PhD(chem), 68. *Prof Exp:* Sr res chemist, 68-73, res specialist, 73-74, res group leader, 74-77, mgr environ sci, 77-80, MGR PROD ACCEPTABILITY, MONSANTO CO, 80- *Mem:* Am Chem Soc; Am Soc Testing Mat; Chem Mfrs Asn. *Res:* Environmental analytical chemistry; environmental fate and aquatic toxicity of chemicals; safety assessment of chemicals. *Mailing Add:* 1242 Chavaniac Dr Manchester MO 63011-3604

MIEYAL, JOHN JOSEPH, b Cleveland, Ohio, Feb 17, 44; m 66; c 4. BIOCHEMISTRY, PHARMACOLOGY. *Educ:* John Carroll Univ, BS, 65; Case Western Reserve Univ, PhD(biochem), 69. *Prof Exp:* NIH fel, Brandeis Univ, 69-71; asst prof pharmacol & biochem, Med Sch, Northwestern Univ, Chicago, 71-76; ASSOC PROF PHARMACOL, MED SCH, CASE WESTERN RESERVE UNIV, 76-, ASSOC PROF CHEM, 81- *Concurrent Pos:* Grants, Res Corp Am, 71-75, Chicago Heart Asn, 74-76, Nat Inst Gen Med Sci, 74-81, Am Heart Asn, 76-79, 87-91, Nat Cancer Inst, 77-81, & Am Cancer Soc, 91. *Mem:* Am Soc Pharmacol & Exp Therapeut; Am Soc Biochem & Molecular Biol; AAAS; Am Chem Soc; Sigma Xi. *Res:* Physicochemical studies of molecular interactions; mechanisms of enzymic reactions; drug metabolism and toxicity; chemistry. *Mailing Add:* Dept Pharmacol Med Sch Case Western Reserve Univ Cleveland OH 44106

MIFFITT, DONALD CHARLES, b Holyoke, Mass, Mar 10, 44; m 67; c 2. ELECTRONICS ENGINEERING. *Educ:* Lowell Technol Inst, BS, 66. *Prof Exp:* Proj engr, US Air Force, Electronic Systs Div, 66-70; sr engr, Bowmar/Ali, Inc, 70-74; prin engr, Gillette Adv Tech Lab, Gillette Co, 74-79; dir elec eng, Parker Bros, 79-84; PRES, VENTURE TECHNOLOGIES, INC, 84- *Mem:* Inst Elec & Electronics Engrs. *Res:* Application of electronics technology to new comsumer products; microcomputers for timing and control; speech recognition/synthesis; temperature measurement; high density memories; video techniques. *Mailing Add:* Venture Technologies Inc 76 Treble Cove Rd North Billerica MA 01862

MIFFLIN, MARTIN DAVID, b Olympia, Wash, Mar 29, 37; m 59; c 4. HYDROGEOLOGY. *Educ:* Univ Wash, Seattle, BS, 60; Mont State Univ, MS, 63; Univ Nev, PhD(hydrogeol), 68. *Prof Exp:* Geologist, Pan Am Petrol Corp, 59, US Geol Surv, 62; res assoc, Water Resouces Ctr, Desert Res Inst, 63-69; assoc prof geol, Univ Fla, 69-75; assoc dir, 75-77, RES PROF, WATER RESOURCES CTR, DESERT RES INST, 77- *Concurrent Pos:* Resident consult, World Bank & chief, Groundwater Planning, Nat Water Plan, Mex, 73-75; chief resident adminr & sr hydrogeologist, UN Develop Prog Proj, Chile, 78-79; consult, Govt Tunisia, USAID, 80. *Mem:* Nat Water Well Asn. *Res:* Groundwater resource assessment: exploration development, management flow system delineation and subsidence; arid and carbonate rock terrains; Quaternary lakes and isostatic rebound in the Great Basin. *Mailing Add:* Mifflin & Assoc Inc 2700 E Sunset Rd Suite B-13 Las Vegas NV 89120

MIFFLIN, THEODORE EDWARD, b Zion, Ill, Aug 4, 46; m 73; c 3. CLINICAL CHEMISTRY. *Educ:* Weber State Univ, BS, 68; Utah State Univ, PhD(biochem), 84. *Prof Exp:* Chemist, Southwest Bioclin Lab, 74-77; res fel clin biochem, Dept Path, Univ Va Med Ctr, 83-86, res asst prof molecular biol, Dept Path & Med, 86-90; ASSOC PROF, DEPT PATH, PRESBY UNIV HOSP PITTSBURGH, 91- *Concurrent Pos:* Instr, dept chem & biochem, Utah State Univ, 80; co-dir, Molecular Probe Lab, Univ Va, 88-90; chief, Molecular Diag Unit, Clin Chem Div, Dept Path, Presby Univ Hosp Pittsburgh, 91- *Honors & Awards:* E Cotlove Award, Am Asn Clin Chem, 85. *Mem:* Am Asn Advan Sci; Am Asn Clin Chem; Am Chem Soc. *Res:* Diagnostic applications of molecular biology in clinical laboratory; regulation of mammalian gene expression; clinical biochemistry of isoenzymes. *Mailing Add:* Presby Univ Hosp Pittsburgh 5845 Main Tower Pittsburgh PA 15213

MIGDALOF, BRUCE HOWARD, b Brooklyn, NY, July 19, 41; m 67; c 4. XENOBIOLOGY, DRUG METABOLISM. *Educ:* Cornell Univ, BA, 62; Purdue Univ, MS, 65; Univ Pittsburgh, PhD(org chem), 69. *Prof Exp:* Sr scientist drug metab, Sandoz Pharmaceuts, Sandoz-Wander Inc, 69-72; sr scientist, McNeil Labs Inc, 72-74, group leader drug disposition, 74-77; dir drug metab, Squibb Inst Med Res, 77-89; DIR, SCI COMMUN & QUAL/PRODUCTIVITY, BRISTOL-MYERS SQUIBB PHARMACEUT RES INST, 89- *Honors & Awards:* Outstanding Achievement Award, Int Soc Study Xenobiotics. *Mem:* Am Chem Soc; NY Acad Sci; Int Soc Study Xenobiotics; Am Asn Pharmaceut Scientists; Sigma Xi. *Res:* Applications of quality/productivity improvement programs in research and development. *Mailing Add:* Bristol-Myers Squibb Pharmaceut Res Inst PO Box 400 Princeton NJ 08543-4000

MIGEON, BARBARA RUBEN, b Rochester, NY, July 31, 31; m 60; c 3. MEDICAL GENETICS. *Educ:* Smith Col, BA, 52; Univ Buffalo, MD, 56. *Prof Exp:* Intern pediat, Johns Hopkins Hosp, 56-57, asst resident, 57-59; fel endocrinol, Med Sch, Harvard Univ, 59-60; fel genetics, 60-62, from instr to assoc prof pediat, 62-79, DIR, PREDOCTORAL PHD PROG IN HUMAN GENETICS, JOHNS HOPKINS HOSP, 80- *Concurrent Pos:* Pediatrician, Johns Hopkins Hosp, 62-,; mem genetics study sect, NIH, 75-79. *Mem:* Am Soc Human Genetics; Am Soc Pediat Res. *Res:* Somatic cell genetics; regulation of expression of X-linked genes; X chromosome inactivation; complementation analysis of human inborn errors. *Mailing Add:* Dept Pediat CMSC 10-04 Johns Hopkins Hosp Baltimore MD 21205

MIGEON, CLAUDE JEAN, b Lievin, France, Dec 22, 23; m 60; c 3. PEDIATRICS, ENDOCRINOLOGY. *Educ:* Lycee de Reims, France, BA, 42; Univ Paris, MD, 50. *Prof Exp:* Asst med biochem, Univ Paris, 47-50; Am Field Serv fel, 50-51; res fel pediat, Johns Hopkins Univ, 51-52; res instr biochem, Univ Utah, 52-54; from asst prof to assoc prof, 54-71, PROF PEDIAT, JOHNS HOPKINS UNIV, 71-, DIR PEDIAT ENDOCRINE CLIN & LABS, 73- *Concurrent Pos:* Fulbright traveling fel, 50; Mayer fel, 51-52; NIH res career award, 64. *Honors & Awards:* Ayerst Award, Endocrine Soc, 82. *Mem:* Endocrine Soc; Soc Pediat Res; Am Soc Clin Invest; Am Physiol Soc; Am Pediat Soc. *Res:* Pediatric endocrinology, particularly steroids biochemistry; abnormalities of human sex differentiation; adrenal function; transplacental passage of steroids from mother to fetus. *Mailing Add:* Dept Pediat Johns Hopkins Hosp Baltimore MD 21205

MIGET, RUSSELL JOHN, b Long Beach, Calif, Oct 22, 42; m 63; c 2. MARINE MICROBIOLOGY. *Educ:* Univ Fla, BS, 64; Fla State Univ, PhD(oceanog), 71. *Prof Exp:* Res assoc marine microbiol, Inst Marine Sci, Univ Tex, 71-75; PRES, TURTLE COVE LAB, INC, 75-; MARINE FISHERIES SPECIALIST, TEX A&M UNIV, 76- *Res:* Marine microbial ecology. *Mailing Add:* PO Box 158 Port Aransas TX 78373

MIGHTON, CHARLES JOSEPH, organic chemistry; deceased, see previous edition for last biography

MIGHTON, HAROLD RUSSELL, b Saskatoon, Sask, Can, Jan 6, 19; nat US; m 43; c 2. CHEMISTRY. *Educ:* Univ Sask, BA, 39, MA, 41; Columbia Univ, PhD(org chem), 45. *Prof Exp:* Res chemist comt med res, Off Sci Res & Develop Proj, Columbia Univ & Rockefeller Inst, 44-45 & Goodyear Tire & Rubber Co, Ohio, 45-48; res chemist rayon dept, E I du Pont de Nemours & Co, Inc, 48-50 & film dept, 50-52, res assoc, 52-53, res supvr, 53-57, res mgr, 57-68, mgr tech liaison, Int Dept, 68-72, planning mgr, Cent Res & Develop Dept, 72-80; RETIRED. *Mem:* Am Chem Soc; Sigma Xi; Chem Inst. *Res:* Surface chemistry; antimalarials; high polymers; thermodynamics of crystallization in high polymers. *Mailing Add:* 711 Ambleside Dr Wilmington DE 19808

MIGLIARO, MARCO WILLIAM, b Brooklyn, NY, March 29, 48; div; c 4. BATTERY TECHNOLOGY, POWER SYSTEMS. *Educ:* Pratt Inst, BEE, 69. *Prof Exp:* Engr, Am Elec Power, 69-78; sr engr, Gibbs & Hill Inc, 78-80, staff eng, 80-81; assoc consult engr, Ebasco Serv Inc, 81-83, Consult engr, 83-87, sr consult, 87-88; staff consult, Impell Corp, 88-90; SR STAFF SPECIALIST, FPL, 90- *Concurrent Pos:* Lectr, Alber Eng Inc, 87- & Dragnetz Technol, 87-89; chmn Standards Coord Comt, Inst Elec & Electronic Engrs, 87-88 & Power Generation Comt, 88, mem, bd dirs & chmn, Standards Bd, 90- *Mem:* Fel Inst Elec & Electronic Engrs; Electrochem Soc; Proj Mgt Inst. *Res:* Battery technology in industrial and utility power systems; numerous publications in other published works. *Mailing Add:* PO Box 9253 Jupiter FL 33468-9253

MIGLIORE, HERMAN JAMES, b Detroit, Mich, July 13, 46. MECHANICAL ENGINEERING, APPLIED MECHANICS. *Educ:* Univ Detroit, BS, 68, MS, 69, PhD(mech eng), 75. *Prof Exp:* Intern, Chrysler Corp, 73-75; res engr, Naval Civil Eng Lab, 75-77; asst prof, 77-80, ASSOC PROF, PORTLAND STATE UNIV, 80- *Concurrent Pos:* Design consult, numerical methods & comput aided design. *Mem:* Am Soc Mech Engrs; Am Soc Eng Educ; Sigma Xi. *Res:* Numerical analysis of nonlinear systems in cable dynamics, structure-fluid interaction and plastic deformation manufacturing processes; investigation of prediction and relief of residual stress; computer aided design procedures. *Mailing Add:* Div Eng PO Box 751 Portland State Univ Portland OR 97207

MIGLIORE, PHILIP JOSEPH, b Pittsburgh, Pa, Dec 18, 31; m 57; c 3. MEDICINE. *Educ:* Univ Pittsburgh, BS, 54, MD, 56. *Prof Exp:* Asst pathologist, Univ Tex M D Anderson Hosp & Tumor Inst, 64-69; ASST PROF PATH, BAYLOR COL MED, 69- *Concurrent Pos:* Asst pathologist, Methodist Hosp, Houston, 69-78, assoc pathologist, 78-82, atten pathologist, 82-87, sr atten pathologist, 87-; head clin chem, Methodist Hosp, 69- *Mem:* NY Acad Sci; AMA; Am Soc Clin Path; Am Asn Clin Chem. *Res:* Gamma globulins; myeloma proteins. *Mailing Add:* Dept Path Baylor Col Med 1200 Moursand Ave Houston TX 77030

MIGNAULT, JEAN DE L, b Sherbrooke, Que, Feb 3, 24; m 53; c 3. MEDICINE, CARDIOLOGY. *Educ:* Univ Montreal, BA, 45, MD, 51; FRCP(C). *Prof Exp:* Clin monitor cardiol, Maisonneuve Hosp, 56-58; asst prof, Univ Montreal & Inst Cardiol Montreal, 58-61, from asst prof to assoc prof, Univ Montreal & Hotel-Dieu Hosp, 62-69, dir cardiac lab, Hosp, 62-69, head dept cardiol, 65-69; chmn dept med, 69-73, dean sch med, 70-73, PROF MED, MED SCH, UNIV SHERBROOKE, 69- *Mem:* Fel Am Col Cardiol; fel Am Col Chest Physicians; Can Med Asn; Can Cardiovasc Soc; fel Am Col Physicians. *Res:* Development of the new cardiac catheterization unit, the Saturn. *Mailing Add:* Fac Med Univ Sherbrooke Sherbrooke PQ J1H 5N4 Can

MIGNEREY, ALICE COX, Brooklyn, NY, Nov 6, 49; m 70. HEAVY-ION REACTION MECHANISMS. *Educ:* Univ Rochester, BS, 71, MS, 73, PhD(nuclear chem), 75. *Prof Exp:* Fel nuclear chem, Univ Rochester, 75-76; res assoc, Chem Div, Argonne Nat Lab, 76-79; from asst prof to assoc prof, 79-89, PROF GEN CHEM & NUCLEAR CHEM, UNIV MD, 89- *Mem:* Am Phys Soc; Am Chem Soc. *Res:* Nuclear reaction mechanisms in heavy-ion induced reactions; mechanisms including preequilibrium emission of light particles, the deep-inelastic reaction, projectile fragmentation and multifragmentation ateneries of million-electron-volt particles. *Mailing Add:* Dept Chem Univ Md College Park MD 20742

MIGNERY, ARNOLD LOUIS, b West Unity, Ohio, Apr 18, 18; m 42; c 4. FORESTRY. *Educ:* Univ Mich, BS, 40, MF, 49. *Prof Exp:* Res forester, Southern Forest Exp Sta, US Forest Serv, 46-56, res ctr leader, 56-64, prin silviculturist & proj leader, 64-75; RETIRED. *Res:* Silviculture and forest management techniques; southern tree species. *Mailing Add:* Running Knob Hollow Rd Sewanee TN 37375

MIGNONE, ROBERT JOSEPH, b Philadelphia, Pa, Apr 21, 45. SET THEORY. *Educ:* Pa State Univ, PhD(math), 79. *Prof Exp:* Vis asst prof math, Univ Tex, Dallas, 79-81; ASST PROF MATH, COL CHARLESTON, 81- *Mem:* Am Math Soc; Asn Symbolic Logic. *Res:* Consequences of the axiom of determinateness on the smallest uncountable cardinals; characterizations of huge cardinals. *Mailing Add:* Math Dept Sci Res Inst 1000 Centennial Dr Berkeley CA 94720

MIHAILOFF, GREGORY A, b Mansfield, Ohio, Aug 17, 45; m 67; c 3. NEUROANATOMY, NEUROBIOLOGY. *Educ:* Ashland Col, BS, 68; Ohio State Univ, MS, 73, PhD(anat), 74. *Prof Exp:* From asst prof to assoc prof cell biol, Health Sci Ctr, Univ Tex, 74-85. *Concurrent Pos:* Prin investr, NSF res grant, Neurobiology Prog, 80-83 & Nat Inst Neurol Commun Disorders Stroke, NIH, 81-84. *Mem:* Soc Neurosci; Am Asn Anatomists; Cajal Club. *Res:* Neurobiology of cerebro-cerebellar interaction; electron microscopy; neurophysiology. *Mailing Add:* 545 Arbor Dr Madison WI 39110

MIHAILOVSKI, ALEXANDER, b Sofia, Bulgaria, Nov 8, 37; US citizen; m 69; c 1. ORGANIC CHEMISTRY, PROCESS RESEARCH. *Educ:* Pa State Univ, BS, 60; Univ Calif, Los Angeles, PhD(org chem), 67. *Prof Exp:* Asst gen & org chem, Univ Calif, Los Angeles, 63-66 & org chem, 66-67; from res chemist to sr res chemist, 67-72, GROUP SUPVR & SECT MGR, WESTERN RES CTR, STAUFFER CHEM CO, 72- *Mem:* AAAS; Am Chem Soc; Royal Soc Chem. *Res:* Acetylene-allene chemistry; organic reaction mechanisms; synthetic organic chemistry in agricultural pest control; fine chemicals process design and optimization. *Mailing Add:* 11 Kensington Ct Kensington CA 94707

MIHAJLOV, VSEVOLOD S, b Kladanj, Yugoslavia, Feb 12, 25; nat US; m 51; c 3. PHYSICAL CHEMISTRY, ORGANIC CHEMISTRY. *Educ:* Univ Munich, BS, 49; Clark Univ, AM, 54, PhD, 56. *Prof Exp:* Sr scientist, Xerox Corp, 56-68, mgr process & mat develop, 68-69, mgr color technol area, 69-74, prin scientist & mgr copy qual technol, 74-81, prin scientist, 81-89; CONSULT, US-USSR TRADE, 89- *Concurrent Pos:* Abstractor, Chem Abstr, 53-70. *Mem:* Am Chem Soc; Soc Photog Sci & Eng; Int-Soc Color Coun. *Res:* Photographic science; graphic arts; photosensitive systems; imaging; xerography; color; color vision. *Mailing Add:* 10-2 Selden St Rochester NY 14605

MIHALAS, BARBARA R WEIBEL, b Berkeley, Calif, Oct 18, 39; m 61, 75; c 3. SOLAR PHYSICS, RADIATION HYDRODYNAMICS. *Educ:* Univ Colo, BA, 61, MS, 77, PhD(astrophysics), 79. *Prof Exp:* Fel, Advan Study Prog, Nat Ctr Atmospheric Res, 79-81; asst astronomer, Sacramento Peak Observ, 81-82; RES SCIENTIST, NAT CTR SUPERCOMPUTING APPLNS, ASST PROF ASTRON, UNIV ILL, 85- *Concurrent Pos:* Vis scientist, high altitude observ, Nat Ctr Atmospheric Res, 82-85; Danforth fel. *Mem:* Am Astron Soc; AAAS. *Res:* Theory and observation of acoustic-gravity waves in the solar atmosphere; theory of spectral line formation in the presence of small-scale fluctuations in velocity and thermodynamic variables, radiation hydrodynamics waves and stability. *Mailing Add:* Nat Ctr Supercomputing Appln-5600 Beckman Drawer 25 405 Mathews Ave Urbana IL 61801

MIHALAS, DIMITRI, b Los Angeles, Calif, Mar 20, 39; m 63, 75; c 2. ASTROPHYSICS. *Educ:* Univ Calif, Los Angeles, AB, 59; Calif Inst Technol, MS, 60, PhD(astron, physics), 64. *Prof Exp:* Mem tech staff, TRW Space Tech Labs, 59; Higgins vis fel astron, Princeton Univ, 63-64, asst prof, 64-67; asst prof physics & astrophys & mem joint inst lab astrophys, Univ Colo, 67-68; from assoc prof to prof astron, Univ Chicago, 68-71; sr scientist, High Altitude Observ, 71-79; astronomer, Sacramento Peak Observ, 79-82; sr scientist, High Altitude Observ, 82-85; GEORGE C MCVITTIE PROF ASTRON, UNIV ILL, 85- *Concurrent Pos:* Alfred P Sloan res fel, 69-71; mem comn 12 & 36, Int Astron Union; mem astron adv panel, NSF, 72-75; assoc ed, Astrophys J, 70-, J Comput Physics, 81-87 & J Quant Spectros Radiative Trans, 84-; adjoint prof depts astrogeophysics & dept physics & astrophysics, Univ Colo, 72-; vis prof dept astrophysics, Oxford Univ, 77-78; sr vis fel, Univ Col London, 78; consult, Los Alamos Nat Lab, 81-; Alexander von Humboldt Found sr US scientist award, 84; mem Theoret & Computational Physics adv panel, Los Alamos Nat Lab, 85- *Honors & Awards:* Helen B Warner Prize, Am Astron Soc, 74. *Mem:* Nat Acad Sci; Am Astron Soc; Int Astron Union. *Res:* Physics of stellar atmospheres and abundances of elements in the stars; theory of radiative transfer; radiation hydrodynamics; solar physics. *Mailing Add:* Dept Astron Univ Ill 1002 W Green St Urbana IL 61801

MIHALCZO, JOHN THOMAS, b Yonkers, NY, May 30, 31; m 52; c 6. NUCLEAR ENGINEERING, PHYSICS. *Educ:* NY Univ, BA, 53, MS, 56; Univ Tenn, Knoxville, PhD(nuclear eng), 70. *Prof Exp:* Physicist, Res Div, Curtiss Wright Corp, 53-58 & Oak Ridge Nat Lab, 58-68; mem res staff physics & nuclear eng, Y-12 Plant, Nuclear Div, Union Carbide Corp, 68-73; MEM SR STAFF, INSTRUMENTATION & CONTROLS DIV, OAK RIDGE NAT LAB, MARTIN MARIETTA ENERGY SYSTS, 73- *Concurrent Pos:* Ford Found prof nuclear eng, Univ Tenn, 71- *Mem:* Fel Am Nuclear Soc. *Res:* Critical and subcritical assemblies; pulse reactors; prompt neutron decay by the pulsed neutron and the Rossi methods; reactor physics; radiation shielding; fusion energy research. *Mailing Add:* 114 Lehigh Lane Oak Ridge TN 37830

MIHALISIN, JOHN RAYMOND, b Passaic, NJ, Dec 18, 24. PROCESS METALLURGY, PRODUCTION ENGINEERING. *Educ:* Yale Univ, BE, 49; Mass Inst Technol, ScD(metall), 53. *Prof Exp:* Chem eng supvr, Curtiss-Wright Corp, 53-55; res metallurgist, Inco, 55-72; TECH DIR, HOWMET CORP, 72- *Mem:* Am Soc Metals; Am Inst Mining Metall & Petrol Engrs; Am Soc Testing & Mat; Asn Res Dirs. *Res:* Development of new materials and processes for high temperature use in the aerospace industry. *Mailing Add:* 545 Mountain Ave North Caldwell NJ 07006

MIHALISIN, TED WARREN, b Houston, Tex, Feb 11, 40; m 61; c 3. LOW TEMPERATURE PHYSICS. *Educ:* Cornell Univ, BA, 61; Univ Rochester, PhD(physics), 67. *Prof Exp:* assoc prof, 69-75, PROF PHYSICS, TEMPLE UNIV, 75- *Concurrent Pos:* Assoc scientist dept physics, Gulf Gen Atomic Inc, Calif, 67-69. *Mem:* Am Phys Soc. *Res:* Study of valence fluctuations, Kondo lattice phenomena, magnetic ordering and superconductivity in rare earth intermetallic compounds. *Mailing Add:* Dept of Physics Temple Univ Philadelphia PA 19122

MIHALOV, JOHN DONALD, b Los Angeles, Calif, Dec 28, 37. SOLAR-TERRESTRIAL RELATIONS, PLANETARY EXPERIMENTAL APPARATUS. *Educ:* Calif Inst Technol, BS, 59, MS, 61; Stanford Univ, 81. *Prof Exp:* Mem tech staff, Space Tech Lab, Inc, 59-60; res asst, Ctr Radiophysics & Space Res, Cornell Univ, 60-61; scientist, Jet Propulsion Lab, NASA, 61; mem tech staff, Aerospace Corp, 61-66; RES SCIENTIST, AMES RES CTR, NASA, 66- *Concurrent Pos:* Guest worker, Lawrence Berkeley Lab, Univ Calif, 85-86. *Mem:* Am Phys Soc; Am Geophys Union; AAAS; Am Astron Soc. *Res:* Author or co-author of over 90 publications dealing with experimental data on the earths radiation belts, the interplanetary medium, the magnetospheres of Jupiter and Saturn, and the solar wind interaction with the moon and with Venus; Galileo Jupiter Atmospheric Probe experiments; Huygens Titan Atmospheric Probe experiments. *Mailing Add:* 761 Garland Dr Palo Alto CA 94303-3604

MIHALYI, ELEMER, b Deva, Rumania, Jan 11, 19; nat US; m 48; c 2. BIOCHEMISTRY. *Educ:* Univ Kolozsvar, Hungary, MD, 43; Cambridge Univ, PhD, 63. *Prof Exp:* Instr med chem, Univ Kolozsvar, 41-44 & biochem, Univ Budapest, 46-48; guest investr, Nobel Inst Med, Stockholm, 48-49; res assoc, Inst Muscle Res, Woods Hole, 49-51; res fel, Harrison Dept Surg Res, Pa, 51-55; chemist, Lab Cell Biol, Nat Heart, Lung & Blood Inst, 55-78; chemist, 78-84, EMER CHEMIST, LAB BIOCHEM PHARMACOL, NAT INST DIABETES & DIGESTIVE & KIDNEY DIS, 84- *Concurrent Pos:* Ed, Thrombosis Res, 72-78; Coordr, US-Spain Coop Proj Biochem, NSF, 74-76. *Mem:* Am Soc Biol Chemists; Am Chem Soc; Int Soc Hemat; Int Soc Thrombosis & Hemostasis. *Res:* Protein chemistry; proteins involved in blood coagulation; fragmentation of proteins into their constituent domains by limited proteolysis; physicochemical and chemical studies of the transformation of fibrinogen into fibrin. *Mailing Add:* 10210 Fleming Ave Bethesda MD 20814

MIHELICH, JOHN L, b Cleveland, Ohio, Oct 10, 37; m 62; c 5. NEW BUSINESS & PRODUCT DEVELOPMENT, ADVANCED COMPOSITES. *Educ:* Case Western Res Univ, Cleveland, PhD(metall), 64. *Prof Exp:* Sr res metallurgist, LTV Steel, 64-70; mgr high-strength low-alloy develop, Climax Molybdenum Div, Amax, 70-76, mgr develop, 76-80, vpres res, 80-86, gen mgr com develop, Mat Res, 86-88; GEN MGR DEVELOP, COMMONWEALTH ALUMINUM TECHNOL, 88- *Concurrent Pos:* Panelist, F&D Div, Bd Rev, Nat Inst Standards & Technol, 83-89; bd mem, Louisville Advan Tech Coun, 88-; mem, Fed Sci & Technol Comt, Indust Res Inst, 90-; vchair, Environ Affairs Comt, Am Soc Metals Int, 90-; chair, AIMMC Task Group, Aluminum Asn, 91- *Honors & Awards:* John Shoemaker Award, Am Soc Metals Int, 86. *Mem:* Am Inst Mining, Metall & Petrol Engrs-Metall Soc; AAAS; Am Soc Metals Int; Indust Res Inst. *Res:* Advanced materials and manufacturing processes; building new business from an advanced materials technology base. *Mailing Add:* Commonwealth Aluminum Unit Comalco Ltd 200 Meidinger Tower Louisville KY 40202

MIHELICH, JOHN WILLIAM, b Colorado Springs, Colo, Jan 2, 22; m 46; c 3. NUCLEAR PHYSICS. *Educ:* Colo Col, AB, 42; Univ Ill, PhD(physics), 50. *Prof Exp:* Assoc physicist, Brookhaven Nat Lab, 50-54; from asst prof to assoc prof, 54-61, PROF PHYSICS, UNIV NOTRE DAME, 61- *Mem:* Fel Am Phys Soc. *Res:* Radioactivity; decay schemes; internal conversion of gamma ray transitions; gamma ray spectra. *Mailing Add:* Dept of Physics Univ of Notre Dame South Bend IN 46556

MIHICH, ENRICO, b Fiume, Italy, Jan 4, 28; m 54; c 1. PHARMACOLOGY. *Educ:* Univ Milan, MD, 51. *Hon Degrees:* Dr, U Marseille, 86. *Prof Exp:* Instr, Inst Pharmacol, Univ Milan, 51, asst prof, 52 & 54-56; from sr cancer res scientist to prin cancer res scientist, Roswell Park Mem Inst, 57-71; assoc prof, 62-68, PROF BIOCHEM PHARMACOL, ROSWELL PARK DIV GRAD SCH, STATE UNIV NY BUFFALO, 68-, CHMN PROG PHARMACOL, 69-; DIR, DEPT EXP THERAPEUT, 71-, ASSOC DIR SPONSORED PROGS, ROSWELL PARK MEM INST & GRACE CANCER DRUG CTR, 87- *Concurrent Pos:* Vis res fel, Sloan-Kettering Inst Cancer Res, 52-54; dir lab pharmacol, Valeas Pharmaceut Indust, Italy, 54-56; docent pharmacol, Univ Milan, 62. *Mem:* AAAS; Am Soc Pharmacol & Exp Therapeut; Soc Exp Biol & Med; Am Asn Cancer Res; NY Acad Sci. *Res:* General and pre-clinical pharmacology; cancer biology and experimental therapy. *Mailing Add:* Dept of Exp Therapeut Roswell Park Mem Inst 666 Elm St Buffalo NY 14263

MIHINA, JOSEPH STEPHEN, b New York, NY, May 4, 18; m 49; c 2. ORGANIC CHEMISTRY. *Educ:* NY Univ, BS, 38; Mich State Col, MS, 48, PhD(chem), 50. *Prof Exp:* Asst foreman, Oil Tempering, Washburn Wire Co, 38-41; field inspector, Chem Warfare Serv, 41-43; fel, Northwestern Univ, 50-51; res chemist, G D Searle & Co, 51-66, mgr chem mfg, 66-77, mgr chem proc optimization, 77-84; RETIRED. *Mem:* Am Chem Soc. *Res:* Emulsion polymerization; tetrazoles steroids. *Mailing Add:* 8959 N Lockwood Ave Skokie IL 60077

MIHM, MARTIN C, JR, b Pittsburgh, Pa. DERMATOLOGY, PATHOLOGY. *Educ:* Duquesne Univ, BA, 55; Univ Pittsburgh, MD, 61; Am Bd Dermat, dipl, 69; Am Bd Path, dipl, 74, cert dermatopath, 75. *Hon Degrees:* MA, Harvard Univ, 89. *Prof Exp:* Clin & res fel dermat, Mass Gen Hosp, 64-67, clin fel path, 69-72, asst pathologist & asst dermatologist, 72-75; res fel dermat, 69-72, asst prof path, 72-75, ASSOC PROF PATH, HARVARD MED SCH, 75-; ASST DERMATOLOGIST, MASS GEN HOSP, 72-, ASSOC PATHOLOGIST, 75- *Concurrent Pos:* Assoc staff, Brigham & Women's Hosp, 75-; consult path, Cambridge City Hosp & Children's Hosp Med Ctr, 75-; consult dermatopath, Addison Gilbert Hosp, 75-; chief dermatopath residence training prog, Mass Gen Hosp, Brigham & Women's Hosp & Children's Hosp Med Ctr, Boston, 77-; consult path, Beth Israel Hosp, Boston, Mass, 78- & Boston Vet Admin Med Ctr, 79- *Mem:* Am Acad Dermat; Am Dermat Asn; Am Soc Dermatopath; Am Soc Clin Oncol; fel Am Col Physicians. *Res:* Biology of malignant melanoma, host response to this tumor and its histology; morphology of delayed hypersensitivity reactions in man; other aspects of cutaneous inflammation. *Mailing Add:* Dermpath Unit Mass Gen Hospital Fruit St Boston MA 02114

MIHRAM, GEORGE ARTHUR, b Norman, Okla, Sept 21, 39; m 65. MATHEMATICAL STATISTICS, SYSTEMIC SCIENCE. *Educ:* Univ Okla, BS, 60; Okla State Univ, MS 62, PhD(statist), 65. *Prof Exp:* Mathematician, Opers Res Inc, 65-66; systs analyst, Orgn Joint Chiefs of Staff, 66-68; asst prof, Univ Pa, 68-74; instr, Univ Southern Calif, 78-79; CONSULT, 79- *Concurrent Pos:* Consult, Hq USAF, 68-69, Off Asst Secy Defense, syst anal, 69, Acad Natural Sci, 70-71 & IBM Corp, 73; NSF res initiation grant, 71, int travel grant, 75, & NATO travel grant, 77; assoc ed, Simulation, 73-75, Int J Gen Syst, 73- & Modeling & Simulation, 74-; res initiation grant, NSF, 70-72, mem peer rev panels, 74-82. *Mem:* AAAS; Asn Comput Mach; Int Asn Statist Comput; Oper Res Soc Am; Sigma Xi; Int Soc Systs Sci; Soc Computer Simulation; Int Soc Cybernet. *Res:* The scientific method is a six-stage model-building process which mimes isomorphically the modelling process (genetic, then neural models) by which all life on earth has to date been assured; tele-cybernetics. *Mailing Add:* PO Box 1188 Princeton NJ 08542-1188

MIHRAN, THEODORE GREGORY, b Detroit, Mich, June 28, 24; m 53, 81; c 3. MICROWAVE ELECTRONICS. *Educ:* Stanford Univ, AB, 44, MS, 47, PhD(elec eng), 50. *Prof Exp:* PHYSICIST, CORP RES & DEVELOP, GEN ELEC CO, 50- *Concurrent Pos:* Lectr, Union Col, NY, 52-53 & 60-61; vis assoc prof, Cornell Univ, 63-64; assoc ed, Inst Elec & Electronics Engrs Trans on Electron Devices, 70-73. *Mem:* Fel, Inst Elec & Electronics Eng; Am Phys Soc. *Res:* Electron physics; microwave tubes and electronics; kylstrons; space charge wave amplification; plasmas; MOSFET modeling; microwave ovens. *Mailing Add:* Res & Develop Ctr Gen Elec Co PO Box Eight Schenectady NY 12301

MIHURSKY, JOSEPH ANTHONY, b Alpha, NJ, May 4, 33; m 55; c 1. ECOLOGY. *Educ:* Lafayette Col, BA, 54; Lehigh Univ, MS, 57, PhD(ecol), 62. *Prof Exp:* Res assoc, 62-67, res asst prof, 67-68, chmn & assoc prof, Dept Environ Res, 68-74, chmn & prof, 74-75, lab head & prof, 75-76, PROF MARINE ECOL, CHESAPEAKE BIOL LAB, UNIV MD, 77- *Concurrent Pos:* Res asst, Moyer-Trembley Consult, 56-62; chmn water working group, Comn Power Plant Siting, Nat Acad Eng, 69-70; panel mem, US Nat Water Comn, 71-72; adv comt mem, Cong Ad-Hoc Comt Environ Matters, US Govt, 68-69; planning comt mem, Inst del a Vie & Acad Sci, France, 73-74; consult adv, US Army Corp Engrs, 73-74 & Calif Marine Rev Comt, 75- *Mem:* Sigma Xi; Estuarine Res Fedn; AAAS; Fedn Am Scientists. *Res:* Determining factors that regulate abundance of estuarine populations and communities; environmental effects of coastal nuclear and fossil fueled power plants; pollution ecology; estuarine fish population dynamics and regional planning. *Mailing Add:* 28 Astor Rd Scientists Cliffs Port Republic MD 20676

MIIKKULAINEN, RISTO PEKKA, b Helsinki, Finland, Dec 16, 61. NEURAL NETWORKS, COGNITIVE SCIENCE. *Educ:* Southwestern Univ, BA, 84; Helsinki Univ, Finland, MS, 86, Univ Calif, Los Angeles, PhD(computer sci), 90. *Prof Exp:* ASST PROF ARTIFICIAL INTEL, DEPT COMPUTER SCI, UNIV TEX AUSTIN, 90- *Concurrent Pos:* Vis res fel, Max Planck Inst Psychol, 91. *Mem:* Am Asn Artificial Intel; Cognitive Sci Soc; Int Neural Networks Soc. *Res:* Neural network models of natural language processing and memory; self-organization and genetic evolution of neural networks. *Mailing Add:* Dept Computer Sci Univ Tex Austin TX 78712-1188

MIJOVIC, JOVAN, b Belgrade, Yugoslavia, Sept 4, 48; m; c 1. ENGINEERING. *Educ:* Univ Belgrade, BS, 72; Univ Wis-Madison, MS, 74, PhD(chem eng), 78. *Prof Exp:* Asst prof, 78-83, ASSOC PROF CHEM ENG, POLYTECH UNIV, 83- *Concurrent Pos:* Prin investr res grants, various insts, 79- *Mem:* Am Inst Chem Engrs; Soc Plastics Engrs; Soc Advan Mat & Process Eng. *Res:* Processing-structure-property-durability correlations in advanced composite materials. *Mailing Add:* Polytech Univ 333 Jay St Brooklyn NY 11201

MIKA, LEONARD ALOYSIUS, b Bay City, Mich, Apr 17, 17; m 43; c 2. MICROBIOLOGY. *Educ:* Univ Mich, BS, 47, MS, 49; George Washington Univ, PhD, 55; Am Bd Microbiol, dipl, 63. *Prof Exp:* Asst bact, Univ Mich, 47-49; sr investr, US Army Sci & Technol Ctr, Ft Detrick, 49-62, staff microbiologist, 62-88, phys sci admin, 63-73, dir, 73-77; res assoc, Sch Med, Univ Va, 77-80; RETIRED. *Mem:* Am Soc Microbiol; Soc Exp Biol & Med; Am Asn Immunol; fel Am Acad Microbiol; NY Acad Sci. *Res:* Research administration. *Mailing Add:* 1207 Oak Island Ct Ft Collins CO 80525-5516

MIKA, THOMAS STEPHEN, b Chicago, Ill, July 16, 41. MINERAL & CHEMICAL ENGINEERING. *Educ:* Stanford Univ, BS, 62; Columbia Univ, MS, 63; Univ Calif, Berkeley, DEng, 71. *Prof Exp:* Asst prof mat sci & eng, Univ Calif, Berkeley, 70-75; scientist & proj leader, 75-79, mgr, Developer Processing Area, 79-91, MGR MARKING/IMAGING ARCHIT, STRATEGY & ARCHIT DIV, XEROX CORP, 91- *Honors & Awards:* Arthur Claudet Prize, Brit Inst Mining & Metall, 76. *Mem:* Am Inst Mining, Metall & Petrol Engrs; Am Inst Chem Engrs; Brit Inst Mining & Metall; Sigma Xi. *Res:* Comminution; mathematical modeling of particulate processing unit operations; flotation; applied colloid chemistry; particulate technology; emulsification; dispersion; polymer processing. *Mailing Add:* Xerox Corp 800 Phillips Rd Bldg 114-20D Webster NY 14580

MIKAMI, HARRY M, b Seward, Alaska, Dec 28, 15; m 55. MINERALOGY, CERAMICS. *Educ:* Univ Alaska, BS, 37; Yale Univ, MS, 42, PhD(petrol), 45. *Prof Exp:* Engr, US Smelting, Ref & Mining Co, Alaska, 37 & Am Creek Operating Co, Alaska, 38-39; geologist, Conn Geol Surv, 43-45; instr mineral, Yale Univ, 45; res scientist, E J Lavino & Co, 45-60, res mgr, 60-65, dir res, 65-67, dir res & develop, Lavino Div, Int Minerals & Chem, Corp, 67-74; res mgr, Basic Refractories, Kaiser Aluminum & Chem Corp, 74-80, prog mgr, 80-82; CONSULT, 82- *Concurrent Pos:* Instr, Pa State Univ, 48-49; consult, Villanova Univ, 57-58. *Honors & Awards:* Theodore J Planje St Louis Refractories Award, 87. *Mem:* Fel Am Ceramic Soc; fel Mineral Soc Am; fel Geol Soc Am; Am Chem Soc; Iron & Steel Soc; Am Inst Mech Engrs. *Res:* Ceramic mineralogy and microstructure; phase equilibria of periclase-chromite-orthosilicate systems; high temperature materials; basic oxygen and electric arc furnace refractories; industrial minerals. *Mailing Add:* 4557 Eull Ct Pleasanton CA 94566

MIKAT, EILEEN M, b Cleveland, Ohio. PATHOLOGY. *Educ:* Case Western Reserve Univ, AB, 52; Duke Univ, MA, 69, PhD(path), 79. *Prof Exp:* Teaching supvr med tech, Metrop Gen Hosp, Cleveland, Ohio, 52-61; res asst, 61-69, assoc res, 69-79, ASST MASTER PROF PATH, DUKE UNIV, 79- *Mem:* Am Asn Pathologists; Am Asn Clin Pathologists; Soc Cardiovasc Pathologists. *Res:* Cardiovascular research in ischemia and atherosclerosis. *Mailing Add:* Dept Path Duke Univ Med Ctr Box 3712 Durham NC 27710

MIKE, VALERIE, b Budapest, Hungary, Aug 20, 34; US citizen. BIOETHICS, TECHNOLOGY ASSESSMENT. *Educ:* Manhattanville Col, BA, 56; NY Univ, MS, 59, PhD(math), 67. *Prof Exp:* From tech asst to mem tech staff, Bell Tel Labs, Inc, 56-67; from res assoc to assoc mem, Sloan-Kettering Inst Cancer Res, 67-78, mem, 78-84, head, Biostatist Lab, 78-83; from res fel to assoc prof biostatist, Cornell Univ Grad Sch Med Sci, 67-78, prof, 79-84, chmn, Biostatist Unit, 79-83; NEH fel, Hastings Ctr, 84-85; CLIN PROF BIOSTATIST PUB HEALTH, CORNELL UNIV MED COL, 79- *Concurrent Pos:* From clin asst prof to clin assoc prof biostatist pub health, Cornell Univ Med Col, 70-78; consult, NIH, 76- & Mayo Clin, 83-87; mem, oncol drugs adv comt, Food & Drug Admin, 78-81; mem, Vis Panel Res, Educ Testing Serv, 83-87. *Mem:* AAAS; Soc Clin Trials; Am Statist Asn; Nat Asn Sci Technol & Soc; Soc Health & Human Values. *Res:* Ethical and value issues in science and technology, with emphasis on the role of statistical evidence in technology diffusion, medical decision-making, and the development of public policy. *Mailing Add:* Dept Pub Health Cornell Univ Med Col 1300 York Ave New York NY 10021

MIKEL, THOMAS KELLY, JR, b E Chicago, Ind, Aug 27,46. MARINE BIOLOGY, MARINE CHEMISTRY. *Educ:* Calif State Univ, San Jose, BA, 73; Univ Calif, Santa Barbara, MA, 75. *Prof Exp:* consult marine biol, US Dept Interior, 73-74; asst dir, Santa Barbara Underseas Found, 75-76; marine biologist, Jacobs Environ, 76-81; lab dir, CRL Environ, 81-88; LAB DIR, ABC LABS, 88- *Concurrent Pos:* Biol coordr, Anacapa Island Underwater Nature Trail, US Nat Park Serv, 75-76; proj mgr, Upper Newport Bay Restoration Proj, 77-78; instr oceanog, Ventura Col, 80-81; res contribr, Third Int Artificial Reef Conf, 83, Tenth Ann Symp on Aquatic Toxicol & Hazard Assessment, Am Soc Testing & Mat, 86. *Mem:* Asn Environ Prof; Soc Pop Ecologists. *Res:* Impact of human activities upon the chemical, physical and biological marine environment. *Mailing Add:* 29 N Olive St Ventura CA 93001

MIKELL, WILLIAM GAILLARD, environmental control, occupational health laboratories, for more information see previous edition

MIKES, JOHN ANDREW, b Budapest, Hungary, Jan 20, 22; m 48; c 1. CHEMISTRY. *Educ:* Pazmany Peter Univ, Budapest, dipl, 45, PhD(org chem), 48; Eotvos Lorand Univ, Budapest, DSc, 68. *Prof Exp:* Asst prof polymer chem, Budapest Tech Univ, 45-48; tech mgr, Hutter & Lever, Co, Budapest, 48-49; dir res & consult serv plastics inst, Nat Polymer Res Ctr, Budapest, 50-69; dir res & develop, Water Treatment Develop Co, Tatabanya, 69-70; consult, Treadwell Corp, NY, 70 & Mocatta Metals Corp, 71; mgr res, Lundy Electronics, Inc, Glen Head, NY, 71-73; mgr develop, Buckman Labs, Inc, Memphis, Tenn, 73-74; mgr res & develop, Ionac Chem Co, Birmingham, NJ, 74-76; sr res fel, Ciba-Geigy Co, Ardsley, 76-87; CONSULT, 87- *Concurrent Pos:* Ed, Ion Exchange & Membranes, 71-76. *Honors & Awards:* Govt Medalist, Technol, Hungary, 60, 65. *Mem:* Am Chem Soc; fel Am Inst Chem. *Res:* Polymer synthesis; water treatment process development; slow release drug form development; ion exchangers; membranes; porosity; cross linking; author of over 100 publications; 38 patents. *Mailing Add:* 12 Rolling Hill Ct Madison NJ 07940

MIKES, PETER, b Prague, Czech, Oct 28, 38; m 69; c 3. POLYMER PHYSICS, INK JET PRINTING. *Educ:* Czech Tech Univ, Ing Phys, 61; Charles Univ, Prague, CSc, 65. *Prof Exp:* Scientist polymers, Inst Macromolecular Chem, Czech, 65-69; fel spectroscopy, Dept Polymer Sci, Case Western Reserve Univ, 69-71; sr scientist syst sci, Res Ctr, Rockland State Hosp, NY, 71-73; scientist rheology, Xerox Corp, 73-86; SCIENTIST, LAWRENCE LIVERMORE NAT LAB, UNIV CALIF, 87- *Concurrent Pos:* Imp Chem Indust fel textiles, Univ Leeds, 66-67. *Mem:* Am Phys Soc; Inst Elec & Electronics Engrs; Comput Soc. *Res:* Mechanical properties of polymers; rheology of elastomers; printing physics; microelectronics fabrication; computational physics. *Mailing Add:* 16641 Cowell St San Leandro CA 94578

MIKESELL, JAN ERWIN, b Macomb, Ill, Feb 19, 43; m 65. PLANT ANATOMY, MORPHOLOGY. *Educ:* Western Ill Univ, BSc, 65, MSc, 66; Ohio State Univ, PhD(bot), 73. *Prof Exp:* Researcher virol & immunol, Viral & Immunol Lab, Sixth US Army Med Labs, 67-69; asst prof, 73-80, ASSOC PROF BOT, GETTYSBURG COL, 80- *Mem:* Bot Soc Am; AAAS; Sigma Xi; Can Soc Plant Physiologists. *Res:* Investigations of anomalous secondary thickening in vascular plants; especially patterns of development and directions of differentiation of anomalous types of cambia in dicotyledonous plants; investigation of flower and fruit development; investigations of anamolous secondary thickening in dicots; seed and fruit development; resource allocation. *Mailing Add:* Dept of Biol Gettysburg Col Gettysburg PA 17325

MIKESELL, JON L, nuclear physics, for more information see previous edition

MIKESELL, SHARELL LEE, b Coshocton, Ohio, Nov 24, 43; m 65. POLYMER CHEMISTRY. *Educ:* Olivet Nazarene Col, AB, 65; Ohio State Univ, MS, 68; Univ Akron, PhD(polymer chem), 71. *Prof Exp:* Prod develop engr polymer chem, Gen Elec Co, 71-72, proj mgr, 72-74, mgr indust mkt develop tech mkt, 74-75, mgr indust prod develop polymer chem, Laminated & Insulating Mat Bus Dept, 75-76; mgr Textile Systs Lab, Owens-Corning Fiberglas Tech Ctr, 76-79, res dir, 79-84, mgr Textile Mat Mkt, 84-85, vpres, 85-90, VPRES TECHNOL, OWENS-CORNING FIBERGLAS, 90- *Mem:* Indust Res Inst. *Res:* Development of high performance thermosetting epoxy and phenolic resins used in high pressure copper clad and multilayer laminates for printed wiring applications; development of fiber glass textile products. *Mailing Add:* Fiberglas Tower Owens-Corning Fiberglas Toledo OH 43659

MIKESKA, EMORY EUGENE, b Abbott, Tex, Aug 24, 27; div; c 2. PHYSICS. *Educ:* Univ Tex, BS, 47, MA, 50. *Prof Exp:* Jr seismic observer, Magnolia Petrol Co, 47-49; res physicist, Defense Res Lab, Univ Tex, Austin, 50-61; sr physicist & proj dir, Tracor, Inc, 61-67; res scientist assoc, Appl Res Lab, Univ Tex, Austin, 67-82; INSTR, AUSTIN COMMUNITY COL, 83- *Concurrent Pos:* Consult, Boner & Lane, 54-59; consult & partner, Lane & Mikesa, 60-61. *Mem:* Fel Acoust Soc Am; Audio Eng Soc. *Res:* Noise control; architectural acoustics; underwater sound. *Mailing Add:* 7613 Rustling Rd Austin TX 78731

MIKHAIL, ADEL AYAD, b Cairo, Egypt, Nov 8, 34; US citizen; m 58; c 2. CARDIOVASCULAR DEVICES RESEARCH & DEVELOPMENT, CLINICAL RESEARCH & REGULATORY AFFAIRS. *Educ:* Univ Alexandria, BPharm, 55, MPharmaceut Chem, 60; Univ Minn, Minneapolis, PhD(med chem), 66. *Prof Exp:* NIH fel & assoc scientist, Cancer Res Lab, Vet Admin Hosp, Minneapolis, Minn, 66; asst prof, Col Pharm, Univ Alexandria, 66-70; USDA Forest Serv fel & res assoc pharm, Ohio State Univ, 70-72; res mgr, Tech Med, 73-78, sr vpres, 78-89; CLIN RES & REGULATORY AFFAIRS; PRES & SR CONSULT, MEDICA NOVA, INC, 89- *Mem:* Am Chem Soc; Am Soc Artificial Internal Organs; Regulatory Affairs Prof Soc. *Res:* Drug design; synthesis; structure elucidation of natural products; biocompatability of polymers; cardiovascular prosthesis and hemodialysis devices; quality control, res and development of medical devices; clinical research and regulatory process of cardiac valve prosthesis. *Mailing Add:* 2332 W 111th St Bloomington MN 55431

MIKHAIL, NABIH N, b Sinai, Egypt, Apr 21, 33; US citizen; m 73; c 1. SAMPLING THEORY, TRANSPORTATION PROBLEMS. *Educ:* Univ Alexandria, Egypt, BSc, 57; Univ Col London, Eng, PhD(statist), 65. *Prof Exp:* Assoc prof math & statist, Univ Assuit, Cairo & High Statist Inst, Egypt, 57-71; assoc prof math & statist, Univ Windsor, Guelph, Western Ont, Simon Fraser & Man, Can, 71-79; PROF MATH & STATIST, LIBERTY UNIV, 79- *Res:* Multivariate analysis; estimation; distribution; sampling theory; operations research. *Mailing Add:* Dept Math Liberty Univ Lynchburg VA 24506

MIKIC, BORA, b Loznica, Yugoslavia, Mar 1, 32; m 57; c 2. HEAT TRANSFER, FLUID MECHANICS. *Educ:* Univ Belgrade, Dipl eng, 57; Mass Inst Technol, ScD(mech eng), 66. *Prof Exp:* Res engr, Aero-Tech Inst, Univ Belgrade, 57-60; asst fluid mech, Univ Belgrade, 60-61; lectr fluid mech & heat transfer, Khartoum Tech Inst, 61-64; res asst heat transfer, 64-66, from asst prof to assoc prof mech eng, 66-73, PROF MECH ENG, MASS INST TECHNOL, 73- *Concurrent Pos:* Nat Res Coun sr fel, Oxford Univ, 75-76. *Honors & Awards:* Heat Transfer Mem Award. *Mem:* Fel Am Soc Mech Engrs; Sigma Xi. *Res:* Contact resistance, dropwise condensation, nucleate boiling; mass transfer; boiling; condensation; fusion nuclear reactor; utilization of geothermal energy; convective heat transfer, cooling of electronic devices, material processing; fluid flore. *Mailing Add:* Dept of Mech Eng Mass Inst Technol 177 Mass Ave Cambridge MA 02139

MIKITEN, TERRY MICHAEL, b New York, NY, June 1, 37; m 60; c 3. NEUROPHYSIOLOGY, COGNITIVE NEUROSCIENCE. *Educ:* NY Univ, BA, 60; Albert Einstein Col Med, PhD(pharmacol), 67. *Prof Exp:* Res asst neurosurg, Mt Sinai Hosp, NY, 59-60; asst pharmacologist, Schering Corp, NJ, 60-61; USPHS fel, Med Sch, Columbia Univ, 67-69; from asst prof to assoc prof, 69-88, PROF PHYSIOL, UNIV TEX MED SCH, SAN ANTONIO, 88- *Concurrent Pos:* Consult, Bexar County Hosp, 69-; regional chmn, Osteogenesis Imperfecta Found, 71-76; chmn, San Antonio Neurosci Group, 76-77; spec asst to pres, Acad Servs, 86-88, assoc dean, Sch Biomed Scis, Health Sci Ctr, San Antonio, 84-; Minnie Piper Stevens Professorship. *Mem:* AAAS; Soc Neurosci; NY Acad Sci. *Res:* Neurophysiology and pharmacology of synaptic transmission; desensitization of cholinergic receptors; electrophysiology of excitable and non-excitable membranes; artificial intelligence, computer-assisted instruction, cognitive science; cognitive theory of learning; intelligent systems for computer-assisted instruction in the neurosciences. *Mailing Add:* Grad Dean's Off Univ Tex Med Sch 7703 Floyd Curl Dr San Antonio TX 78284

MIKKELSEN, DAVID ROBERT, b Ames, Iowa, May 10, 49; m 72; c 2. COMPUTATIONAL PHYSICS. *Educ:* Calif Inst Technol, BS, 71; Univ Wash, PhD(physics), 75. *Prof Exp:* Res fel, Calif Inst Technol, 75-77; res assoc physics, 77-80, res staff physicist, 80-85, RES PHYSICIST, PLASMA PHYSICS LAB, PRINCETON, 85- *Mem:* Am Phys Soc. *Res:* Numerical simulation of magnetically confined plasma experiments. *Mailing Add:* C-Site Forrestal Campus PO Box 451 Princeton NJ 08543

MIKKELSEN, DUANE SOREN, b Payson, Utah, Nov 1, 21; m 43; c 4. SOIL FERTILITY, AGRONOMY. *Educ:* Brigham Young Univ, BS, 46; Rutgers Univ, PhD, 49. *Prof Exp:* From asst prof to prof agron, 49-88, dir int progs, 82-86, EMER PROF AGRON, UNIV CALIF, DAVIS, 88- *Concurrent Pos:* Consult, Rockefeller Found, Colombia, 62, UN, FAO, LATIN AM, 86-87; IRI Res Inst, Brazil, 62, 68-71, Chile-Calif Proj, 63, Int Rice Comn, Manila, Philippines, 64 & 66, Int Atomic Energy Agency, Hong Kong, 66, & Peace Corp Training Progs, 67; vis scientist & Rockefeller Found grant, Int Rice Res Inst, Manila, Philippines, 67-68; chmn, US Rice Tech Working Group, 67-68; mem rice fertilizer adv comt, Tenn Valley Auth, 68-71; consult, Amazon Basin Develop, 70-74 & Malaysian Agr Res Develop, 74; consult agr develop, Govt of Venezuela, 69, Australia, 76, Thailand, 76, Oman, 77, Brazil, Peru & Paraguay, 79-82; vis scientist micronutrient proj, Int Atomic Energy Agency, 74; vis scientist, Int Rice Res Inst, 77-78; dir, Egypt Rice Res & Training Proj, 82-87; consult ed Plant & Soil, J Plant Nutrit, 79-; IRI Int Develop Bank, Guyanna, 87, Indo-US Sci & Technol Initiative, 84-88; consult, UN-FAD, 87-88, Int Inst Tropic Agr, 88; vis prof, McCaughey Mem Inst, 91. *Honors & Awards:* Distinguished Serv Award, USDA, 88, Am Soc Agron, Calif, 88. *Mem:* Fel Soil Sci Soc Am; fel Crop Sci Soc Am; fel Am Soc Agron; Am Soc Plant Physiol. *Res:* Plant-soil interrelations; mineral nutrition of plants; mineral nutrition of rice; chemistry of flooded soils. *Mailing Add:* 617 Oeste Dr Davis CA 95616

MIKKELSEN, WILLIAM MITCHELL, b Minneapolis, Minn, May 25, 23; m 48; c 6. INTERNAL MEDICINE, RHEUMATOLOGY. *Educ:* Univ Mich, MD, 49. *Prof Exp:* From asst prof to assoc prof, 57-69, asst dir health prog, 59-62, actg dir, 62-66, dir, Periodic Health Appraisal Unit, 66-77, PROF INTERNAL MED, MED SCH, UNIV MICH, ANN ARBOR, 69- *Concurrent Pos:* Assoc physician, Rackham Arthritis Res Unit, Univ Mich, Ann Arbor, 55-; attend physician, Vet Admin Hosp, 55-89. *Mem:* Soc Advan Med Syst; Am Soc Clin Pharmacol & Therapeut; Am Col Physicians; Am Fedn Clin Res; Am Geriatrics Soc; Sigma Xi. *Res:* Rheumatic diseases; evaluation and application of periodic health appraisal techniques; geriatric medicine. *Mailing Add:* Univ Mich Turner Geriat Clin 1010 Wall St Ann Arbor MI 48109

MIKKELSON, RAYMOND CHARLES, b Blue Earth, Minn, Mar 22, 37; m 60; c 1. OPTICAL FIBER DESIGN, OPTICAL FIBER SENSORS. *Educ:* St Olaf Col, BA, 59; Univ Ill, MS, 61, PhD(physics), 65. *Prof Exp:* From asst prof to assoc prof, 65-75, head, Dept Physics & Astron, 76-79, dir, Macalester NSF Cause Proj Lab & Graphics, 80-82, PROF PHYSICS, MACALESTER COL, 75-; CONSULT, OPTICAL FIBERS, 86- *Concurrent Pos:* Assoc Cols Midwest physics fac mem, Argonne Nat Lab, 71, vis scientist, Physics Div, 71-72; vis scholar, Univ Minn, 79; vis scientist, 3M Co, 85-86. *Mem:* Am Asn Physics Teachers; Am Phys Soc; Optical Soc Am. *Res:* Interactions of radiation with solids; channeling; environmental radiation levels; teaching of university-college physics; real-time computer instrumentation; fiber optics. *Mailing Add:* Dept Physics Macalester Col St Paul MN 55105

MIKKOLA, DONALD E(MIL), b Champion, Mich, July 30, 38; m 60; c 2. METALLURGICAL ENGINEERING, MATERIALS SCIENCE. *Educ:* Mich Technol Univ, BS, 59; Northwestern Univ, MS, 61, PhD(mat sci), 64. *Prof Exp:* From asst prof to assoc prof, 64-72, PROF METALL & MAT ENG, MICH TECHNOL UNIV, 72- *Concurrent Pos:* Fulbright-Hays res scholar, Helsinki Univ Technol, Finland, 73-74; bd dir, Accreditation Bd Eng & Technol, 86-, mem Eng Accreditation Comn, 81-86; bd dirs, Metall Soc, 89- *Mem:* Fel Am Soc Metals; Metall Soc; Am Crystallog Asn; Sigma Xi. *Res:* Applications of x-ray and electron techniques; structure-property relationships; shock hardening; plastic deformation; order-disorder phenomena; intermetallics; high temperature materials; erosion and abrasion resistance; author of over 60 publications. *Mailing Add:* Dept Metall & Mat Eng Mich Technol Univ Houghton MI 49931

MIKLAVCIC, MILAN, b Ljubljana, Yugoslavia, Sept 1, 54. PARTIAL DIFFERENTIAL EQUATIONS, NUMERICAL ANALYSIS. *Educ:* Univ Ljubljana, BS, 77; Va Polytech Inst & State Univ, PhD(math), 81. *Prof Exp:* Res assoc math, Univ Minn, Minneapolis, 82-83; res assoc math, ISKRA-Ctr Electrooptic, 83-84; res assoc math, Univ Wis, Madison, 84-85; asst prof, 85-90, ASSOC PROF MATH, MICH STATE UNIV, 90- *Concurrent Pos:* Vis asst prof math, Rensselaer Polytechnic Inst, 82. *Mem:* Am Math Soc; Soc Nat Philos. *Res:* Physical mathematics; applied functional analysis; nonlinear partial differential equations; numerical analysis. *Mailing Add:* Dept Math Mich State Univ East Lansing MI 48824

MIKLE, JANOS J, b Kolozsvar, Hungary, Oct 9, 42; Swed citizen. FOOD SCIENCE & TECHNOLOGY. *Educ:* Polytech Inst Yassy, MSC, 71; Can Sch Mgt, MBA, 90. *Prof Exp:* Process engr, Alfa-Laval Ab Tumba, Sweden, 74-78 & Sullivan Eng-US, 78-80, sr process engr, 84-85; group leader & sr res engr, W L Clayton Res Ctr, Tex, 80-84; DIR DEVELOP & TECH SERV, POS PILOT PLANT CORP, 85- *Concurrent Pos:* Consult process eng & mkt support, 85- *Mem:* Am Oil Chemist Soc. *Res:* Process design and optimization in the area of fats and oils technology; degumming, caustic refining, adsorptive bleaching, dewaxing, fractionation, catalytic hydrogenation, interesterification, vacuum deodorization, physical and steam refining; separation technology. *Mailing Add:* 321 Capilano Pl Corp Saskatoon SK S7K 4J7 Can

MIKLOFSKY, HAAREN A(LBERT), b Rochester, NY, Nov 25, 20; m 49; c 2. CIVIL ENGINEERING. *Educ:* George Washington Univ, BCE, 46; Yale Univ, ME, 47, DrEng, 50. *Prof Exp:* Asst instr civil eng, George Washington Univ, 47-48, from asst prof to assoc prof & exec officer dept, 49-55; assoc prof, Rensselaer Polytech Inst, 55-62; prof, Univ SC, 62-66; PROF CIVIL ENG, UNIV ARIZ, 66-, PROF ENG MECH, 81- *Concurrent Pos:* Asst in instr, Yale Univ, 48-49; gen engr, Nat Bur Standards, 50-53. *Mem:* Am Soc Civil Engrs; Am Soc Eng Educ. *Res:* Suspension bridges and arches; finite element analysis. *Mailing Add:* 7442 E 18th Tucson AZ 85710

MIKLOWITZ, JULIUS, b Schenectady, NY, May 22, 19; m 48; c 2. MECHANICS, MECHANICAL ENGINEERING. *Educ:* Univ Mich, BS, 43, MS, 48, PhD(appl mech), 49. *Prof Exp:* Res engr plasticity, Res Labs, Westinghouse Elec Corp, 43-46 & 49; asst prof eng & head dept, NMex Inst Mining & Technol, 49-51; res engr solid mech, US Naval Ord Test Sta, 51-56; from assoc prof to prof, 56-85, EMER PROF APPL MECH, CALIF INST TECHNOL, 85- *Concurrent Pos:* Consult, Space Technol Labs, Ramo-Wooldridge Corp, 56-60, Nat Eng Sci Co, 60-64 & Aerospace Corp, 68-; NSF sr fel, 64-65; W W Clyde prof eng, Univ Utah, 70; mem, US Nat Comn Theoret & Appl Mech, Nat Acad Sci, 76-80. *Mem:* AAAS; fel Am Soc Mech Engrs; Int Soc Interaction, Mech & Math. *Res:* Propagation of waves in elastic and inelastic media; yield phenomena in steel and polymers; fracture of steel. *Mailing Add:* 5255 Vista Miguel Dr La Canada CA 91011

MIKNIS, FRANCIS PAUL, b DuBois, Pa, Jan 31, 40; m 60; c 4. NUCLEAR MAGNETIC RESONANCE. *Educ:* Univ Wyo, BS, 61, PhD(chem), 67. *Prof Exp:* Sr scientist, Aeronutronic Div, Philco-Ford Corp, 66-67; Proj leader, 75-83, RES CHEMIST, LARAMIE ENERGY RES CTR, ENERGY RES & DEVELOP ADMIN, 67-; SR STAFF RES CHEMIST, WESTERN RES INST, 83- *Concurrent Pos:* Chmn, Geochem Div, Am Chem Soc, 83. *Mem:* Am Chem Soc; Sigma Xi. *Res:* Nuclear magnetic resonance of solids, particularly oil shales and coals; elemental analysis of fossil fuels; pyrolysis studies of oil shales and coals. *Mailing Add:* 1819 W Hill Rd Laramie WY 82070

MIKOLAJ, PAUL G(EORGE), b Cleveland, Ohio, Jan 6, 36. CHEMICAL ENGINEERING, ENVIRONMENTAL SCIENCES. *Educ:* Cleveland State Univ, BS, 58; Univ Rochester, MS, 60; Calif Inst Technol, PhD(chem eng), 65. *Prof Exp:* Fel, Calif Inst Technol, 65; res engr, Chevron Res Co, 66-67; asst prof chem eng, Univ Calif, Santa Barbara, 67-73; sr engr, Oceanog Serv, Inc, 73-74 & Dames & Moore, 74-76; DIR REFINING REGULATION, TOSCO CORP, 76- *Concurrent Pos:* Adv Coun, S Coast Air Qual Mgt Dist, 77- *Mem:* AAAS; Am Inst Chem Engrs; Am Chem Soc; Air Pollution Control Asn. *Res:* Environmental policy and regulation; air and water pollution control; toxic and hazardous substances. *Mailing Add:* Tosco Corp Avon Refinery Martinez CA 94553-1486

MIKOLAJCIK, EMIL MICHAEL, b Colchester, Conn, Jan 14, 26; m 53; c 2. DAIRY MICROBIOLOGY. *Educ:* Univ Conn, BS, 50; Ohio State Univ, MS, 51, PhD(dairy microbiol), 59. *Prof Exp:* Prof dairy mfg, Univ PR, 51-61; from asst prof to prof food sci, Ohio State Univ, 61-88, actg chmn dept, 84-85, EMER PROF FOOD SCI, OHIO STATE UNIV, 89- *Concurrent Pos:* NIH grants, 61- *Honors & Awards:* Pfizer Award in Cheese Res, 74. *Mem:* Am Dairy Sci Asn; Int Asn Milk, Food & Environ Sanit; Inst Food Technol. *Res:* Mechanisms of bacteriophage action on lactic organisms; bacterial metabolism of organisms associated with the dairy industry, particularly on lactic streptococci and spore formers; immunoglobulins of bovine milk and colostrum; flat-sour sporeformers in canned thermally processed foods; processing and storage of human milk; psychrotrophic food spoilage organisms; hazard analysis critical control point; sanitation. *Mailing Add:* 1313 Nantucket Ave Columbus OH 43235

MIKOLAJCZAK, ALOJZY ANTONI, b Czestochowa, Poland, Jan 14, 35; m 61; c 3. THERMODYNAMICS, AERODYNAMICS. *Educ:* Cambridge Univ, BA, 57, MA, 61, PhD(magneto-hydrodynamics), 65. *Prof Exp:* Tech asst rocket performance, Rolls-Royce Ltd, 57-60, sect leader jet engines, 60-61; asst prof jet engine & rocket propulsion, Mass Inst Technol, 65-66; consult aerodynamicist, Pratt & Whitney Aircraft, 66-67, head compressor res, 67-72, asst chief engr aerocomponents, 72-77, mgr aerodyn, Thermodyn & Control Systs, 77-79; CORP DIR TECH PLANNING, UNITED TECHNOLOGIES CORP, 79- *Honors & Awards:* Gas Turbine Power Award, Am Soc Mech Engrs, 71 & 72. *Mem:* Am Inst Aeronaut & Astronaut; Am Soc Mech Engrs. *Res:* Turbomachinery, including axial and centrifugal compressors and turbines, combustion, inlets and nozzles, ducts and diffusers, noise and emissions, structures and aeroelasticity, controls and fuel systems for gas turbines. *Mailing Add:* 14 Faber Pl Garfield NJ 07026

MIKOLASEK, DOUGLAS GENE, b Menominee, Mich, Aug 23, 30; m 59; c 3. MEDICINAL CHEMISTRY. *Educ:* Univ Mich, BS, 52, PhD(med chem), 62. *Prof Exp:* Control chemist, Marinette Paper Co, Scott Paper Co, 54; develop chemist, Abbott Labs, 55-56, res chemist, 56-58; sr scientist, 62-68, group leader chem develop, 68-71, prin investr, 71-75, prin res assoc, 75-85, ASSOC DIR, CHEM PROCESS RES, MEAD JOHNSON RES CTR, 85- *Concurrent Pos:* Parke-Davis fel, Univ Kans, 58-60. *Mem:* AAAS; Am Chem Soc; NY Acad Sci. *Res:* Quinoline chemistry and antimalarial research. *Mailing Add:* Mead Johnson Res Ctr 2404 Pennsylvania Ave Evansville IN 47721

MIKSAD, RICHARD WALTER, b Trenton, NJ, Aug 24, 40; m 70; c 3. OCEANOGRAPHY, METEOROLOGY. *Educ:* Bradley Univ, BSME, 63; Cornell Univ, MSc, 64; Mass Inst Technol, ScD(oceanog), 70. *Prof Exp:* Res staff pollution, Ctr Study Responsive Law, 70; res scientist fluid dynamics, Imperial Col, 70-72; asst res prof atmospheric sci, Univ Miami, 72-74; from asst prof to assoc prof eng, 74-84, prof civil eng, 84- 85, assoc dean res, 87-88, CHMN, AEROSPACE ENG & ENG MECH DEPT, UNIV TEX, AUSTIN, 88- *Concurrent Pos:* Nat Res Coun fel, Imperial Col, 70-72; prog dir, NSF, 86-87. *Mem:* Am Phys Soc; Am Geophys Union; Am Meteorol Soc; Am Soc Mech Engrs; Am Soc Civil Engrs. *Res:* Fluid dynamics; non-linear hydrodynamic stability; wave-structure interaction. *Mailing Add:* Dept ASE/EM Univ Tex Austin TX 78712

MIKSCHE, JEROME PHILLIP, b Breckenridge, Minn, June 11, 30; m; c 3. PLANT MORPHOLOGY. *Educ:* Moorhead State Col, BS, 54; Miami Univ, MS, 56; Iowa State Univ, PhD(plant develop, genetics), 59. *Prof Exp:* Instr gen bot, Iowa State Univ, 58-59; fel radiobot, Brookhaven Nat Lab, 59-61, staff position, Biol Dept, 61-65; prof & head, Dept Botany, NC State Univ, 77-84; Nat Prog Staff, USDA Agr Res Serv, 84-90, DIR, PLANT GENOME RES PROG, USDA, 91-; RES BOTANIST CYTOL, INST FOREST GENETICS, US FOREST SERV, 65- *Concurrent Pos:* Asst prof, C W Post Col, Long Island Univ, 60-61; lectr, Adelphi Suffolk Col, 62-65; radiobiol adv, AEC exhibit, Brazil & Lebanon, 61. *Mem:* Bot Soc Am; fel Royal Micros Soc; Am Soc Plant Physiol; AAAS. *Res:* Developmental plant anatomy; plant cytology, genetics and physiology; radiation botany; nucleic acids in connection with genetic relatedness between organisms. *Mailing Add:* 1656 Eton Way Crofton MD 21114-1530

MIKULA, BERNARD C, b Johnstown, Pa, Aug 29, 24; m 51; c 2. GENETICS, TAXONOMY. *Educ:* Col William & Mary, BS, 51; Univ Wash, St Louis, PhD(bot), 56. *Prof Exp:* Asst, Mo Bot Garden, 51-56; proj assoc genetics, Univ Wis, 56-60; from asst prof to assoc prof, 60-67, PROF GENETICS, DEFIANCE COL, 67- *Concurrent Pos:* Vis fel ctr biol of natural syst, Wash Univ, 66-67. *Mem:* AAAS; Genetics Soc Am; Am Inst Biol Sci. *Res:* Mechanisms of allelic variation. *Mailing Add:* Dept of Genetics Defiance Col Defiance OH 43512

MIKULCIK, E(DWIN) C(HARLES), b Glenside, Sask, Dec 10, 36; m 66; c 3. MECHANICAL ENGINEERING. *Educ:* Univ Sask, BE, 60, MSc, 61; Cornell Univ, PhD(mech eng), 68. *Prof Exp:* Instrumentation engr, Du Pont Can, Ltd, 62-63; systs engr, RCA Victor Co, Ltd, 63-65; asst prof mech eng, 68-71, ASSOC PROF MECH ENG, UNIV CALGARY, 71- *Res:* Dynamics, stability and control of ground vehicles; adaptive and optimum control systems; identification and optimal design of dynamic mechanical systems. *Mailing Add:* Dept Mech Eng Univ Calgary 2500 Univ Dr NW Calgary AB T2N 1N4 Can

MIKULCIK, JOHN D, b Ilasco, Mo, July 30, 36; m 61; c 3. AGRONOMY. *Educ:* Univ Mo, BS, 58, MS, 59, PhD(soils), 64. *Prof Exp:* From asst prof to assoc prof, 63-73, PROF AGRON, MURRAY STATE UNIV, 73- *Mem:* AAAS; Am Soc Agron; Soil Sci Soc Am. *Res:* Soil testing; levels of nitrate and phosphate in runoff from rural watersheds; levels of nitrate in soils under barn lot conditions. *Mailing Add:* 1613 Keenland Dr Murray KY 42071

MIKULEC, RICHARD ANDREW, b New Brighton, Pa, Feb 26, 28; m 63; c 1. ANALYTICAL CHEMISTRY. *Educ:* Wayne State Univ, BS, 51; Wash State Univ, MS, 53, PhD(chem), 56. *Prof Exp:* Res fel chem, Univ Minn, 56-57; res investr, G D Searle & Co, 57-77, res anal chemist, 77-83; SR RES INVESTR, NUTRASWEET CO, 83- *Mem:* Am Chem Soc. *Res:* Separation and analytical techniques applied to pharmaceutical and food products. *Mailing Add:* Nutrasweet Co 601 E Kensington Mt Prospect IL 60056

MIKULECKY, DONALD C, b Chicago, Ill, Mar 23, 36; m 57, 73, 81; c 2. NETWORK THERMODYNAMICS, BIOSYSTEMS. *Educ:* Ill Inst Technol, BS, 57; Univ Chicago, PhD(physiol), 63. *Prof Exp:* Fel biophys, Weizmann Inst Sci, Rehovot, Israel, 63-65; asst prof biophys & assoc prof theoret biol, State Univ NY, Buffalo, 65-68; vis scholar physics & math, Philander Smith Col, Little Rock, 68-69; vis lectr biophys, Sch Med, Harvard Univ, 69-71; assoc prof biophys & neurobiol, Meharry Med Col, Nashville, 71-73; prof physiol & biophys, 73-88, PROF PHYSIOL, BIOPHYS & BIOMED ENG, MED COL VA, VA COMMONWEALTH UNIV, 88- *Concurrent Pos:* Vis prof, Col France, Saclay, 77, Univ Rouen, France, 82 & 86, Max Planck Biophys Inst, Frankfurt, 85-86, Weizmann Inst Sci & Univ Padua, 86. *Mem:* Biophys Soc; Biomed Eng Soc; Soc Math Biol; Am Physiol Soc; Int Study Group Biothermokinetics; Am Chem Soc. *Res:* Role of organization in dynamic living systems, using network thermodynamics; computer simulation of biochemical, pharmacological and physiological systems by use of network simulators. *Mailing Add:* Box 551 MCV Sta Richmond VA 23298-0551

MIKULIC, DONALD GEORGE, b Milwaukee, Wis, May 29, 49. PALEOZOIC REEFS, TRILOBITES. *Educ:* Univ Wis-Milwaukee, BA, 75; Ore State Univ, Corvallis, PhD(geol), 79. *Prof Exp:* Cur asst, Dept Geol, Ore State Univ, 75-79; asst geologist, 79-86, ASSOC GEOLOGIST, INST MINERALS & METALS SECT, ILL STATE GEOL SURV, 87-; PROG ASSOC & CUR, GREENE GEOL MUS, UNIV WIS-MILWAUKEE, 83- *Concurrent Pos:* Vol cur asst, Geol Dept, Milwaukee Pub Mus, 65-75; cur asst, Geol Mus, Univ Wis-Milwaukee, 72-75, mus specialist, Greene Geol Mus, 76. *Mem:* Int Palaeontol Asn; Soc Econ Paleontologists & Mineralogists; Geol Soc Am; Paleontol Res Inst. *Res:* Silurian stratigraphy and paleontology of the United States, Canada and Europe with emphasis on triobites and reefs; paleoecology and taphonomy of paleozoic faunas; paleoecology, taphonomy, extinction, function al morphology, biogeography and systematics of Silurian trilobites worldwide; paleoecology and depositional environments of reefs; Silurian stratigraphy of midwestern United States; examination of the only extensive soft-bodied Silurian biota in the world and its related depositional environment at Waukesha, Wisconsin; author and co-author of numerous articles in journals and magazines. *Mailing Add:* Indust Minerals & Metals Sect Ill State Geol Surv 615 E Peabody Dr Champaign IL 61820

MIKULSKI, CHESTER MARK, b Philadelphia, Pa, Nov 26, 46; m 70; c 3. INORGANIC CHEMISTRY. *Educ:* Drexel Univ, BS, 69, PhD(inorg chem), 72. *Prof Exp:* Fel, Dept Chem, Univ Pa, 72-76; ASSOC PROF & CHMN INORG CHEM, BEAVER COL, 76- *Mem:* Am Chem Soc; Am Inst Chemists. *Res:* Synthesis and characterization of metal complexes with organo phosphoryl, nitryl and sulfuryl ligands; decomposition of phosphoryl and thiophosphoryl esters in the presence of metal salts; paramagnetic non-metal silicon and phosphorus compounds; synthesis of polymeric metallic conductors. *Mailing Add:* Dept of Chem & Physics Beaver Col Glenside PA 19038

MIKULSKI, JAMES J(OSEPH), b Chicago, Ill, Feb 18, 34; m 59; c 3. ELECTRICAL ENGINEERING. *Educ:* Fournier Inst Technol, BS, 55; Calif Inst Technol, MS, 56; Univ Ill, Urbana, PhD(elec eng), 59. *Prof Exp:* Mem staff, Lincoln Lab, Mass Inst Technol, 59-65; sr staff engr, Mil Electronics Div, 65-68, sect mgr, Commun Res Lab, Commun Div, 68-76, mgr res, 76-85, VPRES TECH STAFF, MOTOROLA, INC, 85- *Honors & Awards:* Fel, IEEE; Dan Noble Fel, Motorola. *Mem:* Inst Elec & Electronics Engrs. *Res:* Communication theory; signal processing; radar; system and network theory; electromagnetic theory. *Mailing Add:* Dept Res Motorola Inc 1301 E Algonquin Rd Schaumburg IL 60196

MIKULSKI, PIOTR W, b Warsaw, Poland, July 20, 25; m 60; c 1. MATHEMATICAL STATISTICS. *Educ:* Sch Planning & Statist, Warsaw Tech Univ, Dipl, 50, MS, 51; Univ Calif, Berkeley, PhD(statist), 61. *Prof Exp:* Adj statist, Sch Planning & Statist, Warsaw Tech Univ, 50-57, Inst Math, Polish Acad Sci, 52-57; asst prof, Univ Ill, Urbana, 61-62; from asst prof to assoc prof, 62-70, PROF STATIST, UNIV MD, COLLEGE PARK, 70- *Mem:* Inst Math Statist. *Res:* Nonparametric methods in statistics; asymptotic optimal properties of statistical procedures. *Mailing Add:* Dept of Math Univ of Md College Park MD 20742

MILAKOFSKY, LOUIS, b Philadelphia, Pa, Feb 21, 41; m 63; c 2. ANALYTICAL PHARMACOLOGY, CHEMICAL EDUCATION. *Educ:* Temple Univ, BA, 62; Univ Wash, PhD(org chem), 67. *Prof Exp:* Chemist, Dupont Co, 62; asst chem, Univ Wash, 62-67; fel & instr, Ind Univ, 67-68; asst prof, Penn State Univ, Scranton, 68-71, asst prof, 71-86, ASSOC PROF CHEM, PENN STATE UNIV, BERKS CAMPUS, 86- *Mem:* Am Chem Soc; Piaget Soc; Nat Sci Teachers Asn; Soc Col Sci Teaching. *Res:* Chemical education and Piaget: a new paper-pencil inventory to assess cognitive functioning; stress and ammino acids; high pressure liquid chromatography. *Mailing Add:* Dept Chem Pa State Univ Berks Campus RD 5 Tulpehocken Rd Reading PA 19610

MILAM, FRANKLIN D, medicine, for more information see previous edition

MILAM, JOHN D, b Kilgore, Tex, May 22, 33. PATHOLOGY. *Educ:* La State Univ, MD, 60. *Prof Exp:* Assoc dir, 68-85, DIR PATH, ST LUKE'S EPISCOPAL HOSP, HOUSTON, TEX, 85- *Mailing Add:* 11927 Arbordale Houston TX 77024

MILAN, FREDERICK ARTHUR, b Waltham, Mass, Mar 10, 24; m 59; c 3. PHYSICAL ANTHROPOLOGY. *Educ:* Univ Alaska, BA, 52; Univ Wis, MS, 59, PhD(anthrop), 62. *Prof Exp:* Observer meteorol, Mt Wash Observ, NH, 43-44, 46-47, Artic Sect, US Weather Bur, 47-48; res physiologist, US Air Force Arctic Aeromed Lab, 53-54, 56-57, 59-61, chief, Environ Protect Br, 62-67; res physiologist, Oper Deepfreeze, Little Am V, Antarctica, 57-58; assoc scientist-lectr anthrop, Univ Wis-Madison, 67-71; chief behav sci br, Arctic Health Res Ctr, USPHS, 71-73; prof anthrop, 71-73, chem dept, 75-77, prof, 73-86, EMER PROF HUMAN ECOL & ANTHROP, INST ARCTIC BIOL, UNIV ALASKA, 86- *Concurrent Pos:* Dir, Int Study of Eskimos, US Nat Comt Int Biol Prog, US Nat Acad Sci, DC, 67-74; mem, US Man & Biosphere Nat Comt, 77-81, consult panel, Health Prog, Labrador, 79-81. *Res:* Comparative physiology and human biology of aboriginal populations in polar regions; general anthropology of polar regions; human ecology of arctic populations. *Mailing Add:* Inst Arctic Biol Univ Alaska Fairbanks AK 99775

MILANI, VICTOR JOHN, b Mt Vernon, NY, Sept 26, 45; m 73; c 2. MICROBIOLOGY. *Educ:* City Univ New York, BS, 67; NY Univ, MS, 71, PhD(microbiol), 73. *Prof Exp:* Teaching assoc biol, NY Univ, 67-71; asst prof biol, Manhattan Community Col, 71-74; PROF SCI & CHMN DEPT, BAY PATH COL, 74- *Concurrent Pos:* Res assoc, Lab Plant Morphogenesis, Manhattan Col, 73-74. *Mem:* AAAS; Am Inst Biol Sci; Am Soc Microbiol; NY Acad Sci; Sigma Xi. *Res:* The effects of concanavalin A on growth and tumor inducing ability of Agrobacterium tumefaciens. *Mailing Add:* Dept Sci Bay Path Col 588 Longmeadow St Longmeadow MA 01106

MILANOVICH, FRED PAUL, b Rochester, Pa, Nov 22, 44; m 68; c 2. LASER SPECTROSCOPY. *Educ:* US Air Force Acad, BS, 67; Univ Calif, Davis, MS, 68, PhD(appl sci), 74. *Prof Exp:* Proj officer, Air Force Weapons Lab, 68-71; PHYSICIST, LAWRENCE LIVERMORE NAT LAB, 74- *Mem:* Am Chem Soc. *Res:* Application of lasers to biological systems; Raman spectroscopic investigations of macromolecular and membrane structure; macro and micro laser raman spectroscopy with applications in atomic and molecular analysis; fiber-optic sensor development. *Mailing Add:* PO Box 808 L-524 Livermore CA 94550

MILAZZO, FRANCIS HENRY, microbiology, physiology; deceased, see previous edition for last biography

MILBERG, MORTON EDWIN, b New York, NY, July 21, 26; m 62; c 3. SOLID STATE CHEMISTRY. *Educ:* Rutgers Univ, BS, 46; Cornell Univ, PhD(phys chem), 49. *Prof Exp:* Asst chem, Cornell Univ, 46-48; fel, Univ Minn, 49-50; instr, Univ NDak, 50-52; res chemist, Ford Motor Co, 52-59, supvr phys & inorg chem sect, Chem Dept, 59-61, prin res scientist, Sci Lab, 61-88; RETIRED. *Concurrent Pos:* Chmn, Gordon Res Conf Glassy State, 71; prog chmn, Glass Div, Am Ceramic Soc, 74-75. *Mem:* Am Chem Soc; Am Crystallog Asn; Am Ceramic Soc. *Res:* Structure and properties of noncrystalline solids; diffusion in glass; high temperature ceramic materials. *Mailing Add:* 5448 E Placita Apan Tucson AZ 85718-6318

MILBERGER, ERNEST CARL, b Galatia, Kans, Apr 2, 21; m 45; c 2. CHEMISTRY. *Educ:* Univ Mo, AB, 41, MA, 43; Case Western Reserve Univ, PhD(org chem), 57. *Prof Exp:* Chemist, Tex Co, NY, 42-46; sr chemist, Standard Oil Co, 46-60, sect leader, 60-63, sr res assoc, 63-85; RETIRED. *Mem:* Am Chem Soc; Sigma Xi. *Res:* Petrochemical process research; heterogeneous catalysis. *Mailing Add:* 34765 Sherwood Dr Solon OH 44139

MILBERT, ALFRED NICHOLAS, b Great Lakes, Ill, Aug 28, 46; m 67; c 3. MEDICINAL CHEMISTRY, TOXICOLOGY. *Educ:* Seattle Univ, BS, 71; Univ Kans, MS, 74, PhD(med chem), 76. *Prof Exp:* Proj officer, Pub Health Serv, Nat Inst Occup Safety & Health, 75-81; TOXICOLOGIST, FOOD & DRUG ADMIN, 81- *Mem:* AAAS; Am Chem Soc. *Res:* Design and synthesis of biologically active compounds; mechanisms of drug action at the molecular level; product safety evaluation; environmental and occupational health as related to chemical agents; risk analysis; food safety. *Mailing Add:* Div Toxicol HFF-156 200 C St SW Washington DC 20204

MILBOCKER, DANIEL CLEMENT, b Gaylord, Mich, May 25, 31; m 57; c 3. HORTICULTURE. *Educ:* Mich State Univ, BS, 65, MS, 66; Pa State Univ, PhD(hort), 69. *Prof Exp:* Asst prof ornamental hort, Univ Ky, 69-74; plant physiologist, 74-86, ASSOC PROF HORT, VA TRUCK & ORNAMENTALS RES STA, 86- *Mem:* Am Soc Hort Sci; Int Plant Propagators Soc. *Res:* Propagation and container culture of ornamental plants; improvement of ornamental species through genetic and cytological research. *Mailing Add:* Hampton Roads Agr Exp Sta 1444 Diamond Springs Rd Virginia Beach VA 23455

MILBOCKER, MICHAEL, b Detroit, Mich, Mar 10, 62. ELEMENTARY PARTICLE PHYSICS. *Educ:* Mass Inst Technol, BS, 84, MS, 86. *Prof Exp:* SCIENTIST PHYSICS, EYE RES INST, HARVARD, 86- *Concurrent Pos:* Prin investr, Eye Res Inst, 87- *Mem:* Inst Elec & Electronic Engrs; Optical Soc Am; Asn Researchers in Vision & Ophtlal; Int Soc Eye Researchers. *Res:* The application of Laser Doppler Velocimetry in the diagnosis of eye disease; studies of retinal blood flow to characterize surgical procedures; author of various publications. *Mailing Add:* Eye Res Inst 20 Staniford St Boston MA 02114

MILBRATH, GENE MCCOY, b Corvallis, Ore, Feb 15, 41; m 64; c 2. PLANT PATHOLOGY. *Educ:* Ore State Univ, BS, 63; Univ Ariz, MS, 66, PhD(plant path), 70. *Prof Exp:* Asst prof plant path, Univ Hawaii, 70-71; asst plant path, Univ Ill, Urbana, 71-77; plant pathologist, Univ Calif, Salinas, 77-78; PLANT PATHOLOGIST, DEPT AGR, SALEM, ORE, 78- *Mem:* Am Phytopath Soc; Int Soc Plant Path. *Res:* Epidemiology of plant viruses; characterization of plant viruses of economic plants; detection of viruses in shade and fruit trees; virus certification of ornamental and fruit trees; diseases of ornamental plants. *Mailing Add:* Plant Div Ore Dept of Agr Salem OR 97310-0110

MILBURN, GARY L, b Hutchinson, Kans, July 28, 52. MICROBIOLOGY. *Prof Exp:* Fel, Sch Med, Univ Iowa,81-83; SR IMMUNOLOGIST, SYVA CO, 83- *Mem:* Am Asn Immunolochinson, Kans, July 28, 52; Am Asn Pathologists; Am Soc Microbiol. *Mailing Add:* Syva Co 900 Arastradero Rd Palo Alto CA 94303

MILBURN, NANCY STAFFORD, b Syracuse, NY, Sept 7, 27; m 51; c 2. PHYSIOLOGY, ELECTRON MICROSCOPY. *Educ:* Radcliff Col, AB, 49, PhD, 58; Tufts Univ, MS, 50. *Prof Exp:* Asst, Tufts Univ, 49-52; asst, Harvard Univ, 56-58; from instr to assoc prof, 58-71, actg chmn dept, 67-68, dean, Jackson Col, 72-80, PROF PHYSIOL, TUFTS UNIV, 71-, RES ASSOC NEUROPHYSIOL, 58-, DEAN, LIBERAL ARTS & JACKSON COL, 80- *Concurrent Pos:* Mem, Nat Res Coun; chmn, Coun Int Exchange of Scholars, 78-81; trustee, Radcliffe Col, 77- & Corp Woods Hole Oceanog Inst, 81-; coordr, New Eng Region, Am Coun Educ Nat Identification Proj for Women Adminr, 77-80. *Mem:* Am Soc Zool; Am Physiol Soc; fel AAAS; Am Soc Cell Biol; Entom Soc Am; Sigma Xi. *Res:* Neurophysiology, especially synaptic transmission, neurohormones and synaptic transmitters; electron microscopy of insect nervous system, receptor organs and effectors. *Mailing Add:* Dept Biol Tufts Univ Medford MA 02155

MILBURN, RICHARD HENRY, b Newark, NJ, June 3, 28; m 51; c 2. ELEMENTARY PARTICLE PHYSICS. *Educ:* Harvard Univ, AB, 48, AM, 51, PH PhD(physics), 54. *Prof Exp:* Instr physics, Harvard Univ, 54 & 56-57, asst prof, 57-61; assoc prof, 61-65, PROF PHYSICS, TUFTS UNIV, 65- *Concurrent Pos:* Guggenheim fel, Orgn Europ Res Nucleaire, Geneva, 60; Fulbright lectr, India, 84. *Mem:* AAAS; Am Phys Soc; Am Asn Physics Teachers. *Res:* Physics of elementary particles. *Mailing Add:* Dept Physics Tufts Univ Medford MA 02155

MILBURN, RONALD MCRAE, b Wellington, NZ, May 29, 28; m 55; c 2. INORGANIC CHEMISTRY. *Educ:* Victoria Univ, BSc, 49, MSc, 51; Duke Univ, PhD(chem), 54. *Prof Exp:* Demonstr chem, Victoria Univ, NZ, 51-52; asst, Duke Univ, 52-54; lectr, Victoria Univ, NZ, 55; res assoc, Univ Chicago, 56-57; from asst prof to assoc prof, 57-68, PROF CHEM, BOSTON UNIV, 68- *Concurrent Pos:* Fulbright grant, 52; instr & res assoc, Duke Univ, 54 & 56; NIH fel, Oxford Univ, 65-66; vis prof, Univ Basel, 82-83; vis fel, Australian Nat Univ, 74-75 & 91. *Mem:* Am Chem Soc. *Res:* Reactions and stabilities of complex ions; mechanisms of inorganic reactions; reactions of phosphate esters and polyphosphates. *Mailing Add:* Dept Chem Boston Univ Boston MA 02215

MILBY, THOMAS HUTCHINSON, b South Bend, Ind, Feb 7, 31; m 53; c 3. OCCUPATIONAL MEDICINE, TOXICOLOGY. *Educ:* Purdue Univ, BS, 53; Univ Cincinnati, MD, 57, MS, 65; Univ Calif, Berkeley, MPH, 66; Am Bd Prev Med, dipl & cert occup med, 66. *Prof Exp:* Intern med, Ohio State Univ Hosp, 58; med officer, Div Occup Health, USPHS, 59-62; med officer, Bur Occup Health, Calif State Dept Pub Health, 62-66, chief, 66-73; CONSULT OCCUP MED, TOXICOL & EPIDEMIOL, 73- *Concurrent Pos:* Mem comn pesticides & environ health, Secy Health, Educ & Welfare, 69; mem study sect, Nat Inst Occup Safety & Health, 69-72; spec consult, WHO, 70; assoc prof, Sch Pub Health, Univ Calif, Berkeley, 70-; chmn task group on occup exposure to pesticides, Fed Working Group on Pest Mgt, Nat Inst Occup Safety & Health, 72-74. *Mem:* Fel Am Col Occup Med. *Res:* Toxicology and epidemiology and chemical-related diseases. *Mailing Add:* One Aspen Ct Lafayette CA 94549-2302

MILCAREK, CHRISTINE, b Pittsburgh, Pa, Aug 8, 46. MOLECULAR BIOLOGY. *Educ:* Johns Hopkins Univ, PhD(microbiol), 72. *Prof Exp:* ASSOC PROF MICROBIOL, UNIV PITTSBURGH, 83- *Mailing Add:* Dept Microbiol Univ Pittsburgh 720 Scaife Hall Pittsburgh PA 15261

MILCH, LAWRENCE JACQUES, b New York, NY, Sept 5, 18; m 42; c 4. PHARMACOLOGY, BIOCHEMISTRY. *Educ:* Univ Iowa, AB, 40; Rutgers Univ, PhD(physiol, biochem), 50. *Prof Exp:* Biophysicist, US Air Force Sch Aviation Med, 50-55, chief dept pharmacol & biochem, 55-59, dep comdr, 6102 Air Base Wing, Yakota Air Base, 59-61, head space biophys task group, Univ Calif, Berkeley, 61-62; staff res dir, Miles Labs, 62-66; asst res dir, Human Health Res & Develop Div, Dow Chem, USA, 66-71, dir develop, Zionsville, Ind, 72-73; instr physiol & biochem, Butler Univ, 73-76; instr pharmacol, Yavapai Col, Clarkdale, 77-79; INSTR PHARMACOL, NORTHERN ARIZ UNIV, 83- *Mem:* Am Soc Pharmacol & Exp Therapeut. *Res:* Protein chemistry of microscopic airborne proteins. *Mailing Add:* Dept Chem N Ariz Univ Flagstaff AZ 86011

MILCH, PAUL R, b Budapest, Hungary, May 1, 34; US citizen; m 62; c 2. OPERATIONS RESEARCH, STATISTICS. *Educ:* Brown Univ, BS, 58; Stanford Univ, PhD(statist), 66. *Prof Exp:* PROF OPERS RES DEPT, NAVAL POSTGRAD SCH, 63-; ADJ PROF, GOLDEN GATE UNIV, MONTEREY, CA, 80- *Concurrent Pos:* Statistician, Data Dynamics Inc, Calif, 65-66; opers analyst, Mellonics Inc, Litton Industs, 68-70 & BDM, Calif, 71-75. *Mem:* Am Statist Asn; Opers Res Soc Am. *Res:* Manpower modeling; queueing theory; stochastic processes; quantitative; birth and death processes. *Mailing Add:* Dept of Opers Res Naval Postgrad Sch Monterey CA 93943

MILDER, FREDRIC LLOYD, b New York, NY, Nov 16, 49; m 84; c 1. MEDICAL DEVICES, MEDICAL USES OF RADIATION. *Educ:* Mass Inst Technol, BS, 71; Univ Mich, MS, 73, PhD(physics), 76. *Prof Exp:* Res assoc physics, Va Polytech Inst & State Univ, 76-77 & Boston Univ, 77-78; prin scientist radiation physics, Spire Corp, 78-82; prog mgr, comput tomography res & develop, Elscint Inc, 82-84; PRIN SCIENTIST MED DEVICES RES & DEVELOP, APPL BIOMED CORP, 84- *Concurrent Pos:* Consult, Princeton Univ, 83. *Mem:* Am Phys Soc; Sigma Xi. *Res:* Techniques and equipment for measuring wear using radiotracers; hardware and software for transmitting CAT scan images over telephone; cardiac assist devices. *Mailing Add:* 204 Clinton Rd Brookline MA 02146-5814

MILDVAN, ALBERT S, b Philadelphia, Pa, Mar 3, 32; m 57; c 3. BIOPHYSICS, ENZYMOLOGY. *Educ:* Univ Pa, AB, 53; Johns Hopkins Univ, MD, 57. *Prof Exp:* Intern med, Baltimore City Hosps, Md, 57-58; res assoc cell physiol, Geront Br, NIH, 58-60; NIH res fel biochem, Inst Animal Physiol, Cambridge, Eng, 60-62; NIH res fel biophys, 62-64, assoc, 64-65, from asst prof to assoc prof phys biochem, 65-74, assoc mem, Inst, 68-73, mem, Inst Cancer Res, 73-81, PROF PHYS BIOCHEM, JOHNSON FOUND, SCH MED, UNIV PA, 74-, MEM, INST CANCER RES, 73-; JPROF PHYSIOL CHEM & CHEM, SCH MED, JOHNS HOPKINS UNIV, 81- *Concurrent Pos:* Advan fel, Am Heart Asn, 63-65, estab investr, 65-70, mem coun basic sci, 71-; NIH res grant, 65-; NSF res grant, 65; mem adv panel molecular biol, NSF, 71-74; mem, NIGMS, NIH Coun, 86- *Honors & Awards:* Herbert Sober Prize & Lectureship, Am Soc Biochem & Molecular Biol. *Mem:* Am Soc Biochem & Molecular Biol; Brit Biochem Soc; Am Chem Soc. *Res:* mechanisms of enzymes action and metal activation of enzymes; NMR as applied to enzymology; EPR in enzymology. *Mailing Add:* Dept Physiol Chem Sch Med Johns Hopkins Univ 725 N Wolfe St Baltimore MD 21205-2105

MILEDI, RICARDO, b Mexico DF, Mex, Sept 15, 27; Brit citizen; m 55; c 1. NEUROBIOLOGY, ELECTROPHYSIOLOGY. *Educ:* Inst Cientifico Literario, Chihuahua, BSc, 45; Univ Nat Autonoma Mex, MD, 55. *Prof Exp:* Res fel, Nat Inst Cardiol, Mex, 54-56; vis fel, John Curtin Sch Med Res, Canberra, Australia, 56-58; hon res assoc & lectr, Dept Biophys, Univ Col, London, 58-62, reader, 62-65, prof biophys, 65-75, Foulerton res prof, Royal Soc, 75-85, head, Dept Biophys, 78-85; DISTINGUISHED PROF PSYCHOBIOL, UNIV CALIF, IRVINE, 84- *Honors & Awards:* Forbes lectr, Grass Found, 64 & 90; Int Prize for Sci, King Faisal Found, Saudi Arabia, 88. *Mem:* Nat Acad Sci; fel Am Acad Arts & Sci; Europ Molecular Biol Orgn; Physiological Soc UK; Soc Neurosci; NY Acad Sci; Am Soc Cell Biol. *Res:* Transmission of impulses from one nerve cell to another or to a muscle; revealing how the brain transmits information and how drugs and toxic substances affect the nervous system; introduced a new way of studying the brain using the Xenopus oocyte expression system. *Mailing Add:* Dept Psychobiol Univ Calif Irvine CA 92717

MILER, GEORGE GIBBON, JR, b Sumter, SC, Jan 21, 40; m 64; c 3. SYSTEMS DESIGN & SYSTEMS SCIENCE. *Educ:* Clemson Univ, BS, 62; Fla Inst Technol, MS, 68. *Prof Exp:* Res engr, Col Arts & Sci, Clemson Univ, 63-67; engr, Harris Semiconductor, 68-72; PRES, GEORGE MILER, INC, 72- *Concurrent Pos:* Mem, Adv Comt Electronics Eng Technol Prog, Tri-County Tech Col, Pendleton, SC, 82-; inspector var nonfed aircraft navig aids, FAA, 85- *Mem:* Sr mem Inst Elec & Electronics Engrs. *Res:* Instrument package to determine velocity and direction of water currents; custom digital systems; complex data communications systems; computer applications involving real-time data acquisition and/or control, such as cash register communications, process monitoring and control, scientific instrument automation, data communications networks, and/or other applications where computers are required to communicate with other devices. *Mailing Add:* 305 Sasanqua Dr Greenville SC 29615

MILES, CHARLES DAVID, b Kansas City, Mo, Aug 11, 26; m 53; c 2. INVERTEBRATE ZOOLOGY. *Educ:* Univ Kans, AB, 50, MA, 56; Univ Ariz, PhD(zool), 61. *Prof Exp:* Asst prof biol, Eureka Col, 61-64; from asst prof to assoc prof, 64-72, PROF ZOOL, UNIV MO-KANSAS CITY, 72- *Mem:* Am Malacol Union; Sigma Xi. *Res:* Land snail taxonomy and distribution; anatomy; physiology. *Mailing Add:* Dept of Biol Univ of Mo 5100 Rockhill Rd Kansas City MO 64110

MILES, CHARLES DONALD, b Franklin, Ind, Dec 17, 38; m 66; c 1. PLANT PHYSIOLOGY, MOLECULAR BIOLOGY. *Educ:* Franklin Col, AB, 63; Indiana Univ, PhD(bot), 67. *Prof Exp:* NIH fel plant biochem, Cornell Univ, 67-69; from asst prof to assoc prof bot, 69-80, dir, Div Biol Sci, 83-85, PROF BIOL SCI, UNIV MO-COLUMBIA, 80- *Mem:* AAAS; Am Soc Plant Physiol; Bot Soc Am; Am Inst Biol Sci; Soc Plant Molecular Biol. *Res:* Development and control of pigmentation in higher plants; mechanism of photosynthetic phosphorylation and electron transport; chloroplast fluorescence and luminescence; analysis of photosynthesis with genetic mutants of maize; genetics of photosynthesis and chloroplasts; molecular genetics of photosynthesis genes. *Mailing Add:* Div Biol Univ Mo Columbia MO 65211

MILES, CHARLES P, b Chicago, Ill, June 19, 22; m 54; c 3. PATHOLOGY, CYTOLOGY. *Educ:* Univ Calif, Berkeley, BA, 47; Univ Calif, San Francisco, MD, 53; Am Bd Path, dipl, 59. *Prof Exp:* Bank Am-Giannini Found fel, 55-56; Nat Cancer Inst fel, 57-58; asst prof in residence nuclear med & radiation biol, Univ Calif, Los Angeles, 58-59; from instr to asst prof path, Stanford Univ, 59-62; assoc, Sloan-Kettering Inst, 62-66; assoc prof, Univ Calif, San Francisco, 69-70; assoc prof, Univ Utah, 66-69, prof path, 70-87; RETIRED. *Concurrent Pos:* Asst attend pathologist, Mem Hosp, NY, 62-66. *Mem:* Am Soc Exp Path; Am Asn Path & Bact; Am Asn Cancer Res; Soc Human Genetics. *Res:* Cytology analysis in cancer and in tissue culture strains. *Mailing Add:* 1740 E Hubbard Ave Salt Lake City UT 84108

MILES, CORBIN I, b Detroit, Mich, Feb 5, 40; m 60; c 4. ANALYTICAL CHEMISTRY. *Educ:* Northwestern Mo State Col, Bs, 62; Wayne State Univ, MS, 69, PhD(chem), 71. *Prof Exp:* Chemist method develop res & qual control, Detroit, Minneapolis & Kansas City dist labs, 62-72, Consumer Safety Officer, 72-75, chief generally recognized as safe rev br, 75-85, CHIEF, INDIRECT FOOD ADDITIVES BR, CTR FOOD SAFETY & APPL NUTRIT, FOOD & DRUG ADMIN, 85- *Concurrent Pos:* Instr chem, Wayne County Community Col, 71-72; prin investr safety re-eval food ingredients classified as generally recognized as safe, Food & Drug Admin, 75-85, prin investr, food packaging, 85- *Honors & Awards:* Merit Award, Food & Drug Admin, 83. *Mem:* Am Chem Soc; Asn Off Anal Chemists. *Res:* Technical safety of ingredients used in human foods and food packaging in the United States. *Mailing Add:* Food & Drug Admin Ctr Food Safety & Appl Nutrit 200 C St SW Washington DC 20204

MILES, DANIEL S, EXERCISE PHYSIOLOGY, CARDIOVASCULAR EVALUATION. *Educ:* Southern Ill Univ, PhD(physiol), 77. *Prof Exp:* ASSOC PROF CARDIOVASC & RESPIRATORY PHYSIOL, WRIGHT STATE UNIV, 79- *Mailing Add:* Dept Physiol & Biophysics Wright State Univ Dayton OH 45431

MILES, DAVID H, b Price, Utah, Apr 19, 28; m 52; c 11. ORGANIC CHEMISTRY, COMPUTER SCIENCE. *Educ:* Brigham Young Univ, BS, 53; Iowa State Univ, PhD(org chem), 57. *Prof Exp:* Fel dairy chem & gas chromatography, Iowa State Univ, 57-58; res chemist, PPG Ind, Inc, Chem Div, 58-60; comput analyst, Utah County, Utah, 76-77; special instr comput sci, Brigham Young Univ, 73-76; from asst prof to prof chem, 60-73, special instr comput sci, PROF CHEM, BRIGHAM YOUNG UNIV, HAWAII, 77- *Concurrent Pos:* NSF fac fel, 66-67. *Mem:* Am Chem Soc; Asn Develop of Computer-Based Instrnl Systs. *Res:* Reaction of epoxides and organometallic compounds and computers in chemistry. *Mailing Add:* Dept Chem Univ Central Fla PO Box 25000 Orlando FL 32816

MILES, DELBERT HOWARD, b Warrior, Ala, Jan 4, 43; m 63; c 2. SYNTHETIC ORGANIC CHEMISTRY, NATURAL PRODUCTS CHEMISTRY. *Educ:* Birmingham-Southern Col, BS, 65; Ga Inst Technol, PhD(org chem), 70. *Prof Exp:* NIH res fel org chem, Stanford Univ, 69-70; from asst prof to assoc prof, 70-78, coordr chem sci grad prog, 76-81, prof chem, Miss State Univ, 78-88. *Concurrent Pos:* Prog officer, NSF, 79; Fulbright-Hays lectr, Univ Philippines, 83 & 84. *Honors & Awards:* Res Award, Sigma Xi, 73. *Mem:* Am Soc Pharmacog; Phytochem Soc NAm; Am Pharmaceut Asn; Am Chem Soc; Sigma Xi. *Res:* Isolation, structure elucidation and synthesis of natural products which exhibit biological activity of some type or which possess some biosynthetic significance. *Mailing Add:* Chem Dept Univ Cent Fl PO Box 25000 Orlando FL 32816

MILES, DONALD ORVAL, b Callaway, Nebr, May 29, 39; m 88; c 4. CLINICAL MICROBIOLOGY, BORRELIA BURGDORFER. *Educ:* Hastings Col, BA, 64; Univ Nebr, Lincoln, MS, 67, PhD(microbiol), 72. *Prof Exp:* Asst lectr biol, Univ Nebr, Lincoln, 70-71, instr med bact, 72; asst prof microbiol, Sch Health Sci, Grand Valley State Cols, 73-76; clin microbiologist & chief sect microbiol & immunoserol, Dept Path, St Mary's Hosp, 76-81; CLIN MICROBIOLOGIST, ST FRANCIS MED CTR, 81-; ADJ GRAD FAC, SOUTHEAST MO STATE UNIV, 83- *Concurrent Pos:* Res assoc, Dept Microbiol, Univ Nebr, Lincoln, 71 & Dept Oral Biol, Col Dent, 73; clin microbiologist & microbiol consult, Dept Path, Microbiol Lab, St Mary's Hosp, Grand Rapids, 74-76, mem infection control comt & assoc mem med & dent staff, 75-81; mem, infection control pharm & therapeut comt, St Francis Med Ctr, 81-; ed consult, bd abstr/RRM, med & clin microbiol-bact sect, Biol Abstr, 84- *Mem:* Am Soc Microbiol; Sigma Xi; Am Soc Clin Path. *Res:* Amino acid metabolism of gram negative non-spore forming anaerobes; periplasmic enzymes of gram negative bacilli; microbiology of infection control in hospitals and culture of Borrelia Burgdorfer: from human cases of lyme disease. *Mailing Add:* Lab Div St Francis Med Ctr Cape Girardeau MO 63701

MILES, EDITH WILSON, b Dallas, Tex, Feb 22, 37; m 66; c 2. ENZYME CHEMISTRY. *Educ:* Univ Tex, Austin, BA, 57; Univ Calif, Berkeley, PhD(biochem), 62. *Prof Exp:* RES CHEMIST, NIH, 66- *Concurrent Pos:* Mem, biochem & biophys panel B, molecular biol grants panel, NSF, 83; vis prof, Univ Calif, Riverside, 85. *Mem:* Am Chem Soc; Am Soc Biol Chemists; AAAS; Protein Soc. *Res:* Structure and function of enzymes, especially those containing pyridoxal phosphate; studies of tryptophan synthase, a multi-enzyme complex, including spectroscopic studies, chemical modification, substrate analogs, and preparation of crystals for x-ray crystallography; site-directed mutagenesis. *Mailing Add:* NIH Bldg 8 Rm 2A09 Bethesda MD 20892

MILES, EDWARD LANCELOT, b Port-of-Spain, Trinidad, W-I, Dec 21, 39; US citizen; m 63; c 2. MARINES SCIENCES, SCIENCE POLICY. *Educ:* Howard Univ, BA, 62; Univ Denver, PhD(int rel), 65. *Prof Exp:* Instr Int rel, 65-66, from asst prof to assoc prof, Grad Sch Int Studies, Univ Denver, 66-74; PROF MARINES STUDIES & PUB AFFAIRS, INST MARINES STUDIES & GRAD SCH PUB AFFAIRS, UNIV WASH, 74-, DIR, 82- *Concurrent Pos:* Mem adv panel Nat Sea Grant Prog, US Dept Com, 70-72; James P Warburg fel, Ctr Int Affairs, Harvard Univ, 73-74; sr fel, Woods Hole Oceanog Inst, 73-74; chmn ocean policy comt, Nat Res Coun, Nat Acad Sci, 74-79; chief negotiator, micronesian maritime authority, Fed States Micron, 80; consult, Nuclear Energy Agency, 86-87. *Res:* Int regulation of marine resources and marine uses; international science and technology policy. *Mailing Add:* Sch Marine Affairs HF-05 Univ Wash Seattle WA 98195

MILES, ERNEST PERCY, JR, b Birmingham, Ala, Mar 16, 19; m 45; c 2. MATHEMATICS. *Educ:* Birmingham-Southern Col, BA, 37; Duke Univ, MA, 39, PhD(math), 49. *Prof Exp:* Teacher high sch, Ala, 38-39; instr math, NC State Col, 40-41; assoc prof, Ala Polytech Inst, 49-58; mem staff, Nat Sci Found, 58; from assoc prof to prof math, 58-82, dir comput ctr, 61-71, SERV PROF, FLA STATE UNIV, 82-, DIR, MUENCH CTR FOR COLOR GRAPHICS, 81- *Concurrent Pos:* Vis assoc prof & Air Force Off Sci Res contract, Inst Fluid Dynamics & Appl Math, Univ Md, 57-58; participant, Nat Sci Found Training Prog Numerical Anal, Nat Bur Standards, 59; res grants, Air Force Off Sci Res, 60-61, Nat Sci Found, 62-72, 76-80 & indust grants, 79-85; consult, Nat Sci Found, 59-63, 65-70 & US Off Educ, 65-71; coun mem, Conf Bd Math Socs, 74-81; state dir fla, Nat Comput Graphics Asn, 81. *Mem:* Fel AAAS; Am Math Soc; Math Asn Am; Asn Comput Mach; Soc Indust & Appl Math; Nat Comput Graphics Asn. *Res:* Partial differential equations; numerical methods; information retrieval; computer uses in education; computer color graphics and separation. *Mailing Add:* Dept Math Ctr Color Graphics Fla State Univ Tallahassee FL 32306

MILES, FRANK BELSLEY, b Champaign, Ill, May 15, 40; m 66; c 2. MATHEMATICS. *Educ:* Univ Ill, BS, 61; Univ Calif, Berkeley, PhD(chem), 65; Univ Wash, MS, 70, PhD(math), 72. *Prof Exp:* NIH fel, Univ Calif, Los Angeles, 64-65; asst prof chem, Univ Calif, Santa Barbara, 65-68; NSF fel, Univ Wash, 69-71, instr math, 71, res asst, 71-72; asst prof, 72-76, assoc prof, 76-81, PROF MATH, CALIF STATE UNIV, DOMINGUEZ HILLS, 81- *Mem:* Am Math Soc; Math Asn Am. *Res:* Harmonic analysis. *Mailing Add:* Dept Math Calif State Univ 1000 E Victoria St Carson CA 90747

MILES, GEORGE BENJAMIN, b Erin, Tenn, May 14, 26; m 56; c 1. ORGANIC CHEMISTRY. *Educ:* Univ Tenn, BS, 50, PhD(chem), 58. *Prof Exp:* Chemist, US Naval Ord Lab, 53-54; instr chem, Univ Tenn, 57-58; res chemist, Dacron Res Lab, Textile Fibers Dept, E I du Pont de Nemours & Co, 58-61; from asst prof to assoc prof, 61-69, chmn dept chem, 69-77, PROF CHEM, APPALACHIAN STATE UNIV, 77- *Mem:* Am Chem Soc. *Res:* Polymer and steroid chemistry; organic mechanisms. *Mailing Add:* Dept Chem Appalachian State Univ Boone NC 28608

MILES, HARRY TODD, b Maysville, Ky, May 8, 26; m; c 2. MOLECULAR BIOLOGY. *Educ:* Harvard Univ, AB, 47; Northwestern Univ, PhD(chem), 52. *Prof Exp:* USPHS scientist, Enzyme Sect, Lab Cellular Physiol, Nat Heart Inst, 53-58, res chemist, Sect Metab Enzymes, Lab Molecular & Digestive Dis, 59-66, CHIEF, SECT ORG CHEM, LAB MOLECULAR BIOL, NAT INST DIABETES & DIGESTIVE & KIDNEY DIS, NIH, BETHESDA, 66- *Mem:* Am Chem Soc; Soc Biol Chemists. *Res:* Chemistry and biochemistry of polynucleotides and nucleic acids; helix-forming reactions of polynucleotides; infrared spectroscopy; UV and CD spectroscopy; structure and conformation of polynucleotides; specific complexing of alkali metal ions; enzymatic and chemical synthesis of polynucleotides; chemical modifications of nucleic acid bases. *Mailing Add:* NIH Nat Inst Diabetes Digestive & Kidney Dis Lab Molecular Biol Bldg 2 Rm 201 Bethesda MD 20892

MILES, HARRY V(ICTOR), b Columbus, Ohio, Nov 8, 14; wid; c 2. CHEMICAL ENGINEERING. *Educ:* Ohio State Univ, BChE, 36. *Prof Exp:* Appl engr, Airtemp Div, Chrysler Corp, Ohio, 36-39; sales engr, Infilco, Inc, Ill, 39-44; staff engr, Garfield Div, Houdaille-Hershey Corp, Ill, 44-46, mgr res & develop, Honan-Crane Div, Ind, 46-50; mgr res & develop, US Hoffman Mach Corp, NY, 50-53; asst dir res & develop, Oliver United Filters Inc, 53-56, asst dir res, 56-59, filtration develop engr, 59-62, mgr pulp & paper technol, 62-71; dir tech admin, Dorr-Oliver Inc, 71-80; RETIRED. *Res:* Pulp and paper technology; filtration; equipment design; research and development administration; air and water conditioning. *Mailing Add:* 7579 Estrella Circle Boca Raton FL 33433

MILES, HENRY HARCOURT WATERS, b Burnside, La, Sept 18, 15; m 39; c 2. PSYCHIATRY, PSYCHOANALYSIS. *Educ:* Tulane Univ, BS, 36, MD, 39. *Prof Exp:* Res fel psychiat, Harvard Med Sch, 46-48; asst, Harvard Univ & Mass Gen Hosp, 49-52; from asst prof to assoc prof clin psychiat, 52-66, prof psychiat, 66-86, EMER PROF PSYCHIAT, SCH MED, TULANE UNIV, 86- *Concurrent Pos:* Training & supv analyst, New Orleans Psychoanal Inst, 56-86; consult, Family Serv Soc New Orleans, 57-64. *Mem:* Am Psychosom Soc; AMA; fel Am Psychiat Asn; Am Psychoanal Asn. *Res:* Evaluation of psychotherapy; personality factors in cardiovascular diseases. *Mailing Add:* 123 Walnut St No 305 New Orleans LA 70118

MILES, JAMES LOWELL, b Buckhannon, WVa, Aug 15, 37; m 68. CHEMISTRY, BIOCHEMISTRY. *Educ:* WVa Univ, BS, 59, MS, 61, PhD(biochem), 64. *Prof Exp:* Trainee clin chem, Hosp Univ Pa, 64-65; RES BIOCHEMIST, E I DU PONT DE NEMOURS & CO, INC, WILMINGTON, 66- *Mem:* Am Chem Soc; Am Asn Clin Chemists. *Res:* Alpha-chymotrypsin; clinical and analytical chemistry; enzyme assay systems; lipoprotein electrophoresis. *Mailing Add:* 44 Quartz Mill Rd Newark DE 19711-2330

MILES, JAMES S, b Baltimore, Md, Apr 16, 21; m 44; c 4. MEDICINE. *Educ:* Grinnell Col, AB, 42; Univ Chicago, MD, 45; Am Bd Orthop Surg, dipl, 54. *Prof Exp:* Intern, Univ Clins, Sch Med, Univ Chicago, 45-46, resident, 48-51, instr orthop, 51-52; from instr to assoc prof orthop surg, 52-65, chmn div, 58-73, actg chmn dept orthop, 73-74, PROF ORTHOP SURG, SCH MED, UNIV COLO, DENVER, 65-, CHMN DEPT ORTHOP, 74- *Concurrent Pos:* Am Orthop Asn traveling fel, 59; consult, Vet Admin Hosp, Denver, Colo & Fitzsimons Army Hosp, Denver. *Mem:* Orthop Res Soc; fel Am Col Surgeons; Am Acad Orthop Surg; Am Orthop Asn; Clin Orthop Soc. *Res:* Histochemistry of articular cartilage; vascular supply of femoral head. *Mailing Add:* 2635 Kingridge Dr Fallbrook CA 92028

MILES, JAMES WILLIAM, b Henderson, Ky, Sept 19, 18; m 51; c 1. PESTICIDE CHEMISTRY. *Educ:* Western Ky Univ, BS, 40; Univ Ill, MS, 47, PhD(anal chem), 53; Univ Ky, BS, 58. *Prof Exp:* Instr chem, Louisville Col Pharm, 41-42; from asst prof to assoc prof, Univ Ky, 47-55, prof pharmaceut chem & head dept, 55-58; asst chief, Chem Sect, Tech Develop Lab, Ctr for Dis Control, USPHS, 58-64, chief chem sect, Tech Develop Labs, Commun Dis Ctr, 64-74, chief pesticides br, Bur Trop Dis, 74-84; RETIRED. *Concurrent Pos:* Mem expert adv panel on insecticides, WHO, 71-, mem sci & tech adv comt, WHO Onchocerciasis Control Prog, 74- *Honors & Awards:* Superior Serv Award, USPHS, 78. *Mem:* Am Chem Soc; Sigma Xi. *Res:* development of methods of analysis of pesticide residues in the environment; research on pesticide formulations and analysis of formulations. *Mailing Add:* 1228 Bacon Park Dr Savannah GA 31406

MILES, JOHN B(RUCE), b St Louis, Mo, Feb 2, 33; m 58; c 2. MECHANICAL ENGINEERING, FLUID DYNAMICS. *Educ:* Mo Sch Mines, BS, 55, MS, 57; Univ Ill, PhD(mech eng, fluid flow), 63. *Prof Exp:* Instr eng mech, Univ Mo-Rolla, 55-58; from instr to asst prof appl sci, Univ Southern Ill, 58-63; assoc prof, 63-68, PROF MECH ENG, UNIV MO COLUMBIA, 68- *Concurrent Pos:* NSF, NASA & US Air Force res grants, 65-; Ford Found eng residency, Gen Elec Co, 65-66; Nat Res Coun sr res assoc, Ames Res Ctr, NASA, 71; vis prof, Dept Aeronaut & Astronaut, Stanford Univ, 71; sabbatical leave, Solar Energy Res Inst, 81. *Mem:* Am Soc Mech Engrs; Am Soc Eng Educ; Am Inst Aeronaut & Astronaut; Sigma Xi. *Res:* Fluid flow and heat transfer, especially separated flow, jet mixing and turbulence; bio-heat transfer as applied to temperature control in liquid cooled garments; jet mixing studies using a laser velocimeter; aerodynamic holographic interferometry. *Mailing Add:* Dept Mech Eng Univ MO-Columbia Columbia MO 65211

MILES, JOHN WILDER, b Cincinnati, Ohio, Dec 1, 20; m 43; c 3. GEOPHYSICS, APPLIED MATHEMATICS. *Educ:* Calif Inst Technol, BS, 42, MS, 43, PhD(elec eng), 44. *Prof Exp:* Staff mem radiation lab, Mass Inst Technol, 44; res eng, Lockheed Aircraft Corp, Calif, 44-45; from asst prof to prof eng, Univ Calif, Los Angeles, 45-55, prof eng & geophys, 55-61; prof appl math, Inst Adv Studies, Australian Nat Univ, 62-64; chmn appl mech & eng sci, 68-74, chmn, Acad Senate, 77-78, vchancellor acad affairs, 80-83, PROF APPL MECH & GEOPHYS, UNIV CALIF, SAN DIEGO, 65- *Concurrent Pos:* Fulbright lectr, Univ NZ, 51; vis lectr, Univ London, 52; Guggenheim fel, 58-59 & 68-69; Fulbright res fel, Cambridge Univ, 69. *Honors & Awards:* Timoshenko Medal, Am Soc Mech Engrs, 82; Otto Laporte Lectr, Am Phys Soc, 83. *Mem:* Nat Acad Sci; fel Am Acad Arts & Sci; fel Am Acad Mech; fel Am Inst Aeronaut & Astronaut; AAAS; Am Geophys Union. *Res:* Wave propagation and generation; hydrodynamic stability; geophysical fluid dynamics. *Mailing Add:* Inst Geophys & Planetary Physics Univ Calif 9500 Gilman Dr La Jolla CA 92093-0934

MILES, JOSEPH BELSLEY, b Champaign, Ill, June 17, 42; m 70; c 3. MATHEMATICS. *Educ:* Univ Ill, BS, 63; Univ Wis, MS, 64, PhD(math), 68. *Prof Exp:* Res assoc, Cornell Univ, 68-69; from asst prof to assoc prof, 69-79, PROF MATH, UNIV ILL, URBANA, 79- *Concurrent Pos:* Off Naval Res fel, Cornell Univ, 68-69; res assoc, Univ Md, 75-76. *Mem:* Am Math Soc. *Res:* Functions of a complex variable. *Mailing Add:* Dept Math Univ Ill 339 Ill Hall 1409 W Green St Urbana IL 61801

MILES, LINDSEY ANNE, VASCULAR BIOLOGY. *Educ:* Occidental Col, AB, 72; Univ Calif, San Diego, PhD(biol), 82. *Prof Exp:* Res technician, Lab Comp Biochem, San Diego, Calif, 73-74, Dept Allergy & Immunol, Scripps Clin & Res Found, La Jolla, 74-76; grad teaching & res asst, Univ Calif, San Diego, 76-82; postdoctoral res fel, Dept Immunol, Res Inst Scripps Clin, 82-88, sr res assoc, 88, asst mem, 88-89, ASST MEM, COMT STUDY VASCULAR BIOL, RES INST SCRIPPS CLIN, 89- *Concurrent Pos:* Mem, Thrombosis Coun, Am Heart Asn, exec bd, Soc Fels, Scripps Clin, 83-84; numerous invited lect, 87-90. *Mem:* Int Soc Thrombosis & Haemostasis; Am Soc Biol Chemists; Am Fedn Clin Res. *Res:* Author of numerous publications. *Mailing Add:* Scripps Clin & Res Found 10666 N Torrey Pines Rd La Jolla CA 92037

MILES, MARION LAWRENCE, b Columbus, Ga, Sept 5, 29; m 56; c 3. ORGANIC CHEMISTRY. *Educ:* Univ Ga, BS, 57, MS, 59; Univ Fla, PhD(org chem), 63. *Prof Exp:* Fel, Duke Univ, 63-64; asst prof, 65-69, ASSOC PROF ORG CHEM, NC STATE UNIV, 69- & DIR ORG LABS, 77- *Mem:* Am Chem Soc. *Res:* Physical properties of multiple carbanions; mechanisms of condensation reactions. *Mailing Add:* Dept Chem NC State Univ Box 8204 Raleigh NC 27695-8204

MILES, MAURICE HOWARD, b St George, Utah, Nov 20, 33; m 60; c 2. SOLID STATE PHYSICS. *Educ:* Univ Utah, BS, 55, PhD(physics), 63. *Prof Exp:* Res assoc metall, Univ Ill, 63-65; ASST PROF PHYSICS, WASH STATE UNIV, 65- *Mem:* Am Phys Soc. *Res:* Internal friction in metals; electronic properties of dislocations in semiconductors. *Mailing Add:* Dept of Physics Wash State Univ Pullman WA 99164

MILES, MAURICE JARVIS, b St George, Utah, Nov 24, 07; m 31; c 11. ANALYTICAL CHEMISTRY, ENVIRONMENTAL CHEMISTRY. *Educ:* Brigham Young Univ, AB, 30; Univ Utah, MA, 33. *Prof Exp:* Dir phys sci, Dixie Jr Col, 33-53; chief chemist, Titanium Metals Corp Am, 53-70; res assoc, Desert Res Inst, Univ Nev Syst, 70-80; consult, Titanium Metals Corp Am, 80-86; PRES, MILES ASSOCS INC, 86- *Concurrent Pos:* Consult, US Bur Mines, Boulder City, 74-80. *Mem:* AAAS; Am Chem Soc. *Res:* Corrosion rate of selected metals and alloys in Lake Mead water; rapid x-ray ion-exchange analyses of titanium alloys; water pollution analyses related to Lake Mead; trace metals in geothermal waters; trace metals in soils; atmospheric dusts; state of the art in industrial ponding of effluents. *Mailing Add:* 135 Elm St Henderson NV 89015

MILES, MELVIN HENRY, b St George, Utah, Jan 18, 37; m 62; c 4. ELECTROCHEMISTRY, PHYSICAL CHEMISTRY. *Educ:* Brigham Young Univ, BA, 62; Univ Utah, PhD(phys chem), 66. *Prof Exp:* NATO res fel electrochem, Munich Tech, 65-66; res chemist, Naval Weapons Ctr, 67-69; asst prof, Mid Tenn State Univ, 69-72, assoc prof chem, 72-; AT NAVAL WEAPONS CTR, CHINA LAKE, CALIF. *Mem:* Electrochem Soc; Sigma Xi; Am Chem Soc. *Res:* Fast reaction kinetics; electrode kinetics; electrochemical energy conversion; electrode catalysis; fuel cells; water electrolysis; properties of mixed solvents; hydrogen production; oxygen electrode reaction; cold fusion; lithium batteries. *Mailing Add:* Naval Weapons Ctr Code 3853 China Lake CA 93555-6001

MILES, NEIL WAYNE, b River Falls, Wis, June 22, 37; m 59, 86; c 6. PLANT PHYSIOLOGY. *Educ:* Univ Minn, BS, 59, MS, 64, PhD(hort), 65. *Prof Exp:* Exten horticulturist, Univ Minn, St Paul, 65-66; pomologist, Kans State Univ, 66-81; RES SCIENTIST, ONTARIO MINISTRY AGR & FOOD, 81- *Mem:* Am Soc Hort Sci; Am Pomol Soc; Sigma Xi; Int Soc Hort Sci; Can Soc Hort Sci. *Res:* Physiological studies on fruit crops; fruit breeding. *Mailing Add:* Hort Res Inst Ont Vineland Sta ON L0R 2E0 Can

MILES, PHILIP GILTNER, b Olean, NY, Aug 10, 22; m 49; c 3. BOTANY. *Educ:* Yale Univ, BA, 48; Indiana Univ, PhD(bot), 53. *Prof Exp:* Res assoc bot, Univ Chicago, 53-54; res fel, Harvard Univ, 54-56; from asst prof to assoc prof, 56-70, PROF BIOL, STATE UNIV NY BUFFALO, 70- *Concurrent Pos:* Fulbright res scholar, Japan, 63-64; vis scientist, US-China Coop Sci Prog, 70-71; vis prof, Nat Taiwan Univ, 70-71 & 77-78. *Mem:* AAAS; Bot Soc Am; Genetics Soc Am; Mycol Soc Am; Soc Study Evolution; Sigma Xi. *Res:* Genetics and physiology of sexual mechanisms in fungi. *Mailing Add:* Dept of Biol State Univ NY Buffalo NY 14260

MILES, RALPH FRALEY, JR, b Philadelphia, Pa, May 15, 33. PHYSICS, SYSTEMS DESIGNS. *Educ:* Calif Inst Technol, BS, 55, MS, 60, PhD(physics), 63. *Prof Exp:* Sr engr, Jet Propulsion Lab, 63-65, supvr syst eng, 65-69; vis fel econ systs, Stanford Univ, 69-70; vis asst prof aeronaut & environ eng sci, Calif Inst Technol, 70-71; mgr, Mission Analysis & Eng, Outer Planets Missions, 71-75, supvr oper res, 75-82, MEM TECH STAFF, JET PROPULSION LAB, 82- *Mem:* AAAS; Opers Res Soc Soc Am; Sigma Xi. *Res:* Density of cosmic ray neutrons in the atmosphere; systems analysis; design analysis. *Mailing Add:* 3608 Canon Blvd Altadena CA 91001

MILES, RANDALL JAY, b Crawfordsville, Ind, Oct 27, 52; m 75. SOIL MANAGEMENT. *Educ:* Purdue Univ, BS, 74, MS, 76; Tex A&M Univ, PhD(soils), 81. *Prof Exp:* Instr, Tex A&M Univ, 76-81; asst prof soils teaching, Univ Tenn, 81-; AT AGRON DEPT, UNIV MO-COLUMBIA. *Mem:* Am Soc Agron; Soil Conservation Soc Am; Sigma Xi; Nat Asn Col Teachers Agr. *Res:* Use of soil survey for soil management applications with special emphasis on production agriculture; soil erosion; soil varicability; soil genesis. *Mailing Add:* Dept Agron Mumford Hall Univ Mo Columbia MO 65211

MILES, RICHARD BRYANT, b July 10, 43; m 83; c 2. OPTICAL FLOW DIAGNOSTICS, MOLECULAR DYNAMICS. *Educ:* Stanford Univ, BS, 66, MS, 67, PhD(elec eng), 72. *Prof Exp:* Res assoc elec eng, dept elec eng, Stanford Univ, 72; from asst prof to assoc prof, 72-82, PROF MECH & AEROSPACE ENG, DEPT MECH & AEROSPACE, PRINCETON UNIV, 82-, CHMN, ENG PHYSICS PROG, 80- *Concurrent Pos:* Fannie & John K Hertz Found fel, 69-72, mem bd dirs, 89-; trainee, NSF, 72; chmn, Gordon Conf Vibrational Spectros, 86; vchmn, Aerodyn Measurement Technol Tech Comt, Am Inst Aeronaut & Astronaut, 90- *Mem:* Am Phys Soc; Inst Elec & Electronics Engrs; Optical Soc Am; Am Inst Aeronaut & Astronaut. *Res:* Use of lasers as tools studying flowing and nonequilibrium gases, molecular dynamics, and surface phenomena; fluorescing oxygen as a tracer for high speed gas flows; molecular relaxation dynamics; x-ray laser development; optical systems leading to the development of new methods of 3-D data presentation. *Mailing Add:* D-414 Eng Quadrangle Princeton Univ Princeton NJ 08544-5263

MILES, RICHARD DAVID, b Bunkie, La, Oct 29, 47; m 76. EGG SHELL QUALITY, AMINO ACID METABOLISM. *Educ:* Univ Ark, BS, 71, MS, 72; Purdue Univ, PhD(poultry nutrit), 76. *Prof Exp:* From asst prof to assoc prof, 76-85, PROF POULTRY NUTRIT, POULTRY SCI DEPT, UNIV FLA, 85- *Mem:* Poultry Sci Asn; Sigma Xi. *Res:* Dietary mineral interrelationships in laying hens; egg shell quality; influence of antibiotics and probiotics in laying hens, broilers, quail and turkeys; amino acid and nitrogen metabolism in poultry. *Mailing Add:* Poultry Sci Dept Univ Fla Archer Rd Gainesville FL 32611

MILES, ROBERT D(OUGLAS), b Bloomfield, Ind, Dec 23, 24; m 46; c 1. CIVIL ENGINEERING. *Educ:* Purdue Univ, BS, 49, MS, 51. *Prof Exp:* Asst, 49-50, from instr airphoto interpretation to assoc prof civil eng & airphoto interpretation, 50-68, PROF HWY ENG & AIRPHOTO INTERPRETATION, PURDUE UNIV, 68-, RES ENGR, JOINT HWY RES PROJ, 53- *Mem:* Am Soc Eng Educ; Am Soc Photogram; Nat Soc Prof Engrs; Am Soc Civil Engrs. *Res:* Airphoto interpretation and site selection; aerial surveys for planning civil engineering projects and regional planning programs to include natural resources inventories. *Mailing Add:* 1724 Sheridan Rd West Lafayette IN 47906

MILES, WYNDHAM DAVIES, b Wilkes-Barre, Pa, Nov 21, 16; m 52; c 4. HISTORY OF CHEMISTRY. *Educ:* Philadelphia Col Pharm, BS, 42; Pa State Univ, MS, 44; Harvard Univ, PhD(hist of sci), 55. *Prof Exp:* Instr chem, Pa State Univ, 44-50, asst prof, 52-53; historian, US Army Chem Corps, 53-60; specialist in sci, Nat Arch, 60-61; historian, Polaris Proj, 61-62; HISTORIAN, NIH, 62- *Honors & Awards:* Dexter Award in Hist of Chem, 71. *Mem:* Am Chem Soc. *Res:* History of American chemistry. *Mailing Add:* 24 Walker Ave Gaithersburg MD 20877-2704

MILEWICH, LEON, b Buenos Aires, Arg, Mar 26, 27; US citizen; m 59; c 3. ORGANIC CHEMISTRY. *Educ:* Univ Buenos Aires, BS, 56, MS, 58, PhD(org chem), 59. *Prof Exp:* Chemist, Res Inst Armed Forces, Arg, 55-58, res chemist, 60-61; fel, Sch Pharm, Univ Md, 61-64; fel, Sch Med, Johns Hopkins Univ, 64-66, instr gynec & obstet, 66-67; res assoc, Southwest Found Res & Educ, 67-72; asst prof, 72-77, ASSOC PROF, DEPT OBSTET & GYNEC, UNIV TEX SOUTHWESTERN MED SCH DALLAS, 77- *Concurrent Pos:* NIH fel, 63-64. *Mem:* AAAS; Am Chem Soc; Royal Soc Chem; Arg Chem Asn; NY Acad Sci; Sigma Xi. *Res:* Steroids. *Mailing Add:* Obstet & Gynec Dept Univ Tex Health Sci Ctr 5323 Harry Hines Blvd Dallas TX 75235

MILEWSKI, JOHN VINCENT, b Suffern, NY, Nov 4, 28; m 52; c 3. CERAMICS ENGINEERING, MATERIALS SCIENCE. *Educ:* Univ Notre Dame, BSChE, 51; Stevens Inst Technol, MS, 59; Rutgers Univ, PhD(ceramic eng), 72. *Prof Exp:* Sr mat engr, Thiokol Chem Corp, 51-61; proj engr, Curtiss-Wright Corp, 61-62; vpres, Thermokinetic Fibers, 63-68; res assoc mat res, Exxon Res & Eng Co, 68-77; staff engr, Los Alamos Sci Lab, 77-84; PRES, SUPERKINETIC INC, 84-; CONSULT, 84- *Mem:* Am Ceramic Soc; Am Soc Testing & Mat; Soc Plastics Indust. *Res:* High strength materials; growth of single crystal whiskers; development of production processes for whiskers; theory of application of whiskers; short fibers; fiber packing concepts; fillers for plastics; short fiber composites; 25 patents. *Mailing Add:* PO Box 8029 Santa Fe NM 87504-8029

MILEY, G(EORGE) H(UNTER), b Shreveport, La, Aug 6, 33; m 58; c 2. NUCLEAR & FUSION ENGINEERING. *Educ:* Carnegie Inst Technol, BS, 55; Univ Mich, MS, 56, PhD(chem eng), 59. *Prof Exp:* Instr metall, Univ Mich, 57; analyst reactor critical exp, Knolls Atomic Power Lab, 58-61; asst prof nuclear eng & physics, Univ Ill, 61-64, assoc prof nuclear eng, 65-67, chairperson, Nuclear Eng Prog, 75-87, PROF NUCLEAR ENG, UNIV ILL, URBANA, 67-, DIR FUSION STUDIES LAB, 75-; DIR RES, NPL ASSOC, 90- *Concurrent Pos:* Vis prof, Cornell Univ, 69-70, Univ New So Wales, 87, Imperial Col, 87; NATO sr fel, 75; ed, Univ Fusion Asn Newsletter, 80- & J Fusion-Technol, 81-, US ed, Laser & Particle Beams, 88-; Guggenheim fel, 87. *Honors & Awards:* Western Elec Award, Am Soc Eng Educ, 77. *Mem:* fel Am Nuclear Soc; fel Am Phys Soc; Am Soc Eng Educ; fel Inst Elec & Electronic Engrs. *Res:* Fusion plasma engineering; nuclear reactor physics and kinetics; direct energy conversion; fusion technology. *Mailing Add:* Univ Ill 214 NEL 103 S Goodwin Ave Urbana IL 61801

MILEY, JOHN WULBERN, b Lake Charles, La, Apr 23, 42; m 65; c 2. COLOR CHEMISTRY & POLYOLEFIN ADDITIVES. *Educ:* Clemson Univ, BS, 64; Fla State Univ, PhD(phys org chem), 70. *Prof Exp:* Res chemist, Milliken Res Corp, 72-75, mgr anal serv, 75-77; sr develop chemist, 77-79, develop mgr, 79-84, DIR RES & DEVELOP, MILLIKEN SPECIALTY CHEM, 84- *Mem:* Am Chem Soc; Indust Res Inst. *Res:* Discovery, development and commercial deployment of unique performance products such as washable inks, carpet cleaners, corrision inhibitors, polyurethane, colorants, plastic additives. *Mailing Add:* 340 Waldrop Rd Campubello SC 29322

MILFORD, FREDERICK JOHN, b Cleveland, Ohio, July 1, 26; m 51; c 1. PHYSICS. *Educ:* Case Inst Technol, BS, 49; Mass Inst Technol, PhD(physics), 52. *Prof Exp:* Asst physics, Mass Inst Technol, 49-51; instr, Case Western Reserve Univ, 49-51, 52-56, asst prof, 56-59; div consult, 59-64, sr fel & chief theoret physics div, 64-66, dir, Inst Res Phys Sci, 65-73, inst scientist, 73, mgr physics & electronics dept, 73-74, mgr physics electronics & nuclear technol, 74-76, ASSOC DIR, BATTELLE MEM INST, 76- *Mem:* Fel Am Phys Soc; Am Math Soc; Am Nuclear Soc; AAAS; Int Glaciological Soc; Sigma Xi. *Res:* Meson field theory; theoretical nuclear physics; cosmic ray primaries; nuclear magnetic resonance; nuclear moments; electronic structure of solids; magnetism; helium films at low temperature. *Mailing Add:* 1411 London Dr Columbus OH 43221

MILFORD, GEORGE NOEL, JR, b Victoria, PEI, Can, May 4, 24; nat US; m 48; c 3. POLYMER CHEMISTRY. *Educ:* Mt Allison Univ, BSc, 44; Dalhousie Univ, MSc, 48; McGill Univ, PhD(chem), 53. *Prof Exp:* Asst chemist, Best Yeast Co, 45-46; res chemist, Dom Steel & Coal Corp, 48-50; res chemist, E I du Pont De Nemours & Co, 53-60, sr res chemist, 60-85; RETIRED. *Mem:* Am Chem Soc. *Res:* Preparation of monomers and polymers; synthetic fibers, Orlon, Lycra, hollow fibers, Nomex. *Mailing Add:* 1208 Shamrock Lane Waynesboro VA 22980

MILFORD, MURRAY HUDSON, b Honey Grove, Tex, Sept 29, 34; m 61; c 2. SOIL SCIENCE, SOIL MINERALOGY. *Educ:* Tex A&M Univ, BS, 55, MS, 59; Univ Wis, PhD(soil sci), 62. *Prof Exp:* Fel soil chem & res specialist, Cornell Univ, 62-63, from asst prof to assoc prof soil sci, 63-68; assoc prof, 68-74, PROF SOIL SCI, TEX A&M UNIV, 74- *Concurrent Pos:* Soil sci educ award, Soil Sci Soc Am, 88. *Mem:* Fel AAA, 88; fel Am Soc Agron; fel Soil Sci Soc Am; Soil & Water Conserv Soc Am. *Res:* Compacted layers in soils; potassium and magnesium chemistry of soils; movement and degradation of clay minerals in soils in relation to drainage; clay-organic interactions; soil micromorphology. *Mailing Add:* Dept of Soil & Crop Sci Tex A&M Univ College Station TX 77843-2474

MILGRAM, RICHARD JAMES, b South Bend, Ind, Dec 5, 39; m 64; c 2. MATHEMATICS. *Educ:* Univ Chicago, BSc & MSc, 61; Univ Minn, PhD(math), 64. *Prof Exp:* Instr math, Univ Minn, 63-64; instr, Princeton Univ, 64-66; from asst prof to assoc prof, Univ Ill, Chicago, 66-69; PROF MATH, STANFORD UNIV, 69- *Concurrent Pos:* Assoc mem inst advan study, Univ Ill, 67-68; vis prof, Princeton Univ, 69-70, Northwestern Univ, 84, Univ Calif, San Diego, 85. *Res:* Algebraic and differential topology; theory of H-spaces; construction of classifying spaces; structure and classification of manifolds and Poincare duality spaces; structure of the Steenrod algebras. *Mailing Add:* Dept of Math Stanford Univ Stanford CA 94305

MILGROM, FELIX, b Rohatyn, Poland, Oct 12, 19; nat; m 41; c 2. MEDICAL MICROBIOLOGY, IMMUNOLOGY. *Educ:* Wroclaw Univ, MD, 47. *Hon Degrees:* Dr med, Univ Vienna, Austria, 76, Univ Lund, Sweden, 79, Univ Heidelberg, Ger, 79, Univ Bergen, Norway, 80, Univ Med & Dent NJ, 91. *Prof Exp:* From asst prof to assoc prof microbiol, Sch Med, Wroclaw Univ, 46-53, prof & dir in charge, 54; dir in charge, Inst Immunol & Exp Ther, Polish Acad Sci, 54; prof microbiol & head dept, Silesian Med Sch, 54-57; from res assoc to res assoc prof bact & immunol, 58-62, from assoc prof to prof, 62-81, chmn, Dept Microbiol, 67-85, DISTINGUISHED PROF, DEPT MICROBIOL, SCH MED, STATE UNIV NY BUFFALO, 81- *Concurrent Pos:* Ed-in-chief, Int Archives Allergy & Appl Immunol. *Honors & Awards:* Paul Ehrlich & Ludwig Darmstaedter Prize, Frankfurt, Ger, 87. *Mem:* Am Asn Immunol; Soc Exp Biol & Med; Am Acad Microbiol; hon mem Col Int Allergologicum; Transplantation Soc. *Res:* Serology of syphilis and rheumatoid arthritis; natural antibodies; autoimmune processes; transplantation; tissue antigens; tumor immunology. *Mailing Add:* Dept of Microbiol Sch of Med State Univ of NY Buffalo NY 14214

MILGROM, JACK, b Chicago, Ill, May 21, 27; m 46, 48; c 3. POLYMER CHEMISTRY. *Educ:* Univ Chicago, AB, 50, MS, 51, PhD(org chem), 59. *Prof Exp:* Sr chemist, Ninol Labs, Chicago, Ill, 51-56; proj chemist, Standard Oil Co Ind, 56-60; group leader polymerization catalysis, Gen Tire & Rubber Co, 60-66; mgr polymerization & process res, Foster Grant Co, Inc, 66-68; sr staff mem, Arthur D Little Inc, 68-81; sr consult, Sri Int, 81-83; MANAGING DIR, WALDEN RES, INC, 83- *Concurrent Pos:* Adv, Acad Sci, 74-75. *Mem:* Am Chem Soc; NY Acad Sci; Inst Packaging Prof; Soc Plastics Engrs; Soc Plastics Indust. *Res:* Free-radical and coordination chemistry; organometallics; catalysis; environmental studies; radiation chemistry; polymer technology; impact of technology on society; packaging; solid waste management. *Mailing Add:* Five Thornton Lane Concord MA 07142-4107

MILHAM, ROBERT CARR, b Grand Haven, Mich, June 20, 22; m 79. ENVIRONMENTAL CHEMISTRY. *Educ:* Alma Col, BSc, 44; Univ Wis, PhD(inorg chem), 51. *Prof Exp:* Tester, Petrol Lab, Leonard Refining, Mich, 40-42; calculator, Sugar Lab, Hawaiian Sugar Planters Asn, 45-46; asst radiochem, Univ Wis, 46-52; chemist, 52-64, engr, Reactor Eng Div, 64-70, chemist, Radiol Sci Div, 70-78, CHEMIST, ENVIRON EFFECTS DIV, SAVANNAH RIVER LAB, E I DU PONT DE NEMOURS & CO, INC, 73- *Res:* Radiochemistry; radiological physics; trace element analysis in environmental samples; activation analysis; determination of radionuclides in environmental water and air samples; bioassay; plutonium in urine. *Mailing Add:* 1846 Savoy St Augusta GA 30904

MILIAN, ALWIN S, JR, b Tampa, Fla, May 29, 32. FLUORINE CHEMISTRY. *Educ:* Mass Inst Technol, BS, 54; Univ Calif, PhD(chem), 58. *Prof Exp:* Org chemist, plastics dept, Du Pont Exp Sta, 58-67, sr res chemist, 67-84, sr consultant, Comput Div, cent res dept, 84-85; CONSULT, CRIPPEN CONSULT ASSOCS,INC, 85- *Mem:* Am Chem Soc; Sigma Xi. *Res:* Fluorocarbon chemistry; synthetic organic chemistry; analytical chemistry; industrial hygiene; isolation of natural products; organic synthesis and synthesis of fluorinated organic compounds; process analysis and polymer analysis, synthesis; chromatography - liquid, gas, gas chromatography/mass spectrometry, thin layer, ion; spectroscopy; computer hard and software (APL, FORTRAN, BASIC, PASCAL, DBASE). *Mailing Add:* Box 10264 Wilmington DE 19850

MILIC-EMILI, JOSEPH, b Sesana, Yugoslavia, May 27, 31; m 57; c 4. PHYSIOLOGY. *Educ:* Univ Milan, MD, 55. *Hon Degrees:* DSc, Univ Louvain, Belgium, 87; Univ Kunming, China, 88. *Prof Exp:* Asst prof physiol, Univ Milan, 55-58; asst prof, Univ Liege, 59-60; NIH res fel, Sch Pub Health, Harvard Univ, 60-63; from asst prof to assoc prof, 64-70, PROF PHYSIOL, MCGILL UNIV, 70-, CHMN DEPT, 73-; DIR, MEAKINS-CHRISTIE LABS. *Concurrent Pos:* Med Res Coun Can fel, McGill Univ, 63-; prof, Univ Clin, Royal Victoria Hosp, Montreal, 64- *Honors & Awards:* Medalist, Am Col Chest Physicians, 84; Medalist, Australian Thoracic Soc, 88. *Mem:* Am Physiol Soc; Can Physiol Soc; Can Soc Clin Invest; Can Thoracic Soc; fel Royal Soc Can. *Res:* Physiology of respiration. *Mailing Add:* Meakins-Christie Labs 3626 St Urbain St Montreal PQ H2X 2P2 Can

MILICI, ANTHONY J, b New Haven, Conn, Feb 16, 54; m 83. CELL BIOLOGY VASCULAR SYSTEM. *Educ:* Lycoming Col, Williamsport, Pa, 75; Ind Univ, PhD(anat), 81. *Prof Exp:* ASST PROF MED ANAT, SCH MED, YALE UNIV, 81- *Mem:* Am Soc Cell Biol; Am Asn Anat; Sigma Xi; Microcirculatory Soc. *Res:* Electron microscopy; immunocytochemistry; determining and mapping the molecular factors regulating transvascular exchange of macromolecules. *Mailing Add:* Dept Surg Sect Anat Sch Med Yale Univ 333 Cedar St PO Box 3333 New Haven CT 06510

MILICI, ROBERT CALVIN, b New Haven, Conn, Aug 8, 31; m 58; c 2. REGIONAL GEOLOGY, GEOLOGY OF FUELS. *Educ:* Cornell Univ, AB, 54; Univ Tenn, MS, 55, PhD(geol), 60. *Prof Exp:* Instr geol, Univ Tenn, 55-58; geologist, Tenn Div Geol, 58-62; geologist, Va Div Mineral Resources, 62-63; chief geologist res, Tenn Div Geol, 63-79; VA COMNR MINERAL RESOURCES & STATE GEOLOGIST, 79- *Concurrent Pos:* US Geol Surv grant, 75-77; US Dept Energy contract & Eastern Gas Shales Proj, 76-79; US Bur Mines grant & res assoc, Univ Tenn, Chattanooga, 77-78; contract to study coal mine roof falls, Appalachian Regional Comn, 80-82; contract to study struct, oil & gas potential Va, Outer Continental Shelf & Coastal Plain, US Minerals Mgt Serv, 81-82. *Honors & Awards:* Thomas Jefferson Medal, Outstanding Contrib Nat Sci, Va Mus Natural Hist Found, 91. *Mem:* AAAS; fel Geol Soc Am; Am Asn Petrol Geol; Soc Econ Paleont & Mineral; Am Inst Mining Metall & Petrol Engrs; Asn Am State Geologists. *Res:* Geologic mapping; stratigraphy, structural geology, mineral resources and coal reserve studies Virginia and in Tennessee; evaluation of coal, oil and gas resources in Appalachian Basin; coal mine roof fall studies in southwestern Virginia; development and management of geologic and mineral resources and topographic mapping programs for Commonwealth of Virginia; regional stratigraphic synthesis of Appalachian basin; tectonics of Appalachian region; structure of Virginia's continental shelf and coastal plain. *Mailing Add:* 2091 Whippoorwill Rd Charlottesville VA 22901

MILICIC, DRAGAN, b Zagreb, Yugoslavia, Jan 13, 48. SEMI-SIMPLE LIE GROUPS, D-MODULES. *Educ:* Univ Zagreb, BS, 68, MS, 71, PhD(math), 73. *Prof Exp:* Asst math, Univ Zagreb, 69-73, from asst prof to assoc prof math, 73-80; Assoc prof, 80-83, PROF MATH, DEPT MATH, UNIV UTAH, 83- *Concurrent Pos:* Mem, Inst Advan Study, Princeton, 75-76 & 85-86, Math Sci Res Inst, Berkeley, 87-88. *Mem:* Am Math Soc. *Res:* Applications of the theory of D-modules to the representation theory of semi-simple lie groups. *Mailing Add:* Dept Math Univ Utah Salt Lake City UT 84112

MILIONIS, JERRY PETER, b New York, NY, Mar 6, 26; m 48; c 3. ORGANIC CHEMISTRY. *Educ:* Brooklyn Col, BS, 47; Purdue Univ, PhD(chem), 51. *Prof Exp:* Asst, Purdue Univ, 47-49 & 50-51; res chemist, Am Cyanamid Co, 51-54 & new prod develop dept, 54-57, group leader res, 57-63, dir org pigments res, 63-70, mgr agr res & develop, 70-74, sr res chemist, agr res & develop, 74-77, mgr agr, Int Formulation, 83-90; RETIRED. *Mem:* Am Chem Soc. *Res:* Sulfur and heterocyclic chemistry; polymer degradation; pigments; agricultural formulation. *Mailing Add:* 58 Marcy St Somerset NJ 08873

MILIORA, MARIA TERESA, b Somerville, Mass, June 29, 38; div. ORGANIC CHEMISTRY. *Educ:* Regis Col, BA, 60; Tufts Univ, PhD(chem), 65; Boston Univ, MSW, 85. *Prof Exp:* NSF res asst, Tufts Univ, 60-64; res assoc, Tufts Univ. 65-66; assoc prof, 65-71, chmn dept, 72-84, PROF CHEM, SUFFOLK UNIV, 71- *Concurrent Pos:* Adj fac psychol, Suffolk Univ. *Mem:* Am Chem Soc; Sigma Xi. *Res:* Nonchair conformations of 2,5-dialkyl-1, 4-cyclohexanediols; conformational analysis. *Mailing Add:* Dept Chem Suffolk Univ Boston MA 02114

MILJANICH, GEORGE PAUL, b Watsonville, Calif, Feb 2, 50; m 83; c 3. NEUROPHYSIOLOGY, NEUROCHEMISTRY. *Educ:* Univ Calif, Berkeley, BS, 72, Santa Cruz, PhD(chem), 78. *Prof Exp:* Fel biochem, Univ Calif, San Francisco, 78-82; ASST PROF NEUROBIOL, UNIV SOUTHERN CALIF, 82- *Mem:* Am Soc Cell Biol; Soc Neurosci. *Res:* The biochemistry of synaptic transmission and its regulation by ions, hormones, neurotransmitters, intracellular messengers and drugs. *Mailing Add:* Neurex Corp 3760 Haven Ave Menlo Park CA 94025

MILKEY, ROBERT WILLIAM, b Washington, DC, Jan 21, 44; m 65; c 1. ASTRONOMY. *Educ:* Amherst Col, BA, 65; Ind Univ, Bloomington, MA, 67, PhD(astrophys), 70. *Prof Exp:* Res assoc, Los Alamos Sci Lab, 70-71; asst astronr, Kitt Peak Nat Observ, 71-75, mgr comput serv, 75-79, asst dir admin serv, 79-80; asst dir, Inst Astron, Univ Hawaii, 80-82; staff scientist, Aura Corp Hq, 82-84; ASSOC DIR, PROG MGT, SPACE TELESCOPE SCI INST, 84- *Mem:* Int Astron Union; Am Astron Soc; Royal Astron Soc. *Res:* Solar physics; structure of the solar chromosphere; hydromagnetics of the solar atmosphere; radiative transfer and spectral line formation; research administration. *Mailing Add:* Space Telescope Sci Inst 3700 San Martin Dr Baltimore MD 21218

MILKIE, TERENCE H, polymer chemistry, for more information see previous edition

MILKMAN, ROGER DAWSON, b New York, NY, Oct 15, 30; m 58; c 4. MOLECULAR EVOLUTION, POPULATION GENETICS. *Educ:* Harvard Univ, AB, 51, AM, 54, PhD(biol), 56. *Prof Exp:* Asst marine embryol, Marine Biol Lab, 54-55; Nat Sci Found res fel genetics, lab genetics & physiol, Nat Ctr Sci Res, France, 56-57; instr zool, Univ Mich, 57-59, asst prof, 59-60; assoc prof, Syracuse Univ, 60-67, prof, 67-68; PROF BIOL, UNIV IOWA, 68- *Concurrent Pos:* Instr, Marine Biol Lab, 62-64, 88-, investr, 61, 65-72; USPHS res fel, Biol Labs, Harvard Univ, 66-67; assoc ed, Evolution, 74-76, 79-81, ed, 84-86; prin investr grants, NSF & NIH, 80-; vis prof, Grinnell Col, 90. *Mem:* Fel AAAS; Genetics Soc Am; Soc Study Evolution; Am Soc Zool; Am Soc Naturalists (secy, 80-82); Am Soc Microbiol. *Res:* Molecular evolution of the E coli chromosome and its clonal segments; selection theory; genetic structure of species; electrophoretic analysis; Drosophila; genetic basis of natural variation; polygenes; temperature effects. *Mailing Add:* Dept Biol Univ Iowa Iowa City IA 52242-1368

MILKOVIC, MIRAN, b Mar 29, 28; US citizen; m 67; c 1. ELECTRONICS, INSTRUMENTATION. *Educ:* Univ Ljubljana, MSc, 56; Swiss Fed Inst Technol, PhD(electronics), 65. *Prof Exp:* Develop engr, Grundig, Develop Ctr, 56-57; prog mgr, Landis & Gyr Cent Res Lab, 57-69; consult engr, 69-71, RES ENGR, GEN ELEC CO RES & DEVELOP CTR, 71-, SR ENG. *Mem:* Inst Elec & Electronics Engrs. *Res:* Electron devices and circuits; instrumentation; sensors; data conversion; data acquisition; analog microelectronic circuits design. *Mailing Add:* Gen Elec Co Corp Res & Develop 1 River Rd Schenectady NY 12305

MILL, THEODORE, b Hamilton, Ont, Apr 17, 31; nat US; m 57, 75; c 3. ORGANIC CHEMISTRY. *Educ:* Wayne State Univ, BS, 53; Univ Wash, PhD(chem), 57. *Prof Exp:* Res fel chem, Hickrill Res Found, NY, 56-57; res chemist, Org Chem Dept, E I du Pont de Nemours & Co, 57-60; dir, phys org chem dept, 64-85, SR ORGANIC CHEMIST, STANFORD RES INST, 60-, SR SCIENTIST, 85- *Concurrent Pos:* Vis scientist, Oak Ridge Nat Lab, 87. *Mem:* Am Chem Soc; AAAS; Am Geophys Union; Soc Environ Toxicol & Chem. *Res:* Physical organic chemistry; photochemistry; oxidation and free radical chemistry; environmental chemistry. *Mailing Add:* Chem Lab Stanford Res Inst Menlo Park CA 94025

MILLÁN, JOSÉ LUIS, b Buenos Aires, Arg, July 10, 52. EXPERIMENTAL BIOLOGY. *Educ:* Nat Univ Buenos Aires, Arg, Licenciado Analisis Clinicos, 75, Biochemist, 76; Univ Umeå, Sweden, PhD, 83. *Prof Exp:* Head, Clin Enzymol Sect, Inst Analisis Clinicos, Mar del plata, Arg, 76-77; trainee clin enzymol, La Jolla Cancer Res Found, Calif, 77-78, res assoc, 78-81; guest researcher, Dept Physiol Chem, Univ Umeå, 81-83 & State Serum Inst, Denmark, 84; res assoc, 84-86; asst staff scientist, 86-89, STAFF SCIENTIST, CANCER RES CTR, LA JOLLA CANCER RES FOUND, CALIF, 89- *Concurrent Pos:* Adj assoc prof, Univ Umeå, Sweden, 90- *Mem:* Arg Biochem Asn; Int Soc Clin Enzymol; Int Soc Oncodevelop Biol; Am Soc Biol Chemists; Sigma Xi; Protein Soc. *Mailing Add:* La Jolla Cancer Res Found 10901 N Torrey Pines Rd La Jolla CA 92037

MILLAR, C KAY, b Syracuse, NY, Oct 28, 34. INTERNAL MEDICINE, CARDIOLOGY. *Educ:* Syracuse Univ, AB, 56; State Univ NY Upstate Med Ctr, MD, 60. *Prof Exp:* Resident intern med, 61-64, asst prof intern med, State Univ Upstate Med Ctr, 67-68; NIH trainee cardiol, 64-67; asst prof, 68-73, ASSOC PROF INTERNAL MED, MED CTR, UNIV UTAH, 73- *Mem:* Am Heart Asn. *Res:* Electrocardiography; arrhythmias; body surface potentials of cardiac origin. *Mailing Add:* Cardiol Div Bldg 100 Univ Utah Med Ctr Salt Lake City UT 84112

MILLAR, DAVID B, m. NEUROPSYCHIATRICS. *Educ:* City Col New York, BS, 54; Duke Univ, PhD(biochem), 61. *Prof Exp:* Med chemist, US Army, 54-56; res fel, Dept Biochem, Med Ctr, Univ Ky, 59-61; postdoctoral res fel, Nat Cancer Inst, Naval Med Res Inst, 62-63; postdoctoral res assoc, Nat Acad Sci, 63-65; sr scientist, Naval Med Res Inst, 65-67, head, Sect Biophys Chem, 67-70, chief, Lab Phys Biochem, 70-75, dep dir, Environ Sci Dept & head, Biochem Div, 75-79, dep dir, Environ Stress Dept, 79-81, head, Biochem Div, Environ Stress Dept, 79-83, Biochem & Molecular Biol Div, Immunobiol & Transplantation Dept, 83-87, staff mem, Off Dir, 87-88; GUEST SCIENTIST, NEUROPSYCHIAT BR, NEUROSCI CTR ST ELIZABETH'S, NIMH, WASHINGTON, DC, 88- *Concurrent Pos:* Info corresp, NIH, 67-69; conf chmn & head organizing comt, Int Conf Isozymes, St Croix, VI, 76; adj res prof, Dept Biochem, Uniformed Serv Univ Health Sci, 78-83; neurochem consult, Int Symp Behav Med, Greece, 82; consult, Eng Technol, Inc, Md. *Mem:* Am Soc Biochem & Molecular Biol; Am Chem Soc; Environ Mutagen Soc; Excited State Spectros Soc; Undersea Med Soc; fel NY Acad Sci. *Res:* Neuroimmunomodulation and the response of the immune system to stress; fluorescence and phosphorescence of intact cells, proteins and nucleic acids; function and organization of neurogenic membrane constituents and their response to environmental stress; fluorescent probe studies of membranes and cell ion flux; protein structure and function; protein interaction, self-association; protein-ligand interaction; nucleic acid structure and function. *Mailing Add:* Dept Neuropsychiat NIMH Neurosci Ctr St Elizabeth's Washington DC 20032

MILLAR, GORDON HALSTEAD, b Newark, NJ, Nov 28, 23; m 57; c 5. THERMODYNAMICS, SPECTROSCOPY. *Educ:* Univ Detroit, BME, 49; Univ Wis, PhD(mech eng), 52. *Hon Degrees:* DSc, Univ Detroit, 77, Western Mich Univ, 86; LHD, W Coast Univ, 84. *Prof Exp:* Supvr, Mech Eng Dept, Ford Motor Co, 53-57; eng mgr, Meriam Instrument Co, 57-59; dir new prod, McCulloch Corp, 59-63; dir res, Deere & Co, 63-69, asst gen mgr, John Deere Waterloo Tractor Works, 69-71, spec assignment, 71-72, vpres eng, Deere & Co, 72; ENG CONSULT, 84- *Concurrent Pos:* Pres, Accreditation Bd Eng & Technol, 84 & 85. *Mem:* Nat Acad Eng; NY Acad Sci; fel Soc Automotive Engrs (pres, 84); hon fel Am Soc Mech Engrs; Am Soc Agr Engrs; fel Am Soc Eng Educ. *Res:* Combustion thermodynamics; high speed photoelectric electro-optical circuits; engineering; fuels and lubricants. *Mailing Add:* 1840 Wiley Post Trail Daytona Beach FL 32124

MILLAR, JACK WILLIAM, b Ogden, Utah, July 11, 22; m 46; c 4. MEDICINE. *Educ:* Stanford Univ, AB, 45; George Washington Univ, MD, 47; Harvard Univ, MPH, 51, MS, 52; Am Bd Prev Med, dipl, 56. *Prof Exp:* Intern, Naval Hosp, Bethesda, Md, 47-48; med officer in chg, Tinian Leprosarium, Tinian Island, 48-50; med dir & epidemiologist, Am Leprosy Found, Far East, 50-53; instr epidemiol, Naval Med Sch, Md, 54-55, cmndg officer, Naval Med Res Unit 1, Berkeley, Calif, 55-60, dir prev med div, Bur Med & Surg, Navy Dept, 60-67; VIVIAN GILL PROF EPIDEMIOL & ENVIRON HEALTH, SCH MED & HEALTH SCI, GEORGE WASHINGTON UNIV, 67- *Concurrent Pos:* Pres, Gorgas Mem Inst Trop Med; consult epidemiol & trop med, Vet Admin Hosp, Wilmington, Del; sabbatical leave, dir, Pro Tem Gorgas Mem Lab, Panama, 81-82; bd dir, Gorgas Mem Inst; Nat Coun Int Health. *Mem:* Am Soc Trop Med & Hyg; Am Pub Health Asn; AMA; fel Am Col Prev Med; fel Am Col Epidemiol. *Res:* Epidemiology; infectious diseases; leprosy; tropical diseases. *Mailing Add:* Dept Health Sci George Washington Univ 2300 I St Rm 714 Washington DC 20037

MILLAR, JOCELYN GRENVILLE, b London, Eng, May 30, 54; Can citizen. SEMIOCHEMISTRY, PHEROMONE CHEMISTRY. *Educ:* Simon Fraser Univ, BSc, 79, PhD(chem), 84. *Prof Exp:* Res fel, Col Environ Sci & Forestry, State Univ NY, Syracuse, 83-84; res assoc, Nat Res Can Coun, Saskatoon, Sask, 84-86; chief analyst, Prov Toxicol Ctr, Port Coquittam, BC, 87-88; ASST PROF, DEPT ENTOM, UNIV CALIF, RIVERSIDE, 88- *Mem:* Am Chem Soc; Int Soc Chem Ecol; Entom Soc Am. *Res:* Isolation, identification, and synthesis of insect pheromones; application of pheromone technology to integrated pest management; chemical aspects of insect-host plant interactions. *Mailing Add:* Dept Entom Univ Calif Riverside CA 92521

MILLAR, JOHN DAVID, b Dallas Co, Tex, May 24, 21; m 47; c 1. ANALYTICAL CHEMISTRY. *Educ:* Trinity Univ, BA, 47. *Prof Exp:* Chemist, Found Appl Res, 47-49; chemist, Southwest Res Inst, 49-52; org chemist, Celanese Corp Am, 52-53; assoc chemist, Southwest Res Inst, 53-62, sr res chemist, 62-80, staff scientist, 80-83; RETIRED. *Concurrent Pos:* Lab instr, Evening Div, San Antonio Col, 57-59. *Mem:* Am Chem Soc; Sigma Xi. *Res:* Process development; gas chromatography and trace analysis; technical literature and reports; determination of chlorinated pesticides and polychlorinated biphenyls in water. *Mailing Add:* 507 Edgebrook Lane San Antonio TX 78213

MILLAR, JOHN DONALD, b Newport News, Va, Feb 27, 34; m 57; c 3. PREVENTIVE MEDICINE, EPIDEMIOLOGY. *Educ:* Univ Richmond, BS, 56; Med Col Va, MD, 59; London Sch Hyg & Trop Med, dipl trop pub health, 66. *Prof Exp:* Intern, Univ Utah Hosps, 59-60, asst resident med, 60-61; asst chief EIS, Epidemiol Br, 61-62, chief, 62-63, chief smallpox unit, 63-65, chief, Smallpox Eradication Prog, 66-70, dir, Bur State Serv, 70-78, asst dir, Ctr Dis Control, 78-80, dir, Ctr Environ Health, 80-81, DIR, NAT INST OCCUPATIONAL SAFETY & HEALTH, USPHS, 81- *Concurrent Pos:* Mem sci group smallpox eradication, WHO, 67-, mem comn eval smallpox eradication in SAm, 73, consult expanded immunization prog, 74; assoc mem comn immunization, Armed forces Epidemiol Bd, 68-71; mem rural health coord comt, USPHS, 75-81, mem comt maternal & child health, 75-80; chmn, Prog & Policies Adv Comt, Ctr Dis Control, 78; clin assoc, Dept Community Med, Sch Med, Emory Univ, adj prof, Sch Pub Health, 90-; mem, Int Comn to Eval Smallpox Eradication in Somalia, World Health Orgn, 78, adv group on occup health, 88; chmn exec comt, Nat Toxicol Prog, 81; chmn subcomt, Environ Health Risk Assessment, 86-91; chmn, Flag Officers Billet Bd, USPHS Comn Corps, 88- *Honors & Awards:* Surgeon General's Commendation Medal, USPHS, 65, Distinguished Serv Medal, 83 & 88; Joseph Mountain Lectr, Ctr Dis Control, 86, Medal of Excellence, 78; Gorgas Medal, 87; Lucas Lectr, Fac Occup Med, Royal Col Physicians, London, 87; Surgeon General's Exemplary Serv Medal, 89. *Mem:* Am Acad Occup Med; Am Col Epidemiol; Royal Soc Trop Med & Hyg; Asn Pub Health Physicians; Asn Mil Surgeons US; Int Epidemiol Asn; Am Epidemiol Soc. *Res:* Epidemiology of infectious and occupational illness; mass immunization; disease eradication and control methods; occupational disease and injury surveillance; quantitative risk assessment. *Mailing Add:* 3243 Wake Robin Trail Chamblee GA 30341

MILLAR, JOHN ROBERT, b Edinburgh, Scotland, June 3, 27; m 55; c 3. ORGANIC POLYMER CHEMISTRY, ION EXCHANGE. *Educ:* Univ Cambridge, BA, 48, ARIC, 49, MA, 52; FRIC, 59. *Prof Exp:* Res chemist, Howards of Ilford Ltd, 48-50; sr res chemist, Permutit Co Ltd, UK, 50-73; group leader res & develop, Functional Polymers Div, Diamond Shamrock Corp, 73-80; sr res scientist, Duolite Int, 80-82; RETIRED. *Concurrent Pos:* Vchmn, Gordon Res Conf Ion Exchange, 75-77, Chmn, 77-79; consult, 82- *Mem:* Soc Chem Indust; Royal Soc Chem; Royal Inst Chem; Am Chem Soc. *Res:* Preparation, characterization and properties of functional polymers as a function of their chemical and physical structures, including the effect of macroporosity on ion-exchangers and sorbents. *Mailing Add:* Plas-Y-Llyn Three Bell Sq Blagdon Bristol BS186UB England

MILLAR, ROBERT FYFE, b Guelph, Ont, Jan 23, 28; m 76. APPLIED MATHEMATICS. *Educ:* Univ Toronto, BA, 51, MA, 52; Cambridge Univ, PhD, 57. *Prof Exp:* Jr res officer, Microwave Sect, Radio & Elec Div, Nat Res Coun Can, 52-53, asst res officer, 57-60; visitor, Courant Inst Math Sci, NY Univ, 60-61; assoc prof math, Royal Mil Col, Ont, 61-63; sci asst electromagnetic theory lab, Tech Univ Denmark, 63-66; sr res officer, Antenna Eng Sect, Radio & Elec Eng Div, Nat Res Coun Can, 66-73; prof, 73-90, EMER PROF MATH, UNIV ALTA, 90- *Concurrent Pos:* Adj prof math, Univ Victoria, BC. *Mem:* Am Math Soc; Soc Indust & Appl Math; Can Math Asn; Can Appl Math Soc. *Res:* Diffraction and scattering of waves; scattering by periodic structures; complex variable methods in partial differential equations; theory of Hele-Shaw flow. *Mailing Add:* 8654 Sansum Park Dr Sidney BC V8L 5B5 Can

MILLAR, WAYNE NORVAL, b Beverly, Mass, Oct 10, 42; m 65; c 2. MICROBIOLOGY. *Educ:* Bucknell Univ, BS, 64; Pa State Univ, MS, 66, PhD(microbiol), 69. *Prof Exp:* From asst prof to assoc prof bact, WVa Univ, 69-73; sr scientist, Eli Lilly & Co, 73-77, head microbiol & fermentation prod res, 77-81, mgr antibiotic fermentation technol, 81-84, DIR FERMENTATION PROD RES, ELI LILLY & CO, 84- *Mem:* Am Soc Microbiol; Sigma Xi. *Res:* Soil and water microbiology; isolation of antibiotic producing microorganisms; microbial ecology; fermentation technology; natural products chemistry. *Mailing Add:* Lilly Corp Ctr Dept MC930 Indianapolis IN 46285

MILLARD, FREDERICK WILLIAM, b Johnson City, NY, Feb 10, 31; m 53; c 2. PHOTOGRAPHIC CHEMISTRY. *Educ:* Pa State Univ, BS, 53; Mich State Univ, PhD(org chem), 58. *Prof Exp:* Sr chemist, Tex US Chem Co, 57-60; res specialist, Gen Aniline & Film Corp, 60-63, tech assoc silver halide photochemistry, 63-74, mgr Graphic Films Res & Develop, 74-80; MGR, RES & DEVELOP, ANITEC IMAGE CORP, BINGHAMTON, NY, 80- *Concurrent Pos:* Pres, Binghamton Chap, Soc Photog Sci & Eng, 85-86. *Mem:* Am Chem Soc; Soc Photog Sci & Eng; Tech Asn Graphic Arts. *Res:* Photopolymerization theory and adaptation to a photographic system; elucidation of free radical reactions in solution; silver halide technology, particularly in graphic arts; non silver imaging systems; photographic processing solution formulation. *Mailing Add:* PO Box 151 Montrose PA 18801

MILLARD, GEORGE BUENTE, b Kansas City, Kans, Feb 13, 17; m 43; c 4. INORGANIC CHEMISTRY. *Educ:* Wash State Univ, BS, 42, MS, 55. *Prof Exp:* Tech sales agr chem, Sherwin-Williams Co, 46-47, asst to vpres & gen mgr, Calif, 47-49; vpres, Mid-State Chem Co, 49-54; prof, Yakima Valley Col, 51-82; RETIRED. *Concurrent Pos:* Mem bd, Northwest Col & Univ Asn Sci, 71-82. *Mem:* Sigma Xi. *Res:* Investigation of some complex ions of zinc by polarographic methods. *Mailing Add:* 201 N 27th Ave Yakima WA 98902

MILLARD, HERBERT DEAN, b Grayling, Mich, May 22, 24; m 48; c 4. DENTISTRY. *Educ:* Univ Mich, DDS, 52, MS, 56; Am Bd Oral Med, dipl. *Prof Exp:* From instr to assoc prof, 52-64, PROF ORAL DIAG, SCH DENT, UNIV MICH, ANN ARBOR, 64-, CHMN DEPT, 58- *Concurrent Pos:* T C White vis prof, Royal Col Physicians & Surgeons, Glasgow, 89. *Mem:* Am Acad Oral Med; Am Dent Asn; Am Asn Dent Sch; Orgn Teachers Oral Diag; Brit Soc Oral Med. *Res:* Relationships of oral disease to systemic disease; procedures in oral diagnosis. *Mailing Add:* 1205 Glen Leven Rd Ann Arbor MI 48103

MILLARD, RICHARD JAMES, b Peabody, Mass, Oct 3, 18; m 49; c 1. ELECTROCHEMISTRY. *Educ:* Boston Col, BS, 49, MS, 50. *Prof Exp:* Res engr, Sprague Elec Co, 50-65, mgr eng dept, 65-83; RETIRED. *Mem:* Am Chem Soc; Electrochem Soc; Inst Elec & Electronic Engrs. *Res:* Thin films on metals; dielectric breakdown; electrolytic oxidation; identification of phenols; electrolytic capacitor development and engineering. *Mailing Add:* Box 546 Williamstown MA 01267

MILLARD, RONALD WESLEY, b Bridgeport, Conn, Sept 10, 41. CARDIOVASCULAR PHARMACOLOGY. *Educ:* Tufts Univ, BS, 63; Boston Univ, PhD(med sci), 69. *Prof Exp:* Asst prof res, Brown Univ, 75-78; assoc prof, 78-87, PROF, UNIV CINCINNATI, 87- *Concurrent Pos:* Fulbright Hays sr fel, State Dept, 72. *Mem:* Am Physiol Soc; AAAS; Am Heart Asn; Am Pharmacol Soc; Soc Exp Biol & Med. *Res:* Cardiovascular physiology and pharmacology; comparative vertebrate cardiovascular physiology; research administration of cardiovascular physiology, pharmacology and surgery; oxygen deprivation; blood substitutes. *Mailing Add:* Pharmacol & Cell Biophysics Dept Univ Cincinnati Cincinnati OH 45267-0575

MILLARD, WILLIAM JAMES, b Yakima, Wash, Aug 6, 49; m 71; c 2. NEUROENDOCRINOLOGY, ENDOCRINOLOGY. *Educ:* Mount Union Col Alliance Ohio, BS, 71; Univ Toledo, Ohio, PhD(biol), 79. *Prof Exp:* Teaching fel neuroendocrinol, Mass Gen Hosp-Harvard Med Sch, 79-81, res assoc neurol, 81-82, from instr to asst prof physiol & neurol, 82-86; ASSOC PROF PHARM, COL PHARM, UNIV FLA, 86- *Mem:* Endocrine Soc; Soc Neurosci; Int Soc Neuroendocrinol. *Res:* Interaction of the brain in regulating pituitary function. *Mailing Add:* Psychol Bldg Mt Holyoke Col South Hadley MA 01075

MILLEMANN, RAYMOND EAGAN, b New York, NY, Jan 18, 28; m 55; c 3. ZOOLOGY. *Educ:* Dartmouth Col, AB, 48; Univ Calif, Los Angeles, MA, 51, PhD(zool), 54. *Prof Exp:* Teaching asst zool, Univ Calif, Los Angeles, 49-53; from instr to asst prof bact & parasitol, Sch Med, Univ Rochester, 55-63; from asst prof to prof fisheries, Ore State Univ, 63-75; RES BIOLOGIST, OAK RIDGE NAT LAB, 77- *Mem:* Am Soc Parasitol; Soc Protozool; Am Soc Trop Med & Hyg; Am Micros Soc; Wildlife Dis Asn. *Res:* Taxonomy and life cycles of helminths; fish diseases and parasites; marine biology; aquatic toxicology. *Mailing Add:* Environ Sci Div PO Box 2008 Oak Ridge TN 37831

MILLEN, JANE, b Huntsville, Ala, Sept 5, 53. PHARMACEUTICAL CHEMISTRY, BIOCHEMISTRY. *Educ:* Auburn Univ, BS, 76; Univ Miss, PhD(med chem), 83. *Prof Exp:* Postdoctoral fel, Nat Eye Inst, NIH, 83-84; sr scientist, Wyeth-Ayerst Res, 84-88; ASST PROF MED CHEM, UNIV NC, 88- *Mem:* Am Chem Soc; AAAS; Am Diabetes Asn; Am Pharmaceut Asn; Am Asn Cols Pharm. *Res:* Design of novel agents for the treatment of diabetes mellitus and its secondary complications; design of agents modulating collagen metabolism. *Mailing Add:* Sch Pharm Univ NC CB 7360 Chapel Hill NC 27599-7360

MILLENER, DAVID JOHN, b Auckland, NZ, May 2, 44. THEORETICAL NUCLEAR PHYSICS. *Educ:* Univ Auckland, NZ, BSc, 66, MSc, 68; Oxford Univ, Eng, DPhil(nuclear physics), 72. *Prof Exp:* Int Bus Mach res fel nuclear physics, Oxford Univ, Eng, 72-74, res fel, 74-75; from asst physicist to assoc physicist, 76-80, PHYSICIST NUCLEAR PHYSICS, BROOKHAVEN NAT LAB, 80- *Mem:* Am Phys Soc. *Res:* Calculations of the structure and properties of light nuclei and hypernuclei; inelastic scattering reactions. *Mailing Add:* Physics Dept Brookhaven Nat Lab Upton NY 11973

MILLER (GILBERT), CAROL ANN, b Greenville, Ohio, June 27, 43; m 66; c 1. MICROBIOLOGY, IMMUNOLOGY. *Educ:* Defiance Col, BA, 65; Ariz State Univ, MS, 69; Ore State Univ, PhD(microbiol), 70. *Prof Exp:* Clin immunologist, Wilson Mem Hosp, Johnston City, NY, 70-73; res scientist, 73-78, sr res scientist micro-immunol, Ames Res Lab, 78-79, MGR, PROJ MGT, AMES DIV MILES LABS, INC, 79- *Mem:* Am Soc Microbiol. *Res:* Development of immunological and chemical test systems which will be used to detect pathognomonic levels of hormones and microbial products in human biological fluids. *Mailing Add:* Sch Nursing Rm 318 Ind Univ 610 Barnhill Dr Indianapolis IN 46202

MILLER, A EUGENE, b Philadelphia, Pa, Apr 27, 29; m 57; c 2. MATHEMATICS, RESEARCH ADMINISTRATION. *Educ:* Univ Pa, BA, 51, MA, 53. *Prof Exp:* Asst res engr, Burroughs Corp, 51-55, develop engr, 55-58; mem tech staff, Auerbach Corp, Va, 58-62, prog mgr info systs eng,

62-70, dir prog develop, Auerbach Assocs, 70-72; vpres opers, Ins Inst Hwy Safety, 72-80, vpres tech support, 80-85, spec projs, 85-; RETIRED. *Concurrent Pos:* Mayor, Town of Somerset, 82-; trustee, Chesapeake Bay Trust. *Mem:* Sr mem Inst Elec & Electronics Engrs; Sigma Xi; Asn Comput Mach. *Res:* Reducing losses, human and economic, resulting from or associated with the highway transportation system. *Mailing Add:* 6433 Weems Ave Fairhaven MD 20754

MILLER, A(NNA) KATHRINE, b East Orange, NJ, June 8, 13. BACTERIOLOGY. *Educ:* Moravian Col Women, BA, 34; Columbia Univ, MS, 36; Cornell Univ, PhD(bact), 42; Am Bd Med Microbiol, dipl. *Prof Exp:* Asst dir phys educ, Moravian Col Women, 34-35, asst prof biol, 36-41; substitute, Wells Col, 42-43; res fel, Merck Inst, Merck & Co, Inc, 43-74, sr res fel, 74-76, sr investr, 76-78, consult, 79; RETIRED. *Mem:* Fel AAAS; Am Soc Microbiol; fel Am Acad Microbiol; fel NY Acad Sci; Sigma Xi. *Res:* Experimental antibacterial chemotherapy, particularly the development of mouse tests for evaluating new antibiotic agents. *Mailing Add:* 104 B Duncan Hill Westfield NJ 07090-1679

MILLER, ADOLPHUS JAMES, b Patterson, Ark, Mar 8, 12; m 37; c 1. AGRICULTURAL ENGINEERING. *Educ:* Hampton Inst, BS, 37; Mich State Univ, MS, 46, EdD, 56. *Prof Exp:* Instr, jr high sch, Del, 37-40; rural eng, Prairie View State Col, 40-42; from instr to assoc prof agr eng & mech, 42-67, prof, 67-77, EMER PROF AGR MECH EDUC, VA STATE UNIV, 77- *Mem:* Am Voc Educ Res Asn; Am Soc Agr Engrs. *Res:* Agricultural mechanics; fire hazards and fire survival education in Virginia. *Mailing Add:* 2000 Oakland Ave Colonial Heights VA 23834

MILLER, AKELEY, b Phoenix, Ariz, Mar 12, 26; m 49; c 2. MATHEMATICAL PHYSICS, GENERAL PHYSICS. *Educ:* Univ SDak, BA, 50, MA, 52; Univ Mo, Columbia, PhD, 60. *Prof Exp:* Instr physics, Univ SDak, 52-55; asst prof, 60-65, ASSOC PROF PHYSICS, UTAH STATE UNIV, 65- *Concurrent Pos:* Sabbatical leave, Univ NC, Chapel Hill, 78-79. *Mem:* AAAS; Sigma Xi. *Res:* Physics of light sources. *Mailing Add:* Utah State Univ 1423 N 1720 E Logan UT 84321

MILLER, ALAN CHARLES, b Trona, Calif, June 9, 45. MARINE ECOLOGY. *Educ:* Stanford Univ, BA, 67; Univ Ore, MA, 68, PhD(biol), 74. *Prof Exp:* Marine ecologist, Southern Calif Coastal Water Res Proj, 74; ASST PROF ECOL, CALIF STATE UNIV, LONG BEACH, 74- *Mem:* AAAS; Ecol Soc Am; Am Soc Naturalists. *Res:* Species diversity and trophic structure of marine communities. *Mailing Add:* Dept Biol Calif State Univ 1250 Bellflower Blvd Long Beach CA 90840

MILLER, ALAN DALE, b Everett, Wash, May 29, 31; m 56; c 4. CERAMIC ENGINEERING. *Educ:* Univ Wash, BS, 57, PhD(ceramic eng), 67. *Prof Exp:* Proj metallurgist, Pratt & Whitney Aircraft Div, United Aircraft Corp, 58-64; asst prof ceramic eng, 67-75, ASSOC PROF CERAMIC ENG, UNIV WASH, 75- *Mem:* Am Ceramic Soc; Nat Inst Ceramic Engrs; Am Soc Eng Educ. *Res:* High temperature thermochemistry; nature of refractory compounds; processing of electronic ceramics; analysis of brittle fracture processes. *Mailing Add:* Div Ceramic Eng Univ Wash Seattle WA 98195

MILLER, ALAN R(OBERT), b Alameda, Calif, Feb 4, 32; m 57; c 2. MATERIALS SCIENCE, METALLURGY AND PHYSICAL METALLURGICAL. *Educ:* Univ Calif, Berkeley, BS, 53 & 58, MS, 61, PhD(eng), 64. *Prof Exp:* Res asst, Gen Atomic Div, Gen Dynamics Corp, 56-58; phys chemist, Aerojet-Gen Nucleonics Div, Gen Tire & Rubber Co, 58-67; assoc prof, 67-80, PROF METALL & MAT ENG, NMEX INST MINING & TECHNOL, 80- *Mem:* AAAS; Am Chem Soc; Am Soc Metals Int; Sigma Xi. *Res:* Thermodynamics and high temperature chemistry; vapor-pressure determinations; synthesis and compaction of metals and ceramics by explosive shock; published 22 books on computing and engineering. *Mailing Add:* Dept of Metall & Mat Eng Campus Sta Jones Hall Socorro NM 87801

MILLER, ALBERT, b Brooklyn, NY, Mar 18, 11; m 50; c 1. MEDICAL ENTOMOLOGY, MEDICAL & HEALTH SCIENCES. *Educ:* Cornell Univ, BS, 33, MS, 34, PhD(insect embryol), 38. *Prof Exp:* Asst entom & parasitol, Cornell Univ, 34-35, instr, 35-38; instr entom & plant path, Univ Ark, 38-40; guest genetics, Carnegie Inst Washington, 40-41; from instr to asst prof, Sch Med, Tulane Univ, 41-48, assoc prof med entom, 48-76; scientist, Int Ctr Med Res, Cali, Colombia, 77-78; RETIRED. *Mem:* AAAS; emer mem Entom Soc Am; emer mem Am Soc Trop Med & Hyg; Am Mosquito Control Asn. *Res:* Insect morphology, embryology and taxonomy; biology of medically important arthropods; coprophilic fauna. *Mailing Add:* 1600 Green Acres Rd Metairie LA 70003

MILLER, ALBERT EUGENE, b Albion, Nebr, June 22, 38. PHYSICAL METALLURGY. *Educ:* Colo Sch Mines, Engr, 60; Iowa State Univ, PhD(metall), 64. *Prof Exp:* Atomic Energy Comn assoc metall, Ames Lab, 64-66; assoc prof, Univ Alta, 66-67; assoc prof metall, 67-81, PROF METALL ENG & MAT SCI, UNIV NOTRE DAME, 81- *Mem:* Am Soc Metals; Am Chem Soc. *Res:* Alloy theory; magnetism in solids; metallurgical aspects of superconductivity. *Mailing Add:* Dept Metallurgical Univ Notre Dame Notre Dame IN 46556

MILLER, ALBERT THOMAS, b New York, NY, May 7, 39; m 61; c 4. COLLAGEN, PROTEIN CHEMISTRY. *Educ:* Univ Ga, BSA, 62; Univ Mass, MS, 63; Rutgers Univ, PhD, 81. *Prof Exp:* Res technician fats & oils, Lever Brothers Co, 61-62; res asst microbiol, Univ Mass Exp Sta, 62-63; food chemist proteins, Colgate Palmolive Co, 63-66; assoc scientist proteins, Devro Inc/Johnson & Johnson, 66-73, sr scientist prod develop, 75-76, mgr prod develop, 76-77, assoc dir res, Devro Div, 77-81, dir res, 81-85, vpres tech, 85-89, VPRES TECHNOL/SCI AFFAIRS, DEVRO INC/JOHNSON & JOHNSON, 89- *Honors & Awards:* Presidential Achievement Award, 78. *Mem:* Inst Food Technol; Sigma Xi; Am Leather Chemists Asn; AAAS; Am Meat Sci Asn. *Res:* Chemistry of proteins; collagen; meat science; natural polymers; edible packaging materials; nutrition; food packaging; carbohydrates. *Mailing Add:* Devro Inc/Johnson & Johnson PO Box 858 Loeser Ave Somerville NJ 08876

MILLER, ALEX, b Paterson, NJ, Aug 19, 25; m 51; c 2. BIOSTATISTICS, STATISTICS. *Educ:* Purdue Univ, BS, 48; Columbia Univ, MS, 66; Univ Calif, Berkeley, MPH, 70. *Prof Exp:* Sr biostatician, New York Med Col, Flower-Fifth Ave Hosps, 66-69; statistician, Warner-Lambert Co Res Inst, 71-74; RES STATISTICIAN, NABISCO, INC, RES & DEVELOP CTR, 75- *Mem:* Am Statist Asn; Am Pub Health Asn. *Res:* Experimental design; data analysis. *Mailing Add:* 25 Jerome Ave Glen Rock NJ 07452

MILLER, ALEXANDER, b New York, NY, May 16, 28; m 58; c 2. CELLULAR & MOLECULAR IMMUNOLOGY. *Educ:* Univ Wis, BS, 52; Columbia Univ, PhD(biochem), 57. *Prof Exp:* Res fel, div biol, Calif Inst Technol, 57-60, spec res fel, 60-62; asst res zoologist, 62-68, asst res bact, 68-72, lectr bact, 70, assoc res bacteriologist, 72-78, RES MICROBIOLOGIST, UNIV CALIF, LOS ANGELES, 79- *Mem:* Am Asn Immunologists. *Res:* Study of mechanisms involved in immunoregulation at the cellular and molecular level, using carefully characterized proteins, protein derivatives and peptides. *Mailing Add:* Dept Microbiol Univ Calif Life Sci Bldg Los Angeles CA 90024-1489

MILLER, ALFRED CHARLES, b Amsterdam, NY, Sept 10, 47; m 67; c 2. SURFACE PHYSICS, SURFACE CHEMISTRY. *Educ:* Clarkson Col Technol, BS, 68, MS, 70, PhD(physics), 77. *Prof Exp:* Asst prof physics, Hartwick Col, 76-77; adj asst prof, Clarkson Col Technol, 77-78; SR SCIENTIST, ALCOA TECH CTR, 78-; DIR, SURFACE ANAL LAB, ZETTLEMOYER CTR SURFACE SCI, LEHIGH UNIV, 89- *Mem:* Am Vacuum Soc; Sigma Xi; Am Soc Testing & Mat. *Res:* Surface science; reactions at surfaces; properties of metal and metal oxide thin films; ion-surface interaction; ion scattering spectroscopy; sputtering phenomena; surface properties of light metal alloys. *Mailing Add:* 462 W Locust Lane Nazareth PA 18064

MILLER, ALISTAIR IAN, b Edinburgh, Scotland, July 6, 40; m 64; c 4. CHEMICAL ENGINEERING. *Educ:* Univ Glasgow, BSc, 62; Univ London, DIC & PhD(chem eng), 66. *Prof Exp:* Chem engr, 66-80, MGR CHEM ENG BR, CHALK RIVER LABS, ATOMIC ENERGY CAN LTD, 80- *Mem:* Fel Chem Inst Can; Can Soc Chem Eng. *Res:* Research and development of new and established heavy water production processes and other separations of protium, deuterium and tritium. *Mailing Add:* Chem Eng Br Chalk River Labs Chalk River ON K0J 1J0 Can

MILLER, ALLAN STEPHEN, b Arlington Heights, Ill, Feb 21, 28; m 50; c 3. SOLID STATE PHYSICS. *Educ:* Univ Notre Dame, BS, 49, MS, 50; Univ Ill, PhD(physics), 57. *Prof Exp:* Assoc physicist, Res Lab, Int Bus Mach Corp, 57-59, staff physicist, 59-61, develop physicist, IBM Components Div, 61-64; res assoc, Nat Res Corp, 64-66; asst dir res, Norton Res Corp, 66-70; consult solid state physics, Light Emitting Diodes, 71-72; mgr process develop, multigraphics develop ctr, Addressograph Multigraph Corp, 72-77; explor planning specialist, 77-78, mgr physics sect, Babcock & Wilcox, Alliance Res Ctr, 78-82; PHYSICIST, CALIF COMPUT PRODS, 83- *Mem:* Am Phys Soc. *Res:* Photoconductivity in semiconductors and insulators; light emitting diodes; transistors; solid state oxygen sensors; electrographics. *Mailing Add:* 2048 Smokewood Ave Fullerton CA 92631

MILLER, ALLEN H, b Brooklyn, NY, June 23, 32; m 75. PHYSICS. *Educ:* Brooklyn Col, AB, 53; Rutgers Univ, MS, 55, PhD(physics), 60. *Prof Exp:* Physicist, Electronics Corp Am, 55-56; res assoc physics, Univ Ill, 60-62; asst prof, 62-65, ASSOC PROF PHYSICS, SYRACUSE UNIV, 65- *Mem:* Am Phys Soc; Sigma Xi. *Res:* Solid state theory; many-particle problem; thermal physics. *Mailing Add:* Dept Physics Syracuse Univ Syracuse NY 13244-1130

MILLER, ANTHONY BERNARD, b Woodford, Eng, Apr 17, 31; m 52; c 5. EPIDEMIOLOGY. *Educ:* Univ Cambridge, BA, 52, MB & BCh, 55; MRCP, 64, MFCM(UK), 72, FRCP(C), 72, FFCM(UK), 77, FACE, 85, FRCP, 87. *Prof Exp:* House officer, Oldchurch Hosp, Romford, Eng, 55-57; med officer, Royal Air Force, Netheravon, Eng, 57-59; med registr, Luton & Dunstable Hosp, Eng, 59-61; mem sci staff, Med Res Coun Tuberc & Chest Dis Unit, London, 61-71; assoc prof prev med & statist, 72-76, dir epidemiol unit, Nat Cancer Inst Can, 71-86; PROF PREV MED & BIOSTATIST, UNIV TORONTO, 76- *Concurrent Pos:* Mem working cadre, Bladder Cancer Proj, US, 73-75; mem epidemiol comt, Breast Cancer Task Force, US, 73-77, chmn, 75-77; mem, Fed Task Force Cervical Cytol Screening, Can, 74-76 & 80-81, Union Int Contre le Cancer comt, controlled therapeut trials, 78-82, Multidisciplinary Proj, breast cancer, 78-86; mem, Comt Diet, Nutrit, Cancer, US Nat Acad Sci, 80-82, oversight comt, 83-84, Comt Diet & Health, 87-89 & Sci Coun Int Agency Res Cancer, Lyon, 81-85; chmn, Proj Screening for Cancer, US Nat Cancer Inst, 80-, Agency Res Ctr, Lyon, 85, Fel Comt, 87-; chmn, Nat Res Coun Environ Epidemiol, 90- *Mem:* Can Oncol Soc (secy-treas, 75-79, pres, 80-81); Soc Epidemiol Res; Int Epidemiol Asn; Int Asn Study Lung Cancer; Am Soc Prev Oncol (pres, 83-85); Am Col Epidemol. *Res:* Epidemiology of breast, bladder, larynx, lung, gastric colo-rectal cancer and cerebral tumors; radiation, occupation, diet and cancer; monitoring for environmental carcinogenesis; evaluation of screening for cervix and breast cancer; controlled clinical trials in cancer; cancer control. *Mailing Add:* 575 Ave Rd Toronto ON M4V 2S7 Can

MILLER, ARILD JUSTESEN, b Pine City, Minn, May 16, 18; m 43; c 3. PHYSICAL CHEMISTRY. *Educ:* Carleton Col, BA, 39; Purdue Univ, PhD(phys chem), 43. *Hon Degrees:* MS ad eundem, Lawrence Univ, 85. *Prof Exp:* Asst, Carleton Col, 39-41, prof chem & chmn dept, 49-60; asst, Purdue Univ, 41-42; res chemist, Metall Lab, Univ Chicago, 43-45; chemist, Clinton Labs, Tenn, 45-46; asst prof chem, Antioch Col, 46-49; dir admis & assoc dean, 60-83, INSTR, INST PAPER CHEM, FOX VALLEY TECH COL, APPLETON, 85- *Concurrent Pos:* Res chemist, Kettering Found, 46-49. *Mem:* Am Chem Soc. *Res:* Photosynthesis; photochemistry; radiation chemistry; radiochemistry; heats of combustion; solubilities of organic compounds; heats of combustion of some polynitroparaffins. *Mailing Add:* 95 Estherbrook Ct Appleton WI 54915

MILLER, ARNOLD, b New York, NY, May 8, 28; m 50; c 3. TECHNOLOGY MANAGEMENT. *Educ:* Univ Calif, Los Angeles, BS, 48, PhD(chem), 51. *Prof Exp:* Asst, Univ Calif, Los Angeles, 48-49; res phys chemist, William Wrigley Res Lab, 51; res phys chemist, Armour Res Found, Ill Inst Technol, 52-54, supvr phys chem, 55-56; mgr chem, Borg Warner Res Ctr, 56-59; chief mat res, Autonetics Div, NAm Aviation, Inc, 59-62, dir phys res, 62-66, dir cent microelectronics, NAm Rockwell Corp, 66-68; gen mgr res & develop div, Whittaker Corp, 68-69, gen mgr, 69-71; pres, Theta Sensors, Inc, 71-73; dir eng sci, Xerox Corp, 73-78, dir res & adv develop, 78-81, corp vpres, 81-87, PRES TECHNOL STRATEGY GROUP, XEROX CORP, 87- *Concurrent Pos:* Pres, Space Sci, Inc, 68-71; dir, Spectrodiode Labs, 85; consult, World Bank, 87; dir, Merisel, 88- *Honors & Awards:* Armour Res Found Award, 53; Award, Bur Ord, US Navy, 53; Indust Res 100 Awards, 64, 69; Sigma Xi. *Mem:* Am Chem Soc; Inst Elect & Electronics Engrs; Asn Inst Met Engrs. *Res:* Surface physics and chemistry; photochemistry; epitaxial growth; chemical vapor deposition; microelectronics; instrumentation, technology management; published extensively in fields of instrumentation, surface science; electronics and industrial technology management. *Mailing Add:* 505 Westchester Place Fullerton CA 92635

MILLER, ARNOLD I, b New York, NY, Sept 23, 56; m 85; c 2. EVOLUTIONARY PALEOBIOLOGY, PALEOECOLOGY. *Educ:* Univ Rochester, BS, 78; Va Tech, MS, 81; Univ Chicago, PhD(geophys sci), 86. *Prof Exp:* asst prof, 86-91, ASSOC PROF GEOL, UNIV CINCINNATI, 91- *Mem:* Paleont Soc; Sigma Xi; Paleont Res Inst; Soc Econ Paleontologists & Mineralogists. *Res:* Environmental, ecologic and geographic patterns associated with trends in global biotic diversity; formation of fossil assemblages; collection of primary data from the field, literature and numerical modeling of recognized patterns. *Mailing Add:* Dept Geol ML13 Univ Cincinnati Cincinnati OH 45221

MILLER, ARNOLD REED, b Marion, Ind, Dec 25, 44; m 67; c 2. RELIABILITY THEORY. *Educ:* Ind Univ, AB, 69; Univ Ill, Urbana, PhD(chem) 73. *Prof Exp:* Asst dir, Inst Aging, Univ Wis-Madison, 78-79; vis res scientist, Dept Genetics & Develop, 80-82, RES SCIENTIST & VIS ASST PROF, DEPT MED INFO SCI, UNIV ILL, URBANA, 82- *Concurrent Pos:* Assoc ed, Age, J Am Aging Asn, 79-84 & Simulation, J Soc Comput Simulation, 80-85; prin investr, Nat Inst Aging, NIH, 82- *Mem:* Math Asn Am; Reliability Soc; Soc Indust & Appl Math; Soc Math Biol. *Res:* Reliability theory; reliability of evolutionary systems; aging of biological organisms. *Mailing Add:* Counseling Inst 2403 W Springfield Champaign IL 61821

MILLER, ARTHUR, b New York, NY, Apr 3, 30; m 61; c 3. ELECTROMAGNETISM. *Educ:* Polytech Inst Brooklyn, BS, 51; Calif Inst Technol, PhD(chem), 57. *Prof Exp:* Asst chem, Brookhaven Nat Lab, 50-51; asst, Los Alamos Sci Lab, 52; mem tech staff, 56-83, SR MEM TECH STAFF, RCA LABS, 83- *Mem:* Am Phys Soc. *Res:* Magnetic materials; crystal chemistry of spinels; x-ray and neutron diffraction crystallography; crystal optics; radiochemistry; ferroelectrics; nonlinear optics; electro-optic materials; integrated optics; properties of glasses; finite-element analysis; electron optics. *Mailing Add:* David Sarnoff Res Ctr CN 5300 Princeton NJ 08543-5300

MILLER, ARTHUR I, b New York, NY, Feb 6, 40; m; c 2. HISTORY OF SCIENCE, THEORETICAL PHYSICS. *Educ:* City Col New York, BS, 61; Mass Inst Technol, PhD(physics), 65. *Prof Exp:* Asst prof physics, Univ Lowell, 65-70, assoc prof, 70-76; prof philos & hist, Dept Physics, Harvard Univ, 76-90, univ prof, 81-90; RES ASSOC, UNIV CAMBRIDGE, UK, 90- *Concurrent Pos:* Nat Endowment for Humanities fel, Harvard Univ, 72-73, assoc, Physics Dept, 73-, Guggenheim fel, 79-80. *Mem:* Fel Am Phys Soc; Hist Sci Soc Am; Am Asn Physics Teachers; AAAS. *Res:* Interdisciplinary research in the history of 19th and 20th century science, technology and cognitive science. *Mailing Add:* Dept Hist & Philos Sci Univ Cambridge Free School Lane Cambridge CB2 3RH England

MILLER, ARTHUR JAMES, b New York, NY, Apr 7, 50; m 77. FOOD SCIENCE. *Educ:* Kans State Univ, BS, 72; Pa State Univ, MS, 77; Drexel Univ, PhD(food sci), 84. *Prof Exp:* Food technologist, 76-80, res food technologist, 80-85, LEAD SCIENTIST, EASTERN REGIONAL RES CTR, USDA, 85- *Concurrent Pos:* Adj assoc prof, Drexel Univ, 89-; co-ed, J Food Safety, 90- *Mem:* Inst Food Technologists; Am Soc Microbiol; Am Meat Sci Asn; Genetic Toxicol Asn. *Res:* Risks associated with foodborne bacterial pathogens, particularly Clostridium butulinum and Listeria monocytogenes; formation of thermally induced carcinogens in foods. *Mailing Add:* Eastern Regional Res Ctr Agr Res Serv USDA Philadelphia PA 19118

MILLER, ARTHUR JOSEPH, b San Francisco, Calif, Jan 18, 43; m 65; c 3. NEUROPHYSIOLOGY, PHYSIOLOGY. *Educ:* Univ Calif, Los Angeles, PhD(physiol), 70. *Prof Exp:* Trainee, Brain Res Inst, Univ Calif, Los Angeles, 70; asst prof physiol, Univ Ill Med Ctr, 70-75, adj asst prof otolaryngol, 74-75; from asst prof to assoc prof, 75-84, PROF DEPT GROWTH & DEVELOP, SCH DENT & DEPT PHYSIOL, SCH MED, UNIV CALIF, SAN FRANCISCO, 84- *Concurrent Pos:* NIH grants, 71-90; ad hoc, oral biol study groups, NIH; NSF consult. *Mem:* Neurosci Soc; Am Physiol Soc; Int Asn Dent Res; Am Dent Asn; NY Acad Sci; Int Brain Res Orgn. *Res:* Neuromuscular control of craniomandibular muscles; cranial reflexes in developing animals; cranioskeletal growth and development; craniomandibular muscle function in craniomandibular disorders; deglatition; dentistry; craniobiology. *Mailing Add:* Dept Growth & Develop Univ Calif Sch Dent Rm 734C San Francisco CA 94143-0640

MILLER, ARTHUR R, b Boston, Mass, Aug 6, 15; m 41; c 2. PHYSICAL OCEANOGRAPHY. *Prof Exp:* assoc scientist, Woods Hole Oceanog Inst, 46-80; RETIRED. *Concurrent Pos:* Grant from Woods Hole Oceanog Inst, Scripps Inst, Univ Calif, 50; consult, US Weather Bur, 55; mem, Int Comn Bibliog Phys Oceanog, 60-; mem working panel, Int Indian Ocean Exped, 61-63, lectr, US Biol Prog, Bermuda Biol Sta, 62; UN consult, 78; pres, Assoc Scientists, Woods Hole, Inc. *Mem:* AAAS; Am Geophys Union; Am Soc Limnol & Oceanog; Explorers Club; NY Acad Sci. *Res:* Tide and storm surge research; cooperative investigations of the Mediterranean; general ocean research, ocean thermal energy. *Mailing Add:* 175 Lakeview Ave Falmouth MA 02540

MILLER, ARTHUR SIMARD, b Sidney, Mont, Mar 4, 35; m 66. ORAL PATHOLOGY. *Educ:* Mont State Univ, BS, 57; Wash Univ, DDS, 59; Ind Univ, MSD, 63. *Prof Exp:* Instr oral path, Sch Dent, Ind Univ, 63-66; from asst prof to assoc prof, 66-72, prof path, 72-81, PROF ORAL BIOL & PATH, SCH DENT, TEMPLE UNIV, 81-, CHMN DEPT, 68- *Concurrent Pos:* Consult, Cent Am Registry Oral Path, 66-; consult dent aptitude testing, Am Dent Asn, 71- *Mem:* Am Dent Asn; Am Acad Oral Path; Int Asn Dent Res. *Res:* Oral diseases and neoplasms; use of the computer in oral pathology; teaching improvements and innovations. *Mailing Add:* Dept Oral Biol Temple Univ/Health Sci Campus Broad & Ont Philadelphia PA 19140

MILLER, AUDREY, b Danville, Pa, Nov 15, 37. ORGANIC CHEMISTRY. *Educ:* Univ Rochester, BS, 59; Univ Ill, MS, 60; Columbia Univ, PhD(org chem), 62. *Prof Exp:* Res assoc org chem, NMex Highlands Univ, 63-64; from asst prof to assoc prof, 64-81, PROF CHEM, UNIV CONN, 81- *Concurrent Pos:* NSF fel, 63-64, sci fac fel, 71-72. *Mem:* Am Chem Soc. *Res:* Models of biological oxidations and reductions; sulfur and nitrogen compounds. *Mailing Add:* Dept Chem Univ Conn Storrs CT 06269-3060

MILLER, AUGUST, b Isola, Miss, Nov 28, 33; m 63. ATMOSPHERIC PHYSICS. *Educ:* NMex State Univ, BS, 55, PhD(physics), 61; Univ Md, MS, 58. *Prof Exp:* Eng specialist, Ariz Div, Goodyear Aircraft Corp, 61-62; mem res staff, Northrop Space Labs, 62-64; head dept, 79-83, PROF PHYSICS, NMEX STATE UNIV, 64- *Mem:* Am Phys Soc; Am Asn Physics Teachers. *Res:* Scanning electro-optical image sensors; very low pressure radio frequency gaseous discharges; radio frequency plasmoids; atmospheric optics; satellite meteorology. *Mailing Add:* Dept Physics NMex State Univ PO Box 30001 Las Cruces NM 88003-0001

MILLER, AUGUSTUS TAYLOR, JR, b Arlington, Tex, Apr 14, 10; m 38; c 1. PHYSIOLOGY. *Educ:* Emory Univ, BS, 31, MS, 33; Univ Mich, PhD(physiol), 39; Duke Univ, MD, 53. *Prof Exp:* Res assoc, W H Maybury Sanatorium, Mich, 36-39; from instr to prof, 39-82, EMER PROF PHYSIOL, SCH MED, UNIV NC, CHAPEL HILL, 82- *Mem:* Am Physiol Soc. *Res:* Aging. *Mailing Add:* 750 Weaver Dairy Rd No 2110 Chapel Hill NC 27514

MILLER, BARRY, b Passaic, NJ, Jan 22, 33; m 65; c 2. ELECTROCHEMISTRY. *Educ:* Princeton Univ, AB, 55; Mass Inst Technol, PhD(chem), 59. *Prof Exp:* Instr chem, Harvard Univ, 59-62; MEM TECH STAFF, CHEM RES DEPT, BELL TEL LABS, 62- *Mem:* Am Chem Soc; Electrochem Soc. *Res:* Electrochemical kinetics; photoelectrochemistry; electroanalytical chemistry. *Mailing Add:* 54 Fox Run Murray Hill NJ 07974-1206

MILLER, BARRY, b New York, NY, Dec 25, 42. HEALTH PSYCHOLOGY, RESEARCH POLICY. *Educ:* Brooklyn Col, BS, 65; Villanova Univ, MS, 67; Med Col Pa, PhD(psychiat), 71. *Prof Exp:* Sr med res scientist, Eastern Pa Psychiat Inst, 73-80; assoc dean res & assoc prof psychiat, Med Col Pa, 81-90, assoc prof med, 83-90; ALBERT EINSTEIN MED CTR. *Concurrent Pos:* Dir, Pa Bur Res & Training, Off Mental Health, Dept Pub Welfare, 73-81; asst prof psychiat, Univ Pa Med Sch, 75-78, clin asst prof, 78-; adv comt mem, Clin Res Ctr Study Psychopath Elderly, 85-88. *Mem:* Am Psychol Asn; Am Orthopsychiat Asn; AAAS; Asn Mental Health Adminr; Soc Res Adminr. *Res:* Health psychology; research policy. *Mailing Add:* Albert Einstein Med Ctr 5501 Old York Rd Philadelphia PA 19141

MILLER, BERNARD, b Monticello, NY, Sept 1, 30; m 65; c 2. ORGANIC CHEMISTRY. *Educ:* City Col New York, BS, 51; Columbia Univ, MA, 53, PhD(org chem), 55. *Prof Exp:* NIH fel, Univ Wis, 55-56, NSF, 56-57; res chemist, Am Cyanamid Co, 57-60, sr res scientist, 60-67; assoc prof, 67-72, PROF CHEM, UNIV MASS, AMHERST, 72- *Mem:* Am Chem Soc. *Res:* Organic reaction mechanisms; molecular rearrangements; organic phosphorus chemistry. *Mailing Add:* Dept Chem Univ Mass Amherst MA 01002

MILLER, BERNARD, b New York, NY, Apr 9, 27; m 58; c 2. POLYMER SCIENCE, TEXTILES. *Educ:* Va Polytech Inst, BS, 48, MS, 49; McGill Univ, PhD(chem), 55. *Prof Exp:* Jr chemist, Hoffmann-La Roche Inc, 49-52; res chemist, Celanese Corp Am, 55-56; asst prof polymer chem, Lowell Technol Inst, 56-58; Du Pont fel, Textile Res Inst, 58-59; res chemist, E I du Pont de Nemours & Co, 59; asst prof phys chem, Am Univ, 59-61, assoc prof, 61-66; sr scientist, 66-67, assoc dir chem res, 67-69, ASSOC DIR RES, TEXTILE RES INST, 69- *Concurrent Pos:* NASA res grant, 62-64; USPHS res grant, 62-66; consult, NIH, 63-64; Fiber Soc nat lectr, 73-74. *Honors & Awards:* Harold Dewitt Smith Medal, Am Soc Testing & Mat, 77. *Mem:* AAAS; Am Chem Soc; Fiber Soc (pres, 88); Info Coun on Fabric Flammability; NAm Thermal Anal Soc. *Res:* Fiber science; flammability; fiber surface properties; thermal analysis; calorimetry; cellulose chemistry; thermal properties of polymers; fluid flow. *Mailing Add:* Textile Res Inst Prospect Ave PO Box 625 Princeton NJ 08542

MILLER, BETTY M (TINKLEPAUGH), b Corunna, Mich, Apr 23, 30; m 64. PETROLEUM GEOLOGY, RESOURCE ASSESSMENT METHODS. *Educ:* Cent Mich Univ, AB, 52; Mich State Univ, MS, 55, PhD(geol), 57. *Prof Exp:* Cartographer & geologist, McClure Oil Co, 54-55; asst natural sci, Mich State Univ, 56-57, instr, 57-58; res geologist, Pure Oil Co, 58-65; sr res geologist & geostatistician, Sun Oil Co, 66-73; geologist, 73-74, prog chief resource appraisal group, Oil & Gas Resources Br, 74-80, asst dir, Eastern Region, 80-82, staff asst to chief geologist, 82-88, SR RES GEOLOGIST, US GEOL SURV, 88- *Concurrent Pos:* Asst, Mich State Univ, 54-55. *Mem:* Geol Soc Am; Am Asn Petrol Geol; Int Asn Math Geol; Sigma Xi; Comput

Oriented Geol Soc. *Res:* Geostatistical and computer applications as related to petroleum geology, including Geographical Information Systems and Artificial Intelligence-Expert Systems; carbonate petrography and geochemistry; basin analysis petroleum occurrence; oil and gas resource appraisal methodology and appraisals and assessments of the national and worldwide petroleum resources. *Mailing Add:* US Geol Surv Nat Ctr MS 955 12201 Sunrise Valley Dr Reston VA 22092

MILLER, BILLIE LYNN, b Harrodsburg, Ky, Nov 19, 34; m 59; c 2. MEDICINE. *Educ:* Eastern Ky Univ, BS, 54; Univ Chicago, MD, 57; Am Bd Pediat, dipl, 68, cert cardiol, 70. *Prof Exp:* Intern, Hosp, Univ Mich, Ann Arbor, 57-58; resident cardiol, Hammersmith Hosp, London, Eng, 58-59; fel, St Thomas Hosp, 59-61; fel, Nat Heart Hosp, 61-62; res assoc, London Hosp, 62; asst med investr, Nat Inst Cardiol Mex, 62-65; resident pediat, 66-68, spec clin fel pediat cardiol, 68-69, instr pediat, 69-70, asst prof pediat, Col Med, Univ Fla, 70-79; CONSULT, 79. *Concurrent Pos:* Consult, Fla Bur Crippled Children, 68-79. *Mem:* Fel Am Acad Pediat; fel Am Col Cardiol; Am Heart Asn. *Res:* Analysis of disorders of cardiac rhythm in infants and children; application of computer techniques to the interpretation of the body surface manifestations of the electrical activity of the heart in infants and children. *Mailing Add:* 610 SW First Ave Williston FL 32696

MILLER, BOBBY JOE, soil genesis; deceased, see previous edition for last biography

MILLER, BRINTON MARSHALL, b Delaware Co, Pa, Dec 30, 26; m 48; c 3. MICROBIOLOGY, PROTOZOOLOGY. *Educ:* Univ Va, BA, 50, MS, 51; Purdue Univ, PhD(plant sci), 56. *Prof Exp:* Teacher high sch, Va, 51-53; sr microbiologist, Merck Sharp & Dohme Res Labs, 56-66, sect head, 66-73, asst dir basic animal sci, 73-75, dir animal infections, Basic Animal Sci, 75-83; VPRES, SCI, NEOGEN CORP, 84- *Concurrent Pos:* Asst ed, Am Biol Teacher, 53-56, Appl Microbiol, 57-61; vis biologist, Am Inst Biol Sci, 70-73; chmn, Proj Biotech, Am Inst Biol Sci-NSF, 71-75; ed, Indust Microbiol, McGraw Hill; int mem, Mex Prog Dairy Prod in Tropics & mem steering comt biosafety guidelines, Ctr Disease Control, NIH; ed-in-chief, Lab Safety Prin & Pract, Am Soc Microbiol; mem, Hq Adv Comt, Am Soc Microbiol, 85-, Treas Comt, 85- *Honors & Awards:* Merit Award, Soc Indust Microbiol, 70. *Mem:* Soc Indust Microbiol (secy, 58-61, pres, 63); Wildlife Soc; Am Soc Microbiol (treas, 75-84); fel Am Acad Microbiol. *Res:* Pathology of mycoplasma; parasitic Protozoa of man and animals, especially trypanosomes and Coccidia; food science and technology; immunodiagnostics; toxicology; microbiology. *Mailing Add:* Neogen Corp 620 Lesher Pl Lansing MI 48912

MILLER, BRUCE LINN, b Grove City, Pa, Sept 8, 23; m 48; c 2. PHYSICS, MATHEMATICS. *Educ:* SDak State Univ, BS, 47; Univ Kans, MS, 51, PhD(physics), 53. *Prof Exp:* Jr physicist, Univ Iowa, 43-44; res physicist, Sandia Corp, NMex, 53-55; from asst prof to prof physics, Grad Fac, SDak State Univ, 55-87; RETIRED. *Mem:* Am Inst Physics; Am Asn Physics Teachers. *Res:* Geometrical and physical optics; biological effects of nuclear radiation. *Mailing Add:* 1326 Lagaris Dr Brookings SD 57007

MILLER, BRUCE NEIL, b New York, NY, Dec 12, 41; m 66; c 2. STATISTICAL PHYSICS, NON-LINEAR DYNAMICS. *Educ:* Columbia Univ, BA, 63; Univ Chicago, MSc, 65; Rice Univ, PhD(physics), 69. *Prof Exp:* Asst physicist, IIT Res Inst, 65; post doctoral fel, Rice Univ, 69-70; post doctoral fel, State Univ NY Albany, 70-71; from asst prof to assoc prof, 71-85, PROF PHYSICS, TEX CHRISTIAN UNIV, 85- *Mem:* Am Phys Soc; Sigma Xi. *Res:* Influence of gravity on critical opalescence; theory of light scattering from a monuniform fluid; relaxation of a one-dimensional gravitating system; decay of positrons and positronium in fluids. *Mailing Add:* Dept Physics Tex Christian Univ Ft Worth TX 76129

MILLER, BYRON F, b Robinson, Kans, Aug 20, 31; m 52; c 5. POULTRY NUTRITION. *Educ:* Kans State Univ, BS, 53, MS, 60, PhD(animal nutrit), 60. *Prof Exp:* From instr to assoc prof poultry prods, Colo State Univ, 60-88. *Mem:* Poultry Sci Asn; Inst Food Technol. *Res:* Poultry products and food technology. *Mailing Add:* 801 Birky Rd Colo State Univ Ft Collins CO 80526

MILLER, C ARDEN, b Shelby, Ohio, Sept 19, 24; m 48; c 4. PEDIATRICS, PUBLIC HEALTH. *Educ:* Yale Univ, MD, 48. *Prof Exp:* From instr to assoc prof pediat, Med Ctr, Univ Kans, 51-57, from asst dean to dean, Sch Med, 57-66, provost, 65-66, dir med ctr, 60-66; vchancellor health sci, 66-71, PROF MATERNAL & CHILD HEALTH, SCH PUB HEALTH, UNIV NC, CHAPEL HILL, 66- *Concurrent Pos:* Markle scholar med sci, 55-60; vis fel, Cambridge Univ, 86; Chmn bd, Allan Guttmacher Inst, 79-85. *Honors & Awards:* Martha Mae Eliot Award & Sedgwick Medal, Am Pub Health Asn. *Mem:* Inst Med-Nat Acad Sci; Am Pub Health Asn (pres, 74-75); Asn Teachers Maternal & Child Health (pres, 79); Am Soc Pediat Res. *Res:* Health policy; handicapped children; child health. *Mailing Add:* Univ NC Rosenau Hall Chapel Hill NC 27599-7400

MILLER, C DAN, b Salem Ore, Nov 30, 41. GEOMORPHOLOGY & GLACIOLOGY. *Educ:* Univ Wash, BS, 65, MS, 67; Univ Colo, PhD(geol), 71. *Prof Exp:* Res grologist, Standard Oil Co, 70-71; asst prof geol, Colgate Univ, 70-71; RES GEOLOGIST, US GEOL SURVEY, 74- *Mem:* Geol Soc Am; Am Geophys Union. *Res:* Quaternary statigraphy; volcanic statigraphy; volcanic hazards assessment. *Mailing Add:* Geol Survey Cascades Volcano Observ 5400 MacArthur Blvd Vancouver WA 98661

MILLER, C DAVID, b Baltimore, Md, Apr 7, 31; m 61; c 2. ANALYTICAL CHEMISTRY. *Educ:* Columbia Univ, AB, 52; Univ Md, MS, 59; Univ Fla, PhD(analytical chem), 64. *Prof Exp:* Analytical res chemist, E I du Pont de Nemours & Co, 52-53; DIR, RES & DEVELOP & ENG, AM INSTRUMENT CO, 60-61 & 64- *Mem:* Am Chem Soc; Soc Appl Spectroscopy; Electron Micros Soc Am. *Res:* Development of clinical and analytical instrumentation; atomic and molecular spectroscopy, luminescence, electrochemical, neutron activation and enthalpimetric analysis. *Mailing Add:* 44 Lakeside Dr Greenbelt MD 20770

MILLER, C EUGENE, b Buffalo, NY, Sept 25, 28; m 69. ENGINEERING MECHANICS, CIVIL ENGINEERING. *Educ:* Manhattan Col, BA, 53; Fordham Univ, MS, 56; Rensselaer Polytech Inst, MS, 60, PhD(mech), 63. *Prof Exp:* Instr physics & math, De La Salle Inst, 50-55, Christian Bros Acad, 55-57, La Salle Inst, 57-58 & Hillside Hall Jr Col, 58-63; asst prof mech, Manhattan Col, 63-69, res dir rheol, 64-69; dir air pollution res, NY State Dept Health, 69-70; prof environ & biomed eng, 70-77, chmn civil eng dept, 84-85, PROF ENG MECH, UNIV LOUISVILLE, 77- *Concurrent Pos:* Consult rheol, Montefiore Hosp, 63-64; consult air pollution, Chem Construct Co, 65-70; NY State Dept Health res grant & res assoc, Roswell Park Mem Inst, 68-70; res grants, NIH & Hearst Found, 78-82 & Silver Creek Ind Water Dist, 82-84; consult, Rural Water Distrib Systs, 83- *Mem:* Sigma Xi; Soc Rheol; Am Soc Eng Educ; Air Pollution Control Asn; Am Soc Mech Engrs. *Res:* Rheology; vibrations; bionics; biorheology; plates and shells; fetal membranes; mechanics in pregnancy; rural water distribution systems. *Mailing Add:* 202 Wildwood Lane Louisville KY 40223

MILLER, CALVIN F, b Vista, Calif, Aug 6, 47; m; c 2. IGNEOUS PETROLOGY. *Educ:* Pomona Col, BA, 69; George Washington Univ, MS, 73; Univ Calif, Los Angeles, PhD(geol), 77. *Prof Exp:* From asst prof to assoc prof, 77-90, PROF GEOL, VANDERBILT UNIV, 90- *Mem:* Geol Soc Am; Am Geophys Union; Mineral Soc Am; Sigma Xi. *Res:* Origin of granitic magmas & relation to crustal evolution. *Mailing Add:* Vanderbilt Univ Box 6028 Sta B Nashville TN 37235

MILLER, CARL ELMER, b Flint, Mich, Apr 28, 37; m 65. MOLECULAR PHYSICS. *Educ:* Univ Mich, BA, 59, MS, 61, PhD(physics), 67. *Prof Exp:* Instr physics, Flint Jr Community Col, 61-63; from asst prof to assoc prof physics, Mankato State Col, 67-76; sr scientist, 76-79, develop engr, 78-80, SUPVR MATS & APPLN, AC SPARK PLUG CO, 80- *Mem:* Am Phys Soc; Soc Automotive Engrs; Hist Sci Soc. *Res:* Organic materials science and engineering, including organic coatings, plastics and theoretical modeling for automotive applications. *Mailing Add:* Mat Appln AC Spark Plug 1300 N Dort Ave Flint MI 48506

MILLER, CARL HENRY, JR, b Cleveland, Ohio, Sept 18, 20; m 51. PHYSIOLOGICAL CHEMISTRY, ELECTRON MICROSCOPY. *Educ:* Ohio State Univ, BSc, 42, PhD(physiol chem), 67. *Prof Exp:* Asst physics, Case Inst Technol, 45-46, sr technician biochem, 48-56; res asst, Ohio State Univ, 59-67; Nat Inst Ment Health traineeship & assoc res scientist, Med Sch, NY Univ, 67-69; ASSOC PROF BIOL, JERSEY CITY STATE COL, 69- *Mem:* AAAS; Electron Micros Soc Am. *Res:* Biochemistry of mental illness; metabolism of catecholamines. *Mailing Add:* 2396 Johnston Rd Columbus OH 43220

MILLER, CARLOS OAKLEY, b Jackson, Ohio, Feb 19, 23. PLANT PHYSIOLOGY. *Educ:* Ohio State Univ, BSc, 48, MA, 49, PhD, 51. *Prof Exp:* Proj assoc bot, Univ Wis, 51-55, asst prof, 55-57; from asst prof to assoc prof, 57-60, PROF PLANT SCI, IND UNIV, BLOOMINGTON, 63- *Mem:* Bot Soc Am; Am Soc Plant Physiol (secy, 60-61, vpres, 62). *Res:* Plant growth substances, particularly those of kinetin type; chemical control of plant development; control of plant growth by light; plant growth and development. *Mailing Add:* Dept Biol Ind Univ Bloomington IN 47405

MILLER, CAROL RAYMOND, b Asheville, NC, Sept 10, 38; m 59; c 2. PLANT BREEDING, PLANT PATHOLOGY. *Educ:* West Carolina Col, BS, 60; Clemson Univ, MS, 62, PhD(plant path), 65. *Prof Exp:* Res asst, Clemson Univ, 63-64; asst prof plant path, Univ Fla, 64-71; asst dir tobacco res & prod, Coker's Pedigreed Seed Co. 71-72, dir, 72-84; SR STAFF AGRONOMIST RES & DEVELOP, R J REYNOLDS TOBACCO CO, 84- *Mem:* Am Phytopath Soc. *Res:* Tobacco diseases and breeding. *Mailing Add:* 8800 Kingstree Dr Clemmons NC 27012

MILLER, CHARLES A, MEDICAL SCIENCES. *Educ:* Wabash Col, BA, 49; Ind Univ, PhD, 54. *Prof Exp:* Instr & asst prof, Wabash Col, 54-61; health scientist adminr, Res Grants Br, Nat Inst Gen Med Sci, 61-62, exec secy & prog adminr, Res Training Grants Br, 62-66, head, Biophys Sci Sect, Reactor Turbine Generator Br, 66-70, assoc chief, Res Training Grants Br & dir, Minority Access Res Careers Prog, 69-72, chief, Res Training Grants Br & chief, Res Training Grants Br & assoc dir, Nat Inst Gen Med Sci Res Manpower, 72-73, spec asst dir, 73-83, DIR, CELLULAR & MOLECULAR BASIS DIS PROG, NAT INST GEN MED SCI, NIH, 73-, ACTG ASSOC DIR, MINORITY PROG, 89- *Concurrent Pos:* Mem, Contract Rev Panel, Nat Cancer Inst, 74-77. *Honors & Awards:* Super Serv Hon Award, HEW, 75, Merit Award, Sr Exec Serv, 85. *Mem:* Am Soc Biol Chemists; Am Soc Cell Biol; Biophys Soc. *Res:* Training grants. *Mailing Add:* NIH Nat Inst Gen Med Sci Cellular & Molecular Basis Dis Prog Br Westwood Bldg Rm 903 5333 Westbard Ave Bethesda MD 20892

MILLER, CHARLES BENEDICT, b Minneapolis, Minn, Apr 28, 40; m 63, 83; c 3. BIOLOGICAL OCEANOGRAPHY. *Educ:* Carleton Col, BA, 63; Scripps Inst Oceanog, PhD(biol oceanog), 69. *Prof Exp:* NSF fel, Univ Auckland, 69-70; from asst prof to assoc prof, 70-80, PROF BIOL OCEANOG, COL OCEANOG, ORE STATE UNIV, 80- *Concurrent Pos:* Vis scholar, Scripps Inst Oceanog, Univ Tokyo, Univ Maine, Univ Paris VI. *Mem:* Am Soc Limnol & Oceanog; Am Asn Advan Sci; Am Geophys Union; Crustacean Soc; Oceanog Soc. *Res:* Zooplankton ecology; biological oceanography of the subarctic Pacific Ocean. *Mailing Add:* Col Oceanog Ore State Univ Corvallis OR 97331-5503

MILLER, CHARLES EDWARD, b Philadelphia, Pa, Feb 16, 25; m 44; c 1. MYCOLOGY. *Educ:* Furman Univ, BS, 51; Univ NC, MA, 54, PhD(bot), 57. *Prof Exp:* Asst bot, Univ NC, Chapel Hill, 51-57, Am Bact Soc fel, 57-58; instr biol, Emory Univ, 58-59; asst prof bot, Tex A&M Univ, 59-62; assoc prof, Univ Maine, 62-65; assoc prof, 65-70, PROF & CHMN DEPT BOT, OHIO UNIV, 70- *Mem:* Bot Soc Am; Mycol Soc Am; Brit Mycol Soc. *Res:* Taxonomy; morphology; ecology; physiology; ultrastructure of zoosphoric fungi (aquatic Phycomycetes). *Mailing Add:* Psychol Dept Northern Ill Univ De Kalb IL 60115

MILLER, CHARLES FREDERICK, III, b Springfield, Ill, Feb 12, 41; m 66; c 2. MATHEMATICS. *Educ:* Lehigh Univ, BA, 62; NY Univ, MS, 64; Univ Ill, PhD(math), 69. *Prof Exp:* Instr math, Univ Ill, 68-69; mem, Inst Advan Study, Princeton, NJ, 69-70; NSF fel, Oxford Univ, 70-71; asst prof math, Princeton Univ, 71-76; PROF MATH, MELBOURNE UNIV, 76- *Mem:* Am Math Soc; Australian Math Soc; Asn Symbolic Logic. *Res:* Mathematical logic; combinatorial group theory; decision problems in algebra; geometric methods to group theory. *Mailing Add:* Dept Math Melbourne Univ Parkville Victoria 3052 Australia

MILLER, CHARLES G, b Greensburg, Ind, Feb 9, 40; m 65; c 3. MICROBIOLOGY, BIOCHEMISTRY. *Educ:* Ind Univ, Bloomington, AB, 63; Northwestern Univ, PhD(biochem), 68. *Prof Exp:* USPHS fel, Univ Calif, Berkeley, 68-70; from asst prof to assoc prof microbiol, Sch Med, Case Western Reserve Univ, 70-83, prof molecular biol & microbiol, 83-90; PROF & HEAD, DEPT MICROBIOL, UNIV ILL, URBANA-CHAMPAIGN, 90- *Mem:* Am Chem Soc; Am Soc Microbiol; Genetics Soc Am; Am Soc Biochem & Molecular Biol. *Res:* Biochemical genetics. *Mailing Add:* Dept Microbiol 131 Burrill Hall Univ Ill 407 S Goodwin Urbana IL 61801

MILLER, CHARLES LESLIE, b Tampa, Fla, June 5, 29; m 49; c 4. CIVIL ENGINEERING. *Educ:* Mass Inst Technol, BS, 51, MS, 58. *Prof Exp:* Proj engr, Michael Baker, Jr Inc, 51-52, asst to vpres, 52-54, exec engr, 54-55; from asst prof to assoc prof surv, Mass Inst Technol, 55-61, prof civil eng, 61-78, dir photogram lab, 55-60, civil eng systs lab, 60-64 & Inter-Am Prog, 61-65, head dept civil eng, 61-69, dir, Urban Systs Lab, 68-76; PRES, CLM SYSTS INC & C L MILLER CO, INC, 68- *Concurrent Pos:* Chmn, President Elect's Task Force on Transp, 68; consult, Govt Puerto Rico; dir, Spaulding & Slye, Inc; mem adv bd, Ford Found & Latin Am Sci Bd, Nat Acad Sci; chmn bd, CLM-Systs, Inc & Community Assistance Corp; dir, Geo-Transport Found. *Honors & Awards:* George Westinghouse Award, Am Soc Eng Educ. *Mem:* Fel Am Soc Civil Engrs; Am Soc Photogram; Asn Comput Mach; Am Soc Eng Educ; fel Am Acad Arts & Sci. *Res:* Civil engineering systems; information systems and computer methods in civil engineering; man-machine communications and programming systems; surveying and transportation problems. *Mailing Add:* CLM Systs Inc 4023 S Dale Mabry Tampa FL 33611-3769

MILLER, CHARLES WILLIAM, b Quantico, Va, June 7, 42; m 66. BIOLOGY. *Educ:* Purdue Univ, BS, 64; Colo State Univ, MS, 66, PhD(physiol), 69. *Prof Exp:* Res fel, Univ Wash, 68-70; asst prof radiol & radiation biol, 71-75, ASSOC PROF PHYSIOL & BIOPHYSICS, COLO STATE UNIV, 75- *Mem:* Am Physiol Soc; Am Inst Ultrasound in Med. *Res:* Blood flow characteristics at bends and branch points; arterial wall mechanical properties; pulse wave velocity variations with age and blood pressure; noninvasive studies of blood flow in man; investigations of renal function as affected by age and low doses of ionizing radiation; echocardiography; veterinary applications of ultrasound. *Mailing Add:* 35 Crown Blvd Newburgh NY 12550

MILLER, CHRIS H, b Indianapolis, Ind, Feb 28, 42; m 63; c 3. ORAL MICROBIOLOGY. *Educ:* Butler Univ, BA, 64; Univ NDak, MS, 66, PhD(microbiol), 69. *Prof Exp:* Nat Inst Gen Med Sci res fel, Purdue Univ, 69-70; asst prof med & dent microbiol, Med Ctr, Ind Univ-Purdue Univ, Indianapolis, 70-75; assoc prof, 76-81, PROF & CHMN, DEPT ORAL MICROBIOL, IND UNIV, SCH DENT, INDIANAPOLIS, 81- *Mem:* Am Soc Microbiol; Int Asn Dent Res; Am Asn Dent Sch. *Res:* Ecology of oral bacteria; mechanisms of bacterial dental-plaque formation; bacterial extracellular polysaccharides; pathogenicity of actinomycetes. *Mailing Add:* Dept Microbiol Ind Univ Purdue Univ-Med 1100 W Mich St Indianapolis IN 46223

MILLER, CLARENCE A(LPHONSO), b Houston, Tex, Sept 1, 38; m 65; c 1. CHEMICAL ENGINEERING. *Educ:* Rice Univ, BA & BS, 61; Univ Minn, PhD(chem eng), 69. *Prof Exp:* Engr, Div Naval Reactors, US Atomic Energy Comn, Washington, DC, 61-65; asst prof chem eng, Carnegie-Mellon Univ, 69-78, prof, 78-81; PROF CHEM ENG, RICE UNIV, 81- *Mem:* Am Inst Chem Engrs; Am Chem Soc; Am Inst Mining, Metall & Petrol Engrs. *Res:* Interfacial phenomena; enhanced oil recovery. *Mailing Add:* 5614 Mercer Houston TX 77005

MILLER, CONRAD HENRY, b Lowell, WVa, July 28, 26; m 47; c 2. PLANT PHYSIOLOGY. *Educ:* Va Polytech Inst, BS, 54, MS, 55; Mich State Univ, PhD(hort), 57. *Prof Exp:* Asst prof bot & plant path, Mich State Univ, 57; from asst prof to prof hort, NC State Univ, 57-89. *Concurrent Pos:* Mem working group, Plant Growth Regulator. *Mem:* Am Soc Hort Sci. *Res:* Vegetable production research, especially plant nutrition and growth regulators. *Mailing Add:* 4406 Driftwood Dr Raleigh NC 27606

MILLER, CURTIS C, b Shamokin, Pa, Nov 26, 35; m 58; c 2. GENETICS, ANIMAL BREEDING. *Educ:* Iowa State Univ, BS, 61; Mich State Univ, MS, 65, PhD(dairy sci), 68. *Prof Exp:* Exten dairyman, Mich State Univ, 61-64; statistician, 68-71, mgr animal regulatory affairs & statist serv, 71-73, RES MGR ANIMAL REGULATORY AFFAIRS, STATIST SERV & THERAPEUT, ANIMAL HEALTH RES & DEVELOP, 73- *Mem:* Am Dairy Sci Asn; Am Genetic Asn; Crop Sci Soc Am; Am Statist Asn; Biomet Soc. *Res:* Animal breeding, especially response to selection; population genetics, especially effects of selection, linkage, and dominance on genetic parameters. *Mailing Add:* 6324 Meadowview Kalamazoo MI 49004

MILLER, DALE L, b Lebanon, Pa, Nov 21, 41; m 66; c 2. RESEARCH THEORY & NEW PRACTICE IN QUALITY-OF-LIFE TECHNOLOGY, QUALITY & TECHNOLOGY CONSULTING & DOCUMENTATION. *Educ:* Pa State Univ, BSEE, 63. *Prof Exp:* Res engr radio commun, US Naval Res Lab, 63-65; instnl designer elec & signals, Galter & Diehl Inc Consults, 66-67; environ develop engr, H V Precipitation, GE Envirotech Corp, 67; res & develop engr design, evaluation & standards (Teleom), Bus Commun Div, Int Telephone & Telegraph Corp, 68-81; staff qual & reliability engr, Customer Vendor Plant Eng Liaison, AMP Inc, Harrisburg, Pa, 81-89; consult eng, ISO Contracts, BQS Inc, Montville, NJ & San Mateo, Calif, 90; TECH DIR QUAL ASSURANCE MGT & TECH, CHARLES D SNYDER & SON INC, ELECTROPLATERS, 90- *Concurrent Pos:* Div rep, NAm Components Working Group, Int Telephone & Telegraph Corp, 77-81; CQE, qual engr, Am Soc Qual Control, 88; mem, Qual Bd, Charles D Synder & Son Inc, York, Pa, 90-91; reliability engr reliability, Numerous Seminars & RAC Ctr, 77- *Honors & Awards:* Nat Math Asn Award, Nat Math Asn, 59. *Mem:* Inst Elec & Electronic Engrs; Am Defense Preparedness Asn; sr mem Am Soc Qual Control. *Res:* Impact statements and analysis of governmental and industrial science technology trends; economic factors and quality productivity improvement papers; documentation of specifications, managements and operators' work instructions; history of technology and future science. *Mailing Add:* 1419 Ford Ave Harrisburg PA 17109-5617

MILLER, DANIEL NEWTON, JR, b St Louis, Mo, Aug 22, 24; m 50; c 2. GEOLOGY. *Educ:* Mo Sch Mines, MS, 51; Univ Tex, PhD(geol), 55. *Prof Exp:* Jr geologist, Stanolind Oil & Gas Co, 51-52, geologist, 55-57; sr geologist, Pan Am Petrol Corp, 57-59 & Lion Oil Div, Monsanto Chem Co, 59-60; consult geologist, Barlow & Haun, Inc, 60-63; prof geol & chmn dept, Southern Ill Univ, 63-69; state geologist & exec dir, Wyo Geol Surv, 69-81; asst sec, Energy & Minerals, Dept Interior, Washington, DC, 81-83; pres, IWO Exploration Inc, Boise, Idaho, 84-89; DIR, INT ARCH ECON GEOL, UNIV WYO, 89- *Concurrent Pos:* Comnr, Wyo Oil & Gas Conserv Comn; Wyo rep, Dept Interior-Oil Shale Environ Adv Comt, 75-76; Wyo rep, Fed Power Comt-Supply Tech Adv Task Force-Prospective Expor & Develop; chmn, Fed Liason Comt, Asn Am State Geologists, 77-78. *Mem:* Am Asn Petrol Geol; Am Inst Prof Geologists; Asn Am States Geologists (secy-treas, 75-76, pres, 78-79). *Res:* Sedimentation; sedimentary petrology and stratigraphic interpretation of sedimentary rocks and the diagenetic alteration that they have undergone; effect of tectonism on late diagenetic alteration in sedimentary rocks related to petroleum accumulation. *Mailing Add:* Int Arc Econ Univ Wyo Univ Sta PO Box 3924 Laramie WY 82071

MILLER, DANIEL WEBER, b Omaha, Nebr, Jan 24, 26; m 47, 85; c 6. EXPERIMENTAL NUCLEAR PHYSICS. *Educ:* Univ Mo, BS, 47; Univ Wis, PhD(physics), 51. *Prof Exp:* Res assoc, 51-52, from asst prof to assoc prof, 52-62, assoc dean, Col Arts & Sci, 62-64, actg chmn, dept physics, 64-65, assoc dean res & advan studies, 72-73, PROF PHYSICS, IND UNIV, BLOOMINGTON, 62- *Concurrent Pos:* Consult, Los Alamos Nat Lab, 59-64 & 82-86; co-prin investr, Nuclear Reactions Res & 200 MeV Cyclotron Facil, Ind Univ, 63-, co-dir, 79-86; mem bd dirs, Midwest Univs Res Asn, 64-71 & prog adv comt, Los Alamos Meson Physics Facil, 82-85; deleg, Argonne Univ Asn, 66-81; chmn publ comt, Div Nuclear Physics, Am Phys Soc, 75-77. *Mem:* Fel Am Phys Soc; Sigma Xi; AAAS. *Res:* Nuclear reaction mechanisms and nuclear structure investigations utilizing light-ion charged particle beams at intermediate energies; polarization in nuclear reactions; total cross sections of nuclei for fast neutrons. *Mailing Add:* Physics Dept Ind Univ Bloomington IN 47405

MILLER, DARRELL ALVIN, b Lincoln, Ill, Sept 27; 32; m 53; c 4. GENETICS, PLANT BREEDING. *Educ:* Univ Ill, BS, 58, MS, 60; Purdue Univ, PhD(genetics, plant breeding), 62. *Prof Exp:* Asst prof plant breeding & genetics, NC State Univ, 62-64, asst prof in charge crop sci teaching, 64-65, asst dir of instr, Sch Agr & Life Sci, 65-67; assoc prof, 67-71, PROF PLANT BREEDING & GENETICS IN ALFALFA, UNIV ILL, URBANA, 71-, AGRON TEACHING COORD, 69- *Concurrent Pos:* Mem, Crop Sci Writing Conf, 65-74; agron coord for jr cols, 68- *Honors & Awards:* Funk Award. *Mem:* Fel Am Soc Agron; fel Crop Sci Soc Am; Nat Sci Teacher Asn; Sigma Xi; fel Nat Asn Cols & Teachers Agr. *Res:* Inheritance of cytoplasmic male sterility and its interaction in sorghum and alfalfa; alfalfa protein investigations, weevil resistance and physical genetics; allelophathic investigations with alfalfa. *Mailing Add:* Dept Agron Turner Hall Univ Ill Urbana IL 61801

MILLER, DAVID, b Chicago, Ill, Dec 31, 28; div. CHEMICAL ENGINEERING. *Educ:* Ill Inst Technol, BS, 50, MS, 51, PhD(chem eng), 55. *Prof Exp:* Asst purification lard, Ill Inst Technol, 51, asst chem eng, 51-52; res engr, Inst Gas Technol, 52, instr chem eng, 53-55; asst chem engr, Chem Eng Div, Argonne Nat Lab, 54-56, assoc chem engr, Int Inst Nuclear Sci & Eng, 56-64 & Reactor Physics Div, 64-71; TECH DIR, TOTAL SYSTS, 71- *Concurrent Pos:* Lectr, Northwestern Univ, 56; chmn, Nat Heat Transfer Conf, 64, co-chmn, 71; reporter, Int Heat Transfer Confs, 66 & 70; vis prof, Univ Ill, Chicago & Univ Ill, Urbana, 70-71; assoc prof, City Col Chicago, 73-; W Maxwell Reed Sem Mech Eng, Univ Ky, 81. *Mem:* Am Inst Chem Engrs; Am Soc Mech Engrs; Am Soc Eng Educ. *Res:* Thermal and mechanical design problems; physical properties of materials at extreme conditions; air and water pollution control; coal gasification and coal derived chemicals; recovery of materials from scrap tires; detection of plastic weapons and explosives. *Mailing Add:* Total Systs 5522 Main St Downers Grove IL 60516-1335

MILLER, DAVID A B, b Hamilton, UK, Feb 19, 54; m 76; c 2. OPTICAL PROPERTIES OF SEMICONDUCTORS, OPTICAL SWITCHING DEVICES. *Educ:* St Andrews Univ, UK, BSc, 76; Heriot-Watt Univ, UK, PhD(physics), 79. *Prof Exp:* Res assoc physics, Heriot-Watt Univ, UK, 79-80, lectr physics, 80-81; mem tech staff, 81-87, DEPT HEAD, AT&T BELL LABS, HOLMDEL, 87- *Honors & Awards:* Adolph Lomb Medal, Optical Soc Am, 86; RW Wood Prize, Optical Soc Am, 88. *Mem:* Fel Optical Soc Am; Lasers & Electrooptics Soc, IEEE; fel Am Phys Soc. *Res:* Nonlinear optical properties of semiconductors; physics and applications of quantum wells and other quantum-confined semiconductors; optical logic and switching devices, including quantum well self electro-optic effet devices; fundamentals of optical switching. *Mailing Add:* Photonics Switching Device Res Dept Room 4D-401 AT&T Bell Labs Holmdel NJ 07733

MILLER, DAVID ARTHUR, b Marion, Ohio, Jan 7, 42; m 72; c 1. MEDICAL PHYSIOLOGY. *Educ:* Ohio Northern Univ, BSEE, 64; Ohio Univ, MSEE, 66; Ohio State Univ, PhD(physiol), 72. *Prof Exp:* Lectr physiol, Ohio State Univ, 72-73; ASST PROF PHYSIOL, MED COL GA, 73- *Res:* Respiratory mechanics and control. *Mailing Add:* PO Box 11141 Augusta GA 30912

MILLER, DAVID BURKE, b Grand Rapids, Mich, Oct 8, 30; m 54; c 5. ENGINEERING PHYSICS. *Educ:* Colgate Univ, AB, 52, Univ Mich, MS, 53, PhD(elec eng), 61. *Prof Exp:* Physicist, Gen Elec Co, 53-56, res engr, 61-66; res assoc, Univ Mich, 56-61; assoc prof elec eng, Purdue Univ, 66-78; proj mgr, Brown Boveri Elec Co, 78-83; PROF ELEC ENG, MISS STATE UNIV, 83- *Mem:* Inst Elec & Electronics Engrs (secy, 82-83). *Res:* Experimental and theoretical studies of high voltage devices and phenomena; electromagnetic fields, gas discharges and pulsed power. *Mailing Add:* PO Drawer EE Mississippi State MS 39762

MILLER, DAVID CLAIR, entomology, invertebrate zoology, for more information see previous edition

MILLER, DAVID HARRY, b Callington Cornwall, Eng, Mar 3, 39; m 61; c 2. EXPERIMENTAL HIGH ENERGY PHYSICS. *Educ:* Imp Col Univ London, BSc, 60, PhD(high energy physics), 63. *Prof Exp:* Res assoc, 63-65, from asst prof to assoc prof, 65-76, PROF HIGH ENERGY PHYSICS, PURDUE UNIV, 76- *Concurrent Pos:* Guggenheim fel, 72; vis scientist, Europ Orgn Nuclear Res, Geneva, 72-73; res grantee & prin investr, US Dept Energy. *Mem:* Fel Am Phys Soc; AAAS. *Res:* Study of elementary particles using experimental techniques; electron positron annihilations at high energy. *Mailing Add:* Dept Physics Purdue Univ West Lafayette IN 47907

MILLER, DAVID HEWITT, b Russell, Kans, 1918; m 62; c 1. CLIMATOLOGY. *Educ:* Univ Calif, Los Angeles, AB, 39, MA, 44, PhD(geog), Berkeley, 53. *Hon Degrees:* DLitt, Univ Newcastle, 79. *Prof Exp:* Meteorologist, Corps Engrs, US Army, 41-43, forecaster, Transcontinental & Western Air Lines, 43-44; climatologist, Off Qm Gen, 44-46; meteorologist-hydrologist, snow invests, 46-50, asst dir, 50-53; chief environ anal br, Qm Res & Eng Lab, Mass, 53-58; meteorologist, US Forest Serv, Calif, 59-64; prof geog, 64-75, PROF ATMOSPHERIC SCI, DEPT GEOSCI, UNIV WIS-MILWAUKEE, 75- *Concurrent Pos:* NSF res fel, 52-53; vis lectr, Clark Univ, 57-58, Univ Ga, 58, Univ Calif, Berkeley, 61 & 63 & Univ Wis-Madison, 62; Fulbright lectr, Univ Newcastle, Australia, 66; Nat Acad Sci exchange scientist, USSR Acad Sci, 69; Fulbright sr scholar, Univs Newcastle, 71 & 79 & Macquarie, 71; sr acad meterologist, Nat Oceanic Atmospheric Admin, 81-82. *Mem:* Am Meteorol Soc; Inst Australian Geog; Ecol Soc Am; Sigma Xi; Am Geophys Union; Asn Am Geographers; Int Assoc Landscape Ecologists; Int Geog Union Comn Global Monitoring. *Res:* Snow hydrology; climatology; hydrology; energy/mass analysis; landscape structure. *Mailing Add:* Dept Geosci Univ Wis-Milwaukee Milwaukee WI 53201

MILLER, DAVID JACOB, b St Louis, Mo, Sept 20, 10; m 36; c 3. PHARMACEUTICAL CHEMISTRY. *Educ:* Washington Univ, St Louis, BS, 31. *Prof Exp:* Chemist, analysis & res, Food & Drug Admin, 37-58, asst to dir, Bur Field Admin, 58-61, chemist, off comnr, 61-64, dep dir, div cosmetics & colors, 65-69; CONSULT, FOOD, DRUGS & COSMETICS INDUST, 69- *Mem:* fel Am Inst Chemists; fel Royal Soc Chem; Soc Cosmetic Chemists; Inst Food Technologists. *Res:* Methods of analysis for drugs; alkaloids in pharmaceuticals. *Mailing Add:* 6112 32nd Pl NW Washington DC 20015

MILLER, DAVID LEE, solid state physics, for more information see previous edition

MILLER, DAVID LEE, b Knoxville, Tenn, Aug 15, 38; m 61; c 2. PROTEIN CHEMISTRY, MOLECULAR BIOLOGY. *Educ:* Oberlin Col, BA, 60; Harvard Univ, PhD(chem), 66. *Prof Exp:* NIH fel, 65-66; asst prof chem, Oberlin Col, 66-68; staff fel, Nat Heart Inst, 68-69; res assoc, 69-71, asst mem, Roche Inst Molecular Biol, 71-81; HEAD, MOLECULAR BIOL DEPT, NY STATE BASIC RES DEVELOP DISABILITIES, 81- *Mem:* Am Chem Soc; Fedn Am Soc of Exp Biol. *Res:* Structure and function of GTP-regulatory proteins; mechanism of protein biosynthesis; brain protein biochemistry. *Mailing Add:* Molecular Biol NY State Inst Basic Res 1050 Forest Hill Rd Staten Island NY 10314

MILLER, DAVID S, b Brooklyn, NY, July 24, 45; m 78; c 1. MEMBRANE BIOCHEMISTRY, CELL BIOLOGY. *Educ:* Brooklyn Col, BS, 66; Univ Maine, PhD(biochem), 73. *Prof Exp:* Assoc res scientist, Mt Desert Island Biol Lab, 73-78, res scientist, 78-81; chief, Lab Metab Control, 81-85, ADJ SCIENTIST, MICH CANCER FOUND, 85-; expert, Lab Cellular & Molecular Pharmacol, 85-90, RES PHYSIOLOGIST, NAT INST ENVIRON HEALTH SCI, NIH, 90- *Concurrent Pos:* Prin investr, US Pub Health Serv & lectr biochem dept, Univ Maine, 78-81. *Mem:* AAAS; Am Physiol Soc; Sigma Xi. *Res:* Biochemistry and physiology of epithelial membrane transport; hormonal control of cellular metabolism. *Mailing Add:* Lab Pharmacol NIEHS NIH PO Box 12233 Research Triangle Park NC 27709

MILLER, DENNIS DEAN, b Webster, SDak, Feb 26, 45; m 75. FOOD SCIENCE, NUTRITION. *Educ:* Augsburg Col, BA, 67; Univ Wash, MS, 69; Cornell Univ, PhD(nutrit), 78. *Prof Exp:* Instr chem, NDak State Sch Sci, 68-70; asst prof chem, Univ Minn Tech Col, 70-74; NSF fac fel, Univ Wis, 74-75; ASST PROF FOOD SCI, CORNELL UNIV, 78- *Mem:* Inst Food Technologists; AAAS. *Res:* In vitro methodology for assessment of food iron and zinc bioavailabilities; methods development for using iron stable isotopes as biological tracers; study of factors and mechanisms that influence the bioavailability of iron and zinc in human foods. *Mailing Add:* Dept Food Sci Stocking Hall Cornell Univ Ithaca NY 14853

MILLER, DEREK HARRY, b Hull, Eng, Jan 18, 23; m 47; c 3. PSYCHIATRY, PSYCHOANALYSIS. *Educ:* Univ Leeds, MB, ChB, 47, MD, 55. *Prof Exp:* Psychiatrist, Menninger Found, Topeka, Kans, 55-59; dir adolescent unit, Tavistock Clin, London, 59-69; dir adolescent psychiat prog, Med Sch, Univ Mich, Ann Arbor, 69-75, prof psychiat & assoc chmn, 75-76, PROF PSYCHIAT, NORTHWESTERN UNIV, CHICAGO & CHIEF ADOLESCENT PROG, NORTHWESTERN MEM HOSP, 76- *Concurrent Pos:* Assoc mem, Inst Psychoanal, London, 64-68; lectr, Inst Sociol, Bedford Col & Inst Archit, Univ Cambridge, 65-69; WHO training grant, 67-69; consult, W S Hall Psychiat Inst, Columbia, SC, 72-79. *Honors & Awards:* Samuel G Hibbs Award, Am Psychiat Asn. *Mem:* Brit Psychoanal Soc; Am Psychiat Asn; Am Col Psychiat; Int Soc Adolescent Psychiatrists; Royal Col Psychiatrists. *Res:* Adolescent psychiatry, drug abuse, prediction of homicidal behavior and relationship between physical maturation and psychological development. *Mailing Add:* 259 E Erie Chicago IL 60611

MILLER, DON CURTIS, b Oakland, Calif, Apr, 29, 35; m 60; c 2. MARINE ENVIRONMENTAL BIOLOGY. *Educ:* Univ Del, BA, 57; Duke Univ, MA, 62, PhD(zool), 65. *Prof Exp:* Lectr, Queens Col, City Univ New York, 62-65, instr, 65-67; asst prof, Union Col, 67-71; res aquatic biol, 71-73, RES GROUP LEADER, ENVIRON PROTECTION AGENCY, 73- *Concurrent Pos:* Interim dir, I C Darling Marine Ctr, Univ Maine, 65; res assoc, Baruch Coastal Res Inst, Univ SC, 73-; adj prof zool, Univ RI, 75-; US Rep, Int Hydrol Prog, UN Educ Sci & Cult Orgn, 76-79. *Mem:* AAAS; Am Soc Limnol & Oceanog; Ecol Soc Am; New Eng Estuarine Soc. *Res:* Physiological ecology of estuarine animals; crustacean biology; thermal biology; power plant impact assessment; aquatic toxicology; sublethal pollutant effects (development, physiological, behavioral); ocean disposal hazard assessment; marine water quality criteria. *Mailing Add:* Environ Res Lab Environ Protection Agency S Ferry Rd Narragansett RI 02882

MILLER, DON DALZELL, b Menomonie, Wis, Apr 8, 13. SEMIGROUPS. *Educ:* Wayne State Univ, AB, 34, MA, 36; Univ Mich, PhD(math), 41. *Prof Exp:* Instr math, Lawrence Inst Technol, 35-36 & Univ Ohio, 38-42; from assoc prof to prof, 46-78, EMER PROF MATH, UNIV TENN, KNOXVILLE, 78- *Concurrent Pos:* Fulbright lectr, Univ Besancon, 64-65. *Mem:* Am Math Soc; Math Asn Am. *Res:* Semigroups, semirings, binary relations. *Mailing Add:* Lucina 5913 Shell Point Village Ft Myers FL 33908

MILLER, DON WILSON, b Westerville, Ohio, Mar 16, 42; m 66; c 3. NUCLEAR ENGINEERING, PHYSICS. *Educ:* Miami Univ, BSc, 64, MSc, 66; Ohio State Univ, PhD(nuclear eng), 71. *Prof Exp:* Res engr, NAm Aviation, 63-64; from asst prof to assoc prof, 71-80, PROF NUCLEAR ENG, OHIO STATE UNIV, CHMN DEPT & DIR, NUCLEAR REACTOR LAB, 77- *Honors & Awards:* Glenn Murphy Award, Am Soc Eng Educ, 89. *Mem:* Am Nuclear Soc; Am Soc Eng Educ; Inst Elec & Electronics Engrs. *Res:* Nuclear reactor instrumentation and control; nuclear medical instrumentation; application of artificial intelligence; digital x-ray systems. *Mailing Add:* Dept Nuclear Eng Ohio State Univ 206 W 18th Ave 1075 Robinson Columbus OH 43210

MILLER, DONALD ELBERT, b Germano, Ohio, Apr 6, 06; m 31; c 1. BIOLOGY. *Educ:* Thiel Col, AB, 25 & 28; Univ Mich, MS, 29, PhD(zool), 35. *Prof Exp:* Teacher high schs, Pa, 25-28, 30-31; asst zool, Univ Mich, 29-30; instr biol, Gustavus Adolphus Col, 31-33; zool, Univ Idaho, 35-36; asst prof sci, 36-39, from assoc prof to prof, 39-72, EMER PROF BIOL, BALL STATE UNIV, 72- *Res:* Limnology. *Mailing Add:* 5949 Shagway Rd W RR 2 Ludington MI 49431

MILLER, DONALD F, b Laurel, Md, Aug 9, 24; m 77; c 1. FOOD COMPOSITION, FOOD REGULATIONS. *Educ:* Univ Md, BS, 50; Wash State Univ, MS, 52. *Prof Exp:* Tech secy, Comt Feeding Compos, Nat Acad Sci, 52-58; nutrit analyst, US Agr Dept, 59-66; nutrit analyst, Food & Drug Admin, 66-78; CONSULT, 78- *Mem:* Am Inst Nutrit; Am Asn Cereal Chemists; Inst Food Technologists; NY Acad Sci; Am Asn Animal Sci; Nutrit Today Soc. *Res:* Food regulations. *Mailing Add:* Rt Three Box 810 Boone NC 28607

MILLER, DONALD GABRIEL, b Oakland, Calif, Oct 29, 27; m 49; c 2. PHYSICAL CHEMISTRY ELECTROLYTES, DIFFUSION. *Educ:* Univ Calif, BS, 49; Univ Ill, PhD(phys chem), 53. *Prof Exp:* Asst chem, Univ Ill, 49-52; asst prof, Univ Louisville, 52-54; res assoc, Brookhaven Nat Lab, NY, 54-56; CHEMIST, LAWRENCE LIVERMORE NAT LAB, UNIV CALIF, 56- *Concurrent Pos:* Fulbright prof, Univ Lille & Cath Univ Lille, 60-61; vis fel, Australian Nat Univ, 79 & 81; guest prof, Tech High Sch, Aachen, WGermany, 85 & 88; contract prof, Univ Naples, Italy, 83, 87 & 90. *Mem:* Am Chem Soc; Math Asn Am; Asn Symbolic Logic. *Res:* Thermodynamics of irreversible processes; electrolyte solutions; diffusion; history of science. *Mailing Add:* Lawrence Livermore Nat Lab Univ Calif Livermore CA 94550

MILLER, DONALD MORTON, b Chicago, Ill, July 24, 30; div; c 1. COMPARATIVE PHYSIOLOGY. *Educ:* Univ Ill, AB, 60, MA, 62, PhD(physiol), 65. *Prof Exp:* Sci asst org chem, Polymer Res Lab, Univ Ill, 60, asst physiol & biophys, 60-63, 63-64, comp physiol trainee, 63, 65; from asst prof to assoc prof, 66-75, PROF PHYSIOL, SOUTHERN ILL UNIV, CARBONDALE, 76- *Concurrent Pos:* USPHS fel protozool & parasitol, Univ Calif, Los Angeles, 55-66. *Mem:* Biophys Soc; Am Soc Zool; Am Micros Soc; Bot Soc Am; Am Physiol Soc. *Res:* Marine toxins; parasite nervous systems and halotolerant organisms. *Mailing Add:* Dept Physiol Southern Ill Univ Carbondale IL 62903

MILLER, DONALD NELSON, b St Louis, Mo, Aug 25, 23; m 57; c 3. CHEMICAL ENGINEERING. *Educ:* Washington Univ, BSChE, 43, MS, 57; Univ Wis, PhD(chem eng), 55. *Prof Exp:* Consult, 55-70, sr consult chem eng, 70-79, PRIN CONSULT, KINETICS & REACTOR DESIGN, E I DU PONT DE NEMOURS & CO, INC, 80- *Mem:* Am Inst Chem Engrs; Am Chem Soc. *Res:* Chemical kinetics; mass transfer; reactor scaleup. *Mailing Add:* 108 Hitching Post Dr Surrey Park Wilmington DE 19803-1913

MILLER, DONALD PIGUET, b New Orleans, La, Oct 11, 27; m 51; c 3. CHEMICAL PHYSICS, CRYSTALLOGRAPHY. *Educ:* Agr & Mech Col, Tex, BS, 48; Tulane Univ, MS, 52; Polytech Inst Brooklyn, PhD, 62. *Prof Exp:* Asst biophys, Tulane Univ, 51; fel physics, Polytech Inst Brooklyn, 53-54, instr, 54-57; mem tech staff, Cent Res Lab, Tex Instruments, Inc, 57-63; from assoc prof to prof physics, Clemson Univ, 77-91; RETIRED. *Mem:* Am Crystallog Asn; Am Inst Chemists. *Res:* X-ray crystallography; electron diffraction; solid state physical chemistry and chemical physics. *Mailing Add:* 111 A-Coleman Ave PO Box 423 Waveland MS 39576-0423

MILLER, DONALD RICHARD, b Hamilton, Ont, July 4, 36; m 69. MATHEMATICAL BIOLOGY. *Educ:* Univ Toronto, BA, 60, MA, 61, PhD(appl math), 64. *Prof Exp:* Instr math, Univ Toronto, 61-63; asst prof, Univ Western Ont, 63-65, assoc prof appl math, 65-68, res assoc cancer res lab, 68-69; assoc prof indust & systs eng, Univ Fla, Cape Canaveral, 69-71; assoc prof & grad coord, Univ Fla, Gainesville, 71-73; GROUP LEADER, BIOMATH & ECOTOXICOL, DIV OF BIOL SCI, NAT RES COUN CAN 73- *Concurrent Pos:* Mem, Can Comt Man & the Biosphere, MAB Secretariat, Environ Can, Ottawa. *Mem:* Am Col Toxicol; Soc Toxicol Can. *Res:* Environmental studies; pollutant transport; toxic effects and mechanisms of environmental pollutants in agnatic systems; risk assessment. *Mailing Add:* 14 Mt Pleasant Ave Ottawa ON K1S 0L8 Can

MILLER, DONALD SPENCER, b Ventura, Calif, June 12, 32; m 54; c 3. GEOCHEMISTRY. *Educ:* Occidental Col, AB, 54; Columbia Univ, AM, 56, PhD, 60. *Prof Exp:* Asst, Lamont Geol Observ, Columbia Univ, 54-59; res scientist, 59-60; from asst prof to assoc prof, 60-69, chmn dept, 69-76 & 80-90, PROF GEOCHEM, RENSSELAER POLYTECH INST, 69- *Concurrent Pos:* NSF sci fac fel & guest prof, Univ Berne, 66-67, vis prof Isotope Geol Lab, 79; guest res prof, Max-Planck-Inst Nuclear Physics, Heidelberg, 77-78, guest scientist, 79, 80, 81 & 82; partic, NATO Sci Exchange Prog, Demokritos, Greece, 83 & 85; vis res fel, Univ Melbourne, 88. *Mem:* Fel Geol Soc Am; Nat Asn Geol Teachers; Geochem Soc; Am Geophys Union; Sigma Xi. *Res:* Geochronology; fission track techniques in geology; use of quantitative age methods to reveal uplift rates of igneous and sedimentary rocks and to determine their thermal history. *Mailing Add:* Dept of Geol Rensselaer Polytech Inst Troy NY 12181

MILLER, DONALD WRIGHT, b Columbus, Wis, May 28, 27; m 52; c 6. MATHEMATICS. *Educ:* Univ Wis, BS, 50, MS, 51, PhD(math), 57. *Prof Exp:* From instr to assoc prof, 55-69, PROF MATH, UNIV NEBR, LINCOLN, 69- *Mem:* Am Math Soc; Math Asn Am; London Math Soc. *Res:* Algebra, especially the structure of semigroups. *Mailing Add:* Dept Math Univ Nebr Lincoln NE 68588-0323

MILLER, DOROTHEA STARBUCK, b Iowa City, Iowa, Nov 12, 08; m 31. ZOOLOGY. *Educ:* Univ Iowa, BA, 28, MS, 35, PhD(zool), 38. *Prof Exp:* Asst zool, Univ Iowa, 33-38; asst prof, Conn Col, 39-43; instr, Univ Wis, 44-45; res asst, Toxicity Lab, Univ Chicago, 45-46, from instr to asst prof biol sci, 46-63, res assoc zool & asst dean students, Biol Sci Div, 54-63; asst prog dir, NSF, Washington, DC, 63-64; PROG ADMINSTR, GENETICS TRAINING & ANAT SCI TRAINING COMTS, NAT INST GEN MED SCI, 64- *Concurrent Pos:* USPHS res grants, 48-50, 54-63. *Mem:* AAAS; Am Soc Zool; Radiation Res Soc; Genetics Soc Am. *Res:* Audiogenic seizures in mice; mammalian genetics; influence of low-level radiation on audiogenic seizures. *Mailing Add:* Nat Inst Gen Med Sci WB 918 Bethesda MD 20205

MILLER, DOROTHY ANNE SMITH, b New York, NY, 1931; m 54; c 3. GENETICS, CYTOGENETICS. *Educ:* Wilson Col, BA, 52; Yale Univ, PhD(biochem), 57. *Prof Exp:* From res asst to sr res assoc, Col Physicians & Surgeons, Columbia Univ, 63-85, asst prof human genetics, 73-80; PROF PATH & MOLECULAR BIOL & GENETICS, SCH MED, WAYNE STATE UNIV, 85- *Concurrent Pos:* Vis scientist, Med Res Coun, Clin & Pop Cytogenetics Unit, Edinburgh, UK, 84-85; vis prof, Univ La Sapienza, Rome, Italy, 88. *Mem:* Genetics Soc Am; Am Soc Human Genetics. *Res:* Chromosome analysis of human, mouse and interspecific somatic cell hybrids; gene mapping by in situ hybridization. *Mailing Add:* Dept Molecular Biol & Genetics Wayne State Univ Sch Med Detroit MI 48201

MILLER, DOUGLAS CHARLES, b Cincinnati, Ohio, Nov 5, 56; m 81; c 2. BENTHIC ECOLOGY, SEDIMENT TRANSPORT. *Educ:* Univ Notre Dame, BS, 79; Univ Wash, MS, 81, PhD(oceanog), 85. *Prof Exp:* ASST PROF MARINE STUDIES, COL MARINE STUDIES, UNIV DEL, 85- *Mem:* Am Soc Limnol & Oceanog; Am Geophys Union; Crustacean Soc; NAm Benthological Soc. *Res:* Feeding biology of benthic organisms, especially detritivores and deposit-feeding invertebrates; effects of flow and sediment transport on feeding strategies. *Mailing Add:* Col Marine Studies Univ Del Lewes DE 19958-1298

MILLER, DOUGLAS KENNETH, b Devils Lake, NDak, Dec 8, 47; m 78; c 3. BIOCHEMISTRY, CELL PHYSIOLOGY. *Educ:* Univ NDak, BS, 70; Harvard Univ, MA, 71, PhD(biochem), 76. *Prof Exp:* Res fel path, Med Sch, Tufts Univ, 75-78; adj asst prof physiol, Col Med & Dent NJ-Rutgers Med Sch, 78-82; sr res biochemist, 82-85, res fel, 85-90, SR RES FEL, MERCK & CO, 90- *Concurrent Pos:* Teaching fel biochem, Harvard Univ, 71-75. *Mem:* Am Soc Biochem & Molecular Biol; Soc Complex Carbohydrates; Am Soc Cell Biol. *Res:* Glycoprotein biosynthesis and turnover; arachidonic acid metabolism; cell virus interactions; lysosomes; neutrophil activation; inflammation; neutrophil receptors; interleukin 1 synthesis. *Mailing Add:* Dept Immunol & Inflammation Merck Inst Therapeut Res PO Box 2000 Rahway NJ 07065-0900

MILLER, DOUGLASS ROSS, b Monterey Park, Calif, Feb 15, 42; m 64; c 2. ENTOMOLOGY. *Educ:* Univ Calif, Davis, BS, 64, MS, 65, PhD(entom), 69. *Prof Exp:* RES ENTOMOLOGIST, SYST ENTOM LAB, AGR RES SERV, USDA, 69- *Concurrent Pos:* Adj assoc prof entom, Univ Md, 73-; res assoc, Dept Entom, Smithsonian Inst, 78-; res leader, Syst Entom Lab, USDA, 85- *Mem:* Pan-Pac Entom Soc; Entom Soc Am; Soc Syst Zool; Am Inst Biol Sci. *Res:* Systematics of scale insects with emphasis on the families Pseudococcidae and Eriococcidae. *Mailing Add:* 511 Hexton Hill Rd Silver Spring MD 20904

MILLER, DUANE DOUGLAS, b Great Bend, Kans, July 15, 43; m 62; c 3. MEDICINAL CHEMISTRY. *Educ:* Univ Kans, BS Pharm, 66; Univ Wash, PhD(med chem), 69. *Prof Exp:* From asst prof to assoc prof, 69-80, PROF MED CHEM, COL PHARM, OHIO STATE UNIV, 80-, CHMN DEPT, 83- *Mem:* Am Chem Soc; AAAS; Am Asn Physicians & Surgs. *Res:* Drugs used to treat asthma; mechanism of adrenergic and dopaminergic drugs; inhibitors of platelet aggregation. *Mailing Add:* Div Med Chem Col Pharm Ohio State Univ 500 W 12th Ave Columbus OH 43210

MILLER, DWANE GENE, b Cheyenne, Wyo, May 15, 34; m 59; c 2. AGRONOMY. *Educ:* Univ Wyo, BS, 60, MS, 64, PhD(crop sci), 66; Iowa State Univ, PhD, 65. *Prof Exp:* Instr pub sch, Wyo, 60-62; asst prof biol, Southern Ore Col, 66-67; from asst prof to assoc prof agron, Wash State Univ, 67-77; prof plant & soil sci & chmn dept, Tex Tech Univ, 77-; AT DEPT PLANT & SOIL SCI, MONT STATE UNIV. *Mem:* Crop Sci Soc; Nat Asn Cols & Teachers Agr; Am Soc Agron. *Res:* Hybrid wheat, especially artificial induction of male sterility; histological and physiological studies on gametocides and their action; simulated hail studies in wheat and peas; minimum tillage practices in peas; water use studies in sorghum; teaching of agronomy. *Mailing Add:* Dept Plant & Soil Sci Mont State Univ Bozeman MT 59717

MILLER, E(UGENE), b New York, NY, Feb 2, 22; m 47; c 2. CHEMICAL ENGINEERING. *Educ:* City Col New York, BChE, 44; Polytech Inst Brooklyn, MChE, 47; Univ Del, PhD(chem eng), 49. *Prof Exp:* Test engr, Soconv-Vacuum Oil Co, 44; develop engr, MW Kellogg Co Div, Pullman Inc, 44-47, res engr, 49-52; chief res, Army Rocket & Guided Missile Agency, Redstone Arsenal, 52-57; dir res & develop, Olin Mathieson Chem Corp, 57-59; consult, 59-60; asst tech dir, Lockheed Propulsion Co, 60-65, asst to dir res, Lockheed Palo Alto Res Lab, 65-67, mgr propulsion eng, Lockheed Missiles & Space Co, 67-69; pres, Browning Arms Co Inc, 69-73; prof chem eng, 73-85, dept chmn, 81-84, EMER PROF, DEPT CHEM & METAL ENG, MACKAY SCH MINES, UNIV NEV, RENO, 85- *Mem:* Am Chem Soc; Am Inst Chem Engrs. *Res:* Combustion; smoke; thermodynamics. *Mailing Add:* Dept Chem & Metall Eng Univ Nev Reno NV 89557

MILLER, EDMUND K(ENNETH), b Milwaukee, Wis, Dec 24, 35; m 58; c 2. COMPUTER GRAPHICS, SIGNAL PROCESSING. *Educ:* Mich Technol Univ, BS, 57; Univ Mich, MS, 58 & 61, PhD(elec eng), 65. *Prof Exp:* Asst res engr, Radiation Lab, Univ Mich, 58-65, assoc res engr, High Altitude Eng Lab, 66-68; sr staff scientist, MB Assocs, Calif, 68-71; group leader & div leader, Lawrence Livermore Nat Lab, 71-85; Regents distinguished prof elec & comput eng, Univ Kans, 85-87; mgr electromagnetics, Rockwell Int Sci Ctr, 87-88; dir, Electromagnetics Res Oper, Gen Res Corp, 88-89; group leader, Los Alamos Nat Lab, 89. *Concurrent Pos:* Instr physics, Mich Technol Univ, 58-59. *Honors & Awards:* Cert Achievement, Inst Elec & Electronic Engrs, Educ Media Coun Soc, 85. *Mem:* Fel Inst Elec & Electronics Engrs; Am Phys Soc; Sigma Xi; Int Sci Radio Union; Optical Soc Am; Acoust Soc Am; Appl Computational Electromagnetics Soc. *Res:* Computer modeling applications in electromagnetics, and the development of associated numerical methods; time-domain analysis; signal processing and inverse problems; innovative instructional technology. *Mailing Add:* 3225 Calle Celestial Santa Fe NM 87505

MILLER, EDWARD, b Newark, NJ, Mar 10, 22; m 47. MECHANICAL ENGINEERING. *Educ:* Newark Col Eng, BS, 48; Univ Del, MME, 49; Columbia Univ, MA, 51; Stevens Inst Technol, MS, 52; NY Univ, MAeroE, 59. *Prof Exp:* Mech designer, Crucible Steel Corp Am, 41-44; assoc prof mech eng, 48-60, asst exec assoc, 48-63, assoc chmn, 63-70, actg dean eng, 80-81, PROF MECH ENG, NJ INST TECHNOL, 60-, ASSOC DEAN, 81- *Concurrent Pos:* Consult, Savoy Trunk Co, 52-57, Burt Kaplan Asn, 57-58 & Mohawk Refining Co, 58-59; vis lectr, Rutgers Univ, 54-57; adj assoc prof, NY Univ, 59; mem, Nat Bd Gov, Order of Engineer. *Honors & Awards:* Allen R Cullimore Award. *Mem:* Am Soc Mech Engrs; Am Soc Eng Educ; Nat Soc Prof Engrs. *Res:* Experimental stress analysis; dynamic and vibration analysis. *Mailing Add:* NJ Inst Technol Univ Heights Newark NJ 07102

MILLER, EDWARD, b Monticello, NY, July 30, 32; m 61. MATERIALS SCIENCE. *Educ:* City Col New York, BChE, 53; NY Univ, MS, 55, DEngSc(metall), 59. *Prof Exp:* Asst eng sci, NY Univ, 53-59, assoc, 59-61, from asst prof to assoc prof, 61-68; assoc prof mech eng, 68-73, PROF MECH ENG, CALIF STATE UNIV, LONG BEACH, 73- *Honors & Awards:* Ralph R Teetor Award, Am Soc Automotive Engrs, 75; Linback Award, 75. *Mem:* Am Inst Mining, Metall & Petrol Engrs; Am Soc Eng Educ; Soc Plastics Engrs. *Res:* Phase relationships and thermodynamic properties of solids and liquids; mechanical properties of polymers. *Mailing Add:* Dept Mech Eng Calif State Univ 1250 Bellflower Blvd Long Beach CA 90840

MILLER, EDWARD GEORGE, b Columbiana, Ohio, Mar 29, 34; m 57; c 2. ORGANIC CHEMISTRY. *Educ:* Manchester Col, AB, 56; Cornell Univ, PhD(org chem), 61. *Prof Exp:* From instr to assoc prof, 60-74, PROF CHEM & CHMN DEPT, MANCHESTER COL, 74- *Concurrent Pos:* USPHS fel, Princeton Univ, 65-66; vis prof, Silliman Univ, Philippines, 71-72, Ind Univ, Bloomington, 84, 87; dir, Brethren Cols Abroad, Sapporo, Japan, 90-91. *Mem:* AAAS; Am Chem Soc. *Res:* Reaction mechanisms and stereochemistry. *Mailing Add:* Dept Chem Manchester Col North Manchester IN 46962

MILLER, EDWARD GODFREY, JR, b Pittsburgh, Pa, Feb 16, 41; m 64; c 1. BIOCHEMISTRY, ONCOLOGY. *Educ:* Univ Tex, BS, 63, PhD(chem), 69. *Prof Exp:* Fel, McArdle Labs, Univ Wis, 69-72; asst prof, 72-75, assoc prof, 75-88, PROF BIOCHEM & MICROBIOL, BAYLOR UNIV, 88- *Mem:* Am Soc Cell Biol; Sigma Xi; Am Asn Dent Res; Soc Exp Biol Med. *Res:* Nutrition and cancer; modification of nuclear proteins; synthesis of poly ADP-ribose; DNA repair; oral oncology. *Mailing Add:* Dept of Biochem Baylor Col of Dent Dallas TX 75246

MILLER, EDWARD JOSEPH, b Akron, Ohio, Oct 27, 35; m 64; c 3. BIOPHYSICS. *Educ:* Spring Hill Col, BS, 60; Univ Rochester, PhD(radiation biol), 64. *Prof Exp:* Res assoc biochem, Nat Inst Dent Res, 63-71; PROF BIOCHEM, UNIV ALA, BIRMINGHAM, 71- *Honors & Awards:* Award for Basic Res Oral Sci, Int Asn Dent Res, 71; Carol Nachman Prize Rheumatol, 78. *Mem:* AAAS; Am Chem Soc; Am Rheumatol Asn; Am Soc Biochem & Molecular Biol; Protein Soc. *Res:* Chemistry and biology of the genetically distinct collagens. *Mailing Add:* Univ of Ala Med Ctr Univ Sta Birmingham AL 35294

MILLER, ELIZABETH ESHELMAN, b Waxahachie, Tex, Aug 6, 19; m 44. BIOCHEMISTRY, IMMUNOLOGY. *Educ:* Univ Colo, BS, 43; Univ Pa, MS, 54, PhD(med microbiol), 55. *Prof Exp:* Technician plant path, Rockefeller Inst, 43-45; asst chem, Biochem Res Found, 45-46; asst gen biochem, Inst Cancer Res, 46-52; asst med microbiol, Henry Phipps Inst, Univ Pa, 52-55; res assoc, Cancer Res Unit & Dept Biochem, Sch Med, Tufts Univ, 56-57; res assoc protein chem & immunol, Bio-Res Inst, 57-61, res assoc, Univ Pittsburgh, 61-62; ASSOC SURG RES, UNIV PA, 62-, ASST PROF SURG RES, 73- *Mem:* Am Chem Soc; AAAS; Sigma Xi. *Res:* Chemistry and immunology of cancer; primarily with mouse and rat tumors; includes study of tumor antigens, preparation of monoclonal antibodies and lymphokines; testing for antitumor activity in animals or tissue culture. *Mailing Add:* 1333 Prospect Hill Rd Villanova PA 19085

MILLER, ELWOOD MORTON, b Barnesville, Ohio, July 15, 07; m 31; c 1. ZOOLOGY. *Educ:* Bethany Col, BS, 29; Univ Chicago, MS, 30, PhD(zool), 41. *Hon Degrees:* DSc, Bethany Col, 62. *Prof Exp:* From instr to prof, 30-74, EMER PROF ZOOL, UNIV MIAMI, 74- *Concurrent Pos:* Asst to dir biol Century Prog Expos, Chicago, 33; chmn dept zool, Univ Miami, 46-53, dean arts & sci, 53-66; mem, State Bd Exam Basic Sci, Fla, 46-67. *Mem:* Am Soc Zool. *Res:* Termite biology; social insects. *Mailing Add:* Unit 112 Electra Ct 3300 Carpenter Rd SE Olympia WA 98503

MILLER, ELWYN RITTER, b Edon, Ohio, Dec 10, 23; m 51; c 5. ANIMAL NUTRITION. *Educ:* Mich State Univ, BS, 48, PhD(animal nutrit), 56. *Prof Exp:* Instr educ, 51-52, from asst to assoc prof, 52-66, PROF NUTRIT, MICH STATE UNIV, 66- *Honors & Awards:* Am Feed Mfg Nutrit Award, Am Soc Animal Sci, 65, Gustav Bohstedt Trace Mineral Res Award, 69; Calcium Carbonate Mineral Res Award, Nat Feed Ingredients Asn, 66. *Mem:* AAAS; Am Soc Animal Sci; Soc Exp Biol & Med; Am Inst Nutrit; Sigma Xi. *Res:* Hematology, immunology and nutritional requirements of baby pig; mineral nutrition of swine; normal growth and physiological development of swine fetus. *Mailing Add:* Dept Animal Sci Mich State Univ East Lansing MI 48824

MILLER, EMERY B, b Bloomington, Ill, Aug 11, 25; m 48; c 4. ORGANIC CHEMISTRY. *Educ:* Univ Ill, BS, 47; Rice Univ, MS, 49, PhD(org & phys chem), 51. *Prof Exp:* Res chemist, Org Div, Monsanto Chem Co, Mo, 51-53; chem res dir, Maumee Chem Co, Ohio, 53-64; pres, Peninsular Chem Res, Fla, 64-65; mgr res & develop, Houston Res Inst, 65-67; res dir, Tenneco Hydrocarbon Chem Div, 67-69; PRES, EMCHEM CORP, 69- *Mem:* Am Chem Soc; Royal Soc Chem. *Mailing Add:* 5447 Paisley Houston TX 77096-4025

MILLER, EUGENE, b New York, NY, Feb 2, 22; m 47; c 2. COMBUSTION, THERMODYNAMICS. *Educ:* City Col New York, BChE, 44; Brooklyn Polytech Inst, MChE, 47; Univ Del, PhD(chem eng), 49. *Prof Exp:* Chief res, US Army Missile Command, 52-57; dir res & develop, Solid Propellant Orgn, Olin Corp, 57-60; asst tech dir, Lockheed Propulsion Co, 60-65, mgr, Propulsion Mech Systs, Lockheed Missiles & Space Co, 65-69; pres, Browning Arms Co, 69-72; prof, 73-85, EMER PROF CHEM ENG, UNIV NEV, RENO, 85- *Concurrent Pos:* Consult, 52-; gen partner, Miller Assocs, 85- *Mem:* Am Inst Chem Engrs; Am Chem Soc. *Res:* Signatures and visibility of rocket plumes; inhibition of rocket plume after burning; vapor-liquid equilibrium of hydrogen chloride-water-salt below zero degrees centigrade. *Mailing Add:* PO Box 4361 Incline Village NV 89450

MILLER, EUGENE D, b Wilkes Barre, Pa, June 18, 31; m; c 3. PHYSICAL CHEMISTRY. *Educ:* King's Col, Pa, BS, 55; Catholic Univ, PhD(phys chem), 61. *Prof Exp:* Res chemist, Atlantic Ref Co, 60-64; assoc prof phys & anal chem, Cheyney State Col, 64-67; PROF CHEM, LUZERNE COUNTY COMMUNITY COL, 67- *Mem:* Am Chem Soc. *Res:* Free radical reactions and heterogeneous catalysis. *Mailing Add:* Luzerne County Community Col Nanticoke PA 18634

MILLER, FLOYD GLENN, b Chicago, Ill, May 25, 35; m 62; c 3. MAINTENANCE & SAFETY ENGINEERING. *Educ:* Univ Ill, Urbana, BS, 57, PhD(mech eng), 61. *Prof Exp:* Advan mfg engr, Bell & Howell Co, 60-62; tech develop mgr, 3M Co, 62-66; asst mgr systs & planning, Northern Trust Co, 66-69, mgr, 69-70; ASST TO DIR PHYS PLANT & ASST PROF INDUST ENG, UNIV ILL, CHICAGO CIRCLE, 71-, actg dept head, 79-81. *Concurrent Pos:* Mem mgt fac, DePaul Univ, 70- *Mem:* Asn Systs Mgt; Am Inst Indust Engrs. *Res:* Work measurement; work simplification; maintenance systems design; standard data techniques; physical plant systems design; automated reporting systems; safety systems analysis. *Mailing Add:* Dept Mech Eng MC251 Univ Ill Box 4348 Chicago IL 60680

MILLER, FLOYD LAVERNE, physical chemistry, for more information see previous edition

MILLER, FOIL ALLAN, b Aurora, Ill, Jan 18, 16; m 41; c 2. PHYSICAL CHEMISTRY, STRUCTURAL CHEMISTRY. *Educ:* Hamline Univ, BS, 37; Johns Hopkins Univ, PhD(chem), 42. *Prof Exp:* Nat Res Coun fel, Univ Minn, 42-44; asst prof chem, Univ Ill, 44-48; head spectros div, Mellon Inst, 48-58, sr fel fundamental res, 58-67; prof & dir, Spectros Lab, 67-81, EMER PROF CHEM, UNIV PITTSBURGH, 81- *Concurrent Pos:* Lectr, Bowdoin Col, 50-, Univ Pittsburgh, 52-63, adj prof, 63-67; Guggenheim fel, 57-58; ed, Spectrochimica Acta, 57-63; Reilly lectr, Univ Notre Dame, 67; adj sr fel, Mellon Inst, 67-74, mem comn molecular spectros, Int Union Pure & Appl Chem, 67-75, secy, 69-75; vis prof, Tohoku Univ, Sendai, Japan, 77, Brazil, 80. *Honors & Awards:* Pittsburgh Spectros Award, Am Chem Soc, 64, Pittsburgh Award, 65; Hasler Award, Soc for Appl Spectros, 73. *Mem:* Am Chem Soc; fel Optical Soc Am; hon mem Coblentz Soc (pres, 59-60); hon mem Soc Appl Spectros. *Res:* Infrared, Raman and electronic spectra. *Mailing Add:* 960 Lakemont Dr Pittsburgh PA 15243-1816

MILLER, FOREST LEONARD, JR, b Cincinnati, Ohio, June 18, 36; m 61; c 3. EXPERIMENTAL STATISTICS. *Educ:* Purdue Univ, BS, 58, MS, 59; NC State Univ, PhD(statist), 75. *Prof Exp:* Consult statistician, Nuclear Div, Union Carbide Corp, 59-79; SR STATISTICIAN & RES PROF, DESERT RES INST, 79- *Mem:* Am Statist Asn; Biomet Soc; Inst Math Statist; Royal Statist Soc; Health Physics Soc. *Res:* Development of conditional probability integral transformations and their application; development of tests for extreme value distributions. *Mailing Add:* Desert Res Inst 2505 Chandler Ave No 1 Las Vegas NV 89120-4004

MILLER, FOREST R, b York, Pa, Jan 6, 40. MATHEMATICS. *Educ:* Univ Okla, BS, 62; Univ Mass, MA, 67, PhD(math), 68. *Prof Exp:* PROF MATH, KANS STATE UNIV, 84- *Mem:* Am Math Soc. *Res:* Mathematical physics in order to develop models for quantum theory using techniques of global analysis & differential equations. *Mailing Add:* Math Dept Kans State Univ Manhattan KS 66506

MILLER, FRANCIS JOSEPH, b Montgomery, Ala, Aug 6, 17; m 45; c 2. ANALYTICAL CHEMISTRY. *Educ:* Univ Ala, AB, 39. *Prof Exp:* Anal chemist, Oak Ridge Nat Lab, 46-67, tech ed & writer, Isotopes Div, 67-68; asst prof res & admin asst to dir res, Univ Tenn, Mem Res Ctr & Hosp, Knoxville, 68-83; RETIRED. *Mem:* Am Chem Soc; Sigma Xi. *Res:* Instrumental methods of analysis; research administration; technical editing and writing. *Mailing Add:* 8109 Bennington Dr Knoxville TN 37909

MILLER, FRANCIS MARION, b Central City, Ky, Dec 28, 25; m 47; c 3. ORGANIC CHEMISTRY. *Educ:* Western Ky State Col, BS, 46; Northwestern Univ, PhD(chem), 49. *Prof Exp:* Res assoc org chem, Harvard Univ, 48-49; asst prof sch pharm, Univ Md, 49-51, assoc prof, 51-61, prof & chmn dept, 61-68; head dept, 68-77, prof chem, 68-91, EMER PROF CHEM, NORTHERN ILL UNIV, 91- *Concurrent Pos:* Consult, Chem Corps, US Army, 55-69; guest prof, Univ Heidelberg, 58-59, Univ Va, 77. *Mem:* Am Chem Soc. *Res:* Alkaloids; heterocyclics; mechanisms of organic reactions; medicinal chemistry. *Mailing Add:* Dept Chem Northern Ill Univ De Kalb IL 60115

MILLER, FRANK L, b Kansas City, Mo, Aug 31, 30; m 58; c 3. ATOMIC PHYSICS, COSMIC RAY PHYSICS. *Educ:* Univ Okla, BS, 51, MS, 57, PhD(physics), 64. *Prof Exp:* Asst prof physics, Ft Lewis Col, 63-66; assoc prof, 66-73; PROF PHYSICS, US NAVAL ACAD, 73- *Mem:* Am Phys Soc; Am Asn Physics Teachers; AAAS; Sigma Xi. *Res:* Atomic excitation by electron bombardment. *Mailing Add:* Dept of Physics US Naval Acad Annapolis MD 21402

MILLER, FRANK NELSON, b Alexandria, Va, Apr 15, 19; m 54; c 2. PATHOLOGY. *Educ:* George Washington Univ, BS, 43, MD, 48. *Prof Exp:* Asst scientist chem, Allegheny Ballistics Lab, 43-44; asst resident path, Univ Hosp, 49-50, from asst prof to assoc prof, 51-63, assoc dean, 66-73, prof path, 63-85, EMER PROF PATH, GEORGE WASHINGTON UNIV, 85- *Concurrent Pos:* Teaching fel, George Washington Univ, 50-51; attend, Mont Alto Vet Admin Hosp, 56-58; consult, US Air Force Hosp, Washington, DC, 55-58, Baker Vet Admin Hosp, Martinsburg, WVa, 60-68 & Vet Admin Hosp, Washington, DC, 68-85. *Mem:* AAAS; Int Acad Path; AMA; fel Col Am Path; fel Am Soc Clin Pathologists. *Res:* Neoplasms of the breast and female genital tract. *Mailing Add:* Dept Path George Washington Univ 2300 Eye St NW Washington DC 20037

MILLER, FRANKLIN, JR, b St Louis, Mo, Sept 8, 12; m 37; c 1. PHYSICS, SCIENCE EDUCATION. *Educ:* Swarthmore Col, AB, 33; Univ Chicago, PhD(physics), 39. *Hon Degrees:* ScD, Kenyon Col, 81. *Prof Exp:* Asst physics, Univ Chicago, 35-37; from instr to asst prof, Rutgers Univ, 37-48; assoc prof, Kenyon Col, 48-58, chmn dept, 55-66, 69-73, prof physics, 59-81; RETIRED. *Honors & Awards:* Millikan lectr, Am Asn Physics Teachers, 70. *Mem:* AAAS; Am Phys Soc; Am Asn Physics Teachers; Fedn Am Sci; Soc for Social Responsibility in Sci (pres, 53-55). *Res:* Musical acoustics; educational science films; textbook author. *Mailing Add:* PO Box 313 Gambier OH 43022

MILLER, FRANKLIN STUART, b Columbus, Ohio, June 3, 06; m 39; c 1. ECONOMIC GEOLOGY. *Educ:* Williams Col, AB, 28; Harvard Univ, AM, 31, PhD(geol), 34. *Prof Exp:* Geologist, Western Mining Corp, Australia, 34-36; instr geol, Univ Ill, 36-37; consult geologist, Can & Calif, 37-38; instr mineral, Univ Calif, 38-39; field supt, Roseville & Roaring River Gold Dredging Cos, 39-40; consult geologist, Toronto, Ont, 40-41; asst dir, Mining Div, War Prod Bd, 41-45; consult geologist, Columbus, Ohio, 46-47; asst mgr, Pac Tin Consol Corp, Malaya, 48-50, vpres, NY, 50-66, pres, 66-72, chmn, 72-75; vpres, Feldspar Corp, 55-66, pres, 66-72, dir, 55-78; dir, Pac Tin Consol Corp, 50-77; RETIRED. *Concurrent Pos:* Partner, Guggenheim Bros, 73-84; dir, Co-Co Del, Inc, 73-86. *Mem:* Fel AAAS; fel Geol Soc Am; Soc Econ Geol; Am Inst Mining, Metall & Petrol Eng; Can Inst Mining & Metall. *Res:* Petrology of the intrusive rocks; geologic structure of ore deposits; graphs for geologic calculations; production and resources of tin. *Mailing Add:* PO Box 266 Salisbury CT 06068

MILLER, FRED R, b Joplin, Mo, Aug 14, 43; m. IMMUNOLOGY. *Educ:* Kans State Univ, Manhattan, BS, 66; Univ Okla, Norman, MS, 72; Univ Wis-Madison, PhD(med microbiol), 76. *Prof Exp:* Postdoctoral fel, Dept Med, Roger Williams Gen Hosp, Brown Univ, RI, 76-78, res assoc, 78-79; scientist, Dept Immunol, Mich Cancer Found, 79-80, chief, Lab Immunobiol, 80-85, from asst mem to assoc mem, Dept Immunol, 80- 89, RES ADV, ANIMAL

CARE FACIL, MICH CANCER FOUND, 84-, MEM, DEPT IMMUNOL, 89- *Concurrent Pos:* Ad hoc mem, Exp Immunol Study Sect, NIH, 80, consult, Spec Study Sect, 81, 86, 90 & Prog Proj Site Visit, 83 & 84; adj mem, Grad Prog Cancer Biol, Wayne State Univ, 90- *Mem:* Am Asn Cancer Res; Am Soc Cell Biol. *Mailing Add:* Dept Immunol Mich Cancer Found 110 E Warren Ave Detroit MI 48201

MILLER, FREDERICK, b New York, NY, Apr 5, 37; m 62; c 2. IMMUNOPATHOLOGY, RENAL PATHOLOGY. *Educ:* Univ Wis-Madison, BS, 56; NY Univ, MD, 61. *Prof Exp:* USPHS fel, Sch Med, NY Univ, 59-60, instr med, 62-63; clin assoc & attend physician, Nat Inst Arthritis & Metab Dis, NIH, Bethesda, Md, 63-65; resident path, Bellevue & NY Univ Hosps, 65-66, chief resident, 66-67; from asst prof to assoc prof path, Sch Med, NY Univ, 67-70; assoc prof, 70-75, actg chair, 73-83, PROF PATH, STATE UNIV NY, STONY BROOK, 75-, PATHOLOGIST-IN-CHIEF & DIR LABS, UNIV HOSP, 80-, CHMN PATH DEPT, 83- *Concurrent Pos:* Resident med, Bellevue Hosp, 61-63, asst vis pathologist, 67-70; attend pathologist, Med Ctr, NY Univ, 67-70. *Honors & Awards:* Medal, Bausch & Lomb Co, 61; Award, Am Soc Clin Path, 61. *Mem:* Harvey Soc; Am Asn Pathologists; Int Acad Path; NY Acad Sci; Am Asn Immunologists; AAAS; Am Soc Clin Path. *Res:* Chemistry of immunoglobulins; immunopathology of renal glomerular disease; tumor immunology and immunologic markers for neoplasia; clinical immunology; nephropathology. *Mailing Add:* Dept Path HSC State Univ NY Stony Brook NY 11794-8691

MILLER, FREDERICK ARNOLD, b La Crosse, Wis, Dec 1, 26; m 51; c 4. PHYSICAL CHEMISTRY. *Educ:* Luther Col, Iowa, BA, 49; Iowa State Univ, PhD(chem), 53. *Prof Exp:* Assoc scientist, Dow Chem USA, 53-86; RETIRED. *Mem:* Am Chem Soc. *Res:* Synthetic latex; colloid chemistry; organic coatings; emulsion polymerization technology; latex based formulations. *Mailing Add:* Dow Chem Co 1604 Bldg 1113 Glendale Midland MI 48640

MILLER, FREDERICK N, CONTROL OF MACROMOLECULAR MOVEMENT. *Educ:* Univ Cincinnati, PhD(pharmacol), 71. *Prof Exp:* ASSOC PROF CARDIOVASC PHYSIOL, HEALTH & SCI CTR, UNIV LOUISVILLE, 81- *Res:* Fluorescent microscopy. *Mailing Add:* Health & Sci Ctr Univ Louisville Box 35260 Louisville KY 40292

MILLER, FREDERICK POWELL, b Springfield, Ohio, Oct 17, 36; m 65. SOIL SCIENCE, AGRONOMY. *Educ:* Ohio State Univ, BSc, 58, MSc, 61, PhD(agron, soil classification), 65. *Prof Exp:* Res asst soil chem, Ohio State Univ, 59-62, soil classification, 62-63, res asst soil classification, 64-65; soil scientist, Ohio Dept Natural Resources, 63-64; soil & water resource specialist, Univ Md, College Park, 65-69, from asst prof to assoc prof, 65-74, prof soils, 74-; OHIO STATE UNIV. *Mem:* Am Soc Agron; Soil Sci Soc Am; Soil Conserv Soc Am; Int Soc Soil Sci. *Res:* Physical, chemical and mineralogical characterization of soil Fragipans; electrophoretic separation of soil clay minerals; soil survey interpretation. *Mailing Add:* Ohio State Univ 1800 Cannon Dr Columbus OH 43210

MILLER, FREDRIC N, b Chicago, Ill, May 26, 41; m 62; c 2. PAPER CHEMISTRY. *Educ:* Ill Inst Technol, BSc, 62, PhD(org chem), 67. *Prof Exp:* Res chemist, WVa Pulp & Paper Co, 66-69; sr scientist, Am Can Co, 69-72, supvr converting group, 72-74, mgr prod develop, Tissue & Towel, 74-77; dir, Tech Asn Pulp & Paper Indust, 77-78; GEN MGR RES & ENG, TEEPAK INC, 78-, VPRES & GEN MGR, CORIA. *Mem:* Am Chem Soc; Tech Asn Pulp & Paper Indust. *Res:* Cellulose chemistry; collagen chemistry; specialty papers; consumer paper products; paperboard; non-wovens; flame retardancy of cellulosics and cellulosic blends; dry forming of paper; embossing; synthetic and natural binders; injection molding and thermoforming; converting of paper and paperboard. *Mailing Add:* Teepak Inc 1211 W 22nd Ave Oakbrook IL 60521

MILLER, FREEMAN DEVOLD, b Somerville, Mass, Jan 4, 09; m 33. ASTRONOMY. *Educ:* Harvard Univ, SB, 30, MA, 32, PhD(astron), 34. *Prof Exp:* Dir Swasey Observ, Denison Univ, 34-40; assoc prof astron, Univ Mich, Ann Arbor, 46-55; assoc dean Horace H Rackham Sch Grad Studies, 59-66, actg chmn dept astron, 60-61; prof, 55-77, EMER PROF ASTRON, UNIV MICH, ANN ARBOR, 77- *Mem:* Am Astron Soc; Int Astron Union. *Res:* Comets. *Mailing Add:* Dept Astron Dennison Bldg Univ of Mich Ann Arbor MI 48109

MILLER, G(ERSON) H(ARRY), b Philadelphia, Pa, Mar 2, 24; m 61; c 2. MATHEMATICS, MEDICAL STATISTICS. *Educ:* Pomona Col, BA, 49; Temple Univ, MEd, 51; Univ Southern Calif, PhD(ed psychol, math), 57; Univ Ill, ABD, 65. *Prof Exp:* Instr math, Los Angeles Sch Dist, 53-57; assoc prof math & ed, Western Ill Univ, 57-60; prof, Towson State Col, 60-61; prof math, Parsons Col, 61-65; assoc prof, Wis State Univ, Whitewater, 65-66; prof & dir math educ res, Tenn Technol Univ, 66-68; prof math & systs analyst, Comput Ctr, Edinboro State Col, 68-72, asst dir, Off Instnl Res, 73-80, prof math & computer sci dept, 81-89; DIR, STUDIES ON SMOKING, INC & SOS STOP SMOKING CLINIC, 89- *Concurrent Pos:* Dir, Nat Study Math Requirements for Scientists & Engrs, 65-; dir, Studies on Smoking & SOS Stop Smoking Clin, 73-89; bd mem, Nat Interagency Coun on Smoking & Health. *Mem:* AAAS; Math Asn Am; Am Chem Soc; Am Soc Eng Educ; NY Acad Sci; Am Pub Health Asn; Pop Asn; Am Inst Chemists; Int Soc Prev Oncol. *Res:* History of mathematics; comparative mathematics education; retention and deficencies in mathematics; analysis; set theory; curricular improvements in science and engineering; smoking and health research; smoking and longevity; passive smoking; less hazardous cigarette; male-female longevity difference; smoking cessation clinics; smoking and breast and lung cancer. *Mailing Add:* 125 High St Edinboro PA 16412

MILLER, GABRIEL LORIMER, b New York, NY, Jan 18, 28. PHYSICS. *Educ:* Univ London, BSc, 49, MSc, 52, PhD(physics), 57. *Prof Exp:* Physicist, Instrumentation Div, Brookhaven Nat Lab, 57-63; MEM TECH STAFF, 63-, HEAD ROBOTICS RES, BELL TEL LABS. *Mem:* Am Phys Soc; fel Inst Elec & Electronics Eng. *Res:* Nuclear instrumentation; satellite experiments; solid state; electronics. *Mailing Add:* Bell Tel Labs Rm 2D-546 614 Boulevard Westfield NJ 07090

MILLER, GAIL LORENZ, b Pleasant Valley, Iowa, Nov 3, 13; m 44. CANCER. *Educ:* Univ Ill, AB, 33, MA, 34; George Washington Univ, PhD(biochem), 37. *Prof Exp:* Asst plant path, Rockefeller Inst Med Res, 40-45; res chemist, Biochem Res Found, 45-46; assoc mem, Inst Cancer Res, 46-52; supvr biol chem, US Army Natick Lab, 52-61; res assoc, Univ Pittsburgh Sch Med, 61-62; res assoc, Merck Inst Therap Res, 62-64; biochemist, VA Hosp, assoc prof microbiol, Univ Mich, 64-65; head, Tumor Immunol Unit, Lankenau Hosp, 65-68; prof microbiol, Jefferson Med Col, 68-72; prin lit scientist, Franklin Inst Res Lab, 73-76; sr biosci writer, Tracor Jitco Inc, 76-79; BIOMED CONSULT, 79- *Mem:* Am Soc Biol Chem. *Mailing Add:* 1333 Prospect Hill Rd Villanova PA 19085

MILLER, GARY A, b Newark, NJ, Dec 5, 45; m. NUTRITION, FOOD SCIENCE. *Educ:* Rutgers Univ, BS, 68, PhD(food sci), 74. *Prof Exp:* Asst prof food & nutrit, Univ Nebr, Lincoln, 74-76; res scientist, 76-78, group leader, Food Sci Group, 78-80, MGR PHARMACEUT PROD DEVELOP, MCGAW LABS, CALIF, 81- *Concurrent Pos:* Dir, Sci Affairs, Nutrasweet Group, G D Searle & Co, Skokie, Ill, 83-84. *Mem:* Sigma Xi; Inst Food Technologists; NY Acad Sci; AAAS. *Res:* Development and evaluation of biological assays and rapid chemical indices to measure protein nutritive value; product development of medical foods, parenteral solutions, food additives, sweeteners, reduced calorie food technology. *Mailing Add:* Dir Prod Develop McNeil Specialty Prod Co 501 George St New Brunswick NJ 08903-2400

MILLER, GARY ARTHUR, b Mt Clemens, Mich, Nov 7, 50; m 72; c 2. MEDICAL TECHNOLOGY EDUCATION, HEALTH SCIENCES ADMINISTRATION. *Educ:* Spring Arbor Col, BA, 72; St Joseph Sch Med Technol, MT, 72; Cent Mich Univ, MSA, 81; Univ Nebr, PhD(educ admin), 86. *Prof Exp:* Supvr PM shift, Bay Med Ctr, 72-73, blood bank, Midland Hosp Ctr, 73-76; prog coordr lab & pharm technol, Mid Mich Community Col, 76-81; HEAD DEPT MED TECHNOL, NEBR WESLEYAN UNIV, 81- *Concurrent Pos:* Prog consult med technol prog, Creighton Univ, 84- *Mem:* Am Soc Clin Pathologists; Am Soc Med Technol; Asn Study Higher Educ. *Res:* Role of job satisfaction and organizational climate among full-time faculty employed at independent institutions. *Mailing Add:* 4331 Rainer Ct Chino CA 91710-3953

MILLER, GENE WALKER, b Utah, Dec 21, 25; m 53; c 5. PLANT BIOCHEMISTRY. *Educ:* Utah State Univ, BS, 50, MS, 54; NC State Univ, PhD(bot), 57. *Prof Exp:* Plant biochemist, Utah State Univ, 57-69, actg dean col sci, 67; dean, Col Environ Sci, Huxley Col, Western Wash Univ, 69-74; head dept, 74-85, prof, 74-87, EMER PROF BIOL, UTAH STATE UNIV, 87- *Concurrent Pos:* USPHS spec fel, Univ Munster, 61 & 66; mem staff, Univ Melbourne, Australia, 81; isotope res, Japan, 88, 90. *Mem:* AAAS; Biochem Soc; Am Soc Plant Physiol; Japanese Soc Plant Physiol; Am Asn Univ Professors; Am Plant Nutrit; Int Soc Fluoride Res. *Res:* Mineral nutrition of plants, especially the role of metals and nutrients, chlorophyll biosynthesis iron metabolism and plant metabolism; effects of air pollutants on biochemical reactions; environmental study programs; role of iron in plant metabolism is being investigated as well as mechanisms of iron uptake; particular emphasis on iron requirement for chlorophyll biosynthesis; fluoride and its affect on the metabolism with membrane enzymes being emphasized. *Mailing Add:* Dept Biol UMC 53 Utah State Univ Logan UT 84322-5305

MILLER, GEORGE ALFORD, b Madison, Wis, Dec 24, 25; m 63; c 2. PHYSICAL CHEMISTRY. *Educ:* Univ Wis, BS, 50; Univ Mich, PhD(chem), 55. *Prof Exp:* Res assoc, Univ Mich, 55-56; instr, Am Cols Istanbul, 56-57; res assoc, Univ Mich, 57-58; from asst prof to assoc prof, 58-70, PROF CHEM, GA INST TECHNOL, 70- *Mem:* Am Chem Soc. *Res:* Thermodynamics, kinetic theory of gases and light scattering. *Mailing Add:* Sch of Chem Georgia Inst of Technol Atlanta GA 30332

MILLER, GEORGE ARMITAGE, b Charleston, WVa, Feb 1, 20; m; c 2. PSYCHOLOGY. *Educ:* Univ Ala, BA, 40, MA, 41; Harvard Univ, AM, 44, PhD(psychol), 46. *Hon Degrees:* Numerous from US & foreign Univs, 76-84. *Prof Exp:* Instr psychol, Univ Ala, 41-43; assoc prof, Mass Inst Technol, 51-55; res asst, Psycho-Acoust Lab, Harvard Univ, 43-46, res assoc, 46-48, asst prof psychol, 48-51, from assoc prof to prof, 55-68, co-dir, Ctr Cognitive Studies, 60-67, chmn, Dept Psychol, 64-67; prof, Rockefeller Univ, 68-79; prof, 79-82, James S McDonnell distinguished univ prof, 82-90, EMER JAMES S MCDONNELL DISTINGUISHED UNIV PROF PSYCHOL, PRINCETON UNIV, 90- *Concurrent Pos:* Consult ed, J Exp Psychol, 47-50; assoc ed, J Acoust Soc Am, 50-55; Fulbright res prof, Oxford Univ, 63-64; vis prof, Rockefeller Univ, 67-68 & Mass Inst Technol, 76-79; adj prof, Rockefeller Univ, 79-82; chmn, Sect J, AAAS, 81; ed, Psychol Bull, 81-82; Guggenheim fel, 86; William James fel, Am Psychol Soc, 89. *Honors & Awards:* Nat Medal of Sci, 91; Distinguished Sci Contrib Award, Am Psychol Asn, 63; Warren Medal, Soc Exp Psychol, 72; NY Acad Sci Award, 82; Life Achievement in Psychol Sci, Am Psychol Found, 90; Levy Medal, Franklin Inst, 91. *Mem:* Nat Acad Sci; AAAS; Am Philos Soc; Am Psychol Asn (pres, 69); NY Acad Sci; Sigma Xi. *Res:* Cognitive psychology; psycholinguistics. *Mailing Add:* Dept Psychol Princeton Univ Princeton NJ 08544

MILLER, GEORGE C, b Portland, Ore, June 8, 25; m 59; c 2. ICHTHYOLOGY. *Educ:* Univ Wash, BS, 51; Ore State Col, BS, 56, MS, 60. *Prof Exp:* Aquatic biologist, Ore Fish Comn, 51-52, 56, 57-58; fisheries adv to Liberia, US For Opers Admin, 52-54; aquatic biologist, Wash State Dept Fisheries, 54-55 & Ore State Col, 58-59; fisheries biologist, Biol Lab, Bur Com Fisheries, US Fish & Wildlife Serv, Ga, 60-65 & Miami Lab, Southeast

Fisheries Ctr, Nat Marine Fisheries Serv, Fla, 65-80; CONSULT, 81- *Concurrent Pos:* Adj asst prof marine biol, Fla Atlantic Univ. *Mem:* Am Soc Ichthyol & Herpet; Crustacean Soc; Nat Shellfisheries Asn. *Res:* Invertebrate biology; marine ecology; systematic zoology; and zoogeography; biology of marine organisms in relation to technological development. *Mailing Add:* 502 Wesley Oak Dr St Simons Island GA 31522

MILLER, GEORGE E, b Banbury, Eng, May 12, 37; m 64; c 2. PHYSICAL CHEMISTRY, RADIOCHEMISTRY. *Educ:* Oxford Univ, BA, 59, DPhil (chem), 63. *Prof Exp:* Res assoc chem, Univ Kans, 63-65; res assoc, 65-68, lectr chem & reactor supvr, 68-87, SR LECTR, UNIV CALIF, IRVINE, 87- *Concurrent Pos:* NRC Comt on Industry in Sci Educ, Calif Sci Framework Comt. *Mem:* Am Chem Soc; Am Nuclear Soc. *Res:* Reactor utilization in chemistry; activation analysis applications in geochemistry, archeology and medicine; synthesis of labelled molecules and particles. *Mailing Add:* Dept Chem Univ Calif Irvine CA 92717

MILLER, GEORGE EARL, underwater acoustics, for more information see previous edition

MILLER, GEORGE EDWARD, medicine, for more information see previous edition

MILLER, GEORGE MAURICE, b Chicago, Ill, Dec 25, 29; m 62; c 3. COMPOSTING SEWAGE SLUDGE. *Educ:* Univ Ill, BS, 53. *Prof Exp:* Civil engr heavy construct, Raymond Int, 57-63, consult engr, Jacobs Assocs, 64-67, heavy construct, Perini Corp, 68-73; CITY ENGR MUNIC ENG, PLATTSBURGH, NY, 74- *Mem:* Am Soc Civil Engrs. *Res:* Biological degradation rates of various solid wastes that could be used as amendment in composting dewatered sewage sludge. *Mailing Add:* City Hall Plattsburgh NY 12901

MILLER, GEORGE PAUL, b Waiuku, NZ, Sept 9, 51. EXCITATION MECHANISMS IN INDUCTIVELY COUPLED PLASMAS. *Educ:* Univ Waikato, BSc, 77, MSc Hons, 79, DPhil(plasma physics), 87. *Prof Exp:* Res fel physics, Univ Waikato, 87-88; RES FEL CHEM, UNIV ALA, HUNTSVILLE, 88- *Concurrent Pos:* Consult, Solar Physics Corp, 91- *Mem:* Soc Appl Spectros. *Res:* Excitational mechanisms of low temperature plasmas; novel plasma diagnostics; atomic oxygen plasma source for surface chemistry studies; relationships common to both plasma and radiation chemistry. *Mailing Add:* Chem Dept Univ Ala Huntsville AL 35899

MILLER, GERALD ALAN, b New York, NY, Mar 20, 47; m 72; c 2. QUARK & GLUON PHENOMENA IN NUCLEI, FUNDAMENTAL SYMMETRICS. *Educ:* City Col NY, BS, 67; Mass Inst Technol, MS, 68, PhD(physics), 72. *Prof Exp:* Post-doctoral res assoc, Carnegie Mellon Univ, 72-75; from asst prof to assoc prof, 75-86, PROF PHYSICS, UNIV WASH, 86- *Concurrent Pos:* Mem prog, Los Alamos Meson Factory, 79-82, consult, 86; sci assoc, CERN, 82-83; vis scientist, Triumf, 88-89; vis res prof, Univ Ill, 89. *Mem:* Fel Am Phys Soc; Sigma Xi; fel AAAS. *Res:* Theoretical physics. *Mailing Add:* Physics Dept FM-15 Univ Wash Seattle WA 98195

MILLER, GERALD E, b Philadelphia, Pa, Oct 22, 50. ARTIFICIAL ORGANS, BIOMECHANICS. *Educ:* Pa State Univ, BS, 71, MS, 75, PhD(bioeng), 77. *Prof Exp:* Asst dept head indust eng, 85-87, DIR, HUMAN SYSTS ENG, TEX ENG EXP STA, TEX A&M UNIV, 86-, PROF BIOENG, 87-, CHMN DEPT, 89- *Concurrent Pos:* res coordr eng & med, Tex A&M Univ & coordr, Eng Prog, Inst Biosci, 87-, fel, Tex Eng Exp Sta, 88; expert consult, Tex Dept Ment Health & Ment Retardation & Tex Rehab Comn, 89-; mem, Vet Admin Res & Develop Comn & Scott & White Hosp Clin Scientist Comn, 90- *Mem:* Am Soc Mech Engrs; Am Soc Eng Educ; Inst Elec & Electronics Engrs Eng Med & Biol Soc; Biomed Eng Soc. *Res:* Multiple disk, centrifugal, pulsatile artificial heart; early warning detection system for epileptic seizures; hands-free system for automatic control of devices for disabled using voice, electromagnetic gyro and eye tracking; fluid mechanics effects on arterial disease. *Mailing Add:* Bioeng Prog Tex A&M Univ 233 Zachry Eng Ctr College Station TX 77843-3120

MILLER, GERALD R, b McClure, Ill, Dec 4, 34; m 58; c 2. WEED SCIENCE, EXTENSION EDUCATION. *Educ:* Univ Ill, BS, 56, MS, 57; Mich State Univ, PhD(weed control, physiol), 63. *Prof Exp:* Exten agronomist, Purdue Univ, 63-64; exten agronomist, 64-82, ASST DEAN, COL AGR & ASST DIR, MINN EXTEN SERV, UNIV MINN, ST PAUL, 83- *Concurrent Pos:* Pres, N Central Weed Control Conf, 84. *Mem:* Weed Sci Soc Am (secy, 75-76); Crop Sci Soc Am; Am Soc Agron. *Res:* Weed control; crop-weed competition; herbicide development. *Mailing Add:* 400 Harriet Ave St Paul MN 55126

MILLER, GERALD R, b Wellsville, NY, Dec 6, 39; m 61; c 2. MATERIALS SCIENCE, PHYSICS. *Educ:* Cornell Univ, BMetE, 62, MS, 63, PhD, 65. *Prof Exp:* Assoc prof mat sci, 65-76, PROF MAT SCI & ENG, UNIV UTAH, 76-, ADJ ASSOC PROF PHYSICS, 68- *Concurrent Pos:* Sci Res Coun fel, Univ Edinburgh, 71-72. *Mem:* Am Phys Soc. *Res:* Amorphous semiconducting devices; point defects in solids; transport theory. *Mailing Add:* Dept Bio Chem Univ Md Chem Bldg Rm B029 College Park MD 20742

MILLER, GERALD RAY, b Milwaukee, Wis, Nov 13, 36; m 58; c 2. PHYSICAL CHEMISTRY. *Educ:* Univ Wis, BS, 58; Univ Ill, MS, 60, PhD(chem), 62. *Prof Exp:* NSF fel phys chem, Oxford Univ, 61-63; from asst prof to assoc prof, 65-85, assoc chmn, 77-79, actg dean undergrad studies, 86-88, PROF PHYS CHEM, UNIV MD, COLLEGE PARK, 85- *Concurrent Pos:* Assoc dir fels, Nat Acad Sci-Nat Res Coun, 74-76. *Mem:* Am Chem Soc; Royal Soc Chem; Am Phys Soc; AAAS. *Res:* Nuclear magnetic resonance and electron spin resonance spectroscopy; graphite intercalation compounds; molecular motion at interfaces; two-dimensional systems. *Mailing Add:* Dept Chem & Biochem Univ Md College Park MD 20742-2021

MILLER, GLEN A, b Maynaroville, Tenn, Aug 22, 30. GROUND WATER, HYDROLOGY. *Educ:* Univ Tenn, BS, 52, MS, 53. *Prof Exp:* Hydrogeologist, US Geol Surv, 75-86; CONSULT HYDROGEOLOGIST, GLEN A MILLER ASSOC, 87- *Mem:* Am Inst Hydrogeol; AAAS; fel Geol Soc Am; Am Geophys Union. *Mailing Add:* 2264 Willow Wood Rd Grand Junction CO 81503

MILLER, GLENDON RICHARD, b Columbus, Ohio, Oct 28, 38; m 66; c 2. MICROBIOLOGY, BIOCHEMISTRY. *Educ:* Southern Ill Univ, BA, 60, MA, 62; Univ Mo-Columbia, PhD(microbiol), 66. *Prof Exp:* Res microbiologist, Colgate Palmolive Res Ctr, 66-68; asst prof bact & physiol, 68-73, grad coordr dept, 75-80, ASSOC PROF BIOL, WICHITA STATE UNIV, 73-, DIR HAZARDOUS MAT OFF, 90- *Concurrent Pos:* Mallinckrodt microbiol res fel, 62-66; consult, Koch Eng, 71-72, Dold Foods, 79-83 & Cereal Food Processors, 84-86. *Mem:* AAAS; Am Soc Microbiol; Sigma Xi. *Res:* Radiation repair enzymes of yeasts; antibiotics; antibiotic combinations; resistance and cross-resistance to antibiotics; microorganisms in foods (meats and cereal grains); beta-lactamases; Candida albicans; culture collection and transport; clinical microbiology. *Mailing Add:* Dept Biol Sci Wichita State Univ Wichita KS 67208

MILLER, GLENN HARRY, b Pittsburgh, Pa, Feb 10, 22; m 51; c 3. CHEMISTRY. *Educ:* Geneva Col, BS, 43; Brown Univ, PhD(chem), 48. *Prof Exp:* Chemist, Oak Ridge Nat Lab, 44; sr chemist, Butadiene Div, Koppers Co, Inc, 44-45; chemist, Tex Co, 48-49; from instr to assoc prof, 49-63, chmn dept, 60-64, PROF CHEM, UNIV CALIF, SANTA BARBARA, 63- *Concurrent Pos:* Nat Res Coun Can fel, 56-57; Fulbright-Hays lectr, Univ Malaya, 67-68 & Univ Liberia, 74-75. *Mem:* Am Chem Soc. *Res:* Popcorn polymerization; polymers; photolysis. *Mailing Add:* Dept Chem Univ Cal Santa Barbara CA 93106

MILLER, GLENN HOUSTON, physics, for more information see previous edition

MILLER, GLENN JOSEPH, b Crete, Nebr, June 28, 25; m 50; c 3. BIOCHEMISTRY. *Educ:* Doane Col, BA, 51; Purdue Univ, MS, 53, PhD(biochem), 56. *Prof Exp:* Asst, Purdue Univ, 51-56; from asst prof to prof biochem, 56-76, head div, 72-76, PROF FOOD BIOCHEM, DIV ANIMAL SCI, UNIV WYO, 76- *Mem:* AAAS; Am Oil Chem Soc; NY Acad Sci; Am Meat Sci Asn; Inst Food Technol. *Res:* Implication of fats in foods with emphasis upon animal products. *Mailing Add:* Div of Animal Sci Univ Wyo Univ Sta Box 3354 Laramie WY 82071

MILLER, GORDON JAMES, b Utica, NY, May 6, 60; m 90; c 1. SOLID-STATE CHEMISTRY. *Educ:* Univ Rochester, BS, 82; Univ Chicago, PhD(chem), 86. *Prof Exp:* Staff scientist, Max-Planck Inst fur Festkorperforschung, 87-90; ASST PROF INORG CHEM, IOWA STATE UNIV, 90- *Mem:* Am Chem Soc. *Res:* Structure-property relationships in solids; theories of chemical bonding and electronic structure; structural phase transitions; synthesis of transition metal polycompounds. *Mailing Add:* Dept Chem Iowa State Univ Ames IA 50011

MILLER, GORDON LEE, b Milwaukee, Wis, Sept 27, 38. MATHEMATICS. *Educ:* Moorhead State Col, BS, 64; NDak State Univ, MS, 65; Univ Northern Colo, EdD(math), 70. *Prof Exp:* From instr to asst prof, 65-72, assoc prof, 72-80, PROF MATH, UNIV WIS-STEVENS POINT, 80- *Mem:* Math Asn Am; Nat Coun Teachers Math. *Res:* Analysis. *Mailing Add:* Dept of Math Univ of Wis-Stevens Point Stevens Point WI 54481

MILLER, GREGORY DUANE, b Pontiac, Mich, Dec 10, 55; m 90; c 1. NUTRITION RESEARCH, FOOD PRODUCT DEVELOPMENT SUPPORT. *Educ:* Mich State Univ, BS, 78; Pa State Univ, MS, 82, PhD(nutrit toxicol), 86; Ctr Creative Leadership, cert, 89. *Prof Exp:* Instr lab methods, Pa State Univ, 86; res scientist nutrit, 86-89, SR RES SCIENTIST NUTRIT, KRAFT GEN FOODS, 89- *Concurrent Pos:* Prog chmn, Twelfth Symp on Nutrit & Food Technol, 87-88; symp chmn, Diet & Behav: A Workshop on Methodologies, 87-88; comt chmn, Diet & Behav Comt Int Life Sci Inst, Nutrit Found, 89-; co-chair, Planning Comt, Diet & CNS Function Symp, ILSI-NF/Keystone Symp, 90- *Mem:* Sigma Xi; Am Inst Nutrit; Am Col Nutrit. *Res:* Interaction between nutrition and toxicology; mineral metabolism; calcium bioavailability and metabolism; diet and behavior; sports nutrition; diet and cancer; nutritional status. *Mailing Add:* 404 Westmoreland Vernon Hills IL 60061

MILLER, GROVER CLEVELAND, b Jackson, Ky, Jan 23, 27; m 51; c 4. ZOOLOGY. *Educ:* Berea Col, AB, 50; Univ Ky, MS, 52; La State Univ, PhD(zool), 57. *Prof Exp:* Asst parasitologist, La State Univ, 56-57; from instr to assoc prof, 57-69, PROF ZOOL, NC STATE UNIV, 69 - *Mem:* Am Soc Parasitol. *Res:* Invertebrate zoology; parasitology; trematodes of freshwater fishes; helminth parasites in wild animals. *Mailing Add:* Dept Zool NC State Univ Raleigh NC 27695-7617

MILLER, GUTHRIE, b Greensburg, Ind, Oct 16, 42; c 3. HEALTH PHYSICS. *Educ:* Calif Inst Technol, BS, 64; Stanford Univ, PhD(physics), 70. *Prof Exp:* Res asst high energy physics, Stanford Linear Accelerator Ctr, 67-71; res assoc, Univ Wash, 71-74; staff mem Controlled Thermonuclear Div, 74-91, STAFF MEM HEALTH PHYSICS, LOS ALAMOS NAT LAB, 91- *Mem:* Am Phys Soc. *Res:* Confinement of high temperature plasmas, experimental and theoretical magnetohydrodynamics; health physics. *Mailing Add:* Los Alamos Nat Lab Mail Stop K487 Los Alamos NM 87545

MILLER, HAROLD A, b St Paul, Minn, Mar 14, 21; m 46. PHYSICAL CHEMISTRY. *Educ:* Univ Minn, BS, 42; NY Univ, PhD(phys chem), 51. *Prof Exp:* Res chemist, 50-62, asst dir res, 62-72, DIR RES, MEARL CORP, 72-, SR VPRES, 85- *Mem:* Am Phys Soc; Am Chem Soc; Soc Cosmetic Chemists; Electron Micros Soc Am; NY Acad Sci. *Res:* Crystal growth; surface chemistry; nacreous pigments; structure of crystals and molecules; infrared spectroscopy. *Mailing Add:* 37 Meadowbrook Rd White Plains NY 10605-2203

MILLER, HAROLD CHARLES, microbiology, immunology, for more information see previous edition

MILLER, HARRY, b Detroit, Mich, July 11, 15; m 44; c 2. MECHANICAL ENGINEERING. *Educ:* City Col NY, BME, 37. *Prof Exp:* Jr engr, Master Wire & Die Co, 37-38; marine engr, Brooklyn Navy Yard, 38-44; engr, Fed Tel Co, 44-45 & Shirgun Corp, 45-47; head, Dept Eng Transp Flight Controls, 48-67, head, Dept Eng Flight Control Systs, 67-69, mgr, Dept Eng Flight Guide Systs & Instrumentation, 69-80, eng dir commun systs, Sperry Flight Systs Div, 80-85, TECH DIR, COM FLIGHT SYSTS & CONSULT, SPERRY CORP, 85- *Concurrent Pos:* Lectr, City Col NY, 47-50; consult. Sperry Coun Flight Systs, 85-; Pioneer Award, Airlines Avionics Inst, 85. *Mem:* Am Soc Mech Engrs. *Res:* Automatic flight control systems; pressure sensitive devices. *Mailing Add:* 5136 N 68th Pl Scottsdale AZ 85253

MILLER, HARRY BROWN, b Cumberland, Md, May 25, 13; m 41. ORGANIC CHEMISTRY. *Educ:* Univ NC, BS, 36, PhD(chem), 46. *Prof Exp:* With Standard Oil Co (NJ), 36-42; instr chem, Armstrong Jr Col, Ga, 45-47; from asst prof to assoc prof, 47-61, PROF CHEM, WAKE FOREST UNIV, 61- *Mem:* AAAS; Am Chem Soc; Sigma Xi. *Res:* Reactions of organic halogen compounds; fluorine chemistry. *Mailing Add:* Box 7241 Reynolds Sta Winston-Salem NC 27109

MILLER, HARRY GALEN, b Annapolis, Md, May 17, 37; m 57; c 2. PHYSICS. *Educ:* Defiance Col, BS, 59; Ohio State Univ, PhD(nuclear physics), 63. *Prof Exp:* Teacher high sch, 58-59; admissions counsr, 59-60, from asst prof to assoc prof, 63-73, PROF PHYSICS, DEFIANCE COL, 73- *Concurrent Pos:* Resident assoc, Argonne Nat Lab, Ill, 69-70, consult, 70-72. *Res:* Nuclear energy levels; gamma-ray spectroscopy. *Mailing Add:* Dept of Physics Defiance Col 1701 N Clinton St Defiance OH 43512

MILLER, HARVEY ALFRED, b Sturgis, Mich, Oct 19, 28; m 52; c 2. BRYOLOGY, PLANT PATENTS. *Educ:* Univ Mich, BS, 50; Univ Hawaii, MS, 52; Stanford Univ, PhD(biol), 57. *Prof Exp:* Asst bot, Univ Hawaii, 50-53; asst herbarium, Stanford Univ, 53-55; instr bot, Univ Mass, 55-56; from instr to asst prof bot, Miami Univ, Ohio, 56-61, assoc prof & cur herbarium, 61-67; prof biol & bot & chmn, Div Biol Sci & Prog Gen Biol, Wash State Univ, 67-69; chmn, dept biol sci, 70-75, PROF BOT, UNIV CENT FLA, 70- *Concurrent Pos:* Asst herbarium, Univ Mich, 51; grant-in-aid, Sigma Xi, 54; Guggenheim fel, 58-59; prin investr bryophytes, NSF res grant, Pac Islands & Micronesia; hon mem staff, Hattori Bot Lab, Japan; vis prof bot, Univ Ill, Urbana, 69-70; NSF US-Japan Coop Sci prog grant; vis lectr, Col Guam, 65; prin investr, Miami Univ-NSF exped, Micronesia & Philippines, 65 & Univ Cent Fla-NSF exped, Southern Melanesia, 84 & 85; ed, Fla Scientist, 73-78; res assoc, John Young Sci Ctr, Orlando's Mus Sci & Technol, 75-; consult plant patents & land reclamation. *Mem:* Am Bryol & Lichenological Soc (vpres Am Bryol Soc, 62-63, pres, 64-65); Asn Trop Biol; Am Soc Plant Taxon; fel AAAS; Int Asn Plant Taxon. *Res:* Biochemical taxonomy of bryophytes; taxonomy and distribution of Pacific Island bryophytes; world phytogeography; geobotany of bryophytes; ecosystem creation following surface mining. *Mailing Add:* Dept Biol Sci Univ Cent Fla PO Box 25000 Orlando FL 32816

MILLER, HARVEY I, b Brooklyn, NY, May 25, 32; m 54; c 1. PHYSIOLOGY. *Educ:* City Col New York, BS, 55; Hahnemann Med Col, MS, 58, PhD(physiol), 61. *Prof Exp:* Biochemist, Lab & Res Div, State Dept Health, NY, 55-57; asst lipid physiol, Hahnemann Med Col, 58-61; res assoc, Lankenau Hosp, Philadelphia, 61-73; assoc prof, Cardiol Div, Hahnemann Med Col, 72-73; assoc prof, 73-76, PROF PHYSIOL, LA STATE UNIV MED CTR, 76- *Concurrent Pos:* Assoc prof, Jefferson Med Col, 66-74; estab investr, Am Heart Asn, 69-74. *Mem:* Am Physiol Soc; Am Heart Asn; Am Col Sports Med; Sigma Xi. *Res:* Myocardial metabolism; shock; exercise. *Mailing Add:* Dept Physiol La State Univ Med Ctr 1901 Perdido St New Orleans LA 70112

MILLER, HELEN CARTER, b Indianapolis, Ind, Dec 7, 25; m 57; c 3. VERTEBRATE ZOOLOGY, ETHOLOGY. *Educ:* Butler Univ, AB, 48; Cornell Univ, MA, 52, PhD(vert zool), 62. *Prof Exp:* Instr biol, Miami Univ, Ohio, 53-56; from instr to assoc prof biol, Okla State Univ, 63-88; RETIRED. *Res:* Ethological and ecological research on fishes and birds. *Mailing Add:* 2616 Black Oak Dr Stillwater OK 74074

MILLER, HELENA AGNES, b Rudolph, Ohio, Apr 25, 13. BOTANY. *Educ:* Ohio State Univ, BA & BSc, 35, MS, 38; Radcliffe Col, PhD(biol), 45. *Prof Exp:* Teacher high sch, Ohio, 35-37; asst bot, Ohio State Univ, 38-39; lectr biol, Hiram Col, 39; teacher, Milton Acad, 39-41; instr bot, Conn Col, 44-45 & Wellesley Col, 45-48; from assoc prof to prof, Duquesne Univ, 48-66, asst dean arts & sci, 66-78, prof biol, 75-78; RETIRED. *Mem:* Sigma Xi; Bot Soc Am; Soc Develop Biol. *Res:* Developmental anatomy of certain angiosperms; study of growth by culturing embryos of certain angiosperms in vitro. *Mailing Add:* 532 Highview Rd Pittsburgh PA 15234-2414

MILLER, HERBERT CHAUNCEY, b East Orange, NJ, Nov 2, 07; m 34; c 3. PEDIATRICS. *Educ:* Yale Univ, AB, 30, MD, 34. *Prof Exp:* Asst pediat, Sch Med, Yale Univ, 34-37, from instr to asst prof, 37-45; chmn dept, 45-72, prof, 45-78, EMER PROF PEDIAT, MED CTR, UNIV KANS, 78- *Honors & Awards:* Abraham Jacobi Award, Am Acad Pediat, 81. *Mem:* Am Pediat Soc; Soc Pediat Res; fel Am Acad Pediat. *Res:* Diseases of children; fetology; neonatology. *Mailing Add:* PO Box 176 Northford CT 06472

MILLER, HERBERT CRAWFORD, b Lenoir, NC, Mar 30, 44; m 67; c 2. ANALYTICAL CHEMISTRY. *Educ:* Univ Ala, BS, 66, PhD(anal chem), 73. *Prof Exp:* Res chemist, 72-76, head, Analytical Chem Sect, 76-80, HEAD, ANALYTICAL & PHYS CHEM DIV, SOUTHERN RES INST, 80- *Mem:* Am Chem Soc; Am Soc Testing & Mat. *Res:* Environmental chemistry; pollution abatement; chemical analyses for trace constituents. *Mailing Add:* Southern Res Inst PO Box 55305 Birmingham AL 35255

MILLER, HERBERT KENNETH, b New York, NY, Apr 5, 21; m 43; c 2. BIOCHEMISTRY. *Educ:* City Col New York, BS, 40; Univ Ill, MS, 47; Columbia Univ, PhD(biochem), 51. *Prof Exp:* Res assoc infectious diseases, Pub Health Res Inst City New York, Inc, 51-53, assoc, 53-56; biochemist, VA Hosp, Bronx, NY, 56-62; from adj asst prof to assoc prof, 62-72, PROF CHEM, MANHATTAN COL, 72- *Concurrent Pos:* Assoc scientist, Div Cell Biol, Sloan-Kettering Inst Cancer Res, NY, 62-74. *Mem:* Am Chem Soc; Am Soc Microbiol; Sigma Xi. *Res:* Glutamine peptides; influenza virus nucleic acids; adenovirus infected cells; macromolecular methylation; control of DNA synthesis in chick fibroblasts; purine metabolism. *Mailing Add:* Dept Chem Manhattan Col Bronx NY 10471

MILLER, HERMAN LUNDEN, b Detroit, Mich, Apr 23, 24; m 51. NUCLEONICS, OPTICS. *Educ:* Univ Mich, BS, 48, MS, 51. *Prof Exp:* Physicist, Res Labs, Ethyl Corp, 48-49; prof physicist, Rocky Flats Plant, Dow Chem Co, 50-55; mem res staff, Proj Matterhorn, Princeton Univ, 55-65; staff engr, Aerospace Systs Div, Bendix Corp, 65-72; sr nuclear engr, Commonwealth Assocs, Inc, 73-80; RETIRED. *Mem:* Am Phys Soc; Inst Elec & Electronics Engrs; Am Nuclear Soc. *Res:* Nuclear radiation instrumentation; optical instrumentation. *Mailing Add:* 1924 Dunmore Rd Ann Arbor MI 48103-5657

MILLER, HERMAN T, b Syracuse, Mo, Feb 28, 31; m 56. BIOCHEMISTRY, IMMUNOCHEMISTRY. *Educ:* Lincoln Univ, Mo, BS, 53; Kans State Univ, MS, 58; Univ Mo, PhD(biochem), 62. *Prof Exp:* Asst biochem, Kans State Univ, 55-58; from asst to instr, Univ Mo, 58-62; NIH fels, Univ Calif, Davis, 62-64, asst protein biochem, 64-65, res biochemist, 65-66; PROF CHEM, LINCOLN UNIV, MO, 66- *Res:* Immunochemistry as a tool in studying structure and function relationships of protein molecules. *Mailing Add:* Dept Chem Lincoln Univ 820 Chestnut Jefferson City MO 65101

MILLER, HILLARD CRAIG, b Northampton, Pa, Dec 15, 32; m 56; c 4. APPLIED PHYSICS. *Educ:* Lehigh Univ, BA, 54, MS, 55; Pa State Univ, PhD(physics), 60. *Prof Exp:* Physicist, Gen Elec Res Labs, 60-67, PHYSICIST, GEN ELEC CO, 67- *Mem:* AAAS; Am Phys Soc; Inst Elec & Electronics Eng; Am Vacuum Soc; Royal Astron Soc Can; Europ Phys Soc. *Res:* Electrical discharges in vacuum. *Mailing Add:* PO Box 2908 Mail Stop 055 Gen Elec Largo FL 34649-2908

MILLER, HOWARD CHARLES, b Syracuse, NY, Feb 6, 17; m 53; c 2. ENTOMOLOGY. *Educ:* State Univ NY, BS, 41; Cornell Univ, PhD(entom), 51. *Prof Exp:* Entomologist, Forest Insect Div, USDA, 46; asst, Cornell Univ, 47-50; from asst prof to prof biol sci, 54-82, exten entomologist & pathologist, 50-71, emer prof & exten specialist biol sci, State Univ NY Col Environ Sci & Forestry, 82-, assoc pub serv officer, Tree Pest Serv, 71-; RETIRED. *Mem:* AAAS; Entom Soc Am; Soc Am Foresters; Lepidop Soc. *Res:* Insect ecology; forest insect and disease problems; wildlife parasites; medical entomology; biology; urban forestry. *Mailing Add:* 242 Westminster Ave Syracuse NY 13210

MILLER, HUGH HUNT, b Griffin, Ga, Feb 11, 25; m 47; c 5. CHEMISTRY, REACTOR TECHNOLOGY. *Educ:* Univ NC, Chapel Hill, BS, 46. *Prof Exp:* Res chemist, Oak Ridge Nat Lab, 46-55; chief anal chem, Nuclear Power Dept, Curtiss Wright Corp, 55-58; res adminr, US Atomic Energy Comn, 58-61; vpres, Numec Instruments & Controls Corp, 61-65, pres, 65-67; vpres, Waters Assocs Inc, 67-70; dir mkt res & process instrumentation, Esterline Corp, 70-71; exec dir, Nat Acad Eng, 71-81, Off Foreign Secy, 81-82, exec off, 82-87 exec dir, Off External Affairs, 86-87; PRES, JAMES CLARK MAXWELL FOUND, 87. *Concurrent Pos:* Assoc fac mem, Duquesne Univ, 66-67; mem adv bd, int prog, Ga Inst Technol, 76-78, Inst Res & Interactions Technol & Soc, Univ Pittsburgh, 77-79 & Asn Media-Based Continuing Educ for Engrs, Inc, 78-; consult, NSF, 78-80, Congressional Res Serv, 79, USAID, 81, NATO, 81-, State Dept, 81-; sr fel, George Washington Univ Sch Eng & Appl Sci, 82-87. *Honors & Awards:* King Carl Gustaf XVI Gold Medal, Sweden, 79. *Mem:* Am Chem Soc; Am Nuclear Soc; Am Soc Testing & Mat; Am Ceramic Soc; AAAS. *Res:* Instrumental methods of chemical analysis; polarography; coulometry; particle technology liquid chromatography; radioisotope technology; technology and public policy. *Mailing Add:* James Clark Maxwell Found 4109 Great Oak Rd Rockville MD 20853

MILLER, I GEORGE, JR, b Chicago, Ill, Apr 18, 37; m 62; c 3. MICROBIOLOGY. *Educ:* Harvard Univ, AB, 58, MD, 62; Yale Univ, MA, 76. *Prof Exp:* Int & asst resident med, UNiv Hosp Cleveland, 62-66; Epidemic Intel Serv Officer, Commicable Dis Ctr, USPHS, 64-66, chief, Neurotropic Virus Unit, Epid Br, 66; res fel, Children Hosp Med Ctr, Boston & Harvard Med Sch, asst Med, Peter Brent Brigham Hosp, 66-69; from asst prof to prof pediat & epidemiol, 69-79, prof epidemiol, 78-83, JOHN F ENDERS PROF PEDIAT INFECTIOUS DIS, PROF EPIDEMIOL & PROF MOLECULAR BIOPHYS & BIOCHEM, YALE UNIV, 83- *Concurrent Pos:* Investr, Howard Hughes Med Inst, 72-80; exp virol study sect, NIH, 74-77; sci Adv Comt, Leukemia Soc Am,76-81, Mult Sclerosis Soc, 79-85; vaccines & Related Biol Products Adv Comt, FDA,83-86; DNA Virus chmn, Am Soc Microbiol, 84-85; Microbiol & Virol Comt, Am Cancer Soc, 86-, Scholar Award, 90. *Honors & Awards:* Squibb Award, Infectious Dis Soc Am, 82. *Mem:* Sigma Xi; Infectious Dis Soc Am; AAAS; Am Soc Microbiol; Am Soc Virology; Am Pediat Soc; Asn Am Physician. *Res:* Medical virology, especially the oncogenic herpes virus; Epstein-Barr virus. *Mailing Add:* Dept Pediat & Inf Dis Yale Univ New Haven CT 06510

MILLER, INGLIS J, JR, b Columbus, Ohio, Mar 17, 43; m 63; c 2. PHYSIOLOGY, ANATOMY. *Educ:* Ohio State Univ, BS, 65; Fla State Univ, PhD(sensory physiol), 68. *Prof Exp:* USPHS trainee, Univ Pa, 68-71; asst prof anat, 71-77, ASSOC PROF ANAT, BOWMAN GRAY SCH MED, WAKE FOREST UNIV, 77- *Mem:* Am Asn Anat; Soc Neurosci; Assoc Chemoreception Sci. *Res:* Gustatory neurophysiology; neuroanatomy of peripheral gustatory system. *Mailing Add:* Dept Anat Bowman Gray Sch Med 300 Hawthorne Rd SW Winston-Salem NC 27103

MILLER, IRVIN ALEXANDER, b Schellsburg, Pa, Nov 29, 32; m 56; c 3. THEORETICAL PHYSICS. *Educ:* Drexel Inst Technol, BS, 55; Univ Pa, MS, 59; Temple Univ, PhD, 68. *Prof Exp:* From instr to asst prof, 55-70, from asst vpres to assoc vpres acad affairs, 70-82, dir comput servs, 82-85, ASSOC PROF PHYSICS & ATMOS SCI, DREXEL UNIV, 70- *Concurrent Pos:* Actg dean sci, Drexel Univ, 69-70. *Mem:* Am Soc Eng Educ; Am Asn Physics Teachers. *Res:* Electron paramagnetic resonance. *Mailing Add:* Dept Physics & Atmospheric Sci Drexel Univ 32nd & Chestnut St Philadelphia PA 19104

MILLER, IRVING F(RANKLIN), b New York, NY, Sept 27, 34; m 62; c 2. BIOENGINEERING, CHEMICAL ENGINEERING. *Educ:* NY Univ, BChE, 55; Purdue Univ, MSChE, 56; Univ Mich, PhD(chem eng), 60. *Prof Exp:* Res assoc fluid dynamics, Res Inst, Univ Mich, 59; res scientist, Res Labs, United Aircraft Corp, 59-61; from asst prof to prof chem eng, Polytech Inst Brooklyn, 61-72, chmn bioeng prog, 66-72, head dept chem eng, 70-72; prof bioeng & head dept & prof physiol, Med Ctr, 73-79, actg head dept systs eng, 78-79, dean, grad col & assoc vice chancellor res, 79-85, HEAD, CHEM ENG, 86-, PROF PHARMACEUT & CHEM ENG, 85-, PROF MED, UNIV ILL, CHICAGO, 89- *Concurrent Pos:* Consult to various indust, 63- *Mem:* Fel AAAS; Am Chem Soc; Biomed Eng Soc; NY Acad Sci; fel Am Inst Chem Engrs. *Res:* Transport processes; membrane phenomena; biological processes; drug delivery systems; ion exchange; chemical kinetics; fluid dynamics; pulmonary medicine. *Mailing Add:* Chem Eng Dept Univ Ill Chicago Chicago IL 60680

MILLER, IRWIN, b New York City, NY, July 3, 28; m 52; c 3. SURVEY, RESEARCH ADMINISTRATION. *Educ:* Alfred Univ, BA, 50; Purdue Univ, MS, 52; Va Polytech Inst, PhD, 56. *Prof Exp:* Mathematician, Appl Res Lab, US Steel Corp, 56-58; prof statist, Ariz State Univ, 58-65; mem prof staff, 65-73, VPRES, ARTHUR D LITTLE, INC, 73-; CHMN, OPINION RES CORP, 75-, CHIEF EXEC OFFICER, 80- *Concurrent Pos:* Adj prof math & dir comput ctr, Wesleyan Univ, 69-71. *Mem:* AAAS; Am Statist Asn; Inst Math Statist; Biomet Soc. *Res:* Mathematical statistics; continuous stochastic processes; inference. *Mailing Add:* 3480 Princeton Lawrenceville Rd Princeton NJ 08540

MILLER, IVAN KEITH, b Rapid City, SDak, July 28, 21; m 45; c 2. POLYMER CHEMISTRY. *Educ:* SDak State Col, BS, 43; Univ Minn, PhD, 50. *Prof Exp:* Asst, SDak State Col, 42, instr chem, 43; civilian with Off Rubber Res, Univ Minn, 45-50; res chemist, Rayon Dept, E I du Pont de Nemours & Co, Inc, 50-58, res assoc, 58-67, res fel, Textile Fibers Dept, 67-85; RETIRED. *Mem:* AAAS; Am Chem Soc. *Res:* Reaction mechanisms concerning polymerization reactions; rubber; synthetic resins; viscose rayon; nylon. *Mailing Add:* 126 Marcella Rd Webster Farm Wilmington DE 19803

MILLER, JACK CULBERTSON, b Pomona, Calif, Sept 29, 25; m 54; c 2. THEORETICAL PHYSICS. *Educ:* Pomona Col, BA, 47; Univ Calif, MA, 49; Oxford Univ, DPhil, 55. *Prof Exp:* Mathematician, Radiation Lab, Univ Calif, 48-49; from instr to assoc prof, 52-66, PROF PHYSICS, POMONA COL, 66- *Concurrent Pos:* NSF fac fel, 61-62 & 68-69. *Mem:* Am Phys Soc; Am Geophys Union. *Res:* Mathematical physics; physical oceanography. *Mailing Add:* Dept of Physics Pomona Col Claremont CA 91711

MILLER, JACK MARTIN, b Cornwall, Ont, Feb 20, 40; m 80. INORGANIC NMR & MASS SPECTROMETRY. *Educ:* McGill Univ, BSc, 61, PhD(chem), 64; Cambridge Univ, PhD(chem), 66. *Prof Exp:* From asst prof to assoc prof, 66-75, chmn dept, 75-79, PROF CHEM, BROCK UNIV, 75- *Concurrent Pos:* Nat Res Coun Can overseas fel, Cambridge Univ, 64-66; bk rev ed & assoc ed, Can J Spectroscopy; prof, McMaster Univ, Ont, 87- *Mem:* Fel Chem Inst Can; Am Chem Soc; fel Royal Soc Chem; Am Soc Mass Spectrometry; Spectros Soc Can. *Res:* Nuclear magnetic resonance and mass spectra of donor-acceptor complexes; mass spectra of organometallic and coordination compounds; strong hydrogen bonding; fab mass spectrometry; supported catalysis. *Mailing Add:* Dept of Chem Brock Univ St Catharines ON L2S 3A1 Can

MILLER, JACK W, b Knoxville, Tenn, Sept 26, 25; m 52; c 1. PHARMACOLOGY. *Educ:* San Diego State Col, AB, 49; Univ Calif, MS, 52, PhD(pharmacol), 54. *Prof Exp:* Asst pharmacol, Univ Calif, 52-53, lectr, 53-54; from instr to assoc prof, Univ Wis, 54-62; assoc prof, 62-67, PROF PHARMACOL, UNIV MINN, MINNEAPOLIS, 67- *Concurrent Pos:* Mem pharmacol-toxicol comt, Nat Inst Gen Med Sci; Med Curric Course Coordr, Univ Minn, Minneapolis, 76-, interim head pharmacol, 87-89, assoc head, 89. *Mem:* Am Soc Pharmacol & Exp Therapeut; Soc Exp Biol & Med; NY Acad Sci. *Res:* Pharmacology of morphine-type drugs; uterine drugs; adrenergic receptors; catecholamines; drug receptor regulation; posterior pituitary hormones; anticholinesterases. *Mailing Add:* Dept Pharmacol Univ Minn Minneapolis MN 55455

MILLER, JAMES ALBERT, JR, b Peitaiho, China, June 21, 07; US citizen; m 78; c 2. ANATOMY. *Educ:* Col Wooster, AB, 28; Univ Chicago, PhD(zool), 37. *Hon Degrees:* DSc, Wooster Col, 61. *Prof Exp:* Instr biol, Assiut Col, Egypt, 28-31; instr, Ohio Univ, 35-37; instr anat, Med Sch, Univ Mich, 37-42; asst prof, Univ Tenn, 42-46; from assoc prof to prof, Sch Dent, Emory Univ, 46-54, prof div basic health sci, Univ, 54-60; prof anat & chmn dept, 60-72, asst dean basic med sci, 72-73, EMER PROF ANAT, SCH MED, TULANE UNIV, 73- *Concurrent Pos:* NSF sr fel, 57-58; Fulbright fels, Finland, 62 & Germany, 72; Alexander von Humboldt sr res award, 73-74; vis prof anat, Sch Med, Tufts Univ, 77 & 79; vis prof & actg chmn, Univ SFla, Tampa, 78-79, 87-88. *Honors & Awards:* Res Prize, Asn Southeast Biol, 59; Res Citation, Sigma Xi, 59. *Mem:* Am Soc Zool; Soc Cryobiol; Soc Develop Biol; Am Physiol Soc; Am Asn Anat; Sigma Xi. *Res:* Hypothermia in the resuscitation of asphyxiated neonates; cooling and resuscitating animals from zero degrees centigrade; hyperbaric oxygen and blockage of differentiation; physiology and histochemistry of development in coelenterates; effects on heart and brain of hypothermia and metabolic depressants. *Mailing Add:* 307 Shorewood Dr East Falmouth MA 02536

MILLER, JAMES ALEXANDER, b Dormont, Pa, May 27, 15; m 42; c 2. ONCOLOGY, TOXICOLOGY. *Educ:* Univ Pittsburgh, BS, 39; Univ Wis, MS, 41, PhD(biochem), 43. *Hon Degrees:* DSc, Med Col Wis, 82. *Prof Exp:* From instr to prof oncol, 44-85, EMER PROF ONCOL, MED CTR, UNIV WIS-MADISON, 85-; PROF ONCOL, WIS ALUMNI RES FOUND, 80- *Concurrent Pos:* Finney-Howell Found Med res fel, Med Ctr, Univ Wis-Madison, 43-44. *Honors & Awards:* Co-recipient, Teplitz-Langer Award, Ann Langer Cancer Res Found, 63 & Lucy Wortham James Award, James Ewing Soc, 65; G H A Clowes Award, Am Asn Cancer Res, 69; Bertner Award, M D Anderson Hosp & Tumor Inst, 71, Wis Nat Div Award, Am Cancer Soc, 73, Papanicolaou Res Award, Papanicolaou Cancer Res Inst, 75, Lewis S Rosenstiel Award, Brandeis Univ, 76, Nat Award Basic Sci, Am Cancer Soc, 77, Founders' Award, Chem Indust Inst Toxicol, 78, Bristol-Myers Award in Cancer Res, 78 & Int Ann Award, Gairdner Found, 78; Freedman Found Award, NY Acad Sci, 79; 3M Life Sci Award, Fed Am Soc Exp Biol, 79; Mott Award, Gen Motors Cancer Res Found, 80; Res Recognition Award, Samuel Roberts Noble Found. *Mem:* Nat Acad Sci; Am Soc Biol Chemists; Am Acad Arts & Sci; Am Asn Cancer Res; hon mem Japanese Cancer Asn. *Res:* Experimental chemical carcinogenesis; metabolic activation of chemical carcinogens; metabolic activation of carcinogenic chemicals to form electrophilic metabolites that bind covalently to cellular DNA and initiate the formation of tumors. *Mailing Add:* McArdle Lab Univ Wis Med Ctr Madison WI 53706

MILLER, JAMES ANGUS, b Huntington, WVa, Aug 16, 46; m 71; c 2. PHYSICAL CHEMISTRY. *Educ:* Univ Cincinnati, BS, 69; Cornell Univ, MEng, 70, PhD, 74. *Prof Exp:* Mem tech staff, 74-89, DISTINGUISHED MEM TECH STAFF, SANDIA NAT LABS, LIVERMORE, CALIF, 89- *Concurrent Pos:* Vis prof, Stanford Univ, 88. *Honors & Awards:* Silver Combustion Medal, Combustion Inst, 90. *Mem:* Am Phys Soc; Am Chem Soc; Combustion Inst; Sigma Xi. *Res:* Combustion chemistry; theory of flame, detonation and explosion phenomena; applications of quantum mechanics and statistical mechanics in chemistry; theoretical chemical kinetics and dynamics; mathematical modeling and applied mathematics. *Mailing Add:* Combustion Sci Dept Sandia Labs Livermore CA 94551-0969

MILLER, JAMES AVERY, b Los Angeles, Calif, Sept 12, 32; m 54; c 4. AERONAUTICS. *Educ:* Stanford Univ, BSME, 55, MSME, 56; Ill Inst Technol, PhD(mech eng), 63. *Prof Exp:* Consult heat transfer & fluid mech, Stanford Res Inst, 56-57; instr mech eng, Ill Inst Technol, 57-63, consult IIT Res Inst, 60-63; assoc prof aeronaut, Naval Postgrad Sch, 63-90; PRES, LABEX ENG CORP, 90- *Concurrent Pos:* Prin investr, NSF grant, 62-63; consult, Anderson Co, 62-63; chmn, Heat Transfer & Fluid Mech Inst, 65-66. *Mem:* AAAS; Am Soc Mech Engrs; Am Inst Aeronaut & Astronaut. *Res:* Free and forced convection heat and mass transfer; nonsteady boundary layer flows. *Mailing Add:* Box 5898 Carmel CA 93921

MILLER, JAMES EDWARD, b Hanover, Pa, Apr 28, 42; c 2. BIOCHEMISTRY. *Educ:* Shippensburg State Col, BS, 63; Univ NDak, MS, 65, PhD(biochem), 68. *Prof Exp:* USPHS grant, Sch Med, Temple Univ, 68-70; res investr, 70-78, RES SCIENTIST BIOCHEM, G D SEARLE & CO, 78- *Concurrent Pos:* Occupation Safety & Health Admin, US Dept Labor Note, 83- *Mem:* AAAS; Am Chem Soc; Am Indust Hyg Asn; Am Conf Govt Indust Hygienists. *Res:* Enzyme chemistry; protein purification; assay development; lipid metabolism; nutrition; hormonal control of metabolism. *Mailing Add:* 3807 Michael Lane Glenview IL 60025

MILLER, JAMES EDWARD, b Lafayette, La, Mar 21, 40; m 64; c 2. SECURITY, EDUCATION. *Educ:* Univ Southwestern La, 61, PhD (comput sci), 72; Auburn Univ, MS, 63. *Prof Exp:* Systs engr comput applac, IBM, 65-68; asst prof, Univ W Fla, 68-70, chmn comput sci, 72-86; CHMN COMPUT SCI, UNIV S MISS, 86- *Concurrent Pos:* Vis prof, Auburn Univ, 75; comput syst analyst, Environ Protection Agency, 79; ed, Spec Interest Group Comput Sci Educ, Asn Comput Mach, 82-; dir, Data Processing Mgt Asn Educ, 85-86; prog eval, Comput Sci Accreditation Bd, 86- *Mem:* Asn Comput Mach; Data Processing Mgt Asn Educ; Inst Elec Electronics Engrs. *Res:* Computer security and science education. *Mailing Add:* Univ Southern Miss PO Box 5106 Hattiesburg MS 39406

MILLER, JAMES EUGENE, b Loudonville, Ohio, Aug 12, 38; m 63; c 2. BIOCHEMISTRY, MICROBIOLOGY. *Educ:* Denison Univ, BA, 60; Harvard Univ, MA, 63; Amherst Col, PhD(biochem), 65. *Prof Exp:* Fel bact, Univ Calif, Los Angeles, 65-66; res assoc, NASA, 66-69; res biochemist, Corp Res Lab, Allied Chem Corp, 69-71; asst prof, 71-80, ASSOC PROF BIOL & CHMN DEPT, DEL VALLEY COL, 80- *Res:* Bacterial physiology. *Mailing Add:* Dept Biol Del Valley Col Sci & Agr Doylestown PA 18901

MILLER, JAMES FRANKLIN, b Lancaster, Pa, July 18, 12; m 38; c 1. ANALYTICAL CHEMISTRY. *Educ:* Franklin & Marshall Col, BS, 35; Pa State Univ, MS, 37, PhD(anal chem), 39. *Prof Exp:* Asst, Pa State Univ, 35-39, instr chem, Altoona Undergrad Ctr, 39-42; asst prof chem, The Citadel, 42; res fel, Res Found, Purdue Univ, 43-44; fel rubber res, Mellon Inst, 44-46, sr fel insecticides, 46-48, coal tar constituents, 48-49, arsenic, 49-51, head analytical chem sect, 51-59; lab dir gen chem cent res & eng, Div, Continental Can Co, 59-64, mgr appl res, Corp Res & Develop Dept, 64-68. *Concurrent Pos:* Civilian with AEC; Off Sci Res & Develop; US Rubber Reserve Corp. *Mem:* Am Chem Soc; Am Inst Chem Eng. *Res:* Inorganic non-ferrous analysis; analysis of organic halogen compounds and alcohol; butadiene coverter products; physical properties and behavior of emulsions; air elutriation of particular matter; behavior of fractionated particulate matter on falling utilization of arsenic; organic micro-analysis; polarography; spectrophotometry. *Mailing Add:* 6004 Linton Lane Indianapolis IN 46220-5359

MILLER, JAMES FREDERICK, b Davenport, Iowa, Feb 18, 43; m 67; c 2. MICROPALEONTOLOGY, INVERTEBRATE PALEONTOLOGY & STRATIGRAPHY. *Educ:* Augustana Col, Ill, AB, 65; Univ Wis-Madison, MA, 68, PhD(geol), 71. *Prof Exp:* Assoc prof geol, Univ Utah, 70-74; from asst prof to assoc prof, 74-82, PROF GEOL & DIS SCHOLAR, SOUTHWEST MO STATE UNIV, 82- *Mem:* AAAS; Geol Soc Am; Paleont Soc; Int Paleont Asn; Paleont Asn; Sigma Xi; Pander Soc; Soc of Economic Paleontologists & Mineralogists. *Res:* Taxonomy, evolution, and biostratigraphy of Cambrian and Lower Ordovician conodonts; stratigraphic position and international redefinition of Cambrian-Ordovician boundary; Paleozoic crinoids. *Mailing Add:* Geosci Dept Southwest Mo State Univ Springfield MO 65804-0089

MILLER, JAMES GEGAN, b St Louis Mo, Nov 11, 42; m 66; c 1. ULTRASOUND, MEDICAL BIOPHYSICS. *Educ:* St Louis Univ, AB, 64; Washington Univ, MA, 66, PhD(physics), 69. *Prof Exp:* Res assoc physics, 69-70, asst prof, 70-72, assoc prof, 72-77, res asst prof med, 76-81, assoc dir biomed physics, 74-87, res assoc prof, 81-88, PROF PHYSICS, WASHINGTON UNIV, 77-, DIR, 87-, PROF MED, LAB FOR ULTRASONICS, 88- *Concurrent Pos:* Sigma Xi nat lectr, 81-82. *Honors & Awards:* I R 100 Awards, 74 & 78. *Mem:* Inst Elec & Electronic Engrs; Am Phys Soc; fel Am Inst Ultrasound Med; Am Soc Non Destructive Testing; fel Acoust Soc Am. *Res:* Ultrasonics, biomedical physics, ultrasonic tissue characterization; ultrasonic resonators and transducers; ultrasonic materials characterization. *Mailing Add:* Dept of Physics Washington Univ St Louis MO 63130

MILLER, JAMES GILBERT, b El Dorado, Ark, Aug 2, 46; m 69; c 2. GENERAL RELATIVITY. *Educ:* Univ Calif, Berkeley, BA, 68; Princeton Univ, PhD(appl math), 72. *Prof Exp:* Asst prof math, Univ Calif, Los Angeles, 72-74 & asst prof physics, Univ Utah, 74-76; ASST PROF MATH, TEX A&M UNIV, 76- *Mem:* Am Math Soc. *Res:* Global properties of solutions of Einstein's field equations in general relativity, including symmetries, horizons and singularities. *Mailing Add:* Dept Math Tex A&M Univ College Station TX 77843-3368

MILLER, JAMES KINCHELOE, b Elkton, Md, June 16, 32; m 60; c 2. ANIMAL NUTRITION, PHYSIOLOGY. *Educ:* Berry Col, BS, 53; Univ Ga, MS, 59, PhD(animal nutrit), 62. *Prof Exp:* Tech asst dairy nutrit, Univ Ga, 57-58, asst, 58-60; from asst prof to assoc prof dairy physiol, Agr Res Lab, 61-81, assoc prof, Comp Animal Res Lab, Dept Energy, 73-81, PROF ANIMAL SCI, UNIV TENN, 81- *Honors & Awards:* Gustav Bohstedt Mineral & Trace Mineral Award, Am Soc of Animal Sci, 74. *Mem:* Am Dairy Sci Asn; Am Inst Nutrit; Am Soc Animal Sci; fel Am Col Nutrit. *Res:* Nutrition and physiology of the dairy cow; mineral metabolism; periparturient disorders (udder edema, retained placenta, milk fever) in dairy cows. *Mailing Add:* Dept Animal Sci Univ Tenn PO Box 1071 Knoxville TN 37901-1071

MILLER, JAMES L, b Chicago, Ill, May 10, 35; m 58; c 3. ORGANIC CHEMISTRY. *Educ:* Eastern Ill Univ, BS, 57; Univ Iowa, MS, 62, PhD(chem), 63. *Prof Exp:* From asst prof to assoc prof, 63-74, PROF CHEM, EAST TENN STATE UNIV, 74- *Mem:* Am Chem Soc. *Res:* Mechanisms concerning the bromination of stilbene and tolan; cycloaddition reactions which involve benzyne intermediates. *Mailing Add:* Dept of Chem East Tenn State Univ Johnson City TN 37601

MILLER, JAMES MONROE, b Lancaster, Pa, Aug 7, 33; m 55; c 2. ANALYTICAL CHEMISTRY. *Educ:* Elizabethtown Col, BS, 55; Purdue Univ, MS, 58, PhD(analytical chem), 60. *Prof Exp:* From asst prof to assoc prof, 59-69, chmn dept, 71-83 & 87-91, PROF ANALYTICAL CHEM, DREW UNIV, 69-,. *Concurrent Pos:* Vis lectr, Univ Ill, Urbana, 64-65; indust consult, 63-; dir, NSF Col Sci Improv Prog, 67-70; vis prof, Univ Amsterdam, 71; vis scientist, Nat Bur Standards, 85. *Honors & Awards:* Stambaugh Chem Award, Elizabethtown Col, 75. *Mem:* AAAS; Am Asn Clin Chemists; Am Chem Soc; Nat Sci Teachers Asn. *Res:* Gas chromatography; applications in teaching; studies of thermal conductivity detector response; liquid chromatography; detectors; clinical applications; determination of non-ionic detergents by column liquid chromatography. *Mailing Add:* Dept Chem Drew Univ Madison NJ 07940-4037

MILLER, JAMES NATHANIEL, b Detroit, Mich, Mar 16, 26; m 51; c 2. INFECTIOUS DISEASES. *Educ:* Univ Calif, Los Angeles, BA, 50, MA, 51, PhD(infectious dis), 56. *Prof Exp:* Jr res microbiologist, 56-58, asst prof infectious dis, 58-64, from asst prof to assoc prof microbiol & immunol, 66-72, PROF MICROBIOL & IMMUNOL, SCH MED, UNIV CALIF, LOS ANGELES, 72- *Mem:* Am Soc Microbiol; Am Asn Immunologists; Am Veneral Dis Asn. *Res:* Venereal diseases; immunobiology of syphilis. *Mailing Add:* Dept Microbiol & Immunol Univ Calif Sch Med 405 Hilgard Ave Los Angeles CA 90024

MILLER, JAMES PAUL, b Cleveland, Ohio, Sept 10, 46. KAON & MUON PHYSICS, NUCLEAR COMPTON SCATTERING. *Educ:* Carnegie-Mellon Univ, BS, 68, MS, 70, PhD(physics), 75. *Prof Exp:* Physics res assoc, Calif Inst Technol, 74-76; physicist, Lawrence Berkeley Lab, 76-79; from asst prof to assoc prof, 79-91, PROF, BOSTON UNIV, 91- *Mem:* Am Phys Soc; Sigma Xi. *Res:* CP violation using pure Ko or Ko beams; nuclear compton scattering near the delta resonance; precision measurement of the muon g-2. *Mailing Add:* Dept Physics Boston Univ 590 Commonwealth Ave Boston MA 02215

MILLER, JAMES Q, b Lakewood, Ohio, July 6, 26; m 50; c 4. NEUROLOGY, CYTOGENETICS. *Educ:* Haverford Col, BA, 49; Columbia Univ, MD, 53. *Prof Exp:* Nat Inst Neurol Dis & Blindness spec fel neuropath, Harvard Univ, 60-62; asst prof, 62-67, asst dean sch med, 62-70, assoc prof, 67-72, PROF NEUROL, SCH MED, UNIV VA, 72-, NEUROLOGIST, UNIV HOSP, 62- *Mem:* Am Acad Neurol; Am Epilepsy Soc; AMA. *Res:* Chromosome disorders and anomalies of the central nervous system. *Mailing Add:* Dept Neurol Univ Va Hosp Charlottesville VA 22908

MILLER, JAMES RAY, b Lancaster, Pa, Feb 2, 48; m 70; c 2. BIOLOGY, AGRICULTURE. *Educ:* Millersville State Col, BA, 70; Penn State Univ, PhD(entom), 75. *Prof Exp:* Res assoc entom, NY State Agr Exp Sta, 74-77; asst prof, 77-81, PROF ENTOM, MICH STATE UNIV, 81- *Honors & Awards:* Grad Student Recognition Award, Eastern Br Entom Soc Am, 73. *Mem:* AAAS; Entom Soc Am. *Res:* Chemical interactions between plants and insects. *Mailing Add:* 203 Pesticide Res Ctr Mich State Univ East Lansing MI 48824

MILLER, JAMES REGINALD, b Mimico, Ont, Nov 6, 28; m 54; c 5. GENETICS. *Educ:* Univ Toronto, BA, 51, MA, 53; McGill Univ, PhD, 59. *Prof Exp:* Asst develop physiol, Jackson Mem Lab, Maine, 54-56; res assoc genetics, Dept Neurol Res, 58-60, from asst prof to prof pediat, 60-73, head div med genetics, 67-78, PROF MED GENETICS, UNIV BC, 73- *Concurrent Pos:* Biomed consult, 80- *Mem:* Genetics Soc Am; Teratology Soc; Can Col Med Geneticists; Am Soc Human Genetics; Genetics Soc Can. *Res:* Developmental and population genetics of human beings and other mammals. *Mailing Add:* 3744 W 12th Ave Vancouver BC V6R 2N6 Can

MILLER, JAMES RICHARD, b St Louis, Mo, June 11, 22; m 45; c 2. PETROLEUM PRODUCTS CHEMISTRY, APPLIED INDUSTRIAL HYGIENE. *Educ:* Mo Sch Mines, BS, 44; Wash Univ, PhD(chem), 51. *Prof Exp:* Asst chem, Mo Sch Mines, 43; chemist, Ralston-Purina Co, 46-47; asst chem, Washington Univ, 47-50; res chemist, Shell Oil Co, 51-58, sr res chemist, Wood River Res Lab, 58-69; sr res scientist, Shell Res Ltd, Thorton Res Ctr, Chester, Eng, 69-70; mem staff, supv health & environ, Shell Develop Co, Westhollow Res Ctr, Houston, 75-85, chmn, Radiation Safety Comt, 80-85; RETIRED. *Mem:* Am Chem Soc; AAAS; Sigma Xi; fel Am Inst Chem; NY Acad Sci. *Res:* Oxidation and free radical chemistry; electron spin resonance; application of radiochemical techniques to problems in manufacture and use of petroleum; chemistry of porphyrins; corrosion chemistry; interaction of elastomers with organic solutions. *Mailing Add:* 12623 Scouts Lane Cypress TX 77429

MILLER, JAMES ROBERT, b Milford, Nebr, July 2, 22; m 45; c 5. ORGANIC CHEMISTRY, ZOOLOGY. *Educ:* Iowa State Univ, BS, 43; Syracuse Univ, PhD(chem), 50. *Prof Exp:* Res asst & jr res chemist, Parke, Davis & Co, 43-47; from asst prof to prof, 50-84, dept head, 52-65, sr prof chem, 85-88, EMER PROF, HARTWICK COL, 88- *Concurrent Pos:* Consult lab, Fox Hosp, 58-61. *Mem:* AAAS; Am Chem Soc; Ecol Soc Am; Am Soc Naturalists; Am Ornith Union. *Res:* Synthetic organic medicinals and heterocyclic compounds; avian ecology and biogeography. *Mailing Add:* RD 1 Box 1152 Maryland NY 12116

MILLER, JAMES ROBERT, b Holcomb, Mo, Jan 7, 41; m 59; c 3. SOLID STATE PHYSICS. *Educ:* Mo Sch Mines, BS, 62; Tex Christian Univ, MS, 64, PhD(physics), 66. *Prof Exp:* Assoc prof, 66-70, PROF PHYSICS, EAST TENN STATE UNIV, 70- *Res:* Electron spin resonance and nuclear magnetic resonance. *Mailing Add:* Dept Physics ETenn State Univ Johnson City TN 37614

MILLER, JAMES ROLAND, b Millington, Md, May 19, 29; m 54; c 2. SOIL CHEMISTRY, AGRONOMY. *Educ:* Univ Md, BS, 51, MS, 53, PhD(soil chem), 56. *Hon Degrees:* State Farmer Degree, FFA, 60. *Prof Exp:* Asst, Univ Md, 51-56; soil scientist chem, Soils & Plant Relationship Sect, Agr Res Serv, USDA, 56-58; from asst prof to assoc prof soils, 58-63, PROF & HEAD DEPT AGRON, UNIV MD, COLLEGE PARK, 63- *Concurrent Pos:* Consult, NASA, 65-; mem, Potash & Phosphate Inst Adv Coun (vchmn, 77, chmn, 78). *Mem:* Soil Conserv Soc Am; fel Soil Sci Soc Am; fel Am Soc Agron. *Res:* Soil test methods for determining available nutrients in soils; fission products; reactions in soils and uptake by plants. *Mailing Add:* Dept of Agron Univ of Md College Park MD 20742

MILLER, JAMES ROLAND, b Akron, Ohio, Aug 20, 21; m 49; c 2. SYNTHETIC, ORGANIC & NATURAL PRODUCTS CHEMISTRY. *Educ:* Univ Akron, BS, 43, MS & PhD(chem), 48. *Prof Exp:* Instr chem, Univ Akron, 42-43; asst tech mgr, 43-51, tech mgr, 51-56, sr tech mgr, 56-83, RES & DEVELOP ASSOC CHEMIST, BF GOODRICH CO, 83- *Mem:* Am Chem Soc; Am Inst Chemists. *Res:* Styrene-butadiene rubber; development of emulsion polymerization recipes and later colloidal properties with other monomer systems including vinylchloride, acrylate-methacrylate esters, acrylonitrile, vinyl pyridine. *Mailing Add:* 3659 Brecksville Rd Richfield OH 44286-9667

MILLER, JAMES TAGGERT, JR, b Sunbury, Pa, Apr 21, 43. THEORETICAL PHYSICS. *Educ:* Pa State Univ, BS, 65, PhD(physics), 70. *Prof Exp:* Intern radar syst anal, 72-73, mem fac, 73-74, sect supvr, 74-77, sect supvr sensor syst, Appl Physics Lab, 77-79, mem staff, 79-80, PRIN MEM STAFF, COMBAT SYST INTEGRATIONS GROUP, JOHNS HOPKINS HOSP, 80- *Mem:* Am Phys Soc. *Res:* Automated sensor system design, interfacing and data utilization in multisensor environments. *Mailing Add:* 512 Devonshire Lane Severna Park MD 21146

MILLER, JAMES WOODELL, b Detroit, Mich, June 30, 27; m 51; c 2. OCEANOGRAPHY, HUMAN VISION. *Educ:* Mich State Univ, BA, 49, MA, 50, PhD(exp psychol), 56. *Prof Exp:* Res assoc optics, Kresge Eye Inst, 52-60; staff engr human factors, Hughes Aircraft Co, 60-63; dir, eng psychol, Off Naval Res, 63-69; dir, ocean technol, Dept Interior, 69-70; dep dir, undersea sci technol, Nat Oceanic & Atmospheric Admin, 70-80; liaison scientist, psychol & oceanog, Off Naval Res, London, 75-76; assoc dir, 80-85, PROG COORDR, FLA INST OCEANOG, 85-; PRES, WOODELL ENTERPRISES, INC, 83- *Concurrent Pos:* Dep chmn, comt vision, Nat Res Coun, Nat Acad Sci, 54-58; chmn, US-Japan panel diving physiol & technol, 71-82; Nat Acad Sci exchange scientist, Bulgaria, 74; chmn, aquacult interagency coord bd, Fla, 84-87, vchmn, 88- *Honors & Awards:* Arthur S Fleming Award, US Chamber of Commerce, 66; Distinguished Civilian Serv Award, USN, 69; NOGI Award Sci, Underwater Soc Am, 86; Charles Shilling Award for Educ & Res Mgt, Undersea & Hyperbaric Med Soc, 89. *Mem:*

Undersea & Hyperbaric Med Soc; Am Acad Underwater Sci. *Res:* Undersea science and technology, diving and undersea habitation; human vision and perception; development of Florida aquaculture; coral reef ecology. *Mailing Add:* Rte 3 Box 295C Big Pine Key FL 33043

MILLER, JAN DEAN, b Dubois, Pa, Apr 7, 42; m 63; c 3. MINERAL PROCESSING, HYDROMETALLURGY. *Educ:* Pa State Univ, BS, 64; Colo Sch Mines, MS, 66, PhD(metall), 69. *Prof Exp:* Res engr, Anaconda Co, 66; asst prof, 68-72, assoc prof, 72-78, PROF METALL, UNIV UTAH, 78- *Concurrent Pos:* Res engr, Lawrence Livermore Lab, 72; adj prof fuels eng, Univ Utah; assoc dir, Ct Advan Coal Technol; dir, USBM Generic Ctr in Comn; founder & chmn bd, Advan Processing Technol Inc. *Honors & Awards:* Marcus A Grossman Award, 74; Van Diest Gold Medal, 77; Taggart Award, Soc Mining Engrs, 86; Henry Krumb Lectr, Am Inst Mining & Metall Engrs, 87 & Richards Award, 91; Extractive Metall Technol Award, Metall Soc, 89; Stefanko Award, Soc Mining Engrs, 88. *Mem:* Soc Mining Engrs; Metall Soc; Am Chem Soc; Fine Particle Soc. *Res:* Mineral processing; flotation separations including surface chemistry; particle-bubble collision; air-sparged hydrocyclone flotation technology and liberation analysis; hydrometallurgy including the physical chemistry of leaching, cementation, and solvent extraction; technology for gold recovery. *Mailing Add:* Univ Utah 216 WC Browning Bldg Salt Lake City UT 84112

MILLER, JANE ALSOBROOK, b New Orleans, La, Feb 21, 28; c 2. CHEMISTRY. *Educ:* Agnes Scott Col, AB, 48; Tulane Univ, MS, 50, PhD(hist chem), 60. *Prof Exp:* Instr chem, Tulane Univ, 50-52; res asst pharmacol, Wash Univ, 53-54, orthop surg, 63-65, res instr, 65; instr, 65-67, ASST PROF CHEM, UNIV MO-ST LOUIS, 67- *Mem:* Am Chem Soc; Hist Sci Soc; AAAS; Sigma Xi. *Res:* History of chemistry; chemical education. *Mailing Add:* Dept Chem Univ Mo-St Louis St Louis MO 63121

MILLER, JANICE MARGARET LILLY, b McPherson, Kans, Nov 11, 38; m 62; c 2. VETERINARY PATHOLOGY. *Educ:* Kans State Univ, BS, 60, DVM, 62, MS, 63; Univ Wis-Madison, PhD(vet sci), 69. *Prof Exp:* Res assoc animal nutrit, Mass Inst Technol, 64-65; Leukemia Soc Am spec fel, Univ Wis-Madison, 70-72; RES VET PATH, NAT ANIMAL DIS LAB, 72- *Mem:* Am Vet Med Asn; Am Col Vet Path. *Res:* Reproductive diseases of cattle. *Mailing Add:* Box 70 Nat Animal Dis Ctr Ames IA 50010

MILLER, JARRELL E, b San Antonio, Tex, Nov 14, 13; m 39; c 4. MEDICINE. *Educ:* St Mary's Univ, BA, 34; Baylor Univ, MD, 38; Am Bd Radiol, dipl, 42. *Prof Exp:* Intern, Robert B Green Mem Hosp, San Antonio, Tex, 38-39; resident radiol, Cleveland City Hosp, Ohio, 39-42; assoc prof, 47-56, clin prof, 56-66, PROF RADIOL, UNIV TEX HEALTH SCI CTR, DALLAS, 66- *Concurrent Pos:* Radiologist, Parkland Hosp, Dallas, 46-49, Children's Med Ctr, 47-65 & St Paul Hosp, Dallas, 67-; consult, Vet Hosps, Lisbon & McKinney, Tex, 46-57; dir dept radiol, Med Ctr, Baylor Univ, 49-66; lectr, Univ Tex Med Br, Galveston, 59-; deleg, AMA, 71- *Mem:* Fel Am Col Radiol (pres, 67-); Radiol Soc NAm; Am Roentgen Ray Soc (vpres, 63); Nat Tuberc & Respiratory Dis Asn; Am Cancer Soc. *Res:* Hypertrophic phyloric stenosis; angiocardiography; anatomy of the heart and great vessels; childhood malignancies. *Mailing Add:* 6115D Averill Way Dallas TX 75225-3321

MILLER, JERRY K, b Valley City, NDak, Sept 4, 34. ANALYTICAL CHEMISTRY. *Educ:* Univ Minn, BChem, 57, PhD(anal chem), 66; Univ Mo, MA, 63. *Prof Exp:* Res chemist, Am Cyanamid Co, 66-86; RETIRED. *Mem:* Am Chem Soc. *Res:* Polymer characterization. *Mailing Add:* 44 Strawberry Hill Ave Stamford CT 06902

MILLER, JOEL STEVEN, b Detroit, Mich, Oct 14, 44; m 70; c 3. INORGANIC CHEMISTRY. *Educ:* Wayne State Univ, BS, 67; Univ Calif, Los Angeles, PhD(inorg chem), 71. *Prof Exp:* Res assoc inorg chem, Stanford Univ, 71-72; assoc scientist, 72-73, scientist, Webster Res Ctr, Xerox Corp, 73-78; mem tech staff & proj mgr, Sci Ctr, Rockwell Int Corp, 78-79; PRIN SCIENTIST & GROUP LEADER, OCCIDENTAL RES CORP, 79- *Concurrent Pos:* Vis prof chem, Univ Calif, Irvine, 81, Univ Penn, 88. *Mem:* AAAS; Am Phys Soc; Am Chem Soc; fel Royal Soc Chem. *Res:* Synthetic and physical inorganic chemistry; anisotropic inorganic and organic complexes exhibiting unusual magnetic, optical, electrical properties; synthetic metals; homogeneous catalysis; inorganic/organometallic synthesis; structure-function relationships. *Mailing Add:* du Pont Exp Sta E328 Wilmington DE 19880-0328

MILLER, JOHN ALLEN, agricultural engineering, entomology, for more information see previous edition

MILLER, JOHN CAMERON, b Richmond, Va, Aug 12, 49; m 75; c 1. PHYSICAL CHEMISTRY. *Educ:* Ga Inst Technol, BS, 71; Univ Colo, PhD(chem), 75. *Prof Exp:* Instr chem, Colo Col, 76; instr & res assoc, Univ Va, 76-79; RES STAFF MEM, OAK RIDGE NAT LAB, 79-, HEAD, CHEM PHYSICS SECT, 89- *Mem:* Am Chem Soc; Am Phys Soc; Optical Soc Am. *Res:* Atomic and molecular spectroscopy; multiphoton processes; photoionization; laser spectroscopy; nonlinear optics. *Mailing Add:* Health & Safety Res Div Oak Ridge Nat Lab Bldg 4500-5 MS 6125 PO Box 2008 Oak Ridge TN 37831-6125

MILLER, JOHN CLARK, physical chemistry, for more information see previous edition

MILLER, JOHN DAVID, b Todd, NC, Aug 9, 23; m 46, 84; c 3. PLANT BREEDING, PLANT GENETICS. *Educ:* NC State Col, BS, 48, MS, 50; Univ Minn, PhD(plant breeding), 53. *Prof Exp:* Res fel, seedstocks prod, Univ Minn, 53; asst prof cereal breeding, Kans State Univ, 53-57; assoc prof, 57-61, adj prof agron, Agr Exp Sta, Va Polytech Inst & State Univ, 61-78; sr agronomist, Agr Res Serv, USDA, 75-88, adj res assoc, Ga Agr Exp Sta, 78-88; RETIRED. *Concurrent Pos:* Res agronomist, Agr Res Serv, USDA, 57-72, res leader, 72-78. *Mem:* Am Soc Agron; Am Genetics Asn; Sigma Xi. *Res:* Forage breeding; statistical techniques in crops research; improved breeding methods for forages; improved small plot machinery; breeding for tolerance to acid soils. *Mailing Add:* 801 E 12th St Tifton GA 31794-4115

MILLER, JOHN FREDERICK, b Los Angeles, Calif, Mar 24, 28; m 54; c 4. METEOROLOGY. *Educ:* Univ Calif, Los Angeles, AB, 51. *Prof Exp:* Meteorologist, Coop Studies Sect, Hydrologic Serv Div, US Weather Bur, 53-59, Dept Navy, 59-62; asst chief coop studies, Off Hydrol, Nat Weather Serv, Nat Oceanic & Atmospheric Admin, 62-64, chief spec studies br, 64-71, chief water mgt info div, 71-84; CONSULT HYDROMETEOROLOGIST, 85- *Mem:* Am Meteorol Soc; Am Geophys Union. *Res:* Investigation of rainfall with respect to cause, frequency, magnitude and estimating the limiting amounts. *Mailing Add:* 13420 Oriental St Rockville MD 20853

MILLER, JOHN GEORGE, b Philadelphia, Pa, Oct 18, 08; m 40; c 2. PHYSICAL CHEMISTRY. *Educ:* Univ Pa, AB, 29, MSc, 30, PhD(chem), 32. *Prof Exp:* Res chemist, Dermat Res Labs, Pa, 29-30; asst, 30-31, from asst instr to prof, 31-79, EMER PROF CHEM, UNIV PA, 79- *Concurrent Pos:* Proj leader, Thermodyn Res Lab, Univ Pa, 45-51; vis examr, Swarthmore Col, 50-52; consult, Englehard Minerals & Chems Corp, NJ, 44-81, Smith, Kline & French Labs, 46-49, Pennwalt Corp, 49-71 & Eastern Regional Res Labs, USDA, 53-57. *Mem:* Fel AAAS; Am Phys Soc; Clay Minerals Soc; Am Chem Soc; Am Inst Chemists. *Res:* Molecular structure; dielectric constant measurements; homogeneous catalysis; reaction mechanisms; surface chemistry; gas properties; calorimetry. *Mailing Add:* Dept Chem Univ Pa Philadelphia PA 19174

MILLER, JOHN GRIER, b Boston, Mass, Feb 5, 43. MATHEMATICS. *Educ:* Univ Chicago, SB, 63, SM, 64; Rice Univ, PhD(math), 67. *Prof Exp:* Asst prof math, Univ Calif, Los Angeles, 67-69; asst prof, Columbia Univ, 69-72; asst prof, Ill State Univ, 72-76; asst prof, Southern Ill Univ, 76-78; asst prof, 78-83, ASSOC PROF MATH, IND UNIV-PURDUE UNIV INDIANAPOLIS, COLUMBUS, IND, 83- *Mem:* Am Math Soc. *Res:* Topology. *Mailing Add:* Dept of Math Ind Univ-Purdue Univ Columbus IN 47203

MILLER, JOHN H, III, electronic physics, for more information see previous edition

MILLER, JOHN HENRY, b Washington, DC, Mar 16, 33; m 54. PLANT PHYSIOLOGY. *Educ:* Yale Univ, BS, 54, MS, 57, PhD(bot), 59. *Prof Exp:* Nat Cancer Inst fel bot, Yale Univ, 59-60, instr, 60-62; from asst prof to assoc prof, 62-70, PROF BOT, SYRACUSE UNIV, 70- *Mem:* Am Soc Plant Physiol; Bot Soc Am; Scand Soc Plant Physiol. *Res:* Photophysiology; developmental physiology; cell differentiation. *Mailing Add:* 945 Maryland Ave Syracuse NY 13210

MILLER, JOHN HOWARD, b Columbus, Ohio, Oct 13, 43; m 65; c 1. RADIATION PHYSICS. *Educ:* Davidson Col, BS, 66; Univ Va, PhD(physics), 71. *Prof Exp:* Res assoc physics, Univ Fla, 72-74 & Los Alamos Sci Lab, 74-75; STAFF SCIENTIST PHYSICS, PAC NORTHWEST LAB, BATTELLE MEM INST, 75- *Concurrent Pos:* Assoc ed, Radiation Res. *Mem:* Am Phys Soc; Radiation Res Soc; Int Soc Quantum Biol. *Res:* Theory of physical and chemical processes resulting from energy deposition by ionizing radiation and their relationship to biological effects of the radiation exposure. *Mailing Add:* Pac Northwest Lab Battelle Mem Inst PO Box 999 P8-47 Richland WA 99352

MILLER, JOHN JAMES, b Schreiber, Ont, Oct 13, 18; m 51. MICROBIOLOGY. *Educ:* Univ Toronto, BA, 41, PhD(bot), 44. *Prof Exp:* Agr asst, Can Dept Agr, 44-46, asst plant pathologist, 46-47; from asst prof to assoc bot, 47-60, PROF BIOL, MCMASTER UNIV, 60- *Mem:* Can Soc Microbiol; Can Bot Asn. *Res:* Mycology; yeast sporulation and spore germination. *Mailing Add:* Dept Biol McMaster Univ 1280 Main St W Hamilton ON L8S 4K1 Can

MILLER, JOHN JOHNSTON, III, b San Francisco, Calif, Apr 9, 34; m 58; c 4. PEDIATRICS, IMMUNOBIOLOGY. *Educ:* Wesleyan Univ, BA, 55; Univ Rochester, MD, 60; Univ Melbourne, PhD(immunol), 65. *Prof Exp:* Intern pediat, Univ Calif, San Francisco, 60-61, resident, 61-62; resident, 65, clin teaching asst, 65-67, asst prof pediat, 67-73, sr attend physician, 73-77, from assoc prof to prof clin pediat, 77-82, DIR RHEUMATIC DIS SERV, CHILDREN'S HOSP, MED CTR, STANFORD UNIV, 67-, PROF PEDIAT, 82- *Concurrent Pos:* Consult, US Naval Radiol Defense Lab, 67-69. *Mem:* Fel Am Acad Pediat; Am Rheumatism Asn; Am Asn Immunol; Soc Pediat Res; Am Fedn Clin Res. *Res:* Pediatric rheumatology; immunologic abnormalities in juvenile rheumatoid arthritis. *Mailing Add:* Children's Hosp Stanford Univ Med Ctr 520 Sand Hill Rd Palo Alto CA 94304

MILLER, JOHN MICHAEL, b Tulsa, Oklahoma, Apr 29, 54; m 76; c 2. CARDIOVASCULAR DISEASE, ELECTROPHYSIOLOGY. *Educ:* Penn State Univ, BS, 75, MD, 79. *Prof Exp:* Res assoc, HospUniv Penn, 85-86; ASST PROF MED, UNIV PENN, SCHOOL MED, 86- *Mem:* Am Col Cardiology. *Res:* Intracardiac electrophysiology; investigation of mechanisms of arrhythmias in man. *Mailing Add:* Dep Med Cardiovasc Sect Hosp Univ Penn Sch Med Clin Electrophysiol Lab 1598 Salomon Lane Wayne PA 19087

MILLER, JOHN ROBERT, b Berwyn, Ill, June 22, 44; m 67; c 2. PHYSICAL CHEMISTRY, RADIATION CHEMISTRY. *Educ:* Ore State Univ, BS, 66, Univ Wis, PhD(phys chem), 71. *Prof Exp:* Appointee phys & radiation chem, 71-74, asst chemist, 74-76, chemist, 76-86, SR CHEMIST, ARGONNE NAT LAB, 86- *Mem:* Am Chem Soc; AAAS. *Res:* Electron transfer over long distances by quantum mechanical tunneling; picosecond pulse radiolysis; effects of energy, distance and molecular structure on rates of long distance electron transfer between molecules. *Mailing Add:* Chem 200 Argonne Nat Lab Argonne IL 60439

MILLER, JOHN WALCOTT, b Royal Oak, Mich, Nov 10, 30; m 72; c 2. POLYMER CHEMISTRY, ANALYTICAL CHEMISTRY. *Educ:* Wesleyan Univ, BA, 53; Northwestern Univ, PhD(anal chem), 56. *Prof Exp:* Methods develop chemist, 56-60, mgr, Chem Methods Sect, 60-75, supvr, Chromatogr Sect, 75-77, mgr, environ safety Br, 77-83, SR PATENT &

LITIGATION SPECIALIST, PHILLIPS PETROL CO, 83- *Concurrent Pos:* Nat Acad Sci exchange fel, Prague, 68; chmn, Gordon Res Conf Analytical Chem, 75 & Analytical Chem Div, Am Chem Soc, 76-77. *Res:* Coulometric analysis; photometric titrations; organic sulfur functional groups analysis; redox reactions; polarography of coordination compounds; crude oil source identification; water pollution analysis; composition and synthesis of engineering plastics. *Mailing Add:* 306 Autumn Ct Bartlesville OK 74006

MILLER, JOHN WESLEY, JR, b Philadelphia, Pa, Aug 30, 35; m 58; c 2. MARINE ZOOLOGY, ENTOMOLOGY. *Educ:* Dickinson Col, BS, 57; Pa State Univ, MS, 60, PhD(entom), 62. *Prof Exp:* Fel entom & parasitol, Univ Calif, Berkeley, 62-63; from asst prof to assoc prof, 63-74, PROF BIOL, BALDWIN-WALLACE COL, 74- *Concurrent Pos:* Mem, Comt Marine Invertebrates, Nat Res Coun, 76-81. *Mem:* AAAS; Sigma Xi; Am Inst Biol Sci. *Res:* Laboratory culture and use of marine organisms in the undergraduate curriculum; insects as vectors of plant diseases, specifically the relationship of certain plant viruses to their aphid vectors. *Mailing Add:* Geol & Geog Dept Ohio Wesleyan Univ Delaware OH 43015

MILLER, JON PHILIP, b Moline, Ill, Mar 30, 44; m 65; c 2. BIOCHEMISTRY, MOLECULAR PHARMACOLOGY. *Educ:* Augustana Col, Ill, AB, 66; St Louis Univ, PhD(biochem), 70. *Prof Exp:* Fel biophys chem, ICN Nucleic Acid Res Inst, 70-71, biochemist, 72, head molecular pharm, 72-73, head drug metab, 73-74, head biol div, Nucleic Acid Res Inst, 74-76; head, SRI/NCI Liaison Group, 76-78, sr bioorg chemist, 78-80, DIR, MED BIOCHEM PROG, SRI INT, 80-, ASSOC DIR, BIOMED RES LAB, 81- *Res:* Structure-activity relationships and pharmacological activities of c-adenosine monophosphate analogs; development and mechanism of action and metabolism of drugs; control of cellular growth and differentiation. *Mailing Add:* 1147 Blythe St Foster City CA 94404-3646

MILLER, JOSEF MAYER, b Philadelphia, Pa, Nov 29, 37; m 60, 85; c 2. PHYSIOLOGY, PSYCHOLOGY. *Educ:* Univ Calif, Berkeley, BA, 61; Univ Wash, PhD(physiol & psychol), 65. *Hon Degrees:* Univ Goteborg, 87. *Prof Exp:* USPHS fel, Univ Mich, 65-67, res assoc & asst prof psychol, 67-68; from asst prof to prof otolaryngol, physiol & biophys, Univ Wash, 68-84; PROF & DIR, KRESGE HEARING RES INST, UNIV MICH, 84- *Concurrent Pos:* Res affil, Univ Wash, 68-84; Deafness Res Found grant, Univ Wash, 69-71, NIH res grant, 69-73; actg chmn, Univ Mich, 75-76. *Mem:* AAAS; Am Auditory Soc; Acoust Soc Am; Soc Neurosci; Am Otol Soc; Fedn Am Soc Exp Biol; Asn Res Otolaryngol. *Res:* Definitions of the save levels of electrical stimulation; cochlear prosthesis and cochlear blood flow studies; definiton of the mechanisms which underlie control of cochlear blood flow; techniques for measuring cochlear blood flow and experimental manipulations which may be used to influence cochlear blood flow and prove to be useful therapeutically. *Mailing Add:* Kresge Hearing Res Inst Univ Mich 1301 E Ann St Ann Arbor MI 48109-0506

MILLER, JOSEPH, b San Francisco, Calif, Apr 3, 37; m 59; c 3. LASERS & OPTICAL SYSTEMS. *Educ:* Univ Calif, Los Angeles, BS, 57, MS, 58, PhD(eng), 61. *Prof Exp:* Assoc eng, Univ Calif, Los Angeles, 59-61, lectr, 61-69; supvr nuclear reactor eng, Atomics Int Div, N Am Aviation Inc, 61-65; asst mgr, Combustion Systs Lab, TRW Systs Group, 65-74; mgr, Power Systs Technol Lab, 74-79, mgr, High Energy Laser Off, TRW Defense & Space Systs Group, 76-81, mgr, Res & Technol Oper, 81-88, VPRES & GEN MGR, APPL TECHNOL DIV, TRW SPACE & TECHNOL GROUP, REDONDO BEACH, 90- *Mem:* Nat Acad Eng; Am Inst Aeronaut & Astronaut; Optical Soc Am-SPIE; AAAS. *Res:* High energy laser research and development; optical systems; rocket engine and propulsion system development; nuclear reactor physics; heat and mass transfer; thermodynamics; space instruments and sensor systems. *Mailing Add:* 19855 Greenbriar Dr Tarzana CA 91356

MILLER, JOSEPH EDWIN, b Carrollton, Mo, Nov 4, 42; m 62; c 2. PLANT BIOCHEMISTRY, PLANT PHYSIOLOGY. *Educ:* Colo State Univ, BS, 64, MS, 66; Utah State Univ, PhD(plant biochem), 69. *Prof Exp:* Asst prof plant physiol, Univ Colo, Denver, 69-73; res assoc, Univ Ill, 73-75; asst res scientist, Argonne Nat Lab, 75-80, res scientist, 80-84; PLANT PHYSIOLOGIST, AGR RES SERV, USDA, RALEIGH, NC, 84- *Mem:* Am Soc Plant Physiol; Crop Sci Soc Am. *Res:* Plant growth and development; air pollutant effects on plant growth and physiology. *Mailing Add:* 107 Cougar Ct Cary NC 27513

MILLER, JOSEPH HENRY, b Yonkers, NY, May 27, 24; m 48; c 2. MEDICAL PARASITOLOGY. *Educ:* Univ Mich, BS, 48, MS, 49; NY Univ, PhD(biol), 53. *Prof Exp:* Asst biol, NY Univ, 51-53; from instr to prof med parasitol, 53-86, EMER PROF MED PARASITOL, SCH MED, LA STATE UNIV, NEW ORLEANS, 86- *Concurrent Pos:* Fel, China Med Bd, Cent Am, 56; scientist, vis staff, Charity Hosp, New Orleans, 53-; vis prof, Fac Med, Nat Univ Mex, 66, Col Med, Univ Ariz, 75-76; consult, Family Med Proj, Abha Col Med Sci, Univ Riyadh, Saudi Arabia, 81; mem fac, Int Course Trop Med, Fac Med, Univ Costa Rica, 90, course dir, 91. *Mem:* Emer mem Am Soc Trop med & Hyg; emer mem Am Soc Parasitol; emer fel Royal Soc Trop Med & Hyg; hon mem Electron Micros Soc Am (treas, 70-71). *Mailing Add:* Dept Microbiol Immunol & Parasitol La State Univ Sch Med 1901 Perdido St New Orleans LA 70112

MILLER, JOSEPHINE, BIO-AVAILABILITY OF IRON. *Educ:* Ga Col, BS, 46. *Prof Exp:* ASSOC PROF NUTRIT RES, DEPT FOOD SCI, UNIV GA, 46- *Res:* Effects of processing on nutritional quality of foods. *Mailing Add:* Ob/Gyn Dept Univ Ill Med Ctr Chicago IL 60680

MILLER, JOYCE FIDDICK, b Greene, Iowa, Oct 2, 39; div; c 2. GROWTH FACTORS. *Educ:* Univ Iowa, BA, 67; Univ SFla, MA, 70; Southern Ill Univ, PhD(chem), 78. *Prof Exp:* Res assoc biochem, Iowa State Univ, 78-81; res scientist neurol, Univ Iowa, 81-88; ASST PROF CHEM, UNIV WIS, PLATTEVILLE, 88- *Mem:* Am Chem Soc; AAAS. *Res:* Purification and characterization of growth and maturation factors for cells in the nervous system. *Mailing Add:* Univ Wis Platteville WI 53818

MILLER, JOYCE MARY, b Belton, Tex, Mar 12, 45; m 72. BIOCHEMISTRY. *Educ:* Southwest Tex State Univ, BS, 67, MA, 68; Tex A&M Univ, PhD(biochem), 72. *Prof Exp:* Fel lipoprotein chem, Sch Med, Univ Southern Calif, 72-74; res assoc, Sch Med, Wash Univ, 74-75; asst prof, 75-81, PROF BIOL SCI, MARYVILLE COL, MO, 81- *Concurrent Pos:* NIH fel, 73. *Mem:* AAAS; Sigma Xi. *Res:* Determination of qualitative and/or quantitative variations in the peptide composition of various serum lipoprotein classes with increasing age of an individual by protein characterization and immunological techniques. *Mailing Add:* 1706 Bowline Houston TX 77062

MILLER, JUDITH EVELYN, b Rahway, NJ, Dec 30, 51. FERMENTATION. *Educ:* Cornell Univ, BSc, 73; Case Western Reserve Univ, PhD(microbiol), 78. *Prof Exp:* Teaching specialist microbiol, Cornell Univ, 73; asst prof life sci, 78-84, ASSOC PROF BIOL & BIOTECHNOL, WORCESTER POLYTECH INST, 84- *Mem:* Sigma Xi; Soc Indust Microbiol. *Res:* Physiology of immobilized cells. *Mailing Add:* Dept Biol & Biotechnol Worcester Polytech Inst Worcester MA 01609

MILLER, JULIAN CREIGHTON, JR, b Baton Rouge, La, Mar 6, 40; m 65; c 2. PLANT BREEDING, PLANT GENETICS. *Educ:* La State Univ, BS, 65, MS, 67; Mich State Univ, PhD(hort), 72. *Prof Exp:* Res asst hort, La State Univ, 66-67, Univ Wis-Madison, 67-68, Mich State Univ, 68-72; asst prof, Tex Agr Exp Sta, Lubbock, 72-75; from asst prof to assoc prof, 75-82, interim dept head hort sci, 80-83, PROF HORT, TEX A&M UNIV, 82- *Concurrent Pos:* Prin investr, Coop State Res Serv grants, US Agency for Int Develop, 77-80, 78-79, 79-82 & 82-85; grant, Farming Technol, Inc, 90. *Mem:* AAAS; Am Soc Hort Sci; Am Genetic Asn; Am Soc Agron; Am Inst Biol Sci; Sigma Xi; fel, Am Soc Hort Sci (pres-elect, 91). *Res:* Potato variety development; vegetable legume improvement; physiological and biochemical genetics of crop yield under stress. *Mailing Add:* Dept of Hort Sci Tex A&M Univ College Station TX 77843-2133

MILLER, KEITH WYATT, b Chard, UK, Mar 31, 41; m 70; c 2. PHARMACOLOGY. *Educ:* Oxford Univ, BA, 63, DPhil, 67. *Hon Degrees:* AM, Harvard Univ, 84. *Prof Exp:* NSF fel chem, Univ Calif, Berkeley, 67-69; res fel pharmacol, Oxford Univ, UK, 69-71; from instr to assoc prof pharmacol, 71-83, EDWARD MALLINCKRODT JR PROF PHARMACOL ANESTHESIA, HARVARD MED SCH, 83- *Concurrent Pos:* Mem organizing comt, 2nd Int Res Conf, Molecular Mechanisms of Anesthesia, 78-79, co-chmn, 3rd conf, 83-84; mem comt Neurosci & Anesthetic Action, Am Soc Anesthesiologists, 80-88, chmn comt, Exp Neurosci & Biochem, 89-; mem Lamport Award comt, Biophys Soc, 82; chmn, spec rev comt, Nat Inst Alcohol Abuse & Alcoholism, 84; vis fel, St Catherine's Col, Oxford Univ, UK, 85; co-chmn, Conf Molecular & Cellular Mechanism of Alcohol & Anesthetics, NY Acad Sci, 89-90, Calgary, Alta, Can, 90; vis prof, dept biochem, Oxford Univ, 91. *Mem:* Am Soc Pharmacol & Exp Therapeut; Biophys Soc; Undersea Med Soc; Brit Pharmacol Soc; Brit Biophys Soc; Am Soc Anesthesiol; Int Soc Biomed Res Alcoholism; Res Soc Alcoholism. *Res:* Mechanism of action of general anesthetics and their reversal by pressure; thermodynamics of small moleculer-membrane interactions; lipid modulation of protein function; acetylcholine receptors; spectroscopic studies of membrane function. *Mailing Add:* Dept Anesthesia White 4 Mass Gen Hosp Fruit St Boston MA 02114

MILLER, KENNETH JAY, b New York, NY, Oct 12, 24; m 48; c 5. PHYSICAL CHEMISTRY. *Educ:* Eastern Nazarene Col, BS, 49; Johns Hopkins Univ, MA, 50, PhD(phys chem), 52. *Prof Exp:* Develop engr semiconductors, Westinghouse Elec Corp, Pa, 52-55; asst prof chem, Mt Union Col, 55-58; sr scientist phys chem, Res & Advan Develop Div, Avco Corp, 58-59; mem tech staff semiconductors, Bell Tel Labs, 59-68; prof chem, Northeast La Univ, 68-89; RETIRED. *Res:* Thermodynamics and electrochemistry; semiconductors. *Mailing Add:* 3710 College Blvd Monroe LA 71203

MILLER, KENNETH JOHN, b Chicago, Ill, Mar 24, 39; m 63, 75; c 2. THEORETICAL CHEMISTRY. *Educ:* Ill Inst Technol, BS, 60; Johns Hopkins Univ, MA, 64; Iowa State Univ, PhD(chem), 66. *Prof Exp:* Nat Acad Sci-Nat Res Coun resident res assoc, Nat Bur Standards, 66-67; from asst prof to assoc prof, 67-81, PROF THEORET CHEM, RENSSELAER POLYTECH INST, 81- *Concurrent Pos:* Consult & vis scientist, Nat Bur Standards, 71 & 86-; vis prof, Univ Fla, Gainesville, 73-74; Rensselaer Distinguished Teaching Fel. *Mem:* Am Chem Soc; AAAS; Am Asn Univ Prof. *Res:* Theoretical chemistry; interaction of molecules with nucleic acids; design of antitumor agents; characterization of binding sites on nucleic acids; molecular polarizabilities; vibrational spectroscopy; molecular dynamics; conformation and structure of polymers; computer graphics. *Mailing Add:* Dept Chem Rensselaer Polytech Inst 210 Cogswell Troy NY 12180-3590

MILLER, KENNETH L, b Lock Haven, Pa, Aug 14, 43; m 61; c 2. HEALTH PHYSICS, MEDICAL HEALTH PHYSICS. *Educ:* Lock Haven State Col, BS, 65; Univ Pittsburgh, MS, 70; Am Bd Health Physics, cert, 76; Am Bd Med Physics, cert, 90. *Prof Exp:* Teacher math & sci, Pa High Sch, 65-66; health physics asst, Pa State Univ, Univ Park, 66-69; fel radiation safety, Univ Pittsburgh, 69-70; assoc health physicist, Pa State Univ, 70-71; res assoc, Milton S Hershey Med Ctr, Pa State Univ, 71-78, asst prof, 78-81, assoc prof, 81-88, PROF RADIOL, MILTON S HERSHEY MED CTR, PA STATE UNIV, 88- *Concurrent Pos:* Consult, var hosps, clins, physicians & indust; dir, Div Health Physics, Milton S Hershey Med Ctr, Pa State Univ, 71-, bd dirs, 85-88, chair radiol health sect, 87-88; adv panel, Nuclear Regulatory Comn, Decontamination, TMI-2, 84- *Honors & Awards:* Elda E Anderson Award, Health Physics Soc, 82. *Mem:* Health Physics Soc; Am Asn Physicists Med; Am Pub Health Asn; Am Nuclear Soc; AAAS. *Res:* Radioisotope/radiation utilization in research and medicine; equipment design; problem solving. *Mailing Add:* Health Physics Off M S Hershey Med Ctr PO Box 850 Hershey PA 17033

MILLER, KENNETH M, SR, b Chicago, Ill, Nov 20, 21; m 43, 70; c 4. ELECTRONICS. *Educ:* Ill Inst Technol, 40-41; Univ Calif, Los Angeles, 61. *Prof Exp:* Electronics engr, Rauland Corp, 41-48; gen mgr, Learcal Div, Lear, Inc, 48-59; vpres & gen mgr, Motorola Aviation Electronics, Inc, 59-60 & Instrument Div, Daystrom, Inc, 61; gen mgr, Singer Co, 62-65; vpres & gen mgr, Lear Jet Corp, 65-66; pres & dir, Infonics, Inc, Calif, 67-68; vpres & gen mgr, Comput Industs, Inc, Calif, 68-69; vpres & dir, Am Standard Corp, 69-71; pres, Wilcox Elec, Inc & dir, World Wide Wilcox, Inc, 71-73; PRES, CHIEF EXEC OFFICER & DIR, PENRIL CORP, 74- *Concurrent Pos:* Trustee, Park City Hosp, Bridgeport, Conn, 62-64; mem bd assocs, Univ Bridgeport, 62-64. *Mem:* Sr mem Inst Elec & Electronics Engrs; sr mem Instrument Soc Am; Am Soc Nondestructive Testing; Armed Forces Commun Electronics Asn; AAAS. *Res:* Business administration; automatic flight controls; communication and navigation systems; video, audio and radio frequency consumer electronic devices; electromechanical components; precision instruments. *Mailing Add:* 16904 George Washington Dr Rockville MD 20853-1128

MILLER, KENNETH MELVIN, b Indianapolis, Ind, Aug 17, 43; m 69; c 1. FRESH WATER ECOLOGY, FOOD SCIENCE. *Educ:* Ind Univ, AB, 65, AM, 67, PhD(zool), 75. *Prof Exp:* Instr biol, Purdue Univ, NCent Campus, 70-74; lab mgr, 75-76, qual control mgr, 76-81, technical dir, 81-84, Dir Admin, 84-87, DIR QUAL ASSURANCE, AM HOME FOODS, 87- *Concurrent Pos:* Guest lectr, Purdue NCent, 81 & 84, Williamsport Area Community, 87. *Mem:* Am Soc Qual Control; Am Inst Biol Sci; Int Food Technologists. *Res:* Species association and diversity in aquatic coleoptera and odonata; quality assurance technology; popcorn food technology. *Mailing Add:* RD 1 Box 266 P Lewisburg PA 17837-9536

MILLER, KENNETH PHILIP, b Northfield, Minn, Sept 17, 15; m 41; c 4. ANIMAL NUTRITION. *Educ:* Univ Minn, BS, 39, MS, 40; Ohio State Univ, PhD(dairy husb), 56. *Prof Exp:* From instr to prof, 41-85, EMER PROF ANIMAL NUTRIT, SOUTHERN EXP STA, UNIV MINN, WASECA, 85- *Concurrent Pos:* Vis prof, Nepal, 78-79 & Syria, 82; consult animal nutrit, Indonesia, 84-85. *Mem:* AAAS; Am Soc Animal Sci; Am Dairy Sci Asn. *Res:* Animal nutrition; ruminant nutrition; dairy beef production. *Mailing Add:* 808 Fifth Ave SE Waseca MN 56093

MILLER, KENNETH RAYMOND, b Rahway, NJ, July 14, 48; m 72; c 2. CELL BIOLOGY. *Educ:* Brown Univ, ScB, 70; Univ Colo, PhD(biol), 74. *Prof Exp:* Lectr biol, Harvard Univ, 74-76, asst prof, 76-80; assoc prof, 80-86, PROF, BROWN UNIV, 86- *Concurrent Pos:* Ed, J Cell Biol, Advan Cell Biol. *Mem:* AAAS; Am Soc Cell Biol; Am Soc Photobiol. *Res:* Structure, biochemistry, and function of biological membranes; most importantly, the photosynthetic membrane. *Mailing Add:* Div Biol & Med Brown Univ Providence RI 02912

MILLER, KENNETH SIELKE, b New York, NY, June 4, 22; m 53; c 2. MATHEMATICS. *Educ:* Columbia Univ, BS, 43, AM, 47, PhD(math), 50. *Prof Exp:* Lectr, Columbia Univ, 49; from inst to prof math, NY Univ, 50-64; sr staff scientist, Electronics Res Labs, Columbia Univ, 64-89; ADJ PROF MATH, FORDHAM UNIV, 64- *Concurrent Pos:* Mem, Inst Advan Study, 50 & 58-59; consult, Army Res Off, Systs Res Labs & Fed Sci Corp. *Mem:* Am Math Soc; Sigma Xi. *Res:* Differential operators; multivariate distributions; complex stochastic processes; author of a series of text books and research papers in mathematics. *Mailing Add:* 25 Bonwit Rd Rye Brook NY 10573

MILLER, KENT D, b Detroit, Mich, May 9, 25; m 50; c 3. BIOCHEMISTRY. *Educ:* Oberlin Col, AB, 49; Wayne State Univ, MS, 51, PhD, 54; Albany Med Col, MD, 62. *Prof Exp:* Res scientist, Div Labs & Res, NY State Dept Health, 54-57, sr res scientist, 57-62, asst dir, 62-69; PROF MED, SCH MED, UNIV MIAMI, 69-, PROF MICROBIOL, 76- *Mem:* Am Chem Soc; Am Soc Biol Chemists; Am Soc Hemat; Soc Exp Biol & Med. *Res:* Blood proteins; coagulation; bacterial enzymes; immunology. *Mailing Add:* Univ Miami Dept Med PO Box 016960 Miami FL 33101

MILLER, KIM IRVING, b Boone, NC, July 26, 36; m 73. PLANT TAXONOMY. *Educ:* Appalachian State Teachers Col, BS, 58; Purdue Univ, MS, 61, PhD(bot), 64. *Prof Exp:* Vis asst prof biol, Purdue Univ, 64-65, res assoc & cur herbarium, 65-66; asst prof biol, Eastern Ky Univ, 66-67 & Appalachian State Univ, 67-68; asst prof, 68-71, chmn dept, 73-78, assoc prof, 72-78, PROF BIOL & CHMN DIV SCI & MATH & DIR, ENVIRON CTR, JACKSONVILLE UNIV, 78- *Concurrent Pos:* Consult environ res. *Mem:* Bot Soc Am; Am Soc Plant Taxon. *Res:* Evolution and systematics of Euphorbiaceae and related families. *Mailing Add:* Dept of Biol Jacksonville Univ 12800 Univ Blvd N Jacksonville FL 32211

MILLER, KIRK, b New York, NY, June 25, 49; m 72; c 3. PHYSIOLOGICAL ECOLOGY, COMPARATIVE PHYSIOLOGY. *Educ:* Antioch Col, BA, 72; Colo State Univ, MS, 74; Univ Okla, PhD(zool), 78. *Prof Exp:* asst prof, 78-85, ASSOC PROF BIOL, FRANKLIN & MARSHALL COL, 85- *Mem:* AAAS; Am Soc Zoologists. *Res:* Respiration and metabolism of amphibians; metabolism and growth of reptiles. *Mailing Add:* Dept Biol Franklin & Marshall Col Lancaster PA 17604

MILLER, LARRY GENE, b Corsicana, Tex, Dec 30, 46; m 69; c 3. PHARMACEUTICS. *Educ:* Univ Miss, BS, 69, PhD(pharmaceut), 72. *Prof Exp:* Sr scientist prod develop, Mead Johnson & Co, 72-73; sr pharmacist pharmaceut res, A H Robins Co, Inc, 73-76, mgr pharm res prod develop, 76-78, dir prod develop, 78-90. *Concurrent Pos:* Clin instr indust clerkship, Sch Pharm, Med Col Va. *Mem:* Am Pharmaceut Asn; Acad Pharmaceut Sci; Controlled Release Soc. *Res:* Design of dosage forms for pharmaceutical compounds so as to meet the necessary requirements of stability, delivery and bioavailability; new product formula implementation to commercial manufacturing. *Mailing Add:* 2801 Reserve St Richmond VA 23220

MILLER, LARRY O'DELL, b Los Angeles, Calif, Feb 26, 39; m 72; c 2. DEVELOPMENTAL BIOLOGY, NEUROPHYSIOLOGY. *Educ:* Univ Calif, Santa Barbara, BA, 61, MA, 64, PhD(biol), 67. *Prof Exp:* Res biologist, Firestone Tire & Rubber Co, 67-69; lectr biol, Univ Calif, Santa Barbara, 69-71; head dept, 76-78, senate pres, 75-76 & 89-91, INSTR BIOL, MOORPARK COL, 71- *Mem:* AAAS. *Res:* Biochemical control of eukaryotic cellular differentiation. *Mailing Add:* Dept of Biol Moorpark Col Moorpark CA 93021

MILLER, LAURENCE HERBERT, b Newark, NJ, Oct 11, 34; m 59; c 2. DERMATOLOGY. *Educ:* Muhlenberg Col, BS, 56; Univ Lausanne, MD, 61. *Prof Exp:* Intern, Newark Beth Israel Hosp, 62-63; house physician, East Orange Gen Hosp, 63; resident, NY Univ Med Ctr, 63-66; SPEC ADV TO DIR, NAT INST ARTHRITIS, MUSCULOSKELETAL & SKIN DIS, 66- *Concurrent Pos:* Mem, Dermat Found, 68; asst prof, Sch Med, George Washington Univ, 76-; mem, Sci Adv Bd, Nat Psoriasis Found, 71-; tech consult, Skin & Allergy News, 73- *Mem:* Fel Am Acad Dermat; Soc Invest Dermat; Am Dermat Asn. *Res:* Varicella-Zoster virus; cell controls in psoriasis. *Mailing Add:* 5454 Wisconsin Ave Chevy Chase MD 20815

MILLER, LAWRENCE INGRAM, b Jackson Center, Ohio, May 12, 14; m 39; c 2. PHYTOPATHOLOGY, AGRICULTURE. *Educ:* Oberlin Col, AB, 36; Va Polytech Inst, MS, 38; Univ Minn, PhD(plant path), 53. *Prof Exp:* Fel plant path, Freeport Sulphur Co, 38-40,; plant pathologist, 40-79, EMER PROF PLANT PATH, VA POLYTECH INST & STATE UNIV, 80- *Concurrent Pos:* Mem comt biol control of soilborne plant pathogens, Nat Acad Sci-Nat Res Coun, 57-65; NATO res grant plant nematol, 73-74; mem exec coun, Intersoc Consortium for Plant Protection, 76-78. *Honors & Awards:* Golden Peanut Res Award, Nat Peanut Coun, 63; J Shelton Horsley Res Award, Va Acad Sci, 60. *Mem:* Int Soc Plant Path; Soc Europ Nematol; Brit Asn Appl Biol; Am Phytopath Soc; hon mem Soc Nematol (pres, 76-77). *Res:* Plant nematology, particularly Heterodera and Globodera species; diseases of the peanut. *Mailing Add:* McBryde House 800 Price's Fork Rd Blacksburg VA 24060

MILLER, LEE STEPHEN, b Jacksonville, Fla, June 5, 30; m 50; c 2. ELECTRONICS. *Educ:* Ind Inst Technol, BS, 52; Clemson Univ, PhD(eng physics), 67. *Prof Exp:* Engr, Microwave Lab, Int Tel & Tel Co, 52-54; head antenna dept, Radiation Inc, 54-57; vpres, Melbourne Eng Corp, 57-58; sr staff engr, Sperry Rand Corp, 58-62; mem tech staff, Thompson-Ramo-Wooldridge Corp, 62-64; sr div scientist, Res Triangle Inst, 68-80; pres, Appl Sci Assocs, Inc. 80-88; CONSULT, 88- *Concurrent Pos:* Mem countermeasures group, Electronics Warfare Coun, 55-57; NASA grant, Res Triangle Inst, 68-69. *Mem:* AAAS; Inst Elec & Electronics Engrs. *Res:* Radar; electromagnetic scattering; remote sensor systems. *Mailing Add:* 105 Windsor Ct Central SC 29630

MILLER, LEON LEE, b Rochester, NY, Dec 7, 12; m 35, 58; c 6. BIOCHEMISTRY. *Educ:* Cornell Univ, BA & MA, 34, PhD(org chem), 37; Univ Rochester, MD, 45. *Prof Exp:* Res org chemist, Calco Chem Co, NJ, 37-38; instr path & pharmacol, Sch Med & Dent, Univ Rochester, 45-46; asst prof biochem, Jefferson Med Col, 46-48; from assoc prof to prof, 48-78, EMER PROF BIOPHYS & BIOCHEM, UNIV ROCHESTER, 78- *Mem:* AAAS; Am Chem Soc; Am Soc Biol Chem; NY Acad Sci. *Res:* Organic synthesis; blood and liver proteins and their functions; liver injury; enzyme, protein and amino acid metabolism; isolated liver perfusion. *Mailing Add:* Dept Biophys Univ Rochester Sch Med Rochester NY 14642

MILLER, LEONARD DAVID, b Jersey City, NJ, July 8, 30; m 67; c 2. SURGERY. *Educ:* Yale Univ, AB, 51; Univ Pa, MD, 55. *Prof Exp:* Asst instr surg, 56-57 & 59-64, assoc in surg, 64-66, from asst prof to prof surg, 66-78, vchmn dept, 72-78, John Rhea prof surg & chmn dept, 78-81, dir, Harrison Dept Surg Res, 72-81, J WILLIAM WHITE PROF SURG RES, UNIV PA, 70- *Concurrent Pos:* NIH sr clin trainee, Univ Pa, 64-65, NIH grants, 65-, John A Hartford Found, Inc grant, 66 -; intern, Hosp Univ Pa, 55-56, resident surg, 56-57 & 59-64, dir, Shock & Trauma Clin Res Unit, 67-72; exec officer, Harrison Dept Surg Res, Univ Pa, 67-68, dir NIH training grant, 67-72; vis surgeon, Vet Admin Hosp, 71 -; consult, Children's Hosp Philadelphia, 73-81; chmn comt surg educ, Soc Univ Surg, 68-71; actg chmn dept surg, Am Surg Asn. *Mem:* AAAS; Soc Univ Surg; Soc Surg Alimentary Tract; Am Soc Surg Trauma; Am Surg Asn. *Mailing Add:* Dept of Surg 4th Floor Silverstein Hosp of the Univ of Pa Philadelphia PA 19104

MILLER, LEONARD ROBERT, b New York, NY, Oct 31, 33; m 57; c 3. PATHOLOGY, CELL BIOLOGY. *Educ:* Bethany Col, BS, 54; Univ Pittsburgh, MS, 55; Albany Med Col, MD, 59; Am Bd Path, dipl, 65. *Prof Exp:* Intern, Beverly Hosp, 59-60; resident path, Sch Med, Yale Univ, 61-64, instr, 64-67; asst prof path, Sch Med, Univ Ariz, 67-68; assoc prof, Med Ctr, Univ Okla, 68-70, vis assoc prof physiol & biophys, 70-72; CLIN ASSOC PROF PATH, UNIV CALIF, LOS ANGELES & SCH MED, UNIV UTAH, 73- *Concurrent Pos:* Res fel, 60-61; USPHS trainee path, 61-63, spec fel, 63-64; Off Naval Res grant, 68-; consult, Tissue Bank, Naval Med Res Inst, 64-67; sect chief histopath, Armed Forces Radiobiol Res Inst, 64-67; pvt pract path, Sargent, Miller & Morrison Prof Corp, 72- *Mem:* AAAS; Am Asn Pathologists & Bacteriologists; Int Acad Path; Soc Cryobiol. *Res:* In vivo, in vitro mammalian and bacterial cell function; cellular control mechanisms; phage induction; mammalian cell transformation; adaptation; immunology; cryobiology; ultrastructure of immune response; inhibition and enhancement of cell repair. *Mailing Add:* 890 Cherry St Tulare CA 93274

MILLER, LEROY JESSE, b Lebanon, Pa, Aug 12, 33; m 54; c 3. POLYMER CHEMISTRY, PHOTOCHEMISTRY. *Educ:* Elizabethtown Col, BS, 54; Univ Del, MS, 57, PhD(chem), 59. *Prof Exp:* Sr res engr, Atomics Int Div, NAm Aviation Inc, 58-60; sr res chemist, Sundstrand Corp, 60-62; mem tech staff, Hughes Aircraft Co, 62-71, group head, 71-78, staff engr, 71-78, sect head, 78-88, SR SCIENTIST, HUGHES AIRCRAFT CO, 89- *Mem:* AAAS; Am Chem Soc; Soc Imaging Sci & Technol; Electrochem Soc; Sigma Xi; Int Soc Optical Eng. *Res:* Radiation resists; thermally stable aromatic polymers; photopolymerization; liquid crystals; photogalvanic effects; radiation damage; reaction mechanisms. *Mailing Add:* 8313 Hillary Dr West Hills CA 91304-3153

MILLER, LINDA JEAN, b Atlantic City, NJ, Dec 17, 58; m; c 2. LEUKOCYTE ADHESION, HEMATOPOIETIC DIFFERENTIATION. *Educ:* Johns Hopkins Univ, BA, 81; Harvard Univ, PhD(immunol), 87. *Prof Exp:* Postdoctoral fel, Nat Cancer Inst, 87-88; ASSOC ED SCI, SCI, AAAS, 88- *Mem:* Am Asn Immunologists. *Res:* Decisions on and edit most of the immunology, medicine, adhesion, cell biology, and regulation manuscripts. *Mailing Add:* Sci 1333 H St NW Washington DC 20005

MILLER, LLOYD GEORGE, b Brighton, Colo, Dec 15, 34. NUTRITION, BIOCHEMISTRY. *Educ:* Colo State Univ, BS, 62, PhD(nutrit), 67; Univ Nebr, MS, 64. *Prof Exp:* Mgr nutrit, Theracon Co, 71-72; mgr, 67-71, Carnation Co, dir prod develop, 72-82, asst gen mgr, 82-85, VPRES, CALRECO, INC, 85- *Mem:* Am Soc Animal Sci. *Res:* Small animal nutrition; product development of foods for the dog and cat. *Mailing Add:* Carnation Res Labs 8015 Van Nuys Blvd Van Nuys CA 91412

MILLER, LOIS KATHRYN, b Lebanon, Pa, Oct 8, 45; m 74; c 1. BIOCHEMISTRY, VIROLOGY. *Educ:* Upsala Col, BS, 67; Univ Wis-Madison, PhD(biochem), 72. *Prof Exp:* Fel, Calif Inst Technol, 71-74, Imperial Cancer Res Fund, London, 74-76; from asst prof to prof biochem, Univ Idaho, 76-86; PROF ENTOMOL & GENETICS, UNIV GA, 86- *Concurrent Pos:* Adj prof, Wash State Univ, 78-86; NIH grant, 77-; USDA grant, 85- *Honors & Awards:* NIH Merit Award, 86. *Mem:* AAAS; Am Soc Microbiol; Am Soc Biol Chem & Molecular Biol; Soc Invert Path. *Res:* Nucleic acid biochemistry; molecular biology; insect virology; viral genetics; recombinant DNA technology; biological insect pest control, and gene expression vectors. *Mailing Add:* Depts Entomol & Genetics Univ Ga Athens GA 30602

MILLER, LORRAINE THERESA, b Dane, Wis, Mar 12, 31. NUTRITION, BIOCHEMISTRY. *Educ:* Univ Wis, BS, 53, MS, 58, PhD(nutrit, biochem), 67. *Prof Exp:* Dietitian, Med Ctr, Univ Mich, 54-56; instr foods & nutrit, Mich State Univ, 58-63; asst prof, 66-69, ASSOC PROF FOODS & NUTRIT, ORE STATE UNIV, 69- *Mem:* Am Dietetic Asn; AAAS; Sigma Xi. *Res:* Metabolism of tryptophan in vitamin B6 deficiency; metabolism of vitamin B6. *Mailing Add:* Sch Home Economics Ore State Univ Corvallis OR 97331

MILLER, LOUIS HOWARD, b Baltimore, Md, Feb 4, 35; m 59; c 1. TROPICAL MEDICINE, PARASITOLOGY. *Educ:* Haverford Col, BS, 56; Wash Univ, MD, 60; Columbia Univ, MS, 64. *Prof Exp:* Intern, Mt Sinai Hosp, New York, 60-61; resident med, Montefiore Hosp, Bronx, 61-62 & Mt Sinai Hosp, 62-63; res physician, SEATO Med Res Lab, Bangkok, Thailand, 65-67; from asst prof to prof trop med, Col Physicians & Surgeons, Columbia Univ, 67-71; HEAD SECT MALARIA, LAB PARASITIC DIS, NAT INST ALLERGY & INFECTIOUS DIS, 71- *Concurrent Pos:* NIH fel, Cedar-Sinai Med Ctr, 64-65; Gold Medal Lectr Hemat, Calcutta Sch Trop Med, India, 77; vis lectr, Harvard Sch Pub Health, 83-; pres, Am Soc Trop Med & Hyg, 88; comt lectr, Scripps Clin, 88; chmn, adv comt Molecular Parasitology Award, Burroughs-Wellcome Fund, 87-89; mem, steering comt, Special Prog World Health Orgn, Immunization Against Malaria, 79-87, adv comt Molecular Parasitology Award, 81-89, bd dir, Gorgas Mem Inst Trop & Prev Med, 81-84, Sci & Tech Adv Comt, Special Prog Res & Training Trop Dis, 89- *Honors & Awards:* Dyer Lectr, NIH, 85; Paul-Ehrlich & Ludwig-Darmstaedter-Prize, 85; The Sixth Lloyd E Rozeboom Lect in Med Entom, Johns Hopkins Univ, Baltimore, Md, 82; Derrick-Mackerras Lect Queensland Univ, 86; Stoll-Stunkard lectr, Am Soc Parasitology, 89; Shipley lectr, Harvard Med Sch, 90. *Mem:* Nat Acad Sci; Inst Med-Nat Acad Sci; Am Soc Trop Med & Hyg; Infectious Dis Soc Am; Am Soc Clin Invest; hon fel Royal Soc Trop Med & Hyg; hon fel Queensland Inst Med Res; AAAS; Asn Am Physicians; fel Am Col Physicians. *Res:* Malaria; ultrastructure; immunology; physiology; entomology. *Mailing Add:* Lab Parasitic Dis Bldg 4 Rm 126 Nat Inst Allergy & Infect Dis NIH Bethesda MD 20892

MILLER, LOWELL D, b Chicago, Ill, Jan 20, 33; m 59; c 2. RESEARCH & DEVELOPMENT, PHARMACEUTICALS. *Educ:* Univ Mo, BS, 57, MS, 58, PhD(biochem), 60. *Prof Exp:* Dir biochem & toxicol, Neisler Labs, Union Carbide, 60-69; assoc dir biomed res, Warren-Teed Pharmaceuts, 69-70; pres lab exp biol & tech dir clin labs, Smith Klein Inc, 71-73; dir corp sci affairs, Marion Labs Inc, 73-77, corp vpres res & develop, 78-89. *Concurrent Pos:* Mem fac, Millikin Univ, 63-68. *Mem:* Am Chem Soc; NY Acad Sci; Am Asn Clin Chem; Soc Toxicol; fel Am Inst Chemists; fel Am Soc Clin Scientists; Int Soc Forensic Scientists. *Res:* Develops, maufactures and sells pharmaceuticals and products for hospitals and laboratories use. *Mailing Add:* 114 Stone Haven Dr Kansas City MO 64137

MILLER, LYLE DEVON, b Lebanon, Ind, Dec 8, 38; m 62; c 2. VETERINARY PATHOLOGY. *Educ:* Kans State Univ, BS, 61, DVM, 63; Univ Wis-Madison, MS, 68, PhD, 71. *Prof Exp:* Pathologist, Vet Serv, Nat Animal Dis Ctr, USDA, 71-81; PROF COL VET MED, IOWA STATE UNIV, 81- *Mem:* Am Vet Med Asn; Am Asn Vet Lab Diagnosticians; Am Col Vet Pathologists; US Animal Health Asn; Nat Asn Fed Vets. *Mailing Add:* Dept Vet Path Col Vet Med Iowa State Univ Ames IA 50011-1250

MILLER, LYNN, b McCook, Nebr, Nov 6, 32; m 57; c 2. MICROBIAL GENETICS, MOLECULAR BIOLOGY. *Educ:* San Francisco State Col, BS, 57; Stanford Univ, PhD(biol), 62. *Prof Exp:* NIH fel microbiol, Hopkins Marine Sta, Stanford Univ, 62-64; NIH fel genetics, Univ Wash, 64-65; asst prof biol, Am Univ, Beirut, 65-68 & Adelphi Univ, 68-70; assoc prof, 70-74, dean, Sch Natural Sci, 77-78, PROF BIOL, HAMPSHIRE COL, 74- *Concurrent Pos:* Vis scholar, Univ Wash, 75; vis fac, Evergreen State Col, 79-80, Univ Mass, Amherst, 83-84. *Mem:* AAAS; Genetics Soc Am; Am Inst Biol Sci; Am Soc Microbiol. *Res:* Genetics in Saccharomyces; human population genetics; Frankia genetics. *Mailing Add:* Sch of Natural Sci Hampshire Col Amherst MA 01002

MILLER, LYSTER KEITH, b Tonapah, Nev, Aug 8, 32; m 60; c 3. PHYSIOLOGY. *Educ:* Univ Nev, BS, 55, MS, 57; Univ Alaska, PhD(zoophysiol), 66. *Prof Exp:* Res physiologist, Arctic Health Res Ctr, USPHS, Alaska, 60-62; from instr to assoc prof physiol, 62-87, prin res assoc, 88- 89, SR RES ASSOC, INST ARCTIC BIOL, UNIV ALASKA, 89- *Mem:* AAAS; Am Physiol Soc; Entom Soc Am; Soc Cryobiol. *Res:* Temperature adaptation in peripheral nerve; mechanisms of freezing tolerance in insects and other invertebrates; temperature regulation and gross energetics in northern mammals, especially aquatic species; biometeorology of cold regions. *Mailing Add:* Inst Arctic Biol Univ Alaska Fairbanks AK 99775

MILLER, M(URRAY) H(ENRI), b Brooklyn, NY, Oct 1, 28; m 58. ELECTRICAL ENGINEERING. *Educ:* Univ Mich, BSE, 50, MSE, 51, PhD, 59. *Prof Exp:* Res assoc, Eng Res Inst, Univ Mich, Ann Arbor, 51-59, lectr, 52, from asst prof to assoc prof elec eng, 59-69; assoc prof, 69-76, PROF ELEC ENG, UNIV MICH, DEARBORN, 76-, CHMN DEPT, 75- *Concurrent Pos:* Sr scientist, KMS Indust, Mich, 67. *Mem:* Inst Elec & Electronics Engrs; Am Soc Eng Educ; AAAS. *Res:* Electron tubes; physical electronics. *Mailing Add:* 1534 Glastonbury Rd Ann Arbor MI 48103

MILLER, MARCIA MADSEN, b Santa Monica, Calif, May 31, 44; m 66; c 1. BIOLOGY, ELECTRON MICROSCOPY. *Educ:* Univ Calif, Davis, BS, 65, MA, 66; Univ Calif, Los Angeles, PhD(bot), 72. *Prof Exp:* Fel biol, Calif Inst Technol, 72-75; res scientist, City Hope Nat Med Ctr, 75-78, assoc res scientist, 79-87; fel, Biozentrum, Univ Basel, 78-79; ASSOC RES SCIENTIST, BECKMAN RES INST CITY HOPE, 87- *Concurrent Pos:* NSF fel, 65-66; NIH fel, 73-75; Swiss NSF Int fel, 78-79. *Mem:* AAAS; Am Soc Cell Biologists; Sigma Xi; Am Asn Immunol. *Res:* Molecular immuno genetics. *Mailing Add:* Beckman Res Inst City Hope 1450 E Duarte Rd Duarte CA 91010-0269

MILLER, MARTIN WESLEY, b Belden, Nebr, Jan 8, 25; m 48; c 3. MICROBIOLOGY. *Educ:* Univ Calif, AB, 50, MS, 52, PhD(microbiol), 58. *Prof Exp:* Res food technologist, 57-59, from asst prof to assoc prof, 59-70, PROF FOOD TECHNOL, UNIV CALIF, DAVIS, 70- *Concurrent Pos:* Fulbright sr res scientist award, Australia, 64-65; vis prof, Univ NSW, South Wales, Australia, 84. *Mem:* Am Soc Microbiol; fel Inst Food Technologists; fel Am Acad Microbiol; Japanese Mycol Soc. *Res:* Ecology and taxonomy of yeasts; dehydration and drying of fruits; food science and technology educational curricula; food fermentations and microbial (fungal and yeast) food spoilage. *Mailing Add:* Dept Food Sci Technol Univ Calif Davis CA 95616-8598

MILLER, MAURICE MAX, b New Albany, Ind, Feb 18, 29. NUCLEAR PHYSICS. *Educ:* Ind Univ, AB, 48, MS, 50, PhD(nuclear physics), 52. *Prof Exp:* Asst, Off Naval Res, Ind, 48-52; sr nuclear engr, Convair Div, Gen Dynamics Corp, 52-53; mem staff physics, Los Alamos Sci Lab, 53, instr nuclear weapons, Armed Forces Spec Weapons Proj, 53-55; mgr, Nuclear Lab Div, Lockheed Aircraft Corp, 55-62, mgr, Nuclear Aerospace Div, Lockheed, Ga Co, 62-72, consult engr, 72-80, MGR SYSTS ENG DIV, LOCKHEED MISSILES & SPACE CO, 80- *Mem:* Am Phys Soc; Am Nuclear Soc. *Res:* Effects of nuclear radiation on organic and metallic materials; scintillation spectroscopy and measurement of nuclear decay schemes; experimental analysis of beta decay matrix elements; theory of strong interactions. *Mailing Add:* 2400 NASA Rd No One Houston TX 77058

MILLER, MAX K, b Cleburne, Tex, Oct 25, 34; m 58; c 2. MATHEMATICS, GEOPHYSICS. *Educ:* Univ Tex, Austin, BS & BA, 57, MA, 63, PhD(math), 66. *Prof Exp:* Res scientist, Defense Res Lab, Univ Tex, Austin, 60-65; res geophysicist, Tex Instruments, Inc, 65-75; res scientist assoc, Appl Res Lab, Univ Tex, Austin, 75-80; mgr, petrol support group, Cray Res, Inc, 80-84; CONSULT, 84- *Concurrent Pos:* Adj assoc prof, Comput Sci Dept, Univ Tex, Austin, 83-; comput consult, 84- *Mem:* Inst Elec & Electronics Engrs; Acoust Soc Am; Soc Petrol Engrs; Soc Explor Geophys; Sigma Xi. *Res:* Numerical inversion of Laplace transforms; acoustic and seismic wave propagation; numerical analysis; signal processing; vector computers; parallel computers. *Mailing Add:* Max Miller & Assocs 7407 Valburn Dr Austin TX 78731

MILLER, MAYNARD MALCOLM, b Seattle, Wash, Jan 23, 21; m 51; c 2. GEOLOGY. *Educ:* Harvard Univ, SB, 43; Columbia Univ, MA, 48; Cambridge Univ, PhD(geol & geom), 57. *Hon Degrees:* DSc, Univ Alaska, 90. *Prof Exp:* Asst prof navig, Princeton Univ, 46; geologist, Gulf Oil Corp, Cuba, 47; geologist, Off Naval Res, 49-51; vis staff mem, Fed Inst Snow & Avalanche Res, Switz, 52; demonstr phys geog, Cambridge Univ, 53-54; res assoc geol, Lamont Geol Observ, NY, 55-57; sr scientist, Geol Dept, Columbia Univ, 58-59; from asst prof to prof geol, Mich State Univ, 59-75, dir Glaciol Arctic Sci Inst, 60-75; chief & state geologist, Idaho Bur Mines & Geol, 75-83, dir, Idaho Geol Serv, 84-88; dean, Col Mines, 75-78, DIR, GLACIOL & ARCTIC SCI INST, UNIV IDAHO & PROF GEOL, 75-; STATE GEOLOGIST & DIR, IDAHO GEOL SURV, 84- *Concurrent Pos:* Mem var geol & geophys expeds, 40-72; dir & geologist, Juneau Icefield Res Prog, Alaska, 46-; exec dir, Found Glacier & Environ Res, Seattle, Wash, 55-, pres, 75-82, chair, 83-; consult, Boeing Co & US Air Force, 59-60, State of Alaska, 61-63 & US Forest Serv, 63-; res grants from numerous founds and corps, 60-; geologist, Am Mt Everest exped, 63; dir, Alaskan Commemorative Glacier Proj, Nat Geog Soc, 64-; dir, Lemon Glacier Proj, Int Hydrol Decade Prog, Alaska, 65-75; field dir, Mt Kennedy mem mapping proj, 65; dir, Sea Ice Proj, US Navy Oceanog Off, 67-68; chmn & dir, World Explor Ctr Found, 68-71; mem, tech & mineral educ mission, People's Repub China, 81, 85 & 88; leader, Nepalese Langtang Himal geol proj, 84, Manasln-Ganesh Himal geol proj, 87; hon guest prof, China Univ Geosci, Wuhan & Beijing, People's Repub China, 84-, Chongdun Univ, Jilin, 88- *Honors & Awards:* Co-recipient, Hubbard Medal, Nat Geog Soc, 63; Karo Award, Soc Am Mil Engrs, 65; Franklin L Burr Award, Nat Geog Soc, 67. *Mem:* Fel Geol Soc Am; Am Inst Prof Geologists; fel Arctic Inst NAm; Am Asn State Geologists; Am Inst Mining, Metall & Petrol Engrs; Am Geophys Union; Int Glaciol Soc. *Res:* Glaciology and hydrology applications; process geomorphology; mining and environmental geology; volcanology; photogrammetry; inter-disciplinary factors in variation of existing glaciers and sea ice and related problems in global change; Pleistocene and Holocene stratigraphy, arctic climatology and periglacial geology. *Mailing Add:* 514 E First St Moscow ID 83843

MILLER, MELTON M, JR, b Burlington, Vt, Nov 15, 33; m 54; c 4. CIVIL ENGINEERING. *Educ:* Univ Vt, BSCE, 55; Purdue Univ, MSCE, 58, PhD(civil eng), 64. *Prof Exp:* Instr civil eng, Purdue Univ, 57-63; asst prof, 63-68, ASSOC PROF CIVIL ENG, UNIV MASS, AMHERST, 68- *Mem:* Am Soc Eng Educ; Am Soc Civil Engrs; Sigma Xi. *Res:* Model studies for civil engineering structures; analysis and design of structures; computer applications to civil engineering structures. *Mailing Add:* Marston Hall Civil Engr Dept Univ Mass Amherst MA 01002

MILLER, MELVIN J, b St Anthony, Idaho, Sept 11, 40; m 60; c 5. CHEMICAL ENGINEERING, PHYSICS. *Educ:* Univ Utah, BS, 64, PhD(chem eng), 68. *Prof Exp:* Sr res chemist, 68-73, res assoc, 74-77, LAB HEAD, EASTMAN KODAK CO, 77- *Mem:* Am Inst Chem Engrs. *Res:* Surface and bulk rheology; photographic systems; control process equipment; computer stimulation of physical processes; hydrodynamics; transport phenomena; technical management. *Mailing Add:* Eastman Kodak Co Bldg 205 Kodak Park Rochester NY 14652

MILLER, MELVIN P, b Baltimore, Md, May 17, 35; m 64; c 2. PHYSICAL CHEMISTRY. *Educ:* Loyola Col, Md, BS, 57; Princeton Univ, PhD(molten salts), 62. *Prof Exp:* From instr to assoc prof, 60-69, PROF CHEM, LOYOLA COL, MD, 69- *Concurrent Pos:* NSF grant, Conf Surface Colloid & Macromolecular Chem, Lehigh Univ, 65; res partic for col teachers, Boston Univ, 69; sci fac fel, Johns Hopkins Univ & NSF, 71-72, courtesy fel, 72-78, prof eve col, 75; res fel & chemist, Chem Res Develop & Eng Ctr, Aberdeen Proving Ground, 80-90; consult infrared & thermal analysis of polymers. *Honors & Awards:* Distinguished Serv Award, Md Sect, Am Chem Soc, 84. *Mem:* Am Chem Soc; Sigma Xi; Soc Appl Spectros. *Res:* Transport properties of molten salts; heterogeneous catalysis; electroanalytical techniques; surface chemistry; infrared studies of reaction kinetics and molecular structure. *Mailing Add:* Loyola Col Baltimore MD 21210-2699

MILLER, MEREDITH, b Murfreesboro, Tenn, Jan 2, 22; m 46; c 3. PHYSICAL CHEMISTRY. *Educ:* Vanderbilt Univ, BA, 43, MS, 44; Univ Wis, PhD(chem), 50. *Prof Exp:* Instr chem & physics, Middle Tenn State Col, 46-47; res chemist, Film Dept, E I du Pont de Nemours & Co, NY, 50-54, Va, 54-59, staff scientist, 59-63; prin chemist, Morton Thiokol Corp, 63-67, prog mgr, Huntsville Div, 67-76, prin proposal mgr, 76-89; RETIRED. *Res:* Mechanical and physical properties of elastomers; solid propellant rocket motors. *Mailing Add:* 5718 Criner Rd SE Huntsville AL 35802

MILLER, MICHAEL CHARLES, b Brooklyn, NY, Aug 9, 42; m 68. AQUATIC ECOLOGY, BIOLOGICAL LIMNOLOGY. *Educ:* Ind Univ, AB, 64, MS, 66; Mich State Univ, PhD(bot), 72. *Prof Exp:* Asst prof, 70-76, ASSOC PROF BIOL SCI, UNIV CINCINNATI, 76- *Concurrent Pos:* Consult & prin investr, NSF grant, Tundra Biome, Aquatic Prog, 71-74; prin investr, NSF grants, Res Arctic Tundra Environ, 75-77 & Arctic Lake Process Studies, 77-; consult, Dayton Power & Light Co, 71-78. *Mem:* Am Soc Limnol & Oceanog; Int Soc Theoret & Appl Limnol; Phycol Soc Am; Sigma Xi. *Res:* Phytoplankton ecology, especially algal-zooplankton interrelationships; large river and arctic limnology; thermal effects and oil pollution effects on phytoplankton. *Mailing Add:* Dept Biol Sci Univ Cincinnati 1408 Brodie A-2 Cincinnati OH 45221

MILLER, MICHAEL E, immunology, cell biology; deceased, see previous edition for last biography

MILLER, MILLAGE CLINTON, III, b Enid, Okla, Aug 28, 32; m 65. BIOSTATISTICS. *Educ:* Univ Okla, BS, 54, MA, 60, PhD(biostatist), 61. *Prof Exp:* NIH trainee, Med Ctr, Univ Okla, 59-61; grant, Okla State Univ, 61-62; assoc prof prev med, Univ Okla, 62-67; assoc prof biostatist, Tulane Univ, 67-69; PROF BIOMET & CHMN DEPT, MED UNIV SC, 69- *Concurrent Pos:* Consult, Fed & State Govts & various acad & res insts, 59- *Mem:* Biomet Soc; fel Am Statist Asn; Inst Math Statist; Asn Comput Mach; Am Pub Health Asn; fel AAAS. *Res:* Experimental design; multivariate analysis; medical application of statistics; biomedical applications of computers. *Mailing Add:* Dept Biomet Med Univ SC 171 Ashley Ave Charleston SC 29425

MILLER, MILTON H, b Indianapolis, Ind, Sept 1, 27; m; c 3. PSYCHIATRY. *Educ:* Ind Univ, BS, 46, MD, 50. *Prof Exp:* Intern, Indianapolis Gen Hosp, 51; resident, Menninger Sch Psychiat, 53; from instr to prof, Univ Wis, 55-72; prof psychiat & head dept, Univ BC, 72-78; PROF & CHMN DEPT PSYCHIAT, LOS ANGELES COUNTY HARBOR - UNIV CALIF MED CTR, LOS ANGELES, 78- *Concurrent Pos:* Examr, Royal Col Psychiat, Can; mem exec bd, World Fedn Ment Health; dep dir, Ment Health Serv, Coastal Region, Los Angeles County, 78- *Mem:* Fel Royal Col Psychiat; fel Am Psychiat Asn; Can Psychiat Asn. *Mailing Add:* Harbor Calif La Med Ctr 1000 W Carson St Torrance CA 90509

MILLER, MORTON W, b Neptune, NJ, Aug 4, 36; m 68. RADIOBIOLOGY, CYTOGENETICS. *Educ:* Drew Univ, BA, 58; Univ Chicago, MS, 60, PhD(bot), 62. *Prof Exp:* Res assoc radiobiol, Brookhaven Nat Lab, 63-65; second officer, Int Atomic Energy Agency, 65-67; mem staff, Univ Rochester, Dept Radiation Biol & Biophys, 67-75, asst dir atomic energy proj, 69-77, ASSOC PROF RADIATION BIOL & BIOPHYS, SCH MED & DENT, 75-, SR SCIENTIST, UNIV ROCHESTER, 75- *Concurrent Pos:* NATO fel, Oxford Univ, 62-63; planetary quarantine adv panel, Am Inst Biol Sci, 71-73; mem comt, Nat Acad Sci-Nat Res Coun, 75-77, ultrasound comt, Nat Coun Radiation Protection & Measurements, 80-, Bioeffects comt, Am Inst Ultrasound Med, 81-91; chief ed, Environ & Exp Bot, 77- *Mem:* AAAS; Radiation Res Soc; Am Inst Biol Sci; Environ Mutagen Soc; NY Acad Sci; Am Inst Ultrasound Med. *Res:* Effects/mechanisms of action of electric fields & ultrasound; bioelectromagnetics. *Mailing Add:* Dept of Biophys Univ Rochester Sch of Med & Dent Rochester NY 14642

MILLER, MURRAY HENRY, b Ont, Can, July 10, 31; m 54; c 3. SOIL FERTILITY. *Educ:* Ont Agr Col, BSA, 53; Purdue Univ, MS, 55, PhD(agr), 57. *Prof Exp:* From asst prof to assoc prof, 57-66, PROF SOIL SCI, ONT AGR COL, UNIV GUELPH, 66- *Concurrent Pos:* Head dept soil sci, Univ Guelph, 66-71. *Mem:* Fel Can Soc Soil Sci; Agr Inst Can; Can Soc Soil Sci; Int Soc Soil Sci; Soil Conserv Soc Am; Ont Inst Agrologists. *Res:* Soil fertility, especially chemistry of nutrient elements in soils and their absorption by plants; plant nutrients, environmental quality and land productivity; soil physical factors and crop yield. *Mailing Add:* Dept Land Resource Sci Ont Agr Col Univ Guelph Guelph ON N1G 2W1 Can

MILLER, MYRON, b Rochester, NY, Mar 31, 33; m 56; c 3. INTERNAL MEDICINE, ENDOCRINOLOGY. *Educ:* Univ Ill, Urbana, BS, 55; State Univ NY Upstate Med Ctr, MD, 59. *Prof Exp:* Intern & resident internal med, Univ Wis-Madison, 59-63; instr med, State Univ NY Upstate Med Ctr, 65-67, asst prof med & clin investr endocrinol, 67-71, assoc prof med, 71-75, prof med, 75-86; PROF, GERIAT & MED, MT SINAI MED CTR, 86- *Concurrent Pos:* NIH fel endocrinol, Univ Wis-Madison, 62-63; res assoc, Vet Admin Hosp, Syracuse, NY, 65-67, clin investr, 67-71, chief med serv, 72-84; chief, Sect Geriat Med, State Univ NY Upstate Med Ctr, 84-86; Hartford scholar geriat med, 84-85; vchmn, Dept Geriat, Mt Sinai Med Ctr, 86-, chief, Div Geriat Med, 89- *Mem:* Int Soc Neuroendocrinol; Endocrine Soc; Am Col Physicians; Am Fedn Clin Res; Am Physiol Soc; Geront Soc. *Res:* Hypothalamic-pituitary regulation with special interest in the regulation of posterior pituitary function and the role of altered posterior pituitary function in disease states and in aging and aging associated disorders; geriatrics. *Mailing Add:* Mt Sinai Med Ctr One Gustave L Levy Pl, Annenberg 10-14 Box 1070 New York NY 10029

MILLER, NANCY E, b Long Beach, NY, Aug 20, 47; Div. PSYCHOPATHOLOGY, GERIATRICS. *Educ:* NY Univ, BA, 69; Harvard Univ, MA, 70; Univ Chicago, PhD(adult develop & aging), 78. *Prof Exp:* Res assoc geriatric psychiat, Sch Med, Univ Chicago, 72-77; clin psychologist, Div Public Health, City Chicago, 72-77; INSTR GERIATRIC PSYCHIAT, SCH MED, GEORGETOWN UNIV, 78-; CLIN PSYCHOLOGIST, MOBILE MED CARE, 78- *Concurrent Pos:* Exec secy, Aging & Mental Health, Initial Review Group, 77-79, chief, Clin Res Prog Aging, Nat Inst Mental Health, Alcohol Drug Abuse & Mental Health Admin, 77-, rep aging, Int Prog Diagnosis, World Health Prog, 80-, res rep, White House Conf Aging, 81; mem, Task Force Nonenclature & Statist, Am Psychiat Asn, 78-85. *Honors & Awards:* Hon Award, Drug Abuse & Mental Health Admin. *Mem:* AAAS; Am Psychol Asn; Gerontol Soc Am; Int Neuropsychol Soc; Soc Neurosci; Am Psychoanalytic Asn. *Res:* Assessment, cause and treatment of psychopathology in middle and late life, with special emphasis on states of altered brain funciton; Alzheimers Disease, depression and the interaction of medical and psychiatric disease. *Mailing Add:* Ctr Studies Mental Health Aging NIMH 5600 Fishers Lane Rm 11-C-03 Rockville MD 20852

MILLER, NATHAN C, b Winnfield, La, Dec 13, 37; m 67; c 2. ORGANIC CHEMISTRY. *Educ:* Emory Univ, BA, 59; Fla State Univ, PhD(org chem), 64. *Prof Exp:* Res assoc, Univ Vt, 64-65; asst prof, Winthrop Col, 65-66; asst prof, 66-74, ASSOC PROF CHEM, UNIV SOUTH ALA, 74- *Mem:* Am Chem Soc; Sigma Xi. *Res:* Synthesis of small, bicyclic ring systems and study of decarboxylation of bridgehead B-keto acids; synthesis of alkaloids; stereoselective reactions; organometallic synthesis. *Mailing Add:* Dept Chem Univ SAla Mobile AL 36688

MILLER, NEAL ELGAR, b Milwaukee, Wis, Aug 3, 09; m 48; c 2. PSYCHOPHYSIOLOGY. *Educ:* Univ Wash, BS, 31; Stanford Univ, MA, 32; Yale Univ, PhD(psychol), 35. *Hon Degrees:* DSc, Univ Mich, 65, Univ Pa, 68, St Lawrence Univ, 73, Univ Uppsala, 77, La Salle Col, 79 & Rutgers Univ, 85. *Prof Exp:* Fel, Vienna Psychoanal Inst, Austria, 35-36; asst psychol, Inst Human Rels, Yale Univ, 36-41, res assoc, 41-42 & 46-50; officer in chg res, Psychol Res Unit 1, Army Air Corps, Nashville, Tenn, 42-44, dir, Psychol Res Proj, Hq, Flying Training Command, Randolph Field, Tex, 44-46; prof, Yale Univ, 50-52, James Rowland Angell prof, 52-66; prof & head lab, 66-80, EMER PROF PHYSIOL PSYCHOL & HEAD LAB, ROCKEFELLER UNIV, 80-; RES AFFIL, YALE UNIV, 85- *Concurrent Pos:* Mem fel comt, Found Fund Res Psychiat, 56-61; bd sci counrs, NIMH, 57-61; chmn div anthrop & psychol, Nat Res Coun, 58-60, chmn comt brain sci, 69-71; mem bd sci overseers, Jackson Mem Lab, 50-79, chmn, 62-76; Langfield lectr & Sigma Xi lectr, 68; bd sci counrs, Nat Inst Child Health & Human Develop, 69-72; mem clin prog proj res comt, NIMH, 74-78, mem, Bd Mental Health & Behavior Med, NIH, 80- *Honors & Awards:* Warren Medal, Soc Exp Psychol, 54; Cleveland Prize, AAAS, 57; Am Psychol Asn Award, 59; Nat Medal of Sci, 65; Gold Medal, Am Psychol Found, 75. *Mem:* Nat Acad Sci; Soc Neurosci (pres, 71-72); AAAS; Acad Behav Med Res (pres, 78-79); Am Psychol Asn (pres, 60-61 & 80-81); Am Acad Arts & Sci; Biofeedback Soc Am; Am Philos Soc. *Res:* Learning and behavior theory; conflict, fear and stress; mechanisms of psychosomatic effects; physiological and behavioral studies of motivation; electrical and chemical stimulation of the brain; instrumental learning of visceral responses; behavioral medicine; conflict, fear and stress. *Mailing Add:* PO Box 160 Guilford CT 06437

MILLER, NEIL AUSTIN, b Grand Rapids, Mich, Apr 9, 32; m 61; c 2. FOREST ECOLOGY. *Educ:* Mich State Univ, BSF, 58; Memphis State Univ, MS, 64; Southern Ill Univ, PhD(bot), 68; Oak Ridge Radiation Inst, grad, 71. *Prof Exp:* Teacher high sch & chmn, dept biol & physics, Grand Rapids, Mich, 59-62; teacher high sch, Memphis, Tenn, 62-64; instr bot & forestry, Western Ky Univ, 64-65; res asst bot, Southern Ill Univ, 65-66, fel, 66-68; prof, 68-84, DISTINGUISHED PROF BIOL, MEMPHIS STATE UNIV, 84- *Concurrent Pos:* Ill Acad Sci grant, 67-68; consult forestry. *Mem:* Soc Am Foresters; Am Soc Plant Physiol; Ecol Soc Am; Am Inst Biol Sci. *Res:* Wetland management; tree improvement; wildlife enhancement programs. *Mailing Add:* Dept Biol Memphis State Univ Memphis TN 38152

MILLER, NEIL RICHARD, b Wichita Falls, Tex, Nov 11, 45; m 73; c 1. NEURO-OPHTHALMOLOGY, ORBITAL SURGERY. *Educ:* Harvard Col, BA, 67; Johns Hopkins Med Inst, MD, 71. *Prof Exp:* Intern med, 71-72, residency ophthal, 72-75, from asst prof to assoc prof neuro-ophthal, 75-87, PROF OPHTHAL, JOHNS HOPKINS HOSP, 87-, FRANK B WALSH PROF NEURO-OPHTHAL, 87- *Concurrent Pos:* Prof neurol, Johns Hopkins Hosp, 87, prof neurosurg, 87. *Mem:* Am Acad Opht; Am Col Surgeons; Int Neuro-Ophthal Soc; Int Orbital Soc. *Res:* Rewriting the major textbook in neuro-ophthalmology: Walsh and Hoyt's Clinical Neuro-Ophthalmology. *Mailing Add:* Maumenee B-107 Wilmer Inst John Hopkins Hosp Baltimore MD 21205

MILLER, NICHOLAS CARL, b Mason City, Iowa, Aug 9, 42; m 87; c 1. CHEMICAL ENGINEERING, BIOMEDICAL ENGINEERING. *Educ:* Iowa State Univ, BS, 64, MS, 67, PhD(chem eng), 72. *Prof Exp:* Propellant develop engr, Olin Chem Group, Olin Corp, 64-65; res physiologist, Med Sch, Univ Calif, Davis, 72-73; chem engr, Cent Res Lab, 73-81, SUPVR, PHARMACEUT RES & DEVELOP, RIKER LABS, 3M CO, 81- *Mem:* Am Inst Chem Engrs; Am Chem Soc; Fine Particle Soc; Am Asn Pharmaceut Scientists. *Res:* Fine particle size distribution and characterization methods development; generation and processing of fine particulates for medical applications; evaluation of dispersions; pharmaceutical aerosol development and manufacturing. *Mailing Add:* 12746 Ethan Ave N White Bear MN 55110

MILLER, NORMAN E, b Tinley Park, Ill, Aug 14, 31; m 52; c 7. INORGANIC CHEMISTRY. *Educ:* Northern Ill State Teachers Col, BS, 53; Univ Nebr, MS, 55, PhD, 58. *Prof Exp:* Chemist, Cent Res Dept, E I du Pont de Nemours & Co, 58-63; assoc prof, 63-66, PROF CHEM, UNIV S DAK, 66- *Concurrent Pos:* Assoc Western Univs, Los Alamos Sci Lab, 72. *Mem:* Am Chem Soc; Sigma Xi. *Res:* Lewis acid-base phenomena; inorganic synthesis; boron hydride synthesis and reactivity studies; chemistry. *Mailing Add:* Dept of Chem Univ of SDak Vermillion SD 57069

MILLER, NORMAN GUSTAV, b Thermopolis, Wyo, Mar 20, 25; m 54; c 2. MICROBIOLOGY. *Educ:* Western Reserve Univ, BS, 48; Wash State Univ, MS, 50, PhD(bact), 53; Am Bd Microbiol, Dipl. *Prof Exp:* From instr to assoc prof, 55-67, PROF MICROBIOL, COL MED, UNIV NEBR, OMAHA, 67- *Concurrent Pos:* Secy, West-Northcent Interprof Seminar Dis Common to Animals & Man. *Mem:* AAAS; Am Soc Microbiol; Wildlife Dis Asn; Med Mycol Soc Am; Int Soc Human & Animal Mycol; Sigma Xi. *Res:* Infectious diseases of animals transmissible to man; medical mycology. *Mailing Add:* 12323 Izard St Omaha NE 68154

MILLER, NORTON GEORGE, b Buffalo, NY, Feb 4, 42; m 64; c 1. BOTANY. *Educ:* State Univ NY Buffalo, BA, 63; Mich State Univ, PhD(bot), 69. *Prof Exp:* Asst cur, Arnold Arboretum, Harvard Univ, 69-70; vis asst prof bot, Univ NC, Chapel Hill, 70-71, asst prof, 71-74; assoc prof biol, Harvard Univ, 75-80, assoc cur, Arnold Arboretum & Gray Herbarium, 75-80, botanist, 80-81, sr res botanist, 81-82; CHIEF SCIENTIST, NY STATE BIOL SURVEY, 82- *Concurrent Pos:* Fel, Nat Defense Educ Act, 63-66; fel, Mich State Grad Coun, 66-67; mem fac, Univ Mich Biol Sta, 83-84 & mem grad fac, Univ Maine, 84-89; mem Syst Biol Panel, NSF, 83-86; counr, Int Asn Bryologists, 87- *Mem:* AAAS; Am Bryol & Lichenol Soc (pres, 83-87); Bot Soc Am; Am Quaternary Asn; Am Soc Plant Taxon (secy, 89-); Int Asn Bryologists. *Res:* Plant systematics and floristics, especially of bryophytes and seed plants; quaternary paleoecology and palynology; pollen and plant macrofossil analysis; tertiary and quaternary history of the bryophyta. *Mailing Add:* Biol Surv NY State Mus & Sci Serv Albany NY 11230

MILLER, ORLANDO JACK, b Oklahoma City, Okla, May 11, 27; m 54; c 3. HUMAN GENETICS, MOLECULAR CYTOGENETICS. *Educ:* Yale Univ, BS, 46, MD, 50. *Prof Exp:* Intern, St Anthony Hosp, Oklahoma City, Okla, 50-51; asst resident obstet & gynec, Grace-New Haven Community Hosp, 54-57, resident & instr, 57-58; res asst human genetics, Univ Col, Univ London, 58-60; from instr to assoc prof obstet & gynec, Col Physicians & Surgeons, Columbia Univ, 60-70, prof human genetics & develop, obstet & gynec, 70-85; dir ctr molecular biol, 87-90, PROF & CHMN MOLECULAR BIOL & GENETICS & PROF OBSTET & GYNEC, SCH MED, WAYNE STATE UNIV, 85- *Concurrent Pos:* Nat Res Coun fel anat, Yale Univ, 53-54; Pop Coun fel human genetics, Galton Lab, Univ Col, Univ London, 58-60; Josiah Macy, Jr fel obstet & gynec, Columbia Univ, 60-61; NSF sr fel, Oxford Univ, 68-69; career scientist, Health Res Coun, New York, 61-71; basic res comt, March of Dimes Birth Defects Found, 69-; ed, Cytogenetics, 71-72, assoc ed, Cytogenetics & Cell Genetics, 72- & Birth Defects Compendium, 73; consult, J Med Primatol, 78-; mem bd dirs, Am Bd Med Genetics, 82-85, vpres, 83, pres, 84 & 85; vis scientist, Univ Edinburgh, 83-84; adv bd, Human Genetics, 78-; cell & develop biol adv comt, Am Chem Soc, 87-90; ed comm, Genomics, 87-; assoc ed, J Exp Zool, 89-; bd dir, Am Soc Human Genetics, 87-90. *Mem:* Fel AAAS; Am Soc Human Genetics; Am Soc Cell Biol; Genetics Soc Am. *Res:* Cytogenetics; molecular genetics. *Mailing Add:* Dept Molecular Biol & Genetics Wayne State Univ 3216 Scott Hall Detroit MI 48201

MILLER, ORSON K, JR, b Cambridge, Mass, Dec 19, 30; m 54; c 3. MYCOLOGY. *Educ:* Univ Mass, BS, 52; Univ Mich, MF, 57, PhD(bot), 63. *Prof Exp:* Res forester, Northeastern Forest Exp Sta, USDA, 56-57; asst bot, Univ Mich, 58-59, fel, 60-61; plant pathologist, Intermt Forest & Range Exp Sta, USDA, 61-65 & Forest Dis Lab, 65-70; assoc prof, 70-73, PROF BOT, VA POLYTECH INST & STATE UNIV, 73-, CUR FUNGI, 74- *Mem:* AAAS; Mycol Soc Am; Sigma Xi. *Res:* Taxonomy, ecology, genetics and mycorrhizae of Homobasidiomycetes; arctic and northern fungi. *Mailing Add:* Dept Biol Va Polytech Inst & State Univ Blacksburg VA 24060

MILLER, OSCAR LEE, JR, b Gastonia, NC, Apr 12, 25; m 48; c 2. CELL BIOLOGY. *Educ:* NC State Univ, BS, 48, MS, 50; Univ Minn, Minneapolis, PhD, 60. *Prof Exp:* Nat Inst Cancer fel, 60-61; res assoc, Biol Div, Oak Ridge Nat Lab, 61-63, res staff mem, 63-73; prof biol & chmn dept, 73-79, LEWIS & CLARK PROF BIOL, UNIV VA, 78- *Concurrent Pos:* Prof, Univ Tenn-Oak Ridge Grad Sch Biomed Sci, 67-73; mem, Panel Cell Biol, NSF, 79 & Genetics Study Sect, NIH, 80-84; Alexander von Humboldt sr scientist fel, 80; vis prof, Div Biol, Calif Inst Technol & Max Planck Inst Cell Biol, WGer, 80, Univ Calif, Irvine, 84; sr Fulbright scholar, Div Molecular Biol, Commonwealth Sci & Indust Res Orgn, 86; lectr, numerous US & foreign univs. *Mem:* Nat Acad Sci; Am Soc Cell Biol; fel AAAS; Soc Develop Biol; Sigma Xi; Fedn Am Scientists. *Res:* Correlation of fine structure and genetic activity in chromosomes of prokaryotic and eukaryotic cells; identification of specific genes in action. *Mailing Add:* Dept Biol Univ Va Charlottesville VA 22901

MILLER, OWEN WINSTON, b St Louis, Mo, Feb 17, 22; m 50; c 3. INDUSTRIAL ENGINEERING. *Educ:* Washington Univ, St Louis, BS, 50, MS, 58, ScD(indust eng), 66. *Prof Exp:* Asst chief indust eng, Am Steel Foundries, Ill, 50-54; instr, Washington Univ, St Louis, 54-58, asst prof, 59-62, lectr, Spec Bell Tel Prog, 62-63; from asst prof to assoc prof, 64-78, PROF INDUST ENG & DIR, BUS & INDUST PRODUCTIVITY CTR, UNIV MO-COLUMBIA, 78-, DIR, INDUST ENG GRAD PROG, KANSAS CITY, 77-, ASST DIR CREDIT PROGS, 85- *Concurrent Pos:* Consult indust, 55-61, indust & fed govt, 63-; consult, Ellis Fischel State Cancer Hosp, Columbia, Mo, 66-67; co-prin investr, automated patient hist acquisition syst proj, Mo Regional Med Prof, 67-70, consult, stroke intensive care proj, 68-71, co-investr, advan technol proj, 70-71, co-dir, automated physicians' asst proj, 71-72; co-prin investr water pollution models, Mo Water Resources Res Ctr, US Dept Interior, 70-71; mem bd dirs, Community Rehab Ctr, Columbia, 79- *Mem:* Am Inst Indust Engrs; Nat Soc Prof Engrs; fel Am Soc Qual Control; Sigma Xi. *Res:* Quality control and reliability fields; economics of industrial sampling; industrial engineering techniques applied to hospitals and medical activities; industrial engineering analysis methods in surgical operations; measuring direct cost of government regulations on small business; rural and small business productivity improvement; add-on automation; productivity and quality enhancement for Missouri industries. *Mailing Add:* Dept Indust Eng Univ Mo Rm 116 Elec Eng Columbia MO 65211

MILLER, PARK HAYS, JR, b Philadelphia, Pa, Jan 22, 16; m 76; c 3. PHYSICS. *Educ:* Haverford Col, BS, 36; Calif Inst Technol, PhD(physics), 40. *Prof Exp:* Asst physics, Calif Inst Technol, 36-39; from instr to prof, Univ Pa, 39-56; chmn exp physics dept, Gen Atomic Div, Gen Dynamics Corp, 56-62, asst dir lab, 60-69; prof physics & chmn dept, US Int Univ, Calif Western Campus, 69-74; sr tech adv, Gen Atomic Co, 74-81; PRES, EXTRAGALACTIC ENTERPRISES INC, 81- *Concurrent Pos:* Chmn dept physics, Univ Pa, 45-46; consult, US Naval Ord Lab, 51-65, US Dept Defense, 52-55 & 60-62, NASA, 58-60 & Mat Adv Bd, Nat Acad Sci, 56 & 60-61; actg ed, Rev Sci Instruments, Am Inst Physics, 53-55. *Mem:* Fel Am Phys Soc; Soc Explor Geophys; Am Geophys Union; Am Nuclear Soc; Am Asn Physics Teachers. *Res:* Electrical properties of solids; semiconductor devices; radiation effects; x-ray diffraction; optical instruments; energy conversion; design of fusion power reactors. *Mailing Add:* 302 Sea Lane La Jolla CA 92037

MILLER, PAUL, b Philadelphia, Pa, Jan 15, 22. SOLID STATE PHYSICS. *Educ:* George Washington Univ, BS, 43; Univ Pa, MS, 49, PhD(physics), 55. *Prof Exp:* MEM TECH STAFF, BELL LABS, INC, 54- *Mem:* Am Phys Soc. *Res:* Surface and transistor physics; microelectronics; semiconductors; integrated-circuit reliability physics. *Mailing Add:* 221 S Saint George,Apt 3B Allentown PA 18104

MILLER, PAUL DEAN, b Cedar Falls, Iowa, Apr 4, 41; m 65; c 2. ANIMAL BREEDING. *Educ:* Iowa State Univ, BS, 63; Cornell Univ, MS, 66, PhD(animal genetics), 68. *Prof Exp:* Asst prof animal breeding, Cornell Univ, 67-71; DIR BREEDING PROGS, AM BREEDERS SERV, 71- *Concurrent Pos:* Dir, Am Asn Animal Breeders; adj prof, Univ Wis-Madison, 76- *Mem:* Biomet Soc; Am Soc Animal Sci; Am Dairy Sci Asn; Am Asn Animal Breeders. *Res:* Population genetics; statistical estimation and linear models; computer simulation. *Mailing Add:* Am Breeders Serv De Forest WI 53532

MILLER, PAUL GEORGE, b Milwaukee, Wis, Dec, 5, 41. LABORATORY ANIMALS, AVIAN MEDICINE. *Educ:* Marquette Univ, BS, 63; Case Inst Technol, MS, 65, PhD(math), 68; Okla State Univ, vet med, 88- *Prof Exp:* Asst prof math, Okla State Univ, 68-73 & Univ Wis-Whitewater, 76-80; tech rep, Burroughs Corp, 73-76; animal technician, Okla City Zoo, 80-81; assoc prof computer sci, Cent State Univ, Edmond, Okla, 81-88. *Concurrent Pos:* Consult, Pretrus Foods, 74-88; adj herpetologist, Okla State Zoo, 84-88. *Mem:* Am Vet Med Asn; Math Asn Am. *Res:* Veterinary medicine; laboratory animal medicine; avian medicine, birds, poultry. *Mailing Add:* PO Box 183 Stillwater OK 74076-0183

MILLER, PAUL LEROY, JR, b Guthrie, Okla, June 27, 34; m 55; c 3. MECHANICAL ENGINEERING. *Educ:* Kans State Univ, BS, 57, MS, 61; Okla State Univ, PhD(mech eng), 66. *Prof Exp:* From instr to assoc prof, 57-72, head dept, 75-88, PROF MECH ENG, KANS STATE UNIV, 72- *Concurrent Pos:* Consult, Radio Corp Am & Whirlpool Corp, 61-64, Wright-Patterson AFB, 67-69, Air Diffusion Coun, Chicago, 69-, Gen Serv Admin, Washington, DC, 77-84 & US Dept Energy, Washington, DC, 79-81. *Mem:* Am Soc Eng Educ; Am Soc Mech Engrs; fel Am Soc Heat, Refrig & Air Conditioning Engrs. *Res:* Heat transfer; fluid flow; instrumentation; air distribution. *Mailing Add:* Dept of Mech Eng Durland Hall Kans State Univ Manhattan KS 66506

MILLER, PAUL SCOTT, b Brooklyn, NY, Oct 12, 43; m 71. BIO-ORGANIC CHEMISTRY. *Educ:* State Univ NY Buffalo, BA, 65; Northwestern Univ, Ill, PhD(chem), 69. *Prof Exp:* Am Cancer Soc fel nucleic acid chem, 69-73, asst prof biochem & biophys sci, 73-80, assoc prof, 80-88, PROF BIOCHEM, JOHNS HOPKINS UNIV, 88- *Mem:* AAAS; Am Chem Soc. *Res:* Chemical and enzymatic synthesis of nucleic acids and nucleic acid derivatives; interaction of nucleic acids with proteins and nucleic acids; chemical modification of nucleic acids. *Mailing Add:* Dept Biochem Johns Hopkins Univ Baltimore MD 21205

MILLER, PAUL THOMAS, b Atlanta, Ga, May 12, 44; m 77. INORGANIC CHEMISTRY, EDUCATIONAL ADMINISTRATION. *Educ:* Birmingham-Southern Col, BS, 66; Vanderbilt Univ, MA & PhD(chem), 71. *Prof Exp:* Instr, 71-72, asst prof, 72-75, assoc prof chem & phys sci, 75-81, PROF CHEM & CHMN, DIV MATH & SCI, VOL STATE COMMUNITY COL, 81-, COORDR CHMN, 80- *Concurrent Pos:* NSF res grant, 66; res grant, Tulane Univ Scholars & Fels Prog, 66; grad teaching fel, Vanderbilt Univ, 66-69 & 70-71. *Mem:* AAAS; Am Chem Soc; Sigma Xi. *Res:* Structure and properties of coordination compounds; chemical education in the two year college; analytical methods for environmental pollution studies; water pollution by heavy metals. *Mailing Add:* 4001 Sneed Rd Nashville TN 37215

MILLER, PAUL WILLIAM, b Mt Vernon, Ind, May 2, 01; m 28. PLANT PATHOLOGY. *Educ:* Univ Ky, BS, 23, MS, 24; Univ Wis, PhD(plant path), 29. *Prof Exp:* Instr plant path, Univ Wis, 27-29; agent, 29-30, assoc plant pathologist, 30-44, plant pathologist, 44-68, emer res plant pathologist, USDA, Ore State Univ, 68-85; RETIRED. *Res:* Fire blight disease of apples; prune russet; vegetable seed diseases; walnut blight; filbert blight; strawberry root rot; strawberry virus diseases; control of rose mildew and rose rust. *Mailing Add:* 703 NW 30th St Corvallis OR 97330

MILLER, PAULINE MONZ, b Harrisburg, Pa, Apr 2, 31; div. BOTANY, INFORMATION SCIENCE. *Educ:* Pa State Univ, BS, 52; Univ Pa, PhD(bot), 56; Syracuse Univ, MSLS, 76. *Prof Exp:* Instr bot, Wheaton Col, 56-57; instr, Conn Col, 57-59, res assoc, 59-61; res assoc, Yale Univ, 61-62; res assoc, Syracuse Univ, 62-71, ADJ PROF BOT, UNIV COL, SYRACUSE UNIV, 71-, HEAD SCI & TECHNOL LIBR, 78- *Mem:* Am Libr Asn. *Res:* Light effects on plant development; organization and dissemination of science information; economic botany. *Mailing Add:* 3178 Pompey Hollow Rd Cazenovia NY 13035-9507

MILLER, PERCY HUGH, b Yazoo Co, Miss, June 18, 22; m; c 4. AERODYNAMICS. *Educ:* Miss State Univ, BS, 50; Univ Colo, MS, 51; Univ Tex, PhD(aerospace eng), 62. *Prof Exp:* Instr eng, Univ Colo, 50-51; design specialist, Gen Dynamics/Pomona, 51-58; asst prof aerospace eng, Univ Tex, 58-65; assoc prof mech eng, Univ Mo, 65-67; prof mech eng, 67-87, EMER PROF, LA STATE UNIV, BATON ROUGE, 87- *Concurrent Pos:* Consult, Tracor, Inc, 60- & Monsanto Co, 62-; sr assoc, Hanneman Assocs, Inc, 61-; consult, prod liability, safety & design. *Mem:* Assoc fel Am Inst Aeronaut & Astronaut; Am Soc Eng Educ; Am Soc Mech Engrs; Soc Automotive Engrs. *Res:* Aerodynamics and dynamics of aerospace vehicles including analysis, synthesis and preliminary design; safety engineering and design. *Mailing Add:* Consult Engr 18221 Swamp Rd Prairieville LA 70769

MILLER, PHILIP ARTHUR, b Hastings, Nebr, Feb 1, 23; m 45; c 2. PLANT BREEDING. *Educ:* Univ Nebr, BSc, 43, MSc, 47; Iowa State Col, PhD(plant breeding), 50. *Prof Exp:* Res assoc agron, Iowa State Col, 49-50, asst prof, 50-52; from assoc prof to prof agron, NC State Univ, 59-77; MEM STAFF, BELTSVILLE AGR RES CTR-WEST, SCI & EDUC ADMIN, NPS, USDA, 77- *Concurrent Pos:* Dir, NC State Univ Agr Mission, Peru, 59-61. *Mem:* Fel Am Soc Agron; Sigma Xi. *Res:* Genetics; corn; cotton; soybeans; quantitative inheritance. *Mailing Add:* Beltsville Agr Res Ctr-West NPS Agri Res Serv USDA Beltsville MD 20705

MILLER, PHILIP DIXON, b Albuquerque, NMex, June 7, 32; m 64; c 4. ACCELERATOR BASED ATOMIC PHYSICS. *Educ:* Calif Inst Technol, BS, 54; Rice Inst, MA, 56, PhD(physics), 58. *Prof Exp:* physicist, Oak Ridge Nat Lab, 58-88, dir, Vand De Graff Lab, 74-88; RETIRED. *Mem:* Fel Am Phys Soc. *Res:* Atomic collisions physics; low energy physics using Van de Graaff accelerator, including charged particle scattering and reactions and neutron cross sections; measurement of electric and magnetic dipole moments of neutron; properties of the neutron. *Mailing Add:* 12835 Lovelace Rd Knoxville TN 37932

MILLER, PHILIP JOSEPH, b Camden, SC, Mar 20, 41; m 62; c 3. PHYSICAL CHEMISTRY. *Educ:* Johns Hopkins Univ, ScB, 67; Univ Md, PhD(phys chem), 71. *Prof Exp:* Res assoc phys chem, Univ Md, 71-72; instr & res assoc, Brown Univ, 72-74; res scientist, Block Eng, Inc, 74-76; ASST PROF PHYS & ANALYTICAL CHEM, UNIV DETROIT, 76- *Mem:* Am Chem Soc; Am Phys Soc; Soc Appl Spectros; Sigma Xi. *Res:* Molecular spectroscopy; hydrogen bonding in solids; structure and dynamics of disordered materials; laser applications in chemistry. *Mailing Add:* Dept Chem Univ Detroit Detroit MI 48221

MILLER, RALPH ENGLISH, b Hanover, NH, Sept 23, 33; m 62; c 3. NEUROSCIENCE. *Educ:* Dartmouth Col, AB, 58; Harvard Univ, MD, 61, MS, 66, DSc, 70. *Prof Exp:* Intern, Mary Hitchcock Mem Hosp, Hanover, NH, 61-62; NIMH fel, Walter Reed Army Inst Res, 62-64; NASA res & spec fel, Dept Physiol, Harvard Sch Pub Health, 65-69; spec fel physiol, Stanford Univ, 69-70; asst prof, 70-75, assoc prof pharmacol, 75-81, resident med, 81-84, FEL ENDOCRINOL, UNIV KY, 84- *Concurrent Pos:* Chief physiol sect, Dept Neuroendocrinol, Walter Reed Army Inst Res, 64-65. *Mem:* Am Diabetes Asn; Am Physiol Soc; Soc Neurosci; Endocrine Soc. *Res:* Neuro-endocrinol; autonomic neuroendocrinol; role of autonomic nervous system in regulation of hormone secretion from the pancreas and kidney. *Mailing Add:* Dept Med Div Endocrinol Univ Ky Med Sch Rose St Lexington KY 40503

MILLER, RALPH LEROY, b Fountain Hill, Pa, Jan 28, 09; m 39; c 2. ECONOMIC GEOLOGY, STRATIGRAPHY. *Educ:* Haverford Col, BS, 29; Columbia Univ, PhD(geol), 37. *Prof Exp:* Lab asst geol, Columbia Univ, 32-34, lectr, 34-37, instr, 37-46; from assoc geologist to sr geologist, 42-49, US Geol Surv, chief, Navy Oil Unit, 48-51, Fuels Br, 51-57, staff geologist, 58, tech adv, Foreign Geol Br, Afghanistan, 58, Mex, 59-61, Colombia, 62, Cent Am, 63-70, SKorea, 79, Mex 80, Costa Rica, 81 & 84, sr res geologist, 70-79, ANNUITANT, OFF INT GEOL, 80- *Concurrent Pos:* Vis prof, Ohio State Univ, 80. *Mem:* Fel Geol Soc Am; hon mem Am Asn Petrol Geol; Am Inst Mining, Metall & Petrol Engrs. *Res:* Areal geology and stratigraphy of southeast Utah; areal geology, stratigraphy and structure of Appalachian Mountains from southern New York to Tennessee; manganese deposits of Appalachians; oil geology of southern Appalachians; geology and oil resources of the Arctic slope of Alaska; geology of Central America; energy resources, Latin America; geology of northwestern New Mexico. *Mailing Add:* 4932 Sentinel Dr Apt 306 Bethesda MD 20816

MILLER, RAYMOND EARL, b Corning, NY, Jan 26, 56; m. HAZARDOUS WASTE MANAGEMENT, ENVIRONMENTAL COMPLIANCE. *Educ:* Mansfield State Col, BS, 77; WVa Univ, MS, 81. *Prof Exp:* Teaching asst gen & physical chem, dept chem, WVa Univ, 77-78, res asst, phys chem, 78-79 & res fel, 79-81; res chemist phys chem, 81-88, HAZARDOUS WASTE MGR, US ARMY, ABERDEEN PROVING GROUND, MD, 88- *Concurrent Pos:* Supvr of 90 day hazardous waste storage site & 150 satellite storage sites. *Mem:* Am Chem Soc; Sigma Xi. *Res:* Hazardous/RCRA waste management; environmental compliance; Environmental Protection Agency, Dept of Transportation and Occuptational Safety and Health Administration regulations; environmental law; waste disposal technology. *Mailing Add:* 4421 Furley ave Baltimore MD 21206

MILLER, RAYMOND EDWARD, b Bay City, Mich, Oct 9, 28; m; c 4. COMMUNICATION PROTOCOLS, PARALLEL COMPUTATION. *Educ:* Univ Wis-Madison, BS, 50; Univ Ill, Urbana, BS, 54, MS, 55, PhD(elec eng), 57. *Prof Exp:* Design engr, IBM, Endicott & Poughkeepsie, NY, 50-51; Lt, Air Res & Develop Command, Wright-Patterson AFB, Ohio, 51-53; from res asst to res assoc, Digital Comp Lab, Univ Ill, Urbana, 53-57; mem res staff math sci, IBM, Thomas J Watson Res Ctr, 57-81, mgr, Comp Struct Group, 68-71 & Theory Comp Group, 72-79, asst dir, Math Sci Group, 79-80; actg assoc vpres info technol, Ga Inst Technol, 81, dir, Sch Info & Computer Sci, 80-87, prof, 80-89; DIR, CESDIS, NASA/GODDARD SPACE FLIGHT CTR, 88-; PROF, DEPT COMPUTER SCI, UNIV MD, CP, 89-; EMER PROF, SCH INFO & COMPUTER SCI, GA INST TECHNOL, 89- *Concurrent Pos:* Vis asst prof, Digital Computer Lab, Univ Ill, 60-61; vis sr res fel, Elec Eng Dept, Calif Inst Technol, 62-63; lectr, Dept Elec Eng, Univ Conn, Stamford, 65-69; adj prof, Dept Elec Eng & Computer Sci, NY Univ, Bronx, 69-73, Polytech Inst NY, 73-80; vis lectr, Computer Sci Dept, Yale Univ, 73 & 76; coun mem-at-large, Asn Computer Mach, 76-82; mem bd dirs, Comput Res Asn, 83-91; vpres, Comput Sci Accreditation Bd, 84-85, pres 85-87; vis prof, Computer Sci Dept, Univ Tex, Austin, 87, Univ Md, College Park, 88-89; mem bd gov, Inst Elec & Electronics Engrs Computer Soc, 88-91, vpres educ activ, 91. *Honors & Awards:* Mackay lectr, Univ Calif, Berkeley, 69-70; Outstanding Contrib Award, Inst Elec & Electronics Engrs Computer Soc, 85. *Mem:* Fel Inst Elec & Electronics Engrs; fel AAAS; Asn Comput Mach; Inst Elec & Electronics Engrs Computer Soc; Comput Res Asn. *Res:* Formal techniques for specifying, analyzing and designing communication protocols; parallel computation, theory of computation and computer organization. *Mailing Add:* Dept Comput Sci A V Williams Bldg Univ Md College Park MD 20742

MILLER, RAYMOND EDWIN, b Cincinnati, Ohio, July 6, 37; m 61; c 6. PHYSICS. *Educ:* Xavier Univ, Ohio, BS, 59; Johns Hopkins Univ, PhD(physics), 65. *Prof Exp:* Instr physics, Johns Hopkins Univ, 62-65, NASA fel, 65-66; from asst prof to assoc prof, 66-71, PROF PHYSICS & CHMN DEPT, XAVIER UNIV, OHIO, 71- *Concurrent Pos:* NASA grant, Johns Hopkins Univ, 68-71. *Mem:* AAAS; Am Asn Physics Teachers; Optical Soc Am. *Res:* Atmospheric physics; atomic and molecular physics; biophysics. *Mailing Add:* Dept Physics Xavier Univ Cincinnati OH 45207

MILLER, RAYMOND JARVIS, b Claresholm, Alta, Mar 19, 34; m 56; c 3. SOIL CHEMISTRY, PHYSICAL CHEMISTRY. *Educ:* Univ Alta, BS, 57; Washington State Univ, MS, 60; Purdue Univ, PhD(soil chem), 62. *Prof Exp:* From asst prof to assoc prof soil & phys chem, NC State Univ, 62-65; from assoc prof to prof, Univ Ill, Urbana, 65-73; assoc dean, Col Agr & dir, Idaho Agr Exp, Univ Idaho, 73-80, dean, Col Agr, 80-86; vpres agr & dean agr life & sci, 86-90, VCHANCELLOR AGR & NATURAL RESOURCES, UNIV MD SYST, COLLEGE PARK, 90- *Concurrent Pos:* Asst dir, Agr Exp Sta, Univ Ill, Urbana, 69, assoc dir, 70, coordr, Col Agr Coun Environ Qual, 70; chmn exp sta sect, Nat Asn State Univ & Land-Grant Col, 76-77, chmn bd, div agr; mem, US-USSR Joint Comn Coop Agr, 77-79 & 85-86; pres, Idaho Res Found, 81-85. *Mem:* Fel Soil Sci Soc Am; Clay Minerals Soc; Am Soc Plant Physiol; Fel Am Soc Agron; AAAS. *Res:* Clay-water interactions; structure of water in porous media and the effects of water structure on biological activity; membrane transport; agricultural research. *Mailing Add:* Agr & Natural Resources Univ Md Syst 1114 Symons Hall College Park MD 20742

MILLER, RAYMOND MICHAEL, b Chicago, Ill, July 16, 45. MICROBIAL ECOLOGY. *Educ:* Colo State Univ, BS, 69; Ill State Univ, MS, 71, PhD(mycol), 75. *Prof Exp:* Fel microbial ecol, Land Reclamation Prog, 75-76, asst biologist, 76-79, SCIENTIST SOIL ECOLOGIST, ENVIRON RES DIV, ARGONNE NAT LAB, 80- *Concurrent Pos:* Lectr, Comt Evolutionary Biol, Univ Chicago, 86-; pub responsibility rep, Mycological Soc Am & Am Inst Biol Sci, 87- *Mem:* AAAS; Am Soc Microbiol; Mycol Soc Am; Soil Sci Soc Am; Sigma Xi; Bot Soc Am. *Res:* Reestablishment of below-ground ecosystems in relation to disturbance; ecology and physiology of mycorrhiza; aggregate formation in soils. *Mailing Add:* Environ Res Div Argonne Nat Lab Argonne IL 60439-4843

MILLER, RAYMOND SUMNER, analytical chemistry, for more information see previous edition

MILLER, RAYMOND WOODRUFF, b St David, Ariz, Jan 13, 28; m 51; c 5. SOIL FERTILITY, ENVIRONMENTAL SCIENCE. *Educ:* Univ Ariz, BS, 52, MS, 53; Wash State Univ, PhD(agron, soil chem), 56. *Prof Exp:* From asst prof to assoc prof, 56-69, PROF SOILS, UTAH STATE UNIV, 69- *Concurrent Pos:* Consult, Centro Interamericano de Desarollo Integral de Aguas y Tierras, Venezuela, 69-71, USAID, Gambia, Africa, 77, Honduras, 78, Bolivia, 79 & Guinea-Bissau, Africa, 82; Fulbright scholar, India, 84-85,

Israel, 85; consult, FAO in Pakistan, 87-88. *Honors & Awards:* Fulbright lectr, India, 84-85. *Mem:* Am Soc Agron; Soil & Water Conserv Soc Am; Soil Sci Soc Am; Int Soil Sci Soc. *Res:* Soil mineralogy; soil genesis; solid waste management; soil fertility. *Mailing Add:* Dept Plants Soils & Biometeorol Utah State Univ Logan UT 84322-4820

MILLER, REID C, US citizen. CHEMICAL ENGINEERING. *Educ:* Univ Tulsa, BS, 62; Univ Calif, Berkeley, MS, 64, PhD(chem eng), 68. *Prof Exp:* From asst prof to prof chem eng & chem, Univ Wyo, 68-79, assoc dean eng, 79-; AT COL ENG, WASH STATE UNIV. *Concurrent Pos:* NSF, AGA & Exxon res grants, Univ Wyo, 69- *Mem:* Am Inst Chem Engrs; Am Chem Soc; Am Soc Eng Educ. *Res:* Equilibrium and nonequilibrium properties of mixtures of dense fluids; interpretation of the properties of dense fluid mixtures based on the intermolecular potential between unlike species and the resulting fluid structure. *Mailing Add:* 210 SW Skyline Dr Pullman WA 99163-2939

MILLER, RENE H(ARCOURT), b Tenafly, NJ, May 19, 16; c 2. AEROSPACE ENGINEERING. *Educ:* Univ Cambridge, BA, 37, MA, 54. *Prof Exp:* Aeronaut engr, G L Martin Co, Md, 37-39; chief aeronaut & develop, McDonnell Aircraft Corp, Mo, 39-44; assoc prof aeronaut eng, 44-57, prof flight vehicle eng, 57-61, head, dept aeronaut & astronaut, 68-78, slater prof flight transp, 61-86, EMER PROF, MASS INST TECHNOL, 86- *Concurrent Pos:* Vpres eng, Kaman Aircraft Corp, Conn, 52-54; consult, Vertol Div, Boeing Co; mem, US Air Force Sci Adv Bd, 59-70; mem, Comn Aircraft Aerodynamics, NASA, 60-70; mem, tech adv bd, Fed Aviation Agency, 64-66; mem, Army Sci Adv Panel, 66-70; chmn, aviation sci adv group; mem, Aircraft Panel, President's Sci Adv Comn. *Honors & Awards:* Sylvanus Albert Reed Award, Inst Aeronaut & Astronaut, 69; Klemin Award, Helicopter Soc, Nikolsky Lectr. *Mem:* Nat Acad Eng; hon fel Helicopter Soc; hon fel Inst Aeronaut & Astronaut; fel Royal Aeronaut Soc; Nat Acad Air & Space. *Res:* Helicopter and airplane design; jet propulsion; vertical take-off and landing aircraft; space systems engineering. *Mailing Add:* San Jose The Lidden Mass Inst Technol Penzance Cornwall TR18 4PN England

MILLER, RICHARD ALBERT, mathematics; deceased, see previous edition for last biography

MILLER, RICHARD ALLEN, b Johnson City, NY, May 15, 44; m 66. SYSTEMS ENGINEERING. *Educ:* Union Col, BS, 66; Mass Inst Technol, MS, 67; Case Western Reserve Univ, PhD(systs eng), 71. *Prof Exp:* Asst prof, 71-76, assoc prof, 76-80, PROF INDUST, SYSTS ENG, OHIO STATE UNIV, 80-, DIR, MFG SYSTS ENG PROG, 90- *Mem:* Inst Elec & Electronic Engrs; Human Factors Soc; Asn Computing Machinery; Am Asn Artificial Intel; Soc Mfg Engrs. *Res:* Systems theory; systems integration; human-machine systems modeling and design; computer applications in manufacturing. *Mailing Add:* Dept Indust & Systs Eng 1971 Neil Ave Columbus OH 43210

MILLER, RICHARD AVERY, b Erie, Pa, June 23, 11. ENDOCRINOLOGY. *Educ:* Univ Pittsburgh, BS, 32; Univ Iowa, MS, 34, PhD(endocrinol), 37. *Prof Exp:* Res cytologist, Dept Genetics, Carnegie Inst, 37-46; asst prof anat, Univ Minn, 46-48; from asst prof to prof anat, Albany Med Col, 48-76, emer prof, 76-; RETIRED. *Concurrent Pos:* Vis scientist, NIH, 59-60. *Mem:* Am Soc Zool; Am Asn Anat. *Res:* Pituitary adrenal relations; secretory phenomena in cells; neural pathways in brain stem and thalamus. *Mailing Add:* Dept of Anat Albany Med Col Albany NY 12208-1819

MILLER, RICHARD EDWARD, b Hollis, NY, May 31, 37; m 58. PHYSICAL CHEMISTRY, CHEMICAL PHYSICS. *Educ:* Stevens Inst Technol, BE, 59; Univ Wash, PhD(phys chem), 66. *Prof Exp:* Design engr, Boeing Co, 59-62; asst phys chem, Univ Wash, 62-66; res fel, Princeton Univ, 66-67; res fel, Mich State Univ, 67-68, asst prof, 68-69, lab mgr, mem fac, Ohio State Univ, 74-75; SAFETY OFFICER, ERLING RIIS RES LAB, 75- *Concurrent Pos:* Consult, Spectra Physics, Inc, 68-71. *Res:* Molecular and crystal structure using infrared and Raman spectroscopic techniques. *Mailing Add:* 1200 Bristol Bryan TX 77801

MILLER, RICHARD GRAHAM, b St Catharines, Ont, Oct 2, 38; m 63; c 2. CELLULAR IMMUNOLOGY, TOLERANCE. *Educ:* Univ Alta, BSc, 60, MSc, 61; Calif Inst Technol, PhD(physics & biol), 66. *Prof Exp:* From asst prof to assoc prof, 67-76, chmn, Dept Immunol, 84-90, PROF, DEPTS MED BIOPHYS, MED SCI & IMMUNOL, UNIV TORONTO, 76-; SR SCIENTIST, ONT CANCER INST, 67- *Mem:* Am Asn Immunologists; Soc Anal Cytol; Can Soc Immunol (vpres, 89-91, pres, 91-). *Res:* T cell tolerance and immunoregulation. *Mailing Add:* Ont Cancer Inst 500 Sherbourne St Toronto ON M4X 1K9

MILLER, RICHARD HENRY, b Aurora, Ill, Aug 31, 26; m 52. ASTROPHYSICS, COMPUTATIONAL PHYSICS. *Educ:* Iowa State Col, BS, 46; Univ Chicago, PhD(physics), 57. *Prof Exp:* Engr construct cyclotron, Univ Chicago, 47-51, Calif Res & Develop Co div, Standard Oil Co Calif, 52 & Brazilian Nat Res Coun, 52-54; asst physics, 54-57, res assoc, 57-59, asst prof, 59-62, ASSOC PROF ASTROPHYSICS, UNIV CHICAGO, 63- *Concurrent Pos:* Assoc dir, Inst Comput Res, Univ Chicago, 62-63, dir, 63-66, actg chmn, Comt Info Sci, 65-66; mem summer study group, Stanford Linear Accelerator Ctr, 64; Nat Res Coun-NASA sr resident res assoc, Goddard Inst Space Studies, 67 & 68; consult astronr, Kitt Peak Nat Observ, 70 & 71; sr resident res assoc, NASA-Ames Res Ctr, 75-77; co-organizer, Workshop on Computational Astrophysics, NASA- Ames Res Ctr, 80; vis assoc prof, dept phys, Univ Calif, San Diego, 81; joint visitor, Europ Southern Observ-Max Planck Inst Astrophys, Munich, 81-82; vis lectr, Nat Observ Brazil, 83; vis scientist, Max-Planck Inst Physik & Astrophysik, 88. *Mem:* Am Phys Soc; Am Astron Soc; Asn Comput Mach; Int Astron Union. *Res:* Structure, formation, and dynamics of galaxies; large n-body computational studies (numerical experiments) on the dynamics of galaxies; stellar dynamics; formation, evolution, collisions and responses to cluster environment; astronomical instrumentation; interferometry. *Mailing Add:* Astron & Astrophys Ctr 5640 Ellis Ave Chicago IL 60637

MILLER, RICHARD J, b Schenectady, NY, Aug 20, 37; m 89. PHYSICAL CHEMISTRY. *Educ:* Union Col, NY, BS, 59; Lehigh Univ, PhD(phys chem), 64. *Prof Exp:* Asst chem, Lehigh Univ, 59-60 & 63-64; from asst prof to assoc prof, 64-72, dept chair, 79-85, PROF CHEM, STATE UNIV NY COL, CORTLAND, 72-, DEPT CHAIR, 88- *Concurrent Pos:* Vis assoc prof, Tufts Univ, 70-71; res collabr, Brookhaven Nat Lab, 78-79. *Mem:* Am Chem Soc; AAAS. *Res:* Thermodynamics; radiochemistry. *Mailing Add:* Dept Chem State Univ NY Col Cortland NY 13045

MILLER, RICHARD KEITH, b Fresno, Calif, June 12, 49; m 71; c 2. APPLIED MECHANICS, CIVIL ENGINEERING. *Educ:* Univ Calif, Davis, BS, 71; Mass Inst Technol, MS, 72; Calif Inst Technol, PhD(appl mech), 76. *Prof Exp:* asst prof mech eng, Univ Calif, Santa Barabara, 75-79; assoc prof civil eng, 79-85, PROF CIVIL & AEROSPACE ENG, UNIV SOUTHERN CALIF, LOS ANGELES, 85-, ASSOC DEAN ENG, 89- *Concurrent Pos:* Eng consult, Astro Aerospace Corp & Jet Propulsion Lab, Hughes Aircraft Co, Aerospace Corp. *Mem:* Am Soc Mech Engrs; Am Soc Civil Engrs; Am Acad Mech. *Res:* Structural dynamics; nonlinear mechanics. *Mailing Add:* Sch Eng Univ Southern Calif Los Angeles CA 90089-1450

MILLER, RICHARD KEITH, b Clarinda, Iowa, Apr 19, 39; m 63, 84; c 2. MATHEMATICS. *Educ:* Iowa State Univ, BS, 61; Univ Wis, MS, 62, PhD(math), 64. *Prof Exp:* Asst prof math, Univ Minn, Minneapolis, 64-66; from asst prof to assoc prof appl math, Brown Univ, 66-71; from assoc prof to prof math, 72-82, DISTINGUISHED PROF, IOWA STATE UNIV, 82- *Concurrent Pos:* With Collins Radio Co, 61 & 62; Fulbright res fel, Ger, 76. *Mem:* Am Math Soc; Soc Indust & Appl Math; Inst Elec & Electronic Engrs. *Res:* Asymptotic behavior of ordinary differential equations and Volterra integral equations; mathematical control theory. *Mailing Add:* Dept Math Iowa State Univ Ames IA 50011

MILLER, RICHARD KERMIT, b Scranton, Pa, Oct 17, 46. TERATOLOGY. *Educ:* Dartmouth Coll, AB, 68, PhD(pharmacol/toxicol), 73. *Prof Exp:* Fel teratology & develop biol, Jefferson Med Col, 72-74; from asst prof to assoc prof obstet & gynec, pharmacol & toxicol, 74-88, PROF OBSTET & GYNEC & TOXICOL, UNIV ROCHESTER SCH MED & DENT, 88-, DIR DIV RES, 78- *Concurrent Pos:* Prin investr, Nat Cancer Inst, Nat Inst Environ Health Sci & Nat Inst for Drug Abuse grants, 77-; prof reproductive biol, Univ Paris VI, 83,; NIH Fogarty Sr Int Fel, 83; comt mem, Nat Res Coun, Nat Acad Sci, 87-89; Fulbright distinguished prof fel, 88; mem, Bd Sci Counr NIH Nat Toxicol Prog, 88- *Mem:* Nat Acad Sci; Teratology Soc (pres, 91-92); Europ Teratology Soc; Am Soc Pharmacol & Exp Therapeuts; Behav Teratology Soc; Perinatal Res Soc. *Res:* Placental function; reproductive and developmental toxicology; reproductive endocrinology; organ perfusion technology; transplacental carcinogenesis. *Mailing Add:* Dept Obstet-Gynec Sch Med & Dent PO Box 668 Univ Rochester Rochester NY 14642

MILLER, RICHARD LEE, b Glendale, Calif, May 30, 42; m 64; c 2. BIOCHEMISTRY. *Prof Exp:* Res biochemist, 69-72, SR RES BIOCHEMIST, BURROUGHS WELLCOME & CO, 72- *Concurrent Pos:* Vis assoc prof, Univ Calif, San Francisco, 86-87. *Mem:* Am Chem Soc; Am Soc Biol Chemists. *Res:* Purine metabolism; nucleotide interconversion; enzymology; antiviral chemotherapy, antiparasitic chemotherapy. *Mailing Add:* Wellcome Res Labs Burroughs Wellcome & Co Research Triangle Park NC 27709

MILLER, RICHARD LEE, b Boston, Mass, Apr 9, 40. INVERTEBRATE ZOOLOGY, DEVELOPMENTAL BIOLOGY. *Educ:* Univ Chicago, BS, 62, PhD(zool), 65. *Prof Exp:* NIH res fel, Calif Inst Technol & Univ Calif, Berkeley, 65-66; asst prof zool, Ore State Univ, 66-68; asst prof, 68-74, assoc prof, 74-83, PROF BIOL, TEMPLE UNIV, 83- *Concurrent Pos:* NIH career develop award, 71-75. *Mem:* Am Soc Cell Biol; Am Soc Zoologists. *Res:* Chemical mediation of fertilization in the invertebrates, especially sperm attraction; chemical nature of the attractants and their mode of action. *Mailing Add:* Dept Biol Temple Univ Philadelphia PA 19122

MILLER, RICHARD LINN, b Portland, Ore, Sept 7, 33; m 59; c 3. CHEMICAL ENGINEERING. *Educ:* Ore State Univ, BS, 55, MS, 59; Univ Minn, PhD(chem eng), 62. *Prof Exp:* Res engr, Rocketdyne Div, NAm Rockwell Corp, Calif, 55-57; res engr, Chevron Res Corp, 62-64; res engr, USAF Sch Aerospace Med, 64-67, chief biol systs br, Environ Systs Div, 67-71, chief crew systs br, 71-82, dep chief, Crew Technol Div, 82-90, CHIEF CREW TECHNOL DIV, ARMSTRONG LAB, USAF SCH AEROSPACE MED, 90- *Honors & Awards:* Liljencrantz Award, Aerospace Med Asn, 84. *Mem:* AAAS; Am Inst Chem Engrs; fel Aerospace Med Asn; Air Pollution Control Asn; SAFE Asn. *Res:* Research and development on life support apparatus for advanced aerospace systems. *Mailing Add:* Crew Technol Div (VN) Armstrong Lab Brooks AFB TX 78235-5000

MILLER, RICHARD LLOYD, b Mishawaka, Ind, Jan 30, 31; m 52; c 7. ENTOMOLOGY, HORTICULTURE. *Educ:* Purdue Univ, BS, 57; Iowa State Univ, MS, 59, PhD(entom), 62. *Prof Exp:* Exten entomologist, Univ Ky, 62-67; assoc prof, 70-74, PROF ENTOM, OHIO STATE UNIV, 74-, EXTEN ENTOMOLOGIST, 67- *Mem:* Entom Soc Am. *Res:* Insect and mite control on fruits, vegetables, ornamentals, turf, yard and garden. *Mailing Add:* Ohio Coop Exten Serv Ohio State Univ Columbus OH 43210-1090

MILLER, RICHARD LLOYD, b Mesa, Ariz, Aug 22, 31; m 52; c 6. EXTRACTIVE CHEMICAL METALLURGY. *Educ:* Ariz State Univ, BA, 57, MS, 60; Univ Utah, PhD(metall), 68. *Prof Exp:* Chemist, Motorola Inc, Phoenix, 59-61; asst prof chem, Univ Tex, El Paso, 61-65; chemist, Univ Utah, 65-66, res asst metall, 66-68; asst prof metall, Univ Wash, 68-69; SCI SPECIALIST METALL, EG&G INC, IDAHO, 69- *Concurrent Pos:* Affil prof metall, Univ Idaho, 70- *Mem:* Metall Soc Am; Inst Mining, Metall & Petrol Engrs. *Res:* Process development in areas of geothermal minerals and precious metals recovery and supercritical fluid extraction of base metal minerals; corrosion studies and material selection for chemical and metallurgical process applications. *Mailing Add:* 2285 Curlew Dr Idaho Falls ID 83406

MILLER, RICHARD LYNN, b Stevens Point, Wis, Sept 27, 45; m 73; c 3. ANTIVIRAL DRUG RESEARCH, MOLECULAR BIOLOGY. *Educ:* Univ Wis, BS, 68; Univ Minn, PhD(microbiol), 74. *Prof Exp:* Fel, Pa State Med Sch, 75-77; sr microbiologist, 77-79, res specialist, 79-86, sr res specialist, 86-88, MGR BIOL, RIKER LABS, 3M CO, 88- *Concurrent Pos:* Asst prof, Univ Minn, 77-80. *Mem:* Am Soc Microbiol; AAAS; NY Acad Sci. *Res:* Antiviral drug research and virus biochemistry, including herpes virus, DNA polymerases and picornavirus structure and replication. *Mailing Add:* 3M Pharmaceut Bldg 270-2S-06 3M Co St Paul MN 55144

MILLER, RICHARD ROY, b Salt Lake City, Utah, Aug 3, 41. MATHEMATICS, CHEMISTRY. *Educ:* Univ Utah, BS, 63, PhD(math), 69. *Prof Exp:* From asst prof to assoc prof, 69-77, PROF MATH, WEBER STATE COL, 77- *Mem:* Am Math Soc; Math Asn Am. *Res:* Functional analysis. *Mailing Add:* Dept Math Weber State Col 3750 Harrison Blvd Ogden UT 84408

MILLER, RICHARD SAMUEL, animal ecology; deceased, see previous edition for last biography

MILLER, RICHARD WILLIAM, b Moline, Ill, July 24, 47; m 68; c 1. LIMNOLOGY. *Educ:* Col William & Mary, BS, 69; Univ Ga, PhD(zool), 75. *Prof Exp:* Asst prof, 75-80, ASSOC PROF BIOL SCI, BUTLER UNIV, 80- *Concurrent Pos:* Res scientist, The Inst Ecol, 79-81. *Mem:* AAAS; Am Soc Limnol & Oceanog; Ecol Soc Am. *Res:* Effect of acid deposition on freshwater ecosystems. *Mailing Add:* Dept Biol Butler Univ 4600 Sunset Ave Indianapolis IN 46208

MILLER, RICHARD WILSON, b Miami, Fla, May 14, 34; m 57; c 2. ENZYMOLOGY, BIOSPECTROSCOPY. *Educ:* Mass Inst Technol, SB, 56, PhD(biochem), 61. *Prof Exp:* Res assoc enzymol, Mass Inst Technol, 61; res assoc, Sheffield Univ, 62-63 & Univ Mich, 63-64; biochemist, New Eng Inst Med Res, 64-68; sr res scientist, Res Br, Chem & Biol Res Inst, 68-87, SR RES SCIENTIST, PLANT RES CTR, CAN AGR, 87- *Concurrent Pos:* NIH fel, 62-64; USPHS grant, 65-68; vis scientist, AFRC Inst Plant Sci Res Nitrogen Fixation Lab, Univ Sussex, 78-79 & 86-87. *Mem:* AAAS; Am Soc Biol Chemists; Fedn Am Socs Exp Biol. *Res:* Respiratory control over pyramidine biosynthesis; enzymology of pyramidine biosynthesis; role of free radicals and superoxide anion in biosynthesis, biodegradation and toxicity mechanisms in microorganisms and plants; mechanism of metalloprotein catalysis; membrane-bound enzymes; membrane composition and physical structure; membrane factors in microbiol nitrogen fixation; Molybdenum and Vanadium nitrogenase; energy transduction; mechanism and temperature effects; symbiotic nitrogen fixation; bacteroid properties and metabolism. *Mailing Add:* Plant Res Ctr Agr Can Ottawa ON K1A 0C6 Can

MILLER, ROBERT ALAN, b Montclair, NJ, Jan 30, 43; m 71; c 2. PHYSICS, COMPUTER SCIENCE. *Educ:* Univ Ill, BS, 65, MS, 66, PhD(physics), 70. *Prof Exp:* Res assoc physics, Col William & Mary, 70-72; res assoc, Rutgers Univ, 72-74; physicist, Fusion Energy Corp, 74-77; physicist, Princeton Gamma-tech, 77-81; vpres sci & technol, Sci Transfer Assocs, 81-82, pres, 82-83; sr staff scientist, Princeton Gamma-Tech, 83-85; MEM TECH STAFF, A T & T BELL LABS, 85- *Concurrent Pos:* Consult, Fusion Energy Corp, 77- *Mem:* Am Phys Soc; Inst Elec & Electronics Engrs; Sigma Xi. *Res:* Low, medium and high energy nuclear physics; solar energy; energy economics; x-ray fluorescence; materials analysis; fusion; transfer of science and technology to developing countries; telecommunications. *Mailing Add:* 22 Evans Cranbury NJ 08512

MILLER, ROBERT ALON, b Toronto, Ont, Can, Oct 2, 48; m 74; c 3. PHARMACEUTICAL DOSAGE FORM DEVELOPMENT. *Educ:* Univ Toronto, BScPhm, 70; Temple Univ, PhD(pharm), 77. *Prof Exp:* Sr res pharmacist, 78-85, res fel, Merck Frosst Can, Inc, 86-88; DIR, PHARMACEUT DEVELOP, NOVOPHARM, 88- *Concurrent Pos:* Lectr, John Abbott Col, 82-88. *Mem:* Can Pharmaceut Asn; Am Pharmaceut Asn; Acad Pharmaceut Sci; Am Asn Pharmaceut Scientists; Controlled Release Soc. *Res:* Physical characteristics of drugs and pharmaceutical excipients and their influences on the properties of solid dosage forms; stability of pharmaceutical dosage forms; pharmaceutical coating from latex dispersions. *Mailing Add:* 22 Walkerton Dr Markham ON L3P 1H8 Can

MILLER, ROBERT BURNHAM, b Dallas, Tex, Dec 14, 42; m 66; c 3. APPLIED STATISTICS. *Educ:* Univ Iowa, BA, 64, MS, 65, PhD(statist), 68. *Prof Exp:* from asst prof to assoc prof, 68-83, PROF STATIST & BUS, UNIV WIS-MADISON, 83- *Concurrent Pos:* Census fel, Am Statist Asn, 83-84. *Honors & Awards:* Halmstad Prize, Actuarial Educ & Res Fund, 79. *Mem:* AAAS; Am Statist Asn; Am Inst Decision Sci. *Res:* Statistical problems in risk theory; time series analysis applied to business and economic problems. *Mailing Add:* Univ Wis 1210 W Dayton St Madison WI 53706

MILLER, ROBERT CARL, b Chicago, Ill, Oct 26, 38; m 69. EXPERIMENTAL HIGH ENERGY PHYSICS, CRYOGENICS. *Educ:* Ill Inst Technol, BS, 61; Northern Ill Univ, MS, 65, CAS, 72. *Prof Exp:* RESEARCHER, ARGONNE NAT LAB, 61-, SCI ASSOC, HIGH ENERGY PHYSICS DIV. *Mem:* Am Phys Soc; Am Nuclear Soc; Inst Elec & Electronics Engrs; Am Asn Physics Teachers; Instrument Soc Am. *Res:* Spin dependence in proton-proton scattering; phenomenology of pion-proton, proton-proton and proton-neutron scattering including amplitude analysis; polarized proton and deuteron targets. *Mailing Add:* 1105 Elizabeth Ave Naperville IL 60540

MILLER, ROBERT CARMI, JR, b Elgin, Ill, Aug 10, 42; Can citizen; m 64; c 2. MICROBIOLOGY. *Educ:* Trinity Col, BSc, 64; Pa State Univ, MSc, 65; Univ Pa, PhD(molecular biol), 69. *Prof Exp:* US Pub Health Serv trainee, Univ Pa, 66-69; postdoctoral fel, Univ Wis, 69-70 & Am Cancer Soc, 70-71; from asst prof to assoc prof microbiol, 71-82, head, dept microbiol, 82-85, dean sci, 85-88, VPRES RES, UNIV BC, 88- *Concurrent Pos:* Mem, MRC Comt on genetics, 80-82; Nat Cancer Inst, grants panel A, 81-85; NSERC Strategic Grant Comt Biotechnol, 85-87; assoc ed, Virology, 74-85, J Virol, 75-84. *Mem:* Can BiochemSoc; Am Soc Microbiol; Sigma Xi. *Res:* Nucleic acids and molecular genetics; C Fimi cellulases and their genes. *Mailing Add:* Off Pres Univ BC Vancouver BC V6T 2B3 Can

MILLER, ROBERT CHARLES, b State Col, Pa, Feb 2, 25; m 52; c 3. OPTICAL PHYSICS, SOLID STATE PHYSICS. *Educ:* Columbia Univ, AB, 48, MA, 52, PhD(physics), 56. *Prof Exp:* Asst physics, Columbia Univ, 49-51, lectr, 51-53; mem tech staff, Bell Tel Labs, 54-63, head, Solid State Spectros Res Dept, 63-67; mem, Inst Defense Anal, 67-68; head optical electronics res dept, AT&T Bell Labs, 68-77, mem tech staff, 77-84, distinguished mem tech staff, 84-88; RETIRED. *Concurrent Pos:* RCA fel, 53-54. *Honors & Awards:* R W Wood Prize, 86. *Mem:* Fel Am Phys Soc; NY Acad Sci; AAAS; Sigma Xi. *Res:* Ferroelectricity; nonlinear optics; optical spectroscopy of semiconductor heterostructures. *Mailing Add:* 65 Eaton Ct Cotuit MA 02635-2908

MILLER, ROBERT CHRISTOPHER, b Washington, DC, Dec 31, 35; m 65; c 3. CYTOGENETICS, CELL BIOLOGY. *Educ:* Univ Nebr-Lincoln, BSc, 62; Ohio State Univ, MSc, 68, PhD(genetics), 71. *Prof Exp:* Lab technician, King County Cent Blood Bank, Inc, Seattle, 62-63 & Cleveland Clin Found Hosp, 63-65; res asst cytogenetics, Ohio State Univ Hosp, 65-66; grad sch genetics, Ohio State Univ, 66-71; staff scientist, Inst Med Res, Camden, NJ, 72-81; DIR, CYTOGENETICS LAB, PA HOSP, PHILADELPHIA, 81-; ASST PROF OBSTET & GYNEC, MED COL, UNIV PA, 81- *Concurrent Pos:* Prin investr, Nat Inst Environ Health Sci grant, 75-78; coadjutant asst prof, Rutgers Univ, Camden Col, 77. *Res:* Genetics of sex; role of chromosome aberrations in embryo mortality; mutagenesis; chemically induced heritable chromosome defects. *Mailing Add:* Pennsylvania Hosp Eighth & Spruce Philadelphia PA 19107

MILLER, ROBERT CLAY, organic chemistry, for more information see previous edition

MILLER, ROBERT DEMOREST, b Omaha, Nebr, Sept 25, 19; m 41; c 3. SOIL PHYSICS. *Educ:* Univ Mo, BS, 40; Univ Nebr, MS, 42; Cornell Univ, PhD(soil physics), 48. *Prof Exp:* Asst soil physicist, Univ Calif, 48-52; assoc prof, 52-59, prof soil physics, 59-87, EMER PROF, CORNELL UNIV, 87- *Concurrent Pos:* Fulbright res fel, Norway, 65-66; Royal Norweg Coun Sci & Indust Res fel, 65-66; dean fac, Cornell Univ, 67-71. *Mem:* Soil Sci Soc Am; fel Am Soc Agron; Am Geophys Union. *Res:* Soil-water interactions; freezing and heaving of soil; freezing of water in porous media. *Mailing Add:* Dept Agron Cornell Univ Ithaca NY 14853

MILLER, ROBERT DENNIS, b Philadelphia, Pa, Sept 23, 41; m 63; c 3. ORGANIC CHEMISTRY. *Educ:* Lafayette Col, BS, 63; Cornell Univ, PhD(org chem), 68. *Prof Exp:* RES SCIENTIST, IBM CORP, 68- *Mem:* AAAS; Am Chem Soc. *Res:* Organic photochemistry dealing with the production of highly strained, theoretically interesting molecules; high temperature thermal fragmentation reactions; synthetic methods; chemistry and spectroscopy of reactive intermediates; radiation sensitive polymers; nonliner optical materials. *Mailing Add:* IBM Res 650 Harry Rd San Jose CA 95120

MILLER, ROBERT DUWAYNE, b Galesburg, Ill, Nov 27, 12; m 37; c 3. PHYSICS. *Educ:* Knox Col, AB, 34; Univ Ill, MS, 35; Wash Univ, PhD(physics), 37. *Prof Exp:* Res geophysicist, Subterrex, Tex, 37-38; geophysicist, Shell Oil Co, 38-58, mgr tech info, Shell Develop Co, 58-69, consult, 69-90; RETIRED. *Concurrent Pos:* Physicist, Carnegie Inst Dept Terrestrial Magnetism, 42-43; physicist, Appl Physics Lab, Johns Hopkins Univ, 43-45. *Mem:* Am Phys Soc; Soc Explor Geophysicists; Am Geophys Union. *Res:* Diffuse scattering of x-rays; electrical methods of oil exploration. *Mailing Add:* 6150 Cedar Creek Houston TX 77057-1802

MILLER, ROBERT EARL, b Rockford, Ill, Oct 4, 32. MECHANICS. *Educ:* Univ Ill, BS, 54, MS, 55, PhD(theoret & appl mech), 59. *Prof Exp:* From instr to assoc prof, 55-68, PROF THEORET & APPL MECH, UNIV ILL, URBANA, 68- *Mem:* Am Acad Mech; Am Soc Eng Educ; Am Inst Aeronaut & Astronaut; Am Soc Civil Eng. *Res:* Theoretical and applied mechanics. *Mailing Add:* Dept Theoret & Appl Mech 216 Talbot Lab Univ Ill 104 S Wright St Urbana IL 61801-2983

MILLER, ROBERT ERNEST, b Des Moines, Iowa, July 31, 36; m 57; c 3. PLANT PATHOLOGY, SOIL MICROBIOLOGY. *Educ:* Simpson Col, BA, 58; Cornell Univ, MS, 62, PhD(plant path), 63. *Prof Exp:* Res assoc, 63-70, res scientist, Campbell Inst Agr Res, 70-82, VPRES & GEN MGR OPERS, CAMPBELL'S FRESH, INC, 82- *Concurrent Pos:* Res grant plant path, Univ Calif, Berkeley, 65-66. *Mem:* Am Phytopath Soc; Am Soc Microbiol. *Res:* Ecology and physiology of soil-borne plant pathogens; biological control of soil-borne plant pathogens; genetics of fungi. *Mailing Add:* Campbell's Fresh Inc PO Box 169 Blandon PA 19510

MILLER, ROBERT GERARD, b Orange, NJ, Jan 30, 25; m 50; c 4. OPERATIONS RESEARCH, METEOROLOGY. *Educ:* Rutgers Univ, BA, 51; NY Univ, MS, 52; Harvard Univ, PhD(statist), 61. *Prof Exp:* Res staff, Mass Inst Technol, 52-55; dir, Math Statist Div, Travelers Res Ctr, 55-64; mem tech staff, Bell Telephone Labs, 64-65; res fel, Travelers Res Ctr, Inc, 65-71; sr scientist, Life Ins Marketing & Res Asn, 71-76; chief scientist, Air Weather Serv, US Air Force, 76-77; SR SCIENTIST, TECH DEVELOP LAB, NAT WEATHER SERV, 77- *Concurrent Pos:* Consult, Air Force Geophysics Lab, 58-62; expert lectr, World Meterol Orgn, UN, 61; instr, Univ Conn, 62-65; adj prof, St Louis Univ, 77. *Honors & Awards:* USAF Meritorious Civilian Serv Award; Bronze Medal, Dept Com. *Mem:* Fel Am Meterol Soc. *Res:* Techniques development in statistical meteorology; discriminant; analysis in weather prediction; statistical forecasting. *Mailing Add:* 5701 Dun Horse Lane Derwood MD 20855

MILLER, ROBERT GERRY, b Northampton, Mass, Mar 23, 44; m 83; c 2. MECHANICAL ENGINEERING, HEAT TRANSFER. *Educ:* Norwich Univ, BS, 65; Univ Mass, MS, 68; Rutgers Univ, PhD(mech eng), 73. *Prof Exp:* Thermal res engr, Chicago Bridge & Iron Co, 73-75; sr res engr, 75-78, RES ASSOC, JIM WALTER RES CORP, 78- *Concurrent Pos:* Adj instr, Mech Eng Dept, Univ SFla, 81-83, 91-; ed adv bd, J Thermal Insulation, 82-; Accreditation bd eng & technol, 87-90. *Honors & Awards:* Centennial Medallion, Am Soc Mech Engrs, 80. *Mem:* Am Soc Mech Engrs; Am Soc Testing & Mat; Am Soc Eng Educ; Am Soc Heating, Refrigerating & Air Conditioning Engrs. *Res:* Thermal resistance measurements of thermal insulation materials; guarded calibrated hot box testing; thermal contact resistance; fire testing. *Mailing Add:* Jim Walter Res Corp 10301 Ninth St N St Petersburg FL 33716

MILLER, ROBERT H, b Glenside, Pa, Nov 28, 25; m 48; c 3. ELECTRICAL ENGINEERING, INSTRUMENTATION. *Educ:* Va Polytech Inst, BS, 48; Mass Inst Technol, PhD(instrumentation), 64. *Prof Exp:* Engr, Western Union Res Labs, 48-49, Kearfott Co, Inc, 49-50 & Grumman Aircraft Eng Corp, 50-53; chief elec eng, Poly-Sci Corp, 54, chief engr, 55-56, vpres eng, 55-59, vpres res & develop, 60-63; ASSOC PROF ELEC ENG, VA POLYTECH INST & STATE UNIV, 64- *Concurrent Pos:* Consult, Poly-Sci Corp, 64- *Mem:* Inst Elec & Electronics Engrs. *Res:* Electromechanical transducers, especially electromagnetic bearings, differential transformers, motors and sliding contact devices. *Mailing Add:* 1414 Crestview Dr Blacksburg VA 24060

MILLER, ROBERT H(ENRY), b Vassar, Mich, Feb 20, 30; m 53; c 2. CHEMICAL ENGINEERING. *Educ:* Univ Mich, BSE, 52, MSE, 56, PhD(mass transfer), 59. *Prof Exp:* Res engr, Allied Signal Corp, 58-61, res suprv, 61-68, supvr res planning & eval, 68-72, mgr res econ, 72-79, dir planning & admin, 79-80, mgr prog analyst, 80-81, asst to vpres res, 81-88; CONSULT, 89- *Mem:* AAAS; Am Chem Soc; Sigma Xi. *Res:* Economic evaluation of research projects; research planning; research administration. *Mailing Add:* Two Beverly Rd Madison NJ 07940-2817

MILLER, ROBERT HAROLD, b Fremont, Wis, Sept 19, 33; m 57; c 3. SOIL SCIENCE, MICROBIOLOGY. *Educ:* Wis State Univ, River Falls, BS, 58; Univ Minn, MS, 61, PhD(soil microbiol), 64. *Prof Exp:* From asst prof to prof agron, Ohio State Univ, 64-81; head & prof, Dept Soil Sci, NC State Univ, 82-89. *Concurrent Pos:* Fulbright lectr, 74-75; bd dir, Am Soc Agron & Soil Sci Soc Am, 78-81. *Mem:* fel Am Soc Agron; fel Soil Sci Soc Am; Sigma Xi; Am Soc Microbiol. *Res:* Plant rhizosphere microorganisms and their interactions with plants; ecology and physiology of Rhizobium japonicum; chemistry of soil organic matter; recycling of organic wastes in soil; international development. *Mailing Add:* Univ RI Kingston RI 02881-0000

MILLER, ROBERT HAROLD, b Milwaukee, Wis, Nov 25, 57; m 84; c 3. AMINO ACID METABOLISM, METABOLIC REGULATION. *Educ:* Univ Minn, BS, 80; Univ Wis, PhD(nutrit sci), 86. *Prof Exp:* Staff fel, Nat Inst Alchol Abuse & Alcoholism, 86-87; RES ASSOC, ROSS LABS-MED NUTRIT RES, 87- *Mem:* Am Inst Nutrit; Biochem Soc Eng. *Res:* Study of metabolic regulation of energy and amino acid metabolism; investigation of glutamine's effects on cellular growth, proliferation and protein nutriture; protein synthesis and degradation. *Mailing Add:* Med Dept Ross Labs Columbus OH 43215-1754

MILLER, ROBERT JAMES, II, b Dunn, NC, Jan 14, 33; m 59; c 3. PLANT PHYSIOLOGY, ECOLOGY. *Educ:* NC State Univ, BS, 56; Yale Univ, MF, 62, MS, 65, PhD(biol), 67. *Prof Exp:* From assoc prof to prof biol, Radford Col, 65-72; prof biol & dean, St Mary's Col, 73-83, LAWYER, 85- *Concurrent Pos:* Chmn dept biol, Radford Col, 67-68, dean sch natural sci, 68-71, vpres acad affairs, 71-72. *Mem:* Sigma Xi; Am Soc Plant Physiol; Ecol Soc Am; Soc Am Foresters. *Res:* Nitrogen relations of higher plants; ecology of wetlands. *Mailing Add:* 3404 Lake Boone Trail Raleigh NC 27607

MILLER, ROBERT JOSEPH, b Ironton, Ohio, June 10, 41; m 73; c 3. EXPERIMENTAL HIGH ENERGY PHYSICS. *Educ:* Univ Detroit, BS, 63; Purdue Univ, MS, 66, PhD(physics), 69. *Prof Exp:* Fel exp high energy physics, Rutherford Lab, Sch Res Coun UK, 68-72; res assoc, Argonne Nat Lab, 72-75, asst physicist, 75-78; asst prof, 78-80, SR RES ASSOC, DEPT PHYSICS, MICH STATE UNIV, EAST LANSING, 80- *Mem:* Am Phys Soc. *Res:* Form factors of K meson; anti-neutrino interaction with protons; search for fluctuations in the pi meson-proton interaction; direct photon interactions. *Mailing Add:* Dept of Physics Mich State Univ East Lansing MI 48824

MILLER, ROBERT JOSEPH, b Keokuk, Iowa, Nov 17, 39; m 65; c 2. MARINE ECOLOGY. *Educ:* William Jewell Col, AB, 61; Col William & Mary, MA, 64; NC State Univ, PhD(zool), 70. *Prof Exp:* Nat Res Coun Can fel, Marine Ecol Lab, Fisheries Res Bd Can, 69-71, res scientist, St John's Biol Sta, 71-79; RES SCIENTIST, HALIFAX LAB, FISHERIES & OCEANS CAN, 79- *Mem:* Am Soc Limnol & Oceanog; Sigma Xi. *Res:* Energy flow in marine communities; marine fisheries management; efficiency of baited traps. *Mailing Add:* PO Box 550 Halifax NS B3J 2S7 Can

MILLER, ROBERT L, b Chicago, Ill, Jan 26, 26; m 47; c 4. PHYSICAL CHEMISTRY, ACADEMIC ADMINISTRATION. *Educ:* Univ Chicago, PhB, 48, BS, 50, MS, 51; Ill Inst Technol, PhD(chem), 63. *Prof Exp:* From instr to assoc prof chem & assoc dean, Univ Ill, Chicago, 51-68, asst dean, 62-65; prof chem & dean, Col Arts & Sci, 68-85, PROF CHEM, UNIV NC, GREENSBORO, 85-, ACTG DEAN, GRAD SCH & ASSOC PROVOST RES, 89- *Mem:* AAAS; Sigma Xi; Am Chem Soc. *Res:* Applications of quantum mechanics to desription of chemical compounds of biological interest. *Mailing Add:* Dept Chem Univ NC Greensboro NC 27412

MILLER, ROBERT LLEWELLYN, b Chicago, Ill, Jan 19, 29; m 53; c 1. POLYMER SCIENCE. *Educ:* Mass Inst Technol, BS, 50; Brown Univ, PhD(chem), 54. *Prof Exp:* Res chemist, Monsanto Chem Co, 53-59, res specialist, 59-62, group leader, Chemstrand Res Ctr, Inc, Monsanto Co, 62-64, scientist, NC, 64-69, Mo, 69-71; assoc prof, 80-84, SR RES SCIENTIST, MICH MOLECULAR INST, 72-, PROF, 84- *Concurrent Pos:* Sr vis scholar, Univ Manchester, 68-69; adj prof, Case Western Reserve Univ, 78-86, Cent Mich Univ, 81-, Michigan Technol Univ, 87- *Mem:* AAAS; Am Chem Soc; fel Am Phys Soc; Am Crystallog Asn. *Res:* Solid state physics as applied to polymers, particularly semicrystalline polymers. *Mailing Add:* Michigan Molecular Inst 1910 W St Andrews Dr Midland MI 48640

MILLER, ROBERT RUSH, b Colorado Springs, Colo, Apr 23, 16; wid; c 5. ICHTHYOLOGY, PALEOHYDROLOGY. *Educ:* Univ Calif, AB, 38; Univ Mich, MA, 43, PhD(zool), 44. *Prof Exp:* Asst ichthyol surv, Nev, Univ Mich, 38 & Div Fishes, Mus Zool, 39-44; assoc cur fishes, US Nat Mus, Smithsonian Inst, 44-48; from asst prof to prof zool, 48-80, prof biol sci, 80-86, cur fishes, mus zool, 60-86, EMER PROF, BIOL SCI & EMER CUR FISHES, MUS ZOOL, UNIV MICH, ANN ARBOR, 86- *Concurrent Pos:* Mem, Univ Mich expeds, 38-42, 50 & 59-, Mex, 39, 50 & 55-86, Ichthyol Surv, Guatemala, US Dept State, Smithsonian Inst & Govt Guatemala, 46-47, Biol Surv, Arnhem Land, Govt Australia, Nat Geol Soc & Smithsonian Inst, 48; assoc cur, Mus Zool, Univ Mich, 48-59; ichthyol ed, Copeia, Am Soc Ichthyol & Herpet, 50-55; collabr, US Nat Park Serv, 60-; Guggenheim fel, 73-74; ed, Mus Zool publ, 83-85. *Mem:* AAAS; Am Soc Ichthyol & Herpet (vpres, 61, pres, 65); Soc Syst Zool; Soc Study Evolution; Soc Vert Paleont; Am Fish Soc. *Res:* Taxonomy, distribution, variation, hybridization, ecology, life history and evolution of fishes; paleoichthyology. *Mailing Add:* Univ Mus Bldg Univ of Mich Ann Arbor MI 48109-1079

MILLER, ROBERT VANCE, applied physics, for more information see previous edition

MILLER, ROBERT VERNE, b Modesto, Calif, Dec 27, 45; m 68. MICROBIAL GENETICS, MOLECULAR GENETICS. *Educ:* Univ Calif, Davis, BA, 67; Univ Ill, Urbana, MS, 69, PhD(microbiol), 72. *Prof Exp:* Res assoc molecular genetics, Univ Calif, Berkeley, 72-74; asst prof microbiol, Univ Tenn, Knoxville, 74-78, assoc prof, 78-80; assoc prof, 80-85, PROF BIOCHEM, CHICAGO MED SCH, LOYOLA UNIV, 85- *Concurrent Pos:* Dernham fel, Am Cancer Soc, Calif, 72-74; res career develop awardee, NIH-Nat Inst Allergy & Infectious Dis. *Mem:* AAAS; Sigma Xi; Am Soc Microbiol; Genetics Soc Am; Am Soc Biol Chemists. *Res:* Genetic and biochemical mechanisms of DNA transactions in pseudomonas aeruginosa, particularly the molecular mechanisms of recombination, DNA damage repair and gene transfer. *Mailing Add:* Dept Biochem & Biophys Med Ctr Loyola Univ 2160 S First Ave Maywood IL 60153

MILLER, ROBERT W, b Warrensville, NC, Aug 7, 31; m 53; c 2. AGRONOMY. *Educ:* Berea Col, BS, 53; Ohio State Univ, MSc, 60, PhD(agron), 63. *Prof Exp:* Asst county agent, NC State Univ, 55-57; asst prof plant breeding, Cornell Univ, 62-63; from asst prof to assoc prof turfgrass mgt, Ohio State Univ, 63-71, prof agron & turfgrass mgt, 71-; AT DEPT FORESTRY, UNIV WIS-STEVENS POINT. *Mem:* Am Soc Agron; Crop Sci Soc Am. *Res:* Crop physiology; turfgrass management. *Mailing Add:* Dept Forestry Univ Wis Stevens Point WI 54481

MILLER, ROBERT WALKER, JR, b Philadelphia, Pa, Nov 3, 41; m 66; c 2. PLANT PATHOLOGY. *Educ:* Univ Del, BS, 64, PhD(plant path & ecol), 71; Univ Ariz, MS, 70. *Prof Exp:* Exten specialist, 71-72, from asst prof to assoc prof, 72-81, PROF PLANT PATH, CLEMSON UNIV, 81-, PROJ LEADER, 85- *Concurrent Pos:* Consult, plant health policy. *Mem:* Am Phytopath Soc; Soc Nematologists. *Res:* Epidemiology, cultural and chemical controls of the foliar diseases of pecans; diagnosis and control of plant disease, especially fruit. *Mailing Add:* Sue-Craig Rd Six Mile SC 29682

MILLER, ROBERT WARWICK, b Brooklyn, NY, Sept 29, 21; m 55. PEDIATRICS, EPIDEMIOLOGY. *Educ:* Univ Pa, AB, 42, MD, 46; Univ Mich, MPH, 58, DrPH, 61. *Prof Exp:* Mem atomic energy proj, Univ Rochester, 51-53; chief pediat, Atomic Bomb Casualty Comn, Hiroshima, Japan, 53-55, chief pediat child health surv, Hiroshima & Nagasaki, 58-60; prof assoc, Nat Acad Sci, 55-57; chief, Epidemiol Br, 61-75, CHIEF, CLIN EPIDEMIOL BR, NAT CANCER INST, 76- *Mem:* Soc Pediat Res; Am Pediat Soc. *Res:* Epidemiology of cancer; congenital malformations and radiation effects. *Mailing Add:* Clin Epidemiol Br Nat Cancer Inst 400 EPN Bethesda MD 20892

MILLER, ROBERT WITHERSPOON, b Chester, SC, Oct 29, 18; m 43; c 4. ORGANIC CHEMISTRY. *Educ:* Erskine Col, AB, 39; Univ NC, PhD(org chem), 48. *Prof Exp:* Asst instr chem, Clemson Univ, 39-40; chemist, Tenn Eastman Corp, 48-52, sr chemist, Eastman Chem Prod Inc, 52-53, sales rep, 53-56, prod mgr, 56-57, dist sales mgr, 57-58, chief sales develop rep, 58-65, mgr new prod sales, 65-73, res assoc, 73-77, staff asst, Lab Indust Med, Tenn Eastman Co, Eastman Kodak Co, 77-84; RETIRED. *Mem:* Am Chem Soc; Sigma Xi. *Mailing Add:* 4531 Stagecoach Rd Kingsport TN 37664

MILLER, ROGER ERVIN, b Kitchener, Ont, July 23, 52; m 75; c 3. PHYSICAL CHEMISTRY, CHEMICAL DYNAMICS. *Educ:* Univ Waterloo, BSc, 75, MSc, 77, PhD(physics), 80. *Prof Exp:* Res fel physics, Australian Nat Univ, 80-84; assoc prof chem, 85-88, PROF CHEM, UNIV NC, CHAPEL HILL, 88- *Concurrent Pos:* Vis scientist, Max Planck Inst Aerodyn, Göttingen, WGermany, 82 & Hahn-Meitner Inst Nuclear Res, Berlin, 83. *Honors & Awards:* Alfred P Sloan Res Fel. *Mem:* Fel Am Phys Soc; Am Chem Soc. *Res:* Infrared spectroscopy of molecular beams; infrared spectroscopy and photochemistry of weakly bound van der Waals molecules; molecular scattering using crossed beam techniques; differential and inelastic integral scattering. *Mailing Add:* Dept Chem Univ NC Kenan Labs 045A Chapel Hill NC 27599

MILLER, ROGER HEERING, b Dayton, Ohio, June 8, 31; m 57; c 4. PHYSICS, ACCELERATOR PHYSICS. *Educ:* Princeton Univ, AB, 53; Stanford Univ, PhD(physics), 64. *Prof Exp:* Group leader, 61-83, PROF, STANFORD LINEAR ACCELERATOR CTR, STANFORD UNIV, 78- *Concurrent Pos:* Electron optics consult, Haimson Res Corp, 75-, Schonberg Rad Corp, SAIC, Brobeck, Beta Develop Corp. *Mem:* Am Phys Soc; Sigma Xi. *Res:* Accelerator physics; injection and positron production; electron beam optics. *Mailing Add:* Bin 26 SLAC PO Box 4349 Stanford CA 94309

MILLER, RONALD ELDON, b Spokane, Wash, May 13, 41; m 62; c 2. CHEMICAL METALLURGY. *Educ:* Wash State Univ, BS, 64, PhD(metall), 70; Univ Ill, MS, 65. *Prof Exp:* Asst prof metall eng, Univ Wis-Madison, 69-73; sect head, Alcoa Labs, 73-80, mgr, 80-87; MGR, INGOT TECHNOL, 87- *Concurrent Pos:* NSF res initiation grant, 71-72; chmn, Light Metals, Am Inst Mining, Metall & Petrol Engrs, 85-86, mem, bd dirs, 86- *Mem:* Am Inst Mining, Metall & Petrol Engrs; Am Soc Metals; Am Soc Testing & Mat; Metall Soc (pres, 91); Am Foundry Soc. *Res:* Desulfurization of coal; thermodynamics and kinetics of reactions; alloy development; phase transformations; aluminum metal quality; melting technology; energy conservation; recycling. *Mailing Add:* 1000 Riverview Towers 900 S Gay St Knoxville TN 37902

MILLER, RONALD LEE, b Magnolia, Ky, Feb 2, 36; m 59; c 4. BIOCHEMISTRY. *Educ:* Western Ky State Col, BS, 58; Univ Ky, PhD(biochem), 67. *Prof Exp:* Instr biol sci, Cornell Univ, 67-68; fel, Roche Inst Molecular Biol, 68-70, sr investr, 70-72; from asst prof to assoc prof, 72-86, PROF BIOCHEM, MED UNIV SC, 86- *Mem:* AAAS; Am Chem Soc; Am Soc Biol Chem. *Res:* Connective tissue metabolism; protein synthesis, particularly isolation of protein initiation factors from rabbit reticulocytes; purification and properties of lectins. *Mailing Add:* Dept Biochem Med Univ SC 171 Ashley Ave Charleston SC 29425

MILLER, ROSWELL KENFIELD, b Glen Cove, NY, Aug 4, 32; m 55; c 3. FOREST MANAGEMENT. *Educ:* State Univ NY Col Forestry, Syracuse Univ, BS, 58, MF, 59; Univ Mich, Ann Arbor, PhD(forest mgt), 72. *Prof Exp:* Forester, US Forest Serv, Ore, 59-60; forest engr, Crown Zellerbach Corp, Ore, 60-64; logging engr, Navajo Forest Prod Industs, NMex, 64; chief of surv, NMex State Hwy Dept, Gallup, 64-65; asst prof, 65-67 & 69-72, ASSOC PROF FORESTRY, MICH TECHNOL UNIV, 72- *Mem:* Soc Am Foresters; Am Congress Surv & Mapping. *Res:* Cost control; planning natural resource use; small business management; land surveying. *Mailing Add:* Sch Forestry & Wood Prod Mich Technol Univ Houghton MI 49931

MILLER, ROY GLENN, b Columbus, Ohio, Dec 23, 33; m 55; c 4. ORGANIC CHEMISTRY. *Educ:* Ohio Wesleyan Univ, BA, 55; Univ Mich, MS, 60, PhD(chem), 63. *Prof Exp:* Res chemist, Elastomer Chem Dept, Exp Sta, E I du Pont de Nemours & Co, 62-65; from asst prof to prof chem, Univ NDak, 65-82; PROF CHEM & CHMN DEPT, WABASH COL, 82- *Concurrent Pos:* Vis prof, Dartmouth Col, 77-78. *Mem:* Am Chem Soc; Sigma Xi. *Res:* Organic reaction mechanisms; organometallic chemistry and homogeneous catalysis. *Mailing Add:* Dept Chem Wabash Col Crawfordsville IN 47933

MILLER, RUDOLPH J, b Gbely, Czech, Sept 25, 34; US citizen; m 57; c 3. ETHOLOGY, ICHTHYOLOGY. *Educ:* Cornell Univ, BS, 56, PhD(vert zool), 61; Tulane Univ, MS, 58. *Prof Exp:* NIH fel, Univ Groningen, 61-62; from asst prof to assoc prof, 62-69, chmn dept gen & evolutionary biol, 77-79, PROF ZOOL, OKLA STATE UNIV, 69- *Concurrent Pos:* Vis investr, Univ Hawaii, 71-72; vis prof, Cornell Univ, 80 & Univ Wash, 81; Consult, state, fed, private agencies, fish ecol & aquatic ecosystem probs. *Mem:* AAAS; Am Soc Ichthyol & Herpet; Animal Behav Soc. *Res:* Fish behavior; comparative aspects and motivation analysis, primarily on anabantid, centrarchid and cyprinid fishes; fish feeding ecology; correlative studies of brain, sense organs and behavior; streamfish ecology. *Mailing Add:* Dept Zool Okla State Univ Stillwater OK 74078

MILLER, RUPERT GRIEL, statistics; deceased, see previous edition for last biography

MILLER, RUSSELL BENSLEY, b Lake Wales, Fla, Aug 20, 46; c 2. BUMBLEBEE SYSTEMATICS, POLLINATION ECOLOGY. *Educ:* Carleton Col, BA, 68; Yale Univ, PhD(biol), 74. *Prof Exp:* From asst prof to assoc prof, 74-81, RES ASSOC, DEPT BIOL SCI, MT HOLYOKE COL, 81- *Concurrent Pos:* Fac, Rocky Mountain Biol Lab, 74 & Ctr Europ Studies, Tufts Univ, 81; sr investr, Rocky Mountain Biol Lab, 76-79; curatorial affil entom, Peabody Mus Nat Hist, Yale Univ, 84- *Res:* Field studies on the pollination ecology of Angiosperms; systematic relationships of North American bumblebees. *Mailing Add:* Div Entom Peabody Mus Natural Hist New Haven CT 06511

MILLER, RUSSELL BRYAN, b Tyler, Tex, May 31, 40; m 69; c 2. ORGANIC CHEMISTRY, SYNTHETIC CHEMISTRY. *Educ:* Wash & Lee Univ, BS, 62; Rice Univ, PhD(chem), 67. *Prof Exp:* Res fel, Columbia Univ, 66-68; from asst prof to assoc prof, 68-81, chmn chem dept, 85-90, PROF CHEM, UNIV CALIF, DAVIS, 81- *Concurrent Pos:* vis scientist, NSF, 90-91. *Mem:* Am Chem Soc; Sigma Xi. *Res:* Synthetic natural product chemistry; new synthetic methods; conformational analysis. *Mailing Add:* Dept Chem Univ Calif Davis CA 95616

MILLER, RUSSELL LEE, b Cairo, Ga, Dec 23, 22; m 53; c 4. CROP SCIENCE. *Educ:* Univ Ga, BSA, 50, MS, 52; La State Univ, PhD, 58. *Prof Exp:* Instr agron, Univ Ga, 51-52; from instr to assoc prof, 52-68, PROF AGRON, LA STATE UNIV, BATON ROUGE, 68- *Mem:* Am Soc Agron; fel Nat Asn Cols & Teachers Agr (vpres, 80 & pres, 81); Crop Sci Soc Am. *Res:* Agronomic education; teaching improvement. *Mailing Add:* 7033 Menlo Dr Baton Rouge LA 70808

MILLER, RUSSELL LOYD, JR, b Harvey, WVa, June 30, 39; m 63; c 2. IMMUNOPHARMACOLOGY. *Educ:* Howard Univ, Col Liberal Arts, BS, 61, Col Med, MD, 65. *Prof Exp:* Med internship, Med Ctr, Univ Mich, Ann Arbor, 65-66, resident internal med, 66-68; res fel, Dept Internal Med, Div Clin Pharm, Univ Calif, San Francisco, 68-69 & 71-73; vis scientist, Roche Inst Molecular Biol, 73-74; assoc prof, Dept Med & Pharmacol, 74-79, prof clin pharmacol & dean, Col Med, 79-88, VPRES HEALTH AFFIARS, HOWARD UNIV, 88-, SR VPRES, 90- *Concurrent Pos:* Fel, Cardiovasc Res Inst, Univ Calif, San Francisco, 68-69 & 71-73, US Army Med Corps, 69-71; dir, Div Clin Pharmacol, Col Med, Howard Univ, 74-79; vis prof, Dist Columbia Gen Hosp, 74-79; consult, Dept Pharmacol, Univ Miami, 74-79, Med Letter Drugs & Therapeut, 74-; scholar clin pharmacol, Burroughs Wellcome Found, 77; consult ed acad med, 88-89. *Mem:* Am Fedn Clin Res; Am Soc Clin Pharmacol & Therapeut; Am Soc Pharmacol & Exp Therapeut; Am Col Physicians; Nat Med Asn. *Res:* Study of neurotransmitters, neuromodulators, vasoactive peptides, and their relation to clinical conditions; application of techniques of immunopharmacology to improve the understanding of drug actions. *Mailing Add:* Howard Univ Hosp 2041 Georgia Ave NW Washington DC 20060

MILLER, SALLY ANN, b Canton, Ohio, Apr 11, 54; m 76; c 2. AGRICULTURAL DIAGNOSTICS. *Educ:* Ohio State Univ, BS, 76; Univ Wis-Madison, MS, 79, PhD(plant path), 82. *Prof Exp:* Asst chem, Ohio State Univ, 74-76; res asst, Dept Plant Path, Univ Wis, 76-82; res scientist I, DNA Plant Technol Corp, 82-84, res scientist II, 84-85; MGR PLANT PATH, AGRI-DIAG ASSOCS, 85- *Mem:* Am Phytopath Soc; AAAS; Am Soc Plant Physiologists. *Res:* Development of antibody based diagnostics for agriculture. *Mailing Add:* Agri-Diag Assocs Corp 2611 Branch Pike Cinnaminson NJ 08077

MILLER, SANDRA CAROL, b Montreal, Que, Feb 18, 46. IMMUNOLOGY, HEMATOLOGY. *Educ:* Sir George Williams Univ, BSc, 68; McGill Univ, MSc, 71, PhD(immunol), 75. *Prof Exp:* Lectr anat, McGill Univ, 75-76; fel immuno-hemat, Baylor Col Med, 76-78; ASSOC PROF ANAT, MCGILL UNIV, 78- *Honors & Awards:* Murry Barr Award, 85. *Mem:* Am Asn Anatomists; Int Soc Exp Hemat; Can Soc Immunol; Can Asn Anatomists. *Res:* Regulation of cells involved in spontaneous killing of tumor cells in vitro; cells mediating resistance to foreign bone marrow grafts in vivo. *Mailing Add:* Dept Anat 3640 University St Montreal PQ H3A 2B2 Can

MILLER, SANFORD ARTHUR, b Brooklyn, NY, May 12, 31; m 58; c 2. BIOCHEMISTRY, NUTRITION. *Educ:* City Col New York, BS, 52; Rutgers Univ, MS, 56, PhD(physiol, biochem), 57. *Prof Exp:* Chemist, Appl Res Br, Army Chem Ctr, Md, 52; asst, Bur Biol Res, Rutgers Univ, 54-55, Dept Physiol & Biochem, 55-57; res assoc & supvr animal labs, Dept Food Technol, 57-59, from asst prof to assoc prof, 59-70, dir training prog oral sci, 70-78, PROF NUTRIT BIOCHEM, MASS INST TECHNOL, 70-; DIR, BUR FOODS, FDA, 78- *Concurrent Pos:* Mem, Expert Comt Generally Regarded as Safe Substances, Fedn Am Socs Exp Biol, Food & Drug Admin, 72-78 & Comt Maternal & Child Health & Comt Contraceptive Steroids, Nat Inst Child Health Develop, 73- *Honors & Awards:* Conrad Elvehsem Award, Am Inst Nutrit, 81. *Mem:* AAAS; Perinatal Res Soc; Am Soc Pediat Res; Inst Food Technologists; Am Inst Nutrit. *Res:* Nutrition and development; infant nutrition; synthetic dietary energy sources; oral biology. *Mailing Add:* Univ Tex Health Sci Ctr 7703 Floyd Curl Dr San Antonio TX 78284-7819

MILLER, SANFORD STUART, b Paterson, NJ, June 1, 38; m 67; c 2. MATHEMATICAL ANALYSIS. *Educ:* Mass Inst Technol, BS, 60; Wash Univ, MA, 66; Univ Ky, PhD(math), 71. *Prof Exp:* From asst prof to assoc prof, 71-76, PROF MATH, STATE UNIV NY COL BROCKPORT, 76- *Concurrent Pos:* Fel, Int Res Exchange Bd, Poland & Romania, 73-74; Nat Acad Sci exchange scientist, Romania, 74; Fulbright Advan Res Award, Romania, 76; Sigma Xi fac award, 76; exchange scientist, Romania, 77, 79 & res exchange scholar, State Univ NY, 81; fel, Int Res Exchange Bd, Poland, 78, Romania, 80, 85, 88, & 91 Nat Acad Sci, Bulgaria, 81, Romania, 82, Int Res Exchange Bd, Bulgaria, 83, Comt Scholarly Commun, People's Repub China, 84, Fulbright Advan Res Award, Romania, 89; vis prof, Univ Md, 79-80; res assoc, Univ Calif, Berkeley, 86-87. *Mem:* Am Math Soc; Math Asn Am. *Res:* Theory of functions of a complex variable, univalent function theory and differential inequalities in the complex plane. *Mailing Add:* Dept Math State Univ NY Col Brockport NY 14420

MILLER, SCOTT CANNON, b Salt Lake City, Utah, July 3, 47; m 71; c 1. CELL BIOLOGY, ANATOMY. *Educ:* Univ Utah, BS, 67, PhD(anat), 70. *Prof Exp:* Res asst prof, 77-80, RES ASSOC PROF RADIOBIOL, UNIV UTAH, 80- *Concurrent Pos:* Res fel, Harvard Univ, 74-77; mem, OBM study sect, NIH, 81-85. *Mem:* Am Soc Bone & Mineral Res; Radiation Res Soc; Endocrine Soc. *Res:* Mineral metabolism and endocrine effects on bone. *Mailing Add:* Div Radiobiol Univ Utah Salt Lake City UT 84112

MILLER, SHELBY A(LEXANDER), b Louisville, Ky, July 9, 14; m 39, 52; c 1. CHEMICAL ENGINEERING. *Educ:* Univ Louisville, BS, 35; Univ Minn, PhD(chem eng), 43. *Prof Exp:* Asst chemist, Corhart Refractories Co, Ky, 35; asst chem eng, Univ Minn, 35-39; chem engr, Eng Dept, E I du Pont de Nemours & Co, 40-46; from assoc prof to prof chem eng, Univ Kans, 46-55; prof, 55-69, chmn dept, Univ Rochester, 55-68; assoc lab dir, Argonne Nat Lab, 69-74, dir, Ctr Educ Affairs, 69-79, sr chem engr, Chem Eng Div, 79-84, resident sr engr, 84-90; RETIRED. *Concurrent Pos:* Fulbright lectr, King's Col, Durham, 52-53; vis prof, Univ Calif, Berkeley, 67-68; mem, Training Comt, Int Atomic Energy Agency, 75-79; vis lectr, Univ Philippines, 86. *Mem:* Fel AAAS; Am Chem Soc; Am Soc Eng Educ; fel Am Inst Chem Engrs; NY Acad Sci; fel Am Inst Chemists; Filtration Soc. *Res:* Agitation; gas dispersion; filtration; fluidized-bed combustion; coal technology. *Mailing Add:* 825 63rd St Downers Grove IL 60516-1962

MILLER, SHERWOOD ROBERT, b Lamont, Alta, Apr 16, 32; m 55; c 4. POMOLOGY. *Educ:* Univ Alta, BSc, 54, MSc, 56; Cornell Univ, PhD(pomol), 65. *Prof Exp:* RES SCIENTIST HORT, CAN DEPT AGR, 56- *Concurrent Pos:* Supt, Smithfield Exp Farm, Agr Can. *Mem:* Am Soc Hort Sci; Can Soc Hort; Agr Inst Can. *Res:* Use of synthetic and endogenous growth regulators in apple production; spacing trials with emphasis on tree walls and high density plantings. *Mailing Add:* Can Dept Agr PO Box 340 Trenton ON K8V 5R5 Can

MILLER, SIDNEY ISRAEL, b Saskatoon, Sask, May 22, 23; nat US; m 50; c 3. PHYSICAL ORGANIC CHEMISTRY. *Educ:* Univ Man, BSc, 45, MSc, 46; Columbia Univ, PhD(chem), 51. *Prof Exp:* Instr chem, Univ Man, 46 & Univ Mich, 50-51; from instr to assoc prof, 51-64, prof, 64-89, EMER PROF CHEM, ILL INST TECHNOL, 89-; CONSULT, 89- *Concurrent Pos:* NSF sr fel, Univ Col, Univ London, 63-64; vis scientist, Argonne Nat Lab, 71-72 & Japan Soc Prom Sci, 73; vis fel, Latrobe Univ, Australia, 77-78. *Mem:* Am Chem Soc; Am Asn Univ Profs. *Res:* Solution kinetics; mechanisms; stereochemistry; acetylene chemistry; heterocyclics; coal; polymer chemistry. *Mailing Add:* Dept of Chem Ill Inst of Technol Chicago IL 60616

MILLER, SOL, b Akron, Ohio, June 3, 14; m 37; c 2. BIOLOGICAL SAFETY, MICROBIOLOGY. *Educ:* Akron Univ, BA, 36; Ohio State Univ, MSc, 39; Sussex Univ, PhD(microbiol), 75. *Prof Exp:* Bacteriologist, Ohio Dept Health, 39-42; chief chemist, Q O Ordnance Corp, 42-43; res microbiologist, Children's Fund Mich, 44-54; bacteriologist, James Labs, 54-55; group leader & microbiologist, IIT Res Inst, 55-72; staff mem, Corp biohazards Control, Abbott Labs, 72-84; CONSULT BIOSCI, SAFETY SYSTS & SERVS, INC, 84- *Concurrent Pos:* Adj asst prof, Chicago Med Sch, 73-84; training chmn, Res & Develop Sect, Nat Safety Coun, 74-84. *Honors & Awards:* Cameron Award Medalist, 77, 78, 79 & 80; Tanner Shaughnessy Merit Award, 81. *Mem:* Am Soc Microbiol; Am Chem Soc; fel Am Inst Chemists; fel Am Acad Microbiol; Am Soc Indust Microbiol; Am Asn Clin Chemists; Sigma Xi. *Res:* Child nutrition; infectious aerosols; effects of atmospheric pollutants on survival of microorganisms; molecular biology and genetic engineering; industrial hygiene and biological safety. *Mailing Add:* 315 Wayne Pl Apt 401 Oakland CA 94606

MILLER, STANLEY CUSTER, JR, theoretical physics, for more information see previous edition

MILLER, STANLEY FRANK, b Idaho Falls, Idaho, Oct 13, 35; m 60; c 2. RESOURCE ECONOMICS, PRODUCTION ECONOMICS. *Educ:* Brigham Young Univ, BSc, 60; Utah State Univ, MSC, 62; Ore State Univ, PhD(agr econ), 64. *Prof Exp:* Economist, Econ Res Serv, USDA, 61-66 & 68-70; economist, IRI Res Inst Inc, Brazil, 66-68, Venezuela, 70-73; assoc prof agr econ, 73-83, DIR, INT PLANT PROTECTION CTR, ORE STATE UNIV, 73-, PROF AGR ECON DEPT AGR & RESOURCE ECON, 83- *Concurrent Pos:* Dep exec dir, Consortium Int Crop Protection, 85-; dir, Off Int Agr, Ore State Univ, 75-82; consult, Consortium Int Develop, 80-81. *Mem:* Am Agr Econ Asn. *Res:* Pest and pesticide management; soil erosion; international agricultural development; water resource management. *Mailing Add:* Int Plant Protection Ctr Ore State Univ Corvallis OR 97331

MILLER, STANLEY LLOYD, b Oakland, Calif, Mar 7, 30. CHEMISTRY. *Educ:* Univ Calif, BS, 51; Univ Chicago, PhD(chem), 54. *Prof Exp:* Jewett fel chem, Calif Inst Technol, 54-55; instr biochem, Col Physicians & Surgeons, Columbia Univ, 55-58, asst prof, 58-60; from asst prof to assoc prof, 60-68, PROF CHEM, UNIV CALIF, SAN DIEGO, 68- *Honors & Awards:* Oparin Medal, Int Soc Study Origin of Life. *Mem:* Nat Acad Sci; AAAS; Am Chem Soc; Am Soc Biol Chemists. *Res:* Origin of life; natural occurrence of clathrates hydrates; general anesthesia mechanisms. *Mailing Add:* Dept Chem Univ of Calif at San Diego La Jolla CA 92093-0317

MILLER, STEPHEN DOUGLAS, b Greeley, Colo, Mar 27, 46; m 69; c 1. WEED SCIENCE. *Educ:* Colo State Univ, BS, 68; NDak State Univ, MS, 70, PhD(agron), 73. *Prof Exp:* Asst agron, 73-75, asst prof, 75-80, ASSOC PROF AGRON, NDAK STATE UNIV, 80- *Mem:* Weed Sci Soc Am; Agron Soc Am; Crop Sci Soc Am. *Res:* Biology and control of wild oats in field crops; effect of reduced tillage systems on crop yield and crop pests. *Mailing Add:* Plant Sci Div Univ Wyoming Laramie WY 82071

MILLER, STEPHEN DOUGLAS, b Harrisburg, Pa, Jan 22, 48; m 69; c 2. NEUROIMMUNOLOGY, AUTOIMMUNITY. *Educ:* Pa State Univ, BS, 69, MS, 73, PhD(immunol), 75. *Prof Exp:* Postdoctoral fel cellular immunol, Med Sch, Univ Colo, 75-78, from instr to asst prof microbiol-immunol, 78-81; asst prof, 81-85, ASSOC PROF MICROBIOL-IMMUNOL, MED SCH, NORTHWESTERN UNIV, 85- *Concurrent Pos:* Site visitor, Nat Inst Neurol Dis & Stroke, NIH, 88-; NIH Study Sect, Ad Hoc-Immunol Sci Study Sect, 91- *Mem:* Am Asn Immunologists; Int Soc Neuroimmunol; Sigma Xi; AAAS; Am Soc Microbiol. *Res:* Cellular and molecular mechanisms of the immunopathogenesis and immunoregulation of two experimental immune-mediated demyelinating diseases-experimental autoimmune encephalomyelitis and Theiler's virus induced demyelinating disease - which serve as models for the human disease multiple sclerosis. *Mailing Add:* Dept Microbiol-Immunol Med Sch Northwestern Univ 303 E Chicago Ave Chicago IL 60611

MILLER, STEPHEN HERSCHEL, b New York, NY, Jan 12, 41; m 66; c 2. MEDICINE. *Educ:* Univ Calif, Los Angeles, BS, 60, MD, 64. *Prof Exp:* Fel plastic surg, African Med Res Found, 71-72, head & neck surg, 72; asst prof surg & exec head plastic surg, Univ Calif, San Francisco, 73-74; from assoc prof to prof surg, Hershey Med Ctr, Pa State Univ, 74-79, assoc chief plastc surg, 74-79, assoc mem grad fac, 75-79; PROF SURG & CHIEF PLASTIC SURG DIV ORE HEALTH SCI UNIV, 79-; SECT HEAD PLASTIC SURG, VET ADMIN HOSP, 79- *Concurrent Pos:* Chmn written exam, Am Bd Plastic Surg, 85-; chmn, Plastic Surg Res Coun, 84-85; assoc ed, Plastic & Reconstructive Surg, 83-, Yearbk Plastic Surg, 79-; res grant, Am Soc Surg of the Hand, 74, Am Soc Plastic & Reconstructive Surgeons, 76, Ore Health Sci Univ, 80. *Mem:* Am Soc Plastic & Reconstructive Surgeons; Am Soc Plastic & Reconstructive Surgeons (vpres 85-86, treas, 82-85); Am Col Surgeons; Am Asn Plastic Surgeons; Am Soc Aesthetic Plastic Surgeons; Am Col Surgeons; Asn Acad Chmn Plastic Surg (secy & treas, 85-88). *Res:* Effects of trauma on microculation and its alteration by physiological and pharmacological means. *Mailing Add:* Dept Plastic Surg Ore Health Sci Univ 3181 SW Sam Jackson Pk Rd Portland OR 97201

MILLER, STEVE P F, b Albany, NY, Aug 4, 56; m 84; c 1. SYNTHETIC ORGANIC & NATURAL PRODUCTS CHEMISTRY. *Educ:* State Univ NY, BS, 78; Univ Wis-Madison, PhD(chem), 83. *Prof Exp:* Postdoctoral res chemist, Pharmaceut Chem Dept, Univ Wis-Madison, 83-84; res chemist, Nat Heart, Lung & Blood Inst, 84-86, RES CHEMIST, BIOCHEM, NAT INST NEUROL DIS & STROKE, NIH, 86- *Honors & Awards:* Bausch & Lamb Sci Award, 74. *Mem:* Am Chem Soc; Sigma Xi. *Res:* Synthetic organic chemistry geared towards compounds of use in biochemical and biomedical research; targets include enzyme inhibitors and fluorescent substrates for enzymes involved in biosynthesis or catabolism of complex lipids. *Mailing Add:* Bldg 10 Rm 3D-11 NIH Bethesda MD 20892

MILLER, STEVEN RALPH, b Cleveland, Ohio, Feb 26, 36; m 58; c 2. PHYSICAL CHEMISTRY. *Educ:* Case Western Reserve Univ, BS, 58; Mass Inst Technol, PhD(phys chem), 62. *Prof Exp:* Res assoc, Mass Inst Technol, 62; asst prof chem, 62-68, ASSOC PROF CHEM, OAKLAND UNIV, 68- *Mem:* AAAS; Am Chem Soc. *Res:* Nuclear magnetic resonance relaxation phenomena in ferroelectric solids; gas phase reaction kinetics; oxidation of sulfur dioxide; aerosol formation. *Mailing Add:* Dept Chem Oakland Univ Rochester MI 48309

MILLER, STEWART E(DWARD), communications; deceased, see previous edition for last biography

MILLER, SUE ANN, b 1947; m 75; c 2. BIOLOGY. *Educ:* Univ Colo, Boulder, BA, 69, MA, 71, PhD(develop biol), 73. *Prof Exp:* Asst prof cell & develop biol, Dept Biol, Oberlin Col, 73-74; res fel, Dept Anat, Harvard Med Sch, 74-75; asst prof embryol & anat, Kirkland Col, 75-78; asst prof, 78-82, ASSOC PROF EMBRYOL & ANAT, HAMILTON COL, 82- *Mem:* Soc Develop Biol; Am Soc Cell Biol; Am Asn Anatomists; Am Soc Zoologists; Sigma Xi; AAAS. *Res:* Morphogenesis; role of differential growth in morphogenesis of large-scale folds and tubes in avian and mammalian embryos. *Mailing Add:* Dept Biol Hamilton Col Clinton NY 13323

MILLER, SUSAN MARY, b St Louis, Mo, Dec 21, 55; m 79; c 1. ENZYME MECHANISMS. *Educ:* Univ Mo, Columbia, BS, 78; Univ Calif, Berkeley, PhD(org chem), 83. *Prof Exp:* res scholar, 83-87, lectr, Dept Biol Chem, 87-89, ASST PROF, DEPT BIOL CHEM, UNIV MICH, ANN ARBOR, 89- *Mem:* Am Chem Soc; AAAS; Sigma Xi. *Res:* Biological and bio-organic chemistry; enzyme kinetics and mechanism; isotope effects; stopped-flow spectroscopy. *Mailing Add:* 3635 Waldenwood Dr Ann Arbor MI 48105-3042

MILLER, TERRY ALAN, b Girard, Kans, Dec 18, 43; m 66; c 2. CHEMICAL PHYSICS. *Educ:* Univ Kans, BA, 65; Cambridge Univ, PhD(chem), 68. *Prof Exp:* Staff mem & supvr, AT & T Bell Labs,68-84; EMINENT SCHOLAR PROF CHEM, OHIO STATE UNIV, 84- *Concurrent Pos:* Vis prof, Princeton Univ, 68-71 & Stanford Univ, 72; chmn spectros sect, Optical Soc Am, 83-85; vis foreign scholar, Inst Molecular Sci, Okazahi, Japan, 83; counr, Am Chem Soc. *Mem:* Am Chem Soc; fel Optical Soc Am; fel Am Phys Soc. *Res:* Laser spectroscopy of transient molecular species, free radicals,ions, molecular clusters and excited states; chemical reactions and kinetics of atoms and molecules. *Mailing Add:* Dept Chem Ohio State Univ Columbus OH 43210

MILLER, TERRY LEE, b Aberdeen, SDak, Dec 14, 40; m 64; c 2. BIOCHEMISTRY, BIOCHEMICAL TOXICOLOGY. *Educ:* San Diego State Col, AB, 64, MS, 65; Ore State Univ, PhD(biochem), 69. *Prof Exp:* NIH fel, Univ Colo, 68-70; asst prof biochem, 70-72, res assoc, Environ Health Sci Ctr, 73-75, asst prof, 75-81, ASSOC PROF, DEPT AGR CHEM, ORE STATE UNIV, 81- *Mem:* Am Chem Soc; Soc Toxicol. *Res:* Environmental toxicology; environmental chemistry; mechanisms of action of membrane active compounds; effects of selected environmental toxicants on biological and model membranes. *Mailing Add:* Dept of Agr Chem Ore State Univ Corvallis OR 97331

MILLER, TERRY LYNN, b Fulton, Ky, July 9, 45. MICROBIOLOGY. *Educ:* Univ Ky, BS, 67; NC State Univ, MS, 69; Univ Ill, Urbana, PhD(microbiol), 73. *Prof Exp:* Res assoc, Univ Ill, Urbana, 73-74; from res scientist I to III, 74-83, RES SCIENTIST IV, WADSWORTH CTR LABS & RES, NY STATE DEPT HEALTH, 83-; ASSOC PROF, SCH PUB HEALTH, DEPT ENVIRON HEALTH & TOXICOL, STATE UNIV NY, ALBANY & NY STATE HEALTH DEPT, 86- *Concurrent Pos:* Mem, Taxon Subcomt on Methanogenic Bacteria, Int Union Microbiol Soc, 85- *Mem:* Am Soc Microbiol; AAAS; NY Acad Sci; Sigma Xi. *Res:* Physiology, biochemistry and ecology of microorganisms emphasizing anaerobic ecosystems (intestinal tract, waste digestion) including studies of methanogenesis. *Mailing Add:* NY State Dept Health Wadsworth Ctr Labs & Res Albany NY 12201-0509

MILLER, THEODORE CHARLES, b Troy, NY, July 23, 33; m 59; c 3. PHARMACEUTICAL CHEMISTRY, PATENT LAW. *Educ:* Princeton Univ, AB, 55; Univ Ill, PhD(chem), 59. *Prof Exp:* Res chemist, 59-69, patent agent trainee, 69-70, PATENT AGENT, STERLING-WINTHROP RES INST, 70- *Mem:* Am Chem Soc. *Res:* Free radical rearrangements; synthesis of steroid hormones and heterocyclic compounds. *Mailing Add:* Patent Dept Sterling Drug Inc 81 Columbia Turnpike Rensselaer NY 12144-3491

MILLER, THEODORE LEE, b Crab Orchard, WVa, May 25, 40; div; c 2. PHYSICAL CHEMISTRY. *Educ:* Concord Col, BS, 66; Marshall Univ, MS, 70; Univ Cincinnati, PhD(chem), 74. *Prof Exp:* Instr chem, Univ Va, 74-75; asst prof chem, King's Col, Pa, 75-77; asst prof, 77-83, ASSOC PROF CHEM, OHIO WESLEYAN UNIV, 83- *Mem:* Am Chem Soc; Sigma Xi. *Res:* Luminescence spectroscopy; metal ions in biological systems; crocetin chemistry (an atherosclerosis drug); on-line computerized instrumentation. *Mailing Add:* Dept Chem Ohio Wesleyan Univ Delaware OH 43015

MILLER, THOMAS, b Asheville, NC, Apr 17, 32; m 54; c 1. FORESTRY. *Educ:* NC State Univ, BS, 62, MS, 64, PhD(plant path), 72. *Prof Exp:* PLANT PATHOLOGIST, SOUTHEASTERN FOREST EXP STA, US FOREST SERV, 64- *Concurrent Pos:* Adj prof forest pathol, Univ Fla. *Mem:* Am Phytopath Soc; Soc Am Foresters. *Res:* Mechanisms of resistance in southern pines to Cronartium Quercum fsp fusiforme; fungus diseases of pine strobili, cones and seed; mycology-rust fungi; control of tree diseases through silviculture; integrated forest; pest management. *Mailing Add:* SE Forest Exp Sta Sch Forest Resources & Conserv Univ Fla Gainesville FL 32611

MILLER, THOMAS ALBERT, b Sharon, Pa, Jan 5, 40; m 65; c 2. ENTOMOLOGY. *Educ:* Univ Calif, Riverside, BA, 62, PhD(entom), 67. *Prof Exp:* USPHS fel & res assoc insect physiol, Univ Ill, Urbana, 67-68; NATO fel insect physiol, Glasgow Univ, 68-69; from asst prof & asst entomologist to assoc prof & assoc entomologist, 69-76, PROF ENTOM & ENTOMOLOGIST, UNIV CALIF, RIVERSIDE, 76- *Mem:* Am Chem Soc; Am Soc Zoologists; Brit Soc Exp Biol. *Res:* Insect neurophysiology; insect toxicology; mode of action of insecticides; insect cardiac physiology. *Mailing Add:* Entom Dept Univ Calif Riverside CA 92521

MILLER, THOMAS GORE, b Greenfield, Ohio, Nov 3, 24; m 53; c 3. ORGANIC CHEMISTRY. *Educ:* Miami Univ, AB, 48; Univ Ill, MS, 49, PhD(chem), 51. *Prof Exp:* Res chemist, E I du Pont de Nemours & Co, 51-57; from asst prof to assoc prof, 57-69, head dept, 69-79, PROF CHEM, LAFAYETTE COL, 69- *Concurrent Pos:* NSF sci fac fel & vis res fel, Princeton Univ, 70-71. *Mem:* Am Chem Soc; Sigma Xi. *Res:* Molecular rearrangements; clathrate compounds; general organic chemistry. *Mailing Add:* Dept of Chem Lafayette Col Easton PA 18042

MILLER, THOMAS LEE, b Elkhart, Ind, Nov 24, 35; m 62; c 3. BIOCHEMISTRY, MICROBIOLOGY. *Educ:* Ind State Univ, AB, 61; Univ Wis, MS, 64, PhD(biochem), 66. *Prof Exp:* Res asst biochem, Univ Wis, 61-66; res assoc microbiol, 66-67, head microbiol sect, 67-70, res mgr fermentation microbiol, 70-83, ASSOC DIR FERMENTATION OPERS, UPJOHN CO, 83- *Mem:* Am Chem Soc; Am Soc Microbiol; Soc Indust Microbiol. *Res:* Hydrocarbon fermentations; steroid bioconversions; microbiological processes; measurement and control of fermentation variables; antibiotic fermentations. *Mailing Add:* 7599 Orchard Hill Ave Kalamazoo MI 49002

MILLER, THOMAS MARSHALL, b Ft Worth, Tex, Aug 20, 40; m 85; c 1. ATOMIC PHYSICS. *Educ:* Ga Inst Technol, BS, 62, MS, 64, PhD(physics), 68. *Prof Exp:* Asst prof physics, NY Univ, 68-74; physicist, Stanford Res Inst, 74-78; PROF, UNIV OKLA, 78- *Concurrent Pos:* vis fel, Univ Birmingham, 77 & Univ Colo, 84-85. *Mem:* Am Phys Soc. *Res:* Transport properties of low energy ions and electrons in gases; ion-molecule reactions; interactions of low energy electrons with thermal atom beams; low-energy atom-atom collisions; atomic polarizabilities; photodissociation; photodetachment. *Mailing Add:* Dept Physics & Astron Univ Okla Norman OK 73019

MILLER, THOMAS WILLIAM, b Providence, RI, June 12, 29; m 52; c 3. NATURAL PRODUCTS CHEMISTRY. *Educ:* Univ RI, BS, 50. *Prof Exp:* Res chemist, Merck Sharp & Dohme Res Labs, 51-61, sr res chemist, 61-69, res fel, 69-70, asst dir, 70-74, sr res fel, 74- 81, sr investr, 81-89; RETIRED. *Mem:* Am Chem Soc. *Res:* Natural products; isolation of antibiotics, vitamins and other fermentation products. *Mailing Add:* 16 Vermont Ave Carteret NJ 07008

MILLER, TRACY BERTRAM, b Syracuse, NY, Oct 19, 27; m 54; c 3. PHARMACOLOGY. *Educ:* Cornell Univ, AB, 48; Univ Buffalo, MA, 53, PhD, 59. *Prof Exp:* Instr pharmacol, Univ Buffalo, 53-54; res assoc physiol, Harvard Med Sch, 64-66; assoc prof pharmacol & surg res, State Univ NY Upstate Med Ctr, 66-71; PROF PHARMACOL & PHYSIOL, MED SCH, UNIV MASS, 71- *Concurrent Pos:* Am Heart Asn res fel, Col Med, State Univ NY Upstate Med Ctr, 59-64, Am Heart Asn estab investr, 62-67. *Res:* Renal pharmacology; cerebro-spinal fluid physiology. *Mailing Add:* Dept Pharmacol Univ Mass Med Sch 55 Lake Ave N Worcester MA 01655

MILLER, VICTOR CHARLES, geomorphology; deceased, see previous edition for last biography

MILLER, W(ENDELL) E(ARL), b Cissna Park, Ill, June 23, 13; m 41; c 2. ELECTRICAL ENGINEERING. *Educ:* Univ Ill, BS, 36, MS, 46. *Prof Exp:* Trial installation engr, Bell Tel Labs, Inc, NY, 36-39; power sales engr, Northern Ind Pub Serv Co, 39-41; from instr to prof elec eng, 41-70, asst dean eng, 50-55, from asst head dept to assoc head dept, 55-70, EMER PROF ELEC ENG, UNIV ILL, URBANA, 71- *Concurrent Pos:* Dir environ affairs, Ill Power Co, 70-78. *Mem:* Am Soc Eng Educ; Nat Soc Prof Engrs. *Res:* Industrial research and development. *Mailing Add:* 311 Floral Park Savoy IL 61874

MILLER, WADE ELLIOTT, II, b Los Angeles, Calif, Oct 20, 32; m 60; c 3. VERTEBRATE PALEONTOLOGY, GEOLOGY. *Educ:* Brigham Young Univ, BS, 60; Univ Ariz, MS, 63; Univ Calif, Berkeley, PhD(paleont), 68. *Prof Exp:* Instr geol & phys sci, Santa Ana Col, 61-64; instr geol, Fullerton Jr Col, 68-71; ASSOC PROF GEOL, BRIGHAM YOUNG UNIV, 71- *Concurrent Pos:* Geologist & paleontologist, Los Angeles County Mus, 69- *Mem:* Soc Vert Paleont; Paleont Soc; Soc Mammal. *Res:* fossil vertebrates of Utah, especially mammals, late Cenozoic vertebrates of Mexico; jurassic dinosaurs. *Mailing Add:* Dept Zool Brigham Young Univ ESC 258 Provo UT 84602

MILLER, WALTER CHARLES, experimental nuclear physics; deceased, see previous edition for last biography

MILLER, WALTER E, b New York, NY, Jan 28, 14; m 43; c 3. PHYSICAL CHEMISTRY. *Educ:* City Col New York, BS, 35, ChE, 36; NY Univ, PhD(chem), 41. *Prof Exp:* From instr to assoc prof chem, City Col NY, 41-66, prof, 66-; RETIRED. *Concurrent Pos:* Asst instr, NY Univ, 41-42; mem, US Army Chem Corps Adv Coun, 64-69. *Mem:* AAAS; Am Chem Soc; NY Acad Sci. *Res:* Cryogenics; rocket fuels; photo-sensitization; atomic physics; design of radio transmitters and receivers; high polymers; ion exchange; water demineralization; electrolytic treatment of water. *Mailing Add:* 25 Bonnie Ave Bel Air MD 21014-3293

MILLER, WALTER PETER, b Dickinson, NDak, Jan 7, 32; m 60; c 7. ORGANIC POLYMER CHEMISTRY. *Educ:* Univ Minn, BA, 53, PhD, 57. *Prof Exp:* Fel, Max Planck Inst Coal Res, Ger, 57-58; proj chemist, Technol Ucar Emulsion Systs, Union Carbide Chem Co, 58-64, group leader, 65-75, develop assoc, 76-77, from assoc dir to dir, 77-87, MGR HEALTH & PROD SAFETY, UNION CARBIDE, 87- *Concurrent Pos:* Fulbright travel grant, 57-58. *Mem:* Am Chem Soc. *Res:* Emulsion polymerization; organic chemistry. *Mailing Add:* 4504 Yates Pond Rd Raleigh NC 27606

MILLER, WARREN FLETCHER, JR, b Chicago, Ill, Mar 17, 43; m 69; c 2. NUCLEAR ENGINEERING. *Educ:* US Mil Acad, BS, 64; Northwestern Univ, MS, 70, PhD(nuclear eng), 73. *Prof Exp:* Asst prof, Northwestern Univ, 72-74; staff mem, Los Alamos Sci Lab, Los Alamos Nat Labs, 74-75, sect leader, 75, group leader, transport & reactory theory, 75-78, dep assoc dir, 80-81, assoc dir, 81-85 & 88-90, dep dir, 86-88; PROF, UNIV CALIF, 90- *Concurrent Pos:* Consult, Sargent & Lundy Engrs & Argonne Nat Lab, 73-74; NSF res grant, 73-74; vis prof, Howard Univ, 79-80; Goebbel vis prof, Univ Mich, 85. *Mem:* Nat Tech Asn; fel Am Nuclear Soc; Am Phys Soc; AAAS. *Res:* Neutral and charged particle transport theory; numerical analysis; nuclear reactor physics; radiation shielding. *Mailing Add:* Five Erie Lane Los Alamos NM 87544

MILLER, WARREN JAMES, b New Kensington, Pa, Oct 12, 31; m 58; c 3. PHYSICAL CHEMISTRY. *Educ:* Pa State Univ, BS, 57; Fla State Univ, PhD(chem), 62. *Prof Exp:* Res chemist, 62-64, sr res chemist, 64-69, RES ASSOC, RES LABS, EASTMAN KODAK CO, 69- *Mem:* Am Chem Soc. *Res:* Photographic theory; colloid and surface chemistry. *Mailing Add:* 76 Everwild Lane Rochester NY 14616-2056

MILLER, WARREN VICTOR, b Rochester, NY, Oct 4, 44; m 67; c 3. INORGANIC CHEMISTRY, ANALYTICAL CHEMISTRY. *Educ:* Clarkson Col Technol, BS, 66; State Univ NY Binghamton, PhD(chem), 70. *Prof Exp:* Res assoc inorg chem, Clarkson Col Technol, 70-72; chief chemist, Alpha Analysis Labs, 72-74; vpres, Spex Industs Inc, 74-; AT ML LABS. *Mem:* Am Chem Soc. *Res:* Inorganic synthesis; purification of elements and their compounds; platinum group metal chemistry. *Mailing Add:* ML Labs PO Box 370 Three Bridges NJ 08887-0370

MILLER, WATKINS WILFORD, b Hawthorne, Calif, Feb 21, 47; m 69; c 2. SOIL FERTILITY, WATER POLLUTION. *Educ:* Calif Polytech State Univ, BS, 68; Univ Calif, PhD, 73. *Prof Exp:* Agr chemist, Soil Testing Serv, Nelson Labs, Stockton, Calif, 69-70; res asst soil water repellency, Univ Calif, Riverside, 70-73; exten specialist natural resource develop, 73-75, asst prof, 75-78, asst soil & water scientist, 77-81, ASSOC PLANT, SOIL & WATER SCIENTIST, UNIV NEV, RENO, 81- *Mem:* Am Soc Agron; Soil Sci Soc Am; AAAS. *Res:* Soil fertility analysis and calibration; water quality of irrigation return flows; rural development resource inventories; soil and water testing service for Nevada residents; water management, especially Humbolt River system. *Mailing Add:* 1085 Emerson Way Sparks NV 89431

MILLER, WAYNE L(EROY), b Salem, Ore, Dec 6, 24. NUCLEAR & CHEMICAL ENGINEERING. *Educ:* Ore State Univ, BS, 50; Univ Tulsa, MS, 54. *Prof Exp:* Chem engr, US Bur Mines, Ore, 50-52; instr & asst prof chem & petrol refinery eng, Univ Tulsa, 54-57; chem engr, Esso Res & Eng Co, NJ, 57-58; asst prof nuclear eng, 61-75, ASST PROF CHEM ENG, UNIV NEV, RENO, 75- *Mem:* Am Chem Soc; Am Inst Chem Engrs; Am Soc Eng Educ. *Res:* Radiation effects on chemical systems. *Mailing Add:* Dept Chem Eng Univ Nev Reno NV 89557

MILLER, WILBUR HOBART, b Boston, Mass, Feb 15, 15; m 41; c 3. ORGANIC CHEMISTRY, BIOCHEMISTRY. *Educ:* Univ NH, BS, 36, MS, 38; Columbia Univ, PhD(chem), 42. *Prof Exp:* From asst to instr chem, Univ NH, 36-38; asst, Columbia Univ, 38-39, statutory asst, 39-40, Univ fel, 40-41; res chemist, Stamford Res Lab, Am Cyanamid Co, 41-49, tech rep, Washington, DC, 49-53, dir custom sales, Lederle Labs, 53-54, dir indust appln, Fine Chem Div, 54-55, dir food indust develop, Farm & Home Div, 55-57; tech dir prod for agr, Cyanamid Int, 57-60; sr scientist, Dunlap & Assocs, Inc, Conn, 60-66; coordr new prod develop, Celanese Corp, NY, 66-67, mgr commercial res, 67-69, dir diversification develop, 69-78 & dir corp develop, 78-84. *Concurrent Pos:* Consult bus diversification, 84- *Honors & Awards:* Am Design Award, 48. *Mem:* Fel AAAS; fel Am Inst Chemists; Am Chem Soc; Inst Food Technologists; Soc Chem Indust Am (treas, 80-84). *Res:* Agricultural chemicals; animal health and food industry products; chemotherapy; enzyme, organic and general industrial chemistry. *Mailing Add:* 19 Crestview Ave Stamford CT 06907

MILLER, WILLARD, JR, b Ft Wayne, Ind, Sept 17, 37; m 65; c 2. APPLIED MATHEMATICS, MATHEMATICAL PHYSICS. *Educ:* Univ Chicago, SB, 58; Univ Calif, Berkeley, PhD(appl math), 63. *Prof Exp:* NSF fel, Courant Inst, NY Univ, 63-64, vis mem, 64-65; from asst prof to assoc prof, 65-72, head, Sch math, 78-86, PROF MATH, UNIV MINN, MINNEAPOLIS, 72-, ASSOC DIR, INST FOR MATH APPLN, 87- *Concurrent Pos:* Vis mem, Ctr Math Res, Univ Montreal, 73-74; assoc ed, J Math Physics, 73-75 & Applicable Anal, 78-90; managing ed, J Math Anal, 75-81; co-prin investr, Inst Math & Appln, 80- *Mem:* Am Math Soc; Soc Indust & Appl Math. *Res:* Applications of group theory to special functions; separation of variables, q-series. *Mailing Add:* Sch Math Univ Minn 127 Vincent Hall Minneapolis MN 55455-0463

MILLER, WILLIAM, b New York, NY, Sept 1, 22; m 57; c 3. EXPERIMENTAL SOLID STATE PHYSICS. *Educ:* City Col New York, BBA, 43; Univ Pa, PhD(physics), 48. *Prof Exp:* Physicist, Nat Bur Standards, 48-56; from asst prof to assoc prof, 57-67, PROF PHYSICS, CITY COL NEW YORK, 67- *Concurrent Pos:* Opers analyst, Opers Res Off, Johns Hopkins Univ, 52-53; sr physicist & consult, Am Mach & Foundry Co, 56- *Mem:* AAAS; Am Phys Soc; Am Asn Physics Teachers. *Res:* Noise theory; electromagnetic waves; x-rays; solid state. *Mailing Add:* Dept Physics City Col New York 138th St & Convent Ave New York NY 10031

MILLER, WILLIAM ANTON, b Cedar, Mich, Apr 16, 35; m 60; c 4. MATHEMATICS EDUCATION. *Educ:* Mich State Univ, BS, 56, MAT, 61; Univ Ill, Urbana, MA, 63; Univ Wis, Madison, PhD(math educ), 68. *Prof Exp:* Teacher, Sunfield Community Schs, 56-60, Oak Park Schs, 60-61 & Waverly Schs, 61-62; asst prof math educ, Wis State Univ, Whitewater, 65-67, assoc prof math, 67-68; assoc prof, 68-71, PROF MATH, CENT MICH UNIV, 71- *Mem:* Math Asn Am. *Res:* Learning theory as it relates to mathematics. *Mailing Add:* 3909 S Summerton Rd Mt Pleasant MI 48858-8953

MILLER, WILLIAM B, b Tipp City, Ohio, Mar 6, 17; m 39; c 3. MECHANICAL & AERONAUTICAL ENGINEERING. *Educ:* Purdue Univ, BSME, 39. *Prof Exp:* Jr engr, Bendix Aviation Corp, 39-40; jr engr, USAF, 40-42, asst engr, 42-43, assoc engr, 43-44, aeronaut engr, 44-46, sr engr, 46-48, chief test unit struct, Wright Air Develop Ctr, 48-55, asst chief struct, Systs Command, 55-60, chief struct div, Systs Eng Group, 60-76, tech dir airframe subsysts eng, Aeronaut Systs Div, 67-76; CONSULT ENG, 76- *Concurrent Pos:* Mem, NASA Res Adv Comt Aircraft Struct, 52-; mem struct & mat panel, Adv Group Aerospace Res & Develop, NATO, 55-69, dept chmn, 65-66. *Mem:* Assoc fel Am Inst Aeronaut & Astronaut. *Res:* Aerospace structures and their testing, design, requirements, loads and dynamics. *Mailing Add:* 521 Earnshaw Dr Dayton OH 45429

MILLER, WILLIAM BRUNNER, b Bethlehem, Pa, July 27, 23; m 48; c 4. MATHEMATICS. *Educ:* Lehigh Univ, BS, 47, MA, 55, PhD(math), 62. *Prof Exp:* Elec engr, Western Elec Co, Inc, 47-49; engr, Laros Textiles Co, 50-53; prof math, Moravian Col, 53-62; from assoc prof to prof math, 63-89, EMER PROF MATH, WORCESTER POLYTECH INST, 89- *Concurrent Pos:* Instr, Lehigh Univ, 59-60. *Mem:* Math Asn Am; Soc Indust & Appl Math. *Res:* Separation and oscillation theorems of linear differential equations; texts on differential equations. *Mailing Add:* Dept Math Worcester Polytech Inst Worcester MA 01609

MILLER, WILLIAM ELDON, b McAllen, Tex, July 13, 30; m 78; c 6. FOREST ENTOMOLOGY. *Educ:* La State Univ, BS, 50; Ohio State Univ, MS, 51, PhD(entom), 55; Mich State Univ, MS, 61. *Prof Exp:* WGer Govt fel, Univ Gottingen, 56-57; entomologist, US Forest Serv, 56-64, prin insect ecologist, 64-80, chief insect ecologist, 80-82. *Concurrent Pos:* Ed, Forest Sci, 71-75 & J Lepidop Soc, 85-88, Mem Lepidopterist Soc, 89-; adj prof entom, Univ Minn, Twin Cities, 76- *Mem:* Entom Soc Am; Lepidop Soc. *Res:* Population dynamics; reproductive biology; taxonomy of Microlepidoptera; causes of insect outbreaks. *Mailing Add:* Dept Entom Univ Minn St Paul MN 55108

MILLER, WILLIAM EUGENE, audiology; deceased, see previous edition for last biography

MILLER, WILLIAM FRANKLIN, b Stone Creek, Ohio, Jan 16, 20; m 42, 58, 80; c 8. MEDICINE. *Educ:* Wittenberg Univ, BA, 42; Case Western Reserve Univ, MD, 45; Am Bd Internal Med, dipl, 56. *Prof Exp:* Intern, City Hosp, Cleveland, Ohio, 45-46; resident med, neurol & radiol, Dayton Vet Admin Hosp, 46-48; resident med, Dallas Vet Admin Hosp, Tex, 48-51; clin instr, 51-53, from asst prof to assoc prof, 53-67, PROF MED, UNIV TEX SOUTHWESTERN MED CTR DALLAS, 67- *Concurrent Pos:* Dir cardio-respiratory lab, McKinney Vet Admin Hosp, Tex, 51-53; dir pulmonary div, Parkland Mem Hosp & Woodlawn Hosps, 53-67; dir, Pulmonary Div, Methodist Hosp, 67-81; consult, Surgeon Gen, Lackland Air Force Hosp, 59-72, & Brooke Army Hosp, San Antonio, 61-71; consult, US Surgeon Gen, Comt Health Aspects Tobacco, 64-65; Parkland Mem, Vet Admin, St Paul, Methodist & Presby Hosps, Dallas; consult Chronic Pulmonary Dis Sect, NIH, 65-87, task force respiratory dis, Nat Heart & Lung Inst, 71, comt clin training physicians assts, Sch Allied Health Professions, Univ Tex, 72-82; chmn med adv bd, Am Asn Inhalation Ther, 66-68; mem med adv bd, Cystic Fibrosis Found, 69-70; bd trustees, Am Respiratory Care Fedn, 79- *Honors & Awards:* J A Young Medal, Am Asn Respiratory Care, 79 & Forrest M Birid Award, 88; Award of Excellence in Pulmonary Rehab, Am Asn Cardiovasc & Pulmonary Rehab, 89. *Mem:* AAAS; Am Thoracic Soc; fel Am Col Chest Physicians; Am Fedn Clin Res; Sigma Xi; fel Am Col Physicians; Am Asn Respiratory Care; Am Col Qual Med; Am Col Physicians Execs. *Res:* Pulmonary function testing; physiology and therapy in chronic bronchitis, asthma and emphysema; author of 158 publications. *Mailing Add:* Dept Internal Med Univ Tex Southwestern Med Ctr 5323 Harry Hines Blvd Dallas TX 75235-9030

MILLER, WILLIAM FREDERICK, b Vincennes, Ind, Nov 19, 25; m 49; c 1. COMPUTER SCIENCE, ACADEMIC ADMINISTRATION. *Educ:* Purdue Univ, BS, 49, MS, 51, PhD(physics, math), 56. *Hon Degrees:* DSc, Purdue Univ, 72. *Prof Exp:* Assoc physicist, Argonne Nat Lab, 56-59, dir, Appl Math Div, 59-64; vpres & provost, 71-78, PROF COMPUT SCI, STANFORD UNIV, 65-, HERBERT HOOVER PROF PUB & PVT MGT, GRAD SCH BUS, 79- *Concurrent Pos:* Vis prof, Purdue Univ, 62-63; prof lectr, Univ Chicago, 62-64; mem math & comput sci res adv comt, Atomic Energy Comn & US deleg study group digital technol, Europ Nuclear Energy Technol, Europ Nuclear Energy Agency, 62-; consult, Argonne Nat Lab, 65-68 & Comput Usage Corp, 69-72; mem, Comput Sci & Eng Bd, Nat Acad Sci, 68-71; mem sci info coun, NSF, 71-72; dir, Boothe Invest Corp, 72-79, Varian Assocs Inc, 73-, Fireman's Fund Ins Co, 77-91, Ann Rev Inc, 77-91, Resolve Ctr Environ Conflict Resolution, 78-80, First Interstate Bancorp, First Interstate Bank Calif & Pac Gas & Elec Co; mem, educ adv bd, John Simon Guggenheim Mem Found, 76-78, BHP Int Adv Coun, Multifunction Polis, Int Adv Bd, Australia, Computer Sci & Telecommun Bd, Nat Res Coun, Nat Inst Sci & Technol Bd Assessment, Comt Profiting Innovation, Nat Acad Eng & US-China Rels Comt; pres & chief exec officer, SRI Int, 79-90. *Honors & Awards:* Frederic B Whitman Award. *Mem:* Nat Acad Eng; Am Phys Soc; Am Math Soc; Asn Comput Mach; fel Inst Elec & Electronic Engrs; fel AAAS; Am Acad Arts & Sci. *Res:* Computer science and applications; computational physics; nuclear scattering; computing machinery; university administration; computers in management; socio-economic models for planning; policy and planning for science and technology; international science and technology; economic restructuring of developing economies; management science. *Mailing Add:* Dept Pub & Pvt Mgt Stanford Univ Stanford CA 94305-5015

MILLER, WILLIAM HENRY, b Baltimore, Md, Aug 7, 26; m 57; c 3. NEUROSCIENCES. *Educ:* Haverford Col, BA, 49; Johns Hopkins Univ, MD, 54. *Prof Exp:* Intern med, Baltimore City Hosps, 54-55; res assoc, Rockefeller Inst, 55-58, asst prof, 58-64; assoc prof physiol & ophthalmol, 64-69, PROF OPHTHALMOL, PHYSIOL & VISUAL SCI, SCH MED, YALE UNIV, 69- *Res:* Molecular mechanisms of phototransduction. *Mailing Add:* Dept of Ophthal & Vis Sci Yale Univ Sch of Med 333 Cedar St New Haven CT 06510

MILLER, WILLIAM HUGHES, b Kosciusko, Miss, Mar 16, 41; m 66; c 2. THEORETICAL CHEMISTRY. *Educ:* Ga Inst Technol, BS, 63; Harvard Univ, AM, 64, PhD(chem physics), 67. *Prof Exp:* NATO fel, Univ Freiburg, 67-68; Soc Fels jr fel, Harvard Univ, 68-69; from asst prof to assoc prof, 69-74, PROF CHEM, UNIV CALIF, BERKELEY, 74-, CHMN, DEPT CHEM, 89-; PRIN INVESTR, CHEM SCI DIV, LAWRENCE BERKELEY LAB, 69- *Concurrent Pos:* Sloan Found fel, 70; Guggenheim Mem fel, 75-76; fel, Churchill Col, Cambridge Univ, 75-76; Miller prof, Miller Inst Sci, Univ Calif, Berkeley; Alexander von Humboldt Sr Scientist Award, 81. *Honors & Awards:* Int Acad Quantum Molecular Sci Ann Prize, Paris, 74; E O Lawrence Award, US Dept Energy, 85; Irving Langmuis Award Chem Physics, Am Chem Soc, 90. *Mem:* Nat Acad Sci; fel Am Phys Soc; Int Acad Quantum Molecular Sci; fel AAAS. *Res:* Quantum theory of chemical reactions; semiclassical theories and quantum effects in inelastic and reactive scattering of atoms and molecules; collisional transfer of electronic energy. *Mailing Add:* Dept of Chem Univ of Calif Berkeley CA 94720

MILLER, WILLIAM JACK, b Nathans Creek, NC, Feb 7, 27; m 50; c 4. ANIMAL NUTRITION. *Educ:* NC State Col, BS, 48, MS, 50; Univ Wis, PhD(animal nutrit), 52. *Prof Exp:* Wis Alumni Res Found asst, Univ Wis, 50-52; res assoc dairy physiol, Univ Ill, 52-53; from asst prof to prof, 53-64, ALUMNI FOUND DISTINGUISHED PROF DAIRY SCI, UNIV GA, 73- *Honors & Awards:* Nutrit Res Award, Am Feed Mfrs Asn, 63; Excellence in Res Award, Sigma Xi, 69; Borden Award, Am Dairy Sci Asn, 71; Gustav Bohstedt Mineral Res Award, Am Soc Animal Sci, 71; Morrison Award, Am Soc Animal Sci, 80. *Mem:* AAAS; Soc Environ Geochem & Health; Am Inst Nutrit; Am Dairy Sci Asn; fel Am Soc Animal Sci. *Res:* Zinc, manganese, cadmium, nickel and mercury nutrition and metabolism; mineral nutrition of animals; ruminant nutrition; forage evaluation and utilization. *Mailing Add:* Dept Animal & Dairy Sci Univ Ga Athens GA 30602

MILLER, WILLIAM KNIGHT, b Salisbury, NC, Nov 14, 18; m 46; c 2. ANALYTICAL CHEMISTRY. *Educ:* Catawba Col, AB, 40; Univ NC, PhD(anal chem), 50. *Prof Exp:* Chemist, Carbide & Carbon Chem Co, Oak Ridge Nat Lab, 50-52; analysis supvr, S C Johnson & Son, Inc, 52-80; RETIRED. *Mem:* Am Chem Soc; Soc Appl Spectros. *Res:* Infrared spectrophotometry; gas chromatography. *Mailing Add:* 2021 SE 37th Ct Circle Ocala FL 32671

MILLER, WILLIAM LAUBACH, b Charlottesville, Va, Apr 9, 43; m 68; c 2. BIOCHEMISTRY, REPRODUCTIVE ENDOCRINOLOGY. *Educ:* Bucknell Univ, BS, 65; Cornell Univ, M Nut Sci, 67, PhD(biochem), 70. *Prof Exp:* Res biochemist, Walter Reed Army Inst Res, 70-73; Rockefeller fel, Univ Wis-Madison, 73-76; from asst prof to assoc prof, 76-85, PROF BIOCHEM, NC STATE UNIV, 85- *Concurrent Pos:* Prin investr, NIH grant, 77-86, 88-; USDA grant, 83- *Mem:* AAAS; Am Chem Soc; Endocrine Soc. *Res:* Control of gonadotropin synthesis and release; steroid (especially estrogen) hormone action; peptide hormone action, pituitary culture, protein synthesis and RNA characterization. *Mailing Add:* Dept Biochem Box 7622 NC State Univ Raleigh NC 27695-7622

MILLER, WILLIAM LAWRENCE, b Medford, Ore, May 12, 37; m 61; c 2. ENVIRONMENTAL REGULATIONS, MINERALS RESEARCH. *Educ:* Univ Calif, Davis, BS, 63, MS, 67. *Prof Exp:* Assoc chemist nitroplasticizers develop, Aerojet Gen Corp, Gen Tire & Rubber Co, 63-64; asst dir, minerals & mats res, 82-88, CHIEF REGULATORY PROJS COORD, BUR MINES, US DEPT INTERIOR, 88- *Concurrent Pos:* Alt mem coord comt mat res & develop, Fed Coun Sci & Technol, 69. *Mem:* Am Chem Soc; Metall Soc; Mining & Metall Soc Am. *Res:* Emphasis given to regulatory issues affecting the domestic mining and minerals processing industries; attainment of improved information and analysis bases used by regulatory agencies and compliance through the development and use of improved technologies. *Mailing Add:* 2910 N First Rd Arlington VA 22201

MILLER, WILLIAM LLOYD, b Waterman, Ill, Mar 22, 35; m 57; c 4. AGRICULTURAL ECONOMICS. *Educ:* Univ Ill, Urbana, BS, 57, MS, 60; Mich State Univ, PhD(agr econ), 65. *Prof Exp:* from asst prof to assoc prof, Agr Econ, Purdue Univ, 65-76, prof, 76-81; DEPT HEAD, AGR ECON, UNIV NEBR, LINCOLN, 82- *Mem:* Am Agr Econ Asn; Am Econ Asn; Am Water Resources Asn. *Res:* Resource economics. *Mailing Add:* 6410 Monticello Dr Lincoln NE 68570

MILLER, WILLIAM LOUIS, b Springville, Utah, June 28, 25; m 52; c 3. BIOCHEMISTRY. *Educ:* Brigham Young Univ, BS, 48; Univ Wis, MS, 50, PhD(biochem), 52. *Prof Exp:* Asst biochem, Univ Wis, 48-52; res scientist pharmacol, Upjohn Co, 52-59, res assoc endocrinol, 59-68, sr res scientist, Dept Metab Dis Res, 68-77, res assoc, Dept Fertil Res, 77-88; RETIRED. *Mem:* Am Soc Pharmacol & Exp Therapeut; Soc Exp Biol & Med. *Res:* Fate studies on drugs, especially metabolic detoxification of drugs by mammals; skin sterols and azo-dye carcinogenesis; diabetes; hypoglycemic agents; biochemistry of hormone action; lipid metabolism; reproductive physiology; biotelemetry techniques; patent specialist. *Mailing Add:* 6457 Winddrift St Kalamazoo MI 49009

MILLER, WILLIAM RALPH, b Seneca, Pa, Nov 5, 17; m 45. MECHANICAL ENGINEERING. *Educ:* Cleveland State Univ, BME, 49; Case Inst Technol, MSME, 55; Pa State Univ, PhD(eng mech), 65. *Prof Exp:* Instr eng, Evansville Col, 49-50; engr, Trabon Eng Corp, Ohio, 51-54; asst prof mech eng, Case Inst Technol, 56-60; sr eng specialist, TRW, Inc, Ohio, 62-63; chmn dept, 64-70, prof, 64-85, EMER PROF MECH ENG, UNIV TOLEDO, 85- *Concurrent Pos:* Res engr, Thompson Prod, Inc, Ohio, 54-56. *Mem:* Am Soc Mech Engrs; Soc Exp Stress Analysis. *Res:* Metal fatigue. *Mailing Add:* Dept Mech Eng Univ Toledo Toledo OH 43606

MILLER, WILLIAM REYNOLDS, JR, b Philadelphia, Pa, Dec 29, 39; m 76; c 2. POLYMER CHEMISTRY, PHYSICAL CHEMISTRY. *Educ:* Princeton Univ, AB, 61; Columbia Univ, MA, 62, PhD(chem), 65. *Prof Exp:* Chemist & proj leader, Kopper's Co, Inc, 68-73; proj scientist plastics, Arco/Polymers, Inc, 74-78; res asst prof, Dept Metall & Mat Eng, Univ Pittsburgh, 78-79; lectr chem, Chatham Col, 79-80; sr chemist, Tremco, Inc, 80-83; sr res chemist, Protective Treatment, Inc, 83-84; instr chem, Univ Dayton, 85-86; sr chemist, Ohio Sealer, 85-86; CONSULT, 86- *Mem:* Am Chem Soc; Sigma Xi; Soc Plastics Eng. *Res:* Rheology, especially plastics solids and melts; polymer evaluation; dielectric properties of plastics; electron spin resonance spectroscopy of organic free radicals in solution; polymer synthesis and sealant development. *Mailing Add:* 249 Chatham Dr Dayton OH 45429-1407

MILLER, WILLIAM ROBERT, b Norwalk, Ohio, Mar 11, 43; m 64; c 3. GEOCHEMISTRY. *Educ:* Ohio State Univ, BCE, 68; Univ Wyo, MS, 72, PhD(geol), 74. *Prof Exp:* Prod engr, Mobil Oil Corp, 68-69; teaching asst geochem, Univ Wyo, 69-72; GEOLOGIST, BR GEOCHEM, US GEOL SURV, 74- *Mem:* Asn Explor Geochemists; Int Asn Geochem & Cosmochem. *Res:* Low temperature geochemistry, particularly reactions involving natural waters and solid phases; weathering and controls on the partitioning of trace elements between natural waters and solid phases; hydrogeochemical exploration; geochemical exploration. *Mailing Add:* US Geol Surv Mail Stop 973 Fed Ctr Bldg 25 Lakewood CO 80225

MILLER, WILLIAM ROBERT, JR, b Baltimore, Md, June 17, 34; m 58; c 3. ELECTRON PHYSICS, SOLID STATE PHYSICS. *Educ:* Gettysburg Col, BA, 56; Univ Del, MA, 61, PhD(physics), 65. *Prof Exp:* Engr, Westinghouse Elec Corp, 58-59, sr engr, 65-67; teaching asst, Univ Del, 59-61; engr, RCA Corp, 67-68; asst prof, 69-76, ASSOC PROF PHYSICS, PA STATE UNIV, 76- *Concurrent Pos:* Instr physics, York Col Pa, 69-79; guest scientist, Semi conductor Electronics Div, Nat Inst Standards & Technol, Gaithersburg, MD, 90-91. *Mem:* Am Phys Soc; Am Asn Physics Teachers; Math Asn Am. *Res:* Solid state; plasmonexcitation in solids; x-ray studies of semiconductor superlattices. *Mailing Add:* 1029 Preston Rd Lancaster PA 17601-4852

MILLER, WILLIAM ROBERT, b Arlington, Ala, Sept 1, 24; m 47; c 2. MICROBIOLOGY, FOOD HYGIENE. *Educ:* Auburn Univ, DVM, 50, MS, 63; Purdue Univ, PhD(virol), 68; Am Col Vet Prev Med, dipl, 73. *Prof Exp:* From instr to asst prof, Sch Vet Med, Auburn Univ, 60-68, assoc prof microbiol & path, 68-83; chief epidemiol br, 83-86, DIR RESIDUE EVAL & PLANNING DIV, FOOD SAFETY & INSPECTION SERV, USDA, 86- *Concurrent Pos:* Collabr, USDA, 61-65; Nat Inst Neurol Dis & Blindness spec fel, 65-67; consult virol, Kans State Univ-AID, India, 71; Am Vet Med Asn Public Health & Regulatory Vet Med, 78-84, Coun on Educ, 85- *Mem:* Am Vet Med Asn; Sigma Xi; Am Asn Food Hyg Vet; US Animal Health Asn; Nat Asn Fed Vet. *Res:* Pathogenesis of animal virus diseases, specifically respiratory and neurological diseases; epidemiology of food borue diseases. *Mailing Add:* 300 12th SW USDA Food Safety & Inspection Serv Washington DC 20250

MILLER, WILLIAM TAYLOR, b Winston-Salem, NC, Aug 24, 11; m 51. ORGANIC CHEMISTRY, FLUORINE CHEMISTRY. *Educ:* Duke Univ, AB, 32, PhD(org chem), 35. *Prof Exp:* Postdoctoral appt, Stanford Univ, 35-36; from instr to assoc prof, 36-47, prof, 47-77, EMER PROF CHEM, CORNELL UNIV, 77- *Concurrent Pos:* Off investr & consult, Nat Defense Res Comt, 41-43; mem fluorocarbon adv comt, Manhattan Proj, 43-45; head fluorocarbon res, SAM Labs, Manhattan Proj, Columbia Univ & Carbide & Carbon Chem Corp, 43-46; consult var govt & indust labs, 47-; consult, M W Kellogg Co, 47-57, E I du Pont de Nemours & Co, Inc, 57-77. *Honors & Awards:* Award for Creative Work in Fluorine Chem, Am Chem Soc, 74; Moissan Centenary Medal, France, 86. *Mem:* Fel AAAS; Am Chem Soc; Royal Soc Chem. *Res:* Chemistry of carbon-fluorine and related highly halogenated compounds; chemistry of haloorgano-metallic compounds; inventor of the chlorotrifluoroethylene polymers utilized in the first gaseous diffusion plant for the separation of uranium isotopes during world war II; discoverer of the facile reactivity of fluoride ion with fluoroolefins and of the unique importance of fluoride ion to carbon-fluorine chemistry. *Mailing Add:* Baker Lab Dept Chem Cornell Univ Ithaca NY 14853

MILLER, WILLIAM THEODORE, b Belleville, Ill, Feb 8, 25. ORGANIC CHEMISTRY, BIOLOGICAL CHEMISTRY. *Educ:* St Louis Univ, AB, 47, BS, 51; St Mary's Col, Kans, STL, 56; Univ Calif, Berkeley, PhD(chem), 61. *Prof Exp:* Teacher, Marquette Univ High Sch, 51-52; asst prof, 61-66, assoc prof chem, 66-75, chmn dept, 69-75, dir div natural sci & math, 75-77, PROF CHEM, REGIS COL, COLO,75-, CHMN DEPT, 82- *Concurrent Pos:* Res grants, NIH, 62-, Am Chem Soc Petrol Res Fund, 63- & NSF, 64; res fel, Lab Nuclear Med & Radiation Biol, Univ Calif, Los Angeles, 67-69; vis prof, Sogang Univ, Korea, 78-79 & 85-86. *Mem:* AAAS; Am Chem Soc; Sigma Xi. *Res:* Comparative study of lipids isolated from in vitro and in vivo grown tubercle bacillus; synthetic organic chemistry; polynuclear aromatic hydrocarbons. *Mailing Add:* Dept Chem Regis Col W 50th & Lowell Blvd Denver CO 80221

MILLER, WILLIAM WADD, III, b Starkville, Miss, Oct 4, 32; m 57; c 2. REPRODUCTIVE PHYSIOLOGY. *Educ:* Miss State Univ, BS, 54, MS, 58; Auburn Univ, PhD(reprod physiol), 62. *Prof Exp:* Assoc prof biol, Howard Col, 62-67; assoc prof, 67-81, PROF BIOL, NORTHEAST LA UNIV, 81- *Concurrent Pos:* Sigma Xi grant, 65-66. *Mem:* AAAS; Soc Study Reproduction; Am Soc Zoologists. *Res:* Physiology of reproduction; endocrinology. *Mailing Add:* Dept Biol Northeast La Univ 700 University Ave Monroe LA 71203

MILLER, WILLIAM WALTER, b Oakland, Calif, Sept 17, 41; c 2. BIOCHEMISTRY. *Educ:* Univ Calif, Berkeley, BS, 63; Calif Inst Technol, PhD(biochem), 67. *Prof Exp:* Res chemist, E I du Pont de Nemours & Co, 67-73; appln chemist, Beckman Instruments, Inc, 73-76, mgr advan develop, 76-83; DIR FLUORESCENT SYSTS, CARDIOVASC DEVICES, INC, 3M HEALTH CARE, 83- *Mem:* AAAS; Am Chem Soc; Am Asn Clin Chemists; NY Acad Sci. *Res:* Clinical chemistry methods development; fluorescent sensor development. *Mailing Add:* 9782 Sunderland St Santa Ana CA 92705

MILLER, WILLIAM WEAVER, b Winchester, Va, Sept 1, 33; m 58; c 2. PEDIATRICS, CARDIOLOGY. *Educ:* Va Mil Inst, BA, 54; Univ Pa, MD, 58; Am Bd Pediat, dipl, 66, cert pediat cardiol, 70. *Prof Exp:* Intern, Univ Pa Hosp, 59; intern, Children's Hosp Philadelphia, 62-63; instr pediat, Med Sch, Univ Pa, 63-65; instr, Harvard Med Sch, 65-66; assoc, Med Sch, Univ Pa, 66-69, asst prof, 69; assoc prof pediat, Univ Tex Health Sci Ctr, Dallas, & dir pediat cardiol div, Children's Med Ctr, 69-76; mem fac, Med Col Va, 76-; AT DEPT PEDIAT, VA COMMONWEALTH UNIV. *Concurrent Pos:* Fel pediat cardiol, Children's Hosp, Philadelphia, 64-65, res fel, 66-67; assoc cardiologist & assoc dir cardiovasc labs, 67-69; res fel pediat cardiol, Children's Hosp Med Ctr, Boston, 65-66. *Honors & Awards:* Am Acad Pediat Award, 69. *Mem:* Am Acad Pediat; Am Heart Asn. *Res:* Oxygen transport; myocardial chemistry; congenital heart disease. *Mailing Add:* 5855 Bremo Rd Suite 408 Richmond VA 23226

MILLER, WILLIE, b Bolivar, Tenn, Aug 24, 42; m 61; c 1. PLANT GROWTH REGULATOR. *Educ:* Tenn State Univ, BS, 71, MS, 72; Univ Tenn, PhD(plant & soil sci), 76. *Prof Exp:* Plant sci rep, 76-83, RES SCIENTIST, RES & DEVELOP, ELI LILLY & CO 83- *Concurrent Pos:* Res scientist, PowElanco, 89- *Mem:* Soil Sci Soc Am; Weed Sci Soc Am; Coun Agr Sci & Technol; Int Weed Sci Soc. *Res:* Research and development with agricultural chemicals incuding herbicides, insecticides, plant growth regulators and chemical hybridizing agents. *Mailing Add:* 14618 Patrick Circle Omaha NE 68164

MILLER, WILMER GLENN, b Mt Orab, Ohio, Aug 28, 32. POLYMER CHEMISTRY. *Educ:* Capital Univ, BS, 54; Univ Wis, PhD(chem), 58. *Prof Exp:* Res fel chem, Harvard Univ, 58-59 & Univ Minn, 59-60; from asst prof to assoc prof, Univ Iowa, 60-67; assoc prof, 67-70, PROF CHEM, UNIV MINN, MINNEAPOLIS, 70- *Concurrent Pos:* USA-USSR exchange scientist, US Pub Health Serv fel, Guggenheim fel, 64. *Mem:* Am Chem Soc; Sigma Xi. *Res:* Physical chemical studies of synthetic and biological polymers and surfactants; thermodynamics and dynamics of polymer liquid crystals; motion of polymers at or near an interface; surfactant microstructures. *Mailing Add:* Dept of Chem 207 Pleasant St SE Minneapolis MN 55455

MILLER, WILMER JAY, b Lawton, Okla, July 15, 25; m 52; c 2. GENETICS, IMMUNOLOGY. *Educ:* Univ Okla, BA, 48; Univ Wis, PhD(genetics, zool), 54. *Prof Exp:* Proj assoc immunogenetics, Univ Wis, 53-55; assoc specialist, Sch Vet Med, Univ Calif, 55-62, lectr, 56-57; assoc prof, 62-80, PROF GENETICS, IOWA STATE UNIV, 80- *Concurrent Pos:* Collaborating prof, Univ Estadua de Sao Paulo, Brazil. *Honors & Awards:* Willard F Hollander Award, 76. *Mem:* Fel AAAS; Am Inst Biol Sci; Genetics Soc Am; Am Genetic Asn. *Res:* Immunogenetics of birds and bovines. *Mailing Add:* 218 Parkridge Circle Ames IA 50010

MILLER-GRAZIANO, CAROL L, SURGERY. *Educ:* San Diego State Univ, BS, 66; Univ Utah, Salt Lake City, MS, 69, PhD(microbiol-exp path), 72. *Prof Exp:* Technician, Scripps Clin & Res Found, 66-67; res serologist, Epizool Div, Univ Utah, Salt Lake City, 67- 69, res assoc, Dept Path, 69-72; instr, Dept Bact-Immunol, Univ Calif, Berkeley, 72-75; asst prof surg, Univ Calif, San Francisco, 75-80, from asst prof to assoc prof micro-immunol-surg, 80-85; PROF SURG, MOLECULAR GENETICS & MICROBIOL, MED CTR, UNIV MASS, WORCESTER, 85-, CHMN, DIV SURG RES & DIR, GRAD PROG IMMUNOL, 85- *Concurrent Pos:* Mult Sclerosis Soc fel, 75, Pub Health Serv spec fel, 77; travel award, Transplantation Soc, 78; spec res award, Alisa Ruch Burn Found, 78 & 79; mem, Coun Midwinter Conf Immunol, 85-89 & Permanent Study Group Cancer Manpower, NIH, 90-94. *Mem:* Sigma Xi; Am Asn Immunologists; Asn Acad Surg; Transplantation Soc; Am Burn Asn; Leukocyte Biol Soc; Am Soc Microbiologists; Am Fedn Clin Res; Asn Women Sci; Int Burn Asn. *Mailing Add:* Div Surg Res Univ Mass Med Ctr 55 Lake Ave N Worcester MA 01655

MILLERO, FRANK JOSEPH, JR, b Greenville, Pa, Mar 16, 39; m 65; c 3. PHYSICAL CHEMISTRY. *Educ:* Ohio State Univ, BS, 61; Carnegie-Mellon Univ, MS, 64, PhD(phys chem), 65. *Prof Exp:* Asst chem, Carnegie-Mellon Univ, 61-63, asst thermochem, 63-65, fel, 65; phys chemist, Esso Res & Eng Co, 65-66; res scientist, 66-68, from asst prof to assoc prof, 68-73, PROF CHEM, OCEANOG & PHYS CHEM, ROSENSTIEL SCH MARINE & ATMOSPHERIC SCI, UNIV MIAMI, 73-, ASSOC DEAN, 87- *Concurrent Pos:* Mem oceanog panel, NSF, 73-75; mem, Sci Comt Oceanic Res Panel Oceanog Standards, UNESCO, 75; vis prof, Inst di Ricerca sulle Acque,

Rome, 79-80; mem ocean sci bd, Nat Acad Sci, 81-; vis prof, Pontif Catholic Univ, Rio de Janiero, 82, 83, Univ Gothenberg, Swed, 86. *Mem:* AAAS; Am Chem Soc; Am Geophys Union; Geochem Soc; Sigma Xi; Am Soc Limnol & Oceanog. *Res:* Solution kinetics and thermodynamics; electrolyte solutions; thermochemistry; chemical oceanography; physical chemistry of aqueous solutions including seawater. *Mailing Add:* Rosenstiel Sch Marine & Atmospheric Sci Univ Miami Miami FL 33149-1098

MILLER-STEVENS, LOUISE TERESA, b Trinidad, WI, Mar 2, 19; US citizen; m 79; c 1. BIOLOGY. *Educ:* Univ Mich, BA, 49; Syracuse Univ, MS, 51; Yale Univ, MPH, 60; Univ RI, PhD(virol), 70. *Prof Exp:* Teaching & res asst epidemiol, Syracuse Univ, 51-59; res assoc virol, Univ RI, 63-72; assoc prof biol, Spelman Col, 73-81. *Concurrent Pos:* Consult various pvt & pub orgn, 72-78; fel, Ctr Dis Control, 76-77. *Mem:* AAAS; Am Indust Hyg Asn; Am Soc Microbiol; Reticuloendothelial Soc; NY Acad Sci. *Res:* Immunobiology of cancer; viral serology and immunology; virus-cell interaction. *Mailing Add:* Dept Biol Box 186 Spelman Lane Atlanta GA 30314

MILLET, PETER J, b New York, NY, June 28, 40. NUCLEAR MAGNETIC RESONANCE. *Educ:* Rensselaer Polytech Inst, BS, 61; Syracuse Univ, MS, 63, PhD(physics), 69. *Prof Exp:* Asst prof, 68-74, assoc prof, 74-80, PROF PHYSICS, HAMILTON COL, 80- *Concurrent Pos:* Vis assoc prof, Cornell Univ, 80-81, fel, 81; assoc dean & assoc dean students, Hamilton Col, 82-85. *Mem:* Am Phys Soc; Am Asn Physics Teachers; Sigma Xi. *Res:* Theoretical physics including quantum mechanics of three body problem in Heisenberg model of a ferromagnet; experimental studies of luminescence in cadmium sulfide; nuclear magnetic resonance studies of magnetic coupling between liquid helium and susbstrate at low temperature. *Mailing Add:* Dept Physics Hamilton Col Clinton NY 13323

MILLETT, FRANCIS SPENCER, b Madison, Wis, Aug 2, 43; m 68; c 2. BIOCHEMISTRY. *Educ:* Univ Wis, BS, 65; Columbia Univ, PhD(chem physics), 70. *Prof Exp:* NIH fel biochem, Calif Inst Technol, 70-72; from asst prof to assoc prof, 72-80, PROF CHEM, UNIV ARK, FAYETTEVILLE, 80- *Mem:* Am Chem Soc (secy-treas, 73-75); Am Soc Biol Chem; Biophysical Soc. *Res:* Interaction of proteins with biological membranes utilizing nuclear magnetic resonance methods; function of cytochrome C in mitochondria utilizing nuclear magnetic resonance methods. *Mailing Add:* Dept Chem Univ Ark Fayetteville AR 72701

MILLETT, KENNETH CARY, b Hustiford, Wis, Nov 16, 41; m; c 2. MATHEMATICS & BIOMATHEMATICS, THEORETICAL PHYSICS. *Educ:* Mass Inst Technol, BS, 63; Univ Wis-Madison, MS, 64, PhD(math), 67. *Prof Exp:* Instr math, Mass Inst Technol, 67-69; from asst prof to assoc prof, 69-78, PROF MATH, UNIV CALIF, SANTA BARBARA, 78- *Concurrent Pos:* exec dir, Calif Coalition Math. *Honors & Awards:* C B Allendoerfer Award, 89; Chauvenet Prize, 91. *Mem:* Am Math Soc; Math Asn Am; Sigma Xi; AAAS. *Res:* Geometric and algebraic topology. *Mailing Add:* Dept of Math Univ of Calif Santa Barbara CA 93106

MILLETT, MERRILL ALBERT, b Lake Mills, Wis, Nov 17, 15; m 42; c 3. WOOD CHEMISTRY. *Educ:* Univ Wis, BA, 38, MA, 39, PhD(phys chem), 43. *Prof Exp:* Asst chemist, US Forest Prod Lab, 42-43, chemist, 43-77; RETIRED. *Mem:* Sigma Xi. *Res:* Modified woods; cellulose and wood chemistry; molecular properties of celluloses; chromatographic analysis of woods and pulps; cellulose and cellulose esters; characterization and chemical utilization of wood residues; kinetics of aging of wood and cellulose; wood and pulping residues as animal feedstuffs. *Mailing Add:* 322 N Hillside Terr Madison WI 53705

MILLETT, WALTER ELMER, b Hampton, Ill, July 26, 17; m 44. PHYSICS. *Educ:* Univ Fla, BS, 40, MS, 42; Harvard Univ, PhD(physics), 49. *Prof Exp:* Res assoc, Radiation Lab, Mass Inst Technol, 42-45; AEC fel, Calif Inst Technol, 49-50; from instr to asst prof, Univ Fla, 50-52; from asst prof to assoc prof, 52-61, PROF PHYSICS, UNIV TEX, AUSTIN, 61- *Mem:* Fel Am Phys Soc. *Res:* Positron decay; electron and ion optics; scintillation spectrometry. *Mailing Add:* 2301 Lawnmont Ave Austin TX 78706

MILLETTE, CLARKE FRANCIS, b Bridgeport, Conn, Oct 22, 47. REPRODUCTIVE BIOLOGY, IMMUNOLOGY. *Educ:* Johns Hopkins Univ, BS, 69; Rockefeller Univ, PhD(biochem), 75. *Prof Exp:* Fel physiol, 75-77, ASST PROF ANAT, HARVARD MED SCH, 77- *Mem:* Am Soc Cell Biol; Soc Develop Biol. *Res:* Biochemistry and immunology of gametes and gametogenesis. *Mailing Add:* Lab Human Reprod & Reprod Biol Harvard Med Sch 45 Shattuck St Boston MA 02115

MILLETTE, ROBERT LOOMIS, b Rockville Centre, NY, May 17, 33; m 57; c 2. BIOCHEMISTRY, VIROLOGY. *Educ:* Ore State Col, BS, 54; Calif Inst Technol, PhD(biochem), 65. *Prof Exp:* USPHS res fel, Max Planck Inst Biochem, Germany, 64-67; asst prof path, Med Ctr, Univ Colo, Denver, 67-80; vis asst prof, Dept Microbiol & Biophys, Univ Chicago, 74-75; ASSOC PROF, DEPT IMMUNOL & MICROBIOL, SCH MED, WAYNE STATE UNIV, 75- *Mem:* Am Soc Microbiol. *Res:* Herpes virus transcription and gene expression; in vitro transcription of viral eukaryotic genes by RNA polymerase II; eukaryotic gene transfer in vitro. *Mailing Add:* Dept Biol Portland State Univ PO Box 751 Portland OR 97207

MILLHAM, CHARLES BLANCHARD, b Liberal, Kans, Nov 1, 36; m; c 4. COMPUTER GRAPHICS, LINEAR & MIXED INTEGER PROGRAMMING. *Educ:* Iowa State Univ, BS, 58, MS, 61, PhD, 62. *Prof Exp:* From instr to asst prof math, Iowa State Univ, 62-66; from asst prof to assoc prof, Wash State Univ, 66-70, assoc prof math & comput sci, 71-74, prof computer sci, 74-90, PROF MATH, WASH STATE UNIV, 90- *Concurrent Pos:* Assoc environ scientist, Environ Res Ctr, USO, 70-74; vis assoc prof, Univ Wash, 72; Fulbright grant, Univ Jordan, 76-77; prog chair, Environ Sci & Regional Planning, Wash State Univ, 81-83, actg chair comput sci, 85. *Mem:* Asn Comput Mach; Soc Indust & Appl Math; Opers Res Soc Am; Math Prog Soc; Math Asn Am. *Res:* The computer imaging and simulation of human organs; deformations of bicubic spline surfaces; intersection of bicubic spline surfaces; curve fitting and date compression. *Mailing Add:* PO Box 2185 Pullman WA 99165

MILLHEIM, KEITH K, RESEARCH ADMINISTRATION. *Prof Exp:* RES CONSULT, AMOCO PROD CO, 91- *Mem:* Nat Acad Eng. *Mailing Add:* Amoco Prod Co PO Box 3385 Tulsa OK 74135

MILLHOUSE, EDWARD W, JR, b West Hartford, Conn, Oct 28, 22. ELECTRON MICROSCOPY, CYTOCHEMISTRY. *Educ:* Univ Ill, BS, 49, MS, 54, PhD(cytol, anat), 60. *Prof Exp:* NIH fel anat, Sch Med, Ind Univ, 60-61; from instr to asst prof, 61-71, ASSOC PROF ANAT, CHICAGO MED SCH, 71- *Mem:* Electron Micros Soc Am; Am Asn Anatomists. *Res:* Electron microscopy and cyto- and histochemistry of the endocrine organs. *Mailing Add:* 3333 N Green Bay Rd North Chicago IL 60064

MILLHOUSE, OLIVER EUGENE, b Westerville, Ohio, Aug 21, 41. NEUROLOGY, ANATOMY. *Educ:* Ohio State Univ, BS, 63; Univ Calif, Los Angeles, PhD(anat), 67. *Prof Exp:* From instr to assoc prof neurol & anat, Col Med, Univ Utah, 69-90; ED, PUB SECT, MAYO CLINIC, 90- *Concurrent Pos:* Res fel, Dept Anat, Harvard Med Sch, 67-69. *Mem:* Am Asn Anat; Pan-Am Asn Anat; Sigma Xi. *Res:* Structural organization of the mammalian hypothalamus by light and electron microscopy, especially intrinsic connections and relations with other areas of the central nervous system. *Mailing Add:* 2211 Baihly Hills Dr SW Rochester MN 55902

MILLIAN, STEPHEN JERRY, b Okeechobee, Fla, Feb 15, 27; m 56; c 4. VIROLOGY. *Educ:* Brooklyn Col, BS, 49; Ohio State Univ, MS, 50, PhD(bact), 53; Am Bd Microbiol, dipl, 63; Columbia Univ Sch Bus, MDPE, 75. *Prof Exp:* Asst, Ohio State Univ, 50-51; bacteriologist-virologist, Res Div, Armour Pharmaceut Co, 53-57; virologist biol res, Charles Pfizer & Co, Inc, 57-59; assoc cancer res sci, Roswell Park Mem Inst, 60-61, chief virus unit, 61-72; MEM STAFF, BUR LABS, NEW YORK CITY DEPT HEALTH, 72- *Concurrent Pos:* WHO travel-study award, 72; assoc mem, Pub Health Res Inst NY, 61-68; lectr pub health, Hunter Col, 76-78; lab consult, First Army Med Lab, 66-67; res assoc, Mt Sinai Hosp, 66-76; consult, Prof Exam Serv, 66-71; assoc prof clin pediat, NY Univ Med Sch, 73-; dir virol & immunol, Roswell Park Mem Inst. *Mem:* Harvey Soc; fel Am Soc Microbiol; fel Am Pub Health Asn; fel NY Acad Sci. *Res:* Laboratory diagnosis of viral and rickettsial infections; human and veterinary biologics. *Mailing Add:* Bur Labs New York City Dept Health 445 First Ave New York NY 10016

MILLICH, FRANK, b New York, NY, Jan 31, 28; m 60; c 2. POLYMER CHEMISTRY, PHOTOCHEMISTRY. *Educ:* City Col NY, BS, 49; Polytech Inst Brooklyn, MS, 56, PhD(polymer chem), 59. *Prof Exp:* Chemist, Norda Essential Oils & Chem Co, 50-55; Am Cancer Soc res fel, Cambridge Univ, 58-59 & Univ Calif, Berkeley, 59-60; from asst prof to assoc prof, 60-64, res assoc, Syst Space Sci Res Ctr, 66-71, PROF POLYMER CHEM, UNIV MO-KANSAS CITY, 67- *Concurrent Pos:* Consult, Missiles & Space Div, Lockheed Aircraft Corp, Calif, 59-60, Gulf Oil Corp, Kans, 70-75 & Midwest Res Inst, Mo, 70- *Mem:* AAAS; Am Chem Soc; Int Asn Dent Res. *Res:* New polymer synthesis and characterization; chemical evolution; kinetics of nonenzymatic synthesis of polypeptides and nucleic acids; kinetics of dye-sensitive photochemical reactions; synthesis of new chemotherapeutic drugs; interfacial synthesis; luminescence; polyisocyanides; theory of viscosity. *Mailing Add:* Dept Chem Univ Mo-Kans City Kansas City MO 64110

MILLICHAP, J(OSEPH) GORDON, b Wellington, Eng, Dec 18, 18; US citizen; m 46; c 4. NEUROLOGY, PEDIATRIC NEUROLOGY. *Educ:* Univ London, MB, 46, MD, 50; Am Bd Pediat, dipl, 58; Am Bd Psychiat & Neurol, dipl, 60, cert child neurol, 68; FRCP, 71; Am Bd EEG, dipl, 75. *Prof Exp:* House physician, Med & Pediat Units, St Bartholomews Hosp, London 46-47, demonstr physiol, 47-48, chief asst pediat, 51-53; house physician, Great Ormond St Hosp, 50-51; assoc res prof pharmacol, Univ Utah, 54-55; asst prof neurol, George Wash Univ, 55-56; from asst prof to assoc prof pediat & pharmacol, Albert Einsten Col Med, Univ Minn, 56-59; assoc prof pediat neurol & pharmacol, Mayo Grad Sch Med, Univ Minn, 61-63, pediat neurologist, Mayo Clin, 60-63; PROF NEUROL & PEDIAT, NORTHWESTERN UNIV, CHICAGO, 63- *Concurrent Pos:* Traveling fel, Brit Med Res Coun, 53-54; fel pediat, Harvard Univ, 53-54, Nat Inst Neurol Dis spec fel neurol, 58-60; pediat neurologist, Children's Mem Hosp, Wesley Hosp & Passavant Hosp; vis physician, DC Gen Hosp, 55-56; vis scientist, Clin Ctr, Nat Inst Neurol Dis, 55-56; vis pediatrician, Bronx Munic Hosp Ctr, NY, 56-58; resident, Mass Gen Hosp, 58-60; mem, Gov Adv Coun Develop Disabilities, 71-; vis prof, Hosp Sick Children, London, 86-87; vis prof, Southern Ill Univ Sch Med, Springfield, 88-90. *Mem:* AAAS; Soc Pediat Res; Am Soc Pharmacol & Exp Therapeut; Soc Exp Biol & Med; Am Epilepsy Soc. *Res:* Pediatric neurology; neuropharmacology of anticonvulsant drugs; biochemistry of developing nervous system; etiology of neurological disorders of children; behavior and learning disabilities; epilepsy; nutrition; electroencephalography; dyslexia. *Mailing Add:* Northwestern Mem Hosp 250 E Superior St Chicago IL 60611

MILLIER, WILLIAM F(REDERICK), b Skaneateles, NY, Aug 31, 21; m 47; c 4. AGRICULTURAL ENGINEERING. *Educ:* Cornell Univ, BS, 45, PhD, 50. *Prof Exp:* Dist agr engr, State Univ NY Col Agr, Cornell Univ, 42-44, 45-47; res assoc agr eng, Univ Minn, 49-52; from assoc prof to prof, 52-86, EMER PROF AGR ENG, CORNELL UNIV, 86- *Mem:* Am Soc Agr Engrs. *Res:* Farm power and machinery; tree fruit harvesting and post harvest handling; seed coating; pesticide application. *Mailing Add:* Dept Agr Eng Riley-Robb Hall Cornell Univ Ithaca NY 14853-5701

MILLIGAN, BARTON, b Cincinnati, Ohio, Oct 22, 29; m 54; c 1. ORGANIC CHEMISTRY. *Educ:* Haverford Col, AB, 51; Univ NC, MA, 53, PhD(chem), 55. *Prof Exp:* Lectr org chem, Univ Sydney, 55; Fulbright fel chem, Univ Adelaide, 56; assoc prof, Univ Miss, 56-64; from assoc prof to prof, Fla Atlantic Univ, 64-67; res chemist, Air Prod & Chem Inc, 67-75, mgr res nitration prod, 75-77, mgr explor res, Indust Chem Res & Develop, 77-81, res assoc, Indust Chem Technol, 81-86; lab dir, Polymer Dynamics Inc, 87; process develop mgr, Syntex Pharmaceut Int, Bahamas Chem Div, 88; RETIRED. *Concurrent Pos:* NSF fac fel, 63-64. *Mem:* AAAS; Am Chem Soc. *Res:* Free radical reactions; process research; polyurethanes. *Mailing Add:* PO Box 10118 Riviera Beach FL 33419

MILLIGAN, GEORGE CLINTON, b PEI, Can, Sept 6, 19; m 42; c 3. ECONOMIC GEOLOGY, STRUCTURAL GEOLOGY. *Educ:* Dalhousie Univ, MSc, 48; Harvard Univ, AM, 50, PhD(struct geol), 60. *Prof Exp:* Geologist, Man Dept Mines, 50-57; from assoc prof to prof geol, Dalhousie Univ, 57-85; RETIRED. *Concurrent Pos:* Nat Res Coun Can fel & guest prof, Swiss Fed Inst Technol, 64-65; Nat Res Coun Can exchange lectr, Univ Fed Pernambuco, Brazil, 74. *Mem:* Can Inst Mining & Metall; Geol Asn Can; Geol Soc Am. *Mailing Add:* 40 Lyngby Ave Dartmouth NS B3A 3T8 Can

MILLIGAN, JOHN H, b Gary, Ind, Apr 21, 33; m 59; c 1. THERMAL PHYSICS, TECHNOLOGY PLANNING. *Educ:* Purdue Univ, BSME, 59, MSME, 63, PhD, 65. *Prof Exp:* Engr, Lockheed Aircraft Corp, 59-61; phys scientist, 65-76, PHYS SCIENTIST RES & DEVELOP PLANNING, CENT INTEL AGENCY, 77- *Mem:* Sigma Xi; NY Acad Sci. *Res:* Intermolecular potential constants for various gases as determined from their viscosity over wide range of temperature; calculation of thermodynamic properties of dissociating and ionizing alkali-metal vapors; reentry physics and orbital mechanics problems. *Mailing Add:* 3358 Annandale Rd Falls Church VA 22042

MILLIGAN, JOHN VORLEY, b Edmonton, Alta, Feb 21, 36; m 58; c 3. MEDICAL PHYSIOLOGY. *Educ:* Univ Alta, BSc, 58, MSc, 60; Univ Minn, PhD(physiol), 64. *Prof Exp:* Lectr physiol, McGill Univ, 64-66; asst prof, 66-70, ASSOC PROF PHYSIOL, QUEEN'S UNIV, ONT, 70- *Concurrent Pos:* Assoc ed, Can J Physiol & Pharmacol, 73-; fel, Int Brain Res Orgn/UNESCO, 74. *Mem:* Soc Neurosci; Am Physiol Soc; Int Soc Neuroendocrinol; Can Physiol Soc. *Res:* Mechanisms of excitation-contraction coupling in muscle; computer analysis; ion and fluorescent dye interactions with plasma membrane; stimulus-secretion coupling in pituitary. *Mailing Add:* Three Windor St Kingston ON K7M 4K4

MILLIGAN, LARRY PATRICK, b Innisfail, Alta, Dec 12, 40; m 62; c 3. BIOCHEMISTRY, NUTRITION. *Educ:* Univ Alta, BSc, 61, MSc, 63; Univ Calif, Davis, PhD(nutrit), 66. *Prof Exp:* From asst prof to prof animal sci, Univ Alta, 66-88; DEAN RES, ON AGR COL, UNIV GUELPH. *Concurrent Pos:* Consult, UN, 81. *Mem:* Nutrit Soc; Agr Inst Can; Can Soc Animal Sci. *Res:* Nitrogen metabolism in animals, particularly ruminants; energy metabolism in animals. *Mailing Add:* Dean Res On Agr Col Univ Guelph Guelph ON N1G 2W1 Can

MILLIGAN, MANCIL W(OOD), b Shiloh, Tenn, Nov 21, 34; m 56; c 2. MECHANICAL & AEROSPACE ENGINEERING. *Educ:* Univ Tenn, BSME, 56, MS, 58, PhD(eng sci), 63. *Prof Exp:* Res engr, Boeing Co, 56-57, 58-59; instr mech eng, 57-58, from asst prof to assoc prof aerospace eng, 63-73, head, 73-82, PROF MECH & AEROSPACE ENG, UNIV TENN, KNOXVILLE, 82- *Concurrent Pos:* Consult, Oak Ridge Nat Lab, 59-80; pres, Xcel Eng, Knoxville. *Mem:* Am Soc Mech Engrs; Am Soc Eng Educ. *Res:* Low-density gas dynamics; numerical methods in gas dynamics and heat transfer; industrial noise control. *Mailing Add:* Dept of Mech & Aerospace Eng Univ of Tenn Knoxville TN 37916

MILLIGAN, MERLE WALLACE, b Des Moines, Iowa, Mar 7, 22; m 48; c 3. MATHEMATICS. *Educ:* Monmouth Col, BS, 47; Univ Ill, MA, 49; Okla State Univ, EdD(higher educ, math), 60. *Prof Exp:* Asst math, Univ Ill, 47-52; from asst prof to assoc prof, Adams State Col, 53-62; prof, Albion Col, 62-66; dean arts & sci, 66-72, PROF MATH, METROP STATE COL, 66- *Concurrent Pos:* NSF grant, 57, 63, 73. *Mem:* Am Asn Univ Prof; Math Asn Am. *Res:* Analog computation; higher education. *Mailing Add:* Dept Math Sci Metrop State Col 1006 11th St Denver CO 80204

MILLIGAN, ROBERT T(HOMAS), b Taylorville, Ill, Dec 22, 19; m 48; c 4. CHEMICAL ENGINEERING, CHEMISTRY. *Educ:* Univ Ill, BS, 41; Ohio State Univ, PhD(chem eng), 45. *Prof Exp:* Asst instr, Ohio State Univ, 41-44; process engr, Shell Develop Co Div, Shell Oil Co, 44-59, supvr licensing & design eng, 59-70, asst mgr tech dept, Res & Develop, Polymers Div, Shell Chem Co, 70-73; mgr pollution control technol, Bechtel Corp, 73-76, prin engr res & eng, 73-85, mgr org chem process develop, 76-80, consult, Bechtel Group Inc, 85-86; RETIRED. *Mem:* Am Chem Soc; fel Am Inst Chem Engrs. *Res:* Process design of petrochemical plants; plastics and resins research; advanced chemical technology; alternative energy processes; advanced methods of pollution control. *Mailing Add:* 2 Vista del Mar Orinda CA 94563

MILLIGAN, TERRY WILSON, b Hackensack, NJ, Aug 29, 35; m 57; c 3. ORGANIC CHEMISTRY, PHOTOGRAPHIC CHEMISTRY. *Educ:* Marietta Col, BS, 56; Univ Ill, PhD(org chem), 59. *Prof Exp:* Res scientist, Polaroid Corp, 59-64, asst proj mgr, 64-66, mgr color photog res, 66-72, asst to pres, 72-73, mgr, 74-76, sr tech mgr int technol, 76-80, dir appln develop, 80-83, tech dir original equip mfr mkt, 83-86, dir sci mkt, 86-89, TECH DIR CONVENTIONAL FILM, POLAROID CORP, CAMBRIDGE, 89- *Mem:* AAAS; Am Chem Soc. *Res:* Organic synthesis of dyes, hetero-cyclics, photographic developers; spectroscopic interactions between functional groups; reduction-oxidation reactions; coating technology; diffusion transfer photography; polyester films, polymer coatings, paper and plastic laminations; sensitometry and color analysis; medical and electronic imaging systems. *Mailing Add:* 51 Prentiss Lane Belmont MA 02178

MILLIGAN, WILBERT HARVEY, III, b Pittsburgh, Pa, Jan 17, 45; m 71; c 2. VIROLOGY. *Educ:* Washington & Jefferson Col, BA, 66; Univ Pittsburgh, PhD(microbiol), 72; Sch Dent Med, Southern Ill Univ, DMD, 79. *Prof Exp:* Lab instr microbiol & physiol, Washington & Jefferson Col, 65-66, lab instr gen biol, 66; teaching asst microbiol, Univ Pittsburgh, 66-67; lab instr, Allegheny Community Col, 67-69; from vis assoc prof to assoc prof microbiol, Sch Dent Med, 72-80, ADJ ASSOC PROF MICROBIOL-BIOCHEM, SOUTHERN ILL UNIV, EDWARDSVILLE, 87- *Concurrent Pos:* Abstractor, Am Dent Asn, 75-; consult gastroenterol, Sch Med, Washington Univ, 75-79; dent adv, Murray Manor Convalescent Home, 80-; pvt dent practice, 79- *Mem:* Sigma Xi; Am Soc Microbiol; AAAS; Int Asn Dent Res; Am Asn Dent Schs; Am Dent Asn. *Res:* Hepatitis transmission in dentistry; methods for sterilization in dentistry; herpes-simplex latency; immunization against dental caries; immunopathology; viral-mycoplasma interactions; dental caries and peridontal disease; self-instructional learning modules. *Mailing Add:* 3823 Old William Penn Hwy Murrysville PA 15668

MILLIKAN, ALLAN G, b Charleston, WVa, July 31, 27; m 77; c 4. PHOTOGRAPHY, ASTRONOMY. *Educ:* Oberlin Col, AB, 49; Purdue Univ, MS, 51. *Prof Exp:* Photog engr, Color Technol, Eastman Kodak Co, 51-57, res assoc, Res Labs, 58-71; vis scientist astron, Kitt Peak Nat Observ, 71-72; SR RES ASSOC SCI PHOTOG, EASTMAN KODAK RES LABS, 72- *Concurrent Pos:* Mem working group photog mat, Am Astron Soc, 66-; mem organizing comt, Int Astron Union Working Group on photog problems. *Mem:* Am Astron Soc; Int Astron Union. *Res:* Photographic emulsion as a scientific data recorder; improvement of plates and films used in scientific photography; photographic detection of faint objects. *Mailing Add:* 7061 Boughton Hill Rd Victor NY 14564

MILLIKAN, CLARK HAROLD, b Freeport, Ill, Mar 2, 15; m 66; c 3. MEDICINE, NEUROLOGY. *Educ:* Univ Kans, MD, 39; Am Bd Psychiat & Neurol, dipl, 46. *Prof Exp:* Intern, St Luke's Hosp, Cleveland, 39-40, from asst resident to resident med, 40-41; resident neurol, Univ Iowa, 41-44, from instr to asst prof, 44-49; from assoc prof to prof, Mayo Found, Univ Minn, 49-65, prof neurol, Mayo Grad Sch Med, Univ Minn, 65-76; consult, 49-76, head neurol sect, Mayo Clin, 55-65; prof neurol, Sch Med, Univ Utah, 76-86; prof neurol, Univ Miami Sch Med, 86-88; SCHOLAR RESIDENCE, DEPT NEUROL, HENRY FORD HOSP, 88- *Concurrent Pos:* Chmn comt cerebrovascular dis, USPHS & mem adv coun, Nat Inst Neurol Dis & Blindness, 61-65; chmn, Joint Coun Subcomt Cerebrovascular Dis, Nat Inst Neurol Dis & Blindness & Nat Heart Inst, mem nat adv comt regional med progs, NIH; past chmn, Coun Cerebrovascular Dis, Am Heart Asn, ed, Stroke; ed, J Cerebral Circulation, 70- *Honors & Awards:* William O Thompson Gold Medal Award, Am Geriatrics Soc, 71; Gold Heart Award, Am Heart Asn, 76. *Mem:* AMA; Asn Res Nerv & Ment Dis (pres, 61); Am Neurol Asn (pres, 74); fel Am Col Physicians; fel Am Acad Neurol; Sigma Xi. *Res:* Cerebrovascular disease; author or coauthor of over 200 publications. *Mailing Add:* Henry Ford Hosp Jour Off F118A 2799 W Grand Blvd Detroit MI 48202

MILLIKAN, DANIEL FRANKLIN, JR, b Lyndon, Ill, May 31, 18. BOTANY. *Educ:* Iowa State Univ, BS, 47; Univ Mo, PhD, 54. *Prof Exp:* Asst bot, 47-52, instr, 52-54, asst prof hort, 54-58, assoc prof, 58-68, PROF PLANT PATH, UNIV MO-COLUMBIA, 68- *Concurrent Pos:* USDA assignment, Poland, 66; Polish Minister Agr grant, 70 & 72; Nat Acad Sci-Polish Acad Sci grants, 74-77 & 81. *Mem:* Fel AAAS; Am Phytopath Soc; Am Soc Hort Sci; Bot Soc Am; foreign mem Polish Acad Sci; Sigma Xi. *Res:* Pathology of fruit, vegetable and woody ornamental crops; virology; virus diseases of stone and pome fruit crops. *Mailing Add:* Waters Hall Rm 108 Univ Mo Columbia MO 65211

MILLIKAN, LARRY EDWARD, b Sterlin, Ill, May 12, 36; m 62; c 2. DERMATOLOGY, IMMUNOLOGY. *Educ:* Monmouth Col, AB, 58; Univ Mo, MD, 62. *Prof Exp:* Physician intern & med officer, Great Lakes Naval Training Ctr, US Navy, 62, med officer aviation med, Naval Air Sta, Pensacola, Fla, 62-64, flight surgeon, Quonset Point, McGuire AFB, 64-67; resident dermat, Univ Hosp, Ann Arbor, Mich, 67-70; from asst prof to prof dermat Med Ctr, Univ Mo-Columbia, 70-81; PROF & CHMN DERMAT DEPT, MED CTR TULANE UNIV, NEW ORLEANS, LA, 81- *Concurrent Pos:* Consult physician, Student Health Serv, 70-, Vet Admin Hosp, 72- & Ellis Fischell State Cancer Hosp, 72-; mem, Eczema Task Force, Nat Prog Dermat, 73; contrib ed, Int J Dermat, 75-79; consult, Vet Admin Hosp, Pinesville, La & CharityHosp, New Orleans. *Mem:* Sigma Xi; Soc Invest Dermat; AAAS; Am Acad Dermat; Am Dermat Soc; Int Soc Dermat. *Res:* Cellular immunity; immunity in neoplasms; melanoma; immune surveillance. *Mailing Add:* Dermat Dept Tulane Univ Med Ctr 1439 Tulane Ave Suite 3554 New Orleans LA 70112

MILLIKAN, ROGER CONANT, b Tiffin, Ohio, Jan 27, 31; m 53; c 5. CHEMICAL PHYSICS. *Educ:* Oberlin Col, BS, 53; Univ Calif, PhD, 57. *Prof Exp:* Phys chemist, Res Lab, Gen Elec Co, 56-67; PROF CHEM, UNIV CALIF, SANTA BARBARA, 67- *Mem:* Am Chem Soc; fel Am Phys Soc; fel Optical Soc Am. *Res:* Infrared spectroscopy; combustion; high temperature reactions; vibrational relaxation; fluorescence; shock tubes; lasers. *Mailing Add:* Dept of Chem Univ of Calif Santa Barbara CA 93106

MILLIKEN, FRANK R, b Malden, Mass, Jan 25, 14. MINING ENGINEERING, RESEARCH ADMINISTRATION. *Educ:* Mass Inst Technol, BS, 34. *Hon Degrees:* DSc, Univ Utah. *Prof Exp:* Vpres change mining oper, Kennecott Copper Corp, 52-58, exec vpres, 58-61, pres, 61-78, chmn & chief exec officer, 78-79, dir, 58-88; RETIRED. *Concurrent Pos:* Chmn, Fed Reserve Bank, NY, 76-77; dir, Proctor Gamble Co. *Honors & Awards:* Robert H Richards Award, Am Inst Mech Eng. *Mem:* Nat Acad Eng; Mining & Metall Soc Am; Am Inst Mining & Petrol Engrs. *Mailing Add:* Contentment Island Rd Darien CT 06820

MILLIKEN, JOHN ANDREW, b Saskatoon, Sask, May 15, 23; m 46; c 7. INTERNAL MEDICINE. *Educ:* Queen's Univ, Ont, MD, CM, 46; FRCP(C), 54; Am Bd Internal Med, dipl, 60. *Prof Exp:* From asst prof to assoc prof, 56-71, PROF MED QUEEN'S UNIV, ONT, 71- *Concurrent Pos:* Head dept med, Hotel Dieu Hosp, 56-77. *Mem:* Am Col Physicians; NY Acad Sci; fel Am Col Chest Physicians; fel Am Col Cardiol; Can Med Asn; Royal Col Physicians. *Res:* Cardiology. *Mailing Add:* Hotel Dieu Hosp Kingston ON K7L 5G2 Can

MILLIKEN, SPENCER RANKIN, b Dallas, Tex, Dec 5, 24; m 45; c 4. PHYSICAL CHEMISTRY. *Educ:* Ga Inst Technol, BS, 50; Emory Univ, MS, 51; Pa State Univ, PhD, 54. *Prof Exp:* Asst fuel tech, Pa State Univ, 50-53; res engr lubricants, Aluminum Co Am, 53-57, asst chief, 57-58; res & sales coordr, Foote Mineral Co, 58-60, mgr metall res, 60; dir res, Northern Ill Gas Co, 60-65; mgr appl physics, Roy C Ingersoll Res Ctr, Borg-Warner Corp, 65-66; vpres res & develop, Welco Industs Div, Electronic Assistance Corp, 66-70, vpres, Corp, 70-75; vpres eng, Fuel & Energy Consults, 75-79; PRES, S R MILLIKEN & ASSOCS, ENERGY MGT CONSULTS, 79- *Concurrent Pos:* Adj asst prof, Univ Cinn, 67-70. *Mem:* Am Chem Soc; Am Inst Chem; Am Inst Mining, Metall & Petrol Eng; NY Acad Sci; Am Phys Soc. *Res:* Chemical constitution of coal; nature and structure of carbons; mineral preparation; lubricants; high purity metal preparation; nonferrous metals fabrication; metal films. *Mailing Add:* 2724 Chaparral Nacogdoches TX 75961

MILLIKEN, W(ILLIAM) F(RANKLIN), JR, b Old Town, Maine, Apr 18, 11; m 53; c 3. AERONAUTICAL ENGINEERING. *Educ:* Mass Inst Technol, BS, 34. *Prof Exp:* Mem staff, Chance-Vought Aircraft Co, 36-38, Vought-Sikorsky Aircraft, 38-39; asst chief flight test, Boeing Aircraft Co, Wash, 39-43; flight & aerodyn test, Avion, Inc, 43-44; asst head flight res, Cornell Aeronaut Lab, Inc, 44-47, head, 48-56, actg head vehicle dynamics dept, 56-69, dir, Full-Scale Div, 56-77; AT MILLIKEN RES ASSOCS, INC, WILLIAMSVILLE, NY, 77- *Concurrent Pos:* Consult transp res comt, Bur Pub Rds, 64-; mem panel naval vehicles, Nat Acad Sci, 60-61, comt future concepts, Hwy Res Bd, Nat Acad Sci-Nat Res Coun, 64-; adv comt res & develop, Maritime Comn. *Honors & Awards:* Laura Taber Barbour Air Safety Award, 67. *Mem:* Soc Automotive Engrs. *Res:* Flight test organization; crew training and procedures; dynamic stability and control. *Mailing Add:* 245 Brompton Rd Williamsville NY 14221

MILLIMAN, GEORGE ELMER, b New York, NY, Nov 15, 37; m 62; c 2. ANALYTICAL CHEMISTRY. *Educ:* Univ Rochester, BS, 59; Carnegie Inst Technol, PhD(chem), 64. *Prof Exp:* Res chemist, Jersey Prod Res Co, 64 & Esso Prod Res Co, Tex, 64-69; sr res chemist, Exxon Res & Eng Co, 69-86, SR RES CHEMIST, EXXON CHEM CO, 86- *Mem:* Am Chem Soc; Sigma Xi. *Res:* Synthesis; spectroscopy; trace analysis. *Mailing Add:* Four Rainier Rd Fanwood NJ 07023

MILLIMAN, JOHN D, b Rochester, NY, May 5, 38; m 63; c 2. OCEANOGRAPHY, GEOLOGY. *Educ:* Univ Rochester, BS, 60; Univ Wash, MS, 63; Univ Miami, PhD(oceanog), 66. *Prof Exp:* Res asst radiation biol lab, Univ Wash, 61; res asst, Inst Marine Sci, Univ Miami, 63-66, res fel, 66; asst scientist, 66-71, ASSOC SCIENTIST, WOODS HOLE OCEANOG INST, 71- *Concurrent Pos:* Alexander von Humboldt Found scholar, Lab Sedimentology, Univ Heidelberg, 69-70. *Mem:* AAAS; Geol Soc Am; Soc Econ Paleontologists & Mineralogists. *Res:* Deposition and diagenesis of marine sediments; continental shelf sedimentation; Holocene history and shallow structure; submarine precipitation and lithification of ,marine carbonates. *Mailing Add:* Dept Geol & Geophys Woods Hole Oceanog Inst Woods Hole MA 02543

MILLINER, ERIC KILLMON, b Wilmington, Del, June 3, 45; m 73; c 1. PSYCHOTHERAPY, INFANT PSYCHIATRY. *Educ:* Wheaton Col, BS, 66; Hahnemann Med Col, MD, 73. *Prof Exp:* Resident adult psychiat, Mayo Grad Sch Med, 73-74, Hahnemann Med Col, 74-76; staff consult adult psychiat, Mayo Clinic, 77-; DEPT PSYCHIAT, MAYO GRAD SCH MED, ROCHESTER. *Concurrent Pos:* Mayo Found scholar infant psychiat, Irvine Med Ctr, Univ Calif, 81- *Mem:* Am Psychiat Asn; Am Acad Child Psychiat. *Res:* Longitudinal studies of infancy and childhood development to elucidate predictors of adult personality organization and psychopathology; clarifying precursors of adult self and object representations through direct observation of children during separation-individualization (ages 18 through 36 months). *Mailing Add:* Dept Psychiat W-9 Mayo Grad Sch Med Rochester MN 55905

MILLING, MARCUS EUGENE, b Galveston, Tex, Oct 8, 38; m 59; c 1. PETROLEUM GEOLOGY, SEDIMENTOLOGY. *Educ:* Lamar Univ, BS, 61; Univ Iowa, MS, 64, PhD(geol), 68. *Prof Exp:* Res specialist, sr res specialist, sr res geologist & res supvry, Exxon Prod Res Co, 68-77; sr supvry prod geologist, Exxon Co USA, 77-79, dist exploration geologist, 79-80; mgr geol res, Arco Oil & Gas Co, 80-87; ASSOC DIR, BUR ECON GEOL, UNIV TEX, AUSTIN, 87- *Concurrent Pos:* Found trustee, Am Geol Inst; mem bd dirs, Offshore Technol Conf; counr, Geol Soc Am. *Mem:* Fel Geol Soc Am; Am Asn Petrol Geol; Soc Econ Paleontologists & Mineralogists; Soc Petrol Engrs; Soc Explor Geophysicists; Sigma Xi. *Res:* Applied reservoir description; environmental facies analysis; seismic stratigraphy deep-sea fans. *Mailing Add:* Bur Econ Geol Univ Tex PO Box X Univ Sta Austin TX 78713-7508

MILLINGTON, JAMES E, organic chemistry, for more information see previous edition

MILLINGTON, WILLIAM FRANK, b Ridgewood, NJ, June 16, 22; m 47; c 2. PLANT MORPHOGENESIS. *Educ:* Rutgers Univ, BSc, 47, MSc, 49; Univ Wis, PhD(bot hort), 52. *Prof Exp:* Res assoc, Roscoe B Jackson Mem Lab, Maine, 51-53; from instr to asst prof bot, Univ Wis, 53-59; from asst prof to assoc prof, 59-68, PROF BOT, MARQUETTE UNIV, 68- *Mem:* Am Soc Plant Physiol; Am Inst Biol Sci; Bot Soc Am; Phycol Soc Am. *Res:* Plant development; regulation of form and pattern. *Mailing Add:* Dept Biol Marquette Univ Milwaukee WI 53233

MILLION, RODNEY REIFF, b Idaville, Ind, Apr 3, 29; m 55; c 4. RADIOTHERAPY. *Educ:* Ind Univ, BS, 51, MD, 54; Am Bd Radiol, dipl, 63. *Prof Exp:* Intern, Harbor Gen Hosp, Torrance, Calif, 54-55; resident radiol, Ind Univ, 58-60; resident, Univ Tex M D Anderson Hosp & Tumor Inst, 60-62; assoc prof radiol & chief radiother sect, Ind Univ, 62-64; from assoc prof to prof radiol, 64-74, AM CANCER SOC PROF CLIN ONCOL, UNIV FLA, 74- *Res:* Radiation therapy. *Mailing Add:* Div Radiother Univ Fla Gainesville FL 32610

MILLIS, ALBERT JASON TAYLOR, b Philadelphia, Pa, Oct 4, 41; m 65; c 2. CELL BIOLOGY, CELLULAR AGING. *Educ:* Univ Pa, PhD(biol), 71. *Prof Exp:* Instr biol, Univ Pa, 67-69; res assoc pediat, Sch Med, Univ Wash, 71-72, res instr, 73, res asst prof, 74; from asst prof to assoc prof, 74-86, PROF BIOL, STATE UNIV NY, ALBANY, 86-, MEM FAC BIOL, 77- *Concurrent Pos:* Recipient, Res Serv Award, NIH, 81-82. *Mem:* Am Soc Cell Biol. *Res:* Regulation of the mammalian cell cycle and cellular proliferation; smooth muscle cell differentiation; cellular aging. *Mailing Add:* Dept Biol Sci State Univ NY 1400 Washington Ave Albany NY 12222

MILLIS, JOHN SCHOFF, medical education; deceased, see previous edition for last biography

MILLIS, ROBERT LOWELL, b Martinsville, Ill, Sept 12, 41; m 65; c 2. ASTRONOMY, PLANETARY SCIENCE. *Educ:* Eastern Ill Univ, BA, 63; Univ Wis, PhD(astron), 68. *Prof Exp:* Astronr, 67-86, assoc dir, 86-90, DIR, LOWELL OBSERV, 90- *Mem:* Astron Soc Pacific; Am Astron Soc; Int Astron Union; Div Planetary Sci (Secy-Treas, 85-88). *Res:* Planetary satellites and ring systems; occultation studies of solar system objects; comets. *Mailing Add:* Lowell Observ PO Box 1269 Flagstaff AZ 86001

MILLMAN, BARRY MACKENZIE, b Toronto, Ont, Oct 17, 34; m 77; c 8. CONTRACTION MECHANISMS, MUSCLE STRUCTURE. *Educ:* Carleton Univ, BSc, 57; King's Col, Univ London, PhD(biophys), 63. *Prof Exp:* Mem sci staff, Brit Med Res Coun, Biophys Res Unit, King's Col, Univ London, 61-66; from asst prof to prof biol sci, Brock Univ, 67-74, chmn dept, 66-71; chmn biophys interdept group, 75-80, PROF PHYSICS, UNIV GUELPH, 74- *Concurrent Pos:* Res grant, Natural Sci Eng Coun, Cell biol & genetics, 87- 90, chair, 89-90. *Mem:* Biophys Soc; Physiol Soc London; Can Soc Cell Biologists (treas, 68-71); Biophys Soc Can. *Res:* Muscle structure as determined by x-ray diffraction; contraction mechanisms and muscle physiology; intermolecular and inter-particle forces in aqueous gel system of biological interest; tobacco mosaic virus and the muscle filament lattice; electrostatic forces; x-ray diffraction. *Mailing Add:* Dept Physics Univ Guelph Guelph ON N1G 2W1 Can

MILLMAN, GEORGE HAROLD, b Boston, Mass, June 2, 19; m 43; c 1. IONOSPHERIC PHYSICS. *Educ:* Univ Mass, BS, 47; Pa State Univ, MS, 49, PhD(physics), 52. *Prof Exp:* Asst physics, Pa State Univ, 47-50, instr eng res, 50-52; engr, Gen Elec Co, 52-54, res liaison scientist, 54-55, specialist electromagnetic propagation, 55-84, consult physicist, 62-70, sr consult physicist, 70-84; ADJ PROF, DEPT ELEC & COMP ENG, SYRACUSE UNIV, 76- *Concurrent Pos:* Mem, Comn F, wave phenomena in non-ionized media & Comn G, ionospheric radio, Int Union Radio Sci, 58; mem, US Study Group 6, ionospheric propagation, Int Radio Consultative Comt; assoc ed, Radio Sci, 81-85; consult, 85- *Mem:* Am Phys Soc; fel Inst Elec & Electronics Engrs; Am Geophys Union; NY Acad Sci. *Res:* Electromagnetic-atmospheric propagation; ionospheric and space physics; radar and radio astronomy; atmospheric effects on radio wave propagation. *Mailing Add:* 504 Hillsboro Pkwy Syracuse NY 13214

MILLMAN, IRVING, b New York, NY, May 12, 23; m 49; c 2. IMMUNOLOGY, MEDICAL MICROBIOLOGY. *Educ:* City Col New York, BS, 48; Univ Ky, MS, 51; Northwestern Univ, PhD, 54. *Prof Exp:* Res bacteriologist, Armour & Co, 49-52; instr bact, Northwestern Univ, 54-55, asst prof, 55-58; asst, Pub Health Res Inst New York, 58-61; res fel, Merck Inst Therapeut Res, 61-67; mem, Clin Res Unit, 78-87, MEM, POP ONCOL RES UNIT, FOX CHASE CANCER CTR, INST CANCER RES, 88- *Concurrent Pos:* Consult med virol, Am Soc Microbiol. *Mem:* NY Acad Sci; AAAS; Am Soc Microbiol; fel Am Acad Microbiol. *Res:* Immunology, hepatitis and cancer research. *Mailing Add:* 1815 Danforth St Philadelphia PA 19111

MILLMAN, PETER MACKENZIE, astrophysics, planetary sciences; deceased, see previous edition for last biography

MILLMAN, RICHARD STEVEN, b Boston, Mass, Apr 15, 45; c 2. GEOMETRY. *Educ:* Mass Inst Technol, BS, 66; Cornell Univ, MS, 69, PhD(math), 71. *Prof Exp:* Asst prof math, Ithaca Col, 70-71; asst prof math, Southern Ill Univ, Carbondale, 71-74, assoc prof & asst to pres, 74-; AT DEPT MATH, MICH TECHNOL UNIV. *Mem:* Math Asn Am; Am Math Soc. *Res:* Eigenvalues of Laplace operator on Riemannian manifolds; holomorphic connections on fiber bundles. *Mailing Add:* Col Sci & Math Wright St Univ Dayton OH 45435

MILLMAN, ROBERT BARNET, b New York, NY, Aug 25, 39; c 2. PUBLIC HEALTH, PSYCHIATRY. *Educ:* Cornell Univ, BA, 61; State Univ NY, MD, 65. *Prof Exp:* Intern, Bellevue Hosp, NY, 65-66; asst physician, New York Hosp, 68-70; asst prof, Rockefeller Univ, 70-72; asst prof pub health, 70-76, clin asst prof psychiat, 77-87, clin prof pub health, 79-87, DIR, ADOLESCENT DEVELOP PROG, MED CTR, MED COL, CORNELL UNIV, 70-, PROF PSYCHIAT & PUB HEALTH, 87- *Concurrent Pos:* Consult, NY Dept Corrections, 71-72, Manhattan Borough Presidents' Comt Drug Abuse, 71-74; adj asst prof, Rockefeller Univ, 72-80; assoc ed, Millbank Mem Fund Quart, 72-80; mem comnr adv comt, NY Addiction Serv Agency, 74-80; mem adv comt drug abuse, NY Comnr Health, 75-85; mem task force ment health aspects of use & misuse of psychoactive drugs, President's Comn Ment Health, 77; mem adv comt, Gen Pediat Acad Develop Prog, Robert Wood Johnson Found, 78; dir Alcohol & Drug Abuse Prog, Payne Whitney Psychiat Clin, 78; fel, WHO, 79; ed bd, Hosp & Community Psychiat, 80-89; chmn, Adv Comn, NY State Div Substance Abuse Serv. *Mem:* Am Pub Health Asn; Am Med Soc Alcoholism; Sigma Xi; Am Psychiat Asn. *Res:* Pathogenesis and patterns of drug and alcohol abuse; characterization of the addictive process, with particular respect to abstinence syndromes; transmission of acquired immune deficiency syndrome. *Mailing Add:* Med Col Cornell Univ 411 E 69th St New York NY 10021

MILLMAN, SIDNEY, b Dawid Gorodok, USSR, Mar 15, 08; US citizen; m 31; c 1. PHYSICS EDUCATION. *Educ:* City Col New York, BS, 31; Columbia Univ, AM, 32, PhD(physics), 35. *Hon Degrees:* DSc, Lehigh Univ, 74. *Prof Exp:* Tyndall fel physics, Columbia Univ, 35-36, Barnard fel, 36-37, postdoctoral res asst, 37-39; instr physics, City Col New York, 39-41 & Queens Col, NY, 41-42; mem staff radar res, Columbia Radiation Lab, 42-45; exec dir res, Physics Div, Bell Tel Labs, 65-73; RETIRED. *Concurrent Pos:* Consult, Ctr Math Sci & Computer Educ, Rutgers Univ, 85-89, sr fel, 89- *Mem:* Fel Am Phys Soc; fel Inst Elec & Electronic Engrs; fel AAAS. *Res:* Nuclear spins and nuclear magnetic moments; magnetrons for radar; traveling wave tubes for millimeter waves; author of 12 technical publications. *Mailing Add:* 17 Fairview Ave Summit NJ 07901

MILLNER, ELAINE STONE, b Staten Island, NY; c 2. CANCER EPIDEMIOLOGY, HUMAN DEVELOPMENT & BEHAVIOR. *Educ:* Cornell Univ, BS, 53; Kean Col, MA, 69, Prof dipl, 71; Columbia Univ, DrPh(epidemiol), 75. *Prof Exp:* Microbiologist, Res Lab, Wallerstein Labs; chemist, Organon Res Labs, 65; asst prof psychol, Kean Col, 70; psychologist, Caldwell Bd Educ, West Caldwell, NJ, 71-72, consult child develop, 71-73; epidemiologist, Dept Pediat, Col Physicians & Surgeons, Roosevelt Hosp, 75-76, proj dir & co-prin investr, 76-77; sr res scientist & dir, Hepatitis Eval Unit, NY City Dept Health, Bur Preventable Dis, 78-80; actg prog dir epidemiol, Div Cancer Cause & Prev, Spec Progs Br, Nat Cancer Inst, 80-83; dep chief, Epidemiology Div, Am Health Found, 83-85; STAFF, UNION HOSP, 85- *Concurrent Pos:* Vis scientist, Lab Epidemiol, NY Blood Ctr, 78; lectr, Sch Public Health, Columbia Univ, 78-80; pvt pract, med psychotherapy/psychosocial oncology, 85- *Mem:* Soc Epidemiol Res; Am Public Health Asn; NY Acad Sci; Am Pyschol Asn; Sigma Xi. *Mailing Add:* 555 North Ave Apt 16H Ft Lee NJ 07024

MILLS, ALFRED PRESTON, b Fallon, Nev, Jan 8, 22; m 46; c 2. PHYSICAL CHEMISTRY. *Educ:* Univ Nev, BS, 43; Tulane Univ, PhD(phys chem), 49. *Prof Exp:* From instr to asst prof, 49-56, chmn div natural sci & math, 60-61, actg asst dean grad sch, 64-65, ASSOC PROF PHYS CHEM, 56- & ADJ ASST DEAN GRAD SCH FOR CURRIC, UNIV MIAMI, 73- *Mem:* AAAS; Am Chem Soc; Am Phys Soc; Am Inst Chemists; Sigma Xi. *Res:* Thermodynamics of solutions; prediction of physical properties of liquid solutions; dielectric constant and dipole moment, viscosity; physical properties of silicon compounds and nitro compounds. *Mailing Add:* 7540 SW 28th St Miami FL 33155

MILLS, ALLEN PAINE, JR, b Apr 21, 40; US citizen. ATOMIC PHYSICS. *Educ:* Princeton Univ, BA, 62; Brandeis Univ, MA, 64, PhD(physics), 67. *Prof Exp:* From instr to assoc prof physics, Brandeis Univ, 67-75; MEM TECH STAFF, BELL LABS, 75- *Mem:* Am Phys Soc. *Res:* Atomic physics of positronium. *Mailing Add:* 7 Meyersville Rd Chatham NJ 07928

MILLS, ANTHONY FRANCIS, b Cape Town, SAfrica, Jan 21, 38. MECHANICAL ENGINEERING. *Educ:* Univ Cape Town, BSc, 57 & 59, MSc, 60; Imp Col, Univ London, Dipl, 62; Univ Calif, Berkeley, PhD(eng), 66. *Prof Exp:* Lectr eng, Univ Cape Town, 58-61; asst prof, 66-76, assoc prof, 76-80, PROF ENG & APPL SCI, UNIV CALIF, LOS ANGELES, 80- *Concurrent Pos:* Consult, TRW Systs Group, 67- *Mem:* Am Soc Mech Engrs; Am Inst Aeronaut & Astronaut. *Res:* Heat and mass transfer; condensation phenomena. *Mailing Add:* Dept of Energy & Kinetics Univ of Calif Los Angeles CA 90024

MILLS, B(LAKE) D(AVID), JR, b Seattle, Wash, Apr 8, 12; m 45; c 3. HISTORY & PHILOSOPHY OF SCIENCE, MECHANICAL ENGINEERING. *Educ:* Univ Wash, BS, 34, ME, 47; Mass Inst Technol, SM, 35. *Prof Exp:* Test engr, Gen Elec Co, 35-36; instr appl mech & eng mat, Mass Inst Technol, 36-41; ord officer, US Navy Bur Ord, 41-46; from assoc prof to prof 46-77, EMER PROF MECH ENG, UNIV WASH, 77- *Mem:* Am Soc Mech Engrs; Am Soc Metals. *Res:* Properties of engineering materials; vibration; applied mechanics; history of science. *Mailing Add:* 900 Univ St Apt 10-N Seattle WA 98101-2730

MILLS, CLAUDIA EILEEN, b Seattle, Wash, July 27, 50. PLANKTON BIOLOGY, INVERTEBRATE ZOOLOGY. *Educ:* Colo Col, BA, 72; Fla State Univ, MS, 76; Univ Victoria, PhD(biol), 82. *Prof Exp:* Postdoctoral fel, Univ Wash, 83-85, Univ Paris, France, 86; INDEPENDENT SCIENTIST, FRI HARBOR LAB, 86- *Concurrent Pos:* Vis scientist, Sta Biologique, Roscoff, France, 88. *Mem:* Am Soc Limnologists & Oceanogr; Oceanog Soc; Sigma Xi; Western Soc Naturalists. *Res:* Field and laboratory studies of coastal and deep sea medusae, ctenophores and siphonophores including many aspects of individual and population biology. *Mailing Add:* Friday Harbor Lab Univ Wash 620 University Lab Friday Harbor WA 98250

MILLS, DALLICE IVAN, b Endeavor, Wis, July 27, 39; m 61; c 1. BOTANY, GENETICS. *Educ:* Wis State Univ, Stevens Point, BS, 61; Syracuse Univ, MS, 64; Mich State Univ, PhD(bot, plant path), 69. *Prof Exp:* Teacher high schs, Wis & Ariz, 61-65; spec res asst bot & plant path, Mich State Univ, 65-69; NIH res fel genetics, Univ Wash, 69-72; asst prof biol sci, Univ Ill Chicago Circle, 72-76; from asst prof to assoc prof, 76-85, PROF BOT, ORE STATE UNIV, 85-, CHMN GENETICS PROG, 85- *Mem:* Genetics Soc Am; Am Soc Microbiol; AAAS; Am Phytopath Soc. *Res:* Fungal and microbial genetics; genetics and physiology of host-parasite interactions. *Mailing Add:* Dept of Bot Ore State Univ Corvallis OR 97331

MILLS, DAVID EDWARD, b Springfield, Mass, Dec 9, 53. STRESS-REACTIVITY, ESSENTIAL FATTY ACIDS. *Educ:* Purdue Univ, BSc, 75; Indiana Univ, PhD(physiol), 80. *Prof Exp:* Res assoc med, dept med, div endocrinol, Univ NMex Sch Med, 79-81; asst prof, 81-87, ASSOC PROF HEALTH STUDIES, DEPT HEALTH STUDIES, UNIV WATERLOO, 87-, CHAIRPERSON, DEPT HEALTH STUDIES, 89- *Concurrent Pos:* Training grant, NIH, 75. *Mem:* Am Oil Chem Soc; Soc Exp Biol & Med; Soc Behav Med. *Res:* Psychophysiological mediators of stress reactions; nutritional and behavioral modifiers of stress reactivity; stress-related risk factors for disease; lipids and cardiovascular function. *Mailing Add:* Dept Health Studies Univ Waterloo Waterloo ON N2L 3G1 Can

MILLS, DON HARPER, b Peking, China, July 29, 27; US citizen; m 49; c 2. FORENSIC MEDICINE. *Educ:* Univ Cincinnati, BS, 50, MD, 53; Univ Southern Calif, JD, 58. *Prof Exp:* CONSULT FORENSIC MED, 58- *Concurrent Pos:* Fel path, Univ Southern Calif, 54-55, instr path, 58-62, from asst clin prof to assoc clin prof path, 62-69, clin prof, 69-; instr humanities, Loma Linda Univ, 60-66, assoc clin prof, 66-; exec ed, Trauma, 64-88; mem ed bd, J Forensic Sci, 65-80; dep med examr, Off Los Angeles County Coroner, 57-61; mem attend staff, Los Angeles County Hosp, 59-; affil staff, Hosp Good Samaritan, Los Angeles, 67-; res consult, Secy Comn Med Malpract, Dept Health, Educ & Welfare, 72-73; mem adv coun, Assembly Comt Med Malpract, State Calif, 73-75; expert consult, Off Secy, Dept Health, Educ & Welfare, 75-76; consult, Armed Forces Inst Path, Dept Defense, 74-80; Health Resources Admin, Dept Health, Educ & Welfare, 75-76; adminr & prin investr med insurance feasibility study, Calif Med Asn, 76-77; mem adv comt, Joint Legis Comt Tort Reform, State Calif, 77- & mem sci adv bd, Armed Forces Inst Path, Dept Defense, 78-80; bd dirs, Inst Med Risk Studies, Sausalito, Calif, 88-; med dir, Prof Risk Mgt Group of Calif, 89- *Mem:* AAAS; fel Am Acad Forensic Sci (pres, 86-87); AMA; Am Bar Asn; fel Am Col Legal Med (pres, 74-76). *Res:* Forensic medicine, primarily subfield of legal rights and responsibilities of physicians and hospitals. *Mailing Add:* 700 E Ocean Blvd Suite 2606 Long Beach CA 90802

MILLS, DOUGLAS LEON, b Berkeley, Calif, Apr 2, 40; m 61. SOLID STATE PHYSICS. *Educ:* Univ Calif, Berkeley, BS, 61, PhD(physics), 65. *Prof Exp:* NSF fel physics, Paris, 65-66; from asst prof to assoc prof, 66-74, PROF PHYSICS, UNIV CALIF, IRVINE, 74- *Mem:* Am Phys Soc. *Res:* Theoretical investigations of magnetic materials, lattice vibrations, surface effects, light scattering from solids and properties of alloys. *Mailing Add:* Dept Physics Univ Calif Irvine CA 92717

MILLS, EARL RONALD, b Los Angeles, Calif, Mar 28, 43; m 69; c 2. ECOLOGY, ENVIRONMENTAL MANAGEMENT. *Educ:* La State Univ, Baton Rouge, BS, 68, MS, 70; Tex A&M Univ, PhD(biol), 74; Loyola Univ, New Orleans, JD, 86. *Prof Exp:* Teaching asst bot, La State Univ, 68-69; res asst, Tex A&M Univ, 70-74; biologist, US Army Corp Engrs, 74-78; environ specialist, Strategic Petrol Reserve, US Dept Energy, 78-83; PRES, SYNERGISTICS, INC, 83- *Concurrent Pos:* Instr environ mgt & policy, Univ Houston, Clear Lake City, 76; State Admin Law Judge, 87-90. *Mem:* AAAS; Gulf Estuarine Res Soc; Estuarine Res Fedn. *Mailing Add:* 4615 Orleans Blvd New Orleans LA 70121

MILLS, ELLIOTT, b New York, NY, July 17, 35; m 60; c 2. PHARMACOLOGY, PHYSIOLOGY. *Educ:* City Col New York, BS, 57; Columbia Univ, PhD(pharmacol), 64. *Prof Exp:* Nat Inst Neurol Dis & Blindness trainee, Col Physicians & Surgeons, Columbia Univ, 64-65; Nat Heart Inst fel physiol, Middlesex Hosp Med Sch, London, Eng, 65-67; spec fel cardiovasc res inst, Med Ctr, Univ Calif, San Francisco, 67-68; asst prof, 68-74, assoc prof physiol & pharmacol, 74-77, ASSOC PROF PHARMACOL & PHYSIOL, MED CTR, DUKE UNIV, 77- *Concurrent Pos:* Estab investr, Am Heart Asn, 74- *Mem:* Am Phys Soc. *Res:* Chemoreceptor mechanisms; brainstem and reflex control of cardiovascular function; cardiovascular pharmacology. *Mailing Add:* Dept Pharmacol Duke Univ Med Ctr Box 3813 Durham NC 27710

MILLS, ERIC LEONARD, b Toronto, Ont, July 7, 36; m 62; c 2. BIOLOGICAL OCEANOGRAPHY, HISTORY OF OCEANOGRAPHY. *Educ:* Carleton Univ, Can, BSc, 59; Yale Univ, MS, 62, PhD(marine biol), 64. *Prof Exp:* Asst prof biol, Queen's Univ, Ont, 63-67; assoc prof, 67-71, PROF OCEANOG & BIOL, DALHOUSIE UNIV, 71-, CHMN, DEPT OCEANOG, 90- *Concurrent Pos:* Sessional lectr, Carleton Univ, Can, 60; instr, Marine Biol Lab, Woods Hole, Mass, 64-67, mem corp, 65-90; instr, Huntsman Marine Lab, NB, 71-73; vis scholar, Corpus Christi Col, Cambridge Univ, 74-75; Nuffield fel, Univ Edinburgh, 81-82; guest prof, Inst fuer Meereskunde, Univ Kiel, Germany, 84, 88; Ritter Mem fel, Scripps Inst Oceanog, 90. *Mem:* AAAS; Am Soc Limnol & Oceanog; Marine Biol Asn UK; fel Linnean Soc London; Hist Sci Soc. *Res:* History of oceanography; biological oceanography. *Mailing Add:* Dept Oceanog Dalhousie Univ Halifax NS B3H 4J1 Can

MILLS, FRANK D, b Cleveland, Ohio, July 29, 37; m 61; c 3. AGRICULTURAL, FOOD & SYNTHETIC ORGANIC CHEMISTRY. *Educ:* Western Reserve Univ, BA, 60, MS, 61, PhD, 66. *Prof Exp:* Res chemist cereal carbohydrates, Northern Regional Res Ctr, USDA, 66-80, res chemist pesticides, Beltsville Agr Res Ctr E, 80-90; CONSULT, 90- *Honors & Awards:* Citation, USDA, 70. *Mem:* Sigma Xi; Am Chem Soc; Am Asn Meta-Sci; AAAS. *Res:* Carbohydrates, structure and synthesis; pesticide chemistry; mass spectrometry; natural products chemistry. *Mailing Add:* 6591 Castlebay Ct Highland MD 20777

MILLS, FREDERICK EUGENE, b Streator, Ill, Nov 12, 28; m 50; c 3. PHYSICS. *Educ:* Univ Ill, BS, 49, MS, 50, PhD(physics), 55. *Prof Exp:* Res assoc physics, Cornell Univ, 54-56; scientist, Midwestern Univs Res Asn, 56-64, assoc dir sci, 64-65, dir, 65-66; prof physics, Univ Wis Phys Sci Lab, 66-70, dir, 67-70; chmn accelerator dept, Brookhaven Nat Lab, 70-73; SCIENTIST, FERMI NAT ACCELERATOR LAB, 73-; ADJ PROF, PHYSICS DEPT, UNIV WIS, 86- *Concurrent Pos:* Physicist, Saclay Nuclear Res Ctr, France, 61-62; adj prof, Nuclear Eng Dept, Univ Wis, 74- *Mem:* Fel Am Phys Soc. *Res:* Accelerators; high energy and plasma physics; energy loss of fast particles in motion; photoproduction of pi mesons; advanced accelerators. *Mailing Add:* Fermi Nat Accelerator Lab PO Box 500 Batavia IL 60510

MILLS, G(EORGE) J(ACOB), b Philadelphia, Pa, Aug 23, 24; m 48; c 2. METALLURGICAL ENGINEERING, PHYSICS. *Educ:* Pa State Univ, BS, 45, MS, 47; Univ Pa, PhD(metall), 51. *Prof Exp:* Chief res br, Metall Lab, Frankford Arsenal, US Dept of Army, 51-56; res specialist, Chem & Mat Sect, Jet Propulsion Lab, Calif Inst Technol, 56-58; mgr mat dept, Aeronutronic Div, Ford Motor Co, 58-64; prog mgr, Res Anal Corp, 64-68; prin engr plans & progs, Northrop Corp Labs, 68-70, dir mat sci lab, 70-75, asst tech planning,

75-81; RETIRED. *Mem:* Am Soc Metals; Am Inst Mining, Metall & Petrol Engrs; Am Inst Aeronaut & Astronaut; Am Defense Preparedness Asn. *Res:* Physical metallurgy; solid-state physics; materials science; composites; electronic materials; thin film devices. *Mailing Add:* 12142 Sky Lane Santa Ana CA 92705

MILLS, GARY KENITH, b Waynesville, Mo, Jan 27, 46; m 69; c 2. NEUROPSYCHOLOGY, PSYCHOANALYTIC PSYCHOTHERAPY. *Educ:* Univ San Francisco, BA, 66; San Francisco State Univ, MA, 68; Univ Ottawa, PhD(exp psychol), 74. *Prof Exp:* DIR, PAIN REHAB CTR, ST HELENA HOSP, DEER PARK, CALIF, 83- *Concurrent Pos:* Psychologist, pvt clin pract, 76- *Mem:* Am Psychol Asn; Int Asn Study of Pain; Soc Psychophysiol Res; Nat Asn Neuropsychologists; Am Pain Soc. *Res:* Clinical and applied research; psychophysiological research emphasizing self control mechanisms. *Mailing Add:* St Helena Hosp PO Box 399 Deer Park CA 94576

MILLS, GEORGE ALEXANDER, b Saskatoon, Sask, Mar 20, 14; nat US; m 40; c 4. PHYSICAL CHEMISTRY, FUEL SCIENCE. *Educ:* Univ Sask, BSc, 34, MSc, 36; Columbia Univ, PhD(chem), 40. *Prof Exp:* Asst chem, Columbia Univ, 36-39; instr, Dartmouth Col, 39-40; res chemist, Houdry Process & Chem Co, 40-47, asst dir res, 47-52, dir, 52-67; dir res, Houdry Lab, Air Prod & Chem Inc, 67-68; asst dir coal res, US Bur Mines, Washington, DC, 68-70, chief div coal, 70-74, asst dir, Off Coal Res, 74-75; dir fossil energy res, Energy Res & Develop Admin, Dept Energy, 75-77, sr scientist, fossil fuel, 77-81, exec dir, 81-84; SR SCIENTIST, CTR CATALYTIC SCI, UNIV DEL, 84- *Honors & Awards:* Storch Award, Am Chem Soc, 75, Murphree Award, 81; Chem Pioneer Award, Am Inst Chem, 82. *Mem:* Nat Acad Eng; Am Chem Soc; Am Inst Chem Engrs; Am Inst Mining, Metall & Petrol Engrs; Catalysis Soc (pres); AAAS. *Res:* Active hydrogen; separation of isotopes; exchange reactions with oxygen isotopes; reaction kinetics; petroleum refining; clay minerals; mechanism of catalytic reactions; high polymers; organic nitrogen chemicals; polyurethanes; synthetic fuels from coal; materials research; combustion and power. *Mailing Add:* Catalysis Ctr Colburn Lab Univ Del Newark DE 19716

MILLS, GEORGE HIILANI, b Pepeekeo, Hawaii, June 10, 21; m 44; c 4. MEDICINE. *Educ:* Colo Col, BA, 44, MA, 45; Boston Univ, MD, 50. *Prof Exp:* Intern, Queens Med Ctr, Honolulu, 50-51, resident internal med, 51-54; partner, Alsup Clin, Honolulu, 54-70; med dir, Kamehameha Sch, 62-78; mem fac, Sch Med, Univ Hawaii, 78-88; RETIRED. *Concurrent Pos:* Med dir, Manalani Hosp, 60-78; mem nat adv comt, Juv Justice & Delinq Prev, 75-77. *Mem:* Inst of Med of Nat Acad Sci. *Mailing Add:* 53179 Kamehameha Hwy Hauula HI 96717

MILLS, GORDON CANDEE, b Fallon, Nev, Feb 13, 24; m 47; c 3. BIOCHEMISTRY. *Educ:* Univ Nev, BS, 46; Univ Mich, MS, 48, PhD(biochem), 51. *Prof Exp:* Res assoc biochem, Col Med, Univ Tenn, 50-55; PROF BIOCHEM, UNIV TEX MED BR GALVESTON, 55- *Honors & Awards:* John Sinclair Award, Sigma Xi, 87. *Mem:* Am Chem Soc; Am Soc Biol Chem; Sigma Xi; Am Sci Affiliation. *Res:* Erythrocyte metabolism; genetic disorders of erythrocytes; metabolic control mechanisms; biochemistry of immune deficiencies; modified nucleosides of human urine. *Mailing Add:* Div Biochem Univ Tex Med Br Galveston TX 77550

MILLS, HARRY ARVIN, b Paintsville, Ky, Dec 11, 46; m 68; c 2. VEGETABLE CROPS. *Educ:* Univ Ky, BS, 69; Univ Mass, MS, 72, PhD(plant sci, soil sci), 75. *Prof Exp:* Asst prof, 74-76, MEM FAC HORT, UNIV GA, 76- *Mem:* AAAS; Sigma Xi; Am Soc Hort Sci; Am Soc Agron. *Res:* Soil fertility and plant nutrition of vegetable crops; nitrogen utilization. *Mailing Add:* Plant & Soil Sci Dept of Hort Univ of Ga Athens GA 30602

MILLS, HOWARD LEONARD, b Huntington, WVa, May 8, 20. PLANT PHYSIOLOGY, PLANT MORPHOLOGY. *Educ:* Marshall Col, BS, 44, MS, 49; Univ Iowa, PhD(plant physiol), 51. *Prof Exp:* Instr bot, Univ Iowa, 49-51; from asst prof to assoc prof bot, 51-61, PROF BIOL SCI, MARSHALL COL, 61- *Concurrent Pos:* NSF fel, Univ Wyo, 55; NSF-AEC fel, Univ Mich, 59; res assoc, NMex Highlands Univ, 59-60; consult, Div Radiol Health, USPHS, 65-; res consult, Environ Protection Agency, 72- *Mem:* Bot Soc Am; Am Soc Plant Physiol. *Res:* Physiology of growth and floral initiation; anthocyanin production and localization; algal nutrition; bactericides; radiobotany; fission product uptake by plant roots; physiognomic analyses of vegetation. *Mailing Add:* 1234 Ninth St Huntington WV 25701

MILLS, IRA KELLY, b Richmond, Kans, Oct 31, 21; m 43; c 2. PLANT PHYSIOLOGY. *Educ:* Univ Southern Calif, AB, 52, MS, 53; Ore State Col, PhD(bot), 56. *Prof Exp:* Asst plant pathologist, Mont State Univ, 56-62, from asst prof to prof bot, 56-83; RETIRED. *Concurrent Pos:* Fulbright-Hays lectr, Chung Hsing Univ, Taiwan, 66-67; consult, Gen Biol & Genetics Rev Panel, NIH, 68-70. *Mem:* AAAS; Am Soc Plant Physiol. *Res:* Effects of virus infection on host plant physiology; metabolism of aquatic plants and systems. *Mailing Add:* 721 S Tracy Bozeman MT 59717

MILLS, JACK F, b Galesburg, Ill, Feb 3, 28; m 53; c 5. ORGANIC CHEMISTRY. *Educ:* Knox Col, BA, 50; Univ Iowa, PhD(chem), 53. *Prof Exp:* Fel, Univ Ill, 53-54; res chemist, 56-62, sr res chemist, 62-74, sr res specialist, 74-77, RES ASSOC, DOW CHEM USA, 77- *Concurrent Pos:* Instr, Delta Col. *Res:* Halogen and polyhalogen compounds and their reactions; study of basic complexes; disinfection and pollution studies; spectrometric studies; synthesis of biological active compounds. *Mailing Add:* 4524 Andre St Midland MI 48640

MILLS, JAMES HERBERT LAWRENCE, b Guelph, Ont, Jan 14, 33; m 59; c 4. VETERINARY PATHOLOGY. *Educ:* Univ Toronto, BSA, 55, DVM, 61; Univ Conn, MS, 64, PhD(virol), 66; Am Col Vet Path, dipl, 66. *Prof Exp:* Vet practitioner, Ont, 61-62; instr animal dis, Univ Conn, 62-66, assoc prof vet path, 66-67; assoc prof, 67-71, PROF VET PATH, UNIV SASK, 71-, HEAD DEPT, 77- *Mem:* Am Vet Med Asn; US Animal Health Asn; NY Acad Sci; Can Vet Med Asn; Int Acad Path. *Res:* Veterinary virology and pathology, particularly bovine mucosal disease; veterinary parasitology, especially lungworms. *Mailing Add:* Dept Vet Path Col Vet Med Univ Sask Saskatoon SK S7N 0W0 Can

MILLS, JAMES IGNATIUS, b Morganfield, Ky, Nov 11, 44; m 66; c 2. SOLAR PHYSICS, ENERGY POLICY. *Educ:* Univ Okla, BS, 66, PhD(physics), 71. *Prof Exp:* Assoc scientist nuclear physics, Aerojet Nuclear Co, 72-75; vis asst prof & fel teaching & res, Miami Univ, 75-76; SCIENTIST SOLAR PHYSICS, EG&G IDAHO, INC, 76- *Concurrent Pos:* Consult, Nuclear Reactor Safety Anal Consult, EG&G Idaho, Inc & Nuclear Regulatory Comn, 75-76, Alternative Energy Syst Utilization & Develop, 77- *Mem:* Sigma Xi; Int Solar Energy Soc; Am Phys Soc; Am Nuclear Soc. *Res:* Solar energy research and development including total solar energy concepts; alternative energy systems research and development on bio-mass low-head hydropower wind; energy use forecasting and energy policy development. *Mailing Add:* 311 N Placer Ave Idaho Falls ID 83402

MILLS, JAMES LOUIS, b New York, NY, Nov 7, 47; m 74; c 2. PERINATAL EPIDEMIOLOGY, PEDIATRIC ENDOCRINOLOGY. *Educ:* Univ Pa, BA, 69, MS, 79; NY Med Col, MD, 73; Nat Bd Med Examrs, dipl, 74; Am Bd Pediat, cert, 78, cert(pediat endocrinol), 80. *Prof Exp:* Resident pediat, Cornell Univ Med Ctr, 73-75; fel ambulatory pediat, Children's Hosp, Philadelphia, 75-76, endocrinol, 76-79; med staff fel, 79-83, sr investr child health, 83-90, CHIEF, PEDIAT EPIDEMIOL SECT, NAT INST CHILD HEALTH & HUMAN DEVELOP, 90- *Concurrent Pos:* Fel, Dept Pediat, Cornell Univ, 74-75, Dept Res Med, Univ Pa, 76-79; Robert Wood Johnson clin scholar med, 76-79; attend pediat endocrinologist, Clin Ctr, Nat Inst Child Health & Human Develop, 79-, attend pediatrician, 81-84; lectr, dept epidemiol, Johns Hopkins Univ, 82-85, assoc, 85-; consult pediat endocrinologist, Nat Naval Med Ctr, 85- *Mem:* Am Diabetes Asn; Soc Epidemiol Res; Am Pediat Soc; Lawson Wilkins Pediat Endocrine Soc; fel Am Col Epidemiol; Soc Pediat Res; Am Epidemiol Soc. *Res:* Perinatal epidemiology; birth defects, diabetes, causes of early spontaneous abortion, effects of prenatal alcohol exposure and endocrine problems. *Mailing Add:* Nat Inst Health Exec Plaza N Bldg Rm 640 Bethesda MD 20892

MILLS, JAMES WILSON, b Dayton, Ohio, June 7, 42; m 64; c 1. PHYSICAL CHEMISTRY. *Educ:* Earlham Col, AB, 63; Brown Univ, PhD(chem), 68. *Prof Exp:* Res assoc, Joint Inst Lab Astrophys & Univ Colo, 68-69; asst prof chem, Drew Univ, 69-73; asst prof, 73-76, assoc prof, 76-82, chmn dept, 77-80, PROF CHEM, ST LEWIS COL, 82- *Concurrent Pos:* Cotrrell Res Corp grant, Drew Univ, 70-71 & Petrol Res Fund grant, 72-73; consult, Four Corners Environ Res Inst, 74-; vis scientist, Ind Univ, 81-82. *Mem:* Am Chem Soc; Am Phys Soc. *Res:* Molecular spectroscopy; quantum mechanics of small molecules; atomic analytical spectroscopy. *Mailing Add:* Dept Chem Ft Lewis Col Durango CO 81301-3909

MILLS, JANET E, b Bremerton, Wash, Jan 20, 43; m 88. ALGEBRA. *Educ:* Western Wash State Col, BA, 65; Pa State Univ, PhD(math), 70. *Prof Exp:* Asst prof math, Univ Fla, 70-71; from asst prof to prof math, James Madison Univ, 74-84; PROF MATH, SEATTLE UNIV, 84-, CHAIR MATH, 89- *Mem:* Am Math Soc; Math Asn Am; Asn Women Math. *Res:* Algebraic semigroups, particularly in constructions and classifications of inverse and regular semigroups. *Mailing Add:* Math Dept Seattle Univ Seattle WA 98122

MILLS, JERRY LEE, b Midland, Tex, Mar 6, 43. INORGANIC CHEMISTRY. *Educ:* Univ Tex, Austin, BS, 65, PhD(inorg chem), 69. *Prof Exp:* Fel, Ohio State Univ, 69-70; from asst prof to assoc prof, 70-76, assoc to prof, 76-80, PROF CHEM, TEX TECH UNIV, 80- *Mem:* Am Chem Soc; Royal Soc Chem. *Res:* Preparation, reactivity, and structure of non-transition metal compounds. *Mailing Add:* Dept Chem Tex Tech Univ Lubbock TX 79409

MILLS, JOHN BLAKELY, III, b Griffin, Ga, June 15, 39; m 64; c 2. BIOCHEMISTRY. *Educ:* Ga Inst Technol, BS, 61; Emory Univ, PhD(biochem), 65. *Prof Exp:* NSF fel, Cambridge Univ, 65-66; Whitehead fel, 66-67, from instr to asst prof, 67-74, ASSOC PROF BIOCHEM, EMORY UNIV, 74- *Res:* Chemistry of protein hormones. *Mailing Add:* Dept Biochem Emory Univ Atlanta GA 30322

MILLS, JOHN JAMES, b Motherwell, Scotland, May 12, 39; m 71; c 4. MATERIALS SCIENCE, PHYSICS. *Educ:* Glasgow Univ, BSc, 61; Univ Durham, PhD(appl physics), 65. *Prof Exp:* Res fel, Imp Col, Univ London, 64-66; sr scientist, Ill Inst Technol, 66-71; Humbolt res fel, Inst Silicate Res, Ger, 71-73; sect leader, Rosenthal Joint Stock Co, Ger, 73-75; sr scientist, Martin Marietta Labs, Md, 75-79, mgr, Fabric Res & Develop, 79-84, mgr mfg technol, 85-90; DIR AUTOMATION & ROBOTICS RES INST, UNIV TEX, ARLINGTON, 90-, PROF MECH ENG, 90- *Concurrent Pos:* Res fel, Brit Oxygen Co, 62-66; lectr, Ill Inst Technol, 67-69; Von Humboldt sr res fel, 71-73. *Mem:* Am Phys Soc; Brit Inst Physics; Soc Rheology; Am Inst Mining, Metallurg & Petrol Eng; Am Soc Metals; Soc Mech Engrs; Am Soc Mech Engrs. *Res:* Influences of environment on mechanical properties of glasses, ceramics, adhesives, composites and metals; mechanical metallurgy processes; design for manufacturing; precision assembly. *Mailing Add:* Automation & Robotics Res Inst 7300 J Newell Blvd S Ft Worth TX 76118

MILLS, JOHN NORMAN, b Neenah, Wis, Sept 29, 32; m 76; c 4. BIOCHEMISTRY. *Educ:* Wis State Col, BA, 54; Okla State Univ, MS, 56; Univ Okla, PhD, 65. *Prof Exp:* Asst chem, Okla State Univ, 54-58; from instr to asst prof, Okla Baptist Univ, 58-65; sr investr, Okla Med Res Found, 65-67; assoc prof, 67-71, PROF CHEM, OKLA BAPTIST UNIV, 71- *Concurrent Pos:* NIH res grant, 68-74. *Mem:* Am Chem Soc; Sigma Xi; AAAS. *Res:* Human gastric proteolytic enzymes and zymogens; protein structure; enzyme activity. *Mailing Add:* Dept Phys Sci Okla Baptist Univ Shawnee OK 74801

MILLS, JOHN T, b Redhill, Eng, July 31, 37; m 65; c 2. STORAGE MYCOLOGY. *Educ:* Univ Sheffield, BSc, 59; Univ London, PhD(plant path), dipl, Imp Col, 62. *Prof Exp:* Plant pathologist, Tate & Lyle Cent Agr Res Sta, Trinidad, WI, 63-67; SR RES SCIENTIST, RES STA, CAN DEPT AGR, WINNIPEG, 67-, HEAD, STORED PROD SECT, 84- *Mem:* Can Phytopath Soc; Am Phytopath Soc; Am Asn Feed Microscopists. *Res:* Spoilage and heating of stored agricultural commodities: detection, prevention and control; ecology of fungi occurring in stored cereals and oil seeds and their products. *Mailing Add:* Can Dept Agr Res Sta 195 Dafoe Rd Winnipeg MB R3T 2M9 Can

MILLS, KING LOUIS, JR, b Leslie, Ark, Nov 14, 16; m 42, 58; c 1. PETROLEUM CHEMISTRY. *Educ:* Ark State Teachers Col, BS, 38; Univ Ark, MS, 42. *Prof Exp:* Res chemist, 43-54, group leader, 54-65, SECT MGR, RES & DEVELOP DEPT, PHILLIPS PETROL CO, 65- *Mem:* Am Chem Soc; Am Inst Chem Engrs. *Res:* Petroleum refining processes; hydrocarbon conversions; heterogeneous catalysis; petrochemicals; carbon black technology. *Mailing Add:* 4507 SE Bridle Rd Bartlesville OK 74006-5306

MILLS, LEWIS CRAIG, JR, b Chicago, Ill, May 19, 23; m 47; c 4. INTERNAL MEDICINE. *Educ:* Baylor Univ, MD, 46; Am Bd Internal Med, dipl, 57; Endocrinol & Metab Bd, dipl, 75. *Prof Exp:* Intern, John Sealey Hosp, Galveston, Tex, 46-47; resident, Jefferson Davis Hosp, Houston, 49-51 & Methodist Hosp, Houston, 52; from instr to asst prof internal med, Baylor Col Med, 53-57; from asst prof to assoc prof, 57-61, clin prof, 61-64, dir endocrinol & metab dis, Hosp, 69-74, assoc dean affil, 74-77, vpres health affairs, 77-84, PROF MED, HAHNEMANN MED COL, 64-, DIR, ALTERNATIVE HEALTH DELIVERY SYSTS, 84- *Concurrent Pos:* Fel cardiol, Jefferson Davis Hosp, Houston, 51; res fel endocrinol, Peter Bent Brigham Hosp, Boston, 52-53; sr attend physician, Hahnemann Hosp, 58-, assoc vpres med affairs. *Honors & Awards:* Lindback Found Award, 64. *Mem:* Am Soc Pharmacol & Exp Therapeut; Soc Exp Biol & Med; Am Diabetes Asn; Am Fedn Clin Res; fel Am Col Physicians. *Res:* Endocrinology; metabolism; diabetes; vasopressor drugs; shock. *Mailing Add:* Hahnemann Univ & Hosp Broad & Vine St Philadelphia PA 19102

MILLS, MADOLIA MASSEY, b Kansas City, Mo, Oct 21, 19. GEOGRAPHY, CROSS CULTURAL STUDIES. *Educ:* Wayne State Univ, Detroit, BA, 61, MA, 68, MUP, 72. *Prof Exp:* Asst prof earth sci & world geog, Eastern Mich Univ, 68-74; consult & prin investr, Herbert G Whyte & Assocs, Ind, 75-76; asst prof urban planning & design/environ planning, Ill Inst Technol, 74-80 & Res Div, 79-80; res scientist eng, Solar Energy Res Inst, Colo, 80-83; asst prof earth sci, geog Asia, urban planning & glacial geol, Eastern Mich Univ, Upsilanti, Mich, 87-91; RETIRED. *Mem:* Asn Am Geographers. *Res:* Alternative energy; plate tectonics; Asian studies-Middle East, China, Korea and Japan; environmental hazards. *Mailing Add:* Geog Dept Eastern Mich Univ 233 Strong Hall Ypsilanti MI 48197

MILLS, NANCY STEWART, Mar 31, 50; m 77; c 2. CHEMISTRY. *Educ:* Grinnell Col, BA, 72; Univ Ariz, PhD(org chem), 76. *Prof Exp:* Res assoc chem, Ill Inst Technol, 76-77; asst prof org chem, Carleton Col, 77-79; from asst prof to assoc prof, 79-89, PROF ORG CHEM, TRINITY UNIV, 89- *Concurrent Pos:* Prin investr grants, Res Corp, 78-79, Petrol Res Fund, Am Chem Soc, 78-81 & 83-86, 90-, Robert A Welch Found, 80- & NSF, 85-89. *Mem:* Am Chem Soc; Sigma Xi; AAAS; Asn Women Sci; Am Asn Univ Prof. *Res:* Delocalized dianions and dications, including novel aromatic species; preparation and characterization of metallacyclo alkanes, trimethylenemethane metal complexes and catalytically active organometallic complexes. *Mailing Add:* Dept Chem Trinity Univ San Antonio TX 78212

MILLS, NORMAN THOMAS, chemical engineering, for more information see previous edition

MILLS, PATRICK LEO, SR, b Quantico, Va, Sept 24, 52; m 74; c 4. MICROSTRUCTURED MATERIAL RESEARCH. *Educ:* Tri-State Univ, BS, 73; Wash Univ, MSChE, 80, DSc, 80. *Prof Exp:* Process design engr, chem process design, Monsanto Co, 74-75; instr & grad res fel chem eng, Wash Univ, 75-80; staff chem engr res & develop, Gen Elec, 80-81; sr res engr chem eng, Monsanto Co, 81-83, res specialist, 83-88; sr res assoc, dept chem eng, Wash Univ, 88-90; RES SCIENTIST, DU PONT CO, 90- *Concurrent Pos:* Adj prof, dept chem eng, Wash Univ, 82-88; Gen Motors Scholar, 71-73. *Mem:* Am Inst Chem Engr; Int Asn for Math Modelling; Soc for Indust & Appl Math; Sigma Xi; NAm Catalysis Soc; Am Chem Soc. *Res:* Experimental techniques in catalysis and reaction engineering; chemical reaction engineering and applied catalysis; applied mathematics and computer applications in engineering. *Mailing Add:* Cent Res & Develop Exp Sta du Pont Co Wilmington DE 19880-0262

MILLS, ROBERT BARNEY, b Lane, Kans, Feb 10, 22; m 45; c 1. ENTOMOLOGY. *Educ:* Kans State Univ, BS, 49, PhD(entom), 64; Univ Colo, MEd, 53. *Prof Exp:* High sch teacher, Kans, 49-61; from asst prof to assoc prof, 63-76, PROF ENTOM, KANS STATE UNIV, 76- *Mem:* Entom Soc Am; Sigma Xi. *Res:* Stored product entomology. *Mailing Add:* Dept of Entom Kans State Univ Manhattan KS 66506

MILLS, ROBERT GAIL, b Effingham, Ill, Jan 20, 24; m 46; c 2. NUCLEAR PHYSICS. *Educ:* Princeton Univ, BSE, 44; Univ Mich, MA, 47; Univ Calif, Berkeley, PhD(nuclear physics), 52. *Prof Exp:* Instr elec eng, Princeton Univ, 43-44, res assoc elec eng & physics, 45-46; res assoc, Univ Mich, 46-47; Nat Res Coun res fel physics, Univ Zurich, 52-53, instr, 53-54; mem sr tech staff, Princeton Univ, 54-87, prof, 73-87; RETIRED. *Honors & Awards:* Centennial Medal, Inst Elec & Electonics Engrs. *Mem:* Fel Am Nuclear Soc; Am Phys Soc; fel Inst Elec & Electronic Engrs; Sigma Xi; AAAS. *Res:* Controlled thermonuclear research; research machines. *Mailing Add:* 150 Prospect Ave Princeton NJ 08540

MILLS, ROBERT LAURENCE, b Englewood, NJ, Apr 15, 27; m 48; c 5. DISORDERED SYSTEMS. *Educ:* Columbia Univ, AB, 48, PhD(physics), 55; Cambridge Univ, BA, 50, MA, 54. *Prof Exp:* Res assoc physics, Brookhaven Nat Lab, 53-55; mem sch math, Inst Advan Study, 55-56; from asst prof to assoc prof, 56-62, PROF PHYSICS, OHIO STATE UNIV, 62- *Honors & Awards:* Rumford Premium Award, AAAS, 80. *Mem:* Am Phys Soc; Am Asn Univ Professors; Fedn Am Scientists; Arms Control Asn. *Res:* Quantum field theory; many-body theory; theory of alloys. *Mailing Add:* Dept of Physics Ohio State Univ Columbus OH 43210

MILLS, ROBERT LEROY, b Canton, Ohio, June 6, 22; m 45; c 4. SOLID STATE CHEMISTRY. *Educ:* Washington & Jefferson Col, BS, 43; Calif Inst Technol, MS, 48; Stanford Univ, PhD(chem), 50. *Prof Exp:* Asst chem, Calif Inst Technol, 43-48 & Stanford Univ, 48-49; mem staff, 50-75, asst group leader, 75-83, LAB FEL, LOS ALAMOS NAT LAB, UNIV CALIF, 83-, ASSOC, 85- *Mem:* Fel AAAS; Am Chem Soc; fel Am Inst Chem. *Res:* High pressure physics; equation of state; low temperature physics; studies of light molecules to 200 kbar; high pressure solid-state chemistry. *Mailing Add:* Los Alamos Nat Lab MS C345 Los Alamos NM 87545

MILLS, ROGER EDWARD, b Cleveland, Ohio, Nov 19, 30; m 58; c 2. PHYSICS. *Educ:* Ohio State Univ, BSc & MSc, 52, PhD(physics), 63. *Prof Exp:* Physicist, Battelle Mem Inst, 60-63, sr physicist, 63-67, assoc div chief, 67-69; assoc prof, 69-75, asst vpres acad affairs, 74-81, PROF PHYSICS, UNIV LOUISVILLE, 75- *Concurrent Pos:* vis prof, Solar Energy Res Inst, 85-86. *Mem:* Am Phys Soc; Sigma Xi. *Res:* Far-from-equilibrium phenomena; critical phenomena. *Mailing Add:* Dept Physics Univ Louisville Louisville KY 40292

MILLS, RUSSELL CLARENCE, b Milwaukee, Wis, Nov 13, 18; m 40; c 5. MEDICAL EDUCATION. *Educ:* Univ Wis, BS, 40, MS, 42, PhD(biochem), 44. *Prof Exp:* Asst biochem, Univ Wis, 40-44; from asst prof to prof, Univ Kans, 46-51, assoc dean, Grad Sch, 63-70 & Sch Med, 63-72, form assoc vchancellor to asst chancellor, 72-76, univ dir support & serv, 76-79, prof biochem, 51-87, dir geront & assoc to chancellor, 79-87, EMER PROF BIOCHEM, UNIV KANS, 87- *Concurrent Pos:* Alan Gregg traveling scholar med educ, Far East, 70-71; consult, Asn Am Med Cols, 75-77. *Mem:* AAAS; Am Soc Biol Chemists; Geront Soc Am. *Res:* Health and social services for the frail elderly. *Mailing Add:* 2758 Chipperfield Rd Lawrence KS 66047-3183

MILLS, STEVEN HARLON, b Neosho, Mo, Feb 22, 45; m 67; c 2. TEMPERATURE REGULATION. *Educ:* Southwest Mo State Univ, BA, 67; Univ Ill, MS & PhD(physiol), 71. *Prof Exp:* Res assoc physiol, Univ Mo, Columbia, 71-72; from asst prof to assoc prof, 72-84, PROF BIOL, CENT MO STATE UNIV, 84- *Mem:* Sigma Xi; Am Physiol Soc. *Res:* Central nervous system control of temperature regulation in mammalian hibernators; neurotoxic effects on regulatory systems; vasomotor responses to temperature change monitored by telemetry. *Mailing Add:* Dept Biol Cent Mo State Univ Warrensburg MO 64093

MILLS, THOMAS K, b Hartford, Conn, Nov 8, 42; m 70; c 1. ELECTRONICS ENGINEERING. *Educ:* Johns Hopkins Univ, BS, 64; NC State Univ, MEE, 66, PhD(elec eng), 70. *Prof Exp:* Electronics engr, US Army, 70-72; ELECTRONICS ENGR, HARRY DIAMOND LABS, US ARMY ELECTRONICS RES & DEVELOP COMMAND, 72-, SR ELEC ENG. *Mem:* Inst Elec & Electronics Engrs. *Mailing Add:* Harry Diamond Labs Lab 47200 2800 Powder Mill Rd Adelphi MD 20783

MILLS, THOMAS MARSHALL, b Des Moines, Iowa, Nov 2, 38; m 60; c 2. REPRODUCTIVE ENDOCRINOLOGY. *Educ:* Univ Iowa, BA, 61, MS, 64, PhD(zool), 67. *Prof Exp:* Trainee steroid biochem, Ohio State Univ, 67-68; res assoc, Endocrine Lab, Univ Miami, 68-71; from asst prof to assoc prof, 71-82, PROF ENDOCRINOL, MED COL GA, 82- *Mem:* Endocrine Soc; Soc Study Reproduction. *Res:* Control of ovulation; ovarian steroid synthesis; metabolism of ovarian tissues; control penile erection. *Mailing Add:* Dept Physiol & Endocrinol Med Col Ga 1120 15th St Augusta GA 30912-3000

MILLS, WENDELL HOLMES, JR, b Detroit, Mich, July 31, 45; m 69. NUMERICAL ANALYSIS, DIFFERENTIAL EQUATIONS. *Educ:* Univ Mich, BSE, 68, PhD(math), 76; Univ Fla, MS, 72. *Prof Exp:* Mem tech staff, Rockwell Int, 68-72; ASST PROF MATH, PA STATE UNIV, 76- *Concurrent Pos:* Co prin investr, NSF, 81- *Mem:* Am Math Soc; Soc Indust & Appl Math. *Res:* Numerical solution and analysis of and scientific programming for dynamical problems in engineering. *Mailing Add:* Sohio Res 4440 Warrensville Ctr Rd Cleveland OH 44128

MILLS, WILLIAM ANDY, b Lynchburg, Va, Oct 12, 29; m 52; c 4. RADIATION PROTECTION STANDARDS, HEALTH RISK ASSESSMENTS. *Educ:* Lynchburg Col, BS, 51; Vanderbilt Univ, MS, 54; Med Col Va, PhD(biophysics), 64. *Prof Exp:* Health physicist, Oak Ridge Nat Lab, 52-55; div dir, Bureau Radiological Health, US Pub Health Serv, 55-71, Off Radiation Progs, US Environ Protection Agency, 71-81; br chief, US Nuclear Regulatory Comn, 81-85; SR TECH ADVR, OAK RIDGE ASSOC UNIVS, 85- *Concurrent Pos:* Mem, Nat Coun Radiation Protection & Measurements, 81- *Honors & Awards:* Elda E Anderson Award, Health Physics Soc, 69. *Mem:* Health Physics Soc; Soc Risk Anal. *Res:* Scientific and policy assistance on national issues concerning ionizing radiation; compensation for radiation injury; indoor radon; development of radiation protection standards for the workplace and the general environment. *Mailing Add:* 2915 Ascott Lane Olney MD 20832

MILLS, WILLIAM HAROLD, b New York, NY, Nov 9, 21; m 49; c 3. DESIGN THEORY. *Educ:* Swarthmore Col, AB, 43; Princeton Univ, MA, 47, PhD(math), 49. *Prof Exp:* Physicist, Aberdeen Proving Ground, 43-44; asst in instr, math, Princeton Univ, 48-49; from instr to assoc prof, Yale Univ, 49-64; mathematician, 63-90, EMER STAFF MEM, INST DEFENSE ANALYSIS, 90- *Mem:* Am Math Soc. *Res:* Combinatorics; number theory; algebra. *Mailing Add:* Inst Defense Anal Thanet Rd Princeton NJ 08540

MILLS, WILLIAM J, JR, b San Francisco, Calif, July 7, 18; m 52; c 7. COLD INJURY SCOLIOSIS. *Educ:* Univ Calif, Berkeley, BA, 42; Stanford Univ, Calif, MD, 50. *Prof Exp:* Intern, Univ Mich Hosp, Ann Arbor, 49-50, asst resident gen surg, 50-51, clin instr orthop surg, 52-54; United Cerebral Palsy Grant fel, Vanderbilt Univ, 67-68; PVT PRACT ORTHOP SURG, ANCHORAGE, ALASKA, 55-; PROF & DIR COLD RES, UNIV ALASKA, ANCHORAGE, 80-; ASSOC CLIN PROF MED, SCH MED, UNIV WASH, 81- *Concurrent Pos:* Consult, Alaska Native Hosp, Anchorage, 55, med, Armed Forces, 83-; head team physicians, Alaska, 75-; comn mem, Gov Comt Sci & Eng, Alaska. *Honors & Awards:* Gov's Award, State Alaska, 80, Hewitt Mem Award, 82. *Mem:* Fel Am Acad Orthop Surg; Am Orthop Asn; Soc Cryobiol; AMA; fel Am Col Surgeons; Am Col Sports Med; fel Arctic Inst NAm. *Res:* Problems of all aspects of cold injury, including the pathogenesis, etiology and management of frostbite, hypothermia and immersion injury, dehydration in the cold, altitude illness and survival and rescue techniques; cold problems of the winter athlete. *Mailing Add:* Ctr High Latitude Health Res Studies Univ Alaska Anchorage AK 99508

MILLS, WILLIAM RAYMOND, b Dallas, Tex, Feb 14, 30; m 52; c 3. NUCLEAR SCIENCE. *Educ:* Rice Inst, BA, 51; Calif Inst Technol, PhD(physics), 55. *Prof Exp:* Res asst, Knolls Atomic Power Lab, Gen Elec Co, 55-56; sr res technologist, Socony Mobil Oil Co, Inc, 56-63; res assoc, Field Res Lab, 63-83, consult, 83-86, MGR PETROPHYSICS RES, MOBIL RES & DEVELOP CORP, 86- *Concurrent Pos:* Adj prof, Southern Methodist Univ, 69-72. *Mem:* Am Phys Soc; Soc Prof Well Log Analysts. *Res:* Gamma-ray spectroscopy, especially of common earth elements; neutron physics; pulsed neutron phenomena. *Mailing Add:* Mobil Res & Develop Corp PO Box 819047 Dallas TX 75381-9047

MILLS, WILLIAM RONALD, b Clarksville, Tenn, June, 30, 47; m 69; c 1. PLANT BIOCHEMISTRY, PHOTOSYNTHESIS. *Educ:* Austin Peay State Univ, BS, 69, MS, 73; Miami Univ, PhD(bot), 77. *Prof Exp:* Teacher biol, Christian Coun High Sch, Ky, 69-71; teaching asst, Austin Peay State Univ, 71-73; res asst bot, Miami Univ, 73-77; fel biochem, Rothamsted Exp Sta, Eng, 77-78; res assoc, Carleton Univ, Can, 78-79; asst prof, 79-84, ASSOC PROF BIOL, UNIV HOUSTON, CLEAR LAKE, 84-, DEPT CHAIR, 90- *Concurrent Pos:* Bye fel, Robinson Col, 87; vis scholar, Bot Sch, Univ Canbridge, Eng. *Mem:* AAAS; Am Soc Plant Physiologists; Am Inst Biol Sci; Am Chem Soc; Sigma Xi; Int Soc Plant Molecular Biol. *Res:* Biosynthesis of essential amino acids and proteins in leaves of crop plants and the regulation of these processes; isolation and purification of physiologically active organelles from plants; DNA biosynthesis and its regulation in plants. *Mailing Add:* Univ Houston Clear Lake 2700 Bay Area Blvd Houston TX 77058-1098

MILLS, WILLIAM T(ERRELL), agricultural engineering, for more information see previous edition

MILLSAPS, KNOX, applied fluids mathematics, engineering mechanics; deceased, see previous edition for last biography

MILLSTEIN, JEFFREY ALAN, b New Brunswick, NJ, Dec 13, 57; m 84; c 1. ANIMAL POPULATION DYNAMICS, INSECT ECOLOGY. *Educ:* Purdue Univ, BS, 80; Univ Ky, MS, 82; Wash State Univ, PhD(entom), 88. *Prof Exp:* SR RES ASSOC QUANT ECOL, APPL BIOMATH, INC, 89- *Concurrent Pos:* Consult, Ecol Systs Anal, Inc, 88-91. *Honors & Awards:* Comstock Award, Entom Soc Am, 87. *Mem:* Entom Soc Am; Ecol Soc Am; AAAS. *Res:* Ecological risk analysis; mathematical models of population dynamics; epizootiology of insect pathogens and prediction of forest insect outbreaks; ecology of lyme disease; application of qualitative methods to ecological problems. *Mailing Add:* 100 N Country Rd Setauket NY 11733

MILLSTEIN, LLOYD GILBERT, b Brooklyn, NY, Jan 2, 32; m 53; c 2. MEDICAL PHYSIOLOGY, INFORMATION SCIENCE. *Educ:* NY Univ, BA, 53; Rutgers Univ, MS, 60, PhD(physiol, biochem), 64. *Prof Exp:* Microbiologist, Univ Hosp, Bellevue Med Ctr, 53; med writer pharmaceut, Squibb Inst Med Res Div, 55-57, toxicologist, 57-64; sr scientist, Smith Kline & French Labs, 64-67; dir sci info & commun, NcNeil Labs, 67-77; dir, Prescription Drug Labeling Staff, Bur Drugs, 77-81, dep dir, 81-82, actg dir, 82-84, dir, Div Drug Advert & Labeling, 84-86; DIR, ADVERT STANDARD & COMPLIANCE, BURROUGHS WELLCOME, 86- *Concurrent Pos:* Fed Drug Admin liaison, Nat Inst Aging, NIH, 77-; vis scientist, Pharmaceut Mfrs Asn & lectr, clin monitor training prog, Ctr Prof Advan; adj assoc prof, Sch Pharm, Univ NC & adj asst prof, Sch Pharm, Campbell Univ. *Mem:* Sigma Xi. *Res:* Drug regulatory affairs; gerontology; pharmacology and drug evaluation; marketing research drug advertising. *Mailing Add:* 7803 Coach House Lane Raleigh NC 27615

MILMAN, DORIS H, b New York, NY, Nov 17, 17; m 41; c 1. CHILD & ADOLESCENT PSYCHIATRY, PEDIATRICS. *Educ:* Barnard Col, BA, 38; NY Univ, MD, 42; Am Bd Pediat, dipl. *Prof Exp:* From asst prof to assoc prof pediat psychiat, 64-73, actg chmn dept, 73-75, chief-of-serv pediat, Univ Hosp Brooklyn, 73-75, actg chmn dept, 82, chief-of-serv, 82, PROF PEDIAT, STATE UNIV NY, HEALTH SCI CTR, BROOKLYN, 73- *Concurrent Pos:* Chief-of-serv, Kings County Hosp Ctr, 73-75 & 82. *Mem:* Am Pediat Soc; Am Psychiat Asn; Am Acad Pediat; Am Orthopsychiat Asn. *Res:* Minimal brain impairment; group work with parents of handicapped children; school phobia; adolescent phenomena; drug abuse; adolescent suicide. *Mailing Add:* 126 Westminster Rd Brooklyn NY 11218

MILMAN, GREGORY, PATHOGENESIS RESEARCH. *Prof Exp:* CHIEF, PATHOGENESIS BR, DIV AIDS, NAT INST ALLERGY & INFECTIOUS DIS, NIH, 88- *Mailing Add:* NIH Nat Inst Allergy & Infectious Dis Div AIDS Pathogenesis Br Control Data Bldg Rm 242P 6003 Executive Blvd Bethesda MD 20892

MILMAN, HARRY ABRAHAM, b Cairo, Egypt, May 16, 43; US citizen; m 68; c 2. BIOCHEMICAL PHARMACOLOGY. *Educ:* Columbia Univ, BS, 66; St John's Univ, NY, MS, 68; George Washington Univ, PhD(pharmacol), 78. *Prof Exp:* scientist toxicol, Nat Cancer Inst, 70-80; sr toxicol, 80-85, SR SCI ADV, US ENVIRON PROTECTION AGENCY, 80- *Concurrent Pos:* Pub Health Serv, Surgeon Gen Scientist, Prof Adv Comt. *Mem:* Am Soc Pharmacol & Exp Therapeut; Soc Toxicol; Int Asn Comparative Res Leukemia & Related Dis. *Res:* Assessment of the carcinogenicity of chemical compounds and the study of tumor markers; metabolism and homeostasis of asparagine. *Mailing Add:* 14317 Bauer Dr Rockville MD 20853

MILMORE, JOHN EDWARD, b Brooklyn, NY, Oct 31, 43; m 66; c 2. CLINICAL CHEMISTRY, COLLEGE TEACHING. *Educ:* Fordham Univ, BS, 65; Long Island Univ, MS, 68; Rutgers Univ, PhD(animal sci), 74. *Prof Exp:* Res asst pharmacol, US Vitamin Inc, 65-66; res asst endocrinol, Hoffmann-La Roche Inc, 66-68; res assoc pharmacol, Squibb Inst Med Res, 68-75; res assoc nutrit, Am Health Found, 75-78; asst prof pharmacol, NY Med Col, 78-83; asst prof biol, Col Mount St Vincent, 84-88; CLIN CHEMIST, FDR VET ADMIN MED CTR, 88- *Concurrent Pos:* Asst prof biol, Fordham Univ, 88- *Mem:* AAAS; Endocrine Soc; Sigma Xi; Am Asn Clin Chem. *Res:* Endocrine and central nervous system physiology; hypothalamic peptides; prolactin; hormones and mammary cancer; endogenous opioids; pharmacology. *Mailing Add:* 5 Oriole Lane Peekskill NY 10566

MILNE, DAVID BAYARD, b Evanston, Ill, Oct 24, 40; m 66; c 2. NUTRITIONAL BIOCHEMISTRY, BIOINORGANIC CHEMISTRY. *Educ:* Wash State Univ, BS, 62; Ore State Univ, MS, 65, PhD(biochem), 68. *Prof Exp:* Res assoc biochem, NC State Univ, 67-69; res chemist, Vet Admin Hosp, Long Beach, Calif, 69-74; res chemist, Letterman Army Inst Res, 74-79; RES CHEMIST, HUMAN NUTRIT RES CTR, AGR RES SERV, USDA, 79- *Mem:* Am Inst Nutrit; Am Soc Clin Nutrit; Am Chem Soc; Am Asn Clin Chem; Soc Exp Biol & Med. *Res:* Copper metabolism; metabolism and function of new essential trace elements; methods for assessment of nutritional status; clinical chemistry. *Mailing Add:* PO Box 7166 Grand Forks ND 58202

MILNE, DAVID HALL, b Highland Park, Mich, Dec 15, 39; m 64. ENTOMOLOGY. *Educ:* Dartmouth Col, BA, 61; Purdue Univ, PhD(entom), 68. *Prof Exp:* Asst prof gen sci, Ore State Univ, 67-71; MEM FAC, EVERGREEN STATE COL, 71- *Concurrent Pos:* NASA fel, 80-81. *Mem:* AAAS; Pac Estuarine Res Soc; Fel Am Soc Elec Eng; Sigma Xi. *Res:* Computer simulation of predation, competition processes; computer simulation of ecosystem dynamics; crab population biology. processes; computer simulation of ecosystem dynamics. *Mailing Add:* Dept of Biol Evergreen State Col Olympia WA 98505

MILNE, EDMUND ALEXANDER, b Eugene, Ore, Mar 3, 27; m 54; c 2. NUCLEAR PHYSICS. *Educ:* Ore State Col, BA, 49; Calif Inst Technol, MS, 50, PhD(nuclear physics), 53. *Prof Exp:* Asst, Calif Inst Technol, 52-53, res fel, 53-54; asst prof, 54-58, ASSOC PROF PHYSICS, NAVY POSTGRAD SCH, 58- *Mem:* Am Phys Soc; Sigma Xi. *Res:* Energy levels of light nuclei. *Mailing Add:* Dept of Physics Naval Postgrad Sch Monterey CA 93940

MILNE, ERIC CAMPBELL, b Perth, Scotland, Feb 8, 29; Can citizen; m 55; c 5. RADIOLOGY. *Educ:* Univ Edinburgh, MB, ChB, 56, DMRD, 60; Royal Col Radiologists, Eng, FRCR, 62. *Prof Exp:* Intern med surg, Monmouth Med Ctr, NJ, 56-57; resident chest dis, Tulare-Kings Counties Hosp, Calif, 57-58; sr house officer radiol, Royal Infirmary, Univ Edinburgh, 58-60; radiologist, McKellar Gen Hosp, Ft William, Ont, 61-65; asst prof radiol, Univ Western Ont, 65-66 & Peter Bent Brigham Hosp, Harvard Med Sch, 66-68; prof & dir exp radiol, Radiol Res Labs, Univ Toronto, 68-75, consult, Dept Lab Animal Serv, 69-75; chmn dept, 75-80, PROF RADIOL SCI, COL MED, UNIV CALIF, IRVINE, 75- *Concurrent Pos:* UK Med Res Found fel, Depts Radiol & Med, Royal Infirmary, Univ Edinburgh, Edinburgh, 60-61; Ont Cancer Found Gordon Richards fel, Cardiovasc Res Inst, Univ Calif, San Francisco, 65-66; chmn, Nat Adv Comt, Mayo Clinic Biotechnol Resource, 78-81; Fogarty sr int fel, 81. *Mem:* Am Col Radiologists; Soc Photo-Optical Instrument Engrs; Fleischner Soc (pres, 79-80); Asn Univ Radiologists. *Res:* Tumor circulation; radiologic diagnosis of early pulmonary diseases; analysis of x-ray image formation; radiologic magnification techniques; lung water pulmonary microcirculation; x-ray tube construction; three-dimensional image presentation; image analysis. *Mailing Add:* Dept Radiol Sci Univ Calif Col Med Irvine CA 92717

MILNE, GEORGE MCLEAN, JR, b Port Chester, NY, Dec 29, 43; m 65; c 2. MEDICINAL CHEMISTRY, PHARMACOLOGY. *Educ:* Yale Univ, BSc, 65; Mass Inst Technol, PhD(org chem), 69. *Prof Exp:* NIH fel chem, Stanford Univ, 69-70; proj leader, Med Chem Res, 70-73, mgr, Dept Pharmacol, 74-78, exec dir, Dept Immunol & Infectious Dis Res, 81-85, vpres res & develop opers, 85-87, SR VPRES RESEARCHER, PFIZER CENT RES, PFIZER INC, 88- *Concurrent Pos:* Scientific bd, Ventures Med. *Mem:* Am Chem Soc; Sigma Xi; Am Soc Microbiol; NY Acad Sci; Explorers Club. *Res:* Design and pharmacological evaluation of central nervous system drugs; antiemetic and analgesic pharmacology; biologic response modifiers for infectious, chronic degenerative and inflammatory diseases; cancer therapeutants. *Mailing Add:* Pfizer Cent Res Pfizer Inc Groton CT 06340

MILNE, GEORGE WILLIAM ANTHONY, b Stockport, Eng, May, 1937; US citizen. CHEMISTRY. *Educ:* Univ Manchester, BSc, 57, MS, 58, PhD, 60. *Prof Exp:* Res fel, Univ Wis, 60-61; vis fel, Lab Chem, Nat Inst Arthritis & Metab Dis, 62-63, vis assoc, 63-64; chemist, Lab Chem, Nat Heart, Lung & Blood Inst, 65-81; CHIEF INFO TECHNOL BR, DIV CANCER TREATMENT, NAT CANCER INST, 81- *Concurrent Pos:* Adj prof chem, Georgetown Univ, 67-; NIH mgr, NIH/Environ Protection Agency Chem Info Syst, 72-81. *Mem:* AAAS; Am Chem Soc; The Chem Soc. *Res:* Chemistry of steroids, terpenes, alkaloids, amino acids, nucleosides, nucleotides, carbohydrates, and application of nuclear magnetic resonance spectroscopy and mass spectrometry to studies in these fields, particularly biological function of various members of these classes; use of computers to handle data associated with chemicals; use of physical properties data to identify compounds; problems of information retrieval in chemistry and biology. *Mailing Add:* Nat Cancer Inst NIH Landow Bldg 37 Rm 5C-28 Bethesda MD 20805

MILNE, GORDON GLADSTONE, b Deland, Fla, July 13, 16; m 44; c 3. OPTICS. *Educ:* Univ Sask, BA, 38, MA, 39; Univ Rochester, PhD(optics, physics), 50. *Prof Exp:* Res physicist, Inst Optics, Univ Rochester, 42-45, res assoc optics, 45-60, sr res assoc, 60-66; physicist, Tropel, Inc, 66-82; RETIRED. *Concurrent Pos:* Consult optics, 85- *Mem:* AAAS; Optical Soc Am; Soc Photog Scientists & Engrs. *Res:* Optical instrumentation; manufacture of precision optics. *Mailing Add:* 7 Ledgemont Dr Fairport NY 14450

MILNE, LORUS JOHNSON, b Toronto, Ont, Sept 12; nat US. BIOLOGY. *Educ:* Univ Toronto, BA, 33; Harvard Univ, MA, 34, PhD(biol), 36. *Prof Exp:* Asst zool, Harvard Univ, 34-36; prof biol, Southwestern Univ, 36-37; adj prof, Randolph-Macon Woman's Col, 37-39, assoc prof, 39-42; war res, Aviation Med, Johnson Found, Univ Pa, 42-47; assoc prof zool, Univ Vt, 47-48; assoc prof, 48-51, PROF ZOOL, UNIV NH, 51- *Concurrent Pos:* Res grants, Carnegie Corp, Sigma Xi, Am Acad Arts & Sci, Am Philos Soc, Cranbrook Inst Sci & Explorers Club; Fund Advan Educ fac fel, 53-54; exchange lectr, Univs, US-SAfrica Leader Exchange Prog, 59; vis prof environ technol, Fla Int Univ, 74; mem, Marine Biol Lab, Woods Hole. Consult-writer, Biol Sci Curriculum Study, 60-61; Univ NH-Explorers Club deleg, Nairobi Meetings, Int Union Conserv Nature, 63; UNESCO biol consult, NZ, 66. Mem exped, Panama, 51, Cent Am, 53-54, BWI & SAm, 56-57 & 77-78, Equatorial Africa, 59 & 63, NAfrica, Near East, Southeast Africa & Australia, 66 & Semesterat Sea, 85. *Honors & Awards:* Nash Conserv Award, 54. *Mem:* Fel AAAS; Am Soc Zoologists; Animal Behav Soc; Sigma Xi; Explorers Club. *Res:* Behavioral ecology; natural history; author of over 100 articles and 50 books. *Mailing Add:* 1 Garden Lane Durham NH 03824

MILNE, MARGERY (JOAN) (GREENE), b New York, NY. BIOLOGY. *Educ:* Hunter Col, BA, 33; Columbia Univ, MA, 34; Radcliffe Col, MA, 37, PhD(zool), 39. *Prof Exp:* Instr zool, Univ Maine, 36-37; instr biol, Randolph-Macon Woman's Col, 39-40; asst prof biol & bact, Richmond Prof Inst, Col William & Mary, 40-42 & Beaver Col, 42-47; asst prof bot, Univ Vt, 47-48; asst prof zool, Univ NH, 48-50; assoc prof biol, Mass State Teachers Col, Fitchburg, 56; vis prof, Northeastern Univ, 58; consult biologist, Biol Sci Curric Study, Am Inst Biol Sci, Univ Colo, 60; res assoc, Univ NH, 65-74; LECTR, PARKS & RECREATION, UNIV, NH, 75-, PEASE AIR BASE & CONTINUING EDUC, 75- *Concurrent Pos:* Grantee, Am Acad Arts & Sci, Sigma Xi & Am Philos Soc; mem exped, Panama, 51, Cent Am, 53-54, BWI & SAm, 56-57 & 77-78, Equatorial Africa, 63 & NAfrica, Near East, Southeast Asia, Australia & New Zealand, 66; exchangee, US-SAfrica Leader Exchange Prog, 59; UNESCO biol consult, NZ, 66, Nat Geog Soc , 70; vis prof environ technol, Fla Int Univ, 74-; prof, Inst Shipboard Educ, Univ Pitt, 85. *Honors & Awards:* George Westinghouse Award, 47; Nash Conserv Award, 54; Nat Geog Soc Award, 65. *Mem:* AAAS; Nat Audubon Soc; Soc Women Geogrs; Sigma_Xi. *Res:* Plant and animal behavior; ecology; science writing. *Mailing Add:* One Garden Lane Durham NH 03824

MILNE, THOMAS ANDERSON, b Winfield, Kans, Dec 29, 27; m 54; c 3. RESEARCH MANAGEMENT, BIOMASS THERMOCHEMICAL CONVERSION. *Educ:* Univ Kans, AB, 50, PhD(chem), 55. *Prof Exp:* Res engr high-temperature chem, Atomics Int Div, NAm Aviation, Inc, 54-57; sr chemist, Stanford Res Inst, 57-60; sr physicist, Midwest Res Inst, 60-63, prin physicist, 63-68, sr adv chem, 68-77; actg br chief, Biol & Chem Conversion Br, 77-78, prin chemist, 78-80, br chief, Thermochem & Electrochem Res Br, 80-86, PRIN CHEMIST, SOLAR ENERGY RES INST, 86- *Concurrent Pos:* Res assoc, Nat Bur Standards, 83-84. *Mem:* Am Combustion Inst; Am Chem Soc; Am Solar Energy Soc; Biomass Energy Res Asn. *Res:* Molecular beam formation and sampling at high pressures; combustion processes; thermodynamic behavior of systems at high temperature; mass spectroscopic study of high-temperature reactions; coal combustion; inhibition chemistry; biomass pyrolysis chemistry and catalytic upgrading; hazardous waste destruction. *Mailing Add:* Solar Energy Res Inst 1617 Cole Blvd Golden CO 80401

MILNER, ALICE N, b Bay City, Tex, Sept 5, 25. BIOCHEMISTRY. *Educ:* Tex Woman's Univ, BS, 46, MA, 47; Baylor Univ, PhD(biochem), 59. *Prof Exp:* Instr chem, Centenary Col La, 47-50; head clin chem, Baylor Hosp, Dallas, Tex, 50-52; phys scientist, Vet Admin Hosp, Dallas, 52-53; Nat Cancer Inst fel biochem & oncol, Univ Tex M D Anderson Hosp & Tumor Inst, 59-62, res assoc biochem, 62-65; dir biochem, Moody Clin Res Lab, Col Med, Baylor Univ, 65-67; assoc prof nutrit, 67-80, prog dir coord undergrad progs in dietetics, 74-77, prof & actg chmn, 78-80, CHMN, DEPT NUTRIT, TEX WOMAN'S UNIV, 80- *Mem:* AAAS; Am Chem Soc; Am Asn Cancer Res; Am Dietetic Asn. *Res:* Bionutritional interrelationships; bionutritional aspects of mental retardation and obesity. *Mailing Add:* 231 Meadow Lane Lake Dallas TX 75065

MILNER, BRENDA (ATKINSON), b Manchester, Eng, July 15, 18; m 44. NEUROPSYCHOLOGY. *Educ:* Cambridge Univ, BA, 39, MA, 46, ScD, 72; McGill Univ, PhD(psychol), 52. *Hon Degrees:* LLD, Queen's Univ, 80; DSc, Univ Man, 82, Univ Lethbridge & Mt Holyoke Col, 86, Laval Univ & Univ Toronto, 87, McGill Univ & Wesleyan Univ, 91; LHD, Mt St Vincent Univ, 88; Dr, Univ Montreal, 88. *Prof Exp:* Exp officer, Ministry of Supply, UK, 41-44; asst prof psychol, Univ Montreal, 45-52; res assoc psychol, McGill Univ, 52-53, lectr, 53-60, from asst prof to assoc prof, 60-70, PROF NEUROL & NEUROSURG, MCGILL UNIV, 70- *Honors & Awards:* Karl Spencer Lashely Award, Am Philos Asn, 79. *Mem:* Am Psychol Asn; Am Acad Neurol; Brit Exp Psychol Soc; fel Royal Soc London; Royal Soc Can; Sigma Xi. *Res:* Brain function; perception and learning in human patients undergoing brain operation for focal cortical epilepsy. *Mailing Add:* 3553 Durocher Apt 802 Montreal PQ H2X 2B4 Can

MILNER, CLIFFORD E, b Concord, NH, Aug 10, 28; m 51; c 5. PHYSICAL CHEMISTRY. *Educ:* Wesleyan Univ, BA, 50, MA, 52; Yale Univ, PhD(phys chem), 55. *Prof Exp:* From res chemist to sr res chemist, E I du Pont de Nemours & Co, Inc, 55-77, res assoc, 77-90, sr res assoc, 90-91; RETIRED. *Mem:* Am Chem Soc; Soc Photog Scientists & Engrs (vpres). *Res:* Pressure dependence of the dielectric constant of water; photographic emulsions and processing solutions; diffusion transfer processes. *Mailing Add:* 1763 Winton Rd N Rochester NY 14609

MILNER, DAVID, b Birkenhead, Eng, July 23, 38; US citizen; m 61; c 2. ANALYTICAL CHEMISTRY. *Educ:* LRIC, 75 Univ Santa Clara, MBA, 80. *Prof Exp:* Tech asst chem, Distiller Co Ltd, Eng, 56-61; tech asst, Eli Lilly Ltd, Eng, 61-62, control chemist, 62-64; sect head pharmaceut anal, Syntex Corp, 64-81; TECH MGR, CHEM INDUST, NZ, 81- *Mem:* Royal Inst Chem; fel Royal Soc Chem. *Res:* Development of analytical methods for pharmaceutical dosage forms. *Mailing Add:* Chem Indust 46 Ben Lomond Crescent Pakuranga New Zealand

MILNER, ERIC CHARLES, b London, Eng, May 17, 28; m 54; c 4. MATHEMATICS. *Educ:* Univ London, BSc, 49, MSc, 50, PhD(math), 63. *Prof Exp:* Lectr math, Univ Malaya, 52-61 & Univ Reading, 61-67; PROF MATH, UNIV CALGARY, 67-, CHMN DIV PURE MATH, DEPT MATH, STATIST & COMPUT SCI, 74- *Mem:* Am Math Soc; Math Asn Am; Can Math Cong; London Math Soc. *Res:* Set theory; combinatorics; graph theory. *Mailing Add:* Dept Math Statist & Comput Sci Univ Calgary 2500 University Dr Calgary AB T2N 1N4 Can

MILNER, JOHN AUSTIN, b Pine Bluff, Ark, June 11, 47; c 2. NUTRITION. *Educ:* Okla State Univ, BS, 69; Cornell Univ, PhD(nutrit), 74. *Prof Exp:* Res assoc animal sci, Cornell Univ, 74-75; from asst prof to assoc prof, 75-84, PROF NUTRIT, UNIV ILL, URBANA-CHAMPAIGN. *Concurrent Pos:* Young Investr Award, Nutrit Found, 77-78; Dir Div Nutrit Sci, 81. *Mem:* Sigma Xi; Nutrit Today Soc; Inst Food Technol; Am Asn Animal Sci; Am Inst Nutrit; Am Soc Clin Nutrit; Am Asn Cancer Res. *Res:* Diet and Cancer; selenium metabolism; form and distribution of selenium in biological tissues. *Mailing Add:* Dept Nutrit Pa State Univ 126 Henderson Bldg S University Park PA 16802

MILNER, MAX, b Edmonton, Alta, Jan 24, 14; nat US; m 42; c 2. FOOD SCIENCE, NUTRITION. *Educ:* Univ Sask, BSc, 38; Univ Minn, MS 41, PhD(biochem), 45. *Hon Degrees:* LLD, Univ Sask, 79. *Prof Exp:* Res chemist, Pillsbury Mills, Inc, Minn, 41-42; res assoc, Univ Minn, 45-46; prof cereal chem, Kans State Univ, 47-59; sr food technologist, UNICEF, 59-71; dir secretariat, Protein-Calorie Adv Group UN Syst, 71-75; coordr, NSF-Mass Inst Technol Protein Resources Study, 75; nutrit coordr, Off Technol Assessment, US Cong, 76; assoc dir int nutrit prog, Mass Inst Technol, 76-78; exec officer, Am Inst Nutrit, 78-84; RETIRED. *Concurrent Pos:* Consult, Food & Agr Orgn, UN, 54-58; adj prof, Columbia Univ, 64-; sr lectr, Mass Inst Technol, 75-; mem panel world food supply, President's Sci Adv Comt, 68-; chmn, Gordon Res Conf Food & Nutrit, 68; consult, food & nutrit sci, 84- *Honors & Awards:* Inst Food Technologists Int Award. *Mem:* Fel AAAS; Am Chem Soc; fel Am Inst Nutrit; Am Asn Cereal Chem; Inst Food Technologists. *Res:* Cereal chemistry; nutrition, protein technology and development. *Mailing Add:* 10401 Grosvenor Pl Apt 721 Rockville MD 20852-4635

MILNER, PAUL CHAMBERS, b Washington, DC, Aug 23, 31. PHYSICAL CHEMISTRY. *Educ:* Haverford Col, BS, 52; Princeton Univ, MA, 54, PhD(chem), 56. *Prof Exp:* Mem tech staff, Bell Labs, 57-69, head, electrochem & contamination res dept, 69-87; RETIRED. *Mem:* Am Chem Soc; fel Electrochem Soc. *Res:* Electrochemistry; kinetics; thermodynamics. *Mailing Add:* 9832 W Escuda Dr Peoria AZ 85382-4139

MILNER, REID THOMPSON, b Carbondale, Ill, Aug 13, 03; m 28, 39; c 1. FOOD SCIENCE & TECHNOLOGY. *Educ:* Univ Ill, BS, 24, MS, 25; Univ Calif, PhD(phys chem), 28. *Prof Exp:* From asst to assoc chemist, US Bur Mines, 29-30; assoc chemist, 30-36, sr chemist, Regional Soybean Indust Prod Lab, Bur Agr Chem & Eng, 36-39, dir, 39-41, head anal & phys chem div, Northern Regional Res Lab, 41-48, dir, USDA, 48-54; prof food sci & head dept, 54-71, EMER PROF FOOD SCI, UNIV ILL, URBANA, 71- *Mem:* AAAS; Am Chem Soc; Am Oil Chem Soc (vpres, 46, pres, 47); Inst Food Technologists (pres, 74). *Res:* Low temperature specific heats; microanalysis; gas analysis; agricultural and food chemistry. *Mailing Add:* 101 W Windsor Rd Urbana IL 61801

MILNES, ARTHUR G(EORGE), b Heswall, Eng, July 30, 22; nat US; m 55; c 3. ELECTRICAL ENGINEERING. *Educ:* Bristol Univ, BSc, 43, MSc, 47, DSc(elec eng), 56. *Prof Exp:* Engr, Royal Aircraft Estab, Eng, 43-54; vis assoc prof elec eng, Carnegie Inst Technol, 54-55; engr, Royal Aircraft Estab, Eng, 56-57; assoc prof, 57-59, assoc head dept, 66-69, BUHL PROF ELEC ENG, CARNEGIE-MELLON UNIV, 60- *Concurrent Pos:* Bd mem, Sensormatic Electronics Corp, 68- *Honors & Awards:* J J Ebers Award, Inst Elect & Electronics Engrs, 82. *Mem:* Fel Inst Elec & Electronics Engrs, Electron Devices Soc, 82; fel Am Phys Soc; Electrochem Soc; fel Brit Inst Elec Engrs. *Res:* Semiconductor device studies; heterojunctions; deep impurities; solar cells. *Mailing Add:* Dept of Elec Eng Carnegie-Mellon Univ Pittsburgh PA 15213

MILNES, DALE J, ozone systems, for more information see previous edition

MILNOR, JOHN WILLARD, b Orange, NJ, Feb 20, 31; m 54 & 68; c 3. TOPOLOGY, DYNAMICAL SYSTEMS. *Educ:* Princeton Univ, AB, 51, PhD(math), 54. *Hon Degrees:* ScD, Syracuse Univ, 65; DSc, Univ Chicago, 67. *Prof Exp:* Higgins res asst math, Princeton Univ, 53-54, Higgins lectr, 54-55, from asst prof to prof, 55-62, Henry Putnam Univ prof, 62-67, chmn dept, 63-66; prof, Mass Inst Technol, 68-70; prof math, Inst Advan Study, 70-90; PROF & DIR, INST MATH SCI, STATE UNIV NY, STONY BROOK, 89-

Concurrent Pos: Alfred P Sloan fel, 55-59; ed, Ann Math, 62-68 & 73-79; vis prof, Univ Calif, Berkeley, 59-60 & Univ Calif, Los Angeles, 67-68; mem, Inst Advan Study, 63-70; Alexander von Humboldt-Stiftung sr US scientist award, 79. *Honors & Awards:* Fields Medal, Int Math Union, 62; Page-Barbour Lectr, Univ Va, 63; Hedrick Lectr, Math Asn Am, 65; Nat Medal Sci, 67; Steele Prize, Am Math Soc, 82; Roever Lectr, Wash Univ, St Louis, 83; Procelli Lectr, La State Univ, 84; Pitcher Lectr, Lehigh Univ, 87; Wolf Prize, Israel, 89. *Mem:* Nat Acad Sci; Int Cong Math; Am Math Soc (vpres, 75-76). *Res:* Topology of manifolds; differential geometry; dynamical systems. *Mailing Add:* Inst Math Sci State Univ NY Stony Brook NY 11794-3660

MILNOR, TILLA SAVANUCK KLOTZ, b New York, NY, Sept 29, 34; div; c 2. MATHEMATICS. *Educ:* NY Univ, BA, 55, MS, 56, PhD(math), 59. *Prof Exp:* NSF fel, 58-59; instr math, Univ Calif, Los Angeles, 59-60, lectr, 60-61, from asst prof to assoc prof, 61-69; assoc prof, Boston Col, 69-70; chmn dept, 70-73, PROF MATH, DOUGLASS COL, RUTGERS UNIV, 70-, CHMN DEPT, 78- *Concurrent Pos:* Vis mem, Courant Inst Math Sci, NY Univ, 64-65 & 77. *Mem:* Math Asn Am; Am Math Soc. *Res:* Differential geometry of immersed surfaces, especially questions involving ordinary or nonstandard conformal structures on surfaces; geometric application of methods form Riemann surface theory. *Mailing Add:* Dept Math Rutgers Univ New Brunswick NJ 08903

MILNOR, WILLIAM ROBERT, b Wilmington, Del, May 4, 20; m 44; c 2. MEDICAL PHYSIOLOGY. *Educ:* Princeton Univ, AB, 41; Johns Hopkins Univ, MD, 44. *Prof Exp:* From instr to assoc prof med, 51-69, dept dir, 81-85, PROF PHYSIOL, SCH MED, JOHNS HOPKINS UNIV, 69- *Concurrent Pos:* Nat Heart Inst res fel, 49-51; physician, Johns Hopkins Hosp, 52-, physician-in-chg heart sta, 51-60. *Mem:* Fel Am Col Physicians; Am Physiol Soc; Am Fedn Clin Res. *Res:* Cardiovascular physiology; hemodynamics; vascular and cardiac mechanics; control of pulsatile blood flow. *Mailing Add:* Dept Physiol Johns Hopkins Univ Sch Med Baltimore MD 21205

MILO, GEORGE EDWARD, b Montpelier, Vt, Nov 6, 32; m 56; c 4. CARCINOGENESIS, BIOCHEMISTRY. *Educ:* Univ Vt, BA, 58, MS, 61; State Univ NY Buffalo, PhD(virol), 68. *Prof Exp:* Instr biol, Rosary Hill Col, 63-65; NIH fel, Roswell Park Mem Inst, 67-69; sr res virologist, Battelle Mem Inst, 69; asst prof vet pathobiol, 69-75, assoc prof, 75-80, PROF PHYSIOL CHEM, COL MED, OHIO STATE UNIV, 80-, SR RES ASSOC, CAMPUS CANCER CTR, 79- *Concurrent Pos:* Nat chmn, Health Res Effects Grants Rev Panel, Environ Protection Agency. *Mem:* Tissue Cult Asn; Am Asn Cancer Res; Am Soc Biol Chemists; Soc Toxicol. *Res:* Chemical toxicology and in vitro chemical carcinogenesis; carcinogen, carcinogen and steroid administration to human cell systems in vitro. *Mailing Add:* Dept Physiol Chem 314 Hamilton Hall Ohio State Univ Col Med 1645 Neil Ave Columbus OH 43210

MILO, HENRY L(OUIS), JR, aeronautical & mechanical engineering, for more information see previous edition

MILONE, CHARLES ROBERT, b Uhrichsville, Ohio, Feb 13, 13; m 40; c 2. ORGANIC POLYMER CHEMISTRY. *Educ:* Mass Inst Technol, BS, 36, PhD(org chem), 39. *Prof Exp:* Res chemist, Goodyear Tire & Rubber Co, 39-45, sect head, 45-52; supt develop lab, Goodyear Atomic Corp, 52-57, mgr tech div, 57-60, dep gen mgr, 60-67; dir gen prod develop, Goodyear Tire & Rubber Co, 67-68, dir res & gen prod develop, 68-70, vpres, 70-77; RETIRED. *Mem:* AAAS; Am Chem Soc. *Res:* Synthetic rubber; polymerization; new plastics and their applications; gaseous diffusion; atomic energy. *Mailing Add:* 3068 Kent Rd 509C Stow OH 44224-4420

MILONE, EUGENE FRANK, b New York, NY, June 26, 39; m 59; c 2. ASTRONOMY, ASTROPHYSICS. *Educ:* Columbia Univ, AB, 61; Yale Univ, MS, 63, PhD(astron), 67. *Prof Exp:* From instr to asst prof physics, Gettysburg Col, 66-71; dir, Hatter Planetarium, 66-71; astronr, US Naval Res Lab, 67-79; from asst prof to assoc prof, 71-81, PROF, DEPT PHYSICS & ASTRON, UNIV CALGARY, 81-; CO-DIR, ROTHNEY ASTROPHYS OBSERV, 75- *Concurrent Pos:* Lutheran Church Am & Gettysburg Col res & creativity grant, 67-68, Gettysburg Col fac fel, 68-69 & 71-72; Can Nat Res Coun grants, 71-; mem, infrared & optical astron subcomt, Nat Res Coun, 74-77; NATO res grant, 79-81; Can Nat Sci Eng Res Coun Major Equip grant, 81-82, & Observ Infrastructure grant, 85-87; Killam resident res fel, Univ Calgary, 82 & 88; exec mem, Calgary Coun Lutheran Churches, 82-; exec mem, Calgary Inst Humanities, 82-; mem comns 42, 27 & 25, Int Astron Union, 85-, mem organizing comt com, 25, 85-88; vis prof, Univ Fla, 86. *Mem:* AAAS; Am Astron Soc; Sigma Xi; Int Astron Union; Can Astron Soc. *Res:* Optical and infrared photometry and spectroscopy of variable stars, especially interacting binaries; eclipsing binaries; ultraviolet solar and stellar limb darkening; rapid alternate detection system (RADS) for differential variable star photometry; computer modeling; author of 3 books. *Mailing Add:* Dept Physics & Astron Univ Calgary Calgary AB T2N 1N4 Can

MILSOM, WILLIAM KENNETH, b Toronto, Ont, June 24, 47. COMPARATIVE PHYSIOLOGY, RESPIRATORY PHYSIOLOGY. *Educ:* Univ Alta, BSc, 69; Univ Wash, MSc, 74; Univ BC, PhD(zool), 78. *Prof Exp:* Res assoc, 78-79, from asst prof to assoc prof, 79-91, PROF ZOOL, UNIV BC, 91- *Mem:* Can Soc Zoologists; Am Physiol Soc; Soc Exp Biol; Can Physiol Soc; Am Soc Zoologists. *Res:* Peripheral and central control of respiratory and cardiovascular systems in vertebrates with an emphasis on physiological adjustments to exercise, diving and hibernation. *Mailing Add:* Dept Zool Univ BC Vancouver BC V6T 2A9 Can

MILSTED, AMY, b Zanesville, Ohio, Nov 5, 44; m 79. GENE REGULATION. *Educ:* Ohio State Univ, BScEd, 67; City Univ New York, PhD(cell biol), 77. *Prof Exp:* Fel cell biol, Carnegie-Mellon Univ, 76-77, Muscular Dystrophy Asn fel, 78-79; res assoc, Case Western Reserve Univ, 79-82; res chemist, Vet Admin Med Ctr, Cleveland, Ohio, 82-88; ASSOC STAFF, CLEVELAND CLIN FOUND, 88- *Concurrent Pos:* Co-prin investr, Vet Admin Merit Rev grant, 86- *Mem:* Am Soc Cell Biol; Am Soc Microbiol; Asn Women Sci; AAAS. *Res:* Regulation of gonadotropin gene expression in cultured cells and in tumors; eutopic and ectopic hormone production; mechanisms of action of sodium butyrate and ccyclic adenosine monophosphate. *Mailing Add:* Brain & Vasc Res Dept Cleveland Clin Found Res Inst 1 Clin Ctr 9500 Euclid Ave Cleveland OH 44195-5070

MILSTEIN, CESAR, b Bahia Blanca, Arg, Oct 8, 27. MOLECULAR BIOLOGY. *Prof Exp:* Prof microbiol, Nat Inst Microbiol, Buenos Aires, 61-63; head, Div Protein & Nucleic Acid Chem, 63-83, DEP DIR, LAB MOLECULAR BIOL, MED RES COUN, 83- *Concurrent Pos:* Fel, Darwin Col, Cambridge. *Honors & Awards:* Nobel Prize in Med, 84; Ciba Medal & Prize, 78. *Mailing Add:* Med Res Coun Lab Molecular Biol Hills Rd Cambridge CB2 2QH England

MILSTEIN, FREDERICK, b New York, NY, May 14, 39; m 60; c 3. ENGINEERING, MATERIALS SCIENCE. *Educ:* Univ Calif, Los Angeles, BSc, 62, MS, 63, PhD(eng & appl sci), 66. *Prof Exp:* Nat Ctr Sci Res grant, Electrostatic & Solid State Physics Lab, France, 66-67; res scientist mat sci, Rand Corp, Calif, 67-69; actg asst prof energy & kinetics, Sch Eng & Appl Sci, Univ Calif, Los Angeles, 69-70; from asst prof to assoc prof mech eng, 70-78, assoc dean Col Eng, 73-75, chmn, dept mech eng, 81-82, PROF MAT SCI & MECH ENG, UNIV CALIF, SANTA BARBARA, 78- *Concurrent Pos:* Lectr, Univ Calif, Los Angeles, 67-69; consult, Rand Corp, Calif, 69-71, Civil Eng Lab, US Navy, Calif, 78-85, Mission Res Corp, 85-; vis fel, Clare Hall, Univ Cambridge, 75, Guggenheim fel, 75; sr fel electronics, Weizmann Inst Sci, Israel, 75-76, sr fel, NATO, 76; prin investr grants, NSF, Mission Res Corp & EG&G, Inc, 76; distinguished sr fel, Soc Eng Educ, 88. *Mem:* Am Phys Soc. *Res:* Crystal elasticity; mechanical behavior and phase transformations in solids; physical metallurgy; solid-state physics; theoretical elasticity. *Mailing Add:* 456 Braemar Ranch Lane Santa Barbara CA 93109

MILSTEIN, JAIME, mathematics, for more information see previous edition

MILSTEIN, STANLEY RICHARD, b Brooklyn, NY, Nov 27, 44; m; c 1. ORGANIC CHEMISTRY, MEDICINAL CHEMISTRY. *Educ:* Rensselaer Polytech Inst, BS, 66; Adelphi Univ, PhD(chem), 74; Univ Cincinnati, Col Pharm, MS, 86. *Prof Exp:* Res assoc, Dept Chem, Pomona Col, 72-73; asst prof, Dept Chem, Adelphi Univ, 73-74; res assoc, Col Pharm, Univ Cincinnati, 74-78, vis asst prof org chem, 78-79; SR MGR, TECH SERV, ANDREW JERGENS CO, 80-; ADJ ASST PROF ORGANIC CHEM, UNIV CINCINNATI, 80- *Concurrent Pos:* Soc Cosmetic Chemists grant, 78-79; adj asst prof cosmetic sci, Univ Cincinnati Col Phar, 87- *Honors & Awards:* Merit Award, Soc Cosmetic Chemists, 89. *Mem:* AAAS; Am Chem Soc; Soc Cosmetic Chemists (vpres, 91, pres, 92); Derm Clin Eval Soc; Sigma Xi. *Res:* Organic synthesis of biologically active compounds; computer-assisted development of quantitative structure-activity relationships between biological activity and chemical structure; Hansch approach; synthesis of new cosmetic raw materials; claim substantiation; N-nitrosamine assessment and inhibition. *Mailing Add:* Andrew Jergens Co 2535 Spring Grove Ave Cincinnati OH 45214

MILSTOC, MAYER, b Iasy, Rumania, Dec 14, 20; US citizen; m 45. MEDICINE, PATHOLOGY. *Educ:* Univ Bucharest, MD, 52; State Univ NY, MD, 65; Am Bd Path, cert anat & clin path, 69. *Prof Exp:* Asst in res, Inst Res Antibiotics, Bucharest, 51-55, chief lab, 55-57; chief lab clin path, Colentina Hosp, 57-61; asst path, Montefiore-Morrisania Hosp, New York, 66; asst prof path, 67-73, ASSOC PROF CLIN PATH, MED CTR, NY UNIV, 73-; DIR LABS, GOLDWATER MEM HOSP, 69- *Concurrent Pos:* Asst prof microbiol, Inst Medico Pharmaceut, Sch Med, Bucharest. *Mem:* Col Am Path; Am Soc Clin Path. *Res:* Biology of microorganisms, especially antibiotic problems; enzymes, especially cholinesterase in the normal and the diseased. *Mailing Add:* 370 E 76th St New York NY 10021

MILSUM, JOHN H, b Sussex, Eng, Aug 15, 25; Can citizen; m 55; c 2. HEALTH & HEALTH CARE SYSTEM. *Educ:* Univ London, BSc, 45; Mass Inst Technol, SM, 55, ME, 56, ScD(control eng), 57. *Prof Exp:* Proj engr, Nat Res Coun Can, 50-54, head anal sect, 57-61; Abitibi prof control eng, McGill Univ, 61-72, dir Biomed Eng Unit, 66-72; Imp Oil prof gen systs, Univ BC, 72-77, dir, Div Health Systs, 72-85, prof & chmn, Div Health Prom, Dept Health Care & Epidemiol, 72-90, actg dir, Inst Health Prom Res, 90-91; RETIRED. *Concurrent Pos:* Consult health care systs, Sci Coun Can, 72; assoc ed, Automatica, Behav Sci, Kubernetes, Methods Info & Med & J Gen Systs; adv ed ser biomed eng & health systs, John Wiley & Sons, Inc. *Mem:* Fel AAAS; sr mem Inst Elec & Electronic Engrs; Sigma Xi; Soc Prospective Med; Soc Gen Systs Res. *Res:* Health dynamics and promotion; general systems theory; control; homeostasis; hierarchy; optimization; health and spirituality. *Mailing Add:* Dept Health Car & Epidemiol Univ of BC 5804 Fairview Crescent Vancouver BC V6T 1W5 Can

MILTON, ALBERT FENNER, b New York, NY, Oct 16, 40; m 76. SOLID STATE PHYSICS. *Educ:* Williams Col, BA, 62; Harvard Univ, MA, 63, PhD(appl physics), 68. *Prof Exp:* Staff mem, Inst Defense Anal, 68-71; res physicist, Naval Res Lab, 71-76, dep head, Optical Techniques Br, 76-77, head Electro Optical Technol Br, 77-84; vpres, Roosevelt Ctr Am Policy Studies, 84-85; mgr, Electro Optics Lab, Electronics Lab, Gen Elec Co, 85-90; DIR TECHNOL, OFF ASST SECY OF THE ARMY (RDA), 90- *Mem:* Am Phys Soc; Optical Soc Am. *Res:* Integrated optics; fiber optics; IR focal plane arrays; infrared detection. *Mailing Add:* 5200 Reno Rd NW Washington DC 20015

MILTON, CHARLES, geology; deceased, see previous edition for last biography

MILTON, DANIEL JEREMY, geology; deceased, see previous edition for last biography

MILTON, JAMES E(DMUND), b Florala, Ala, May 12, 34; m 56; c 2. ENGINEERING SCIENCES, APPLIED PHYSICS. *Educ:* Univ Fla, BANE, 60, PhD(physics), 66. *Prof Exp:* Asst prof aerospace eng, Univ Fla, 66-73, resident dir & assoc engr, Grad Ctr, 73-77, RESIDENT DIR & ENGR, GRAD CTR, UNIV FLA, 77- *Concurrent Pos:* Consult, US Air Force Armament Lab, 71-72 & 79-81. *Mem:* Am Inst Aeronaut & Astronaut; Am Soc Eng Educ. *Res:* Flight dynamics; aerodynamics; terradynamics. *Mailing Add:* 265 S Bayshore Dr Valparaiso FL 32580

MILTON, JOHN CHARLES DOUGLAS, b Regina, Sask, June 1, 24; m 53; c 4. NUCLEAR PHYSICS. *Educ:* Univ Man, BSc, 47; Princeton Univ, MA, 49, PhD(physics), 51. *Prof Exp:* From asst res officer to sr res officer physics, Chalk River Nuclear Labs, 51-67, actg dir res, 85-86, dir physics div, 83-85, dir res, 85-86, vpres physics & health sci, 86-90, HEAD NUCLEAR PHYSICS, CHALK RIVER NUCLEAR LABS, ATOMIC ENERGY CAN, LTD, 67-, EMER RESEARCHER, 90- *Concurrent Pos:* Vis physicist, Lawrence Radiation Lab, Univ Calif, 60-62; dir res, Ctr Nuclear Res, Strasbourg, 75; vis physicist, Ctr Study, Bruyeres-le-Chatel, France, 75-76; chmn, Nuclear Physics Grants Selection Comt, Nat Sci & Eng Res Coun Can, 77- *Mem:* Royal Soc Can; Am Phys Soc; Can Asn Physicists (vpres). *Res:* Fission physics; directional correlation of radiations in radioactive decay; production of very high thermal neutron fluxes; intermediate energy physics; high voltage electrostatic accelerators. *Mailing Add:* Phys Sci Chalk River Nuclear Lab 53 Chalk River ON K0J 1J0 Can

MILTON, KIMBALL ALAN, b La Grande, Ore, Nov 29, 44; m 78; c 1. QUANTUM FIELD THEORY. *Educ:* Univ Wash, BS, 67; Harvard Univ, AM, 68, PhD(physics), 71. *Prof Exp:* Asst res physicist, Univ Calif, Los Angeles, 71-79, assoc res physicist, 79-81; from assoc prof to prof physics, Okla State Univ, 81-86; PROF PHYSICS, UNIV OKLA, 86- *Concurrent Pos:* Vis assoc prof physics, Ohio State Univ, 79-81; co-prin investr, US Dept Energy, 81-86, prin investr, 86-; collabr, Los Alamos Nat Lab, 83-; vis prof physics, Univ Okla, 85-86, Ohio State Univ, 89; assocs distinguished lectr, Univ Okla, 88. *Honors & Awards:* Regents' Award for Super Res, Univ Okla, 91. *Mem:* Am Phys Soc; AAAS; NY Acad Sci. *Res:* Theoretical high-energy physics; Casimir effect; nonperturbative aspects of quantum field theory; author of over 100 technical papers. *Mailing Add:* Dept Physics & Astron Univ Okla Norman OK 73019

MILTON, KIRBY MITCHELL, b St Joseph, Mich, May 4, 23; m 45, 68; c 8. CHEMISTRY, BIOCHEMISTRY. *Educ:* Harvard Univ, SB, 43; Univ Mich, MS, 48, PhD(org chem), 51. *Prof Exp:* Res chemist, Am Cyanamid Co, Conn, 43-44 & Manhattan Dist, 44-46; res & develop chemist, Eastman Kodak Co, 50-59, res assoc, 59-85; RETIRED. *Mem:* AAAS; Am Chem Soc; Am Inst Chemists; Soc Photog Scientists & Engrs. *Res:* Photographic emulsions and supports; gelatin; plasticizers; polymers. *Mailing Add:* 309 Fishers Rd Pittsford NY 14534

MILTON, NANCY MELISSA, b Salem, Ore, July 17, 42; c 1. GEOBOTANY, REMOTE SENSING. *Educ:* Howard Univ, BS, 73; Johns Hopkins Univ, PhD(geog), 81. *Prof Exp:* Botanist, Johns Hopkins Univ, 73-75; BOTANIST, US GEOL SURV, RESTON, VA, 75- *Concurrent Pos:* Adv & tutor, Wash Int Col, 74-77; vis prof, Mackay Sch Mines, Univ Nev-Reno, 88-89. *Mem:* AAAS; Inst Elec & Electronics Engrs; Asn Women Geoscientists; Am Soc Photogram & Remote Sensing. *Res:* Geobotanical remote sensing for geologic applications; biogeochemistry; biophysics. *Mailing Add:* MS 927 Nat Ctr US Geol Surv Reston VA 22092

MILTON, OSBORNE, b Denver, Colo, Oct 10, 20; m 45; c 4. ELECTRICAL & METALLURGICAL ENGINEERING. *Educ:* Univ Mo, BSEE, 45, MetEng, 56. *Prof Exp:* Trainee, Westinghouse Elec Corp, 45-46; researcher, Labs. Denver & Rio Grande Western RR Co, 47; mem staff, Los Alamos Sci Lab, 47-49; mem staff, Sandia Nat Lab, 49-56, sect supvr, 56-65, mem tech staff, 65-85; RETIRED. *Mem:* AAAS; sr mem Inst Elec & Electronics Engrs. *Res:* Electrostatic fields; dielectrics; x-ray diffraction; materials science; electromagnetic fields; lightning phenomena. *Mailing Add:* 9024 Los Arboles Ave NE Albuquerque NM 87112

MILTON, ROBERT MITCHELL, b St Joseph, Mich, Nov 29, 20; m 46; c 3. PRODUCT SAFETY & LIABILITY, ZEOLITE SYNTHESIS. *Educ:* Oberlin Col, AB, 41; Johns Hopkins Univ, MA, 43, PhD(phys chem), 44. *Prof Exp:* Res assoc, dept chem, Johns Hopkins Univ, 44-46; res scientist, Linde Div, 46-51, res supvr, 51-54, mgr develop, 54-58, asst mgr new prod, 58-59, asst res dir, 59-64, dir res, 64-73, exec vpres Showa Unox-Showa Union Gosei, joint venture, 73-77, dir agr bus develop, 77-79, vpres, Keystone Seed Co, subsid Union Carbide Corp, 78-85, assoc corp dir prod safety & liability, 80-85, CONSULT, UNION CARBIDE CORP, 86-; PRES, R MILTON ASSOCS, INC, 86- *Concurrent Pos:* Consult anti-submarine warfare, US Navy, 44; mem adv bd hyperbaric med, State Univ NY, 66-73; rep, Chem Indust Inst Toxicol, 81-85; mem subcomt environ & harzardous wastes, Nat Res Coun, 85-87; chmn bd trustees, Am Inst Chemists Found, 91; mem coun, Sci Soc Presidents. *Honors & Awards:* Jackob F Schoelkopf Medal, Am Chem Soc, 63; Pioneer Award, Am Inst Chemists, 80. *Mem:* Am Chem Soc; Am Inst Chemists (pres, 90-91); Am Inst Chem Engrs; AAAS. *Res:* Linde molecular sieve adsorbents and catalysts; Linde hi-flux boiling tubes; argon-oxygen decarburization of stainless steel; zeolite synthesis, adsorption and catalysis; cryogenic blood preservation; superconducting bolometers & infra-red detection; air separation and purification; boiling heat transfer; rare gas recovery and purification; cryobiology; gas physiology; superconductivity; waste water treatment; hybrid seed development. *Mailing Add:* 5991 Set N Sun Pl Jupiter FL 33458

MILTON, ROY CHARLES, b St Paul, Minn, Mar 10, 34; m 55; c 2. STATISTICAL COMPUTING, OPHTHALMOLOGY. *Educ:* Univ Minn, BA, 55, MA, 63, PhD(statist), 65. *Prof Exp:* Statistician, Atomic Bomb Casualty Comn, 64-66; sr scientist, Comput Ctr, Univ Wis-Madison, 66-70; statistician, Atomic Bomb Casualty Comn, 70-71; RES MATH STATISTICIAN, NAT EYE INST, 72- *Mem:* Int Statist Inst; Am Statist Asn; Biomet Soc; Asn Res Vision Ophthal. *Res:* Application of digital computers to applied and theoretical statistical research; biostatistical applications in ophthalmology. *Mailing Add:* Biomet & Epidemiol Prog Nat Eye Inst Bldg 31/6A18 Bethesda MD 20892

MILZ, WENDELL COLLINS, b Bedford, Ohio, Feb 21, 18; m 42; c 3. PHYSICAL CHEMISTRY. *Educ:* Hiram Col, BA, 40. *Prof Exp:* Asst chief chemist, Cleveland Plant, Aluminum Co Am, 40-48, res engr, Alcoa Res Labs, 48-57, asst chief lubricants div, 57-67; sci assoc, Alcoa Tech Ctr, 67-79; RETIRED. *Mem:* Sigma Xi. *Res:* Lubricants; friction; wear; forging. *Mailing Add:* 1725 Pleasant Ave New Kensington PA 15068

MILZOFF, JOEL ROBERT, b New York, NY, Mar 23, 44; m 71; c 1. ANALYTICAL CHEMISTRY, MEDICINE. *Educ:* Columbia Univ, BS, 66; Ind Univ Sch Med, MS, 69, PhD(toxicol), 72; Am Bd Forensic Toxicol, dipl. *Prof Exp:* TOXICOLOGIST, TOXICOL SECT, DEPT HEALTH LABS, CONN, 72- *Concurrent Pos:* Clin assoc, Dept Lab Med, Univ Conn Sch Med, 77- *Mem:* Fel Acad Forensic Sci; Soc Forensic Toxicologists; Sigma Xi. *Res:* Forensic toxicology. *Mailing Add:* Conn Dept Health Labs 10 Clinton St Hartford CT 06106

MIMMACK, WILLIAM EDWARD, b Eaton, Colo, Aug 22, 26. OPTICS. *Educ:* Univ Colo, BA, 56; Univ Rochester, PhD(optics), 73. *Prof Exp:* Res physicist, White Sands Missile Range, Dept of Army, 51-73; STAFF OPTICAL ENGR, KEUFFEL & ESSER CO, 74- *Honors & Awards:* Karl Fairbanks Award, Soc Photo-Optical Instrumentation Engrs, 61. *Mem:* Optical Soc Am; Am Phys Soc; Soc Photo-Optical Instrumentation Engrs. *Res:* Geometrical optics; lens design; lens design methods; optimization. *Mailing Add:* 4927 Wichita Cir El Paso TX 79904

MIMNAUGH, MICHAEL NEIL, b Detroit, Mich, Apr 15, 49; m 76; c 3. MEDICINAL CHEMISTRY. *Educ:* Univ Mich, Ann Arbor, BS, 71; Purdue Univ, PhD(med chem), 75. *Prof Exp:* Asst prof med chem, Univ Ill, Chicago, 76-82, assoc res scientist biochem, Med Sch, 82-83; ASSOC PROF CHEM, CHICAGO STATE UNIV, 83-, CHMN DEPT CHEM & PHYSICS, 88- *Mem:* Am Chem Soc. *Res:* Organic synthesis and in-vitro evaluation of potential beta adrenergic blockers; high performance liquid chromatography investigation of red blood cell membrane protein phosphorylation. *Mailing Add:* Dept Chem & Physics Chicago State Univ Chicago IL 60628

MIMS, CHARLES WAYNE, b Waukegan, Ill, May 3, 44; m 68; c 1. MYCOLOGY. *Educ:* McNeese State Col, BS, 66; Univ Tex, Austin, PhD(bot), 69. *Prof Exp:* Asst prof, 69-77, ASSOC PROF BIOL, STEPHEN F AUSTIN STATE UNIV, 77- *Mem:* Bot Soc Am; Mycol Soc Am. *Res:* Morphogenesis in fungi, ultrastructure. *Mailing Add:* Dept Biol Austin Bldg Stephen F Austin State Univ North St Nacogdoches TX 75962

MIMS, WILLIAM B, b Mansfield, Eng, Feb 13, 22. PHYSICS, BIOPHYSICS. *Educ:* Oxford Univ, Eng, MA, 52, PhD(nuclear physics), 52. *Prof Exp:* Sr sci officer nuclear physics, UK Atomic Energy Authority, 55-56; mem tech staff solid state physics, Bell Labs, Murray Hill, NJ, 56-85; consult, Exxon Res & Eng Co, 87-90; RETIRED. *Concurrent Pos:* Prof physics, Univ Calif, Los Angeles, 67-68; vis prof physics, Univ Konstanz, WGer, 72. *Mem:* Fel Am Phys Soc. *Res:* Solid-state physics; biophysics. *Mailing Add:* RD Two Box 31 Rocky Run Rd Glen Gardner NJ 08826

MIN, KONGKI, b Seoul, Korea, May 24, 31; m 58; c 1. NUCLEAR PHYSICS. *Educ:* Amherst Col, BA, 57; Univ Ill, PhD(physics), 63. *Prof Exp:* Res assoc physics, Univ Va, 62-63, asst prof, 63-68; assoc prof, 68-77, PROF PHYSICS, RENSSELAER POLYTECH INST, 77- *Mem:* Am Phys Soc. *Res:* Photonuclear reactions; nuclear spectroscopy. *Mailing Add:* Dept of Physics Rensselaer Polytech Inst Troy NY 12180

MIN, KWANG-SHIK, b Seoul, Korea, Sept 25, 27; m 56; c 2. MATHEMATICAL PHYSICS, NUCLEAR SCIENCE. *Educ:* Seoul Nat Univ, BS, 51; Univ Minn, MS, 59, PhD(physics), 61. *Prof Exp:* Instr physics, Seoul Nat Univ, 55-57; res assoc reactor physics, Argonne Nat Lab, 61-62; asst prof physics, Seoul Nat Univ, 62-64; assoc prof, 64-71, PROF PHYSICS, ETEX STATE UNIV, 71- *Concurrent Pos:* Res subcontract, E-Systs, Greenville, 83-; univ fac res grant, Etex State Univ, 65-; vis scientist, A F Armament Lab, 89-91. *Mem:* Am Phys Soc; Korean Phys Soc; Inst Elec & Electronic Engrs. *Res:* Digital signal processing; thermal neutron scattering; application of stochastic processes in physics. *Mailing Add:* 2802 Rix Commerce TX 75428

MIN, KYUNG-WHAN, b Seoul, Korea, May 5, 37; m; c 2. ELECTRON MICROSCOPY, VIRAL ONCOGENESIS. *Educ:* Seoul Nat Univ, MD, 62. *Prof Exp:* Asst instr path, Baylor Col Med, 65-68, asst prof, 68-70 & 71-78; assoc prof, Chosun Univ, 70-71; clin assoc prof path, Creighton Univ, 78-87; staff pathologist, Mercy Hosp Med Ctr, Des Moines, Iowa, 78-87, dir, EM Lab, 80-87; ASSOC PROF, OKLA UNIV HEALTH SCI CTR, 87-, DIR SURG PATH & ELECTRON MICROS, 87- *Concurrent Pos:* Staff pathologist, Vet Admin Med Ctr, Houston, Tex, 71-72, asst chief, 72-78. *Mem:* Fel Am Soc Clin Pathologists; Int Acad Path; Am Asn Pathologists; fel Am Col Physicians. *Res:* Pathology of cancer. *Mailing Add:* Dept Pathology Okla Univ Health Sci Ctr PO Box 26901 Oklahoma City OK 73190

MIN, TONY C(HARLES), b Shanghai, China, Jan 5, 23; nat US; m 68; c 1. MECHANICAL ENGINEERING, ENGINEERING MECHANICS. *Educ:* Chiao Tung Univ, BS, 47; Univ Tenn, MS, 53, PhD(eng sci), 69. *Prof Exp:* Instr mech eng, Univ Tenn, 50-51; mech engr, Div Design, Tenn Valley Authority, 51-54; res engr, Res Lab, Am Soc Heating, Refrig & Air-Conditioning Engrs, 54-57; assoc prof mech eng, Auburn Univ, 57-64; instr eng mech, Univ Tenn, 64-68; prof eng mech, Mich Technol Univ, 68-82; prog mgr, Res & Develop Br, Off Asst Secy Conserv & Solar Appln, Dept Energy, 78-79; PROF & CHMN MECH ENG, NC AGR TECHNOL STATE UNIV, GREENSBORO, 81- *Concurrent Pos:* NSF sci fac fel, Univ Minn, 61-63;

consult, Oak Ridge Nat Lab, 63-68, 88, Argonne Nat Lab, 62, 75, 82, Babcock & Wilcox, 69, Beloit Corp, 76-, Tenn Eastman, 89, Gen Atomics, 90; mem exec comt, Solar Energy Div, Am Soc Mech Engrs, 79-83 & 83-87, chmn, 81-82; mem exec comt, Energy Res Bd, 82-86, Bd Commun, 86-; assoc ed, J Solar Energy Eng, Am Soc Mech Engrs, 82-84; consult prof, Northwestern Polytech Univ, Xian, China, 87-, Huazhong Univ Sci & Technol, Wuhan, China, 87- *Mem:* Fel Am Soc Mech Engrs; Am Soc Eng Educ; Int Solar Energy; Am Acad Mech; NY Acad Sci; Sigma Xi. *Res:* Fluid mechanics; heat transfer; energy conversion and conservation; solar energy; environmental conditioning. *Mailing Add:* Dept Mech Eng NC Agr Technol State Univ Greensboro NC 27411

MINAH, GLENN ERNEST, b Providence, RI, Mar 15, 39; m 67; c 4. ORAL MICROBIOLOGY. *Educ:* Duke Univ, AB, 61; Univ NC, DDS, 66; Univ Mich, MS, 70, PhD(microbiol), 76. *Prof Exp:* Asst prof, 76-81, ASSOC PROF ORAL MICROBIOL & PEDIAT DENT, DENT SCH, UNIV MD, 81- *Concurrent Pos:* Consult, Children's Oral Health Prog, & Nat Caries Prog, Nat Inst Dent Res, NIH, 76- *Mem:* Am Soc Microbiol; Int Asn Dent Res; Acad Oral Med. *Res:* Microbial etiology; pathogenesis, diagnosis and treatment of oral diseases; dental caries; peridontal diseases; oral infections of immunosuppressed cancer patients. *Mailing Add:* Dept Microbiol Dent Sch Univ Md 666 W Baltimore St Baltimore MD 21201

MINARD, FREDERICK NELSON, neurochemistry, for more information see previous edition

MINARD, ROBERT DAVID, b Buffalo, NY, Mar 21, 41; m 66; c 2. MASS SPECTROMETRY. *Educ:* St Olaf Col, BA, 63; Univ Wis, PhD, 68. *Prof Exp:* Asst prof chem, Col of the Virgin Islands, 68-69; fel, Univ Ill, Chicago Circle, 69-70, asst prof, 70-73; LECTR CHEM, PA STATE UNIV, 73- *Mem:* Am Chem Soc; Am Soc Mass Spectrometry. *Res:* Mass spectrometry; pesticide degradation; coal structure. *Mailing Add:* Dept Chem Penn State Univ Univ Park PA 16802

MINASSIAN, DONALD PAUL, b New York, NY, Dec 8, 35; m 64; c 2. ALGEBRA, ACTUARIAL STUDIES. *Educ:* Fresno State Col, BA, 57; Brown Univ, MAT, 64; Univ Mich, Ann Arbor, MS, 65, EdD(math), 67; Ind Univ, JD, 86. *Prof Exp:* Assoc prof, 67-73, PROF MATH, BUTLER UNIV, 73- *Concurrent Pos:* Actuarial consult; teaching fel, Butler Univ. *Mem:* Assoc Soc Actuaries; Math Asn Am; Asn Study Grants Econ; Am Acad Actuaries. *Res:* Ordered algebraic structures, particularly ordered groups of modern algebra; foreign aid; industrial concentration; expository papers; acturial science; law. *Mailing Add:* Dept Math Butler Univ 4600 Sunset Ave Indianapolis IN 46208

MINATOYA, HIROAKI, b Japan, Nov 8, 11; US citizen; m 45; c 2. PHARMACOLOGY, PHYSIOLOGY. *Educ:* Univ Ore, BA, 38; Univ Ill, MS, 42; Nara Med Col, Japan, PhD(pharmacol), 64. *Prof Exp:* Res asst physiol, Col Med, Univ Ill, 41-42 & 44-48; pharmacologist, Sterling-Winthrop Res Inst, 48-71, sr res pharmacologist, 71-76; RETIRED. *Mem:* Am Soc Pharmacol & Exp Therapeut; Am Chem Soc; Sigma Xi. *Res:* Conditioned reflex and insulin hypoglycemia; experimental hypertension in dogs; diuretics; antihypertensive compounds in the renal hypertensive rat; catecholamines, especially absorption, metabolism and elimination; bronchodilators in dogs. *Mailing Add:* Ten Van Buren Ave East Greenbush NY 12061

MINC, HENRYK, b Lodz, Poland, Nov 12, 19; m 43; c 3. LINEAR ALGEBRA, COMBINATORIAL MATRIX THEORY. *Educ:* Univ Edinburgh, MA, 55, PhD(math), 59. *Prof Exp:* Lectr math, Dundee Tech Col, Scotland, 56-58; lectr, Univ BC, 58-59, asst prof, 59-60; assoc prof, Univ Fla, 60-63; prof, 63-90, EMER PROF MATH, UNIV CALIF, SANTA BARBARA, 90- *Concurrent Pos:* Prin investr, Air Force Off Sci Res, 60-83, Off Naval Res, 85-88; vis prof, Israel Inst Technol, 69-80; assoc ed, Linear & Multilinear Algebra, 75-; Lady Davis fel, 75, 78; mem adv bd, Inst Antiquity & Christianity. *Honors & Awards:* Ford Award, Math Asn Am, 66. *Mem:* Am Math Soc; Int Linear Algebra Soc; Inst Antiquity & Christianity. *Res:* Linear and multilinear algebra; matrix theory; combinatorial matrix theory; nonnegative matrices; theory of permanents. *Mailing Add:* Dept of Math Univ of Calif Santa Barbara CA 93106

MINCER, ALLEN I, b New York, NY, June 12, 57. ELEMENTARY PARTICLE PHYSICS. *Educ:* Brooklyn Col, BS, 78; Univ Md, PhD(high energy cosmic ray physics), 84. *Prof Exp:* Lady Davis fel, Technion-Israel Inst Technol, 84-86; res fel, Caltech, 86-89; ASST PROF, NY UNIV, 89- *Mem:* Am Phys Soc. *Res:* Experimental particle physics using accelerators and cosmic rays. *Mailing Add:* Dept Physics NY Univ Four Washington Pl New York NY 10003

MINCH, EDWIN WILTON, b Warren, Ohio, Apr 6, 51. AGRONOMY, ENTOMOLOGY. *Educ:* Cornell Univ, BS, 73; Ariz State Univ, PhD(zool), 77. *Prof Exp:* Asst prof anat & physiol, Ariz State Univ, 77, asst prof biol, 78-81; biologist, Western Sod Co, 81-83; asst in exten, Univ Ariz, 83-84; pesticide specialist, 84-88, ENVIRON SPECIALIST, ARIZ COMN AGR & HORT, 88- *Concurrent Pos:* Consult, Western Sod Co, 78-81; vis fac, Phoenix Col, 79- *Mem:* Asn Study Animal Behav; Am Soc Zoologists; Am Arachnological Soc; Brit Arachnological Soc. *Res:* Annual and daily activity patterns of arthropods, especially Araneae and Coccoidea; general turfgrass culture; pesticides. *Mailing Add:* 2207 W Main St Mesa AZ 85201

MINCH, MICHAEL JOSEPH, b Klamath Falls, Ore, Apr 7, 43; m; c 1. PHYSICAL ORGANIC CHEMISTRY. *Educ:* Ore State Univ, BS, 65; Univ Wash, PhD(chem), 70. *Prof Exp:* NIH fel chem, Univ Calif, Santa Barbara, 70-72; asst prof chem, Tulane Univ, 72-74; from asst prof to assoc prof, 74-82, PROF CHEM, UNIV PAC, 82- *Mem:* Am Chem Soc; Sigma Xi. *Res:* Protein-nucleic acid interactions; micellar catalysis; nuclear magnetic resonance spectroscopy of biochemically significant complexes. *Mailing Add:* Dept of Chem Univ of the Pac 3061 Pacific Ave Stockton CA 95211

MINCHAK, ROBERT JOHN, b Cleveland, Ohio, Sept 14, 29; m 57; c 5. COMPUTER SCIENCE. *Educ:* Case Tech, BS, 51; Univ Akron, MS, 62. *Prof Exp:* Chem engr, Hercules Powder Co, 51-53; chem engr, B F Goodrich, 55-60, sr chemist, 60-85, res & develop assoc, 86-88, SR RES & DEVELOP ASSOC, B F GOODRICH, 89- *Mem:* Am Chem Soc. *Res:* Polymerization; Ziegler Natta; metathesis; dienes, olefins and cyclic olefins; granted 20 patents; polymer synthesis including rubbers and plastics, thermosets. *Mailing Add:* Res & Develop Ctr B F Goodrich Corp Brecksville Rd Brecksville OH 44141

MINCHER, BRUCE J, b Cohoes, NY, Dec 8, 57. SOLVENT EXTRACTION STUDIES OF ACTINIDES & LANTHANIDES, ACTINIDE SEPARATIONS. *Educ:* State Univ NY, Albany, BS, 79; Univ Idaho, MS, 89. *Prof Exp:* Instr chem/radcon, Knolls Atomic Pur Lab, 80-82; sr tech, Pub Serv Co New Hampshire, 82-84; SR SCIENTIST, IDAHO NAT ENG LAB, EG&G IDAHO INC, 84- *Mem:* Am Nuclear Soc. *Res:* Gamma ray decomposition of organochlorines, the use of isotope sources of x-rays to decompose persistent, toxic halogenated hydrocarbons; decomposition mechanisms decomposition products with a goal of producing a waste treatment system. *Mailing Add:* EG&G Idaho Inc PO Box 1625 Idaho Falls ID 83415-7111

MINCK, ROBERT W, b Defiance, Ohio, Sept, 26, 34; m 58; c 4. ELECTRICAL ENGINEERING. *Educ:* Univ Notre Dame, BS, 56; Univ Wis-Madison, MS, 58, PhD(elec eng), 60. *Prof Exp:* Res scientist, Sci Lab, Ford Motor Co, 60-76, staff scientist, Sci Res Lab, 76-80; TECH SPECIALIST, ADVANCED DEVELOP OPER, FORD AEROSPACE & COMMUN CORP, 80-, STAFF SCIENTIST, AERONUTRONICS DIV. *Mem:* Inst Elec & Electronics Engrs. *Res:* Nonlinear optics; battery development; laser development; energy conversion. *Mailing Add:* Aeronutronic Div FACC Ford Motor Co Ford Rd Newport Beach CA 92663

MINCKLER, JEFF, b Knox, NDak, June 4, 12; m 33; c 6. PATHOLOGY. *Educ:* Univ Mont, AB, 37; Univ Minn, MA & PhD(neuroanat), 39; St Louis Univ, MD, 44. *Prof Exp:* Asst anat, Univ Minn, 37-39; from instr to asst prof, Creighton Univ, 39-41; instr, St Louis Univ, 41-43, instr path, 43-45; asst prof, Med Sch, Univ Ore, 45-46, prof gen path & actg head dept, Sch Dent, 49-59; assoc clin prof path, Med Sch, Univ Colo, 60-71; dir labs, Century City Hosp, 73-75; ASSOC RES PROF NEUROSURG, MED CTR, LOMA LINDA UNIV, 71-; DIR LABS, MAD RIVER COMMUNITY HOSP, ARCATA, CALIF, 75- *Concurrent Pos:* Dir labs, Gen Rose Mem Hosp, 60-71 & Eisenhower Med Ctr, 71-73; lectr radiol, Univ Calif, Los Angeles; adj prof, Univ Denver, 65-71 & speech & hearing, Humboldt State Univ, Arcata, Calif, 76- *Mem:* AAAS; Am Asn Anat; Am Asn Neuropath; Am Soc Clin Path; Int Acad Path. *Res:* General and speech pathology; neuropathology; neuroanatomy. *Mailing Add:* 3276 Buttermilk Lane Arcata CA 95521

MINCKLER, LEON SHERWOOD, JR, b Lockport, NY, Apr 4, 30; m; c 2. ORGANIC CHEMISTRY, PROCESS IMPROVEMENT. *Educ:* Univ Southern Ill, BA, 51; Northwestern Univ, PhD(org chem), 55. *Prof Exp:* Asst org chem, Northwestern Univ, 51-53; sr chemist, Esso Res & Eng Co, 55-67, res assoc, 67-74, res assoc, Exxon Chem Co, 74-84; RETIRED. *Mem:* Am Chem Soc; Am Inst Chemists. *Res:* Polymers; butyl rubber; cationic polymerization; stereochemistry; polymer modification; oil additives; process improvement. *Mailing Add:* 989 E Silver Lake Rd Oak Harbor WV 98277-8517

MINCKLER, TATE MULDOWN, b Kalispell, Mont, Apr 1, 34; m 56; c 5. PATHOLOGY. *Educ:* Reed Col, BA, 55; Univ Ore, MD, 59. *Prof Exp:* Pathologist, Nat Cancer Inst, Washington, DC, 63-65, head tissue path unit, 64-65; asst prof path & med systs analyst, Univ Tex M D Anderson Hosp & Tumor Inst, Houston, 65-69, chief sect med info mgt systs, Dept Biomath, 67-69; head dept med automation, Presby Med Ctr, 69-71; assoc prof lab med & dir comput div, Sch Med, Univ Wash, 71-75; assoc pathologist & adminr, Mad River Community Hosp, Arcata, Calif, 75-76, pathologist, 76-80; chief path, Al Hada Hosp, TAIF, Saudi Arabia, 81-84; PATHOLOGIST & MED DIR LABS, GEN HOSP, EUREKA, CALIF, 84- *Concurrent Pos:* Med dir, N Coast Community Blood Bank, Eureka, Calif, 85- *Mem:* AAAS; Am Soc Clin Path; fel Col Am Path; NY Acad Sci; AMA. *Res:* Development of medical information systems, especially pathology data automation, application of computers to medical problems; utilization of human tissues for research and patient care; clinical and anatomical pathology; cryobiology; medical records. *Mailing Add:* Gen Hosp 2200 Harrison Ave Eureka CA 95501

MINCKLEY, WENDELL LEE, b Ottawa, Kans, Nov 13, 35; m 56; c 4. ICHTHYOLOGY, AQUATIC ECOLOGY. *Educ:* Kans State Univ, BS, 57; Univ Kans, MA, 59; Univ Louisville, PhD, 62. *Prof Exp:* Asst prof zool, Western Mich Univ, 62-63; from asst prof to assoc prof, 63-77, PROF ZOOL, ARIZ STATE UNIV, 77-, DIR LOWER COLO RIVER BASIN RES LAB, 72- *Concurrent Pos:* Fac res grant, Western Mich, 62-63; NSF grant, 63-65; Fac Res Comt awards, Ariz State Univ, 63-65; Sport Fishery Inst res grant, 65-66. *Mem:* Am Soc Ichthyologists & Herpetologists; Am Fisheries Soc; Wildlife Soc; Am Soc Limnol & Oceanog; Am Inst Biol Sci. *Res:* Systematic and ecological ichthyology; radiation, stream, crustacean and algal ecology; crustacean taxonomy. *Mailing Add:* Dept of Zool Ariz State Univ Tempe AZ 85287

MINDA, CARL DAVID, b Cincinnati, Ohio, Nov 5, 43; m 73; c 3. MATHEMATICAL ANALYSIS. *Educ:* Univ Cincinnati, BS, 65, MS, 66; Univ Calif, San Diego, PhD(math), 70. *Prof Exp:* Asst prof math, Univ Minn, 70-71; from asst prof to assoc prof, 71-80, PROF MATH, UNIV CINCINNATI, 80-, HEAD, 90- *Concurrent Pos:* Vis assoc prof, Univ Calif, San Diego, 80-81 & 83, vis prof, 88-89. *Mem:* Am Math Soc; Math Asn Am. *Res:* Complex analysis; Riemann surfaces. *Mailing Add:* Dept of Math Sci Univ of Cincinnati Cincinnati OH 45221-0025

MINDAK, ROBERT JOSEPH, b Chicago, Ill, Mar 1, 25; m 53; c 6. MECHANICAL ENGINEERING. *Educ:* Northwestern Univ, US, 46, MS, 48. *Prof Exp:* Lectr mech eng, Northwestern Univ, 46-48; asst res engr, Standard Oil Co, Ind, 48-52; assoc res engr, Armour Res Found, Ill Inst Technol, 52-55; phys sci adminr, Off Naval Res, 55-68, staff asst to asst chief res, 68-74, mech engr, 74-80; pres & owner, Independent Nursing Serv, 85-90, CHMN BD, INDEPENDENT NURSING SERV, INC, 90- *Concurrent Pos:* Lectr, Ill Inst Technol, 50-56. *Mem:* Am Soc Mech Engrs. *Res:* Heat transfer; magnetohydrodynamics; gas dynamics; air conditioning; refrigeration; temperature measurement; thermodynamics; scientific and technical information; mechanized information systems; research and development management. *Mailing Add:* 3714 Forest Grove Dr Annandale VA 22003

MINDE, KARL KLAUS, b Leipzig, Ger, Dec 27, 33; Can citizen; c 3. PSYCHIATRY. *Educ:* Columbia Univ, MA, 60; Munich Univ, MD, 57. *Prof Exp:* Staff physician, Mont Children's Hosp & Queen Elizabeth Hosp, 65-71; WHO sr lectr psychiat, Makerere Univ, Uganda, 71-73; staff physician, Hosp Sick Children, 73, assoc prof, 73-78, dir psychiat res, 73-86; from asst prof to prof pediat, Univ Toronto, 75-86, prof psychiat, 78-86; prof & chair, dept psychiat, Queen's Univ, 86-89, prof pediat & psychol, 86-89; asst prof psychiat, 66-73, PROF PEDIAT & CHAIR, DIV CHILD PSYCHIAT, MCGILL UNIV, 89-; PROF PSYCHIAT & DIR, DEPT CHILD PSYCHIAT, MONTREAL CHILDREN'S HOSP, 89- *Concurrent Pos:* Staff physician, Montreal Childrens Hosp, 65-73; consult psychiat, Baird Residential Treatment Ctr, Burlington, Vt, 65-67. *Mem:* Can Psychiat Asn; Am Psychiat Asn; Am Acad Child Psychiat; Can Psychoanal Asn; Soc Res Child Develop. *Res:* Bonding between premature babies and their mothers, and a follow-up study of aggressive kindergarten children. *Mailing Add:* Montreal Children's Hosp 2300 Tupper St Montreal PQ H3H 1P3 Can

MINDEL, JOSEPH, b New York, NY, May 3, 12; m 34, 75; c 1. HISTORY OF SCIENCE, SCIENCE EDUCATION. *Educ:* City Col New York, BS, 32; Columbia Univ, MA, 37; NY Univ, PhD(chem), 43. *Prof Exp:* Instr high sch, NY, 32-46, head dept sci, 46-60; secy to pres, Bd Ed, NY, 60-61; staff mem, Lincoln Lab, Mass Inst Technol, 61-69, Secy steering comt, 69-75, educ dir, , 71-75, consult, 75-76; CONSULT, DEFENSE COMMUN AGENCY, US DEPT DEFENSE, 76- *Concurrent Pos:* Consult, Beth Israel Hosp, Boston, 75-76; consult & mem teaching staff, Wash Sch Psychiat, Wash,DC, 80-81. *Mem:* AAAS; NY Acad Sci. *Mailing Add:* 527 Grant Pl Frederick MD 21702

MINDELL, EUGENE R, b Chicago, Ill, Feb 24, 22; m 45; c 4. ORTHOPEDIC SURGERY. *Educ:* Univ Chicago, BS, 43, MD, 45. *Prof Exp:* prof orthop surg & head dept, Sch Med, State Univ NY, Buffalo, 64-89; RETIRED. *Concurrent Pos:* Nat Res Coun fel orthop surg, Univ Chicago Clin, 48-49; Orthop Res & Educ Found res grant, 58-61; NIH res grant, 63-; examr bone path, Am Bd Orthop Surg, 57-, mem, 78-84, pres, 83-84; residing, Orthop Surg Rev Comt, 86- *Mem:* Am Orthop Asn; Am Soc Surg Trauma; Orthop Res Soc (pres-elect, 71); Am Acad Orthop Surg; fel Am Col Surg; Muscoloskeletal Tumor Soc (pres, 90-91). *Res:* Mechanisms by which chondrogenesis occurs; bone pathology; structure and function of cartilage; chemotherapy in bone sarcoma; experimental studies on strength of remodeled bone; fate of bone grafts. *Mailing Add:* Dept Orthop Surg Sch Med State Univ NY 3435 Main St Buffalo NY 14214-3098

MINDEN, HENRY THOMAS, b New York, NY, Aug 23, 23; m 53; c 3. CHEMICAL PHYSICS. *Educ:* Johns Hopkins Univ, BA, 43; Columbia Univ, PhD(chem), 52. *Prof Exp:* Mem tech staff, Bell Tel Labs, Inc, 48 & RCA Labs, 50-52; sr physicist, Midway Labs, Chicago, 52-56; eng specialist, Res Labs, Sylvania Elec Prod, Inc, 56-60; physicist, Semiconductor Prod Dept, Gen Elec Co, 60-62; tech staff mem, Sperry Rand Res Ctr, 62-83; sr mem tech staff, Northrop Corp Precision Prods Div, Norwood Mass, 83-89; RETIRED. *Mem:* Am Phys Soc; Inst Elec & Electronic Engrs; Optical Soc Am. *Res:* Semiconductor materials and devices; laser optics; ring laser gyro mirror fabrilation and design; ion beam coating technology. *Mailing Add:* 31 Loring Rd Concord MA 01742

MINDESS, SIDNEY, b Winnipeg, Man, Aug 29, 40; m 81; c 2. FIBRE REINFORCED CONCRETE, IMPACT LOADING. *Educ:* Univ Man, BA, 64, BSc, 65; Stanford Univ, MS, 66, PhD(civil eng), 70. *Prof Exp:* PROF CIVIL ENG, UNIV BC, 69- *Concurrent Pos:* Vis scholar, Univ Ill, Urbana, 75-76; Lady Davis fel, Technion Israel Inst Technol, 81-82. *Mem:* Am Concrete Inst; fel Am Ceramic Soc; Can Soc Civil Engrs; Am Soc Testing & Mat; Int Union Testing & Res Lab Mat. *Res:* Mechanical properties of cementitious materials, and in particular fibre reinforced concrete; impact behaviour of concrete. *Mailing Add:* 1406 W 40th Ave Vancouver BC V6M 1V6 Can

MINDICH, LEONARD EUGENE, b New York, NY, May 24, 36; m 59; c 3. MICROBIAL PHYSIOLOGY. *Educ:* Cornell Univ, BS, 57; Rockefeller Univ, PhD, 62. *Prof Exp:* Asst virol, 62-64, assoc, 64-70, ASSOC MEM MICROBIOL, PUB HEALTH RES INST OF CITY OF NEW YORK, INC, 70- *Concurrent Pos:* Mem microbial chem study sect, NIH, 72-76. *Mem:* AAAS; Am Soc Microbiol. *Res:* Genetics of bacteriocin production, fractionation of desoxyribonucleic acid and the synthesis of bacterial membranes. *Mailing Add:* Dept Microbiol NY Univ Sch Med New York NY 10016

MINDLIN, HAROLD, b Bethlehem, Pa, Nov 23, 30; m 58; c 2. MATERIALS SCIENCE. *Educ:* Lehigh Univ, BS, 56, MS, 60. *Prof Exp:* Assoc stress analyst, Lockheed Aircraft Corp, 56-57; engr, Metall Res Div, Reynolds Metals Co, 60-63, supvr mech testing sect, 63-65; res engr, Battelle Mem Inst, 65-77, assoc div chief, struct mat eng, 67-69, sect mgr, 69-77, struct mat eng & prog mgr, Metals & Ceramics Info Ctr, Columbus Div, 77-90; CONSULT, 91- *Mem:* Am Soc Testing & Mat; Am Inst Aeronaut & Astronaut. *Res:* Fatigue, crack propagation and fracture behavior of materials and structures; materials information; numeric databases. *Mailing Add:* 171 S Cassingham Rd Bexley OH 43209-1846

MINDLIN, RAYMOND D(AVID), applied mechanics; deceased, see previous edition for last biography

MINDLIN, ROWLAND L, b New York, NY, Jan 30, 12; m 40, 76; c 2. PEDIATRICS, COMMUNITY HEALTH. *Educ:* Harvard Univ, BS, 33, MD, 37, MPH, 62. *Prof Exp:* Dir maternal & child health, Boston Dept Health & Hosps, 71-74; dir ambulatory care, St Mary's Hosp, Brooklyn, 74-78; vpres & med dir, Mile Sq Health Ctr, Chicago, 78-87; RETIRED. *Concurrent Pos:* Chmn coun child & adolescent health, Am Acad Pediat; mem comt med in soc, NY Acad Med. *Mem:* Am Acad Pediat; NY Acad Med; Am Pub Health Asn. *Res:* Medical care in maternal and child health. *Mailing Add:* 1924 Harbourside Dr Longboat Key FL 34228

MINEAR, ROGER ALLAN, b Seattle, Wash, June 16, 39; m 66; c 2. ENVIRONMENTAL CHEMISTRY, CHEMICAL LIMNOLOGY. *Educ:* Univ Wash, BS, 64, MSE, 66, PhD(civil eng), 71. *Prof Exp:* Lectr environ chem, Dept Civil Eng, Ore State Univ, 66-67; asst prof environ eng & sci, Dept Environ Eng, Ill Inst Technol, 70-73; from assoc prof to prof environ eng & sci, Dept Civil Eng, Univ Tenn, 73-82, Armour T Granger Prof, 83-84; sr scientist, Radian Corp, Austin, Tex, 80-81; DIR INST ENVIRON STUDIES, UNIV ILL URBANA-CHAMPAIGN, 85- & INTERDISCIPLINARY ENVIRON TOXICOL PROG & OFF SOLID WASTE RES, 87- *Concurrent Pos:* Mem, Bd Dirs, Asn Environ Eng Prof, 78-80; mem, Environ Studies Bd, Bd Environ Sci & Toxicol, Nat Res Coun, 83-86 & Tech Comt, Int Conf Chem Environ, Lisbon, Portugal, 86; invited lectr, Nankai Univ, Tianjin, People's Repub China, 88; mem, Bd Sci Counr, Agency Toxic Substances & Dis Registry, Dept Health & Human Serv, 88-93 & Sci Adv Comt, Hazardous Waste Res Ctr, La State Univ, 89-91; treas, Environ Chem Div, Am Chem Soc, 77-82, chmn, 84-85, counr, 89-91. *Honors & Awards:* 1st Distinguished Serv Award, Asn Environ Eng Prof, 84; Distinguished Serv Award, Am Chem Soc, 85. *Mem:* Asn Environ Eng Prof (vpres, 79, pres, 80); Am Chem Soc; Am Water Works Asn; Am Soc Limnol & Oceanog; Am Soc Civil Engrs; Water Pollution Control Fedn. *Res:* Nature, origin, transport, and transformation of organic and inorganic compounds in natural and wastewaters; chemistry of aqueous solutions and chemical processes of water and wasterwater treatment; trace and environmental analysis. *Mailing Add:* Inst Environ Studies Univ Ill 1101 W Peabody Dr Urbana IL 61801

MINEHART, RALPH CONRAD, b Mitchell, SDak, Jan 25, 35; m 59; c 4. EXPERIMENTAL NUCLEAR PHYSICS. *Educ:* Yale Univ, BS, 56; Harvard Univ, MA, 57, PhD(physics), 62. *Prof Exp:* Res assoc & lectr physics, Yale Univ, 62-66; from asst prof to assoc prof, 66-81, PROF, UNIV VA, 81- *Concurrent Pos:* Vis staff mem, Swiss Inst Nuclear Res, 75-76. *Mem:* Am Phys Soc; Sigma Xi. *Res:* Electroproduction of nucleon resonances; medium energy nuclear physics; interactions of pi mesons with nucleons and nuclei; meson decay and neutrino mass; electro-nuclear interactions. *Mailing Add:* Dept Physics Univ Va Charlottesville VA 22901

MINER, BRYANT ALBERT, b Moroni, Utah, Aug 9, 34; m 60, 87; c 10. PHYSICAL CHEMISTRY. *Educ:* Univ Utah, BA, 61, PhD(phys chem), 65. *Prof Exp:* From asst prof to assoc prof, Weber State Col, 64-73, PROF CHEM, WEBER STATE UNIV, 73- *Mem:* Am Chem Soc. *Res:* Significant structure theory of liquids and electrochemistry; reactions and kinetics. *Mailing Add:* Dept Chem 2503 Weber State Univ Ogden UT 84408-0002

MINER, ELLIS DEVERE, JR, b Los Angeles, Calif, Apr 16, 37; m 61; c 7. UNMANNED EXPLORATION OF PLANETS. *Educ:* Utah State Univ, BS, 61; Brigham Young Univ, PhD(physics & astron), 65. *Prof Exp:* Sr scientist, 65-77, asst proj scientist, 77-90, C/C PROJ SCI MGR, JET PROPULSION LAB, 90- *Concurrent Pos:* Mem, Speaker's Bur, Jet Propulsion Lab, 77- *Mem:* Am Astron Soc. *Res:* Science planning and analysis on Mariners 6, 7, 9 to Mars, Marine, 10 to Venus and Mercury, Viking 1 and 2 to Mars, Voyagers 1 and 2 to Jupiter and Saturn, Voyager 2 to Uranus and Neptune; comet rendezvous asteroid flyby mission, Cassini Saturn orbiter mission. *Mailing Add:* 11335 Sunburst St Lake View Terrace CA 91342

MINER, FREND JOHN, b Loveland, Colo, Dec 8, 28; m 58; c 2. ANALYTICAL CHEMISTRY. *Educ:* Univ Colo, BA, 50; Ore State Univ, MS, 52, PhD(chem), 55. *Prof Exp:* Asst chem, Ore State Univ, 50-54; sr res chemist, 54-68, assoc scientist, 68-77, MGR CHEM RES & DEVELOP, ROCKY FLATS PLANT, ROCKWELL INT, 77- *Mem:* Am Chem Soc; Am Soc Test & Mat; Sigma Xi. *Res:* Ion exchange separations; complex ions; chemistry of plutonium. *Mailing Add:* 2445 Dartmouth Ave Boulder CO 80303

MINER, GARY DAVID, b Waseca, Minn, Dec 17, 42; m 69; c 2. BEHAVIORAL GENETICS, NEUROSCIENCE. *Educ:* Hamline Univ, BS, 64; Univ Wyo, MS, 66; Univ Kans, PhD(genetics), 70. *Prof Exp:* Res assoc psychiat genetics, Univ Minn, 70-72, asst prof, 72-77; prof biol & behav genetics, Northwest Nazarene Col, 77-82, actg chmn, Dept Biol, 78-82; prof biol, Eastern Nazarene Col, 82-85; FOUNDER & DIR, ALZHEIMERS FOUND, 85- *Concurrent Pos:* Prin investr, Nichiman Res Found grant, 77-78, Res Corp grant, 78-80; fel psychiat epidemiol, Univ Iowa, 80-82. *Mem:* AAAS; Soc Neurosci; Behav Genetics Asn; Int Soc Neurochem; Am Acad Neurol. *Res:* Brain proteins in schizophrenia, dementia and old age; Alzheimer's disease; brain proteins and aluminum; genetics of behavior. *Mailing Add:* 8177 S Harvard M-C-114 Tulsa OK 74137

MINER, GEORGE KENNETH, b Asheville, NC, Dec 16, 36; m 62; c 3. MAGNETIC RESONANCE, TRANSPORT PROPERTIES. *Educ:* Thomas More Col, AB, 58; Univ Notre Dame, MS, 60; Univ Cincinnati, PhD(physics), 65. *Prof Exp:* From instr to prof physics, Thomas More Col, 64-76, chmn dept, 66-76; assoc prof, 76-83, PROF PHYSICS, UNIV DAYTON, 83- *Concurrent Pos:* NSF sci fac fel, Univ Dayton, 71; Nat Coun Soc Physics Students, 80-93; dir, Univ Dayton, NSF Chautauqua Field Ctr, 80- *Mem:* Am Phys Soc; Am Asn Physics Teachers; Sigma Xi. *Res:* Radiation damage; electron paramagnetic resonance of rare earths in fluorites. *Mailing Add:* Dept Physics Univ Dayton Dayton OH 45469-2314

MINER, GORDON STANLEY, b Howell, Mich, Apr 25, 40; m 64; c 2. SOIL FERTILITY, AGRONOMY. *Educ:* Mich State Univ, BS, 62, MS, 64; NC State Univ, PhD(soil sci), 69. *Prof Exp:* Res agronomist, Rockefeller Found, 69-71; regional dir, Int Soil Fertil Eval & Improv Proj, Costa Rica & Nicaragua, 71-74; from asst prof to assoc prof, 74-86, PROF SOIL SCI, NC STATE UNIV, 86- *Mem:* Sigma Xi; Am Soc Agron. *Res:* Plant bed and field management of tobacco for mechanized production. *Mailing Add:* Box 7619 NC State Univ Raleigh NC 27695-7619

MINER, J RONALD, b Scottsburg, Ind, July 4, 38; m 63; c 3. LIVESTOCK WASTE MANAGEMENT, ODOR CONTROL. *Educ:* Univ Kans, BSE, 59; Univ Mich, MSE, 60; Kans State Univ, PhD(chem eng), 67. *Prof Exp:* Sanitary engr, Kans State Dept Health, 59-64; asst prof agr eng, Iowa State Univ, 67-71, assoc prof, 71-72; from assoc prof to prof agr eng, Ore State Univ, 72-86, head, 76-86, actg assoc dean, Col Agr Sci, 83-84, assoc dir int res & develop, 86-90, WATER QUAL SPECIALIST, ORE STATE UNIV, 90- *Concurrent Pos:* Environ engr, UN Develop Prog, Singapore, 80-81; expert witness, Various law firms & livestock enterprises faced with pollution complaints; consult, Am Nat Cattlemen's Asn, 77-; vis prof agr eng, Univ Fla, 90-91. *Mem:* Am Soc Agr Engrs; Water Pollution Control Fedn; Am Soc Engrs Educ; Sigma Xi. *Res:* Minimize or eliminate adverse environmental impacts of livestock and poultry production, particular emphasis on animal waste management and odor control; author or co-author of over 100 technical publications, a two volume textbook and three books of children's sermons. *Mailing Add:* Dept Bioresource Eng Ore State Univ Corvallis OR 97331

MINER, JAMES JOSHUA, b Waldo, Ark, June 11, 28; m 57; c 4. NUTRITION, BIOCHEMISTRY. *Educ:* Univ Ark, Fayetteville, BS, 53, MS, 54; La State Univ, Baton Rouge, PhD(nutrit), 62. *Prof Exp:* Poultry husbandman, Poultry Res Br, Agr Res Serv, USDA, 56-58; dir res, Ala Flour Mills, Nebr, 60-62; nutritionist, Loret Mills, Seed Feed Supply Co, 62-65; exten serv poultry man, Univ Ark, 65-66; vpres, Pilgrim Industs, Inc, 66-90, VPRES FARM PROD, PILGRIM'S PRIDE CORP, 90- *Mem:* Am Soc Animal Sci; Poultry Sci Asn. *Res:* Biological evaluation of protein supplements for non-ruminants. *Mailing Add:* Pilgrim Industs Inc 110 S Texas St Pittsburgh TX 75686

MINER, KAREN MILLS, b Callicoon, NY; m; c 2. INFORMATION RETRIEVAL, MEDICAL WRITING. *Educ:* State Univ NY, Plattsburgh, BA, 69; Univ Md, PhD(biochem), 74. *Prof Exp:* NIH fel microbiol, Case Western Reserve Univ, 74-75; vis asst prof biol, Kent State Univ, 75-76; asst prof chem & biol, Lake Erie Col, 76-77; asst prof biol, Sam Houston State Univ, 77-79; fel molecular biol, Univ Calif, Irvine, 79-82; sr res immunologist, Merck Sharp & Dohme Res Labs, 82-84; sr lit scientist, Am Cyanamid Co, 84-87; PRES BIOMED, BIOMED INFO CTR, INC, 87- *Concurrent Pos:* NIH res fel, Univ Calif, 80-82; inst res grant, Am Cancer Soc, 80-81. *Mem:* Am Soc Cell Biol; Drug Info Asn. *Res:* Medical author in areas of biochemistry, molecular biology, immunology and medicine. *Mailing Add:* Biomed Info Ctr Inc PO Box 1298 Ridgewood NJ 07451-1298

MINER, MERTHYR LEILANI, b Honolulu, Hawaii, Apr 13, 12; m 37; c 4. VETERINARY PATHOLOGY. *Educ:* Utah State Univ, BS, 37; Iowa State Univ, DVM, 41. *Prof Exp:* Asst vet bact, Mich State Univ, 41-43; from asst prof to assoc prof, 43-54, head dept, 54-73, prof, 54-77, EMER PROF VET SCI, UTAH STATE UNIV, 77- *Concurrent Pos:* Fel, Ralston Purina Co, Univ Minn, 53-54. *Mem:* Am Vet Med Asn; US Animal Health Asn; Am Asn Avian Path. *Res:* Microbiology; staphylococci; salmonellae; avian skeletal diseases; bovine coccidia. *Mailing Add:* 996 Sumac Dr Logan UT 84321

MINER, ROBERT SCOTT, JR, b Chicago, Ill, June 16, 18; m 42; c 3. ORGANIC CHEMISTRY. *Educ:* Univ Chicago, SB, 40; Polytech Inst Brooklyn, MS, 53; Princeton Univ, MA, 55, PhD(org chem), 56. *Prof Exp:* Develop chemist, Merck & Co, Inc, NJ, 40-44; chief res chemist, Tung-Sol Lamp Works, Inc, 45-47; asst mfg chemist, Ciba Pharmaceut Co, 47-49, mfg chemist, 49-58, mgr chem mfg div, 58-59, dir, 59-69; asst to chmn, Dept Physics, Princeton Univ, 70-74, res assoc fac chem, 74-75, mem prof res staff, 75-81; CONSULT CHEMIST ENGR, 70- *Concurrent Pos:* Fel & eve instr, Union Jr Col, 56-58, guest lectr, 60; guest lectr, Westfield High Sch, 60-; mem, Westfield Bd Educ, 62-65, pres, 64-65; guest lectr, MacMurray Col, 63, trustee, 65-70; chem consult, Haemodialysis Team, Overlook Hosp, 64-; NSF fel chem catalysis, USSR, 75, 76, & 78; consult chemist/engr, UN, 77-79, Int Exec Corp, Turkey & Brazil, 83-; lectr chem safety, 79- *Honors & Awards:* Honor Scroll Award, Phi Beta Kappa, 40, Sigma Xi, 56 & Am Inst Chem, 71. *Mem:* AAAS; Am Chem Soc; Royal Soc Chem; fel Am Inst Chem (treas, 70-75); Am Inst Chem Eng. *Res:* Synthesis of carbohydrates; vitamins; sulfonamides; steroid hormones; pharmaceuticals; chemical education; catalysis and electrode phenomena involving precious metals; large-scale pharmaceutical and intermediate synthesis. *Mailing Add:* 1139 Lawrence Ave Westfield NJ 07090

MINERBO, GERALD N, b Alexandria, UAR, Nov 21, 39; US citizen; m 66; c 2. THEORETICAL PHYSICS. *Educ:* Polytech Inst Brooklyn, 60; Cambridge Univ, PhD(theoret physics), 65. *Prof Exp:* Res assoc theoret physics, Atomic Energy Res Estab, Harwell, Eng, 64-65; proj physicist, Vitro Lab, Vitro Corp Am, 65-66; asst prof physics, Adelphi Univ, 66-71; STAFF MEM, LOS ALAMOS SCI LAB, 71- *Concurrent Pos:* Consult, Vitro Corp Am, 66-68; vis prof, Univ Buenos Aires, 68-69. *Mem:* Am Phys Soc. *Res:* Theory of elementary particles; formal theory of scattering; atomic physics; radiation transfer; laser theory. *Mailing Add:* 2314 Canyon Meadows Dr Missouri City TX 77489

MINERBO, GRACE MOFFAT, m; c 2. PATHOLOGY INTERNAL MEDICINE. *Educ:* City Univ NY, BS, MA; NY Univ, PhD; Univ Autonoma, Juarez, MD. *Prof Exp:* Res scientist mammalian radiobiol, Brookhaven Nat Lab, Upton, Long Island, NY, 61-63; res scientist radiobiol, Strangeways Res Lab, Cambridge, England, 63-65; instr anat, Dept Anat, Downstate Med Ctr, State Univ NY, Brooklyn, 65-71; adj asst prof cell physiol, Univ NMex, Los Alamos Grad Ctr, NMex, 72-75; asst prof, Depts Anat & Physiol, Sch Med, Univ NMex, Albuquerque, 72-75; med clerk, New Hyde Park, NY, Long Island Jewish Hillside Med Ctr, New Hyde Park, NY & Queens Hosp Ctr, Jamaica, NY, 77-78; clin med, Sante Fe, NMex, 78-80; fel, Dept Path, Sch Med, Univ Tex Health Sci Ctr, 81-83; VIS ASSOC PROF, SCH PHYS THER, TEX WOMAN'S UNIV HOUSTON, 83- *Concurrent Pos:* Teaching fel gen biol, physiol & comp anat, Grad Sch Arts & Sci, NY Univ, 61-62, instr gen biol, Wash Square Col, 62; Brit Empire Cancer Campaign res grant, Strangeways Res Lab, Cambridge, Eng, 63-65; Radiation Safety Comn, Texas Woman's Univ, 84-86; mem PhD Adv Comt, 84-90, Human Subj Rev Comt, Houston, 85-89; chair, AIDS Task Force, Houston Ctr, Texas Woman's Univ, 88-90, sem ser, 87-90; fel, phys chem, NSF & radiobiol, Atomic Energy Comn. *Mem:* Sigma Xi; NY Acad Sci; Am Med Asn. *Res:* Radiobiology; cell population kinetics; nucleic acid metabolism and protein synthesis; histopathology of musculoskeletal systems; pathophysiology therapeutic exercise; Parkinsonism; AIDS; atherosclerosis; aneurisms; author of numerous publications. *Mailing Add:* 2314 Canyon Meadows Dr Missouri City TX 77489

MINES, ALLAN HOWARD, b New York, NY, Apr 11, 36; m 59; c 2. PHYSIOLOGY. *Educ:* Univ Ill, Urbana, BS, 61, MS, 62; Univ Calif, San Francisco, PhD(physiol), 68. *Prof Exp:* From lectr to asst prof, 68-74, ASSOC PROF PHYSIOL, MED CTR, UNIV CALIF, SAN FRANCISCO, 74- *Mem:* Am Physiol Soc. *Res:* Comparative physiology; regulation of respiration, particularly the mechanism responsible for ventilatory acclimatization to altitude. *Mailing Add:* Dept Physiol Univ Calif Med Ctr 513 Parnassus Ave San Francisco CA 94143

MINET, RONALD G(EORGE), b New York, NY, Aug 13, 22; m 77; c 1. CHEMICAL ENGINEERING. *Educ:* City Col New York, BChE, 43; Stevens Inst Technol, MS, 50; NY Univ, DSc(chem eng), 59. *Prof Exp:* Res engr, Foster Wheeler Corp, 47-50, group leader, 50-52; chief process engr, United Engrs & Constructors, Inc, 52-61; mgr carbon dept, FMC Corp, 61-63; vpres process eng, Comp Tecnica Ind Petroli, 65-69; managing dir, Chem Projs Int, 69-72; pres, 72-78, CHMN BD, KINETICS TECHNOL INT CORP, 78-; CHMN BD, PYROTEC, 81- *Mem:* Am Chem Soc; Am Inst Chem Engrs; Am Inst Mining, Metall & Petrol Engrs. *Res:* Coal processing; catalytic processes; fluidized bed processes; petroleum and petrochemical processing; methanol. *Mailing Add:* 592 Garfield Ave South Pasadena CA 91030-2211

MINFORD, JAMES DEAN, b Clairton, Pa, Feb 27, 23; m 44; c 2. ADHESIVE JOINING, METALLURGICAL CHEMISTRY. *Educ:* Carnegie Inst Technol, BS, 43; Univ Pittsburgh, MLitt, 48, PhD(chem), 51. *Prof Exp:* Res chemist org chem, Goodyear Tire & Rubber Co, 44; asst chem, Univ Pittsburgh, 46-50, res assoc biochem, Sch Pub Health, 50-52, instr biochem & nutrit, 52-53; sect head, Alcoa Labs, 53-72, group leader, Chem Metall Div, 72-76, sci assoc alloy technol, 76-78, sci assoc, Joining Div, 78-83; OWNER, MINFORD CONSULT, 83- *Concurrent Pos:* Consult, Ky Steel Corp, 83, Andrews Corp, 83, Sea-Land Corp, 84- & Loctite Corp, 84-85; lectr, Univ Gothenburg, Sweden, 84, Granges Aluminium, Sweden, 84; auth. *Honors & Awards:* Adhesives Age Award, Am Soc Testing & Mat, 82. *Mem:* Sigma Xi; Adhesion Soc; Am Soc Testing & Mat; fel Am Inst Chemists. *Res:* Porphyrin synthesis in microorganisms; cancer in experimental animals; nutritional factors in disease; corrosive action of cooling waters on metals; reactions of aluminum and halogenated hydrocarbons; adhesives; high polymers; surface cleaning aluminum; metal joint durability; joining dissimilar materials. *Mailing Add:* One Harleston Green Hilton Head Island SC 29928

MING, LI CHUNG, b Shantung, China, July 19, 45; m 72. CRYSTAL CHEMISTRY, X-RAY DIFFRACTION. *Educ:* Nat Taiwan Univ, BS, 67; Univ Rochester, MS, 71, PhD(geol), 74. *Prof Exp:* Teaching asst, Nat Taiwan Univ, 68-69; teaching asst, Univ Rochester, 70-73, res asst, 73-74, res assoc, 74-75; asst geophysicist, 76-80, assoc geophysicist, 80-86, GEOPHYSICIST UNIV HAWAII, 86-, PROF, 88- *Mem:* Am Geophys Union; Sigma Xi; AAAS. *Res:* Laboratory simulation of the physical conditions, temperature and pressure, of the earth's deep interior and postulation of the most likely chemical constituents and the stable mineral assemblages. *Mailing Add:* Hawaii Inst Geophysics 2525 Correa Rd Honolulu HI 96822

MING, SI-CHUN, b Shanghai, China, Nov 10, 22; US citizen; m 57; c 6. PATHOLOGY. *Educ:* Nat Cent Univ, China, MD, 47. *Prof Exp:* Resident path, Mass Gen Hosp, Boston, 52-56; from instr to asst prof, Harvard Med Sch, 56-67; assoc prof, Med Sch, Univ Md, 67-71; PROF PATH, MED SCH, TEMPLE UNIV, 71- *Concurrent Pos:* Assoc pathologist, Beth Israel Hosp, Boston, 56-67; Nat Cancer Inst sr fel, Dept Tumor Biol, Karolinska Inst, Sweden, 64-65. *Mem:* Int Acad Path; Am Asn Pathologists; NY Acad Sci; AAAS. *Res:* Digestive tract disease, gastrointestinal oncology, carcinogenesis and tumor-host relationship. *Mailing Add:* Dept of Path Temple Univ Med Sch Philadelphia PA 19140

MINGES, MERRILL LOREN, b Denver, Colo, Sept 25, 37; m 60; c 4. MATERIALS SCIENCE, CHEMICAL ENGINEERING. *Educ:* Mass Inst Technol, BSc, 59, MSc, 60; Ohio State Univ, PhD(chem eng), 68; Stanford Univ, Sloan Exec Fel, Dipl, 72. *Prof Exp:* Proj engr, 60-63, group leader thermophys res, 64-65, tech mgr thermal protection systs, 66-71, br chief, Elastomers & Coatings Br, 73-75, asst div chief, Electromagnetic Mats Div, 75-77, asst chief, Systs Div, 77-79, chief, Electromagnetics Div, 79-85, DIR, NON-METALLIC MATS DIV, US AIR FORCE MATS LAB, 86- *Concurrent Pos:* Consult, Adv group Aerospace Res & Develop, NATO, 66-; consult, Adv Res Projs Agency, 67-; consult math adv bd, Nat Acad Sci, 68-; Sloan exec fel, Stanford Univ, 71-72; mem adv bd, Int J High Temperatures-High Pressures; Sloan Exec Fel, Stanford Univ, 72; task group chmn, Comt Data Sci & Technol, Int Coun Sci Union, 78- *Honors & Awards:* US Air Force Sci Achievement, 68, 69; Int Thermal Conductivity Award, 84. *Mem:* Assoc fel Am Inst Aeronaut & Astronaut; Sigma Xi. *Res:* High temperature thermophysical properties; high temperature heat transfer; materials applications for advanced reentry vehicles, space and propulsion systems; thermo-optical properties. *Mailing Add:* 2071 Beaver Valley Rd Beaver Creek OH 45385-9521

MINGLE, JOHN O(RVILLE), b Oakley, Kans, May 6, 31; m 57; c 2. NUCLEAR ENGINEERING, CHEMICAL ENGINEERING. *Educ:* Kans State Univ, BS, 53, MS, 58; Northwestern Univ, PhD(chem eng), 60; Washburn Univ, JD, 80. *Hon Degrees:* JD, Washburn Univ, 80. *Prof Exp:* Training engr, Gen Elec Co, 53-54; instr chem eng, 56-58, from asst prof to assoc prof nuclear eng, 60-65, Black & Veatch distinguished prof, 73-77, PROF NUCLEAR ENG, KANS STATE UNIV, 65-, DIR INST COMPUTATIONAL RES ENG, 69-, EXEC VPRES, RES FOUND, 83- *Concurrent Pos:* Res grants, NSF, 62-65, 65-67, Petrol Res Fund, 63-66, Dept Defense, 68-71 & AEC, 69-71; vis prof, Univ Southern Calif, 67-68; consult, Gulf-Gen Atomic, 68 & Wilson & Co, Engrs, 72-81; resident res assoc, Argonne Nat Lab, 76-77; patent atty, 84. *Mem:* Am Nuclear Soc; Am Inst Chem Engrs; Am Soc Eng Educ; Nat Soc Prof Engrs; Sigma Xi. *Res:* Nuclear heat transfer; applied mathematics; numerical analysis; nuclear transport theory; engineering law; patent law. *Mailing Add:* Res Found Fairchild Hall Kans State Univ Manhattan KS 66506

MINGORI, DIAMOND LEWIS, b Jersey City, NJ, June 7, 38; m 60. DYNAMICS, CONTROL SYSTEMS. *Educ:* Univ Calif, Berkeley, BS, 60, Los Angeles, MS, 62; Stanford Univ, PhD(aeronaut, astronaut), 66. *Prof Exp:* Mem tech staff, Aerospace Corp, 66-68; from asst prof to assoc prof, 68-79, PROF APPL MECH, UNIV CALIF, LOS ANGELES, 79- *Concurrent Pos:* Consult, Aerospace Corp, Calif, 68-70, 77-; consult, Comsat Corp, Md, 69-70, Jet Propulsion Lab, Calif Inst Technol, 76-, Aerojet Elec Optical Systs, 77, TRW Systs, 78-, Acurex Corp, 78, NASA Adv Comn, 78-83; Fulbright sr scholar, 80-81. *Mem:* Am Inst Aeronaut & Astronaut; Inst Elec & Electronic Engrs Control Systs Soc. *Res:* Dynamics and control with applications in aerospace vehicle attitude control; attitude dynamics and control of gravity gradient, dual-spin and large flexible craft. *Mailing Add:* Mech Aerospace & Nuclear Eng Dept Univ Calif 38-137L Engr IV Los Angeles CA 90024-1597

MINGRONE, LOUIS V, b Pittsburgh, Pa, Jan 27, 40; m 64; c 3. BOTANY. *Educ:* Slippery Rock State Col, BS, 62; Ohio Univ, MS, 64; Wash State Univ, PhD(bot, taxon), 68. *Prof Exp:* Assoc prof, 68-73, PROF BIOL & BOT, BLOOMSBURG STATE COL, 73-, asst chmn, 78-85, CHAIRPERSON, 86- *Mem:* Bot Soc Am; Am Soc Plant Taxon; Am Inst Biol Sci; Sigma Xi. *Res:* Plant systematics; cytotaxonomy and phytochemistry of the secondary plant constituents; floristic studies. *Mailing Add:* Dept Biol Allied Health Bloomsburg Univ Bloomsburg PA 17815

MINIATS, OLGERTS PAULS, b Besagola, Lithuania, Sept 13, 23; Can citizen; m 47; c 3. VETERINARY MEDICINE. *Educ:* Univ Toronto, DVM, 55; Univ Guelph, MSc, 66. *Prof Exp:* Clinician, High River Vet Clin, 55-64; asst prof, 66-72, assoc prof clin res, 72-81, PROF, DEPT CLIN STUDIES, ONT VET COL, UNIV GUELPH, 81- *Mem:* Can Vet Med Asn; Asn Gnotobiotics; Asn Advan Baltic Studies; Am Asn Swine Practitioners; Latvian Am Asn Univ Teachers & Professors. *Res:* Development of techniques for procurement and rearing of gnotobiotic research animals; investigation of the characteristics of gnotobiotic swine; application of gnotobiotic animals for the investigation of infectious diseases. *Mailing Add:* 30 Eleanor Ct Guelph ON N1E 1S8 Can

MINICH, MARLIN, b Orange, Calif, Aug 13, 38; m 60; c 5. CIVIL ENGINEERING, ENGINEERING MECHANICS. *Educ:* Fenn Col, BCE, 61; Case Western Reserve Univ, MSEM, 64, PhD(civil eng), 68. *Prof Exp:* Asst eng mech, Case Inst Technol, 61-67; assoc prof civil eng & eng mech, Cleveland State Univ, 67-79; PROF CIVIL ENG, 79-, CHMN CIVIL ENGR DEPT, OHIO NORTHERN UNIV, 87- *Mem:* Am Soc Civil Eng; Am Soc Eng Educ. *Res:* Design and analysis of structures and their related behavior. *Mailing Add:* Dept Civil Eng Ohio Northern Univ Ada OH 45810

MINICK, CHARLES RICHARD, b Sheridan, Wyo, Feb 28, 36; m 57; c 3. PATHOLOGY. *Educ:* Univ Wyo, BS, 57; Cornell Univ, MD, 60. *Prof Exp:* Intern path, New York Hosp, 60-61; USPHS trainee, Cornell Univ, New York Hosp Med Center, 61-65, asst res path, 62-63, chief resident, 63-64, asst attend pathologist, 64-70, assoc attend pathologist, 70-76; from asst to assoc prof, 65-76, PROF PATH, MED COL, CORNELL UNIV, 76-; A HENDING PATHOLOGIST, CORNELL UNIV, NEW YORK HOSP MED CTR, 76-, DIR, SURG PATH, 81- *Concurrent Pos:* Fel, coun arteriosclerosis, Am Heart Asn; dir, Surg Path, 81- *Mem:* AAAS; Harvey Soc; Am Asn Path; Am Heart Asn. *Res:* Pathology of arteriosclerosis and hypertension; immunology; culture of cells derived from blood vessels; scanning and transmission electron microscopy. *Mailing Add:* Dept of Path Cornell Univ Med Col New York NY 10021

MINISCALCO, WILLIAM J, b Homewood, Ill, Apr 28, 46; m 79; c 3. OPTICAL INVESTIGATION OF SOLIDS, LASER MATERIALS. *Educ:* Univ Ill, Urbana, BS, 68, MS, 72, PhD(physics), 77. *Prof Exp:* Res asst, Physics Dept, Univ Ill, 72-77; res assoc, Physics Dept, Univ Wis, 77-79; sr mem tech staff, 79-84, PRIN MEM TECH STAFF, GTE LABS INC, 84- *Concurrent Pos:* Prin investr, new mat solid state lasers, 82-84, radiation hardened silica-based optical fibers, 85-87, photoluminescent tech ultra-trace anal, 87- & optically pumped fiber amplifiers, 89-; mem, Tech Prog Comt, Integrated Photonics Res Conf, 90- & Optical Amplifiers & their Applications Conf, 91-; co-chair, Fiber Laser Sources & Amplifiers Conf, Soc Photo-Optical Instrumentation Engrs, 90- *Mem:* Am Phys Soc; Sigma Xi; Optical Soc Am; Inst Elec & Electronics Engrs Lasers & Electro-Optic Soc. *Res:* New materials for lasers and optical amplifiers; excited state dynamics of optical ions; electronic defects in solids; electronic structure of magnetic semiconductors and doped insulators; nonlinear magneto-optics; author of over 60 publications. *Mailing Add:* GTE Lab Inc 40 Sylvan Rd Waltham MA 02254

MINK, GAYLORD IRA, b Lafayette, Ind, Sept 23, 31; m 54; c 4. PLANT VIROLOGY. *Educ:* Purdue Univ, BS, 56, MS, 59, PhD(plant path), 62. *Prof Exp:* Asst prof plant path, Purdue Univ, 62; from asst to assoc plant pathologist, 62-72, PROF, IRRIGATED AGR RES & EXTEN CTR, WASH STATE UNIV, 73-, PLANT PATHOLOGIST, 73- *Concurrent Pos:* Sr ed, Plant Dis, 83-85. *Honors & Awards:* Lee Hutchins Award, Am Phytopath Soc, 88. *Mem:* Am Phytopath Soc. *Res:* Identification, purification and serology of plant viruses; nature of inactivation of plant virus. *Mailing Add:* Dept Plant Path Wash State Univ Prosser WA 99350-9687

MINK, IRVING BERNARD, b New York, NY, Sept 23, 27; m 50; c 4. HEMATOLOGY. *Educ:* Univ Buffalo, BA, 48; State Univ NY Buffalo, MS, 69. *Prof Exp:* Asst cancer res scientist pharmacol, 61-62, cancer res scientist hematol, 62-69, SR CANCER RES SCIENTIST HEMATOL & HEMOSTASIS, ROSWELL PARK MEM INST, 69- *Concurrent Pos:* Instr pharmacol, Sch Pharm, Univ Buffalo, 56-62; mem bd, Oper Comt & 2nd vpres, Western NY Br, Nat Hemophilia Ctr, 69-76, mem bd, 76-; asst res prof, Roswell Park Div, Grad Sch, State Univ NY Buffalo, 72-; clin instr dept path, Med Sch, State Univ NY, Buffalo; res prof, Niagara Univ. *Mem:* AAAS; Int Soc Thrombosis & Hemostasis. *Res:* Hemostatic mechanism; cancer chemotherapy; bleeding and clotting disorders associated with malignancy; platelet function; development of hemostatic mechanism methodological procedures. *Mailing Add:* 665 N Forest Rd Amherst NY 14221

MINK, JAMES WALTER, b Elgin, Ill, Apr 23, 35; m 59; c 1. MILLIMETER WAVES, CONFORMAL ANTENNAS. *Educ:* Univ Wis-Madison, BS, 61, MS, 62, PhD(elec eng), 64. *Prof Exp:* Res assoc elec eng, Univ Wis-Madison, 61-64; res physical scientist electromagnetic res, Inst Explor Res, 64-71, Commu-Automatic Data Processing Lab, 71-76; elec engr, 76-84, prog mgr & assoc dir, 84-90, DIR, ELECTRONICS DIV, US ARMY RES OFF, 90. *Concurrent Pos:* Bd dir & secy, Raleigh Acad, 78-; adj assoc prof, NC State Univ, 79- *Honors & Awards:* Issai Lefkowitz Award, 87. *Mem:* Fel Inst Elec & Electronic Engrs; Int Sci Radio Union; Soc Photo-Optical Instrumentation Engrs. *Res:* Areas of conformal micro and millimeter wave antennas; millimeter wave integrated circuits and devices for high speed optical communications. *Mailing Add:* US Army Res Off Box 12211 Research Triangle Park NC 27709

MINK, LAWRENCE ALBRIGHT, b Birmingham, Ala, Nov 10, 36; m 60; c 1. EXPERIMENTAL PHYSICS, PROGRAMMING. *Educ:* NC State Univ, BE, 58, MS, 61, PhD(nuclear physics), 67. *Prof Exp:* Instr physics, NC State Univ, 65-66; from asst prof to assoc prof, 66-88, PROF PHYSICS, ARK STATE UNIV, 88- *Mem:* AAAS; Am Phys Soc; Sigma Xi; Asn Comput Mach. *Res:* Thin liquid films; solar energy conversion. *Mailing Add:* Div Comput Sci Math & Physics Ark State Univ State University AR 72467

MINKER, JACK, b Brooklyn, NY, July 4, 27; wid; c 2. MATHEMATICS. *Educ:* Brooklyn Col, BA, 49; Univ Wis, MS, 50; Univ Pa, PhD(math), 59. *Prof Exp:* Dynamics engr, Bell Aircraft Corp, 51-52; engr, Radio Corp Am, 52-57, mgr info tech, Data Systs Ctr, 57-63; tech dir & acting off mgr, Auerbach Corp, 63-67; chmn dept, 74-79, PROF COMPUT SCI, UNIV MD, 67- *Concurrent Pos:* Invited lectr, Gordon Res Conf Info Storage & Retrieval, 60; co-chmn Nat Conf Info Storage & Retrieval Asn Comput Mach, 60; mem, Nat Acad Sci-US Nat Comt Fedn Info Documentationalists, 69-72; mem adv comt, Annual Rev Info Sci & Technol, 70-; vchmn, Jerusalem Conf Info Technol, 71; co-chmn conf & prog, Nat Info Storage & Retrieval Conf, 71; mem adv comt, Encycl Comput Sci, 71-; mem staff grad sch, NIH; mem, NASA Study Group Mach Intel & Robotics, 78-81; mem adv comt computing, NSF, 79-, chmn, 80- *Honors & Awards:* Outstanding Contrib Award, Asn Comput Mach, 85. *Mem:* Asn Comput Mach; fel AAAS; fel Inst Elec & Electronic Engrs; Soc Indust & Appl Math; fel Am Asn Artificial Intel. *Res:* Computer application; logic programming; nonmonatonic reasoning; operations research; artificial intelligence. *Mailing Add:* Dept Comput Sci Univ Md College Park MD 20742

MINKIEWICZ, VINCENT JOSEPH, b Shenandoah, Pa, Oct 24, 38; m 65; c 2. SOLID STATE PHYSICS. *Educ:* Villanova Univ, BS, 60; Univ Calif, Berkeley, PhD(physics), 65. *Prof Exp:* Assoc scientist solid state physics, Brookhaven Nat Lab, 65-72; assoc prof, Univ Md, College Park, 72-74; RES STAFF MEM, IBM CORP, 75- *Mem:* Am Phys Soc. *Res:* Magnetic materials; magnetic bubbles. *Mailing Add:* 13760 Camino Rico Saratoga CA 95070

MINKIN, JEAN ALBERT, b Philadelphia, Pa, Nov 17, 25; m 47; c 2. MINERALOGY OF FINE-GRAINED ROCKS, TRACE ELEMENTS IN MINERALS. *Educ:* Bryn Mawr Col, BA, 47. *Prof Exp:* Res engr, Franklin Inst Lab, 47-51; physicist, Nat Bur Standards, 51-52; res assoc, Inst Cancer Res, 60-68; RES PHYSICIST, US GEOL SURV, 68- *Mem:* Am Crystallog Asn; Mineral Soc Am; Am Asn Petrol Geologists; Microbeam Analytical Soc. *Res:* Investigation of the mineralogy and petrology of fine grained rocks associated with mineral deposits; determination of trace-element characteristics of minerals. *Mailing Add:* 3440 Round Table Ct Annandale VA 22003

MINKOFF, ELI COOPERMAN, b New York, NY, Sept 5, 43; m 68; c 2. EVOLUTIONARY BIOLOGY, COMPARATIVE ANATOMY. *Educ:* Columbia Univ, AB, 63; Harvard Univ, AM, 67, PhD(biol), 69. *Prof Exp:* Res asst primate anat, New Eng Regional Primate Res Ctr, Southboro, Mass, 67-68; asst prof, 68-75, actg chmn dept, 76-77, ASSOC PROF BIOL, BATES COL, 76-, ACTG CHMN DEPT, 81- *Mem:* Soc Study Evolution; Am Soc Zoologists; Soc Vert Paleont; Soc Syst Zool; Am Soc Mammalogists. *Res:* Anatomy and evolutionary biology of primates and other mammals; neuromuscular anatomy, especially of facial nerve and facial muscles, vertebrate paleontology; evolutionary theory. *Mailing Add:* Dept Biol Bates Col Lewiston ME 04240

MINKOWITZ, STANLEY, b Brooklyn, NY, July 1, 28; m 57; c 3. PATHOLOGY, DERMATOLOGICAL PATHOLOGY. *Educ:* City Col New York, BS, 48; Univ Colo, MS, 50; Univ Geneva, MD, 56. *Prof Exp:* Chief surg path, Kings County Hosp Med Ctr, 62-71; CLIN ASSOC PROF PATH, STATE UNIV NY DOWNSTATE MED CTR, 71-; DIR LABS, MAIMONIDES MED CTR, 71- *Concurrent Pos:* Fel pediat med, Jewish Chronic Dis Hosp, 57; asst prof path, State Univ NY Downstate Med Ctr, 63-

67, assoc prof, 67-71, consult vet pathologist, 64-; consult pathologist, Brooklyn Women's Hosp, 63- & Unity Hosp, Brooklyn, 65-; consult, Brooklyn State Hosp, 67- *Res:* Organic chemistry, especially amino acids synthesis; experimental study of the use of gold leaf in skin graft; heterotopic liver transplantation; coxsackie B-virus infection in adult mice; research in aging; homologous lung transplantation. *Mailing Add:* 85 Buckingham Rd Brooklyn NY 11226

MINKOWSKI, JAN MICHAEL, b Zurich, Switz, Mar 7, 16; nat US; m 51; c 4. SOLID STATE PHYSICS. *Educ:* Swiss Fed Inst Technol, Dipl Physicist, 49; Johns Hopkins Univ, PhD(physics), 63. *Prof Exp:* Res physicist, Inst Tehoret Physics, Switz, 49-50; res physicist solid-state physics, Erie Resistor Corp, 50-52; head phys res, Carlyle Barton Lab, 52-63, assoc prof, 63-80, PROF ELEC ENG, JOHNS HOPKINS UNIV, 80- *Mem:* Am Phys Soc. *Res:* Quantum electronics, solid state. *Mailing Add:* Dept Elec Eng 210 Barton Hall Johns Hopkins Univ 34th & Charles Sts Baltimore MD 21218

MINKOWYCZ, W J, b Libokhora, Ukraine, Oct 21, 37; US citizen; m 73; c 1. MECHANICAL ENGINEERING, HEAT TRANSFER. *Educ:* Univ Minn, BS, 58, MS, 61, PhD(mech eng), 65. *Prof Exp:* From asst prof to assoc prof, 66-79, PROF MECH ENG, UNIV ILL, CHICAGO, 79- *Concurrent Pos:* Co-ed, Int J Heat & Mass Transfer, 67-; coord ed, Lett in Heat & Mass Transfer, 74-82; ed-in-chief, Numerical Heat Transfer J, 78-; coord ed, Int Commun Heat & Mass Transfer, 83- *Honors & Awards:* Ralph Coats Road Award, Am Soc Eng Educ. *Mem:* Fel Am Soc Mech Engrs. *Res:* Heat transfer; two-phase flows; non-similar boundary layers; porous media heat transfer; numerical methods. *Mailing Add:* Dept Mech Eng Univ Ill Chicago Box 4348 Chicago IL 60680

MINN, FREDRICK LOUIS, b Waukegan, Ill, Aug 9, 35. MEDICINE, PHYSICAL CHEMISTRY. *Educ:* Univ Ill, AB, 57, PhD(chem), 64; Univ Miami, MD, 73. *Prof Exp:* Asst prof chem, Columbia Univ, 63-65; assoc prof, George Washington Univ, 65-74, asst med, 73-74; asst dir clin pharmacol, Squibb Inst Med Res, 74-76; asst dir clin invest, 76-80, CLIN RES FEL, MCNEIL PHARMACEUT, JOHNSON & JOHNSON, 80- *Concurrent Pos:* Consult engr, Lockheed Electronics Co, Lockheed Aircraft Corp & Taag Designs, Inc, 67-70. *Mem:* Am Chem Soc; AMA; Am Col Clin Pharmacol; Am Soc Clin Pharmacol & Therapeut. *Res:* Clinical pharmacology; theoretical chemistry. *Mailing Add:* 601 Midway Lane Blue Bell PA 19422

MINNA, JOHN, ONCOLOGY. *Prof Exp:* PROF INTERNAL MED & DIR, SIMMONS COMPREHENSIVE CANCER CTR, SOUTHWESTERN MED CTR, UNIV TEX, 91- *Mailing Add:* Simmons Comprehensive Cancer Ctr Southwestern Med Ctr Univ Tex 5323 Harry Hines Blvd Dallas TX 75235-8590

MINNE, RONN N, b Menominee, Mich, Oct 3, 24. INORGANIC CHEMISTRY. *Educ:* Northwestern Univ, BS, 50, AM, 51; Harvard Univ, PhD(chem), 60. *Prof Exp:* Teacher high sch, Ill & NMex, 51-54; teacher chem, Culver Mil Acad, 54-56, chmn dept, 60-65; chmn, Sci Div, 72-80, INSTR CHEM, PHILLIPS ACAD, 65- *Concurrent Pos:* Off Naval Res grant, 62-65; vis scholar, Cambridge Univ, 71-72; instr chem, Martha Cochran Found, 80- *Honors & Awards:* Aula Laudis Award, Am Chem Soc, Northeastern Sect, 85. *Mem:* AAAS; Am Chem Soc; Am Inst Chem; NY Acad Sci. *Res:* Inorganic polymers; silicon chemistry. *Mailing Add:* Dept of Chem Phillips Acad Andover MA 01810

MINNEAR, WILLIAM PAUL, b Pittsburgh, Pa, July 17, 46; m 70; c 2. OPTICAL & ELECTRONIC CERAMICS, MANUFACTURING. *Educ:* Pa State Univ, BS, 68, MS, 69 & 76, PhD(metall), 75. *Prof Exp:* Mem staff, Cutting & Wear Resistant Mat Prog, GE Corp Res & Develop, 75-82, mgr, process eng, Advan Process Develop, Qual & Product Eng, GE Carboloy Systs Dept, 82-86, metallurgist/ceramist, 86-89, MGR OPTICAL & ELEC CERAMICS PROG, GE CORP RES & DEVELOP, 89- *Mem:* Am Ceramic Soc; Electrochem Soc. *Res:* Materials and process development of transparent, scintillating and piezoelectric ceramics. *Mailing Add:* Gen Elec Co PO Box Eight Schenectady NY 12301

MINNEMAN, KENNETH PAUL, b Sacramento, Calif, Sept 1, 52; m 81; c 3. NEUROPHARMACOLOGY, NEUROCHEMISTRY. *Educ:* Mass Inst Technol, BS, 74; Univ Cambridge, Eng, PhD(pharmacol), 77. *Prof Exp:* Fel, Med Ctr, Univ Colo, 77-80; from asst prof to assoc prof, 80-90, PROF PHARMACOL, EMORY UNIV, 90- *Honors & Awards:* John Jacob Abel Award, Am Soc Pharmacol & Exp Therapeut, 89. *Mem:* Am Soc Pharmacol & Exp Therapeut; AAAS; Int Soc Neurochem; Soc Neurosci. *Res:* Biochemical studies of neurotransmitter receptors. *Mailing Add:* Dept Pharmacol Emory Univ Atlanta GA 30322

MINNEMAN, MILTON J(AY), b Brooklyn, NY, July 31, 23; m 55; c 1. ELECTRONICS ENGINEERING. *Educ:* Cooper Union, BEE, 43; Univ Pa, MSEE, 49; Polytech Univ, Brooklyn, PhD(elec eng), 66. *Prof Exp:* Elec develop engr, Radio Corp Am, NJ, 43-47; sect head, Martin Co, Md, 47-52; chief engr, Utility Electronics Corp, NJ, 52; chief fuel gage engr, Avien, Inc, NY, 52-54; chief electronics engr, Bulova Res & Develop Labs, Inc, 54-56; chief electrophys res, Repub Aviation Corp, 56-64; prog mgr space systs, Fairchild Hiller Corp, 64-65; tech dir, Systs & Instruments Div, Bulova Watch Co, Inc, 65-67; eng consult to group vpres, Airborne Instruments Lab Div, Cutler-Hammer, Inc, 67-68, tech asst to exec vpres, 68-69; dir, Eng Electronic Systs Div, Gen Instrument Corp, 69-73; spec asst, 73-81, DIR, MOBILITY & SPEC PROJ, OFF SECY DEFENSE, 81- *Concurrent Pos:* Adj inst, Polytech Inst NY, 66-72; pres, Wash Soc Engrs. *Mem:* Sr mem Inst Elec & Electronics Engrs; Sigma Xi. *Res:* Inertial navigation and guidance systems; fuel gages; magnetics; nuclear magnetic resonance; space craft systems; communications; radar; data processing; nuclear systems; airlift; sealift. *Mailing Add:* 8815 Hidden Hill Lane Potomac MD 20854-4230

MINNEMEYER, HARRY JOSEPH, b Buffalo, NY, July 12, 32; div; c 3. ORGANIC CHEMISTRY. *Educ:* Univ Buffalo, BA, 58, PhD(chem), 62. *Prof Exp:* Res assoc chem, State Univ NY Buffalo, 61-66, lectr, Millard Fillmore Col, 63-64; chemist, Starks Assocs, Inc, 66-67, supvr, 67-68, sect mgr, 68-69; sr res chemist, Lorillard Res Ctr, 70, supvr org chem, 70-75, mgr res, 75-80, DIR RESEARCH, LORILLARD, DIV LOEWS THEATERS, INC, 80- *Mem:* Am Chem Soc. *Res:* Organic synthesis; heterocyclic and medicinal chemistry; reaction mechanisms; tobacco science. *Mailing Add:* Lorillard Res Ctr PO Box 21688 Greensboro NC 27420-1688

MINNERS, HOWARD ALYN, b Rockville Centre, NY, Sept 1, 31; m 58; c 2. MEDICINE. *Educ:* Princeton Univ, AB, 53; Yale Univ, MD, 57; Harvard Univ, MPH, 60; Am Bd Prev Med, cert aerospace med, 65. *Prof Exp:* Intern, Wilford Hall US Air Force Hosp, 57-58, flight surgeon, Langley Air Force Base, Va, 61-62; head, Flight Med Br, Manned Spacecraft Ctr, NASA, 62-66; spec asst to chief, Off Int Res, NIH, 66-68; chief, Geog Med Br, Nat Inst Allergy & Infectious Dis, NIH, 68-72, assoc dir collab res, 72-76, assoc dir int res, 76-77; responsible officer, Off Res Prom & Develop, WHO, Geneva, Switz, 77-80; dep dir, Off Int Health, US Pub Health Serv, 80-81; SCI ADV, AID, 81- *Mem:* AAAS; Am Pub Health Asn; fel World Acad Arts & Sci; fel Am Col Prev Med. *Res:* Ecology, distribution and determinants of disease prevalence in man; career motivation for international research; tropical medicine. *Mailing Add:* AID/SCI Rm 320 SA-18 Washington DC 20523-1818

MINNICH, JOHN EDWIN, b Long Beach, CAlif, Oct 23, 42; m 68; c 2. ENVIRONMENTAL PHYSIOLOGY, VERTEBRATE ZOOLOGY. *Educ:* Univ Calif, Riverside, AB, 64; Univ Mich, Ann Arbor, PhD(zool), 68. *Prof Exp:* Asst prof, 68-72, ASSOC PROF ZOOL, UNIV WIS-MILWAUKEE, 72- *Concurrent Pos:* Wis Alumni Res Found fel physiol, 69-71; environ consult, Environ Analysts, Inc, 74-; textbook consult physiol, MacMillan & Co, Inc, 74- & Wadsworth Pub Co, 75-; reviewer, Reg Biol Div, NSF, 78-; reviewer, Am J Physiol & J Morphol, 79- *Mem:* Sigma Xi; Am Soc Ichthyol & Herpet; Am Soc Zool; Ecol Soc Am; Soc Study Amphibians & Reptiles. *Res:* Environmental physiology of animals; water, electrolyte, energy and nitrogen metabolism of terrestrial vertebrates; excretion and osmoregulation of vertebrates; ecological changes, as seen through rephotography. *Mailing Add:* Box 11527 Milwaukee WI 53211

MINNICK, DANNY RICHARD, b Export, Pa, July 17, 37; m 71. EDUCATIONAL ADMINISTRATION. *Educ:* Southern Col, BA, 61; Univ Fla, MS, 67, PhD(entom), 70. *Prof Exp:* Asst entom, Univ Fla, 65-70; asst prof biol, Fla Keys Community Col, 70-71; from asst prof to assoc prof entom, Univ Fla, 71-90; HEAD TRAINING CTR, INT RICE RES INST, MANILLA, PHILIPPINES, 90- *Concurrent Pos:* Danforth assoc. *Honors & Awards:* Outstanding Grad Studies Entom Soc Am, 70; Serv Award, Am Beekeeping Fedn. *Mem:* Am Soc Training & Develop; Nat Asn Col Teachers Agr; Intergovernmental Sci Eng & Tech Adv Panel; Sigma Xi. *Res:* Cognitive psychology; courseware development; HR and institute development technology trans. *Mailing Add:* IRRI PO Box 933 Manila Philippines

MINNICK, ROBERT C, b Houston, Tex, Feb 7, 26. ELECTRICAL ENGINEERING, PHYSICS. *Educ:* Johns Hopkins Univ, BA, 50; Harvard Univ, AM, 51, PhD(appl math), 53. *Prof Exp:* From instr to asst prof appl math, Harvard Univ, 53-57; sr physicist, Burroughs Corp, 57-60; sr res engr, Stanford Res Inst, 60-65; prof elec eng, Mont State Univ, 66-70; prof elec eng & computer sci & dir, Lab Comput Sci & Eng, Rice Univ, 71-73; PRES, MINNICK ENG CO, 73-; PRES & CHIEF EXEC OFFICER, MARK FIVE SYSTS INC, 79- *Mem:* Sr mem Inst Elec & Electronics Engrs; Asn Comput Mach. *Res:* Switching theory; design of digital computer subsystems and systems; cellular logic components. *Mailing Add:* 2115 N Grand Ave Pueblo CO 81003

MINNIFIELD, NITA MICHELE, b Jefferson Co, Ala, May 3, 60. LIVESTOCK INSECTS. *Educ:* Stillman Col, BS, 82; Purdue Univ, PhD(cell biol), 89. *Prof Exp:* Teaching asst, Dept Biol Sci, Purdue Univ, 82-83, res asst, Dept Med Chem & Pharmacog, 86-89; POSTDOCTORAL FEL, LIVESTOCK INSECTS LAB, AGR RES SERV, USDA, 89- *Concurrent Pos:* Mem, Comt Inst Coop Fel. *Mem:* Am Soc Cell Biol; AAAS. *Res:* Decreasing-controlling the insect population in the agricultural arena without the use of harmful chemicals; antibody production; immunoprecipitation; protein and receptor purification; cellular fractionation; membrane separation by free-flow electrophoresis; gel electrophoresis; immunocytochemistry; high performance liquid chromatography; general electron microscopy. *Mailing Add:* Agr Res Serv Livestock Insect Lab USDA Rm 120 Bldg 307 Barc-E 10300 Baltimore Ave Beltsville MD 20705

MINNIX, RICHARD BRYANT, b Salem, Va, June 20, 33; m 55; c 3. PHYSICS. *Educ:* Roanoke Col, BS, 54; Univ Va, MS, 57; Univ NC, Chapel Hill, PhD(physics), 65. *Prof Exp:* From instr to assoc prof, 56-69, head dept, 74-79, PROF PHYSICS, VA MIL INST, 69-, HEAD DEPT, 89- *Honors & Awards:* Distinguished Serv Award, Am Asn Physics Teachers, 88. *Mem:* Am Phys Soc; Am Asn Physics Teachers. *Res:* Lecture demonstrations as method of instruction and/or stimulation of public interest in science. *Mailing Add:* Dept Physics & Astron Va Mil Inst Lexington VA 24450

MINOCA, SUBHASH C, b New Delhi, India, Aug 22, 47; US citizen; m 76; c 2. PLANT CLONING, GENETIC ENGINEERING. *Educ:* Punjab Univ, Chandigarh, India, BS, 68, MS, 69; Univ Wash, Seattle, PhD(bot), 74. *Prof Exp:* Teaching asst, Univ Notre Dame, 70-71; teaching asst & res asst, Univ Wash, Seattle, 71-74; from asst prof to assoc prof bot & plant path, 74-85, assoc prof genetics, 82-85, PROF BOT & PLANT PATH, UNIV NH, 85-, PROF GENETICS, 85-, CHMN, DEPT BOT & PLANT PATH, 82- *Honors & Awards:* Alexander von Humboldt fel, 81-82. *Mem:* Am Soc Plant Physiologists; Int Plant Tissue Cult Asn; Sigma Xi; Int Plant Growth Substance Asn; Int Plant Molecular Biol Orgn; Int Union Forest Res Orgn. *Mailing Add:* Dept Bot & Plant Path Univ NH Durham NH 03824

MINOCHA, HARISH C, b Aug 31, 32; m 55; c 4. MICROBIOLOGY, VIROLOGY. *Educ:* Punjab Univ, India, BVSc, 55; Kans State Univ, MS, 63, PhD(virol), 67. *Prof Exp:* Res asst microbiol, Indian Vet Res Inst, 55-61; res asst, Kans State Univ, 61-66, res assoc, 66-67; asst prof, NC State Univ, 67-69; assoc prof microbiol, 69-77, asst head, dept lab med, 82-88, PROF VIROL, KANS STATE UNIV, 77-, ASSOC DEAN RES & GRAD AFFAIRS, 89- *Concurrent Pos:* Nat Cancer Soc res grant, 69-72; res grant, USDA, 72- & animal health res grant, 80-81; sr scientist, Mid Am Cancer Ctr, 75-; mem, Ctr Basic Cancer Res, 80-; spec res grant, 87-; competitive res grant, 88- *Honors & Awards:* Beecham Award for Excellence in Res. *Mem:* Am Soc Microbiol; Asn Am Vet Med Cols; Am Vet Med Asn; fel Am Soc Microbiol. *Res:* Biochemical studies on polyoma and shope fibroma virus infected tissue cultures; immunological studies on bovine Herpesvirus-1; antigenic analysis of influenza viruses; monoclonal antibodies. *Mailing Add:* Dept Lab Med Kans State Univ Manhattan KS 66502

MINOCK, MICHAEL EDWARD, b Los Angeles, Calif, Dec 6, 37. VERTEBRATE ECOLOGY. *Educ:* Stanford Univ, AB, 60; Calif State Univ, Northridge, MA, 65; Univ Nebr, Omaha, MS, 66; Utah State Univ, PhD(zool), 70. *Prof Exp:* Teacher biol, Los Angeles City Schs, 62-65; instr, State Univ NY Col, Brockport, 69-70; NIH fel behav biol & ecol, Univ Minn, 70-71; from asst prof to assoc prof, 71-84, PROF BIOL, UNIV WIS CTR-FOX VALLEY, 84- *Mem:* Sigma Xi; Am Ornithologists Union; Wilson Ornith Soc; AAAS; Cooper Ornith Soc. *Res:* Bird social organizations and their ecological significance. *Mailing Add:* Univ Wis Ctr-Fox Valley PO Box 8002 Menasha WI 54952

MINOR, CHARLES OSCAR, b Churdan, Iowa, June 22, 20; m 43; c 3. FORESTRY. *Educ:* Iowa State Col, BS, 41; Duke Univ, MF, 42, DF, 58. *Prof Exp:* Asst prof forestry, La State Univ, 46-54; admin forester, Kirby Lumber Corp, 54-56; assoc prof forestry, Clemson Col, 58; dean, sch Forestry, Northern Ariz Univ, 58-79, prof forestry, 58-84; RETIRED. *Concurrent Pos:* Consult forester, 79- *Mem:* Soc Am Foresters; Am Forestry Asn; Am Soc Photogrammetry. *Res:* Forest management and mensuration, including aerial photo interpretation. *Mailing Add:* 2140 Edgewood Dr Sedona AZ 86336

MINOR, JAMES E(RNEST), b Davenport, Wash, Apr 10, 19; m 50; c 3. PHYSICAL CHEMISTRY, METALLURGY. *Educ:* Wash State Univ, BS, 41; Univ Wash, PhD(phys chem), 50. *Prof Exp:* Phys chemist, Procter & Gamble Co, 50-52; mgr fuel design & fuel fabrication, Hanford Labs, Gen Elec Co, 52-65; mgr metal fabrication develop, 65-68, sr res assoc, 68-72, mgr major proj develop, 72-75, mgr Exxon nuclear prog, 75-78, mgr nuclear waste vitrification prog, 78-80, mgr seasonal thermal energy storage, 80-82, MGR HANFORD NUCLEAR WASTE VITRIFICATION, PAC NORTHWEST LABS, BATTELLE MEM INST, 82- *Mem:* Am Chem Soc; Am Soc Metals; Am Inst Chem Engrs. *Res:* Physical and fabrication metallurgy of fissionable metals, high strength alloys and refractory metals; defense programs; nuclear fuel development; nuclear waste vitrification; seasonal thermal energy storage. *Mailing Add:* 2105 Symons St Richland WA 99352

MINOR, JOHN THREECIVELOUS, organic chemistry, information science, for more information see previous edition

MINOR, JOHN THREECIVELOUS, b Fulton, Mo, Nov 17, 50. MECHANICAL DEDUCTION, RULE-BASED EXPERT SYSTEMS. *Educ:* Rice Univ, BA, 73; Univ Tex, Austin, PhD(computer sci), 79. *Prof Exp:* Asst prof computer sci, Univ Okla, 79-85; assoc prof computer sci, 85-90, CHMN COMPUTER SCI, UNIV NEV, LAS VEGAS, 90- *Concurrent Pos:* Res fel, USAF Summer Res Prog, 84; prin investr, Air Force Off Sci Res grant, 85 & Army Res Off grant, 86-91; consult, Nev Gaming Control Bd, 90. *Mem:* Asn Automated Reasoning; Am Asn Artificial Intel; Asn Comput Mach. *Res:* Symbolic logic and mechanical deduction techniques in applications of articficial intelligence and expert systems; modified versions of logic, such as fuzzy logic. *Mailing Add:* 4071 Grasmere Ave Las Vegas NV 89121

MINOR, RONALD R, b Donora, Pa, Sept 13, 36; m 58; c 4. DEVELOPMENTAL ANATOMY, EXPERIMENTAL PATHOLOGY. *Educ:* Univ Pa, VMD, 66, PhD(path), 71. *Prof Exp:* Assoc, Univ Pa, 71-72, asst prof anat, med, path & vet med, 72-76; ASSOC PROF DEPT PATH, CORNELL UNIV, 76- *Mem:* Am Vet Med Asn; Soc Develop Biol; Soc Cell Biol; AAAS. *Res:* Regulation of mesodermal differentiation and connective tissue development; basement membrane-laminin and procollagen; inheritable diseases of connective tissue; toxicity of intracellular helium 3, carbon 14 and sulfer 35. *Mailing Add:* Dept Vet Cornell Univ Vet Sch Ithaca NY 14853

MINORE, DON, b Chicago, Ill, Oct 31, 31; m 63; c 2. ECOLOGY, FOREST REGENERATION. *Educ:* Univ Minn, BS, 53; Univ Calif, Berkeley, PhD(bot), 66. *Prof Exp:* Res forester, Pac Northwest Forest & Range Exp Sta, US Forest Serv, 55, 57-60; asst bot, Univ Calif, Berkeley, 61-63, teaching fel, 63-65; PLANT ECOLOGIST, FORESTRY SCI LAB, US FOREST SERV, 65- *Mem:* Ecol Soc Am; Am Forestry Asn; Northwest Sci Asn. *Res:* Species-site relationships in the mixed conifer forests of Oregon and Washington; shrub ecology; natural regeneration of conifers. *Mailing Add:* Forestry Sci Lab 3200 Jefferson Way Corvallis OR 97331

MINOT, MICHAEL JAY, b Apr 4, 46; m 73; c 2. CHEMICAL ENGINEERING. *Educ:* New York Univ, BA, 68; Johns Hopkins Univ, PhD(chem), 73. *Prof Exp:* Sr chemist elec mat tech, 74-76, sr scientist surface chem res, 76-78, plant mfg engr, 78-83, PROJ MGR, OPTICAL WAVEGUIDES, CORNING GLASS WORKS, INC, 83- *Concurrent Pos:* Adj prof modern glass technol, Clarkson Col, 80. *Mem:* Am Chem Soc; Am Ceramics Soc. *Res:* Surface properties of glass; physical-inorganic chemistry. *Mailing Add:* American Superconductor Corp 149 Grove St Watertown MA 02172-2828

MINOTTI, PETER LEE, b Burlington, Vt, Aug 20, 35; m 62; c 2. VEGETABLE CROP NUTRITION. *Educ:* Univ Vt, BS, 57, MS, 62; NC State Univ, PhD(soil sci, plant physiol), 65. *Prof Exp:* Crop physiologist, Int Minerals & Chem Corp, 65-66; ASSOC PROF VEG CROPS, CORNELL UNIV, 66- *Mem:* Weed Sci Soc Am; Am Soc Hort Sci. *Res:* Vegetable crop nutrition. *Mailing Add:* Dept of Veg Crops Cornell Univ Ithaca NY 14853

MINOWADA, JUN, b Kyoto, Japan, Nov 5, 27; m 59; c 3. PATHOLOGY, IMMUNOLOGY. *Educ:* Mie Med Col, Japan, BS, 48; Kyoto Univ, MD, 52, DMedSci, 59. *Prof Exp:* Sr res assoc. 61-64, sr res scientist, 66-67, assoc res scientist, 68-72, prin res scientist, 73-74, ASSOC CHIEF SCIENTIST, ROSWELL PARK MEM INST, 75-; DIR, FUJISAKI CELL CTR, HAYASHABIRA BIOCHEM LABS, INC,; AT VET ADMIN MED CTR, HINES, ILL. *Concurrent Pos:* Int Atomic Energy Agency res fel, Roswell Park Mem Inst, 60; vis fel, Karolinska Inst, Sweden, 64-66; res prof, Roswell Park Div, State Univ NY, Buffalo, 69- *Mem:* Am Asn Cancer Res; Am Soc Microbiol; Am Soc Clin Oncol. *Res:* Leukemia-lymphoma; immunology. *Mailing Add:* Fujisaka Cell Ctr Hayashabira Biochem Labs Inc 675-1 Fujisaki Okayama 702 Japan

MINSAVAGE, EDWARD JOSEPH, b Nanticoke, Pa, June 1, 18; m 51. MEDICAL MICROBIOLOGY. *Educ:* Univ Scranton, BS, 49; Univ Pa, MS, 51, PhD(microbiol), 55. *Prof Exp:* Ed asst, Biol Abstracts, 50-51; res bacteriologist, Eaton Labs, Inc, 51-52; instr, Med Sch, Univ Pa, 54-55, res assoc, 55-57; from asst prof to prof biol, King's Col, Pa, 57-83; RETIRED. *Concurrent Pos:* USPHS res grant bact physiol, 63-66. *Mem:* AAAS; Am Soc Microbiol. *Res:* Cytology and cytochemistry; bacterial respiration and oxygen transport systems; analysis and biochemistry of nucleic acids and their derivatives; electrophoresis and chromatographic analysis; microphotography; organ deficiency in experimental animals; loss of essential biochemical intermediates by normal cells. *Mailing Add:* 221 S Hanover St Nanticoke PA 18634

MINSHALL, GERRY WAYNE, b Billings, Mont, Aug 30, 38; m 63; c 3. AQUATIC ECOLOGY. *Educ:* Mont State Univ, BS, 61; Univ Louisville, PhD(zool), 65. *Prof Exp:* NATO fel, Freshwater Biol Asn, Eng, 65-66; from asst prof to assoc prof, 66-74, PROF ECOL & ZOOL, IDAHO STATE UNIV, 74- *Concurrent Pos:* Fac res grants, Idaho State Univ, 66, 68, 70, 73, 75, 78, 82, 83, 84, 86, 87 & 89, US Dept Health, Educ & Welfare Res grant, 68-69, NSF-Int Biol Prog grant, 70-76, NSF-Anal Ecosyst Prog grant, 74-81, NSF-Ecosyst Prog grant, 89-91, US Forest Serv grants, 75, 77, 78, 79, 80, 84, US Corps Eng grants, 72 & 77, US Soil Conserv Serv grant, 76-78, US Dept Energy grant, 84, Smithsonian Inst grant, 84, US Fish & Wildlife Serv grants, 86-87, 89-92, Nat Park Serv grants, 89-92, Id Dept Environ Qual grant, 90; vis res scientist, Freshwater Biol Asn, Eng, 74, Stroud Water Res Ctr, Philadelphia Acad Nat Sci, 81-82; guest lectr, Nordic Coun Ecol, Sweden, 79; mem numerous Adv Panels. *Honors & Awards:* R S Campbell Mem lectr, Univ Mo, 84; Tom Wallace Chair Lectr, Univ Louisville, 90. *Mem:* NAm Benthol Soc; Fresh Water Biol Asn; Ecol Soc Am. *Res:* Stream ecosystem dynamics; biotic production; wildfire; disturbance; water pollution; benthic invertebrates. *Mailing Add:* Dept Biol Sci Idaho State Univ Pocatello ID 83209

MINSHALL, WILLIAM HAROLD, b Brantford, Ont, Dec 6, 11; m 39; c 2. PLANT PHYSIOLOGY. *Educ:* Ont Agr Col, BSA, 33; McGill Univ, MSc, 38, PhD(plant physiol), 41. *Prof Exp:* Asst bot, Can Dept Agr, 33-41, from jr botanist to botanist, 41-51, sr plant physiologist, 51-75; RETIRED. *Concurrent Pos:* Hon lectr, Univ Western Ont, 52-76. *Mem:* Fel AAAS; Am Soc Plant Physiol; Bot Soc Am; fel Weed Sci Soc Am; Can Soc Plant Physiol; fel Agr Inst Can; fel Can Pest Mgt Soc. *Res:* Physiology of herbicidal action; metabolic root pressure and uptake of solutes; plant phenology. *Mailing Add:* 91 Huron St London ON N6A 2H9 Can

MINSINGER, WILLIAM ELLIOT, b Quincy, Mass, Aug 3, 50; m 80; c 3. HISTORIC WEATHER EVENTS, CLIMATE STUDY & CHANGE. *Educ:* Northwestern Univ, BA, 73; Boston Univ, MD, 78. *Prof Exp:* MED CLIN DIR, HITHERAL ASSOCS RANDOLPH, 84- *Concurrent Pos:* Pres, Blue Hill Observ, 80- *Mem:* Am Med Soc; Am Meteorol Soc. *Res:* Climate, weather and historic storms. *Mailing Add:* PO Box 101 East Milton MA 02186

MINSKER, DAVID HARRY, b Huntingdon, Pa, Jan 17, 38; m 65; c 1. PHYSIOLOGY, PHARMACOLOGY. *Educ:* Juniata Col, BS, 63; Univ Wis, PhD(physiol), 67. *Prof Exp:* Res pharmacologist thrombosis, 67-69, res pharmacologist hypertension, 69-71, res fel thrombosis & teratology, 71-77, RES FEL TERATOLOGY, MERCK SHARP & DOHME RES LABS, 77- *Mem:* Am Heart Asn; Int Soc Thrombosis & Hemostasis. *Res:* Development of anti-thrombotic drugs; teratology; toxicology; embryology. *Mailing Add:* Minehill Rd & Hillcrest Ave Schwenksville PA 19473

MINSKY, MARVIN LEE, b New York, NY, Aug 9, 27; m 52; c 3. MATHEMATICS, COMPUTER SCIENCE. *Educ:* Harvard Univ, BA, 50; Princeton Univ, PhD(math), 54. *Hon Degrees:* Dr, Free Univ Brussels & Pine Manor Col, 86. *Prof Exp:* Asst math, Princeton Univ, 50-53; res assoc, Tufts Univ, 53-54; jr fel, Soc Fels, Harvard Univ, 54-57; mem staff, Lincoln Lab, Mass Inst Technol, 57-58, asst prof math, 58-61, assoc prof elec eng, 61-64, PROF ELEC ENG & COMPUTER SCI, MASS INST TECHNOL, 64-, TOSHIBA PROF MEDIA ARTS & SCI, MEDIA LAB, 89- *Concurrent Pos:* Dir, Artificial Intel Lab, 64-73, Donner prof sci, 74-89. *Honors & Awards:* Turing Award, Asn Comput Mach, 70; Doubleday Lectr, Smithsonian Inst, 78; Messenger Lectr, Cornell Univ, 79; Killian Lectr, Mass Inst Technol, 89. *Mem:* Nat Acad Sci; Nat Acad Eng; fel Am Acad Arts & Sci; fel Inst Elec & Electronics Engrs; fel NY Acad Sci. *Res:* Artificial intelligence; theory of computation; psychology; engineering. *Mailing Add:* Media Lab Mass Inst Technol 20 Ames St Cambridge MA 02139

MINTA, JOE ODURO, b Fomena, Ghana, Mar 16, 42; m 70; c 1. IMMUNOLOGY. *Educ:* Univ Ghana, BSc, 66; Univ Guelph, MSc, 68; Univ Toronto, PhD(biochem), 71. *Prof Exp:* Res asst chem, Univ Guelph, 66-68; res asst biochem, Univ Toronto, 68-71; fel immunol, Med Res Coun Can, 71-73; asst prof, 73-78, ASSOC PROF PATH, UNIV TORONTO, 78-, MEM INST IMMUNOL, 74-, ASSOC PROF MED, 79- *Concurrent Pos:* Res scholar, Can Heart Found, 73-77; clin immunopathologist, Toronto Western Hosp, 77-79; res assoc, Can Arthritis & Rheumatism Soc, 79- *Mem:* AAAS; NY Acad Sci; Can Soc Immunol; Am Asn Exp Path; Am Asn Immunologists. *Res:* Immunochemistry, immunogenetics and immunopathology of the complement system. *Mailing Add:* Dept Path Univ Toronto Toronto ON M5S 1A8 Can

MINTER, JERRY BURNETT, b Ft Worth, Tex, Oct 31, 13; m 40; c 5. PHOTOGRAPHY, AUDIO ENGINEERING. *Educ:* MIT, BS, 34. *Prof Exp:* Engr, Boouton Radio Corp, 35-36; engr instr, Ferris Inst Co, 36-39; vpres, Measurements Corp, 39-53; PRES, COMPONENTS CORP, 53- *Honors & Awards:* Armstrong Medal, Radio Club Am, 68. *Mem:* Fel Inst Elec & Electronic Engrs; fel Radio Club Am; Am Soc Metals; fel Audio Eng Soc; Soc Motion Picture & TV Eng. *Mailing Add:* Components Corp Six Kinsey Pl Denville NJ 07834

MINTHORN, MARTIN LLOYD, JR, b Grand Rapids, Mich, Aug 8, 22; m 55, 64; c 4. ENVIRONMENTAL HEALTH, RESEARCH ADMINISTRATION. *Educ:* Univ Nebr, BS, 44, MS, 49; Univ Ill, PhD(biochem), 53. *Prof Exp:* Jr chemist, Nat Bur Stand, 44-46; from res assoc to instr biochem, Univ Ill, 52-55; instr, Wash Univ, 53-54; res assoc, Univ Tenn, 55-56, asst prof, 57-64; biochemist, asst br chief, dep prog mgr & div dir, Health & Environ Res, AEC, US Energy Res & Develop Admin, US Dept Energy, 64-88; RETIRED. *Concurrent Pos:* USPHS sr res fel, 57-61; prof lectr, George Washington Univ, 75-81. *Mem:* AAAS; Sigma Xi; Am Chem Soc; fel Am Inst Chem; Biolectromagnetics Soc. *Res:* Biochemistry of amino acids; stereochemistry; isotopic tracers; health impacts of energy technologies. *Mailing Add:* 15715 Ancient Oak Dr Gaithersburg MD 20878

MINTON, ALLEN PAUL, b Takoma Park, Md, July 5, 43; m 70; c 2. BIOPHYSICAL CHEMISTRY. *Educ:* Univ Calif, Los Angeles, BS, 64, PhD(phys chem), 68. *Prof Exp:* Guest scientist, polymer dept, Weizmann Inst Sci, 68-70, polymer & biophysics dept, 78-79, staff fel, 70-74, sr staff fel, 74-75; res chemist, Lab Biophys Chem, 75-78, RES CHEMIST, LAB BIOCHEM PHARMACOL, NAT INST ARTHRITIS, DIABETES, DIGESTIVE & KIDNEY DIS, 78- *Concurrent Pos:* Vis prof, Sci Univ, Tokyo, 88. *Honors & Awards:* Inventor's Award, US Dept Com, 87. *Mem:* Am Soc Biol Chemists; Sigma Xi; Biophys Soc. *Res:* self- and hetero-association of biological macromolecules in solution; biochemical equilibria and kinetics in concentrated protein solutions; new techniques and applications of analytical ultracentrifugation. *Mailing Add:* Lab Biochem Pharmacol Bldg 8 Rm 226 NIADDK, NIH Bethesda MD 20892

MINTON, NORMAN A, b Spring Garden, Ala, Oct 12, 24; m 44; c 1. PLANT NEMATOLOGY. *Educ:* Auburn Univ, BS, 50, MS, 51, PhD(zool), 60. *Prof Exp:* Asst county agr agent, Coop Exten Serv, Ala, 51-53; horticulturist, Berry Col, 53-55; nematologist, US Dept Agr, Auburn Univ, 55-64; NEMATOLOGIST, US DEPT AGR, COASTAL PLAIN EXP STA, 64- *Concurrent Pos:* Assoc ed, Nematropica, 77-; mem comt common names of nematodes, 74-75, mem crop loss comt, Soc Nematologists, 74-85, mem sustaining assocs comt, 84-87, chmn, 86-87; chmn, Sustaining Assocs Comm, Soc Nematologists, 86-87. *Mem:* Am Phytopath Soc; Soc Nematologists; Orgn Trop Am Nematologists; Am Peanut Res & Educ Soc, Inc; Sigma Xi. *Res:* Host-parasite relationship of nematodes to plants; nematode-fungus relationship to plant diseases; nematode population dynamics as influenced by crops, cultural practices, and chemicals; development of nematode resistant varieties; economic nematode control. *Mailing Add:* US Dept Agr Coastal Plain Exp Sta Tifton GA 31793-0748

MINTON, PAUL DIXON, b Dallas, Tex, Aug 4, 18; m 43; c 2. STATISTICS. *Educ:* Southern Methodist Univ, BSc, 41, MSc, 48; NC State Col, PhD(exp statist), 57. *Prof Exp:* Instr math, Univ NC, 48, asst, Inst Statist, 49-52, statist asst, Inst Res Social Sci, 51-52; asst prof math, Southern Methodist Univ, 52-55; assoc prof statist, Va Polytech Inst, 55-56; assoc prof math & dir comput lab, Southern Methodist Univ, 57-61, prof statist & chmn dept, 61-72; dean, Sch Arts & Sci, Va Commonwealth Univ, 72-79, dir, Inst Statist, 79-87; RETIRED. *Concurrent Pos:* Consult statist, qual control, indust & med res groups. *Mem:* Biomet Soc; Inst Math Statist; fel Am Statist Asn; fel Am Soc Qual Control; Math Asn Am. *Res:* Distribution theory; experimental statistics. *Mailing Add:* 2626 Stratford Rd Richmond VA 23225

MINTON, SHERMAN ANTHONY, b New Albany, Ind, Feb 24, 19; m 43; c 3. HERPETOLOGY, MICROBIOLOGY. *Educ:* Ind Univ, AB, 39, MD, 42. *Prof Exp:* From asst prof to assoc prof microbiol, Sch Med, Ind Univ, Indianapolis, 47-48; vis prof, Postgrad Med Ctr, Karachi, Pakistan, 58-62; from assoc prof to prof, 62-84, EMER PROF MICROBIOL, SCH MED IND UNIV, 84- *Concurrent Pos:* Res assoc, Am Mus Natural Hist, 57-64, 64- *Honors & Awards:* Redi Award, Int Soc Toxinol, 85. *Mem:* Am Soc Trop Med & Hyg; Am Soc Ichthyol & Herpet; Int Soc Toxinology (pres, 66-68); Soc Study Amphibians & Reptiles (pres, 85); NY Acad Sci. *Res:* Venomous animals and the injuries they cause; geographic distribution and taxonomy of amphibians and reptiles; serological techniques in taxonomy; arthropods as human parasites and disease vectors. *Mailing Add:* Dept of Microbiol Ind Univ Med Ctr Indianapolis IN 46202-5120

MINTZ, A AARON, b Houston, Tex, July 29, 22; m 47; c 3. MEDICINE. *Educ:* Rice Inst, BA, 48; Univ Tex, MD, 48. *Prof Exp:* From instr to assoc prof pediat, 52-73, PROF PEDIAT & COMMUNITY MED, BAYLOR COL MED, 73- *Res:* Pediatrics. *Mailing Add:* Dept Pediat Baylor Col Med One Baylor Plaza Houston TX 77030

MINTZ, BEATRICE, b New York, NY, Jan 24, 21. BIOLOGY. *Educ:* Hunter Col, AB, 41; Univ Iowa, MS, 44, PhD(zool), 46. *Hon Degrees:* DSc, NY Med Col, 80, Med Col Pa, 80, Northwestern Univ, 82, Hunter Col, 86; LHD, Holy Family Col, 88. *Prof Exp:* Res asst, Guggenhiem Dent Clin, New York, 41-42; res asst develop biol, Univ Iowa, 42-46, instr, 46; from instr to assoc prof biol sci, Univ Chicago, 46-60; assoc mem, 60-65, SR MEM, INST CANCER RES, PHILADELPHIA, 65- *Concurrent Pos:* Fulbright res scholar, Univ Paris, 51 & Univ Strasbourg, 51; Harvey Soc lectr, 76; mem bd adv, Jane Coffin Children's Mem Fund, 77-; NIH lectr, 78. *Honors & Awards:* Bertner Found Award, 77; Award Biol & Med Sci, NY Acad Sci, 79; Papanicolaou Award, Papanicolaou Cancer Res Inst, 79; Medal, Genetics Soc Am, 81; Amory Prize, Am Acad Arts & Sci, 88; Ernst Jung Medal, 90. *Mem:* Nat Acad Sci; fel AAAS; Sigma Xi; Genetics Soc Am; Am Inst Biol Sci; fel Am Acad Arts & Sci; hon fel Am Gynec & Obstet Soc; Am Philos Soc. *Res:* Gene control of differentiation and disease in mammals. *Mailing Add:* Inst Cancer Res Fox Chase Cancer Ctr 7701 Burholme Ave Philadelphia PA 19111

MINTZ, DANIEL HARVEY, b New York, NY, Sept 16, 30; c 3. DIABETES. *Educ:* St Bonaventure Col, Olean, NY, BS, 51; NY Med Col, MD, 56. *Hon Degrees:* PhD, St Bonaventure Univ, Olean, NY, 85. *Prof Exp:* Chief med officer, Georgetown Univ Med Div, DC Gen Hosp, Washington, DC, 61-64, chief serv, 63-64; chief med, Magee Women's Hosp, Pittsburg, Pa, 64-69; co-dir, Clin Res Unit, Univ Miami Hosps & Clinics, 69-77, chief, Div Endocrinol & Metab, Univ Miami Sch Med, Jackson Mem Hosp, 69-80, SCI DIR, DIABETES RES UNIT, SCH MED, UNIV MIAMI, 80- , MARY LOU HELD PROF MED, 80- *Concurrent Pos:* Instr med, Sch Med, Georgetown Univ, Washington, DC, 61-63, asst prof med, 63-64; assoc prof med, Sch Med, Univ Pittsburgh, 64-69; prof med, Sch Med, Univ Miami, 69-80, interim chmn, Dept Med, 79-80; invited prof, Biochem Inst, Sch Med, Univ Geneve, Geneva, Switz, 76-77. *Mem:* Fel Am Col Physicians; Endocrine Soc; Am Diabetes Asn; Transplantation Soc; Europ Asn Study Diabetes; Am Asn Physicians. *Res:* Diabetes research in islet cell transplantation. *Mailing Add:* Diabetes Res Inst PO Box 016960 (R-134) Miami FL 33101

MINTZ, ESTHER URESS, b New York, NY, May 18, 07; m 29; c 1. PHYSICS. *Educ:* Hunter Col, BA, 28; Columbia Univ, MA, 31; NY Univ, PhD, 36. *Prof Exp:* Instr physics, Hunter Col, 28-33; res asst biophys, Columbia Univ, 36-39; instr physics, Hunter Col, 39-42; physicist, Signal Corps Labs, US Dept Army, NJ, 42-44; instr math, Fieldston Sch, 54-57; from asst prof to prof, 57-73, EMER PROF PHYSICS & ASTRON, BARUCH COL, CITY UNIV NEW YORK, 73- *Mem:* Am Phys Soc. *Res:* Optics; spectroscopy; astronomy for non-science student. *Mailing Add:* 2025 E 71st St Tulsa OK 74136

MINTZ, FRED, b New York, NY, June 30, 18; m 42, 76; c 2. MECHANICAL ENGINEERING, ACOUSTICS. *Educ:* George Washington Univ, BS, 40, MechEng, 46; Univ Calif, Los Angeles, ME, 65. *Prof Exp:* Physicist, David Taylor Model Basin, US Dept Navy, 40-46 & Bur Ships, 46-49; asst supvr acoust res sect, Armour Res Found, Ill Inst Technol, 49-51, supvr mech instrumentation & vibrations sect, 51-55; group engr, Eng Lab, Lockheed-Calif Co, 55-57, mgr physics res dept, 57-60, Phys & Chem Sci Dept, 60-64 & Vehicle Systs Lab, 64-70, sr res & develop engr, 70-73; prog mgr, Off Noise Abatement, US Environ Protection Agency, 73-82; sr scientist, Noise Control Technol Div, Underwater Systs, Inc, 82- 84; NOISE CONSULT, EPA, 85- *Concurrent Pos:* Mem subpanel acoust noise reduction, Panel Acoust, Res & Develop Bd, Dept Defense, 47-51; mem comt hearing & bioacoust, Nat Acad Sci-Nat Res Coun, 62-70. *Mem:* AAAS; fel Acoust Soc Am; Inst Elec & Electronics Engrs. *Res:* Shock and vibration engineering, especially measurement and control; acoustical engineering, especially noise control. *Mailing Add:* 4601 N Park Ave Chevy Chase MD 20815

MINTZ, LEIGH WAYNE, b Cleveland, Ohio, June 12, 39; m 62; c 2. PALEONTOLOGY, STRATIGRAPHY. *Educ:* Univ Mich, BS, 61, MS, 62; Univ Calif, Berkeley, PhD(paleont), 66. *Prof Exp:* From asst prof to assoc prof, 65-75, assoc dean instr, 69-70, assoc dean sci, 71-72, actg dean instr, 72-73, dean undergrad studies, 73-79, PROF GEOL SCI, CALIF STATE UNIV, HAYWARD, 75-, ASSOC VPRES ACAD PROG, 79- *Mem:* Paleont Soc; Geol Soc Am; Sigma Xi. *Res:* Geology of western United States; historical geology, particularly the significance of plate tectonics in earth history; fossil and recent irregular echinoids; Paleozoic crinozoan and Mesozoic echinozoan echinoderms. *Mailing Add:* Assoc Vpres Acad Prog Calif State Univ Hayward CA 94542-3011

MINTZ, STEPHEN LARRY, b Washington, DC, June 28, 43; m 71; c 1. NEUTRINO REACTIONS. *Educ:* John Hopkins Univ, BA, 65, PhD (physics), 72;Columbia Univ, MA, 67. *Prof Exp:* Res assoc physics, Ctr Theoret Studies, Univ Miami, 72-73, vis asst prof, 73-74; from asst prof to assoc prof, 74-84, PROF PHYSICS, FLA INT UNIV, 85-, CHMN DEPT, 86- *Concurrent Pos:* Prin investr, NSF grant, 79-81. *Mem:* Am Phys Soc; Sigma Xi. *Res:* Weak semileptomic processes in nuclei, particularly moon capture, neutrino reactions and electron-induced weak processes and corresponding strong processes with view to understanding nuclear structure. *Mailing Add:* 17220 SW 84th Ct Fla Int Univ Miami FL 33157

MINTZER, DAVID, b New York, NY, May 4, 26; m 49; c 2. GAS DYNAMICS, UNDERWATER ACOUSTICS. *Educ:* Mass Inst Technol, BS, 45, PhD(physics), 49. *Prof Exp:* Res assoc, US Navy Opers Eval Group, Mass Inst Technol, 46-48, res assoc, Acoust Lab, 48-49, mem staff, 49; asst prof physics, Brown Univ, 49-55; res assoc, Yale Univ, 55-56, assoc prof & dir lab marine physics, 56-62; assoc dean, Northwestern Univ, 70-73, actg dean, 71-72, vpres res & dean sci, 73-86, spec asst to pres, 86-87, PROF MECH ENG & ASTRONAUT SCI, NORTHWESTERN UNIV, 62-, PROF PHYSICS & ASTRON, 68- *Concurrent Pos:* Mem mine adv comt, Nat Acad Sci-Nat Res Coun, 63-73; mem bd trustees, EDUCOM, 75-81, chmn, 78-81; mem, Appl Res Lab Adv Bd, Pa State Univ, 76-81, chmn, 80-81; mem bd trustees, Adler Planetarium, Chicago, Ill, 76-; mem bd dirs, Res Park, Inc, Evanston, Ill, 86-; mem bd trustees, Ill Math & Sci Acad, Aurora, Ill, 87-; chmn bd dirs, Heartland Venture Capital Network, Inc, Evanston, Ill, 87-

Mem: Fel Am Phys Soc; fel Acoust Soc Am; Am Soc Mech Eng; Am Soc Eng Educ; Am Astron Soc. *Res:* Kinetic theory; plasma physics; acoustic wave propagation. *Mailing Add:* Northwestern Univ Technol Inst 2145 Sheridan Rd Evanston IL 60208

MINYARD, JAMES PATRICK, b Greenwood, Miss, May 11, 29; m 56; c 5. CONSUMER AGROCHEMICALS REGULATION, SCIENTIFIC LABORATORY ADMINISTRATION. *Educ:* Miss State Univ, BS, 51, PhD(org chem), 67. *Prof Exp:* From asst chemist to chemist, Miss State Chem Labs, 58-64; res chemist, Boll Weevil Res Lab, Agr Res Serv, US Dept Agr, 64-67; PROF CHEM, MISS STATE UNIV, 67-; STATE CHEMIST, MISS STATE CHEM LABS, 67- *Concurrent Pos:* Instr, Miss State Univ, 59-64; mem bd dirs, Am Chem Soc, 67-69, Asn Off Analytical Chemists, 78-83, Asn Am Feed Control Off, 71-77, Asn Am Plant Food Control Off, 84-89; adv & consult numerous couns & univs, 72-; mem, Coun Agr Sci & Technol Task Force Feeding Animal Waste, 78, Environ Protection Agency Off Enforcement Anal & Testing Activities Subcomt, 79-81, peer rev panel, 83, J Asn Off Analytical Chemists Rev & Eval Task Force, 86-88; reviewer, prog & lab capabilities & orgn, TVA-Nat Fertilizer Develop Ctr, 84, USDA-Agr Res Serv Ann Coop Res Prog, 85. *Honors & Awards:* Fel, Div Agrochem, Am Chem Soc; Fel, Asn Off Analytical Chemists; Res Group Award of Merit, USDA, 75; Comnr's Spec Citation, US Food & Drug Admin, 87; Outstanding Chemist Award, Am Chem Soc, 90. *Mem:* AAAS; fel Am Chem Soc; fel Asn Off Anal Chemists (pres, 81-82); Asn Am Feed Control Off (pres, 76); Asn Am Plant Food Control Off (pres, 87-88); Soc Environ Toxicol & Chem; Newcomen Soc NAm; Plant Growth Regulator Soc Am; Am Oil Chemists Soc; Asn Am Pesticide Control Off; Asn Food & Drug Off; Am Soc Testing & Mat; Coun Agr Sci & Technol. *Res:* Instrumental analysis; pesticide residues; natural products; organic reaction mechanisms; pesticide photochemistry; insect pheremones; environmental analytical chemistry and toxicology of agricultural chemicals to humans, plants and animals; author & manager of US national databases on toxic chemicals in animal feeds and human foods. *Mailing Add:* Box 2198 Miss State Univ Mississippi State MS 39762

MINZNER, RAYMOND ARTHUR, b Lawrence, Mass, June 9, 15; m 40; c 4. AERONOMY, METEOROLOGY. *Educ:* Mass State Col, BS, 37, MS, 40. *Prof Exp:* Lab instr physics, Mass State Col, 38-41; instr, Univ Ariz, 41-42; res assoc electronics, Mass Inst Technol, 42-45; electronic engr, Air Force Cambridge Res Ctr, 46-50, physicist & unit chief, 50-53, sect chief atmospheric standards, 53-56, chief compos sect, 56-58; staff physicist, Geophys Corp Am, 59-64, prin scientist, GCA Tech Div, GCA Corp, Mass, 65-67; aerospace technologist, Electronic Res Ctr, NASA, 67-70 & Goddard Space Flight Ctr, 70-76, sr res scientist, 76-80; environ scientist, Space Div, Gen Elec Co, 80-81; consult, 81-85; RETIRED. *Concurrent Pos:* Mem comt for exten, US Standard Atmosphere, 53-, co-ed, 75. *Mem:* Am Meteorol Soc; Am Geophys Union; Sigma Xi. *Res:* Statistical studies and uncertainty of inferred meteorological parameters; stratospheric climatology; cloud motion vs wind relationships; cloud height from aircraft photos and satellite imagery; standard and model atmospheres. *Mailing Add:* 215 Rio Villa Dr No 3263 Punta Gorda FL 33950

MIONE, ANTHONY J, nuclear engineering, physics, for more information see previous edition

MIOVIC, MARGARET LANCEFIELD, b Salem, Ore, Apr 15, 43; m 64; c 3. MICROBIOLOGY. *Educ:* Radcliffe Col, BA, 65; Univ Pa, PhD(microbiol), 70. *Prof Exp:* Instr microbiol, Cornell Univ, 70-72 & biochem, 73-74; asst prof, Swarthmore Col, 75-80, assoc prof biol, 80-81; LECT, DEPT MED, UNIV PA SCH MED, 87- *Mem:* Am Soc Microbiol; Sigma Xi. *Res:* Metabolic regulation in photosynthetic bacteria. *Mailing Add:* 3400 Spruce St Nine Penn Tower Philadelphia PA 19104

MIQUEL, JAIME, b Agres, Spain, Jan 7, 29; US citizen; m 55; c 2. NEUROSCIENCES, ANIMAL PHYSIOLOGY. *Educ:* Univ Granada, MSc, 50, PhD(pharmacol), 52. *Prof Exp:* NIH res assoc, Nat Inst Neurol Dis & Blindness, 58-61; res scientist, 61-65, chief br exp path, Ames Res Ctr, NASA, 65-81; RES ASSOC, PAULING INST, PALO ALTO, 84- *Concurrent Pos:* Span Inst Pharmacol fel, Span Ministry of Educ, Madrid, 52-54; Cajal Inst fel, Span Ministry of Educ, Valencia, 55-58; prof neurogerontol, Univ Sch Med, Alicante Spain, 83- *Honors & Awards:* Cosmos Awards, NASA, 75 & 77; Biocore Award, 75. *Mem:* Geront Soc; Am Asn Neuropath; Span Soc Geront; Soc Invert Path. *Res:* Experimental gerontology; experimental neuropathology; radiobiology; space biomedicine; pharmacology; cellular and molecular mechanisms of aging (with emphasis in the study of free radical and peroxidation processes) and on the biological effects of space flight. *Mailing Add:* Linus Pauling Inst Sci & Med 440 Page Mill Rd Palo Alto CA 94306

MIR, GHULAM NABI, b Srinagar, Kashmir, June 26, 39; US citizen; div; c 2. PHARMACOLOGY, PHYSIOLOGY. *Educ:* Univ Jammu & Kashmir, BSc, 58; Punjab Univ, India, B Pharm, 63, M Pharm, 65; Univ Miss, PhD(pharmacol), 71; Temple, MBA, 78. *Prof Exp:* Anal chemist, Drug Res Labs, Kashmir, 58-59; res pharmacologist, Inst Hist Med & Med Res, New Delhi, India, 65-66; group leader toxicol, 72-73 & gastrointestinal pharmacol, 73-75, sect head, 75-84, dept pharmacol, William H Rorer, Inc, Pa, 84-88; asst dir, dept pharmacol, Ayerst Labs Res Inc, Princeton, NJ, 84-88; PRES, DRUG DEVELOP INC, BUCKINGHAM, PA, 88- *Mem:* Soc Toxicol; Drug Info Asn; Am Soc Pharmacol & Exp Therapeut; Am Gastroenterol Asn. *Res:* Drug effects on gastric secretion; peptic and duodenal ulcers; intestinal motility; intestinal absorption and secretion; uterine contraction and general toxicity of drugs; experimental therapy. *Mailing Add:* PO Box 394 Buckingham PA 18912

MIR, LEON, b Krystynopol, Poland, July 13, 38; US citizen; m 66; c 1. CHEMICAL ENGINEERING. *Educ:* Columbia Univ, AB, 58, BS, 59; Mass Inst Technol, MS, 61, ScD(chem eng), 63. *Prof Exp:* Engr, Esso Res & Eng Co, NJ, 63-65; proj leader & actg prog mgr large scale gas chromatog, Abcor, Inc, 65-71; sr consult scientist, Avco Systs Div, Avco Corp, 71-77; SR CONSULT SCI, ABCOR, INC, NEWTON, MASS, 77- *Mem:* Am Chem Soc; Am Inst Chem Engrs. *Res:* Physical chemistry of polyelectrolytes; chromatographic processes; applied mathematics. *Mailing Add:* 15 Hobart Rd Newton MA 02159-1312

MIRABELLA, FRANCIS MICHAEL, JR, b Dec 27, 43; m 67. POLYMER CHEMISTRY. *Educ:* Univ Bridgeport, BA, 66; Univ Conn, MS, 74, PhD(polymer chem), 75. *Prof Exp:* Res chemist, Technichem Co, 66-69; analyst, Olin Corp, 69-72; teaching asst anal chem, Univ Conn, 72-75; res scientist polymer eval, Arco/Polymers Inc, Atlantic Richfield Co, 75-77; res scientist polymer eval, Northern Petrochem Co, 77-85; sr res assoc, Norchem, Inc, 85-; AT USI CHEMICALS CO. *Mem:* Am Chem Soc. *Res:* Theory of composition of copolymers and relationship of copolymer composition to molecular weight. *Mailing Add:* USI Chem Co 31000 Golf Rd Rolling Meadows IL 60008

MIRABELLI, CHRISTOPHER KEVIN, EXPERIMENTAL BIOLOGY. *Educ:* State Univ NY, Fredonia, BS, 77; Baylor Col Med, PhD(pharmacol), 81. *Prof Exp:* Res assoc, Dept Biol, Brookhaven Nat Lab, 77-78; assoc sr investr, Dept Molecular Pharmacol, Smith Kline & French Labs, 81-82, sr investr, 82-83, prog head, Transition Metals Prog, 82-85, from asst dir to assoc dir, Dept Molecular Pharmacol, 83-88, biol dir, Antineoplastic Prog, 85-88, feasability study head, Azaspirane-Suppressor Cell Feasability Study, 87-88, dir, Dept Molecular Pharmacol, 88; vpres res, 89-90, CO-FOUNDER & CORP OFFICER, ISIS PHARMACEUT, 89-, SR VPRES RES & PRECLIN DEVELOP, 91- *Concurrent Pos:* Numerous invited lect, 83-91; consult, Develop Therapeut Contracts Rev Comt, Nat Cancer Inst, NIH, 84 & mem, Prog Proj Spec Rev Comt, 88; adj asst prof, Dept Pharmacol & mem, Grad Group Pharmacol Sci, Univ Pa, 84- *Mem:* AAAS; Am Asn Cancer Res; Am Soc Pharmacol & Exp Therapeut. *Res:* Author of numerous publications. *Mailing Add:* ISIS Pharmaceut 2280 Faraday Ave Carlsbad CA 92008

MIRABILE, CHARLES SAMUEL, JR, b Hartford, Conn, Jan 10, 37; m 60; c 3. PSYCHIATRY. *Educ:* Yale Univ, BA, 59; McGill Univ, MDCM, 63. *Prof Exp:* RES PSYCHIATRIST, INST LIVING, 67- *Concurrent Pos:* Mem med staff, Hotchkiss Sch, 70-, Indian Mountain Sch, 78, Berkshire Sch, 79 & Salisbury Sch, 81-; pvt pract psychiatrist, Sharon Clin, 69- *Mem:* AAAS; Am Psychiat Asn. *Res:* Orienting mechanisms; sensory function and perception; neurophysiology of individual difference; neurophysiologic correlates of mental illness. *Mailing Add:* Upper Main St PO Box 683 Sharon CT 06069

MIRABITO, JOHN A, b Somerville, Mass, May 16, 17; m 39. METEOROLOGY. *Educ:* Wake Forest Col, BS, 41; Mass Inst Technol, cert, 43. *Prof Exp:* Meteorologist, Pac Fleet & Naval Air Reserve Training Prog, US Navy, 42-52, personnel & training officer meteorol, Off Chief Naval Oper, 52-54, staff meteorologist & oceanogr, Naval Support Forces, Antarctica, 54-59, exec officer, Fleet Weather Cent, DC, 59-61, staff meteorologist & oceanogr, 6th Fleet, 61-63; marine serv coordr, Environ Sci Serv Admin, 63-70; res prog mgr, Nat Marine Fisheries Serv, Nat Oceanic & Atmospheric Admin, 71-72, prog analyst, 72-81; RETIRED. *Concurrent Pos:* Exec secy working group VIII, US-USSR Bilateral on Protection of Environ, 75-81. *Honors & Awards:* Personal Commendation Medal, Secy Navy, 56. *Mem:* AAAS; Am Meteorol Soc; Am Geophys Union. *Res:* Polar meteorology in Antarctica; oceanographic forecasts; marine science, air-sea interaction; fisheries research management. *Mailing Add:* 4713 Jasmine Dr Rockville MD 20853

MIRACLE, CHESTER LEE, b Barbourville, Ky, Apr 30, 34; m 57; c 3. MATHEMATICS. *Educ:* Berea Col, 54, BA; Auburn Univ, MS, 56; Univ Ky, PhD(math), 59. *Prof Exp:* From instr to asst prof, 59-63, ASSOC PROF MATH, UNIV MINN, MINNEAPOLIS, 63- *Mem:* Am Math Soc; Math Asn. *Res:* Analytical continuation and summability of series. *Mailing Add:* Dept of Math Univ of Minn Minneapolis MN 55455

MIRAGLIA, GENNARO J, b Italy, Jan 18, 29; US citizen; m 54; c 3. MEDICAL MICROBIOLOGY, PHYSIOLOGY. *Educ:* St Bonaventure Univ, BS, 51; Univ NH, MS, 57; Univ Tenn, PhD(bact), 60. *Prof Exp:* Fel bact, Bryn Mawr Col, 61-62; asst prof bact, Seton Hall Col Med & Dent, 62-63; RES FEL, SQUIBB INST MED RES, 63- *Mem:* AAAS; Am Soc Microbiol; NY Acad Sci; Sigma Xi. *Res:* Host-bacteria interrelationships; gram-negative infections; effect of low ambient temperature on infections; chemotherapy of a number of bacterial infections. *Mailing Add:* Two Tompkins Rd East Brunswick NJ 08816

MIRALDI, FLORO D, b Lorain, Ohio, Mar 21, 31; m 57; c 1. BIOENGINEERING, NUCLEAR MEDICINE. *Educ:* Col Wooster, AB, 53; Mass Inst Technol, SB, 53, SM, 55, ScD(nuclear eng), 59; Case Western Reserve Univ, MD, 70. *Prof Exp:* Engr, Atomic Power Div, Westinghouse Elec Corp, 55; asst prof elec eng, Purdue Univ, 58-59; from asst prof to assoc prof nuclear eng, 59-69, assoc prof biomed eng, 69-72, PROF BIOMED ENG, SCHS ENG & MED, CASE WESTERN RESERVE UNIV, 72-; DIR, NUCLEAR RADIOL, UNIV HOSP CLEVELAND, 84- *Concurrent Pos:* Asst radiologist, Univ Hosps of Cleveland, 71-72, radiologist, 72-; chief nuclear med, Cleveland Metrop Gen Hosp, 73-75; mem adv comt biotechnol & human resources, NASA, 63-65; pres, Am Radiation Res Corp, 63-70; mem res adv bd, Euclid Clin Found, 66-69; assoc prof radiology, Nuclear Scanners & Radiation Diag, 81- *Mem:* AAAS; Am Nuclear Soc; Soc Nuclear Med; Sigma Xi; Radiol Soc NAm; Roentgen Ray Soc; Am Asn Univ Radiologists. *Res:* Radiological health; applications of radiation and radioisotopes; nuclear reactor physics; nuclear medical instrumentation. *Mailing Add:* Dept Radiol Univ Hosp Cleveland Cleveland OH 44106

MIRAND, EDWIN ALBERT, b Buffalo, NY, July 18, 26. BIOLOGY. *Educ:* Univ Buffalo, BA, 47, MA, 49; Syracuse Univ, PhD, 51. *Hon Degrees:* DSc, Niagara Univ, 70; DSc, D'Youville Col, 74. *Prof Exp:* Asst, Univ Buffalo, 47, instr biol, 48; instr, Utica Col, Syracuse, 50; assoc cancer res scientist, 51-60, asst to Inst Dir, 60-67, PRIN CANCER RES SCIENTIST, DIR CANCER

RES, ROSWELL PARK MEM INST, 67-, ASSOC DIR & DEAN INST, 67- *Concurrent Pos:* Res prof, Grad Sch, State Univ NY Buffalo, 54-, prof & dir grad studies, 67-, dean, Roswell Park Grad Div, Grad Sch, 67-; prof & dean, Roswell Park Grad Div, Niagara Univ, 68-, coun mem, 71-; mem comt tech guid, Nat Acad Sci-Nat Res Coun, Human Cancer Virus Task Force, NIH, 63-67; prof clin cancer comt mem, Nat Cancer Inst, 74-82; chmn, USA Nat Comt, Int Union Against Cancer, 80-86; secy-gen, 13th Int Cancer Cong, 78-82; NY State adv coun on AIDS, 82- *Honors & Awards:* Billings Silver Medal, AMA, 60. *Mem:* Fel AAAS; Radiation Res Soc; Am Soc Zool; Soc Exp Biol & Med; Am Asn Cancer Inst (secy-treas, 68-); Int Soc Gnotobiol (pres, 83-84); fel NY Acad Sci; Sigma Xi; Int Soc Hemat; Pub Health Asn Am; Am Soc Hemat; Am Soc Prev Oncol; Am Asn Cancer Educ; Int Union Health Educ. *Res:* Abnormality of iron metabolism in iron deficiency anemias; erythropoietin; relationships of viruses to cancer; gnotobiology; author of 493 publications in the fields of cancer research, endocrinology, hematology, virology and cancer education. *Mailing Add:* Roswell Park Mem Inst 666 Elm St Buffalo NY 14263

MIRANDA, ARMAND F, b Paramaribo, Suriname, SAfrica, Nov 10, 35. PATHOLOGY. *Educ:* Columbia Univ, BSc, 69, MPh, 73, PhD(pathobiol), 74. *Prof Exp:* Muscular Dystrophy Asn fel, 74; instr path, Columbia Univ, 75, res assoc, 75-77, asst prof, 77-86, PROF CLIN PATH, COLUMBIA UNIV, 86-; DIR, TISSUE CULT LAB, H HOUSTON MERRITT CLIN RES CTR MUSCULAR DYSTROPHY & RELATED DIS, 75- *Mem:* Sigma Xi; Am Soc Cell Biol; Tissue Cult Asn. *Res:* Muscle development and regeneration; cytopathology. *Mailing Add:* Dept Path Col Physicians & Surgeons Columbia Univ 630 W 168th St New York NY 10032

MIRANDA, CONSTANCIO F, b Raia-Goa, India, Dec 4, 26; m 57; c 4. STRUCTURAL & CIVIL ENGINEERING. *Educ:* Univ Bombay, BE, 49; Univ Notre Dame, MS, 62; Ohio State Univ, PhD(struct eng), 64. *Prof Exp:* Asst lectr civil eng, Col Eng, Poona, 50; asst engr & exec engr, Govt Bombay, India, 50-60; from teaching asst to instr, Univ Notre Dame, 60-62; instr & res assoc, Ohio State Univ, 62-64; assoc res engr, Air Force Shock Tube Facility, Univ NMex, 64-65; chmn dept, Univ Detroit, 65-80, prof civil eng, 65-89, dir, Prof Adv Serv Ctr, 76-89; CONSULT ENGR, 89- *Concurrent Pos:* Indust engr, Messrs Ibcon Ltd, India, 50; instr nuclear defense design, Off Civil Defense, 65-89; res grant & tech proj dir, Army Tank Automotive Ctr, 66-89; prin, Miranda, Baker & Assocs, Consult Engrs, Mich, 67. *Mem:* Am Soc Civil Engrs; Am Soc Eng Educ; assoc mem Indian Inst Eng. *Res:* Structural engineering in general, particularly stability; design in the inelastic range; matrix formulation of structural problems; soil mechanics; nuclear defense and hydraulic engineering. *Mailing Add:* 100 Silvercliff Terr Cary NC 27513

MIRANDA, FRANK JOSEPH, b Erie, Pa, June 30, 46; m 76; c 2. DENTAL CLINIC ADMINISTRATION. *Educ:* Univ Calif, Los Angeles, DDS, 71; Cent State Univ, MEd, 76, MBA, 79. *Prof Exp:* Dentist, Lynwood Children's Found, 71-72; assoc dentist, Sch Dent, Univ Calif, Los Angeles, 71-74; res dentist, Proj Acorde, Dept Health, Educ & Welfare, 72-74; from asst prof to assoc prof, 74-87, PROF DENT, COL DENT, UNIV OKLA, 87-, ASST DEAN, CLIN AFFAIRS, 89- *Concurrent Pos:* Consult, State Okla Chief Med Examr, 75-80, Cent Regional Dent Testing Serv, Inc, 82-88, Food & Drug Admin Dent Devices & Radiol Health Panel, 84-; adj fac dent hyg, Rose State Col, 80-; assoc ed, J Okla Dent Asn, 83-89; mem comt dent accreditation, Am Dent Asn, 86-, joint comn nat dent exam, 87- *Honors & Awards:* Assoc Distinguished Lectrship, 88. *Mem:* Fel Acad Gen Dent; fel Acad Dent Int; fel Acad Int Dent Studies; fel Am Col Dentists; Am Dent Asn; Am Asn Dent Schs; Acad Oper Dent. *Res:* Improvements in dental materials, particularly amalgam and composite, and techniques of clinical usage. *Mailing Add:* 6645 Whitehall Dr Oklahoma City OK 73132

MIRANDA, GILBERT A, b Los Angeles, Calif, Oct 21, 43; div; c 4. DETONATION PHYSICS, CRYOGENICS. *Educ:* Calif State Col, Los Angeles, BS, 65; Univ Calif, Los Angeles, MS, 66, PhD(solid state physics), 72. *Prof Exp:* APPL PHYSICIST, LOS ALAMOS SCI LAB, UNIV CALIF, 72- *Res:* Critical current measurements of 10-50 kiloampere superconducting cables, braids and monolithic wires; measurements of the effects of colliding detonation waves on metals; nuclear magnetic resonance; initiation of insensitive high explosives and generation of shaped detonation waves. *Mailing Add:* Los Alamos Sci Labs MS-960 Los Alamos NM 87545

MIRANDA, HENRY A, JR, b New York, NY, Sept 20, 24; m 50; c 2. GEOENVIRONMENTAL SCIENCE. *Educ:* Iona Col, BS, 52; Fordham Univ, MS, 53, PhD(physics), 56. *Prof Exp:* Sr scientist, Hudson Labs, Columbia Univ, 56-60; sr consult engr, Braddock, Dunn & McDonald, Inc, Tex, 60-62; sr scientist & group leader radiation physics, Lowell Tech Inst Res Found, 62-65; sr scientist & dept mgr atmospheric sci, GCA Technol Div, 65-67, dir space sci lab, 67-72; vpres & dir res, Epsilon Labs, Inc, 72-86; PROPRIETOR, MIRANDA LABS, 86- *Mem:* Am Geophys Union; Am Phys Soc; Inst Elec & Electronics Eng; Sigma Xi. *Res:* Earth and atmospheric radiation; turbulent oceanic diffusion; scintillation counting; optical and infrared tracking systems; nuclear effects; spectroscopy; densitometry; optical instrumentation; laser development; aeronomy; upper-atmospheric winds and diffusion; stratospheric aerosols; aerosol physics; imaging systems; photogrammetry. *Mailing Add:* 13 Old Stage Coach Rd Bedford MA 01730

MIRANDA, QUIRINUS RONNIE, b Bombay, India, June 4, 39; US citizen; m 76. VIROLOGY, MICROBIOLOGY. *Educ:* Univ Bombay, BSc, 61, MSc, 66; Univ Manchester, PhD(virol), 71. *Prof Exp:* Sr scientist & virologist, Schering-Plough Corp, 71-74; asst res microbiol, Univ Calif, Los Angeles, 74-75; sr proj leader, Organon Diag, Akzona, Inc, 75-78; MGR RES & DEVELOP, INT DIAG TECHNOL, INC, 78- *Concurrent Pos:* Adj asst prof, Univ Southern Calif, 77- *Mem:* AAAS; Am Soc Microbiol; Soc Gen Microbiol; Tissue Culture Asn; NY Acad Sci. *Res:* Rapid diagnosis of pathogens using IFA and EIA; interferon; viral chemotherapy and viral vaccines. *Mailing Add:* 2097 Jonthan Ave San Jose CA 95125

MIRANDA, THOMAS JOSEPH, b Ewa Mill, Hawaii, Nov 18, 27; m 53; c 5. ORGANIC POLYMER CHEMISTRY. *Educ:* San Jose State Col, AB, 51, MA, 53; Univ Notre Dame, PhD(org chem), 59; Ind Univ, MBA, 80. *Prof Exp:* Instr chem, San Jose State Col, 52-53; chemist, Eitel-McCullough Inc, 53; dir res chem, O'Brien Corp, 59-69; sr res mat scientist, 69-71, STAFF SCIENTIST, WHIRLPOOL CORP, 71-, ADJ ASST PROF CHEM, UNIV IND, SOUTH BEND, 80- *Concurrent Pos:* Tech ed, J Coatings Technol; pres, Paint Res Inst, 83-84. *Honors & Awards:* Ernest T Trigg Award, Fedn Socs Coatings Technol, 67 & George Baugh Heckel Award, 80; JJ Mattiello Mem Lectr, 84. *Mem:* Sigma Xi; Am Soc Testing & Mat; Fedn Soc Coatings Technol. *Res:* Stereospecific polymerization of olefins; radiation polymerization of olefins; rocket fuels; emulsion polymerization; thermosetting acrylics; epoxy resins; water soluble polymers; thermal analysis; polymer stabilization; coatings technology; semiorganic foams. *Mailing Add:* Elisha Gray II Res & Eng Ctr Whirlpool Corp Monte Rd Benton Harbor MI 49022

MIRANKER, WILLARD LEE, b Brooklyn, NY, Mar 8, 32; m 52; c 3. MATHEMATICS. *Educ:* NY Univ, BA, 52, MS, 53, PhD(math), 57. *Prof Exp:* Asst math, NY Univ, 53-56; mem staff, Bell Tel Labs, Inc, 56-58; staff mathematician, Res Ctr, Int Bus Mach Corp, 59-63; sr res fel, Calif Inst Technol, 63-64; asst to dir res, 65, STAFF MATHEMATICIAN, RES CTR, IBM CORP, 65-, ASST DIR, MATH SCI DEPT, 72- *Concurrent Pos:* Adj prof, City Univ New York, 66-67; vis prof dept math, Hebrew Univ, Israel, 68-69; adj prof, Dept Math, NY Univ, 70-73; vis prof, Univ Paris, 74-75; vis lectr, Yale Univ, 73-74; adj prof, State Univ NY, Purchase, 77. *Mem:* Fel AAAS; Am Math Soc; Soc Indust & Appl Math. *Res:* Applied mathematics; numerical analysis. *Mailing Add:* Res Ctr IBM PO Box 218 Yorktown Heights NY 10598

MIRARCHI, RALPH EDWARD, b Mt Carmel, Pa, Jan 30, 50; m 77; c 2. WILDLIFE BIOLOGY. *Educ:* Muhlenberg Col, BS, 71; Va Polytech Inst & State Univ, MS, 75, PhD(wildlife biol), 78. *Prof Exp:* ASSOC PROF WILDLIFE, DEPT ZOOL-ENTOM, AUBURN UNIV, 78- *Mem:* Wildlife Soc; Wilson Ornith Soc; Am Ornith Union; Wildlife Mgt Inst; Am Soc Mammalogists. *Res:* Effects of biotic and abiotic factors upon behavior and reproductive physiology of wildlife species. *Mailing Add:* 403 Dunlop Dr Opelika AL 36801

MIRCETICH, SRECKO M, b Skela, Yugoslavia, Sept 2, 26; US citizen; m 56; c 2. PLANT PATHOLOGY. *Educ:* Univ Sarajevo, BS, 52; Univ Belgrade, MS, 54; Univ Calif, Riverside, PhD(plant path), 66. *Prof Exp:* Teaching asst plant path, Sch Agr, Univ Sarajevo, 49-52; asst plant pathologist, Exp Sta Subtrop Agr, Bar, Yugoslavia, 52-54, head plant protection dept, 54-57; res assoc plant path, Univ Calif, Riverside, 58-66; res plant pathologist, Plant Sci Res Div, 66-73, PLANT PATHOLOGIST, AGR RES SERV, US DEPT AGR, PACIFIC WEST AREA, 73- *Honors & Awards:* CIBA-Geigy Award & Lee M Hutchins Award, Am Phytopath Soc. *Mem:* Am Phytopath Soc; Int Soc Plant Path; fel Am Pytopath Soc. *Res:* Soilborne pathogens and diseases; soil ecology of Phytophthora and Pythium; soil populations; diseases of fruit and nut trees; Phytophthora root and crown rot of deciduous fruit and nut trees. *Mailing Add:* Dept of Plant Path Univ of Calif Davis CA 95616

MIRELS, HAROLD, b New York, NY, July 29, 24; m 53; c 3. AERONAUTICAL ENGINEERING. *Educ:* Cooper Union, BME, 44; Case Inst Technol, MS, 49; Cornell Univ, PhD(aeronaut eng), 53. *Prof Exp:* Aeronaut res engr, NASA, 44-61, chief plasma physics br, 60-61; head adv propulsion & fluid mech dept, 61-64, head aerodyn & heat transfer dept, 64-76, asst dir, 76-78, assoc dir, 78-84, PRIN SCIENTIST, AEROPHYSICS LAB, AEROSPACE CORP, 84- *Concurrent Pos:* Spec lectr, Case Inst Technol, 50, Fenn Col, 54-61 & Univ Calif, Los Angeles, 72-77; mem fluid mech comt, NASA, 60-61, mem adv subcomt fluid dynamics, 67-70. *Honors & Awards:* Fluid & Plasma Dynamics Award, Am Inst Aeronaut & Astronaut, 88. *Mem:* Nat Acad Eng; fel Am Phys Soc; Fel Am Inst Aeronaut & Astronaut. *Res:* Aerodynamics; gas dynamics; heat transfer; chemical lasers. *Mailing Add:* 3 Seahurst Rd Rolling Hills Estates CA 90274

MIRES, RAYMOND WILLIAM, b Mansfield, Tex, Mar 16, 33; div; c 3. PHYSICS. *Educ:* Tex Tech Univ, BS, 55, MS, 60; Univ Okla, PhD(physics), 64. *Prof Exp:* Mathematician, Holloman Air Develop Ctr, Holloman AFB, NMex, 55-56; engr, Martin Co, 56-57; instr physics, Tex Tech Univ, 57-60; asst, Univ Okla, 60-64; from asst prof to assoc prof, 64-71, PROF PHYSICS, TEX TECH UNIV, 71- *Concurrent Pos:* Physicist, LTV Electrosysts, Inc, Tex, 66-67; Advan Res Projs Agency res grant, Tex Tech Univ, 67-72; consult forensic physicist, 67- *Mem:* AAAS; Am Phys Soc; Am Asn Physics Teachers. *Res:* Optical and magnetic properties of solids; atomic structure; forensic applications of engineering physics. *Mailing Add:* Dept of Physics Tex Tech Univ Lubbock TX 79409

MIRHEJ, MICHAEL EDWARD, b Dhour-Shweir, Lebanon, Dec 25, 31; US citizen; div; c 2. TEXTILE TECHNOLOGY. *Educ:* McGill Univ, BSc, 58; Univ Western Ont, MSc, 59; Univ BC, PhD(chem oceanog, chem physics), 62. *Prof Exp:* Fel physics, Univ BC, 62-63; sr res chemist, Polymer Corp Ltd, Ont, 63-66; res chemist, E I du Pont de Nemours & Co Inc, 66-75, sr res chemist, 75-81, res assoc, Chattanooga Nylon Plant, 81-82, mkt mgr, Wilmington, 82-91; RETIRED. *Res:* Nuclear magnetic resonance spin echo technique to study influence of paramagnetic oxygen on proton relaxation; polymer synthesis and characterization; auto regenerating dehydrogenation catalysts; development of new textile products. *Mailing Add:* 20 Hanover Dr Westchester PA 19382

MIRKES, PHILIP EDMUND, b Oshkosh, Wis, May 8, 43; m 68; c 2. DEVELOPMENTAL BIOLOGY, DEVELOPMENTAL BIOCHEMISTRY. *Educ:* St Norbert Col, BS, 65; Univ Mich, MS, 67, PhD(zool), 70. *Prof Exp:* Trainee zool, Univ Wash, 70-71, fel genetics, 71-73; asst prof biol, Univ SC, 73-79; res asst prof, 79-83, RES ASSOC PROF, UNIV WASH, 83- *Concurrent Pos:* Prin investr, Res Corp grant, 75-78 & NSF grant, 78-81; mid career develop award-toxicol, Nat Inst Environ Health Sci, 79-81; Nat Inst Child Health & Human Develop, 81- *Mem:* AAAS; Soc Teratology; Soc Develop Biol. *Res:* Molecular mechanisms of teratogenesis. *Mailing Add:* Dept Pediat RD-20 Univ Wash Seattle WA 98195

MIRKIN, BERNARD LEO, b New York, NY, Mar 31, 28; m 54, 86; c 2. PEDIATRICS, CLINICAL PHARMACOLOGY. *Educ:* NY Univ, AB, 49; Yale Univ, PhD(pharmacol), 53; Univ Minn, MD, 64. *Prof Exp:* Instr physiol & pharmacol, State Univ NY Downstate Med Ctr, 52-54, 56-57, asst prof pharmacol, 57-58, res assoc, 58-60; from asst prof to assoc prof pediat & pharmacol, 66-72, dir, Div Clin Pharmacol, 66-89, PROF PEDIAT & PHARMACOL, MED SCH, UNIV MINN, MINNEAPOLIS, 72-; HEAD DIR RES, CHILDREN'S MEM INST EDUC RES, CHILDREN'S MEM HOSP, 89- *Concurrent Pos:* Consult, NIH & Nat Res Coun-Nat Acad Sci; fel, Jesus Col, Oxford Univ & Ford Found. *Mem:* Am Soc Pharmacol & Exp Therapeut; Am Soc Clin Pharmacol & Therapeut; Soc Pediat Res; Am Asn Cancer Res; Am Acad Pediat. *Res:* Chemotherapy of neural crest tumors; developmental pharmacology. *Mailing Add:* Dept Pediat Children's Mem Inst Educ & Res Children's Hosp 2300 Children's Plaza Chicago IL 60614

MIRKIN, L DAVID, b Buenos Aires, Arg; US citizen. PATHOLOGY. *Educ:* Mariano Moreno Col, BS, 48; Univ Buenos Aires, MD, 57. *Prof Exp:* Asst dir path, Hosp Ninos, Buenos Aires, Arg, 67-70; dept head path, Hosp Roberto Rio, Santiago, Chile, 70-76; dir pediat path, J W Riley Hosp Children, Indiana Univ, Indianapolis, 78-87; PROF PATH & PEDIAT, WRIGHT STATE UNIV, 87- *Concurrent Pos:* Consult, Pan Am Health Orgn, 70 & Fundacion Maximo Castro, Buenos Aires, Arg, 70-76; vis prof, Nat Univ Chile, Santiago, 71-76. *Mem:* Am Asn Pathologists; Soc Pediat Path; Am Soc Clin Pathologists; NY Acad Sci; Am Med Asn; Col Am Pathologists. *Res:* Electron microscopy of pediatric tumors; pathology of pediatric diseases. *Mailing Add:* Children's Med Ctr One Children's Plaza Dayton OH 45404-1815

MIROCHA, CHESTER JOSEPH, b Cudahy, Wis, Feb 7, 30; m 52; c 6. PLANT PATHOLOGY. *Educ:* Marquette Univ, BS, 55; Univ Calif, PhD(plant path), 60. *Prof Exp:* Lab technologist, Univ Calif, 57-60; res plant pathologist, Union Carbide Chem Co, 60-63; from asst prof to assoc prof plant path & physiol, 63-66, PROF PLANT PATH, UNIV MINN, ST PAUL, 72- *Concurrent Pos:* Chmn comt mycotoxicol, Int Soc Plant Path. *Mem:* AAAS; Am Phytopath Soc; Sigma Xi. *Res:* Mycotoxicology; physiology of fungi; mass spectroscopy; analytical chemistry. *Mailing Add:* Plant Path Dept 221 Univ Minn St Paul MN 55108

MIRONESCU, STEFAN GHEORGHE DAN, cell biology; deceased, see previous edition for last biography

MIROWITZ, L(EO) I(SAAK), b Mannheim, Ger, Dec 11, 23; nat US; m 47; c 3. STRUCTURAL DYNAMICS. *Educ:* Wash Univ, BS, 44, MS, 57. *Prof Exp:* From asst res engr to sr res engr flutter & vibrations, McDonnell Aircraft Corp, 46-51, from design engr to sr design engr, 51-54, proj dynamics engr, 54-57, chief dynamics engr, 57-58, chief struct dynamics engr, 58-64, eng mgr, Space & Missile Systs, 64-66, Voyager prog mgr, 66-69; dir advan technol, McDonnell Douglas Astronaut Co, 69-75, vpres diversification, McDonnell Douglas Corp, 75-84, pres, McDonnell Douglas Health Systs Co, 84-87; RETIRED. *Concurrent Pos:* Mem subcomt vibration & flutter, NASA, 56-58; dir, Microdata, Inc, 79-84, Coaliquid Inc, 81-84 & Repub Health Corp, 84-87; pres, Vitek Systs, Inc, 79-86. *Mem:* Assoc fel Am Inst Aeronaut & Astronaut. *Res:* Applied mechanics; influence of dynamics, flutter and vibrations on design; reliability and performance of aircraft, missiles and control systems; design of space and missile systems. *Mailing Add:* 40 County Fair St Louis MO 63141

MIRSKY, ALLAN FRANKLIN, b New York, NY, Feb 2, 29; m 51, 86; c 2. NEUROPSYCHOLOGY, CLINICAL PSYCHOLOGY. *Educ:* City Col NY, BS, 50; Yale Univ, MS, 52, PhD(psychol), 54. *Prof Exp:* Res psychologist, Lab Psychol, Nat Inst Mental Health, 54-61; prof neuropsychol, Sch Med, Boston Univ, 61-80; CHIEF, LAB OF PSYCHOL & PSYCHOPATH, NAT INST MENTAL HEALTH, 80- *Concurrent Pos:* Consult, Nat Inst Mental Health, Nat Sci Found & Nat Inst Neurological & Communicative Disorders & Stroke, 55-80; adj prof, Johns Hopkins Univ, 86- *Mem:* Nat Acad Sci; Am EEG Soc; Soc Neurosci; Am Psychol Asn; Am Col Neuropsychopharm. *Res:* Neuropsychology and neurophysiology of attention and attention impairment. *Mailing Add:* Lab Psychol & Psychopath Nat Inst Mental Health-IRP Bldg 10 Rm 4C110 Bethesda MD 20892

MIRSKY, ARTHUR, b Philadelphia, Pa, Feb 8, 27; m 61; c 1. ENVIRONMENTAL GEOLOGY, URBAN GEOLOGY. *Educ:* Univ Calif, Los Angeles, BA, 50; Univ Ariz, MS, 55; Ohio State Univ, PhD(stratig geol), 60. *Prof Exp:* Field geologist, Atomic Energy Comn, 51-53; consult uranium geologist, 55-56; asst dir Inst Polar Studies, Ohio State Univ, 60-67, asst prof geol, 64-67; from asst to assoc prof, 67-72, PROF GEOL, IND UNIV-PURDUE UNIV, INDIANAPOLIS, 73-, CHMN DEPT, 69- *Concurrent Pos:* NSF res grant, 62-63. *Mem:* Geol Soc Am; Soc Econ Paleont & Mineral; Nat Asn Geol Teachers; Am Inst Prof Geologists; AAAS. *Res:* Environmental and urban geology; geologic factors in urban planning; history of man's use of geologic resources; medical geology; archaeological geology; stratigraphy-sedimentation; geologic hazards in urbanization. *Mailing Add:* Dept Geol Ind Univ-Purdue Univ Indianapolis IN 46202

MIRSKY, W(ILLIAM), b Poland, July 10, 22; nat US; m 50; c 2. MECHANICAL ENGINEERING. *Educ:* Univ Conn, BS, 44; Univ Mich, MS, 51, PhD(mech eng), 56. *Prof Exp:* Assoc eng ballistic comput, Westinghouse Elec Corp, 46-50; from instr to assoc prof mech eng, 55-70, PROF MECH ENG, UNIV MICH, ANN ARBOR, 70- *Mem:* Am Soc Mech Engrs. *Res:* Combustion of fuel drops and sprays; effect of ultrasonic energy on combustion; internal combustion engines; air pollution. *Mailing Add:* 3950 Waldenwood Dr Ann Arbor MI 48105

MIRVISH, SIDNEY SOLOMON, b Cape Town, SAfrica, Mar 12, 29; m 60; c 2. CANCER. *Educ:* Univ Cape Town, BSc, 48, MSc, 50; Cambridge Univ, PhD(org chem), 55. *Prof Exp:* Lectr physiol, Univ Witwatersrand, 55-60; res assoc chem carcinogenesis, Weizmann Inst Sci, 61-69; assoc prof, 69-72, PROF CHEM CARCINOGENESIS, EPPLEY INST RES CANCER, UNIV NEBR MED CTR, OMAHA, 72- *Concurrent Pos:* Res fel med, Hadassah Med Sch, Hebrew Univ, Israel, 60-61; Eleanor Roosevelt Inst Cancer fel, McArdle Labs, Med Sch, Univ Wis-Madison, 65-66, res fel, Univ, 65-67. *Mem:* Am Chem Soc; Am Asn Cancer Res. *Res:* Chemical carcinogenesis; metabolism, formation and carcinogenic action of urethane and N-nitroso compounds; biochemistry of carcinogenesis in esophagus and stomach. *Mailing Add:* Eppley Inst for Res in Cancer Univ of Nebr Med Ctr 42nd & Dewey Omaha NE 68104

MIRVISS, STANLEY BURTON, b Minneapolis, Minn, Sept 15, 22; m 49; c 2. SYNTHETIC ORGANIC CHEMISTRY. *Educ:* Univ Wis, BS, 44, PhD(org chem), 50. *Prof Exp:* Asst org chem, Univ Wis, 46-49; from res chemist to proj leader, Chems Res Div, Esso Res & Eng Co, 50-60, res assoc, 60-63; supvr org chem res, Eastern Res Ctr, Stauffer Chem Co, 63-83, res assoc, 83-87, sr res assoc, 87-89; pioneering res, Lonza Chem Co, 89-90; SR RES SCIENTIST, AKZO CHEM CO, 90- *Mem:* AAAS; Am Chem Soc; NY Acad Sci; Am Inst Chem. *Res:* Organometallic chemistry; polymerization; catalysis; high-pressure reactions; organic synthesis; organophorus chemistry; phosphorus. *Mailing Add:* Akzo Res Lab Akzo Chem Co Dobbs Ferry NY 10522

MIRZA, JOHN, b Baghdad, Iraq, Dec 1, 22; m 45; c 2. ORGANIC CHEMISTRY. *Educ:* DePauw Univ, AB, 44; Univ Ill, PhD(org chem, biochem & physiol), 49. *Prof Exp:* Res chemist, Rohm & Haas Co, 49-51; RES CHEMIST, MILES LABS, INC, 51-, VPRES, 70- *Mem:* Am Chem Soc. *Res:* Syntheses from 2-vinylpyridine; applications of dialdehyde polysaccharides. *Mailing Add:* Box 3032 Rancho Santa Fe CA 92067-3032

MISCH, DONALD WILLIAM, b Providence, RI, Jan 1, 29; m 59; c 5. INSECT CELL BIOLOGY, PHYSIOLOGY & TOXICOLOGY. *Educ:* Northeastern Univ, BS, 53; Univ Mich, MS, 58, PhD(zool), 63. *Prof Exp:* Trainee electron micros, Biol Labs, Harvard Univ, 62-63; asst prof, 63-68, dir Electron Micros Lab, 67-84, ASSOC PROF ZOOL, UNIV NC, CHAPEL HILL, 68-; ASST DEAN, COL ARTS & SCI, 87- *Concurrent Pos:* Vis scholar, Dept Zool, Duke Univ, 75; vis scientist, US Army Biomedical Res & Develop Lab, Ft Detrick, Frederick, MD, 66. *Mem:* AAAS; Am Soc Cell Biol; Sigma Xi; Entom Soc Am; Int Soc Invert Pathol. *Res:* Intestinal structure and function (insects); bacterial control of insect pests and disease vectors; mucociliiary clearance. *Mailing Add:* Dept Biol Univ NC Chapel Hill NC 27599

MISCH, HERBERT LOUIS, b Sandusky, Ohio, Dec 7, 17; m 39; c 2. AUTOMOTIVE ENGINEERING. *Educ:* Univ Mich, BS, 41. *Prof Exp:* Chief engr, Packard Motor Car Co, 41-56; dir advan planning, Cadillac Div, Gen Motors Corp, 56-57; mem staff, Ford Motor Co, 57-70, vpres eng & mfg, 70-72, vpres environ safety eng, 72-82; RETIRED. *Mem:* Nat Acad Eng; Soc Automotive Engrs; Coord Res Coun; Am Soc Body Engrs. *Mailing Add:* 1411 Lochridge Bloomfield Hills MI 48302

MISCH, PETER, geology; deceased, see previous edition for last biography

MISCHKE, CHARLES R(USSELL), b Queens, NY, Mar 2, 27; m 51; c 2. MECHANICAL ENGINEERING. *Educ:* Cornell Univ, BSME, 47, MME, 50; Univ Wis, PhD(mech eng), 53. *Prof Exp:* From asst prof to assoc prof mech eng, Univ Kans, 53-57; prof & chmn dept, Pratt Inst, 57-64; PROF MECH ENG, IOWA STATE UNIV, 64- *Concurrent Pos:* Consult, John Deere Ankeny Works, Iowa, Deere & Co, Ill & FMC Corp, Iowa, John Deere Prod Eng Ctr, Iowa, Union Carbide, Iowa & Ford Res, Mich. *Honors & Awards:* Ralph R Teetor Award, Soc Automotive Engrs, 77; Machine Design Award, Am Soc Mech Engrs, 90. *Mem:* Fel Am Soc Mech Engrs; Am Soc Eng Educ; Soc Automotive Engrs; Am Gear Mfgrs Asn. *Res:* Engineering analysis; computer-aided design; mechanical engineering design; originator of Iowa Cadet algorithm for computer-aided design; author of books on mechanical analysis, mathematical model building, computer-aided design and co-author of a book on mechanical engineering design; co ed-in-chief of Standard Handbook of Machine Design, Mechanical Designers' Workbooks, also many papers. *Mailing Add:* 3029 Mech Eng Bldg Iowa State Univ Ames IA 50011

MISCHKE, RICHARD E, b Bristol, Va, Aug 19, 40; m 62; c 2. PARTICLE PHYSICS, NUCLEAR PHYSICS. *Educ:* Univ Tenn, BS, 61; Univ Ill, MS, 62, PhD(physics), 66. *Prof Exp:* From instr to asst prof physics, Princeton Univ, 66-71; staff mem, 71-75, assoc group leader, 75-78, dept group leader, 78-86 STAFF MEM, MP DIV, LOS ALAMOS NAT LAB, 86- *Concurrent Pos:* Mem bd dirs users group, Los Alamos Meson Physics Facil, 77-79; guest scientist, Swiss Inst Nuclear Res, 78-79; mem, organizing comt, Int Conf Intersections between Particle & Nuclear Physics, 84- & Int Conf Spin Physics, 88; prog monitor, US Dept Energy, 87-88. *Mem:* Fel Am Phys Soc; Sigma Xi. *Res:* Experimental high-energy physics; experimental research in particle and nuclear physics at medium energies. *Mailing Add:* 2172 Loma Linda Dr Los Alamos NM 87544

MISCHKE, ROLAND A(LAN), b New York, NY, May 28, 30; m 61; c 2. CHEMICAL ENGINEERING. *Educ:* Pratt Inst, BChE, 50; Northwestern Univ, PhD(chem eng), 61. *Prof Exp:* Jr chem engr, Chem Construct Corp, 50-53; chem eng asst, Army Chem Ctr, Md, 53-55; process engr, Chem Construct Corp, 55-58; assoc prof chem eng, Va Polytech Inst & State Univ, 61-85. *Concurrent Pos:* Consult, Mfg Res Lab, IBM Corp, 69-70. *Mem:* Am Inst Chem Engrs; Am Soc Eng Educ. *Res:* Heat transfer; reaction kinetics; educational methods. *Mailing Add:* Dept Chem Eng Va Polytech Inst & State Univ Blacksburg VA 24061

MISCONI, NEBIL YOUSIF, b Baghdad, Iraq, Dec 8, 39; m 72; c 1. ASTRONOMY. *Educ:* Istanbul Univ, BS, 65; State Univ NY Albany, PhD(astron & space sci), 75. *Prof Exp:* Lab instr physics, Univ Baghdad, 66-70; res asst astron, Dudley Observ, NY, 70-75; res assoc astron, Space Astron Lab, State Univ NY Albany, 75-80; from asst res scientist to assoc res scientist, 80-87, ASSOC DIR, SPACE ASTRON LAB, UNIV FLA, 87-

Concurrent Pos: Co-prin investr, Laser Levitation Exp, Goddard Space Flight Ctr, 78-79; co-investr, Joint NASA/Eng Study Authorization Div Solar Polar Mission, 79-81; scientific collabr, Study Interplanetary Dust, NASA, 75-82, prin investr, small particles-laser beams interaction, 84- *Mem:* Am Astron Soc; Sigma Xi; Royal Astron Soc; Int Astron Union. *Res:* Interplanetary medium; zodiacal light; dynamics of cosmic dust; cometary physics; background starlight. *Mailing Add:* 1810 NW Sixth St Univ Fla Gainesville FL 32601

MISEK, BERNARD, b New York, NY, Mar 29, 30; m 56; c 4. PHARMACEUTICAL CHEMISTRY, COSMETIC CHEMISTRY. *Educ:* Columbia Univ, BS, 51; Univ Md, MS, 53; Univ Conn, PhD(pharm), 56. *Prof Exp:* Pharmaceut chemist, Nepera Chem Co, 53-54; asst prof pharm, Univ Houston, 56-57; head pharm res & develop, Lloyd Brothers Inc, 57-59; dir med prod develop, Nopco Chem Co, 59-60; sr scientist explor res, Richardson-Merrell Inc, 60-64, asst dir, 64-66, dir, 66-72; mgr res & develop, Beecham Prod, 72-80, res mgr, 80-84, res dir, 84-90; CONSULT, 90- *Mem:* Am Pharmaceut Asn; Soc Cosmetic Chem. *Res:* Pharmaceutical, toiletry and cosmetic product development; dermatological products; dentifrices; proprietary drugs; feminine hygiene; hair care; personal products. *Mailing Add:* 4718 Mill Village Rd Raleigh NC 27612-3795

MISELIS, RICHARD ROBERT, b Boston, Mass, Mar 13, 45; m 65; c 2. NEUROBIOLOGY. *Educ:* Tufts Univ, BS, 67; Univ Pa, VMD, 73 & PhD(biol), 73. *Prof Exp:* Fel, Dept Neurophysiol, Col France, 73-75; asst prof, 75-80, ASSOC PROF ANIMAL BIOL, SCH VET MED, UNIV PA, 80- *Concurrent Pos:* Alred P Sloan Found fel award, 78-80, Fogarty Sr Int fel, Howard Florey Inst, Univ Melbourne, Australia, 83-84. *Mem:* AAAS; Soc Neurosci; Am Asn Anatomists; Sigma Xi. *Res:* Neurological and physiological basis of feeding and drinking behavior. *Mailing Add:* 19 Colwyn Lane Bala Cynwyd PA 19004

MISENHIMER, HAROLD ROBERT, obstetrics & gynecology, for more information see previous edition

MISFELDT, MICHAEL LEE, b Davenport, Iowa, June 15, 50; m 73; c 2. MICROBIAL SUPERANTIGENS, BACTERIAL DERIVED IMMUNOMODIFIERS. *Educ:* Univ Ill, Urbana, BS, 72; Univ Iowa, PhD(microbiol), 77. *Prof Exp:* Staff fel res, Lab Molecular Genetics, Nat Inst Child Health Develop, NIH, 77-81; ASSOC PROF IMMUNOL, DEPT MOLECULAR, MICROBIOL & IMMUNOL, UNIV MO-COLUMBIA, 81-, PRIN INVESTR, 83- *Concurrent Pos:* Ad hoc grant rev, NIH, USDA. *Mem:* Am Soc Microbiol; AAAS; NY Acad Sci; Am Asn Immunologists; Reticuloendothelial Soc. *Res:* Microbial Superantigens; T-cell activation; examination of the mechanisms by which bacterial products act as immunomodifiers; Psudomonas exotoxin A represents a unique microbial superantigen; swine T lymphocytes. *Mailing Add:* Dept Molecular Microbiol & Immunol Univ Mo, M-642 Med Sci Bldg Columbia MO 65212

MISH, LAWRENCE BRONISLAW, b Stamford, Conn, Feb 27, 23; m 50; c 4. BOTANY. *Educ:* Univ Conn, AB, 50; Harvard Univ, AM, 50, PhD(bot), 53. *Prof Exp:* From instr to asst prof biol, Wheaton Col, 53-59; prof bot, Bridgewater State Col, 59-85; CONSULT, 85- *Concurrent Pos:* Res asst, United Fruit Co, 60-63; ecol consult, 71-; mem, SShore Nature Ctr. *Mem:* AAAS; Bot Soc Am. *Res:* Biology of lichens, their algae and fungi; marine and fresh water plankton; taxonomy of flowering plants. *Mailing Add:* Box 1245 East Orleans MA 02643

MISHELOFF, MICHAEL NORMAN, b Brooklyn, NY, Feb 15, 44; m 71; c 3. SEMICONDUCTOR DEVICE PHYSICS. *Educ:* Calif Inst Technol, BS, 65; Univ Calif, Berkeley, PhD(physics), 70. *Prof Exp:* Instr physics, Princeton Univ, 70-73; res assoc, Univ Nebr, Lincoln, 73-76; exp physicist, Lawrence Livermore Lab, 76-80; mem res staff, Fairchild Camera & Instrument Advan Res & Develop Lab, 80-82; MGR, DEVICE CHARACTERIZATION & MODELLING, VLSI TECHNOL INC, 82- *Mem:* Inst Elec & Electronics Engrs. *Res:* Metal oxide semiconductor device physics. *Mailing Add:* VLSI Technol Inc 1101 McKay Dr San Jose CA 95131

MISHKIN, ELI ABSALOM, b Poland, Apr 25, 17; m 47; c 2. APPLIED PHYSICS. *Educ:* Israel Inst Technol, Ingenieur, 42, ScD, 52. *Prof Exp:* Lectr elec eng, Israel Inst Technol, 48-53; asst prof, Mass Inst Technol, 53-55; assoc prof, 55-60, PROF APPL PHYSICS, POLYTECH INST NEW YORK, 60- *Concurrent Pos:* Res fel physics, Harvard Univ, 68; vis prof, Inst Sci, Rehorot Univ, 70, Dept Physics & Astron, Univ Tel Aviv, Israel, 71; consult, Lawrence Livermore Lab, 73-76; Judge Mac fel, Stein Club, Mass Inst Technol; dept physics, Harvard Univ, 68-69. *Mem:* Am Phys Soc; NY Acad Sci. *Res:* Electromagnetic theory; automatic control systems; quantum and nonlinear optics; laser induced fusion. *Mailing Add:* Dept of Elec Eng & Appl Physics Polytech Inst New York 333 Jay St Brooklyn NY 11201

MISHKIN, MORTIMER, b Fitchburg, Mass, Dec 13, 26. NEUROBIOLOGY. *Educ:* Dartmouth Col, AB, 46; McGill Univ, MA, 49, PhD(psychol), 51. *Prof Exp:* Res psychologist, Inst Living, 51-55; res psychologist, 55-80, CHIEF, LAB NEUROPSYCHOL, NIMH, 80- *Concurrent Pos:* Consult ed, J Comp & Physiol Psychol, 63-73, Neuropsychologia, 63-, Exp Brain Res, 65-, Brain Res, 74-78 & Sci, 85-; mem, Psychol Servs Panel, NIH, 59-61, Exp Psychol Study Sect, 65-69, adv panel, Fogarty Int Scholar in Residence, 85-89, adv bd, McDonnell-Pew Prog Cognitive Neurosci. *Mem:* Nat Acad Sci; Am Psychol Asn; AAAS; Int Brain Res Orgn; Soc Neurosci (pres, 86-87); Int Neuropsychol Soc; Inst Med. *Res:* Cerebral mechanisms underlying basic mental processes (perception, memory, affect, volition) in primates. *Mailing Add:* NIMH Bldg 9 Rm 1N107 Bethesda MD 20892

MISHLER, JOHN MILTON, IV, b Cairo, Ill, Sept 25, 46; m 81; c 1. HEMATOLOGY. *Educ:* Univ Calif, San Diego, AB, 69, ScM, 71, Univ Oxford, Ophil, 78; Royal Col Pathologists, MRCPath, 81. *Prof Exp:* Assoc clin coordr biol, McGaw Labs, 72-74, clin coordr, 74-75; res fel hematol, Radcliffe Infirmary, Oxford, 75-78, Med Res Coun Leukemia Unit, Royal Postgrad Med Sch, 77-78; sr res fel & co-dir, Lab Tumorimmunol, Med Univ Clin, Cologne, 78-80; chief, Blood Res Br, Nat Heart, Lung & Blood Inst, NIH, 80-82; assoc vchancellor res & prof basic life scis, Med & Pharmacol, Univ Mo-Kans City, 83-89; DEAN GRAD STUDIES & RES & PROF NATURAL SCIS, UNIV MD EASTERN SHORE, 89- *Concurrent Pos:* Session co-chmn, Second Int Symp Leucocyte Separation & Transfusion, London, 76, Int Soc Hemat, Europ & African Div, Hamburg, 79, Int Soc Blood Transfusion, Bethesda, 80, 33rd Ann Meeting Am Asn Blood Banks, Washington, DC, 80, Int Symp Viral Hepatitis, NY, 81, Int Grad Educ, Nat Asn State Univs & Land- Grant Cols, Hilton Head, 90, Ann Meeting Soc Res Adminrs, 88-90; referee, NSF, Transfusion, Blood, JAMA, Am J Hosp Pharm. *Honors & Awards:* Excellance Award, Soc Res Adminrs, 89. *Mem:* Am Soc Hemat; NY Acad Sci; Int Soc Hemat; Am Asn Blood Banks; Int Soc Blood Transfusion; Sigma Xi; Ger Soc Hemat; Soc Res Adminrs; Nat Coun Univ Res Adminrs; Nat Asn State Univs & Land-Grant Cols. *Res:* Pharmacology of the hydroxyethyl starches; Various aspects surrounding the use of leukacytapheresis and plateletpheresis in medical applications; Management studies in the field of research administration. *Mailing Add:* Off Grad Studies & Res Univ Md Eastern Shores Princess Anne MD 21853-1299

MISHMASH, HAROLD EDWARD, b Richmond, Calif, Nov 8, 42; m 64; c 3. ANALYTICAL CHEMISTRY. *Educ:* Iowa State Univ, BS, 64; Kans State Univ, PhD(anal chem), 68. *Prof Exp:* RES SPECIALIST, CENT RES LABS, 3M CO, 68-, MGR INORG ANAL RES, 81- *Mem:* Am Vacuum Soc. *Res:* Electron microprobe analysis; scanning electron microscopy; emission spectroscopy; ion beam surface studies using ion scattering and secondary ion mass spectroscopy; x-ray photoelectron spectroscopy (ESCA) investigations of surface defects and modifications. *Mailing Add:* Cent Res Labs 3M Co PO Box 33221 St Paul MN 55133-3221

MISHOE, LUNA I, mathematics, mathematical physics; deceased, see previous edition for last biography

MISHRA, DINESH S, b Deoria, India, Sept 15, 61; m 84; c 1. PHARMACEUTICAL CHEMISTRY, PHYSICAL CHEMISTRY. *Educ:* Univ Bombay, India, BS, 82, MS, 84; Univ Ariz, PhD(pharm sci), 89. *Prof Exp:* Asst prof pharmaceut, Univ Ariz, 89-90; ASST PROF PHARMACEUT, UNIV PITTSBURGH, 90- *Mem:* Am Chem Soc; Am Asn Pharmaceut Scientists. *Res:* Physical and chemical properties of drugs as it relates to development of pharmaceutical dosage forms. *Mailing Add:* Univ Pittsburgh 724 Salk Hall Pittsburgh PA 15261

MISHRA, RAM K, b Jaipur, India, Aug 2, 45; Can citizen; m 67; c 2. NEUROPSYCHOPHARMACOLOGY, BIOTECHNOLOGY. *Educ:* Udaipur Univ, BS, 65; La State Univ, New Orleans, MS, 70; Mem Univ, Nfld, PhD(biochem), 75. *Prof Exp:* Fel med sci, Albert Einstein Col Med, 73-76; from asst prof to assoc prof, 76-83, PROF MED SCI, HEALTH SCI CTR, MCMASTER UNIV, 84- *Concurrent Pos:* Prin investr grants, NIH, Med Res Coun Can, Ont Ment Health Found & Parkinson's Found, 77-; Res Career Develop Award, Med Res Coun Can, 77-82 & Ont Ment Health Found, 82-87; vis scientist, Roche Inst Molecular Biol, NJ, 83-84. *Honors & Awards:* Meller Award, 76; John Dewan Award, 84. *Mem:* Am Soc Pharmacol & Exp Therapeut; Soc Neurosci; Can Col Neuropsychopharmacol. *Res:* Neurotransmitter (dopamine) and neuropeptide receptors in health and disease. *Mailing Add:* Dept Soc Work McMaster Univ 1280 Main St W Hamilton ON L8N 3Z5 Can

MISHRA, SATYA NARAYAN, b Varanasi, India, Feb 2, 47; m 72; c 2. SELECTION THEORY, STATISTICAL INFERENCE. *Educ:* Univ Gorakhpur, BSc, 66; Benares Hindu Univ, MSc, 69; Univ Mass, MA, 74; Ohio State Univ, MS & PhD(statist), 82. *Prof Exp:* Lectr math, Benares Hindu Univ, 69-72; teaching assoc, Univ Mass, 72-74; instr & teaching assoc, Univ Cincinnati, 74-77; teaching assoc statist, Ohio State Univ, 77-82; asst prof, 82-87, ASSOC PROF STATIST, UNIV S ALA, 87- *Concurrent Pos:* Mem, bd referees, Commun Statist, 80-, J Educ Statist, 82- & Prog Math, 86-; consult, US Naval Air Sta, Pensacola, 85 & 89, Statist Lab, Univ S Ala, 85 & 87 & ICI Am, 88- *Mem:* Am Statist Asn; Int Statist Inst; Bernoulli Soc Math Statist & Probability. *Res:* Selection theory, selecting extreme populations; statistical inference; co-author of one book. *Mailing Add:* Dept Math & Statist FCS No 3 Univ S Ala Mobile AL 36688

MISHUCK, ELI, b Buenos Aires, Arg, July 27, 23; nat US; m 44; c 2. PHYSICAL CHEMISTRY. *Educ:* Brooklyn Col, BA, 44; Polytech Inst Brooklyn, MS, 46, PhD(chem), 50. *Prof Exp:* Res assoc chem, Manhattan Proj, SAM Labs, Columbia Univ, 44-46; chief chemist, Bingham Bros Co, NJ, 46-48; res chemist polymers, Polytech Inst Brooklyn, 48-50; head dept solid propellant res, Aerojet-Gen Corp, Gen Tire & Rubber Co, 50-63, sr mgr chem & biol opers, Space-Gen Corp, 63-66, dir chem & biol systs, 66-70, pres, Aerojet Med & Biol Systs, 70-74; gen mgr, Organon Diagnostics, Inc, 74-77; ASST V PRES, SCI APPLN, INC, 77- *Concurrent Pos:* Mem tech adv bd to pres, Gen Tire & Rubber Co. *Mem:* Sigma Xi. *Res:* Biological and chemical detection; medical diagnostic equipment; life sciences; environmental sciences; oceanography; biomedical instrumentation; cultural resources managements. *Mailing Add:* 7605 Hillside Dr La Jolla CA 92037

MISIASZEK, EDWARD T, b Utica, NY, May 23, 28; m 52; c 2. SOIL MECHANICS, ENGINEERING GEOLOGY. *Educ:* Clarkson Col Technol, BCE, 52, MCE, 54; Univ Ill, PhD(soil mech), 60. *Prof Exp:* Instr mech, Lehigh Univ, 54-55; instr mech, Univ Ill, 55-60, asst prof civil eng, 60-62; assoc prof, 62-66, asst dean eng, 66-68, ASSOC DEAN ENG, CLARKSON COL TECHNOL, 68- *Concurrent Pos:* Chmn, NY State 4 year-2 year Eng Col Curric Study Comt, 69-; mem, NY State Bd Eng & Land Surv, 71. *Mem:* Am Soc Civil Engrs; fel Am Soc Eng Educ; Nat Soc Prof Engrs; Sigma Xi. *Res:* Soil foundations, compaction and consolidation; physical properties of soils. *Mailing Add:* Eng Sch Clarkson Univ Potsdam NY 13699-5702

MISIEK, MARTIN, b Buffalo, NY, Sept 6, 19; m 45; c 2. MICROBIOLOGY. *Educ:* Univ Buffalo, BA, 43; Syracuse Univ, MS, 49, PhD(microbiol), 55. *Prof Exp:* Assoc dir microbiol dept, Prod & Develop, Bristol-Myers Co, Syracuse, 43-86; RETIRED. *Mem:* Am Soc Microbiol. *Res:* In vitro evaluation of antibacterial agents. *Mailing Add:* 70 Maple Rd Syracuse NY 13201

MISKEL, JOHN ALBERT, b San Francisco, Calif, Aug 21, 19; m 49. CHEMISTRY. *Educ:* Univ Calif, BS, 43; Washington Univ, PhD(chem), 49. *Prof Exp:* Jr scientist, Manhattan Proj, Univ Calif, 43-46; asst phys chem, Washington Univ, 46-48, AEC fel phys sci, 48-49; from res assoc to chemist, Brookhaven Nat Lab, 49-55; CHEMIST, LAWRENCE LIVERMORE LAB, UNIV CALIF, 55- *Concurrent Pos:* mem comt disarmament, US Deleg, 81-82, 84. *Mem:* Am Phys Soc. *Res:* Chemistry of heavy elements; fission fragments; ranges of fission fragments; neutron cross sections; absolute beta counting; nuclear excitation functions; radiochemistry. *Mailing Add:* 23 Castledown Rd Pleasanton CA 94566

MISKEL, JOHN JOSEPH, JR, b Brooklyn, NY, Aug 2, 33; m 55; c 7. POLYMER CHEMISTRY, ORGANIC CHEMISTRY. *Educ:* Univ Notre Dame, BS, 55; Univ Pa, PhD(org chem), 60. *Prof Exp:* Polymer chemist, Atlantic Ref Co, 59-60; supvr polymer develop, Rexall Chem Co, 60-68; MGR TECHNOL, COATINGS & INK DIV, HENKEL CORP, 68- *Mem:* Am Chem Soc; Soc Chem Indust; Am Oil Chem Soc. *Res:* Polyolefin process and product research; specialty polymers and surfactants. *Mailing Add:* 17 Valley Way Mendham NJ 07945

MISKIMEN, CARMEN RIVERA, b Mayaguez, PR, Mar 15, 33; m 63; c 4. PLANT PATHOLOGY, VIROLOGY. *Educ:* Univ PR, Mayaguez, BS, 53; Univ Wis-Madison, PhD(plant path), 62. *Prof Exp:* Instr sci, Commonwealth of PR Dept Educ, 53-55; technician III, US Dept Agr, PR, 55-58, res plant pathologist, Crops Res Div, 62-63; from asst prof to prof biol, Univ PR, Mayaguez, 63-83, coordr med technol, 73-83; CONSULT, 83- *Concurrent Pos:* USDA Hatch grant, Univ PR, 64-67, NIH res grant, 73-86, NSF grant, 76-78. *Mem:* Am Phytopath Soc. *Res:* Mosaic and other viruses of economically important crops; morphology and histology of insect visual structures. *Mailing Add:* PO Box 210 Crystal River FL 32629

MISKIMEN, GEORGE WILLIAM, b Appleton, Wis, May 21, 30; m 63; c 4. ENVIRONMENTAL SCIENCE, ECOLOGY. *Educ:* Ohio Univ, BS, 53, MS, 55; Univ Fla, PhD(biol, entom), 66. *Prof Exp:* Entomologist, VI Agr Prog, US Dept Agr, 58-61, invests leader, Grain & Forage Insects, Agr Res Serv, PR, 62-66; dir entom pioneering res lab, Univ PR, 66-72, prof biol, 66-87, dir, NIH Biomed Res Prog, 73-86; environ specialist, Citrus County, Fla, 89-90; ENVIRON CONSULT, 90- *Concurrent Pos:* USDA Hatch grant, Univ PR, 64-67; USDA res grant, 72-73; NIH res grant, 73-87 & NSF grants, 74 & 76-78; adj prof ophthal, Sch Med, Univ Fla, 81- *Mem:* Soc Neurosci; Sigma Xi; Entom Soc Am; Asn Trop Biol; Int Orgn Biol Control; Coleopterists Soc. *Res:* Population dynamics and ecology of insects; electrophysiology and anatomy of insect visual receptors; relationships between agriculturally important pests, climate, and biological control organisms. *Mailing Add:* PO Box 210 Crystal River FL 32623-0210

MISKOVSKY, NICHOLAS MATTHEW, b Passaic, NJ, Oct 4, 40; m 71; c 1. PHYSICS. *Educ:* Rutgers Univ, New Brunswick, AB, 62; Pa State Univ, PhD(physics), 70. *Prof Exp:* ASSOC PROF PHYSICS, PA STATE UNIV, 70- *Mem:* Am Phys Soc. *Res:* Theoretical solid-state physics; optical properties and band structure of solids; physics of nanosize devices. *Mailing Add:* Dept Physics Penn State Univ Altoona PA 16603

MISLEVY, PAUL, b Scranton, Pa, May 5, 41; m 64; c 1. AGRONOMY. *Educ:* Pa State Univ, BS, 66, MS, 69, PhD(agron), 71. *Prof Exp:* From asst prof to assoc prof, 71-82, PROF AGRON, UNIV FLA, 82- *Mem:* Am Soc Agron; Am Forage & Grassland Coun; Weed Sci Soc Am; Am Soc Surface Mining & Reclamation. *Res:* Forage management; crop physiology; pasture herbicides; multicropping; phosphate spoil bark reclamation; grazing studies; biomass extension. *Mailing Add:* Agr Res Ctr Univ Fla Ona FL 33865

MISLIVEC, PHILIP BRIAN, b Danville, Ill, Feb 12, 39. MYCOLOGY, PLANT PATHOLOGY. *Educ:* St Meinrad Col, BS, 61; Ind State Univ, Terre Haute, MA, 65; Purdue Univ, PhD(plant path), 68. *Prof Exp:* Asst prof bot, Ind State Univ, Terre Haute, 68-69; RES MYCOLOGIST, US FOOD & DRUG ADMIN, 69- *Concurrent Pos:* Prof, US Dept Agr Grad Sch, DC, 70- *Mem:* Mycol Soc Am; Inst Food Technologists; Am Soc Microbiol. *Res:* Effects of environment and microbial competition on growth and toxin production by mycotoxin-producing mold species; mold systematist. *Mailing Add:* Food & Drug Admin 200 C St SW Washington DC 20204

MISLOVE, MICHAEL WILLIAM, b Washington, DC, Feb 8, 44; m; c 2. MATHEMATICS, THEORETICAL COMPUTER SCIENCE. *Educ:* Univ of the South, BA, 65; Univ Tenn, Knoxville, PhD(math), 69. *Prof Exp:* Asst prof math, Univ Fla, 70; asst prof, 70-75, assoc prof, 75-79, PROF MATH, TULANE UNIV, LA, 79- *Concurrent Pos:* NSF res grant, Tulane Univ, La, 71-80, sabbatical support grant, 84; Alexander von Humboldt res fel, 76, 78 & 82; Fulbright Scholar, 76; guest prof, Univ Tuegingen, 76; vis prof, Technische Hochschule Darmstadt, 78, 82; sr vis, Maths Inst, Oxford Univ, 84, vis prof prog res group, 91; Off Naval Res Contracts, 88-91, 91-94. *Mem:* AAAS; Am Math Soc; Sigma Xi. *Res:* Topological algebra, ordered structures, spectral theory and stone duality; theoretical computer science: semantics of high level programming languages, models of concurrency. *Mailing Add:* Dept of Math Tulane Univ 6823 St Charles Ave 400C Gibson Hall New Orleans LA 70118

MISLOW, KURT MARTIN, b Berlin, Germany, June 5, 23; nat US; m 66; c 2. CHEMISTRY. *Educ:* Tulane Univ, BS, 44; Calif Inst Technol, PhD(org chem), 47. *Hon Degrees:* Dr, Free Univ Brussels, 74 & Univ Uppsala, 77; DSc, Tulane Univ, 75. *Prof Exp:* From instr to prof chem, NY Univ, 47-64; chmn dept, 68-74, H S TAYLOR PROF CHEM, PRINCETON UNIV, 64- *Concurrent Pos:* Guggenheim fel, 56, 75; Sloan fel, 59; mem adv panel, NSF, 63-66; mem med & org chem panel, NIH, 63-66; univ lectr, Univ London, 65; Frontiers chem lectr, Wayne State Univ, 66; distinguished lectr, Mich State Univ, 70; lectr, Nobel Inst, Royal Swed Acad Sci, 74; Churchill fel, Univ Cambridge, 75; distinguished vis speaker, Univ Calgary, 78; indust-sci lectr, Ramapo Col, 80; hon lectr, Ariz State Univ, 81; res scholar lectr, Drew Univ, 83; Morris S Kharasch vis prof, Univ Chicago, 89. *Honors & Awards:* FMC lectr, Princeton, 63; J A McRae mem lectr, Queen's Univ, 67; Solvay Medal, Univ Brussels, 72; H A Iddles lectr, Univ NH, 72; E C Lee lectr, Univ Chicago, 73; James Flack Norris Award, Am Chem Soc, 75; A A Vernon lectr, Northeastern Univ, 76; PPG Indust lectr, Ohio Univ, 77; J Musher Mem lectr, Hebrew Univ Jerusalem, 78; E Ritchie Mem lectr, Univ Sydney, 83; R C Fuson lectr, Univ Nev, 83; J F McGregory lectr, Colgate Univ, 84; R B Sandin lectr, Univ Alta, 84; C B Purvis lectr, McGill Univ, 85; Arnold lectr, Southern Ill Univ, 85; W Bergman lectr, Yale Univ, 86; Prelog Medal, ETH Zurich, 86; W H Nichols, Medal, Am Chem Soc, 87; H C Brown lectr, Purdue Univ, 88. *Mem:* Nat Acad Sci; fel AAAS; Am Chem Soc; fel Am Acad Arts & Sci. *Res:* Stereochemistry. *Mailing Add:* Dept Chem Princeton Univ Princeton NJ 08544

MISNER, CHARLES WILLIAM, b Jackson, Mich, June 13, 32; m 59; c 4. THEORETICAL PHYSICS, COSMOLOGY. *Educ:* Univ Notre Dame, BS, 52; Princeton Univ, MA, 54, PhD(physics), 57. *Prof Exp:* Instr, Princeton Univ, 56-59, asst prof, 59-63; assoc prof physics & astron, 63-66, prof physics & astron, 66-76, PROF PHYSICS, UNIV MD, COLLEGE PARK, 76- *Concurrent Pos:* Sloan res fel, 58-62; lectr, Les Houches, Univ Grenoble, 63; NSF sr fel, Univ Cambridge, 66-67 & Niels Bohr Inst, Copenhagen, 67; vis prof, Princeton Univ, 69-70 & Calif Inst Technol, 72; Guggenheim fel, 72; vis fel, All Souls Col, Oxford, 73; mem, NSF Adv Comt Res, 73-75; vis scientist, Inst Theoret Phys, Univ Calif, Santa Barbara, 80-81. *Honors & Awards:* Notre Dame Sci Centennial Award, 65. *Mem:* AAAS; Am Phys Soc; Am Math Soc; Royal Astron Soc; Philos Sci Asn; Am Asn Phys Teachers. *Res:* Physics education; general relativity; relativistic astrophysics; philosophy of physics. *Mailing Add:* Dept Physics Univ of Md College Park MD 20742-4111

MISNER, ROBERT DAVID, b Waynesville, Ill, May 1, 20; m 49; c 2. PHYSICS. *Educ:* George Washington Univ, BS, 46. *Prof Exp:* Physicist, US Naval Res Lab, 42-44, electronic engr, 45-54, sect head, data processing, 54-66, br head, signal exploitation, 66-85, CONSULT, US NAVAL RES LAB, 85- *Concurrent Pos:* Pres, Memre, 84- *Honors & Awards:* Silver Technol Medal, Asn Old Crows, 70. *Mem:* Sigma Xi; Asn Old Crows; Inst Elec & Electronics Engrs. *Res:* Radio frequency intercept and data handling, especially data storage and processing techniques; magnetic tape storage. *Mailing Add:* 7107 Sussex Pl Alexandria VA 22307

MISNER, ROBERT E, b Yonkers, NY, May 25, 41; m 68. ORGANIC CHEMISTRY. *Educ:* Manhattan Col, BS, 63; Fordham Univ, PhD(org chem), 68. *Prof Exp:* Teaching asst org chem, Fordham Univ, 63-65, res asst, 65-67; SR RES CHEMIST, AM CYANAMID CO, 67- *Mem:* Am Chem Soc. *Res:* Organic Synthesis of basic organic chemicals; heterocyclic synthesis; research and development of dyes and organic intermediates. *Mailing Add:* 40 Winding Way RD 3 Flemington NJ 08822-9490

MISONO, KUNIO SHIRAISHI, b Hiroshima, Japan, Nov 21, 46; US citizen; m; c 2. BIOCHEMISTRY. *Educ:* Saitama Univ, BS, 69; Osaka State Univ, MS, 71; Vanderbilt Univ, PhD(biochem), 78. *Prof Exp:* Res assoc microbiol, Duke Univ Sch Med, 78-80; from res instr to asst prof biochem, Vanderbilt Univ Sch Med, 80-87; prin scientist microbiol, Res Div, Scherring-Plough Corp, 86-87; ASSOC STAFF, HYPERTENSION, CLEVELAND CLIN FOUND RES INST, 87- *Mem:* Am Soc Biochem & Molecular Biol; Protein Soc; Sigma Xi; Am Chem Soc; Am Heart Asn. *Res:* Structure and function of proteins, and peptides; chemistry and molecular biology of peptide hormones, receptors and their signal transduction mechanisms; proteolytic enzymes and biological control; protein sequencing, peptide synthesis chemistry. *Mailing Add:* Cleveland Clin Found Res Inst 9500 Euclid Ave Cleveland OH 44195-5071

MISRA, ALOK C, b Kanpur, India, Sept 29, 50; m 81. REAL TIME CONTROL SYSTEMS, INSTRUMENTATION & DIGITAL SIGNAL PROCESSING. *Educ:* Worcester Polytech Inst, MS, 78; Kanpur Univ, BS, 69; Univ Allahabad, BS, 73. *Prof Exp:* Computer engr software eng, Continental Group Inc, 78-79; software engr, Computer Controls Corp, 79-81; CONSULT SOFTWARE ENG, ACM SOFTWARE, SALEM, NH, 81- *Mem:* Asn Comput; Inst Elec & Electronic Engrs; Math Asn Am. *Res:* Real time distributed systems; instrumentation and control; digital image processing. *Mailing Add:* 31 Barron Ave Salem NH 03079

MISRA, ANAND LAL, b Kanpur, India, June 5, 28; US citizen. DRUG ABUSE, DRUG DISPOSITION & METABOLISM. *Educ:* Allahabad Univ, India, BSc, 46, MSc, 48, PhD(plant prod chem), 50; Heriot-Watt Col, Edinburgh, Scotland, FH-WC, 55. *Prof Exp:* Res fel structure activity, chem dept, Heriot-Watt Col, Edinburgh, UK, 53-55; jr sci officer res, Nat Chem Lab, Poona, India, 56-57; res fel pharmacol, Univ Mich, Ann Arbor, 57-60; sci officer res biophys, Cent Drug Res Inst, Lucknow, India, 60-61; res chemist, CIBA Ltd, Basel, Switz, 61-64, head, dept prod develop, CIBA India Ltd, Bombay, 64-67; res assoc pharmacol, Univ Iowa, Iowa City, 69-70; RES SCIENTIST V, RES LAB, NY STATE DIV SUBSTANCE ABUSE SERV, 71- *Concurrent Pos:* Adj assoc prof, Dept Psychiat, Health Sci Ctr, State Univ NY, Brooklyn, 80-89. *Mem:* Fel Royal Soc Chem; fel Am Inst Chemists; Am Soc Pharmacol & Exp Therapeut; NY Acad Sci; Int Soc Study Xenobiotics; AAAS. *Res:* Radioactive tracers studies on the biological disposition, metabolism, pharmacokinetics and interactions of drugs subject to human abuse; investigation of the mechanisms of the development of tolerance, psychic and physiological dependence on such psychoactive drugs. *Mailing Add:* 8646 Ft Hamilton Pkwy Brooklyn NY 11209

MISRA, DHIRENDRA N, b India, Mar 1, 36; US citizen; m 65; c 3. IMMUNO-CHEMISTRY, BIOPHYSICS. *Educ:* Univ Calcutta, India, PhD(biophys), 66. *Prof Exp:* ASSOC PROF PATH, SCH MED, UNIV PITTSBURGH, 85- *Mem:* Am Asn Immunologists; Am Soc Biochem & Molecular Biol. *Res:* Molecular biology of rat histocompatibility complex. *Mailing Add:* Dept Path Sch Med Univ Pittsburgh Pittsburgh PA 15261

MISRA, DWARIKA NATH, b Sarai-Miran, India, Mar 17, 33; m 54; c 3. PHYSICAL CHEMISTRY, SURFACE CHEMISTRY. *Educ:* Univ Lucknow, BS, 51, MSc, 53; Howard Univ, PhD(phys chem), 63. *Prof Exp:* Res asst phys chem, Regional Res Lab, Hyderabad, India, 54-58; NSF fel, Pa State Univ, 63-66; sr scientist, Itek Corp, Mass, 66-68; lectr, Howard Univ, 70-72; RES ASSOC, AM DENT ASN HEALTH FOUND PAFFENBARGER RES CTR, 72- *Mem:* Am Chem Soc; Sigma Xi; NY Acad Sci. *Res:* Adsorption and catalysis on heterogeneous surfaces; surface chemistry of semiconducting oxides and hydroxyapatite, behavior and kinetics of adsorbed organic molecules; chemical bond between bone mineral and resins. *Mailing Add:* Am Dent Asn Health Found Paffenbarger Res Ctr Nat Bur of Standards Gaithersburg MD 20899

MISRA, HARA PRASAD, b Khallikote, Orissa, India, June 1, 40; m 62; c 3. BIOCHEMISTRY, VETERINARY MEDICINE. *Educ:* Utkal Univ, BVSc & AH, 62; Va Polytech Inst & State Univ, MS, 68, PhD(biochem), 70. *Prof Exp:* Vet asst surgeon, Orissa Govt, India, 62-64; instr biochem, Orissa Univ Agr & Technol, 64-66; res asst poultry sci, Va Polytech Inst & State Univ, 66-68, res asst biochem, 68-70; res assoc, Duke Univ, 70-73 & 75-78; asst prof microbiol, Univ Ala, Birmingham, 73-75; asst res biochemist & lectr physiol sci, Univ Calif, Davis, 78-84; asst mem, Okla Med Res Found, 84-86; assoc prof, 86-87, PROF & HEAD BIOMED SCI, COL VET MED, VA TECH, 87- *Concurrent Pos:* Assoc scientist, Cancer Res Ctr, Univ Ala, Birmingham, 73-75; asst adj prof, Univ Calif, Davis, 80-; dir, Univ Ctr Toxicol, Va Tech, 89- *Mem:* Am Soc Biol Chemists; Tissue Cult Asn; NY Acad Sci; Soc Toxicol; Free Radicals Biol; Am Vet Med Asn; Am Soc Photobiol. *Res:* Superoxide and superoxide dismutase; molecular biology; environmental toxicology; oxidative damage and cell dysfunction. *Mailing Add:* Col Vet Med Va Tech Blackburg VA 24061

MISRA, PRABHAKAR, b Lucknow, India, May 7, 55; m 81; c 1. LASER SPECTROSCOPY, CHEMICAL PHYSICS. *Educ:* Univ Calcutta, India, BSc Hons, 75, MSc, 78; Carnegie-Mellon Univ, MS, 81; Ohio State Univ, PhD(physics), 86. *Prof Exp:* Res fel biophys, Bose Inst, 78, Indian Inst Chem Biol, 79; grad asst physics, Carnegie-Mellon Univ, 79-81; grad assoc physics, Ohio State Univ, 81-86, postdoctorate laser spectros, 86-88; ASST PROF PHYSICS, HOWARD UNIV, 88- *Concurrent Pos:* Res incentive award, Howard Univ, 89-90, prin investr, 90-; vis scholar, Northwestern Univ, 90. *Mem:* Am Phys Soc; Am Mensa. *Res:* Detection and spectroscopic characterization of stable molecules, free radicals and molecular ions in a supersonic jet expansion using laser-induced fluorescence; ion-molecule reactions of importance to combustion and atmospheric processes; laser-induced release of dyes and drugs from liposomes. *Mailing Add:* Dept Physics Howard Univ 2355 Sixth St NW Washington DC 20059

MISRA, RAGHUNATH P, b Feb 1, 28; c 4. ANATOMIC NEPHROPATHOLOGY, CLINICAL IMMUNOPATHOLOGY. *Educ:* Sci Col, Calcutta Univ, BSc Hons, 48, Med Col, MBBS, 54; McGill Univ, PhD(med), 65. *Prof Exp:* Instr med, Sch Med, Univ Louisville, 66-68, asst prof, 68; assoc investr kidney dis, Mt Sinai Hosp, Cleveland, 68-73; asst prof pathol, Sch Med, Case Western Res Univ, 73-76; from asst prof to assoc prof, 76-86, PROF PATHOL & OPHTHAL, SCH MED, LA STATE UNIV, 86- *Concurrent Pos:* Vis prof, Sch Med, Univ Ky, 86, Kanazahwa Med Sch, Japan, 90. *Honors & Awards:* Jean-Tallisman Award, Mt Sinai Hosp, Cleveland, 71-73. *Mem:* Am Soc Nephrol; Col Am Pathologists; Am Soc Clin Pathologists; Am Med Asn. *Res:* Molecular pathoibiology of renal disease and tumor markers. *Mailing Add:* Dept Pathol Sch Med La State Univ PO Box 33932 Shreveport LA 71130-3932

MISRA, RAJ PRATAP, b Chhaterpur, India, Dec 23, 19; US citizen; m 48; c 4. RELIABILITY MANAGEMENT, PRODUCT IMPROVEMENT & COST REDUCTION. *Educ:* Mass Inst Technol, SB, 41; Cornell Univ, MEE, 45, PhD(elec eng & indust mgt), 55. *Prof Exp:* Gen mgr & chief engr, Hamara Radio & Gen Industs Ltd, Delhi, India, 47-50; instr elec eng, Cornell Univ, 50-52; mgr reliability & high frequency, Philco Corp, Lansdale, Pa, 52-58; mgr reliabilty res & develop, Tex Instruments, Dallas, Tex, 58-62; EMER PROF RELIABILITY, NJ INST TECHNOL, 62- *Concurrent Pos:* Chmn, Ref Planar Diode Task Force, Am Soc Testing & Mat, 55-59, & Reliability Group, Inst Elec & Electronic Engrs, NJ, 65-68; consult, Tex Instruments, 62-93, Soletron Inc, NY & Fla, 66-69 & Kertron, Reriera Beach, Fla, 69-80; consult reliability, Astro Electronics Div, Westinghouse, Calif, 62-65, & Autonetics, Calif, 65-66; vis prof from US Acad Sci to Romanian Acad, 88-89; Fulbright scholar, Fulbright fel, Coun Int Exchange of Scholars, 91-92. *Mem:* Sigma Xi; fel Indian Asn Engrs; Inst Elec & Electronic Engrs. *Res:* Increasing life and reducing failure rates by two orders of magnitude for semiconductors and increasing life of more than 6 times of cathode ray and other electron tubes; oxide cathodes; failure of dielectrics; author of over 50 publications. *Mailing Add:* Ctr Reliability Res NJ Inst Technol 323 Martin Luther King Jr Blvd Newark NJ 07102

MISRA, RENUKA, b India, July 3, 40; Can citizen. ANTITUMOR ANTIBIOTICS, NATURAL PRODUCTS. *Educ:* Agra Univ, India, BSc, 57, MSc, 59, PhD(org chem), 66. *Prof Exp:* Res fel chem, Nat Chem Lab, India, 61-66; NIH fel org chem, Univ Nebr, Lincoln, 66-68; res assoc, NC State Univ, Raleigh, 68-69; Nat Res Coun fel, Queen's Univ, Can, 69-70; sci pool officer, Nat Chem Lab, India, 71-72; lectr, Univ Toronto, 73-72; SCIENTIST NATURAL PROD, FREDERICK CANCER RES FACIL, NAT CANCER INST, NIH, 81- *Concurrent Pos:* Vis scientist, Univ Ill, 77 & 80 & Malti Chem Res Ctr, India, 80. *Mem:* Am Chem Soc; Am Soc Pharmacog; Sigma Xi. *Res:* Isolation and structural studies of biologically active compounds; anticancer and antitumor drugs; natural products; antibiotics isolation and structural elucidation produced in fermentation broths. *Mailing Add:* 1405 Key Pkwy E Frederick MD 21702

MISRA, SUDHAN SEKHER, b Puri, India, Sept 2, 38; US citizen; m 69; c 2. SECONDARY & PRIMARY BATTERIES. *Educ:* Utkal Univ, BSc, 56, MSc, 58; Indian Inst Sci, PhD(electrochem), 64. *Prof Exp:* Res chemist, Oak Ridge Nat Lab, Tenn, 64-67; res scientist, Cent Fuel Res Inst, New Delhi, India, 68-70; sr res chemist, Graner & Weil Ltd, Bombay, India, 70-72; from asst mgr to res & develop mgr, Estrela Batteries Ltd, 72-77; staff scientist, Corp Res Labs, Gould Inc, Rolling Meadows, Ill, 77-84; eng mgr, C&D Power Systs Inc, 84-87, DIR ENG, C&D CHARTER POWER SYSTS INC, 87- *Mem:* Electrochem Soc; Am Chem Soc. *Res:* Developed new alloy plating interpretations and brighteners; worked on several aspects of corrosion, including post-seal corrosion and grid corrosion in lead-acid batteries; developed recombinant sealed lead-acid batteries and molten salt lithium batteries; Electroplating, electrowinning & electrochemical corrosion. *Mailing Add:* C & D Power Syst Inc 3043 Walton Rd Plymouth Meeting PA 19462

MISRA, SUSHIL, b Budaun, India, Sept 5, 40; c 3. SOLID STATE PHYSICS. *Educ:* Agra Univ, BSc, 58; Gorakhpur Univ, MSc, 60; St Louis Univ, PhD(physics), 64; Concordia Univ, BA Hons, 91. *Prof Exp:* Jr res scholar physics, Indian Asn Cultivation Sci, Calcutta, 60-61; fel, Univ Toronto, 64-67; from asst prof to assoc prof, 67-77, PROF PHYSICS, CONCORDIA UNIV, 77- *Concurrent Pos:* Consult, Electronics & Equip Div, McDonnell Aircraft Corp, Mo, 64; Chinese Ministry Educ, appraise teaching & res, lectr-electron paramagnetic resonance, Nanjing Univ, 85. *Mem:* Am Phys Soc. *Res:* Electron paramagnetic resonance; reorientation of oriented nuclei; Mossbauer detection of dynamically oriented nuclei; low-temperature magnetic ordering; spin-lattice relaxation. *Mailing Add:* Dept Physics Concordia Univ Montreal PQ H3G 1M8 Can

MISSEN, R(ONALD) W(ILLIAM), b St Catharines, Ont, Feb 26, 28; m 51; c 4. CHEMICAL ENGINEERING. *Educ:* Queen's Univ, Ont, BSc, 50, MSc, 51; Univ Cambridge, PhD(phys chem), 56. *Prof Exp:* Chem engr, Polysar Corp, Ont, 51-53; from asst prof to assoc prof chem eng, Univ Toronto, 56-68, assoc dean grad studies, 76-77, vprovost prof faculties, 77-81, PROF CHEM ENG, UNIV TORONTO, 68- *Concurrent Pos:* Dir, Chem Eng Res Consult, Ltd. *Honors & Awards:* Plummer Medal, 62. *Mem:* Am Inst Chem Engrs; fel Chem Inst Can; Can Soc Chem Eng; Asn Prof Engrs Ont. *Res:* Chemical reaction engineering; chemical equilibrium analysis and properties of solutions. *Mailing Add:* Dept Chem Eng & Appl Chem Univ of Toronto Toronto ON M5S 1A4 Can

MISTLER, RICHARD EDWARD, b New York, NY, Mar 17, 35; m 59; c 2. CERAMIC ENGINEERING. *Educ:* Alfred Univ, BS, 59; Rensselaer Polytech Inst, MMetallEng, 61; Mass Inst Technol, ScD(ceramics), 67. *Prof Exp:* Ceramic engr, Knolls Atomic Power Lab, Gen Elec Co, 59-63; sr res engr, Western Elec Co, 67-69, mem res staff & res leader ceramics processing, 69-78, dir res & develop, Plessey, Frenchtown Div, 78-81, vpres, 81-84; PRES, KERAMOS INDUST, INC, 85- *Mem:* AAAS; fel Am Ceramic Soc; Int Soc Hybrid Microelectronics. *Res:* Research and development of ceramic materials and processes. *Mailing Add:* 1038 Lafayette Dr Yardley PA 19067

MISTREE, FARROKH, b Ponna, India. COMPUTER AIDED DESIGN, OCEAN ENGINEERING. *Educ:* Indian Inst Technol, BTech, 68; Univ Calif, Berkeley, MS, 70, PhD(eng), 74. *Prof Exp:* Res fel, Univ New South Wales, Australia, 74-76, lectr appl mech, 76-80, sr lectr, 80-81; ASSOC PROF MECH ENG, UNIV HOUSTON, 81- *Honors & Awards:* Walter Atkinson Mem Award, Royal Inst Naval Architects, Sydney, 77. *Mem:* Royal Inst Naval Architects; Soc Naval Architects & Marine Engrs; Am Soc Mech Engrs; Am Soc Eng Educ; Oper Res Soc Am. *Res:* Developing design methods and the necessary decision support software for enhancing the effectiveness of designers operating in a computer-assisted design environment; design of damage-tolerant systems. *Mailing Add:* Dept Mech Eng Univ of Houston 4800 Calhoun Rd Houston TX 77004

MISTRETTA, CHARLOTTE MAE, b Washington, DC, Feb 16, 44; m 68. NEUROPHYSIOLOGY, DEVELOPMENTAL PHYSIOLOGY. *Educ:* Trinity Col, Washington, DC, BA, 66; Fla State Univ, MS, 68, PhD(sensory physiol), 70. *Prof Exp:* Am Asn Univ Women fel, Nuffield Inst Med Res, Oxford Univ, 70-72; sr res assoc, 72-74, asst res scientist, 74-76, ASSOC RES SCIENTIST, DEPT ORAL BIOL, SCH DENT, UNIV MICH, ANN ARBOR, 76-, ASSOC RES SCIENTIST, CTR HUMAN GROWTH & DEVELOP, 75-, ASSOC PROF RES, SCH NURSING, GRAD STUDIES, 75- *Concurrent Pos:* Lectr, Ctr Human Growth & Develop, Univ Mich, Ann Arbor, 73-75; mem int comn olfaction & taste, Int Union Physiol Sci, 74-; Res Career Develop Award, Nat Inst Dent Res, 78-83. *Mem:* AAAS; Am Physiol Soc; Europ Chemoreception Res Orgn; Asn Chemoreception Sci; Soc Neurosci; Sigma Xi. *Res:* Investigation of the development of the sense of taste using anatomical, behavioral and electrophysiological techniques, including study of fetal swallowing activity; study of nerve-epithelium tissue interactions in development. *Mailing Add:* Dept of Oral Biol Univ of Mich Sch of Dent Ann Arbor MI 48109

MISTRY, NARIMAN BURJOR, b Bombay, India, Oct 21, 37; US citizen; m 64; c 4. PHYSICS. *Educ:* Univ Bombay, BS, 56; Columbia Univ, MA, 60, PhD(neutrino physics), 63. *Prof Exp:* Res asst high-energy physics, Columbia Univ, 63-64; instr & res assoc, 64-67, SR PHYSICIST, HIGH-ENERGY PHYSICS, CORNELL UNIV, 67-, ADJ PROF PHYSICS, 88- *Concurrent Pos:* Vis prof, Phys Inst, Aachen, WGer, 73-74. *Mem:* Fel Am Phys Soc. *Res:* High-energy particle physics; cosmic rays; spark chambers; high-energy neutrino interactions; weak interaction physics; high-energy photon and electron physics; electron-positron storage ring design; colliding beam experiments; instrumentation and electronics for detectors. *Mailing Add:* 214 Newman Lab Cornell Univ Ithaca NY 14853-5003

MISTRY, SORAB PIROZSHAH, b Bombay, India, Dec 18, 20; nat US; m 53; c 2. BIOCHEMISTRY, NUTRITION. *Educ:* Univ Bombay, BSc, 42; Indian Inst Sci, Bangalore, MSc, 46; Cambridge Univ, PhD(biochem), 51. *Prof Exp:* Sr res asst, Indian Inst Sci, Bangalore, 45-47; instr, Dunn Nutrit Lab, Cambridge Univ, 48-50, Med Res Coun fel, 50-52; fel, 52-54, res assoc, 54, from asst prof to assoc prof animal nutrit, 54-62, assoc prof, 62-66, PROF BIOCHEM, UNIV ILL, URBANA, 66- *Concurrent Pos:* Sreenivasaya res award, 45-46; vis prof, Univ Amsterdam, 56 & Univ Zurich, 56-57, 63-64;

NIH spec fel, 63-64; vis prof, Ind Inst Sci, Bangalore, 70 & Univ Madrid, 71; prof, Autonoma Univ, Madrid, 79 & Swiss Fed Inst Technol, Zurich, 79, 81 & 83-86. *Mem:* Am Soc Biochem & Molecular Biol. *Res:* Nutritional and comparative biochemistry; regulation of metabolic processes. *Mailing Add:* Dept Animal Sci 124 Animal Sci Lab Univ Ill 1207 W Gregory Dr Urbana IL 61801

MISUGI, TAKAHIKO, b Nishinomiya, Japan, Feb 17, 27; m 58; c 2. SEMICONDUCTOR, VACUUM TUBES. *Educ:* Osaka Univ, BS, 50. *Hon Degrees:* PhD, Osaka Univ, 62. *Prof Exp:* Develop engr, Kobe Kogyo, 51-70; mgr, Fujitsu Labs, 70-82, dir, 83-88, managing dir, 89-91, PRES, FUJITSU TECHNO RES, 91- *Mem:* Fel Inst Elec & Electronic Engrs. *Res:* Development of microwave tubes such as klystron, triode and TWT; development of compound semiconductor devices and materials. *Mailing Add:* Fujitsu Techno Res 1812-10 Shimonumabe Nakalara-ku Kawasaki 211 Japan

MITACEK, EUGENE JAROSLAV, b Hluk, Moravia, Czech, Apr 22, 35; US citizen; m 64; c 1. BIOCHEMISTRY, ENVIRONMENTAL HEALTH. *Educ:* Palacky Univ, Czech, MA, 58; Charles Univ, Prague, PhD(chem & chem educ), 66. *Prof Exp:* Instr chem, State Lyceum, Veseli na Morave, 58-60, asst prof chem & res assoc chem educ, Inst Educ, Charles Univ, Prague, 60-66, sr res assoc prof chem & chem educ, 66-67; head, dept earth sci, UN Int Sch, NY, 67-70; from asst prof to assoc prof biochem, Dept Life Sci, NY Inst Technol, 70-87; RES ASSOC PROF PREV MED, SCH MED, STATE UNIV NY, STONY BROOK, 86-; PROF BIOCHEM, NY INST TECHNOL, 88- *Concurrent Pos:* Lectr, Inst Educ, Univ Warsaw, 61 & 64, Univ Sofia, 63 & Univ Zagreb, 65; consult, Prague Br, UNESCO, Paris, 64-67; Nuffield Found fel, Univ London, 67; res fel, Inst Environ Med, NY Univ Med Sch, 74-75; hon prof, San Marcos Univ, Lima, Univ Buenos Aires & Univ Rio de Janeiro, 78, Med Sch Univ Haiti, 79-81, Med Sch Mahidol, Univ Bangkok, 86-89, Med Sch Univ Chiang Mai, 90. *Mem:* AAAS; Am Chem Soc; NY Acad Sci; fel Am Inst Chemists; Am Asn Univ Prof. *Res:* Geochemistry; polarographic analysis; structural chemistry; comparative chemical education; effectiveness of teaching methods in chemistry; motivation and learning in chemistry; biochemistry of cancer; clinical chemistry; environmental chemistry; nutrition; dietary carcinogens; cancer epidemiology; carcinogenesis; preventive medicine. *Mailing Add:* Eight Lendale Pl Huntington NY 11743-9998

MITACEK, PAUL, JR, b Johnson City, NY, Jan 1, 32; m 67. PHYSICAL CHEMISTRY. *Educ:* Oberlin Col, AB, 54; Pa State Univ, PhD(chem), 62. *Prof Exp:* Res asst chem, Cryogenic Lab, Pa State Univ, 61-62; instr, Bucknell Univ, 62-63; resident res assoc radiation physics, Argonne Nat Lab, 63-67; from asst prof to assoc prof, 67-77, PROF CHEM, ST JOHN FISHER COL, 77-, CHMN DEPT, 70- *Mem:* Am Chem Soc. *Res:* Cryogenics; radiation dosimetry of mixed radiations of neutrons and gamma rays; mini and microprocessor computers; interfacing of laboratory equipment. *Mailing Add:* 164 Willowbend Rd Rochester NY 14618-4049

MITALA, JOSEPH JERROLD, b Trenton, NJ, June 21, 47; m 70; c 3. TOXICOLOGY, TERATOLOGY. *Educ:* Temple Univ Sch Pharm, BS, 70, MS, 73, PhD(pharmacol & toxicol), 77; Am Bd Toxicol, cert. *Prof Exp:* Asst prof pharmacotherapeut, Ferris State Col Sch Pharm, 76-79, assoc prof pharm/pharmacol, 79-81; res scientists reprod toxicol & sr res scientist, Safety Eval Dept, Adria Labs, Inc, 81-87; group leader, reproductive & genetic toxicol, Exxon Biomed Sci, Inc, 87-88; PRIN SCIENTIST, REPRODUCTIVE & INVESTIGATIVE TOXICOL, R W JOHNSON PRI, 88- *Concurrent Pos:* Pharmacol instr, Temple Univ Sch Dent Hyg, 72-76, Camden County Col, 73-76, Community Col Philadelphia, 75-76; teaching asst, Temple Univ Sch Pharm, 71-76; asst examr, Pa State Bd Pharm, 72-76. *Honors & Awards:* Pharm Res Award, Lunsford-Richardson, 70. *Mem:* Teratol Soc; Sigma Xi; Neurobehavioral Teratology Soc. *Res:* General toxicology; developmental toxicology; safety evaluation of new investigational drugs intended for human use. *Mailing Add:* 16 Fawn Dr Flemington NJ 08822

MITALAS, ROMAS, b Kaunas, Lithuania, Feb 28, 33; nat Can; m 79. THEORETICAL PHYSICS, ASTROPHYSICS. *Educ:* Univ Toronto, BA, 57, MA, 58; Cornell Univ, PhD(physics), 64. *Prof Exp:* Asst prof physics, 64-68, asst prof astron, 68-72, ASSOC PROF ASTRON, UNIV WESTERN ONT, 73- *Concurrent Pos:* Res assoc, Yale Univ, 73-74, Cambridge Univ, 79-80. *Mem:* Am Astron Soc; Can Astron Soc. *Res:* Stellar structure and evolution. *Mailing Add:* Dept Astron Univ of Western Ont London ON N6A 3K7 Can

MITCH, FRANK ALLAN, b State College, Pa, Apr 21, 20; m 67. ORGANIC CHEMISTRY. *Educ:* Pa State Col, AB, 41, MS, 48. *Prof Exp:* Researcher, Photo Prod, E I du Pont de Nemours & Co, Inc, 41-45; control, Floridin Co, 49-51, researcher, 51-58; mem staff res & develop, Ariz Chem Co, 58-61; MGR TECH SERV, SCM-ORGANIC CHEM, 61- *Mem:* AAAS; Am Chem Soc; Soc Soft Drink Technol. *Res:* Soft drinks; tall oil; terpenes; essential oils. *Mailing Add:* 4134 Rogero Rd Jacksonville FL 32211-2159

MITCH, WILLIAM EVANS, b Birmingham, Ala, July 22, 41; m 65; c 2. NEPHROLOGY. *Educ:* Harvard Univ, BA, 63, MD, 67. *Prof Exp:* Intern med, Peter Bent Brigham Hosp, 67-68, asst resident, 68-69; clin assoc oncol, Nat Cancer Inst, 69-71; fel med, Johns Hopkins Univ, 71-73; chief resident, Peter Bent Brigham Hosp, 73-74; asst prof med & pharmacol, Johns Hopkins Univ, 74-78, assoc prof, 78-79; assoc prof med, Harvard Med Sch, 79-87; AT PETER BENT BRIGHAM HOSP, BOSTON, MA, 87-; GARLAND HERNDON PROF MED & DIR, RENAL DIV, EMORY UNIV SCH MED, 87- *Concurrent Pos:* Secy treas renal dis, Int Soc Nutrit & Metab; pres, Region II, Nat Kidney Found; vchmn, Coun Kidney, Am Heart Asn. *Mem:* Am Soc Nephrology; Am Inst Nutrit; Am Soc Clin Nutrit; Am Physiol Soc; Am Soc Clin Invest; Sigma Xi. *Res:* Metabolism in chronic renal insufficiency. *Mailing Add:* Renal Div Emory Univ Sch Med 1364 Clifton Rd NE Atlanta GA 30322

MITCHAM, DONALD, b Hazlehurst, Miss, Nov 15, 21. PHYSICS. *Educ:* Tulane Univ, BS, 48. *Prof Exp:* Physicist, 48-80, RES PHYSICIST, SOUTHERN REGIONAL RES LAB, US DEPT AGR, 68- *Mem:* Am Crystallog Asn; Am Asn Textile Chem & Colorists; Soc Appl Spectros; Am Oil Chem Soc; Sigma Xi. *Res:* X-ray diffraction of cotton cellulose; physical properties of textiles; structure of fatty acids by x-ray diffraction. *Mailing Add:* 3326 Marigny New Orleans LA 70122

MITCHELL, A RICHARD, b Pine Bluff, Ark, Feb 27, 39; m 59; c 3. APPLIED MATHEMATICS. *Educ:* Southern Methodist Univ, BS, 60; NMex State Univ, MS, 62, PhD(math), 64. *Prof Exp:* Assoc prof math, Hendrix Col, 64-65; asst prof, Univ Tex, Arlington, 65-67, assoc prof, 67-77, prof math, 77-80; PROF MATH, UNIV TEX, TYLER, 87- *Mem:* Am Math Soc; Math Asn Am. *Res:* Differential equations. *Mailing Add:* Tylor 3900 University Blvd Tyler TX 75701

MITCHELL, ALEXANDER REBAR, b San Pedro, Calif, June 13, 38; m 63; c 1. BIOCHEMISTRY. *Educ:* Univ Calif, Berkeley, BA, 61; Ind Univ, PhD(biochem), 69. *Prof Exp:* Asst biochem, Edgewood Arsenal, US Army, 62-63; res assoc, Rockefeller Univ, 69-75, asst prof biochem, 75-77; CHEMIST, LAWRENCE LIVERMORE LAB, UNIV CALIF, 77- *Mem:* Am Chem Soc; AAAS; NY Acad Sci. *Res:* Chemical synthesis of peptides; solid phase peptide synthesis; cells; synthesis of dense energetic materials. *Mailing Add:* Lawrence Livermore Nat Lab L282 PO Box 808 Livermore CA 94550-0622

MITCHELL, ANN DENMAN, b Nashville, Tenn, Oct 29, 39; m 68; c 1. CELL BIOLOGY, BIOCHEMICAL GENETICS. *Educ:* Univ Tex, Austin, BA, 60, PhD(zool), 71. *Prof Exp:* Teacher biol & chem, Tex Independent Sch Dists, Ft Worth & Austin, 61-66; instr biol, Huston-Tillotson Col, 68; res assoc path, Stanford Univ, Sch Med, 70-72, NIH fel develop biol, 72-73; cell biologist & prog mgr biochem cytogenetics, SRI Int, 73-79, dir, Cellular & Genetic Toxicol Dept, 79-85; PRES, GENESYS RES, INC, 85- *Mem:* AAAS; Soc Toxicol; Environ Mutagen Soc. *Res:* Mutagenesis; carcinogenesis; toxicology; DNA repair; cytogenetics; metabolic activation systems; cell synchronization; drug screening; in vitro mutagenesis and cell transformation. *Mailing Add:* Genesys Research Inc 2300 Englert Dr PO Box 14165 Research Triangle Park NC 27709

MITCHELL, BRIAN JAMES, b Minneapolis, Minn, July 25, 36; m 60; c 2. GEOPHYSICS. *Educ:* Univ Minn, BA, 62, MS, 65; Southern Methodist Univ, PhD(geophys), 70. *Prof Exp:* Physicist rock properties, US Bur Mines Res Ctr, 62-65; res fel seismol, Calif Inst Technol, 71-72; geophysicist mining res, Newmont Explor Ltd, 72-73; from asst prof to assoc prof, 74-80, PROF GEOPHYS & CHMN, DEPT EARTH & ATMOSPHERIC SCI, ST LOUIS UNIV, 80- *Mem:* Am Geophys Union; Seismol Soc Am; Soc Explor Geophysicists; Royal Astron Soc. *Res:* Crust and upper mantle structure; seismic surface wave propagation; arctic research. *Mailing Add:* Dept Earth & Atmospheric Sci St Louis Univ 221 N Grand Blvd St Louis MO 63103

MITCHELL, CARY ARTHUR, b Woodstock, Ill, May 28, 43; m 68; c 3. PLANT STRESS PHYSIOLOGY. *Educ:* Univ Ill, BS, 65; Cornell Univ, MS, 68; Univ Calif, Davis, PhD(plant physiol), 72. *Prof Exp:* From asst prof to assoc prof, 72-84, PROF PLANT PHYSIOL, DEPT HORT, PURDUE UNIV, 84- *Concurrent Pos:* Consult, Am Inst Biol Sci; prin investr, Controlled Ecol Life Support Syst & Space Biol, grants from Nat Aeronaut & Space Admin, 76- *Mem:* Am Soc Plant Physiologists; Am Soc Hort Sci; Am Inst Biol Sci; AAAS; Am Soc Gravitational & Space Biol; Sigma Xi. *Res:* Mechanical stress regulation of plant growth-effects of wind, shaking, and touching on plant hormones and metabolism; physiology of flood tolerance in woody plants; mechanisms of seed dormancy; controlled environment agriculture; physiology of plant stress hardening. *Mailing Add:* Dept Hort Purdue Univ West Lafayette IN 47907

MITCHELL, CHARLES ELLIOTT, b Newark, NJ, Apr 14, 41; c 3. MECHANICAL ENGINEERING. *Educ:* Princeton Univ, BSE, 63, MA, 65, PhD(aerospace & mech sci), 67. *Prof Exp:* Res asst combustion physics, Princeton Univ, 63-67; from asst prof to assoc prof, 67-80, PROF MECH ENG, COLO STATE UNIV, 80- *Concurrent Pos:* Consult, RDA, Inc, 77-79; fac fel, Laramie Energy Tech Ctr, 79-80; fac fel, Southeastern Ctr Elec Eng Educ, US Air Force Rocket Propulsion Lab, 84; rep, Assoc Western Univs, 84-86; consult, Rockwell Int, 86-87, SRS Technologies, 85-88, Battelle Mem Inst, 86-87 & Aerojet Propulsion Div, 90-91. *Mem:* Combustion Inst; Am Inst Aeronaut & Astronaut. *Res:* Combustion instability; nonlinear wave oscillations; combustion of fossil fuels; acoustics; gas dynamics. *Mailing Add:* Dept Mech Eng Colo State Univ Ft Collins CO 80523

MITCHELL, CLIFFORD L, b Ottumwa, Iowa, Dec 7, 30; m 54; c 4. PHARMACOLOGY. *Educ:* Univ Iowa, BA, 52, BS, 54, MS, 58, PhD(pharmacol), 59. *Prof Exp:* Asst prof, Col Med, Univ Iowa, 59-60 & 62-66, from assoc prof to prof pharmacol, 66-73; sr res scientist, Riker Labs, 73-74, mgr cent nerv syst & cardiopulmonary pharmacol, 74-76; head behav toxicol prog, 76-77, chief lab behav & neurol toxicol, 77-84, HEAD MEMBRANE PHYSIOL SECT, LAB MOLECULAR & INTEGRATIVE NEUROSCI, NAT INST ENVIRON HEALTH SCI, 84- *Concurrent Pos:* Res fel, Stanford Univ, 60-62; adj prof pharmacol, Univ NC, 77- *Mem:* AAAS; Am Soc Pharmacol & Exp Therapeut; Soc Neurosci; Soc Exp Biol & Med. *Res:* Neuropharmacology; neurotoxicology; primarily analgesic fields, neurobehavioral toxicology and epilepsy. *Mailing Add:* Nat Inst of Environ Health Sci Box 12233 Research Triangle Park NC 27709

MITCHELL, DAVID FARRAR, dentistry; deceased, see previous edition for last biography

MITCHELL, DAVID HILLARD, b Philadelphia, Pa, May 17, 45. GERONTOLOGY, GENETICS. *Educ:* Stanford Univ, BA, 67; Harvard Univ, MA, 70, PhD(biochem), 75. *Prof Exp:* Res fel geront, 75-78, STAFF SCIENTIST GERONT, BOSTON BIOMED RES INST, 78- *Concurrent Pos:* Prin investr, Nat Inst Aging res grant, 78-81. *Mem:* Geront Soc. *Res:* Genetics and biochemistry of aging. *Mailing Add:* 78 Gilman Boston MA 02145

MITCHELL, DAVID WESLEY, b Columbia, SC, Jan 28, 13; m 43; c 2. METALLURGY, PHYSICAL CHEMISTRY. *Educ:* Univ Calif, Berkeley, BS, 38; Univ Utah, MS, 40; Univ Calif, PhD(metall), 47. *Prof Exp:* From instr to assoc prof metall, Dept Mineral Technol, Univ Calif, Berkeley, 42-57; mgr dir res develop minerals, chem, metall & ceramics, Foote Mineral Co, 57-62; prof metall, NMex Inst Mining & Technol, 62-73; vpres & tech dir elec & magnetic separation solids, Carpco Inc, 74-80; CONSULT, 80- *Concurrent Pos:* Vpres & tech dir, Oil Shale Corp, 55-57; expert in metall, UNESCO, 70-72. *Mem:* Am Inst Mining, Metall & Petrol Engrs. *Res:* Electrostatic and magnetic separation of minerals and waste products. *Mailing Add:* 7000 Arroyo Del Oso NE Albuquerque NM 87109

MITCHELL, DEAN LEWIS, b Montour Falls, NY, Apr 9, 29; m 51, 89; c 4. SOLID STATE PHYSICS. *Educ:* Syracuse Univ, BS, 52, PhD(physics), 59. *Prof Exp:* Asst prof physics, Utica Col, Syracuse Univ, 58-61; res physicist, Naval Res Lab, 61-70, head solid-state appln br, 70-74; prog dir solid-state physics, NSF, 74-81, sect head, condensed matter sci, 81-82, staff assoc nat facil, 82-85; vis scientist, Lawrence Berkeley Lab, 85; liaison scientist, Europ Off, Off Naval Res, 88-90; RES CONSULT, MITCHELL ASSOC, 86-88 & 90- *Concurrent Pos:* Nat Acad Sci-Acad Sci USSR exchange prog fel, A F Ioffe Phys Tech Inst, Leningrad, 67-68. *Mem:* AAAS; Am Phys Soc. *Res:* Optical and magneto-optical properties; electronic band structure of solids; semiconductors; infrared sources and detectors. *Mailing Add:* PO Box 2337 Reston VA 22090

MITCHELL, DENNIS KEITH, b Hollywood, Calif, Apr 17, 46. INORGANIC CHEMISTRY, GENERAL CHEMISTRY. *Educ:* Univ Calif, Los Angeles, BS, 68; Univ Calif, Santa Barbara, PhD(chem), 72. *Prof Exp:* Asst prof chem, Univ Andes, Venezuela, 73-76; res asst chem, Univ Calif, Berkeley, 76-77; lectr chem, Calif State Univ, Hayward, 77-80; from asst prof to assoc prof, 80-85, PROF CHEM, LOS ANGELES CITY COL, 85- *Concurrent Pos:* Res grant, Nat Coun Sci & Tech Res, Venezuela, 75-76. *Mem:* Am Chem Soc. *Res:* Organometallic synthesis and catalysis. *Mailing Add:* 15480 Antioch St No 204 Pacific Palisades CA 90272-4304

MITCHELL, DONALD GILMAN, b Somerville, Mass, Sept 17, 17; m 42; c 2. FOOD TECHNOLOGY. *Educ:* Mass Inst Technol, SB, 38. *Prof Exp:* Chemist, Walter Baker & Co, Inc, Gen Foods Corp, Dover, 38-41, res chemist, 46-50, asst to res dir, 50-54, qual control mgr, 54-59, mgr chocolate develop, 59-62, mgr tech serv, Baker's Chocolate & Coconut Div, 62-81; RETIRED. *Honors & Awards:* Stroud Jordan Award, Am Asn Candy Technologists, 80; Res & Educ Award, Nat Confectioners Asn, 80. *Mem:* Am Chem Soc; Inst Food Technologists; Am Asn Candy Technologists; Am Soc Bakery Engrs. *Res:* Quality control and research and development on chocolate products and their processing. *Mailing Add:* 706 North Shore Dr Milford DE 19963

MITCHELL, DONALD J, b New Castle, Pa, May 12, 38; m 68. PHYSICAL CHEMISTRY, CRYSTALLOGRAPHY. *Educ:* Westminster Col, Pa, BS, 60; Vanderbilt Univ, PhD(phys chem), 64. *Prof Exp:* Res chemist with Dr Jerome Karle, Naval Res Labs, US Govt, DC, 64-67; ASST PROF CHEM, JUNIATA COL, 67-, CHMN DEPT, 77- *Concurrent Pos:* NSF fel, 75. *Mem:* Am Chem Soc; Am Crystallog Asn. *Res:* Crystal structure analysis by x-ray diffraction, particularly gas bearing shales, coals and polymers. *Mailing Add:* Dept Chem Juniata Col 1700 Moore St Huntingdon PA 16652-2196

MITCHELL, DONALD JOHN, plant physiology, for more information see previous edition

MITCHELL, DONALD W(ILLIAM), b New York, NY, May 24, 23; m 51; c 6. MINING ENGINEERING, COMBUSTION. *Educ:* Pa State Univ, BS, 48; Columbia Univ, MS, 51. *Prof Exp:* Engr, H C Frick Coke Co, Pa, 41-42; mining engr, Hudson Coal Co, 46-50; mining engr, US Bur Mines, 51-59, chief mine exp sect, 59-74, asst chief, Br Dust Explosions, 61-65, proj coordr eng appln, 65-74, staff engr, Pittsburgh Mining & Safety Res Ctr, 71-74; prin mining engr, Mining Safety & Health Admin, 74-78; chief engr, Gates Eng Co & Foster Miller Assoc, 78-82; CHIEF ENGR, MITCHELL ENG, 82- *Concurrent Pos:* Chmn eng comt & gen chmn, Coal Mining Sect, Nat Safety Coun, 63-; chmn task force underground storage of oil, gas & nuclear waste, 76-78. *Honors & Awards:* Secy Labor Spec Recognition Award, US Dept of Labor, 78. *Res:* Dust explosions and fires; strata control. *Mailing Add:* 5858 Horseshoe Dr Bethel Park PA 15102

MITCHELL, EARL BRUCE, b Louisville, Miss, Sept 1, 27; m 50; c 3. ENTOMOLOGY. *Educ:* Miss State Univ, BS, 50, MS, 51, PhD(entom), 71. *Prof Exp:* Entomologist, La State Univ, 51-52 & Food Mach Corp, 53-54; farmer, 54-62; ENTOMOLOGIST, BOLL WEEVIL RES LAB, AGR RES SERV, USDA, MISSISSIPPI STATE, 63- *Mem:* Entom Soc Am. *Res:* Use of pheromones, traps, insecticides, sterile males, and combinations of these for the purpose of suppressing or eradicating the boll weevil. *Mailing Add:* Rte 2 Box 284 Louisville MS 39339

MITCHELL, EARL DOUGLASS, JR, b New Orleans, La, May 16, 38; m 59; c 3. BIOCHEMISTRY. *Educ:* Xavier Univ La, BS, 60; Mich State Univ, MS, 63, PhD(biochem), 66. *Prof Exp:* Res assoc, Okla State Univ, 67-69, from asst prof to assoc prof, 69-78, prof & asst dean grad sch, 78-82, PROF BIOCHEM, OKLA STATE UNIV, 82- *Concurrent Pos:* Res chemist, NIH, 78-79. *Mem:* Am Chem Soc; Sigma Xi; Am Soc Biochem & Molecular Biol. *Res:* DNA fingerprinting and restriction fragment length polymorphism analysis of plants and animals; plant cell culture and plant host/pathogen studies. *Mailing Add:* Dept Biochem Okla State Univ Stillwater OK 74078

MITCHELL, EARL NELSON, b Centerville, Iowa, Aug 30, 26; m 55. PHOTOGRAPHIC SCIENCE. *Educ:* Univ Iowa, BA, 49, MS, 51; Univ Minn, PhD(physics), 55. *Prof Exp:* Res physicist, Univac Div, Sperry Rand Corp, 55-58; from asst prof to assoc prof physics, Univ NDak, 58-62; vis assoc prof, 62-65, assoc prof, 65-69, asst chmn dept, 68-76, PROF PHYSICS, UNIV NC, CHAPEL HILL, 69- *Mem:* Am Phys Soc; Am Soc Enol & Viticult; Am Asn Physics Teachers. *Res:* Electric and magnetic properties of thin metal films; writing (photographic science) history of technology (photography and viticulture); winter hardiness of grape vines; harvesting of grapes; cosmic ray physics. *Mailing Add:* Dept Physics & Astron Univ NC Chapel Hill NC 27514

MITCHELL, EVERETT ROYAL, b Itasca, Tex, Sept 17, 36; m 57; c 1. ENTOMOLOGY. *Educ:* Tex Tech Univ, BS, 59; NC State Univ, MS, 61, PhD(entom, bot), 63. *Prof Exp:* Res entomologist, Cotton Insects Br, USDA, Florence, SC, 63-65; asst prof entom & head dept, Coastal Plain Exp Sta, Univ Ga, Tifton, 65-67; prof biol sci & chmn div sci & math, Tarrant County Community Col, Ft Worth, Tex, 67-69; sr res biologist, Monsanto Co, St Louis, 69-70; res entomologist, 71, actg dir lab, 71-72, RES LEADER BEHAV ECOL & REPRODUCTION UNIT, INSECT ATTRACTANTS LAB, USDA, 72- *Concurrent Pos:* Assoc prof entom & nematol, Univ Fla, 71-77, prof, 77-; pres, Southeastern Br, Entom Soc Am, 90-91. *Mem:* Entom Soc Am; Int Soc Chem Ecol. *Res:* Biology, ecology, behavior and control of economic pests of field and vegetable crops, particularly chemical messengers produced by insects and plants. *Mailing Add:* Insect Attractants Lab USDA PO Box 14565 Gainesville FL 32604

MITCHELL, GARY EARL, b Louisville, Ky, July 5, 35; m 57; c 1. PHYSICS. *Educ:* Univ Louisville, BS, 56; Duke Univ, MA, 58; Fla State Univ, PhD(physics), 62. *Prof Exp:* Res assoc physics, Columbia Univ, 62-64, asst prof, 64-68; assoc prof, 68-74, PROF PHYSICS, NC STATE UNIV, 74- *Concurrent Pos:* Humboldt sr scientist award, 75-76. *Mem:* AAAS; fel Am Phys Soc. *Res:* Nuclear structure physics; fundamental symmetry tests and statistical properties of nuclear states. *Mailing Add:* Dept Physics Box 8202 NC State Univ Raleigh NC 27695-8202

MITCHELL, GEORGE ERNEST, JR, b Duoro, NMex, June 7, 30; m 52; c 3. ANIMAL SCIENCE. *Educ:* Univ Mo, BS, 51, MS, 54; Univ Ill, PhD(animal sci), 56. *Prof Exp:* Asst animal sci, Univ Ill, 54-56, asst prof, 56-60; assoc prof animal husb, 60-67, PROF ANIMAL SCI, UNIV KY, 67 - *Concurrent Pos:* Sr Fulbright Res Scholar, NZ, 73-74; mem, subcomt beef cattle nutrit, Nat Res Coun, subcomt vitamin tolerance, comt animal nutrit; AID consult, Morocco & Indonesia, John Lee Pratt Nutrit Prog Consult; adj prof, Hassan II Agron Vet Sci Inst, Morocco, 85 - *Mem:* AAAS; Am Soc Animal Sci; Am Dairy Sci Asn; Am Inst Nutrit; Am Register Prof Animal Scientists. *Res:* Vitamin A metabolism; digestive physiology; carbohydrate utilization; rumen function; ruminant-nonruminant comparisons; effect of hormones on nutritive requirements; beef cattle feeding and management. *Mailing Add:* 814 Blue Ridge Ave Culpeper VA 22701

MITCHELL, GEORGE JOSEPH, b Vancouver, BC, Nov 21, 25; m 56; c 4. POPULATION ECOLOGY, BIOLOGY OF REPRODUCTION. *Educ:* Univ BC, BA, 50, MA, 52; Wash State Univ, PhD(zool), 65. *Prof Exp:* Wildlife biologist, Alta Dept Lands & Forests, 52-64, chief wildlife biologist, 64-66; assoc prof zool, Univ Sask, Regina, 66-73; prof zool, Univ Regina, Can, 73-88; RETIRED. *Concurrent Pos:* Can mem-at-large & Wildlife Soc Can coun rep, Wildlife Soc, 79-88. *Mem:* Wildlife Soc; Can Soc Zoologists; Wildlife Soc; Am Soc Mammalogists. *Res:* Distribution, ecology, growth, regulation and demography of some wild ungulate, carnivore and avian populations of the Canadian Plains in Western Canada. *Mailing Add:* 4322 Castle Rd Regina SK S4S 4W3 Can

MITCHELL, GEORGE REDMOND, JR, organic polymer chemistry; deceased, see previous edition for last biography

MITCHELL, GERALDINE VAUGHN, b Woodland, NC, Oct 23, 40; m 66; c 2. NUTRITIONAL SCIENCES, BIOCHEMISTRY. *Educ:* Va Union Univ, BS, 62; George Washington Univ, MS, 68; Univ Md, PhD(nutrit sci), 78. *Prof Exp:* Biochemist, Walter Reed Army Res Inst, 62-64; RES CHEMIST, FOOD & DRUG ADMIN, 64- *Mem:* Am Inst Nutrit. *Res:* Amino acid and protein metabolism with emphasis on protein quality; metabolic role of vitamin A. *Mailing Add:* 413 Van Buren St NW Washington DC 20012

MITCHELL, H REES, b New London, Conn, Oct 13, 08; m 37; c 4. PHYSICS. *Educ:* Trinity Col, Conn, BS, 31; Johns Hopkins Univ, PhD(physics), 38. *Prof Exp:* Instr math, Gettysburg Col, 37-38; asst prof physics, The Citadel, 38-44; physicist, Appl Physics Lab, Johns Hopkins Univ, 44-48; asst prof, Georgetown Univ, 48-51; assoc prof, 51-58, prof 58-76, EMER PROF PHYSICS, MICH TECHNOL UNIV, 76- *Mem:* AAAS; Am Phys Soc; Am Asn Physics Teachers. *Res:* Thin metallic films servomechanisms; computing mechanism of gun fire control and guided missiles. *Mailing Add:* PO Box 215 Manset ME 04656-0215

MITCHELL, HAROLD HUGH, b New York, NY, Apr 10, 16. MEDICINE. *Educ:* Univ Ariz, BS, 36; Univ Southern Calif, MS, 38; Wash Univ, MD, 45. *Prof Exp:* Lectr bact, Univ Southern Calif, 38-41; asst med dir, Calif Physicians Serv, 48-52; from assoc scientist to sr scientist, Physics Div, Rand Corp, 52-71; sr scientist, R&D Assocs, 71-85; RETIRED. *Concurrent Pos:* Chmn, Gov's Adv Comt Radiol Defense, Calif, 62-68. *Mem:* AAAS; AMA. *Res:* Biological and environmental consequences of nuclear war radiation effects and radiobiology; mass casualty; civil defense. *Mailing Add:* 426 S Spalding Dr Beverly Hills CA 90212

MITCHELL, HELEN C, PHARMACOLOGY. *Educ:* Washington Univ, St Louis, MD, 48. *Prof Exp:* RES ASST PROF, UNIV TEX HEALTH SCI CTR, 80- *Mailing Add:* Dept Pharmacol & Internal Med Univ Tex Health Sci Ctr 5323 Harry Hines Blvd Dallas TX 75235

MITCHELL, HENRY ANDREW, b Joplin, Mo, Oct 6, 36; m 61; c 3. MAMMALIAN PHYSIOLOGY. *Educ:* Southwest Mo State Col, BS, 58; Univ Ariz, MS, 60, PhD(zool), 63. *Prof Exp:* Asst zool, Univ Ariz, 58-63; asst prof biol, 63-68, asst dean, Sch Grad Studies, 69, assoc prof biol & assoc dean, Col Arts & Sci, 68-73, lectr med & assoc dean, Sch Med, 72-73, actg dean, Col Arts & Sci, 74-75, assoc provost health sci, 73-80, asst vchancellor, 80-81, assoc vchancellor acad affairs, 81-84, PROF BIOL, MED & PHARM, UNIV MO-KANSAS CITY, 73-, ASSOC VCHANCELLOR ACAD & INT AFFAIRS & HEAD CTR INT AFFAIRS, 84- *Concurrent Pos:* Mem, Am Conf Acad Deans; field ctr coordr, NSF Chautauqua-Type Short Courses Prog for Col Teachers, 71-80; proj adminr, Western Mo Area Health Educ Ctr, 75-77; hon coun mem, Smedley, Strong & Snow Soc, China, 85-91; mem & chair, Discipline Screening Comt for Fulbright Scholar Awards in Life Sci, coun, Int Exchange Scholars, 84-87, chair, Fulbright East Asian Scholars in Residence Rev Comt, 85-88; mem adv coun, People to People Int Goodwill Collegiate Ambassador's Prog, 86-; acad assoc, Atlantic Coun US, 87-; US Adv Ningbo Univ, China, 90-; mem, bd trustees, People to People Int, 90, hon coun mem, China Soc People's Friendship Studies, 91- *Mem:* Fel AAAS; Am Soc Mammal; Am Inst Biol Sci; Mex Natural Hist Soc. *Res:* Hematological studies of bats. *Mailing Add:* Off Acad Affairs Univ Mo 5100 Rockhill Rd Kansas City MO 64110

MITCHELL, HERSCHEL KENWORTHY, b Los Nietos, Calif, Nov 27, 13; m 34; c 5. BIOCHEMISTRY, GENETICS. *Educ:* Pomona Col, AB, 36; Ore State Col, MS, 38; Univ Tex, PhD(org chem, biochem), 41. *Prof Exp:* Res assoc chem, Ore State Col, 38-39; res biochemist, Univ Tex, 41-43; res assoc biochem, Stanford Univ, 43-46; sr res fel, 46-48, assoc prof biol, 48-53, prof, 53-84, EMER PROF BIOL, CALIF INST TECHNOL, 84- *Concurrent Pos:* NSF sr fel zool, Univ Zurich, 59-60. *Mem:* Am Soc Biol Chemists; Am Chem Soc; Genetics Soc Am; Sigma Xi; Am Asn Univ Professors; AAAS. *Res:* Growth factors for microorganisms; microchemical methods; biochemical genetics in neurospora; synthesis of compounds of biological significance; biochemistry of Drosophila; moleculr basis of differentiation. *Mailing Add:* Dept Biol Calif Inst Technol Pasadena CA 91109

MITCHELL, HUGH BERTRON, b Bolton, Miss, Dec 8, 23; m 45; c 4. RADIOBIOLOGY, AEROSPACE MEDICINE. *Educ:* La State Univ, MD, 47; Univ Calif, Berkeley, MBioradiol, 65. *Prof Exp:* Physician, Tulsa Clin, Okla, 48-49; pvt pract, Richton, Miss, 49-50; med corps, US Navy, 50-52; physician, Baton Rouge Clin, La, 52-57; med corps, US Air Force, 57-75, med officer, Barksdale Air Force Base Hosp, 57-58, med officer, Sch Aviation Med, Randolph Air Force Base, 58, med officer spec weapons prog & radiobiol, Off Surgeon, Strategic Air Command Hq, 59-62, from dep dir to dir, Armed Forces Radiobiol Res Inst, Nat Naval Med Ctr, Md, 65-71, surgeon, 314th Air Div & comdr, Osan Air Force Base Hosp, Korea, 71-72, staff, Radiobiol Div, US Air Force Sch Aerospace Med, 72-73, comdr, US Air Force Hosp, Plattsburgh Air Force Base, NY, 73-75; area med dir, Springs Mills, Inc, Ft Mills, SC, 75-84; RETIRED. *Mem:* Asn Mil Surg US. *Res:* General medicine and surgery; occupational medicine; medical aspects of special weapons and of disaster operations; combined effects of radiation plus other stresses. *Mailing Add:* 2079 Marquesas Ft Mill SC 29715

MITCHELL, I(DA) MERLE, mathematics, for more information see previous edition

MITCHELL, JACK HARRIS, JR, b Auburn, Ala, Sept 15, 11; m 44; c 4. BIOCHEMISTRY. *Educ:* Clemson Col, BS, 33; Purdue Univ, PhD(biochem), 41. *Prof Exp:* Asst chemist, State Chem Lab, SC, 33-36; asst, Columbia Univ, 36-37; asst, Purdue Univ, 37-40, asst chemist, 40-41; res chemist, Am Meat Inst Found, Univ Chicago, 41-42 & Petrol Chem Co, Md, 46-47; head biochem sect, Southern Res Inst, Ala, 47-50; asst chief, Stability Div, Qm Food & Container Inst, Chicago Ill, 50-55, chief chem & microbiol div, 55-57; head dept food tech & human nutrit, 57-64, prof food sci, 64-77, EMER PROF FOOD SCI, CLEMSON UNIV, 77- *Concurrent Pos:* Capt, Edgewood Arsenal, Chem Warfare Res, 42-46; mem bd dirs, Res & Develop Assoc for Military Food & Packaging Systs, Inc, 64-67. *Mem:* Fel Inst Food Technologists. *Res:* Research administration; food biochemistry; lipids; food processing and product development; new peanut processing technology; new peanut products. *Mailing Add:* 101 Bradley St Clemson SC 29631

MITCHELL, JAMES EMMETT, b Triadelphia, WVa, July 18, 39; m 62; c 2. CHEMICAL ENGINEERING. *Educ:* WVa Univ, BSChE, 61; Univ Ill, Urbana, MS, 63, PhD(chem eng), 65. *Prof Exp:* From res engr to sr res engr, Esso Res & Eng Co, 65-73, sect head, 73-75, lab dir, 75-78; vpres Res & Develop, Albany Int Corp, 79-83; res mgr, 83-90, LICENSING MGR, ARCO OIL & GAS CO, 91- *Mem:* Am Chem Soc; Am Inst Chem Engrs; Am Phys Soc; AAAS; Soc Petrol Engrs; AAAS. *Res:* Fluid mixing and mechanics; electrical aspects of combustion; adsorption and other separation processes; petroleum recovery processes. *Mailing Add:* 7616 Dunleer Way Dallas TX 75248-1639

MITCHELL, JAMES GEORGE, b Kitchener, Ont, Apr 25, 43; c 3. COMPUTER SCIENCE. *Educ:* Univ Waterloo, BSc, 66; Carnegie-Mellon Univ, PhD(comput sci), 70. *Prof Exp:* Programmer, Berkeley Comput Corp, 70-71; mem res staff comput sci, 71-79, prin scientist, Palo Alto Res Ctr, Xerox Corp, 79-84; pres, Acorn Res Ctr, Inc, 84-88, dir Res & Develop, Acorn Computers plc, UK, 86-88; DIR TECHNOL, SUN MICROSYSTS, INC, 88- *Concurrent Pos:* Sr vis fel, Comput Lab, Cambridge Univ, 80-81. *Mem:* Asn Comput Mach; Inst Elec & Electronic Engrs. *Res:* Programming systems and methodologies; reliable software; programming languages; personal computer systems; computer networks; computer architecture; distributed computing. *Mailing Add:* Sun Microsystems 2550 Garcia Ave MS 10-21 Mountain View CA 94043

MITCHELL, JAMES K(ENNETH), b Manchester, NH, Apr, 19, 30; m 51; c 5. GEOTECHNICAL ENGINEERING. *Educ:* Rensselaer Polytech Inst, BCE, 51; Mass Inst Technol, SM, 53, ScD(civil eng), 56. *Prof Exp:* Asst soil mech, Mass Inst Technol, 51-55; from asst prof & asst res engr to assoc prof & assoc res engr, 58-68, chmn dept, 79-84, prof civil eng & res eng, 68-88, EDWARD G & JOHN R CAHILL PROF CIVIL ENG, UNIV CALIF, BERKELEY, 89- *Concurrent Pos:* Mem, Transp Res Bd, Nat Res Coun, 82-85; mem, US Nat Comn, Int Soc Soil Mech & Found Engrs; prin investr, Soil Mech Exp, Apollo 14-17; geotech consult, 60-; geotech bd, Nat Res Coun, 88, chmn, 90- *Honors & Awards:* Thomas A Middlebrooks Award, Am Soc Civil Engrs, 62, 70 & 73, Walter L Huber Prize, 65; Norman Medal, 72; Terzaghi Lectr, 84; Terzaghi Award, 85; Rankin Lectr, Brit Geotech Soc, 91. *Mem:* Nat Acad Eng; fel Am Soc Civil Engrs; Clay Minerals Soc; Am Soc Eng Educ. *Res:* Soil behavior; soil and site improvement; in-situ measurement of soil properties; environmental geotechnology. *Mailing Add:* Dept Civil Eng Univ Calif 440 Davis Hall Berkeley CA 94720

MITCHELL, JAMES WINFIELD, b Durham, NC, Nov 16, 43; m 65; c 3. ANALYTICAL CHEMISTRY. *Educ:* NC A&T State Univ, BS, 65; Iowa State Univ, PhD(anal chem), 70. *Prof Exp:* SUPVR ANAL CHEM, BELL LABS, 70-, HEAD, ANAL CHEM DEPT, 75- *Honors & Awards:* Pharmacia Prize; Percy L Julian Res Award; IR 100 Award, 82,. *Mem:* Nat Acad Engrs; Am Nuclear Soc; Am Inst Chem; NY Acad Sci; Am Chem Soc. *Res:* Quantitative analysis of submicrogram amounts of inorganic species by radiochemical methods; ultrapurification of reagents for trace analysis; microwave plasma induced chemistry, spectroscopy. *Mailing Add:* MH 1D 239 AT&T Bell Labs 600 Mountain Ave Murray Hill NJ 07974

MITCHELL, JERALD ANDREW, b Dallas, Tex, May 8, 41; m 71. REPRODUCTIVE PHYSIOLOGY, NEUROENDOCRINOLOGY. *Educ:* Southern Methodist Univ, BS, 63; Univ Kans, PhD(physiol), 69. *Prof Exp:* Asst prof, 72-77, ASSOC PROF ANAT, SCH MED, WAYNE STATE UNIV, 77- *Concurrent Pos:* NIH fel, Dept Anat, Baylor Col Med, 69-72. *Mem:* AAAS; Am Asn Anat; Soc Study Reproduction; Am Asn Hist Med. *Mailing Add:* Dept Anat Sch Med Wayne State Univ 540 E Canfield Ave Detroit MI 48201

MITCHELL, JERE HOLLOWAY, b Longview, Tex, Oct 17, 28; m 60; c 3. CARDIOVASCULAR PHYSIOLOGY. *Educ:* Va Mil Inst, BS, 50; Univ Tex Southwestern Med Sch, MD, 54. *Prof Exp:* From intern to resident med, Parkland Mem Hosp, Dallas, 54-56; from sr asst surgeon to surgeon, Lab Cardiovasc Physiol, Nat Heart Inst, 58-62; from asst prof to assoc prof med & physiol, 62-69, PROF MED & PHYSIOL, SOUTHWESTERN MED SCH, UNIV TEX HEALTH SCI CTR, DALLAS, 69- *Concurrent Pos:* Nat Heart Inst cardiac trainee, Univ Tex Southwestern Med Sch, Dallas, 56-57, res fel, 57-58, USPHS career develop award, 68-73; estab investr, Am Heart Asn, 62-67, chmn coun basic sci, 69-71, fel, Coun Clin Cardiol & Coun Circulation; vis sr scientist, Lab Physiol, Oxford Univ, 70-71; dir, Pauline & Adolph Weinberger Lab Cardiopulmonary Res, 66-; dir, Harry S Moss Heart Ctr, 76-; vis sr scientist, August Krogh Inst, Univ Copenhagen, 76, 81, 86-87; consult, Appl Physiol & Orthop Study Sect, Nat Heart Lung Blood Inst, 79-81 & Respiration & Appl Physiol Study Sect, 81-82. *Honors & Awards:* Young Investr Award, Am Col Cardiol, 61; Donald W Seldin Res Award, Univ Tex Southwestern, 78; Citation Award, 83 & Honor Award, Am Col Sports Med, 88. *Mem:* Am Physiol Soc; Am Soc Clin Invest; fel Am Col Cardiol; Am Asn Physicians; Sigma Xi; Am Col Sports Med. *Res:* Cardiovascular and exercise physiology. *Mailing Add:* Univ Tex Southwestern Med Ctr Harry S Moss Heart Ctr 5323 Harry Hines Blvd Dallas TX 75235-9034

MITCHELL, JERRY R, b Detroit, Mich, July 27, 41; m 65; c 2. METABOLIC PROBLEMS. *Educ:* Univ Ky, BA, 63; Vanderbilt Univ, MD, 68, PhD(pharmacol), 69. *Prof Exp:* Intern, internal med, Vanderbilt Univ Hosp, 68-69; asst resident, internal med, Cornell Med Ctr, 69-70; res assoc pharmacol-toxicol, Nat Heart & Lung Inst, NIH, 70-72, sr clin invest, Exp Therapeut, 72-77; prof chem toxicol lipid res, 77-86, PROF MED, BAYLOR COL MED, 77-, CHIEF, DEPT MED, 77-, DIR EXP THERAPEUT, CTR EXP THERAPEUT, 86- *Concurrent Pos:* Pfizer vis prof clin pharmacol, 78-87; chmn, select comt, Am Soc Pharmacol & Exp Therapeut, 80; prin investr, prog proj grant, NIH, 79-; Willian N Creasy-Burroughs Wellcome vis prof clin pharm, 76 & 78 & 87; chmn, Pharmacol Study Sect, NIH, 87-89. *Honors & Awards:* John J Abel Award, Am Soc Pharmacol & Exp Therapeut, 77. *Mem:* Am Fedn Clin Res; Am Soc Pharmacol & Exp Therapeut; Am Asn Study Liver Diseases; Am Soc Clin Invest; Asn Am Physicians. *Res:* Molecular and clinical basis for human tissue injury from oxygen radicals, lipid derived chemotactics factors, and reactive drug metabolites. *Mailing Add:* Ctr Exp Therapeut Baylor Col Med One Baylor Plaza Houston TX 77030

MITCHELL, JOAN LAVERNE, b Palo Alto, Calif, May 24, 47. DATA COMPRESSION, PHYSICS. *Educ:* Stanford Univ, BS, 69; Univ Ill, Urbana, MS, 71, PhD(physics), 74. *Prof Exp:* Res staff appl res, T J Watson Res Ctr, 74-88, mgr, 79-88, CONSULT, IBM IMAGE TECH, IBM CORP, 89- *Concurrent Pos:* Comt mem, Inst Sci Orgn, Int Electronics Corp, JTC 1, WG10, JPEG Standards, 87- *Mem:* Am Phys Soc; Sigma Xi; Inst Elec & Electronic Engrs. *Res:* Data compression algorithm development; display of images; image processing. *Mailing Add:* IBM Res Ctr PO Box 704 Yorktown Heights NY 10598

MITCHELL, JOHN, JR, b Catonsville, Md, Oct 5, 13; m 35; c 2. ANALYTICAL CHEMISTRY. *Educ:* Johns Hopkins Univ, BE, 35; Univ Del, MS, 45. *Prof Exp:* Control chemist, Consol Edison Co, NY, 35; chem engr, E I du Pont de Nemours & Co, Inc, 35-36, chemist, 36-37, anal chemist, 37-45, res supvr, 45-51, sr supvr, 51-66, res mgr, 66-78; RETIRED. *Concurrent Pos:* Consult, analytical chem & polymer sci, 79-; ed, Hanser Verlag, Munich, Fed Repub Germany. *Honors & Awards:* Fisher Award Anal Chem, 64; Anachem Award, Asn Anal Chemists, 74. *Mem:* Am Chem Soc. *Res:* Organic and physical chemistry; chemical engineering. *Mailing Add:* 3 Meadows Lane The Meadows Wilmington DE 19807

MITCHELL, JOHN ALEXANDER, b Providence, RI, Aug 5, 45; m 67; c 2. PARASITOLOGY, MICROBIOLOGY. *Educ:* Okla State Univ, BS, 67; Univ Mont, PhD(zool), 75. *Prof Exp:* Vector control officer, Tulsa City-County Health Dept, 69-70; lab coordr physiol, Univ Mont, 72-73; res asst statist analyst, Mont Coop Wildlife Res Unit, 73-75; asst prof & chmn dept biol, Pac Univ, 75-80; asst dir, Wildlife-Wildlands Inst, 80-83; EDUC COORDR, ST PATRICK HOSP, 84- *Concurrent Pos:* NDEA fel, 67-68 & 70-72; vis prof microbiol, Univ Mont, 83-84. *Mem:* AAAS; Sigma Xi. *Res:* Disease causation and epidemiology/epizootiology; protozoan parasites; medical mycology. *Mailing Add:* 3851 Beverly Ridge Dr Sherman Oaks CA 91403

MITCHELL, JOHN CHARLES, b Jourdanton, Tex, April 23, 32; m 53; c 2. RADIATION PHYSICS. *Educ:* St Marys Univ, Tex, BS, 58. *Prof Exp:* Radiation proj eng, Ling Temco Vought, 58-64; mgr, res prog dir & chief radiation physics, 64-88, res dir radiation sci, 88-90, DIR OCCUP & ENVIRON HEALTH, ARMSTRONG LAB, AFSC, USAF AEROSPACE MED, 91- *Concurrent Pos:* Chmn, Tri-Service Electromagnetic Radiation Panel, 78, 81 & 85 & Panel VIII Res Study Group 2, NATO, 79-82. *Mem:* Inst Elec & Electronic Engrs; Aerospace Med Asn; Am Conf Govt Indust Hygienists; Bioelectromagnetic Soc. *Res:* Biological effects of radio frequency electromagnetic radiation; electromagnetic interference; electromagnetic compatibility; occupational and environmental health. *Mailing Add:* Armstrong Lab/OE Brooks AFB San Antonio TX 78235

MITCHELL, JOHN CLIFFORD, b Palo Alto, Calif, Dec 20, 55. PROGRAMMING LANGUAGES, MATHEMATICAL LOGIC. *Educ:* Stanford Univ, BS, 78; Mass Inst Technol, MS, 82, PhD(computer sci), 84. *Prof Exp:* Mem tech staff, AT&T Bell Labs, 84-88; asst prof, 88-90, ASSOC PROF COMPUTER SCI, STANFORD UNIV, 90- *Mem:* Asn Comput Mach. *Res:* Theory and practice of programming language design; applications of mathematical logic to computer science. *Mailing Add:* Dept Computer Sci Stanford Univ Stanford CA 94305

MITCHELL, JOHN DOUGLAS, b Bennington, Vt, Dec 16, 44; m 66; c 2. INSTRUMENTATION, ELECTROMAGNETICS. *Educ:* Pa State Univ, BS, 66, MS, 68, PhD(elec eng), 73. *Prof Exp:* From asst prof to assoc prof elec eng, Univ Tex, El Paso, 73-80; ASSOC PROF ELEC ENG, PA STATE UNIV, 80- *Mem:* Inst Elec & Electronics Engrs; Am Geophys Union; AAAS. *Res:* Electrical properties of the middle atmosphere; designing probes for measuring atmospheric electrical parameters, planning and conducting field experiments to obtain data, and analyzing and interpreting the measurement results. *Mailing Add:* Dept Elec Eng Pa State Univ University Park PA 16802

MITCHELL, JOHN EDWARDS, b San Francisco, Calif, Mar 27, 17; m 42; c 3. PLANT PATHOLOGY. *Educ:* Univ Minn, BS, 39; Univ Wis, PhD(biochem), 48. *Prof Exp:* Asst plant path, Univ Minn, 39-40 & 41-42; asst plant path, Univ Wis, 46-48; plant pathologist, Camp Detrick, Md, 48-51, chief biol br, C Div, 51-56; assoc prof plant path, 56-63, PROF PLANT PATH, UNIV WIS-MADISON, 63-, CHMN DEPT, 75- *Mem:* Am Phytopath Soc; Am Soc Plant Physiol; Sigma Xi. *Res:* Ecology and control of soil borne plant pathogens; resistance of roots to disease; soil environment and plant disease. *Mailing Add:* Rocky Mountain Forest Range Exp Sta 240 W Prospect St Ft Collins CO 80526

MITCHELL, JOHN JACOB, b Schenectady, NY, Mar 4, 17; m 42; c 3. PHYSICAL CHEMISTRY. *Educ:* Johns Hopkins Univ, PhD(chem), 41. *Prof Exp:* Asst chem, Johns Hopkins Univ, 40-41; phys chemist, Tex Co, 41-53, staff chemist, Texaco Inc, 53-57, res assoc, 57-60, sr res assoc, 60-82; RETIRED. *Concurrent Pos:* Supvr lab, Manhattan Eng Dist, Fercleve Corp, Oak Ridge, Tenn, 44-45. *Mem:* AAAS; Am Chem Soc; Am Phys Soc; Health Physics Soc. *Res:* Mass spectroscopy; use of radioactive tracers; radiation chemistry; use of isotope tracers in studying reaction mechanisms; research planning. *Mailing Add:* 226 Main St Fishkill NY 12524

MITCHELL, JOHN LAURIN AMOS, b Lincoln, Nebr, July 18, 44; m 68; c 2. CELL PHYSIOLOGY, ENZYME REGULATION. *Educ:* Oberlin Col, BA, 66; Princeton Univ, PhD(biol), 70. *Prof Exp:* Fel cancer res, McArdle Labs Cancer Res, Univ Wis-Madison, 70-73; from asst prof to assoc prof, 73-84, PROF BIOL & CHEM, NORTHERN ILL UNIV, 84- *Mem:* Am Soc Cell Biol; Am Soc Biol Chemists; Sigma Xi. *Res:* Regulation of polyamine synthesis in mammalian tissues; post-translational enzyme modifications; biochemistry of cell cycle; rapid protein turnover. *Mailing Add:* Dept Biol Sci Northern Ill Univ De Kalb IL 60115

MITCHELL, JOHN MURRAY, JR, meteorology; deceased, see previous edition for last biography

MITCHELL, JOHN PETER, b Toronto, Ont, June 28, 32; m 56; c 2. PHYSICS. *Educ:* Univ Toronto, BA, 55, MA, 57, PhD(physics), 60. *Prof Exp:* Engr, Bell Tel Co, Can, 55-56; teacher math & physics, York Mem Col Inst, Toronto, 60-63; mem tech staff, Bell Labs, Inc, 63-89; RETIRED. *Mem:* Am Phys Soc; Inst Elec & Electronics Engrs. *Res:* Surface effects of radiation on semiconductor devices; resistivity of dilute alloys of lead at low temperatures; thin insulating films for cryogenic use; materials used for printed circuits. *Mailing Add:* 17 Warwick Rd Summit NJ 07901

MITCHELL, JOHN RICHARD, microbiology, public health, for more information see previous edition

MITCHELL, JOHN TAYLOR, b Buffalo, NY, Aug 16, 31; m 62; c 3. DEVELOPMENTAL BIOLOGY, EXPERIMENTAL EMBRYOLOGY. *Educ:* Amherst Col, BA, 53; NY Univ, MS, 64, PhD(biol), 66. *Prof Exp:* Asst cancer res scientist, Roswell Park Mem Inst, 56-60; res asst biol, NY Univ, 60-62, NIH fel, dept biol, 62-66; NIH staff fel, Nat Cancer Inst, Md, 66-68; asst prof anat, State Univ NY Upstate Med Ctr, 68-75; PROF BIOL, COLGATE UNIV, 75- *Concurrent Pos:* State Univ NY Res Found grant, 68-70. *Mem:* AAAS; Soc Develop Biol; Am Asn Anat. *Res:* Developmental biology, especially cytogenetic, cytological and in vitro approaches; effect of chemical compounds on early developing mammalian, avian and reptilian embryos; teratology. *Mailing Add:* Dept of Biol Colgate Univ Hamilton NY 13346

MITCHELL, JOHN WESLEY, b Christchurch, NZ, Dec 3, 13; m 76. PHYSICS. *Educ:* Univ NZ, BSc, 33, MSc, 34; Oxford Univ, DrPhil, 38, DSc, 60. *Prof Exp:* Res physicist, Brit Ministry Supply, 39-45; reader exp physics, Bristol Univ, 45-59; prof, 59-65, William Barton Rogers prof, 65-79, EMER PROF PHYSICS & SR RES FEL, UNIV VA, 79- *Honors & Awards:* Boys Prize, Brit Inst Physics & Phys Soc, 55; Renwick Mem Medal, Royal Photog Soc, 56; Cult Prize, German Soc Photog, 81; Lieven Gevaert Medal, Soc Photog Scientists and Engrs, 83. *Mem:* Hon mem Soc Photog Sci & Eng; fel Am Phys Soc; fel Royal Soc; fel Royal Inst Chem; fel Royal Photog Soc; hon mem Soc Photog Sci & Technol Japan. *Res:* Physics of crystals; surface properties of crystalline materials; dislocations and plastic deformation of ionic crystals and metals; theory of photographic sensitivity. *Mailing Add:* Dept of Physics Univ of Va Charlottesville VA 22901

MITCHELL, JOHN WRIGHT, b Palo Alto, Calif, Mar 17, 35; m 55; c 5. MECHANICAL ENGINEERING. *Educ:* Stanford Univ, BS, 56, MS, 57, Engr, 59, PhD(mech eng), 63. *Prof Exp:* From asst prof to assoc prof, 62-71, chmn dept, 83-87, PROF MECH ENG, UNIV WIS-MADISON, 71- *Concurrent Pos:* Consult, Dept Energy, Solar Energy Res Inst, Oscar Mayer, New Zealand; vis assoc prof, Yale Univ, 70; vis scientist, CSIRO, Australia, 81, Ecole des Mines, Paris, 87, Univ de Liege, Belgium, 88. *Mem:* Am Soc Mech Engrs; Am Soc Heating Ventilation Air Conditioning Engrs; Int Solar Energy Soc. *Res:* Heat transfer and natural convection; solar cooling systems; desiccant air conditioning systems; building energy systems and control. *Mailing Add:* Dept Mech Eng 1513 University Ave Madison WI 53706

MITCHELL, JOSEPH CHRISTOPHER, b Albany, Ga, Oct 8, 22; m 45; c 3. PARASITOLOGY & ENTOMOLOGY. *Educ:* Ft Valley State Col, BS, 43; Atlanta Univ, MS, 49; Princeton Univ, cert, 59; Univ Mich, cert, 60; Duke Univ, cert, 60. *Hon Degrees:* ScD, London Inst Appl Res, 72. *Prof Exp:* Chmn dept sci high schs, Ga, 43-45, 47 & 48-49 & Ala, 46-49; scholar, Atlanta Col Mortuary Sci, 45-46, instr physiol, 47-48; asst prof biol, Ft Valley State Col, 49-52 & Albany State Col, 54-63; chmn dept phys sci & zool, Ala State Col, 52-54, assoc prof biol & chmn div natural sci & math, Mobile Ctr, 63-66, interim pres, 81; DIR MORTUARY SCI & CHMN DIV SCI & MATH, BISHOP STATE COMMUNITY COL, 78-; interim pres, 81, ASSOC PROF BIOL & CHMN DIV NATURAL SCI & MATH, S D BISHOP STATE COL, 70-, DIR MORTUARY SCI, 78- *Concurrent Pos:* Carnegie res grant-in-aid & sci fac grant, 50-52; res technician, Cornell Univ, 70; Elem & Sec Educ Act title III field reader, US Off Educ; consult sec sci pub schs, Southwest Ga; proj dir sci improv, NSF, NSF fel, Cornell Univ, 61-62, Ford Found fel, 69-70, res assoc Entom & Med Entom, 70; sci consult, Vis Comt Southern Asn Cols & Schs, 80, 82 & 83. *Mem:* AAAS; Am Soc Microbiol; Nat Sci Teachers Asn; Nat Asn Biol Teachers; Nat Inst Sci; Am Educ Asn; Nat Educ Asn. *Res:* Mammalian physiology; medical entomology and parasitology; anatomy and physiology. *Mailing Add:* S D Bishop State Community Col 351 N Broad St Mobile AL 36690

MITCHELL, JOSEPH SHANNON BAIRD, b Pittsburgh, Pa, July, 24, 59; m 87; c 2. COMPUTATIONAL GEOMETRY, ANALYSIS OF ALGORITHMS. *Educ:* Carnegie-Mellon Univ, BS & MS, 81; Stanford Univ, PhD(opers res), 86. *Prof Exp:* Mem tech staff, Hughes Res Lab, 81-86; ASST PROF OPERS RES, CORNELL UNIV, 86-; ASSOC PROF APPL MATH, STATE UNIV NY, STONY BROOK, 91- *Mem:* Asn Comput Mach; Opers Res Soc Am; Inst Elec & Electronic Engrs. *Res:* Computational geometry applied to robotics, computer vision, route planning, and optimization. *Mailing Add:* Appl Math Dept State Univ NY Stony Brook NY 11794-3600

MITCHELL, JOSEPHINE MARGARET, b Edmonton, Alta; nat US; m 53. MATHEMATICS. *Educ:* Univ Alta, BS, 34; Bryn Mawr Col, MA, 41, PhD(math), 42. *Prof Exp:* Teacher pub sch, Alta, 35-38; instr math, Hollins Col, 42-44; instr, Conn Col, 44-45; assoc prof, Winthrop Col, 45-46; assoc prof, Tex State Col Women, 46-47; asst prof, Okla State Univ, 47-48; asst prof, Univ Ill, 48-54; with adv electronics lab, Gen Elec Co, 55-56; with res lab, Westinghouse Elec Corp, 56-57; assoc prof math, Univ Pittsburgh, 57-58; assoc prof, Pa State Univ, 58-61, prof, 61-69; prof, 69-82, EMER PROF MATH, STATE UNIV NY BUFFALO, 82- *Concurrent Pos:* Grantee, Am Philos Soc, 48; NSF grant, 52-53; sr fel, 64-65; mem, Inst Adv Study & Am Asn Univ Women fel, 54-55; Air Force contract, 60-63; mem, US Army Math Res Ctr, Univ Wis, 64-65; Math Asn Am Lectr, 65-72; NSF contract, 68-71; sabbatical leave, Univ Mich & Calif, Berkeley, 74 & vis fel, Australian Nat Univ, 81. *Mem:* AAAS; Am Math Soc; Math Asn Am. *Res:* Multiple Fourier and orthogonal series; bounded symmetric domains in space of several complex variables; solutions of partial differential equations. *Mailing Add:* 1701 W River Rd Grand Island NY 14072

MITCHELL, KENNETH FRANK, b Hornchurch, Eng, Apr 18, 40; m 61; c 2. IMMUNOBIOLOGY. *Educ:* Heriot Watt Univ, BSc, 69; Univ Pa, PhD(immunol), 75. *Prof Exp:* Technician biochem, Animal Health Trust, Eng, 56-59; May & Baker, Eng, 59-62 & Brit Vitamin Prod Ltd, 62-66; instr, Univ Pa, 69-72, res specialist immunol, 72-75, fel immunol, 75-77, asst prof path, 77-81; asst prof, Wistar Inst, 78-82; STAFF SCIENTIST, E I DU PONT DE NEMOURS & CO, INC,, 82- *Mem:* Fel Royal Soc Arts; Brit Inst Biol. *Res:* Structure of human, normal & tumor cell antigens; structure of guinea pig normal and tumor antigens; immunobiology and immunogenetics of the laboratory rat; immunochemistry of receptors on T and B cells. *Mailing Add:* Box 4049 Elwyn PA 19063-0588

MITCHELL, KENNETH JOHN, b Bralorne, BC, Sept 8, 38; m 68; c 2. FOREST GROWTH & YIELD. *Educ:* Univ BC, BSF, 61; Yale Univ, MF, 64, PhD(forest mensuration), 67. *Prof Exp:* Res officer forest mensuration, Can Forestry Serv, Govt of Can, 61-67, res scientist, 67-70; asst prof forest mgt, Yale Univ, 70-75; vis prof forestry, Univ BC, 75-76; assoc prof forest resources, Univ Idaho, 76-80; tech adv biomet, Res Br, 80-87, LEADER, STAND MODELLING RES, BC MINISTRY OF FORESTS, 88- *Mem:* Soc Am Foresters; Can Inst Forestry; Sigma Xi. *Res:* Dynamics and simulated yield of commercial coniferous species in British Columbia. *Mailing Add:* Res Br BC Ministry of Forests 1450 Government St Victoria BC V8W 3E7

MITCHELL, LAWRENCE GUSTAVE, b West Chester, Pa, Oct 15, 42; m 87; c 2. FISH PARASITOLOGY & PATHOLOGY, INVERTEBRATE ZOOLOGY. *Educ:* Pa State Univ, BS, 64; Univ Mont, PhD(zool), 70. *Prof Exp:* NIH fel, Stella Duncan Mem Res Inst, Univ Mont, 70, asst prof zool, 70-71; asst prof zool & biol, Iowa State Univ, 71-78, assoc prof zool, 78-88, prof zool & animal ecol, 88-91; AFFIL PROF BIOL SCI, UNIV MONT, 89-

Concurrent Pos: Instrnl sci equip prog grant, NSF, 74; coinvestr, NSF Comprehensive Assistance to Undergrad Sci Educ Prog grant, 76-79; res contracts, Fish & Wildlife Serv, US Dept Interior, 84- *Mem:* Am Inst Biol Sci; AAAS; Am Soc Zoologists; Am Fisheries Soc; Sigma Xi. *Res:* Host-parasite relationships of freshwater fish parasites; systematics, ultrastructure, and transmission ecology of myxozoa; ecology of freshwater mussels and host-parasite relationships of mussel glochidia; author zoology & biology textbooks. *Mailing Add:* Div Biol Sci Univ Mont Missoula MT 59812

MITCHELL, MADELEINE ENID, b Jamaica, WI, Dec 14, 41; US citizen. NUTRITION. *Educ:* McGill Univ, BSc, 63; Cornell Univ, MS, 65, PhD(nutrit), 68. *Prof Exp:* ASSOC PROF NUTRIT, WASH STATE UNIV, 69- *Concurrent Pos:* Nutrit scientist, USDA-CSRS, 80-81; asst dir, Agr Exp Sta, Wash State Univ, 84-86. *Mem:* Am Dietetic Asn; Nutrit Today Soc. *Res:* Protein and amino acid nutrition; carnitine; nutritional status of human subjects; zinc and alcohol in pregnancy. *Mailing Add:* NE 1010 Alfred Lane Pullman WA 99163

MITCHELL, MALCOLM STUART, b New York, NY, May 6, 37; m 59, 76; c 4. ONCOLOGY, TUMOR IMMUNOLOGY. *Educ:* Harvard Univ, AB, 57; Yale Univ, MD, 62. *Prof Exp:* From instr to assoc prof med & pharmacol, Sch Med, Yale Univ, 68-78; dir clin invests, 78-83, chief med oncol, Univ Southern Calif Cancer Ctr, 78-84, PROF MED & MICROBIOL, SCH MED, UNIV SOUTHERN CALIF, 78- *Concurrent Pos:* Scholar, Leukemia Soc, 68-73; res career develop, Nat Cancer Inst, 74-79; ed, Int Encyclopedia Pharmacol & Therapeut, 75-85, Yale J Biol & Med, 76-78 & Directions in Oncol, 84-86; assoc ed, J Immunol 75-79, J Immunopharmacol, 78-, Cancer Res, 80-, Cancer Invest, 81-, J Biol Resp Modif, 81-, J Clin Oncol, 82-86, Cancer Immunol Immunotherapy, 83-, Hubridoma, 87-, J Nat Cancer Inst, 88-, Am J Clin Oncol, 90-, Vaccine Res, 90-; chmn, res adv comt, Nat Cancer Cytol Ctr, 81-86; chmn, NIH conf on Hybridomas in Cancer Treatment, 82 & Immunity to Cancer, 84 & 87. *Mem:* Am Asn Immunologists; Am Soc Clin Invest; Am Asn Cancer Res; Am Soc Clin Oncol; Am Fedn Clin Res; Soc Biol Ther; Clin Immunol Soc. *Res:* Development of active specific immunotherapy for melanoma; clinical immunotherapy; mechanisms of action of biomodulators in man. *Mailing Add:* Univ Southern Calif Cancer Ctr 2025 Zonal Ave Los Angeles CA 90033

MITCHELL, MAURICE MCCLELLAN, JR, b Lansdowne, Pa, Nov 27, 29; m 52. PHYSICAL CHEMISTRY. *Educ:* Carnegie Inst Technol, BS, 51, MS, 57, PhD(phys chem), 60. *Prof Exp:* Group leader process chem group, Coal Coke & Coal Chem Div, Appl Res Lab, US Steel Corp, 51-56; asst, Carnegie Inst Technol, 56-58; sr technologist & group leader phys chem group, Coal Chem Div, Appl Res Lab, US Steel Corp, 60-61, actg head process chem sect, 61; supvr phys chem br, Res Div, Melpar, Inc, 61-64; group leader, Res & Develop Dept, Atlantic-Richfield Co, 64-73; dir res & develop, Houdry Div, 73-81, dir res planning & eng & sr res scientist, 81-84, dir res & develop dept, 84-86, VPRES, RES & DEVELOP, ASHLAND OIL INC, 86- *Concurrent Pos:* Instr eve sch, Carnegie Inst Technol, 60-61; consult, Blaw-Knox Co, 52-53 & Melpar, Inc, 64-66; ed newsletter, Catalysis Soc, 70-77. *Mem:* AAAS; Am Chem Soc; Sigma Xi; Catalysis Soc NAm (vpres, 81-85, pres, 85-89); fel Am Inst Chemists; Am Soc Testing & Mat; Am Inst Chem Engrs; Asn Res Dir; Indust Res Inst. *Res:* Chemical kinetics; heterogeneous and homogeneous catalysis; catalytic processes; preparation of catalysts; transition metal chemistry; coal chemistry. *Mailing Add:* Ashland Oil PO Box 391 Ashland KY 41114

MITCHELL, MICHAEL A, b Austin, Tex, Dec 23, 41; m 65. PHYSICS. *Educ:* Univ Calif, Riverside, BA, 64; Univ Conn, MS, 66, PhD(physics), 70. *Prof Exp:* Res physicist, US Naval Ord Lab, 70-74, RES PHYSICIST, NAVAL SURFACE WEAPONS CTR, WHITE OAK LAB, 74- *Concurrent Pos:* Air Force Off Sci Res-Nat Res Coun fel, 70-72. *Mem:* AAAS; Am Phys Soc. *Res:* Transport and magnetic properties of metal alloys; transport properties and structure of metallic materials; amorphous metals. *Mailing Add:* 2104 Essex Rd Minnetonka MN 55343

MITCHELL, MICHAEL ERNST, b Montclair, NJ, Apr 11, 43; m 67; c 4. UROLOGY, PEDIATRIC UROLOGY. *Educ:* Princeton Univ, BA, 65; Harvard Univ, MD, 69. *Prof Exp:* Clin asst prof urol, Mass Gen Hosp, Harvard, 77-78; PROF UROL PEDIAT UROL RILEY CHILDREN'S HOSP IND UNIV SCH MED, 78- *Concurrent Pos:* clin dir res lab, dept urol, Ind Univ Sch Med, 87- *Mem:* Am Urol Asn; fel Am Col Sci; Am Asn Genito-Urinary Surgeons; Soc Pediat Urol Surgeons; fel Am Acad Pediat; Am Med Asn. *Res:* The physiologic and metabolic efforts of stomach in the lower urinary tract; reconstructive pediatric urology. *Mailing Add:* 3120 E Laurelhurst Dr NE Seattle WA 98105-5333

MITCHELL, MICHAEL ROGER, b Detroit, Mich, Jan 24, 41; wid; c 2. MECHANICAL METALLURGY, FATIGUE. *Educ:* Lawrence Inst Technol, BS, 63; Wayne State Univ, MS, 69; Univ Ill, Urbana, PhD(theoret & appl mech), 76. *Prof Exp:* Design engr, Eng & Foundry Div, Ford Motor Co, Mich, 63-65, staff mem sci res, Metall Dept, 65-71; res & teaching asst, Univ Ill, Urbana, 71-76, vis asst prof metall eng, 76-77; mem tech staff, 77-79, mgr phys metall, 79-83, MEM TECH STAFF, ROCKWELL INT SCI CTR, 83- *Concurrent Pos:* Consult, Deere & Co, Ill & Structural Dynamics Res Corp, Ohio, 76-77; instr, statics & strength mats, Moorpark Col, Calif, 88- *Honors & Awards:* Colwell Award, Soc Automotive Engrs, 75; Vanadium Award, Brit Inst Metals, 84. *Mem:* Am Soc Testing & Mat; Sigma Xi; Soc Auto Engrs. *Res:* Cyclic deformation and fracture behavior of materials; casting technology; laser surface alloying, cumulative fatigue damage analysis; hydrogen-assisted fatigue crack initiation and propagation; threshold stress intensity; statistics of fatigue; environmental effects; use of ultrasonics in material processing. *Mailing Add:* Rockwell Int Sci Ctr 1049 Camino Dos Rios Thousand Oaks CA 91360

MITCHELL, MYRON JAMES, b Denver, Colo, Apr 11, 47; m 73; c 1. ECOLOGY. *Educ:* Lake Forest Col, BA, 69; Univ Calgary, PhD(soil ecol), 74. *Prof Exp:* Res asst soil ecol, Cornell Univ, 69-70 & Univ Calgary, 70- 74; Nat Res Coun Can fel math modelling, Univ BC, 74-75; asst prof invert zool & ecol energetics, 75-80, PROF ECOL, STATE UNIV NY COL ENVIRON SCI & FORESTRY, 80- *Mem:* Ecol Soc Am; AAAS; Brit Ecol Soc; Sigma Xi. *Res:* Study of floral-faunal interactions and how these interactions affect decomposition, mineral cycling, and energy transformation in terrestrial and aquatic ecosystems; sulfur dynamics. *Mailing Add:* Fac Environ & Forest Biol State Univ of NY Col of Environ Sci & Forestry Syracuse NY 13210

MITCHELL, NORMAN L, b Jamaica, WI, Nov 21, 28; m 68. PLANT PATHOLOGY. *Educ:* Univ London, BSc, 62; Univ Western Ont, PhD(bot), 67. *Prof Exp:* Instr biol, WI Col, Jamaica, 62-64; asst prof, 67-69, assoc prof, 69-81, PROF BIOL, LOMA LINDA UNIV, 81- *Mem:* AAAS; Bot Soc Am; Am Soc Microbiol. *Res:* Electron microscopic studies on powdery mildews mainly Sphaerotheca macularis of strawberry. *Mailing Add:* Dept of Biol Loma Linda Univ Riverside CA 92515

MITCHELL, OLGA MARY MRACEK, b Montreal, Can, Aug 5, 33; m 56; c 2. TELECOMMUNICATIONS. *Educ:* Univ Toronto, BA, 55, MA, 58, PhD(physics), 62; Pace Univ, MS, 82. *Prof Exp:* Res asst textile & metal physics, Ont Res Found, 55-57, res assoc metal physics, 62-63; mem tech staff & supvr, Bell Labs, 63-77; asst eng mgr, 77-80, dist mgr, Am Tel & Tel Co, 80-83, DIST MGR, BELLCORE, 84- *Concurrent Pos:* Ed Info Serv Data Network Vol, Inst Elec & Electronic Engrs, frontiers in Commun. *Mem:* Am Phys Soc; Acoust Soc Am; Inst Elec & Electronic Engrs; Soc Women Engrs. *Res:* Integrated services digital networks, planning and applications; acoustics and signal processing; interaction of ultrasonic waves in crystals with dislocations, phonons, electrons; nuclear photodisintegration; digital transmission systems including optical fiber systems. *Mailing Add:* Bellcore Rm 1E204 290 W Mt Pleasant Ave Livingston NJ 07039

MITCHELL, ORMOND GLENN, b Long Beach, Calif, Sept 17, 27. HUMAN ANATOMY, HISTOLOGY. *Educ:* San Diego State Univ, AB, 49; Univ Southern Calif, MS, 53, PhD(zool), 57. *Prof Exp:* Instr zool, Univ Southern Calif, 57; asst prof biol, Calif State Polytech Col, 57-62; res assoc, Univ Southern Calif, 62-63; from asst prof to assoc prof anat, Col Dent, NY Univ, 63-69; prof biol, Adelphi Univ, 69-71; assoc prof, 71-75, PROF ANAT, COL DENT, NY UNIV, 75- & CHMN DEPT, 71- *Mem:* AAAS; Am Soc Mammal; Int Asn Dent Res; NY Acad Sci; Harvey Soc. *Res:* Growth and development of skin and hair, glands and control of secretion. *Mailing Add:* Dept of Anat NY Univ Col Dent 345 E 24th St New York NY 10010

MITCHELL, OWEN ROBERT, b Beaumont, Tex, July 4, 45; m 68; c 3. IMAGE PROCESSING, COMPUTER VISION. *Educ:* Lamar Univ, BSEE, 67; Mass Inst Technol, SMEE, 68, PhD(elec eng), 72. *Prof Exp:* From asst prof to assoc prof, 72-81, PROF ELEC ENG, PURDUE UNIV, 81-, ASST DEAN ENG, 84- *Concurrent Pos:* Expert engr, White Sands Missle Range, 77 & 79; prin investr, Army Res Off, 81-85. *Mem:* Inst Elec & Electronics Engrs. *Res:* Digital image processing for bandwidth compression and information extraction; shape and texture analysis for computer vision systems used in inspection, cartography, and guidance. *Mailing Add:* Three Castellan Dr Lafayette IN 47905

MITCHELL, PETER, b Surrey, Eng, Sept 29, 20; m 58; c 4. OSMOCHEMISTRY, HUMAN COMMUNICATION. *Educ:* Jesus Col, Cambridge, BA, PhD. *Hon Degrees:* Numerous from US and foreign univs, 77-90. *Prof Exp:* Biochemist, Univ Cambridge, 43-55, demonstr, 50-55; dir, Chem Biol Unit, Univ Edinburgh, 55-63, sr lectr, 61-62, reader, 62-63; dir res, 64-86, CHMN & HON DIR, GLYNN RES FOUND LTD, 87- *Concurrent Pos:* Mem, Econ Res Coun, 75-, comt Royal Inst, Davy Farady Lab, 82-85; hon ed adv, Biosci Reports, 85-; vis prof, Dept Biochem, Univ London, 87-89. *Honors & Awards:* Nobel Prize in Chem, 78; CIBA Medal & Prize, Biochem Soc, 73; Louis & Bert Freedman Found Award, NY Acad Sci, 74; Wilhelm Feldberg Found Prize, Feldberg Found Anglo/Ger Sci Exchange, 76; Sir Hans Krebs lectr & Medal, Fedn Europ Biochem Soc, 78; Humphry Davy Mem lectr, Royal Inst Chem, 80; James Rennie Bequest lectr, Univ Edinburgh, 80; Copley Medal, Coun Royal Soc, 81; Croonian lectr, Royal Soc, 87. *Mem:* Foreign assoc Nat Acad Sci; NY Acad Sci; Soc Gen Microbiol; Japanese Biochem Soc. *Res:* The chemiosmotic theory of transport and metabolism; the mechanisms by which chemical action can be organized in space so as to drive the spatially-directed movement of chemical substances and physical structures. *Mailing Add:* Glynn Res Found Ltd Bodmin Cornwall PL30 4AU England

MITCHELL, RALPH, b Dublin, Ireland, Nov 26, 34; m 57; c 3. MICROBIOLOGY. *Educ:* Trinity Col, Dublin, BA, 56; Cornell Univ, MS, 59, PhD, 61. *Hon Degrees:* AM, Harvard Univ, 70. *Prof Exp:* Res assoc, Cornell Univ, 61-62; sr scientist, Weizmann Inst, 62-65; from asst prof to assoc prof appl microbiol, 65-70, GORDON MCKAY PROF APPL BIOL, HARVARD UNIV, 70- *Mem:* Am Soc Microbiol; Brit Soc Gen Microbiol. *Res:* Microbial ecology; applied microbiology; microbial predator-prey systems; chemoreception in microorganisms; marine fouling, biological control of disease; microbial corrosion; water pollution; energy production. *Mailing Add:* Div of Appl Sci Harvard Univ Cambridge MA 02138

MITCHELL, REGINALD EUGENE, b Houston, Tex, May 16, 47; m; c 1. COMBUSTION RESEARCH, COMPUTATIONAL FLUID DYNAMICS. *Educ:* Univ Denver, BSChe, 68; NJ Inst Technol, MSChe, 70; Mass Inst Technol, ScD, 75. *Prof Exp:* mem tech staff combustion sci, Sandia Nat Labs, Calif, 75-90; ASSOC PROF MECH ENG, STANFORD UNIV, CALIF, 91- *Concurrent Pos:* Distinguished mem tech staff, Sandia Nat Labs, 89; Am Assoc Western Univs distinguished lectr, Dept Energy Lab, 90-91. *Mem:* Combustion Inst; Nat Orgn Black Chemist & Chem Engrs; Am Inst Chem Engrs; Sigma Xi. *Res:* Combustion science; physical and chemical processes governing the combustion of hydrocarbons and coals; characterizing the effects which influence the formation of atmospheric pollutants during the combustion process. *Mailing Add:* 107 Sunnyside Ave Piedmont CA 94611

MITCHELL, REGINALD HARRY, b Woking, Eng, Sept 1, 43; Can citizen; m 65, 87; c 2. ORGANIC CHEMISTRY, ENVIRONMENTAL CHEMISTRY. *Educ:* Univ Cambridge, BA, 65, PhD(chem), 68, MA, 69. *Prof Exp:* Res fel org chem, Fitzwilliam Col, Univ Cambridge, 67-68; res assoc, Univ Ore, 68-70; sr scientist, Formica Res Div, Eng, 70-71, mgr res & develop, 71-72; from asst to assoc prof, 72-82, PROF ORG CHEM, UNIV VICTORIA, BC, 82- *Concurrent Pos:* Fel, Univ Ore, 68-70; course tutor technol, Open Univ, Eng, 72; actg chmn dept chem, Univ Victoria, 81-82. *Mem:* Am Chem Soc; Royal Soc Chem; Chem Inst Can. *Res:* Synthesis of novel aromatic hydrocarbons; new synthetic reactions; hydrocarbons and pesticides in the environment; synthesis and transformations. *Mailing Add:* Dept Chem Univ Victoria Box 3055 Victoria BC V8W 3P6 Can

MITCHELL, RICHARD LEE, b Cleveland, Ohio, Sept 23, 38; m 63; c 3. ELECTRICAL ENGINEERING. *Educ:* Purdue Univ, BS, 60, MS, 61, PhD(elec eng), 64. *Prof Exp:* Sect head elec eng, Aerospace Corp, 64-68; sr scientist, Technol Serv Corp, 68-74; VPRES, MARK RESOURCES INC, 74- *Mem:* Inst Elec & Electronics Engrs. *Res:* Radar system analysis and simulation; signal processing. *Mailing Add:* Mark Resources Inc 2665 30th St Suite 200 Santa Monica CA 90405

MITCHELL, RICHARD SCOTT, b Longmont, Colo, Jan 28, 29. MINERALOGY, CRYSTALLOGRAPHY. *Educ:* Univ Mich, BS, 50, MS, 51, PhD(mineral), 56. *Prof Exp:* From asst prof to prof geol, 53-59, chmn dept, 64-69, PROF ENVIRON SCI, UNIV VA, 69- *Concurrent Pos:* Exec ed, Rocks & Minerals mag, 76- *Mem:* AAAS; fel Mineral Soc Am; fel Geol Soc Am; Sigma Xi; Mineral Asn Can; Am Crystallog Asn. *Res:* Morphology and crystal structures of inorganic compounds; structural polytypism; metamict state; mineralogy of coal ash; mineralogical compositions of archaeological ceramics; mineral and rock nomenclature. *Mailing Add:* Dept of Environ Sci Clark Hall Univ Va Charlottesville VA 22903

MITCHELL, RICHARD SHEPARD, b Indianapolis, Ind, Mar 30, 38; c 2. SCIENTIFIC BOTANICAL EDITING. *Educ:* Fla State Univ, BS, 60, MS, 62; Univ Calif, Berkeley, PhD(bot), 67. *Prof Exp:* Asst prof biol, Va Polytech Inst & State Univ, 67-75; STATE BOTANIST, NY STATE BIOL SURV, NY STATE MUS, 75- *Concurrent Pos:* Prin investr, NY State Flora Proj, 75-; founder & co-dir, NY Flora Asn, 90-91. *Honors & Awards:* Hort Award, Garden Club Am, 90. *Mem:* Am Soc Plant Taxon. *Res:* Flora of New York State; taxonomy of vascular plants; systematics and morphology of the Smartweed family (Polygonaceae). *Mailing Add:* Biol Surv 3132 CEC NY State Mus Albany NY 12230

MITCHELL, RICHARD SIBLEY, b Barnsdall, Okla, Sept 19, 32; m 59; c 4. ANALYTICAL CHEMISTRY. *Educ:* Austin Col, AB, 58; Univ Okla, MS, 63, PhD(chem), 64. *Prof Exp:* Res chemist, Lion Oil Co Div, Monsanto Chem Co, 57 & 59-61; from asst prof to assoc prof chem, 64-71, PROF CHEM, ARK STATE UNIV, 71- *Mem:* Am Chem Soc; Sigma Xi; Col Sci Teachers Asn. *Res:* Gas chromatography and electroanalytical methods. *Mailing Add:* Box 700 Dept Chem Ark State Univ State University AR 72467

MITCHELL, RICHARD WARREN, b Lynchburg, Va, Aug 15, 23; m 45 & 62; c 3. PHYSICS. *Educ:* Lynchburg Col, BS, 44; Agr & Mech Col, Tex, MS, 53, PhD(physics), 60. *Prof Exp:* Instr math & physics, Lynchburg Col, 45-47; from instr to asst prof physics, Agr & Mech Col, Tex, 47-62, res assoc, Res Found, 54-60; assoc prof physics, Univ SFla, 62-86; RETIRED. *Mem:* Am Asn Physics Teachers. *Res:* Raman and infrared spectroscopy; nuclear magnetic resonance studies relating relaxation times of protons to physical properties of solutions. *Mailing Add:* 2502 Victarra Circle Tampa FL 33549

MITCHELL, ROBERT A, b Oakland, Calif, Sept 11, 22; m 51; c 4. PHYSIOLOGY. *Educ:* Univ Calif, Berkeley, BS, 47; Creighton Univ, MS, 52, MD, 53. *Prof Exp:* Intern Hosp, Univ Nebr, 53-54, resident internal med, 54-55; resident, Hosp, 55-57, asst res physician, Univ, 59-61, from instr to assoc prof med, 60-74, PROF PHYSIOL & MED, UNIV CALIF, SAN FRANCISCO, 74-, ASSOC STAFF MEM, UNIV CALIF HOSP, 60- *Concurrent Pos:* Bank of Am Giannini fel pulmonary physiol, Cardiovasc Res Inst, Univ Calif, San Francisco, 57-59; USPHS res career develop award, 63- *Mem:* Am Physiol Soc. *Res:* Regulation of respiration; chemoreceptors. *Mailing Add:* Dept Physiol HSE 1386 Med Ctr Univ Calif San Francisco CA 94143

MITCHELL, ROBERT ALEXANDER, b Belfast, Northern Ireland, Apr 3, 35. BIOCHEMISTRY. *Educ:* Queen's Univ Belfast, BSc, 55, PhD(chem), 60. *Prof Exp:* Teacher chem, Col Technol, Belfast, 57-60; asst prof, 65-72, ASSOC PROF BIOCHEM, WAYNE STATE UNIV, 72- *Concurrent Pos:* Fel biochem, Okla State Univ, 60-62, Univ Minn, 63 & Univ Calif, Los Angeles, 63-65. *Mem:* Am Chem Soc; Biophys Soc; Am Soc Biol Chem. *Res:* Mitochondrial energy metabolism; enzyme kinetics; use of substrate analogs to study enzyme catalysis and regulation; metabolic alterations in Reye's Syndrome. *Mailing Add:* Dept Biochem Wayne State Univ Sch Med 540 E Canfield Detroit MI 48201-1908

MITCHELL, ROBERT BRUCE, b Rochester, Pa, Sept 24, 42; c 1. PHYSIOLOGY. *Educ:* Denison Univ, BS, 64; Ohio Univ, MS, 66; Pa State Univ, PhD(physiol), 69. *Prof Exp:* ASSOC PROF BIOL, PA STATE UNIV, 69- *Mem:* AAAS; Geront Soc. *Res:* Quantitative histochemistry; cytophotometric and interferometric analyses of nucleic acids and protein in aging cells biology of aging. *Mailing Add:* Dept Biol Pa State Univ University Park PA 16802

MITCHELL, ROBERT CURTIS, b Ft Dodge, Iowa, Mar 29, 28; m 49; c 3. ASTRONOMY. *Educ:* NMex State Univ, BS, 49, PhD(physics), 66; Univ Wash, MS, 52. *Prof Exp:* Physicist ultrasonics, Anderson Labs, West Hartford, Conn, 52-53; physicist atmosphere, Univ Conn, 53; solar observer, Harvard Col Observ, 53-54; teacher math & sci, Colo Rocky Mountain Sch, 54-56 & Gadsden High Sch, Anthony, NMex, 56-62; chmn dept geol & physics, 78-82, PROF PHYSICS, CENT WASH UNIV, 66- *Mem:* Am Asn Physics Teachers; Am Asn Variable Star Observers. *Res:* Wide field astrophotography for instructional use. *Mailing Add:* Dept of Physics Cent Wash Univ Ellensburg WA 98926

MITCHELL, ROBERT DALTON, b Bellingham, Wash, June 2, 23; m 45; c 3. MEDICINE. *Educ:* La Sierra Col, BS, 44; Col Med Evangelists, MD, 47, MSc, 61; Am Bd Internal Med, dipl & cert gastroenterol. *Prof Exp:* Asst clin prof, 56-64, from asst prof to assoc prof, 64-81, PROF MED, LOMA LINDA UNIV, 81- *Mem:* AMA; fel Am Col Physicians; Am Gastroenterol Asn. *Res:* Gastroenterology. *Mailing Add:* Loma Linda Univ Med Ctr Loma Linda CA 92350

MITCHELL, ROBERT L(YNNE), b Floresville, Tex, Oct, 25, 23; m 52; c 3. CHEMICAL ENGINEERING. *Educ:* Tex Col Arts & Industs, BS, 43; Mass Inst Technol, SM, 47. *Prof Exp:* Pilot plant engr, Celanese Corp Am, 47-49, group leader, 49-52, chief chem engr, 52-55, dir eng res, 55-57, dir tech & econ eval dept, 57-60; vpres planning, 60-64, vpres com develop & int, 64-66, vpres tech & mfg, 66-67, vpres & mgr, 67-69, exec vpres, 67-71, pres, 71-76, exec vpres, 76-80, VCHMN, CELANESE CORP, 80- *Mem:* Am Chem Soc; Am Inst Chem Eng. *Res:* Hydrocarbon oxidation; petrochemicals. *Mailing Add:* 35 Salem Rd Weston CT 06883-1720

MITCHELL, ROBERT W, biospeleology, ecology, for more information see previous edition

MITCHELL, RODGER (DAVID), b Wheaton, Ill, July 22, 26; m 89. POPULATION BIOLOGY. *Educ:* Univ Mich, PhD(zool), 54. *Prof Exp:* Instr zool, Univ Vt, 54-57; asst prof biol, Univ Fla, 57-61, assoc prof zool, 61-69; PROF ZOOL, OHIO STATE UNIV, 69- *Concurrent Pos:* NSF fac fel, Univ Calif, Berkeley, 59-60; Fulbright res fel, Ibaraki Univ, Japan, 65-66; Fulbright prof, Univ Agr Sci, Bangalore, India, 76-77. *Mem:* AAAS; Ecol Soc Am; Soc Study Evolution; Am Soc Naturalists. *Res:* Morphology and biology of water mites; population biology of bruchid beetles; analysis of agro-ecosystems. *Mailing Add:* Dept Zool Ohio State Univ Columbus OH 43210

MITCHELL, ROGER HAROLD, b Englewood, NJ, Nov 27, 46; m 71; c 2. EXPERIMENTAL PATHOLOGY, IMMUNOPATHOLOGY. *Educ:* Lafayette Col, AB, 68; Med Col Va, PhD(exp path), 76. *Prof Exp:* Res staff mem, Dept Microbiol & Immunol, Bowman Gray Sch Med, Wake Forest Univ, 76-78; res assoc, Tumor-Host Sect, Biomed Div, Samuel Roberts Noble Found, 78-84; DIR NEW BUS DEVELOP, AM MED SYSTS, 90- *Concurrent Pos:* Res grant, Dept Microbiol & Immunol, Bowman Gray Sch Med, Wake Forest Univ, 77-78. *Mem:* Reticuloendothelial Soc; Soc Exp Biol & Med; Sigma Xi. *Res:* Immunological and pathological mechanisms involved in host defenses against carcinogenesis; immunopathology of infectious diseases; leukocytic endogenous mediator; endogenous pyrogen. *Mailing Add:* Am Med Systs 11001 Bren Rd E Minnetonka MN 55343

MITCHELL, ROGER L, b Grinnell, Iowa, Sept 13, 32; m 55; c 4. AGRONOMY, CROP PHYSIOLOGY. *Educ:* Iowa State Univ, BS, 54, PhD(agron), 61; Cornell Univ, MS, 58. *Prof Exp:* From asst prof to prof agron, Iowa State Univ, 61-69; chmn dept agron, Univ Mo-Columbia, 69-72, dean exten div, 72-75; vpres agr, Kans State Univ, 75-80; chmn dept agron, 81-83, DEAN COL AGR & DIR AGR EXP STA, UNIV MO-COLUMBIA, 83- *Concurrent Pos:* Prof-in-chg farm oper curric, 62-66; Am Coun Educ acad admin intern, Univ Calif, Irvine, 66-67. *Mem:* Fel AAAS; fel Am Soc Agron (pres-elect, 78-79, pres, 79-80, past pres, 80-81); Crop Sci Soc Am (pres, 75-76). *Res:* Soybean physiology; rooting patterns under field conditions; sorghum physiology. *Mailing Add:* 2-69 Agr Univ Mo Columbia MO 65211

MITCHELL, ROGER W, b Ft Worth, Tex, Oct 20, 37; m 59; c 2. MATHEMATICS. *Educ:* Hendrix Col, AB, 59; Southern Methodist Univ, MS, 61; NMex State Univ, PhD(math), 64. *Prof Exp:* From asst prof to prof math, Univ Tex, Arlington, 64-81; CTA, QED COMMODITIES, 81- *Mem:* Am Math Soc; Math Asn Am; Soc Indust & Appl Math; Nat Coun Teachers Math. *Res:* Infinite Abelian groups; algebra. *Mailing Add:* Qed Commodities 120 E South Town Dr Tyler TX 75703

MITCHELL, ROY ERNEST, b Ft Worth, Tex, Aug 27, 36; m 58; c 1. INORGANIC CHEMISTRY, ENOLOGY. *Educ:* Tex A&M Univ, BS, 58; Purdue Univ, PhD(inorg chem), 64. *Prof Exp:* Fel, Tex A&M Univ, 64-65; asst prof chem, Purdue Univ, 65-66; asst prof, 66-79, ASSOC PROF CHEM, TEX TECH UNIV, 79- *Mem:* Am Chem Soc; Am Soc Enologists. *Res:* Vinification procedures for Texas grapes; buffer nature of wine; wine analysis. *Mailing Add:* Dept Chem Tex Tech Univ Lubbock TX 79409

MITCHELL, STEPHEN KEITH, b Houston, Tex, Aug 15, 42; m 75; c 2. UNDERWATER ACOUSTICS. *Educ:* Univ Tex, BS, 64, MA, 66, PhD(physics), 76. *Prof Exp:* RES SCIENTIST ASSOC, APPL RES LABS, UNIV TEX, 64- *Mem:* Acoust Soc Am; Sigma Xi. *Res:* Ocean acoustics, particularly measurement and analysis of ocean acoustic parameters and signal propagation modeling and measurements. *Mailing Add:* 4307 Wildridge Circle Austin TX 78759

MITCHELL, TERENCE EDWARD, b Haywards Heath, Eng, May 18, 37; m 59; c 2. PHYSICAL METALLURGY, CERAMICS. *Educ:* Univ Cambridge, BA, 58, MA & PhD(physics), 62. *Prof Exp:* asst prof matall, Case Inst Technol, 62-66; assoc prof, 66-75, PROF METALL, CASE WESTERN RESERVE UNIV, 75- *Concurrent Pos:* Res fel, Cavendish Lab, Cambridge, 62-63; mem, Ctr Student Mat, Case Inst Technol, 62-66, assoc dir, Mat Ctr, 69-74; dir, High Voltage Electron Micros Fac, 69-82, phys Sci, Micros Soc Am, 84-; vis scientist, NASA Ames Lab, Stanford Univ & Elec Power Res Inst, Palo Alto, Calif, 75-76; chemn, Agronne HVEM-Tandem Steering Comt, 80-82, Dept Matell & Mat Sci, Case Western Reserv Univ, 83-; scientist comt, Electron Micros Soc, France, 81-; prog chmn, Electron Micros Soc Am, 82; comt, Am Soc Testing & Mats; Metallog comt, Am Soc Metals; co-dir Mat Res Lab, Case Western Univ, 82-83. *Mem:* Metall Soc of Asn Inst Mech Engrs; Metals Soc; Am Ceramic Soc; Am Soc Testing & Mat; Electron Micros Soc Am. *Res:* Physical metallurgy and cermics; mechanical properties; electron microscopy; dislocation theory; radiation damage; phase transformation; oxidation. *Mailing Add:* Dept Metall Case Western Reserve Univ 2040 Adelbert Rd Cleveland OH 44106

MITCHELL, TERENCE EDWARD, b Sussex, Eng, May 18, 37; US citizen; m 59; c 2. ELECTRON MICROSCOPY, CERAMIC SCIENCE. *Educ:* Univ Cambridge, Eng, BA, 58, MA & PhD(physics), 62. *Prof Exp:* Res fel physics, Cavendish Lab, Univ Cambridge, 62-63; asst prof metall, Case Inst Tech, 63-66, assoc prof metall, Case Western Reserve Univ, 66-75, prof mat sci, 76-87, chmn, 83-86; STAFF MEM MAT SCI, LOS ALAMOS NAT LAB, 87-, FEL, 91- *Concurrent Pos:* Vis prof mat sci, Stanford Univ, 75-76; prog chair, Electron Micros Soc Am, 81-82, dir, 84-86; mem, Univ Mat Coun, 83-; ed, J Electron Micros Technol, 86-; adj prof mat sci, Case Western Reserve Univ, 87-; mem mat sci rev comt, Ames Lab, 87-90; assoc ed, J Am Ceramic Soc, 88-; mem sci adv comt, Sci & Technol Ctr Superconductivity, 89- *Mem:* Fel Am Soc Metals; fel Am Ceramic Soc; Minerals Metals & Mat Soc; Mat Res Soc; Electron Micros Soc Am; Am Phys Soc. *Res:* Relationship between structure and properties of metals, ceramics and semiconductors; electron microscopy; dislocations and mechanical properties; radiation damage; phase transformations; oxidation; interfaces; superconductivity. *Mailing Add:* Ctr Mat Sci MS K705 Los Alamos Nat Lab NM 87545

MITCHELL, THEODORE, b Chicago, Ill, Apr 30, 26; m 57; c 2. ABSRACT HARMONIC ANALYSIS. *Educ:* Ill Inst Technol, BS, 51, MS, 57, PhD(math), 64. *Prof Exp:* Jr mathematician inst air weapons res, Univ Chicago, 52-54, mathematician, 54-58, sr staff mem labs appl sci, 59-62; staff mathematician, Weapons Systs Eval Group, Inst Defense Anal, 58-59; sr mathematician, Acad Intersci Methodology, 63-64; lectr math, State Univ NY Buffalo, 64-65, assoc prof, 65-67; assoc prof, 67-69, PROF MATH, TEMPLE UNIV, 69- *Mem:* AAAS; Am Math Soc; Math Asn Am. *Res:* Functional analysis; invariant means; fixed point theorems. *Mailing Add:* Dept Math 038-16 Temple Univ Broad & Montgomery Sts Philadelphia PA 19122

MITCHELL, THOMAS GEORGE, b Philadelphia, Pa, Feb 27, 27; m 46; c 6. PHYSIOLOGY, NUCLEAR MEDICINE. *Educ:* St Joseph's Col, Pa, BS, 50; Univ Rochester, MS, 56; Georgetown Univ, PhD(physiol), 63; Am Bd Radiol, dipl, 58. *Prof Exp:* Med Serv Corpsman, US Navy, 50-69, asst supvr radioisotope lab, Naval Hosp, St Albans, NY, 51-55, med nuclear physicist, Bethesda Naval Hosp, Md, 56-60, physiologist & med nuclear physicist, Naval Med Res Unit 2, Taipei, Taiwan, 63-65, med nuclear physicist, Radiation Exposure Eval Lab, Naval Hosp, 65-69; assoc prof radiol sci, Sch Hyg & Pub Health, Johns Hopkins Univ, 69-74; dir radiation control, assoc prof physiol & biophys & assoc prof radiol, Med Ctr, Georgetown Univ, 74-; AT DEPT RADIOL, JOHNS HOPKINS UNIV. *Concurrent Pos:* Lectr, Naval Med Sch, Bethesda, 56-63 & 65-; adj acad staff, Children's Hosp, DC, 59-63; lectr, Sch Hyg & Pub Health, Johns Hopkins Univ, 65-70; instr, Sch Med & Dent, Georgetown Univ, 65-69; consult, Naval Med Res Inst, 65-70. *Mem:* Am Col Radiol; Soc Nuclear Med; Health Physics Soc; Am Asn Physicists in Med. *Res:* Medical nuclear physics; clinical applications of radioisotopes; regional blood flow. *Mailing Add:* Dept Radiol Johns Hopkins Univ 615 N Wolfe St Rm 2001 Baltimore MD 21205-2179

MITCHELL, THOMAS GREENFIELD, b New York, NY, Mar 30, 41; div; c 2. MEDICAL MYCOLOGY. *Educ:* NTex Univ, BA, 63; Tulane Univ, PhD(microbiol & immunol), 71. *Prof Exp:* Res assoc, microbiol dept, La State Univ Med Ctr, La State Univ, 71-72; res specialist pediat, Univ Minn, 72-74; asst prof, 74-80, ASSOC PROF, DUKE UNIV MED CTR, 80-, DIR CLIN MYCOL LAB, 74-, DIR MYCOBACTERIOL LAB, 76- *Concurrent Pos:* Prin investr, Minn Med Found res grant, 72-73 & Nat Inst Allergy & Infectious Dis res grant, 76-79 & 87-95; co-investr, Univ Minn res grant, 73-74 & Nat Inst Allergy & Infectious Dis res grant, 74-77; dir serol lab, Duke Univ Med Ctr, 74-82; guest reviewer, numerous sci journals, 76-; co-founder, Triangle Area Med Mycol Asn, 81-; chmn, Div F Med Mycol, Am Soc Microbiol, 87-89; mem, Bact & Mycol II Study Sect, Div Res Grants, NIH, 90-94. *Mem:* Am Soc Microbiol; Med Mycol Soc Am; Sigma Xi; Int Soc Human Animal Mycol; Mycol Soc Am. *Res:* Host defense mechanisms against mycotic infections; fungal morphogenesis; serodiagnosis of fungal disease. *Mailing Add:* Dept Microbiol & Immunol Box 3803 Duke Univ Med Ctr Durham NC 27710

MITCHELL, THOMAS OWEN, b New York, NY, Oct 18, 44; m 77. ORGANIC CHEMISTRY, PETROLEUM CHEMISTRY. *Educ:* Trinity Col, Conn, BS, 66; Northwestern Univ, Ill, PhD(org chem), 70. *Prof Exp:* Res chemist, Mobil Res & Develop Corp, 70-73, sr res chemist, 74-79, assoc, 80-90, RES ASSOC, MOBIL RES & DEVELOP CORP, 90- *Mem:* Am Chem Soc. *Res:* Enhanced oil recovery; catalysis in petrochemistry; catalytic chemistry; organosilicon chemistry; coal chemistry and conversion; organic geochemistry. *Mailing Add:* Mobil Res & Develop Corp PO Box 1025 Princeton NJ 08540

MITCHELL, TOM M, US citizen. EXPERT SYSTEMS, ARTIFICIAL INTELLIGENCE. *Educ:* Mass Inst Technol, BS, 73; Stanford Univ, MS, 75, PhD(elec eng & comput sci), 78. *Prof Exp:* asst prof, 78-82, assoc prof comupt sci, Rutgers Univ, 80-86; PROF COMPUT SCI, CARNEGIE-MELLON UNIV, 86- *Concurrent Pos:* Consult, Defense Res Estab Atlantic, Halifax, Nova Scotia, 80-81; prin investr, NSF grants, 80- *Mem:* Inst Elec & Electronic Engrs; Asn Comput Mach. *Res:* Applications of artificial intelligence and expert systems; artificial intelligence approaches to problems in curcuit design and medicine; automated improvement of problem-solving strategies through practice. *Mailing Add:* Comput Sci Dept Carnegie-Mellon Univ 5000 Forbes Ave Pittsburgh PA 15213

MITCHELL, VAL LEONARD, b Salt Lake City, Utah, Sept 9, 38; m 61; c 4. CLIMATOLOGY. *Educ:* Univ Utah, BS, 64; Univ Wis-Madison, MS, 67, PhD(meteorol), 69. *Prof Exp:* Res asst meteorol, Univ Wis-Madison, 64-69; from asst prof to assoc prof earth sci, Mont State Univ, 69-74; asst prof meteorol, Univ Wis-Madison, 74-78; STATE CLIMATOLOGIST, WIS GEOL & NATURAL HIST SURV, UNIV WIS-EXTEN, 74-, ASSOC PROF METEOROL, UNIV WIS-MADISON, 78- *Mem:* Am Meteorol Soc; Asn Am Geogr. *Res:* Applied climatology; interaction between the atmosphere and the biosphere; climatology of mountainous regions. *Mailing Add:* 6105 Monticello Madison WI 53719-1514

MITCHELL, WALLACE CLARK, b Ames, Iowa, Nov 12, 20; m 58; c 3. ENTOMOLOGY. *Educ:* Iowa State Col, BS, 47, PhD(zool, entom), 55. *Prof Exp:* Sr engr aide & entomologist malaria control in war areas, USPHS, Fla, 42-43; asst to state entomologist, Nursery Inspection, State Dept Agr, Iowa, 47-49; instr zool & entom & asst entom, Agr Exp Sta, Univ Hawaii, 49-53; field entomologist midwestern area, Geigy Agr Chem Inc, 53-54; asst prof zool & entom, SDak State Col, 55-56; entomologist fruit insect sect, Agr Res Serv, USDA, 56-62; assoc prof entom & assoc entomologist, 62-69, actg dir, Hawaii Agr Exp Sta, 69-70, prof entom, 78-85, chmn dept & entomologist, 68-77, actg dean, Col Trop Agr & Human Resources, 75-76, act assoc dean, Col Trop Agr & Human Resources, 78-85, EMER PROF ENTOM, COL TROP AGR & HUMAN RESOURCES, UNIV HAWAII, 85- *Concurrent Pos:* Bd dirs, Consortium for Int Crop Protection, 79-, treas, 80-82, vchmn, 83-87. *Honors & Awards:* Ezra Taft Bensen Award, USDA, 59, Orville Freeman Award, 64. *Mem:* Entom Soc Am; Am Registry Cert Entomologists; Sigma Xi. *Res:* Economic entomology; field evaluation of fruit fly lures, attractants, insecticide sprays, sterilization and eradication procedures for control of fruit and vegetable insects; tropical economic entomology; insect behavior, pest management; insect ecology. *Mailing Add:* Dept Entomol CTAHR 3050 Maile Way Honolulu HI 96822

MITCHELL, WALTER EDMUND, JR, b Franklin, Mass, Nov 16, 25. ASTRONOMY. *Educ:* Tufts Univ, BS, 49; Univ Va, MS, 51; Univ Mich, PhD, 58. *Prof Exp:* Asst astron, Univ Mich, 51-52, observer infrared proj, Mt Wilson Observ, 53-55; vis asst prof astron, Brown Univ, 56-57; from instr to assoc prof, 57-69, PROF ASTRON, OHIO STATE UNIV, 69- *Mem:* Am Astron Soc; Royal Astron Soc; Am Asn Physics Teachers. *Res:* Solar spectroscopy; solar-terrestrial relations; planetarium science. *Mailing Add:* Dept Astron Ohio State Univ 5062 Smith Lab Columbus OH 43210

MITCHELL, WILLIAM ALEXANDER, b Raymond, Minn, Oct 21, 11; m 38; c 7. FOOD SCIENCE, CARBOHYDRATE CHEMISTRY. *Educ:* Nebr Wesleyan Univ, BA, 35; Univ Nebr, MS, 38. *Hon Degrees:* DrSci, Nebr Wesleyan Univ, 80. *Prof Exp:* Teacher high sch, Nebr, 35-36; asst chem, Univ Nebr, 37-38; org res chemist, Eastman Kodak Co, NY, 38-41; head biochem sect, Gen Foods Corp, 41-53, head biocolloids sect, 53-56, sect head chem res, 56-62, res specialist, 62-68, sr res specialist, 68-76; FOOD CONSULT, 76- *Mem:* Am Chem Soc. *Res:* Synthetic organic biochemistry; wheat flour; starch; proteins; fats and oils; emulsifiers; gas reactions; enzymes; gelatin; food processing and products; sugars; pectins; gums; colloid chemistry; coffee; food flavors; cake mixes; eggs; carbonation systems; Kodachrome couplers. *Mailing Add:* RR#4 Gardenside F-1 Shelburne VT 05482

MITCHELL, WILLIAM COBBEY, b Rochester, NY, Aug 2, 39; m 62; c 4. OPTICAL RECORDING, ENERGY CONVERSION. *Educ:* Oberlin Col, AB, 61; Wash Univ, PhD(physics), 67. *Prof Exp:* Sr res physicist, 3M Co, 67-69; Nat Bur Standards-Nat Res Coun fel, Nat Bur Standards, 69-71; consult, ThermoelecTRIC Syst Sect, 71-72, sr res physicist, 72-75, res specialist, 75-76, mat res supvr, 76-78, mgr, 79-88, STAFF SCIENTIST, 3M CO, 88- *Mem:* Am Phys Soc. *Res:* Non-equilibrium statistical mechanics; solid-state transport theory; thermoelectric materials; energy storage; optical and magnetic recording. *Mailing Add:* 3568 Siems St St Paul MN 55112

MITCHELL, WILLIAM H, b Acworth, NH, Dec 12, 21; m 46; c 2. AGRONOMY, BOTANY. *Educ:* Univ NH, BS, 46, MS, 49; Pa State Univ, PhD(agron, bot), 60. *Prof Exp:* Headmaster & teacher high sch, NH, 45-47; from asst res prof to assoc res prof agron, 49-71, PROF PLANT SCI, UNIV DEL, 71- *Concurrent Pos:* Dir, Int Coop Improv Asn, 53-66; chmn collabrs regional pasture res lab, Pa State Univ, 60-66; tech comt, Northeastern Regional Res Group, 66-68; mem, Am Grassland Coun. *Mem:* Soil Sci Soc Am; Am Soc Agron; Crop Sci Soc Am. *Res:* Management of forage crops and turf; soil fertility. *Mailing Add:* 506 Briar Lane Newark DE 19711

MITCHELL, WILLIAM JOHN, b Minneapolis, Minn. SET THEORY. *Educ:* Univ Wis, BA, 65; Berkeley Sch, PhD(math), 70. *Prof Exp:* Instr math, Univ Chicago, 70-72; asst prof math, Rockefeller Univ, 72-77; mem staff math, Inst Advan Studies in Humanities, New York, 77-78; ASSOC PROF MATH, PA STATE UNIV, 79- *Res:* Inner models for large cardinals. *Mailing Add:* 5621 Pelham Rd Durham NC 27713

MITCHELL, WILLIAM WARREN, b Butte, Mont, Mar 10, 23; m 55; c 2. AGRONOMY, BOTANY. *Educ:* Univ Mont, BA, 57, MA, 58; Iowa State Univ, PhD(bot), 62. *Prof Exp:* Instr biol sci, Western Mont Col, 58-59; asst prof, Chadron State Col, 62-63; prof agron & head dept, 63-86, prof agron & asst dir, 87, EMER PROF AGRON, AGR & FORESTRY EXP STA, UNIV ALASKA, 88- *Mem:* Am Soc Plant Taxon; Sigma Xi; Soc Range Mgt; Am Agron Soc; Am Forage & Grassland Coun. *Res:* Applications of introduced and indigenous grass taxa; varietal selection and development of grasses; revegetation; biosystematics of grasses. *Mailing Add:* Agr & Forestry Exp Sta 533 E Fireweed Palmer AK 99645

MITCHEM, JOHN ALAN, b Sterling, Colo, July 28, 40; m 62; c 2. MATHEMATICS. *Educ:* Univ Nebr, Lincoln, BS, 62; Western Mich Univ, MA, 67, PhD(math), 70. *Prof Exp:* Asst prof math, 70-72, assoc prof, 72-77, PROF & CHMN MATH, SAN JOSE STATE UNIV, 77- *Mem:* Math Asn Am; Am Math Soc. *Res:* Graph theory, especially partioning problems. *Mailing Add:* Dept Math San Jose State Univ Washington Square San Jose CA 95192

MITCHNER, HYMAN, b Vancouver, BC, Nov 23, 30; m 53; c 4. PHARMACEUTICAL CHEMISTRY. *Educ:* Univ BC, BA, 51, MSc, 53; Univ Wis, Madison, PhD(pharmaceut chem), 56. *Prof Exp:* Asst prof pharmaceut chem, Sch Pharm, Univ Wis, Madison, 56-59; group leader analytical res, Miles Lab, 59-61, sect head, 61-63; dir qual control, Barnes-Hind Labs, 63-66; dir, 66-69, vpres qual control, 69-78, vpres qual assurance & tech serv, 78-82, VPRES CORP QUAL ASSURANCE, SYNTEX USA LABS, INC, 82- *Mem:* Am Pharmaceut Asn; Am Chem Soc; Am Soc Qual Control; fel Acad Pharmaceut Sci. *Res:* Analytical research on drug products; physical chemical studies in drug systems. *Mailing Add:* 270 Yerba Buena Pl Los Altos CA 94022-2153

MITCHNER, MORTON, b Vancouver, BC, Jan 17, 26; US citizen; m 60; c 2. MAGNETOHYDRODYNAMICS. *Educ:* Univ BC, BA, 47, MA, 48; Harvard Univ, PhD(physics), 52. *Prof Exp:* Sheldon traveling fel, Harvard Univ, 52-53, res fel appl sci, 53-54; staff mem opers res, Arthur D Little, Inc, 54-58; staff scientist, Lockheed Missiles & Space Co, 58-63; vis lectr mech eng, 63-64, assoc prof, 64-69, prof, 69-84, PROF EMER MECH ENG, STANFORD UNIV, 84- *Concurrent Pos:* Vis prof eng sci, Columbia Univ, 61-62; consult, United Tech Ctr, United Aircraft Corp, 65-68; mem & chmn steering comt, Eng Appln Magnetohydrodynamics, 67-; consult, Nat Comt Nuclear Energy, Italy, 70-71. *Mem:* Am Phys Soc; Am Inst Aeronaut & Astronaut; Opers Res Soc Am. *Res:* High-temperature gas dynamics; plasma physics; magnetohydrodynamics; kinetic theory; turbulence; combustion aerodynamics; energy conversion; operations research; electrostatic precipitation. *Mailing Add:* 44 Palm Court Menlo Park CA 94025

MITCHUM, RONALD KEM, b Elk City, Okla, Dec 2, 46; m 65; c 3. MASS SPECTROMETRY. *Educ:* Southwestern Okla State Univ, BS, 68; Okla State Univ, PhD(chem), 73. *Prof Exp:* Fel mass spectrometry, Univ Houston, 72-74 & Univ Warwick, 74-75; fel mass spectrometry, Univ Nebr-Lincoln, 75-76; dir chem, Nat Ctr Toxicol Res, HEW/Food & Drug Admin, 76-84; dir, Qual Assurance Div, US Environ Protection Agency, Las Vegas, 84-88; dir methods develop, Battelle Mem Inst, Columbus, Ohio, 88-90; PRES, TRIANGLE LABS, 90- *Concurrent Pos:* Consult, instrument design. *Mem:* Am Chem Soc; Am Soc Mass Spectrometry; Asn Off Anal Chemists. *Res:* Atmospheric pressure, field desorption and development of analytical methodology for the analysis of toxicants in environmental media; analytical chemistry management. *Mailing Add:* 6385 Schier Rings Rd Dublin OH 43017

MITESCU, CATALIN DAN, b Bucarest, Romania, May 2, 38; Can citizen; m 72; c 3. PHYSICS, PHASE TRANSITIONS. *Educ:* McGill Univ, BEng, 58; Calif Inst Technol, PhD(physics), 66. *Prof Exp:* From instr to assoc prof, 65-77, PROF PHYSICS, POMONA COL, 77- *Concurrent Pos:* Vis prof, Univ Paris Sud, 71-72, Univ Provence, Aix-Marseille I, 76, Univ Provence, 77, 79, 83, 86 & 87, Ecole Super Phys Chim Indust, Paris, 79, 90. *Mem:* Am Phys Soc. *Res:* Low-temperature physics-superconductivity, thin films; liquid helium; phase transitions; liquid crystals; fractals; electrical properties of disordered matter; flows through porous media; Rayleigh Taylor instabilities. *Mailing Add:* Dept Physics Pomona Col 610 N College Ave Claremont CA 91711-6359

MITHCELL, WILLIAM MARVIN, b Atlanta, Ga, Mar 3, 35; m 59; c 3. PATHOLOGY. *Educ:* Vanderbilt Univ, BA, 57, MD, 60; Johns Hopkins Univ, PhD(biol, biochem), 66. *Prof Exp:* Mem house staff med, Johns Hopkins Hosp, 60-61; from asst prof to assoc prof microbiol, 66-74, asst prof med, 69-77, assoc prof, 74-78, PROF PATH, VANDERBILT UNIV, 78- *Concurrent Pos:* USPHS fel biol, McCullom-Pratt Inst, Johns Hopkins Univ, 61-66; USPHS grant & Vanderbilt Inst award, 66-67; planning dir, Vanderbilt Cancer Ctr, 71-73; Nat Inst Arthritis & Metab Dis grant, 66-81; Nat Cancer Inst grant, 71-73; staff physician, Vet Admin Hosp, 66-67; consult, NIH, 71-, AIDS study sect, 89-92; vis prof, Inst Anat Path, Univ Lausanne, 76-77; Eleanor Roosevelt Int Cancer fel, 76-77. *Honors & Awards:* Borden Award Med Res, 60. *Mem:* AAAS; Am Asn Pathologists; Int Acad Path; Am Soc Biol Chem; Am Chem Soc; Sigma Xi. *Res:* Molecular pathology; AIDS and anti-HIV drugs; HIV vaccines; chronic fatigue syndrome; biological response modifiers; molecular modeling; HIV vaccines. *Mailing Add:* Dept Path Vanderbilt Univ Nashville TN 37232

MITLER, HENRI EMMANUEL, b Paris, France, Oct 26, 30; US citizen; m 86; c 3. MATHEMATICAL MODELING, ASTROPHYSICS. *Educ:* City Col New York, BS, 53; Princeton Univ, PhD(nuclear structure), 60. *Prof Exp:* Jr physicist, Nuclear Develop Assocs, 53; instr physics, Princeton Univ, 57-58; res assoc physics & adj lectr, Brandeis Univ, 59-60; sr staff physicist, Smithsonian Astrophys Observ, 61-75; res assoc, Harvard Col Observ, 62-75, lectr astron, Harvard Univ, 63-75, res assoc, Div Appl Sci, 75-83; HEAD, FIRE DYNAMICS GROUP, BLDG & FIRE RES LAB, NAT INST STANDARDS & TECHNOL, 83- *Concurrent Pos:* Lectr, Brandeis Univ, 67-68 & 71-72; consult. *Mem:* Am Phys Soc; Am Astron Soc; Sigma Xi; Combustion Inst; Int Asn Fire Safety Sci. *Res:* Quantum theory; nuclear structure and reactions; origin of the elements; radiochemical production in meteoroids; nucleosynthesis by cosmic rays; electron-ion screening in dense plasmas; origin of the moon; evolution of fires in compartments; dynamics of fires in various environments. *Mailing Add:* 1517 Columbia Ave Rockville MD 20850

MITOFF, S(TEPHAN) P(AUL), b San Francisco, Calif, May 17, 24; m 48; c 3. CERAMICS, SYSTEMS ENGINEERING. *Educ:* Univ Calif, BS, 50, PhD(eng sci), 56. *Prof Exp:* Asst, Univ Calif, 50-55; res assoc & ceramist, Res Labs, Gen Elec Co, 55-73, prin investr high temperature battery proj, Res & Develop Ctr, 73-79, proj tech leader, 79-82, staff ceramist, Integrated Circuit Packaging, 82-87, consult, Elec Properties Ceramics, 87-88; RETIRED. *Mem:* Fel Am Ceramic Soc; Am Phys Soc. *Res:* Electrical conductivity; defect structure; thermodynamics of oxides and salts; fuel cells; solid electrolytes; batteries; battery system design; ceramic engineering. *Mailing Add:* 6114 43rd St W Bradenton FL 34210

MITOMA, CHOZO, b San Francisco, Calif, July 21, 22; m 50; c 4. BIOCHEMICAL PHARMACOLOGY. *Educ:* Univ Calif, BA, 48, PhD(biochem), 51. *Prof Exp:* Asst, Univ Calif, 48-51; biochemist, NIH, 52-59; sr biochemist, 59-69, dir, Biomed Res Lab, 69-85, SR STAFF SCIENTIST, SRI INT, 85- *Concurrent Pos:* Hite fel, Univ Tex, 51-52. *Mem:* Am Soc Biol Chem; Am Soc Pharmacol & Exp Therapeut; AAAS; Int Soc Study Xenobiotics. *Res:* Mechanism of action of drugs; drug metabolism; toxicokinetics. *Mailing Add:* Life Sci Div SRI Int Menlo Park CA 94025

MITRA, GRIHAPATI, b Oct 29, 27; m 58; c 1. INORGANIC CHEMISTRY. *Educ:* Univ Calcutta, BSc, 47, MSc, 49, DSc, 54. *Prof Exp:* Fel chem, Univ Wash, Seattle, 55-57; res officer, AEC, India, 58; lectr chem, Dum Dum Col, 59-60; fel physics, Pa State Univ, 60-61; asst prof chem, 61-64, chmn dept, 64-67, PROF CHEM, KING'S COL, PA, 67- *Concurrent Pos:* Vis lectr, Sylvania Elec Prods, Inc, Pa, 62; dir rocket fuel res, Adv Res Projs Agency, 63-67; res dir, Beryllium Corp, 67. *Mem:* AAAS; Am Chem Soc; Indian Chem Soc. *Res:* Preparation and properties of compounds containing halogens. *Mailing Add:* Dept Chem King's Col 133 N River St Wilkes-Barre PA 18711

MITRA, JYOTIRMAY, b Calcutta, India, Nov 25, 21; m 70. CYTOGENETICS, CYTOTAXONOMY. *Educ:* Univ Calcutta, BA Hons, 42, MA, 44; Cornell Univ, PhD(cytogenetics), 55. *Prof Exp:* Res fel plant taxon, Bot Surv India, 45-47; lectr plant sci, Univ Calcutta, 48-51, ICI fel tissue cult, 48-51; res asst cytogenetics, Cornell Univ, 54-55, res assoc, 57-61; cytogeneticist, Beth Israel Med Ctr, NY, 61-63; assoc prof, 63-66, PROF BIOL, NY UNIV, 67- *Concurrent Pos:* Chief cytogeneticist, Beth Israel Med Ctr, NY, 63-77; chmn develop genetics, Fifteenth Int Bot Cong, 69. *Honors & Awards:* Fulbright Award & Smith-Mundt Award, US State Dept, 52. *Mem:* Genetics Soc Am; Am Genetic Asn; AAAS. *Res:* Cytogenetics and cytotaxonomy of eukaryotic organisms involving mechanisms of alteration of somatic and meiotic chromosomes and their genetic consequences; cytogenetical studies on experimental leukemia in mice. *Mailing Add:* Dept Biol NY Univ 1009 Main Bldg Washington Sq New York NY 10003

MITRA, SANJIT K, b Calcutta, India, Nov 26, 35; US citizen. SIGNAL & IMAGE PROCESSING, COMPUTER-AIDED DESIGN. *Educ:* Utkal Univ, BSc, 53; Calcutta Univ, MS, 56; Univ Calif-Berkeley, MS, 60, PhD (elec eng), 62. *Hon Degrees:* Dr Tech, Tampere Univ Technol, 87. *Prof Exp:* Asst prof elec eng, Cornell Univ, 62-65; mem tech staff, AT & T Bell Telephone Labs, 65-77; chmn, 79-82, PROF ELEC ENG, UNIV CALIF, SANTA BARBARA, 77- *Concurrent Pos:* Prin investr, NSF res grants, 69-, Off Naval Res Contract, 86-, Univ Calif Micro Res Grants, 82-; hon prof, Northern Jiaotong Univ, Beijing, China, 85-; vis prof, Indian Inst Technol, New Delhi, 72-73 & Univ Erlangen-Nuernberg, Erlangen, W Ger, 75; vis fel, Australian Nat Univ, Canberra, 82; vis prof, Fed Univ Rio De Janeiro, Brazil, 84, Tech Univ Zagreb, Yugoslavia, 86, Istanbul Tech Univ & Bilkent Univ, Turkey, 88. *Honors & Awards:* Terman Award, Am Soc Eng Educ, 73; AT&T Found Award, Am Soc Eng Educ, 85; Educ Award, Inst Elec & Electronic Engrs Circuits & Systs Soc, 88; Distinguished Fulbright Prof Award, Brazil, 84, Yugoslavia, 86, Turkey, 88; Distinguished Sr Scientist Award, Alexander von Humboldt Found, Germany, 89. *Mem:* Fel Inst Elec & Electronic Engrs; fel AAAS; Inst Elec & Electronic Engrs Circuits & Systs Soc (pres, 86); Am Soc Eng Educ; Europ Asn Signal Processing; Sigma Xi; Soc Photo-optical Instrumentation Engrs. *Res:* All aspects of analog and digital signal processing which is concerned with the representation of signals and its transformation by linear and nonlinear systems for the purpose of improving the quality of signal, and extract pertinent features, etc. *Mailing Add:* Dept Elec & Comput Eng Univ Calif Santa Barbara CA 93106

MITRA, SANKAR, b Calcutta, India, July 7, 37; m 66; c 2. MOLECULAR BIOLOGY, BIOCHEMISTRY. *Educ:* Univ Calcutta, BSc, 57, MSc, 59; Univ Wis-Madison, PhD(biochem), 64. *Prof Exp:* Res asst biochem, Univ Calcutta, 59-60; res asst, Univ Wis-Madison, 60-62, res fel, 62-63; res assoc, Stanford Univ, 64-65; Indian Govt Coun Sci & Indust Res sci officer, Bose Inst, India, 66-67, sr res fel, 67-70, reader, 71; BIOCHEMIST, BIOL DIV, OAK RIDGE NAT LAB, 71- *Concurrent Pos:* Mem panel V, Int Cell Res Orgn, 69-73; lectr sch biomed sci, Univ Tenn, 72- *Mem:* Indian Soc Biol Chem; Am Soc Biol Chemists. *Res:* Molecular biology of viruses and nucleic acids; synthesis of nucleic acids in vivo and in vitro; repair and miscoding properties of simple alkylated bases in DNA. *Mailing Add:* Biol Div Bldg 9211 Oak Ridge Nat Lab PO Box 2009 Oak Ridge TN 37831-8080

MITRA, SHASHANKA S, b Calcutta, India, May 20, 32; m 63; c 2. PROBABILITY. *Educ:* Univ Calcutta, BS, 52, MS, 54; Univ Wash, PhD(math), 61. *Prof Exp:* Teaching asst math, Univ Wash, 56-61; asst prof, Univ Idaho, 61-62, Univ Ariz, 62-64, Clarkson Col Technol, 64-67, Western Wash State, 67-69 & Wilkes Col, 69-72; ASST PROF MATH, PA STATE UNIV, 72- *Mailing Add:* Dept Math Pa State Univ College Place Dubois PA 15801

MITRA, SUNANDA, b Bengal, Feb 5, 36; m 60; c 2. VISUAL CONTRAST SENSITIVITY, CONTRAST DETECTION PERIMETERY. *Educ:* Calcutta Univ, BS, 55, MS, 57; Phillipps Univ, WGer, PhD(physics), 66. *Prof Exp:* Res assoc nuclear physics, Saha Inst Nuclear Physics, India, 58-59; lectr physics, Lady Brabourne Col, India, 59-64; res asst, Phillipps Univ, WGer, 64-66; res assoc, plasma physics, 69-73, biomed inst, 74-75, res assoc visual sci, dept ophthalmol, Sch Med, 77-, AT DEPT ELEC ENG & COMPUT SCI, TEX TECH UNIV. *Mem:* Optical Soc Am; Asn Res Vision & Ophthalmol. *Res:* Development of a method of determining spatial contrast sensitivity function loss in maculopathy with an aim to detect and classify macular disorders at an early stage. *Mailing Add:* Dept Elec Eng Comput Sci Tex Tech Univ Lubbock TX 79409

MITRIUS, JOAN C, neuropharmacology, for more information see previous edition

MITRUKA, BRIJ MOHAN, b Hanuman Garh Town, Rajasthan, India, May 12, 37; US citizen; wid; c 3. MICROBIOLOGY, VETERINARY MEDICINE. *Educ:* Rajasthan Vet Col, BVSc & AH, 59; Mich State Univ, MS, 62, PhD(microbiol, pub health), 65. *Prof Exp:* Vet asst surg, Vet Hosp, Hanamangarh County, Rajasthan, 59-60; USPHS fel & res assoc microbiol, Mich State Univ, 65-66; res assoc, Cornell Univ, 66-68; clin pathologist, Sch Med, Yale Univ, 68-69, asst prof lab animal sci & lab med, 69-74; assoc prof, Sch Med, Univ Pa, 74-78; dir clin labs & res & develop, BHP Inc, West Chester, Pa, 78-; AT DEPT MICROBIOL, TEMPLE UNIV. *Concurrent Pos:* USPHS grants, Yale Univ, 69-73; prof microbiol & head dept, Punjab Agr Univ, 74-78; dir qual control, BHP Inc; res prof microbiol, Temple Univ,

78- *Mem:* AAAS; Am Soc Microbiol; Am Asn Clin Chem; Am Vet Med Asn. *Res:* Bases of microbial pathogenicity; microbial metabolites detection and identification in tissues and body fluids; microbial metabolism in vitro and in vivo; study of pathogenic mechanisms involved in infectious diseases of man and animals; rapid, automated diagnosis in infectious and non-infectious diseases; manufacturing and development of in vitro diagnostic products. *Mailing Add:* Sch Vet Med Ross Univ PO Box 266 Roseau Dominica West Indies

MITSCH, WILLIAM JOSEPH, b Wheeling, WVa, Mar 29, 47; m 70; c 3. WETLAND ECOLOGY. *Educ:* Univ Notre Dame, BS, 69; Univ Fla, ME, 72, PhD(environ eng sci), 75. *Prof Exp:* Asst prof, Ill Inst Technol, 75-79; from assoc prof to prof systs ecol, Univ Louisville, 79-85; PROF, SCH NATURAL RESOURCES, OHIO STATE UNIV COLUMBUS, 86- *Concurrent Pos:* Consult var pub & pvt clients, 75-, auth var books & publ, 76; res grant & consult, Argonne Nat Lab, 76-78; var grants, 76-; chmn & ed, Energy Ecol Modelling Int Conf, 81; exec dir, Ohio River Basin Consortium, 85-; Fulbright fel, Copenhagen, Denmark, 86-87; chmn, Intecol Wetlands Conf, 92. *Mem:* Ecol Soc Am; Sigma Xi; Am Soc Limnol & Oceanog; fel AAAS; Int Soc Ecol Modelling (secy, pres). *Res:* Wetland and freshwater ecology; ecological modelling; ecological engineering; water quality; management and restoration of wetlands. *Mailing Add:* Sch Natural Resources Ohio State Univ Columbus OH 43210

MITSCHER, LESTER ALLEN, b Detroit, Mich, Aug 20, 31; m 53; c 3. BIO-ORGANIC CHEMISTRY. *Educ:* Wayne State Univ, BS, 53, PhD(chem), 59. *Prof Exp:* Spec instr pharm, Wayne State Univ, 56-58; res scientist bio-org chem, Lederle Labs, Am Cyanamid Co, 58-61, group leader fermentation biochem, 61-67; prof pharmacog & natural prod, Col Pharm, Ohio State Univ, 67-75; UNIV DISTINGUISHED PROF MEDICINAL CHEM & CHMN DEPT, UNIV KANS, 75- *Concurrent Pos:* Intersearch prof, Victorian Col Pharm, Melbourne, Australia. *Honors & Awards:* Ernst Volwieler Award, Am Asn Col Pharm, 85; Res Achievement Award, Nat Prod Chem, Am Pharm Asn, 80; Higuchi-Simons Award, Biomed Sci, Kans Univ, 86; Smissman Award, Med Chem Div, Am Chem Soc, 88. *Mem:* Am Chem Soc; Am Soc Pharmacog; The Chem Soc; Am Soc Microbiol; Japanese Antibiotics Asn. *Res:* Chemistry of organic compounds of natural origin, especially alkaloids, terpenes, steroids and antibiotics. *Mailing Add:* Col Pharm Univ Kans Malott Hall Lawrence KS 66044

MITSOULIS, EVAN, b Athens, Greece, Sept 17, 54; Can citizen. NUMERICAL METHODS, RHEOLOGY. *Educ:* Nat Tech Univ, Athens, Greece, BScE, 77; Univ NB, MScE, 79, McMaster Univ, PhD(chem eng), 84. *Prof Exp:* Asst prof, 84-89, ASSOC PROF CHEM ENG, UNIV OTTAWA, 89- *Concurrent Pos:* Vis prof chem eng, Paris Sch of Mines, France, 90-91, Nat Tech Univ, Athens, Greece, 91- *Honors & Awards:* Award of Appeciation, Soc Plastics Engrs, 90. *Mem:* Chem Inst Can; Can Soc Chem Engrs; Am Inst Chem Engrs; Soc Plastics Engrs; Soc Rheology; Polymer Processing Soc. *Res:* Computer applications in polymer processing and rheology of polymer solutions and melts, coextrusion and other processing operations; computer-aided design/computed-aided manufacturing; finite element method. *Mailing Add:* Dept Chem Eng Univ Ottawa Ottawa ON K1N 6N5 Can

MITSUI, AKIRA, b Japan, Jan 25, 29; m 64; c 3. BIOLOGICAL OCEANOGRAPHY. *Educ:* Univ Tokyo, BS, 51, MA, 55, PhD(plant physiol), 58. *Prof Exp:* PROF BIOL OCEANOG, SCH MARINE & ATMOSPHERIC SCI, UNIV MIAMI, 72- *Mem:* Am Soc Microbiol; Am Soc Plant Physiologists; Int Asn Hydrogen Energy. *Res:* Marine biochemistry; bioenergetics; bioconversion of solar energy; hydrogen energy. *Mailing Add:* Univ Miami 4600 Rickenbacker Causeway Miami FL 33149

MITSUTOMI, T(AKASHI), b Honolulu, Hawaii, Dec 20, 23; m 48. SYSTEMS ENGINEERING. *Educ:* Mass Inst Technol, 52, MS, 53. *Prof Exp:* Machinist-draftsman, Am Can Co, Hawaii, 42-45, 48-49; res engr, Autonetics Div, NAm Aviation, Inc, 53-55, eng supvr, 55-59, group leader, 59-62, res mgr adv tech, 62-64, staff sci adv to vpres, 64-65, res mgr appl res, 65-67, dir adv tech, NAm Rockwell Corp, 67-71; vpres, Datapet Corp, 71-72; PRES, HYCOM, INC, 72- *Concurrent Pos:* First Lieutenant, US Army, 45-48; Instr, Univ Calif, Los Angeles, 57-58, Lectr, 59-68. *Mem:* Inst Elec & Electronics Engrs. *Res:* Research and advanced development of microelectronics, solid-state devices and space system controls; velocity meters and platform dynamic analysis and synthesis; microelectronics large-scale integration; data communication systems; microcomputer developments. *Mailing Add:* Hycom 16851 Armstrong Ave Irvine CA 92714

MITSUYA, HIROAKI, b Sasebo, Nagosaki, Japan, Aug 8, 50; m. ONCOLOGY. *Educ:* Kumamoto Univ, MD, 75, PhD(med sci), 82. *Prof Exp:* Clin staff, Sec Div Internal Med, Med Sch, Kumamoto Univ, 75-77, res fel, 77-80, instr, 80-82; vis fel, Metab Br, Nat Cancer Inst, 82-83, Clin Oncol Prog, 83-84, cancer expert, 84-88, vis scientist, 88-89, SR INVESTR, CLIN ONCOL PROG, NAT CANCER INST, BETHESDA, 89- *Res:* AIDS therapy. *Mailing Add:* Clin Oncol Prog Bldg 10 Rm 13N248 NIH Nat Cancer Inst 9000 Rockville Pike Bethesda MD 20892

MITTAG, THOMAS WALDEMAR, b Pecs, Hungary, Mar 14, 37; m 59; c 2. PHARMACOLOGY, EXPERIMENTAL EYE RESEARCH. *Educ:* Univ Cape Town, BS, 59, Hons, 61, PhD(org chem), 64. *Prof Exp:* Jr lectr chem, Univ Cape Town, 63-65; staff scientist, Worcester Found Exp Biol, 66-67; instr pharmacol, New York Med Col, Flower & Fifth Ave Hosps, 68-69, asst prof, 69-71; assoc prof, 71-78, PROF PHARMACOL, MT SINAI SCH MED, 78-, RES PROF OPHTHAL, 82- *Concurrent Pos:* Res fel biochem, Purdue Univ, 65-66; res fel pharmacol, Georgetown Univ, 67-68; NIH grants, 69-95. *Mem:* AAAS; Asn Res Vision Opthal; NY Acad Sci; Am Soc Pharmacol & Exp Therapeut; Int Soc Eye Res. *Res:* Molecular pharmacology of neurohormone receptors and their effector systems in the eye; hormone regulation of epithelial transport in the eye. *Mailing Add:* Dept Pharmacol Mt Sinai Sch Med Box 1215 New York NY 10029

MITTAL, BALRAJ, b Gohana, India, Jan 8, 51; m 81; c 2. CELL BIOLOGY, FLUORESCENT ANALOG CYTOCHEMISTRY. *Educ:* Panjab Univ, Chandigarh, India, BSc, 72, MSc, 75; Indian Inst Sci, Bangalore, India, PhD(biochem), 81. *Prof Exp:* Res assoc biochem, dept animal biol, Sch Vet Med, 81-82, res assoc cell biol, 82-85, RES BIOCHEMIST CELL BIOL, DEPT ANAT, SCH MED, UNIV PA, PHILADELPHIA, 85- *Mem:* Am Soc Cell Biol. *Res:* Analysis of cell motility; purification, fluorescent-labeling and microinjection of contractile proteins into living muscle and non-muscle tissue culture cells; immunofluorescence; cell biology of cytoskeleton. *Mailing Add:* Sarjay Gandhi Inst Med Sci PO Box 375 Raebareli Rd Lucknow 226001 India

MITTAL, GAURI S, b June 1, 47; Can citizen; m 72; c 3. FOOD SCIENCE & TECHNOLOGY, INTELLIGENT SYSTEMS. *Educ:* Punjab Agr Univ, India, BS, 69; Univ Manitoba, Winnipeg, Canada, MS, 76; Ohio State Univ, Columbus, PhD (food eng), 79. *Prof Exp:* Doctoral fel food eng, 80-82, asst prof food eng, 82-86, ASSOC PROF FOOD ENG, UNIV GUELPH, 86- *Mem:* Am Soc Agr Engrs; Inst Food Technologists; Can Soc Agr Eng; Can Inst Food Sci & Technol; Can Meat Res Coun. *Res:* Food process engineering with emphasis on thermal processing, meat processing, food-biotechnology, dairy processing, and fruits and vegetable processing; food and grain drying; energy management, recovery and conservation; simulation and modelling; expert systems and sensor development. *Mailing Add:* Sch Eng Univ Guelph Guelph ON N1G 2W1 Can

MITTAL, KAMAL KANT, b Pihani, India, July 1, 35; m; c 2. IMMUNOGENETICS, TRANSPLANTATION IMMUNOLOGY. *Educ:* Agra Univ, BS, 54, DVM, 58; Univ Ill, Urbana, MS, 62, PhD(immunogenetics), 65. *Prof Exp:* Res assoc animal sci, Univ Ill, Urbana, 65-66; res fel biol, Calif Inst Technol, 66-67; res geneticist II, Dept Surg, Univ Calif, Los Angeles, 67-69; asst prof microbiol, Baylor Col Med, Houston, 69-70; asst prof immunogenetics, Univ Calif, Los Angeles, 70-72; asst prof surg & physiol, Northwestern Univ, Chicago, 72-75; res microbiologist, Dept Health & Human Serv, NIH, 75-84, chief, Lab Transplantation Biol, Ctr Biol Eval & Res, Food & Drug Admin, 84-88, exec secy, Allergy, Immunol & Transplantation Res Comt, Prog & Proj Rev Br, Div Extramural Activ,88-90, SR SCIENTIST, OFF SCI INTEGRITY, OFF DIR, DEPT HEALTH & HUMAN SERV, NIH, 90- *Concurrent Pos:* ASHI Workshop & Repository Comt, Am Soc for Histocompatibility & Immunogenics, 84, 85 & 88; chmn, Orthoclone OKT*3 Rev Comt, Food & Drug Admin, 84-88; Hybridoma Comt, Off Biol Res & Rev, Ctr Drugs & Biol, Food & Drug Admin, Dept Health & Human Serv, 84-88, Recombinant DNA Comt, 85-88; Transplantation Res Coord Comt, NIH, 90-; reviewer, J Immunol, J Transplantation, J Tissue Antigens, J Sci, J Clin Invest. *Mem:* AAAS; Transplantation Soc; Am Asn Immunol; Am Soc Histocompatibility & Immunogenetics; Am Asn Blood Banks; Indian Soc Human Genetics. *Res:* Immunogenetics, population genetics, transplantation genetics; transplantation of organs and tissues, transplantation immunology; mammalian major histocompatibity complex; HLA Region-serology, genetics, clinical applications; immunopathology, genetic diseases, autoimmunity allergy; hybridomas, monoclonal antibodies, differentiation antigens; tumor immunology, HLA-restriction, interferon, lymphokines; recombinant DNA technology: RFLPS gene therapy; complement component deficiencies; immunoreproduction, fertility, infertility. *Mailing Add:* Off Sci Integrity Off Dir NIH Bldg 31 Rm B1-C39 9000 Rockville Pike Bethesda MD 20892

MITTAL, KASHMIRI LAL, b Kilrodh, India, Oct 15, 45; m 70; c 4. PHYSICAL CHEMISTRY. *Educ:* Panjab Univ, Chandigarh, BSc, 64; Indian Inst Technol, New Delhi, MSc, 66; Univ Southern Calif, PhD(phys chem), 70. *Prof Exp:* Res assoc, Pa State Univ, 70-71; fel chem, Univ Pa, 71-72; fel, IBM Corp, San Jose, Calif, 72-74, staff engr, Poughkeepsie, 74-77, STAFF ENGR, IBM CORP, HOPEWELL JUNCTION, 77- *Mem:* Am Chem Soc; Electrochem Soc; Am Vacuum Soc; Adhesion Soc; fel Am Inst Chemists; Sigma Xi. *Res:* Surface, colloid, polymer and electrochemistry; surface properties of materials; adhesion and corrosion. *Mailing Add:* Corp Tech Inst 500 Columbus Ave Thornwood NY 10594

MITTAL, YASHASWINI DEVAL, b Poona, India, Oct 1, 41. STATISTICS. *Educ:* Poona Univ, BSc, 61; Univ Ill, Urbana, MS, 66; Univ Calif, Los Angeles, PhD(math), 72. *Prof Exp:* Teaching asst math, Univ Ill, Urbana, 64-66 & Univ Calif, Los Angeles, 66-71; asst prof, Northwestern Univ, Evanston, 72-73; vis mem, Inst Advan Study, Princeton, 73-74; asst prof statist, Stanford Univ, 74-80; ASSOC PROF STATIST, VA POLYTECH INST, 80- *Concurrent Pos:* Prog dir statist & probability, NSF, 86-88. *Mem:* Am Statist Asn; Inst Math Statist. *Res:* Convergence properties of maxima of stationary Gaussian processes. *Mailing Add:* Dept Statist Va Tech Blacksburgh VA 24061

MITTELMAN, ARNOLD, b New York, NY, Dec 21, 24; m 56; c 2. SURGERY, MEDICINE. *Educ:* Columbia Univ, AB, 49, MD, 54; Am Bd Surg, dipl, 66. *Prof Exp:* Instr surg, Columbia-Presby Med Ctr, 59-61; assoc cancer res surgeon, 61-65, ASSOC CHIEF RES SURGEON & DIR SURG DEVELOP ONCOL, ROSWELL PARK MEM INST, 65-, ASSOC RES PROF BIOCHEM, 69- *Concurrent Pos:* Asst attend & asst vis surgeon, Presby Hosp, NY, 61. *Mem:* AAAS; Am Inst Chem; Am Asn Cancer Res. *Res:* Nucleic acid and steroid biochemistry; acid base physiology. *Mailing Add:* Surg Develop Oncol Roswell Park Mem Inst 666 Elm St Buffalo NY 14203

MITTELSTAEDT, STANLEY GEORGE, b Connell, Wash, Oct 15, 09; m 40; c 4. PHARMACEUTICAL CHEMISTRY, PHARMACY. *Educ:* Northwest Nazarene Col, 34; State Univ Wash, BS & MS, 38; Purdue Univ, PhD(pharm, pharmaceut chem), 48. *Prof Exp:* Asst, State Univ Wash, 37-38; asst, Purdue Univ, 38-40; actg head dept chem, Boise Jr Col, 40-42; assoc prof, Univ Tex, 48-51; asst dean, Sch Pharm, 51-53, prof pharm & pharmaceut chem, 51-77, dean, 53-77, EMER PROF PHARM & PHARMACEUT CHEM, SCH PHARM, UNIV ARK, & EMER DEAN, 77- *Concurrent Pos:* Co-ed, Vet

Drug Encyclop; mem nat adv coun health educ, Dept Health, Educ & Welfare, Washington, DC, 72-76; exec vpres, Sounds of Music Found, Inc, 77-; consult Health Sci & Serv. *Mem:* AAAS; Am Chem Soc; Am Pharmaceut Asn. *Res:* Iodo-radio opaques. *Mailing Add:* 2000 Magnolia Little Rock AR 72202

MITTEN, LORING G(OODWIN), b Danville, Ill, Dec 29, 20; m 42; c 3. INDUSTRIAL ENGINEERING. *Educ:* Drexel Inst, BS, 42; Mass Inst Technol, SM, 47; Ohio State Univ, PhD(indust eng), 52. *Prof Exp:* Prof indust eng, Ohio State Univ, 48-57; prof indust eng & mgt sci, Northwestern Univ, 57-70, chmn dept, 63- 65; PROF MGT SCI & CHMN DIV, UNIV BC, 70- *Mem:* Fel AAAS; fel Am Soc Qual Control; Inst Mgt Sci; Am Inst Indust Engrs; Soc Indust & Appl Math; Sigma Xi. *Res:* Operations research and management science with emphasis on mathematical models of sequencing problems, optimal design and operation of industrial processes, dynamic programming and sequential decision processes. *Mailing Add:* 2501 Panorama Dr North Vancouver BC V7G 1V4 Can

MITTENTHAL, JAY EDWARD, b Boston, Mass, July 28, 41; m 68; c 2. MORPHOGENESIS, PATTERN FORMATION. *Educ:* Amherst Col, BA, 62; Johns Hopkins Univ, PhD(biophys), 70. *Prof Exp:* Fel neurobiol, Stanford Univ, 70-72; asst prof biol, Purdue Univ, 73-79; res assoc, Univ Ore, 79-81; ASSOC PROF CELL & STRUCT BIOL, UNIV ILL, 81- *Mem:* Soc Develop Biol; Soc Math Biol. *Res:* Mechanics of morphogenesis; principles of organization in organisms. *Mailing Add:* Dept Cell & Struct Biol 505 S Goodwin St Urbana IL 61801

MITTER, SANJOY, b Calcutta, India, Dec 9, 33; US citizen. ESTIMATION & STATISTICAL SIGNAL PROCESSING. *Educ:* Calcutta Univ, India, BS, 54; Imp Col Sci Technol, London, BSc, 57, PhD(elec eng), 65. *Prof Exp:* Develop engr, Brown Boveri & Co, Ltd, Baden, Switz, 57-61; res engr, Battelle Mem Inst, Geneva, 61-62; from asst prof to assoc prof & mem, Systs Res Ctr, Case Western Reserve Univ, 65-69; vis assoc prof, Mass Inst Technol, 69-70, assoc prof elec eng, 70-73, dir, Lab Info & Decision Systs, 81-86, PROF ELEC ENG, MASS INST TECHNOL, 73-, CO-DIR, LAB INFO & DECISION SYSTS, 86-, DIR, CTR INTEL CONTROL SYSTS, 86- *Concurrent Pos:* Vis prof, Inst Res Info & Automation, Versailles, 70, Imp Col, London, 72, Technische Univ, Berlin, 73-74, Univ Groningen, Holland, 76, Dept Systs Sci & Math, Wash Univ, 77, Dept Elec Eng, Univ Md, 77, Sch Math, Tata Inst Fundamental Res, Bombay, 79, Math Inst, Univ Florence, 81, Scuola Normale Superiore, Italy, 83-88; Consult, Cleveland Elec Illuminating Co, Ohio, 66-70, Charles Stark Draper Lab, Cambridge, Mass, 72-76, IBM Watson Res Lab, NY, 79-80, Sci Systs Inc, Cambridge, Mass, 76-; mem, Panel on Futore Directions in Control Theory, NSF, 86-87; mem, Comt Recommendations US Army Basic Res, 86-88; lectr, Markov Random Fields & Prob Comput Visions, Nagoya, Japan, 85, Info Processing Conf, Tokyo, Japan, 87; dir, Softron, Inc, 88- *Mem:* Nat Acad Eng; fel Inst Elec & Electronics Engrs; Am Math Soc; AAAS; Soc Appl Math. *Res:* Contributions in the area of control systems and theory estimation and statistical signal processing, computational methods in optimal control; author or co author of over 80 publications; control of delay and distributed parameter systs; optimisation, linear syst theory. *Mailing Add:* Mass Inst Technol Rm 35-308 77 Massachusetts Ave Cambridge MA 02139

MITTERER, RICHARD MAX, b Lancaster, Pa, Sept 8, 38; m; m 72; c 4. GEOCHEMISTRY. *Educ:* Franklin & Marshall Col, BS, 60; Fla State Univ, PhD(geol), 66. *Prof Exp:* Fel geophys lab, Carnegie Inst, 66-67; asst prof geosci, Southwest Ctr Advan Studies, 67-69; from asst prof to assoc prof, 69-80, head dept, 75-85, PROF GEOSCI, UNIV TEX, DALLAS, 80- *Concurrent Pos:* Mem bd dir, Geol Info Library, Dallas, 80-85; mem adv bd, Petrol Res Fund (ACS), 87-90. *Mem:* AAAS; Geol Soc Am; Soc Econ Paleont & Mineral; Geochem Soc; Am Geophys Union. *Res:* Amino acid diagenesis; carbonate geochemistry; sedimentary geochemistry. *Mailing Add:* Progs for Geosci Univ Tex Dallas PO Box 830688 Richardson TX 75083-0688

MITTLEMAN, JOHN, b Roslyn Heights, NY. UNDERWATER NONDESTRUCTIVE TESTING. *Educ:* Cornell Univ, BS, 69; Mass Inst Technol, MS, 70. *Prof Exp:* MEM, ENG & TEST-EVAL DEPT, NAVAL COASTAL SYSTS CTR, 74- *Concurrent Pos:* Navy weapons control systs fel, Mass Inst Technol, 73-74. *Honors & Awards:* Photogrammetry Award, Bausch & Lamb, 69; Solberg Award, Am Soc Naval Eng, 81. *Mem:* Am Soc Photogrammetry; Marine Technol Soc; Am Soc Testing & Mat; Am Soc Nondestructive Testing; Inst Diving. *Res:* Underwater stereophotographic, ultrasonic, and magnetic particle inspection systems. *Mailing Add:* Code S130 Naval Coastal Syst Lab Panama City FL 32407

MITTLEMAN, MARVIN HAROLD, b New York, NY, Mar 13, 28; m 55; c 3. PHYSICS. *Educ:* Polytech Inst Brooklyn, BS, 49; Mass Inst Technol, PhD(physics), 53. *Prof Exp:* Instr physics, Columbia Univ, 52-55; staff scientist, Lawrence Radiation Lab, Livermore, Calif, 55-65; staff scientist space sci lab, Univ Calif, Berkeley, 65-68; assoc prof physics, 68-69, exec officer PhD prog, 70-74, PROF PHYSICS, CITY COL NEW YORK, 69- *Concurrent Pos:* Div Sci & Indust Res, Brit Govt fel, Univ Col, Univ London, 62-63; NASA grant, Univ Calif, Berkeley, 69-70; consult, Lockheed Aircraft Corp, Calif, Convair, Inst Defense Analysis, DC & Goddard Space Flight Ctr, NASA. *Mem:* Fel Am Phys Soc. *Res:* Atomic scattering and structure; quantum optics. *Mailing Add:* Dept of Physics City Col of New York 138th St New York NY 10031

MITTLER, ARTHUR, b Paterson, NJ, July 15, 43; m 66. PHYSICS. *Educ:* Drew Univ, BA, 65; Univ Ky, MS, 67, PhD(physics), 70. *Prof Exp:* Asst prof, 69-77, ASSOC PROF PHYSICS, UNIV LOWELL, 77- *Mem:* AAAS; Am Phys Soc; Am Nuclear Soc. *Res:* Low-energy nuclear physics; neutron cross section measurements. *Mailing Add:* 18 Abbott Lane Chelmsford MA 01824

MITTLER, JAMES CARLTON, b Denver, Colo, Nov 7, 35. PHYSIOLOGY, ENDOCRINOLOGY. *Educ:* Univ NMex, BS, 56; Univ Ill, Urbana, MS, 61; Mich State Univ, PhD(physiol), 66. *Prof Exp:* Instr physiol, Rutgers Univ, 65-67; res fel med, Sch Med, Tulane Univ, 67-69; res physiologist, Vet Admin, 69-70; res fel endocrinol, Coney Island Hosp, Maimonides Med Ctr, 70-72; SUPV CHEMIST, VET ADMIN, 72- *Concurrent Pos:* Adj assoc prof med, NJ Med Sch, 82- *Mem:* Endocrine Soc; Am Soc Andrology; Int Soc Psychoneuroendocrinol. *Res:* Neuroendocrinology; physiology of reproduction; hormone assays. *Mailing Add:* Med Dept US Dept Veterans Affairs Med Ctr East Orange NJ 07019

MITTLER, ROBERT S, EXPERIMENTAL BIOLOGY, IMMUNOLOGY. *Educ:* State Univ NY, Stony Brook, BS, 72; NY Univ, MS, 74, PhD(immunol), 77. *Prof Exp:* Adj asst prof biol, Nassau Community Col, Garden City, NY, 72-78; NIH fel, Sloan-Kettering Inst Cancer Res, 77-78; prin investr, Becton Dickinson Res Ctr, NC, 78-80; group leader, Molecular Immunol Lab, Immunobiol Div, Ortho Pharmaceut Corp, NJ, 80-84; sr res scientist, 84-88, RES FEL, DEPT IMMUNOL, BRISTOL-MYERS SQUIBB CO, 88- *Concurrent Pos:* Adj asst prof immunol, Sch Pub Health, Univ NC, Chapel Hill, 78-80 & Dept Microbiol & Immunol, NY Med Col, Valhalla, 89-; adj prof immunol, Dept Molecular & Cell Biol, Univ Conn, Storrs, 88-; ad hoc reviewer, J Immunol. *Mem:* Sigma Xi; Am Soc Microbiol; Am Asn Immunologists; Soc Anal Cytol. *Mailing Add:* Pharmaceut Res & Develop Div Bristol-Myers Squibb Co Five Research Pkwy Wallingford CT 06492-7660

MITTMAN, BENJAMIN, b Chicago, Ill, Dec 24, 28; m 50; c 2. COMPUTER SCIENCE. *Educ:* Ill Inst Technol, BS, 50; Univ Calif, Los Angeles, MA, 51. *Prof Exp:* Mathematician, Boeing Airplane Co, 51-53; sci rep, Remington Rand Univac Div, Sperry Rand Corp, 56-58; mathematician, Armour Res Found, Ill Inst Technol, 58-65, mgr comput appln, Ill Inst Technol Res Inst, 65-66; PROF COMPUT SCI, NORTHWESTERN UNIV, 66-, DIR VOGELBACK COMPUT CTR, 66- *Concurrent Pos:* Mem panel comput applns in res, NSF. *Mem:* AAAS; Asn Comput Mach; Am Soc Info Sci; Inst Elec & Electronics Engrs; Int Comput Chess Asn. *Res:* Information retrieval; computing center management; computer graphics; computer chess. *Mailing Add:* Dept Comput Sci Northwestern Univ 633 Clark St Evanston IL 60208

MITTON, JEFFRY BOND, b Glen Ridge, NJ, Mar 16, 47; m 69. POPULATION GENETICS. *Educ:* Univ Conn, BA, 69; State Univ NY Stony Brook, PhD(ecol, evolution), 73. *Prof Exp:* NIH fel genetics, Univ Calif, Davis, 73-74; asst prof, 74-79, ASSOC PROF BIOL, DEPT ENVIRON POP & ORGANISMIC BIOL, UNIV COLO, BOULDER, 79-, RES ASSOC, INST ARCTIC ALPINE RES, 80- *Concurrent Pos:* Vis res scientist, Marine Biol Lab, Woods Hole, Mass; John Simon Guggenheim Fel, 83. *Mem:* Genetics Soc Am; Soc Study Evolution (secy, 82-84); AAAS; Sigma Xi; Soc Syst Zool. *Res:* Processes of natural selection resulting in population structuring and geographic variation of gene frequencies; protein polymorphisms; multi locus systems. *Mailing Add:* Dept Environ Pop & Org Biol Univ Colo Boulder CO 80301

MITUS, WLADYSLAW J, b Zywiec, Poland, May 14, 20; US citizen; m 52; c 2. PATHOLOGY, HEMATOLOGY. *Educ:* Univ Edinburgh, MB, ChB, 46. *Prof Exp:* Intern med, Weymouth & Dist Hosp, Eng, 48-49; registr, Kilton Hosp Workshop, 49-52; res pathologist, Children's Hosp, Sheffield, 52-53; res assoc, New Eng Ctr Hosp, Boston, 57-69; ASSOC PROF MED, TUFTS UNIV, 65-; CHIEF HEMAT & DIR RES, CARNEY HOSP, 69- *Concurrent Pos:* Fel path, City Hosp, Cleveland, Ohio, 54-55; fel hemat, Blood Res Lab, New Eng Ctr Hosp, Boston, 55-57; USPHS grant, 57-58; asst prof, Tufts Univ, 60-65; consult, Med Found, Boston, 61-63. *Mem:* Am Soc Exp Path; Am Soc Hemat; sr mem Am Fedn Clin Res. *Mailing Add:* Dept Hemat, Carney Hospital 2100 Dorchester Ave Dorchester MA 02124

MITYAGIN, BORIS SAMUEL, b Voronevzh, USSR, Aug 12, 37. MATHEMATICS. *Educ:* Moscow Univ, PhD(math & physics), 61, DS, 63. *Prof Exp:* PROF MATH, OHIO STATE UNIV, 79- *Honors & Awards:* Moscow Math Soc Prize, 60. *Mem:* Am Math Soc; Soc Indust & Appl Math. *Mailing Add:* Dept Math Ohio State Univ 231 W 18th Ave Columbus OH 43210

MITZNER, KENNETH MARTIN, b Brooklyn, NY, May 7, 38; m 68; c 3. ELECTROMAGNETIC SCATTERING. *Educ:* Mass Inst Tech, BS, 58, Calif Tech Inst, MS, 59, PhD(elec eng), 64. *Prof Exp:* Mem tech staff, Hughes Aircraft, 59-64; PRIN ENGR, NORTHROP CORP, 64- *Concurrent Pos:* Instr, Univ Calif, Santa Barbara, 64-65; mem, Comm B, US Nat Comt, Int Union Radio Sci, 79-; deleg, 20th Gen Assembly, Int Union Radio Sci, 81; mem, Electromagnetics Acad, 90- *Mem:* Fel Inst Elec & Electronics Engrs. *Res:* Research in the theory of electromagnetic scattering and its applications. *Mailing Add:* Northrop Corp B-2 DivMS W944/AP 8900 E Wash Blvd Pico Rivera CA 90660-0848

MIURA, CAROLE K MASUTANI, b Hilo, Hawaii, June 8, 38; m 62; c 2. MATHEMATICAL STATISTICS. *Educ:* Cornell Univ, BA, 60; Univ Hawaii, MA, 62; Boston Univ, PhD(math), 73. *Prof Exp:* Teaching asst, Dept Math, 61-62, instr, 62-65, ASST PROF MATH DISCIPLINE, UNIV HAWAII, HILO, 73- *Mem:* Am Statist Asn. *Res:* Theoretical studies of inverse gaussian distribution; statistical studies of remedial education. *Mailing Add:* Univ Hawaii of Hilo 1175 Mawono St Hilo HI 96720

MIURA, GEORGE AKIO, b Honolulu, Hawaii, Aug 6, 42. PLANT HORMONES, ALGAL TOXINS. *Educ:* Univ Hawaii, BS, 64; Ind Univ, PhD(plant physiol), 68. *Prof Exp:* Res biologist, Univ Calif, San Diego, 68-70; res fel, McMaster Univ, Hamilton, Ont, 70-72; biol sci asst, Med Res Inst Infect Dis, US Army, 79-90; INSTR, ACAD HEALTH SCI, FT SAM HOUSTON, TEX, 90- *Mem:* Sigma Xi. *Res:* Plant hormes; algal toxins; plant tissue culture. *Mailing Add:* Lab Sci Div Acad Health Sci Ft Sam Houston TX 78234

MIURA, ROBERT MITSURU, b Selma, Calif, Sept 12, 38; m 87; c 4. NEUROBIOLOGY, NONLINEAR WAVE. *Educ:* Univ Calif, Berkeley, BS, 60, MS, 62; Princeton Univ, MA, 64, PhD(aerospace eng), 66. *Prof Exp:* Res assoc nonlinear wave propagation, Plasma Physics Lab, Princeton Univ, 65-67; asst prof math, NY Univ, 68-71; assoc prof, Vanderbilt Univ, 71-75; vis assoc prof, 75-76, assoc prof, 76-78, PROF MATH, INST APPL MATH, PHARMACOL & THERAPEUTS, UNIV BC, 78- *Concurrent Pos:* Assoc res scientist, Courant Inst Math, NY Univ, 67-69; John Simon Guggenheim fel, 80-81; sr Killam hon fel, Univ BC, 80-81; vis prof, Kyoto Univ, 81, Univ Wash & Univ Calif, Los Angeles, 85-87; assoc ed, Can J Math, 81-85, Japan J Industrial & Appl Math, 84-; chmn, comt math in life sci, Am Math Soc-Soc Indust & Appl Math, 81-84; mem bd dirs, Can Math Soc, 83-85; mem adv bd, J Math Biol, 82-; mem steering comt, Math Res Ctr, Univ Montreal, 90- *Mem:* AAAS; Am Math Soc; Soc Indust & Appl Math; Can Math Soc; Can Appl Math Soc; Sigma Xi; Soc Math Biol; Can Soc Theoret Biol. *Res:* Nonlinear partial differential equations; nonlinear wave propagation; fluid mechanics; asymptotic methods; kinetic theory; Mathematical neurophysiology; reaction-diffusion problems; singular perturbation boundary-value problems of differential-difference equations. *Mailing Add:* Dept Math Univ BC Vancouver BC V6T 1Y4 Can

MIURA, TAKESHI, medical entomology, for more information see previous edition

MIWA, GERALD T, b Poston, Ariz, June 14, 45. BIOCHEMICAL TOXICOLOGY. *Educ:* Univ Calif, Los Angeles, PhD(pharmacol), 75. *Prof Exp:* From res fel to sr res fel, Animal Drug Metab, Merck, Sharp & Dohme, 82-87; VPRES, DRUG SAFETY & METAB, RES INST, GLAXO INC *Mailing Add:* Dept Drug Metabology Glaxo Inc Five Moore Dr Research Triangle Park NC 27709

MIX, DWIGHT FRANKLIN, b Fayetteville, Ark, Feb 18, 32; m 54; c 2. ELECTRICAL ENGINEERING. *Educ:* Univ Ark, BS, 56, MS, 61; Purdue Univ, PhD(elec eng), 66. *Prof Exp:* Design engr, Gen Dynamics/Convair, Tex, 56-59 & Tex Instruments, Inc, 59; instr elec eng, Univ Ark, 59-61 & Purdue Univ, 61-65; asst prof, 65-70, ASSOC PROF ELEC ENG, UNIV ARK, FAYETTEVILLE, 70- *Mem:* Inst Elec & Electronic Engrs; Am Soc Eng Educ. *Res:* Statistical communication theory. *Mailing Add:* Dept Elec Eng Univ Ark Fayetteville AR 72701

MIX, MICHAEL CARY, b Deer Park, Wash, June 27, 41; m 62; c 2. INVERTEBRATE PATHOLOGY, ENVIRONMENTAL CARCINOGENS. *Educ:* Wash State Univ, BS, 63; Univ Wash, PhD(fisheries), 70. *Prof Exp:* Asst prof biol, 70-74, ASSOC PROF BIOL, ORE STATE UNIV, 74- *Concurrent Pos:* Consult, Shapiro & Assoc, Nalco Environ, Inc. *Mem:* AAAS; Nat Shellfisheries Asn; Soc Invert Path; NY Acad Sci; Sigma Xi. *Res:* Chemical carcinogens in the marine environment; histopathological effects of irradiation on higher invertebrates; experimental invertebrate pathobiology; diseases of invertebrates; cell renewal systems of mollusks; invertebrate oncology. *Mailing Add:* Dept of Gen Sci Ore State Univ Corvallis OR 97331

MIXAN, CRAIG EDWARD, b Berwyn, Ill, July 21, 46; m 69; c 2. ORGANIC CHEMISTRY. *Educ:* Holy Cross Col, BA, 68; Northwestern Univ, PhD(org chem), 72. *Prof Exp:* Res assoc, Dow Chem Co, 72-89; SR PATENT AGENT, DOWELANCO, 89- *Mem:* Am Chem Soc; Sigma Xi. *Res:* Synthesis of bioactive compounds; agricultural products; process research and development. *Mailing Add:* 3558 Hawthorne W Dr Carmel IN 46032-9287

MIXON, AUBREY CLIFTON, b Tifton, Ga, Sept 20, 24; m 43. AGRONOMY, PLANT BREEDING. *Educ:* Univ Ga, BSA, 49; NC State Univ, MSA, 53; Auburn Univ, PhD(plant path), 66. *Prof Exp:* Asst agronomist, Fla Agr Exten Serv, 53-57; res agronomist, Coop Ala Agr Exp Sta & USDA, 57-73, res agronomist, Col Agr, Coop Univ Ga Coastal Plain Sta & USDA, 73-81. *Concurrent Pos:* Res agronomist, Auburn Agr Exp Sta, 57-73; coordr, USDA Nat Winter Peanut Nursery, Mayaguez, PR, 72-; adj res assoc, Col Agr, Univ Ga, 73-81; recorder, Nat Peanut Prod Workshop, 74. *Honors & Awards:* Twenty-Five Year Serv Award, USDA, 77. *Mem:* Am Peanut Res & Educ Asn; Am Soc Agron; AAAS; Am Inst Biol Sci; Sigma Xi. *Res:* Breeding, agronomic, physiological, ecological and pathological investigations associated with developing peanut varieties that are resistant to toxin-producing fungi. *Mailing Add:* PO Box 748 Tifton GA 31793

MIXON, FOREST ORION, chemical engineering, mathematics; deceased, see previous edition for last biography

MIXTER, RUSSELL LOWELL, b Williamston, Mich, Aug 7, 06; m 31; c 4. ANATOMY. *Educ:* Wheaton Col, Ill, AB, 28; Mich State Col, MS, 30; Univ Ill, PhD(anat), 39. *Prof Exp:* From instr to prof zool, Wheaton Col, Ill, 28-79; RETIRED. *Concurrent Pos:* Instr, Univ Ill, 35-36; ed jour, Am Sci Affil, 64-68; vis prof, Trinity Col, 74-77, Barat Col, 77-78, Judson Col, 82-83 & 84. *Mem:* Am Sci Affil (pres, 51-54). *Res:* Macrophages of connective tissue; flexed tail in mice; evolution; spiders of Black Hills. *Mailing Add:* 120 Windsor Park Dr #206 Carol Stream IL 60188

MIYA, TOM SABURO, b Hanford, Calif, Apr 6, 23; m 48; c 1. PHARMACOLOGY. *Educ:* Univ Nebr, BSc, 47, MSc, 48; Purdue Univ, PhD(pharmacol), 52. *Hon Degrees:* DSc, Univ Nebr, 86. *Prof Exp:* Asst, Univ Nebr, 47-48; from instr to asst prof pharmacol, Purdue Univ, 48-56; chmn dept, Univ Nebr, 56-57; prof, Purdue Univ, West Lafayette, 58-76, head dept, 64-76; PROF PHARMACOL, SCH MED & DEAN SCH PHARM, UNIV NC, CHAPEL HILL, 77-, CHMN TOXICOL PROG, SCH MED, 80- *Concurrent Pos:* Mem rev comt, US Pharmacopoeia, 70-80 & pharmacol-toxicol prog comt, Nat Inst Gen Med Sci; assoc ed, Toxicol & Appl Pharmacol; chmn chem & biol info handling panel, Res Resources Div, HEW, NIH, 75-76; nat adv coun, Nat Inst Environ Health, 87-90. *Honors & Awards:* Award, Am Pharmaceut Asn, 64; Merit Award, Soc Toxicol, 84; Toxicology Educ Award, 86. *Mem:* AAAS; Am Soc Pharmacol & Exp Therapeut; Soc Toxicol (pres, 79); Am Chem Soc; Am Asn Cols Pharm (pres, 75-76). *Res:* Hormonal determinants of drug metabolism; factors modifying the normal disposition of drugs. *Mailing Add:* Sch of Pharm Univ NC CBN 7360 Chapel Hill NC 27599

MIYADA, DON SHUSO, b Oceanside, Calif, May 21, 25; m 60; c 4. BIOCHEMISTRY. *Educ:* Univ Calif, Los Angeles, BS, 49; Mich State Univ, PhD, 53. *Prof Exp:* Res assoc dept food tech, Univ Calif, 53-55; res assoc dept chem, Ohio State Univ, 55-56; res assoc, McArdle Mem Lab, Univ Wis, 56-57; asst res biochemist dept med, Univ Calif, Los Angeles, 57-61; biochemist, Long Beach Vet Admin Hosp, 61-67; biochemist, Biochem Procedures, Inc, 67-69; biochemist, Orange County Med Ctr & adj asst prof, 69-76, ADJ ASSOC PROF DEPTS PATH & BIOCHEM, UNIV CALIF, IRVINE, 76- *Mem:* AAAS; Am Chem Soc; Am Asn Clin Chem; NY Acad Sci. *Res:* Clinical chemistry. *Mailing Add:* Univ Calif Irvine Med Ctr 101 City Dr S Orange CA 92668

MIYAGAWA, ICHIRO, b Hiratsuka, Japan, Mar 5, 22; m 49; c 3. CHEMICAL PHYSICS. *Educ:* Nagoya Univ, BS, 45; Univ Tokyo, DrS(chem physics), 54. *Prof Exp:* Res assoc chem, Nagoya Univ, 48-49; res assoc chem, Univ Tokyo, 50-55, asst prof, Inst Solid-State Physics, 60-62; res assoc, Duke Univ, 56-59, vis asst prof, 63-64; from asst prof to assoc prof, 65-71, PROF PHYSICS, UNIV ALA, 71- *Mem:* Am Phys Soc. *Res:* Dielectric constant of liquids; electron spin resonance of irradiated molecular crystals. *Mailing Add:* Box 870324 Tuscaloosa AL 35487-0324

MIYAI, KATSUMI, b Yokosuka, Japan, Oct 17, 31; m 66; c 3. PATHOLOGY. *Educ:* Keio Univ, Japan, MD, 56; Univ Toronto, PhD(path), 67; Am Bd Path, dipl, 62. *Prof Exp:* Intern med, US Naval Hosp, Yokosuka, Japan, 56-57; intern surg, Barnes Hosp, St Louis, Mo, 57-58; resident path, Jewish Hosp St Louis, 58-60; fel path, Johns Hopkins Univ Sch Med, 60-63; Nat Cancer Inst Can res fel path, Banting Inst, Fac Med, Univ Toronto, 63-67, from lectr to asst prof path, Fac Med, 68-70; from asst prof to assoc prof, 70-82, PROF PATH, SCH MED, UNIV CALIF, SAN DIEGO, 82- *Concurrent Pos:* Asst res pathologist, Univ Hosp San Diego, 70-, assoc pathologist, 76-82, pathologist, 82- *Mem:* Am Soc Cell Biol; Int Acad Path; AAAS; Electron Micros Soc Am; Can Asn Path; Am Asn Pathologists; Am Asn Study Liver Dis. *Res:* Pathology of hepatobiliary and gastrointestinal system with emphasis on the mechanism of cholestatic diseases. *Mailing Add:* Dept Path Sch Med Univ Calif San Diego M-012 La Jolla CA 92093

MIYAKODA, KIKURO, b Yonago City, Japan, Nov 7, 27; m 54; c 1. ATMOSPHERIC DYNAMICS, OCEANOGRAPHY. *Educ:* Univ Tokyo, BS, 56, PhD(geophys), 61. *Prof Exp:* GROUP LEADER & RES METEOROLOGIST, GEOPHYS FLUID DYNAMICS LAB, PRINCETON UNIV, 65-, VIS PROF METEOROL, 68- *Honors & Awards:* Gold Medal, US Dept Com, 72; Fujiwava Award, Japan Metereol Soc, 83; Gold Medal, US Dept Com, 88; Carl-Gustaf Rossby Medal, Am Meteorol Soc, 91. *Mem:* Am Geophys Union; fel Am Meteorol Soc; Japan Meteorol Soc; Sigma Xi. *Res:* Numerical weather prediction; long-range weather forecast; meteorological dynamics; feasibility of the seasonal forecast involving air-sea interactions. *Mailing Add:* Geophys Fluid Dynamics Lab Princeton Univ Princeton NJ 08542

MIYAMOTO, MICHAEL DWIGHT, b Honolulu, Hawaii, Apr 22, 45; m 73; c 2. NEUROPHARMACOLOGY. *Educ:* Northwestern Univ, Evanston, BA, 66, PhD(biol), 71. *Prof Exp:* Instr pharmacol, Rutgers Med Sch, Col Med & Dent, NJ, 70-72; asst prof, Health Ctr, Univ Conn, 72-78; assoc prof, 78-87, PROF PHARMACOL, COL MED, E TENN STATE UNIV, 87- *Concurrent Pos:* Prin investr, Pharm Mfrs Asn Found grant, 75; USPHS grant neurol dis & stroke, 75-79 & 88-91; Epilepsy Found Am award, 76; consult, NIH, NSF, Nat Inst Aging. *Mem:* Am Soc Pharmacol & Exp Therapeut; Soc Neurosci. *Res:* Neuromuscular transmitter release; mathematical modeling. *Mailing Add:* Dept of Pharmacol ETenn State Univ Col of Med Johnson City TN 37614-0002

MIYAMOTO, MICHAEL MASAO, b Gardena, Calif, July 2, 55; m. MOLECULAR EVOLUTION, SYSTEMATICS. *Educ:* Calif State Univ, Dominquez Hills, BA, 77; Univ Southern Calif, PhD(biol sci), 82. *Prof Exp:* Res assoc, dept biol, Univ Miami, 82-84; res assoc, dept anat, Wayne State Univ, 85-86; ASST PROF ZOOL, UNIV FLA, 87- *Concurrent Pos:* Asst cur, div herpet, dept biol, Univ Miami, 82-84; asst cur, Fla Mus Natural Hist, 87- *Honors & Awards:* Presidential Young Investr, Nat Sci Found, 88- *Mem:* AAAS; Am Soc Ichthyologists & Herpetologists; Am Soc Mammologists; Soc Study Evolution; Molecular Biol & Evolution Soc; Soc Syst Zool. *Res:* Molecular evolution and systematics of mammals and other vertebrates, as ascertained from DNA sequences. *Mailing Add:* Dept Zool Univ Fla Gainesville FL 32611

MIYAMOTO, SEIICHI, b Nagasaki, Japan, Oct 1, 44; US citizen; c 3. AGRONOMY, HORTICULTURE. *Educ:* Gifu Univ, BS, 67; Kyushu Univ, MS, 69; Univ Calif, Riverside, PhD(soil sci), 71. *Prof Exp:* Res assoc soil sci, Univ Ariz, 71-75; res assoc & asst prof soil sci, NMex State Univ, 75-77; assoc prof, 77-87, PROF SOIL & WATER SCI, EL PASO RES CTR, TEX A&M UNIV, 87- *Concurrent Pos:* Consult, World Bank. *Mem:* Soil Sci Soc Am; Am Soc Agron. *Res:* Soil salinity, irrigation and drainage; soils and soil science; water quality. *Mailing Add:* 11417 Dean Refram Dr El Paso TX 79936

MIYANO, KENJIRO, b Okayama, Japan, May 25, 47; m 75; c 1. LIQUID INTERFACES, MONOMOLECULAR FILMS. *Educ:* Univ Tokyo, BS, 70; Northwestern Univ, PhD(physics), 75. *Prof Exp:* Res assoc physics, Lawrence Berkeley Lab, 75-76; res assoc, Argonne Nat Lab, 76-78, asst physicist, 78-81, physicist, 81-; AT RES INST ELEC COMMUN, TOHOKU UNIV, JAPAN. *Mem:* Am Phys Soc; Phys Soc Japan; Am Chem Soc. *Res:* Ultrasonics and light scattering in liquid crystals; surface acoustic waves; liquid interfaces and monomolecular films. *Mailing Add:* Res Inst Elec Commun Tohoku Univ Katahira Bunkyo-Ku 112 Tokyo Japan

MIYASAKA, KYOKO, b Utsunomiya City, Japan, Mar 15, 50; m 74; c 2. GASTRO INTESTINE PHYSIOLOGY, GERONTOLOGY. *Educ:* Tokyo Med & Dent Univ, MD, DMes Sci. *Prof Exp:* Clin fel internal med, Tokyo Med & Dent Univ, 74-78; postdoctoral physiol, Univ Calif, San Francisco, 79-81; res instr gastrenterol, Univ Tex, San Antonio, 81-82; ASSOC PROF PHYSIOL, TOKYO METRO INST GERONT, 83- *Concurrent Pos:* Lectr, Itabashi Nursing Sch, 84-90; ed, Japan Soc Biomed Geront, 88- *Mem:* Am Gastroenterol Asn; Am Physiol Soc; Am Aging Soc; NY Aging Soc; Fedn Am Socs Exp Biol. *Res:* Regulation of pancreatic exocrine secretion in conscious rats, mainly neurohormonal control of pancreas; aging on pancreatic function. *Mailing Add:* Dept Clin Physiol Tokyo Metro Inst Geront 35-2 Sakaecho Itabashiku Tokyo 173 Japan

MIYASHIRO, AKIHO, b Okayama, Japan, Oct 30, 20. PETROLOGY. *Educ:* Univ Tokyo, BSc, 43, PhD(petrol), 53. *Prof Exp:* Asst instr petrol, Univ Tokyo, 46-58, assoc prof, 58-67; vis prof, Lamont-Doherty Geol Observ, Columbia Univ, 67-70; prof, 70-90, EMER PROF GEOL, STATE UNIV NY, ALBANY, 91- *Concurrent Pos:* Vis int scientist, Am Geol Inst, 65; vis prof, Univ Kyoto, 80; NSF res grant, 71-90. *Honors & Awards:* Prize, Geol Soc Japan, 58; A L Day Medal, Geol Soc Am, 77; Paul Fourmarier Medal, Royal Acad Sci, Belgium, 81; P Base Mem Medal, Asiatic Soc, India, 84. *Mem:* Mineral Soc Am; Geol Soc Am; Geol Soc France; Geol Soc London. *Res:* Metamorphic and igneous petrology; earth science, especially geology. *Mailing Add:* 14 Stonehenge Dr Albany NY 12203

MIYOSHI, KAZUHISA, b Kobe, Japan, Feb 15, 46; US citizen; m 73; c 4. AERONAUATICAL, ASTRONAUTICAL & CERAMICS ENGINEERING. *Educ:* Osaka Inst Technol, BS, 68; Osaka Univ, MS, 70, PhD(eng), 75. *Prof Exp:* Asst prof eng, Kanazawa Univ, 70-78; res assoc tribol, Nat Res Coun, 76-78; res scientist, 79-82, SR RES SCIENTIST SURFACE SCI & TRIBOL, LEWIS RES CTR, NAT AERONAUT & SPACE ADMIN, 82- *Mem:* Soc Tribologists & Lubrication Engrs; Am Vacuum Soc; Mat Res Soc. *Res:* Pioneered tribology research in magnetic recording systems; tribology and surface science of structural ceramics and composites, ceramic coatings and advanced solid lubricants. *Mailing Add:* 5541 Quail Run North Olmsted OH 44070

MIZE, CHARLES EDWARD, b Smithville, Tex, Mar 3, 34; m 62; c 1. PEDIATRICS, BIOCHEMISTRY. *Educ:* Rice Inst, BA, 55; Johns Hopkins Univ, PhD(biochem), 61, MD, 62. *Prof Exp:* From intern pediat to resident, Johns Hopkins Hosp, 62-64; staff assoc metab, Nat Heart Inst, 64-67; asst prof, 67-73, ASSOC PROF PEDIAT & BIOCHEM, UNIV TEX HEALTH SCI CTR DALLAS, 74- *Mem:* Soc Pediat Res; Am Soc Neurochem; Am Soc Human Genetics; Am Fedn Clin Res; NY Acad Sci; Sigma Xi. *Res:* Biochemistry of nutrition in growth and development; metabolic disorders of childhood. *Mailing Add:* Dept Pediat-Univ Tex Health Sci Ctr 5323 Harry Hines Blvd Dallas TX 75235-9063

MIZE, JACK PITTS, b Kansas City, Mo, July 27, 23; m 49; c 3. NUCLEAR PHYSICS. *Educ:* Duke Univ, BS, 47; Univ Rochester, MS, 49; Iowa State Col, PhD(physics), 53. *Prof Exp:* Res assoc physics, Inst Atomic Res, Iowa State Col, 53; mem staff, Los Alamos Sci Lab, 53-60; TECH STAFF MEM PHYSICS, TEX INSTRUMENTS, INC, DALLAS, 60- *Mem:* Fel Am Phys Soc. *Res:* Nuclear spectroscopy; plasma and solid-state physics. *Mailing Add:* 918 Beechwood Richardson TX 75080

MIZE, JOE H(ENRY), b Colorado City, Tex, June 14, 34; m 66; c 1. COMPUTER SIMULATION MODELING, PRODUCTION CONTROL. *Educ:* Tex Tech Col, BS, 58; Purdue Univ, MS, 63, PhD(indust eng), 64. *Prof Exp:* Indust engr, White Sands Missile Range, NMex, 58-61; assoc prof eng, Auburn Univ, 64-65, assoc prof indust eng, 66-69, dir comput ctr, 65-69; prof, indust eng, Ariz State Univ, 69-72; prof & head, dept indust eng & mgt, 72-80, dir, Inst Energy Analysis, 80-83 REGENTS PROF, INDUST ENG, OKLA STATE UNIV, 80- *Concurrent Pos:* Chmn, Tech Adv Coun, Southern Growth Policies Bd, 76-78; consult, various orgns, 64-; prin investr, several res projs, NSF, Dept Energy; ed indust eng, Prentice Hall Int Series, 72- *Honors & Awards:* H B Maynard Innovative Achievement Award, Am Inst Indust Engrs, 77; Frank & Lillian Gilbreath Indust Eng Award, 90. *Mem:* Fel Am Inst Indust Engrs (exec vpres, 78-80, pres, 81-82); fel Am Soc Eng Educ; Nat Soc Prof Engrs; Nat Acad Engr; Asn Comput Mach; Soc Computer Simulation. *Res:* Development of new modeling and simulation methodologies based upon sound mathematical and statistical foundations; modeling and analysis of socio-economic systems, such as energy systems, state econometric models. *Mailing Add:* 1511 N Glenwood Stillwater OK 74075

MIZEJEWSKI, GERALD JUDE, b Pittsburgh, Pa, Aug 1, 39; m 65; c 6. DEVELOPMENTAL PHYSIOLOGY, IMMUNOLOGY. *Educ:* Duquesne Univ, BS, 61; Univ Md, MS, 65, PhD(zool), 68. *Prof Exp:* From asst zool to res asst immunol, Univ Md, 61-68; from res assoc to lectr, Med Sch, Univ Mich, 68-71; asst prof physiol & immunol, Univ SC, 71-74; sr res scientist-IV, Wadsworth Ctr Labs & Res, 74-78, SR RES SCIENTIST-IV, BIRTH DEFECTS INST, DIV LABS & RES, NY STATE DEPT HEALTH, 78-, NEWBORN HYPOTHYROID SCREENING, 90- *Concurrent Pos:* Am Cancer Soc grant, Univ Mich, 69-70; Upjohn res grant, 69-70; Abbott radiopharmaceut res gift, 69-71 & Cal-Biochem res gift, 78-80; res assoc prof, Dept Pediat, Albany Med Col, 78-, adj assoc prof, Dept Obstet & Gynec, 80-; assoc prof, Grad Sch Pub Health, State Univ NY, Albany, 85-; assoc prof, Biol Dept, Union Col, Schnectady, NY. *Mem:* AAAS; Am Soc Zool; Am Inst Biol Sci; Reticuloendothelial Soc; NY Acad Sci; Am Fedn Clin Res. *Res:* Alpha-fetoprotein bioassay and perinatal and neonatal biology; MCF7 breast carcinoma; carcinoembryonic antigen, tumor transplantation; radiolabeled antibodies and antigens; hepatoma and lymphoma cell culture; cytotoxic antibodies; estradiol binding assays; sexual differentiation of gonads; study onset of puberty, immunoadsorbants; affinity chromatography; immunochemistry; cystic fibrosis; newborn screening; physiology of alpha-fetoprotein, alpha-fetoprotein receptors; tumor cell surface antigens; radioimmuno and enzyme-immuno assay; immuno-reactive trypsin; serine protease; alpha-fetoprotein fluorescent ligand. *Mailing Add:* Wadsworth Ctr Labs & Res NY State Dept Health Albany NY 12201

MIZEL, STEVEN B, b San Francisco, Calif, Oct 1, 47; m; c 3. MICROBIOLOGY. *Educ:* Univ Calif, Berkeley, BA, 69; Stanford Univ, PhD(pharmacol), 73. *Prof Exp:* Pharmaceut Mfrs Found postdoctoral fel, Dept Biochem, Colo State Univ, 73-74; NIH postdoctoral fel, Dept Biochem, Weizmann Inst Sci, Israel, 74-76; sr staff fel, Cellular Immunol Sect, Nat Inst Dent Res, 76-80; assoc prof, Pa State Univ, University Park, 80-85; PROF & CHMN, DEPT MICROBIOL & IMMUNOL, BOWMAN GRAY SCH MED, 85- *Concurrent Pos:* Assoc ed, J Immunol, 80-82 & sect ed, 82-86; chair, Allergy & Immunol Study Sect, NIH, 91- *Mem:* AAAS; Am Asn Immunologists; Asn Med Sch Dept Chairmen. *Res:* Cytokines; interleukin 1 and macrophage function; regulation of immune and inflammatory responses; transcription factors. *Mailing Add:* Dept Microbiol & Immunol Bowman Gray Sch Med 300 S Hawthorne Rd Winston-Salem NC 27103

MIZELL, LOUIS RICHARD, b Gettysburg, Pa, Jan 25, 18; m 43; c 4. TEXTILE CHEMISTRY. *Educ:* Gettysburg Col, AB, 38; Georgetown Univ, MS, 42. *Prof Exp:* Res assoc, Textile Found, Inc, 39-42; mem staff, Harris Res Labs, Inc, 46-58, asst dir, 58-67; wool mgr new mkt outlets, 67-80, WOOL DIR NEW MKT OUTLETS, INT SECRETARIAT, WOODBURY, 80- *Mem:* Am Chem Soc; Fiber Soc; Am Asn Textile Chem & Colorists; Am Inst Chem; World Future Soc. *Res:* Chemical and engineering research on fibrous materials; conception and development of new products and processes and taking them to commercial fruition on a world-wide scale. *Mailing Add:* 8122 Misty Oaks Blvd Sarasota FL 34243-3615

MIZELL, MERLE, b Chicago, Ill, Apr 25, 27; m 58; c 2. DEVELOPMENTAL GENETICS, ONCOLOGY. *Educ:* Univ Ill, Urbana, BS, 50, MS, 52, PhD(zool), 57. *Prof Exp:* Asst, Univ Ill, Urbana, 54-57; instr zool, 57-60, from asst prof to assoc prof biol, 60-70, PROF ANAT & BIOL, MED SCH, TULANE UNIV, 70-, DIR, CHAPMAN H HYAMS | LAB TUMOR CELL BIOL, 69. *Concurrent Pos:* Am Cancer Soc & Cancer Asn Greater New Orleans grant, Tulane Univ; NSF, NIH & Damon Runyon Mem Fund grants; consult, Spec Virus Cancer Prog, Nat Cancer Inst & proj site visitor, Cancer Res Centers, 69-; corp mem, Marine Biol Lab, Woods Hole, Mass; vis lectr, Tokyo, Cambridge Univ, workshop genetic eng, Berlin, Univ Bristol & numerous univs in the US; mem bd dir, Cancer Asn Greater New Orleans. *Mem:* AAAS; Am Soc Zool; Am Inst Biol Sci; Soc Exp Biol & Med (secy, 69-71); Soc Develop Biol; Sigma Xi; Am Asn Cancer Res; Am Soc Microbiol. *Res:* Oncogene expression during normal and malignant differentiation; mechanism of limb regeneration in spontaneous regeneration and induced regeneration; role of viruses as agents of normal and neoplastic differentiation; effects of the regeneration environment on neoplastic growths; tumor biology; genetic engineering; oncogenic herpes viruses; biology of natural occuring tumors. *Mailing Add:* Lab Tumor Cell Biol Tulane Univ New Orleans LA 70118

MIZELL, SHERWIN, b Chicago, Ill, Apr 27, 31; m 57; c 3. GROSS ANATOMY. *Educ:* Univ Ill, BS, 52, MS, 54, PhD(physiol), 58. *Prof Exp:* Res fel, Med Col SC, 58-59, from instr to asst prof anat, 59-64; assoc prof physiol sch med, Creighton Univ, 64-65; head sect, 76-84, assoc prof anat & physiol, 65-75, PROF ANAT, MED SCI PROG, IND UNIV, BLOOMINGTON, 75- *Mem:* Am Asn Anat; Am Physiol Soc. *Res:* Biological rhythms; synchronization and mechanisms responsible for changes in physiology and behavior. *Mailing Add:* Med Sci Prog Ind Univ Bloomington IN 47405

MIZERES, NICHOLAS JAMES, b Pittsburgh, Pa, Nov 13, 24; m 52; c 1. ANATOMY. *Educ:* Kent State Univ, BS, 48; Mich State Univ, MS, 51; Univ Mich, PhD(anat), 54. *Prof Exp:* From instr to assoc prof, 54-66, PROF ANAT, SCH MED, WAYNE STATE UNIV, 66- *Honors & Awards:* Lamp Award, 59. *Mem:* AAAS; Asn Am Med Cols; Am Asn Anat. *Res:* Human anatomy; cardiovascular research, especially heart coronary circulation and the autonomic nervous system; descriptive anatomy related to surgery. *Mailing Add:* Dept Anat 212 Scott Wayne State Univ 540 E Canfield Detroit MI 48201

MIZIOLEK, ANDRZEJ WLADYSLAW, b Hannover, Ger, Feb 17, 50; US citizen; m 74; c 2. SPECTROSCOPY & SPECTROMETRY. *Educ:* Wayne State Univ, BS, 71; Univ Calif, Berkeley, PhD(chem), 76. *Prof Exp:* Fel chem, Univ Calif, Irvine, 76-77; asst res chemist, Scripps Inst Oceanog, Univ Calif, San Diego, 77-81; RES PHYSICIST, US ARMY BALLISTIC RES LAB, 81- *Concurrent Pos:* Sci adv, Off Naval Res, 89. *Mem:* Am Chem Soc; Sigma Xi; Optical Soc Am; Soc Appl Spectros; Combustion Inst; Am Inst Aeronaut & Astronaut. *Res:* Application of laser spectroscopy and photochemistry to combustion and chemical analysis research. *Mailing Add:* Ignition & Combustion Br US Army Ballistic Res Lab Aberdeen Proving Ground MD 21005-5066

MIZIORKO, HENRY MICHAEL, b Philadelphia, Pa, Oct 11, 47; m 71; c 2. ENZYMOLOGY. *Educ:* St Joseph's Col, Pa, BS, 69; Univ Pa, PhD(biochem), 74. *Prof Exp:* Fel physiol chem, Sch Med, Johns Hopkins Univ, 74-77; from asst prof to assoc prof, 81-86, PROF BIOCHEM, MED COL WIS, 87-, VCHMN, 89-, INTERIM CHMN, 90- *Concurrent Pos:* Mellon Found fel, Sch Med, Johns Hopkins Univ, 76-77; NIH res career develop award, 79-84; vis prof, Max Planck Inst Exp Med, Gottingen, Fed Repub Germany, 83; Alexander von Humboldt fel, 83; mem, NIH Biomed Sci Study Sect, 87-90, NIH Reviewers Reserve, 90- *Mem:* Am Chem Soc; Am Soc Biol Chemists; Protein Soc. *Res:* Mechanism of enzyme action; regulation in biological systems; ketogenesis; photosynthetic carbon assimilation; cholesterogenesis. *Mailing Add:* Dept Biochem 8701 Watertown Plank Rd Milwaukee WI 53226

MIZMA, EDWARD JOHN, b Rochester, NY, Mar 29, 34; m 60, 84; c 3. CLINICAL CHEMISTRY, CHEMICAL ENGINEERING. *Educ:* Bucknell Univ, ScB, 55; Cornell Univ, PhD(chem eng), 59. *Prof Exp:* Tech assoc, Mfg Exp Div, 58-71, supvr Eng Div, 71-72, supvr Mfg Tech Div, 72-76, asst dir Health Safety & Human Fac Lab, 80-84, dir Clin Chem Tech Ctr, 80-84, dir Clin Prod Tech Ctr, 84-85, UNIT DIR PROJ MGT DIV, EASTMAN KODAK CO, 85- *Res:* Biochemical engineering; continuous electrophoresis equipment; rheology of viscous fluids especially polymer solutions; direction of pilot plant and semi-plant coating machines; technical supervision. *Mailing Add:* 35 Overlook Dr Hilton NY 14468

MIZUKAMI, HIROSHI, b Otaru-Shi, Japan, Oct 11, 32; m 59; c 1. BIOPHYSICS, HEMATOLOGY. *Educ:* Int Christian Univ, Tokyo, BA, 57; Univ Ill, PhD(biophys), 63. *Prof Exp:* Res fel phys chem, Univ Minn, 62-65, res fel med, Univ Hosps, 64-65; from asst prof to assoc prof, 65-74, PROF BIOL, WAYNE STATE UNIV, 74- *Concurrent Pos:* Vis res prof, Tokyo Med & Dent Univ, 73-74. *Mem:* AAAS; Biophys Soc; Am Chem Soc. *Res:* Structure and function of proteins, membrane proteins of erythrocytes; sickle cells, NMR of biological molecules. *Mailing Add:* Dept Biol Wayne State Univ Detroit MI 48202

MIZUNO, NOBUKO S(HIMOTORI), b Oakland, Calif, Apr 20, 16; m 42. BIOLOGY. *Educ:* Univ Calif, Berkeley, BA, 37, MA, 39; Univ Minn, PhD(biochem), 56. *Prof Exp:* Res asst, Inst Exp Biol, Univ Calif, 39-41; instr med technol, Macalester Col, 43-51; res assoc, Col Vet Med, Univ Minn, 56-62, res assoc, Dept Surg, 63-78; res biochemist, Vet Admin Med Ctr, Minneapolis, 62-78, prin investr, 67-78; RETIRED. *Mem:* Am Soc Biol Chemists; Am Inst Nutrit; Am Chem Soc; Am Asn Cancer Res; NY Acad Sci; Soc Exp Biol & Med. *Res:* Metabolism and pharmacological effects of cancer chemotherapy drugs, particularly as they affect DNA; biochemical studies in hypoplastic anemia; prevention of muscular dystrophy with vitamin E. *Mailing Add:* 3628 Loma Way San Diego CA 92106-2034

MIZUNO, SHIGEKI, b Tsinan, China, Aug 17, 36; Japanese citizen; m 64; c 1. ANIMAL GENE EXPRESSION, CHROMOSOME & CHROMATIN STRUCTURE. *Educ:* Univ Tokyo, BS, 59, MS, 61, DrAgrSc, 64. *Prof Exp:* Res assoc develop biol, Dept Microbiol & Friday Harbor Lab, Univ Wash, 70-72; res assoc molecular cytogenetics, Dept Zool, Univ Leicester, UK, 72-75; res assoc molecular biol, Dept Molecular Med, Mayo Clin, 75-77; asst microbiol, Fac Agr & Inst Appl Microbiol, Univ Tokyo, 64-70, asst prof molecular biol, Dept Appl Biol Sci, 77-82; assoc prof, 82-84, PROF BIOCHEM, DEPT AGR CHEM, TOHOKU UNIV, 84-, PROF MOLECULAR BIOL, GENE RES CTR, 86- *Concurrent Pos:* Ed, J Biochem, Tokyo, 88-90, Agr Biol Chem, Tokyo, 90-92. *Mem:* Am Soc Microbiol; Am Soc Cell Biol; Int Soc Develop Biologists. *Res:* Structure and function of fibroin 1-chain; mechanisms of secretion-deficient mutations of fibroin; coordinate control of fibroin h- and 1-chain gene expression; molecular structure of chicken w-heterochromatin; gene expressions during sex differentiation of chicken embryos. *Mailing Add:* Dept Agr Chem Tohoku Univ 1-1 Tsutsumidori-Amamiyamachi Aoba-ku Sendai 981 Japan

MIZUNO, WILLIAM GEORGE, b Ocean Falls, BC; nat US; m 48; c 3. BACTERIOLOGY. *Educ:* Univ Minn, BA, 48, MS, 50, PhD, 56. *Prof Exp:* Bacteriologist, 43-47, res bacteriologist, 48-65, sr res scientist, 65-71, mgr corp tech serv, 71-77, ASSOC DIR CORP SCI & TECHNOL, ECON LAB, INC, 77- *Concurrent Pos:* Hon fel, Univ Minn, 58; consult, 88- *Mem:* Am Soc Microbiol; Am Pub Health Asn; Sigma Xi. *Res:* Endamoeba histolytica; biocidal agents; cysticides; bactericides; fungicides; sanitation; bacterial metabolism; radioactive tracers; surface active agents; detergents; enzymes. *Mailing Add:* 2925 Regent Ave N Golden Valley MN 55422-2732

MIZUSHIMA, MASATAKA, b Tokyo, Japan, Mar 30, 23; US citizen; m 55; c 5. MOLECULAR PHYSICS, GRAVITATIONAL WAVES. *Educ:* Univ Tokyo, BA, 46, DrSc, 51. *Prof Exp:* Res assoc physics, Duke Univ, 52-55; from asst prof to assoc prof, Univ Colo, Boulder, 55-60, prof physics, 60-89; RETIRED. *Concurrent Pos:* Mem staff, Nat Bur Standards, 55-69; vis prof univ & inst solid-state physics, Univ Tokyo, 62-63, fac sci, Univ Rennes, 64, inst atomic physics, Univ Bucharest, 69-70, Cath Univ Nijmegen, 72 & Univ Electro-Comm, Tokyo, 80-81; vis prof, Inst Molecular Sci, Japan 82, Nagoya Univ, 86, ATR Res Int, Osaka, Japan, 87 & 88. *Res:* Theory of microwave and laser spectroscopy; molecular structure, particularly hyperfine structure and Zeeman effect; theory of radiation processes; propagation of radiation through earth's atmosphere; spectral-line shape; isotope separation; gravitational waves. *Mailing Add:* 523 Theresa Dr Boulder CO 80303

MIZUTANI, SATOSHI, b Yokohama, Japan, Nov 19, 37; m 66. VIROLOGY. *Educ:* Tokyo Univ Agr & Tech, BS, 62; Univ Kans, PhD(microbiol), 69. *Prof Exp:* Res scientist antibiotics, Nippon Kayaku Co, Ltd, 62-65; instr tumor virol, McArdle Lab Cancer Res, Univ Wis-Madison, 71-72, asst scientist, 72-75, assoc scientist, 75-80; sr scientist, Abbott Lab, North Chicago, 80-81; DIR, CELL & MOLECULAR BIOL GENETICS DIV, BETHESDA RES LAB, INC, 81- *Concurrent Pos:* Scholar, Leukemia Soc Am, Inc, 73-78. *Mem:* Am Soc Microbiol. *Res:* Molecular mechanism of replication of RNA tumor viruses and their relatives; mechanism of tumor formation by RNA tumor viruses; gene structure and expression in eukaryotes; manipulation of genes of eukaryotes. *Mailing Add:* 854 Village Circle Blue Bell PA 19422

MJOLSNESS, RAYMOND C, b Chicago, Ill, Apr 22, 33; m 58; c 3. FLUID DYNAMICS, ATOMIC PHYSICS. *Educ:* Reed Col, BA, 53; Oxford Univ, BA, 55; Princeton Univ, PhD(math physics), 63. *Prof Exp:* Asst physics, Los Alamos Sci Lab, 58-61; asst prof math, Reed Col, 61-62; theoret physicist space sci lab, Gen Elec Co, 62-64, consult, 64; staff mem, Los Alamos Sci Lab, 64-67; assoc prof astron, Pa State Univ, 67-69; STAFF MEM, LOS ALAMOS SCI LAB, 69- *Concurrent Pos:* State secy, Rhodes Scholar Trust, 74-76. *Mem:* Am Phys Soc. *Res:* Plasma stability; collisional relaxation of plasmas; scattering of electrons on atoms and molecules; cosmology and galaxy formation; laser energy absorption; fluid dynamics and hydrodynamic turbulence theory; low gravity flows. *Mailing Add:* Los Alamos Sci Lab T3 PO Box 1663 Los Alamos NM 87545

MLODOZENIEC, ARTHUR ROMAN, b Buffalo, NY, Mar 29, 37. PHYSICAL CHEMISTRY, PHARMACY. *Educ:* Fordham Univ, BS, 59; Univ Wis, PhD, 64. *Prof Exp:* Res assoc prod develop, Upjohn Co, 64-68; sr phys chemist, Solid Surfaces Lab, Hoffmann-La Roche, Inc, 68-74, group leader appl sci, 74-76, mgr qual control, 77-78, dir res & diag prod, 78-80; EXEC DIR INT RES, MERCK, SHARP & DOHME RES LAB, 81-; AT COL PHARM, UNIV KY, LEXINGTON. *Concurrent Pos:* Nat chmn, Indust Pharmaceut Technol, 74-78; adj prof, Sch Pharm, Univ Ky, 77- & pharmaceut chem, Univ Kans, 81-; mem exec comt, Acad Pharmaceut Sci, 78-81 & Nat Nominating Comt, 79-82. *Honors & Awards:* Cosmetic Soc Award, Soc Cosmetic Chem, 79. *Mem:* AAAS; Soc Cosmetic Chem; Am Chem Soc; Sigma Xi; fel Acad Pharmaceut Sci. *Res:* Thermal analysis; molecular organic solid physics; scanning electron microscopy and surface analysis; small particle technology; particle flow and cohesion; phase transitions; liquid crystal behavior; microencapsulation of drugs; drug specifications; dosage form design. *Mailing Add:* 652 Sand Hill Circle 3401 Hillview Ave Menlo Park CA 94025

MO, CHARLES TSE CHIN, b China 43;US citizen; m; c 3. PLASMA PHYSICS, PROBABILITY THEORY. *Educ:* Calif Inst Technol, MS, 66, PhD(elec eng & physics), 69. *Prof Exp:* Res asst elec eng, Calif Inst Technol, 65-69, Sloan res fel, 69-70, res fel, 70-72, sr res fel, 72-73; MEM SR TECH STAFF & PROF MGR, R & D ASSOCIATES LOS ANGELES, CALIF, 73- *Concurrent Pos:* Anthony fel, Calif Inst Technol, 66. *Mem:* Am Phys Soc; Int Union Radio Sci; Sigma Xi; Am Statist Asn. *Res:* Electrodynamics; relativity; probability and mathematical statistics; waves in plasma; foundation of quantum mechanics. *Mailing Add:* 782 Radcliffe Ave Pacific Palisades CA 90272

MO, LUKE WEI, b Shantung, China, June 3, 34; m 60; c 2. PHYSICS. *Educ:* Nat Taiwan Univ, BS, 56; Tsing Hua Univ, Taiwan, MS, 59; Columbia Univ, PhD(physics), 63. *Prof Exp:* Res assoc physics, Columbia Univ, 63-64; res physicist linear accelerator ctr, Stanford Univ, 65-69; asst prof physics, Univ Chicago, 69-76; assoc prof, 76-78, PROF PHYSICS, VA POLYTECH INST & STATE UNIV, 78 - *Concurrent Pos:* Guggenheim fel, 81. *Mem:* Fel Am Phys Soc. *Res:* Electromagnetic and weak interactions in high-energy physics; experiments on conserved-vector current theorem; electron scatterings; time-reversal invariance; muon-nucleon and neutrino-electron scatterings at Fermilab; search for axion-like particles; electron-proton collisions at DESY (Hamburg, Germany). *Mailing Add:* Dept Physics Va Polytech Inst & State Univ Blacksburg VA 24061

MOAD, M(OHAMED) F(ARES), b Damascus, Syria, Sept 23, 28; m 58; c 2. ELECTRICAL ENGINEERING. *Educ:* Ga Inst Technol, BS & MS, 57, PhD(elec eng), 61. *Prof Exp:* Instr elec eng, Ga Inst Technol, 57-61; dir dept of studies, Syrian Broadcasting Serv, 61-63; asst prof elec eng, 63-68, ASSOC PROF ELEC ENG, GA INST TECHNOL, 68- *Res:* Network theory; communication; systems. *Mailing Add:* Dept Elec Eng Ga Inst Technol 225 North Ave NW Atlanta GA 30332

MOAK, CHARLES DEXTER, b Marshall, Tex, Feb 24, 22; m 43; c 2. PHYSICS. *Educ:* Univ Tenn, BS, 43, PhD(physics), 54. *Hon Degrees:* DSc, Univ Witwatersrand, S Africa. *Prof Exp:* Asst, Univ Chicago, 44; PHYSICIST, OAK RIDGE NAT LAB, 44- *Mem:* Fel Am Phys Soc. *Res:* Alpha particles accompanying fission; slow neutron cross-sections; neutron capture gamma-ray studies; high-voltage accelerator research on light element charged-particle reactions; interactions of heavy particles with matter. *Mailing Add:* 332 Louisiana Ave Oak Ridge TN 37830-8550

MOAK, JAMES EMANUEL, b Norfield, Miss, Oct 26, 16; m 43; c 3. FOREST ECONOMICS. *Educ:* Univ Fla, BSF, 52; Ala Polytech Inst, MSF, 53; State Univ NY Col Forestry, Syracuse, PhD(forestry econ), 65. *Prof Exp:* From instr to prof forestry, Miss State Univ, 53-84; RETIRED. *Concurrent Pos:* Sci fac fel, 59. *Mem:* Soc Am Foresters. *Res:* Economics of private forests. *Mailing Add:* 34 Hillcrest Circle Starkville MS 39762

MOAT, ALBERT GROOMBRIDGE, b Nyack, NY, Apr 23, 26; m 49; c 3. MICROBIOLOGY. *Educ:* Cornell Univ, BS, 49, MS, 50; Univ Minn, PhD(bact), 53; Am Bd Microbiol, dipl. *Prof Exp:* From asst prof to prof bact, microbiol & immunol, Hahnemann Med Col, 52-78; PROF MICROBIOL & CHMN DEPT, SCH MED, MARSHALL UNIV, 78- *Concurrent Pos:* USPHS spec res fel & vis prof, Cornell Univ, 71-72. *Mem:* AAAS; Am Soc Microbiol; Am Chem Soc; Am Soc Biochem & Molecular Biol; fel Am Acad Microbiol. *Res:* Nutrition, metabolism and genetics of microorganisms. *Mailing Add:* Dept Microbiol Marshall Univ Sch of Med Huntington WV 25755-9330

MOATES, ROBERT FRANKLIN, b Birmingham, Ala, May 16, 38; m 62; c 2. TOBACCO CHEMISTRY, TECHNOLOGY. *Educ:* Duke Univ, BS, 60; Univ SC, PhD(org chem), 66. *Prof Exp:* Res chemist, R J Reynolds Tobacco Co, 65-89; CONSULT, 89- *Mem:* Am Chem Soc; The Chem Soc; Sigma Xi. *Res:* Alkaloid isolation; synthesis of alkaloid systems; synthetic organic chemistry; synthesis of natural products; isolation and identification of natural products. *Mailing Add:* 717 Lankashire Rd Winston-Salem NC 27106

MOATS, WILLIAM ALDEN, b Des Moines, Iowa, Nov 30, 28; m 58; c 2. FOOD BIOCHEMISTRY. *Educ:* Iowa State Univ, BS, 50; Univ Md, PhD(chem), 57. *Prof Exp:* Asst chem, Univ Md, 50-55; res chemist, Field Crops & Animal Prod Br, Mkt Qual Res Div, 57-72, res chemist, Agr Mkt Res Inst, Sci & Educ Admin-Agr Res, 72-80, RES CHEMIST, MEAT SCI RES LAB, ANIMAL SCI INST, AGR RES SERV, USDA, 80- *Mem:* Am Chem Soc; Am Soc Microbiol; Poultry Sci Asn; Am Dairy Sci Asn. *Res:* Quality tests for dairy products; staining of bacteria for microscopic examination; determination of pesticide residues in foods; heat resistance of bacteria; improved media for salmonella detection; egg washing and sanitizing procedures; physicochemical methods for detecting antibiotic residues. *Mailing Add:* Agr Res Ctr USDA Bldg 201-East Beltsville MD 20705

MOAVENZADEH, FRED, b Rasht, Iran, Oct 14, 35; m 61; c 2. CIVIL ENGINEERING, MATERIALS SCIENCE. *Educ:* Univ Tehran, BS, 58; Cornell Univ, MS, 60; Purdue Univ, PhD(civil eng), 62. *Prof Exp:* Field engr, Kampsax Overseas, Inc, 56; res asst civil eng, Purdue Univ, 60-62; asst prof, Ohio State Univ, 62-65; assoc prof, 65-72, PROF CIVIL ENG, MASS INST TECHNOL, 72-, DIR TECHNOL ADAPTATION PROG, 75- *Concurrent Pos:* Comt mem, Hwy Res Bd, Nat Acad Sci-Nat Res Coun; vis prof civil eng, Grad Sch Design, Harvard Univ, 70- *Honors & Awards:* Sanford E Thompson

Award, Am Soc Testing & Mat. *Mem:* Bldg Res Inst-Nat Acad Eng; Am Concrete Inst; Am Soc Civil Engrs; AAAS; Sigma Xi. *Res:* Mechanics of materials; viscoelasticity; highway engineering and construction. *Mailing Add:* Mass Inst Technol Rm 1-171 77 Massachusetts Ave Cambridge MA 02139

MOAWAD, ATEF H, b Dec 2, 35; Can citizen; m 66; c 2. OBSTETRICS & GYNECOLOGY, PHARMACOLOGY. *Educ:* Cairo Univ, MD, 58; Jefferson Med Col, MS, 63; Am Bd Obstet & Gynec, dipl, 68; FRCS(C), 69. *Prof Exp:* Fel obstet & gynec, Case Western Reserve Univ, 64-65; vis investr, Univ Lund, 65-66; from lectr pharmacol to assoc prof obstet & gynec & pharmacol, Univ Alta, 66-72; prof obstet & gynec & pharmacol, 72-75, PROF OBSTET & GYNEC, DIV BIOL SCI & PRITZKER SCH MED, UNIV CHICAGO, 75-, PROF PEDIAT, 76- *Concurrent Pos:* Brush Found scholar, 66-67. *Mem:* Fel Am Col Obstet & Gynec; Soc Gynec Invest; Pharmacol Soc Can; NY Acad Sci; Can Med Asn. *Res:* Reproductive physiology and pharmacology, chiefly of the structure and function of uterine and fallopian tube smooth muscle. *Mailing Add:* Dept Med Box 446 Univ Chicago 5841 S Maryland Ave Chicago IL 60637

MOAZED, K L, b Meshed, Iran, Sept 14, 30; nat US; m 53; c 4. PHYSICAL METALLURGY. *Educ:* Rensselaer Polytech Inst, BS, 53, MMetEng, 56; Carnegie Inst Technol, MS, 58, PhD(metall), 59. *Prof Exp:* Res assoc metall res, Rensselaer Polytech Inst, 53-56; proj engr, Carnegie Inst Technol, 56-59; from asst prof to assoc prof metall, Ohio State Univ, 59-68; PROF MAT SCI & ENG, NC STATE UNIV, 68- *Mem:* Sigma Xi. *Res:* Surface science; thermodynamics and kinetics of surface reactions; metallization of semiconductors, nucleation and growth of thin films. *Mailing Add:* Dept Mat Sci & Eng NC State Univ Raleigh NC 27695-7916

MOBARHAN, SOHRAB, b Kerman, Iran, Nov 19, 41. GASTROINTESTINAL DISEASES. *Educ:* Rome Univ, MD, 65. *Prof Exp:* PROF MED, LOYOLA UNIV, ILL, 89- *Concurrent Pos:* Assoc chief gastroenterol & dir clin nutrit unit, Loyola Univ, Ill. *Mem:* Am Col Physicians; Am Col Nutrit; Am Gastroenterol Asn. *Mailing Add:* Sect Gastroenterol Rm 25 Bldg 117 Loyola Univ Med Ctr 2160 S First Ave Maywood IL 60153

MOBERG, GARY PHILIP, b Monmouth, Ill, Feb 14, 41; m 67; c 2. STRESS PHYSIOLOGY, NEUROENDOCRINOLOGY. *Educ:* Monmouth Col, Ill, BA, 63; Univ Ill, Urbana, MS, 65, PhD(physiol), 68. *Prof Exp:* NIH fel med ctr, Univ Calif, San Francisco, 68-70; from asst prof to assoc prof, 70-82, res physiologist, Calif Primate Rest Ctr, 79-86, PROF ANIMAL SCI & ANIMAL PHYSIOL, UNIV CALIF, DAVIS, 82- *Concurrent Pos:* USDA res grants, 70-, NIH res grants, Univ Calif, Davis, 71-76, 79- & Calif sea grants, 88-; vis prof, Utrecht Med Sch, Neth & NATO sr scientist fel, 77; affil, Calif Primate Res Ctr, 77-79. *Mem:* Soc Study Reproduction; Am Physiol Soc; Endocrine Soc; Am Soc Animal Sci. *Res:* Neural control of anterior pituitary function; effects of stress on animals; stress biology; endocrine control of fish reproduction. *Mailing Add:* Dept Animal Sci Univ Calif Davis CA 95606

MOBERG, WILLIAM KARL, b Fargo, NDak, July 1, 48; m 70; c 2. SYNTHETIC ORGANIC & NATURAL PRODUCTS CHEMISTRY, ARGRICULTURAL & FOOD CHEMISTRY. *Educ:* Bowdoin Col, AB, 69; Harvard Univ, AM, 71, PhD(chem), 74. *Prof Exp:* NIH fel, Mass Inst Technol, 75-77; mem staff, Cent Res Dept, 77-80, MEM STAFF, 80-, SR RES ASSOC, AGR PROD DEPT, 80- *Concurrent Pos:* Chmn, Gordon Res Conf on Chem & Heterocyclic Compounds, 84. *Mem:* Am Chem Soc; Am Phytopath Soc; Royal Soc Chem. *Res:* Organic synthesis; fungicide chemistry and biochemistry; organic chemistry; phytopathology. *Mailing Add:* DuPont Agr Prod Stine-Haskell 300-312C PO Box 30 Newark DE 19714

MOBERLY, RALPH M, b St Louis, Mo, Apr 17, 29; m 54; c 2. MARINE GEOLOGY. *Educ:* Princeton Univ, AB, 50, PhD(geol), 56. *Prof Exp:* Geologist, Standard Oil Co, Calif, 56-59; from asst prof to assoc prof, 59-70, PROF GEOL, UNIV HAWAII, 70-, CHMN DEPT GEOL & GEOPHYSICS, 75-80 & 90- *Concurrent Pos:* Vis prof Univ Calif, Berkeley & Santa Barbara, 65-66, Bryn Mawr Col, 73 & Univ Calif, Davis, 82; vis fel, US Nat Mus, 66, Lamont-Doherty Geol Observ & Scripps Inst Oceanog, 73 & Oxford Univ, 81; assoc dir, Hawaii Inst Geophysics, Univ Hawaii, 83-90, geologist, 62- *Mem:* Fel AAAS; fel Geol Soc Am; Am Asn Petrol Geologists; Am Geophys Union; Soc Econ Paleontologists & Mineralogists; Int Asn Sedimentologists. *Res:* Marine geology, from the coastal zone into the deep sea; sedimentology, tectonics and ocean drilling; Pacific basin and its margin. *Mailing Add:* Dept Geol & Geophysics Univ Hawaii 2525 Correa Rd Honolulu HI 96822

MOBLEY, BERT A, ELECTROPHYSIOLOGY, MUSCLE MECHANICS. *Educ:* Univ Chicago, PhD(physiol), 70. *Prof Exp:* PROF PHYSIOL, UNIV OKLA COL MED, 79- *Mailing Add:* Dept Physiol Col Med Univ Okla Oklahoma City OK 73190

MOBLEY, CARROLL EDWARD, b Baltimore, Md, Oct 22, 41; m 64; c 2. MATERIALS SCIENCE. *Educ:* Johns Hopkins Univ, BA, 63, PhD(mech), 68. *Prof Exp:* Sr researcher metall, Battelle Columbus Labs, 67-78; assoc prof, 78-81, PROF METALL, OHIO STATE UNIV, 81- *Concurrent Pos:* Vis prof mech & mat sci, Johns Hopkins Univ, 76-77 & 87. *Honors & Awards:* IR-100 Award, Indust Res Mag, 75. *Mem:* Am Soc Metals; Am Inst Mining, Metall & Petrol Engrs; Sigma Xi; Am Foundrymen's Soc. *Res:* Solidification of metals; structure and properties of rapidly solidified materials; process metallurgy of iron and steel production; high velocity deformation of metals. *Mailing Add:* 2753 Cranford Rd Columbus OH 43221

MOBLEY, CURTIS DALE, b Canyon, Tex, June 15, 47; m 79. HYDROLOGIC OPTICS. *Educ:* Univ Tex, Austin, BS, 69; Univ Md, College Park, PhD(meteorol), 77. *Prof Exp:* Nat Res Coun res assoc, Pac Marine Environ Lab, 77-79, inst scientist, Joint Inst, Study of Atmosphere & Ocean, Univ Wash & Pac Marine Environ Lab, Nat Oceanic & Atmospheric Admin, Seattle, 79-88; ASSOC PROF PHYSICS, PACIFIC LUTHERAN UNIV, TACOMA, 88- *Mem:* Am Meteorol Soc; Soc Photo-Optical Inst Eng. *Res:* Numerical studies of light in the ocean (hydrologic optics); computational fluid dynamics, especially free-surface fluid flows. *Mailing Add:* 20137 53rd Ave NE Seattle WA 98155

MOBLEY, JEAN BELLINGRATH, b Norfolk, Va, Mar 13, 27; m 49; c 2. MATHEMATICS, MATHEMATICS EDUCATION. *Educ:* Duke Univ, AB, 48; Univ NC, MA, 54, PhD, 70. *Prof Exp:* Teacher, Pub Schs, NC, 48-56; asst prof math, Flora MacDonald Col, 56-59, assoc prof & head dept, 59-61; assoc prof & head, dept math, St Andrews Presby Col, 61-63; assoc prof, 63-70, prof math, 70-76, prof math & educ & head, Sec Educ Dept, 76-80, prof math, 80-83, HEAD, DEPT MATH SCI, PFEIFFER COL, 83- *Mem:* Nat Coun Teachers Math; Nat Educ Asn; Math Asn Am. *Res:* Geometry; mathematics education. *Mailing Add:* 227 Camelot Dr Salisbury NC 28144

MOBLEY, RALPH CLAUDE, b Buffalo, NY. NUCLEAR PHYSICS. *Educ:* Univ Wis, PhD(physics), 50. *Prof Exp:* Res assoc physics, Univ Wis, 50-51; res assoc, Duke Univ, 51-53; from asst prof to assoc prof, La State Univ, 53-61; chmn dept, 61-72, PROF PHYSICS, OAKLAND UNIV, 61- *Mem:* Fel Am Phys Soc; Inst Elec & Electronics Engrs. *Res:* Photoneutron thresholds; charged-particle scattering; neutron scattering; ion buncher and accelerator development and neutron scattering by time-of-flight method; mass spectrometry; monopole and macromolecule mass spectrometer development. *Mailing Add:* Dept Physics Oakland Univ Rochester MI 48309

MOBRAATEN, LARRY EDWARD, b Fergus Falls, Minn, Sept 6, 38; m 67; c 2. GERMPLASM PRESERVATION, CRYOBIOLOGY. *Educ:* Univ Calif, Berkeley, AB, 62; Univ Maine, Orono, PhD(zool), 72. *Prof Exp:* Curie Found fel immunogenetics, 72-74; assoc staff scientist immunogenetics, 74-79, assoc staff scientist genetics, 80-82, STAFF SCIENTIST, JACKSON LAB, 82-, DIR GENETIC RESOURCES, 89- *Concurrent Pos:* Lectr zool, Univ Maine, Orono, 75-76; mem, Comt Genetic Standards, Inst Lab Animal Resources, Nat Res Coun, 75-76; Mem, Comm on Preserv Lab Animal Resources, Inst Lab Animal Resources, 85- *Mem:* Assoc Sigma Xi; Soc Cryobiol; Am Genetic Asn; AAAS. *Res:* Cryopresevation of mouse embryos; analysis of histocompatibility mutations in mice. *Mailing Add:* Jackson Lab 600 Main St Bar Harbor ME 04609

MOCELLA, MICHAEL THOMAS, catalysis, organometallic chemistry, for more information see previous edition

MOCH, IRVING, JR, b New York, NY, Jan 28, 27; m 51; c 1. CHEMICAL ENGINEERING. *Educ:* Columbia Univ, AB, 47, BS, 49, MS, 50, PhD(chem eng), 56. *Prof Exp:* Res engr, E I du Pont de Nemours & Co, Inc, La Porte, Tex, 54-63, supt pilot plant, 63-68, tech supt, Electrochem Dept, 68-72, res supt, Polymer Prod Dept, 72-78, develop & tech serv mgr, 78-85, APPL TECHNOL MGR, POLYMER PROD DEPT, DU PONT DE NEMOURS & CO, INC, 85- *Concurrent Pos:* Dir, Int Desalination Asn; chmn, NSF Standard, 61, Mech Devices & Am Soc Testing & Mat D-19 RO Task Group. *Mem:* AAAS; Am Inst Chem Engrs; Am Chem Soc; Sigma Xi. *Res:* Process and products development. *Mailing Add:* Polymer Prod Dept Permasep Bldg E I du Pont de Nemours & Co Inc Wilmington DE 19898

MOCHAN, EUGENE, BIOCHEMISTRY. *Prof Exp:* Instr, Dept Chem, Pa State Univ, Pottsville, 62-63, res assoc, Dept Biochem, 71-72; lab instr biochem, State Univ NY, Buffalo, 62-67; conf instr, Univ Pa, 68-72; from asst prof to assoc prof, Dept Physiol Chem, Philadelphia Col Osteop Med, 72-81, asst prof, Dept Family Med, 78-81; assoc prof family med & biochem, 81-83, actg chmn, 83-85, CHMN, DEPT FAMILY MED, SCH OSTEOP MED, UNIV MED & DENT NJ, CAMDEN, 85-, PROF, 86- *Concurrent Pos:* Chmn, Human Inst Rev Bd, Sch Osteop Med, Univ Med & Dent NJ, 86-88, Biochem Sect & Basic Sci Sect, Nat Bd Med Examnr, 86-89. *Mem:* Sigma Xi; Am Osteop Asn; Am Col Rheumatology; Am Soc Cell Biol; AAAS; Am Osteop Col Gen Pract; Am Osteop Col Rheumatology; Soc Teachers Family Pract; Am Acad Family Pract; Nat Bd Med Examnr (vpres, 86-89, pres, 89-). *Mailing Add:* Biochem & Family Med Dept Univ Med & Dent NJ 401 Haddon Ave Camden NJ 08103

MOCHARLA, RAMAN, b Mattigiri, Karnataka, India, Sept 19, 53; nat US; m 83; c 2. MICROBIOLOGY, MEDICINE. *Educ:* Univ Agra, BS, 72; GBP Univ, India, MS, 76; Univ Kurukshetra, India, PhD(microbiol), 80. *Prof Exp:* Res asst, GBP Univ, 72-74; scientist S-1, Indian Coun Agr Res, New Delhi, 77-80; UNESCO/WHO fel, Czech Acad Sci, Prague, 80-81; postdoctoral fel, Univ Okla, Norman, 81-83; assoc res scientist, Okla Med Res Found, Oklahoma City, 83-84 & S R Noble Found, Ardmore, Okla, 84-86; res assoc, Sch Med, Ind Univ, Indianapolis, 86-89. *Mem:* AAAS; Am Soc Biol Chemists; Nat Geog Soc; Am Acad Family Physicians; Am Soc Indust Microbiol; Am Soc Microbiol; Am Asn Immunologists; AMA; Am Med Student Asn; NY Acad Sci. *Mailing Add:* Indiana Univ Sch Med 3138 Shadow Brook Dr Indianapolis IN 46214

MOCHEL, JACK MCKINNEY, b Boston, Mass, Jan 27, 39; m 62; c 3. PHYSICS. *Educ:* Cornell Univ, BA, 61; Univ Rochester, PhD(physics), 65. *Prof Exp:* Fel physics, Univ Rochester, 65-66; from asst prof to assoc prof, 66-72, PROF PHYSICS, UNIV ILL, 72- *Concurrent Pos:* A P Sloan fel, 68-74; assoc, Ctr Advan Study, 73-74; consult, Bell Labs, Murray Hill, 77-78. *Mem:* Am Phys Soc; Sigma Xi. *Res:* Low-temperature physics; properties of helium in two and three dimensions; superconductivity and phase transitions. *Mailing Add:* Dept Physics 1110 W Green St Urbana IL 61801

MOCHEL, MYRON GEORGE, b Fremont, Ohio, Oct 9, 05; m 30; c 3. MECHANICAL ENGINEERING. *Educ:* Case Inst Technol, BS, 29; Yale Univ, MS, 30. *Prof Exp:* Develop engr, Socony Mobil Oil Co, Inc, 31-37; mech design & develop engr, Gearing Eng Div, Westinghouse Elec Corp, 37-43; mech design engr, Underwater Sound Lab, Harvard Univ, 43-45; supvr training, Steam Turbine Div, Worthington Corp, 45-49; assoc prof eng graphics, 49-55, prof mech eng, 55-71, EMER PROF MECH ENG, CLARKSON UNIV, 71- *Concurrent Pos:* Lectr, Eve Div, Univ Pittsburgh, 38-43 & NY State Adult Educ Prog, 46-49; partic, Conf Eng Graphics Sci Eng, NSF, 60; adv mgr, J Eng Graphics, 63-66. *Mem:* Am Soc Mech Engrs; Nat Soc Prof Engrs; Am Soc Eng Educ. *Res:* Fundamentals of engineering graphics. *Mailing Add:* Three Castle Dr Potsdam NY 13676-1610

MOCHEL, VIRGIL DALE, b Woodland, Ind, Oct 29, 30; m 51; c 3. PHYSICAL CHEMISTRY, PHYSICS. *Educ:* Purdue Univ, BS, 52, MS, 54; Univ Ill, PhD(nuclear magnetic resonance), 60. *Prof Exp:* Chief chemist, Globe Am Corp, 53-54; res chemist, US Army Chem Ctr, 54-56; sr chemist, Corning Glass Works, 59-64; from res chemist to sr res chemist, 64-67, RES ASSOC POLYMER NUCLEAR MAGNETIC RESONANCE, FIRESTONE TIRE & RUBBER CO, AKRON, 67- *Mem:* Am Chem Soc (Rubber & Polymer divs); Am Phys Soc. *Res:* High resolution nuclear magnetic resonance; kinetics; polymer structure studies; electroluminescence; photoluminescence; semiconduction; photoconduction; glass composition; rubber curing optimization; tire construction. *Mailing Add:* 1804 Wall Rd Wadsworth OH 44281

MOCHIZUKI, DIANE YUKIKO, b Feb 17, 52; m; c 1. CELLULAR BIOCHEMISTRY, CELL BIOLOGY. *Educ:* Univ Calif, Irvine, PhD(biol sci), 81. *Prof Exp:* SR STAFF SCIENTIST & HEAD CELL BIOLOGY LAB, DEPT BIOCHEM, IMMUNEX CORP, 84- *Mem:* Am Asn Immunologists; Asn Women Sci; AAAS. *Mailing Add:* Immunex Corp 2644 Bridle Lane Walnut Creek CA 94596

MOCHRIE, RICHARD D, b Lowell, Mass, Feb 17, 28; m 50; c 3. DAIRY SCIENCE. *Educ:* Univ Conn, BS, 50, MS, 53; NC State Col, PhD(rumen nutrit), 58. *Prof Exp:* Res technician, Univ Conn, 50-53; asst animal sci, 53-54, res instr, 54-58, from asst prof to assoc prof, 58-72, fac senate vchmn, 80-81, chmn, 81-82, prof animal sci, 72-88, EMER PROF, NC STATE UNIV, 89- *Concurrent Pos:* Mem comt abnormal milk control & adv to Coun I, Nat Conf Interstate Milk Shipments, 71-78; US organizer & co-editor, Proceedings of Joint US-Australian Forage Workshop, Nat Sci Found, Armidale, New South Wales, Aug, 80. *Honors & Awards:* Distinguished Serv Award, Nat Mestitis Coun, 78. *Mem:* Am Dairy Sci Asn; Nat Mastitis Coun (vpres, 75-76, pres, 76-77); Int Asn Milk, Food & Environ Sanitarians, Inc; Asn Off Analytical Chemists; Assoc referee, Am Forage & Grass Coun; Sigma Xi; Dairy Shrine Club. *Res:* Dairy cattle nutrition, physiology of lactation and forage utilization. *Mailing Add:* 505 S Dixon Ave Cary NC 27511

MOCK, DAVID CLINTON, JR, b Redlands, Calif, May 6, 22; m 52. INTERNAL MEDICINE. *Educ:* Univ Southern Calif, AB, 44; Hahnemann Med Col, MD, 48; Am Bd Internal Med, dipl, 58. *Prof Exp:* Intern, Hahnemann Hosp, Philadelphia, 48-49; resident, Community Hosp of San Mateo County, Calif, 49-50, resident med, 50-51 & 54; chief med serv, Navajo Base Hosp, Ft Defiance, Ariz, 51-53, med officer-in-chg, 52-53; resident med, Vet Admin Hosp, Oklahoma City, Okla, 54-55; pvt pract, Calif, 55-56; from clin asst to assoc prof, 56-72, assoc dean med student affairs, 69-76, PROF MED, COL MED, UNIV OKLA, 72-, ASSOC DEAN POSTDOCTORAL EDUC, 76-, DIR, CONTINUING MED EDUC, 80- *Concurrent Pos:* Res fel exp therapeut, Col Med, Univ Okla, 56-57; Upjohn fel, 57-59; attend physician, Vet Admin Hosp, Oklahoma City, 56- & Univ Okla Hosp, 59-; dir therapeut, Univ Okla, 58-60, asst dean student affairs, 66-69. *Mem:* Am Fedn Clin Res; fel Am Col Physicians; NY Acad Sci. *Res:* Clinical drug investigation. *Mailing Add:* 570 Alameda Blvd Coronado CA 92118

MOCK, DOUGLAS WAYNE, b New York, NY, July 4, 47; m 69. ANIMAL BEHAVIOR, ECOLOGY. *Educ:* Cornell Univ, BS, 69; Univ Minn, MS, 72, PhD(ecol), 76. *Prof Exp:* Fel ethol, Smithsonian Inst, 76-77; vis scientist, Percy Fitzpatrick Inst African Ornith, Univ Cape Town, 77; ASST PROF ZOOL, UNIV OKLA, 78- *Honors & Awards:* Alexander P Wilson Prize, Wilson Ornith Soc, 75; A Brazier Howell Prize, Cooper Ornith Soc, 76. *Mem:* Animal Behav Soc; Ecol Soc Am; Wilson Ornith Soc; Cooper Ornith Soc; Am Ornithologists Union. *Res:* Evolutionary biology, especially theoretical and field studies of ecology and behavior of vertebrates; evolution of social organization, mating systems and communication in birds. *Mailing Add:* Dept Zool Univ Okla 660 Parrington Oval Norman OK 73019

MOCK, GORDON DUANE, b Bloomington, Ill, Oct 21, 27; m 54; c 2. MATHEMATICS. *Educ:* Univ Ill, BS, 50, MS, 51; Univ Wis, PhD(educ), 59. *Prof Exp:* Teacher high sch, Ill, 51-53; teacher lab sch, Univ Wis-Madison, 54-55; teacher lab sch, Univ Southern Ill-Carbondale, 55-56; teacher lab sch, Univ Wis-Madison, 56-58; teacher math, State Univ NY Col Oswego, 58-60; assoc prof, 60-67, PROF MATH, WESTERN ILL UNIV, 67- *Concurrent Pos:* NSF sci fac fel, 63-64. *Mem:* Am Math Soc; Math Asn Am; Nat Coun Teachers Math. *Res:* Mathematics education. *Mailing Add:* Dept Math Western Ill Univ 711 E Franklin Macomb IL 61455-7097

MOCK, JAMES JOSEPH, b Geneseo, Ill, Feb 15, 43; m 66; c 3. AGRONOMY, PLANT BREEDING. *Educ:* Monmouth Col, Ill, BA, 65; Iowa State Univ, PhD(agron), 70. *Prof Exp:* From asst prof to prof plant breeding, Iowa State Univ, 70-78; dir corn res, 78-82, mgr Western Prod Unit, 82-84, asst vpres res, 84-86, vpres, Int Res Prod, 86-87, VPRES, AGRI-PROD RES, NORTHRUP KING CO, 88- *Mem:* Am Soc Agron; Crop Sci Soc Am. *Res:* Physiological corn breeding; breeding maize genotypes that will efficiently intercept and convert solar energy into grain; seed production research for corn, soybeans, sorghum and sunflowers. *Mailing Add:* 1336 Liberty Ct Northfield MN 55057

MOCK, JOHN E(DWIN), b Altoona, Pa, Sept 29, 25; m 47; c 2. ENERGY TECHNOLOGY, SCIENCE POLICY. *Educ:* US Mil Acad, BS, 47; Purdue Univ, BS & MS, 50, PhD(nuclear eng), 60; Ohio State Univ, MS, 53; George Washington Univ, MS, 65, MBA, 68, MA(econ), 76, LLM (patent law), 82. *Prof Exp:* US Air Force, 47-68, proj scientist, Wright Air Develop Ctr, Ohio, 49-54, staff scientist, Hq Ger, 55-58, sci dir, Defense Atomic Support Agency, Washington, DC, 60-64, staff scientist, Nuclear Res Assocs, NY, 64-66, sci dir, Advan Res Projs Agency, Washington, DC, 66-68; dir, Ga Sci & Technol Comn, 68-74, Ga Ctr Technol Forecasting & Technol Assessment, 69-74 & Ga Inst Biotechnol, 69-74, chmn Ga Energy Comn, 73-74; SR TECHNOL ADV, US ENERGY RES & DEVELOP ADMIN, WASHINGTON, DC, 76- *Concurrent Pos:* Assoc prof, George Washington Univ, 60-64; consult, Kaman Nuclear Div, 64-65; assoc prof US Air Force Acad, 64-65; consult, Off Secy Defense, 68-71; adv, Gulf Univs Res Corp, 68-71; Southern Interstate Nuclear Bd, 68-74; dir, Coastal State Orgn, 69-70; mem coun, State Govts Comt Sci & Technol, 69-74; chmn, Nat Gov Coun Sci & Technol, 70-71; consult comt cities of future, Nat Acad Eng, 70-73. *Honors & Awards:* Mark Mills Award, Am Nuclear Soc, 60. *Mem:* AAAS; Am Nuclear Soc; Am Phys Soc; Opers Res Soc Am; Am Soc Eng Educ. *Res:* Nuclear effects; geosciences; geothermal research; heat transfer; hydrodynamics; energy policy; technology transfer; theoretical analysis; game theory; civil defense; military strategy; oceanography; science policy. *Mailing Add:* 1326 Round Oak Ct McLean VA 22101

MOCK, ORIN BAILEY, b Elmer, Mo, Oct 22, 38; m 67; c 3. REPRODUCTIVE PHYSIOLOGY, MAMMALOGY. *Educ:* Northeast Mo State Col, BS, 60, MA, 65; Univ Mo-Columbia, PhD(zool), 70. *Prof Exp:* Teacher high schs, Mo, 59-60 & 61-62; vol, Peace Corps, Philippines, 62-64; from asst prof to assoc prof, Northeast Mo State Univ, 69-76; ANAT, KIRKSVILLE COL OSTEOP MED, 76- *Mem:* AAAS; Am Soc Mammalogists. *Res:* Reproduction in the least shrew, Cryptotis parva; taxonomy of North American shrews. *Mailing Add:* Dept Anat Kirksville Col Osteo Med 204 W Jefferson Kirksville MO 63501

MOCK, STEVEN JAMES, b Philadelphia, Pa, July 3, 34; m 87; c 5. GEOLOGY, GLACIOLOGY. *Educ:* Antioch Col, BA, 57; Dartmouth Col, MA, 65; Northwestern Univ, PhD(geol), 75. *Prof Exp:* Geologist glaciol, Snow, Ice & Permafrost Res Estab, 59-61, res geologist, Cold Regions Res & Eng Lab, 61-77, CHIEF TERRESTRIAL SCI BR, US ARMY RES OFF, 77- *Concurrent Pos:* Adj prof, dept marine, earth & atmospheric sci, NC State Univ. *Mem:* Am Geophys Union. *Res:* Glacier-climate interactions; quantitative geomorphology; climatic trends. *Mailing Add:* Army Res Off PO Box 12211 Research Triangle Park NC 27709

MOCK, WILLIAM L, b Los Angeles, Calif, Aug 5, 38. ORGANIC CHEMISTRY. *Educ:* Calif Inst Technol, BS, 60; Harvard Univ, PhD(org chem), 65. *Prof Exp:* PROF CHEM, UNIV ILL CHICAGO, 89- *Concurrent Pos:* Fel, A P Sloan Found, 72. *Mem:* Am Chem Soc. *Res:* Synthetic methods; reaction mechanisms; molecular recognition; enzymology. *Mailing Add:* Dept Chem M/C 111 Univ Ill Chicago Chicago IL 60680-4348

MOCKETT, PAUL M, b San Francisco, Calif, Apr 9, 36; m 65; c 3. EXPERIMENTAL HIGH ENERGY PHYSICS, PARTICLE PHYSICS. *Educ:* Reed Col, BA, 59; Mass Inst Technol, PhD(physics), 65. *Prof Exp:* Res assoc physics, Mass Inst Technol, 65-67; asst physicist, Brookhaven Nat Lab, 67-70, assoc physicist, Div Particles & Fields, 70-72; sr res assoc, 72-75, res assoc prof, 75-81, RES PROF PHYSICS, UNIV WASH, 81- *Mem:* Am Phys Soc. *Res:* Experimental particle physics. *Mailing Add:* Dept Physics Univ Wash Seattle WA 98195

MOCKFORD, EDWARD LEE, b Indianapolis, Ind, June 16, 30. TAXONOMY, ENTOMOLOGY. *Educ:* Ind Univ, AB, 52; Univ Fla, MS, 54; Univ Ill, PhD, 60. *Prof Exp:* Asst limnol, Lake & Stream Surv, Ind, 48-52; asst biol, Univ Fla, 52-54; tech asst entom, Nat Hist Surv, Ill, 56-60; from asst prof to prof, 60-84, DISTINGUISHED PROF BIOL SCI, ILL STATE UNIV, 84- *Concurrent Pos:* Coop scientist, USDA, 59-; res assoc, Fla Dept Agr & Consumer Serv, 60-; vis prof, Inst Technol & Higher Studies, Monterrey, Mexico, 63-64. *Mem:* Soc Syst Zool; Wilson Ornith Soc; Entom Soc Am; Asn Trop Biol. *Res:* Taxonomic entomology; taxonomy and evolution of insects, especially order Psocoptera, psocids or bark-lice; life history. *Mailing Add:* Dept Biol Sci Ill State Univ Normal IL 61761

MOCKRIN, STEPHEN CHARLES, b Columbus, Ohio, Apr 17, 46; m 87; c 2. CARDIOVASCULAR RESEARCH. *Educ:* Univ Mich, Ann Arbor, BS, 68; Univ Calif, Berkeley, PhD(biochem), 73. *Prof Exp:* Res fel biochem, Med Ctr, Univ Calif, San Francisco, 73-77; sr staff fel biochem, 77-83, health sci adminr, 83-87, dep chief, 88-90, CHIEF, HYPERTENSION & KIDNEY DIS BR, NAT HEART LUNG & BLOOD INST, NIH, 90- *Honors & Awards:* Merit Award, NIH, 89. *Mem:* AAAS; Am Soc Cell Biol. *Res:* Application of molecular and cellular technologies to cardiovascular research; high blood pressure and blood pressure regulation. *Mailing Add:* Nat Heart Lung & Blood Inst/NIH Fed Bldg Rm 4C10 Bethesda MD 20892

MOCKROS, LYLE F(RED), b Chicago, Ill, July 19, 33; m 60; c 3. FLUID MECHANICS, BIOMECHANICS. *Educ:* Northwestern Univ, BS, 56, MS, 57; Univ Calif, Berkeley, PhD(fluid mech), 62. *Prof Exp:* Asst res engr, Inst Eng Res, Univ Calif, Berkeley, 59-62; asst prof civil eng, 62-66, assoc prof civil eng & eng sci, 66-70, PROF CIVIL ENG & BIOL ENG, TECHNOL INST, NORTHWESTERN UNIV, 70- *Concurrent Pos:* Chmn, biomed eng, Northwestern Univ,76-83 & assoc chmn, civil eng, 78-80. *Mem:* Biomed Eng Soc; Am Soc Eng Educ; Am Soc Civil Engrs; Am Soc Artificial Internal Organs. *Res:* Biological fluid mechanics; artificial lungs; transport processes; blood coagulation; separation in filtering devices. *Mailing Add:* Dept Civil Eng Technol Inst Northwestern Univ Evanston IL 60208

MOCZYGEMBA, GEORGE A, b Panna Maria, Tex, Jan 7, 39; m 66; c 3. OLEFIN & DIENE POLYMERIZATION, CATALYSIS. *Educ:* Univ Tex, Austin, BS, 62, PhD(chem), 69. *Prof Exp:* Instr chem, St Edward's Univ, 65-67; SR CHEMIST, PHILLIPS PETROL CO, 68- *Mem:* Am Chem Soc; Sigma Xi; AAAS. *Res:* Polymerization mechanisms, kinetics and catalysis; polymer research for rubber and plastic applications; studies involving anionic polymerization of dienes and vinyl aromatics. *Mailing Add:* 824 SE Crestland Bartlesville OK 74006

MODABBER, FARROKH Z, b Rasht, Iran, Feb 27, 40; m 67; c 3. IMMUNODIAGNOSTICS, RESEARCH GRANT EVALUATION & ADMINISTRATION. *Educ:* Univ Calif, Los Angeles, BA, 64, PhD(microbiol), 68. *Prof Exp:* Postdoctoral fel immunol, Harvard Med Sch, 68-71, asst prof, Sch Pub Health, 71-76; head, Dept Pathobiol, Tehran Univ Sch Pub Health, 74-79; dir, Pasteur Inst, Tehran, 79-80; head, Dept Immunol, Syntex Res Inst, 82-84; SCIENTIST, WHO, 84- *Concurrent Pos:* Lectr, Harvard Sch Pub Health, 76-82; pres, bd dirs, Asn Control Leprosy, 76-79; vis scientist, Pasteur Inst, Paris, 80-82. *Mem:* Am Asn Immunologists; Am Soc Microbiol; NY Acad Sci; Brit Soc Immunol. *Res:* Immunology of parasitic diseases, particularly leishmaniasis; administration of research grants on leishmaniasis, diagnostic tests, vaccine development; author of over 70 publications; awarded one US patent. *Mailing Add:* WHO 1211 Geneva 27 Switzerland

MODAFFERI, JUDY HALL, b Chicago, Ill, July 15, 43; m 77; c 1. CELL BIOLOGY. *Educ:* Ill Wesleyan Univ, BA, 64; Smith Col, MA, 66; Purdue Univ, PhD(biol), 71. *Prof Exp:* Electron microscopist, Dept Path & Toxicol, Dow Chem Co, 71-72; postdoctoral res fel, Dept Cell Physiol, Boston Biomed Res Inst, 72-74; asst prof, Dept Biol Sci, State Univ NY, Binghamton, 74-81; ELECTRON MICROSCOPIST, DEPT PATH, ROBERT PACKER HOSP, 81- *Mem:* AAAS; Am Soc Cell Biol; Electron Micros Soc Am; Sigma Xi. *Mailing Add:* Dept Path Robert Packer Hosp Guthrie Sq Sayre PA 18840

MODAK, ARVIND T, b Bombay, India; US citizen; m 65; c 2. NEUROCHEMISTRY, TOXICOLOGY. *Educ:* Univ Bombay, BSc, 61, BSc, 63, MSc, 65; Univ Tex, Austin, MS, 68, PhD(pharmacol), 70. *Prof Exp:* Demonstr pharmacol, Univ Bombay, 63-64, lectr, 64-65; teaching asst pharm, Univ Tex, Austin, 65-70; fel pharmacol, Health Sci Ctr, Univ Tex, San Antonio, 70-72; pharmacologist & toxicologist, Pharma Corp, 72-74; res coordr, 73-75, ASST PROF PHARMACOL, HEALTH SCI CTR, UNIV TEX, 75- *Concurrent Pos:* Adj instr pharmacol, Health Sci Ctr, Univ Tex, 72-74. *Mem:* Am Soc Neurochem. *Res:* Study of labile metabolites in the central nervous system; acetylcholine, cyclic nucleotides, indoleamines and catecholamines through the use of microwave irradiation technique; modification of these metabolites by drugs and heavy metals. *Mailing Add:* Brooks AFB San Antonio TX 78284

MODAK, ASHOK TRIMBAK, b Pune, India, Apr 11, 46; m 71. MECHANICAL ENGINEERING, PHYSICAL CHEMISTRY. *Educ:* Indian Inst Technol, BTech, 68; Univ Calif, Berkeley, MS, 69, PhD(mech eng), 73. *Prof Exp:* Prod design engr air pollution control, Ford Motor Co, 69-71; res specialist combustion, Factory Mutual Res Corp, 73-80; sr proj mgr, Combustion Res & Develop, Northern Res & Eng Corp, 80-83; prin res scientist, lasers, Laser Diag Combustion Systs, Phys Sci Inc, 83-91; CONSULT, 91- *Concurrent Pos:* Consult fire safety & combustion in gas turbine engines. *Mem:* Am Soc Mech Engrs; Combustion Inst; Sigma Xi; NY Acad Sci. *Res:* Combustion problems; heat and mass transfer in flames and fires; gaseous radiation; diffusion flame structure and chemical kinetics; soot formation and burnout in flames; flammability of plastics; combustion in gas turbine engines and in fluidized beds. *Mailing Add:* 22 Standish Rd Watertown MA 02172

MODDEL, GARRET R, b Dublin, Ireland, Feb 7, 54; US citizen. AMORPHOUS SILICON, OPTO ELECTRONICS. *Educ:* Stanford Univ, BS, 76; Harvard Univ, MS, 78, PhD(appl physics), 81. *Prof Exp:* Teaching asst solid-state physics, Harvard Univ, 77-78, res asst, 77-81; staff scientist, SERA Solar Corp, 81-85; asst prof, 85-90, ASSOC PROF, DEPT ELEC & COMPUT ENG, UNIV COLO, 85-90. *Concurrent Pos:* Prog mgr, Univ Colo Optoelectronic Comput Systs Ctr, 90- *Mem:* Am Phys Soc; Inst Elec & Electronics Engrs; Mat Res Soc. *Res:* Deposition, processing and characterization of semiconductor materials and basic devices; amorphous silicon, particularly its optoelectronic properties; development of state-of-the-art solar cells; high speed optically addressed spatial light modulators. *Mailing Add:* Dept Elec & Comput Eng Univ Colo Campus Box 425 Boulder CO 80309-0425

MODDERMAN, JOHN PHILIP, b Grand Rapids, Mich, Dec 4, 44; m 79; c 1. FOOD ADDITIVE CHEMISTRY. *Educ:* Calvin Col, AB, 67; Wayne State Univ, PhD(analytical chem), 71. *Prof Exp:* Fel chemiluminescence res, Univ Ga, 71-73; CHEMIST, DIV CHEM & PHYSICS, FOOD & DRUG ADMIN, 73- *Concurrent Pos:* Consult, Joint Food & Agr Orgn/WHO Expert Comt Food Additives; mem deleg, Codex Alimentanus Comt Food Additives. *Mem:* Am Chem Soc; AAAS. *Res:* Analytical chemistry of additives and contaminants in food. *Mailing Add:* Keller & Heckman 1150 17th St NW Suite 1100 Washington DC 20036-4614

MODE, CHARLES J, b Bismarck, NDak, Dec 29, 27; m 60; c 1. MATHEMATICS, POPULATION STUDIES. *Educ:* NDak State Univ, BS, 52; Kans State Univ, MS, 53; Univ Calif, Davis, PhD(genetics), 56. *Prof Exp:* Res fel statist, NC State Univ, 56-57; prof math, Mont State Univ, 57-66; assoc prof statist, State Univ NY Buffalo, 66-70; PROF MATH, DREXEL UNIV, 70- *Concurrent Pos:* Mem, Inst Pop Studies, Drexel Univ. *Mem:* Biomet Soc; Inst Math Statist; Am Math Soc; Pop Asn Am; Int Union Sci Study Pop. *Res:* Probability theory; application of mathematics and statistics to biology and medicine, particularly in family planning evaluation; stochastic processes; branching processes, models of population growth; mathematical demography; computer simulation. *Mailing Add:* Dept of Math Drexel Univ 32nd & Chestnut Sts Philadelphia PA 19104

MODE, VINCENT ALAN, b Gilroy, Calif, May 25, 40; m 76; c 1. APPLIED RESEARCH MANAGEMENT, COMPUTER SIMULATION. *Educ:* Whitman Col, AB, 62; Univ Ill, Urbana, PhD(inorg chem), 65; Golden Gate Univ, MBA, 80. *Prof Exp:* Res chemist, Lawrence Livermore Lab, Univ Calif, 65-71, group leader, 71-75, sect leader, 75-77, assoc div leader, 77-80; exec dir, BC Res, 80-84; facil mgr, 84-85, dep assoc dir, chem dept, 85-89, DEP ASSOC DIR, LASER PGM ASSURANCES, LAWRENCE LIVERMORE LAB, 89- *Concurrent Pos:* Pres, Techwest Enterprises, Ltd, 81-84; vchmn, Western Regional Coun, Soc Comput Simulation, 85-86. *Mem:* Sigma Xi; Soc Comput Simulation. *Res:* Fluorescence of rare earth chelates; high-pressure liquid chromatography of inorganic species; development of interactive computer programs and large data base analysis; computer simulation of management problems. *Mailing Add:* Lawrence Livermore Nat Lab PO Box 808 Livermore CA 94551

MODEL, FRANK STEVEN, b New York, NY, May 5, 42; m 65; c 2. CHEMISTRY, POLYMER SCIENCE. *Educ:* Mass Inst Technol, BS, 63; Harvard Univ, MA, 65, PhD(chem), 68. *Prof Exp:* Res chemist, Celanese Res Co, Summit, 67-72, proj leader, 72-75, sr res chemist, 72-76; dir res, Gelman Sci Inc, Ann Arbor, 76, vpres res & develop, 76-79; sr res assoc, 79-80, lab mgr, 80-82, assoc dir, 82-86, ASSOC DIR RES & DEVELOP, PALL CORP, GLEN COVE, 86- *Mem:* AAAS; Am Chem Soc; Sigma Xi. *Res:* Physical chemistry of formed polymers; polymer processing; membrane science and technology; microfiltration processes and devices; ultrafiltration; reverse osmosis; hemodialysis; polymer characterization; biomedical applications of polymers; electrophoresis and immunodiffusion; inorganic photochromism. *Mailing Add:* 28 Buttonwood Dr Dix Hills NY 11746

MODEL, PETER, b Frankfurt, Ger, May 17, 33; US citizen; m 81; c 1. MOLECULAR BIOLOGY, GENETICS. *Educ:* Stanford Univ, AB, 53; Columbia Univ, PhD(biochem), 65. *Prof Exp:* Res assoc biochem, Columbia Univ, 65-67; NSF fel, 67-69, from asst prof to assoc prof, 69-87, PROF BIOCHEM ROCKEFELLER UNIV, 87- *Mem:* Am Soc Microbiol; Am Soc Virol. *Res:* Genetics and physiology of prokaryotic membrane protein biosynthesis; in vitro protein synthesis; bacteriophage genetics and physiology. *Mailing Add:* Dept Genetics Rockefeller Univ 1230 York Ave New York NY 10021

MODELL, JEROME HERBERT, b St Paul, Minn, Sept 9, 32; m 52; c 3. ANESTHIOSOLOGY. *Educ:* Univ Minn, BA, 54, BS & MD, 57; Am Bd Anesthesiol, dipl, 64. *Prof Exp:* Jr scientist internal med, Sch Med, Univ Minn, 57-59; from instr to assoc prof anesthesiol, Sch Med, Univ Miami, 63-69; PROF ANESTHESIOL & CHMN DEPT, COL MED, UNIV FLA, 69- *Concurrent Pos:* NIH res career develop award, 67-69. *Mem:* AAAS; Am Soc Anesthesiol; Asn Univ Anesthetists. *Res:* Pathophysiology and treatment of near-drowning; physiologic applications of liquid breathing; intensive pulmonary therapy. *Mailing Add:* PO Box J-254 Gainesville FL 32610

MODELL, WALTER, b Waterbury, Conn, July 18, 07; m 33; c 1. PHARMACOLOGY. *Educ:* City Col New York, BS, 28; Cornell Univ, MD, 32. *Prof Exp:* From instr clin pharmacol to prof pharmacol, 32-73, dir clin pharmacol, 56-73, EMER PROF PHARMACOL, MED COL, CORNELL UNIV, 73- *Concurrent Pos:* Ed-in-chief, Clin Pharmacol & Therapeut, 60-; mem bd dirs, US Pharmacopoeia. *Mem:* Am Soc Pharmacol & Exp Therapeut; Am Soc Clin Pharmacol & Therapeut; fel AMA; fel Am Col Physicians; NY Acad Sci. *Res:* Clinical pharmacology. *Mailing Add:* Cornell Univ Med Col New York NY 10021

MODER, JOSEPH J(OHN), b St Louis, Mo, Dec 12, 24; m 51; c 6. OPERATIONS RESEARCH, STATISTICS. *Educ:* Washington Univ, BS, 47; Northwestern Univ, MS, 49, PhD(chem eng), 50. *Prof Exp:* Chem engr, Petrolite Corp, 47; asst, Northwestern Univ, 47-50; from assoc prof to prof indust eng, 50-64, res assoc, Exp Sta, Ga Inst Technol, 50-57; chmn dept, 66-78, PROF MGT SCI, UNIV MIAMI, 64- *Concurrent Pos:* NSF fel, Iowa State Univ, 60-61 & Stanford Univ, 61-62; Joseph Lucas vis prof eng prod, Univ Birmingham, 73-74. *Mem:* Am Statist Asn; Opers Res Soc Am; Inst Mgt Sci; Proj Mgt Inst. *Res:* Project management methodology and network techniques; work sampling; operations research and systems analysis. *Mailing Add:* Dept Mgt Sci Univ Miami University Station Coral Gables FL 33124

MODESITT, DONALD ERNEST, b Richmond, Ind, Oct 14, 36; m 59; c 3. ENVIRONMENTAL & SANITARY ENGINEERING. *Educ:* Mo Sch Mines, BS, 58; Univ Mo-Rolla, MS, 66; Okla State Univ, PhD(bioenviron eng), 70. *Prof Exp:* Civil engr, State of Ill Div Hwy, 58-60; from instr to asst prof, 60-73, ASSOC PROF CIVIL ENG, UNIV MO-ROLLA, 73- *Concurrent Pos:* Consult to indust, Munic & individuals with sanit-environ problems. *Mem:* Water Pollution Control Fedn; Am Water Works Asn; Am Soc Civil Engrs; Nat Soc Prof Engrs; Am Acad Environ Engrs; Am Soc Eng Educ. *Res:* Bioenvironmental engineering; water quality, wastewater and water treatment; environmental health. *Mailing Add:* Dept Civil Eng Univ Mo-Rolla Rolla MO 65401

MODEST, EDWARD JULIAN, b Boston, Mass, Sept 9, 23; m 47; c 3. CANCER CHEMOTHERAPY, TOXICOLOGY. *Educ:* Harvard Univ, BA, 43, MA, 47, PhD(org chem), 49. *Prof Exp:* Res asst path, The Children's Med Ctr, Boston, Mass, 49-52, res assoc, 52-80; head, Lab Bio Org Chem, The Children's Cancer Res Found, Mass, 52-74; res assoc path, Sch Med, Harvard Univ, Mass, 56-69, assoc, 69-77, lectr pharmacol, 77-80; head, Clin Chem Lab, Sidney Farber Cancer Inst, Mass, 72-80, scientist, 74-80, head, Div Med Chem & Pharmacol, 74-80; assoc dir pharmacol & exp therapeut, oncol res ctr, 80-83, PROF CHEM, DEPT BIOCHEM, BOWMAN GRAY SCH MED, WINSTON-SALEM, NC, 80-, DIR, PHARMACOL CORE LAB, ONCOL RES CTR, 80-, ASSOC DIR PRECLIN STUDIES, 83-; CONSULT PHARMACOL, SIDNEY FARBER CANCER INST, MASS, 80- *Concurrent Pos:* Spec consult, Nat Cancer Inst, NIH, 59-62, consult, 65-, div res grants, 68-; mem, chem panel, Cancer Chemother Nat Serv Ctr, Nat Cancer Inst, NIH, 59-62, liaison mem to biochem comt, drug eval panel, 61-62, mem, med & org chem B fel rev comt, 68-70, pharmacol B study sect, 71-73, exp therapeut study sect, 73-75 & 86-, antifolate adv group, div cancer treatment, 76-; radiation safety comt, The Children's Hosp Med Ctr, 63-80, chmn, 70-80, mem, pharm comt, 71-80; adj prof pharmaceut chem, Univ RI, Kingston, 67-; mem classification & pharm & therapeut comts, Sidney Farber Cancer Inst, 73-80; consult, Cognitive Info Processing Group, dept elec eng, Mass Inst Technol, 74-78, toxicol, Bioassays Systs Corp, Woburn, 79-, infusion technol, Infusaid Corp, Norwood, 80- & Pharmacia Nu-Tech Co, Med field, 82-; scientific consult, Grace Cancer Drug Ctr, Roswell Park Mem

Inst, Buffalo, NY, 77-78 & Cancer Ctr, Univ Miami, Fla, 78-; vis lectr pharmacol, dept pharmacol, Sch Med, Harvard Univ, 80-; mem, biosafety comt, Bowman Gray Sch Med, 80- *Mem:* Am Chem Soc; Am Asn Cancer Res; AAAS; Am Soc Clin Oncol. *Res:* Biochemical and cancer pharmacology; preclinical and clinical pharmacokinetics and metabolism; mechanisms of drug resistance in human malignant cells; interaction of antitumor drugs with malignant cell membranes, enzymes and DNA; experimental and clinical cancer chemotherapy; drug development and mechanistic studies on antifolates, antitumor proteins and intercalating agents; preclinical toxicology; new approches to drug delivery. *Mailing Add:* Dept Biochem Boston Univ Sch Med 80 Concord St Boston MA 02118-2394

MODI, V J, b Bhavnagar, India, Dec 15, 30; m 59; c 1. AERONAUTICAL ENGINEERING. *Educ:* Univ Bombay, BE, 53; Indian Inst Sci, Bangalore, dipl, 55; Univ Wash, MS, 56; Purdue Univ, PhD(aeronaut eng), 59. *Prof Exp:* Bombay Govt fel, 53-55; res asst, Univ Wash, 55-56 & Purdue Univ, 56-57; res specialist dynamics, Cessna Aircraft Co, 59-61; from asst prof to assoc prof mech eng, 61-67, PROF MECH ENG, UNIV BC, 67- *Mem:* Sr mem Am Astronaut Soc; assoc fel Can Aeronaut & Space Inst; Am Soc Mech Engrs; Am Inst Aeronaut & Astronaut. *Res:* Aeroelasticity; separated flows; industrial aerodynamics; dynamics of offshore structures; bioengineering; satellite dynamics. *Mailing Add:* Dept Mech Eng Univ BC 2075 Wesbroook Pl Vancouver BC V6T 1W5 Can

MODIC, FRANK JOSEPH, b Cleveland, Ohio, Sept 20, 22; m 56; c 4. PHYSICAL CHEMISTRY, ORGANIC CHEMISTRY. *Educ:* Case Western Reserve Univ, BS, 43; Iowa State Col, PhD(phys org chem), 51. *Prof Exp:* Jr chem engr, Kellex Corp, 43-45; prod supvr, Carbide & Carbon Chem Co, 45-46; instr chem, Iowa State Col, 46-48; CHEMIST, SILICONE PROD DEPT, GEN ELEC CO, 51- *Mem:* Am Chem Soc; Royal Soc Chem. *Res:* Physical-organic research in organosilicon chemistry; reaction mechanisms and synthesis of silicones and silicone copolymers. *Mailing Add:* Gen Elec Co Six Lillian Dr Scotia NY 12302-3908

MODINE, FRANKLIN ARTHUR, b Sault Ste Marie, Mich, Feb 8, 36; m 63; c 2. PHYSICS. *Educ:* Mich State Univ, BS, 60; Univ Southern Calif, MA, 66; Univ Ore, PhD(physics), 71. *Prof Exp:* Engr, NAm Aviation, 61-66; physicist, Aerojet Gen, 66-67; PHYSICIST, OAK RIDGE NAT LAB, 71- *Mem:* Am Phys Soc. *Res:* Optical and magnetic properties of solids. *Mailing Add:* Solid State Div Bldg 3025 MS 30 Oak Ridge Nat Lab Oak Ridge TN 37831

MODISETTE, JERRY L, b Minden, La, July 28, 34; m 59; c 1. FLUID FLOW SIMULATIONS, ENERGY CONVERSION. *Educ:* La Tech Univ, BS, 56; Va Polytech Inst, MS, 60; Rice Univ, PhD(space sci), 67. *Prof Exp:* Aerospace technologist, NASA Langley Res Ctr, 56-62, chief, Space Physics Div, Johnson Space Ctr, 62-69; dean sci, Houston Baptist Univ, 69-80; pres, Modisette, Inc, 80-87; EXEC VPRES, ADVAN PIPELINE TECHNOLOGIES, 87- *Mem:* Am Inst Aeronaut & Astronaut; Am Geophys Union. *Res:* Fluid dynamics-thermodynamics applied to pipeline operations. *Mailing Add:* 18323 Hereford Lane Houston TX 77058

MODLIN, HERBERT CHARLES, b Chicago, Ill, Jan 12, 13; m 33; c 1. PSYCHIATRY. *Educ:* Univ Nebr, BSc, 35, MD, 38, MA, 40. *Hon Degrees:* DSc, Univ Nebr, 85. *Prof Exp:* Intern, Univ Nebr Hosp, 38-39; resident neuropsychiat, Clarkson Hosp, 39-40; fel psychiat, Pa Hosp, Philadelphia, 40-41 & Adams House, Boston, 41-42; resident neurol, Montreal Neurol Inst, 42-43; chief neuropsychiat serv, Winter Vet Admin Hosp, 46-49; SR PSYCHIATRIST, MENNINGER FOUND, 49-; NOBLE PROF FORENSIC PSYCHIAT, SCH MED, UNIV KANS, 80- *Concurrent Pos:* Lectr, Sch Law, Univ Kans, 47-; assoc prof forensic psychiat, Sch Med, Univ kans, 60-80. *Honors & Awards:* Golden Apple Award, Am Acad Phychiat & Law. *Mem:* Am Psychiat Asn; Sigma Xi. *Res:* Forensic and social psychiatry; psychiatric education. *Mailing Add:* Menninger Found Topeka KS 66601

MODLIN, IRVIN M, b Cape Town, SAfrica, Mar 14, 46; US citizen. SURGERY. *Educ:* Univ Cape Town, MB, 68, PhD, 89; Yale Univ, MA, 87; FCS(SAfrica), 75; FRCS, 75; FACS, 86. *Prof Exp:* Asst prof surg, Univ Calif, Los Angeles, 77-79; asst prof, Downstate Med Ctr, Brooklyn, 79-83; assoc prof, 84-87, VCHMN, DEPT SURG, SCH MED, YALE UNIV, 84- , PROF, 87- *Concurrent Pos:* Assoc dir, Gastrointestinal Peptide Physiol Res Labs, 80-83; chief surg, Vet Admin, Dept Surg, Sch Med, Yale Univ, 84-88, dir, Gastrointestinal Surg Res Group, 84-, Surg Endoscopy Div, 84- & Gastrointestinal Surg Pathobiol Res Unit, 89-; examnr, Royal Col Surgeons, Edinburgh, 85-; mem, Educ & Training Comt, Am Gastroenterol Asn, 85-89; Fulbright scholar, 89. *Honors & Awards:* Eisenberg Mem Lectr, Brigham Hosp, Harvard Med Sch, 88. *Mem:* Royal Col Surgeons Edinburgh; SAfrican Col Surgeons; Asn Acad Surgeons; Am Gastroenterol Asn; Soc Univ Surgeons; Royal Soc Med Eng; NY Acad Sci; Am Fedn Clin Res; Am Soc Endocrine Surgeons; Soc Surg Alimentary Tract. *Res:* Pancreatitis; pepsinogen section; parietal cells; mastomys ECL cells; colonic physiology; enterocyte migration; intracellular pH regulation; author of numerous publications. *Mailing Add:* Dept Surg Sch Med Yale Univ 333 Cedar St PO Box 3333-109 FMB New Haven CT 06510-8060

MODLIN, RICHARD FRANK, b Toledo, Ohio, Nov 16, 37; m 76. LIMNOLOGY, ZOOLOGY. *Educ:* Univ Wis-Milwaukee, BS, 67, MS, 69; Univ Conn, PhD(marine ecol), 76. *Prof Exp:* Teaching asst zool, Univ Wis-Milwaukee, 65-67; aquatic biologist, Wis Dept Nat Resources, 68; res asst limnol, Ctr Great Lakes Studies, 68-69, res assoc, 69-70; fisheries supvr, Conn Dept Environ Protection, 70-72; teaching asst biol, Univ Conn, 72-76; ASST PROF BIOL, UNIV ALA, HUNTSVILLE, 76- *Concurrent Pos:* Prin investr, Proj-77-14, Univ Ala, Huntsville, 76-77 & Proj-78-16, 77-78; vis prof marine sci, Dauphin Island Sea Lab, 79-; res grant, Tenn Valley Authority, 79-81. *Mem:* Am Soc Limnol & Oceanog; Int Asn Theoret & Appl Limnol; NAm Benthological Soc; Estuarine Res Fedn; Sigma Xi. *Res:* Limnology of estuaries, reservoirs and temporary waters; ecology, physiology and taxonomy of benthic invertebrates, crustacea; fisheries biology; southern estuaries and Gulf of Mexico. *Mailing Add:* Dept Biol Univ Ala Box 1247 Huntsville AL 35899

MODRAK, JOHN BRUCE, b New Britain, Conn, June 14, 43; m 68; c 2. PHARMACOLOGY. *Educ:* Univ Conn, BS, 67, PhD(pharmacol), 75. *Prof Exp:* ASST PROF PHARMACOL, COL PHARM, UNIV NEBR, 75- *Mem:* NY Acad Sci. *Res:* Role of connective tissue, collagen, elastin and mucopolysaccharides in the progression and regression of atherosclerosis. *Mailing Add:* Dept Med Commun Glaxo Inc Five Moore Dr Research Triangle Park NC 27709

MODRESKI, PETER JOHN, b New Brunswick, NJ, Dec 22, 46; m 68; c 1. EXPERIMENTAL PETROLOGY, MINERAL CHEMISTRY. *Educ:* Rutgers Univ, BA, 68; Pa State Univ, MS, 71, PhD(geochem), 72. *Prof Exp:* Res chemist laser chem, Air Force Weapons Lab, Kirtland AFB, 72-75; mem tech staff geochem, Sandia Labs, 75-79; GEOCHEMIST, US GEOL SURV, 79- *Mem:* Geol Soc Am; Mineral Soc Am. *Res:* Chemistry and phase relations of minerals and hydrothermal solutions, relating to the genesis of ore deposits; chemistry of magnetite; geochemistry of cobalt; luminescence of minerals; shock metamorphism. *Mailing Add:* US Geol Surv Box 25046 Fed MS 922 Denver CO 80225

MODREY, JOSEPH, b New York, NY, Jan 29, 16; m 43; c 2. MECHANICAL ENGINEERING. *Educ:* Columbia Univ, BS, 37, MME, 38; Rensselaer Polytech Inst, DEng, 63. *Prof Exp:* Res engr gas turbine design, Wright Aeronaut Corp, 38-47; prof mach design, Polytech Inst Brooklyn, 47-55; prof mech eng & head dept, Union Col, 55-65; PROF MECH ENG, PURDUE UNIV, 65- *Concurrent Pos:* Sr scientist, Midwest Appl Sci Corp. *Mem:* Am Soc Mech Engrs; Soc Automotive Engrs; Am Soc Eng Educ; Sigma Xi. *Res:* Very high-speed machinery, vibration and lubrication. *Mailing Add:* Mech Eng Purdue Univ Lafayette IN 47907

MODRICH, PAUL L, b Raton, NMex, June 13, 46. BIOCHEMISTRY, GENETICS. *Educ:* Mass Inst Technol, BS, 68; Stanford Univ, PhD(biochem), 73. *Prof Exp:* Postdoctoral fel, Dept Biol Chem, Harvard Univ, 73-74; asst prof chem, Univ Calif, Berkeley, 74-76; from asst prof to prof biochem, 76-88, mem, Duke Comprehensive Cancer Ctr, 77-80, JAMES B DUKE PROF BIOCHEM, DUKE UNIV, 88-, DIR, PROG GENETICS, 89- *Concurrent Pos:* Henry & Camille Dreyfus teacher scholar award, 77; NIH res career develop award, 78-83; mem, Biochem Study Sect, NIH, 80-84; vis comt biol, Corp Mass Inst Technol, 81-84 & 84-87; counr, Am Soc Biochem & Molecular Biol, 89-92. *Honors & Awards:* Pfizer Award in Enzyme Chem, 83. *Res:* Mechanisms of protein-nucleic acid interactions. *Mailing Add:* Dept Biochem Med Ctr Duke Univ Box 3711 Durham NC 27710

MODRZAKOWSKI, MALCOLM CHARLES, b Enid, Okla, Mar 12, 52; m 73; c 2. MICROBIOLOGY. *Educ:* Univ Mass, Amherst, BS, 74; Univ Ga, PhD(microbiol), 77. *Prof Exp:* Fel, Univ NC, Chapel Hill, 77-79; ASST PROF MICROBIOL, DEPT ZOOL & MICROBIOL, OHIO UNIV, 79-, ASST PROF BIOMED SCI, COL OSTEOPATRIC MED, 79- *Concurrent Pos:* Prin investr, Pub Health Serv grants, 80- *Mem:* Am Soc Microbiol. *Res:* Inter-relationships that exist between intrusive gram-negative bacteria and host non-specific leukocyte defense mechanisms. *Mailing Add:* Dept Basic Sci & Zool Ohio Univ Athens OH 45701

MOE, AARON JAY, b Duluth, Minn, Aug 15, 55. PEDIATRICS. *Educ:* Univ Minn, BS, 79; Va Polytech Inst & State Univ, MS, 82, PhD(animal sci-ruminant nutrit), 84. *Prof Exp:* Res asst, Dept Dairy Sci, Va Polytech Inst & State Univ, Blacksburg, 79-84; res assoc, Dept Physiol, Med Ctr, George Washington Univ, 84-87; res assoc, 87-90, RES INSTR, DEPT PEDIAT, SCH MED, WASH UNIV, ST. LOUIS, MO, 90- *Mem:* Am Physiol Soc; AAAS; Tissue Cult Asn. *Res:* Author of numerous publications. *Mailing Add:* Dept Pediat Sch Med Wash Univ 400 S Kingshighway Blvd St Louis MO 63110

MOE, CHESNEY RUDOLPH, b Rainy River, Ont, Oct 6, 08; US citizen; m 35, 51; c 2. PHYSICS. *Educ:* Stanford Univ, AB, 29, AM, 31; Univ Southern Calif, PhD(physics), 41. *Prof Exp:* Asst, 29-30, instr, 31-35, from asst prof to prof, 35-73, assoc chmn div phys sci, 56-62, chmn dept physics, 62-65, EMER PROF PHYSICS, SAN DIEGO STATE UNIV, 73- *Concurrent Pos:* Mem bd rev, Marine Phys Lab, Univ Calif, 46-51; consult, US Navy Electronics Lab, 53-56, Tracor Corp, Tex & Calif, 67-69 & Jet Propulsion Lab, 69 & 70. *Mem:* Fel Acoust Soc Am. *Res:* Acoustics. *Mailing Add:* 4669 E Talmadge Dr San Diego CA 92116-4829

MOE, DENNIS L, b SDak, Apr 1, 17; m 43; c 4. AGRICULTURAL ENGINEERING. *Educ:* SDak State Univ, BS, 48, MS, 49; Augustana Col, PhD, 71. *Prof Exp:* Field supvr, Mutual Benefit Life Ins Co, 40; asst chemist, Am Chem & Potash Corp, 41; from instr to assoc prof, 46-56, from asst agr engr to assoc agr engr, 46-56, PROF AGR ENG & HEAD DEPT & AGR ENGR, AGR EXP STA, SDAK STATE UNIV, 56-, DIR, INST IRRIGATION TECHNOL, 68- *Concurrent Pos:* Consult, State Eng Bd Dirs & indust firms; NSF nat grant panel, mem, Review Comn, Eng Coun Prof Develop; Am consult, foreign univs, Western Europe & Scandinavia; consult struct eng; mem, Eng Accreditation Teams. *Mem:* Fel Am Soc Agr Eng; Am Soc Eng Educ; Soil Conserv Soc Am. *Res:* Farm structures, power and machinery; development of lightweight aggregate from native shales and clays; low-pressure gas in farm tractors; silage and structures; world food production and energy used in food chain production; review of research institutes and universities. *Mailing Add:* RR 3 No 23 Brookings SD 57006

MOE, GEORGE, b Portland, Ore, Dec 31, 25; m 52; c 2. PHYSICAL CHEMISTRY, INORGANIC CHEMISTRY. *Educ:* Reed Col, BA, 47; Univ Rochester, PhD(chem), 50. *Prof Exp:* Asst chem, Univ Rochester, 47-49; instr, Univ Buffalo, 50-51; from res chemist to head astronaut dept, Aerojet-Gen Corp Div, Gen Tire & Rubber Co, 51-60; vpres res, Astropower, Inc, Douglas Aircraft Corp, 60-64, dir, Astropower Lab, Missiles & Space Syst Div, 64-70, dep dir res & develop, 70-72, dir res & develop, 72-76, DIR, ENERGY SYSTS, MCDONNELL-DOUGLAS ASTRONAUT CO, 76- *Mem:* Am Chem Soc; Am Phys Soc; Am Inst Aeronaut & Astronaut. *Res:* Aerospace technology; electrochemical systems; advanced computers and advanced materials; synthesis and evaluation of high-energy propellants; photochemistry and radiation chemistry of free radicals; solid-state reactions. *Mailing Add:* 12231 Afton Lane Santa Ana CA 92705-3052

MOE, GEORGE WYLBUR, b Opportunity, Wash, Apr 24, 42. ENGINEERING PHYSICS. *Educ:* Gonzaga Univ, BS, 66; Univ Wash, PhD(physics), 74. *Prof Exp:* res assoc physics, Columbia Radiation, Columbia Univ, 73-78; PRIN MEM RES STAFF, RIVERSIDE RES INST, 78- *Mem:* Inst Elec & Electronics Engrs; Am Phys Soc; Am Inst Physics; Soc Photo-Optical Instrumentation Engrs. *Res:* Electro-optical data acquisition systems; image to signal processing; real-time computer applications; experimental atomic physics. *Mailing Add:* Riverside Res Inst 330 W 42nd St New York NY 10036

MOE, GREGORY ROBERT, b Granite Falls, Minn, Dec 17, 56; m 88; c 1. BIOCHEMISTRY. *Educ:* Univ Minn, BChem, 80; Univ Chicago, PhD(chem), 85. *Prof Exp:* Postdoctoral fel biochem, Univ Calif, Berkeley, 85-88; ASST PROF BIOCHEM, UNIV DEL, 88- *Mem:* Protein Soc. *Res:* Protein folding; rational design of peptides and proteins; mechanisms of transmembrane signaling by receptor proteins. *Mailing Add:* Dept Chem & Biochem Univ Del Newark DE 19716

MOE, JAMES BURTON, b Hayfield, Minn, Oct 4, 40; c 4. VETERINARY PATHOLOGY, MICROBIOLOGY. *Educ:* Univ Minn, BS, 62, DVM, 64; Univ Calif, PhD(comp path), 78. *Prof Exp:* Staff pathologist, US Army Med Res Inst, 69-75, microbiologist infectious dis, 78-80, dir, Div Path, 80-85; consult, 80-86; res pathologist, 86-88, dir, Path/Toxicol, 88-89, EXEC DIR, DRUG SAFETY RES, UPJOHN, 89- *Concurrent Pos:* Vis consult, Armed Forces Inst Path, 73-75; consult, Physicians Clin Lab, 76-78; Calif Lung Asn grant, 77-78. *Mem:* Am Col Vet Pathologists; Int Acad Path. *Res:* Immunological, pathological and virological studies of infectious diseases affecting animals and man; pathological studies of neoplastic diseases of various animal species; toxicologic pathology. *Mailing Add:* 301 Henrietta St Kalamazoo MI 49001

MOE, MAYNARD L, b Lake Mills, Iowa, Jan 5, 35; m 57; c 3. ELECTRICAL ENGINEERING. *Educ:* Iowa State Univ, BS, 57; Northwestern Univ, MS, 59, PhD(elec eng), 61. *Prof Exp:* From asst prof to prof elec eng, Univ Denver, 61-75; PRES, DENCOR, INC, 74- *Concurrent Pos:* Res engr, Denver Res Inst, 61-75. *Mem:* Inst Elec & Electronics Engrs. *Res:* Feedback control system design; application of artificial intelligence to control systems; biomedical control systems; energy control systems. *Mailing Add:* Biol Dept Calif State Col Bakersfield CA 93309

MOE, MICHAEL K, b Milwaukee, Wis, Nov 17, 37; m 61; c 1. ELEMENTARY PARTICLE PHYSICS. *Educ:* Stanford Univ, BS, 59; Case Western Reserve Univ, MS, 61, PhD(physics), 65. *Prof Exp:* Res fel cosmic ray physics, Calif Inst Technol, 65-66; asst res physicist, 66-68, asst prof physics, 68-73, asst res physicist, 73-75, assoc res physicist, 75-80, RES PHYSICIST, UNIV CALIF, IRVINE, 80- *Res:* Low-level scintillation spectrometry; cosmic rays; experimental search for double beta decay. *Mailing Add:* One Mann St Irvine CA 92715

MOE, MILDRED MINASIAN, b Philadelphia, Pa, Nov 18, 29; m 51; c 2. ATMOSPHERIC PHYSICS, MECHANICS OF SATELLITE MOTIONS. *Educ:* Univ Calif, Los Angeles, BA, 51, MA, 53, PhD(physics), 57. *Prof Exp:* Mem tech staff, Ramo-Wooldridge Corp, Calif, 56-60; asst prof physics, Loyola Univ Los Angeles, 67-70, res assoc, 70-72; asst res physicist, Physics Dept, Univ Calif, Irvine, 73-78; consult, 79-81; LECTR, DEPT PHYSICS, UNIV CALIF, IRVINE, 82- *Mem:* Am Phys Soc; Am Asn Physics Teachers. *Res:* Quantum mechanical scattering theory; supersonic jet flows; satellite orbital and attitude motions; gas-surface interactions; measurement of upper atmospheric properties; atmospheric modeling. *Mailing Add:* 1520 Sandcastle Dr Corona del Mar CA 92625

MOE, OSBORNE KENNETH, b Los Angeles, Calif, Dec 29, 25; m 51; c 2. PHYSICS. *Educ:* Univ Calif, Los Angeles, BA, 51, MA, 53, PhD(planetary & space sci), 66. *Prof Exp:* Analyst airborne radar facil, Radio Corp Am, 54-55; mem tech staff, Ramo-Wooldridge Corp, 56-59, consult satellite orbits & model atmospheres, Space Tech Labs, 60-63, consult orbital predictions & model atmospheres, TRW, Inc, 63-67; sr res scientist, McDonnell Douglas Astronaut Co, 67-75; CONSULT, 75- *Concurrent Pos:* Asst res geophysicist, Univ Calif, Los Angeles, 66-67; consult, Aerospace Corp, 65-67 & Northrop Corp, 67; adj prof, Calif State Univ, Fullerton, 83-86; geophys scholar, Geophys Labs, US Air Force, 84-85; engr, Air Force Plant Rep Off/TRW, 87-89; staff meteorologist, Space Syst Div, AFSC, 89- *Mem:* Am Geophys Union; Am Meteorol Soc; Am Inst Aeronaut & Astronaut. *Res:* Theoretical analysis of the various types of density and composition measurements in the thermosphere; adsorption and energy accommodation at artificial satellite surfaces; thermospheric models; solar-terrestrial relationships; improvement of ozone measurements. *Mailing Add:* 1520 Sandcastle Dr Corona del Mar CA 92625

MOE, ROBERT ANTHONY, b Jersey City, NJ, Mar 26, 23; m 49; c 3. PHYSIOLOGY. *Educ:* Seton Hall Univ, BS, 48; Fordham Univ. MS, 49, PhD(physiol), 51. *Prof Exp:* Asst biol, Grad Sch, Fordham Univ, 48-51; head physiol testing sect, Toxicol Res Div, Inst Med Res, E R Squibb & Sons, 51-59; head cardiovasc sect, Pharmacol Dept, Hoffmann-La Roche, Inc, 59-65, from asst dir pharmacol dept to asst dir biol res div, 65-70; dir biol res, Searle Labs, 70-72, vpres res & develop, 72-73, sr vpres sci affairs, 73-76, exec vpres, 76-79; CONSULT SCI AFFAIRS & VPRES RES & DEVELOP, AYERST LABS, 79- *Mem:* AAAS; Am Soc Pharmacol & Exp Therapeut; assoc fel Am Soc Clin Pharmacol & Therapeut. *Res:* Toxicology and physiology of biological preparations; cardiovascular and autonomic systems. *Mailing Add:* Res & Develop Ayerst Labs 685 Third Ave New York NY 10017

MOECKEL, W(OLFGANG) E(RNST), b Ger, Feb 11, 22; nat US; m 50; c 1. AEROSPACE ENGINEERING. *Educ:* Univ Mich, BS, 44. *Prof Exp:* Aeronaut res scientist, Nat Adv Comt Aeronaut, Lewis Res Ctr, NASA, 44-47, sect head, Spec Projs Br, 47-50, chief, 50-55, asst chief, Propulsion Aerodyn Div, 55-58, chief, Electromagnectic Propulsion Div, 58-71, chief, Phys Sci Div, 71-77, chief scientist, 77-80; CONSULT, 80- *Honors & Awards:* Arthur S Flemming Award, 60; NASA Exceptional Sci Achievement Award, 69. *Mem:* AAAS; assoc fel Am Inst Aeronaut & Astronaut; Am Phys Soc. *Res:* Supersonic and hypersonic aerodynamics and propulsion; fluid mechanics; astronautics; space propulsion systems. *Mailing Add:* 29033 Lincoln Rd Bay Village OH 44140

MOEDRITZER, KURT, b Prague, Czech, June 20, 29; nat US; m 57; c 3. ORGANOMETALLIC CHEMISTRY, ORGANIC SYNTHESIS. *Educ:* Univ Munich, dipl, 53, PhD(inorg chem), 55. *Prof Exp:* Asst inorg chem, Univ Munich, 53-56, asst & instr, 57-59; Fulbright fel, Univ Southern Calif, 56-57; instr, Munich Br, Univ Md, 58-59; res chemist, Monsanto Co, 59-70, sci fel, 70-77, sr sci fel, 77-89, DISTINGUISHED FEL, MONSANTO CO, 89- *Concurrent Pos:* Ed, Synthesis & Reactivity in Inorg & Metal-org Chem. *Mem:* Am Chem Soc. *Res:* Organic chemistry, inorganic and organic phosphorus compounds; inorganic polymers; organometallics; metal hydrides; nuclear magnetic resonance; agricultural chemistry. *Mailing Add:* Monsanto Co St Louis MO 63167

MOEHLMAN, PATRICIA DES ROSES, b Washington, DC, Oct 5, 43. BEHAVIORAL ECOLOGY. *Educ:* Wellesley Col, BA, 65; Univ Tex, Austin, MA, 68, Univ Wis-Madison, PhD(zool), 74. *Prof Exp:* Instr biol, State Univ Calif, Chico, 73-74; fel zool, Univ Wis-Madison, 75-80; vis scientist animal behav, Univ Cambridge, 80-81; ASST PROF ANIMAL ECOL, YALE UNIV, 81- *Concurrent Pos:* Scientist behav ecol, Serengeti Res Inst, 76-82. *Mem:* Sigma Xi; Animal Behav Soc; Am Soc Mammalogists; Royal Geog Soc. *Res:* Behavioral ecology of feral asses (Equus asinus) in Death Valley National Monument and siverbacked (Canis mesomelas) and golden jackals (Canis aureus) in the Serengeti, Tanzania. *Mailing Add:* 91 High Ridge Road Guilford CT 06437

MOEHLMAN, ROBERT STEVENS, geology; deceased, see previous edition for last biography

MOEHRING, JOAN MARQUART, b Orchard Park, NY, Sept 23, 35; m 74. CELL BIOLOGY. *Educ:* Syracuse Univ, BS, 61; Rutgers Univ, MS, 63, PhD(microbiol), 65. *Prof Exp:* Fel microbiol, Stanford Univ, 65-68; res assoc cell & tissue cult, Dept Med Microbiol, Univ Vt, 68-70, res assoc, Dept Med, 70-71, res assoc, Dept Path, 71-73, res assoc, asst prof cell & tissue cult, consult & res assoc spec ctr res pulmonary fibrosis, Col Med, 73-77, res assoc prof, 77-80, RES PROF CELL & TISSUE CULT, DEPT MICROBIOL & MOLECULAR GENETICS, UNIV VT, 80- *Concurrent Pos:* Mem, Vermont Regional Cancer Ctr, 80- *Mem:* Am Soc Microbiol; Am Soc Cell Biol; Sigma Xi. *Res:* Application of cell and tissue culture to biomedical research; use of cultured mammalian cells to study the molecular action of bacterial toxins and the genetics of resistance to toxins. *Mailing Add:* Dept Microbiol & Molecular Genetics Univ Vt Burlington VT 05405

MOEHRING, THOMAS JOHN, b New York, NY, Aug 15, 36; m 64. MICROBIOLOGY. *Educ:* Fairleigh Dickinson Univ, BS, 61; Rutgers Univ, MS, 63, PhD(microbiol), 66. *Prof Exp:* Fel, Stanford Univ, 65-68; from asst prof to prof med microbiol, 68-86, prof microbiol, 86-89, PROF MICROBIOL & MOLECULAR GENETICS, UNIV VT, 89- *Concurrent Pos:* Fac, cell biol prog, fac grad col, Univ Vt; mem, Vt Regional Cancer Ctr. *Mem:* Sigma Xi; Am Soc Microbiol; Tissue Cult Asn; AAAS. *Res:* Cell culture in biomedical research; mechanisms of pathogenesis; in vitro action of microbial toxins; biochemical genetics of cultured cells; protein synthesis; replication of animal viruses. *Mailing Add:* Dept Microbiol & Molecular Genetics Univ Vt Col Med & Col Agr & Life Sci Burlington VT 05405

MOEHS, PETER JOHN, b Yonkers, NY, Apr 4, 40; m 65; c 2. CHEMICAL EDUCATION, ANALYTICAL CHEMISTRY. *Educ:* Norwich Univ, BS, 62; Univ NH, PhD(inorg chem), 67. *Prof Exp:* From assst prof to assoc prof, 69-84, chmn div sci, 71-72, PROF CHEM, SAGINAW VALLEY COL, 84- *Mem:* Am Chem Soc; Sigma Xi. *Res:* Chemistry of Group IV organometallics. *Mailing Add:* Dept Chem Saginaw Valley Univ 2250 Pierce Rd University Center MI 48710

MOELLER, ARTHUR CHARLES, b Cleveland, Ohio, Dec 14, 19. ELECTRICAL ENGINEERING, MATHEMATICS. *Educ:* Western Reserve Univ, BS, 41; Mich State Univ, MS, 48; Marquette Univ, BEE, 51; Univ Wis, PhD(elec eng), 65. *Prof Exp:* Asst math, Mich State Univ, 41-43; instr, Denison Univ, 43-44; instr, Marquette Univ, 44-51, elec eng, 51-52, asst prof, 52-58; res asst, Univ Wis, 58-60; from asst prof to assoc prof, Marquette Univ, 60-70, chmn dept, 61-63, actg dean col, 63-64, dean, 64-65, vpres acad affairs, 65-72, PROF ELEC ENG, MARQUETTE UNIV, 70- *Concurrent Pos:* Bd dir, Nat Eng Consortium, 63- *Mem:* AAAS; Math Asn Am; Inst Elec & Electronics Engrs; Sigma Xi. *Res:* Calculation of propagation constants in waveguides loaded with dielectric slabs; numerical solution of elliptic boundary value problems. *Mailing Add:* Marquette Univ 1515 W Wisconsin Ave Milwaukee WI 53233

MOELLER, CARL WILLIAM, JR, b Carroll, Iowa, Mar 2, 24; m 52; c 2. INORGANIC CHEMISTRY. *Educ:* Harvard Univ, BS, 49; Univ Southern Calif, PhD(chem), 54. *Prof Exp:* Asst chem, Univ Southern Calif, 49-53; Fulbright fel, Univ Tübingen, 54-55; dep dept head br, 54-80, from instr to assoc prof inorg chem, 55-78, PROF INORG CHEM, UNIV CONN, 78-, ACTG DEPT HEAD, 80- *Mem:* AAAS; Am Chem Soc. *Res:* Magnetochemical studies of free radicals and solid ternary oxides; photochemistry of coordination compounds; organoboron chemistry; chemical bonding. *Mailing Add:* Dept Chem Univ Conn Storrs CT 06269-3060

MOELLER, D(ADE) W(ILLIAM), b Grant, Fla, Feb 27, 27; m 49; c 5. ENVIRONMENTAL ENGINEERING, RADIATION PROTECTION. *Educ:* Ga Inst Technol, BCE & MS, 48; NC State Univ, PhD, 57. *Hon Degrees:* AM, Harvard Univ, 69. *Prof Exp:* Asst radioactive waste, Johns Hopkins Univ, 48-49; res engr, Los Alamos Sci Lab, 49-52; staff asst, Radiol

Health Br, USPHS, 52-54; res assoc reactor cooling systs, Oak Ridge Nat Lab, 56-57; chief radiol training, Robert A Taft Sanit Eng Ctr, USPHS, 57-61, officer-in-chg, Northeastern Radiol Health Lab, 61-66; assoc dir, Kresge Ctr Environ Health, 66-83, head dept Environ Health Sci, 68-83, dir, Off Cont Educ, Sch Pub Health, 82-84, PROF ENG ENVIRON HEALTH, SCH PUB HEALTH, HARVARD UNIV, 82-, ASSOC DEAN CONT EDUC, 84- *Concurrent Pos:* Chmn, Am Bd Health Physics, 67-70; mem, Nat Coun Radiation Protection & Measurements, 67-; mem, Adv Comt Reactor Safeguards, US Nuclear Regulator Comn, 73-88, chmn, 76; mem, Comt 4, Int Comn Radiol Protection, 78-85; chmn, Adv Comt Nuclear Waste, US Nuclear Regulatory Comn, 88- *Mem:* Nat Acad Eng; AAAS; fel Am Nuclear Soc; Am Pub Health Asn; Health Physics Soc (pres, 71-72). *Res:* Analyses of reportable events at nuclear power plants; control of airborne radon decay products inside buildings; environmental radiation surveillance. *Mailing Add:* 27 Wildwood Dr Bedford MA 01730

MOELLER, HENRY WILLIAM, b Woodbury, NJ, Aug 4, 37; c 2. MARINE BIOLOGY. *Educ:* Drew Univ, AB, 59; Rutgers Univ, MS, 65, PhD(bot), 69. *Prof Exp:* Physiologist, Wallace Pharmaceut Corp, NJ, 61-62; from instr to asst prof marine sci, Southampton Col, 65-69; from asst prof to assoc prof, 69-80, PROF BIOL, DOWLING COL, 81-, BIOL DISCIPLINE COORDR, 88- *Concurrent Pos:* Pres, Hydro Bot Co, Shelter Island, NY, 81-84. *Res:* Marine botany; mass culture of marine algae in a water charged atmosphere for energy, food and biochemicals. *Mailing Add:* Dept Biol Dowling Col Oakdale NY 11769

MOELLER, THEODORE WILLIAM, b Cincinnati, Ohio, Jan 26, 43; m 66; c 2. FOOD SCIENCE. *Educ:* Ohio State Univ, BS, 65; Mich State Univ, MS, 67, PhD(food sci), 71. *Prof Exp:* Group leader food res, 71, MGR PET FOODS RES, JOHN STUART RES LAB, QUAKER OATS CO, 71- *Concurrent Pos:* Mgr/dir qual assurance, Int Foods Res & Develop, 76-81, dir, 81-, vpres, Food Res & Develop, 82. *Mem:* Inst Food Technologists; Am Soc Cereal Chem. *Mailing Add:* Quaker Oats Co 617 W Main St Barrington IL 60010

MOELLMANN, GISELA E BIELITZ, b Dessau, Ger, Feb 15, 29; US citizen; c 1. CELL BIOLOGY, CYTOCHEMISTRY. *Educ:* Univ Dayton, BS, 53; Yale Univ, PhD(anat), 67. *Prof Exp:* Am Cancer Soc fel, 66-68, res assoc anat, 68-69, res assoc dermat & lectr anat, 69-73, asst prof, 73-78, LECTR CELL BIOL, YALE UNIV, 78-, ASSOC PROF RES DERMAT (ADJ), SCH MED, 83- *Mem:* AAAS; Am Soc Cell Biol; Soc Invest Dermat; Pan Am Soc Pigment Cell Res; Int Pigment Cell Soc. *Res:* Cell biology, especially the fine structure and cytochemistry of mammalian and amphibian pigment cells; intracellular transport of particles; topography of peptide hormone receptor sites; fine-structural changes in vitiligo; pigment cell aging. *Mailing Add:* LCI 500 Dept Dermat Yale Univ Sch of Med New Haven CT 06510

MOELTER, GREGORY MARTIN, b Dover, NJ, Aug 14, 19; m 42; c 2. CHEMICAL ENGINEERING. *Educ:* Newark Col Eng, BS, 41; NY Univ, MS, 44. *Prof Exp:* Anal chemist, Barrett Chem Co, 40; from sr res engr to sect head fiber spinning res, Celanese Corp Am, 41-63; develop mgr, Celanese Fibers Co, 63-65, tech mgr, 65-72, tech admin, 72-75, proj mgr, 75-83; RETIRED. *Concurrent Pos:* Consult, 83- *Mem:* AAAS; Am Chem Soc; Am Inst Chem Engrs; Am Soc Eng Educ. *Res:* Technology of synthetic fibers productions. *Mailing Add:* 3637 Henshaw Rd Charlotte NC 28209

MOEN, ALLEN LEROY, b Badger, Minn, June 15, 33; m 59; c 3. SURFACE PHYSICS. *Educ:* Pac Lutheran Univ, BA, 55; Wash State Univ, MS, 61, PhD(physics), 68. *Prof Exp:* Asst physics, 55-57, 59-60 & 61-63; from asst prof to assoc prof, 63-82, chmn dept, 68-82, chmn, Natural Sci Div, 81-85, PROF PHYSICS, CENT COL IOWA, 82- *Concurrent Pos:* vis assoc prof, Univ Mo, Rolla, 79-80; vis prof, Zhejiang Univ, Hangzhow, China, 88-89. *Mem:* Am Asn Physics Teachers; Sigma Xi. *Res:* Ion bombardment of metals; sputtering; surface physics; ultra-high vacuum physics; surface area measurement of Mt St Helens ash aerosols using Bruauer-Emmett-Teller method. *Mailing Add:* Dept of Physics Cent Col Pella IA 50219

MOEN, WALTER B(ONIFACE), b Rockville Centre, NY, Feb 13, 20; m 46; c 2. MECHANICAL ENGINEERING. *Educ:* Pratt Inst, BME, 40; Columbia Univ, MS, 46. *Prof Exp:* Jr engr, Combustion Eng Co, NY, 40-41; instr mech eng, Pratt Inst, 41-47; asst engr, Thatcher Furnace Co, NJ, 47-48; res engr, Res Lab, Air Reduction Co, Inc, 48-52, head metall res sect, 52-56, asst dir, 56-59, eng mgr, Spec Prod Dept, 59-60, mgr, Cryogenics Eng Dept, 60-62, asst chief engr, Cent Eng Dept, 63-67, asst to group vpres, 67-69; dir res & develop, Am Sterilizer Co, 69-71; mgr auto arc div, Nat Standard Co, 71-73; dir eng, Consol Energy Prod, Inc, 73-75; managing dir tech progs, Am Soc Mech Engrs, 75-83, staff exec, 83-85; CONSULT, 85- *Concurrent Pos:* Consult, 85- *Mem:* Am Soc Mech Engrs; Am Soc Eng Educ; Am Welding Soc. *Res:* Industrial combustion processes; arc and gas welding; propellant evaluation; propulsion; inorganic chemical manufacture; ocean engineering. *Mailing Add:* Consult 244 Deerwood Dr Huddleston VA 24104

MOENCH, ROBERT HADLEY, b Boston, Mass, Oct 23, 26; m 55; c 3. GEOLOGY. *Educ:* Boston Univ, AB, 50, AM, 51, PhD(geol), 54. *Prof Exp:* geologist, 52-88, EMER SCIENTIST, US GEOL SURV, 88- *Mem:* AAAS; Geol Soc Am; Soc Econ Geol; Sigma Xi. *Res:* Stratigraphy, structure and tectonics of metamorphic rocks, mineral resources and environmental geology, especially in New England. *Mailing Add:* 902 Grant Pl Boulder CO 80302-7117

MOENS, PETER B, b Neth, May 15, 31; Can citizen; m 53; c 5. CELL BIOLOGY, CYTOGENETICS. *Educ:* Univ Toronto, BScF, 59, MA, 61, PhD(biol), 63. *Prof Exp:* Assoc prof, 64-72, chmn dept natural sci, Atkinson Col, 74-77 & dept biol, 81-84, PROF BIOL, YORK UNIV,72- *Concurrent Pos:* Assoc ed, Genetic Soc Can, 70; ed, Genome, formerly Can J Gen Cytol, 82-; managing ed, Chromosoma (BERL), 88- *Honors & Awards:* Award of Merit, Genetics Soc Can. *Mem:* Fel Royal Soc Can; Genetics Soc Can (pres, 78-79); Genetics Soc Am; Can Soc Cell Biol; Am Cell Biol Soc. *Res:* Electron microscopy; meiosis; development of germ cells in fungi, plants and animals with emphasis on genetically significant aspects; specifically electron microscopy of the synaptonemal complex; immuno cytochemical in situ hybridization analysis of synaptonemal complex proteins and associated DNA. *Mailing Add:* Dept Biol York Univ Downsview ON M3J 1P3 Can

MOERMOND, TIMOTHY CREIGHTON, b Sioux City, Iowa, Apr 4, 47; m 68; c 1. ECOLOGY. *Educ:* Univ Ill, Urbana-Champaign, BS, 69; Harvard Univ, PhD(biol), 74. *Prof Exp:* Asst prof, 73-81, ASSOC PROF ZOOL, UNIV WIS-MADISON, 81- *Mem:* Ecol Soc Am; Am Ornithologists Union; Soc Study Evolution; Am Soc Ichthyologists & Herpetologists; Wilson Ornith Soc. *Res:* Field and theoretical studies of foraging strategies and habitat use patterns. *Mailing Add:* Dept Zool 226 Russell Labs Univ Wis Birge Hall 1630 Linden Dr Madison WI 53706

MOERNER, WILLIAM ESCO, b Pleasanton, Calif, June 24, 53; m 83; c 1. LASER SPECTROSCOPY, NONLINEAR ORGANIC MATERIALS. *Educ:* Wash Univ, BS & BSEE & AB, 75; Cornell Univ, MS, 78, PhD(physics), 82. *Prof Exp:* Res asst, dept physics, Wash Univ, 72-75; grad res asst, Lab Atomic & Solid State Physics, Cornell Univ, 75-81; res staff mem, 81-88, mgr, 88-89, PROJ LEADER, LASER MAT INTERACTIONS, IBM RES DIV, ALMADEN RES CTR, 89- *Concurrent Pos:* Prin investr, US Off Naval Res contract, 83-90; asst treas, Inst Elec & Electronic Engrs, Lasers & Electro-Optics Soc, 88, mem prog comt, 88, treas, 89; gen chair, Opt Soc Am Topical Mtg, 90-91. *Mem:* AAAS; Am Chem Soc; Am Phys Soc; sr mem Inst Elec & Electronic Engrs; Optical Soc Am. *Res:* Utilizing linear and nonlinear laser spectroscopy, precision detection and systems analysis to study mechanisms for persistent spectral hole-burning, organic photorefrativity and fundamental properties of defects in solids as well as possible future technologies for optical data storage. *Mailing Add:* IBM Res Div Almaden Res Ctr K95/801 650 Harry Rd San Jose CA 95120-6099

MOERTEL, CHARLES GEORGE, b Milwaukee, Wis, Oct 17, 27; m 52; c 4. GASTROINTESTINAL CANCER, CLINICAL PHARMACOLOGY. *Educ:* Univ Ill, BS & MD, 53; Univ Minn, MS, 57. *Hon Degrees:* Dr, Univ Grenoble. *Prof Exp:* Internship, Los Angles County Gen Hosp, 53-54; resident internal med, Mayo Found, Rochester, Minn, 54-57, asst to staff Mayo Clin, 57-58, prof med & dir Mayo Comprehensive Ctr, 75-86, Purvis & Roberta Tabor prof, 81-87, PROF ONCOL, MAYO MED SCH, 76-, CONSULT, MAYO CLIN, 58- *Concurrent Pos:* Chmn, Eastern Coop Oncol Group, Gastrointestinal Cancer Comt, 72-74; co-chmn, Gastrointestinal Tumor Study Group, 73-79, Oncol Drugs Adv Comt, Food & Drug Admin, 74-89; Coun Cancer, Am Med Asn, 78-81; chmn, Nat Adv Comt Colorectal, Am Cancer Soc, 85-, N Cent Cancer Treat Group, 85-, Bd Sci Coun Div, Cancer Prev Control, Nat Cancer Inst, 81-86; Walter Lawrence vis prof, Univ Calif, Los Angeles. *Honors & Awards:* Walter Hubert lectureship, British Asn Cancer Res, 76; Gold Medal, Swed Surg Asn, 78; Karnotsky Award, Am Soc Clin Oncol, 86; Clin Res Award, Asn Community Cancer Ctr, 87. *Mem:* Am Asn Cancer Res; Am Soc Clin Oncol (pres, 79-80). *Res:* Diagnosis and treatment of gastrointestinal cancer; clinical pharmacology and clinical trial methodology. *Mailing Add:* 1009 Skyline Lane SW Rochester MN 55902

MOESCHBERGER, MELVIN LEE, b Berne, Ind, June 26, 40; m 62; c 3. APPLIED STATISTICS, PUBLIC HEALTH. *Educ:* Taylor Univ, BS, 62; Ohio Univ, MS, 65; NC State Univ, PhD(statist), 70. *Prof Exp:* Instr math, Taylor Univ, 62-63; res assoc biostatist, Univ NC, Chapel Hill, 65-70; asst prof, Univ Mo-Columbia, 70-76; assoc prof statist, Ohio State Univ, 76-80; assoc prof, 80-86, PROF PREV MED, OHIO STATE UNIV, 87- *Concurrent Pos:* NIH res grant, 74; math statistician, Nat Ctr Toxicol Res, 76-78; AFOSR res grant. *Mem:* Am Statist Asn; Biomet Soc. *Res:* Survival analyses and competing risk theory. *Mailing Add:* Dept Prev Med Col Med Ohio State Univ 320 W Tenth Ave Columbus OH 43210

MOFFA, DAVID JOSEPH, b Fairmont, WVa, Dec 6, 42; m 64; c 2. BIOCHEMISTRY, CLINICAL CHEMISTRY. *Educ:* WVa Univ, AB, 64, MS, 66, PhD(biochem), 68. *Prof Exp:* Res asst biochem, 64-68, instr, 68-70, ASST PROF BIOCHEM, SCH MED, WVA UNIV, 70- *Concurrent Pos:* NIH res fel, WVa Univ, 68-70; dir, BioPreps Labs, 69- *Mem:* Am Asn Clin Chem; Am Soc Clin Pathologists; Am Med Technologists. *Res:* Metabolism of vitamin A; lipid metabolism; enzymology; clinical methodology. *Mailing Add:* 501 Locust Ave Fairmont WV 26554

MOFFAT, ANTHONY FREDERICK JOHN, b Toronto, Ont, Jan 30, 43; m 66; c 2. ASTROPHYSICS. *Educ:* Univ Toronto, BSc, 65, MSc, 66; Univ Ruhr, W Germany, Dr rer nat, 70, Dr habil(astron), 76. *Prof Exp:* Instr astron, Univ Bonn, 67-69; sci aid, Univ Ruhr, 70-76; sci asst, 77-80, PROF ASTRON, UNIV MONTREAL, 81- *Concurrent Pos:* Imperial Oil fel, Univ Bonn, 66-69; Alexander von Humboldt fel, 82-83, 89; mem, Natural Sci & Eng Res Coun Can grant selection comt, Space & Astron, 85-88, pres, 88-89; dir, Ctr Observatoire der mont Mégantic, 90- *Honors & Awards:* Gold medal of Royal Astron Soc Can, 65. *Mem:* W German Astron Soc; Am Astron Soc; Can Astron Soc; Royal Astron Soc Can; Int Astron Union. *Res:* Nature and evolution of massive stars; structure and dynamics of the galaxy; cataclysmic variable stars. *Mailing Add:* Dept de Physique Univ Montreal CP6128 Succ A Montreal PQ H3C 3J7 Can

MOFFAT, JAMES, b Turtle Creek, Pa, Feb 24, 21; m 46; c 1. CHEMISTRY. *Educ:* Allegheny Col, BS, 42; Northwestern Univ, PhD(org chem), 48. *Prof Exp:* Asst prof chem, Univ Miami Fla, 47-49; res fel, Calif Inst Technol, 49-50; sr res chemist, Nepera Chem Co, 51; res assoc, Northwestern Univ, 51-53; res asst prof chem, Univ Louisville, 53-58; from asst prof to assoc prof chem, Univ Mo, Kansas City, 58-88; RETIRED. *Mem:* AAAS; Am Chem Soc. *Res:* Heterocyclic compounds; organic isocyanides; hydrogen bonding; tautomerism. *Mailing Add:* 950 E Grayson St San Antonio TX 78208

MOFFAT, JOHN BLAIN, b Owen Sound, Ont, Aug 7, 30; m 56; c 3. HETEROGENEOUS CATALYSIS, SURFACE CHEMISTRY. *Educ:* Univ Toronto, BA, 53, PhD(phys chem), 56. *Prof Exp:* Res chemist, Res & Develop Lab, Du Pont Can Ltd, 56-61; from asst prof to assoc prof, 61-74, PROF CHEM, UNIV WATERLOO, 74- *Concurrent Pos:* dir, Can Soc Chem, 87-; vpres, Can Catalysis Found, 90- *Honors & Awards:* Catalysis Award, Chem Inst Can, 88. *Mem:* Am Chem Soc; Royal Soc Chem; Am Phys Soc; fel Chem Inst Can. *Res:* Heterogeneous catalysis; surface chemistry; surface, structural and catalytic properties of heteropoly oxometalates and inorganic stoichiometric and nonstoichiometric phosphates; partial oxidation and oxidative coupling of methane; conversion of methanol to hydrocarbons. *Mailing Add:* Dept Chem Univ Waterloo Waterloo ON N2L 3G1 Can

MOFFAT, JOHN KEITH, b Edinburgh, Scotland, Apr 3, 43. BIOPHYSICS. *Educ:* Univ Edinburgh, BSc, 65; Cambridge Univ, PhD(protein crystallog), 70. *Prof Exp:* Sci staff mem protein crystallog, Med Res Coun Lab Molecular Biol, 68-69; res assoc reaction kinetics, 69-70, from asst prof to assoc prof, 70-84, PROF BIOCHEM & MOLECULAR BIOL, CORNELL UNIV, 85- *Concurrent Pos:* John Simon Guggenheim fel, 85; Royal Soc guest res fel, 85. *Honors & Awards:* Res Career Develop Award, NIH, 78. *Mem:* Am Crystallog Asn; Biophys Soc; Am Soc Biol Chemists; AAAS. *Res:* Protein structure determination by physico-chemical techniques; relation between structure and function in calcium binding proteins and polypeptide hormones; protein crystallography; synchrotron radiation techniques. *Mailing Add:* Sect Biochem Molecular & Cell Biol 207 Biotech Bldg Cornell Univ Ithaca NY 14853

MOFFAT, JOHN WILLIAM, b Copenhagen, Denmark, Dec 24, 32; Can citizen; m 85; c 2. PARTICLE THEORY, GRAVITATION. *Educ:* Cambridge Univ, UK, PhD(gravitation), 58. *Hon Degrees:* DSc, Univ Winnipeg, Man, 89. *Prof Exp:* Sr scientist physics, Res Inst Advan Studies, 59-64; assoc prof, 64-67, PROF PHYSICS, UNIV TORONTO, 67- *Concurrent Pos:* Fel, Cambridge Philos Soc, 59-; vis scientist, Cern Geneva, Switz, 60-61; vis prof, Univ Dijon & Univ Paris, France, 84-85 & Orsay Lab, Univ Paris, France, 87-88. *Mem:* NY Acad Sci; Int Union Astron. *Res:* Particle theory-symmetry breading and electroweak theory; nonsymmetric gravitation theory; astronomy-gravitational tests binary observations; cosmology-large scale structure; astrophysics-neutron stars etc. *Mailing Add:* Dept Physics Univ Toronto Toronto ON M5S 1A7 Can

MOFFAT, ROBERT J, b Grosse Pointe, Mich, Nov 29, 27; div; c 1. MECHANICAL ENGINEERING. *Educ:* Univ Mich, BS, 52; Wayne State Univ, MS, 61; Stanford Univ, MS, 66, PhD(heat transfer), 67. *Prof Exp:* Sr res engr, Res Labs, Gen Motors Corp, 52-62; assoc prof, 66-71, PROF MECH ENG & CHMN, THERMOSCI DIV, STANFORD UNIV, 71- *Mem:* Fel, Am Soc Mech Engrs; Instrument Soc Am. *Res:* Heat transfer, mass transfer, temperature measurement and systems or devices related to these fields. *Mailing Add:* Dept of Mech Eng Stanford Univ Stanford CA 94305

MOFFATT, DAVID JOHN, b Staffordshire, Eng, July 23, 39; m 64; c 2. COMPUTER-AIDED INSTRUCTION, HEMATOLOGY. *Educ:* Bristol Univ, Eng, BSc, 61, MB & ChB, 64. *Prof Exp:* Demonstr anat, Bristol Univ, 65-67, res assoc, 67-68; from asst prof to assoc prof anat, Univ Iowa, 68-79; prof & chmn, Dept Anat, 79-85, PROF, UNIV MO-KANSAS CITY, 85- *Concurrent Pos:* Vis prof anat, Univ Geissen, Ger, 75; consult, Asn Med Schs Mid East, 75- & 4th Saudi Med Conf, 79; vis prof anat, Cairo Univ, 76; mem learning mat panel, Nat Libr Med, 76-; ed, Health Sci News, 76-78; vis prof, El Minya Univ, Egypt & King Faisal Univ, Saudi Arabia, 79 & King Abdulaziz Univ, 80; external examr anat, King Abdulaziz Univ, Saudi Arabia, 80, 81 & 85. *Mem:* Am Asn Anatomists; Nat Soc Performance & Instr; Anat Soc Gt Brit & Ireland; Int Exp Hemat Soc; Soc Exp Biol Med; Brit Asn Clin Anatomists. *Res:* Use of mediated learning systems in medical education; computer aided instruction; experimental hematology; kinetics of stem cell proliferation; medical education systems. *Mailing Add:* Dept Anat Univ Mo 2411 Holmes St Kansas City MO 64108

MOFFATT, DAVID LLOYD, b Wheeling, WVa; m 50; c 2. ELECTRICAL ENGINEERING, ELECTROMAGNETISM. *Educ:* Ohio State Univ, BS, 58, MSc, 61, PhD(elec eng), 67. *Prof Exp:* Res assoc, 58-67, from asst prof to assoc prof, 69-83, PROF ELEC ENG, OHIO STATE UNIV, 83-, ASSOC SUPVR, ELECTRO-SCI LAB, 67- *Mem:* Sigma Xi; Int Union Radio Sci. *Res:* Radar target identification; time domain electromagnetics; radar cross-section. *Mailing Add:* Dept Elec Eng Ohio State Univ Columbus OH 43210

MOFFATT, JOHN GILBERT, b Victoria, BC, Sept 19, 30; m 53; c 4. ORGANIC CHEMISTRY. *Educ:* Univ BC, BA, 52, MSc, 53, PhD(org chem), 56. *Prof Exp:* Tech officer, Defence Res Bd Can, 53-54; res assoc, BC Res Coun, 56-60; group leader, Calif Corp Biochem Res, 60-61; head org chem, 61-65, assoc dir, 65-67, dir, Inst Molecular Biol, 67-77, dir synthetic chem, Inst Org Chem, 77-81, dir, Inst Bio-Org Chem, Syntex Res, 81-87; RETIRED. *Mem:* Am Chem Soc; Chem Soc; Am Soc Biochemists; AAAS. *Res:* Chemistry of nucleosides, nucleotides, nucleoside polyphosphates, sugar phosphates, carbohydrates and nucleic acids. *Mailing Add:* 22 Mirada Rd Half Moon Bay CA 94019-1780

MOFFATT, WILLIAM CRAIG, b Owen Sound, Ont, Apr 19, 33; m 56; c 2. FLUID MECHANICS, AERODYNAMICS. *Educ:* Queen's Univ, Ont, BSc, 56, MSc, 58; Mass Inst Technol, ScD(mech eng), 61. *Prof Exp:* Lectr mech eng, Royal Mil Col, Ont, 56; asst prof, Mass Inst Technol, 61-65; proj engr, Northern Res & Eng Corp, 65-66; assoc prof, 66-67, head dept, 69-78, PROF MECH ENG, ROYAL MIL COL CAN, 67-, DEAN ENG, 84- *Concurrent Pos:* Consult, Northern Res & Eng Corp, 61-65 & 66-; vis prof, Nat Defence Col, 74-75, Van Karman Inst, Brussels, 78-79. *Mem:* Am Soc Mech Engrs; Am Inst Aeronaut & Astronaut; assoc fel Can Aeronaut & Space Inst; Sigma Xi. *Res:* Gas turbine propulsion; turbomachinery; combustion. *Mailing Add:* Dean Eng Royal Mil Col Can Kingston ON K7K 5L0 Can

MOFFA-WHITE, ANDREA MARIE, b Marlboro, Mass, May 19, 49; m 78; c 3. POPULATION GENETICS. *Educ:* Fordham Univ, BS, 71; Univ RI, MS, 73, PhD(biol), 76. *Prof Exp:* Asst prof biol, Wheeling Col, 76-77; asst prof biol, WVa Univ, 77-81; asst pediat, Univ Fla, Gainsville, 81-84; PVT GENETIC COUNSELLOR, 84- *Concurrent Pos:* NIH grant, Univ Fla, 83-85. *Mem:* AAAS; Sigma Xi; Am Soc Zoologists; Genetics Soc Am; Am Soc Human Genetics; Teratology Soc. *Res:* Etiology of neural tube defects in animal models; demographic and genetic changes in laboratory populations. *Mailing Add:* 133 Shady Branch Trail Ormond Beach FL 32074

MOFFEIT, KENNETH CHARLES, b Vinson, Okla, Dec 24, 39; c 2. PHYSICS. *Educ:* Univ Calif, Riverside, BA, 65; Univ Calif, Berkeley, PhD(physics), 70. *Prof Exp:* Res assoc physics, Stanford Linear Accelerator Ctr, Stanford Univ, 70-74; staff physicist, Deutches, 74-76; STAFF PHYSICIST PHYSICS, STANFORD LINEAR ACCELERATOR CTR, STANFORD UNIV, 76- *Concurrent Pos:* Fel, Atomic Energy Comn, 65-66. *Mem:* Am Phys Soc. *Res:* High-energy physics. *Mailing Add:* Stanford Linear Accelerator Ctr PO Box 4349 Stanford CA 94305

MOFFET, ALAN THEODORE, radio astronomy; deceased, see previous edition for last biography

MOFFET, HUGH L, b Monmouth, Ill, Jan 6, 32; m 54, 84; c 3. PEDIATRICS, INFECTIOUS DISEASES. *Educ:* Harvard Univ, AB, 53; Yale Univ, MD, 57. *Prof Exp:* Instr pediat, Bowman Gray Sch Med, 62-63; from asst prof to assoc prof, Med Sch, Northwestern Univ, Chicago, 63-71; assoc prof, 71-74, PROF PEDIAT, MED SCH, UNIV WIS-MADISON, 74- *Concurrent Pos:* Nat Inst Allergy & Infectious Dis fel, 60-63. *Mem:* Fel Infectious Dis Soc Am. *Res:* Diagnostic microbiology; epidemiology of pediatric infections. *Mailing Add:* Dept Pediat Univ Wis Madison WI 53706

MOFFET, ROBERT BRUCE, PROTEIN CHEMISTRY, MOLECULAR GENETICS. *Educ:* Purdue Univ, PhD(molecular biol), 74. *Prof Exp:* PROJ SCIENTIST, CLEVELAND CLIN FOUND, 82- *Res:* Molecular biology of hypertension. *Mailing Add:* Dept Brain & Vasc Res Res Inst Cleveland Clin Found 9500 Euclid Ave Cleveland OH 44106

MOFFETT, BENJAMIN CHARLES, JR, b Spring Lake, NJ, Oct 28, 23; m 49, 78; c 3. DENTAL RESEARCH. *Educ:* Syracuse Univ, BA, 48; NY Univ, PhD(anat), 52. *Prof Exp:* From asst prof to assoc prof anat, Med Col, Univ Ala, 52-63; assoc prof, Col Med, Wayne State Univ, 63-64; assoc prof, 64-67, PROF ORTHOD, SCH DENT, UNIV WASH, 67- *Concurrent Pos:* Res fel anat, Med Sch, Gothenburg Univ, 59-60; res fel anat, Armed Forces Inst Path, 60-61; vis prof, Sch Dent, Cath Univ Nijmegen, 71-72; mem, Cranio-facial Biol Group, Int Asn Dent Res. *Honors & Awards:* Jerome Schweitzer Res Award, NY Acad Prosthodont, 69. *Mem:* Am Asn Anat; Int Asn Dent Res; hon mem Am Acad Craniomandular Disorders. *Res:* Arthrology; history of anatomy; cranio-facial morphogenesis. *Mailing Add:* Dept Orthod SM/46 Univ Wash Sch Dent Seattle WA 98195

MOFFETT, DAVID FRANKLIN, JR, b Raleigh, NC, Sept 4, 47; m 70; c 3. COMPARATIVE PHYSIOLOGY, EPITHELIAL TRANSPORT. *Educ:* Duke Univ, BS, 69; Univ Miami, PhD(biol), 73. *Prof Exp:* Fel, Dept Physiol, Duke Univ, 72-73; fel, Zoophysiol Prog, 74-75, asst prof, Dept Zool, 75-82, ASSOC PROF PHYSIOL, DEPT ZOOL, WASH STATE UNIV, 82- *Concurrent Pos:* Vis assoc prof, Dept Physiol/Biophys, Univ Tex Med Br, Galveston, 89-90. *Mem:* Biophys Soc; Soc Gen Physiologists; Am Soc Zoologists. *Res:* Cellular mechanisms of solute and water transport in animals; ion channels and pumps in native and artificial lipid membranes; transport by insect intestinal epithelia. *Mailing Add:* Dept Zool Wash State Univ Pullman WA 99164-4236

MOFFETT, JOSEPH ORR, entomology, for more information see previous edition

MOFFETT, ROBERT BRUCE, b Madison, Ind, June 8, 14; c 2. ORGANIC CHEMISTRY, MEDICINAL CHEMISTRY. *Educ:* Hanover Col, AB, 37, Univ Ill, AM, 39, PhD(org chem), 41. *Hon Degrees:* DSc, Hanover Col, 59. *Prof Exp:* Asst, Univ Ill, 37-39; Abbott, Upjohn & Glidden fel, Northwestern Univ, 41-43; sr res chemist, George A Breon & Co, Mo, 43-44, asst dir labs, 44-47; res chemist, Upjohn Co, 47-77; RETIRED. *Mem:* Am Chem Soc; Sigma Xi. *Res:* Chemiluminescence; benzopyrylium salts; steroids; analgesics; antispasmodics; drugs for mental diseases. *Mailing Add:* 2895 Bronson Blvd Kalamazoo MI 49008

MOFFETT, STACIA BRANDON, INVERTEBRATE BEHAVIOR, REGENERATION. *Educ:* Univ Miami, PhD(neurophysiol), 73. *Prof Exp:* ASSOC PROF ZOOL, WASH STATE UNIV, 74- *Mailing Add:* Dept Zool Wash State Univ Pullman WA 99164

MOFFITT, EMERSON AMOS, b McAdam, NB, Sept 9, 24; US citizen; m 51; c 3. ANESTHESIOLOGY. *Educ:* Dalhousie Univ, MD, CM, 51; Univ Minn, MS, 58; FRCP(C), 58; Am Bd Anesthesiol, dipl, 60. *Prof Exp:* Pvt pract, 51-54; resident anesthesiol, Mayo Grad Sch Med, Univ Minn, 54-57, from instr to assoc prof, 59-72; prof anesthesia & head dept, Med Sch Dalhousie Univ, 73-80, assoc dean, Clin Affairs, 80-86, develop officer, 86-89; RETIRED. *Concurrent Pos:* Consult, Mayo Clin, 57-72, sect head anesthesiol, 66-72, NIH grant gen med sci, Mayo Found, 67-72, NS Heart Found, 81- *Honors & Awards:* Can Anesthetists' Soc Medal, 90. *Mem:* Am Soc Anesthesiol; Int Anesthesia Res Soc; Can Anesthetists Soc; Sigma Xi. *Res:* Cardiovascular and metabolic effects of anesthetic state and cardiac surgery with whole body perfusion; metabolism in acute stress and shock. *Mailing Add:* Dept Anesthesia Victoria Gen Hosp Halifax NS B3H 2Y9 Can

MOFFITT, HAROLD ROGER, b Ukiah, Calif, Aug 8, 34; m 54; c 3. AGRICULTURAL ENTOMOLOGY. *Educ:* Univ Calif, Davis, BS, 57, Univ Calif, Riverside, MS, 63, PhD(entom), 67. *Prof Exp:* Lab technician entom, Univ Calif, Riverside, 57-63, res asst, 63-67; res leader & tech adv, 72-80, RES

ENTOMOLIGIST, AGR RES SERV, USDA, 67. *Concurrent Pos:* Vis scientist, Entomology Div, Dept Sci & Indust Res, New Zealand, 80-81. *Mem:* AAAS; Entom Soc Am; Entom Soc Can. *Res:* Biology and control of insects and mites of agricultural importance; integrated control and applied ecology; postharvest entomology of horticultural crops. *Mailing Add:* Yakima Agr Res Lab USDA 3706 W Nob Hill Blvd Yakima WA 98902

MOFFITT, ROBERT ALLAN, b Gillette, Wyo, June 17, 18; m 44; c 4. CHEMISTRY, BIOCHEMISTRY. *Educ:* Univ Calif, Los Angeles, BA, 40; Univ Southern Calif, MS, 54. *Prof Exp:* Lab mgr, Fernando Valley Milling & Supply Co, 40-44 & Ralston Purina Co, 46-60; head analysis lab, Res Labs, 60-67, MGR ANALYTICAL SERV, CARNATION CO, 67- *Mem:* Am Chem Soc; Am Asn Cereal Chem; Inst Food Technol. *Res:* Accurate analysis of chemical components of food products; contamination of foods; analytical chemistry; instrumentation. *Mailing Add:* 1960 Escarpa Dr Los Angeles CA 90041-3015

MOG, DAVID MICHAEL, b Cleveland, Ohio, Oct 21, 42; m 68; c 2. BIOCHEMISTRY. *Educ:* Case Inst Technol, BS, 64; Calif Inst Technol, PhD(chem), 70. *Prof Exp:* Asst prof chem, Muskingum Col, 68-69; res fel Brazil, Nat Acad Sci, 70-73; asst prof chem, Oberlin Col, 73-78; admin officer, Chem Dept, Princeton Univ, 78-81; sr prog officer, Off Int Affairs, Nat Res Coun, 81-89; SCI TEACHER, SIDWELL FRIENDS SCH, 90- *Concurrent Pos:* Mem, Peace Corps Rev Comt, Nat Acad Sci, 73-74. *Mem:* Am Chem Soc; AAAS; Sigma Xi. *Res:* Science and technology in Third World Development; chemistry and biotechnology in agriculture, health and environment; resource conserving systems; science education. *Mailing Add:* Sci Dept Sidwell Friends Sch 3825 Wisconsin Ave NW Washington DC 20016

MOGAB, CYRIL JOSEPH, electronic materials, device processing, for more information see previous edition

MOGABGAB, WILLIAM JOSEPH, b Durant, Okla, Nov 2, 21; m; c 7. INTERNAL MEDICINE, INFECTIOUS DISEASES. *Educ:* Tulane Univ, BS, 42, MD, 44; Am Bd Internal Med, dipl, 51; Am Bd Microbiol, dipl. *Prof Exp:* Intern, Charity Hosp La, New Orleans, 44-45; from asst to instr med, Sch Med, Tulane Univ, 46-49, instr, Div Infectious Dis, 49-51; Nat Found Infantile Paralysis fel, 51-52; asst prof, Baylor Col Med, 52-53; head virol div, Naval Med Res Unit 4, Great Lakes, Ill, 53-55; assoc prof, 56-62, PROF MED, SCH MED, TULANE UNIV, 62- *Concurrent Pos:* From resident to chief resident, Tulane Serv, Charity Hosp, La, 46-49, vis physician, 49-51 & sr vis physician, 61-75, consult, 76-; vis investr & asst physician, Hosp, Rockefeller Inst, 51-52; chief infectious dis, Vet Admin Hosp, Houston, 52-53; consult, Vet Admin Hosp, New Orleans, 56-; mem, Comn Influenza, Armed Forces Epidemiol Bd, 59-71; mem, Mayor's Health Adv Comt, New Orleans, 83-; mem, Orphan Prod Develop Initial Rev Group, Food & Drug Admin, 84- *Mem:* Am Col Physicians; Infectious Dis Soc Am; Soc Epidemiol Res; Am Soc Clin Invest; Am Soc Virol; Am Acad Microbiol. *Res:* Virology; tissue culture; bacteriology; microbiology; mycoplasma; vaccines for respiratory infections; new antibiotics; sexually transmitted diseases. *Mailing Add:* Dept Med Tulane Univ Sch of Med 1430 Tulane Ave New Orleans LA 70112

MOGENSEN, HANS LLOYD, b Price, Utah, Dec 16, 38; m 58; c 2. PLANT ANATOMY, PLANT MORPHOLOGY. *Educ:* Utah State Univ, BS, 61; Iowa State Univ, MS, 63, PhD(plant anat), 65. *Prof Exp:* Assoc prof, 65-74, PROF BOT, NORTHERN ARIZ UNIV, 74- *Concurrent Pos:* Vis prof, Carlsberg Res Lab, Copenhagen, Denmark, 76-77. *Mem:* Bot Soc Am; Int Soc Plant Morphologists. *Res:* Ultrastructure of fertilization in flowering plants; plant development. *Mailing Add:* Dept Biol Box 5640 Northern Ariz Univ Flagstaff AZ 86011

MOGENSON, GORDON JAMES, b Delisle, Sask, Jan 24, 31; m 54; c 2. NEUROPHYSIOLOGY, BEHAVIOR NEUROSCIENCE. *Educ:* Univ Sask, BA, 55, MA, 56; McGill Univ, PhD(neurosci), 59. *Prof Exp:* From asst prof to assoc prof psychol, Univ Sask, 58-65; assoc prof psychol & physiol, 65-68, chmn dept, 76-84, PROF PHYSIOL, UNIV WESTERN ONT, 68- *Concurrent Pos:* Ed, Can J Psychol, 69-74; res prof, Med Res Coun Can, 81-82, vpres, 85-86. *Mem:* AAAS; Am Physiol Soc; Animal Behav Soc; Can Psychol Asn; Can Physiol Soc (pres, 81-82); fel Royal Soc Can; Soc Neurosci. *Res:* Neurophysiology of the limbic system; the mechanism of limbic-motor integration. *Mailing Add:* Dept Physiol Univ Western Ont London ON N6A 5C1 Can

MOGFORD, JAMES A, b McCamey, Tex, Dec 6, 30; m 53; c 4. PHYSICS. *Educ:* Tex Tech Col, BS, 56; Univ Wis, MS, 59. *Prof Exp:* Physicist, Midwestern Univs Res Asn, 59-61; staff mem, Sandia Labs, NMex, 61-67, supvr, Anal Div, Sandia Labs, Livermore, 67-74, supvr, Inductive Energy Storage Div, 74-75, mem mgt staff, Sandia Nat Labs, Albuquerque, 75-88; foreign affairs specialist, Off Arms Control, Dept Energy, 88-90; spec assignment arms control, 90, SUPVR, TECH PROJ REV DIV, SANDIA NAT LABS, 91- *Concurrent Pos:* Coordr, Sandia interactions with UK, 84-88; mem, US Deleg to Negotiate Arms Control Agreement, 89 & 90. *Res:* Theoretical studies of orbit dynamics in particle accelerators; effects of radiation energy deposition; effects of photon-electron irradiation of materials and systems; design and development of pulsed-power systems. *Mailing Add:* Star Rte Box 1238 Corrales NM 87048

MOGGIO, MARY VIRGINIA, b Baton Rouge, La, Mar 10, 47; div. ENVIRONMENTAL EPIDEMIOLOGY, ENVIRONMENTAL SCIENCES. *Educ:* Univ Pittsburgh, BS, 68; Univ NC, Chapel Hill, MSPH, 72. *Prof Exp:* Environ epidemiologist, 71-79, EPIDEMIOLOGIST, PEDIAT INFECTIOUS DIS, MED CTR, DUKE UNIV, 79- *Mem:* Am Pub Health Asn; Nat Environ Health Asn; Asn Practitioners Infection Control; Sigma Xi. *Res:* Environmental and nosocomial infection epidemiology in health care institutions; preventative disease epidemiology in pediatric population; risk for defined bacterial complication of previous underlying disease states; participation in preventative vaccine trials for pediatric populations. *Mailing Add:* 130 Murdock Rd Millsborough NC 27278

MOGHADAM, OMID A, b Teheran, Iran, Jan 27, 68. IMAGE-PROCESSING & PATTERN RECOGNITION, BIO-MEDICAL IMAGING SYSTEMS DESIGN. *Educ:* State Univ NY Buffalo, BS, 89, MS, 90. *Prof Exp:* RES ENGR THERMOGRAPHIC IMAGING, DEPT BIOPHYS, SCI, SCH MED & BIOMED SCI, STATE UNIV NY BUFFALO, 89- *Mem:* Int Soc Optical Eng; Inst Elec & Electronic Engrs; Bio-med Eng Soc. *Res:* Image processing and opto-electronic aspects of various clinical diagnostic systems. *Mailing Add:* 69 Framingham Lane Pittsford NY 14534

MOGHISSI, KAMRAN S, b Tehran, Iran, Sept 11, 25; US citizen; m 52; c 2. OBSTETRICS & GYNECOLOGY. *Educ:* Univ Geneva, MB & ChB, 51, MD, 52; Am Bd Obstet & Gynec, dipl, 67, cert, 80, dipl reprod endocrinol, 75. *Prof Exp:* Intern & resident obstet & gynec var hosps, Eng, 52-56; assoc prof, Med Sch & Nemazee Hosp, Univ Shiraz, 56-59; res assoc biochem, Sch Med, Wayne State Univ, 59-61; sr resident obstet & gynec, Detroit Gen Hosp, 61; from asst to assoc prof, 62-71, PROF OBSTET & GYNEC & CHIEF DIV REPROD ENDOCRINOL & INFERTILITY, SCH MED, WAYNE STATE UNIV, 71- *Concurrent Pos:* Attend obstetrician & gynecologist, Detroit Gen Hosp, 62; sr attend obstet & gynec, Hutzel Hosp, 62; surgeon obstet & gynec, Harper Grace Hosp, 63; gynecologist, Children's Hosp Mich; examr, Am Bd Obstet & Gynec; consult, NIH & WHO. *Mem:* Fel Am Col Obstet & Gynec; fel Am Col Surg; Am Fertil Soc; Am Gynec & Obstet Soc; Am Soc Androl; Am Med Asn. *Res:* Human reproduction and reproductive endocrinology; infertility and conception control. *Mailing Add:* Dept Obstet & Gynec Wayne State Univ Sch Med Detroit MI 48201

MOGIL, H MICHAEL, b New York, NY, July 9, 45; m 88; c 2. METEOROLOGY, TEACHER TRAINING. *Educ:* Fla State Univ, BS, 67, MS, 69. *Prof Exp:* Forecaster, Nat Severe Storms Forecast Ctr, Kansas City, Mo, 72-74 & Nat Meteorol Ctr, Camp Springs, Md, 74-75; warning meteorologist, Nat Oceanic & Atmospheric Admin, Nat Weather Serv Hq, Silver Springs, Md, 75-80; lead forecaster, N O A A, Forecast Off, Nat Weather Serv, Redwood City, Calif, 80-81, meteorologist-in-charge, Ft Worth, Tex, 81-83, res meteorologist, Nat Weather Serv, 83-85; EDUC METEOROLOGIST, WEATHER, HOW WEATHER WORKS, ROCKVILLE, MD, 79-; CHIEF, TRAINING BR, NAT OCEANIC & ATMOSPHERIC ADMIN, NESDIS, CAMP SPRINGS, MD, 85- *Concurrent Pos:* TV meteorologist, AM weather, Pub TV, Owings Mills, Md, 77; teacher trainer, Nat Sci Teachers Asn, 78-91; contrib ed, Weatherwise Mag, Washington, DC, 80-91; Pocono Environ Educ Ctr, Dingman's Ferry, Pa, 88-91; column ed, Sci & Children Mag, Nat Sci Teachers Asn, Washington, DC, 89-91; adj prof, Univ Mo, Columbia, 89-91; chmn, Nat Weather Asn Conf, 90; educ consult. *Mem:* Nat Weather Asn; Am Meteorol Soc; Nat Sci Teachers Asn; Nat Asn Sci, Technol & Soc; Am Asn Weather Observers; Nat Earth Sci Teachers Asn. *Res:* Educational consultant-develops and documents innovative and hands-on inter- and multidisciplinary weather activities for students and teachers grades K-12; conducts workshops and courses; writes articles. *Mailing Add:* 1522 Baylor Ave Rockville MD 20850

MOGREN, EDWIN WALFRED, b Minn, Sept 16, 21; m 44; c 2. FORESTRY. *Educ:* Univ Minn, BSF, 47, MF, 48; Univ Mich, PhD(forest ecol), 55. *Prof Exp:* Asst, Lake States Forest Exp Sta, US Forest Serv, 41; asst, Univ Minn, 47-48; from instr to asst prof forestry, 48-55, from assoc prof to prof, 55-86, EMER PROF FOREST MGT, COLO STATE UNIV, 86- *Concurrent Pos:* Collabr, Rocky Mt Forest & Range Exp Sta, 52-60. *Mem:* Fel AAAS; fel Soc Am Foresters; Ecol Soc Am; Sigma Xi. *Res:* Forest management; ecology; siliviculture. *Mailing Add:* Dept Forest & Wood Sci Col Forest & Natural Resources Colo State Univ Ft Collins CO 80523

MOGRO-CAMPERO, ANTONIO, b Liverpool, Eng, Aug 25, 40; US citizen; m 66; c 3. HIGH TEMPERATURE SUPERCONDUCTOR THIN FILMS. *Educ:* Columbia Univ, BS, 63; Univ Chicago, MS, 67, PhD(physics), 71. *Prof Exp:* Res engr, Lab Cosmic Physics, Univ San Andres, Bolivia, 63-65; res assoc physics, Enrico Fermi Inst, Univ Chicago, 70-74; res physicist & lectr, Univ Calif, San Diego, 74-75; PHYSICIST, GEN ELEC RES & DEVELOP CTR, 75- *Mem:* Am Phys Soc; Mat Res Soc. *Res:* Amorphous metals; gas flow in the Earth; earthquake prediction; Jupiter's and Earth's radiation belts; solar and galactic cosmic rays; lifetime control in semiconductor devices; silicon-on-insulator structures; deposition, properties and applications of high temperature superconductor thin films. *Mailing Add:* Gen Elec Res & Develop Ctr PO Box 8 Schenectady NY 12301

MOGUS, MARY ANN, b Greensburg, Pa. BIOPHYSICS, APPLIED PHYSICS. *Educ:* Seton Hill Col, BA, 65; Pa State Univ, MS, 67, PhD(biophys), 70. *Prof Exp:* Res assoc bioacoust lab, Eye & Ear Hosp, Univ Pittsburgh, 70-71, NIH res fel, 71-72; res fel biophys lab, Carnegie-Mellon Univ, 72-74; from asst prof to assoc prof, 74-83, PROF PHYSICS, EAST STROUDSBURG UNIV, 83- *Concurrent Pos:* Res fel, Carnegie-Mellon Univ, 81- *Mem:* Am Soc Photobiol; NY Acad Sci; Sigma Xi; Biophys Soc. *Res:* Biophysics of sensory systems; vision and auditory brain organization and information processing; physics applied to archaeology, infrared analysis of sites and artifacts; artifact composition studies; developing vision device for the low-visioned; microcomputers and education; History of Science and Technology. *Mailing Add:* Dept Physics East Stroudsburg Univ East Stroudsburg PA 18301

MOHACSI, ERNO, b Zalaegerszeg, Hungary, Jan 26, 29; US citizen. SYNTHETIC CHEMISTRY, MEDICINAL CHEMISTRY. *Educ:* Eotvos Lorand Univ, Budapest, dipl org chem, 56; Columbia Univ MA, 60, PhD(org chem), 62. *Prof Exp:* Fel, Columbia Univ, 62-64 & Harvard Univ, 64-66; SR RES CHEMIST, 66-, RES LEADER, HOFFMANN-LA ROCHE, INC, 87- *Mem:* Am Chem Soc. *Res:* Synthetic organic chemistry, especially design and synthesis of potential medicinal agents. *Mailing Add:* 133 Summit Ave Apt 7 Summit NJ 07901

MOHAMED, ALY HAMED, b Cairo, Egypt, Aug 29, 24; m 43; c 2. GENETICS. *Educ:* Univ Alexandria, BS, 46; Univ Minn, MS, 53, PhD(genetics), 54. *Prof Exp:* Asst genetics, Univ Alexandria, 46-48, lectr, 54-60, assoc prof, 60-63; res assoc range sci, Tex A&M Univ, 63-64, res fel genetics & air pollution trainee, 64-66; assoc prof, 66-69, PROF BIOL, UNIV MO-KANSAS CITY, 69- *Concurrent Pos:* Dept Health, Educ & Welfare res grant, 67-71. *Mem:* AAAS; Bot Soc Am; Genetics Soc Am; Am Genetic Asn; Int Soc Fluoride Res (vpres). *Res:* Plant cytogenetics and the inheritance of quantitative characters; studying the effect of some air pollutants on chromosomes. *Mailing Add:* Dept Biol Univ Mo 5100 Rockhill Rd Kansas City MO 64110

MOHAMED, FARGHALLI ABDELRAHMAN, b Assuit, Egypt, Sept 25, 43; US citizen; m 69; c 2. MECHANICAL ENGINEERING. *Educ:* Cairo Univ, Egypt, BS, 65; Univ Calif, Berkeley, MS, 70, PhD(mat sci eng), 72. *Prof Exp:* Res assoc mat, dept mat sci, Univ Southern Calif, Los Angeles, 72-75, asst prof, 76-80; from asst prof to assoc prof, 80-84, PROF MAT, DEPT MECH ENG, UNIV CALIF, IRVINE, 84- *Concurrent Pos:* Prin investr, Univ Southern Calif, Los Angeles, 78-80 & Univ Calif, Irvine, 80- *Mem:* Am Soc Metals; Sigma Xi. *Res:* Engineering materials with specialization in mechanical behavior; high-temperature deformation of materials; superplasticity; correlation of mechanical behavior with microstructure. *Mailing Add:* 20131 Swansea Lane Huntington Beach CA 92646

MOHAMMAD, SYED FAZAL, b Faizabad, India, Sept 2, 42; m 69; c 2. HEMOSTASIS & THROMBOSIS, COAGULATION BIOCHEMISTRY. *Educ:* Univ Lucknow, India, BSc, 61, MSc, 63; All India Inst Med Sci, PhD(biophys), 72. *Prof Exp:* Fel path, Univ NC, 72-75; instr clin path, Brown Univ, 75-77; asst prof, Univ SFla, Tampa, 77-79; asst prof, 79-80, RES ASSOC PROF PATH, UNIV UTAH, SALT LAKE CITY, 80-, DIR HEMATOL, ARTIFICIAL HEART RES LAB, 86- *Concurrent Pos:* Consult, Div Artificial Organs, Univ Utah, 80-, mem & consult, Ctr Artificial Hearts & Med Devices, 85-, Ctr Biopolymers at Interfaces, 85-; NIH res grant, 78-; mem, Coun Thrombosis & Hemostasis, Am Heart Asn, 85-; res assoc prof pharmaceut, 87- *Mem:* AAAS; Am Asn Pathologists; Int Thrombosis & Hemostasis Soc; NY Acad Sci; Am Heart Asn; Am Soc Artificial Internal Organs; Int Soc Artificial Organs; Biomat Soc. *Res:* Role of platelets, endothelium and coagulation factors in hemostasis and thrombosis; pathophysiology of blood vessels; interaction of blood with artificial surfaces. *Mailing Add:* Dept Path Med Ctr Univ Utah Sch Med 5c-239 Salt Lake City UT 84132

MOHAMMED, AUYUAB, b Trinidad, WI, Jan 11, 28; m 55; c 4. APPLIED MATHEMATICS. *Educ:* Univ Man, BSc, 54, MSc, 56; Univ BC, PhD(appl math), 65. *Prof Exp:* Sci officer, 54-60, HEAD APPL MATH SECT, DEFENCE RES ESTAB ATLANTIC, DEFENCE RES BD CAN, 61- *Mem:* Fel Acoust Soc Am; Asn Comput Mach; Inst Elec & Electronics Engrs; Brit Comput Soc. *Res:* Underwater acoustics; scientific applications of computers; non-linear and linear control theory; communication theory. *Mailing Add:* 5794 Atlantic St Halifax NS B3H 1H2 Can

MOHAMMED, KASHEED, b Trinidad, WI, Apr 27, 30; m 64; c 2. NUTRITIONAL BIOCHEMISTRY, HUMAN PHYSIOLOGY. *Educ:* Univ Ariz, BS, 62, MS, 63, PhD(biochem), 67. *Prof Exp:* Scientist, Angostura Bitters, Trinidad, 52-56; NIH fel, Univ Ill Med Ctr, 67-69, res fel, 69; nutrit biochemist, Pharmaceut Div, Johnson & Johnson Res Ctr, 69-80; SR NUTRITIONAL SCIENTIST, ROSS LABS, CLEVELAND, OHIO, 80- *Concurrent Pos:* Sci adv, State Univ NY Agr & Tech Col, Canton, 70- *Mem:* AAAS; Am Chem Soc; Am Dietetic Asn; Sigma Xi. *Res:* Research and development of enteral and parenteral products for therapeutic usages. *Mailing Add:* Grandview Rd POBox 3000 Skillman NJ 08558-3000

MOHAMMED, M HAMDI A, dental materials, prosthodontics, for more information see previous edition

MOHAN, ARTHUR G, b Trenton, NJ, Mar 26, 35; m 59; c 4. ANALYTICAL CHEMISTRY. *Educ:* St Bonaventure Univ, BS, 57, MS, 59; Seton Hall Univ, PhD(org chem), 66. *Prof Exp:* Chemist, Nopco Chem Co, NJ, 59-65; res chemist, Chem Res Div, Am Cyanamid Co, Bound Brook, 66-73, sr res chemist, 73-75, assoc res fel, 75-82; sr res assoc, org chem, Electro-Nucleonics Inc, Fairfield, NJ, 83-84; RES SCIENTIST, ENVIRON & CHEM LABS, NJ DEPT HEALTH, TRENTON, 85- *Concurrent Pos:* Adj assoc prof chem, Seton Hall Univ, 72-80, adj prof, 80-85; adj prof sci, Raritan Valley Community Col; pvt consult, 88- *Mem:* Am Chem Soc. *Res:* Organic reaction mechanisms; catalysis and organic process research; photochemistry and chemiluminescence; environmental analysis. *Mailing Add:* 34 Windy Willow Way Somerville NJ 08876

MOHAN, CHANDRA, b Lucknow, India, Aug 3, 50; US citizen; m 78; c 2. DIABETES, AGING. *Educ:* Bangalore Univ, India, BS, 70, MA, 72, PhD(biochem & physiol), 76. *Prof Exp:* Res assoc, pharmacol, 77-83, ASST PROF, PHARMACOL & NUTRIT, SCH MED, UNIV SOUTHERN CALIF, 83- *Concurrent Pos:* Assoc ed, Biochem Med & Metab Biol, Acad Press, 87- *Mem:* Am Diabetes Asn; Am Instit Nutrit; AAAS; NY Acad Sci; Soc Exp Biol & Med. *Res:* Metabolic basis of diabetic stress and the regulation of gluconeogenesis; insulin regulation of intermediary metabolism. *Mailing Add:* 13638 E Dicky St Whittier CA 90605

MOHAN, J(OSEPH) C(HARLES), JR, b Philadelphia, Pa, Nov 2, 21; m 47; c 2. CHEMICAL ENGINEERING. *Educ:* Pa State Univ, BS, 46. *Prof Exp:* Jr chem engr, Sinclair Oil Co, 46-48; sr chem engr, Pennsalt Chem Co, 48-57; sect leader res & develop, film opers, Am Viscose Div, FMC Corp, 57-66, mfg mgr, Indust Packaging Dept, 66-69; SECT LEADER POLYMER & PLASTICS RES & DEVELOP, AMOCO CHEM CORP, 69- *Mem:* Am Inst Chem Engrs. *Res:* Industrial chemicals; elemental fluorine development; benzene hexochloride; chlorofluorohydrocarbons; cellophane; polyethylene; polypropylene; polyvinyl-chloride and other plastic films; polymer stabilization; rheology applications. *Mailing Add:* 8053 E Via De La Escuela St Scottsdale AZ 85258

MOHAN, NARENDRA, b India, Oct 5, 46; m 73. ELECTRICAL ENGINEERING, NUCLEAR ENGINEERING. *Educ:* Indian Inst Technol, India, BS, 67; Univ NB, MS, 69; Univ Wis-Madison, MS, 72, PhD(elec eng), 73. *Prof Exp:* Proj assoc elec eng, Univ Wis, 73-75; asst prof elec eng, 75-80, ASSOC PROF ELEC ENG, UNIV MINN, MINNEAPOLIS, 80- *Mem:* Inst Elec & Electronics Engrs. *Res:* High-voltage direct current transmission; transients in power systems; power conditioning; solar and wind energy conversion systems. *Mailing Add:* Dept Elec Eng Univ Minn 123 Church St SE Minneapolis MN 55455-0113

MOHAN, PREM, b Colombo, Sri Lanka, Aug 3, 54. DRUG DESIGN, PRECLINICAL DEVELOPMENT. *Educ:* Banares Hindu Univ, BS, 77; Mass Col Pharm, MS, 80; Purdue Univ, PhD(med chem), 84. *Prof Exp:* Postdoctoral res assoc med chem, Univ Iowa, 84-87; ASST PROF MED CHEM, UNIV ILL CHICAGO, 87- *Concurrent Pos:* Scholar award, Am Found AIDS Res, 88. *Mem:* Am Chem Soc; AAAS; Am Soc Phamacog. *Res:* Search design synthesis and development of potential anti-AIDS agents; design and synthesis of new antitumor agents; reaction mechanisms and the study of biological structure-activity relationships. *Mailing Add:* Dept Med Chem & Pharmacog (M/C 781) Col Pharm Univ Ill Chicago CHI 833 S Wood Chicago IL 60680

MOHANAKUMAR, THALACHALLOUR, m; c 2. IMMUNOLOGY. *Educ:* Duke Univ, PhD(microbiol & immunol), 74. *Prof Exp:* prof surg & immunol, Med Col Va, Va Commonwealth Univ, 83-87; PROF SURG & PATH, WASH UNIV SCH MED, DIR HISTOCOMPATIBILITY & IMMUNOGENETICS, 88- *Mem:* Am Asn Immunologists; Am Asn Cancer Res. *Res:* Tumor immunology; transplantation immunology. *Mailing Add:* Dept Surg Box 8109 Wash Univ Sch Med 4939 Audubon Ave St Louis MO 63110

MOHANDAS, THULUVANCHERI, b Guruvayur, India, Feb 26, 46; US citizen; m 82; c 1. HUMAN GENETICS, CYTOGENETICS. *Educ:* Univ Kerala, India, BS, 66; Ind Agr Res Inst, MS, 69; McGill Univ, PhD(genetics), 72. *Prof Exp:* Fel human cytogenetics, Dept Pediat, Univ Man, 72-75; from asst prof to assoc prof, 75-86, PROF PEDIAT, HUMAN CYTOGENTICS, HARBOR-UNIV CALIF LOS ANGELES MED CTR, 86- *Mem:* AAAS; Am Soc Human Genetics. *Res:* Human cytogenetics; human gene mapping using somatic cell hybrids; mechanism of X chromosome inactivation. *Mailing Add:* Div Med Genetics E-4 Harbor-Univ Calif Los Angeles Med Ctr Torrance CA 90509

MOHANTY, GANESH PRASAD, b Cuttack, India, Mar 11, 34; m 69; c 1. MATERIALS SCIENCE. *Educ:* Utkal Univ, India, BS, 54; Mich Col Mining & Technol, MS, 58; Ill Inst Technol, PhD(metall), 61. *Prof Exp:* Res scientist, Res Div, A O Smith Corp, Wis, 60-63; from asst prof mat sci to prof eng sci, 63-76; mem staff, 76-81, PROF ENG SCI, UNIV NC, CHARLOTTE, 81- *Concurrent Pos:* NSF res grant, 64-70; partic fac, Chem Physics Prog, Fla State Univ, 70- *Mem:* Am Inst Mining, Metall & Petrol Engrs. *Res:* X-ray diffraction; imperfections in crystals; phase equilibria; structures of deformed intermetallics and amorphous substances; temperature diffusion and short-range order scattering; diffusion in metallic systems; x-ray and electron optics and crystallography. *Mailing Add:* Dept Eng Univ NC Univ Sta Charlotte NC 28223

MOHANTY, NIRODE C, filtering, control systems, for more information see previous edition

MOHANTY, SASHI B, b India, Sept 4, 32; nat US; m 57; c 4. MICROBIOLOGY, VIROLOGY. *Educ:* Univ Bihar, BVSc & AH, 56; Univ Md, MS, 61, PhD(microbiol), 63. *Prof Exp:* Vet asst surgeon, Civil Vet Dept, Govt Orissa, India, 56-60; asst microbiol, 60-63, asst prof vet sci & microbiol, 63-69, assoc prof vet sci, 69-74, PROF & ASSOC DEAN VET MED, UNIV MD, 74- *Concurrent Pos:* NIH grant, 63-65; mem, Md State Proj Bovine Respiratory viruses, 63-; head working team on bovine, equine and porcine picorna viruses, WHO, Food & Agr Orgn, UN, 73- *Mem:* Am Soc Microbiol; Electron micros Soc Am; Soc Exp Biol & Med; Am Vet Med Asn; fel Am Acad Microbiol. *Res:* Animal virus diseases; experimental infection of cattle with viruses; viral growth; electron microscopy; interferon induction; cellular immunity, prevention and control of animal diseases; anti-viral drugs. *Mailing Add:* 4306 Kenny St Beltsville MD 20705

MOHANTY, SRI GOPAL, b Soro, India, Feb 11, 33; m 63; c 3. COMBINATORICS & FINITE MATHEMATICS. *Educ:* Utkal Univ, India, BA, 51; Indian Coun Agr Res, dipl agr & animal husb statist, 57; Panjab Univ, MA, 57; Univ Alta, PhD(math statist), 61. *Prof Exp:* Tech asst, Ministry Food & Agr, Govt of India, 54-56; res fel, Indian Coun Agr Res, 56-58, asst statistician, 58-59; teaching asst math, Univ Alta, 59-61, sessional lectr, 61-62; from asst prof to assoc prof, State Univ NY Buffalo, 62-64; assoc prof, 64-72, PROF MATH, MCMASTER UNIV, 72- *Concurrent Pos:* Asst prof, Indian Inst Technol, New Delhi, 66-68; vis prof, Univ Bonn, 66; vis prof, Indian Statist Inst, 74-75, Stanford Univ, 81-82, & Univ Delhi, 82. *Mem:* Am Statist Asn; Inst Math Statist; Can Statist Soc; Int Statist Inst. *Res:* Combinatorial probability; random walk; discrete probability distributions; nonparametric methods in inferences; theory of queues; fluctuation theory; enumeration of trees and certain finite structures. *Mailing Add:* Dept Math & Statist McMaster Univ Hamilton ON L8S 4K1 Can

MOHAPATRA, PRAMODA KUMAR, b Cuttack, Orissa, India, May 15, 55; m 83; c 2. SUPER STRING PHENOMENOLOGY. *Educ:* Ravenshaw Col, India, MSc, 79; Univ Md, PhD (physics), 87. *Prof Exp:* POSTDOCTORAL, BARTOL RES INST, UNIV DEL, NEWARK, 87- *Mem:* Am Phys Soc; Indian Physics Asn. *Res:* Phenomenology of superstring models. *Mailing Add:* Bartol Res Inst Univ Del Newark DE 19716

MOHAPATRA, RABINDRA NATH, b Musagadia, India, Sept 1, 44; m 69; c 2. HIGH-ENERGY PHYSICS. *Educ:* Utkal Univ, India, BSc, 64; Delhi Univ, MSc, 66; Univ Rochester, PhD(physics), 69. *Prof Exp:* Res assoc physics, Inst Theoret Physics, State Univ NY Stony Brook, 69-71 & Univ Md, College Park, 71-74; from asst prof to prof physics, City Col NY, 74-82; PROF PHYSICS, UNIV MD, COL PARK, 83- *Concurrent Pos:* Alexander von Humboldt Found fel, 80-81. *Mem:* Fel Am Phys Soc. *Res:* Gauge theories of weak, electromagnetic and strong interactions; approximate hadronic symmetries; quark models; neutrino interactions; selection rules in weak and strong interactions; field theories; radiative corrections to weak transitions; mass differences among elementary particles; cosmology; particle statistics; neutrino physics; CP violation. *Mailing Add:* Dept Physics & Astron Univ Md College Park MD 20742

MOHAT, JOHN THEODORE, b El Paso, Tex, Apr 8, 24; m 45; c 2. MATHEMATICS. *Educ:* Tex Western Col, BA, 50; Univ Tex, PhD(math), 55. *Prof Exp:* Instr math, Univ Tex, 51-55 & Duke Univ, 55-57; chief math br, Math Sci Div, Off Ord Res, US Dept Army, 57-59; from asst prof to assoc prof, 59-64, dir dept, 65-69, actg chmn dept, 75-77, PROF MATH, N TEX STATE UNIV, 64- *Concurrent Pos:* Vis asst prof, Duke Univ, 58-59. *Mem:* Am Math Soc; Math Asn Am. *Res:* General topology; points sets and transformations. *Mailing Add:* 609 Pennsylvania Dr Denton TX 76205

MOHBERG, JOYCE, b Britton, SD, Apr 28, 31. BIOCHEMISTRY OF THE CELL CYCLE. *Educ:* Univ Wis-Madison, PhD(biochem), 62. *Prof Exp:* PROF BIOCHEM & PHYSIOL, GOV STATE UNIV, 78- *Mailing Add:* Gov State Univ University Park IL 60466

MOHBERG, NOEL ROSS, b Britton, SDak, Dec 16, 39; m 63; c 2. BIOSTATISTICS. *Educ:* NDak State Univ, BS, 61; Va Polytech Inst, MS, 62; Univ NC, PhD(biostatist), 72. *Prof Exp:* Statistician, NIH, 63-65 & Sandia Corp, 65-68; biostatistician, 72-78, res head, 78-89, DIR, RES SUPPORT BIOSTATIST, UPJOHN CO, 89- *Mem:* Biomet Soc; Am Statist Asn; Clin Trials Soc; Drug Info Asn. *Res:* Methods of analysis of categorized data; design and analysis of clinical trials. *Mailing Add:* Upjohn Co 301 Henrietta St Kalamazoo MI 49001

MOHILNER, DAVID MORRIS, electrochemistry, physical chemistry; deceased, see previous edition for last biography

MOHIUDDIN, SYED M, b Hyderabad, India, Nov 14, 34; m 61; c 3. CARDIOLOGY, INTERNAL MEDICINE. *Educ:* Osmania Univ, MD, 60; Creighton Univ, MS, 67; Laval Univ, DSc(med), 70. *Prof Exp:* Fel cardiol, Sch Med, Creighton Univ, 65-67; fel res cardiol, Laval Univ, 68-70, adj prof med, 69-70; asst prof, 70-74, assoc prof, 74-78, PROF MED, SCH MED, CREIGHTON UNIV, 78-, DIR CARDIAC GRAPHIC LAB, 73-, ASSOC DIR, DIV CARDIOL. *Mem:* Can Cardiovasc Soc; Am Fedn Clin Res; fel Am Col Cardiol; fel Am Col Physicians; fel, Coun Clin Cardiol. *Res:* Clinical cardiology; coronary flow and cardiac metabolism; cardiomyopathies; graphic methods in cardiology; cardiac pharmacology. *Mailing Add:* Dept Med Creighton Univ 2500 Cal St Omaha NE 68178

MOHLA, SURESH, b Calcutta, WBengal, India, May 5, 43; US citizen; m 69; c 1. HORMONES & BREAST CANCER, CARCINOGEN METABOLISM. *Educ:* Univ Delhi India, BSc, 63, MSc, 65, PhD(endocrinol), 68. *Prof Exp:* Fel encocrinol, Ministry Family Planning, India & Ford Found, US, 68-70; res assoc & asst prof, Ben May Lab Cancer Res, Univ Chicago, 70-76; asst prof oncol, Col Med, Howard Univ, Washington, DC, 76-82, dir, Hormone Receptor Lab, 76-90, assoc prof oncol & pharmacol, 82-90, assoc prof human genetics, Grad Sch Arts & Sci, 82-90, assoc dir, Div Cancer Educ, Cancer Ctr, 85-90, prof oncol, Col Med, 90-91; HEALTH SCI ADMINR, GRANTS REV BR, NAT CANCER INST, 90- *Concurrent Pos:* Prin investr, res grants, NIH & US Environ Protection Agency, 80-90. *Mem:* Endocrine Soc; Am Physiol Soc; Am Asn Cancer Res; Am Soc Cell Biol; Am Asn Cancer Educ. *Res:* Hormonal control of growth in Normal and Neoplastic Breast tissue; hormonal control of carcinogen metabolism; steroid hormone receptors; reproductive endocrinology; cell biology. *Mailing Add:* Div Extramural Activ Grants Rev Br Westwood Bldg 822 Nat Cancer Inst 5333 Westbard Ave Bethesda MD 20892

MOHLENBROCK, ROBERT H, JR, b Murphysboro, Ill, Sept 26, 31; m 57; c 3. SYSTEMATIC BOTANY. *Educ:* Southern Ill Univ, BS, 53, MA, 54; Wash Univ, PhD(bot), 57. *Prof Exp:* From asst prof to prof, 57-85, chmn dept, 66-79, DISTINGUISHED PROF BOT, SOUTHERN ILL UNIV, 85- *Mem:* Nature Conservancy. *Res:* Flora of Midwest and Illinois; tropical legumes; national forests. *Mailing Add:* Dept Bot Southern Ill Univ Carbondale IL 62903

MOHLENKAMP, MARVIN JOSEPH, JR, b Louisville, Ky, Apr 22, 40; m 63; c 4. FOOD CHEMISTRY. *Educ:* Univ Notre Dame, BS, 62; Univ Wis, MS, 65, PhD(biochem), 68. *Prof Exp:* CHEMIST, PROCTER & GAMBLE CO, 68- *Mem:* Am Chem Soc; Inst Food Technologists. *Res:* Chemical and organoleptic aspects of food flavors with emphasis on thermally induced flavors. *Mailing Add:* 9113 Zoellner Rd Cincinnati OH 45251-3049

MOHLER, IRVIN C, JR, b Lancaster, Pa, Nov 4, 25; m 56; c 2. BACTERIOLOGY. *Educ:* Franklin & Marshall Col, BS, 49; Pa State Univ, MS, 52. *Prof Exp:* With US govt, 52-55; asst exec dir biol, Am Inst Biol Sci, 55-59; exec off, McCollum Pratt Inst, Johns Hopkins Univ, 59-61; asst to dir, Am Type Cult Collection, 61-67; asst dir biol sci commun proj, George Washington Univ Med Ctr, 67-75, asst res prof, dept med & pub affairs, 75-78, dir, Off Sponsored Res, 78-88; RETIRED. *Concurrent Pos:* Managing ed, Environ Biol & Med. *Honors & Awards:* NASA Group Achievement Award. *Mem:* Am Soc Microbiol. *Res:* Research adminstration. *Mailing Add:* 6 Stratton Ct Potomac MD 20854

MOHLER, JAMES DAWSON, b Liberal, Mo, June 2, 26; m 51; c 3. GENETICS. *Educ:* Univ Mo, AB, 49, AM, 50; Univ Calif, PhD(zool), 55. *Prof Exp:* Instr zool, Univ Mo, 50-51; asst, Univ Calif, 51-53; from asst prof to assoc prof, Ore State Univ, 55-66; from assoc prof to prof, 66-90, EMER PROF BIOL, UNIV IOWA, 90- *Concurrent Pos:* USPHS trainee, Syracuse Univ, 63-64; vis scholar, Univ Ariz, 75; vis prof zool, Ariz State Univ, 75-76; vis scientist, Ind Univ, 79; vis prof, Univ Wash, 75-88. *Mem:* Genetics Soc Am; AAAS. *Res:* Developmental genetics with drosophila. *Mailing Add:* PO Box 17 Waldport OR 97394

MOHLER, JOHN GEORGE, b Los Angeles, Calif, May 25, 32; m 58; c 3. MEDICINE, PULMONARY PHYSIOLOGY. *Educ:* Col Osteop Physicians & Surgeons, DO, 60; Univ Calif, MD, 62. *Prof Exp:* Instr med, Calif Col Med, 60-62; from instr to asst prof, 62-69, ASSOC PROF MED, UNIV SOUTHERN CALIF, 69-, MED DIR PULMONARY PHYSIOL LABS, LOS ANGELES COUNTY-UNIV SOUTHERN CALIF MED CTR, 69- *Concurrent Pos:* Fel pulmonary med, Los Angeles County-Univ Southern Calif Med Ctr, 64-66, assoc med dir pulmonary lab, 68-70; spec fel physiol, Sch Med, Univ Southern Calif, 66-67; instr med & spec fel physiol, Univ Colo, 68-69; co-dir, Univ Southern Calif air pollution res fel, 68-70; med dir pulmonary serv, Alhambra Community Hosp, 68-, chief med, 70-76, chief staff, 78-79, bd trustees, 81- *Mem:* Fel Am Col Med; fel Am Col Chest Med; Am Physiol Soc; Am Thoracic Soc; Am Fedn Clin Res; fel Am Col Sports Med. *Res:* Air pollution effects on human health, lung reactions specifically; study of exercise physiology generally, transients from rest to exercise to rest specifically; distribution of ventilation as measured by nitrogen clearance; dyspnea, cause and quantitation. *Mailing Add:* Pulmonary Physiol Labs Univ Southern Calif 1200 N State St Rm 11720 Los Angeles CA 90033

MOHLER, ORREN (CUTHBERT), astronomy; deceased, see previous edition for last biography

MOHLER, RONALD RUTT, b Ephrata, Pa, Apr 11, 31; m 50; c 8. ELECTRICAL & SYSTEMS ENGINEERING. *Educ:* Pa State Univ, BS, 56; Univ Southern Calif, MS, 58; Univ Mich, PhD(systs), 65. *Prof Exp:* Designer, Textile Mach Works, Pa, 49-56; staff mem systs, Hughes Aircraft Co, Calif, 56-58; Los Alamos Sci Lab, Univ Calif, 58-65; assoc prof elec eng, Univ NMex, 65-69; vis assoc prof eng, Univ Calif, Los Angeles, 68-69; prof elec, aero, mech & nuclear eng & dir systs res ctr, Univ Okla, 69-70, prof info & comput sci & chmn dept, 70-71; head dept, 72-79, PROF ELEC & COMPUT ENG, ORE STATE UNIV, 72- *Concurrent Pos:* Adj prof, Univ NMex, 59-65; indust consult, 65-; scientific visitor, Nat Acad Sci, USSR & China, People's Repub, 80; vis prof, Univ Rome, 77, Imp Col, London, 78-79, Australian Nat Univ, 88 & Int Inst Appl Syst Anal, 88-; sr fel, N Atlantic Treaty Orgn, 78-79. *Mem:* Am Soc Eng Educ; fel Elec & Electronic Engrs. *Res:* Systems; control theory; optimization; nuclear systems; biological engineering; immune process; random processes; power systems; aerospace control. *Mailing Add:* Dept Elec & Comput Eng Ore State Univ Corvallis OR 97331-3211

MOHLER, STANLEY ROSS, b Amarillo, Tex, Sept 30, 27; m 53; c 3. AEROSPACE MEDICINE. *Educ:* Univ Tex, BA & MA, 53, MD, 56. *Prof Exp:* Chem analyst, Longhorn Tin Smelter, Tex, 49-50; intern, USPHS, 56-57, med officer, Div Gen Med Sci, Ctr Aging Res, NIH, 57-61; dir, Civil Aeromed Res Inst, Oklahoma City, Okla, 61-65; chief, Aeromed Appln Div, Fed Aviation Admin, 65-78; PROF COMMUN HEALTH, V CHMN DEPT & DIR AEROSPACE MED, SCH MED, WRIGHT STATE UNIV, 78- *Concurrent Pos:* Assoc prof prev med & pub health, Univ Okla, 61-65. *Honors & Awards:* Boothby Award, 66, Moseley Award, 74, Lysler Award, 84, Aerospace Med Asn; Sharples Award, Aircraft Owners & Pilots Asn, 84. *Mem:* Aerospace Med Asn; Soc Air Safety Investrs; AMA. *Res:* Gerontology; general medicine; blood clotting; aviation medicine; aircraft accident research. *Mailing Add:* Dept Community Health PO Box 927 Dayton OH 45401

MOHLER, WILLIAM C, b Bridgeton, NJ, Nov 16, 27; m 56; c 3. COMPUTER SCIENCE, MEDICINE. *Educ:* Yale Univ, BA, 49; Columbia Univ, MD, 53. *Prof Exp:* Intern & resident, Presby Hosp, New York, 54-55; investr, Nat Cancer Inst, 56-65, asst dir labs & clins, 65-67, DEP DIR, DIV COMPUT RES & TECHNOL, NIH, 67- *Res:* Computing in support of biomedical research and clinical medicine. *Mailing Add:* Comput Res & Technol, Bldg 12A Rm 3033 Nat Inst of Health Bethesda MD 20892

MOHLKE, BYRON HENRY, b Valparaiso, Ind, Mar 4, 38; m 62; c 2. SPACE SCIENCE, COMPUTER SCIENCE. *Educ:* Purdue Univ, BSME, 59; Univ Wis, MS, 61. *Prof Exp:* PRIN PHYSICIST, ARVIN CO, CALSPAN CORP, 61- *Mem:* Sigma Xi. *Res:* Radar data analysis techniques; space object identification; system analysis and testing. *Mailing Add:* 26 Old Spring Lane Williamsville NY 14221

MOHN, JAMES FREDERIC, b Buffalo, NY, Apr 11, 22; m 45; c 4. IMMUNOLOGY, MEDICAL BACTERIOLOGY. *Educ:* Univ Buffalo, MD, 44. *Prof Exp:* From instr to assoc prof, 45-55, prof bact & immunol, 55-76, PROF MICROBIOL & DIR, CTR IMMUNOL, SCH MED, STATE UNIV NY BUFFALO, 76- *Concurrent Pos:* Buswell fel, 59-; asst bacteriologist & serologist, Niagara Sanitorium, NY, 46-48, bacteriologist & dir lab, 48-53; asst bacteriologist, serologist & asst dir blood bank, Buffalo Gen Hosp, 47-58, assoc bacteriologist, serologist & assoc dir blood bank, 58-; consult, Blood Bank, Deaconess Hosp, Buffalo, 57- & Walter Reed Army Inst Res, 58-; mem, Subcomt Transfusion Probs, Nat Acad Sci-Nat Res Coun, 58- *Mem:* AAAS; Soc Exp Biol & Med; Am Soc Hemat; fel Inst Soc Hemat; Am Asn Immunologists. *Res:* Investigation of Rh substances and antibodies; blood group specific substances; characterization of blood group isoagglutins; immunologic aspects of hemolytic agents anemia; immunohematologic blood transfusion studies. *Mailing Add:* Witebsky Ctr Immunol State Univ NY 233 Sherman Hall Buffalo NY 14214

MOHN, MELVIN P, b Cleveland, Ohio, June 19, 26; m 52; c 2. HISTOLOGY, EMBRYOLOGY. *Educ:* Marietta Col, AB, 50; Brown Univ, ScM, 52, PhD(biol), 55. *Prof Exp:* From instr to asst prof anat, State Univ NY Downstate Med Ctr, 55-63; from asst prof to prof, 63-88; EMER PROF ANAT, UNIV KANS MED CTR, KANSAS CITY, 88- *Concurrent Pos:* Vis prof, Nat Med AV Ctr, 72. *Mem:* Fel AAAS; Am Asn Anat; Am Soc Zool; Am Inst Biol Sci; Sigma Xi. *Res:* Structure, function and embryology of skin and its appendages, particularly hair, nail and ceruminous glands; histochemistry; electron microscopy. *Mailing Add:* Dept Anat Univ Kans Med Ctr 39th & Rainbow Kansas City KS 66103

MOHN, WALTER ROSING, b Fairmont, WVa, Mar 20, 48; m 77; c 2. HEAT TREATMENT DEVELOPMENT OF HIGH STRENGTH STEELS, COMPOSITE MATERIALS DESIGN & APPLICATIONS. *Educ:* Tex A&M Univ, BS, 72; Univ Conn, MS, 74. *Prof Exp:* Mat engr gas turbine superalloys, Pratt & Whitney Aircraft, 74-76; res metallurgist mat & processing, res & develop, Gen Elec Corp, 76-78; res engr rapid solidification, Allied-Signal Inc, 78-83; dir advan technol, Advan Composite Mat Corp, 83-90; RES SPECIALIST PROCESS DEVELOP, RES & DEVELOP DIV, BABCOCK & WILCOX, 90- *Concurrent Pos:* Secy, Old South Chap, Am Soc Metals Int, 87-88, vchmn, 88-89, chmn, 89-90. *Honors & Awards:* R & D 100 Award, Res & Develop Mag, 87. *Mem:* Am Soc Metals Int; Am Soc Mech Engrs; Metall Soc; Soc Mfg Engrs. *Res:* High performance steels; metallic and ceramic composites; rapidly solidified materials; associated processes; granted 2 patents. *Mailing Add:* 7836 Campton Circle NW North Canton OH 44720

MOHNEN, VOLKER A, b Stuttgart, WGer, Mar 11, 37; m 63; c 2. ATMOSPHERIC SCIENCES, PHYSICS. *Educ:* Univ Karlsruhe, BS, 59; Univ Munich, MS, 63, PhD(physics, meteorol astrophys), 66. *Prof Exp:* Res assoc, Univ Munich, 62-67; sr res assoc, Atmospheric Sci Res Ctr, State Univ NY Albany, 67-75, assoc dir, 72-75, assoc prof, 67-77, SR RES ASSOC, ATMOSPHERIC SCI RES CTR, STATE UNIV NY ALBANY, 75-, RES PROF, 77- *Mem:* Fel AAAS; Am Chem Soc; fel NY Acad Sci; Deutsche Physikalische Gesellschaft; Am Inst Aeronaut & Astronaut. *Res:* Air pollution; aerosol physics; solar energy. *Mailing Add:* Atmospheric Sci Res Ctr State Univ NY 100 Fuller Rd Albany NY 12205

MOHOS, STEVEN CHARLES, b Sopron, Hungary, Jan 20, 18; div; c 2. PATHOLOGY, IMMUNOLOGY. *Educ:* Pazmany Peter Univ, Budapest, MD, 41; Am Bd Path, dipl, 56. *Prof Exp:* Res fel exp path, Med Sch, Pazmany Peter Univ, 41-43, asst prof path, 42-43; med staff chief serv, Hosp of Int Red Cross, Ger, 46-50; resident path, Polyclin Hosp, Harrisburg, Pa, 51-52, 53-54, intern, 52-53; instr & res assoc, Med Col, Cornell Univ, 54-56; asst prof path, State Univ NY Downstate Med Ctr, 56-63; assoc prof, 63-78, PROF PATH, NY MED COL, 78- *Concurrent Pos:* Asst attend pathologist, New York Hosp, 54-56; Life Ins Res Fund, Off Naval Res & NIH res grants. *Honors & Awards:* Chinoin Award, 42. *Mem:* AMA; Col Am Path; Am Soc Exp Path; NY Acad Sci; Int Acad Path. *Res:* Tissue immunology; filtration membrane; immunological aspects of cancer research; transplantation immunity; experimental nephritis; in vivo effects of complement and application of electron microscopy to these problems. *Mailing Add:* Dept Path Basic Sci Bldg New York Med Col Valhalla NY 10595

MOHR, JAY PRESTON, b Philadelphia, Pa, Mar 5, 37; m 62; c 2. NEUROLOGY. *Educ:* Haverford Col, AB, 58; Univ Va, MS & MD, 63. *Prof Exp:* USPHS fel pharmacol, Univ Va, 60-63; from intern to asst resident internal med, Mary Imogene Bassett Hosp, Cooperstown, NY, 63-65; asst resident neurol, NY Neurol Inst, Columbia-Presby Med Ctr, New York, 65-66; fel neurol, Mass Gen Hosp, Boston, 66-68, Nat Inst Neurol Dis & Stroke fel, 67-69; instr, Univ Md Hosp, Baltimore, 69-71; asst prof neurol, Harvard Med Sch, 72-78; PROF NEUROL & CHMN DEPT, UNIV S ALA, 78- *Concurrent Pos:* Asst neurologist, Johns Hopkins Hosp, 69-71 & Mass Gen Hosp, 72-75; assoc neurologist, Mass Gen Hosp, 75-78. *Mem:* Acad Aphasia; Am Acad Neurol. *Res:* Behavioral neurology; cerebrovascular disease; aphasiology. *Mailing Add:* Box 131 NYNI 710 W 168th St New York NY 10032

MOHR, JOHN LUTHER, b Reading, Pa, Dec 1, 11; m 39; c 2. MARINE BIOLOGY, PROTOZOOLOGY. *Educ:* Bucknell Univ, AB, 33; Univ Calif, PhD(zool), 39. *Prof Exp:* Asst zool, Univ Calif, 34-38, technician, 38-42; res assoc, Pac Islands, Stanford Univ, 42-44; asst prof & res assoc, Allan Hancock Found, 44-47, vis asst prof zool, 47-48, asst prof, 48-54, assoc prof biol, 54-57, head dept, 59-62, prof, 57-77, EMER PROF BIOL, UNIV SOUTHERN CALIF, 77-, EMERITI COL, 91- *Concurrent Pos:* Vis prof, Univ Wash Friday Harbor Labs, 56-57; Guggenheim fel, Plymouth Lab, Marine Biol Asn UK; chief, Marine Zool Group, Antarctic Ship Eltanin, 62 & 65; mem gen invert comt, Smithsonian Oceanog Sorting Ctr, 63-66; res assoc, Los Angeles County Mus Natural Hist, 64-; vpres, Biol Stain Comn, 76-80, dir emer, 81; mem bd dirs, Calif Natural Areas Coord Coun. *Mem:* Am Soc Parasitologists; Am Soc Zoologists; Ecol Soc Am; Soc Protozoologists; Marine Biol Asn UK; Am Micros Soc. *Res:* Protozoology and parasitology, especially opalinida, chonotrichs, chilotrichs and ciliates of elephants; effects of drilling slurries on marine populations; irrationalities in marine bioassays; biology of polar seas; philosophy and folkways of biologists; biological stains. *Mailing Add:* 3819 Chanson Dr Los Angeles CA 90043-1601

MOHR, MILTON ERNST, b Milwaukee, Wis, Apr 9, 15; m 38; c 2. ELECTRICAL ENGINEERING. *Educ:* Univ Nebr, BS, 38. *Hon Degrees:* DrEng, Univ Nebr, 59. *Prof Exp:* Mem tech staff, Bell Tel Labs, 38-50; dept head, Radar Lab, Hughes Aircraft Co, Calif, 50-54; vpres & gen mgr, TRW Comput Div, Thompson-Ramo-Wooldridge, Inc, 54-64; vpres, Bunker-Ramo Corp, 64-66, pres, 66-70; pres & chief exec officer, Quotron Systs Inc, 70-89; RETIRED. *Mem:* Am Inst Aeronaut & Astronaut; fel Inst Elec & Electronics Engrs. *Mailing Add:* Quotron Systs Inc Malibu CA 90265

MOHR, RICHARD ARNOLD, chemical engineering, for more information see previous edition

MOHR, SCOTT CHALMERS, b Jamestown, NY, Aug 30, 40; m 64; c 3. BIOCHEMISTRY. *Educ:* Williams Col, BA, 62; Harvard Univ, MA, 66, PhD(chem), 68. *Prof Exp:* NIH fel, Cornell Univ, 68-69; asst prof, 69-75, ASSOC PROF CHEM, BOSTON UNIV, 75- *Mem:* AAAS; Am Chem Soc; Sigma Xi. *Res:* Fast kinetics in biochemical systems; allosteric proteins; transfer RNA; nucleic acid-protein interactions; protein synthesis elongation factor I (Tu); compact states of nucleic acids; carcinogen-nucleic acid interactions. *Mailing Add:* 40 Calvin Rd Wellesley MA 02181

MOHRAZ, BIJAN, b Tehran, Iran, May 3, 37. CIVIL ENGINEERING. *Educ:* Univ Ill, BS, 61, MS, 62, PhD(civil eng), 66. *Prof Exp:* Res asst civil eng, Univ Ill, Urbana, 62-66; proj engr, Agbabian-Jacobsen Assocs, 66-67; asst prof civil eng, Univ Ill, Urbana, 67-74; assoc prof, 74-81, PROF CIVIL ENG, SOUTHERN METHODIST UNIV, 81-, ASSOC DEAN, SCH ENG & APP SCI, 84- *Concurrent Pos:* Consult, govt agencies & private indust; NSF grant, 80-82. *Mem:* Am Soc Civil Engrs; Seismol Soc Am; Sigma Xi; Earthquake Eng Res Inst; Sigma Xi. *Res:* Structural analysis and design; earthquake engineering. *Mailing Add:* 4027 University Dallas TX 75205

MOHRENWEISER, HARVEY WALTER, b Mora, Minn, Oct 12, 40; m 61; c 2. BIOCHEMISTRY. *Educ:* Univ Minn, BS, 62, MS, 66; Mich State Univ, PhD(biochem), 70. *Prof Exp:* NIH fel, McArdle Lab, Univ Wis, 70-73; res chemist mutagenesis, Nat Ctr Toxicol Res, Food & Drug Admin, HEW, 73-76; asst prof human genetics, Sch Med, Univ Mich, Ann Arbor, 76-84, assoc res scientist, Dept Human Genetics, 84-87; SR BIOMED SCIENTIST, BIOMED SCI DIV, LAWRENCE LIVERMORE NAT LAB, 87- *Concurrent Pos:* Asst prof biochem, Univ Ark, 73-76. *Res:* Biochemical mechanisms of mutagenesis; structure, function and metabolic significance of variant enzymes; biochemical/molecular genetics. *Mailing Add:* Biomed Sci Div L452 Lawrence Livermore Nat Lab Livermore CA 94550

MOHRIG, JERRY R, b Grand Rapids, Mich, Feb 24, 36; m 60; c 2. BIOCHEMISTRY. *Educ:* Univ Mich, BS, 57; Univ Colo, PhD(chem), 63. *Prof Exp:* Asst prof, Hope Col, 64-67; from asst prof to assoc prof, 67-75, PROF CHEM, CARLETON COL, 75-, L M GOULD PROF NAT SCI, 90- *Concurrent Pos:* Vis prof. Univ Calif, Berkeley, 82-83; pres, Coun Undergrad Res, 83-86, chmn, 85-87; mem PRF adv bd, Am Chem Soc, 84-86; adv comt chem, NSF, 85-88; pres, Midwestern Asn Chem Teachers Lib Arts Col, 87-88. *Honors & Awards:* Catalyst Award, Chem Mfg Asn, 78. *Mem:* Am Chem Soc; Sigma Xi. *Res:* Organic reaction mechanisms; stereospecific synthesis; elimination and addition reactions; enzymic catalysis. *Mailing Add:* Dept Chem Carleton Col Northfield MN 55057

MOHRLAND, J SCOTT, b Walla Walla, Wash, Dec 3, 50; m 74; c 3. PHARMACEUTICAL RESEARCH, DRUG DEVELOPMENT. *Educ:* Western Wash State Univ, BA, 73; Wash State Univ, PhD(pharmacol), 77. *Prof Exp:* Asst prof pharmacol, Col Pharm, Wash State Univ, 76-77; postdoctoral fel pain res, Col Med, Univ Iowa, 77-79; res scientist, Upjohn Co, 79-81, sr clin res scientist, 81-88, assoc dir, 88-90, DIR, CLIN PHARMACOL, UPJOHN CO, 90- *Concurrent Pos:* From adj asst prof to adj assoc prof med, Col Health & Human Serv, Western Mich Univ, 79-89, adj prof, 89-; consult, Onsite Systs, Inc, 86-87. *Mem:* Fedn Exp Scientists & Biologists; Int Asn Study Pain. *Res:* Clinical trials on new cardiovascular, endocrine and cancer drugs; angina, hypertension, heart failure and thrombosis; pain and analgesia. *Mailing Add:* Clin Res 7216-BRN-5 Upjohn Co Kalamazoo MI 49001

MOHRMAN, HAROLD W, b Quincy, Ill, Oct 1, 17; m 39; c 4. CHEMISTRY. *Educ:* Univ Ill, BS, 39. *Prof Exp:* Res chemist, Monsanto Co, 39, group leader, 40-46, asst res dir, 46-50, dir res, Plastics Div, 50-59, assoc interests, 59-60 & overseas div, 60-63, dir polymer sect & overseas res, 63-68, sect dir, Corp Res Dept, 68-79; INDEPENDENT CONSULT, 79- *Mem:* AAAS; Am Chem Soc; NY Acad Sci; Soc Am Archeol. *Res:* Condensation resin and vinyl polymers; manufacture of phenolic and melaminealdehyde condensation products; polyelectrolytes and research administration. *Mailing Add:* 46 Ballas Ct St Louis MO 63131

MOHS, FREDERIC EDWARD, b Burlington, Wis, Mar 1, 10; m 34; c 3. SURGERY. *Educ:* Univ Wis, BS, 31, MD, 34. *Prof Exp:* Brittingham asst cancer res, zool dept, Univ Wis, 29-34, Bowman cancer res fel, 35-38, assoc cancer surg, 67-80, EMER CLIN PROF SURG, MED SCH, UNIV WIS-MADISON, 80-, EMER DIR MOHS SURG CLIN, UNIV HOSPS, 80- *Honors & Awards:* Lila Gruber Award for Cancer Res, Am Acad Dermat, 77, Int Facial Plastic Surg Award, 79. *Mem:* AAAS; AMA; Am Asn Cancer Res; Am Col Micrographic Surg & Cutaneous Oncol. *Res:* Micrographic surgery for the microscopically controlled excision of cancer of the skin, lip, parotid gland and other external structures; inventor & developer. *Mailing Add:* 2880 University Madison WI 53705

MOHSENIN, NURI N, b Tehran, Iran, Sept 15, 23; m 52; c 6. FOOD MATERIALS. *Educ:* Okla State Univ, BS, 51; Mich State Univ, MS, 53, PhD(agr eng), 56. *Prof Exp:* Prof agr eng, Pa State Univ, 60-79; CONSULT FOOD PHYSICS, 79- *Concurrent Pos:* Alexander von Humbolt US sr scientist award, 80. *Mem:* Am Soc Agr Engrs; Soc Rheol; Inst Food Technol. *Res:* Physical properties of plant and animal materials; thermal properties of foods and agricultural materials; electromagnetic radiation properties of foods and agricultural products. *Mailing Add:* Food Physics Info Systs 390 Sixth Ave Santa Cruz CA 95062

MOHTADI, FARHANG, b Tehran, Iran, Jan 6, 26; m 52; c 3. CHEMICAL ENGINEERING, THERMODYNAMICS. *Educ:* Univ Tehran, BEng, 45; Univ Birmingham, BSc, 48, PhD(chem eng), 51. *Prof Exp:* Imp Chem Indust fel chem eng, Univ Birmingham, 52-55, lectr, 55-64; Bwisa prof & head dept, Univ WI, 64-67; head dept chem eng, Univ Calgary, 68-71, prof, 67-87; Albert Baumann prof chem eng, Univ Natal, 87-89; RETIRED. *Concurrent Pos:* Royal Norweg Coun fel, 62-63; mem, Royal Comn Sugar Indust Trinidad-WI, 66-67. *Mem:* Fel Brit Inst Chem Engrs; fel Brit Inst Petrol; Am Inst Chem Engrs; fel Eng Inst Can; fel Chem Inst Can. *Res:* Electrokinetic phenomena in non-aqueous dispersions; flow and mass transfer from drops; physical behavior of particulate systems; degradation of oil in soil; combustion in a flowing stream of water; flow in the cardiovascular system; downhole steam generation. *Mailing Add:* 3243 Alfege St SW Calgary AB T2T 3S4 Can

MOINUDDIN, JESSIE FISCHER, nutrition, biochemistry, for more information see previous edition

MOIOLA, RICHARD JAMES, b Reno, Nev, Oct 5, 37; m 70; c 2. SEDIMENTOLOGY. *Educ:* Univ Calif, Berkeley, AB, 59, PhD(geol), 69. *Prof Exp:* Geologist, Shell Oil Co, 60; res asst geol, Univ Calif, Berkeley, 61-63; res geologist, Field Res Lab, Mobil Oil Corp, Tex, 63-66, sr res geologist, Mobil Res & Develop Corp, 67-74, res assoc, 74-81, geol consult, 81-83, mgr geol res, 83-87, mgr geol & geochem res, 87-90, SR RES SCIENTIST, MOBIL RES & DEVELOP CORP, 91- *Concurrent Pos:* Lectr, Univ Tex, Dallas, 75, Continuing Educ Prog, Am Asn Petrol Geologists, 77-88; assoc ed, Bull Am Asn Petrol Geologists, 77-, Bull Geol Soc Am, 84-, J Sedimentart Petrol, 84-88; sedimentology counr, Soc Econ Paleontologists & Mineralogists, 83-85. *Mem:* Fel Geol Soc Am; Soc Econ Paleontologists & Mineralogists; Am Asn Petrol Geologists; Int Asn Sedimentologists. *Res:* Sedimentology of modern and ancient sand bodies; sequence stratigraphy; tectonics and sedimentation. *Mailing Add:* Mobil Res & Develop Corp Dallas Res Lab PO Box 819047 Dallas TX 75381

MOIR, DAVID CHANDLER, b Globe, Ariz, Dec 9, 47; m 74; c 2. MATERIAL SCIENCE, RADIOGRAPHY. *Educ:* NMex State Univ, BS, 69; Ariz State Univ, MS, 70, PhD(physics), 75. *Prof Exp:* Res assoc fel nuclear & particle physics, 75-77, staff mem Mat Sci & Radiography, 77-80, STAFF MEM INTENSE ELECTRON BEAM GENERATORS ACCELERATORS & TRANSPORT, LOS ALAMOS SCI LAB, 80- *Mem:* Am Phys Soc; Am Asn Physics Teachers. *Res:* Material response at high strain rates; quantitative flash radiography of explosive systems; electron beam accelerators. *Mailing Add:* MS 940 M-4 Los Alamos Nat Lab Los Alamos NM 87545

MOIR, RALPH WAYNE, b Bellingham, Wash, Jan 21, 40; m 63; c 3. FUSION BREEDERS, DIRECT ENERGY CONVERSION. *Educ:* Univ Calif, Berkeley, BS, 62; Mass Inst Technol, ScD(nuclear eng), 67. *Prof Exp:* Jolliot-Curie fel fusion res, French Atomic Energy Comn, Fontenay-aux-Roses, France, 67-68; RES PHYSICIST, GROUP LEADER & PROJ MGR FUSION ENERGY RES, LAWRENCE LIVERMORE NAT LAB, 68- *Mem:* Am Phys Soc; Am Nuclear Soc. *Res:* Direct energy conversion; fusion breeders; magnet design. *Mailing Add:* Lawrence Livermore Nat Lab L-644 PO Box 5511 Livermore CA 94550

MOIR, ROBERT YOUNG, b Estevan, Sask, Oct 30, 20; m 46; c 3. ORGANIC CHEMISTRY. *Educ:* Queen's Univ, Ont, BA, 41, MA, 42; McGill Univ, PhD(org chem), 46. *Prof Exp:* Jr chemist, Inspection Bd, Can, 41; res chemist, Indust, Org Chem, Res Labs, Dom Rubber Co, Ltd, 43-44, 46-49; from asst prof to assoc prof, 49-64, PROF CHEM, QUEEN'S UNIV, ONT, 64- *Mem:* Am Chem Soc; Chem Inst Can. *Res:* Steric effects in cyclohexanes and in diphenyl ethers; general synthesis. *Mailing Add:* 223 Victoria St Kingston ON K7L 3Y9 Can

MOIR, RONALD BROWN, JR, b Romulus, NY, Dec 18, 53; m 81; c 3. ECOLOGY. *Educ:* Hampshire Col, BA, 77; Antioch Univ, MST, 79. *Prof Exp:* Asst prof natural sci, Hampshire Col, 76-77; naturalist field biol, Habitat Inst Environ, 77-78; field biologist, Antrim Conserv Comn, NH, 78-79; asst scientist marine sci, Sea Educ Asn, 79-80; teacher gen sci, Pike Sch, Andover, Mass, 81-84; CUR NATURAL HIST, PEABODY MUS, SALEM, MASS, 84- *Concurrent Pos:* Naturalist-consult, New England Aquarium, 81-82; naturalist, Mass Audubon Soc, 81- & Dirigo Cruises, Schooner Harvey Gamage, 82-83; consult, Whale Proj, Mem Univ, St Johns, Nfld, 82- *Honors & Awards:* James Centorino Award, Nat Marine Educr Asn, 88. *Mem:* Am Nature Study Soc; AAAS. *Res:* Fauna and flora abundance and distribution in northeastern Massachusetts; marine ecology; ecological inventorying. *Mailing Add:* Peabody Mus E India Sq Salem MA 01970

MOISE, EDWIN EVARISTE, b New Orleans, La, Dec 22, 18; div; c 2. MATHEMATICS. *Educ:* Tulane Univ La, BA, 40; Univ Tex, PhD(math), 47; Harvard Univ, MA, 60. *Prof Exp:* Instr math, Univ Tex, 46-47; from instr to prof, Univ Mich, 47-60; prof math & educ, Harvard Univ, 60-71; distinguished prof math, 71-87, EMER PROF, MATH DEPT, QUEENS COL, NY, 88- *Concurrent Pos:* Nat Res Coun fel, Inst Advan Study, 49-50, asst, 50-51, Guggenheim fel, 56-57; vis prof, Res Ctr, Nat Polytech Inst, Mex, 70-71; Hudson prof, Auburn Univ, 80-81. *Mem:* Fel Am Acad Arts & Sci; Math Asn Am (pres, 67-68); Am Math Soc (vpres, 73-74). *Res:* Topology. *Mailing Add:* 77 Bleeker St Apt 323 E New York NY 10012

MOISE, NANCY SYDNEY, b Houston, Tex, Sept 27, 54; m 82. VETERINARY CARDIOLOGY, PULMONARY DISEASE. *Educ:* Tex A&M Univ, BS, 76, DVM, 77; Cornell Univ, MS, 85. *Prof Exp:* Med resident, 79-81, instr, 81-82, ASST PROF VET MED, NY STATE COL VET MED, 85- *Mem:* Am Vet Med Asn; Am Col Vet Internal Med. *Res:* Insulin therapy in diabetic cats; ectocardiography of cats with cardiomyopathies; mycoplasma arthritis in cats; bronchitis in cats. *Mailing Add:* Dept Clin Sci NY State Col Vet Med Cornell Univ Ithaca NY 14853

MOISEYEV, ALEXIS N, b Paris, France, June 27, 32; US citizen. GEOCHEMISTRY, GEOLOGY. *Educ:* Sorbonne, Lic es Sci, 55, Univ Paris, Dr, 59; Stanford Univ, PhD(geochem), 66. *Prof Exp:* Mining geologist, Compagnie Royale Asturienne des Mines, 56-58 & 60-61; lectr geol, Univ Calif, Davis, 66-67; asst prof, San Jose State Col, 67-68; asst prof, 68-71, ASSOC PROF GEOL, CALIF STATE COL, HAYWARD, 71- *Mem:* Am Geophys Union; Geol Soc Am. *Res:* Geochemistry of hydrothermal processes and low temperature sedimentary deposits; plate tectonics. *Mailing Add:* Dept Geol Calif State Col Univ Hayward Hayward CA 94542

MOISON, ROBERT LEON, b Fitchburg, Mass, Mar 15, 29; m 55; c 2. CHEMICAL ENGINEERING. *Educ:* Worcester Polytech Inst, BS, 50, MS, 51. *Prof Exp:* Res engr, Eng Res Lab, E I du Pont de Nemours & Co, 51-53 & 55-58, sr res eng, 58-61, res supvr, 61-63; proj leader, T L Daniels Res Ctr, Archer Daniels Midland Co, Minn, 63-64, group leader, 64-67; sect mgr, Ashland Oil & Refining Co, 67-70; consult engr, David R Conkey & Assocs, 70-71; CONSULT ENGR, ROBERT L MOISON & ASSOCS, 71- *Mem:* Am Chem Soc; Am Inst Chem Engrs; Am Sci Affiliation; Am Oil Chemists Soc; Consult Engrs Coun Am. *Res:* Chemical & food processing; process development and design; mass transfer; reactor design; pollution abatement. *Mailing Add:* 112 S Surrey Trail Apple Valley MN 55124

MOISSIDES-HINES, LYDIA ELIZABETH, b Newton, Mass, Mar 11, 48; m 74; c 3. MEDICINAL CHEMISTRY, INDUSTRIAL CHEMISTRY. *Educ:* Aurora Col, BS, 67; Univ Ill, Urbana-Champaign, MS, 69, PhD(org chem), 71. *Prof Exp:* Sr scientist med chem, Mead Johnson & Co, 71-75; mem staff, Patent Liaison Dept, Upjohn Co, 75-78, mem sr staff, Tech Intel Dept, 78-85. *Concurrent Pos:* Chmn, Comt Copyrights Nat, Am Chem Soc, 90-; teacher sci, Orgn Hands-on Sci Progs for Lay Pub, particularly children. *Mem:* AAAS; Am Chem Soc; Metric Asn; Sigma Xi. *Mailing Add:* 5596 Parkview Ave Kalamazoo MI 49009

MOIZ, SYED ABDUL, b Hyderabad, India, Feb 3, 37; US citizen; m 68; c 1. EDUCATION ADMINISTRATION, SOLID STATE PHYSICS. *Educ:* Osmania Univ, India, BSc, 56; Univ Houston, MS(physics), 62, ABD(physics); Clarkson Col Technol, MS(eng sci), 71. *Prof Exp:* Res fel physics, Univ Houston, 60-65; asst prof math & physics, Jarvis Col, 65-67; chmn math & sci, 67-85, ASST DEAN OF INSTRUCTION, GALVESTON COL, 85- *Concurrent Pos:* Adj instr, Tex Southern Univ, 61-62; prin investr, Proj Curriculum Improvement through Comput Utilization, Galveston Col, 77-80; adj instr, Tex A&M Univ, Galveston, 80-82. *Res:* Copper crystal growing; x-ray diffraction tecniques to determine crystal structure; studies of the motion of dislocations by bombarding single crystals with high energy argon ions. *Mailing Add:* 2642 Gerol Galveston TX 77551

MOJICA-A, TOBIAS, b Soata, Colombia, Mar 1, 43; m 74; c 1. MICROBIAL GENETICS, MOLECULAR BIOLOGY. *Educ:* Brandeis Univ, BA, 68; McGill Univ, MSc, 72; Polish Acad Sci, PhD(biochem), 74. *Prof Exp:* Res assoc radiobiol, Atomic Ctr, Belg, 74-76; res assoc biochem, Hunter Col, 76-77; ASST PROF MICROBIOL, COL MED & DENT NJ-NJ SCH OSTEOPATH MED, 77- *Concurrent Pos:* Mem, Grad Prog Microbiol, Rutgers Univ, 78- *Mem:* Sigma Xi; Am Soc Microbiol; Genetics Soc Am; Belg Biochem Soc; Colombian Soc Genetics. *Res:* Genetics and molecular biology of DNA processing systems in gram positive bacteria, membrane and cell wall in gram negative bacteria, phage DNA injection. *Mailing Add:* Apartado Aero 059708 Bogata De 6 Colombia

MOK, MACHTELD CORNELIA, b De Bilt, Neth, Aug 27, 47; m 72. GENETICS OF PLANT DEVELOPMENT. *Educ:* Univ Wageningen, Neth, BS, 69; Univ Wis-Madison, MS, 73, PhD(plant breeding & genetics), 75. *Prof Exp:* Asst prof, 75-80, ASSOC PROF HORT GENETICS, ORE STATE UNIV, CORVALLIS, 80- *Mem:* Genetics Soc Am; Am Soc Plant Physiologists; Tissue Cult Asn; Bot Soc Am; Am Soc Hort Sci; Int Asn Plant Tissue Cult. *Res:* Genetic regulation of cytokinin metabolism is studied through a combination of callus culture bioassays, determination of radiolabeled cytokinin metabolism and genetic analyses; tissue culture techniques (anther culture, protoplast culture and regeneration techniques) are devised and applied for plant improvement; plant hormone metabolism. *Mailing Add:* Dept Hort Ore State Univ Corvallis OR 97331

MOKADAM, RAGHUNATH G(ANPATRAO), b Akola, India, Oct 13, 23; m 53; c 4. THERMODYNAMICS, HEAT TRANSFER. *Educ:* Benares Hindu Univ, BSc, 46; Univ Louisville, MME, 49; Univ Minn, PhD(mech eng), 53. *Prof Exp:* Jr engr, Govt Elec Dept, Univ Nagpur, 46-48; from asst prof to prof mech eng, Indian Inst Technol, Khragpur, 55-67; sr mech engr, Inst Gas Technol, Ill, 68-71; consult, Chicago, Ill, 71-72; prin engr, Res Dept, 73-87, RES ADV ENGR, SUNSTRAND CORP, 87- *Concurrent Pos:* Vis assoc prof, Ill Inst Technol, 66-68, from adj assoc prof to adj prof, 69-71. *Mem:* Am Soc Mech Engrs. *Res:* Nonreversible thermodynamics applied to fluid flow in porous media; turbulent steady-state flow of coal and gas mixtures in long transmission lines; energy systems; heat transfer; fluid flow and power; fluidics; refrigeration and airconditioning; properties of binary mixtures of fluids; magnetic refrigeration and heat pump systems. *Mailing Add:* Res Dept 4747 Harrison Ave Rockford IL 61101

MOKE, CHARLES BURDETTE, b Pittsburgh, Pa, Mar 13, 10; m 50; c 2. MINERALOGY-PETROLOGY, STRUCTURAL GEOLOGY. *Educ:* Col Wooster, BA, 31; Harvard Univ, MA, 35 & PhD(geol), 48. *Prof Exp:* Instr geol, Col Wooster, 36-48, from assoc prof to prof, 48-72, dept chmn, 54-72, EMER PROF GEOL, COL WOOSTER, 72- *Concurrent Pos:* Vis lectr, Univ Ill Summer Field Camp, Sheridan, Wyo, 56-58 & Rio Salado Community Col, Sun City, AZ, 78. *Mem:* Geol Soc Am; Am Asn Petrol Geologists; Nat Asn Geol Teachers; Sigma Xi. *Mailing Add:* 13373 N Plaza Del Rio Blvd No 5553 Peoria AZ 85381-4873

MOKLER, BRIAN VICTOR, b Los Angeles, Calif, May 1, 36; m 68; c 2. AEROSOL SCIENCE, INDUSTRIAL HYGIENE. *Educ:* Pomona Col, BA, 58; Mass Inst Technol, SM, 60; Harvard Univ, SM, 68, ScD(environ health sci), 73; Am Bd Indust Hyg, dipl & cert air pollution, 75. *Prof Exp:* Chem engr, Arthur D Little Inc, 60-67; teaching asst environ health sci, Sch Pub Health, Harvard Univ, 69-70, teaching fel, 70-71; consult environ sci & air pollution, 73; mem sr staff aerosol sci, Lovelace Biomed & Environ Res Inst, 74-82; CONSULT INDUST HYG & AEROSOL SCI, 82- *Concurrent Pos:* Ed consult, Biol Abstrs & Biol Abstrs/RRM, 84-91. *Mem:* Am Chem Soc; Am Indust Hyg Asn; Am Acad Indust Hyg; Am Asn Aerosol Res. *Res:* Chemistry and physics of disperse systems; aerosol generation and characterization methodology; condensation aerosols; toxicology of inhaled materials; industrial hygiene; characteristics of aerosols emitted from energy production and transportation sources. *Mailing Add:* 7800 Phoenix NE Suite E Albuquerque NM 87110-1462

MOKLER, CORWIN MORRIS, b Forsythe, Ill, Dec 10, 25; m 50; c 2. CARDIOVASCULAR PHYSIOLOGY, PHARMACOLOGY. *Educ:* Colo Col, BA, 50; Univ Nev, MS, 52; Univ Ill, PhD(physiol), 58. *Prof Exp:* Technician virol, Harvard Med Sch, 52-54; biologist, NIH, 54; asst anat, Univ Ill, 54-58; investr cardiac pharmacol & physiol, G D Searle & Co, Ill, 58-61; asst prof pharmacol, Univ Fla, 61-67; assoc prof pharmacol, Col Pharm, Univ Ga, 67-88; RETIRED. *Mem:* Am Soc Pharmacol & Exp Therapeut; Sigma Xi; Am Heart Asn; Int Soc Heart Res. *Res:* Cardiovascular physiology and pharmacology; cardiac excitability; anti-arrhythmic drugs. *Mailing Add:* 509 Ponderosa Dr Athens GA 30605

MOKMA, DELBERT LEWIS, b Holland, Mich, Sept 21, 42; m 77; c 1. SOIL SCIENCE. *Educ:* Mich State Univ, BS, 64, MS, 66; Univ Wis-Madison, PhD(soil sci), 71. *Prof Exp:* From res asst to res assoc soil sci, 71-75, asst prof, 75-80, assoc prof crop & soil sci, 80-87, PROF, MICH STATE UNIV, 87- *Mem:* Soil Sci Soc Am; Am Soc Agron; Soil Water Conserve Soc; Int Soc Soil Sci; Soil Class Asn Mich. *Res:* Soil genesis and classification; use of remote sensing and soil surveys in land use planning; soil mineralogy; soil erosion. *Mailing Add:* Dept Crop & Soil Sci Mich State Univ 584D Plant Soil Sci Bldg East Lansing MI 48824

MOKOTOFF, MICHAEL, b Brooklyn, NY, Jan 23, 39; m 67; c 3. MEDICINAL CHEMISTRY, PEPTIDE CHEMISTRY. *Educ:* Columbia Univ, BS, 60; Univ Wis, MS, 63, PhD(med chem), 66. *Prof Exp:* NIH staff fel med chem, Lab Chem, Nat Inst Arthritis & Metab Dis, 66-68; from asst prof to assoc prof Med Chem, 68-85, ASSOC PROF PHARMACEUT SCI, SCH PHARM, UNIV PITTSBURGH, 85- *Concurrent Pos:* Health Res Serv Found grant, 69-70; Nat Cancer Inst grant, 70-73; res grant, Am Cancer Soc, 73-75; NIH grant, 76-79, contract, 80-83; Black Athletes Found grant, 77; Am Cancer Soc Inst grant, 79, 89; vis sci, Weizmann Inst Sci, Israel, 78; vis lectr, Pharmaceut Univ China, Nanjing, 84; sci expert, drug related trials; mem, Pa Drug, Device & Cosmetic Bd. *Mem:* Am Chem Soc; Am Asn Col Pharm; AAAS; NY Acad Sci; Am Asn Cancer Res. *Res:* Azabicyclo chemistry; potential inhibitors of asparagine biosynthesis; peptide synthesis using polymeric reagents; antagonists of gastrin releasing peptide; laminin peptide-like fragments as inhibitors of cancer metastases; radiolabeled peptides for diagnostic imaging. *Mailing Add:* Sch Pharm 529 Salk Hall Univ Pittsburgh Pittsburgh PA 15261

MOKRASCH, LEWIS CARL, b St Paul, Minn, May 9, 30. BIOCHEMISTRY. *Educ:* Col St Thomas, BS, 52; Univ Wis, PhD(physiol chem), 55; Am Bd Clin Chem, cert. *Prof Exp:* Res assoc psychiat & neurochem, Sch Med, La State Univ, 56-57; sr res assoc biochem & instr med, Univ Kans Med Ctr, Kansas City, 57-62, dir neurochem lab & assoc med, 59-62; from assoc to asst prof biol chem, Harvard Med Sch, 60-71; assoc prof, 71-76, actg head dept biochem, 78-79, PROF BIOCHEM, LA STATE UNIV MED CTR, NEW ORLEANS, 76- *Concurrent Pos:* Nat Inst Neurol Dis & Blindness spec fel, 60-62 & 81-82; assoc biochemist, McLean Hosp, Mass, 60-71; resident scientist, Neurosci Res Prog, Mass Inst Technol, 70-71; vis prof, Div Neurol, Duke Med Ctr, Durham, 81-82. *Mem:* Soc Exp Biol & Med; Am Asn Biol Chem; Am Soc Neurochem; AAAS; Am Asn Univ Professors. *Res:* Neurochemistry; clinical chemistry; metabolism and chemistry of brain proteins in relation to development; neurochemistry of aging; biochemical methods of analysis. *Mailing Add:* Dept of Biochem La State Univ Sch Med 1100 Florida Ave New Orleans LA 70119

MOLAISON, HENRI J(EAN), engineering, for more information see previous edition

MOLAU, GUNTHER ERICH, b Leipzig, Ger, Oct 15, 32; m 58; c 3. POLYMER CHEMISTRY. *Educ:* Carolo-Wilhelmina Inst Technol, BS, 56, MS, 59, PhD(chem), 61. *Prof Exp:* Chemist, Dow Chem Co, 61-63, res chemist, 63-65, sr res chemist, 65-72, assoc scientist, Dow Chem USA, 69-86; RETIRED. *Concurrent Pos:* Computer consult, 86- *Res:* Polymer chemistry; membranes; colloidal and heterogeneous polymers; computer simulation of polymer and membranes processes. *Mailing Add:* 30 Mount Tamalpais Ct Clayton CA 94517

MOLD, CAROLYN, MEMBRANE PROTEINS. *Educ:* Univ Minn, PhD(microbiol), 77. *Prof Exp:* ASSOC PROF, RUSH MED COL, 77- *Mailing Add:* Univ NMex 2211 Lomas NE Albuquerque NM 87131

MOLD, JAMES DAVIS, b Carlton, Minn, Sept 26, 20; m 46; c 3. BIO-ORGANIC CHEMISTRY. *Educ:* Univ Minn, BCh, 42; Northwestern Univ, MS, 44, PhD(org chem), 47. *Prof Exp:* Asst, Northwestern Univ, 42-44, asst chem, 44-46, res assoc, 46-47; biochemist, Parke, Davis & Co, 47-49; org chemist, Allied Sci Div, Biol Lab, US Army Chem Corps, 49-55; chief org chem res, Liggett & Myers Tobacco Co, 55-64, asst dir res, 64-79; RETIRED. *Concurrent Pos:* Consult, Liggett Group, Inc, 79-84. *Mem:* Sigma Xi. *Res:* Isolation of natural organic substances; synthesis of organic chemicals; degradation and structure proof of natural organic products; chemistry of tobacco and smoke. *Mailing Add:* 4901 Whitfield Rd Durham NC 27707

MOLDAVE, KIVIE, b Kiev, Russia, Oct 22, 23; nat US; m 49; c 2. BIOCHEMISTRY. *Educ:* Univ Calif, AB, 47; Univ Southern Calif, MS, 50, PhD(biochem), 52. *Prof Exp:* USPHS res fels, Univ Wis, 52-53 & Fac Sci, Univ Paris, 53-54; from asst prof to prof biochem, Sch Med, Tufts Univ, 54-66; prof & chmn dept, Sch Med, Univ Pittsburgh, 66-70; chmn dept biol chem, 70-80, prof biochem, Col Med, Univ Calif, Irvine, 80- 84; acad vice chancellor & prof biochem, 84-87, PROF BIOCHEM, UNIV CALIF, SANTA CRUZ, 87- *Concurrent Pos:* Mem physiol chem study sect, NIH, 67-71 & res serv merit rev bd basic sci, Vet Admin, 72-76; mem adv coun res & clin invest awards, Am Cancer Soc, 76-81, Aging Review Comt, NIH, 77-81; ed, Methods Enzymol, Nucleic Acids & Protein Synthesis, 67-, co ed, Progress Nucleic Acids & Molecular Biol, 81-; sci adv comt acad personnel, Am Cancer Soc, 81-84. *Mem:* Am Soc Biol Chem; Am Chem Soc; fel AAAS. *Res:* Nucleic acid and protein biosynthesis. *Mailing Add:* Dept Biol Univ Calif Santa Cruz CA 95064

MOLDAWER, MARC, b Philadelphia, Pa, June 4, 22; m 63; c 3. MEDICINE, ENDOCRINOLOGY. *Educ:* Univ Pa, 39-42; Harvard Univ, MD, 50. *Prof Exp:* Intern med, 50-51, resident, Presby Hosp, Columbia Univ, 51-52; clin res fel endocrinol, Mass Gen Hosp, 52-55; res fel biochem, Cambridge Univ, 55-56; res fel endocrinol, Harvard Med Sch, 56-57; asst prof, 57-63, ASSOC PROF MED, BAYLOR COL MED, 63-; dir med endocrine sect, Methodist Hosp, Houston, 63-78; CLIN PRACT, 78. *Concurrent Pos:* Consult, Tex Inst Rehab & Res, Ben Taub Gen, Methodist & Vet Admin Hosps, Houston, 58- *Mem:* AAAS; Endocrine Soc; Am Fedn Clin Res. *Res:* Gynecomastia and estrogen metabolism in the male; human growth hormone; immunology and physiology. *Mailing Add:* 6410 Fannin #1528 Houston TX 77030

MOLDAY, ROBERT S, b New York, NY, Oct 27, 43; m 72; c 2. BIOCHEMISTRY, CELL BIOLOGY. *Educ:* Univ Pa, BSc, 65, PhD(biochem), 71; Georgetown Univ, MSc, 67. *Prof Exp:* Res assoc biochem & fel, Calif Inst Technol, 72-75; asst prof, 75-80, ASSOC PROF BIOCHEM, UNIV BC, 80- *Concurrent Pos:* Am Cancer Soc fel, 72-74. *Mem:* Can Biochem Soc. *Res:* Protein and membrane biochemistry; electron microscopy; cell surface receptors; biochemistry of vision. *Mailing Add:* Dept Biochem Univ BC Vancouver BC V6T 1W5 Can

MOLDENHAUER, RALPH ROY, b Detroit, Mich, Apr 8, 35; m 59; c 2. PHYSIOLOGY, ECOLOGY. *Educ:* Mich State Univ, BS, 60; Ore State Univ, MS, 65, PhD(zool), 69. *Prof Exp:* ASSOC PROF BIOL, SAM HOUSTON STATE UNIV, 68- *Concurrent Pos:* Soc Sigma Xi res grant-in-aid, Sam Houston State Univ, 69-71, Am Mus Natural Hist Frank M Chapman Mem Fund grant, 70-72 & Am Philos Soc res grant, 71-72. *Mem:* Am Inst Biol Sci; Am Ornith Union; Cooper Ornith Soc; Ecol Soc Am. *Res:* Physiology and ecology of birds; salt and water balance; thermoregulation; behavioral adjustments in stressful environments. *Mailing Add:* Dept Biol Sam Houston State Univ Huntsville TX 77340

MOLDENHAUER, WILLIAM CALVIN, b New Underwood, SDak, Oct 27, 23; m 47; c 5. SOIL SCIENCE, SOIL CONSERVATION. *Educ:* SDak State Univ, BS, 49; Univ Wis-Madison, MS, 51, PhD(soil sci), 56. *Prof Exp:* Asst agronomist, SDak State Univ, 49-54; soil scientist, Agr Res Serv, USDA, Tex, 54-57 & Iowa, 57-72, soil scientist & res leader, Minn, 72-75, Nat Soil Erosion Lab, West Lafayette, 75-85; EXEC SECY, WORLD ASN SOIL/WATER CONSERV, 85- *Honors & Awards:* Hugh Hammond Bennett Award, Soil & Water Conserv Soc. *Mem:* Fel Soil Conserv Soc (pres, 79); fel Am Soc Agron; fel Soil Sci Soc Am; Am Soc Agr Engrs; World Asn Soil & Water Conserv (pres, 82-85). *Res:* Soil erosion research; soil management research; soil erosion and tillage and their relationship. *Mailing Add:* 317 Marvin Ave Volga SD 57071

MOLDER, S(ANNU), b Tallin, Estonia, Sept 1, 35; Can citizen; c 3. AERONAUTICAL ENGINEERING. *Educ:* Univ Toronto, BASc, 58, MASc, 61, MEng, 78. *Prof Exp:* Res asst aerophys, Inst Aerophys, Univ Toronto, 58-61; res assoc propulsion, McGill Univ, 63-64, from asst prof to assoc prof gas dynamics, 64-70, chmn mech eng dept, 69-70; chmn mech technol dept, 71-77, PROF MECH TECHNOL, RYERSON POLYTECH INST, 77- *Concurrent Pos:* Defense Res Bd Can grant propulsion, 63; consult, Appl Physics Lab, Johns Hopkins Univ, 64 & 85; mem assoc comt aerodyn, Nat Res Coun Can, 64; vis assoc prof, Univ Sydney, 70-71; consult, CANADAIR, Ltd, 81. *Mem:* Can Soc Mech Engrs; assoc fel Am Inst Aeronaut & Astronaut; fel Can Aeronaut & Space Inst. *Res:* Hypersonic propulsion; supersonic combustion ramjets; hypersonic air inlets; supersonic aerodynamics of wings; aerodynamics of curved shockwaves and their interactions. *Mailing Add:* Dept of Mech/Aerosp Engr, Ryerson Polytechnical Inst 350 Victoria St Toronto ON M5B 2K3 Can

MOLDOVER, MICHAEL ROBERT, b New York, NY, July 19, 40. PHYSICS. *Educ:* Rensselaer Polytech Inst, BS, 61; Stanford Univ, MS, 62, PhD(physics), 66. *Prof Exp:* Res assoc physics, Stanford Univ, 66; asst prof, Univ Minn, Minneapolis, 67-72; from physicist to sr physicist, 72-88, GROUP LEADER, NAT INST STANDARDS & TECHNOL, 89- *Honors & Awards:* Stratton Award, Dept Com, Gold Medal, Silver Medal, Bronze Medal. *Mem:* Sigma Xi; Acoust Soc Am; Am Phys Soc; Am Asn Physics Teachers. *Res:* Thermodynamic properties of liquids and solids, especially near phase transitions; properties of alternative refrigerants; low temperature physics; phase equilibria in fluid mixtures; physical acoustics; dynamics of phase changes in fluids; fluid-fluid and fluid-solid interface properties; temperature scales and thermometry. *Mailing Add:* Nat Inst Standards & Technol Bldg 221 Rm A105 Gaithersburg MD 20899

MOLE, JOHN EDWIN, b Macon, Ga, May 14, 44; m 72. IMMUNOCHEMISTRY, BIOCHEMISTRY. *Educ:* Berry Col, BA, 66; NC State Univ, PhD(biochem), 72. *Prof Exp:* Fel immunol, 72-74, asst prof comp med, 74-75, asst prof microbiol, Univ Ala, Birmingham, 75-, assoc scientist, Comprehensive Cancer Ctr, 76-; AT DEPT BIOCHEM, MED CTR, UNIV MASS, WORCESTER. *Concurrent Pos:* NIH fel, 75-76, grants, 75-76 & 78-81; Nat Cancer grant, 76-79; Am Heart Asn grant, 78-80, prin investr, 79-84. *Mem:* Sigma Xi; Am Chem Soc; NY Acad Sci; Am Asn Immunologists. *Res:* Structure-function of biologically-active macromolecules. *Mailing Add:* Dept Biochem Dept Biochem Rm 56315 Worcester MA 01655

MOLE, PAUL ANGELO, b Jamestown, NY, Mar 13, 38; m 57; c 3. MUSCULAR PHYSIOLOGY, EXERCISE PHYSIOLOGY. *Educ:* Univ Ill, Urbana, BS, 60, MS, 62, PhD(physiol), 69. *Prof Exp:* NIH trainee nutrit, Sch Med, Wash Univ, 69-71; asst prof phys educ, Temple Univ, 71-74; asst prof physiol, La State Univ Med Ctr, 74-77; ASSOC PROF PHYS EDUC & FAC GRAD PHYSIOL, UNIV CALIF, DAVIS, 77- *Mem:* AAAS; NY Acad Sci; fel Am Col Sports Med; Am Physiological Soc. *Res:* Modeling energetics and NMR spectroscopy; biochemical and contractile properties of skeletal and heart muscle. *Mailing Add:* Muscle Lab Univ Calif Davis CA 95616

MOLECKE, MARTIN A, b Cleveland, Ohio, Sept 14, 45; m 68; c 2. PHYSICAL CHEMISTRY. *Educ:* Bowling Green State Univ, BS, 67; Carnegie-Mellon Univ, MS, 70, PhD(nuclear chem), 72. *Prof Exp:* Fel, Carnegie-Mellon Univ, 72-73; sr scientist radiochem, Bettis Atomic Power Lab, Westinghouse Elec Co, 73-75; SR MEM TECH STAFF NUCLEAR SCI, SANDIA NAT LABS, AT&T, 76- *Concurrent Pos:* Chemistry Hon Prog-Argonne Nat Lab, 67. *Mem:* Am Chem Soc; Am Nuclear Soc; Mat Res Soc; Sigma Xi. *Res:* Nuclear waste technology; high-level waste laboratory and in situ experimentation; waste package materials performance research; transuranic waste experimental characterization. *Mailing Add:* 12100 St Marys NE Albuquerque NM 87111

MOLENDA, JOHN R, b Scranton, Pa, Apr 12, 31; m 62; c 3. FOOD POISONING, FOOD TECHNOLOGY. *Educ:* Univ Scranton, BS, 52; Utah State Univ, MS, 57, PhD(bact), 65; Johns Hopkins Univ, MPH, 63. *Prof Exp:* Med lab technician, US Army Med Corps, 53-56; res asst bact & pub health, Utah State Univ, 56-61; USPHS trainee, Sch Pub Health, Johns Hopkins Univ, 61-63, lab supvr, 63-65, from instr to asst prof epidemiol, 65-68; asst chief, Microbiol Div, Bur Labs, State of Md, 68-72, chief, Salisbury Pub Health Lab, Dept Health, 72-76; prof biol sci, 76-81, chmn dept, 79-84, actg vpres acad affairs, 81-82, dean sci, 84-90, PROF BIOL SCI, SALISBURY STATE COL, 90- *Concurrent Pos:* Scientist dir, USPHS, 57-; consult lab & epidemiol, County Health Dept, 76-, infectious dis, Sable Mink Ranch, 79-80, blood typing, Court Syst, Worcester County, Md, 79-81, food poisoning, legal firms, 80- & Ceva Co, 82-83; coordr, Md Acad Sci, 80- & epidemiol & food poisoning home study courses, Ctrs Dis Control, 82- *Mem:* Am Soc Microbiol; Epidemiol Soc Southern Africa. *Res:* Bacterial food poisoning in food service and canning industries. *Mailing Add:* Sch Sci Salisbury State Col Salisbury MD 21801

MOLENKAMP, CHARLES RICHARD, b San Francisco, Calif, Aug 26, 41; m 67; c 2. ATMOSPHERIC PHYSICS, POLLUTANT TRANSPORT & DEPOSITION. *Educ:* Calvin Col, BS, 63; Univ Ariz, MS, 68, PhD(atmospheric physics), 72. *Prof Exp:* PHYSICIST ATMOSPHERIC SCI, LAWRENCE LIVERMORE LAB, UNIV CALIF, 72- *Mem:* Am Meteorol Soc; Am Sci Affil. *Res:* Cloud physics; numerical modeling; mesoscale atmospheric modeling; precipitation scavenging; regional and climatic effects of nuclear weapons. *Mailing Add:* Lawrence Livermore Lab L-262 PO Box 808 Livermore CA 94550

MOLER, CLEVE B, b Salt Lake City, Utah, Aug 17, 39; m 60; c 3. NUMERICAL ANALYSIS, MATHEMATICAL SOFTWARE. *Educ:* Calif Inst Technol, BS, 61; Stanford Univ, PhD(math), 65. *Prof Exp:* Instr comput sci, Stanford Univ, 65; from asst prof to assoc prof math, Univ Mich, Ann Arbor, 66-72; assoc prof math, Univ NMex, 72-74, prof, 74-80, chmn Dept Comput Sci, 80-84; MGR APPLNS RES, INTEL SCI COMPUT, 84- *Concurrent Pos:* Off Naval Res assoc, Swiss Fed Inst Technol, 65-66; vis assoc prof, Stanford Univ, 70-71 & 78-79. *Mem:* Am Math Soc; Asn Comput Mach; Soc Indust & Appl Math. *Res:* Numerical analysis; computer science; linear algebra; partial differential equations. *Mailing Add:* 325 Linfield Pl Menlo Park CA 94025

MOLGAARD, JOHANNES, b Kunming, China, Apr 23, 36; m 65; c 3. MATERIALS ENGINEERING, TRIBOLOGY. *Educ:* Queen's Univ Belfast, BS, 57; Univ Leeds, PhD(textiles), 66. *Prof Exp:* Tech officer, Brit Nylon Spinners Ltd, 57-63; Nat Res Coun Can fel, McMaster Univ, 66-68; from asst prof to assoc prof eng, 68-78, PROF ENG, MEM UNIV NFLD, 78- *Concurrent Pos:* Mem assoc comt on tribology, Nat Res Coun Can, 71-, chmn, 81-89; Alexander von Humboldt Found fel, Ger, 75-76. *Mem:* Can Soc Mech Engrs. *Res:* Wear of metals; friction of ice; system theory. *Mailing Add:* Fac Eng Mem Univ Nfld St John's NF A1B 3X5 Can

MØLHAVE, LARS, b Saeby, Denmark, Aug 5, 44; m 69; c 2. INDOOR CLIMATE & AIR QUALITY, SOURCES OF INDOOR AIR POLLUTION. *Educ:* Univ Aarhus, Denmark, Master, 70; Inst Hyg, Arhus, PhD(med), 83; Inst Environ & Occup Med, Arhus, PhD(med), 88. *Prof Exp:* Res asst med, Inst Hyg, Arhus, Denmark, 72-78, asst prof, 78-86; ASSOC PROF MED, INST ENVIRON & OCCUP MED, UNIV AARHUS, 86- *Concurrent Pos:* Res consult, US Environ Protection Agency, Research Triangle Park, NC, 86-; appointee, Comt Indoor Climate, Ministry Health, 86-, Ministry Buildings, 88-; vis prof, Inst Environ Med, Beijing, China, 87; vchmn, Cost-613, EEC Comt Environ & Health, 88-; vis scientist, Dept Epidemiol, Yale Univ, New Haven, 90- *Honors & Awards:* Nevins Award, Am Soc Heating, Refrig & Air Conditioning Engrs, 86. *Mem:* Am Soc Testing & Mat; Am Soc Heating Refrig & Air Conditioning Engrs. *Res:* Human response, comfort and health effects caused by indoor air pollution by volatile pollutants or particulates in non- industrial or industrial buildings. *Mailing Add:* Inst Environ & Occup Med Univ Aarhus Bld 180 Universitetsparken Arhus DK 8000 Denmark

MOLINA, JOHN FRANCIS, b Jamaica, NY, Jan 4, 50; m 73; c 2. ANALYTICAL, PHYSICAL & ORGANIC ENVIRONMENTAL CHEMISTRY. *Educ:* Northeastern Univ, BS, 73; Univ New Orleans, PhD(chem), 77. *Prof Exp:* Chemist qual control drug anal, Fougera, Inc, 73; sr res chemist, Bristol-Myers Co, 77-78; res chemist & group leader spectrochem, Celanese Res Co, 78-81; inst anal group leader, Apollo Technol Inc, 81-83; lab mgr, AT-SEA Incineration, 83-85; mgr field chem, OH Mats, 85-86; vpres opers & dir, Hager Labs, 87-89; VPRES, ENVIRON SERV, LAB TESTING SERV INC, 90- *Mem:* Am Chem Soc; Soc Appl Spectros; NY Acad Sci; Am Indust Hyg Asn; Am Inst Chemists; Col Am Pathologists. *Res:* Trace analysis of both inorganic and organic substances via spectroscopic techniques, especially atomic absorption, atomic emission and x-ray fluorescence; wet chemical analysis and separation techniques, especially gas chromatography and high pressure liquid chromatography. *Mailing Add:* 36 Melbourne St Oyster Bay NY 11771-1607

MOLINA, MARIO JOSE, b Mar 19, 43; m 73; c 1. ATMOSPHERIC CHEMISTRY, CHEMICAL KINETICS. *Educ:* Univ Nat Autonoma De Mex, BS, 65; Univ Calif, Berkeley, PhD(phys chem), 72. *Prof Exp:* Asst prof chem eng, Univ Nat Autonoma De Mex, 67-68; res assoc phys chem, Univ Calif, Berkeley, 72-73; res assoc phys chem, Univ Calif, Irvine, 73-75, from asst prof to assoc prof, 75-82; SR RES SCIENTIST, JET PROPULSION LAB, CALIF INST TECHNOL, 83- *Concurrent Pos:* Fel, Alfred P Sloan Found, 76-78; teacher & scholar, Camille & Henry Dreyfus Found, 78-82; Mem, Panel Chem Kinetic & Photochem Data Eval, NASA, 78-, Comn Human Resources, Nat Res Coun, Nat Acad Sci, 78-81. *Honors & Awards:* Esselen Award, Am Chem Soc, 87; Soc Hispanic Prof Engrs Award, 83. *Mem:* Am Chem Soc; Am Phys Soc; Am Geophys Union; Photochem Soc; Sigma Xi; AAAS; Sci Res Soc. *Res:* Chemistry of the stratosphere particularly as it can be affected by man-made perturbations such as the release of chlorofluorocarbons; various aspects of the theory of stratospheric ozone depletion by chlorofluorocarbons. *Mailing Add:* Dept EAPS 54-1312 Mass Inst Technol Cambridge MA 02139

MOLINA, RANDOLPH JOHN, b Los Angeles, Calif, May 13, 51; m 71; c 2. MYCOLOGY. *Educ:* Univ Calif, Santa Barbara, BA, 73; Ore State Univ, PhD(bot), 81. *Prof Exp:* Botanist, 73-80, RES BOTANIST, PAC NORTHWEST FOREST & RANGE EXP STA, FOREST SERV, USDA, 81- *Concurrent Pos:* Assoc prof, Dept Forest Sci, Ore State Univ, 81- *Mem:* Sigma Xi. *Res:* Specificity and compatibility between ectomycorrhizal fungus and host symbionts; practical application of beneficial mycorrhizal relationships in forestry; interactions between non-leguminous nitrogen fixing plants; nodule symbionts; mycorrhizal fungi. *Mailing Add:* Foresty Sci Lab 3200 Jefferson Way Corvallis OR 97331

MOLINARI, JOHN A, IMMUNOLOGY. *Educ:* Univ Pittsburgh, PhD(microbiol), 70. *Prof Exp:* PROF MICROBIOL & BIOCHEM & CHMN DEPT, SCH DENT, UNIV DETROIT, 77- *Mailing Add:* Dept Microbiol & Biochem Univ Detroit Sch Dent Detroit MI 48207

MOLINARI, PIETRO FILIPPO, b Mestre-Venice, Italy, Sept 9, 23; US citizen; m 56; c 2. ENDOCRINOLOGY, HEMATOLOGY. *Educ:* Univ Milan, DVM, 52, PhD(clin path), 60. *Prof Exp:* From asst prof clin vet med to assoc prof clin methodology, Univ Milan, 52-61, dir res lab, 61-64; endocrinologist, Mason Res Inst, 64-67, sr investr endocrinol, 67-71; asst dir hemat, St Vincent Hosp, 71-79; ASSOC PROF, DEPT MED, VET SCH, TUFTS UNIV, 79- *Concurrent Pos:* Ital Res Coun fel physiol, Vet Sch, Cornell Univ, 61; asst prof med, Med Sch, Univ Mass. *Mem:* Int Soc Exp Hemat; NY Acad Sci; Endocrine Soc; Sigma Xi. *Res:* Study of erythropoietic activity of hormones, particularly steroid hormones in laboratory animals and human cells in vitro. *Mailing Add:* Seven Hancock Hill Dr Worcester MA 01609

MOLINARI, ROBERT JAMES, b Peckville, Pa, May 30, 52. PHYSICAL POLYMER CHEMISTRY, BIOPHYSICS. *Educ:* Dartmouth Col, AB, 74, MBA, 79; Brown Univ, PhD(chem), 77. *Prof Exp:* At Res Div, W R Grace & Co, 78; staff scientist, 79, group leader, 80, mkt develop mgr, Raychem Corp, 81-85; consult, McKinsey Co, 85-88; PRES & CHIEF EXEC OFFICER, AT BIOCHEM, 88- *Concurrent Pos:* Consult, Schlumberger Corp, Ridgefield, Conn & Houston, Tex, 80. *Mem:* NY Acad Sci. *Res:* Electrical and dielectric properties of polymers, particularly industrial and biological macromolecules; polymer radiation chemistry and physics, biophysics and separations science. *Mailing Add:* 30 Spring Mill Dr Malvern PA 19355

MOLINDER, JOHN IRVING, b Erie, Pa, June 14, 41; m 62; c 2. COMMUNICATION SYSTEMS, SIGNAL PROCESSING & CONTROL. *Educ:* Univ Nebr, BS, 63; Air Force Inst Technol, MS, 64; Calif Inst Technol, PhD(elec eng), 69. *Prof Exp:* Sr engr, commun systs, Jet Propulsion Lab, 69-70; from asst prof to assoc prof, 70-80, PROF ENG, HARVEY MUDD COL, 80- *Concurrent Pos:* Lectr elec eng, Calif State Univ, Los Angeles, 70-74; mem tech staff, Jet Propulsion Lab, NASA, 74- & rep, NASA Hq, 79-80; vis prof, Calif Inst Technol, 82-83. *Mem:* Inst Elec & Electronics Engrs. *Res:* Analysis and design of communiciation systems. *Mailing Add:* Dept Eng Harvey Mudd Col Claremont CA 91711

MOLINE, HAROLD EMIL, b Frederic, Wis, Nov 13, 39; m 65; c 2. PHYTOPATHOLOGY, PLANT VIROLOGY. *Educ:* Univ Wis-River Falls, BSc, 67, MS, 69; Iowa State Univ, PhD(plant path), 72. *Prof Exp:* Plant pathologist, Northern Grain Insect Res Lab, Brookings, SDak, 72-73, RES PLANT PATHOLOGIST, HORT CROPS QUAL LAB, HORT SCI INST, AGR RES SERV, USDA, BELTSVILLE, MD, 74- *Concurrent Pos:* Adj prof, Bot Dept, Howard Univ, Washington, DC. *Mem:* Am Phytopath Soc; Bot Soc Am; Sigma Xi; Am Soc Plant Physiologists; Electron Micros Soc Am. *Res:* Epidemiology of post harvest diseases of fresh market fruits and vegetables; ultrastructural and histochemical modifications of host cells invaded by bacteria, fungi or viruses; physiological disorders. *Mailing Add:* Prod Qual & Devel Inst Hort Crops Qual Lab Beltsville Agr Res Ctr-W Beltsville MD 20705

MOLINE, SHELDON WALTER, b Chicago, Ill, Feb 15, 31; m 52; c 4. BIOCHEMISTRY, IMMUNOLOGY. *Educ:* Roosevelt Univ, BS, 52; Univ Chicago, PhD, 58; State Univ NY Buffalo, MBA, 69. *Prof Exp:* Res technician, Argonne Nat Lab, 52-56; res chemist, Linde Div, Union Carbide Corp, 58-63, sr staff biochemist, 64-67, proj scientist chem & plastics, 67-68, tech mgr fermentation, Chem Div, 68-71, dir res, Creative Agr Systs, 71-74, sr group leader, Corp Res Dept, 74-79; mgr med diagnostics, 79-83, corp projs dir, 83-88, PRES, MOLINE BIOTECHNOL RESOURCES, STAUFFER CHEM CO, 88- *Mem:* AAAS; Am Chem Soc; Inst Food Technol; Sigma Xi; NY Acad Sci. *Res:* Thermal properties and metabolic processes of biological systems at low temperatures; cryosurgery; cell culture; preservation of foods at low temperatures; shipment of produce and meats; seed and plant physiology; medical and health sciences. *Mailing Add:* 101 Greenridge Ave White Plains NY 10605

MOLINE, WALDEMAR JOHN, b Fredric, Wis, Oct 29, 34; m 57; c 3. AGRONOMY. *Educ:* Wis State Univ, Riverfalls, BS, 59; Univ Minn, MS, 61; Iowa State Univ, PhD(agron), 65. *Prof Exp:* Res asst agron, Univ Minn, 59-61; res assoc, Iowa State Univ, 61-65; asst prof, Univ Md, 65-66; from assoc prof to prof agron, Univ Nebr, Lincoln, 66-76; prof agron & head dept, Univ Ark, 76-; DIR COOPER EXTEN SERV & ASSOC DEAN, CNAR, MICHIGAN STATE UNIV. *Concurrent Pos:* Chmn, mem comt & mem exec comt, Am Forage & Grassland Coun, 60-70; chmn pub rels & info comt, Am Soc Agron, Crop Sci Soc Am & Soil Sci Soc Am, 75. *Mem:* Am Soc Agron; Soc Range Mgt; Am Forage & Grassland Coun (pres, 75). *Res:* Forage crops management, production and utilization. *Mailing Add:* Mich State Univ Rm 101 Agriculture E Lansing MI 48824-1039

MOLINOFF, PERRY BROWN, b Smithtown, NY, June 3, 40; m 63; c 2. NEUROPHARMACOLOGY. *Educ:* Harvard Univ, BS, 62, MD, 67. *Hon Degrees:* MS, Univ Penn. *Prof Exp:* Intern med, Univ Chicago Hosps & Clins, 67-68; res assoc, NIMH, 68-70; vis fel biophys, Univ Col, London, 70-72; from asst prof to prof, Dept Pharmacol, Univ Colo, Health Sci Ctr, 72-81; A N RICHARDS PROF PHARMACOL & CHMN, PHARMACOL DEPT, MED SCH, UNIV PENN, 81- *Concurrent Pos:* Guggenheim fel; estab investr, Am Heart Asn. *Honors & Awards:* Grass Lectr neurosci. *Mem:* Am Soc Pharmacol & Exp Therapeut; Soc Neurosci; Am Soc Neurochem; Am Heart Asn; AAAS; Am Soc Biol Chemists. *Res:* Effects of in vivo manipulations on receptors for catecholamines; receptor properties and coupling to effector systems in vitro. *Mailing Add:* Dept Pharmacol Univ Pa 36th & Hamilton Walk Philadelphia PA 19104-6084

MOLITCH, MARK E, b Ft Knox, Ky, Dec 10, 43. PITUITARY DISEASE. *Educ:* Univ Pa, MD, 69. *Prof Exp:* Asst prof endocrinol, metab & nutrit, Tufts Univ, 75-84; ASSOC PROF ENDOCRINOL, METAB & NUTRIT, SCH MED, NORTHWESTERN UNIV, 84- *Mem:* Endocrine Soc; Am Diabetes Asn; fel Am Col Path. *Res:* Pathogenesis and treatment of pituitary tumors; pathogenisis and complications of diabetes. *Mailing Add:* Ctr Endocrinol Metab & Nutrit Sch Med Northwestern Univ 303 E Chicago Ave Chicago IL 60611

MOLL, ALBERT JAMES, b Vergennes, Ill, Feb 2, 37; m 64; c 2. CHEMICAL ENGINEERING. *Educ:* Univ Ill, BS, 59, Univ Wash, Seattle, MS, 61, PhD(chem eng), 66. *Prof Exp:* Chem engr res & develop, Union Carbide Corp, WVa, 66-70; sr staff engr, Chem Systs Inc, 70-72; sr engr & economist, SRI Int, 72-74, dir energy technol dept, 74-80; PRES, SFA PAC INC, 80- *Mem:* Am Chem Soc; Am Inst Chem Engrs; Sigma Xi. *Res:* Synthetic fuels technology and economics; underground coal gasification; liquefied natural gas. *Mailing Add:* 3571 Cambridge Lane Mountain View CA 94040

MOLL, EDWARD OWEN, b Peoria, Ill, Nov 30, 39; m 60; c 2. HERPETOLOGY. *Educ:* Univ Ill, BS, 61, MS, 63; Univ Utah, PhD(zool), 68. *Prof Exp:* From asst prof to assoc prof, 68-78, PROF ZOOL, EASTERN ILL UNIV, 78- *Concurrent Pos:* Environ consult, Westinghouse Elec Corp, 73-74; consult turtle conserv, W Malaysian Dept Wildlife, 75-76, 86-87, 89-90; res grants, NY Zool Soc, 75-78, 81 & 86, World Wildlife Fund, 75-76, 81, 89-90, Am Philos Soc, 75-76 & Fauna Preserv Soc, 75-76 & 78, Ill Dept Conserv, 86, 87 & 88; mem, Freshwater Chelonian Specialist Group, Int Union Conserv Nature; mem, Marine Turtle Specialist Group, Int Union Conserv Nature; Fulbright fel, India, 82-83; field assoc, Field Mus Natural Hist, Chicago, 83-; affil prof scientist, Ill Natural Hist Surv, 90- *Mem:* AAAS; Am Soc Ichthyologists & Herpetologists; Soc Study Amphibians & Reptiles (pres, 91-); Herpetologists League; Sigma Xi; Malayan Nat Soc; Int Union Conserv Nature. *Res:* Reptilian ecology; taxonomy; distribution and management of river turtles. *Mailing Add:* Dept Zool Eastern Ill Univ Charleston IL 61920

MOLL, HAROLD WESBROOK, b Detroit, Mich, Apr 2, 14; m 38; c 3. ORGANIC CHEMISTRY, CHEMICAL ENGINEERING. *Educ:* Andrews Univ, BS, 37. *Hon Degrees:* DSc, Andrew's Univ, 82. *Prof Exp:* Res chemist, Dow Chem, USA, 37-48, supt latex pilot plant, 48-58, supvr, Instrument Lab, 59-63, tech expert, E C Britton Res Lab, 63-71, tech expert, Phys Res Lab, 72-79; consult, 80-88. *Concurrent Pos:* Adj prof chem, Andrew's Univ, Berrien Springs, Mich, 78-89. *Mem:* AAAS; Am Chem Soc; Instrument Soc Am. *Res:* Organic synthesis; polymers; latexes; chemical engineering; instrumentation science. *Mailing Add:* 4220 Isabella Rd Midland MI 48640-8362

MOLL, JOHN L, b Ohio, Dec 21, 21; m 44; c 3. ELECTRICAL ENGINEERING. *Educ:* Ohio State Univ, BSc, 43, PhD(elec eng), 52. *Hon Degrees:* Dr, Cath Univ Leuven, Belgium, 83. *Prof Exp:* Develop engr magnetrons, Radio Corp Am, 44-45; asst math, Ohio State Univ, 46-49, instr & res assoc, 50-52; mem tech staff transistor develop, Bell Tel Labs, Inc, 53-56 & solid state electronics res, 56-58; from assoc prof to prof elec eng, Stanford Univ, 58-69; tech dir eng, Fairchild Microwave & Opto-Electronics Div, Fairchild Camera & Instrument Corp, 69-74; tech dir, 74-81, dir, Integrated Circuits Lab, 81-87, ASSOC DIR, SUPERCONDUCTIVITY LAB, NEWLETT-PACKARD CO, INC, 87 - *Concurrent Pos:* NSF sr fel, Tech Univ Denmark, 64-65; consult, HP Assocs, Hewlett-Packard Co; Guggenheim fel, 64. *Honors & Awards:* Howard N Potts Medal, Franklin Inst, 67; Ebers Award, Inst Elec & Electronics Engrs, 71; Benjamin G Lamme Medal, Ohio State Univ, 88. *Mem:* Nat Acad Sci; Nat Acad Eng; fel Inst Elec & Electronics Engrs; Am Phys Soc. *Res:* Semiconductor electronics; high frequency oscilators; transistors. *Mailing Add:* 4111 Old Trace Rd Palo Alto CA 94306

MOLL, KENNETH LEON, b Jackson, Mo, Oct 16, 32; m 57; c 2. SPEECH SCIENCE. *Educ:* Southeast Mo State Col, BS, 54; Univ Iowa, MA, 59, PhD(speech path), 60. *Prof Exp:* Res assoc speech sci, 59-61, res asst prof, 61-64, assoc prof, 64-68, chmn, Dept Speech Path & Audiol, 68-76, assoc dean fac, 76-81, actg vpres acad affairs, 81-82, assoc vpres acad affairs, 82-89, PROF SPEECH SCI, UNIV IOWA, 68-,. *Concurrent Pos:* Nat Inst Neurol Dis & Blindness spec fel, Univ Mich, 65-66. *Mem:* Am Speech-Lang-Hearing Asn; Acoust Soc Am. *Res:* Physiological aspects of human speech production through use of x-ray techniques, electromyography and air pressure and air flow recordings. *Mailing Add:* Dept Speech Path & Audiol Univ Iowa Iowa City IA 52242

MOLL, MAGNUS, b East Orange, NJ, July 23, 28; m 58. ELECTRICAL & SYSTEMS ENGINEERING. *Educ:* Purdue Univ, BS, 53; Univ Ill, MS, 56; Ohio State Univ, PhD(elec eng), 62. *Prof Exp:* Prin elec engr, Battelle Mem Inst, 56-57, sr elec engr, 57-62, group consult, 62; mem prof staff, Arthur D Little Inc, Mass, 63-71; SR SCIENTIST, BOLT BERANEK & NEWMAN INC, 72- *Mem:* Inst Elec & Electronics Engrs; Sigma Xi. *Res:* Systems engineering and analysis; stochastic processes; sonar systems; signal processing. *Mailing Add:* 4460 Dexter St NW Washington DC 20007

MOLL, PATRICIA PEYSER, b New Rochelle, NY, Oct 29, 46; m 69. HUMAN & POPULATION GENETICS. *Educ:* Univ Vt, BA, 68; State Univ NY, Stony Brook, PhD(biol), 75. *Prof Exp:* Systs analyst, Int Bus Mach, 68-69; math assoc, Brookhaven Nat Lab, 69-70; res asst ecol & evolution, State Univ NY, Stony Brook, 71-75; scholar, 75-78, ASST PROF EPIDEMIOL & ASST RES SCIENTIST, HUMAN GENETICS, UNIV MICH, ANN ARBOR, 79- *Concurrent Pos:* NIH fel, 76-78. *Mem:* Genetics Soc Am; Am Soc Human Genetics; Biomet Soc; Soc Epidemiol Res. *Res:* Genetic epidemiology. *Mailing Add:* Dept Pub Health Univ Mich Main Campus Ann Arbor MI 48104-2029

MOLL, RICHARD A, b Chicago, Ill, Sept 2, 35; m 55; c 6. METALLURGICAL & MECHANICAL ENGINEERING. *Educ:* Ill Inst Technol, BS, 62; Lehigh Univ, MS, 64, PhD(metall, mat sci), 66. *Prof Exp:* Asst experimentalist metall, IIT Res Inst, 55-62; instr, Lehigh Univ, 62-66; asst prof metall & mech eng, 66-74, assoc prof metall eng, 74-80, PROF METALL ENG, UNIV WIS-MADISON, 80- *Concurrent Pos:* Consult, prod liability & expert witness. *Honors & Awards:* F L Plummer lectr, Am Welding Soc, 81. *Mem:* Am Soc Metals; Soc Mfg Engrs; Nat Acad Forensic Engrs. *Res:* Product safety and liability prevention, as it relates to design, manufacturing, warnings, and current legal-engineering aspects of products. *Mailing Add:* Dept Metall Univ Wis Madison 1505 University Ave Madison WI 53706

MOLL, ROBERT HARRY, b Lackawanna, NY, July 17, 27; m 50; c 3. QUANTITATIVE GENETICS. *Educ:* Cornell Univ, BS, 51; Univ Idaho, MS, 53; NC State Col, PhD(plant breeding), 57. *Prof Exp:* From asst prof to assoc prof, 57-65, PROF GENETICS, NC STATE UNIV, 65- *Mem:* AAAS; Am Soc Agron; Genetics Soc Am; Am Soc Naturalists; Am Genetics Assoc; fel Crop Sci Soc Am. *Res:* Quantitative genetics; statistics; plant breeding; horticulture. *Mailing Add:* Dept Genetics NC State Univ Raleigh NC 27695-7614

MOLL, RUSSELL ADDISON, b Bound Brook, NJ, Aug 12, 46; m 69. LIMNOLOGY. *Educ:* Univ Vt, BA, 68; Long Island Univ, MS, 71; State Univ NY Stony Brook, PhD(biol), 74; Univ Mich, MS, 83. *Prof Exp:* Jr res assoc, Dept Biol, Brookhaven Nat Lab, 72-73; res innvestr limnol, 74-76, asst res scientist, 76-81, ASSOC RES SCIENTIST, GREAT LAKES RES DIV, 81-, ASST DIR, MICH SEA GRANT, UNIV MICH, 85- *Mem:* Am Soc Limnol & Oceanog; AAAS; Phycol Soc Am; Ecol Soc Am; Int Asn Great Lakes Res; Int Asn Theoret & Appl Limnol. *Res:* Ecology and community structure of aquatic ecosystems; phytoplankton and bacterial distribution and productivity; data analysis and biostatistics applied to ecosystems. *Mailing Add:* Great Lakes Res Div IST Bldg Univ Mich Ann Arbor MI 48109-2099

MOLL, WILLIAM FRANCIS, JR, b Jacksonville Fla, Feb 20, 31; m 60; c 1. GEOCHEMISTRY, MINERALOGY. *Educ:* Univ Fla, BS, 54; Ind Univ, Bloomington, MA, 58; Wash Univ, PhD(geol), 63. *Prof Exp:* Instr mineral, Wash Univ, 59-60; technician, Emerson Elec Mfg Co, 61-62; fel mat res, Pa State Univ, 63-65; technologist, Baroid Div, Nat Lead Co, 65-67; mineralogist, Collodial Minerals Lab, Georgia Kaolin Co, Elizabeth, 67-74, head, 74-79; mgr tech serv & com develop, Cab-O-Sil Div, Cabot Corp, 79-83; tech dir, 83-86, VPRES RES & DEVELOP, OIL-DRI CORP AM, 86- *Honors & Awards:* Clarence E Earle Award, Nat Lubricating Grease Inst, 83; Cert Appreciation, NASA, 86. *Mem:* Am Chem Soc; Mineral Soc Am; Mineral Soc Gt Brit & Ireland; Clay Minerals Soc; Soc Cosmetic Chemists. *Res:* Characterization and application of fine-particle inorganic materials, including surface chemistry and rheology; paragenesis and chemistry of layer silicates; surface modification of fine-particle materials; industrial mineralogy. *Mailing Add:* Oil-Dri Corp Am 520 N Mich Ave Chicago IL 60611

MOLLARD, JOHN D, b Regina, Sask, Jan 3, 24; m 52; c 3. AIRPHOTO INTERPRETATION, SPACE IMAGERY ANALYSIS. *Educ:* Univ Sask, BE, 45; Purdue Univ, MSCE, 47; Cornell Univ, PhD(transp eng & eng geol), 52. *Prof Exp:* Resident engr, Sask Dept Hwys, 45; grad asst res, Purdue Univ, 45-47; air surveys engr, Prairie Farm Rehab Admin, Agr Can, 47-48; res engr, Cornell Univ, 50-52; chief air surveys & engr geol div, Prairie Farm Rehab Admin, Agr Can, 52-56; PRES J D MOLLARD & ASSOC LTD, 56- *Concurrent Pos:* Tech adv, Colombo Plan Overseas, govt's Can, West Pakistan & Ceylon (Sri Lanka); vis lectr, Harvard Univ & Hawaii Univ; guest lectr, Univ Calif, Berkeley, Univ Wis-Madison & Wash State Univ, Pullman; distinguished lectr, Can Geotech Soc, 69; First Allocution R M Hardy Mem Keynote Address. *Honors & Awards:* Keefer Medal, Eng Inst Can, 48; Pioneering Contrib, Am Soc Photogram & Remote Sensing, 79; Massey Medal, Royal Can Geol Soc, 89; Thomas Roy Award Eng Geol, Can Geotech Soc, 89. *Mem:* Fel Am Soc Civil Engrs; fel Geol Soc Am; fel Geol Asn Can; fel Am Soc Photogram; Eng Inst Can; Assoc Consult Eng Can; fel Int Explorers Club; Can Soc Petrol Geol. *Res:* Applications of airphoto interpretation and remote sensing to resource exploration development and environmental management; civil engineering; geology, applied, economic and engineering; geomorphology and glaciology. *Mailing Add:* 2960 Retallack ST Regina SK S4S 1S9 Can

MOLLENAUER, LINN F, b Washington, Pa, Jan 6, 37; m 62; c 2. PHYSICS. *Educ:* Cornell Univ, BEngPhys, 59; Stanford Univ, PhD(physics), 65. *Prof Exp:* Asst prof physics, Univ Calif, Berkeley, 65-72; RES STAFF MEM, BELL LABS, 72- *Honors & Awards:* R W Wood Prize, Optical Soc Am, 82. *Mem:* Am Phys Soc; Inst Elec & Electronics Engrs; fel Optical Soc Am; fel AAAS. *Res:* Optical spectroscopy of solids; lasers; solitons in optical fibers; infrared tunable lasers; color centers. *Mailing Add:* 11 Carriage Hill Rd Colts Neck NJ 07722

MOLLER, KARLIND THEODORE, b Chisago City, Minn, May 25, 42; m 65; c 1. SPEECH PATHOLOGY. *Educ:* Univ Minn, Minneapolis, BS, 64, MA, 67, PhD(speech path), 70. *Prof Exp:* Nat Inst Dent Res spec res fel, 70-72, ASSOC PROF SPEECH PATH, SCH DENT UNIV MINN, MINNEAPOLIS, 70-, DIR CLEFT PALATE MAXILLOFACIAL CLIN, 77- *Mem:* Am Speech & Hearing Asn; Am Cleft Palate Asn. *Res:* Oral physiology; speech production; speech in persons with orofacial anomalies; modification of speech and oral structures. *Mailing Add:* Dept Commun Disorders 115 Shevlin Univ Minn 164 Pillsbury Dr SE Minneapolis MN 55455-0209

MOLLER, PETER, b Hamburg, Ger, Nov 19, 41; m 67; c 1. ETHOLOGY. *Educ:* Free Univ Berlin, dipl biol, 64, PhD(zool), 67. *Prof Exp:* From asst prof to assoc prof, 70-81, PROF PSYCHOL, HUNTER COL, 82-; RES ASSOC, AM MUS NATURAL HIST, 72- *Concurrent Pos:* NATO res fel, Nat Ctr Sci Res, Paris, 68-69, Dr Carl Duisberg fel, 69-70; City Univ New York fac res grant, Hunter Col, 72-73, 78-90, NIMH, 80 & Nat Geog Soc, 76, 77 & 85. *Mem:* AAAS; NY Acad Sci; Ger Zool Soc; Sigma Xi; Soc Neurosci; Int Soc Neuroethology. *Res:* Sensory and behavioral physiology; ethology and ecology of electric fish; arthropod orientation. *Mailing Add:* 42-23 215 Pl Bayside NY 11361

MOLLER, PETER C, b Nov 26, 38; c 2. CANCER RESEARCH. *Educ:* Rice Univ, PhD(biol), 71. *Prof Exp:* ASST PROF CELL BIOL, UNIV TEX BR, 75- *Mem:* Am Soc Cell Biol; Am Thoracic Soc. *Res:* Structure and function of the respiratory tract. *Mailing Add:* Dept Human Biol Chem & Genetics CF-43 Univ Tex Med Br Galveston TX 77550

MOLLER, RAYMOND WILLIAM, b Brooklyn, NY, Jan 28, 20; m 48; c 8. MATHEMATICS. *Educ:* Manhattan Col, BS, 41; Cath Univ, PhD(math), 51. *Prof Exp:* Instr math, Trinity Col, 43-45; from instr to asst prof, 46-57, head dept, 57-70, ASSOC PROF MATH, CATH UNIV AM, 57- *Mem:* Am Math Soc; Math Asn Am; Sigma Xi. *Res:* Congruences; primitive roots; cyclotomic polynomials. *Mailing Add:* 1609 Michigan Ave NW Washington DC 20017

MOLLERE, PHILLIP DAVID, b New Orleans, La, Dec 18, 44; m 68; c 3. SOLVENT EXTRACTION, HYDROMETALLURGY. *Educ:* Washington & Lee Univ, BS, 66; La State Univ, PhD(chem), 71. *Prof Exp:* Res assoc chem, Inst Inorg Chem II, Goethe Univ, Frankfurt, 71-72 & Cornell Univ, 72-73; vis asst prof, Univ Mo, Kansas City, 73-74; res assoc, La State Univ, 74-75; res chemist prod develop, 75-78, supt chem res & develop, 79-82, mgr chem process develop, 82-85, ASST DIR, FREEPORT MINERALS RES & DEVELOP, 85- *Mem:* Am Chem Soc. *Res:* Chemical process development related to the recovery and/or grade improvement of mineral values (uranium, phosphate, gold, nickel, cobalt, sulfur, silver), from natural resources. *Mailing Add:* 1185 Robert E Lee Blvd New Orleans LA 70124-4332

MOLLES, MANUEL CARL, JR, b Gustine, Calif, July 30, 48; m 77; c 2. STREAM ECOLOGY, REEF FISH ECOLOGY. *Educ:* Humboldt State Univ, BS, 71; Univ Ariz, PhD(zool), 76. *Prof Exp:* ASSOC PROF ECOL & CUR FISHES, UNIV NMEX, 84- *Mem:* N Am Benthological Soc; Ecol Soc Am; Soc Conserv Biol; AAAS. *Res:* Riparian Ecology; stream ecology. *Mailing Add:* Dept Biol Univ NMex Albuquerque NM 87131

MOLLICA, JOSEPH ANTHONY, b Providence, RI, Oct 24, 40; m 64; c 3. PHARMACEUTICAL CHEMISTRY, ANALYTICAL CHEMISTRY. *Educ:* Univ RI, BS, 62; Univ Wis, MS, 65, PhD, 66. *Prof Exp:* Sr chemist, Ciba Pharmaceut Co, 66-68, mgr phys chem, 68-70, asst dir anal res & develop, 70-75, dir, 75-78, sr dir anal res & develop & advan drug delivery systs, 78-80, EXEC DIR RES & DEVELOP, CIBA-GEIGY CORP, 80- *Mem:* NY Acad Sci; Am Chem Soc; Am Pharmaceut Asn; Acad Pharmaceut Sci. *Res:* Development of analytical methods for pharmaceuticals; investigation of rates and mechanisms of organic reactions; pharmaceutical product development. *Mailing Add:* Dupont Co Barley Mill Plaza P252232 Wilmington DE 19880

MOLLIN, RICHARD ANTHONY, b Kingston, Ont, Can, Dec 12, 47. ALGEBRA. *Educ:* Univ Western Ont, BA, 71, MA, 72; Queen's Univ, PhD(math), 75. *Prof Exp:* Res assoc comput sci, Concordia Univ, 75-76; asst prof math, Univ Victoria, 76-77, Univ Toronto, 77-78, McMaster Univ, 78-79 & Univ Lethbridge, 79-80; NSERC res fel, Queen's Univ, 81-82; from asst prof to assoc prof, 82-86, PROF MATH, UNIV CALGARY, 86- *Concurrent Pos:* Reviewer, Am Math Soc, 84-; asst ed, Fibonacci Quart, 84- *Honors & Awards:* NATO ASI Award. *Mem:* Am Math Soc; Can Math Soc. *Res:* Use of tools of algebraic number theory to investigate various open questions, in algebra, number theory and computing. *Mailing Add:* Math Dept Univ Calgary Calgary AB T2N 1N4 Can

MOLLO-CHRISTENSEN, ERIK LEONARD, b Bergen, Norway, Jan 10, 23; nat US; m 48; c 3. FLUID DYNAMICS, OCEANOGRAPHY. *Educ:* Mass Inst Technol, SB, 48, SM, 49, ScD, 54. *Prof Exp:* Fel, Norweg Defense Res Estab, 48-49, sci officer, 49-51; res assoc, Mass Inst Technol, 54-55, asst prof aeronaut eng, 54-57; sr res fel, Calif Inst Technol, 57-58; from assoc prof to prof aeronaut, Mass Inst Technol, 58-64, prof meteorol, 64-76, prof oceanog, 76-; AT NASA GODDARD SPACE FLIGHT CTR. *Honors & Awards:* Von Karman Award, Am Inst Aeronaut & Astronaut, 70. *Mem:* Am Inst Aeronaut & Astronaut; Am Meteorol Soc; fel Am Phys Soc; Am Acad Arts & Sci; Am Geophys Union. *Res:* Fluid mechanics. *Mailing Add:* NASA Goddard Space Flight Ctr Code 970 Greenbelt MD 20771

MOLLOW, BENJAMIN R, b Trenton, NJ, Nov 26, 38; div. THEORETICAL PHYSICS. *Educ:* Cornell Univ, AB, 60; Harvard Univ, PhD(physics), 66. *Prof Exp:* NSF fel physics, Brandeis Univ, 66-68, asst prof, 68-69; from asst prof to assoc prof, 69-83, PROF PHYSICS, UNIV MASS, BOSTON, 83- *Res:* Quantum optics; light scattering; parametric processes. *Mailing Add:* Dept of Physics Univ of Mass Harbor Camus Boston MA 02125

MOLLOY, ANDREW A, b New York, NY, Mar 19, 30; m 66; c 4. ENVIRONMENTAL CHEMICAL ANALYSIS. *Educ:* Marist Col, BA, 51; Cath Univ, PhD(chem), 61. *Prof Exp:* Teacher parochial schs, 51-56; asst prof chem & chmn dept, Marist Col, 60-66; dir career serv, Elmira Col, 74-77, assoc prof chem, 66-80, dean grad & advan studies, 77-80; vpres & prof chem, 80-85, PROF CHEM, MARIST COL, 85- *Mem:* Sigma Xi; Am Chem Soc; AAAS. *Res:* Pesticide analysis in water & soil. *Mailing Add:* 46 Spacken Kill Rd Poughkeepsie NY 12603

MOLLOY, CHARLES THOMAS, b New York, NY, Nov 22, 14; m 36; c 2. PHYSICS. *Educ:* Cooper Union, BS, 35; NY Univ, MS, 38, PhD(physics), 48. *Prof Exp:* Res chemist, Muralo Co, NY, 35-38; res engr, Johns Manville Co, NJ, 38-42; physicist, Brooklyn Navy Yard, 42-45 & Bell Tel Labs, Inc, 45-50; head, Physics Res Dept, Labs Div, Vitro Corp Am, 50-54; staff engr sound & vibration, Lockheed Aircraft Corp, 54-59; sr staff engr, 59-70, SR STAFF ENGR ENGR, WASH OPERS, TRW SYSTS GROUP, 73- *Concurrent Pos:* Instr, Long Island Univ, 42-45 & Brooklyn Col, 39-42; adj prof, Polytech Inst Brooklyn, 50-53, Univ Southern Calif, 57-70 & Univ Calif, Los Angeles, 59-; prof lectr, George Washington Univ, 73-82; prog mgr, US Environ Protection Agency, 73-82; consult, 82-85; mem, Comt Hearing & Bio-Acoust. *Mem:* Fel Acoust Soc Am; Am Math Soc; Math Asn Am; Audio Eng Soc. *Res:* Underwater acoustics; applied mathematics; propagation of sound in lined tubes; radiation theory; loud speaker design; hearing; sound absorbing materials; transients; sound and vibration problems associated with aircraft and missiles. *Mailing Add:* 2400 Claremont Dr Falls Church VA 22043

MOLLOY, MARILYN, b Caney, Kans, Apr 24, 31. MATHEMATICS. *Educ:* Our Lady of the Lake Col, BS, 55; Univ Tex, MA, 62, PhD(math) 66. *Prof Exp:* From instr to asst prof, 59-70, ASSOC PROF MATH, OUR LADY OF THE LAKE COL, 70-, ASSOC ACAD DEAN, 72-, PLANNING OFFICER, 77- *Mem:* Math Asn Am; Am Math Soc. *Res:* Mathematical analysis; Stieltjes integral. *Mailing Add:* Dept Math Our Lady of the Lake Col 411 SW 24 St San Antonio TX 78207-4666

MOLMUD, PAUL, b Brooklyn, NY, May 29, 23; m 57; c 2. PHYSICS. *Educ:* Brooklyn Col, AB, 43; Ohio State Univ, PhD(physics), 51. *Prof Exp:* Jr engr, Nat Union Radio Corp, NJ, 43-45; asst prof physics, Clarkson Col Technol, 51-53; sr res scientist, TRW Systs Inc, 54-68, mgr, Theoret Phyiscs Dept, 68-74, sr res scientist, 74-86; CONSULT, 86- *Concurrent Pos:* Consult, 86- *Mem:* Sr mem Inst Elec & Electronics Eng; Am Phys Soc. *Res:* Gaseous electronics; ionospheric disturbances; plasma physics; electromagnetic properties of rocket exhausts; electromagnetic fields from nuclear explosions; rarefied gas dynamics. *Mailing Add:* 324 McCarty Dr Beverly Hills CA 90212

MOLNAR, CHARLES EDWIN, b Newark, NJ, Mar 14, 35; m 57; c 2. NEUROPHYSIOLOGY, COMPUTER SCIENCE. *Educ:* Rutgers Univ, BS, 56, MS, 57; Mass Inst Technol, ScD(elec eng), 66. *Prof Exp:* Staff assoc, Lincoln Lab, Mass Inst Technol, 57-61; res electronics engr, Air Force Cambridge Res Labs, 64-65; assoc prof physiol & biophys, 65-71, assoc prof elec eng, 67-71, assoc dir comput systs lab, 67-72, PROF PHYSIOL & ELEC ENG, WASH UNIV, 71-, DIR COMPUT SYSTS LAB, 72-, DIR INST BIOMED COMPUT, 84- *Concurrent Pos:* Vis prof, Int Brain Res Orgn Vis Sem, Univ Chile, 67; mem comput & biomath study sect, NIH, 71-74, chmn, 74-75; regent, Nat Libr Med, 80-84; consult, IBM Fed Systs, 85-; dir cert, NIH, 83. *Mem:* Sigma Xi. *Res:* Peripheral auditory system; models for neural networks; design of computer systems for biological research; parallel computer engineering. *Mailing Add:* 471 Toft Lane Webster Groves MO 63119

MOLNAR, GEORGE D, b Szekesfehervar, Hungary, July 30, 22; Can citizen; m 47; c 2. ENDOCRINOLOGY, INTERNAL MEDICINE. *Educ:* Univ Alta, BSc, 49, MD, 51; Univ Minn, PhD(med), 56; Am Bd Internal Med, dipl, 59. *Prof Exp:* From instr to prof, Mayo Grad Sch Med, 56-73, prof med, Mayo Med Sch, Univ Minn, 73-75; chmn dept, 75-86, prof, 75-90, EMER PROF MED, UNIV ALTA, 90-, CO-DIR, MUTTART DIABETES RES & TRAINING CTR, 81- *Mem:* AMA; Am Diabetes Asn; fel Royal Col Physicians & Surgeons; Endocrine Soc; fel Am Col Physicians; Can Soc Clin Invest. *Res:* Unstable diabetes; endocrine correlates of the diabetic state; applied physiologic, biochemical and clinical aspects; hepatic metabolism of insulin; pancreatic islet transplantation. *Mailing Add:* 458 Heritage Med Res Ctr CW Univ Alta Edmonton AB T6G 2S2 Can

MOLNAR, GEORGE WILLIAM, b Detroit, Mich, Feb 14, 14; m 37; c 2. PHYSIOLOGY. *Educ:* Oberlin Col, AB, 36; Yale Univ, PhD(zool), 40. *Prof Exp:* Asst biol, Yale Univ, 36-39; instr zool, Miami Univ, 40-42; asst prof, RI State Col, 42-44; instr physiol, Sch Med & Dent, Univ Rochester, 44-46; physiologist, Army Med Res Lab, Ft Knox, Ky, 46-62; physiologist, Res Lab, Kerrville, Tex, 62-66, coord prof serv, Southern Res Support Ctr, Little Rock, 66-68, res physiologist, Vet Admin Med Ctr, N Little Rock Div, 68-80; RETIRED. *Mem:* Am Physiol Soc; NY Acad Sci. *Res:* Thermal physiology of man. *Mailing Add:* 10801 Bainbridge Dr Little Rock AR 72212

MOLNAR, IMRE, b Budapest, Hungary, Oct 25, 06; nat US;. TELECOMMUNICATIONS ENGINEERING. *Educ:* Univ Berlin, Dipl, 30; Northwestern Univ, PhD, 50. *Prof Exp:* Installer, Int Tel & Tel Corp, 21-24; engr, Automatic Elec Co, Gen Tel & Electronics Corp, 30-56, dir develop labs, 56-59, tech dir, Automatic Elec Int, Inc, 59-63; sr res engr, Stanford Res Inst, 63-66; CONSULT, STANDDARD OIL CO CALIF, BECHTEL, & TRW, 66- *Concurrent Pos:* Lectr, Univ Ill, San Jose State Col, Mich State Univ & Univ Santa Clara; exchange grant, Nat Acad Sci; US deleg,

Int Telecommunication Union, 60-67. *Mem:* Inst Elec & Electronics Engrs; Am Numis Soc. *Res:* Wire and radio communications; control and electronics; probability and mathematical statistics; chemistry; numismatics. *Mailing Add:* 12832 Lamaida St North Hollywood CA 91607

MOLNAR, JANOS, b Budapest, Hungary, Nov 28, 27; US citizen; m 55; c 2. BIOCHEMISTRY. *Educ:* Eotvos Lorand Univ, Budapest, dipl, 53; Northwestern Univ, PhD(biochem), 60. *Prof Exp:* Instr chem, Med Univ Budapest, 53-56; res asst biochem, Northwestern Univ, 60-62; res asst, 62-65, from asst prof to assoc prof, 65-74, PROF BIOCHEM, UNIV ILL COL MED, 74- *Mem:* AAAS; Am Soc Biol Chemists; Reticuloendothelial Soc; Am Chem Soc. *Res:* Metabolism of glycoproteins; mechanism of photocytosis; cell membranes topography and surface antigens as related to cancer cells. *Mailing Add:* Dept Biochem Univ Ill Col Med 1853 W Polk St Chicago IL 60612

MOLNAR, MICHAEL ROBERT, b Passaic, NJ, Sept 15, 45; m 76. ASTROPHYSICS, SPACE PHYSICS. *Educ:* Bucknell Univ, BS, 67; Univ Wis-Madison, PhD(astron), 70. *Prof Exp:* Univ Wis proj assoc satellite opers, OAO Prog, Goddard Space Flight Ctr, 70-71; res assoc satellite data anal, Lab Atmospheric & Space Physics, Univ Colo, 71-73; asst prof astron, Univ Toledo, 73-77, assoc prof, 77-70; mem staff, Hycel, Inc, 80-; AT AT&T BELL LABS. *Concurrent Pos:* Guest investr & researcher grants, NASA, 73- *Mem:* Am Astron Soc; Int Astron Union. *Res:* Ultraviolet satellite observations of magnetic and chemically peculiar stars; development of astronomical instrumentation, especially echelle spectrographs, image tubes and solid state devices with computer interfacing and software development. *Mailing Add:* Rm 3C-618 AT&T Bell Labs Crawfords Corner Rd Holmdel NJ 07733

MOLNAR, PETER HALE, b Pittsburgh, Pa, Aug 25, 43; m; c 1. GEOPHYSICS. *Educ:* Oberlin Col, AB, 65; Columbia Univ, PhD(geol & seismol), 70. *Prof Exp:* Res scientist seismol, Lamont-Doherty Geol Observ, Columbia Univ, 70-71; asst res scientist, Scripps Inst Oceanog, Univ Calif, San Diego, 71-73; exchange scientist, Acad Sci, USSR, 73; from asst prof to prof earth sci, 74-76, SR RES ASSOC, MASS INST TECHNOL, 86- *Concurrent Pos:* Res assoc, Univ Grenoble, France, 86-87; fel Royal Soc, Oxford Univ, 88-89. *Mem:* Am Geophys Union; Seismol Soc Am Geol; Royal Astron Soc. *Res:* Geophysics; large scale continental tectonics. *Mailing Add:* Dept Earth & Planetary Sci Mass Inst Technol 54-712 Cambridge MA 02139

MOLNAR, STEPHEN P, b Toledo, Ohio, July 8, 35; div; c 2. ORGANIC CHEMISTRY, INORGANIC CHEMISTRY. *Educ:* Univ Toledo, BS, 57; Purdue Univ, MS, 63; Univ Cincinnati, PhD(chem), 67. *Prof Exp:* Res chemist, Owens Ill Glass Co, 59-60; teaching asst chem, Purdue Univ, 60-63; res fel, Univ Cincinnati, 63-65, teaching asst, 65-67; asst prof, Miami Univ, 67-75; res chemist, Armco Inc, 75-78, sr res chemist, 78-83; mgr chem serv, Nuclear Consult Serv Inc, Columbus, Ohio, 87-88; SR TECH STAFF, EDISON WELDING INST, COLUMBUS, OHIO, 88- *Concurrent Pos:* Sr res assoc & mgr, adhesives & sealants lab, Franklin Int, Columbus, Ohio, 84-85; consult, 86-87. *Mem:* AAAS; Am Chem Soc; Sigma Xi. *Res:* Chemical kinetics; complexes of Group VIII elements; electronic and chemical spectroscopy; polymer chemistry; thermal analysis of polymers; chemistry and physics of absorbents; mathematical modeling; computer applications in chemistry; materials joining technology. *Mailing Add:* 1921 Hillside Dr Columbus OH 43221

MOLNAR, ZELMA VILLANYI, b Komotau, Czech, Jan 29, 31; US citizen; div; c 2. PATHOLOGY. *Educ:* Med Univ Budapest, MD, 56; Univ Chicago, PhD(path), 73. *Prof Exp:* Res asst, Dept Surg, Univ Chicago, 57-58 & Dept Anat, 58, from res asst to res assoc, Depts Anat & Physiol, 58-61, resident trainee, Dept Path, 61-65, from instr to asst prof path, 65-73; assoc prof, 73-79, PROF PATH, LOYOLA UNIV, 79- *Concurrent Pos:* Lectr, Cook County Grad Sch Med, 74- *Honors & Awards:* Hektoen Award, Chicago Path Soc, 65. *Mem:* AAAS; Col Am Path; Sigma Xi; Am Soc Hemat. *Mailing Add:* Vet Admin Hosp Box 1216 Hines IL 60141

MOLNIA, BRUCE FRANKLIN, b Bronx, NY, Oct 17, 45; m 78; c 4. MARINE GEOLOGY. *Educ:* State Univ NY Binghamton, BA, 67; Duke Univ, MA, 69; Univ SC, PhD(geol), 72. *Prof Exp:* Geophys res asst, Lamont Geol Observ, 65-67; res fel geol, Duke Univ, 67-69 & Univ SC, 70-72; teaching fel, Cornell Univ, 69-70; asst prof, Amherst Col & Mt Holyoke Col, 72-73; chief geologist & vpres, Marine Environ Sci Assoc, 82-83; sr prof officer, Polar Res Bd, Nat Acad Sci, Nat Res Coun, 85-88; geol oceanogr, US Bur Land Mgt, 73-74; marine geologist, US Geol Surv, 74-82, supvry phys scientist, 83-87, CHIEF INT POLAR PROG, OFF INT GEOL, US GEOL SURV, US DEPT INTERIOR, 87- *Concurrent Pos:* Sci ed, SC Educ TV Network, 71-73; consult, Environ Defense Fund, 73; res fac, Juneau Icefield Res Proj, 78-; adj prof, Calif State Univ, Northridge, 83- *Mem:* Fel Geol Soc Am; Am Geophys Union; Soc Econ Paleontologists & Mineralogists; Am Asn Petrol Geologists; Sigma Xi; Antarctican Soc. *Res:* Geology and geophysics of the marine environment; glacial-marine sedimentation; glacial, arctic and antarctic processes; coastal systems; remote sensing for geological purposes. *Mailing Add:* US Geol Surv 917 Nat Ctr Reston VA 22092

MOLOF, ALAN H(ARTLEY), b Vineland, NJ, Dec 18, 28; m 52; c 4. ENVIRONMENTAL & SANITARY ENGINEERING. *Educ:* Bucknell Univ, BS, 49; Univ Mich, MSE, 51, MSE, 53, PhD(sanit & civil eng), 61. *Prof Exp:* Res & develop engr, Dorr-Oliver, Inc, Conn, 57-58; div sanit engr, Lederle Labs Div, Am Cyanamid Co, NY, 58-62; assoc prof civil eng, NY Univ, 62-73; ASSOC PROF, CIVIL/ENVIRON ENG, POLYTECH UNIV, 73- *Mem:* Water Pollution Control Fedn; Am Water Works Asn; Am Soc Civil Engrs; Am Chem Soc; Int Asn Water Pollution Res & Control; Asn Environ Eng Professors. *Res:* Wastewater treatment; stream and estuarine pollution; effluent and river monitoring; water treatment; industrial waste treatment; hazardous waste. *Mailing Add:* 32 London Terr New City NY 10956-4035

MOLONEY, JOHN BROMLEY, b Lowell, Mass, Jan 18, 24; m 49; c 2. VIRAL ONCOLOGY. *Educ:* Tufts Col, BS, 47; George Washington Univ, MS, 53, PhD, 59. *Prof Exp:* Biologist, Nat Cancer Inst, 47-60, supvry res biologist, 60-64, head viral leukemia sect, 64-66, assoc chief viral biol br, 66-67, chief, Viral Leukemia & Lymphoma Br, 67-70, assoc sci dir viral oncol & chmn, Spec Virus Cancer Prog, 70-80; CONSULT BIOMED RES, 80- *Mem:* AAAS; Am Asn Cancer Res. *Res:* Biological and biochemical properties of tumor viruses, especially the leukemia agents, a murine sarcoma virus and the Rous virus. *Mailing Add:* 6814 Greyswood Rd Bethesda MD 20817

MOLONEY, MICHAEL J, b Albany, NY, Nov 16, 36; m 64; c 3. MICROWAVE PHYSICS. *Educ:* Ill Inst Technol, BS, 58; Univ Md, PhD(physics), 66. *Prof Exp:* Asst prof physics, Rose-Hulman Inst Technol, 66-68 & Lafayette Col, 68-70; assoc prof physics, 70-76, chmn dept, 70-72, PROF PHYSICS, ROSE-HULMAN INST TECHNOL, 76- *Concurrent Pos:* NSF res partic, Univ Md, 68; electronics engr, Naval Weapons Support Ctr, Crane, Ind, 72- *Mem:* Am Asn Physics Teachers. *Res:* Computer assisted analysis of microwave spectra involving quadropole or centrifugal distortion effects. *Mailing Add:* Dept Physics Rose-Hulman Inst Technol Terre Haute IN 47803

MOLONEY, THOMAS W, MEDICAL ADMINISTRATION. *Prof Exp:* SR VPRES, COMMONWEALTH FUND. *Mem:* Inst Med-Nat Acad Sci. *Mailing Add:* Commonwealth Fund One E 75th St New York NY 10021

MOLONEY, WILLIAM CURRY, b Boston, Mass, Dec 19, 07; c 4. HEMATOLOGY. *Educ:* Tufts Col, MD, 32. *Hon Degrees:* DSc, Col of the Holy Cross, 61. *Prof Exp:* From asst to clin prof med, Med Sch, Tufts Univ, 34-67; clin prof, 67-71, prof, 71-74, EMER PROF MED, HARVARD MED SCH, 74-; EMER PHYSICIAN & CHIEF HEMAT DIV, PETER BENT BRIGHAM HOSP, 67- *Concurrent Pos:* Consult, Boston Hosps, 38-; consult, Boston City Hosp, 48-, dir clin labs; dir res, Atomic Bomb Casualty Comn, Hiroshima, Japan, 52-54. *Mem:* AAAS; fel Am Fedn Clin Res; fel AMA; fel Am Col Physicians; Asn Am Physicians. *Res:* Leukemia. *Mailing Add:* Brigham & Women's Hosp 75 Francis St Boston MA 02115

MOLOTSKY, HYMAN MAX, b Russia, Nov 1, 19; nat US; m 52; c 3. CARBOHYDRATE CHEMISTRY. *Educ:* Univ Man, BSc, 43; Univ Mo, AM, 49, PhD(chem), 53. *Prof Exp:* Asst chem, McGill Univ, 45-46 & Univ Mo, 47-52; sr res chemist, Velsicol Corp, 52-55, proj leader, 55-57; res chemist, Richardson Co, 57-58, proj leader, 58-59; res chemist, Corn Prod Co, 59-64, sect leader basic & appl res, Moffett Tech Ctr, 64-66, patent liaison & tech coordr, Moffett Res, 66-74, asst to exec dir res & develop, Indust Div, 74-78, ASST TO VPRES RES & DEVELOP, CORN PRODS, MOFFETT TECH CTR, CPC INT INC, 78- *Mem:* Am Chem Soc; Sigma Xi; AAAS. *Res:* Derivatives of sugars and starch for industrial applications; intermediates for urethane foams; synthetic organic intermediates; intermediates for resins and plasticisers. *Mailing Add:* 2735 W Birchwood Ave Chicago IL 60645

MOLT, JAMES TEUNIS, b Plainfield, NJ, Mar 31, 47; m 75; c 2. NEUROSCIENCES, NEUROPHYSIOLOGY. *Educ:* Colgate Univ, AB, 69; Cornell Univ, PhD(physiol), 74. *Prof Exp:* Instr physiol, Albany Med Col, 74-76, asst prof, 76-80; asst prof physiol, Hahnemann Med Col, Philadelphia, 80-83; DIR REGULATORY AFFAIRS, MERCK, SHARP & DOHME RES LABS, 90- *Concurrent Pos:* Fel, Albany Med Col, 74-76; Nat Inst Neurol & Commun Dis & Stroke grant, 76-79. *Mem:* Soc Neurosci; AAAS; Am Physiol Soc. *Res:* Sensory processing in the mammalian nervous system; experimental spinal cord trauma. *Mailing Add:* Merck Sharp & Dohme Res Labs Bldg 30 Sumay Town Pike West Point PA 19486

MOLTENI, AGOSTINO, b Como, Italy, Nov 12, 33; US citizen; c 2. EXPERIMENTAL PATHOLOGY. *Educ:* Univ Milan, MD, 57; Ital Bd Internal Med, cert, 63; State Univ NY Buffalo, PhD(path), 70. *Prof Exp:* Asst prof med, Univ Milano, 58-62; sr investr med res, Farmitalia, Milan, Italy, 63-65; Henry C & Bertha Buswell fel & res asst prof path, State Univ NY Buffalo, 70-72; assoc prof path, Univ Kans Med Ctr, Kansas City, 73-76; PROF PATH, NORTHWESTERN UNIV SCH MED, CHICAGO, 76- *Concurrent Pos:* NIH res career develop award, 72; consult path, Vet Admin Hosp, Kansas City, 73-; dir, Endocrine Pathol Labs, Northwestern Mem Hosp, Chicago, 76-; vis prof pathol, Harvard Med Sch, 83-84. *Honors & Awards:* Sharer, Albert E Lasker Award, 80. *Mem:* Am Asn Clin Chem; Am Soc Exp Path; Endocrine Soc; Am Soc Exp Biol & Med; Nat Acad Clin Biochem; Int Acad Path. *Res:* Cardiovascular diseases, especially systemic and pulmonary hypertension; hormone receptor mechanisms and control. *Mailing Add:* Dept of Path Northwestern Univ Med Sch Chicago IL 60611

MOLTENI, RICHARD A(LOYSIUS), b Teaneck, NJ, Nov 11, 44; m 68; c 2. PEDIATRICS, NEONATOLOGY. *Educ:* Fairfield Univ, BS, 66; NJ Col Med, MD, 70. *Prof Exp:* Intern pediat, Denver Children's Hosp, 70-71; resident, Univ Colo Med Ctr, 71-73; major, US Army Med Corps, 73-75; chief resident, Univ Colo Med Ctr, 75-76, instr, 75-78, fel neonatology, 76-78; asst prof pediat, Johns Hopkins Hosp, 78-83, med dir, Newborn Intensive Care Unit, 79-82, asst prof obstet, 78-86, assoc prof pediat, 83-86; ASSOC PROF PEDIAT, MED CTR, UNIV UTAH, 86-; DIR NEONATOLOGY, PRIMARY CHILDREN'S MED CTR, 86- *Concurrent Pos:* Chief pediat, Baltimore City Hosp, 82-84; neonatal consult, Project Hope, 84-86. *Mem:* Am Acad Pediat; Nat Perinatal Soc. *Res:* Blood flow distribution with microspheros in chronically catheterized fetal sheep; pulmonary blood flow and fetal breathing. *Mailing Add:* Div Neonatology Primary Children's Hosp 100 N Medical Dr Salt Lake City UT 84103

MOLTER, LYNNE ANN, b Pittsburgh, Pa, 1957. ELECTRICAL ENGINEERING. *Educ:* Swarthmore Col, BS, 79, BA, 79; Mass Inst Technol, SM, 83, ScD, 87. *Prof Exp:* Prod engr, Hewlett Packard, 79-81; res asst elec engr, Mass Inst Technol, 81-87; ASST PROF ENG, SWARTHMORE COL, 87- *Concurrent Pos:* Consult, US Army Electronic Devices & Technol, 87-; reviewer, NSF, 87- & Inst Elec & Electronics Engrs

J Quantum Electronics, 87-; prin investr, NSF, 89- *Mem:* Inst Elec & Electronics Engrs; Optical Soc Am; Sigma Xi. *Res:* Integrated optical waveguiding for signal processing applications; optics; photonics; optoelectronics; quantum electronics; author of several publications. *Mailing Add:* Dept Eng Swarthmore Col Swarthmore PA 19081

MOLTYANER, GRIGORY, b USSR, Apr 26, 43; Can citizen; m 67; c 3. HYDROLOGY & WATER RESOURCES. *Educ:* State Univ, Tashkent, MSc, 66; Inst Cybernet, Acad Sci, PhD(phys & math sci), 72. *Prof Exp:* Res officer contaminant hydrogeol, Res Inst Hydrogeol, 67-72; assoc prof advan math, Inst Agr Eng, 75-79; res assoc contaminant hydrogeol, Univ Waterloo, Can, 80-81; RES OFFICER CONTAMINANT HYDROGEOL, ATOMIC ENERGY CAN, 81- *Concurrent Pos:* Adj prof contaminant hydrogeol, Univ Toronto, 84- *Mem:* Am Geophys Union; Int Asn Hydraul Res. *Res:* Mass transport phenomena in porous media and groundwater hydrology; experimental and numerical methods for determining fluid-medium properties; finite-element codes; large-scale field studies of environmental processes; uncertainty associated with measurements; physical mathematics. *Mailing Add:* PO Box 746 Deep River ON K0J 1P0 Can

MOLTZAN, HERBERT JOHN, b Chicago, Ill, May 26, 33; m 55, 76; c 5. PROCESS CHEMISTRY. *Educ:* Ill Inst Technol, BS, 56; St Louis Univ, MS, 60. *Prof Exp:* Res chemist, Monsanto Co, 56-61, Vulcan Mat Corp, 61-65; sr engr, 65-66, EPI Eng, 66-68, mem tech staff, 68-76, supvr plastics lab, 76-84, MGR ANAL SERV LABS, TEX INSTRUMENTS, 85- *Mem:* Am Chem Soc; Electrochem Soc. *Res:* Electronic grade silicon; circuit boards; quartz; chloromethylations; plasticizer synthesis; catalytic hydrogenations; use and characterization of polymers for electronics. *Mailing Add:* 6935 Northaven Dallas TX 75230-3502

MOLVIK, ARTHUR WARREN, b L'Anse, Mich, Mar 19, 43; m 68; c 3. PHYSICS. *Educ:* Concordia Col, Moorhead, Minn, BA, 64; Univ Wis-Madison, MA, 66, PhD(physics), 71. *Prof Exp:* PHYSICIST, LAWRENCE LIVERMORE NAT LAB, 72- *Mem:* Am Phys Soc. *Res:* Magnetic confinement of thermonuclear plasmas; neutral beam and radio frequency heating of plasmas; plasma stability; compact torus acceleration. *Mailing Add:* Lawrence Livermore Nat Lab L-637 PO Box 5511 Livermore CA 94550

MOLYNEUX, JOHN ECOB, b Philadelphia, Pa, May 19, 35; m 62; c 2. APPLIED MATHEMATICS, MECHANICAL ENGINEERING. *Educ:* Univ Pa, BS, 57, MS, 61, PhD(mech eng), 64. *Prof Exp:* Res scientist, Courant Inst Math Sci, NY Univ, 64-65; from asst prof to prof mech & aerospace sci, Univ Rochester, 65-84; PROF MECH ENG, WIDENER UNIV, 84- *Concurrent Pos:* Vis prof, Tel Aviv Univ, 72-73; res sci, Oak Ridge Nat Lab, 79-80. *Mem:* Soc Indust & Appl Math; Sigma Xi; Am Soc Mech Engrs; Math Asn Am. *Res:* Inverse scattering, free surface problems. *Mailing Add:* Dept Mech Eng Widener Univ Chester PA 19013

MOLYNEUX, RUSSELL JOHN, b Luton, Eng, Aug 3, 38; m 74; c 2. ORGANIC CHEMISTRY. *Educ:* Univ Nottingham, BSc, 60, PhD(org chem), 63. *Prof Exp:* NSF fel org chem, Univ Ore, 63-65; asst prof forest prod chem, Ore State Univ, 65-67; res assoc, 67-74, RES CHEMIST, WESTERN REGIONAL RES CTR, AGR RES SERV, USDA, 74- *Concurrent Pos:* Consult, US Brewers Asn, 68-74. *Mem:* Am Chem Soc; Royal Soc Chem; Phytochem Soc; AAAS; Am Soc Pharmacog. *Res:* Naturally occurring quinones; naturally occurring phenolic compounds; constituents of toxic range plants; pyrrolizidine alkaloids; indolizidine alkaloids; glycosidase inhibitors. *Mailing Add:* Western Regional Res Ctr Agr Res Serv 800 Buchanan St Albany CA 94710

MOLZ, FRED JOHN, III, b Mays Landing, NJ, Aug 13, 43; m 66; c 2. HYDROLOGY, SOIL PHYSICS. *Educ:* Drexel Univ, BS, 66, MSCE, 68; Stanford Univ, PhD(hydrol), 70. *Prof Exp:* From asst prof to prof civil eng, Auburn Univ, 70-84, dir, Eng Exp Sta & asst dean res, 79-84, Feagin prof, 84-89, HUFF EMINENT SCHOLAR, AUBURN UNIV, 90- *Concurrent Pos:* Off Water Resources res grants, Auburn Univ, 71-77; consult various private & govt agencies, 72-; res grants, Fed Hwy Admin, 74-76, US Geol Surv, 75-77, Dept Energy, 77-79, Batelle Aquifer Storage, 80-83 & US Environ Agency, 83-93; vis prof, Univ Ill, 77 & US Geol Surv, 89-91. *Mem:* Am Geophys Union; Am Soc Agron; Soil Sci Soc Am; Nat Water Well Asn. *Res:* Transport process in the groundwater-soil-plant-atmosphere system including physical, mathematical, biological and engineering aspects. *Mailing Add:* Civil Eng Dept Auburn Univ Auburn AL 36849

MOMBERG, HAROLD LESLIE, b Sedalia, Mo, Mar 24, 29. ZOOLOGY, HISTOLOGY. *Educ:* Cent Mo State Col, BS, 51; Univ Mo, MA, 55, PhD(zool), 61. *Prof Exp:* Head dept biol, Hannibal-LaGrange Col, 55-57; instr zool, Univ Mo, 57-58; from assoc prof to prof biol, histol & embryol, William Jewell Col, 60-67, NSF res grant, 64-66; assoc prof biol, Cent Mo State Col, 67-69; assoc prof biol, 70-75, head dept biol & geol, 74-85, PROF BIOL, CENT METHODIST COL, 75- *Mem:* AAAS; Am Soc Zoologists; Soc Study Reproduction. *Res:* Mammalian implantation patterns and cycles; histochemistry and physiology of reproductive tract. *Mailing Add:* Dept Biol Cent Methodist Col Fayette MO 65248

MOMENT, GAIRDNER BOSTWICK, developmental biology; deceased, see previous edition for last biography

MOMMAERTS, WILFRIED, b Broechem, Belg, Mar 4, 17; nat US; m 44. PHYSIOLOGY. *Educ:* State Univ Leiden, BA, 37, MA, 39; Kolozsvar Univ, Hungary,. *Prof Exp:* Vis assoc prof biochem, Am Univ Beirut, 45-46,; res assoc biochem, Duke Univ, 48-53; assoc prof, Case Western Reserve Univ, 53-56; from prof to emer prof med & physiol, Sch Med, Univ Calif, Los Angeles, 56-87, chmn dept physiol, 66-87; RETIRED. *Concurrent Pos:* Estab investr, Am Heart Asn, 49-59; dir res lab, Los Angeles Heart Asn, 55-; conseiller exceptionel, Nat Inst Health & Med Res, France; Humboldt Award, Max Planck Inst, Heidelberg, Ger, 87-88. *Honors & Awards:* DHc, Dijon Med Sch, France, 76. *Mem:* Biophys Soc; Am Physiol Soc; Am Heart Asn; fel Am Acad Arts & Sci. *Res:* Molecular physiology and biochemistry of contractile tissues; physical chemistry of tissue proteins and cellular processes; functional influences upon differential gene expression. *Mailing Add:* Dept Physiol 53-170 Chs Univ Calif 405 Hilgard Ave Los Angeles CA 90024-1751

MOMOT, WALTER THOMAS, b Hamtramck, Mich, Oct 12, 38; m 66; c 3. FISH BIOLOGY. *Educ:* Wayne State Univ, BS, 60; Univ Mich, MS, 61, PhD(fisheries), 64. *Prof Exp:* Asst prof zool, Univ Okla, 64; from instr to assoc prof, Ohio State Univ, 64-75; PROF ZOOL, LAKEHEAD UNIV, 75- *Concurrent Pos:* Vis prof, La State Univ, 81-82, Univ Hawaii & Western Australian Marine Res Lab, 90-91. *Mem:* Int Asn Astocology; Am Fisheries Soc; Can Soc Zoologists; NAm Benthological Soc; fel Am Inst Fishery Res Biologists. *Res:* Production, population dynamics, trophic ecology of fish and crayfish populations. *Mailing Add:* Dept Biol Lakehead Univ Thunder Bay ON P7B 5E1 Can

MOMPARLER, RICHARD LEWIS, b New York, NY, Jan 6, 35; m 66. PHARMACOLOGY, BIOCHEMISTRY. *Educ:* Mich State Univ, BS, 57; Univ Vt, PhD(pharmacol), 66. *Prof Exp:* USPHS fel, Yale Univ, 64-65 & Int Lab Genetics, Italy, 66; asst prof biochem, McGill Univ, 67-74; assoc prof pharmacol, Sch Med, Univ Southern Calif, 74-77; PROF PHARMACOL, SCH MED, UNIV MONTREAL, 77- *Mem:* AAAS; Am Asn Cancer Res; Cell Kinetics Soc; Int Asn Comparitive Res Leukemia; Can Soc Clin Pharmacol. *Res:* Cancer chemotherapy; enzymes; cell cycle; DNA methylation. *Mailing Add:* Ctr Pediat Res Hosp Ste-Justine 3175 Chemin Ste-Catherine Montreal PQ H3T 1C5 Can

MONACELLA, VINCENT JOSEPH, b Erie, Pa, May 2, 26; m 61; c 4. HYDRODYNAMICS. *Educ:* Gannon Col, BS, 48. *Prof Exp:* Mathematician, David Taylor Model Basin, 53-54, physicist, 54-67, physicist, Naval Ship Res & Develop Ctr, 67-70, head, Submarine Dynamics Br, 70-71, asst res, David W Taylor Naval Ship Res & Develop Ctr, 71-74, asst res & develop, 74-81, ASST DESIGN SUPPORT & RES, 81-, DAVID TAYLOR RES CTR, BETHESDA, MD, 87- *Mem:* Soc Naval Archit & Marine Eng; Am Soc Naval Engrs. *Res:* Ship hydrodynamics; water waves; analysis of motions, forces and moments of arbitrary bodies in proximity of a fluid surface. *Mailing Add:* 2908 Hideaway Rd Fairfax VA 22031-1327

MONACK, A(LBERT) J(AMES), b Charleroi, Pa, Dec 30, 04; wid. MECHANICS. *Educ:* WVa Univ, BS, 27; Univ Ill, MS, 29; NY Univ, PhD(eng ed), 62. *Prof Exp:* Asst, Univ Ill, 29-32; consult engr, Pa, 32-35; develop engr, Western Elec Co, NY, 35-39; at Radio Corp Am, NJ, 39-42; vpres & dir eng, Mycalex Corp Am, 42-46; asst prof physics & chmn dept, Fairleigh Dickinson Col, 47-50, asst prof sci & eng, 50-54; asst prof physics, Newark Col Eng, 54-60; from assoc prof to prof, 60-70, EMER PROF APPL MECH, NJ INST TECHNOL, 70- *Concurrent Pos:* Instr, Cooper Union, 39-42; consult engr, 46- *Res:* Inorganic electrical insulations; glass-metal seals; machinable ceramic composition of high dielectric constant; engineering materials; applied mechanics; properties of materials. *Mailing Add:* 287 Mortimer Ave Rutherford NJ 07070

MONACO, ANTHONY PETER, b Philadelphia, Pa, Mar 12, 32; m 60; c 4. SURGERY, IMMUNOLOGY. *Educ:* Univ Pa, BA, 52; Harvard Med Sch, MD, 56. *Prof Exp:* From intern to resident surg, Mass Gen Hosp, 56-63; Am Cancer Soc fel, 63-67, from instr to assoc prof, 63-76, prof surg & chief transplantation div, Harvard Med Sch, 76-; AT CANCER RES INST, BOSTON, MASS. *Concurrent Pos:* Lederle fac award, 67-70; ed, Transplantation, 69-; chief transplantation unit, Harvard Surg Serv, New Eng Deaconess Hosp; pres, Interhosp Organ Bank New Eng, 73- *Mem:* Am Asn Immunol; Transplantation Soc (treas, 70, secy, 74-78, vpres, 78-). *Res:* Transplantation immunology and immunobiology; experimental and clinical transplantation. *Mailing Add:* Cancer Res Inst New Eng Deacononess Hosp 185 Pilgrim Boston MA 02115

MONACO, LAWRENCE HENRY, b Philadelphia, Pa, Mar 31, 25; m 54; c 5. ZOOLOGY. *Educ:* LaSalle Col, BA, 49; Univ Notre Dame, MS, 52, PhD(parasitol), 54. *Prof Exp:* Instr biol sci, Del Mar Col, 54-57 & Villa Madonna Col, 57-58; from instr to assoc prof, 58-61, head dept, 61-66, dean, 66-82, PROF BIOL SCI, DUTCHESS COMMUNITY COL, 61- *Concurrent Pos:* Consult, Culinary Inst Am, 82- *Mem:* Am Inst Biol Sci; Sigma Xi. *Res:* Gill parasites of fish. *Mailing Add:* Six Thorndale Ave Poughkeepsie NY 12603

MONACO, PAUL J, b Boston, Mass, Sept 6, 52; m 84. EVOLUTIONARY & POPULATION BIOLOGY. *Educ:* Merrimack Col, BA, 74; Marquette Univ, MS, 77, PhD(biol), 82. *Prof Exp:* Instr biophys, 81-84, ASST PROF CELL BIOL, COL MED, E TENN STATE UNIV, 84- *Mem:* AAAS; Am Inst Biol Sci; Am Soc Cell Biol; Histochem Soc; Sigma Xi. *Res:* Reproductive biology, with emphasis on cellular and developmental events relating to speciation and evolution; molecular biology; organization and evolution of eukaryotic DNA. *Mailing Add:* Dept Biophys Col Med E Tenn State Univ Johnson City TN 37614-0002

MONAGHAN, PATRICK HENRY, b Memphis, Tenn, July 25, 22; m 43; c 2. GEOCHEMISTRY, PETROLEUM ENGINEERING. *Educ:* La Polytech Inst, BS, 43; La State Univ, MS, 49, PhD(chem), 50. *Prof Exp:* Field engr, Sperry Gyroscope Co, 46-47; instr chem, La State Univ, 49-50; asst res engr, Prod Res Div, Humble Oil & Ref Co, 50-51, from res engr to sr res engr, 51-59, from res specialist to sr res specialist, 59-65; res assoc, Exxon Prod Res Co, 65-73, res adv, 73-81, sr res adv, 81-86; RETIRED. *Mem:* AAAS; Am Asn Petrol Geologists; Am Chem Soc; Soc Econ Paleontologists & Mineralogists; Am Inst Mining, Metall & Petrol Eng. *Res:* Oil well drilling and completion techniques; geochemistry; instrumental analysis; organic geochemistry; petroleum geology; environmental management. *Mailing Add:* 13627 Queensbury Houston TX 77079

MONAGLE, DANIEL J, b Eddystone, Pa, Oct 14, 36; m 61; c 6. CELLULOSE CHEMISTRY, WATER SOLUBLE POLYMERS. *Educ:* Mt St Mary's Col Md, BS, 58; Duquesne Univ, MS, 60; Univ Del, PhD(org polymer chem), 67. *Prof Exp:* Chemist, Hercules Res Ctr, Del, 60-65; lectr anal & gen chem, Univ Del, 65-66; res chemist, 66-70, res supvr coatings & spec prod, 70-77, mgr mat sci, 77-78, dir technol water soluble polymers, 78-85, MGR ANAL SCI DIV, HERCULES RES CTR, 85- *Mem:* Am Chem Soc; Sigma Xi. *Res:* Synthesis process development, characterization and applications of synthetic water-soluble polymers, particularly polyelectrolytes; coatings; pigments; magnetic materials; cellulose and water soluble cellulose derivatives. *Mailing Add:* 605 Halstead Rd Sharpley Wilmington DE 19803

MONAGLE, JOHN JOSEPH, JR, b Chester, Pa, Feb 2, 29; m 54; c 5. ORGANIC CHEMISTRY. *Educ:* Villanova Univ, BS, 50; Polytech Inst Brooklyn, PhD(chem), 54. *Prof Exp:* Res chemist, Sinclair Res Labs, Inc, 54-56 & Jackson Lab, E I du Pont de Nemours & Co, 56-61; from asst prof to assoc prof chem, NMex State Univ, 61-63, prof & head dept, 63-67; prof & head dept, Univ Ala, Tuscaloosa, 67-68; dean col arts & sci, 68-70, assoc dir res ctr arts & sci, 70-71, PROF CHEM, NMEX STATE UNIV, 68-, ASSOC DEAN ARTS & SCI & DIR, ARTS RES CTR, 75- *Concurrent Pos:* Consult, Melpar Corp, 64; sci & technol adv, US State Dept Agency for Int Develop, Cairo, Egypt, 84-86. *Mem:* Am Chem Soc; Nat Coun Univ Res Adminrs; Soc Res Adminrs; fel AAAS. *Res:* Synthetic and physical organic chemistry; polymer chemistry; synthesis and properties of organophosphorus compounds; organic chemistry of polymers. *Mailing Add:* HRD/USAID PSC Box 4 APO New York NY 09614

MONAHAN, ALAN RICHARD, b Schenectady, NY, June 17, 39; m 62; c 6. PHYSICAL CHEMISTRY, POLYMER CHEMISTRY. *Educ:* Rensselaer Polytech Inst, BChE, 61, PhD(phys chem), 64. *Prof Exp:* Sr chemist gaseous electronics res br, 64-66, scientist chem physics res br, 66-69, sr scientist org solid state physics res br, 69-71, prin scientist, Corp Physics Lab, 71-72, mgr org solid state physics area, 72-75, mgr-developer, Mat Technol Ctr, 75-78, MGR EXPLOR MAT AREA, XEROX CORP, 78- *Mem:* Am Chem Soc. *Res:* Thermal and photochemical conversions in polymers; photochromics; organic photoconductors; analytical chemistry; molecular spectroscopy; physical properties of dyes and pigments; research and development of of xerographic developer materials (toners and carriers); coating technology; small particle processing. *Mailing Add:* Xerox Corp 800 Phillips Rd Webster NY 14580

MONAHAN, EDWARD CHARLES, b Bayonne, NJ, July 25, 36; m 60; c 3. OCEANOGRAPHY. *Educ:* Cornell Univ, BEP, 59; Univ Tex, Austin, MA, 61; Mass Inst Technol, PhD(oceanog), 66; Nat Univ Ireland, DSc, 84. *Prof Exp:* Res asst, Woods Hole Oceanog Inst, 64-65; asst prof physics, Northern Mich Univ, 65-68; asst prof oceanog, Hobart & William Smith Cols, 68-69; from asst prof to assoc prof oceanog, Univ Mich, Ann Arbor, 69-75; dir educ & res, Sea Educ Asn, Woods Hole, 75-76; statutory lectr phys oceanog, Dept Oceanog, Univ Col, Galway, Ireland, 76-86; DIR, CONN SEA GRANT PROG, 86-; PROF MARINE SCI & ASSOC DIR, MARINE SCI INST, UNIV CONN, AVERY POINT, 86- *Concurrent Pos:* Adj assoc prof, Boston Univ, 75-76; Haltiner res chair prof, Naval Postgrad Sch, 81-82. *Mem:* Royal Meteorol Soc; Int Asn Theoret & Appl Limnol; Am Meteorol Soc; Am Geophys Union; Am Soc Limnol & Oceanog. *Res:* Air-sea interaction; marine aerosols; physical limnology; oceanographic instrumentation; design of drogues and drifters; remote sensing of oceanic whitecaps. *Mailing Add:* Marine Sci Inst Univ Conn Avery Pt Groton CT 06340

MONAHAN, EDWARD JAMES, b Bayonne, NJ, Sept 18, 31; m 70. FORENSIC CONSULTING. *Educ:* Newark Col Eng, BSCE, 58, MSCE, 61; Okla State Univ, PhD(civil eng), 68. *Prof Exp:* From instr to prof, Newark Col Eng, 58-84, head, Geotech Group, 68-84, EMER PROF CIVIL ENG, NJ INST TECHNOL, 84- *Concurrent Pos:* Consult, Raamot Assocs, 68-70, pvt, 70- *Mem:* Am Soc Civil Engrs. *Res:* Author of book dealing with the practical aspects of earthwork construction; two patents dealing with novel methods of foundation construction; authored a number of instructional manuscripts dealing with laboratory procedures in soils and various aspects of engineering report writing, excerpts of which have been published. *Mailing Add:* 85 Newark Ave Bloomfield NJ 07003

MONAHAN, HUBERT HARVEY, b Oshkosh, Wis, Jan 14, 22; m 53. METEOROLOGY, OPERATIONAL WEATHER FORECASTING. *Educ:* Wis State Col, Oshkosh, BS, 47; Univ Wis-Madison, BBA, 59. *Prof Exp:* Weather officer forecasting, US Air Force, Dept Defense, 50-57; meteorologist, Atmospheric Sci Off, US Army Electronics Command, 59-61; staff weather officer, US Air Force, Dept Defense, 62-65; supvr meteorologist, US Weather Bur, Nat Oceanic & Atmospheric Admin, 66-67; meteorologist res, Atmospheric Sci Lab, US Army Electronics Command, 67-77; meteorologist climat, Atmospheric Sci Lab, US Army Electronics Res & Develop Command, NMex, 77-80; RETIRED. *Res:* Mesoscale time and space variability of low-level meteorological phenomena and short-term weather predictions for application of effects to Army tactical weapons systems. *Mailing Add:* 4723 Larkspur Ct El Paso TX 79924-3021

MONAHAN, JAMES EMMETT, b Kansas City, Mo, Jan 10, 25; m 48. THEORETICAL PHYSICS. *Educ:* Rockhurst Col, BS, 48; St Louis Univ, MS, 50, PhD(physics, math), 53. *Prof Exp:* SR PHYSICIST, ARGONNE NAT LAB, 51- *Concurrent Pos:* Weizmann fel, 66-67; prof lectr, St Louis Univ, 66-; vis scientist, Univ Ohio, 71- *Mem:* Am Phys Soc. *Res:* Theoretical nuclear physics. *Mailing Add:* 701 South Dr Hinsdale IL 60521

MONALDO, FRANCIS MICHAEL, b Wash, DC, Jan 4, 56; m 77; c 3. OCEANOGRAPHY, GENERAL PHYSICS. *Educ:* Catholic Univ Am, BA, 77, MS, 78. *Prof Exp:* SR PHYSICIST, APPL PHYSICS LAB, JOHN HOPKINS UNIV, 77- *Mem:* Am Geophys Union. *Res:* Full integration of remotely sensed spacecraft data into the global geo-physical data base; sensor design and configuration to the assimilation of this data base into global forecast models. *Mailing Add:* Appl Phys Lab John Hopkins Univ John Hopkins Rd Laurel MD 20707

MONAN, GERALD E, b Klamath Falls, Ore, Sept 14, 33; m 53; c 2. ELECTRONIC SURVEILLANCE TECHNIQUES ELECTRONIC TAGS FISH PASSAGE DAMS. *Educ:* Univ Wash, BA, 55. *Prof Exp:* Res biologist, 58-65, spec planning biologist, 65-69, subtask mgr, 69-76, dep dir, 76-88, DIR FISHERIES RES, COASTAL ZONE & ESTUARINE STUDIES DIV, NAT MARINE FISHERIES SERV, 88- *Concurrent Pos:* Second lieutenant to Colonel, US Air Force & US Air Force Reserves, 55-86. *Mem:* Am Inst Fishery Res Biologists; Am Fisheries Soc. *Res:* Behavior of anadromous salmonids during migrations as applied to fish passage problems in the Columbia River Basin. *Mailing Add:* 2725 Montlake Blvd E Seattle WA 98112

MONARD, JOYCE ANNE, b Bethlehem, Pa, Nov 5, 46; c 2. SALES SUPPORT, NEW PRODUCT DEVELOPMENT. *Educ:* Bryn Mawr Col, BA, 68; Univ Tenn, Knoxville, PhD(physics), 72. *Prof Exp:* Postdoctoral res fel, Lawrence Berkeley Lab, Univ Calif, 72-75; tech leader, Gen Elec, 75-78, sr prog mgr, 78-83, new prod mgr, 83-87; new prod mgr, 88-89, GROUP MGR, SUN MICROSYSTS, 90- *Mailing Add:* 1757 Erinbrook Pl San Jose CA 95131

MONATH, THOMAS P, b Hewlett Harbor, NY, Aug 13, 40; m 64, 88; c 2. ARBOVIROLOGY. *Educ:* Harvard Univ, AB, 62, MD, 66. *Prof Exp:* Intern med, Peter Bent Brigham Hosp, 66-67, resident, 67-68 & 73-74; res fel virol, Univ Ibadan, Nigeria, 70-72; med officer, Ctrs Dis Control, 68-70, chief arbovirus sect, 72-73, dir, Div Vector-Borne Viral Dis, 74-88; CHIEF, VIROL DIV, US ARMY MED RES INST INFECTIOUS DIS, 88- *Concurrent Pos:* Consult, WHO, 75-; chmn, Am Comt Arthropod-Borne Viruses, 77-80 & adv comt viral & rickettsial dis, US Army, 85-; assoc ed, Am J Trop Med Hyg, J Virol Meth, & Acta Trop, 78-; vis res fel, Harvard Med Sch, 84-85. *Honors & Awards:* Nat Young Mem Award Arborirology. *Mem:* Am Soc Virol; Am Soc Trop Med & Hyg; Infectious Dis Soc; Royal Soc Trop Med & Hyg. *Res:* Clinical features, pathophysiology, diagnosis, epidemiology and ecology of arthropod-borne viruses; yellow fever; mosquito-borne viral encephalitides. *Mailing Add:* US Army Med Res Inst Infectious Dis Ft Detrick Frederick MD 21702-5011

MONCE, MICHAEL NOLEN, b Honolulu, Hawaii, Apr 13, 52. ION-MOLECULE & ION ATOM COLLISIONS. *Educ:* Univ Colo, BA, 74; Colo State Univ, MS, 76; Univ Ga, PhD(physics), 81. *Prof Exp:* ASSOC PROF PHYSICS, CONN COL, 81- *Mem:* Am Phys Soc; Am Asn Physics Teachers. *Res:* Accelerator-based atomic and molecular physics; photon emission from ion-molecule collisions. *Mailing Add:* Dept Physics & Astron Conn Col New London CT 06320

MONCHAMP, ROCH ROBERT, b Manchester, NH, Sept 27, 31; m 69; c 3. SOLID STATE CHEMISTRY, CRYSTAL GROWTH. *Educ:* St Anselms Col, AB, 53; Mass Inst Technol, PhD(chem), 59. *Prof Exp:* Sr res chemist, Res Lab, Merck & Co, 59-63; mgr crystal growth res & develop, Airtron Div, Litton Precision Prod, Inc, NJ, 64-67 & Raytheon Co, 67-73; chief engr, Saphikon Div, Tyco Labs, Inc, 73-75; mgr crystal growth dept, Adolf Meller Co, 75-82; PROJ MGR, EG&G/ENERGY MEASUREMENTS, 87- *Concurrent Pos:* Mgr, Mat Res & Develop, New Eng Res Ctr, 82-87. *Mem:* AAAS; Am Asn Crystal Growth; fel Am Inst Chemists; Sigma Xi. *Res:* Semiconductors; lasers and acoustic crystals; crystal growth; hydrothermal synthesis. *Mailing Add:* 124 San Milano Goleta CA 93117

MONCHICK, LOUIS, b Brooklyn, NY, Dec 27, 27; m 66. CHEMICAL PHYSICS. *Educ:* Boston Univ, AB, 48, MA, 51, PhD(chem), 54. *Prof Exp:* Cloud physicist, Air Force Cambridge Res Ctr, 53-54; fel radiation chem, Univ Notre Dame, 54-56; res assoc fused salts, Knolls Atomic Power Lab, Gen Elec Co, 56-57; CHEMIST, APPL PHYSICS LAB, JOHNS HOPKINS UNIV, 57-, LECTR DEPT CHEM ENG, 80- *Concurrent Pos:* Assoc prof chem, Johns Hopkins Univ, 68-69, vis prof chem, 75-76; vis prof chem eng, Univ Bielefeld, 82-83; vis scientist, Univ Leiden, 79, vis scientist, 80. *Mem:* Am Chem Soc; Am Phys Soc. *Res:* Diffusion controlled reactions; kinetic theory of gases; intermolecular forces; molecular collisions. *Mailing Add:* Johns Hopkins Univ Appl Phys Lab Johns Hopkins Rd Laurel MD 20707

MONCRIEF, EUGENE CHARLES, b Washington, DC, July 7, 32; m 55; c 1. ENGINEERING MANAGEMENT. *Educ:* Va Polytech Inst, BS, 54, MS, 55, PhD(chem eng), 57. *Prof Exp:* Res assoc chem eng, Sterling Forest Lab, Union Carbide Nuclear Co, 57-58; develop specialist, Oak Ridge Nat Lab, 58-63; tech adv, Res & Develop Div, Nuclear Develop Ctr, 63-64, sect chief, 64-69, mgr process develop commercial nuclear fuel, 69-71, mgr fuel contract, Fuel Dept, 61, mgr qual & tech serv, Numec, 71-73, mgr mfg, 73-76, gen mgr, Indust & Marine Div, 76-77 & vpres, 77-81, SR VPRES, FOSSIL POWER & CONSTRUCT GROUP, BABCOCK & WILCOX CO, 81- *Concurrent Pos:* Vis lectr, Darden Grad Sch Bus Admin, Univ Va, 78- *Mem:* Am Nuclear Soc; Am Inst Chem Engrs; Soc Naval Architects & Marine Engrs; Tech Asn Pulp & Paper Indust. *Res:* Uranium and tungsten separation technology; nuclear fuel reprocessing and cycle development; nuclear chemical engineering; utility and industrial boiler manufacture and development. *Mailing Add:* 874 Merriman Rd Akron OH 44303

MONCRIEF, JOHN WILLIAM, b Brunswick, Ga, Jan 23, 41; m 63; c 3. CHEMISTRY. *Educ:* Emory Univ, BS, 63; Harvard Univ, PhD(phys chem), 66. *Prof Exp:* Asst prof chem, Amherst Col, 66-68; from asst prof to prof chem, Emory Univ, Atlanta, 68-86, dean & div exec, Oxford Col, 76-86; VPRES ACAD AFFAIRS, DEAN OF FAC, PROF CHEM, PRESBY COL, 86- *Concurrent Pos:* Res grants, NIH, 66-70, Eli Lilly, 68-69, NSF, 71-74, Sloan fel, Emory Univ, 68-70. *Mem:* Am Chem Soc; Am Crystallog Asn; Sigma Xi. *Res:* X-ray crystallography; structures of organic and inorganic molecules; relation of mechanism to structures. *Mailing Add:* Vpres Acad Affairs Presby Col Clinton SC 29325

MONCRIEF, NANCY D, b Memphis, Tenn, Nov 20, 57. SYSTEMATICS, EVOLUTIONARY BIOLOGY. *Educ:* Memphis State Univ, BS, 78; Ft Hays State Univ, MS, 81; La State Univ, PhD(zool), 87. *Prof Exp:* Fel, dept biol, Univ Va, 87-89; CUR, MAMMAL, VA MUS NAT HIST, 89- *Mem:* AAAS; Am Inst Biol Sci; Am Soc Mammal. *Res:* Evolutionary biology and systematics of North American mammals expecially insectivores and rodents. *Mailing Add:* Va Mus Natural Hist 1001 Douglas Ave Martinsville VA 24112

MONCTON, DAVID EUGENE, b New York, NY, Oct 7, 48; m 75; c 2. SOLID STATE PHYSICS. *Educ:* Cornell Univ, BS, 70; Mass Inst Technol, MS, 73, PhD(physics), 75. *Prof Exp:* Mem tech staff physics res, Bell Labs, 75-83; physicist, Brookhaven Nat Lab, 82-85; sr res assoc, Exxon Corp Res, 85-90; ASSOC LAB DIR, ARGONNE NAT LAB, 87- *Honors & Awards:* E O Lawrence Award, Dept Energy, 87. *Mem:* Fel Am Phys Soc. *Res:* X-ray and neutron scattering studies of condensed matter systems. *Mailing Add:* Argonne Nat Lab Argonne IL 60439

MONCURE, HENRY, JR, b Stafford, Va, Feb 16, 30; m 55; c 4. ORGANIC POLYMER CHEMISTRY. *Educ:* Univ Va, BS, 51, PhD(org chem), 58. *Prof Exp:* Instr, US Naval Acad, 53-55; NIH fel, Cambridge Univ, 58-59; from chemist to sr res chemist, 59-68, from res supvr to sr supvr, 68-74, res mgr, Plastics Prod & Resins Dept, 74-80, tech mgr, polymer prod dept, 80-84, tech dir Europe, E I du Pont de Nemours & Co, Inc, 84-86; PRIN CONSULT, MONCURE ASSOC, 86- *Mem:* Am Chem Soc. *Res:* Ring chain tautomerism; aromatic polymides; polymer stability; composite polymers; fluorocarbon polymers, plastics for use in harsh environments; engineering plastics. *Mailing Add:* 1601 Woodsdale Rd Bellvue Manor Wilmington DE 19809-2248

MOND, BERTRAM, b New York, NY, Aug 24, 31; m 57; c 3. MATHEMATICS. *Educ:* Yeshiva Univ, BA, 51; Bucknell Univ, MA, 59; Univ Cincinnati, PhD(math), 63. *Prof Exp:* Mech comput analyst, Gen Elec Co, 59-60; res assoc biomet, Col Med & instr math, Univ Cincinnati, 62-63; res mathematician, Aerospace Res Labs, Wright-Patterson AFB, Ohio, 63-69; dean, Sch Phys Sci, 76-78, chmn dept, 70-81, PROF MATH, LA TROBE UNIV, 69-, DEAN, SCH MATH INFO SCI, 90- *Concurrent Pos:* Lectr eve col, Univ Cincinnati, 61-64; ed, J Australian Math Soc, 69-74, mem coun, 69-74. *Mem:* Am Math Soc; Australian Math Soc; Math Prog Soc; Australian Soc Opers Res. *Res:* Operations research; linear and nonlinear programming; approximation theory. *Mailing Add:* Dept Math La Trobe Univ Bundoora 3083 Melbourne Australia

MOND, JAMES JACOB, b New York, NY, Jan 5, 46; m 71; c 3. RHEUMATOLOGY. *Educ:* Yeshiva Univ, BS, 67; NY Univ Med Ctr, MD, 73, PhD(immunol), 73. *Prof Exp:* Intern med, Mt Sinai Hosp, New York, 73-74, resident, 74-75; res assoc immunol, immunol lab, NIH, 75-80; asst prof, 80-83, ASSOC PROF MED & RHEUMATOLOGY, UNIFORMED SERVS UNIV HEALTH SCI, MD, 83- *Concurrent Pos:* Staff ed, J Immunol, 84- *Mem:* Asn Immunologists; Am Rheumatism Asn. *Res:* Exploration of the mechanisms of B lymphocyte activation using biochemical and molucular biological approaches. *Mailing Add:* Dept Med A3041 Uniformed Serv Univ Health Sci 4301 Jones Bridge Rd Bethesda MD 20814

MONDAL, KALYAN, b Calcutta, India, Aug 17, 51; m 81; c 2. DIGITAL SIGNAL PROCESSING, COMPUTER-AIDED DESIGN. *Educ:* Univ Calcutta, India, BSc, 69, BTech, 72, MTech, 74; Univ Calif, Santa Barbara, PhD(elec eng), 78. *Prof Exp:* Res asst, Univ Calif, Davis, 75-77, Santa Barbara, 77-79; asst prof, Lehigh Univ, 80-81; mem tech staff, 82-87, DISTINGUISHED MEM TECH STAFF, AT&T BELL LABS, ALLENTOWN, PA, 87- *Concurrent Pos:* Teaching asst, Univ Calif, Santa Barbara, 78, lectr, 79; consult, AT&T Bell Labs, Allentown, Pa, 80-81; adj asst prof, Lehigh Univ, 82-83. *Mem:* Inst Elec & Electronics Engrs; Sigma Xi; Asn Comput Mach; AAAS. *Res:* Development of silicon compilers for digital signal processing; very-large-scale integrated circuits. *Mailing Add:* AT&T Bell Labs 555 Union Blvd Allentown PA 18103-1285

MONDER, CARL, b US, Aug 24, 28; m 59, 83; c 2. BIOCHEMISTRY. *Educ:* City Col New York, BS, 50; Cornell Univ, MS, 52; Univ Wis, PhD(biochem), 56. *Prof Exp:* Assoc technologist, Res Lab, Gen Foods Corp, 52-54; USPHS fel, Sch Med, Tufts Univ, 56-57; from instr to asst prof biochem, Albert Einstein Col Med, 58-69; res assoc prof, 69-78, prof biochem, Mt Sinai Sch Med, 78-81; head, Sect Steroid Studies, Res Inst Skeletomuscular Dis, Hosp for Joint Dis, 64-81; SR SCIENTIST, POPULATION COUN, NY, 81- *Concurrent Pos:* Vis prof chem, Stevens Inst Technol, 63-69; career develop award, USPHS, 69-73; adj prof biochem & adj prof pediat, Cornell Med Sch, 81- *Mem:* AAAS; Am Soc Biol Chem; Am Chem Soc; Endocrine Soc; NY Acad Sci. *Res:* Steroid chemistry and metabolism; hormone action; perinatal metabolism and development; tumor metabolism. *Mailing Add:* Population Coun 1230 York Ave New York NY 10021

MONDOLFO, L(UCIO) F(AUSTO), b Senigallia, Italy, Aug 20, 10; nat US; m 35; c 2. METALLURGY, MATERIALS SCIENCE. *Educ:* Univ Bologna, BS, 29; Milan Polytech Inst, Dr Sc(indust eng), 33. *Prof Exp:* Metallurgist, Isotta Fraschini, Italy, 35-38; metallurgist, US Reduction Co, Ill, 39; chemist, R Lavin Sons, 39-40; res metallurgist, Reynolds Metals Co, Ky, 40-41, asst chief res metallurgist, NY, 43-45; chief metallurgist, Howard Foundry, Ill, 42-43; assoc prof metall, Ill Inst Technol, 46-50, prof, 51-65, head dept, 55-65; sr res metallurgist, Revere Copper & Brass, Inc, 66-76; RETIRED. *Concurrent Pos:* Consult, 76-; adj prof, Rensselaer Polytech Inst, 78- *Mem:* AAAS; Am Soc Metals; Am Inst Mining, Metall & Petrol Engrs; Brit Metals Soc. *Res:* Aluminum alloys; crystallization. *Mailing Add:* RR 4 Box 432 Clinton NY 13323

MONDON, CARL ERWIN, DIABETES, EXERCISE PHYSIOLOGY. *Educ:* Univ Southern Calif, PhD(med physiol), 62. *Prof Exp:* RES PHYSIOLOGIST, GERIAT RES EDUC CLIN CTR, 73- *Mailing Add:* Dept Med Geriat Res Educ Clin Ctr 3810 Miranda Ave Palo Alto CA 94304

MONDY, NELL IRENE, b Pocahontas, Ark, Oct 27, 21. AGRICULTURAL BIOCHEMISTRY. *Educ:* Ouachita Univ, BS & BA, 43; Univ Tex, MA, 45; Cornell Univ, PhD(biochem), 53. *Prof Exp:* Asst prof chem, Ouachita Univ, 43-44; asst biochem, Univ Tex, 44-45; res assoc, Cornell Univ, 45-46; from instr to asst prof chem, Sampson Col, 46-48; from instr to assoc prof, 48-57, PROF FOOD & NUTRIT, CORNELL UNIV, 81- *Concurrent Pos:* Prof food chem, Fla State Univ, 69-70 & R T French Co, 66-67; consult, Environ Protection Agency; supvry food Specialist, USDA, 60-61; vis scholar, Gadjak Mada Univ, Indonesia, 89; consult, Frito-Lay, Procter & Gamble & Gen Mills. *Honors & Awards:* Centennial Award, Ouachita Univ, 1986. *Mem:* Fel AAAS; Am Chem Soc; fel Am Inst Chemists; fel Inst Food Technol; NY Acad Sci; hon mem Potato Asn Am; hon mem Grad Women Sci. *Res:* Vitamin B-6 group; choline and betaine aldehyde dehydrogenase in rats; enzymes; phenols; ascorbic acid; glycoalkaloids, nitrates, nitrites and lipids in potatoes; protein minerals sensory evaluation. *Mailing Add:* Div Nutrit Sci N231 Van Rensselaer Cornell Univ Ithaca NY 14853

MONER, JOHN GEORGE, b Bayonne, NJ, Sept 4, 28; m 87; c 4. CELL BIOLOGY, BIOCHEMISTRY. *Educ:* Johns Hopkins Univ, AB, 49; Princeton Univ, MA, 51, PhD(biol), 53. *Prof Exp:* From instr to assoc prof 55-73, PROF ZOOL, UNIV MASS, AMHERST, 73- *Concurrent Pos:* NSF sci fac fel, Biol Inst, Carlsberg Found, Copenhagen, 61-62; pub health spec fel biol, Univ Calif, San Diego; actg ed, J Protozool, 73-74; vis prof, Mass Inst Technol, 84-85. *Mem:* Soc Protozool; Am Soc Cell Biol. *Res:* Physiology and biochemistry of cell division, with emphasis on synchronized tetrahymena; nutrient uptake and cytoskeleton in tetrahymena. *Mailing Add:* Dept Zool Univ Mass Amherst MA 01003

MONESTIER, MARC, b La Tronche, France, Jan 8, 59. AUTOIMMUNE DISEASES, CANCER VACCINES. *Educ:* Univ Lyon, France, MD, 83, PhD(immunol), 88. *Prof Exp:* Fel, Blood Ctr, Lyon, France, 83-84 & Mt Sinai Med Ctr, NY, 86-87; ASST MEM IMMUNOL, CTR MOLECULAR MED & IMMUNOL, NEWARK, 87- *Concurrent Pos:* Adj asst prof, Microbiol & Immunol Dept, NY Med Ctr, Vahalla, 89- *Mem:* Am Asn Immunologists; Am Asn Cancer Res. *Res:* Molecular genetics of autoantibodies; immunopathology of lupus nephritis; anti-histone antibodies; anti-cardiolipin antibodies; idiotype vaccines. *Mailing Add:* Dept Immunol Ctr Molecular Med & Immunol One Bruce St Newark NJ 07103

MONET, MARION C(RENSHALL), information science, for more information see previous edition

MONETI, GIANCARLO, b Rome, Italy, Nov 2, 31; m 55; c 5. ELEMENTARY PARTICLE PHYSICS. *Educ:* Univ Rome, Dr, 54, Libero Docente, 63. *Prof Exp:* Asst prof physics, Univ Rome, 54-62, Nat Res Coun Italy fel, 54-55; physicist nat comt nuclear energy, Frascati Labs, Italy, 57-60; vis assoc physicist, Brookhaven Nat Lab, 61-62; assoc prof physics, Univ Rome, 62-64 & hist of physics, 64-68; PROF PHYSICS, SYRACUSE UNIV, 68- *Concurrent Pos:* Mem bd dirs, Nat Agency Nuclear Physics, Italy, 66-68; vis scientist, Europ Orgn Elem Particle Res, 74-75, 81-82. *Mem:* Ital Phys Soc; Europ Phys Soc; fel Am Phys Soc. *Res:* History of physics; experimental elementary particle physics; heavy quark meson and baryon resonances; electron-positron annihilation; quark and gluon fragmentation. *Mailing Add:* Dept Physics 323 Physics Bldg Syracuse Univ Syracuse NY 13244-1130

MONETTE, FRANCIS C, b Lowell Mass, Aug 9, 41; m 68; c 2. HEMATOLOGY, CELL PHYSIOLOGY & BIOLOGY. *Educ:* St Anselm's Col, BA, 62; NY Univ, MS, 65, PhD(biol), 68. *Prof Exp:* Teaching fel biol, NY Univ, 62-63, res asst, 63-65, asst res scientist, 65-68; NIH trainee hemat, St Elizabeth's Hosp-Tufts Med Sch, 68-71; from asst prof to assoc prof biol, 71-82, asst prof health sci, 74-77, PROF BIOL & HEALTH SCI, BOSTON UNIV, 82- *Concurrent Pos:* Lectr, Iona Col, 66-67; NIH res career develop award, 77-82; rev panels, Hemat Study Sect, NIH, 88 & 90; ad hoc rev, NSF, 86- *Mem:* AAAS; Am Soc Hemat; Am Soc Cell Biol; Int Soc Exp Hemat; Sigma Xi; Am Phys Soc. *Res:* Experimental hematology; regulation of cell production and differentiation; erythropoietic physiology and biochemistry; cell growth factors; interleukin-3; serum-free cell growth. *Mailing Add:* Dept Biol Sci Boston Univ 2 Cummington St Boston MA 02215

MONEY, JOHN WILLIAM, b Morrinsville, NZ, July 8, 21, US citizen. GENETICS, PEDIATRICS. *Educ:* Univ NZ, MA, 42, dipl, 44; Harvard Univ, PhD, 52. *Prof Exp:* Instr psychiat, 51-55, from asst prof to assoc prof med psychol, 55-72, assoc prof pediat, 59-86, prof med psychol, 72-76, EMER PROF, DEPT PSYCHIAT & BEHAV SCI, JOHNS HOPKINS HOSP, 86- *Concurrent Pos:* NIH res career award, 62-72; mem bd dirs, Sex Info & Educ Coun US, 65-68; mem task force homosexuality, NIMH, 67-69; mem bd dirs, Neighborhood Family Planning Ctr, Inc, 70-82; mem develop & behav sci study sect, NIH, 70-74; Grant Found grant, 72-; dir, Psychohormonal Res Unit, Johns Hopkins Hosp, 51-, psychologist, 55. *Honors & Awards:* Hofheimer Prize, Am Psychiat Asn, 56; Gold Medal Award, Children's Hosp Philadelphia, 66; Awards, Soc Sci Study Sex, 72 & 76; Harry Benjamin Medal Honor, Erickson Educ Found, 76; Lindemann lectr, Cornell Univ, 83; Bernadin lectr, Univ Mo, 85; Distinguished Sci Award, Am Psychol Asn, 85; Outstanding Res Award, Nat Inst Child Health & Human Develop, 87. *Mem:* Soc Sci Study Sex (pres, 75-76); Am Found Gender & Genital Med & Sci (pres, 78-); fel AAAS; Soc Pediat Psychol; Int Acad Sex Res; Int Soc Psychoneuroendocrinol. *Res:* Medical psychology; sexology; behavioral endocrinology and genetics; gender identity; paraphilias; abuse dwarfism. *Mailing Add:* Dept Psychiat & Behav Sci Johns Hopkins Univ Hosp Baltimore MD 21205

MONEY, KENNETH ERIC, b Toronto, Ont, Jan 4, 35; m 58; c 1. PHYSIOLOGY, BIOLOGY. *Educ:* Univ Toronto, BA, 58, MA, 59, PhD(physiol), 61, Nat Defense Col, 72. *Prof Exp:* Res scientist, Defence Res Med Labs, 61-66, sect head vestibular physiol, 66-76, dir biosci, 76-79, SR SCIENTIST, DEFENCE & CIVIL INST ENVIRON MED, 79-; ASSOC PROF PHYSIOL, UNIV TORONTO, 72- *Concurrent Pos:* Res assoc, Univ Toronto, 61-62, lectr, 62-68, asst prof, 69-72; assoc prof, Can Astronaut Corp,

84- *Mem:* Can Physiol Soc; fel Royal Soc Health; fel Areospace Med Asn; Int Acad Astronuntics. *Res:* Vestibular physiology; motion sickness; histology of the inner ear; eye movements; alcohol; pilot disorientation. *Mailing Add:* Dept Physiol Univ Toronto Toronto ON M5S 1A1 Can

MONEY, LLOYD J(EAN), b Lawton, Okla, Sept 14, 20; m 44; c 3. ELECTRICAL ENGINEERING. *Educ:* Rice Univ, BS, 42; Purdue Univ, MS, 50, PhD(elec eng), 52. *Prof Exp:* Instr elec eng, Rice Univ, 42-43, 47-49, Tulane Univ, 46-47 & Purdue Univ, 49-52; mgr, Interceptor Systs Dept, Systs Analysis Lab, Hughes Aircraft Co, 52-60, assoc mgr, Tactical Systs Lab, 60-62, mgr, Europ Opers Aeronaut Systs Div, 62-64, assoc mgr, Advan Projs Labs, Calif, 64-68; staff mgr advan progs, TRW, Inc, 68-71; asst dir, Off Systs Eng, Off of the Secy, Dept Transp, 71-77, actg dir, Off Univ Res, 72-74, actg assoc adminr res & develop, Urban Mass Transp Admin, 74, actg asst secy systs develop & tech, 77-78, dir, Transp Progs Bur, Res & Special Progs Admin, 78-79, dir off univ res, 81-84; SR STAFF ENGR, TRW, INC, 84- *Concurrent Pos:* Vis scholar, Univ Calif, 80-81. *Mem:* Inst Elec & Electronics Engrs; NY Acad Sci. *Res:* Airborne weapons design; automatic control systems; system planning, analysis and coordination, including radar computers, navigation, missiles and automatic controls; transportation engineering; research and development administration. *Mailing Add:* 904 21st St Hermosa Beach CA 90254

MONFORE, GERVAISE EDWIN, b Waverly, Kans, Feb 26, 10; m 36; c 1. INSTRUMENTATION. *Educ:* Col of Emporia, AB, 32; Univ Denver, MS, 50. *Prof Exp:* Jr physicist, Nat Bur Stand, 36-41; res physicist, B F Goodrich Co, 42-45; res engr, US Bur Reclamation, 45-55; res physicist, Portland Cement Asn, 55-60, from sr res physicist to prin res physicist, 67-73; RETIRED. *Mem:* Sigma Xi. *Res:* Properties and behavior of portland cement and concrete, including strength, stress, elasticity, creep, shrinkage, electrical and thermal conductivity, heat of hydration and corrosion of embedded metals, fiber reinforcement; ice pressure. *Mailing Add:* 4083 S Eaton Ave Springfield MO 65807

MONFORTON, GERARD ROLAND, b Windsor, Ont, July 21, 38; m 60; c 4. CIVIL ENGINEERING. *Educ:* Assumption Univ, BASc, 61, MASc, 62; Case Western Reserve Univ, PhD(civil eng), 70. *Prof Exp:* Lectr civil eng, Univ Windsor, 62-64; res asst solid mech, Case Western Reserve Univ, 64-68; from asst prof to assoc prof civil eng, 68-76, PROF CIVIL ENG, UNIV WINDSOR, 76- *Mem:* Eng Inst Can; Can Soc Civil Engrs. *Res:* Solid mechanics; structural design. *Mailing Add:* Dean Eng Univ Windsor Windsor ON N9B 3P4 Can

MONG, SEYMOUR, cellular pharmacology, molecular pharmacology, for more information see previous edition

MONGAN, EDWIN LAWRENCE, JR, b Cincinnati, Ohio, June 27, 19; m 49; c 8. PROCESS DESIGN, THERMODYNAMICS. *Educ:* Univ Cincinnati, ChE, 42. *Prof Exp:* Chemist soap mfg, DuBois Co, 46; instr chem eng, Univ Cincinnati, 47-51; engr & prin consult econ chem thermodyn, E I Du Pont de Nemours & Co, Inc, 51-85; PRES, CHEMCODE INC, 85- *Concurrent Pos:* Adj prof, Univ Louisville, 51-53 & Univ Del, 73-76; mem adv bd coal res, City Univ New York, 72-; consult chem eng, 85- *Res:* Commercial applications of research; chemical process conception and development; chemical thermodynamics; economics. *Mailing Add:* Six Calgary Rd Newark DE 19711

MONGER, JAMES WILLIAM HERON, b Reading, Eng, Sept 26, 37; Can citizen; m 60; c 4. STRUCTURAL GEOLOGY. *Educ:* Univ Reading, Eng, BSc, 59; Univ Kans, MS, 61: Univ BC, PhD(geol), 66. *Prof Exp:* GEOLOGIST, GEOL SURV CAN, 65- *Concurrent Pos:* Distinguished lectr, Am Asn Petrol Geologists, 81-82; assoc ed, Can J Earth Sci, 81- *Mem:* Fel Royal Soc Can; fel Geol Soc Am; fel Geol Asn Can. *Res:* Cordilleran tectonics; upper Paleozoic, lower Mesozoid stratigraphy of cordillera; paleobiogeography, paleomagnetism applied to cordillera; structural geology. *Mailing Add:* Geol Survey Can 100 W Pender St Vancouver BC V6B 1R8

MONGINI, PATRICIA KATHERINE ANN, b Cottonwood, Ariz, Dec 13, 50; c 2. IMMUNOLOGY. *Educ:* Northern Ariz Univ, BS, 72; Stanford Univ, PhD(med microbiol), 76. *Prof Exp:* Fel immunol, Sch Med, Tufts Univ, 76-78; fel immunol, lab immunol, Nat Inst Allergy & Infectious Dis, 78-81; asst prof, Hosp Joint Dis, Mt Sinai Sch Med, New York, NY, 82-86; ASST PROF, HOSP JOINT DIS, SCH MED, NY UNIV, NY, 87- *Concurrent Pos:* Prin investr, Monoclonal anti-Igm regulation of Human B cell function, NIH grant, 84- *Mem:* AAAS; Am Asn Immunologists. *Res:* regulation of B lymphocyte activation. *Mailing Add:* Dept Rheumatic Dis 301 E 17th St Hosp Joint Dis New York NY 10003

MONHEIT, ALAN G, b Philadelphia, Pa, Apr 5, 49; m 75; c 3. OBSTETRICS & GYNECOLOGY, MATERNAL-FETAL MEDICINE. *Educ:* Muhlenberg Col, BSc, 71; Univ Pa, MD, 75. *Prof Exp:* Resident obstet & gynecol, Univ Calif, San Diego, 76-79; fel fetal med, 79-81, ATTENDING PHYSICIAN, DEPT OBSTET & GYNECOL, DIV MATERNAL FETAL MED, HEALTH SCI CTR, STATE UNIV NY, STONY BROOK, 81- *Honors & Awards:* Poster Prize, Am Col Obstet & Gynecol, 85-87. *Mem:* Am Col Obstet & Gynecol; Soc Perinatal Obstetricians. *Res:* Pathophysiology of uteroplacental insufficiency and fetal distress; fetal adaptation to stress. *Mailing Add:* Dept Obstet & Gynecol Div Maternal Fetal Med SUNY Stony Brook Health Sci Ctr T-9 080 Stony Brook NY 11794

MONIE, IAN WHITELAW, b Paisley, Scotland, May 24, 18; nat US; m 42; c 4. EMBRYOLOGY. *Educ:* Glasgow Univ, MB, ChB, 40, MD, 72. *Prof Exp:* From demonstr to lectr anat, Glasgow Univ, 42-47; from asst prof to assoc prof, Univ Man, 47-52; from asst prof to prof anat, 52-70, vchmn dept, 58-63, chmn dept, 63-70, prof, 70-87, EMER PROF ANAT & EMBRYOL, UNIV CALIF, SAN FRANCISCO, 87- *Concurrent Pos:* Mem study sect human develop, NIH, 65-69; Guggenheim fel, 67-68. *Mem:* Am Asn Anat; Anat Soc Gt Brit & Ireland; Teratology Soc (pres, 64-65). *Res:* Mammalian embryology and teratology; human gross anatomy; comparative embryology. *Mailing Add:* Dept Anat Univ Calif San Francisco CA 94143

MONIER, LOUIS MARCEL, b La Seyne, France, Mar 21, 56. VERY LARGE SCALE INTEGRATION DESIGN, CAD TOOLS. *Educ:* Univ Paris, Orsay, PhD(math & computer sci), 80. *Prof Exp:* Vis scientist, Carnegie-Mellon Univ, Pittsburgh, 80-83; mem res staff, Xerox Palo Alto Res Ctr, 83-89; PRIN ENGR, DEC, WESTERN RES LAB, 89- *Concurrent Pos:* Lectr, Ceris, France, 90- *Res:* Design of high-performance very large scale integration circuits; cad tools for very large scale integration; synthesis tools for bipolar circuits; algorithms and data structures. *Mailing Add:* DEC Western Res Lab 250 University Ave Palo Alto CA 94301

MONIOT, ROBERT KEITH, b Butler, Pa, Nov 26, 50. METEORITICS. *Educ:* State Univ NY, Fredonia, BS, 72; Univ Calif, Berkeley, PhD(physics), 79; NY Univ, MS, 88. *Prof Exp:* Res fel, dept physics, Rutgers Univ, 79-82; ASST PROF, DIV SCI & MATH, FORDHAM UNIV, 82- *Mem:* Am Phys Soc; AAAS; Asn Comp Mach. *Res:* Measurement of beryllium-10 in meteorites and terrestrial materials by technique of accelerator-based mass spectrometry; mass-spectrometric analysis of noble gases in meteorites. *Mailing Add:* Div Sci & Math Fordham Univ Box 122 New York NY 10023

MONISMITH, CARL L(EROY), b Harrisburg, Pa, Oct 23, 26; m 49; c 2. CIVIL ENGINEERING. *Educ:* Univ Calif, BS, 50, MS, 54. *Prof Exp:* From instr to assoc prof, 52-66, asst res engr, 52-61, assoc res engr, 61-66, chmn dept, 74-79, PROF CIVIL ENG & RES ENGR, INST TRANSP STUDIES, UNIV CALIF, BERKELEY, 66- *Concurrent Pos:* Consult, Chevron Res Co, Woodward Clyde Consults, Corps Eng, US Army, Waterways Exp Sta, Are Inc & Bechtel Corp, San Francisco Calif; assoc, Transp Res Bd, Nat Acad Sci-Nat Res Coun, chmn sect B, Group 2, Div A, 73-79. *Honors & Awards:* Emmons Award, Asn Asphalt Paving Technologists, 61, 65 & 85; K B Woods Award, Transp Res Bd, 72; State of the Art Award, Am Soc Civil Eng, 78; James Laurie Prize, Am Soc Civil Engrs, 88. *Mem:* Nat Acad Eng; fel Am Soc Civil Engrs; mem Am Soc Testing & Mat; Am Soc Eng Educ; hon mem, Asn Asphalt Paving Technologists (pres, 68). *Res:* Physical behavior of asphalts and asphalt paving mixtures; behavior, design and rehabilitation of pavements for highways and airfields; highway engineering. *Mailing Add:* Dept Civil Eng Univ Calif 115 McLaughlin Hall Berkeley CA 94720

MONIZ, ERNEST JEFFREY, b Fall River, Mass, Dec 22, 44; m 73; c 1. THEORETICAL PHYSICS, NUCLEAR STRUCTURE. *Educ:* Boston Col, BS, 66; Stanford Univ, PhD(physics), 71. *Prof Exp:* NSF res fel, Ctr Nuclear Energy Res, Saclay, Belg, 71-72; res assoc, Univ Pa, 72-73; from asst prof to assoc prof, 73-83, PROF PHYSICS, MASS INST TECHNOL, 83-; DIR ADMIN, BATES LINEAR ACCELERATOR CTR, 83- *Concurrent Pos:* Vis mem staff, Los Alamos Nat Lab, 76, consult, 76- *Mem:* Am Phys Soc. *Res:* The role of subnuclear degrees of freedom (mesons and quarks) in nuclear structure and dynamics and development of quantum mechanical scattering theory. *Mailing Add:* Dept Physics 26-403 Mass Inst Technol 77 Mass Ave Cambridge MA 02139

MONIZ, WILLIAM B, b New Bedford, Mass, Feb 12, 32; m 55; c 6. POLYMER CHEMISTRY. *Educ:* Brown Univ, BS, 53; Pa State Univ, PhD(org chem), 60. *Prof Exp:* Petrol Res Fund fel, Pa State Univ, 60-61; NIH vis scientist, Univ Ill, 61-62; head nuclear magnetic resonance spectros sect, Chem Div, 62-74, head polymer diag sect, 74-83, HEAD, POLYMERIC MAT BR, CHEM DIV, NAVAL RES LAB, 84- *Concurrent Pos:* Mem admis comt, Am Chem Soc, 65-66. *Mem:* Am Chem Soc; Sigma Xi. *Res:* Materials degradation; polymer characterization. *Mailing Add:* 7104 Block Rd Ft Washington MA 20744

MONJAN, ANDREW ARTHUR, b New York, NY, Feb 9, 38; m 69; c 2. NEUROSCIENCES, IMMUNOPATHOLOGY. *Educ:* Rensselaer Polytech Inst, BS, 60; Univ Rochester, PhD(psychol), 65; Johns Hopkins Univ, MPH, 71. *Prof Exp:* USPHS res fel, Univ Rochester, 64-66; asst prof psychol & physiol, Univ Western Ont, 66-69; USPHS community health trainee, Sch Hyg & Pub Health, Johns Hopkins Univ, 69-70, asst prof epidemiol, 71-75, assoc prof, 75-83; spec expert, Epidemiol Br, Nat Cancer Inst, NIH, 83-85; CHIEF, NEUROBIOL & NEUROPSYCHOL BR, NAT INST AGING, NIH, 85- *Concurrent Pos:* Exec secy, Nat Comn Sleep Disorders Res. *Mem:* Soc Neurosci; Am Asn Immunologists. *Res:* Effects of viruses upon the nervous system; developmental pathogenesis, mechanisms of immunopathology and long-term psychological sequela; stress and the immune systems; neurobiology of aging. *Mailing Add:* Neurosci & Neuropsychol Aging Bldg 31C/5C35 Nat Inst Aging NIH Bethesda MD 20892

MONK, CARL DOUGLAS, b Hurdles Mill, NC, July 28, 33; m 57; c 2. PLANT ECOLOGY. *Educ:* Duke Univ, AB, 55; Rutgers Univ, MS, 58, PhD(bot), 59. *Prof Exp:* Asst prof bot, Univ Fla, 59-64; from asst prof to assoc prof, 64-71, PROF BOT, UNIV GA, 71- *Mem:* Bot Soc Am; Ecol Soc Am; Torrey Bot Club; Brit Ecol Soc; Sigma Xi. *Res:* Vegetation analysis; mineral cycling. *Mailing Add:* Dept of Bot Univ Ga Athens GA 30601

MONK, CLAYBORNE MORRIS, b Atlantic City, NJ, Dec 4, 38; m 65; c 2. BIOPHARMACEUTICS, PHARMACOKINETICS. *Educ:* Howard Univ, BS, 60; Univ Calif, San Francisco, PhD(pharmaceut chem), 75. *Prof Exp:* Asst prof pharm, Sch Pharm, Univ Southern Calif, 72-75; asst prof, 76-80, ASSOC PROF PHARM, SCH PHARM, TEX SOUTHERN UNIV, 80- *Concurrent Pos:* Vis assoc prof pharmaceut, Sch Pharm, Philadelphia Col Pharm & Sci, Pa, 91. *Mem:* AAAS; Am Pharmaceut Asn; Nat Pharmaceut Asn; Am Asn Col Pharm. *Res:* Biopharmaceutics and pharmacokinetics with special interest in drug disposition within the maternal-placental-fetal system. *Mailing Add:* Col Pharm & Health Sci Tex Southern Univ 3100 Cleburne Ave Houston TX 77004

MONK, JAMES DONALD, b Childress, Tex, Sept 27, 30; m 53; c 2. MATHEMATICS. *Educ:* Univ Chicago, AB, 51; Univ NMex, BS, 56; Univ Calif, Berkeley, MA, 59, PhD(math), 61. *Prof Exp:* Math analyst, Los Alamos Sci Labs, 51-53; instr math, Univ Calif, Berkeley, 61-62, asst res mathematician, 63-64; from asst prof to assoc prof, 62-67, PROF MATH,

UNIV COLO, BOULDER, 67- *Concurrent Pos:* NSF res grants, 63-; vis prof, Univ Calif, Berkeley, 67-68. *Mem:* Am Math Soc; Asn Symbolic Logic. *Res:* Algebraic logic; general algebra; model theory; foundations of set theory. *Mailing Add:* Dept Math Univ Colo 2135 Goddard Pl Boulder CO 80309-0426

MONK, MARY ALICE, epidemiology, public health, for more information see previous edition

MONKE, EDWIN J, b Ill, June 7, 25; m 63; c 3. AGRICULTURAL ENGINEERING. *Educ:* Univ Ill, BS, 50, MS, 53, PhD(civil eng), 59. *Prof Exp:* Instr, Univ Ill, 51-58; from asst prof to assoc prof, 58-67, PROF AGR ENG, PURDUE UNIV, 67- *Mem:* Am Geophys Union; fel Am Soc Agr Engrs; Nat Soc Prof Engrs; Am Soc Eng Educ; Soil Conserv Soc Am. *Res:* Groundwater hydraulics; watershed hydrology; erosion and sedimentation control; water quality. *Mailing Add:* Dept Agr Eng Purdue Univ West Lafayette IN 47907

MONKEWITZ, PETER ALEXIS, b Jegenstorf, Switz, Nov 9, 43; m 72; c 3. HYDRODYNAMIC INSTABILITIES. *Educ:* Fed Inst Technol, Switz, dipl, 67, PhD(natural sci), 77. *Prof Exp:* Software consult, Sperry Rand Univac, 68; res assoc aerospace eng, Univ Southern Calif, Los Angeles, 77-80; PROF FLUID MECH-AEROSPACE ENG, MANE DEPT, UNIV CALIF, LOS ANGELES, 80- *Concurrent Pos:* Reviewer, J Fluid Mech, Physics Fluids, NSF, 79-; prin investr grants, NASA, NSF, ARO, Air Force Off Sci Res, Off Naval Res, 82-; consult, Anco Eng, Pac Sierra Res Corp, 85-; vis sr scientist, Ger Aeronaut & Astronaut Res Estab, DLR, Berlin, 87; vis prof, HF-Inst, Tech Univ Berlin & DLR, 89-90; maitre de conf, Univ Aix, Marseille, France, 89 & 90; co-ed, J Appl Math & Physics, 91- *Honors & Awards:* Sr US Scientist Award, Alexander von Humboldt Found, Ger, 88. *Mem:* Am Phys Soc; Am Inst Aeronaut & Astronaut. *Res:* Fluid mechanics, in particular the theoretical and experimental study of hydrodynamic instabilities in shear flows such as mixing layers, jets and wakes; flow control; transition to turbulence and acoustics. *Mailing Add:* MANE Dept Univ Calif Los Angeles CA 90024-1597

MONKHORST, HENDRIK J, b Kampen, Neth, Oct 13, 38; US citizen; m 64; c 2. MUON CATALYZED FUSION, NEUTRINO MASS DETERMINATION. *Educ:* Univ Groningen, Neth, PhD(theoret chem), 68. *Prof Exp:* Res asst physics, Univ Utah, Salt Lake City, 68-73, asst res prof, 75-78; vis prof chem, Univ Aarhus, Denmark, 74; assoc prof, 78-82, PROF PHYSICS & CHEM, UNIV FLA, GAINESVILLE, 82- *Concurrent Pos:* Res fel chem, Inst Rundjer Boskovic, Zagreb, Yugoslavia, 71-72. *Honors & Awards:* Hercules Award, Hercules Powder Co. *Mem:* Am Phys Soc. *Res:* Computational quantum chemistry; muon catalyzed fusion; neutrino mass determinations; polymer superconductivity. *Mailing Add:* Dept Physics Univ Fla Gainesville FL 32611

MONLUX, ANDREW W, b Algona, Iowa, Jan 29, 20; m 50; c 2. VETERINARY PATHOLOGY. *Educ:* Iowa State Univ, DVM, 42, MS, 47; George Washington Univ, PhD(comp path), 51. *Prof Exp:* Pvt pract, Iowa, 42; asst, Iowa State Univ, 46-47; vet, USDA, Colo & DC, 51-56; head dept, 56-72, PROF VET PATH, OKLA STATE UNIV, 56-, REGENTS PROF, 72- *Mem:* Am Vet Med Asn; US Animal Health Asn; Conf Res Workers Animal Dis; Vet Cancer Soc; Vet Urol Soc. *Res:* Pathology of neoplastic and kidney diseases; lead and photosensitivity syndromes of animals. *Mailing Add:* 2202 Black Oak Dr Stillwater OK 74074

MONMONIER, MARK, b Baltimore, Md, Feb 2, 43; m 65; c 1. CARTOGRAPHY, INFORMATION SYSTEMS. *Educ:* Johns Hopkins Univ, BA, 64; Pa State Univ, MS, 67, PhD(geog), 69. *Prof Exp:* Asst prof geog, Univ RI, 69-70 & State Univ NY, Albany, 70-73; assoc prof, 73-79, PROF GEOG, SYRACUSE UNIV, 79- *Concurrent Pos:* Consult, Syracuse Police Dept, 76-79, Nat Geog Soc, 88; mem, US Nat Comt, Int Cartog Asn, 77-81, assoc ed, Am Cartogr, 77-82, ed, 82-84; res geographer, US Geol Surv, 79-84; Guggenheim fel, 84-85; contrib ed, Cartographica, 84-; assoc ed, Mapping Sci & Romote Sensing, 87-; dep dir, NYS Prog Geog Info & Anal, 88-90. *Mem:* Am Soc Photogram & Remote Sensing; Asn Am Geogrs; Am Statist Asn; Can Cartog Soc; Sigma Xi; Am Cartog Asn (pres, 83). *Res:* Automated mapping; especially computer-assisted map design, pattern recognition and cartographic generalization; information graphic and electronic publishing; information policy and geographic data bases; statistical graphics and data visualization; mapping environmental hazards and risk; dynamic cartography and animation. *Mailing Add:* Dept of Geog Syracuse Univ Syracuse NY 13244-1160

MONN, DONALD EDGAR, b Chambersburg, Pa, June 21, 38; m 63; c 1. ANALYTICAL CHEMISTRY. *Educ:* Elizabethtown Col, BS, 59; Univ Del, PhD(anal chem), 64. *Prof Exp:* Analytical chemist & group leader, Data Mgt Group & tech asst, 64-80, mgr, res serv div, 80-84, DIR, ELECTRONIC & COMPUT SERV & DIR, TECH SERV, CONOCO INC, 84- *Mem:* Am Chem Soc; Am Soc Testing & Mat. *Res:* Interfacing analytical instruments to computers and data acquisition equipment; writing computer programs to perform analytical calculations; techniques of data evaluation; evaluation of data base systems; coordination of office automation planning and implementation; evaluation of personal computer hardware and software; electronic publishing; networking personal computers with mainframe computers. *Mailing Add:* Continental Oil Co PO Drawer 1267 Ponca City OK 74603-0002

MONOPOLI, RICHARD V(ITO), b Providence, RI, Nov 22, 30; m 56; c 2. ELECTRICAL ENGINEERING. *Educ:* US Naval Acad, BSc, 52; Brown Univ, MSc, 60; Univ Conn, PhD(elec eng), 65. *Prof Exp:* Design engr, Int Bus Mach Corp, NY, 56-57; proj engr, Indust Div, Speidel Corp, 57-62; asst automatic control, Univ Conn, 62-65; assoc prof elec eng, 65-69, PROF ELEC ENG, UNIV MASS, AMHERST, 69- *Mem:* Inst Elec & Electronics Engrs; Am Soc Eng Educ. *Res:* Automatic control; application of magnetohydrodynamic principles to gyroscope design; development of tape recorders for missile use; application of Liapunov's direct method to control system design. *Mailing Add:* 55 Huntington Rd Hadley MA 01035

MONOS, EMIL, b Csabdi, Hungary, Jan 4, 35; m 62; c 2. CARDIOVASCULAR PHYSIOLOGY, BIOMECHANICS OF ARTERIES & VEINS. *Educ:* Med Univ Budapest, MD, 59; Hungarian Acad Sci, PhD(physiol), 70, DmSc, 82. *Prof Exp:* Asst prof, 59-83, PROF HUMAN PHYSIOL, EXP RES DEPT & SECOND INST PHYSIOL, SEMMELWEIS UNIV MED, BUDAPEST, 83-, DIR, 90- *Concurrent Pos:* Prin investr, Ministry Health, Hungary, 86-, Hungarian Acad Sci, 87-, NSF, 88-; adj prof human physiol, Dept Physiol, Med Col Wis-Milwaukee, 90-; mem, Comt Theoret Med, Hungarian Acad Sci, 90-; mem coun, Int Soc Pathophysiol, 91- *Honors & Awards:* Medal of Honor, Hungarian Ministry Health & Social Affiars, 89. *Mem:* Am Physiol Soc; Int Soc Pathophysiol. *Res:* Biomechanical aspects of normal and pathological vascular functions; cellular mechanisms; neural and humoral control of blood vessels; physiological control of smooth muscle activity in female reproductive organs; computer modelling of cardiovascular processes. *Mailing Add:* Semmelweis Med Univ Ulloi ut 78-a 1146 Budapest PO Box 448 Budapest H-1082 Hungary

MONOSON, HERBERT L, b Chicago, Ill, Dec 23, 36; m 65; c 2. MYCOLOGY. *Educ:* Western Ill Univ, BS, 58, MS, 60; Univ Ill, Urbana, PhD(bot), 67. *Prof Exp:* From asst prof to assoc prof, 66-77, PROF BIOL, BRADLEY UNIV, 77- *Concurrent Pos:* Res Corp grant, Bradley Univ, 71-72. *Mem:* Mycol Soc Am. *Res:* Phycology; nematology; hormonal regulation of fungi; rust fungi taxonomy. *Mailing Add:* Dept of Biol Bradley Univ 1501 West Bradley Ave Peoria IL 61625

MONOSTORI, BENEDICT JOSEPH, b Kovagoors, Hungary, July 4, 19; US citizen. PHYSICS. *Educ:* Pazmany Peter Univ, Budapest, BS, 45; Pontifical Univ, St Anselm, Rome, MA, 51; Fordham Univ, PhD(physics), 64. *Prof Exp:* Instr math & physics, St Stephen's Acad, Hungary, 45-48; prof philos, St Bernard's Col, Hungary, 48-50; instr math, Acad Mary Immaculate, Wichita Falls, Tex, 55-56; instr math & philos, Univ Dallas, 56-60, asst prof physics, 60-64, chmn dept, 71-86, assoc prof physics, 64-89, EMER ASSOC PROF PHYSICS, UNIV DALLAS, 89- *Mem:* Am Asn Physics Teachers; Sigma Xi. *Res:* Molecular spectra and structure; Raman spectroscopy; philosophy of science. *Mailing Add:* One Cistercian Rd Irving TX 75039-4501

MONRAD, DITLEV, b Copenhagen, Denmark, Aug 2, 49; m 72; c 1. MATHEMATICS, STATISTICS. *Educ:* Copenhagen Univ, BA, 72; Univ Calif, Berkeley, PhD(math), 76. *Prof Exp:* Asst prof math, Univ Southern Calif, 76-78; ASST PROF MATH, UNIV ILL, URBANA, 78- *Res:* Probability theory. *Mailing Add:* Dept Math Univ Ill 725 S Wright St Urbana IL 61802

MONROE, BARBARA SAMSON GRANGER, anatomy; deceased, see previous edition for last biography

MONROE, BRUCE MALCOLM, b Indianapolis, Ind, July 7, 40; m 68. ORGANIC CHEMISTRY. *Educ:* Wabash Col, AB, 62; Univ Ill, Urbana, MS, 64, PhD(org chem), 67. *Prof Exp:* NIH fel, Calif Inst Technol, 67-69; res chemist, Explosives Dept, 69-71, res chemist, Cent Res & Develop Dept, Exp Sta, 71-81, sect supvr, Haskell Lab Toxicol & Indust Med, 81-85, sr res chemist, Photosysts & Electronic Prod Dept, Exp Sta, 85-87, SPECIALIST, INTELLECTUAL PROPERTY, ELECTRONICS DEPT, E I DU PONT DE NEMOURS & CO INC. *Mem:* Am Chem Soc; Royal Soc Chem; Sigma Xi. *Res:* Organic photochemistry; Photoimaging. *Mailing Add:* 3030 Maple Shade Lane Wilmington DE 19810

MONROE, BURT LEAVELLE, JR, b Louisville, Ky, Aug 25, 30; m 60. ORNITHOLOGY. *Educ:* Univ Louisville, BS, 53; La State Univ, PhD(zool), 65. *Prof Exp:* Vis instr zool, La State Univ, 65; from asst prof to assoc prof, 65-74, PROF VERT ZOOL, UNIV LOUISVILLE, 74-, CHMN DEPT BIOL, 71- *Mem:* Soc Syst Zool; Am Soc Ichthyologists & Herpetologists; Am Ornith Union (treas, 68-75, vpres, 83, pres, 90-92); fel AAAS; Cooper Ornith Soc; Wilson Ornith Soc. *Res:* Distribution and systematics of birds, especially neotropical; entomology, especially Lepidoptera and Coleoptera; herpetology; zoogeography. *Mailing Add:* Dept Biol Univ Louisville Louisville KY 40292

MONROE, ELIZABETH MCLEISTER, b Pittsburgh, Pa, Dec 11, 40; m 68. ORGANIC CHEMISTRY. *Educ:* Bucknell Univ, BS, 62; Univ Ill, Urbana, MS, 64, PhD(org chem), 68. *Prof Exp:* Info chemist, 69-71, SR INFO CHEMIST, E I DU PONT DE NEMOURS & CO, INC, 71- *Mem:* Am Chem Soc. *Res:* Chemical information. *Mailing Add:* 3030 Maple Shade Lane Wilmington DE 19810-3424

MONROE, EUGENE ALAN, b Kansas City, Kans, May 31, 34; m 54; c 3. MINERALOGY, CRYSTALLOGRAPHY. *Educ:* Univ Wis, BS, 55; Univ Ill, MS, 59, PhD(geol-mineral), 61; Columbia Univ, DDS(dent), 73. *Prof Exp:* From asst prof to assoc prof crystallog, Col Ceramics, State Univ NY, Alfred Univ, 61-88; CO-DIR, CM RES, 88- *Mem:* Soc Biomat; Sigma Xi; NY Acad Sci; Am Dental Asn. *Res:* Electron microscopy and crystallography of materials; biomedical material development; mineralogical studies; dental research. *Mailing Add:* 2039 Civic Ctr Dr No F North Las Vegas NV 89030-6311

MONROE, GORDON EUGENE, agricultural engineering, for more information see previous edition

MONROE, ROBERT JAMES, b Dysart, Iowa, Dec 28, 18; m 86. STATISTICS. *Educ:* Iowa State Univ, BS, 39; NC State Col, PhD, 49. *Prof Exp:* Res collabr & supvr, Statist Lab, USDA, Iowa State Univ, 39-41; instr mach comput, NC State Col, 41-42, instr statist, 46-49, asst prof statist & plant sci statistician, 48-52, assoc prof exp statist, 52-53; chief, Oper Anal Off, Air Force Missile Test Ctr, 53-54; prof, 54-82, EMER PROF EXP STATIST, SCH PHYS & MATH SCI, NC STATE UNIV, 82- *Concurrent Pos:* Agent, USDA, 41-42; vis prof, Med Col Va, 61-62. *Mem:* Biomet Soc; fel Am Statist Asn; Sigma Xi. *Res:* Estimation of nutrition requirements; statistical methodology; biometry; operations research. *Mailing Add:* Dept Statist NC State Univ Raleigh NC 27650

MONROE, RONALD EUGENE, b Porterville, Calif, Jan 17, 33; m 58; c 1. ENTOMOLOGY. *Educ:* Fresno State Col, BA, 56; Ore State Col, MS, 58; Kans State Univ, PhD(entom), 64. *Prof Exp:* Jr vector control officer, Bur Vector Control, State Dept Pub Health, Calif, 55-56; med entomologist, Agr Res Serv, USDA, Ore, 57-58, gen entomologist, Insect Physiol Lab, 58-61, from asst prof to prof entom, Mich State Univ, 64-73; PROF BIOL, SAN DIEGO STATE UNIV, 73- *Mem:* Entom Soc Am. *Res:* Biochemistry and physiology of insects, chiefly lipid, carbohydrate and amino acid metabolism and insect nutrition and reproduction. *Mailing Add:* Dept of Biol San Diego State Univ 5300 Campanile Dr San Diego CA 92182-0057

MONROE, RUSSELL RONALD, b Des Moines, Iowa, June 7, 20; m 45; c 3. MEDICINE. *Educ:* Yale Univ, BS, 42, MD, 44. *Prof Exp:* Asst med, Yale Univ, 45-46; from asst prof to assoc prof psychiat, Sch Med, Tulane Univ, 50-60; chmn dept psychiat & dir, Inst Psychiat & Human Behav, 76-85, PROF PSYCHIAT, SCH MED, UNIV MD, BALTIMORE, 60- *Mem:* Am Psychiat Asn; Am Acad Psychoanal. *Res:* Episodic behavioral disorders. *Mailing Add:* Inst of Psychiat & Human Behav Univ of Md Sch of Med Baltimore MD 21201

MONROE, STUART BENTON, b Manassas, Va, Oct 26, 34; m 60; c 3. ORGANIC CHEMISTRY, POLYMER CHEMISTRY. *Educ:* Randolph-Macon Col, BS, 56; Univ Fla, PhD(org chem), 62. *Prof Exp:* Res chemist, Hercules Res Ctr, Del, 61-65; assoc prof, Randolph-Macon Col, 65-67, chmn, Dept Chem, 75-78 & 84-88, chmn area sci & math, 78-81, PROF ORG CHEM, RANDOLPH-MACON COL, 67- *Mem:* Am Chem Soc. *Mailing Add:* Dept Chem Randolph-Macon Col Ashland VA 23005

MONROE, WATSON HINER, b Parkersburg, WVa, Dec 1, 07; m 33; c 1. GEOLOGY. *Prof Exp:* From jr geologist to geologist, US Geol Surv, 30-49, chief, Eastern Field Invests Sect, Fuels Br, 49-53, staff geologist, 53-55, chief, P R Coop Invest, Gen Geol Br, 55-66, res geologist, Atlantic Environ Br, 66-81; RETIRED. *Concurrent Pos:* Lectr, Univ PR, 60-65 & 68-69. *Mem:* Fel & hon mem Geol Soc Am. *Res:* Stratigraphy of Cretaceous and Tertiary rocks of the Gulf Coast Plain; karst geology; stratigraphy, structure, economic geology and geomorphology of Puerto Rico. *Mailing Add:* 218 North St NE Leesburg VA 22075

MONSE, ERNST ULRICH, b Bautzen, Ger, Jan 10, 27; m 70; c 2. PHYSICAL CHEMISTRY. *Educ:* Univ Mainz, MS, 53, PhD(phys chem), 57. *Prof Exp:* Res assoc phys chem, Columbia Univ, 57-59; res assoc, 59-64, assoc prof, 64-74, PROF PHYS CHEM, RUTGERS UNIV, NEWARK, 74- *Mem:* Sigma Xi. *Res:* Isotope effects and their correlation with molecular structures and force fields. *Mailing Add:* Dept Chem Rutgers Univ 73 Warren St Newark NJ 07102

MONSEES, JAMES E, b Sedalia, Mo, Mar 27, 37; m 61; c 2. UNDERGROUND STRUCTURES. *Educ:* Univ Mo, BS, 60, MS, 61; Univ Ill, PhD(civil eng), 70. *Prof Exp:* Res assoc geotech eng, Univ Ill, 66-69; sr vpres underground energy, A A Mathews Inc, 69-80; dept mgr high level nuclear waste depository, Battelle Mem Inst, 80-82; vpres & engr geotech consult, Lachel Hanson & Assocs, 82-83; chief tunnel engr, Los Angeles Subway, Metro Rail Transit Consults, 83-90; PROJ MGR & COLLIDER, SSC DESIGN, PB/MK TEAM, 90- *Concurrent Pos:* Course coordr & lectr, Tunneling Short Course, Univ Calif, Los Angeles, 73; lectr, Soft Ground Tunnel Design, Univ Colo, 83, Univ Wis, 89; ed, Rapid Excavation & Tunneling Conf, 91-93. *Mem:* Nat Acad Eng; fel Am Soc Civil Engrs; US Nat Comt Rock Mech; Int Soc Rock Mech. *Res:* Application of soil-structure interaction to design of underground structures; seismic criteria for underground structures; improved analysis and design methods for tunnels; author of several books. *Mailing Add:* 7220 S Westmoreland Rd Suite 200 Dallas TX 75237

MONSEN, ELAINE R, b Oakland, Calif, June 6, 35. LIPID METABOLISM, NUTRIENT COMPOSITION OF FOODS. *Educ:* Univ Calif, Berkeley, PhD(nutrit), 61. *Prof Exp:* PROF MED, UNIV WASH, 76- *Mem:* Am Inst Nutrit; Am Dietetic Asn; Am Soc Clin Nutrit. *Mailing Add:* DL-10 Univ Wash Seattle WA 98195

MONSEN, HARRY, b Trondheim, Norway, Aug 24, 24; nat US; m 50; c 2. ANATOMY. *Educ:* Univ Minn, MS, 51; Univ Ill, PhD(anat), 54. *Prof Exp:* Asst anat, Univ Minn, 49-51; asst cancer biol, 51-54, from instr to assoc prof, 54-70, PROF ANAT, UNIV ILL COL MED, 70- *Mem:* Sigma Xi. *Res:* Cancer biology; carcinogenesis; pituitary-adrenal-gonadal interrelationships. *Mailing Add:* Dept Anat M/C 512 Univ Ill Col Med PO Box 6998 Chicago IL 60680

MONSIMER, HAROLD GENE, b Las Vegas, NMex, Feb 5, 28; m 63; c 2. ORGANIC CHEMISTRY. *Educ:* Univ Calif, BS, 52; Wayne State Univ, MS, 54, PhD(chem), 57. *Prof Exp:* Sr res chemist, Nat Drug Co, 56-65; head org chem, MacAndrews & Forbes Co, 65-66; sr res chemist, 66-73, PROJ LEADER, PENNWALT CORP, 73-, SUPVR, 84- *Mem:* Am Chem Soc. *Res:* Medicinal chemistry; organic synthesis. *Mailing Add:* 2916 Toll Gate Dr Norristown PA 19403

MONSON, FREDERICK CARLTON, b Philadelphia, Pa, Aug 3, 39; m 65; c 3. REPRODUCTIVE PHYSIOLOGY. *Educ:* Lehigh Univ, BA, 65, MS, 67, PhD(biol), 71. *Prof Exp:* ASST PROF BIOL, ST JOSEPH'S COL, PA, 71- *Concurrent Pos:* Consult, microscopy, 77. *Mem:* Soc Study Reproduction; Soc Develop Biol; Electron Micros Soc Am; Sigma Xi. *Res:* Physiology of contractility; metabolism and histochemistry of the seminiferons tubule; development of the Sertoli cell population in rat and mouse testes; electron and light microscopy of Myoiol and Sertoli cells in the testes. *Mailing Add:* Div Urol Univ Pa Hosp 3400 Spruce St Philadelphia PA 19104

MONSON, HARRY O, b Feb 21, 19; m 52; c 2. NUCLEAR ENGINEERING. *Educ:* Purdue Univ, BS, 40, MS, 47, PhD(eng), 50. *Prof Exp:* Group leader, Argonne Nat Lab, 50-60, nuclear researcher & sr engr, 60-80; consult, 80-88; RETIRED. *Concurrent Pos:* Mem, Reactor Safeguard, 66-74; sci attache, Geneva Conf. *Mem:* Nat Acad Sci; Am Nuclear Soc; Nat Acad Eng. *Res:* Experimental breeder reactor II; all phases of fast neutron reactor work; decomposition of water by radiation. *Mailing Add:* 171 Eggleston Ave Elmhurst IL 60126

MONSON, JAMES EDWARD, b Oakland, Calif, June 20, 32; m 54; c 3. ELECTRICAL ENGINEERING. *Educ:* Stanford Univ, BS, 54, MS, 55, PhD(elec eng), 61. *Prof Exp:* Mem tech staff, Bell Tel Labs, 55-56; develop engr, Hewlett-Packard Co, 56-61; ROBERT C SABINI PROF ENG, HARVEY MUDD COL, 61- *Concurrent Pos:* Consult, Bell & Howell Res Labs, 61-73 & Spin Physics, Inc, 74-87; Ford Found residency in eng practice, Western Elec Co, 65-66; dir, Pac Measurements, Inc, 66-76; vis prof, Trinity Col, Dublin, 71-72; Fulbright res grant & sr lectr, Tehnichki Fakultet, Titograd, Yugoslavia, 75-76 & 80; Japan Soc Prom Sci fel, Tohoku Univ, Sendai, Japan, 84; vis scientist, TRL, IBM, Japan Ltd, 87; consult, Kodak, 88- *Mem:* Inst Elec & Electronic Engrs; Am Soc Eng Educ; Sigma Xi; Magnetics Soc Japan. *Res:* Magnetic recording. *Mailing Add:* Dept of Eng Harvey Mudd Col Claremont CA 91711-5990

MONSON, PAUL HERMAN, b Fargo, NDak, Sept 29, 25; m 50; c 3. PLANT TAXONOMY. *Educ:* Luther Col, BA, 50; Iowa State Univ, MS, 52, PhD(plant taxon), 59. *Prof Exp:* Instr biol, Luther Col, 52-55; instr bot, Iowa State Univ, 57-58; from asst prof to assoc prof, Univ Minn, Duluth, 58-68, prof biol, 68-90, CUR, OLGA LAKELA HERBARIUM, 74- *Concurrent Pos:* Partic, NSF Acad Year Inst, Brown Univ, 65-66. *Mem:* Nat Asn Biol Teachers; Soc Biol Lab Educ. *Res:* Flora of the midwest and aquatic plants. *Mailing Add:* Dept of Biol Univ Minn 2400 Oakland Ave Duluth MN 55812

MONSON, RICHARD STANLEY, b Los Angeles, Calif, May 28, 37; m 66; c 2. CHEMISTRY. *Educ:* Univ Calif, Los Angeles, BS, 59; Univ Calif, Berkeley, PhD(chem), 64. *Prof Exp:* From asst prof to assoc prof, 63-72, PROF CHEM, CALIF STATE UNIV, HAYWARD, 72- *Concurrent Pos:* Fulbright-Hays fel & lectr, Univ Sarajevo, Yugoslavia, 72-73. *Mem:* Am Chem Soc. *Res:* Organic reaction mechanisms; novel methods of organic synthesis; reactions of organophosphorus intermediates. *Mailing Add:* Dept of Chem Calif State Univ Hayward CA 94542

MONSON, WARREN GLENN, b Clay Center, Nebr, Dec 24, 26; m 58; c 2. AGRONOMY. *Educ:* Univ Nebr, BSc, 51, MSc, 55, PhD, 58. *Prof Exp:* Asst agron, Univ Nebr, 57-58; res agronomist, NY, 58-66, RES AGRONOMIST, ARS, USDA. *Mem:* Fel Am Soc Agron; Crop Sci Soc Am; Am Forage & Grassland Coun. *Res:* Management and quality of pasture and forage crops. *Mailing Add:* 1905 N Atlantic Blvd 8-D Ft Lauderdale FL 33305-3706

MONSON, WILLIAM JOYE, b Menomonie, Wis, Jan 12, 27; m 49; c 1. POULTRY NUTRITION, ANIMAL NUTRITION. *Educ:* Beloit Col, BS, 49; Univ Wis, MS, 51, PhD(biochem), 53. *Prof Exp:* Tech adv, Nutrit Res Lab, Chem Div, Borden Inc, 53-58, dir tech serv, Feed Suppl Div, 58-62, tech dir, Nutrit Res Lab, Chem Div, 62-81, gen mgr, Pet-Ag Div, 81-86; VPRES RES & DEVELOP, PET-AG, INC, 86- *Mem:* Poultry Sci Asn; Animal Nutrit Res Coun (pres, 74); Am Inst Nutrit. *Res:* General nutrition of poultry, large animals and pets with particular emphasis on unknown factors and minerals; critical care nutrition. *Mailing Add:* Vpres Res & Develop Pet-Ag Inc 30W432 Rte 20 Elgin IL 60120

MONSOUR, VICTOR, b Shreveport, La, Aug 28, 22; m 50; c 2. MICROBIOLOGY. *Educ:* La State Univ, Baton Rouge, BS, 48, MS, 50; Univ Tex, Austin, PhD(microbiol), 54. *Prof Exp:* Bacteriologist, Shreveport Charity Hosp, La, 50-51; microbiologist, Confederate Med Ctr, Shreveport, La, 54-57; asst dir bur of lab, Div Health Mo, 57-59; prof microbiol & head dept, McNeese State Univ, 59-84, emer prof, 85; RETIRED. *Concurrent Pos:* Consult, Ark-La-Tex area hosps & clins, 54-56; on loan from Div Health Mo to Sch Med, Wash Univ, 58-59; consult, area industs, La, 68-86. *Mem:* Am Soc Microbiol; Am Chem Soc; Sigma Xi. *Res:* Microbial metabolism; environmental science; chemical and/or biological pollution. *Mailing Add:* 206 McVay St Lake Charles LA 70605

MONT, GEORGE EDWARD, b New Bedford, Mass, Aug 6, 35; m 57; c 4. POLYMER CHEMISTRY. *Educ:* Brown Univ, BS, 57; Clark Univ, MA, 59, PhD(chem), 64. *Prof Exp:* Res chemist, Shawiningan Resins Corp, 61-63; res specialist chem, 63-76, GROUP LEADER, MONSANTO CO, 76- *Res:* Polymer structure-property relationships, poly(vinyl butyral) chemistry. *Mailing Add:* 30 Brentwood Dr Wilbraham MA 01095

MONTAG, MORDECHAI, b Czech, Oct 30, 25; US citizen; m 53; c 4. EQUIPMENT FOR ATOMIC PHYSICS RESEARCH, HIGH VACUUM EQUIPMENT. *Educ:* Israel Inst Technol, Ingeneur, 50. *Prof Exp:* Mech engr, Soltam Ltd, Haifa, Israel, 51-55; design engr, Induction Heating Corp, 55-56; sr engr, Singmaster & Bryer, 56-57; chief mech engr, Semiconductor Div, Gen Instrument Corp, 57-64; mech engr, CBS Labs, 64-65; PROJ ENGR, BROOKHAVEN NAT LAB, 65- *Mailing Add:* Three Malton Rd Plainview NY 11803

MONTAGNA, WILLIAM, b Roccacasale, Italy, July 6, 13; nat US; m 39, 80; c 4. BIOLOGY. *Educ:* Bethany Col, WVa, AB, 36; Cornell Univ, PhD(histol, embryol), 44. *Hon Degrees:* DSc, Bethany Col, WVa, 60; DBiolSci, Univ Sardinia, 64. *Prof Exp:* Instr zool, Cornell Univ, 44-45; from instr to asst prof anat, Long Island Col Med, 45-47; from asst prof to Herbert L Ballou prof biol, Brown Univ, 48-63; dir, Ore Regional Primate Res Ctr & prof exp biol & head div, Med Sch, Univ Ore, 63-82; PROF DERMAT, ORE HEALTH SCI UNIV, 82- *Concurrent Pos:* Spec lectr, Univ London, 53; vis prof, Univ Cincinnati, 58; mem sci comt, Int Cong Dermat, 62; consult, Nat Inst Child Health & Human Develop, 65-; counr, Japan Monkey Ctr, Aichi, 65-; mem adv comt, Washington County Child Develop Prog, Ore, 65-; mem comn natural sci, Nat Bd Fels, Bethany Col, 66-; sci consult, Inst Clin Dermat, Univ Cattolica Sacro Cuore, Rome, Italy, 81-; Harold Cummins Memorial lectr, Sch Med, Tulane Univ, New Orleans, 80; Louis A Duhring lectr, Pa Acad Dermat, Hershey, 81; William Montagna annual lectr, State Univ NY Downstate Med Ctr, 81; Frederick G Novy lectr, Univ Calif, Davis, 81. *Honors & Awards:* Soc Cosmetic Chemists Award, 57; Gold Award, Am Acad Dermat, 58; Gold Medal, Decorated Cavaliere, Ital Repub, 63; Cavaliere Ufficiale, Italian Repub, 69, Commendatore, 75; Stephen Rothman Award in Dermat, 72; Aubrey R Watzek Award, Lewis & Clark Col, 77; Hans Schwarzkopf Res Award, German Dermat Soc, 81. *Mem:* Soc Investigative Dermat; Am Acad Dermat; Geront Soc; NY Acad Sci; Sigma Xi (vpres, 57, pres, 60). *Res:* Cytophysiology; histophysiology and comparative anatomy of the skin; skin of primates; prosimians; primate reproduction. *Mailing Add:* 505 NW 185th Ave Beaverton OR 97005

MONTAGNE, JOHN M, b White Plains, NY, Apr 17, 20; m 42; c 2. PHYSICAL GEOLOGY, ENVIRONMENTAL GEOLOGY. *Educ:* Dartmouth Col, BA, 42; Univ Wyo, MA, 51, PhD(geol), 55. *Prof Exp:* From instr to asst prof geol, Colo Sch Mines, 53-57; from asst prof to prof, 57-83, EMER PROF GEOL, MONT STATE UNIV, 83- *Concurrent Pos:* Supply instr, Univ Wyo, 53-54; mem, Int Field Inst Geol, Italy, 64; mem curric panel, Am Geol Inst; chmn, ad comt, Mont Bur Mines & Geol, 79-82; chmn res comn, Am Asn Avalanche Prof, 86-88, Educ comt, 88-90, pres, 91-92; chmn, Int Snow Sci Workshop, 82; expert witness, snow & geol hazard litigation. *Mem:* Geol Soc Am; Am Inst Prof Geol; Glaciol Soc; Am Quarternary Asn (treas, 70-76); Am Asn Petrol Geol. *Res:* Cenozoic history of the Rocky Mountain region, particularly structural, stratigraphic and geomorphologic aspects; Pleistocene glacial geology and geomorphology; field geology in undergraduate geological education; snow dynamics; geology applied to land use planning. *Mailing Add:* Dept Earth Sci Mont State Univ Bozeman MT 59715

MONTAGUE, BARBARA ANN, b Hagerstown, Md, Aug 29, 29. INFORMATION SCIENCE. *Educ:* Randolph-Macon Woman's Col, AB, 51. *Prof Exp:* Anal chemist, Res Div, Plastics Dept, 51-60, info chemist, 60-61, head plastic dept info syst, 61-64, develop coord, Cent Report Index, Info Syst Dept, 64-67, supvr tech opers, 67-74, mgr info serv photo prods, Int Opers Div, 74-76, mgr info systs develop, 76-77, compensation supvr, 77-80, COMPENSATION CONSULT, CORP HUMAN RESOURCES, E I DU PONT DE NEMOURS & CO, INC, 80- *Concurrent Pos:* Mem adv bd, J Chem Info & Comput Sci, Am Chem Soc, 75-79; chmn, Div Chem Info & Div Activ Comt, Am Chem Soc, mem, Bd Trustees Group Ins Progs, Bd Comt Pensions, Eng Manpower Comn. *Mem:* Am Chem Soc. *Res:* Quantitative organic analysis; spectra-structure correlations qualitative and quantitative using infrared spectroscopy; designing, installation, operation and testing of coordinate indexing systems for storage and retrieval of scientific and technical information; professional pay and progression systems. *Mailing Add:* 668 Meeting House Rd Hockessin DE 19707

MONTAGUE, DANIEL GROVER, b Yakima, Wash, July 7, 37; m 57; c 4. PHYSICS. *Educ:* Ore State Col, BS, 59; Univ Wash, MS, 63; Univ Southern Calif, PhD(physics), 66. *Prof Exp:* Reactor physicist, Gen Elec Co, Wash, 59-61; sr sci officer, Rutherford High Energy Lab, Sci Res Coun, Chilton, Eng, 66-69; from asst prof to assoc prof, 69-81, PROF PHYSICS, WILLAMETTE UNIV, 81- *Concurrent Pos:* Vis scientist, Univ Kent, England, 80-81; scientist-in-residence, Argonne Nat Lab, 87, guest fac res partic, 84, 85, 88- *Mem:* Am Phys Soc; Am Asn Physics Teachers. *Res:* Charged particle scattering and induced reactions for projectiles in the 10 to 50 million electron volts range; neutron and x-ray. *Mailing Add:* Dept of Physics Willamette Univ 900 State St D184 Salem OR 97301

MONTAGUE, DROGO K, b Alpena, Mich, Dec 11, 42; m 67; c 2. IMPOTENCE, GENITOURINARY PROSTHESES. *Educ:* Univ Mich, MD, 68. *Prof Exp:* Intern gen surg, Cleveland Clin, 68-69, resident, 69-70, urol resident, 70-73, HEAD SECT PROSTHETIC SURG, DEPT UROL, CLEVELAND CLIN FOUND, 73-, DIR CTR SEXUAL FUNCTION, 87- *Concurrent Pos:* Mem & consult Exam Comt, Am Bd Urol, 75-87; consult, Am Med Systs, 75-; mem Lithotripsy Training Site Rev Comt, Am Urol Asn, 86-; dir residency training prog, Cleveland Clin Found, 85-; mem ad hoc Impotence Adv Comn, Am Urol Asn, 87-; NAm chmn, Urol Sect, Pan Am Med Asn, 87-88; trustee, Am Bd Urol, 89. *Mem:* Am Col Surgeons; Am Urol Asn; Am Fertility Soc; Int Soc Impotence Res; Pan Am Med Asn; Int Urol Soc. *Res:* Impotence and genitourinary prostheses (various types of penile prostheses and the artificial urinary sphincter). *Mailing Add:* Desk A-100 Cleveland Clin Found One Clin Ctr 9500 Euclid Ave Cleveland OH 44195-5041

MONTAGUE, ELEANOR D, b Genoa, Italy, Feb 11, 26; US citizen; m 53; c 4. RADIOTHERAPY. *Educ:* Univ Ala, BA, 47; Med Col Pa, MD, 50. *Prof Exp:* Resident path, Kings County Hosp, Brooklyn, 52-53; resident radiol, Columbia-Presby Med Ctr, 53-55; radiologist, 6160th US Air Force Hosp, Japan, 55-56; staff physician, Am Tel & Tel Co, NY, 56-57; Am Cancer Soc fel radiother, Univ Tex M D Anderson Hosp & Tumor Inst, 59-61, from asst radiotherapist to radiotherapist, 61-69, assoc prof radiother, 66-69; assoc clin prof radiol, Baylor Col Med, 69-72; prof radiation ther, M D Anderson Hosp & Tumor Inst, Univ Tex, 73-90; RETIRED. *Mem:* Am Col Radiol; Am Radium Soc; Radiol Soc NAm; Am Soc Therapeut Radiol; AMA. *Res:* Clinical use of radiation therapy for treatment of neoplasia. *Mailing Add:* 5230 Yarwell Dr Houston TX 77096

MONTAGUE, FREDRICK HOWARD, JR, b Lafayette, Ind, May 31, 45; m 68; c 1. WILDLIFE ECOLOGY. *Educ:* Purdue Univ, BS, 67, PhD(vert ecol), 75. *Prof Exp:* From res asst wildlife ecol to teaching asst, 70-75, ASST PROF WILDLIFE ECOL & DIR OFF STUDENT SERV, PURDUE UNIV, 75- *Mem:* Wildlife Soc; Am Soc Mammalogists. *Res:* Ecology of wild canids in Midwest; urban wildlife and disease; farm game management; farm habitat improvement projects. *Mailing Add:* 2404 E 800 N Rd Battleground IN 47920-9446

MONTAGUE, HARRIET FRANCES, b Buffalo, NY, June 9, 05. MATHEMATICS. *Educ:* Univ Buffalo, BS, 27, MA, 29; Cornell Univ, PhD(math), 35. *Prof Exp:* Asst, 27-29, from instr to prof, 29-73, dir, NSF Inst Math, 57-70, actg chmn dept math, 61-64, dir undergrad studies, Dept Math, 70-73, EMER PROF MATH, STATE UNIV NY BUFFALO, 73- *Mem:* Am Math Soc; Math Asn Am. *Res:* Mathematics education. *Mailing Add:* 236 Fayette Ave Kenmore NY 14223

MONTAGUE, JOHN H, b Winnipeg, Man, July 16, 25; m 55; c 2. SCIENCE-TECHNOLOGY PLANNING & STRATEGY. *Educ:* Univ Man, BSc, 46; Univ Chicago, SM, 48, PhD(physics), 50. *Prof Exp:* Fel nuclear physics, Nat Res Coun Can, Chalk River, Ont, 50-52; physicist, Assoc Elec Industs, Eng, 52-54; lectr physics, Queen's Univ, Ont, 54-55; sr Harwell fel nuclear physics, Atomic Energy Res Estab, Eng, 55-58, prin scientist, 58-66; prof physics, Queen's Univ, Kingston, Ont, 66-73; prin scientist, Atomic Energy Res Estab, Harwell, Eng, 74-76, head, Bus Develop Group, 76-86; head, Bus Strategy Group, AEA Technol, Harwell, Eng, 86-90; CONSULT, 90- *Concurrent Pos:* Prin res fel, Atomic Energy Res Estab, Harwell, Eng, 71-73; attache to Chief Scientist's Off, Dept of Energy, London, 74-76. *Mem:* Am Phys Soc; Sigma Xi; Royal Soc Arts; Can Asn Physicists. *Res:* Nuclear physics, mainly at low energies with electrostatic generators; research and development management and planning. *Mailing Add:* Seven Surley Row Reading Berks RG48ND England

MONTAGUE, L DAVID, b Washington, DC; m; c 1. DEFENSIVE MISSILE SYSTEMS. *Educ:* Cornell Univ, Bacheloris, 56. *Prof Exp:* Assoc engr, Lockheed, 56-65, chief, Poseidon Missile Develop & proj engr & mgr, Poseidon Systs Eng & mgr, Advan Defense Systs, prog mgr & asst chief engr, Missile Systs Div, vpres, Tactical & Defense Systs, vpres, Missile Systs Div, PRES, MISSILE SYSTS DIV, LOCKHEED MISSILES & SPACE CO, 88- *Concurrent Pos:* Adv, Dept Defense; mem, Defense Sci Bd Task Force. *Mem:* Nat Acad Eng; fel Am Inst Aeronaut & Astronaut. *Res:* Author of various publications; granted 1 patent. *Mailing Add:* 80-01 B/181 Lockheed Missiles & Space Co PO Box 3504 Sunnyvale CA 94088-3504

MONTAGUE, MICHAEL JAMES, b Flint, Mich, Dec 25, 47. PLANT PHYSIOLOGY, CELL BIOLOGY. *Educ:* Univ Mich-Flint, AB, 70, Univ Mich, Ann Arbor, MS, 72, PhD(cell & molecular biol), 74. *Prof Exp:* Res assoc plant physiol, Stanford Univ, 74-75; res group leader, 75-84, mgr res oper, 84-89, RES OPER DIR, MONSANTO CO, 90- *Mem:* AAAS. *Res:* Investigate biochemical events involved in plant development, including the biochemical action of plant hormones and other processes which lead to differentiation. *Mailing Add:* Monsanto Co 800 N Lindbergh Mail Zone 02A St Louis MO 63166

MONTAGUE, PATRICIA TUCKER, b Emporia, Kans, Nov 4, 37; m 65; c 2. MATHEMATICS, COMPUTER SCIENCE. *Educ:* Kans State Univ, BS, 57; Univ Wis-Madison, MS, 58, PhD(math), 61. *Prof Exp:* From instr to asst prof math, Univ Ill, Urbana, 61-67; assoc prof, Univ Tenn, Knoxville, 67-75; assoc prof, Univ Nebr, Omaha, 77-84; ASSOC PROF MATH SCI, METROP STATE COL, DENVER, COLO, 85- *Concurrent Pos:* Vis assoc prof comput sci, Univ Colo, Denver, 84-85; mem, Comput Soc, Inst Elec & Electronics Engrs. *Mem:* Am Math Soc; Math Asn Am; Asn Comput Mach; Inst Elec & Electronics Engrs Comput Soc. *Res:* Algebra; representations of finite groups. *Mailing Add:* 6251 S Leyden St Englewood CO 80111

MONTAGUE, STEPHEN, b Los Angeles, Calif, July 17, 40; m 65; c 2. MATHEMATICS, OPERATIONS RESEARCH. *Educ:* Pomona Col, BS, 62; Univ Ill, Urbana, PhD(math), 67; Univ Tenn, Knoxville, MS, 75. *Prof Exp:* Asst prof math, Univ Tenn, Knoxville, 67-74; asst prof math, Univ Nebr, Omaha, 75-80; tech analyst, TRW Corp, 80-84; STAFF ENGR, MARTIN MARIETTA DENVER AEROSPACE, 84- *Mem:* Asn Comput Mach; Math Asn Am; Oper Res Soc Am. *Res:* Algebra; transitive extentions of finite permutation groups; operations research; optimization techniques; non-linear programming. *Mailing Add:* 6251 S Leyden St Englewood CO 80111

MONTALBETTI, RAYMON, b Cranbrook, BC, Can, Feb 7, 24; m 49; c 3. PHYSICS. *Educ:* Univ Alta, BSc, 46; Univ Sask, PhD(nuclear physics), 52. *Prof Exp:* Res officer physics, Nat Res Coun Can, 46-49; sci officer upper atmospheric physics, Defense Res Bd, 52-64, officer-in-charge, Defense Res North Lab, 58-60; from asst prof to assoc prof, 60-68, PROF PHYSICS, UNIV SASK, 68-, HEAD PHYSICS DEPT, 76- *Mem:* Can Asn Physicists. *Res:* Photonuclear reactions; auroral and upper atmospheric physics. *Mailing Add:* Dept of Physics Univ of Sask Saskatoon SK S7N 0W0 Can

MONTALVO, JOSE MIGUEL, b Cali, Colombia, June 30, 28; m 51; c 5. MEDICINE, PEDIATRICS. *Educ:* Univ Tenn, BS, 51, MD, 57; Am Bd Pediat & Am Bd Pediat Endocrinol, dipl. *Prof Exp:* From intern pediat to chief resident, Frank T Toby Children's Hosp, Univ Tenn, 57-59, Rockefeller Found fel, 58-59; asst prof, Univ Valle, Colombia, 59-60; chief resident, Frank T Toby Children's Hosp, Univ Tenn, 60-61; from instr to prof pediat, Med Ctr, Univ Miss, 62-86, chief div pediat endocrinol, 80-86; PVT PRACT, 86- *Concurrent Pos:* Consult, Miss State Ment Hosp, Whitfield & USPHS Indian Hosp, Philadelphia, 61-86. *Mem:* Endocrine Soc; Am Soc Pediat Nephrology; Am Fedn Clin Res; Int Soc Nephrology; fel Am Acad Pediat; Lawson Wilkins Pediat Endocrine Soc. *Res:* Pediatric endocrine and metabolic disorders. *Mailing Add:* 848 Adams Ave Suite 103 Memphis TN 38103

MONTALVO, JOSEPH G, JR, b Cottonport, La, Oct 30, 37; m 61; c 2. ANALYTICAL CHEMISTRY, SPECTROSCOPY. *Educ:* Univ Southwestern La, BS, 59, MS, 61; La State Univ, New Orleans, PhD(analytical chem), 68. *Prof Exp:* Analytical chemist, Shell Oil Co, 61-64; NSF res fel, La State Univ, New Orleans, 65-68, res fel, 68-69; mgr, Dept Analytical Chem, Gulf South Res Inst, 71-77, staff scientist, 77-79; res leader, 79-83, RES CHEMIST, USDA, AGR RES SERV, SOUTHERN REGIONAL RES CTR, 83-; ANALYTICAL CHEMIST, GULF SOUTH RES INST, 69- *Mem:* Am Chem Soc; Soc Appl Spectros. *Res:* Ion-selective electrodes; dipsticks; personnel badges; dioxin methods development; total organic chlorine instrumentation. *Mailing Add:* SRRR PO Box 19687 New Orleans LA 70179

MONTALVO, RAMIRO A, b Monterrey, Mex, Dec 18, 37; US citizen; m 67; c 2. SOLID STATE PHYSICS. *Educ:* Ill Inst Technol, BS, 60; Northwestern Univ, PhD(physics), 67. *Prof Exp:* Asst physics, Northwestern Univ, 60-66; physicist, Aerospace Res Labs, Off Aerospace Res, US Air Force, 66-69; res scientist, Geomet Inc, 69-73; prin staff mem, Opers Res Inc, 73-76; ELECTRONIC ENGR, US NAVY, 76- *Mem:* Am Phys Soc. *Res:* Magnetic properties of metals at low temperatures; operations research; test and evaluation; data and voice communications; underwater acoustics; computer modeling. *Mailing Add:* Three Simms Court Kensington MD 20895

MONTANA, ANDREW FREDERICK, b Oil City, Pa, Jan 15, 30; m 67. ORGANIC CHEMISTRY. *Educ:* Seattle Pac Col, BS, 51; Univ Wash, PhD(org chem), 57. *Prof Exp:* Asst prof chem, Seattle Pac Col, 55-61 & Univ Hawaii, 61-63; assoc prof & dept chmn, 63-70, PROF CHEM, CALIF STATE UNIV, FULLERTON, 70- *Mem:* AAAS; Am Chem Soc; NY Acad Sci. *Res:* Pseudoaromatics. *Mailing Add:* Dept of Chem Calif State Univ 800 N State College Blvd Fullerton CA 92634

MONTANA, ANTHONY J, b Brooklyn, NY, Apr 1, 50. NUCLEAR MAGNETIC RESONANCE SPECTROSCOPY. *Educ:* York Col, BS, 72; Columbia Univ, MA, 73, PhD(phys chem), 76; Fairleigh Dickinson Univ, MBA, 84. *Prof Exp:* Sr res chemist compositional, GAF Corp, 76-78, group leader polymer characterizations, 79-82; mgr, Dept Analytical Res & Develop, Diamond Shamrock Chem Co, 83-85, & Analysis & Info Serv, 85-88; DIR, CENT ANALYSIS DEPT, ATOCHEM N A, 88- *Concurrent Pos:* Mem bd dirs, Am Oil Chemists Soc, 83-86. *Honors & Awards:* Sect Award, Am Oil Chemists Soc, 90. *Mem:* Am Oil Chemists Soc (pres, 87-89); Am Soc Testing & Mat; Am Chem Soc; Am Mgt Asn; Am Lab Mgrs Asn; AAAS. *Res:* Use of smectic liquid crystals as orienting solvents in nuclear magnetic resonance studies; determination of polymer molecular weight distribution using gel permeation chromatography. *Mailing Add:* Atochem N A PO Box 1295 Somerville NJ 08876-1295

MONTANI, JEAN-PIERRE, b Morat, Switz, July 9, 51; m 79; c 3. CARDIOVASCULAR PHYSIOLOGY, COMPUTER SIMULATION OF BIOLOGICAL SYSTEMS. *Educ:* Med Sch Geneva, Switz, MD, 77. *Prof Exp:* Res asst physiol, Cardiovasc Inst, Univ Fribourg, Switz, 77-79; from intern resident, dept internal med , Univ Hosp Geneva, Switz, 79-82; vis asst prof, 83-85, asst prof physiol, Dept Physiol & Biophys, 86-88, ASSOC PROF PHYSIOL, DEPT PHYSIOL & BIOPHYS, UNIV MISS MED CTR, 88- *Concurrent Pos:* Fed expert, Swiss Fed Comn Experts for Revision of Swiss Fed Regulations for Med Studies Exam, 75-80. *Mem:* Am Soc Hypertension; AAAS; Swiss Fedn Physicians. *Res:* Long-term regulation of arterial blood pressure; sodium excretion and cardiac output in various forms of experimental hypertension; computer simulation of biological systems. *Mailing Add:* Dept Physiol Univ Miss Med Ctr 2500 N State St Jackson MS 39126-4505

MONTANO, PEDRO ANTONIO, b Havana, Cuba, Feb 26, 40; m 62; c 2. SOLID STATE PHYSICS, PHYSICAL CHEMISTRY. *Educ:* Israel Inst Technol, BSc, 67, MSc, 68, DSc(physics), 72. *Prof Exp:* Res assoc, Univ Calif, Santa Barbara, 72-75; asst prof, 75-77, assoc prof solid state physics, 78-80, PROF PHYSICS, WVA UNIV, 80- *Concurrent Pos:* NSF grant, 78; US Airforce, 86; Dept of Energy, 78-87. *Mem:* Am Phys Soc; Mat Res Soc; Am Chem Soc. *Res:* Matrix isolation techniques; characterization of metal clusters; mineral matter in coal; synthetic fuels; catalysis; surface science; synchrotron radiation applications. *Mailing Add:* Dept Physics Brooklyn Col CUNY Brooklyn NY 11210

MONTAZER, G HOSEIN, b Abadan, Iran, Dec 12, 61. HIGH SPEED SIGNAL PROCESSING COMPONENTS, NEW & INNOVATIVE CONVERSION TECHNIQUES. *Educ:* Ga Inst Technol, BSc, 83; Northeastern Univ, MSc, 80. *Prof Exp:* Design engr, Dymel Div, BBF Inc, 83-86; DESIGN ENGR ELECTRONICS, HYBRID SYSTS DIV, SIPEX INC, 86- *Mem:* Inst Elec & Electronics Engrs. *Res:* New product development in data acquisition and conversion area. *Mailing Add:* 62 N Lexden St Brockton MA 02401

MONTEFUSCO, CHERYL MARIE, b New Kensington, Pa, May 14, 48. DONOR ORGAN PROCUREMENT. *Educ:* St Francis Col, Loretto, Pa, BS, 70; NJ Col Med & Dent, PhD(physiol), 75. *Prof Exp:* From asst prof to assoc prof surg, Albert Einstein Col Med, 77-83; coordr, Montefiore Med Ctr, 77-84, proj leader, USPHS-Nat Heart, Lung & Blood Inst, 80-, co-dir lung transplantation, 84-, ASSOC PROF & COORDR DIR, NON-INVASIVE VASC LAB, DEPT SURG, MONTEFIORE MED CTR. *Concurrent Pos:* Guest lectr, Univ Utrecht, Neth, 76. *Mem:* Fel Int Col Angiol; fel Am Col Angiol; Am Physiol Soc; NY Acad Sci; Am Transplant Soc; AAAS. *Res:* Solid organ procurement, preservation and transplantation; hemodynamics of vascular prosthetics; non-invasive vascular diagnostic testing. *Mailing Add:* Dept Surg Montefiore Med Ctr 111 E 210th St Bronx NY 10467

MONTEITH, LARRY KING, b Bryson City, NC, Aug 17, 33; m 52; c 3. ELECTRICAL ENGINEERING, SOLID STATE ELECTRONICS. *Educ:* NC State Univ, BS, 60; Duke Univ, MS, 62, PhD(elec eng), 65. *Prof Exp:* Mem tech staff, Bell Tel Labs, 60-62; res engr, Res Triangle Inst, 62-66, sr scientist, 66-67, group leader mat & devices, 67-68; assoc prof solid state electronics, 68-72, prof, 72-74, prof elec eng & head dept, 74-78, DEAN ENG, NC STATE UNIV, 78- *Concurrent Pos:* Dir, Res Triangle Inst; mem, Am Soc Eng Educ, Prof Engrs NC; chmn, bd dirs, Water Resources Res Inst, 78-; mem, bd trustees, NC Sch Sci & Math, 78-85; mem, Coun NC Inst Transp Res & Educ, 81-87; mem, bd dirs, Nat Driving Ctr, 81-87, Microwave Labs, 82-, NCSU Res Corp, 84-, Res Triangle Inst, 84-; mem, NC Sch Sci & Math Educ Adv Coun. *Mem:* AAAS; Inst Elec & Electronics Engrs; Am Soc Elec Engrs; Nat Soc Prof Engrs. *Res:* Charge transport in organic polymers; metal-organic compounds and silicon oxides; electronic properties and device applications of silicon; electronic materials. *Mailing Add:* Dean Sch Eng NC State Univ Raleigh NC 27695-7901

MONTELARO, JAMES, b Melville, La, Mar 3, 21; m 55; c 1. HORTICULTURE. *Educ:* Southwest La Inst, BS, 41; La State Univ, MS, 50; Univ Fla, PhD(hort), 52. *Prof Exp:* Asst hort, Univ Fla, 52-55; horticulturist, Minute Maid Corp, 55-57; assoc veg specialist, 58-65, VEG CROPS SPECIALIST, UNIV FLA, 65-, PROF, INST FOOD & AGR SCI, 74- *Mem:* Am Soc Hort Sci. *Res:* Nutrition and physiology of vegetable crops. *Mailing Add:* 1605 SW 56th Pl Gainesville FL 32608

MONTELL, CRAIG, b New York, NY, Oct 4, 55; m 87. SIGNAL TRANSDUCTION, VISION. *Educ:* Univ Calif, Berkeley, BA, 78; Univ Calif, Los Angeles, PhD(microbiol), 83. *Prof Exp:* Postdoctoral fel molecular neurobiol, Univ Calif, Berkeley, 83-88; ASST PROF, DEPT BIOL CHEM & DEPT NEUROSCI, JOHNS HOPKINS SCH MED, 88- *Concurrent Pos:* NSF presidential young investr, 89; US-Israel Binat Sci Found Bergmann Mem award, 89; Am Cancer Soc jr fac award, 90. *Res:* Molecular basis of vision; phototransduction cascade of the fruitfly, Drosophila melanogaster. *Mailing Add:* Dept Biol Chem Johns Hopkins Med Sch 725 N Wolfe St Baltimore MD 21205

MONTEMURRO, DONALD GILBERT, b North Bay, Ont, May 27, 30; m 54; c 2. NEUROANATOMY, NEUROENDOCRINOLOGY. *Educ:* Univ Western Ont, BA, 51, MSc, 54, PhD(physiol), 57. *Prof Exp:* Sr res asst, Med Sch, Univ Western Ont, 57-58; Brit Empire Cancer Campaign exchange fel, Chester Beatty Res Inst, Royal Cancer Hosp, London, 58-60; Cancer Inst Can res fel, Sch Med, Yale Univ, 60-61; assoc prof physiol, 61-68, anat, Health Sci Centre, 68-72, chmn dept anat, 73-78, PROF ANAT, HEALTH SCI CENTRE, UNIV WESTERN ONT, 72- *Concurrent Pos:* vis prof, Charing Cross Hosp Med Sch, Dept Anat, London, 78-79; vis prof, Hon Res Fel, Dept Anat, Univ Col, London, Eng, 88-89. *Mem:* Am Asn Anatomists; Can Asn Anat. *Res:* role of the hypothalamus in water and energy metabolism; scanning and transmission electron microscopy; neuroanatomy and neuroendocrine function of the hypothalamus; producer of teaching videotapes in anatomical & neurological sciences. *Mailing Add:* Dept of Anat Univ Western Ont Health Sci Ctr London ON N6A 5C1 Can

MONTENYOHL, VICTOR IRL, b Akron, Ohio, Mar 18, 21; m 46; c 3. CHEMISTRY. *Educ:* Stanford Univ, AB, 42; Princeton Univ, MA, 47, PhD(chem), 50. *Prof Exp:* Res assoc, Princeton Univ, 42-46; res chemist, Pigments Dept, E I du Pont de Nemours, Co, Inc, 46-50, Atomic Energy Div, Explosion Dept, 50-53, res supvr, 53- 74, chief supvr, 74-79; RETIRED. *Mem:* Fel AAAS; Am Chem Soc. *Res:* Surface chemistry; corrosion; non-destructive testing; metallurgy; research planning and funding. *Mailing Add:* 155 Inwood Dr Aiken SC 29803

MONTES, MARIA EUGENIA, b Mexico, DF, Dec 30, 52; m 81; c 2. MATHEMATICS. *Educ:* Univ PR, Mayaguez, BS, 74, Rio Piedras, MA, 79. *Prof Exp:* Instr math, Inter-Am Univ, 78-81, Univ PR, 81-83, CHAIRPERSON, DEPT MATH, AGUADILLA REGIONAL COL, UNIV PR, 83- *Concurrent Pos:* Coordr, activity No 3, Aguadilla Regional Col, Univ PR, 85- *Mem:* Puerto Rican Asn Math Teachers. *Res:* Effectivity of competency based teaching in the basic mathematics courses. *Mailing Add:* PO Box 199 Ramey Br Aguadilla PR 00604

MONTET, GEORGE LOUIS, b Ventress, La, Dec 10, 19; m 46; c 2. CHEMICAL PHYSICS. *Educ:* La State Univ, BS, 40; Univ Chicago, MS, 49, PhD(chem), 51. *Prof Exp:* Plant engr, Bird & Son, Inc, 40-42 & 46; shift supt, Huntsville Arsenal, 42; assoc chemist, Argonne Nat Lab, 51-74, proj leader, 72-78, chemist, 74-81, prog mgr, Environmental Impact Studies, 78-81; RETIRED. *Concurrent Pos:* Fulbright lectr, Univ Ankara, 62-63; adj prof, Northern Ill Univ, 68-71. *Mem:* AAAS; Am Phys Soc; Sigma Xi. *Res:* Solid state physics; quantum chemistry; statistical mechanics; environmental assessment. *Mailing Add:* 52 Mark Twain Rd Rte Two Asheville NC 28805

MONTGOMERIE, ROBERT DENNIS, b Toronto, Ont, Jan 22, 47; m 79; c 3. BEHAVIORAL ECOLOGY, EVOLUTIONARY BIOLOGY. *Educ:* Univ Guelph, BSc, 72; McGill Univ, PhD(biol), 79. *Prof Exp:* Lectr ecol, Univ Victoria, 78-79; fel, McGill Univ, 79-80; asst prof, 80-86, res fel, 80-90, ASSOC PROF BIOL, QUEEN'S UNIV, 86- *Mem:* Int Soc Behav Ecol; Am Soc Naturalists; Soc Study Evolution; Am Ornithologists Union; Animal Behav Soc; Artic Inst N Am. *Res:* Evolutionary ecology; sexual selection; behavioral ecology; territoriality; reproductive strategies; animal coloration; arctic and tropical ecology; birds and insects; pollination ecology; parental care; mating systems. *Mailing Add:* Dept Biol Queen's Univ Kingston ON K7L 3N6 Can

MONTGOMERY, ANTHONY JOHN, optics, for more information see previous edition

MONTGOMERY, CHARLES GRAY, b Philadelphia, Pa, Apr 9, 37; m 66; c 2. THEORETICAL PHYSICS. *Educ:* Yale Univ, BA, 59; Calif Inst Technol, MS, 61, PhD(physics), 65. *Prof Exp:* Vis lectr physics, Hollins Col, 65; from asst prof to assoc prof, 65-75, PROF PHYSICS, UNIV TOLEDO, 75- *Concurrent Pos:* Consult, Owens-Ill, Inc, 65-70 & GTE Labs, Inc, 75-76. *Mem:* Am Phys Soc; Am Asn Physics Teachers. *Res:* Statistical and solid-state physics. *Mailing Add:* Dept Physics Univ Toledo Toledo OH 43606

MONTGOMERY, DANIEL MICHAEL, b Indianapolis, Ind, Nov 2, 43; m 66; c 1. RADIATION ECOLOGY, RADIOCHEMISTRY. *Educ:* St Martin's Col, BS, 65; Purdue Univ, Lafayette, PhD(nuclear chem), 69. *Prof Exp:* Res assoc nuclear chem, Univ Marburg; NSF res assoc, Carnegie-Mellon Univ, 70-71; res chemist, Nat Environ Res Ctr, Environ Protection Agency, 71-74, head radioecol sect, 74-78; RADIATIONS SPECIALIST ENVIRON & SPEC PROJ SECT, US NUCLEAR REGULATORY COMN, 78- *Concurrent Pos:* Vis scientist, Europ Orgn Nuclear Res, Geneva, Switz, 69-70. *Res:* Radiochemical analysis of environmental samples; radioecology; neutron activation analysis; radiological surveillance at nuclear facilities. *Mailing Add:* Analytics Inc 1380 Seaboard Industrial Blvd Atlanta GA 30318-5454

MONTGOMERY, DAVID CAMPBELL, b Milan, Mo, Mar 5, 36; m 57; c 2. PLASMA PHYSICS, STATISTICAL MECHANICS. *Educ:* Univ Wis, BS, 56; Princeton Univ, MA, 58, PhD(physics), 59. *Prof Exp:* Assoc, Proj Matterhorn, Princeton Univ, 59-60; res assoc physics, Univ Wis, 61, instr, 61-62; res asst prof, Univ Md, 62-64; vis researcher, Utrecht (Netherlands), 64-65; from assoc prof to prof physics, Univ Iowa, 65-77; prof physics, Col William & Mary, 77-84; prof, 84-88, ELEANOR & A KELVIN SMITH PROF PHYSICS, DARTMOUTH COL, 88- *Concurrent Pos:* Consult, Oak Ridge Nat Lab, 62-70, Goddard Space Flight Ctr, NASA, 63-64 & Los Alamos Sci Lab, 69-71; vis prof, Univ Colo, 66, Univ Alaska, 68, Univ Calif, Berkley, 69-70, Univ Nagoya, Japan, 83, Columbia Univ, 85, Univ Wis, 89; assoc ed, Physics of Fluids, 71, 72 & 73; assoc ed, Int J Eng Sci, 71-; vis prof, Hunter Col, 73-74, adj prof, 74-75; consult, NASA, 77-; vis res prof, Univ Md, College Park, 77-; consult, NASA Hq, 77-; vis staff mem, Los Alamos Nat Lab, 77-78, 80-81, 82, 86, 87-88. *Mem:* Fel Am Phys Soc; Sigma Xi; Am Geophys Union. *Res:* Theoretical plasma physics; kinetic theory; strongly magnetized plasmas; turbulence theory; non-equilibrium statistical mechanics and transport theory. *Mailing Add:* Dept Physics & Astron Dartmouth Col Hanover NH 03755

MONTGOMERY, DAVID CAREY, b Elmhurst, Ill, Aug 21, 38; m 68, 87. INSTITUTIONAL RESEARCH, PLANNING. *Educ:* Mass Inst Technol, BS, 60; Univ Ill, MS, 61, PhD(physics), 67. *Prof Exp:* From instr to asst prof physics, Oberlin Col, 66-71, registrar & asst provost, 71-73, dir, Inst Res & Planning, 73-74; coordr, State Univ Syst Fla, 75-76, dir planning & analysis, 77-82; sr res scientist, Ctr Nuclear Studies, Memphis State Univ, 82-84; dir, Inst Anal & Studies, Cent Admin, Univ Md, 84-87; VPRES, ACAD AFFAIRS, UNIV MD UNIV COL, 87- *Concurrent Pos:* Consult admin. *Mem:* Am Asn Higher Educ. *Res:* Evaluation of educational programs. *Mailing Add:* Acad Affairs Univ Md Univ Col Univ Blvd Adelphi Rd College Park MD 20742-1600

MONTGOMERY, DEANE, b Weaver, Minn, Sept 2, 09; m 33; c 2. TOPOLOGY. *Educ:* Hamline Univ, BA, 29; Univ Iowa, MS, 30, PhD(math), 33. *Prof Exp:* Nat Res Coun fel, Harvard Univ, 33-34 & Inst Adv Study, 34-35; from asst prof to prof math, Smith Col, 35-46; assoc prof, Yale Univ, 46-48; Guggenheim fel, 41-42, mem, Nat Defense Res Comt Proj, 45-46, mem, Inst Advan Study, 48-51, prof, 51-80, EMER PROF MATH, INST ADVAN STUDY, 80- *Concurrent Pos:* Vis assoc prof, Princeton Univ, 43-45. *Mem:* Nat Acad Sci; Int Math Union (pres, 74-75); Am Philos Soc; Am Math Soc (pres, 61-62). *Res:* Topology; topological groups. *Mailing Add:* Carolina Meadows Apt I-207 Whippoorwill Lane Chapel Hill NC 27514

MONTGOMERY, DONALD BRUCE, b Hartford, Conn, July 1, 33; m 57; c 2. PHYSICS, ELECTRICAL ENGINEERING. *Educ:* Williams Col, BA, 57; Mass Inst Technol, BS & MS, 57; Univ Lausanne, DSc, 68. *Prof Exp:* Mem staff eng, Arthur D Little, Inc, 57-59; mem staff magnet develop, Lincoln Lab, 59-61, group leader magnet develop, 61-78, ASSOC DIR TECHNOL, PLASMA FUSION CTR, MASS INST TECHNOL, 78- *Res:* High field magnet design; fusion engineering and superconductivity. *Mailing Add:* Plasma Fusion Ctr Mass Inst Technol Cambridge MA 02139

MONTGOMERY, DONALD JOSEPH, b Cincinnati, Ohio, June 11, 17; m 42; c 4. SOLID STATE PHYSICS. *Educ:* Univ Cincinnati, ChE, 39, PhD(theoret physics), 45. *Prof Exp:* Instr, Univ Cincinnati, 42-44; res assoc, Princeton Univ, 45-46, asst prof, 46-47; physicist, London Br, Off Naval Res, 47-48, Ballistic Res Labs, 48-50 & Textile Res Inst, 50-53; assoc prof physics, Mich State Univ, 53-56, prof, 56-60, res prof, 60-65, prof metall, mech & mat sci & chmn dept, 66-71, prof physics, 66-88, RES PROF ENG, MICH STATE UNIV, 71-, EMER PROF PHYSICS, 88- *Concurrent Pos:* Fulbright lectr, Univ Grenoble, 59-60, Guggenheim fel, 60; spec asst to dir, Off Grants & Res Contracts, NASA, Washington, DC, 64-65; vis res physicist, Space Sci Lab, Univ Calif, Berkeley, 65-66; mem, Int Inst for Empirical Socioecon, Augsburg-Leiterhofen, 74-75; Fulbright sr researcher & vis prof, Dept Macroecon, Univ Augsburg, Ger; mem, Textile Res Inst; vis scholar, dept polit sci, Univ Ill, 84-86. *Mem:* AAAS; Am Phys Soc; Soc Social Studies Sci; Policy Studies Orgn; Acad Polit Sci; Am Nuclear Soc. *Res:* Materials science, chemical physics; technology and public policy; sociotechnical assessment. *Mailing Add:* Col Eng Mich State Univ East Lansing MI 48824-1226

MONTGOMERY, DOUGLAS C(ARTER), b Roanoke, Va, June 5, 43; m 65; c 3. INDUSTRIAL ENGINEERING. *Educ:* Va Polytech Inst, BS, 65, MS, 67, PhD(indust eng), 69. *Prof Exp:* Instr indust eng, Va Polytech Inst, 65-69; from asst prof to prof indust & systs eng, Ga Inst Technol, 69-84-; John M Fluke Distinguished Prof Mfg Eng, Dir Indust Eng, Prof Mech Eng, Univ Wash, 84-88; PROF ENG, ARIZ STATE UNIV, 88- *Concurrent Pos:* Consult to over 50 US corp. *Mem:* Am Inst Indust Engrs; Am Statist Asn; Opers Res Soc Am; Inst Mgt Sci; fel Am Soc Qual Control. *Res:* Engineering statistics, including experimental design; operations research; inventory theory; forecasting and time series analysis. *Mailing Add:* Dept Indust & Mgt Systs Eng Ariz State Univ Tempe AZ 85044

MONTGOMERY, EDWARD HARRY, b Houston, Tex, July 8, 39; m 63; c 3. PHARMACOLOGY, PHYSIOLOGY. *Educ:* Univ Houston, BS, 61, MS, 63; Univ Tex, PhD(pharmacol), 67. *Prof Exp:* Teaching fel, Univ Houston, 61-63, Nat Inst Dent Res training grant, 63-67; from asst prof to assoc prof pharmacol, Dent Sch, Univ Ore, 67-72; ASSOC PROF PHARMACOL, UNIV TEX DENT BR HOUSTON, 72- *Concurrent Pos:* USPHS gen res serv fund grant, Univ Ore, 67-69, Nat Inst Dent Res grant, 69- *Honors & Awards:* Lehn & Fink Award, 61. *Mem:* AAAS; Sigma Xi; Int Asn Dent Res. *Res:* Role of vasoactive polypeptides as inflammatory mediators; mechanism of action of anti-inflammatory drugs; release of catecholamines by bradykinin and other vasoactive polypeptides; studies of the mediator systems involved in gingival inflammation. *Mailing Add:* Dept Pharmacol Health Sci Ctr Univ Tex PO Box 20036 Houston TX 77225

MONTGOMERY, ERROL LEE, b Roseburg, Ore, May 1, 39; m 60; c 2. HYDROGEOLOGY. *Educ:* Ore State Univ, BS, 62; Univ Ariz, MS, 63, PhD(hydrogeol), 71. *Prof Exp:* Groundwater geologist, Wyo State Engrs Off, 63-65; geohydrologist, Wright Water Engrs, Colo, 65-67; asst prof geol, Northern Ariz Univ, 70-77; prin, Hargis & Montgomery, Inc, 79-84; PRES, ERROR L MONTGOMERY & ASSOC, INC, CONSULTS HYDROGEOL, 84- *Concurrent Pos:* Hydrogeologist, Harshbarger & Assocs, 68-79; Fulbright-Hays scholar, Lisbon, Portugal, 78. *Mem:* Am Geophys Union; Am Water Resources Asn; Asn Eng Geol; Sigma Xi. *Res:* Hydrogeology; applications of geophysics to hydrogeology; aquifer and aquifer systems analysis; engineering geology. *Mailing Add:* 447 E Canyon View Place Tucson AZ 85704

MONTGOMERY, G(EORGE) FRANKLIN, b Oakmont, Pa, May 1, 21; m 67; c 1. ELECTRICAL ENGINEERING. *Educ:* Purdue Univ, BS, 41. *Prof Exp:* Radio engr, Naval Res Lab, Washington, 41-44; electronic scientist, Nat Bur Standards, 46-58, chief, Electronic Instrumentation Div, 58-60, Instrumentation Div, 60-64 & Measurement Eng Div, 64-75, sr eng adv, 75-78; CONSULT ENGR, 78- *Concurrent Pos:* Tech ed, J Audio Eng Soc, 80- *Mem:* Fel Inst Elec & Electronics Engrs; Audio Eng Soc. *Res:* Instrumentation and communication circuit design; modulation theory; product performance. *Mailing Add:* 2806 Kanawha St NW Washington DC 20015

MONTGOMERY, GEORGE PAUL, JR, b Atlanta, Ga, Jan 5, 43; m 68; c 2. ELECTRO-OPTICAL PHYSICS, LIQUID CRYSTALS. *Educ:* Loyola Col, Md, BS, 64; Univ Ill, Urbana, MS, 66, PhD(physics), 71. *Prof Exp:* Nat Acad Sci-Nat Res Coun resident res assoc, US Naval Res Lab, 70-72; STAFF RES SCIENTIST, GEN MOTORS RES LABS, 72- *Mem:* Optical Soc Am; Sigma Xi; Am Phys Soc; Soc Appl Spectros; Soc Info Display. *Res:* Liquid crystal materials and display. *Mailing Add:* Gen Motors Res Labs Physics Dept Warren MI 48090-9055

MONTGOMERY, HUGH, b Austin, Tex, Apr 17, 04; m 30; c 3. MEDICINE. *Educ:* Haverford Col, BS, 25; Harvard Univ, MD, 30. *Prof Exp:* Res, Harvard Med Sch & Marine Biol Lab, Woods Hole, 27-28; intern, Mass Gen Hosp, 31-32; res fel pharmacol, 32-35, instr clin med, 35-41, Heckscher fel, 37-38, Thompson fel, 38-39, assoc med, 41-47, asst prof clin med, 47-52, from assoc prof to prof med, 52-60, EMER PROF MED, SCH MED, UNIV PA, 72- *Mem:* AAAS; Am Soc Clin Invest; Am Med Asn; Asn Am Physicians; fel Am Col Physicians. *Res:* Peripheral circulation; metabolism and oxygen tension of tissue; chemical constitution of glomerular and tubular fluids. *Mailing Add:* Waverly Heights B-220 1400 Waverly Rd Gladwyne PA 19035

MONTGOMERY, HUGH LOWELL, b Muncie, Ind, Aug 26, 44; div; c 2. MATHEMATICS. *Educ:* Univ Ill, BS, 66; Cambridge Univ, PhD(math), 72. *Prof Exp:* From asst prof to assoc prof, 72-75, PROF MATH, UNIV MICH, ANN ARBOR, 75- *Concurrent Pos:* Fel math, Trinity Col, Cambridge Univ, 69-73; Sloan Found fel, 74-77. *Honors & Awards:* Salem Prize, Fr Math Soc, 74. *Mem:* Am Math Soc; Math Asn Am. *Res:* Number theory; analytic number theory. *Mailing Add:* Dept Math Univ Mich Ann Arbor MI 48109

MONTGOMERY, ILENE NOWICKI, b Detroit, Mich, Aug 20, 42; m 68; c 2. NEUROGENETICS, NEUROIMMUNOLOGY. *Educ:* Univ Detroit, BS, 64; Univ Mo, MS, 67; Univ Ill, Urbana, PhD(genetics), 73. *Prof Exp:* Fel molecular biol, Univ, 73-75, fel immunol, 75-77, RES ASSOC IMMUNOL, SCH MED, WAYNE STATE UNIV, 77- *Mem:* Sigma Xi; Genetics Soc Am; Am Genetic Asn. *Res:* Genetics of neuroimmunolgy, with particular emphasis on the genetics of neurologic disorders of immunologic etiology. *Mailing Add:* 1674 Crestline Dr Troy MI 48083-5531

MONTGOMERY, JAMES DOUGLAS, b Morristown, NJ, July 28, 37. BOTANY, TAXONOMY. *Educ:* Bucknell Univ, BS, 59; Rutgers Univ, MS, 61, PhD(bot), 64. *Prof Exp:* From instr to assoc prof biol, Upsala Col, 64-74; res biologist, Ichthyol Assocs, 74-84; ENVIRON STUDIES DIR, ECOLOGY III, 84- *Concurrent Pos:* Pres NY chap Am Fern Soc, 86-88, mem coun, 86- *Mem:* Am Soc Plant Taxonomists; Am Fern Soc; Ecol Soc Am; Torrey Bot Club; Sigma Xi. *Res:* Ecology and effects of power plants; floristics; hybridization and distribution of ferns; fern ecology. *Mailing Add:* Ecology III RR 1 Berwick PA 18603

MONTGOMERY, JOHN ATTERBURY, b Greenville, Miss, Mar 29, 24; m 47; c 4. ORGANIC CHEMISTRY, MEDICINAL CHEMISTRY. *Educ:* Vanderbilt Univ, BA, 46, MS, 47; Univ NC, PhD(chem), 51. *Prof Exp:* Fel, Univ NC, 51-52; chemist, 52-56, head org div, 56-62, dir org chem res, 61-74, vpres, 74-80, SR VPRES & DIR KETTERING MEYER LAB, SOUTHERN RES INST, 81- *Concurrent Pos:* Adj prof, Birmingham Southern Col, 57-62; mem chem adv panel, 60-61, consult, Cancer Chemother Nat Serv Ctr, Nat Cancer Inst, 62-70; consult, 62-63, mem med chem study sect, 64-68 & 71, mem exp therapeut study sect, 75-79, mem biol org & natural prod chem study sect, Health Res Facilities Br, NIH, 81-83; mem bd sci adv, Sloan Kettering Inst, 76-85; adj sr scientist, Comprehensive Cancer Ctr, Univ Ala, 78-; mem ed bds, J Heterocyclic Chem, 65, J Med Chem, 72-74, 74-76, Cancer Treatment Reports, 76-79, J Org Chem, 81-85, Nucleosides & Nucleotides, 82, Anticancer Drug Design, 84-, Pteridines, 88-, Antiviral Res, 88-, Cancer Res Assoc Ed, 84-89; hon prof, Shanghai Inst Materia Medica, Chinese Acad Sci, 87. *Honors & Awards:* Herty Award, Am Chem Soc, 74; T O Soine Award, Unif Minn, 79; Southern Chemist Award, Am Chem Soc, 80 & Alfred Burger Award, 86; Cain Award, Am Asn Cancer Res, 82; Machson Award, Am Chem Soc, 86. *Mem:* AAAS; Am Chem Soc; Sigma Xi; Am Soc Pharmacol & Exp Therapeut; fel NY Acad Sci; Am Asn Cancer Res; Int Soc Heterocyclic Chem. *Res:* Organic syntheses; biochemistry; chemotherapy; pharmacology. *Mailing Add:* Southern Res Inst PO Box 55305 Birmingham AL 35255-5305

MONTGOMERY, JOHN R, b Burnsville, Miss, Oct 24, 34; m 65; c 2. PEDIATRIC INFECTIOUS DISEASES, TRANSPLANTATION. *Educ:* Univ Ala, BS, 55; Med Col Ala, MD, 58. *Prof Exp:* From asst prof to assoc prof pediat, Baylor Col Med, 66-75; PROF & CHIEF PEDIAT, UNIV ALA, HUNTSVILLE, 75- *Mem:* Am Acad Pediat; Soc Pediat Res; Infectious Dis Soc Am; Am Asn Immunologists. *Res:* Immunodeficiency diseases and congenital viral infections of the newborn; care of the "Bubble Baby" over a 12 year period; viral effects on the immunological development of newborn infants, primarily effects of the rubella virus and cytomegalovirus. *Mailing Add:* 201 Governors Dr Huntsville AL 35801

MONTGOMERY, LAWRENCE KERNAN, b Denver, Colo, May 6, 35; m 58; c 3. SOLID STATE PHYSICS. *Educ:* Colo State Univ, BS, 57; Calif Inst Technol, PhD(chem), 61. *Prof Exp:* Fel, Harvard Univ, 60-62; from instr to asst prof, 62-67, ASSOC PROF CHEM, IND UNIV, BLOOMINGTON, 67- *Mem:* Am Chem Soc. *Res:* Reaction mechanisms; organic superconductors. *Mailing Add:* Dept of Chem Ind Univ Bloomington IN 47405

MONTGOMERY, LESLIE D, b Otterbein, Ind, Sept 4, 39; m 71; c 3. BIOMEDICAL ENGINEERING, CARDIOVASCULAR PHYSIOLOGY. *Educ:* Monmouth Col, BA, 61; Iowa State Univ, MS, 63; Univ Calif, Los Angeles, PhD(eng), 72. *Prof Exp:* Mem tech staff III nuclear eng, NAm Rockwell Int, 63-73; sr res engr biomed eng, SRI Int, 80-85; SR ENGR BIOMED ENG, LDM ASSOC, 73- *Concurrent Pos:* Postdoctoral fel, Nat Res Coun, Ames Res Ctr, NASA, 73-75; sr postdoctoral fel, Nat Res Coun, Wright Patterson AFB, 87-89. *Mem:* Aerospace Med Asn; Biomed Eng Soc. *Res:* Noninvasive monitoring of physiologic responses to stressful environments; cerebral hemodynamic and neurologic responses to cognition. *Mailing Add:* 1764 Emory St San Jose CA 95126

MONTGOMERY, M SUSAN, b Tampa, Fla, Apr 2, 43; m 83. ALGEBRA. *Educ:* Univ Mich, Ann Arbor, BA, 65; Univ Chicago, MS, 66, PhD(math), 69. *Prof Exp:* Asst prof math, DePaul Univ, 69-70; asst prof to assoc prof, 70-82, PROF MATH, UNIV SOUTHERN CALIF, 82- *Concurrent Pos:* Vis asst prof, Hebrew Univ Jerusalem, 73; vis assoc prof, Univ Chicago, 78; vis res assoc, Univ Leeds, 81; John Simon Guggenheim mem fel, 84-85. *Mem:* Am Math Soc; Asn Women Math; Math Asn Am; London Math Soc. *Res:* Group actions on rings. *Mailing Add:* Dept Math Univ Southern Calif Los Angeles CA 90089-1113

MONTGOMERY, MABEL D, mathematics, for more information see previous edition

MONTGOMERY, MAX MALCOLM, medicine, for more information see previous edition

MONTGOMERY, MICHAEL DAVIS, b San Luis Obispo, Calif, June 4, 36; m 58; c 4. PHYSICS. *Educ:* Stanford Univ, BS, 58, MS, 59; Univ NMex, PhD(physics), 67. *Prof Exp:* Staff physicist, Los Alamos Sci Lab, 62-75; mem staff, Max Planck Inst Extraterrestrial Physics, 75-77; staff physicist, Los Alamos Sci Lab, 77-83; VPRES RES & DEVELOP, MAXWELL LABS INC, 83- *Concurrent Pos:* Assoc ed, J Geophys Res, Am Geophys Union, 72-74; Alexander v Humboldt Prize, 72-73. *Mem:* Am Phys Soc; Am Geophys Union; NY Acad Sci; Sigma Xi. *Res:* Space plasma physics; laser fusion; electron beam plasma heating. *Mailing Add:* 13545 Mira Montana Dr Del Mar CA 92014

MONTGOMERY, MONTY J, b Longview, Miss, Dec 26, 39; m 61; c 3. DAIRY SCIENCE. *Educ:* Miss State Univ, BS, 61; Univ Wis, MS, 63, PhD(dairy sci), 65. *Prof Exp:* From asst prof to assoc prof, 65-76, PROF ANIMAL SCI, UNIV TENN, KNOXVILLE, 76- *Mem:* Am Dairy Sci Asn. *Res:* Dairy cattle nutrition especially feed intake regulation and forage evaluation; applied dairy cattle feeding. *Mailing Add:* Dept Animal Sci Univ Tenn 208B Brehn Sci Bldg Knoxville TN 37996

MONTGOMERY, MORRIS WILLIAM, b Fargo, NDak, Mar 24, 29; m 50; c 4. FOOD SCIENCE, BIOCHEMISTRY. *Educ:* NDak State Univ, BS, 51, MS, 57; Wash State Univ, PhD(dairy sci, biochem), 61. *Prof Exp:* Qual control supvr dairy tech, Nat Dairy Prod Corp, Wis, 53-54, asst prod mgr, 54-55, prod mgr, 55-57; res assoc food sci, 61-63, asst prof, 63-69, ASSOC PROF FOOD SCI, ORE STATE UNIV, 69- *Concurrent Pos:* US Dept Health, Educ & Welfare grant, 65-67; vis prof, State Univ Campinas, Brazil, 73-74. *Mem:* Inst Food Technologists; Am Chem Soc. *Res:* Enzymic browning of fruits; polyphenol oxidase; fish muscle enzymes. *Mailing Add:* Sch Food Eng Univ Campinas CT 61214 13081 Campinas S P Brazil

MONTGOMERY, PAUL CHARLES, b Philadelphia, Pa, Jan 29, 44; m 64; c 3. IMMUNOLOGY, MICROBIOLOGY. *Educ:* Dickinson Col, BSc, 65; Univ Pa, PhD(microbiol), 69. *Prof Exp:* Smith Kline & French traveling fel, Nat Inst Med Res, London, 69-70; asst prof microbiol, Sch Dent Med, Univ Pa, 70-73, assoc prof, 74-81, chmn immunol grad group & prof microbiol, 81-83; PROF & CHMN, DEPT IMMUNOL & MICROBIOL, SCH MED, WAYNE STATE UNIV, DETROIT, MICH, 83- *Concurrent Pos:* Fogarty sr int fel, Int Inst of Cellular & Molecular Path, Cath Univ Louvain, 78-79. *Mem:* Am Soc Microbiol; Brit Soc Immunol; Am Asn Immunol; AAAS; Asn Med Sch Microbiol Immunol Chairs. *Res:* Mucosal immunity; regulation of secretory IDA antibody induction; mechanisms governing lymphocyte traffic; ocular immunobiology; immune responses to mucosal pathogens. *Mailing Add:* Dept Immunol & Micrbiol Sch Med Wayne State Univ 540 E Canfield Ave Detroit MI 48201-9960

MONTGOMERY, PETER WILLIAMS, b Denver, Colo, May 27, 35; m 60; c 3. PHYSICAL CHEMISTRY. *Educ:* Univ Colo, BA, 57; Univ Calif, Berkeley, PhD(phys chem), 61. *Prof Exp:* Sr res chemist, Cent Res Labs, Minn Mining & Mfg Co, 61-67 & Isotope Power Lab, 67-69; asst prof chem, St Cloud State Col, 69-71; consult & itinerant lectr, 71-74; dir, HMO feasibility study, 74, dir resource develop, Ramsey Action Progs, 74-76; CONSULT HUMAN SERV, HEALTH & SOCIAL SERV PLANNING, 76- *Mem:* AAAS; Sigma Xi. *Res:* High-pressure physics and chemistry; thermodynamics. *Mailing Add:* 1477 Goodrich Ave St Paul MN 55105

MONTGOMERY, PHILIP O'BRYAN, JR, b Dallas, Tex, Aug 16, 21; m 53; c 4. PATHOLOGY. *Educ:* Southern Methodist Univ, BS, 42; Columbia Univ, MD, 45; Am Bd Path, dipl. *Prof Exp:* Intern, Mary Imogene Bassett Hosp, 46; fel path, Univ Tex Southwest Med Sch Dallas, 50-51; asst path & cancer, Cancer Res Inst, New Eng Deaconess Hosp, 51-52; from asst prof to assoc prof, 52-61, NIH career develop award, 62-68, assoc dean, 68-70, PROF PATH, UNIV TEX HEALTH SCI CTR DALLAS, 61- *Concurrent Pos:* Consult path var hosps, 52-; mem sci adv comt, Damon Runyan Mem Fund Cancer Res, 66-72, mem bd dir, 74-; pres bd dirs, Damon Runyon-Walter Winchell Cancer Fund, 74-; spec asst to chancellor, Univ Tex Syst, 71-75; mem bd regents, Uniformed Serv Univ of Health Sci, 74-; pres, Biol Humanics Found, Dallas, 74- *Honors & Awards:* Astronauts' Silver Snoopy. *Mem:* Fel Am Soc Clin Path; AMA; fel Col Am Path; fel NY Acad Sci; fel Royal Micros Soc. *Res:* Pathological aspects of medicolegal cases; ultraviolet irradiation and microscopy; time-lapse photography; cell ultrastructure; carcinogenesis and nucleolar structure and function. *Mailing Add:* Path Dept Univ Tex Southwestern Med Ctr Dallas TX 75235

MONTGOMERY, RAYMOND BRAISLIN, b Philadelphia, Pa, May 5, 10; m 44; c 4. PHYSICAL OCEANOGRAPHY. *Educ:* Harvard Univ, AB, 32; Mass Inst Technol, SM, 34, ScD(oceanog), 38. *Prof Exp:* Asst meteorol, Mass Inst Technol, 35-36, mem staff, 44-45; jr meteorol statistician, Bur Agr Econ, USDA, 36-37; jr oceanogr, Woods Hole Oceanog Inst, 38-40, phys oceanogr, 40-42 & 45-49; assoc prof meteorol, NY Univ, 43-44; vis prof oceanog, Brown Univ, 49-54; from assoc prof to prof, 54-75, EMER PROF OCEANOG, JOHNS HOPKINS UNIV, 75- *Concurrent Pos:* Nat Res Coun fel, Univ Berlin, Ger & Helsinki, Finland, 38-39; vis prof, Scripps Inst, Univ Calif, San Diego, 48; Fulbright res scholar, Commonwealth Sci & Indust Res Orgn, Australia, 58; mem corp, Woods Hole Oceanog Inst, 70-80, hon mem corp, 80-; vis prof, Univ Hawaii, 71; prof, State Univ NY, Stony Brook, 78. *Honors & Awards:* Sverdrup Gold Medal Award, Am Meteorol Soc, 77. *Mem:* Oceanog Soc Japan; hon mem Royal Meteorol Soc; Am Geophys Union. *Res:* Analysis of water characteristics and oceanic flow patterns; oceanic leveling. *Mailing Add:* 44 Whitman Rd Woods Hole MA 02543

MONTGOMERY, REX, b Birmingham, Eng, Sept 4, 23; nat US; wid; c 4. MEDICAL SCIENCES. *Educ:* Univ Birmingham, BSc, 43, PhD(chem), 46, DSc(chem), 63. *Prof Exp:* Res chemist, Colonial Prod Res Coun, Eng, 43-46 & Dunlop Rubber Co, 46-47; sci off, Ministry of Supply, Brit Govt, 47-48; fel, Ohio State Univ, 48-49; Sugar Res Found fel, USDA, 49-51; res assoc, Univ Minn, 51-55; from asst prof to assoc prof, 55-64, PROF BIOCHEM, UNIV IOWA, 64-, ASSOC DEAN, COL MED, 74-, HEAD DIV ASSOC MED SCI, 80- *Concurrent Pos:* Mem staff, Physiol Chem Study Sect, NIH, 68-72, Drug Develop Contract Rev Comt, Nat Cancer Inst, 75-77, Develop Therapeut Comt, Nat Cancer Inst, 77-87; USPHS sr fel, Australian Nat Univ, 69-70; prog dir, Physician's Asst Prog, Col Med, Univ Iowa, 73-76. *Mem:* Am Chem Soc; Am Soc Biol Chem; Royal Soc Chem. *Res:* Carbohydrates; protein-carbohydrate complexes; natural products; glycoproteins; carbohydrases; membrane biochemistry; polypeptide antitumor agents; serospecific antigens of Legionella sp. *Mailing Add:* Dept of Biochem Univ of Iowa Iowa City IA 52242

MONTGOMERY, RICHARD A(LAN), b Vancouver, BC, Jan 11, 19; nat US; m 44; c 6. ELECTRICAL ENGINEERING. *Educ:* Univ BC, BA, 40; Calif Inst Technol, MS, 46, PhD(elec eng), 48. *Prof Exp:* Jr res physicist, Nat Res Coun Can, 41-43; asst, Calif Inst Technol, 46-48; develop engr, Gen Elec Co, 48-51, Boeing Airplane Co, 51-62, Off Secy Defense, 62-64 & Boeing Co, 64-74; mem staff, R & D ASSOCS, 74-; dir, Arroyo Ctr, 83-85; CHMN, MONTGOMERY & ASSOC, 85- *Concurrent Pos:* Consult, Off Secy Defense, 64- & Ballistic Syst Div, US Air Force, 66-69; mem, Army Sci Adv Panel & Army Sci Bd, 68-86. *Mem:* Inst Elec & Electronics Engrs. *Res:* Guided missiles; systems engineer; electronics engineer. *Mailing Add:* 1398 Avenida Decortez Pacific Palisades CA 90272

MONTGOMERY, RICHARD C, geology, for more information see previous edition

MONTGOMERY, RICHARD GLEE, b Grayslake, Ill, Feb 9, 38; m 61; c 3. MATHEMATICS. *Educ:* San Francisco State Univ, AB, 60; Brown Univ, MAT, 65; Clark Univ, MA, 68, PhD(math), 69. *Prof Exp:* Asst prof math, Humboldt State Col, 69-70; assoc prof, 70-81, PROF MATH, SOUTHERN ORE STATE COL, 81- *Mem:* Math Asn Am. *Res:* Algebraic, topological and categorical structures of rings of continuous functions. *Mailing Add:* Dept Math Southern Ore State Col Ashland OR 97520

MONTGOMERY, RICHARD MILLAR, b Cleveland, Ohio, Apr 19, 41; div; c 2. FORENSIC DRUG ANALYSIS, COSMETIC PRODUCT ANALYSIS. *Educ:* Carnegie Inst Technol, BS, 68; Univ Pittsburgh, MS, 73, PhD(chem), 74. *Prof Exp:* Criminalogist, Pittsburgh & Allegheny County Crime Lab, 72-73; dir, lab serv, MSP, Mylan Parmaceut, 73-75; dir, Characterization Ctr, Mellon Inst, 75-80; US mgr, Varian MAT, Varian Assocs, 80-81; mgr, anal dept, Avon Prods, Inc, 81-89; SN MGR, ANALYTICAL CHEM, UNILEVER RES US INC, 89- *Concurrent Pos:* Consult, Am Water Works Serv Co & NUS, Cyrus Rice, 77. *Mem:* Fel Am Inst Chemists; Sigma Xi; Am Soc Mass Spectoscopy; Soc Appl Spectros; Asn Analytical Chem; Forensic Sci Soc London. *Res:* Industrial problem solving through creative analytical chemistry; synthesis and characterization of natural products; problem solving with gas chromatography-mass spectrometry. *Mailing Add:* 17 Old Dutch Pl Bedminster NJ 07921-9716

MONTGOMERY, ROBERT L, b 1935; c 4. SPACECRAFT CONTAMINATION CONTROL. *Educ:* Univ Calif, Berkeley, BS, 56; Okla State Univ, PhD(phys chem), 75. *Prof Exp:* Tech data engr, M W Kellogg Co, Houston, 77-82; sr res assoc, dept chem, Rice Univ, 82-84; STAFF ENGR, MARTIN MARIETTA DENVER AEROSPACE, 84- *Concurrent Pos:* US Calorimetry Conf, Nat Soc Prof Engrs. *Mem:* Sigma Xi; AAAS; Am Chem Soc; Am Soc Met. *Res:* Effects of rocket exhaust, outgassed materials and vented waste products on instruments and systems of spacecraft. *Mailing Add:* Martin Marietta Astronaut Group PO Box 179 Denver CO 80201

MONTGOMERY, RONALD EUGENE, b Rural Valley, Pa, Feb 17, 37; m 61; c 4. ORGANIC CHEMISTRY, PESTICIDE CHEMISTRY. *Educ:* Waynesburg Col, BS, 59; Duke Univ, MA, 61, PhD(org chem), 63. *Prof Exp:* Res assoc, Duke Univ, 63-64; from res chemist to sr res chemist, FMC Corp, 64-69, mgr org synthesis, Agr Chem Div, 69-76, mgr insecticide/nematicide res, 76-80, mgr process res, 80-81, dir, process res & eng, 81-88, dir, Develop Chem, 88-90, DIR, RES & DEVELOP, FMC CORP, 90- *Mem:* Am Chem Soc. *Res:* Biological laboratory evaluation, field testing, process research and chemical manufacturing pesticide residue and metabolism chemistry; structure-activity relationships of organic chemicals. *Mailing Add:* 297 Cinnabar Lane Yardley PA 08543

MONTGOMERY, ROYCE LEE, b Hartsville, Tenn, Nov 8, 33; m 67; c 1. GROSS ANATOMY, NEUROANATOMY. *Educ:* Univ Va, BA, 55; WVa Univ, MS, 60, PhD(gross anat), 63. *Prof Exp:* Instr gross anat & neuroanat, Med Sch, WVa Univ, 63-65; from instr to asst prof, 68-72, ASSOC PROF GROSS ANAT & NEUROANAT, SCH MED, UNIV NC, CHAPEL HILL, 72- *Mem:* Am Asn Anat. *Res:* Morphological changes in aging intervertebral disks. *Mailing Add:* Dept Anat Univ NC Sch Med Chapel Hill NC 27514

MONTGOMERY, STEWART ROBERT, b Pottsville, Pa, July 16, 24; m 60. INDUSTRIAL CHEMISTRY, ORGANIC CHEMISTRY. *Educ:* Pa State Univ, BS, 49; Univ Rochester, PhD(org chem), 55. *Prof Exp:* Res chemist org synthesis, Cent Res Lab, Allied Chem Corp, 49-51; res chemist petrochem, Res Dept, Lion Oil Co, Monsanto Co, 55-61; sr res chemist, Indust Catalysts Res Dept, 61-73 & 78-84, Indust Chem Tech Ctr, 73-78, INDUST CATALYSTS RES DEPT, DAVIDSON DIV, W R GRACE CO, 84- *Mem:* Am Chem Soc; Catalysis Soc NAm; Sigma Xi. *Res:* Synthesis of amino acids, organosilicon compounds and aromatic hydrocarbons; heterogeneous, vaporphase catalysis; synthetic resins; chemistry of asphalt and bitumens; oxidation of hydrocarbons; Raney nickel; preparation and evaluation of industrial catalysts. *Mailing Add:* 17943 Pond Rd Ashton MD 10861-9756

MONTGOMERY, THEODORE ASHTON, b Los Angeles, Calif, Oct 27, 23. PUBLIC HEALTH, PEDIATRICS. *Educ:* Univ Southern Calif, MD, 46; Harvard Univ, MPH, 52; Am Bd Pediat & Am Bd Prev Med, dipl. *Prof Exp:* Consult pediat, Calif Dept Pub Health, 52-53; pvt pract, 52-54; consult child health, Calif Dept Pub Health, 54-58, chief maternal & perinatal health, 58-61 & bur maternal & child health, 61-62, from asst chief to chief div prev med serv, 62-72, dep dir, 69-72, chief div prev med, Alameda County Health Agency, 73-74, med consult, Calif Dept Health Serv, 74-83. *Concurrent Pos:* Mem, Surg Gen Adv Comt Immunization Practices, 64-68, President's Adv Comt Ment Retarded, 65 & Ment Retarded Proj Rev Comt, USPHS, 65-66; chmn, Calif Interdept Coun on Food & Nutrit, 77- *Mem:* Fel Am Acad Pediat; fel Am Pub Health Asn. *Res:* Maternal and child health morbidity and mortality causation and procedures for reducing these factors, including designing and implementing approaches to preventive measures. *Mailing Add:* 85 Wildwood Gardens Piedmont CA 94611

MONTGOMERY, WILLIAM WAYNE, b Proctor, Vt, Aug 20, 23. OTOLARYNGOLOGY. *Educ:* Middlebury Col, AB, 44; Univ Vt, MD, 47; Am Bd Otolaryngol, dipl, 56. *Prof Exp:* Intern, Mary Fletcher Hosp, Burlington, Vt, 47-48; physician pvt pract, Vt, 48-50; resident otolaryngol, Mass Eye & Ear Infirmary, 52-55, asst, 55-56; asst otol, 56-58, from instr to assoc prof, 59-70, PROF OTOLARYNGOL, HARVARD MED SCH, 70-; SR SURGEON, DEPT OTOLARYNGOL, MASS EYE & EAR INFIRMARY, 69- *Concurrent Pos:* Asst surgeon, Mass Eye & Ear Infirmary, 58-60, assoc surgeon, 60-69. *Honors & Awards:* Harris P Mosher Award, 63. *Mem:* Fel Am Col Surgeons; AMA; Am Acad Ophthal & Otolaryngol; Am Broncho-Esophagol Asn; Am Laryngol Asn. *Res:* Dysfunctions of the human larynx; reconstruction of the cervical respiratory areas; carcinoma of the head and neck; radical surgery of the nose and sinuses; surgery of the upper respiratory system. *Mailing Add:* Dept Otolaryngol Harvard Med Sch 100 Charles River Plaza Boston MA 02114

MONTGOMERY, WILLSON LINN, b Detroit, Mich, May 8, 46; m 73. BEHAVIORAL ECOLOGY, ICHTHYOLOGY. *Educ:* Univ Calif, Berkeley, BA, 68; Univ Calif, Los Angeles, MA, 73; Ariz State Univ, PhD(zool), 78. *Prof Exp:* ACTG ASST PROF ZOOL, ARIZ STATE UNIV, 79-;; AT BIOL SCI DEPT, NORTHERN ARIZ UNIV. *Mem:* Am Soc Ichthyologists & Herpetologists; Sigma Xi; Am Soc Zoologists. *Res:* Physiological and behavioral bases of ecological patterns in fishes; impact of herbivorous fishes on marine algal communities; interspecific feeding associations of fishes. *Mailing Add:* Biol Sci Dept Box 5640 Northern Ariz Univ Nau Box 4084 Flagstaff AZ 86011-5640

MONTH, MELVIN, b Montreal, Que, June 20, 36; m 57; c 2. ACCELERATOR PHYSICS, MANAGEMENT. *Educ:* McGill Univ, BSc, 57, MSc, 61, PhD(physics), 64; Hofstra Univ, MBA, 73. *Prof Exp:* Res assoc physics, Univ Ill, Urbana, 64-66; PHYSICIST ACCELERATOR PHYSICS, BROOKHAVEN NAT LAB, 66- *Concurrent Pos:* Consult, High Energy Physics Off, Dept of Energy, 79- *Mem:* AAAS; fel Am Phys Soc. *Res:* Particle beams in accelerators; analysis of beam stability at high currents; limitations in colliding beam performance; analysis of facility development. *Mailing Add:* Accelerator Dept Brookhaven Nat Lab Upton NY 11973

MONTI, JOHN ANTHONY, b Birmingham, Ala, Aug 7, 49; m 72; c 2. PHYSIOLOGY, BIOPHYSICS. *Educ:* Birmingham-Southern Col, BS, 71; Univ Ala, Birmingham, PhD(physiol, biophys), 75. *Prof Exp:* Res assoc, Univ Ala, Birmingham, 76-79, res asst prof neurobiol, Neurosci Prog, 79-85. *Mem:* Soc Neurosci; Biophys Soc. *Res:* Membrane biochemistry and biophysics; central nervous system metabolism; fluorescence spectroscopy, theory and applications. *Mailing Add:* Sparks Ctr Univ Ala Birmingham AL 35294

MONTI, STEPHEN ARION, b Ross, Calif, Nov 23, 39; m 81; c 2. ORGANIC CHEMISTRY. *Educ:* Univ Calif, Berkeley, BS, 61; Mass Inst Technol, PhD(org chem), 64. *Prof Exp:* Asst prof chem, Mich State Univ, 64-66; NSF fel, Harvard Univ, 66-67; from asst prof to assoc prof, 67-79, asst to the pres, 74-81, assoc vpres Acad planning 81-84, assoc vpres Acad Affairs Res, 84-86, PROF CHEM, UNIV TEX, AUSTIN 79-, VICE PROVOST, 86- *Mem:* Am Chem Soc. *Res:* Synthetic organic chemistry; natural products and related areas. *Mailing Add:* Off the Exec VPres & Provost MAI 201 Univ of Tex Austin TX 78712

MONTIE, THOMAS C, b Cleveland, Ohio, Oct 8, 34; m 56, 79; c 2. BIOCHEMISTRY, MICROBIOLOGY. *Educ:* Oberlin Col, AB, 56; Univ Md, MS, 58, PhD(microbial physiol), 60. *Prof Exp:* Asst bot, Univ Md, 56-58, fungus physiol, 59-60; res assoc, Inst Cancer Res, 60-62; asst mem, Albert Einstein Med Ctr, 62-67, assoc mem, 67-69; assoc prof, 69-78, PROF MICROBIOL, UNIV TENN, KNOXVILLE, 78- *Concurrent Pos:* Prin investr, Cystic Fibrosis Grant, 80-81; Army contract, Pseudomonas vaccine, 82-87; indust consult, 83, 90 & 91; sci alliance awards, Univ Tenn, Knoxville, 85-88. *Mem:* AAAS; Am Soc Microbiol; Am Soc Biochem & Molecular Biol. *Res:* Location, synthesis, regulation, mode of action and chemistry of toxic proteins from bacteria; microbial physiology; amino acid transport and metabolism, chemotaxis; pseudomonas pathogenicity; vaccines; vaccine patents foreigh and United States. *Mailing Add:* Dept Microbiol Univ Tenn Knoxville TN 37996-0845

MONTIERTH, MAX ROMNEY, b Phoenix, Ariz, Dec 12, 38; m 59; c 5. CERAMIC & CHEMICAL ENGINEERING. *Educ:* Brigham Young Univ, BES, 63; Univ Utah, PhD(ceramic eng), 67; Harvard Sch Bus, PMD, 72. *Prof Exp:* Sr ceramic engr, Tech Staffs, Corning Glass Works, 66-70, mgr glass-ceramic prod develop, 70-72, mgr mkt & prod develop, Electronic Mat, TV Div, 72-74, mgr special prod develop, 74-75, mgr prod develop, Optical Waveguide Technol, 75-79, tech leader, indust heat exchanges, 79-80, composite develop, 85-86, tech leader, Diesel Particulate Filter Tech Staffs Div, 80-88, MGR, TECHNOL & DEVELOP, CORNING KERAMIK GMBH & COKG, EUROPE, 88- *Concurrent Pos:* Mem staff, Elmira Col, 68-70. *Mem:* Am Ceramic Soc; Soc Automotive Engrs. *Res:* Effects of crystal morphology and composition on physical properties of glass-ceramics; glass, glass-ceramic and ceramic materials for electronic applications; glass composition and fiber forming optical waveguides; polymer coatings for OWG fibers; high temperature low-expansion ceramics, heat exchange materials and mechanisms; diesel particulate filter; durability and life testing of ceramics. *Mailing Add:* Corning Keramik GmbH & COKG Abraham Lincoln Str 30 Wiesbaden 6200 Germany

MONTIETH, RICHARD VOORHEES, b Indianapolis, Ind, Aug 13, 13; m 41; c 4. CHEMISTRY. *Educ:* Butler Univ, AB, 35. *Prof Exp:* Chemist, Allison Div, 40-44, supvr bearing chem lab, 44-54, gen supvr prod metall lab, 54-58, supvr prod metall sect, 58-64, sr exp metallurgist foundry technol, 64-66, sr exp metallurgist, Detroit Diesel Allison Div, Gen Motors Corp, 66-74; RETIRED. *Concurrent Pos:* Nutrit consult, 74- *Mem:* Am Soc Metals. *Res:* Metallurgical, nonferrous and ferrous and high-temperature alloys, selection and properties. *Mailing Add:* 5475 Happy Hollow Indianapolis IN 46268

MONTJAR, MONTY JACK, b Latrobe, Pa, Aug 12, 24. PHYSICAL CHEMISTRY. *Educ:* St Vincent Col, BS, 49; Univ Notre Dame, MS, 50; Carnegie Inst Technol, PhD(phys chem), 55. *Prof Exp:* Chemist, Callery Chem Co, 54-57; asst prof chem, Pa State Univ, 57-60; from asst prof to assoc prof, St Vincent Col, 60-65; assoc scientist, Brookhaven Nat Lab, 66-68; assoc prof chem, Mt Sinai Sch Med, 68-70; ASSOC PROF CHEM, PA STATE UNIV, HAZLETON, 70- *Mem:* Sigma Xi. *Res:* RNA; protein biosynthesis; tissue transplantation; fractionation; peptide synthesis. *Mailing Add:* 324 Chestnut St Latrobe PA 15650

MONTO, ARNOLD SIMON, b Brooklyn, NY, Mar 22, 33; m 58; c 4. EPIDEMIOLOGY, INFECTIOUS DISEASES. *Educ:* Cornell Univ, BA, 54, MD, 58. *Prof Exp:* Intern & asst resident med, Vanderbilt Univ Hosp, 58-60; USPHS fel, Med Sch, Stanford Univ, 60-62; mem virus dis sect, Nat Inst Allergy & Infectious Dis, 62-65; from res assoc to assoc prof, 65-75, PROF EPIDEMIOL, SCH PUB HEALTH, UNIV MICH, ANN ARBOR, 75- *Concurrent Pos:* NIH career develop award, 68; adv coun, Nat Inst Allergy & Infectious Dis. *Mem:* Am Epidemiol Soc; Soc Exp Biol & Med; Am Asn Immunologists; Infectious Dis Soc Am; Soc Epidemiol Res. *Res:* Epidemiology of respiratory and enteric disease in the community; viral diseases and diagnosis; evaluations of vaccines. *Mailing Add:* Sch Pub Health Univ Mich Ann Arbor MI 48109

MONTOURE, JOHN ERNEST, b Shawano, Wis, May 10, 27; m 48; c 4. DAIRY SCIENCE, AGRICULTURAL BIOCHEMISTRY. *Educ:* Univ Wis, BS, 54, MS, 55; Wash State Univ, PhD(dairy sci), 61. *Prof Exp:* Asst prof dairy sci, 61-70, head dept food sci, 70-75, ASSOC PROF FOOD SCI, UNIV IDAHO, 70-, FOOD SCIENTIST, 73- *Mem:* Am Dairy Sci Asn; Inst Food Technol. *Res:* Pesticide detection and analytical methods; fate of pesticides in dairy products; enzyme isolation; bacteriology. *Mailing Add:* 1121 E 7th St Moscow ID 83843

MONTROSE, CHARLES JOSEPH, b Pittsburgh, Pa, Jan 3, 42; m 63; c 3. PHYSICS. *Educ:* John Carroll Univ, BS, 62, MS, 64; Cath Univ Am, PhD(physics), 67. *Prof Exp:* Asst prof, 67-70,chmn dept, 83-86 ASSOC PROF PHYSICS, CATH AM UNIV, 70-,. *Mem:* Am Phys Soc; Bioelectromagnetics Soc. *Res:* Study of the structure and dynamics of liquids and glasses using computer simulation and optical techniques; interaction of electromagnetic fields with biological molecules and systems. *Mailing Add:* Dept Physics Cath Univ Am Washington DC 20064

MONTS, DAVID LEE, b Taylorville, Ill, Apr 14, 51; m; c 3. CHEMISTRY, PHYSICS. *Educ:* Univ Ill, Urbana, BS, 73; Columbia Univ, MA, 74, PhD(chem), 78. *Prof Exp:* Fel chem, William Marsh Rice Univ, 77-79; lectr, Princeton Univ, 79-81; asst prof chem & prin investr, Univ Ark, Fayetteville, 81-88; ASST PROF PHYSICS, MISS STATE UNIV, 88- *Mem:* Am Chem Soc; Am Phys Soc; Sigma Xi; Soc Appl Spectros. *Res:* Identification and characterization of molecules and atoms by laser spectroscopy; development of new experimental techniques for identification and characterization of molecular species, including laser optogalvanic spectroscopy and differential absorption laser spectroscopyl; investigation of combustion processes. *Mailing Add:* Dept Physics & Astron Miss State Univ PO Box 3574 Mississippi State MS 39762-3574

MONTVILLE, THOMAS JOSEPH, b Somerville, NJ, Jan 10, 53; m 76; c 3. FOOD SAFETY MICROBIOLOGY, FOOD BIOTECHNOLOGY. *Educ:* Rutgers Univ, BS, 75; Mass Inst Technol, PhD(food sci), 79. *Prof Exp:* Res microbiologist, Eastern Regional Res Ctr, USDA, 80-84; assoc prof, 84-91, PROF FOOD MICROBIOL, COOK COL, RUTGERS UNIV, 91- *Concurrent Pos:* Consult & expert witness, 85-; course dir, Ctr Prof Advan, 86-; co-ed, J Food Safety, 88-; dir, Grad Prog Food Sci, Rutgers Univ, 91-; chmn, Biotechnol Div, Inst Food Technologists, 91-92. *Mem:* Inst Food Technologists; Am Soc Microbiol; AAAS; Soc Indust Microbiol; Soc Appl Bact; Int Asn Food Dairy & Environ Sanitarians. *Res:* Food safety and fermentation microbiology; physiology of lactic acid bacteria; novel antimicrobials; bacteriocins; application of biotechnology to food science. *Mailing Add:* Dept Food Sci Rutgers Univ New Brunswick NJ 08903-0231

MONTY, KENNETH JAMES, b Sanford, Maine, Sept 11, 30; m 52; c 2. BIOCHEMISTRY, CELL BIOLOGY. *Educ:* Bowdoin Col, BA, 51; Univ Rochester, PhD(biochem), 56. *Prof Exp:* Fel biochem, McCollum-Pratt Inst, Johns Hopkins Univ, 55-57, asst prof biol, 57-63; head dept, 63-75, 77-79, coordr biol sci, 73-84, PROF BIOCHEM, UNIV TENN, KNOXVILLE, 63- *Concurrent Pos:* Dir, Tenn Govs Sch Sci, 89 & 90, Univ Tenn Acad Teachers Sci & Math, 91. *Mem:* AAAS; Am Chem Soc; Am Soc Biochem & Molecular Biol; NY Acad Sci. *Res:* Cellular physiology; sulfur metabolism; microbial genetics and metabolic control; biological adaptation. *Mailing Add:* Dept Biochem Univ Tenn Knoxville TN 37996-0840

MONTZINGO, LLOYD J, JR, b Chicago, Ill, Nov 29, 27; m 48; c 6. MATHEMATICS. *Educ:* Houghton Col, AB, 49; Univ Buffalo, MA, 51, PhD, 61. *Prof Exp:* From instr to asst prof math, Roberts Wesleyan Col, 51-56; from instr to asst prof, Univ Buffalo, 56-62; assoc prof, 62-66, chmn dept, 62-73, dir, Sch Sci, 73-78, PROF MATH, SEATTLE PAC UNIV, 66- *Concurrent Pos:* NSF sci fac fel, Univ Wash, 70-71. *Mem:* Nat Coun Teachers Math; Math Asn Am (assoc secy, 58-62). *Res:* Mathematical statistics. *Mailing Add:* Sch Sci Seattle Pac Univ Seattle WA 98119

MONZYK, BRUCE FRANCIS, b Washington, Mo. HIGH OXIDATIONS STATES OF TRANSITION METALS. *Educ:* Univ Mo, Columbia, BS, 72, MS, 76; Duke Univ, PhD(bio-inorg chem), 80. *Prof Exp:* Res chemist, 80-81, sr res chemist, 82-83, RES SPECIALIST RES & DEVELOP, MONSANTO CO, 84-, GRAD RECRUITER, 83- *Concurrent Pos:* Mem, adv comt, dept chem, Univ Mo, Columbia, 84- *Mem:* Am Chem Soc; Am Inst Mining, Metall & Petrol Engrs. *Res:* Coordination chemistry; hydrometallurgy; bio-inorganic chemistry; reinforced plastic composites; biological sequestration and transport of iron and copper; kinetics and thermodynamics of ligand substitution processes involving transition metals; biomass utilization. *Mailing Add:* 2066 Pheasant Run Maryland Heights MO 63043

MOOBERRY, JARED BEN, b Pekin, Ill, Mar 6, 42; m 75. SYNTHETIC ORGANIC CHEMISTRY. *Educ:* Univ Ill, BS, 64; Cornell Univ, PhD(chem), 69. *Prof Exp:* Fel chem, Swiss Fed Inst Technol, 69-70; res assoc, Cornell Univ, 70-71; RES CHEMIST, EASTMAN KODAK RES LABS, 71- *Mem:* Am Chem Soc. *Res:* Synthesis of organic chemicals of photographic utility. *Mailing Add:* 175 Huntington Hills Rochester NY 14622-1135

MOOD, ALEXANDER MCFARLANE, b Amarillo, Tex, May 31, 13; m 36; c 3. APPLIED MATHEMATICS, PUBLIC POLICY. *Educ:* Univ Tex, BA, 34; Princeton Univ, PhD(math statist), 40. *Prof Exp:* Instr appl math, Univ Tex, 40-42; statistician, Bur Labor Statist, 42-44; res assoc, Princeton Univ, 45; prof math statist, Iowa State Col, 45-48; dep chief, Math Div, Rand Corp, 48-55; pres, Gen Anal Corp, 55-60; vpres-in-charge, Los Angeles Res Ctr, C-E-I-R, Inc, 60-64; asst commissioner ed, US Off Ed, 64-67; dir, Pub Policy Res Orgn, 67-73, prof, 73-77, EMER PROF ADMIN & POLICY ANALYST, UNIV CALIF, IRVINE, 73- *Concurrent Pos:* Guest lectr, Army War Col & Air War Col; consult, US Dept Defense; del, NATO Opers Res Conf, Paris, 58, Int Opers Res Conf, Aix-en-Provence, 60 & Oslo, 63. *Mem:* Am Statist Assoc; Opers Res Soc Am (pres, 63); fel Inst Math Statist (pres, 56). *Res:* Mathematical statistics; operations research; game theory; administration; policy analysis. *Mailing Add:* 4705 Royce Rd Irvine CA 92715

MOODERA, JAGADEESH SUBBAIAH, b Bangalore, India, Dec 3; m 82. SOLID STATE PHYSICS. *Educ:* Mysore Univ, India, BSc, 71, MSc, 73; Indian Inst Technol, Madras, India, PhD(solid state physics), 78. *Prof Exp:* Fel, physics, Indian Inst Technol, India, 78; res assoc, magnetism, Physics Dept, WVa Univ, 79-81; MEM RES STAFF, SUPERCONDUCTIVITY, TUNNELING & MAGNETISM, NAT MAGNET LAB, MASS INST TECHNOL, 81- *Concurrent Pos:* consult. *Mem:* Mat Res Soc; Am Phys Soc. *Res:* Electrical and magnetic properties of thin film; magnetic properties of antiferromagnets and semiconductors; electron spin polarized tunneling into ferromagnets from superconductors; STM, high Tc superconductors (bulk and films). *Mailing Add:* NW14-2114 Mass Inst Technol 170 Albany St Cambridge MA 02139

MOODY, ARNOLD RALPH, b Augusta, Maine, Oct 8, 41; m 65. PLANT PATHOLOGY. *Educ:* Univ Maine, BS, 63; Univ NH, MS, 65; Univ Calif, Berkeley, PhD(plant path), 71. *Prof Exp:* Asst res plant pathologist, Univ Calif, Berkeley, 71-72; res plant pathologist, Changins Fed Agr Exp Sta, Nyon, Switz, 72-74; assoc prof plant path, VA State Univ, 74-84; RES LAB, WATER PLANT, APPOMATOX RIVER WATER AUTHORITY, 90- *Mem:* Am Phytopath Soc. *Res:* Plant diseases; control of plant disease through natural resistance mechanisms; physiology of disease resistance. *Mailing Add:* 967 Edinborough Dr Colonial Heights VA 23834-2611

MOODY, CHARLES EDWARD, JR, b Portsmouth, NH, Feb 14, 48. MICROBIOLOGY. *Educ:* Providence Col, BA, 70; Univ RI, MSc, 73, PhD(microbiol), 76. *Prof Exp:* Res fel immunol, dept med, Med Col, Cornell Univ, 76-79, asst prof, 79-82; ASSOC PROF IMMUNOL, DEPT MICROBIOL, UNIV ME, 82- *Concurrent Pos:* Adj prof, dept microbiol, Cornell Med Col, 81-82; prin investr, NIH grant, 82-85 & 87-, Nat Oceanic & Atmospheric Admin, 85-87. *Mem:* Am Asn Immunologists; NY Acad Sci; Harvey Soc; Int Soc Comparative & Develop Immunol; Sigma Xi; Am Soc Microbiol. *Res:* Immunology of teleost fish, primarily salmonids; role of carbohydrates in modulation of immune response and cancer; syngeneic mixed lymphocyte reaction. *Mailing Add:* Dept Microbiol Univ Maine Rm 279 Hitcher Hall Orono ME 04469

MOODY, DAVID BURRITT, b Rochester, NY, June 20, 40; m 64; c 2. PSYCHOACOUSTICS. *Educ:* Hamilton Col, BA, 62; Columbia Univ, MA, 64, PhD(psychol), 67. *Prof Exp:* Dept HEW res fel, Kresge Hearing Res Inst, Med Sch, 67-68, res assoc psychoacoustics, 68-71, asst prof otorhinolaryngol, 71-79, asst prof psychol, Univ, 69-80, assoc res scientist otorhinilaryngol, 79-88, assoc prof psychol, 80-89, RES SCIENTIST OTORHINILARYNGOL, KRESGE HEARING INST, 88-, PROF PSYCHOL, MED SCH, UNIV MICH, ANN ARBOR, 89- *Mem:* Acoust Soc Am; Asn Res Otorhinolaryngol; Sigma Xi. *Res:* Psychoacoustics; hearing in non-human subjects; experimentally produced hearing loss; computers in psychology. *Mailing Add:* Kresge Hearing Res Inst Univ Mich Ann Arbor MI 48109

MOODY, DAVID COIT, III, b Florence, SC, Dec 21, 48; m 68; c 2. INORGANIC CHEMISTRY. *Educ:* Univ SC, BS, 67; Ind Univ, PhD(chem), 75. *Prof Exp:* Res asst chem, Ind Univ, 72-75; fel chem, 75-77, MEM STAFF, LOS ALAMOS NAT LAB, 77- *Mem:* Am Chem Soc; Sigma Xi. *Res:* Organometallic actinide chemistry; waste treatment; nuclear medicine. *Mailing Add:* 402 Pine Tree Lane Boulder CO 80304

MOODY, DAVID EDWARD, b Oak Lawn, Ill, Mar 30, 50; m 72; c 1. XENOBIOTIC-METABOLISM, FORENSIC TOXICOLOGY. *Educ:* Univ Kans, BA, 72, PhD(exp path), 77. *Prof Exp:* Postdoctoral fel exp chem path, Dept Path, Univ Calif, San Francisco, 77-80, asst res pathologist, 80-83; asst res toxicologist, Dept Entom & Dept Environ Toxicol, Univ Calif, Davis, 82-86; res asst prof, 86-89, ASSOC DIR TOXICOL, DEPT PHARMACOL & TOXICOL, CTR HUMAN TOXICOL, UNIV UTAH, SALT LAKE CITY, 86-, RES ASSOC PROF, 89- *Concurrent Pos:* Gianina Found fel, Bank Am, 77 & Monsanto Found fel toxicol, 78. *Mem:* Am Soc Cell Biol; Am Asn Pathologists; Soc Toxicol; Soc Forensic Toxicol; Int Soc Study Xenobiotics. *Res:* Metabolism of abused drugs and other xenobiotics, with an emphasis on environmental and genetic factors which regulate specific enzymes or organelles involved in xenobiotic-metabolism. *Mailing Add:* Ctr Human Toxicol Univ Utah 417 Wakara Way Rm 290 Salt Lake City UT 84108

MOODY, DAVID WRIGHT, b Boston, Mass, Nov 23, 37; m 63; c 1. HYDROLOGY. *Educ:* Harvard Univ, AB, 60; Johns Hopkins Univ, PhD(geog), 64. *Prof Exp:* Hydrologist, Water Resources Div, 64-74, prog analyst, Off Dir, 74-77, staff scientist, 77-83, chief, Off Nat Water Summary & Long-Range Planning, 83-87, ASST CHIEF HYDROLOGIST, WATER ASSESSMENT & DATA COORDR, US GEOL SURV, 88- *Concurrent Pos:* Am Polit Sci Asn Cong Fel, 76-77; mem geophysics study comt, Panel Geophysics Data & Pub Policy, Nat Res Coun, 81-83. *Mem:* AAAS; Geol Soc Am; Am Geophys Union; Am Water Resources Asn (treas, 87-88, vpres, 89-90, pres-elect, 90-91); Asn Am Geogr. *Res:* Information systems; long-range planning; water resources assessment; water policy analysis. *Mailing Add:* Nat Ctr Stop 407 US Geol Surv Reston VA 22092

MOODY, EDWARD GRANT, b Thatcher, Ariz, Dec 23, 19; m 43; c 7. ANIMAL NUTRITION. *Educ:* Univ Ariz, BS, 41; Kans State Col, MS, 47; Purdue Univ, PhD(nutrit), 51. *Prof Exp:* Asst dairy husb, Kans State Col, 46-47; asst, Purdue Univ, 47-49, instr, 49-51; from asst prof to assoc prof, 51-63, PROF ANIMAL SCI, ARIZ STATE UNIV, 63- *Mem:* Am Soc Animal Sci; Am Dairy Sci Asn; Soc Nutrit Educ. *Res:* Fat metabolism in ruminants; dairy herd management and production. *Mailing Add:* 10225 Kensington Pkwy No 911 Kensington MD 20895

MOODY, ELIZABETH ANNE, b Portland, Maine, Oct 29, 48. INFRARED & OPTICAL SEEKER SYSTEMS, GIMBAL STABILIZATION. *Educ:* Simmons Col, AB, 71. *Prof Exp:* Data analyst, Smithsonian Astrophys Observ, 69; res consult, Mass Inst Technol Instrumentation Lab, 73-74, Independent Consults, 75-76; res scientist, Aerodyne Res, 77-78, Sci Applications Int, 79-80, USAF Geophys Lab, 81; DESIGN & DEVELOP ENGR, RAYTHEON CO, 82- *Concurrent Pos:* Pres, Independent Consults, 75-; liaison officer, Harvard Kirkland Cosmology Club, 80- *Mem:* Am Astron Soc; Am Phys Soc; Asn Women Sci. *Res:* Detection systems; signal and image processing; astronomical instrumentation; dynamical astronomy. *Mailing Add:* PO Box 5549 Beverly Farms MA 01915-0520

MOODY, ERIC EDWARD MARSHALL, b Neath, Gt Brit, Dec 15, 38; m 67; c 2. MOLECULAR GENETICS, VENEREAL DISEASES. *Educ:* Univ London, BSc, 64; Univ Edinburgh, PhD(molecular biol), 71. *Prof Exp:* Res scientist molecular genetics, Southwest Ctr Advan Studies, 65-68; tech officer, Med Res Coun Gt Brit, 68-71; asst prof, 72-77, ASSOC PROF MOLECULAR GENETICS, UNIV TEX HEALTH SCI CTR, 77- *Concurrent Pos:* Fel, Col Med, Univ Ariz, 71-72. *Mem:* Soc Gen Microbiol; Am Venereal Dis Asn. *Res:* Molecular genetics of plasmids; genetics of virulence and antibiotic resistance in Candida albicans. *Mailing Add:* Dept Microbiol Univ Tex Health Sci Ctr 7703 Floyd Curl Dr San Antonio TX 78284

MOODY, FRANK GORDON, b Franklin, NH, May 3, 28; m 64; c 3. SURGERY. *Educ:* Dartmouth Col, BA, 52; Cornell Univ, MD, 56. *Prof Exp:* Surg intern, NY Hosp-Cornell Med Ctr, 56-57, surg residency, 57-63; clin instr surg, Med Ctr, Univ Calif, San Francisco, 63-65, asst prof, 65-66; assoc prof, Univ Ala, 66-69, prof, 69-71; AT DEPT SURG, MED SCH, UNIV TEX, HOUSTON. *Concurrent Pos:* Cardiovasc Res Inst fel, 63-65; Am Heart Asn advan res fel, 63-65; consult, Vet Admin Hosp, Birmingham, Ala, 66-; mem surg training comt, Nat Inst Gen Med Sci, 68-72; mem med sect A, NIH, 72-75 & surg sect B, 75-80. *Mem:* Biophys Soc; Am Col Surg; Am Gastroenterol Asn; Am Surg Asn; Soc Surg Alimentary Tract. *Res:* Gastric acid secretion; pancreatic function; portal hypertension. *Mailing Add:* Dept Surg Univ Tex Health Sci Ctr 6431 Fannin Suite 4020 Houston TX 77030

MOODY, HARRY JOHN, b Birsay, Sask, May 8, 26; m 52; c 3. MICROWAVE PHYSICS, COMMUNICATIONS ENGINEERING. *Educ:* Univ Sask, BEng, 48, MS, 50; McGill Univ, PhD(physics), 55. *Prof Exp:* Asst, Univ Ill, 50-51; res asst, Nat Res Coun Can, 51-55; sr physicist, Can Marconi Co, 55-61; sr mem sci staff, RCA Ltd, Montreal, 61-73, res fel, 73-77; STAFF SCIENTIST, SPAR AEROSPACE LTD, 77- *Mem:* Can Asn Physicists. *Res:* Microwave; millimeter wave; infrared; characterizing and optimizing a space system for SCPC communications. *Mailing Add:* 4470 Coronation Montreal PQ H4B 2C4 Can

MOODY, JOHN ROBERT, b Richmond, Va, Feb 6, 42. ANALYTICAL CHEMISTRY. *Educ:* Univ Richmond, BS, 64; Univ Md, College Park, MS, 67, PhD(anal chem), 70. *Prof Exp:* Asst anal chem, Univ Md, 65-70; RES CHEMIST, US NAT BUR STANDARDS, WASHINGTON, DC, 71- *Mem:* Am Chem Soc; Am Soc Testing & Mat; Soc Appl Spectros. *Res:* General analytical procedures as applied to isotope dilution mass spectrometry for trace metals; high accuracy trace analysis; analytical applications of clean rooms, preparation of ultra pure reagents. *Mailing Add:* 2846 E Weyburn Rd Richmond VA 23235

MOODY, JUDITH BARBARA, b Detroit, Mich, Oct 14, 42. APPLIED ECONOMIC & ENGINEERING GEOLOGY, SCIENCE EDUCATION. *Educ:* Univ Mich, Ann Arbor, BS, 64, MS, 67; McGill Univ, PhD(geol), 74. *Prof Exp:* Geochemist, Anaconda Am Brass Ltd, Bathurst, NB, 67; vis scientist geochem, Univ Liege, 68-70; asst prof geol, Univ NC, Chapel Hill, 74-80; proj mgr, Off Nuclear Waste Isolation, Battelle Mem Inst, 80-88; PRES, J B MOODY & ASSOCS, COLUMBUS, OHIO, 88- *Concurrent Pos:* Mem women geoscientist comt, Am Geol Inst, 72-77; NSF grant, 75-80; consult, Pedco Environ, 77 & Vulcan Mat Co, 78; ed, Geol, 84-; vpres & pres, Asn Women Sci Cent Ohio, 84-86; chief judge, Asn Women Geoscientists Found Int Sci & Eng Fair, 84, 88, 89 & 90. *Honors & Awards:* Howley Award, Mineralog Asn Can, 86; Harriet Parker Award for Sci & Technol, 87. *Mem:* AAAS; Am Geophys Union; Geochem Soc; Geol Soc Am; Can Mineral Soc; Asn Women Geoscientists; Asn Women in Sci. *Res:* Mineral stability as a function of temperature, pressure and composition; mechanisms of geochemical processes; plate tectonic cyclicity; applied geoengineering problems eg high-level nuclear waste; radar and asbestos environmental problems. *Mailing Add:* J B Moody & Assocs 25 W Washington St Suite 10 Athens OH 45701-2447

MOODY, KENTON J, b Vallejo, Calif, Dec 5, 54; m 82; c 2. NUCLEAR CHEMISTRY, ANALYTICAL CHEMISTRY. *Educ:* Univ Calif, Santa Barbara, BS, 77; Univ Calif, Berkeley, PhD(nuclear chem), 83. *Prof Exp:* Staff scientist nuclear chem, Lawrence Berkeley Lab, 83; staff scientist nuclear chem, Gesellschaft für Schwerionen forsch, 83-85; NUCLEAR CHEMIST NUCLEAR CHEM, LAWRENCE LIVERMORE NAT LAB, 85- *Mem:* Am Phys Soc; AAAS. *Res:* Decay properties and production techniques of the heavy elements; spontaneous fission of transfermium isotopes. *Mailing Add:* Lawrence Livermore Nat Lab PO Box 808 L-232 Livermore CA 94551

MOODY, LEROY STEPHEN, b Elyria, Ohio, Nov 18, 18; m 45; c 3. PHYSICAL CHEMISTRY. *Educ:* Wesleyan Univ, BA, 41; Univ Wis, PhD(phys chem), 44. *Prof Exp:* Res chemist, Mass Inst Technol, 44-45; res chemist, Gen Elec Co, 45-51, supvr prod eval, 51-52, specialist mkt res & prod planning, 52-54 & proj eval, 55-57, mgr, New Prod Develop Lab, 57-59, & polycarbonate eng, 59-64, managing dir, N V Polychemie, AKU-GE, Neth, 64-67, gen mgr irradiation processing oper, Vallecitos Nuclear Ctr, Calif, 67-69 & reactor fuels & reprocessing dept, 69-71, mgr strategy planning oper, 71-74, res & develop mgr mat sci & eng, 74-78, Corp Res & Develop Ctr, 78-82; RETIRED. *Mem:* Fel AAAS; Am Chem Soc; Am Nuclear Soc. *Res:* General management; administration of scientific programs; technology transfer; international technical programs. *Mailing Add:* 603 S Owl Dr Sarasota FL 34236

MOODY, MARTIN L(UTHER), b Corpus Christi, Tex, Apr 14, 25; m 50; c 4. CIVIL & STRUCTURAL ENGINEERING. *Educ:* Univ Mo, BS, 49; Univ Colo, MS, 57; Stanford Univ, PhD(civil eng), 65. *Prof Exp:* Inspector construct, Pa Dept Hwy, 49-50; eng aide, Tex Hwy Dept, 50; civil engr, Brown & Root, 50-51; field engr, Denver Water Bd, Colo, 51-55; from instr to assoc prof civil eng, Univ Colo, Boulder, 55-69; assoc chmn dept, 68-72, prof civil eng, 69-85, assoc dean eng, 72-73, vchancellor admin, 74-79, chmn dept, 81-83, EMER PROF CIVIL ENG, UNIV COLO, DENVER, 85- *Res:* Mechanical properties of materials; dynamic response of structures. *Mailing Add:* 3150 Fifth St Boulder CO 80204

MOODY, MAX DALE, b Onaga, Kans, Sept 29, 24; m 50; c 3. MEDICAL BACTERIOLOGY, QUALITY ASSURANCE. *Educ:* Univ Kans, AB, 48, MA, 49, PhD(bact), 53; Am Bd Med Microbiol, dipl. *Prof Exp:* Asst med bact, Univ Kans, 46-53; in chg spec projs lab, Nat Commun Dis Ctr, USPHS, 53-62, chief staphylococcus & streptococcus unit, 62-67, dir, Nat Streptococcal Dis Ctr, 67-70, chief reagents eval unit, Ctr Dis Control, 70-71; dir sci & tech affairs, Wellcome Diag Div, Burroughs Wellcome Co, 71-89; CONSULT, MED LAB DIAG, 89- *Concurrent Pos:* Mem expert adv panel coccal infections, WHO, 67-79; secy subcomt streptococci & pneumococci, Int Cong Microbiol; mem comt on rheumatic fever, Am Heart Asn; assoc mem, Comn on Staphylococcal & Streptococcal Dis, Armed Forces Epidemiol Bd; area chmn microbiol, Nat Comt Clin Lab Standards, 78-81; chmn, Conf Pub Health Lab Dirs, 80-81; mem bd gov, Am Acad Microbiol; mem bd dir, Nat Comt Clin Lab Standards, 83-; mem bd gov, Am Acad Microbiol, 79-82, 86-; mem, Pan Am Group, Rapid Viral Diagnosis. *Honors & Awards:* Kimble Methodology Res Award, 67. *Mem:* Am Soc Microbiol; Sigma Xi; NY Acad Sci; Am Pub Health Asn; AAAS; fel Am Acad Microbiol. *Res:* Pathogenesis and immunity of Tularemia; rapid detection and identification of pathogenic microorganisms; fluorescent antibody identification of bacterial pathogens; rapid identification of streptococci; rapid antigen detection. *Mailing Add:* 115 Dunedin Ct Cary NC 27511

MOODY, MICHAEL EUGENE, b Murray, Ky, May 23, 52; m 74; c 1. POPULATION GENETICS, EVOLUTIONARY MODELING. *Educ:* Univ Calif, San Diego, BA, 75; Univ Chicago, PhD(theoret biol), 79. *Prof Exp:* USPHS postdoctoral genetics, Dept Genetics, Univ Wis-Madison, 79-81; asst prof, 81-88, ASSOC PROF MATH & GENETICS, WASH STATE UNIV, 88- *Concurrent Pos:* Prin investr, var NSF grants, 88-90, 89-90 & 91-92; vis prof & Fulbright fel, Inst Math, Univ Vienna, 90-91. *Mem:* Genetics Soc Am; Soc Math Biol. *Res:* Theoretical population genetics, with an emphasis on evolution in geographically structured populations. *Mailing Add:* Dept Pure & Appl Math Wash State Univ Pullman WA 99164-2930

MOODY, PAUL AMOS, zoology, for more information see previous edition

MOODY, ROBERT ADAMS, b Swampscott, Mass, Oct 1, 34; m; m; c 5. NEUROSURGERY. *Educ:* Univ Chicago, BA, 55 & 56, MD, 60. *Prof Exp:* Intern, Royal Victoria Hosp, Mon, 60-61; resident surg, Univ Vt, 61-62 & resident neurosurg, 62-66; asst prof neurosurg, Univ Chicago, 66-71; asst prof, Sch Med, Tufts Univ, 71-74; chmn div neurosurg, Cook County Hosp, 74-81, assoc chmn, dept surg, 76-81; prof neurosurg, Abraham Lincoln Sch Med, Univ Ill Med Ctr, 74-81; CHIEF NEUROSURG, GUTHRIE CLINIC, SAYRE, PA, 81-; CLIN PROF NEUROSURG, STATE UNIV NY, 83- *Concurrent Pos:* Fel neurosurg, Lahey Clin, 63-64; vis prof neurosurg, Okayama, 85. *Mem:* Am Asn Neurol Surg; Am Col Surgeons; AMA; Sigma Xi. *Res:* Head injury; blood brain barrier; brain scanning; microvascular surgery. *Mailing Add:* Guthrie Clinic Guthrie Square Sayre PA 18840

MOODY, WILLIAM GLENN, b Jeffersontown, Ky, Dec 19, 33; m 64. PHYSIOLOGY, ANATOMY. *Educ:* Univ Ky, BS, 56, MS, 57; Univ Mo, PhD(animal sci), 63. *Prof Exp:* From asst prof to assoc prof, 63-73, PROF MEAT SCI, UNIV KY, 73- *Concurrent Pos:* Vis prof, Univ Wis, 66-67 & Meat Res Inst, Bristol, Eng, 74-75; dir, Am Meat Sci Asn, 69-72. *Mem:* Am Meat Sci Asn (pres-elect, 85, pres, 86); Sigma Xi; Am Soc Animal Sci; Inst Food Technologists; Coun Agr Sci & Technol. *Res:* Growth and development of meat animals; differentiation of muscle fiber types; quantitative evaluation of subcutaneous and intramuscular fat cells. *Mailing Add:* Agr Sci South Rm 206 Univ Ky Lexington KY 40546

MOODY, WILLIS E, JR, b Raleigh, NC, Mar 30, 24; m 47; c 5. CERAMIC ENGINEERING. *Educ:* NC State Univ, Raleigh, BS, 48, MS, 49, PhD(ceramic eng), 56; Woodrow Wilson Col of Law, JD, 79. *Prof Exp:* Ceramic engr, Elec Auto-Lite Co, Ohio, 49-50; ceramic engr, Lab Equip Corp, Mich, 50-51; instr ceramics & metall, NC State Col, 51-56; PROF CERAMIC ENG & RES ASSOC, ENG EXP STA, GA INST TECHNOL, 56- *Concurrent Pos:* Chmn, Mat div, Am Soc Eng Educ, 79. *Mem:* AAAS; fel Am Ceramic Soc; Nat Inst Ceramic Engrs (vpres, 78, pres, 80); Ceramic Educ Coun (pres, 63); Am Soc Eng Educ; Clay Mineral Soc. *Res:* High temperature ceramics and metals; dental materials; clays; application of materials to nuclear reactors; electrical ceramics; solar energy. *Mailing Add:* 6607 Park Ave Atlanta GA 30342

MOOERS, CALVIN NORTHRUP, b Minneapolis, Minn, Oct 24, 19; m 45; c 2. MATHEMATICS, SOFTWARE SYSTEMS. *Educ:* Univ Minn, AB, 41; Mass Inst Technol, MS, 48. *Prof Exp:* Physicist, Naval Ord Lab, Md, 41-46; PROPRIETOR & DIR RES, ZATOR CO, 47-; PRES & DIR RES, ROCKFORD RES, INC, 61- *Mem:* Int Fedn Doc; Asn Comput Mach; Am Soc Info Sci; Inst Elec & Electronics Eng. *Res:* Programming languages; information retrieval and processing, including theory, systems and machines; reactive typewriter. *Mailing Add:* Zator Co 13 Bowdoin St Cambridge MA 02138

MOOERS, CHRISTOPHER NORTHRUP KENNARD, b Hagerstown, Md, Nov 11, 35; m 60; c 2. MESOSCALE OCEAN SCIENCE, PHYSICAL OCEANOGRAPHY. *Educ:* US Naval Acad, BS, 57; Univ Conn, MS, 64; Ore State Univ, PhD(phys oceanog), 69. *Prof Exp:* NATO fel, Univ Liverpool, 69-70; from asst prof to assoc prof phys oceanog, Sch Marine & Atmospheric Sci, Univ Miami, 70-76; from assoc prof to prof phys oceanog, Col Marine Studies, Univ Del, 76-79; prof phys oceanog & chmn dept, Naval Postgrad Sch, 79-86; dir, Inst Naval Oceanog, 86-89; res prof oceanog, Inst Study Earth Oceans in Space, Univ NH, 89-91; PROF & CHMN, DIV APPL MARINE PHYSICS, RSMAS, UNIV MIAMI, 91- *Mem:* Challenger Soc; Oceanog Soc; Am Geophys Union (secy, 78-80, pres, 82-84); Am Meteorol Soc; Sigma Xi; AAAS. *Res:* Mesoscale ocean prediction, fluid dynamics of coastal margins; oceanic fronts, coastal upwelling, transient circulations due to storms, tidal and other long waves and wave-mean current interactions; analysis of observations; geophysical fluid dynamics. *Mailing Add:* Div Appl Marine Physics RSMAS/Univ Miami 4600 Rickenbacker Causeway Miami FL 33149-1098

MOOERS, HOWARD T(HEODORE), b Minneapolis, Minn, May 19, 21; m 47; c 4. SYSTEMS ANALYSIS, INFLUENCE FUZING. *Educ:* Univ Minn, BEE, 43, MS, 49. *Prof Exp:* Design engr, Zator Co, Mass, 47-48; chief elec res engr, Minneapolis-Honeywell Regulator Co, 50-52, proj engr, Solid State Res, 52-54, chief appln engr transistor circuitry, 54-56, asst dir eng transistor design, 56-57, staff asst transistor sales, 57-59; mgr, Appl Res Am Electronics Co, 59-61; mgr mil elec dept, Rosemount Eng Co, 61-62; proj supvr, Govt & Aeronaut Prod Div, 63-66, staff engr, Defense Systems Div, Honeywell, Inc, 66-86; TEACHER, BASIC ENGR, 86- *Mem:* Sr mem Inst Elec & Electronics Engrs. *Res:* Semiconductor devices; circuitry; power conditioning and influence fuzing; specifically magnetic influence fuzing for scatterable mines; variation of magnetic signatures of military vehicles and their charges with compass heading and geomagnetic latitudes; detection and localizing algorithms, countermeasures and counter counter measures. *Mailing Add:* 3983 Heathcote Rd Wayzata MN 55391

MOOK, DEAN TRITSCHLER, b Ridgeway, Pa, Aug 3, 35; m 54; c 4. ENGINEERING MECHANICS. *Educ:* Va Polytech Inst & State Univ, BS, 58, MS, 61; Univ Mich, Ann Arbor, PhD(eng mech), 66. *Prof Exp:* Instr, 58-60, ASST PROF ENG MECH, VA POLYTECH INST & STATE UNIV, 66- *Concurrent Pos:* Instr, Univ Mich, Ann Arbor, 60. *Mem:* Am Inst Aeronaut & Astronaut; Am Soc Mech Engrs; Sigma Xi. *Res:* Aerodynamics; nonlinear oscillations. *Mailing Add:* 212 Preston Forest Dr Blacksburg VA 24060

MOOK, DELO EMERSON, II, b Cleveland, Ohio, Apr 28, 42; m 65, 72; c 2. ASTRONOMY. *Educ:* Case Inst Technol, BS, 64; Univ Mich, PhD(astron), 69. *Prof Exp:* Instr, Lawrence Univ, 67; res assoc astron, Univ Chicago, 69-70; from asst prof to assoc prof, 70-83, PROF PHYSICS & ASTRON, DARTMOUTH COL, 83- *Mem:* Am Astron Soc; Int Astron Union. *Res:* Astronomical photometry; polarimetry; celestial x-ray sources. *Mailing Add:* Dept of Physics & Astron Dartmouth Col Hanover NH 03755

MOOK, HERBERT ARTHUR, JR, b Meadville, Pa, Apr 17, 39; m 65; c 1. APPLIED PHYSICS, SOLID STATE PHYSICS. *Educ:* Williams Col, BA, 60; Harvard Univ, MA, 61, PhD(physics), 65. *Prof Exp:* PHYSICIST, OAK RIDGE NAT LAB, 65- *Mem:* Fel Am Phys Soc; Sigma Xi. *Res:* Neutron diffraction studies of solid state physics, especially magnetic materials; polarized and inelastic neutron scattering investigations of the spin density and magnetic excitations in the rare earths and transition metals. *Mailing Add:* 112 Wingate Rd Oak Ridge TN 37830

MOOKERJEA, SAILEN, b Jamalpur, India, Sept 1, 30; m 56; c 2. BIOCHEMISTRY, PHYSIOLOGY. *Educ:* Univ Calcutta, BSc, 49, MSc, 51, PhD(physiol), 56. *Prof Exp:* Lectr biochem, Univ Nagpur, India, 53-61; lectr physiol, Univ Calcutta, 61-62; res assoc, Banting & Best Dept Med Res, Univ Toronto, 62-65, from asst prof to assoc prof med, 65-76; chmn dept, 78-86, PROF BIOCHEM, MEM UNIV NFLD, 76-, DISTINGUISHED UNIV RES PROF, 87- *Concurrent Pos:* Nat Res Coun Can fel, 56-57; res fel, Banting & Best Dept Med Res, Univ Toronto, 57-58; Med Res Coun scholar, 65-70; mem grants various comt, Med Res Coun Can, 78-, mem grants comt, Can Heart Found, 85- *Mem:* Can Biochem Soc; US Soc Biol Chemists; Soc Complex Carbohydrates; fel Am Heart Asn. *Res:* Lipoprotein biosynthesis; role of soluble and membrane glycosyltransferases; membrane function; role of choline in membrane function and lipoprotein biogenesis; control of dolicholphosphate and glycoprotein biosynthesis; phosphorylcholine binding protein and serum lipoproteins; Rat C-Reactive protein, structure and function. *Mailing Add:* Dept Biochem Mem Univ Nfld St Johns NF A1B 3X9 Can

MOOKHERJI, TRIPTY KUMAR, b Calcutta, India, Sept 15, 38; m 65; c 1. SOLID STATE PHYSICS. *Educ:* Agra Univ, BSc, 57, MSc, 59; Univ Burdwan, DPhil(physics), 66. *Prof Exp:* Res assoc, Marshall Space Flight Ctr, NASA, 66-68; sr physicist solid state, 68-79, SR SYSTS ANALYST, TELEDYNE-BROWN ENG, 80- *Concurrent Pos:* Mem, Indian Asn Cultivation of Sci; mgr, space station MMPF Prog Requirements. *Honors & Awards:* New Technol Utilization Award, NASA, 72; NASA Recognition, 75. *Mem:* Am Phys Soc; Optical Soc Am; Am Asn Crystal Growth. *Res:* Electronic excitation; color center; optical band structure; surface physics; growing and characterization of metallic and nonmetallic crystals; manufacturing in space; photovoltaic solar energy conversion; amniotic fluid study; materials processing in the low gravity of space. *Mailing Add:* 1013 Toney Dr SE Huntersville AL 35802

MOOLENAAR, ROBERT JOHN, b DeMotte, Ind, Nov 30, 31; m 55; c 3. PHYSICAL CHEMISTRY, ENVIRONMENTAL SCIENCES. *Educ:* Hope Col, BA, 53; Univ Ill, PhD, 57. *Prof Exp:* Instr phys chem, Univ Ill, 56-57; chemist, 57-65, sr res chemist, 65-69, assoc scientist, 69-74, dir environ sci res, 74-83, PROJ DIR, HEALTH & ENVIRON SCI, DOW CHEM CO, 83- *Mem:* Am Chem Soc; Soc Environ Toxicol & Chem; Sigma Xi. *Res:* Chemical carcinogenesis; biodegradation; photodegradation; toxicology; effects of chemicals on human health and the environment; fate and distribution of man made chemicals in the environment and the impact of low level chronic exposure on humans and the environment. *Mailing Add:* Dow Chem Co 1803 Bldg Midland MI 48674

MOOLGAVKAR, SURESH HIRAJI, b Bombay, India, Jan 3, 43; m 68; c 2. MATHEMATICAL BIOLOGY, EPIDEMIOLOGY. *Educ:* Univ Bombay, MB & BS, 65; Johns Hopkins Univ, PhD(math), 73. *Prof Exp:* Instr math, Johns Hopkins Univ, 72-73; asst prof math, Ind Univ, Bloomington, 73-77; from asst prof to assoc prof res med, Univ Pa, 77-84; epidemiologist & res physician, Fox Chase Cancer Ctr, 77-84; MEM, FRED HUTCHINSON CANCER CTR, 84-; PROF EPIDEMIOL & BIOSTATIST, UNIV WASH, 84- *Concurrent Pos:* Sr fel, Univ Wash, 76-77. *Honors & Awards:* Lester R Ford Award, Math Asn Am, 77; Founder's Award, Chem Indust Inst Toxicol, 90. *Res:* Application of mathematical and statistical techniques in the biological sciences especially epidemiology. *Mailing Add:* Fred Hutchinson Cancer Res Ctr 1124 Columbia St Seattle WA 98104

MOOLTEN, FREDERICK LONDON, b New York, NY, Oct 11, 32; m 60; c 4. CANCER. *Educ:* Harvard Col, BA, 53; Harvard Med Sch, MD, 63. *Prof Exp:* Internship med, Mass Gen Hosp, Boston, 63-64, residency, 64-65 & 67-68, res fel, 65-67, 68-69; res assoc, 69-71, asst prof, 71-74, ASSOC PROF MICROBIOL, SCH MED, BOSTON UNIV, 74-; ASSOC CHIEF STAFF, RES & DEVELOP, VET ADMIN MED CTR, BEDFORD, MASS, 86- *Concurrent Pos:* Fel, Am Cancer Soc, 65-67 & NIH, 69-71. *Honors & Awards:* Cancer Res Scholar Award, Am Cancer Soc, 71. *Mem:* Am Asn Cancer Res. *Res:* Use of gene insertion techniques to enhance tumor curability; immunotherapy of tumors with immunotoxins; immunization against carcinogens. *Mailing Add:* Sch Med Boston Univ 80 E Concord St Boston MA 02118

MOOLTEN, SYLVAN E, b New York, NY, Sept 1, 04; m 28; c 2. INTERNAL MEDICINE, PATHOLOGY. *Educ:* Columbia Univ, AB, 24, MD, 28; Am Bd Path, dipl, 38; Am Bd Internal Med, dipl, 41. *Prof Exp:* Atten physician, Mt Sinai Hosp, New York, 34-42; dir lab, Middlesex Gen Hosp, 46-70, dir lab res, 56-70; assoc prof, Col Physicians & Surgeons, Columbia Univ, 63-67; dir lab, 50-83, DIR MED EDUC, ROOSEVELT HOSP, 83- *Concurrent Pos:* Consult, St Peter's Gen Hosp, 41-, dir lab, 46-56; res grants, USPHS, 50-55 & 61-66, NJ Heart Asn, 56-58 & Wesson Fund for Med Res, 61; consult, Middlesex Gen Hosp, 56-; clin prof, Rutgers Med Sch, Col Med & Dent NJ, 63- *Mem:* AAAS; Col Am Path; Am Col Physicians; Am Soc Clin Oncol; Asn Hosp Med Educ. *Res:* Lung pathology; blood platelets and thrombosis; Hodgkin's disease; pulmonary lymphatics; adsorption of virus-like agents to erythrocytes; systemic lupus erythematous. *Mailing Add:* 103 N 4th Pl Highland Park NJ 08904

MOOMAW, WILLIAM RENKEN, b Kansas City, Mo, Feb 18, 38; m 64; c 2. ENVIRONMENTAL SCIENCE, MOLECULAR SPECTROSCOPY. *Educ:* Williams Col, BA, 59; Mass Inst Technol, PhD(phys chem), 65. *Prof Exp:* Asst prof chem, Williams Col, 64-65; researcher, Univ Calif, Los Angeles, 65-66; from asst prof to prof chem, Williams Col, 66-90, dir, Ctr Environ Studies, 82-88; DIR, CTR ENVIRON MGT, TUFTS UNIV, 89- *Concurrent Pos:* Froman distinguished prof, Russell Sage Col, 71; vis scholar, Univ Calif, Los Angeles, 71-72 & Tech Univ Munich, 86-87; Dreyfus teacher scholar, 72-78; AAAS cong sci fel, 75-76; W R Kenan Jr prof chem, Williams Col, 81-86, E Fitch prof, 86-90; dir, Climate Energy & Pollution Prog, World Resources Inst. *Mem:* AAAS; Am Phys Soc; Am Chem Soc; Sigma Xi. *Res:* Electronic spectra of small organic molecules; nature of the triplet states of these molecules; excited states and structure of molecular complexes; application of physical science to environmental problems. *Mailing Add:* PO Box 379 Williamstown MA 02167

MOON, BYONG HOON, b Seoul, Korea, Jan 15, 26; US citizen; m 57; c 4. PHARMACOLOGY, MEDICINAL CHEMISTRY. *Educ:* Seoul Nat Univ, BSc, 51; Univ Nebr, BPharm, 60, MS, 64; Wash State Univ, PhD(pharmacol), 69. *Prof Exp:* Res assoc pharmacol, Univ Ill Col Med, 69-72; instr pharmacol, 72-73, asst prof & asst scientist, 73-80, ASSOC PROF PHARMACOL & ASSOC SCIENTIST INTERNAL MED, RUSH MED COL, 81-, COURSE DIR MED PHARMACOL, 74- *Concurrent Pos:* Nursing Pharmacol, Col Nursing, 75- *Mem:* Am Soc Pharmacol & Exp Therapeut. *Res:* The synthesis of 3-aminopiperidone derivatives to provide a basis for study of structure versus potential anticonvulsant activity relationship; norepinephrine-adenosine triphosphate complex formation, spectroscopical proof of a complex in vitro and in adrenal medulary granules; role of dopaminergic mechanisms in noloxone-induced inhibition of apomorphine-induced sterotyped behavior; pharmacological analysis of electric foot-shock-induced antinociception and role of pituitary opioid peptides in physostigmine-induced antinociception; cyclosporin induced hypotension in rabbits. *Mailing Add:* Rush-Presby-St Luke's Med Ctr 1725 W Harrison Chicago IL 60612

MOON, DEUG WOON, b Korea; US citizen; c 3. MATERIALS SCIENCE ENGINEERING. *Educ:* Seoul Nat Univ, BS, 61; State Univ NY, 70; Rensselaer Polytech Inst, PhD(mat sci eng), 75. *Prof Exp:* Fel mat, Rensselaer Polytech Inst, 75-77; MEM STAFF MAT SCI, NAVAL RES LAB, 77- *Mem:* Am Soc Metals; Am Welding Soc. *Res:* High-energy laser applications in materials processing; fundamentals in solid state bonding; solidification; interaction between laser beam and solids; heat flow structure, property and process parameters relationships; surface science. *Mailing Add:* Code 6324 Naval Res Lab Washington DC 20375-5000

MOON, DONALD W(AYNE), b Herrick, Ill, Apr 1, 34; div; c 2. MATERIALS SCIENCE, MECHANICAL ENGINEERING. *Educ:* Univ Ill, BS, 56, MS, 57; Calif Inst Technol, PhD(mat sci), 66. *Prof Exp:* Proj engr eng staff, Gen Motors Corp, 58-61 & defense systs, 61; asst prof mech eng, Univ Calif, Davis, 65-72, group leader, 72-74, from assoc div leader,to div leader, 78-81, assoc prog leader, Precision eng prog, 81-83, mgr, QA Off, 83-86, LASER PROG, LAWRENCE LIVERMORE NAT LAB, UNIV CALIF, 86- *Concurrent Pos:* NSF fes initiation grant, 67-69. *Mem:* AAAS; Am Soc Metals; Nat Soc Prof Engrs; Am Inst Mining, Metall & Petrol Engrs; Sigma Xi. *Res:* Dislocation theory; observation and application to the deformation of crystalline solids; mechanical deformation of biological materials. *Mailing Add:* Lawrence Livermore Lab L197 Univ Calif Livermore CA 94550

MOON, FRANCIS C, b Brooklyn, NY, May 1, 39; m 62; c 3. ENGINEERING MECHANICS. *Educ:* Pratt Inst, BSME, 62; Cornell Univ, MS, 64, PhD(theoret & appl mech), 67. *Prof Exp:* Asst prof mech & aerospace eng, Univ Del, 66-67; asst prof aerospace & mech sci, Princeton Univ, 67-74; assoc prof, 75-81, PROF & CHMN, DEPT THEORET & APPL MECH, CORNELL UNIV, 81- *Concurrent Pos:* Consult, Rand Corp, Calif, 67-69, Boeing Vertol, 71-72, Int Tel & Tel Corp, 78, SCM Corp, 79-80 & Argonne Nat Lab, 80- *Mem:* Soc Eng Sci; Am Soc Mech Engrs; Am Acad Mech. *Res:* Nonlinear chaotic dynamics, magneto-solid mechanics, wave propagation in solids, structural dynamics, mechanics of superconducting magnets. *Mailing Add:* Dept Theoret & Appl Mech Cornell Univ Thurston Hall 105 Upson Ithaca NY 14853

MOON, GEORGE D(ONALD), JR, b Dayton, Ohio, Apr 15, 27; m 53; c 4. CHEMICAL ENGINEERING. *Educ:* Univ Dayton, BChE, 49. *Prof Exp:* Asst chem eng, Univ Cincinnati, 49-51, from instr to assoc prof, 51-73; partner & sr process engr, 73-80, vpres, Raphael Katzen Assocs Int, Inc, 80-85. *Concurrent Pos:* Consult, Raphael Katzen Assocs, 57-73; adj assoc prof, 73-84, adj prof chem eng, Univ Cincinnati, 85- *Honors & Awards:* Chem Engr of Year Award, Am Inst Chem Engrs, 78. *Mem:* Am Soc Eng Educ; fel Am Inst Chem Engrs; Am Chem Soc; Sigma Xi. *Res:* Distillation-perforated tray contractors; computers in engineering design. *Mailing Add:* 1181 Hawkstone Dr Cincinnati OH 45230

MOON, HARLEY W, b Tracy, Minn, Mar 1, 36; m 56; c 4. VETERINARY PATHOLOGY. *Educ:* Univ Minn, BS, 58, DVM, 60, PhD(vet path), 65. *Prof Exp:* Instr vet path, Univ Minn, 60-62, NIH res fel, 62-65; asst scientist, Brookhaven Nat Lab, 65-66; assoc prof vet path, Univ Sask, 66-68; res veterinarian, Nat Animal Dis Lab, Agr Res Serv, USDA, 68-73; PROF VET PATH, IOWA STATE UNIV, 74-; DIR, NAT ANIMAL DIS CTR, USDA-ARS, 88- *Concurrent Pos:* Res collabr, Brookhaven Nat Lab, 66- *Mem:* Nat Acad Sci; Am Col Vet Path; Am Vet Med Asn. *Res:* Infectious diseases of the intestinal tract of animals; role of the lymphocyte in the immune response. *Mailing Add:* 211 Valley View Nevada IA 50201

MOON, JOHN WESLEY, b Hornell, NY. MATHEMATICS. *Educ:* Bethany Nazarene Col, BA, 59; Mich State Univ, MS, 60; Univ Alta, PhD(math), 62. *Prof Exp:* Nat Res Coun Can fel, Univ Col, Univ London, 62-64; from asst prof to assoc prof, 64-69, PROF MATH, UNIV ALTA, 69- *Concurrent Pos:* Vis prof, Univ Cape Town, 70-71; vis, Math Inst, Oxford Univ, 71 & Univ Witwatersrand, 77-78; vis prof, Univ Stirling, 84-85. *Mem:* Am Math Soc; Can Math Soc. *Res:* Graph theory; combinatorial analysis. *Mailing Add:* Dept Math Univ Alta Edmonton AB T6G 2G1 Can

MOON, MILTON LEWIS, b New Providence, Iowa, Dec 16, 22; m 47; c 4. PHYSICS, TECHNICAL MANAGEMENT. *Educ:* Iowa State Teachers Col, BA, 43; Univ Iowa, MS, 48, PhD(physics), 51. *Prof Exp:* Physicist, Johns Hopkins Univ, 51-72, power plant site eval prog mgr, Appl Physics Lab, 72-87; RETIRED. *Concurrent Pos:* Mem, Environ Res Guid Comt, 72-75, lectr mgt. *Res:* Communications and navigation systems; environmental effects; cooling systems; air pollution; impact modelling. *Mailing Add:* 10510 Greenacres Dr Silver Spring MD 20903

MOON, PETER CLAYTON, b St Louis, Mo, May 17, 40; m 82; c 2. DENTAL MATERIALS SCIENCE, ADHESION SCIENCE. *Educ:* Univ Toledo, BS, 64; Univ Va, MS, 67, PhD(mat sci), 71. *Prof Exp:* Mat scientist, Texaco Exp, Inc, 66-68; ASSOC PROF DENT MAT, SCH DENT, MED COL VA, VA COMMONWEALTH UNIV, 71- *Concurrent Pos:* Adj prof biomed eng, Va Commonwealth Univ. *Mem:* Int Asn Dent Res; Am Asn Dent Res; Am Chem Soc; fel Acad Dent Mat; Adhesion Soc. *Res:* Mechanical properties; fracture; mineralization of bone and teeth; metallurgy; virginia resin bonded bridge; interfacial bonding. *Mailing Add:* Dept Restorative Dent Box 566 Med Col Va Sch of Dent Richmond VA 23298

MOON, RALPH MARKS, JR, b Bombay, India, Oct 11, 29; US citizen; m 50; c 4. PHYSICS. *Educ:* Univ Kans, BA, 50, MA, 52; Mass Inst Technol, PhD(physics), 63. *Prof Exp:* Physicist, Naval Ord Test Sta, Calif, 52-54; PHYSICIST, OAK RIDGE NAT LAB, 63- *Mem:* Am Phys Soc. *Res:* Neutron diffraction; magnetism. *Mailing Add:* Solid State Div Oak Ridge Nat Lab Oak Ridge TN 37831

MOON, RICHARD C, b Beech Grove, Ind, July 25, 26; m 48; c 3. ENDOCRINOLOGY. *Educ:* Butler Univ, BS, 50; Univ Cincinnati, MS, 52, PhD(zool), 55. *Prof Exp:* Instr physiol, Col Pharm, Univ Cincinnati, 53-54, instr zool, 54-55; asst prof biol, Duquesne Univ, 55-58; asst prof physiol, Med Units, Univ Tenn, Memphis, 59-63, from assoc prof to prof physiol & biophys, 63-73; head endocrinol res, 73-74, sr sci adv, 74-77, HEAD PATHOPHYSIOL, RES INST, ILL INST TECHNOL, 77- *Concurrent Pos:* Nat Cancer Inst res fel endocrinol, Univ Mo, 58-59; Lederle med fac award, 63-66; Nat Cancer Inst spec res fel, Imp Cancer Res Fund, London, Eng, 65-66; guest investr, Ben May Lab Cancer Res, Univ Chicago, 61. *Mem:* AAAS; Am Physiol Soc; Endocrine Soc; Soc Exp Biol & Med; Am Asn Cancer Res; Soc Study Reproduction. *Res:* Hormonal control of mammary gland growth and hormonal influence upon mammary tumor formation; chemical carcinogenesis. *Mailing Add:* Pathophysiol Sect Ill Inst Technol Res Inst Ten W 35th St Chicago IL 60616

MOON, ROBERT JOHN, b Carbondale, Pa, Aug 6, 42; m 66; c 1. MICROBIOLOGY. *Educ:* Eastern Col, AB, 64; Bryn Mawr Col, PhD(biol), 69. *Prof Exp:* From instr to asst prof, 68-75, ASSOC PROF MICROBIOL, MICH STATE UNIV, 75- *Concurrent Pos:* NIH fel, Harvard Univ, 72-73. *Mem:* AAAS; Am Soc Microbiol. *Res:* Medical microbiology and immunology; host responses to infectious agents. *Mailing Add:* 1428 25th Ave Moline IL 61265-5205

MOON, SUNG, b Seoul, Korea, Oct 6, 28; US citizen; m 61; c 2. ORGANIC CHEMISTRY. *Educ:* Univ Ill, BS, 56; Mass Inst Technol, PhD(org chem), 59. *Prof Exp:* Res assoc org chem, Mass Inst Technol, 59-62; from asst prof to assoc prof, 62-71, PROF ORG CHEM, ADELPHI UNIV, 71- *Concurrent Pos:* Am Chem Soc Petrol Res Fund grant, 65-68. *Mem:* Am Chem Soc. *Res:* Organic synthesis and reaction mechanisms. *Mailing Add:* Dept Chem Adelphi Univ Garden City NY 11530-1299

MOON, TAG YOUNG, b Seoul, Korea, May 11, 31; US citizen; m 59; c 4. PHYSICAL CHEMISTRY, STATISTICAL MECHANICS. *Educ:* Seoul Nat Univ, BS, 59; Yale Univ, PhD(chem), 65. *Prof Exp:* Lectr & NSF grants, Yale Univ, 65-66; res chemist, 66-86, PATENT ANALYST, CHEM ABSTRACTS SERV, DOW CHEM USA, 86- *Concurrent Pos:* Oriental lang transl; comput terminal design. *Mem:* Am Chem Soc; Sigma Xi; Asn Comput Machin; Asn Comput Ling; NY Acad Sci; Chinese Lang Comput Soc. *Res:* Statistical thermodynamics of surfaces; interaction of colloidal particles; membrane device design for separation of liquids and gases, pervaporation and membrane distillation; corrosion of steel in concrete; decontamination of nuclear power plants; sinitic language keyboard. *Mailing Add:* 1050 Putney Dr Worthington OH 43085-2903

MOON, TESSIE JO, b Butler, Pa, Dec 7, 61. FLUID MECHANICS, MANUFACTURING. *Educ:* Grove City Col, BS, 83; Univ Ill, Urbana-Champaign, MS, 86, PhD(eng mech), 89. *Prof Exp:* Develop engr, AMP Inc, Winston-Salem, NC, 83-84; grad teaching res asst eng mech, Univ Ill, Urbana-Champaign, 84-89, res assoc, ONR-URI Nat Ctr Composite Mat Res, 89; ASST PROF MECH ENG, UNIV TEX, AUSTIN, 89- *Concurrent Pos:* Prin investr, Fusion Power Prog, Dept Energy, Argonne Nat Lab, 90. *Mem:* Am Soc Mech Engrs; Am Acad Mech; Am Welding Soc; Am Soc Eng Educ. *Res:* Analytical and numerical modeling of physical systems, primarily manufacturing processes such as gas-metal-arc welding and composites processing; liquid-metal magnetohydrodynamics; development of numerical methods to treat manufacturing processes. *Mailing Add:* Dept Mech Eng Univ Tex Austin TX 78712-1063

MOON, THOMAS CHARLES, b Detroit, Mich, Apr 1, 40; m 63; c 2. SCIENCE EDUCATION, ECOLOGY. *Educ:* Kalamazoo Col, BA, 62; Oberlin Col, MAT, 63; Mich State Univ, PhD(sci educ, biol), 69. *Prof Exp:* Teacher, Detroit Country Day Sch, 63-64 & Farmington High Sch, 64-66; sr lectr sci educ, Mich State Univ, 66-69; dir environ studies, 74-77, PROF SCI EDUC & BIOL, CALIF UNIV OF PA, 69- *Concurrent Pos:* Sci consult, Rand McNally & Co, Ill, 68-72; vpres, Aquatech, Inc, 72- *Mem:* AAAS; Nat Asn Res Sci Teaching. *Res:* Interdisciplinary environmental science and its effects on science education endeavors; stress factors in aquatic ecosystems. *Mailing Add:* Dept Biol Calif Univ Pa California PA 15419

MOON, THOMAS EDWARD, b Pontiac, Mich, Aug 16, 43; m 64; c 1. DISEASE ETIOLOGY & PREVENTION. *Educ:* Northern Ill Univ, BS, 65; Univ Chicago, MS, 67; Univ Calif, Berkeley, PhD(biostatist & epidemiol), 73. *Prof Exp:* Mathematician, US Army Aberdeen Proving Grounds, 67; statistician, NIH, 67-70; biostatistician, Gorgas Mem Inst, Mid Am Res Unit, 73-74; asst biometrician, M D Anderson Hosp & Tumor Inst, 74-77; res assoc prof, Dept Family & Community Med, Univ Ariz, 77-83, head biomet, Comput & Epidemiol Share Resource, Cancer Ctr Div, 77-86, res prof, Dept Internal Med & Family & Community Med, 83-87, DIR EPIDEMIOL RES PROG, UNIV ARIZ, 87-, PROF, DEPT FAMILY & COMMUNITY MED, 87-; DIR, ARIZ DIS PREV CTR, 90- *Concurrent Pos:* Asst biometrician, Southwest Oncol Group, 74-77; asst prof biomet, Grad Sch Biomed Sci, Univ Tex, 74-77; adj assoc prof, dept family & community med, 81-83; chief epidemiol & biomet sect, dept family & community med, Univ Ariz, 87-; chmn Cancer Control Intervention Rev Comt, Nat Cancer Inst, 85-86; mem, Sci Adv Comt Prev Diag & Ther, Am Cancer Soc, 90-; res grants, Nat Cancer Inst, Centers Dis Control & Nat Dairy Coun. *Mem:* Am Statist Asn; Sigma Xi; Am Soc Clin Oncol; Biomet Soc; Soc Epidemiol Res; Am Asn Cancer Res; Am Soc Prev Oncol (pres, 91-); Am Col Epidemiol; Int Epidemiol Asn. *Res:* Disease prevention; epidemology; nutrition; biometry; health promotion. *Mailing Add:* Dept Family & Community Med Univ of Ariz Tucson AZ 85724

MOON, THOMAS WILLIAM, b Portland, Ore, June 10, 44; m 65; c 2. COMPARATIVE PHYSIOLOGY, BIOCHEMISTRY. *Educ:* Ore State Univ, BSc, 66, MA, 68; Univ BC, PhD(zool), 71. *Prof Exp:* Fel, Marine Sci Res Lab, Mem Univ, 71-72; from asst prof to assoc prof, 72-82, PROF BIOL & CHMN DEPT, UNIV OTTAWA, 82- *Concurrent Pos:* Partic exped, Galapagos Island, 70, Kona Coast, Hawaii, 73 & Amazon, 76; vis prof zool, Univ Toronto, 75, Odense Univ, Denmark, 79, St Andrews Univ, Scotland, 79, 89 & Nat Marine Fisheries, Honolulu, 81 & 84; exec dir, Huntsman Marine Lab, St Andrews, NB, 82-85; Carnegie Trust fel, Univ Scotland, 89. *Mem:* Am Phys Soc; Can Soc Zoologists. *Res:* Organism-environment interactions using the tools of biochemistry to probe molecular strategies associated with the adaptation process; enzymes, hormones and hepatocytes in lower vertabrates; role of toxicants in vertebrates. *Mailing Add:* Dept Biol Univ Ottawa Ottawa ON K1N 6N5 Can

MOON, WILCHOR DAVID, b Rose Bud, Ark, June 2, 33; m 53; c 2. MATHEMATICS. *Educ:* State Col Ark, BS, 58; Univ Tenn, Knoxville, MA, 62; Okla State Univ, PhD(math), 67. *Prof Exp:* Instr math, Univ Tenn, Knoxville, 60-62; asst prof, Southern State Col, Ark, 62-63; assoc prof, Ark Col, 63-65, prof math & chmn, Sci Div, 66-68; PROF MATH, UNIV CENT ARK, 68-, DEAN UNDERGRAD STUDIES, 72- *Concurrent Pos:* Consult, Health Physics Div, Union Carbide Nuclear, Oak Ridge Nuclear Co, Tenn, 60-62; vis lectr, Ark Acad Sci, 66-68. *Mem:* Math Asn Am. *Res:* Topology; geometry. *Mailing Add:* Off Dean Undergrad Studies Univ Cent Ark Conway AR 72032

MOONEY, DAVID SAMUEL, b Albany, NY, Oct 13, 28; m 60; c 2. ORGANIC CHEMISTRY. *Educ:* State Univ NY Albany, BA, 50, MA, 51; Rensselaer Polytech Inst, PhD(org chem), 55. *Prof Exp:* Asst org chem, Rensselaer Polytech Inst, 51-55; from asst prof to assoc prof, 55-65, PROF ORG CHEM, WASHINGTON & JEFFERSON COL, 65- *Mem:* Am Chem Soc; Sigma Xi. *Res:* Organo lithium compounds; organic reaction mechanisms. *Mailing Add:* Dept Chem Washington & Jefferson Col Washington PA 15301

MOONEY, EDWARD, JR, b New Castle, Pa, Aug 11, 42; m 70; c 2. NUCLEAR SCIENCE, GENERAL PHYSICS. *Educ:* Youngstown State Univ, BS, 64; Cornell Univ, MS, 66; Va Polytech Inst, PhD(nuclear sci), 71. *Prof Exp:* PROF PHYSICS, YOUNGSTOWN STATE UNIV, 66- *Concurrent Pos:* Prof res physics, Col Med, Northeastern Ohio Univ, 76- *Mem:* Am Inst Physics; Sigma Xi. *Res:* Neutron activation analysis studies of environmental pollutants; elementary particle structure. *Mailing Add:* 2907 N Mercer St New Castle PA 16105

MOONEY, HAROLD A, b Santa Rosa, Calif, June 1, 32; m 74; c 3. PLANT ECOLOGY. *Educ:* Univ Calif, Santa Barbara, AB, 57; Duke Univ, MA, 58, PhD, 60. *Prof Exp:* From instr to assoc prof biol, Univ Calif, Los Angeles, 60-68; assoc prof biol, 68-75, prof environ biol, 75- 76, PAUL F ACHILLES PROF ENVIRON BIOL, STANFORD UNIV, 76- *Honors & Awards:* Mercer Award, Ecol Soc Am; Merit Award, Bot Soc Am. *Mem:* Nat Acad Sci; Ecol Soc Am; AAAS; Am Acad Arts & Sci. *Res:* Physiological adaptation of plants; global ecology. *Mailing Add:* Dept Biol Sci Stanford Univ Stanford CA 94305

MOONEY, JOHN BERNARD, b Coalinga, Calif, June 9, 26; m 83; c 2. ANALYTICAL CHEMISTRY, MATERIALS RESEARCH. *Educ:* Univ Santa Clara, BS, 50; Stanford Univ, MS, 53. *Prof Exp:* Jr develop chemist, Int Minerals & Chem Corp, 51-54; instr chem, Univ Santa Clara, 54-57; res chemist, Kaiser Aluminum & Chem Corp, 57-59; chief chemist, Thermatest Labs, Inc, 59-60; sr engr chem, Varian Assocs, 60-69; mgr mat technol & corp vpres, Photophys, Inc, 69-75; staff scientist, SRI Int, 76-86; vpres, Solitec Inc, 86-87; PROG MGR, SRI INT, 87- *Concurrent Pos:* Vis assoc, Calif Inst Technol, 75-76. *Res:* Emission spectroscopy; gas chromatography; materials research; chemical vapor deposition. *Mailing Add:* Stanford Res Inst Menlo Park CA 94025

MOONEY, JOHN BRADFORD, JR, b Portsmith, NH, Mar 26, 31. OCEAN ENGINEERING. *Educ:* Naval Acad, BS, 53. *Prof Exp:* Navy officer, US Navy, 53-87; PRES & MANAGING DIR, HARBOR BR OCEANOG INST, 88- *Concurrent Pos:* Oceanogr & Navy dep, Nat Oceanic & Atmospheric Admin, 81-83; chief naval res, USN, 83-87 & exec dir, Naval Res Adv Comt, 83-87; chair, US Space Oceanog Comt, 84; vis prof oceanog eng, NH Univ, 87-; res assoc elec eng, Tex A&M Univ, 87-; consult, Fla Atlantic Univ, 87- *Mem:* Nat Acad Eng; Marine Technol Soc (pres, 91-92). *Mailing Add:* Harbor Br Oceanog Inst Inc 5600 Old Dixie Hwy Ft Pierce FL 34946

MOONEY, LARRY ALBERT, b Idaho Falls, Idaho, Jan 28, 36; m 58; c 7. CLINICAL CHEMISTRY. *Educ:* Willamette Univ, BS, 58; Univ Wash, PhD(biochem), 64; Am Bd Clin Chem, dipl. *Prof Exp:* Fel lipid biochem, Va Mason Res Ctr, 64-66; res fel, 66-68; clin biochemist, Mason Clin, 68-69; ASSOC DIR CHEM LAB, SACRED HEART HOSP, 69- *Concurrent Pos:* Biomed consult, Athletics W Running Club, 78- *Mem:* AAAS; assoc Am Soc Clin Path; Am Asn Clin Chem; NY Acad Sci; Am Chem Soc. *Res:* Biosynthesis of essential fatty acids and biological elongation of fatty acids; development of analytical methods for clinical chemistry; application of biochemical knowledge to problems in medicine; applications of microprocessors in the clinical laboratory. *Mailing Add:* Sacred Heart Hosp FCU PO Box 10905 Eugene OR 97401

MOONEY, MARGARET L, b Stinnett, Tex, July 29, 29. METEOROLOGY. *Educ:* Univ Calif, Los Angeles, BA, 62. *Prof Exp:* Meteorologist, Dept Army, Dugway Proving Ground, 62-66; meteorologist, 66-71, supv meteorologist, 71-81, DIR METEOROL SERV, PAC GAS & ELEC CO, 81- *Mem:* Am Meteorol Soc; Air Pollution Control Asn; Weather Modification Asn; Am Soc Testing & Mat; Am Wind Energy Asn. *Res:* Techniques for prediction of dispersion of airborne materials; structure of low level atmospheric turbulence and diffusion processes; micrometeorological studies. *Mailing Add:* Pac Gas & Elec Co Rm 328 San Francisco CA 94106

MOONEY, PATRICIA MAY, b Bryn Mawr, Pa, July 12, 45. SOLID STATE PHYSICS. *Educ:* Wilson Col, AB, 67; Bryn Mawr Col, MA, 69, PhD(physics), 72. *Prof Exp:* Asst prof physics, Hiram Col, 72-74; asst prof physics, Vassar Col, 74-80; RES MEM STAFF, WATSON RES CTR, IBM CORP, 80- *Concurrent Pos:* Res assoc, Physics Dept, State Univ NY Albany, 77-78; vis sci, Group Physique Solides, l'Ecole Normale Superieure, Univ Paris VII, 79-80; vis scientist, Fraunhoffer IAF, Freiburg, W Ger, 87-88. *Mem:* Am Phys Soc; Mat Res Soc; Sigma Xi. *Res:* Point defects in semiconductors; transport properties of semiconductors; semiconductor devices for opto-electronics and high speed digital applications. *Mailing Add:* IBM Watson Res Ctr PO Box 218 Yorktown Heights NY 10598

MOONEY, RICHARD T, b New York, NY, Jan 12, 25; m 50; c 2. RADIOLOGICAL HEALTH. *Educ:* Pratt Inst, BS, 44; NY Univ, MS, 52; Am Bd Radiol, dipl. *Prof Exp:* Jr physicist, Physics Lab, New York, City Dept Hosps, 50-51; dir physics serv, New York City Health & Hosps Corp, 67-83; staff physicist, Lenox Hill Hosp & St Clare's Hosp, 70-83; SR PHYSICIST, FRANCIS DELAFIELD HOSP, 51-; CHIEF PHYSICIST, LONG ISLAND JEWISH MED CTR, QUEENS HOSP, 83- *Concurrent Pos:* Staff scientist, AEC contracts, Columbia Univ; staff physicist, Westchester County Med Ctr, 69-; clin assoc, NY Med Col; consult, Vet Admin, hosps & indust; mem, Sci Comt Number 7 & 9, mem task group M-2 & consult subcomt M-2, Nat Coun Radiation Protection. *Mem:* Assoc fel Am Col Radiol; Radiol Soc NAm; Am Asn Physicists in Med; Am Phys Soc; Health Physics Soc. *Res:* Radiation protection and dose distribution in patients. *Mailing Add:* four Edgemont Circle Scarsdale NY 10583

MOONEY, RICHARD WARREN, b Lynn, Mass, Aug 2, 23; m 44; c 3. PHYSICAL CHEMISTRY, INORGANIC CHEMISTRY. *Educ:* Tufts Univ, BS, 44; Cornell Univ, PhD(phys chem), 51. *Prof Exp:* Engr, Bakelite Corp, NJ, 46-47; sr engr, Lighting Div, Sylvania Elec Prod Inc, Gen Tel & Electronics Corp, 53-56, engr-in-chg, Chem & Metall Div, 56-58, sect head, 58-59, eng mgr, 59-64, mgr res & develop, 65-66, chief engr, 66-69, vpres & gen mgr, 69-73, pres, Wilbur B Driver Div, 73-79, vpres prod improv, 79-83; pres, Aavid Eng, 84-85; RETIRED. *Concurrent Pos:* Ed, J Electrochem Soc, 63-64; Fulbright-Hays res scholar, Tech Univ Norway, 64-65. *Mem:* Am Chem Soc; fel Am Inst Chemists. *Res:* Phosphors; phosphates; tungsten and molybdenum; x-ray crystallography; theoretical spectroscopy. *Mailing Add:* Rte one Box 235 Deering Hillsborough NH 03244

MOONEY, WALTER D, b Floral Park, NY,Nov 17, 51. GEOPHYSICS IN CHINA, COMPUTER MODELLING OF GEOPHYSICAL DATA. *Educ:* Cornell Univ, BS, 73; Univ Wis, PhD(geophys),79. *Prof Exp:* Res asst geophys, Univ Wis, 73-78; GEOPHYSICIST CRUSTAL SEISMOL, US GEOL SURV, 78- *Concurrent Pos:* Res asst geophys , Univ Karlsruhe, WGer, 76; consult prof, Stanford Univ, 84-; vis prof, Univ Kiel, WGer, 88. *Mem:* Am Geophys Union; Geol Soc Am; Seismol Soc Am; Am Asn Petrol Geologists; Soc Explor Geophysicists. *Res:* Investigations of the structure, composition and evolution of the earth,s crust and lithosphere using seismological and other geophysical data. *Mailing Add:* 1095 Sherman Ave Menlo CA 94025

MOOR, WILLIAM C(HATTLE), b St Louis, Mo, Jan 17, 41; m 64; c 2. INDUSTRIAL ENGINEERING, ORGANIZATION THEORY. *Educ:* Washington Univ, BS, 63, MS, 65; Northwestern Univ, PhD(indust eng), 69. *Prof Exp:* Asst prof, 68-73, ASSOC PROF INDUST ENG, ARIZ STATE UNIV, 73- *Concurrent Pos:* Consult, Allied Signal Corp, Garrett Turbine Engine Co, Ariz St Legis, Ariz Auditor Gen Off. *Mem:* Am Inst Indust Engrs; Sigma Xi; Am Soc Eng & Mgt. *Res:* Study of information-related behavior in research and development; study of organizational structure and process; application of industrial engineering techniques to the analysis of complex human and man/machine systems; human factors/ergonomics. *Mailing Add:* Dept Indust Eng Ariz State Univ Tempe AZ 85287

MOORADIAN, ARSHAG DERTAD, b Aleppo, Syria, Aug 20, 53; US citizen; m 85; c 2. GERONTOLOGY, ENDOCRINOLOGY. *Educ:* Am Univ Beirut, BS, 76, MD, 80. *Prof Exp:* Post-doctoral fel, endocrinol, Univ Minn, 82-85; asst prof med, Univ Calif, Los Angeles, 85-88; ASSOC PROF MED, UNIV ARIZ, 88- *Concurrent Pos:* Staff physician, VA Admin Med Ctr, Los Angeles, 85-88; prin investr, Aging Res Lab, VA Med Ctr, Los Angeles, 85-88; dir, Geront Res, Univ Ariz, 88- *Mem:* Endocrine Soc; Am Fedn Clin Res; AAAS. *Res:* Use of diabetes mellitus as a model of premature aging, e.g. establishing a parallelism between diabetes and aging in various physiological parameters; studies on biomarkers of aging. *Mailing Add:* Univ Ariz 1821 E Elm St Tucson AZ 85719

MOORCROFT, DONALD ROSS, b Toronto, Ont, June 2, 35; m 64; c 3. PHYSICS. *Educ:* Univ Toronto, BASc, 57; Univ Sask, MSc, 60, PhD(physics), 62. *Prof Exp:* NATO sci fel & Rutherford fel, Radiophysics Lab, Stanford Univ, 62-63; from asst prof to assoc prof, 63-74, PROF PHYSICS, UNIV WESTERN ONT, 74- *Mem:* Can Asn Physicists; Am Geophys Union. *Res:* Physics of the upper atmosphere; propagation and scattering of radio waves in ionized media; small-scale ionospheric structure. *Mailing Add:* Dept of Physics Univ of Western Ont London ON N6A 3K7 Can

MOORCROFT, WILLIAM HERBERT, b Detroit, Mich, Feb 1, 44; m 71; c 3. PSYCHOBIOLOGY, SLEEP. *Educ:* Augustana Col, BA, 66; Princeton Univ, PhD(psychol), 70. *Prof Exp:* USPHS fel ment retardation, Nebr Psychiat Inst & instr psychol, Col Med, Univ Nebr, 70-71; ASSOC PROF PSYCHOL, LUTHER COL, IOWA, 71- *Concurrent Pos:* Vis res psychologist, Ment Retardation Ctr, Univ Calif, Los Angeles, 72; fel, Sleep Disorders Serv, Rush-St Lukes-Presby Med Ctr, Chicago, Ill, 80. *Mem:* Asn Study Dreaming; Sleep Res Soc. *Res:* Sleep, dreaming, and sleep disorders. *Mailing Add:* Dept of Psychol Luther Col Decorah IA 52101

MOORE, A(RTHUR) DONALD, b Assiniboia, Sask, Apr 28, 23; m 53; c 2. ELECTRICAL ENGINEERING, COMMUNICATIONS. *Educ:* Queen's Univ, Ont, BSc, 45, MSc, 50; Stanford Univ, PhD(elec eng), 53. *Prof Exp:* Jr engr, Turbo Res, Ltd, Ont & Power Jets, Ltd, Eng, 45-46; r design engr, Gas Turbine Div, Avro Can, 46-47; instr elec eng, Queen's Univ, Ont, 47-49; instr, Univ BC, 49-50; asst, Electronics Res Lab, Stanford Univ, 50-52; from asst prof to assoc prof, 52-61, head dept, 70-81, PROF ELEC ENG, UNIV BC, 61-,. *Concurrent Pos:* Nat Res Coun Can sr res fel, 64-65. *Mem:* Can Soc Elec Eng; Inst Elec & Electronics Engrs. *Res:* Network theory; communication theory; electronic instrumentation; computer-aided design. *Mailing Add:* Dept Elec Eng Univ BC 2075 Wesbrook Pl Vancouver BC V6T 1W5 Can

MOORE, AIMEE N, b Conway, SC, Nov 8, 18. NUTRITION. *Educ:* Univ NC, BS, 39; Columbia Univ, MA, 47; Mich State Univ, PhD(higher educ), 59. *Prof Exp:* Intern hosp dietetics, Univ Mich, 40; hosp dietitian, Miss State Sanatorium & Stuart Circle Hosp, Richmond, Va, 40-43; teacher inst mgt, Col Human Ecol, Cornell Univ, 47-61; dir dept nutrit & dietetics, 61-80, dir dietetics, 61-80, dir dietetic educ, health sci ctr, 80-83, prof, 61-84, EMER PROF FOOD SYSTEMS MGT, MED CTR, UNIV MO-COLUMBIA, 84- *Concurrent Pos:* Rose fel, Am Dietetic Asn, 56 & 58; Hacettepe Univ, 67, Ramathibide Hosp Bangkok, 65, Pan Am Health Orgn, 75, Title VII Prog, US VI, 79; NIH educ grant, 71-77. *Honors & Awards:* Medallion Award, Am Dietetic Asn, 79, Copher Award, 86. *Mem:* Am Dietetic Asn; Am Home Econ Asn; Am Inst Decision Sci; Asn Schs Allied Health Prof; Am Mgt Asn. *Res:* Applications of computer technology in hospital departments of nutrition and dietetics; labor productivity in hospital food service. *Mailing Add:* 700 Morningside Dr Columbia MO 65201

MOORE, ALEXANDER MAZYCK, b Charleston, SC, Nov 6, 17; m 45; c 4. ORGANIC CHEMISTRY, INFORMATION SCIENCE. *Educ:* Col Charleston, BS, 38; Johns Hopkins Univ, PhD(org chem), 42. *Prof Exp:* Lab asst biol, Col Charleston, 36-38; lab asst chem, Johns Hopkins Univ, 38-42; res chemist, Socony-Vacuum Oil Co, Inc, NJ, 42-44 & Surv Antimalarial Drugs, Md, 44-46; res chemist, Parke, Davis & Co, 46-52, lab dir org chem, 52; admin fel, Mellon Inst, 52-56; asst to vpres, Parke, Davis & Co, 56-66, dir res info, 66-78; RETIRED. *Mem:* Am Chem Soc; NY Acad Sci. *Res:* Chemicals from petroleum; classification and nomenclature of organic compounds; chemotherapy; pharmaceuticals; science information; use of chemical structure and biological activity to develop new and useful medicinal products. *Mailing Add:* 805 Heatherway Ann Arbor MI 48104-2731

MOORE, ALLEN MURDOCH, b Ithaca, NY, Mar 15, 40; m 69; c 2. ECOLOGY, ZOOLOGY. *Educ:* Cornell Univ, AB, 61; Univ Tex, Austin, PhD(zool), 68. *Prof Exp:* Fel, Univ NC, 68-69; asst prof, 68-76, ASSOC PROF BIOL, WESTERN CAROLINA UNIV, 76- *Concurrent Pos:* Cooperator, Southeastern Forest Exp Sta, US Forest Serv, 73-; co-investr, Paleolimnol proj in Yunnan, China. *Mem:* AAAS; Ecol Soc Am; Fedn Am Scientists; Am Soc Limnol Oceanogr. *Res:* Systems ecology, especially ecological modelling; energy flow in ecosystems; human impact on ecosystems. *Mailing Add:* Dept Biol Western Carolina Univ Cullowhee NC 28723

MOORE, ALTON WALLACE, b Lewiston, Idaho, Sept 16, 16; m 45; c 3. DENTISTRY. *Educ:* Univ Calif, DDS, 41; Univ Ill, MS, 48; Am Bd Orthod, dipl, 55. *Prof Exp:* Intern dent, Univ Chicago, 41-42; instr oral path & diag, Univ Louisville, 42-43, asst prof, 43-44; from instr to asst prof, Univ Ill, 45-48; assoc prof, 48-50, head dept orthod, 48-66, dir grad dent educ, 51-58, actg asst dean, Sch Dent, 55-56, head dept dent sci & lit, 66-70, assoc dean Sch Dent, 66-77, prof orthod, 50-80, dean, 77-80, EMER PROF ORTHOD & EMER DEAN SCH DENT, UNIV WASH, 80- *Concurrent Pos:* Grieve Mem lectr, 56; dir, Am Bd Orthod, 59-60, secy, 60-65, pres, 65-66; Wylie Mem lectr, 67; mem dent study sec, Nat Inst Dent Res, 70-74, chmn, 71-74; Fogarty Int sr fel, Roman Cath Univ Nijmegen, 76-77. *Honors & Awards:* Award, Am Asn Orthod, 49; Albert H Ketcham Mem Award, Am Bd Orthod, 73. *Mem:* Am Asn Orthod; Am Dent Asn; fel Am Col Dent; Int Asn Dent Res. *Res:* Orthodontics; human facial growth and cephalometric appraisal of orthodontic treatment. *Mailing Add:* 5724 Princeton Ave NE Seattle WA 98105-2132

MOORE, ANA MARIA, b Buenos Aires, Arg, Apr 14, 42; US citizen; m 68; c 2. PHOTOCHEMISTRY, PHOTOBIOLOGY. *Educ:* Nat Univ Plata, BPharm, 64; Fed Univ Rio de Janeiro, MSc, 66; Tex Tech Univ, PhD(chem), 72. *Prof Exp:* Res assoc, Univ Wash, 73-76; vis asst prof, 77-82, RES ASSOC CHEM, ARIZ STATE UNIV, 82- *Concurrent Pos:* Vis scientist chem, lab biophys, Nat Mus Natural Hist, Paris, France, 82-83, lab phys chem systs polyphasd, Nat Ctr Sci Res, Montpellia, 84-85. *Mailing Add:* Dept Chem Ariz State Univ Tempe AZ 85287

MOORE, ANDREW BROOKS, pigment chemistry, risk analysis, for more information see previous edition

MOORE, ARNOLD ROBERT, solid state physics; deceased, see previous edition for last biography

MOORE, ARTHUR WILLIAM, b Windsor, Nfld, July 22, 37; m 60; c 4. CHEMICAL ENGINEERING. *Educ:* Mem Univ, BSc, 57; Mass Inst Technol, SM, 59; Univ London, PhD(chem eng), 63. *Prof Exp:* Res engr, Textile Fibres Div, Can Indust Ltd, 59-60; res scientist, Res Labs, Philips' Gloeilampenfabrieken, Neth, 63-66; SR RES SCIENTIST ADVAN CERAMICS, UNION CARBIDE CORP, 66- *Honors & Awards:* Indust Res-100 Award, 68. *Mem:* Am Inst Chem Engrs; Am Ceramic Soc; Mat Res Soc. *Res:* Carbon and graphite; chemical vapor deposition; high temperature technology; semiconductors; textile fibers. *Mailing Add:* Parma Tech Ctr Advan Ceramics Union Carbide Corp 12900 Snow Rd Parma OH 44130

MOORE, BARRY NEWTON, b San Antonio, Tex, Jan 27, 41; m 87; c 2. PLASMA PHYSICS. *Educ:* Univ Tex, Austin, BS, 62, MA, 69, PhD(physics), 72. *Prof Exp:* Engr, Missiles & Space Div, LTV Aerospace Corp, 62-67; res assoc plasma physics, Fusion Res Ctr, Univ Tex, Austin, 73-76; PHYSICIST, AUSTIN RES ASSOCS, 76- *Mem:* Sigma Xi; Am Phys Soc; Inst Elec & Electronic Engrs. *Res:* Feasibility studies of fuel cycles for advanced fusion reactors; applications of radio-frequency heating to plasmas; plasma simulation; plasma waves; free electron lasers. *Mailing Add:* 7501 Bluff Springs Rd No 31 Austin TX 78744

MOORE, BENJAMIN LABREE, b Louisville, Ky, Feb 10, 15; m 43; c 2. NUCLEAR PHYSICS. *Educ:* Davidson Col, AB, 34; Vanderbilt Univ, MA, 35; Cornell Univ, PhD(physics), 40. *Prof Exp:* Asst, Cornell Univ, 35-40; physicist, Bur Ord, Navy Dept, 40-42 & Naval Mine Warfare Sch, 42-44; sr physicist, Tenn Eastman Corp, 44-46; res fel, Harvard Univ, 46-49, asst dir comput lab, 49-50; mem staff, Los Alamos Sci Lab, Univ Calif, 50-51, asst div leader, 51-60, assoc div leader, 60-73; SCI CONSULT, 73- *Concurrent Pos:* Extra-mural prof, Washington Univ, 57-59; consult, Los Alamos Sci Lab, 73-83. *Mem:* Fel AAAS; Am Phys Soc. *Res:* Nuclear physics; isotope separation by electromagnetic means; design of large scale digital computing machines; atomic weapons. *Mailing Add:* 3801 Pased Del Prado Boulder CO 80301-1521

MOORE, BERRIEN, III, b Atlanta, Ga, Nov 12, 41; m 67. MATHEMATICS. *Educ:* Univ NC, BS, 63; Univ Va, PhD(math), 69. *Prof Exp:* Asst prof, 69-75, ASSOC PROF MATH, UNIV NH, 75- *Mem:* Am Math Soc. *Res:* Hilbert space; operator theory. *Mailing Add:* Dept Math Univ NH Durham NH 03824

MOORE, BETTY CLARK, b Dedham, Mass, Jan 1, 15; m 38; c 1. EMBRYOLOGY. *Educ:* Radcliff Col, AB, 36; Columbia Univ, MA, 37, PhD(zool), 49. *Prof Exp:* Asst zool, Manhattanville Col, 37-40; res assoc embryol, Columbia Univ, 49-50; instr biol, Queens Col, 51-52; res assoc cytochem, Columbia Univ, 54-69, lectr, 64-69; RES ASSOC BIOL, UNIV CALIF, RIVERSIDE, 69- *Mem:* Am Soc Zool; Am Soc Cell Biol; Am Soc Nat. *Res:* Biological sciences; population genetics of drosophila. *Mailing Add:* Dept of Biol Univ of Calif Riverside CA 92521

MOORE, BILL C, b Kansas City, Mo, Sept 12, 07; m 42; c 2. MATHEMATICS, OPERATIONS RESEARCH. *Educ:* Univ Kans, AB, 29; Princeton Univ, AM, 37. *Prof Exp:* Assoc prof math, Tex A&M Univ, 48-74; RETIRED. *Concurrent Pos:* Consult, Humble Oil Co, 54 & 55 & Dow Chem Co, 56; guest worker, Nat Bur Standards, 59. *Mem:* AAAS; Math Asn Am; Sigma Xi. *Res:* Optimization; digital computers. *Mailing Add:* 1000 Munson College Station TX 77840

MOORE, BLAKE WILLIAM, b Ohio, Sept 18, 26; m 55. BIOCHEMISTRY. *Educ:* Univ Akron, BS, 48; Northwestern Univ, PhD(biochem), 52. *Prof Exp:* Asst biochem, Northwestern Univ, 48-52; asst prof, Sch Dent, 52-57, instr cancer res, 57-58, asst prof biochem in cancer res, 58-59, from asst prof to assoc prof, 59-70, PROF BIOCHEM IN PSYCHIAT, SCH MED, WASHINGTON UNIV, 70- *Mem:* AAAS; Am Soc Biol Chemists. *Res:* Neurochemistry; proteins; biochemistry of cancer; protein chromatography; enzymes. *Mailing Add:* Dept Biochem Washington Univ Sch Med 660 S Euclid Ave St Louis MO 63110

MOORE, BOBBY GRAHAM, b Fulton, Miss, June 22, 40; m 60; c 2. CELL BIOLOGY. *Educ:* Miss State Univ, BS, 62, MS, 64; Auburn Univ, PhD(biochem), 68. *Prof Exp:* Res technologist microbiol, Oak Ridge Nat Lab, 64-65; instr, Auburn Univ, 65-68; asst prof, 68-72, ASSOC PROF BIOL, UNIV ALA, TUSCALOOSA, 72- *Concurrent Pos:* Grant-in-aid, Univ Ala, 69-71. *Mem:* AAAS. *Res:* Biochemistry of nucleic acid methylation in diverse organisms and environmental toxicology of naturally occurring chemical substances. *Mailing Add:* 1901 Alexander Dr SE Huntsville AL 35801

MOORE, C BRADLEY, b Boston, Mass, Dec 7, 39; m; c 2. CHEMISTRY. *Educ:* Harvard Univ, BA, 60; Univ Calif Berkeley, PhD(chem), 63. *Prof Exp:* Harvard Nat Scholarships, 58-60; NSF predoc fel, 60-63; from asst prof to assoc prof, 63-72, vchmn, 71-75, chmn, 82-86, PROF CHEM, UNIV CALIF BERKELEY, 72-, DEAN, COL CHEM, 88-, FAC SR SCIENTIST, CHEM SCI DIV, LAWRENCE BERKELEY LAB, 74- *Concurrent Pos:* Alfred Sloan Found fel, 68-72; John Simon Guggenheim Mem fel, 69-70; Miller res prof, 72-73 & 87-88; assoc prof, Fac Sci, Paris, 70; vis prof, Inst Molecular Sci, Okazaki, Japan & Fudan Univ, Shanghai, Peoples Repub China, 79, adv prof, 88-; consult, AVCO Everett Res Labs, Exxon Nuclear & Corp Res Labs & Mallinckrodt Chem; vis fel, Joint Inst Lab Astrophys, Univ Colo, 81-82. *Honors & Awards:* Coblentz Award, 73; E O Lawrence Mem Award, 86; Lippincott Award, 87. *Mem:* Nat Acad Sci; Fel AAAS; fel Am Phys Soc; mem Am Chem Soc. *Res:* Molecular energy transfer; chemical reaction dynamics; photochemistry. *Mailing Add:* Dept Chem Univ Calif Berkeley CA 94720

MOORE, C FRED, b Louisville, Ky, Mar 18, 36; m 61; c 5. NUCLEAR PHYSICS, ATOMIC PHYSICS. *Educ:* Univ Notre Dame, BS, 59; Univ Louisville, MS, 61; Fla State Univ, PhD(physics), 64. *Prof Exp:* Asst, Univ Louisville, 60-61; asst, Fla State Univ, 61-63, instr, 63-64, res assoc, 64-65; res scientist, Univ Tex, Austin, 65, asst prof physics, 65-68; vis prof, Univ Heidelberg, 68-69; assoc prof, 69-71, PROF PHYSICS, UNIV TEX, AUSTIN, 71- *Mem:* Am Phys Soc. *Res:* Isobaric analogue states as compound nuclear resonances; fission studies; low energy electron scattering on ions; atoms and molecules; meson-nuclear physics. *Mailing Add:* Dept of Physics RLM 5 208 Univ of Tex Austin TX 78712-1081

MOORE, CALVIN C, b New York, NY, Nov 2, 36; m 74. MATHEMATICS. *Educ:* Harvard Univ, AB, 58, MA, 59, PhD(math), 60. *Prof Exp:* From asst prof to assoc prof, 61-66, dean phys sci, 71-76, dir, Ctr Pure & Appl Math, 77-80, Miller res prof, 78-79, PROF MATH, UNIV CALIF, BERKELEY, 66-, DIR, MATH SCI RES INST, 81- *Concurrent Pos:* Mem, Inst Adv Study, 64-65; Alfred P Sloan Found fel, 65-67; mem bd trustees, Am Math Soc, 71-79; assoc ed, Pac J Math, 77-; mem, Pres Comt Nat Medal Sci, 79-81. *Mem:* Am Math Soc; fel Am Acad Arts & Sci. *Res:* Group representations. *Mailing Add:* Dept Math Univ Calif 2120 Oxford St Berkeley CA 94720

MOORE, CARL, b Cleveland, Ohio, Aug 28, 19; m 56; c 2. POLYMER CHEMISTRY, COLLOID CHEMISTRY. *Educ:* Ohio State Univ, BA, 41; Oberlin Col, MA, 43; Case Inst, PhD(phys chem), 51. *Prof Exp:* Chemist, Standard Oil Co Ohio, 41; asst gen chem, Oberlin Col, 41-43; res assoc colloid sci, Govt Synthetic Rubber Prog, Case Inst, 43-52; chemist, 52-62, sr res chemist, 62-73, RES SPECIALIST, DOW CHEM CO, 73- *Mem:* Am Chem Soc; Sigma Xi. *Res:* Physical chemistry of latexes; emulsion polymerization; colloid science; radiation grafting of polymers; organic basic research; complexing resins; plastics and latex foam; rheology of polymer solutions; latex paints; adhesion of coatings; organic coatings. *Mailing Add:* 1105 Evamar Dr Midland MI 48640

MOORE, CARL EDWARD, b Frankfort, Ky, Sept 25, 15; m 40; c 4. ANALYTICAL CHEMISTRY. *Educ:* Eastern Ky State Col, BS, 39; Univ Louisville, MS, 47; Ohio State Univ, PhD(anal chem), 52. *Prof Exp:* Chemist, Nat Distillers Prod Corp, 39-41 & E I du Pont de Nemours & Co, 41-45; instr, Univ Louisville, 47-50; PROF CHEM, LOYOLA UNIV CHICAGO, 52- *Mem:* Am Chem Soc; Soc Appl Spectros; AAAS. *Res:* Organic reagents for inorganic analysis; analytical methods for detection and determination of atmospheric contaminants. *Mailing Add:* Dept of Chem Loyola Univ 6525 N Sheridan Rd Chicago IL 60626

MOORE, CARLA JEAN, b Mt Morris, Ill, July 10, 50; m 79; c 1. GEOLOGY, MARINE GEOLOGY. *Educ:* Univ Mo-Columbia, BS, 72, MA, 75. *Prof Exp:* Geol field asst basin analysis, US Geol Surv, Denver, 75-76; GEOLOGIST, MARINE GEOL & GEOPHYSICS DIV, NAT GEOPHYS DATA CTR, NAT ENVIRON SATELLITE DATA & INFO SERV, NAT OCEANIC & ATMOSPHERIC ADMIN, 76- *Mem:* Geol Soc Am; Sigma Xi. *Res:* Data Base management applications in marine geology; systems design. *Mailing Add:* Nat Geophys Data Ctr Nat Oceanic & Atmospheric Admin E/GC3 325 Broadway Boulder CO 80303

MOORE, CARLETON BRYANT, b New York, NY, Sept 1, 32; m 59; c 2. GEOCHEMISTRY, METEORITES. *Educ:* Alfred Univ, BS, 54; Calif Inst Technol, PhD(chem, geol), 60. *Hon Degrees:* DSc, Alfred Univ, 77. *Prof Exp:* Asst prof geol, Wesleyan Univ, 59-61; asst prof geol, 61-66, assoc prof, 66-70, prof chem & geol, 70-, DIR CTR METEORITE STUDIES, 61-, REGENTS PROF CHEM & GEOL, ARIZ STATE UNIV, 88- *Concurrent Pos:* Prin investr, Apollo 11-17, mem preliminary exam team, Apollo 12-17; ed, Meteoritics. *Mem:* Geochem Soc; Geol Soc Am; Am Chem Soc; Am Geophys Union; Mineral Soc Am; Meteoritical Soc (pres, 66). *Res:* Chemistry and mineralogy of meteorites; analytical geochemistry. *Mailing Add:* Ctr Meteorite Studies Ariz State Univ Tempe AZ 85287-2504

MOORE, CAROL WOOD, b Columbus, Ohio, Sept 23, 43; m 66; c 1. GENETICS. *Educ:* Ohio State Univ, BS, 65; Pa State Univ, MS, 69, PhD(genetics), 70. *Prof Exp:* Teacher pub schs, NY, 65-67; asst human genetics, 70-71, res assoc, 71-74, asst prof biol, 74-77, ASSOC PROF RADIATION BIOL & BIOPHYS, SCH MED & DENT, UNIV ROCHESTER, 77- *Mem:* Genetics Soc Am; AAAS. *Res:* Recombination and mutation in yeast cells. *Mailing Add:* Microbiol Dept City Univ New York Med Sch 138th St & Convent Ave New York NY 10031

MOORE, CHARLEEN MORIZOT, b Shreveport, La, July 29, 44; div. CYTOGENETICS. *Educ:* Northeast La Univ, BS, 66; Ind Univ, Bloomington, MA, 67; Univ Tenn, Knoxville, PhD(zool), 71. *Prof Exp:* Fel med cytogenetics, John F Kennedy Inst, Sch Med, Johns Hopkins Univ, 71-73; ASST PROF PEDIAT & DIR MED CYTOGENETICS LAB, UNIV TEX HEALTH SCI CTR HOUSTON, 74- *Concurrent Pos:* Assoc prof human genetics, Univ Tex Health Sci Ctr, San Antonio. *Mem:* Am Soc Human Genetics; Genetics Soc Am; Tissue Cult Asn; Sigma Xi. *Res:* Chromosome behavior. *Mailing Add:* 7703 Floyd Curl Dr San Antonio TX 78284

MOORE, CHARLES B(ERNARD), b Cincinnati, Ohio, May 1, 27; m 49; c 3. ELECTRICAL ENGINEERING, PHYSICS. *Educ:* Univ Cincinnati, EE, 51; Univ Pa, MS, 60. *Prof Exp:* Jr engr, Cincinnati Gas & Elec Co, 50-51; res engr, Eng Res Lab, 51-60, field engr, Wash Works, 60-62, sr res engr, 62-65, res supvr, 65-70, res mgr, 70-77, eng mgr, Instrument Prod, Sci & Process Div, 77-79, MGR APPL PHYSICS, ENG DEPT, E I DU PONT DE NEMOURS & CO, INC, 79-; SR CONSULT, DU PONT ENG, 84- *Mem:* Inst Elec & Electronic Engrs; Sigma Xi. *Res:* Laboratory and process instruments; mass spectrometers; liquid chromatographs; process photometer systems; use of computing techniques in process control; instruments for non destructive assay for plutonium and uranium. *Mailing Add:* Box 7145 Wilmington DE 19803

MOORE, CHARLES GODAT, b Linn, Mo, July 13, 27; m 57; c 3. MATHEMATICS. *Educ:* Cent Mo State Col, BS, 51, MS, 54; Univ Mich, MA, 60, PhD, 67. *Prof Exp:* Teacher high schs, Mo, 51-53, 54-59 & Kemper Mil Sch, 53-54; from asst prof to assoc prof, 60-74, PROF MATH, NORTHERN ARIZ UNIV, 74- *Res:* Mathematics education; continued fractions. *Mailing Add:* Dept Math Fac Box 5717 Northern Ariz Univ Flagstaff AZ 86001

MOORE, CLARENCE L, b Britton, SDak, Dec 6, 31; m 55; c 3. DAIRY SCIENCE. *Educ:* SDak State Univ, BS, 53, MS, 57, PhD(dairy sci), 59. *Prof Exp:* Area livestock specialist, Univ Hawaii, 59-61; from assoc prof to prof, 61-89, EMER PROF DAIRY SCI, ILL STATE UNIV, 90- *Concurrent Pos:* Sabbatical, Ruakura Animal Res Sta, Hamilton, NZ, 72. *Mem:* Am Dairy Sci Asn. *Res:* Dairy cattle nutrition; physiology of milk secretion; dairy cattle management; animal behavior; embryo transfer. *Mailing Add:* Dept Agr Ill State Univ Normal IL 61761-6901

MOORE, CLAUDE HENRY, b Greensboro, Ala, May 8, 23; m 56; c 5. POULTRY GENETICS. *Educ:* Auburn Univ, BS, 47; Kans State Col, MS, 48; Purdue Univ, PhD(poultry genetics), 52. *Prof Exp:* Asst, Kans State Col, 46-48 & Purdue Univ, 48-50; asst nat coordr, Poultry Div, Agr Res Serv, USDA, 50-56; from assoc prof to prof poultry sci, 56-86, head dept, 59-86, ASSOC DIR, ALA AGR EXP STA, AUBURN UNIV, 86- *Mem:* AAAS; Poultry Sci Asn; Sigma Xi. *Res:* Poultry breeding and physiology. *Mailing Add:* 510 S Dean Rd Auburn Univ Auburn AL 36830

MOORE, CLYDE H, JR, b Jacksonville, Fla, June 10, 33; m 53; c 3. GEOLOGY. *Educ:* La State Univ, BS, 55; Univ Tex, MS, 59, PhD(geol), 61. *Prof Exp:* Res geologist, Shell Develop Co, 61-66; from asst prof to assoc prof, 66-73 chem dir,83-87, PROF GEOL & DIR APPL CARBONATE RES PROG,LA STATE UNIV,BATON ROUGE,77-,RES DIR,BASIN RES INST, 88- *Concurrent Pos:* Grants, NSF & Am Chem Soc; consult geologist. *Mem:* Am Asn Petrol Geol; Soc Econ Paleont & Mineral; Int Asn Sedimentologists. *Res:* Carbonate petrology; recent carbonate sedimentation and stratigraphy in Caribbean and Gulf Coastal plain, Central America and Canada; petroleum geology; porosity in carbonates; carbonate diagenesis. *Mailing Add:* Dept Geol La State Univ Baton Rouge LA 70803

MOORE, CONDICT, b Essex Fells, NJ, Apr 29, 16; m 43; c 2. SURGERY. *Educ:* Princeton Univ, BA, 38; Columbia Univ, MD, 42. *Prof Exp:* Resident path, St Luke's Hosp, 46-47; resident surg, Methodist Hosp, Brooklyn, 47-49; resident, Mem Hosp, New York, 49-52; from instr to asst prof, 52-62, clin assoc prof, 62-65, assoc prof, 65-69, PROF SURG, SCH MED, UNIV LOUISVILLE, 69- *Concurrent Pos:* Fel surg, Mem Hosp, New York, 49-51. *Mem:* Fel Am Col Surgeons; James Ewing Soc; Am Radium Soc. *Res:* Cancer research; carcinogenesis; tobacco and oral cancer. *Mailing Add:* Dept Surg Univ Louisville Sch Med 529 S Jackson St Louisville KY 40202

MOORE, CRAIG DAMON, b Youngstown, Ohio, July 13, 42; div. ACCELERATOR PHYSICS, EXPERIMENTAL HIGH ENERGY PHYSICS. *Educ:* Ohio Univ, BS, 64; Univ Wis, PhD(physics), 70. *Prof Exp:* Res assoc physics, State Univ NY Stony Brook, 70-73; PHYSICIST, FERMI NAT LAB, 73- *Mem:* Am Phys Soc. *Res:* Radiation shielding experiments; high energy muon scattering. *Mailing Add:* Fermi Nat Lab Cross Gallery PO Box 500 Batavia IL 60510

MOORE, CYRIL L, b Trinidad, WIndies, Feb 14, 28; US citizen; m 54; c 7. BIOCHEMISTRY, BIOPHYSICS. *Educ:* Brooklyn Col, BA, 53, MA, 58; Albert Einstein Col Med, PhD(biochem), 63. *Prof Exp:* Johnson Res Found fel, Univ Pa, 63-65; univ fel, Albert Einstein Col Med, 65-66; asst prof, 66-69, assoc prof biochem & neurol, Albert Einstein Col Med, 70-76; CHMN BIOCHEM, MOREHOUSE COL, 77- *Mem:* Soc Exp Biol & Med; Int Soc Neurochem; Biophys Soc; Am Chem Soc; AAAS; Am Soc Biol Chem; Can Biochem Soc. *Res:* Ion transport; energy metabolism; development of the central nervous system; mitochondrial biogenesis; neurologic diseases, their origin and causes. *Mailing Add:* 2352 Miriam Lane Decatur VA 30032

MOORE, DAN HOUSTON, b Elk Creek, Va, Apr 7, 09; m 41, 75; c 2. BIOPHYSICS. *Educ:* Duke Univ, AB, 32, MA, 33; Univ Va, PhD(physics), 36. *Hon Degrees:* DS, Emory & Henry Col, 74. *Prof Exp:* Instr physics, Univ Columbia Univ, 36-38; assoc anat, Col Physicians & Surgeons, Columbia Univ, 39-43, from asst prof to assoc prof, 43-53, assoc prof microbiol, 53-58; assoc mem & prof biophys, Rockefeller Inst, 58-66; mem & head dept biophys cytol, Inst Med Res, Camden, NJ, 66-77; AM CANCER SOC PROF, DEPT MICROBIOL & IMMUNOL, HAHNEMANN MED COL & HOSP, 77- *Concurrent Pos:* Mem div war res, Off Sci Res & Develop, 40-45; sci liaison officer, Off Naval Res, London, 49-51; mem virol & rickettsiology study sect, USPHS, 59-63; assoc ed, Cancer Res, 70-74; Am Cancer Soc res prof, 73. *Mem:* Soc Exp Biol & Med; Am Physiol Soc; Harvey Soc; Electron Micros Soc. *Res:* Tumor viruses; electron microscopy; characterization of proteins by electrophoresis and ultracentrifugation. *Mailing Add:* Dept Microbiol & Immunol Hahnemann Univ230 N Broad & Vine Philadelphia PA 19102

MOORE, DAN HOUSTON, II, b New York, NY, Sept 24, 41; m 66; c 1. BIOSTATISTICS. *Educ:* Univ Calif, Santa Barbara, BA, 63; Univ Calif, Berkeley, PhD(biostatist), 70. *Prof Exp:* Biostatistician, Univ Pa, 70-73; BIOSTATISTICIAN, LAWRENCE LIVERMORE LAB, UNIV CALIF, 73- *Mem:* Biomet Soc; Sigma Xi; Am Statist Asn. *Res:* Application of statistical methods to mutation research; mixture decomposition; epidemiology of melanoma; discriminant analysis. *Mailing Add:* Biomed Div L-452 Livermore Lab Univ of Calif Livermore CA 94550

MOORE, DANIEL CHARLES, b Cincinnati, Ohio, Sept 9, 18; m 45; c 4. ANESTHESIOLOGY. *Educ:* Amherst Col, BA, 40; Northwestern Univ, BM & MD, 44. *Prof Exp:* Clin assoc prof, 56-64, CLIN PROF ANESTHESIOL, SCH MED, UNIV WASH, 65- *Concurrent Pos:* Mem consult staff, Children's Orthop Hosp, 62-; dir anesthesiol, Mason Clin & chmn dept, Virginia Mason Hosp, 47-72, sr consult, 72-; NIH grant anesthesiol, 67-71. *Honors & Awards:* Distinguished Serv Award, Am Soc Anesthesiologists, Inc, 76; Labat Award, Am Soc Regional Anesthesia, 77. *Mem:* AMA; Am Soc Anesthesiol (1st vpres, 53-54, 2nd vpres, 54-55, pres-elect, 57-58, pres, 58-59); Acad Anesthesiol; Pan-Am Med Asn. *Res:* Local anesthetic agents and their distribution and fate in man. *Mailing Add:* 1100 Ninth Ave PO Box 900 Seattle WA 98111

MOORE, DAVID A, b Medford, Ore, Jan 12, 17; m 40; c 1. STRATIGRAPHY & SEDIMENTATION, APPLIED GEOLOGY. *Educ:* Univ Calif, Los Angeles, BA, 42. *Prof Exp:* Explor geologist, Superior Oil Co, Calif, 43-49, div explor mgr, Northern Rocky Mountains, 49-51; explor mgr, Hancock Oil Co, Calif, 51-57; CONSULT EXPLOR GEOL, ROCKY MOUNTAINS, 57- *Mem:* Fel AAAS; Am Asn Petrol Geologists. *Res:* Pennsylvanian and Permian stratigraphy; oil and gas production of northern Rocky Mountains. *Mailing Add:* 10707 W Ctr Ave Denver CO 80226

MOORE, DAVID JAY, b Cecil, Pa, Sept 8, 36; m 58; c 2. ANIMAL ECOLOGY. *Educ:* Clarion State Col, BS, 59; Ohio Univ, MS, 61; NC State Univ, PhD(animal ecol), 69. *Prof Exp:* Asst prof biol, SMacomb Community Col, Mich, 62-63; from asst prof to assoc prof, 63-73, dean, Sch Natural Sci, 71-72, PROF BIOL, RADFORD UNIV, 73-, VPRES ACAD AFFAIRS, 72- *Concurrent Pos:* Teacher pub schs, Mich, 61-63. *Mem:* Sigma Xi. *Res:* Water pollution, especially effects of fluoride pollution on the blue crab. *Mailing Add:* Off VPres Acad Affairs Radford Univ Box 5792 Radford VA 24142

MOORE, DAVID LEE, b Springfield, Mo, June 25, 40; m 66; c 2. ORAL & MAXILLOFACIAL SURGERY. *Educ:* Drury Col, AB, 62; Wash Univ, DDS, 66; Univ Mo-Kansas City, MD, 75. *Prof Exp:* Resident oral surg, Geisinger Med Ctr, Danville, Pa, 66-68; asst dent surgeon, Barnes Hosp, St Louis, Mo, 69-70; asst prof, 74-81, PROF ORAL SURG, SCH MED & DENT, UNIV MO-KANSAS CITY, 81- *Mem:* Am Asn Oral & Maxillofacial Surgeons; Am Dent Asn; AMA. *Res:* Surgical correction of mandibular retrusion with mandibular alveolar protrusion and extrusion; use of silastic chin implants; contraindications and complications of silastic chin implants. *Mailing Add:* 4400 Broadway Suite 400 Kansas City MO 64111

MOORE, DAVID SHELDON, b Plattsburg, NY, Jan 28, 40; m 64; c 2. STATISTICS. *Educ:* Princeton Univ, AB, 62; Cornell Univ, PhD(math), 67. *Prof Exp:* Asst prof math & statist, 67-71, assoc prof, 71-76, PROF STATIST, PURDUE UNIV, 77- *Concurrent Pos:* Assoc ed, J Am Statist Asn, 74-77; asst dean, Grad Sch, Purdue Univ, 77-80; prog dir, NSF, 80-81; assoc ed, Technometrics, 90- *Mem:* Inst Math Statist; Math Asn Am; Am Statist Asn; Am Soc Qual Control. *Res:* Tests of fit, nonparametric and categorical data analysis, statistical quality control. *Mailing Add:* 1399 Math Sci Purdue Univ Lafayette IN 47907-1399

MOORE, DAVID WARREN, b Bennettsville, SC, July 7, 1939. GEOMORPHOLOGY, GLACIOLOGY. *Educ:* Col Wooster, BA, 61; Univ NC, Chapel Hill, MS, 72; Univ Ill, Urbana, PhD(geol), 81. *Prof Exp:* Consult geologist, Am Dredging Co, Philadelphia, 68; explor geologist, United Nuclear Corp, New York, 69; planning consult, Laz, Edwards & Dankert Architects, Champaign, Ill, 73-74; GEOLOGIST, US GEOL SURV, 75- *Mem:* Geol Soc Am; Am Quaternary Asn; Sigma Xi. *Res:* Bedrock mapping in the thrust belt at Idaho-Wyoming border determining bedrock structures for economic mineral appraisal; quaternary deposits of southern San Juan Mountains, Colorado; glacial deposits in central Illinois. *Mailing Add:* US Geol Survey MS 913 Fed Ctr PO Box 25045 Denver CO 80225

MOORE, DONALD R, b Reading, Pa, Dec 29, 33; m 72; c 4. ORGANIC CHEMISTRY, POLYMER CHEMISTRY. *Educ:* Lafayette Col, BS, 54; Harvard Univ, AM, 55, PhD(org chem), 58, Harvard Bus Sch, AMP, 75. *Prof Exp:* Chemist, Trubek Labs, 58-62; group leader, Cent Res Lab, J P Stevens & Co, Inc, 62-63, sect leader, 63-68, mgr, 68-69; mgr org fiber chem, Burlington Industs Res Ctr, 70-72; mgr tech, Abrasive Technol Ctr, Carborundum Co, 72-74, dir, 75-77, gen mgr, Activated Carbon Div, 77-84; PRES, MOORE CONSULT, 84- *Mem:* Am Chem Soc; Am Inst Chem; Am Asn Textile Chemists & Colorists. *Res:* Organic synthesis; reaction mechanisms. *Mailing Add:* 5401 NW 110th St Oklahoma City OK 73162-5906

MOORE, DONALD RICHARD, b West Palm Beach, Fla, Feb 16, 21; m 70; c 1. MALACOLOGY. *Educ:* Univ Miami, BS, 54, PhD, 64; Miss Southern Col, MS, 60. *Prof Exp:* Field biologist, Oyster Div, State Bd Conserv, Fla, 53 & Shell Develop Co, 54; res scientist marine invert, Inst Marine Sci, Univ Tex,

55; asst marine biologist, Gulf Coast Res Lab, Miss, 55-60; res instr, Inst Marine Sci, 60-64, asst prof, 64-73, ASSOC PROF MARINE INVERT, UNIV MIAMI, 73- *Concurrent Pos:* Consult, Shell Oil Co, Tex, 55-64 & Freeport Sulphur Co, La, 58-60. *Mem:* Am Malacol Union (pres, 74-75); Paleont Soc; Soc Syst Zool. *Res:* Ahermatypic corals; distribution of marine invertebrates, especially mollusks; systematics; ecology and zoogeography; marine geology; systematics, distribution and biology of marine micromollusca. *Mailing Add:* Dept Maine Geol-Geophys Sch Marine Sci Univ Miami 4600 Rickenbacker Causeway Miami FL 33149

MOORE, DONALD VINCENT, b York, Nebr, Dec 29, 15; m 42; c 2. PARASITOLOGY. *Educ:* Hastings Col, BA, 37; Univ Nebr, MA, 39; Rice Inst, PhD(parasitol), 42. *Prof Exp:* Asst, Univ Nebr, 37-39; sr parasitologist, Bur Labs, State Health Dept, Texas, 42-46; asst prof prev med, Col Med, NY Univ, 46-52, assoc prof, 52-55; asst prof microbiol, 55-74, ASSOC PROF PATH, MICROBIOL & MED LAB SCI, SOUTHWESTERN MED SCH, UNIV TEX HEALTH SCI CTR, DALLAS, 74- *Concurrent Pos:* Lectr, Univ Tex, 43-46; asst parasitologist, Bellevue Hosp, 47-55; lectr, Shelton Col, 50; assoc attend microbiologist, Univ Hosp, Post-Grad Med Sch, NY Univ, 50-55; mem, Nat Res Coun-NSF biol sci fel panel, 61-63, Nat Res Coun res associateships panel, 72-74, 83 & 85. *Honors & Awards:* Spec Award, Am Soc Parasitol, 77. *Mem:* Am Micros Soc; Am Soc Parasitol (secy-treas, 66-77); Am Soc Trop Med & Hyg; Sigma Xi. *Res:* Acanthocephala life histories; schistosomiasis; helminthology; trypanosomiasis medical parasitology; drug resistance in malaria; diagnostic medical parasitology. *Mailing Add:* Dept Path Univ Tex Health Sci Ctr & Southwestern Med Sch Dallas TX 75235

MOORE, DOUGLAS HOUSTON, b Los Angeles, Calif, Apr 22, 20; m 47; c 2. APPLIED MATHEMATICS. *Educ:* Univ Calif, Berkeley, AB, 42; Univ Calif, Los Angeles, MA, 48, PhD(eng), 62. *Prof Exp:* Instr math, West Coast Univ, 49-53; res engr, NAm Aviation, Inc, 53-54 & Hughes Aircraft Co, 55-56; instr math, West Coast Univ, 56-58; from asst prof to prof, Calif State Polytech Col, 58-68; assoc prof, 68-70, actg dir concentration in environ control, 69-70, prof math, Univ Wis-Green Bay, 70-75; assoc prof, Northrop Univ, Inglewood, Calif, 78-87; RETIRED. *Mem:* Math Asn Am; Inst Elec & Electronics Engrs; Sigma Xi. *Res:* Heaviside operational calculus; operational calculus of sequences; scalar dimensional analysis. *Mailing Add:* 900 E Harrison Ave No G3 Pomona CA 91767-2045

MOORE, DUANE GREY, b Barron, Wis, Nov 18, 29; m 54; c 4. FOREST SOILS. *Educ:* Univ Wis, BS, 53, MS, 55, PhD(soils), 60. *Prof Exp:* Proj assoc bot & chem, Univ Wis, 59-62; asst prof soil sci, Univ Hawaii, 62-65; SOIL SCIENTIST, FORESTRY SCI LAB, PAC NORTHWEST FOREST & RANGE EXP STA, 65- *Concurrent Pos:* Soil-herbicide chemist, Hawaiian Sugar Planter's Asn Exp Sta, Honolulu, 63. *Mem:* AAAS; Am Soc Agron; Soil Sci Soc Am; Int Soc Soil Sci. *Res:* Forest fertilization and water quality; pesticide residues in the forest-soil ecosystem; micronutrient fertility of range forage and soil herbicide interactions as related to water quality. *Mailing Add:* 1065 SW Stamm Pl Corvallis OR 97330

MOORE, DUANE MILTON, b Rochelle, Ill, Apr 17, 33; m 53, 90; c 3. MINERALOGY, GEOCHEMISTRY. *Educ:* Univ Ill, BA, 58, MS, 61, PhD(geol), 63. *Prof Exp:* Asst geologist, Ill State Geol Surv, 62-63; asst prof geol, Marshall Univ, 63-64; asst prof, Knox Col, Ill, 64-70, assoc prof geol, 70-87; CLAY MINEROLOGIST, ILL STATE GEOL SURV, 87- *Concurrent Pos:* Vis assoc prof, Univ Iowa, 72-73, Univ Chicago, 79; vis res prof, Univ Ill-Urbana, 80; Fulbright lectureship, Pakistan, 84-85. *Mem:* Mineral Soc A; Geol Soc Am; Sigma Xi; Clay Minerals Soc. *Res:* Mineralogical and geochemical investigations of sedimentary rocks, especially distribution of trace metallic elements; man in geologic perspective; clay minerology; co-authored book. *Mailing Add:* Ill Geol Surv 615 E Peabody Dr Champaign IL 61820

MOORE, DUNCAN THOMAS, b Biddeford, Maine, Dec 7, 46; m 69. OPTICS, OPTICAL ENGINEERING. *Educ:* Univ Maine, Orono, BA, 69; Univ Rochester, MS, 71, PhD(optics), 74. *Prof Exp:* Optical engr, Western Elec Co, Inc, Princeton, NJ, 69-71; asst prof, 74-78, ASSOC PROF OPTICS, INST OPTICS, UNIV ROCHESTER, 78-; PRES, GRADIENT LENS CORP, 80- *Concurrent Pos:* Consult optical eng, 71-; vis scientist, Nippon Schlumberger KK, Tokyo, 83. *Mem:* Fel Optical Soc Am; Am Ceramic Soc; fel Soc Photog & Instrumentation Engrs. *Res:* Design of optical systems with gradient index materials; design of optical instruments for metrology; medical optical instrumentation. *Mailing Add:* Inst Optics Univ Rochester Wilson Blvd Rochester NY 14627

MOORE, E(ARL) NEIL, b Morgantown, WVa, Dec 19, 32; m 59; c 2. PHYSIOLOGY, CARDIOLOGY. *Educ:* Cornell Univ, DVM, 56; State Univ NY, PhD, 62; Univ Pa, MA, 71. *Prof Exp:* Pvt pract, 56; from asst prof to assoc prof animal biol, 62-70, PROF PHYSIOL, SCH VET MED & GRAD SCH ARTS & SCI, UNIV PA, 70-, PROF PHYSIOL MED, SCH MED, 71- *Concurrent Pos:* William Stroud Estab investr, Am Heart Asn, 66-71; fel, Am Col Cardiol, 72; vis prof med, Div Cardiol, Dept Med, Johns Hopkins Sch Med, 86-; adj prof med, Div Cardiol, Dept Med, Hahnemann Med Sch, Hahnemann Univ, 89- *Mem:* Am Physiol Soc; Am Heart Asn; Cardiac Muscle Soc; Soc Gen Physiol. *Res:* Cardiac arrhythmias; cardiac electrophysiology; His' bundle electrocardiology; electrocardiology of congenital heart disease; Wolf-Parkinson-White syndrome; cardiac pharmacology. *Mailing Add:* Physiol Dept Univ Pa Sch Vet Med 3800 Spruce St Philadelphia PA 19104

MOORE, EARL NEIL, veterinary medicine; deceased, see previous edition for last biography

MOORE, EARL PHILLIP, industrial chemistry, for more information see previous edition

MOORE, EDGAR TILDEN, JR, b Middlesboro, Ky, Jan 18, 37; m 58; c 3. APPLIED PHYSICS. *Educ:* Univ Ky, BS, 58. *Prof Exp:* Physicist, Lawrence Radiation Lab, 59-62; physicist, 63-67, mgr shock dynamics dept, 67-70, dep dir appl sci div, 70-72, dir appl sci div, 72-75, vpres, 75-81, SR VPRES, BUS OPERS, PHYSICS INT CO, SAN LEANDRO, 81- *Mem:* Soc Petrol Engrs; Soc Explosive Engrs. *Res:* Design and development of new methods to enhance the recovery of oil, gas and other natural resources using chemical explosives; use of one and two-dimensional, finite-differencing computer programs. *Mailing Add:* Physics Int Co 2700 Merced St San Leandro CA 94577

MOORE, EDWARD FORREST, b Baltimore, Md, Nov 23, 25; m 50; c 3. COMPUTER SCIENCE. *Educ:* Va Polytech Inst, BS, 47; Brown Univ, MS, 49, PhD(math), 50. *Prof Exp:* Asst prof math, Electronic Digital Comput Proj, Univ Ill, 50-51; mem tech staff, Bell Tel Labs, Inc, 51-66; prof comput sci & math, Univ Wis-Madison, 66-85; RETIRED. *Concurrent Pos:* Vis lectr, Harvard Univ, 61-62; vis prof, Mass Inst Technol, 61-62 & Stevens Inst Technol, 65-66. *Mem:* AAAS; Am Math Soc; Inst Elec & Electronic Engrs; Soc Indust & Appl Math; Math Asn Am; Nat Speleological Soc. *Res:* Logical design of switching circuits; automata theory; data base management; binary codes; graph theory; diagnostic identification problems. *Mailing Add:* 4337 Keating Terr Madison WI 53711

MOORE, EDWARD LEE, b Springfield, Mo, Dec 12, 29; m 55; c 3. INORGANIC CHEMISTRY. *Educ:* Drury Col, BS, 51; Washington Univ, MS, 56. *Prof Exp:* From res chemist to sr res chemist, Monsanto Co, 56-64, res group leader nitrogen chem, 64-67, mgr mfg technol, Inorg Div, 67-69, mgr mfg, Spec Chem Systs, 69-71; mgr mfg technol, 71-77, mgr mfg serv specialists, Chem Div, 77-80, MGR TOLL MFG, MONSANTO INDUST CHEM CO, 80- *Mem:* Am Chem Soc; Synthetic Org Chem Mfg Asn. *Res:* Chlorinated cyanurates; inorganic phosphates; sulfuric acid and sulfuric acid catalyst. *Mailing Add:* Monsanto Chem Co Box 526 St Louis MO 63166-0526

MOORE, EDWARD T(OWSON), b Wytheville, Va, Feb 26, 37; m 65. ELECTRICAL ENGINEERING. *Educ:* Va Polytech Inst, BS, 58; Duke Univ, PhD(elec eng), 63. *Prof Exp:* Engr, Sperry Farragut Co Div, Sperry Rand Corp, 58-59; res assoc elec eng, Duke Univ, 61-64; PRES, WILMORE ELECTRONICS CO, INC, 64- *Concurrent Pos:* Consult, Duke Power Co & Disston Div, H K Porter Co, Va, 63 & NC Res Triangle Inst, 63-; res fel, Duke Univ, 63- *Mem:* Inst Elec & Electronics Engrs. *Res:* Energy conversion; power conditioning; feedback control systems; nonlinear circuits. *Mailing Add:* Wilmore Electronics Inc PO Box 2973 W Durham Sta Durham NC 27705

MOORE, EDWARD WELDON, b Madisonville, Ky, July 6, 30; m 63; c 5. GASTROENTEROLOGY, ELECTROCHEMISTRY. *Educ:* Vanderbilt Univ, BA, 52, MD, 55. *Prof Exp:* Intern med, Harvard Med Serv, Boston City Hosp, 55-56; resident, Lemuel Shattuck Hosp, 56-57; clin assoc cancer, Nat Cancer Inst, 57-59; resident med, Harvard Med Serv, Boston City Hosp, 59-60; from asst prof to assoc prof, Med Sch, Tufts Univ, 64-70; PROF MED, MED COL VA, VA COMMONWEALTH UNIV, 70-, PROF PATH & PHYSIOL, 70- *Concurrent Pos:* USPHS res fel, Harvard Med Sch, 60-62; Med Found Boston fel, Med Sch, Tufts Univ, 62-65 & NIH res career develop award, 65-70; consult, Gen Med Study Sect, NIH, 70-; informal consult, Electrochem Sect, Nat Bur Stand, 70-; consult, NASA, 71-; consult, Rev Panel Over-the-counter Antacids, Food & Drug Admin, 72-; surg & bioeng study sect, NIH, 78-82. *Mem:* AAAS; Am Fedn Clin Res; Am Gastroenterol Asn; NY Acad Sci; Am Soc Clin Invest; Am Asn Study of Liver Dis; Asn Am Physicians. *Res:* Gastrointestinal physiology; ion-exchange electrodes in biomedical research. *Mailing Add:* Dept Internal Med Health Sci Ctr Med Col Va-Va Commonwealth Univ Box 711 Mcv Sci Richmond VA 23298

MOORE, EDWIN FORREST, economic statistics; deceased, see previous edition for last biography

MOORE, EDWIN GRANVILLE, b Joliet, Ill, Apr 12, 50; m 70; c 2. PHYSICAL BIOCHEMISTRY, ENZYMOLOGY. *Educ:* Univ Ill, BS, 72; Cornell Univ, PhD(biochem), 76. *Prof Exp:* Scholar biochem, Univ Mich, 76-78; res scientist biochem, 78-82, RES MGR DIAG DIV, ABBOTT LABS, 82- *Mem:* AAAS; Am Asn Clin Chemists; Am Diabetes Asn. *Res:* Physical chemical characterization of hemoglobin and flavoenzyme, lipoamide dehydrogenase; rapid kinetic and thermodynamic investigations of ligand binding and details of enzyme mechanisms; medical diagnostic enzyme and immune assays. *Mailing Add:* 512 E Sunnyside Ave Libertyville IL 60048

MOORE, EDWIN LEWIS, b Springfield, Mass, May 26, 16; m. FOOD SCIENCE. *Educ:* Mass State Col, BS, 38, MS, 40, PhD(food tech), 42. *Prof Exp:* Fel, Fla Citrus Comn & collabr, Bur Agr & Indust Chem, USDA, 42-48; Fla Citrus Comn fel, Citrus Exp Sta, 47-67, res chemist, 67-77, RES SCIENTIST III & ADJ PROF, FLA DEPT CITRUS, UNIV FLA, 77- *Honors & Awards:* Award, USDA, 52. *Mem:* Am Chem Soc; Inst Food Technol. *Res:* Nutrition; citrus by-products and processing. *Mailing Add:* Citrus Res & Educ Ctr Univ Fla 700 Exp Sta Rd Lake Alfred FL 33850-2299

MOORE, EDWIN NEAL, b Dallas, Tex, Aug 14, 34; m 62; c 2. ATOMIC PHYSICS. *Educ:* Southern Methodist Univ, BS, 57; Yale Univ, MS, 58, PhD(physics), 62. *Prof Exp:* Lectr physics, Univ Calif, Santa Barbara, 61-62; asst prof, 62-67, chmn dept, 69-73, ASSOC PROF PHYSICS, UNIV NEV, RENO, 67-, CHMN DEPT, 86- *Concurrent Pos:* NASA res grant, 65-71. *Mem:* Am Phys Soc; Am Asn Physics Teachers. *Res:* Theoretical calculations of accurate atomic wave functions and photoionization cross-sections and other continuum problems; atomic quantum mechanics. *Mailing Add:* Dept Physics Univ Nev Reno NV 89557

MOORE, ELLIOTT PAUL, b Ninnekah, Okla, May 24, 36. ASTROPHYSICS. *Educ:* Univ Chicago, BA, 56, BS, 57; Univ Ariz, PhD(astron), 68. *Prof Exp:* actg chmn, Dept Physics, 77-78, ASSOC PROF ASTROPHYS, NMEX INST MINING & TECHNOL, 68- *Concurrent Pos:* Co-dir, Joint Observ Cometary Res, 74- *Mem:* Int Astron Union; Am Astron Soc; Astron Soc Pac. *Res:* Co-dir, Joint Observ Cometary Res, 74- Mem: Int Astron Union; Am Astron Soc; Astron Soc Pac. Res: Comets; plasma taildynamics; stellar content of galaxies; instrumentation. *Mailing Add:* Dept Physics NMex Inst Mining Technol Socorro NM 87801

MOORE, EMMETT BURRIS, JR, b Bozeman, Mont, June 14, 29; m 60; c 2. CHEMICAL PHYSICS. *Educ:* Wash State Univ, BS, 51; Univ Minn, PhD(phys chem), 56. *Prof Exp:* Asst prof physics, Univ Minn, Duluth, 57-59; staff mem solid-state physics, Boeing Sci Res Labs, 59-73; lectr phys chem, Seattle Univ, 73; dir power plant siting, Minn Environ Qual Bd, 73-76; gen mgr, Richland Div, Olympic Eng Corp, 76-78; SR RES SCIENTIST, PAC NORTHWEST LABS, BATTELLE MEM INST, 78- *Concurrent Pos:* Lect environ sci, Wash State Univ, 90. *Mem:* Fel AAAS; Am Phys Soc; Am Chem Soc. *Res:* Energy and environmental policy; molecular and crystal structure; electron paramagnetic resonance in biological systems; decommissioning of nuclear facilities. *Mailing Add:* Battelle Northwest Box 999 Richland WA 99352

MOORE, ERIN COLLEEN, b Arlington, Tex, Nov 16, 24. ONCOLOGY, NUCLEOTIDES. *Educ:* Univ Tex, BA, 45, MA, 50; Univ Wis, PhD(oncol), 58. *Prof Exp:* Analytical chemist, Univ Tex, 45-50; asst biochem, Dept Genetics, Carnegie Inst, 50-52; jr res chemist, Int Minerals & Chem Corp, 52-54; asst biochem & oncol, Univ Wis, 55-58; sr fel biochem, M D Anderson Hosp & Tumor Inst, Univ Tex, Houston, 58-59, asst, 59-66, assoc biochemist, 66-81, biochemist, 81-87, assoc prof biochem, Grad Sch Biomed Sci, 66-87; RETIRED. *Concurrent Pos:* USPHS fel, Univ Uppsala, 61-63. *Mem:* AAAS; Am Chem Soc; Am Asn Cancer Res; Am Soc Biol Chem. *Res:* Biosynthesis of deoxyribonucleotides; intermediary metabolism of nucleotides; biochemical effects of antitumor agents. *Mailing Add:* 6501 Brush Country No 141 Austin TX 78749

MOORE, ERNEST J(ULIUS), b Stuttgart, Ger, Oct 15, 19; nat US; m 42; c 3. ELECTRICAL ENGINEERING. *Educ:* Calif, MS, 43, PhD(elec eng), 50. *Prof Exp:* Asst elec eng, Univ Calif, 41; electronics engr antenna develop, US Naval Electronics Lab, Calif, res engr systs studies, Stanford Res Inst, 50-56, mgr systs eval dept, 56-68, exec dir, Eng Systs Div, 68-74, vpres off res opers, 74-77, vpres progs & admin, 77-80, VPRES HEALTH & SOCIAL SYSTS, SRI INT, 80- *Mem:* Sigma Xi; sr mem Inst Elec & Electronics. *Res:* Electromagnetic boundary value problems; requirements for and evaluation of large systems and system complexes, especially communication nets, air defense weapons and environmental systems. *Mailing Add:* 731 Josina Ave Palo Alto CA 94306

MOORE, EUGENE ROGER, b Saginaw, Mich, Oct 20, 33; m 58; c 4. POLYMER CHEMISTRY, PROCESS DEVELOPMENT. *Educ:* Mich Tech Univ, BS(chem) & BS(chem eng), 56; Case Inst, PhD(chem eng), 62. *Prof Exp:* Chem engr, Dow Chem Co, 56-58; asst, Case Western Reserve Univ, 58-59; proj leader, Nuclear & Basic Res Lab, Dow Chem Co, 61-67, group leader, 67-70, sr res engr phys res, 70-72, res specialist styrene molding polymers res & develop, 72-74, mfg rep process develop, 74-81, res assoc styrene molding polymers, 81-89, MICH ENG, DOW CHEM CO, 89- *Honors & Awards:* Union Carbide Corp Award, Am Chem Soc, 61. *Mem:* Am Chem Soc; Sigma Xi. *Res:* Chemistry and physics of alkyd resins; exploration of the chemistry of reactive copolymers; process design and development. *Mailing Add:* 5600 Woodview Pass Midland MI 48640

MOORE, FENTON DANIEL, b Cleveland, Ohio, Nov 4, 38; m 61; c 5. ANATOMY, CELL BIOLOGY. *Educ:* John Carroll Univ, BS, 64, MS, 66; Case Western Reserve Univ, PhD(anat), 71. *Prof Exp:* NSF fel, 71-72, asst prof, 72-77, ASSOC PROF BIOL, JOHN CARROLL UNIV, 77- *Mem:* Am Soc Cell Biol; AAAS; Int Fedn Cell Biol; Int Res Group Acetabularia. *Res:* Chloroplast biogenesis and intraplastidal biosynthesis; biochemical correlates of fish distribution. *Mailing Add:* Dept Biol John Carroll Univ 20700 N Park Blvd University Heights OH 44118

MOORE, FLETCHER BROOKS, b Heiberger, Ala, June 15, 26; m 54; c 2. ELECTRICAL & ELECTRONICS ENGINEERING. *Educ:* Ala Polytech Inst, BS, 48; Ga Inst Technol, MSEE, 49. *Prof Exp:* Physicist, Navy Exp Sta, Panama City, Fla, 49-50, electronic scientist, 50-52, elec engr, Mixing Comput Unit, 52-53, chief, Simulator Unit, 53-54, dep chief, Analog Comput Sect, 54-56, chief control sect, Army Ballastic Missile Agency, Redstone, Ala, 56-58, chief, Navig Br, Redstone Arsenal, Ala, 59-62; chief, guid & control br, Marshall Space Flight Ctr, NASA, 62-63, chief, guid & Control div, 63-69, dir, astrionics Lab, 69-74, dir, electronics & Control Lab, sci & eng, 74-81; chief, missile system sec, Teledyne Brown Eng, Huntsville, 81-83; dir opers, 83-86, VPRES, CONTROL DYNAMICS CO, HUNTSVILLE, 86- *Concurrent Pos:* Mem, eng & res couns, Auburn Univ, Ala Indust Coun Engr Educ, Univ Ala-Huntsville Adm Sci Coun. *Honors & Awards:* Leadership Medal, NASA; Except Serv Medal, NASA; Spec Achievement Award, NASA. *Mem:* Assoc fel Am Inst Aeronaut & Astronaut; Inst Navig. *Res:* Guidance and control; instrumentation; power systems; solar energy; optical systems; communications; automated systems. *Mailing Add:* Control Dynamics Co 600 Blvd South Huntsville AL 35802

MOORE, FRANCIS BERTRAM, b Des Moines, Iowa, July 31, 05. PHYSICAL CHEMISTRY. *Educ:* Des Moines Univ, BA, 26; Iowa State Col, PhD(phys chem), 40. *Prof Exp:* Instr & asst coach high sch, Iowa, 26-27, 29-32, prin & coach, 27-29; instr chem, Iowa State Col, 33-40; assoc prof & head dept, Phillips Univ, 40-43; prof, Southeast Mo State Col, 43-52; assoc prof, 52-61, prof, 61-74, head dept, 54-72, EMER PROF CHEM, UNIV MINN, DULUTH, 74- *Mem:* Am Chem Soc; Sigma Xi. *Res:* Electron sharing ability of organic radicals; qualitative separation of copper and cadmium; condensation of mercaptans with chloral in the gaseous phase; chemical composition of western Lake Superior waters; extension of studies of the bathophanthroline determination of iron. *Mailing Add:* 800 Chester Park Dr Duluth MN 55812

MOORE, FRANCIS DANIELS, b Evanston, Ill, Aug 17, 13; m 35; c 5. SURGERY. *Educ:* Harvard Univ, AB, 35, MD, 39. *Hon Degrees:* MCh, Nat Univ Ireland, 61; LLD, Glasgow Univ, 65; DSc, Suffolk Univ, 66, Harvard Univ, 82; FRCS, 67, FRCS(E), 68, FRCS(C), 70 & FRCS(I), 72; MD, Univ Goteberg, Sweden, 75, Univ Edinburgh, 76, Univ Paris, 76, Univ Copenhagen, 79. *Prof Exp:* Nat Res Coun fel med, Harvard Univ, 41-42; instr, 43-46, assoc, 46-47, asst prof & tutor, 47-48, Moseley prof surg, 48-76, Elliott Carr Cutler prof surg, 76-80, EMER MOSELEY PROF SURG, HARVARD MED SCH, 80-; EMER SURGEON-IN-CHIEF, PETER BENT BRIGHAM HOSP, 80- *Concurrent Pos:* Asst resident surgeon, Mass Gen Hosp, Boston, 42-43, resident surgeon, 43, asst surg, 43-46, asst surgeon, 46; surgeon-in-chief, Peter Bent Brigham Hosp, 48-76, surgeon, 76-; consult, Surgeon Gen, Korea, 51; vis prof, Univ Edinburgh, 52 & Univ London, 55; chmn surg study sect, USPHS, 56-59; vis prof, Univ Colo, 58 & Univ Otago, NZ, 67; mem, Exec Comt, Nat Res Coun; chmn, Adv Comt Metab Trauma, Off Surgeon Gen; consult, Surgeon Sidney Farber Cancer Inst, 76- & Health Resources Admin, HEW, 76; mem bd regents, Uniformed Serv Univ Health Sci, 76; pres, Mass Health Data Consortium, Inc, 81-87; consult, Life Sci, NASA, 68-71 & 86- *Honors & Awards:* Lister Medal, Royal Col Surgeon, 78; Gross Medal, Am Surgeon Asn, 78. *Mem:* Nat Acad Sci; Soc Univ Surgeons (pres, 58-); hon mem Polish Acad Sci; Am Surg Asn (pres, 71-; AMA. *Res:* Clinical surgery as related to gastrointestinal tracts; biochemistry of cellular changes in surgery as revealed by radioactive and stable isotopes; metabolic care in trauma; transplantation of tissues and organs; cancer of the breast; surgical manpower and health care delivery. *Mailing Add:* Countway Libr Ten Shattuck St Boston MA 02115

MOORE, FRANK ARCHER, b Tribune, Kans, Mar 30, 20; m 48; c 3. BIOCHEMISTRY. *Educ:* Ft Hays Kans State Col, BS, 42; Kans State Univ, MS, 52, PhD(chem), 60. *Prof Exp:* Jr inspector powder & explosives, Weldon Spring Ord Works, 42-43; chemist & analyst, Phillips Petrol Co, 47-49; asst chem, Univ Calif, 51-52, lab technician food technol, 52; asst chem, Kans State Univ, 52-56; instr, NMex State Univ, 56-59; from asst prof to prof, 59-85, EMER PROF CHEM, ADAMS STATE COL, 85- *Mem:* Am Chem Soc; Sigma Xi. *Res:* Interaction of proteins with small molecules; organic chemistry; blood lipids; history of chemistry. *Mailing Add:* 82 Monterey Ave Alamosa CO 81101

MOORE, FRANK DEVITT, III, b Philadelphia, Pa, June 16, 31; m 59; c 2. HORTICULTURE. *Educ:* Pa State Univ, BS, 58; Univ Del, MS, 60; Univ Md, PhD(hort), 64. *Prof Exp:* Asst hort, Univ Del, 58-60 & Univ Md, 60-64; asst horticulturist, 64-70, ASSOC PROF HORT, COLO STATE UNIV, 70- *Concurrent Pos:* Sr res horticulturist, 3M Co, 69; mem, Inst Soc Hort Sci. *Mem:* AAAS; Am Soc Hort Sci; Am Inst Biol Sci. *Res:* Effect of environment and nutrition on plant growth and physiology; vegetable crops. *Mailing Add:* Dept of Hort Plant Sci Bldg Colo State Univ Ft Collins CO 80523

MOORE, FRANK LUDWIG, b Fremont, Ohio, Mar 22, 45; m 68; c 1. ENDOCRINOLOGY. *Educ:* Col Wooster, BA, 67; Univ Colo, MA & PhD(biol), 74. *Prof Exp:* Teacher biol, Pub Schs, Lakewood, Ohio & Denver, Colo, 67-71; instr, Univ Colo, 71-72; ASST PROF BIOL, ORE STATE UNIV, 73- *Mem:* Am Soc Zoologists; Sigma Xi; Am Inst Biol Sci. *Res:* Reproductive endocrinology of amphibia. *Mailing Add:* Dept Zool Ore State Univ Corvallis OR 97331

MOORE, FRANKLIN K(INGSTON), b Milton, Mass, Aug 24, 22; m 46; c 6. AERODYNAMICS, HEAT TRANSFER. *Educ:* Cornell Univ, BS, 44, PhD(aeronaut eng), 49. *Prof Exp:* Aeronaut res scientist, Nat Adv Comt Aeronaut, 49-55; dir aerodyn div, Cornell Aeronaut Lab, Inc, 55-65; head dept thermal eng, 65-73, JOSEPH C FORD PROF MECH ENG, CORNELL UNIV, 65- *Concurrent Pos:* Vis scientist, NASA, 80-81, 88-89; mem, Aeronaut & Space Eng Bd, Nat Res Coun, 86- *Honors & Awards:* Except Sci Achievement Award, NASA. *Mem:* Nat Acad Eng; fel Am Soc Mech Engrs; fel Am Inst Aeronaut & Astronaut; Am Phys Soc. *Res:* Gas-turbine dynamics; unsteady fluid dynamics; thermal engineering; boundary layer theory. *Mailing Add:* 240 Upson Hall Cornell Univ Ithaca NY 14853

MOORE, FRED EDWARD, geology; deceased, see previous edition for last biography

MOORE, GARRY EDGAR, b Washington, DC, Aug 29, 48; m 76. NUCLEAR PHYSICS, SPACE SCIENCES. *Educ:* Va Polytech Inst & State Univ, BS, 70; Fla State Univ, MS, 73, PhD(physics), 75. *Prof Exp:* Res assoc nuclear physics & fel, Univ Pa, 75-77; ANALYST SPACE SCI, ANAL SERV, INC, 77- *Concurrent Pos:* Fulbright fel, W Ger. *Mem:* Am Phys Soc; Am Inst Aeronaut & Astronaut. *Res:* Advanced space systems concepts and technology planning, utility of crews in space, comprehensive study of tritium induced nuclear reactions, nuclear structure of light nuclei and lithium induced transfer reactions. *Mailing Add:* 13102 Rounding Run Circle Herndon VA 22071

MOORE, GARY T, b Calgary, Alta, Mar 13, 45; m 88; c 2. ARCHITECTURE RESEARCH, ENVIRONMENTAL PSYCHOLOGY. *Educ:* Univ Calif, Berkeley, BArch, 68; Clark Univ, MA, 73, PhD(environ psychol), 82. *Prof Exp:* Instr geog, Clark Univ, 74-75; from asst prof to assoc prof, 76-88, PROF ARCHIT RES, UNIV WIS-MILWAUKEE, 88- *Concurrent Pos:* Vis lectr, Sydney Univ & Univ NSW, Australia, 75; prin, Gary T Moore & Assoc, 83-; vis prof archit res, Godjah Mada Univ, Indonesia, 85-86; vis assoc prof, Univ Ore, 87; dir, Advan Design Prog, NASA, 89- & Wis Space Grant Consortium, 91- *Mem:* Sr mem Am Inst Aeronaut & Astronaut; fel Am Psychol Asn. *Res:* Environment-behavior relationships, with special focuses on child-environment relations; environmental cognition and aging and environment; applications to architecture, urban design and planning. *Mailing Add:* Ctr Archit & Urban Planning Res Univ Wis Milwaukee WI 53201-0413

MOORE, GARY THOMAS, b Dallas, Tex, Apr 6, 42; m 61; c 3. VETERINARY MEDICINE, LABORATORY ANIMAL SCIENCE. *Educ:* Tex A&M Univ, BS, 65, DVM, 66. *Prof Exp:* Res veterinarian, Div Microbiol, US Army, Vet Corps, 66-68; veterinarian, Yerkes Regional Primate Ctr, Emory Univ, 68-69; chief res support br, Southwest Found Res & Educ, 69-72, dir, Dept Animal Resources & Facil, 72-80; DIR, LAB ANIMAL SCI, SOUTHWEST RES INST, SAN ANTONIO, TEX, 80- *Concurrent Pos:* Mem, Nat Comt Conserv Nonhuman Primates, Inst of Lab Animal Resources, 73-75; prin investr, NIH contract, 74-78, grant, 78-83. *Mem:* Am Asn Lab Animal Sci. *Res:* Management of laboratory and animal medicine colonies; experimental surgery; reproductive physiology and domestic breeding of nonhuman primates. *Mailing Add:* 6435 Ridge Circle San Antonio TX 78233

MOORE, GEORGE A(NDREW), b New York, NY, Feb 14, 13; m 39; c 5. MICROSCOPY, METALLURGY. *Educ:* Union Col, BS, 34; Harvard Univ, MA, 35; Princeton Univ, PhD(chem), 39. *Prof Exp:* Asst instr chem & metall, Princeton Univ, 35-39, instr, 39-40; res engr, Battelle Mem Inst, 40-47; asst prof metall eng, Univ Pa, 48-51; metallurgist, Nat Bur Standards, 51-78; RETIRED. *Concurrent Pos:* Lectr, Univ Md, 54-62; chmn organizing comt, Fourth Int Cong Stereology, 75; consult, var indust & govt, 78- *Honors & Awards:* Electrochem Soc, 39. *Mem:* Am Soc Metals; Electrochem Soc; fel Am Soc Testing & Mat; Am Inst Mining, Metall & Petrol Engrs; Int Soc Stereology; Sigma Xi. *Res:* Hydrogen in metals; absorption and evolution; analysis; effect on microstructure and properties; hydrogen in steel brittle failures; low-temperature properties; high purity metals; superconducting alloys; quantitative microscopy; computer analysis of micrographs; automated image analysis. *Mailing Add:* 1108 Agnew Dr Rockville MD 20851

MOORE, GEORGE EDWARD, b Evanston, Ill, Sept 28, 22; m 49; c 2. CONTINUING EDUCATION, PERSONNEL DEVELOPMENT. *Educ:* Northwestern Univ, BS, 43; Univ Tenn, PhD(chem), 61. *Prof Exp:* Asst chem, Univ Minn, 43-44 & Univ Chicago, 44; res chemist, Clinton Labs, Oak Ridge Nat Lab, 44-48 & Union Carbide Nuclear Co, 48-71; res chemist, Chem Div, 71-74, res staff dir admin, 74-78, HEAD, UNIV RELS & PERSONNEL DEVELOP, OAK RIDGE NAT LAB, 78- *Concurrent Pos:* Chemist, Metall Lab, Manhattan Eng Dist, 44. *Mem:* AAAS; Am Chem Soc; Sigma Xi; Am Soc Eng Educ. *Res:* Heterogeneous catalysis; radiation chemistry; ion exchange; corrosion; chemistry of actinide elements; molecular beams; gas-surface interactions; water pollution research; radiological protection. *Mailing Add:* 7108 Wellington Dr Knoxville TN 37919

MOORE, GEORGE EMERSON, JR, b Lebanon, Mo, Jan 2, 14; m 39; c 3. GEOLOGY. *Educ:* Univ Mo, AB, 36, MA, 38; Harvard Univ, MA, 41, PhD(geol), 47. *Prof Exp:* Instr geol, Univ Mo, 38-39; rodman, US Geol Surv, Washington, DC, 39; asst, Harvard Univ, 40-42; geologist, A P Green Fire Brick Co, Mo, 42-46; asst, Harvard Univ, 46-47; from instr to prof, 47-84, EMER PROF GEOL, OHIO STATE UNIV, 84- *Mem:* Geol Soc Am; Sigma Xi. *Res:* Structure and metamorphism of southwestern New Hampshire, bedrock geology of southeastern New England. *Mailing Add:* 58 Mulberry Dr Wakefield RI 02879

MOORE, GEORGE EUGENE, b Minneapolis, Minn, Feb 22, 20; m 44; c 5. SURGERY, ONCOLOGY. *Educ:* Univ Minn, BA, 42, MA, 43, BS, 44, BM, 46, MD, 47, PhD, 50; Am Bd Surg, dipl. *Prof Exp:* Lab asst physiol, Univ Minn, 41-42, asst histol & zool, 42-43, intern surg, Univ Hosps, 46-47, clin instr, Med Sch, 48-50, asst prof & cancer coordr, 51, assoc prof, 51-52; res prof biol & dir, Roswell Park Div, Grad Sch, State Univ NY Buffalo, 53-69, clin prof surg, Sch Med, 62-73; PROF SURG, UNIV COLO, DENVER, 73-; CHIEF DIV SURG ONCOL, DENVER GEN HOSP, 73- *Concurrent Pos:* Markle Fund scholar, 48-53; inst dir, 53-68, dir & chief surg, Roswell Park Mem Inst, 53-69; consult, Nat Cancer Chemother Ctr, NIH, 56; dir pub health res, NY State Dept Health, 69- *Honors & Awards:* Gross Award, 50; Chilean Iodine Educ Bur Award, Am Pharmaceut Asn, 51. *Mem:* Soc Exp Biol & Med; Soc Univ Surgeons; Am Surg Asn; NY Acad Sci; fel Am Col Surgeons. *Res:* Localization of brain tumors; carcinogenesis; experimental biology and oncology; cell culture chemotherapy; cancer surgery. *Mailing Add:* 3247 Locust St Denver CO 80207

MOORE, GEORGE WILLIAM, b Palo Alto, Calif, June 7, 28; m 60; c 2. GEOLOGY. *Educ:* Stanford Univ, BS, 50, MS, 51; Yale Univ, PhD(geol), 60. *Prof Exp:* Geologist, US Geol Surv, 51-66; geologist-in-chg, La Jolla Marine Geol Lab, 66-75; GEOLOGIST, US GEOL SURV, 75- *Concurrent Pos:* Res assoc, Scripps Inst Oceanog, 70-75; partic, Deep Sea Drilling Proj, Leg 57, 77; arctic chmn, Circum-Pac Map Proj, 79-; invited lectr, USSR Acad Sci, 80, Indonesian Marine Geol Inst, 86; courtesy prof geol, Ore State Univ, 87- *Honors & Awards:* 00080976x. *Mem:* Fel AAAS; fel Geol Soc Am; hon mem Nat Speleol Soc (pres, 63); Am Asn Petrol Geologists; Am Geophys Union. *Res:* Stratigraphy and geochemistry of sedimentary rocks; structural geology; cave mineralogy; marine geophysics. *Mailing Add:* Dept Geosci Ore State Univ Corvallis OR 97331-5506

MOORE, GERALD L, b Cincinnati, Ohio, Apr 24, 39; m. BIOCHEMISTRY, ANALYTICAL CHEMISTRY. *Educ:* Univ Cincinnati, BS, 61, PhD(biochem), 67; Xavier Univ, Ohio, MS, 63. *Prof Exp:* Res biochemist blood, US Army Med Res Lab, Ft Knox, Ky, 67-74, RES BIOCHEMIST BLOOD, LETTERMAN ARMY INST RES, 74- *Concurrent Pos:* Asst prof chem, Univ Ky, Ft Knox Exten, 71-74 & Bowling Green State Univ, Ft Knox Exten, 73-74. *Mem:* Am Chem Soc; Am Asn Blood Banks; AAAS; Sigma Xi; Soc Armed Forces Med Lab Scientists. *Res:* Metabolism and membrane structure and function of red blood cells; improved methods of blood storage and blood banking techniques; hemoglobin biochemistry. *Mailing Add:* Blood Res Div Letterman Army Inst of Res Presidio Blvd Presidio of San Francisco CA 94129

MOORE, GLENN D, b Barron, Wis, Apr 9, 23; m 47; c 3. ENTOMOLOGY, PLANT BREEDING. *Educ:* Univ Wis, BS, 50, MS, 51; Univ Minn, PhD, 68. *Prof Exp:* Res agronomist & asst to mgr, Res Serv Dept, Northrup King Co, 52-80, sr entomologist & dir entom, 70-82; RETIRED. *Concurrent Pos:* Consult & prof witness, 82- *Res:* Insect resistance in crop plants; alfalfa pollination for seed production; fungicide and fungicide-insecticide seed treatments; herbicides and their application in field crops and vegetables; forage crop management studies. *Mailing Add:* 4007 61 Ave N Brooklyn MN 55429

MOORE, GORDON EARLE, b San Francisco, Calif, Jan 3, 29; m 50. PHYSICAL CHEMISTRY. *Educ:* Univ Calif, BS, 50; Calif Inst Technol, PhD(chem), 54. *Prof Exp:* Asst chem, Calif Inst Technol, 50-52; res chemist phys chem, Johns Hopkins Univ, 53-56; mem tech staff, Shockley Semiconductor Labs, Beckman Instruments Corp, 56-57; head eng, Fairchild Semiconductor Corp, 57-58, dir res & develop, 58-68; exec vpres, 68-75, pres, 75-79, CHMN BD, INTEL CORP, 79- *Honors & Awards:* Nat Medal Technol, 90. *Mem:* Nat Acad Eng; Electrochem Soc; Am Phys Soc; Inst Elec & Electronics Engrs. *Res:* Molecular spectroscopy and structure; semiconductors; transistors; microcircuits. *Mailing Add:* Intel Corp 3065 Bowers Ave Santa Clara CA 95051

MOORE, GORDON GEORGE, b Des Moines, Iowa, Mar 18, 35; m 56; c 2. ORGANIC CHEMISTRY. *Educ:* Iowa State Univ, BS, 56; Yale Univ, MS, 58, PhD(org chem), 62. *Prof Exp:* Res assoc, Brookhaven Nat Lab, 60-62; asst prof org chem, Marshall Univ, 62-65; from asst prof to assoc prof, 65-77, PROF ORG CHEM, PA STATE UNIV, OGONTZ CAMPUS, 77- *Concurrent Pos:* Res chemist, USDA, Pa, 68-83, 87-88 & 90, Warminster Naval Air Devel Ctr, 84-85. *Mem:* Am Chem Soc. *Res:* Organic mechanisms; organic synthesis; compounds of biochemical interest; corrosion. *Mailing Add:* Dept Chem Pa State Univ 1600 Woodland Ave Abington PA 19001

MOORE, GRAHAM JOHN, b Bristol, Eng, Feb 13, 46; Can citizen; m 78; c 2. PEPTIDE SYNTHESIS, PROTEIN CHEMISTRY. *Educ:* Univ Exeter, UK, BSc, 67, MSc, 69; Univ Ottawa, PhD(biochem), 72. *Prof Exp:* Fel, Max Planck Inst Molecular Genetics, 72-73; fel, 73-76, ASST PROF PHARMACOL & BIOCHEM, UNIV CALGARY, CAN, 76- *Mem:* Soc Endocrinol; Can Biochem Soc; Pharmacol Soc Can; Western Pharmacol Soc. *Res:* Synthesis and evaluation of peptide hormones and analogues; biosynthesis of neurohypophysial hormones. *Mailing Add:* 1762 First Ave NW Calgary AB T2N 0B1 Can

MOORE, GREGORY FRANK, b Bismarck, NDak, Sept 25, 51; m 74. GEOLOGY. *Educ:* Univ Calif, Santa Barbara, BA, 73; Johns Hopkins Univ, MA, 74; Cornell Univ, PhD(geol), 77. *Prof Exp:* Res asst, Cornell Univ, 74-77, fel, 77; asst res geologist structural geol, Scripps Inst Oceanog, Univ Calif, 78-82; res geologist, Res Lab, Cities Serv Oil Co, 82-83; assoc prof, Univ Tulsa, 84-88; ASSOC PROF, DEPT GEOL GEOPHYS, UNIV HAWAII, 88- *Concurrent Pos:* Lectr, Univ Calif, San Diego, 80-82. *Mem:* Geol Soc Am; Am Geophys Union; Soc Explor Geophysicists; Soc Econ Paleontologists & Mineralogists; Int Asn Sedimentologists. *Res:* Structural geology; marine geophysics applied to the study of convergent plate margins in an attempt to understand subduction processes. *Mailing Add:* Dept Geol Geophys Univ Hawaii 2525 Correa Rd Honolulu HI 96822

MOORE, HAL G, b Vernal, Utah, Aug 14, 29; m 56; c 3. MATHEMATICS. *Educ:* Univ Utah, BS, 52, MS, 57; Univ Calif, Santa Barbara, PhD(math), 67. *Prof Exp:* Jr high sch teacher, Utah, 52-53; instr math, Carbon Jr Col & Carbon High Sch, 53-55; instr, Purdue Univ, 57-61, admin asst to chmn dept, 60-61; exec asst to chmn dept, Brigham Young Univ, 61-64, from asst prof to prof math, 61-89, assoc dep chmn, 86-89; RETIRED. *Concurrent Pos:* Mem bd dirs, Sigma Xi, 74-90; chmn, Nat lectureships, 82-90; mem bd gov, Math Asn Am, 89- & Math Coalition, Utah, 90- *Mem:* Sigma Xi; Am Math Soc; Math Asn Am. *Res:* Structure of rings. *Mailing Add:* Dept Math Brigham Young Univ 316 TMCB Provo UT 84602

MOORE, HAROLD ARTHUR, b Brackettville, Tex, Feb 4, 25. NUCLEAR PHYSICS. *Educ:* St Mary's Univ, Tex, BS, 44; Wash Univ, MS, 48; Univ Fla, PhD, 56. *Prof Exp:* Asst, Wash Univ, 46-48; physicist, Mo Res Labs, Inc, 48-51; asst, Univ Fla, 51-52, 55-56, instr phys sci, 52-55; from asst prof to assoc prof, 56-62, PROF PHYSICS, BRADLEY UNIV, 62- *Concurrent Pos:* Consult, Metric Photo Br, US Naval Ord Test Sta, Calif, 57-58. *Mem:* Am Phys Soc; Am Asn Physics Teachers; Sigma Xi. *Res:* Fast resolving time coincidence circuits; deuteron stripping reactions on carbon and oxygen; unified field theory. *Mailing Add:* 904 W Armstrong Peoria IL 61606

MOORE, HAROLD BEVERIDGE, b Alix, Ark, Sept 11, 28; m 50; c 2. MEDICAL BACTERIOLOGY. *Educ:* San Diego State Col, AB, 51; Univ Calif, Los Angeles, MA, 55, PhD(microbiol), 57; Am Bd Med Microbiol, dipl pub health & med, 65. *Prof Exp:* Chief microbiologist, Donald N Sharp Mem Community Hosp, San Diego, 57-60; from asst prof to assoc prof, 60-67, PROF MICROBIOL, SAN DIEGO STATE UNIV, 67- *Concurrent Pos:* Consult, Donald N Sharp Mem Community Hosp, San Diego, 60-86, Palomar Hosp, Escondido, 64- *Mem:* AAAS; Am Soc Microbiol; Sigma Xi. *Res:* Isolation, identification, taxonomy and clinical significance of poorly defined gram negative rods; clincally significant anaerobic bacteria. *Mailing Add:* Dept of Biol San Diego State Univ San Diego CA 92182

MOORE, HAROLD W, b Ft Collins, Colo, May 21, 36; m 59; c 2. ORGANIC CHEMISTRY. *Educ:* Colo State Univ, BS, 59; Univ Ill, PhD(chem), 63. *Prof Exp:* Asst prof, 65-69, assoc prof, 69-74, chmn dept, 70-74, PROF CHEM, UNIV CALIF, IRVINE, 74- *Mem:* Am Chem Soc. *Res:* Organic chemistry, especially synthesis and mechanistic studies. *Mailing Add:* Dept of Chem Univ of Calif Irvine CA 92717

MOORE, HARRY BALLARD, JR, b Rocky Mount, NC, Sept 28, 28; m 55; c 5. WOOD PRODUCTS INSECTS. *Educ:* E Carolina Col, AB, 51; Purdue Univ, MS, 55; NC State Univ, PhD(entom), 64. *Prof Exp:* Chief inspector, NC Struct Pest Control Comn, 58-60; from instr to prof, 60-88, EMER PROF ENTOM, NC STATE UNIV, 89-; CONSULT, DM ASSOCS, 89- *Mem:* Entom Soc Am; Nat Pest Control Asn. *Res:* Wood-destroying insects. *Mailing Add:* DM Assocs 3725 Eakley Ct Raleigh NC 27606

MOORE, HENRY J, II, b Albuquerque, NMex, Sept 2, 28; m 59; c 3. GEOLOGY. *Educ:* Univ Utah, BS, 51; Stanford Univ, MS, 59, PhD(geol), 65. *Prof Exp:* Prospector-geologist, 54-55; geologist, US Geol Surv, 55-57; asst geol, Stanford Univ, 57-59; GEOLOGIST ASTROGEOL, US GEOL SURV, 60- *Concurrent Pos:* Team leader, Phys Properties Viking Mars. *Honors & Awards:* Spec Commendation for Astronaut Training, Geol Soc Am, 73; Exceptional Serv Viking, NASA, 77; Meritorius Serv Award, US Dept of Interior, 85. *Mem:* AAAS; Sigma Xi; Geol Soc Am; Am Inst Mining, Metall & Petrol Engrs; Am Geophys Union; AAAS. *Res:* Application of the principles of geology to lunar, Martian, and Venusian problems; application of remote sensing data to lunar, Martian, and Venusian problems. *Mailing Add:* 528 Jackson Dr Palo Alto CA 94303

MOORE, HOWARD FRANCIS, b Louisville, Ky, Jan 12, 48; m 75; c 2. REFINING PROCESS DEVELOPMENT, PETROLEUM CATALYST EVALUATION. *Educ:* Univ Ky, BS, 74, MS, 75. *Prof Exp:* Res engr, refining res & develop, 75-77, group leader, synthetic fuels, 77-83, MGR, PROCESS RES & DEVELOP, ASHLAND OIL INC, 83- *Mem:* Am Inst Chem Engrs; Am Chem Soc; N Am Catalysis Soc. *Res:* Evaluation of presently available petroleum refining catalysts and technologies; prediction of future directions for development of justifiable research and development programs; new and improved refining technology. *Mailing Add:* PO Box 391 Ashland KY 41114

MOORE, J STROTHER, b Seminole, Okla, Sept 11, 47; m 68; c 1. COMPUTER SCIENCE. *Educ:* Mass Inst Technol, BS, 70; Univ Edinburgh, PhD(comput logic), 73. *Prof Exp:* Mem res staff comput sci, Xerox Palo Alto Res Ctr, 73-76; RES MATHEMATICIAN COMPUT SCI, STANFORD RES INST, 76-;; AT DEPT COMPUT SCI, UNIV TEX. *Honors & Awards:* McCarthy Prize. *Res:* Automatic computer program verification and advanced debugging aids. *Mailing Add:* 1717 W Sixth St Suite 290 Austin TX 78703

MOORE, JAMES ALEXANDER, b Johnstown, Pa, Nov 8, 23; m 51; c 3. ORGANIC CHEMISTRY. *Educ:* Washington & Jefferson Col, BS, 43; Purdue Univ, MS, 44; Pa State Col, PhD(chem), 49. *Prof Exp:* Res chemist, Parke, Davis & Co, 49-55; from asst prof to assoc prof, Univ Del, 55-63, prof chem, 63-88; RETIRED. *Concurrent Pos:* Asst, Univ Basel, 52-53; sr ed, J Org Chem, 63-88; NIH spec fel, 64-65; consult ed, Encycl Sci & Tech, 88- *Mem:* Am Chem Soc. *Res:* Novel heterocyclic systems; new forms of chemical publications. *Mailing Add:* Dept Chem Univ Del Newark DE 19716

MOORE, JAMES ALFRED, b Brooklyn, NY, Aug 30, 39; m 66; c 2. ORGANIC CHEMISTRY, POLYMER CHEMISTRY. *Educ:* St John's Univ, NY, BS, 61; Polytech Inst Brooklyn, PhD(chem), 67. *Prof Exp:* NIH fel, Univ Mainz, 67-68; res assoc, Univ Mich, Ann Arbor, 68-69; from asst prof to assoc prof, 69-84, coordr polymer sci & eng, 73-77, PROF CHEM, RENSSELAER POLYTECH INST, 84- *Concurrent Pos:* Assoc ed, Org Preparations & Procedures Int, 73- *Honors & Awards:* Am Cyanamide Award, 85. *Mem:* Am Chem Soc. *Res:* Synthesis, characterization and reactions of novel polymers; structure-property relationships; polymers in microelectronics. *Mailing Add:* Dept Chem Rensselaer Polytech Inst Troy NY 12180-3590

MOORE, JAMES ALLAN, b Fresno, Calif, Mar 11, 39; m 60; c 5. WATER QUALITY, LIVESTOCK WASTE MANAGEMENT. *Educ:* Calif State Polytech, BS, 62; Univ Ariz, MS, 64; Univ Minn, PhD(agr eng), 75. *Prof Exp:* Res engr, dept agr eng, Univ Calif, Davis, 64-68; from instr to asst prof eng, dept agr eng, Univ Minn, 68-79; assoc prof, 79-85, actg head dept, 83-84, PROF AGR ENG, ORE STATE UNIV, 85-, EXTEN AGR ENGR, 79- *Concurrent Pos:* Ed, Trans Am Soc Agr Engrs, 72-75. *Honors & Awards:* G B Gunlogson Countryside Eng Award, Am Soc Agr Engrs, 85. *Mem:* Am Soc Agr Engrs; Coun Agr Sci & Technol; Soil Sci Soc Am. *Res:* Impact of livestock and processing operations on water quality (both nutrient and microbiological); modeling water quality; parameters. *Mailing Add:* Dept Agr Eng Ore State Univ Corvallis OR 97331-3906

MOORE, JAMES CARLTON, b Harrodsburg, Ky, May 31, 23; m 46; c 3. PHYSIOLOGY. *Educ:* Univ Ky, BS, 44; Univ Louisville, MD, 46. *Prof Exp:* Commonwealth Fund fel, Mass Inst Technol, 49-50; from asst prof to assoc prof physiol, 53-70, asst dean student affairs, 73-74, PROF PHYSIOL & BIOPHYS, SCH MED, UNIV LOUISVILLE, 70-, ASSOC DEAN ADMIS, 74- *Mem:* AAAS; Biophys Soc; Am Physiol Soc; Soc Exp Biol & Med; Sigma Xi. *Res:* Blood volume measurement; mechanics of respiration; pulmonary edema; pulmonary circulation; microcirculation. *Mailing Add:* Health Sci Ctr Sch Med Univ Louisville Louisville KY 40292

MOORE, JAMES FREDERICK, JR, b Barbourville, Ky, July 19, 38; m 65; c 2. PLANT PATHOLOGY. *Educ:* Western Ky Univ, BS, 65; Clemson Univ, MS, 68; Univ Ariz, PhD(plant path), 76. *Prof Exp:* Res asst plant path, Clemson Univ, 65-69; plant pathologist, Dept Agr, Guam, 69-72; res assoc, Univ Ariz, 72-75; res plant pathologist, 75-80, assoc dir res, 80-85, assoc res prof, 80-87, RES PROF, STATE FRUIT EXP STA, SOUTHWEST MO STATE UNIV, 87-, DIR, 85- *Mem:* Am Phytopath Soc; Am Hort Sci; Am Soc Enol & Viticult; Sigma Xi. *Res:* Disease of fruit and nut crops; ecology of soil-borne plant pathogens and soil microorganisms; antibiosis; chemical, cultural and biological control; epiphytology. *Mailing Add:* Rte 2 Box 246 Mountain Grove MO 65711

MOORE, JAMES GREGORY, b Palo Alto, Calif, Apr 30, 30; m 52, 86; c 3. GEOLOGY. *Educ:* Stanford Univ, BS, 51; Univ Wash, MS, 52; Johns Hopkins Univ, PhD(geol), 54. *Prof Exp:* Instr geol, Johns Hopkins Univ, 52-53; GEOLOGIST, US GEOL SURV, 56- *Concurrent Pos:* Scientist-in-charge, Hawaiian Volcano Observ, US Geol Surv, 62-64 & chief br field geochem & petrol, 70-74. *Mem:* Geol Soc Am; Geochem Soc; Am Geophys Union. *Res:* Petrology and geochemistry of igneous rocks; structural geology; volcanology; submarine geology. *Mailing Add:* US Geol Surv 345 Middlefield Rd Menlo Park CA 94025

MOORE, JAMES MENDON, b Winchester, Mass, Apr 25, 25; m 50; c 4. INDUSTRIAL ENGINEERING. *Educ:* Rensselaer Polytech Inst, BME, 50; Cornell Univ, MS, 56; Stanford Univ, PhD(indust eng), 65. *Prof Exp:* Prod control supvr, Stanley Tools, 50-52; instr indust eng, Cornell Univ, 52-56; asst prof mech eng, Clarkson Tech Col, 56-60; prof indust eng & chmn dept, Northeastern Univ, 61-73; head dept, 73-75, prof, 73-81, ADJ PROF, INDUST ENG & OPERS RES, VA POLYTECH INST & STATE UNIV, 81-; PRES, MOORE PRODUCTIVITY SOFTWARE, 81- *Concurrent Pos:* Fulbright lectr, Tech Univ Helsinki, 68-69; vis Lucas prof eng prod, Univ Birmingham, Eng, 75-76; indust consult, 60- *Mem:* Fel Am Inst Indust Engrs; Am Soc Eng Educ; fel The Royal Soc. *Res:* Application of computers to solve problems involving the location of economic activities in an optimal manner. *Mailing Add:* Dept Indust Eng & Opers Res Va Polytech Inst & State Univ Blacksburg VA 24061

MOORE, JAMES NORMAN, b Vilonia, Ark, June 10, 31; m 53; c 2. HORTICULTURE. *Educ:* Univ Ark, BSA, 56, MS, 57; Rutgers Univ, PhD(hort), 61. *Prof Exp:* Instr hort, Univ Ark, 57; res assoc pomol, Rutgers Univ, 57-61; res horticulturist, USDA, 61-64; from assoc prof to prof, 64-85, univ prof, 85-89, DISTINGUISHED PROF HORT, UNIV ARK, FAYETTEVILLE, 89- *Concurrent Pos:* Hon asst prof, Rutgers Univ, 61-64; assoc ed, Am Soc Hort Sci, 80-85. *Honors & Awards:* Outstanding Res Award, Am Soc Hort Sci, 90; Distinguished Serv Award, USDA, 87. *Mem:* Am Soc Hort Sci (vpres, 82-83, pres, 87-88); Am Genetic Asn; Am Pomol Soc (pres, 83); Int Soc Hort Sci; NY Acad Sci. *Res:* Fruit breeding and genetics; physiology of fruit crops. *Mailing Add:* 316 Plant Sci Bldg Univ Ark Fayetteville AR 72701

MOORE, JAMES THOMAS, b Mineola, NY, Feb 9, 52; m 78. ATMOSPHERIC SCIENCES. *Educ:* New York Univ, BS, 74; Cornell Univ, MS, 76, PhD(atmospheric sci), 79. *Prof Exp:* Teaching res asst meteorol, Dept Agron, Cornell Univ, 74-78; asst prof meteorol, Earth Sci Dept, State Univ NY, Oneonta, 78-80; asst prof, 80-85,ASSOC PROF METEOROL, DEPT EARTH SCI & ATMOSPHERIC SCI, ST LOUIS UNIV, 85- *Concurrent Pos:* Fac fel, NASA, 79 & Am Soc Eng Educ, 80, 82; prin investr NSF grant, 83-85. *Mem:* Am Meteorol Soc; Nat Weather Asn. *Res:* Dynamics of jet stream propagation, especially with respect to the diagnosis and forecasting of severe local storms including thunderstorms, hailstorms, strong surface winds and tornadoes. *Mailing Add:* Dept Earth & Atmospheric Sci St Louis Univ 3507 Laclede Ave St Louis MO 63103

MOORE, JAMES W(ALLACE) I, b Birmingham, Ala, Feb 19, 23; m 48; c 3. MECHANICAL ENGINEERING. *Educ:* Tenn Polytech Inst, BS, 51; Univ Ky, MS, 52; Purdue Univ, PhD(mech eng), 62. *Prof Exp:* Proj engr, Carbide & Carbon Chem Co, 52-55; proj engr, Allison Div, Gen Motors Corp, 55-56, sr proj engr, 56-58; res asst mech eng, Purdue Univ, 59-62; assoc prof, 62-67, sesquicentennial fel to Barcelona, Spain, 71-72, PROF MECH ENG, UNIV VA, 67- *Concurrent Pos:* NASA res grants, 64-71; Dept of Defense res grant, 67-72. *Mem:* Am Soc Mech Engrs; Sigma Xi. *Res:* Automatic control; algebraic techniques in control design; automatic control in aerodynamic drag instrumentation and in a wind tunnel force balance system; automatic control systems. *Mailing Add:* 3409 Indian Springs Rd Charlottesville VA 22901

MOORE, JAY WINSTON, b Madison, Wis, Apr 20, 42; m 65. DEVELOPMENTAL GENETICS, POULTRY GENETICS. *Educ:* Cedarville Col, BS, 64; Univ Nebr, MS, 67; Univ Mass, PhD(genetics), 70. *Prof Exp:* From asst prof to assoc prof, 70-77, PROF BIOL, EASTERN COL, 77- *Mem:* AAAS; Genetics Soc Am; Sigma Xi; Am Soc Affil; Am Inst Biol Sci. *Res:* Plumage color inheritance; genetic factors controlling pigment and keratin development. *Mailing Add:* 1209 Guildford Ct Iowa City IA 52240

MOORE, JERRY LAMAR, b Anderson, SC, Feb 28, 42; m 66; c 2. NUTRITION. *Educ:* Clemson Univ, BS, 63; Univ Wis, MS, 65, PhD(food sci, biochem), 68. *Prof Exp:* Clin lab officer, Wilford Hall US Air Force Med Ctr, 68-69, biomed lab officer, Nutrit Br, US Air Force Sch Aerospace Med, 69-72; assoc dir nutrit, Pillsbury Co, 72-74, assoc dir corp res & develop, 74-77; vpres res & develop, Mead Johnson Nutrit Div, 78-85; VPRES RES & DEVELOP, BRISTOL MYERS US NUTRIT GROUP, 85- *Concurrent Pos:* Mem, Food & Nutrit Bd, Nat Res Coun, Nat Acad Sci, 82-84. *Mem:* Am Dietetic Asn; Sigma Xi; Inst Food Technol; Soc Nutrit Educ (secy, 74-77 & pres, 82-83); Am Inst Nutrit. *Res:* Management of applied human nutrition and product development leading to new or nutritionally improved foods and special dietary products, new health related services and more effective informational and educational materials. *Mailing Add:* 700 Cedar Hill Dr Evansville IN 47710

MOORE, JOANNE IWEITA, b Greenville, Ohio, July 23, 28. CARDIOVASCULAR PHARMACOLOGY. *Educ:* Univ Cincinnati, AB, 50; Univ Mich, PhD(pharmacol), 59. *Prof Exp:* Asst, Christ Hosp Inst Med Res, Cincinnati, 50-55; asst pharmacol, Univ Mich, 55-57; fel cardiovasc pharmacol, Basic Sci Div, Emory Univ, 59-61; from asst prof to assoc prof, 61-71, actg chmn dept, 69-70, interim chmn dept, 71-73, PROF PHARMACOL, COL MED, UNIV OKLA, 71-, CHMN DEPT, 73- *Concurrent Pos:* Mem, Biomed Study Sect, NIH, Div Res Grants, 86-90; Fogarty Sr Int fels. *Mem:* Am Soc Pharmacol & Exp Therapeut; Soc Exp Biol & Med; Am Heart Asn; Asn Med Sch Pharmacol; NY Acad Sci; Sigma Xi. *Res:* Cardiovascular pharmacology; mechanism of action of drugs upon the contractile and/or conduction system of the heart. *Mailing Add:* Dept Pharmacol 753 BMSB Univ Okla Col Med Oklahoma City OK 73190-3042

MOORE, JOHN ALEXANDER, b Charles Town, WVa, June 27, 15; m 38; c 1. EVOLUTIONARY BIOLOGY. *Educ:* Columbia Univ, AB, 36, MS, 39, PhD(zool), 40. *Prof Exp:* Asst zool, Columbia Univ, 36-39; tutor biol, Brooklyn Col, 39-41; instr, Queens Col, NY, 41-43; from asst prof to prof zool, Barnard Col & Columbia Univ, 43-69, chmn dept, Col, 48-52, 53-54 & 60-66, chmn dept, Univ, 49-52; prof, 69-82, EMER PROF BIOL, UNIV CALIF, RIVERSIDE, 82- *Concurrent Pos:* Res assoc, Am Mus Natural Hist, 42-, Fulbright res scholar, Australia, 52-53; managing ed, J Morphol, 55-60; Guggenheim fel, 59-60; partic, Biol Sci Curric Study, 59-76; mem, US Nat Comn, Int Union Biol Sci, 76-; mem, Comn Human Resources, Nat Res Coun, 79-82. *Honors & Awards:* Gold Medal, XVI Int Cong Zool. *Mem:* Nat Acad Sci; Am Acad Arts & Sci; Am Soc Zoologists (pres, 74); Am Soc Naturalists (pres, 72); hon mem Australian Soc Herpet; Genetic Soc Am; Soc Study Evolution; Soc Col Sci Teachers; AAAS. *Res:* Evolution of amphibians and Drosophila. *Mailing Add:* Dept Biol Univ Calif Riverside CA 92521

MOORE, JOHN ARTHUR, b Salem, Mass, Feb 9, 39; m 62; c 3. TOXICOLOGY. *Educ:* Mich State Univ, BS, 61, DVM, 63, MS, 67; Am Col Lab Animal Med, dipl, 68; Am Bd Toxicol, dipl, 80. *Prof Exp:* Asst dir lab animal med, Sch Med, Western Reserve Univ, 63-65; asst prof vet med, Mich State Univ, 65-69; chief animal sci, 69-71, assoc dir res resources, 75-80, CHIEF ENVIRON BIOL CHEM, NAT INST ENVIRON HEALTH SCI, 71-; at off pesticides & toxic substances, Environ Protection Agency, Washington, DC; PRES, INST EVALUATING HEALTH RISKS, 89- *Concurrent Pos:* Exec secy comt coord toxicol & related progs, HEW, 74-76, chmn subcomt health effects polybrominated biphenyl, 77; consult, Am Asn Accreditation Lab Animal Care, 74-; chmn comt health effects dibenzodioxins & dibenzofurans, Int Agency Res Cancer, 78. *Honors & Awards:* Super Serv Award, HEW, 75. *Mem:* Soc Toxicol; Teratology Soc; Am Vet Med Asn. *Res:* Chemical toxicology research and methods development; functional teratogenic defects; effects of chemicals on immune response; development and use of experimental animals for research. *Mailing Add:* Inst Evaluating Health Risks 1101 Vermont Ave Suite 608 Washington DC 20005

MOORE, JOHN BAILY, JR, b Hattiesburg, Miss, Apr 26, 44; m 67; c 3. IMMUNOLOGY. *Educ:* Wake Forest Univ, BS, 66; Purdue Univ, MS, 68; Johns Hopkins Univ, PhD(biochem), 71. *Prof Exp:* NIH fel biochem, Stanford Univ, 71-73; res scientist, Lederle Labs, Am Cyanamid Co, 73-75; ASST PROF BIOCHEM, TEX A&M UNIV, 75-; RES FEL, ORTHO PHARM CORP. *Mem:* Am Chem Soc. *Res:* Protein chemistry and enzymology; immunopharmacology; early enzyme events in T-cell activation; tyrosine kinases. *Mailing Add:* Ortho Pharm Corp Rte 202 Raritan NJ 08869

MOORE, JOHN CARMAN GAILEY, b Belleville, Ont, Aug 15, 16; m 52; c 3. ECONOMIC GEOLOGY. *Educ:* Univ Toronto, BA, 38; Cornell Univ, MS, 40; Harvard Univ, PhD, 55. *Prof Exp:* Field asst, Ont Dept Mines, 37-39; geologist, Int Petrol Co, Ecuador, 40-41, Labrador Mining & Explor Co, 46 & Oliver Iron Mining Co, Venezuela, 47; tech officer, Geol Surv Can, 48-51; geologist, Photog Surv Corp, Ltd, 51-52, Am Metal Co, Ltd, 52-54 & Dom Gulf Co, 54-56; div supvr, McIntyre Porcupine Mines, Ltd, 56-61; from asst prof to prof, 62-82, head dept, 66-73, EMER PROF GEOL, MT ALLISON UNIV, 82- *Concurrent Pos:* Vis lectr geol, Univ NB, 61-62. *Honors & Awards:* Coleman Gold Medal Univ of Toronto, 38. *Mem:* Geol Asn Can; Can Inst Mining & Metall; Soc Econ Geologists. *Res:* Dispersion haloes around ore deposits. *Mailing Add:* One Bennett St PO Box 471 Sackville NB E0A 3C0 Can

MOORE, JOHN COLEMAN, b Staten Island, NY, May 27, 23. MATHEMATICS. *Educ:* Mass Inst Technol, BS, 48; Brown Univ, PhD(math), 52. *Prof Exp:* Fine instr, Princeton Univ, 53, NSF fel, 53-55, from asst prof to prof, 55-89, EMER PROF MATH, PRINCETON UNIV, 89-; ADJ PROF MATH, UNIV ROCHESTER, 89- *Mem:* Am Math Soc; Math Asn Am; London Math Soc. *Res:* Algebraic topology; homological algebra. *Mailing Add:* Dept Math Univ Rochester Rochester NY 14620

MOORE, JOHN DOUGLAS, b Austin, Tex, Oct 5, 39; m 63. MATHEMATICS. *Educ:* Idaho State Univ, BS, 61, MS, 63; Syracuse Univ, PhD(math), 69. *Prof Exp:* Asst prof math, State Univ NY Col Oswego, 67-69; asst prof, 69-77, ASSOC PROF MATH, ARIZ STATE UNIV, 77- *Mem:* Am Math Soc; Math Asn Am. *Res:* Theory of Abelian groups. *Mailing Add:* Dept Math Ariz State Univ Tempe AZ 85287

MOORE, JOHN DOUGLAS, b Chicago, Ill, July 11, 43; div; c 2. MATHEMATICS. *Educ:* Univ Calif, Berkeley, BA, 65, PhD(math), 69. *Prof Exp:* From asst prof to assoc prof, 69-81, PROF MATH, UNIV CALIF, SANTA BARBARA, 81- *Mem:* Am Math Soc. *Res:* Isometric immersions of Riemannian manifolds; minimal surfaces and their applications to curvature and topology of Riemannian manifolds. *Mailing Add:* Dept Math Univ Calif Santa Barbara CA 93106

MOORE, JOHN DUAIN, b Lancaster, Pa, Dec 11, 13; m 40; c 2. PLANT PATHOLOGY. *Educ:* Pa State Univ, BS, 39; Univ Wis, PhD(plant path), 45. *Prof Exp:* From asst prof to assoc prof, 45-54, prof, 54-80, EMER PROF PLANT PATH & EMER DIR UNIV EXP FARMS, UNIV WIS-MADISON, 80- *Concurrent Pos:* Plant pathologist, USDA, 56-65; prof plant sci & head dept, Fac Agr, Univ Ife, Nigeria, 68-70, dean grad studies, 69, dean agr, 69-70; consult, Univ Calabar, Nigeria & Inst Pertanian Bogor, Indonesia, 81- *Mem:* AAAS; Am Phytopath Soc; Bot Soc Am; Am Inst Biol Sci. *Res:* Fungus, bacterial and virus diseases of sour cherries and apples; effects of fungicides on quality of cherries and apples; greenhouse energy conservation. *Mailing Add:* Dept Plant Path Univ Wis-Madison Madison WI 53706

MOORE, JOHN EDWARD, b Kirkersville, Ohio, Mar 7, 35; m 57; c 1. RUMINANT NUTRITION. *Educ:* Ohio State Univ, BS, 57, MS, 58, PhD(animal sci), 61. *Prof Exp:* PROF ANIMAL SCI, UNIV FLA, 61- *Mem:* Am Dairy Sci Asn; Am Soc Animal Sci; Am Forage & Grassland Coun. *Res:* Forage evaluation and utilization. *Mailing Add:* Nutrit Lab Univ Fla Gainesville FL 32611

MOORE, JOHN EZRA, b Columbus, Ohio, Jan 25, 31; m 56. HYDROLOGY, GROUNDWATER GEOLOGY. *Educ:* Ohio Wesleyan Univ, BA, 53; Univ Ill, MS, 58, PhD(geol), 60. *Prof Exp:* Geologist, Am Zinc Co, Mo, 53-54; teaching asst geol, Univ Ill, 57-60; geologist, Nev Test Site, Nev & Colo, 60-63, proj chief, Groundwater Br, Water Resources Div, 63-67, supvry hydrologist, Colo Dist, 67-71, staff hydrologist, Rocky Mt Region, 71-75, chief, Southwest Fla Subdist Tampa, 75-78; staff hydrologist, Water Resources Div, US Geol Surv, 78-79, dep asst chief hydrologist Sci Info Mgt, 79-89; SR HYDROLOGIST, ENVIRON STRATEGIES CORP, 89- *Mem:* Int Asn Hydrogeologists; fel Geol Soc Am; Am Water Res Asn; Asn Earth Sci Ed. *Res:* Clay mineralogy; sedimentation; groundwater hydrology; application of electrical analog and digital models to the study of groundwater and surfacewater systems; evaluations of water resources for planning and management. *Mailing Add:* 2302 Foxfire Ct Reston VA 22091

MOORE, JOHN FITZALLEN, b London, Eng, Feb 23, 28; US citizen; m 48, 63; c 7. RADIOLOGICAL PHYSICS. *Educ:* Mass Inst Technol, BS, 49; Harvard Univ, MS, 52; Columbia Univ, PhD(optical physics), 71. *Prof Exp:* Microwave engr, Raytheon Mfg Co, Inc, 48-52; syst mgr, Hycon Eastern Inc, 53-57; new prod mgr, Appl Sci Corp, Princeton, 57-60; sci adv to pres, Div Lockheed Aircraft, Lockheed Electronics Co, 60-68; dir res, Computone Systs, Inc, 68-70; dir prod develop, Spex Industs, Inc, 71-76; vpres eng, E M I Med, Inc, 76-80; PRES, BIO-IMAGING RES, INC, 80- *Concurrent Pos:* Pres, Princeton Custom Serv, 68-80; chmn, Surface Weapons Bd Naval Res Adv Comt, 74-77 & Working Group Specif, ACR-NEMA Digital Standards Comt, 82-87; assoc prof, Chicago Med Sch, 88- *Mem:* Sigma Xi; AAAS; Inst Elec & Electronic Engrs. *Res:* Synthesis of multiple technologies to create new instrumentation in spectroscopy, military sensors, computed tomography, etc. *Mailing Add:* 28328 Ivy Lane Libertyville IL 60048

MOORE, JOHN HAYS, b Pittsburgh, Pa, Nov 6, 41; m 63; c 2. PHYSICAL CHEMISTRY, MOLECULAR PHYSICS. *Educ:* Carnegie Inst Technol, BS, 63; Johns Hopkins Univ, MA, 65, PhD(chem), 67. *Prof Exp:* Res assoc chem, Johns Hopkins Univ, 67-69; from asst prof to assoc prof, 69-78, PROF CHEM, UNIV MD, COLLEGE PARK, 78- *Concurrent Pos:* Vis fel, Joint Inst Lab Astrophys, Colo, 75-76; prog officer, Nat Sci Found, 80-81. *Mem:* Fel Am Phys Soc; Sigma Xi. *Res:* Valence electronic structure of molecules; electron transmission spectroscopy; (e, ze) spectroscopy; atomic physics; ionospheric physics. *Mailing Add:* 3905 Commander Dr Hyattsville MD 20782

MOORE, JOHN JEREMY, b Stourbridge, Eng, Mar 28, 44; US citizen; m 68; c 3. PROCESS METALLURGY, CHEMICAL SYNTHESIS & PROCESSING OF MATERIALS. *Educ:* Univ Surrey, Eng, BSc, 66; Univ Birmingham, Eng, PhD(metall), 69. *Prof Exp:* Mgr indust eng, Birmingham Aluminum Casting Co, Ltd, Eng, 69-73; sr lectr chem metall, Dept Metals Technol, Sandwell Col Further & Higher Educ, 73-79; from assoc prof to prof process metall, Mineral Resources Res Ctr, Dept Civil & Mineral Eng, Univ Minn, 79-86, assoc dir, 82-86; prof & head mat processing, Dept Chem & Mat Eng, Univ Auckland, NZ, 86-89; PROF & HEAD MAT PROCESSING, DEPT METALL & MAT ENG, COLO SCH MINES, 89- *Concurrent Pos:* Steering Comt, Process Technol Div, Iron & Steel Soc, 83-86; hon prof mat eng, Dept Chem & Mat Eng, Univ Auckland, NZ, 89- *Mem:* Metall Soc; Am Soc Metals; Am Ceramic Soc. *Res:* Chemical processing of materials, materials synthesis and process metallurgy; combustion and plasma synthesis of ceramics, intermetallics and composite materials; pyrometallurgical and high temperature reactions; continuous casting of steel; fused salt electrolysis; iron and steel making; wear in ferrous materials. *Mailing Add:* 15216 W Ellsworth Pl Golden CO 80401

MOORE, JOHN MARSHALL, JR, b Winnipeg, Man, Aug 14, 35; m 58; c 3. GEOLOGY, PETROLOGY. *Educ:* Univ Man, BSc, 56; Mass Inst Technol, PhD(petrol), 60. *Prof Exp:* Queen Elizabeth II fel, 60-62, from asst prof to assoc prof, 62-75, chmn dept, 75-78, PROF GEOL, CARLETON UNIV, 75- *Concurrent Pos:* Guest prof, Inst Mineral, Univ Geneva, 68-69; adv, Can Int Develop Agency, Ethiopia, 72-74; consult regional planning & resource develop, Y Bojard Assocs, Vancouver, 78- *Mem:* Geol Asn Can; Soc Int Develop; Mineral Asn Can; Mineral Soc Am; Sigma Xi. *Res:* Precambrian geology; metamorphic petrology; resources and development. *Mailing Add:* Dept of Geol Carleton Univ Ottawa ON K1S 5B6 Can

MOORE, JOHN R(OBERT), b St Louis, Mo, July 5, 16; m 41; c 3. MECHANICAL ENGINEERING. *Educ:* Washington Univ, BS, 37. *Hon Degrees:* DSc, W Coast Univ, 63. *Prof Exp:* Asst chief engr, Mech Sect, Aeronaut & Marine Eng, Gen Elec Co, 40-45 & chief, Theoret Sect, Proj Hermes, 45-46; assoc prof mech & dir, Dynamical Control Lab, Washington Univ, 46-48; group leader, Electromech Eng Dept, NAm Aviation, Inc, 48-53, dir dept, 53-55, gen mgr, Autonetics Div, 55-57, pres, 60-66, vpres, NAm Aviation, 57-66, exec vpres, mem bd, 66-67, pres, Aerospace Systs Group, Rockwell Corp, 67-70; vpres subsid, McDonald-Douglas, 70-78, vpres, ACTRON, 78; vpres, Bus Develop Electronic Systs Group, Northrop, 78-89, gen mgr, Electro Mech Div, 89; RETIRED. *Concurrent Pos:* Vis assoc prof, Univ Calif, Los Angeles, 48-56; mem space technol panel, Air Force Sci Adv Bd, 59-60; mem adv bd, Joint Task Force Two, Dept Defense, 60-89; mem adv comt, Presidents Roundtable. *Mem:* Nat Acad Eng; Am Inst Aeronaut & Astronaut; Am Inst Navig; fel Inst Elec & Electronics Engrs. *Res:* Airborne controls; computers; missile guidance systems. *Mailing Add:* 15980 Meadowcrest Rd Sherman Oaks CA 91403

MOORE, JOHN ROBERT, b Bridgeville, Pa, Mar 3, 34; m 55; c 2. PLANT ECOLOGY. *Educ:* Clarion State Col, BS, 57; Univ Pittsburgh, 62, PhD(biol), 65. *Prof Exp:* Pub sch teacher, Pa, 57-62; asst biol, Univ Pittsburgh, 62-65; assoc prof, 65-67, PROF BIOL, UNIV PA, CLARION, 67- *Mem:* Ecol Soc Am. *Res:* Primary production of vascular aquatic plants. *Mailing Add:* Dept Biol Univ Pa Clarion Clarion PA 16214

MOORE, JOHN WARD, b Lancaster, Pa, July 17, 39; m 61. COMPUTERS & CHEMISTRY, CHEMISTRY EDUCATION. *Educ:* Franklin & Marshall Col, AB, 61; Northwestern Univ, PhD(phys chem), 65. *Prof Exp:* NSF fel phys chem, Univ Copenhagen, 64-65; asst prof inorg chem, Ind Univ, Bloomington, 65-71; assoc prof chem, Eastern Mich Univ, 71-75, prof, 76-89; PROF CHEM, UNIV WIS-MADISON, 89- *Concurrent Pos:* Fel, Ctr Study Contemp Issues, Eastern Mich Univ, 74-75; mem, Mich Environ Rev Bd, 79-84; vis prof, Univ Wis-Madison, 81-82, Univ Nice, France, 87-88; dir, NSF, Proj Seraphim, 82-; specialist, Chinese Univ Develop Proj, 85; ed, J Chem Educ Software, 88-; dir, Inst Chem Educ, 89- *Honors & Awards:* Catalyst Award, Chem Mfrs Asn, 82; George C Pimentel Award, Chem Educ, Am Chem Soc, 91. *Mem:* AAAS; Nat Sci Teachers Asn; Sigma Xi; Am Chem Soc. *Res:* Coordination chemistry; molecular orbital theory; chemical education; environmental chemistry; computer/instrument interfacing; computer graphics; computer-aided instruction. *Mailing Add:* Dept Chem Univ Wis-Madison Madison WI 53706

MOORE, JOHN WILLIAMSON, physical chemistry, for more information see previous edition

MOORE, JOHN WILSON, b Winston-Salem, NC, Nov 1, 20; m 44, 78; c 4. NEUROPHYSIOLOGY. *Educ:* Davidson Col, BS, 41; Univ Va, MA, 42, PhD(physics), 45. *Prof Exp:* Researcher, RCA Labs, 45-46; asst prof physics, Med Col Va, 46-50; biophysicist, Naval Med Res Inst, 50-54; biophysicist, Nat Inst Neurol Dis & Blindness, 54-56, assoc chief lab biophys, 56-61; from assoc prof to prof physiol, 61-90, PROF NEUROBIOL, DUKE UNIV, 90- *Concurrent Pos:* Nat Neurol Res Found fel, 61-66; trustee, Marine Biol Lab, Woods Hole, 71-79, 81-85, exec comm, 78-79; Dupont fel, Nat Neurol Res Found Scientist. *Honors & Awards:* K S Cole Award, Biophys Soc, 81. *Mem:* AAAS; Am Physiol Soc; Biophys Soc; Inst Elec & Electronic Engrs; Soc Neurosci. *Res:* Neurophysiological experiments, computation and instrumentation. *Mailing Add:* Dept of Neurosci Duke Univ Med Ctr Durham NC 27710

MOORE, JOHNES KITTELLE, b Washington, DC, Apr 20, 31; m 87; c 3. MARINE ECOLOGY, MARINE EDUCATION. *Educ:* Bowdoin Col, AB, 53; Univ RI, PhD(oceanog), 67. *Prof Exp:* Asst acct exec, Batton, Barton, Durstine & Osborn, Inc, 55-60; res asst oceanog, Narragansett Marine Lab, Univ RI, 61-66; from asst prof to assoc prof, 66-74, PROF BIOL, SALEM STATE COL, 74- *Concurrent Pos:* Consult, NSF, 67-69; Woods Hole Oceanog Inst sr fel marine policy, 78-79; sr scientist, Gulf of Maine Res Ctr, Inc, Salem, Mass, 85- *Mem:* New Eng Estuarine Res Soc. *Res:* Estuarine ecology; marine science education. *Mailing Add:* Dept of Biol Salem State Col Salem MA 01970

MOORE, JOHNNIE NATHAN, b San Fernando, Calif, Mar 25, 47. STRATIGRAPHY, TECTONICS. *Educ:* San Fernando Valley Col, BS, 70; Univ Calif, Los Angeles, MS, 73, PhD(geol), 76. *Prof Exp:* Lectr geol, Calif State Univ, Fresno, 76-77; ASST PROF GEOL, UNIV MONT, 77- *Concurrent Pos:* Mgr uranium eval contract, Bendix Field Eng, 78-79. *Mem:* Geol Soc Am; Soc Econ Paleontologists & Mineralogists; Sigma Xi. *Res:* Precambrian-Cambrian depositional environments; controls of sedimentation on stable margins; lacustrine delta sedimentation. *Mailing Add:* Dept Geol Univ Mont Missoula MT 59812

MOORE, JOSEPH B, economic entomology, for more information see previous edition

MOORE, JOSEPH B, b Tuscaloosa, Ala, Jan 19, 26. MATERIALS ENGINEERING. *Educ:* Univ Ala, BS, 48, MS, 55. *Prof Exp:* Supvr mat develop, Gen Elec, Evendale, Ohio, 54-57; mgr metall develop, Wyman Gordon Co, Grafton, Mass, 57-60; sr scientist, Southern Res Inst, Birmingham, Ala, 60-61; dir mat eng, Pratt & Whitney Aircraft, 61-87, DIR MAT ENG, PRATT & WHITNEY GOVT ENGINE BUSINESS, WEST PALM BEACH, FLA, 87- *Concurrent Pos:* Consult numerous univs. *Honors & Awards:* George Mead Gold Medal for Eng Achievement, United Technol Corp. *Mem:* Nat Acad Eng; Sigma Xi. *Res:* Metals and alloys: application, compositional development, economics of use, fabrication, processing/working, property development, and structural and compositional characterization; holder of seven patents; author or co-author of 11 publications and numerous presentations. *Mailing Add:* Pratt & Whitney PO Box 109600 Mail Stop M-L 706-04 West Palm Beach FL 33410-9600

MOORE, JOSEPH CURTIS, b Washington, DC, June 5, 14; m 40; c 3. MAMMALOGY. *Educ:* Univ Ky, BS, 39; Univ Fla, MS, 42, PhD(biol), 53. *Prof Exp:* Asst bot, Univ Ky, 37-39; asst biol, Univ Fla, 39-41; park biologist, Everglades Nat Park, 49-55; res fel, Dept Mammals, Am Mus Natural Hist, 55-61; cur mammals, Field Mus Natural Hist, 62-71, res assoc, 71-73; mem staff, Ecol & Systematics, Fla Southern Col, 72-80; RETIRED. *Concurrent Pos:* Hon lectr, Dept Anat, Univ Chicago, 65-71. *Mem:* AAAS; Am Soc Mammal; Zool Soc London. *Res:* Systematics and zoogeography of mammals, world-wide in Cetacea and Sciuridae; ecology of vertebrates, especially Sciuridae and Sirenia in Florida; local mid-peninsular Florida ecology, especially vertebrate, terrestrial and volant, emphasizing animal and plant relationships to each other, the seasons, temperature, and rainfall. *Mailing Add:* 4210 Mossy Oak Dr Lakeland FL 33805

MOORE, JOSEPH HERBERT, b Spartanburg, SC, May 13, 22; m 43; c 2. STRUCTURAL ENGINEERING. *Educ:* The Citadel, BS, 43; Pa State Univ, MS, 49; Purdue Univ, PhD, 61. *Prof Exp:* Civil engr, The Citadel, 46-48, asst prof, 49-50; asst prof, Pa State Univ, 48-55, assoc prof civil eng, 56-62, prof eng, 67-72; prof & head, Civil Eng Dept, Clemson Univ, 62-67; prof civil eng, Va Polytech Inst, 72-87; RETIRED. *Concurrent Pos:* Consult, Gannett, Fleming, Corddry & Carpenter, Inc, 53-55, Hwy Res Bd, 72-, Van Note-Harvey Asn, 76-, Olver Inc, 75-, Hatcher-Sayre, 88-; mem, Eng Manpower Comn, Eng Joint Coun, 77-80; prof bridge engr, Joseph K Knoerle & Assocs, 55-56; prof & head dept, Clemson Univ, 62-67; head progs eng & Technol, Capitol Campus, Va Polytech Inst, 67-70, head, Div Eng & Technol, 70-72, dir, 72-78. *Mem:* Am Soc Civil Engrs; Am Soc Eng Educ; Sigma Xi. *Res:* Structures and transportation. *Mailing Add:* 506 Forest Hill Dr Blacksburg VA 24060

MOORE, JOSEPH NEAL, b New York, NY, Jan 21, 48. GEOLOGY. *Educ:* City Col NY, BS, 69; Pa State Univ, MS, 72, PhD(geol), 75. *Prof Exp:* Staff geologist, Anaconda Co, 75-77; GEOLOGIST, EARTH SCI LAB, UNIV UTAH RES INST, 77- *Mem:* Geol Soc Am. *Mailing Add:* Res Inst Earth Sci Lab Univ Utah 391 Chipeta Way Suite A Salt Lake City UT 84108

MOORE, JOSEPHINE CARROLL, b Ann Arbor, Mich, Sept 20, 25. NEUROANATOMY. *Educ:* Univ Mich, BA, 47, MS, 59, PhD(anat), 64; Eastern Mich Univ, BS, 54. *Hon Degrees:* DSc, Eastern Mich Univ, Health & Human Serv, 82. *Prof Exp:* From instr to asst prof occup ther, Eastern Mich Univ, 55-60; instr anat, Med Sch, Univ Mich, 64-66; from asst prof to assoc prof, 66-74, PROF ANAT, MED SCH, UNIV S DAK, 74-, VCHM DEPT ANAT, 78- *Mem:* AAAS; Am Asn Anat; World Fedn Occup Ther; Am Occup Ther Asn. *Res:* Rehabilitation of handicapped individuals, especially dealing with neuroanatomical and neurophysiological concepts; neurobehavioral sciences and rehabilitation. *Mailing Add:* Dept of Anat Univ of SDak Med Sch Vermillion SD 57069

MOORE, KEITH LEON, b Brantford, Ont, Oct 5, 25; m 49; c 5. EMBRYOLOGY. *Educ:* Univ Western Ont, BA, 49, MSc, 51, PhD(micro anat), 54. *Prof Exp:* Lectr anat, Univ Western Ont, 54-56; from asst prof to prof, Univ Man, 56-76; prof anat & chmn, 76-84, ASSOC DEAN, BASIC SCI, UNIV TORONTO, 84- *Concurrent Pos:* Consult, Children's Hosp, Winnipeg, Man, 59-76; Nat Bd Med Examiners, 84-88. *Honors & Awards:* JCB Grant Award. *Mem:* Am Asn Anat; Anat Soc Gt Brit & Ireland; Can Asn Anat (secy, 62-65, pres, 66-68); Can Fedn Biol Socs; Can Cytol Coun; Am Asn Clin Anat. *Res:* Embryology and teratology. *Mailing Add:* Dept Anat Dept Toronto Med Sci Bldg Toronto ON M5S 1A8 Can

MOORE, KENNETH BOYD, psychiatry, for more information see previous edition

MOORE, KENNETH EDWIN, b Edmonton, Alta, Aug 8, 33; m 53; c 3. BIOCHEMICAL PSYCHOPHARMACOLOGY, NEUROENDOCRINOLOGY. *Educ:* Univ Alta, BS, 55, MS, 57; Univ Mich, PhD(pharmacol), 60. *Prof Exp:* Instr pharmacol, Dartmouth Med Sch, 60-61, asst prof, 62-65; assoc prof, 66-69, PROF PHARMACOL, MICH STATE UNIV, 70-, CHMN, 87- *Concurrent Pos:* Mem rev comt pharmacol & endocrinol, NIH, 68-70, mem pharmacol sci study sect, 75-79 & 82-86; vis scholar, Cambridge Univ, Eng, 74. *Honors & Awards:* Merit Award, Alcohol, Drug Abuse & Mental Health Admin. *Mem:* Am Soc Pharmacol & Exp Therapeut; Soc Exp Biol & Med; Soc Neurosci; Am Col Neuropsychopharmacol; Endocrine Soc. *Res:* Biochemical pharmacology, neuropharmacology and toxicology related to the role of endocrine and nervous systems; central nervous system transmitters; catecholamines. *Mailing Add:* Dept Pharm Mich State Univ E Lansing MI 48824

MOORE, KENNETH HOWARD, physics; deceased, see previous edition for last biography

MOORE, KENNETH J, b Phoenix, Ariz, June 6, 57; m 79; c 2. AGRONOMY, ANIMAL SCI & NUTRITION. *Educ:* Ariz State UNiv, BS, 79; Purdue Univ, MS, 81, PhD(agron), 83. *Prof Exp:* Res asst, Dept Agron, Purdue Univ, 79-83; asst prof, Dept Agron, Univ Ill, 83-87; from asst prof to assoc prof, NMex State Univ, 87-89; RES AGRONOMIST & ADJ ASSOC PROF, UNIV NEBR, 89- *Mem:* Am Soc Agron; Crop Sci Soc Am; Am Soc Animal Sci; Am Dairy Sci Asn; Am Forage & Grassland Coun. *Res:* Forage quality and utilization; identification of factors which limit utilization of plant fiber by ruminant animals; post-harvest physiology and preservation of forage crops; chemical regulation of the growth and quality of forage crops. *Mailing Add:* USDA Agr Res Serv Univ Nebr 336 Keim Hall Lincoln NE 68583

MOORE, KENNETH VIRGIL, b Emporia, Kans, May 16, 33; m 52; c 2. NUCLEAR SCIENCE. *Educ:* Kans State Teachers Col, BA, 54; Stanford Univ, MS, 64. *Prof Exp:* Jr geophysicist, Gulf Oil Corp, 56; nuclear analyst, Phillips Petrol Co, Atomic Energy Div, 57-63, group leader nuclear res, 64-69; group leader thermal hydraulics, Idaho Nuclear Corp, 69-71, Aerojet Nuclear Corp, 71-74; vpres anal, 74-81, SR VPRES, ENERGY INC, 81-; PRES, COMPUTER SIMULATION & ANALYSIS CORP. *Concurrent Pos:* US AEC fel, 64. *Mem:* Am Soc Mech Engrs; Soc Automotive Engrs. *Res:* Major director and author of thermal hydraulic (transient, two-phase water) computer codes used by the governments and the nuclear reactor industry for safety analysis and experimental predictions. *Mailing Add:* Computer Simulation & Anal Corp PO Box 51596 Idaho Falls ID 83405

MOORE, KRIS, b Frost, Tex, Apr 11, 41; m 64; c 4. STATISTICS, COMPUTER SCIENCE. *Educ:* Univ Tex, Austin, BA, 64, MA, 66; Tex A&M Univ, PhD(statist), 74. *Prof Exp:* Engr scientist res, Tracor Inc, 64-67; instr math, Tex Tech Univ, 67-69; ASSOC PROF STATIST, BAYLOR UNIV, 70- *Concurrent Pos:* Consult, Tex Educ Res Corp, 76-78, Word Inc, 77-78, Joe L Ward's, 78- & Cherry Law Firm, 81- *Mem:* Am Statist Asn; Decision Sci. *Res:* Nonparametric multivariate statistics. *Mailing Add:* Dept Comput Sci Info Syst Baylor Univ PO Box 6367 Waco TX 76798

MOORE, LARRY WALLACE, b Menan, Idaho, Aug 24, 37; m 61; c 5. PLANT PATHOLOGY. *Educ:* Univ Idaho, BS, 62, MS, 64; Univ Calif, Berkeley, PhD(plant path), 70. *Prof Exp:* ASSOC PROF PLANT PATH, ORE STATE UNIV, 69- *Mem:* Am Phytopath Soc; Am Soc Microbiol; Sigma Xi; AAAS. *Res:* Biological disease control; biology of phytopathogenic bacteria; bacterial biology. *Mailing Add:* 4737 NW Elmwood Dr Corvallis OR 97330

MOORE, LAURENCE DALE, b Danville, Ill, July 12, 37; m 58; c 4. PLANT PATHOLOGY. *Educ:* Univ Ill, BS, 59; Pa State Univ, MS, 61, PhD(plant path), 65. *Prof Exp:* Grad asst, Pa State Univ, 59-65; from asst prof to assoc prof, 65-84, PROF PLANT PATH & HEAD DEPT, VA POLYTECH INST & STATE UNIV, 85- *Mem:* Am Phytopath Soc; Am Soc Plant Path; Am Inst Biol Sci. *Res:* Physiology of disease; role of enzymes and plant nutrition in disease development; air pollution effects on the physiology and biochemistry of plants. *Mailing Add:* Dept Plant Path Physiol & Weed Sci Va Polytech Inst & State Univ Blacksburg VA 24061-0331

MOORE, LAWRENCE EDWARD, b Conway, SC, May 28, 38; m 64; c 2. INORGANIC CHEMISTRY, ANALYTICAL CHEMISTRY. *Educ:* Davidson Col, BS, 60; Univ Tenn, PhD(chem), 64. *Prof Exp:* Asst prof chem, Wofford Col, 66-70; from asst prof to assoc prof, 70-77, PROF CHEM, UNIV SC, SPARTANBURG, 77- *Mem:* Am Chem Soc; AAAS; Sigma Xi. *Res:* Coordination compounds of nickel in which the perchlorate ion is a ligand; aromatic substituent effects in coordination compounds of nickel with substituted pyridines. *Mailing Add:* Univ SC 800 University Way Spartanburg SC 29303

MOORE, LEE E, b Chelsea, Okla, Mar 6, 38; m 64; c 2. NEUROPHYSIOLOGY, BIOPHYSICS. *Educ:* Univ Okla, BS, 60; Duke Univ, PhD(physiol), 66. *Prof Exp:* Asst prof physiol, Ohio Univ, 66-68; from asst prof to assoc prof physiol, Med Sch, Case Western Reserve Univ, 68-76; PROF PHYSIOL & BIOPHYS, MED BR, UNIV TEX, 76- *Concurrent Pos:* Nat Inst Neurol Dis & Stroke res grant, 67-75. *Mem:* Neurosci Soc; Biophys Soc; Soc Gen Physiol; Am Physiol Soc. *Res:* Cellular neurophysiology; biophysics of membrane transport; effect of low temperature on the ionic conductance of myelinated nerve; ion permeability of skeletal muscle; ion conduitance fluituation analysis of nerve; optical fluituation spectroscopy of nerve and muscle. *Mailing Add:* Dept Physiol & Biophys Med Br Univ Tex 301 Univ Blvd Galveston TX 77550

MOORE, LEON, b Waldron, Ark, Oct 2, 31; m 51; c 4. ENTOMOLOGY. *Educ:* Univ Ark, BSA, 57, MS, 59; Kans State Univ, PhD, 72. *Prof Exp:* Surv entomologist, Agr Exp Sta, 59-62, assoc prof entom & assoc entomologist, 74-77, PROF ENTOM, UNIV ARIZ, 77-, EXTEN ENTOMOLOGIST, AGR EXP STA, 62- *Mem:* Entom Soc Am. *Res:* Extension entomology with emphasis on educational programs on cotton and vegetable insects, cereal and forage crop insects and pesticides. *Mailing Add:* Dept Entom Univ Ariz Tucson AZ 85721

MOORE, LEONARD ORO, b Payson, Utah, Sept 20, 31; m 54; c 2. RESEARCH ADMINISTRATION, ORGANIC CHEMISTRY. *Educ:* Brigham Young Univ, BS, 53; Iowa State Univ, PhD(chem), 57. *Prof Exp:* Europ Res Assocs fel, Swiss Fed Inst Technol, 57-58; res chemist, Union Carbide Corp, 58-64, proj leader, 64-72; mgr res, Ansul Co, 72-76, mgr res & develop, 76-83; vpres & dir Technol, 83-88, VPRES ENVIRON SPECIALTY CHEM PROD CORP, 88- *Concurrent Pos:* Chmn adv comt, Kanawha Valley Grad Ctr. *Mem:* Am Chem Soc; fel Am Inst Chemists; Sigma Xi; AAAS. *Res:* Halogen chemicals; fluorocarbons; pesticides; free radicals; kinetics; environmental chemistry; organometallics; glycol derivatives. *Mailing Add:* Specialty Chem Prod Corp Two Stanton St Marinette WI 54143

MOORE, LILLIAN VIRGINIA HOLDEMAN, b Moberly, Mo, Aug 8, 29; m 85. BACTERIOLOGY. *Educ:* Duke Univ, AB, 51; Mont State Univ, PhD(bact), 64. *Prof Exp:* Med technologist serol lab, Commun Dis Ctr, USPHS, 51-55, bacteriologist, Diag Bact Lab, 55-56, asst chief microbiol reagents unit, 57-62, microbiologist-in-charge, Anaerobe Lab, 62-66; assoc prof bact, 66-68, prof microbiol, 68-78, UNIV DISTINGUISHED PROF BACTERIOL, VA POLYTECH INST & STATE UNIV, 78- *Concurrent Pos:* Bd dir & exec comm, Am Type Culture Collection. *Honors & Awards:* Kimble Methodology Res Award, Conf Pub Health Lab Directors, 73-; Becton-Dickinson Award, Am Soc Microbiol, 85. *Mem:* Am Soc Microbiol; Am Pub Health Asn; Sigma Xi; Int Asn Dent Res; Am Asn Dent Res. *Res:* Isolation and identification of anaerobic bacteria of medical importance; anaerobes in the normal flora. *Mailing Add:* Dept Anaerobic Microbiol Va Polytech Inst & State Univ Blacksburg VA 24061-0305

MOORE, LORNA GRINDLAY, MATERNAL OXYGEN TRANSPORT. *Educ:* Univ Mich, PhD(anthrop), 73. *Prof Exp:* ASSOC PROF ANTHROP, UNIV COLO, 76- *Res:* Human adaptation of high altitude. *Mailing Add:* Univ Colo Box B133 4200 E Ninth Ave Univ Colo 1100 14th St Denver CO 80262

MOORE, LOUIS DOYLE, JR, b Royston, Ga, Sept 16, 26; wid; c 2. POLYMER CHEMISTRY. *Educ:* Duke Univ, BS, 47; Mass Inst Technol, PhD(phys chem), 51. *Prof Exp:* Res chemist, Tenn Eastman Co, 51-57, sr res chemist, 57-67, res assoc, 67-71, sr res assoc, 71-86; RETIRED. *Mem:* Am Chem Soc; Sigma Xi. *Res:* Molecular characterization of polymers; polymer rheology; polymer morphology. *Mailing Add:* 4411 Leedy Rd Kingsport TN 37664

MOORE, MALCOLM A S, b Edinburgh, Scotland, Jan 18, 44. HEMATOPOIETIC GROWTH FACTORS, EXPERIMENTAL LEUKEMIA RESEARCH. *Educ:* Univ Oxford, BM, 63, BA, 64, DPhil, 67 & MA, 70. *Prof Exp:* Prize fel, Magdalen Col, Oxford Univ, 65-70; head, Lab Develop Biol, Walter & Eliza Hall, Inst Med Res, 70-74; PROF BIOL, SLOAN-KETTERING DIV, CORNELL GRAD SCH MED SCI, MEM & HEAD, LAB DEVELOP HEMATOPOIESIS, CANCER RES, 74-, ATTEND BIOLOGIST, DIV MED ONCOL, MEM SLOAN-KETTERING CANCER CTR, 87- *Concurrent Pos:* Dir, Gar Reichmon Lab Advan Cancer, 75-; mem review comt, Int Cancer Res Workshop Proj, Int Union Against Cancer, 76-, Immunol Sci Study Sect, NIH, 76-80, adv comt res needs in hemat, Nat Inst Arthritis, Metab & Digestive Dis, 79-84; class B trustee, Leukemia Soc Am, 84- *Honors & Awards:* Armand Hammer Cancer Prize, 87; Dr Kenny Award, 87. *Mem:* Am Asn Cancer Res; Am Asn Immunologists; Int Soc Exp Hemat; Reticuloendothelial Soc; Leukemia Soc Am; Am Asn Hematologists. *Res:* Hematopoietic growth factors, biomolecular, molecular and functional characterization; use of these factors in vivo and clinical situations to protect against chemotherapy or radiation induced myelosuppression. *Mailing Add:* Mem Sloan-Kettering Cancer Ctr 1275 York Ave New York NY 10021

MOORE, MARION E, b Boise City, Okla, May 22, 34; m 53; c 1. MATHEMATICS. *Educ:* W Tex State Univ, BS, 57; Tex Tech Col, MS, 60; Univ NMex, PhD(math), 68. *Prof Exp:* Instr math, W Tex State Univ, 58-61 & Univ NMex, 61-66; asst prof, 66-70, ASSOC PROF MATH, UNIV TEX, ARLINGTON, 70- *Concurrent Pos:* Vis prof, Tex Tech Univ, 81; grad adv, Univ Tex, Arlington, 87- *Mem:* Math Asn Am; Am Math Soc; Sigma Xi. *Res:* Ring theory. *Mailing Add:* Dept of Math Univ of Tex Arlington TX 76019

MOORE, MARTHA MAY, b Durham, NC, Feb 17, 49; m 82. GENETIC TOXICOLOGY, MUTAGENICITY. *Educ:* Western Md Col, BS, 71; Univ NC, Chapel Hill, PhD(genetics), 80. *Prof Exp:* Biologist, Nat Inst Environ Health Sci, NIH, 71-72; res biologist, 77-84, CHIEF, MUTAGENESIS & CELLULAR TOXICOL BR, US ENVIRON PROTECTION AGENCY, 84- *Mem:* Environ Mutagen Soc; AAAS. *Res:* Use of short-term mammalian mutagenesis assays to predict genotoxicity from environmental exposure; development and characterization of such assays, including their genetic basis. *Mailing Add:* 3200 S Walnut Pkwy Raleigh NC 27606

MOORE, MARVIN G, b Harrison, Ark, Dec 8, 08; m 35; c 1. MATHEMATICS. *Educ:* Univ Ark, AB, 31; Univ Ill, MA, 32, PhD(math), 38. *Prof Exp:* Instr math, Ala Polytech Inst, 36-37 & Ind Univ, 37-40; prof, Tri-State Col, 40-43; from asst prof to prof & chmn dept, 43-74, EMER PROF MATH, BRADLEY UNIV, 74- *Mem:* Am Math Soc; Math Asn Am; Soc Indust & Appl Math. *Res:* Generalizations of Fourier series in complex plane; relativity; expansions in series of exponential functions; residual stresses. *Mailing Add:* 708 Henryetta Springdale AR 72764-5112

MOORE, MARY ELIZABETH, b New London, Conn, Dec 7, 30; m 74. RHEUMATOLOGY. *Educ:* Douglass Col, AB, 52; Rutgers Univ, MS, 58, PhD(psychol), 60; Temple Univ, MD, 67. *Prof Exp:* Res assoc sociol, Rutgers Univ, 52-60; assoc psychol & psychiat, Univ Pa, 60-63; asst prof, 72-75, assoc prof, 76-81, PROF MED, SCH MED, TEMPLE UNIV, 81- *Concurrent Pos:* Head, Div Rheumatology, Albert Einstein Med Ctr, Philadelphia, Pa; dir, Einstein-Moss Arthritis Ctr, Philadelphia, Pa; physician ed, Med Knowledge Self-Assessment Tests, Am Col Physicians. *Honors & Awards:* Nellie Westerman Prize for Res in Ethics, Am Fedn Clin Res. *Mem:* Fel Am Col Physicians; Am Col Rheumatology. *Res:* Rheumatic diseases, methods of pain relief. *Mailing Add:* 701 John Barry Dr Bryn Mawr PA 19010

MOORE, MAURICE LEE, b Laurel Hill, Fla, Sept 11, 09; m 33; c 3. MEDICINAL CHEMISTRY. *Educ:* Univ Fla, BS, 30, MS, 31; Northwestern Univ, PhD(org chem), 34. *Prof Exp:* Asst chem, Univ Fla, 27-31; asst instr gen chem, Dental Sch, Northwestern Univ, 31-34; Smith fel org chem, Yale Univ, 34-36; res chemist, Sharp & Dohme, Inc, Pa, 36-43; dir org res, Frederick Stearns & Co, Mich, 43-45, asst dir res, 45-47; dir, Smith, Kline & French Labs, Pa, 47-51; vpres, Vick Chem Co, 51-59; dir new prod develop, Sterling Drug Inc, 59-65, exec vpres, Winthrop Labs Div, 59-70, corp sci officer, 70-74; RETIRED. *Honors & Awards:* Hon Award, Am Inst Chemists, 79. *Mem:* AAAS; Am Chem Soc; Soc Chem Indust; Am Pharmaceut Asn; fel Am Inst Chemists. *Res:* Research and development of new therapeutic agents. *Mailing Add:* Seven Brookside Circle Bronxville NY 10708-5618

MOORE, MELITA HOLLY, b Joplin, Mo, Sept 12, 09; m 46, 75; c 1. STATISTICS. *Educ:* Wellesley Col, BA, 32; Univ Mich, MA, 34. *Prof Exp:* Instr math, pvt sch & Wellesley Col, 34-41; admin asst math, Navel Ord Lab, 41-45; chief statist serv, C/L Br, Bur Labor Statist, 45-46; chief res sect, UN Statist Off, 46-55, statist consult, 57-76; statist consult, Nat Bur Econ Res, 77-79. *Mem:* Am Statist Asn. *Res:* International Economic Indicators; coauthor of book with G H Moore. *Mailing Add:* 1171 Valley Rd New Canaan CT 06840

MOORE, MICHAEL STANLEY, b Grass Creek, Wyo, Oct 24, 30; m 54; c 4. NUCLEAR PHYSICS. *Educ:* Rice Univ, BA, 52, MA, 53, PhD(physics), 56. *Prof Exp:* Physicist, Atomic Energy Div, Phillips Petrol Co, 56-66 & Idaho Nuclear Corp, 66-68; STAFF MEM, LOS ALAMOS SCI LAB, 68- *Mem:* Fel Am Phys Soc; fel Am Nuclear Soc; Sigma Xi. *Res:* Neutron cross sections by time of flight; fission physics. *Mailing Add:* 104 Tesuque Los Alamos NM 87544

MOORE, MORTIMER NORMAN, b Los Angeles, Calif, Mar 16, 27; m 58; c 2. MATHEMATICAL PHYSICS. *Educ:* Univ Calif, Los Angeles, BA, 50; Univ London, PhD(math physics), 55. *Prof Exp:* Sr res scientist, Atomics Int, NAm Aviation, 55-57; lectr theoret physics, Birkbeck Col, London, 57-61; assoc prof physics, 61-66, PROF PHYSICS, CALIF STATE UNIV, NORTHRIDGE, 66-, PROF ASTRON, 74- *Concurrent Pos:* Consult, Inst Nuclear Res, Stuttgart, Ger, 58, Atomics Int, 59-64 & Fr AEC fast reactor prog, 64-65; vis prof, Univ Ariz, 69. *Mem:* Am Phys Soc; Am Nuclear Soc; Brit Inst Physics. *Res:* Stochastic processes; transport processes; partial coherence theory; low-energy neutron physics; non-equilibrium processes; laser physics. *Mailing Add:* Dept Physics Calif State Univ 18111 Nordoff St Northridge CA 91330

MOORE, NADINE HANSON, b Idaho Falls, Idaho, Feb 10, 41; m 63. MATHEMATICS. *Educ:* Idaho State Univ, BS, 63; Syracuse Univ, MA, 65, PhD(math), 69. *Prof Exp:* Instr math, State Univ NY Col Oswego, 68-69; asst prof math, Ariz State Univ, 69-75; TEACHER MATH, GLENDALE COMMUNITY COL, ARIZ, 83- *Mem:* Am Math Soc; Asn Women Math. *Res:* Homological algebra; structure of projective modules. *Mailing Add:* 5223 E Tamblo Dr Phoenix AZ 85044

MOORE, NELSON JAY, b Greenville, Ohio, Aug 29, 41; m 64; c 2. ETHOLOGY, ORNITHOLOGY. *Educ:* Manchester Col, BA, 63; Ohio State Univ, MS, 68; Univ Ariz, PhD(zool), 72. *Prof Exp:* chmn biol dept, 80-85, PROF BIOL, OHIO NORTHERN UNIV, 72- *Mem:* Am Ornithologists Union; Asn Field Ornithologists; Inland Bird Banding Asn; Sigma Xi. *Res:* Description and analysis of the behaviors of the Yellow-eyed Junco; raptors of central Ohio. *Mailing Add:* Dept of Biol Ohio Northern Univ Ada OH 45810

MOORE, NOEL E(DWARD), b Ft Wayne, Ind, Dec 23, 34; m 57; c 3. CHEMICAL ENGINEERING. *Educ:* Purdue Univ, West Lafayette, BSChE, 56, PhD(chem eng), 67; Mass Inst Technol, MS, 58. *Prof Exp:* Asst prof chem eng, Univ Ky, 64-68; assoc prof, 68-72, PROF CHEM ENG, ROSE-HULMAN INST TECHNOL, 72-, CHMN, 87- *Concurrent Pos:* Consult, Humble Oil & Refining Co, 67, Dow Chem Co, 68 & Eli Lilly & Co, 73-75, 90; dir div air pollution control, Vigo County Health Dept, 69-71, E

I Du Pont, 82-83. *Mem:* Am Soc Eng Educ; Am Inst Chem Engrs; Air Pollution Control Asn; Instrument Soc Am. *Res:* Process control; air pollution. *Mailing Add:* Dept Chem Eng Rose-Hulman Inst of Technol Terre Haute IN 47803-3999

MOORE, PAUL BRIAN, b Stamford, Conn, Nov 24, 40. MINERALOGY, CRYSTALLOGRAPHY. *Educ:* Mich Technol Univ, BS, 62; Univ Chicago, SM, 63, PhD(geophys), 65. *Prof Exp:* NSF fel, Swed Natural Hist Mus, Stockholm, 65-66; from instr to assoc prof, 66-71, PROF MINERAL & CRYSTALLOG, UNIV CHICAGO, 72- *Honors & Awards:* Dreyfus Found Award, 71; Mineral Soc Am Award, 73; Alexander von Humboldt Sr Sci Award, 76. *Mem:* Am Crystallog Asn; Mineral Soc Am; Mineral Asn Can; Am Geophys Union; Explorers Club; Sigma Xi; Lepodopterus Soc. *Res:* Descriptive, paragenetic and taxonomic mineralogy; crystal structure analysis of silicates, phosphates and arsenates; crystallochemical classifications; dense-packed structures; polyhedra theory; butterfly pigments. *Mailing Add:* 5559 S Blackstone Chicago IL 60637

MOORE, PAUL HARRIS, b Trenton, Mo, Apr 6, 38; m 65; c 2. PLANT PHYSIOLOGY, BIOCHEMISTRY. *Educ:* Calif State Univ, Long Beach, BS, 61; Univ Calif, Los Angeles, MA, 65, PhD(plant physiol), 66. *Prof Exp:* PLANT PHYSIOLOGIST, USDA ARS EXP STA, HAWAIIAN SUGAR PLANTERS ASN, 67- *Concurrent Pos:* Fel Univ Calif, Los Angeles, 66-67. *Mem:* Am Soc Plant Physiologists; AAAS; Crop Sci Soc Am; Int Soc Sugar Cane Technologists; Plant Growth Regulator Working Group. *Res:* Developmental biology; regulatory role of phytohormones in growth and development; cellular water relations; growth regulation; plant cell and tissue culture; flowering; stress physiology. *Mailing Add:* 99-193 Aiea Heights Dr Aiea HI 96701

MOORE, PETER BARTLETT, b Boston, Mass, Oct 15, 39; m 66; c 2. BIOCHEMISTRY. *Educ:* Yale Univ, BS, 61; Harvard Univ, PhD(biophys), 66. *Prof Exp:* NSF fel, Univ Geneva, 66-67; US Air Force Off Sci Res fel, Med Res Coun Lab, Cambridge, Eng, 67-69; from asst prof to assoc prof molecular biophysics & biochem, 69-76, assoc prof, 76-79, chmn dept chem, 87-90, PROF CHEM & MOLECULAR BIOPHYSICS & BIOCHEM, YALE UNIV, 76- *Concurrent Pos:* NIH & NSF res grants, 69-; Guggenheim fel, Univ Oxford, 79-80. *Mem:* AAAS; Sigma Xi; Am Chem Soc; Am Soc Biol Chem; Biophys Soc. *Res:* Structure and function of ribosomes; application of neutron scattering to study of quaternary structure; structure of RNA and RNA nucleoproteins; NMR. *Mailing Add:* PO Box 6666 Dept Chem Yale Univ New Haven CT 06511

MOORE, PETER FRANCIS, b New York, NY, July 24, 36; m 60; c 5. IMMUNOLOGY, PHYSIOLOGY. *Educ:* Fordham Univ, BS, 56; Purdue Univ, MS, 59, PhD(pharmacol), 61. *Prof Exp:* Mem staff, 61-73, SR RES INVESTR, PHARMACOL DEPT, PFIZER, INC, 73- *Mem:* AAAS; Am Heart Asn. *Res:* Pulmonary-asthma, allergy, immediate hypersensitivity, bronchodilators, leukotrienes, inflammation, NSAID, immune function, interleukins, mucolytics; cyclic mononucleotides; hypolipemics; diuretics. *Mailing Add:* Pharmacol Dept Eastern Point Rd Groton CT 06340

MOORE, RALPH BISHOP, b Carbonear, Nfld, Mar 23, 41; m 65; c 2. INDUSTRIAL CHEMISTRY, POLYMER PHYSICS. *Educ:* Mem Univ Nfld, BSc, 62; Univ Alta, PhD(phys org chem), 67. *Prof Exp:* Res chemist, 67-69, process chemist, 69-74, sr process chemist, 74-79, sr supervisor, 79-81, supt, Dept Org Chem, 81-84, SR TECH CONSULT, E I DU PONT DE NEMOURS & CO INC, 84 - *Mem:* Am Chem Soc; AAAS; Am Hort Soc. *Res:* Organolead chemistry; photochemistry; reaction kinetics; organofluorine chem; polymer dynamics. *Mailing Add:* Dupont Chemicals PO Box 80709 Wilmington DE 19880-0709

MOORE, RALPH LESLIE, b Wilmington, Del, May 7, 24; m 48; c 2. PETROCHEMICAL PLANTS, MATHEMATICAL MODELING PETROCHEMICAL PROCESSES. *Educ:* Univ Del, BME, 50; Drexel Univ, MSME, 54; Case Western Reserve Univ, MSInstE, 59. *Prof Exp:* Instruments engr, E I du Pont de Nemours & Co, 50-56, Control Systs consult, 60-85; instr instruments, Case Western Reserve Univ, 57-59; RETIRED. *Honors & Awards:* C R Otto Tech Award, Instrument Soc Am, 80, D P Eckman Educ Award, 84, Golden Achievement Award, 86. *Mem:* Instrument Soc Am; Nat Soc Prof Engrs. *Res:* Written books on the subjects of process measurements, the dynamics of automatic control, neutralization of wastewater by pH control and the control of centrifugal compressors. *Mailing Add:* Five Colesbery Dr New Castle DE 19720

MOORE, RANDY, b Columbus, Tex, June 21, 54. CALCIUM & GROWTH EFFECTORS. *Educ:* Tex A&M Univ, BS, 75; Univ Ga, MS, 77; Univ Calif, Los Angeles, PhD(plant anat), 80. *Prof Exp:* From asst prof to assoc prof biol, Baylor Univ, 80-88; PROF & CHMN, DEPT BIOL, WRIGHT STATE UNIV, 88- *Concurrent Pos:* Fulbright scholar, Thialand, 87; vis prof, 85, 87, corresp prof, Pontificia Univ, Catolica de Chile, 87- *Honors & Awards:* Presidential Award, Scanning Electron Microscopy, 82. *Mem:* Bot Soc Am; Am Soc of Plant Physiologists; Am Inst Biol Sci; AAAS; Sigma Xi; Nat Asn Biol Teachers. *Res:* How plants perceive and respond to gravity with structural and functional approaches centering in the involvement of calcium and growth effectors such as auxin mediate differential growth characteristic of gravicurvature. *Mailing Add:* Rm 235A Biol Sci Bldg Wright State Univ Dayton OH 45435

MOORE, RAYMOND A, b Britton, SDak, Nov 16, 27; m 51; c 4. AGRONOMY, PLANT PHYSIOLOGY. *Educ:* SDak State Univ, BS, 51, MS, 56; Purdue Univ, PhD(agron), 63. *Prof Exp:* High sch voc agr instr, SDak, 51-56; instr agron, 56-58, asst agronomist field crops, 58-62, assoc prof, 63-67, head dept plant sci, 69-73, PROF FIELD CROPS, S DAK STATE UNIV, 67-, DIR S DAK EXP STA, 73- *Mem:* Am Soc Agron; Soc Range Mgt. *Res:* Administration; university teaching; pasture crops. *Mailing Add:* Agr Exp Sta SDak State Univ Brookings SD 57007

MOORE, RAYMOND F, JR, b Fishersville, Va, Dec 17, 27; m 51; c 3. ENTOMOLOGY. *Educ:* Bridgewater Col, BS, 51; Univ Richmond, MA, 56; Rutgers Univ, PhD(entom), 59. *Prof Exp:* Entomologist, USDA, 59-65; asst prof biol, Univ SC, 65-66; entomologist, USDA, 66-90; RETIRED. *Mem:* AAAS; Entom Soc Am; Am Chem Soc; Sigma Xi. *Res:* Insect physiology and nutrition. *Mailing Add:* 851 Championship Dr Florence SC 29501

MOORE, RAYMOND H, b Spokane, Wash, May 29, 18; m 41; c 2. INORGANIC CHEMISTRY, CHEMICAL ENGINEERING. *Educ:* Gonzaga Univ, BS, 40. *Prof Exp:* Asst analytical chem, Gonzaga Univ, 40-41; asst org chem, Univ Pittsburgh, 41-42; asst chemist, Explosives Div, US Bur Mines, 42-44, analytical chemist, Electrometall Sta, 44-47; sr scientist, Hanford Atomic Prods Dept, Gen Elec Co, 47-65; staff scientist, Pac Northwest Lab Div, Battelle Mem Inst, 65-85; RETIRED. *Mem:* Am Chem Soc; Am Inst Mining Engrs. *Res:* Instrumental methods of analysis; infrared spectroscopy; inorganic and physical chemistry; fused salt chemistry; radiochemical separations process development; kinetics of diffusion in metals; chemical process design-development; molten salt chemistry; coal liquefaction and gasification; hot fuel gas cleaning. *Mailing Add:* 2041 Greenbrook Blvd Richland WA 99352-9621

MOORE, RAYMOND KENWORTHY, b Meriden, Conn, Jan 9, 42; m 61; c 4. GEOCHEMISTRY, MINERALOGY. *Educ:* Univ Fla, BS, 63, MS, 65; Pa State Univ, PhD(geochem), 69. *Prof Exp:* Res assoc, Mat Res Labs, Pa State Univ, 69-70; assoc prof, 70-81, chmn dept, 79-81, PROF GEOL, RADFORD UNIV, 82- *Mem:* Nat Asn Geol Teachers; Nat Sci Teachers Asn. *Res:* Petrology; spectroscopic studies of minerals; geomorphological processes; spectroscopic properties of naturally occurring solids; landform development. *Mailing Add:* Dept Geol Radford Univ Radford VA 24142

MOORE, RAYMOND KNOX, b Orlando, Fla, Apr 15, 44; m 71; c 2. GEOTECHNICAL ENGINEERING, PAVEMENT ENGINEERING. *Educ:* Okla State Univ, BSCE, 66, MS, 68; Univ Tex, Austin, PhD(civil eng), 71. *Prof Exp:* From asst prof to assoc prof, Dept Civil Eng, Auburn Univ, 71-84; PROF CIVIL ENG, DEPT CIVIL ENG, UNIV KANS, 84- *Concurrent Pos:* Mem, Transp Res Bd, Nat Acad Sci; chmn comt, A2J03, lime-lime/Fly Ash Stabilization, mem comt, A2B01, (pavement mgt systs), A2J02, (chem stabilization of soil & rock); mem publ comt, Highway Div, Am Soc Civil Engrs; comts D-4 (road & paving mat) & D-18 (soil & rock), Am Soc Testing & Mat; consult, Kans Dept Transp; res assoc, Transp Ctr, Kans Univ, 84- *Mem:* Fel Am Soc Civil Engrs; Am Soc Testing & Mat; Int Soc Soil Mech & Found Eng; Asn Asphalt Paving Tech. *Res:* Physical and engineering properties of soils; soil stabilization; geotechnical engineering soil testing. *Mailing Add:* Dept Civil Eng Univ Kansas Lawrence KS 66045-2225

MOORE, REGINALD GEORGE, b Brantford, Ont, Mar 16, 32; m 57; c 3. INVERTEBRATE PALEONTOLOGY. *Educ:* Univ Western Ont, BSc, 54; Univ Mich, MS, 55, PhD(geol), 60. *Prof Exp:* Instr geol, Oberlin Col, 57-58; assoc prof, 60-71, PROF GEOL, ACADIA UNIV, 71- *Mem:* Sigma Xi. *Res:* Non-marine invertebrate paleontology and paleoecology; malacology. *Mailing Add:* Dept of Geol Acadia Univ RR 2 Wolfville NS B0P 1X0 Can

MOORE, RICHARD, b Los Angeles, Calif, Jan 19, 27; m 57, 69; c 2. MEDICAL PHYSICS, BIOMEDICAL ENGINEERING. *Educ:* Univ Mo, BS, 49; Univ Rochester, PhD(biophys), 56; George Washington Univ, DSc(bioeng), 70; Am Bd Health Physics, cert; Am Bd Radiol, cert diag radiol physics, 75. *Prof Exp:* Asst biophys, Univ Rochester, 52-53, res assoc, 53-56; biophysicist, Metab Dis Br, Nat Inst Arthritis & Metab Dis, 57-60; res biophysicist, Blood Prog Res Lab, Am Nat Red Cross, 60-69; assoc prof radiol, Sch Med, Univ Minn, Minneapolis, 69-82; prof, Univ Witwatersrand, Johannesburg, 82-87; HEALTH PHYSICIST, IDAHO NAT ENG LAB, 87- *Concurrent Pos:* Sr asst engr, Div Sanit Eng Serv, USPHS, 55-57, lectr, R A Taft Sanit Eng Ctr, 56-57; vis prof, Howard Univ, 58-; consult, Comt Data Logging Systs, NIH, 59-60 & Dept Biophys, Div Nuclear Med, Walter Reed Army Inst Res, 61-65; from vis asst prof to vis prof, Sch Med, George Washington Univ, 64-69; assoc ed, Pattern Recognition, 68- & Comput in Biol & Med, 69-; contrib ed, Med Electronics & Data, 70-; consult ed, Measurements & Data, 71-; mem coun cardiovasc radiol, Am Heart Asn; sr scientist fel, NATO; German acad exchange fel. *Honors & Awards:* President's Prize, Soc Radiol Eng. *Mem:* Fel AAAS; fel Soc Advan Med Systs; Am Col Radiol; Am Asn Physicists in Med; Radiol Soc NAm; fel Am Heart Asn; Am Col Med Physicist. *Res:* Radioisotopic tracing; compartmental analysis; radiological physics; ultracentrifugation; radiation biology; diagnostic radiologic sciences; membrane permeability; mathematical modeling; computer simulation; dose assessment; math modelling. *Mailing Add:* 812 Garden Dr Chubbuck ID 83202

MOORE, RICHARD ALLAN, b Mansfield, Ohio, Jan 11, 24; m 49; c 3. MATHEMATICS. *Educ:* Wash Univ, AB, 48, AM, 50, PhD, 53. *Prof Exp:* Instr math, Univ Nebr, 53-54 & Yale Univ, 54-56; from asst prof to prof, 56-87, EMER PROF MATH, CARNEGIE-MELLON UNIV, 87- *Concurrent Pos:* Assoc head dept, Carnegie-Mellon Univ, 65-71 & 75-85, chmn dept, 71-75. *Mem:* Am Math Soc; Math Asn Am. *Res:* Ordinary differential equations, especially second order equations. *Mailing Add:* Dept of Math Carnegie-Mellon Univ Pittsburgh PA 15213

MOORE, RICHARD ANTHONY, b Pittsburgh, Pa, Oct 29, 28; m 55; c 8. ORGANIC CHEMISTRY. *Educ:* Duquesne Univ, BS, 51; Purdue Univ, MS, 56; Univ Pittsburgh, PhD(org chem), 62. *Prof Exp:* Proj officer, Wright Air Develop Ctr, Ohio, 51-53; teaching asst chem, Purdue Univ, 53-56; res chemist, Develop Dept, Koppers Co, Inc, Pa, 56-57; teaching asst chem, Univ Pittsburgh, 57-58; res chemist, Wyandotte Chem Corp, 61-72; sr res chemist, BASF Corp, 72-89; RETIRED. *Mem:* Am Chem Soc; fel Am Inst Chem; Sigma Xi. *Res:* Heterocyclic and organic fluorine chemistry; polyethers research; vitamin research. *Mailing Add:* 4042 Longmeadow Trenton MI 48183

MOORE, RICHARD ARTHUR, nuclear engineering, for more information see previous edition

MOORE, RICHARD DANA, b Battle Creek, Mich, Feb 11, 26. ANATOMY. *Educ:* Olivet Col, BS, 48; Mich State Univ, MS, 52, PhD(anat), 56. *Prof Exp:* Asst biol, Olivet Col, 48-49; asst anat, Mich State Univ, 51-55; from asst prof to prof & actg head dept biol, Hardin-Simmons Univ, 55-66; assoc prof, Albright Col, 66-67; chmn dept, McMurry Col, 68-70, from assoc prof to prof biol, 67-88; RETIRED. *Concurrent Pos:* Am Physiol Asn fel, NTex State Col, 59-88. *Mem:* NY Acad Sci. *Res:* Histology of urinary system of domestic animals; human histology. *Mailing Add:* 2121 N Sixth St Apt 116 Abilene TX 79603

MOORE, RICHARD DAVIS, b Salina, Kans, Mar 17, 32. BIOPHYSICS. *Educ:* Purdue Univ, PhD(oxytocin action), 63; Ind Univ, MD, 57. *Prof Exp:* NIH fel biophys, Sch Med, Univ Md, 63-64, res asst prof, 64; assoc prof, 64-67, PROF BIOPHYSICS, STATE UNIV NY COL PLATTSBURGH, 67- *Mem:* AAAS; Biophys Soc; Sigma Xi. *Res:* Cellular control mechanisms, especially ionic regulation; role of ions in hormone action and gene activation; electrical interactions between hormones and receptor sites; active transport. *Mailing Add:* 138 Dumbarton Rd Baltimore MD 21212

MOORE, RICHARD DONALD, b Spokane, Wash, Mar 7, 24; m 46; c 4. PATHOLOGY. *Educ:* Western Reserve Univ, MD, 47; Am Bd Path, dipl, 55. *Prof Exp:* Resident path, Univ Hosps, Western Reserve Univ, 49-52, from instr to sr instr, Sch Med, 52-55, asst prof, 55-56; asst prof, Sch Med, Univ Rochester, 56-57; assoc prof, Sch Med, Western Reserve Univ, 57-67, prof, 67-69; prof path & chmn dept, Med Sch, Univ Ore, 69-85; RETIRED. *Mem:* Am Soc Exp Path; Reticuloendothelial Soc; Am Asn Path & Bact. *Res:* Structure and function of connective tissue and the reticuloendothelial system. *Mailing Add:* 53236 E Marmot Rd Sandy OR 97055

MOORE, RICHARD E, b San Francisco, Calif, July 30, 33; m 60; c 4. ORGANIC CHEMISTRY. *Educ:* Univ San Francisco, BS, 57, MS, 59; Univ Calif, Berkeley, PhD(org chem), 62. *Prof Exp:* Res assoc, 63-66, from asst prof to assoc prof org chem, 66-75, PROF CHEM, UNIV HAWAII, HONOLULU, 75- *Mem:* Am Chem Soc. *Res:* Isolation, structure determination and biosynthesis of natural products from microalgae and marine organisms. *Mailing Add:* Dept Chem Univ Hawaii Manoa 2545 The Mall Honolulu HI 96822

MOORE, RICHARD K(ERR), b St Louis, Mo, Nov 13, 23; m 44; c 2. MICROWAVE REMOTE SENSING, RADIO WAVE PROPAGATION. *Educ:* Washington Univ, BSEE, 43; Cornell Univ, PhD(elec eng), 51. *Prof Exp:* Test equip engr, Victor Div, Radio Corp Am, 43-44; instr elec eng & res engr, Washington Univ, 47-49; res assoc, Cornell Univ, 49-51; res engr & sect supvr, Sandia Corp, 51-55; assoc prof elec eng, Univ NMex, 55-56, prof, 56-72, chmn dept, 55-62; dir, Remote Sensing Lab, 64-74, BLACK & VEATCH PROF ELEC ENG, UNIV KANS, 62-, DIR, RADAR SYSTS & REMOTE SENSING LAB, 84- *Concurrent Pos:* Pres, Cadre Corp, 68-90. *Honors & Awards:* Outstanding Tech Achievement Award, Coun Oceanic Eng, Inst Elec & Electronic Engrs, 78, Distinguished Achievement Award, Geosci & Remote Sensing Soc, 82. *Mem:* Nat Acad Eng; Am Soc Eng Educ; Am Geophys Union; fel Inst Elec & Electronics Engrs; Int Sci Radio Union. *Res:* Submarine, tropospheric and ionospheric wave propagation; radar scattering; submarine communication; traveling wave analogies; application of radar and radio propagation to earth sciences; radar systems. *Mailing Add:* Radar Systs & Remote Sensing Labs Univ Kans 2291 Irving Hill Rd Lawrence KS 66045-2969

MOORE, RICHARD LEE, b Philadelphia, Pa, June 20, 18; m 42; c 3. PLASMA PHYSICS, THEORETICAL PHYSICS. *Educ:* Univ Calif, Los Angeles, AB, 41, MA, 42; Ohio State Univ, PhD(physics), 53. *Prof Exp:* Meteorologist & res physicist, US Army & US Air Force, 41-51; res asst physics, Los Alamos Sci Lab, 51-53; mem staff, 54, staff asst, Weapon Systs Anal Dept, Northrop Aircraft, Inc, Calif, 54-56; sr assoc, Planning Res Corp, Calif, 56-57; consult physicist, R L Moore Consults, 57-62; head appl & theoret physics, Douglas Aircraft Co, Inc, Calif, 62-66, mgr phys sci & math res, 66-70; phys scientist, US Army Weapons Command, 71-73, phys scientist, US Army Armament Command, 73-78, OPERS RES ANALYST, US ARMY ARMAMENT RES & DEVELOP COMMAND, 78- *Mem:* Am Phys Soc; assoc fel Am Inst Aeronaut & Astronaut. *Res:* Operations analysis; Gauss-Hertz principle in continuum theory; data reduction principles; clear air turbulence; tropical cyclone theory. *Mailing Add:* 14724 107th Ave Bothell WA 98011

MOORE, RICHARD NEWTON, b Bernice, La, Mar 4, 26; m 60; c 3. ORGANIC CHEMISTRY. *Educ:* Miss Col, BS, 49. *Prof Exp:* Chemist, Southern Regional Res Lab, USDA, 49-57; res chemist, Monsanto Co, 57-59, sr res chemist, 59-67, res specialist, 67-68, res group leader, 68-70, sr res specialist, Hydrocarbons & Polymers Div, 70-76, fel petrochem res, 76-82; RETIRED. *Mem:* Am Chem Soc; Sigma Xi. *Res:* Chemistry of hydrocarbons, free radical reactions and process research. *Mailing Add:* 302 Northwood Pl Ruston LA 71270

MOORE, RICHARD OWEN, biochemistry, for more information see previous edition

MOORE, ROBERT ALONZO, b Indianapolis, Ind, June 15, 31; m 58; c 3. MATHEMATICS. *Educ:* Hanover Col, BA, 53; Ind Univ, Bloomington, PhD(math), 60. *Prof Exp:* Asst prof math, Pa State Univ, 61-64; ASSOC PROF MATH, SOUTHERN ILL UNIV, CARBONDALE, 64- *Mem:* Am Math Soc; Math Asn Am. *Res:* Class field theory. *Mailing Add:* Dept of Math Southern Ill Univ Carbondale IL 62901

MOORE, ROBERT AVERY, b Cullman, Ala, Aug 12, 32; m 56; c 3. ELECTRICAL ENGINEERING, ELECTROMAGNETISM. *Educ:* Univ Ala, BS, 54; Northwestern Univ, MS, 56, PhD(elec eng), 60. *Prof Exp:* Sr engr, Appl Physics Group, 58-68, mgr Solid-State Microwaves, Westinghouse Defense & Electronics Systs Ctr, Baltimore, 68-89, MGR RES & DEVELOP, WESTINGHOUSE ELECTRONICS SYSTS CTR, BALTIMORE, 89- *Mem:* Inst Elec & Electronics Engrs. *Res:* Research and development on antennas; solid-state microwaves, and ultrasonic technology. *Mailing Add:* 1243 Balfour Dr Arnold MD 21012

MOORE, ROBERT B, b Windsor, Nfld, June 28, 35; m 58. NUCLEAR PHYSICS. *Educ:* McGill Univ, BEngPhys, 57, MSc, 59, PhD(physics), 61. *Prof Exp:* Lectr, 61-63, from asst prof to assoc prof, 63-77, PROF PHYSICS, McGILL UNIV, 77- *Mem:* Am Phys Soc; Can Asn Physicists; Eng Inst Can; Am Asn Physics Teachers; Sigma Xi; Prof Engrs Quebec. *Res:* On-line isotope separators; nuclear spectroscopy; fission and short-lived isomeric states; ion trapping; proton and heavy ion irradiations; radioactive ion acceleration. *Mailing Add:* Dept Physics McGill Univ 3610 University st Montreal PQ H3A 2B2 Can

MOORE, ROBERT BLAINE, MEMBRANE PEROXIDATION, SICKLE-CELL ANEMIA. *Educ:* Univ Toronto, PhD(biochem), 78. *Prof Exp:* ASSOC PROF PEDIAT & BIOCHEM, COL MED, UNIV S ALA, 82- *Mem:* Fel Am Inst Chemists. *Mailing Add:* Dept Pediat Univ S Ala Col Med 2451 Fillingim St Mobile AL 36617

MOORE, ROBERT BYRON, b Bangkok, Siam, Nov 11, 29; US citizen; m 55; c 2. CHEMICAL ENGINEERING. *Educ:* Univ Mich, BS, 52. *Prof Exp:* Process engr, Phillips Petrol Co, 52-59; eng supvr, 59-61, opers supt, 61-63, proj engr, 63-67, proj mgr, 67-71, ECON EVALUATOR, AIR PROD & CHEM, INC, 71- *Res:* Optimization of cryogenic separation processes; development of gas separation techniques; production and utilization of hydrogen and carbon monoxide synthesis gases. *Mailing Add:* 2951 Edgemont Ct Allentown PA 18103

MOORE, ROBERT E(DWARD), b Winsted, Conn, July 29, 23; m 46; c 3. MECHANICAL ENGINEERING. *Educ:* Univ Wis, BSME, 48. *Prof Exp:* Mech engr, Wis Axle Div, Rockwell Int, 48-51, chief inspector, Ohio Axle & Gear Div, 51-53; mech engr, John I Thompson & Co, 53-57, vpres, 57-65; PRES, POTOMAC RES, INC, 65- *Mem:* Am Soc Mech Engrs. *Res:* Engineering and research management in the field of military hardware development; system planning of technical, scientific or socio-economic efforts. *Mailing Add:* 3610 Bent Branch Court Falls Church VA 22041

MOORE, ROBERT EARL, b South Bend, Ind, Dec 26, 23; m 47, 56; c 2. ATMOSPHERIC CHEMISTRY & PHYSICS, ENVIRONMENTAL SCIENCES. *Educ:* Purdue Univ, BS, 45; Univ Chicago, SM, 48, PhD(chem), 50. *Prof Exp:* Chemist, Navy Inorg Chem Res Proj, Chicago, 47-50; chemist, Oak Ridge Nat Lab, 50-80; mem staff, Lockheed Missiles & Space Co, Oak Ridge, Tenn, 80-84; PRES, MOORE TECH ASSOCS, INC, 84- *Mem:* AAAS; Am Chem Soc. *Res:* Physical chemistry; chemical processes; computer-implemented modeling of atmospheric dispersion and deposition of air pollutants and radionuclides; environmental radiation dose calculations; radiation damage; solvent extraction processes; high-temperature phase equilibria. *Mailing Add:* 60 Webster Rd Ellington CT 06029

MOORE, ROBERT EDMUND, b Orange, NJ, Dec 12, 25; m 51; c 1. DESIGN FOR RELIABILITY, MICROELECTRONIC MANUFACTURING. *Educ:* NJ Inst Technol, BS, 48; Rensselaer Polytech Inst, MS, 50. *Prof Exp:* Teaching asst electrochem, Rensselaer Polytech Inst, 49-50; develop chemist metal finishing, MacDermid Inc, 50-52; eng sales engr metal finishing, Frederick Gumm Chem Co, 52-56; engr, Semiconductor Prods, Gen Elec Co, 56-60, consult engr, 60-66, consult engr microelectronics, Space Systs Div, 66-68, prog mgr microelectronics, Ord Systs Div, 68-75, consult mat eng, 75-88; CONSULT, MICROELECTRONICS-INT EXEC SERV CORPS, 89- *Concurrent Pos:* Instr semiconductor technol, Semiconductor Prod Div, Gen Elec Co, 60-66, instr microelectronics, Ord Systs, 70-88, instr reliability, Eng Serv, 85-88. *Mem:* Int Soc Microelectronics. *Res:* Microelectronics packaging, assembly, reliability and failure analysis; development of semiconductor, hybrid and printed wire board processes; development and assembly of semiconductor imaging arrays; development of high resolution photolithography and microlithography. *Mailing Add:* 68 Gravesleigh Terr Pittsfield MA 01201

MOORE, ROBERT EMMETT, b Wichita Falls, Tex, Nov 17, 31; m 53; c 3. VERTEBRATE ECOLOGY. *Educ:* N Tex State Univ, BA, 52; Ore State Univ, MS, 59; Univ Tex, PhD(mammalian speciation), 62. *Prof Exp:* From asst prof to assoc prof, 62-72, PROF ZOOL, MONT STATE UNIV, 72- *Mem:* Am Soc Naturalists; Ecol Soc Am; Am Soc Mammal; Sigma Xi. *Res:* Isolating mechanisms in the speciation of small mammals; population biology of small mammals. *Mailing Add:* Dept Zool Mont State Univ Bozeman MT 59717

MOORE, ROBERT H, b Baltimore, Md, Dec 19, 30; m 58, 85; c 4. MATHEMATICS. *Educ:* Univ Md, BS, 53, MA, 55; Univ Mich, PhD(math), 59. *Prof Exp:* NSF fel, Munich Tech Univ, 59-60; res mathematician, Math Res Ctr, US Army, Univ Wis, 60-63; asst prof, 63-68, ASSOC PROF MATH, UNIV WIS-MILWAUKEE, 68- *Mem:* Am Math Soc; Math Asn Am; Soc Indust & Appl Math. *Res:* Functional analytic study of methods in numerical analysis, especially integral equations, collectively compact operators and Newton's method, generalized inverses of linear operators; numerical solution of hyperbolic partial differential equations. *Mailing Add:* Dept Math Univ Wis PO Box 413 Milwaukee WI 53201

MOORE, ROBERT J(AMES), b Aberdeen, Wash, May 31, 17; m 41; c 2. PHYSICAL CHEMISTRY. *Educ:* Univ Wash, BS, 37, MS, 40; Univ Utah, MS, 51. *Prof Exp:* Asst chief chem, Fibreboard Prod, Inc, 37-38; res chemist, Shell Develop Co Div, 41-55; group leader phys chem, Shell Oil Co Inc, 55-57; spec assignment, Shell Res Ltd Eng, 57-58; asst chief res chemist, Shell Oil Co, 59-60, chief res chemist, 61-63; asst dept mgr, Shell Develop Co, 64; dept mgr, Shell Chem Co, 64-70; sr engr, Petro-Chem Div, Shell Chem Co, 70-82; CONSULT, M D ANDERSON CANCER CTR, 90- *Concurrent Pos:* Pvt consult, R J Moore, Inc, Houston, Tex, 82-87. *Mem:* Am Chem Soc. *Res:* Engine lubricants and fuels; waxes; resins and polymers applications; liquid viscosity. *Mailing Add:* 5110 San Felipe 371 W Tower Houston TX 77056-3615

MOORE, ROBERT LAURENS, b Lake City, Fla, Nov 25, 45. MATHEMATICS. *Educ:* Mass Inst Technol, BA, 68; Ind Univ, PhD(math, physics), 74. *Prof Exp:* PROF MATH, UNIV ALA, 78- *Mem:* Am Math Soc; Sigma Xi. *Mailing Add:* Math Dept Univ Ala Box 1416 Tuscaloosa AL 35487-0350

MOORE, ROBERT LEE, b Gainesville, Tex, Dec 28, 20; m 53. PHYSICAL CHEMISTRY. *Educ:* N Tex State Univ, BA, 42; Univ Tex, MA, 44, PhD(phys chem), 47. *Prof Exp:* Instr math, N Tex State Univ, 41-42; instr chem, Univ Tex, 42-45; chemist, Hanford Atomic Prod Oper, Gen Elec Co, 47-56, mgr fission prod chem, 56-65; mgr, Pac Northwest Lab, Battelle Mem Inst, 65-70; MGR APPL CHEM & ANAL, WESTINGHOUSE HANFORD CORP, 70- *Mem:* Am Chem Soc; Am Nuclear Soc; fel Am Inst Chemists; Sigma Xi. *Res:* Nuclear fuel processing; separations chemistry; waste treatment; analytical chemistry. *Mailing Add:* Rte 1 Box 5238 Richland WA 99352

MOORE, ROBERT STEPHENS, b Dubuque, Iowa, Sept 12, 33; m 56; c 3. POLYMER SCIENCE, POLYMER CHEMISTRY. *Educ:* Univ Wis, BS, 55, PhD(phys chem), 62. *Prof Exp:* Mem tech staff, Bell Tel Labs, Inc, 62-69; res assoc, 69-71, head, Polymer Phys Chem Lab, 71-81, HEAD, CHEMIPHOTOGRAPHIC SYSTS LAB, EASTMAN KODAK CO, 81- *Mem:* Am Chem Soc; fel Am Phys Soc; Soc Rheol; Sigma Xi. *Res:* Viscoelastic properties of polymers; dilute solution properties of polymers; rheological and rheo-optical properties of polymers; spectroscopy of polymers; light scattering from polymeric systems; light sensitive polymers for lithographic and microresist applications; physical chemistry; technical management. *Mailing Add:* Res Lab Bldg 81 Eastman Kodak Co Rochester NY 14650

MOORE, ROBERT YATES, b Harvey, Ill, Dec 5, 31; m 69; c 4. NEUROLOGY, PEDIATRIC NEUROLOGY. *Educ:* Lawrence Univ, BA, 53; Univ Chicago, MD, 57, PhD(psychol), 62; Am Bd Psychiat & Neurol, dipl & cert neurol, 66. *Hon Degrees:* MD, Univ Lund, 74. *Prof Exp:* Intern, Univ Mich Hosp, 58-59; instr anat & res neurol, Sch Med, Univ Chicago, 59-64, from asst prof to prof pediat, neurol & anat, 64-74; prof neurosci, Univ Calif, San Diego, 74-79; PROF & CHAIR, DEPT NEUROL, STATE UNIV NY, STONY BROOK, NY. *Concurrent Pos:* Markle scholar, 65-70; consult, NIH, 71-75, 76-80 & Food & Drug Admin, 72-76; assoc, Neurosci Res Prog, 74-83. *Mem:* Am Acad Neurol; Am Neurol Asn; Child Neurol Soc; Soc Neurosci. *Res:* Organization and function of monoamine neuron systems in the mammalian brain; central neural regulation of circadian rhythms. *Mailing Add:* Dept Neurol Health Sci Ctr SUNY Stony Brook NY 11794

MOORE, RONALD LEE, b Ft Wayne, Ind, Mar 30, 42; m 72; c 3. SOLAR PHYSICS, ASTROPHYSICS. *Educ:* Purdue Univ, BS, 64; Stanford Univ, MS, 65, PhD(aeronaut & astronaut sci), 72. *Prof Exp:* Res fel solar physics, Calif Inst Technol, 72-75, sr res fel, 75-80; MEM TECH STAFF, MARSHALL SPACE FLIGHT CTR, NASA, HUNTSVILLE, 80- *Concurrent Pos:* Mem Solar Physics Mgt & Oper Working Group, NASA, 79-80 & 84-86, co-chmn & study scientist, Advan Solar Observ Sci Study Team, 81-, co-investr, Photometric Filtergraph Instrument High Resolution Solar Observ Sci Working Group, 87-, US Sci & Tech Working Group for large earth-based solar telescope, 88-, Solar Physics Panel Astron & Astrophys Survey Co Nat Acad Sci, 89-90 & Solar Physics Panel of NASA Space Physics Strategy Implementation Study, 90. *Mem:* Am Astron Soc; Int Astron Union. *Res:* Physics of solar flares; fine-scale structure and dynamics of quiet solar atmosphere, especially oscillations, spicules, macrospicules, and ephemeral active regions; energy balance of solar corona; solar magnetic cycle. *Mailing Add:* Space Sci Lab ES52 Marshall Space Flight Ctr NASA Huntsville AL 35812

MOORE, ROSCOE MICHAEL, JR, b Richmond, Va, Dec 2, 44; m 69; c 2. EPIDEMIOLOGY. *Educ:* Tuskegee Inst, BS, 68, DVM, 69; Univ Mich, MPH, 70; John Hopkins Univ MHS, 82, PhD, 85. *Hon Degrees:* DSc, Tuskegee Univ, 90. *Prof Exp:* Researcher gnotobiotics, NIH, 70-71; epidemiologist, Ctr Dis Control, 71-73 & Food & Drug Admin, 73-74; sr epidemiologist, Nat Inst Occup Safety & Health, 74-81; SR EPIDEMIOLOGIST, CTR DEVICES & RADIOL HEALTH, FOOD & DRUG ADMIN, 81- *Concurrent Pos:* Consult, Dade County, Fla Health Dept & Ala State Health Dept, 72-73; adv, Washington Tech Inst, 74-77; assoc dir epidemiol & biostatist, Cancer Ctr, Col Med, Howard Univ, 80-81; chief, Epidemiol Br, Food & Drug Admin, 87-; chief vet med officer, USPHS, 89- *Mem:* Am Pub Health Asn; Soc Occup & Environ Health; fel Am Col Epidemiol; Soc Epidemiol Res; AAAS. *Res:* Comparative medicine and dentistry; growth and development among human fetuses during different stages of gestation. *Mailing Add:* Ctr Devices & Radiol Health Food & Drug Admin 12200 Wilkins Ave Rockville MD 20857

MOORE, RUFUS ADOLPHUS, b Brackettville, Tex, Feb 8, 23. PHYSICS. *Educ:* St Mary's Univ, Tex, BS, 43; Univ Tex, MA, 49, PhD(physics), 58. *Prof Exp:* Asst math & physics, Univ Tex, 48-50, res scientist physics, Defense Res Lab, 51-52, asst math & physics, Univ, 53; mathematician, Sch Aviation Med, 54-56; asst math & physics, Univ Tex, 56-58; mathematician, Personnel Lab, Wright Air Develop Ctr, 58-59; asst prof physics, St Mary's Univ, Tex, 59-62; from assoc prof to prof physics, State Univ NY Col Oswego, 62-83; RETIRED. *Mem:* AAAS; Am Math Soc; Inst Elec & Electronics Engrs; Am Phys Soc; Am Asn Physics Teachers; Sigma Xi. *Res:* Molecular theory; electronic communications. *Mailing Add:* PO Box 430 Brackettville TX 78832-0430

MOORE, RUSSELL THOMAS, b Las Animas, Colo, Dec 30, 43; m 64; c 2. PLANT ECOLOGY, RANGE SCIENCE & MANAGEMENT. *Educ:* Univ Idaho, BS, 66; Utah State Univ, PhD(plant ecol), 72. *Prof Exp:* Fel plant ecophysiol, San Diego State Univ, 71-73; proj mgr plant ecol, ENSR Consult & Eng, 73-75, sect mgr Mining & Reclamation, 75-79, proj mgr, Mine Permitting Projs, 79-81, Multidisciplinary Projs, 81-89, GEN MGR, ENSR CONSULT & ENG, 89- *Mem:* Ecol Soc Am; Soc Range Mgt; Northwest Mining Asn; Colo Mining Asn. *Res:* Environmental impact studies associated with large-scale development of oil shale, coal, and hard rock minerals throughout the western states. *Mailing Add:* ENSR 1716 Heath Pkwy Ft Collins CO 80524

MOORE, SEAN, b Belfast, Northern Ireland, Nov 24, 26; m 57; c 3. PATHOLOGY, EDUCATION ADMINISTRATION. *Educ:* Queen's Univ, Belfast, MB, BCh & BAO, 50; FRCP(C), 74. *Prof Exp:* Resident path, Montreal Gen Hosp & St Mary's Hosp, 54-55, asst pathologist, Montreal Gen Hosp, McGill Univ, 58-61, assoc pathologist, Montreal Gen Hosp, 61-69; pathologist-in-chief, Jewish Gen Hosp, Montreal, 69-71; coordr anat path, Med Ctr, McMaster Univ, 72, chmn dept path, 72-78, dir labs, 72-84, univ prog dir, residency educ lab med, 78-83; PROF PATH & CHMN DEPT, MCGILL UNIV, 84-, STRATHCONA PROF, 85-; PATHOLOGIST-IN-CHIEF, ROYAL VICTORIA HOSP, 84- *Concurrent Pos:* Mem Coun Thrombosis & Coun Arteriosclerosis, Am Heart Asn; managing ed, Exp & Molecular Path, 87- *Mem:* Fedn Am Socs Exp Biol; Int Acad Path; Can Asn Pathologists; Am Heart Asn; Int Soc Thrombosis & Haemostasis. *Res:* Injury mechanisms in atherosclerosis; observation of animal lesions closely resembling human disease in response to endothelial injury; proteoglycan alterations in response to endothelial injury and lipid accumulation. *Mailing Add:* Dept Path Lyman Duff Med Sci Bldg McGill Univ 3775 Univ St Montreal PQ H3A 2B4 Can

MOORE, SEAN BREANNDAN, b Dublin, Ireland, Aug 4, 44; m 69; c 2. TRANSFUSION MEDICINE. *Educ:* Univ Col, Dublin, MB, BCh, BAO 68, DCH, 70; FRCP(I). *Prof Exp:* CONSULT & PHYSICIAN, MAYO CLIN BLOOD BANK & TRANSFUSION SERV, 76-; DIR MAYO CLIN HISTOCOMPATIBILITY LAB, 77- *Concurrent Pos:* From instr to assoc prof, 76-88, prof lab med, Mayo Grad Sch Med, 88-; histocompatibility comt, United Network Organ Sharing, 85-88; Comt Clin Affairs, Am Soc Histocompatibility & Immunogenetics, 82- *Mem:* Am Asn Blood Banks; Am Soc Histocompatibility & Immunogenetics; Transplantation Soc; AMA. *Res:* Immunological aspects of transfusion reactions; transfusion-induced immune modulation; pathology. *Mailing Add:* Mayo Clin Blood Bank 200 First St SW Rochester MN 55905

MOORE, STEVENSON, III, entomology, for more information see previous edition

MOORE, THEODORE CARLTON, JR, b Kinston, NC, Feb 16, 38; m 60; c 2. OCEANOGRAPHY, MARINE GEOLOGY. *Educ:* Univ NC, BS, 60; Univ Calif, San Diego, PhD(oceanog), 68. *Prof Exp:* Res assoc oceanog, Ore State Univ, 68-69, asst prof, 69-75; FROMASSOC PROF TO PROF OCEANOG, UNIV RI, 75,; AT EXXON PROD RES CO. *Mem:* AAAS; Am Geophys Union. *Res:* Stratigraphy and sedimentation in the deep-sea; micropaleontological studies of Radiolaria and calcareous nanoplankton. *Mailing Add:* Dept Geol Sci 1006 C C Little Bldg Univ Mich Ann Arbor MI 48109

MOORE, THERAL ORVIS, b Emerson, Ark, Oct 16, 27; m 62; c 2. MATHEMATICS. *Educ:* Univ Ark, BA, 49, MA, 51; Univ Mo, PhD(math), 55. *Prof Exp:* Asst math, Univ Mo, 51-55; asst prof, 55-66, ASSOC PROF MATH, UNIV FLA, 66- *Mem:* Am Math Soc; Math Asn Am. *Res:* Topology; lattice theory; abstract algebra. *Mailing Add:* Dept of Math Univ of Fla Gainesville FL 32611

MOORE, THERON LANGFORD, b Mayo, Va, Nov 24, 34; m 59; c 3. ORGANIC CHEMISTRY. *Educ:* Yale Univ, BS, 56; Univ Calif, Los Angeles, PhD(org chem), 61. *Prof Exp:* Chemist, Procter & Gamble Co, 61-66; assoc prof, 66-68, PROF CHEM, NORFOLK STATE UNIV, 68- *Mem:* AAAS; Am Chem Soc. *Mailing Add:* Dept of Chem Norfolk State Univ Norfolk VA 23504

MOORE, THOMAS ANDREW, b Pensacola, Fla, June 3, 44; m 68; c 2. BIOCHEMISTRY, BIOPHYSICS. *Educ:* Tex Tech Univ, BA, 68, PhD(chem), 75. *Prof Exp:* Res assoc chem, Univ Wash, 73-76; from asst prof to assoc prof, 76-85, PROF CHEM, ARIZ STATE UNIV, 85- *Mem:* Am Chem Soc; Am Soc Photobiol. *Res:* Biochemistry; photobiology; molecular spectroscopy. *Mailing Add:* 5721 E Edgemont Scottsdale AZ 85257-1027

MOORE, THOMAS CARROL, b Sanger, Tex, Sept 22, 36; m 56; c 3. PLANT ECOLOGY. *Educ:* N Tex State Univ, BA, 56; Univ Colo, MA, 58, PhD(bot), 61. *Prof Exp:* Instr biol, Univ Colo, 58-59; asst prof bot, Ariz State Col, 61-63; from asst prof to assoc prof, 63-71, chmn dept bot & plant path, 73-86, PROF BOT, ORE STATE UNIV, 71- *Concurrent Pos:* Ed-in-chief, J Plant Growth Regulation, 81-89. *Mem:* Am Soc Plant Physiologists; Bot Soc Am; Int Plant Growth Substances Asn; Sigma Xi; Plant Growth Regulator Soc Am. *Res:* Hormonal regulation of growth and flowering in angiosperms, including metabolic control mechanisms; modes of action of plant growth-regulating chemicals; physiological ecology of seed plants. *Mailing Add:* Dept Bot & Plant Path Ore State Univ Corvallis OR 97331-2902

MOORE, THOMAS EDWIN, b Amarillo, Tex, Jan 15, 18; m 42; c 4. INORGANIC CHEMISTRY. *Educ:* Univ Tex, BA, 40, MS, 42, PhD(chem), 46. *Prof Exp:* Res assoc, Radio Res Lab, Harvard Univ, 43-45; US Signal Corps fel, Northwestern Univ, 46-47; from asst prof to assoc prof, 47-57, prof, 57-82, EMER PROF CHEM, OKLA STATE UNIV, 82- *Mem:* Am Chem Soc. *Res:* Solvent extraction of inorganic salts; nonaqueous solutions; thermodynamics of electrolyte mixtures. *Mailing Add:* 1124 W Knapp St Stillwater OK 74075-2711

MOORE, THOMAS EDWIN, b Champaign, Ill, Mar 10, 30; m 51; c 2. SCIENCE ADMINISTRATION, ACOUSTICS. *Educ:* Univ Ill, BS, 51, MS, 52, PhD, 56. *Prof Exp:* Asst, Ill Natural Hist Surv, 52-56; from instr to assoc prof, 56-67, chmn Comt Trop Studies, 74-79, chmn dept ecol & evolutionary biol, 77-78, 81-82, PROF BIOL, UNIV MICH, ANN ARBOR, 67-, CUR INSECTS, MUS ZOOL, 56-, DIR, EXHIBIT MUSEUM, 88- *Concurrent*

Pos: Mem, Orgn Trop Studies, vis prof, 70 & 72; mem, Nat Acad Sci-Nat Res Coun panel on NSF grad fels, 71-73. *Mem:* AAAS; Entom Soc Am; Soc Study Evolution; Royal Entom Soc London; Asn Trop Biol (pres, 73-76). *Res:* Acoustical behavior of insects; evolution and systematics of cicadas; forensic entomology; insect nerve and muscle physiology; insect skeletal structure and function; mitochondrial DNA analysis of isolated species populations. *Mailing Add:* Exhib Mus Univ Mich 4502 Museums Bldg Ann Arbor MI 48109-1079

MOORE, THOMAS FRANCIS, geochemistry; deceased, see previous edition for last biography

MOORE, THOMAS MATTHEW, US citizen. MATERIALS SCIENCE ENGINEERING, ENGINEERING PHYSICS. *Educ:* Univ Va, BA, 76, MS, 78, PhD(eng mat sci), 80. *Prof Exp:* Mem tech staff, 80-89, SR MEM TECH STAFF, TEX INSTRUMENTS, 89- *Mem:* Am Phys Soc; Electron Micros Soc Am; Inst Elec & Electronics Engrs. *Res:* Electron beam analysis of semiconductor materials and devices; scanning acoustic inspection of integrated circuit packages. *Mailing Add:* Tex Instruments Cent Res MS-147 PO Box 655936 Dallas TX 75265

MOORE, THOMAS STEPHEN, JR, b Kennett, Mo, Aug 27, 42; m 68; c 1. MEMBRANE SYNTHESIS, LIPID METABOLISM. *Educ:* Univ Ark, Fayetteville, BS, 64, MS, 66; Ind Univ, Bloomington, PhD(plant physiol), 70. *Prof Exp:* Res assoc bot, Univ Calif, Santa Cruz, 70-73; asst prof, Univ Wyo, Laramie, 73-78, assoc prof & chmn dept, 78-82; prof bot & chmn, 82-88, interim chmn microbiol, 86-88, DIR PLANT PHYSIOL PROG, LA STATE UNIV, BATON ROUGE, 85- *Concurrent Pos:* Alexander von Humboldt fel, Univ Munich. *Mem:* Am Soc Plant Physiologists; AAAS; Sigma Xi. *Res:* Regulation of membrane lipid synthesis in plants with emphasis on phospholipid biosynthesis in the endoplasmic reticulum and mitochondria; solubilization and reconstruction of activity of phospholipid synthesizing enzymes. *Mailing Add:* Dept Bot La State Univ Baton Rouge LA 70803-1705

MOORE, THOMAS WARNER, b New Haven, Conn, Mar 22, 28; m 51; c 3. SOLID-STATE PHYSICS. *Educ:* Calif Inst Technol, BS, 50; Univ Calif, Berkeley, PhD(physics), 61. *Prof Exp:* Staff mem, Gen Elec Res & Develop Ctr, 61-67; assoc prof, 67-75, chmn dept, 76-80, PROF PHYSICS, MT HOLYOKE COL, 75- *Mem:* Am Phys Soc; Am Asn Physics Teachers. *Res:* Metal physics; transport phenomena in metals at cryogenic temperatures. *Mailing Add:* 156 Amherst Rd South Hadley MA 01075

MOORE, VAUGHN CLAYTON, b Osawatomie, Kans, May 16, 34; m 56; c 3. RADIOLOGICAL PHYSICS, BIOPHYSICS. *Educ:* Univ Kans, BS, 56, MS, 57; Univ Minn, PhD(biophys), 68; Am Bd Radiol, dipl, 64. *Prof Exp:* Health physicist, Univ Chicago, 57-58, asst dir health physics, 58-59; asst prof radiol, Health Sci Ctr, Univ Minn, 68-70, asst prof therapeut radiol, 70-75; dir radiation physics, Neuropsychiat Inst, Fargo, 75-86, St Lukes Hosp, 86-87; PRES, MIDWEST RADIATION PHYSICISTS, INC, 75- *Concurrent Pos:* Attend, Vet Admin Hosp, Minneapolis, 68-75; consult, St Joseph's Hosp, St Paul, 69-75; adj prof physics & bionucleonics, NDak State Univ, 75-, clin assoc prof, Radiol, 78-; physics instr, St Luke's Sch Radiol Technol, Fargo, NDak, 81-; radiation physicist, Dakota Clin, 88-89. *Mem:* Am Asn Physicists Med. *Res:* Applications of electron linear accelerators and computers to radiation therapy. *Mailing Add:* 3601 Evergreen Rd Fargo ND 58102

MOORE, VERNON LEON, b Albion, Okla, Oct 21, 36; m 70; c 1. IMMUNOLOGY, PATHOLOGY. *Educ:* Okla State Univ, BS, 63; Tripler Army Med Ctr Sch Med, MT, 64; Bowman Gray Sch Med, MS, 70, PhD(immunol & path), 72. *Prof Exp:* From instr to asst prof, 72-77, ASSOC PROF MED, MED COL WIS, 77-, ASSOC PROF PATH, 76-; AT IMMUNOL & INFLAMMATION RES, MERCK, SHARP & DOHME RES LABS, RAHWAY, NJ, 82- *Concurrent Pos:* Res assoc, Vet Admin Ctr, 72-75, res health scientist, 75-; Nat Heart Lung & Blood Inst fel, 77-80; NC Tuberc & Respiratory Dis Asn, 71-72. *Mem:* Am Asn Immunologists; Reticuloendothelial Soc; Fedn Am Soc Exp Biol; Am Soc Clin Pathologists; Sigma Xi. *Res:* Cellular immunology; immunology of the respiratory tract; immunogenetics of chronic granulomatous inflammation; biology of cartilage. *Mailing Add:* Merck Sharp & Dohme Res Labs PO Box 2000 Rahway NJ 07065

MOORE, WALTER CALVIN, b Oklahoma City, Okla, Oct 21, 10; m 31; c 3. CHEMICAL ENGINEERING. *Educ:* State Univ NY, BS, 86. *Prof Exp:* Develop engr, Oak Ridge Gaseous Diffusion Plant, Union Carbide Corp, 44-45, eng dept head, 45-51, asst chief engr, 51-54, proj engr, Oak Ridge Nat Lab, 54, asst plant supt, Oak Ridge Y-12 Plant, 54-58, proj mgr plastics appln packaging, Union Carbide Corp, NY, 58-59; proj mgr nuclear reactor develop, Gen Atomic Div, Gen Dynamics Corp, Calif, 59-62; vpres eng & res, York Div, Borg-Warner Corp, 62-76; RETIRED. *Concurrent Pos:* Consult mgt & res, 76- *Mem:* Am Inst Chem Engrs; Am Soc Heating, Refrig & Air-Conditioning Engrs; Am Nuclear Soc; Int Solar Energy Soc; Int Inst Refrig; NY Acad Sci. *Res:* Chemical and metallurgical processes and related process equipment; air conditioning, refrigeration and ice making equipment; statistical analysis of data. *Mailing Add:* 360 Tri-Hill Dr York PA 17403

MOORE, WALTER EDWARD C, b Rahway, NJ, Oct 12, 27; m; c 3. MEDICAL MICROBIOLOGY, BACTERIOLOGY. *Educ:* Univ NH, BS, 51; Univ Wis, MS, 52, PhD(dairy husb, bact), 55. *Prof Exp:* Asst, Alumni Res Found, Univ Wis, 51-52; assoc prof, 54-61, dept head, Anaerobe Lab, 71-85, PROF BACT, VA POLYTECH INST & STATE UNIV, 61-, UNIV DISTINGUISHED PROF, 81- *Concurrent Pos:* Vchmn judicial comn, Int Comn Bact Nomenclature, 75-; ed-in-chief, Int J Syst Bact. *Honors & Awards:* Kimble Methodology Res Award, Conf Pub Health Lab Dirs, 73; Bergey Award, 84; Porter Award, 88. *Mem:* AAAS; Am Soc Microbiol; Am Acad Microbiol. *Res:* Intestinal, periodontal and anaerobic microbiology. *Mailing Add:* Anaerobe Lab Va Polytech Inst & State Univ Blacksburg VA 24061

MOORE, WALTER GUY, b Detroit, Mich, June 21, 13; m 39; c 6. AQUATIC BIOLOGY, ECOLOGY. *Educ:* Wayne State Univ, BA, 34; Univ Minn, MA, 38, PhD(zool), 40. *Prof Exp:* Instr, Wayne State Univ, 34-35; asst, Univ Minn, 35-40; from instr to assoc prof, 40-51, actg chmn depts biol & med technol, 42-44, prof, 51-79, EMER PROF BIOL, LOYOLA UNIV, LA, 79- *Mem:* AAAS; Ecol Soc Am; Am Micros Soc; Am Soc Limnol & Oceanog; Sigma Xi. *Res:* Limnology; ecology of Anostraca; temporary ponds. *Mailing Add:* 5231 S Derbigny St New Orleans LA 70125

MOORE, WALTER JOHN, b New York, NY, Mar 25, 18; m 43; c 3. PHYSICAL CHEMISTRY. *Educ:* NY Univ, BS, 37; Princeton Univ, PhD(phys chem), 40. *Prof Exp:* Nat Res Coun fel, Calif Inst Technol, 40-41; from instr to assoc prof chem, Cath Univ Am, 41-51; Guggenheim & Fulbright fels, Bristol Univ, 51-52; prof chem, Ind Univ, Bloomington, 52-63, res prof, 63-74; prof, 74-84, EMER PROF CHEM, UNIV SYDNEY, 84- *Concurrent Pos:* Manhattan proj engr, US Army, 42-46; NSF sr fel, Paris, 58-59; vis prof, Harvard Univ, 60 & Univ Brasil, 62 & 63; chmn comn phys chem, Nat Acad Sci-Nat Res Coun, 64-66; Austalian Am Educ Found prof, Univ Queensland, 66, vis prof, 68-; adj prof, Ind Univ, 90- *Honors & Awards:* James F Norris Award, Am Chem Soc, 65. *Mem:* Am Chem Soc; Biophys Soc; Am Soc Biol Chemists; Am Soc Neurochem; Int Soc Neurochem. *Res:* Solid-state chemistry; neurochemistry; biophysical chemistry of brain function. *Mailing Add:* 916 S Mitchell St Bloomington IN 47401

MOORE, WALTER L(EON), b Estrella, Calif, Mar 12, 16; m 42; c 4. WATER RESOURCES ENGINEERING, HYDRAULICS. *Educ:* Calif Inst Technol, BS, 37, MS, 38; Univ Iowa, PhD(mech, hydraul), 51. *Prof Exp:* Asst, Soil Conserv Serv Coop Lab, Calif Inst Technol, 39-40; res engr & analyst, Lockheed Aircraft Corp, 40-47; assoc prof, 47-53, chmn dept, 58-65, PROF CIVIL ENG, UNIV TEX, AUSTIN, 53- *Concurrent Pos:* US deleg, Working Group Educ & Training, Int Hydraul Decade, UNESCO, 65-75; coordr water resources, Hydraul Div, Am Soc Civil Engrs, 72-74. *Honors & Awards:* Collingwood Prize, Am Soc Civil Engrs, 44. *Mem:* AAAS; Am Soc Civil Engrs; Am Soc Eng Educ; Am Geophys Union; Am Water Resources Asn; Sigma Xi. *Res:* Stilling basins; diffusion; fluid mechanics; hydrology; hydraulic structures. *Mailing Add:* Civil Eng ECJ 4-610 Univ Tex Austin TX 78712

MOORE, WALTER P, JR, STRUCTURAL DESIGN. *Educ:* Rice Univ, Houston, Tex, BA, 59, BS, 60; Univ Ill, Urbana, MS, 62, PhD(civil eng), 64. *Prof Exp:* PRES & CHMN BD, WALTER P MOORE & ASSOCS INC. *Concurrent Pos:* Activ chmn, Am Soc Civil Engrs, Houston, 68-69, secy, Struct Group, Tex Sect, 76-77, vchmn, 77-78, chmn, 78-79; mem bd dirs, Consult Engrs Coun Tex, 74-76, vpres, 76-77, pres, 77-78, alt dir, 79-81, dir, 81-83; adj prof archit, Rice Univ, 75-82; mem bd dirs, Rice Ctr Commun Design & Res, 77-82, Houston Eng & Sci Soc, 84-87 & Rice Design Alliance, 76-79; trustee, Am Consult Engrs, 79-85, secy, 85-87, chmn, 87-89, vpres, 90-92; mem, Nat Specif Comt, Am Inst Steel Construct, 83; chmn architects & engrs, United Way, 85. *Honors & Awards:* Spec Citation Award, Am Inst Steel Construct, 85. *Mem:* Nat Acad Eng; Am Consult Engrs Coun; Am Soc Civil Engrs; Soc Am Mil Engrs; Int Asn Bridge & Struct Engrs; Am Concrete Inst; Nat Soc Prof Engrs; Coun Tall Bldgs & Urban Habitat; Post Tensioning Inst. *Res:* Author of various publications. *Mailing Add:* Walter P Moore & Assocs Inc 3131 Eastside 2nd Floor Houston TX 77098

MOORE, WARD WILFRED, b Cowden, Ill, Feb 12, 24; m 49; c 4. PHYSIOLOGY. *Educ:* Univ Ill, AB, 48, MS, 51, PhD(physiol), 52. *Prof Exp:* Asst animal physiol, Univ Ill, 50-52, res assoc, 52-54; asst prof physiol, Okla State Univ, 54-55; from asst prof to assoc prof physiol, 55-66, actg chmn dept anat, 71-73, assoc dean basic med sci, 71-76, PROF PHYSIOL, IND UNIV, INDIANAPOLIS, 66-, ASSOC DEAN & DIR MED SCI PROG, 76- *Concurrent Pos:* Vis prof, Jinnah Postgrad Med Ctr, Karachi, 63-64; mem staff, Rockefeller Found & vis prof & chmn dept physiol, Fac Sci, Mahidol Univ, Thailand, 68-71. *Mem:* AAAS; Am Physiol Soc; Am Asn Anat; Endocrine Soc; Am Soc Nephrology; Soc Study Reproduction. *Res:* Neuroendocrinology; stress; regulation of antidiuretic hormone secretion. *Mailing Add:* Med Sci Prog Meyers Hall 203 Ind Univ Sch of Med Bloomington IN 47401

MOORE, WARREN KEITH, b Wellington, Kans, Feb 11, 23; m 44; c 5. MATHEMATICS. *Educ:* Southwestern Col, Kans, AB, 47; Univ Kans, MA, 48, PhD(math), 51. *Prof Exp:* Cottrell asst, Univ Kans, 48-49, instr math, 51-52; from asst prof to assoc prof, 52-63, chmn dept, 61-79, PROF MATH, ALBION COL, 63- *Mem:* Am Math Soc; Math Asn Am; Sigma Xi. *Res:* Mathematical analysis. *Mailing Add:* 1201 Jackson Albion MI 49224

MOORE, WAYNE ELDEN, b McLeansboro, Ill, Sept 2, 19; m 44; c 4. PETROLEUM GEOLOGY. *Educ:* Univ Ill, BS, 46; Cornell Univ, MS, 48, PhD(geol), 50. *Prof Exp:* Assoc prof geol, Va Polytech Inst, 50-56; paleontologist, Chevron Oil Co, Stand Oil Co, Calif, 56-70; chmn dept, 72-81, prof, 71-86, EMER PROF GEOL, CENT MICH UNIV, 87- *Mem:* Am Asn Petrol Geologists; Am Asn Stratig Palynologists; AAAS. *Res:* Origin, maturation and migration of Paleozoic petroleum and subsurface brines, especially in Michigan; micropaleontology of the source beds. *Mailing Add:* Dept Geol Cent Mich Univ Mt Pleasant MI 48859

MOORE, WESLEY SANFORD, b San Bernardino, Calif, Aug 1, 35; m 60; c 2. VASCULAR SURGERY. *Educ:* Univ Southern Calif, BS, 55; Univ Calif, San Francisco, MD, 59. *Prof Exp:* Intern surg, Univ Calif Hosp, Univ Calif, San Francisco, 59-60, asst resident surg, 60-63, chief resident, 63-64, from clin instr to assoc prof, Sch Med. 66-77; prof surg & chief vascular surg sect, Univ Ariz, 77-80; PROF SURG & CHIEF SECT VASCULAR SURG, CTR HEALTH SCI, UNIV CALIF, LOS ANGELES, 80- *Concurrent Pos:* NIH fel cerebrovascular insufficiency, 66-67; mem comt prosthetics & orthotics, Nat Acad Sci; chief vascular surg sect, Vet Admin Hosp, San Francisco, 66-77; chief vascular surg serv, Vet Admin Hosp, Tucson, Ariz, 77-80; staff surgeon, Wadsworth Vet Admin Hosp, Los Angeles, 80-85 & Sepulveda Vet Admin Hosp, 84- *Mem:* Soc Vascular Surg (pres elect, 85-86, pres, 86-87);

Int Cardiovascular Soc; Am Heart Asn; fel Am Col Surg; Soc Univ Surgeons. *Res:* Vascular surgery; circulation research; stroke prevention; amputation surgery and rehabilitation; development of infection-resistant vascular grafts. *Mailing Add:* Ctr Health Sci Univ Calif, Los Angeles Los Angeles CA 90024-1749

MOORE, WILLARD S, b Jackson, Miss, Sept 27, 41. MARINE GEOCHEMISTRY, NATURAL RADIOISOTOPES. *Educ:* Millsaps Col, BS, 62; Columbia Univ, MA, 65; State Univ NY Stony Brook, PhD(earth & space sci), 69. *Prof Exp:* Oceanogr, Ocean Floor Anal Div, US Naval Oceanog Off, 69-76; assoc prof geol & marine sci, 76-81, chmn dept geol, 81-85, PROF GEOL & MARINE SCI, UNIV SC, 81- *Concurrent Pos:* Vis asst prof, State Univ NY Stony Brook, 70; vis res fel, Tata Inst Fundamental Res, India, 71; vis adj prof, Univ SC, 70 & 71; assoc ed, Geol Soc Am Bull. *Mem:* AAAS; Am Chem Soc; Sigma Xi; Am Geophys Union; Geol Soc Am. *Res:* Radioisotopes in the ocean; manganese nodules; radium in groundwaters; paleo sea levels; ocean mixing. *Mailing Add:* Dept Geol Univ SC Columbia SC 29208

MOORE, WILLIAM EARL, protein chemistry, for more information see previous edition

MOORE, WILLIAM MARSHALL, b Lincoln, Nebr, Dec 25, 30; m 58, 68; c 4. PHYSICAL CHEMISTRY. *Educ:* Colo Col, BA, 52; Iowa State Univ, PhD(phys chem), 59. *Prof Exp:* Res chemist, Monsanto Chem Co, 52-53; NIH fel, Cambridge Univ, 59-60; from asst prof to assoc prof, 60-69, PROF CHEM, UTAH STATE UNIV, 69- *Mem:* Am Chem Soc; Am Soc Photobiol. *Res:* Mechanisms for photochemical reactions; atmospheric chemistry and photochemistry; emission spectroscopy. *Mailing Add:* Dept Chem & Biochem UMC0300 Utah State Univ Logan UT 84322

MOORE, WILLIAM ROBERT, b Minneapolis, Minn, July 18, 28; m 56; c 3. ORGANIC CHEMISTRY. *Educ:* Univ Calif, Los Angeles, BS; Univ Minn, PhD, 54. *Prof Exp:* Res assoc org chem, Mass Inst Technol, 54-55, from instr to assoc prof, 55-72; PROF CHEM & CHMN DEPT, WVA UNIV, 72- *Mem:* AAAS; Am Chem Soc. *Res:* Mechanisms of organic reactions; unsaturated cyclic hydrocarbons; highly-strained compounds; carbenes. *Mailing Add:* Dept Chem WVa Univ Box 6045 Morgantown WV 26506

MOORE, WILLIAM SAMUEL, b Manhattan, Kans, June 8, 42; m 63; c 2. POPULATION BIOLOGY. *Educ:* Mich State Univ, BS, 66; Univ Conn, PhD(biol), 71. *Prof Exp:* Asst prof, 71-75, ASSOC PROF BIOL, WAYNE STATE UNIV, 75- *Mem:* Soc Study Evolution; Ecol Soc Am; Genetics Soc Am. *Res:* Evolution and adaptive strategies of various genetic systems. *Mailing Add:* Dept Biol Wayne State Univ 5950 Cass Ave Detroit MI 48202

MOORE, WILLIAM W, b Pasadena, Calif, Jan 24, 12. FOUNDATION ENGINEERING, EARTHQUAKE ENGINEERING. *Educ:* Calif Inst Technol, BS, 33, MS, 34. *Prof Exp:* Teaching asst, Calif Inst Technol, 33-34; staff mem, US Coast & Geodetic Surv, Los Angeles, 34-35; jr engr, R V Labarre Soil Mech Consult, 35-38; FOUNDING PARTNER, DAMES & MOORE, 38- *Concurrent Pos:* Vpres, Earthquake Eng Res Inst, 65; pres, Appl Technol Coun, 73-74; chmn, Bldg Seismic Safety Coun, 78-81. *Honors & Awards:* Golden Beavers Award for Eng, 78. *Mem:* Nat Acad Eng; Am Soc Civil Engrs; Int Fedn Consult Engrs; fel Inst Civil Engrs UK; NY Acad Sci; Sigma Xi. *Res:* Foundation engineering; soil mechanics; earthquake engineering. *Mailing Add:* Dames & Moore 221 Main St Suite 600 San Francisco CA 94105

MOORE, WILLIS EUGENE, pharmacy, for more information see previous edition

MOORE-EDE, MARTIN C, b London, Eng, Nov 22, 45; m 79; c 2. CIRCADIAN PHYSIOLOGY, AEROSPACE PHYSIOLOGY. *Educ:* Univ London, BSc, 67; Guy's Hosp Med Sch, MB, BS, 70; Harvard Univ, PhD(physiol), 74. *Prof Exp:* Instr, physiol, Guy's Hosp Med Sch, 70; intern, Toronto East Gen Hosp, 70-71; res fel, Peter Bent Brigham Hosp, 71-74; assoc in surg, 74-75, asst prof, 75-81, ASSOC PROF PHYSIOL, HARVARD MED SCH, 81-; DIR, INST CIRCADIAN PHYSIOL, 87- *Concurrent Pos:* Frank Knox fel, Harvard Univ, 71; consult surg, Peter Bent Brigham Hosp, 74- & comt biosphere effects of extremely low frequency radiation, Nat Res Coun, 77; mem, NASA Sci Working Group, primate res, 80- & Nat Acad Sci Subcomt, space biol & med, 82-; chmn, Int Sci Adv Bd, Work & Sleep Schedules, 81- & Int Comn on Circadian Rhythm & Sleep Physiol, 82-; guest ed, Photochem & Photobiol, 81; assoc ed, Am Jour Physiol. *Honors & Awards:* Bowditch lectr, Am Physiol Soc, 85. *Mem:* Am Physiol Soc; Aerospace Med Soc; Asn Psychophysiol Study of Sleep; Soc Neurosci; Endocrine Soc. *Res:* Anatomy and physiology of the circadian timing system; applications of circadian theory to clinical medicine and occupational schedules; human performance in round-the-clock operations. *Mailing Add:* Inst Circadian Physiol 677 Beacon St Boston MA 02215

MOOREFIELD, HERBERT HUGHES, b Baltimore, Md, July 25, 18; m 53; c 1. TOXICOLOGY. *Educ:* Univ Md, BS, 51; Univ Ill, MS, 52, PhD, 53. *Prof Exp:* Res assoc, Univ Ill, 53-54; entomologist, Boyce Thompson Inst, 54-60; dir agr res, Res Sta, Union Carbide Chem Co, 60-64; asst dir res & develop, Union Carbide Corp, NC, 64-67, mgr agr prod mkt develop, Calif, 67-72, technol mgr agr prod, 72-77, corp res fel, Res & Develop Dept, Tech Ctr, 77-80, Res Triangle Park, 80- 86; RETIRED. *Res:* Pesticides; insect physiology-toxicology. *Mailing Add:* 2301 Tamarack Ct Raleigh NC 27612

MOOREHEAD, THOMAS J, b Jersey City, NJ, Nov 1, 47; m 86; c 1. METABOLISM, BIO-ORGANIC CHEMISTRY. *Educ:* St Peter's Col, NJ, BS, 69; Univ Notre Dame, PhD(chem), 75. *Prof Exp:* Res asst bio-org chem, Pa State Univ, University Park, 74-77; RES SCIENTIST, NORWICH EATON PHARMACEUT, NORWICH, 77- *Mem:* Am Chem Soc; Int Soc Study Xenobiotics. *Res:* Metabolism of developmental drugs; pharmaceutical research and development. *Mailing Add:* Norwich Eaton Pharmaceut Inc PO Box 191 Norwich NY 13815

MOOREHEAD, WELLS RUFUS, b Hickory Flat, Miss, July 20, 31; m 62; c 1. CLINICAL CHEMISTRY, BIOCHEMISTRY. *Educ:* Miss State Univ, BS, 53, MS, 60; Univ Tenn, PhD(biochem), 65. *Prof Exp:* Assoc prof, 70-76, PROF CLIN PATH, MED CTR, IND UNIV, INDIANAPOLIS, 76-, ASSOC DIR CLIN CHEM, 70- *Mem:* AAAS; Am Asn Clin Chemists; NY Acad Sci; Sigma Xi. *Res:* Serum enzyme activity levels in different disease conditions; sports medicine. *Mailing Add:* Dept Clin Path Sch Med Ind Univ Purdue Univ 926 W Michigan St Indianapolis IN 46223

MOORES, ELDRIDGE MORTON, b Phoenix, Ariz, Oct 13, 38; m 65; c 3. PETROLOGY, STRUCTURAL GEOLOGY. *Educ:* Calif Inst Technol, BS, 59; Princeton Univ, MA, 61, PhD(geol), 63. *Prof Exp:* Teaching asst geol, Princeton Univ, 62-63, NSF vis res fel, 63-65, res assoc, 65-66; lectr, 66-67, from asst prof to assoc prof, 67-75, chmn dept, 71-72, 73-76 & 88-89, PROF GEOL, UNIV CALIF, DAVIS, 75- *Concurrent Pos:* Ed, Geol, 81-88, GSA Today, 91- *Honors & Awards:* Distinguished Serv Award, Geol Soc Am, 88. *Mem:* Geol Soc Am; Mineral Soc Am; Am Geophys Union. *Res:* Ophiolites and plate tectonics; plate tectonics of deformed belts; history of plate interactions. *Mailing Add:* Dept of Geol Univ of Calif Davis CA 95616

MOORES, RUSSELL R, b St Louis, Mo, Feb 25, 35; m 57; c 7. MEDICINE. *Educ:* Ark State Univ, BS, 55; Univ Ark, MD, 58; Am Bd Internal Med, dipl, 65. *Prof Exp:* From intern to resident med, Strong Mem Hosp, Rochester, NY, 58-60; resident, Barnes Hosp, St Louis, Mo, 60-61; NIH fel hemat, 61-63; staff hematologist, US Naval Hosp, Oakland, Calif, 63-65; from asst prof to assoc prof med, 65-71, assoc dean curric, 72-74, prof humanties & med, 71-84, assoc dean spec progs, 74-84, PROF MED, MED COL GA, 85- *Mem:* Am Soc Hemat; AMA; fel Am Col Physicians; Am Fedn Clin Res; fel Int Soc Hemat. *Res:* Erythropoiesis. *Mailing Add:* Off Humanities Med Col Ga Augusta GA 30912

MOORHEAD, EDWARD DARRELL, b Massillon, Ohio, May 9, 30; m 54; c 2. ELECTROCHEMISTRY, CHEMICAL INSTRUMENTATION & COMPUTER CONTROL. *Educ:* Ohio State Univ, BSc, 54, PhD(chem), 59. *Prof Exp:* Fel chem, Princeton Univ, 59-60; instr chem, Harvard Univ, 60-63; asst prof chem, Rutgers Univ, 63-69; assoc prof, 69-84, PROF CHEM ENG, UNIV KY, 84- *Concurrent Pos:* Consult & auth, D Van Nosrand Pub, 59-62 & Marcel Dekker, 78; prin investr, Res Corp, 64-65, Alfred P Sloan Found, 65-66, Nat Sci Found, IBM, NIH & Am Can Soc. *Mem:* Sigma Xi; Am Chem Soc; NY Acad Sci. *Res:* Various aspects of electrochemical charge transfer processes and electroanalytical measurements, particularly as these topics are related to electrode hydrodynamics, catalyzed charge transfer, ionic equilibria, computer data acqusition and control and numerical modelling. *Mailing Add:* Dept Chem Eng Univ Ky Lexington KY 40506

MOORHEAD, JOHN WILBUR, b Neodesha, Kans, Aug 7, 42; m 70; c 3. CELLULAR IMMUNOLOGY. *Educ:* Univ Kans, BA, 64; Mich State Univ, MS, 66; State Univ NY, Buffalo, PhD(immunol), 70. *Prof Exp:* Fel immunol, 70-73, from instr to assc prof, 73-90, PROF IMMUNOL, UNIV COLO MED SCH, 90- *Concurrent Pos:* Spec fel, Leukemia Soc Am, 73-75; NIH grant, 76-79; NIH res grant, 76-91. *Mem:* Am Asn Immunol. *Res:* Immunoregulation of the immune response. *Mailing Add:* Dept Med Div Immunol B164 Univ Colo Health Med Ctr 4200 E Ninth Ave Denver CO 80220

MOORHEAD, PAUL SIDNEY, b El Dorado, Ark, Apr 18, 24; m 49; c 3. CYTOGENETICS. *Educ:* Univ NC, AB, 48, MA, 50; Univ Tex, PhD(zool), 54. *Prof Exp:* Res assoc cytol, Med Br, Univ Tex, 54-56; res assoc, Sch Med, Univ Pittsburgh, 56-58; assoc mem, Wistar Inst Anat & Biol, 59-69; assoc prof, 69-85, EMER PROF HUMAN GENETICS & PEDIAT, SCH MED, UNIV PA, 85- *Concurrent Pos:* Damon Runyon Mem fel, 55; NIH Career Develop Awards, 62-67 & 68-69. *Mem:* Sigma Xi; Am Soc Human Genetics; Environ Mutagen Soc; NY Acad Sci; Tissue Cult Asn (pres, 80-82). *Res:* Mammalian chromosomes. *Mailing Add:* 55 E Greenwood Ave Lansdowne PA 19050

MOORHEAD, PHILIP DARWIN, b Pratt, Kans, Nov 21, 33; m 56; c 3. VETERINARY PATHOLOGY. *Educ:* Kans State Univ, BS & DVM, 57; Purdue Univ, MS, 64, PhD(vet path), 66. *Prof Exp:* Vet, gen pract, 57-62; instr vet path, Col Vet Med, Purdue Univ, 63-66; ASSOC PROF VET PATH, OHIO AGR RES & DEVELOP CTR, 66- *Concurrent Pos:* Secy, NC-65 Comt Poultry Respiratory Dis, Coop Res Serv, USDA, 69- *Mem:* Am Vet Med Asn; Am Asn Avian Path; Conf Res Workers Animal Dis. *Res:* Sporadic toxocologic environmental infections and non-specific problems in chickens and turkeys. *Mailing Add:* 816 Quinby Ave Wooster OH 44691

MOORHEAD, WILLIAM DEAN, b Youngstown, Ohio, Nov 2, 36; m 65; c 1. THEORETICAL PHYSICS. *Educ:* Ohio Wesleyan Univ, BA, 58; Ohio State Univ, PhD(physics), 68. *Prof Exp:* Asst prof physics & astron, Youngstown State Univ, 68-72, assoc prof, 68-78; staff res physicist, Shell Develop Co, 78-90; SR CONSULT SCIENTIST, THINKING MACHINES CORP, 90- *Mem:* Am Phys Soc. *Res:* Theories of angular momentum, magnetism, and the many body problem; statistical physics. *Mailing Add:* 3816 Northwestern Houston TX 77005

MOORHOUSE, DOUGLAS C, SCIENCE ADMINISTRATION. *Educ:* Univ Calif, Berkeley, BS, 50. *Prof Exp:* Res & resident engr, Div Hwys, State Calif, 50-54; chief hwy & airport engr, Woodward-Clyde & Assocs, Oakland, 54-59, br mgr, San Diego, 59-62, pres, chief exec officer & prin, Woodward-Moorhouse & Assocs, Inc, NY & NJ, 62-73, pres, Woodward- Clyde Consults, 73-76, pres & chief exec officer, 76-87, CHMN & CHIEF EXEC OFFICER, WOODWARD-CLYDE GROUP, INC, 87- *Concurrent Pos:* Mem adv bd of the built environ, Comn Eng & Tech Systs, Nat Res Coun, 80; mem eng adv bd, Univ Calif, Berkeley; pres bd dirs, Hazardous Waste Action Coalition, 88; mem planning cabinet, Am Consult Engrs Coun, 89; chmn, Task Comt on Int Competitiveness, Am Soc Civil Engrs, 89. *Honors & Awards:* Wesley W Horner Award, Am Soc Civil Engrs, 69. *Mem:* Nat Acad Eng; Am Soc Civil Engrs. *Res:* Civil and geotechnical engineering. *Mailing Add:* Woodward-Clyde Group 30th Floor 600 Montgomery St San Francisco CA 94111

MOORHOUSE, JOHN A, b Winnipeg, Man, Oct 4, 26; m 80; c 1. MEDICINE, ENDOCRINOLOGY. *Educ:* Univ Man, MD, 50, MSc, 55; FRCP(C), FACP. *Prof Exp:* Fel path, Winnipeg Gen Hosp, Man, 50-51, asst resident med, 51-53, chief resident, 53-54; res fel metab, Res & Educ Hosp & Presby Hosp, Univ Ill, 54-56; res fel & clin asst endocrinol & metab, Univ Mich Hosp, 56-58; res assoc, Clin Invest Unit, Winnipeg Gen Hosp, 58-61; from asst prof to assoc prof, 60-72, PROF PHYSIOL, UNIV MAN, 72-, ASSOC PROF MED, 70- *Concurrent Pos:* Dir endocrine & metab lab, Health Sci Ctr, 61-77; asst physician, Health Sci Ctr, 63- *Mem:* Fel Am Col Physicians; Endocrine Soc; Am Diabetes Asn; Can Soc Clin Invest. *Res:* Carbohydrate and fat metabolism, particularly as related to diabetes mellitus. *Mailing Add:* Health Sci Ctr NA511-700 McDermot Ave Winnipeg MB R3E 0T2 Can

MOORING, FRANCIS PAUL, b Stokes, NC, Feb 6, 21; m 48; c 3. NUCLEAR PHYSICS. *Educ:* Duke Univ, BA, 44; Univ Wis, PhD(physics), 51. *Prof Exp:* Instr physics, Duke Univ, 44-46; from assoc physicist to physicist, Argonne Nat Lab, 51-83; adj prof, 66-83, ASSOC ED, APPL PHYSICS LETT, ST LOUIS UNIV, 83. *Concurrent Pos:* Fulbright res fel, Univ Helsinki, Finland, 62-63; ed, Am Inst Physics, 83- *Mem:* Am Phys Soc; AAAS. *Res:* Particle accelerators; nuclear physics-nuclear cross sections from 0.1-3.0 MeV. *Mailing Add:* 295 Abbotsford Ct Glen Ellyn IL 60137

MOORING, JOHN STUART, b Long Beach, Calif, July 14, 26; m 55; c 4. BOTANY. *Educ:* Univ Calif, Santa Barbara, AB, 50; Univ Calif, Los Angeles, PhD, 56. *Prof Exp:* Instr biol, Occidental Col, 55; from instr to asst prof bot, Wash State Univ, 55-61; vis asst prof, Univ Calif, Riverside, 61-62, lectr, 62-63; from asst prof to assoc prof, 63-69, PROF BOT, UNIV SANTA CLARA, 69- *Concurrent Pos:* NSF res grants, 60-63, 65-66 & 66-69. *Mem:* Bot Soc Am; Soc Study Evolution. *Res:* Biosystematics. *Mailing Add:* Dept of Biol Univ of Santa Clara Santa Clara CA 95053

MOORING, PAUL K, b Houston, Tex, Sept 16, 23; m 62; c 2. PEDIATRIC CARDIOLOGY, PEDIATRICS. *Educ:* NY Univ, BS, 55; Columbia Univ, MD, 57. *Prof Exp:* Intern med, Bellevue Hosp, NY, 57-58; res pediat, Presby Med Ctr-Babies Hosp, 58-60; NIH vis fel pediat cardiol, Presby Med Ctr, NY, 60-62; from asst prof to assoc prof pediat, 62-71, dir pediat cardiol & cardiovasc labs, 64-71, PROF PEDIAT & MED, SCH MED, UNIV NEBR, OMAHA, 71- *Mem:* Am Acad Pediat; AMA; fel Am Col Cardiol; Am Heart Asn. *Res:* Radiotelemetry electrocardiology at maximum stress; educational television as a technique for postgraduate physician education; heart disease in the retarded; teenage heart disease. *Mailing Add:* Dept Pediat Univ Nebr 42nd & Dewey Omaha NE 68105

MOORJANI, KISHIN, b Apr 9, 35; US citizen; m 62. HIGH TEMPERATURE SUPERCONDUCTIVITY, DISORDERED MATERIALS. *Educ:* Univ Delhi, BSc, 55, MSc, 57; Cath Univ Am, PhD(physics), 64. *Prof Exp:* Res assoc, 64-65, asst prof physics, Cath Univ Am, 65-66; sr staff physicist, 67-74, prog supvr, 84-90, PRIN STAFF PHYSICIST, APPL PHYSICS LAB, JOHNS HOPKINS UNIV, 74-, GROUP SUPVR, 91- *Concurrent Pos:* Vis prof physics, Univ Grenoble, 74-75 & 82; Parsons prof physics, 77-78, chmn appl physics, Johns Hopkins Univ, 86-; consult physicist, Melpar Inc, 61-67 & Goddard Space Flight Ctr, NASA, 64-65; vis scientist, Nat Ctr Sci Res, Paris & Grenoble, 74-75, Free Univ, WBerlin, 77, Univ Delhi, 81. *Mem:* Am Phys Soc. *Res:* High temperature superconductivity; thin film phenomena; intense photon beam processing of materials; magnetic glasses; amorphous semiconductors; physics of disorder; spin glasses; magnetic resonance; non-linear effects; nanostructures; quasicrystals; author of 105 papers. *Mailing Add:* Appl Physics Lab Johns Hopkins Univ Johns Hopkins Rd Laurel MD 20723-6099

MOOR-JANKOWSKI, J, b Czestochowa, Poland, Feb 5, 24; US citizen. IMMUNOGENETICS, PRIMATOLOGY. *Educ:* Univ Bern, MD, 54. *Prof Exp:* Prin investr, Population Genetics, Swiss Nat Fund Sci Res, 53-57; res asst, Inst Human Genetics, Univ Geneva, 55-57; res assoc, Inst Study Human Variation, Columbia Univ, 57-58; res fel, Blood Group Res Unit, Lister Inst Prev Med, London & Dept Path, Cambridge Univ, 59-62; vis scientist, Human Genetics Br, NIH, 62-64; chief div exp immunogenetics oncol, Yerkes Primate Ctr, Emory Univ, 64-65; res assoc prof forensic med, 65-69, RES PROF FORENSIC MED, MED SCH, NY UNIV, 69-, DIR, LAB EXP MED & SURG PRIMATES, 69-; DIR, WHO COLLAB CTR HEMAT PRIMATE ANIMALS, 75- *Concurrent Pos:* Ed-in-chief, Primates in Med & J Med Primatol. *Honors & Awards:* Trumpeldor Medal, Life Sci Div, Prime Minister's Off, Israel, 71; Medal, Inst Exp Path & Ther, USSR Acad Med Sci, 78; G Budé Medal, Col France, Paris, 79; Vis Prof, Col de France, 79; Consult, USSA Acad Scis, 79. *Mem:* AAAS; Am Soc Human Genetics; Am Soc Immunol; Tissue Cult Asn; Soc Cryobiol; Brit Soc Immunol; Swiss Genetic Soc; Swiss Soc Nat Sci; Int Primatology Soc; Transplantation Soc; Int Soc Biol Standardization. *Res:* Immunological specificity of tissue cells in culture; primate immunogenetics; experimental medicine in nonhuman primates. *Mailing Add:* LEMSIP NY Univ Med Ctr 550 First Ave New York NY 10016

MOORMAN, GARY WILLIAM, b Albany, NY, Jan 30, 49. COOPERATIVE EXTENSION SERVICE. *Educ:* Univ Maine, BS, 71; Univ Vt, MS, 74; NC State Univ, PhD(plant path), 78. *Prof Exp:* Electron microscopist, W Alton Jones Cell Sci Ctr, 73-75; asst prof dis floricult crops, Suburban Exp Sta, 78-82; ASST PROF DIS ORNAMENTAL CROPS, PA STATE UNIV, 83- *Mem:* Am Phytopath Soc; Mycological Soc Am; Brit Mycol Soc. *Res:* The effects of current plant growing practices on root rots of ornamental crops caused by soil borne organisms; detection and management of fungicide resistance in Botrytis and Pythium in greenhouse. *Mailing Add:* Dept Path 211 Buckhout Lab Univ Park PA 16802

MOORMAN, ROBERT BRUCE, b Chadron, Nebr, Oct 21, 16; m 43; c 3. FISHERIES. *Educ:* Iowa State Col, BS, 39, MS, 42, PhD(fisheries mgt), 53. *Prof Exp:* Asst, Iowa State Univ, 39-42, exten wildlife conservationist, 48-53; asst prof zool, Kans State Univ, 53-56; from asst prof to prof zool & entom, Iowa State Univ, 56-86; RETIRED. *Concurrent Pos:* With Iowa Conserv Comn, 46-48; exten wildlife conservationist, 56- *Mem:* Am Fisheries Soc; Wildlife Soc. *Res:* Fisheries management in farm ponds; bobwhite quail. *Mailing Add:* 1223 Ninth St Ames IA 50011

MOORREES, COENRAAD FRANS AUGUST, b Hague, Neth, Oct 23, 16; nat US; m 39; c 2. ORTHODONTICS. *Educ:* State Univ Utrecht, dipl, 39; Univ Pa, DDS, 41. *Hon Degrees:* AM, Harvard Univ, 59; Dr Med, Univ Utrecht, 71. *Prof Exp:* Intern, Eastman Dent Dispensary, 41; sr fel orthod, 47-48, actg chief dept orthod, 48-56, assoc prof, 59-64, PROF ORTHOD, FORSYTH DENT CTR, HARVARD UNIV, 64-, CHIEF DEPT, 56- *Concurrent Pos:* Mem sci exped, Aleutian Islands, 48; res fel ondontol, Peabody Mus, 51-66, res assoc odontol, 66-68, hon assoc, 68-; mem, Nat Adv Dent Res Coun, Nat Inst Dent Res, 83-86; hon acad mem, Charles H Tweed Found Orthod Res. *Honors & Awards:* Northcroft Mem lectr, Brit Soc Study Orthod, 76; Albert H Ketcham Mem Award, Am Asn Orthodontists, 77; Sheldon Friel Mem lectr, Europ Orthod Soc, 78; Medal of Paris, Europ Orthod Soc, 80. *Mem:* Am Asn Orthod; Am Asn Phys Anthrop; Int Asn Dent Res; Neth Soc Study Orthod; Neth Dent Soc. *Res:* Child growth and development, especially face and dentition; dental anthropology; evolutionary and racial aspects of the dentition. *Mailing Add:* Four Peacock Farm Rd Lexington MA 02173

MOOS, ANTHONY MANUEL, physical chemistry; deceased, see previous edition for last biography

MOOS, CARL, b New York, NY, Mar 3, 30; m 51; c 3. BIOCHEMISTRY, CELL PHYSIOLOGY. *Educ:* Mass Inst Technol, SB, 50; Columbia Univ, PhD(biophys), 57. *Prof Exp:* Res assoc chem, Northwestern Univ, 55-57; res assoc physiol, Col Med, Univ Ill, 57-58, instr, 58-59; assoc biophys, Sch Med, State Univ NY Buffalo, 59-61, asst prof, 61-66; assoc prof biol sci, 66-69, ASSOC PROF BIOCHEM, STATE UNIV NY STONY BROOK, 69- *Concurrent Pos:* Vis scientist, King's Col, Univ London, 70-71 & 77-78, VA Med Ctr, Albany, NY, 85. *Mem:* Fel AAAS; Biophys Soc; Soc Gen Physiologists; Am Soc Biol Chemists; Am Soc Cell Biol. *Res:* Muscle contraction; molecular mechanisms of contraction and relaxation; role of muscle proteins. *Mailing Add:* Dept of Biochem State Univ of NY Stony Brook NY 11794-5215

MOOS, GILBERT ELLSWORTH, cancer, teaching; deceased, see previous edition for last biography

MOOS, HENRY WARREN, b New York, NY, Mar 26, 36; m 57; c 4. PHYSICS. *Educ:* Brown Univ, ScB, 57; Univ Mich, MA, 59, PhD(physics), 62. *Prof Exp:* Res assoc physics, Stanford Univ, 61-63, actg asst prof, 63-64; from asst prof to assoc prof, 64-71, PROF PHYSICS, JOHNS HOPKINS UNIV, 71-; DIR, CTR ASTRO PHYSICAL SCI, Johns Hopkins Univ, 88- *Concurrent Pos:* Sloan Found fel, 65-67; vis fel, Joint Inst Lab Astrophys & Lab Atmospheric & Space Physics, Univ Colo, 72-73 & 80-81; mem, NASA & Dept Energy Comts; consult indust. *Mem:* Fel Am Phys Soc; Am Astron Soc. *Res:* Astrophysics; ultraviolet astronomy; ultraviolet spectroscopy of fusion plasmas. *Mailing Add:* Dept Physics & Astron Johns Hopkins Univ Baltimore MD 21218

MOOS, WALTER HAMILTON, b Canton, NY, July 25, 54; m 79; c 1. MEDICINAL CHEMISTRY, COMPUTER ASSISTED DRUG DESIGN. *Educ:* Harvard Univ, AB, 76; Univ Calif, Berkeley, PhD(chem), 82. *Prof Exp:* Scientist, 82-84, sr scientist, 84, from res assoc to sr res assoc, 84-87, sect dir, 87-89, dir, 89, sr dir, 90, VPRES, NEUROSCI & BIOL CHEM, PARKE-DAVIS, PHARMACEUT RES DIV, WARNER-LAMBERT CO, 90-; ADJ ASSOC PROF, MED CHEM, UNIV MICH, ANN ARBOR, 90- *Mem:* Am Chem Soc; Sigma Xi; AAAS. *Res:* Medicinal organic chemistry directed toward novel drug discovery and development, particularly involving neurosciences and computer-assisted drug design; adenosine receptors; cardiotonics; cognition activators; hypnotics; quantitative structure activity relationships; nucleosides; nitrogen heterocycles; morphine analogues; molecular modeling. *Mailing Add:* 3635 Waldenwood Dr Ann Arbor MI 48105-3042

MOOSE, LOUIS F, engineering, for more information see previous edition

MOOSMAN, DARVAN ALBERT, b Toledo, Ohio. ANATOMY, SURGERY. *Educ:* Bowling Green Univ, AB, 34; Univ Mich, MD, 37, MS, 47; Am Bd Surg, dipl, 52. *Prof Exp:* Intern & asst resident surg, Univ Mich Hosp, Ann Arbor, 37-40; resident, St Joseph Mercy Hosp, Pontiac, 40-42; clin instr surg, Hosp, 50-52, prof, 52-82, EMER PROF ANAT & GEN SURG, MED CTR, UNIV MICH, ANN ARBOR, 82- *Concurrent Pos:* Vis prof clin anat, Med Sch, Univ Hawaii, 70 & Med Ctr, Univ Ala, Birmingham, 73. *Mem:* Am Soc Abdominal Surg. *Res:* Biliary duct system; blood supply liver; thyroid gland. *Mailing Add:* Dept Anat Univ Mich Med Ctr 2610 Devonshire Rd Ann Arbor MI 48104

MOOSSY, JOHN, b Shreveport, La, Aug 24, 25; m 51; c 2. NEUROPATHOLOGY, NEUROLOGY. *Educ:* Tulane Univ, MD, 50. *Prof Exp:* USPHS fel neuropath, Columbia Univ, 53-54; lectr, Sch Med, Tulane Univ, 54-60; from asst prof to prof neurol & path, Sch Med, La State Univ, 57-65; prof path & neurol, Sch Med, Univ Pittsburgh, 65-67, prof, Bowman Gray Sch Med, Winston Salem, NC, 67-72; PROF PATH & NEUROL, SCH MED, UNIV PITTSBURGH, 72- *Concurrent Pos:* Ed-in-chief, J Neuropath Exp Neurol, 81- *Mem:* Am Asn Neuropath (pres, 74-75); Am Neurol Asn (vpres, 77-78); Am Acad Neurol (secy-treas, 63-65); Coun Biol Ed; Asn Res Nervous & Mental Dis; Am Asn Homeop Med. *Res:* Neuropathology and neurology, especially vascular disease in the nervous system. *Mailing Add:* Div Neuropath Sch Med Univ Pittsburgh Pittsburgh PA 15261

MOO-YOUNG, MURRAY, b Lucea, Jamaica; Can citizen. BIOCHEMICAL ENGINEERING, INDUSTRIAL BIOTECHNOLOGY. *Educ:* Univ London, BSc, 54, PhD(biochem eng), 60; Univ Toronto, MASc, 55. *Prof Exp:* Sci officer, Ministry Trade & Indust, UK, 57-59; res fel chem eng, Univ Edinburgh, UK, 61; lectr, Univ Toronto, 62-63; asst prof, Univ Western Ont, 63-66; assoc prof, 66-71, PROF CHEM ENG, UNIV WATERLOO, 71-, DIR, INST BIOTECHNOL RES, 83- *Concurrent Pos:* Vis prof, Imperial Col, London, 73, Univ Calif, Berkeley, 76 & 81, Mass Inst Technol, 81 & Univ Karlsruhe, 81; ed, Biotechnol Advances, 82-; ed-in-chief, Comprehensive Biotechnol, 85. *Honors & Awards:* Erco Award, Can Soc Chem Engrs, 73; Eng Medal, Asn Prof Engrs, 83. *Mem:* Can Soc Chem Eng; Am Chem Soc; Am Inst Chem Engrs; Soc Indust Microbiol. *Res:* Development and characterization of multiphase contacting devices for use as heterogeneous chemical or biochemical reactors; raw materials pretreatment and product recovery techniques in bioreactor operations; computer-aided design and technoeconomic sensitivity analyses of bioprocesses; industrial biotechnology and biochemical engineering. *Mailing Add:* 63-50 Blue Springs Dr Waterloo ON N2J 4M4 Can

MOOZ, ELIZABETH DODD, b Middletown, Conn, Nov 22, 39; m 64; c 2. BIOCHEMISTRY. *Educ:* Hollins Col, BA, 61; Tufts Univ, PhD(biochem), 67. *Prof Exp:* Instr res surg, Univ Pa Grad Hosp, 67-68; fel biochem, Univ Del, 68-71, asst prof health sci, 71-73; res assoc chem, Bowdoin Col, 73-76; asst prof biochem, 77-78, res fel, Med Col VA, 78-79; RES SCIENTIST, PHILIP MORRIS, USA, 79- *Concurrent Pos:* Mem, New Eng Bd Higher Educ, 74-75; gov comt to select Outstanding Scientist, VA, 85. *Mem:* AAAS; Sigma Xi; Am Chem Soc; Nat Women Chemist Comt. *Res:* Enzyme purification, substrate specificity, immunology; biochemical modification of tobacco; tobacco process development; new product development. *Mailing Add:* Moody Found 704 Moody Nat Bank Bldg Galveston TX 77550-1994

MOOZ, WILLIAM ERNST, b Staten Island, NY, Feb 28, 29. RESOURCE ANALYSIS, METALLURGICAL ENGINEERING. *Educ:* Mass Inst Technol, SB, 50. *Prof Exp:* Res engr/develop engr/asst to plant mgr, Titanium Metals Corp Am, 50-60; econ analyst, US Borax & Chem Co, 60-62; exec vpres, G B Smith Chem Works Inc, 62-64; sr staff mem resource analyst, Rand Corp, 64-89; PRES, MET-L-CHEK CO, 64- *Concurrent Pos:* Chmn bd dir, Met-L-Chek Corp, 64-, NDT Europa BV, 66-68; mem, Calif Gov Energy Seminar, 73; Southern Gov Energy Comt, 74. *Honors & Awards:* Earl P L Apfelbaum Mem Award, 85. *Mem:* Am Inst Mining, Metall & Petrol Engrs; Am Inst Non-Destructive Testing. *Res:* Capital costs of light water reactor and breeder reactor power plants; sectoral demand for electrical energy; transportation model energy intensiveness; peacetime military aircraft attrition; economics of regulating chlorofluorocarbons. *Mailing Add:* Box 1714 Santa Monica CA 90406

MOPPER, KENNETH, b Philadelphia, Pa, Jan 4, 47; m 77; c 2. MARINE ORGANIC CHEMISTRY. *Educ:* Queens Col, City Univ New York, BA, 68; Mass Inst Technol, MS, 71, Woods Hole Oceanog Inst, PhD(oceanog), 73. *Prof Exp:* Fel, Dept Eng Chem, Univ Gothenburg, Swed, 73-74; staff researcher, Geol Inst, Univ Hamburg, Ger, 75-77; guest researcher, Dept Anal & Marine Chem, Univ Gothenburg, Swed, 77-80; asst prof marine chem, marine studies, Univ Del, 80-; prof, Univ Miami; PROF, WASHINGTON STATE UNIV. *Mem:* AAAS; Sigma Xi. *Res:* Cycling of organic compounds in marine environments with emphasis on the formation of humics; trace analysis of carbonyls and carbohydrates; amino acids, carboxylic acids; phenols, thiols and flavins. *Mailing Add:* Chem Dept Wash State Univ Pullman WA 99164-4630

MOPPETT, CHARLES EDWARD, b London, Eng, Sept 23, 41; m 65; c 2. ORGANIC CHEMISTRY. *Educ:* Univ London, BSc & ARCS, 63, PhD(org chem) & dipl, Imp Col, 66. *Prof Exp:* Res grant org chem, Res Inst Med & Chem, Cambridge, Mass, 66-67 & Harvard Univ, 67-69; res chemist, Dow Chem USA, 69-73; DIR, PFIZER, USA, 73- *Mem:* The Chem Soc; Am Chem Soc. *Res:* Organic chemistry, especially structure determination of natural products and their synthesis; general synthetic chemistry; new synthetic methods; discovery isolation; physico-chemical characterization; chemistry and biology of microbial metabolites. *Mailing Add:* 148 Cedar Rd Groton CT 06355

MOPSIK, FREDERICK ISRAEL, b New York, NY, May 20, 38; m 65; c 2. PHYSICAL CHEMISTRY, DIELECTIVES MEASUREMENT. *Educ:* Queen's Col, NY, BS, 59; Brown Univ, PhD(chem), 64. *Prof Exp:* PHYS CHEMIST, POLYMERS DIV, NAT BUR STANDARDS, 63- *Mem:* Am Phys Soc. *Res:* Dielectrics research; equations of state; polymer research. *Mailing Add:* Polymers Div Voc Inst Standards & Technol Gaithersburg MD 20899

MORA, EMILIO CHAVEZ, b Valedon, NMex, Aug 14, 28; m 52; c 4. BACTERIOLOGY. *Educ:* Univ NMex, BS, 51; NMex State Univ, MS, 54; Kans State Univ, PhD(bact), 59. *Prof Exp:* Assoc prof, 58-70, PROF POULTRY SCI, AUBURN UNIV, 70- *Mem:* Am Soc Microbiol. *Res:* Chemistry of virus-host cell relationships; virus vaccines; electron microscopy. *Mailing Add:* Dept of Poultry Sci Auburn Univ Auburn AL 36849

MORA, JOSE O, HEALTH CARE, COMMUNITY RESEARCH. *Educ:* Harvard Univ, MS, 81. *Prof Exp:* Sr med nutritionist, Off Int Health, Dept Health & Human Serv, 84-90; SR ASSOC, INT SCI & TECHNOL INST, 90- *Mailing Add:* Intl Sci & Technol Inst 1129 20th St NW Suite 206 Washington DC 20036

MORA, PETER T, Szolnok, Hungary, July 18, 24; US citizen; m 51. MOLECULAR BIOLOGY. *Educ:* Univ Budapest, Hungary, PhD(chem), 48. *Prof Exp:* Fel org chem & res assoc, dept chem, Princeton Univ, 48-50; res chemist polymer chem, E I DuPont de Nemours & Co, Inc, 50-54; vis scientist biochem, Inst Cancer Res, Philadelphia, 54-55; Brit-Am exchange fel cancer res, dept biochem, Oxford Univ, Eng, 55-56; chemist, sr investr, supvry chemist & chief molecular biol sect, 56-90, EMER SCIENTIST, NAT CANCER INST, NIH, 90- *Concurrent Pos:* Vis scientist virol, Virus Res Unit, Cambridge, Eng, 62-63; vis scientist biochem, Oxford Univ, Eng, 78-79. *Mem:* Am Chem Soc; Am Soc Biol Chemists. *Res:* Molecular controls of cell division, differentiation and transformation to malignancy; tumor specific transplantation antigens after transformation of cells by tumorigenic viruses. *Mailing Add:* Nat Cancer Inst NIH Bldg 36 Rm 1D28 Bethesda MD 20892

MORABITO, BRUNO P, b Monticella, Italy, Feb 10, 22; US citizen; m 45; c 1. HEATING & AIR CONDITIONING, DISTRICT HEATING & COOLING PLANTS. *Educ:* Syracuse Univ, BCE, 45. *Prof Exp:* Eng & mgt, Carrier MSD United Technologies, 45-81, mgr mkt, 76-81; group vpres, 81-84, PRES, BPM PLANNING & CONSULT, 86- *Concurrent Pos:* Bd dirs, Am Soc Heating, Refrig & Air Conditioning Engrs, 70-73, officer, 73-77. *Honors & Awards:* Wolverine Diamond Key Award, Am Soc Heating, Refrig & Air Conditioning Engrs, 61 & Distinguished Serv Award, 75; Silver Knight of Mgt Award, Nat Mgt Asn, 78. *Mem:* Fel Am Soc Heating, Refrig & Air Conditioning Engrs (pres, 77-78). *Res:* Heating; ventilating; air conditioning; marketing; management. *Mailing Add:* 302 Saltmakers Rd Liverpool NY 13088

MORABITO, JOSEPH MICHAEL, b Asbury Park, NJ, Feb 26, 41; m 68; c 4. ELECTRONIC SPECTROSCOPY. *Educ:* Univ Notre Dame, BSMet, 63; Univ Pa, PhD(mat sci), 67. *Prof Exp:* Fel low-energy electron diffraction, Univ Calif, Berkeley, 68-69; vis scientist thin film res & develop, Phillips Res, Neth, 69-70; mem tech staff, 70-74, SUPVR THIN FILM RES & DEVELOP, BELL LABS, AM TEL & TEL CO, 75- *Concurrent Pos:* Consult, NSF, Dept Energy, Elec Power Res Inst, 74- *Mem:* Am Vacuum Soc. *Res:* Thin film research and development using auger electron spectroscopy and secondary ion mass spectrometry; bonding techniques; interdiffusion studies; solar cells. *Mailing Add:* 1236 Moffit Ave Bethleham PA 18018

MORACK, JOHN LUDWIG, b Schenectady, NY. PHYSICS. *Educ:* Union Col, NY, BS, 61; Ore State Univ, PhD(physics), 67. *Prof Exp:* From asst prof to assoc prof, 67-78, PROF PHYSICS, UNIV ALASKA, 78- *Mem:* Am Phys Soc; Am Asn Physics Teachers. *Res:* Acoustics; geophysics. *Mailing Add:* Dept Physics Univ Alaska Fairbanks AK 99775

MORAFF, HOWARD, b New York, NY. COMPUTER SCIENCE. *Educ:* Columbia Col, BA; Columbia Univ, BS, MS, Cornell Univ, PhD(neurophysiol). *Prof Exp:* Dir vet med comp resources, Cornell Univ, 67-82; dir comp resources, Merck Sharp & Dohme Res Labs, 82-84; prog dir automation, 84-87, eng res ctr, 87-90, PROG DIR ROBOTICS, NSF, 90- *Mem:* Sr mem Inst Elec & Electronic Engrs; NY Acad Sci; Sigma Xi; Am Asn Artificial Intel. *Res:* Robotic manipulation, coordination and control; computer vision; machine intelligence; neural networks; granted one patent. *Mailing Add:* NSF Rm 310 Washington DC 20550

MORAHAN, PAGE SMITH, b Newport News, Va, Jan 7, 40; m. VIROLOGY, IMMUNOLOGY. *Educ:* Agnes Scott Col, BA, 61; Hunter Col, MA, 64; Marquette Univ, PhD(microbiol), 69. *Prof Exp:* Res technician, Rockefeller Univ, 61-65; NIH trainee, Med Col Va, 69-70, A D Williams Jr acad fel, 70, from asst prof to prof microbiol, 71-82; PROF & CHMN MICROBIOL, MED COL PA, 82- *Concurrent Pos:* NIH res career development award, 74-79; mem, Nat Cancer Inst manpower rev comt, 77-81; vis researcher, Harvard Med Sch, 78-79; mem cancer ctrs rev comt, Nat Cancer Inst, 86-90; pres-elect, Asn Med Microbiol, Dept Chair, 88- *Mem:* AAAS; Am Asn Immunologists; Reticuloendothelial Soc; Am Asn Cancer Res; Sigma Xi; Soc Toxicol; Am Soc Microbiol; Am Soc Virol; Antiviral Soc. *Res:* Host resistance to viruses and tumors; interferon inducers and antitumor drugs; macrophages; immunomodulators. *Mailing Add:* Dept Microbiol Med Col Pa 3300 Henpry Ave Philadelphia PA 19129

MORAIS, REJEAN, b Montreal, Que, Oct 26, 38; m 63; c 2. BIOCHEMISTRY. *Educ:* Univ Montreal, BSc, 60, MSc, 62, PhD(biochem), 65. *Prof Exp:* Res fel med, Harvard Med Sch, 65-67; from asst prof to assoc prof, 67-79, PROF BIOCHEM, UNIV MONTREAL, 79-; RES ASST CANCER BIOCHEM, INST CANCER MONTREAL, NOTRE-DAME HOSP, 67- *Mem:* Am Asn Cancer Res; AAAS. *Res:* Role of mitochondrial genes on vertebrate cell phenotype. *Mailing Add:* Dept Biochem Univ Montreal CP 6228 Succ A Montreal PQ H3C 3T4 Can

MORALES, DANIEL RICHARD, b San Francisco, Calif, Jan 1, 29; m 57. CLINICAL BIOCHEMISTRY. *Educ:* Univ San Francisco, BS, 55; Calif Inst Technol, MS, 56; Univ Calif, PhD(biochem), 62. *Prof Exp:* Fel pharmacol, Yale Univ, 62-65, res assoc, 65-66; asst prof, Sch Med, Univ Kans, 66-68; res clin chemist, 68-72, CHIEF CLIN CHEM LAB, CALIF DEPT HEALTH SERV, 72- *Concurrent Pos:* Mem var ad comt. *Res:* Forensic alcohol analysis; drug analysis; nutritional surveys; laboratory tests for pesticide toxicity; genetic disease testing; newborn screening tests; prenatal diagnostic testing. *Mailing Add:* 7141 Mound Ave El Cerrito CA 94530

MORALES, GEORGE JOHN, b Havana, Cuba, Nov 24, 45; US citizen. PLASMA PHYSICS. *Educ:* Fairleigh-Dickinson Univ, BS, 67; Univ Calif, San Diego, MS, 71, PhD(physics), 73. *Prof Exp:* Asst res physicist & adj asst prof physics, 73-78, assoc res physicist & adj assoc prof physics, 78-82, assoc prof, 82-83, PROF PHYSICS, UNIV CALIF, LOS ANGELES, 83- *Mem:* Fel Am Phys Soc. *Res:* Nonlinear phenomena in plasmas theory experiment and computer simulation; heating of fusion plasmas; ionosphere modification. *Mailing Add:* Dept Physics 174 Knudsen Hall Univ Calif 405 Hilgard Ave Los Angeles CA 90024

MORALES, MANUEL FRANK, b San Pedro, Honduras, July 23, 19; US citizen. CONTRACTILITY. *Educ:* Univ Calif, AB, 39, PhD(physiol), 42; Harvard Univ, AM, 41. *Prof Exp:* Teaching fel physiol, Univ Calif, 41-42; instr physics, Western Reserve Univ, 42-43; instr math biophys & asst prof physiol, Univ Chicago, 46-48; head phys biochem div, Naval Med Res Inst, 48-57; prof biochem & chmn dept, Dartmouth Med Sch, 57-60; prof biochem,

60, prof biophys, Dept Biochem & Biophys & Cardiovasc Res Inst, Sch Med, Univ Calif, San Francisco, 60-90; ADJ PROF PHYSIOL & BIOPHYS, UNIV OF PACIFIC, SAN FRANCISCO, 90- *Concurrent Pos:* Mem panel physiol, Comt Undersea Warfare, Nat Res Coun, 49; mem US cultural mission, Honduras, 51 & physiol study sect, USPHS, 52-; mem, Nat Adv Res Resources Coun, 67-, mem sr fel selection comt; mem, molecular biol panel, NSF, bd sci coun, Nat Inst Aging; career investr, Am Heart Asn, 60-89; ed, Annual Rev Biophys & Bioeng; Fogarty Scholar-in-Residence, NIH, 90- *Honors & Awards:* Flemming Award, US Fed Serv; Merit Award, Nat Heart Lung Blood Inst, 89. *Mem:* Nat Acad Sci; Am Soc Biol Chem; Soc Gen Physiol; Biophys Soc (pres, 68). *Res:* Biochemical thermodynamics and kinetics; physical chemistry of muscle contraction. *Mailing Add:* Univ Pac 2155 Webster St San Francisco CA 94115

MORALES, RAUL, b San Pedro Sula, Honduras, Sept 27, 35; US citizen; m 59; c 4. ANALYTICAL CHEMISTRY. *Educ:* Univ Southwestern La, BS, 61; La State Univ, PhD(anal chem), 66. *Prof Exp:* Asst prof chem, Nicholls State Col, 65-66; res chemist, Indust & Biochem Dept, E I du Pont de Nemours & Co, 66-75; MEM STAFF, LOS ALAMOS NAT LAB, 75- *Mem:* Am Chem Soc. *Res:* Precipitations from homogeneous solution; titrations in non-aqueous solvents; metabolism and analytical chemistry of agricultural chemicals; sampling and trace analytical methods for carcinogenic substances in the occupational environment; analytical chemistry of actinide coordination compounds. *Mailing Add:* 116 La Senda Rd Los Alamos NM 87544-3820

MORAN, DANIEL AUSTIN, b Chicago, Ill, Feb 17, 36; m 68; c 2. TOPOLOGY. *Educ:* St Mary's Univ, Tex, BS, 57; Univ Ill, MS, 58, PhD(math), 62. *Prof Exp:* Res instr math, Univ Chicago, 62-64; from asst prof to assoc prof, 64-76, assoc chmn dept, 69-70, PROF MATH, MICH STATE UNIV, 76- *Concurrent Pos:* Vis scholar, Cambridge Univ, 70-71 & Univ Col NWales, 78. *Mem:* Am Math Soc; Math Asn Am. *Res:* Theory of topological manifolds. *Mailing Add:* Dept Math Mich State Univ East Lansing MI 48823

MORAN, DAVID DUNSTAN, US citizen. NAVAL ARCHITECTURE, OCEAN ENGINEERING. *Educ:* Mass Inst Technol, SB, 65, SM, 67; Univ Iowa, PhD(hydrodyn), 71. *Prof Exp:* Prog element adminr, Naval Mat Command, 79-80; prof res, US Naval Acad, 80-81; res naval architect, 71-78, br head, 81-83, RES COORDR, D TAYLOR NAVAL SHIP RES & DEVELOP CTR, 83- *Concurrent Pos:* Adj prof, George Washington Univ, 73-86. *Mem:* Sigma Xi; Soc Naval Architects & Marine Engrs; Am Soc Civil Engrs; Am Soc Mech Engrs. *Res:* Naval architecture of large amplitude ship motions and hovercraft hydrodynamics. *Mailing Add:* David W Taylor Naval Ship Res & Develop Ctr Code 1A Bethesda MD 20084

MORAN, DAVID TAYLOR, b New York, NY, June 30, 40; m 63; c 2. NEUROBIOLOGY, CELL BIOLOGY. *Educ:* Princeton Univ, AB, 62; Brown Univ, PhD(biol), 69. *Prof Exp:* NIH fel, Harvard Univ, 69-70; NIH fel, 70-71, ASST PROF ANAT, MED SCH, UNIV COLO, DENVER, 71- *Mem:* Am Soc Cell Biol; AAAS; Am Soc Zoologists; Sigma Xi. *Res:* Neurobiology of sensory transduction in mechanoreceptors. *Mailing Add:* Dept Anat B-111 Univ Colo Med Ctr 4200 E Ninth St Denver CO 80262

MORAN, DENIS JOSEPH, b New York, NY, Aug 1, 42; m 70; c 2. DEVELOPMENTAL BIOLOGY. *Educ:* City Univ NY, BA, 67; NY Univ, PhD(biol), 72. *Prof Exp:* Asst prof biol, NY Univ, 71-72; res assoc develop biol, Columbia Col Physicians & Surgeons, 72-73; assoc prof, 73-82, PROF BIOL, STATE UNIV NY, NEW PALTZ, 82- *Concurrent Pos:* Consult, Orthopedic Res Labs, Columbia Univ, 73-, vis scientist, 76 & 78; NSF grant, 74-75. *Mem:* Am Soc Cell Biologists; Sigma Xi; Bioelectrochem Soc; AAAS. *Res:* Roles of extracellular materials and of microfilaments and microtubules in morphogenesis; calcium. *Mailing Add:* Develop Biol Lab State Univ NY New Paltz NY 12562

MORAN, EDWARD FRANCIS, JR, b Lowell, Mass, July 1, 32; m 58; c 4. INORGANIC CHEMISTRY. *Educ:* Villanova Univ, BS, 54; Univ Pa, PhD(inorg chem), 61. *Prof Exp:* Asst, Yale Univ, 60-62; RES ASSOC INORG RES, EXP STA, E I DUPONT DE NEMOURS & CO, 62- *Mem:* Am Chem Soc. *Res:* Synthesis and structure determination of inorganic polymers; low-temperature spectroscopy; hydrometallurgy of copper concentrates and ores; synthesis of zeolite catalysts; electrochemical processing. *Mailing Add:* Exp Sta E I Du Pont de Nemours & Co PO Box 80336 Wilmington DE 19880-0336

MORAN, EDWIN T, JR, FOOD SCIENCES. *Educ:* Wash State Univ, PhD(animal sci), 65. *Prof Exp:* Prof poultry sci, Univ Guelph, Can, 65-86; PROF POULTRY SCI, AUBURN UNIV, 86- *Mailing Add:* Dept Poultry Sci Auburn Univ Auburn AL 36849-4201

MORAN, JAMES MICHAEL, JR, b Plainfield, NJ, Jan 3, 43; m 74; c 2. ASTRONOMY, ELECTRICAL ENGINEERING. *Educ:* Univ Notre Dame, BS, 63; Mass Inst Technol, SM, 65, PhD(elec eng), 68. *Prof Exp:* Staff scientist radio physics, Mass Inst Technol, Lincoln Lab, 68-70; staff scientist astron, 70-86, ASSOC DIR, SMITHSONIAN ASTROPHYS OBSERV, 87- *Concurrent Pos:* Prof astron, Harvard Univ, 88- *Honors & Awards:* Rumford Prize, Am Acad Arts & Sci, 71; Newton L Pierce Prize, Am Astron Soc, 78. *Mem:* Am Astron Soc; AAAS; Int Astron Union; Inst Elec & Electronics Engrs. *Res:* Very long baseline interferometry; early stellar evolution and molecular cosmic masers. *Mailing Add:* Smithsonian Astrophys Observ 60 Garden St Cambridge MA 02138

MORAN, JEFFREY CHRIS, b Van Nuys, Calif, Aug 18, 54. NEUROPHYSIOLOGY. *Educ:* Univ Ore, BS, 79, MS, 81, PhD(psychol), 85. *Prof Exp:* Res asst psychol, dept psychol, Univ Ore, 78-79 & 80-82, teaching asst, 79-80; psychol technician, Nat Inst Mental Health, 82-85, psychologist, 85-86; postdoctoral assoc, Yale Univ Sch Med, 86-87; STAFF FEL LAB CLIN STUDIES, NAT INST ALCOHOL ABUSE & ALCOHOLISM, 87- *Mem:* Soc Neuroscience. *Res:* Neural mechanisms of higher cognitive functions such as attention and perception. *Mailing Add:* 11511 Aldburg Way Germantown MD 20876

MORAN, JOHN J, b Scranton, Pa, Jan 11, 27; m 52; c 3. MEDICINE. *Educ:* Univ Scranton, BS, 48; Jefferson Med Col, MD, 52. *Prof Exp:* Assoc path, Univ Pa & assoc pathologist, Univ Hosp, 57-61; from asst prof to assoc prof path, Jefferson Med Col, 61-77, asst dir clin labs, Hosp, 61-77; CHMN & DIR DIV LAB MED, GEISINGER MED CTR, 77- *Concurrent Pos:* Consult, Philadelphia Vet Admin Hosp, 57-59. *Mem:* AMA; Col Am Path; Am Soc Clin Path; Int Acad Path. *Res:* Pathology. *Mailing Add:* Geisinger Med Ctr Danville PA 17822

MORAN, JOHN P, b Queens Village, NY, Apr 24, 34; m 55; c 5. AERODYNAMICS. *Educ:* Cornell Univ, BME, 59, MAeroE, 60, PhD(aerospace eng), 65. *Prof Exp:* Propulsion engr, Grumman Aircraft Eng Co, 56-58, consult, 58-59; staff scientist, Therm Adv Res Div, Inc, 60-65; asst prof theoret mech, Cornell Univ, 65-68; asst prof, 68-69, ASSOC PROF AEROSPACE ENG & MECH, UNIV MINN, MINNEAPOLIS, 69- *Concurrent Pos:* Consult, United Technol Res Ctr, 78-80. *Res:* Numerical methods in fluid mechanics. *Mailing Add:* Dept Aerospace Eng & Mech Univ Minn 126 Aerospace Bldg Minneapolis MN 55455

MORAN, JOSEPH MICHAEL, b Boston, Mass, Feb 14, 44; m 71, 85. CLIMATOLOGY, PALEOCLIMATOLOGY. *Educ:* Boston Col, BS, 65, MS, 67; Univ Wis-Madison, PhD(meteorol), 72. *Prof Exp:* From instr to assoc prof earth sci, Univ Wis-Green Bay, 69-75; vis assoc prof climat, Univ Ill, Urbana, 75-76; assoc prof, 76-80, chmn dept, 83-86, PROF EARTH SCI, UNIV WIS-GREEN BAY, 80- *Concurrent Pos:* Hon fel, Univ Wis-Madison, 86. *Mem:* Am Meteorol Soc; AAAS; Sigma Xi; Nat Asn Geol Teachers; Asn Am Geographers; Am Quaternary Asn. *Res:* Pleistocene climatology; nature of climatic change and environmental impact; environmental science education; co-author of textbooks on environmental science; co-author of textbooks on meteorology and climatology. *Mailing Add:* Col Environ Sci Univ Wis Green Bay WI 54311-7001

MORAN, JULIETTE MAY, b New York, NY, June 12, 17. SCIENCE ADMINISTRATOR TECHNICAL MANAGEMENT. *Educ:* Columbia Univ, BS, 39; NY Univ, MS, 48. *Prof Exp:* Asst chem, Columbia Univ, 41; jr engr, Signal Corps Lab, US Army, 42-43; jr chemist, Process Develop Dept, Gen Aniline & Film Corp, 43-44, tech asst, 44-48, tech asst to dir, Cent Res Lab, 49-52 & com develop, 52-55, supvr tech serv, Com Develop Dept, 55-59, sr develop specialist, Develop Dept, 59-60, mgr planning, 61, asst to pres, 62-67, vpres, GAF Corp, 67-71, sr vpres, 71-74, exec vpres, 74-80, vchmn, GAF Corp, 80-82; MEM BD NY STATE SCI & TECHNOL FOUND, 79- *Mem:* Fel AAAS; Am Chem Soc; Com Develop Asn; fel Am Inst Chem. *Res:* Structure of hemocyanine; synthesis and application of dyestuffs, detergents and acetylene derivatives; technical information systems; commercial development; research administration. *Mailing Add:* Ten W 66th St New York NY 10023-6206

MORAN, NEIL CLYMER, b Phoenix, Ariz, Oct 12, 24; wid; c 3. PHARMACOLOGY. *Educ:* Stanford Univ, AB, 49, MD, 50. *Prof Exp:* Irving fel physiol, Stanford Univ, 50-51; med officer, USPHS Hop, Savannah, Ga, 51-52; head sect pharmacodynamics, Nat Heart Inst, Bethesda, Md, 54-56; pharmacologist, Nat Heart Inst, 52-54, assoc prof pharmacol, 56-62, prof & chmn dept, 62-80, CHARLES HOWARD CANDLER PROF PHARMACOL, EMORY UNIV, 80- *Concurrent Pos:* USPHS sr res fel, 57-60 & res career develop award, 60-62; asst ed, J Pharmacol & Exp Therapeut, 58-60 & ed, 61-65; vis scientist, Karolinska Inst, Sweden, 60-61; mem res career award comt, Nat Inst Gen Med Sci, 64-68; mem res comt, Am Heart Asn, 68-73, chmn, 72-73, Am Nat Heart & Lung Inst, 75-78. *Mem:* AAAS; Am Soc Pharmacol & Exp Therapeut; Soc Exp Biol & Med; Am Heart Asn; Am Asn Univ Prof. *Res:* Cardiovascular and autonomic pharmacology and physiology; allergy. *Mailing Add:* Dept of Pharmacol Emory Univ Atlanta GA 30322

MORAN, PAUL RICHARD, b Buffalo, NY, June 1, 36; m 58; c 4. SOLID-STATE PHYSICS. *Educ:* Univ Notre Dame, BS, 58; Cornell Univ, PhD(physics), 63. *Prof Exp:* Staff consult scientist, Kaman Nuclear Co, Colo, 63; NSF fel physics, Univ Ill, Urbana, 63-65; from asst prof to assoc prof, 65-73, PROF PHYSICS, UNIV WIS-MADISON, 73- *Mem:* AAAS; Am Phys Soc. *Res:* Magnetic resonance and optical studies of defect states in insulating solids. *Mailing Add:* Dept Physics Sterling Hall Univ Wis Sch Clin Ctr Madison WI 53792

MORAN, REID (VENABLE), b Los Angeles, Calif, June 30, 16; c 1. PLANT TAXONOMY. *Educ:* Stanford Univ, AB, 39; Cornell Univ, MS, 42; Univ Calif, PhD(bot), 51. *Prof Exp:* Botanist, Santa Barbara Bot Garden, 47-48; instr bot, Bailey Hortorium, Cornell Univ, 51-53; lectr Far East prog, Univ Calif, 53-56; cur, 57-82, EMER CUR BOT, SAN DIEGO MUS NATURAL HIST, 82- *Concurrent Pos:* Ed, San Diego Mus Natural Hist, 57-62, actg dir, 65-66. *Mem:* Am Fern Soc; fel Cactus & Succulent Soc Am. *Res:* Taxonomy of Crassulaceae and Cactaceae; vascular flora of Baja California; flora of Isla de Guadalupe; Dudleya; Graptopetalum; Pachyphytum. *Mailing Add:* PO Box 13901 San Diego CA 92112

MORAN, THOMAS FRANCIS, b Manchester, NH, Dec 11, 36; m 60; c 4. PHYSICAL CHEMISTRY. *Educ:* St Anselm's Col, BA, 58; Univ Notre Dame, PhD(chem), 62. *Prof Exp:* Asst instr chem, Univ Notre Dame, 58-59, res asst, 59-62; AEC fel, Brookhaven Nat Lab, 62-64, assoc scientist, 64-66; from asst prof to assoc prof, 66-72, PROF CHEM, GA INST TECHNOL, 72- *Concurrent Pos:* Danforth fel, 71- *Mem:* AAAS; Am Chem Soc; Am Phys Soc; Sigma Xi; Am Soc Mass Spectros. *Res:* Collisions of electrons and ions with molecules; mass spectrometry; energy transfer processes; kinetics of chemical reactions. *Mailing Add:* Sch Chem Ga Inst Technol Atlanta GA 30332

MORAN, THOMAS IRVING, b Amsterdam, NY, Nov 8, 30; m 53; c 4. PHYSICS. *Educ:* Union Univ, NY, BS, 53; Yale Univ, PhD, 57. *Prof Exp:* Res physicist, Gen Elec Co, 58-59; res assoc physics, Syracuse Univ, 59, asst prof, 60; res assoc, Brookhaven Nat Lab, 61-63; Fulbright sr scholar, Univ

Heidelberg, 63-64; asst prof, 64-74, ASSOC PROF PHYSICS, UNIV CONN, 74- *Mem:* Am Phys Soc. *Res:* Thermal diffusion; atomic beams; atomic physics; optical biophysics. *Mailing Add:* Dept of Physics U-46 Univ of Conn Storrs CT 06268

MORAN, THOMAS J, b St Paul, Minn, Aug 19, 38; m 61; c 6. FAST REACTOR STRUCTURES. *Educ:* Univ Notre Dame, BS, 60; Univ Minn, MS, 65, PhD(eng mech), 69. *Prof Exp:* From instr to asst prof mech, Ill Inst Technol, 68-73; MECH ENGR, ARGONNE NAT LAB, 73- *Mem:* Am Soc Mech Engrs; Sigma Xi; AAAS. *Res:* seismic and structural analysis of fast reactor cores. *Mailing Add:* Argonne Nat Lab 9700 S Cass Ave Argonne IL 60439

MORAN, THOMAS JAMES, b Rennerdale, Pa, Oct 14, 12; m 41; c 4. PATHOLOGY, MEDICINE. *Educ:* Univ Pittsburgh, BS & MD, 36; Am Bd Path, dipl, 45. *Prof Exp:* Dir labs, City Hosp, Pittsburgh, Pa, 39-42; pathologist, Welborn Hosp, Ind, 46; mem staff, Hosp, Danville, Va, 46-50 & St Margaret Hosp, 50-54; from asst prof to prof path, Univ Pittsburgh, 54-62; dir labs, 62-84, EMER DIR LABS, MEM HOSP, DANVILLE, 84- *Mem:* Am Soc Clin Path; Am Asn Pathologists; Int Acad Path; Col Am Path; Path Soc Gt Brit & Ireland. *Res:* Aspiration pneumonia; cortisone effects. *Mailing Add:* Creekside Danville VA 24541

MORAN, THOMAS PATRICK, b Detroit, Mich, Nov 6, 41; m 71; c 2. COMMUNICATION SCIENCE. *Educ:* Univ Detroit, BArch, 65; Carnegie-Mellon Univ, PhD(comput sci), 74. *Prof Exp:* scientist comput sci & psychol, 74-83, mgr, user-Systs Res, 84-86, PRIN SCIENTIST, PALO ALTO RES CTR, XEROX CORP, 90- *Concurrent Pos:* Dir, Europspace Res Lab, Rank Xerox, Cambridge, Eng, 86-90. *Mem:* Asn Comput Mach; Inst Elec & Electronic Engrs; Human Factors Soc. *Res:* Human-computer interaction; User-oriented system design; design processes and techniques. *Mailing Add:* Xerox Palo Alto Res Ctr 3333 Coyote Hill Rd Palo Alto CA 94304

MORAN, WALTER HARRISON, JR, b Grand Forks, NDak, Nov 16, 30; m 52; c 2. SURGERY, PHYSIOLOGY. *Educ:* Univ NDak, BA, 52, BS, 53; Harvard Univ, MD, 55; Am Bd Surg, dipl, 63. *Prof Exp:* Intern, Dept Surg, Univ Minn Hosps, 55-56; from instr to assoc prof surg, 60-70, dir surg res labs, 60-73, coordr div & actg dir emergency room, 73-81, PROF PHYSIOL, SURG & BIOPHYS, SCH MED, WVA UNIV, 70-, CHIEF EMERGENCY SERV, 81- *Concurrent Pos:* Med fel, Dept Surg, Univ Minn Hosps, 56-58, med fel specialist, 58-59, Nat Heart Inst fel, 59-60; USPHS fel, Sch Med, WVa Univ, 60-63; Nat Inst Arthritis & Metab Dis res career develop award, 62-67; co-dir metab unit, Univ Hosp, WVa Univ, 61-70. *Mem:* Endocrine Soc; AMA; Soc Univ Surgeons; Am Physiol Soc; Sigma Xi. *Res:* Biophysics; surgical endocrinology, especially vasopressin physiology; burn surgery. *Mailing Add:* Rt One Box 107 Swanton MD 21561

MORAN, WILLIAM RODES, b Los Angeles, Calif, July 29, 19. GEOLOGY. *Educ:* Stanford Univ, AB, 42. *Prof Exp:* Geologist, Union Oil Co, Calif, 43-46, Paraguay, 46-50, sr geologist, Nev, 50-51, Cia Petrol de Costa Rica, 51-52, mem spec explor staff, 52-59, actg mgr, Union Oil Develop Corp, Australia, 59-60, opers, Los Angeles, 60-63, vpres & mgr, Minerals Explor Co, Union Oil Co, Calif, 63-85, vpres, Australia Div, 68-85, vpres, Explor Molycorp, 79-85; RETIRED. *Concurrent Pos:* Consult, Stanford Univ Archive Recorded Sound, hon cur, 70-; dir, Quebec Columbium Ltd & Geosat Inc, 80-85. *Mem:* Fel AAAS; fel Am Geog Soc; Am Inst Mining, Metall & Petrol Engrs; fel Geol Soc Am. *Res:* Historical sound recording; author and discographer. *Mailing Add:* 1335 Olive Lane La Canada CA 91011

MORAND, JAMES M(CHURON), b Cincinnati, Ohio, Jan 27, 38; m 60; c 2. SANITARY ENGINEERING. *Educ:* Univ Cincinnati, CE, 60, MS, 61; Univ Wis, PhD(sanit eng), 64. *Prof Exp:* From asst prof to prof, 64-89, EMER PROF CIVIL ENG, UNIV CINCINNATI, 89- *Honors & Awards:* Fulbright lectr, Wroctaw, Poland, 91. *Mem:* Water Pollution Control Fedn; Am Soc Civil Engrs; Am Acad Environ Eng. *Res:* Water and waste treatment; biological waste treatment. *Mailing Add:* Dept Civil Eng Univ Cincinnati Cincinnati OH 45221-0071

MORAND, PETER, b Montreal, Can, Feb 11, 35; m 57; c 2. ORGANIC CHEMISTRY, RESEARCH ADMINISTRATION. *Educ:* Bishop's Univ, Can, BSc, 56; McGill Univ, PhD(org chem), 59. *Prof Exp:* NATO fel, Imp Col, Univ London, 59-61; sr chemist, Ayerst Labs, Can, 61-63; asst prof, 63-67, asst vrector acad, 68-71, assoc prof org chem, 67-81, dean fac sci & eng, 76-81, PROF CHEM, UNIV OTTAWA, 81-; VRECTOR, UNIV REL & DEVELOP, 87- *Concurrent Pos:* Res grants, Nat Sci Eng Res Coun Can, 77-; Nat Sci Eng Res Coun Can, 77- *Mem:* Am Chem Soc; Royal Soc Chem; fel Chem Inst Can. *Res:* Chemistry of natural products; cell membrane fluorescent probes; biosynthesis of estrogens; plant defence substances; phototoxic biocides. *Mailing Add:* Dept of Chem Univ of Ottawa Ottawa ON K1N 6N5 Can

MORARI, MANFRED, b Graz, Austria, May 13, 51; m 84; c 1. PROCESS CONTROL. *Educ:* Swiss Fed Inst Technol, dipl, 74; Univ Minn, PhD(chem eng), 77. *Prof Exp:* From asst prof to assoc prof chem eng, Univ Wis-Madison, 77-83; PROF DEPT CHEM ENG, CALIF INST TECHNOL, 83-, EXEC OFFICER, 90- *Concurrent Pos:* Dir Comput & Syst Technol Div, Am Inst Chem Eng, 87-90; Gulf vis prof, Carnegie Mellon Univ, 87. *Honors & Awards:* D P Eckman Award, Am Automatic Control Coun, 80; Alan P Colburn Award, Am Inst Chem Eng, 84; Curtis W McGraw Res Award, Am Soc Eng Educ, 89. *Mem:* Am Inst Chem Engrs; Am Chem Soc; Inst Elec Electronic Engrs; Am Soc Eng Educ. *Res:* Process design; process control. *Mailing Add:* Chem Eng 210-41 Calif Inst Technol Pasadena CA 91125

MORASH, RONALD, b Boston, Mass, Feb 5, 45. MATHEMATICS. *Educ:* Boston Col, AB, 65, MA, 67; Univ Mass, Amherst, PhD(math), 71. *Prof Exp:* PROF MATH, UNIV MICH, 72- *Mem:* Sigma Xi; Am Math Soc; Math Asn Am. *Res:* Writing about the foundations of mathematics & linear algebra. *Mailing Add:* 1315 W Huron St Ann Arbor MI 48103

MORATH, RICHARD JOSEPH, b St Paul, Minn, July 13, 25; wid; c 2. ORGANIC CHEMISTRY. *Educ:* Univ Minn, BChem, 49; Wash State Univ, MS, 51, PhD(chem), 54. *Prof Exp:* Res assoc, Univ NC, 53-54 & Univ Iowa, 54-55; asst prof org chem, Univ Dayton, 55-57; from asst prof to assoc prof, 57-70, PROF CHEM, COL ST THOMAS, 70- *Mem:* Am Chem Soc; The Chem Soc. *Res:* Organic syntheses; reaction mechanisms. *Mailing Add:* 2128 Stanford Ave St Paul MN 55105

MORAVCSIK, MICHAEL JULIUS, b Budapest, Hungary, June 25, 28; US citizen; m 56; c 2. HIGH-ENERGY PHYSICS, SCIENCE POLICY. *Educ:* Harvard Col, AB, 51; Cornell Univ, PhD(theoret physics), 56. *Prof Exp:* Res assoc theoret physics, Brookhaven Nat Lab, 56-58; physicist & head elem particle & nuclear theory group, Lawrence Radiation Lab, Univ Calif, 58-67; dir inst, 69-72, RES ASSOC, INST THEORET SCI & PROF PHYSICS, UNIV ORE, 67- *Concurrent Pos:* Vis prof, Purdue Univ, 57, Int Atomic Energy Agency, Atomic Energy Ctr, Pakistan AEC, Lahore, 62-63, Osaka Univ, 72, Nat Lab High Energy Physics Japan, 72 & Physics Dept & Sci Policy Res Univ, Sussex Univ, 75-76; vis lectr, Harvard Univ, 66-67 & Univ Chile, 78; mem spec study group role sci & technol in develop in 70's, Nat Acad Sci, Woods Hole, Mass; Scientists & Engrs Econ Develop grant, Nigeria, 74; mem adv comt East Asia, Comt Int Exchange Persons, 71-74; SEED grant, Nigeria, 75; NATO sr fel sci, Sci Policy Res Unit, Brighton, Eng, 74; consult, Los Alamos Meson Physics Facil, NMex; mem, Grants Adv Panel, Fund Overseas Res Grants & Educ & Coun Sci & Technol Develop, 78-; ed-in-chief, Scientometrics. *Mem:* Sigma Xi; Am Asn Physics Teachers; fel Am Phys Soc. *Res:* Theoretical elementary particle physics, especially as it is related to experiments; photoproduction processes; nuclear forces; spin structure of particle reactions; assignments of intrinsic quantum numbers; problems of science policy, organization and management in developing countries; nuclear and elementary particle physics. *Mailing Add:* Inst of Theoret Sci Univ of Ore Eugene OR 97403

MORAVEC, HANS PETER, b Kautzen, Austria, Nov 30, 48; Can citizen; m. COMPUTER SCIENCE. *Educ:* Acadia Univ, NS, BS, 69; Univ Western Ont, London, Can, MS, 71; Stanford Univ, PhD(computer sci), 80. *Prof Exp:* Res asst computer sci, Univ Western Ont, London, Can, 69-71; res asst computer sci, Stanford Univ, 71-80; res scientist 80-85, SR RES SCIENTIST ROBOTICS, CARNEGIE MELLON UNIV, 85- *Concurrent Pos:* Consult for var bus, 68- *Mem:* Asn Comput Mach; Inst Elec & Electronic Engrs; AAAS; Nat Space Soc. *Res:* Intelligent robots in outer space; mobile robots; computer vision; supercomputers; robot manipulators; novel space travel schemes; computational models of physics; computer animation; three dimensional graphics; author of various publications. *Mailing Add:* Robotics Inst Carnegie-Mellon Univ Pittsburgh PA 15213-3876

MORAWA, ARNOLD PETER, b Detroit, Mich, Feb 14, 40; m 65; c 3. PEDODONTICS, ORAL BIOLOGY. *Educ:* Univ Mich, DDS, 64, MS, 66 & 68, PhD(anat), 73. *Prof Exp:* Asst prof, 73-77, ASSOC PROF DENT & HOSP DENT, UNIV MICH, ANN ARBOR, 77- *Concurrent Pos:* Consult, Hawthorne Ctr, State Mich, 66-72, Plymouth State Home, 68-72 & Ctr Study Ment Retardation, Univ Mich, 75-; pedodontist, pvt pract, Ann Arbor, 66- *Mem:* Int Asn Dent Res; Sigma Xi. *Res:* Biochemical and ultrastructural evaluation of dilute formocresol upon dental pulp; concurrent studies of clinical application of this medicament; biochemical and ultrastructural effects of hypoxia upon protein secreting cells; clinical trial DNA probes for detection of periodontal disease. *Mailing Add:* 3920 Waldenwood Ann Arbor MI 48105

MORAWETZ, CATHLEEN SYNGE, b Toronto, Ont, May 5, 23; nat US; m 45; c 4. APPLIED MATHEMATICS. *Educ:* Univ Toronto, BS, 44; Mass Inst Technol, MS, 46; NY Univ, PhD(math), 51. *Hon Degrees:* Dr, Eastern Mich Univ, 80; Smith Col, 82; Brown Univ, 82; Princeton Univ, 86; NJ Inst Tech, 86; Duke Univ, 88. *Prof Exp:* Res assoc, NY Univ, 51-57, from asst prof to assoc prof, 57-66, dir, 84-88,PROF MATH, COURANT INST MATH SCI, NY UNIV, 66- *Concurrent Pos:* Guggenheim fels, 66-67 & 78-79; term trustee, Princeton Univ, 74-78, dir, NCR, 78-, trustee Sloan Found, 80- *Honors & Awards:* Josiah Willard Gibbs lectr, Am Math Soc, 81. *Mem:* fel Nat Acad Sci; Am Math Soc; Soc Indust & Appl Math; fel AAAS; fel Am Acad Arts & Sci; Math Asn Am. *Res:* Applications of partial differential equations. *Mailing Add:* Courant Inst Math Sci NY Univ 251 Mercer St New York NY 10012

MORAWETZ, HERBERT, b Prague, Czech, Oct 16, 15; nat US; m 45; c 4. PHYSICAL CHEMISTRY. *Educ:* Univ Toronto, BASc, 43, MASc, 44; Polytech Inst Brooklyn, PhD(chem), 50. *Prof Exp:* Res chemist, Bakelite Co, 45-49; fel, NIH, 50-51; from asst prof to assoc prof polymer chem, 51-58, dir, Polymer Res Inst, 71-81, inst prof, 81-86, PROF POLYMER CHEM, POLYTECH UNIV, NY, 58-, EMER INST PROF, 86- *Concurrent Pos:* NIH fel, 67-68; Centennial scholar, Case Western Reserve Univ,80. *Honors & Awards:* Award in Polymer Chem, Am Chem Soc, 85; Hegrovsky Medal, Czech Acad Sci, 90. *Mem:* AAAS; Am Chem Soc. *Res:* Molecular association of polymers; reaction kinetics in polymer solutions; polyelectrolytes; solid-state polymerization; polymer compatibility; application of fluorescence techniques to polymer studies. *Mailing Add:* Dept Chem Polytech Univ Brooklyn NY 11201

MORAWITZ, HANS, b Neustadt, Austria, Feb 6, 35; m 63; c 2. THEORETICAL PHYSICS. *Educ:* Stanford Univ, BS, 56, PhD(physics), 63. *Prof Exp:* Staff physicist, IBM Res Lab, 63-65; Australian Dept Supply fel & sr lectr physics, Monash Univ, Australia, 65-66; Austrian Ministry Educ res assoc, Univ Vienna, 66-67; STAFF PHYSICIST, IBM RES LAB, 67- *Concurrent Pos:* Vis prof, Ulm Univ, Germany, 82 & Inst Advan Study, Australian Nat Univ, Canberra, 84, Bayreuth Univ, Germany, 86. *Mem:* Am Phys Soc; Europ Phys Soc. *Res:* Theoretical atomic, molecular and solid-state physics; surface physics; cooperative radiative processes; phase transition in organic solids and magnetic systems; macroscopic quantum states; high temperature superconductivity. *Mailing Add:* Dept K31 IBM Almaden Res Ctr 650 Harry Rd San Jose CA 95120-6099

MORAY, NEVILLE PETER, b London, Eng, May 27, 35; Brit & Can citizen; m; c 2. HUMAN FACTORS. *Educ:* Oxford Univ, BA, 57, DPhil, 60. *Prof Exp:* Lectr psychol, Sheffield Univ, 59-70; assoc prof, Univ Toronto, 70-74, prof indust eng, 81-87; prof psychol, Stirling Univ, 74-81; PROF MECH & INDUST ENG & PROF PSYCHOL, UNIV ILL, URBANA-CHAMPAIGN, 87- *Concurrent Pos:* Consult. *Mem:* Inst Elec & Electronics Engrs; fel Human Factors Soc; Ergonomics Soc. *Res:* Human-machine systems; human error; workload; role of human operators in automated manufacturing. *Mailing Add:* 1109 S Orchard Urbana IL 61801

MORBEY, GRAHAM KENNETH, b Birmingham, Eng, Apr 5, 35; m 60; c 2. TEXTILE CHEMISTRY, CHEMICAL ENGINEERING. *Educ:* Univ Birmingham, BSc, 56; Univ Toronto, MASc, 57; Princeton Univ, MA, 58, PhD(chem eng), 61. *Prof Exp:* Res scientist, Dunlop Res Ctr, Can, 60-61, asst tech dir, Dunlop Tyre Co, Eng, 61-63; group leader, Celanese Fibers Mkt Co, 63-64, develop mgr, 64-67, develop dir, 67-69; tech dir, Hoechst Fibers Inc, 69-72; dir appl res, Personal Prod Co, 72-77, dir develop, 77-79; vpres res & develop, Texon Inc, 79-86; PROF MGMT, UNIV MASS, 86- *Mem:* Textile Res Inst; Tech Asn Pulp & Paper Indust; Prod Develop Mgmt Asn; Fiber Soc; AAAS; Sigma Xi. *Res:* Polymer chemistry; statistics; fiber physics and textiles; research and production management. *Mailing Add:* 81 Blossom Lane Amherst MA 01002

MORCK, ROLAND ANTON, b Crookston, Minn, July 11, 13; m 39; c 4. BIOCHEMISTRY. *Educ:* St Olaf Col, BA, 35; Pa State Univ, MS, 37, PhD(biochem), 39. *Prof Exp:* Nutrit specialist, R B Davis Co, 39-41, chemist, 41-43, chief chemist, 44-55; res chemist, Nat Biscuit Co, 55-58, asst dir res, 58-72, dir res, Nabisco, Inc, 72-75, vpres res, 75-78; RETIRED. *Mem:* AAAS; Am Chem Soc; Am Oil Chem Soc; Am Asn Cereal Chem; Inst Food Technol; Sigma Xi. *Res:* Food research. *Mailing Add:* PO Box 2 Allendale NJ 07401-0002

MORCK, TIMOTHY ANTON, b Glenn Ridge, NJ, May 11, 49; m 72; c 2. NUTRITION. *Educ:* Pa State Univ, BS, 71; Cornell Univ, MS, 75, PhD(nutrit), 78. *Prof Exp:* Res assoc, State Univ NY Upstate Med Ctr, 78; instr med, Univ Kans Med Ctr, 78-81; CHIEF NUTRIT PHYSIOL SECT, VET ADMIN MED CTR, HAMPTON, VA, 81- *Concurrent Pos:* Asst prof biochem & instr med, Eastern Va Med Sch, 81- *Mem:* Sigma Xi. *Res:* Interaction of dietary components that affect the bioavailability of food ion in humans; assessment of nutritional status and repletion of malnourished elderly hospitalized patients using various protein to calorie ratios. *Mailing Add:* ILSI Nutrit Found 1126 Sixteenth St NW Washington DC 20036

MORCOCK, ROBERT EDWARD, b Washington, DC, Aug 26, 38; m 67; c 2. PHYSIOLOGICAL ECOLOGY, ENVIRONMENTAL TOXICOLOGY. *Educ:* NC Wesleyan Col, BA, 66; Wake Forest Univ, MA, 70, PhD(physiol ecol), 74. *Prof Exp:* Res assoc parasitol, Univ Mass, Amherst, 74-76; asst prof biol, Hood Col, Frederick, Md, 76-80; MEM STAFF, US ENVIRON PROTECTION AGENCY, 80- *Mem:* AAAS; Sigma Xi; Soc Environ Toxicol & Chem. *Res:* Physiology and biochemistry of symbiotic helminths; carbon dioxide fixation by developing cestodes; effects of malnutrition on host-parasite relations; toxic effects of chemicals on biologic systems. *Mailing Add:* 8226 Morning Dew Ct Frederick MD 21702-3009

MORDES, JOHN PETER, b New Haven, Conn, May 12, 47. ENDOCRINOLOGY, DIABETES. *Educ:* Harvard Univ, AB, 69, MD, 73. *Prof Exp:* Asst prof, 81-85, ASSOC PROF MED, SCH MED, UNIV MASS, 85- *Honors & Awards:* Balodemos prize, Am Diabetes Asn. *Mem:* Am Diabetes Asn; Endocrine Soc; Am Fedn Clin Res; AAAS. *Res:* Immunology of insulin dependent diabetes; regulation of appetite in obesity. *Mailing Add:* Dept Med Univ Mass 55 Lake Ave N Worcester MA 01605

MORDESON, JOHN N, b Council Bluffs, Iowa, Apr 22, 34; m 60; c 5. MATHEMATICS. *Educ:* Iowa State Univ, BS, 59, MS, 61, PhD(math), 63. *Prof Exp:* PROF MATH, CREIGHTON UNIV, 63- *Mem:* Am Math Soc. *Res:* Field theory and ring theory. *Mailing Add:* Dept Math Creighton Univ 2500 Cal St Omaha NE 68178

MORDFIN, LEONARD, b Brooklyn, NY, June 23, 29; m 54; c 3. MECHANICAL TEST METHODS, FAILURE ANALYSIS. *Educ:* Cooper Union, BME, 50; Univ Md, MS, 54, PhD, 66. *Prof Exp:* Engr, Nat Bur Standards, 50-67; phys sci adminr, Off Aerospace Res, US Air Force, 67-69; aerospace engr, 69-77, dep chief, Off Nondestructive eval, 77-90, GROUP LEADER, MECH PROPERTIES & PERFORMANCE, NAT INST STANDARDS & TECHNOL, 90- *Concurrent Pos:* Tech consult, Forensic Technologies, Inc, Int Atomic Energy Agency, Int Trade Admin, Minn Pub Radio, Soc Automotive Engrs; US deleg to Int Orgn for Standardization, tech comt on non-destructive testing, 84- *Honors & Awards:* Bronze Medal, US Dept Com, 83; Award Merit, Am Soc Testing & Mat, 89; E B Rosa Award, Nat Inst Standards & Technol, 89. *Mem:* fel Am Soc Testing & Mat; Soc Exp Mech; Am Soc Mech Engrs; Am Soc Nondestructive Testing; Am Soc Microbiol Int. *Res:* Experimental and analytical studies of the mechanical behavior of advanced materials and structures under severe environmental and operating conditions including creep and fatigue; development and promulgation of standard test methods; metal-matrix composites; nondestructive evaluation; residual stresses. *Mailing Add:* Mech Properties & Performance Nat Inst Standards & Technol Gaithersburg MD 20899

MORDUCHOW, MORRIS, b Rogachev, Russia, Sept 25, 21; US citizen; m 75. FLUID DYNAMICS, SOLID DYNAMICS. *Educ:* Brooklyn Col, BA, 42; Polytech Inst Brooklyn, BAEE, 44, MAEE, 45, PhD(aeronaut eng), 47. *Prof Exp:* Res assoc aero eng, Polytech Inst Brooklyn, 44-47, instr, 47-50, from asst prof to prof appl mech, 50-76, prof mech & aerospace eng, Polytech Inst New York, 76-87, PROF AEROSPACE ENG, POLYTECH UNIV, 88- *Concurrent Pos:* Mem nat tech comt struct, Am Inst Aeronaut & Astronaut, 72-73; adj prof, Brooklyn Col, 70-72; res grants Nat Adv Comt Aeronaut, Off Naval Res, 45-57. *Honors & Awards:* I B Laskowit Award, NY Acad Sci, 80- *Mem:* Assoc fel Am Inst Aeronaut & Astronaut; Am Math Soc; Math Asn Am; Soc Indust & Appl Math; AAAS; Am Soc Mech Engrs; NY Acad Sci; Am Soc Eng Educ. *Res:* Fluid dynamics; solid dynamics; numerical analysis. *Mailing Add:* 48 Skyline Dr Keene NH 03431

MORDUCHOWITZ, ABRAHAM, b New York, NY, Aug 17, 33; m 56; c 3. ORGANIC POLYMER CHEMISTRY. *Educ:* Yeshiva Col, BA, 54; Univ Chicago, MS, 58, PhD(org photochem), 62. *Prof Exp:* Chemist, Beacon Res Lab, Texaco, Inc, 62-63, sr chemist, 63-68, res chemist, 68-75, campus tech recruiter, 75-83, res assoc, 83- 89, SR RES CHEMIST, BEACON RES LAB, TEXACO, INC, 75-, GROUP LEADER, 80- *Concurrent Pos:* Sr assoc-prod safety & coordr, Chem Regulation Compliance, Beacon Res Lab, Texaco, Inc, 89- *Mem:* Am Chem Soc. *Res:* Synthesis and study of solution properties of macromolecules; solar energy conversion; enhanced oil recovery; electrochemistry; polymeric lubricant additives. *Mailing Add:* Three Crown Rd Monsey NY 10952

MORDUE, DALE LEWIS, b Colchester, Ill, Apr 26, 33; m 60; c 3. PHYSICS, MATHEMATICS. *Educ:* Western Ill Univ, BS, 54; Univ Ill, Urbana, MS, 59; Tex A&M Univ, PhD(physics), 65. *Prof Exp:* Teacher high sch, Ill, 56-58; instr phys sci & chem, Evansville Col, 59-61; asst physics, Tex A&M Univ, 61-63; from asst prof to assoc prof, 65-71, PROF PHYSICS, MANKATO STATE UNIV, 71- *Concurrent Pos:* Dir, Energy Efficiency Sch Bldg Prog, Dept Energy, 78. *Res:* Biophysics using electrophysiological methods to determine spectral sensitivity of insects; conservation and solar energy. *Mailing Add:* Dept Physics Mankato State Univ S Roadland Ellis Ave Mankato MN 56001

MORDUKHOVICH, BORIS S, b Moscow, USSR, Apr 8, 48; m 69; c 2. OPTIMIZATION, SYSTEMS CONTROL. *Educ:* Byelorussian State Univ, Minsk, Russia, MSc, 71, PhD(appl math), 73. *Prof Exp:* Res engr, Res Inst Autom Control, Minsk, 71-73; lectr math, Byelorussian Polytech Inst, Minski, USSR, 71-73; sr scientist, Res Inst Water Mgt, Minsk, 73-88; PROF MATH, WAYNE STATE UNIV, DETROIT, MICH, 89- *Concurrent Pos:* adj prof math, Byelorussian State Univ, Minsk, USSR, 73-88; vis prof math, Univ Montreal, Que, Can, 89 & Univ Pau, France, 90; NSF res grant, Washington, DC, 90. *Mem:* Am Math Soc; Soc Indust & Appl Math; Math Asn Am; Math Prog Soc. *Res:* Applied mathematics; optimization; nonlinear analysis; control systems; developing new mathematical tools for solving such problems and applications to engineering systems. *Mailing Add:* Math Dept Wayne State Univ Detroit MI 48202

MORE, KENNETH RIDDELL, b Vancouver, BC, Jan 9, 10; nat US; m 42; c 2. PHYSICS. *Educ:* Univ BC, BA, 29, MA, 31; Univ Calif, PhD(physics), 34. *Prof Exp:* Asst physics, Univ BC, 29-31; asst, Univ Calif, 31-34, fel, 34-35; Royal Soc Can fel, Mass Inst Technol, 35-36; Sterling fel, Yale Univ, 36-38; instr physics, Ohio State Univ, 38-44, asst prof, 44-45; res physicist, Phillips Petrol Co, Okla, 46-47; prof physics, Univ BC, 47-50; opers analyst, US Dept Air Force, 50-61; sr physicist, Stanford Res Inst, 61-63; opers res scientist, Ctr Naval Analysis, 63-64, dir naval objectives analysis group, 64-67, naval implications technol & tech group, Naval Res Lab, 67-68; sci & eng adv, US Bur Mines, 68-70, chief off opers res, 70-72, staff asst educ & training, 72-73; staff asst educ & training, Mining Enforecement & Safety Admin, Dept Interior, 73-75, coordr, Res & Planning Off, Nat Mine Health & Safety Acad, 75-80; RETIRED. *Concurrent Pos:* Mem staff, Radiation Lab, Mass Inst Technol, 42-45. *Mem:* AAAS; fel Am Phys Soc; Opers Res Soc Am. *Res:* Atomic and molecular spectroscopy; microwave magnetron design; nuclear physics; operations research. *Mailing Add:* 9901 E Bexhill Dr Kensington MD 20895

MORE, RICHARD MICHAEL, b Kenosha, Wis, July 27, 42; m 60; c 2. THEORY OF ATOMIC PROCESSES IN LASER PLASMAS. *Educ:* Univ Calif, Riverside, BA, 63; Univ Calif, San Diego, MS, 64, PhD(physics), 68. *Prof Exp:* Res assoc, Univ Pittsburgh, 68-69, asst prof physics, 69-76, assoc prof physics & astron, 76-77; MEM STAFF, LAWRENCE LIVERMORE LAB, UNIV CALIF, 77- *Mem:* Am Phys Soc. *Res:* Laser-produced plasmas; atomic physics of dense plasmas; laser and particle-beam effects; fast-ion energy deposition and ion implantation; statistical mechanics and large-scale computational physics. *Mailing Add:* Lawrence Livermore Lab PO Box 808 L-321 Livermore CA 94550

MOREAU, JEAN RAYMOND, b Village des Aulnaies, Que, June 27, 24; m 50; c 4. FOOD SCIENCE. *Educ:* Laval Univ, BA, 44, BSc, 48; Mass Inst Technol, PhD(food sci & eng), 57. *Prof Exp:* Res asst biol, Dept Fisheries, Que, 48-49; supt fisheries plant, 49-51; sr res scientist, Can Packers Ltd, 57-64; prof food sci, Univ Toronto, 64-65; PROF FOOD SCI & BIOCHEM ENG, LAVAL UNIV, 65- *Concurrent Pos:* Sci consult to govt & indust, 67- *Honors & Awards:* Knight Mark Twain Outstanding Contrib Mod Educ. *Mem:* Sigma Xi; fel Brit Royal Soc Arts; Int Platform Asn; Can Inst Food Sci & Technol; Soc Am Inventors. *Res:* Food science and engineering of meat, milk products and protein foods; hygiene, quality and nutritive value of meat, meat products, vitamins and beverages; microbial biomass; forestry biomass. *Mailing Add:* Dept Chem Eng Laval Univ Ste-Foy PQ G1K 7P4 Can

MOREAU, ROBERT ARTHUR, b North Adams, Mass, Aug 10, 52; m 78; c 3. BIOCHEMISTRY, PHYSIOLOGY. *Educ:* Boston Univ, BA, 74; Univ SC, PhD(biochem), 79. *Prof Exp:* Postdoctoral fel, Univ Calif, Davis, 79-81; RES CHEMIST, AGR RES SERV, USDA, 81- *Concurrent Pos:* Lectr, Thomas Jefferson Univ, 87-88. *Mem:* Am Soc Plant Physiologists; Am Oil Chemists Soc. *Res:* Lipid metabolism in higher plants and in microbes; developed HPLC methodology for the analysis of lipids in plants and microbes. *Mailing Add:* Eastern Regional Res Ctr USDA 600 E Mermaid Lane Philadelphia PA 19118

MOREHART, ALLEN L, b Williamsport, Pa, Apr 1, 33; m 57; c 5. PHYTOPATHOLOGY, MEDICAL MYCOLOGY. *Educ:* Lycoming Col, AB, 59; Univ Del, MS, 61, PhD(biol sci), 64. *Prof Exp:* Res asst plant path, Univ Del, 59-61, res assoc, 61-64; fel med mycol & res assoc microbiol, Univ

Okla, 64-65; asst prof, WVa Univ, 65-68; assoc prof biol, Lycoming Col, 68-71, chmn dept, 69-71; assoc prof plant path, 71-81, actg chmn dept plant sci, 72-76, PROF PLANT PATH, UNIV DEL, 81-, CHMN DEPT PLANT SCI, 84- *Concurrent Pos:* Danforth Assoc, 69. *Mem:* AAAS; Mycol Soc Am; Am Soc Microbiol; Int Soc Human & Animal Mycol; Am Phytopath Soc; Sigma Xi. *Res:* Mechanism of fungicidal action; fungal physiology; study of deciduous forest tree tissue cultures for resistance to nectria galligena. *Mailing Add:* Dept of Plant Sci Univ of Del Newark DE 19711

MOREHEAD, FREDERICK FERGUSON, JR, b Roanoke, Va, July 30, 29; m 54; c 5. PHYSICAL CHEMISTRY, SOLID-STATE PHYSICS. *Educ:* Swarthmore Col, BA, 50; Univ Wis, MS, 51, PhD(phys chem), 53. *Prof Exp:* Asst prof chem, Union Col, NY, 53-54; mem staff, Lamp Develop Lab, Gen Elec Co, Ohio, 54-59; MEM STAFF, T J WATSON RES CTR, IBM CORP, 59- *Mem:* Am Phys Soc; Sigma Xi. *Res:* Luminescence and photoconductivity in II-IV compounds; electroluminescence; ion implantation; diffusion modeling. *Mailing Add:* 1719 Summit St Yorktown Heights NY 10598

MOREHOUSE, ALPHA L, b Lafayette, Ind, Sept 27, 23; m 47; c 4. BIOCHEMISTRY. *Educ:* Purdue Univ, BS, 48; Pa State Univ, MS, 50, PhD(biochem), 52. *Prof Exp:* Mgr, Spec Prods Res, 80-90, SR BIOCHEMIST, GRAIN PROCESSING CORP, 52-, RES FEL, 90- *Mem:* Am Chem Soc; Am Asn Cereal Chemists. *Res:* Poultry nutrition; recovery of vitamins and antibiotics from fermentation liquors; chemical modification of corn starch; enzymatic hydrolysis of starch; biological value of proteins; extraction and purification of plant proteins. *Mailing Add:* Tech Dept Grain Processing Corp Muscatine IA 52761

MOREHOUSE, CHAUNCEY ANDERSON, b Palatka, Fla, June 7, 26; m 53; c 3. BIOMECHANICS OF SPORT, PROTECTIVE EQUIPMENT FOR SPORTS. *Educ:* Springfield Col, BS, 49; Pa State Univ, MS, 54, PhD(phys educ), 64. *Prof Exp:* Dir athletics, Perkiomen Sch, Pennsburg, Pa, 49-53 & Detroit Country Day Sch, Detroit, Mich, 53-55; asst prof health educ, Panzer Col Phys Educ & Hyg, East Orange, NJ, 55-58; asst prof phys educ, Lock Haven State Col, Pa, 58-62; instr phys educ prep, 62-65, from asst prof to assoc prof statist & measurement, 65-72, prof statist, Measurement & Res Methodology, 72-87, EMER PROF PHYS EDUC, PA STATE UNIV, UNIV PARK, 87- *Concurrent Pos:* Vis prof, Inst Sport & Sport Sci, Johann Wolfgang Goethe Univ, Frankfurt-on-Main, WGermany, 72; vis researcher, Lab Biomech, Ger Sports Univ, Cologne, WGermany, 79; Fulbright grant, 79; archivist, Int Soc Biomech, 91- *Honors & Awards:* E B Cottrell Award, Pa, 74; Award of Merit, Am Soc Testing & Mat, 83. *Mem:* Am Alliance Health Phys Educ, Recreation & Dance; Int Soc Biomech (treas, 81-89); fel Am Col Sports Med; fel Am Soc Testing & Mat. *Res:* Development of protective equipment for sports; statistics; experimental design. *Mailing Add:* Biomech Lab Pa State Univ University Park PA 16802

MOREHOUSE, CLARENCE KOPPERL, b Boston, Mass, Apr 8, 17; m 42; c 2. ELECTROCHEMISTRY. *Educ:* Tufts Col, BS, 39; McGill Univ, MS, 40; Mass Inst Technol, PhD(chem), 47. *Prof Exp:* Res chemist, Naval Ord Lab, Washington, DC, 42-43 & Nat Bur Standards, 43-45; leader battery res & develop, Olin Industs, Inc, Ill, 47-49, asst dir res, Elec Div, Conn, 49-52, mgr, Res & Develop, 52-53; chemist, Res Labs, Radio Corp Am, 53-58, mgr battery & capacitor develop & eng, Semiconductor & Mat Div, 58-59, component develop, 59-60; vpres eng, Globe Battery Co Div, Globe Union Inc, 60-67, vpres res & eng, 62-70, vpres & gen mgr, Int Div, 70-77, vpres technol, 77-79; vpres & dir appl res, Johnson Controls, Inc, 79-80; RETIRED. *Concurrent Pos:* Consulting, 84- *Mem:* Am Chem Soc; Electrochem Soc. *Res:* Inorganic and physical chemistry, specifically batteries and capacitors; corrosion and chemistry of less familiar elements. *Mailing Add:* 569 Village Place Longwood FL 32779

MOREHOUSE, LAURENCE ENGLEMOHR, b Danbury, Conn, July 13, 13; m 73; c 2. PHYSIOLOGY. *Educ:* Springfield Col, BS, 36, MEd, 37; Univ Iowa, PhD(phys ed), 41. *Prof Exp:* Asst, Univ Iowa, 37-40; head dept health & phys ed, Univ Wichita, 41-42; asst prof ed, Univ Kans, 42; fel, Harvard Univ, 45-46; assoc prof phys educ & assoc aviation med, Univ Southern Calif, 46-54; PROF KINESIOL, UNIV CALIF, LOS ANGELES, 54- *Concurrent Pos:* Mem exped, Nat Acad Sci, NH, 46; researcher, Off Naval Res, 47-49; chief performance physiol sect, Care of Flyer Dept, Randolph Air Force Sch Aviation Med, 49; mem port study comt, Nat Res Coun, 58-; specialist, US State Dept, 60-61, consult to minister sci res, UAR, 63; consult, Henry Dryfuss, 60-63, Douglas Aircraft Co, 63- & Mayor's Space Adv Comt, 65-; Nat Res Coun vis scientist, Manned Spacecraft Ctr, NASA, 68-69; adv, Fitness Systs Inc, 74-82, Los Robles Regional Med Ctr, 79-80. *Mem:* Am Physiol Soc; Ergonomics Res Soc; Human Factors Soc; Aerospace Med Asn; Am Asn Health, Phys Educ & Recreation. *Res:* Physiology of exercise; industrial physiology; sports medicine; aerospace medicine; kinesiology; fatigue. *Mailing Add:* Dept of Kinesiology Univ of Calif Los Angeles CA 90024

MOREHOUSE, LAWRENCE G, b Manchester, Kans, July 21, 25; c 2. VETERINARY PATHOLOGY, VETERINARY MICROBIOLOGY. *Educ:* Kans State Univ, BS & DVM, 52; Purdue Univ, MS, 56, PhD(animal path), 60. *Prof Exp:* Supvr brucellosis labs, USDA, Purdue Univ, 53-60, staff vet, Agr Res Serv, 60-61, discipline leader path & toxicol, Animal Health Div, Nat Animal Dis Lab, Iowa, 61-64; chmn dept, 64-71, prof vet path & dir, Vet Med Diag Lab, 71-88, EMER PROF VET PATH, UNIV MO, COLUMBIA, 88- *Concurrent Pos:* Consult, Agr Res Serv, USDA, 64-; mem, Comt Salmonellosis, NCent US Poultry Dis Conf, 62-68, studies enteric dis in young swine, NCent State Tech Comt, 64-88 & Nat Conf Vet Lab Diagnosticians; chmn Accreditation Bd, Am Asn Vet Lab Diagnosticians; mem, Animal Health Spec Res Grant Peer Panel, Coop State Res Serv/USDA, 83; mem, Med Res Develop Adv Comt, US Army, 83-84; consult, Am Inst Biol Sci-Nat Acad Sci. *Honors & Awards:* E P Pore Award, Am Asn Vet Diagnosticians, 75. *Mem:* AAAS; Am Vet Med Asn; Am Asn Avian Path; Am Asn Vet Lab Diagnosticians (pres, 78-79); fel Royal Soc Health; Conf Res Workers Animal Dis. *Res:* Virus-host cell relationships in tissue culture systems; respiratory diseases of poultry; brucellosis and tuberculosis in domestic animals; enteric diseases of young swine; rabies in swine; streptococcic lymphadenitis of swine; mycotoxins-mycotoxicoses. *Mailing Add:* Vet Med Diagn Lab Sch Vet Med Univ Mo-Columbia Columbia MO 65201

MOREHOUSE, MARGARET GULICK, b Champaign, Ill, Aug 22, 04; m 33. BIOCHEMISTRY. *Educ:* Univ Calif, AB, 27; Univ Southern Calif, PhD(biochem), 39. *Prof Exp:* Asst biochem, 29-33 & 36-40, from instr to assoc prof, 40-72, EMER ASSOC PROF BIOCHEM, SCH MED, UNIV SOUTHERN CALIF, 73- *Mem:* Am Soc Biol Chem; Soc Exp Biol & Med; Am Oil Chem Soc. *Res:* x-radiation effects upon absorption of a labeled dioleoylstearin; doubly labeled glyceride absorption and synthesis of intestinal mucosa; absorption of lipids linked to fatty acid composition and placement; dietary fat-level effects; rat heart lipids under normal and stress conditions together with changes in diet and age. *Mailing Add:* 842 E Villa No 246 Pasadena CA 91106

MOREHOUSE, NEAL FRANCIS, poultry science; deceased, see previous edition for last biography

MOREIRA, ANTONIO R, b Porto, Portugal, Oct 27, 50; m 73; c 2. BIOCHEMICAL ENGINEERING, BIOTECHNOLOGY. *Educ:* Univ Porto, Portugal, BS, 73; Univ Pa, Philadelphia, MS, 75, PhD(chem & biochem eng), 77. *Prof Exp:* Res assoc biochem eng, Univ Waterloo, 77-78; from asst prof to assoc prof biochem eng, Colo State Univ, 78-82; group leader, Int Flavors & Fragrances, Inc, 82-84; assoc dir, Schering-Plough Corp, 84-90; DIR CHEM & BIOCHEM ENG, UNIV MD, 90- *Concurrent Pos:* Consult, Solar Energy Res Inst, 79-81, EG&G, Idaho, 80-81, Arco Coal Co, 81-82; counr, Am Chem Soc, 83-85; affil prof, Colo State Univ, 83-85; actg co-dir, Ctr Biotechnol Mfg, 90-; NATO sr guest fel, 91. *Honors & Awards:* Award Excellence, Halliburton Educ Found, 80. *Mem:* Am Inst Chem Engrs; AAAS; Soc Indust Microbiol; Am Soc Microbiol; Am Soc Eng Educ. *Res:* Recombinant DNA processes; cell culture; bioisolation technology; antibiotics; steroids; biotransformations; regulatory-engineering issues in bioprocessing; bioprocess scale-up. *Mailing Add:* Univ Md Baltimore County Baltimore MD 21228

MOREIRA, JORGE EDUARDO, b Porono, Arg, Apr 5, 43; c 2. CELL BIOLOGY, ELECTRON MICROSCOPY. *Educ:* Escuela Normal, Paraná, Arg, BA, 60; Northeast Univ, Arg, MVD, 66; Univ Sao Paulo, Brazil, MVD, 78, PhD(cell biol), 82. *Prof Exp:* Asst prof histol, Northeast Univ, Arg, 68-76 & State Univ Sao Paulo, Brazil, 77-86; lab instr, Univ Cordoba Med Sch, Arg, 73-74 & Georgetown Univ Med Sch, Wash, DC, 87; vis assoc cell biol, Clin Invest & Patient Care Br, Nat Inst Dent Res, 86-89, VIS SCIENTIST, LAB NEUROBIOL, NAT INST NEUROL DIS & STROKE, NIH, BETHESDA, 89- *Honors & Awards:* Renato Locchi Award, Brazilian Soc Anat, 84. *Mem:* Am Soc Cell Biol; Electron Micros Soc Am; Histochem Soc. *Res:* Immunocytochemistry; immunoelectron microscopy; localization of secretory proteins; cytoskeletal proteins; cytoplasmic proteins; cell differentiation, secretion. *Mailing Add:* Lab Neurobiol NINDS NIH Bethesda MD 20892

MOREJON, CLARA BAEZ, b Matanzas, Cuba, Nov 30, 40; US citizen; m 65; c 2. CHEMISTRY. *Educ:* Univ Miami, BS, 64, MS, 67. *Prof Exp:* Scientist II spec chem, 66-70, supvr spec chem & chem res & develop, 70-72, SR RES SCIENTIST IMMUNOCHEM RES & DEVELOP, BAXTER DADE DIV, 72-, RADIATION SAFETY OFFICER, 70- *Mem:* Am Chem Soc. *Res:* Immunology for compounds of clinical significance. *Mailing Add:* Baxter Dade Div 1851 Delaware Pkwy Miami FL 33152

MOREL, THOMAS, b New York, NY, Sept 26, 43; m 70; c 2. INTERNAL COMBUSTION ENGINES, FLUID MECHANICS. *Educ:* Czech Tech Univ, Praque, dipl eng, 65; Ill Inst Technol, MS, 69, PhD(mech eng), 72. *Prof Exp:* Eng, Res Inst Transp Mechanisms, 66-67; sr res eng, Res Labs, Gen Motors, 72-81; mgr, Int Harvester, 81-82; res assoc, Univ Minn, 82-83; SR VPRES, RICARDO N AM, 83- *Honors & Awards:* R T Knapp Award, Am Soc Mech Engrs, 79; R T Colwell Award, Soc Automotive Engrs, 88. *Mem:* Fel Am Soc Mech Engrs; Soc Automotive Engrs; Sigma Xi. *Res:* Heat transfer, combustion, tribology, and internal combustion engines; energy systems; vehicle aerodynamics; basic fluid mechanics; turbulence modeling of momentum and heat transfer. *Mailing Add:* Ricardo N Am 645 Blackhawk Dr Westmont IL 60559

MORELAND, ALVIN FRANKLIN, b Morven, Ga, Sept 5, 31; m 55; c 3. VETERINARY & COMPARATIVE MEDICINE, NUTRITION & IMMUNOLOGY. *Educ:* Ga Teachers Col, BS, 51; Univ Ga, MS, 52, DVM, 60; Am Col Lab Animal Med, Dipl. *Prof Exp:* NIH fel lab animal med, Bowman Gray Sch Med, 60-62; asst prof exp path, Univ Va, 62-63; from asst prof to prof, lab animal & wildlife med, Col Vet Med, 63-81, head div, 63-75, prof comp med, Col Med, 72-90, prof & chmn, dept spec clin sci, Col Vet Med, 82-88, PROF SMALL ANIMAL CLIN SCI, COL VET MED, UNIV FLA, 88- *Concurrent Pos:* Vchmn, Coun Accreditation, Am Asn Accreditation of Lab Animal Care, 68-72, chmn, 72-74; consult field lab animal med; vis fel, Australian Nat Univ, 85-86. *Mem:* Am Vet Med Asn; Am Asn Lab Animal Sci; Int Asn Aquatic Animal Med. *Res:* Atherosclerosis; subhuman primate medicine; animal nutrition; organ transplantation. *Mailing Add:* Dept Small Animal Clin Sci Univ Fla Col Vet Med Gainesville FL 32601

MORELAND, CHARLES GLEN, b St Petersburg, Fla, Nov 24, 36; m 60; c 3. PHYSICAL CHEMISTRY. *Educ:* Univ Fla, BS, 60, MS, 62, PhD(phys chem), 64. *Prof Exp:* From asst prof to assoc prof, 64-76, PROF CHEM, NC STATE UNIV, 76- *Honors & Awards:* Sigma Xi Res Award. *Mem:* Am Chem Soc. *Res:* Nuclear magnetic resonance; thermodynamics and kinetics of redistribution reactions of organo-arsenic (V) and antimony (V) derivatives. *Mailing Add:* Dept Chem NC State Univ PO Box 8204 Raleigh NC 27650-8204

MORELAND, DONALD EDWIN, b Enfield, Conn, Oct 12, 19; m 54; c 3. PLANT PHYSIOLOGY, WEED SCIENCE. *Educ:* NC State Univ, BS, 49, MS, 50, PhD(plant physiol), 53. *Prof Exp:* Asst, 50, res asst prof field crops, 53-61, assoc prof crop sci, 61-65, PROF CROP SCI, BOT & FORESTRY, NC STATE UNIV, 65-; PLANT PHYSIOLOGIST, S ATLANTIC AREA, AGR RES SERV, USDA, 53- *Concurrent Pos:* Asst, State Univ NY Col Forestry, Syracuse, 52-53; mem toxicol study sect, NIH, 63-67; res leader, S Atlantic Area, Agr Res SERV, USDA, 73-, sr exec, 79-, location coordr, 84-87. *Honors & Awards:* Outstanding Res Award, Weed Soc Am, 73; Chem Indust CIBA-GeigyAgr Recognition Award, 73; Super Serv Award, USDA, 76. *Mem:* Fel AAAS; Am Soc Plant Physiol; Bot Soc Am; fel Weed Sci Soc Am; Am Chem Soc; Plant Growth Regulator Soc Am. *Res:* Biochemical mechanisms of action of herbicides and growth regulators; photosynthesis; respiration; enzymology; translocation. *Mailing Add:* Dept Crop Sci NC State Univ Raleigh NC 27695-7620

MORELAND, FERRIN BATES, b Portland, Ore, Aug 12, 09; m 37; c 2. FORENSIC TOXICOLOGY, CLINICAL CHEMISTRY. *Educ:* Ore State Univ, BS, 30; Rice Univ, MA, 32; Vanderbilt Univ, PhD(biochem), 36; Am Bd Clin Chem, dipl clin chem, 51, dipl toxicol chem, 72; Am Bd Forensic Toxicol, dipl, 76. *Prof Exp:* Asst biochem, Vanderbilt Univ, 32-36; instr, Tulane Univ, 36; chemist-bacteriologist, Tenn Dept Pub Health, 36; instr biochem, Univ Iowa, 36-42; chief chemist, Kans Dept Pub Health Labs, 45-47; assoc prof biochem, Col Med, Baylor Univ, 47-65; biochemist & dir clin lab, Tex Inst Rehab & Res, 59-65; chief, Biomed Support Sect, Crew Systs Labs, Brown & Root-Northrop, Manned Spacecraft Ctr, Tex, 65-67, mgr, Bio-Med Support Labs, 67-68; dir clin chem & qual control, Metab Res Found, 68-70; chief toxicologist, Off Harris County Med Examr, 70-79; consult toxicologist, 79-90; RETIRED. *Concurrent Pos:* biochemist, Methodist Hosp, Houston, 47-52 & US Vet Admin Hosp, 49-65. *Mem:* Am Chem Soc; Am Asn Clin Chemists (vpres, 57-58, pres, 58-59); Am Acad Forensic Sci; Int Asn Forensic Toxicol; Forensic Sci Soc. *Res:* Biochemical effects of anoxia; methods for clinical chemistry and analytical toxicology; gallbladder activity. *Mailing Add:* 4141 S Braeswood No 366 Houston TX 77025-3330

MORELAND, PARKER ELBERT, JR, b Ft Worth, Tex, Nov 5, 31; m 55; c 3. APPLIED PHYSICS, ENGINEERING MANAGEMENT. *Educ:* Baylor Univ, BS, 54; Harvard Univ, AM, 55, PhD(physics), 62. *Prof Exp:* Asst physicist, Argonne Nat Lab, 62-70; sr physicist, Packard Instrument Co, Ill, 70-73; vpres-tech dir, Sperry Div, Automation Industs, Inc, 73-84; VPRES, TECH DEVELOP, QUALCORP, 84- *Mem:* Am Phys Soc; Acoust Soc Am; Am Soc Nondestructive Testing. *Res:* Mass spectroscopy; atomic masses; ion-molecule reactions; nuclear instrumentation; x-ray fluorescence spectrometry; ultrasonic transducers and instrumentation; materials evaluation. *Mailing Add:* 13 Strawberry Hill Rd Danbury CT 06810

MORELAND, SUZANNE, b Honolulu, Hawaii, Nov 23, 51; m 85. CONTRACTILE PROTEINS. *Educ:* Univ SFla, BA, 73, PhD(med sci), 79. *Prof Exp:* res investr, Squibb Inst Med Res, 82-86; SR RES INVESTR, BRISTOL-MYERS SQUIBB PHARMACEUT RES INST, 86- *Mem:* Am Physiol Soc; Biophys Soc; Am Heart Asn; NY Acad Sci; AAAS. *Res:* Vascular-smooth muscle reactivity; contractile protein regulation; discovery and development of antihypertensive agents. *Mailing Add:* Dept Pharmacol Bristol-Myers Squibb Pharmaceut Res Inst PO Box 4000 Princeton NJ 08543-4000

MORELAND, WALTER THOMAS, JR, b New London, NH, Apr 30, 26; m 48; c 4. MEDICINAL CHEMISTRY, PHARMACOLOGY. *Educ:* Univ NH, BS, 48, MS, 50; Mass Inst Technol, PhD(chem), 52. *Prof Exp:* Res chemist, Chas Pfizer & Co, Inc, Groton, 52-59, proj leader, 59-61, group supvr, 61-63, sect mgr, 63-65, asst dir med chem res, 65-68, dir, 68-72, exec dir med chem res, 72-76, exec dir med sci, 76-80, vpres med sci, 80-84, sr vpres med prod Res, C Res, Pfizer, Inc, 84-86; CONSULT,86- *Mem:* Am Chem Soc; AAAS; NY Acad Sci. *Res:* Cardiovascular drugs; adrenergic agents; medicinal products; antibiotics; central nervous system; cardiopulmonary and metabolic diseases. *Mailing Add:* 45 Westwood Dr Waterford CT 06385-3826

MORELL, PIERRE, b Dominican Repub, Dec 10, 41; US citizen; m 65; c 2. BIOCHEMISTRY. *Educ:* Columbia Univ, AB, 63; Albert Einstein Col Med, PhD(biochem), 68. *Prof Exp:* Asst prof biochem in neurol, Albert Einstein Col Med, 69-72; assoc prof, 72-76, dir curriculum neurobiol, 77-78, RES SCIENTIST, BIOL SCI RES CTR, UNIV NC, 72-, PROF BIOCHEM & BIOPHYS, 76- *Concurrent Pos:* Nat Inst Ment Health fel, Ment Health Res Inst, Univ Mich, Ann Arbor, 68-69. *Mem:* Am Soc Neurochem; Int Soc Neurochem; Am Soc Biochem & Molecular Biol. *Res:* Neurochemistry including metabolism of myelin, axonal transport, neurotoxicology and second messenger systems. *Mailing Add:* Biol Sci Res Ctr 220H Univ NC Chapel Hill NC 27599-7250

MORELLO, EDWIN FRANCIS, b Marseilles, Ill, Jan 12, 28. ORGANIC CHEMISTRY, POLYMER CHEMISTRY. *Educ:* Univ Ill, BS, 48; Univ Minn, PhD(org chem), 52. *Prof Exp:* Res chemist, Standard Oil Co, Ind, 52-57; group leader, 58-65, res chemist, 65-72, RES ASSOC, AMOCO CHEM CORP, 72- *Mem:* Am Chem Soc. *Res:* Engineering polymers; high-temperature polymers; organic chemicals; solid propellants; catalysis. *Mailing Add:* 19 Olympus Dr Apt 1A Naperville IL 60540

MORELLO, JOSEPHINE A, b Boston, Mass, May 2, 36; m 71. MEDICAL MICROBIOLOGY. *Educ:* Simmons Col, BS, 57; Boston Univ, AM, 60, PhD(microbiol), 62; Am Bd Microbiol, Cert med microbiol. *Prof Exp:* Instr microbiol, Boston Univ, 62-64; res assoc, Rockefeller Univ, 64-66; resident med microbiol, Col Physicians & Surgeons, Columbia Univ, 66-68, asst prof microbiol, 68-69; asst prof, 70-73, assoc prof path & med, 73-78, PROF PATH & MED, UNIV CHICAGO, 78-, DIR CLIN MICROBIOL, 70- *Concurrent Pos:* Dir microbiol, Harlem Hosp Ctr, 68-69, ed, Clin Microbiol Rev; ed Clin Microbiol Newsletter. *Honors & Awards:* Sonnenwirth Award, Am Soc Microbiol, 91. *Mem:* Acad Clin Lab Physicians & Scientists; Am Soc Microbiol; fel Am Acad Microbiol; Am Soc Clin Path; Sigma Xi. *Res:* Improved methods of clinical microbiology; epidemiology and characteristics of pathogenic neisseria. *Mailing Add:* 425 Luthin Rd Oak Brook IL 60521-2770

MORELOCK, JACK, b Houston, Tex, Nov 27, 28; m 58. OCEANOGRAPHY. *Educ:* Univ Houston, BS, 50 & 53; Tex A&M Univ, PhD(oceanog), 67. *Prof Exp:* Geologist, Magnolia Petrol Co, Tex, 53; teaching asst geol, Univ Kans, 53-55; explor geologist, Continental Oil Co, Wyo, 55-59 & Oasis Oil Co, Libya, 59-62; teaching asst oceanog, Tex A&M Univ, 62-66; prof & head dept, Fla Inst Technol, 66-69 & Inst Oceanog, Univ Oriente, Venezuela, 69-72; assoc prof, 71-80, PROF MARINE SCI, UNIV PR, MAYAGUEZ, 80- *Concurrent Pos:* Consult to secy natural resources, Dept Natural Resources, PR, 74-, US Geol Surv, 80. *Mem:* Soc Econ Paleont & Mineral; Explorers Club. *Res:* Geological oceanography in areas of marine sedimentation; beach and coastal processes; estuarine geology; coral reef studies. *Mailing Add:* Dept Marine Sci Univ PR PO Box 5000 Mayaguez PR 00709-5000

MOREL-SEYTOUX, HUBERT JEAN, b Calais, France, Oct 6, 32; US citizen; m 60; c 4. HYDROLOGY, WATER RESOURCES. *Educ:* Nat Sch Civil Eng, Paris, MS, 56; Stanford Univ, PhD(eng), 62. *Prof Exp:* Res engr, Chevron Res Co, 62-66; assoc prof, 66-74, PROF CIVIL ENG, ENG RES CTR, COLO STATE UNIV, 74- *Concurrent Pos:* Chief researcher, Nat Ctr Sci Res, Univ Grenoble, France, 72-73; organizer & ed of proc of hydrol days, 81-91; master in res, Sch Mines, Paris, 82; mem, US Nat Comt Hydrol; invited prof, Fed Polytech Sch Lausanne, Switz, 87. *Mem:* Am Soc Civil Engrs; Am Geophys Union; Soil Sci Soc Am; Am Inst Mining, Metall & Petrol Engrs; Int Asn Hydraul Res; Int Asn Hydrol Sci. *Res:* Floods and droughts; infiltration; recharge; drainage; groundwater models; stream-aquifer models for management; oil reservoir recovery simulation on computers; watershed models; river basin planning. *Mailing Add:* B20 Eng Colo State Univ Ft Collins CO 80523

MORENG, ROBERT EDWARD, b New York, NY, Jan 29, 22; m 50; c 7. POULTRY PHYSIOLOGY. *Educ:* Univ Md, BS, 44, MS, 48, PhD(poultry physiol), 50. *Prof Exp:* Asst poultry husb, Univ Md, 47-50; asst prof, NDak Agr Col, 50-55; prof avian sci, 55-72, asst dir, 72-88, EMER PROF EXP STA, COLO STATE UNIV, 88- *Concurrent Pos:* Sr res fel, Univ Sidney, Australia, 89-90. *Mem:* AAAS; Soc Exp Biol & Med; Poultry Sci Asn; NY Acad Sci; Food Technol & Am Genetics Asn. *Res:* Effects of low temperature on the chicken and the chicken embryo; x-ray irradiation and avian embryo; environment and growth; physiological genetics of reproductive adaptation to high altitude in turkeys. *Mailing Add:* 6221 N County Rd 15 Ft Collins CO 80524

MORENO, CARLOS JULIO, b Sevilla, Colombia, July 30, 46; US citizen; m 72; c 2. MATHEMATICS. *Educ:* NY Univ, BA, 68, PhD(math), 71. *Prof Exp:* From asst prof to prof math, Univ Ill, Urbana-Champaign, 71-85; PROF MATH, CITY UNIV NY, BARUCH GRAD CTR, 85- *Concurrent Pos:* Assoc mem, Ctr Advan Study, Univ Ill, 75-76; vis mem, Inst Advan Study, Princeton, 75-76 & Inst Hautes Etudes Scientifique, Paris, 79-80; Fulbright sr visitor, Colombia, 80. *Mem:* Am Math Soc; Math Asn Am; NY Acad Sci. *Res:* Applications of group representations and automorphic forms to problems in number theory and algebraic geometry; analytic properties of Euler products; algebraic curves and error correcting codes. *Mailing Add:* Box 545 North Salem NY 10560

MORENO, ESTEBAN, b San Juan, PR, Aug 3, 26; US citizen. MEDICINE, PATHOLOGY. *Educ:* Columbia Univ, BA, 48; Temple Univ, MD, 51. *Prof Exp:* Attend pathologist, Univ Hosp, PR, 57-63; assoc prof, 61-73, PROF PATH, SCH MED, UNIV PR, SAN JUAN, 73- *Concurrent Pos:* Attend path, Oncol Hosp, PR Med Ctr, 66-73. *Res:* Dermatopathology; training of physicians in pathology. *Mailing Add:* Dept of Path GPO Box 5067 Univ of Pr Sch of Med San Juan PR 00936

MORENO, HERNAN, b Medellin, Colombia, Sept 9, 39; m 64; c 3. PEDIATRICS, HEMATOLOGY. *Educ:* Berchmans Col, Colombia, BS, 56; Univ Valle, Colombia, MD, 63; Am Bd Pediat, dipl, 67; Educ Coun for Med Grad, cert. *Prof Exp:* Intern, Univ Hosp, Cali, Colombia, 61-62, jr pediat resident, Dept Pediat, 62-63, sr pediat resident, 63-64; pediat resident, Children's Hosp, Birmingham, Ala, 64-65; instr pediat, Col Med, Univ Cincinnati, 65-67; from instr to assoc prof pediat, Sch Med, Univ Ala, Birmingham, 67-76, dir pediat hemat & oncol div, 69-76; CHIEF PEDIAT CLIN & CHIEF PEDIAT, CARRAWAY METHODIST MED CTR, 78- *Concurrent Pos:* Fel pediat hemat, Children's Hosp Res Found, Cincinnati, Ohio, 65-67; consult, Children's Hosp, Birmingham, 67-; mem hosps & clins staff, Sch Med, Univ Ala, 68- *Mem:* Am Acad Pediat; Am Soc Hemat; NY Acad Sci. *Res:* Cancer chemotherapy. *Mailing Add:* 101 Payne Rd Gardendale AL 35071

MORENO, OSCAR, b Camaguey, Cuba, Jan 5, 46. ALGEBRA, COMBINATORICS. *Educ:* Univ PR, BA, 67; Univ Calif, Berkeley, MA, 68, PhD(math), 73. *Prof Exp:* Instr gen studies, Univ PR, 68-69; teaching asst, Dept Math, Univ Calif, 69-72, teaching & res assoc, Dept Math & Electronics Lab, 72-73; asst prof, 74-78, ASSOC PROF MATH, UNIV PR, 78- *Concurrent Pos:* Consult, NSF Miss Proj, Univ PR, 77-78; Univ PR fel, 77-79; NSF grant, 78-80. *Mem:* NY Acad Sci; Am Math Soc; Inst Elec & Electronics Engrs. *Res:* Finite fields and combinatorial theory; its application to coding theory; goppa codes. *Mailing Add:* Dept Math Univ PR Rio Piedras PR 00931

MORENO-BLACK, GERALDINE S, b Brooklyn, NY, Aug 25, 46. PHYSICAL ANTHROPOLOGY. *Educ:* State Univ NY Buffalo, BA, 67; Univ Ariz, MA, 70; Univ Fla, PhD(anthrop), 74. *Prof Exp:* ASSOC PROF ANTHROP, UNIV ORE, 74- *Concurrent Pos:* Tech consult breastfeeding, cult pract, Univ Indonesia; teaching & consult, Univ Mahidal, Bangkok, Thailand. *Mem:* Am Asn Phys Anthropologists; Soc Med Anthrop; Am Anthrop Asn; AAAS. *Res:* Women's issues in nutrition; human ecology; human nutrition; growth and development. *Mailing Add:* Dept Anthrop Univ Ore Eugene OR 97403

MORENZONI, RICHARD ANTHONY, b Sonoma, Calif, Sept 30, 46; m 74. MICROBIOLOGY. *Educ:* Univ Calif, Davis, BS, 68, PhD(microbiol), 73. *Prof Exp:* RES & STAFF MICROBIOLOGIST, E & J GALLO WINERY, 73- *Mem:* Sigma Xi. *Res:* Industrial research related to microbiological practices or problems encountered in the production of wines and spirits. *Mailing Add:* Res Microbiol Dept E & J Gallo Winery PO Box 1130 Modesto CA 95353

MOREST, DONALD KENT, b Kansas City, Mo, Oct 4, 34; m 63; c 2. OTHER MEDICAL & HEALTH SCIENCES. *Educ:* Univ Chicago, BA, 55; Yale Univ, MD, 60. *Prof Exp:* Sr asst surgeon, NIH, 60-63; asst prof anat, Univ Chicago, 63-65; assoc, 65-67, from asst prof to assoc prof anat, Harvard Med Sch, 67-77; PROF ANAT, UNIV CONN HEALTH CTR, 77-, PROF COMMUN SCI, 81-, DIR CTR NEUROL SCI, 86- *Concurrent Pos:* Res assoc otolaryngol, Mass Eye & Ear Infirmary, 65-77; vis scientist, Salk Inst, La Jolla, Calif. *Honors & Awards:* Herrick Award Comp Neurol, 66; Javits Award, NIH, 84, Pepper Award, 90. *Mem:* AAAS; Am Asn Anatomists; Soc Neurosci; Asn Res Otolaryngol; Sigma Xi. *Res:* Neuroembryology; neuroanatomy; connections of central neural pathways and neurocytology; autonomic and sensory systems; auditory system; neurobiology. *Mailing Add:* Dept of Anat Univ of Conn Health Ctr Farmington CT 06030

MORETON, JULIAN EDWARD, b Mobile, Ala, July 5, 43; m 63; c 3. NEUROPHARMACOLOGY, BEHAVIORAL PHARMACOLOGY. *Educ:* Univ Miss Sch Pharm, BS, 66, PhD(pharmacol), 71. *Prof Exp:* NIMH fel, Psychiat Res Univ, Sch Med, Univ Minn, 71-73; res assoc, Sch Med, Univ Cincinnati, 73-74; from asst prof to assoc prof, 74-88, PROF PHARMACOL & TOXICOL, SCH PHARM, UNIV MD, 88- *Concurrent Pos:* Co-prin investr, res grants, Nat Inst Drug Abuse, 74-82; prin investr res grants, Nat Inst Drug Abuse, 84-, grad prog dir, 82-88; mem drug abuse adv comt, Food & Drug Admin, 88-; sr res fel, Nat Acad Sci/Nat Res Coun, 90-91. *Mem:* Am Soc Pharmacol & Exp Therapeut; Soc Neurosci; Am Asn Col Pharm; AAAS. *Res:* Neuropharmacology and behavioral pharmacology of drug abuse and drug dependence using EEG, behavioral and self-administration techniques in the rat and monkey. *Mailing Add:* Dept Pharmacol & Toxicol Sch Pharm Univ Md 20 N Pine St Baltimore MD 21201

MORETON, ROBERT DULANEY, b Brookhaven, Miss, Sept 24, 13; m 45. RADIOLOGY. *Educ:* Millsaps Col, BS, 34; Univ Miss, cert, 36; Univ Tenn, MD, 38; Am Bd Radiol, dipl, 43. *Prof Exp:* Intern, Lloyd Noland Mem Hosp, Fairfield, Ala, 38-39; instr, Sch Med, Univ Miss, 40; lectr, Univ Tex Med Br, Galveston, 45-50; from instr to assoc prof clin radiol, Univ Tex Southwestern Med Sch, Dallas, 51-65; PROF RADIOL, UNIV TEX M D ANDERSON HOSP & TUMOR INST, HOUSTON, 65-, VPRES PROF & PUB AFFAIRS, 69- *Concurrent Pos:* Fel radiol, Mayo Found, 40-42; staff radiologist, Scott & White Hosp & Clin & Santa Fe Hosp, Temple, Tex, 42-50; chmn & partner, Bond Radiol Group, Tex, 50-65; consult var hosps, companies & railroads, Tex, 52-65; chmn & dir dept radiol, Harris Hosp, Ft Worth & Ft Worth Children's Hosp, 61-65; vpres, Univ Cancer Found, Tex, 65-; founding mem bd, Carter Blood Bank & Bd Radiation & Res Found Southwest; mem bd & exec comt, Radiation Ctr, Ft Worth; chmn, Tex Bd Health, 75-80. *Mem:* Am Col Radiol; AMA; Am Roentgen Ray Soc; Am Geriat Soc; Indust Med Asn. *Mailing Add:* Dept Radiol Health Univ Tex Health Sci Ctr 1515 Holcombe Blvd Houston TX 77030

MORETTI, G(INO), b Turin, Italy, Jan 2, 17; US citizen; m 41; c 5. APPLIED MATHEMATICS, FLUID DYNAMICS. *Educ:* Univ Turin, PhD(math), 39. *Prof Exp:* Asst prof mech, Turin Polytech Inst, 45-48; prof fluid mech, Univ Cordoba, 48-55; prof classical & fluid mech, Inst Physics, Bariloche, 55-58; sci supvr, Gen Appl Sci Labs, Inc, 58-67; prof, 67-86, EMER PROF AEROSPACE ENG, POLYTECH UNIV, 87-; PRES, GMAF INC, 81- *Concurrent Pos:* Adv, Aerotech Inst Cordoba, 48-55; res scientist, Atomic Energy Comn, Argentina, 55-58; consult, Grumman Aerospace. *Honors & Awards:* 00081286x; Medal Except Sci Achievement, NASA, 85. *Mem:* Fel Am Inst Aeronaut & Astronaut. *Res:* Numerical gas dynamics; time-dependent, three-dimensional, inviscid and viscous flows; real gas and chemical effects. *Mailing Add:* PO Box 184 Freeport NY 11520

MORETTI, PETER M, b Zurich, Switz, Apr 13, 35; US citizen; m 77; c 5. MECHANICAL ENGINEERING. *Educ:* Calif Inst Technol, BS, 57, MS, 58; Stanford Univ, PhD(mech eng), 65. *Prof Exp:* Proj mgr nuclear eng, Interatom, Bensberg, Ger, 64-68; sr engr, Westinghouse ARD, 68-70; PROF MECH ENG, OKLA STATE UNIV, 70- *Concurrent Pos:* Fulbright grant, Technische Hochschule, Darmstadt, 58-59; prog mgr, Solar Tech Div Wind Syst Br, US Dept Energy, 77-78; consult flow-induced vibrations; alternative energy systems. *Mem:* Am Soc Mech Engrs. *Res:* Heat-exchanger vibrations; wind-power applications; stratified lakes; flutter of high-speed belts and webs. *Mailing Add:* Sch Mech & Aerospace Eng Okla State Univ Stillwater OK 74078-0545

MORETTI, RICHARD LEO, b Ft Collins, Colo, Feb 15, 29; m 54; c 3. CARDIOVASCULAR PHYSIOLOGY. *Educ:* Univ Calif, Riverside, AB, 57, Berkeley, MA, 60, PhD(zool), 64. *Prof Exp:* Lectr embryol, Univ Calif, Riverside, 64-65, asst prof embryol & asst res zoologist, Air Pollution Res Ctr, 65-73; res biologist, Bruce Lyon Mem Res Lab, Children's Hosp, Oakland, 73-80, res physiologist, 80-; RES PHYSIOLOGIST, BIOMED RES UNIT. *Concurrent Pos:* Consult, May, Ecker, Iverson & Young Cardiac & Thoracic Surgery Med Group. *Mem:* Am Physiol Soc; Soc Exp Biol Med. *Res:* Mechanisms regulating coronary blood flow; mechanisms controlling platelet aggregation; vascular and hemostatic factors in sickle cell disease. *Mailing Add:* Biomed Res Unit 5711 Columbia Richmond CA 94804

MORETTO, LUCIANO G, b Bisuschio, Italy, Feb 18, 40; m 65; c 2. NUCLEAR CHEMISTRY, NUCLEAR PHYSICS. *Educ:* Univ Pavia, Italy, Laurea, 64, Libera Docenza(nuclear chem), 70. *Prof Exp:* Teaching assoc, Univ Pavia, 64-65, asst prof chem, 68-71; fel nuclear chem, Lawrence Berkeley Lab, 65-68; from asst prof to assoc prof, 71-77, PROF CHEM, UNIV CALIF, BERKELEY, 77- *Concurrent Pos:* Sr scientist, Lawrence Berkeley Lab, 71-; assoc ed, Nuclear Physics, 76- *Mem:* Am Chem Soc. *Res:* Transport properties in heavy ion induced reactions; fission-statistical properties of excited atomic nuclei. *Mailing Add:* 5570 Harbord Dr Oakland CA 94618-2624

MORETZ, ROGER C, b Warren, Ohio, July 4, 42; m 67; c 2. CELL BIOLOGY, ULTRASTRUCTURAL PATHOLOGY. *Educ:* Greenville Col, Ill, AB, 64; Rutgers Univ, MS, 65; State Univ NY, Buffalo, PhD(biophys), 73. *Prof Exp:* Res assoc, TAM Div, Nat Lead Co, 66-67; asst res scientist, Roswell Park Mem Inst, 67-68, res scientist I, Electron Optics Lab, 69-72; NIH res assoc fel, dept molecular, cellular & develop biol, Univ Colo, 72-74; res assoc, dept path, Peter Bent Brigham Hosp & depts anat & path, Med Sch, Harvard Univ, 74-76; res scientist III, Dept Path Neurobiol, NY State Inst Basic Res Develop Disabilities, Staten Island, 76-82, res scientist IV, 82-86; SR PRIN SCIENTIST, BOEHRINGER INGELHEIM PHARMACEUT, INC, RIDGEFIELD, CONN, 86- *Concurrent Pos:* Lectr histol, Dept Anat, Rutgers Med Sch, Univ Med & Dent NJ, 78-81. *Mem:* Am Soc Cell Biol; Electron Micros Soc Am; NY Acad Sci; Microbeam Anal Soc; AAAS; Int Soc Stereology. *Res:* Application of electron and confocal laser scanning microscopy to the discovery and development of pharmaceuticals. *Mailing Add:* Dept Anal Sci Boehringer Ingelheim Pharmaceut Inc PO Box 368 Ridgefield CT 06877

MORETZ, WILLIAM HENRY, b Hickory, NC, Oct 23, 14; m 47; c 6. SURGERY. *Educ:* Lenoir-Rhyne Col, BS, 35; Harvard Univ, MD, 39. *Hon Degrees:* DSc, Lenoir-Rhyne Col, 61. *Prof Exp:* Instr surg, Sch Med, Univ Rochester, 44-47; from asst prof to assoc prof, Sch Med, Univ Utah, 47-55; prof & chmn dept, 55-72, PRES, MED COL GA, 72- *Concurrent Pos:* Attend surgeon, North Bench Vet Admin Hosp & assoc surgeon, Salt Lake Gen Hosp, Salt Lake City, Utah, 47-55; consult surgeon, Tooele Army Hosp, 51-55; chief surg, Eugene Talmadge Mem Hosp, Ga, 55-72; consult, US Vet Admin Hosp, 55- *Mem:* Soc Univ Surgeons; Am Surg Asn; Int Cardiovasc Soc; Int Soc Surgeons; Soc Surg Alimentary Tract. *Res:* Thrombo-embolism and arterial diseases. *Mailing Add:* 1120 15th St Augusta GA 30912

MOREWITZ, HARRY ALAN, b Newport News, Va, June 2, 23; m 48; c 2. REACTOR SAFETY, AEROSOL PHYSICS. *Educ:* Col William & Mary, BS, 43; Columbia Univ, AM, 49; NY Univ, PhD(physics), 53. *Prof Exp:* Res assoc cosmic rays, NY Univ, 49-53; supvr reactor physics, Westinghouse Elec Corp, 53-59; proj mgr, Rockwell Int Corp, 59-80, sr staff engr, Atomics Int DIv, 80-82; PRIN, HM ASSOC LTD, 82- *Concurrent Pos:* Proj engr, Metavac, Inc, 52-53; lectr, Dept Radiol, Ctr Health Sci, Univ Calif, Los Angeles, 68-72; mem, Adv Comt Reactor Physics, US AEC, 69-73; consult. *Mem:* Am Phys Soc; Am Nuclear Soc; Sigma Xi; NY Acad Sci; Health Physics Soc; Am Geophysical Union. *Res:* Reactor physics; reactor safety; aerosol physics; transient liquid metal heat transfer; radiation dosimetry. *Mailing Add:* 5300 Bothwell Rd Tarzana CA 91356-2923

MOREY, BOOKER WILLIAMS, b Rochester, NY, May 12, 41; div. SEPARATION TECHNOLOGY. *Educ:* Pa State Univ, BS, 63, MS, 65; Stanford Univ, PhD(mineral eng), 69. *Prof Exp:* Sr res engr, Garrett Res & Develop Co, Inc, 69-72, group leader, Occidental Res Corp, 72-78; dir advan technol, Envirotech Corp, 78-81; vpres, Waste Energy Technol, 82-84; consult, 81-84, SR CONSULT, SRI INT, 84- *Mem:* Am Inst Mining, Metall & Petrol Engrs (vpres, 81-84); Filtration Soc; Int Precious Metals Inst; Air & Waste Mgt Asn. *Res:* Technology and innovation management; equipment and process development for plastics, scrap and municipal refuse recycling; pollution control technology; development of separation equipment for solid-solid, liquid-solid and gas-solid separations. *Mailing Add:* SRI Int 333 Ravenswood Ave Menlo Park CA 94025

MOREY, DARRELL DORR, b Manhattan, Kans, Dec 6, 14; m 40; c 6. AGRONOMY. *Educ:* Kans State Col, BS, 37; Tex Tech Col, MS, 38; Iowa State Col, PhD(agron), 47. *Prof Exp:* Asst, Tex Tech Col, 37-38, instr crops, 38-39; jr supvr grain inspection, Grain Br, USDA, Minn, 39-44; assoc, Exp Sta, Iowa State Col, 44-47; assoc prof agron, Univ Ga, 48-49; assoc agronomist, Fla Agr Exp Sta, 49-53; assoc plant breeder, Wheat Breeding Invests, Ga Coastal Plain Exp Sta, Crop Res Div, Agr Res Serv, USDA, 53-65, plant breeder rye triticale, 65-85; prof agron, Univ GA, 65-85; RETIRED. *Concurrent Pos:* Mem, Nat Oat Conf Comt, 51-52 & Nat Wheat Improv Coun, 73-75; coordr, Uniform Southern Soft Wheat Exps, 73-76. *Mem:* Crop Sci Soc Am; Am Soc Agron. *Res:* Breeding disease resistance in small grains; genetics and cytology of cereals; rye breeding; triticale breeding; wheat breeding. *Mailing Add:* 207 W 18th St Tifton GA 31794

MOREY, DONALD ROGER, b Buffalo, NY, Mar 27, 07; m 32; c 1. OPTICAL MATERIALS, INSTRUMENTATION. *Educ:* Univ Buffalo, BS, 27; Cornell Univ, PhD(physics), 31. *Prof Exp:* Sr res fel, Textile Found, Washington, DC, 32-35; res physicist, Tenn Eastman Div, Eastman Kodak Co, 35-45, sr res assoc, Navy Ordnance Div, 50-70; WRITER & CONSULT, 78- *Concurrent Pos:* Chmn, NY Sect, Am Phys Soc, 61-63. *Mem:* Optical Soc Am; fel Am Phys Soc. *Res:* Mechanical and optical properties of polymers and fibers; photoconductive and infrared properties of binary compounds; military instrumentation; limitations of physical laws in relation to the philosophic problem of determinism. *Mailing Add:* Box 127 Naples NY 14512

MOREY, GLENN BERNHARDT, b Duluth, Minn, Oct 17, 35; m 67; c 2. HISTORY OF SCIENCE. *Educ:* Univ Minn, Duluth, BA, 53, Minneapolis, MS, 62, PhD(geol), 65. *Prof Exp:* Instr geol, Univ Minn, Minneapolis, 64-65, asst prof, Minn Geol Surv, 65-69, prin geologist Precambrian, Minn Geol Surv, 73-76, assoc prof geol, 69-86, ASSOC DIR, MINN GEOL SURV, UNIV MINN, ST PAUL, 76-, CHIEF GEOLOGIST, 79-, PROF GEOL, 86- *Concurrent Pos:* Group leader, Int Geol Correlation Prog & Int Union Geol Sci, 75-83; co-leader, problems of iron formations, Int Union Geol Sci, 75-84; mem, DNAG prog, Geol Soc Am, 83-88; coordr, Proj Correlation Stratig Units, NAm, 82-84, adv bd, Nat Resources Res Inst, Duluth, 86-; field trip leader, Int Geol Cong, Great Lakes Region, 87-88; mem steering comt,

Midcontinent Rift Syst Sci Drilling Prog, Deep Observation & Sampling Earth's Continental Crust-NSF, 87-89; mem, NAM Comn Stratig Nomenclature, 91- *Honors & Awards:* Goldich Medal, Inst Lake Super Geol, 86. *Mem:* Fel Geol Soc Am; Hist Earth Sci Soc; Soc Mining Metall & Explor. *Res:* Field and laboratory studies of Precambrian rocks, primarily iron-formations, and sedimentary manganese deposits in Minnesota and the Lake Superior Region; history of geology. *Mailing Add:* Minn Geol Surv 2642 Univ Ave St Paul MN 55114-1057

MOREY, ROBERT VANCE, b Sturgis, Mich, Jan 19, 45; m 65; c 2. AGRICULTURAL ENGINEERING. *Educ:* Mich State Univ, BS, 67; Purdue Univ, PhD(agr eng), 71. *Prof Exp:* From asst prof to assoc prof, 70-81, PROF AGR ENG, UNIV MINN, 81- *Mem:* Am Soc Engrs; Soc Comput Simulation; Sigma Xi. *Res:* Grain drying; storage and handling; computer modeling and simulation. *Mailing Add:* 206 Agr Eng Univ of Minn St Paul MN 55108

MOREY-HOLTON, EMILY RENE, b Kirksville, Mo, Dec 9, 36; m 67; c 1. PHARMACOLOGY, PHYSIOLOGY. *Educ:* W Va Univ, BA, 58, BS, 61, PhD(pharmacol), 64. *Prof Exp:* Technician surgical res, Peter Bent Brigham Hosp, 58-59; instr pharmacol, Sch Med, Univ Pittsburgh, 63-67; asst prof pharmacol, Med Sch, Ind Univ, 67-68; aerospace technologist biomed, NASA-Wallops Flight Ctr, 68-74; SR RES SCIENTIST BIOMED, NASA-AMES RES CTR, 74- *Mem:* Am Physiol Soc; Am Soc Bone & Mineral Res; fel AAAS. *Res:* Calcium homeostasis; bone changes produced by space flight and simulated space flight. *Mailing Add:* M/S 236-Seven Ames Res Ctr NASA Moffett Field CA 94035-1000

MORFOPOULOS, VASSILIS C(ONSTANTINOS) P, b Athens, Greece, Oct 22, 37; US citizen; m 62. METALLURGICAL & MATERIALS ENGINEERING. *Educ:* Purdue Univ, BS, 58, MS, 61; Columbia Univ, EngScD(eng sci), 64. *Prof Exp:* Res assoc metall eng, Purdue Univ, 57-60 & Columbia Univ, 60-61; res engr, US Steel Corp, 61; instr chem, City Univ New York, 61-63; res engr, Argonne Nat Lab, AEC, 63; Am Iron & Steel res fel metall eng, Columbia Univ, 64-65; sr metall scientist, 65-66, TECH DIR RES & DEVELOP TESTING, AM STANDARDS TESTING BUR, 66- *Concurrent Pos:* Consult, govt & indust, 66-; mem, Int Comn Chem Thermodyn & Kinetics; mem, Transp Res Bd, Nat Res Coun. *Mem:* AAAS; Am Inst Mining, Metall & Petrol Engrs; Am Soc Eng Educ; Asn Consult Chemists & Chem Engrs; NY Acad Sci. *Res:* Research and consulting in fields of corrosion and oxidation phenomena; low and high-temperature thermodynamics; liquid metals and compounds; surface phenomena; electrometallurgy and electrode phenomena; electrical and magnetic properties of matter; failure and stress analysis; metal finishing, joining and working. *Mailing Add:* Am Standards Testing Bur 40 Water St New York NY 10004-2672

MORGAL, PAUL WALTER, b Council Bluffs, Iowa, Dec 10, 11; c 3. CHEMICAL ENGINEERING, INDUSTRIAL CHEMISTRY. *Educ:* Iowa State Univ, BS, 33, MS, 35, PhD(chem), 37. *Prof Exp:* Chemist rayon plant, Du Pont Rayon Co, 33-34; develop engr inorg salts, Gen Chem Co, 37-38; Rackham Found fel indust agr prod, Mich State Univ, 38-42; eng assoc hydrogenation coal, Ind Gas Assoc, Purdue Univ, 42-43; process engr petrol refining, Union Oil Calif, 43-47, proj mgr petrol refining construct, 48-77; CONSULT CHEM ENGR, RES DEPT, UNION OIL CO CALIF, BREA, 78- *Concurrent Pos:* Instr, Univ Calif, 48-63. *Mem:* Am Chem Soc; Am Inst Chem Engrs. *Res:* Industrial uses of agricultural products; hydrogenation of coal. *Mailing Add:* 4446 Rutgers Ave Long Beach CA 90808-2398

MORGALI, JAMES R, b Salem, Ore, Sept 7, 32; m 57; c 3. CIVIL ENGINEERING. *Educ:* Willamette Univ, BA, 56; Stanford Univ, BS, 56, MS, 57, PhD(civil eng), 64. *Prof Exp:* PROF CIVIL ENG, UNIV OF THE PAC, 61- *Mem:* Am Soc Civil Engrs; Am Soc Eng Educ. *Res:* Synthesis of runoff hydrographs from small drainage areas. *Mailing Add:* Univ Pac 3601 Pacific Ave Stockton CA 95211

MORGAN, ALAN RAYMOND, b Tredeger, Gwent, UK, July 22, 55; m 80; c 1. HEME PROTEINS, HETEROCYCLIC CHEMISTRY. *Educ:* Exeter Univ, BSc, 76; Univ London, PhD(chem), 79. *Prof Exp:* Teaching fel, Univ Sussex, UK, 79-81; res assoc, Univ BC, Can, 81-83; from asst prof to assoc prof, 83-89, PROF CHEM & BIOCHEM, UNIV TOLEDO, 90- *Concurrent Pos:* Treas, Toledo Sect, Am Chem Soc, 85-86; Sigma Xi res award, Univ Toledo, 90. *Mem:* Am Chem Soc; Sigma Xi; Royal Soc Chem. *Res:* Synthesis and properties of petroporphyrins; the biological role of iron; the use of photosensitizers in cancer control. *Mailing Add:* Dept Chem Univ Toledo Toledo OH 43606

MORGAN, ALVIN H(ANSON), b Highmore, SDak, July 8, 08; m 37; c 4. RADIO ENGINEERING. *Educ:* Kans State Univ, BS, 37; Georgetown Univ, MS, 52. *Prof Exp:* Electronic engr, Shell Geophys Div, Okla, 37-41; asst chief field opers sect, Nat Bur Standards, 46-51, proj leader in charge frequency & time standards, High Frequency Standards Sect, 51-54, chief frequency & time dissemination, Boulder Lab, 54-63; chief frequency-time dissemination res, 63-68, sci consult, 68; prof, 69-76, EMER PROF ELECTRONICS, METROP STATE COL, 76- *Concurrent Pos:* Lectr, George Washington Univ, 53-54; instr, Exten Ctr, Univ Colo, 63; supvr design & construct, radio stas WWVB & WWVL. *Mem:* AAAS; Inst Elec & Electronics Engrs; Sigma Xi; NY Acad Sci. *Res:* Design of radio equipment for radio propagation measurement; frequency measurements; oscillators; transistors. *Mailing Add:* Salina Star Rte Boulder CO 80302

MORGAN, ANTONY RICHARD, b Mombasa, Kenya, Jan 5, 40. MOLECULAR BIOLOGY. *Educ:* Cambridge Univ, BA, 61; Univ Alta, PhD(chem), 64. *Prof Exp:* Proj assoc DNA chem & enzymol, Enzyme Inst, Univ Wis-Madison, 65-67, asst prof, 67-69; asst prof DNA chem & enzymol, 69-77, PROF DNA CHEM & ENZYMOL, UNIV ALTA, 77- *Mem:* AAAS; Can Fedn Biol Socs. *Res:* Chemical mechanism of DNA replication in vitro; three-stranded complexes between DNA and RNA and their biological implications for transcription; multistranded nucleic acid complexes and DNA replication. *Mailing Add:* Dept Biochem Univ Alta Edmonton AB T6G 2E2 Can

MORGAN, ARTHUR I, JR, b Berkeley, Calif, May 21, 23; m 48, 83; c 2. CHEMICAL ENGINEERING. *Educ:* Univ Calif, BS, 43, MS, 48; Swiss Fed Inst Technol, PhD(chem eng), 52. *Prof Exp:* Chem engr, USDA, 52-58, head unit opers invest, 58-62, chief eng & develop lab, 62-69, ctr dir, Western Regional Res Ctr, 69-84, NAT PROG LEADER, USDA, 84- *Concurrent Pos:* Adj prof nutritional sci & lectr chem eng, Univ Calif, Berkeley, 67-84; Consult, ITT, 80. *Honors & Awards:* Babcock-Hart Award, Inst Food Technol, 68; Food & Bioeng Award, Am Inst Chem Engrs, 71. *Mem:* Am Chem Soc; Am Inst Chem Engrs; Inst Food Technol. *Res:* Heat transfer, drying, fluid flow and evaporation relative to agricultural products. *Mailing Add:* Bldg 005 BARC-W Beltsville MD 20705

MORGAN, BERNARD S(TANLEY), JR, b Brooklyn, NY, June 30, 27; m 51; c 5. ENGINEERING. *Educ:* US Naval Acad, BS, 51; Univ Mich, MS(aeronaut eng) & MS(instrumentation eng), 57, PhD(instrumentation eng), 63. *Prof Exp:* US Air Force, 51-79, missile officer, 3499 mobil training wing, 51-55, asst aeronaut eng, US Air Force Inst Technol, 58-60, proj scientist, Air Force Off Sci Res, 60-65, dir, Aerospace Mech Div, F J Seiler Res Lab, Colo, 65-69, vcomdr aerospace res labs, Wright-Patterson AFB, 69-70, chief, Command Control & Reconnaissance Div & asst chief staff studies & anal, hq, 71-73, comdr, Air Force Cambridge Res Labs, 74-76, comdr, Air Force Geophysics Lab, 76-79; vpres systs technol & integration, BDM Corp, 79-82; PROG MGR, TELEDYNE BROWN ENGINEERING. *Mem:* Fel Inst Elec & Electronics Engrs; Am Inst Aeronaut & Astronaut; Am Soc Mech Engrs; Am Automatic Control Coun (pres, 78-79); Sigma Xi. *Res:* Multivariable control systems; high-speed aerodynamics; operations research; applied mathematics. *Mailing Add:* 2808 Bentley St RD SE Huntsville AL 35801

MORGAN, BEVERLY CARVER, b New York, NY, May 29, 27; m 54; c 3. PEDIATRIC CARDIOLOGY. *Educ:* Duke Univ, MD, 55; Am Bd Pediat, dipl, 60, cert cardiol, 61. *Prof Exp:* Intern & asst resident pediat, Stanford Univ, 55-56; trainee pediat cardiol, Babies Hosp, Columbia Univ, 57-60; dir cardiol, Heart Sta, Robert Green Hosp, San Antonio, Tex, 60-62; from instr to assoc prof pediat, Sch Med, Univ Wash, 62-73, prof & chmn dept, 73-80; PROF PEDIAT & CHMN DEPT, SCH MED, UNIV CALIF, IRVINE, 80- *Concurrent Pos:* Clin fel, Babies Hosp, Columbia Univ, 56-57, res fel, Cardiovasc Lab, Presby Hosp, 59-60; NIH res career prog award, 66-71; consult to Surgeon Gen, Brooke Army Med Ctr, Tex, 60-62; clin lectr, Sch Med, Univ Tex, 60-62; mem pulmonary training comt, Nat Heart & Lung Inst, 72, mem pulmonary acad award panel, 72-; mem comt NIH training & fel progs, Nat Res Coun, 72-74; mem, Grad Med Educ Nat Adv Comt, 76-80. *Mem:* Am Pediat Soc; Asn Med Sch Pediat Dept Chmn (secy-treas, 81-); Am Acad Pediat; Soc Pediat Res; Am Col Cardiol. *Res:* Clinical research; effects of respiration on circulation. *Mailing Add:* Dept Pediat Sch Med Univ Calif 101 City Dr S Orange CA 92668

MORGAN, BRUCE HARRY, b Sharon, Pa, Sept 30, 31; m 58; c 2. PHYSICS. *Educ:* Harvard Univ, AB, 53; Calif Inst Technol, MS, 54; George Washington Univ, JD, 68. *Prof Exp:* Assoc engr systs anal, Westinghouse Elec Corp, 56-57; ASSOC PROF GEN PHYSICS, US NAVAL ACAD, 57- *Mem:* Am Asn Physics Teachers; Am Solar Energy Soc; Sigma Xi. *Res:* General physics. *Mailing Add:* Dept of Physics Michelson Hall US Naval Acad Annapolis MD 21402

MORGAN, BRUCE HENRY, food microbiology, for more information see previous edition

MORGAN, BRYAN EDWARD, b Lamesa, Tex, Jan 15, 19; m 40. PETROLEUM ENGINEERING, CHEMISTRY. *Educ:* W Tex State Univ, BS, 40. *Prof Exp:* Chemist, Oil Refining Lab, Phillips Petrol Co, 41-42; plant chemist, J S Abercrombie Co, 43-44, lab supvr, 44-46; res engr, Humble Oil & Refining Co, 46-50, sr res engr, 50-60, sr res specialist, 60-64, res assoc, Esso Prod Res Co, 64-77; SR ENGR ASSOC, EXXON PROD RES CO, 77- *Concurrent Pos:* Vchmn nat comt standardization oil well cements, Am Petrol Inst, 62- *Mem:* Soc Petrol Engrs. *Res:* Drilling; logging; formation evaluation; formation stimulation for production of oil and gas. *Mailing Add:* Box 528 Grandview TX 76050

MORGAN, CARL WILLIAM, b Long Island, Kans, Feb 16; m 40; c 2. CIVIL ENGINEERING. *Educ:* Kans State Univ, BS, 38; Univ Tex, MS, 51, PhD(civil eng, math), 58. *Prof Exp:* Instr surv, Univ Tex, Austin, 46-47, from asst prof to assoc prof civil eng, 47-77, prof civil eng & dir eng, Career Asst Ctr & Coop Eng Progs, 77-83, RES ENGR, STRUCT MECH RES LAB, UNIV TEX, AUSTIN, 70-, EMER PROF CIVIL ENG, 83- *Mem:* Am Soc Civil Engrs; Am Soc Eng Educ; Am Geophys Union; Int Asn Hydraul Res; Sigma Xi. *Res:* Fluid mechanics; hydrology. *Mailing Add:* 4611 Edgemont Dr Austin TX 78731

MORGAN, CHARLES D(AVID), b Spring Valley, NY, Nov 7, 34; m 56; c 4. MECHANICAL ENGINEERING. *Educ:* Stevens Inst Technol, ME, 56; Rensselaer Polytech Inst, MS, 60; Lehigh Univ, PhD(mech eng), 65. *Prof Exp:* Engr, Nuclear Div, Combustion Eng Co, 56-60; instr mech eng, Lehigh Univ, 60-64; prin engr, Atomic Energy Div, Babcock & Wilcox Co, 64-67, unit mgr thermal-hydraul methods, Nuclear Power Generation Dept, 67-73, mgr, tech staff, 73-80; TECH CONSULT, 80- *Mem:* Am Soc Mech Engrs. *Res:* Film boiling from vertical surfaces; two phase flow studies; effects of turbulent mixing on burnout; applications of thermal/hydraulics to nuclear reactor design and safety. *Mailing Add:* 1444 Northwood Circle Lynchburg VA 24503

MORGAN, CHARLES ROBERT, b Kingston, Pa, July 18, 34; m 68; c 2. PHYSICAL ORGANIC CHEMISTRY. *Educ:* Pa State Univ, BS, 56; Mass Inst Technol, PhD(org chem), 63. *Prof Exp:* Res assoc, Mass Inst Technol, 63-65; res chemist, Org Res Dept, 65-73, sr res chemist, Photopolymer Systs, 73-74, mgr prod develop, 74-79, mgr res, 79-87, SR RES ASSOC, W R GRACE & CO, 87- *Mem:* Am Chem Soc; Inter-Am Photochem Soc; NAm Membrane Soc. *Res:* Kinetics and mechanisms of organic reactions; sulfur chemistry; polymers for electronics; sealants and adhesives; membrane technology; photochemistry and photopolymers; conductive polymers; packaging technology. *Mailing Add:* W R Grace & Co Columbia MD 21044

MORGAN, DAVID ZACKQUILL, b Fairmont, WVa, Mar 27, 25; m 48; c 2. INTERNAL MEDICINE, GERIATRICS. *Educ:* WVa Univ, AB, 48, BS, 50; Med Col Va, MD, 52; Am Bd Internal Med, dipl, 64. *Prof Exp:* From instr to assoc prof, WVa Univ, 63-73, asst dean, 65-72, assoc dean, 72-86, PROF MED, SCH MED, WVA UNIV, 73-, DIR, GRAD PROG, 86- *Concurrent Pos:* Pres, WVa State Med Asn, 85-86. *Mem:* Fel Am Col Physicians. *Res:* Cardiology; geriatrics. *Mailing Add:* Dept Med Sch Med WVa Univ Morgantown WV 26506

MORGAN, DENNIS RAYMOND, b Cincinnati, Ohio, Feb 19, 42; div; c 2. ELECTRONICS ENGINEERING. *Educ:* Univ Cincinnati, BSEE, 65; Syracuse Univ, MSEE, 68, PhD(elec eng), 70. *Prof Exp:* Sr engr, Gen Elec Co, Electronics Lab, 65-84; DISTINGUISHED MEM TECH STAFF AT&T BELL LABS, 84- *Mem:* Inst Elec & Electronic Engrs; Sigma Xi. *Res:* Signal processing; communications; probability and statistics; applied math; authored over 35 papers in technical journals and one patent. *Mailing Add:* AT&T Bell Labs Rm 2D-331 600 Mountain Ave Murray Hill NJ 07974-2070

MORGAN, DONALD E(ARLE), b Regina, Sask, Sept 1, 17; US citizen; m 41; c 5. INDUSTRIAL ENGINEERING, APPLIED MATHEMATICS. *Educ:* Ore State Univ, BS, 40; Stanford Univ, MS, 62, PhD(indust eng), 64. *Prof Exp:* Mgr appln eng, Electronics Div, Westinghouse Elec Corp, 40-46; owner & gen mgr, Intermt Surg Supply Co, 46-60; prof indust eng, Stanford Univ, 63-65; owner & sr staff mem, Decision Studies Group, 65-68; HEAD DEPT INDUST ENG, CALIF STATE POLYTECH UNIV, SAN LUIS OBISPO, 68- *Concurrent Pos:* Prin investr, US Air Force Systs Command, 62-64, US Navy & Exec Br, US Govt, 64-66. *Mem:* Am Inst Indust Engrs; Inst Elec & Electronics Engrs; Am Soc Qual Control. *Res:* Management decision processes using applied mathematical, statistical, and decision theoretic aids implanted by electronic data processing equipment based on economic and human factors. *Mailing Add:* Dept Indust Eng Polytech Univ Calif State San Luis Obispo CA 93407

MORGAN, DONALD LEE, b Huntingburg, Ind, Dec 19, 46; m 65; c 2. MANUFACTURING ENGINEERING. *Prof Exp:* Chief engr, King Indust Corp, 66-75; dir eng, NAm Prod, 75-82; plant maintenance supt, Curon Div, Reeves Bros, Inc, 82-89; PLANT GEN AFFAIRS MGR, ASMO NC, INC, 89- *Concurrent Pos:* Mfg mgr, Precision Disc Corp, 79-80; instr, Mitchell Community Col, ASMO NC, 90-91. *Honors & Awards:* Dale Carnegie Achievement Award, 73. *Mem:* Soc Mfg Engrs; Precision Metalforming Asn. *Mailing Add:* 630 Danbury Lane Statesville NC 28677

MORGAN, DONALD O'QUINN, b Star, NC, Mar 24, 34; m 59; c 2. VETERINARY IMMUNOLOGY. *Educ:* NC State Univ, BS, 55, MS, 63; Univ Ga, DVM, 59; Univ Ill, PhD(physiol), 67. *Prof Exp:* Res asst microbiol, NC State Univ, 59-62, instr animal sci, 62-63; diagnostician, Dept Agr, NC, 63-64; NIH fel, Univ Ill, 64-67, asst prof vet physiol, 67; vet med officer, Plum Island Animal Dis Lab, Agr Res Serv, USDA, 67-69; from asst prof to assoc prof vet sci, Univ Ky, 69-74; assoc prof vet med, Col Vet Med, Univ Ill, Urbana, 74-76; HEAD SCIENTIST MOLECULAR BIOL, PLUM ISLAND ANIMAL DIS CTR, AGR RES SERV, US DEPT AGR, 76- *Honors & Awards:* Newcomb-Cleveland Prize, AAAS, 81-82. *Mem:* NY Acad Sci; Am Soc Vet Physiol & Pharmacol; Am Vet Med Asn; Am Soc Microbiol; Conf Res Workers Clin Dis; Am Soc Virol; Am Soc Trop Vet Med; Am Asn Vet Immunologists. *Res:* Exotic viral diseases of food animals, research and development of foot and mouth diseases vaccines; participation in international vaccine field trials in South America. *Mailing Add:* 2930 Little Neck Rd Cutchogue NY 11935

MORGAN, DONALD PRYSE, b Indianapolis, Ind; m 52; c 3. PREVENTIVE MEDICINE. *Educ:* Ind Univ, MD, 47; Northwestern Univ, PhD(physiol), 53. *Prof Exp:* Asst prof physiol, Sch Med, Ind Univ, 53-54; asst prof, Med Sch, Northwestern Univ, 54-60; pvt pract, Ariz, 60-67; epidemiologist, Community Pesticide Proj, Univ Ariz, 67-73; asst prof prev med, Med Sch, Univ Iowa, 73-77, assoc prof, 77-88; RETIRED. *Mem:* Soc Toxicol; Am Physiol Soc; Sigma Xi. *Res:* Pesticide toxicology; environmental toxicology. *Mailing Add:* 326 Highland Dr Iowa City IA 52246

MORGAN, DONALD R, b Boston, Mass, May 26, 33; m 61; c 2. SOLID-STATE PHYSICS. *Educ:* Boston Col, BS, 55, MS, 57; Univ Notre Dame, PhD(physics), 61. *Prof Exp:* Res engr, Melpar Inc, Mass, 56-57; instr physics, Univ Notre Dame, 60-61; from asst prof to assoc prof, 61-70, PROF PHYSICS, ST MARY'S COL, MINN, 70-, CHMN DIV NATURAL SCI & MATH, 62- *Mem:* Am Phys Soc; Am Asn Physics Teachers. *Res:* Physical electronics; thermionic and field emission with special emphasis on the effects of absorbed molecules on the work function. *Mailing Add:* Dept Physics St Mary's Col Winona MN 55987

MORGAN, EDDIE MACK, JR, ECONOMIC DEVELOPMENT. *Educ:* Univ Miss, DPA, 74. *Prof Exp:* Dir, Indust & Community Develop Dept, 75-82, EXEC VPRES, DELTA COUN, 82- *Res:* Agricultural, forestry and wildlife; flood protection measures in the Mississippi Delta. *Mailing Add:* Delta Council PO Box 257 Stoneville MS 38776

MORGAN, EVAN, b Spokane, Wash, Feb 26, 30; m 59; c 1. ANALYTICAL CHEMISTRY. *Educ:* Gonzaga Univ, BS, 52; Univ Wash, MS, 54, PhD(anal chem), 56. *Prof Exp:* Staff chemist, Int Bus Mach Corp, 56-60; group supvr, Olin Mathieson Chem Corp, 60-64; assoc prof chem, High Point Col, 64-65; sr chemist, Metall Res Dept, Reynolds Metals Co, Va, 65-72; group supvr, 72, sr res chemist, 72-81, PRIN ENGR, BABCOCK & WILCOX, 81- *Concurrent Pos:* Cent Va Air Pollution Control Comt. *Mem:* Am Chem Soc; Am Inst Chemists. *Res:* Nuclear fuels; gas chromatography; emmision spectrometry; mass spectrometry. *Mailing Add:* PO Box 11165 Lynchburg VA 24506-1165

MORGAN, FRANK W, b Green Bay, Wis, Jan 9, 15; m 39; c 2. CHEMICAL ENGINEERING, MATHEMATICS. *Educ:* Purdue Univ, BSChE, 36; Union Univ, NY, MS, 70. *Prof Exp:* Chem engr, Texaco Res Ctr, Texaco, Inc, NY, 36-52, Tex, 52-55 & Beacon, 55-77; RETIRED. *Concurrent Pos:* Instr, Eve Div, Dutchess Community Col, 59-; mem comt info retrieval, Am Petrol Inst, 62-63; mkt assoc, Matt Jordan Realty, Inc, 77-; energy adv, NY State Energy Off, 81- *Mem:* Am Chem Soc; Am Statist Asn; Am Inst Chem Engrs; Am Soc Qual Control; Asn Comput Mach. *Res:* Gasoline middle distillate and lube oil processing; motor and diesel oil development; gasoline research; digital and analog computing; information retrieval. *Mailing Add:* 31 Losee Rd Wappingers Falls NY 12590-2332

MORGAN, GEORGE L, b Cleveland, Ohio, Aug 8, 37. INORGANIC CHEMISTRY. *Educ:* Case Inst, BS, 58; Univ Ill, MS, 60, PhD(inorg chem), 63. *Prof Exp:* From asst prof to assoc prof, 62-72, PROF INORG CHEM, UNIV WYO, 72- *Mem:* Am Chem Soc; Royal Soc Chem. *Res:* Electron-deficient bonding; structures of organometallic compounds; infrared and nuclear magnetic resonance spectroscopy. *Mailing Add:* Dept of Chem Univ of Wyo Laramie WY 82071

MORGAN, GEORGE WALLACE, b Shreveport, La, Aug 14, 41; m 62; c 4. POULTRY PHYSIOLOGY. *Educ:* Miss State Univ, BS, 64, MS, 66, PhD(animal physiol), 70. *Prof Exp:* Staff fel immunol, Div Biologics Standards, 70-73; physiologist, Bur Biologics, Food & Drug Admin, 73-74; ASST PROF IMMUNOL & PHYSIOLOGIST, DEP POULTRY SCI,NC STATE UNIV, 74-;; AT DEPT POULTRY SCI, MISS STATE UNIV. *Mem:* Poultry Sci Asn; Sigma Xi. *Res:* Physiological regulation of immune responsiveness. *Mailing Add:* Dept Poultry Sci Box 5188 Miss State Univ PO Box 5328 Mississippi Station MS 39762

MORGAN, H KEITH, b Indianapolis, Ind, June 6, 42. NUCLEAR PHARMACY. *Educ:* Purdue Univ, BS, 65, MS, 70, PhD(bionucleonics), 70. *Prof Exp:* Adv sr fel nuclear pharm, M D Anderson Hosp & Tumor Inst, 70-71; radiopharm instr, Univ Tex Med Br, 71-76; ASST PROF PHARMACEUT, UNIV HOUSTON COL PHARM, 76- *Mem:* Soc Nuclear Med; Am Pharmaceut Asn; Soc Health Physics; Am Soc Hosp Pharmacists; Am Asn Col Pharm. *Res:* Synthesis, development and evaluation of radiopharmaceuticals; use of radiation in pharmacy and biological research, radiation health physics. *Mailing Add:* 5533 Aspen Houston TX 77081

MORGAN, HARRY CLARK, b Kalamazoo, Mich, Dec 17, 16; c 1. PHYSICS. *Educ:* Mich State Univ, BS, 38, MS, 39. *Prof Exp:* Physicist, Moraine Prod Div, Gen Motors Corp, 39-41; partner, Optron Labs, Ohio, 41-42; proj engr, Curtis Eng Co, 42-47; nuclear physicist, Mound Labs, Monsanto Chem Co, 47-49; owner, Morgan Instruments Co, 49-52; proj engr, Electronics Div, Century Metalcraft Corp, 52-54; lead engr instrumentation data handling, Rocketdyne Div, NAm Aviation Inc, 54-57; sr engr, Res Div, Cohu Electronics, Inc, 57-58; mem tech staff, Ramo Wooldridge Div, TRW, Inc, 58-60; mem tech staff, Autonetics Div, NAm Rockwell Corp, 60-70, space div, 70-71, sr staff assoc, Sci Ctr, 71-74; mem staff, Eon Instrumentation Inc, 74-77; consult & audio visual design, Litton Indust, Inc, 79-81, sr engr 2nd Foundan Converters Med Ultrasonic Scanners, 79-81, eng specialist, Guidance & Control Systs Div, 81-83; CONSULT, 83- *Concurrent Pos:* Mem staff, McDonnel Astronauts, 83-85. *Mem:* Am Phys Soc; Inst Elec & Electronics Engrs; Am Inst Aeronaut & Astronaut. *Res:* Computers and digital data handling systems; optics and nuclear physics; analog electronic systems. *Mailing Add:* 1710 W Hillcrest Dr No 79 Newbury Park CA 91320

MORGAN, HERBERT R, b St Paul, Minn, June 19, 14; m 48; c 3. CANCER. *Educ:* Univ Calif Los Angeles, BS, 36, MA, 38; Harvard Med Sch, MD, 42; Am Bd Prev Med & Pub Health, dipl; Am Bd Microbiol, dipl. *Prof Exp:* Assoc prof med, Univ Mich, 48-50; assoc prof med & prof microbiol, Med Sch, Univ Rochester, 50-79; VIS PROF MICROBIOL, UNIFORMED SERV MED SCH, 79- *Concurrent Pos:* Mem fac, Univ Calif, Harvard Med Sch, Harvard Sch Pub Health, Univ Mich, 48-50; head, Dept Microbiol, Univ Rochester Sch Med & Dent, 50-68, dir, Indust Studies Prog Med, 72-79; dir, Rochester Health Bur Labs, 50-59, Bact Labs, Strong Mem Hosp, 50-68, Microbiol Labs, Monroe County Health Dept, 58-68; consult, Span Ministry Health, 56; consult to surgeon gen, Spain, 56, Polish Ministry Health, 77; vis prof virol, fac med, Univ Oslo; consult, WHO, 56-, Nat Cancer Inst, 62, 68-76; adv virol, Govt Norway, 59-60; mem, Virol & Rickettsiology Study Sect, NIH, 59-63, Microbiol Training Comt, USPHS, 64-68, Comt Res Etiology Cancer, Am Cancer Soc, 65-71, Comt Virol & Cell Biol, Tumor Virol, 67, Instnl Grant Comt, 72-76; assoc ed, J Immunol, 59-63; Fulbright scholar, 60; chmn, Med Sect, Am Soc Microbiol, 61-62; mem, USA-USSR Exchange Group on Tumor Virol, 67; prof fel, Inst Cancer Res, Paris, 67-68; mem, Am Bd Med Microbiol, 70-73, adv panel microbiol, Off Naval Res, 52-60; mem, Certifying Comt, Am Asn Immunol, 70-73; ed, Infection & Immunity, 70-79; med officer, Food & Drug Admin, 79-84; guest investr, Naval Med Res Inst, 80-; vis prof, Ministry Def Hosps, Saudi Arabia, 81. *Mem:* Sigma Xi; AAAS; Am Soc Microbiol; Am Asn Immunol. *Res:* Bacterial endotoxins; chlamydia; virology; tumor viruses; cancer cell biology. *Mailing Add:* Uniformed Serv Univ Health Sci Med Sch 9728 Shadow Oak Dr Gaithersburg MD 20879

MORGAN, HORACE C, b Piedmont, Ala, July 3, 28; m 56; c 2. CLINICAL PATHOLOGY. *Educ:* Auburn Univ, DVM, 55, MS, 58. *Prof Exp:* Instr physiol, Auburn Univ, 55-58, asst prof, 58-59; asst prof clin path, Univ Ga, 60-61, from asst prof to assoc prof physiol, 61-65, assoc prof clin path, 65-70; prof & dir continuing educ & learning resources, 70-73, asst dean, 73-81, ASSOC DEAN, SCH VET MED, AUBURN UNIV, 81- *Concurrent Pos:* Consult, WHO, 73-81, Am Animal Hosp Asn, 80- *Mem:* Am Vet Med Asn; Am Soc Vet Physiol & Pharmacol; Am Soc Vet Clin Path (pres, 67); Am Physiol Soc; Am Heartworm Soc (exec secy, 77-85). *Res:* Liver function; heartworm disease; renal function; medical photography; hematology of domestic animals; fetal electrocardiography. *Mailing Add:* Dept of Admin Col of Vet Med Auburn Univ Auburn AL 36849

MORGAN, HOWARD E, b Bloomington, Ill, Oct 8, 27; m 86. PHYSIOLOGY, BIOCHEMISTRY. *Educ:* Johns Hopkins Univ, MD, 49. *Hon Degrees:* DSc, Susquehanna Univ, 90. *Prof Exp:* Intern obstet & gynec, Vanderbilt Univ, 49-51, asst resident, 51-53, instr, 53-55, fel med res, Howard Hughes Instr, 54-55, investr, 57-67, from asst prof to prof, Dept Physiol, 59-67; prof & chmn, Dept Physiol, Milton S Hershey Med Ctr, Pa State Univ, 67-87, assoc dean res, 73-87, Evan Pugh prof physiol, 74-87, J Lloyd Huck prof, 86-87, EMER EVAN PUGH PROF & J LLOYD HUCK PROF PHYSIOL, PA STATE UNIV, 87-; SR VPRES RES, WEIS CTR RES GEISINGER CLIN, 87- *Concurrent Pos:* Vis scientist, Dept Biochem, Cambridge Univ, Eng, 60-61; mem, Metab Study Sect, NIH, 67-71, Cardiol Adv Comt, Nat Heart, Lung & Blood Inst, 75-78, adv coun, 79-83; ed, Physiol Rev, 73-78, Cell Physiol Sect, Am J Physiol, 81-84 & Heart & Cardiovasc Syst, 84-; mem, Physiol Chem A Res Study Group, Am Heart Asn, 73-75, chmn, 76-79, mem res comt, 74-79, vpres res, chmn res comt & mem bd dirs, 77-78; coord prob area 3, Myocardial Metab, US-USSR Exchange Prog, 74-83 & 86-; assoc ed, Endocrinol & Metab Sect, Am J Physiol, 79-81, J Molecular & Cellular Cardiol, 79-; consult ed, Circulation Res, 82-, Fedn Am Socs Exp Biol J, 86-; Hughes scholar, Howard Hughes Med Inst, 82; mem, US Nat Comt, Int Union Physiol Sci, 84-87; sect ed, Ann Rev Physiol, 86-; mem, Comt Responsible Conduct Res, Inst Med-Nat Acad Sci, 88-89; coun deleg, AAAS, 90, bd dirs, 91- *Honors & Awards:* Merit Award, Am Heart Asn, 79, Distinguished Achievement Award, 88; Carl J Wiggers Award, Cardiovasc Sect, Am Physiol Soc, 84; Citation Distinguished Serv, Am Soc Biol Chemists, 86. *Mem:* Inst Med-Nat Acad Sci; Am Physiol Soc (pres, 85-86); Int Soc Heart Res (pres-elect, 80-83, pres, 83-86); Biophys Soc; Brit Biochem Soc; Am Heart Asn (pres-elect, 86-87, pres, 87-88); hon fel Am Col Cardiol; fel AAAS; Am Soc Biol Chemists. *Res:* Regulation of glucose and glycogen metabolism; membrane transport; regulation of protein and RNA turnover. *Mailing Add:* Geisinger Clin-Weis Ctr Res N Academy Ave Danville PA 17822-2601

MORGAN, IRA LON, b Ft Worth, Tex, Aug 3, 26; m 48; c 3. NUCLEAR PHYSICS. *Educ:* Tex Christian Univ, BA, 49, MA, 51; Univ Tex, PhD(physics), 54. *Prof Exp:* Instr physics, Tex Christian Univ, 48-51; res scientist, Univ Tex, 51-56; vpres nuclear physics, Tex Nuclear Corp, 56-60, exec vpres & dir res, 60-66, pres, 66-68; prof physics & dir ctr nuclear studies, Univ Tex, Austin, 68-73; pres, CSI Corp, 68-75; pres, Sci Measurement Systs, Inc, Austin, Tex, 75-79, vpres, 79-82, sr vpres, 82-87; prof, Univ N Tex, Denton, Tex, 88; PRES, IDMCORP, 87- *Concurrent Pos:* AEC fel, 54-56; vpres, Nuclear-Chicago Corp, 65-78; consult, Los Alamos Sci Lab; mem bd dirs, Allan Shivers Cancer Ctr; chmn, Adv Comt Isotopes & Radiation Develop, AEC, 70-72; mem, Comn Nuclear Sci, Nat Res Coun & Tech Electronic Prod Radiation Safety Standards Comn, Food & Drug Admin, 74-77. *Mem:* Fel Am Phys Soc; fel Am Nuclear Soc; Am Inst Mgt; Sigma Xi. *Res:* Inelastic neutron scattering; activation analysis; radiation interaction in biological systems; photon tomography; nucleon-nucleon interactions; 3D computer analysis. *Mailing Add:* 3800 Palomar Lane Austin TX 78727

MORGAN, JAMES FREDERICK, b Gretna, La, Oct 21, 15; m 43; c 6. INDUSTRIAL HYGIENE. *Educ:* George Washington Univ, BS, 39. *Prof Exp:* Asst chemist, US Food & Drug Admin, 39-42, biochemist, 46-47; head chem hygienist, Indust Hyg Found, Inc, 47-55; dir indust hyg, Pa RR Co, 55-61; asst to dir, Haskell Labs, E I du Pont de Nemours & Co Inc, 61-73, consult indust hyg, 73-80; CONSULT, 80- *Concurrent Pos:* Lectr, Grad Sch Pub Health, Univ Pittsburgh, 50-55; consult, Threshold Limit Values Comt, Am Cong Govt Indust Hyg, 72-80. *Mem:* Am Chem Soc; Am Indust Hyg Asn. *Res:* Physical and chemical environmental health factors; industrial toxicology; design of company operated industrial hygiene services; direction of company-wide industrial hygiene program; composition, analysis and health effects of diesel exhaust; scientific information storage and retrieval; industrial chemical carcinogenic agents. *Mailing Add:* 612 Merion Ave Havertown PA 19083

MORGAN, JAMES FREDERICK, b Minneapolis, Minn, June 20, 41; m 68; c 4. NUCLEAR PHYSICS. *Educ:* St Mary's Col, Minn, BA, 63; Univ Minn, MA, 66, PhD(physics), 68. *Prof Exp:* Res fel, Calif Inst Technol, 68-70; res assoc, Ohio State Univ, 70-74, asst prof nuclear physics, 74-77; MEM STAFF, LAWRENCE LIVERMORE LAB, UNIV CALIF, 77- *Mem:* Am Phys Soc. *Res:* Nuclear astrophysics; understanding of the origin and abundance of the elements, nuclear spectroscopy via analog resonances; understanding of extreme states of matter. *Mailing Add:* 5463 Kathy Way Livermore CA 94550

MORGAN, JAMES JOHN, b New York, NY, June 23, 32; m 57; c 6. ENVIRONMENTAL SCIENCE, WATER CHEMISTRY. *Educ:* Manhattan Col, BCE, 54; Univ Mich, MSE, 56; Harvard Univ, AM, 62, PhD(water chem), 64. *Hon Degrees:* PhD, Manhattan Col, 89. *Prof Exp:* Instr, Univ Ill, 56-60; assoc prof chem & eng, Univ Fla, 63-65; assoc prof eng, 65-69, dean students, 72-75, exec officer, 74-80, actg dean grad studies, 81-84, prof environ eng sci, 69-87, vpres student affairs, 80-89, MARVIN L GOLDBERGER PROF, CALIF INST TECHNOL, 87- *Concurrent Pos:* Ed, Environ Sci & Technol, Am Chem Soc, 66-74. *Honors & Awards:* Water Purification Award, Am Water Works Asn, 63; Award, Am Chem Soc, 80. *Mem:* Nat Acad Eng; AAAS; Am Chem Soc; Sigma Xi; Am Soc Limnol & Oceanog. *Res:* Chemistry of natural water systems; coagulation processes in aqueous systems; water purification. *Mailing Add:* Div Eng & Appl Sci Calif Inst Technol Pasadena CA 91125

MORGAN, JAMES PHILIP, b Cincinnati, Ohio, Jan 13, 48; m 73; c 2. CLINICAL PHARMACOLOGY, CARDIOLOGY. *Educ:* Univ Cincinnati, BS, 70, MD, 76, PhD(pharmacol), 74. *Prof Exp:* Instr pharmacol, Univ Cincinnati, 76; clin fel internal med, Mayo Clin, 76-79, fel, cardiovasc dis & internal med & dir, advan cardiac life support prog, 79-83, asst prof pharmacol & instr med, Mayo Med Sch, 80-83; asst med, Beth Israel Hosp, Boston, 83-88, asst prof, 83-88, ASSOC PROF MED, HARVARD MED SCH, 88- *Concurrent Pos:* Dir, Advanced Cardiac Life Support Prog, Mayo Clin, 79-; affil fac, Minn Heart Asn, 80-83 & Mass Heart Asn, 83-; res career develop award, Nat Heart Lung & Blood Inst, 85- *Honors & Awards:* Balfour Award, 83. *Mem:* Am Med Asn; Am Pharmaceut Asn; Am Col Physicians; Fedn Am Soc Exp Biol; Biophys Soc. *Res:* Cardiovascular and autonomic pharmacology; intracellular ionic calcium handling and excitation-contraction coupling in the heart. *Mailing Add:* 56 Norwood Ave Boston MA 02159

MORGAN, JAMES RICHARD, b Tacoma, Wash, Oct 23, 53; m 86. EARTHQUAKE ENGINEERING, STRUCTURAL DYNAMICS. *Educ:* Univ Ill, BSCE, 75, MSCE, 77, PhD(civil eng), 79. *Prof Exp:* Res asst civil eng, Univ Ill, Urbana-Champaign, 75-79; asst prof, Univ Houston, 79-81; asst prof Civil Eng, 81-87, ASSOC PROF CIVIL ENG, TEX A&M UNIV, 87- *Concurrent Pos:* Teaching asst, Univ Ill, Urbana-Champaign, 77-78; prin investr, NSF, 80-82; mem, Transp Res Bd. *Mem:* Am Soc Civil Eng; Am Soc Eng Educ; Am Concrete Inst; Sigma Xi; Earthquake Eng Res Inst; Am Soc Testing & Mat. *Res:* Structural mechanics and dynamics including earthquake ground motion effects on structures, foundation compliance and soil structure interaction; dynamic analysis of tension-leg offshore platforms; behavior and design of window glass; design of highway safety appurtenances. *Mailing Add:* Civil Eng Dept Tex A&M Univ College Station TX 77843

MORGAN, JOE PETER, b Olds, Iowa, July 30, 31; m 52; c 4. VETERINARY RADIOLOGY. *Educ:* Iowa State Teachers Col, BA, 52; Colo State Univ, DVM, 60, MS, 62; Royal Vet Col, Sweden, Vet med dr, 67. *Prof Exp:* Instr radiol, Colo State Univ, 60-63, asst prof, 63-64 & 67-68; chmn dept radiol sci, 74-77, PROF RADIOL, SCH VET MED, UNIV CALIF, DAVIS, 68- *Res:* Comparative orthopedics. *Mailing Add:* Dept Radiol Sci Univ Calif Sch Vet Med Davis CA 95616

MORGAN, JOHN CLIFFORD, II, b Darby, Pa, Oct 29, 38; m 61; c 4. GENERAL MATHEMATICS, TOPOLOGY. *Educ:* San Diego State Col, AB, 64; Univ Calif, Berkeley, MA, 68, PhD(statist), 71. *Prof Exp:* From asst prof to assoc prof, 78-82, PROF MATH, CALIF STATE POLYTECH UNIV, POMONA, 82- *Mem:* Polish Math Soc. *Res:* Methods of classifying point sets, including Baire category, measure theory and dimension theory; history of modern mathematics. *Mailing Add:* Dept Math Calif State Polytech Univ 3801 W Temple Ave Pomona CA 91768

MORGAN, JOHN D(AVIS), b Newark, NJ, Feb 14, 21; m 53; c 2. MINERAL ENGINEERING, RESOURCE ECONOMICS. *Educ:* Pa State Col, BS, 42, MS, 47, PhD(mining eng), 48, EM, 50; Indust Col Armed Forces, dipl, 53. *Prof Exp:* Asst for mat & stockpile policies, Nat Security Resources Bd, 48-51; dir mat rev div, Defense Prod Admin, 51-53; strategic mat expert, Off Defense Mobilization, 53-56; head, Dept Sci & Math, Daytona Beach Community Col, 61-71; asst dir, 71-74, actg dir, 73-74 & 77-78, assoc dir, 74-78, CHIEF STAFF OFFICER, US BUR MINES, 79- *Concurrent Pos:* Rep, interdept stockpile comt, Munitions Bd, Dept Defense, 48-53; US Govt rep, UN Sci Conf Conserv & Utilization of Resources, 49; mem staff, President's Cabinet Mineral Policy Comt, 53-54; mem Nat Defense Exec Reserve, 56-; consult, Off Defense Mobilization, 56-61, mem spec stockpile adv comt to dir, 57-58; minerals indust consult, 56-71; mem comt scope & conduct of mat res, Nat Acad Sci, 59-60, mem comt mineral sci & technol & chmn panel on mineral econ & resources, 66-70; Dept Interior liaison, Coun Int Econ Policy staff, 73-77, Econ Policy Bd staff, 74-77, Fed Preparedness Agency-Fed Emergency Mgt Agency, Dept Defense, stockpile activities, 76, Dept Defense Mat Steering Group, 75- & Winter Energy Emergency Planning Group, Dept Energy, 77-; mem, Adv Comt Mining & Mineral Resources Res, 77-; hon prof, Indust Col Armed Forces, Nat Defense Univ, Washington, DC, 83. *Honors & Awards:* Legion of Honor, Am Inst Mining, Metall & Petrol Engrs, 89. *Mem:* Fel Soc Am Mil Engrs; Sigma Xi; Am Defense Preparedness Asn; Am Inst Mech Engrs; Mining & Metall Soc Am. *Res:* Geopolitics of strategic and critical materials; mineral economics and national mineral policy. *Mailing Add:* 5013 Worthington Dr Bethesda MD 20816

MORGAN, JOHN DAVIS, III, b Washington, DC, April 10, 55. ATOMIC & MOLECULAR THEORY, MATHEMATICAL PHYSICS. *Educ:* George Washington Univ, BS, 74; Univ Oxford, MSc, 78; Univ Calif, Berkeley, PhD(chem), 78. *Prof Exp:* Dr Chaim Weizmann fel, Dept Physics, Princeton Univ, 78-80, Nat Sci Found Nat Needs fel, 80-81, lectr physics, Princeton Univ, 80-81; asst prof, 81-86, ASSOC PROF PHYSICS, UNIV DEL, 86- *Concurrent Pos:* Assoc ed, J Math Physics, 85-87; visitor, Dept Chem, Univ Chicago, 85; Precision Measurement Grant, Nat Inst Standards & Technol, 87-90; Dept Chem, Harvard, 88-90; vis assoc prof, Dept Physics, Harvard, 90; vis scientist, Harvard-Smithsonian Ctr Astrophysics, 89-90. *Mem:* Am Phys Soc. *Res:* Analytic structure of atomic and molecular wave functions; theory of rates of convergence of Rayleigh-Ritz variational calculations; high-precision calculation of properties of small atoms and molecules; large-order perturbation theory. *Mailing Add:* Dept Physics Univ Del Newark DE 19716

MORGAN, JOHN DERALD, b Hays, Kans, Mar 15, 39; m 62; c 4. ELECTRICAL ENGINEERING. *Educ:* La Tech Univ, BSEE, 62; Univ Mo-Rolla, MSEE, 65; Ariz State Univ, PhD(eng), 68. *Prof Exp:* Elec engr, Tex Eastman Div, Eastman Kodak Co, 62-63; instr elec eng, Univ Mo-Rolla, 63-65 & Ariz State Univ, 65-68; assoc prof, Univ Mo-Rolla, 68-72, Alcoa Found Prof, 72-75, Emerson Elec Prof, 76-85, chmn dept, elec eng, 78-85; DEAN ENG, NEW MEX STATE UNIV, 85- *Concurrent Pos:* Consult, Ariz Pub Serv Co, 68, A B Chance Co, 69-, Westinghouse Corp, 70-72, Emerson Elec, 76- & others; mem planning comt elec div, Am Power Conf, 68-; vis prof, Carnegie-Mellon Univ, 70 & Univ Pittsburgh, 70; Nat Acad Sci exchange prof, Romania, 71. *Honors & Awards:* Centennial Award, Inst Elec & Electronics Engrs. *Mem:* Fel Inst Elec & Electronic Engrs; Am Soc Eng Educ; Am Soc Testing & Mat; Nat Soc Prof Engrs(dir); fel Nat Acad Forensic Engrs. *Res:* Power systems control, analysis, and design; power apparatus testing techniques; high voltage operation of power systems; dielectric materials analysis and design. *Mailing Add:* Col Eng NMex State Univ Box 30001 Dept 3449 Las Cruces NM 88003-3001

MORGAN, JOHN P, b Cincinnati, Ohio, Jan 14, 40; m 64; c 2. CLINICAL PHARMACOLOGY, DRUG MISUSE & ABUSE. *Educ:* Univ Cincinnati, MD, 65. *Prof Exp:* prof med & dir, Dept Pharmacol, 82-90, PROF PHARMACOL, CITY UNIV NEW YORK, 90- *Concurrent Pos:* Assoc prof pharmacol & med, Mt Sinai Sch Med. *Mem:* Am Soc Clin Pharmacol & Therapeut; Am Soc Pharmacol & Exp Therapeut. *Res:* Pharmacology of alcohol and drug misuse. *Mailing Add:* CUNY Med Sch City Col New York 138th St & Convent Ave New York NY 10031

MORGAN, JOHN WALTER, b Walsall, Eng, Jan 27, 32; div; c 2. GEOCHEMISTRY, RADIOCHEMISTRY. *Educ:* Univ Birmingham, BSc, 55; Australian Nat Univ, PhD(geochem), 66. *Prof Exp:* Sci asst electronics, Atomic Energy Res Estab, UK, 48-51, asst exp off, 51-52, anal chem, 55-59; exp off, Australian Atomic Energy Comn Res Estab, 59-61, sr res scientist, 66-68; res asst geochem, Australian Nat Univ, 61-66, res fel, 66; res assoc, Univ Ky, 68-69, chem, 69-70; sr res assoc, Enrico Fermi Inst, Univ Chicago, 70-76; assoc prof geochem, Univ Tex, San Antonio, 76-77; res chemist, 77-80, chief chemist, 80-85, RES CHEMIST, US GEOL SURV, RESTON, VA, 85- *Concurrent Pos:* Prin investr, NASA, 77-81. *Mem:* AAAS; Int Asn Geochem & Cosmochem; Am Geophys Union; Meteoritical Soc. *Res:* Application of analytical chemistry, particularly those aspects involving isotope and radiochemistry to the problems of earth and space science. *Mailing Add:* US Geol Surv MS-981 Nat Ctr Reston VA 22092

MORGAN, JOSEPH, b Kiev, Russia, Mar 4, 09; nat US; m 42; c 2. PHYSICS, X-RAY CRYSTALLOGRAPHY. *Educ:* Temple Univ, AB, 31, MA, 33; Mass Inst Technol, PhD(physics), 37. *Prof Exp:* Asst physics, Temple Univ, 29-31; supvr, Wave Length Proj, Mass Inst Technol, 37-38; from instr to asst prof physics, Tex Agr & Mech Col, 38-41; from asst prof to prof physics, 41-78, dir res coord, 69-76, EMER PROF PHYSICS, TEX CHRISTIAN UNIV, 78- *Concurrent Pos:* Dir eng, Tex Christian Univ, 52-69, chmn, Div Natural Sci, 53-55 & Dept Physics, 58-59, vpres, Univ Res Found, 66-73. *Mem:* Sigma Xi; Am Phys Soc; Am Asn Physics Teachers. *Res:* Optical spectroscopy and physical optics; x-ray diffraction; neutron diffraction; nuclear physics. *Mailing Add:* 10153 Oakton Terrace Rd Oakton VA 22124

MORGAN, JULIET, b Clacton-on-Sea, Eng, Jan 30, 37; US citizen; m 59; c 2. CELL BIOLOGY, ZOOLOGY. *Educ:* Australian Nat Univ, BSc, 66; MacQuarie Univ, MS, 68; Univ Ky, PhD(cell biol), 70. *Prof Exp:* Asst zool, Australian Nat Univ, 64-66; exp officer, Australian AEC, 66-68; res asst cell biol, T H Morgan Sch Biol Sci, Univ Ky, 68-70; asst prof, 78-81, ASSOC PROF CARDIOL, DEPT MED, UNIV CHICAGO, 81-, RES ASSOC, 78- *Concurrent Pos:* Proj dir & co-prin investr, Multiple Risk Factor Intervention Trial, 78-82; asst dir, Cardiol Outpatient Clin, 84-89; fel, pharmacol & morphol, Pharmaceut Mfr Asn Found; commonwealth scholar, Australia. *Honors & Awards:* Pricilla Fairfield Bok Prize. *Mem:* Tissue Culture Asn; Soc Exp Biol & Med; Sigma Xi. *Res:* Basic research in muscle disease; cell structure and function; basic research in treatment of cardiac; cell ultrastructure; basic research in cardiac arrhythmias. *Mailing Add:* Dept Med Box 401 Univ Chicago 5841 S Maryland Ave Chicago IL 60637

MORGAN, KAREN J, b Feb 22, 45; m; c 2. DIETARY ASSESSMENT, SURVEY METHODOLOGY. *Educ:* Purdue Univ, BS, 67; Mich State Univ, MS, 68; Univ Mo, PhD(food econ), 77. *Prof Exp:* Asst prof, Mich State Univ, 77-81; assoc prof, Univ Mo, 81-86; SR DIR, NUTRIT & CONSUMER AFFAIRS, NABISCO BRANDS INC, 87- *Concurrent Pos:* Consult, Food Indust, 77-86. *Mem:* Am Dietetic Asn; Am Inst Nutrit; Inst Food Technologists; Nat Food Processors Asn. *Mailing Add:* Nabisco Brands Inc 100 De Forest Ave East Hanover NJ 07936

MORGAN, KARL ZIEGLER, b Enochsville, NC, Sept 27, 07; m 36; c 4. GENERAL PHYSICS. *Educ:* Univ NC, AB, 29, MA, 30; Duke Univ, PhD(physics), 34. *Hon Degrees:* DSc, Lenoir Rhyne Col, 67. *Prof Exp:* Chmn, Dept Physics, Lenoir Rhyne Col, 34-43; sr scientist health physics, Manhattan Dist & Univ Chicago, 43; dir, Health Physics Div, Oak Ridge Nat Lab, 44-72; adj prof health physics, Vanderbilt Univ, 60-72 & Univ Tenn, 65-72; prof health physics, Ga Tech, 72-83; RETIRED. *Concurrent Pos:* Mem, Int Comn Radiol Protection, 45-71, Nat Coun Radiation Protection; ed-in-chief, Health Physics J, 55-78; vis prof health physics, Appalachian State Univ, 83-86. *Honors & Awards:* Gold Medal for Radiation Protection, Royal Acad Sci Sweden, 62; First Distinguished Serv Award, Health Physics Soc, 68. *Mem:* Hon mem Sigma Xi; fel Am Phys Soc; assoc fel Am Col Radiol; AAAS; fel Am Nuclear Soc; Radiation Res Soc; Int Radiation Protection Asn (pres, 68); Health Physics Soc (pres, 56). *Mailing Add:* 1984 Castleway Dr Atlanta GA 30345

MORGAN, KATHLEEN GREIVE, b Dayton, Ohio, Sept 22, 50; m 73; c 2. PHARMACOLOGY, BIOPHYSICS. *Educ:* Col Mt St Joseph, BS, 72; Univ Cincinnati, PhD(pharmacol), 76. *Prof Exp:* Grass fel, 75, fel, 76-79, ASST PROF PHYSIOL, MAYO FOUND, 79-, ASST PROF PHARMACOL, 81-;; ASSOC PROF PHYSIOL MED, BETH ISRAEL HOSP, HARVARD MED SCH, BOSTON, MASS, 86- *Concurrent Pos:* Minn Heart Asn fel, 77-; estab investr, Am Heart Asn, 84-89. *Mem:* AAAS; Am Physiol Soc; Sigma Xi; Biophys Soc; Am Heart Asn. *Res:* Vascular and gastrontestinal smooth muscle; smooth muscle; pharmacology and electrophysiology of microcirculation; putative peptide neurotransmitters; calcium; protein kinase C. *Mailing Add:* Beth Israel Hosp Harvard Med Boston MA 02215

MORGAN, KATHRYN A, b Riverside, Calif, Nov 18, 22. MATHEMATICS. *Educ:* Stanford Univ, AB, 42, AM, 43, PhD(math), 46. *Prof Exp:* From instr to assoc prof math, Syracuse Univ, 46-82; RETIRED. *Res:* Number theory. *Mailing Add:* Box 222 Creston CA 93432

MORGAN, KENNETH ROBB, b Oxford Eng, June 27, 52; US citizen; m 80; c 1. ECOLOGICAL PHYSIOLOGY, COMPARATIVE PHYSIOLOGY. *Educ:* Case Western Reserve Univ, BA, 77; Univ Calif, Los Angeles, MA, 79, PhD(biol), 84. *Prof Exp:* Fel, Univ Vt, 85, fel biol, Univ Calif, Riverside, 85-87; fel Univ Calif, Los Angeles, 87, FEL, UNIV WASH, 87- *Mem:* AAAS; Am Soc Zoologists; Ecol Soc; Am Ornithologists Union. *Res:* Comparative physiology of vertebrates and insects. *Mailing Add:* Dept Zool Univ Wash Seattle WA 98195

MORGAN, KEVIN THOMAS, b Bristol, Eng, June 29, 43; US citizen; m 68; c 3. EXPERIMENTAL PATHOLOGY, EXPERIMENTAL PHYSIOLOGY. *Educ:* Bristol Univ, BVSc, 67; Edinburgh Univ, PhD, 75; Am Col Vet Path, dipl, 79. *Prof Exp:* Sr vet res officer, Moredun Res Inst, 70-75; researcher path, Battelle Res Labs, 75-78; pathologist, Nat Ctr Toxicol, 78-80; researcher path, Battelle Res Labs, 80-81; SCIENTIST EXP PATH, CHEM INDUST INST TOXICOL, 81- *Concurrent Pos:* Consult, Nat Inst Environ Health Sci, 81-; adj asst prof, Univ NC, Chapel Hill, 85-88. *Mem:* Am Col Vet Pathologists; Brit Neuropath Soc; Royal Col Pathologists; Soc Toxicol Pathologists; Royal Col Vet Surgeons; Fr Soc Toxicol Pathologists. *Res:* Determination of mechanisms of nasal toxicity in rodents and primates to determine the appropriateness of the rodent as a model for inhalation toxicology studies designed for human risk estimation; nasal cancer and olfactory damage. *Mailing Add:* Chem Ind Inst Toxicol PO Box 12137 Research Triangle Park NC 27514

MORGAN, LEE ROY, JR, b New Orleans, La, Nov 5, 36; m 57; c 4. PHARMACOLOGY, CHEMOTHERAPY. *Educ:* Tulane Univ, BS, 58, MS, 59, PhD, 60; Imp Col, Univ London, DIC, 61; La State Univ, MD, 71. *Prof Exp:* Instr, 61-63, asst prof, 63-72, assoc prof, 72-73, PROF PHARMACOL, SCH MED, LA STATE UNIV, MED CTR, 73-, CHMN DEPT, 80- *Concurrent Pos:* Res assoc, Imp Col, Univ London, 60-61; asst prof, Loyola Univ, 61-63; vis prof pharmacol, Sch Med, Univ Costa Rica, 63; consult clin biochemist, Charity Hosp, New Orleans, 67- *Honors & Awards:* Sigma Xi Award, 58. *Mem:* AAAS; Am Chem Soc; Royal Soc Chem; Soc Exp Med & Biol; Am Asn Cancer Res. *Res:* Drug profiles; enzyme kinetics; immunology and chemotherapy of cancer. *Mailing Add:* King Found Cancer Res 3839 Ulloa St New Orleans LA 70119

MORGAN, LEON OWEN, b Oklahoma City, Okla, Oct 25, 19; m 42; c 4. CHEMISTRY. *Educ:* Oklahoma City Univ, BA, 41; Univ Tex, MS, 43; Univ Calif, PhD(chem), 48. *Prof Exp:* Anal chemist, Okla Gas & Elec Co, 41; instr chem, Univ Tex, 41-44; res assoc, Metall Lab, Univ Chicago, 44-45; chemist, Radiation Labs, Univ Calif, 45-47; from asst prof to assoc prof, 47-61, PROF CHEM, UNIV TEX, AUSTIN, 62- *Concurrent Pos:* Consult, Los Alamos Nat Lab, 61- *Mem:* Am Chem Soc; Am Phys Soc; Sigma Xi. *Res:* Transuranium elements; nuclear and electron paramagnetic resonance and relaxation; fast reaction rate processes; radiation effects; inorganic biochemistry. *Mailing Add:* 3307 River Rd Austin TX 78703-1028

MORGAN, LUCIAN L(LOYD), b Wichita Falls, Tex, Dec 22, 28. AEROSPACE ENGINEERING, CHEMICAL ENGINEERING. *Educ:* Tex A&M Univ, BS, 49; Southern Methodist Univ, MS, 59; Univ Southern Calif, MS, 75. *Prof Exp:* Process engr, US Gypsum Co, 49-50; process analyst, Gen Dynamics/Ft Worth, 50-52, sr chemist, 52-55, res chemist, 55-56, asst chief chemist, 56-58, sr eng chemist, 58-60, design specialist supvr, Gen Dynamics/Astronautics, 60-62; res specialist, 62-63, staff engr, 63-67, sr staff engr, 67-80, mgr syst eng, 80-84, CHIEF SYST ENG, LOCKHEED MISSLES & SPACE CO, 84- *Honors & Awards:* Pub Serv Award, NASA, 84. *Res:* Laser systems; vehicle design; project engineering; high-energy laser systems. *Mailing Add:* 2029 Kent Dr Los Altos CA 94022

MORGAN, M(ILLETT) GRANGER, b Hanover, NH, Mar 17, 41; m 63; c 2. SCIENCE POLICY, ENVIRONMENTAL MANAGEMENT. *Educ:* Harvard Univ, BA, 63; Cornell Univ, MS, 65; Univ Calif, San Diego, PhD(appl physics), 69. *Prof Exp:* Lectr appl physics & info sci, Univ Calif, San Diego, 70-71, actg asst prof info sci, 71-72, dir comput jobs through training proj, 69-72; prog dir, Comput Impact on Soc, NSF, 72-74; assoc res physicist, Biomed & Environ Assessment Group, Brookhaven Nat Lab, 74-75; asst prof elec eng & pub affairs & coordr elec eng & pub affairs grad prog, 75-77, PROF ENG & PUB POLICY & ELEC ENG, CARNEGIE-MELLON UNIV, 80-, HEAD DEPT ENG & PUB POLICY, 80-, DIR PROG INT PEACE & SECURITY, 89- *Concurrent Pos:* Mem comts, EPA Sci Adv Bd, Nat Sci Found, Cong Off Technol Assessment, Nat Acad Sci & Inst Elec & Electronic Engrs. *Mem:* AAAS; Am Geophys Union; fel Inst Elec & Electronic Engrs; Air Pollution Control Asn; Soc Risk Anal; Bioelectromagnetics Soc. *Res:* Problems in technology and public policy in which technical issues play a central role; techniques for dealing with uncertainty in quantitative policy analysis; risk assessment and communication; social impacts of eletrical technologies. *Mailing Add:* Dept Eng & Pub Policy Carnegie-Mellon Univ 5000 Forbes Ave Pittsburgh PA 15213

MORGAN, MARJORIE SUSAN, b Philadelphia, Pa, July 3, 56. PROTEIN BIOCHEMISTRY. *Educ:* Juniata Col, Huntingdon, Pa, BS, 77; Miami Univ, Oxford, Ohio, PhD(chem), 81. *Prof Exp:* Postdoctoral res assoc, Fels Res Inst, Wright State Univ Sch Med, Dayton, Ohio, 81-84; from asst to assoc prof chem & health sci, Salem Col, WVa, 84-88; CHEM LAB DIR, CANCER BIOLOGICS AM INT INC, LEXINGTON, KY, 88- *Concurrent Pos:* Chair, Dept Health Sci, Salem Col, WVa, 86-87; consult, Cancer Biologics Am, Stonewood, WVa, 87-88. *Mem:* Am Soc Biochem & Molecular Biol; Am Chem Soc; Am Women Sci. *Res:* Develop monoclonal antibodies for cancer diagnosis and therapy; protein purification; immunoglobulin fragmentation; radiolabeling. *Mailing Add:* Cancer Biologics Am Int Inc 600 Perimeter Dr Lexington KY 40517

MORGAN, MARK DOUGLAS, b Stuttgart, Germany, Apr 4, 52; US citizen; m 75; c 2. LIMNOLOGY, BIOGEOCHEMISTRY. *Educ:* Univ Calif, Davis, BS, 74, PhD(ecol), 79; Univ Wis-Milwaukee, MS, 76. *Prof Exp:* Teaching asst zool, Univ Wis-Milwaukee, 74-75, res asst, 75-76; res asst, Univ Calif, Davis, 76-79; fel, Univ Tex Marine Sci Inst, 79-81, prof, 81-87, ASSOC PROF ZOOL, RUTGERS UNIV, 87- *Mem:* AAAS; Am Soc Limnol & Oceanog; Ecol Soc Am; Soc Int Limnologie; Sigma Xi. *Res:* Dynamics of freshwater plant and animal populations, particularly as influenced by stress, such as low food and low pH; ecosystems effects of acid rain; biogeochemistry. *Mailing Add:* Dept Biol Rutgers Univ Camden NJ 08102

MORGAN, MARVIN THOMAS, b Nashville, Tenn, Jan 27, 21; m; c 3. NUCLEAR ENGINEERING. *Educ:* Tenn Technol Univ, BS, 50. *Prof Exp:* Jr physicist irradiation & post irradiation exp on high temperature gas-cooled reactor fuel, Oak Ridge Nat Lab, 51-53, assoc physicist, 53-56, physicist, 56-70, res assoc, 70-76, develop assoc nuclear waste disposal, 76-77, develop staff mem nuclear waste disposal, 77-82; RETIRED. *Mem:* Sigma Xi. *Res:* High temperature gas-cooled reactor fuels; nuclear waste fixation in concrete; thermal properties of rocks and concrete; properties of concrete related to borehole plugging. *Mailing Add:* 109 N Purdue Ave Oak Ridge TN 37830

MORGAN, MEREDITH WALTER, b Kingman, Ariz, Mar 22, 12; m 37; c 1. OPTOMETRY, VISUAL PHYSIOLOGY. *Educ:* Univ Calif, AB, 34, MA, 39, PhD(physiol), 41. *Hon Degrees:* DOS, Ill Col Optom, 68; DOS, Southern Calif Col Optom, 75; DSc, Pa Col Optom, 76; DSc, State Univ NY, 89. *Prof Exp:* Clin asst optom, 36-42, from instr to assoc prof, 42-51, dean sch optom, 60-73, prof physiol optics & optom, 51-75, EMER PROF PHYSIOL OPTICS & OPTOM, UNIV CALIF, BERKELEY, 75- *Concurrent Pos:* Pvt pract, 34-60; mem ed coun, Am J Optom, 55-74; mem, Coun Optom Educ, 60-73; mem rev comt construct schs optom, USPHS, 64-65, nat adv coun med, dent, optom, podiatric & vet educ, 65-67; mem adv coun, Nat Eye Inst, 69-71; vis prof, Univ Waterloo, Can, 74 & Univ Ala, 77. *Honors & Awards:* Prentice Medal, Am Acad Optom, 67; Apollo Award, Am Optom Asn, 75; Berkeley Citation, 75. *Mem:* Am Optom Asn; AAAS; Geront Soc; Am Acad Optom (pres, 53-54). *Res:* Accommodation and convergence; binocular vision. *Mailing Add:* 1217 Skycrest 4 Walnut Creek CA 94595

MORGAN, MERLE L(OREN), b Whittier, Calif, May 28, 19; m 52; c 2. PRECISION ELECTRICAL MEASUREMENTS. *Educ:* Calif Inst Technol, BS, 49, MS, 50, PhD(elec eng), 54. *Prof Exp:* Res engr, McCullough Tool Co, Calif, 46-48; dir res & eng, Electro-Sci Industs, 54-78; consult, 78-85; RETIRED. *Mem:* Inst Elec & Electronics Engrs. *Res:* Precision electrical measurements. *Mailing Add:* 1480 NW 138th Ave Portland OR 97229

MORGAN, MICHAEL ALLEN, b Los Angeles, Calif, July 8, 48; m 66; c 2. ELECTRICAL ENGINEERING, APPLIED MATHEMATICS. *Educ:* Calif State Polytech Univ, BS, 71; Univ Calif, Berkeley, MS, 73, PhD(elec eng, comput sci), 76. *Prof Exp:* Staff scientist, Sci Appl, Inc, 75; res asst, Univ Calif, Berkeley, 71-76; res eng in radar systs, Stanford Res Inst, 76-77; asst prof elec eng, Univ Miss, 77-80; PROF, DEPT ELEC ENG, NAVAL POST GRAD SCH, 80- *Concurrent Pos:* Consult, Stanford Res Inst, 77-78 & Dept of Bioeng, Univ Utah, 78-; consult, Arnold Eng Develop Ctr, 78-79; prog mgr, Off Naval Res, 85-86. *Mem:* Inst Elec & Electronics Engrs, 76- *Res:* Electromagnetics; antennas; microwave techniques; communications theory and signal processing; numerical analysis; applied mathematics and statistics. *Mailing Add:* Dept of Elec Eng Code 62MW Naval Post Grad Sch Monterey CA 93943

MORGAN, MICHAEL DEAN, b Marion, Ind, Oct 27, 41; m 66; c 1. PLANT ECOLOGY, BIOCLIMATOLOGY. *Educ:* Butler Univ, BA, 63; Univ Ill, Urbana-Champaign, MS, 65, PhD(bot), 68. *Prof Exp:* Asst prof, 68-72, ASSOC PROF BIOL, UNIV WIS-GREEN BAY, 72- *Concurrent Pos:* Wis Alumni Res Fund grant, Univ Wis-Green Bay, 72. *Mem:* Ecol Soc Am; Am Inst Biol Sci; AAAS; Sigma Xi; Nat Sci Teachers Asn. *Res:* Relationships between climatic variability and plant distribution and production; reproductive ecology of herbacous plants; management endangered plant species; science education for nonscience majors. *Mailing Add:* 2249 Hillside Lane Green Bay WI 54302

MORGAN, MONROE TALTON, SR, b Mars Hill, NC, June 29, 33; m 60; c 2. ENVIRONMENTAL HEALTH, PUBLIC HEALTH. *Educ:* ETenn State Univ, BA & CPHS, 60; Univ NC, Chapel Hill, MSPH, 62; Tulane Univ, La, DrPh(environ & pub health), 69. *Prof Exp:* Sanitarian, Fairfax County Health Dept, Va, 60-61; training officer pub health, Va State Health Dept, 62-63; PROF ENVIRON HEALTH & CHMN DEPT, E TENN STATE UNIV, 63- *Concurrent Pos:* Consult ed, Nat J Environ Health, 69-; consult environ health educ, Nat Environ Health Asn, 70-; mem, Pub Health Rev Comt, 71-75. *Mem:* Am Pub Health Asn; Nat Environ Health Asn (2nd vpres, 71-). *Res:* Survival of aerobic sporeformers and mycobacterium tuberculosis vas hominis; human ecology as it relates to environmental stresses. *Mailing Add:* Dept of Environ Health ETenn State Univ Johnson City TN 37614

MORGAN, MORRIS HERBERT, b Atlanta, Ga, Feb 10, 50; m 69; c 2. CHEMICAL ENGINEERING. *Educ:* Vanderbilt Univ, BS, 69; Univ Dayton, BS, 73; Rensselaer Polytech Inst, PhD(chem eng), 78. *Prof Exp:* Develop engr chem eng, Inland Mfg Div Gen Motors Corp, 69-72; safety engr chem eng, Mound Labs Monsanto Co, 72-73; staff chem engr, Gen Elec Res & Develop Ctr, 73-82; asst prof, 82-88, ASSOC PROF, RENSSELAER POLYTECH INST, 88- *Concurrent Pos:* Mem comt high technol, NY State, 78-79; mem comt Educ & Utilization Engrs, Nat Acad Engrs, 81. *Mem:* Am Inst Chem Engr. *Res:* Fluid and particle mechanics of spouted bed systems; fluid bed reactor modeling. *Mailing Add:* 1233 Viewmont Dr Schenectady NY 12309

MORGAN, NANCY H, b Hartwell, Ga, Nov 8, 33; m 58; c 2. RELATIONAL DATA BASES, COMPUTER SOLUTIONS FOR SMALL BUSINESSES. *Educ:* Wash Univ, BS, 70, MS, 77. *Prof Exp:* Grad teaching asst computer sci, Wash Univ, 70-73, computer specialist, 73-78; sr systs analyst, Monsanto Co, 78-84; database adminr, Bridge Info Systs, Inc, 84-89; consult, Booz-Allen & Hamilton Inc, 90; PRES, NHM COMPUTER CONSULTS, 90- *Concurrent Pos:* Lectr, Computer Sci Dept, Wash Univ, 74-78; Nat Bd, Asn Women Comput, 83-84. *Mem:* Inst Elec & Electronic Engrs; Asn Computer Mach; Digital Equip Computer Users Soc; Soc Women Engrs; Asn Women Comput. *Res:* User friendly computer systems for small business applications; relational data bases; operating systems; software systems testing; indexing of small data bases. *Mailing Add:* 7282 Princeton Ave St Louis MO 63130

MORGAN, NEAL O, medical entomology, ecology, for more information see previous edition

MORGAN, OMAR DRENNAN, JR, b Shelbyville, Mo, Mar 5, 13; m 42; c 2. PLANT PATHOLOGY. *Educ:* Ill State Univ, BEd, 40; Univ Ill, PhD(plant path), 50. *Prof Exp:* Teacher high sch, 41-42; asst plant path, Univ Ill, 48-49; from asst prof to prof plant path, Univ Md, 49-81; RETIRED. *Mem:* Am Phytopath Soc; Sigma Xi. *Res:* Soil fumigation for control of soil borne diseases; developing tobacco resistance to major diseases. *Mailing Add:* Plant Path Dept Univ Md College Park MD 20742

MORGAN, ORA BILLY, JR, b Kershaw, SC, March 24, 30; m 53; c 2. NUCLEAR ENGINEERING, ENGINEERING PHYSICS. *Educ:* NC State Univ, BS, 56, MS, 58; Univ Wis, PhD(nuclear eng), 70. *Prof Exp:* Res staff mem fusion energy, Oak Ridge Nat Lab, 58-62, assoc group leader, 62-69, group leader, 69-74, assoc div dir, 74-77, DIV DIR FUSION ENERGY, OAK RIDGE NAT LAB, 77- *Concurrent Pos:* Oak Ridge Nat Lab fel, Univ Wis, 66-68. *Mem:* Fel Am Phys Soc; Am Nuclear Soc. *Res:* Developing fusion reactors as a long-term energy source with special emphasis on developing, applying and understanding ion sources and neutral injection plasma heating systems. *Mailing Add:* Oak Ridge Nat Lab Fusion Energy Div Oak Ridge TN 37830

MORGAN, PAGE WESLEY, b Phoenix, Ariz, Apr 3, 33; m 55; c 3. PLANT HORMONES, DEVELOPMENTAL PHYSIOLOGY. *Educ:* Tex A&M UNiv, BS, 55, MS, 58, PhD(plant physiol), 61. *Prof Exp:* Asst plant physiol & range mgt, 56-58, Anderson-Clayton fel plant physiol, 58-60, from asst prof to assoc prof, 60-69, PROF PLANT PHYSIOL, TEX A&M UNIV, 69- *Concurrent Pos:* Assoc ed, Plant Physiol, 78- *Honors & Awards:* Distinguished Serv Award, Southern Sect, Am Soc Plant Physiol, 90. *Mem:* Am Soc Plant Physiol (pres, 80-81); Am Soc Agron; Sigma Xi; Plant Growth Regulator Soc Am (pres, 77-78). *Res:* Genetic regulation of plant development; phytohormones; plant responses to stress; synthesis of ethylene; interaction of phytohormones; floral initiation; apical dominance abscission root growth. *Mailing Add:* Dept Soil & Crop Sci Tex A&M Univ College Station TX 77843-2474

MORGAN, PAUL E(MERSON), b Hudson, Iowa, June 18, 23; m 45; c 3. CIVIL ENGINEERING. *Educ:* Iowa State Univ, BS, 44, MS, 56. *Prof Exp:* City engr, Algona, Iowa, 46-48; consult, Estherville, Iowa, 48-50; sanit engr, State Dept Health, Iowa, 50-53; from asst prof to prof civil eng, Iowa State Univ, 53-89, from asst dean to assoc dean, Col Eng, 64-89, dir eng exten, 81-82, actg head 85-86; RETIRED. *Mem:* Am Soc Civil Engrs; Nat Soc Prof Engrs; Am Waterworks Asn; Am Soc Eng Educ; Water Pollution Control Fedn. *Res:* Water. *Mailing Add:* 1428 Northwestern Dr Ames IA 50011

MORGAN, PAUL WINTHROP, b West Chesterfield, NH, Aug 30, 11; m 39; c 2. TEXTILE FIBERS, POLYMER SYNTHESIS. *Educ:* Univ Maine, BS, 37; Ohio State Univ, PhD(org chem), 40. *Prof Exp:* Asst chem, Ohio State Univ, 37-40, Du Pont fel, 40-41; res chemist, E I Du Pont de Nemours & Co, 41-46, res assoc, 46-50, Pioneering Res Labs, Textile Fibers Dept, Exp Sta, E I Du Pont de Nemours & Co, 50-57, res fel, 57-73, sr res fel, 73-76; CONSULT, 76- *Concurrent Pos:* Mem, Polymer Nomenclature Comt, Am Chem Soc, 70-75; chmn, Gordon Conf Polymers, 74, Polymer Div Awards Comt, 75-76, alt coun, 76-77, coun, 78. *Honors & Awards:* Am Chem Soc Award, 60, 77, 86, Polymer Chem Award, 76 & Midgely Award, 79; Howard N Potts Medal, Franklin Inst, 76, 78; Swinburne Medal, Plastics & Rubber Inst, London, 78; Citation, Am Soc Metals, 78; Whitby Lectr Award, Univ Akron, 87; World Mat Cong Award, 88; Carothers Award. *Mem:* Nat Acad Eng; Franklin Inst; Am Chem Soc; hon mem Fiber Soc; Int Union Pure Appl & Chem; Friends of Mineral. *Res:* Cellulose derivatives; low temperature and interfacial polycondensations; condensation polymers; thermally stable polymers; extended chain polymers and their liquid crystalline solutions; high strength-high modulus fibers; tire cord and reinforcement fibers. *Mailing Add:* 822 Roslyn Ave West Chester PA 19382

MORGAN, R JOHN, b Fairfax, Okla, Nov 28, 23. BIOMEDICAL ENGINEERING. *Educ:* Colo Sch Mines, EM, 49; Colo State Univ, MS, 60; Iowa State Univ, PhD(biomed eng), 69. *Prof Exp:* Design engr, Aeronaut Electronics, Inc, 49-52 & Sandia Corp, 52-54; pres, Scintillonics, Inc, 54-62; from instr to asst prof elec eng, 60-67, asst prof bioeng, 67-80, ASSOC PROF BIOENG, COLO STATE UNIV, 80- *Concurrent Pos:* Consult instrumentation probs, TV stas & indust co, 58- *Mem:* AAAS; Inst Elec & Electronics Engrs; Instrument Soc Am. *Res:* Data processing of neuroelectric signals from animals and man. *Mailing Add:* Dept of Physiol & Biophys Colo State Univ Ft Collins CO 80523

MORGAN, RAYMOND P, b Frederick, Md, Nov 24, 43; div; c 2. ECOLOGY, PHYSIOLOGY. *Educ:* Frostburg State Col, BS, 66; Univ Md, PhD(zool), 71. *Prof Exp:* Res assoc, Univ Md, 71-74, res asst prof, 74-77; prin res scientist, Columbus Div, Battelle Mem Inst, 77-79; asst prof, 79-85, ASST LAB HEAD, UNIV MD, 85- *Mem:* Am Fisheries Soc; AAAS; Am Inst Fishery Res Biologist; Am Soc Zoologists; Estuarine Res Fedn. *Res:* Biochemical systematics; pollution ecology; physiological ecology; ecology of fishes. *Mailing Add:* PO Box 3464 La Vale MD 21504-3464

MORGAN, RAYMOND VICTOR, JR, b Brownwood, Tex, May 10, 42; m 67; c 2. MATHEMATICS. *Educ:* Howard Payne Col, BA, 64; Vanderbilt Univ, MA, 65; Univ Mo, PhD(math), 69. *Prof Exp:* Instr math, Univ Mo, 69; asst prof math, Southern Methodist Univ, 69-75; assoc prof math & chmn dept, 75-85, dir, sci div, 79-86, PRES & PROF MATH, SUL ROSS STATE UNIV, ALPINE, TEX, 85- *Mem:* Nat Coun Teachers Math; Math Asn Am. *Res:* Non-associative algebra; generalizations of alternative algebras and their structures relative to Peirce decompositions and Wedderburn decompositions. *Mailing Add:* Box C-114 Sul Ross State Univ Alpine TX 79832

MORGAN, RELBUE MARVIN, b Oxford, Miss, May 7, 39; m 58. PHYSICS. *Educ:* Christian Bros Col, BS, 62; Iowa State Univ, PhD(physics), 67. *Prof Exp:* Asst prof, 67-69, ASSOC PROF PHYSICS, CHRISTIAN BROS COL, 69- *Mem:* Am Phys Soc; Am Asn Physics Teachers. *Res:* Electronic properties of metals and alloys; electrical and optical properties of thin metal films. *Mailing Add:* 3731 Woodland Dr Memphis TN 38111

MORGAN, RICHARD C, applied mathematics, for more information see previous edition

MORGAN, ROBERT LEE, organic chemistry, for more information see previous edition

MORGAN, ROBERT P, b Brooklyn, NY, Feb 26, 34; m 58; c 2. TECHNOLOGY & HUMAN AFFAIRS, ENERGY & ENVIRONMENTAL POLICY. *Educ:* Cooper Union, BChE, 56; Mass Inst Technol, MS, 59, NuclE, 61; Rensselaer Polytech Inst, PhD(chem eng), 65. *Prof Exp:* Asst dir, Oak Ridge Eng Pract Sch, Mass Inst Technol, 58-59; instr chem eng, Rensselaer Polytech Inst, 60-64; asst prof nuclear & chem eng, Univ Mo-Columbia, 64-68; vis assoc prof eng & actg dir int develop technol prog, Wash Univ, 68-69, assoc prof eng, 69-76, chmn dept technol & human affairs, 71-83, prof technol & human affairs, 76-87, DIR, CTR DEVELOP TECHNOL, WASH UNIV, 69-, ELVERA & WILLIAM STUCKENBERG PROF TECHNOL & HUMAN AFFAIRS, 87- *Concurrent Pos:* Nat lectr, Sigma Xi, 80-82; sci & pub policy fel, Brookings Inst, 82-83; vis sr analyst, Off Technol Assessment, US Cong, 89-90. *Honors & Awards:* Chester F Carlson Award, Am Soc Eng Educ, 78. *Mem:* AAAS; Fedn Am Sci; Am Soc Eng Educ; Am Asn Univ Professors. *Res:* Technology for international development; science and technology policy; energy technology and policy; issues in engineering education; radioactive waste management. *Mailing Add:* Dept Eng & Policy Wash Univ Box 1106 St Louis MO 63130

MORGAN, ROGER JOHN, b Manchester, Eng, Nov 2, 42; US citizen; m 67; c 3. POLYMERIC COMPOSITES, POLYMER PHYSICS. *Educ:* Univ London, BSc, 65; Univ Manchester, PhD(polymer physics), 68. *Prof Exp:* Scientist polymer physics, McDonnell Douglas Res Lab, 72-78; group leader polymer composites, Lawrence Livermore Nat Lab, 78-85; mem tech staff polymer composites, Rockwell Int Sci Ctr, 85-86; HEAD COMPOSITES, POLYMER COMPOSITES, MICH MOLECULAR INST, 86- *Concurrent Pos:* Assoc ed, J Composites Technol & Res, 81-83, ed adv bd, 83-, assoc ed, Advan Composites Bull, 88-; consult, Lawrence Livermore Nat Lab, 85-, Dow Chem, 87-; ed adv bd, Jour Composite Mat, 85-; adj prof, Mich Technol Univ, 86-, Cent Mich Univ, 87- *Mem:* Am Chem Soc; Am Phys Soc; Soc Advan Mat & Process Engr. *Res:* Ninety-two publications on the relations between the chemical and physical structure, deformation and failure process and mechanical response of polymers and high performance fibrous composites and how such relations are modified by environmental factors. *Mailing Add:* Mich Molecular Inst 1910 W St Andrews Rd Midland MI 48640

MORGAN, SAMUEL POPE, b San Diego, Calif, July 14, 23; m 48; c 4. MATHEMATICAL PHYSICS, COMPUTER SCIENCE. *Educ:* Calif Inst Technol, BS, 43, MS, 44, PhD(physics), 47. *Prof Exp:* Asst physics, Univ Calif, 43-44 & Calif Inst Technol, 44-47; mem tech staff, Bell Tel Labs, 47-59, head math physics dept, 59-67, dir comput technol, 69-70, dir comput sci res, 67-82, DISTINGUISHED MEM TECH STAFF, BELL LABS, INC, NJ, 82- *Mem:* AAAS; Am Phys Soc; Soc Indust & Appl Math; Asn Comput Mach; Sigma Xi; fel Inst Elec & Electronic Engrs. *Res:* Electromagnetic theory; mechanics of continua; wave propagation; queueing theory. *Mailing Add:* AT&T Bell Labs Rm 2C-555 PO Box 263 Murray Hill NJ 07974-2070

MORGAN, STANLEY L, b Sandyville, Ohio, Jan 28, 18; m 41; c 3. PHARMACEUTICS, CHEMICAL ENGINEERING. *Educ:* Case Inst Technol, BSChE, 39. *Prof Exp:* Chem engr, Ben Venue Labs Inc, 40-42, mgr, Blood Plasma Labs, 42-43, gen mgr, 43-63, vpres, 63-64, exec vpres, 64-85, pres, 85-87; CONSULT, 87- *Mem:* Am Inst Chem Engrs; Am Chem Soc; NY Acad Sci; Am Soc Testing & Mat; fel Am Inst Chem. *Res:* Pharmaceuticals under aseptic conditions; ethylene oxide sterilization; freeze drying; racing chemistry; preservation of viable organisms; cryobiology. *Mailing Add:* 31051 Northwood Dr Pepper Pike OH 44124

MORGAN, THOMAS EDWARD, b Bay Shore, NY, Dec 27, 43. ASTRONOMY, ARCHAEOASTRONOMY. *Educ:* Hofstra Univ, BA, 65; Ind Univ, Bloomington, MA, 67, PhD(astrophys), 71. *Prof Exp:* Instr, 70-71, asst prof astron, State Univ NY Col Oswego, 71-78; TECH INFOSPEC & SR LAWS SPEC, LIBR CONGRESS, 78- *Mem:* Am Astron Soc; Sigma Xi. *Res:* Line blanketing in stellar atmospheres, primarily Epsilon Virginis and Arcturus; ancient solar cycle; astronomy education; archaeoastronomy. *Mailing Add:* 7205 Auburn St Annandale VA 01003

MORGAN, THOMAS HARLOW, astrophysics, planetary sciences, for more information see previous edition

MORGAN, THOMAS JOSEPH, b Brooklyn, NY, Oct 20, 43; m 68; c 2. ATOMIC PHYSICS. *Educ:* Carroll Col, BA, 65; Mont State Univ, BSc, 66; Univ Calif, Berkeley, MSc, 68, PhD(eng physics), 71. *Prof Exp:* Univ fel pure & appl physics, Queen's Univ Belfast, 71-73; from asst prof to assoc prof, 73-80, PROF PHYSICS, WESLEYAN UNIV, 86-, DEAN SCI, 90- *Concurrent Pos:* Consult, Oak Ridge Nat Lab, 76-, mem, Atomic Data Ctr; vis scientist, Lawrence Berkeley Lab, 74, Oak Ridge Nat Lab, 75 & Univ Mex, 79; fac res award, Asn Western Univ, 82; vis fel, Joint Inst Lab Astrophys, 82; Moshinsky vis chair physics, Univ Mex, 84; vis prof, Univ Paris-Orsay, 86 & Huazhong Univ, Peoples Repub China, 88; vis fel, Centre Chem Physics, Univ Western Ont, 90- *Mem:* Am Phys Soc; Sigma Xi. *Res:* Heavy particle collision phenomena; atomic beams; collisional properties of excited states; laser-atomic collisions; synchrotron radiation photoionization. *Mailing Add:* Dept of Physics Wesleyan Univ Middletown CT 06457

MORGAN, THOMAS KENNETH, JR, b Upper Darby, Pa, Oct 25, 49; m 71; c 4. ORGANIC CHEMISTRY. *Educ:* Villanova Univ, BS, 71; Univ Del, PhD(chem), 75. *Prof Exp:* Res assoc org chem, State Univ NY Binghamton, 75-77; sr chemist org chem, Cooper Labs, Inc, 77-79; sr org chemist, Berlex Labs, 79-82, proj leader, 82-87, sr scientist, 88-89, SECT HEAD, BERLEX LABS, 89- *Concurrent Pos:* NIH fel, 76-77. *Mem:* AAAS; Am Chem Soc. *Mailing Add:* Berlex Labs Inc 110 E Hanover Ave Cedar Knolls NJ 07927

MORGAN, WALTER CLIFFORD, b Ledyard, Conn, Dec 22, 21; m 48; c 3. ANIMAL GENETICS. *Educ:* Univ Conn, BSc, 46, PhD(genetics), 53; George Washington Univ, MSc, 49. *Prof Exp:* Animal geneticist, Nat Cancer Inst, 46-49; res assoc mammalian genetics, Columbia Univ, 50-53; asst prof poultry genetics, Univ Tenn, 53-54; assoc prof, 54-58, prof genetics & physiol, SDak State Univ, 58-85; RETIRED. *Concurrent Pos:* Vis scientist, Radiobiol Dept, Nuclear Energy Ctr, Mol, Belg, 68-69 & Animal Genetics Unit, Commonwealth Sci & Indust Res Orgn, North Ryde, Australia, 75-76. *Honors & Awards:* Outstanding Poultry Genetic Res Award, Czechoslovakian Govt, 79. *Mem:* Am Genetic Asn; World Poultry Sci Asn; Am Inst Biol Sci; Radiation Res Soc; NY Acad Sci. *Res:* Developmental genetics of poultry and mice; physiology of reproduction; new mutations; lethals; irradiation effects on chick embryos and on wheat. *Mailing Add:* 1610 First St Brookings SD 57006

MORGAN, WALTER L(EROY), b Passaic, NJ, Dec 20, 30; m 65; c 1. ENGINEERING ECONOMICS, SYSTEMS ENGINEERING. *Educ:* Carnegie-Mellon Univ, BSEE, 54. *Prof Exp:* Res engr, David Sarnoff Res Labs, RCA(GE), 54-55, sr systs engr, Defense Prod Div, 55-59, Astro-Electronic Div, 59-79; mem tech staff long-range planning, Comsat Labs, Commun Satellite Corp, 70-75, sr staff exec, 75-80; CONSULT & PRES, COMMUN CTR, 80- *Concurrent Pos:* Lectr, Univ Calif, Los Angeles, 78; consult ed, Satellite Commun Mag, 80- *Honors & Awards:* Aerospace Commun Award, Am Inst Aeronaut & Astronaut, 82. *Mem:* Nat Acad Sci; sr mem Inst Elec & Electronics Engrs; fel Am Inst Aeronaut & Astronaut; Int Acad Astronaut; AAAS; fel Brit Interplanetary Soc. *Res:* Planning and management of telecommunications applications; operations analysis of communication via satellite and fiber; international and domestic telecommunication system analysis; market research; superconductivity applications. *Mailing Add:* Commun Ctr 2723 Green Valley Rd Clarksburg MD 20871-8599

MORGAN, WILLIAM BRUCE, b Fairfield, Iowa, Dec 20, 26; m 50; c 2. HYDRODYNAMICS, NAVAL ARCHITECTURE. *Educ:* US Merchant Marine Acad, BS, 50; Univ Iowa, MS, 51; Univ Calif, Berkeley, DEng(naval archit), 61. *Prof Exp:* Hydraul engr, Propeller Br, David Taylor Model Basin, 51-52, naval architect, 52-58, supv naval architect propeller res & design, Supercavitating & Design Sect, 58-62, supv naval architect ship propulsion & head propeller br, 62-70, head Naval Hydromech Div, Naval Ship Res & Develop Ctr, 70-80, ASSOC TECH DIR SHIP HYDROMECH, DAVID TAYLOR RES CTR, 80- *Honors & Awards:* William Froude Gold Medal, Contrib Naval Archit, Rotal Inst Naval Architects, 89; Davidson Medal for Sci Accomplishment in Ship Res, 86; Presidential Rank. *Mem:* Fel Soc Naval Archit & Marine Engrs; fel Am Soc Mech Engrs; German Soc Naval Archit; Am Soc Naval Engrs; Sigma Xi. *Res:* Ship hydrodynamics, especially propeller design theories for conventional, contrarotating, supercavitating and ducted propellers; cavitation and propulsion in general. *Mailing Add:* 110 Upton Rockville MD 20850

MORGAN, WILLIAM JASON, b Savannah, Ga, Oct 10, 35; m 59; c 2. GEOPHYSICS. *Educ:* Ga Inst Technol, BS, 57; Princeton Univ, PhD(physics), 64. *Prof Exp:* Res assoc, 64-66, from asst prof to assoc prof, 66-77, PROF GEOPHYS, PRINCETON UNIV, 77- *Mem:* Nat Acad Sci; Am Geophys Union; Europ Union Geol Sci. *Res:* Mantle convection; heat flow; plate tectonics; marine geophysics. *Mailing Add:* Dept Geol & Geophys Sci Princeton Univ Princeton NJ 08544

MORGAN, WILLIAM KEITH C, b July 1, 29; US citizen; m 53; c 3. INTERNAL MEDICINE. *Educ:* Univ Sheffield, MB, ChB, 53, MD, 61; MRCP(D), 58, FRCP(E), 71, FACP, 72, FRCP(C), 78. *Prof Exp:* From instr to assoc prof, Univ Md, 59-67; from assoc prof to prof med, Med Ctr, WVa Univ, 67-78, head div pulmonary dis, 74-78; PROF MED, UNIV WESTERN ONT, 78-, DIR CHEST DIS SERV, UNIV HOSP, 78- *Concurrent Pos:* Chief, Appalachian Lab Occup Respiratory Dis, Nat Inst Occup Safety & Health, 67-71, dir, 71-74. *Mem:* Am Thoracic Soc; Brit Med Asn; Am Col Chest Physicians; Can Thoracic Soc (pres, 85-86); Brit Thoracic Soc; Can Soc for Clinical Invest. *Res:* Chest diseases and occupational medicine. *Mailing Add:* Univ Hosp Rm 6-0F6 Univ Western Ont Box 5339 Postal Sta A London ON N6A 5A5 Can

MORGAN, WILLIAM L, JR, b Honolulu, Hawaii, Nov 18, 27; m 54; c 2. MEDICINE, CARDIOVASCULAR DISEASES. *Educ:* Yale Univ, BA, 48; Harvard Med Sch, MD, 52; Am Bd Internal Med, dipl, 62. *Prof Exp:* Instr med, Harvard Univ, 56-58, tutor med sci, 57-58; assoc, Div Cardiovasc Dis, Henry Ford Hosp, 58-62; assoc prof, 62-66, PROF MED, SCH MED & DENT, UNIV ROCHESTER, 66-, ASSOC CHMN DEPT MED & DIR EDUC PROGS, 69- *Concurrent Pos:* Clin assoc, Nat Heart Inst; mem, Am Bd Internal Med, 73-, mem, Residency Rev Comt, 74- *Mem:* Fel Am Col Physicians. *Res:* Cardiovascular hemodynamics; cardiac arrhythmias; medical education; physical diagnosis. *Mailing Add:* 601 Elmwood Ave Rochester NY 14642

MORGAN, WILLIAM R(ICHARD), b Cambridge, Ohio, Mar 27, 22; m 46; c 2. MECHANICAL ENGINEERING, TECHNICAL MANAGEMENT. *Educ:* Ohio State Univ, BS, 44; Purdue Univ, MS, 50, PhD, 51. *Prof Exp:* Asst mech, Ohio State Univ, 43-44; design engr, Curtiss-Wright Corp, 46-47; instr, Purdue Univ, 47-51; group leader heat transfer & thermodyn analysis unit, Aircraft Nuclear Propulsion Dept, Gen Elec Co, 51-53, supvr exp mech eng unit, 53-55, supvr controls analysis unit & mgr controls analysis subsect, Flight Propulsion Lab Dept, 55-57, mgr controls develop design subsect, 57-59, mgr vertical take off & landing aircraft prog, 59-63, prog dir high bypass propulsion engines, 63-67, mgr acoustic eng, 67-69, mgr quiet engine prog, Advan Technol Prog Dept, 69-72; vpres & gen mgr, SDRC Int, Inc, 72-80. *Concurrent Pos:* Consult Eng & Mgt, 80- *Mem:* Am Soc Mech Engrs; Acoust Soc Am; Inst Noise Control Eng; Sigma Xi. *Res:* Heat transfer; thermodynamics; fluid dynamics. *Mailing Add:* 312 Ardon Lane Cincinnati OH 45215-4102

MORGAN, WILLIAM T, b Pittsburgh, Pa, Nov 3, 41. HEME TRANSPORT, MEMBRANE PROTEINS. *Educ:* Univ Pittsburgh, BS, 65; Univ Calif, Santa Barbara, PhD(biochem), 70. *Prof Exp:* PROF BIOCHEM & MOLECULAR BIOL, LA STATE UNIV, 79-, DIR BIOTECHNOL UNIT, 87- *Mem:* AAAS; Am Soc Biochem & Molecular Biol; Am Chem Soc; Protein Soc; Sigma Xi. *Res:* Plasma protein and membrane protein structure and function; heme and metal metabolism; hemopexin; histidine-rich glycoprotein. *Mailing Add:* Dept Biochem & Molecular Biol La State Univ New Orleans LA 70112

MORGAN, WILLIAM WILSON, b Bethesda, Tenn, Jan 3, 06; m 28, 66; c 2. ASTRONOMY. *Educ:* Univ Chicago, BS, 27, PhD, 31. *Hon Degrees:* Dr, Nat Univ Cordoba; DSc, Yale Univ, 78. *Prof Exp:* Asst astron, Yerkes Observ, 26-32, from instr to prof, 32-66, chmn dept, 60-66, dir, Yerkes & McDonald Observs, 60-64, Bernard E & Ellen C Sunny distinguished serv prof, 66-74, EMER PROF ASTRON, UNIV CHICAGO, 74- *Honors & Awards:* Bruce Gold Medal Award, Astron Soc Pac, 58; Henry Draper Medal, Nat Acad Sci, 80; Herschel Medal, Royal Astron Soc, 83. *Mem:* Nat Acad Sci; Am Acad Arts & Sci; Royal Danish Acad; Royal Soc Sci Liege; corresp mem Arg Nat Acad Sci. *Res:* Stellar spectroscopy; galaxies. *Mailing Add:* Yerkes Observ PO Box 258 Williams Bay WI 53191-0258

MORGAN, WINFIELD SCOTT, b Takoma Park, Md, Jan 9, 21; m 48; c 5. PATHOLOGY. *Educ:* Albright Col, BS, 42; Temple Univ, MD, 45. *Prof Exp:* Resident path, Mass Gen Hosp, 48-51, asst, 52-53, from asst pathologist to assoc pathologist, 53-62; dir path, Cleveland Metrop Gen Hosp, 62-67; dir labs, Aultman Hosp, Canton, Ohio, 67-74; PROF PATH & DIR SURG PATH, MED CTR, WVA UNIV, 74- *Concurrent Pos:* Fel med, Guthrie Clin, Sayre, Pa, 46; Am Cancer Soc-Brit Empire Cancer Campaign exchange fel, Oxford Univ, 51-52; Nat Cancer Inst spec res fel cellular physiol, Wenner-Gren Inst, Univ Stockholm, 58-60; consult, Health Res Facil Br, NIH, 65-67; instr, Med Sch, Tufts Univ, 52-60; Harvard Med Sch, 60-61, assoc, 61-62; prof, Sch Med, Case Western Reserve Univ, 62-67. *Mem:* Am Asn Path & Bact; Am Soc Exp Path; NY Acad Med; Int Acad Path. *Res:* Biochemical pathology. *Mailing Add:* Dept Path WVa Univ Med Ctr Morgantown WV 26506

MORGAN, WM LOWELL, b Royal Oak, Mich, Sept 27, 46; m 82; c 2. COMPUTATIONAL PHYSICS, PLASMA CHEMISTRY. *Educ:* Univ Mich, BSE, 69; Wayne State Univ, MA, 72; Univ Windsor, Can, PhD(physics), 76. *Prof Exp:* Asst res engr, Willow Run Labs, Univ Mich, 68-74; res assoc, Joint Inst Astrophys, Univ Colo, 76-78; physicist, Theoret Atomic & Molecular Physics Group, Lawrence Livermore Nat Lab, Univ Calif, 78-87; SR SCIENTIST, KINEMA RES, 89- *Concurrent Pos:* Res asst, Res Inst Eng Sci, Wayne State Univ, 72-76; lectr statist physics, Dept Appl Sci, Univ Calif, Davis-Livermore, 83; adj prof physics, Univ Denver, Colo. *Mem:* Am Phys Soc; Mats Res Soc. *Res:* Applied and computational physics; simulation of gas-surface interaction processes; use of neural networks and cellular automata in applied physics; charged particle transport in gases, plasmas, and liquids; plasma chemistry modeling; kinetics of excimer lasers; atomic & molecular collision processes; development of PC based scientific data bases. *Mailing Add:* Kinema Res 18720 Autumn Way Monument CO 80132

MORGAN, WYMAN, b Russellville, Ala, Jan 7, 41; m 59. INORGANIC CHEMISTRY. *Educ:* Florence State Col, BS, 62; Univ Fla, PhD(chem), 67. *Prof Exp:* Res chemist, Inorg Div, 67-71, sr res chemist, 71-75, sr res specialist, 75-77, mgr Manpower & Personnel Develop, Corp Res & Develop, 77-80, mgr Chem Res, Corp Res & Develop Biomed Prog, 80-81, mgr res & develop, 81-84, RES DIR SOMATOTROPIN PROJ, NUTRIT CHEM DIV, MONSANTO AGR PROD CO, 84- *Mem:* Am Chem Soc. *Res:* Chemistry of metallic elements; exploratory process development; development of biotechnology pharmaceutical products for animal agriculture. *Mailing Add:* 220 Monroe Mill Dr St Louis MO 63164

MORGANE, PETER J, b Atlanta, Ga, May 14, 27. NEUROPHYSIOLOGY, PSYCHOPHARMACOLOGY. *Educ:* Tulane Univ, BS, 48; Northwestern Univ, MS, 57, PhD(physiol), 59. *Prof Exp:* Resident res, Northwestern Univ, 56-59; instr physiol, Col Med, Univ Tenn, 59, asst prof, 59-61; sr physiologist, Life Sci Br, Goodyear Aircraft Corp, Ohio, 61-62; sr vis scientist, Brain Res Unit, Mex, 62-63; chmn div neurol sci, Commun Res Inst, 63-68; SR SCIENTIST, WORCESTER FOUND EXP BIOL, 68- *Concurrent Pos:* Affil prof, Worcester Polytech Inst, Clark Univ & Boston Univ, 69-; mem panel neurobiol, NSF, 71-74; mem study sect, Nat Inst Neurol Dis & Stroke, 74-78; dir post-doctoral training prog neurobiol, NIMH, 70-74; chmn, Neurol B Study Sect, NIH, 76-78 & mem, 84-88. *Mem:* Am Physiol Soc; Am Asn Anatomists; fel Am Col Neuropsychopharmacol; Am Psychol Asn; NY Acad Sci. *Res:* Comparative morphology of the brains of mammals, especially whale brain anatomy; neural regulation of food and water intake; hypothalamic-limbic interactions in behavior; anatomical, physiological, pharmacological and neurobiochemical studies on the sleep states; quantitative electroencephalographic analysis following pharmacological manipulation of the biogenic amines; neurophysiological role of serotonin; effects of protein malnutrition on developing brain; electro-ontogenesis of the brain; evolutionary morphology of brains of aquatic mammals; electrophysiological studies of hippocampal formation, especially long term potentiation following protein malnutrition insult. *Mailing Add:* Worcester Found for Exp Biol Nine Heatherwood Dr Shrewsbury MA 01545

MORGAN-POND, CAROLINE G, b Morristown, NJ. DEFECTS IN SEMICONDUCTORS, MAGNETIC MATERIALS. *Educ:* Swarthmore Col, BA; Princeton Univ, PhD(physics), 80. *Prof Exp:* Res assoc, NY Univ, 80-83; mem res staff, Riverside Res Inst, 83-84; asst prof, 84-89, ASSOC PROF PHYSICS, WAYNE STATE UNIV, 89- *Concurrent Pos:* Consult, Riverside Res Inst, 84, Energy Conversion Devices, 85-86 & Lockheed, 86. *Mem:* Am Phys Soc; Sigma Xi. *Res:* Solid state theory: work on electronic and structural properties of narrow gap semiconductors and defects in semiconductors; spin glasses and other random magnetic systems. *Mailing Add:* Physics Dept Wayne State Univ Detroit MI 48202

MORGANROTH, JOEL, b Detroit, Mich, Oct 29, 45; m 72; c 3. CARDIOLOGY, INTERNAL MEDICINE. *Educ:* Univ Mich, BS, 68, MD, 70; Am Bd Internal Med, dipl, 73, Am Bd Cardiovasc Dis, dipl, 75. *Prof Exp:* From intern to resident, Beth Israel Hosp, Harvard Med Sch, 70-72; clin assoc cardiol, Nat Heart & Lung Inst, 72-74; clin instr med, Georgetown Univ, 74; clin fel, 74-75, asst prof med, Med Sch, Univ Pa, 75-78, dir, Ecg-Exercise Lab & assoc dir, Non-Invasive Lab, Univ Hosp, 75-78; assoc prof med, Thomas Jefferson Med Col, Philadelphia, 78-82; assoc dir cardiovasc sect, Lankenan Hosp, 78-82; PROF MED & PHARMACOL, SCH MED, HAHNESMANN ANN UNIV, 82- *Concurrent Pos:* Fel, Coun on Clin Cardiol & Artherosclerosis; dir, Sudden Death Prev Prog, Likoff Cardiovasc Inst, 82-, Nat Cardiovasc Res Ctr, 82-; dir cardiol res & develop, Grad Hosp, 87; clin prof med, Sch Med, Univ Pa, 88. *Honors & Awards:* Physician Recognition Award, AMA, 73-76, 76-79 & 79-82. *Mem:* Am Fedn Clin Res; fel Am Col Physicians; fel Am Col Cardiol; Am Heart Asn; fel Am Col Chest Physicians. *Res:* Pathophysiology of clinical ischemic heart disease utilizing primarily non-invasive techniques; echocardiography and exercise testing with particular attention to computerized models; evaluation of pathogenesis and treatment of sudden cardiac death. *Mailing Add:* Cardiac R&D Grad Health Syst One Grad Plaza Philadelphia PA 19146

MORGANS, LELAND FOSTER, b Stuttgart, Ark, June 30, 39; m 60; c 2. HISTOLOGY, ZOOLOGY. *Educ:* Tex Lutheran Col, BS, 61; Univ Ark, Fayetteville, MS, 65; Okla State Univ, PhD(zool), 68. *Prof Exp:* Asst prof biol, Phillips Univ, 68-69; asst prof, 69-77, ASSOC PROF BIOL, UNIV ARK, LITTLE ROCK, 77- *Mem:* AAAS; Sigma Xi; Am Asn Anatomists. *Res:* Comparative vertebrate physiology. *Mailing Add:* Dept of Biol Univ of Ark 33 & Univ Ave Little Rock AR 72204

MORGENSTERN, ALAN LAWRENCE, b Brooklyn, NY, Dec 21, 33; div; c 2. PSYCHIATRY. *Educ:* Cornell Univ, BA, 54; Duke Univ, MD, 59. *Prof Exp:* Intern internal med, Med Sch, Duke Univ, 59-60; resident psychiat, Med Sch, Univ Colo, 60-63; from instr to assoc prof, 65-81, PROF CLIN PSYCHIAT, MED SCH, UNIV ORE, 81- *Concurrent Pos:* WHO travel-study fel, Inst Psychiat, London, 72-73; consult, Training Br, NIMH & prog develop, Good Samaritan Hosp, Portland, 74-87, chmn dept psychiat. *Honors & Awards:* Alumni Lectr, Tavistock Clin, London, 84. *Mem:* Fel Am Psychiat Asn. *Res:* Psychotherapy of borderline states; marital psychotherapy. *Mailing Add:* 7929 SW Ruby Terr Portland OR 97219

MORGENSTERN, MATTHEW, b New York, NY. COMPUTER SCIENCE. *Educ:* Columbia Univ, BS, 68, MS, 70; Mass Inst Technol, SM, 75, PhD(comput sci), 76. *Prof Exp:* Tech staff comput sci, Bell Telephone Labs, 68-70; res asst, Mass Inst Technol, 70-76; ASST PROF COMPUT SCI, RUTGERS UNIV, 76- *Concurrent Pos:* Vis fac, Goddard Space Flight Ctr, NASA, 80-81. *Mem:* Asn Comput Mach; Inst Elec & Electronic Engrs; AAAS; Sigma Xi. *Res:* Database systems research, design, and development: schema design to support multiple data model views for users of distrubuted information systems; file and index structure implementation; intelligent systems and user interfaces; automatic programming: computer aided design of information systems. *Mailing Add:* 825 Menlo Ave Menlo Park CA 94025

MORGENSTERN, N(ORBERT) R, b Toronto, Ont, May 25, 35; m 60; c 3. SOIL MECHANICS, ENGINEERING GEOLOGY. *Educ:* Univ Toronto, BASc, 56; Univ London, DIC & PhD(soil mech), 64. *Hon Degrees:* DEng, Univ Toronto, 83, DSc, Queen's Univ, 89. *Prof Exp:* Res asst soil mech, Imp Col, Univ London, 58-60, lectr civil eng, 60-68; prof civil eng, 68-83, UNIV PROF, UNIV ALTA, 83- *Concurrent Pos:* Consult to var eng firms & govt agencies, 60-; hon res fel, Inst Water Conservancy & Hydroelec Res, Beijing. *Honors & Awards:* Brit Geotech Soc Prizes, 61 & 66; Huber Res Prize, Am Soc Civil Engrs, 72; Can Geotech Soc Prize, 77, Legget Award, 79; Rankine lectr, Brit Geotech Soc, 81; Haultain Award, Govt Alta, 87; Roger Brown Award, 87; Thomas Roy Award, 87; Rocha Lectr, Portugese Soc Geotech, 89. *Mem:* Fel Royal Soc Can; Can Geotech Soc; Am Soc Civil Engrs; Asn Eng Geol; Brit Geotech Soc; fel Eng Inst Can; fel Can Acad Eng. *Res:* Mechanics of landslides and other geomorphological processes; properties of natural materials; design and construction of dams; relation between mechanics and geological structures; permafrost engineering; underground excavations; rock mechanics; offshore engineering. *Mailing Add:* Dept Civil Eng Univ Alta Edmonton AB T6G 2G7 Can

MORGENTHALER, FREDERIC R(ICHARD), b Cleveland, Ohio, Mar 12, 33; m 58; c 2. ELECTRICAL ENGINEERING. *Educ:* Mass Inst Technol, SB & SM, 56, PhD(elec eng), 60. *Prof Exp:* From asst prof to assoc prof, 60-65, PROF ELEC ENG, MASS INST TECHNOL, 68- *Concurrent Pos:* Ford Found fel, Mass Inst Technol, 60-62; res fel appl physics, Harvard Univ, 60-61, vis lectr, 61, Cecil H Green Prof Elec Eng, 84-86. *Mem:* Fel Inst Elec & Electronics Engrs; Sigma Xi; Am Phys Soc. *Res:* Microwave magnetics; ultrasonics; electromagnetic theory, hyperthermia. *Mailing Add:* 71 Abbott Rd Wellesley MA 02181

MORGENTHALER, GEORGE WILLIAM, b Chicago, Ill, Dec 16, 26; m 49; c 4. ASTRODYNAMICS, STATISTICAL ANALYSIS. *Educ:* Concordia Col, BS, 46; Univ Chicago, MS, 48, PhD, 53; Univ Denver, MS, 63; Mass Inst Technol, MS, 70. *Hon Degrees:* LLD, Concordia Col, 83. *Prof Exp:* Mathematician, Inst Air Weapons Res, Chicago, 51-55, group leader, 55-58; assoc prof math & head dept, Chicago Undergrad Div, Univ Ill, 58-60; mgr electronics & math res dept, Martin Co, 60-65, dir res & develop, Martin Marietta Corp, 66-69, corp dir res & develop, 70-74, vpres tech opers, 74-76, vpres & gen mgr, Martin Marietta Aerospace, Baltimore, 76-78, vpres, Martin Marietta Aluminum Co, 79-85; prof & chmn Aerospace Eng Dept & Dir Eng Res Ctr, 86-89, PROF AEROSPACE ENG DEPT & ASSOC DEAN, COL ENG, UNIV COLO, BOULDER, 89- *Concurrent Pos:* Instr, Ill Inst Technol, 51-58; consult, Caywood-Schiller Assocs, 58-59; adj prof, Univ Colo, 60-65 & vis prof 70-76; bk rev ed, J Astronaut Sci, 63-78; mem bd dirs, Columbia Aluminum Co, Exp Fabricators & Ctr Sapce & Advan Technol. *Mem:* Fel AAAS; Sigma Xi; Opers Res Soc Am; assoc fel Inst Math Statist; fel Am Astronaut Soc (pres, 64-66); Am Math Soc; fel Am Inst Aeronaut & Astronaut. *Res:* Statistics; operations research; applied mathematics; astronautics. *Mailing Add:* Col Eng & Appl Sci Univ Colo Eng Ctr Campus Box 0429 Boulder CO 80309-0429

MORGERA, SALVATORE DOMENIC, b Providence, RI, Aug 5, 46; m 83. TELECOMMUNICATIONS, DIGITAL SIGNAL PROCESSING. *Educ:* Brown Univ, BSc, 68, MSc, 70, PhD(elec eng), 75. *Prof Exp:* Sr res scientist, advan technol, Submarine Signal Div, Raytheon Co, 68-78; prof, 78-85, ADJ PROF ELEC ENG, FAC ENG & COMPUT SCI, CONCORDIA UNIV, 85-; PROF ELEC ENG, FAC ENG, MCGILL UNIV, 85- *Concurrent Pos:* Scientist, Nat Security Indust Agency, 70-78; sr consult & prin investr, Dept Fisheries & Oceans, Bedford Inst Oceanog, 78-82; responsible adminr, Commun Circuits & Systs Group, Concordia Univ, 78-85; sr consult, Le Groupe Videoway, 84-85, Sigma Securities, 84-85, Comput Res Inst, Montreal, 85- & Bell Can Ottawa, 86-; fel, Sci Col, Concordia Univ, 86-; chmn, Commun Chap, Montreal Sect, Inst Elec & Electronic Engrs, 88-; pres, eng grant selection comt, FCAR, Que, 88-; maj proj leader & res prof, Can Inst Telecommun Res, 90-; vis expert, Bell Can Ottawa, 90. *Mem:* Fel Inst Elec & Electronic Engrs; Sigma Xi; Statist Soc Can; Am Statist Asn; European Soc Signal Processing; Can Inst Fifth Generation Res; Can Soc Elec Eng; Soc Indust & Appl Math; NY Acad Sci. *Res:* Detection and estimation theory; pattern recognition and image processing; information theory and data compression; complexity theory; computer communications networks; algebraic coding theory; digital signal processing. *Mailing Add:* Dept Elec Eng Mc Connell Eng Bldg McGill Univ 3480 Univ St Montreal PQ H3A 2A7 Can

MORI, ERIK JUN, b Kobe, Japan, June 12, 62. SEMICONDUCTOR RESEARCH, ENVIRONMENTAL. *Educ:* Univ Hawaii, Manoa, BS, 84; Univ Tex, Arlington, DSc, 90. *Prof Exp:* Res technician, Hitco Corp, 85; analytical chemist, Radian Corp, 85-86; res asst, Univ Tex, Arlington, 86-90; ANALYTICAL SCIENTIST, MEMC ELECTRONIC MAT, INC, 90- *Mem:* Am Chem Soc; Electrochem Soc; Soc Appl Spectros. *Res:* Semiconductor fabrication and characterization; solar cell technology; metrology; trace element and chemical analyses; environmental monitoring and control. *Mailing Add:* 209 Trails of Sunbrook St Charles MO 63301

MORI, KEN, pathology, for more information see previous edition

MORI, PETER TAKETOSHI, b Montebello, Calif, Feb 16, 25; m 57; c 3. ORGANIC CHEMISTRY. *Educ:* Park Col, BS, 45; Okla State Univ, MS, 48; Purdue Univ, PhD, 54. *Prof Exp:* Res chemist, Dearborn Chem Co, 52-55, Am Potash & Chem Corp, 55-57, Turco Prod, Inc, 57 & Stuart Co, 58-62; RES CHEMIST, ICI UNITED STATES, INC, 62- *Mem:* AAAS; Am Chem Soc; Sigma Xi. *Res:* Medicinals, insecticides; detergents and corrosion inhibitors. *Mailing Add:* 1309 Quincy Dr Green Acres Wilmington DE 19803-5131

MORI, RAYMOND I, b Hawaii, Oct 7, 26; m 51; c 4. ORGANIC CHEMISTRY. *Educ:* Univ Hawaii, BA, 49; Northwestern Univ, PhD(chem), 55. *Prof Exp:* Org chemist, Hawaiian Pineapple Co, Ltd, 54-61; asst dir qual assurance, Dole Co, 61-64, dir qual assurance, 64-69; corp dir qual assurance, Castle & Cooke Foods, Inc, 69-87; dir qual assurance, Dole Packaged Foods, 82-87; RETIRED. *Mem:* Am Chem Soc; Inst Food Technologists; Asn Offs Agr. *Res:* Quality control; enzymes. *Mailing Add:* 188 Paseo Del Rio Moraga CA 94556

MORI, SCOTT ALAN, b Janesville, Wis, Oct 13, 41. PLANT TAXONOMY. *Educ:* Univ Wis-Stevens Point, BS, 64; Univ Wis-Madison, MS, 68, PhD(bot), 74. *Prof Exp:* Instr bot, Univ Wis-Marshfield, 69-74, taxonomist, Mo Bot Garden, 74-75; taxonomist, 75-77, assoc cur, 77-90, CUR, NY BOT GARDEN, 90- *Concurrent Pos:* taxonomist, Cocoa Res Ctr, Itabuna, Bahia, Brazil, 78-80; Mellon fel, Smithsonian Inst. *Mem:* Org Trop Biol; Bot Soc Am; Brazilian Bot Soc. *Res:* Taxonomy of neotropical lecythidaceae (Brazil Nut Family); phenology and floral biology of tropical trees; taxonomy of vochysiacene. *Mailing Add:* 3065 Sedgwick Ave Bronx NY 10468

MORIARTY, C MICHAEL, b Schenectady, NY, Apr 12, 41; m 66, 75; c 3. PHYSIOLOGY, BIOPHYSICS. *Educ:* Carnegie Mellon Univ, BS, 62; Cornell Univ, MS, 65; Univ Rochester, PhD(biophys), 68. *Prof Exp:* Instr physiol & biophys, Sch Med, Univ Iowa, 68-70; from asst prof to assoc prof physiol & biophys, Univ Nebr Med Ctr, Omaha, 70-80, assoc dean res & develop, 81-83, asst exec vpres & provost, 87-88, PROF PHYSIOL & BIOPHYS, UNIV NEBR MED CTR, OMAHA, 80-; ASSOC VPRES RES, UNIV GA, 88- *Concurrent Pos:* USPHS fel, Sch Med, Univ Iowa, 68-70; NSF res grant, 71-74, NIH res grants, 74-, Am Cancer Soc grant, 75-77 & Environ Protection Agency grant, 91- *Mem:* AAAS; Am Physiol Soc; Am Soc Cell Biol. *Res:* Cell physiology, membrane transport; calcium; hormone regulation; heavy metal toxicology. *Mailing Add:* Off Vpres Res Univ Ga 616 Boyd Grad Studies Res Ctr Athens GA 30602

MORIARTY, DANIEL DELMAR, JR, b New Orleans, La, June 16, 46; m 69; c 1. ANIMAL BEHAVIOR. *Educ:* La State Univ, New Orleans, BA, 68; Tulane Univ, MS, 72, PhD(psychol), 73. *Prof Exp:* Asst prof, 73-77, ASSOC PROF EXP PSYCHOL, UNIV SAN DIEGO, 77- *Mem:* Animal Behav Soc; Am Psychol Asn. *Res:* Genetic analysis of the behavioral response of frustration in rats and mice; effects of food deprivation and partial reinforcement on estrus cycling in female rats; effects of irrelevant drive on aversively motivated instrumental responses; Hippocampal chemical stimulation and delay of reinforcement. *Mailing Add:* Psychol Dept Univ San Diego Alcala Park San Diego CA 92110

MORIARTY, DAVID JOHN, b Chicago, Ill, Feb 25, 48; m 74. ECOLOGY, BIOSTATISTICS. *Educ:* Univ Ill, Urbana, BS, 70, MS, 72, PhD(zool), 76. *Prof Exp:* asst prof, 76-80, ASSOC PROF BIOL SCI, CALIF STATE POLYTECH UNIV, 80- *Mem:* Ecol Soc Am; AAAS; Am Inst Biol Sci; Biomet Soc; Am Ornithologists Union; Sigma Xi. *Res:* Application of multivariate statistics to investigations of structure of vertebrate communities; adaptive value of flocking in birds. *Mailing Add:* Dept of Biol Sci Calif State Polytech Univ Pomona CA 91768

MORIARTY, JOHN ALAN, b Chicago, Ill, Jan 17, 44; m 69; c 2. THEORETICAL SOLID-STATE PHYSICS. *Educ:* Univ Calif, Berkeley, AB, 65; Stanford Univ, PhD(appl physics), 71. *Prof Exp:* Res assoc, Los Alamos Sci Lab, NMex, 71-73; res assoc, Univ Cambridge, 73-74; res assoc physics, Col William & Mary, 74-79; asst prof, Univ Cincinnati, 79-82; PHYSICIST, LAWRENCE LIVERMORE NAT LAB, 82- *Concurrent Pos:* Contractor, NASA Langley Res Ctr, 74-79; prin investr, NASA grants, 79-82; consult, Lawrence Livermore Nat Lab, 81-82. *Mem:* Am Phys Soc; Sigma Xi; Mat Res Soc. *Res:* Electronic structure, interatomic forces, and the physical properties of metals; pseudopotential, tight-binding, and resonance methods. *Mailing Add:* L299 Lawrence Livermore Lab Livermore CA 94550

MORIARTY, JOHN LAWRENCE, JR, b Fayette, Iowa, Nov 26, 32; m. CHEMICAL & MECHANICAL ENGINEERING, INDUSTRIAL MANUFACTURING ENGINEERING. *Educ:* St Ambrose Col, BA, 55; Univ Iowa, MS, 57, PhD(chem), 60. *Prof Exp:* Resident res assoc phys chem, Argonne Nat Lab, 60-61; dir chem & metall res, Lunex Co, 61-68; sr scientist, Davenport Div, Bendix Corp, 68-70; independent consult, metallurgist & chemist, 71-73; prof physics, Bettendorf Sch, 73-83; SR METALLURGIST, ROCK ISLAND ARSENAL, 83- *Mem:* Am Chem Soc; fel Am Soc Metals Int; Am Crystallog Asn; Soc Mfg Engr; Metall Soc; Soc Carbide & Tool Engrs; Sigma Xi. *Res:* Physical chemistry, metallurgy of rare earth, alloys; metallurgical chemistry of steelmaking; wear life testing of cutting tools; flexible manufacturing systematic tool management and in-process control sensors. *Mailing Add:* 1917 Perry St Davenport IA 52803

MORIARTY, KEVIN JOSEPH, b Halifax, NS, June 28, 40; m 86. QUANTUM CHROMODYNAMICS, LATTICE GAUGE THEORIES. *Educ:* St Mary's Univ, BSc, 61; Dalhousie Univ, MSc, 63; Univ London, DIC, 65, PhD(math physics), 71. *Prof Exp:* Lectr math, Royal Holloway Col, Univ London, 71-74, reader appl math, 74-83; asst dean comput resources, 83-86, PROF COMPUT, DALHOUSIE UNIV, 83-, DIR, DIV COMPUT SCI, 90- *Concurrent Pos:* Staff tutor math, Open Univ, Eng, 75-77; sr applications engr, John Von Neumann Nat Super Ctr, 86-88; mem, Inst Advan Study, Princeton, 87-89; consult, ETA Systs Inc, 86-88 & Alliant Computer Systs, 91-; pres, Scotia High End Comput Ltd, 88-; res fel, NEC Supercomputers, 90- *Mem:* Royal Inst Gt Brit; Inst Math & Its Appications; Inst Physics. *Res:* Theoretical physics in elementary particle physics and condensed matter physics; parallel algorithm design on parallel supercomputers; computer-aided software environments for programming; author of 254 technical publications. *Mailing Add:* Dept Math Statist & Comput Sci Dalhousie Univ Halifax NS B3H 3J5 Can

MORIARTY, ROBERT M, b New York, NY, Oct 9, 33; m 57; c 3. ORGANIC CHEMISTRY. *Educ:* Fordham Univ, BS, 55; Princeton Univ, PhD(org chem), 59. *Prof Exp:* Res worker org chem, Merck, Sharpe & Dohme, 55-56; NSF, NATO, Fulbright & NIH fels, Harvard Univ & Univ Munich, 59-61; from assoc prof to prof org chem, Cath Univ Am, 61-68; PROF ORG CHEM, UNIV ILL, CHICAGO CIRCLE, 68- *Mem:* Am Chem Soc; Royal Soc Chem. *Res:* Biochemistry; physical methods. *Mailing Add:* Dept of Chem Univ of Ill at Chicago Circle Box 4348 Chicago IL 60680

MORIBER, LOUIS G, b New York, NY, July 12, 17; m 41; c 2. CELL BIOLOGY. *Educ:* Brooklyn Col, AB, 38; Columbia Univ, MA, 52, PhD(zool, cytol), 56. *Prof Exp:* Jr aquatic physiologist, US Fish & Wildlife Serv, 41-43; from assoc prof to prof, 50-86, chmn dept, 81-83, EMER PROF BIOL, BROOKLYN COL, 86- *Concurrent Pos:* Exec officer, PhD prog biol, City Univ New York, 71-81. *Res:* Cytochemistry; cellular endocrinology; neurophysiology. *Mailing Add:* 2239 Troy Ave Brooklyn NY 11234

MORIE, GERALD PRESCOTT, b St Louis, Mo, Mar 20, 39; m 61; c 3. ANALYTICAL CHEMISTRY. *Educ:* Cent Mo State Col, BS, 61; Ohio State Univ, MS, 63, PhD(analytical chem), 66. *Prof Exp:* Asst instr chem, Ohio State Univ, 65; res chemist, Eastman Kodak Co, 66-67, sr res chemist, 68-74, res assoc, Tenn Eastman Co, 74-84, dir phys & anal chem res, 84-85, DIR TECH SERV & DEVELOP, TENN EASTMAN CO, EASTMAN KODAK CO, 85- *Honors & Awards:* Philip Morris Award in Tobacco Sci, 76. *Mem:* Am Chem Soc; Sigma Xi. *Res:* Chemical separations especially liquid chromatography; selective ion electrodes; gas chromatography; analysis of tobacco smoke; technical management. *Mailing Add:* 4522 Mitchell Rd Kingsport TN 37664

MORIN, CHARLES RAYMOND, b Burlington, Vt, Apr 8, 47; m 68; c 4. ENGINEERING. *Educ:* Ohio State Univ, BMetE & MSE, 72. *Prof Exp:* From res asst to res assoc, Ohio State Univ, 68-72; staff consult, Packer Engrs, 72-79, vpres safety eng, 79-81, vpres, 81-85, pres, 85-87; CO-FOUNDER & CHMN BD, ENG SYSTS, INC, 87- *Concurrent Pos:* Mem, Pulp & Paper Comt, Nat Asn Corrosion Engrs, 80-, Corrosion & Mat Eng Comt, Tech Asn Pulp & Paper Indust, 82-, Fatigue Comt, Am Soc Testing & Mat, 83-; chmn, Failure Anal Comt, Am Soc Metals, 83-; lectr, Transp Safety Inst, OKC, 84- *Mem:* fel Am Soc Metals Int; Am Soc Testing & Mat; Nat Asn Corrosion Engrs; Tech Asn Pulp & Paper Indust; Metall Soc; Nat Soc Prof Engrs. *Res:* Material behavior and selection for engineered products; failure mechanisms; reliability investigations. *Mailing Add:* 833 Shiloh Crele Naperville IL 60540

MORIN, DORNIS CLINTON, b Detroit, Mich, Oct 9, 23; m 48; c 3. PLASMA PHYSICS, ELECTRICAL ENGINEERING. *Educ:* Wayne State Univ, BS, 48, MSE, 51; Univ Wis, PhD(physics), 62. *Prof Exp:* Instr elec eng, Wayne State Univ, 48-51; res assoc, Univ Mich, 50-52; test engr, Vickers, Inc, 52-54; res assoc, Univ Mich, 54-55; mgr res dept, Dynex, Inc, 55-58; engr, Midwestern Univs Res Asn, 59-62, physicist, 62-63; res assoc, Univ Wis, 63-64, asst prof physics, Univ Wis-Milwaukee, 64-68; assoc prof, Univ Wis-Whitewater, 68-71; proj assoc, Plasma Physics Group, Univ Wis-Madison, 71-76; comput specialist, Systs Group, 76-80, ENGR NEWS LOG INT, SYSTS BY GRABER, 80- *Concurrent Pos:* Consult, Dynex, Inc, 58-60. *Mem:* AAAS; Am Phys Soc; Inst Elec & Electronics Eng; Asn Comput Mach; Sigma Xi. *Res:* Mechanics; fluid mechanics; electromagnetic theory; electronics; automatic control systems; applied mathematics. *Mailing Add:* 622 Jacobson Ave Madison WI 53714

MORIN, FRANCIS JOSEPH, b Laconia, NH, Oct 10, 17; m 46; c 7. SOLID-STATE PHYSICS. *Educ:* Univ NH, BS, 39, MS, 40. *Prof Exp:* Mem tech staff, Bell Tel Labs, Inc, 41-62; assoc dir, Sci Ctr, NAm Aviation, Inc, 62-67, dir, Space Sci, Space & Info Systs Div, 65-67, dir, Sci Ctr, NAm Rockwell Corp, 67-70, mem tech staff, Sci Ctr, 70-74, DISTINGUISHED FEL, SCI CTR, ROCKWELL INT CORP, 74- *Concurrent Pos:* Vis prof physics, Univ Mo, 75. *Mem:* Am Chem Soc; Am Phys Soc. *Res:* Semiconductors; low temperature; electronic transport and energy-band structure in solids; transition metal oxides; diffusion and dislocations in solids; transition metal superconductors; surface physics; mechanism of catalysis of d-band surface states. *Mailing Add:* 1870 Fearn St Los Osos CA 93402-2516

MORIN, GEORGE CARDINAL ALBERT, b Natick, Mass, June 7, 43; m 72; c 1. ALTERNATE ENERGY, SOIL PHYSICS. *Educ:* Univ Ariz, Tuscon, BS, 65, MS, 68, PhD(soil physics), 77. *Prof Exp:* Geologist, Tex Water Develop Bd, 68-69; res assoc agr eng, Univ Ariz, 70-76; asst prof agr eng, Univ Nebr, Lincoln, 76-80; ASST PROF AGR ENG, NMEX STATE UNIV, 80- *Mem:* Soil Sci Soc Am; Am Soc Agr Engrs. *Res:* Alternate energy, including solar, wind and biomass fuel; soil and water management including irrigation scheduling and water harvesting. *Mailing Add:* 2207-B N French Dr Hobbs NM 88240

MORIN, JAMES GUNNAR, b Minneapolis, Minn, Sept 13, 42. INVERTEBRATE ZOOLOGY, INVERTEBRATE BEHAVIOR. *Educ:* Univ Calif, Santa Barbara, BA, 65; Harvard Univ, MA, 67, PhD(biol), 69. *Prof Exp:* Asst prof, 69-75, assoc prof zool, 75-79, PROF BIOL, UNIV CALIF, LOS ANGELES, 79- *Mem:* AAAS; Am Soc Zoologists; Soc Exp Biol. *Res:* Bioluminescence in marine invertebrate, especially ostracodes and cnidaria; physiology of primitive nervous systems; community structure of subtidal sand bottoms. *Mailing Add:* Dept of Biol Univ Calif 405 Hilgard Ave Los Angeles CA 90024

MORIN, JOHN EDWARD, b Margretville, NY, Aug 12, 51. REGULATION OF GENE EXPRESSION, VACCINE DESIGN & DEVELOPMENT. *Educ:* Univ NH, BS, 73; Purdue Univ, PhD(biochem), 82. *Prof Exp:* Postdoctoral, Med Sch, Univ Vt, 82-83 & Metab Res Unit, Univ Calif, San Francisco, 83-84; sr res scientist, Wyeth-Ayerst Res, Am Home Corp, 84-89; SR RES SCIENTIST, MED RES DIV, AM CYANAMID, 89- *Mem:* AAAS; Fedn Am Socs Exp Biol. *Res:* Design and develop Eukaryotic expression systems which have been used as: Research tools for study of gene expression and protein processing and live recombinant vaccines. *Mailing Add:* Am Cyanamid Med Res Div Pearl River NY 10965

MORIN, LEO GREGORY, b Berlin, NH, May 9, 41; m 68; c 3. CLINICAL BIOCHEMISTRY. *Educ:* Spring Hill Col, BS, 65; Boston Col, PhD(molecular biol), 68. *Prof Exp:* Asst prof biol, Rollins Col, 68-70; res biochemist, Sunland Hosp, Orlando, Fla, 70-71 & Kiess Instruments, Inc, 71-73; CLIN BIOCHEMIST & DIR CLIN CHEM, VET ADMIN HOSP, ATLANTA, 73- *Concurrent Pos:* Assoc prof, Sch Med, Emory Univ, 76- *Mem:* AAAS; Am Chem Soc; Am Asn Clin Chem. *Res:* Metabolic regulation and disorders; analytical methodology; creatine kinase enzymology. *Mailing Add:* 3421 Donegal Way Lithonia GA 30058-8635

MORIN, NANCY RUTH, b Albuquerque, NMex, Feb 16, 48; US citizen. TAXONOMY, FLORISTICS. *Educ:* Univ Calif, Berkeley, AB, 75, PhD(bot), 80. *Prof Exp:* Postdoctoral fel, Bot Dept, Smithsonian Inst, 80-81; ADMIN CUR, HERBARIUM, MO BOT GARDEN, 81- *Concurrent Pos:* Ed, Ann Mo Bot Garden, Mo Bot Garden, 81-87, convening ed, Flora NAm, 83-; Adj fac, Biol Dept, Univ Mo, St Louis, 83- *Mem:* Am Soc Plant Taxonomists (coun mem, 88-); Am Inst Biol Sci; Bot Soc Am; Int Asn Plant Taxon; Sigma Xi; AAAS. *Res:* Evolution and classification of Campanulaceae; north temperate floristics; systematics of Githopsis, Campanula, and Triodanis; breeding systems; seed and pollen morphology; cytology. *Mailing Add:* Mo Bot Garden PO Box 299 St Louis MO 63166-0299

MORIN, PETER JAY, b New Britain, Conn, Sept 8, 53; m 76. COMMUNITY ECOLOGY, POPULATON ECOLOGY. *Educ:* Trinity Col, BS, 76; Duke Univ, PhD(zool), 82. *Prof Exp:* Asst prof biol, Col Holy Cross, 82-83; asst prof, 83-89, ASSOC PROF BIOL, RUTGERS UNIV, 89- *Honors & Awards:* George Mercer Award, Ecol Soc Am, 85. *Mem:* Ecol Soc Am; Am Soc Naturalists; Soc Study Evol; Brit Ecol Soc; Am Soc Limnologists & Oceanographers; Am Soc Ichthyologists & Herpetologists. *Res:* Experimental studies of population dynamics and community organization. *Mailing Add:* Biol Sci Nelson Labs Rutgers Univ Piscataway NJ 08854-1059

MORIN, RICHARD DUDLEY, b Quincy, Ill, Oct 5, 18; m 42; c 4. MEDICINAL CHEMISTRY. *Educ:* Univ Mich, BS, 40, MS, 42, PhD(org chem), 43. *Prof Exp:* Res chemist, Battelle Mem Inst, 43-52, asst div chief, 52-62, res assoc, 62-63; from assoc prof to prof med chem, 70-89, EMER PROF, SCH MED, UNIV ALA, BIRMINGHAM, 89- *Mem:* AAAS; Am Chem Soc. *Res:* Alkaloid synthesis; synthesis and biochemical and psychopharmacological action of psychotomimetic agents related to mescaline and psilocybin; structure-activity relationships among excitant and depressant compounds. *Mailing Add:* Prog Sch of Med Neuropsychiat Res Prog Univ of Ala Birmingham AL 35294

MORIN, THOMAS LEE, b Rahway, NJ, Aug 27, 43; m 65; c 1. OPERATIONS RESEARCH. *Educ:* Rutgers Univ, BS, 65; Univ NMex, MS, 67; Case Western Reserve Univ, MS, 70, PhD(opers res), 71. *Prof Exp:* Spec lectr opers res, Case Western Reserve Univ, 69-70; opers res analyst, Univ Assocs, Inc, 70-71; asst prof opers res, Dept Indust Eng & Mgt Sci, Technol Inst, Northwestern Univ, Evanston, 71-77; ASSOC PROF OPERS RES, SCH INDUST ENG, PURDUE UNIV, 77- *Concurrent Pos:* NSF grant, Urban Systs Eng Ctr, Northwestern Univ, Evanston, 71-72. *Mem:* Opers Res Soc Am; Inst Mgt Sci; Am Soc Civil Engrs. *Res:* Dynamic programming; optimization of stochastic processes; application of operations research to urban and water resources systems. *Mailing Add:* Sch of Indust Eng Purdue Univ West Lafayette IN 47907

MORIN, WALTER ARTHUR, b Salem, Mass, Oct 31, 33; m 57; c 7. NEUROPHYSIOLOGY. *Educ:* Merrimack Col, AB, 58; Boston Col, MS, 60; Clark Univ, PhD(physiol), 66. *Prof Exp:* Assoc prof, 61-74, PROF ZOOL, BRIDGEWATER STATE COL, 74- *Concurrent Pos:* NIH res grants, 66-67 & 69-71; Nat Res Coun Can Fel, Univ Toronto, 67-68; Sigma Xi res grant, 69-70. *Mem:* Am Soc Zool. *Res:* Excitatory and inhibitory synaptic vesicles in crustaceans. *Mailing Add:* Dept of Zool Bridgewater State Col Junction Rte 18-28-104 Bridgewater MA 02324

MORINIGO, FERNANDO BERNARDINO, b Parana, Arg, June 1, 36; m 63; c 2. PHYSICS, OPERATIONS RESEARCH. *Educ:* Univ Southern Calif, BS, 57; Calif Inst Technol, PhD(physics), 63. *Prof Exp:* Asst prof physics, Calif State Col, Los Angeles, 63-64; vis prof, Univ Freiburg, 64-65; from asst prof to prof physics, Calif State Univ, Los Angeles, 65-90; CHIEF SCIENTIST, HUGHES AIRCRAFT CO, 82- *Concurrent Pos:* Res fel, Calif Inst Technol, 63-64, sr res fel, 65; res physicist, Lab Nuclear Spectrometry, Nat Ctr Sci Res, Strasbourg, France, 68-69; res fel, Niels Bohr Inst, Copenhagen, 73; vis prof, Calif Inst Technol, 77-78. *Mem:* Am Phys Soc. *Res:* Nuclear beta decay; experimental nuclear spectroscopy; theoretical nuclear physics; nuclear reaction theory; mathematical modelling of macrosystems. *Mailing Add:* 8834 Glider Ave Los Angeles CA 90045

MORINO, LUIGI, b Rome, Italy, July 21, 38; m 84; c 2. AEROELASTICITY. *Educ:* Univ Rome, DrMechEng, 63, DrAeroEng, 66. *Prof Exp:* Asst prof aerospace eng, Univ Rome, 65-68; sr res engr, Mass Inst Technol, 68-69; from assoc prof to prof aerospace & mech eng, Boston Univ, 69-87, dir, Ctr Comput & Appl Dynamics, 77-87; PROF AEROELASTICITY, UNIV ROME LA SAPIENZA, 87- *Concurrent Pos:* NATO fel, Mass Inst Technol, 67-68; adj assoc prof, Boston Univ, 68-69; consult, Mass Inst Technol, 69-72 & Aerospace Syst Inc, 72-77; prin investr, NASA grant, 72-, NSF, 75-77, 79-81, Dept Energy, 76-78, Army Res Off, 79-; pres, Anal & Comput Sci Inc, 77-81 & Inst Comput Appln Res & Utilization in Sci, 79-85; NATO fac fel, 85. *Mem:* Am Inst Aeronaut & Astronaut; Am Helicopter Soc. *Res:* Computational steady, oscillatory and unsteady subsonic and supersonic aerodynamics of complex aircraft configurations; extensions to helicopters, propellers and windmills; applications to aeroelasticity; extension to aeroelasticity; transonic flows; boundary layers and separated flows; vortex flows for inviscid and viscous fluids. *Mailing Add:* Univ Rome La Sapienza V Eudossiana 18 Rome 00184 Italy

MORISAWA, MARIE, b Toledo, Ohio, Nov 2, 19. GEOLOGY, GEOMORPHOLOGY. *Educ:* Hunter Col, AB, 41; Union Theol Sem, MA, 45; Univ Wyo, MA, 52; Columbia Univ, PhD(geol), 60. *Prof Exp:* Instr geol, Bryn Mawr Col, 55-59; asst prof, Mont State Univ, 59-61; assoc prof, Antioch Col, 63-69; assoc prof, State Univ NY Binghamton, 69-75, prof geol, 75-90; RETIRED. *Concurrent Pos:* Geologist, US Geol Serv. *Honors & Awards:* Fulbright Lectureship to India, 87-88. *Mem:* Geol Soc Am; Am Quaternary Asn; Am Geophys Union; Sigma Xi; Am Water Resources Asn. *Res:* Fluvial geomorphology; environmental geomorphology, geology and planning; geological aesthetics. *Mailing Add:* Dept of Geol State Univ of NY Binghamton NY 13902-6000

MORISHIMA, AKIRA, b Tokyo, Japan, Apr 18, 30; US citizen; m 61; c 2. PEDIATRICS, CYTOGENETICS. *Educ:* Keio Univ, Japan, MD, 54, PhD(med), 61. *Prof Exp:* Fulbright travel grant, 55-57; fel pediat endocrinol, Col Physicians & Surgeons, Columbia Univ, 58-61; instr pediat, 61-63, assoc, 63-65, asst prof pediat endocrinol, 65-66; asst prof, Med Ctr, Univ Calif, San Francisco, 66-68; ASSOC PROF PEDIAT ENDOCRIOL, COL PHYSICIANS & SURGEONS, COLUMBIA UNIV, 68-, DIR DIV PEDIAT ENDOCRINOL, 68- *Concurrent Pos:* From asst attend to assoc attend pediatrician, Babies Hosp, New York, 63-; NIH res career develop award, 66-68; vis prof, Keio Univ, Tokyo, 77; keynote speaker, 27th Ann Meeting, Japan Soc Pediat Neurol, 85. *Mem:* NY Acad Sci; Environ Mutagen Soc; Am Soc Human Genetics; Soc Pediat Res; Lawson Wilkins Pediat Endocrine Soc; Am Pediat Soc; Endocrine Soc. *Res:* Human cytogenetics; endocrinology. *Mailing Add:* Dept Spec Northeastern Ill Univ 5500 N St Louis Ave Chicago IL 60625

MORISHIMA, HISAYO ODA, b Ito-shi, Japan, July 27, 29; m 61; c 2. MEDICINE. *Educ:* Toho Univ, MD, 51; Tokyo Univ, PhD(med), 60. *Prof Exp:* Resident anesthesiol, Sch Med, Tokyo Univ, 54-56; dir anesthesiol, Izu Teishin Hosp, Shizuoka-ken, Japan, 56-59; resident, DC Gen Hosp, 59-60 & Washington Hosp Ctr, DC, 60-61; res assoc, Col Physicians & Surgeons, Columbia Univ, 61-66; clin instr, Med Ctr, Univ Calif, San Francisco, 66-68; from asst prof to assoc prof, 68-84, PROF ANESTHESIOL, COL PHYSICIANS & SURGEONS, COLUMBIA UNIV, 84- *Concurrent Pos:* Prin investr, NIH res grant, 81-, Nat Inst Drug Abuse res grant, 88-; vis prof, Yamanashi Med Col, Japan, 82; mem, NIH Pharmacol Sci Rev Comt, 83- *Mem:* Am Soc Anesthesiol; Soc Obstet Anesthesia & Perinatology; NY Acad Sci; Asn Univ Anesthesiologists. *Res:* Fetal and maternal physiology and pharmacology; obstetric anesthesia; perinatal drug abuse. *Mailing Add:* Dept of Anesthesiol Columbia Univ New York NY 10032

MORISON, IAN GEORGE, b Perth, Australia, Jan 31, 28; m 72; c 4. FORESTRY. *Educ:* Univ Western Australia, BScFor, 51; Australian Forestry Sch, DipFor, 52; Univ Wash, PhD(forestry), 70. *Prof Exp:* Sr scientist soils, 69-72, ASSOC PROF FORESTRY, COL FOREST RESOURCES, UNIV WASH, DIR INST FOREST PROD & DIR DIV CONTINUING EDUC FOREST RESOURCES, 72- *Concurrent Pos:* Proj mgr regional forest nutrit res prog & land use res appl nat needs, NSF, 69-72. *Mem:* Am Soc Agron; AAAS; Soc Am Foresters; Am Soc Plant Physiologists; Int Soc Soil Sci; Sigma Xi. *Res:* Growth response of Douglas fir to application of nitrogenous fertilizers and the relationship of response to stand, soil and foliar characteristics. *Mailing Add:* 1021 Arlington Blvd #830 Arlington VA 22209-2216

MORISSET, PIERRE, b Quebec, Que, Nov 17, 38; m 70; c 2. PLANT TAXONOMY, PLANT ECOLOGY. *Educ:* Laval Univ, BA, 58, BScA, 62; Univ Cambridge, PhD(bot), 67. *Prof Exp:* Botanist plant taxon, Bot Garden Montreal, 67-68; res assoc, Smithsonian Inst, Washington, DC, 68. PROF BIOL, LAVAL UNIV, 68- *Concurrent Pos:* Univ Col N Wales, Bangor, UK, 76-77; chmn, comt prog eval Que univs, 86-; ed, Natural Can, 87- *Mem:* Can Bot Asn; Genetics Soc Can; Inst Asn Plant Taxon. *Res:* Biosystematic studies of vascular plants; ecology and distribution of Arctic-Alpine species in eastern Canada; population biology of weedy plant species; flora of Northern Quebec. *Mailing Add:* Dept Biol Laval Univ Quebec PQ G1K 7P4 Can

MORITA, HIROKAZU, b Steveston, BC, July 17, 26; m 65. ORGANIC CHEMISTRY. *Educ:* Univ Man, BSc, 48, MSc, 49; Univ Notre Dame, PhD(org chem), 51. *Prof Exp:* US Off Naval Res fel, Northwestern Univ, 51-52; agr res off, Chem Div, Sci Serv, Can Dept Agr, 52-57, res off chem, Soil Res Inst, 67, res scientist chem, 67-85; RES ASSOC, PALUDICAN INT, 85- *Concurrent Pos:* Goodyear res fel, Princeton Univ, 56-57; Commonwealth Sci & Indust Res Orgn vis scientist, Univ Melbourne, 66-67. *Mem:* Fel AAAS; Phytochem Soc NAm; Am Chem Soc; fel Chem Inst Can; Can Soc Soil Sci. *Res:* Natural products in peat; organic mass spectrometry; gas and high-pressure liquid chromatography; synthesis and properties of polyphenols; differential thermal analysis of organic polymers; organic geochemistry. *Mailing Add:* 2957 Marcel St Ottawa ON K1V 8H6 Can

MORITA, RICHARD YUKIO, b Pasadena, Calif, Mar 27, 23; m 53; c 3. MICROBIOLOGY, OCEANOGRAPHY. *Educ:* Univ Nebr, BS, 47; Univ Southern Calif, MS, 49; Univ Calif, PhD(microbiol, oceanog), 54. *Prof Exp:* Asst, Univ Southern Calif, 47-49; asst, Univ Calif, 49-54, res microbiologist, Scripps Inst Oceanog, 54-55; asst prof bact, Univ Houston, 55-58; assoc prof, Univ Nebr, 58-62; prof, 62-89, EMER PROF MICROBIOL & OCEANOG, ORE STATE UNIV, 89- *Concurrent Pos:* Microbiologist, Mid-Pac Exped, 50 & Trans-Pac Exped, 53; vis investr, Danish Galathea Deep-Sea Exped, 52 & Dodo Exped, 64; prog dir biochem, NSF, 68-69, mem, Panel Molecular Biol, 69-70; consult, Nat Inst Gen Med Sci, 68-71; proj reviewer, Environ Protection Agency, 72; mem, Panel Biol Oceanog, NSF, 73; Queen Elizabeth II sr fel, Govt Australia, 73-74; Japan Soc Promotion Sci fel, 78; distinguished vis prof, Kyoto Univ, Japan, 89; Glaser vis prof, Fla Int Univ, 90. *Honors & Awards:* Fisher Award, Am Soc Microbiol. *Mem:* Fel AAAS; Soc Gen Microbiol; Oceanog Soc Japan; Can Soc Microbiol; fel Am Acad Microbiol; hon mem Am Soc Microbiol; hon mem Chilean Soc Microbiol. *Res:* Effect of hydrostatic pressure on the physiology of microorganisms; study of Beggiatoa, marine psychophilic bacteria; eutrophication; starvation-survival of microbes. *Mailing Add:* Dept Microbiol Ore State Univ Corvallis OR 97331-3804

MORITA, TOSHIKO N, b Los Angeles, Calif, May 29, 26; m 53; c 3. FOOD MICROBIOLOGY. *Educ:* Univ Calif, Los Angeles, BS, 48, MA, 50, PhD(zool), 52. *Prof Exp:* Res assoc chemother, Barlow Sanitarium, 51-52; serologist, Vet Admin, 52-53; res assoc biochem, Scripps Metab Clin, 53-55; asst prof biol, Univ Houston, 57-58; asst prof immunol, Ore State Univ, 65-67; res assoc biol, Georgetown Univ, 68-69; res assoc food & nutrit, Ore State Univ, 70-88; RETIRED. *Res:* Staph toxins. *Mailing Add:* 1515 NW 14th St Corvallis OR 97330

MORITSUGU, TOSHIO, b Honolulu, Hawaii, Apr 2, 25; m 59; c 2. SUGAR CHEMISTRY. *Educ:* Univ Louisville, BA, 49; Ohio State Univ, MS, 51, PhD(org chem), 54. *Prof Exp:* Res fel, Ohio State Univ, 54-55; assoc technologist sugar cane res, 55-68, SUGAR TECHNOLOGIST, EXP STA, HAWAIIAN SUGAR PLANTERS' ASN, 68- *Concurrent Pos:* Mem, US Nat Comt Sugar Analysis & Int Comt Uniform Methods Sugar Analysis. *Mem:* AAAS; fel Am Inst Chemists; Am Chem Soc; Int Soc Sugar Cane Technol; Sigma Xi. *Res:* Carbohydrate chemistry; crystallization, clarification and ion exchange in sugar cane processing; steric effects in organic chemistry; cane factory analysis, control and calculations. *Mailing Add:* 3664 Loala Honolulu HI 96822

MORITZ, BARRY KYLER, b Elizabeth, NJ, Apr 16, 41; m 65; c 2. PHYSICS, COMPUTER SCIENCE. *Educ:* Calif Inst Technol, BS, 63; Univ Md, PhD(physics), 69. *Prof Exp:* Asst physics, Univ Md, 63-69; NSF fel, E O Hulburt Ctr Space Res, US Naval Res Lab, Washington, DC, 69-71; sr assoc, PRC Info Sci Co, 71-75, prin, 75-81; PRES, MEASUREMENT CONCEPT CORP, 81- *Concurrent Pos:* Lectr, The Heights, Washington, DC, 69-70. *Mem:* Asn Comput Mach. *Res:* General relativity, astrophysics and cosmology; solar corona and streamers; digital image analysis and pattern recognition; distributed logic systems; associative and parallel techniques; computer information and science. *Mailing Add:* 124 Cavalier Dr Virginia Beach VA 23451-2528

MORITZ, CARL ALBERT, b Bellevue, Ky, Jan 27, 14; m 43; c 5. GEOLOGY. *Educ:* Univ Kans, AB, 40; Harvard Univ, MA, 48, PhD(geol), 50. *Prof Exp:* Jr geologist, St Louis Smelting & Refining, 41; topog engr, US Coast & Geod Surv, 42-43; jr geologist, Union Mines Develop Corp, 44; sr geologist, Phillips Petrol Co, 44-46; instr, Univ Tenn, 46; asst prof, Dartmouth Col, 48-53; CONSULT GEOLOGIST, ALEX W McCOY ASSOCS, INC, 53- *Mem:* Fel Geol Soc Am; Paleont Soc; Am Asn Petrol Geologists; Am Inst Mining, Metall & Petrol Engrs; Sigma Xi. *Res:* Stratigraphy; sedimentation and petroleum geology. *Mailing Add:* 5223 S Birmingham Rd Tulsa OK 74105

MORITZ, ROGER HOMER, b Cleveland, Ohio, Mar 11, 37; m 86; c 9. EGYPTIAN FRANCTIONS, INFINITE SERIES. *Educ:* Valparaiso Univ, BS, 59; Univ Pittsburgh, MS, 61, PhD(math), 64. *Prof Exp:* Asst math, Univ Pittsburgh, 59-61; sr engr, Goodyear Atomic Corp, 62-64; res mathematician, Cornell Aeronaut Lab, Inc, Cornell Univ, 64-70; assoc prof, 70-80, chmn dept, 70-85, PROF MATH, ALFRED UNIV, 80-, COLE PROF APPLIED MATH, 85- *Concurrent Pos:* Instr math, State Univ NY Buffalo, 65-69. *Mem:* Math Asn Am. *Res:* infinite series; probability; Egyptian fractions; Canchy-Euler differential equation. *Mailing Add:* Dept Math Alfred Univ Alfred NY 14802

MORIWAKI, MELVIN M, b Lihue, Hawaii, May 10, 52. SOLID STATE PHYSICS. *Educ:* Univ Hawaii, BS, 74, PhD(physics), 83. *Prof Exp:* MEM, RES TECH STAFF, RES TECH CTR, 84- *Mem:* Am Phys Soc; Sigma Xi. *Mailing Add:* 1215 Front Hill Ave Torrance CA 90503-6009

MORIYAMA, IWAO MILTON, b San Francisco, Calif, Jan 26, 09; m 46; c 2. PUBLIC HEALTH. *Educ:* Univ Calif, BS, 31; Yale Univ, MPH, 34, PhD(pub health statist), 37. *Prof Exp:* Sanit engr, George Williams Hooper Found Med Res, Calif, 31-32; asst, Pierce Lab Hyg, Conn, 33-39; tech secy comt hyg housing, Am Pub Health Asn, 39-40; jr biometrician, US Bur Census, Washington, DC, 40-46; chief mortality analysis sect, Nat Off Vital Statist, 47-61, dir off health statist anal, Nat Ctr Health Statist, 61-71, chief statist dept, Atomic Bomb Casualty Comn, 71-73, assoc dir int statist, Nat Ctr Health Statist, USPHS, 74-75, chief, dept epidemiol & statist, Radiation Effects Res Found, 75-78, dep exec dir, 78-86, EXEC DIR, INT INST FOR VITAL REGISTRATION & STATISTICS, 86- *Concurrent Pos:* WHO expert comt on health statist; secy, US Nat Comt, Vital & Health Statist. *Mem:* Fel AAAS; fel Am Statist Asn; fel Am Pub Health Asn; Sci Pop Asn Am. *Res:* Vital and health statistics; demography. *Mailing Add:* 7120 Darby Rd Bethesda MD 20817

MORIYASU, KEIHACHIRO, experimental high-energy physics, for more information see previous edition

MORIZUMI, S JAMES, b San Francisco, Calif, Nov 13, 23; m 55; c 1. APPLIED SCIENCES, THERMODYNAMICS. *Educ:* Univ Calif, Berkeley, BS, 55; Calif Inst Technol, MS, 57; Univ Calif, Los Angeles, PhD(appl math), 70. *Hon Degrees:* Mech Eng, Kumamoto Univ. *Prof Exp:* Aerodynamicist aerodyn & performance group, Douglas Aircraft, 55-60; mem tech staff aerothermal dept, TRW Defense & Space Syst, 60-68; res engr IR radiation heat transfer, Univ Calif, Los Angeles, 68-69; mem tech staff, Electro-Optical Sensor Dept, TRW Defense & Space Syst Group, 69-71, Space Guid Software Design Dept, 71-73, Mission Design Dept, 73-76, HEL Optical Dept, 76-77, staff engr, Syst Survivability Dept, 77-78, staff scientist opers syst, 78-81; dir guid & control/space applns, Hydraulic Res Textron, 81-82; sr staff engr, Electro-Optics & Data Systs Group, Hughes Aircraft Co, El Segundo, Calif, 82-84, Radar Syst Group, 84-87, Space & Commun Syst Group, 87-88; INSTR MATH, EL CAMINO COL & CALIF STATE UNIV, LONG BEACH, 88- *Concurrent Pos:* Invited lectr, Univ Ariake, 75, Univ Kumamoto, 75 & 81 & Univ Kyushu, 81. *Mem:* Sigma Xi; Am Inst Aeronaut & Astronaut. *Res:* Molecular gas non-grey thermal radiation, vibration rotation bands; applied mathematics, mathematical logic, methodology and optimization; laser, electro-optic sensing, guidance and control, trajectory analysis, detection/pointing/tracking; author of 30 technical papers in US, Japan, West Germany and Sweden. *Mailing Add:* 29339 Stadia Hill Lane Palos Verde Peninsula CA 90274

MORK, DAVID PETER SOGN, b Thief River Falls, Minn, Sept 25, 42; m 66; c 2. BIOLOGY, PHYSIOLOGY ANIMAL. *Educ:* Moorhead State Univ, BS, 64; Purdue Univ, MS, 66, PhD(bionucleonics), 69. *Prof Exp:* Asst prof, 68-75, assoc prof, 75-79, lectr, Sch Nursing, 73-79, 83-84, PROF BIOL, ST CLOUD STATE COL, 79- *Concurrent Pos:* Sabbitical, Univ Oslo & Radium Hosp, Oslo, Norway, 86-87. *Mem:* AAAS; Am Soc Zoologists; Sigma Xi. *Res:* Annual bat census. *Mailing Add:* Dept Biol Sci St Cloud State Univ St Cloud MN 56301

MORKEN, DONALD A, b Crookston, Minn, Feb 2, 22; m 47; c 2. BIOPHYSICS. *Educ:* Cornell Univ, BEE, 49; Univ Rochester, PhD(biophys), 54. *Prof Exp:* From instr to assoc prof radiation biol, Sch Med & Dent, Univ Rochester, 54-84, assoc prof biophys, 77-84; RETIRED. *Mem:* Radiation Res Soc; Inst Elec & Electronics Engrs; Biophys Soc; Health Physics Soc. *Res:* Effects of ionizing radiation on physiological functions of organisms. *Mailing Add:* 218 Newton Rd Rochester NY 14626

MORKOC, HADIS, b Erzurum, Turkey, Oct 2, 47; US citizen; m 76. SEMICONDUCTORS. *Educ:* Istanbul Tech Univ, BS, 68, MS, 69; Cornell Univ, PhD(elec eng), 76. *Prof Exp:* Instr elec eng, Istanbul Tech Univ, 69-71; fel, Cornell Univ, 76; mem tech staff, Varian Assoc, Inc, 76-78; asst prof, 78-82, ASSOC PROF ELEC ENG, UNIV ILL, 82- *Concurrent Pos:* Consult, Gen Dynamics, 80-81, Hughes, 81, Int Bus Mach, 81-, Gould, Perkin Elmer, TRW & Bendix & AT&T Bell Labs. *Mem:* Fel Int Elec & Electronic Engrs; AAAS; Am Phys Soc; Electrochem Soc; Optical Soc Am; Mat Res Soc. *Res:* Compound semiconductor materials prepared by molecular beam opitaxy; high-speed devices; semiconductor physics. *Mailing Add:* 1803 Golfview Dr Urbana IL 61801

MORKOVIN, MARK V(LADIMIR), b Prague, Czech, July 28, 17; nat US; m 40; c 2. FLUID DYNAMICS, AEROMECHANICAL ENGINEERING. *Educ:* Univ Southern Calif, AB, 37; Syracuse Univ, MA, 38; Univ Wis, PhD(appl math), 42. *Prof Exp:* Instr, Mich State Col, 41-42; Rockefeller fel, Brown Univ, 42-43, instr civil eng, 43; res aerodynamicist, Bell Aircraft Corp, NY, 43-46 & Off Naval Res, Washington, DC, 46-47; asst & assoc prof aeronaut eng, Univ Mich, 47-51; res scientist, Johns Hopkins Univ, 51-58; prin staff scientist, Res Dept, Baltimore Div, Martin Co, 58-67; prof, 67-82, EMER PROF MECH & AEROSPACE ENG, ILL INST TECHNOL, 82- *Concurrent Pos:* Lectr, Johns Hopkins Univ, 51-67; consult in the field; mem subcomt high-speed aerodyn, Nat Adv Comt Aeronaut, 46-49; mem adv res comt fluid dynamics, NASA, 68-70; mem tech comt fluid dynamics, Am Inst Aeronaut & Astronaut, 73-74; assoc mem fluid mech comt, Am Soc Mech Engrs; mem, US Transition Study Group, 75-; sr exchange scientist with USSR Acad Sci, Nat Acad Sci, 79; mem adv comt, Ctr Turbulence Res, NASA, 90- *Honors & Awards:* First Fluid & Plasmadynamics Award, Am Inst Aeronaut & Astronaut, 76; Alexander von Humboldt Sr Award, 77, 78; Fluids Eng Award, Am Soc Mech Eng, 87. *Mem:* Nat Acad Eng; Sigma Xi; fel Am Soc Mech Engrs; fel Am Inst Aeronaut & Astronaut; fel Am Phys Soc; AAAS. *Res:* Experimental and theoretical fluid dynamics; transition and turbulent flow; stability and transition to turbulence; unsteady flows; separated flows; coupled fluid-elastic instabilities; simulation of wind effects in the atmospheric boundary layer. *Mailing Add:* 1104 Linden Ave Oak Park IL 60302

MORLANG, BARBARA LOUISE, b Conn. NUTRITION. *Educ:* Brigham Young Univ, BS, 60; Columbia Univ, MS, 64; Univ Mass, PhD(food & nutrit), 69. *Prof Exp:* Therapeut dietitian, Yale-New Haven Community Hosp, 61-62 & St Luke's Hosp, New York, 62-64; nutrit consult, Bur Nutrit, City New York, 64-65; dir & nutritionist, Springfield Dairy Coun, Mass, 65-66; STATE NUTRIT CONSULT, VA DEPT HEALTH, 70- *Mem:* Am Dietetic Asn; Am Home Econ Asn; Am Pub Health Asn. *Res:* Nutritional status of local populations. *Mailing Add:* 6828 Ardmore Dr Hollins VA 24019

MORLEY, COLIN GODFREY DENNIS, b Sittingbourne, Eng, Nov 12, 41; div; c 2. BIOCHEMISTRY. *Educ:* Univ Nottingham, BSc, 63; Australian Nat Univ, PhD(biochem), 69. *Prof Exp:* Res scientist, Imp Chem Industs, 63-64; asst lectr chem, Woolwich Col Further Educ, London, 64-65; NIH fel, Nat Heart & Lung Inst, 68-70; asst prof, Dept Med, Univ Chicago, 70-76; asst prof, 76-81, ASSOC PROF DEPT BIOCHEM, RUSH PRESBYTERIAN-ST LUKE'S MED CTR, CHICAGO, 81- *Concurrent Pos:* Guest lectr, Purdue Univ, Calumet Campus, 74-75, 77-78 & 80-81. *Mem:* AAAS; Royal Soc Chem; Tissue Culture Asn; Am Chem Soc. *Res:* Cell biology, particularly control of cell growth in mammalian systems, hormonal aspects of such control; relationship between normal and cancer cells for growth control. *Mailing Add:* Dept Biochem Rush Univ 1753 W Harrison Chicago IL 60612

MORLEY, GAYLE L, b Moroni, Utah, Feb 29, 36; m 64; c 3. THEORETICAL SOLID-STATE PHYSICS. *Educ:* Brigham Young Univ, BS, 58; Univ Calif, Los Angeles, MS, 60; Iowa State Univ, PhD(physics), 67; Univ Idaho, ME, 83. *Prof Exp:* Microwave engr, Hughes Aircraft Co, 58-60; fel, Tex A&M Univ, 67-68; assoc prof, Mankato State Univ, 68-77, prof physics & electronics eng technol, 77-79; assoc prof, 79-81, PROF PHYSICS, COL IDAHO, 81- *Res:* Theory of lattice vibrations in solids. *Mailing Add:* Dept Math/Physics Col Idaho 2112 Cleveland Blvd Caldwell ID 83605

MORLEY, GEORGE W, b Toledo, Ohio, June 6, 23; m 46; c 3. MEDICINE. *Educ:* Univ Mich, BS, 44, MD, 49, MS, 55; Am Bd Obstet & Gynec, dipl, 53, cert gynec oncol, 74. *Prof Exp:* Intern, 49-50, from asst resident to resident, 50-54, from instr to assoc prof, 56-70, PROF OBSTET & GYNEC, MED CTR, UNIV MICH, ANN ARBOR, 70- *Concurrent Pos:* Consult, US Vet Admin Hosp, Ann Arbor, Mich, 56- & Wayne County Gen Hosp, Eloise, 60- *Mem:* Fel Am Col Surg; fel Am Col Obstet & Gynec; Soc Gynec Oncol; Soc Pelvic Surg; Int Soc Study Vulvar Dis. *Res:* Malignancy of the female genital tract. *Mailing Add:* Dept of Obstet & Gynec Univ of Mich Med Ctr Ann Arbor MI 48109

MORLEY, HAROLD VICTOR, b Buenos Aires, Arg, July 21, 27; m 53; c 1. PESTICIDE CHEMISTRY, ENVIRONMENTAL CHEMISTRY. *Educ:* Univ London, BSc, 48, PhD(org chem), 55. *Prof Exp:* Asst lectr chem, Royal Free Hosp, Sch Med, Univ London, 51-57; fel, Nat Res Coun Can, 57-59 & McMaster Univ, 59-60; res scientist, Anal Chem Res Serv, Can Dept Agr, 60-71, Chem Biol Res Inst, 71-72, res coordr, Environ & Resources, Res Br, Can Dept Agr, Ottawa, 72-78; DIR LONDON RES INST, 78- *Concurrent Pos:* Mem, Reference Group Land Use Activities, Int Joint Comn, 73-77. *Mem:* The Chem Soc; Chem Inst Can. *Res:* Determination and function of ergothioneine in biological fluids; goitrogenic compounds, especially imidazole-2-thiols; porphyrins, especially structure of chlorophylls; pesticide residue chemistry. *Mailing Add:* 231 Univ Cresent Agr Can Univ Sub PO London ON N6A 2L7 Can

MORLEY, JOHN E, b SAfrica, June 13, 46; c 3. GERIATRICS, ENDOCRINOLOGY. *Educ:* Univ Witwatersrand, MD, 72. *Prof Exp:* Assoc prof endocrinol, Univ Minn, 78-85; prof geriat, Med Ctr, Univ Calif, Los Angeles, 85-89; dir grecc, St Louis Vet Admin Ctr, 85-89; DAMMERT PROF GERONT, ST LOUIS UNIV, 89- *Concurrent Pos:* Assoc ed, J Am Geriat Soc, ed, Yrbk Endocrinol & ed-in-chief, Geriat Med Today. *Honors & Awards:* Mead Johnson Award, Am Inst Nutrit, 85. *Mem:* Geriat Soc Am; Endocrine Soc; Am Soc Clin Invest; Am Fed Clin Res; Soc Neurosci; Am Soc Pharmacol & Exp Therapeut. *Res:* Neuropeptides and regulation of appetite and memory; diabetes mellitus in the elderly; nutrition and the elderly; home health care; health service research; sexual dysfunction. *Mailing Add:* Dept Med/Geriat St Louis Univ Sch Med 1402 S Grand Rm M-239 St Louis MO 63104

MORLEY, LAWRENCE WHITAKER, b Toronto, Ont, Feb 19, 20; m 50; c 4. GEOPHYSICS. *Educ:* Univ Toronto, BA, 46, MA, 49, PhD(geophys), 52. *Hon Degrees:* DSc, York Univ, 74. *Prof Exp:* Geophysicist, Fairchild Aerial Surv, Inc, Calif, 46-48; chief geophysicist, Dom Gulf Co, 48-49; chief geophys div, 52-69, dir, Can Ctr Remote Sensing, Dept Energy, Mines & Resources, Geol Surv Can, 69-80; sci counr, Can High Comn, London, England, 80-82; PROF, FAC SCI, YORK UNIV, TORONTO, 85- *Concurrent Pos:* Lectr, Carleton Univ, 57 & 59; consult exp geophysics & remote sensing, 82- *Honors & Awards:* McCurdy Medal, Can Aeronaut & Space Inst, 74; Tuzo Wilson Prize, Can Geophys Union, 80. *Mem:* Soc Explor Geophys; Am Geophys Union; Can Inst Mining & Metall; fel Can Aeronaut & Space Inst; fel Royal Soc Can. *Res:* Paleomagnetic research; geophysical instrumentation and interpretation; aerogeophysics; remote sensing; theory of magnetic imprinting of reversing earths magnetic field in rocks of ocean basin. *Mailing Add:* 20 Wellesley Ave Toronto ON M4X 1V3 Can

MORLEY, LLOYD ALBERT, b Provo, Utah, Oct 28, 40; m 75; c 1. MINING ENGINEERING, ELECTRICAL ENGINEERING. *Educ:* Univ Utah, BS, 68, PhD(mining eng), 72. *Prof Exp:* Technician & dept mgr elec, Strevell Paterson Co, 61-67; from asst prof to prof mining eng, Pa State Univ, 71-85; PROF & HEAD MINERAL ENG, UNIV ALA, 85- *Concurrent Pos:* Consult, Cumberland Coal Div, US Steel Corp, 75-78, Dept Mine Safety, Pa Dept Environ Resources, 77-, Skelley & Loy, 80, IMC, 81, Standard Oil Co, Ohio, 84 & MET, 85-; res grants Bur Mines, US Dept Interior, 72-; res grants Bur Mines, US Dept Interior, 72-; secy, Indust Appln Soc, 86. *Mem:* Am Inst Mining Metall & Petrol Engrs; Soc Mining Engrs; sr mem Inst Elec & Electronics Engrs; Inst Elec & Electronics Engrs Indust Applns Soc (secy, 86). *Res:* Mine electrical systems; protective relaying, electrocution prevention, transients, monitoring and simulation, power equipment design; coal mining; materials handling. *Mailing Add:* Univ Ala PO Box 870207 Tuscaloosa AL 35487

MORLEY, MICHAEL DARWIN, b Youngstown, Ohio, Sept 29, 30; m 54. MATHEMATICAL LOGIC. *Educ:* Case Inst Technol, BS, 51; Univ Chicago, MS, 53, PhD, 62. *Prof Exp:* Sr mathematician, Labs Appl Math, Univ Chicago, 55-61; instr math, Univ Calif, Berkeley, 62-63; asst prof, Univ Wis-Madison, 63-66; assoc prof, 67-70, PROF MATH, CORNELL UNIV, 70- *Mem:* Am Math Soc; Asn Symbolic Logic. *Res:* Foundations of mathematics. *Mailing Add:* Dept Math Cornell Univ Ithaca NY 14853

MORLEY, ROBERT EMMETT, JR, b St Louis, Mo, April 20, 51; c 1. COMMUNICATIONS SYSTEMS. *Educ:* Washington Univ, St Louis, BS, 73, MS, 75, DSc, 77. *Prof Exp:* Staff engr, Lincoln Lab, Mass Inst Technol, 75; vpres engr, Micro-Term Inc, 76-81; ASSOC PROF ELEC ENG, WASH UNIV, ST LOUIS, 81- *Honors & Awards:* Young Prof Award, Inst Elec & Electronics Engrs, 81. *Mem:* Inst Elec & Electronics Engrs. *Res:* Design of very large scale integration systems for digital signal processing; massively parallel computer architectures. *Mailing Add:* Washington Univ Lindell-Skinker Blvd Box 1127 St Louis MO 63130

MORLEY, THOMAS PATERSON, b Manchester, Eng, June 13, 20; Can citizen; m 43; c 3. NEUROSURGERY. *Educ:* Oxford Univ, BA, 41, BM, BCh, 43; FRCS, 49; FRCPS(C), 53. *Prof Exp:* Consult, Sunnybrook Hosp, Dept Vet Affairs, Govt of Can, 54-60; chmn, Div Neurosurg, Univ & Toronto Gen Hosp, 64-79, prof surg, Univ Toronto, 64-; AT TORONTO GEN HOSP. *Concurrent Pos:* Ont Cancer Treatment & Res Found res fel, 55-65; consult, Toronto Gen Hosp, 62-69, Queen Elizabeth Hosp, Toronto, 63-, Princess Margaret Hosp, 63-, Ont Cancer Inst, 63- & Wellesley Hosp, Toronto, 64-69. *Mem:* Am Asn Neurol Surg; Soc Neurol Surg; Neurosurg Soc Am; Can Neurosurg Soc (secy, 60-64, pres, 71-72); Soc Brit Neurol Surg. *Res:* Diagnostic use of radio-isotopes in neurosurgery; echoencephalography; recovery of tumor cells from blood in glioma cases; radiotherapy in gliomas. *Mailing Add:* PO Box 58 Claremont ON L0H 1E0 Can

MORLOCK, CARL G, b Crediton, Ont, Sept 11, 06; nat US; m 37; c 2. MEDICINE, GASTROENTEROLOGY. *Educ:* Univ Western Ont, BA, 29, MD, 32; Univ Minn, MS, 37. *Prof Exp:* Intern, Victoria Hosp, London, Ont, 32-33, resident physician, 33-34; res med, Mayo Found, 34-39, consult, Mayo Clin, 39-75, from instr to prof clin med, Mayo Grad Sch Med, 39-72, prof med, 72-75, EMER PROF MED, MAYO MED SCH, UNIV MINN, 75-; consult, Mayo Clin, 39-75, from instr to prof clin med, Mayo Grad Sch Med, 39-72, prof med, 72-75, EMER PROF MED, MAYO MED SCH, UNIV MINN, 75- *Mem:* AMA; Am Gastroenterol Asn; fel Am Col Physicians; Sigma Xi. *Res:* Peptic ulcer and its complications; gastric carcinoma; anorexia nervosa; regional enteritis; arteriolar pathology of hypertension; blood pressure in renal tumors; liver disease; hemochromatosis; suprarenal insufficiency. *Mailing Add:* Mayo Clin 200 First St SW Rochester MN 55905

MORMAN, KENNETH N, b Philadelphia, Pa, July 17, 40; m 65. COMPUTER-AIDED ENGINEERING, APPLIED MATHEMATICS. *Educ:* Pa State Univ, BS, 62, BS, 62; NJ Inst Technol, MS, 69; Columbia Univ, Eng ScD, 73. *Prof Exp:* Munic engr, Michael Baker Jr, Inc, 62-66; design engr, Gibbs & Hill, Inc, 66-68; struct mech engr, Grumman Aerospace Corp, 68-73; prin res engr assoc, 73-77, prin staff engr, 77-91, PRIN RES ENGR, FORD MOTOR CO, 91- *Concurrent Pos:* Lectr, Battele Sem Ser, design of rubber parts, London, Eng, Columbus, Ohio, 84-85, Columbia Univ, Univ Mich, Wayne State Univ; adj prof mech eng, Wayne State Univ, 87-91; chmn, Soc Auto Engrs, Computer Appln Comm, 88-91; Ford rep, NSF comt to increase number of women & minorities sci & eng, 90. *Mem:* Sigma Xi; Soc Indust & Appl Math; Am Soc Mech Eng; Soc Automotive Engrs; NY Acad Sci; Am Inst Aeronaut & Astronaut. *Res:* Development of constitutive models for rubber-like materials; characterization of the fatigue and fracture properties of polymers; computational mechanics; computational methods for the finite element analysis of elastomeric components; applied mathematics including method for the identification of model parameters. *Mailing Add:* 2572 Wickfield Rd West Bloomfield MI 48323

MORMAN, MICHAEL T, b Wausau, Wis, Sept, 28, 45; m 67. POLYMER PROCESSING, MATERIAL INNOVATION. *Educ:* Univ Wis, Madison, BS, 69, Rice Univ, PhD(chem eng), 73. *Prof Exp:* Captain, US Army Reserve, 69-77; res engr, E I Dupont de Nemours, 73-76; res scientist, 76-80, sr res scientist, 80-85, res fel, 85-89, SR RES FEL, KIMBERLY-CLARK CORP, 89- *Res:* 15 US Patents. *Mailing Add:* 1400 Holcomb Bridge Rd Roswell GA 30076-9702

MORNEWECK, SAMUEL, b Meadville, Pa, Sept 3, 39; m 61. ORGANIC BIOCHEMISTRY. *Educ:* Allegheny Col, BS, 60; Case Inst Technol, PhD(org chem), 65. *Prof Exp:* Chemist, Agr Prod Labs, Esso Res & Eng Co, 65-67, res chemist, 67-70; asst prof, 70-75, ASSOC PROF CHEM, ST PETER'S COL, NJ, 75- *Res:* Determination of steric substituent constants. *Mailing Add:* Dept of Chem St Peter's Col 2641 Kennedy Blvd Jersey City NJ 07306

MOROI, DAVID S, b Tokyo, Japan, Oct 15, 26; m 59; c 3. THEORETICAL PHYSICS. *Educ:* St Paul's Univ, Tokyo, BSc, 53; Johns Hopkins Univ, PhD(physics), 59. *Prof Exp:* Res assoc physics, Iowa State Univ, 59-61, instr, 60-61; res assoc, Univ Notre Dame, 61-63; asst prof, 63-68, assoc prof, 68-73, PROF PHYSICS, KENT STATE UNIV, 74- *Concurrent Pos:* US Air Force Off Sci Res grant, 64-69. *Mem:* Am Phys Soc; Sigma Xi. *Res:* Quantum electrodynamics; field theory; electromagnetic and other physical properties of liquid crystals; elementary particles in intense laser beams. *Mailing Add:* Dept Physics Kent State Univ 207 Smith Hall Kent OH 44242

MORONI, ANTONIO, b Barga, Italy, Oct 25, 53. COMPUTER CHEMISTRY & MOLECULAR DESIGN, CHARACTERIZATION OF POLYMERS. *Educ:* Univ Pisa, PhD(chem), 79. *Prof Exp:* Teaching fel polymers, Univ Mass, Amherst, 80-81, Univ Pisa, 81-83, Polytech Inst NY, 83-85; sr res chemist polymers, Pennwalt Corp, 85-88; SR RES CHEMIST POLYMERS, PHARMACEUT DOSAGE FORMS, WARNER-LAMBERT CO, 88- *Mem:* Am Chem Soc; Sigma Xi; Controlled Release Soc. *Res:* Polymer synthesis and modification; vinyl acrylic epoxies; radical anionic and photopolymerization synthesis of monomers and organic synthesis; characterization of polymers; kinetics of polymerization; computer modeling and molecular design; polymer applied to controlled release; polymers applied to the manufacturing of dosage forms; evaluation of dosage forms. *Mailing Add:* 43 Burnham Rd Morris Plains NJ 07950

MORONI, ENEO C, b Fiume, Italy, Feb 6, 23; US citizen; m 60; c 1. INDUSTRIAL ORGANIC CHEMISTRY. *Educ:* Univ Milan, PhD(phys org chem), 49. *Prof Exp:* Res chemist, Ledoga SpA, Italy, 49-55; prod chemist, Industria Saccarifera Parmense, 55-56; prod chemist, Pitt-Consol Chem Co Div, Consol Coal Co, 56-57, res chemist, 57-64; sr res chemist, Jones & Laughlin Steel Corp, 64-67; res chemist, Pittsburgh Energy Res Ctr, US Bur Mines, 67-75; prog mgr, Energy Res & Develop Admin, 75-78; PROG MGR, DEPT ENERGY, 78- *Mem:* Am Chem Soc. *Res:* Activated carbons; phenolic and cresylic acids resins; catalytic alkylation, dealkylation and isomerization of phenols and thiophenols; corrosion and coating surface phenomena; organometallics; coal desulfurization and liquefaction. *Mailing Add:* Apt 334N 1600 S Eads St Arlington VA 22202-2916

MOROS, STEPHEN ANDREW, b New York, NY, July 29, 28; m 57; c 2. ANALYTICAL CHEMISTRY. *Educ:* Polytech Inst Brooklyn, BS, 48, PhD(electroanalytical chem), 61; Cornell Univ, MS, 50. *Prof Exp:* Proj group leader, Foster D Snell, Inc, NY, 50-53; res chemist, Am Cyanamid Co, NJ, 58-71; sr chemist, Hoffmann-La Roche, Inc, 71-73, group leader, 73-78, mgr, 78-81, asst dir, 81-85; mgr chem develop, Photomedica, Inc, 85-89; SR TECH ADV, ORTHO PHARMACEUT CORP, 89- *Mem:* Am Chem Soc; Am Pharmaceut Asn; Am Soc Testing & Mat. *Res:* Electroanalytical chemistry, including coulometry and potentiometry; organic electrosynthesis; instrumentation; thermal methods of analysis; spectroscopic methods; radiotracers; instrumental methods of analysis; spectrochemical analysis; chromatography. *Mailing Add:* 144 Konner Ave Pine Brook NJ 07058

MOROSIN, BRUNO, b Klamath Falls, Ore, Feb 10, 34; m 58; c 4. X-RAY & NEUTRON DIFFRACTION, STRUCTURAL PROPERTIES. *Educ:* Univ Ore, BA, 56; Univ Wash, PhD, 59. *Prof Exp:* Asst, Univ Wash, 56-59; mem, tech staff, Hughes Aircraft Co, 60-61; staff mem, Sandia Labs, 61-67, supvr, chem physics div, 67-73, supvr, solid-state mat, 73-80, supvr, Shock Wave & Explosive Physics Div, 80-89, DISTINGUISHED MEM TECH STAFF, SANDIA NAT LABS, 90- *Concurrent Pos:* Co-ed, J Appl Crystallog, 86- *Honors & Awards:* Mat Sci Award, Dept Energy, 88 & 89. *Mem:* Am Phys Soc; Am Crystallog Asn; Mat Res Soc. *Res:* Diffraction studies; structural properties related to magnetic and dielectric behavior; shock-loaded materials and superconductors. *Mailing Add:* Organ 1150 Sandia Nat Labs Kirkland Air Force Base Albuquerque NM 87185

MOROSON, HAROLD, radiation immunology, radiobiology, for more information see previous edition

MOROWITZ, HAROLD JOSEPH, b Poughkeepsie, NY, Dec 4, 27; m 49; c 5. BIOENERGETICS, ORIGIN OF LIFE. *Educ:* Yale Univ, BS, 47, MS, 50, PhD(biophys), 51. *Prof Exp:* Biophysicist, Nat Bur Standards, 51-53 & Nat Heart Inst, 53-55; from asst prof to prof biophys & biochem, Yale Univ, 67-87; ROBINSON PROF BIOL & NATURAL PHILOS, GEORGE MASON UNIV, 87- *Concurrent Pos:* Mem planetary biol subcomt, NASA, 66-72; assoc ed, J Biomed Computing, 69-; mem eval panel phys chem, Nat Bur Standards, 69-74; columnist, Hosp Pract, 74-; mem bd biol, Nat Res Coun, 87-; mem coun, NCRR, NIH, 88- *Mem:* Explorers Club; Biophys Soc. *Res:* Energy transduction in biological systems; prebiotic chemistry; thermodynamic foundations of biology. *Mailing Add:* Ox Bow Lane Woodbridge CT 06525

MOROZ, LEONARD ARTHUR, b Winnipeg, Man, Dec 9, 35; c 3. IMMUNOLOGY, BIOCHEMISTRY. *Educ:* Univ Man, MD, 59; FRCP(C), 64. *Prof Exp:* Res fel, Mass Gen Hosp & Harvard Med Sch, 61-63; guest investr, Rockefeller Univ, 64-67; asst prof, 67-73, assoc prof, 73-85, PROF, MCGILL UNIV, 86- *Concurrent Pos:* Med res scholar, Med Res Coun Can, 67-72; Helen Hay Whitney Found fel, 64-67. *Mem:* Am Asn Immunologists; Am Fedn Clin; Am Rheumatism Asn; Am Acad Allergy; Can Soc Clin Invest. *Res:* Proteolytic enzymes; fibrinolysis; inflammation; immunologically mediated lung disease; connective tissue diseases; immunochemistry. *Mailing Add:* Royal Victoria Hosp 687 Pine Ave W Montreal PQ H3A 1A1 Can

MOROZ, WILLIAM JAMES, b Toronto, Ont, Aug 29, 27; m 54; c 3. MECHANICAL ENGINEERING, METEOROLOGY. *Educ:* Univ Toronto, BASc, 49, MASc, 51; Univ Mich, MS, 62, PhD(meteorol), 64. *Prof Exp:* Combustion engr, Empire Hanna Coal Co, 50-57; lectr mech eng, Univ Toronto, 57-61, asst prof, 65-67, assoc prof, 67-68; prof mech eng & dir, Ctr Air Environ Studies, Pa State Univ, 68-76; vpres, Jas F MacLaren Consult Engrs, 77-80; supvr sci & field studies, Environ Dept, Ont Hydro, 80-86; PRES, WM J MOROZ ASSOC CONSULT ENGRS, 87- *Concurrent Pos:* Consult, Air Pollution Control Serv, Ont, 65-68, actg head meteorol & air qual sect, 67-68; consult, James F McLaren, Ltd, 65-67; mem adv comt, Nat Air Pollution Control Tech, Environ Protection Agency, 73-76; adj prof mech eng, Univ Toronto, 79-86. *Honors & Awards:* Centennial Medal, Am Soc Mech Engrs, 79. *Mem:* fel Royal Meteorol Soc; Can Inst Combustion & Fuel Technol (pres, 67 & 68); Am Soc Mech Engrs; fel NY Acad Sci; Air & Waste Mgt Asn. *Res:* Air pollution control; atmospheric diffusion and dispersion; particle technology; combustion. *Mailing Add:* RR Two Hastings ON K0L 1Y0 Can

MOROZOWICH, WALTER, b Irwin, Pa, Oct 27, 33; m 69; c 3. MEDICINAL CHEMISTRY. *Educ:* Duquesne Univ, BS, 55; Ohio State Univ, MS, 56, PhD(pharmaceut chem), 59. *Prof Exp:* SR RES SCIENTIST, UPJOHN CO, 59- *Honors & Awards:* Upjohn Award, 70. *Mem:* Am Pharmaceut Asn; Am Chem Soc; Acad Pharm Sci. *Res:* Design of biophysical properties of drugs, prodrugs, soft drugs, drug delivery and membrane transport. *Mailing Add:* Upjohn Co 301 Henrietta St Kalamazoo MI 49001

MORR, CHARLES VERNON, b Ashland, Ohio, Oct 7, 27; m 51; c 2. FOOD CHEMISTRY. *Educ:* Ohio State Univ, BS, 52, MS, 55, PhD(dairy technol), 59. *Prof Exp:* Res assoc, Carnation Res Lab, Calif, 59-61; asst prof Ohio State Univ, 61-64; from asst prof to prof, Univ Minn, St Paul, 64-73; dir protein res, Ralston Purina Co, 73-76; prof & chmn depts Food & Nutrit & Food Technol, Tex Tech Univ, 76-78; STENDER PROF, DEPT FOOD SCI, CLEMSON UNIV, 78- *Honors & Awards:* Dairy Res Award, Am Dairy Sci Asn, 73. *Mem:* Am Chem Soc; Am Dairy Sci Asn; Inst Food Technologists; Am Oil Chem Soc. *Res:* Chemistry and functional properties of milk and plant proteins; effects of fractionation and processing treatments on the composition, physichemical, functional and nutritional properties of whey protein concentrates, caseinates and soy protein isolates. *Mailing Add:* 390 Ridgecrest Dr Clemson SC 29631-1849

MORRAL, F(ACUNDO) R(OLF), b Chemnitz, Germany, June 19, 07; nat US; m 34; c 5. METALLURGY, METALLURGICAL ENGINEERING. *Educ:* Mass Inst Technol, BS, 32; Purdue Univ, PhD(metall), 40. *Prof Exp:* Mgr, Textile Mill, Martin Morral, Sabadell, Spain, 34-36; res metallurgist, Continental Steel Corp, 38-41; asst prof metall, Pa State Univ, 41-43; indust fel, Mellon Inst Sci, 43-44; group leader, Metal Trades Lab, Am Cyanamid Co, 44-48; assoc prof mat eng, Syracuse Univ, 48-51; x-ray sect head, Dept Metall Res, Kaiser Aluminum & Chem Corp, 51-56; head, Cobalt Info Ctr Proj & consult, Battelle Mem Inst, 56-72; assoc dir Emilio Jimeno Inst Technol & Metall, Univ Barcelona, 72-73; CONSULT, 73- *Concurrent Pos:* Foreign affil, Royal Acad Exact Physical & Natural Sci. *Mem:* Fel AAAS; Am Inst Mining, Metall & Petrol Engrs; Am Soc Metals. *Res:* Cobalt; metallurgy; engineering education; materials engineering. *Mailing Add:* 2075 Arlington Ave Columbus OH 43221-4313

MORRAL, JOHN ERIC, b Kokomo, Ind, Aug 3, 39; c 2. PHYSICAL METALLURGY. *Educ:* Ohio State Univ, BMetE, 64, MS, 65; Mass Inst Technol, PhD, 69. *Prof Exp:* Asst prof metall, Univ Ill, Urbana, 68-71; assoc prof, 71-86, PROF METALL, UNIV CONN, 86- *Mem:* Minerals Metals & Mat Soc; Am Soc Metals Int. *Res:* Kinetics and thermodynamics of phase transformations. *Mailing Add:* Dept of Metall Univ of Conn Storrs CT 06268

MORRE, D JAMES, b Drake, Mo, Oct 20, 35; m 56; c 3. BIOCHEMISTRY. *Educ:* Univ Mo, BS, 57; Purdue Univ, MS, 59; Calif Inst Technol, PhD(biochem), 63. *Hon Degrees:* Dr, Univ Geneva, Switz, 85. *Prof Exp:* From asst prof to assoc prof, 63-71, PROF MED CHEM & DIR PURDUE CANCER CTR, PURDUE UNIV, LAFAYETTE, 71- *Honors & Awards:* Purdue Cancer Res Award, 78. *Mem:* Am Soc Cell Biol; Am Soc Biol Chem; Am Asn Cancer Res. *Res:* Cell growth; cancer; Golgi apparatus structure-function; membrane biogenesis; cell surfaces; secretion. *Mailing Add:* Dept Med Chem Purdue Univ West Lafayette IN 47907

MORRÉ, DOROTHY MARIE, b Bonnots Mill, Mo, Jan 18, 35; m 56; c 3. NUTRITION, CELL BIOLOGY. *Educ:* Univ Mo, BS, 58; Purdue Univ, PhD(nutrit), 77. *Prof Exp:* Lab technician, Ger Cancer Res Ctr, 76; vis asst prof, 78, from asst prof to assoc prof, 78-89, PROF NUTRIT, DEPT FOODS & NUTRIT, PURDUE UNIV, 89- *Concurrent Pos:* Prin investr, Am Cancer Soc, 78-79; co-prin investr, NIH, 78-81, 90-94. *Mem:* Sigma Xi; Am Dietetics Asn; Am Inst Nutrit; Am Soc Cell Biologists; Am Diabetes Asn. *Res:* Chemopreventative effect of vitamin A on tumorigenesis and metastasis; hematological profile; cytotoxicity of megadoses, lysosomal stability and liability; ultrastructural changes. *Mailing Add:* Dept Foods & Nutrit 109 Stone Hall West Lafayette IN 47907

MORREL, BERNARD BALDWIN, b Lynchburg, Va, Nov 28, 40; m 64; c 2. MATHEMATICS. *Educ:* Univ Va, BA, 62, MA, 66, PhD(math), 68. *Prof Exp:* Teaching asst math, Johns Hopkins Univ, 62-64; from teaching asst appl math to jr instr math, Univ Va, 65-68; asst prof math, Univ Ga, 68-75; vis asst prof, Ind Univ, Bloomington, 75-77; ASSOC PROF MATH SCI, IND UNIV-PURDUE UNIV, INDIANAPOLIS, 77- *Mem:* Am Math Soc; Math Asn Am; Soc Indust & Appl Math. *Res:* Functional analysis; theory of operators in Hilbert Space. *Mailing Add:* Dept Math Sci Ind Univ-Purdue Univ 1125 E 38th St Indianapolis IN 46205-2810

MORRELL, FRANK, b New York, NY, June 4, 26; c 4. NEUROLOGY, NEUROPHYSIOLOGY. *Educ:* Columbia Univ, AB, 48, MD, 51; McGill Univ, MSc, 55; Am Bd Psychiat & Neurol, dipl, 58. *Prof Exp:* Med intern, Montefiore Hosp, New York, 51-52, chief resident neurol, 53-54; from instr to assoc prof neurol, Med Sch, Univ Minn, 55-61; prof & chmn dept, Sch Med, Stanford Univ, 61-69; prof neurol & psychiat, New York Med Col, 69-72; PROF NEUROL SCI, MED COL, RUSH UNIV, 72- *Concurrent Pos:* Rosenthal fel, Nat Hosp, London, Eng, 52-53; fel neurophysiol, Montreal Neurol Inst, 54-55; consult, Epilepsy Found Am, 59-; consult, NIH & NSF, 61-; assoc neurosci res prog, Mass Inst Technol, 62-; mem brain sci comt, Nat Acad Sci, 66-70; Wall Mem lectr, Children's Hosp, DC, 67. *Mem:* Fel Royal Soc Health; Am Electroencephalog Soc (pres, 77-78); Am Epilepsy Soc; Am Acad Neurol; Soc Neurosci. *Res:* Pathophysiology of epilepsy and neural mechanisms of learning. *Mailing Add:* 1753 W Congress Pkwy Chicago IL 60612

MORRELL, GEORGE, clinical toxicology, biochemistry, for more information see previous edition

MORRELL, WILLIAM EGBERT, b Logan, Utah, July 30, 09; m 33; c 3. CHEMISTRY, SCIENCE EDUCATION. *Educ:* Utah State Univ, BS, 33; Univ Calif, PhD(phys chem), 38. *Hon Degrees:* LHD, Suffolk Univ, 73. *Prof Exp:* Instr phys sci, Chicago City Col, 38-42; from instr to prof phys sci, Univ Ill, 42-59; prog dir summer study, NSF, 59-76; RETIRED. *Concurrent Pos:* Asst prog dir summer study, NSF, 58-59. *Honors & Awards:* Meritorious Serv Award, NSF, 75. *Mem:* AAAS; Am Chem Soc. *Mailing Add:* 3440 S Jefferson St#817 Falls Church VA 22041-3127

MORREY, JOHN ROLPH, b Joseph, Utah, May 30, 30; m 52; c 5. PHYSICAL CHEMISTRY, INORGANIC CHEMISTRY. *Educ:* Brigham Young Univ, BA, 53; Univ Utah, PhD(phys chem), 58. *Prof Exp:* Sr scientist, Gen Elec Co, 58-63, tech specialist, 63-64; res assoc chem, 65-77, STAFF SCIENTIST, PAC NORTHWEST LABS, BATTELLE MEM INST, 77- *Concurrent Pos:* Adj assoc prof, Wash State Univ, 67- *Mem:* AAAS; Am Chem Soc; NY Acad Sci. *Res:* Boron hydride fused salt and actinide element chemistry; chemical kinetics; molecular spectroscopy; computer applications to chemistry; laser chemistry; thermodynamics and materials; chemistry of coal; general earth sciences. *Mailing Add:* 1408 Westwood Ct Richland WA 99352

MORRICAL, DANIEL GENE, b Hartford City, Ind, Apr 23, 55; m 78; c 2. LAMB & WOOL PRODUCTION. *Educ:* Purdue Univ, BS, 77; NMex State Univ, MS, 82, PhD(animal sci), 84. *Prof Exp:* Res asst animal sci, NMex State Univ, 79-82, grad asst, 82-84; ASST PROF ANIMAL SCI, IOWA STATE UNIV, 84- *Res:* Development of management practices to improve the production of lamb and wool. *Mailing Add:* 109 Kildee Hall Iowa State Univ Ames IA 50011

MORRILL, BERNARD, b Boston, Mass, May 31, 10; m 40; c 1. MECHANICAL ENGINEERING. *Educ:* Mass Inst Technol, BSME, 47; Univ Del, MME, 49; Univ Mich, PhD, 59. *Prof Exp:* From instr to prof, 47-60, actg chmn dept, 59-60, chmn, 60-64, chmn engr dept, 72-73, Henry C & J Archer Turner prof, 67-75, EMER PROF MECH ENG, SWARTHMORE COL, 75- *Concurrent Pos:* NSF sci fac fel, Univ Mich, 58-59, Imp Col, Univ London, 64-65. *Mem:* Am Soc Mech Engrs; Am Soc Eng Educ; Soc Eng Sci; NY Acad Sci. *Res:* Mechanical vibration, linear and nonlinear; thermodynamics. *Mailing Add:* Strath Haven Condos 1014 Swarthmore PA 19081

MORRILL, CALLIS GARY, b Tridell, Utah, Nov 6, 38; m 64; c 5. PHYSIOLOGY. *Educ:* Brigham Young Univ, BA, 64; Univ Calif, San Francisco, PhD(physiol), 70. *Prof Exp:* NIH res fel physiol, Univ Colo Med Ctr, 71-73; RES ASSOC PHYSIOL, NAT ASTHMA CTR, DENVER, 73- *Concurrent Pos:* Am Lung Asn fel, Nat Asthma Ctr, 74-75; NIH res grant, 75. *Mem:* Am Thoracic Soc. *Res:* Ventilatory control of asthmatic children; effects of low levels of carbon monoxide on exercise tolerance. *Mailing Add:* 1205 Clayview Dr Liberty MO 64068-3408

MORRILL, CHARLES D(UNCKER), b St Louis, Mo, Oct 31, 19; m 42; c 5. ELECTRONICS. *Educ:* Univ Ill, BS, 41. *Prof Exp:* Engr, Am Tel & Tel Corp, 46-48; develop engr, Goodyear Aerospace Corp, 48-50, group leader electronic comput, 50-53, sect head, 53-56, dept mgr, 56-60, div mgr mil guid & data handling systs, 60-70, chief engr, Electronics Div, 70-80, ASST CHIEF ENGR, DEFENSE SYSTS DIV, GOODYEAR AEROSPACE CORP, 80-, CHIEF ENG, 83- *Mem:* Fel Inst Elec & Electronics Engrs. *Res:* Military electronics; terminal guidance; trainers and simulators; intelligence and command control; associative array processor development; undersea weapons. *Mailing Add:* 2248 16th St Cuyahoga Falls OH 44223

MORRILL, GENE A, b Bend, Ore, Aug 5, 31; m 60; c 2. BIOCHEMISTRY, DEVELOPMENTAL BIOLOGY. *Educ:* Univ Portland, BS, 54; Univ Utah, PhD(biochem), 59. *Prof Exp:* Res asst prof, 63-64, from asst prof to assoc prof, 65-76, PROF PHYSIOL, ALBERT EINSTEIN COL MED, 76- *Concurrent Pos:* Fel biochem, Albert Einstein Col Med, 58-60, sr fel physiol, 62; fel, Inst Training Res Behav & Neurol Sci, 60-61; City of New York Health Res Coun career scientist award, 69- *Mem:* Biophys Soc; Am Physiol Soc; Am Soc Biol Chemists; Soc Develop Biol. *Res:* Cell physiology; developmental biology; biophysics. *Mailing Add:* Dept Physiol Albert Einstein Col Med 1300 Morris Park Ave New York NY 10461

MORRILL, JAMES LAWRENCE, JR, b Graves Co, Ky, Nov 23, 30; m 52; c 5. DAIRY SCIENCE. *Educ:* Murray State Col, BS, 58; Univ Ky, MS, 59; Iowa State Univ, PhD(dairy cattle nutrit), 63. *Prof Exp:* From instr to assoc prof, 62-78, PROF & DIARY CATTLE RES NUTRITIONIST, AGR EXP STA, KANS STATE UNIV, 78- *Mem:* Am Dairy Sci Asn; Am Soc Animal Sci. *Res:* Dairy cattle nutrition, especially nutrition of young. *Mailing Add:* Dept of Animal Sci Kans State Univ Manhattan KS 66506

MORRILL, JOHN BARSTOW, JR, b Chicago, Ill, Nov 20, 29; m 53; c 2. DEVELOPMENTAL BIOLOGY. *Educ:* Grinnell Col, BA, 51; Iowa State Col, MS, 53; Fla State Univ, PhD(zool), 58. *Prof Exp:* From instr to asst prof biol, Wesleyan Univ, 58-65; assoc prof, Col William & Mary, 65-67; assoc prof, 67-69, coordr Environ Studies Prog, 74-81, PROF BIOL, NEW COL FLA, 69- *Concurrent Pos:* NSF fel, 58; NIH special fel, 64; mem corp, Marine Biol Lab; AAAS fel, 81. *Mem:* Am Soc Zoologists; Soc Develop Biol; Am Soc Cell Biol; Int Soc Develop Biol; fel AAAS. *Res:* Development of mollusk eggs; experimental analyses of molluscan development; effects of chemicals on mollusk eggs; SEM of sea urchin development. *Mailing Add:* Div Natural Sci New Col-USF Sarasota FL 34243-2197

MORRILL, JOHN ELLIOTT, b Oak Park, Ill, Nov 4, 35; m 58; c 3. MATHEMATICS. *Educ:* DePaul Univ, BA, 57; Univ Mich, MA, 60, PhD(math), 64. *Prof Exp:* From asst prof to assoc prof math, DePaul Univ, 64-70; vis assoc prof, Univ Mich, 70-71; assoc prof, 71-76, PROF MATH, DePAUL UNIV, 76- *Concurrent Pos:* Acad guest, Res Inst Math, ETH, Zurich, 72-73. *Mem:* Assoc Soc Actuaries; Am Math Soc; Math Asn Am. *Res:* Actuarial mathematics; mathematical economics. *Mailing Add:* Dept Math DePaul Univ Locust St Greencastle IN 46135-0037

MORRILL, LAWRENCE GEORGE, b Tridell, Utah, July 21, 29; m 49; c 4. SOIL CHEMISTRY, FERTILITY. *Educ:* Utah State Univ, BS, 55, MS, 56; Cornell Univ, PhD(soil chem), 59. *Prof Exp:* From assoc chem to sr chemist, Thiokol Chem Corp, Utah, 59-66; res specialist, Cornell Univ, 60-61; assoc prof, 66-75, PROF SOIL CHEM & FERTIL, OKLA STATE UNIV, 75- *Mem:* Am Soc Agron; Soil Sci Soc Am. *Res:* Soil chemistry and fertility research and teaching. *Mailing Add:* Dept of Agron Okla State Univ Stillwater OK 74078

MORRILL, TERENCE CLARK, b Albany, NY, Mar 1, 40; m 65; c 2. ORGANIC CHEMISTRY, SPECTROMETRY. *Educ:* Syracuse Univ, BS, 61; San Jose State Col, MS, 64; Univ Colo, PhD(org chem), 66. *Prof Exp:* NSF fel org chem, Yale Univ, 66-67; from asst prof to assoc prof, 67-75, PROF ORG CHEM, ROCHESTER INST TECHNOL, 75- *Concurrent Pos:* Vis assoc prof, Univ Rochester, 75, vis prof, 81; sci fac fel, NSF, 75-77, fac fel, Rochester Inst Technol, 76-78; vis prof, Univ NC, Chapel Hill, 86, Univ Rochester, 87-90. *Mem:* Am Chem Soc; Sigma Xi; AAAS. *Res:* Stereochemistry of addition and solvolysis reactions; mechanisms of additions, solvolyses and oxidation reactions: alkylations; use of Wilkonson's Reagent with boraxes; lanthanide shift reagents; reactions of organoboranes and organomercurials; relaxation reagents. *Mailing Add:* Dept Chem Rochester Inst Technol Rochester NY 14623

MORRILL, WENDELL LEE, b Madison, SDak, May 22, 41; m 65; c 2. ENTOMOLOGY, AGRICULTURE. *Educ:* SDak State Univ, BS, 67, MS, 68; Univ Fla, PhD(entom), 71. *Prof Exp:* Res assoc fire ants, Univ Fla, 71-73; asst prof grassland entom, Univ Ga, 73-78; ASSOC PROF CROPPING SYSTS ENTOM, MONT STATE UNIV, 78- *Mem:* Entom Soc Am. *Res:* Insect/plant relationships; effects of changing farming practices on insect populations. *Mailing Add:* Dept Entom Mont State Univ Bozeman MT 59717

MORRIN, PETER ARTHUR FRANCIS, b Dublin, Ireland, Oct 8, 31; m 60; c 3. MEDICINE, NEPHROLOGY. *Educ:* Nat Univ Ireland, MB, BCh & BAO, 54, BSc, 55; FRCP(C), 61. *Prof Exp:* Instr med, Wash Univ, 60-61; lectr, 62-63, from asst to assoc prof, 63-76, PROF MED, QUEEN'S UNIV, ONT, 76- *Mem:* Am Fedn Clin Res; Can Med Asn; Can Soc Clin Invest; Am Soc Nephrol; Am Soc Artificial Internal Organs; Int Soc Nephrol; Soc Nephrol France. *Res:* Nephrology. *Mailing Add:* Renal Unit Kingston Gen Hosp Kingston ON K7L 3N6 Can

MORRIS, ALAN, b Baltimore, Md, Jan 4, 31; m 51; c 3. AUTOMOTIVE ENGINEERING, ENVIRONMENTAL SCIENCE & ENGINEERING. *Educ:* Johns Hopkins Univ, BE, 51, DrEng, 55. *Prof Exp:* Res asst, Inst Coop Res, Johns Hopkins Univ, 51-55; res engr, Opers Res, Inc, Md, 58-61; assoc prof physics, Am Univ, 61-65, dir opers res, Ctr Technol & Admin, 65-69; SR PARTNER, MORRIS & WARD ENGRS, 69- *Concurrent Pos:* Aeronaut Eng Officer, LT(jg), US Naval Reserve, 55-58. *Mem:* Am Consult Engrs Coun; Am Soc Mech Engrs; Am Soc Civil Engrs; Inst Elec & Electronic Engrs; Soc Automotive Engrs. *Res:* Forensic science and engineering; hazardous waste management technologies. *Mailing Add:* 5817 Plainview Rd Bethesda MD 20817

MORRIS, ALLAN J, biochemistry, for more information see previous edition

MORRIS, ALVIN LEONARD, b Detroit, Mich, July 2, 27; wid; c 3. DENTISTRY, ORAL MEDICINE. *Educ:* Univ Mich, DDS, 51; Univ Rochester, PhD(path), 57. *Prof Exp:* Asst prof oral med & actg head dept oral diag, Sch Dent, Univ Pa, 57-60, asst prof, Grad Sch Med, 58-61, assoc prof & head dept, 60-61; dean, Sch Dent, Univ Ky, 61-68, asst vpres, Med Ctr, 68-69, spec asst to pres admin, 69-70, prof oral diag & oral med, 61-75, vpres admin, 70-75; exec dir, Asn Acad Health Ctrs, 75-79; prof dent med & vpres health affairs, 79-87, EMER PROF DENT MED, UNIV PA, 87- *Concurrent Pos:* Consult, Vet Admin, Lexington, Ky, 62; consult, USPHS, Ky, 63, chmn dent study sect, Nat Inst Dent Res, 65-67; mem army med serv adv comt prev dent, Off Surgeon Gen, 67-74; mem nat adv coun educ health professions, NIH, 68-72; mem dent adv comt, Dept Defense, 70-73; pres, Am Fund Dent Educ, 70-74. *Mem:* Nat Inst Med; AAAS; Int Asn Dent Res; Am Dent Asn; Am Acad Oral Med. *Res:* Experimental oral cancer with emphasis on the histochemistry and biochemistry of carcinogenesis; normal and abnormal keratinization of oral mucosa. *Mailing Add:* River Club 6-31 Tall Pines Pawleys Island SC 29585

MORRIS, ARTHUR EDWARD, b Billings, Mont, Apr 26, 35; m 57; c 2. METALLURGY. *Educ:* Mont Col Mineral Sci & Technol, BS, 56; Purdue Univ, MS, 59; Pa State Univ, PhD(metall), 65. *Prof Exp:* Extractive metallurgist, Boulder City Sta, US Bur Mines, 59-61; from asst prof to assoc prof, 65-74, PROF METALL, UNIV MO-ROLLA, 74- *Concurrent Pos:* Metallurgist, Lawrence Livermore Lab, 73; extractive metallurgist, Anaconda Minerals Res Lab, Tucson, 83. *Mem:* Am Soc Metals; Am Inst Mining, Metall & Petrol Engrs. *Res:* Chemical metallurgy; phase equilibria, thermodynamics and kinetics of reactions in high-temperature metallurgical systems; software for modeling pyrometallurgical processes. *Mailing Add:* 809 Bray Ave Rolla MO 65401-0249

MORRIS, BROOKS T(HERON), b Pasadena, Calif, June 11, 13; m 39; c 5. AERONAUTICAL & CIVIL ENGINEERING. *Educ:* Stanford Univ, AB, 34, CE, 38. *Prof Exp:* Jr civil engr, US Bur Reclamation, Colo, 34-36; asst hydraul mach, Calif Inst Technol, 38-39; assoc hydraul engr, Co-op Lab, Soil Conserv Serv, USDA, Calif Inst Technol, 39-42; res engr, C F Braun & Co, 42-43; proj engr & asst dir res, Aerojet Eng Corp, Calif, 43-46; chief propulsion sect, Aircraft Res & Develop Div, Willys-Overland Motors, Inc, 46-47; chief res & develop div, Gen Tire & Rubber Co, Calif, 48-49; proj engr, pulse-jet & subsonic ramjet engines, Marquardt Corp, 49-50, admin engr, 51-52, chief proj engr, 53-54, exec engr, 54-59, dir propulsion div, 60-61; mgr, Qual Assurance & Reliability Off, Jet Propulsion Lab, Calif Inst Technol, 61-81. *Concurrent Pos:* Consult engr, 81- *Honors & Awards:* Except Serv Medal, NASA, 72. *Mem:* Am Soc Civil Engrs; assoc fel Am Inst Aeronaut & Astronaut; Am Geophys Union; Earthquake Eng Res Inst. *Res:* Reliability of space exploration systems; development of pulsejet engines, ramjet engines and rocket motors. *Mailing Add:* 3745 Normandy Dr Flintridge CA 91011

MORRIS, CATHERINE ELIZABETH, b Toronto, Ont, Dec 24, 49; c 2. ION CHANNELS, ELECTROPHYSIOLOGY. *Educ:* Univ Canterbury, Christchurch, NZ, BSc, 73; Univ Cambridge, Eng, PhD(zool), 77. *Prof Exp:* Postdoctoral fel, Lab Biophysics, Nat Inst Neurol & Commun Dis & Strokes, NIH, Bethesda, 78-81; ASSOC PROF, DEPT BIOL, UNIV OTTAWA, 81- *Mem:* Biophys Soc; Soc Neurosci. *Res:* Stretch-sensitive ion channels; alkaloid pumps; P-glycoprotein. *Mailing Add:* Neurosci Unit Loeb Inst Ottawa Civic Hosp 1053 Carling Ave Ottawa ON K1Y 4E9 Can

MORRIS, CECIL ARTHUR, SR, b Jacksonville, Fla, May 20, 43; m 66; c 2. SOFTWARE METRICS, EXPERT SYSTEMS. *Educ:* Pfeiffer Col, Misenheimer, NC, BA, 65; Appalachian State Univ, Boone, MA, 67; Va Polytech Inst & State Univ, Blacksburg, PhD(math), 74. *Prof Exp:* Asst prof math, Lynchburg Col, 71-73; systs analyst, Cannon Mills Co, Kanaapolis, NC, 73-74; assoc prof math, Barber-Scotia Col, Concord, NC, 74-75; assoc prof, Cumberland Col, Williamsburg, Ky, 75-81, prof math & data processing, 81-84; info resources specialist, Va Power Co, Richmond, 86-87; proj mgr, Cestaro & Co, Inc, Colonial Heights, 87-89; SECT MGR, COMPUTER SCI CORP, 84-86 & 89- *Concurrent Pos:* Chmn, Math Sci Dept & dir, Computer Serv, Cumberland Col, Williamsburg, Ky, 81-84. *Mem:* Data Processing Mgt Asn. *Res:* Measurement and metrics of software development; software engineering; algebraic topology, deleted products; expert systems; artificial intelligence. *Mailing Add:* 4203 Montreal Ave Prince George VA 23875

MORRIS, CHARLES EDWARD, b Detroit, Mich, Feb 17, 41; m 68. SOLID-STATE PHYSICS. *Educ:* Iowa State Univ, BS, 63, PhD(physics), 68. *Prof Exp:* Mem staff, 68-84, SECT LEADER MG, LOS ALAMOS NAT LAB, 84- *Mem:* Am Phys Soc. *Res:* Shock wave physics; mechanical properties of solids and liquids; optical properties of solids. *Mailing Add:* Los Alamos Nat Lab MSJ 970 Los Alamos NM 87545

MORRIS, CHARLES ELLIOT, b Denver, Colo, Mar 30, 29; m 51; c 2. NEUROLOGY. *Educ:* Univ Denver, BA, 50, MA, 51; Univ Colo, MD, 55. *Prof Exp:* Teaching fel neurol, Harvard Med Sch, 56-59; from asst prof to prof neurol & med, Sch Med, Univ NC, Chapel Hill, 61-76; PROF NEUROL & CHMN DEPT, UHS/CHICAGO MED SCH, 76-; ACTG CHIEF, NEUROL SERV, VET ADMIN MED CTR, NORTH CHICAGO, 76- *Concurrent Pos:* Consult coun drugs, AMA, 66; dir, Nat Insts Neurol Dis & Stroke Res Ctr, Agana, Guam, 70-71. *Mem:* Fel Am Acad Neurol; Asn Res Nerv & Ment Dis; AMA; Am Epilepsy Soc; Sigma Xi. *Res:* Neuroimmunology, especially investigations into the pathogenesis and diagnosis of autoimmune diseases of the central and peripheral nervous system; diseases of muscle; Parkinsonism and neurogenic amines; amyotropic lateral sclerosis; movement disorders. *Mailing Add:* Dept Neurol Chicago Med Sch 3333 Green Bay Rd North Chicago IL 60064

MORRIS, CLAUDE C, b Come by Chance, Nfld, Sept 23, 58; m 88. SCIENCE EDUCATION. *Educ:* Mem Univ, Nfld, BSc, 80, MSc, 83; Univ Cambridge, PhD(zool), 88. *Prof Exp:* Postdoctoral, 89-90, RES ASSOC, MEM UNIV NFLD, 91- *Mem:* Am Soc Zoologists. *Res:* Physiology of biomineralization in cephalopods, particularly growth increment formation in statoliths; fluid dynamics of mucoid structures of larvaceans; ultrastructure of larvacean houses. *Mailing Add:* Ocean Sci Ctr Mem Univ Nfld St John's NF A1C 5S7 Can

MORRIS, CLETUS EUGENE, b Alcorn County, Miss, Jan 30, 35; m 62; c 1. TEXTILE CHEMISTRY. *Educ:* Auburn Univ, BS, 59, PhD(org chem), 66. *Prof Exp:* RES CHEMIST, SOUTHERN REGIONAL RES CTR, AGR RES SERV, USDA, 65- *Mem:* Am Chem Soc; Sigma Xi; Am Asn Textile Chemists & Colorists. *Res:* Heterocyclic compounds; chemical modification and finishing of cotton; chemistry of crosslinking agents for cellulose. *Mailing Add:* Southern Regional Res Ctr USDA PO Box 19687 New Orleans LA 70179

MORRIS, DANIEL JOSEPH, b Minneapolis, Minn, Oct 19, 51. GAMMA RAY ASTRONOMY. *Educ:* Mass Inst Technol, SB, 73; Univ Md, PhD(physics), 82. *Prof Exp:* Res fel, Nat Res Coun, 82-84; RES SCIENTIST, UNIV NH, 84- *Mem:* Sigma Xi; Am Phys Soc. *Res:* Construction and calibration of the ComCompton Imaging Telescope, experiment and preparations for data analysis; Monte Carlo simulations of gamma ray production and scattering; search of existing data for low level gamma ray bursts. *Mailing Add:* Sci & Eng Res Bldg Univ NH Durham NH 03824

MORRIS, DANIEL LUZON, b Newtown, Conn, July 29, 07; m 29, 61, 75. CHEMISTRY. *Educ:* Yale Univ, AB, 29, PhD(org chem), 34. *Prof Exp:* Field dir speed surv, State Hwy Dept, Conn, 33-34; fel physiol, Sch Med, Yale Univ, 34-35; teacher, Putney Sch, 35-44; res chemist, Mead Johnson & 44-48; lab supvr, Food, Chem & Res Labs, Inc, 48-51; head dept sci, 51-69, teacher math & sci Lakeside Sch, 51-83; RETIRED. *Concurrent Pos:* Fulbright exchange teacher, Eng, 58-59. *Mem:* AAAS; Am Chem Soc; Sigma Xi. *Res:* Descriptive geometry of four dimensions; lipids from animal sources; effects of colloids on crystallization; isolation, determination and physiological effects of carbohydrates; isolation of tryptophane; Christian theology. *Mailing Add:* 1202 Eighth Ave W Seattle WA 98119

MORRIS, DAVID, b Jackson, Miss, May 9, 33; m 59; c 2. CIVIL ENGINEERING. *Educ:* Clemson Col, BS, 55; NC State Univ, MS, 60; Rensselaer Polytech Inst, DrEngSc, 64. *Prof Exp:* Jr engr, Am Bridge Co, 55-56; asst engr, Southern Prestressed Concrete, 60-61; instr struct, Rensselaer Polytech Inst, 62-64; struct designer, Gorgwer & Kraas, Consults, 64-66; ASSOC PROF CIVIL ENG, UNIV VA, 66- *Mem:* Am Soc Civil Engrs. *Res:* Civil engineering systems and structural analysis and design. *Mailing Add:* Dept Civil Eng Thorton Hall A-122 Univ Va Charlottesville VA 22903

MORRIS, DAVID ALBERT, b Marietta, Ohio, July 30, 36; m 58; c 2. GEOCHEMISTRY, EXPLORATION GEOLOGY. *Educ:* Marietta Col, BS, 58; Univ Kans, MS, 61, PhD(geol), 67. *Prof Exp:* From res geologist to sr res geologist, Phillips Petrol Co, 67-69, mgr appl projs sect, 70-75, mgr geochem br, Explor Prod Res Div, 75-78, dir, Geol Europe-Africa Div, London, 78-80, mgr explor & develop, Norway Div, 80-85, chief geologist, Basin Analysis Group, Bartlesville, Okla, 85-88; petrol geol consult, 88-90; MGR EXPLOR, YATES CO INT, 90- *Mem:* AAAS; Am Asn Petrol Geologists; Soc Econ Paleontologists & Mineralogists; Geol Soc Am. *Res:* Stratigraphy, sedimentation and depositional environments of Pennsylvania rocks in North America and Europe; paleogeomorphology of coal swamps and formation of coal splits; origin and geology of petroliferous source rocks and migration of oil; diagenesis and compaction of rocks. *Mailing Add:* Yates Co Int 2 Sunwest Centre PO Box 1798 Roswell NM 88202

MORRIS, DAVID ALEXANDER NATHANIEL, b Jamaica, WI, May 13, 44; m 70; c 2. PHYSICAL CHEMISTRY, CLINICAL CHEMISTRY. *Educ:* Interam Univ Puerto Rico, BS, 66; Univ Wis-Milwaukee, MS, 71; Univ Notre Dame, PhD(phys chem), 77. *Prof Exp:* Asst res scientist, Miles Labs, Inc, 70-73, assoc res scientist, 73-76, res scientist clin chem, 76-79, supvr res & develop, 79-87; res scientist, 87-89, SR STAFF SCIENTIST, BAYER AG, WGER, 89- *Concurrent Pos:* Spec assignment, Notre Dame Radiation Lab, 74-76. *Mem:* Am Chem Soc; Am Asn Clin Chem. *Res:* Methods for clinical analysis of body fluids; chemical reactions in paper matrix; chemical reactions of material applied to solid surfaces; instrumental measurement of color; color theory. *Mailing Add:* Ames Miles Labs Div PO Box 70 Elkhart IN 46515

MORRIS, DAVID JULIAN, b Ramsgate, Eng, May 17, 39; m 65; c 2. ORGANIC CHEMISTRY, ENDOCRINOLOGY. *Educ:* Oxford Univ, BA, 60, MA & DPhil(org chem), 63. *Prof Exp:* Fel with Prof F W Barnes, Brown Univ, 63-66; sect leader med chem, Beecham Res Labs, 66-68; chief biochemist, Dept Med, 68-80, CHIEF BIOCHEMIST, DEPT LAB MED, MIRIAM HOSP, PROVIDENCE, 68-; PROF BIOCHEM PHARMACOL, BROWN UNIV, PROF PATH, 81- *Concurrent Pos:* Chief biochemist, Dept Med, Miriam Hosp, Providence, 68-80; asst prof biochem pharmacol, Brown Univ, 68-75, assoc prof, 75-80. *Mem:* Endocrine Soc; Am Asn Clin Chem; Am Chem Soc. *Res:* Mechanism of action of aldosterone in kidney; primary receptors of steroid hormones in target tissues; characterization and physiological role of aldosterone metabolites in the kidney; steroid metabolism. *Mailing Add:* Dept Lab Med Miriam Hosp 164 Summit Ave Providence RI 02990

MORRIS, DAVID ROBERT, b Whittier, Calif, June 25, 39; m 61, 79; c 3. BIOCHEMISTRY. *Educ:* Univ Calif, Los Angeles, BA, 61; Univ Ill, PhD(chem), 64. *Prof Exp:* NIH fel, 64-66; asst prof biochem, 66-70, assoc prof, 70-77, PROF BIOCHEM, UNIV WASH, 77- *Concurrent Pos:* John Simon Guggenheim fel, 71-72; assoc ed, Biochem, 89- *Mem:* AAAS; Am Soc Biol Chem; Am Soc Microbiol. *Res:* Regulation of cell growth and division; biological function of polyamines. *Mailing Add:* Dept Biochem Univ Wash Seattle WA 98195

MORRIS, DAVID ROWLAND, b London, Eng, June 18, 30; m 56; c 3. CHEMICAL ENGINEERING & METALLURGY. *Educ:* Univ Birmingham, BSc, 51; Univ London, PhD(chem eng), 54. *Prof Exp:* Sci officer, Atomic Energy Res Estab, Eng, 54-57, sr officer, 57-59; lectr chem eng, Univ London, 59-66; assoc prof, 66-73, PROF CHEM ENG, UNIV NB, 73- *Honors & Awards:* Corecipient, Extractive Metall Technol Award, Am Inst Mining, Metall, & Petrol Engrs, 81. *Mem:* Can Inst Mining & Metall; assoc Brit Inst Chem Engrs; Metall Soc Am Inst Mech Engrs. *Res:* Graphite technology; fluidization; thermodynamics of alloy systems; high temperature electrochemistry; metallurgical processes; corrosion; electrochemical sensors; exergy analysis. *Mailing Add:* Dept Chem Eng Univ NB Col Hill Box 4400 Fredericton NB E3B 5A3 Can

MORRIS, DEREK, b Hove, Eng, May 6, 30; m 53; c 3. METROLOGY, QUANTUM PHYSICS. *Educ:* London Univ, BSc, 50, PhD(physics), 53. *Prof Exp:* Fel, Div Pure Physics, Nat Res Coun, Can, 53-55; sci officer, Guided Weapons Dept, Royal Aircraft Estab, Eng, 55-57; from asst res officer to assoc res officer div appl physics, 57-75, SR RES OFFICER, NAT RES COUN, CAN, 75- *Concurrent Pos:* Assoc ed, Transactions Instrumentation & Measurement, Inst Elec & Electronics Engrs, 80- *Mem:* Can Asn Physicists; sr mem Inst Electric & Electronics Engrs. *Res:* Atomic frequency standards. *Mailing Add:* Inst Nat Measurement Standards Nat Res Coun Montreal Rd Ottawa ON K1A 0R6 Can

MORRIS, DONALD EUGENE, b Tulsa, Okla, July 9, 40; m 63; c 4. ORGANOMETALLIC CHEMISTRY, HOMOGENEOUS CATALYSIS. *Educ:* Univ Tulsa, BS, 63; Northwestern Univ, PhD(inorg chem), 67. *Prof Exp:* NSF res grant, Stanford Univ, 67-68; res chemist, 68-73, res specialist, 73-77, sr res specialist, 77-81, SR GROUP LEADER CORP RES LABS, MONSANTO CO, 81- *Mem:* Am Chem Soc. *Res:* Metal complexes and their application in homogeneous catalysis. *Mailing Add:* Corp Res Labs Monsanto Co 800 N Lindbergh Blvd St Louis MO 63166

MORRIS, EDWARD C, chemistry; deceased, see previous edition for last biography

MORRIS, ELLIOT COBIA, b Ely, Nev, June 24, 26; m 50; c 3. ASTROGEOLOGY, MINERALOGY. *Educ:* Univ Utah, BS, 50, MS, 53; Stanford Univ, PhD(geol), 62. *Prof Exp:* Seismic comput, Seismic Explor, Inc, Wyo, 53-54; asst explor geologist, Phillips Petrol Co, Utah, 54-56; geologist & coordr Surveyor TV invests, Astrogeol Br, US Geol Surv, 61-69, geologist, Viking Mars Lander Imaging Team, 69-78 & Astrogeol Br, 69-87; RETIRED. *Concurrent Pos:* Staff scientist planetary progs, NASA hq, 70-71. *Res:* Structural and stratigraphic geology of Western Uinta and Wasatch Mountains, Utah and southern Alaskan areas; sedimentary mineralogy of central California; astrogeologic studies. *Mailing Add:* 95 Lakeview Stansbury UT 84074

MORRIS, EUGENE RAY, b Albion, Nebr, Aug 26, 30; m 52; c 2. HUMAN NUTRITION, TRACE ELEMENTS. *Educ:* Univ Mo, BS, 52, MS, 56, PhD(agr chem), 62. *Prof Exp:* Assoc chemist, Midwest Res Inst, 62-64; instr agr chem, Univ Mo, 64-68; RES CHEMIST, HUMAN NUTRIT RES CTR, USDA, BELTSVILLE, 68- *Mem:* Am Chem Soc; Am Inst Nutrit; Am Asn Cereal Chemists; Soc Environ Geochem & Health; Sigma Xi. *Res:* Animal and human nutrition; physiological chemistry of magnesium; unidentified growth factors; iron and zinc bioavailability. *Mailing Add:* 12215 Valerie Lane Laurel MD 20708

MORRIS, EVERETT FRANKLIN, b Bellmont, Ill, May 23, 24; m 46, 74; c 2. BOTANY. *Educ:* Eastern Ill Univ, BS, 50; Univ Wyo, MS, 52; Univ Iowa, PhD(bot), 55. *Prof Exp:* Asst prof biol, Millikin Univ, 55-57; assoc prof, Martin Br, Univ Tenn, 57-58; asst mycologist, Ill Natural Hist Surv, 58; from asst prof to assoc prof, 58-66, prof biol, Western Ill Univ, 66-78, co-adminr, A L Kibbe Life Sci Sta, 65-70, chmn dept biol sci, 69-78; prof biol & dean, Sch Sci & Nursing, Purdue Univ, Calumet Campus, 78-80; PROF BIOL & DEAN ACAD AFFAIRS, LA STATE UNIV, ALEXANDRIA, 80- *Mem:* Mycol Soc Am. *Res:* Taxonomy of myxomycetes and Fungi Imperfecti. *Mailing Add:* La State Univ Alexandria LA 71302

MORRIS, FRED(ERICK) W(ILLIAM), b Los Angeles, Calif, Feb 28, 22; m 49. ELECTRONICS & COMMUNICATIONS. *Educ:* Calif Inst Technol, BSEE, 44. *Hon Degrees:* DSc, Capital Inst Technol, 75. *Prof Exp:* asst prof elec eng, Univ Southern Calif, 47-50; chief res studies, Electronic Defense, Signal Corps, US Army, 50-54; pres, electronic eng & mgt consult, Fred W Morris & Assoc, 54-64; exec vpres & dir Electromagnetic Technol Corp, Calif, 61-64; assoc dir telecommun mgt, Exec Off of the President, White House, 64-66; vpres corp planning, Radiation Inc, 66-69; pres & chief exec officer, TRT Telecommun Corp, Washington, DC, 72-75; vpres corp devel & consult, Comsat Gen Corp, Washington, DC, 75-78; pres & chief exec officer, 69-86, CHMN, TELE-SCI ASSOCS, 69- *Concurrent Pos:* Nat Aeronaut & Space Coun, 64-65; vpres & dir, Electromagnetic Tech Corp, Calif; dir, Astro Tech Corp; mem res & develop bd, Dept Defense, 50-53; mem, President Johnson's Task Force on Commun Policy, 67-69; consult, Stanford Res Inst, Mass Inst Technol, & Inst Defense Analysis; assoc, Calif Inst Technol. *Honors & Awards:* Officier de l'Ordre Grand-Ducal de la Couronne de Chene, Grand Duchy of Luxembourg. *Mem:* Sr mem Inst Elec & Electronic Engrs; fel Am Inst Aeronaut & Astronaut; Armed Forces Commun-Electronics Asn; fel East-West Ctr. *Res:* Electronic defense; telecommunications systems engineering, policy and management; technology investment advisor/consultant. *Mailing Add:* Tele-Sci Assocs 137 Ash Lane Portola Valley CA 94028

MORRIS, GENE FRANKLIN, b Cedar Rapids, Iowa, Nov 22, 34; m 59; c 4. ORGANIC CHEMISTRY. *Educ:* Iowa State Univ, BS, 55; Kans State Univ, PhD(org chem), 61. *Prof Exp:* USPHS fel, Duke Univ, 59-61, Iowa State Univ, 61-63, instr chem, 63-66; from asst prof to assoc prof, Wis State Univ, Eau Claire, 66-69; vis prof, Univ Calgary, 69; ASSOC PROF CHEM, WESTERN CAROLINA UNIV, 69- *Mem:* Am Chem Soc; Sigma Xi. *Res:* Physical organic chemistry; mechanisms of reactions, epoxidations, hydrogenations. *Mailing Add:* 870 Tilley Creek Rd Cullowhee NC 28723-9708

MORRIS, GENE RAY, b Joplin, Mo, Oct 7, 27; m 50, 70; c 5. PAVEMENT DESIGN, EMULSIFIED ASPHALT. *Educ:* Univ Ariz, BSCE, 51. *Prof Exp:* Field engr construct, Portland Cement Asn, 54-59; sr resident, Ariz Dept Transp, 62-65, asst dist engr, 65-68, qual control engr, 68-70, res engr, Res & Develop, 70-78; dir res & develop, Ariz Transp Res Ctr, 78-83; prin consult, Western Technologies Inc, 83-88; TECH DIR RES & DEVELOP, INT SURFACING INC, 88- *Concurrent Pos:* Chmn, Transp Res Bd, 72-78. *Honors & Awards:* W J Emmons Award, Asn Asphalt Paving Technologists, 82; John C Parks Award, Am Soc Civil Engrs, 88. *Mem:* Asn Asphalt Paving Technologists; fel Am Soc Civil Engrs; Nat Soc Prof Engrs; Am Mil Engrs; Nat Asn County Engrs. *Res:* Use of ground tire rubber in asphalt; control of expansive clay soils; sub-bituminous coal ash in concrete and soil stabilization; polymer modified asphalt paving systems. *Mailing Add:* 8140 N First Dr Phoenix AZ 85021

MORRIS, GEORGE COOPER, JR, b Evanston, Ill, Feb 15, 24; m 46; c 5. SURGERY. *Educ:* Univ Pa, MD, 48. *Prof Exp:* Instr surg, Sch Med, Univ Pa, 49-50; from instr to assoc prof, 50-68, PROF SURG, BAYLOR COL MED, 68- *Concurrent Pos:* Markle scholar. *Mem:* Soc Vascular Surg; Soc Thoracic Surgeons; Am Col Chest Physicians; Am Asn Surg Trauma; Am Surg Asn. *Res:* Cardiovascular research. *Mailing Add:* Dept Surg Baylor Med 1200 Moursund Ave One Baylor Plaza Houston TX 77030

MORRIS, GEORGE MICHAEL, b Tulsa, Okla, Dec 12, 52; m 74; c 2. STATISTICAL OPTICS, HOLOGRAPHY. *Educ:* Univ Okla, BS, 75; Calif Inst Technol, MS, 76, PhD(elec eng), 79. *Prof Exp:* Scientist, 79, asst prof optics, 82-86, ASSOC PROF OPTICS, UNIV ROCHESTER, 86- *Concurrent Pos:* Consult, Eastman Kodak Co, 85-, 3M Co, 85-, Loral Electro-Optical Syst, 86-87; mem, Optical Soc Am Educ Coun, 88-90; prin invest, US Army Res Office, 82-, US Army Missle Command, 83-86, DARPA, MIT Lincoln Labs, 86-, Nat Sci Found, 82-83. *Honors & Awards:* Faculty Develop Award, IBM Co, 84 & 85. *Mem:* Optical Soc Am; Soc Photo-Optical & Instrumentation Eng. *Res:* Statistical optics, optical data processing and holography; quantum-limited imaging and image processing, optical coherence theory and measurement, diffractive optics, and real-time holography. *Mailing Add:* Inst Optics Univ Rochester Rochester NY 14627

MORRIS, GEORGE RONALD, atomic physics, molecular physics, for more information see previous edition

MORRIS, GEORGE V, b Providence, RI, Nov 18, 30; m 59; c 2. PHYSICAL & ANALYTICAL CHEMISTRY, MATERIALS ENGINEERING. *Educ:* Providence Col, BS, 52; Univ RI, MS, 57, PhD(phys chem), 62. *Prof Exp:* Asst chem, Univ RI, 55-56; res chemist, Nat Res Corp, Mass, 56-57 & Eltex Res Corp, RI, 57-59; asst chem, Univ RI, 59-61; res chemist, US Naval Underwater Ord Sta, 62-65; from instr to assoc prof, 63-72, PROF PHYS CHEM, SALVE REGINA COL, 72- *Concurrent Pos:* Sr engr, Raytheon Co, RI, 66-; fel, Oregon State Univ, 68-69; pres, Riverside Res Corp, 70- *Mem:* Am Chem Soc. *Res:* Corrosion of metals in sea water; physical and mechanical properties of materials, coatings and composites; development of stable, long deployment life, dissolved oxygen probe; thermal decomposition of inorganic superoxides. *Mailing Add:* Dept Physics & Chem Salve-Regina Col Newport RI 02840

MORRIS, GEORGE WILLIAM, b Granite, Okla, Apr 23, 21; m 59; c 2. MATHEMATICS, MATHEMATICAL PHYSICS. *Educ:* Southest Inst Technol, BA, 42; Univ Okla, MA, 48; Univ Calif, Los Angeles, PhD(math), 57. *Prof Exp:* Instr math, Univ Tulsa, 47-48; engr, Northrop Aircraft, Inc, 51-53; sr engr, NAm Aviation, Inc, 53-58; proj engr, Aerolab Develop Co, 58-60; mem tech staff, Land-Air, Inc, Point Mugu, 60-62 & Douglas Aircraft Co, 62-68; PROF MATH, NAVAL POSTGRAD SCH, 68- *Mem:* Soc Indust & Appl Math. *Res:* Numerical analysis; celestial mechanics; exterior ballistics; elasticity; aerodynamics; electromagnetic wave propagation; thermodynamics; operations research. *Mailing Add:* Dept Math Naval Postgrad Sch Code 0223 Monterey CA 93940

MORRIS, GERALD BROOKS, b Decatur, Tex, July 2, 33; m 59. SEISMOLOGY, UNDERWATER ACOUSTICS. *Educ:* Tex A&M Univ, 56, MS, 62; Univ Calif, San Diego, PhD(earth sci), 69. *Prof Exp:* Res engr, Res Lab, Carter Oil Co, 57-58 & Jersey Prod Res Co, 58-63; asst res geophysicist, Marine Phys Lab, Univ Calif, San Diego, 69-70; asst prof geophys, Univ Hawaii, 70-72; res geophysicist, Marine Phys Lab, Univ Calif, San Diego, 72-77; supvr res physicist, 77-82, res geophysicist, Off Naval Res, 82-86, PROG MGR, UNDERWATER ACOUST, NAVAL OCEAN RES & DEVELOP ACTIV, 86- *Concurrent Pos:* Dir, Off Naval Res Detachment, Bay St Louis, Miss. *Mem:* Soc Explor Geophysicists; Am Geophys Union; Acoust Soc Am. *Res:* Underwater acoustics and sound propagation; explosion seismology, particularly marine seismic refraction studies; elastic properties of earth materials; low frequency ambient noise in the ocean. *Mailing Add:* Naval Ocean Res & Develop Activ Code 211 NSTL Station MS 39529

MORRIS, GERALD PATRICK, b Edmonton, Alta, June 13. 39; m 64; c 2. CELL BIOLOGY, GASTROENTEROLOGY. *Educ:* Univ BC, BSc, 64, MSc, 66; Queens Univ, Belfast, PhD(zool), 68. *Prof Exp:* Lectr zool, Univ BC, 65-66; NIH fel electron micros, Univ Kans, 68-69; from asst prof to assoc prof, 69-84, PROF BIOL, QUEEN'S UNIV, ONT, 84- *Mem:* Can Asn Gastroenterol; Micros Soc Can. *Res:* Ultrastructure and physiology of gastrointestinal ulcers; prostaglandins and cytoprotection; roles of gastrointestinal mucus, inflammatory bowel disease. *Mailing Add:* Dept Biol Queens Univ Kingston ON K7L 3N6 Can

MORRIS, HAL TRYON, b Salt Lake City, Utah, Oct 24, 20; m 42; c 3. GEOLOGY. *Educ:* Univ Utah, BS, 42, MS, 47. *Prof Exp:* Geologist, 46-69, res geologist, 69-86, EMER GEOL, US GEOL SURV, 86- *Honors & Awards:* Meritorious Serv Award, US Dept Interior, 76. *Mem:* Fel Geol Soc Am; Soc Econ Geologists. *Res:* Mineral deposits; structural geology; stratigraphy; geochemical prospecting; detection and discovery of concealed ore deposits. *Mailing Add:* 2220 Camino A Los Cerros Menlo Park CA 94025

MORRIS, HALCYON ELLEN MCNEIL, b Delphos, Kans, May 24, 27; m 59. GEOPHYSICS, OCEANOGRAPHY. *Educ:* Kans State Univ, BS, 51; Univ Tulsa, MS, 58. *Prof Exp:* Mathematician, Carter Oil Co, 51-58, Jersey Prod Res Co, 58-59, Pan Am Petrol Co, 60-63 & Naval Electronics Lab, 64-68; div head & supvry mathematician, Naval Ocean Systs Ctr, Naval Ocean Res & Develop Activ, Nat Space Technol Lab Sta, Miss, 68-77, mem staff, 77-81, asst dir ocean sci, 81-88; DIR BASIC RES PROG, NAVAL OCEANOG & ATMOSPHERIC RES LAB, STENNIS SPACE CTR, MISS, 88- *Mem:* Am Geophys Union; fel Acoust Soc Am. *Res:* Detection of submarines by surveillance systems; theory development for new sonar systems; underwater acoustics sound propagation; manage basic research program in ocean sciences; manage academic graduate programs. *Mailing Add:* Code 114 Naval Oceanog & Atmospheric Res Lab Stennis Space Center MS 39529-5004

MORRIS, HAROLD HOLLINGSWORTH, JR, b Shanghai, China, Sept 23, 17; US citizen; m 44; c 2. MEDICINE, PSYCHIATRY. *Educ:* Haverford Col, BS, 39; Tulane Univ, MD, 43. *Prof Exp:* ASSOC PROF PSYCHIAT, SCH MED, UNIV PA, 56- *Concurrent Pos:* Clin dir psychiat, Inst Pa Hosp, 48-56; consult, Episcopal Diocese Pa, 54-, Vet Admin Hosp, Coatesville, 56-, Am Friends Serv Comt, 62 & Peace Corps, 64; dir, Mercy-Douglass Hosp, Philadelphia, 56-63 & Misericordia Hosp, 66- *Mem:* Am Psychiat Asn. *Res:* Effect of phenothiazines on schizophrenia; long-term follow-up studies on 3, 000 patients, including 225 children; effect of psychological mileu on organic illness. *Mailing Add:* 362 Devon Way West Chester PA 19380

MORRIS, HENRY MADISON, JR, b Dallas, Tex, Oct 6, 18; m 40; c 6. HYDRAULIC ENGINEERING, HYDROGEOLOGY. *Educ:* Rice Inst, BS, 39; Univ Minn, MS, 48, PhD(civil eng), 50. *Hon Degrees:* LLD, Bob Jones Univ, 66; DLitt, Liberty Univ, 89. *Prof Exp:* Mem State Hwy Dept, Tex, 38-39; from jr engr to asst engr, Int Boundary Comn, 39-42; instr civil eng, Rice Inst, 42-46; from instr to asst prof & proj supvr, St Anthony Falls Hydraul Lab, Univ Minn, 46-51; prof civil eng & head dept, Southwestern La Inst, 51-56; prof appl sci, Univ Southern Ill, 57; prof civil eng & head dept, Va Polytech Inst & State Univ, 57-70; vpres acad affairs, Christian Heritage Col, 70-78, pres, 78-80; dir, 70-80, PRES, INST CREATION RES, SAN DIEGO, 80- *Concurrent Pos:* Pres, Trans Nat Asn Christian Schs, 84- *Mem:* Fel AAAS; fel Am Soc Civil Engrs; Am Geophys Union; Geol Soc Am; Am Asn Petrol Geol; Sigma Xi. *Res:* Conduit and engineering hydraulics; design of hydraulic structures; hydraulic studies; hydro-morphology; correlation of science and christianity. *Mailing Add:* Inst for Creation Res 10746 Woodside Ave N Santee CA 92071

MORRIS, HERBERT ALLEN, b Okla, Sept 15, 19; m 45; c 3. MATHEMATICS. *Educ:* Southeastern State Col, BA, 46; Univ Tex, MA, 51. *Prof Exp:* Asst prof math, Lamar State Col, 51-S5 & Colo Sch Mines, 55-57; mathematician, Atomic Energy Div, Phillips Petrol Co, 57-59, sr mathematician, Comput Dept, 59-65; dir, Comput Ctr, 65-74, ASSOC PROF COMPUT SCI, BRADLEY UNIV, 74-, CHMN DEPT, 65- *Concurrent Pos:* Adj prof, Okla State Univ, 60-65. *Mem:* Soc Indust & Appl Math; Asn Comput Mach. *Res:* Numerical analysis; point set topology; operations research. *Mailing Add:* Comput Ctr Bradley Univ 1501 W Bradley Ave Peoria IL 61625

MORRIS, HERBERT COMSTOCK, b Dayton, Ohio, Apr 18, 17; m 43; c 5. CHEMICAL ENGINEERING, CHEMISTRY. *Educ:* Univ Dayton, BCE, 42. *Prof Exp:* Chem engr, Beacon Res Labs, Texaco, Inc, 42-49, asst to dir res, NY, 49-55, supvr, Port Arthur Res Labs, 55-67, dir res, Beacon Res Labs, 67-70, asst mgr petrol res, 70-77; mgr, Texaco, Inc, 77-79, BUS MGR, TEXACO CHEM CO, 79- *Mem:* Am Chem Soc; Am Inst Chem Engrs; Sci Res Soc Am. *Res:* Petroleum research; process and product development; research administration. *Mailing Add:* Texaco Chem Co 610 Santa Maria Sugar Land TX 77478

MORRIS, HORTON HAROLD, b Post, Tex, May 26, 22; m 45; c 3. CLAY MINERALOGY, PULP AND PAPER TECHNOLOGY. *Educ:* Tex Tech Col, BS, 49; Univ Maine, MS, 52. *Prof Exp:* From instr to assoc prof chem, Univ Maine, 52-57; res dir, South Clays, Inc, 57-63; vpres res & develop, Freeport Kaolin Co, 63-74; PRES, SSI CONSULTS, 79- *Mem:* AAAS; Am Chem Soc; Tech Asn Pulp & Paper Indust; Am Soc Test & Mat; NY Acad Sci. *Res:* Molecular rearrangements; Glycidic esters; 2, 3-dihydroxy esters and halohydrins; kaolin clay studies; mineral benefaction; delaminated clays; calcined clays. *Mailing Add:* 478 Island Ave Peaks Island ME 04108

MORRIS, HOWARD ARTHUR, b Draper, Utah, Feb 9, 19; m 41; c 4. FOOD SCIENCE. *Educ:* Univ Minn, MS, 49, PhD(dairy technol), 52. *Prof Exp:* Asst & instr, 46-51, asst prof dairy technol, 52-55, assoc prof dairy indust, 55-60, PROF FOOD SCI, UNIV MINN, ST PAUL, 60- *Honors & Awards:* Pfizer Award, Am Diary Sci Asn, 81. *Mem:* Am Soc Microbiol; Inst Food Technologists; Am Dairy Sci Asn; Sigma Xi. *Res:* Chemistry, bacteriology and enzymology applied to food processing. *Mailing Add:* Dept Food Sci & Nutrit Univ Minn St Paul MN 55108

MORRIS, HUGHLETT LEWIS, b Big Rock, Tenn, Mar 18, 31; m 50; c 3. SPEECH PATHOLOGY. *Educ:* Univ Iowa, BA, 52, MA, 57, PhD(speech path), 60. *Prof Exp:* Clinician speech & hearing, Pub Schs, Iowa, 54-56; res assoc otolaryngol & maxillofacial surg, 58-61, coordr cleft palate clin, 59-64, res asst prof otolaryngol & maxillofacial surg, 61-64, assoc prof otolaryngol, maxillofacial surg, speech path & audiol, 65-67, PROF OTOLARYNGOL & MAXILLOFACIAL SURG, SPEECH PATH & AUDIOL, UNIV IOWA, 68-, DIR DIV SPEECH & HEARING, DEPT OTOLARYNGOL & MAXILLOFACIAL SURG & DIR CLEFT PALATE RES PROG, 65-, CHMN DEPT SPEECH PATH & AUDIOL, 76- *Concurrent Pos:* Ed, Cleft Palate J, Am Cleft Palate Asn, 64-70; prin investr, Proj Res Grant, Nat Inst Dent Res Prog, 65-; mem, Am Bds Examrs Speech Path & Audiol, 73-76; vpres, Am Cleft Palate Educ Found, 74-75. *Mem:* AAAS; fel Am Speech & Hearing Asn; Am Cleft Palate Asn (pres, 73-74); Sigma Xi. *Res:* Cleft lip and palate; disorders of the voice. *Mailing Add:* Dept Otolaryngol-Maxillofacial Surg Univ Iowa Hosps Iowa City IA 52241

MORRIS, J(AMES) WILLIAM, b Clarksville, Tex, June 25, 18; m 49; c 3. CHEMICAL ENGINEERING, COOPERATIVE EDUCATION. *Educ:* Univ Tex, BS, 40, MS, 41, PhD(chem eng), 44. *Prof Exp:* Instr chem eng, Univ Tex, 40-44; chem engr, E I du Pont de Nemours & Co, Inc, 44-83, chem engr, Tenn, 44, supvr, Wash, 44-45, Del, 45-47 & Ohio, 47-49, process engr, Del, 49-50, from asst tech supt to tech supt, Ind, 50-53, sect dir, Savannah River Lab, SC, 53-83; RETIRED. *Mem:* Am Nuclear Soc; Am Chem Soc; Am Inst Chem Engrs; Am Soc Eng Educ. *Res:* Cooperative engineering education; cooperative education; technical personnel placement; environmental assessment; chemical engineering process development. *Mailing Add:* 3418 Meadow Dr Aiken SC 29801

MORRIS, JAMES ALBERT, b Crawfordsville, Ind, June 4, 42; c 2. ENZYMOLOGY, MICROBIAL BIOCHEMISTRY. *Educ:* Wabash Col, BA, 64; Purdue Univ, PhD(microbiol), 69. *Prof Exp:* Res assoc gustation, Monell Chem Senses Ctr, Univ Pa, 70-73; sr microbiologist, 73-80, proj biochemist, 80-86, SR PROJ MICROBIOLOGIST, INT FLAVORS & FRAGRANCES, INC, 86- *Mem:* Soc Indust Microbiol. *Res:* Production or modification of flavors by microbial and/or enzymatic processes. *Mailing Add:* 1515 Hwy 36 Union Beach NJ 07735

MORRIS, JAMES ALLEN, b Vienna, Ga, Jan 18, 29; m 50; c 3. MECHANICAL ENGINEERING. *Educ:* Mass Inst Technol, BSME, 52. *Prof Exp:* Teacher, high sch, Ga, 49-50; gen engr, Warner Robins AFB, 50-51; test engr, US Naval Eng Exp Sta, Md, 52-55; proj engr, Ford Sci Lab, Mich, 55-57 & Arnold Eng Develop Ctr, Tenn, 57-68; proj mgr gen consult, Planning Res Corp, 68-73; mgr printer mechanisms, Sci Systs Inc, Ala, 73-80; SR STAFF SYSTS ENGR, SCI APPLN INC, 80- *Concurrent Pos:* Instr, Tenn State Univ, 62-68. *Mem:* Assoc fel Am Inst Aeronaut & Astronaut. *Res:* Ultra low pressure research; cryogenic studies; propulsion research and development; transportation; environmental simulation; heat transfer; design and development of high-speed printers. *Mailing Add:* Rte 2 Box 271A Somerville AL 35670

MORRIS, JAMES F, b New York, NY, Mar 22, 22. PULMONARY DISEASE. *Educ:* Ohio Wesleyan Univ, AB, 43; Univ Rochester, MD, 48. *Prof Exp:* Fel med bact, Sch Med, Univ Rochester, 50-51; instr med, Col Med, Univ Utah, 53-54; assoc prof, 57-71, PROF MED, ORE HEALTH SCI UNIV, 71- *Concurrent Pos:* Chief pulmonary & infectious dis, Vet Admin Med Ctr, Portland, 57- *Mem:* Am Thoracic Soc; Am Col Chest Physicians; Am Col Physicians. *Res:* Clinical pulmonary physiology, primarily obstructive airway diseases. *Mailing Add:* Dept Med Vet Admin Med Ctr 3181 SW Sam Jackson Park Rd Portland OR 97201

MORRIS, JAMES G, b Parkersburg, WVa, Mar 20, 28; m 53; c 1. METALLURGICAL ENGINEERING, MATERIALS SCIENCE. *Educ:* Purdue Univ, BS, 51, PhD(metall eng), 56. *Prof Exp:* Instr metall, Purdue Univ, 53-56; res engr, Dow Chem Co, 56-57; proj engr, Kaiser Aluminum & Chem Corp, 57-58; sr res metallurgist, Olin Mathieson Chem Corp, 58-59; asst prof metall, 59-60, dir honors prog, 64-65, chmn dept, 65-67, dir grad studies, 67-68, DIR LIGHT METALS RES LAB, UNIV KY, 65- *Concurrent Pos:* NSF fac partic res award, 75 & 76; sr res fel & consult, United Technol Corp, 76-; numerous res contracts with industs & govt orgn; res consult, Adv Technol Group & Continental Packaging Co, Inc, Stamford, Conn; res consult, ARCO Metals, Louisville, Ky, Alumax, Inc, Atlanta, Ga. *Honors & Awards:* Tarr Award, 51. *Mem:* AAAS; Am Inst Mining, Metall & Petrol Engrs; Am Soc Eng Educ; NY Acad Sci; Sigma Xi; Am Soc Metal; Inst Metals. *Res:* Thermomechanical processing; deformation dynamics; warm and hot working; softening kinetics; fracture toughness; mechanical anisotropy of Al alloys. *Mailing Add:* Col of Eng Anderson Hall Univ of Ky Lexington KY 40506

MORRIS, JAMES GRANT, b Brisbane, Australia, Aug 30, 30; US citizen; m 59; c 3. ANIMAL NUTRITION. *Educ:* Univ Queensland, BAgrSc, 53, Hons, 55, BSc, 58, MAgrSci, 59; Utah State Univ, PhD(nutrit & biochem), 61. *Prof Exp:* Dir husb res, Animal Res Inst, Brisbane, Australia, 65-69; assoc prof ruminant nutrit & assoc nutritionist, 69-75, PROF VET MED & PHYSIOL CHEM & NUTRITIONIST, EXP STA, UNIV CALIF, DAVIS, 75- *Concurrent Pos:* Full Bright Scholar, 58-61. *Honors & Awards:* FA Brodie Award, 69. *Mem:* Brit Nutrit Soc; Australian Soc Animal Prod; Am Inst Nutrit; Am Soc Animal Sci; Brit Soc Animal Prod. *Res:* Feline and canine nutrition; nutrition of ruminants, particularly grazing cattle and sheep; mineral nutrition. *Mailing Add:* Dept Physiol Sci Sch Vet Med Univ Calif Davis CA 95616

MORRIS, JAMES JOSEPH, JR, b Jersey City, NJ, Aug 16, 33; m 54; c 3. INTERNAL MEDICINE, CARDIOLOGY. *Educ:* Hofstra Univ, BA, 55; State Univ NY, MD, 59. *Prof Exp:* Intern, 59-60, instr, 60-61, resident, 61-62, instr, 62-63, chief resident, 63-64, assoc, 64-66, asst prof, 66-70, assoc prof med, 70-80, PROF MED CARDIOL, DUKE UNIV, 80- *Concurrent Pos:* USPHS fels, 60-61, 62-63 & spec fel, 64-67. *Mem:* Am Heart Asn; Am Col Physicians. *Res:* Electrocardiology; arrhythmia; cardiac catherization; hemodynamics. *Mailing Add:* Dept Med Box 2993 Duke Univ Med Ctr Durham NC 27706

MORRIS, JAMES RUSSELL, b Turlock, Calif, Nov 21, 41; m 69. PHYSICS, OPTICS. *Educ:* Univ Calif, Berkeley, AB, 64, MA, 68, PhD(physics), 77. *Prof Exp:* Comput programmer, US Air Force, 65-67; math programmer, 68, COMPUTATIONAL PHYSICIST, LAWRENCE LIVERMORE NAT LAB, 73- *Mem:* Optical Soc Am; Am Phys Soc. *Res:* Computational modeling of non-linear optical propagation effects; thermal blooming, self-focusing, and coherent pulse propagation. *Mailing Add:* Lawrence Livermore Nat Lab PO Box 808 L-495 Livermore CA 94550

MORRIS, JAMES T, b Langdale, Ala, April 16, 50; c 2. NUMERICAL MODELING, WETLANDS ECOLOGY. *Educ:* Univ Va, BA, 73; Yale Univ, MS, 75, PhD(environ sci), 79. *Prof Exp:* Fel, Ecosyst Ctr, Marine Biol Lab, Woods Hole, 79-81; asst prof, 81-87, ASSOC PROF, DEPT BIOL, UNIV SC, 87- *Concurrent Pos:* Prin investr, Nat Sci Found & US Geol Surv sponsored res. *Mem:* Ecol Soc Am; AAAS; Am Soc Limnol & Oceanog; Bot Soc Am, Estuarine Res Fedn. *Res:* Plant physiological ecology, plant nutrition and nutrient cycling within wetland ecosystems. *Mailing Add:* Dept Biol & Marine Sci Univ SC Columbia SC 29208

MORRIS, JERRY LEE, JR, b Charlotte, NC, Feb 27, 52; m 84; c 4. ELECTROCHEMISTRY, INSTRUMENTATION. *Educ:* NC State Univ, BS, 74; Univ Ill, Urbana-Champaign, PhD(chem), 78. *Prof Exp:* Scientist assoc res electrochem, Lockheed Palo Alto Res Lab, 78-80; staff scientist, Energy Systs Lab, The Continental Group, Inc, 80-82; DIR PROGS, PINNACLE RES INST, 82- *Mem:* Am Chem Soc; Electrochem Soc; AAAS. *Res:* Electrochemistry of battery systems; electrochemical corrosion and passivation; electrochemistry of semiconductors; chemiluminescence; electrochemical kinetics; analytical methods development. *Mailing Add:* 207 Vineyard Dr San Jose CA 95119

MORRIS, JOHN EDWARD, b Pasadena, Calif, July 9, 36; m 58; c 2. DEVELOPMENTAL BIOLOGY, CELL BIOLOGY. *Educ:* Stanford Univ, BA, 58; Univ Hawaii, MS, 60; Univ Calif, Los Angeles, PhD(zool), 66. *Prof Exp:* Investr, Wenner-Gren Inst, Sweden, 65-67 & Univ Chicago, 67-68; from asst prof to assoc prof, 68-82, PROF ZOOL,Ore State Univ, 82-, ASSOC DEAN COL SCI, 87- *Concurrent Pos:* Vis asst prof pediat, Univ Chicago, 74-75; vis investr, NIH, 83-84. *Mem:* AAAS; Am Soc Zoologists; Am Soc Cell Biol; Soc Develop Biol. *Res:* Tissue and cell interactions during embryonic differentiation and growth; mammalian embryo implantation; proteglycans. *Mailing Add:* Dept Zool Ore State Univ Corvallis OR 97331

MORRIS, JOHN EMORY, b Takoma Park, Md, June 15, 37. BIOCHEMISTRY, ONCOLOGY. *Educ:* Cornell Univ, BA, 59; Univ Wis, MS, 62, PhD(oncol), 66. *Prof Exp:* Res asst clin oncol, Univ Wis-Madison, 61-65; Am Peace Corps vis asst prof biochem, Fac Med, Pahlavi Univ, Iran, 66-67; asst prof, 67-70, assoc prof, 70-76, PROF CHEM, STATE UNIV NY COL BROCKPORT, 76- *Concurrent Pos:* Vis scientist biol & med, Argonne Nat Lab, 73-74 & 81- *Mem:* Am Chem Soc; Sigma Xi. *Res:* nitrogen metabolism and its regulation in gerbils; protein structure; mechanism of enzyme regulation. *Mailing Add:* Dept of Chem State Univ of NY Col Brockport NY 14420

MORRIS, JOHN F, b Kansas City, Mo, Oct 23, 28; m 52; c 3. BLOWN FILM COEXTRUSION, FLEXOGRAPHIC PRINTING. *Educ:* Kansas Univ, BS, 51. *Prof Exp:* Process engr, Hercules PWD Co, 51-56; res assoc, Spencer Chem Co, 56-63; tech serv supvr, Gulf Oil Corp, 63-67; tech mgr, Gulf Plastic Prod Co, 67-73; tech sales supvr, Crown Zellerbach Corp, 73-77, tech mgr, 77-86; TECH MGR, JAMES RIVER CORP, 86- *Mem:* Soc Plastic Engrs; Flexible Packaging Asn. *Res:* Polymer development and flexible packaging in the various food industries. *Mailing Add:* PO Box 439 Greensburg IN 47240-0439

MORRIS, JOHN LEONARD, b Des Moines, Iowa, Dec 12, 29; m 52; c 2. PLANT BREEDING, PLANT PATHOLOGY. *Educ:* Iowa State Univ, BS, 61; Utah State Univ, PhD(plant breeding & path), 67. *Prof Exp:* Fieldman, 61-63, res asst seed develop, 63-68, DIR, PEA & BEAN RES, ROGERS BROS CO, 68- *Mem:* Am Soc Agron; Sci Res Soc Am. *Res:* Breeding, development and research of snap beans and garden peas; seed quality research. *Mailing Add:* 130 Pierce Twin Falls ID 83301

MORRIS, JOHN MCLEAN, b Kuling, China, Sept 1, 14; m 51; c 5. SURGERY, GYNECOLOGY. *Educ:* Princeton Univ, AB, 36; Harvard Univ, MD, 40; Am Bd Surg, dipl, 50; Am Bd Obstet & Gynec, dipl, 58. *Hon Degrees:* MA, Yale Univ, 62. *Prof Exp:* Asst surg, Mass Gen Hosp, Boston, 47-52; assoc prof, 52-61, prof, 61-69, JOHN SLADE ELY PROF GYNEC, SCH MED, YALE UNIV, 69- *Concurrent Pos:* Am Cancer Soc fel, Radiumhemmet, Stockholm, Sweden, 51-52; consult gynecologist, Hosps, Conn, 52-; chief obstet & gynec, Yale-New Haven Hosp, 65-66; vis prof gynec & obstet, Stanford Univ, 66-67; vis prof, Univ Tex M D Anderson Hosp & Tumor Inst, 70, Univ Calif, Los Angeles, 78 & Univ Tenn, Memphis, 83; consult, Walter Reed Hosp, Gorgas Hosp, Tripler Gen Hosp & Armed Forces Hosp, Khamis, Jeddah, Tabuk, Saudi Arabia, 84 & Taiwan, 87-88; med corps, US Navy, 42-46. *Mem:* Fel Am Col Surg; fel Am Col Obstet & Gynec; Am Gynec Soc (vpres, 81-82); Am Fertil Soc; Soc Pelvic Surg (pres, 74-75). *Res:* Gynecology; surgery; endocrinology; radiation biology; intersexuality; agents affecting ovum development. *Mailing Add:* Dept Obstet & Gynec Yale Univ New Haven CT 06510-8063

MORRIS, JOHN WILLIAM, JR, b Birmingham, Ala, June 7, 43; div; c 1. ALLOY DESIGN, THERMODYNAMICS & PROSE TRANSFORMATIONS. *Educ:* Mass Inst Technol, BS, 64, ScD(mat sci), 69. *Prof Exp:* Res scientist, Bell Aerospace Co, Textron, 68-71; from asst prof to assoc prof, 71-77, PROF MAT SCI, UNIV CALIF, BERKELEY, 77- *Concurrent Pos:* Fac Sr Scientist, Mat & Chem Sci Div, Lawrence Berkeley Lab, 71-; chmn, Chem & Phys of Metal Comt, 78-80, publ comt, TMS-Aime, 78-79, Golden Gate Chapter Am Soc Metals, 79-80, Eng Sci Prog, Col Eng, 79-80, 5th Int Cryogenic Mat Conf, 81, Comt Aeronaut, Col Eng, 87-, Int Cryogenic Metals Conf, Los Angelos, 88- *Honors & Awards:* Happy Gold Medal, Am Inst Metalurg Engrs, 72. *Mem:* Am Inst Metalurg Engrs; fel Am Soc Metals; Mat Res Soc; Iron & Steel Inst Japan. *Res:* Materials science and laboratory development of metal alloys with exceptional and useful properties including lightweight structural alloys, formable steels, electrical contacts; structural alloys and conductors for high field superconducting magnets. *Mailing Add:* Dept Mat Sci & Mineral Eng Univ Calif Berkeley CA 94720

MORRIS, JOHN WOODLAND, b Princess Anne, Md, Sept 10, 21; m 47; c 2. CONSTRUCTION ENGINEERING & MANAGEMENT. *Educ:* US Mil Acad, BS, 43; Univ Iowa, MS, 48. *Prof Exp:* Var mil positions, US Army, 43-76, chief Corp Engrs, 76-80; pres, J W Morris, Ltd, 81-86; mem staff, PRC Eng, McLean, Va, 86-87; PRES, J W MORRIS, LTD, 87- *Concurrent Pos:* Chair prof construct eng & mgt, Univ Md. *Honors & Awards:* Schwab lectr, Am Iron & Steel Inst, 76; Palladium Medal, Audubon Soc. *Mem:* Nat Acad Eng; hon mem Soc Am Value Engrs; fel Am Soc Civil Engrs; Permanent Int Asn Navig Congs; Soc Am Mil Engrs. *Mailing Add:* 3800 N Fairfax Dr Suite 5 Arlington VA 22102

MORRIS, JOSEPH ANTHONY, b Prince Georges Co, Md, Sept 6, 18; m 42; c 4. BACTERIOLOGY. *Educ:* Cath Univ Am, BS, 40, MS, 42, PhD(bact), 47. *Prof Exp:* Bacteriologist, Josiah Macy Jr Found, NY, 43-44, US Dept Interior & USDA, 44-47, Walter Reed Army Inst Res, DC, 47-56 & US Army Med Command, Japan, 56-59; virologist, NIH, 59-72; dir, Slow, Latent & Temperate Virus Br, Bur Biologics, Food & Drug Admin, 72-76; vchmn, 76-82, CHMN, BELL OF ATRI, INC, 83- *Concurrent Pos:* Instr, Am Univ, 43-46. *Mem:* Am Soc Microbiol; Am Soc Trop Med & Hyg; Soc Exp Biol & Med; Am Asn Immunol. *Res:* Virus and rickettsial diseases. *Mailing Add:* Bell of Atri Inc PO Box 40 College Park MD 20740

MORRIS, JOSEPH BURTON, b Del, Jan 25, 25. ANALYTICAL CHEMISTRY. *Educ:* Howard Univ, BS, 49, MS, 51; Pa State Univ, PhD(anal chem), 56. *Prof Exp:* Instr chem, Howard Univ, 51-53; asst, Pa State Univ, 53-56; res & develop chemist, E I du Pont de Nemours & Co, 56-57; from asst prof to assoc prof, 57-70, PROF ANAL CHEM, HOWARD UNIV, 70-, CHMN DEPT CHEM, 75- *Concurrent Pos:* Fel, Univ Brussels, 67-68. *Mem:* AAAS; Am Chem Soc. *Res:* Polarography; chronopotentiometry; voltammetry at solid electrodes; coulometry; instrumental methods of analysis; solid electrode voltammetry; trace analysis by anodic stripping voltammetry; differential pulse polarography. *Mailing Add:* Dept of Chem Howard Univ Washington DC 20059

MORRIS, JOSEPH RICHARD, b Richmond, Va, Aug 3, 35; m 59; c 2. TOPOLOGY. *Educ:* Va Polytech Inst, BS, 57, MS, 60; Univ Ala, MA, 65, PhD(math), 69. *Prof Exp:* Asst prof math, Samford Univ, 59-64; asst prof, 69-78, ASSOC PROF MATH, VA COMMONWEALTH UNIV, 78- *Mem:* Am Math Soc; Math Asn Am. *Res:* The existence of invariant means of Banach spaces and the common fixed point property for a family of functions. *Mailing Add:* Va Commonwealth Univ 11413 Homestead Lane Va Commonwealth Univ 923 W Franklin St Richmond VA 23233

MORRIS, JUSTIN ROY, b Nashville, Ark, Feb 20, 37; m 56; c 2. HORTICULTURE, PLANT PHYSIOLOGY. *Educ:* Univ Ark, BSA, 57, MS, 61; Rutgers Univ, PhD(hort), 64. *Prof Exp:* Res asst, Univ Ark, 57-61; instr pomol, Rutgers Univ, 61-64; exten horticulturist & assoc prof food sci, 64-75, prof hort & food sci, 75-85, UNIV PROF, UNIV ARK, FAYETTEVILLE, 85- *Concurrent Pos:* Consult indust, Int Exten, Fed Exten Serv, USDA, 71-72 & US AID, 74. *Honors & Awards:* Gourley Award, Am Soc Hort Sci, 79, Raw Prod Award, 82; White Outstanding Res Award, 83; Ware Outstanding Res Award, 83; Hort Soc Serv Appreciation Award, 83. *Mem:* Fel Am Soc Hort Sci; Inst Food Technol; Am Soc Enol. *Res:* Viticulture and preharvest production and handling of mechanically harvested fruits; enology and production processing systems. *Mailing Add:* Dept Food Sci Univ of Arkansas 272 Young Ave Fayetteville AR 72703

MORRIS, LARRY ARTHUR, b Hamilton, Ont, Sept 8, 37; m 58; c 4. FAILURE ANALYSIS. *Educ:* McMaster Univ, BASc, 60, MSc, 62, PhD(metall), 65. *Prof Exp:* Mgr phys metall, Falconbridge Metall Labs, 72-78; dir mkt develop, 78-90, DIR PROD ENVIRON SERV, FALCONBRIDGE LTD, 90- *Concurrent Pos:* Lectr, Ryerson Polytech Inst, 68-72; trustee, Am Soc Metals Int, 86-89. *Mem:* Fel Am Soc Metals Int; Am Powder Metall Inst; Iron & Steel Soc; Am Soc Testing & Mat; Am Electroplaters & Surface Finishing Soc. *Res:* Oxidation, sulfidation and corrosion studies on stainless steels and high nickel alloys; wear resistant alloys; stainless steels for high temperature service; improved nickel electroplating anodes. *Mailing Add:* Falconbridge Ltd PO Box 40 Com Ct W Toronto ON M5L 1B4 Can

MORRIS, LAWRENCE ROBERT, b Toronto, Ont, Apr 7, 42; m 65; c 1. COMMUNICATION, COMPUTER SCIENCE. *Educ:* Univ Toronto, BASc, 65; Univ London, DIC & PhD(speech commun), 70. *Prof Exp:* Vis res assoc elec eng, Univ Rochester, 69-70; asst prof eng, 70-77, ASSOC PROF ENG & COMPUT SCI, CARLETON UNIV, 77- *Concurrent Pos:* Ont Dept Univ Affairs res grant, Carleton Univ, 71-72. *Mem:* Inst Elec & Electronics Engrs; Acoust Soc Am; assoc mem Brit Inst Elec Engrs. *Res:* Speech analysis, synthesis and perception; digital signal processing; interactive computer graphics; computer-aided instruction. *Mailing Add:* Dept Systs Eng Carleton Univ Colonel By Dr Ottawa ON K1S 5B6 Can

MORRIS, LEO RAYMOND, b South Whitley, Ind, June 19, 22; m 45; c 3. INDUSTRIAL ORGANIC CHEMISTRY. *Educ:* Manchester Col, BA, 47; Univ Wis, PhD(org chem), 52. *Prof Exp:* Asst chemist, Univ Wis, 47-50; from org chemist to sr res chemist, Dow Chem USA, 51-71, res specialist, 71-77, sr res specialist, 77-82; RETIRED. *Mem:* Am Chem Soc. *Res:* Dehydrohalogenation, synthesis of new monomers and bioactive compounds; free-radical additions; ag-chemical process development; 30 US patents. *Mailing Add:* 405 Britton Creek Hendersonville NC 28739

MORRIS, LEONARD LESLIE, b Terre Haute, Ind, Aug 5, 14; m 40. POSTHARVEST PHYSIOLOGY OF VEGETABLES, POSTHARVEST TECHNOLOGY. *Educ:* Purdue Univ, BS, 37; Cornell Univ, MS, 39, PhD(veg crops), 41. *Prof Exp:* Res asst veg crops, Cornell Univ, 37-41; instr truck crops, 41-45, from asst prof to prof, 45-82, EMER PROF VEG CROPS, UNIV CALIF, DAVIS, 82- *Honors & Awards:* C W Hauck Award, Produce Parkaging Asn Am, 56; L H Vaughn Award, Am Soc Hort Sci, 58. *Mem:* Fel Am Soc Hort Sci; Int Soc Hort Sci; Am Inst Biol Sci; fel AAAS. *Res:* Postharvest physiology of vegetables in relation to deterioration during handling, transportation, storage and marketing; chilling injury, senescence and ripening; postharvest technology to control deterioration of vegetables. *Mailing Add:* Dept Veg Crops Mann Lab 16 Willowbark Rd Davis CA 95616

MORRIS, LUCIEN ELLIS, b Mattoon, Ill, Nov 30, 14; m 42; c 5. ANESTHESIOLOGY. *Educ:* Oberlin Col, AB, 36; Western Reserve Univ MD, 43; Am Bd Anesthesiol, dipl, 49. *Hon Degrees:* FFARCS, Royal Col Surgeons, 78; FFARACS, Royal Australian Col Surgeons, 89. *Prof Exp:* Intern, Grasslands Hosp, Valhalla, NY, 43; resident anesthesia, Wis Gen Hosp, Madison, 46-48; instr anesthesiol, Univ Wis, 48-49; from asst prof to assoc prof, Univ Iowa, 49-54; prof, Univ Wash, 54-60, clin prof, 61-68; prof anaesthesia, Fac Med, Univ Toronto, 68-70; prof anesthesia & chmn dept, 70-80, prof, 80-86, EMER PROF ANESTHESIOL, MED COL OHIO, 87- *Concurrent Pos:* Mem traveling med fac, WHO & Unitarian Serv Comt, Israel & Iran, 51; mem subcomt anesthesia, Nat Res Coun, 56-61; dir anesthesia res labs, Providence Hosp, Seattle, 60-68, dir med educ & res, 65-68; chief anaesthetist, St Michael's Hosp Unit, Toronto, 68-70; vis prof, Anaesthetics Unit, London Hosp Med Col, 80-81. *Mem:* Am Soc Anesthesiol; Am Soc Pharmacol & Exp Therapeut; Can Anaesthetists Soc; Fac Anaesthetists, Royal Col Surgeons; Asn Anaesthetists Gt Brit & Ireland; Anaesthesia Res Soc (UK); hon mem Australian Soc Anaesthetists; Sigma Xi; Anesthesia Hist Asn (secy-treas, 89-). *Res:* Cardiac conduction; placental transmission of drugs; anesthetic apparatus; fundamental neurophysiologic mechanism in anesthesia; carbon dioxide homeostasis; cardiac output; acid-base status with cardiopulmonary bypass and hypothermia; liver function with various anesthetic agents; medical education. *Mailing Add:* 15670 Point Monroe Dr NE Bainbridge Island WA 98110-1116

MORRIS, MANFORD D, b Kamiah, Idaho, Apr 18, 26; m 51; c 3. BIOCHEMICAL LIPID METABOLISM. *Educ:* Univ San Francisco, BS, 49, MS, 51; Univ Calif, PhD(biochem), 58. *Prof Exp:* Asst res biochemist, Sch Med, Univ Calif, 58-61; asst prof, 61-65, assoc prof, 65-72, PROF BIOCHEM, SCH MED SCI, UNIV ARK, LITTLE ROCK, 72-, PROF PEDIAT, 77- *Concurrent Pos:* Asst res biochemist, Clin Invest Ctr, US Naval Hosp, Oakland, 58-61. *Res:* Cholesterol metabolism; sterol methodology; primate lipid and lipoprotein metabolism; heritable diplipoproteinemia; lipoprotein structure. *Mailing Add:* Dept Pediat Sch Med Sci Univ Ark 4301 W Markham Little Rock AR 72205

MORRIS, MARILYN EMILY, b Winnipeg, Man. HEPATIC TRANSPORT & METABOLISM, INORGANIC SULFATE. *Educ:* Univ Man, BSc, 73; Univ Ottawa, MSc, 76; State Univ NY, PhD(pharmaceut), 84. *Prof Exp:* Lectr pharm, Dalhousie Univ, 76-77, asst prof pharm, 77-78; postdoctoral fel pharmaceut, Univ Toronto, 84-85; ASST PROF PHARMACEUT, STATE UNIV NY, BUFFALO, 85- *Mem:* Am Asn Pharmaceut Scientists; Am Pharmaceut Asn; Soc Exp Biol & Med; AAAS; Am Asn Cols Pharm. *Res:* Inorganic sulfate homeostasis; hepatic transport; sulfate conjugation. *Mailing Add:* Dept Pharmaceut State Univ NY Buffalo 527 Hochstetler Hall Amherst NY 14260

MORRIS, MARION CLYDE, b Akron, Ohio, Oct 14, 32; m 54, 78; c 3. POLYMER CHEMISTRY. *Educ:* Univ Akron, BS, 54, MS, 60, PhD(polymer chem), 63. *Prof Exp:* Sr res chemist, Goodyear Tire & Rubber Co, 62-71, sect head phys chem, Res Div, 71-78; SR RES SCIENTIST, KIMBERLY CLARK, 78- *Mem:* Am Chem Soc. *Res:* Rubber-like elasticity; physical properties of polymers; characterization of polymers by gel permeation chromatography and other physical methods; rheology and processing properties and product development. *Mailing Add:* 16 Silver Spur Lane Appleton WI 54915-2313

MORRIS, MARK ROOT, b Aberdeen, Wash, Sept 2, 47; m 76; c 2. RADIO ASTRONOMY. *Educ:* Univ Calif, Riverside, BA, 69; Univ Chicago, PhD(physics), 75. *Prof Exp:* Res fel radio astron, Owens Valley Radio Observ, Calif Inst Technol, 74-77; asst prof dept astron, Columbia Univ, 77-82; assoc prof, 83-84, PROF, DEPT ASTRON, UNIV CALIF, LOS ANGELES, 84- *Concurrent Pos:* Vis prof, Group d'Astrophys, Univ Sci et Med de Grenoble, 81-82,& 85. *Mem:* Am Astron Soc; Int Astron Union; Int Union Radio Sci. *Res:* Mass loss from red giant stars; the galactic center; interstellar molecular clouds. *Mailing Add:* Dept Astron Math-Sci Bldg Univ Calif Los Angeles 405 Hilgard Ave Los Angeles CA 90024

MORRIS, MARLENE COOK, b Washington, DC, Dec 20, 33; m 61; c 3. CRYSTALLOGRAPHY, PHYSICAL CHEMISTRY. *Educ:* Howard Univ, BS(chem), 55. *Prof Exp:* Res assoc, Off Ord Res, US Army, Howard Univ, 53-55; res assoc, Nat Bur Standards, 55-75, dir, res assoc, Int Ctr Diffractional Data, 75-86; STAFF CHEMIST, JCPDS INT CTR DIFFRATION DATA, 86- *Mem:* Am Crystallog Asn; Am Chem Soc; AAAS; Am Inst Physics; Sigma Xi. *Res:* Crystallography; x-ray powder diffraction. *Mailing Add:* CPDS Int Ctr Diffaction Data 1601 Park Lane Swarthmore PA 19081

MORRIS, MARY ROSALIND, b Ruthin, Wales, May 8, 20; US citizen. GENETICS & CYTOGENETICS. *Educ:* Ont Agr Col, BSA, 42; Cornell Univ, PhD(plant breeding), 47. *Prof Exp:* Asst agron, Univ Nebr, Lincoln, 47-51, from asst prof to prof cytogenetics, 51-90, EMER PROF CYTOGENETICS, UNIV NEBR, LINCOLN, 90- *Concurrent Pos:* Univ Nebr Johnson fel, Calif Inst Technol, 49-50; Guggenheim fel, Sweden & Eng, 56-57. *Mem:* Fel AAAS; fel Crop Sci Soc Am; Genetics Soc Can; fel Am Soc Agron. *Res:* Wheat cytogenetics; assignment of genes for important wheat characters to specific chromosomes by use of aneuploids and chromosome substitutions; chromosome studies on wheat varieties and germplasm. *Mailing Add:* Dept Agron Univ Nebr Lincoln NE 68583-0915

MORRIS, MELVIN L, b Cincinnati, Ohio, Mar 27, 29. INORGANIC CHEMISTRY. *Educ:* Ohio State Univ, BSc, 51, MSc, 55, PhD(chem), 58. *Prof Exp:* Fel, Northwestern Univ, 58-59 & Ohio State Univ, 59-60; asst prof, Tex Tech Col, 60-61; from asst prof to assoc prof, NDak State Univ, 63-68; educ sci adminr, NSF, 68-69; assoc prof, 69-74, PROF CHEM, NDAK STATE UNIV, 74- *Concurrent Pos:* Res chemist, Wright Patterson AFB, Ohio, 61. *Mem:* Am Chem Soc. *Res:* Mass spectrometry and synthesis of inorganic compounds; inorganic synthesis of beta-diketone complexes. *Mailing Add:* Dept of Chem NDak State Univ Fargo ND 58102

MORRIS, MELVIN LEWIS, b New York, NY, Nov 28, 14; m 43; c 3. DENTISTRY, PERIODONTICS. *Educ:* City Col New York, BS, 34; Columbia Univ, MA, 37, DDS, 41; Am Bd periodont, dipl, 51. *Prof Exp:* From instr to clin prof, 48-85, adj prof, 70-72, EMER CLIN PROF DENT, SCH DENT & ORAL SURG, COLUMBIA UNIV, 85- *Concurrent Pos:* Consult, Vet Admin Hosp, Castle Point, 53-56 & Franklin Delano Hosp, 53-59; NIH res grant, 66-68; Emer consult dent, Presby Hosp, 86- *Honors & Awards:* Hirschfeld Mem Award, 80; Townsend Harris Medal, City Col NY, 84. *Mem:* Am Dent Asn; Int Asn Dent Res; Am Acad Periodont; Sigma Xi. *Res:* Experimental wound healing of periodontal tissues. *Mailing Add:* Columbia Univ Sch of 630 W 168th St New York NY 10032

MORRIS, MELVIN SOLOMON, range science; deceased, see previous edition for last biography

MORRIS, MICHAEL D, b New York, NY, Mar 27, 39; m 61; c 4. ANALYTICAL CHEMISTRY. *Educ:* Reed Col, BA, 60; Harvard Univ, MA, 62, PhD(chem), 64. *Prof Exp:* Asst prof chem, Pa State Univ, 64-69; assoc prof, 69-82, PROF CHEM, UNIV MICH, ANN ARBOR, 82- *Concurrent Pos:* Res grants, USPHS, 76-79, 80-83 & 87-92, NSF, 78-79, 79-82, 82-86, Dept Educ, 89-92. *Mem:* Soc Appl Spectros; Am Chem Soc; AAAS; Optical Soc Am. *Res:* Applications of Raman spectroscopy and laser spectroscopy to analytical chemistry; gel and capillary electrophoresis. *Mailing Add:* Dept Chem Univ Mich Ann Arbor MI 48109-1055

MORRIS, N RONALD, b New York, NY, July 22, 33; m 57; c 2. CELL BIOLOGY. *Educ:* Yale Univ, BS, 55, MD, 59. *Prof Exp:* Asst prof pharmacol, Sch Med, Yale Univ, 63-67; asst prof, 67-68, assoc prof, 68-72, PROF PHARMACOL, RUTGERS MED SCH, COL MED & DENT NJ 72-; PROF PHARMACOL, ROBERT WOOD JOHNSON MED SCH, UNIT MED & DENT NJ, 72- *Concurrent Pos:* Sabbatical molecular biol, Med Res Ctr Lab, Cambridge, Eng. *Mem:* Am Soc Pharmacol & Exp Therapeut; Am Soc Cell Biol; Genetics Soc Am. *Res:* Biochemical genetics of mitosis; molecular genetics of tubulin. *Mailing Add:* Dept Pharmacol Robert Wood Johnson Univ NJ Piscataway NJ 08854

MORRIS, NANCY MITCHELL, b Griffin, Ga, Jan 9, 40; m 62; c 1. SPECTROSCOPY, TEXTILE CHEMISTRY. *Educ:* LaGrange Col, AB, 60; Auburn Univ, MS, 64. *Prof Exp:* Chemist textiles anal, Res Div, West Point-Pepperell Mfg Co, 64-65; chemist vesicular photog, Kalvar Corp, 66; chemist, 66-72, RES CHEMIST TEXTILES ANAL, SOUTHERN REGIONAL RES CTR, USDA, 72- *Mem:* Am Chem Soc; Am Asn Textile Chemists & Colorists; Soc Appl Spectros; Sigma Xi; Coblentz Soc. *Res:* Trace metals in foods and fibers; effect of chemical modification on structure of cellulose by infrared and Fourier transform infrared spectroscopy; determination of endotoxins in cotton dust; characterization and quantification of polycarboxylic acids on cellulose. *Mailing Add:* Southern Regional Res Ctr PO Box 19687 New Orleans LA 70179

MORRIS, OWEN G, b Shawnee, Okla, Feb 3, 27; m 48; c 2. AERODYNAMICS, AEROSPACE TECHNOLOGY. *Educ:* Univ Okla, BS, 47, MS, 48. *Prof Exp:* Instr mech, Univ Okla, 47-48; aeronaut res scientist, Nat Adv Comt Aeronaut, NASA, 48-49, aerospace technologist, 57-61, asst chief test div, 61-63, chief, Reliability & Qual Assurance Div, Apollo Spacecraft Prog Off, 63-65, chief, Lunar Module Proj Eng Div, 65-69, mgr lunar module, 69-72, mgr Apollo Spacecraft Prog, 72-73, mgr systs integration space shuttle, 73-80; PRES, EAGLE AEROSPACE, 86- *Concurrent Pos:* Mgt consult. *Mem:* Assoc fel Am Inst Aeronaut & Astronaut; Am Astronaut Soc. *Res:* Design, development and operation of the lunar module, the first manned lunar landing spacecraft; supersonic aerodynamics. *Mailing Add:* 14914 Timberland Ct Houston TX 77062-2922

MORRIS, PETER ALAN, b Oakland, Calif, Oct 6, 45; m 69; c 1. OPERATIONS RESEARCH, SYSTEMS ANALYSIS. *Educ:* Univ Calif, Berkeley, BS, 68; Stanford Univ, MS, 70, PhD(eng-econ syst), 71. *Prof Exp:* Opers res analyst, Off Systs Analysis, Dept Defense, 71-74, opers res mgr, Modeling & Analysis Off, Manpower & Reserve Affairs, 74; res scientist anal res group, Palo Alto Res Ctr, Xerox Corp, 74-79; PRIN, APPL DECISION ANALYST, INC, 79- *Concurrent Pos:* Assoc ed, Mgt Sci, 71-; consult assoc prof eng-econ syst, Stanford Univ, 75- *Mem:* Inst Mgt Sci; Opers Res Soc Am; Inst Elec & Electronics Engrs. *Res:* Decision analysis; systems modeling; probabilistic models; use of experts. *Mailing Add:* 213 Cape Fear Blvd Carolina Beach NC 28428

MORRIS, PETER CRAIG, b Kansas City, Mo, Sept 5, 37; m 60; c 3. MATHEMATICS. *Educ:* Southern Ill Univ, BA, 59; Univ Iowa, MS, 61; Okla State Univ, PhD(math), 67. *Prof Exp:* Asst prof math, State Col Iowa, 63-65; Belg-Am Educ Found fel, 67-68; asst prof math, Fla State Univ, 68-72; assoc prof, 72-79, PROF MATH, SHEPHERD COL, 79-, HEAD DEPT, 72- *Concurrent Pos:* Sabbatical leave, Southern Ill Univ, Carbondale, 80-81. *Mem:* Am Math Soc; Math Asn Am. *Mailing Add:* PO Box 681 Shepherdstown WV 25443

MORRIS, RALPH DENNIS, b Humboldt, Sask, Feb 13, 40; m 63; c 2. BEHAVIORAL ECOLOGY. *Educ:* Univ Sask, BSc, 63, PhD(ecol), 69; Univ Colo, Boulder, BEd, 63. *Prof Exp:* Postdoctoral fel biol, McGill Univ, 69-70; from asst prof to assoc prof, 70-85, PROF BIOL, BROCK UNIV, 85- *Concurrent Pos:* Ed, Colonial Waterbids, 89-91. *Mem:* Am Orinthol Union; Can Soc Orinthol; Wilson Ornithol Soc; Am Soc Mammalogists; Cooper Ornithol Soc; Colonial Waterbird Soc. *Res:* Mate choice criteria and parental care activities in colonial nesting seabirds; foraging and movement patterns in gulls and terns; overwintering ecology and behaviour in common terns; parental care behaviour of brown noddies. *Mailing Add:* Dept Biosci Brock Univ St Catharines ON L2S 3A1 Can

MORRIS, RALPH WILLIAM, b Cleveland Heights, Ohio, July 30, 28; c 5. PHARMACOLOGY, CHRONOBIOLOGY. *Educ:* Ohio Univ, BA, 50, MS, 53; Univ Iowa, PhD(pharmacol), 55. *Prof Exp:* Asst pharmacol, Univ Iowa, 52-53; instr, Col Med, 55-56, from asst prof to assoc prof, Col Pharm, 56-69, PROF PHARMACOL, COL PHARM, UNIV ILL, CHICAGO, 69- *Concurrent Pos:* Adj prof educ, Col Educ, Univ Ill, Chicago, 75-83. *Mem:* AAAS; Drug Info Asn; Am Soc Pharmacol & Exp Therapeut; Am Pharmaceut Asn; Int Soc Chronobiol. *Res:* Chronopharmacology; chronopathology; drug abuse; drug education; chronometrics. *Mailing Add:* Dept Pharmacodynamics MC 865 Univ Ill Col Pharm Chicago IL 60680

MORRIS, RANDAL EDWARD, b Shoemaker, Calif, Dec 10, 45; m 70; c 2. IMMMUNOLOGY, VIROLOGY. *Educ:* Aurora Col, BS, 69; Northern Ill Univ, MS, 71; Emory Univ, PhD(microbiol), 74. *Prof Exp:* Fel microbiol, Univ Mich, 74-76; ASST PROF MICROBIOL, UNIV CINCINNATI, 76- *Mem:* Am Soc Microbiol. *Res:* Demonstration and quantitation of cell associated antigens at the ultrastructural level. *Mailing Add:* Dept Microbiol Univ Cincinnati 231 Bethesda Ave Cincinnati OH 45267

MORRIS, RICHARD HERBERT, b Oakland, Calif, Nov 22, 28; m 56; c 2. ELECTROMAGNETICS. *Educ:* Univ Calif, Berkeley, AB, 50, PhD(nuclear physics), 57. *Prof Exp:* Asst physics, Univ Calif, Berkeley, 50-53, asst nuclear physics, Lawrence Radiation Lab, 53- 55; instr physics, Sacramento State Col, 56-57; from asst prof to assoc prof, 57-66, PROF PHYSICS, SAN DIEGO STATE COL, 66- *Concurrent Pos:* Consult, Naval Electronics Lab, 62-63. *Mem:* Am Asn Physics Teachers; Am Phys Soc; Sigma Xi; Optical Soc Am. *Res:* Theoretical physics; modern optics; teaching of physics. *Mailing Add:* 1775 Granite Hills Dr El Cajon CA 92019

MORRIS, ROBERT, b Akron, Ohio, Nov 21, 10; m 40. GERONTOLOGY. *Educ:* Univ Akron, BA, 31; Western Reserve Univ, MSc, 35. *Hon Degrees:* DSW, Columbia Univ, 59; DHL, Brandeis Univ, 84. *Prof Exp:* Kirstein prof, policy, Brandeis Univ, 59-79; dir geront, Levinson Policy Inst, Brandeis Univ, 70-78; consult, health policy, Univ Consortium, Brandeis, Mass Inst Technol, Boston Univ, 79-81; prof, health, policy, Inst Health Prof, Mass Gen Hosp, 80-84; lectr Pub Health Policy, Harvard Univ, Sch Pub Health, 70-85; CARDINAL MEDEIROS SR RES & LECTR, GERONT, UNIV MASS, BOSTON, 83- *Concurrent Pos:* Prin investr, US Pub Health Serv, res grant, 57-60, co-prin investr, US Vet Admin on Nursing Homes, 65-68; vis prof, health policy, Univ Calif, LA, 67; consult, Vet Admin spec med adv bd geriatric res ctrs, 74-78; prin investr, Nat Sci Found, 75-77; vis prof, Case Western Reserve Univ, 76 Univ Wis, 77, Univ Tex, Arlington, 80; mem, Harvard Working Group on Health Policy & Aging, 84-89. *Honors & Awards:* Donald Kent Award, Gerontol Soc Am, 87. *Mem:* Fel AAAS; Gerontol Soc Am (pres 66-67); Am Pub Health Asn. *Res:* Chronic illness and long term health care; organization of health services; national and international trends in health and social policy; aging in national societies and health and social organization adaptation. *Mailing Add:* 80 Park St No 24 Brookline MA 62146-6340

MORRIS, ROBERT, US citizen. VISUAL PSYCHOPHYSICS, IMAGE PROCESSING. *Educ:* Reed Col, BA, 65; Cornell Univ, MA, 67, PhD(math), 70. *Prof Exp:* From instr to asst prof math, State Univ NY, Albany, 69-75; assoc prof, Univ Okla, 75-78; assoc prof, 78-86, PROF MATH, UNIV MASS, BOSTON, 86- *Concurrent Pos:* Mem, Inst Advan Study, 73-75. *Mem:* Asn Comput Mach; Inst Elec & Electronics Engrs Computer Soc. *Res:* Application of image processing and human vision science to digital typography. *Mailing Add:* Dept Math & Computer Sci Univ Mass Harbor Campus Boston MA 02125

MORRIS, ROBERT ALAN, b Patchogue, NY, Mar 2, 58; m 84; c 1. ION CHEMISTRY, CHEMICAL KINETICS & DYNAMICS. *Educ:* Bates Col, BS, 80; Boston Col, PhD(phys chem), 87. *Prof Exp:* Geophysics scholar, Air Force Geophysics Lab, 87-89; sr scientist, Systs Integration Eng, Inc, 89-91; RES CHEMIST, AIR FORCE PHILLIPS LAB, GEOPHYSICS DIRECTORATE, 91- *Mem:* Am Chem Soc; Am Phys Soc; Am Geophys Union; Am Soc Mass Spectrometry; AAAS; Sigma Xi. *Res:* Radical-radical reactions; ion-molecule reaction kinetics; internal energy effects in ion-molecule reactions; plasma modification chemical physics; vibrational quenching of ions; tropospheric ion chemistry; chemical kinetics modeling. *Mailing Add:* Phillips Lab Geophysics Directorate-LID Hanscom AFB MA 01731-5000

MORRIS, ROBERT CARTER, b Richmond, Va, Oct 3, 43; m 63; c 1. EXPERIMENTAL SOLID STATE PHYSICS. *Educ:* Hampden-Sydney Col, BS, 66; Univ Va, PhD(physics), 70. *Prof Exp:* Res assoc physics, Univ Va, 70-71, asst prof, 71-73; from asst prof to assoc prof physics, Fla State Univ, 77-85; PHYS SCIENTIST, CENT INTEL AGENCY, 85- *Mem:* Am Phys Soc. *Res:* Experimental studies of the normal and superconducting state properties of layer-structure, transition-metal dichalcogenide compounds and tungsten bronze compounds, both pure and doped with impurity atoms. *Mailing Add:* C/O ORD Washington DC 20505

MORRIS, ROBERT CLARENCE, geology; deceased, see previous edition for last biography

MORRIS, ROBERT CRAIG, b Hemet, Calif, Jan 7, 44; m 79. SINGLE CRYSTAL GROWTH, SOLID STATE LASER MATERIALS. *Educ:* Rensselaer Polytech Inst, BS, 66, PhD(mat sci), 71. *Prof Exp:* staff scientist, Corp Res & Develop, 71-82, mgr, electronic & optical physics group, 82-84, DIR ELECTRONIC MATS & DEVICES LAB, ALLIED CORP, 84- *Mem:* Am Asn Crystal Growth; AAAS; Sigma Xi. *Res:* New single crystal solid state laser materials. *Mailing Add:* Mat Lab Allied Corp PO Box 1021R Morristown NJ 07960

MORRIS, ROBERT GEMMILL, b Des Moines, Iowa, July 20, 29; m 55; c 3. SOLID STATE PHYSICS. *Educ:* Iowa State Univ, BS, 51, PhD(physics), 57; Calif Inst Technol, MS, 54. *Prof Exp:* Am-Swiss Found Sci exchange fel, Swiss Fed Inst Technol, 57-58; from asst prof to prof physics & head dept, SDak Sch Mines & Technol, 58-68; physicist, Physics Prog, Off Naval Res, Va, 68-72, actg dir, Electronics Prog, 72-73, dir, Electronics Prog, 73-74; dep dir, Off Technol Policy & Space Affairs, Bur Oceans & Int Environ & Sci Affairs, US Dept State, 74-76, actg dir, 76-77, dir, Off Soviet & East Europ Sci & Technol Affairs, 77-78, counr sci & technol affairs, US Mission Orgn Econ Coop & Develop, US Embasssy Paris, 78-82, US Embassy Bonn, 82-85, dep asst Secy of State Sci & Technol Affairs, US Dept State, 85-87; counr, Buenos Aires, 87-90, COUNR SCI & TECHNOL AFFAIRS, US EMBASSY, MADRID, 90- *Concurrent Pos:* Vis prof, Swiss Fed Inst Technol, 63-64. *Mem:* Fel Am Phys Soc; Inst Elec & Electronic Engrs. *Res:* Electrical, thermal and magnetic properties of semiconductor elements and compounds; technology transfer. *Mailing Add:* Sci & Technol US Embassy APO New York NY 09285

MORRIS, ROBERT WHARTON, b Liberal, Mo, Aug 27, 20; m 45; c 2. ICHTHYOLOGY. *Educ:* Wichita State Univ, AB, 42; Ore State Col, MS, 48; Stanford Univ, PhD(biol), 54. *Prof Exp:* Biologist agr exp sta, Ore State Col, 48-49 & US Fish & Wildlife Serv, 51-55; from instr to assoc prof, 55-68, PROF BIOL, UNIV ORE, 68- *Concurrent Pos:* Guggenheim fel, 62-63. *Res:* Biology of fishes and lower vertebrates. *Mailing Add:* PO Box 348 Moraga CA 94556

MORRIS, ROBERT WILLIAM, b Staten Island, NY, Sept 28, 41; m 68; c 2. GEOLOGY, INVERTEBRATE PALEONTOLOGY. *Educ:* Duke Univ, AB, 63; Columbia Univ, MA, 65, PhD(geol), 69. *Prof Exp:* Teaching asst geol, Columbia Univ, 63-66; asst instr, Rutgers Univ, 66-67; from asst prof to assoc prof, 68-81, PROF GEOL, WITTENBERG UNIV, 81- *Mem:* Sigma Xi; Soc Econ Paleontologists & Mineralogists; Int Paleont Union. *Res:* Paleoecology; micropaleontology. *Mailing Add:* Dept Geol Wittenberg Univ Springfield OH 45501

MORRIS, ROSEMARY SHULL, b Los Angeles, Calif, Aug 11, 29. SCIENCE ADMINISTRATION, NUTRITION. *Educ:* Univ Calif, Berkeley, 50; Univ Southern Calif, BS, 53, MS, 56, PhD(biochem, nutrit), 59. *Prof Exp:* Res assoc biochem, Univ Southern Calif, 59; jr res biochemist, Univ Calif, Los Angeles, 59-61; res biochemist, Eastern Utilization Res & Develop Div, Agr Res Serv, USDA, Washington, DC, 61-66; res chemist, Div Nutrit, US Food & Drug Admin, 66-67 & Human Nutrit Res Div, Agr Res Serv, USDA, Md, 67-72; health scientist adminr, Nat Heart & Lung Inst, NIH, 72-75, asst chief res, Referral Br, Div Res Grants, 75-77, exec secy, Cardiovasc Renal Study Sect, 77-91, ASST CHIEF, REFERRAL SECT, DIV RES GRANTS, NIH, 91- *Mem:* Am Inst Nutrit. *Res:* Lipid and cholesterol metabolism; vitamin E; essential fatty acids. *Mailing Add:* Div Res Grants Nat Inst Health Bethesda MD 20892

MORRIS, ROY OWEN, b Kingston-on-Thames, Eng, May 24, 34; m 63; c 2. BIOCHEMISTRY. *Educ:* Univ London, BSc, 55, PhD(chem), 59. *Prof Exp:* Asst lectr chem & biochem, St Thomas Hosp Med Sch, London, Eng, 59-61; res assoc biochem, Sci Res Inst, 61-64, asst prof agr chem, 64-70, ASSOC PROF AGR CHEM, ORE STATE UNIV, 70-, ASSOC PROF CHEM, 74- *Mem:* AAAS; Am Soc Plant Physiol. *Res:* Biochemistry of plant development; protein and nucleic acid biosynthesis. *Mailing Add:* 117 Schwitzer Hall Univ Mo Columbia MO 65211

MORRIS, SAMUEL CARY, III, b Summit, NJ, Dec 16, 42; m 66; c 3. HEALTH & ENVIRONMENTAL RISK ANALYSIS. *Educ:* Va Mil Inst, BS, 65; Rutgers Univ, MS, 67; Univ Pittsburgh, ScD, 73. *Prof Exp:* Environ health engr serv div, US Army Med Lab, 66-67, chief div, 67-68; asst prof environ health sci, Ill State Univ, 71-72; res assoc, Grad Sch Pub Health, Univ Pittsburgh, 72-73; from asst to assoc scientist, 73-77, SCIENTIST BIOMED & ENVIRON ASSESSMENT GROUP, BROOKHAVEN NAT LAB, 77-, DEP DIV HEAD, ANAL SCI DIV, 90- *Concurrent Pos:* Asst prof dept appl math sci, State Univ NY, Stony Brook, 76-78; adj asst prof dept eng & pub policy, Carnegie Mellon Univ, 76-79, adj assoc prof, 80-; consult, Power Authority State NY, 77-79, Columbus Div, Battelle Mem Inst, 76-77, US Army Construct Eng Res Lab, 73-75 & Occup & Environ Analysts, Inc, 79-88; mem adv group health effects reactor safety study, US Nuclear Regulatory Comn, 75; sanit engr mobilization design, US Army, 69-74; chmn task force energy, Am Pub Health Asn, 76-77; mem Air Resources Task Group, Environ Div, Am Soc Civil Engrs, 76-79, Environ Effects Comt, Energy Div, 85-; mem Comt Epidemiol Invest Air Pollutants, Comn Life Sci, Nat Res Coun, 84-85; mem Gov Coun, Soc Risk Anal, 84-87; mem, Risk Panel, Technol Forecasting & Assessment Comn Pub Affairs Coun, Am Asn Eng Soc, 86-88. *Mem:* Soc Risk Anal; Am Soc Civil Engrs; Air & Waste Mgt; Inst Mgt Sci. *Res:* Energy systems analysis and health and environmental risks of energy systems. *Mailing Add:* Dept Appl Sci Brookhaven Nat Lab Upton NY 11973

MORRIS, SIDNEY MACHEN, JR, b Tyler, Tex, Mar 31, 46; m 88; c 2. MOLECULAR ENDOCRINOLOGY. *Educ:* Univ Tex, Austin, BS, 68; Univ Calif, Berkeley, PhD(biochem), 75. *Prof Exp:* NIH fel physiol chem, Univ Wis Med Ctr, 75-78; res assoc, Dept Pharmacol, Case Western Reserve Univ, 78-92; asst prof, 83-89, ASSOC PROF DEPT MOLECULAR GENETIC & BIOCHEM, UNIV PITTSBURGH, 89- *Concurrent Pos:* Mem, ad hoc, NIH Study Sect, 87-90, 91- *Mem:* AAAS; Am Soc Biochem & Molecular Biol; Am Soc Microbiol; Endocrine Soc. *Res:* Eucaryotic gene regulation and development; urea cycle enzymes; hormone action; liver-specific gene expression. *Mailing Add:* Dept Molecular & Genetic Biochem Univ Pittsburgh Pittsburgh PA 15261

MORRIS, STANLEY P, b Montreal, Que, Nov 23, 37; m 63; c 4. THEORETICAL SOLID-STATE PHYSICS. *Educ:* McGill Univ, BSc, 58, PhD(physics), 64. *Prof Exp:* Lectr physics, Loyola Col, Can, 63-64; asst prof, Sir George Williams Univ, 64-69; chmn dept, 74-79, ASSOC PROF PHYSICS, CONCORDIA UNIV, 69- *Mem:* Am Phys Soc; Am Asn Physics Teachers; Can Asn Physicists. *Res:* Electron interactions in the presence of a uniform magnetic field; Bloch electrons in a magnetic field. *Mailing Add:* Dept Physics Concordia Univ SGW Campus 1455 de Maisonneuve Montreal PQ H3G 1M8 Can

MORRIS, STEPHEN JON, structural biophysics, neurobiology, for more information see previous edition

MORRIS, THOMAS JACK, b Montreal, Que, Apr 28, 47; m 69; c 3. PLANT VIROLOGY & PATHOLOGY, INSECT VIROLOGY. *Educ:* MacDonald Col McGill Univ, BSc, 68; McGill Univ, MSc, 70; Univ of Nebr-Lincoln, PhD(plant virol), 73. *Prof Exp:* Asst prof biol, Univ NB, 74-76; prof plant path, Univ Calif, Berkeley, 76-90; DIR, UNIV NEBR, LINCOLN, 90- *Concurrent Pos:* Nat Res Coun Can fels, 70-74. *Mem:* Am Phytopath Soc; Soc Gen Microbiol; Am Soc Virol. *Res:* Role of viruses and viroids in plant disease; comparative virology of plant and invertebrate viruses; applied emphasis on the detection, diagnosis and control of virus and virus-like diseases of plants; molecular genetics and comparative virology of small RNA viruses of plants and invertebrates; defective interfering RNAs. *Mailing Add:* Sch Biol Scis Univ Nebr Lincoln NE 68588-0118

MORRIS, THOMAS WENDELL, b Emory, Ga, Jan 31, 30; m 54; c 2. PHYSICS. *Educ:* Duke Univ, BS, 51; Yale Univ, MS, 53, PhD(physics), 55. *Prof Exp:* PHYSICIST, BROOKHAVEN NAT LAB, 55- *Concurrent Pos:* Vis physicist, Saclay Nuclear Res Ctr, France, 59-60. *Mem:* Am Phys Soc. *Res:* Particle physics; computer applications; nuclear instrumentation. *Mailing Add:* Dept Radiol/Physiol Univ Rochester Med Ctr Rochester NY 14627

MORRIS, THOMAS WILDE, b Bay Shore, NY, Jun 25, 43; m 66; c 2. CARDIOVASCULAR PHYSIOLOGY, DIAGNOSTIC IMAGERY. *Educ:* Univ Rochester, BS, 65; Univ Mich, MS, 67 & PhD(bio eng), 72. *Prof Exp:* Captain bioinstrumentation, US Air Force Sch Aerospace Med, 67-71; fel, bioeng & physiol, Univ Mich, 72-73; assoc physiol Univ Penn, 73-75, asst prof physiol & radiol, Univ Rochester, 75-81, ASSOC PROF PHYSIOL & RADIOL, 81-, DEPT RES DIR, 88- *Concurrent Pos:* Prin investr, 15 Nat Inst Health & Indust Grants, 76-; reviewer radiol, 76-, Am J Roentgenog, 77-; ed, Invest Radiol, 84- *Honors & Awards:* Picker Found Scholar Award. *Mem:* Asn Univ Radiologists; AAAS; Sigma Xi. *Res:* Application and testing of diagnostic agents for imaging disease; hemodynamics; blood vessel structure; development of techniques for clinical measurements of blood flow. *Mailing Add:* Radiology Box 648 Univ Rochester Rochester NY 14642

MORRIS, WILLIAM GUY, b Great Falls, Mont, Jan 4, 40. PHYSICAL METALLURGY. *Educ:* Univ Calif, BS, 61; Mass Inst Technol, SM, 63, ScD(metall), 65. *Prof Exp:* Scientist, Adv Metals Res Corp, 63-65; res scientist, 65-78, proj mgr, 78-80, UNIT MGR, GEN ELEC CO, 80- *Mem:* Am Inst Mining, Metall & Petrol Engrs; Am Soc Metals; Microbeam Anal Soc; Sigma Xi. *Res:* Electron beam microprobe; scanning electron microscopy; electrical properties of polycrystalline ceramics. *Mailing Add:* 668 Riverview Rd Rexford NY 12148

MORRIS, WILLIAM JOSEPH, b Baltimore, Md, Oct 14, 23; m 45; c 2. GEOLOGY. *Educ:* Syracuse Univ, BA, 48; Princeton Univ, MA & PhD(geol), 51. *Prof Exp:* Prof geol, Agr & Mech Col, Tex, 51-55; chmn dept, 71-73, PROF GEOL, OCCIDENTAL COL, 55- *Concurrent Pos:* Res assoc, Mus Natural Hist, Los Angeles. *Honors & Awards:* Arnold Guyot Mem Award, Nat Geog Soc, 68. *Mem:* Fel Geol Soc Am; Soc Vert Paleont; AAAS; Soc Study Evolution. *Res:* Vertebrate paleontology; sedimentary petrology; invertebrate paleontology. *Mailing Add:* 11167 Tujunga Canyon Blvd Tujunga CA 91042

MORRIS, WILLIAM LEWIS, b Hamilton, Ohio, Aug 19, 31; m 56; c 7. COMPUTATIONAL LINEAR ALGEBRA, DIFFERENTIAL EQUATIONS. *Educ:* Univ Cincinnati, AB, 58, AM, 60; Univ Tenn, PhD(appl math), 67. *Prof Exp:* Sr engr res, Gen Dynamics Corp, 60-62; from instr to asst prof, Univ Tenn, 62-67; res staff, Oak Ridge Nat Lab, 67-68; prof, Univ Houston, 68-77; DIR, NUMERICAL ANAL GROUP, 77- *Concurrent Pos:* Lectr, Oak Ridge Grad Prog, 67-68; consult, MRI, Inc, 70-75. *Res:* Matrix theory and its applications including computational linear algebra, numerical solutions of differential equations and approximation theory. *Mailing Add:* Rte 4 Box 963 Center TX 75935

MORRISETT, JOEL DAVID, b Winston-Salem, NC, May 2, 42; m 67; c 7. BIOPHYSICS, BIOCHEMISTRY. *Educ:* Davidson Col, BS, 64; Univ NC, Chapel Hill, PhD(org chem), 68. *Prof Exp:* NIH fel biophys, Stanford Univ, 70-71; from asst prof to assoc prof, 71-82, PROF EXP MED, BAYLOR COL MED, 82- *Concurrent Pos:* Estab investr, Am Heart Asn, 74-79; actg dir, Mag Res Ctr, Baylor Col Med. *Mem:* Am Heart Asn; Biophys Soc; AAAS; Fedn Am Soc Exp Biol; NY Acad Sci; Am Chem Soc; Soc Mag Res Med. *Res:* Correlation and determination of structure and function of proteins, particularly enzymes and lipoproteins using biophysical and chemical methods. *Mailing Add:* Methodist Hosp A601 Baylor Col of Med Houston TX 77030

MORRISH, ALLAN HENRY, b Winnipeg, Man, Apr 18, 24; m 52, 89; c 2. PHYSICS. *Educ:* Univ Manitoba, BSc, 43; Univ Toronto, MA, 46; Univ Chicago, PhD(physics), 49. *Prof Exp:* Asst physics, Univ Chicago, 48-49; lectr, Univ BC, 49-50, res assoc 51-52; physicist, McGill Univ, 52-53; from res assoc to prof elec eng, Univ Minn, Minneapolis, 53-64; head dept, 66-87, PROF PHYSICS, UNIV MAN, 64-, DISTINGUISHED PROF, 84- *Concurrent Pos:* Nat Res Coun Can fel, 50-51; Guggenheim fel, 57-58. *Honors & Awards:* Gold Medal Achievement in Physics, Can Asn Physicists, 77. *Mem:* Fel Am Phys Soc; Royal Soc Can; fel Brit Inst Physics; Can Asn Physicists(pres, 74-75). *Res:* Particle accelerators; elementary and small magnetic particles; nuclear emulsions; ferromagnetism; magnetic resonance; low temperatures; Mossbauer effect in magnetic materials. *Mailing Add:* Dept Physics Univ Man Winnipeg MB R3T 2N2 Can

MORRIS HOOKE, ANNE, b Sydney, Australia, July 15, 39. MICROBIOLOGY. *Educ:* George Mason Univ, BS, 72; Georgetown Univ, PhD(microbiol), 79. *Prof Exp:* From res assoc to asst prof pediat, Georgetown Univ, 78-87; ASSOC PROF MICROBIOL, MIAMI UNIV, 87- *Concurrent Pos:* Vis prof, Fla Int Univ, 87. *Mem:* Am Soc Microbiol; AAAS; Asn Women Sci. *Res:* Temperature-sensitive bacterial vaccines against influenzal meningitis, typhoid fever, Pseudomonas aeruginosa, avian salmonellosis and air sacculitis. *Mailing Add:* 325 W High St Oxford OH 45056

MORRISON, ADRIAN RUSSEL, b Philadelphia, Pa, Nov 5, 35; m 58; c 5. NEUROANATOMY, NEUROPHYSIOLOGY. *Educ:* Cornell Univ, DVM, 60, MS, 62; Univ Pa, PhD(anat), 64. *Prof Exp:* NIH spec fel neurophysiol, Univ Pisa, 64-65; asst prof, 66-70, ASSOC PROF ANAT, SCH VET MED, UNIV PA, 70- *Mem:* Am Vet Med Asn; Am Asn Vet Anat; Am Asn Anat; Asn Psychophysiol Study Sleep. *Res:* Neuroanatomical and neurophysiological bases of mammalian behavior. *Mailing Add:* Dept Anat Univ Pa Philadelphia PA 19104

MORRISON, ASHTON BYROM, b Belfast, Ireland, Oct 13, 22; m 50; c 1. PATHOLOGY. *Educ:* Queens Univ Belfast, MB, 46, PhD, 50; Duke Univ, MD, 46. *Prof Exp:* Asst lectr biochem, Queens Univ Belfast, 47-50 & anat, 50-51; mem sci staff exp med, Univ Cambridge, 52-55; assoc path, Duke Univ, 55-58; asst prof, Sch Med, Univ Pa, 58-61; assoc prof, Sch Med, Univ Rochester, 61-65; dean, Eastern Va Med Sch, Norfolk, 80-83; prof path & chmn dept, 65-80, DEAN, ASSOC-IN-CHARGE, RUTGERS MED SCH, CAMDEN, NJ, 83- *Concurrent Pos:* Markle scholar, 56-61. *Mem:* Am Asn Path; Am Fedn Clin Res; Am Physiol Soc; Soc Exp Biol & Med. *Res:* Experimental chronic renal insufficiency; experimental nephropathies; islet cell tumor of pancreas (Verner-Momson Syndrome). *Mailing Add:* UMDNJ Robert Wood Johnson Med Sch 401 Haddon Ave Camden NJ 08103

MORRISON, CHARLES FREEMAN, JR, b Yakima, Wash, Sept 24, 29; m 52; c 4. THIN FILM PROCESSES, VACUUM TECHNOLOGY. *Educ:* Univ Puget Sound, BS, 53; Mass Inst Technol, PhD(anal chem), 57. *Prof Exp:* From instr to asst prof anal chem, Wash State Univ, 57-62; res scientist, Granville-Phillips Co, Colo, 62-70; sr scientist, Univ Instruments Corp, 70-71; eng & opers mgr, Valleylab, Inc, 71-75, chief res, 75-76; gen mgr, Vac-Tec Systs, 76-80, tech vpres, 80-84; sr scientist, Granville-Phillips Co, Colo, 84-89; CHIEF SCIENTIST, TURBULENCE PREDICTION SYSTS, COLO, 89- *Concurrent Pos:* NSF fel, 61-63. *Mem:* Am Chem Soc; Am Vacuum Soc. *Res:* Magnetically trapped plasmas are applied to produce thin film coating; specific processes are developed and fundamentals are re-explored for more effective methods of sputter coating and cleaning; analytical instruments; vacuum instrumentation; low pressure measurement methods; calibration techniques; biomedical instrumentation; electrosurgical mechanisms and methods; detection instrumentation for microbursts and clear air turbulence; atmospheric dynamics. *Mailing Add:* 4790 Sioux Dr Boulder CO 80303

MORRISON, CLARENCE C, b June 19, 32; m; c 2. MATHEMATICS. *Educ:* Davidson Col, BS, 54; Univ NC, Chapel Hill, MA, 56, PhD, 64. *Prof Exp:* Asst prof, Univ Va, 64-67; assoc prof, Univ Ga, 67-70; PROF ECON, IND UNIV, BLOOMINGTON, 70- *Concurrent Pos:* Vpres, Atlantic Econ Soc, 79-80. *Mem:* Am Econ Soc. *Res:* General equilibrium with monopoly firms and public goods. *Mailing Add:* Econ Dept Ind Univ Ballantine Hall Bloomington IN 47405

MORRISON, DAVID CAMPBELL, b Stoneham, Mass, Sept 1, 41; m 66, 80; c 2. IMMUNOLOGY. *Educ:* Univ Mass, BS, 63; Yale Univ, MS, 66, PhD(molecular biol), 69. *Prof Exp:* Fel, Lab Biochem Pharm, Nat Inst Allergy & Infectious Dis, NIH, 69-71; fel, Dept Exp Path, Scipps Clin & Res Found, 71-74, asst, 74-75, assoc, Dept Immunopath, 75-78, assoc mem, 78-80; assoc prof, Emory Univ, Atlanta, 80-81, prof, dept microbiol & immunol, 81-85, William PTimmie Chair, 80-85; PROF & CHMN DEPT MICROBIOL, KANSAS UNIV MED CTR, 85- *Concurrent Pos:* Res career develop award, Nat Inst Allergy & Infectious Dis, NIH, 75-80. *Mem:* Am Asn Immunol; Am Soc Exp Path; Am Soc Microbiol; NY Acad Sci; Reticuloendomelitis Soc; Int Endotoxin Soc. *Res:* Interaction of bacterial lipopolysaccharides with cellular and humoral mediation systems. *Mailing Add:* Dept Microbiol Kansas Univ Med Ctr 39 St and Rainbow Blvd Kansas City KS 66103

MORRISON, DAVID DOUGLAS, astronomy, planetary science, for more information see previous edition

MORRISON, DAVID LEE, b Butler, Pa, Jan 25, 33; m 54; c 2. NUCLEAR CHEMISTRY, PHYSICAL CHEMISTRY. *Educ:* Grove City Col, BS, 54; Carnegie Inst Technol, MS, 60, PhD(chem), 61. *Prof Exp:* Chemist, Callery Chem Co, 54; sr chemist, Battelle Mem Inst, 61-65, from assoc chief to chief chem physics div, 65-70, mgr environ systs & processes sect, 70-74, mgr energy & environ prog off, 74-75, dir prog develop & mgt, 75-77; exec vpres, 77-79, PRES, ITT RES INST, CHICAGO, 80- *Concurrent Pos:* Mem, Nat Mat Adv Bd, 82- *Mem:* AAAS; Sigma Xi; Am Chem Soc; Am Nuclear Soc. *Res:* Energy and environmental research; technology and environmental impact assessment; research and development planning; nuclear reactor safety analysis; radiochemistry; research management. *Mailing Add:* 8400 Martingale Dr McLean VA 22102

MORRISON, DONALD ALLEN, b Mt Forest, Ont, July 20, 36; US citizen; m 69; c 2. PETROLOGY, GEOLOGY. *Educ:* State Univ NY Buffalo, BS, 62; Univ Alaska, MS, 64; Univ Idaho, PhD(geol), 68. *Prof Exp:* PLANETARY SCIENTIST, JOHNSON SPACE CTR, NASA, 68- *Mem:* AAAS; Am Geophys Union; Geol Soc Am. *Res:* Petrology; petrology of lunar rocks; meteorite and micrometeorite studies as related to lunar surface processes; Precambrian geology. *Mailing Add:* Mail Code SN-4 NASA Johnson Space Ctr Houston TX 77058

MORRISON, DONALD FRANKLIN, b Stoneham, Mass, Feb 10, 31; m 67; c 2. MATHEMATICAL STATISTICS. *Educ:* Boston Univ, BS, 53, AM, 54; Univ NC, MS, 57; Va Polytech Inst, PhD, 60. *Hon Degrees:* MA, Univ Pa, 71. *Prof Exp:* Asst math, Boston Univ, 53-54 & Univ NC, 54-56; res math statistician, Biomet Br, NIMH, 56-63; chmn dept, 77-85, PROF STATIST, WHARTON SCH, UNIV PA, 63- *Concurrent Pos:* Instr, Found Advan Educ in Sci, 60-63; mem staff, Lincoln Lab, Mass Inst Technol, 56, consult, 56-57; div comput sci, NIH, 63-65; mem tech staff, Bell Tel Labs, NJ, 67; ed, Am Statistician, 72-75; assoc ed, Biometrics, 72-75. *Mem:* fel Am Statist Asn; fel Inst Math Statist; Biomet Soc; Psychomet Soc; fel Royal Statist Soc; Sigma Xi; Int Statist Inst. *Res:* Statistical theory and methodology; multivariate analysis. *Mailing Add:* Dept of Statist Wharton Sch Univ Pa Philadelphia PA 19104-6302

MORRISON, DONALD ROSS, b Tacoma, Wash, May 3, 22; m 43; c 3. MATHEMATICS, COMPUTER SCIENCE. *Educ:* Northern Ill State Teachers Col, BE, 42; Univ Wis, PhM, 46, PhD(math), 50. *Prof Exp:* Instr math, Univ Wis, 50; asst prof, Tulane Univ, 50-55; from mem staff to mgr, Sandia Corp, 55-71; prof math & comput sci, Univ NMex, 71-89; RETIRED. *Concurrent Pos:* Consult, Los Alamos Sci Lab, Univ Calif, 71-74, 88-, Sandia Corp, 77-79. *Mem:* Am Math Soc; Math Asn Am; Asn Comput Mach. *Res:* Pattern recognition; information retrieval; graph theory; abstract algebra; cryptology. *Mailing Add:* 712 Laguayra Dr NE Albuquerque NM 87108

MORRISON, DOUGLAS WILDES, b Schenectady, NY, Nov 8, 47; m 76; c 1. ZOOLOGY. *Educ:* Univ Rochester, AB, 69; Cornell Univ, PhD(behav & ecol), 75. *Prof Exp:* Asst prof, 75-80, ASSOC PROF ZOOL, RUTGERS UNIV, NEWARK, 80- *Mem:* Animal Behav Soc; Ecol Soc Am; Asn Trop Biol; Am Soc Naturalists. *Res:* Radio-tracking studies of social, foraging and roosting behavior of neotropical fruit bats and New Jersey black birds. *Mailing Add:* Dept Zool & Physiol Rutgers Univ Newark NJ 07102

MORRISON, ESTON ODELL, b Sabinal, Tex, Sept 18, 32; m 58; c 3. ENTOMOLOGY, PARASITOLOGY. *Educ:* Tex Col Arts & Indust, BS, 57; Tex A&M Univ, MS, 60, PhD(entom), 63. *Prof Exp:* Instr biol, Tex Col Arts & Indust, 60-61; asst prof, Lamar State Col, 63-66; assoc prof, 66-68, PROF BIOL, TARLETON STATE COL, 68- *Concurrent Pos:* Lamar Res Ctr grant, 65-66. *Mem:* Am Soc Parasitol. *Res:* Lung flukes of salientia; helminthology. *Mailing Add:* Dept Biol Sci Tarleton State Univ Tarleton Station Stephenville TX 76402

MORRISON, FRANK ALBERT, JR, mechanical engineering; deceased, see previous edition for last biography

MORRISON, GEORGE HAROLD, b New York, NY, Aug 24, 21; m 52; c 3. ANALYTICAL CHEMISTRY. *Educ:* Brooklyn Col, BA, 42; Princeton Univ, MA & PhD, 48. *Prof Exp:* Instr chem, Rutgers Univ, 48-50; head inorg & analytical chem, Gen Tel & Electronics Labs, 51-61; PROF CHEM, CORNELL UNIV, 61- *Concurrent Pos:* Res chemist, US AEC, 49-51; mem chem adv panel, NSF, 62-65; chmn comt anal chem, Nat Acad Sci, Nat Res Coun, 66-75; NSF sr fel, Univ Calif, San Diego, 67-68; Guggenheim fel, Univ Paris, Orsay, 74-75; NIH sr fel, Harvard Med Sch, Boston, 82-83; ed, Anal Chem, 80-91. *Honors & Awards:* Anal Chem Award, Am Chem Soc, 71; Medal Soc Appl Spectros, 75; Benedetti-Pichler Award, Am Microchem Soc, 77. *Mem:* Fel AAAS; Am Chem Soc; Soc Appl Spectros. *Res:* Ion microprobe; microscopy; mass spectroscopy; radiochemistry; atomic spectroscopy; trace and microanalysis. *Mailing Add:* Dept of Chem Cornell Univ Ithaca NY 14853

MORRISON, GLENN C, b New Haven, Conn, Mar 24, 33; m 55; c 4. ORGANIC CHEMISTRY. *Educ:* Brown Univ, ScB, 54; Univ Rochester, PhD(org chem), 58. *Prof Exp:* Res chemist org chem, Am Cyanamid Co, 57-60; sr scientist, 60-70, SR RES ASSOC ORG CHEM, WARNER-LAMBERT/PARKE-DAVIS PHARMACEUT RES DIV, 70- *Mem:* Am Chem Soc. *Res:* Synthetic organic medicinals. *Mailing Add:* 1445 Arlington Ave Ann Arbor MI 48106

MORRISON, HARRY, b New York, NY, Apr 25, 37; m 58; c 3. ORGANIC CHEMISTRY, PHOTOBIOLOGY. *Educ:* Brandeis Univ, BS, 57; Harvard Univ, PhD(org chem), 61. *Prof Exp:* NSF-NATO fel, Swiss Fed Inst Technol, 61-62; res fel org chem, Univ Wis, 62-63; from asst prof to assoc prof, 63-76, PROF, PURDUE UNIV, 76-, DEPT HEAD CHEM, 87- *Concurrent Pos:* Mem bd fels, Brandeis Univ, 65-; vis scientist, Weizmann Inst, Rehovot, Israel, 72; consult, Sun Chem Corp, 72-77, Eli Lilly Co, 77- 79 & Great Lakes Chem Co, 77-87; vis prof, Oxford Univ, 79 & Univ Calif, Berkeley, 87. *Mem:* Interam Photochem Soc (pres, 86-90); Sigma Xi; Am Soc Photobiol; Am Chem Soc. *Res:* Organic photochemistry; organic reaction mechanisms; bio-organic photochemistry. *Mailing Add:* Dept Chem 1393 Brwn Purdue Univ West Lafayette IN 47907

MORRISON, HARRY LEE, b Arlington Co, Va, Oct 7, 32; m 62; c 1. QUANTUM LIQUID THEORY, STATISTICAL PHYSICS. *Educ:* Cath Univ Am, AB, 55, PhD(chem), 60. *Prof Exp:* Chemist, NIH, 55-56; physicist, Nat Bur Standards, 60-61; physicist, asst prof & first lieutenant, US Air Force Acad, 61-64; physicist, Lawrence Livermore Nat Lab, 64-72; assoc prof physics, 72-77, PROF PHYSICS, UNIV CALIF, BERKELEY, 77- *Concurrent Pos:* Theoret physicist, Denver Res Inst, 62-64; assoc dir, Lawrence Hall Sci, 69-75; vis prof, Howard Univ, 72, Univ Colo, 73-74, Mass Inst Technol, 75-76; asst dean, Col Lett & Sci, Univ Calif, Berkeley, 85- *Mem:* Fel Am Phys Soc; Sigma Xi; Am Math Soc; Int Asn Math Physics. *Res:* Quantum liquid theory and quantum statistical mechanics; the phenomenon of superfluidity and also two dimensional phase transition theory. *Mailing Add:* Dept Physics Univ Calif Berkeley CA 94720

MORRISON, HUGH MACGREGOR, b Liverpool, Eng, Aug 28, 36; m 63; c 3. PHYSICS. *Educ:* Univ Edinburgh, BSc, 59, PhD(physics), 65. *Prof Exp:* Asst lectr physics, Univ Edinburgh, 62-65; asst prof, 65-70, ASSOC PROF PHYSICS, UNIV WATERLOO, 70- *Mem:* Brit Inst Physics. *Res:* Dislocation enhanced diffusion at relatively low temperatures; diffusion in metals. *Mailing Add:* 123 Renaud Dr Waterloo ON N2J 3T3 Can

MORRISON, HUNTLY FRANK, b Montreal, Que, May 16, 38; m 70; c 2. GEOPHYSICS. *Educ:* McGill Univ, BSc, 59, MSc, 61; Univ Calif, Berkeley, PhD(eng geosci), 67. *Prof Exp:* From asst prof to assoc prof, 67-77, PROF GEOPHYS ENG, UNIV CALIF, BERKELEY, 77- *Mem:* Am Geophys Union; Soc Explor Geophys; Europ Asn Explor Geophys. *Res:* Applied geophysics; electromagnetic and electrical prospecting methods. *Mailing Add:* Dept Mat Sci & Eng Univ Calif 2120 Oxford St Berkeley CA 94720

MORRISON, IAN KENNETH, b Barrie, Ont, Aug 5, 39; m 69; c 2. FOREST SOILS, FOREST ECOLOGY. *Educ:* Univ Toronto, BScF, 62, MScF, 64, PhD(forestry, bot), 69. *Prof Exp:* RES SCIENTIST FORESTRY, FORESTRY CAN, ONT REGION, 68- *Honors & Awards:* J A Bothwell Award, 80. *Mem:* Can Inst Forestry; Soil Sci Soc Am; Ont Forestry Asn. *Res:* Production ecology and biogeochemical cycling of elements, in both conifer and hardwood forest, including effects of fertilizers on growth; effects of both timber harvesting and acid precipitation on site productivity. *Mailing Add:* Forestry Can Ont Region Box 490 Sault Ste Marie ON P6A 5M7 Can

MORRISON, JAMES ALEXANDER, physical chemistry; deceased, see previous edition for last biography

MORRISON, JAMES BARBOUR, b Glasgow, Scotland, Aug 16, 43; Can citizen; m 69; c 2. ERGONOMICS, DIVING PHYSIOLOGY. *Educ:* Univ Glasgow, BSc, 64; Univ Strathclyde, PhD(bioeng), 67. *Prof Exp:* Res assoc bioeng, Mass Inst Technol, 67-68; sr scientist underwater physiol, Royal Naval Physiol Lab, Ministry Defence, UK, 68-74; assoc prof, 74-82, PROF KINESIOLOGY, SIMON FRASER UNIV, 82- *Concurrent Pos:* Vis lectr, Dept Ship Bldg & Naval Archit, Univ Strathclyde, Glasgow, 80-81 & Bioeng Unit, 87-88. *Mem:* Undersea & Hyperbaric Med Soc. *Res:* Rehabilitation biomechanics (locomotion function and lower limb prosthetics; human factors engineering (whole body vibration and ergonomic design); environmental physiology (thermoregulation, respiratory mechanics and underwater breathing apparatus. *Mailing Add:* Sch Kinesiology Simon Fraser Univ Burnaby BC V5A 1S6 Can

MORRISON, JAMES DANIEL, b Bryn Mawr, Pa, Mar 28, 36; m 58; c 3. ORGANIC CHEMISTRY, UNIVERSITY ADMINISTRATION. *Educ:* Franklin & Marshall Col, BS, 58; Northwestern Univ, PhD(org chem), 63. *Prof Exp:* From teaching asst to teaching assoc gen chem, Northwestern Univ, 58-62; NSF fel, Stanford Univ, 62-63; asst prof org chem, Wake Forest Col, 63-65; from asst prof to assoc prof, 65-72, dir indust res, 82-85, PROF ORG CHEM, UNIV NH, 72-, ASSOC VPRES RES, 85- *Concurrent Pos:* NSF sci fac fel, Univ NC, 71-72; vis prof, Univ Hawaii, 80-81. *Mem:* Am Chem Soc; Urban Land Inst; Sigma Xi. *Res:* Asymmetric organic reactions; novel peptides. *Mailing Add:* Assoc Vpres Res Univ NH Consult Ctr-Horton Durham NH 03824

MORRISON, JOHN AGNEW, b Ridgefield Park, NJ, Mar 13, 32; m 54; c 3. CHEMISTRY. *Educ:* Fairleigh Dickinson Univ, BS, 59; Rutgers Univ, PhD(org chem), 70. *Prof Exp:* From chemist to res chemist, 57-73, group leader, Lederle Labs, 73-84, SCI DEVELOP COORDR, MED RES DIV, AM CYANAMID CO, 85- *Concurrent Pos:* Lectr, Fairleigh Dickinson Univ, 60-88. *Mem:* AAAS; Am Chem Soc. *Res:* Pharmacokinetics; drug metabolism. *Mailing Add:* c/o Lederle Labs Pearl River NY 10965

MORRISON, JOHN ALBERT, b Wichita, Kans, Dec 1, 24; m 50; c 3. WILDLIFE MANAGEMENT, ECOLOGY. *Educ:* Mont State Univ, BS, 55, MS, 57; Wash State Univ, PhD(zool), 65. *Prof Exp:* Wildlife biologist, Idaho Fish & Game Dept, 57-61; res asst zool, Wash State Univ, 61-65; res biologist, lab perinatal physiol & chief sect primate ecol, US Dept Health, Educ & Welfare, PR, 65-67; leader, Okla Coop Wildlife Res Unit, Okla State Univ, 67-75; terrestrial ecologist, Western Energy & Land Use Team, 75-78; team leader regional info & technol transfer, US Fish & Wildlife Serv, Anchorage, Alaska, 78-85; environ interpretation consult, 85-89; BIOLOGIST, ALASKA DEPT FISH & GAME, ANCHORAGE, 89- *Concurrent Pos:* Staff ecology & natural history, Alaska Pac Univ & Univ Alaska-Anchorage, 86-88. *Honors & Awards:* Superior Service Award, US Dept of Interior, 85. *Mem:* Wildlife Soc; Sigma Xi; Ecol Soc Am. *Res:* Behavior, reproductive and nutritional physiology and general ecology of birds and mammals; environmental protection and reclamation of disturbed energy development locations. *Mailing Add:* Environ Interpretation 12651 Mariner Dr Anchorage AK 99515

MORRISON, JOHN ALLAN, b Beckenham, Eng, June 10, 27; nat US; m 55. APPLIED MATHEMATICS, QUEUEING THEORY. *Educ:* Univ London, BSc, 52; Brown Univ, ScM, 54, PhD(appl math), 56. *Prof Exp:* Asst appl math, Brown Univ, 52-56; mem tech staff, Math Sci Res Ctr, 56-83, DISTINGUISHED MEM TECH STAFF, AT&T BELL LABS, 83- *Concurrent Pos:* Vis prof mechanics, Lehigh Univ, 68; managing ed, Soc Indust & Appl Math Rev, 82-83. *Mem:* Am Math Soc; Soc Indust & Appl Math; Sigma Xi. *Res:* Mathematical physics; nonlinear oscillations; methods of averaging; stochastic differential equations; queueing theory; singular perturbations. *Mailing Add:* AT&T Bell Labs Rm 2C-378 Murray Hill NJ 07974

MORRISON, JOHN B, b White Plains, NY, Apr 6, 38; m 70; c 2. CARDIOVASCULAR RESEARCH. *Educ:* St Lawrence Univ, BS, 60; Cornell Univ, MD, 64. *Prof Exp:* Fel cardiol, 69-71, from instr to asst prof, 71-76, ASSOC PROF MED CARDIOL, MED COL, CORNELL UNIV, 76-; ATTEND PHYSICIAN , NORTH SHORE UNIV HOSP, 73- *Concurrent Pos:* Prin investr, Nassau Chap Grant-in-Aid, Am Heart Asn, 76- & NIH, 86- *Mem:* Am Heart Asn; Am Fedn Clin Res; Inst Elec & Electronics Engrs; Harvey Soc. *Res:* Clinical research designed to understand the pathophysiology of evolving myocardial infarction in man. *Mailing Add:* North Shore Univ Hosp 300 Community Dr Manhasset NY 11030

MORRISON, JOHN COULTER, b Hickman, Ky, Sept 11, 43; m 67; c 3. OBSTETRICS & GYNECOLOGY, BIOCHEMISTRY. *Educ:* Memphis State Univ, BS, 65, Univ Tenn, MD, 68. *Prof Exp:* Res asst biochem, Univ Tenn, Memphis, 65-66, res assoc, 67-68; intern, City of Memphis Hosps, 68-69, resident, 69-72; from instr to asst prof obstet & gynec, Univ Tenn, Memphis, 71-75, assoc prof, 75-80; MEM FAC, MED CTR, UNIV MISS, 80- *Concurrent Pos:* Chief resident, City of Memphis Hosps, 71-72, asst prof, 72- *Mem:* Am Chem Soc; Am Col Obstet & Gynec; AMA; Am Fertil Soc; Sigma Xi. *Res:* Clinical research in obstetrics and gynecology; basic research in carcinogenesis and protein biosynthesis. *Mailing Add:* Dept Obstet-Gynec Univ Miss Med Ctr 2500 N State St Jackson MS 39216

MORRISON, JOHN EDDY, JR, b Ionia, Mich, Aug 1, 39; m 62; c 3. ENGINEERING, AGRICULTURE. *Educ:* Mich State Univ, BS, 61; Univ Mich, MS, 68; Univ Ky, PhD(agr engr), 78. *Prof Exp:* Engr res, Massey-Ferguson Inc, 61-66 & Eaton Corp, 66-68; ENGR RES, SCI & EDUC ADMIN, AGR RES, USDA, 68- *Honors & Awards:* Young Designer Award, Am Soc Agr Engrs, 74. *Mem:* Am Soc Agr Eng; Sigma Xi. *Res:* Tillage; planting; microenvironment; seeding; seed germination; plant stress; system design; physical properties of agricultural materials. *Mailing Add:* 4517 Chestnut Temple TX 76501

MORRISON, JOHN STUART, b St Louis, Mo, Jan 22, 47; m 71; c 3. SOFTWARE REUSE, LARGE-SCALE SYSTEM INTEGRATION. *Educ:* Wash Univ, St Louis, Mo, BA, 68; Naval Postgrad Sch, Monterey, Calif, MS, 80. *Prof Exp:* Air intel officer, 388 Tactical Fighter Wing, 70-71; res officer, Res Div, 544 ARTW(SAC), 71-74; chief air intel, 32 Tactical Reconnaissance Squadron, 75-78; interoperability proj officer, Tactical Air Forces Interoperability Group, 80-83; joint C3 adv to Saudi Arabian Moda, US Mil Training Mission, 83-85; dir plans, 85-88, interoperability, 88-89, DIR SYST ENG & DEVELOP, NAT TEST BED, 89- *Mem:* Armed Forces Commun & Electronics Asn; Soc Computer Simulation; Am Inst Aeronaut & Astronaut; Air Force Asn. *Res:* Large scale simulation; software test and integration; software reuse. *Mailing Add:* Technol Transfer Int Inc 6736 War Eagle Pl Colorado Springs CO 80919-1634

MORRISON, KENNETH JESS, b Rudy, Ark, Feb 14, 21; m 46; c 2. AGRONOMY. *Educ:* Kans State Univ, BS, 48; Purdue Univ, Lafayette, MS, 50, PhD(agr), 67. *Prof Exp:* Assoc agronomist, 66-68, AGRONOMIST, WASH STATE UNIV, 68-, EXTEN AGRONOMIST, 52- *Mem:* Am Soc Agron; Soc Range Mgt; Crop Sci Soc Am; Sigma Xi. *Res:* Effect of environment on cultivars of wheat and barley; interrelation of emergence in dry soils and cold hardiness in common winter wheats. *Mailing Add:* NW 1300 Orion Dr Pullman WA 99163

MORRISON, MALCOLM CAMERON, b Pittsburgh, Pa, Apr 12, 42; m 74; c 2. CHEMICAL ENGINEERING. *Educ:* Calif Inst Technol, BS, 64, PhD(chem eng), 69. *Prof Exp:* Sr scientist, Havens Int, 69-70; group leader, Calgon Corp, 70-72; chief eng, Chem Systs Inc, 72-76, vpres, 76-79; vpres, Puropore, 79-84; CONSULT, 84- *Mem:* Am Inst Chem Engrs; Am Soc Qual Control; Sigma Xi; Filtration Soc; Parenteral Drug Asn. *Res:* Engineering development in membrane technology, primarily in manufacturing techniques for flat sheet and hollow fiber phase inversion membranes. *Mailing Add:* PO Box 765 Silverado CA 92676

MORRISON, MARTIN, biochemistry; deceased, see previous edition for last biography

MORRISON, MARY ALICE, PROTEIN & AMINO ACID UTILIZATION. *Educ:* Univ Wis, PhD(biochem & nutrit), 60. *Prof Exp:* PROF NUTRIT SCI, NY STATE COL HUMAN ECOL, CORNELL UNIV, 60- *Mailing Add:* Cornell Univ 118 Savage Hall Ithaca NY 14850

MORRISON, MILTON EDWARD, b Sigourney, Iowa, July 18, 39; m 61; c 3. CHEMICAL ENGINEERING. *Educ:* Iowa State Univ, BS, 61; Calif Inst Technol, MS, 62, PhD(chem eng), 65. *Prof Exp:* Res engr, E I du Pont de Nemours & Co, Inc, 65-66; chem res engr, Chem Res Lab, Aerospace Res Labs, US Air Force, 66-69; sect head, Res Lab, Am Enka Corp, 69-70, mgr melt spinning res, 70-72, tech mgr, 72-74, plant mgr, 74-75; mgr new venture develop, Akzona, Inc, 75-77; GEN MGR CATALYST DIV, ARMAK CO, 77- *Concurrent Pos:* Adj asst prof, Univ Dayton, 66-69. *Mem:* Am Chem Soc. *Res:* Refining and petrochemical catalysts; polymers and man-made fibers. *Mailing Add:* Akzo Chemei Amer 13000 Baypark Rd Pasadena TX 77507-1104

MORRISON, NANCY DUNLAP, b Schenectady, NY, Dec 14, 46; div. ASTRONOMY. *Educ:* Radcliffe Col, BA, 67; Univ Hawaii, MS, 71, PhD(astron), 75. *Prof Exp:* Res assoc astron, Joint Inst Lab Astrophys, Univ Colo, Boulder, 75-78; asst prof, 78-83, ASSOC PROF ASTRON, DEPT PHYSICS & ASTRON, UNIV TOLEDO, 83- *Mem:* Am Astron Soc; AAAS; Int Astron Union; Astron Soc Pac. *Res:* Photometry and spectroscopy of O-type stars, utilizing these techniques to determine the masses of stars in binary systems; spectroscopic study of mass loss in hot, luminous stars. *Mailing Add:* Dept Physics & Astron Univ Toledo 2801 W Bancroft St Toledo OH 43606

MORRISON, NATHAN, b RI, Dec 4, 12; m; c 1. MATHEMATICAL STATISTICS. *Educ:* Brooklyn Col, AB, 32. *Prof Exp:* Prin actuary, State Dept Labor, NY, 44-61; exec assoc, Assoc Hosp Serv NY, 61-73; CONSULT, 74- *Concurrent Pos:* Consult, US War Dept, 43-46; exec secy, State Adv Coun Employ & Unemploy Ins, NY, 43-70; vis prof, Cornell Univ, 56-; lectr, Teachers Col, Columbia Univ, 59-61, State Univ NY, Stoneybrook, 77-80, Pace Univ, 81-84 & Tours Col, 85-; adj prof, Grad Div, Brooklyn Col, 65-67. *Mem:* Fel Am Statist Asn; Economet Soc; Am Math Soc; fel AAAS; Inst Math Statist. *Res:* Labor market analysis and unemployment insurance; mathematical physics; social insurance; hospital and medical care. *Mailing Add:* 196 Elm St New Rochelle NY 10805

MORRISON, PETER REED, b Washington, DC, Nov 11, 19; m 45; c 6. COMPARATIVE & ENVIRONMENTAL PHYSIOLOGY, BIOCHEMISTRY. *Educ:* Swarthmore Col, BA, 40; Harvard Univ, PhD(biol), 47. *Prof Exp:* Asst physiol, Harvard Univ, 42; asst phys chem, Harvard Med Sch, 42-46; from asst prof to prof zool & physiol, Univ Wis, 47-64; prof zoo physiol, 63-80, dir Inst, 66-74, EMER PROF ZOO PHYSIOL, INST ARCTIC BIOL, UNIV ALASKA, 81- *Concurrent Pos:* Guggenheim & Fulbright fel, Australia, 54-55; NSF sr fel, SAm, 59-60. *Honors & Awards:* Petersen Found Prize Contrib Animal Biometerol, 75. *Mem:* Am Soc Zoologists; Am Soc Biol Chemists; Am Physiol Soc; Am Soc Mammal; Int Soc Biometeorol. *Res:* Energy metabolism and temperature regulation in mammals; fibrinogen and blood coagulation; comparative cold and high altitude physiology; hibernation; protein amino acid sequences. *Mailing Add:* PO Box 787 Friday Harbor WA 98250

MORRISON, PHILIP, b Somerville, NJ, Nov 7, 15; m 38, 64. PHYSICS. *Educ:* Carnegie Inst Technol, BS, 36; Univ Calif, PhD(theoret physics), 40. *Hon Degrees:* DSc, Case Western Reserve Univ, Rutgers Univ & Denison Univ. *Prof Exp:* Instr physics, San Francisco State Col, Calif, 41 & Univ Ill, 41-42; physicist, Metall Lab, Univ Chicago, 43-44; physicist & group leader, Los Alamos Sci Lab, Univ Calif, 44-46; from assoc prof to prof physics, Cornell Univ, 46-65; prof, 65-76, inst prof physics, 76-86, EMER PROF PHYSICS, MASS INST TECHNOL, 86- *Concurrent Pos:* Ed, Sci Am; consult, 86- *Honors & Awards:* Pregel Prize, 55; Babson Prize, 57; Oersted Medal, 65; Killin Award, 85; Gemante Award, 88. *Mem:* Nat Acad Sci; Am Phys Soc; Am Astron Soc; Fedn Am Scientists (chmn, 72-76); AAAS. *Res:* Applications of physics in astronomy. *Mailing Add:* Dept Physics Mass Inst Technol Cambridge MA 02139

MORRISON, RALPH M, b Annapolis, Md, June 23, 32; m 60. PLANT PHYSIOLOGY, MICROBIOLOGY. *Educ:* Col William & Mary, BS, 55; Ind Univ, PhD(bot), 60. *Prof Exp:* From instr to asst prof biol, 60-66, ASSOC PROF BIOL, UNIV NC, GREENSBORO, 66- *Mem:* AAAS; Mycol Soc Am; Bot Soc Am; Am Soc Plant Physiologists. *Res:* Mathematical method for studying the rate of seed germination; botany; mycology. *Mailing Add:* Dept of Biol Univ NC Greensboro NC 27412

MORRISON, RICHARD CHARLES, b Lowell, Mass, Jan 24, 38; m 60; c 4. FIBER OPTICS. *Educ:* Princeton Univ, AB, 59; Yale Univ, MS, 61, PhD(physics), 65. *Prof Exp:* From instr to asst prof physics, New Haven Col, 63-67; asst prof, Iowa State Univ, 67-74; PROF PHYSICS, UNIV NEW HAVEN, 74- *Concurrent Pos:* Assoc physicist, Ames Lab, AEC, 67-74; sr partner, Enercon Assocs, 75-81; vis prof, Bldg Res Sta, Watford, England, 77-78; dean prof studies, Univ New Haven, 80-83; res fac fel, US Navy Underwater Systs Ctr, 84-85. *Mem:* Am Nuclear Soc; Am Phys Soc; Am Asn Physics Teachers; Optical Soc Am. *Res:* Energy production, conversion, and consumption; alternative energy systems and fiber optics. *Mailing Add:* Dept Physics Univ New Haven West Haven CT 06516

MORRISON, ROBERT DEAN, b Wetumka, Okla, Sept 27, 15; m 39; c 1. BIOSTATISTICS. *Educ:* Okla State Univ, BS, 38, MS, 42; NC State Col, PhD(exp statist), 57. *Prof Exp:* From asst prof to assoc prof math, 46-61, statistician, Agr Exp Sta, 57-81, PROF STATIST, OKLA STATE UNIV, 61- *Mem:* Biomet Soc; Am Statist Asn; Sigma Xi. *Res:* Technometrics; experimental statistics; statistics for agricultural research. *Mailing Add:* 1802 N Washington Stillwater OK 74075

MORRISON, ROBERT W, JR, b Columbia, SC, Dec 5, 38; m 61; c 2. ORGANIC CHEMISTRY. *Educ:* Davidson Col, BS, 60; Princeton Univ, MA, 62, PhD(org chem), 64. *Prof Exp:* Res chemist, Chemstrand Res Ctr, Inc, Monsanto Co, 64-69; sr res chemist, Burroughs Wellcome Co, USA, 69-73, group leader, 73-85, assoc head org chem, 85-88, DIR, DIV ORG CHEM, BURROUGHS WELLCOME CO, USA, 88- *Concurrent Pos:* From vis asst prof to adj assoc prof chem, NC State Univ, 68-71. *Mem:* Am Chem Soc; NY Acad Sci; Am Soc Microbiol; AAAS; Int Soc Heterocyclic Chem. *Res:* Synthesis of nitrogen heterocyclic compounds; enzyme inhibitors; transport in microorganisms; antivirals and antitumor compounds. *Mailing Add:* Dir Div Org Chem Burroughs Wellcome Co 3030 Cornwallis Rd Research Triangle Park NC 27709

MORRISON, ROGER BARRON, b Madison, Wis, Mar 26, 14; m 41; c 3. SOILS & SOIL SCIENCE, EARTH & MARINE SCIENCES. *Educ:* Cornell Univ, BA, 33, MS, 34; Univ Nev, Reno, PhD(geol), 64. *Prof Exp:* Geologist, US Geol Surv, 39-76; CONSULT GEOLOGIST, MORRISON & ASSOCS, 78- *Concurrent Pos:* Assoc ed, CATENA, 72-; vis prof, 76-77, adj prof, geosci dept, Univ Ariz, 77-81; adj prof, Makay Sch Mines, Univ Nev, Reno, 84-86; ed, Quaternary Nonglacial Geol: Conterminous US, Decade N Am Geol, Geol Soc Am; consult, Sci Applications Int Corp, 86- *Mem:* Fel Geol Soc Am; AAAS; Int Asn Quaternary Res; Am Soc Photogram; Soil Sci Soc Am; Int Soil Sci Soc. *Res:* Quaternary geology and geomorphology; environmental geology; remote sensing of earth resources; geologic writing; paleoclimatology; hydrogeology; soil stratigraphy. *Mailing Add:* Morrison Assocs 13150 W Ninth Ave Golden CO 80401

MORRISON, ROLLIN JOHN, b Akron, Ohio, Oct 8, 37; m 64; c 2. HIGH ENERGY PHYSICS. *Educ:* Ohio Wesleyan Univ, BA, 59; Univ Ill, MA, 61, PhD(physics), 64. *Prof Exp:* Volkswagen fel, Ger Electron Syncrotron, 64-66; res physicist, 67, from asst prof to assoc prof, 67-78, PROF PHYSICS, UNIV CALIF, SANTA BARBARA, 78- *Mem:* Fel Am Phys Soc. *Res:* High energy experimental physics with an emphasis on heavy quark physics; use of precision vartex detectors to determine charm particle lifetimes and other properties of weak decays. *Mailing Add:* Dept Physics Univ Calif Santa Barbara CA 93106

MORRISON, SHAUN FRANCIS, m; c 2. ELECTROPHYSIOLOGY. *Educ:* Univ Vt, PhD(physiol & biophysics), 80. *Prof Exp:* Asst prof neurobiol, Med Col, Cornell Univ, 84-90; ASSOC PROF, DEPT PHYSIOL, NORTHWESTERN UNIV MED SCH, 90- *Mem:* Soc Neurosci; Am Physiol Soc. *Res:* Central nervous system control of cardiovascular function; single cell electrophysiology; sympathetic nervous system recording. *Mailing Add:* Dept Physiol Northwestern Univ Med Sch 303 E Chicago Ave Chicago IL 60611-3010

MORRISON, SHERIE LEAVER, b New Eagle, Pa, July 5, 42; m 64; c 2. IMMUNOLOGY. *Educ:* Stanford Univ, BA, 63, PhD(biol), 66. *Prof Exp:* Res assoc, Dept Biol, Columbia Univ, 66-70, Cold Spring Harbor Lab Quant Biol, 70 & Dept Molecular Biol Virus Labs, Univ Calif, Berkeley, 70-71; res fel, Dept Cell Biol, Albert Einstein Col Med, 71-74; from asst prof to assoc prof, 74-84, prof microbiol, Col Physicians & Surgeons, Columbia Univ, 84-88; PROF MICROBIOL, MOLECULAR BIOL INST, UNIV CALIF, LOS ANGELES, 88- *Concurrent Pos:* Mem, Allergy & Immunol Study Sect, 78-82; prin investr, Nat Cancer Inst grant, 75-82; exec ed, Analytical Biochem, 86-88; grant, Nat Cancer Inst, 75-; assoc ed, J Immunol, 84-; assoc ed, J Molecular & Cellllular Immunol, 84- *Honors & Awards:* Career Scientist Award, Irma T Hirschl Trust, 75. *Mem:* Am Asn Immunologists; Sigma Xi. *Res:* Genetic biochemistry of immunoglobulin production by isolating and characterizing mouse myeloma cells mutant in their production of immunoglobulin. *Mailing Add:* 258 Denslow Ave Los Angeles CA 90049

MORRISON, SIDONIE A, b Rocford, Eng, July 27, 47; m; c 1. HEMATOLOGY, BLOOD COAGULATION. *Educ:* Oxford Univ, DPh, 73. *Prof Exp:* asst prof, 76-86, ASSOC PROF MED, STATE UNIV NY, STONY BROOK, 86- *Mem:* Am Soc Biol Chemists; Am Soc Hemat; Asn Women Sci. *Res:* Participation of cellular elements (platelets, monocytes/ macrophaces, endothelial cells) in coagulation; role of these in immune disease (lupus erythematosus, AIDS). *Mailing Add:* Div Hemat Health Sci Ctr T-15 Rm 040 State Univ New York Stony Brook NY 11794

MORRISON, SPENCER HORTON, b Madison, Wis, Apr 17, 19; m 46; c 6. VETERINARY MEDICINE, ANIMAL NUTRITION. *Educ:* Cornell Univ, BS, 39, MS, 46, PhD(animal nutrit & physiol, biochem), 49; Univ Ga, DVM, 54. *Prof Exp:* Asst animal husb, Cornell Univ, 46-48; assoc, Univ Calif, 48-49; assoc prof dairying, Univ Ga, 49-54; tech sales dir, Feed & Soy Div, Pillsbury Mills, Inc, 54-57, res dir, 57-58; ed & mgr, Morrison Pub Co, 58-78, partner, 40-78; DIR, AGRICON, 58- *Mem:* Fel AAAS; Am Soc Animal Sci; Poultry Sci Asn; Am Dairy Sci Asn; Am Vet Med Asn. *Res:* Animal physiology; biochemistry; feeding value of concentrates and roughages; antibiotics, vitamins; physiology of reproduction; protein synthesis in ruminants. *Mailing Add:* Box 729 Wiarton ON N0H 2T0 Can

MORRISON, STANLEY ROY, b Saskatoon, Sask, Sept 24, 26; m 49; c 3. SURFACE CHEMISTRY, SURFACE PHYSICS. *Educ:* Univ BC, BA, 48, MA, 49; Univ Pa, PhD(physics), 52. *Prof Exp:* Res assoc solid state physics, Univ Ill, 52-54; sr scientist, Sylvania Elec Prod Inc, 54-55; from staff scientist to asst dir res, Res Ctr, Minneapolis-Honeywell Regulator Co, 54-64; sr physicist, Stanford Res Inst, 64-82; AT DEPT PHYS, SIMON FRASER UNIV. *Concurrent Pos:* Guest prof, Inst Physics & Chem, Gottingen Univ, 71-72; guest scientist, Device Lab, Ft Monmouth, 68. *Honors & Awards:* Cert of Recognition, NASA, 73 & 74. *Mem:* Am Phys Soc. *Res:* Adsorption and catalysis; semiconductor electrochemistry; electrical and chemical properties of surfaces; influence of adsorbed gases and other surface imperfections on the electrical properties of solid state materials and devices. *Mailing Add:* Dept Phys Simon Fraser Univ Burnaby BC V5A 1S6 Can

MORRISON, THOMAS GOLDEN, b Chicago, Ill, Sept 28, 18; m 50; c 2. STRUCTURAL ENGINEERING, ENGINEERING PHYSICS. *Educ:* Univ Ill, BS, 49, CE, 58; NMex State Univ, ScD(appl mech), 75. *Prof Exp:* Asst resistance of mat, Univ Ill, 47-49; bridge designer, Alfred Benesch & Assocs, 49-52; pvt consult engr, 52-53; sr tech adv mech, Am Mach & Foundry Co, 53-63; sr engr, Cosmic Ray Observ, Univ Chicago, 63-68, mgr & sr engr, High Altitude Cosmic Ray Observ, 68-73; instr, NMex State Univ, 73-75; assoc prof, 75-83, PROF, CHRISTIAN BROTHERS UNIV, 83-

Concurrent Pos: Consult, Dept Physics & Astron, Univ Md, 73- & NASA, 85- *Honors & Awards:* NASA Group achievement Award, Ballon Prog-Prog Recovery Team, 88. *Mem:* Fel Am Soc Civil Engrs. *Res:* Applied mechanics; mechanics of dynamically loaded shell structures; transient vibrations; buckling of structure. *Mailing Add:* Christian Brothers Univ 650 E Parkway S Memphis TN 38104

MORRISON, WILLIAM ALFRED, b Chicago, Ill, Mar 27, 48; div. INORGANIC CHEMISTRY. *Educ:* Ill Wesleyan Univ, BA, 70; Univ Kans, PhD(inorg chem), 74. *Prof Exp:* Asst prof chem, Monmouth Col, 74-75; vis asst prof, 75-76, asst prof, 76-81, ASSOC PROF CHEM, UNIV EVANSVILLE, 81- *Mem:* Am Chem Soc; AAAS. *Res:* Transition metal organometallics; carbenes; metal-metal bonds; nonaqueous solvents. *Mailing Add:* 1800 Lincoln Ave Evansville IN 47722

MORRISON, WILLIAM D, b Provost, Alta, Oct 16, 27; m 49; c 4. ANIMAL NUTRITION, ANIMAL PHYSIOLOGY. *Educ:* Ont Agr Col, Toronto, BSA, 49; Univ Ill, MSc, 54, PhD(animal nutrit), 55. *Prof Exp:* Territory mgr, Master Feeds Div, Maple Leaf Mills Ltd, 49-52, nutritionist, 55-57, dir nutrit & res, 57-71; chmn dept, 71-82, PROF ANIMAL & POULTRY SCI, UNIV GUELPH, 82- *Mem:* AAAS; Am Soc Animal Sci; Poultry Sci Asn; Can Soc Animal Sci. *Res:* Amino acid utilization by the chick, with special emphasis on the D isomer; protein and energy requirements of chickens and turkeys; environmental physiology of pigs and poultry. *Mailing Add:* Dept Animal & Poultry Sci Univ Guelph Guelph ON N1G 2W1 Can

MORRISON, WILLIAM JOSEPH, b Plainfield, NJ, Feb 17, 42. GENETICS. *Educ:* Clemson Univ, BS, 65; Pa State Univ, PhD(genetics), 69. *Prof Exp:* NIH fel genetics, Hershey Med Ctr, Pa State Univ, 69-70; NIH fel, Cornell Univ, 70-73; ASSOC PROF BIOL, SHIPPENSBURG UNIV, PA, 73- *Concurrent Pos:* vis investr, Waksman Inst, Rutgers Univ, 82 & 83. *Mem:* AAAS; Genetics Soc Am; Sigma Xi; Am Genetic Asn. *Res:* Biochemical and development genetics; Drosophila genetics. *Mailing Add:* Dept Biol Shippensburg Univ Shippensburg PA 17257

MORRISS, FRANK HOWARD, JR, b Birmingham, Ala, Apr 20, 40; m 68; c 2. PEDIATRICS, NEONATOLOGY. *Educ:* Univ Va, BA, 62; Duke Univ, MD, 66. *Prof Exp:* Resident pediat, Duke Univ Med Ctr, 66-68, fel neonatology, 70-71; lt comdr med, USN Med Corps, 68-70; fel neonatal res, Univ Colo Med Ctr, 71-73; from asst prof to prof pediat & obstet-gynec, Univ Tex Med Ctr, Houston, 73-86, vchair dept, 82-86; PROF & HEAD, DEPT PEDIAT, COL MED, UNIV IOWA, 87- *Concurrent Pos:* Assoc dean clin affairs, Univ Tex Med Sch, Houston, 82-84; mem bd, Am Bd Pediat Sub-bd Neonatal-Pediat Med, 89- *Mem:* Am Pediat Soc; Soc Pediat Res; Soc Gynec Invest; Am Acad Pediat; Am Soc Clin Nutrit. *Res:* Perinatal physiology; developmental nutrition and gastroenterology. *Mailing Add:* Dept Pediat Univ Iowa Hosps & Clins Iowa City IA 52242

MORRISSETTE, MAURICE CORLETTE, b Clyde, Kans, Aug 27, 21; m 45; c 3. PHYSIOLOGY, PHARMACOLOGY. *Educ:* Kans State Univ, BS & DVM, 54; Okla State Univ, MS, 56, PhD(reprod physiol), 64. *Prof Exp:* Instr physiol & pharmacol, Sch Vet Med, Okla State Univ, 54-56, asst prof, 56-57; asst prof physiol, Kans State Univ, 57-59; from asst prof to assoc prof physiol & pharmacol, Okla State Univ, 59-69; PROF & HEAD DEPT, SCH VET MED, LA STATE UNIV, BATON ROUGE, 69- *Mem:* Am Soc Vet Physiologists & Pharmacologists (pres, 72-73); Am Vet Med Asn; Am Fertil Soc; Soc Study Reproduction; Sigma Xi. *Res:* Reproductive physiology; veterinary medicine; female reproduction in swine and cattle; veterinary physiology, pharmacology and toxicology. *Mailing Add:* 8956 Tallyho Ave Baton Rouge LA 70806

MORRISSEY, BRUCE WILLIAM, b Danbury, Conn, Apr 18, 42; m 67; c 3. PHYSICAL CHEMISTRY, BIOCHEMISTRY. *Educ:* Rensselaer Polytech Inst, BS, 64, PhD(theoret chem), 70; Yale Univ, MS, 66; George Washington Univ, JD, 81. *Prof Exp:* Nat Res Coun res assoc, Nat Bur Standards, 70-72, res chemist, Polymer Div, 72-76, prog anal, Dir Off, 76-78, prog mgr, Water Pollution Measurements, 78-81; atty patent law, 81-84, pharmaceut regulatory dir, 84-86, SR COUN, PATENT LAW BIOTECH, E I DUPONT, 86- *Mem:* Am Chem Soc; Am Soc Artificial Internal Organs. *Res:* Polymer surface chemistry; determination of the conformation and conformational changes of adsorbed proteins and their reactivity and function at surfaces. *Mailing Add:* 3302 Coachman Rd Wilmington DE 19803

MORRISSEY, DAVID JOSEPH, b White Plains, NY, Dec 7, 53; m 75; c 2. NUCLEAR CHEMISTRY. *Educ:* Pa State Univ, BS, 75; Univ Calif, Berkeley, PhD(chem), 78. *Prof Exp:* Fel nuclear chem, Nuclear Sci Div, Lawrence Berkeley Lab, 78-81; ASST PROF CHEM, MICH STATE UNIV, 81- *Mem:* Am Chem Soc; Am Phys Soc; AAAS; Sigma Xi. *Res:* Reaction mechanisms operating in heavy-ion reactions; detection of reaction products, their kinematic properties and other characteristics such as internal excitation and intrinsic spin. *Mailing Add:* Chem Dept Chem Bldg Mich State Univ East Lansing MI 48824

MORRISSEY, J EDWARD, b Grinnell, Iowa, Aug 7, 32; m 57; c 2. ZOOLOGY, PHYSIOLOGY. *Educ:* St Ambrose Col, BA, 56; Northwestern Univ, Ill, MS, 58; Univ Mo-Columbia, PhD(zool), 68. *Hon Degrees:* MHL, Ottawa Univ, 72. *Prof Exp:* Instr biol, Stevens Col, 60-65; asst prof, MacMurray Col, 68; from asst prof to assoc prof, 68-77, PROF BIOL, OTTAWA UNIV, 77- *Mem:* AAAS. *Res:* Immunological development; studies of immune responses in chickens. *Mailing Add:* Dept Biol Ottawa Univ Tenth Cedar Ottawa KS 66067

MORRISSEY, JAMES HENRY, b Tucson, Ariz, May 27, 53; m 73; c 1. BLOOD COAGULATION, THROMBOSIS. *Educ:* Univ Calif, Irvine, BS, 75, BA, 75, MS, 75; Univ Calif, San Diego, PhD(biol), 80. *Prof Exp:* Postdoctoral fel, Univ Calif, San Diego, 80-81, Univ Oxford, UK, 82-83, Res Inst Scripps Clin, 83-84, asst mem, 84-89; ASSOC MEM, OKLA MED RES FOUND, 89-; OKLA MED RES FOUND ASSOC PROF PATH, HEALTH SCI CTR, UNIV OKLA, 91- *Mem:* Am Heart Asn; AAAS; Int Soc Thrombosis & Hemostasis; Am Soc Cell Biol; Am Soc Microbiol. *Res:* Regulation of the blood clotting cascade in hemostasis, thrombosis and inflammation; tissue factor (thromboplostin) structure, function and gene activity. *Mailing Add:* Okla Med Res Found 825 NE 13th St Oklahoma City OK 73104

MORRISSEY, JOHN F, b Brookline, Mass, June 16, 24; m 50; c 2. MEDICINE, GASTROENTEROLOGY. *Educ:* Dartmouth Col, AB, 46; Harvard Univ, MD, 49. *Prof Exp:* Asst prof med, Univ Wis, 56-60; asst prof, Univ Wash, 60-62; from asst prof to prof, 62-89, EMER PROF MED, UNIV WIS-MADISON, 89- *Honors & Awards:* Schindler Award, Am Soc Gastrointestinal Endoscopy. *Mem:* Am Gastroenterol Asn; Am Col Physicians; Am Soc Gastrointestinal Endoscopy. *Res:* Evaluation of new instruments for digestive tract endoscopy; effects of drugs on gastrointestinal mucosa. *Mailing Add:* Dept of Med H6/516 600 Highland Ave Madison WI 53792

MORRISSEY, PHILIP JOHN, LYMPHOKINE RESEARCH, IMMUNE REGULATION. *Educ:* Tufts Univ, PhD(physiol), 79. *Prof Exp:* STAFF SCIENTIST, IMMUNEX CORP, 84- *Mailing Add:* Immunex Corp 51 University St Seattle WA 98101

MORRISSEY, RICHARD EDWARD, teratology, reproductive toxicology, for more information see previous edition

MORRISSEY, ROBERT LEROY, toxicological pathology, nutrition, for more information see previous edition

MORRONE, TERRY, b New York, NY, May 30, 36; m 62; c 2. PLASMA PHYSICS. *Educ:* Columbia Univ, BS, 57, MS, 58; Polytech Inst Brooklyn, PhD(electrophys), 64. *Prof Exp:* From asst prof to assoc prof, 64-76, PROF PHYSICS, ADELPHI UNIV, 76- *Mem:* Am Phys Soc. *Mailing Add:* Eight Bowdon Rd Greenlawn NY 11740

MORROW, ANDREW GLENN, b Indianapolis, Ind, Nov 3, 22; m 45; c 2. SURGERY. *Educ:* Wabash Col, AB, 43; Johns Hopkins Univ, MD, 46. *Prof Exp:* Asst, 50-51, from instr to asst prof, 51-60, ASSOC PROF SURG, JOHNS HOPKINS UNIV, 60-; CHIEF CLIN SURG, NAT HEART & LUNG INST, 53- *Mem:* Soc Vascular Surg (pres, 71-72); Soc Univ Surg; Am Asn Thoracic Surg; fel Am Col Surg; Am Fedn Clin Res. *Res:* Cardiovascular surgery; diagnostic methods and allied physiology. *Mailing Add:* 809 N Calvert St Baltimore MD 21202

MORROW, BARRY ALBERT, b Regina, Sask, Aug, 39. SURFACE CHEMISTRY, CATALYSIS. *Educ:* Univ BC, BSc, 61, MSc, 62; Cambridge Univ, PhD(chem), 65. *Prof Exp:* Nat Res Coun fel, 65-66; lectr chem, Univ West Indies, 66-67; PROF CHEM, UNIV OTTAWA, 67- *Mem:* Am Chem Soc; Royal Soc Chem. *Res:* Spectroscopic studies of adsorption and catalysis; spectroscopic studies of small molecules. *Mailing Add:* Dept of Chem Univ of Ottawa Ottawa ON K1N 6N5 Can

MORROW, CHARLES T(ERRY), b Clarksburg, WVa, June 20, 41; c 2. AGRICULTURAL ENGINEERING. *Educ:* Univ WVa, BSAE, 63; Pa State Univ, MSAE, 65, PhD(eng mech), 69. *Prof Exp:* Physicist, Eastern Utilization Lab, USDA, 65; PROF AGR ENG, PA STATE UNIV, 67- *Mem:* Am Soc Agr Engrs; Am Soc Eng Educ. *Res:* Mechanical characterization of biological tissue; mechanical harvesting and landing of fruit and vegetable; bioinstrumentation; computer application; horticultural engineering. *Mailing Add:* Dept Agr & Biol Eng Col Agr Pa State Univ University Park PA 16802

MORROW, CHARLES TABOR, b Gloucester, Mass, May 3, 17; m 49; c 2. ACOUSTICS. *Educ:* Harvard Univ, AB, 37, SM, 38, ScD, 46. *Prof Exp:* Engr, Harvard Univ, 37-40, res assoc physics, 40-41, spec res assoc underwater sound, 41, instr radio & radar officers training, 42-44; lectr indust electronics, Northeastern Univ, 44-45; sr proj engr, Sperry Gyroscopy Co, 46-51; res physicist, Hughes Aircraft Co, Inc, 51-55; mem sr staff, Space Technol Labs, Inc, 55-60 & Aerospace Corp, Calif, 60-67; from mem staff to staff scientist, Western Div, LTV Res Ctr, 67-69; staff scientist, Advan Technol Ctr, Inc, 69-76; CONSULT, 76- *Honors & Awards:* Vigness Award, Inst Environ Sci, 71. *Mem:* Fel Inst Environ Sci; fel Acoust Soc Am; Am Inst Aeronaut & Astronaut; Inst Elec & Electronics Eng; Am Soc Eng Educ. *Res:* Vibratory gyroscopes; silencing of diving masks; study of speech in gas oxygen and diving masks and in helium atmospheres; development of microphone for divers; shock and vibration analysis; prediction of structural response to aerodynamic turbulence and rocket noise; vibration instrumentation. *Mailing Add:* 1345 Cherry Tree Ct Encinitas CA 92024

MORROW, DAVID AUSTIN, b Tyrone, Pa, Jan 14, 35; m 65; c 3. THERIOGENOLOGY. *Educ:* Pa State Univ, BS, 56; Cornell Univ, DVM, 60, PhD(theriogenol), 67. *Prof Exp:* Pvt pract vet med, 60-61; intern, Cornell Univ, 61-62, asst, 62-64, NIH fel, 64-67, res assoc, 67-68; from assoc prof to prof vet med, Mich State Univ, 68-90; CONSULT, 90- *Honors & Awards:* Borden Award, 80; Vet Med Res Award, AFMA, 82. *Mem:* Soc Study Reprod; Am Col Theriogenologists; Am Dairy Sci Asn; Am Soc Animal Sci; Am Asn Bovine Practitioners; Soc Theriogenology; Am Vet Med Asn. *Res:* Bovine reproductive physiology; bovine theriogenology; herd health. *Mailing Add:* 1060 Haymaker Rd State College PA 16801

MORROW, DEAN HUSTON, b Indianapolis, Ind, June 11, 31; m 53, 86; c 6. ANESTHESIOLOGY, CARDIOVASCULAR PHYSIOLOGY. *Educ:* Butler Univ, BS, 53; Ind Univ, MD, 56; Am Bd Anesthesiol, dipl, 63. *Prof Exp:* Staff anesthesiologist, Clin Ctr, NIH, Md, 59-61, res anesthesiologist, 62-64; assoc prof anesthesiol, Col Med, Univ Ky, 64-66, prof anesthesiol & dir dept res, 66-73, prof pharmacol, 71-73; assoc dir clin res, Travenol Labs, 73-75; med dir, Dept Biomed Instrumentation, Monitoring Dept, Methodist Hosp, Houston, 76-81, chmn, Dept Anesthesiol, 79-88; vchmn, Dept Anesthesiol, 78-79, PROF ANESTHESIOL, BAYLOR COL MED, 75-;

ASSOC CHIEF STAFF/EDUC, VET ADMIN MED CTR, HOUSTON, 90- *Concurrent Pos:* Nat Inst Gen Med Sci res career award, 67-71; res affil, Nat Heart Inst, 61; mem comt anesthesia, Nat Res Coun-Nat Acad Sci, 63-65. *Mem:* Am Soc Anesthesiol; Int Anesthesia Res Soc; Asn Univ Anesthetists; fel Am Col Anesthesiol; Am Soc Pharmacol & Exp Therapeut. *Res:* Pharmacology. *Mailing Add:* Assoc Chief Staff Educ (141) Vet Admin Med Ctr 2002 Holcombe Blvd Houston TX 77030

MORROW, DUANE FRANCIS, b Detroit, Mich, June 23, 33; m 58; c 3. PHARMACEUTICAL CHEMISTRY, SYNTHETIC ORGANIC & NATURAL PRODUCTS CHEMISTRY. *Educ:* Wayne State Univ, BS, 54; Univ Ill, PhD(chem), 57. *Prof Exp:* From assoc res chemist to sr res chemist, Parke, Davis, & Co, 57-69; from group leader to sr investr, Mead Johnson Co, 69-74, prin investr, Res Ctr, 74-83; div Pharm Res, Bristol-Myers Co, 83-88, ASSOC DIR, BRISTOL-MYERS SQUIBB US PHARMACEUT GROUP, 88- *Mem:* Am Chem Soc; Regulatory Affairs Prof Soc; Drug Info Asn. *Res:* Pharmaceutical chemistry; steroids; anti-fertility. *Mailing Add:* Bristol-Myers Squibb US Pharmaceut Group 2400 W Lloyd Expressway B305 Evansville IN 47721-0001

MORROW, GRANT, III, b Pittsburgh, Pa, Mar 18, 33; m 60; c 2. NEONATAL-PERINATAL MEDICINE, CLINICAL GENETICS. *Educ:* Haverford Col, BA, 55; Univ Pa, MD, 59. *Prof Exp:* Intern, Univ Colo, 59-60; resident pediat, Univ Pa, 60-62, fel neonatology, 62-63, assoc prof, 64-65, assoc prof metab dis, 65-71; from assoc prof to prof neonatology & metab, Univ Ariz, 72-78, assoc chmn dept, 76-78; PROF NEONATOLOGY & METAB & CHMN DEPT, OHIO STATE UNIV, 78-; MED DIR, COLUMBUS CHILDREN'S HOSP, 78- *Mem:* Am Pediat Soc; Soc Pediat Res; Am Soc Clin Nutrit. *Res:* Children suffering inborn errors of metabolism, mainly amino and organic acids; patients on total parental nutrition. *Mailing Add:* Columbus Children's Hosp 700 Children's Dr Columbus OH 43205

MORROW, JACK I, b New York, NY, Jan 30, 33; m 58; c 1. PHYSICAL INORGANIC CHEMISTRY. *Educ:* NY Univ, BA, 54, PhD(surface chem), 59. *Prof Exp:* From asst prof to assoc prof chem, 57-73, PROF CHEM, CITY COL, NEW YORK, 73- *Mem:* Fel Am Inst Chemists; Am Chem Soc. *Res:* Inorganic reaction mechanisms and instrumentation for the study of fast reactions. *Mailing Add:* Dept Chem City Col New York Convent Ave & 138th St New York NY 10031

MORROW, JAMES ALLEN, JR, b Little Rock, Ark, Sept 14, 41; m 68; c 1. MATHEMATICS. *Educ:* Calif Inst Technol, BSc, 63; Stanford Univ, PhD(math), 67. *Prof Exp:* Teaching asst math, Stanford Univ, 63-67; instr, Univ Calif, 67-68, lectr, 68-69; from asst prof to assoc prof, 69-78, PROF MATH, UNIV WASH, 78- *Mem:* Am Math Soc; Math Asn Am. *Res:* Complex manifolds; singularities. *Mailing Add:* Dept Math Univ Wash Seattle WA 98195

MORROW, JANET RUTH, b Santa Monica, Calif, Feb 18, 57; M 85; c 1. BIOINORGANIC CHEMISTRY, NUCLEIC ACID CHEMISTRY. *Educ:* Univ Calif, Santa Barbara, BS, 80; Univ NC, Chapel Hill, PhD(chem), 85. *Prof Exp:* NSF postdoctoral inorg chem, Univ Bordeaux, France, 85-86; postdoctoral bioinorg, Univ Calif, San Diego, 86-88; ASST PROF CHEM, STATE UNIV NY, BUFFALO, 88- *Concurrent Pos:* Postdoctoral researcher, Industrialized Countries exchange of scientists & engrs, NSF, 85; prin investr, several res grants, NIH, NSF & Petrol Res Fund, 88- *Mem:* Am Chem Soc. *Res:* Synthesis of molecules that site-specifically cleave RNA for use in the selective suppression of gene expression. *Mailing Add:* Dept Chem State Univ NY Buffalo NY 14214

MORROW, JODEAN, b Woodbine, Iowa, Oct 16, 29; m 50; c 4. ENGINEERING MECHANICS. *Educ:* Rose-Hulman Inst Technol, BS, 50; Univ Ill, MS, 54, PhD(theoret & appl mech), 57. *Prof Exp:* Asst proj engr, State Hwy Dept, Ind, 50-51; asst theoret & appl mech, 53-54, res assoc, 54-57, from asst prof to assoc prof, 57-64, PROF THEORET & APPL MECH, UNIV ILL, URBANA, 64-, GRAD PROG COORDR, 80- *Concurrent Pos:* Vis prof, Kyoto Univ, 69. *Mem:* Am Soc Testing & Mat; Soc Mat Sci Japan. *Res:* Mechanics of materials; flow and fracture of metals; fatigue; residual stresses; cyclic stress-strain behavior; contact fatigue. *Mailing Add:* 201 Draper Ave Champaign IL 61821

MORROW, JOHN CHARLES, III, physical chemistry; deceased, see previous edition for last biography

MORROW, JON S, MOLECULAR CELL BIOLOGY. *Educ:* Ind Univ, BS, 65, PhD(biochem & biophys), 74; Yale Univ, MD, 76. *Prof Exp:* PROF PATH, YALE UNIV MED SCH, 89-, CHMN DEPT, 90-, CHIEF PATH, YALE-NEW HAVEN HOSP, 90- *Mem:* Am Asn Pathologists; Am Soc Cell Biol; Am Soc Biochem & Molecular Biol; Int Acad Path; Col Am Pathologists. *Res:* Mechanisms of polarized cytoskeletal assembly in eukasyotic cells; role of the membrane cytoskeleton in signal transduction; molecular pathology of disorders involving the cytoskeleton. *Mailing Add:* 333 Cedar St New Haven CT 06510

MORROW, KENNETH JOHN, JR, b Wallace, Idaho, Nov 2, 38; m 82. GENETICS, IMMUNOLOGY. *Educ:* Whitman Col, AB, 60; Univ Wash, MS, 62, PhD(genetics), 64. *Prof Exp:* Fulbright fel, 64; NIH fel, Inst Genetics, Univ Pavia, 64-66; res assoc biol, Inst Cancer Res, Philadelphia, 66-68; asst prof physiol & cell biol, Univ Kans, 68-73; ASSOC PROF BIOCHEM, SCH MED, TEX TECH UNIV, 73- *Mem:* Am Soc Cell Biol; Am Tissue Cult Asn. *Res:* Somatic cell genetics; genetics of mammalian somatic cell cultivated in vitro; isolation of variants and their characterization. *Mailing Add:* Dept Biochem Tex Tech Univ Health Sci Ctr Fourth St & Indiana Ave Lubbock TX 79430

MORROW, LARRY ALAN, b Boise, Idaho, Oct 3, 38; m 57; c 4. WEED SCIENCE. *Educ:* Utah State Univ, BS, 65; Univ Nebr, MS, 71, PhD(weed sci), 74. *Prof Exp:* Biologist, 66-74, res agronomist, Sci & Educ Admin-Agr Res, 74-80 & agr res serv, USDA, 80-; AT DEPT AGRON, WASH STATE UNIV. *Mem:* Am Soc Agron; Crop Sci Soc Am; Coun Agr Sci & Technol; Weed Sci Soc Am. *Res:* Annual grass weed research in small grains. *Mailing Add:* PO Box 11712 Pullman WA 99163

MORROW, NORMAN LOUIS, b Brooklyn, NY, Feb 1, 42; m 64; c 2. PHYSICAL CHEMISTRY, ANALYTICAL CHEMISTRY. *Educ:* Stevens Inst Technol, BS, 63; Univ Conn, PhD(chem), 67. *Prof Exp:* Res chemist, Exxon Res & Eng Co, NJ, 67-70; group head analytical chem, Exxon Chem Co, 70-73, chief chemist, 73-75, staff chemist, 75-77, sect supvr, 78-86, sr staff engr, 87-90, ENVIRON ASSOC, EXXON CHEM CO, 90- *Mem:* Am Chem Soc. *Res:* Project development in petrochemical and environmental control areas. *Mailing Add:* Exxon Chem Co Basic Chem Group PO Box 241 Baton Rouge LA 70821-0241

MORROW, NORMAN ROBERT, b Barking, Eng, Mar 16, 37; US citizen; m 59; c 4. POROUS MEDIA, FLUID FLOW & CAPILLARITY. *Educ:* Univ Leeds, BSc, 59, PhD(mineral eng), 62. *Prof Exp:* Res asst mineral eng, Univ Leeds, 59-62; res assoc, Columbia Univ, 63-64; sr res scientist, Esso Prod Res Co, Standard Oil, NJ, 64-69 & Petrol Recovery Res Inst, Univ Calgary, 69-77; HEAD, PETROPHYSICS & SURFACE CHEM, PETROL RECOVERY RES CTR, NMEX INST, 77- *Honors & Awards:* Bert Cohen Award, Petrol Soc Can Inst Mining, Metall & Petrol Engrs, 76. *Mem:* Petrol Soc Can Inst Mining, Metall & Petrol Engrs; Am Chem Soc; Soc Petrol Engrs; Am Inst Mining, Metall & Petrol Engrs; Sigma Xi; Soc Core Analysts. *Res:* Capillarity of porous materials; wettability; thermodynamics; structure of porous media; stability of interfaces; adsorption; enhanced oil recovery; low interfacial tension systems; residual oil; low permeability gas reservoirs. *Mailing Add:* Petrol Recovery Res Ctr NMex Inst Socorro NM 87801

MORROW, PAUL EDWARD, b Fairmont, WVa, Dec 27, 22; m 47; c 2. TOXICOLOGY. *Educ:* Univ Ga, BS, 42, MS, 47; Univ Rochester, PhD(pharmacol), 51; Am Col Toxicol, dipl. *Prof Exp:* Indust hygienist, Holston Ord Works, Tenn Eastman Corp, 42-43; asst, Univ Ga, 46-47; res assoc, 47-52, from instr to prof radiation biol & pharmacol, 52-67, prof radiation biol, pharmacol, toxicol & biophys, 67-83, prof toxicol radiation biol & biophys, 83-85, EMER PROF TOXICOL BIOPHYS, SCH MED & DENT, UNIV ROCHESTER, 85- *Concurrent Pos:* NIH spec res fel, Univ Göttingen, 59-60; mem, Int Comn Radiol Protection, 65-77 & Nat Coun Radiation Protection & Measurements, 69-; mem subcomt health effects fossil fuel combustion, Nat Res Coun-Nat Acad Sci; mem toxicol design rev comt, Nat Inst Environ Health & Sci. *Honors & Awards:* Inhalation Toxicol Sect Achievement Award, Soc Toxicol, 85; Award Outstanding Contrib to Aerosol Res, Int Soc Aerosols Med, 88; Founder's Award in Toxicol, Chem Indust Inst Toxicol, 89. *Mem:* Health Physics Soc; Radiation Res Soc; fel NY Acad Sci; Am Indust Hyg Asn; Am Thoracic Soc; Sigma Xi; Soc Toxicol; Am Acad Indust Hyg; AAAS; Int Soc Aerosol Med; Reticuloendothelial Soc; Am Asn Aerosol Res. *Res:* Radiation toxicology; deposition and retention of inhaled dusts; dust clearance mechanisms in the lung; radioactive dust hazards; aerosols; lung models, health effects of air pollutants. *Mailing Add:* Dept Biophys Univ Rochester Med Ctr Box EHSC Rochester NY 14627

MORROW, RICHARD ALEXANDER, b Powassan, Ont, Apr 19, 37; m 64; c 2. SEMICONDUCTOR PHYSICS. *Educ:* Queens Univ, Ont, BSc, 58; Univ BC, MSc, 59; Princeton Univ, PhD(physics), 63. *Prof Exp:* Instr physics, Princeton Univ, 63-64; asst prof, Dartmouth Col, 64-70; assoc prof, 70-78, PROF PHYSICS, UNIV MAINE, ORONO, 78- *Mem:* Am Phys Soc; Can Asn Physicists; Am Asn Physics Teachers. *Res:* Theoretical semiconductor physics. *Mailing Add:* Dept Physics Univ Maine Orono ME 04469

MORROW, RICHARD JOSEPH, b Portland, Ore, Aug 28, 28; m 58; c 5. NUCLEAR CHEMISTRY, PHYSICAL CHEMISTRY. *Educ:* Reed Col, BA, 52; Univ Idaho, MS, 58. *Prof Exp:* Res chemist, Reed Inst, 52 & Gen Elec Co, Hanford, Wash, 53-55, opers chemist, 56-58; nuclear chemist, Lawrence Livermore Lab, Univ Calif, 58-71; staff scientist atmospheric res, Tech Appln Ctr, US Air Force, 71-80; PHYS SCI OFFICER, NUCLEAR TESTING, CHEMICAL WEAPONS, US ARMS CONTROL & DISARMAMENT AGENCY, 80- *Mem:* AAAS; Am Chem Soc; NY Acad Sci; Sigma Xi. *Res:* Nuclear decay schemes, actinide and rare earth separation chemistry; Raman and infrared studies of metal oxide systems; fast neutron reactions; radioactivity in the environment; seismology of attenuation of plate boundaries. *Mailing Add:* 5115 Holden St Fairfax VA 22032

MORROW, RICHARD M, b WVa, 1926; m; c 3. PETROLEUM ENGINEERING. *Educ:* Ohio State Univ, BS. *Prof Exp:* Staff, Amoco Prod Co, 48-62, chief engr, Prod Dept, 62-65, mgr, Denver Region, 65-66, exec vpres, Amoco Int Oil, 66-70, exec vpres, Amoco Chem, 70-74, pres, 74-76, dir, Amoco Corp, 76-78, pres, 78-83, chmn & chief exec officer, 83-91; RETIRED. *Mem:* Nat Acad Eng; Am Petrol Inst; Nat Petrol Coun. *Mailing Add:* 200 E Randolph Dr Suite 7909 Chicago IL 60601

MORROW, ROY WAYNE, b Hopkinsville, Ky, Sept 28, 42; m 66; c 2. ANALYTICAL CHEMISTRY. *Educ:* Murray State Univ, BS, 64; Univ Tenn, MS, 67, PhD(chem), 70. *Prof Exp:* Develop chemist, Union Carbide Corp, Oak Ridge, 70-75, sect head, Anal Develop, Y-12 Plant, 76-79, mgr spec serv dept, 79-80, mgr, Anal Chem Dept, Gaseous Diffusion Plant, 80-85; MGR, ANAL CHEM DEPT, GASEOUS DIFFUSION PLANT, MARTIN-MARIETTA ENERGY SYSTS CORP, OAK RIDGE, 85- *Mem:* Am Chem Soc; Soc Appl Spectros; Sigma Xi. *Res:* Flame emission and atomic absorption spectroscopy; gas chromatography. *Mailing Add:* 102 E Morningside Dr Oak Ridge TN 37830

MORROW, SCOTT, b Oklahoma City, Okla, Sept 11, 20; m 45; c 3. LIGHT & ELECTRON MICROSCOPY. *Educ:* Case Western Reserve Univ, BS, 47, MS, 49, PhD(inorg chem), 51. *Prof Exp:* Res chemist, Mound Lab, Monsanto Chem Co, Ohio, 51-53, Mass, 53-54 & Socony Mobil Co, NJ, 54-56; proj chemist, Thiokol Chem Co, 56-66; res chemist, US Army Armament Res & Develop & Eng Ctr, Armament Eng Directorate, Dover, NJ, 66-89; RETIRED. *Mem:* Am Chem Soc. *Res:* Solid state propellants and explosives; catalysis and combustion research; mixed crystals and solid solutions; microscopy of thin films; thermal analysis by microscopy; ignition and combustion of gun propellants; electric ignition of guns; microscopical characterization of nitrocellulose and explosives; scanning electron microscopy of energetic materials; inorganic chemistry, phase rule studies; chemistry of explosives and propellants. *Mailing Add:* 36 East Shore Rd Denville NJ 07834

MORROW, TERRY ORAN, b Latrobe, Pa, May 24, 47; m 75; c 1. MICROBIAL GENETICS. *Educ:* Grove City Col, BS, 69; Bowling Green State Univ, MA, 71, PhD(microbial genetics), 73. *Prof Exp:* Asst prof biol, Univ Wis-River Falls, 73-75; ASST PROF BIOL, CLARION STATE STATE UNIV, 75- *Mem:* Am Soc Microbiol; Sigma Xi; Am Inst Biol Sci; Sigma Xi. *Res:* Genetics of Staphylococcus aureus. *Mailing Add:* Dept of Biol Clarion State Col Clarion PA 16214

MORROW, THOMAS JOHN, b Kew Gardens, NY, Mar 11, 46; m; c 3. NEUROPHYSIOLOGY, PHYSIOLOGY. *Educ:* Fordham Univ, BS, 67; Long Island Univ, MS, 69; Univ Mich, PhD(physiol), 76. *Prof Exp:* Res assoc II neurophysiol, Dept Physiol, 75-78, asst res scientist, 78-80, adj instr, Dept Physiol, 84-88, ADJ ASST PROF, DEPT PHYSIOL, UNIV MICH, 88-, ASST RES SCIENTIST, DEPT NEUROL, 89-; RES HEALTH SCI SPECIALIST, VET ADMIN MED CTR, ANN ARBOR, MICH, 80- *Concurrent Pos:* Adj instr, dept physiol, Univ Mich, 80- *Mem:* NY Acad Sci; Int Asn Study Pain; Am Pain Soc; Soc Neurosci; AAAS; Am Acad Clin Neurophysiology. *Res:* Neurophysiology of pain and somesthesis. *Mailing Add:* Vet Admin Med Ctr Res 151 2215 Fuller Rd Ann Arbor MI 48105

MORROW, WALTER E, JR, b Springfield, Mass, July 24, 28; m 51; c 3. ELECTRICAL ENGINEERING, SPACE PHYSICS. *Educ:* Mass Inst Technol, SB, 49, SM, 51. *Prof Exp:* Asst elec eng, 49-51, staff mem, 51-55, asst group leader, 55-56, group leader, 56-64, assoc div head, 64-66, div head, 66-68, from asst dir to assoc dir, 68-77, DIR, LINCOLN LAB, MASS INST TECHNOL, 77-, PROF ELEC ENG & COMPUT SCI, 78- *Concurrent Pos:* Mem comn VI, Int Union Radio Sci, 62-; consult, Dept Navy, 71-; mem, NASA Space Appln Comt, 72-; mem, Nat Res Coun, Assembly Eng Space Appln Bd Comt on Satellite Commun, 75-; mem, Defense Commun Agency Sci Adv Group, 76-; mem, Air Force Sci Adv Bd, 78, mem, Defense Sci Bd, 87- *Honors & Awards:* Edwin Howard Armstrong Achievement Award, Inst Elec & Electronics Engrs, 76. *Mem:* Fel Nat Acad Eng; fel Inst Elec & Electronics Engrs. *Res:* Ionospheric and tropospheric radio communication and propagation; orbital scatter communication; orbital dipole experiment; communication satellites; lunar and planetary radar studies. *Mailing Add:* Lincoln Lab Mass Inst of Technol Lexington MA 02173

MORROW, WILLIAM JOHN WOODROOFE, Brit citizen; m 80; c 1. AUTOIMMUNITY, VIROLOGY. *Educ:* Univ Col, Cardiff, UK, BSc, 74; Plymouth Polytech, UK, PhD(immunol), 78. *Prof Exp:* Postdoctoral res fel, Middlesex Hosp Med Sch, London, 77-82; vis scientist, Sch Med, Univ Calif, San Francisco, 82-85, asst res immunologist, 85-87; SR SCIENTIST, IDEC PHARMACEUT CORP, 87- *Mem:* Am Asn Immunologists. *Res:* Pathology and immunotherapy for infectious disease, autoimmunity and cancer. *Mailing Add:* IDEC Pharmaceut Corp 11099 N Torrey Pines Rd No 160 La Jolla CA 92037

MORROW, WILLIAM SCOT, b New York, NY, Jan 26, 31; m 70; c 1. BIOCHEMISTRY, ANALYTICAL CHEMISTRY. *Educ:* Philadelphia Col Pharm & Sci, BS, 59; St Joseph's Col, MS, 64; Univ NC, Chapel Hill, PhD(biochem), 69. *Prof Exp:* Res technician high temperature res, Res Inst Temple Univ, 55-58; assoc res chemist electronics res, Int Resistance Co, 59-60; res biochemist ophthal res, Wills Eye Hosp, 60-63; asst prof biol, Concord Col, 68-70; ASST PROF CHEM, WOFFORD COL, 70- *Mem:* Am Chem Soc; AAAS. *Res:* Chromosomal proteins, insect biochemistry; abiogenesis: early cellular organization; development of techniques in analytical chemistry. *Mailing Add:* Dept Chem Wofford Col N Church St Spartanburg SC 29303-3840

MORS, WALTER B, natural products chemistry, for more information see previous edition

MORSE, BERNARD S, b New York, NY, Sept 11, 34; m 60; c 2. HEMATOLOGY, INTERNAL MEDICINE. *Educ:* NY Univ, BA, 55, BS, 56; Seton Hall Univ, MD, 60. *Prof Exp:* Asst prof med, Tufts Univ, 67-69; from assoc prof to prof med, 69-87, CLIN PROF MED, UNIV MED & DENT, NJ, 87- *Mem:* Am Soc Hemat; Int Soc Hemat; Am Fedn Clin Res. *Res:* Erythropoiesis; ferrokinetics; platelet immunology. *Mailing Add:* 2787 Kennedy Blvd Jersey City NJ 07306

MORSE, BURT JULES, b New York, NY, June 17, 26. APPLIED MATHEMATICS. *Educ:* City Col New York, BS, 49; Columbia Univ, AM, 51; NY Univ, PhD(math), 63. *Prof Exp:* Mathematician, Vitro Corp Am, 52-54; mathematician, Int Bus Mach Corp, 54-58; res assoc electromagnetic theory, NY Univ, 58-63; asst prof math, St John's Univ, NY, 63 & Univ NMex, 63-66; res mathematician, Gen Elec Co, 66-67 & Philco-Ford Corp, Pa, 67-68; mathematician, Nat Hurricane Res Lab, Nat Oceanic & Atmospheric Admin, 68-73, Satellite Res Lab, 73-90; RETIRED. *Mem:* Am Math Soc; Soc Indust & Appl Math. *Res:* Geophysical fluid dynamics; numerical analysis; meteorology; wave propagation. *Mailing Add:* 6129 Leesburg Pike Apt 1005 Falls Church VA 22041

MORSE, DANIEL E, b New York, NY, May 20, 41. MOLECULAR REGULATION OF DEVELOPOMENT. *Educ:* Harvard Unv, BA, 63; Albert Einstein Col Med, PhD(molecular biol), 67. *Prof Exp:* Fel molecular genetics, Stanford Univ, 67-69; from Silas Arnold Houston asst prof to Silas Arnold Houston assoc prof, Harvard Univ Med Sch, 69-73; PROF MOLECULAR GENETICS & BIOCHEM, UNIV CALIF, SANTA BARBARA, 73-, CHMN, MARIE BIOTECHNOL CTR, 86- *Concurrent Pos:* Chmn, Sect Molecular Biol & Biochem, Dept Biol Sci, Univ Calif, Santa Barbara, 81-85; mem, Nat Res Coun, US Nat Comt Int Union Biol Sci, 86-; chmn, Task Force Biotechnol in Ocean Sci, NSF, 87- *Mem:* Fel AAAS; Am Soc Molecular Biol & Biochem; NY Acad Sci; Int Soc Chem Ecol; Am Soc Limnol & Oceanog; Am Soc Microbiol; Am Soc Zool. *Res:* Molecular mechanisms controlling reproduction, larval metamorphosis, development & gene expression; signal molecules, receptors, & transducers; molecular marine biology; molecular neurobiology; molecular chemosensory mechanisms. *Mailing Add:* Marine Sci Inst Univ Calif Santa Barbara CA 93106

MORSE, DENNIS ERVIN, b Loup City, Nebr, Mar 21, 47; m 67, 83; c 6. HUMAN ANATOMY. *Educ:* Hastings Col, BA, 69; Univ NDak, MS, 71, PhD(anat), 73. *Prof Exp:* Asst prof anat, George Washington Univ, 73-76; asst prof, 76-79, ASSOC PROF ANAT, MED COL OHIO, 79- *Mem:* Am Asn Anatomists; Am Soc Cell Biol; Sigma Xi; Electron Micros Soc Am; Teratology. *Res:* Developmental biology of cardiovascular system; extracellular connective tissue ultrastructure; ocular development. *Mailing Add:* Dept Anat PO Box 10008 Toledo OH 43699-0008

MORSE, EDWARD EVERETT, b Gardner, Mass, June 7, 32; m 54; c 4. MEDICINE, HEMATOLOGY. *Educ:* Harvard Univ, AB, 54, MD, 58; Am Bd Path, dipl, 69, cert, 73. *Prof Exp:* Intern & resident, Johns Hopkins Hosp, 58-60; asst surgeon, Nat Cancer Inst, 60-62; USPHS spec fel hemat, Johns Hopkins Hosp, 62-63, from instr asst prof med, Johns Hopkins Univ, 63-68; assoc prof, 68-74, PROF LAB MED, 74-, DIR HEMAT DIV, SCH MED, UNIV CONN, FARMINTON, 68-, PROF MED, 80- *Concurrent Pos:* Med dir, Conn Red Cross Blood Prog, 68-74, consult, 74-; consult, Newington Vet Admin Hosp, Hartford & Bristol Hosps. *Mem:* Fel Am Col Physicians; fel Am Soc Clin Path; fel Asn Clin Scientists; Am Soc Hemat; Int Soc Blood Transfusion; Int Soc Hemat. *Res:* Laboratory medicine; cell preservation; stem cell and granulocyte transfusion therapy; platelet physiology; cell markers and flow cytometry. *Mailing Add:* Dept Lab Med Univ Conn Health Ctr Farmington CT 06030

MORSE, ERSKINE VANCE, b Peoria, Ill, June 25, 21; m 45; c 4. VETERINARY MICROBIOLOGY, VETERINARY PUBLIC HEALTH. *Educ:* Cornell Univ, DVM, 44, MS, 48, PhD(vet bact), 49; Am Col Vet Prev Med, dipl & cert vet pub health, 75. *Prof Exp:* Asst pathogenic bact, Cornell Univ, 47-48, Am Vet Med Asn fel, 48-49; from asst prof to assoc prof vet sci, Univ Wis, 49-55; from assoc prof to prof microbiol & pub health, Mich State Univ, 55-58; prof & assoc dir vet med res inst, Iowa State Univ, 58-60; prof vet sci, head dept & dean sch agr exp sta, 60-70, H W Handley prof vet med & environ health & assoc dir Environ Health Inst, Sch Vet Med, 70-86, EMER PROF & DEAN, PURDUE UNIV, 86- *Concurrent Pos:* Consult, Nat Asn Stand Med Vocab, 62-, Vet Admin, 66-74, Surgeon Gen, US Air Force, 68-70, USPHS & AID; alt chmn US deleg, Conf Vet Med Educ, Food & Agr Orgn-WHO, Copenhagen, 65; judge, Int Sci Fair, 65, 68, 69, 74-79; mem, Nat Bd Vet Med Examr, 65-74; nat counr, Purdue Res Found, 65-; mem bd trustees, Am Asn Accreditation of Lab Animal Care, 68-75, vchmn bd, 72-73, chmn bd, 74-75; mem nat coun health prof educ assistance, Dept Health, Educ & Welfare, 69-72; chmn comt animal health, Nat Res Coun-Nat Acad Sci, 69-72; evaluator-consult, NCent Asn Cols & Schs-Comn Insts Higher Educ, 72-; Purdue Univ liaison rep for vet serv to Asst Surgeon Gen, US Army, 73-; reviewer, Jour, Am Vet Med Asn, 74- & Am J Vet Res, 75-80. *Mem:* Soc Exp Biol & Med; Conf Res Workers Animal Dis; Am Vet Med Asn (secy, 66-69); US Animal Health Asn; Am Asn Lab Animal Sci. *Res:* College administration; environmental health; pathogenic bacteriology; infectious diseases of animals; salmonellosis, leptospirosis, vibriosis, brucellosis and corynebacterial infections; laboratory animal medicine; epidemiology, microbiology and treatment of zoonotic diseases; monitoring environmental quality and water pollution. *Mailing Add:* 345 Leslie Ave West Lafayette IN 47906

MORSE, FRANCIS, b Pittsfield, Mass, Sept 8, 17; m 45; c 4. AEROSPACE ENGINEERING. *Educ:* Yale Univ, BE, 39; Calif Inst Technol, MS, 40. *Prof Exp:* Aeronaut engr, Lockheed Aircraft Corp, 40-42 & Goodyear Aircraft Corp, 42-45; ASSOC PROF AEROSPACE ENG, COL ENG, BOSTON UNIV, 62-, ASSOC PROF THERMODYN & ENG MECH, 63- *Concurrent Pos:* Consult eng, 46- *Mem:* Am Soc Eng Educ; Am Inst Aeronaut & Astronaut. *Res:* Lighter-than-air craft design; aircraft propulsion; engineering graphics. *Mailing Add:* Dept Aerospace & Mech Eng Boston Univ Boston MA 02215

MORSE, FRED A, b Colorado Springs, Colo, Jan 11, 37; m 57; c 3. SPACE PHYSICS. *Educ:* Univ Idaho, BS, 58; Univ Mich, MS, 60, PhD(phys chem), 62. *Prof Exp:* Fel, Univ Mich, 62; instr phys chem, Univ Del, 63-64; mem tech staff, Space Physics Lab, Aerospace Corp, 64-63, staff scientist, 68-72, assoc dept head atmospheric physics, 72-75, head atmospheric physics, 75-80; group leader, neutron measurement group P-15, Los Alamos Nat Lab, 80-82, assoc div leader, physics div, 82-84, physics div leader, 84-86, ASSOC DIR RES, LOS ALAMOS NAT LAB, 87- *Mem:* AAAS; Am Geophys Union; Am Phys Soc. *Res:* Molecular and ion beam collision; atomic spectroscopy; atmospheric and auroral physics; chemiluminescent reaction rate; aeronomy; ionospheric physics. *Mailing Add:* 181 San Ildefonso Rd Los Alamos NM 87544

MORSE, GARTH EDWIN, physics, for more information see previous edition

MORSE, HELVISE GLESSNER, b Frederick, Md, Sept 17, 25; m 49; c 2. CYTOGENETICS OF SOMATIC CELLS, RESEARCH TUMOR CYTOGENETICS. *Educ:* Hood Col, Frederick, Md, BA, 46; Univ Ky, Lexington, MS, 49; Univ Colo, Denver, MS, 63, PhD(biophyics & genetics), 66. *Prof Exp:* Teaching fel, 66-69, instr, 70-73, cytogeneticist, 73-79, asst prof, 79-87, ASSOC PROF BIOCHEM, BIOPHYS & GENETICS, UNIV COLO, DENVER, 88-, FEL ELEANOR ROOSEVELT, INST CANCER RES, 79- *Concurrent Pos:* Mem, Cytogenetics Comt Children's Cancer Study Group, 83-86; leukemia cytogeneticist, Children's Hosp, Denver, Colo, 76-86; dir, cytogenetics core lab, Univ Colo Cancer Ctr, 88- *Mem:* Am Soc Human Genetics. *Res:* Diagnostic cytogenetics of malignancy in patients; premature chromosome condensation as a means of studying direct cytogenetics in tumor specimens; cytogenetic interpretations of human material in Chinese hamster and human hybrids; comparisons of cytogenetics of malignant melanoma derived from various metastatic sites. *Mailing Add:* Biochem/Biophys & Genetics Dept Univ Colo Health Sci Ctr 4200 E Ninth Ave Denver CO 80262

MORSE, HERBERT CARPENTER, III, b Washington, DC, May 7, 43; m 69; c 2. IMMUNOLOGY. *Educ:* Oberlin Col, BA, 65; Harvard Univ, MD, 70. *Prof Exp:* From intern to resident, Peter Bent Brigham Hosp, 70-72; res assoc, 72-75, SR INVESTR IMMUNOL, NAT INST ALLERGY & INFECTIOUS DIS, 75- *Concurrent Pos:* Attend physician, Arthritis & Rheumatism Br, Nat Inst Arthritis, Metab & Digestive Dis, 76- *Mem:* Am Asn Immunologists; Sigma Xi. *Res:* Interactions of viruses and the immune response; regulation of antibody formation. *Mailing Add:* Bldg Seven Rm 302 NIAID NIH 9000 Rockville Pike Bethesda MD 20892

MORSE, IVAN E, JR, b Fountain, Mich, Dec 9, 25; m 51; c 3. MECHANICAL ENGINEERING. *Educ:* Mich State Univ, BS, 50, MS, 54; Purdue Univ, PhD(mech eng), 61. *Prof Exp:* From instr to asst prof mech eng, Mich State Univ, 50-61; assoc prof, 61-64, prof-in-chg dept, 67-68, PROF MECH & INDUST ENG, UNIV CINCINNATI, 64-, HEAD DEPT, 79- *Concurrent Pos:* Mem sci adv bd & consult, KDI Corp, 66-70, Nelson Industs, Inc, 74- *Mem:* Am Soc Eng Educ; Am Soc Mech Engrs; Acoust Soc Am. *Res:* Machine tool vibrations; signature analysis; testing and design of mechanical equipment. *Mailing Add:* Dept Mech Eng Univ Cincinnati Cincinnati OH 45221

MORSE, J(EROME) G(ILBERT), b New York, NY, Oct 22, 21; m 49; c 2. ENVIRONMENTAL RADIATION. *Educ:* City Col New York, BS, 42; Univ Pa, MS, 47; Ill Inst Technol, PhD(chem), 52. *Prof Exp:* Dir small power systs depts, Martin Co, 55-64, tech dir, Hispano-Martin, Paris, 65-67; mgr space sci, Martin Marietta Corp, Colo, 67-74; dep dir, Colo Energy Res Inst, 74-77; PRES, MORSE ASSOCS, INC, 77-; RES PROF PHYSICS, COLO SCH MINES, 84- *Concurrent Pos:* Staff consult, Oak Ridge Inst Nuclear Studies, 58-64; lectr, Colo Sch Mines, adj prof, dept of physics, 76-86; consult, Colo Energy Res Inst, 77-83 & US Dept Energy, 78-81. *Mem:* Health Physics Soc; fel Am Nuclear Soc; Sigma Xi. *Res:* Environmental radiation; materials science. *Mailing Add:* 5066 Tule Lake Dr Littleton CO 80123-2758

MORSE, JANE H, b Grosse Pointe, Mich, Aug 27, 29; wid; c 2. IMMUNOLOGY, RHEUMATOLOGY. *Educ:* Smith Col, BA, 51; Columbia Univ, MD, 55. *Prof Exp:* Res assoc immunol, Rockefeller Univ, 60-62; instr & asst prof med, 62-75, assoc prof, 75-86, PROF CLIN MED, COL PHYSICIANS & SURGEONS, COLUMBIA UNIV, 86- *Concurrent Pos:* Mem, Allergy & Immunol Study Sect, USPHS, 73-77; mem, Allergy & Clin Immunol Res Comt, NIAID, NIH, 79-82. *Mem:* Am Asn Immunologists; Am Rheumatism Asn. *Res:* Immune complexes and complement in autoimmune diseases; the role of inhibitors in the immune response; immunoregulatory factors found in crude human chorionic gonadotropin and pregnancy urine. *Mailing Add:* 630 W 168th St Dept Med New York NY 10032

MORSE, JOHN THOMAS, b Oakland, Calif, Apr 30, 35; m 60; c 2. ENVIRONMENTAL PHYSIOLOGY. *Educ:* Ore State Univ, BS, 56; Univ Calif, Davis, PhD(physiol), 68. *Prof Exp:* Pvt indust res grant, 68-69, asst prof, 68-72, ASSOC PROF PHYSIOL, CALIF STATE UNIV, SACRAMENTO, 72- *Mem:* Am Physiol Soc; Am Inst Biol Sci. *Res:* Physiological adaptations favoring physical work performance in varying environments. *Mailing Add:* Dept Biol Sci Calif State Univ 6000 J St Sacramento CA 95819

MORSE, JOHN WILBUR, b Ft Dodge, Iowa, Nov 11, 46; m; c 1. GEOCHEMISTRY. *Educ:* Univ Minn, BS, 69; Yale Univ, MPhil, 71, PhD(geol), 73. *Prof Exp:* Asst prof oceanog, Fla State Univ, 73-76; assoc prof Marine & atmospheric Chem, Univ Miami, 76-81; PROF OCEANOG, TEX A & M UNIV, 81- *Concurrent Pos:* Fulbright Traveling Scholar, 87. *Mem:* Am Geophys Union; Int Asn Geochemists & Cosmochemists; Am Chem Soc; AAAS; Geochem Soc Am; Geol Soc Am; Oceanog Soc. *Res:* Application of chemical kinetics and surface chemistry to diagenetic reactions in marine sediment; transuranic element chemistry in environment; modelling nonequilibrium process. *Mailing Add:* 3910 Old Oaks Dr Bryan TX 77802

MORSE, JOSEPH GRANT, b Colorado Springs, Colo, Oct 16, 39; m 63; c 2. INORGANIC CHEMISTRY. *Educ:* SDak State Col, BS, 61; Univ Mich, MS, 63, PhD(inorg chem), 67. *Prof Exp:* Lectr, Univ Mich, 65-66; asst prof, 68-74, ASSOC PROF CHEM, UTAH STATE UNIV, 74-, DIR, UNIV HONORS PROG, 86- *Concurrent Pos:* Vis assoc prof, Univ NC, 80-81. *Mem:* Fel AAAS; Am Chem Soc; Sigma Xi. *Res:* Synthesis and properties of new ligands, especially of phosphorus group; chemistry of pi-acid chelates; photochemistry of fluorophosphines; reactions of coordinated liquids. *Mailing Add:* Dept Chem & Biochem Utah State Univ Logan UT 84322-0300

MORSE, JOSEPH GRANT, b Ithaca, NY, May 26, 53. ENTOMOLOGY. *Educ:* Cornell Univ, BS, 75; Mich State Univ, MS(systs sci), 77, MS, 78, PhD(entom), 81. *Prof Exp:* Asst prof, 81-88, ASSOC PROF ENTOM, UNIV CALIF, RIVERSIDE, 88- *Mem:* Entom Soc Am. *Res:* Utilization of quantitative research techniques in integrated pest management research programs; pest management of citrus and subtropical fruit crops. *Mailing Add:* Dept Entom Univ Calif Riverside PO Box 112 Riverside CA 95251-0314

MORSE, KAREN W, b Monroe, Mich, May 8, 40; m 63; c 2. INORGANIC CHEMISTRY. *Educ:* Denison Univ, BSc, 62; Univ Mich, MSc, 64, PhD(chem), 67. *Hon Degrees:* DSc, Denison Univ, 90. *Prof Exp:* Res scientist, Ballistic Res Inst, 67-68; lectr, Utah State Univ, 68-69, from asst prof to assoc prof, 69-83, dept head, 81-88, dean sci, 88-89, PROF CHEM, UTAH STATE UNIV, 83-, PROVOST, 89- *Concurrent Pos:* Mem, Women's Chemist Comt, Am Chem Soc, 76-84, Comt Prof Training, 84- & adv bd, Petrol Res Fund, 84. *Mem:* Am Chem Soc; fel AAAS; Asn Women Sci; Sigma Xi. *Res:* Synthesis and behavior of new hydroborates; boron analogs of biologically active compounds. *Mailing Add:* Dept Chem & Biochem Utah State Univ Logan UT 84322

MORSE, LEWIS DAVID, b Brooklyn, NY, Oct 29, 24; m 46; c 1. BIOCHEMISTRY, SYNTHETIC POLYMER RESEARCH. *Educ:* NY Univ, BA, 48; Brooklyn Col, MA, 52. *Prof Exp:* Res chemist, Col Physicians & Surgeons, Columbia Univ, 50-51 & Stein Hall Corp, 51-55; chief chemist & prod supvr, Myer 1890 Beverages, Inc, 55-59; proj leader, Am Sugar Refining Co, Inc, 59-63 & Nat Cash Register Co, 63-64; mgr prod res, Ionac Chem Co, NJ, 64-67; prod dev fel, Calgon Corp, Merck & Co Inc, Pittsburgh, PA, 67, group leader fine chem, Prod Develop, Rahway, 67-80, sr res assoc, 80-89; CONSULT, MICROENCAPSULATION, 89- *Mem:* AAAS; Am Chem Soc; Inst Food Technologists; NY Acad Sci. *Res:* Polymer chemistry and synthesis; microencapsulation organic synthesis; ion exchange; surfactant technology; flavor compounding; enzyme, sugar, starch, adhesive and microbiological chemistry; statistical design and analysis of experiments; vitamin technology in foods and medicinals; nutrition; antiseptics. *Mailing Add:* 307 S Dithridge St No 705 Pittsburgh PA 15214

MORSE, M PATRICIA, b Hyannis, Mass, Aug 29, 38. MALACOLOGY. *Educ:* Bates Col, BS, 60; Univ NH, MS, 62, PhD(zool), 66. *Hon Degrees:* DSc, Plymouth State Col, NH, 88. *Prof Exp:* Instr biol, Suffolk Univ, 62-63; from instr to asst prof, 64-79, PROF BIOL, NORTHEASTERN UNIV, 79-; AT MARINE LABS, UNIV WASHINGTON. *Concurrent Pos:* Trustee, Bates Col; res assoc malacol, Harvard Univ, 74-83; Brasilian grant to study interstitial mollusks, 75; Fulbright fel, Fiji Islands, 78-79; researcher, Univ Wash Marine Labs, 85-86. *Mem:* Am Soc Zoologists (pres, 85); Sigma Xi (pres, 89). *Res:* Functional cytomorphology of bivalve molluscs; ultrastructure of molluscan blood cells and blood forming tissues; biology and systematics of interstitial molluscs and priapulids. *Mailing Add:* Marine Sci Ctr Northeastern Univ Nahant MA 01908

MORSE, MELVIN LAURANCE, b Hopkinton, Mass, Feb 23, 21; m 49; c 2. GENETICS. *Educ:* Univ NH, BS, 44; Univ Ky, MS, 47; Univ Wis, PhD(genetics), 55. *Prof Exp:* Jr biologist radiation res, Biol Div, Oak Ridge Nat Lab, 47- 51; res assoc genetics, Univ Wis, 55-56; res microbiologist, Webb-Waring Lung Inst, 56-58, asst dir inst, 59-71, actg dir, 71-72, from asst prof to assoc prof, Med Ctr, 56-66, vchmn dept, 71-73, dean grad sch, 76-84, PROF BIOPHYS & GENETICS, MED CTR, UNIV COLO, 66-, JAMES J WARING CHAIR BIOL, 60- *Concurrent Pos:* USPHS sr res fel, 61, career develop award, 62-70; consult, Army Med Res & Nutrit Lab, Fitzsimons Gen Hosp, 60-70; foreign res, Inst Molecular Biol, Univ Geneva, 62-63. *Mem:* Am Soc Microbiol; Am Genetic Asn; Am Soc Human Genetics. *Res:* Biochemical genetics; biophysics; microbiology. *Mailing Add:* Webb-Waring Lung Inst Univ of Colo Health Sci Ctr 4200 E 9th Ave Denver CO 80262

MORSE, N(ORMAN) L(ESTER), b Bethlehem, Pa, Nov 25, 17; m 45; c 3. CHEMICAL ENGINEERING. *Educ:* Lehigh Univ, BS, 40; Univ Rochester, PhD(phys chem), 43. *Prof Exp:* Petrol technologist, Shell Oil Co, Inc, 43-60, asst chief res technologist, Houston Res Lab, Shell Oil Co Refinery, Wood River, Ill, 60-70, tech Adv, 70-76; tech adv, Shell Develop C0, 76-83; RETIRED. *Mem:* AAAS; Am Chem Soc; Am Inst Chem Engrs; Sigma Xi. *Res:* Mechanism; kinetics of catalytic reactions of hydrocarbons. *Mailing Add:* Shell Develop Co 1211 Briar Park Dr Houston TX 77042

MORSE, PHILIP DEXTER, II, molecular biology, for more information see previous edition

MORSE, R(ICHARD) A(RDEN), b Independence, Kans, Feb 18, 20; m 42; c 1. PETROLEUM ENGINEERING. *Educ:* Univ Okla, BS, 42; Pa State Univ, MS, 47. *Prof Exp:* Petrol Engr, Halliburton Oilwell Cementing Co, 42-43; staff engr, Superior Oil Co, 43-44 & West Edmond Eng Asn, 44-45; asst, Pa State Col, 45-47; res sect head, Stanolind Oil & Gas Co, 47-55; assoc dir res, Gulf Res & Develop Co, Pa, 55-64; mgr, Mene Grande Oil Co, Venezuela, 64-70; PROF PETROL ENG, TEX A&M UNIV, 70- *Mem:* Am Inst Mining, Metall & Petrol Engrs (vpres, 60-). *Res:* Methods of recovery of oil and gas from natural reservoirs; oil well drilling, completion and stimulation. *Mailing Add:* 1001 Goode Bryan TX 77803

MORSE, RICHARD STETSON, physics; deceased, see previous edition for last biography

MORSE, ROBERT MALCOLM, b Haverhill, Mass, Dec 21, 38; m 61; c 2. HIGH ENERGY PHYSICS. *Educ:* San Jose State Col, BA, 63; Univ Wis-Madison, MA, 65, PhD(physics), 69. *Prof Exp:* Res assoc physics, Univ Wis, 69-70; res assoc physics, Univ Colo, Boulder, 70-76, asst prof physics & astrophys, 74-76; asst prof, Univ Pa, 76-77; ASSOC SCIENTIST, UNIV WIS, 77- *Mem:* Am Phys Soc. *Res:* Weak interactions of strange particles; electron-positron storage ring physics; hadron calorimetery. *Mailing Add:* 5130 Minocqua Crescent Madison WI 53705-1320

MORSE, ROBERT WARREN, b Boston, Mass, May 25, 21; m 43; c 3. PHYSICS. *Educ:* Bowdoin Col, BS, 43; Brown Univ, ScM, 47, PhD(physics), 49. *Hon Degrees:* ScD, Bowdoin Col, 66. *Prof Exp:* From asst prof to prof physics & head dept, Brown Univ, 49-62, dean col, 62-64; asst secy Navy for Res & Develop, 64-66; pres, Case Inst Technol, 66-67; pres, Case Western Reserve Univ, 67-71; dir res, 71-73, assoc dir & dean, 73-79, sr scientist, 79-83, EMER SCIENTIST, WOODS HOLE OCEANOG INST, 83- *Concurrent Pos:* Howard Found fel, Cambridge Univ, 54-55; mem comt undersea warfare, Nat Acad Sci, 58-64, chmn, 62-64, chmn bd human resources, 70-74, chmn ocean affairs bd, 71-75, chmn comt emergency mgt, 81-82; chmn interagency comt oceanog, Fed Coun Sci & Technol, 64-66; mem, Naval Res Adv Comt, 71-74; dir, Res Corp, 69-, PPG Indust Inc, 69- *Mem:* Fel Acoust Soc Am (pres, 65-66); fel Am Phys Soc; fel Am Acad Arts & Sci; fel AAAS. *Res:* Ultrasonics and underwater sound; superconductivity and properties of electrons in metals. *Mailing Add:* Box 574 North Falmouth MA 02556

MORSE, ROGER ALFRED, b Saugerties, NY, July 5, 27; m 51; c 3. ENTOMOLOGY, APICULTURE. *Educ:* Cornell Univ, BS, 50, MS, 53, PhD(entom), 55. *Hon Degrees:* Dr, Univ Warclaw, Poland. *Prof Exp:* Entomologist, State Plant Bd, Fla, 55-57; asst prof hort, Univ Mass, 57; from asst prof to assoc prof apicult, 57-72, chmn, Dept Entom, 86-89, PROF APICULT, CORNELL UNIV, 72- *Concurrent Pos:* Vis prof, Col Agr, Univ Philippines, 68, Univ Sao Paulo, 78, Univ Helsinki, 88. *Mem:* Fel AAAS; fel Entom Soc Am; Bee Res Asn; Sigma Xi. *Res:* Evolution of the Apoidea; toxicity of insecticides to honey bees, honey wine, honey production and handling; social structure of honey bee colony; diseases of honey bees. *Mailing Add:* Dept Entom Comstock Hall Cornell Univ Ithaca NY 14850

MORSE, RONALD LOYD, b Kearney, Nebr, May 15, 40; m 63; c 2. INDUSTRIAL ORGANIC CHEMISTRY. *Educ:* Univ Nebr, BS, 63; Univ Wis, PhD(org chem), 68. *Prof Exp:* Sr res chemist, 67-80, SR RES SPECIALIST, MONSANTO INDUST CHEM CO, 80- *Mem:* Am Chem Soc. *Mailing Add:* 876 Holly Ridge Rd Ballwin MO 63011-3554

MORSE, ROY E, b Boston, Mass, Nov 3, 16; m 46; c 3. RESEARCH-MANAGEMENT RELATIONS, EXERCISE PHYSIOLOGY. *Educ:* Univ Mass, BS, 40, MS, 41, PhD(food technol), 48. *Prof Exp:* Control food technologist, Hills Bros Co, NY, 41; instr food technol, Ore State Col, 41-42; packaging researcher, Owens-Ill Co, Calif, 42-43; prod mgr, Featherweight Foods, Maine, 43-44; instr food technol, Univ Mass, 46-48; assoc prof, Univ Ga, 48-49; group leader, Food Lab, Monsanto Chem Co, 49-51; dir res, Kingan & Co, Ind, 51-53 & Wm J Stange Co, 53-55; chmn dept food sci, Rutgers Univ, 55-59, prof, 69-80; vpres res, Thomas J Lipton, Inc, 59-66, Pepsico Co, Inc, NY, 66-69, R J Reynolds Tobacco Co, 80-84 & R J Reynolds Indust, 84-85; RETIRED. *Concurrent Pos:* Mem nutrit adv comt, Govt Tunisia; ed, Biol Abstracts; vis prof, Univ New South Wales, 75 & Univ Iceland, 76; mem adv comt, Radford Univ, 83-87; adj prof, Va Polytech & State Univ, 85-, NC State Univ, 86-; consult, food res, self employed, 85-; interim pres, NC Biotechnol Ctr, 87-89; chair, Mil Foods Comt, Nat Res Coun. *Honors & Awards:* Carl R Fellers Award, Inst Food Technologists. *Mem:* Fel AAAS; Am Chem Soc; fel Inst Food Technologists; Am Mgt Soc; Sigma Xi. *Res:* Food industry; bakers yeast, tobacco, tea, carbonated beverages and food packaging. *Mailing Add:* 7522 Lasater Rd Clemmons NC 27012

MORSE, STEARNS ANTHONY, b Hanover, NH, Jan 3, 31; m 60; c 3. PETROLOGY, GEOCHEMISTRY. *Educ:* Dartmouth Col, AB, 52; McGill Univ, MSc, 58, PhD(geol), 62. *Prof Exp:* Mem, Blue Dolphin Labrador Exped, 49, 51, 52 & 54; petrologist, Brit Nfld Explor Ltd, Can, 59-61; geologist, Cold Regions Res & Eng Lab, US Army, 62; from asst prof to assoc prof geol, Franklin & Marshall Col, 62-71; assoc prof, 71-74, PROF GEOL, UNIV MASS, AMHERST, 74- *Concurrent Pos:* Carnegie Corp fel, Carnegie Inst Geophys Lab, Washington, DC, 67-68; res fel, Mineralogist Mus, Oslo, Norway, 77-78 & Univ Cambridge, 84-85; assoc ed, J Geophys Res, 81-84. *Honors & Awards:* Peacock Mem Prize, 61. *Mem:* AAAS; Geochem Soc; Mineral Soc Am; Am Geophys Union; Mineral Asn Can; Europ Union Geosci; Europ Asn Geochem. *Res:* Geochemistry; igneous and metamorphic petrology; layered intrusions and magma evolution; feldspars; anorthosites; phase equilibria; magma dynamics; planetary crusts. *Mailing Add:* Dept of Geol & Geog Univ of Mass Amherst MA 01003

MORSE, STEPHEN ALLEN, b Los Angeles, Calif, Apr 11, 42; m 74, 88; c 2. MICROBIOLOGY. *Educ:* San Jose State Col, BA, 64; Univ NC, MSPH, 66, PhD(microbiol), 69. *Prof Exp:* NSF fel, Univ Ga, 69-70; asst prof biol, Southeastern Mass Univ, 70-71; res assoc microbiol, Sch Pub Health, Harvard Univ, 71-72, asst prof, 72-74; from asst prof to prof microbiol, Ore Health Sci Univ, 74-84; CONSULT. *Concurrent Pos:* Vis lectr, Sch Pub Health, Univ Calif, Los Angeles, 72; sr instr clin bact, Harvard Med Sch, 72-74; adj prof microbiol, Emory Univ; Res Career Develop Award, USPHS, 76-81; Tartar res fel, 82. *Honors & Awards:* Mary Poston Award, Am Soc Microbiol, 65. *Mem:* AAAS; Am Soc Microbiol; Soc Gen Microbiol; Am Venereal Dis Asn; Soc Exp Biol Med; Infectious Dis Soc Am. *Res:* Physiology and metabolism of infectious agents. *Mailing Add:* Ctrs for Disease Control Div STD Lab Res Bldg One Rm 6040 Atlanta GA 30333

MORSE, STEPHEN SCOTT, b New York, NY, Nov 22, 51; m 91. VIROLOGY, VIRAL IMMUNOLOGY. *Educ:* City Col, City Univ NY, BS, 71; Univ Wis, Madison, MS, 74, PhD(bacteriol & virol), 77. *Prof Exp:* NSF trainee, Dept Bacteriol, Univ Wis-Madison, 71-72, res asst, Dept Bacteriol, Virol & Slow Virus Lab, Dept Vet Sci, 72-77; res fel microbiol & infectious dis, Med Col, Va Commonwealth Univ, 77-79, Nat Cancer Inst fel, 79-80, instr microbiol, 80-81; asst prof microbiol, Rutgers Univ, 81-85; FAC MEM, ROCKEFELLER UNIV, 85-, ASST PROF, 88- *Concurrent Pos:* Reader, Marine Biol Lab, Woods Hole Mass, 81-84; fac mem, Bur Biol Res & fel, Rutgers Col, Rutgers Univ, 81-85; corp mem Marine Biol Lab, Woods Hole; chmn, Conf Emerging Viruses, NIH, 89; mem, Comt Microbiol, Threats to Health, Inst Med, Nat Acad Sci, 91- *Mem:* Am Soc Microbiol; Am Asn Pathologists; Sigma Xi; Am Soc Virol. *Res:* Lymphotropic viruses (T-lymphocytolytic viruses); viral immunology; emerging viral diseases; virus effects on developing T-lymphocytes and thymus; virus-macrophage interactions especially herpes viruses; role of macrophages in host resistance; viral evolution; pathogenesis of infectious disease. *Mailing Add:* Rockefeller Univ 1230 York Ave Box 2 New York NY 10021-6399

MORSE, THEODORE FREDERICK, b New York, NY, Feb 28, 32; m 55; c 2. ENGINEERING. *Educ:* Duke Univ, BA, 53, MA, 54; Hartford Univ, BSc, 58; Rensselaer Polytech Inst, MSc, 59; Northwestern Univ, PhD(eng), 61. *Prof Exp:* Res engr, Pratt & Whitney Aircraft Div, United Aircraft Corp, 55-59; sr res engr, Aeronaut Res Assocs Princeton, Inc, 61-63; from asst prof to assoc prof eng, 64-67, PROF ENG, BROWN UNIV, 67- *Concurrent Pos:* Sr Fulbright res fel, Ger Exp Estab Air & Space Res, Porz-Wahn, Ger, 69-70. *Mem:* Am Phys Soc. *Res:* Kinetic theory of gases; fluid mechanics; laser isotope separation; gas laser theory. *Mailing Add:* Div of Eng Brown Univ Providence RI 02912

MORSE, WILLIAM HERBERT, b Yorktown, Va, May 30, 28; m 58; c 4. PHARMACOLOGY. *Educ:* Univ Va, BA, 50, MA, 52; Harvard Univ, PhD(psychol), 55. *Prof Exp:* Res fel psychol, 55-58, from instr to assoc prof psychol, Med Sch, 58-76, PROF PSYCHOBIOL, MED SCH, HARVARD UNIV, 76- *Mem:* Am Soc Pharmacol & Exp Therapeut. *Res:* Behavioral pharmacology and physiology. *Mailing Add:* Dept Psychiat Lab Psychobiol Harvard Med Sch 220 Longwood Ave Boston MA 02115

MORSE, WILLIAM M, b Portland, Maine, May 16, 47; m 73; c 4. RARE K MESON DECAYS. *Educ:* State Univ NY, Stony Brook, BS, 69; Purdue Univ, PhD(physics), 76. *Prof Exp:* PHYSICIST, BROOKHAVEN NAT LAB, 76- *Concurrent Pos:* Co-spokesman, Alternating Gradient Synchrotron exp 845, Brookhaven Nat Lab, 84- *Mem:* Am Phys Soc. *Res:* Flavor changing neutral currents; muon g-2; charge conjugation-parity violations. *Mailing Add:* Dept Physics Brookhaven Nat Lab Upton NY 11973

MORT, ANDREW JAMES, b Kent, Eng, Apr 28, 51; m 80; c 1. PLANT POLYSACCHARIDE, POLYSACCHARIDE CHEMISTRY. *Educ:* McGill Univ, Montreal, BSc, 71; Mich State Univ, PhD(biochem),78. *Prof Exp:* Res asst, C F Kettering Res Lab, 77-81; from asst prof to assoc prof, 81-89, PROF, DEPT BIOCHEM, OKLA STATE UNIV, 89- *Mem:* Am Chem Soc; Am Soc Plant Physiologists; Am Phytopathol Soc. *Res:* Development of new methods for the determination of polysaccharide structure; determination of the structure of plant cell wall polysaccharides and glycoproteins. *Mailing Add:* Dept Biochem Okla State Univ Stillwater OK 74078

MORT, JOSEPH, b Oldham, Eng, Sept 21, 36; m 64; c 2. SCIENCE POLICY, SCIENCE COMMUNICATIONS. *Educ:* Univ Leicester, BSc, 59, PhD(physics), 62. *Prof Exp:* Res assoc physics, Univ Wis, 65-66; RES MGR PHYS SCI, XEROX CORP, 66-, RES FEL, 85- *Concurrent Pos:* Fulbright fel, Univ Ill, Urbana, 62-64; mem solid state sci panel, Nat Acad Sci, 78- *Mem:* Fel Am Phys Soc; Electrochem Soc. *Res:* Amorphous solids including polymers with principal interests in photoelectronic and transport properties, electrography and electronic materials and devices; amorphous inorganic and organic solids; amorphous silicon; diamond and diamond-like thin films. *Mailing Add:* Xerox Webster Res Ctr Bldg 114-41D Xerox Corp 800 Philips Rd Rochester NY 14580

MORTADA, MOHAMED, b Alexandria, Egypt, Mar 14, 25; US citizen; m 58; c 1. SIMULATED OIL & GAS PROCESSES. *Educ:* Univ Cairo, Egypt, BSc Hons, 46; Univ Calif, Berkeley, MSc, 49, PhD(petrol eng), 52. *Prof Exp:* Instr petrol refining, Univ Cairo, Egypt, 46-47; petrol engr, Mobil Oil Corp, 52-54, sr res engr, 54-56, res assoc reservoir eng, 56-59, sr res assoc opers res, 59-63; sr consult oil reserves, Ministry Finances & Oil, Kuwait, 63-67; PRES, MORTADA INT INC, 68- *Concurrent Pos:* Distinguished lectr, Soc Petrol Engrs, 71-72. *Honors & Awards:* Alfred Noble Prize, 56; Rossiter W Raymond, Am Inst Mining, Metall & Petrol Engrs, 57. *Mem:* Soc Petrol Engrs; Am Asn Petrol Geologists; Res Soc Am. *Res:* Fluid flow through porous media; use of simulation method to study fluid flow in oil and gas reservoir. *Mailing Add:* 4820 Mill Creek Rd Dallas TX 75244

MORTARA, LORNE B, b Chicago, Ill, Sept 29, 32; m 57; c 4. HIGH ENERGY PHYSICS. *Educ:* Purdue Univ, BS, 53, PhD(physics), 63. *Prof Exp:* Asst physics, Purdue Univ, 56-63; res assoc, Univ Ariz, 63-65; sr scientist physics instrumentation, Albuquerque Lab, Edgerton, Germeshausen & Grier, Inc, 65-68; SR ENG PHYSICIST, AURA, INC, 68- *Mem:* AAAS; Am Phys Soc. *Res:* Instrumentation design; application of computer systems. *Mailing Add:* 7865 N 86th St Milwaukee WI 53224

MORTEL, RODRIGUE, b Saint-Marc, Haiti, Dec 3, 33; m 71. OBSTETRICS & GYNECOLOGY. *Educ:* Univ Haiti, MD, 60; Am Bd Obstet & Gynec, dipl, 70. *Prof Exp:* Asst clin instr obstet & gynec, Hahnemann Med Col & Hosp, 67-68, instr, 68-70, sr instr, 70-71, asst prof, 71-72; from asst prof to assoc, 72-83, PROF OBSTET & GYNEC & CHMN DEPT, HERSHEY MED CTR, PA STATE UNIV, 83- *Concurrent Pos:* USPHS grant gynec oncol, Hahnemann Med Col & Hosp, 68-69; USPHS grant, Mem Hosp Cancer & Allied Dis, New York, 69-70. *Mem:* Fel Am Col Obstet & Gynec; fel Am Col Surgeons; James Ewing Soc; NY Acad Sci. *Res:* Clinical and basic research in gynecologic oncology. *Mailing Add:* Dept of Obstet & Gynec Hershey Med Ctr Pa State Univ Hershey PA 17033

MORTENSEN, EARL MILLER, b Salt Lake City, Utah, June 25, 33; m 62; c 4. PHYSICAL CHEMISTRY. *Educ:* Univ Utah, BA, 55, PhD(chem), 59. *Prof Exp:* NSF fel chem, Univ Calif, Berkeley, 59-60, lectr, 60-61, chemist, Radiation Lab, 60-62; asst prof chem, Univ Mass, Amherst, 62-69; ASSOC PROF CHEM, CLEVELAND STATE UNIV, 69- *Mem:* Am Phys Soc; Am Chem Soc; Sigma Xi. *Res:* Theoretical reaction kinetics; computer applications to chemical education. *Mailing Add:* Dept of Chem Cleveland State Univ Cleveland OH 44115

MORTENSEN, GLEN ALBERT, b Moscow, Idaho, Dec 1, 33; m 59; c 4. NUCLEAR ENGINEERING. *Educ:* Univ Idaho, BS, 55; Oak Ridge Sch Reactor Technol, dipl, 56; Univ Calif, Berkeley, PhD(nuclear eng), 63. *Prof Exp:* Engr, USAEC, Washington, DC, 56-58, Phillips Petrol Co, Idaho, 63-69, Idaho Nuclear Corp, 69-70 & Aerojet Nuclear Corp, 70-75; engr, Intermountain Technologies Inc, 75-86; PRIN ENGR, EG&G, IDAHO, 86- *Concurrent Pos:* Affil prof, Nat Reactor Testing Sta Educ Prog, Univ Idaho, 63-75. *Mem:* Am Nuclear Soc; Asn Comput Mach. *Res:* Database management systems; numerical methods; fluid flow; reactor kinetics. *Mailing Add:* EG&G Idaho PO Box 1625 Idaho Falls ID 83415

MORTENSEN, HARLEY EUGENE, b Albuquerque, NMex, Mar 1, 31; m 52; c 5. ORGANIC MOLECULAR BIOLOGY. *Educ:* Regis Col, BS, 54; Kans State Univ, PhD(org chem), 61. *Prof Exp:* Res chemist, Benger Lab, E I du Pont de Nemours & Co, 61-67; from asst prof to prof chem, 67-84, actg head biomed sci, 84-86, PROF BIOMED SCI, SOUTHWEST MO STATE UNIV, 84-, HEAD DEPT, 86- *Concurrent Pos:* NSF res partic, Acad Year Exten, Southwest Mo State Univ, 70-72. *Mem:* Am Chem Soc; AAAS; Am Soc Microbiol. *Res:* Enzymology; molecular biology. *Mailing Add:* Biomed Sci Southwest Mo State Univ Springfield MO 65804

MORTENSEN, JOHN ALAN, b San Antonio, Tex, May 11, 29; m 57; c 1. HORTICULTURE, GENETICS. *Educ:* Tex A&M Univ, BS, 50, MS, 51; Cornell Univ, PhD(plant breeding), 58. *Prof Exp:* Plant breeder, Birds Eye Div, Gen Foods Corp, 57-60; asst geneticist, Fla Agr Exp Sta, Univ Fla, 60-68, assoc prof, 68-77, assoc geneticist, 68-77, prof & geneticist, Inst Food & Agr Sci, Cent Fla Res & Educ Ctr, 77-89, asst ctr dir, 89-91; RETIRED. *Mem:* Am Soc Hort Sci; Am Pomol Soc. *Res:* Disease and insect resistance in grapes; development of improved varieties of scions and rootstocks in grapes through breeding and testing; nutritional and inheritance studies in grapes. *Mailing Add:* Cent Fla Res & Educ Ctr 5336 Univ Ave Leesburg FL 34748

MORTENSEN, KNUD, b Outrup, Denmark, Mar 22, 39; Can citizen; m 67; c 2. PLANT PATHOLOGY. *Educ:* Univ Copenhagen, BSc, 69, PhD(plant path), 74. *Prof Exp:* Plant pathologist, Sask Agr, 78-80; RES SCIENTIST, AGR CAN, 81- *Mem:* Am Phytopath Soc; Can Phytopath Soc. *Res:* Biological control of weeds using plant pathogens. *Mailing Add:* Agr Can Res Station Box 440 Regina SK S4P 3A2 CAN

MORTENSEN, RICHARD E, b Denver, Colo, Sept 29, 35; div. SYSTEMS SCIENCE. *Educ:* Mass Inst Technol, BSEE & MSEE, 58; Univ Calif, Berkeley, PhD(stochastic optimal control), 66. *Prof Exp:* Coop engr, Gen Elec Co, 55-57; mem tech staff guid & control, Space Technol Labs, Inc, 58-61; teaching fel elec eng, Univ Calif, Berkeley, 61-65; asst prof eng, 65-70, assoc prof eng & appl sci, 70-90, PROF ELEC ENG, UNIV CALIF, LOS ANGELES, 90- *Concurrent Pos:* Vis asst prof, Univ Colo, 66-67; consult, TRW Systs, Inc, Calif, 66-80. *Mem:* Union Concerned Scientists; Inst Elec & Electronics Engrs. *Res:* Electric power systems analysis and optimization; stochastic processes; optimal control theory. *Mailing Add:* Dept Elec Eng Univ Calif Los Angeles CA 90024

MORTENSON, KENNETH ERNEST, b Melrose, Mass, Dec 14, 26; m 49. MICROWAVE SOLID STATE DEVICES, SEMICONDUCTOR PHYSICS & DEVICES. *Educ:* Rensselaer Polytech Inst, BS, 47, BEE, 48, MEE, 50, PhD(appl physics), 54. *Prof Exp:* From instr to asst prof elec eng & physics & res assoc, Rensselaer Polytech Inst, Troy, NY, 47-56; res physicist, Gen Elec Res Lab, Schenectady, NY, 56-60; dir res & develop, Microwave Assocs, Inc, Burlington, Mass, 60-63; chmn & prof, elec eng dept, 63-67, assoc dean, Sch Eng, Rensselaer Polytech Inst, 67-69; chmn, pres & chief exec officer, RRC Int, Inc, Latham, NY, 69-77; sr vpres planning & technol, Am District Tel Co, New York, 77-79; exec vpres technol & dir, Nat Micronetics, Inc, 80-81; exec consult mgt, bus develop & technol, 81-87; SR VPRES CORP DEVELOP, & DIR SECURITY TAG SYSTS, INC, 87- *Concurrent Pos:* Dir, Appl Robotics Inc, 83-89, Nat Micronetics, Inc, 74-80, ADT Energy Systs, Ltd, 78-80; exec vpres, chief oper officer & dir, Nat Micronetics, Inc, Kingston, NY & San Diego, Calif, 77. *Mem:* Fel Inst Elec & Electronic Engrs; AAAS; Am Soc Eng Educ; NY Acad Sci. *Res:* Microwave solid state devices; semiconductor physics & devices; electro-optics devices; RF devices; circuits & transmission; security devices & systems; author of 26 technical papers & 2 books. *Mailing Add:* 13 Ross Ct Loudonville NY 12211

MORTENSON, LEONARD EARL, b Melrose, Mass, June 24, 28; m 52; c 5. BIOCHEMISTRY. *Educ:* RI State Col, BS, 50; Univ Wis, MS, 52, PhD(bact, biochem), 54. *Prof Exp:* Asst bact, Univ Wis, 50-52, asst, Enzyme Inst, 52-53, NSF fel, 53-54; res biochemist, E I du Pont de Nemours & Co, 54-61; assoc prof biol, Purdue Univ, West Lafayette, 62-66, prof, 66-81; sr res assoc, Exxon Res & Eng Co, Linden, NJ, 81-85; CHMN, DIV BIOL SCI, UNIV GA, 85-, CALLAWAY PROF BIOCHEM, 85- *Concurrent Pos:* Lectr, Found Microbiol, 70-71; vis scientist, Chem Dept, Stanford Univ, 75-76 & CNRS, Marseille, France, 78; Lembert traveling fel, Australian Acad Sci, 80; chmn & organizer, Int Hydrogenase Conf, 88; dir, Biochem Prog, NSF, 88-89. *Honors & Awards:* Hoblitzelle Nat Award, 65. *Mem:* Corresp mem French Acad Sci; Am Soc Microbiol; Am Soc Biol Chemists; Am Chem Soc; Sigma Xi. *Res:* Biological nitrogen fixation; electron transport; energy and carbohydrate metabolism; biosynthetic reactions; ferredoxin biochemistry; hydrogenase. *Mailing Add:* Univ Ga #400 Biosci Bldg Athens GA 30602

MORTENSON, THEADORE HAMPTON, b Miami, Fla, Apr 27, 34; div; c 1. ECOLOGICAL PLANT ANATOMY. *Educ:* Youngstown Univ, BS, 60; Whittier Col, MEd, 64; Univ Mont, MS, 67; Claremont Men's Col, PhD(bot), 70. *Prof Exp:* Teacher sci, Brookfield Jr High Sch, 59-60, Tallmadge High Sch, 60-61; teacher sci & math, Newton Jr High Sch, 61-63; teacher sci & biol, La Puente High Sch, 63-66; from asst prof to assoc prof biol & bot, 70-80, chmn dept biol & health sci, 79-87, PROF BIOL & BOT, CHAPMAN COL, 86- *Mem:* Bot Soc Am. *Res:* Ecological plant anatomy involving native trees and shrubs of the more arid portions of the western United States. *Mailing Add:* Dept Biol & Health Sci Chapman Col 333 N Glassell St Orange CA 92666

MORTER, RAYMOND LIONE, b Arlington, Wis, Sept 7, 20; m 46, 76; c 2. FOOD ANIMAL MEDICINE, CELLULAR IMMUNOLOGY. *Educ:* Iowa State Univ, BS, 54, DVM, 57; Mich State Univ, MS, 58, PhD(microbiol & path), 60. *Prof Exp:* NSF fel vet microbiol, Mich State Univ, 57-59; asst vet anat, Iowa State Univ, 55-57, asst prof, Vet Med Res Inst, 59-60; assoc prof vet microbiol, path & public health, 60-64, prof vet microbiol, Sch Vet Sci & Med, 64-66, head dept vet sci & assoc dean res, 66-76, PROF VET MED, SCH VET MED, PURDUE UNIV, WEST LAFAYETTE,76- *Concurrent Pos:* AID staff, Philippines, 66; consult, Food & Agr Orgn, UN. *Mem:* AAAS; Am Asn Lab Animal Sci; Am Soc Microbiol; Am Vet Med Asn; Am Soc Exp Path; Sigma Xi; US Animal Health Asn; Nat Cattlemen's Asn; Am Asn Bovine Practitioners. *Res:* Mechanism and course of infectious diseases and alveolar macrophyage functions; preventive medicine in food producing animals. *Mailing Add:* Lynn Hall Purdue Univ Sch Vet Med West Lafayette IN 47907

MORTIMER, CHARLES EDGAR, b Allentown, Pa, Nov 21, 21; m 60; c 1. ORGANIC CHEMISTRY, HISTORY OF SCIENCE. *Educ:* Muhlenberg Col, BS, 42; Purdue Univ, MS, 48, PhD(chem), 50. *Hon Degrees:* ScD, Muhlenberg Col, 88. *Prof Exp:* Line shift supvr, Hercules Powder Co, Va & Kans, 42-44; res assoc, Manhattan Project, NY, 44-46; from asst prof to prof chem, Muhlenberg Col, 50-83, head dept, 76-83, EMER PROF CHEM, MUHLENBERG COL, 83- *Mem:* Am Chem Soc. *Res:* Chemical education; author of editions in chemical textbooks. *Mailing Add:* 4762 Parkview Dr S Emmaus PA 18049-1212

MORTIMER, CLIFFORD HILEY, b Somerset, Eng, Feb 27, 11; m 36; c 2. LIMNOLOGY-PHYSICAL, INTERACTIONS HYDRODYNAMICS & BIODYNAMICS. *Educ:* Univ Manchester, Eng, BSc, 32, DSc, 46; Univ Berlin, Germany, DPhil, 35. *Hon Degrees:* DSc, Univ Wis, Milwaukee, 85 & Ecole Polytech Fed, Lausanne, Switz, 87. *Prof Exp:* Res fel, Kaiser Wilhelm Inst Biol, 32-35; staff mem, Freshwater Biol Asn 35-41 & 46-56; mem oceanog group, Admiralty res lab, Royal Naval Sci serv, 41-46; secy & dir marine sta Millport, Scottish Marine Biol Asn, 56-66; distinguished prof zool & dir, Ctr Great Lakes Studies, 66-81, DISTINGUISHED EMER PROF BIOL SCI, UNIV WIS-MILWAUKEE, 81- *Concurrent Pos:* External examr oceanog, Univ Liverpool & Wales, 60; Brittingham vis prof, Univ-Wis, Madison, 62-63; hon lectr oceanog, Univ Glasgow, Scotland, 63-66. *Honors & Awards:* Naumann Medal, Int Asn Limnol, 65. *Mem:* Fel Royal Soc London; Freshwater Biol Asn, (vpres, 66-); Am Soc Limnol & Oceanog (pres, 69); Int Asn Great Lakes Res (pres, 73). *Res:* Hydrodynamics of large lakes and enclosed seas, particularly internal wave response to wind impulses; physical and chemical factors controlling biological production, with particular focus on Laurentian Great Lakes. *Mailing Add:* Dept Biol Sci Univ Wis Milwaukee WI 53201

MORTIMER, EDWARD ALBERT, JR, b Chicago, Ill, Mar 22, 22; m 44; c 3. MEDICINE. *Educ:* Dartmouth Col, AB, 43; Dartmouth Med Sch, dipl, 44; Northwestern Univ, MD, 47. *Prof Exp:* Resident pediat, Boston Children's Hosp, 50-52; from sr instr to prof, Case Western Reserve Univ, 52-66; prof & chmn dept, Sch Med, Univ NMex, 66-75; prof community health & pediat & chmn dept community health, 75-85, PROF PEDIAT & EPIDEMIOL, SCH MED, CASE WESTERN RESERVE UNIV, 85- *Concurrent Pos:* Markle scholar, Case Western Reserve Univ, 61-66; asst dir dept pediat, Cleveland Metrop Gen Hosp, 52-66; chief pediat, Bernalillo County Med Ctr, 66-75; mem comn streptococcal & staphylococcal dis, Armed Forces Epidemiol Bd, 69-72; mem epidemiol & dis control study sect, NIH, 69-73; chmn, Joint Coun Nat Pediat Soc, 72-74; vis prof epidemiol, Sch Pub Health, Harvard Univ, 73; consult, Off Biol Res & Rev, FDA, 73-; USPHS adv comt on immunization pract, 83- *Mem:* Soc Pediat Res; Am Pediat Soc; Am Epidemiol Soc. *Res:* Pediatrics; epidemiology; rheumatic fever; streptococcal and staphylococcal diseases and infections. *Mailing Add:* Dept Community Health Case Western Reserve Univ 2119 Adelbert Rd Cleveland OH 44106

MORTIMER, J(OHN) THOMAS, b Las Vegas, NMex, Oct 12, 39; m 67; c 3. BIOMEDICAL ENGINEERING. *Educ:* Tex Technol Col, BSEE, 64; Case Western Reserve Univ, MS, 65, PhD(eng), 68. *Prof Exp:* Swed Bd Technol Develop grant, Chalmers Univ Technol, Sweden, 68-69; from asst prof to assoc prof, 69-81, PROF BIOMED ENG, CASE WESTERN RESERVE UNIV, 81-, DIR APPL NUEROL CONTROL LAB, 72- *Concurrent Pos:* Vis prof, Univ Karlsruhe, WGermany, 77-78. *Honors & Awards:* Humboldt Award, Alexander von Humboldt Found, 77. *Mem:* Biomed Eng Soc; Soc Neurosci. *Res:* Clinical application of electrical stimulation to the neuro-muscular system; applied neural control (electrical excitation of the nervous system). *Mailing Add:* 2885 Fairfax Rd Cleveland OH 44118

MORTIMER, JAMES ARTHUR, b Boston, Mass, May 13, 44; m 65; c 1. NEUROPSYCHOLOGY, EPIDEMIOLOGY OF DEMENTIA. *Educ:* Tufts Univ, BS, 65; Univ Mich, Ann Arbor, MS, 67, PhD(comput commun sci), 70. *Prof Exp:* Res fel, biomed eng unit, McGill Univ, 68, Logic Comput Group, Ann Arbor, 68-69; staff fel, Lab Appl Studies, NIH, 70-73; asst prof neurosurg, 74-76, asst prof, 77-90, ASSOC PROF NEUROL, MED SCH, UNIV MINN, 90-; ASSOC DIR, GERIAT RES, EDUC & CLIN CTR, VET AFFAIRS MED CTR, MINNEAPOLIS, 77- *Concurrent Pos:* Lectr comput & commun sci, Univ Mich, Ann Arbor, 70; guest work, lab neuropysiol, NIMH, 70-73; prin investr, Longitudinal Study of Dementing Illness, 86-90; co-chmn, Nat Task Force on Health Prom & Dis Prev in the Elderly, 87-90; vis prof, Univ Quebec, 88-; mem, Life Course & Prevention Res Review, Comt, NIMH, 89-93; vchmn, clin med, Geront Soc Am, 89-90. *Honors & Awards:* Jennie & Lillian Fischer Mem Lectr, 88. *Mem:* Fel Geront Soc Am; Soc Neurosci; Int Neuropsychol Soc; Soc Epidemiol Res; Inst Psychogeriat Asn; Am Acad Neurol. *Res:* Epidemiology and etiology of dementia; neuropsychology of aging and dementia; motor system neurophysiology. *Mailing Add:* Geriat Res, Educ & Clin Ctr Vet Affairs Med Ctr Minneapolis MN 55417

MORTIMER, KENNETH, b Aberdeen, Scotland, Apr 22, 22; nat US; m 80; c 4. ENGINEERING MECHANICS. *Educ:* Ill Inst Technol, BS, 47, MS, 49. *Prof Exp:* Instr mech eng, Ill Inst Technol, 48-50; from asst prof to prof civil eng, Valparaiso Univ, 50-87; RETIRED. *Concurrent Pos:* Consult, Chicago South Shore & South Bend RR & US Army Corps Engrs. *Mem:* Am Soc Eng Educ. *Res:* Computers; structural analysis. *Mailing Add:* 352 Green Acres Dr Valparaiso IN 46383

MORTIMER, RICHARD W(ALTER), b Philadelphia, Pa, Dec 7, 36; m 57; c 4. SOLID MECHANICS. *Educ:* Drexel Univ, BS, 62, MS, 64, PhD(appl mech), 67. *Prof Exp:* Chmn, Dept Mech Eng & Mech, 76-85, assoc vpres acad affairs & dir res develop, 85-90, from instr to PROF MECH ENG, DREXEL UNIV, 65- *Concurrent Pos:* Am Soc Mech Engrs-NASA fac fel, Drexel Univ, 67, NASA grant, 67-73; Air Force Mat Lab grant, 73-76; chmn, Nat Mech Engr Dept Heads' Comt, Am Soc Mech Engrs, 84-85, mem bd eng educ, 84-; mem Eng Acreditation Comn, 86- *Honors & Awards:* Achievement Award, Am Soc Nondestructive Testing, 73. *Mem:* Am Soc Mech Engrs; Am Soc Eng Educ; Soc Mfg Engrs. *Res:* Dynamic elasticity; stress analysis; fluid mechanics; response of structures to impact loadings; wave propagation; mechanics of composite materials. *Mailing Add:* Prof Mech Eng Drexel Univ 32nd & Chestnut Sts Philadelphia PA 19104

MORTIMER, ROBERT GEORGE, b Provo, Utah, Aug 25, 33; m 60; c 5. PHYSICAL CHEMISTRY, THEORETICAL CHEMISTRY. *Educ:* Utah State Univ, BS, 58, MS, 59; Calif Inst Technol, PhD(chem), 63. *Prof Exp:* Res chemist, Univ Calif, San Diego, 62-64; asst prof chem, Ind Univ, Bloomington, 64-70; from asst prof to assoc prof, 70-81, PROF CHEM, RHODES COL, 81- *Res:* Statistical mechanics; irreversible thermodynamics; experimental study of transport processes in liquids. *Mailing Add:* Dept Chem Rhodes Col 2000 N Parkway Memphis TN 38112

MORTIMER, ROBERT KEITH, b Didsbury, Alta, Nov 1, 27; nat US; m 49; c 4. GENETICS, BIOPHYSICS. *Educ:* Univ Alta, BSc, 49; Univ Calif, Berkeley, PhD(biophys), 53. *Prof Exp:* From instr to assoc prof, 53-66, PROF MED PHYSICS, UNIV CALIF, BERKELEY, 66-, CHMN MED, 72- *Mem:* AAAS; Genetics Soc Am; Radiation Res Soc; Sigma Xi. *Res:* Genetics and radiation biology of microorganisms. *Mailing Add:* 600 Vicente Ave Berkeley CA 94707

MORTIMORE, GLENN EDWARD, b Portland, Ore, Apr 13, 25; m 59; c 2. PHYSIOLOGY, BIOCHEMISTRY. *Educ:* Ore State Col, BS, 49; Univ Ore, MD, 52. *Prof Exp:* NSF fel, 57-58; sr investr, Nat Inst Arthritis & Metab Dis, 58-67; assoc prof, 67-71, PROF PHYSIOL, MILTON S HERSHEY MED CTR, PA STATE UNIV, 71- *Mem:* AAAS; Endocrine Soc; Am Fedn Clin Res; Am Physiol Soc; Am Soc Biol Chemists. *Res:* Mechanism of hormone action; effect of insulin of liver metabolism; regulation of metabolism and protein turnover; lysosomes. *Mailing Add:* Dept Physiol Hershey Med Ctr Pa State Univ Hershey PA 17033

MORTLAND, MAX MERLE, b Streator, Ill, Mar 30, 23; m 47; c 4. SOIL CHEMISTRY. *Educ:* Univ Ill, BS, 46, MS, 47 & 50, PhD(agron), 51. *Prof Exp:* Asst prof soils, Univ Wyo, 51-53; from asst prof to assoc prof, Mich State Univ, 53-69, prof soil sci, 69-89, crop sci, 74-89; RETIRED. *Concurrent Pos:* Fulbright sr res scholar, Cath Univ Louvain, 61-62. *Mem:* Am Soc Agron; Soil Sci Soc Am. *Res:* Physical chemical reactions of soils; reactions of ammonia in soils; rate controlling processes in potassium release from minerals; clay-organic complexes. *Mailing Add:* 1188 Chartway Carriageway N East Lansing MI 48823

MORTLOCK, ROBERT PAUL, b Bronxville, NY, May 12, 31; m 54; c 3. MICROBIAL PHYSIOLOGY. *Educ:* Rensselaer Polytech Inst, BS, 53; Univ Ill, PhD(bact), 58. *Prof Exp:* Bacteriologist, US Army Chem Corps Res & Develop Labs, 59-61; res assoc biochem, Mich State Univ, 61-63; from asst prof to prof, Univ Mass, Amherst, 68-78, head dept, 72-78; chmn dept microbiol, 78-88, PROF MICROBIOL, CORNELL UNIV, 78- *Concurrent Pos:* USPHS fel, 61-63. *Mem:* AAAS; Am Soc Microbiol; Am Acad Microbiol. *Res:* Physiological bacteriology; microbial physiology and metabolism; cellular regulatory mechanisms; carbohydrate metabolism and enzyme regulation in microorganisms; the utilization of uncommon and unnatural carbohydrates by microorganisms. *Mailing Add:* Sect Microbiol Cornell Univ Ithaca NY 14853-7201

MORTOLA, JACOPO PROSPERO, b Milan, Italy, Nov 7, 49; m 76; c 3. RESPIRATORY PHYSIOLOGY, NEONATAL RESPIRATION. *Educ:* Univ Milan, Italy, MD, 73. *Prof Exp:* Med officer, Italian Army, 75-76; res assoc physiol, Univ Tex Med Br, 76-78; asst prof, 78-84, ASSOC PROF PHYSIOL, MCGILL UNIV, MONTREAL, QUE, 84- *Honors & Awards:* Lepetit Award, 74. *Mem:* Ital Med Asn; Ital Physiol Soc; Am Physiol Soc; Can Physiol Soc; NY Acad Sci; Can Soc Clin Invest. *Res:* Respiratory adaptation at birth; mechanics of breathing and repiratory control in the perinatal period; development of the lung and factors controlling it. *Mailing Add:* Dept Physiol McIntyre Med Sci Bldg McGill Univ 3655 Drummond St Montreal PQ H3G 1Y6 Can

MORTON, BRUCE ELDINE, b Loma Linda, Calif, May 9, 38; m 60, 76; c 3. HUMAN BRAIN RESEARCH,. *Educ:* La Sierra Col, BS, 60; Univ Wis, MS, 63, PhD(biochem), 65. *Prof Exp:* Fel, Inst Enzyme Res, Univ Wis, 65-66; NIH fel, Mass Inst Technol, 66-67; Harvard Univ res fel med, Beth Israel Hosp, 67-69; from asst prof to assoc prof, 69-83, dir, Brain Syst Res Lab, 87, PROF BIOCHEM, UNIV HAWAII, MANOA, 84- *Concurrent Pos:* Consult, New Eng Mem Hosp, Stoneham, Mass, 67-69 & St Francis Hosp, Honolulu, 70-71; vis prof, Dept Biochem, Univ Southern Calif, 75-76, Dept Neurol, Univ Mich, 85, Dept Psychol, Stanford Univ, 85-86. *Mem:* Soc Neurosci; Am Soc Biol Chemists; Int Soc Res Agression; AAAS; Am Soc Neurochem. *Res:* Mode of action of psychoactive compounds; neurotransmitter receptor autoradiographic analysis of human brain in health and disease; development of neurophilosophic framework for analysis of molecular data; synaptic transmission and emotional illness. *Mailing Add:* Dept of Biochem & Biophys Univ Hawaii Sch Med 2538 The Mall Honolulu HI 96822

MORTON, DONALD CHARLES, b Kapuskasing, Ont, June 12, 33; m 70; c 2. ASTROPHYSICS. *Educ:* Univ Toronto, BA, 56; Princeton Univ, PhD, 59. *Prof Exp:* Astronr, US Naval Res Lab, 59-61; res assoc, Princeton Univ, 61-63, res staff mem, 63-65, res astronr, 65-68, sr res astronr & lectr astrophys sci, 68-76; dir, Anglo-Australian Observ, 76-86; DIR-GEN, HERZBERG INST ASTROPHYS, 86- *Honors & Awards:* Pawsey Mem lectr, Australia, 85. *Mem:* Int Astron Union; Am Astron Soc; Assoc Royal Astron Soc; Royal Astron Soc Can; Astron Soc Australia (pres, 81-83); fel Australian Acad Sci; hon mem Astron Soc Australia; Australian Inst Phys; Canada Astron Soc; Can Asn Physicist. *Res:* spectroscopy of stars, interstellar gas, galaxies, quasars; stellar mass loss; interstellar abundances; instrumentation for space and ground-based telescopes. *Mailing Add:* Herzberg Inst Astrophys Nat Res Coun 100 Sussex Dr Ottawa ON K1A 0R6 Can

MORTON, DONALD JOHN, b Brooklyn, NY, Jan 11, 31; m 53; c 3. INFORMATION SCIENCE. *Educ:* Univ Del, BS, 52; La State Univ, MS, 54; Univ Calif, Berkeley, PhD(plant path), 57; Simmons Col, MLS, 69, DA(libr sci), 76. *Prof Exp:* Asst prof plant nematol, NMex State Univ, 57-58; asst plant path, NDak State Univ, 59-61; sr res plant pathologist, USDA, Ga, 61-65; assoc prof plant path, Univ Del, 65-68; dir sci libr, Northeastern Univ, 69-70; asst prof hist med, 70-74, ASSOC PROF LIBR SCI, MED SCH, UNIV MASS, 74-, LIBR DIR, 70- *Concurrent Pos:* Prof libr sci, Worcester State Col, 74- *Mem:* Mycol Soc Am; Med Libr Asn; Am Soc Info Sci; Am Libr Asn; Spec Libr Asn. *Res:* Air pollution effects on plants; serological studies of plant pathogens; organization and retrieval of scientific information. *Mailing Add:* Libr Univ of Mass Med Sch Worcester MA 01605

MORTON, DONALD LEE, b Richwood, WVa, Sept 12, 34; m 57; c 4. SURGERY, ONCOLOGY. *Educ:* Univ Calif, BA, 55, MD, 58; Am Bd Surg, dipl, 67; Am Bd Thoracic Surg, dipl, 69. *Prof Exp:* Intern med, Med Ctr, Univ Calif, 58-59, resident surg, 59-60; clin assoc, Nat Cancer Inst, 60-62; resident surg, Med Ctr, Univ Calif, 62-66; sr surgeon, Nat Cancer Inst, 66-69, head tumor immunol sect, 69-71; assoc prof surg, Sch Med, Johns Hopkins Univ, 70-71; PROF SURG & CHIEF DIV ONCOL & GEN SURG, SCH MED, UNIV CALIF, LOS ANGELES, 71- *Concurrent Pos:* Fel, Cancer Res Inst, Med Ctr, Univ Calif, 62-66; immunol adv mem, Spec Virus Cancer Prog, Nat Cancer Inst, 69-71, mem bd sci counr, 74-; mem comt for objective 6, Nat Cancer Plan, 71-; chief surg, Sepulveda Vet Admin Hosp, Calif, 71-74, chief oncol sect, Surg Serv, 74-; mem sci adv coun, Cancer Res Inst, Inc, 74- *Honors & Awards:* Langer Award, 78. *Mem:* Am Asn Cancer Res; Am Surg Asn; Am Soc Clin Oncol; Soc of Surg Oncol; Soc Univ Surgeons. *Res:* Immunologic and virologic aspects of neoplastic disease, including immunotherapy of melanoma, skeletal and soft tissue sarcoma and mammary carcinoma; surgical oncology; thoracic surgery. *Mailing Add:* Div Surg Oncol Rm 9-260 Louis Factor Bldg Univ Calif Sch Med Los Angeles CA 90024

MORTON, DOUGLAS M, b Hemet, Calif, June 15, 35. GEOLOGY. *Educ:* Univ Calif, BA, 58, PhD(geol), 66. *Prof Exp:* GEOLOGIST, US GEOL SURV, RIVERSIDE, CALIF, 72- *Concurrent Pos:* Chief, Off Regional Geol, US Geol Surv, 78-80, chief western region geol, 80-83. *Mem:* Fel Geol Soc Am. *Mailing Add:* Dept Earth Sci Univ Calif Riverside CA 92521

MORTON, G A, b New Hartford, NY, Mar 24, 03; m; c 4. ELECTRICAL ENGINEERING. *Educ:* Mass Inst Technol, BS, 26, MS, 28, PhD(physics), 32. *Prof Exp:* res asst & instr, Mass Inst Technol, 33-68; consult, RCA, Princeton, 68-70; consult, Lawrence Berkeley Lab, 70-89; RETIRED. *Honors & Awards:* Inst Elec & Electronic Engrs Award, 37; V K Zworyhim Award, Inst Radio Engrs, 62; David Richardson Award, Optical Soc, 67; Nuclear & Plasma Soc/Inst Elec & Electronic Engrs Award, 74. *Mem:* Fel Inst Elec & Electronic Engrs; Sigma Xi; fel Am Phys Soc. *Res:* Electronic imaging; nuclear radiation detectors. *Mailing Add:* 1122 Skycrest Dr Apt 6 Walnut Creek CA 94595

MORTON, HARRISON LEON, b St Paul, Minn, Oct 19, 38; m 62; c 5. URBAN FORESTRY. *Educ:* Univ Minn, BS, 61, MS, 64, PhD(plant path), 67. *Prof Exp:* From asst prof to assoc prof forest path, 72-78, PROF FOREST PATH, FORESTRY & WILDLIFE PROG, SCH NATURAL RESOURCES, UNIV MICH, ANN ARBOR, 78-; OWNER, AM THREE CONSULT, 80- *Concurrent Pos:* Chmn fisheries, forestry & wildlife, 72-75; prof, forestry & wildlife, 78-; dir, Nichol's Arboneturn, 87- *Mem:* Int Soc Arboriculture; Am Phytopath Soc; Soc Am Foresters; Am Asn Botanical Gardens & Arboreta; Forestry & Park Asn. *Res:* Diseases of forest and shade trees; value of - litigation - tree losses and product liability; construction damage; foliage diseases. *Mailing Add:* Sch Nat Resources Dana Bldg Ann Arbor MI 48109-1115

MORTON, HARRY E, microbiology; deceased, see previous edition for last biography

MORTON, HOWARD LEROY, b Moscow, Idaho, Dec 13, 24; m 50; c 2. PLANT PHYSIOLOGY, WEED SCIENCE. *Educ:* Univ Idaho, BS, 50, MS, 52; Agr & Mech Col, Tex, PhD(plant physiol), 61. *Prof Exp:* Asst agronomist, Univ Idaho, 52-57; res agronomist, Tex, 57-66, plant physiologist, 66-68, PLANT PHYSIOLOGIST, CROPS RES DIV, AGR RES SERV, USDA, ARIZ, 68- *Mem:* Weed Sci Soc Am; Am Soc Plant Physiol; Soc Range Mgt; Sigma Xi; Bot Soc Am. *Res:* Absorption, translocation and metabolism of herbicides; weed control on range lands; range revegetation; poisonous weeds; pesticide residues; plant growth and development; controlled environment systems; effects of herbicides on honey bees. *Mailing Add:* Agr Res Serv USDA 2000 E Allen Rd Tucson AZ 85719

MORTON, JEFFREY BRUCE, b Chicago, Ill, Apr 25, 41; m 63; c 3. FLUID DYNAMICS. *Educ:* Mass Inst Technol, BS, 63; Johns Hopkins Univ, PhD(fluid mech), 67. *Prof Exp:* Res scientist, 67-68, asst prof to assoc prof, 68-80, PROF AEROSPACE ENG, UNIV VA, 80- *Honors & Awards:* Am Soc Eng Educ Res Unit Award Outstanding Contrib Res, 81. *Mem:* AAAS; Am Inst Aeronaut & Astronaut; Am Phys Soc; Sigma Xi. *Res:* Turbulence; boundary layers; fluid mechanics. *Mailing Add:* Dept Mech & Aerospace Eng Univ Va Charlottesville VA 22901

MORTON, JOHN DUDLEY, b Southampton, Eng, July 25, 14; US citizen; m 62; c 3. ENVIRONMENTAL HEALTH. *Educ:* Cambridge Univ, BA, 36, MA, 40. *Prof Exp:* Sect leader appl chem, Exp Sta, Eng, 36-47; asst dir aerobiol, Microbiol Res Estab, Eng, 47-62; lab mgr meteorol, Melpar Inc, 63-70; PRIN SCIENTIST ENVIRON HEALTH, DYNAMAC CORP, 70- *Concurrent Pos:* Consult, Nat Acad Eng, 69-70. *Res:* Studies of human health in relation to environmental and occupational exposure to toxic substances. *Mailing Add:* Dynamac Corp 2275 Research Blvd Suite 500 Rockville MD 20850-3268

MORTON, JOHN HENDERSON, b New Haven, Conn, Jan 15, 23; m 49; c 4. SURGERY. *Educ:* Amherst Col, BA, 45; Yale Univ, MD, 46. *Prof Exp:* Intern surg, gynec & obstet, Strong Mem Hosp & Rochester Munic Hosp, 46-47, asst resident surg, 47 & 49-52, resident, 53; from instr to assoc prof surg, 53-69, from instr to asst prof surg anat, 55-67, PROF SURG, SCH MED & DENT, UNIV ROCHESTER, 69- *Concurrent Pos:* From asst surgeon to assoc surgeon, Med Ctr, Univ Rochester, 54-62, sr assoc surgeon, 62-67, surgeon, 67- *Mem:* AMA; fel Am Col Surg; Am Asn Surg of Trauma; Am Burn Asn; Sigma Xi; Am Surg Asn. *Res:* Liver and gastrointestinal tract. *Mailing Add:* Dept Surg Univ Rochester Sch of Med Rochester NY 14642

MORTON, JOHN KENNETH, b Tamworth, Eng, Jan 3, 28; m 51; c 2. BOTANY. *Educ:* Univ Durham, BSc, 44, PhD(bot), 53; Univ Newcastle Upon Tyne, DSc(bot), 87. *Prof Exp:* Lectr bot, Univ Ghana, 51-60, sr lectr & cur, Ghana Herbarium, 60-61; lectr, Birkbeck Col, London, 61-63; prof & chmn dept, Fourah Bay Col, Sierra Leone, 63-67; chmn dept, 74-80, PROF BIOL, UNIV WATERLOO, 68- *Mem:* Can Bot Asn (pres, 74-75); Bot Soc Brit Isles; Linnean Soc London; Am Soc Plant Taxonomists; Int Asn Plant Taxon. *Res:* Experimental taxonomy and biogeography of North American and tropical African vascular plants; palynology; evolution. *Mailing Add:* Dept Biol Univ Waterloo Waterloo ON N2L 3G1 Can

MORTON, JOHN ROBERT, III, b Palestine, Tex, June 5, 29; m 53; c 2. NUCLEAR SCIENCE. *Educ:* Univ Ala, BS, 50; Univ Calif, Berkeley, PhD(chem), 61. *Prof Exp:* Tech grad, Hanford Atomic Prod Oper, Gen Elec Co, 50-51, supvr health physics, 51-56; PHYSICIST, LAWRENCE LIVERMORE LAB, UNIV CALIF, 61- *Mem:* AAAS; Am Phys Soc; Am Nuclear Soc. *Res:* Nuclear weapons test diagnostic techniques; pinhole imagery; reactor and critical assembly physics; seeking evidence for neutrino decay from fission explosions. *Mailing Add:* 4384 E Ave Livermore CA 94550

MORTON, JOHN WEST, JR, b Dallas, Tex, Mar 3, 25; m 50; c 2. ORGANIC CHEMISTRY. *Educ:* Southern Methodist Univ, BS, 46; Iowa State Univ, PhD(org chem), 52. *Prof Exp:* Res chemist, Procter & Gamble Co, Ohio, 52-54; from assoc prof to prof chem, La Polytech Inst, 54-62; assoc prof chem, 62-74, PROF CHEM, WESTERN NMEX UNIV, 74-, CHMN DEPT, 77- *Mem:* Am Chem Soc. *Res:* Organolithium compounds. *Mailing Add:* Dept Phys Sci Western NMex Univ 1404 Fla St Silver City NM 88061-5199

MORTON, JOSEPH JAMES PANDOZZI, b Hartford, Conn, May 9, 41; m 68; c 2. PHARMACOLOGY, ENVIRONMENTAL HEALTH. *Educ:* Univ Hartford, BS, 63; Univ Conn, MS, 66, PhD(pharmacol), 68; Am Bd Toxicol, cert, 80. *Prof Exp:* Dir pharmacol, Amazon Natural Drug Co, 67-69; toxicologist, Gillette Co, 69-70, med rev officer, 70-73, chief, Off Med Rev, 74-82, chief, Med Rev Off, 82-88, SR STAFF TOXICOLOGIST, GILLETTE CO, 89- *Mem:* Soc Toxicol; Soc Cosmetic Chemists; Am Indust Hyg Asn; Inst Food Technologists. *Res:* Safety and toxicity evaluation of drugs, cosmetics, foods, writing instruments, household and industrial chemical products; screening of natural products for potential therapeutic activity; environmental and industrial hygiene safety; expertise in technical regulatory affairs. *Mailing Add:* Gillette Med Eval Labs 401 Professional Dr Gaithersburg MD 20879

MORTON, MARTIN LEWIS, b Tony, Wis, May 1, 34; m 53; c 4. ZOOLOGY, PHYSIOLOGY. *Educ:* San Jose State Col, BA, 59, MA, 61; Wash State Univ, PhD(zoophysiol), 66. *Prof Exp:* Res asst zool, Wash State Univ, 63-65; res assoc, Univ Wash, 65-66, asst prof, 66-67; from asst prof to assoc prof, 67-77, PROF BIOL, OCCIDENTAL COL, 77- *Mem:* Am Ornith Union; Am Soc Mammal; Am Soc Zoologists; Ecol Soc Am; Am Inst Biol Sci; Sigma Xi. *Res:* Bioenergetics, orientation, phenology, endocrinology, annual cycles and biological clocks of migratory birds. *Mailing Add:* Dept of Biol Occidental Col 1600 Campus Rd Los Angeles CA 90041

MORTON, MAURICE, b Latvia, June 3, 13; nat US; m 33; c 3. POLYMER CHEMISTRY. *Educ:* McGill Univ, BSc, 34, PhD(chem), 45. *Prof Exp:* Chief chemist, Johns-Manville Co, Can, 36-41; chemist, Congoleum of Can, Ltd, 41-44; from asst prof to prof chem, Sir George Williams Col, 45-48; prof polymer chem, Univ Akron, 53-78, dir, Inst Polymer Sci, 56-78, head dept, 67-78, regents prof, 69-78, EMER REGENTS PROF POLYMER CHEM, UNIV AKRON, 78- *Concurrent Pos:* Lectr, McGill Univ, 46-48; chmn, comt macromolecular chem, Nat Acad Sci-Nat Res Coun, 63-67. *Honors & Awards:* Colwyn Medal, Eng, 79; Goodyear Medal, Am Chem Soc, 85; Polymer Education Award, Am Chem Soc, 88; Medaille de la Ville de Paris, 90. *Mem:* Am Chem Soc. *Res:* Polymerization kinetics; emulsion and anionic polymerization; synthetic rubber. *Mailing Add:* Inst Polymer Sci Univ Akron Akron OH 44325

MORTON, NEWTON ENNIS, b Camden, NJ, Dec 21, 29; m 49; c 5. POPULATION GENETICS. *Educ:* Univ Hawaii, BA, 51; Univ Wis, MS, 52, PhD, 55. *Hon Degrees:* MD, Univ Umea, 76. *Prof Exp:* Geneticist, Atomic Bomb Casualty Comn, Japan, 52-53; fel, Nat Cancer Inst, 55-56; asst prof med genetics, Univ Wis, 56-60, assoc prof, 60-61; dir genetics res proj, 58-59, prof genetics, 61-69, chmn dept, 62-65, dir pop genetics lab, 69-85, prof, sch pub health, Univ Hawaii, Honolulu, 75-85; MEM & CHMN DEPT EPIDEMIOL & BIOSTATIST, MEM SLOAN-KETTERING CANCER CTR, 85- *Concurrent Pos:* Consult, NIH, 59 & genetics training comt, 61-65; mem expert adv comt human genetics, WHO, 61-; dir med genetics proj, Immigrants Hosp, Sao Paulo, Brazil, 62-63. *Honors & Awards:* Lederle Award, 58; Allan Award, 63. *Mem:* Nat Acad Sci; Genetics Soc Am; Am Soc Human Genetics; Am Soc Naturalists; Brazilian Acad Sci; AAAS. *Res:* Human and population genetics. *Mailing Add:* Comm Med, Univ Southampton Southampton Gen Hosp Southampton S09 4XY England

MORTON, PERRY WILKES, JR, b Strong, Ark, Jan 19, 23; m 58; c 3. PHYSICS. *Educ:* Rice Univ, BS, 47; Miss State Univ, MS, 51; Duke Univ, PhD(physics), 57. *Prof Exp:* Asst math & instr physics, Miss State Univ, 52-53; asst, Duke Univ, 53-57; from assoc prof to prof, Miss State Univ, 57-63; prof physics & math & chmn div natural sci, Ky Southern Col, 63-69; CHMN DEPT PHYSICS, SAMFORD UNIV, 69- *Mem:* Am Phys Soc. *Res:* Nuclear and classical physics; beta-ray spectroscopy. *Mailing Add:* Dept of Physics Samford Univ Birmingham AL 35229

MORTON, PHILLIP A, m 78; c 3. CELL BIOLOGY, RECEPTOR BIOLOGY. *Educ:* Univ Ky, PhD(microbiol), 85. *Prof Exp:* Fel, Univ Va, 85-87; FEL, WASH UNIV, 87- *Mem:* Sigma Xi; Am Soc Cell Biology; Am Soc Microbiol. *Res:* Receptor biology. *Mailing Add:* Box 8116 Children's Hosp Wash Univ Sch Med 400 S Kings Highway St St Louis MO 63110

MORTON, RANDALL EUGENE, b Portland, Ore, May 4, 50; m 79; c 3. FIBER OPTICS, INSTRUMENTATION. *Educ:* Univ Wash, BS, 72, MS, 74, PhD(nuclear eng), 79. *Prof Exp:* Exec consult nuclear & qual eng, Holloran & Assocs, 77-81; corp consult eng recruitment, AGA Consults, 81-82; sr mfg syst anal, 82-83, sr eng res & develop, Fiber Optics & Instrumentation, 83-85, MGR, NEW PROD DEVELOP, ELDEC CORP, 85- *Concurrent Pos:* Consult, Innovative Concepts, 84-88. *Mem:* Soc Automotive Engrs; Inst Elec & Electronics Engrs; Soc Photo-Optical Instrumentation Engrs; Optical Soc Am. *Res:* Development of fiber optic sensors and systems and advanced aerospace instrumentation systems; management of research and development group developing aerospace electronic and electromechanical products such as flat panel displays space vehicle data acquisition systems, fiber optic sensors, power supplies and electric motor actuator controllers. *Mailing Add:* 10320 181 St NE Redmond WA 98052

MORTON, RICHARD ALAN, b Chicago, Ill, Dec 14, 38; m 62; c 4. BIOPHYSICS. *Educ:* Univ Chicago, SB, 61, SM, 62, PhD(biophys), 65. *Prof Exp:* Res assoc biophys, Johns Hopkins Univ, 65-67; NSF res fel, Univ Calif, Santa Barbara, 68, res assoc, 69; asst prof, 69-73, ASSOC PROF BIOL, MCMASTER UNIV, 73- *Mem:* Sigma Xi. *Res:* Structure and function of proteins, especially c and acetylchotinesterase; evolution of proteins and population genetics. *Mailing Add:* Dept Biol McMaster Univ 1280 Main St W Hamilton ON L8S 4L8 Can

MORTON, ROBERT ALEX, b Cincinnati, Ohio, Oct 17, 42; m 68; c 2. SEDIMENTOLOGY. *Educ:* Univ Chattanooga, BA, 65; WVa Univ, MS, 66, PhD(geol), 72. *Prof Exp:* Petrol geologist, Chevron Oil Co, 66-69; assoc res scientist geol, Univ Texas, Austin, 72-76, res scientist, 76-80 & 85-87, lectr dept marine studies, 78-80, assoc dir, 80-85, SR RES SCIENTIST, UNIV TEX, AUSTIN, 87- *Concurrent Pos:* Assoc ed, J Sed Petrology, 88-; consult, Conoco Inc, 85-86, BP Expl, 88-91, CNG Prod, 89, Mobil Corp, 88-90. *Mem:* Am Asn Petrol Geologists; Int Asn Sedimentologists; Geol Soc Am; Soc Econ Paleontologists & Mineralogists. *Res:* Coastal processes; ancient and modern clastic depositional systems; marine geology; environmental geology. *Mailing Add:* Bur Econ Geol University Sta-Box X Austin TX 78713-7508

MORTON, ROGER DAVID, b Nottingham, Eng, Oct 20, 35; m 61; c 2. MINERALOGY. *Educ:* Univ Nottingham, BSc, 56, PhD(geol), 59. *Prof Exp:* Sci asst, Univ Oslo, 59-61; lectr geol, Univ Nottingham, 61-66; assoc prof, 66-73, PROF GEOL, UNIV ALTA, 73- *Concurrent Pos:* G V Hobson Bequest Fund, Brit Inst Mining & Metall, 62; consult, Can Int Develop Agency, 75- *Mem:* Fel Geol Asn Can; Can Inst Mining & Metall; Soc Econ Geologists; Mineral Soc Am. *Res:* Investigation of uranium deposits in northwest Canada and mineral resources of Indonesia. *Mailing Add:* Rm 366 Dept Geol Univ Alta Edmonton AB T6G 2M7 Can

MORTON, ROGER ROY ADAMS, b Melbourne, Australia, June 1, 41; m 65; c 2. ELECTRICAL ENGINEERING. *Educ:* Royal Melbourne Inst Technol, Assoc dipl elec eng, 61; Univ Melbourne, BEE, 63; Monash Univ, Australia, PhD(elec eng), 66. *Prof Exp:* Exp officer comput res, Commonwealth Sci & Indust Res Orgn, Australia, 63-64; sr scientist image analysis develop, 67-73, Image Analysis Systs, Analysis Syst Div, Bausch & Lomb Inc, dir res & develop, 73-84; MGR, ELECTRONIC IMAGING RES LABS, IMAGE ELECTRONICS, 84- *Concurrent Pos:* Vis prof, Rochester Inst Technol, 77-78. *Honors & Awards:* Indust Res 100 Award, 71. *Mem:* Inst Elec & Electronic Engrs; Am Soc Testing & Mat. *Res:* Image analysis; instrument development; research and applications; electronic imaging; imaging science; 25 US patents. *Mailing Add:* Electronic Imaging & Res Labs Eastman Kodak MC 35712 Rochester NY 14653-5712

MORTON, STEPHEN DANA, b Madison, Wis, Sept 7, 32. WATER CHEMISTRY, PHYSICAL CHEMISTRY. *Educ:* Univ Wis, BS, 54, PhD(chem), 62. *Prof Exp:* Asst prof chem, Otterbein Col, 62-66; fel water chem, Univ Wis, 66-67; res chemist, Warf Inst, 67-73, head environ qual dept, 73-76; mgr qual assurance, Raltech Sci Serv, 77-82; PRES, SDM CONSULTS, 82- *Mem:* Am Chem Soc; Am Soc Limnol & Oceanog; Am Water Works Asn; Water Pollution Control Fedn. *Res:* Water pollution; lake and stream studies; waste treatment; toxicology. *Mailing Add:* SDM Consults 1202 Ann St Madison WI 53713

MORTON, THOMAS HELLMAN, b Los Angeles, Calif, Feb 10, 47; m 75. ORGANIC CHEMISTRY. *Educ:* Harvard Univ, AB, 68; Calif Inst Technol, PhD(chem), 73. *Prof Exp:* Asst prof chem, Brown Univ, 72-80; vis asst prof chem, Brandeis Univ, 80-81; asst prof, 81-85, ASSOC PROF CHEM, UNIV CALIF, RIVERSIDE, 85- *Concurrent Pos:* Vis scholar, Interdisciplinary Progs Health, Harvard Univ, 78-79 & 86. *Mem:* Am Chem Soc; Am Soc Mass Spectrometry; Asn Chemorecept Sci; Soc Neurosci; NY Acad Sci. *Res:* Neutral products from gas phase ionic reactions; molecular mechanisms of the sense of smell. *Mailing Add:* Univ Calif Chem Dept Riverside CA 92521

MORTON, WILLIAM EDWARDS, b Boston, Mass, June 30, 29; m 56; c 3. EPIDEMIOLOGY. *Educ:* Univ Puget Sound, BS, 52; Univ Wash, MD, 55; Univ Mich, MPH, 60, DrPH, 62. *Prof Exp:* Intern med, Doctors Hosp, Seattle, Wash, 55-56; USPHS heart dis control officer, Colo Dept Pub Health, 56-58; sr resident med, San Mateo County Hosp, Calif, 58-59; trainee epidemiol, Sch Pub Health, Univ Mich, 59-62; res epidemiologist, Colo Heart Asn & asst clin prof prev med, Med Sch, Univ Colo, 62-67; assoc prof, 67-70, PROF PUB HEALTH & PREV MED, MED SCH, UNIV ORE, 70-, HEAD DIV ENVIRON MED, 72- *Concurrent Pos:* Med res consult, Selective Serv, Colo, 64-67 & Ore, 70-76; consult, Environ Health Sci Ctr, Ore State Univ, 72-79; contrib ed, Am J Indust Med, 79- *Mem:* Am Col Prev Med; Am Pub Health Asn; Am Col Epidemiol; Am Occup Med Asn; Asn Teachers Prev Med; Soc Epidemiol Res. *Res:* Cancer epidemiology; job related chronic encephalopathy; cardiocvascular disease epidemiology; screening method evaluation; environmental and occupational health hazards. *Mailing Add:* Environ Med Div Ore Health Sci Univ Med Sch Portland OR 97201

MORTVEDT, JOHN JACOB, b Dell Rapids, SDak, Jan 25, 32; m 55; c 3. SOIL CHEMISTRY. *Educ:* SDak State Col, BS, 53, MS, 59; Univ Wis, PhD(soil chem), 62. *Prof Exp:* Soil chemist, Soils & Fertil Res Br, 62-80, Agr Res Br, 80-86, SR SCIENTIST, AGR RES DEPT, NAT FERTIL & ENVIRON RES CTR, TENN VALLEY AUTHORITY, 86- *Concurrent Pos:* Ed-in-chief, Soil Sci Soc Am J, 82-87. *Mem:* Fel Am Soc Agron; fel Soil Sci Soc Am (pres, 89); Int Soil Sci Soc; fel AAAS; hon mem Colombian Soil Sci Soc. *Res:* Chemistry of micronutrients and their soil-plant-fertilizer relationships; heavy metal contaminants added to soils. *Mailing Add:* 213 Forest Hills Dr Florence AL 35630

MOSAK, RICHARD DAVID, b Washington, DC, Oct 8, 45; m 67; c 4. MATHEMATICAL ANALYSIS, SOFTWARE SYSTEMS. *Educ:* Columbia Univ, AB, 66, PhD(math), 70. *Prof Exp:* Instr math, Yale Univ, 70-72 & Univ Chicago, 72-73; from asst prof to assoc prof math, Univ Rochester, 73-83; PROF MATH & COMPUT SCI, LEHMAN COL, 83- *Concurrent Pos:* Vis scientist, ctr appl math, Nat Bur Standards, 81-82. *Mem:* Am Math Soc; Asn Comput Mach. *Res:* Lie groups. *Mailing Add:* Dept Math & Comput Sci Lehman Col Bronx NY 10468

MOSBACH, ERWIN HEINZ, b Ger, Feb 18, 20; nat US; m 44; c 1. BIOCHEMISTRY, NUTRITION. *Educ:* Columbia Univ, BA, 43, MA, 48, PhD(chem), 50. *Prof Exp:* Tutor chem, Brooklyn Col, 42-46; asst, Columbia Univ, 46-50; biochemist, Biol Div, Oak Ridge Nat Lab, 50-51; res assoc biochem, Col Physicians & Surgeons, Columbia Univ, 51-54, from asst prof to assoc prof, 54-61; assoc mem & chief biochem sect, Dept Lab Diag, Pub Health Res Inst New York, 61-71, mem & chief, Dept Lipid Res, 72-78; DIR, SURG LIPID LAB, BETH ISRAEL MED CTR, 78- *Concurrent Pos:* Lectr, Hunter Col, 51-54; mem, Metab Study Sect, NIH, 67-71, lipid metab adv comt, 74-78, chmn, 78; adj assoc prof med, Med Sch, NY Univ, 61-78; asst dir bur labs, Dept Health, New York, 61-68; assoc ed, J Lipid Res, 68-72, Alcoholism, 85-, & ed, J Lipid Res, 76-78; consult, Manhattan Vet Admin Hosp, 71- 90; fel coun arteriosclerosis, Am Heart Asn; chmn, Search Comt, Vet Admin Spec Alcoholism Res Prog, 78; res prof surg, Mount Sinai Sch Med, 78- *Honors & Awards:* Windaus Prize, 82. *Mem:* Am Soc Biol Chemists; Soc Exp Biol & Med; Am Inst Nutrit; Am Asn Study Liver Dis; Am Gastroenterol Asn; Am Soc Clin Nutrit. *Res:* Biochemistry of sterols, bile acids, and gallstone disease. *Mailing Add:* Beth Israel Med Ctr First Ave at 16th St New York NY 10003

MOSBERG, ARNOLD T, b Cleveland, Ohio, Mar 4, 46; m 67. BIOMEDICAL ENGINEERING, FLUID MECHANICS. *Educ:* Ohio State Univ, BAAE, 69, MS, 73, PhD(biomed eng), 78. *Prof Exp:* Aeronaut eng combustion, Gen Elec Co, 69-70, aeronaut eng thermodyn diag, 70-71; RES SCIENTIST BIOMED ENG, COLUMBUS DIV, BATTELLE MEM INST, 77- *Mem:* Asn Advan Med Instrumentation; Aerospace Med Asn; Am Inst Aeronaut & Astronaut; Int Lung Sounds Asn. *Res:* Biomedical engineering; pulmonary function studies; cardiovascular physiology; biofluid mechanics; biomedical device development and design; medical instrumentation design and evaluation. *Mailing Add:* 7989 Peak Rd Clemmons NC 27012

MOSBO, JOHN ALVIN, b Davenport, Iowa, June 11, 47; m 68; c 2. INORGANIC CHEMISTRY. *Educ:* Univ Northern Colo, BA, 69; Iowa State Univ, PhD(inorg chem), 73. *Prof Exp:* Prof chem, Ball State Univ, 73-86; PROF CHEM & HEAD DEPT, JAMES MADISON UNIV, 86- *Mem:* Am Chem Soc. *Res:* Macrocylic compounds; stereochemistry of phosphorus-containing heterocycles; semi-empirical and empirical calculations of organophosphorus compounds. *Mailing Add:* Dept Chem James Madison Univ Harrisonburg VA 22807

MOSBORG, ROBERT J(OHN), b Chicago, Ill, Dec 20, 24; m 51; c 2. STRUCTURAL ENGINEERING. *Educ:* Univ Ill, BS, 46, MS, 49. *Prof Exp:* Eng aide, Bridge Dept, Ill Cent RR, 46-47; res asst, 47-49, res assoc, 49-52, ASST PROF, DEAN OF ENG, UNIV ILL, URBANA, 52- *Concurrent Pos:* NSF fac fels, 59, 60-62. *Mem:* Am Soc Civil Engrs; Am Soc Eng Educ; Nat Soc Prof Engrs. *Res:* Structural mechanics; material behavior. *Mailing Add:* 1104 Civil Eng Bldg Univ Ill Urbana IL 61803

MOSBURG, EARL R, JR, b Frederick, Md, Jan 23, 28; m 58; c 1. ATOMIC PHYSICS. *Educ:* Yale Univ, BS, 52, PhD(physics), 56. *Prof Exp:* PHYSICIST, NAT BUR STANDARDS, 56- *Concurrent Pos:* Mem subcomt neutron standards & measurements, Comt Nuclear Sci, Nat Res Coun, 59-61. *Mem:* Am Phys Soc; Sigma Xi. *Res:* Plasma physics; gas discharges; gaseous electronics. *Mailing Add:* 1525 Sunset Blvd Boulder CO 80302

MOSBY, JAMES FRANCIS, b Owensville, Ind, Nov 8, 37; m 77; c 1. FUEL TECHNOLOGY & PETROLEUM ENGINEERING. *Educ:* Purdue Univ, BS, 59, PhD(chem eng), 64; Univ Chicago, MBA, 74. *Prof Exp:* Chem engr, Am Oil Co, 64-68, proj mgr, Res & Develop Dept, 68-83, SR RES ASSOC, AMOCO OIL CO, 83- *Mem:* Am Chem Soc; Am Inst Chem Engrs. *Res:* Reaction mechanisms and reaction kinetics in multiphase systems; petroleum technology, particularly hydrotreating of petroleum distillates and residues. *Mailing Add:* Amoco Oil Co Res & Develop PO Box 3011 Naperville IL 60566-3011

MOSBY, WILLIAM LINDSAY, b Rockford, Ill, Nov 30, 21; m 49. ORGANIC CHEMISTRY. *Educ:* Harvard Univ, BSc, 43; Ohio State Univ, PhD(org chem), 49. *Prof Exp:* Res chemist, Gen Aniline & Film Corp, 49-52; res chemist, Res Dept, 52-54, group leader, 55-58, res assoc, 59-76, PRIN RES SCIENTIST, AM CYANAMID CO, 76- *Concurrent Pos:* Am Cyanamid fel, Univ Munich, 64-65. *Mem:* Am Chem Soc. *Res:* Synthetic and theoretical organic chemistry; intermediates for dyes and pharmaceuticals; light stabilizers; vat dyes; polyester dyes; antioxidants; aromatic and polycyclic compounds and heterocyclic system with bridgehead nitrogen atoms. *Mailing Add:* 66-B Heritage Village Southbury CT 06488-1653

MOSCATELLI, DAVID ANTHONY, b Stockton, Calif, July 26, 49. ANGIOGENESIS, GROWTH FACTORS. *Educ:* Univ Calif Davis, BS, 71, Univ Calif Berkeley, PhD(molecular biol), 77. *Prof Exp:* res asst prof, 81-87, RES ASSOC PROF CELL BIOL, MED CTR, NY UNIV, 87- *Mailing Add:* Dept Cell Biol Med Ctr NY Univ 550 First Ave New York NY 10016

MOSCATELLI, EZIO ANTHONY, b New York, NY, Nov 17, 26; c 1. BIOCHEMISTRY, NEUROCHEMISTRY. *Educ:* Columbia Univ, AB, 48; Univ Ill, MS, 49, PhD(biochem), 58. *Prof Exp:* Assoc chemist, Merck & Co, Inc, 49-55; chemist, Nat Heart Inst, 58-59; sr chemist, Merck, Sharp & Dohme Res Labs, 59-62; asst prof biochem, Univ Tex Southwest Med Sch, Dallas, 62-70; assoc prof psychiat & biochem, Mo Inst Psychiat, 70-74; ASSOC PROF BIOCHEM, SCH MED & COL AGR, UNIV MO-COLUMBIA, 74- *Mem:* Am Oil Chem Soc; Am Soc Biol Chemists; Am Soc Neurochem; Int Soc Neurochem; Sigma Xi. *Res:* Biochemistry of brain membrane lipids in alcohol abuse, cold adaptation and spinal cord injury. *Mailing Add:* Biochem Dept Univ Mo-Columbia M121 Med Sci Columbia MO 65212

MOSCATELLI, FRANK A, b New York, NY, July 2, 51; m 81. DISCHARGE PHYSICS, OPTOGALVANIC EFFECT. *Educ:* C W Post Col, BS, 72; NY Univ, MS, 74, PhD(physics), 80. *Prof Exp:* Res assoc physics, Clarendon Lab, Oxford Univ, Eng, 80-82; ASST PROF PHYSICS, SWARTHMORE COL, 82- *Concurrent Pos:* Vis scientist, Atomic Spectros Lab, Univ Caen, France, 86; mem, Coun Undergrad Res. *Mem:* Am Phys Soc; Sigma Xi. *Res:* High resolution atomic spectroscopy using laser Doppler-free methods; discharge diagnosis using optogalvanic effect. *Mailing Add:* Dept Physics Swarthmore Col Swarthmore PA 19081

MOSCHANDREAS, DEMETRIOS J, b Thessaloniki, Greece, Feb 3, 43; US citizen. EMISSION RATES, CHARACTERIZATION INDOOR AIR. *Educ:* Stetson Univ, BS, 66; Univ Cincinnati, PhD(physics), 73. *Prof Exp:* Dir environ sci, Geomet Technol, Inc, 73-81; sr sci adv, Ill Inst Technol Res Inst, 81-91, PROF ENVIRON ENG, ILL INST TECHNOL, 90- *Concurrent Pos:* Mem, Comt Indoor Pollutants, Nat Acad Sci, 77-81, Bd, Ctr Hazardous Waste, 84-86, Sci Bd, Ctr Indoor Air Res, 89- & Bd, J- Indoor Air, 90-; co-chmn, IAQ-81, 79-81; consult, World Health Orgn, 84. *Mem:* Am Phys Soc; Am Soc Heating, Refrigerating & Air Conditioning Engrs; Water Pollution Control Fedn; Air & Waste Mgt Asn. *Res:* Total exposure to air pollutants; measures public exposure to pollutants using specially designed, sampling devices and questionnaires. *Mailing Add:* Ill Inst Technol Chicago IL 60616

MOSCHEL, ROBERT CARL, b Cincinnati, Ohio. ORGANIC CHEMISTRY, BIOCHEMISTRY. *Educ:* Ohio State Univ, BSc, 68, PhD(biochem), 73. *Prof Exp:* Res assoc org chem, Univ Ill, 73-75; CHEMIST, CHEM CARCINOGENESIS, FREDERICK CANCER RES CTR, 76- *Mem:* Am Chem Soc. *Res:* Chemical carcinogenesis; reactivity of chemical carcinogens with nucleic acids. *Mailing Add:* Frederick Cancer Ctr PO Box B Bldg 538 Frederick MD 21701-1230

MOSCHOPEDIS, SPEROS E, b Piraeus, Greece, June 1, 26; Can citizen; m 56; c 3. ORGANIC CHEMISTRY. *Educ:* Nat Univ Athens, BSc, 54, PhD(chem), 69. *Prof Exp:* Teacher chem, Archimides Inst Technol, Greece, 54-56; chemist, Sherritt Gordon Mines, Ltd, Alta, 56-57; RES OFFICER ORG CHEM, RES COUN ALTA, 57- *Mem:* Am Chem Soc; Greek Chem Asn. *Res:* Humic acids, lignites, coals and asphaltic type bituminous materials; water-soluble derivatives of humic acids; synthesis of polypeptides. *Mailing Add:* 9640 Austin OBrien Rd Edmonton AB T6B 2C2 Can

MOSCHOVAKIS, JOAN RAND, b Glendale, Calif, Dec 24, 37; m 63. MATHEMATICS. *Educ:* Univ Calif, Berkeley, AB, 59; Univ Wis, MS, 61, PhD(math), 65. *Prof Exp:* Instr math, Oberlin Col, 63-64; asst prof, 65-67, 69-74, assoc prof, 74-86, PROF MATH, OCCIDENTAL COL, 86- *Mem:* Asn Symbolic Logic; Am Math Soc. *Res:* Foundations of mathematics; formal and symbolic logic; intuitionism. *Mailing Add:* 721 24th St Santa Monica CA 90402

MOSCHOVAKIS, YIANNIS N, b Athens, Greece, Jan 18, 38; m 63; c 2. MATHEMATICS. *Educ:* Mass Inst Technol, SB & SM, 60; Univ Wis, PhD(math), 63. *Hon Degrees:* PhD, Univ Athens, 87. *Prof Exp:* Actg instr math, Univ Wis, 62-63; Benjamin Peirce instr, Harvard Univ, 63-64; from asst prof to assoc prof, 64-74, PROF MATH, UNIV CALIF, LOS ANGELES, 74- *Mem:* Am Math Soc; Asn Symbolic Logic. *Res:* Foundations of mathematics; recursive functions; foundations of computer science. *Mailing Add:* Dept of Math Univ of Calif 405 Hilgard Ave Los Angeles CA 90024

MOŚCICKI, EVE KARIN, b Jönköing, Sweden, May 1, 48; US citizen; m 73; c 2. EPIDEMIOLOGY OF SUICIDAL & SUICIDAL BEHAVIOR, PSYCHIATRIC EPIDEMIOLOGY. *Educ:* Michigan State Univ, BA, 69, MA, 71; Johns Hopkins Univ, ScD, 80, MPH, 82. *Prof Exp:* Speech-language pathologist ther, Montgomery County Pub Sch, 71-74; instr, Johns Hopkins Univ Eve Col, 77-78; scientist, Off Biomet & Field Studies, Nat Inst Neurol

& Commun Dis & Stroke, NIH, 81-84; epidemiologist, Epidemiol & Psychopath Res Br, NIMH & Alcohol, Drug Abuse & Ment Health Admin, 84-87; chief res/sci admin, Adult Epidemiol Prog, Epidemiol & Psychopath Res Br, 87-91, br asst chief, 87-91, CHIEF RES/SCI ADMIN, PREV RES BR, NIMH, 91- *Concurrent Pos:* Ed consult, Am J Epidemiol, Am J Pub Health, Archives Gen Psychiat, Ear & Hearing, Gen Hosp Psychiat, J Am Med Asn, J Appl Psycholing, J Nervous & Ment Dis, J Speech & Hearing Dis, Psychosom, Pub Health Reports, 81-; postdoctoral fel, Epidemiol Training Prog, USPHS, 81-84; co-chair, Epidemiol Training Prog Sem Ser, USPHS, 82-83; mem, publ comt, Nat Health & Nutrit Exam Surv, Epidemiol Followup Study, 82-87, Scientist Category Promo Bd, USPHS, 88, Task Force on Nat Strategic Res Plan, Nat Inst Deafness & Other Commun Dis, 89, Panel on Violence Prev, Nat Agenda Injury Control, 90-; alt rep, Alcohol, Drug Abuse & Ment Health Admin, Off Surgeon Gen, 88-90, mem, Liaison Comt Revitalizatin Comn Corps, 87-; Collabr, Nat Health & Nutrit Exam Surv-III Res Consortium, 87-; chair, Suicide Consortium, NIMH, 87-; consult ed, Suicide & Life-Threatening Behav, 89-; assoc ed, Am J Epidemiol, 91- *Mem:* AAAS; Am Asn Suicidology; Am Col Epidemiol; Soc Epidemiol Res; Am Pub Health Asn. *Res:* Epidemiology, especially suicide and suicidal behavior; cross cultural issues in psychiatric epidemiology; epidemiology of intentional injuries; survey research and epidemiologic methods; epidemiology of communication disorders. *Mailing Add:* 5600 Fishers Lane Rm 10-85 Rockville MD 20857

MOSCONA, ARON ARTHUR, b Israel, July 4, 22; m 55; c 1. NEUROBIOLOGY. *Educ:* Hebrew Univ Jerusalem, MSc, 47, PhD(zool), 50. *Prof Exp:* Res fel, Strangeways Res Lab, Cambridge Univ, 50-52; assoc prof physiol, Sch Med, Hebrew Univ Jerusalem, 53-55; vis investr Rockefeller Univ, 55-57; from assoc prof to prof, 58-74, chmn, Comt Develop Biol, 69-76, LOUIS BLOCK PROF BIOL SCI, UNIV CHICAGO, 74- *Concurrent Pos:* Vis prof, Col France, Stanford Univ, 59, Univ Palermo, 66, Univ Jerusalem, Columbia Univ, Harvard Univ, Tel Aviv Univ, Univ NC, Univ Helsinki, Kyoto Univ, 80, Academia Sinica & Univ Fla; Claude Bernard vis prof, Univ Montreal, 60; Lillie fel, Marine Biol Lab, Woods Hole, 60; founder and co-ed, Current Topics Develop Biol, 66-89; mem, President's Biomed Res Panel, 75; chmn, Bd Sci Counselors, Nat Inst Child Health & Human Develop, 82-86. *Honors & Awards:* Claude Bernard Medal; Alcon Prize; Azabu Gold Medal. *Mem:* Nat Acad Sci; Int Soc Develop Biol (pres, 76-81); fel NY Acad Sci; fel Am Acad Arts & Sci; fel AAAS. *Res:* Mechanisms of embryonic development and cell differentiation. *Mailing Add:* Cummings Life Sci Ctr Univ Chicago 920 E 58th St Chicago IL 60637

MOSCONY, JOHN JOSEPH, b Philadelphia, Pa, Aug 26, 29. CHEMISTRY. *Educ:* St Joseph's Col, Pa, BS, 51, MS, 58; Univ Pa, PhD(chem), 65. *Prof Exp:* Chemist, Elec Storage Battery Co, 51-54 & Waterman Prod Co, 54-57; ENG LEADER, RCA CORP, 57- *Concurrent Pos:* Lectr, RCA Corp Eng Serv, 71- *Mem:* Am Chem Soc; Am Vacuum Soc. *Res:* High pressure synthesis of silicon fluorides; thermoelectric and thermionic energy conversion; materials and processes related to vacuum and color television picture tubes. *Mailing Add:* 1860 Beverly Dr Lancaster PA 17601-4102

MOSCOVICI, CARLO, b 1925; m 55; c 2. VIROLOGY. *Educ:* Univ Rome, PhD(microbiol), 52. *Prof Exp:* Asst prof pediat, Med Sch, Univ Colo, 57-67; assoc prof, 67-71, PROF IMMUNOL & MICROBIOL, MED SCH, UNIV FLA, 71-, RES CAREER SCIENTIST, 78-; CHIEF VIROL RES LAB, VET ADMIN HOSP, 67- *Honors & Awards:* USPHS-Carrier Award, 61. *Res:* Tumor virology; avian tumor viruses; RNA tumor viruses; cell differentiation. *Mailing Add:* Dept Pathol/Immunol Univ Fla Med Col J Hillis Miller Health Ctr Gainesville FL 32601

MOSCOVICI, HENRI, b Tecuci, Romania, May 5, 44; m; c 1. GEOMETRY & ANALYSIS. *Educ:* Univ Bucharest, MS, 66, PhD(math), 71. *Prof Exp:* PROF MATH, OHIO STATE UNIV, 84- *Concurrent Pos:* Vis prof, Collège de France, IHES, Paris, IAS, Princeton. *Honors & Awards:* G Tzitzeica, Nat Acad Sci/Romania, 74. *Mem:* Am Math Soc. *Res:* Research in index theory & operator algebra; non commutative differential geometry. *Mailing Add:* Dept Math Ohio State Univ Columbus OH 43210

MOSCOVICI, MAURICIO, anatomy, surgery, for more information see previous edition

MOSCOWITZ, ALBERT, b Manchester, NH, Aug 20, 29. PHYSICAL CHEMISTRY. *Educ:* City Col New York, BS, 50; Harvard Univ, MA, 54, PhD, 57. *Prof Exp:* Nat Res Coun-Am Chem Soc fel petrol res, Harvard Univ, 57-58 & Wash Univ, 58-59; from asst prof to assoc prof, 59-65, PROF PHYS CHEM, UNIV MINN, MINNEAPOLIS, 65- *Concurrent Pos:* Fulbright lectr, Copenhagen Univ, 61-62, vis prof, 61-62 & 67-68; Alfred P Sloan Found fel, 62-66; mem nat screening comt, Fulbright Awards to Scandinavia, 65, chmn, 66; Seydel-Woolley vis prof, Ga Inst Technol, 66; adv ed, Chem Physics Lett, 67-; vchmn, Gordon Conf Theoret Chem, 68, chmn, 70; assoc ed, J Chem Physics, 70-73. *Mem:* AAAS; Am Chem Soc; Royal Soc Chem; The Chem Soc; fel NY Acad Sci; foreign mem Royal Danish Soc Sci & Letters; fel Am Phys Soc. *Res:* Electronic structure of molecules; optical activity; stereochemistry. *Mailing Add:* Dept Chem Univ Minn Minneapolis MN 55455

MOSE, DOUGLAS GEORGE, b Chicago, Ill, July 18, 42; m 69. GEOCHEMISTRY, GEOCHRONOLOGY. *Educ:* Univ Ill, Urbana, BS, 65; Univ Kans, MS, 68, PhD(geol), 71. *Prof Exp:* Asst prof geol, Brooklyn Col, 71-75; ASSOC PROF GEOL, GEORGE MASON UNIV, 75- *Mem:* Geol Soc Am; Geol Soc Can; Nat Asn Geol Teachers; Sigma Xi. *Res:* Evolution of igneous and metamorphic rocks in North American Precambrian and Paleozoic terranes. *Mailing Add:* Dept Geol George Mason Univ Fairfax VA 22030

MOSELEY, HARRISON MILLER, b Dundee, Tex, Dec 14, 21. PHYSICS. *Educ:* Tex Christian Univ, AB, 43; Univ NC, PhD(physics), 50. *Prof Exp:* From asst prof to assoc prof, 50-65, PROF PHYSICS, TEX CHRISTIAN UNIV, 65- *Mem:* Am Phys Soc; Am Asn Physics Teachers; Sigma Xi. *Res:* Thermal diffusion; fundamental particle theory. *Mailing Add:* 6016 Wrigley Way Ft Worth TX 76133

MOSELEY, HARRY EDWARD, b New Iberia, La, Oct 18, 29; m 55; c 2. CHEMISTRY. *Educ:* La State Univ, BS, 51, MS, 52, PhD(chem), 69. *Prof Exp:* Res chemist, Monsanto Chem Co, 54-61; from instr to assoc prof, 61-75, PROF CHEM, LA TECH UNIV, 75- *Mem:* AAAS; Am Chem Soc; Sigma Xi. *Res:* Separation and determination of the platinum metals. *Mailing Add:* 2606 Cypress Springs Ave Ruston LA 71270

MOSELEY, JOHN TRAVIS, b New Orleans, La, Feb 26, 42; m 61, 79; c 4. ATOMIC PHYSICS, MOLECULAR PHYSICS. *Educ:* Ga Inst Technol, BS, 64, MS, 66, PhD(physics), 69. *Prof Exp:* Asst res physicist, Eng Exp Sta, Ga Inst Technol, 64-65; asst prof physics, Univ West Fla, 68-69; physicist, SRI Int, 69-75, sr physicist, 75-77, prog mgr, 77-79; assoc prof, Univ Ore, 79-82, dir, 81-84, head, Dept Physics, 84-85, PROF, DEPT PHYSICS, UNIV ORE, 82-, VPRES RES, CHEM PHYSICS INST, 85- *Concurrent Pos:* Vis scientist, Univ Paris, 75-76, vis prof, 77, 78, 80 & 82; mem, Comt Atomic & Molecular Sci, Nat Res Coun. *Mem:* Fel Am Phys Soc; Am Chem Soc; AAAS; Sigma Xi. *Res:* Laser techniques are used to study small molecules, principally ions, in the gas phase; spectroscopy, reactions, multi-photon ionization, and photodissociation; photodissociation and photodetachment of ions. *Mailing Add:* Univ Oregon 110 Johnson Hall Eugene OR 97405

MOSELEY, LYNN JOHNSON, b Washington, DC, July 7, 48; m 72; c 2. ANIMAL BEHAVIOR, ORNITHOLOGY. *Educ:* Col William & Mary, BS, 70; Univ NC, Chapel Hill, PhD(zool), 76. *Prof Exp:* Instr biol, Elon Col, 75-77; asst prof, 77-85, ASSOC PROF BIOL, GUILFORD COL, 85- *Mem:* Sigma Xi; Animal Behav Soc; Am Ornithologists Union; Wilson Ornith Soc. *Res:* Behavior and communication in vertebrates, specifically auditory communication in birds. *Mailing Add:* Dept Biol Guilford Col Greensboro NC 27410

MOSELEY, MAYNARD FOWLE, b Boston, Mass, July 15, 18; m 49; c 2. PLANT ANATOMY. *Educ:* Univ Mass, BS, 40; Univ Ill, MS, 42, PhD(bot), 47. *Prof Exp:* Instr bot, Cornell Univ, 47-49; from instr to assoc prof, Univ Calif, Santa Barbara, 49-63, prof bot, 63-84; RETIRED. *Concurrent Pos:* Clin lab tech, US Army, 42-47. *Mem:* Bot Soc Am; Int Soc Plant Morphol; Int Asn Wood Anat. *Res:* Determination of the wood and floral anatomy of certain plant families; use of data for phylogenetic considerations. *Mailing Add:* Dept Biol Sci Univ Calif Santa Barbara CA 93106

MOSELEY, PATTERSON B, b Holland, Mo, May 27, 18; m 42; c 3. CHROMATOGRAPHY. *Educ:* Ouachita Col, BS, 43; La State Univ, MS, 49, PhD, 51. *Prof Exp:* Res chemist, Hercules Powder Co, Del, 51-57, res supvr, 57-64; assoc prof, 64-69, PROF CHEM, LA TECH UNIV, 69-, DIR RES, COL ARTS & SCI & DIR DIV ALLIED HEALTH, 68-, ASSOC DEAN COL ARTS & SCI, 70- *Mem:* AAAS; Am Chem Soc; Sigma Xi. *Res:* Adsorption chromatography; naval stores chemistry; ion exchange. *Mailing Add:* 1005 Maple Ruston LA 71270

MOSELEY, SHERRARD THOMAS, b Roanoke, Va, May 16, 21; m 41; c 1. ELECTRICAL ENGINEERING. *Educ:* Va Polytech Inst, BSEE, 42; Syracuse Univ, MEE, 52. *Prof Exp:* Elec engr, Gen Elec Co, 46-48; instr & res assoc elec eng, Syracuse Univ, 48-54; assoc prof & actg head dept, Univ SC, 57-62; pres, Wytheville Community Col, 62-67; res dir, Fla Jr Col Syst, 67-69; mgr duplicating prod div, Caldwell-Sites Co, 69-74; mgr safety & training, Electro-Optical Prod Div, ITT Corp, 74-88; RETIRED. *Mem:* Inst Elec & Electronics Engrs. *Res:* Microwave antennas and propagation; ionospheric propagation at low frequencies. *Mailing Add:* 2310 Stanley Ave SE Roanoke VA 24014

MOSELEY, WILLIAM DAVID, JR, b Cleveland, Ohio, Nov 27, 3. PHYSICAL CHEMISTRY. *Educ:* Williams Col, BA, 58; Wash State Univ, PhD(chem), 63. *Prof Exp:* Jr scientist quantum chem, Univ Uppsala, 63-66; asst prof chem, Howard Univ, 66-69; asst prof, 69-78, ASSOC PROF CHEM, WASH STATE UNIV, 78- *Res:* Quantum theory; valence theory; collisions. *Mailing Add:* NE 405 Colorado Wash State Univ Pullman WA 99164

MOSEMAN, JOHN GUSTAV, b Oakland, Nebr, Dec 7, 21; m 48; c 3. PLANT PATHOLOGY, AGRONOMY. *Educ:* Univ Nebr, BS, 43; Wash State Univ, MS, 48; Iowa State Univ, PhD(agron, plant path), 50. *Prof Exp:* Res plant pathologist, NC, Agr Res Serv, USDA, 50-54, res plant pathologist, Cereal Crops Res Br, 54-69, leader barley invest, 69-72, chmn, Plant Genetics & Germplasm Inst, 72-81, res plant pathologist, 81-86; RETIRED. *Mem:* Am Phytopath Soc; Am Soc Agron; fel AAAS. *Res:* Develop improved wheat, barley germplasm by studying host pathogen interactions, resistance biotic, abiotic stresses, such as rusts, powdery mildew, salts and drouth; coevolution, and genetics of host pathogens. *Mailing Add:* 1918 Blackbriar Silver Springs MD 20903

MOSEMAN, ROBERT FREDRICK, b Indianapolis, Ind, Dec 18, 41; m 65; c 3. PESTICIDE CHEMISTRY. *Educ:* Marian Col, BS, 63; Univ NC, MS, 65; Univ Mo, PhD(agr chem), 71. *Prof Exp:* Chemist, NC State Bd Health, 65-66; chemist, USPHS, 66-68; res chemist, US Environ Protection Agency, 71-79; DEPT HEAD, ANAL CHEM DIV, RADIAN CORP, 79-, AT RADIAN CORP PROGRESS CTR. *Mem:* Sigma Xi; Am Chem Soc. *Res:* Development of analytical methodology for the determination of trace levels of pesticides and other toxic substances and their transformation products in various types of biological and environmental samples. *Mailing Add:* 4505 Sterling Pl Raleigh NC 27612-3660

MOSEN, ARTHUR WALTER, b Bemidji, Minn, July 11, 22; m 46; c 2. ANALYTICAL CHEMISTRY. *Educ:* Ore State Col, BS, 49, MS, 51. *Prof Exp:* Instr chem, San Diego State Col, 50-51; res asst analytical chem, Los Alamos Sci Lab, 51-53, staff mem, 53-55; chemist, Rohr Corp, 55-56; staff mem, John Jay Hopkins Lab Pure & Appl Sci, Gen Atomic Div, Gen Dynamics Corp, 56-59, group leader analytical chem, 59-66, asst chmn chem dept, 66-70; mgr analytical chem br & sr staff mem, Mat Sci Dept, Gulf Gen Atomic Co, 70-73; mgr analytical chem dept, Gen Atomic Co, 73-78; chief chemist, Industrial Marine, 78-81, prin chemist, AVX Mat Div, 79-81; RETIRED. *Concurrent Pos:* Lectr, San Diego State Col, 55-57 & 59-60. *Mem:* Am Chem Soc; Am Soc Testing & Mat. *Res:* Analytical chemistry methods development applied particularly to nuclear reactor materials; gases in metals; rare earth elements. *Mailing Add:* 323 Alpine Ave Chula Vista CA 91910

MOSER, ALMA P, b Auburn, Wyo, Aug 9, 35; m 57; c 4. MECHANICAL ENGINEERING, APPLIED MECHANICS. *Educ:* Utah State Univ, BS, 61, MS, 63; Univ Colo, PhD(civil eng), 67. *Prof Exp:* Assoc engr, Thiokol Chem Corp, 61; from instr to assoc prof mech eng, 61-76, PROF MECH ENG & HEAD DEPT & DIR, PIPING SYSTS INST, UTAH STATE UNIV, 76- *Concurrent Pos:* Res assoc, Johns-Manville Corp, 72-75. *Mem:* Am Soc Mech Engrs; Am Water Works Asn; Sigma Xi. *Res:* Engineering elasticity; viscoelasticity; crack propagation and fracture analysis; soil-structure interaction mechanics. *Mailing Add:* Dept Mech Eng Utah State Univ Logan UT 84322-4130

MOSER, BRUNO CARL, b Elmhurst, Ill, Mar 31, 40; m 62; c 3. HORTICULTURE. *Educ:* Mich State Univ, BS, 62, MS, 64; Rutgers Univ, PhD(hort), 69. *Prof Exp:* From asst prof to assoc prof hort, Rutgers Univ, New Brunswick, 69-75; PROF HORT & HEAD DEPT, PURDUE UNIV, WEST LAFAYETTE, 75- *Honors & Awards:* Kenneth Post Award, Am Soc Hort Sci, 69. *Mem:* Fel Am Soc Hort Sci; Int Plant Propagation Soc; Sigma Xi; Am Soc Plant Physiologists. *Res:* Physiology of root regeneration; tuberization; salt stress. *Mailing Add:* Dept Hort Purdue Univ West Lafayette IN 47907

MOSER, CHARLES R, b Woodland, Calif, Oct 8, 39; m 60; c 1. DEVELOPMENTAL BIOLOGY. *Educ:* Humboldt State Col, AB, 61; State Univ NY Buffalo, PhD(biol), 67. *Prof Exp:* From asst prof to assoc prof, 66-77, PROF BIOL SCI, CALIF STATE UNIV, SACRAMENTO, 77- *Mem:* Sigma Xi. *Res:* DNA, RNA, protein synthesis and their interrelationships in developing frog embryos. *Mailing Add:* 8986 La Riviera Dr Sacramento CA 95826

MOSER, DONALD EUGENE, b Steubenville, Ohio, Jan 22, 25; m 46; c 3. MATHEMATICS. *Educ:* Amherst Col, AB, 47; Brown Univ, AM, 49; Univ Pittsburgh, PhD(math), 56. *Prof Exp:* From instr to asst prof math, Univ Mass, 49-60; assoc prof, 60-70, PROF MATH, UNIV VT, 70- *Concurrent Pos:* Dept chair, Univ Vt, 77-84, 86-88. *Mem:* Math Asn Am; Sigma Xi. *Mailing Add:* Dept Math Univ Vt 16 Colchester Ave Burlington VT 05405

MOSER, FRANK, b Winnipeg, Man, Sept 5, 27; US citizen; c 3. SOLID STATE PHYSICS. *Educ:* Univ Man, BSc hons, 49; Univ Minn, MSc, 52. *Prof Exp:* Res physicist, Res Labs, Eastman Kodak Co, 52-83; TEL AVIV UNIV, ISRAEL, 84- *Concurrent Pos:* Vis scientist, Oxford Univ, 65-66 & Israel Inst Technol, 73-74. *Mem:* Am Phys Soc. *Res:* Electronic properties of semiconductors; photoeffects and imaging phenomena in semiconductors and other solids and thin films. *Mailing Add:* PO Box 18098 Jerusalem 91180 Israel

MOSER, FRANK HANS, b Chicago, Ill, Aug 4, 07; m 30, 69; c 2. PIGMENTS, GLASS COATING. *Educ:* Hope Col, AB, 28; Univ Mich, MS, 29, PhD(chem), 31. *Prof Exp:* Asst, Univ Mich, 28-31, chief analyst, Eng Res Dept, 31-32; sr res chemist, Nat Aniline Co, NY, 32-38; supt, Intermediate Dept, Standard Ultramarine Co, 38-53; supt, Intermediate & Phthalocyanine Depts, Standard Ultramarine & Color Co, 53-59, res dir, 59-64; res dir, Holland Suco Color Co, 65-68; tech dir, Pigments Div, Chemetron Corp, 68-72; CONSULT, PHTHALOCYANINE COMPOUNDS, 72- *Concurrent Pos:* Consult, Donnelly Corp, 79. *Mem:* Am Chem Soc; fel Am Inst Chemists; NY Acad Sci. *Res:* Phthalocyanine compounds; pigments; intermediates for pigments and dyes; pigments application; process, product and raw material safety; glass coatings. *Mailing Add:* 3373 Lakeshore Dr Holland MI 49424

MOSER, GENE WENDELL, b Hawley, Pa, July 25, 30; c 2. INFORMATION THEORY, ORIGIN OF LIFE EVOLUTION. *Educ:* E Stroudsburg State Univ, BS, 53; Columbia Univ, MS, 55; Cornell Univ, PhD(sci educ & evolution), 63. *Prof Exp:* Asst prof sci educ, State Univ NY, Oswego, 57-58; asst prof biol & sci educ, Univ Rochester, 62-64; asst prof, sci educ & natural sci, Colgate Univ, 64-67; assoc prof, 67-69, PROF SCI EDUC, UNIV PITTSBURGH, 69- *Concurrent Pos:* Dir higher educ, sci, NY State Bd Regents, 60-62; assoc dean & dir sci, Univ Rochester, 62-64; asst to provost, Colgate Univ, 64-67; sr inventor, H-Inventor Group, 70- *Mem:* Fel AAAS; Nat Asn Res Sci Teaching; Sigma Xi. *Res:* Quantifying sequences of behaviors and biological events; computing and information processing; poikiothermic nerve stain. *Mailing Add:* 9230 Perry Hwy Pittsburgh PA 15237

MOSER, GLENN ALLEN, b West Reading, Pa, Sept 19, 43. ORGANOMETALLIC CHEMISTRY, INORGANIC CHEMISTRY. *Educ:* Lebanon Valley Col, BSchem, 65; Bucknell Univ, MS, 68; Univ Mass, MS, 69, PhD(inorg chem), 72. *Prof Exp:* Sci Res Coun fel inorg chem & organometallic chem, Inorg Chem Lab, Oxford Univ, Eng, 72-74; res assoc inorg chem, Univ Wis-Madison, 74-76; SR RES CHEMIST, RESINS & POLYMERS, HERCULES INC, 76- *Mem:* Am Chem Soc; Sigma Xi. *Res:* Organometallic research; homogeneous and heterogeneous catalysis; metathesis; cationic polymerization; hydrogenation; graphic arts and adhesive application of resins. *Mailing Add:* 2521 Eaton Rd Chalfonte Wilmington DE 19810-3557

MOSER, HERBERT CHARLES, b Camp Verde, Ariz, Mar 5, 29; m 51; c 2. PHYSICAL CHEMISTRY. *Educ:* San Jose State Col, BS, 52; Iowa State Univ, PhD(chem), 57. *Prof Exp:* From asst prof to assoc prof, 57-69, PROF CHEM, KANS STATE UNIV, 69- *Mem:* Am Chem Soc; Sigma Xi. *Res:* Atomic and free radical reactions; radiation chemistry; use of radioisotopes as tracers. *Mailing Add:* Dept of Chem Kans State Univ Manhattan KS 66502

MOSER, JAMES HOWARD, b Santa Rosa, Calif, Apr 29, 28; m 50; c 3. AIR DISPERSION MODELING & MEASUREMENTS, CHEMICAL ENGINEERING. *Educ:* Univ Calif, BS, 50; Ore State Col, MS, 52, PhD(chem), 54. *Prof Exp:* Asst food technol, Ore State Col, 50-51, asst chem, 51-52; res chemist & technologist, Shell Oil Co, Calif, 54-60, sr res chemist, Houston Res Lab, 60-68, staff res engr, Shell Develop Co, Houston, 68-86, sr staff res engr, 86-88; PRES, SOFTSKILLS, INC, CYPRESS, TEX, 88- *Mem:* Am Inst Chem Eng; Sigma Xi; Am Meteorol Soc; Air & Waste Mgr Asn. *Res:* Air quality modeling and analysis; computer science and applications; mathematical simulation of chemical and petroleum processes; control systems, environmental engineering; operations research; analytical and physical chemistry; reaction kinetics; chemical and petroleum process design and development; applied mathematics and statistics. *Mailing Add:* Softskills Inc PO Box 898 Cypress TX 77429

MOSER, JOHN WILLIAM, JR, b Hagerstown, Md, Oct 8, 36; m 64; c 3. FOREST BIOMETRY, SOFTWARE DEVELOPMENT. *Educ:* WVa Univ, BS, 58; Pa State Univ, MS, 61; Purdue Univ, PhD(forest biomet), 67. *Prof Exp:* Forester, US Forest Serv, 61-63 & Coop Exten Serv, WVa Univ, 63-64; asst, 64-66, from instr to assoc prof forestry & conserv, 66-78, PROF FORESTRY, PURDUE UNIV, WEST LAFAYETTE, 78- *Concurrent Pos:* Assoc ed, Forest Sci, 84-85; chmn, Forest Sci & Technol Bd, 85-88 & Soc Am Forester Coun, 89- *Honors & Awards:* Fel, Soc Am Foresters. *Mem:* Biomet Soc; Soc Am Foresters. *Res:* Modeling the dynamics of forest stands; computer applications to forest management; geographic information systems; management information systems. *Mailing Add:* Dept of Forestry & Natural Resources Purdue Univ West Lafayette IN 47907

MOSER, JOSEPH M, b Spring Hill, Minn, Apr 13, 30; wid; c 1. MATHEMATICAL STATISTICS. *Educ:* St John's Univ, Minn, BA, 54; St Louis Univ, MA, 55, PhD(math statist), 59. *Prof Exp:* Statistician, Aberdeen Proving Ground, Md, 55-56; from asst prof to assoc prof, 59-69, PROF MATH STATIST, SAN DIEGO STATE UNIV, 69- *Concurrent Pos:* Consult, Navy Electronics Lab, Calif, 64-71. *Mem:* Inst Math Statist; Math Asn Am; Am Statist Asn. *Res:* Distribution-free statistics. *Mailing Add:* Dept of Math San Diego State Univ San Diego CA 92182

MOSER, JURGEN (KURT), b Königsberg, Ger, July 4, 28; nat US; m 55; c 3. DYNAMICAL SYSTEMS, CELESTIAL MECHANICS. *Educ:* Univ G06ttingen, Dr rer nat, 52. *Prof Exp:* Asst math, Univ Göttingen, 53; res assoc, NY Univ, 53-54; asst, Univ Göttingen, 54-55; res assoc, NY Univ, 55-56, asst prof, 56-57; from asst prof to assoc prof, Mass Inst Technol, 57-60; prof math, NY Univ, 60-80, dir, Courant Inst Math Sci, 67-70; PROF MATH, EIDGENOSSICHE TECHNISCHE, 80- *Concurrent Pos:* Sloan fel, 62 & 63; Am Acad Arts & Sci fel, 64; Guggenheim fel, 70. *Honors & Awards:* G B Birkhoff Prize, 68; J Craig Watson Medal, 69. *Mem:* Nat Acad Sci; Am Math Soc; corresp mem Int Astron Union; Royal Swed Acad Sci. *Res:* Ordinary and partial differential equations; spectral theory; celestial mechanics. *Mailing Add:* Dept Math E T H Ctr 8092 Zurich Switzerlanda

MOSER, KENNETH BRUCE, b Malverne, NY, Mar 27, 33; m 58; c 5. CARBOHYDRATE CHEMISTRY. *Educ:* Tusculum Col, BS, 54; Duke Univ, PhD(chem), 59. *Prof Exp:* Res chemist, 58-60, sr res chemist, 60-70, lab head, 70-73, group leader, 73-85, SECT MGR, A E STALEY MFG CO, 85- *Mem:* Am Chem Soc; Sigma Xi. *Res:* Synthesis of polycyclic aromatic systems containing quaternary nitrogen at the bridgehead position; preparation of acrylic polymer emulsions; carbohydrates; nitrogen heterocyclics; starch modification. *Mailing Add:* 3054 Mac Arthur Rd Decatur IL 62526

MOSER, KENNETH MILES, b Baltimore, Md, Apr 12, 29; m 51; c 4. MEDICINE. *Educ:* Haverford Col, AB, 50; Johns Hopkins Univ, MD, 54. *Prof Exp:* From instr to assoc prof med, Georgetown Univ, 58-68, chief pulmonary div, Univ Hosp, 61-68; assoc prof, 68-73, PROF MED, UNIV CALIF, SAN DIEGO, 73-, DIR PULMONARY & CRITICAL CARE DIV, 68-, DIR PULMONARY SPEC CTR RES, 70- *Concurrent Pos:* Dir head, chest & contagious dis div, Nat Naval Med Ctr, Md, 59-61; chief pulmonary sect, Georgetown Clin Res Inst, Bur Aviation Med, Fed Aviation Agency, 61-66; consult, US Naval Hosp, Md, 61 & NIH, 65- *Mem:* Am Thoracic Soc; Am Heart Asn; Am Fedn Clin Res; fel Am Col Physicians; fel Am Col Chest Physicians. *Res:* Pulmonary and cardiac physiology; blood coagulation; pathogenesis and therapy of thromboembolism; clinical pulmonary disease. *Mailing Add:* Dept Med Univ Calif 225 W Dickinson St San Diego CA 92103

MOSER, LOUISE ELIZABETH, b Racine, Wis, July 24, 43. MATHEMATICS. *Educ:* Univ Wis-Madison, BS, 65, MS, 66, PhD(math), 70. *Prof Exp:* Asst prof, 70-74, assoc prof, 74-80, PROF MATH, CALIF STATE UNIV, HAYWARD, 80- *Mem:* Am Math Soc; Math Asn Am. *Res:* Topology of 3-manifolds; knot theory; group theory; ring theory. *Mailing Add:* Dept Math Calif State Univ Santa Barbara CA 93106

MOSER, LOWELL E, b Akron, Ohio, Mar 19, 40; m 64; c 2. AGRONOMY. *Educ:* Ohio State Univ, BS, 62, PhD(agron), 67; Kans State Univ, MS, 64. *Prof Exp:* Asst prof agron, Ohio State Univ, 67-70; assoc prof, 70-75, PROF AGRON, UNIV NEBR-LINCOLN, 75- *Mem:* Fel Am Soc Agron; fel Crop Sci Soc Am; Soc Range Mgt. *Res:* Forage physiology; physiological and morphological investigation into cool and warm season grasses in Nebraska. *Mailing Add:* Dept Agron E Campus Univ Nebr Lincoln NE 68583-0915

MOSER, PAUL E, b Auburn, Wyo, Jan 18, 42; m 63; c 4. PLANT BREEDING, PLANT PATHOLOGY. *Educ:* Utah State Univ, MS, 68, PhD(plant path & breeding), 71. *Prof Exp:* Fel plant path, McGill Univ, 71-72; PATHOLOGIST & PLANT BREEDER, GALLATIN VALLEY SEED CO, 72- *Mem:* Am Phytopath Soc; Sigma Xi. *Res:* Breeding of new pea and bean varieties; indexing and screening for disease resistance of peas and beans. *Mailing Add:* 5417 N Fieldcrest Ave Boise ID 83704

MOSER, PAUL H, b Matto Grosso, Brazil, July 23, 31; US citizen; m 54; c 1. KARST HYDROLOGY, HYDROGEOLOGY. *Educ:* Berea Col, BA, 54; Univ Ky, Lexington, MS, 61. *Prof Exp:* Eng geologist, Stokley & Assoc, 60-62; petrol geologist, Texaco Inc, 62-64; geologist, 64-72, eng geologist, 72-80, HYDROGEOLOGIST, GEOL SURV ALA, 80- *Mem:* Am Inst Prof Geologists; Sigma Xi; Am Inst Hydrol. *Res:* Hydrogeologic, hydrologic and geologic research in the relationship and interconnection of ground water and surface water, including recharge areas, potential contaminating tracer dye work, time of travel and fluctuation. *Mailing Add:* 2709 Inland St Northport AL 35476

MOSER, ROBERT E, bio-organic chemistry; deceased, see previous edition for last biography

MOSER, ROBERT HARLAN, b Trenton, NJ, June 16, 23; m 48; c 2. CARDIOLOGY, MEDICAL EDUCATION. *Educ:* Loyola Univ, BS, 44; Georgetown Univ, MD, 48; Am Bd Internal Med, dipl, 74. *Prof Exp:* Intern, Washington, DC Gen Hosp, 48-49, fel pulmonary dis, 49-50; surgeon, Korea, 50-51; asst resident, Georgetown Univ Hosp, 51-52, chief resident, 52-53; chief med serv, US Army Hosp, Salzburg, Austria, 53-55; chief med serv, Wurzburg, Ger, 55-56; resident cardiol, Brooke Gen Hosp, 56-57, asst chief dept med, 57-59; fel hemat, Col Med, Univ Utah, 59-60; asst chief, US Army Tripler Gen Hosp, 60-64; chief, William Beaumont Gen Hosp, 65-67, Brooke Gen Hosp, 67-68 & Walter Reed Gen Hosp, 68-69; chief staff, Maui Mem Hosp, 72-73; exec vpres, Am Col Physicians, 77-87; VPRES, MED AFFAIRS, THE NUTRA SWEET CO, 87- *Concurrent Pos:* Assoc prof med, Col Med, Baylor Univ, 58-59; flight controller, Proj Mercury, 59-62; consult & mem med eval team, Proj Gemini, 62-66; ed, Med Opinion & Rev, 66-; chief dept med, Brooke Gen Hosp, 67-68; consult, Proj Apollo, 67-73; clin prof med, Col Med, Univ Hawaii, 69-, Col Med, Wash Univ, 70- & Abraham Lincoln Sch Med, 74-; ed, J AMA, 73-75. *Mem:* Inst of Med of Nat Acad Sci; master Am Col Physicians; Am Col Cardiol; Am Therapeut Soc; AMA; AAAS; Am Clin & Climat Asn, 78; hon fel Col Physicians & Surgeons Can. *Mailing Add:* The Nutra Sweet Co 1751 Lake Cook Rd PO Box 730 Dearfield IL 60015

MOSER, RONNY LEE, animal science & nutrition, for more information see previous edition

MOSER, ROY EDGAR, b Steubenville, Ohio, Sept 30, 22; m 42; c 2. FOOD TECHNOLOGY. *Educ:* Univ Mass, Amherst, BS, 47, MS, 49. *Prof Exp:* Food technologist, USDA, DC, 48-50; exten food technologist, Va Polytech Inst, 50-57; assoc prof food technol, Ore State Univ, 58-67, Univ Hawaii, Manoa Campus, 67-68 & Ore State Univ, 68-69; chmn, dept food sci & technol, 70-73, PROF FOOD TECHNOL, UNIV HAWAII, MANOA CAMPUS, 69-, CHMN DEPT FOOD SCI & TECHNOL, 78- *Mem:* Fel Inst Food Technol. *Res:* Food processing. *Mailing Add:* 557 Paokano Place Kailua HI 96734

MOSER, ROYCE, JR, b Versailles, Mo, Aug 21, 35; m 58; c 2. AEROSPACE MEDICINE & PHYSIOLOGY. *Educ:* Harvard Univ, BA, 57, MD, 61, MPH, 65. *Prof Exp:* Dir aerospace med, Schilling AFB Hosp, Salina, Kans, 62-64; resident aerospace med, Sch Pub Health, Harvard Univ, & US Air Force Sch Aerospace Med, Brooks AFB, Tex, 64-67; chief aerospace med, Aerospace Defense Command, Colo Springs, 67-70; comdr, US Air Force Clin, Phan Rang AB, Repub Viet Nam, 70-71; chief, aerospace med br, Sch Aerospace Med, Brooks AFB, 71-77; comdr, US Air Force Hosp, Tyndall AFB, Fla, 77-79; chief clin sci div, US Air Force Sch Aerospace Med, Brooks AFB, Tex, 79-81, chief educ div, 81-83, comdr, 83-85; PROF PREV & OCCUP MED VCHMN & COORDR PREV MED PROGS, DEPT FAMILY & PREV MED, SCH MED, UNIV UTAH, 85-, DIR, ROCKY MOUNTAIN CTR OCCUP & ENVIRON HEALTH, DEPT FAMILY & PREV MED, 87- *Concurrent Pos:* Assoc prof, dept family pract, Univ Tex Health Sci Ctr, San Antonio, 79-85 & dept environ sci, Sch Pub Health, Houston, 81-85; assoc prof, Uniformed Serv Univ Health Sci, Bethesda, 83- *Honors & Awards:* H G Moseley Award, Aerospace Med Asn, 81; T C Lyster Award, Aerospace Mod Asn, 88. *Mem:* Fel Am Col Prev Med; fel Am Acad Family Physicians; fel Am Col Occup Med; Am Bd Prev Med; fel Aerospace Med Asn (pres 89-90). *Res:* Managing aeromedical and biotechnology research involving ionizing and nonionizing radiation, aircrew performance, life support equipment and disease detection and prevention in flyers; programs in epidemiology, biostatistics, preventive medicine and environmental and occupational medicine. *Mailing Add:* Dept Family & Prev Med Sch Med Univ Utah 50 N Medical Dr Salt Lake City UT 84132

MOSER, STEPHEN ADCOCK, b Artesia, Calif, Mar 24, 46; m 69; c 3. MICROBIOLOGY, MEDICAL MYCOLOGY. *Educ:* Calif State Univ, Long Beach, BS, 69, MS, 72; Ohio State Univ, PhD(med microbiol), 76. *Prof Exp:* Trainee med mycol, 77-79, asst prof microbiol & immunol mycol, Sch Med, Tulane Univ, 80-; postdoctoral trainee, Pub Health & Med Lab Microbiol, Mt Sinai Med Ctr, Milwaukee, Wis, 80-82; assoc dir, 82-84, DIR MICRO & SEROLOGY, JEWISH HOSP, WASH UNIV MED CTR, 84-, ASST PROF, DEPT PATH, SCH MED, 82- *Mem:* Am Soc Microbiol; Med Mycological Soc Am; AAAS; NY Acad Sci; Southwestern Asn Clin Microbiol. *Res:* Host-parasite relationships; chemistry of fungal antigens; immune response in mycotic diseases. *Mailing Add:* Dept of Microbiol/ Serology, Jewish Hosp, Washington Univ Med Ctr PO Box 14109 St Louis MO 63178

MOSER, VIRGINIA CLAYTON, b Apr 23, 54; m 80; c 2. NEUROTOXICOLOGY, NEUROPHARMACOLOGY. *Educ:* Univ NC, BS, 77; Med Col Va, PhD(pharmacol), 83. *Prof Exp:* Res assoc, US Environ Protection Agency, 83-85; RES SCIENTIST, NORTHROP SERV INC, 85- *Concurrent Pos:* Consult, Nat Toxicol Prog; dipl, Am Bd Toxicol, 88. *Mem:* Neurosci Soc; Behavioral Toxicol Soc; Am Soc Pharmacol Exp Therapeut; Soc of Toxicol. *Res:* Develop and validate methods for neurotoxicity testing. *Mailing Add:* ManTech Environ Tech, Inc PO Box 12313 Res Triangle Park NC 27709

MOSER, WILLIAM O J, b Winnipeg, Man, Sept 5, 27; m 53; c 3. DISCRETE GEOMETRY. *Educ:* Univ Man, BSc, 49; Univ Minn, MA, 51; Univ Toronto, PhD(math), 57. *Prof Exp:* From lectr to asst prof math, Univ Sask, 55-59; assoc prof, Univ Man, 59-64; assoc prof, 64-66, PROF MATH, McGILL UNIV, 66- *Concurrent Pos:* Ed, Can Math Bull, 61-69; assoc ed, Can J Math, 82-85. *Mem:* Am Math Soc; Math Asn Am; Can Math Soc (pres, 75-77); Soc Indust & Appl Math. *Res:* Finite groups; combinatorial mathematics. *Mailing Add:* Dept Math McGill Univ 805 Sherbrooke St W Montreal PQ H3A 2K6 Can

MOSER, WILLIAM RAY, b Old Hickory, Tenn, Aug 3, 35; m 60; c 3. CATALYTIC SCIENCES. *Educ:* Mid Tenn State Univ, BS, 59; Mass Inst Technol, PhD(org chem), 64. *Prof Exp:* Res chemist, Monsanto Res SA, Switz, 66-67, staff scientist, 67-69; sr res chemist, Corp Res Labs, Esso Res & Eng Co, NJ, 69-75; res assoc, Badger Co, 75-81; PROF CHEM ENG, WORCESTER POLYTECH INST, 81- *Mem:* Am Chem Soc; NY Acad Sci; NAm Catalysis Soc; Am Inst Chem Engrs; Sigma Xi. *Res:* Homogeneous and heterogeneous catalysis in areas of chemical and petrochemical processes; organometallic synthesis and reaction mechanisms; metal oxide and superconductor synthesis. *Mailing Add:* Dept Chem Eng Worcester Polytech Inst 100 Inst Rd Worcester MA 01609

MOSER-VEILLON, PHYLIS B, b Cincinnati, Ohio, June 23, 47; m 69; c 1. NUTRITION. *Educ:* Univ Md, BS, 69, MS, 73, PhD(nutrit sci), 76. *Prof Exp:* Res nutritionist, Dept Pediat, Georgetown Med Ctr, 77-78; biologist, Carbohydrate Lab, Nutrit Inst, Agr Res Serv, USDA, 69-72; teaching asst nutrit, 72-74, res asst, 75, asst prof, 75-77, from asst prof to assoc prof nutrit, 78-89, PROF NUTRIT, DEPT HUMAN NUTRIT, UNIV MD, 89- *Concurrent Pos:* Fel, Am Col Nutrit. *Mem:* Sigma Xi; Am Dietetic Asn; AAAS; Am Women in Sci; Am Inst Nutrit; Am Soc Clin Nutrit. *Res:* Nutritional status of women and children especially in regard to trace minerals. *Mailing Add:* Dept Human Nutrit Marie Mount Hall Univ Md College Park MD 20742

MOSES, CAMPBELL, JR, b Pittsburgh, Pa, Feb 12, 17; m 40; c 4. MEDICINE. *Educ:* Univ Pittsburgh, BS, 39, MD, 41. *Prof Exp:* Instr physiol & pharmacol, Sch Med, Univ Pittsburgh, 41-46, from asst prof to assoc prof med, 46-48, dir, Addison H Gibson Lab, 48-68, dir postgrad educ, 60-68; med dir, Am Heart Asn, 68-73; vpres med affairs, Medicus Comn, 74-85; CONSULT, HEALTH COMMUN, 85- *Concurrent Pos:* Mem coun arteriosclerosis, Am Heart Asn. *Mem:* Am Physiol Soc; Soc Exp Biol & Med; fel AMA; fel Am Col Physicians; Am Diabetes Asn. *Res:* Thrombosis and embolism; anticoagulants; liver and kidney function tests; nutrition, liver and kidney function in arteriosclerosis; interactive videodisc role in medical education. *Mailing Add:* Medicus Intercon Inc 1675 Broadway New York NY 10019-5809

MOSES, EDWARD JOEL, b Newark, NJ, Oct 9, 38; m 65; c 2. UNDERWATER ACOUSTICS, OPERATIONS RESEARCH. *Educ:* Rensselaer Polytech Inst, BS, 60; Johns Hopkins Univ, PhD(physics), 67. *Prof Exp:* Res assoc physics, Vanderbilt Univ, 67-68, instr, 68-71; res scientist, Raff Assocs, Inc, Md, 71-75; dept head, Ocean Systs Dept, Gen Res Co, 75; sr proj staff, Opers Res, Inc, 75-78, prog dir, 78-80, assoc div dir, 80-81, EXEC SCIENTIST, ORI, INC, 81- *Mem:* Am Phys Soc. *Res:* Sound propagation and noise background in the oceans; statistical character of oceanic noise; statistics of signal detection; tactical analysis and operations research for naval operations and systems. *Mailing Add:* 14506 Woodcrest Dr Rockville MD 20853

MOSES, FRANCIS GUY, b Baltimore, Md, Nov 15, 37. PHYSICAL ORGANIC CHEMISTRY. *Educ:* Univ Del, BA, 59; Calif Inst Technol, PhD(chem), 67. *Prof Exp:* Fel chem, Iowa State Univ, 64-65; supvry chemist, 66-86, CONSULT, E I DU PONT DE NEMOURS & CO, INC, 89-; RETIRED. *Res:* High pressure chemistry; reaction mechanisms; organic photochemistry. *Mailing Add:* Three E Kenmore Dr Hyde Park Wilmington DE 19808

MOSES, GERALD ROBERT, speech pathology, for more information see previous edition

MOSES, GREGORY ALLEN, b Kalamazoo, Mich, Apr 7, 50; m 83; c 2. COMPUTATIONAL ENGINEERING, ENGINEERING RESEARCH ADMINISTRATION. *Educ:* Univ Mich, BSE, 72, MSE, 74, PdD(nuclear eng), 76. *Prof Exp:* From asst prof to assoc prof, 76-84, PROF NUCLEAR ENG, UNIV WIS-MADISON, 84-, ASSOC DEAN RES ENG, 89- *Mem:* Am Nuclear Soc; Am Phys Soc; Inst Elec & Electronic Engrs. *Res:* Computational engineering; radiation hydrodynamics; fusion technology; fission reactor analysis; particle transport. *Mailing Add:* 1500 Johnson Dr Madison WI 53706-1687

MOSES, HAL LYNWOOD, b Goldston, NC, Oct 12, 34; m 69; c 2. FLUID MECHANICS, TURBOMACHINERY. *Educ:* Va Polytech Inst, BS, 60; Mass Inst Technol, MS, 61, PhD(mech eng), 64. *Prof Exp:* Apprentice design, Newport News Shipbuilding & Dry Dock Co, 52-59; fel & asst prof mech eng, Mass Inst Technol, 64-66; engr, Corning Glass Works, NC, 66-69; assoc prof, 69-75, PROF MECH, VA POLYTECH INST & STATE UNIV, 75- *Concurrent Pos:* Consult, Foster-Miller Assocs, Mass, 65-66; adj asst prof, NC State Univ, 66-69; adj prof, Naval Postgrad Sch, 83-85. *Mem:* Am Soc Mech Engrs; Am Inst Aeronaut & Astronaut. *Res:* Turbulent boundary layer; fluidics and fluid mechanics; gas turbines. *Mailing Add:* Dept Mech Eng Va Polytech Inst & State Univ Blacksburg VA 24061

MOSES, HARRY ELECKS, b Canton, Ohio, Aug 30, 22; m 58; c 1. THEORETICAL PHYSICS. *Educ:* Univ Mich, BS, 44, MS, 47; Columbia Univ, PhD(physics), 50. *Prof Exp:* Res scientist aerodyn, Nat Adv Comt Aeronaut, 44-46 & Univ Mich, 46-47; asst physics, Columbia Univ, 47-49; asst wave propagation, NY Univ, 49-50, res assoc upper atmosphere res, 50-60; assoc prof physics, Polytech Inst Brooklyn, 60-61; staff mem, Geophys Corp Am, Mass, 61-62 & Lincoln Lab, Mass Inst Technol, 62-69; staff mem, Air Force Cambridge Res Labs, 69-76; mem staff, 76-80, res prof physics, Ctr Atmospheric Res, Univ Lowell, 80-87; RES PROF ELEC ENG, BOSTON UNIV, 87-; PRES, HARRY E MOSES-APPLIMATH CO, 87- *Mem:* Fel Am Phys Soc; NY Acad Sci. *Res:* Aerodynamics and fluid flow; electromagnetic, acoustic and quantum direct and inverse scattering; soliton theory and nonlinear equations; information theory; applications of group theory to quantum theory; meteorology and atmospheric physics. *Mailing Add:* 150 Tappan St Brookline MA 02146

MOSES, HENRY A, b Gastonia, NC, Sept 8, 39. BIOCHEMISTRY, PHYSIOLOGY. *Educ:* Livingstone Col, BS, 59; Purdue Univ, MS, 62, PhD(biochem), 64. *Prof Exp:* From asst prof to assoc prof, 64-74, PROF BIOCHEM, MEHARRY MED COL, 74-, CONTINUING EDUC & ASST VPRES ACAD SUPPORT, 82- *Concurrent Pos:* Tenn Heart Asn res grant, 65; vis lectr, Wheaton Col, Mass, 65; consult biochemist, Vet Admin Hosp, Tuskegee, Ala, 65; asst prof, Tenn State Univ, 65-66; consult, George W Hubbard Hosp, Nashville; vis prof biochem, Fisk Univ, 67-; Environ Protection Agency grant, 74-77; HHS grants, 77-83; provost for internal affairs, Meharry Med Col, 78-82. *Mem:* AAAS; Am Chem Soc. *Res:* Zinc metabolism; cadmium and lead toxicity; toxic metals in the environment. *Mailing Add:* Dept Biochem Meharry Med Col 17th Ave North Nashville TN 37208

MOSES, HERBERT A, b Hartford, Conn, July 22, 29; m 54; c 3. ATOMIC PHYSICS. *Educ:* Mich State Univ, BS, 51, MS, 53; Univ Conn, PhD(physics), 63. *Prof Exp:* From asst to instr physics, Mich State Univ, 51-56; res asst, Univ Conn, 56-59; from asst prof to assoc prof, 59-66, PROF PHYSICS, TRENTON STATE COL, 66- *Mem:* Am Asn Physics Teachers. *Res:* Experimental work in nuclear magnetic resonance in liquid crystals total cross sections for multiple electron stripping in atomic collisions at one hundred kiloelectron volts and five billion electron volts c-muon interactions; theoretical calculations of atomic wave functions for cesium and electron distribution in a deuterium plasma; nuclear magnetic resonance signals in liquid crystals. *Mailing Add:* Dept of Physics Trenton State Col Hillwood Lakes Trenton NJ 08625

MOSES, JOEL, b Petach Tikvah, Israel, Nov 25, 41; US citizen; m 70; c 2. COMPUTER SCIENCE. *Educ:* Columbia Univ, BA, 62, MA, 63; Mass Inst Technol, PhD(math), 67. *Prof Exp:* From asst prof to assoc prof comput sci, 67-76, assoc dir lab comput sci, 74-78, assoc head dept, 78-81, prof comput sci & eng, 77-89, head dept elec eng & comput sci, 81-89, D C JACKSON PROF COMPUT SCI & ENG, MASS INST TECHNOL, 89- *Concurrent Pos:* Vis prof, Harvard Grad Sch Bus, 89-91. *Mem:* Nat Acad Eng; Sigma Xi; Asn Comput Mach; fel Inst Elec & Electronic Engrs; fel Am Acad Arts & Sci, 87. *Res:* Design large systems symbolic formula manipulation; symbolic integration algorithms; artificial intelligence. *Mailing Add:* Lab Computer Sci Bldg NE43-407 Mass Inst of Technol Cambridge MA 02139

MOSES, LINCOLN ELLSWORTH, b Kansas City, Mo, Dec 21, 21; m 42, 68; c 5. STATISTICS. *Educ:* Stanford Univ, AB, 41, PhD(statist), 50. *Prof Exp:* Asst prof educ, Teachers Col, Columbia Univ, 50-52; from asst prof to assoc prof, 52-59, exec head dept, 64-68, assoc dean humanities & sci, 65-68 & 85-86, dean grad div, 69-75, PROF STATIST & BIOSTATIST, UNIV & SCH MED, STANFORD UNIV, 59- *Concurrent Pos:* Guggenheim fel, 60-61; fel, Ctr Advan Study in Behav Sci, 75-76, consult, 81-; adminr, Energy Info Admin, US Dept Energy, 78-80; mem, Comt Nat Statist coun, 75-77 & 81-85, (chmn, 84-85) & Inst Med Nat Acad Sci, 75- *Mem:* Inst Med; Biomet Soc; Am Statist Asn; Inst Math Statist; AAAS; Amer Acad Arts & Sci. *Res:* Health care technology assessment; biological and psychological applications of statistical methods; data analysis. *Mailing Add:* Dept of Statist Sequoia Hall Stanford Univ Stanford CA 94305

MOSES, MONTROSE JAMES, b New York, NY, June 26, 19; m 49; c 2. CELL BIOLOGY, ELECTRON MICROSCOPY. *Educ:* Bates Col, BS, 41; Columbia Univ, AM, 42, PhD(zool), 49. *Prof Exp:* Assoc cytochemist, Brookhaven Nat Lab, 48-52, cytochemist, 52-55; assoc prof anat, 59-66, PROF ANAT, SCH MED, DUKE UNIV, 66- *Concurrent Pos:* Vis investr, Rockefeller Inst, 54-55, asst, 55-, assoc, 55-56, asst prof cytol, 56-59; mem, Nat Res Coun, 62-64 & 70-74; mem molecular biol study sect, NIH, 66-69; adv ed, Int Rev Cytol, 71-76. *Mem:* Am Soc Zoologists; Genetics Soc Am; Am Soc Naturalists; Am Asn Anatomists; Am Soc Cell Biol (secy, 61-67, pres, 68-69); Sigma Xi. *Res:* Cytology; fine structure and cytochemistry of nucleus and chromosomes; synaptonemal complex in meiosis; microtubules in motility and cell differentiation; light and electron microscopic techniques for investigating cell structure and function. *Mailing Add:* Dept of Anat Duke Univ Sch of Med Durham NC 27710

MOSES, RAY NAPOLEON, JR, b Clinton, NC, Jan 23, 36; m 62; c 1. ASTRONOMY. *Educ:* Ga Inst Technol, BS, 64; Ohio State Univ, PhD(astron), 74. *Prof Exp:* Astronaut engr, Boeing Co, 64-67 & Lockheed Missiles & Space Co, Inc, 67-68; systs analyst, US Air Force, Wright-Patterson AFB, Ohio, 69-70; ASST PROF PHYSICS, FURMAN UNIV, 74- *Mem:* Am Astron Soc. *Res:* Solar system studies, especially those related to interplanetary exploration. *Mailing Add:* 2117 Laverne Dr Hunstville AL 35810

MOSES, RONALD ELLIOT, b Chelsea, Mass, Dec 29, 30; m 52; c 2. ORGANIC CHEMISTRY. *Educ:* Harvard Univ, AB, 52; Northeastern Univ, MS, 59. *Prof Exp:* From jr res chemist to sr res chemist, Atlantic Gelatin Div, Gen Foods Corp, Mass, 54-60; sr prof chemist, Gillette Safety Razor Co, 60-66; proj chemist, Gillette Co, 66-71, sr mgr res, 71-72, dir prod develop, 72-75, dir tech & admin serv, Toiletries Div, 75-77, dir prod develop-skin care prod, Personal Care Div, 77-79, dir prod develop-hair care prod, 79-83, vpres res & develop, Personal Care Div, 83-87, vpres res & develop, Personal Care Group, 87-89, DIR RES & DEVELOP, NATLANTIC GROUP, GILLETTE CO, 89- *Mem:* Fel Am Inst Chemists; Am Chem Soc; Soc Cosmetic Chemists. *Res:* Gelatin, processing and properties; organic synthesis, particularly of heterocyclic compounds, especially pyridines, quinolines and pyrimidines. *Mailing Add:* 1039 Shirley St Winthrop MA 02152

MOSES, SAUL, b Pittsburgh, Pa, Dec 31, 21; m 49; c 3. CHEMICAL ENGINEERING. *Educ:* Carnegie Inst Technol, BS, 42; Johns Hopkins Univ, PhD(chem eng), 48. *Prof Exp:* Chem engr, Calvert Distilling Co, Md, 42, Lasting Prod Co, 42-43 & Westinghouse Elec & Mfg Co, 43-44; chem engr-physicist, Naval Res Lab, 47-50; chem engr, 50-59, PRES, DENTOCIDE CHEM CO, 59- *Concurrent Pos:* Consult chem engr. *Mem:* AAAS; Am Chem Soc; Am Inst Chem Engrs; NY Acad Sci; Am Soc Safety Engrs. *Res:* Protective coatings; ultrasonics; process operations-unit phase changes; surface chemistry and plastics; evaluation of adhesion of organic coatings by ultrasonic vibrations. *Mailing Add:* 3525 Barton Oaks Rd Baltimore MD 21208-4332

MOSESSON, MICHAEL W, b New York, NY, Dec 31, 34; m 67; c 3. BIOCHEMISTRY, INTERNAL MEDICINE. *Educ:* Brooklyn Col, BS, 55; State Univ NY, Downstate Med Ctr, MD, 59. *Prof Exp:* Intern, II & IV Med Serv, Boston City Hosp, Mass, 59-60; asst resident, Ward Med Serv, Barnes Hosp, St Louis, Mo, 63-64, instr, Dept Med, 65-67; from asst prof to assoc prof med, 67-75, prof med, Col Med, State Univ NY Downstate Med Ctr, 75-81; PROF MED, UNIV WIS MED SCH, 81- *Concurrent Pos:* NIH res career develop award, 67; Josiah Macy Jr Found Scholar, 77. *Mem:* Am Fedn Clin Res; Int Soc Hemat; Am Soc Clin Invest; Am Soc Biol Chemists; Int Soc Thrombosis & Haemostasis; Asn Am Physicians. *Res:* Structure of fibrinogen and related proteins; structure and function of coagulation proteins; hemostasis. *Mailing Add:* Sinai Samaritan Med Ctr 950 N 12th St PO Box 342 Milwaukee WI 53201

MOSHELL, ALAN N, b New York, NY, Mar 17, 47. DERMATOLOGY. *Educ:* NY Univ, BA, 67, MD, 71. *Prof Exp:* PROG DIR, SKIN DIS BR, NAT INST ARTHRITIS, MUSCULOSKELETAL & SKIN DIS, NIH, 80- *Mem:* Am Acad Dermat; Am Dermat Asn; Soc Investigative Dermat; Am Soc Photobiologists; AMA. *Mailing Add:* Skin Dis Br Nat Inst Arthritis Musculoskeletal & Skin Dis NIH Westwood Bldg Rm 405 5333 Westbard Ave Bethesda MD 20892

MOSHER, CAROL WALKER, b Loveland, Colo, June 23, 21; m 44; c 3. BIO-ORGANIC CHEMISTRY. *Educ:* Colo State Col, BS, 42; Pa State Col, MS, 43, PhD(chem), 47. *Prof Exp:* RES CHEMIST, STANFORD RES INST, 47- *Mem:* Am Chem Soc. *Res:* Organic syntheses; drug-DNA interactions. *Mailing Add:* 713 Mayfield Stanford CA 94305-1016

MOSHER, CLIFFORD COLEMAN, III, b Lima, Ohio, Aug 18, 27; m 54; c 2. ELECTRICAL ENGINEERING. *Educ:* Mich State Univ, BS, 53; Univ Tex, MS, 55; Stanford Univ, PhD(elec eng), 65. *Prof Exp:* Consult, Magnetohydrodyn Proj, Dept Mech Eng, Stanford Univ, 63-65; asst prof elec eng, Univ Mo, 65-66; consult, Alternating & Direct Current Transmission Proj, Edison Elec Inst, 66-67; assoc prof elec eng & dir energy systs div, Drexel Univ, 67-77; POWER PROF ELEC ENG, WASH STATE UNIV, 77- *Concurrent Pos:* Chmn, Western Protective Relay Conf, 74- *Mem:* Inst Elec & Electronics Engrs; Nat Soc Prof Engrs; Am Soc Eng Educ. *Res:* Electrical power systems; technical and economic analysis, direct current transmission. *Mailing Add:* Dept Elec Eng Wash State Univ Pullman WA 99164-2752

MOSHER, DEANE FREMONT, JR, b Newport, Vt, July 10, 43; m 68; c 2. MEDICINE. *Educ:* Dartmouth Col, AB, 65; Dartmouth Med Sch, BMS, 66; Harvard Med Sch, MD, 68. *Prof Exp:* Intern med, Duke Univ Med Ctr, 68-69; resident, Beth Israel Hosp, Boston, Mass, 69-70; fel biol chem, Harvard Med Sch, 70-72; physician med, US Army Med Res Inst, Frederick, Md, 72-75; vis scientist virol, Univ Helsinki, Finland, 75-76; from asst prof to assoc prof, 76-83, PROF MED, UNIV WIS, MADISON, 83-, HEAD, SECT HEMATOL, 84-, PROF PHYSIOL CHEM, 85- *Concurrent Pos:* Fel, NIH, 70-72; dir, hemophilia prog, Univ Hosp, Madison, Wis, 76-82, consult hemat, 76-, dir, spec coagulation lab, 76-; estab investr, Am Heart Asn, 78-83; mem grad fac, dept anat, Univ Wis, Madison, 79-, dept path & lab med, 80-, dept physiol chem, 80-85, Romnes fel, 82; mem, path res comt, coun Arteriosclerosis & Thrombosis, Am heart Asn, 83-86; chmn, subcomt hemostasis, Am Soc Hemat, 81-83. *Mem:* Am Fedn Clin Res; Am Heart Asn; Am Soc Cell Biol; Am Soc Clin Invest; Am Soc Biol Chemists; Am Soc Hemat. *Res:* Biochemistry of blood coagulation and the connective tissue matrix; cell-matrix and cell-cell adhesion. *Mailing Add:* 600 Highland Ave Madison WI 53792

MOSHER, DON R(AYMOND), b Davenport, Iowa, Aug 7, 21; m 42; c 4. PHYSICAL METALLURGY. *Educ:* St Ambrose Col, BS, 42; Univ Denver, MS, 60, PhD, 69. *Prof Exp:* Res engr, Res Lab, Westinghouse Elec Corp, 42-48; res scientist, Nat Adv Comt Aeronaut, 48-53; from instr to asst prof, 54-60, assoc prof, 60-80, EMER PROF METALL, UNIV COLO, BOULDER, 80- *Mem:* Am Soc Metals; Am Inst Mining, Metall & Petrol Engrs. *Res:* Internal friction; refractory metals; mechanical behavior; failure analysis. *Mailing Add:* 1025 Paragon Dr Boulder CO 80303

MOSHER, DONNA PATRICIA, b BC, Oct 5, 41; US citizen; m 63; c 3. BACTERIOLOGY, MEDICAL TECHNOLOGY. *Educ:* Univ Wash, Seattle, BS, 63; Marshall Univ, MS, 74. *Prof Exp:* Lab supvr forest dis res, Intermountain Forest & Range Exp Sta, Moscow, Idaho, 65-67; teaching asst microbiol, Marshall Univ, Huntington, WVa, 72- 74; SUPVR, CHEM DEPT, OAK HILL HOSP, JOPLIN, MO, 76- *Concurrent Pos:* Clin chem instr, Northeast Okla, A&M, 86- *Mem:* Am Med Technologists. *Res:* Development of tissue culture medium for Chronartium ribicola, causative organism of white pine blister rust; differentiate subspecies of Rhynichthes atratulus by isoenzyme electrophoresis. *Mailing Add:* Rte 7 Box 776 Joplin MO 64801

MOSHER, HAROLD ELWOOD, b Sterling, Mass, Aug 6, 20; m 46; c 3. HORTICULTURE. *Educ:* Mass State Col, BS, 42; Univ Mass, BLA, 47, MLA, 57. *Prof Exp:* Landscape architect & supt grounds, Lake Placid Club, NY, 47-50; from instr to asst prof hort, Univ Mo, 50-58; assoc exten prof, Univ Mass, Amherst, 58-66, from assoc prof to prof landscape archit, 66-90; RETIRED. *Res:* Ornamental plants and their uses in landscape architecture; ecological determinants in land use and landscaping. *Mailing Add:* 102 Hills North Univ of Mass Amherst MA 01003

MOSHER, HARRY STONE, b Salem, Ore, Aug 31, 15; m 44; c 3. ORGANIC CHEMISTRY. *Educ:* Willamette Univ, AB, 37; Ore State Col, MS, 39; Pa State Col, PhD(org chem), 42. *Hon Degrees:* DSc, Willamette Univ, 81. *Prof Exp:* Asst prof chem, Willamette Univ, 39-40; from instr to asst prof, Pa State Col, 43-46; from asst prof to emer prof, 47-81, PROF CHEM, STANFORD UNIV, 81- *Concurrent Pos:* US sr res fel, Univ London, 59-60; Am Chem Soc fel, Univ Zurich, 67-68; vis prof, Univ Amsterdam, 75. *Mem:* Fel AAAS; Am Chem Soc. *Res:* Stereochemistry; synthetic drugs; mechanisms of organic reactions; chemistry of pyridine compounds; peroxides; animal toxins. *Mailing Add:* Dept Chem Stanford Univ Stanford CA 94305

MOSHER, JAMES ARTHUR, b Green, NY, Oct 25, 42; m 68; c 2. PHYSIOLOGICAL ECOLOGY, ORNITHOLOGY. *Educ:* Utica Col, BS, 65; State Univ NY Col Environ Sci & Forestry, MS, 73; Brigham Young Univ, PhD(zool), 75. *Prof Exp:* Res assoc physiol ecol, Naval Arctic Res Lab, Univ Alaska, 75-76; fel, Univ Md, 76, asst prof, Appalachian Environ Lab, CEES, 77-83; HEAD, SAVAGE RIVER CONSULT, 83- *Concurrent Pos:* Mem bd, Raptor Res Found. *Mem:* Ecol Soc Am; AAAS; Am Ornithologists Union; Cooper Ornith Soc; Wilson Ornith Soc; Raptor Res Found. *Res:* Study of avian ecology, especially birds of prey; physiological ecology of birds; thermoregulatory and metabolic adaptations of vertebrates. *Mailing Add:* Appalachian Environ Lab Gunter Hall Frostburg State Col Campus Frostburg MD 21532

MOSHER, JOHN IVAN, b Waterloo, NY, Sept 26, 33; m 60; c 4. HUMAN ECOLOGY. *Educ:* Hobart Col, BA, 56; Western State Col Colo, MA, 61; Utah State Univ, PhD, 72. *Prof Exp:* Chem analyst, NY Agr Sta, Cornell Univ, 58-59; instr biol, Lyndon Inst, Vt, 59-60; asst & preparator, Univ Rochester, 60-61; assoc prof, 61-77, PROF BIOL SCI, STATE UNIV NY COL BROCKPORT, 77- *Concurrent Pos:* Sr res fel & vis lectr, Univ Manchester, England; vis prof, Univ Calif, Davis. *Mem:* Ecol Soc Am; Am Soc Zoologists; Zool Soc London; Bio-Dynamic Farming & Gardening Asn; Sigma Xi. *Res:* Human ecology, connected with understanding the practical applications of basic ecological principles to living a life style harmonious with the environment; experimentations with shelter, biodynamic food production, and human reactions to employing environmentally sound living practices. *Mailing Add:* Dept of Biol State Univ of NY Brockport NY 14420

MOSHER, LOREN CAMERON, b Phoenix, Ariz, June 20, 38; m 63; c 5. PETROLEUM GEOLOGY, PALEONTOLOGY. *Educ:* Calif Inst Technol, BS, 60; Univ Wis, MS, 64, PhD(micropaleont), 67. *Prof Exp:* Asst prof geol, Fla State Univ, 67-71; assoc prof geosci, Univ Ariz, 71-75; res geologist, Phillips Petrol Co, 75-77, sect supvr, 77-79; TRAINER/FACILITATOR, HIGH CHALLENGE INC, 84- *Concurrent Pos:* NSF grant, Fla State Univ, 68-71; adj prof, Brigham Young Univ, 79- *Mem:* Asn Experiential Educ; Am Soc Training & Develop. *Res:* Petroleum and alternate energy; stratigraphic and zoologic studies of conodonts. *Mailing Add:* High Challenge Inc 1255 N 500 E Pleasant Grove UT 84062

MOSHER, LOREN RICHARD, b Monterey, Calif, Sept 3, 33; div; c 3. PSYCHIATRY. *Educ:* Stanford Univ, BA, 56; Harvard Med Sch, MD, 61. *Prof Exp:* Resident psychiat, Mass Ment Health Ctr, 62-64; clin assoc twin studies, NIMH, 64-66; USPHS ment health spec res fel, Tavistock Clin, London, 66-67; asst prof psychiat, Conn Ment Health Ctr, Sch Med, Yale Univ, 67-68; chief, Ctr Studies Schizophrenia, NIMH, 68-81; prof psychiat, uniformed servs Univ Health Sci, 81-88; ASSOC DIR ADDICTION VICTIM, MENT HEALTH SERV, 88- *Concurrent Pos:* Consult, Nat Naval Med Ctr, Bethesda, Md, 72-, Arbours Assoc, London, Eng, 72-, Woodley House, Washington, DC, 73-, Green Door, 77-, Crossing Place, 77- & Va Dept Ment Health & Ment Retardation, 84- *Mem:* Fel Am Psychiat Asn; Am Family Ther Asn; Am Orthopsychol Assoc; Soc Psychother Res. *Res:* Psychosocial treatment of mental illness, specifically, schizophrenia; twin studies, schizophrenia; labeling of mental patients; deinstitutionalism of mental patients and their proper care in the community. *Mailing Add:* Addiction Victim & Mental Health Serv 401 Hungerford Dr #500 Rockville MD 20850

MOSHER, MELVILLE CALVIN, b Port Chester, NY, Aug 1, 45. APPLIED MATHEMATICS. *Educ:* Univ Calif, Berkeley, BA, 69, PhD(math), 79. *Prof Exp:* Asst prof math, Univ Miami, Coral Gables, 79-81; RES MATHEMATICIAN, NUMERICAL FLUID MECH DIV, D W TAYLOR NAVAL SHIP RES & DEVELOP CTR, 81- *Mem:* Computional and mathematical fluid mechanics. *Mailing Add:* 451 E 83rd St - 12D New York NY 10028-6139

MOSHER, MELVYN WAYNE, b Palo Alto, Calif, June 10, 40; m 63; c 3. PHYSICAL ORGANIC CHEMISTRY. *Educ:* Univ Washington, BA, 62; Univ Idaho, MS, 64, PhD(org chem), 68. *Prof Exp:* Fel, Univ Alta, 67-69; asst prof chem, Marshall Univ, 69-74; PROF CHEM & ASST DIR, REGIONAL CRIME LAB, MO SOUTHERN STATE COL, 74- *Mem:* Am Chem Soc. *Res:* Free radical reactions and mechanisms; polar effects in free radical reactions; free radical introduction of functional groups into alkanes. *Mailing Add:* Dept of Chem Mo Southern State Col Joplin MO 64801

MOSHER, RICHARD ARTHUR, b Troy, NY, Aug 27, 46; m 87. ELECTROPHORETIC SEPARATIONS. *Educ:* SUNY Oneonta, BS, 68; Univ Ariz, PhD(biochem), 78. *Prof Exp:* From res asst to res assoc, biophys tech lab, Univ Ariz, 79-85, RES SCIENTIST, CTR SEPARATION SCI, 85-ASSOC DIR, 87- *Mem:* Electrophoresis Soc; Am Chem Soc. *Res:* Computer simulation of electrophoretic processes; development of new buffer systems for electrophoretic separations; large scale protein purification by isoelectric focusing or isotachophoresis; capillary electrophoresis; theory of electrophoresis. *Mailing Add:* 4473 E Haven Lane Tucson AZ 85721-5426

MOSHER, ROBERT EUGENE, b Detroit, Mich, Sept 27, 20; m 43, 79; c 6. ANALYTICAL CHEMISTRY. *Educ:* Wayne State Univ, BS, 42, MS, 49, PhD(analytical chem), 50. *Prof Exp:* Chemist rubber develop, US Rubber Co, 42-45 & 46-47; instr, Wayne State Univ, 47-50; dir dept physiol & res, 50-83, MB RES ASSOC, PROVIDENCE HOSP, 83-, TECH ENG & CONSULT, 86- *Concurrent Pos:* Assoc prof geol, Wayne State Univ, 70-80, adj prof, 80- *Mem:* Fel AAAS; Am Chem Soc; Am Fedn Clin Res; Am Asn Clin Chemists; Sigma Xi; Soc Econ Paleontologists & Mineralogists. *Res:* Application of analytical chemistry to research in clinical and geochemical research. *Mailing Add:* 3907 Ravena Royal Oak MI 48072

MOSHER, SHARON, b Freeport, Ill, Jan 6, 51; m 80; c 2. STRUCTURAL GEOLOGY, METAMORPHIC PETROLOGY. *Educ:* Univ Ill, Urbana, BS, 73, PhD(geol), 78; Brown Univ, MS, 75. *Prof Exp:* asst prof, 78-84, ASSOC PROF GEOL, UNIV TEX, AUSTIN, 84- *Concurrent Pos:* Chmn, Structure & Tectonics Div, Geol Soc Am, 81-82, Div Struct Geol; prin investr, 3 NSF grants, 82-84, US comt on geol, Int Geologic Correlation Prog subcomt vchmn, 87-90. *Mem:* Am Geophys Union; Geol Soc Am. *Res:* Structural geology and metamorphic petrology with an emphasis on deformation mechanisms, especially pressure solution, strain analysis, and field mapping of highly deformed terrains; crystal contraction mechanisms in the brittle ductile transition zone. *Mailing Add:* Dept of Geol Sci Univ of Tex Austin TX 78712

MOSHEY, EDWARD A, US citizen. VACUUM & CRYOGENIC ENGINEERING, X-RAY & GAMMA RAY SPECTROSCOPY-SPECTOMETRY. *Educ:* Rochester Inst Technol, BS, 64. *Prof Exp:* Engr, Astro-Electronics Div, RCA Corp, 66-79; proj engr, Plasma Physics Lab, Princeton Univ, 79-89; MGR ENG, PRINCETON GAMMA-TECH, INC, 90- *Concurrent Pos:* Consult, E A Moshey, Consult, 85- *Res:* Design of x-ray spectrometers; design of vacuum & erygenic systems; opto-mechanical design; laser optics instrumentation; closed circuit television cameras for aerospace. *Mailing Add:* 743 Prospect Ave Princeton NJ 08540

MOSHIRI, GERALD ALEXANDER, b Shiraz, Iran, June 1, 29; US citizen. LIMNOLOGY, PHYSIOLOGICAL ECOLOGY. *Educ:* Oberlin Col, BA, 52, MA, 54; Univ Pittsburgh, PhD(biol), 68. *Prof Exp:* Instr biol sci, Cent Fla Jr Col, 64-66; asst prof, 69-73, assoc prof, 73-80, PROF BIOL, UNIV WEST FLA, 80- *Concurrent Pos:* Fed Water Pollution Control Admin sr res fel, Inst Ecol, Univ Calif, Davis, 68-69; mem, Gov's Comt on Northwest Fla Coastal Resource Mgt & Planning. *Honors & Awards:* Sigma Xi. *Mem:* AAAS; Brit Freshwater Biol Asn; Ecol Soc Am; Am Soc Limnol & Oceanog; Int Asn Theoret & Appl Limnol. *Res:* Aquatic ecology; energetics of aquatic ecosystems, including problems involving cycling of nutrients and eutrophication of inland waters and estuaries. *Mailing Add:* Dept Biol Univ Fla Pensacola FL 32504

MOSHMAN, JACK, b Richmond Hill, NY, Aug 12, 24; m 47; c 4. STATISTICS, OPERATIONS RESEARCH. *Educ:* NY Univ, BA, 46; Columbia Univ, MA, 47; Univ Tenn, PhD(math), 53. *Prof Exp:* Tutor math, Queens Col, NY, 47; instr, Univ Tenn, 47-50; sr statistician, Oak Ridge Nat Lab, 50-54; mem tech staff, Bell Tel Labs, Inc, 54-57; vpres & gen mgr appl res & mgt sci div, C-E-I-R Inc, Washington, DC, 57-66; managing dir, EBS Mgt Consults, Inc, Washington, DC, 66-67; sr vpres, Leasco Systs & Res Corp, Md, 67-69; PRES, MOSHMAN ASSOCS, INC, 70- *Concurrent Pos:* Statistician, AEC, 48-50; lectr & mem adv comt math & statist, USDA Grad Sch, 59-75; prof lectr, George Washington Univ, 60-61; vis prof, Eagleton Inst Polit, Rutgers Univ, adj prof, 72-74; exec secy, Comt to Evaluate Nat Ctr Health Statist, 71-73; mem, Comt Nat Info Syst in Math Sci; mem adv comt statist policy, Off Mgt & Budget, 74-77; chmn, Inst Safety Analysis, 75-88; exec secy, Comt Eval Nat Ctr Social Statist, 76-77 & Expert Panel Eval Coop Health Statist Systs, 78-79; mem, Exec Comt, Am Fed Info Processing Soc, 82-88, pres, 86-87. *Mem:* Opers Res Soc Am; Inst Mgt Sci; Asn Comput Mach (secy, 56-60, vpres, 60-62); fel Am Statist Asn; Inst Math Statist; Europ Group Oper Res in Health. *Res:* Health services; Monte Carlo methods; operations research; mathematical models of political behavior; information systems; statistical problems of compliance with safety and other regulatory requirements. *Mailing Add:* Moshman Assocs Inc Suite 410N 7315 Wisconsin Ave Bethesda MD 20814

MOSHY, RAYMOND JOSEPH, b Brooklyn, NY, Aug 12, 25; m 48; c 5. FOOD SCIENCE. *Educ:* St John's Univ, NY, BS, 48; Fordham Univ, MS, 49, PhD(chem), 53. *Prof Exp:* Chemist res & develop, Am Lecithin Co, 50-52; synthetic org chem, Heyden Chem Corp, 52-55; proj leader, Res Ctr, Gen Foods Corp, 55-59; group mgr chem lab, Am Mach & Foundry Co, 59-60, sect mgr, 60-65, mgr chem develop lab, Res & Develop Div, 65-66, staff vpres & dir res div, 66-70; vpres res & develop, Hunt-Wesson Foods, Inc, 70-75, vpres & group exec, 75-89; VPRES, ESCAGENETICS CORP, 89- *Mem:* Am Chem Soc; Sigma Xi; Am Inst Chemists; Inst Food Technologists; Indust Res Inst. *Res:* Agricultural chemicals; proteins; starches; food processing; nutrition; agricultural research. *Mailing Add:* ESCAgenetics Corp 830 Bransten Rd San Carlos CA 94070

MOSIER, ARVIN RAY, b Olney Springs, Colo, June 11, 45; m 65; c 2. AGRICULTURAL CHEMISTRY. *Educ:* Colo State Univ, BS, 67, MS, 68, PhD(soil sci), 74. *Prof Exp:* RES CHEMIST, AGR RES SERV, USDA, 67- *Concurrent Pos:* Vis scientist, Common Wealth Sci & Indust Res Orgn, Griffith, Australia, 83-84, CSIRO, Canberra, Australia, 87, FAL Braunschweig, WGer, 78, 83 & 85; assoc ed, Geoderma, Fertilizer Res. *Mem:* AAAS; fel Am Soc Agron; Soil Sci Soc Am; Int Soc Soil Sci; Sigma Xi. *Res:* Distribution of nitrogen and organic compounds emanating from agricultural sources and the effect of these chemicals on soil, water and plant systems;

nitrogen metabolism and nutrition of algae and higher plants; assess importance of volatile nitrogen losses from native and fertilized soils and determine factors that control soil nitrification, denitrification and interaction with soil c and n trace gas flux. *Mailing Add:* Agr Res Serv PO Box E Ft Collins CO 80522

MOSIER, BENJAMIN, b Corsicana, Tex, July 15, 26; m 54; c 4. CHEMISTRY. *Educ:* Tex A&M Univ, BS, 49, MS, 52; Univ Ill, PhD(chem), 57. *Prof Exp:* Instr chem, Kilgore Col, 49-50; asst, Tex A&M Univ, 50-51; res chemist, Gen Dynamics Corp, 51-52; res scientist, Humble Oil & Refining Co, 57-60; owner & res dir, 60-69, PRES, INST RES, INC, 69-, DIR, 60- *Concurrent Pos:* Lectr, Univ Houston, Rice Univ, Baylor Col Med, Univ Tex M D Anderson Hosp & Tumor Inst & Tex Res Inst Ment Sci; mem, Am Coun Independent Labs; adj res asst prof path, Baylor Col Med, 74-, res asst prof, 75-; sr res assoc, Dept Chem, Rice Univ, 75-85; adj prof, Dept Chem, Kans State Univ, Manhattan, 77-85. *Honors & Awards:* Apollo & Gemini-Apollo Awards, NASA. *Mem:* Am Chem Soc; Nat Asn Corrosion Eng; Am Inst Mining, Metall & Petrol Engrs; fel Am Inst Chemists; Royal Soc Chem; Union Pure & Appl Chem; Nat Asn Corrosion Engrs; Sigma Xi; NY Acad Sci. *Res:* Electrochemistry; surface and colloidal phenomena; microencapsulation; instrumental methods including polarography, infrared, ultraviolet and visible spectrometry, gas chromatography, nuclear radiation methods, x-ray diffraction and fluorescence; microprobe analysis; electron microscopy; mass spectrometry; nuclear magnetic spectroscopy; differential thermal, thermogravimetry, emission spectrographic analysis. *Mailing Add:* Inst for Res Inc 8330 Westglen Dr Houston TX 77063

MOSIER, H DAVID, JR, b Topeka, Kans, May 22, 25. PEDIATRICS, ENDOCRINOLOGY. *Educ:* Notre Dame Univ, BS, 48; Johns Hopkins Univ, MD, 52; Am Bd Pediat, dipl, 57. *Prof Exp:* Intern pediat, Johns Hopkins Hosp, 52-53; asst path, Univ Southern Calif, 54-55; fel pediat endocrinol, Johns Hopkins Hosp, 55-57; from asst prof to assoc prof pediat, Sch Med, Univ Calif, Los Angeles, 57-63; assoc prof, Univ Ill Col Med, 63-67; PROF PEDIAT & HEAD DIV ENDOCRINOL & METAB, UNIV CALIF, IRVINE-CALIF COL MED, 67- *Concurrent Pos:* Asst resident, Los Angeles Children's Hosp, Calif, 53-54, resident pediat path, 54-55; consult, Pac State Hosp, Pomona, Calif, 57-; dir res, Ill State Pediat Inst, 63-67; mem staff, Childrens Hosp Med Ctr, Long Beach & Med Ctr, Univ Calif, Irvine, Orange. *Mem:* Endocrine Soc; Soc Pediat Res; Soc Exp Biol & Med; Lawson Wilkins Pediat Endocrine Soc; Am Pediat Soc; Sigma Xi. *Res:* Mechanisms controlling physical growth; the control of catch-up growth. *Mailing Add:* PO Box 4504 Irvine CA 92716-4504

MOSIER, JACOB EUGENE, b Hoxie, Kans, Feb 5, 24; m 45; c 4. VETERINARY MEDICINE. *Educ:* Kans State Univ, DVM, 45, MS, 48; Am Col Vet Internal Med, dipl. *Prof Exp:* Instr anat, surg & med, Kans State Univ, 45-47, instr surg & med, 47-48, asst prof large animal med, 48-49; asst prof large animal surg, Univ Ill, 49-50; assoc prof surg & med, 50-54, prof small animal med, 54-61, prof surg & med & head dept, 61-81, PROF VET MED, KANS STATE UNIV, 81- *Concurrent Pos:* Mem nat adv comt, Food & Drug Admin, 71-76, consult, Bur Vet Med, 74-76; consult, Tex Coord Bd Higher Educ, 71, Col Vet Med, Miss State Univ, 74-85, La State Bd Educ, 84, Whittle Publ, 82- & Norden Labs, 84- *Honors & Awards:* Am Animal Hosp Asn Award, 73; Intermoutain Vet Med Asn Award, 77. *Mem:* Am Vet Med Asn (pres, 81-82); NY Acad Sci; Am Animal Hosp Asn; World Vet Asn (vpres, 77-80). *Res:* Canine pediatrics; internal medicine of the dog and cat; cause and effect of perinatal disease in the cat and dog; canine and feline geratrics; pet population dynamics. *Mailing Add:* Dept Vet Med Kans State Univ Manhattan KS 66506

MOSIER, STEPHEN R, b San Rafael, Calif, Nov 14, 42; m 64; c 2. INTERNATIONAL SCIENCE, TECHNOLOGY & EDUCATIONAL AFFAIRS. *Educ:* Col William & Mary, BS, 64; Univ Iowa, MS, 67, PhD(physics), 70. *Prof Exp:* Res assoc physics, Univ Iowa, 69-70; res assoc radio astron, NASA Goddard Space Flight Ctr, 70-71, space scientist physics & astron, 71-78; prog dir, US-Japan coop sci prog, NSF, 78-81 & US-France & US-Belgium, 81-83; assoc vpres int affairs, Univ Houston Syst Admin, 83-86; DIR RES SERV, UNIV NC-GREENSBORO, 86- *Concurrent Pos:* Vchmn bd dirs, NC Asn Biomed Res. *Honors & Awards:* Spec Achievement Award, NASA, 76, Group Achievement Award, 73, 77. *Mem:* Am Geophys Union; AAAS; Nat Coun Univ Res Admin; Soc Res Admin. *Res:* Space physics; magnetospheric radio physics; solar radio physics; international science and technology policy. *Mailing Add:* Off of Res Servs Univ NC-Greensboro 100 McIver Bldg Greensboro NC 27412

MOSIG, GISELA, b Schmorkau, Ger, Nov 29, 30. GENETICS. *Educ:* Univ Cologne, Dr.rer nat(bot), 59. *Prof Exp:* Res assoc phage genetics, Vanderbilt Univ, 59-62; NIH fel, Carnegie Inst Genetics Res Unit, 62-63, res assoc, 63-65; from asst prof to assoc prof, 65-71, PROF MOLECULAR BIOL, VANDERBILT UNIV, 71- *Concurrent Pos:* Humboldt Award for sr scientist, Max Planck Inst Plant Breeding, Cologne, W Germany, 76-77. *Mem:* AAAS; Genetics Soc Am; Am Soc Microbiol; NY Acad Sci. *Res:* Mechanism of genetic recombination and replication of DNA in bacteriophage; control of transcription in chloroplasts. *Mailing Add:* Dept Molecular Biol Vanderbilt Univ Nashville TN 37240

MOSIMANN, JAMES EMILE, b Charleston, SC, Oct 26, 30; m 53; c 8. BIOSTATISTICS, MATHEMATICAL STATISTICS. *Educ:* Univ Mich, BA, 52, MS, 53, PhD(zool), 56; Johns Hopkins Univ, MS, 61. *Prof Exp:* Res assoc, Willow Run Labs, Univ Mich, 55; asst prof biol, Univ Montreal, 55-61; NIH res fel, 61-62; res assoc, Univ Ariz, 62-63; math statistician, 63-75, CHIEF LAB STATIST & MATH METHODOLOGY, DIV COMPUT RES & TECHNOL, NIH, 75- *Concurrent Pos:* NIH res fel, 59-60. *Mem:* Am Statist Asn; Biomet Soc; Royal Statist Soc. *Res:* Discrete probability models in biology; statistical distribution theory; biometry; ecological and population statistics. *Mailing Add:* Five Balmoral Ct Rockville MD 20850

MOSKAL, JOSEPH RUSSELL, b Saginaw, Mich, June 10, 50; m 80. DEVELOPMENTAL NEUROBIOLOGY. *Educ:* Univ Notre Dame, BSc, 72, PhD(chem), 77. *Prof Exp:* Res assoc, Dept Biochem, Mich State Univ, 77-79; staff fel, NIH, 79-81, sr staff fel, Lab Biochem Genetics, Nat Heart, Lung & Blood Inst, 82-90; DIR RES, CINN MED GROUP, 90- *Res:* Identification, purification and regulation of cell surface glycoconjugates involved in the formation of synapses. *Mailing Add:* CINN Med Group 428 W Deming Pl Chicago IL 60614

MOSKALYK, RICHARD EDWARD, b Hafford, Sask, Apr 17, 36; m 57; c 3. MEDICINAL CHEMISTRY. *Educ:* Univ Sask, BS, 56, MSc, 59; Univ Alta, PhD(pharmaceut chem), 65. *Prof Exp:* Control chemist, Merck, Sharp & Dohme Ltd, Can, 56-57; res chemist, Food & Drug Directorate, Dept Nat Health & Welfare, 58-61; from asst prof to prof, Univ Alta, 63-75, asst dean, 79-80, assoc dean, 81-89, actg dean, 89, PROF MED CHEM, FAC PHARM & PHARMACEUT SCI, UNIV ALTA, 75-, DEAN, 90- *Concurrent Pos:* Invited prof, Univ Geneva, 71-72; pres, Pharm Exam Bd Can, 85-86; bd dir, Can Pharmaceut Asn, 85-88; vpres, Western, 88-89. *Mem:* Can Pharmaceut Asn; Am Chem Soc; Chem Inst Can; Asn Fac Pharm Can (pres, 79-80); Alta Pharmaceut Asn (pres, 83-84). *Res:* The relationship of physicochemical properties to the absorption, distribution, metabolism and elimination of drugs; the arachidonic acid cascade and its significance to future drug development. *Mailing Add:* Fac Pharm & Pharmaceut Sci Univ Alta Edmonton AB T6G 2N8 Can

MOSKO, SIGMUND W, b Philadelphia, Pa, June 9, 36; m 63; c 3. PARTICLE ACCELERATORS. *Educ:* Drexel Univ, BSEE, 58. *Prof Exp:* Res staff mem, 58-80, SR RES STAFF MEM, OAK RIDGE NAT LAB, 80- *Mem:* Sr mem Inst Elec & Electronic Engrs; Inst Elec & Electronic Engrs Nuclear & Plasma Physics Soc; Nat Soc Prof Engrs. *Res:* Design and development of particle accelerators for nuclear physics research; radio frequency powered acceleration systems and peripheral radio frequency electronics; direct current supplies for ultra stable electromagnet components. *Mailing Add:* Oak Ridge Nat Lab PO Box 2008 MS 6368 Oak Ridge TN 37831-6368

MOSKOVITS, MARTIN, b Apr 13, 43; Can citizen; m; c 3. SPECTROCHEMISTRY, SURFACE CHEMISTRY. *Educ:* Univ Toronto, BSc, 65, PhD(chem), 70. *Prof Exp:* Res scientist chem, Alcan Int, 70-71; from asst prof to assoc prof, 72-82, PROF CHEM, UNIV TORONTO, 82- *Concurrent Pos:* Guggenheim fel, 87. *Mem:* Am Chem Soc; Optical Soc Am; AAAS; Mat Res Soc. *Res:* Chemistry and spectroscopy of metal surfaces and metal particles, both bulk and isolated in rare gas solids, raman spectroscopy, surface photochemistry. *Mailing Add:* Dept of Chem Univ Toronto Toronto ON M5S 1A1 Can

MOSKOWITZ, GERARD JAY, b Yonkers, NY, June 17, 40; m 63; c 1. BIOCHEMISTRY, ENZYMOLOGY. *Educ:* Univ Buffalo, BA, 62; State Univ NY, Buffalo, PhD(biochem), 68. *Prof Exp:* Res assoc oncol, Med Sch, Baylor Univ, 67-69; asst mem biochem, Albert Einstein Med Ctr, 69-70; mgr res & develop, Wallerstein Co, Div Baxter, 70-76; sr res scientist, Baxter-Travenol Labs Inc, 76-77; TECH DIR, DAIRYLAND FOOD LABS INC, 77- *Mem:* Am Chem Soc; Inst Food Technologists; AAAS; Sigma Xi; NY Acad Sci. *Res:* Preparation, isolation, purification, characterization of enzymes and their application in commercial processes; microbiology; propagation of microorganisms and the study of their enzyme systems and metabolic pathways; application of biotechnology to food systems. *Mailing Add:* PO Box 406 620 Progress Ave Waukesha WI 53187

MOSKOWITZ, GORDON DAVID, b New York, NY, Dec 25, 34; m 60; c 2. BIOMECHANICS, MECHANICAL ENGINEERING. *Educ:* City Col New York, BME, 60; Princeton Univ, MSE, 62, MA, 63, PhD(mech eng), 64. *Prof Exp:* Res engr plastics, E I du Pont de Nemours & Co, Inc, 63-64; asst prof mech eng, Univ Maine, 65-67; from asst prof to assoc prof, 67-77, PROF MECH ENG, DREXEL UNIV, 77- *Concurrent Pos:* Assoc dir, Rehab Eng Ctr, Mass Hosp, 77- *Honors & Awards:* Bausch & Lomb Award, 63. *Mem:* AAAS; Inst Elec & Electronics Engrs; Am Soc Testing & Mat; Am Soc Eng Educ; Sigma Xi. *Res:* Biomechanics of pulmonary and cardiovascular function; bioengineering design of medical support equipment; basic mechanisms of locomotion, design and development of electromyographic control of prosthetic limbs. *Mailing Add:* 4600 Marvine Ave Drexel Hill PA 19026

MOSKOWITZ, JULES WARREN, b Newark, NJ, June 11, 34; m 57; c 1. PHYSICAL CHEMISTRY. *Educ:* Princeton Univ, AB, 56; Mass Inst Technol, PhD(chem), 61. *Prof Exp:* From asst prof to assoc prof, 63-72, PROF CHEM, NY UNIV, 72- *Concurrent Pos:* Consult, Bell Tel Labs, NJ, 65- *Mem:* Am Phys Soc; Sigma Xi. *Res:* Quantum mechanics of solid state and molecular systems; application of digital computers to problems of chemical interest. *Mailing Add:* Dept of Chem NY Univ-Washington Sq New York NY 10003

MOSKOWITZ, MARK LEWIS, b Brooklyn, NY, Dec 5, 25; m 49; c 1. IMAGING CHEMISTRY. *Educ:* City Col New York, BS, 50; Syracuse Univ, PhD(org chem), 54. *Prof Exp:* Org chemist, 54-59, supvr, Prod Develop Lab, 59-65, mgr reproduction prod res & develop, 65-69, tech dir, Off Systs Div, 69-74, mgr adv reproduction mat res & develop, 74-80, mgr explor, res & develop, 80-83, SR TECH ASSOC, GAF CORP, 83- *Mem:* AAAS; Am Chem Soc; Soc Photog Sci & Eng; Am Soc Test & Mat. *Res:* Diazotype coatings chemistry and sensitometry; light sensitive coatings; electrophotography; electron beam imaging. *Mailing Add:* 1636 Brookhouse Dr Unit 130 Sarasota FL 34231-8948

MOSKOWITZ, MARTIN A, b New York, NY, June 25, 35; m 58; c 2. MATHEMATICS. *Educ:* Brooklyn Col, BA, 57; Univ Calif, Berkeley, MA, 59, PhD(math), 64. *Prof Exp:* Instr math, Univ Chicago, 64-66; asst prof, Columbia Univ, 66-69; assoc prof, 69-76, dept chmn, 84-86, PROF MATH, GRAD CTR, CITY UNIV NY, 76-, DEPT CHMN, 89- *Concurrent Pos:* Prin investr, NSF contract, 74-; vis prof, Univ Paris, 76, NSF sr fel, Paris, 76, Univ Rome, La Sapienzo, 90. *Mem:* Am Math Soc. *Res:* Topological groups, lie groups and representation theory. *Mailing Add:* Dept of Math City Univ of NY Grad Ctr 33 W 42 St New York NY 10036

MOSKOWITZ, MERWIN, b New York, NY, May 26, 21; m 44; c 3. MICROBIOLOGY, CELL BIOLOGY. *Educ:* Univ Mich, BS, 44; Univ Calif, PhD(bact), 49. *Prof Exp:* Asst chem, Univ Calif, 46-49; fel, USPHS, 49-50; instr microbiol, Yale Univ, 50; from asst prof to assoc prof bact, 51-61, prof biol, 61-73, PROF BIOL SCI, PURDUE UNIV, 73- *Mem:* AAAS; Am Soc Microbiol; Am Soc Cell Biol; Am Asn Immunol. *Res:* Medical microbiology; cell biology. *Mailing Add:* Dept Biol Sci Purdue Univ West Lafayette IN 47907

MOSKOWITZ, MICHAEL ARTHUR, b New York, NY, May 26, 42; div; c 1. NEUROSCIENCE. *Educ:* Johns Hopkins Univ, AB, 64; Tufts Univ, MD, 68; Am Bd Internal Med, cert, Am Bd Psychiat & Neurosurg, cert. *Prof Exp:* Chief resident, Dept Neurol, 72-73, instr, 74-76, asst prof, 76-80, ASSOC PROF NEUROL, SCH MED, HARVARD UNIV, 80- *Concurrent Pos:* Res assoc, Mass Inst Technol, 73-, lectr neurosci, Dept Nutrit, 75-, asst prof neursci, 77-; res fel, Found Fund Res Psychiat, 73-75, Alfred P Sloan Found, 76; estab investr, Am Heart Asn. *Honors & Awards:* Enrico Greppi Prize. *Mem:* Neurosci Soc; Am Acad Neurol. *Res:* Biogenic amines and neurological diseases; neuropeptides. *Mailing Add:* Dept Neurol & Neurosurg Mass Gen Hosp 32 Fruit St Boston MA 02114

MOSKOWITZ, NORMAN, b Trenton, NJ, Jan 25, 22. ANATOMY. *Educ:* Rutgers Univ, BS, 43; Univ Pa, MS, 47, PhD(zool), 51. *Prof Exp:* Asst instr zool, Univ Pa, 48-50; lectr biol, Rutgers Univ, 58-59; from asst prof to assoc prof, 62-74, PROF ANAT, JEFFERSON MED COL, 74- *Concurrent Pos:* Fel neuroanat, Col Physicians & Surgeons, Columbia Univ, 59-62. *Mem:* Am Asn Anat; Harvey Soc. *Res:* Neuroanatomy; central auditory system. *Mailing Add:* 1025 Walnut St Philadelphia PA 19107

MOSKOWITZ, RONALD, b New York, NY, Feb 15, 39; m 57; c 4. ELECTRICAL ENGINEERING. *Educ:* City Col New York, BEE, 61; Rutgers Univ, MSc, 63, PhD(elec eng), 67. *Prof Exp:* Assoc engr, Norden-Ketay Div, United Aircraft Corp, 58-59; engr, Radio Corp Am, 61-65; instr elec eng, Rutgers Univ, 65-67; prof, Univ Miss, 67; sr consult engr, Space Systs Div, Avco Corp, 67-68; exec vpres, 68-72, PRES, FERROFLUIDICS CORP, 72- *Concurrent Pos:* Mem tech staff, RCA Labs, David Sarnoff Res Ctr, 66. *Mem:* Inst Elec & Electronics Engrs; Am Soc Lubrication Engrs; Am Soc Mech Engrs; Am Inst Physics. *Res:* Ferromagnetic fluids; magnetic fine particles; demagnetization and depolarization phenomena; magnetic recording and resonance; spacecraft magnetism, dynamics and control. *Mailing Add:* Ferrofluidics Corp 40 Simon Street Nashua NH 03061

MOSLEY, JAMES W, b Temple, Tex, Aug 8, 29; div; c 3. EPIDEMIOLOGY. *Educ:* Univ Tex, BA, 50; Cornell Univ, MD, 54. *Prof Exp:* Intern med, New York Hosp, 54-55; chief hepatitis unit, Commun Dis Ctr, USPHS, Ga, 57-58; resident med, Peter Bent Brigham Hosp, 58-59; fel virol, Harvard Med Sch, 59-61; resident, New Eng Ctr Hosp, 61-62; chief hepatitis unit, Commun Dis Ctr, 62-66, chief viral dis sect, 66-70; assoc prof med, 70-75, PROF MED, UNIV SOUTHERN CALIF, 75-, DIR, TRANSFUSION SAFETY STUDY, 84- *Concurrent Pos:* Dir, Transfusion-Transmitted Viruses Study, 74-80; mem, World Health Orgn Expert Adv panel, virus dis, 62, comt Plasma & Plasma Substitutes, Nat Res Coun, 64-69, Expert Group Viral Hepatitis, Lister Hill Nat Ctr Biomed Commun, Nat Libr Med, 67-69; chmn policy bd, Vet Admin Coop Hepatitis Study, 68-72; Fogarty Sr Int fel, Int Ctr Med Res & Training, San Jose, Costa Rica, 78-79; assoc ed, Am J Epidemiol, 77-, J Acquired Immunodeficiency Syndromes, 88- *Mem:* Fel Infectious Dis Soc Am; fel Am Col Epidemiol; Int Soc Blood Transfusion; Am Asn Blood Banks; Int Asn Study Liver. *Res:* Infectious diseases; transfusion safety; epidemiology of hepatitis viruses; epidemiology of AIDS and other retrovirus infections; clinical trials and cooperative studies; water and food transmission of enteric viruses; infections due to blood transfusion and plasma products. *Mailing Add:* Dept Med 1840 N Soto St EDM 108 Los Angeles CA 90032

MOSLEY, JOHN ROSS, b Wichita, Kans, Oct 18, 22; m 50; c 2. PHYSICAL CHEMISTRY. *Educ:* Stanford Univ, AB, 44, PhD(chem), 49. *Prof Exp:* Mem staff, Los Alamos Nat Lab, 48-68, assoc group leader, 68-73, alt group leader, 73-84; CONSULT, 84- *Mem:* Am Phys Soc; Am Chem Soc. *Res:* Materials science; chemical kinetics; high vacuum. *Mailing Add:* 1124 E Grandview Rd Phoenix AZ 85022-2627

MOSLEY, RONALD BRUCE, solid state physics, for more information see previous edition

MOSLEY, STEPHEN T, b Detroit, Mich, May 14, 49; m 86; c 1. CELL BIOLOGY, CHOLESTEROL METABOLISM. *Educ:* Univ Tex, PhD(molecular genetics), 83. *Prof Exp:* Postdoctoral fel, cell & develop biol, Harvard Univ, 83-86; sr res investr, Squibb Inst Med Res, Princeton, 86-88; RES LEADER, BRISTOL-MYERS SQUIBB INST PHARM RES, PRINCETON, NJ, 88- *Mem:* Am Soc Cell Biol; Am Heart Asn. *Mailing Add:* Bristol-Myers Squibb Inst Pharm Res PO Box 4000 Princeton NJ 08543

MOSLEY, WILBUR CLANTON, JR, b Birmingham, Ala, Oct 30, 38; m 60; c 2. SOLID STATE PHYSICS, MATERIALS SCIENCE. *Educ:* Auburn Univ, BEP, 60, MS, 62; Univ Ala, Tuscaloosa, PhD(physics), 65. *Prof Exp:* Res physicist, E I du Pont de Nemours & Co Inc, 65-80, staff physicist, Nuclear Mat div, 80-88; RES STAFF PHYSICIST, HYDROGEN TECHNOL SECT, SAVANNAH RIVER LAB, WESTINGHOUSE SAVANNAH RIVER CO, 88- *Mem:* Am Soc Metals; Microbeam Analysis Soc. *Res:* Chemical and radiation stability of compounds of actinides; hydrogen and helium effects; hydride technology. *Mailing Add:* 202 Fairway Dr New Ellenton SC 29809

MOSS, ALFRED JEFFERSON, JR, b Little Rock, Ark, Nov 22, 40; m 65. BIOPHYSICS, PHYSIOLOGY. *Educ:* Univ Ark, Little Rock, BS, 62, MS, 64, PhD(biophys, physiol), 70. *Prof Exp:* Res chemist, Dow Chem Co, 65-68; asst prof, Med Ctr, Little Rock, 70-77, assoc prof radiol & physiol, 77-, AT DEPT PHYSIOL & BIOPHYSICS, UNIV ARK-FAYETTEVILLE. *Mem:* Biophys Soc; Radiation Res Soc. *Res:* Radiation biophysics; molecular biology. *Mailing Add:* 8312 Linda Lane Little Rock AR 72207

MOSS, BERNARD, b New York, NY, July 26, 37; m 60; c 3. BIOCHEMISTRY, VIROLOGY. *Educ:* NY Univ, BA, 57, MD, 61; Mass Inst Technol, PhD(biochem), 67. *Prof Exp:* Intern med, Children's Hosp Med Ctr, Boston, Mass, 61-62; USPHS basic sci training fel, 62-66; investr, lab biol viruses, 66-71, head macromolecular biol sect, Lab Biol Viruses, 71-83, CHIEF, LAB VIRAL DIS, NIH, 84- *Honors & Awards:* Commendation Medal, USPHS, Meritorious Serv Medal, Distinguished Serv Medal. *Mem:* Nat Acad Sci; Am Soc Microbiol; Am Soc Biol Chemists; AAAS; Am Soc Virol. *Res:* Animal viruses; proteins, enzymes; DNA replication; RNA synthesis; vaccines. *Mailing Add:* Lab Viral Dis NIH Bethesda MD 20892

MOSS, BUELON REXFORD, b Columbia, Ky, Oct 24, 37; m 59; c 3. ANIMAL NUTRITION, DAIRY SCIENCE. *Educ:* Berea Col, BS, 60; Univ Tenn, Knoxville, PhD(animal sci), 68. *Prof Exp:* Instr high sch, Ky, 61-63; res technician, Univ Tenn-AEC Agr Res Lab, Oak Ridge, 67-68; res assoc animal sci, Univ Tenn, Knoxville, 68-69; from asst prof to prof animal sci & dairy nutrit, Mont State Univ, 69-83; PROF DAIRY SCI, ANIMAL & DAIRY SCI DEPT, AUBURN UNIV, 83- *Mem:* Am Dairy Sci Asn. *Res:* Forage evaluation; pasture for lactating cows; by-product feeds. *Mailing Add:* Animal & Dairy Sci Dept Auburn Univ Auburn AL 36849

MOSS, CALVIN E, b Richmond, Va, Nov 27, 39; m 61. EXPERIMENTAL NUCLEAR PHYSICS. *Educ:* Univ Va, BS, 61; Calif Inst Technol, MS, 63, PhD(physics), 68. *Prof Exp:* Res assoc nuclear physics, Duke Univ, 67-69 & Univ Colo, 69-71; res fel, Australian Nat Univ, 71-72; res assoc, Univ Colo, 72-73; STAFF MEM, LOS ALAMOS SCI LAB, 73- *Mem:* Am Phys Soc. *Res:* Nuclear spectroscopy of light nuclei and accelerator development. *Mailing Add:* Group EES-9 LANL MS D436 Los Alamos Sci Lab PO Box 1663 Los Alamos NM 87545

MOSS, CLAUDE WAYNE, b Rural Hall, NC, Mar 20, 35; m 58; c 4. MICROBIOLOGY, BIOCHEMISTRY. *Educ:* NC State Univ, BS, 57, MS, 62, PhD(microbiol), 65. *Prof Exp:* Res assoc microbiol, NC State Univ, 63-65; RES MICROBIOLOGIST, CTR DIS CONTROL, USPHS, 65- *Mem:* Sci Res Soc Am; Am Soc Microbiol; Sigma Xi. *Res:* Physiology and metabolism of pathogenic bacteria; application of chromatographic techniques to diagnostic bacteriology and clinical chemistry; cryobiology of microorganisms and animal cells; biochemistry of drug metabolism. *Mailing Add:* CDC L Div Mic BR4-112 1600 Clifton Rd NE Atlanta GA 30333

MOSS, DALE NELSON, b Thornton, Idaho, Mar 27, 30; m 53; c 8. AGRONOMY, PLANT PHYSIOLOGY. *Educ:* Ricks Col, BS, 55; Cornell Univ, MS, 55, PhD(crop physiol), 59. *Prof Exp:* Asst prof chem, Ricks Col, 55-56; asst agr scientist, Conn Agr Exp Sta, 59-61, assoc agr scientist, 61-63, agr scientist, 63-67; prof crop physiol, Univ Minn, St Paul, 67-77; head dept, 77-84, PROF CROP SCI, ORE STATE UNIV, CORVALLIS, 84- *Concurrent Pos:* Vis prof, Agr Univ, Wagening The Netherlands, 66; mem, US AID Res Adv Comm, 75-83. *Mem:* Fel Am Soc Agron (pres-elect, 84-85, pres, 85-86); fel Crop Sci Soc Am (pres-elect, 75-76, pres, 76-77, past pres, 77-78). *Res:* Effects on photosynthesis, respiration and transpiration of higher plants of light intensity, temperature, carbon dioxide concentration, nutrition, removal of storage organs, air turbulence, planting patterns, and age of leaves; effect of acid rain on crop yields; modeling crop growth and yield; leaf growth in cereals. *Mailing Add:* Dept Crop & Soil Sci Ore State Univ Corvallis OR 97331-3002

MOSS, DONOVAN DEAN, b Bunker Hill, Ind, Feb 28, 26; m 48; c 2. FISH BIOLOGY, AQUACULTURE. *Educ:* Auburn Univ, BS, 49, MS, 50; Univ Ga, PhD(zool), 62. *Prof Exp:* Fisheries biologist, Ala State Dept Conserv, 51-56, asst chief fisheries, 56-57; res asst zool, Univ Ga, 57-61; from assoc prof to prof, Univ Ky, 62-65; assoc prof biol, Tenn Technol Univ, 65-67; from assoc prof to prof fisheries, 67-89, assoc dir, Int Ctr Aquacult, 70-89, EMER PROF FISHERIES, AUBURN UNIV, 89- *Mem:* Am Fisheries Soc; World Mariculture Soc. *Res:* Fish management and biology of fishes including ecological requirements of fish species. *Mailing Add:* Fisheries Bldg Auburn Univ Auburn AL 36849

MOSS, FRANK EDWARD, b Paris, Ill, Feb 10, 34; m 62; c 1. LOW TEMPERATURE PHYSICS. *Educ:* Univ Va, BEE, 56, MNE, 61, PhD(physics), 64. *Prof Exp:* Res engr electronics, Univ Va, 56-61, sr scientist cryog eng, 67-71; assoc prof, 71-76, PROF PHYSICS, UNIV MO-ST LOUIS, 76- *Concurrent Pos:* NSF fel, Univ Rome, 65-66, vis researcher, 66-67. *Mem:* Sigma Xi; AAAS; Fed Am Scientists; Am Phys Soc. *Res:* Mechanics of superfluids; phase transitions; phonon interactions in solids and liquids. *Mailing Add:* Dept of Physics Univ of Mo 8001 Natural Bridge Rd St Louis MO 63121

MOSS, GERALD, b New York, NY, Feb 1, 31; m 63; c 1. BIOMEDICAL ENGINEERING, SURGERY. *Educ:* NY Univ, BA, 51, MS, 56; Union Univ, NY, PhD(biochem), 61; Albany Med Col, MD, 61. *Prof Exp:* Res phys chemist, Vitro Labs, Inc, 56; resident surg, Albany Med Ctr Hosp, 61-68; PROF BIOMED ENG, RENSSELAER POLYTECH INST, 68- *Concurrent Pos:* Clin asst prof surg, Albany Med Col, 68- *Res:* Biochemistry; postoperative protein metabolism and wound healing; cerebral glucose metabolism; circulatory physiology; gastrointestinal physiology. *Mailing Add:* ORR One Box 295 West Sand Lake NY 12196

MOSS, GERALD ALLEN, b Milwaukee, Wis, Jan 24, 40. NUCLEAR PHYSICS. *Educ:* Univ Wis, BS, 61; Univ Ore, MS, 63, PhD(physics), 66. *Prof Exp:* Fel nuclear physics, Univ Man, 67-69; from asst prof to assoc prof, 69-80, PROF NUCLEAR PHYSICS, UNIV ALTA, 80-, ASSOC DIR SCI, SCI WORLD BRITISH COLUMBIA. *Concurrent Pos:* Vis scientist, Saclay Nuclear Res Ctr, France, 75-76. *Mem:* Am Asn Physics Teachers. *Res:* Nuclear reactions; spectroscopy; intermediate energy experimental nuclear physics. *Mailing Add:* 3700 Willingdon Ave Burnaby BC V5G 3H2 Can

MOSS, GERALD S, b Cleveland, Ohio, Mar 4, 35; m; c 3. SURGERY, EXPERIMENTAL SURGERY. *Educ:* Ohio State Univ, BA, 56, MD, 60. *Prof Exp:* Teaching fel anat, Sch Med, Harvard Univ, 62; tutor surg, Manchester Royal Infirmary, Eng, 64; head exp surg, US Naval Res Inst, 66-68; from asst prof to assoc prof, 68-72, PROF SURG, UNIV ILL, CHICAGO, 73-; CHMN DEPT SURG, MICHAEL REESE HOSP & MED CTR, 77- *Concurrent Pos:* Asst chief surg, Vet Admin West Side Hosp, Chicago, 68-70; attend surg, Cook County Hosp, 70-72, chmn dept, 72-; dir surg res, Hektoen Inst Med Res, Chicago, 72- *Mem:* Soc Univ Surgeons; fel Am Col Surgeons; Asn Acad Surg (pres, 76-); Nat Soc Med Res; Int Cardiovasc Soc. *Res:* Shock and resuscitation; blood preservation; blood component therapy and blood substitutes; cardiopulmonary physiology. *Mailing Add:* 1042 Saxong Dr Highland Park IL 60035

MOSS, HERBERT IRWIN, b Brooklyn, NY, Mar 8, 32; m 60; c 3. MAGNETIC MATERIALS, CERAMICS. *Educ:* Univ Louisville, BS, 53; Indiana Univ, PhD(chem), 60. *Prof Exp:* Mem staff, David Sarnoff Res Ctr, RCA Labs, 59-87, CONSULT, MAT CERAMIC PROCESSING, DAVID SARNOFF RES CTR/SRI, 87- *Mem:* Am Chem Soc; Sigma Xi; Am Ceramic Soc. *Res:* Pressure sintering of magnetic, electronic and optically active materials; synthesis and properties of electronically active materials; thin films; materials for magnetic recording heads. *Mailing Add:* PO Box 569 Point Pleasant PA 18950-0569

MOSS, J ELIOT B, b Staunton, Va, Jan 1, 54; m 76. PROGRAMMING LANGUAGES, DATABASES. *Educ:* Mass Inst Technol, SB, 75, SM, 78, PhD(elec eng & computer sci), 81. *Prof Exp:* Staff programmer, US Army War Col, 81-85; ASST PROF COMPUTER SCI, UNIV MASS, 85- *Mem:* Asn Comput Mach; Inst Elec & Electronic Engrs; Sigma Xi. *Res:* Object-oriented programming languages, systems, databases, and architectures; emphasis on system design, prototypes, and performance evaluation. *Mailing Add:* Dept Computer & Info Sci Univ Mass Amherst MA 01003

MOSS, JOEL, b Brooklyn, NY, Nov 27, 46. PULMONARY CRITICAL CARE. *Educ:* NY Univ, MD, 72, PhD(biochem), 72. *Prof Exp:* HEAD, SECT MOLECULAR MECHANISMS, NAT HEART, LUNG & BLOOD INST, NIH, 79- *Honors & Awards:* Young Investr Award, Am Fedn Clin Res, 87. *Mem:* Am Soc Biol Chemists; Am Soc Clin Investrs; Am Fedn Clin Res. *Res:* Signal transduction by eukaryotic cells. *Mailing Add:* Sect Molecular Mechanism Nat Heart Lung & Blood Inst NIH Bldg Ten Rm 5N307 Bethesda MD 20892

MOSS, LEO D, b Berlin-Spandau, Germany, June 25, 11; US citizen; m 35; c 3. ANATOMIC PATHOLOGY, CLINICAL PATHOLOGY. *Educ:* Med Sch, Univ Berne, Switz, MD, 34. *Prof Exp:* ATTEND PATHOLOGIST, OLEAN GEN HOSP & ST FRANCIS HOSP, 42-; LAB DIR, CATTARAUGUS COUNTY LAB, OLEAN, NY, 60-; MED DIR, MDS REGIONAL LAB, OLEAN, NY, 81- *Concurrent Pos:* Adj prof, dept biol, St Bonaventure Univ; health comnr, Cattaraugus County, 72-84. *Mem:* Fel Am Soc Clin Pathologists; fel Am Col Physicians; Int Acad Path; fel Col Am Pathologists. *Res:* Hypertension and Goldblatt mechanism; experimental arteriosclerosis; multiple case reports. *Mailing Add:* 477 Vermont St Olean NY 14760

MOSS, LLOYD KENT, b Los Angeles, Calif, Aug 8, 24; m 50; c 4. BIOCHEMISTRY. *Educ:* Univ Calif, Los Angeles, BS, 50; Stanford Univ, PhD(chem), 57. *Prof Exp:* Chemist, Aerojet-Gen Corp, Gen Tire & Rubber Co, 50-52; assoc chemist, Stanford Res Inst, 56-58, biochemist, 58-66; from asst prof to prof chem, Foothill Col, 66-87; consult chem, Tanzania, EAfrica, 87-89; SCI TEACHER, VERDE VALLEY SCH, 90- *Mem:* AAAS; Am Chem Soc; NY Acad Sci. *Res:* Protein characterization; preparative chromatography; continuous flow electrophoresis; periodate oxidations; photosynthetic energy transfer mechanisms; allergens, hemagglutinins and toxins. *Mailing Add:* Verde Valley Sch 3511 Verde Valley Sch Rd Sedona AZ 86336

MOSS, MARVIN, b New York, NY, Dec 8, 29; m 57; c 2. SOLID STATE PHYSICS. *Educ:* Queens Col, NY, BS, 51; Cornell Univ, PhD(eng physics), 63. *Prof Exp:* Jr engr, Sylvania Elec Prod, Inc, 51-54; asst physics, Cornell Univ, 54-62; MEM TECH STAFF, SANDIA NAT LABS, 63- *Mem:* AAAS. *Res:* Thermophysical properties of solids. *Mailing Add:* Div 4051 Sandia Nat Labs Albuquerque NM 87185

MOSS, MELVIN LANE, b Deerfield, Ohio, July 3, 15; m 52; c 1. BIOCHEMISTRY, SCIENCE ADMINISTRATION. *Educ:* Mt Union Col, BS, 38; Purdue Univ, MS, 40, PhD(anal chem), 42. *Prof Exp:* Asst instr chem, Purdue Univ, 38-41; res chemist, Hercules Powder Co, 42-48; asst div head, Alcoa Res Labs, Aluminum Co Am, 48-62; NIH fel, Inst Neurobiol, Univ Gothenburg, 63-64; sr scientist staff mem, Oak Ridge Nat Lab, 69-73; assoc mem, Inst Muscle Dis, Inc, 64-69, actg dir, 73-74, dir res, Muscular Dystrophy Asn, Inc, 73-87; RETIRED. *Mem:* AAAS; Am Chem Soc; Am Microchem Soc. *Res:* Analytical methods and instrumentation; chemical and metallurgical process control; analysis of isolated nerve, muscle and amniotic cells; biochemistry of muscle disease. *Mailing Add:* Nine Ridgemead Fields Verona PA 15147

MOSS, MELVIN LIONEL, b New York, NY, Jan 3, 23; m 70; c 2. ANATOMY. *Educ:* NY Univ, AB, 42; Columbia Univ, DDS, 46, PhD, 54. *Prof Exp:* From asst anat to assoc prof, 52-67, dean, Sch Dent & Oral Surg, 68-73, PROF ANAT, COLUMBIA UNIV, 68-, PROF ORAL BIOL, COL PHYSICIANS & SURGEONS, 67- *Concurrent Pos:* Lederle med fac award, 54-56; mem int comt standardization human biol. *Honors & Awards:* Craniofacial Biol Award, Int Asn Dent Res. *Mem:* AAAS; Am Soc Zoologists; Am Asn Anatomists; Am Asn Phys Anthrop; Int Asn Dent Res. *Res:* Skeletal morphology and physiology; physical anthropology; bioengineering; information sciences. *Mailing Add:* Dept Anat Columbia Univ 630 W 168th St New York NY 10032

MOSS, RANDY HAYS, b Searcy, Ark, Aug 7, 53; m 78; c 2. PATTERN RECOGNITION, IMAGE PROCESSING. *Educ:* Univ Ark, BS, 75, MS, 77; Univ Ill, PhD(elec eng), 81. *Prof Exp:* asst prof, 81-86, ASSOC PROF ELEC ENG, UNIV MO-ROLLA, 86- *Concurrent Pos:* Vis instr, Univ Ill, 79, vis lectr, 80; ed bd, Pattern Recognition; assoc ed, Computerized Med Imaging & Graphics. *Honors & Awards:* Lindbergh Award, Am Inst Aeronaut & Astronaut; Tenth Ann Award, Pattern Recognition Soc; Ralph R Teetor Educ Award, Soc Automotive Engrs, 88; Outstanding Young Mfg Engr Award, Soc Mfg Engrs, 87. *Mem:* Inst Elec & Electronics Engrs; Sigma Xi; Pattern Recognition Soc. *Res:* Vision systems for industrial robots; medical applications of machine vision; pattern recognition; image processing; digital systems, including microprocessor systems; computer aided design and manufacturing. *Mailing Add:* Elec Eng Dept Univ Mo-Rolla Rolla MO 65401-0249

MOSS, RICHARD, b 1947; m 68; c 1. CALCIUM REGULATION OF CONTRACTION. *Educ:* Univ Vt, PhD(physiol), 75. *Prof Exp:* from asst prof to assoc prof, 79-87, PROF, UNIV WIS, 87-, CHMN, DEPT PHYSIOL, 88- *Mem:* Int Soc Heart Res; Biophys Soc; Am Physiol Soc. *Res:* Developmental aspects of muscle; muscle contraction. *Mailing Add:* Dept Physiol Univ Wis 1300 University Ave Madison WI 53706

MOSS, RICHARD WALLACE, b Charlotte, NC, June 12, 41; m 62; c 2. ELECTRONICS ENGINEERING, COMMUNICATIONS. *Educ:* Ga Inst Technol, BEE, 65, MSEE, 68. *Prof Exp:* From res engr to prin res engr, 65-79, CHIEF COMMUN SYSTS DIV, GA INST TECHNOL, 74- *Concurrent Pos:* Consult engr, 73- *Mem:* Sigma Xi; Inst Elec & Electronics Engrs; Nat Soc Prof Engrs. *Res:* Communications systems and technology; signal monitoring systems, spectrum utilization management; simulation modeling. *Mailing Add:* 3855 Mceachern Farms Dr SW Power Springs GA 30073

MOSS, ROBERT ALLEN, b Brooklyn, NY, May 27, 40; m 67; c 2. ORGANIC CHEMISTRY. *Educ:* Brooklyn Col, BS, 60; Univ Chicago, MS, 62, PhD(chem), 63. *Prof Exp:* Nat Acad Sci-Nat Res Coun res fel chem, Columbia Univ, 63-64; from asst prof to assoc prof, 64-73, PROF CHEM, RUTGERS UNIV, 73- *Concurrent Pos:* Vis scientist, Mass Inst Technol, 71-72; Alfred P Sloan Found fel, 71-73; acad vis, Univ Oxford, 76-77; assoc dean natural sci, Rutgers Univ, 81-87; michael vis prof, Weizmann Inst, Israel, 84. *Mem:* Am Chem Soc; Royal Soc Chem. *Res:* Organic and bioorganic chemistry in aggregate phases; reactive intermediates; kinetics in the micro-nanosecond regime; carbenes and carbenoids. *Mailing Add:* Dept of Chem Rutgers Univ New Brunswick NJ 08903

MOSS, ROBERT HENRY, physical inorganic chemistry, solid state chemistry; deceased, see previous edition for last biography

MOSS, ROBERT L, b Brooklyn, NY, Aug 24, 40; m 62; c 2. NEUROPHYSIOLOGY, NEUROENDOCRINOLOGY. *Educ:* Villanova Univ, BS, 62; Claremont Grad Sch & Univ Ctr, MS, 67; PhD(neurophysiol, neuropsychol), 69. *Prof Exp:* Res asst, Inst Behav Res, Silver Spring, Md, 63-64; res asst grade I operant behav, Patton State Hosp, Calif, 64-65; res assoc anat, Med Sch, Univ Bristol, 69-71; from asst prof to prof physiol, 71-83, chmn physiol grad prog, 83-87, PROF PHYSIOL & NEUROL, UNIV TEX SOUTHWESTERN MED CTR, 83- *Concurrent Pos:* NIMH fel, Dept Anat, Med Sch, Univ Bristol, 69-71; Instnl grant, Dept Physiol, Univ Tex Southwestern Med Ctr, Dallas, 71-72, NIH grant, 72-78, NSF grant, 74-76 & Ayerst Labs Inc grant, 74-, Arcopharma Ltd Pharmaceut grant, 84-, Texas-Salk grant, 85-, & NIH grant, 87-; NIH career develop award, 76-81. *Honors & Awards:* Young Scientist Award, Am Psychol Asn, 69. *Mem:* AAAS; Am Physiol Soc; Endocrine Soc; Int Soc Neuroendocrinol; Soc Neurosci; Soc Experimental Biol & Med. *Res:* Neural cellular, molecular, and biochemical mechanisms involved in hypothalamic control over pituitary function(s) and reproductive behavior. *Mailing Add:* Dept Physiol Univ Tex Southwestern Med Ctr Dallas TX 75235-9040

MOSS, RODNEY DALE, b Oakdale, Nebr, Apr 9, 27; m 50; c 2. ORGANIC CHEMISTRY, ANALYTICAL CHEMISTRY. *Educ:* Univ Nebr, BS, 48, MS, 49; Indiana Univ, PhD(chem), 51. *Prof Exp:* Res chemist, 51-60, proj leader, 60-63, group leader, 63-65, head pharm chem res dept, 65-68, dir chem, 68-76, dir prod develop, Agr Prod Ctr, Dow Chem, USA, 76-81, dir agr chem res & develop, Dow Chem Pac Ltd, 81-84, DIR GLOBAL AGR PROD LICENSING & PATENTS, DOW CHEM USA, 84- *Mem:* Am Chem Soc. *Res:* Organic sulfur and phosphorus chemistry; residue analysis and environmental studies. *Mailing Add:* PO Box 3866 Prescott AZ 86302-3866

MOSS, SIMON CHARLES, b Woodmere, NY, July 31, 34; m 58; c 4. MATERIALS SCIENCE, SOLID STATE PHYSICS. *Educ:* Mass Inst Technol, SB, 56, SM, 59, ScD(metall), 62. *Prof Exp:* Mem res staff metall, Res Div, Raytheon Mfg Co, 56-57; from asst prof to assoc prof, Mass Inst Technol, 62-70; dir sci dept, Energy Conversion Devices, Inc, 70-72; PROF PHYSICS, UNIV HOUSTON, 72- *Concurrent Pos:* Ford fel eng, Mass Inst Technol, 62-64; Guggenheim fel, 68-69; Alexander von Humboldt sr scientist grant, Univ Munich, 79. *Mem:* Fel Am Phys Soc. *Res:* X-ray and neutron diffraction; structure of disordered and defective solids; crystallography and thermodynamics of phase transformations; amorphous semiconductors; hydrogen in metals; biological structures. *Mailing Add:* Dept of Physics Univ of Houston 4800 Calhoun Rd Houston TX 77004

MOSS, STEVEN C, b Ann Arbor, MI, Mar 19, 48. ULTRA FAST PHENOMENA, NONLINEAR OPTICS. *Educ:* Ark A&M Col, BS, 70; Purdue Univ, MS, 72; N Tex State Univ, PhD(Physics), 81. *Prof Exp:* Postdoctoral res assoc, Nat Res Coun-Naval Res Lab, Washington, DC, 81-92; res assoc, Dept Physics & Ctr Appl Quantum Electronics, N Tex State Univ, Denton, 82-84; MTS, OPTICAL PHYSICS DEPT, AEROSPACE CORP, 84- *Mem:* Am Phys Soc; Optical Soc Am; Inst Elec & Electronic Engrs, Mat Res Soc; Int Soc Optical Eng; Am Asn Physics Teachers. *Res:* Ultrafast optoelectronics; ultrafast phenomena; nonlinear optical properties of materials. *Mailing Add:* Aerospace Corp MS M2-253 PO Box 92957 Los Angeles CA 90009

MOSS, WILLIAM WAYNE, b Toronto, Ont, Mar 14, 37; US citizen; m 67; c 2. ACAROLOGY. *Educ:* Carleton Univ, Bsc, 59, MSc, 61; Univ Kans, PhD(entom), 66. *Prof Exp:* Asst cur entom, Acad Natural Sci, Philadelphia, 66-70, assoc cur, 70-82; biol adminr syst biol, NSF, Washington, DC, 78-81; ADJ SR SCIENTIST ENTOM, UNIV KANS, LAWRENCE, 81-; ASSOC PROF BIOL, DEPT BIOL, NEUMANN COL, ASTON, PA, 84- *Concurrent Pos:* Adj asst prof biol, Univ Pa, Philadelphia, 66-75; Miller fel entom, Univ Calif, Berkeley, 68-70, lectr entom, 69-70; prin investr res grants, NSF, 67-88, consult, 76-78. *Mailing Add:* Dept Biol Neumann Col Aston PA 19014-1297

MOSSBERG, THOMAS WILLIAM, b Minneapolis, Mn, June 28, 51; c 2. QUANTUM OPTICS, OPTICAL OPTICS. *Educ:* Columbia Univ, MS, 75, PhD(physics), 78. *Prof Exp:* Res assoc, Columbia Univ, 78-80, asst prof physics, 80-81; asst prof, 81-85, assoc prof physics, Harvard Univ, 85-87; PROF PHYSICS, UNIV OREGON, 87- *Mem:* Am Phys Soc; Optical Soc Am. *Res:* Interaction of light with matter concentration on regimes in which coherence and/or quantum optical effects are important. *Mailing Add:* Dept Physics Univ Oregon Eugene OR 97403

MOSSER, JOHN SNAVELY, b Canton, Ohio, Apr 7, 28. PHYSICAL CHEMISTRY, ENVIRONMENTAL MANAGEMENT. *Educ:* Case Western Reserve Univ, BS, 50, MS, 52, PhD(phys chem), 63; Univ Akron, MBA, 77. *Prof Exp:* Res engr, Indust Rayon Corp, 52-54; sr res chemist, Gen Tire & Rubber Co, 62-70; WASTEWATER QUAL COORDR, CITY OF AKRON, 72- *Mem:* Am Chem Soc; Sigma Xi; Water Pollution Control Fedn. *Res:* Cryogenic measurement and calculation of thermodynamic properties; preparation and characterization of polymers and latices. *Mailing Add:* Water Pollution Control 2460 Akron Peninsula Rd Akron OH 44313

MOSSMAN, ARCHIE STANTON, b Madison, Wis, Feb 5, 26; m 73; c 4. WILDLIFE ECOLOGY, GAME RANCHING. *Educ:* Univ Wis, BA, 49, PhD(zool, wildlife mgt), 55; Univ Calif, MA, 51. *Prof Exp:* Biologist, Dept Fish & Game, Alaska, 55-57; instr, Exten Div, Univ Wis, 57-58; asst prof, Univ Wyo, 58-59; Fulbright res scholar, Nat Mus Southern Rhodesia, 59-61; from asst prof to prof, 61-80, EMER PROF WILDLIFE ECOL, COL NATURAL RESOURCES, HUMBOLDT STATE UNIV, 80- *Concurrent Pos:* Sr lectr, Univ Col Rhodesia & Nyasaland, 63-65; Food & Agr Orgn consult, Malawi, 69; prin investr evaluation game ranching in southern Africa, Int Union Conserv Nature & Natural Resources & World Wildlife Fund Joint Proj, 74-75; consult, Int Union Conserv Nature & Natural Resources, Vienña, Pêru, 79. *Mem:* Ecol Soc Am; Wildlife Soc; Am Soc Mammal; Cooper Ornith Soc; Asn Study Animal Behavior. *Res:* Animal behavioral response to environment; food production from wildlife; predation; natural landscape production. *Mailing Add:* Col Natural Resources Humboldt State Univ Arcata CA 95521

MOSSMAN, BROOKE T, b Schenectady, NY, Aug 2, 47. CANCER RESEARCH. *Educ:* Univ Vt, PhD(cell biol), 77. *Prof Exp:* ASSOC PROF PATH, UNIV VT, 83-, CHAIR CELL BIOL PROG, 84- *Mem:* Am Asn Cancer Res; Am Soc Cell Biol; Am Soc Path. *Mailing Add:* Dept Path Col Med Univ Vt 85 S Prospect St Burlington VT 05401

MOSSMAN, DAVID JOHN, b Mar 9, 38; Can citizen; m 63; c 3. MINERALOGY, GEOLOGY. *Educ:* Dalhousie Univ, BSc, 59, MSc, 63; Univ Otago, NZ, PhD(geol), 70. *Prof Exp:* Field geologist, Anglo Am Corp, SAfrica, 59-62; party chief groundwater res, NS Govt, Can, 64; lectr appl geol, Univ Otago, NZ, 70; geologist, Dept Natural Resources, NB, Can, 71; from asst prof to assoc prof econ geol, Univ Sask, 71-82; assoc prof, 82-86, PROF ECON GEOL, MT ALLISON UNIV, 86- *Mem:* Geol Soc NZ; Mineral Asn Can; Mineral Soc Am. *Res:* Economic geology; petrology of basic and ultrabasic rocks; mineralogy; vertebrate trace fossils; geology of gold with focus on Proterozoic paleoplacer deposits in Witwatersrand (South Africa) and the Huronian (Canada); nature and origin of Kerogens in these deposits; study of biomineralization particularly as it applies to the ability of prokaryotic micro bacteria to dissolve and/or concentrate gold. *Mailing Add:* Dept Geol Mt Allison Univ Sackville NB E0A 3C0 Can

MOSSMAN, KENNETH LESLIE, b Windsor, Ont, Apr 14, 46; US citizen; m 70. RADIATION BIOLOGY, HEALTH PHYSICS. *Educ:* Wayne State Univ, BS, 68; Univ Tenn, MS, 70, PhD(radiation biol), 73; Univ Md, MEd, 88. *Prof Exp:* Radiation biologist & asst prof radiol, Med Ctr, Georgetown Univ Grad Sch, 73-79, from assoc prof to prof radiation med, 79-90, chmn, Dept Radiation Sci, 85-90; ASST VPRES RES & PROF DEPT MICROBIOL, ARIZ STATE UNIV, TEMPE, 90- *Concurrent Pos:* Asst prof, Howard Univ, 76-90; mem bd dirs, Health Physics Soc, 87-90. *Honors & Awards:* Elda E Anderson Award, Health Physics Soc, 84. *Mem:* Radiation Res Soc; Sigma Xi; Health Physics Soc; AAAS; NY Acad Sci. *Res:* Human radiobiology; radiation and public policy; radon; radiation carcinogenesis; radiation and pregnancy; risk perception and assessment. *Mailing Add:* 8046 E Kalil Dr Scottsdale AZ 85260

MOSSOP, GRANT DILWORTH, b Calgary, Alta, Apr 15, 48; m 69; c 3. SEDIMENTOLOGY, STRATIGRAPHY. *Educ:* Univ Calgary, BSc, 70, MSc, 71; Univ London, PhD(sedimentary geol), & DIC, 73. *Prof Exp:* Res fel geol, Univ Calgary, 74; res officer oil sand geol, 75-80, head, 80-84, SR RES OFFICER, ALTA GEOL SURV, ALTA RES COUN, 85- *Concurrent Pos:* Acad vis, Oxford Univ, 84-85; distinguished lectr, Can Soc Petrol Geologists, 79 & 88, Am Asn Petrol Geologists, 81; pres, Can Geosci Coun, 89. *Honors & Awards:* Link Award, Can Soc Petrol Geologists, 78. *Mem:* Can Soc Petrol Geologists; Geol Asn Can (pres, 87); Soc Econ Paleontologists & Mineralogists; Int Asn Sedimentologists. *Res:* Synthesis stratigraphy of the Western Canadian Sedimentary Basin. *Mailing Add:* Alta Res Coun PO Box 8330 Sta F Edmonton AB T6H 5X2 Can

MOSS-SALENTIJN, LETTY, b Amsterdam, Netherlands, Apr 14, 43; m 70. ANATOMY, DENTAL RESEARCH. *Educ:* State Univ Utrecht, DDS, 67, PhD(anat), 76. *Prof Exp:* Chief instr oral hist growth & develop, Holland Lab Hist & Micros Anat, State Univ Utrecht, 67-68; asst prof anat, 68-74, assoc prof oral biol, 74-78, assoc prof orofacial growth & develop, 78-86, dir dent radiol unit, 80-86, PROF OROFACIAL GROWTH & DEVELOP, COLUMBIA UNIV, 86-, DIR GRAD PROG IN DENT SCI, 86- *Mem:* AAAS; Int Asn Dent Res; Am Soc Zool; Int Soc Stereology; fel Royal Micros Soc. *Res:* Growth and development of skeletal tissues; growth of cartilages; orofacial embryology; comparative odontology. *Mailing Add:* 560 Riverside Dr Apt 20K New York NY 10027

MOST, DAVID S, b Boston, Mass, Feb 7, 29; m 52; c 4. PAPER CHEMISTRY. *Educ:* Boston Univ, AB, 52; Lawrence Col, MS, 54, PhD(chem), 57. *Prof Exp:* Group leader, Res Dept, Albemarle Paper Co Div, Ethyl Corp, 57-60; dept mgr appl res, Itek Corp, 60-62; vpres & gen mgr, New Eng Labs, Inc, Rahn Corp, 62-65; consult paper chem, 65-70; pres, M/K Systs, Inc, 70-75; CONSULT, 75- *Mem:* AAAS; Am Chem Soc; Tech Asn Pulp & Paper Indust; Soc Photog Sci & Eng. *Res:* Reproduction technology; development of special papers for use in recording and communications applications including copy papers and facsimile. *Mailing Add:* 114 Waters Edge Dr Jupiter FL 33477-4001

MOST, JOSEPH MORRIS, b New York, NY, Apr 24, 43; m 65; c 2. GENERAL & PHYSICAL CHEMISTRY, COMPUTER SCIENCE. *Educ:* Rutgers Univ, AB, 64, PhD(inorg chem), 74; NJ Inst Technol, MS, 89. *Prof Exp:* Chemist, NL Industs, Inc, 64-66; instr chem, Rutgers Col, Rutgers Univ, 71-74; asst prof chem, 74-82, dir hons prog, 76-79, assoc prof chem & chmn chem, 82-89, PROF CHEM & COMPUTER SCI, UPSALA COL, 89- *Concurrent Pos:* Danforth Found Assoc, 80-86, sci educ consult, sci educ consult, E Orange Sch Dist, 90-; special asst pres, computer appln, 90- *Mem:* Am Chem Soc; Cognitive Sci Soc; Asn Comput Mach; Inst Elec & Electronic Engrs Computer Soc. *Res:* Chemical education; artificial intelligence and expert systems; computer applications in instructional and administrative environments. *Mailing Add:* Upsala PO Box 1186 East Orange NJ 07019-1186

MOSTAFAPOUR, M KAZEM, b Dezful, Iran, Feb 14, 37; US citizen; m 63; c 2. PROTEIN CHEMISTRY, HORMONAL REGULATION. *Educ:* Am Univ Beirut, Lebanon, BSc, 62; Wayne State Univ, Mich, PhD(chem), 72. *Prof Exp:* NIH res fel biochem, Univ Wash, Seattle, 72-73; Nat Cancer Inst res assoc cell biol, Mass Inst Technol, 73-75; asst prof biochem, Free Univ, Tehran, Iran, 75-76; res assoc, Wayne State Univ, Detroit, Mich, 76-77; Nat Eye Inst fel, Inst Biol Sci, Oakland Univ, Rochester, Mich, 77-78, asst prof biomed sci, 78-83; ASSOC PROF CHEM & BIOCHEM, UNIV MICH, DEARBORN, 83- *Concurrent Pos:* Prin investr, Nat Eye Inst grant, 80-83, Rackham res grant, Univ Mich, 85-86 & NIH, 86-889. *Mem:* NY Acad Sci; Am Chem Soc; Asn Res Vision & Ophthal; Sigma Xi. *Res:* Structure and topography of lens proteins; cell cycle regulation; adrenocorticotrophic hormone-adrenal axis and regulation of steroid biosynthesis; role of glutathione and its mixed-disulfides in enzymatic regulation. *Mailing Add:* Dept Natural Sci Univ Mich 4901 Evergreen Dearborn MI 48128

MOSTARDI, RICHARD ALBERT, b Bryn Mawr, Pa, July 1, 38; m 62; c 4. PHYSIOLOGY. *Educ:* Kent State Univ, BS, 60, MEd, 64; Ohio State Univ, PhD(physiol), 68. *Prof Exp:* Res asst physiol, Aviation Med Lab, Ohio State Univ, 66-68; PROF PHYSIOL, UNIV AKRON, 68- *Concurrent Pos:* NIH fel, Milan, Italy, 72-73. *Mem:* NY Acad Sci; Sigma Xi. *Res:* Acoustic diagnosis of arthritis; effects of drag reducing polymers in the vertebrate system; exercise in humans. *Mailing Add:* Dept Biol Univ Akron Akron OH 44325

MOSTELLER, C FREDERICK, b Clarksburg, WVa, Dec 24, 16; m 41; c 2. MATHEMATICAL STATISTICS. *Educ:* Carnegie Inst Technol, BS, 38, MS, 39; Princeton Univ, AM, 42, PhD(math), 46. *Hon Degrees:* DSc, Univ Chicago, 73, Carnegie-Mellon Univ, 74 & Wesleyan Univ, 83; DSSc, Yale Univ, 81. *Prof Exp:* Instr math, Princeton Univ, 42-44; res mathematician, Statist Res Group, 44-45; chmn dept statist, 57-76, chmn dept biostatist, 77-81, DIR, TECHNOL ASSESSMENT GROUP, 88-, EMER PROF MATH STATIST, 88- *Concurrent Pos:* Fund Advan Educ fel, Univ Chicago, 54-55; mem staff probability & statist, NBC's Continental Classroom TV Course, 60-61; fel, Ctr Advan Study Behav Sci, 62-63; chmn bd dirs, Soc Sci Res Coun, 66-68; Guggenheim fel, 69-70; vchmn, President's Comn Fed Statist, 71; mem fac, Dept Social Rels, Harvard Univ, 46; prof math statist, 51; mem fac, JFK Sch Govt, 70; mem fac, Med Sch, 77. *Honors & Awards:* Myrdal Prize Sci Res, 78; Lazarsfeld Prize, Appl Social Sci Res, 79; Samuel S Wilks Award, Am Statist Asn, 86; R A Fisher Award & Lectr, Comt Presidents Statist Soc, 87. *Mem:* Nat Acad Sci; Inst of Med of Nat Acad Sci; AAAS (pres, 80); Am Philos Soc; Am Acad Arts & Sci; Int Statist Inst (vpres, 85). *Res:* Theoretical statistics and its applications to social science, medicine, public policy and industry. *Mailing Add:* Rm 603 Sci Ctr Harvard Univ One Oxford St Cambridge MA 02138

MOSTELLER, HENRY WALTER, b Detroit, Mich, Nov 6, 32; m 55; c 2. POWER SYSTEMS ENGINEERING, OPTIMIZATION THEORY. *Educ:* Univ Mich, BS(math) & BS(elec eng), 55; Cornell Univ, MS, 59, PhD(elec eng), 63. *Prof Exp:* Anal engr, 63-68, systs engr, Avionics Control Dept, 68-72, SR ENGR, POWER SYSTS ENG DEPT, GEN ELEC CO, 72- *Concurrent Pos:* Asst prof, Elec Eng Dept, Union Col, Schenectady, NY, 63-

68 & Sch Advan Technol, State Univ NY, Binghamton, NY, 68-72; reviewer, Asn Comput Mach. *Mem:* Sigma Xi; sr mem Inst Elec & Electronic Engrs. *Res:* System Identification and torsional fatigue evaluation of turbine-generator torsional systems; developed new algorithms to calculate torsional and uniaxial fatigue for nonlinear shaft systems. *Mailing Add:* 24 Cedar Lane Scotia NY 12302

MOSTELLER, RAYMOND DEE, b Austin, Tex, Dec 30, 41; m 75; c 2. BIOCHEMISTRY, MOLECULAR BIOLOGY. *Educ:* Univ Tex, Austin, BA, 64, PhD(biochem), 68. *Prof Exp:* Asst prof, 70-77, ASSOC PROF BIOCHEM, SCH MED, UNIV SOUTHERN CALIF, 77- *Concurrent Pos:* Fel molecular biol, Stanford Univ, 68-70; NSF res grants, 71-73, 73-75 & 78-80; NIH res grant, 73-76, 76-79, 80-83; vis scientist, Calif Inst Technol, 89-90. *Mem:* Am Soc Microbiol; AAAS; Am Soc Biol Chem; Protein Soc. *Res:* Mechanism of protein biosynthesis; control of gene expression; regulation of protein degradation; post-translational modification of proteins; structure-function relationships in proteins; protein-protein interactions. *Mailing Add:* Dept Biochem Univ Southern Calif Sch Med Los Angeles CA 90033

MOSTELLER, ROBERT COBB, b Lynchburg, Va, Oct 14, 38; m 60; c 2. BIOMETRICS. *Educ:* Randolph-Macon Col, BA, 61; Emory Univ, MS, 67, PhD(statist), 76. *Prof Exp:* Res assoc biostatist, Biomet Unit, Dept Plant Breeding, Cornell Univ, 65; instr biometry, Univ Kans Med Ctr, 68-70; assoc, dept biometry & statist, Emory Univ, 70-73, chief statist, unit biometry, mammography sect, dept radiol, 73-80, asst prof, Sch Med, 76-80; PROF, DEPT BUS ADMIN, MORRIS BROWN COL, 76- *Res:* Statistical evaluation of breast cancer data aimed at the identification of women with a high risk of either current or future breast cancer. *Mailing Add:* 2469 Helmsdale Drive Atlanta GA 30345

MOSTER, MARK LESLIE, b Toronto, Ont, Dec 15, 53; US citizen; m 75; c 3. NEURO-OPHTHALMOLOGY. *Educ:* State Univ NY, BS, 75, MD, 79; Am Bd Psychiat & Neurol, dipl, 84. *Prof Exp:* Fel neuro-ophthal, Wills Eye Hosp, Philadelphia, 83-84; asst prof neurol & ophthal, 84-89, ASSOC PROF NEUROL, TEMPLE UNIV SCH MED, 89- *Concurrent Pos:* Dir neurol residency training prog, Temple Univ Hosp, 85-; examr, Am Bd Psychiat & Neurol, 86; NIH res grant, 87; Nat Inst Neurol Dis & Stroke res grant, 89-94; Clin Invest Support Award, 89-94. *Mem:* Clin Eye Movement Soc; Asn Res Vision & Ophthal; Asn Res Neurol & Ment Dis; Am Acad Neurol; NAm Neuro-Ophthalmology Soc. *Res:* Neurology; ophthalmology. *Mailing Add:* Dept Neurol Temple Univ Hosp 3401 N Broad St Philadelphia PA 19140

MOSTERT, PAUL STALLINGS, b Morrilton, Ark, Nov 27, 27; m 47, 90; c 4. MATHEMATICS. *Educ:* Southwestern at Memphis, BS, 50; Univ Chicago, MS, 51; Purdue Univ, PhD(math), 53. *Prof Exp:* Asst math, Purdue Univ, 51-53; res instr, Tulane Univ La, 53-54, from asst prof to prof, 54-70, chmn dept, 68-70; chmn dept, 70-73, PROF MATH, UNIV KANS, 70- *Concurrent Pos:* Vis prof, Univ Tubingen, 62-63 & 66, Univ Ky, 84-85; NSF sr fel & mem, Inst Advan Study, 67-68; mem selection of postdoctoral fels panel, Nat Res Coun, 69-71; co-founder, Semigroup Forum, managing ed, 68-84, exec ed, 74-84; chmn comt acad freedom, tenure and employment security, Am Math Soc, 72-77, mem coun, 72-75; pres, Equix, Inc, 84-85, Pennfield Biomechanics Corp, 85-89 & Equix Biomechanics, 89- *Mem:* Am Math Soc; Soc Photo-Opt Instr Eng; Int Neural Network Soc. *Res:* Topological semigroups; transformation groups; category theory; applied mathematics; biomathematics. *Mailing Add:* 3298 Roxburg Dr Lexington KY 40503

MOSTOFI, F KSH, GENITOURINARY PATHOLOGY, CANCER. *Educ:* Harvard Univ, MD, 39. *Prof Exp:* CHMN, DEPT GENITOURINARY PATH, ARMED FORCES INST PATH, 48-, CTR ADVAN PATH, 76- *Mailing Add:* Armed Forces Inst Path Washington DC 20306

MOSTOLLER, MARK ELLSWORTH, b Somerset, Pa, Sept 13, 41; m 63; c 2. SOLID STATE PHYSICS. *Educ:* Harvard Col, AB, 62; Harvard Univ, SM, 63, PhD(appl physics), 69. *Prof Exp:* MEM RES STAFF, OAK RIDGE NAT LAB, 69- *Mem:* Am Phys Soc. *Res:* Lattice dynamics; surface excitations; defects in solids; electronic structure and properties. *Mailing Add:* 6032 Solid State Div Oak Ridge Nat Lab PO Box 2008 Mail Stop 6032 Oak Ridge TN 37831-6032

MOSTOV, GEORGE DANIEL, b Boston, Mass, July 4, 23; m 47; c 4. LIE GROUP THEORY. *Educ:* Harvard Univ, BA, 43, MA, 46, PhD(math), 48. *Prof Exp:* Mem, Inst Adv Study, 47-49; instr math, Princeton Univ, 47-48; asst prof math, State Univ NY, Syracuse, 48-51; assoc prof math, Johns Hopkins Univ, 51-61; prof math, 61-83, HENRY FORD 2ND PROF, YALE UNIV, 83- *Concurrent Pos:* Vis prof, Inst Math Applications, 54-55, Univ Paris, 66, Hebrew Univ, 67, Tata Inst Fundamental Res, Bombay, 75; pres, Am Math Soc, 87-89; bd trustee, Inst Adv Study, Princeton, 83-, Weiznien Inst, 88- *Mailing Add:* Beechwood Rd Woodbridge CT 06525

MOSTOW, GEORGE DANIEL, b Boston, Mass, July 4, 23; m 47; c 4. MATHEMATICS. *Educ:* Harvard Univ, BA, 43, MA, 46, PhD(math), 48. *Prof Exp:* Instr math, Princeton Univ, 47-48; asst prof, Syracuse Univ, 49-52; from asst prof to prof, Johns Hopkins Univ, 52-61; prof, 61-63, chmn dept, 71-74, James E English prof, 63-83, HENRY FORD II PROF MATH, YALE UNIV, 83- *Concurrent Pos:* Mem staff, Inst Advan Study, 47-49, 56-57, 75; vis prof, Inst Pure & Appl Math, Brazil, 53-54; Guggenheim fel & Fulbright res scholar, State Univ Utrecht, 57-58; ed, Am J Math, 63-67, assoc ed, 67-; exchange prof, Univ Paris, 66; vis prof, Hebrew Univ, Israel, 67; Tata Inst Fundamental Res, India, 69; Inst Advan Study Sci, 71 & 75; chmn, US Nat Comt Math, 73-74; chmn, Off Math Sci, Nat Acad Sci-Nat Res Coun, 75-78; mem bd trustees, Inst Advan Study. *Mem:* Nat Acad Sci; Am Acad Arts & Sci; Am Math Soc (pres-elect, 86, pres, 87,88). *Res:* Lie groups; discrete subgroups of algebraic groups. *Mailing Add:* Dept of Math Yale Univ New Haven CT 06520

MOSZKOWSKI, STEVEN ALEXANDER, b Berlin, Ger, Mar 13, 27; nat US; m 52, 78; c 3. THEORETICAL PHYSICS. *Educ:* Univ Chicago, BS, 46, MS, 50, PhD(physics), 52. *Prof Exp:* Jr physicist, Argonne Nat Lab, 50-51; res asst, Columbia Univ, 52-53; FROM ASST PROF TO PROF PHYSICS, UNIV CALIF, LOS ANGELES, 53- *Concurrent Pos:* Consult, Rand Corp, 53-71, Oak Ridge Nat Lab, 62-68 & Lawrence Livermore Lab, 73-; Guggenheim fel, 61-62. *Mem:* Fel Am Phys Soc. *Res:* Nuclear shell structure; many-body problem. *Mailing Add:* Dept Physics Univ Calif Los Angeles CA 90024

MOTA DE FREITAS, DUARTE EMANUEL, b Funchal, Portugal, Aug 20, 57. NUCLEAR MAGNETIC RESONANCE, MANIC DEPRESSION-LITHIUM. *Educ:* Imp Col, Univ London, BS, 79; Univ Calif, Los Angeles, PhD(chem), 84. *Prof Exp:* Tutor chem, Davies' Col, London, UK, 79-80; instr chem, Univ Coimbra, Portugal, 80-81; res assoc bioinorg chem, Univ Calif, Los Angeles, 81-82 & 84, teaching assoc chem, 82-83; asst prof, 84-90, ASSOC PROF BIOINORG CHEM, LOYOLA UNIV, CHICAGO, 90- *Concurrent Pos:* Prin investr, Res Corp Cottrell grant, 86-88, Am Heart Asn, Metrop Chicago, 87-89, First Award, NIMH, 90- *Mem:* Am Chem Soc; Magnetic Resonance Med; Int Soc Magnetic Resonance; Biophys Soc. *Res:* Biological action of lithium in manic depression and hypertension; biological applications of nuclear magnetic resonance spectroscopy; enzyme mechanism of copper, zinc-superoxide dismutase; role of metal ions in biology and medicine. *Mailing Add:* Chem Dept Loyola Univ 6525 N Sheridan Rd Chicago IL 60626

MOTARD, R(ODOLPHE) L(EO), b Ottawa, Ont, May 26, 25; m 47; c 4. CHEMICAL ENGINEERING. *Educ:* Queen's Univ, Ont, BSc, 47; Carnegie Inst Technol, MS, 48, DSc(chem eng), 52. *Prof Exp:* Res engr, Shell Oil Co, 51-56, group leader, 56-57; from assoc prof to prof chem eng, Univ Houston, 57-78; assoc prof chem eng, 57-78, chmn dept, 78-91, PROF CHEM ENG, WASH UNIV, 78- *Concurrent Pos:* Nat Acad Sci-NASA sr fel, 66-67. *Mem:* AAAS; Am Inst Chem Engrs; Am Chem Soc; Am Soc Eng Educ. *Res:* Computer science; systems engineering; computer aided design; engineering information systems. *Mailing Add:* Dept of Chem Eng Washington Univ St Louis MO 63130

MOTCHENBACHER, CURTIS D, electronics engineering, for more information see previous edition

MOTE, C(LAYTON) D(ANIEL), JR, b San Francisco, Calif, Feb 5, 37; m 62; c 2. ENGINEERING MECHANICS, MECHANICAL ENGINEERING. *Educ:* Univ Calif, Berkeley, BSc, 59, MSc, 60, PhD(eng mech), 63. *Prof Exp:* Asst mech engr & lectr mech eng, Univ Calif, Berkeley, 62-63; NSF res fel mech eng, Univ Birmingham, 63-64; asst prof, Carnegie Inst Technol, 64-67; from asst prof to assoc prof, 67-72, vchmn, PROF UNIV CALIF, BERKELEY, 72-, CHMN, DEPT MECH ENG, 87- *Concurrent Pos:* Consult, Westinghouse Elec Corp, Calif Cedar Prod Co, Teknekron & Gen Motors Res. *Honors & Awards:* Blackall Award, Am Soc Mech Engrs, 75; Sr Sci Award, Royal Norweg Coun Sci Res, 76 & Alexander von Humboldt Found, 88; Frederick W Taylor Res Medal, 91. *Mem:* Nat Acad Eng; AAAS; fel Am Soc Mech Engrs; Am Testing & Mats; fel Int Acad Wood Sci; Acoustical Soc Am; fel Am Soc Mec; Orthop Res Soc. *Res:* Mechanics of solids, dynamics and dynamic systems; vibration and stability; biomechanics; human injury. *Mailing Add:* Dept of Mech Eng Univ Calif Berkeley CA 94720

MOTE, JIMMY DALE, b Sheridan, Ark, Nov 4, 30; m 51; c 1. METALLURGY. *Educ:* Univ Calif, BS, 57, PhD(metall), 64. *Prof Exp:* Res metallurgist, Lawrence Radiation Lab & Inst Eng Res, Univ Calif, Berkeley, 57-64; tech staff mem, Sandia Corp, NMex State Univ, 64-65; assoc res scientist, Denver Div, Martin Co, 65-68, chief mat sci, 66-68, dir ctr high energy forming, Martin Marietta Corp, 68-71; vpres opers, E F Industs Inc, Univ Louisville, 71-77; SR RES METALL, METALL & MAT SCI DIV, DENVER RES INST, UNIV DENVER, 77- *Mem:* Am Inst Mining, Metall & Petrol Engrs; Am Soc Metals; Am Soc Testing & Mat; Soc Mining Engrs; Sigma Xi. *Res:* Explosive forming and composite materials; fundamental deformation mechanisms of deformation in metals and alloys under static and dynamic loading. *Mailing Add:* 12011 N Antelope Trail Parker CO 80134

MOTE, MICHAEL ISNARDI, b San Francisco, Calif, Feb 5, 35; m 65; c 2. NEUROPHYSIOLOGY. *Educ:* Univ Calif, Berkeley, AB, 58; San Francisco State Col, MA, 63; Univ Calif, Los Angeles, PhD(zool), 68. *Prof Exp:* NIH fel biol, Yale Univ, 68-70; from asst prof to assoc prof, 70-81, dept chmn, 81-84, PROF BIOL, TEMPLE UNIV, 81- *Concurrent Pos:* Res fel, Swiss Nat Sci Found, 77 & Roche Res Found, Basel, 77; vis prof, Univ Zurich, 77; NIH fel, Eye Inst & NSF Fel psychobiol, 71-; vis prof, Univ Zurich, 77; res fel, Swiss Nat Sci Found, 77 & Roche Res Found, Basel, 77. *Mem:* Fel AAAS; Am Soc Zool; Soc Gen Physiol; Asn Res Vision & Opthal. *Res:* Integrative neurophysiology of invertebrate nervous systems with emphasis on vision in arthropods. *Mailing Add:* Dept Biol Temple Univ Broad & Montgomery Sts Philadelphia PA 19122

MOTES, DENNIS ROY, b Beloit, Kans, Dec 17, 50; m 77; c 2. VEGETABLE RESEARCH. *Educ:* Kans State Univ, BS, 72; Univ Ark, MS, 78. *Prof Exp:* RES ASST, VEG SUB STA, UNIV ARK, 78-, RESIDENT DIR, 78- *Res:* Vegetable and agronomic crops. *Mailing Add:* Veg Sub Sta Univ Ark PO Box 2608 Alma AR 72921

MOTHERSILL, JOHN SYDNEY, b Ottawa, Ont, Mar 24, 31; m 79; c 1. GEOLOGY. *Educ:* Carleton Univ, Can, BSc, 53; Queen's Univ, Ont, BSc, 56, PhD(geol), 67. *Prof Exp:* Geologist, Esso Standard Turkey, Inc, Stand Oil Co NJ, 56-58; geologist, Mobil Explor Nigeria, Inc, Mobil Int Oil Co, Inc, 58-61, sr geologist, 62-64; asst prof, Lakehead Univ, 66-70, assoc prof stratig & sedimentation, 70-75, prof geol & dean sci, 75-84; PRIN DIR STUDIES, ROYAL ROADS MIL COL, 84- *Concurrent Pos:* Reader, Univ Nigeria, 72-73; mem staff, Limnol Surv Lake Superior, Can Ctr Inland Waters; sect head admin stratig, Columbia Petrol Co, 61-62. *Mem:* Asn Prof Eng(Ontario); Geol Asn Can; Am Geophys Union. *Res:* Stratigraphy; sedimentation; paleomagnetic studies; nearshore clastic sedimentation. *Mailing Add:* Royal Roads Mil Col Victoria BC V0S 1B0 Can

MOTHERWAY, JOSEPH E, b Providence, RI, Jan 28, 30; m 55; c 9. MECHANICAL ENGINEERING. *Educ:* Brown Univ, ScB, 55; Univ Conn, MS, 61, PhD, 70. *Prof Exp:* Mech engr, Esso Bayway Refinery, 55-56; design engr, Elec Boat Div, Gen Dynamics Corp, 56-58, design group leader, 58-59; sr proj engr, Remington Rand Univac Labs, 59-60; chief engr res & develop, CHI Div, Speidel Corp, 60-63; chief engr, CHI Div, Textron, Inc, 63-64; from asst prof to assoc prof mech eng, 64-70, head dept, 71-75, Bullard Prof Eng Design, Univ Bridgeport, 70-82; PROF MECH ENG, UNIV MASS, 82- *Concurrent Pos:* Lectr, Brown Univ, 62-64; indust consult, 64- *Mem:* Am Soc Mech Engrs; Soc Exp Mech; Sigma Xi. *Res:* Machinery dynamics; computer aided engineering; expert system software. *Mailing Add:* Dept Mech Eng Univ Mass Amherst MA 01003

MOTICKA, EDWARD JAMES, b Oak Park, Ill, May 21, 44; div; c 3. CELLULAR IMMUNOLOGY, IMMUNOREGULATION. *Educ:* Kalamazoo Col, BA, 66; Univ Ill, PhD(anat), 70. *Prof Exp:* Vis scientist immunol, Czech Acad Sci, Inst Microbiol, 71-72; asst prof cell biol, Univ Tex Health Sci Ctr, Dallas, 72-78; assoc prof, Sch Med, Southern Ill Univ, Carbondale, 78-80; assoc prof, 80-91, PROF, SCH MED, SOUTHERN ILL UNIV, SPRINGFIELD, 91- *Mem:* Am Asn Immunologists; Am Soc Microbiol; Asn Res Vision & Ophthal. *Res:* Network regulation of autoantibody synthesis and secretion; regulation of corneal graft rejection. *Mailing Add:* PO Box 19230 Southern Ill Univ Springfield IL 62794-9230

MOTIFF, JAMES P, b Green Bay, Wis, Sept 3, 43; m 81; c 1. BEHAVIORAL MEDICINE, CLINICAL & HEALTH PSYCHOLOGY. *Educ:* St Norbert Col, BS, 65; Univ SDak, MA, 67, PhD (psychol), 69. *Prof Exp:* From asst prof psychol to assoc prof psychol, 69-85, PROF PSYCHOL, HOPE COL, 85- *Concurrent Pos:* Consult, 75- *Mem:* Am Psychol Asn; Biofeedback Soc Am (pres, 85); Soc Behavioral Med; Sigma Xi. *Res:* Stress, biofeedback, behavioral medicine, bulemia. *Mailing Add:* 73 W 21st St Holland MI 49423

MOTILL, RONALD ALLEN, theoretical physics, cosmology, for more information see previous edition

MOTLEY, DAVID MALCOLM, b Long Beach, Calif, June 4, 29; m 52; c 4. COMMUNICATIONS SCIENCE. *Prof Exp:* Staff engr, Collins Radio Co, 52-63; supvr commun res group, Advan Technol Dept, Autonetics Div, N Am Rockwell Corp, Anaheim, 63-; CHIEF SCIENTIST, HYCOM. *Mem:* Inst Elec & Electronics Engrs. *Res:* Digital communications; adaptive equalization for data modems; high frequency transmission; communication systems. *Mailing Add:* 18182 Romelle Ave Santa Ana CA 92705

MOTOYAMA, ETSURO K, b Japan, Apr 11, 32; US citizen. ANESTHESIOLOGY, RESPIRATORY PHYSIOLOGY. *Educ:* Chiba Univ, Japan, BS, 53; Chiba Univ Med Sch, Japan, MD, 57. *Prof Exp:* Res assoc pediat, Med Sch, Yale Univ, 64-66, asst prof anesthesiol & pediat, 66-70, assoc prof, 70-79; PROF ANESTHESIOL & PEDIAT, SCH MED, UNIV PITTSBURGH, 79-, VCHMN FOR SCI & RES, DEPT ANESTHESIOL, CCM, 85-; DIR, PULMONOLOGY PROG & ATTEND ANESTHESIOLOGIST & PEDIATRICIAN, CHILDREN'S HOSP, PITTSBURGH, 79- *Concurrent Pos:* Fel respiratory physiol, Harvard Med Sch, 62-64; attend anesthesiologist & pediatrician, Yale-New Haven Hosp, 66-; prin investr lung res ctr, Yale Univ, 71-77. *Mem:* Am Physiol Soc; Am Pediat Soc; Soc Pediat Res; Am Soc Anesthesiol; Asn Univ Anesthetists. *Res:* Developmental physiology of lung in fetus, infants and children; pediatric respiratory function in health and disease; pediatric anesthesiology. *Mailing Add:* Dept Anesthesiol & Pediat Sch Med Univ Pittsburgh One Children's Plaza Pittsburgh PA 15213

MOTSAVAGE, VINCENT ANDREW, b Scranton, Pa, May 10, 34; m 57; c 5. PHARMACY, PHYSICAL CHEMISTRY. *Educ:* Philadelphia Col Pharm, BS, 55; Temple Univ, MS, 57, PhD(pharm), 62. *Prof Exp:* Res assoc pharmaceut develop, Merck Sharp & Dohme Res Labs, Merck & Co, Inc, 57-59, pharm res, 61-63; asst instr pharm, Temple Univ, 59-60, res asst, 60-61; sr phys chemist, Avon Prod, Inc, 64-67, sect head chem res, 67-69; head, Dept Cosmetic & Toiletries Develop, 69-80, ASSOC DIR RES & DEVELOP, MENLEY & JAMES LABS, INC, SMITH KLINE & FRENCH LABS, 80- *Mem:* Am Chem Soc; Am Pharmaceut Asn; Soc Cosmetic Chemists. *Res:* Physical and colloid chemistry as applied to the research and development of cosmetics and pharmaceuticals. *Mailing Add:* 151 Gwynedd Manor Rd N Wales PA 19454

MOTSINGER, RALPH E, plant pathology, for more information see previous edition

MOTT, HAROLD, b Harris, NC, June 16, 28; m 59; c 1. ELECTRICAL ENGINEERING. *Educ:* NC State Univ, BEE, 51, MSEE, 53, PhD(elec eng), 60. *Prof Exp:* Asst elec eng, NC State Univ, 51-53, instr, 54-60; engr, Wright Mach Co, NC, 53-54; assoc prof, 60-64, PROF ELEC ENG, UNIV ALA, TUSCALOOSA, 64- *Concurrent Pos:* Consult, Troxler Elec Labs, NC, 56-59 & US Army Missile Res & Develop Command. *Mem:* Inst Elec & Electronics Engrs; Am Soc Eng Educ. *Res:* Antennas; radio wave propagation; electromagnetic theory; radar. *Mailing Add:* PO Box 870286 Tuscaloosa AL 35487-0286

MOTT, JACK EDWARD, b Hammond, Ind, May 4, 37; m 59; c 3. ENGINEERING PHYSICS. *Educ:* Univ Chicago, MS, 60; Northwestern Univ, PhD(physics), 67. *Prof Exp:* Assoc scientist, Bettis Atomic Power Div, Westinghouse Elec Co, 60-63; res asst physics, Northwestern Univ, Evanston, 63-67, res assoc, 67-68; from asst prof to assoc prof, Indiana Univ, 68-75; sr physicist, Energy Systs & Technol Div, Gen Elec Co, 75-84; sect mgr, Los Alamos Tech Asn, 84-85; vpres, Saratoga Eng Consults, 85-86; technol dir, EI Int, 86-90; PRES, ADVAN MODELING TECHN CORP, 90- *Mem:* Am Phys Soc; Sigma Xi; Am Nuclear Soc. *Res:* Fundamental particle research and teaching; nuclear physics research and teaching; nuclear reactor and energy systems research. *Mailing Add:* 191 11th St Idaho Falls ID 83404

MOTT, JOE LEONARD, b Linden, Tex, Apr 1, 37; m 60; c 2. ALGEBRA. *Educ:* E Tex Baptist Col, BS, 58; La State Univ, MS, 60, PhD(math), 63. *Prof Exp:* Asst prof math, Univ Kans, 63-65; assoc prof, 65-75, PROF MATH, FLA STATE UNIV, 75- *Concurrent Pos:* Vis prof, Mich State Univ, 73-74 & Univ Tenn, 78- *Mem:* Am Math Soc; Am Math Asn. *Res:* Ideal theory of commutative rings; partially ordered Abelian groups. *Mailing Add:* Dept of Math Fla State Univ Tallahassee FL 32306

MOTT, JULIAN EDWARD, b Lebanon, Mo, Aug 22, 29; m 54; c 2. NUCLEAR ENGINEERING. *Educ:* Univ Tenn, BS, 56; Univ Minn, PhD(aeronaut eng), 66. *Prof Exp:* Test engr, Pratt & Whitney Aircraft Co, 56-57; engr, Oak Ridge Nat Lab, 57-59; AT TECHNOL ENERGY CORP, KNOXVILLE, TENN. *Concurrent Pos:* Consult, Oak Ridge Nat Lab, 59-76, Oak Ridge Assoc Univs, 66-67, Oak Ridge Gaseous Diffusion Plant, 67-74; mgr mech eng, Technol for Energy Corp, 76- *Mem:* Am Nuclear Soc; Am Soc Eng Educ; Am Soc Mech Engrs. *Res:* Two-phase heat transfer and fluid mechanics; nuclear reactor safety and loss of coolant accidents; scientific use of mini-computers. *Mailing Add:* One Energy Ctr Lake Dr PO Box 22996 Knoxville TN 37933-0996

MOTT, NEVILL FRANCIS, b Leeds, Eng, Sept 30, 05. EXPERIMENTAL PHYSICS. *Educ:* Cambridge Univ, PhD(physics), 29. *Prof Exp:* Prof physics, Univ Bristol, 33-54; dir, H H Wills Phys Lab, 48-54; chmn bd dir, 70-75, PRES TAYLOR & FRANCIS LTD, 76- *Honors & Awards:* Nobel Prize in Physics, 77. *Mem:* Am Acad Arts & Sci. *Mailing Add:* Physics Dept Univ Cambridge Cambridge CB3 UHE England

MOTT, RALPH LIONEL, b Tooele, Utah, Nov 14, 37; m 57; c 1. PLANT TISSUE CULTURE, PLANT MORPHOGENESIS. *Educ:* Univ Utah, BS, 62, MS, 64; Cornell Univ, PhD(plant physiol & gen biochem), 69. *Prof Exp:* Teaching & res asst, plant physiol, Cornell Univ, Ithaca, NY, 63-66, NIH fel, 66-69, res assoc, 69-72; res assoc, 72-73, assoc prof, 73-79, PROF BOT, NC STATE UNIV, RALEIGH, 79- *Concurrent Pos:* Assoc ed, Plant Cell, Tissue & Organ Cult Int J, 82-85; chmn, Forest Biol Subcomt Biotechnol, Technol Asn Pulp & Paper Indust, 84-; panel mem, USDA Competitive Res Grant Off, SIBR, 85; dir, plant cell & tissue cult lab. *Mem:* Sigma Xi; Am Soc Plant Physiologists; Am Soc Hort Sci; Int Asn Plant Tissue Cult; Int Conifer Tissue Cult Asn (pres, 85-); Soc Gen Physiologists. *Res:* Fundamental morphogenesis and embryogenesis from cultured cells and application of tissue culture methods to crop improvement, conifer mass propagation and disease resistance. *Mailing Add:* Dept Bot Box 7612 NC State Univ Raleigh NC 27650

MOTT, THOMAS, b Oswego, NY, Feb 14, 26; m 77; c 7. MATHEMATICS. *Educ:* Union Col, AB, 50; Univ Pa, AM, 52; Pa State Univ, PhD, 67. *Prof Exp:* Jr mathematician, Cornell Aeronaut Lab, Buffalo, NY, 52-53; instr math, Pa State Univ, Erie, 53; instr, Clarkson Inst Technol, 53-55; instr, Pa State Univ, Hazleton, 55-57, University Park, 57-62; assoc prof, State Univ NY Col Fredonia, 62-67; PROF MATH, STATE UNIV NY COL BUFFALO, 67- *Concurrent Pos:* Math Asn Am lectr, 70- *Mem:* Math Asn Am. *Res:* Analysis; integration theory; limit theorems. *Mailing Add:* Dept Math State Univ NY Col 1300 Elmwood Ave Buffalo NY 14222

MOTTA, JEROME J, b Los Angeles, Calif, July 6, 33. MYCOLOGY. *Educ:* San Francisco State Col, AB, 58, MA, 64; Univ Calif, Berkeley, PhD(bot), 68. *Prof Exp:* Res plant pathologist, Univ Calif, Berkeley, 68-69; ASSOC PROF BOT, UNIV MD, COLLEGE PARK, 69- *Mem:* AAAS; Mycol Soc Am; Bot Soc Am. *Res:* Cytology and ultrastructure of fungi. *Mailing Add:* Dept of Bot Univ of MD College Park MD 20742

MOTTELER, ZANE CLINTON, b Wenatchee, Wash, July 4, 35; m 60; c 4. COMPUTER SCIENCE. *Educ:* Stanford Univ, BS, 57, MS, 62, PhD(math), 64; Mich State, MS, 81. *Prof Exp:* NSF fel, Univ Minn, 57-58 & Univ NMex, 58-59; res asst math, Los Alamos Sci Lab, Univ Calif, 58-60, staff mem, 60-65; from asst prof to assoc prof math, Gonzaga Univ, 65-72, chmn dept, 66-71; prof, dept math & comput sci, Mich Technol Univ, 72-82, chmn dept, 72-80; PROF COMPUT SCI, CALIF POLYTECH STATE UNIV, 82- *Concurrent Pos:* NSF vis scientist prog lectr, 60-61, 65-66; consult, Northwest Col & Univ Asn Sci, 70; nat lectr, Soc Indust & Appl Math, 70-72; vis lectr comp sci, Mich State Univ, 80-81; NASA-Am Soc Eng Educ fel, Langley Res Ctr, NASA, 82; sr analyst, Chevron Corp, 83 & IBM Corp, 84-85, Lawrence Livermore Nat Lab, 86- *Mem:* Inst Elec & Electronics Engrs Comput Soc; Asn Comput Mach. *Res:* Existence theory for non-linear elliptic partial differential equations of second order; quantitative and qualitative behavior of polynomials near roots; computer systems; compilers and programming languages; automata theory. *Mailing Add:* Dept Comput Sci Calif Polytech State Univ San Luis Obispo CA 93407

MOTTET, NORMAN KARLE, b Renton, Wash, Jan 8, 24; m 52; c 3. TERATOLOGY, ENVIRONMENTAL PATHOLOGY. *Educ:* Wash State Univ, Pullman, BS, 47; Yale Univ, New Haven, MD, 52. *Prof Exp:* Instr physiol, Yale Univ Sch Med, 51-52, instr path, 55-59; from asst prof to assoc prof path, 59-66, PROF PATH, UNIV WASH, SCH MED, 66-, PROF ENVIRON HEALTH, SCH PUB HEALTH & COMMUNITY MED, 82- *Concurrent Pos:* James Hudson Brown fel, Yale Univ, 49-50; fel Nat Found Infantile Paralysis, Strangeways Res Lab, 52-53; vis scientist, Strangeways Res Lab, Cambridge, 69-70, UN Environ Monitoring & Assessment Res Ctr, Univ London, Eng, 84-85; spec res fel, US Pub Health Serv, 69-70; mem, Int Comt Trace Metals, 85-, Int Working Groups WHO, 88- *Mem:* Am Asn Pathologists; Am Soc Clin Pathologist; Soc Toxicol; Teratology Soc. *Res:* To investigate the embryopathic effects of methylmercury, inorganic mercury and cadmium particularly as they affect the developing nervous system; also study of the brain biotransformation of methylmercury in the primate. *Mailing Add:* Dept Path SM-30 Univ Wash Seattle WA 98195

MOTTINGER, JOHN P, b Detroit, Mich, Nov 28, 38. CYTOGENETICS. *Educ:* Ohio Wesleyan Univ, BA, 61, Indiana Univ, PhD(cytogenetics), 68. *Prof Exp:* Instr, 67-68, asst prof, 68-75, ASSOC PROF BOT, UNIV RI, 75- *Mem:* AAAS; Genetics Soc Am. *Res:* Cytogenetics of maize and study of endosperm mutants in tissue culture. *Mailing Add:* Dept of Bot Univ RI Kingston RI 02881

MOTTLEY, CAROLYN, b Palestine, Tex, Oct 29, 47. PHYSICAL CHEMISTRY. *Educ:* Wayland Col, BS, 69; Univ NC, Chapel Hill, PhD(chem), 73. *Prof Exp:* Res assoc chem, Univ Ala, Tuscaloosa, 74-75; from asst prof to assoc prof, 75-81, PROF CHEM, LUTHER COL, 87- *Concurrent Pos:* Res chemist, Nat Inst Environ Health Sci, 80-81, 85-86 & 88; vis assoc prof, Med Col Wis. *Mem:* Am Chem Soc; Am Phys Soc. *Res:* Spin trapping of free radical metabolites of various xenobiotics. *Mailing Add:* Dept Chem Luther Col Decorah IA 52101-1042

MOTTMANN, JOHN, b Alsfeld, WGer, Apr 6, 44; US citizen. ASTRONOMY. *Educ:* Univ Calif, Los Angeles, BA, 66, MS, 67, PhD(astron), 72. *Prof Exp:* Astron, Aerospace Corp, 70-74; lectr astron, Santa Monica Community Col, 72-74; ASSOC PROF ASTRON, CALIF POLYTECH UNIV, 74- *Mem:* Am Astron Soc; Astron Soc Pac; AAAS. *Res:* Galaxies; radio properties of extra-galactic sources. *Mailing Add:* Dept of Physics Calif Polytech State Univ San Luis Obispo CA 93407

MOTTO, JEROME (ARTHUR), b Kansas City, Mo, Oct 16, 21. PSYCHIATRY. *Educ:* Univ Calif, AB, 48, MD, 51. *Prof Exp:* Intern, San Francisco Gen Hosp, 51-52; resident psychiat, Henry Phipps Psychiat Clin, Johns Hopkins Hosp, 52-55; sr resident, Langley Porter Neuropsychiat Inst, 55-56; from instr to asst prof, 56-64, lectr, 64-67, assoc clin prof, 67-69, assoc prof, 69-73, PROF PSYCHIAT, SCH MED, UNIV CALIF, SAN FRANCISCO, 73- *Concurrent Pos:* Attend psychiatrist, Langley Porter Neuropsychiat Inst, 56-; assoc dir psychiat consult serv, Univ Calif, San Francisco Med Ctr. *Res:* Clinical psychiatry. *Mailing Add:* Dept Psychiat Univ Calif Med Ctr San Francisco CA 94143

MOTTOLA, HORACIO ANTONIO, b Buenos Aires, Arg, Mar 22, 30; m 58; c 2. ANALYTICAL CHEMISTRY. *Educ:* Indust Tech, Indust Nat Sch, Arg, 49; Univ Buenos Aires, MS, 57, PhD(chem), 62. *Prof Exp:* Teaching asst chem, Univ Buenos Aires, 56-57, instr, 57-58, 60-63; res assoc, Univ Ariz, 63-64; asst prof, Elbert Covell Col, Univ of the Pac, 64-67; from asst prof to assoc prof, 67-75, PROF CHEM, OKLA STATE UNIV, 75- *Concurrent Pos:* Lectr, Univ Ariz, 66-67. *Mem:* Am Chem Soc; Sigma Xi. *Res:* Kinetic aspects of several analytical techniques; flow injection analysis; metal chelate photochromism (reactors for continuous flow analysis); enzyme and metal chelate immobilization; chemically modified electrodes. *Mailing Add:* Dept Chem Okla State Univ Stillwater OK 74075

MOTTS, WARD SUNDT, b Cleveland, Ohio, Oct 31, 24; m 51; c 2. GEOLOGY. *Educ:* Columbia Univ, BA, 49; Univ Minn, MS, 51; Univ Ill, PhD(geol), 57. *Prof Exp:* Geologist, US Bur Reclamation, 51-53; geologist, US Geol Surv, 53-60 & Okla Geol Surv, 60-61; ASSOC PROF GEOL, UNIV MASS, AMHERST, 61- *Concurrent Pos:* Asst, Univ Ill, 55-57. *Mem:* Fel Geol Soc Am; Am Geophys Union. *Res:* Hydrogeology; environmental geology; geomorphology; engineering geology. *Mailing Add:* Dept Geol Univ Mass Amherst MA 01002

MOTTUR, GEORGE PRESTON, b New York, NY, Jan 5, 50; m 68; c 2. NEW PRODUCT DEVELOPMENT, CEREAL CHEMISTRY. *Educ:* Long Island Univ, BS, 73; Univ Ore, PhD(molecular biol), 77. *Prof Exp:* Res assoc, dept biochem, Univ Va Med Sch, 78-81; res scientist, basic res dept, Frito-Lay, Inc, 81-82, proj mgr, new prod dept, 82-83; DIR RES & DEVELOP, SNACK FOODS DIV, BORDEN INC, 83- *Mem:* Am Asn Cereal Chemists; Inst Food Technologists; AAAS. *Res:* Development of new processes for snack food manufacture; new product formulation and consumer testing; potato cultivation, breeding and biochemistry; chemistry and functionality of cereal-derived ingredients. *Mailing Add:* Res Dept Borden Snacks-Wise Foods 228 Raseley St Berwick PA 18603

MOTTUS, EDWARD HUGO, b Eckville, Alta, June 12, 22; m 45; c 3. ORGANIC CHEMISTRY. *Educ:* Univ Alta, BSc, 49; Univ Ill, PhD(chem), 52. *Prof Exp:* Teacher pub schs, Can, 41-42; asst, Univ Ill, 49-51; fel, Nat Res Coun Can, 52-53; res chemist, 53-67, scientist, 67-71, SR SCI FEL, MONSANTO CO, 71- *Mem:* AAAS; Am Chem Soc; The Chem Soc; Sigma Xi. *Res:* Electrolytic reduction of bicyclic aminoketones; polarographic reduction of diketones; lycopodine; protopine; polyethylene, ionic ringopening polymerizations; polyethers; polyamides; catalysis; polymer syntheses. *Mailing Add:* 850 Claymont Dr Baldwin MO 63011

MOTULSKY, ARNO GUNTHER, b Fischhausen, Ger, July 5, 23; nat US; m 45; c 3. INTERNAL MEDICINE, MEDICAL GENETICS. *Educ:* Univ Ill, BS, 45, MD, 47. *Hon Degrees:* DSc, Univ Ill, 82. *Prof Exp:* Res assoc internal med, Sch Med, George Washington Univ, 52-53; from instr to assoc prof, 53-61, PROF MED & GENETICS, UNIV WASH, 61- *Concurrent Pos:* Clin investr, Army Med Serv Grad Sch, Walter Reed Army Med Ctr, DC, 52-53; attend physician, King County & Vet Admin Hosps, Seattle, 54-70; consult, Madigan Army Hosp, Tacoma, 55-74; Commonwealth Fund fel, Univ London, 57-58; Markle scholar, 57-62; mem subcomt transfusion probs, Nat Res Coun, 58-63; attend physician, Univ Wash Hosp, 59-; mem human ecol study sect, NIH, 61-65 & hemat study, 69-72; mem, US Panel Methods Eval Environ Mutagenesis & Carcinogenesis, Nat Inst Allergy & Infectious Dis, 72-76; mem, Nat Heart, Lung & Blood Inst Lipid Metab Adv Comt, 76-80; mem, Nat Res Coun Comt on Study of Nat Needs for Biomed & Behav Res Personnel, 77-80; elected mem founding bd, Am Bd Med Genetics, 79-82; mem President's Comn Study Ethical Probs Med & Biomed & Behav Res, 79-83; sci counr, Radiation Effects Res Found, 83-; fel, Inst Advan Study, Berlin, 84; chmn, OTA Adv Panel Determining Mutation Frequencies in Human Beings, 85; Recombinant DNA Adv Comt Working Group on Human Gene Ther, Nat Inst Allergy & Infectious Dis, 85-87; pres, VII Int Cong Human Genetics, Berlin, 86; chmn, Food & Nutrit Bd Comt on Diet & Health, Nat Acad Sci-Nat Res Coun, 86-89; ed, Am J Human Genetics, 69-75, Human Genetics, 69-, Progress in Med Genetics, 74- & Oxford Monographs on Med Genetics, 79-; assoc ed, Arteriosclerosis, 80-90. *Honors & Awards:* William M Allan Mem Award, Am Soc Human Genetics, 70; Alexander von Humboldt Award, 84; Sam Reino Int Prize Genetic Res, 88. *Mem:* Nat Acad Sci; Inst Med-Nat Acad Sci; Asn Am Physicians; fel Am Col Physicians; Am Fedn Clin Res; Am Acad Arts & Sci; Am Soc Human Genetics (pres, 77-78). *Res:* Role of genetic factors in disease etiology; genetics of coronary heart disease; hereditary hemolytic anemias; abnormal hemoglobins; genetics of drug reaction and response; human population genetics; molecular genetics of color vision abnormalities. *Mailing Add:* Div Med Genetics Univ Wash Sch Med Seattle WA 98195

MOTWANI, NALINI M, US citizen; m 67; c 2. HYBRIDOMA, MOLECULAR BIOLOGY. *Educ:* Wayne State Univ, PhD(microbiol), 76. *Prof Exp:* SR RES SCIENTIST, STROH-TECH, 83- *Concurrent Pos:* Vis scientist, Karolinskee Inst, 82. *Res:* Cell biology and immunology; expression of heterologous proteins; gene expression in mammalian cell culture (liver cells); blood substitute (recombinant). *Mailing Add:* Res & Develop Strohtech Detroit MI 48207

MOTZ, HENRY THOMAS, b St Louis, Mo, June 10, 23; m 47; c 3. NUCLEAR PHYSICS. *Educ:* Yale Univ, BS, 44, MS, 48, PhD, 49. *Prof Exp:* Physicist, Brookhaven Nat Lab, 49-56; physicist, 56-61, group leader res reactor group, 61-65, assoc physics div leader, 65-71, PHYSICS DIV LEADER, LOS ALAMOS SCI LAB, 71- *Concurrent Pos:* Res fel, Univ Zurich, 53-54; guest lectr, Netherlands-Norweg Reactor Sch, 63; secy & chmn nuclear cross sect adv comt & controlled thermonuclear res standing comt, USAEC; mem, Europ-Am Nuclear Data Comt; adv to US mem, Int Nuclear Data Comt, Int Atomic Energy Agency & US mem, Nuclear Energy Agency-Nuclear Data Comt, 73-; Univ Calif contractor to Energy Res & Develop Admin. *Mem:* Fel Am Phys Soc; NY Acad Sci. *Res:* Cyclotron bombardment; si reactions; slow neutron capture gamma rays. *Mailing Add:* Los Alamos Nat Lab MS A103 Los Alamos NM 87545

MOTZ, JOSEPH WILLIAM, b Binghamton, NY, Nov 11, 18; m 45; c 2. PHSYICS. *Educ:* Univ Wis, BS, 41; Cornell Univ, MS, 42; Ind Univ, PhD(physics), 49. *Prof Exp:* Physicist, Armour Res Found, 43-46; PHYSICIST, X-RAY DIV, NAT BUR STANDARDS, 49- *Mem:* Fel Am Phys Soc. *Res:* Radiation physics; photon and electron scattering processes. *Mailing Add:* 9701 Fields Rd Apt 1605 Gaithersburg MD 20878

MOTZ, KAYE LA MARR, b Bluffton, Ind, Aug 10, 32; m 59; c 2. INDUSTRIAL ORGANIC CHEMISTRY. *Educ:* Univ Colo, BA, 54; Univ Ill, PhD(chem), 58. *Prof Exp:* Instr chem, Mich State Univ, 58-59; res assoc, Univ Mich, 59-60; RES ASSOC, CONOCO INC, 60- *Mem:* Am Chem Soc. *Res:* oil additives; properties of petroleum fluids; tertiary oil recovery; surfactants; oil field chemicals. *Mailing Add:* Res & Develop Dept Conoco Inc Drawer 1267 Ponca City OK 74603

MOTZ, LLOYD, b Susquehanna, Pa, June 5, 10; m 34; c 2. ASTROPHYSICS, NUCLEAR PHYSICS. *Educ:* City Col NY, BS, 30; Columbia Univ, PhD(physics), 36. *Prof Exp:* Instr physics, City Col New York, 31-40; dir res & optical design, Dome Precision Corp, NY, 42-46; dir, Park Instrument Co, NJ, 46-49; from asst prof to assoc prof, 50-62, PROF ASTRON, COLUMBIA UNIV, 62- *Concurrent Pos:* Lectr, Columbia Univ, 35-49; adj prof, Polytech Inst Brooklyn, 50-; consult, AMF, Inc, Grumman Aerospace Corp, Polarad Electronics Corp & Razdow Labs; mem bd dirs, Geosci Instrument Corp & Thexon Corp. *Honors & Awards:* Award, Gravity Res Found, 60; Boris Pregel Award in Astron & Physics, NY Acad Sci, 72. *Mem:* AAAS; fel Am Phys Soc; Am Astron Soc; Royal Astron Soc; NY Acad Sci (pres, 70-71). *Res:* Internal constitution of stars; unified field theory; design of optical instruments; geometrical optics; structure of elementary particles; cosmology. *Mailing Add:* 140 Cabrini Blvd, # 86 New York NY 10033-3437

MOTZ, ROBIN OWEN, b New York, NY, Mar 9, 39; m 59, 90; c 3. PLASMA PHYSICS, INTERNAL MEDICINE. *Educ:* Columbia Univ, AB, 59, AM, 60, PhD(physics), 65, MD, 75. *Prof Exp:* Asst astron, Columbia Univ, 58-63; lectr physics, City Col New York, 63-65; asst prof, Stevens Inst Technol, 65-71; lectr, Columbia Univ, 71-75; med house staff mem, Presby Hosp, New York, 75-78; ASST PROF CLIN MED, COLUMBIA UNIV, 78- *Concurrent Pos:* Astron ed, Am Oxford Encyclop, 62-63; assoc ed, Am J Physics, 69-72; vchmn, Sect Phys Sci, NY Acad Sci, 67-69. *Mem:* Fel NY Acad Sci; AAAS; Am Phys Soc; AMA; Am col Physicians; Sigma Xi. *Res:* Magnetohydrodynamic drag; Alfven waves; plasma radiation; optical diagnostics; nuclear magnetic resonance of cancer cells; biological effects of electromagnetic waves; pharmacology; cholesterol. *Mailing Add:* 404 Tenafly Rd Tenafly NJ 07670

MOTZKIN, SHIRLEY M, b New York, NY, Jan 12, 27; m 52; c 3. ANATOMY, DEVELOPMENTAL BIOLOGY. *Educ:* Brooklyn Col, BS, 47; Columbia Univ, AM, 49; NY Univ, PhD(anat), 58. *Prof Exp:* Instr biol, Brooklyn Col, 47-52; instr histol, NY Univ, 51-59, asst prof, 59-66; assoc prof, 66-78, PROF BIOL, POLYTECH INST NEW YORK, 81-, DIR LIFE SCI, 66- *Concurrent Pos:* Adj instr & prof, Brooklyn Col, 52-; guest lectr, Guggenheim Dent Clin, 60-; partic, interdisciplinary prog, NIH basic res prog, 62-66. *Mem:* Sigma Xi; NY Acad Sci; Int Asn Dent Res. *Res:* Interactive effects of millimeter waves on living systems; effects of ionizing radiation on development; developmental teratologic studies of bone formation, palate and tooth development and tissue interactions using biochemical, embryological, histological, cytological, histochemical, cytochemical and radioisotopic techniques. *Mailing Add:* Dept Biol Polytech Inst NY 333 Jay St Brooklyn NY 11201

MOU, DUEN-GANG, b China, Nov 5, 48; m 72; c 4. BIOCHEMICAL ENGINEERING, INDUSTRIAL & ENVIRONMENTAL MICROBIOLOGY. *Educ:* Nat Taiwan Univ, BS, 70; Univ RI, MS, 75; Mass Inst Technol, PhD(biochem eng), 79. *Prof Exp:* Res engr, Food Indust Res Develop Inst, 71-72; develop engr, Indust Div, Bristol-Myers Co, 79-80; sr staff, Res Labs, Eastman Kodak Co, 80-86; DIR MICROBIOL, DEV CTR BIOTECHNOL, 87- *Concurrent Pos:* Adj prof, dept chem eng, Univ Rochester, 84-86; adj prof, dept biol, Rochester Inst Technol, 86; vis comt, dept chem eng, Univ Calif, Davis, 86; managing ed, Bioindust, Taiwan, Repub of China. *Mem:* Am Chem Soc; Am Inst Chem Engrs; Soc Indust Microbiol.

Res: Fermentation process development and scale-up; mass culture screening and strain improvement; lab instrumentation, automation and computer control; environmental biotechnology; fermentation production of antibiotics and amino acids; recombinant DNA fermentation. *Mailing Add:* Develop Ctr Biotechnol 81 Chang Hsing St Taipei Taiwan

MOUDGIL, BRIJ MOHAN, b Pataudi, Haryana, India, Aug 4, 45; m 73; c 3. APPLIED SURFACE COLLOID CHEMISTRY, MINERAL PROCESSING. *Educ:* Panjab Univ, India, 65; Indian Inst Sci, BEng, 68; Columbia Univ, MS, 72, ScD(mineral eng), 81. *Prof Exp:* Prod engr, Hindustan Steel Ltd, 68-70; proj engr, Ctr Res Mining & Mineral Resources, Univ Fla, 73-76; res engr, Occidental Res Corp, 76-80; assoc prof, 81-85, PROF MAT SCI & ENG, UNIV FLA, 85-; DIR, MINERAL RESOURCES RES CTR, 82- *Concurrent Pos:* Consult, var indust co, 74-; prin investr, Fla Inst Phosphate Res, 82-, NSF, 84- & Eng Found, 84-85. *Honors & Awards:* US Presidential Young Investr Award, 84. *Mem:* Am Inst Mining Metall & Petrol Engrs; Am Chem Soc; Am Inst Chem Engrs; Am Ceramics Soc; Sigma Xi; Fine Particles Soc. *Res:* Fine particles processing; applied colloid and surface chemistry; mineral engineering; solid-solid and solid-liquid separation technology; processing of energy minerals; hydrometallurgy. *Mailing Add:* 2101 NW 20th St Gainesville FL 32605

MOUDGIL, VIRINDER KUMAR, b Ludhiana, India. BIOCHEMICAL ENDOCRINOLOGY. *Educ:* Panjab Univ, BSc, 67; Banaras Hindu Univ, MSc, 69, PhD(zool), 72. *Prof Exp:* Res fel biochem, Banaras Hindu Univ, 69-71, asst res officer, 71-73, sr res fel, 73; fel molecular med, Mayo Clin, 73-76; from asst prof to assoc prof, 76-87, PROF BIOL SCI, OAKLAND UNIV, 87- *Concurrent Pos:* Prin investr, NIH, 78-; mem, Am Deleg Adult Endocrinol, People's Repub China, 83; chmn, Int Conf Steroid Receptors in Health and Dis; mem, NIH Study Sect; consult, UN Develop Prog, 90. *Honors & Awards:* Miriam Wilson Award, 86. *Mem:* Am Physiol Soc; Am Soc Biol Chemists; Endocrine Soc; Gerontological Soc Am; Sigma Xi; Am Asn Appl Psychol. *Res:* Investigating the mode of action of steroid hormones with particular emphasis on the role of nucleotides and phosphorylation in the process of activation of steroid hormone-receptor complexes. *Mailing Add:* Dept Biol Sci Oakland Univ Rochester MI 48309-4401

MOUFTAH, HUSSEIN T, b Alexandria, Egypt, Jan 1, 47; Can citizen; m 71; c 3. BROADBAND PACKET SWITCHING ARCHITECTURES. *Educ:* Univ Alexandria, BSc, 69, MSc, 72; Laval Univ, DSc(elec eng), 75. *Prof Exp:* Instr computer sci, Univ Alexandria, 69-72; teaching & res asst, Laval Univ, 73-75; postdoctoral fel digital systs, Univ Toronto, 75-76; sr digital systs engr, Adaptive Microelectronics Ltd, Toronto, 76, chief engr, 76-77; mem sci staff, Bell Northern Res, Ottawa, 77-79; from asst prof to assoc prof, 79-88, PROF ELEC ENG, QUEEN'S UNIV, KINGSTON, 88- *Concurrent Pos:* Assoc prof, Royal Mil Col, Kingston, 85; consult, Bell Northern Res, Ottawa, 86-87; prin investr, Telecommun Res Inst, Ont, 89-; vis expert, Transp Develop Ctr, Govt Can, 90. *Honors & Awards:* Eng Medal In Res & Develop, Asn Prof Engrs Ont, 89. *Mem:* Fel Inst Elec & Electronic Engrs; Can Soc Elec & Comput Eng; Can Soc Univ Teachers. *Res:* Computer engineering; high speed computer communication networks; digital systems and fault tolerant computing; computer aided modeling, analysis and design of computer and communication systems and networks. *Mailing Add:* Dept Elec Eng Queens Univ Kingston ON K7L 3N6 Can

MOUK, ROBERT WATTS, b Trenton, NJ, June 23, 40; m 65. POLYMER CHEMISTRY, SYNTHETIC ORGANIC CHEMISTRY. *Educ:* Wittenberg Univ, BS, 63; Bowling Green State Univ, MA, 67; Mich State Univ, PhD(chem), 70. *Prof Exp:* DEVELOP CHEMIST, ASHLAND CHEM CO, ASHLAND OIL INC, 69- *Mem:* Am Chem Soc. *Res:* Reaction injection molding; new polymers and intermediates; adhesives; emulsion polymerization. *Mailing Add:* 251 Chinkaobin Way Westerville OH 43081

MOULD, RICHARD A, b Reading, Pa, Mar 4, 27. THEORETICAL PHYSICS. *Educ:* Lehigh Univ, BS, 51; Yale Univ, MS, 55, PhD(physics), 57. *Prof Exp:* Asst prof, 57-64, master, Learned Hand Col, 70-78, ASSOC PROF PHYSICS, STATE UNIV NY STONY BROOK, 64- *Mem:* Am Asn Physics Teachers. *Res:* General relativity quantum theory of measurements. *Mailing Add:* Dept of Physics State Univ of NY Stony Brook NY 11794

MOULDEN, TREVOR HOLMES, b Leicester, Eng, Oct 13, 39. ENGINEERING. *Educ:* Imperial Col, London, BSc, 61; Univ London, MPhil, 68; Univ Tenn, PhD(aeronaut eng), 73. *Prof Exp:* Sci officer aeronaut eng, Nat Phys Lab, Eng, 61-66; res engr aeronaut eng, Lockheed-Georgia Co, 66-69; res fel, 69-73, asst prof, 73-78, ASSOC PROF AERONAUT ENG, SPACE INST, UNIV TENN, 78- *Concurrent Pos:* Vis prof, Chung-Cheng Inst, Repub China, 89. *Mem:* Am Acad Mech; Soc Indust Appl Math. *Res:* Fluid motion with special reference to viscous flow and transonic flows. *Mailing Add:* Univ Tenn Space Inst Tullahoma TN 37388

MOULDER, JAMES WILLIAM, b Burgin, Ky, Mar 28, 21; m 42; c 4. MICROBIOLOGY. *Educ:* Univ Chicago, SB, 41, PhD(biochem), 44. *Prof Exp:* Res assoc malaria, Off Sci Res & Develop Proj, 44-45, Logan fel, 46, instr biochem, Dept Microbiol & Biochem, 46-47, from asst prof to prof microbiol, 47-84, chmn dept, 60-69, PROF MOLECULAR GENETICS & CELL BIOL, UNIV CHICAGO, 84- *Concurrent Pos:* Fulbright scholar & Guggenheim fel, Oxford Univ, 52-53; Ciba lectr microbial biochem, 63; ed, J Infectious Dis, 57-68; vis prof microbiol, Western Ky Univ, 84. *Honors & Awards:* Lilly Award, 54. *Mem:* AAAS; Am Soc Biol Chem; Am Soc Microbiol; Am Acad Microbiol. *Res:* Biochemistry of intracellular parasitism. *Mailing Add:* 10349 E Edna Pl Tucson AZ 85748

MOULDER, JERRY WRIGHT, b Bowling Green, Ky, Sept 2, 42; m 67; c 1. PHYSICS. *Educ:* Western Ky Univ, BS, 64; Univ Tenn, PhD(physics), 70. *Prof Exp:* Asst prof physics, WVa Inst Technol, 70-75; asst prof, 75-77, assoc prof physics, Tri State Col, 77-; PROGRAM DIR & ASSOC PROF PHYSICS, TRANSYLVANIA UNIV. *Concurrent Pos:* NSF Acad Yr exten grant, WVa Inst Technol, 71-73. *Mem:* Am Asn Physics Teachers. *Res:* The interaction of low energy K mesons with nuclei. *Mailing Add:* Transylvania Univ Lexington KY 40508

MOULDER, PETER VINCENT, JR, b Jackson, Mich, Jan 26, 21; m 46; c 4. CARDIOTHORACIC VASCULAR SURGERY. *Educ:* Univ Notre Dame, BS, 42; Univ Chicago, MD, 45; Am Bd Surg, dipl, 54; Am Bd Thoracic Surg, dipl, 56. *Hon Degrees:* MS, Univ Pa, 72. *Prof Exp:* Intern surg, Univ Chicago Clins, 45-46, resident gen surg, 48-51; resident, Univ Ill, 52; from instr to prof surg, Univ Chicago Clins, 52-68; prof thoracic & cardiovasc surg & dir dept, Pa Hosp, Sch Med, Univ Pa, 68-72; prof thoracic & cardiovasc surg, Univ Fla, 72-80; PROF, THORACIC & CARDIOVASC SURG, SCH MED, DIR SURG RES LAB, MEM STAFF, TULANE MED CTR, TULANE UNIV SCH MED, 80- *Concurrent Pos:* Resident thoracic surg, Univ Chicago Clins, 52-53, chief resident surgeon, 53-54, secy, dept surg, 59-64; consult & lectr, Great Lakes Naval Hosp, Chicago, 59-68; consult, Philadelphia Naval Hosp, 68-; med investr, Vet Admin Hosp, 73-74 & 76-82. *Honors & Awards:* Alexander Vishneusky Medal, USSR, 66. *Mem:* Am Physiol Soc; Am Asn Thoracic Surg; Am Surg Asn; Am Col Surg; Soc Clin Surg. *Res:* Biochemical and physiological studies on the heart; pulmonary hypertension; myocardial hypertrophy; time series analysis of cardiovascular phenomena; computer science; ventricular assist device. *Mailing Add:* Div Thoracic & Cardiovasc Surg Sch Med Tulane Univ 1430 Tulane Ave New Orleans LA 70112-2699

MOULDS, WILLIAM JOSEPH, b Newton, Kan, Mar 7, 33; m 55; c 6. VEHICLE ACCIDENT INVESTIGATION & RECONSTRUCTION. *Educ:* Univ NMex, BS, 57; NMex State Univ, MS, 70. *Prof Exp:* Design engr, Sandia Labs, Albuquerque, NMex, 57-58; aeronaut & mech engr, Test Directorate, Air Force Spec Weapons Ctr, Kirtland AFB, 58-61, supvry aeronaut engr, Res Directorate, 61-63, sr aeronaut res engr, Electronics Div, Air Force Weapons Lab, 63-69, sr aeronaut res engr, Radiation Div & asst chief, Applications Br, 69-72, res gen engr & tech adv, Survivability & Vulnerability Br, 72, res gen engr & chief, spec study group, 73-74, chief, Eng Br & sr staff engr to Comdr, Air Force Weapons Lab, 74-77, dep div chief & sr staff engr to Lab Dir/Comdr, 77-79, chief, Tech Serv Div & Staff engr to Lab Comdr, 79-89; PVT CONSULT ENGR, 89- *Concurrent Pos:* Guest lectr, Dept Physics, USAF Acad, 70-71. *Mem:* Nat Soc Prof Engrs; Am Inst Aeronaut & Astronaut; Res Soc Am. *Res:* Six degree of freedom trajectory computer code that has the capability to determine the effects of an exoatmospheric nuclear explosion encounter on the vehicle motions (translational and rotational); code is not restricted to the principal axes (center of gravity offset) as it employs the complete inertia tensor. *Mailing Add:* 1401 Cardenas NE Albuquerque NM 87110

MOULE, DAVID, b Hamilton, Ont, Nov 17, 33; m 62; c 3. PHYSICAL CHEMISTRY, MOLECULAR SPECTROSCOPY. *Educ:* McMaster Univ, BSc, 58, PhD(chem), 62. *Prof Exp:* Asst res officer chem, Atomic Energy Comn Can, 64-66; from asst prof to assoc prof, 66-76, PROF CHEM, BROCK UNIV, 76- *Concurrent Pos:* NATO fel, 62-64. *Res:* spectroscopy of polyatomic molecules in excited electronic states; isotope equilibria. *Mailing Add:* Dept Chem Brock Univ St Catharines ON L2S 3A1 Can

MOULIS, EDWARD JEAN, JR, b Natchitoches, La, July 26, 40; m 67; c 2. MATHEMATICS. *Educ:* Harvard Univ, BA, 62; Univ Del, MS, 67, PhD(math), 71. *Prof Exp:* Instr math, US Navy Nuclear Power Sch, Md, 62-66; asst prof, Frostburg State Col, 71-75; ASST PROF MATH, US NAVAL ACAD, 75- *Mem:* Math Asn Am; Am Math Soc. *Res:* Complex analysis; univalent function theory; conformal mapping. *Mailing Add:* Dept Math US Naval Acad Annapolis MD 21402

MOULTON, ARTHUR B(ERTRAM), b Rochester, NH, Nov 29, 20. ELECTRICAL ENGINEERING. *Educ:* Univ Maine, BS, 43. *Prof Exp:* Electronic scientist, US Dept Defense, 46-52; sr electronic engr, Gen Dynamics/Convair, 52-55; assoc engr, Scripps Inst, Univ Calif, 55-56; asst prof elec eng, Univ Maine, 56-59; sr electronic engr, Litton Industs, 59; sr commun engr, Holmes & Narver, 60-61; engr, Lawrence Radiation Lab, Univ Calif, 61-63 & ACF Industs Inc & Link Group, Gen Precision Inc, 64-67, field eng dept, Systs Div, 67-68, engr, Link Div, Singer-Gen Precision, 68-81; engr, Nesea, 81-86; RETIRED. *Mem:* Inst Elec & Electronic Engrs. *Res:* Communications; digital computers; navigation; holder of two US patents. *Mailing Add:* Rte 5 PO Box 3 Scotland MD 20687-0003

MOULTON, BRUCE CARL, b Oneida, NY, Nov 6, 40; m 81; c 3. ENDOCRINOLOGY, BIOCHEMISTRY. *Educ:* Hamilton Col, AB, 62; Cornell Univ, MS, 65, PhD(endocrinol), 68. *Prof Exp:* Fel, Reprod Biol Training Prog, Col Med, Univ Nebr, Omaha, 68-70, res asst prof obstet-gynec & biochem, 70; asst prof obstet-gynec & biol chem, 70-75, assoc prof res obstet-gynec, biol chem & physiol, 75-84, PROF RES OBSTET-GYNEC, COL MED, UNIV CINCINNATI, 84- *Mem:* AAAS; Endocrine Soc; Soc Study Reprod; Am Physiol Soc; Soc Gynecol Invest. *Res:* Biochemical mechanisms of hormone action in reproductive physiology. *Mailing Add:* Dept Obstet & Gynec Univ Cincinnati Col Med Cincinnati OH 45267-0526

MOULTON, EDWARD Q(UENTIN), b Kalamazoo, Mich, Nov 16, 26; m 54; c 4. CIVIL ENGINEERING, EDUCATION ADMINISTRATION. *Educ:* Mich State Univ, BS, 47; La State Univ, MS, 48; Univ Calif, Berkeley, PhD(environ eng), 56. *Hon Degrees:* DSc, Univ Wittenberg, 79; LLD, Wilmington Col, 83, Xavier Univ, 83. *Prof Exp:* Instr civil eng, Mich State Univ, 47; asst prof, Auburn Univ, 48-50; lectr, Univ Calif, 51-54; from asst prof to assoc prof civil eng, Ohio State Univ, 54-64, prof eng mech, 64-66, asst dean, Grad Sch, 58-62, assoc dean, 62-64, assoc dean arts & sci & chmn, Dept Geod Sci, 62-64, assoc dean acad affairs & dean off campus educ, 64-66; pres, Univ SDak, 66-68; prof civil eng, Ohio State Univ, 68-79, secy bd trustees, 68-79, exec asst to pres, 68-69, vpres admin opers, 69-70, exec vpres, 70-73, vpres bus & admin, 73-79; chancellor, Ohio Bd Regents, 79-84, pres, Lake Erie Col, 85-86; RETIRED. *Concurrent Pos:* Exec vpres, Granston Securities, 84-85; pres & gen mgr, Columbus Symphony Orchestra, 86-88. *Mem:* Fel Am Soc Civil Engrs. *Res:* Water resources; effects of toxic ions of waste water recovery procedures; application of electrodialysis to the removal of saline and waste waters; acid mine drainage. *Mailing Add:* 1303 London Dr Columbus OH 43221

MOULTON, GRACE CHARBONNET, b New Orleans, La, Nov 1, 23; m 47; c 2. BIOPHYSICS, EXPERIMENTAL SOLID STATE PHYSICS. *Educ:* Tulane Univ, BA, 44; Univ Ill, MS, 48; Univ Ala, PhD(physics), 62. *Prof Exp:* Asst biophys, Univ Ill, 50-52; physicist, Argonne Cancer Res Hosp, Ill, 52; asst physics, Univ Ala, 59-61, asst prof, 61-65; from asst prof to assoc prof, 65-80, PROF PHYSICS, FLA STATE UNIV, 80- *Mem:* Sigma Xi; Am Phys Soc. *Res:* Radiation effects in materials of biological importance with emphasis on the mechanisms involved, as studied by electron spin resonance and electron nuclear double resonance. *Mailing Add:* Dept Physics Fla State Univ Tallahassee FL 32306

MOULTON, JAMES FRANK, JR, b Wash, DC, Nov 9, 21; m 44; c 2. NUCLEAR EXPLOSIONS & EFFECTS, TECHNICAL WRITING. *Educ:* Georgetown Univ, BS, 43. *Prof Exp:* Res physicist, Underwater & Air Explosion Effects, US Dept Navy, 43-46, sr res assoc, Shock Wave Phenomena in Air, Naval Ord Lab, 46-58, chief, Air-Ground Explosions Div, Naval Ord Lab, 58-65; chief, Naval Effects Br, Defense Atomic Support Agency, 65-67; chief, Aerospace Systs Div, Defense Nuclear Agency, 67-79; CONSULT, 79- *Concurrent Pos:* Sci consult, Energy Res & Develop Admin, Dept Navy, Armed Serv Explosives Safety Bd & Nat Mat Adv Bd; mem working group, S2-54 Atmospheric Blast Effects, Am Nat Standards Comt, 71- *Honors & Awards:* Newmann Award, 40. *Mem:* AAAS; Am Phys Soc; Am Inst Aeronaut & Astronaut. *Res:* Detection and measurement of blast and shock phenomena in high explosive and nuclear explosion environments; impulsive irradiation and response of aerospace systems materials and structures; research and development resource management; technical writing, editing and updating manuals for Department of Defense. *Mailing Add:* 4105 Glenrose St Kensington MD 20895-3718

MOULTON, PETER FRANKLIN, b Springfield, Mass, May 27, 46; m 70; c 2. SOLID STATE LASERS, SOLID STATE SPECTROSCOPY. *Educ:* Harvard Univ, BA, 68; Mass Inst Technol, MS, 71, PhD(elec eng), 75. *Prof Exp:* Physicist, Arthur D Little, Inc, 68-69; MEM STAFF, LINCOLN LAB, MASS INST TECHNOL, 75-; VPRES, SHWARTS ELECTRO-OPTICS INC. *Res:* Quantum electronics. *Mailing Add:* Schwarts Electro-Optics Inc 142 The Valley Rd Concord MA 01742

MOULTON, ROBERT HENRY, b Springfield, Mo, Oct 14, 25; m 54; c 4. TOXICOLOGY, ORGANS EFFECTED BY PHARMACEUTICALS. *Educ:* Drury Col, BS(chem) & BS(biol), 50; Mo Univ, MS, 52. *Prof Exp:* Assoc scientist, Colgate-Palmolive Co, 52; toxicologist, State Mo, 52-57; assoc prof physiol, Wash Univ, 57-60; VPRES, SCI ASSOCS, INC, 60- *Mem:* Am Inst Chemists; Soc Clin Toxicologists; Soc Toxicol; Am Acad Forensic Toxicol. *Res:* Transport across cell membranes; kidney tubule reabsorption; single organ toxicity. *Mailing Add:* 15 St Georges Dr St Louis MO 63123

MOULTON, RUSSELL DANA, b New York, NY, Feb 1, 65. THERMODYNAMICS & MATERIAL PROPERTIES. *Educ:* Univ NH, BA, 87. *Prof Exp:* STAFF SCIENTIST ELECTROCHEM, EIC LABS, 88- *Mem:* Electrochem Soc; Space Studies Inst. *Res:* Primary and secondary lithium batteries. *Mailing Add:* EIC Labs 111 Downey St Norwood MA 02062

MOULTON, WILLIAM G, b Waverly, Ill, Jan 4, 25; m 47; c 2. SOLID STATE PHYSICS, LOW TEMPERATURE PHYSICS. *Educ:* Western Ill State Col, BS, 46; Univ Ill, MS, 48, PhD(physics), 52. *Prof Exp:* Asst physics, Univ Ill, 46-51; from instr to asst prof, Univ Ill, Chicago, 51-56; from asst prof to prof, Univ Ala, 56-65; PROF PHYSICS, FLA STATE UNIV, 65-, ASSOC DIR, CTR MAT RES & TECHNOL, 86- *Mem:* Am Phys Soc. *Res:* Superconductivity; magnetic resonance; magnetic ordered states; solid state and low temperature physics. *Mailing Add:* Dept Physics Fla State Univ Tallahassee FL 32306

MOULY, RAYMOND J, b France, Nov 15, 22; US citizen; m 69; c 2. INTERNATIONAL TECHNOLOGY TRANSFER, GLASS & CERAMICS. *Educ:* Ecole Superieure D'Electricite, Ing Dipl, 47; Univ Paris, France, Lices Sci, 45. *Prof Exp:* Develop engr indust electronics, S A Phillips, France, 47-53; res engr electro, Que Iron & Titanium Corp, Can, 53-55; dept mgr glass, Corning Inc, 55-71; staff engr glass, PPG Industs Inc, 72-90; PRIN CONSULT, INT TECHNOL TRANSFER, MOULY ASSOCS, 90- *Mem:* Sr mem Inst Elec & Electronics Engrs; Nat Soc Prof Engrs; fel Instrument Soc Am. *Res:* Process, product and instrumentation and control systems development with particular emphasis on the glass and ceramics industries. *Mailing Add:* 4230 Centre Ave Pittsburgh PA 15213

MOUNCE, TROY G(ASPARD), chemical engineering, chemistry, for more information see previous edition

MOUNT, BERTHA LAURITZEN, b Valparaiso, Ind, Mar 26, 40; m 66; c 2. INTELLIGENT SYSTEMS, SOFTWARE SYSTEMS. *Educ:* Carleton Col, BA, 62; Northwestern Univ, MA, 64, PhD(topology), 70. *Prof Exp:* Coordr, Proj SEED, 75-80; instr computer sci, DePaul Univ, 81-83; partner, ACCESS Computer Clubs, 81-83; sr prog & analyst, MRCA Info Serv, 83-90; SYSTS OFFICER, FIRST CHICAGO CORP, 90- *Mailing Add:* 2705 Noyes St Evanston IL 60201

MOUNT, DAVID WILLIAM ALEXANDER, b Bromley, Eng, Jan 15, 38; m 60; c 2. GENETICS, MOLECULAR BIOLOGY. *Educ:* Univ Alta, BSc, 60; Univ Toronto, MA, 63, PhD(med biophys), 66. *Prof Exp:* Asst prof, 69-81, PROF MICROBIOL, COL MED, UNIV ARIZ, 81- *Concurrent Pos:* Fel genetics, Univ Alta, 67-68; USPHS fel molecular biol, Univ Calif, Berkeley, 68-69; lectr, Univ Ottawa, 66, Carleton Univ, 66-67 & Univ Alta, 67-68. *Mem:* AAAS; Genetics Soc Am; Am Soc Microbiol. *Res:* Genetics of bacteria and bacterial viruses; radiation biology; biophysics. *Mailing Add:* Dept Molecular & Cell Biol Univ Ariz Biosci W Tucson AZ 85721

MOUNT, DONALD I, b Miamisburg, Ohio, Sept 20, 31; div; c 2. FISH BIOLOGY. *Educ:* Ohio State Univ, BS, 53, MS, 57, PhD(zool, fish physiol), 60. *Prof Exp:* Fisheries res biologist, R A Taft Sanit Eng Ctr, Fed Water Pollution Control Admin, US Environ Protection Agency, 60-67, dir, Nat Water Qual Lab, 67-79, res aquatic biologist, 79-90; DIR AQUATIC TOXICOL PROGS, ASCI CORP, 90- *Mem:* AAAS; Soc Environ Toxicol & Chem. *Res:* Fish toxicology and physiology; effects of water pollution on fishes, especially the chronic effects of pollutants. *Mailing Add:* Asci Corp 112 E Second St Duluth MN 55805

MOUNT, ELDRIDGE MILFORD, III, b Springfield, Pa, Jan 22, 50; m 72; c 2. ORGANIC CHEMISTRY. *Educ:* West Chester State Col, BA, 72; Rensselaer Polytech Inst, ME, 76, PhD(chem eng), 79. *Prof Exp:* Asst res chem, Sterling Winthrop Res Inst, 72-74; process develop engr, Films Res & Develop, ICI Americas, 78-81; SR RES ASSOC, FILMS RES & DEVELOP, MOBIL CHEM, 81- *Concurrent Pos:* Mem bd dir, Extrusion Div, Soc Plastics Engrs, 81-, chmn, 90-91. *Honors & Awards:* Acad Achievement Medal, Am Inst Chemists, 72. *Mem:* Soc Plastics Engrs; Soc Rheology. *Res:* Mechanism of polymer melting in single screw extruders, both theoretical and experimental; film properties versus processing conditions and polymer composition; screw design. *Mailing Add:* Box 798 Macedon NY 14502-0798

MOUNT, GARY ARTHUR, b Bristow, Okla, Oct 8, 36; c 3. MEDICAL & VETERINARY ENTOMOLOGY, COMPUTER MODELING & SIMULATION. *Educ:* Okla State Univ, BS, 58, MS, 60, PhD(entomol), 63. *Prof Exp:* RES ENTOMOLOGIST, AGR RES SERV, USDA, 63- *Mem:* Entom Soc Am; Soc Vector Ecologists; Am Mosquito Control Asn. *Res:* Author of more than 100 scientific publications on control of arthropod pests and vectors of disease; computer modeling and simulation of population dynamics, disease transmission, and integrated management. *Mailing Add:* Med & Vet Entomol Res Lab USDA-ARS PO Box 14568 Gainesville FL 32604

MOUNT, KENNETH R, b Champaign, Ill, Apr 29, 33; m 66; c 2. MATHEMATICS. *Educ:* Univ Ill, BA, 54, MA, 55; Univ Calif, Berkeley, PhD(math), 60. *Prof Exp:* From instr to asst prof, 60-66, PROF MATH, NORTHWESTERN UNIV, ILL, 66- *Concurrent Pos:* NSF res grant, France, 64-65; NATO grant, 74. *Mem:* Am Math Soc; Econ Soc. *Res:* Algebraic geometry and commutative algebra; economics. *Mailing Add:* Dept Math Northwestern Univ Evanston IL 60208

MOUNT, LLOYD GORDON, b Central Square, NY, Mar 29, 16; m 41; c 3. INDUSTRIAL ORGANIC CHEMISTRY. *Educ:* Cornell Univ, AB, 37; Yale Univ, PhD(org chem), 40. *Prof Exp:* Res chemist, Calco Chem Div, Am Cyanamid Co, NJ, 40-41, group leader process develop, 45-51; chief, Pyrotech Res & Develop Sect, Picatinny Arsenal, 51-52; tech liaison mkt res, Chemstrand Corp, 52-55; commercial develop, Food Mach & Chem Corp, 55-58; head chem res & develop dept, Vitro Labs, 58-60; pres, Carnegies Fine Chem Kearny, NJ, 60-62; dir bus res, Thiokol Chem Corp, 62-65; vpres planning & develop, Clarkson Col Technol, 65-67; dir res & develop, Ruetgers-Nease Chem Co, Inc, 67-68, tech dir, 68-71, dir mfg, 71-72, vpres prod, 72-81, vpres regulatory affairs, 81-86; PRES, LGM INDUST CONSULTS, 86- *Concurrent Pos:* Lectr, Rutgers Univ, 46-47; vchmn, Pa Chem Indust Coun, 80-86. *Mem:* Am Chem Soc; Am Inst Chem; Am Soc Safety Engrs. *Res:* Sulfa compounds; vat dyes; military explosives; synthetic fibers; planning, development and marketing research; industrial health/safety. *Mailing Add:* L G M Indust Consults 1647 Cherry Hill Rd State College PA 16803-3217

MOUNT, MARK SAMUEL, b Crawfordsville, Ind, Nov 18, 40; m 63; c 2. PLANT PATHOLOGY, MOLECULAR BIOLOGY. *Educ:* Ill Wesleyan Univ, BS, 63; Mich State Univ, MS, 65, PhD(bot, plant path), 68. *Prof Exp:* Res assoc, Cornell Univ, 68-69; from asst prof to assoc prof, 69-82, PROF & DEPT HEAD PLANT PATH, UNIV MASS, AMHERST, 82- *Concurrent Pos:* NSF res grant, Univ Mass, Amherst, 70-72, NIH res grant, 73-76; Dept of Defence Res grant, 76-79, Allied Chem Co grant, 82-85, USDA competitive grant, 85-87; Gloeckner grant, 90. *Mem:* Am Phytopath Soc. *Res:* Physiology of plant disease development; nucleic acid metabolism in diseased plants; molecular genetics of host-pathogen interactions; transgenic plant development. *Mailing Add:* Dept of Plant Path Univ of Mass Amherst MA 01003

MOUNT, RAMON ALBERT, b Lohrville, Iowa, May 4, 39; m 62; c 2. ORGANIC CHEMISTRY. *Educ:* Ariz State Univ, BS, 61; Mich State Univ, PhD(org chem), 67; St Louis Univ, MBA, 72. *Prof Exp:* Chemist, Dow Chem Co, 61-64; res specialist, 67-75, sr res specialist, 75-80, sr res group leader, 81-83, MGR TECHNOL PLANNING, MONSANTO CO, 84- *Mem:* AAAS; Am Chem Soc; Sigma Xi. *Res:* Catalytic oxidation of hydrocarbons; heterogeneous catalysis; industrial process research in organic chemistry. *Mailing Add:* 15815 Kersten Ridge Ct Chesterfield MO 63017-8724

MOUNT, ROBERT HUGHES, b Lewisburg, Tenn, Dec 25, 31; m 61; c 2. VERTEBRATE ZOOLOGY. *Educ:* Auburn Univ, BS, 54, MS, 56; Univ Fla, PhD(biol), 61. *Prof Exp:* From asst prof to assoc prof biol, Ala Col, 61-66; from assoc prof to prof, 66-87, EMER PROF ZOOL, AUBURN UNIV, 87- *Honors & Awards:* W Kelly Mosley Environ Award, 81; Gopher Tortoise Coun Award, 87. *Mem:* Am Soc Ichthyologists & Herpetologists; Soc Study Amphibians & Reptiles. *Res:* Herpetology; natural history of reptiles and amphibians of southern United States. *Mailing Add:* Dept Zool & Wildlife Sci Auburn Univ Auburn AL 36849-5414

MOUNTAIN, CLIFTON FLETCHER, b Toledo, Ohio, Apr 15, 24; m 45; c 3. THORACIC SURGERY. *Educ:* Harvard Col, AB, 47; Boston Univ, MD, 54; Am Bd Surg, dipl, 62. *Prof Exp:* Dir dept statist res, Univ Boston, 47-50; consult & res analyst, Mass Dept Pub Health, 51-53; resident surgeon, Univ Chicago Clins, 54-58; instr surg, Univ Chicago, 58-59; from asst prof to assoc prof, 60-76, chmn, Dept Thoracic Surg, 76-86, PROF SURG, UNIV TEX SYST CANCER CTR, M D ANDERSON HOSP & TUMOR INST, 76-,

PROF SURG, UNIV TEX MED SCH AT HOUSTON, 87- *Concurrent Pos:* Fel surg physiol, Univ Chicago, 55-58; sr fel thoracic surg, Univ Tex M D Anderson Hosp & Tumor Inst, 59-60; prin investr & mem solid tumor study group, Cancer Chemother Nat Serv Ctr, NIH, 61-72, prin investr, Cancer Res Progs, Nat Cancer Inst, 63-; sect ed chest dis, Yearbk Cancer, 61-; sr investr & chmn prog biomath & comput sci, Univ Tex, 62-64; consult med sci adv comt, Systs Develop Corp, Calif, 63-68; mem, Am Joint Comt Cancer Staging & End Result Reporting; chmn, Task Force Lung & Esophageal Cancer & Nat Working Party Lung Cancer; chmn, Task Force Surg; many vis prof, lectr & consult worldwide. *Honors & Awards:* Kelsey-Leary Res Award, 80; Presidential Leadership Award, Int Asn Study Lung Cancer, 80; Medal, Nat Cancer Inst Japan, 85; Enrique M Garcia Mem Medal, Phillipine Col Surgeons. *Mem:* Am Asn Thoracic Surg; Am Col Chest Physicians; Am Radium Soc; Soc Surg Oncol; Soc Thoracic Surgeons; Sigma Xi; Int Asn Study Lung Cancer. *Res:* Thoracic malignant diseases; surgical techniques and adjunctive therapeutic programs in cancer chemotherapy and supervoltage irradiation; quantitative biology through biomathematics and computer sciences; study of radiant energy and its potential use in detecting and treating malignant disease. *Mailing Add:* M D Anderson Hosp & Tumor Inst Univ Tex 1515 Holcombe Dr Houston TX 77030

MOUNTAIN, DAVID CHARLES, JR, b Boston, Mass, Oct 3, 46. NEUROPHYSIOLOGY, BIOMEDICAL INSTRUMENTATION. *Educ:* Mass Inst Technol, BS, 68; Univ Wis, Madison, MS, 73, PhD(elec eng), 78. *Prof Exp:* Med engr, Mass Gen Hosp, 68-70; asst prof & asst res prof, 79-84, ASSOC PROF BIOMED ENG & ASSOC RES PROF OTOLARYNGOL, BOSTON UNIV, 84- *Mem:* AAAS; Asn Res Otolaryngol. *Res:* Physiology of the auditory system and development of related instrumentation. *Mailing Add:* Boston Univ 44 Cummington St Boston MA 02215

MOUNTAIN, RAYMOND DALE, b Great Falls, Mont, Mar 28, 37; m 61; c 3. THEORETICAL PHYSICS. *Educ:* Mont State Col, BS, 59; Case Western Reserve Univ, MS, 61, PhD(physics), 63. *Prof Exp:* Physicist, Nat Bur Standards, 63-68, chief statist physics sect, 68-82, physicist, 82-86; FEL, NAT INST STANDARDS & TECHNOL, 86- *Concurrent Pos:* Nat Acad Sci-Nat Res Coun fel, 63-65; John Simon Guggenheim Mem Found fel, 74. *Mem:* Am Phys Soc; Sigma Xi. *Res:* Statistical mechanics; physics of liquids; computer simulation. *Mailing Add:* Nat Inst Standards & Technol Gaithersburg MD 20899

MOUNTCASTLE, VERNON BENJAMIN, b Shelbyville, Ky, July 15, 18; m 45; c 3. NEUROSCIENCE. *Educ:* Roanoke Col, BS, 38; Johns Hopkins Univ, MD, 42. *Hon Degrees:* DSc, Roanoke Col, 68, Univ Pa, 76, Zurich, 83, Siena, 84, Northwestern, 85, Santiago, 91. *Prof Exp:* House officer surg, Johns Hopkins Hosp, 42-43; lieutenant jr grade, US Naval Amphibious Force, 43-46; res fel physiol, 46-48, from asst prof to prof, 48-84, dir dept physiol, 64-80, UNIV PROF NEUROSCI, SCH MED, JOHNS HOPKINS UNIV, 80-, DIR BARD RES LABS NEUROPHYSIOL, DEPT NEUROSCI, 80- *Concurrent Pos:* physiol study sect, NIH, 57-61, chmn, 58-61, physiol training comt, 57-61; vis lectr, Col France, 59; mem, vis comt, dept, physiol, Mass Inst Technol, 66-75, Neurosci Res Prog, 66-75, adv coun, Nat Eye Inst, 71-74, adv comt to dir, NIH, 84; dir, Neurosci Res Prog, Rockefeller Univ, 81-84, pres, Neurosci Res Found, 81-85; mem bd biol & med, NSF, 70-73; mem coun, Nat Eye Inst, 71-74; chmn sect physiol, Nat Acad Sci, 71-74; mem comn neurophysiol, Int Union Phys Sci. *Honors & Awards:* Lashley Prize, Am Philos Soc, 74; F O Schmitt Prize & Medal, 75; Shrrington Gold Medal, Royal Soc Med, Great Brit, 77; Horwitz Prize, 78; Gerard Prize, Soc Neurosci, 80; Helmholtz Prize & Medal, Cognitive Neurosci Inst, 82; Fyssen Int Prize, 83; Lasker Prize, 83; Nat Medal of Sci, 86; Zotterman Prize & Medal, Swedish Physiol Soc, 89; Fidia-Georgetown Medal & Award in Neurosci, 90; McGovern Medal & Prize, AAAS, 90. *Mem:* Nat Acad Sci; Nat Inst Med; Am Neurol Asn; Am Asn Neurol Surg; Soc Neurosci (pres, 70-71); Am Philosophical Soc. *Res:* Central nervous mechanisms in sensation and perception; study of brain mechanisms of higher functions. *Mailing Add:* Dept Neurosci Johns Hopkins Univ Sch Med Baltimore MD 21205

MOUNTCASTLE, WILLIAM R, JR, b Smyrna, Ga, Oct 31, 21; m 50; c 2. PHYSICAL CHEMISTRY, ANALYTICAL CHEMISTRY. *Educ:* Ga Inst Technol, BS, 43; Univ Ala, MS, 56, PhD(chem), 58. *Prof Exp:* Rubber chemist, Goodyear Tire & Rubber Co, Ala, 43-53; chem engr, Southeastern Exp Sta, US Bur Mines, 54-56; instr chem, Univ Ala, 57-58; assoc prof analytical & phys chem, Birmingham-Southern Col, 58-65, prof, 65-66; ASST PROF CHEM, AUBURN UNIV, 66- *Concurrent Pos:* Consult res ctr, Med Col, Univ Ala, 58-60; participation contract, Oak Ridge Assoc Univs, 61-; res assoc, Union Carbide Nuclear Corp, 63; dir, NSF-Undergrad Res Chem, 64-66. *Mem:* Am Chem Soc; Sigma Xi. *Res:* Solvent extraction using phenyl phosphate diester and applications of electrochemical and spectrographic methods to this study; coulometry; spectroscopy. *Mailing Add:* 776 Heaod Auburn AL 36830

MOUNTFORD, KENT, b Plainfield, NJ, July 23, 38; m 71. MARINE ECOLOGY. *Educ:* Rutgers Univ, BS, 60, MS, 69, PhD(bot), 71. *Prof Exp:* Res asst bur biol res, Rutgers Univ, 67-71, teaching asst bot & zool, 69-70; asst cur, Benedict Estuarine Lab, Acad Natural Sci, Philadelphia, 71-78; res assoc, Chesapeake Biol Lab, Univ Md, 78-79; mgr, Aquatic Toxicol Biospherics, Inc, 79-80; environ scientist qual assurance, Washington, DC Govt, 80-84; coordr environ monitoring, 84-87, SR SCIENTIST, CHESAPEAKE BAY PROG, US ENVIRON PROTECTION AGENCY, 87- *Concurrent Pos:* Lectr, var cols, univs & co; proj officer, state & fed govt. *Mem:* Am Soc Limnol & Oceanog; Am Littoral Soc; Sigma Xi; Estuarine Res Fedn; Atlantic Estuarine Res Soc. *Res:* Human estuary interactions, field and lab studies on ecology of estuarine plankton systems, multi-media public information programs. *Mailing Add:* Barnegat Assocs 10200 Breeden Rd Lusby MD 21403

MOUNTJOY, ERIC W, b Calgary, Alta, Nov 28, 31; m 58. GEOLOGY, STRUCTURAL. *Educ:* Univ BC, BASc, 55; Univ Toronto, PhD(stratig, struct geol), 60. *Prof Exp:* Tech officer field geol, Geol Surv Can, 57-60, geologist, 60-63; from asst prof to assoc prof, 63-74, PROF SEDIMENTATION STRATIG, McGILL UNIV, 74- *Concurrent Pos:* Assoc ed, Geol Soc Am, 84-89, Can Soc Petrol Geologists, 89-; assoc ed, Sedimentary Geol, 90- *Honors & Awards:* Douglas Medal, Can Soc Petrol Geol, 86. *Mem:* Soc Econ Paleont & Mineral (vpres, 78-79); Am Asn Petrol Geol; fel Geol Soc Am,; fel Geol Asn Can; Int Asn Sedimentologists; Can Soc Petrol Geologists. *Res:* Sedimentation; structural and regional geology of Canadian Rocky Mountains; devonian reef complexes; carbonate sedimentology and diagenesis, dolomitization; recent carbonates; Proterozoic and early Paleozoic sedimentation. *Mailing Add:* Dept of Geol Sci McGill Univ 3450 University St Montreal PQ H3A 2A7 Can

MOUNTS, RICHARD DUANE, b San Diego, Calif, Nov 15, 41; m 75; c 2. ANALYTICAL CHEMISTRY. *Educ:* Wheaton Col, Ill, BS, 64; Ariz State Univ, MS, 68; Univ Ariz, PhD(analytical chem), 74. *Prof Exp:* Res chemist, J T Baker Chem Co, 67-69 & Grefco, Inc, 69-70; asst prof chem, Wake Forest Univ, 74-75; ASST PROF CHEM, FLA INST TECHNOL, 75- *Mem:* Am Chem Soc. *Res:* Voltammetric methods of analysis and electronic modules for instruction in analytical chemistry instrumentation. *Mailing Add:* Dept of Chem Fla Inst of Technol 150 W Univ Blvd Melbourne FL 32901

MOUNTS, TIMOTHY LEE, b Peoria, Ill, Sept 14, 37; m 58; c 3. AGRICULTURAL CHEMISTRY, RADIOCHEMISTRY. *Educ:* Bradley Univ, BS, 59, MS, 68. *Prof Exp:* Res chemist, 57-75, leader edible oils prod & processes, 75-80, chief, Oilseed Crops Lab, 80-85, res leader, Vegetable Oil Res, Northern Regional Res Ctr, 85-90, RES LEADER, FOOD QUAL & SAFETY RES, NAT CTR AGR RES, ARS/USDA, 90- *Concurrent Pos:* Air Nat Guard, 66-90. *Honors & Awards:* Bond Award, Am Oil Chemists Soc, 69 & 71; Chevreul Medal, French Oil Chemists Soc, 85; Res Award, Am Soybean Asn, 87; Bailey Award, Am Oil Chemists Soc. *Mem:* Am Oil Chemists Soc(pres, 88-89); Am Chem Soc. *Res:* Development of improved edible oil products and processes so as to maintain a safe and nutritious food supply; hydrogenation and refining of edible oils, organoleptic evaluation of oils for flavor and stability, oils from damaged soybeans; assess genetically modified vegetable oil for end-use performance. *Mailing Add:* 6802 N Wilshire Dr Peoria IL 61614

MOURAD, A GEORGE, b Bludan, Syria, Nov 6, 31; US citizen; m 58; c 1. GEODESY, PETROLEUM ENGINEERING. *Educ:* Ohio State Univ, BSc, 57, MSc, 59. *Prof Exp:* Res asst gravity & geod, Inst Geod, Ohio State Univ Res Found, 56-59, res assoc, 59-62; sr engr, NAm Aviation, 62-64; res geodesist, 64-66, SR GEODESIST, BATTELLE MEM INST, 66-, PROG DIR MARINE GEOD, 68-, PROJ MGR GEOD & OCEAN PHYSICS, 72- *Concurrent Pos* Vchmn navig subcomt antisubmarine warfare, Nat Security Indust Asn, 70-; chmn spec study group on marine geod, Int Asn Geod, 70-, chmn comt marine geod, 75-79. *Mem:* Marine Technol Soc; Am Geophys Union; Am Inst Navig. *Res:* Satellite applications to earth and ocean dynamics disciplines; satellite altimetry for determining mean sea level; radar techniques for sea state measurements; satellite interferometry techniques for navigation; traffic control; data transfer; search and rescue applications. *Mailing Add:* 3680 Dublin Rd Columbus OH 43221

MOURITSEN, T(HORVALD) EDGAR, b Waukegan, Ill, Aug 13, 26; m 60; c 3. MECHANICAL ENGINEERING. *Educ:* Univ Ill, BS, 48; Univ Tex, MS, 54. *Prof Exp:* Propulsion & flight test engr, Gen Dynamics Corp, 50-53; res asst, Univ Tex, 53-54; propulsion engr, Chance Vought Aircraft, Inc, 54-59, res specialist, Vought Res Ctr, 59-63, sr specialist, Vought Corp, LTV Corp, 63-85; LEAD ENGR, FT WORTH DIV, GEN DYNAMICS, 86- *Mem:* Am Soc Mech Engrs. *Res:* Aerospace propulsion and environmental control; heat transfer and fluid mechanics; infrared counter measures. *Mailing Add:* 4739 Harvest Hill Rd Dallas TX 75244-6516

MOURNING, MICHAEL CHARLES, b Jerseyville, Ill, Oct 6, 40; m 67; c 2. CHEMISTRY. *Educ:* Univ Ill, Urbana, BS, 63; Univ NC, Chapel Hill, PhD(chem), 68. *Prof Exp:* Chemist, GAF Corp, 68-80; CHEMIST, ANITEC IMAGE CORP, 80- *Mem:* Am Chem Soc; Soc Photog Sci & Eng. *Res:* Organic synthesis. *Mailing Add:* 713 Imperial Woods Dr Vestal NY 13850-2513

MOURSUND, DAVID G, b Eugene, Ore, Nov 3, 36; m 61, 89; c 4. COMPUTER SCIENCE, MATHEMATICS EDUCATION. *Educ:* Univ Ore, BA, 58; Univ Wis, MS, 60, PhD(math), 63. *Prof Exp:* From asst prof to assoc prof math, Mich State Univ, 63-67; res assoc, Comput Ctr, Univ Ore, 67-70, head dept computer sci, 69-75, from assoc prof to prof computer sci, 67-86, PROF, COL EDUC, UNIV ORE, 86- *Mem:* Int Soc Technol Educ; Asn Comput Mach. *Res:* Computers in education, with major emphasis upon uses of computers in pre-college education; computer literacy. *Mailing Add:* Int Soc Technol Educ 1287 Agate St Eugene OR 97403

MOURY, JOHN DAVID, b Johnson City, Tenn, Sept 21, 60. VERTEBRATE MORPHOGENESIS, COMPARATIVE VERTEBRATE ANATOMY. *Educ:* East Tenn State Univ, BS, 81, MS, 83; Univ Tex, Austin, PhD(zool), 88. *Prof Exp:* Instr biol & anat, Dept Biol Sci, ETenn State Univ, 90-91; instr anat, 91, POSTDOCTORAL FEL, DEPT ENVIRON POP & ORGANISMIC BIOL, UNIV COLO, 91- *Concurrent Pos:* Lectr biol & physiol, Dept Zool, Univ Tex, 88-90. *Mem:* AAAS; Am Soc Zool; Soc Develop Biol; Am Asn Anatomists. *Res:* Examination of cell movements and interactions during vertebrate neurulation; formation migration and derivatives of the neural crest; tissue interactions during morphogenesis; amphibian skeletal anatomy. *Mailing Add:* Dept EPO Biol Univ Colo Boulder Boulder CO 80309-0334

MOUSCHOVIAS, TELEMACHOS CHARALAMBOUS, b Famagusta, Cyprus, Dec 29, 45. THEORETICAL ASTROPHYSICS, MAGNETOHYDRODYNAMICS. *Educ:* Yale Univ, BS, 68; Univ Calif, Berkeley, PhD(physics), 75. *Prof Exp:* Res assoc astrophys, Princeton Univ, 75-76; spec proj scientist solar physics, Nat Ctr Atmospheric Res, 76-77; ASST PROF PHYSICS & ASTRON, UNIV ILL, URBANA-CHAMPAIGN, 77- *Concurrent Pos:* Prin investr, NSF, 78-80; proj dir, NASA, 79-81. *Honors & Awards:* Trumpler Award, Astron Soc Pac, 77. *Mem:* Am Astron Soc; Int Astron Union. *Res:* Interstellar gas dynamics; star formation; galactic structure; dynamics of solar corona. *Mailing Add:* Dept Physics & Astron Univ Ill 1011 W Springfield Urbana IL 61801

MOUSHEGIAN, GEORGE, b Detroit, Mich, Jan 19, 23; m 52; c 3. PHYSIOLOGICAL PSYCHOLOGY. *Educ:* Wayne State Univ, BS, 47, MA, 51; Univ Tex, PhD, 57. *Prof Exp:* Res scientist, Defense Res Lab, Univ Tex, 56-59; res fel hearing, neurophysiol & psychol, Walter Reed Army Inst Res, 59-64; prof physiol psychol, Lab Sensory Commun, Syracuse Univ, 64-68; actg dean & actg dir, Sch Human Develop & Callier Ctr, 76-77, DIR RES, CALLIER CTR COMMUN DIS, UNIV TEX, DALLAS, 68-, PROF HUMAN DEVELOP, 75-, DEAN & DIR, SCH HUMAN DEVELOP & CALLIER CTR, 77- *Concurrent Pos:* Adj prof, Dept Physiol, Univ Tex Southwest Med Br, 69- *Mem:* AAAS; Am Psychol Asn; Acoust Soc Am; Am Physiol Soc; Soc Neurosci. *Res:* Electrophysiological study of responses from the brain stem to acoustic stimulation, using micro and macro electrodes; study of human responses to sounds; neurophysiology; psychophysics. *Mailing Add:* Sch Human Develop Callier Ctr Univ Tex 1966 Inwood Rd Dallas TX 75235

MOUSSOURIS, HARRY, CYTOPATHOLOGY. *Educ:* State Univ NY, Downstate Col Med, MD, 75. *Prof Exp:* CHIEF CYTOPATHOLOGY, ST VINCENT'S HOSP, 84- *Res:* Thin needle aspiration biopsy interpretation. *Mailing Add:* Dept Path St Vincent's Hosp 153 W 11th St New York NY 10011

MOUSTAKAS, THEODORE D, b Greece, Jan 28, 40; m 74; c 2. SOLID STATE PHYSICS, MATERIAL SCIENCE. *Educ:* Univ Salonika, Greece, BS, 64; Columbia Univ, NY, MPhil, 74, PhD(solid state sci), 74. *Prof Exp:* Res fel amorphous semiconductors, Harvard Univ, 74-77; SR PHYSICIST SOLAR ENERGY, EXXON RES & ENG CO, 77- *Concurrent Pos:* Fel, IBM Corp, 74-75. *Mem:* Am Phys Soc; Sci Res Soc NAm. *Res:* Optical and electronic properties of amorphous semiconductors (chalcogenide glasses and tetrahedrally corrdinated); photovoltaic studies of amorphous silicon and related materials. *Mailing Add:* Exxon Res Co PO Box 101 Florham Park NJ 07932-0101

MOUW, DAVID RICHARD, b Carlisle, Pa, Aug 22, 42; c 2. PHYSIOLOGY. *Educ:* Hope Col, BA, 64; Univ Mich, Ann Arbor, PhD(physiol), 69. *Prof Exp:* Instr biol, Hampton Inst, 67-68; NIH fel exp physiol, Howard Florey Labs Exp Physiol, Melbourne, Australia, 70-71; from asst prof to prof physiol, Univ Mich, Ann Arbor, 71-81. *Concurrent Pos:* Fogarty Int fel, Howard Floney Inst, Melbourne, 77-78. *Mem:* Am Physiol Soc; AAAS. *Res:* Central nervous system control of electrolyte metabolism. *Mailing Add:* 104 Keller Heights Waynesville NC 28786

MOVAT, HENRY ZOLTAN, b Temesvar, Romania, Aug 11, 23; nat Can; m 56; c 3. PATHOLOGY. *Educ:* Innsbruck Univ, MD, 48; Queen's Univ, Ont, MSc, 54, PhD, 56; Royal Col Physicians & Surgeons Can, cert path, 59; FRCP(C), 67. *Prof Exp:* from asst prof to assoc prof path, 57-65, mem, Inst Immunol, 71-84, head, Div Exp Path, 68-80, PROF PATH, UNIV TORONTO, 65-, PROF IMMUNOL, 84- *Concurrent Pos:* Career investr, Med Res Coun, 60- *Mem:* Am Asn Path; Am Asn Immunol; Soc Exp Biol & Med; Int Acad Path; Am Soc Pharmacol Exp Ther; Can Soc Immunol. *Res:* Acute inflammatory reaction; chemical mediators of acute inflammation and hypersensitivity; disseminated intravascular coagulation. *Mailing Add:* Dept Path Med Sci Bldg Univ Toronto Toronto ON M5S 1A8 Can

MOVIUS, WILLIAM GUST, b Portland, Ore, Jan 15, 43. INORGANIC CHEMISTRY. *Educ:* Univ Ore, BA, 65; Pa State Univ, PhD(chem), 68. *Prof Exp:* Fel, Univ Calif, San Diego, 68-69; ASST PROF CHEM, KENT STATE UNIV, 70- *Res:* Oxidation-reduction reactions, especially those involving uncommon oxidation states; coordination compounds in nonaqueous electrolyte solutions, especially those incompatible with water. *Mailing Add:* Dept of Chem Kent State Univ Kent OH 44242

MOVSHON, J ANTHONY, b New York, NY, Dec 10, 50; m 75; c 2. NEUROPHYSIOLOGY, PSYCHOPHYSICS. *Educ:* Univ Cambridge, BA, 72, MA, 76, PhD(psychol), 75. *Prof Exp:* From asst prof to assoc prof pyschol, 75-84, dir, Ctr Neural Sci, 87-91, PROF NEURAL SCI & PSCYHOL, NY UNIV, 84-, INVESTR, HOWARD HUGHES MED INST, 91- *Honors & Awards:* Young Investr Award, Soc Neurosci, 85. *Mem:* Soc Neurosci; Asn Res Vision Opthamol. *Res:* Neurophysiology and psychophysics of vision and visual development. *Mailing Add:* Ctr Neural Sci NY Univ Six Washington Pl New York NY 10003

MOW, C(HAO) C(HOW), b Nanking, China, Apr 28, 30; US citizen; m 54; c 4. APPLIED MECHANICS, MATHEMATICS. *Educ:* Rensselaer Polytech Inst, BME, 53, MS, 56, PhD(appl mech), 59. *Prof Exp:* Teaching asst appl mech, Rensselaer Polytech Inst, 53-56, instr, 56-59; chief stress analysis unit, Watervliet Arsenal, 59-60; mem staff, Mitre Corp, 61-62, sub-dept head appl mech, 62-63; mem staff, Rand Corp, 63-65, mech group leader eng sci dept, 65-72, dep dept head, 72-77, sr staff mem, 77-78; CHMN BD & CHIEF EXEC OFFICER, CENTURY WEST DEVELOP, INC, 78- *Concurrent Pos:* Lectr, Univ Calif, Los Angeles, 64 & Univ Southern Calif, 68. *Mem:* Am Soc Mech Engrs; NY Acad Sci. *Res:* Shell stability; stress wave propagation and scattering phenomena; bio-mechanics; environmental impact of transportation systems; electricity demand forecast methodology. *Mailing Add:* 20541 Roca Chica Malibu CA 90265

MOW, MAURICE, b China, June 24, 40; US citizen; m 66; c 2. CIVIL ENGINEERING, APPLIED MECHANICS. *Educ:* Rensselaer Polytech Inst, BCE, 63, MS, 64, PhD(appl mech), 68. *Prof Exp:* Tech staff struct mech, TRW Inc, 68-71; vpres, Doch Corp, 71-73; partner, Design & Planning Assoc, 73-76; asst prof eng technol, Ulster Co Community Col, 75-76; asst prof civil technol, Univ Maine, 76-78; ASSOC PROF CIVIL ENG, CALIF STATE UNIV, CHICO, 78- *Mem:* Am Soc Civil Engrs; Am Soc Eng Educrs; Sigma Xi. *Res:* Marketing freshwater resources. *Mailing Add:* Dept of Civil Eng Calif State Univ Chico CA 95929

MOW, VAN C, b China, Jan 10, 39; US citizen; m 76; c 2. MECHANICS, BIOMEDICAL ENGINEERING. *Educ:* Rensselaer Polytech Inst, BAE, 62, PhD(mech), 66. *Prof Exp:* Asst prof mech, Rensselaer Polytech Inst, 66-67; vis mem electromagnetic div, Courant Inst Math Sci, NY Univ, 67-68; mem tech staff, Bell Tel Labs, 68-69; from assoc prof to prof mech, Rensselaer Polytech Inst, 69-82, dir, Biomech Res Lab, 71-86, Clark & Crossan prof eng, 82-86; PROF MECH ENG & ORTHOP BIOENG, COLUMBIA UNIV, 86-; DIR, NY ORTHOP HOSP RES LAB, COLUMBIA-PRESBY MED CTR, 86- *Concurrent Pos:* Consult, Mech Technol, Inc, 66-67; various grants, NIH & NSF, 70-; NATO sr fel, 76 & 78; adj prof orthop, Albany Med Col, 77-89; lectr, Harvard Med Col, 78-86; chmn prog comt, Bioeng Div, Am Soc Mech Engrs, 79-80, chmn div, 84-85; assoc ed, J Biomech Eng, 79-86; chmn, First Gordon Res Conf Bioeng & Orthop Sci, 80, Orthop Musculo Dis Study Sect, NIH, 82-84, rev comt, Bioeng Rehab Disabled Prog, NSF, 88, subcomt Govt Rels, US Coun Biomech, 88- & GMA-1 AHR-MA Spec Study Sect, 90; mem, Appl Physiol Orthop Study Sect, NIH, 80-82, Res Grt Rev Comt, Orthop Res Ed Found, 85-, Int Steering Comt, First World Cong Biomech, 86-90, Sec World Cong Biomech, 91-, Nat Adv Bd Mat, 90-91 & spec study sects, Nat Inst Aging & NIAMDS, 91; hon prof, Chengdu Univ Sci & Technol, 81, Shanghai Univ Sci & Technol, China, 83 & Shanghai Jiao Tong Univ, 87; consult ed, Springer-Verlag, 85-; consult, Hong Kong Poly Bioeng Prog, 89-; chair, US Nat Comt Biomech, 91-94. *Honors & Awards:* Melville Medal, Am Soc Mech Engrs, 82 & H R Lissner Award, 87; Alza Distinguished Lectr, Biomed Eng Soc, 87. *Mem:* Nat Acad Sci; Am Asn Univ Prof; AAAS; fel Am Soc Mech Engrs; Am Soc Biomech; Am Acad Mech; Am Acad Orthop Surgeons; Am Phys Soc; Sigma Xi; Orthop Res Soc (pres, 82-83). *Res:* Continuum mechanics; classical elasticity and thermoelasticity theory; fluid mechanics; applied mathematics; lubrication biomechanics; biomechanics of synovial joints; study of mechanical processes in various degenerative arthritic diseases; mixture theory; finite element analysis; analytical stereophotogrammetry; biomacromolecules; experimental mechanics; computer aided surgery; experimental orthopaedic surgery. *Mailing Add:* Ortho Res Lab Rm 1412 Columbia Univ 630 W 168th St New York NY 10030

MOWAT, J RICHARD, b Honolulu, Hawaii, Aug 29, 43; US citizen; m 63; c 2. EXPERIMENTAL ATOMIC PHYSICS. *Educ:* Univ Calif, Berkeley, AB, 66, PhD(physics), 69. *Prof Exp:* Res assoc physics, Brandeis Univ, 69-72; res assoc physics, Univ Tenn, 72-73, asst prof, 73-74; asst prof physics, City Col New York, 74-76; from asst prof to assoc prof, 76-89, PROF PHYSICS, NC STATE UNIV, 89- *Concurrent Pos:* Invited assignee, Oak Ridge Nat Lab, 72-74; res collabr, Brookhaven Nat Lab, 74-76; staff scientist, Lawrence Berkeley Lab, 84-85, 88 & 89; vis scientist, Manne Siegbahn Inst, Sweden, 86-87. *Mem:* Am Phys Soc; Sigma Xi. *Res:* Experimental atomic physics; inner shell phenomena; x-ray and auger electron spectroscopy of highly ionized atoms; excited state lifetimes; ion-atom collisions. *Mailing Add:* Dept Physics NC State Univ PO Box 8202 Raleigh NC 27695-8202

MOWATT, THOMAS C, b Orange, NJ, Apr 24, 36; m 59; c 1. DIAGENESIS, PETROPHYSICS. *Educ:* Rutgers Univ, BA, 59; Univ Mont, PhD(geol), 65. *Prof Exp:* Res scientist, Amoco Prod Co, Okla, 65-67; asst prof geol, Winona State Col, 67-68 & Univ SDak, 68-70; supvr minerals analysis & res, Alaska Geol Surv, 70-74; geologist, Bur Land Mgt, 74-75, Bur Mines, US Dept Interior, 75-82; consult petrologist, Resevoirs Inc, 82-83, gen proj mgr, 83-84; GEOLOGIST, BUR LAND MGT, US DEPT INTERIOR, 84- *Concurrent Pos:* Lectr, Univ Tulsa, 67 & Univ Alaska, Juneau, 80-82; res assoc, SDak State Geol Surv, 68-70 & Alaska Pac Univ, 87-; adj assoc prof, Inst Marine Sci, Univ Alaska, 73- *Mem:* Geochem Soc; Mineral Soc Am; Clay Minerals Soc; Mineral Asn Can; Soc Econ Paleontologists & Mineralogists; Sigma Xi. *Res:* Petrology; clay mineralogy; environmental studies; economic geology; geochemistry, mineralogy and petrology in the contexts of petroleum geology, economic geology, environmental science and marine science; petrophysics; diagenesis; Petroleum science engineering. *Mailing Add:* 9905 Amchitka Circle Eagle River AK 99577-8728

MOWBRAY, DONALD F, b Duluth, Minn, July 29, 37; m 62. ENGINEERING MECHANICS. *Educ:* Univ Minn, BS, 60, MS, 62; Rensselaer Polytech Inst, PhD(mech), 68. *Prof Exp:* Asst mech, Univ Minn, 60-62; engr, Knolls Atomic Power Lab, 62-68, engr mat & processes lab, 68-71, mgr, Gas Turbine, 71-74, mgr mech mat, Mat & Processes Lab, 74-79, MGR SOLID MECHS, RES LAB, GEN ELEC CO, 79- *Mem:* Am Soc Testing & Mat; Am Soc Mech Engrs. *Res:* Material damping; metal fatigue; fracture of metals; dynamic thermoelasticity. *Mailing Add:* Gen Elec Co Bldg K1-3A19 Schenectady NY 12301

MOWBRAY, THOMAS BRUCE, b Duluth, Minn, Mar 1, 40; m 66. BOTANY, PLANT ECOLOGY. *Educ:* Univ Minn, Duluth, BA, 62; Duke Univ, MA, 64, PhD(bot), 67. *Prof Exp:* Instr biol, Duke Univ, 67-68; from asst prof to assoc prof biol & chairperson pop dynamics, Univ Wis-Green Bay, 68-78; mem fac, 78-80, FULL PROF BIOL, SALEM COL, 80- *Concurrent Pos:* Fulbright scholar, 89-90. *Mem:* Ecol Soc Am; Am Inst Biol Sci. *Res:* Plant community analysis; vegetation gradient analysis; environmental impact analysis. *Mailing Add:* Dept of Biol Salem Col PO Box 10548 Winston-Salem NC 27108

MOWER, HOWARD FREDERICK, b Chicago, Ill, Aug 25, 29; m; c 2. ORGANIC CHEMISTRY. *Educ:* Calif Inst Technol, BS, 51, PhD(org chem), 56. *Prof Exp:* Res chemist, Cent Res Dept, E I du Pont de Nemours & Co, Del, 56; assoc prof, 65-69, PROF BIOCHEM, UNIV HAWAII, 69- *Res:* Ferredoxins; hydrogenase enzymes; biological nitrogen fixation. *Mailing Add:* Dept Biochem Univ Hawaii Manoa 1960 East-West Rd Honolulu HI 96822

MOWER, LYMAN, b Berkeley, Calif, June 15, 27; m 48; c 3. PHYSICS. *Educ:* Univ Calif, BS, 49; Mass Inst Technol, PhD(physics), 53. *Prof Exp:* Eng specialist, Sylvania Elec Prod Inc, 53-57; from asst prof to assoc prof, 57-64, PROF PHYSICS, UNIV NH, 64- *Concurrent Pos:* Vis fel, Joint Inst Lab Astrophys, 64-65. *Mem:* Am Phys Soc. *Res:* Atomic and plasma physics; quantum electronics. *Mailing Add:* Dept Physics Univ NH Durham NH 03824

MOWER, ROBERT G, b Gasport, NY, Sept 27, 28. FLORICULTURE, ORNAMENTAL HORTICULTURE. *Educ:* Cornell Univ, BS, 56, MS, 59, PhD(turf dis), 61. *Prof Exp:* From asst prof to assoc prof, 61-77, PROF WOODY ORNAMENTALS, CORNELL UNIV, 77- *Mem:* Am Soc Hort Sci; Int Soc Hort Sci. *Res:* Taxonomy, evaluation of woody plants for landscape use. *Mailing Add:* Dept Floriculture Cornell Univ 33 Plant Sci Bldg Ithaca NY 14853

MOWERY, DWIGHT FAY, JR, b Moorehead, Minn, May 1, 15; m 43. CARBOHYDRATE CHEMISTRY, CHEMICAL KINETICS. *Educ:* Harvard Univ, AB, 37; Mass Inst Technol, PhD(org chem), 40. *Prof Exp:* Res chemist, E I du Pont de Nemours & Co, Del, 40-42; res chemist, Hercules Powder Co, 42-43; head chem dept, Elms Col, 43-46; head dept, Franklin Tech Inst, 46-49; asst prof, Trinity Col, Conn, 49-53; chmn dept chem, Ripon Col, 53-57; prof chem & dir grad prog, New Bedford Inst Tech, 57-64; chmn dept, 64-70, COMMONWEALTH PROF CHEM, SOUTHEASTERN MASS UNIV, 65- *Concurrent Pos:* Researcher, J B Williams Co, Conn, 52-53, WTM Mfg Co, Wis, 54-56, Aerovox Corp, Mass, 62, Acushnet Process Corp, Mass, 65-67 & Tibbetts Eng Corp, Mass, 78- *Mem:* Am Chem Soc. *Res:* Seed disinfectants and bactericides; carbohydrate chemistry; chromatographic adsorption; gas chromatography; organic microanalysis; chemical kinetics and computer programming; environmental analysis. *Mailing Add:* Dept Chem North Dartmouth MA 02747

MOWERY, RICHARD ALLEN, JR, b Newboston, Ohio, June 2, 38. ANALYTICAL CHEMISTRY. *Educ:* Univ Calif, Los Angeles, BS, 63; Univ Southern Calif, MA, 69; Ariz State Univ, PhD(chem), 74. *Prof Exp:* Sr measurement chemist, Appl Automation Inc, 74-84, sr chromatographer, 84-87, prin chromatographer, 87-88; SR RES CHEM, HERCULES RES CTR, 88- *Mem:* Am Chem Soc. *Mailing Add:* 24 Cannon Run Newark DE 19702-2446

MOWITZ, ARNOLD MARTIN, b New York, NY, Jan 14, 23; m 46; c 2. ANALYTICAL CHEMISTRY, TOXICOLOGY. *Educ:* Univ Buffalo, MA, 53. *Prof Exp:* Supvr control analysts, Nat Aniline Div, Allied Chem & Dye Corp, 46-48, chief analyst res & develop, 48-50, analytical res chemist, 50-53, chief analytical res, 53-55; group leader, Analytical Dept, Interchem Corp, 55-65, prog mgr, 66-67, mgr, Res Serv Dept, 67-70, mgr, Opers Dept, 70-77, corp mgr, Indust Toxicol & Prod Safety, Cent Res Labs, Inmont Corp, 77-88; INDEPENDENT CONSULT, BASF CORP, 89- *Concurrent Pos:* Consult, health comt, Nat Asn Printing Ink Mfrs; mem, task forces, Toxic Substances Control Act & Occup Health & Prod Safety, Nat Paint & Coatings Asn. *Mem:* Am Chem Soc; Am Microchem Soc; Soc Appl Spectros; sr mem Am Soc Test & Mat; NY Acad Sci. *Res:* Instrumental analysis; infrared and ultraviolet absorption analysis; spectrographic and microchemical analysis; x-ray diffraction; gas chromatography; light and electron microscopy; physical testing; nuclear magnetic resonance spectroscopy; research management; environmental chemistry; toxicology. *Mailing Add:* Chemsultants PA Environ Health 1111 Falmouth Ave Teaneck NJ 07666-1907

MOWLE, FREDERIC J, b Orange, NJ, Aug 4, 37; m 63; c 3. ELECTRICAL ENGINEERING. *Educ:* Univ Notre Dame, BS, 59, MS, 61, PhD(elec eng), 66. *Prof Exp:* PROF ELEC ENG, PURDUE UNIV, WEST LAFAYETTE, 66- *Concurrent Pos:* Pres, Unicorn Tech Consults, Inc, 82- *Mem:* Inst Elec & Electronics Engrs; Asn Comput Mach; Am Soc Eng Educ. *Res:* Software engineering with emphasis on planning and testing; computer networking; application of computers for industry; reliable software for robotic application. *Mailing Add:* 2215 Rainbow Dr W Lafayette IN 47906

MOWLES, THOMAS FRANCIS, b Boston, Mass, Feb 26, 34; m 56; c 4. PHARMACOLOGY, IMMUNOLOGY. *Educ:* Boston Univ, BA, 55; NY Univ, MS, 64; Rutgers Univ, PhD(zool), 68. *Prof Exp:* Lab supvr, Ciba Pharmaceut Co, 56-68; sr res biochemist, 68-74, res group chief, 75-81, RES SECT HEAD, HOFFMANN-LA ROCHE INC, 81- *Mem:* AAAS; Am Chem Soc; Int Soc Psychoneurol Endocrinol; NY Acad Sci; Sigma Xi. *Res:* Mechanism of hormone action; psychoneuroendocrinology; biochemical pharmacology; steroid biosynthesis; peptide hormones. *Mailing Add:* 266 Changebridge Rd Pine Brook NJ 07058-9559

MOWREY, GARY LEE, b Carlisle, Pa, June 17, 47; m 75. COAL-ROCK INTERFACE DETECTION, ADAPTIVE SIGNAL DISCRIMINATION. *Educ:* Pa State Univ, BS, 69, MS, 77, PhD(mining eng), 80. *Prof Exp:* Geophys engr, Schlumberger, Ltd, 69-72; asst prof geomech, Pa State Univ, 77-80; res engr, Shell Develop Co, 80-86; RES COORDR, US BUR MINES, 86- *Concurrent Pos:* Prin investr, US Bur Mines, 86- *Honors & Awards:* Stefanko Award, Soc Mining Engrs, 91. *Mem:* Soc Petrol Engrs. *Res:* Coal-rock interface detection techniques including mining machine vibrations; adaptive signal discrimination; infrared, natural gamma radiation; image processing; ground-penetrating radar; x-ray fluorescence; neural networks; geomechanics; microseismics; borehole stability. *Mailing Add:* US Bur Mines PO Box 18070 Pittsburgh PA 15236

MOWRY, DAVID THOMAS, b Pyengyang, Korea, Mar 11, 17; US citizen; m 38; c 3. INDUSTRIAL CHEMISTRY. *Educ:* Col Wooster, BS, 38; Ohio State Univ, MSc, 40, PhD(org chem), 41. *Prof Exp:* Chemist, Ohio State Univ, 38-41; res mgr, Cent Res Labs, Monsanto Co, 41-52, mgr chem develop, Phosphate Div, 52-53, mgr fine chem, Org Div, Develop Dept, 54-57, asst dir, 57-58, dir res & eng div, 58-61, mgr plastics div, 61-64, mgr planning East Asia, Int Div, 64-68, dir, Monsanto Japan Ltd & Ryoko Chemstrand Ltd, 68-74; prin engr, Int Opers Div, Nus Corp, 74-77; ASSOC DIR, NAT TECH INFO SERV, US DEPT COM, 77- *Mem:* Am Chem Soc; Com Develop Asn; Licensing Exec Soc; Technol Transfer Soc. *Res:* Structure of natural products; synthesis and reactions of nitriles; exploratory organic synthesis; high polymers; agricultural chemicals; commercial development; patent and know-how licensing; energy economics. *Mailing Add:* PO Box 1080 St Michaels MD 21663-1080

MOWRY, ROBERT WILBUR, b Griffin, Ga, Jan 10, 23; m 49; c 3. PATHOLOGY. *Educ:* Birmingham Southern Col, BS, 44; Johns Hopkins Univ, MD, 46. *Prof Exp:* Intern, Med Col Ala, 46-47, asst resident path, 47-48; sr asst surgeon, NIH, 48-52; asst prof path, Sch Med, Wash Univ, 52-53; from asst prof to prof path, Med Ctr, Univ Ala, Birmingham, 53-89, dir grad progs path, 64-72, sr scientist, Inst Dent Res, 67-72, prof health serv admin, 76-84, EMER PROF PATH, MED CTR, UNIV ALA, BIRMINGHAM, 89- *Concurrent Pos:* Fel, Mallory Inst Path, Boston Univ, 49-50; dir, Anat Path Lab, Univ Ala Hosp, 60-64 & 75-79; assoc ed, Stain Technol, 65-; mem, Path A Study Sect, USPHS, 64-68, trustee, Biol Stain Comn, 66-, vpres, 74-76, pres, 76-81; consult, Food & Drug Admin, 77-81; vis scientist, dept path, Cambridge Univ, 72-73. *Mem:* Am Asn Path; Am Asn Path; Biol Stain Comn; Int Acad Path. *Res:* Histochemistry and its applications to pathology; histopathologic technic; methods for detection and characterization of complex carbohydrates, microbial agents, amyloids and insulin in cells and tissues. *Mailing Add:* Dept Path Univ Ala Med Ctr Birmingham AL 35294

MOWSHOWITZ, ABBE, b Liberty, NY, Nov 13, 39; m 64; c 2. COMPUTER SCIENCE. *Educ:* Univ Chicago, SB, 61; Univ Mich, Ann Arbor, MA, 65, MS, 66, PhD(comput sci), 67. *Prof Exp:* Res assoc methodology, Human Sci Res, Inc, 62-63; res asst appl math, Ment Health Res Inst, Univ Mich, 63-67, asst res mathematician, 67-68; asst prof comput sci & indust eng, Univ Toronto, 68-69; asst prof comput sci, Univ BC, 69-74, assoc prof, 74-79; vis prof, Grad Sch Mgt, Delft, Netherlands, 79-80; dir, Croton Res Group, Inc, 80-82; prof & res dir sci & technol studies, Rensselaer Polytech Inst, 82-84; PROF COMPUT SCI, CITY COL, CITY UNIV NEW YORK, 84- *Concurrent Pos:* Res assoc, Inst Social Res & lectr, Dept Commun & Comput Sci, Univ Mich, 67-68; vis res assoc, Dept Comput Sci, Cornell Univ, 75-76; pres, Technol Impact Res, Larchmont, NY, 83-; Tinbergen chair, Erasmus Univ-Rotterdam, Neth, 90-91. *Mem:* Asn Comput Mach; Sigma Xi; Int Fed Info Process. *Res:* Economics of information; effects of information technology on economic and political organization, decision-making, and ethics; science and technology policy. *Mailing Add:* 41 Mayhew Ave Larchmont NY 10538-2740

MOXLEY, JOHN H, III, b Elizabeth, NJ, Jan 10, 35; div; c 3. MEDICAL ADMINISTRATION. *Educ:* Colo Univ, MD, 61. *Hon Degrees:* DSc, Hahnemann Univ, 86. *Prof Exp:* Intern, 61-62, jr asst res, 62-63, sr asst res, Peter Bent Brigham Hosp, Boston, MA, 65-66; instr med & asst to dean, Harvard Univ, 66-69; assoc prof med & Dean Sch Med, Univ MD, 69-73; VChancellor Health Scis & Dean Sch Med, Univ Calif San Diego, 73-80; asst secy, Dept Defense, 80-81; sr vpres, Am Med Int, Beverly Hills, 81-88; VPRES/PARTNER, KORN/FERRY INT, 88- *Concurrent Pos:* Surg USPHS, Nat Cancer Inst, 63-65. *Honors & Awards:* Secy Defense Medal Distinguished Public Serv. *Mem:* Am Med Asn; Fel Am Col Physicians; Inst Med-Nat Acad Sci; distinguished fel Am Col Physician Execs. *Mailing Add:* 8180 Manitoba St No 210 Playa del Rey CA 90293

MOXNESS, KAREN, CLINICAL NUTRITION, CLINICAL MANAGEMENT. *Educ:* Univ Minn, MS, 63. *Prof Exp:* ASST DIR CLIN DIETETICS & EDUC, ST MARY'S HOSP, ROCHESTER, MINN, 72- *Mailing Add:* 114 36th Ave NW Rochester MN 55901

MOY, DAN, b Chicago, Ill, July 11, 55; m 78; c 2. ELECTRONICS ENGINEERING, SOLID STATE PHYSICS. *Educ:* Mass Inst Technol, BS, 78; Univ Ill, Champaign-Urbana, MS, 80, PhD(physics), 83. *Prof Exp:* Sr engr, Intel Corp, 83-84; staff engr, 84-89, adv engr, 89-90, MGR, IBM CORP, 90- *Mem:* Am Vacuum Soc; Int Soc Optical Eng; Mat Res Soc. *Res:* Demonstration and use of state-of-art microlithographic technology for fabrication of leading edge devices and circuits primarily of the ultralarge-scale integration bipolar, cmos or bicmos type. *Mailing Add:* IBM Corp PO Box 218 Yorktown Heights NY 10598

MOY, JAMES HEE, b Canton, China, Feb 20, 29; US citizen; m 67; c 2. FOOD SCIENCE, CHEMICAL ENGINEERING. *Educ:* Univ Wis-Madison, BSChE, 57, MSChE, 58; Rutgers Univ, New Brunswick, PhD(food sci), 65. *Prof Exp:* Chem engr, Esso Res & Eng Co, 58-60 & Lipton Co, 60-61; from asst prof to assoc prof, 65-80, PROF FOOF ENG, UNIV HAWAII, 80- *Concurrent Pos:* USAEC res grant, 68-71; Int Sugar Res Found res grant, 72-76; US Dept Energy res grant, 76-82; US NSF sea grant, 77-81, US Dept Energy, 81-85; adv & consult, S Pac Comn, USDA, Food & Agr Orgn & Int Atomic Energy Agency. *Mem:* Inst Food Technologists; Am Inst Chem Engrs; Am Soc Agr Engrs. *Res:* Food engineering; tropical foods and root crops processing including solar dehydration, freeze dehydration, freezing and ionizing radiation. *Mailing Add:* Dept Food Sci & Human Nutrit Univ Hawaii Manoa 1920 Edmondson Rd Honolulu HI 96822

MOY, MAMIE WONG, b San Antonio, Tex, Sept 4, 29; wid; c 1. SCIENCE EDUCATION. *Educ:* Univ Tex, Austin, BA, 50; Univ Houston, MS, 52. *Prof Exp:* From instr to asst prof, 52-76, ASSOC PROF CHEM, UNIV HOUSTON, CENT CAMPUS, 76-, ASSOC CHAIR, 83- *Mem:* Am Chem Soc; AAAS; Nat Sci Teachers Asn. *Mailing Add:* Dept Chem Univ Houston Univ Park 4800 Calhoun Rd Houston TX 77004-5641

MOY, RICHARD HENRY, b Chicago, Ill, Feb 2, 31; m 54; c 2. MEDICINE, HEALTH SCIENCE. *Educ:* Univ Chicago, BA, 53, BS, 54, MD, 57; Am Bd Internal Med, dipl, 66. *Prof Exp:* Intern med, Univ Chicago, 57-68; clin assoc endocrinol, USPHS, 58-60; res internal med, Univ Chicago Clins, 60-62, instr & chief res med, 62-63, from instr to assoc prof, 63-70; PROF MED, DEAN & PROVOST, SCH MED, SOUTHERN ILL UNIV, 70- *Concurrent Pos:* Nat Cancer Inst spec fel, 63-64; res assoc, Univ Chicago, 64-68; dir, Univ Health Serv, Univ Chicago, 64-70; consult, HEW, 77-78; consult, Vet Admin, Washington, DC, 77-78. *Mem:* Fel Am Col Physicians; Am Med Asn; Soc Health Human Values; Sigma Xi. *Mailing Add:* 25 Wildwood Rd Springfield IL 62704

MOY, WILLIAM A(RTHUR), b St Paul, Minn, June 16, 31; m 54; c 2. INDUSTRIAL ENGINEERING, OPERATIONS RESEARCH. *Educ:* Univ Minn, BIE, 54, MS, 56; Northwestern Univ, PhD(indust eng, mgt sci), 65. *Prof Exp:* Dept mgr, Proctor & Gamble Mfg Co, 56-57, methods engr, 57-58; from asst prof to assoc prof mech eng, Univ Wis-Madison, 58-69, prof indust eng, 69-74, chmn dept, 69-71; dean sch mod indust, 72-76, PROF INDUST ENG, UNIV WIS-PARKSIDE, 73- *Mem:* Opers Res Soc Am; Inst Mgt Sci; Am Inst Indust Engrs; Am Soc Qual Control; Am Prod & Inventory Control Soc. *Res:* Operations research and management science especially digital computer simulation methods. *Mailing Add:* Dept of Indust Eng Univ of Wis-Parkside Box 2000 Kenosha WI 53141

MOYE, ANTHONY JOSEPH, b McAdoo, Pa, Oct 15, 33; m 57; c 3. ACADEMIC ADMINISTRATION, PHYSICAL ORGANIC CHEMISTRY. *Educ:* Upsala Col, BS, 55; Iowa State Univ, MS, 57, PhD(org chem), 62. *Prof Exp:* Prof chem & dean acad planning & grad studies, Calif State Col, Los Angeles, 62-71; prof chem & vpres acad affairs, Quinnipiac Col, 71-72; state univ dean, 72-79, asst vchancellor educ progs & resources, 79-84, ASSOC VCHANCELLOR, EDUC PROGS & RESOURCES, CALIF STATE UNIV, 84-, DEPUTY VCHANCELLOR, ACAD AFFAIRS, RESOURCES, 88- *Mem:* AAAS; Am Chem Soc; Royal Soc Chem. *Res:* Free radicals in solution; chemiluminescence. *Mailing Add:* Calif State Univ 400 Golden Shore Suite 314 Long Beach CA 90802

MOYE, HUGH ANSON, b Mobile, Ala, Oct 18, 38. ANALYTICAL CHEMISTRY. *Educ:* Spring Hill Col, BS, 61; Univ Fla, PhD(chem), 65. *Prof Exp:* Asst prof, 65-75, assoc prof, 75-78, ASSOC CHEMIST PESTICIDE RES, PESTICIDE RES LAB, UNIV FLA, 75-, PROF CHEM, 78- *Mem:* Am Chem Soc; Asn Offs Analytical Chem; Sigma Xi. *Res:* Analytical methods for pesticides; reaction gas chromatography of pesticides; gas chromatography detectors. *Mailing Add:* 2208 SW 43rd Pl Gainesville FL 32608

MOYED, HARRIS S, b Philadelphia, Pa, May 15, 25; m 54; c 2. BACTERIOLOGY, BIOCHEMISTRY. *Prof Exp:* Nat Found res fel biochem, Mass Gen Hosp, 54-55 & bact, Harvard Med Sch, 55-57; from instr to asst prof bact, Harvard Med Sch, 57-63; Hastings prof microbiol, Sch Med, Univ Southern Calif, 63-69; PROF MICROBIOL, COL MED, UNIV CALIF, IRVINE, 69- *Concurrent Pos:* Lederle award, 58-60; vis prof, Rockefeller Univ, 81-82. *Mem:* Am Soc Microbiol; Am Soc Biol Chem; Sigma Xi. *Res:* Biochemistry of bacteria; regulation of biosynthetic reactions; action of plant auxin. *Mailing Add:* Dept of Med Microbiol Univ of Calif Irvine CA 92717

MOYER, BRUCE A, b Harrisburg, Pa, Nov 15, 52; m 82. SEPARATIONS CHEMISTRY, COORDINATION CHEMISTRY. *Educ:* Duke Univ, BS, 74; Univ NC, Chapel Hill, PhD(inorg chem), 79. *Prof Exp:* Res chemist, Chem Technol Div, 79-83 & Chem Div, 83-87, GROUP LEADER, CHEM DIV, OAK RIDGE NAT LAB, 87- *Mem:* Am Chem Div. *Res:* Solvent extraction and ion exchange of metal complexes; separations, actinides, transition metals, equilibrium analysis, coordination reactions, chemical recognition and redox process. *Mailing Add:* Oak Ridge Nat Lab Bldg 4500S MS6119 PO Box 2008 Oak Ridge TN 37831-6119

MOYER, CALVIN LYLE, b Philadelphia, Pa, Nov 2, 41; m 63; c 2. ORGANIC CHEMISTRY. *Educ:* Ursinus Col, BS, 63; Harvard Univ, MA, 65, PhD(chem), 68. *Prof Exp:* Res chemist, Benger Lab, E I du Pont de Nemours & Co, Inc, 68-72, sr res chemist end use res, 72-73, col recruiter, 74-75, prog coordr, Col Rels, 76-77, col rels supvr, 77-78, staff asst, 78-81, employee rels supt, 82-84; admin & planning asst, Du Pont Polymers, 85-87, training & develop mgr, 88-89, human rels mgr, 89-90; CONSULT, 90- *Mem:* Am Chem Soc; Am Soc Training & Develop. *Mailing Add:* Skyline Orchards 367 Skyline Orchard Dr Hockessin DE 19707

MOYER, CARL EDWARD, b Dayton, Ohio, Dec 24, 26; m 50; c 4. PHYSIOLOGICAL CHEMISTRY. *Educ:* Univ Dayton, BS, 53; Ohio State Univ, MS, 57, PhD, 59. *Prof Exp:* Clin biochemist & head clin lab, Res Labs, Parke Davis & Co, 63-77, clin biochemist & head clin lab, Pharmaceut Res Div, 77-81, MGR, CLIN PATHOL LAB, DEPT PATHOL & EXP TOXICOL, WARNER-LAMBERT/PARKE-DAVIS, 81- *Concurrent Pos:* Supvr clin labs, Riverside Methodist Hosp, Columbus, Ohio, 59-63. *Mem:* Am Chem Soc; Am Asn Clin Chem. *Res:* Clinical biochemistry. *Mailing Add:* 1810 Crestland Dr Ann Arbor MI 48104

MOYER, DEAN LA ROCHE, b Pa, Mar 17, 25; m 53; c 5. PATHOLOGY. *Educ:* Lehigh Univ, BA, 48; Univ Rochester, MD, 52. *Prof Exp:* From instr to prof path, Med Ctr, Univ Calif, Los Angeles, 56-69; head, Exp Path Sect, 69-76, PROF PATH, OBSTET & GYNEC, MED SCH, UNIV SOUTHERN CALIF, 69-, CHIEF PATHOLOGIST, WOMEN'S HOSP LOS ANGELES, COUNTY-UNIV SOUTHERN CALIF MED CTR, 76- *Concurrent Pos:* Fel oncol, Mass Gen Hosp, 55-56; dir labs, Harbor Gen Hosp, Torrance, 61-69. *Res:* Early reproduction. *Mailing Add:* Women's Hosp Los Angeles County Univ Southern Calif Med Ctr 14120 Magnolia Blvd Sherman Oaks CA 91423

MOYER, GEOFFREY H, b Cleveland, Ohio, Feb 7, 38. PATHOLOGY. *Educ:* Univ Wis, MD. 63, PhD(oncol), 75. *Prof Exp:* DIR, DAMON REF LAB, NEWBURY PARK, CALIF, 81- *Mailing Add:* 720 Lachman Lane Pacific Palisades CA 90272

MOYER, H(ALLARD) C(HARLES), b Crawford, Nebr, Dec 27, 18; m 47; c 2. PETROLEUM ENGINEERING & TECHNOLOGY. *Educ:* Univ Nebr, BS, 40. *Prof Exp:* Chem engr, Sinclair Refining Co, 41-42; exp engr, S C Johnson & Son, Inc, 47-52; sr res engr, Sinclair Res, Inc, 52-69; dir indust prod res, Atlantic Richfield Co, 69-71, mgr tech support, Asphalt & Process Oils, 71-75, mgr indust lubricants & asphalt res, Develop & Tech Serv, 75-80; RETIRED. *Res:* Processing and functionality of refined petroleum oils and waxes; composition and physical properties of asphalts. *Mailing Add:* 1537 W 186th Pl Homewood IL 60430

MOYER, JAMES ROBERT, b Kitchener, Ont, June 28, 42; m 74; c 2. AGRICULTURE. *Educ:* Univ Waterloo, BSc, 66; Univ Guelph, MSc, 68; Univ Sask, PhD(soil sci), 72. *Prof Exp:* Res assoc soil sci, Univ Sask, 72 & Univ Guelph, 72-75; RES SCIENTIST WEED SCI, AGR CAN, 75- *Mem:* Int Weed Sci Soc; Can Soc Soil Sci; Weed Sci Soc Am. *Res:* Control of weeds in rangeland and pastures and the effect of soil physical properties on the persistence and efficacy of herbicides. *Mailing Add:* Lethbridge Res Sta Agr Can Lethbridge AB T1J 4B1 Can

MOYER, JOHN ALLEN, b Lebanon, Pa, Oct 20, 51; m 77; c 1. NEUROPSYCHOPHARMACOLOGY, PSYCHOBIOLOGY. *Educ:* Albright Col, BS, 73; Bucknell Univ, MS, 75; Temple Univ, PhD(psychobiol), 78. *Prof Exp:* Res asst psychol, Bucknell Univ, 73-75; res asst psychobiol, Temple Univ, 75-78; fel neuropharmacol, 78-80, res assoc, Dept Psychiat, Univ Pa Sch Med, 80-82; supvr psychopharmacol, 80-85, res scientist psychopharmacol, 86, mgr psychopharmacol, 86-88, SECT HEAD PSYCHOPHARMACOL, WYETH-AYERST RES, 88- *Concurrent Pos:* Vis scientist, Lab Clin Sci, NIMH, 75-76; instr, Dept Psychol, Villanova Univ, 84; adj asst prof, Dept Psychiat, Univ Pa Sch Med, 82- *Mem:* Soc Neurosci; Int Soc Psychoneuroendocrinol; AAAS; Behav Pharmacol Soc; Int Brain Res Orgn. *Res:* Neuropsychopharmacology of central nervous system disorders; behavioral pharmacology, neurochemistry, and neuropharmacology. *Mailing Add:* Psychopharmacol Sect Wyeth-Ayerst Res CN-8000 Princeton NJ 08543-8000

MOYER, JOHN CLARENCE, b Chicago, Ill, Jan 9, 46; m 75. MATHEMATICS. *Educ:* Christian Bros Col, BS, 67; Northwestern Univ, MS, 69, PhD(math educ), 74. *Prof Exp:* Teacher math, St Patrick High Sch, Chicago, 67-69 & St Joseph High Sch, Chicago, 69-72; asst prof, 74-80, ASSOC PROF MATH, MARQUETTE UNIV, 80- *Mem:* Math Asn Am; Am Math Soc; Nat Coun Teachers Math. *Res:* Problem solving research with children ages 9-14. *Mailing Add:* Dept of Math & Statist Marquette Univ 1515 W Wisconsin Ave Milwaukee WI 53233

MOYER, JOHN HENRY, b Hershey, Pa, Apr 1, 19; m; c 7. MEDICINE. *Educ:* Lebanon Valley Col, BS, 39; Univ Pa, MD, 43; Am Bd Internal Med, dipl. *Hon Degrees:* DSc, Lebanon Valley Col, 68. *Prof Exp:* Intern, Pa Hosp, 43; resident, Belmont Hosp, Worcester, Mass, 44-45; asst instr tuberc & contagious dis, Univ Vt, 44-45; chief resident med, Brooke Gen Hosp, 47; fel pharmacol & med, Sch Med, Univ Pa, 48-50; from asst prof to prof pharmacol, Col Med, Baylor Univ, 50-57; prof med, Hahnemann Med Col & Hosp, 57-74, chmn dept med, 57-71, vpres acad affairs, 71-73; VPRES, DIR PROF & EDUC AFFAIRS, CONEMAUGH VALLEY MEM HOSP, 74-; PROF MED, SCH MED, TEMPLE UNIV, 76- *Concurrent Pos:* From attend physician to sr attend, Jefferson Davis Hosp, Houston, 50-57; consult, Vet Admin Hosp, Houston & Houston Tuberc Hosp, 50-57; vis prof, Sch Med, La State Univ, 52; consult, Vet Admin Hosp, Philadelphia, 58-68, Philadelphia Naval Hosp, 58-, Bd Vet Appeals, 63- & comn drugs, AMA, 68-; deleg at large, AMA, 70-75; adv & consult, Hypertension Info & Educ Adv Comt, US Dept Health, 72-75; pres bd trustees, US Pharmacopeia, 70-75; adv, Gov Task Force Hypertension, State of Pa, 74-; ed consult, Am J Cardiol; ed cardiovasc sect, Cyclopedia Med, Surg & Specialties; Milliken lect, Pa Hosp, 58; chmn adv group, Pa High Blood Pressure Control Prog, 79-81. *Honors & Awards:* Hunter Award, Am Therapeut Soc, 59; Clyde M Fish Mem Lect, 60; Susan & Theodore Cummings Humanitarian Award, 62, 65 & 66. *Mem:* Fel Am Col Cardiol; Am Soc Clin Pharmacol & Therapeut (pres, 65); Am Acad Tuberc Physicians (pres, 61); fel Am Col Clin Pharmacol & Chemother (pres, 64-66); fel NY Acad Sci; Sigma Xi. *Res:* Hypertension and pharmacodynamics of the cardiovascular system; renal function. *Mailing Add:* Conemaugh Valley Hosp 1086 Franklin St Johnstown PA 15547

MOYER, JOHN RAYMOND, b Buffalo, NY, June 9, 31; m 52; c 4. INORGANIC CHEMISTRY, CERAMIC ENGINEERING. *Educ:* Eastern Mich Univ, AB, 52; Univ Mich, PhD(phys inorg chem), 58. *Prof Exp:* Res chemist, Electro-inorg Res Lab, Dow Chem, USA, 59-63, sr res chemist, 63-68, assoc scientist, 68-70, environ res lab, 70-74, assoc scientist, 74-76, res scientist, Cent Res-inorg Mat & Catalysis Lab, 76-86, SR RES SCIENTIST, CTR RES, ADVAN CERAMICS LAB, DOW CHEM, USA, 86- *Mem:* Am Chem Soc; Mat Res Soc. *Res:* Chemistry of halogens; synthesis of ceramic precursors. *Mailing Add:* Dow Chem USA 1776 Bldg Midland MI 48674

MOYER, KENNETH EVAN, b Chippewa Falls, Wis, Nov 19, 19; m 43; c 2. PHYSIOLOGICAL PSYCHOLOGY. *Educ:* Park Col, AB, 43; Wash Univ, MA, 48, PhD(psychol), 51. *Prof Exp:* Dir phys educ, Park Col, 42-43; instr psychol & phys educ, Pearl River Col, 46-47; vet counr psychol, Wash Univ, 47-49; instr psychol, Carnegie Inst Technol, 49-50, assoc prof, 54-61; actg head dept, Carnegie-Mellon Univ, 61, from prof to emer prof, 61-87; RETIRED. *Concurrent Pos:* Consult higher educ, Govt Norway, 54; ed-in-chief, Aggressive Behav, 74-78. *Honors & Awards:* Carnegie Found Award, 54. *Mem:* Fel Am Psychol Asn; Psychonomic Soc; fel AAAS. *Res:* Physiology of aggressive behavior. *Mailing Add:* 625 Escambia Loop Lillian AL 36549

MOYER, KENNETH HAROLD, b Poughkeepsie, NY, Sept 30, 29; m 51; c 7. METALLURGICAL ENGINEERING. *Educ:* Polytech Inst Brooklyn, BS, 59, MS, 62. *Prof Exp:* Jr staff mem metall, US Hoffman Mach Corp, 53-54 & Int Nickel Res Lab, 54-55; proj engr, Grumman Aircraft Eng Corp, 55-58, Sylvania Corning Nuclear Corp, 58-60 & Beryllium Corp Am, 60-62; mgr beryllium opers, Gen Astrometals Corp, 62-66; mgr spec alloys res & develop,

66-80, NEW PROD ENG, HOEGANAES CORP, 80- *Concurrent Pos:* Consult; instr, Spring Garden Col Eve Div, 73- *Mem:* Fel Am Soc Metals; Am Inst Mining, Metall & Petrol Engrs; Am Ord Asn; Inst Elec & Electronics Engrs. *Res:* Fabrication of beryllium; high alloy iron powders. *Mailing Add:* Magna-Tech P/M Labs Four Green Briar Lane Cinnaminson NJ 08077

MOYER, LEROY, b Erie, Pa, June 7, 42. PHYICS, MATHEMATICS. *Educ:* Univ Rochester, PhD(physics), 72. *Prof Exp:* ANALYST, ARMY FOREIGN SCI & TECHNOL CTR, 72- *Mailing Add:* Ducky Oaks Rte Nine Box 236 Charlottesville VA 22901

MOYER, MARY PAT SUTTER, b Arlington, Mass, Apr 27, 51; m 74; c 1. ONCOLOGY, GASTROINTESTINAL CELL BIOLOGY. *Educ:* Fla Atlantic Univ, BS, 72, MS, 74; Univ Tex, Austin, PhD, 81. *Prof Exp:* Dir tissue culture res virol, Equine Res Inst, 70-73; cancer res scientist viral oncol, Thorman Cancer Res Lab, Trinity Univ, 74-81; instr, 81-82, res asst prof, 82-85, RES ASSOC PROF, DEPT SURG, UNIV TEX HEALTH SCI CTR, 85- *Concurrent Pos:* Consult, tumor biol div, Smith Kline French & Beckman, Philadelphia, Pa, 84-; invited lectr, var socs, univs & corps in US, Eng & Austria. *Mem:* Am Asn Cancer Res; Am Soc Microbiol; AAAS; Tissue Culture Asn; Am Soc Cell Biol; Sigma Xi. *Res:* Experimental oncology; biological activity of simian virus 40 DNA fragments; persistent virus infections in vitro; animal tumorigenesis models; tumor immunology; biological effect of DNAs in vivo. *Mailing Add:* Univ Tex Health Sci Ctr 7703 Floyd Curl Dr Univ Tex Health Sci Ctr San Antonio TX 78284-7842

MOYER, MELVIN ISAAC, b Newton, Kans, June 30, 21; m 61. ORGANIC CHEMISTRY. *Educ:* Bethel Col, AB, 42; Univ Okla, MS, 44; Univ Kans, PhD(chem), 52. *Prof Exp:* Asst, Univ Okla, 42-44; res chemist, Cities Serv Oil Co, 46-48; asst instr, Univ Kans, 48-50; develop chemist, Am Cyanamid Co, 52-57, chief chemist mfg, 57-62, sr chemist, NJ , 62-73, sr chemist, 73-87; RETIRED. *Mem:* AAAS; Am Chem Soc. *Res:* Manufacturing. *Mailing Add:* 111 Lynn Lane No 9 Newton KS 67114

MOYER, PATRICIA HELEN, b Greensboro, NC, Sept 30, 27; m 50; c 3. ORGANIC CHEMISTRY. *Educ:* Northwestern Univ, BA, 49; Univ Wis, PhD(chem), 54. *Prof Exp:* Res chemist, Phillips Petrol Co, 53; sr chemist, Clevite Corp, 55-56; sr res chemist, Res Ctr, B F Goodrich Co, 56-63; head biochem lab, Midwest Med Res Found, 65-66; sr res chemist, Frontier Chem Co, 66-67; chem div, Vulcan Mat Co, 67-68, group leader chem res, 68-73; instr vis staff, 74-78, RES FAC, DEPT CHEM, PHOENIX COL, 78- *Mem:* AAAS; Am Asn Univ Women; Am Chem Soc; Sigma Xi. *Res:* Rates and mechanisms of organic reactions; polymerization; organometallics. *Mailing Add:* 8102 N Sixth St Phoenix AZ 85020

MOYER, RALPH OWEN, JR, b New Bedford, Mass, May 19, 36. INORGANIC CHEMISTRY. *Educ:* Southeastern Mass Univ, BS, 57; Univ Toledo, MS, 63; Univ Conn, PhD(inorg chem), 69. *Prof Exp:* Develop engr, Union Carbide Corp, 57-64; from asst prof to assoc prof, 69-86, chmn dept, 85-88, PROF CHEM, TRINITY COL, CONN, 86- *Concurrent Pos:* Vis asst prof, Wesleyan Univ, 75; res collabr, Brookhaven Nat Lab, 77-80; vis lectr, Univ West Indies, 85. *Mem:* Am Chem Soc; Sigma Xi; NY Acad Sci. *Res:* Preparation, structure, magnetic and electrical properties of metal hydrides. *Mailing Add:* Clement Chem Lab Trinity Col Hartford CT 06106

MOYER, REX CARLTON, b Elkhart, Ind, Dec 8, 35; m 58; c 5. CANCER, MICROBIOLOGY. *Educ:* Purdue Univ, BS, 57; Univ Nebr, MS, 61; Univ Tex, PhD(microbiol), 65. *Prof Exp:* Asst bacteriologist, Miles-Ames Res Labs, 57-58; lab instr gen microbiol, Univ Nebr, 58-61; trainee molecular biol, Univ Tex, 61-65; Nat Acad Sci-Nat Res Coun res fel microbial genetics & bacteriophagy, Ft Detrick, 65-66; res microbiologist, Ft Detrick, 66-69; co-dir, Bettye Thorman Cancer Res Lab, Trinity Univ, 70-75, from asst prof to assoc prof biol, 69-86, DIR, BETTYE THORMAN CANCER RES LAB, TRINITY UNIV, 75-, PROF BIOL, 86- *Concurrent Pos:* Assoc ed, Tex J Sci; owner, Moyer Tree Farm; Air Force Off Sci Res univ resident fel, 87; mem, bd dirs, San Antonio Bot Ctr & Am Cancer Soc, chair, Sci Educ Comt & Mahncke Arboretum Comt. *Honors & Awards:* Outstanding Serv Award, Am Cancer Soc, 86. *Mem:* Am Soc Microbiol; Nitrogen Fixing Tree Asn; Sigma Xi; Am Asn Cancer Res; Int Asn Vitamin & Nutrit Oncol. *Res:* Cancer virology and nutrition; air purification by higher plants; safety of recombinant DNAs in biologicals; agroforestry; phototaxis in algae. *Mailing Add:* Dept Biol Trinity Univ 715 Stadium Dr San Antonio TX 78212

MOYER, RICHARD W, b Sept 25, 40; m; c 2. ORTHODOX VIRUS GENES EXPRESSION. *Educ:* Univ Calif, Los Angeles, PhD(chem), 67. *Prof Exp:* Prof microbiol, Vanderbilt Univ, 76-90; PROF & CHMN, DEPT IMMUNOL & MED MICROBIOL, UNIV FLA, 90- *Mem:* Am Soc Microbiol; Am Soc Biol Chemists. *Res:* Virus-cell interactions. *Mailing Add:* Dept Immunol & Med Microbiol JHMHC Univ Fla PO Box J266 Gainesville FL 32610

MOYER, ROBERT (FINDLEY), b New York, NY, May 12, 37; c 3. RADIATION PHYSICS. *Educ:* Pa State Univ, BS, 59, MS, 61; Univ Calif, Los Angeles, PhD(med physics), 65. *Prof Exp:* Inst radiol physics, State Univ NY Upstate Med Ctr, 65-70, asst prof radiol, 70-81; CHIEF PHYSICIST & RADIATION SAFETY OFFICER, READING HOSPITAL, 81- *Concurrent Pos:* Consult medical radiation physicist. *Mem:* Am Asn Physicists in Med; Am Col Medical Physics; Health Physics Soc. *Res:* Radiologic physics and biology. *Mailing Add:* Dept Radiol Reading Hosp & Med Ctr Reading PA 19603

MOYER, SAMUEL EDWARD, b Hershey, Pa, Oct 5, 34; m 59; c 2. POPULATION GENETICS. *Educ:* Pa State Univ, BS, 56; Univ NH, MS, 59; Univ Minn, PhD(genetics), 64. *Prof Exp:* Asst geneticist, NC State Univ, 64-65; asst prof genetics, Northeastern Univ, 66-71; ASSOC PROF BIOL, BURLINGTON COUNTY COL, 71- *Mem:* Genetics Soc Am; Am Genetic Asn; Sigma Xi. *Res:* Genetic traits of economic importance in poultry; effects of linkage on survival; genetic loads of populations. *Mailing Add:* CA 171 Burlington County Col Pemberton NJ 08068

MOYER, VANCE EDWARDS, b Orwigsburg, Pa, Nov 22, 14; m 53; c 2. METEOROLOGY. *Educ:* Pa State Univ, BS, 50, MS, 51, PhD(meteorol), 54. *Prof Exp:* Res asst meteorol, Pa State Univ, 51, res assoc, 52-54; asst prof, Univ Tex, 54-58; assoc prof, 58-61, chmn instruct meteorol, 60-66, actg head dept, 66-67, head dept, 67-75, prof, 61-80, EMER PROF METEOROL, TEX A&M UNIV, 80- *Concurrent Pos:* NSF lectr, 58-64; mem earth sci curriculum proj, 65. *Mem:* Am Meteorol Soc; Am Geophys Union. *Res:* Cloud and precipitation physics; physical and radar meteorology; satellite determination of atmospheric structure. *Mailing Add:* 210 W North Ave Bryan TX 77801

MOYER, WALTER ALLEN, JR, b Philadelphia, Pa, Nov 16, 22; m 46; c 2. ORGANIC CHEMISTRY. *Educ:* Philadelphia Col Pharm, BSc, 43; Middlebury Col, MSc, 48; Univ Del, PhD(org chem), 51. *Prof Exp:* From instr to assoc prof, 51-67, PROF CHEM, MIDDLEBURY COL, 67-, ASSOC DEAN, COL INST RES & SPEC ADMIS, 71- *Concurrent Pos:* Dir career coun & placement, Middleburg Col, 74-, assoc dean sci, 81- *Mem:* Nat Asn Adv Health Professions; Am Chem Soc. *Res:* Carbohydrates; natural products; organic synthesis. *Mailing Add:* Nine Adirondark View Middlebury VT 05753-1301

MOYER, WAYNE A, b Brooklyn, NY, Mar 9, 30; m 54; c 1. CELL ADHESION. *Educ:* Bucknell Univ, BS, 52; Syracuse Univ, MS, 55; Brown Univ, ScM, 64; Princeton Univ, PhD(develop biol), 73. *Prof Exp:* Dept chmn biol, East Brunswick Bd Educ, 55-69; asst prof develop biol, City Univ NY, 73-76, Trenton State Col, 76-77 & Seton Hall Univ, 77-79; EXEC DIR, NAT ASN BIOL TEACHERS, 79- *Concurrent Pos:* Teacher gen sci, Fair Lawn Bd Educ, NJ, 55-58. *Mem:* AAAS; Am Inst Biol Sci; Nat Sci Teachers Asn; Soc Develop Biol; Nat Asn Biol Teachers. *Res:* Sequestered mRNA in sea urchin embryos as revealed by actinomycin block to RNA snynthesis; rate studies of cell adhesion between various cell types derived from embryonic chick tissues; strength of adhesion as measured by hierarchical envelopment, interpreted by the differential adhesion hypothesis. *Mailing Add:* 1547 Scandia Circle Reston VA 22090

MOYER, WILLIAM C, JR, b Dallas, Tex, Apr 5, 37. UNDERWATER ACOUSTICS. *Educ:* Southern Methodist Univ, BS, 59; Univ Tex, PhD(mech eng), 66. *Prof Exp:* Mem tech staff, Hughes Aircraft Co, 59; sr scientist, 66-68, asst dir res dept, 68-69, dir anal dept, 69-71, asst vpres appl technol div, 71-74, vpres, Anal & Appl Res Div, 74-76, VPRES APPL SCI, TRACOR INC, 76- *Res:* Sonar system performance analysis; radiation associated with underwater arrays; transducer and baffle interactions; specialized underwater sensor systems. *Mailing Add:* 4308 Sinclair Dr Austin TX 78721

MOYERMAN, ROBERT MAX, b Atlantic City, NJ, Sept 14, 25; m 51; c 3. ORGANIC CHEMISTRY, ANALYTICAL CHEMISTRY. *Educ:* Rutgers Univ, BS, 49; Univ Ala, MS, 51. *Prof Exp:* Chemist high temperature nuclear reactors, Nuclear Develop Assocs, Inc, 51-52; assoc chemist carbohydrate & analytical chem, Johns Hopkins Univ, 52-53, chem kinetics, Appl Physics Lab, 53-55; res investr, Am Smelting & Ref Co, 55-58; res chemist analytical chem & org separations, Ansul Co, Wis, 58-63, sr res chemist process res, Org Res & Sect Head Analytical Dept, 63-65; sr chemist, Scholler Bros, Inc, Pa, 65-68; group leader, Chem Div, Sun Chem Corp, RI, 68-70; mgr res & develop, Hydrolabs, Inc, Paterson, 74-75; OWNER & NEW PROD MGR, WARWICK LABS, 70- *Concurrent Pos:* Tech serv dir, Org Chem Corp, 70-72. *Mem:* Am Chem Soc; Am Microchem Soc; Am Asn Textile Chemists & Colorists; Sigma Xi. *Res:* Organic syntheses; organic arsenic. *Mailing Add:* 118 Edmond Dr Warwick RI 02886-8545

MOYERS, JACK, b Sidney, Iowa, Dec 7, 21; m 45, 84; c 2. ANESTHESIOLOGY. *Educ:* Univ Iowa, BS, 43, MD, 45; Am Bd Anesthesiol, dipl, 53. *Prof Exp:* Intern med, Mt Carmel Mercy Hosp, Detroit, 45-46; resident anesthesiol, Col Med, Univ Iowa, 48-50; instr, WHO Anesthesiol Training Ctr, Univ Copenhagen, 50-51; instr, 51-52, assoc, 52-53, from asst prof to assoc prof, 53-66, actg head dept, 67-68, head dept, 68-77, PROF ANESTHESIA, COL MED, UNIV IOWA, 66-; ATTEND ANESTHESIOLOGIST, VET ADMIN HOSP, 52- *Honors & Awards:* Husfeldt lectr, Univ Copenhagen. *Mem:* Am Soc Anesthesiol; Asn Univ Anesthetists; fel Am Col Anesthesiol; Scand Soc Anesthesiol; NY Acad Sci. *Res:* Clinical and laboratory investigation in field of anesthesiology. *Mailing Add:* Dept of Anesthesia Univ of Iowa Col of Med Iowa City IA 52240

MOYERS, JARVIS LEE, b Houston, Tex, Sept 7, 43; m 66; c 1. CHEMISTRY. *Educ:* Marshall Univ, BS, 65; Univ Hawaii, PhD(chem), 70. *Prof Exp:* Lab dir, Dept Chem, Univ Ariz, 71-82; PROG DIR, ATMOSPHERIC CHEM PROG, NSF, 82- *Mem:* AAAS; Am Chem Soc; Am Geophys Union; Am Soc Testing & Mat. *Res:* Analytic environmental chemistry; atmospheric chemistry. *Mailing Add:* NSF ATmer Sci 1800 G St NW Washington DC 20550

MOYLE, DAVID DOUGLAS, b Wilkes-Barre, Pa, Dec 10, 42; m 64; c 2. BIOMECHANICS, BIOMATERIALS. *Educ:* Wilkes Col, BS, 64; Rensselaer Polytech Inst, PhD(physics), 69. *Prof Exp:* Fel biomech, Rensselaer Polytech Inst, 69-71; asst prof, 71-75, ASSOC PROF BIOENG, CLEMSON UNIV, 75- *Concurrent Pos:* Vis scientist, Emery Univ Med Sch, 85-86; adj prof surg, Univ SC Med Sch, 84- *Mem:* Sigma Xi; AAAS. *Res:* Mechanics of skeletal tissue including properties of bone and bone replacement materials. *Mailing Add:* 216 Heather Dr Central SC 29630

MOYLE, PETER BRIGGS, b Minneapolis, Minn, May 29, 42; m 66; c 2. ICHTHYOLOGY, CONSERVATION BIOLOGY. *Educ:* Univ Minn, BA, 64, PhD(zool), 69; Cornell Univ, MS, 66. *Prof Exp:* Asst prof biol, Fresno State Col, 69-72; from asst prof to assoc prof, 72-83, chair, 82-88, PROF BIOL, DEPT WILDLIFE & FISHERIES BIOL, UNIV CALIF, DAVIS, 83- *Mem:* AAAS; Ecol Soc Am; Am Fisheries Soc; Am Soc Ichthyol & Herpet. *Res:* Ecology of freshwater and estuarine fishes; effects of species introductions; conservation of fishes of California and Sri Lanka. *Mailing Add:* Dept of Wildlife & Fisheries Biol Univ Calif Davis CA 95616

MOYLE, RICHARD W, b American Fork, Utah, Mar 22, 30; m 57; c 2. ENVIRONMENTAL, EARTH & MARINE SCIENCES, EARTH SCIENCES TEACHING. *Educ:* Brigham Young Univ, BS, 52, MS, 57; Univ Iowa, PhD(gen geol), 63. *Prof Exp:* Instr geol, Western State Col Colo, 61-63, asst prof, 63-65; assoc prof, Weber State Univ, 65-71, actg head dept geol & geog, 68-69, chmn dept, 69-72, & 85-87, PROF GEOL, WEBER STATE UNIV, 71- *Concurrent Pos:* Geologist mat eng, Region 4, Regional Off, US Forest Serv, 78-; Am Fedn of Mineralogical Soc Scholarship Award, 90. *Mem:* Geol Soc Am; Nat Asn Geol Teachers; Soc Econ Paleont & Mineral; Paleont Asn; Sigma Xi. *Res:* Mississippian and Pennsylvanian sponges; paleoecology of Upper Mississippian and Lower Pennsylvanian sediments in west central Utah; ammonoids of Wolfcampian from the Glass Mountains of west Texas and continguous areas; microcrystals and photography of same; Mississippian Blastoids of Utah. *Mailing Add:* Dept Geol Weber State Univ Ogden UT 84408-2507

MOYLS, BENJAMIN NELSON, b Vancouver, BC, May 1, 19; m 42, 76; c 2. ALGEBRA. *Educ:* Univ BC, BA, 40, MA, 41; Harvard Univ. AM, 42, PhD(math), 47. *Prof Exp:* From instr to prof, 47-84, asst dean grad studies, 67-76, head math dept, 78-83, EMER PROF MATH, UNIV BC, 84- *Mem:* Am Math Soc; Math Asn Am; Soc Indust & Appl Math; Can Math Soc (vpres, 81-83). *Res:* Linear algebra. *Mailing Add:* Dept Math Univ BC Vancouver BC V6T 1W5 Can

MOYNE, JOHN ABEL, b Yezd, Iran, July 6, 20; m 63; c 3. COMPUTERS, COMPUTATIONAL LINGUISTICS. *Educ:* Georgetown Univ, BA, 59, MA, 60; Harvard Univ, PhD(ling), 70. *Prof Exp:* Res assoc mach transl, Georgetown Univ, 56-63; mgr appl ling, IBM Corp, 63-71; assoc prof, 71-75, prof comput sci & chmn dept, 75-81, exec officer, PhD prog ling, 83-88, PROF LING & COMPUT SCI, GRAD SCH, CITY UNIV NY, 88- *Concurrent Pos:* Free lance writer, Brit Broadcasting Corp, London, 52-53; translr foreign broadcast, Cyprus, 53-56; researcher, Europ AFC, Italy, 61-63; teaching fel, Harvard Univ, 69-70; chmn info & technol comt, Ling Soc Am, Comput Sci Dept, Queens Col, 75-81, Div Math & Natural Sci, 79-81, univ comt instrnl comput, 73-77, res comt grad coun & various comt comput sci & ling progs; pres, Univ Consult Serv Inc, 84-85; adv comt ling, NY Acad Sci. *Mem:* Ling Soc Am; Asn Computational Ling; Brit Inst Eng Technol. *Res:* Linguistic theory; formal languages automata; programming languages and compilers; computational linguistics; author of over 100 publications and 9 books in linguistics, computer science and poetry. *Mailing Add:* PhD Prog Ling City Univ NY Grad Ctr 33 W 42nd St New York NY 10036-8099

MOYNIHAN, CORNELIUS TIMOTHY, b Inglewood, Calif, Feb 2, 39; div; c 2. GLASS SCIENCE, PHYSICAL CHEMISTRY. *Educ:* Univ Santa Clara, BS, 60; Princeton Univ, MA, 62, PhD(chem), 65. *Prof Exp:* From asst prof to assoc prof chem, Calif State Col Los Angeles, 64-69; assoc prof mat sci, Cath Univ Am, 69-75, prof mat sci & chem, 75-81; PROF MAT ENG, RENSSELAER POLYTECHNIC INST, 81- *Concurrent Pos:* Res assoc, Purdue Univ, 68-69. *Mem:* Fel Am Chem Soc; Am Ceramic Soc; Electrochem Soc. *Res:* Materials engineering of glasses; physical chemistry of molten salts, electrolyte solutions and glasses. *Mailing Add:* Dept Mat Eng Rennselaer Polytech Inst Troy NY 12181

MOYNIHAN, MARTIN HUMPHREY, b Chicago, Ill, Feb 5, 28. ANIMAL BEHAVIOR. *Educ:* Princeton Univ, AB, 50, DPhil(zool), Oxford Univ, 53. *Prof Exp:* Vis fel, Cornell Univ, 53-55; res fel, Harvard Univ, 55-57; DIR SMITHSONIAN TROP RES INST, 57- *Mem:* Soc Study Evolution; Am Ornith Union; Asn Trop Biol. *Res:* Behavior, ecology and evaluation. *Mailing Add:* Smithsonian Trop Res Inst PO Box 2072 Balboa Panama

MOZELL, MAXWELL MARK, b Brooklyn, NY, May 20, 29; m 68; c 5. SENSORY PHYSIOLOGY, PSYCHOPHYSIOLOGY. *Educ:* Brown Univ, AB, 51, MSc, 53, PhD(phsiol psychol), 56. *Prof Exp:* Fel physiol, Fla State Univ, 59-61; from asst prof to assoc prof, 61-70, assoc dean, 71-77, prof physiol, State Univ NY Upstate Med Ctr, 70-; PROJ DIR, SUNY CLIN OLFACTORY RES CTR, 83-, DEAN, COL GRAD STUDIES, 90- *Concurrent Pos:* Mem sensory physiol & perception study panel, NSF, 74-78; mem commun disorders rev comt, NIH, 81-85; chairperson, Int Comn Olfaction & Taste, 84-; mem adv bd, Nat Inst Deafness & other Commun Dis, NIH, 89- *Honors & Awards:* Manheimer Award, 89. *Mem:* Am Physiol Soc; Am Psychol Asn; Soc Neurosci; Sigma Xi; Asn Chemoreception Sci. *Res:* Sensory psychophysiology; olfaction; electrophysiology and voltage sensitive dyes; determine the physical, chemical, physiological mechanisms basic to olfactory discriminations. *Mailing Add:* Col Grad Studies State Univ NY Health Sci Ctr Syracuse Syracuse NY 13210

MOZER, BERNARD, b Denver, Colo, Dec 2, 25; m 57; c 3. PHYSICS. *Educ:* Univ Denver, BS, 50; Univ Colo, MS, 52; Carnegie Inst Technol, PhD(physics), 60. *Prof Exp:* Res assoc physics, Univ Denver, 52-53; jr physicist, Brookhaven Nat Lab, 53-55; res asst physics, Carnegie Inst Technol, 55-59; res assoc, Brookhaven Nat Lab, 59-61, from asst physicist to assoc physicist, 61-67; PHYSICIST, NAT BUR STANDARDS, 67- *Mem:* Am Phys Soc. *Res:* Theoretical plasma physics; theoretical and experimental aspects of Mossbauer effect; inelastic neutron scattering experiments on liquids, metals and alloys; vibrational and electronic effects of impurities in solids. *Mailing Add:* Reactor Bld Nat Bur Stand Gaithersburg MD 20899

MOZER, FORREST S, b Lincoln, Nebr, Feb 13, 29; m 61; c 3. SPACE PHYSICS. *Educ:* Univ Nebr, BS, 51; Calif Inst Technol, MS, 53, PhD(physics), 56. *Prof Exp:* Fel, Calif Inst Technol, 56-57; res scientist, Lockheed Res Lab, 57-61 & Aerospace Corp, 62-63; res dir space physics, Univ Paris, 63-66; from asst prof to assoc prof, 66-70, PROF PHYSICS, UNIV CALIF, BERKELEY, 70- *Mem:* Am Geophys Union; fel Am Phys Soc. *Mailing Add:* Space Sci Lab Univ of Calif 2120 Oxford St Berkeley CA 94720

MOZERSKY, SAMUEL M, b Sask, Can, Sept 19, 24; US citizen; m 55; c 2. FIELD FLOW FRACTIONATION & SUB CELLULAR PARTICLES. *Educ:* Univ Calif, Los Angeles, BA, 47; Univ Southern Calif, PhD(biochem), 57. *Prof Exp:* Sr lab asst turnover of plasma proteins, Col Med, Univ Ill, 52-55; grad res physiol chemist, Med Ctr, Univ Calif, Los Angeles, 56-57, asst res physiol chemist, 57-61; RES CHEMIST, EASTERN REGIONAL LAB, USDA, WYNDMOOR, 63- *Concurrent Pos:* USPHS spec fel, Dept Biol, Brookhaven Nat Lab, 61-63. *Mem:* Biophys Soc; Am Chem Soc; NY Acad Sci; Sigma Xi; Am Soc Biochem Molecular Biol; Fedn Am Soc Molecular Biol. *Res:* Biosynthesis of plasma proteins; purification and characterization of hyaluronidases; protein modification; mechanisms of enzyme action; structure and enzymatic properties of contractile proteins; size distribution of micro particles of biological origin; protein micelles. *Mailing Add:* 7916 Heather Rd Elkins Park PA 19117

MOZINGO, HUGH NELSON, b Monongahela, Pa, Apr 23, 25; m 49. BOTANY. *Educ:* Univ Pittsburgh, BS, 46, MS, 47; Columbia Univ, PhD(bot), 50. *Prof Exp:* Asst biol, Univ Pittsburgh, 46-47; asst bot, Columbia Univ, 47-50; instr, Univ Tenn, 50-51; assoc prof biol & bot & chmn sci div, Fla Southern Col, 51-55; asst prof, Mich State Univ, 55-59; from assoc prof to prof biol, 59-85, Univ Nev, Reno, chmn dept, 69-76, prof bot, 85; RETIRED. *Concurrent Pos:* Herbarium Cur, 59, emer herbarium cur. *Mem:* AAAS; Bot Soc Am; Am Soc Plant Physiol; Am Bryol & Lichenological Soc; Electron Micros Soc Am. *Res:* Plant morphogenesis; plant systematics; Nevada bryophytes. *Mailing Add:* Dept Biol Univ Nev Reno NV 89507

MOZLEY, JAMES MARSHALL, JR, b Marion, Ill, Nov 1, 22; m 44; c 1. BIOMEDICAL ENGINEERING. *Educ:* Washington Univ, BS, 43, MS, 47, PhD(chem & elec eng), 50. *Prof Exp:* Instr chem eng, Washington Univ, 47-49; assoc chemist sec oil recovery, Atlantic Refining Co, Tex, 49-51; sr instr chem eng, Polytech Inst Brooklyn, 51-52; res engr, automatic control, E I du Pont de Nemours & Co, Del, 51-57; assoc prof radiol, Johns Hopkins Univ, 57-65, dir div radiation chem, Sch Pub Health & Hyg, 59-65; dir, Div Radiol Physics & Eng, Univ Hosp, State Univ NY Upstate Med Ctr, 75-78, prof radiol, 65-85; PROF CHEM ENG, SYRACUSE UNIV, 67- *Concurrent Pos:* Pres, Radiation Assocs Md, Inc, 59-84; ed, Trans, Instrument Soc Am, 62-68; vis prof biomed eng, Wash Univ, St Louis, 74-75; adj prof elec eng, Syracuse Univ, 80-, mech & aero eng, 81- *Mem:* AAAS; Am Inst Chem Eng; Am Chem Soc; Am Soc Mech Eng; sr mem Instrument Soc Am. *Res:* Automatic control of chemical processes; radiological instrumentation; data processing; computation; nuclear medicine. *Mailing Add:* 126 Windcrest Camillus NY 13031-1924

MOZLEY, ROBERT FRED, b Boston, Mass, Apr 18, 17; c 2. ELEMENTARY PARTICLE PHYSICS, ARMS CONTROL. *Educ:* Harvard Univ, AB, 38; Univ Calif, MS & PhD(physics), 50. *Prof Exp:* Elec engr radar, Sperry Gyroscope Co, 41-45; asst radiation lab, Univ Calif, 45-50; from instr to asst prof physics, Princeton Univ, 50-53; assoc prof, 53-62, PROF PHYSICS, STANFORD UNIV, 62- *Mem:* Fel Am Phys Soc. *Res:* Elementary particle physics. *Mailing Add:* Stanford Linear Accelerator Ctr Stanford Univ Stanford CA 94305

MOZLEY, SAMUEL CLIFFORD, b Atlanta, Ga, Aug 13, 43; m 64, 85; c 2. AQUATIC ECOLOGY. *Educ:* Emory Univ, BS, 64, MS, 66, PhD(animal ecol), 68. *Prof Exp:* NATO fel, Max Planck Inst Limnol, Ger, 68-69, NSF fel, 69-70; Nat Res Coun Can fel, Univ Toronto, 70; from asst res scientist to assoc res scientist, Great Lakes Div, Univ Mich, Ann Arbor, 76-77; ASSOC PROF ZOOL, NC STATE UNIV, 77- *Mem:* Am Soc Limnol & Oceanog; NAm Lake Mgt Soc; Ecol Soc Am. *Res:* Zooplankton community ecology; reservoir limnology; freshwater invertebrate biology. *Mailing Add:* Dept Zool NC State Univ Raleigh NC 27695-7617

MOZUMDER, ASOKENDU, b Baherok, India, June 2, 31; m 61. RADIATION CHEMISTRY, RADIATION PHYSICS. *Educ:* Univ Calcutta, BSc, 50, MSc, 53; Indian Inst Technol, Kharagpur, PhD(physics). 61. *Prof Exp:* From assoc lectr to lectr physics, Indian Inst Technol, Kharagpur, 54-62; assoc, 62-65, from assoc res scientist to res scientist, 65-69, assoc fac fel, 69-86, FAC FEL RADIATION LAB, UNIV NOTRE DAME, 86- *Concurrent Pos:* Vis prof, Kyoto Univ, 73; vis fel, Wolfson Col, Oxford Univ, 86; vis scholar, Waseda Univ, 88; vis prof, Univ Paris, 86, Latvian State Univ, USSR, 89. *Mem:* Fel Am Phys Soc; Radiation Res Soc. *Res:* Theoretical radiation chemistry; application of the methods of theoretical physics to problems involving interaction of radiation with matter. *Mailing Add:* Radiation Lab Univ Notre Dame Notre Dame IN 46556

MOZURKEWICH, GEORGE, b Blakely, Pa, Apr 3, 53. CHARGE DENSITY WAVES, ELASTIC PROPERTIES. *Educ:* Muhlenberg Col, Allentown, Pa, BS, 75; Wash Univ, St Louis, PhD(physics), 81. *Prof Exp:* Res assoc physics, Univ Calif, Los Angeles, 81-84; ASST PROF PHYSICS, UNIV ILL, URBANA-CHAMPAIGN, 84- *Honors & Awards:* Presidential Young Investr Award, NSF, 85. *Mem:* Am Phys Soc. *Res:* Thermal, transport and electrical properties of quasi-one dimensional materials containing sliding charge density waves. *Mailing Add:* Dept Physics Univ Ill 1110 W Green St Urbana IL 61801

MOZZI, ROBERT LEWIS, b Meriden, Conn, Dec 8, 31; m 56; c 1. EXPERIMENTAL SOLID STATE PHYSICS. *Educ:* Villanova Univ, BS, 53; Univ Pittsburgh, MS, 56; Mass Inst Technol, PhD(physics), 68. *Prof Exp:* Physicist, Pratt & Whitney Aircraft Div, United Aircraft Corp, 55-57; CONSULT SCIENTIST, RES DIV, RAYTHEON CO, 57- *Mem:* Am Phys Soc. *Res:* X-ray diffraction studies of the structure of glass and imperfections in crystals; ion implantation in semiconductors; gallium arsenide microwave device technology; high resolution lithography. *Mailing Add:* Res Div Raytheon Co 131 Spring St Lexington MA 02173

MRAW, STEPHEN CHARLES, b Trenton, NJ, Jan 27, 50; m 71; c 5. THERMODYNAMICS. *Educ:* Fordham Univ, BS, 70; Univ Calif, Berkeley, PhD(phys chem), 74. *Prof Exp:* NATO fel chem, Inorg Chem Lab, Oxford Univ, 74-75; fel res assoc chem eng, Rice Univ, 75-76; SECT HEAD, EXXON RES & ENG CO, 76- *Concurrent Pos:* Dir, Calorimetry Conf, 80-82, prog chmn, 85, chmn, 86. *Honors & Awards:* Sunner Mem Award, 83. *Mem:* Am Chem Soc; Am Inst Chem Eng. *Res:* Heavy hydrocarbon chemistry, coal liquefaction, oil shale retorting and combustion thermodynamics, principally calorimetry of pure compounds and complex systems at high temperatures as related to fossil fuel technology; study of the structure of coal as related to other porous systems. *Mailing Add:* Exxon Res & Eng Co Rte 22E Annandale NJ 08801

MRAZEK, ROBERT VERNON, b Chicago, Ill, Jan 15, 36; m 59; c 3. CHEMICAL ENGINEERING. *Educ:* Purdue Univ, BS, 57; Rensselaer Polytech Inst, PhD(chem eng), 60. *Prof Exp:* From asst prof to assoc prof, 60-67, PROF CHEM ENG, ORE STATE UNIV, 67- *Concurrent Pos:* Consult, Albany Metall Res Sta, US Bur Mines, 63- *Mem:* Am Inst Chem Engrs; Sigma Xi. *Res:* Thermodynamics; applied mathematics. *Mailing Add:* 1525 N 14th Pl Corvallis OR 97330

MRAZEK, RUDOLPH G, b Chicago, Ill, May 23, 22; m 44; c 3. SURGERY. *Educ:* Univ Ill, BA, 41, MD, 44, MS, 45. *Prof Exp:* Resident surg, MacNeal Hosp, 45-46 & Hines Vet Admin Hosp, 49-52; from instr to clin assoc prof, 52-73, PROF CLIN SURG, UNIV ILL COL MED, 73-; DIR MED EDUC, MACNEAL HOSP, 71- *Mem:* AMA; Am Asn Cancer Res; fel Am Col Surg; Soc Surg Alimentary Tract; Soc Surg Oncol. *Res:* Cancer chemotherapy. *Mailing Add:* 3416 S Harlem Ave Riverside IL 60546

MROCZKOWSKI, STANLEY, b Poland, Mar 29, 25; US citizen; m 54; c 1. SOLID STATE CHEMISTRY. *Educ:* Adam Mickiewicz Univ, Poznan, MPH, 52; Univ Warsaw, PhD (inorg chem), 56. *Hon Degrees:* LLD, Jagiellonian Univ, Poland, 73. *Prof Exp:* Group leader rare earth compounds, Inst Electron Technol, Polish Acad Sci, Warsaw, 56-58; head analytical sect, Weitzman Inst, Albar Kvar-Saba, 59-61; res assoc chem, Columbia Univ, 61-63; SR RES ASSOC & LECTR APPL PHYSICS, CHEM RARE EARTH COMPOUND & SEMICONDUCTORS MAT, YALE UNIV, 63- *Concurrent Pos:* Tech consult, David Sarnoff Res Ctr, RCA Corp, 67- & Autoclave Engrs, Inc, Erie, Pa, 74-; mem staff Du Pont Exp Sta, 80- *Mem:* Am Chem Soc; Inst Elec & Electronics Engrs; Am Crystal Growth Asn. *Res:* Crystallization processes under high pressure and high temperature; crystal growth from high temperature solution; vapor transport reaction. *Mailing Add:* Yale Univ Dept of Eng & Appl Sci 427 Becton Ctr New Haven CT 06520

MROTEK, JAMES JOSEPH, b Loyal, Wis, Mar 19, 39; m 67. CELL BIOLOGY, ENDOCRINOLOGY. *Educ:* Univ Wis-Madison, BS, 64, MS, 65; Clark Univ, PhD(biol), 73. *Prof Exp:* Res technician qual control, Barley & Malt Lab, USDA, 60-61; res scientist adrenal physiol, Worcester Found Exp Biol, 68-69; fel adrenal physiol, Dept Physiol, Calif Col Med, Univ Calif, Irvine, 73-76; ASST PROF CELLULAR & MOLEC BIOL, DEPT BIOL SCI, N TEX STATE UNIV, 76- *Mem:* Soc Study Reprod; AAAS. *Res:* Cellular and molecular mechanisms of steroidogenesis; age-related changes in adrenal physiology; biochemistry of microfilaments; toxicological effects of cigarettes on cultured cells; cellular changes occurring in neoplastic cells. *Mailing Add:* Dept Biol Sci Meharry Med Col 100 S D B Todd Blvd Nashville TN 37208

MROWCA, JOSEPH J, b Taylor, Pa, Jan 25, 39; m 60; c 3. ORGANOMETALLIC CHEMISTRY. *Educ:* Univ Scranton, BS, 60; Columbia Univ, MA, 62, PhD(chem), 65. *Prof Exp:* SR RES ASSOC, AGRICHEM DEPT, E I DU PONT DE NEMOURS & CO, INC, 65- *Mem:* Am Chem Soc. *Res:* Transition metal catalysis. *Mailing Add:* 1513 Forsythia Ave Wilmington DE 19810

MROZIK, HELMUT, b Habelschwerdt, Germany, Oct 23, 31; m 57; c 2. ORGANIC CHEMISTRY. *Educ:* Univ Basel, PhD(chem), 58. *Prof Exp:* Asst chem, Columbia Univ, 59-60; SR SCIENTIST, MERCK & CO, 60- *Mem:* Am Chem Soc. *Res:* Chemotherapy of parasitic diseases; medicinal chemistry; natural product chemistry. *Mailing Add:* 159 Idlebrook Lane Matawan NJ 07747

MROZINSKI, PETER MATTHEW, b Chicago, Ill, Apr 22, 47; m 70; c 1. APPLIED PHYSICS, LOW TEMPERATURE PHYSICS. *Educ:* St Mary's Col, Minn, BA, 69; Ohio State Univ, MS, 72, PhD(physics), 75- *Prof Exp:* RES PHYSICIST, E I DU PONT DE NEMOURS & CO, INC, 74- *Mem:* Am Phys Soc. *Res:* Applications of x-ray fluorescence. *Mailing Add:* Exp Sta Bldg 357 E I du Pont de Nemours & Co Inc Wilmington DE 19898

MRTEK, ROBERT GEORGE, b Oak Park, Ill, Sept 2, 40; m 66. PHARMACY, HISTORY OF PHARMACY. *Educ:* Univ Ill, BS, 62, PhD(pharm), 67. *Prof Exp:* Resident res assoc chem, Argonne Nat Lab, 64-66; from asst prof to assoc prof pharm, 67-73, coordr educ res develop, 70-73, asst dean educ develop, 73-79, spec asst, Chancellor's off, Med Ctr, 80-81, PROF PHARM, COL PHARM, UNIV ILL, CHICAGO, 73-,. *Concurrent Pos:* Consult, Walter Reed Army Inst Res, 66-69 & Health Sci Educ Planning, Off Pres, Univ Calif, Berkeley, 79; res bd grants, Univ Ill, 68-69; consult & examr, Civil Serv Comn, City of Chicago, 69; res grant, US Vitamin & Pharmaceut Co, 70; spec proj prog, Health Professions, Bur Health Manpower Educ grant, 70; vis prof, Health Sci Ctr, Univ Wis, Madison, 75-76. *Honors & Awards:* C P van Schaak Chem Award, Lehn & Fink Gold Medal Award & Elich Prize, 62; Kremers Award, Am Inst Hist Pharm, 81; Lyman Award, Am Asn Col Pharm, 77; Fischelis Scholar, Am Inst Hist Pharm, 83. *Mem:* AAAS; Am Pharmaceut Asn; Am Inst Hist Pharm; Am Asn Col Pharm. *Res:* History of pharmacy in Illinois; history of pharmaceutical education and of American pharmacy practice; computer application in pharmacy. *Mailing Add:* 901 S Plymouth Ct Chicago IL 60605

MUAN, ARNULF, slag-refractory chemistry, high-temperature thermodynamics; deceased, see previous edition for last biography

MUCCI, JOSEPH FRANCIS, b Southington, Conn, Apr; m 53; c 3. PHYSICAL CHEMISTRY. *Educ:* Cent Conn State Col, BS, 50; Wesleyan Univ, MA, 53; Yale Univ, PhD, 57. *Prof Exp:* Asst, Yale Univ, 54-57; from instr to assoc prof, 57-66, PROF CHEM, VASSAR COL, 66-, CHMN DEPT, 75- *Mem:* AAAS; Am Chem Soc; NY Acad Sci; fel Am Inst Chem. *Res:* Complexions in various media by spectrophotometric, conductometric, polarographic, ion exchange, extraction and radiochemical methods; quantum chemical studies employing self consistent field-linear combination of atomic orbitals-molecular orbitals; statsitical mechanics. *Mailing Add:* Dept of Chem Vassar Col Poughkeepsie NY 12601

MUCENIEKS, PAUL RAIMOND, b Riga, Latvia, Feb 3, 21; US citizen; m 56; c 1. PHYSICAL CHEMISTRY, ELECTROCHEMISTRY. *Educ:* Johns Hopkins Univ, MA, 61, PhD(phys chem), 64. *Prof Exp:* Prin physicist, Litton Systs, Litton Indust, Inc, 64; sr res chemist, FMC Corp, 64-80, res assoc, 81-85. *Concurrent Pos:* Consult, 86- *Mem:* Am Chem Soc; Electrochem Soc; Am Inst Chem Engrs. *Res:* Studies of reaction kinetics and mechanisms; x-ray diffraction in heavy metal salt solutions; synthesis of stable free radicals; electrochemical synthesis and electrodialysis; chemical instrumentation; elimination of industrial pollutants; physical properties of solids; mineral processing. *Mailing Add:* 338 Glenn Ave Lawrenceville NJ 08648-3255

MUCHMORE, HAROLD GORDON, b Ponca City, Okla, Mar 8, 20; m 54; c 4. INTERNAL MEDICINE. *Educ:* Rice Univ, BA, 43; Univ Okla, MD, 46, MS, 56; Am Bd Internal Med, dipl, 62. *Prof Exp:* Intern, Jersey City Med Ctr, 46-47; univ fel, Univ Okla, 47-48, instr pharmacol, Sch Med, 48-49, instr med, 49-52, resident, Univ Hosps, 54-56, asst prof, 57-62; assoc prof, Med Sch, Univ Minn, 62-66; assoc prof med, microbiol & immunol, 66-70, chief infectious dis sect, 66-68, Carl Puckett assoc prof pulmonary dis, 68-70, prof microbiol & immunol, 71-79, Carl Puckett prof pulmonary dis & prof med, Med Sch, 70-86, adj prof, 80-87, EMER PROF MICROBIOL & IMMUNOL, UNIV OKLA, 87- *Concurrent Pos:* Clin investr, Vet Admin, 57-60, chief tuberc & infectious dis sect, Vet Admin Hosp, Oklahoma City, 60-62 & 66-; chief infectious dis sect, Ancker Hosp, St Paul, Minn, 62-66. *Mem:* Fel Am Col Physicians; Am Fedn Clin Res; Am Thoracic Soc; Med Mycol Soc Americas; Sigma Xi; fel Infectious Dis Soc Am; fel Int Soc Human & Animal Mycol. *Res:* Infectious diseases, especially fungus disease. *Mailing Add:* Vet Admin Infectious Dis & Med 921 NE 13th St Rm A542 Oklahoma City OK 73104

MUCHMORE, ROBERT B(OYER), b Augusta, Kans, July 8, 17; m 44; c 2. ELECTRONICS ENGINEERING. *Educ:* Univ Calif, BS, 39; Stanford Univ, EE, 42. *Prof Exp:* Proj engr, Sperry Gyroscope Co, 42-46; sr mem tech staff, Res & Develop Labs, Hughes Aircraft Co, 46-54; mem tech staff, TRW Systs, 54-60, dir electronics div, 60-61, dir phys res div, 61-65, vpres & assoc dir syst labs, 65-69, vpres & gen mgr, 69-71, vpres & chief scientist, Software & Info Systs Div, 71-73; CONSULT, 73- *Concurrent Pos:* Lectr, Univ Calif, Los Angeles, 53-58. *Mem:* Sigma Xi; fel Inst Elec & Electronics Engrs. *Res:* Microwave; stochastic processes; system analysis; radio propagation. *Mailing Add:* 4311 Grove St Sonoma CA 95476

MUCHMORE, WILLIAM BREULEUX, b Cincinnati, Ohio, July 7, 20; m 43; c 2. ARACHNOLOGY, PSEUDOSCORPIONIDA. *Educ:* Oberlin Col, AB, 42; Washington Univ, PhD(zool), 50. *Prof Exp:* From instr to prof, 50-85, EMER PROF BIOL, UNIV ROCHESTER, 85- *Concurrent Pos:* Res assoc, Fla State Collection Arthropods, 74- *Mem:* Am Micros Soc; Am Arachnolog Soc; Brit Arachnolog Soc. *Res:* Systematics and biogeography of pseudoscorpions. *Mailing Add:* Dept of Biol Univ of Rochester Rochester NY 14627-0211

MUCHOVEJ, JAMES JOHN, b Elizabeth, NJ, June 3, 53; m 75; c 2. MYCOLOGY, PHYSIOLOGY OF PARASITISM. *Educ:* Purdue Univ, BSc, 75, MSc, 76; Va Polytech Inst & State Univ, PhD(plant path), 84. *Prof Exp:* Lectr plant path, Fac Agr Sci, Para, Brazil, 77; lectr, 78-80, asst prof, 80-85, ASSOC PROF PLANT PATH, FED UNIV VICOSA, BRAZIL, 85- *Mem:* Am Phytopath Soc; Asn Appl Biologists; Mycol Soc Am; Brit Soc Plant Path; Brit Mycol Soc. *Res:* Interaction of phylloplane inhabiting microorganisms and the plant surfaces on which they reside; etiology of plant diseases. *Mailing Add:* Departamento de Fitopatologia Universidade Fed de Vicosa Vicosa MG 36570 Brazil

MUCHOW, GORDON MARK, b Evanston, Ill, June 15, 21; m 44; c 3. PHYSICAL CHEMISTRY. *Educ:* Northwestern Univ, Ill, BS, 42, MS, 51; St Louis Univ, PhD(chem), 54. *Prof Exp:* Res chemist, Pure Oil Co, 46-48; chemist, Graymills Corp, 48; res chemist, Monsanto Co, 53-62; from res scientist to sr res scientist, Owens-Ill, Inc. 62-73; staff technologist, Brush Wellman, Inc, 73-75; sr mat engr, 75-85, SCI WRITER, PRESTOLITE CO, 85- *Mem:* Sigma Xi. *Res:* Silicates; x-ray crystallography; crystal chemistry. *Mailing Add:* 5923 Winding Way Sylvania OH 43560

MUCHOWSKI, JOSEPH MARTIN, b Odessa, Sask, Jan 30, 37; m 65; c 3. ORGANIC CHEMISTRY. *Educ:* Univ Sask, BSc, 58, MSc, 59; Univ Ottawa, PhD(org chem), 59. *Prof Exp:* Nat Res Coun Can overseas fel with Prof A Eschenmoser, Swiss Fed Inst Technol, 62-63; sr res chemist, Bristol Labs of Can, Que, 63-71; sr chemist, Syntex, SA, Mexico City, 71-72, asst dir res, 72-73, DIR CHEM RES, SYNTEX, SA, MEXICO CITY, 73-, ASST DIR RES, SYNTEX RES CENTRE, PALO ALTO, 75- *Concurrent Pos:* Prof extraordinary, Iberoamerican Univ, Mex, 74-76; tour speaker, Atlantic Provinces Coun Sci, Can, 84. *Mem:* Am Chem Soc; Chem Soc Mex; Mex Acad Sci Res; Chem Inst Can. *Res:* Mechanistic and synthetic organic chemistry; medicinal chemistry. *Mailing Add:* Syntex Res Inst Org Chem 3401 Hillview Ave Palo Alto CA 94304

MUCK, DARREL LEE, b Larned, Kans, Jan 26, 38; m 60, 81; c 2. PHYSICAL ORGANIC CHEMISTRY. *Educ:* Wichita State Univ, BS, 59, MS, 62; Univ Fla, PhD(phys org chem), 65. *Prof Exp:* Res chemist, Procter & Gamble Co, 65-71 & Pfizer, Inc, 71-72; tech mgr detergents res, 72-74, prod mgr detergent chem, 74-77, mgr polymer additives, 77-80, bus mgr zeolite group,

Philadelphia Quartz Co, Valley Forge, Pa, 81-87; mgr technol, Global Detergent Chem, 88-89; com develop dir, 89-91, CONSULT, FMC CORP, PHILADELPHIA, 91- *Mem:* Am Oil Chemists Soc; Am Chem Soc; Soc Plastic Engrs. *Res:* Chelating and/or sequestering tendencies of organic hydroxy; acids and inorganic polymers to heavy metal ions. *Mailing Add:* Box 87 RD 5 Malvern PA 19355

MUCK, GEORGE A, b Fillmore, Ill, Sept 28, 37; m 59; c 5. FOOD SCIENCE, BIOCHEMISTRY. *Educ:* Univ Ill, BS, 59, MS, 61, PhD(dairy tech), 62. *Prof Exp:* Res asst food tech, Univ Ill, 59-62; head prod develop sect, Res Dept, Dean Foods Co, 62-67, dir res, 67-70, VPRES, RES & DEVELOP, RES DEPT, DEAN FOODS CO, 70- *Mem:* Am Dairy Sci Asn; Inst Food Technol; Am Chem Soc; Am Oil Chemist Soc. *Res:* Dairy technology; isolation and identification of flavors; effect of high heat treatment on model milk systems; development of new dairy and food products. *Mailing Add:* Res Dept Dean Foods Co 1126 Kilburn Ave Rockford IL 61103

MUCKENFUSS, CHARLES, b Cleveland, Ohio, May 2, 27; m 54; c 2. THEORETICAL CHEMISTRY. *Educ:* Univ Wis, PhD(chem), 57. *Prof Exp:* Nat Res Coun fel, Nat Bur Standards, Washington, DC, 57-58; res assoc, Gen Elec Res & Develop Ctr, NY, 58-67; ASSOC PROF CHEM ENG, RENSSELAER POLYTECH INST, 67- *Mem:* Am Phys Soc. *Res:* Kinetic theory; statistical mechanics; irreversible thermodynamics; transport phenomena. *Mailing Add:* Div Chem Eng Rensselaer Polytech Inst Troy NY 12180

MUCKENHOUPT, BENJAMIN, b Newton, Mass, Dec 22, 33; m 64; c 2. MATHEMATICAL ANALYSIS. *Educ:* Harvard Univ, AB, 54; Univ Chicago, MS, 55, PhD(math), 58. *Prof Exp:* From instr to asst prof math, DePaul Univ 58-60; from asst prof to assoc prof, 60-70, PROF MATH, RUTGERS UNIV, 70- *Concurrent Pos:* Vis assoc prof, Mt Holyoke Col, 63-65; visitor, Inst Advan Study, 68-69 & 75-76; vis prof, Stat Univ NY, Albany, 70-71. *Mem:* Am Math Soc; Math Asn Am. *Res:* Singular transformations; Fourier series. *Mailing Add:* Dept Math Rutgers Univ New Brunswick NJ 08903

MUCKENTHALER, FLORIAN AUGUST, b McFarland, Kans, July 31, 33; m 69; c 3. GENETICS, DEVELOPMENTAL BIOLOGY. *Educ:* Spring Hill Col, BS, 59; Catholic Univ, PhD(zool), 64. *Prof Exp:* USPHS fel cell biol, Johns Hopkins Univ, 64-65; asst prof biol, State Univ NY Albany, 65-71; from asst prof to assoc prof biol, 71-79, PROF BIOL, BRIDGEWATER STATE COL, 79-, DEPT CHMN, 90- *Concurrent Pos:* Vis prof med sci, Brown Univ, 80. *Mem:* AAAS; Am Genetics Asn; Am Inst Biol Sci; Am Soc Zool. *Res:* Developmental genetics; cell biology; mechanisms in meiosis and development. *Mailing Add:* Dept Biol Sci Bridgewater State Col Bridgewater MA 02324

MUCKERHEIDE, ANNETTE, MICROBIOLOGY, IMMUNOLOGY. *Educ:* Col Mt St Joseph, AB, 63; Drake Univ, MA, 73; Univ Cincinnati, PhD(microbiol), 78. *Prof Exp:* Teacher biol & chem, Sec Schs, NMex & Ohio, 65-73; teaching asst microbiol, Col Med, Univ Cincinnati, 74-78; instr biol, Col Mt St Joseph, 73-74, from asst prof to assoc prof, 78-88, chairperson dept, 79-85, PROF BIOL, COL MT ST JOSEPH, 88- *Concurrent Pos:* Adj prof microbiol, Col Med, Univ Cincinnati, 85-89; NIH acad res enhancement award, 87; Col Mt St Joseph res grants, 90 & 91. *Mem:* Am Asn Immunologists; AAAS; Nat Sci Teachers Asn. *Res:* Author of various publications on microbiology and immunology. *Mailing Add:* Dept Biol Col Mt St Joseph Mt St Joseph OH 45051

MUDD, J GERARD, b St Louis, Mo, Apr 7, 21; m 46; c 5. CARDIOLOGY. *Educ:* Col Holy Cross, BS, 43; St Louis Univ, MD, 45. *Prof Exp:* Intern med, 45-46, resident, 48-50, instr, 51-55, sr instr, 55-56, from asst prof to assoc prof med, 56-76, PROF INTERNAL MED, ST LOUIS UNIV, 76- *Concurrent Pos:* Fel, Johns Hopkins Univ, 50-51; St Louis Heart Asn fel cardiol, 51-52; Nat Heart Inst fel, 52-54. *Mem:* Am Heart Asn; Am Col Cardiol; Am Fedn Clin Res. *Res:* Cardiac catheterization, including right heart, left, retrograde and coronary arteriography. *Mailing Add:* Dept Internal Med 1325 S Grand Blvd St Louis MO 63104

MUDD, JOHN BRIAN, b Darlington, Eng, Aug 31, 29; m 74; c 1. BIOCHEMISTRY, ENVIRONMENTAL SCIENCES. *Educ:* Cambridge Univ, BA, 52; Univ Alta, MSc, 55; Univ Wis, PhD(biochem), 58. *Prof Exp:* Jane Coffin Childs Mem Fund Med Res fel, Univ Calif, Davis, 59-60; group leader, 81-86, vpres res, Plant Cell Res Inst, 87-90; from asst prof to prof biochem, 61-80, PROF BOT & PLANT SCI, UNIV CALIF, RIVERSIDE, 90-, DIR, STATEWIDE AIR POLLUTION RES CTR, 90- *Concurrent Pos:* Vis prof, Plant Res Lab, Mich State Univ, 79-80; sabbatical, Univ Utredrt, Neth, 67-68, Biozentrum, Basil, Switz, 74-75 & INRA, Bordeaux, France, 75. *Mem:* Am Chem Soc; Am Soc Plant Physiol; Brit Biochem Soc; Am Soc Biol Chem. *Res:* Lipid metabolism in plants; function of membranes in plants; biochemical basis for air pollution toxicity. *Mailing Add:* Statewide Air Pollution Res Ctr Univ Calif Riverside Riverside CA 92521

MUDD, STUART HARVEY, b Bryn Mawr, Pa, Apr 29, 27; m 55; c 3. BIOCHEMISTRY. *Educ:* Harvard Univ, BS, 49, MD, 53. *Prof Exp:* Intern med, Mass Gen Hosp, 53-54; MED DIR, LAB GEN & COMP BIOCHEM, NAT INST MENT HEALTH, 56- *Concurrent Pos:* NSF res fel, Biochem Res Lab, Mass Gen Hosp, Boston, 54-56. *Mem:* Am Soc Biol Chemists. *Res:* Oxidative phosphorylation; transmethylation; plant metabolism; mechanism of enzyme action. *Mailing Add:* Nat Inst Mental Health PHS 9507 Wadsworth Dr Bethesda MD 20892

MUDGE, GILBERT HORTON, b Brooklyn, NY, Apr 19, 15; m 41; c 4. PHYSIOLOGY, MEDICINE. *Educ:* Amherst Col, BA, 36; Columbia Univ, MD, 41, Med Sci Dr, 45. *Prof Exp:* Instr med, Columbia Univ, 48-49, assoc, 49-51, from asst prof to assoc prof, 51-55; prof pharmacol & exp therapeut & dir dept, Johns Hopkins Univ, 55-62, dean, 62-65, actg chmn dept med, 65-66, chmn dept, 66-67; prof, Dartmouth Med Sch, 65-81; RETIRED. *Concurrent Pos:* Mem pharmacol study sect, USPHS, 57-60, Life Ins Med Res Fund, 60-64 & Nat Res Coun, 60-62; assoc dean, Johns Hopkins Univ, 60-62, prof exp therapeut, 62-66; mem regulatory biol panel, NSF, 61-65. *Mem:* Soc Exp Biol & Med; Am Physiol Soc; Am Soc Clin Invest; Am Soc Pharmacol & Exp Therapeut; Asn Am Physicians. *Res:* Renal function; electrolyte physiology. *Mailing Add:* RR 1 Box 154 Lyme NH 03768

MUDHOLKAR, GOVIND S, b Aurangabad, India, Jan 5, 34. MATHEMATICAL STATISTICS. *Educ:* Univ Poona, BSc, 56, MSc, 57 & 58; Univ NC, PhD(statist), 63. *Prof Exp:* Lectr math & statist, SP Col, Poona, 57-60; from asst prof to assoc prof, 63-75, PROF STATIST & BIOSTATIST, MED SCH, UNIV ROCHESTER, 75- *Concurrent Pos:* Vis prof, Stanford Univ, 68-69 & KUL, Belgium, 78-79. *Mem:* Fel Inst Math Statist; Am Statist Asn; Am Math Soc; Math Asn Am; Biomet Soc. *Res:* Multivariate analysis; inequalities; goodness of fit problems; robust inference; statistical inference and methodology; biostatistics; quantal response problems. *Mailing Add:* Dept Biostatist Univ Rochester Med Ctr Rochester NY 14647

MUDREY, MICHAEL GEORGE, JR, b Pinebluff, Ark, Sept 22, 45; m 70; c 2. PRECAMBRIAN GEOLOGY, ORE DEPOSITS. *Educ:* Princeton Univ, AB, 67; Northern Ill Univ, MS, 69; Univ Minn, PhD(geol), 73. *Prof Exp:* Geologist precambrian geol, Minn Geol Surv, 69-72; proj scientist, Dry Valley Drilling Proj, Northern Ill Univ, 73-75, asst prof environ geol, 75; sect head geol, 79-81, GEOLOGIST, WIS GEOL SURV, 76- *Concurrent Pos:* Geologist mineral res, US Geol Surv, 74; adj prof, Dept Geol, Northern Ill Univ, 76-; asst prof, Dept Environ Sci, Univ Wis, Extension, 76-; asst prof, Dept Geol Sci, Univ Wis, Madison, 76- *Mem:* Fel Geol Asn Can; Soc Econ Geologists; Am Polar Soc. *Res:* Geologic evolution of the Lake Superior region, and the relationship of that history to mineral deposits; radiologic evaluation of geologic problems. *Mailing Add:* Wis Geol & Natural Hist Surv 3817 Mineral Pt Rd Madison WI 53705

MUDROCH, ALENA, b Prague, Czech, Nov 4, 30; Can citizen; m 50; c 2. GEOCHEMISTRY, LIMNOLOGY. *Educ:* State Col, Prague, BSc, 48; McMaster Univ, MSc, 73. *Prof Exp:* Lab asst chem, State Res Inst Metal Finishing & Corrosion, Prague, 48-52 & Acad Sci, Prague, 64-68; lab asst limnol, Can Ctr Inland Waters, 68-73, phys scientist geochem, 73-82; RES SCIENTIST, NAT WATER RES INST, BURLINGTON, ONT, 82- *Res:* Effects of toxic substances originated from dredge spoil disposal on existing aquatic ecosystem; evaluation the release and/or retention of nutrients and toxic metals by different type of marshes; pathways of environmental contaminants in aquatic ecosystem. *Mailing Add:* Nat Water Res Inst Burlington ON L7R 4A6 Can

MUDRY, KAREN MICHELE, b Philadephia, Pa, Nov 3, 48; m 72. BIOENGINEERING, BIOMEDICAL ENGINEERING. *Educ:* Villanove Univ, BEE, 70; Johns Hopkins Univ, MS, 72; Cornell Univ, PhD(bioeng), 78. *Prof Exp:* Elec engr, Dept Defense, 70-72; fel neurophysiol, Eye & Ear Hosp, Pittsburgh, Pa, 78-79; ASST PROF ELEC ENG, UNIV AKRON, 79- *Concurrent Pos:* Prin investr grants, Neurol & Commun Disorders & Stroke, NIH, 81-, Deafness Res Found, 81- *Mem:* Soc Neurosci; Sigma Xi; AAAS; Acoustical Soc Am; Inst Elec & Electronic Engrs. *Res:* Processing of sensory information in the peripheral and central nervous system of vertebrates; determination of organizational principles for central nervous system analysis of sensory information through a combination of neurophysiological, anatomical, and behavioral studies. *Mailing Add:* Dept Elec Eng Univ Akron Akron OH 44325

MUECKE, HERBERT OSCAR, b Kenedy, Tex, Jan 14, 40; m 61. MATHEMATICS. *Educ:* Univ Tex, Austin, BS & MA, 62, PhD(math), 68. *Prof Exp:* Asst prof, 68-74, ASSOC PROF MATH, SAM HOUSTON STATE UNIV, 74- *Res:* Complex analysis; symbolic logic; history and philosophy of science. *Mailing Add:* Dept Math Sam Houston State Univ 125 Hickory Huntsville TX 77341

MUECKLER, MIKE MAX, b Racine, Wis, Dec 27, 53; m; c 2. CELL BIOLOGY, PHYSIOLOGY. *Educ:* Univ Wis-Madison, BA, 76, PhD(oncol), 82. *Prof Exp:* NIH postdoctoral trainee, McArdle Lab Cancer Res, Univ Wis-Madison, 82-83; Damon Runyon-Walter Winchell Cancer Fund postdoctoral fel, Mass Inst Technol & Whitehead Inst Biomed Res, 83-86; asst prof, 86-91, ASSOC PROF, DEPT CELL BIOL & PHYSIOL, MED SCH, WASH UNIV, 91- *Concurrent Pos:* Ad hoc reviewer, J Biol Chem, J Cell Biol, J Clin Invest & J Cellular Physiol; mem, NIH Site Visit Comt, Albert Einstein Col Med, 88, Spec Reviewer Pool, NIH Physiol Sci Stud Sect, 89-, Prog Proj Rev Comt, Nat Inst Diabetes & Digestive & Kidney Dis, 90, Med Sci Rev Comt, Juv Diabetes Found, 91- & NIH Site Visit Comt, Boston Univ, 91; assoc ed, Diabetes, 92- *Honors & Awards:* Inbusch Award, 83. *Mem:* AAAS; Am Soc Cell Biologists. *Res:* Membrane transport, especially the structure, function and regulation of glucose transporters; insulin and growth hormone action; glucose metabolism, the genetics of obesity, and diabetes mellitus; membrane structure and biogenesis; intracellular protein trafficking; author of 37 technical publications. *Mailing Add:* Dept Cell Biol & Physiol Sch Med Wash Univ 660 S Euclid Ave St Louis MO 63110

MUEGGLER, WALTER FRANK, b Enterprise, Ore, May 22, 26; m 58; c 6. PLANT ECOLOGY, RANGE SCIENCE. *Educ:* Univ Idaho, BS, 49, Univ Wis, MS, 53; Duke Univ, PhD, 61. *Prof Exp:* Plant ecologist, 49-74, PROJ LEADER, INTERMOUNTAIN FOREST & RANGE EXP STA, US FOREST SERV, 74- *Mem:* Ecol Soc Am; Soc Range Mgt. *Res:* Range and wildlife habitat research; plant synecology and autecology. *Mailing Add:* 1541 E 1630 N N Logan UT 84321

MUEHLBERGER, WILLIAM RUDOLF, b New York, NY, Sept 26, 23; m 49; c 2. GEOLOGY, EARTH SCIENCES. *Educ:* Calif Inst Technol, BS & MS, 49, PhD(geol), 54. *Prof Exp:* From asst prof to prof geol, Univ Tex Austin, 54-84, chmn dept geol sci, 66-70, William Stamps Farish Chair, 85-88, PETER T FLAWN CENTENNIAL CHAIR GEOL, UNIV TEX, AUSTIN, 88- *Concurrent Pos:* Geologic field asst, US Geol Surv, 48-49, geologist, 49

& 71-; geologist, State Bur Mines & Mineral Resources, NMex, 53-61, dir crustal studies lab, 61-66; prin investr, Apollo Field Geol Invests, Apollo 16 & 17, 71-74; dir, Tectonic Map NAm Proj, 80-88. *Honors & Awards:* Matson Award, Am Asn Petrol Geologists, 64; Medal Except Sci Achievement, NASA, 73. *Mem:* Geol Soc Am; Am Asn Petrol Geol; Am Geophys Union. *Res:* Structural, areal and lunar geology; analysis of global tectonics; tectonics of North America. *Mailing Add:* Dept Geol Sci Univ Tex Austin TX 78713-7909

MUEHLING, ARTHUR J, b Cissna Park, Ill, June 3, 28; m 56; c 2. PLANNING SWINE SYSTEMS. *Educ:* Univ Ill, BS, 50; Univ Mo, MS, 51. *Prof Exp:* Res assoc agr eng, 56-59, from asst prof to assoc prof, 59-74, PROF AGR ENG, UNIV ILL, 74- *Concurrent Pos:* Weather officer, USAF, 52-56; vis res fel, Max Planck Inst, Bad Kreuznach, Ger, 69-70; sabbatical, Australian Depts Agr, 80-81, Europ Bldg Res Insts, 80. *Honors & Awards:* Paul A Funk Award, Col Agr, Univ Ill, 79. *Res:* Assisting producers plan intensified livestock systems, including function and environmental planning; livestock waste management systems. *Mailing Add:* 1501 W John St Champaign IL 61821

MUEHRCKE, ROBERT C, b Cincinnati, Ohio, Aug 4, 21. NEPHROLOGY, NUTRITION. *Educ:* Univ Ill, MSc, 52. *Prof Exp:* DIR MED EDUC, W SUBURBAN HOSP, 65-; PROF MED, RUSH MED COL, 78- *Mem:* Am Fedn Clin Res; Am Heart Asn. *Mailing Add:* Muehroke Med Inc PO Box 391256 Cleveland OH 44139

MUELLER, ALFRED H, b Chicago, Ill, June 9, 39. PHYSICS, THEORETICAL PHYSICS. *Educ:* Univ Ohio, BS, 61; Mass Inst Technol, PhD(physics), 65. *Prof Exp:* Physicist, Brookhaven Lab, 65-72; PROF PHYSICS, COLUMBIA UNIV, 72- *Mem:* Am Phys Soc. *Mailing Add:* Columbia Univ New York NY 10027

MUELLER, AUGUST P, b Fargo, NDak, July 30, 33; m 58; c 2. IMMUNOLOGY, SEROLOGY. *Educ:* Moorhead State Col, BSc, 55; Univ Wis, MSc, 57, PhD(zool, chem), 60. *Prof Exp:* Res assoc zool, Univ Wis, 60 & 62; fel, Univ Edinburgh, 61-62; asst prof, 62-67, ASSOC PROF BIOL, STATE UNIV NY BINGHAMTON, 67- *Concurrent Pos:* NIH res grant, 63-66. *Mem:* AAAS; Am Soc Zool; Reticuloendothelial Soc; Genetics Soc Am; Am Inst Biol Sci; Sigma Xi. *Res:* Immune unresponsiveness in juvenile and adult animals; development of the immune system; bursa of Fabricus in chickens and the thymus in mammals; genetics; biometry. *Mailing Add:* Dept Biol State Univ NY Binghamton NY 13901

MUELLER, BRUNO J W, b Bad Emn, Ger, Jul 10, 34. ALGEBRA. *Educ:* Univ Maiaz, Ger, PhD(math), 59, Hobiliation, 64. *Prof Exp:* PROF MATH, MCMASTER UNIV, 66- *Mem:* Am Math Soc. *Res:* Non commulative xoetherian rings. *Mailing Add:* Math Dept McMaster Univ Hamilton ON L8S 4K1 Can

MUELLER, CHARLES CARSTEN, b Glenwood Springs, Colo, July 11, 37; m 59, 73. GEOTECHNICAL FOUNDATIONS ENGINEERING. *Educ:* Calif State Polytech Col, San Luis Obispo, BS, 63; Mich State Univ, MS, 65, PhD(agr eng), 69. *Prof Exp:* Res asst agr eng, Mich State Univ, 63-67; asst prof agr eng & asst agr engr, Wash State Univ, 68-73; lectr, 73-74, assoc prof, 74-90, PROF CIVIL ENG, CIVIL ENG DEPT, CALIF STATE UNIV, CHICO, 78- *Mem:* Am Soc Civil Engrs; Am Soc Testing & Mat; Am Soc Eng Educ. *Res:* Building foundations, irrigation and drainage; soil mechanics. *Mailing Add:* Dept Civil Eng Calif State Univ Chico CA 95929-0930

MUELLER, CHARLES FREDERICK, b Sharon, Pa, Oct 3, 39; m 64; c 2. ECOLOGY. *Educ:* Ind Univ Pa, BSEd, 62; Ohio Univ, MS, 65; Mont State Univ, PhD(zool), 67. *Prof Exp:* Assoc prof, 67-76, PROF BIOL, SLIPPERY ROCK UNIV, 76-, CHMN DEPT, 75- *Concurrent Pos:* NSF res grant, Slippery Rock State Col, 69-71; consult herpetol, Aquatic Ecol Assocs, Pittsburgh; res asst zool, Univ Bristol, Eng, 84. *Mem:* Soc Study Amphibians & Reptiles; Am Soc Ichthyologists & Herpetologists; Sigma Xi. *Res:* Temperature and energy characteristics; bioenergetics of poikilotherms. *Mailing Add:* Dept of Biol Slippery Rock Univ Slippery Rock PA 16057

MUELLER, CHARLES RICHARD, theoretical chemistry, physical chemistry; deceased, see previous edition for last biography

MUELLER, CHARLES W(ILLIAM), b New Athens, Ill, Feb 12, 12; m 41; c 2. ELECTRONICS. *Educ:* Univ Notre Dame, BS, 34; Mass Inst Technol, MS, 36, ScD(physics), 42. *Prof Exp:* Prod engr, Raytheon Prod Corp, Mass, 36-37, develop engr, 37-38; res asst, Off Sci Res & Develop Proj, Mass Inst Technol, 40-41; res physicist, RCA Labs, 42-85; RETIRED. *Honors & Awards:* David Sarnoff Sci Award, 66; J J Ebers Award, Inst Elec & Electronics Engrs, 72. *Mem:* Am Phys Soc; fel Inst Elec & Electronics Engrs. *Res:* Development of transistors; single-crystal silicon films on insulators; silicon-target image and storage tubes; RCA Corp. *Mailing Add:* 381 Riverside Dr Princeton NJ 08540

MUELLER, DALE M J, b Grand Forks, NDak, June 24, 39; m 64; c 2. BOTANY, BRYOLOGY. *Educ:* Sch Forestry, NDak State, AS, 60, BS, 62; Okla State Univ, Stillwater, MS, 64; Univ Calif, Berkeley, PhD(bot), 70. *Prof Exp:* Mus technician res, Dept Bot, Univ Calif, Berkeley 70-71, actg asst prof & fel, 71; vis asst prof, Univ Minn, 71-72; asst prof, 72-78, ASSOC PROF BIOL, TEX A&M UNIV, COLLEGE STATION, 78- *Concurrent Pos:* Ed, Bryologist, Am Bryological & Lichenological Soc, 75-79. *Mem:* Sigma Xi; Am Bryological & Lichenological Soc (secy-treas, 85-89); Bot Soc Am; Int Asn Bryologists. *Res:* Studies of the anatomy-morphology (including ultrastructure) and phylogeny of bryophytes, specifically, sporophyte development, sporogenesis, mechanisms of spore dispersal and methods of asexual propagation (reproductive biology). *Mailing Add:* Dept Biol Tex A&M Univ College Station TX 77843-3258

MUELLER, DELBERT DEAN, b Claremore, Okla, Oct 22, 33; m 59; c 3. PHYSICAL BIOCHEMISTRY. *Educ:* Univ Okla, BS, 62, PhD(phys chem), 66. *Prof Exp:* Res assoc phys biochem, Northwestern Univ, 66-68; asst prof, 68-75, ASSOC PROF BIOCHEM, KANS STATE UNIV, 75- *Concurrent Pos:* Vis staff mem, Los Alamos Sci Lab, 74- *Mem:* AAAS; Am Soc Biol Chemists; Am Chem Soc; Sigma Xi. *Res:* Physical studies on biopolymers with emphasis on the applications of carbon-13 and phosphorus-31 nuclear magnetic resonance spectroscopy and hydrogen exchange techniques to conformational change and binding problems. *Mailing Add:* 2116 McDowell Ave Manhattan KS 66502

MUELLER, DENNIS, MAGNETICALLY CONFINED PLASMA, PLASMA-WALL INTERACTIONS. *Educ:* MacMurray Col, Ill, BA, 68; Mich State Univ, MS, 74, PhD(physics), 76. *Prof Exp:* Instr physics, Princeton Univ Cyclotron Lab, 76-78; mem res staff, 78-83, res physicist, 83-91, PRIN RES PHYSICIST, PRINCETON PLASMA PHYSICS LAB, 91-, BR HEAD, TOROIDAL KAMERA MAGNETIC FUSION TEST REACTOR PHYSICS OPERS, 87- *Mem:* Am Phys Soc; AAAS. *Res:* Magnetically-confined controlled thermonuclear fusion; confinement; auxilliary heated tokamalc plasmas; plasma-wall interactions and conditioning. *Mailing Add:* Princeton Plasma Physics Lab PO Box 451 L0B-258 Princeton NJ 08544

MUELLER, DENNIS W, b Abock, Nebr, May 11, 49. PHYSICS, SOLID STATE PHYSICS. *Educ:* Univ Nebr, MA, 76, PhD(physics), 82. *Prof Exp:* Res assoc, Univ Colo, 82-84; res assoc, Univ Bielefeld, 84-85; asst prof physics, La State Univ, 85-87; ASST PROF PHYSICS, N TEX STATE UNIV, 87- *Mem:* Am Phys Soc. *Res:* Study of atom physics and molecular physics. *Mailing Add:* Dept Physics NTex State Univ PO Box 5368 Denton TX 76203-5368

MUELLER, DONALD SCOTT, plastics chemistry, polymer chemistry, for more information see previous edition

MUELLER, EDWARD E(UGENE), b Wood River, Ill, July 26, 24; m 46; c 3. CERAMICS. *Educ:* Univ Mo, BS, 48; Rutgers Univ, MS, 52, PhD(ceramics), 53. *Prof Exp:* Instr, Rutgers Univ, 48-53; assoc prof ceramic eng, Univ Wash, 53-59; dir ceramics res, Chem Group, Glidden Co, Md, 59-65; prof ceramics & dean, Alfred Univ, 65-73, prof ceramic eng, 73-90, dir, Placement & Co-op Prog, 82-90, EMER PROF CERAMIC ENG, NY STATE COL CERAMICS, ALFRED UNIV, 90- *Concurrent Pos:* Consult, Boeing Airplane Co, United Control Corp, US Army Corps Engrs, Gladding, McBean & Co, Inc, Gen Elec Co, Du Pont, Union Carbide & Xerox; dir, Study Abroad Prog, State Col Ceramics, Alfred Univ, 88-90. *Mem:* Fel Am Ceramic Soc; Nat Inst Ceramic Engrs (pres, 65-66); Am Soc Eng Educ. *Res:* Ceramic engineering, science education. *Mailing Add:* NY State Col Ceramics Alfred Univ Alfred NY 14802

MUELLER, FRED(ERICK) M(ARION), electrical engineering, for more information see previous edition

MUELLER, FRED MICHAEL, b Chicago, Ill, Oct 8, 38; div; c 2. SOLID STATE PHYSICS. *Educ:* Univ Chicago, SB, 61, SM, 62, PhD(physics), 66. *Prof Exp:* Assoc physicist, Argonne Nat Lab, 66-73, sr scientist, 73-75; prof, physics lab, Nijmegen, Neth, 75-85; STAFF MEM, LOS ALAMOS NAT LAB, 85- *Concurrent Pos:* Consult, Northwestern Univ, 69-, Univ Chicago, 69-, Stanford Univ, 70- & Am Photo Copying Equip Co, 71-; prof physics, Northern Ill Univ, 69-75. *Mem:* Fel Am Phys Soc. *Res:* Electronic structure of transition metal compound; phonon spectra; superconductivity. *Mailing Add:* Ctr Mat Sci Los Alamos Nat Lab Los Alamos NM 87545

MUELLER, GEORGE E(DWIN), b St Louis, Mo, July 16, 18. COMMUNICATIONS, ELECTRONICS. *Educ:* Univ Mo, BS, 39; Purdue Univ, MS, 40; Ohio State Univ, PhD(physics), 51. *Hon Degrees:* PhD(eng), Univ Mo & Wayne State Univ, 64, Purdue Univ, 65; PhD(laws), NMex State Univ, 64; LLD, Pepperdyne Univ, 79. *Prof Exp:* Mem tech staff, Bell Tel Labs, Inc, 40-46; prof elec eng, Ohio State Univ, 46-57; vpres & dir space systs, Space Technol Labs, Calif, 47-63; assoc adminr manned space flight, NASA, DC, 63-69; sr vpres, Gen Dynamics Corp, NY, 69-71; chmn bd & pres, Syst Develop Corp, 71-80, chmn & chief exec officer, 81-83; sr vpres, Burroughs Corp, 81-83; PRES, JOJOBA PROPAGATION LABS, 81-; PRES, GEORGE E MUELLER CORP, 84- *Concurrent Pos:* Vpres, Int Astronaut Fedn; pres, Next Peripherals Inc, 87-88. *Honors & Awards:* Nat Medal Sci, 70; Sterry Award, 86; Eugen Sanger Award, 70; Goddard Medal, Am Inst Aeronaut & Astronaut, 83; Fono Albert Emlekerem Space Award, Hungary, 83; Nat Transp Award, Am Soc Mech Engrs, 79; Gaugarin Medal, Acad Sci, USSR, 89. *Mem:* Nat Acad Eng; hon fel Am Inst Aeronaut & Astronaut (pres, 79-80); Inst Elec & Electronics Engrs; fel Am Geophys Union; NY Acad Sci; fel AAAS; fel Am Astronaut Soc; fel Royal Aeronaut Soc; Inst Adv Eng; Sigma Xi; Am Phys Soc; Nat Acad Aeronaut (pres, 83-). *Mailing Add:* PO Box 5856 Santa Barbara CA 93150

MUELLER, GEORGE PETER, b Atchison, Kans, Aug 7, 18; m 46; c 1. ORGANIC CHEMISTRY. *Educ:* Univ Nebr, BSc, 40, MSc, 41; Univ Ill, PhD(org chem), 43. *Prof Exp:* Chemist, State Hwy Testing Lab, Nebr, 38-40 & Eastman Kodak Co, 42; asst, Univ Ill, 44; res assoc, Harvard Univ, 44-46; chemist, Wyeth Inst Appl Biochem, Pa, 46-47; assoc prof chem, Univ Tenn, 47-52; res supvr, G D Searle & Co, 52-62, coordr, Searle Chem Inc, Ill & Mex, 59-62; dir res, Marine Colloids, Inc, Maine, 62-68; CONSULT, 68- *Concurrent Pos:* Fel, Univ Ill, 44-45; consult, Oak Ridge Nat Lab & Oak Ridge Inst Nuclear Studies, 49-52. *Mem:* AAAS; Am Chem Soc; NY Acad Sci. *Res:* Synthetic and steroidal estrogens; barbiturates; insecticides; alicyclic synthesis; androgens; corticoids; isolation, pharmacology; structure of botanical isolates; sources, chemistry steroidal sapogenins; seaweed sources; structure, modification, utilization of seaweed colloids; polysaccharides. *Mailing Add:* Five Cedar St Camden ME 04843

MUELLER, GERALD CONRAD, b Centuria, Wis, May 22, 20; m 44; c 3. BIOCHEMISTRY. *Educ:* Univ Wis, BS, 43, MD, 46, PhD(biochem, physiol), 50. *Prof Exp:* Intern, Med Col, Va Hosp, 47; from instr to assoc prof oncol, 50-58, asst prof acad affairs, 63-67, PROF ONCOL, MCARDLE LAB CANCER RES, UNIV WIS-MADISON, 58-, PROF HEALTH SCI MED, 77- *Concurrent Pos:* Schering scholar, Max Planck Inst Virus Res, 58; mem drug eval panel, Cancer Chemother Nat Serv Ctr, 60-61, chmn, 61-62, mem biochem comt, 60-62; USPHS res career award, Univ Wis-Madison, 62-; mem bd sci coun, Nat Cancer Inst, 65-69, mem organizational task force, 68, mem chemother adv comt, 69-; vis prof, Univ Sao Paulo, 71. *Mem:* Am Soc Biol Chem; Am Asn Cancer Res. *Res:* Mechanism of action of estrogenic hormones; molecular processes regulating animal cell replication; intermediate metabolism of growth regulation. *Mailing Add:* McArdle Lab Cancer Res Univ Wis 450 N Randall Ave Madison WI 53706

MUELLER, GERALD SYLVESTER, chemical engineering, automatic control systems, for more information see previous edition

MUELLER, HELMUT, b Schneeberg, Ger, Jan 2, 26; US citizen; m 53; c 2. BIOCHEMISTRY. *Educ:* Univ W rzburg, MD, 52; Univ Birmingham, PhD(biochem), 61. *Prof Exp:* Univ fel, Univ Colo, 53-54; resident physician, Northwestern Univ, 54-56; Am Heart Asn res fel, Inst Muscle Res, Mass, 57-58, adv res fel, 58-60; asst res prof biochem, Univ Pittsburgh, 61-64, assoc res prof, 64-65; sr fel, Mellon Inst, 65-68; Health Sci Adminr & Med Officer Drug Surveillance, Ctr Pop Res, Nat Inst Child Health & Human Develop, 68-86; RETIRED. *Concurrent Pos:* Am Heart Asn estab investr, Univ Birmingham, 60-61 & Univ Pittsburgh, 61-65. *Mem:* AAAS; Am Soc Biol Chem; Soc Gen Physiol; Biophys Soc. *Res:* Structure and function of contractile proteins; cardiovascular disease; epidemiology of adverse drug effects; drug-nutrient interactions. *Mailing Add:* 319 Brown St Box 36 Washington Grove MD 20880

MUELLER, HELMUT CHARLES, b Milwaukee, Wis, Mar 20, 31; m 59; c 1. ZOOLOGY, ANIMAL BEHAVIOR. *Educ:* Univ Wis-Madison, BS, 53, MS, 58, PhD(zool), 62. *Prof Exp:* Res assoc zool, Univ Wis-Madison, 62-65, lectr, 66; from asst prof to assoc prof, 66-77, PROF BIOL, UNIV NC, CHAPEL HILL, 77- *Concurrent Pos:* Vis prof, Univ Vienna, 75. *Mem:* Fel AAAS; fel Am Ornithologists Union; Wilson Ornith Soc; Ecol Soc Am; Animal Behav Soc. *Res:* Behavioral aspects of the predator-prey interaction; hawk behavior; bird migration; behavioral ecology. *Mailing Add:* Dept Biol Univ NC Chapel Hill NC 27514

MUELLER, HILTRUD S, b Amorbach-Heidelberg, Ger, July 4, 26; US citizen. CARDIOLOGY, INTERNAL MEDICINE. *Educ:* Univs Jena & Heidelberg, MD, 50. *Prof Exp:* Fac mem, USPHS training prog, St Vincent's Hosp, NY, 67-68, dir intensive care & shock unit, Hosp & Med Ctr, 69-73; dir div cardiol, St Vincent's Hosp, Mass, 73-75; chief div cardiol & prof med, Sch Med, St Louis Univ, 76-81; ASSOC CHIEF, DIV CARDIOL, MONTEFIORE HOSP & PROF MED, ALBERT EINSTEIN COL, BRONX, NY, 81- *Concurrent Pos:* Fel cardiol, Univ Hosp Minn, 65-66 & St Vincent's Hosp & Med Ctr, NY, 66-67; mem cardiol adv comt, Nat Inst Heart, Lung & Blood Inst, 76-80, Cardiovasc & Renal Study Sect, Div Res Grants; consult, Ischemic Sci Ctr Res Comt, Nat Inst Heart, Lung & Blood Inst, 76- *Mem:* Fel Am Col Chest Physicians; fel Am Col Cardiol; fel Am Col Physicians; fel NY Acad Sci; Fedn Clin Res. *Res:* Acute ischemic heart disease; influence of sympathetic nervous systems on ischemia; myocardial metabolism; coronary blood flow; beta adrenergic blockade. *Mailing Add:* Dept Med Albert Einstein Col Med 1300 Morris Park Ave Bronx NY 10461

MUELLER, IRENE MARIAN, b St Libory, Nebr, July 12, 04. PLANT ECOLOGY. *Educ:* Nebr Cent Col, AB, 27; Univ Nebr, AM, 37, PhD(plant ecol), 40. *Prof Exp:* Teacher high schs, Nebr, 27-35; asst bot, Univ Nebr, 36-39; instr biol sci, Wis State Teachers Col, Platteville, 40-43; from assoc prof to prof, 43-75, EMER PROF BIOL, NORTHWEST MO STATE UNIV, 75- *Res:* Rhizomes of prairie plants; drought resistance in prarie plants. *Mailing Add:* 728 West Third St Maryville MO 64468

MUELLER, IVAN I, b Budapest, Hungary, Jan 9, 30; US citizen; m 50; c 2. GEODESY, GEOPHYSICS. *Educ:* Budapest Tech Univ, dipl eng, 52; Ohio State Univ, PhD(geod sci), 60. *Prof Exp:* Asst prof geod, Budapest Tech Univ, 52-56; design engr, C H Sells Consult Engr, NY, 57-58; from instr to assoc prof, 59-66, PROF GEOD, OHIO STATE UNIV, 66-, CHMN, 84- *Concurrent Pos:* Mem geod-cartog working group, Manned Space Sci Coord Comt, NASA, 65-66, prin investr, Nat Geod Satellite Prog, 65-74, mem adv group satellite geod, 66-67, geod & cartog subcomt & space sci & appln steering comt, 67-68; panel solid earth geophys & earthquake eng, comt adv to Environ Sci Serv Admin, Nat Acad Sci-Nat Acad Eng, 67-69; assoc ed, J Geophys Res, 67-74; consult, UN Develop Prog, Ctr Survey Training & Map Prod, Hyderabad, India, 71 & 72; ed in chief, Bull Geodesique, Paris, 75-87; vis prof, Univ Berlin, 65, 69, Univ Parana, Brazil, 72-74, 78 & Univ Stuttgart, 80; chmn subcomt, Int Asn Geodesy, mem, US Nat Comm, Int Union of Geodesy & Geophys, 76-, vchmn, 80-84; mem comt geod, Nat Acad Sci, 76-81, chmn, 78-81; corres mem, Ger Geod Comn, Bavarian Acad Sci, Munich, 80; pres, Int Union Geod & Geophysics, Int Asn Geod, vchmn, Joint Working Group Determination Rotation of Earth. *Honors & Awards:* Alexander von Humboldt Prize, Bonn-Munich, 77. *Mem:* Am Soc Photogram; fel Am Geophys Union (pres/geod, 74); Am Astron Soc; Int Astron Union; Int Cong Surv & Mapping; foreign mem Austrian Nat Acad Sci; Int Asn Geod (vpres, 83-87, pres, 87-). *Res:* Geodetic astronomy; gravimetric and satellite geodesy. *Mailing Add:* Dept Geod Sci & Surv Ohio State Univ 405 Cockins Columbus OH 43210

MUELLER, JAMES LOWELL, remote sensing, ocean optics, for more information see previous edition

MUELLER, JOHN FREDERICK, b Goshen, Ind, June 15, 22; m 45; c 4. INTERNAL MEDICINE. *Educ:* Capital Univ, BA, 44; Univ Cincinnati, MD, 46. *Prof Exp:* Intern, King's County Hosp, Brooklyn, NY, 46-47; resident internal med & fel hemat & nutrit, Univ Cincinnati, 47-50; fel hemat, Western Reserve Univ, 50-51; from instr to assoc prof med, Sch Med, Univ Cincinnati, 51-62, assoc dir lab hemat & nutrit, 55, co-dir, 57-62; prof med, Univ Colo, 62-64; prof, State Univ NY Downstate Med Ctr, 64-73; PROF MED, UNIV COLO MED CTR, DENVER, 73- *Concurrent Pos:* Chief clinician, Outpatient Dept, Cincinnati Gen Hosp, 57-62; chief med, Denver Vet Admin Hosp, 62-64; physician-in-chief, Brooklyn-Cumberland Med Ctr, 64-73; consult, Surgeon Gen, NIH; dir internal med, St Lukes Hosp, 73-81; dir acad affairs, Presbyterian/St Lukes Med Ctr, 81- *Mem:* Am Soc Clin Invest; Am Fedn Clin Res; Am Soc Clin Nutrit (secy-treas, 63-66); Am Heart Asn; Am Col Physicians. *Res:* General field of lipid chemistry of blood and tissues as related to various disease states; nutrition and hematology. *Mailing Add:* 3333 E Florida Ave No 88 Denver CO 80210

MUELLER, JUSTUS FREDERICK, b Baltimore, Md, Nov 20, 02. ZOOLOGY. *Educ:* Johns Hopkins Univ, AB, 23; Univ Ill, MA, 26, PhD(zool), 28. *Prof Exp:* Sci asst, Bur Fisheries, 23-24; asst zool, Univ Ill, 24-28; from instr to assoc prof zool, State Univ NY Col Forestry, Syracuse Univ, 28-42, assoc prof parasitol, Col Med, Univ, 42-50; assoc prof parasitol, 50-56, actg chmn dept microbiol, 54-57, prof microbiol, 56-72, EMER PROF, STATE UNIV NY UPSTATE MED CTR, 72- *Concurrent Pos:* Field naturalist, Roosevelt Wild Life Forest Exp Sta, 28-35; mem trop med & parasitol study sect, NIH, 62-66; leader Ore Biol Colloquium, 65; fel trop med, Cent Am Prog, La State Univ, 65; consult Merck Inst Therapeut Res, 68-69; lectr, Col Med, Syracuse Univ, 30-42, Med Sch, Marquette Univ, 56-60, Univ Pittsburgh, 65 & Yale Univ, 67; ed, J Parasitol, 62-78. *Mem:* Am Soc Parasitol (pres elect, 72, pres, 73); Am Micros Soc; Am Soc Trop Med & Hyg. *Res:* Invertebrates; fish and human parasites; pseudophyllidean tapeworms; sparganosis; visual education; models; in vitro culture of cestodes; growth-promoting substances in cestodes; parasite-induced obesity. *Mailing Add:* 770 James Ave Syracuse NY 13210

MUELLER, KARL ALEXANDER, b Chur & Schaffhausen, Switz, Apr 20, 27. CERAMIC PHYSICS. *Educ:* Swiss Fed Inst Technol, PhD(physics), 58. *Hon Degrees:* DSc, Univ Genera, 87 & Tech Univ Munich, 87. *Prof Exp:* Proj mgr, Battelle Int, Geneva, 58-63; res solid state physics, 63-73, mgr physics dept, 73-82, IBM fel, 82-85, RES FEL, IBM ZURICH RES LAB, 85- *Concurrent Pos:* Lect, Univ Zurich, 62, prof, 70. *Honors & Awards:* Nobel Prize in Physics, 87; Marcel Benoist Prize Physics, 86; Fritz London Mem Award, 87; Dannie Heineman Prize, Minna James Heineman Stiftung Acad Sci, 87; Robert Wichard Pohl Prize, 88; Hewlett Packard Europhysics Prize, 88. *Mem:* Fel Am Physic Soc; Swiss Physics Soc. *Mailing Add:* IBM Zurich Res Lab Saemerstrasse Four 8803 Ruschlikon Zurich Switzerland

MUELLER, KARL HUGO, JR, b Ft Worth, Tex, May 27, 43; m 65; c 2. SOLID STATE PHYSICS. *Educ:* Rice Univ, BA, 65; Duke Univ, PhD(physics), 72. *Prof Exp:* Res assoc physics, Duke Univ, 71-73; staff mem, Kernforschungsanlage Julich GMBH, 73-75; STAFF MEM PHYSICS, LOS ALAMOS SCI LAB, 75- *Mem:* Am Phys Soc. *Res:* Shock wave data of cryogenic fluids; high pressure phase transitions that are shock induced. *Mailing Add:* MS P940 Los Alamos Nat Lab PO Box 1663 Los Alamos NM 87545

MUELLER, MARVIN MARTIN, b Broken Arrow, Okla, Sept 29, 28; m 53, 67. FUSION PHYSICS. *Educ:* Univ Okla, BS, 51, MS. 54, PhD(physics), 59. *Prof Exp:* RES STAFF MEM, LOS ALAMOS NAT LAB, UNIV CALIF, 59- *Mem:* AAAS; Am Asn Physics Teachers; Am Phys Soc. *Res:* Laser-generated plasmas; x-ray plasma diagnostics; radiation physics; laser-fusion research; hybrid-drive fusion research. *Mailing Add:* 409 Estante Way Los Alamos NM 87544

MUELLER, MELVIN H(ENRY), b Spencer, Iowa, Feb 22, 18; m 42; c 4. CHEMISTRY, METALLURGY. *Educ:* Univ Northern Iowa, BA, 40; Univ Ill, PhD(chem), 49. *Prof Exp:* Asst chem, Univ Northern Iowa, 41; chemist, Deere & Co, 42; chemist & metallurgist, US Rubber Co, 43-45; asst, Univ Ill, 45-49; assoc metallurgist, 49-60, SR METALLURGIST, ARGONNE NAT LAB, 60-; RETIRED. *Mem:* Am Chem Soc; Am Phys Soc; Am Crystallog Asn; Mat Res Soc. *Res:* X-ray and neutron diffraction; solid state physics; chemical and magnetic structure determination. *Mailing Add:* 465 S Kenilworth Elmhurst IL 60126-3930

MUELLER, NANCY SCHNEIDER, b Wooster, Ohio, Mar 8, 33; m 59; c 1. DEVELOPMENTAL BIOLOGY, IMMUNOLOGY. *Educ:* Col Wooster, AB, 55; Univ Wis, MS, 57, PhD(zool), 62. *Prof Exp:* Hon fel, Univ Wis, Madison, 65-66, instr develop biol, 66; vis asst prof zool, NC State Univ, 68 & zool & poultry sci, 68-71; vis prof, 71-72, assoc prof, 72-79, PROF BIOL, NC CENT UNIV, 79- *Concurrent Pos:* Vis scientist, Univ Wien, Vienna, Austria, 75. *Mem:* AAAS; Am Soc Zool; Am Ornith Union; Sigma Xi; Wilson Ornith Soc; Cooper Ornith Soc. *Res:* Hematology and immunology of birds; genetic and hormonal influences on sexual differentiation of birds; sex determination and differentiation in bird embryos. *Mailing Add:* Dept of Biol NC Cent Univ Durham NC 27707

MUELLER, PAUL ALLEN, b Anniston, Ala, Sept 9, 45. PRECAMBRIAN GEOLOGY. *Educ:* Washington Univ, AB, 67; Rice Univ, MA & PhD(geol), 71. *Prof Exp:* Advan res projs agency res assoc geochem, Mat Res Ctr, Univ NC, Chapel Hill, 71-73; from asst prof to assoc prof, 73-88, PROF GEOL, UNIV FLA, 88- *Concurrent Pos:* Sr res assoc, Nat Res Coun, 81-83. *Honors & Awards:* Nininger Prize, 69. *Mem:* Geol Soc Am; Geochem Soc; Am Geophys Union. *Res:* Petrology, geochemistry, isotopic geochemistry and geochronology of igneous and metamorphic rocks; Archean geology. *Mailing Add:* Dept Geol Univ Fla Gainesville FL 32611

MUELLER, PETER KLAUS, b Hanover, Ger, Dec 30, 26; US citizen; m 50; c 3. ENVIRONMENTAL SCIENCE, BIOCHEMISTRY. *Educ:* George Washington Univ, BS, 50; Rutgers Univ, MS, 53, PhD(sanit biochem), 55. *Prof Exp:* Res asst environ sci, Rutgers Univ, 52-55; res chemist, Komline-Sanderson Eng Corp, NJ, 55-57; res chemist, Calif State Dept Pub Health, 57-63, chief, Air & Indust Hyg Lab, 63-77; mem staff, Environ Res & Technol Inc, 77-80; MEM STAFF, ENVIRON ASSESSMENT DEPT, ELEC POWER RES INST, INC, 80- *Concurrent Pos:* Mem tech adv comt instrumentation, Calif Motor Vehicle Control Bd, 65-; tech adv comt motor vehicle emissions, Calif Div Indust Safety, 66; consult, NIH, 66 & Nat Inst Occup Safety & Health, 71; air pollution res grants adv comt mem, Environ Protection Agency, 67-71; Nat Acad Sci measurements for air qual eval panel, 71-72 & air chem panel, 74-76; mem, Calif Air Resources Bd, 75. *Mem:* AAAS; Am Chem Soc; Air Pollution Control Asn. *Res:* Air pollution composition and analysis; aerosols; effects on man and animals. *Mailing Add:* Environ Assessment Dept Elec Power Res Inst Inc PO Box 10412 Palo Alto CA 94304

MUELLER, RAYMOND KARL, b East St Louis, Ill, Oct 18, 41; m 67. APPLIED MATHEMATICS. *Educ:* Washington Univ, BS, 63, MS, 65, DSc(appl math), 67. *Prof Exp:* Mem tech staff, Bell Tel Labs, 67-70; asst prof, 70-78, assoc prof math, 78-80, PROF METALL ENG & HEAD DEPT, COLO SCH MINES, 80- *Mem:* Opers Res Soc Am; Inst Math Statist. *Res:* Applied probability; probability and stochastic processes; operations research. *Mailing Add:* 2053 Crestview Circle Golden CO 80401

MUELLER, ROBERT ANDREW, b Rockville Center, NY, July 29, 52; m; c 1. COMPUTER SCIENCE. *Educ:* Colo State Univ, BS, 74, MS, 76; Univ Colo, PhD(comput sci), 80. *Prof Exp:* Res asst, Colo State Univ, 74-76; sci programmer, Tex Instruments Inc, 76-77; mem staff algorithms, Univ Colo, 77-80; asst prof comput sci, 80-85, ASSOC PROF COMPUT SCI, COLO STATE UNIV, 85-; DIR, RES, QUANT TECHNOL CORP, 87- *Concurrent Pos:* Consult, Pinon Syst & Technol, 81-; NSF grant prin investr, 81-88. *Mem:* Asn Comput Mach; Inst Elec & Electronics Engrs. *Res:* Microprogramming; algorithms; logic programming; formal methods; retargetable compilers for parallel architectures. *Mailing Add:* 1812 Seminole Dr Ft Collins CO 80525-1536

MUELLER, ROBERT ARTHUR, b Fond du Lac, Wis, July 24, 38; m 62; c 4. ANESTHESIOLOGY, PHARMACOLOGY. *Educ:* Univ Wis-Madison, BS, 60, MS, 63; Univ Minn, Minneapolis, MD, 65, PhD(pharmacol), 66. *Prof Exp:* Am Cancer Soc fel, Univ Minn, Minneapolis, 65-66, intern surg, Hosps, 66-67, resident anesthesiol, 67; res assoc pharmacol & toxicol, Lab Clin Sci, NIMH, 67-69; resident anesthesia, Med Sch, Northwestern Univ, Ill, 69-70; assoc prof anesthesiol & asst prof pharmacol, Med Sch, Univ NC, Chapel Hill, 70-75, prof anesthesiol & assoc prof pharmacol, prof anesthesiol & pharmacol, 78-81. *Mem:* Am Soc Pharmacol & Exp Therapeut; Fedn Am Socs Exp Biol; Soc Neurosci; Am Soc Anesthesiol; Sigma Xi. *Res:* Adrenergic pharmacology. *Mailing Add:* Rm 223 Burnett-Womack Bldg Univ NC Chapel Hill NC 27514

MUELLER, ROBERT KIRK, b St Louis, Mo, July 25, 13; m 40; c 3. CHEMICAL ENGINEERING. *Educ:* Washington Univ, BS, 34; Univ Mich, MS, 35; Harvard AMP, 40. *Prof Exp:* Plant chemist, Monsanto Co, 35-38, gen supt, Plastics Div, 40-42; chmn, Shawinigan Resins Corp, Mass, 38-40; supt, Longhorn Ord Works, Tex, 42-44, plant mgr, 44-46; prod sput, Plastics Div, Monsanto Co, 46-48 from asst prod mgr to prod mgr, Arthur D Little, Inc, 48-50, from asst gen mgr to gen mgr, 50-54, vpres, 54-68; vpres, 73-77, chmn, Arthur D Little, Inc, 77-86, DIR, ARTHUR D LITTLE, LTD, 68-; DIR, HEC CORP, DECISION RESOURCES, INC. *Concurrent Pos:* Mem bd dirs exec comt, Monsanto Co, 61-68; dir, Mass Mutual Life Ins Co, 68-86, BayBanks Inc, Boston, 69-86 & Nat Asn Corp Dirs; chmn fac & lectr, Salzburg Sem Am Studies, Austria, 70; chmn, C F Systs Inc; trustee, Am Austrian, Acad Advan Corp Governance, Fordham Univ. *Mem:* Fel AAAS; Am Chem Soc; Am Inst Chem Engrs; Am Mgt Asn; NY Acad Sci; fel Int Acad Mgt. *Res:* Chemicals and plastics; fibers; international business; management consulting and writing. *Mailing Add:* Arthur D Little Inc Acorn Park Cambridge MA 02140

MUELLER, ROLF KARL, b Zurich, Switz, Aug 30, 14; nat US; m 42; c 3. APPLIED PHYSICS, ACOUSTIC IMAGING. *Educ:* Munich Tech Univ, Dipl phys, 39, Dr rer nat, 42, habil, 50. *Prof Exp:* Asst prof appl physics, Univ Jena, 39-45; asst prof theoret physics, Stuttgart Tech Univ, 47-48; dozent, Munich Tech Univ, 48-52; consult, Air Force Cambridge Res Ctr, Mass, 52-55; sr tech specialist & head solid state res sect, Gen Mills, Inc, 56-63; mgr, Gen Sci & Technol Lab, Bendix Corp, 63, Lab dir, Bendix Ctr, 63-74; prof elec eng, 74-85, EMER PROF, UNIV MINN, 85 - *Mem:* AAAS; fel Am Phys Soc; Soc; sr mem Inst Elec & Electronics Engrs. *Res:* Acoustic imaging and image processing applied to biomedical diagnostic instrumentation; nondestructive materials evaluation; sonar processing and acoustic microscopy. *Mailing Add:* 9707 Manning Ave Stillwater MN 55082

MUELLER, SABINA GERTRUDE, b Binghamton, NY, Apr 29, 40; m 70; c 2. BOTANY. *Educ:* Swarthmore Col, BA, 61; Univ NC, Chapel Hill, PhD(bot), 68. *Prof Exp:* From asst prof to assoc prof bot, Shippensburg State Col, 66-70; staff mem plant records, Cox Arboretum, Dayton, 72-73, botanist, Educ Serv, 73-75; botanist, Fullmer's Landscape Serv, Dayton, 75-76; prof bot, Wilberforce Univ, 77-80; pres & dir, Ivy Res Ctr, Cox Abore, 80-85, DIR, IVY RES CTR, DAYTON, OH, 85- *Concurrent Pos:* Vis prof, Miami Univ, Middletown, 78, 79 & 80; registrar, Ivy Soc, 77-; adj prof, Wright State Univ, 85-; Natural Resource Specialist, Natural Heritage Prog, Ohio Dept Natural Resources, 88. *Mem:* Am Daffodil Soc; Am Ivy Soc; Am Hort Soc; Am Hosta Soc. *Res:* Plant nomenclature; Hedera cultivars. *Mailing Add:* 5512 Woodbridge Lane Dayton OH 45429

MUELLER, SISTER RITA MARIE, b Cincinnati, Ohio, July 16, 18. PHYSICAL CHEMISTRY, THEORETICAL CHEMISTRY. *Educ:* Villa Madonna Col, BA, 41; Cath Univ Am, MS, 42; Univ Cincinnati, PhD(chem), 62. *Prof Exp:* From asst prof to assoc prof, 42-52, PROF CHEM, THOMAS MORE COL, 52-, CHMN DEPT, 73- *Concurrent Pos:* Argonne Nat Lab fel; DuPont fel. *Mem:* Am Chem Soc; Sigma Xi. *Res:* Physics, mathematics & institutional research. *Mailing Add:* Dept of Chem Thomas More Col 333 Thomas More Pkwy Crestview Hills KY 41017-3428

MUELLER, STEPHEN NEIL, b New York, NY, Mar 17, 47; m 79; c 1. CELL BIOLOGY. *Educ:* LeMoyne Col, BS, 69; Syracuse Univ, PhD(biol), 76. *Prof Exp:* Fel, Wistar Inst Anat & Biol, 76-78, res investr, 78-80, res assoc, 80-84, asst prof, 84-86; ASSOC PROF, CORIELL INST MED RES, 87- *Mem:* Tissue Cult Asn; AAAS; Am Soc Cell Biol. *Res:* Studies of the regulation of growth, differentiated function and senescence of vascular endothelial cells in vitro. *Mailing Add:* Coriell Inst Med Res 401 Haddon Ave Camden NJ 08103

MUELLER, THEODORE ARNOLD, b St Louis, Mo, Jan 29, 38; m 63; c 2. BIOPHYSICS. *Educ:* Cent Methodist Col, AB, 59; NMex Highlands Univ, MS, 62, PhD(biophys chem), 65. *Prof Exp:* PROF SCI, ADAMS STATE COL, 65-, HEAD DEPT PHYSICS & ASTRON, 77- *Mem:* Asn Physics Teachers; Biophys Soc; Air Pollution Control Asn. *Res:* Cellular and photochemical ultraviolet effects; photobiology; environmental pollution; pollution monitoring. *Mailing Add:* Div Sci & Math Adams State Col Alamosa CO 81102

MUELLER, THEODORE ROLF, b Ft Wayne, Ind, Dec 14, 28; m 51; c 3. ANALYTICAL CHEMISTRY, INSTRUMENTATION. *Educ:* Valparaiso Univ, BA, 50; Univ Kans, PhD(analytical chem), 63. *Prof Exp:* Instr math & sci, Lutheran High Sch, Houston, Tex, 50-51; instr chem, Concordia Col Inst, 51-53, Callery Chem Co, 57-59; CHEMIST, OAK RIDGE NAT LAB, 61- *Mem:* AAAS; Am Chem Soc; Sigma Xi; Am Soc Mass Spectrometry. *Res:* Application of electroanalytical techniques to environmental problems; research and development in analytical procedures; instruments for automated analyses. *Mailing Add:* 218 East Dr Oak Ridge TN 37830

MUELLER, THOMAS J, b Chicago, Ill, May 25, 34; m 61; c 5. GAS DYNAMICS, FLUID MECHANICS. *Educ:* Ill Inst Technol, BS, 56; Univ Ill, MS, 58, PhD(gas dynamics), 61. *Prof Exp:* Res assoc thermodyn & gas dynamics, Univ Ill, 60-61, asst prof gas dynamics, 61-63; sr res scientist, United Aircraft Res Labs, 63-65; assoc prof gas dynamics, 65-69, dir eng res & grad studies, 85, PROF FLUID MECH, UNIV NOTRE DAME, 69-, CHMN, AEROSPACE & MECH ENG, 88. *Concurrent Pos:* Vis res prof, von Karman Inst Fluid Dynamics, 73-74; consult, US Army, Md, 76-77, Coachmen Indust, Inc, 77 & Lockheed-Georgia Co, 80-81, AGARD NATO, 83-84. *Mem:* Am Inst Aeronaut & Astronaut; fel Am Soc Mech Engrs; Am Soc Eng Educ. *Res:* Compressible and incompressible laminar and turbulent separated or wake flows; propulsive exhaust nozzles; low reynolds number aerodynamics. *Mailing Add:* Dept Aerospace & Mech Eng Univ Notre Dame Notre Dame IN 46556

MUELLER, THOMAS JOSEPH, b Evanston, Ill, June 2, 46. NEUROBIOLOGY. *Educ:* Loyola Univ, BS, 69; Univ Southern Calif, PhD(biol), 81. *Prof Exp:* ASST PROF BIOL, HARVEY MUDD COL, 81- *Concurrent Pos:* Consult comput & statist, Univ Southern Calif, 78- *Mem:* Inst Elec & Electronics Engrs; Soc Neurosci; Sigma Xi. *Res:* Mauthner cell; neurobiology and behavior of lower vertebrates and invertebrates. *Mailing Add:* Dept Biol Harvey Mudd Col Claremont CA 91711

MUELLER, W(HEELER) K(AY), mechanical engineering, for more information see previous edition

MUELLER, WALTER CARL, b Newark, NJ, Nov 29, 34; m 56; c 3. PLANT PATHOLOGY. *Educ:* Rutgers Univ, BS, 56; Cornell Univ, PhD(plant path), 61. *Prof Exp:* From asst prof to assoc prof, 61-74, PROF PATH-ENTOM, UNIV RI, 74- *Mem:* AAAS; Am Phytopath Soc; Electron Micros Soc Am; Sigma Xi. *Res:* Viruses and virus diseases of plants; electron microscopy. *Mailing Add:* Dept of Plant Path-Entom Univ of RI Kingston RI 02881

MUELLER, WAYNE PAUL, b Evansville, Ind, July 27, 33; m 57; c 2. ECOLOGY. *Educ:* Univ Evansville, BA, 56; Ind Univ, MA, 61, PhD(zool), 62. *Prof Exp:* From asst prof to assoc prof, 62-72, PROF BIOL, UNIV EVANSVILLE, 72-, HEAD DEPT BIOL, 77- *Mem:* Nat Asn Biol Teachers. *Res:* Protozoan ecology. *Mailing Add:* 1800 Lincoln Ave Evansville IN 47702

MUELLER, WENDELIN HENRY, III, b St Louis, Mo, Feb 12, 41; m 66; c 2. CIVIL & STRUCTURAL ENGINEERING. *Educ:* St Louis Univ, BS, 62; Univ Mo-Rolla, MS, 66, PhD(civil eng), 72. *Prof Exp:* Engr flood struct, US Army Corps Engrs, St Louis, 62-65; asst civil eng, Univ Mo-Rolla, 65-66; res engr aerospace, Boeing Airplane Co, Seattle, 66-68; engr-programmer civil mech eng, Nat Comput Serv, St Ann, Mo, 68-69; instr civil eng, Univ Mo-Rolla, 69-72; struct engr flood struct, US Army Corps Engrs, Portland, 72-73; from asst prof to assoc prof, 73-80, PROF CIVIL STRUCT, PORTLAND STATE UNV, 80- *Mem:* Am Soc Civil Engrs; Am Soc Eng Educrs; Sigma Xi; fel Am Inst Steel Construct. *Res:* Investigation of non-linear performance of steel structures using computer analysis; bio-medical engineering, particularly mechanical improvement of artificial joint systems and monitoring and control using micro-processors. *Mailing Add:* Dept of Eng PO Box 751 Portland OR 97207

MUELLER, WILLIAM M(ARTIN), b Denver, Colo, Jan 14, 17; m 42; c 2. PHYSICAL METALLURGY. *Educ:* Colo Sch Mines, MetE, 40, MS, 49, DSc(phys metall), 52. *Prof Exp:* Metallurgist, Aluminum Co Am, 40-45; metall engr, Gates Rubber Co, 45-47; instr, Colo Sch Mines, 47-52; metallurgist, Dow Chem Co, 52-57; mgr phys metall lab, Denver Res Inst, 57-59, head metall div, 59-65, prof metall & chmn dept, Univ Denver, 61-65; dir educ, Am Soc Metals, 65-74; prof metall eng & head dept, 74-79, vpres, 79-83, EMER VPRES, ACAD AFFAIRS, COLO SCH MINES, 83- *Concurrent Pos:* Consult, Climax Molybdenum Co & Res Found, Colo Sch Mines, 49-52, Sundstrand Aviation, 60-61, Gordon & Breach Sci Publ, 62-65 & Western Forge Corp, 75-83; mem, US-Indonesian Workshop, Nat Acad Sci, 71; vis, Beijing Univ Iron & Steel Technol, China, 80; leader, People to People Deleg, China, 84, Southeast Asia, 86, Soviet Union, 90. *Mem:* Am Soc Metals; Mining & Metall Soc Am; Am Inst Mining, Metall & Petrol Engrs; Am Soc Eng Educ; Am Soc Testing & Mat. *Res:* Spectrochemical analysis of ores; eutectic aluminum-silicon alloys; synthetic mica; radioactive metals; metal hydrides; metallurgical and industrial development; engineering education. *Mailing Add:* Emer Vpres Colo Sch Mines Golden CO 80401

MUELLER-DOMBOIS, DIETER, b Bethel, Ger, July 26, 25; nat Can; m 51; c 5. PLANT ECOLOGY. *Educ:* Univ Gottingen, BS, 51; Hohenheim Agr Univ, dipl, 51; Univ BC, BS, 55, PhD, 60. *Prof Exp:* Asst biol & bot, Univ BC, 55-57; forest ecologist, Res Div, Can Dept Forestry, 58-63; forest ecologist, Dept Bot, Univ Hawaii, 63-66; prin field investr, Smithsonian-Ceylon Ecol Proj, Univ Ceylon, 67-69; PRIN FIELD INVESTR, UNIV HAWAII, 69-, PROF BOT & ECOL, 71- *Concurrent Pos:* Co-dir & sci coordr, Island Ecosysts Integrated Res Prog, Int Biol Prog, 70-71, dir, 72- *Mem:* Fel AAAS; Asn Trop Biol; Ecol Soc Am; Int Soc Trop Ecol. *Res:* Agriculture; botany; soil science; climatology; forest site classification; vegetation and environmental studies; synecology and autecology; tree physiology; soil water-plant growth relations; ecology of vegetation on recent volcanic matter; tropical and ecosystems ecology; animal-vegetation interactions. *Mailing Add:* Dept Bot St John 505 Univ Hawaii-Manoa 2500 Campus Rd Honolulu HI 96822

MUENCH, DONALD LEO, b Rochester, NY, Jan 31, 34; m 60; c 4. MATHEMATICS. *Educ:* St John Fisher Col, BS, 55; St John's Univ, NY, MS, 60; Idaho State Univ, DA(math), 74. *Prof Exp:* Asst prof math, US Naval Acad, 60-66; assoc prof, 66-81, chmn math dept, 68-80, dir comput sci prog, 81-82, PROF MATH, ST JOHN FISHER COL, 81-, DIR COMPUT SCI PROG, 85- *Concurrent Pos:* Asst dir, IFRICS, 87 & 88. *Mem:* Math Asn Am. *Res:* Linear Algebra; matrix theory; computer science education; mathematics education. *Mailing Add:* Dept of Math St John Fisher Col Rochester NY 14618

MUENCH, KARL HUGO, b St Louis, Mo, May 3, 34; m 76; c 5. BIOCHEMISTRY, GENETICS. *Educ:* Princeton Univ, AB, 56; Washington Univ, MD, 60. *Prof Exp:* Intern med, Barnes Hosp, St Louis, 60-61; USPHS fel biochem, Stanford Univ, 61-65; from instr to assoc prof med & biochem, 65-73, PROF MED, SCH MED, UNIV MIAMI, 73-, CHIEF GENETIC MED, 68- *Concurrent Pos:* Am Cancer Soc fac res assoc, Univ Miami, 65-70; Markle scholar acad med, 69-74; Leukemia Soc Am scholar, 71-76; USPHS res career develop award, 71-76; NSF adv, Panel Molecular Biol, 78-81. *Mem:* Am Soc Biol Chem; Am Chem Soc; Am Soc Human Genetics; fel Am Col Physicians. *Res:* Human genetics. *Mailing Add:* Dept Med Univ Miami Sch Med Miami FL 33101

MUENCH, NILS LILIENBERG, b Houston, Tex, Feb 27, 28; m 50; c 1. PHYSICS. *Educ:* Rice Univ, BA, 49, MA, 50, PhD(physics), 55; S Tex Col, LLB, 59. *Prof Exp:* Sr res engr, Exxon Co, 55-59; chief scientist, Army Rocket & Guided Missile Agency, 59-62; mem Inst Defense Anal, 62-63; head physics dept, 63-69, tech dir, 69-87, EXEC DIR, GEN MOTORS RES LABS, 88- *Concurrent Pos:* Mem bd dirs, Teknowledge, Inc, Palo Alto, Calif, 1986-89. *Mem:* Am Phys Soc; Am Inst Aeronaut & Astronaut. *Res:* Solid state physics; management of research. *Mailing Add:* Gen Motors Res Labs Warren MI 48090

MUENCH, ROBIN DAVIE, b N Conway, NH, Sept 16, 42; m 66; c 2. PHYSICAL OCEANOGRAPHY. *Educ:* Bowdoin Col. AB. 64; Dartmouth Col, MA, 66; Univ Wash, PhD(oceanog), 70. *Prof Exp:* From res asst oceanogr to oceanogr, Univ Wash, 68-70; from asst prof to assoc prof oceanog, Inst Marine Sci, Univ Alaska, 75-76; res oceanogr, Pac Marine Environ Lab, Nat Oceanog & Atmospheric Admin, 76-79; SR RES OCEANOGR, SCI APPLN INC, 79- *Concurrent Pos:* Vis scholar, Cambridge Univ, 86; chmn, Div River Lake & Sea Ice, Int Comn Snow & Ice, 83-87; prin investr on numerous progs & pres, Comn Sea Ice, IAPSO, 87- *Mem:* AAAS; Am Geophys Union; Am Meteorol Soc; Int Glaciological Soc. *Res:* Physical oceanography of the Polar Regions; air-sea interaction; sea ice research; journal of geophysical research. *Mailing Add:* Sci Appln Inc-Northwest 13400B Northrup Way #36 Bellevue WA 98005

MUENDEL, CARL H(EINRICH), b New York, NY, July 27, 30; m 61; c 3. CHEMICAL ENGINEERING. *Educ:* Columbia Univ, BS, 52, MS, 54, Eng ScD, 59. *Prof Exp:* Instr chem eng, Columbia Univ, 55-56; chem engr, 57-66, sr res engr, 66-68, res supvr, Res Div, 68-78, tech serv supvr, 78-86, TECH SERV MGR, CHEM & PIGMENTS DEPT, E I DU PONT DE NEMOURS & CO, INC, 86- *Mem:* Am Inst Chem Engrs; Am Chem Soc. *Res:* Pigments; industrial chemicals. *Mailing Add:* E I du Pont de Nemours Co Inc Chestnut Run Plaza PO Box 80709 Wilmington DE 19880-0709

MUENDEL, HANS-HENNING, b Kosten, Poland, Mar 31, 42; Can citizen; m 68; c 3. PLANT BREEDING. *Educ:* Univ BC, BScAg, 64; Univ Calif, Davis, MS, 66; Univ Man, PhD(plant breeding), 73. *Prof Exp:* Farm mgr tribal rehab farm colony, Nilgiris Adivasi Welfare Asn, India, Can Univ Serv Overseas, 66-67, res agronomist safflower breeding & field experimentation, Nimbkar Agr Res Inst, Phaltan, India, 67-69; plant breeder wheat breeding, Can Int Develop Agency, Kenya, Univ Man, 72-74; proj mgr plant genetic resources ctr, Ger Agency Tech Coop, Ethiopia Proj, 75-77; RES SCIENTIST NEW CROPS, BREEDING & MGT SOYBEANS, SAFFLOWER, WHEAT & FIELD BEANS, LETHBRIDGE RES STA, AGR CAN, 78- *Concurrent Pos:* Chmn, Int Safflower Continuing Comt, 81-85. *Mem:* Agr Inst Can; Am Soc Agron; Crop Sci Soc Am; Can Soc Agron. *Res:* Breeding of early-maturing saflower, soybeans of high levels of nitrogen fixation support, and sawfly resistant spring wheats and upright field beans; crop genetic resources worldwide. *Mailing Add:* Agr Can Res Sta Lethbridge AB T1J 4B1 Can

MUENOW, DAVID W, b Chicago, Ill, May 28, 39. PHYSICAL CHEMISTRY, GEOCHEMISTRY. *Educ:* Carleton Col, BA, 61; Purdue Univ, PhD(chem), 67. *Prof Exp:* Welch fel mass spectrometry, Rice Univ, 67-70; asst prof, 70-74, ASSOC PROF CHEM, UNIV HAWAII, 75- *Mem:* Am Chem Soc; Am Geophys Union. *Res:* High temperature mass spectrometry; thermodynamics; silicate chemistry; geochemistry; magmatic volatiles. *Mailing Add:* Dept Chem Univ Hawaii-Manoa 2500 Campus Rd Honolulu HI 96822

MUENTENER, DONALD ARTHUR, b Pigeon, Mich, Apr 24, 26; m 52; c 2. BACTERIOLOGY, FOOD SCIENCE & TECHNOLOGY. *Educ:* Alma Col, BSc, 50; Mich State Univ, MS, 53. *Prof Exp:* Instr biol, Alma Col, 50; asst, Mich State Univ, 53-55; microbiologist, Mich Dept Agr, 56-60, asst dir, 61-75, actg dir, 75-76, state analyst & dir lab div, 76-84; RETIRED. *Mem:* Regist Nat Registry Microbiol; Am Soc Microbiol; Brit Soc Appl Bact; Asn Food & Drug Off; Asn Off Anal Chemists; Sigma Xi; Inst Food Technologists. *Res:* Laboratory administration; agricultural, industrial, food, dairy and sanitation microbiology. *Mailing Add:* 3312 Inverary Dr Lansing MI 48911

MUENTER, ANNABEL ADAMS, b New York, NY, Dec 3, 44; m 68; c 1. PHOTOGRAPHIC & PHYSICAL CHEMISTRY. *Educ:* Univ Mich, BSChem, 66; Harvard Univ, PhD(chem physics), 72. *Prof Exp:* Sr res chemist, 70-78, res assoc, 78-90, SR RES ASSOC, EASTMAN KODAK RES LABS, 90- *Concurrent Pos:* Mem, NSF Ctr Photoinduced Charge Transfer, Univ Rochester, 89- *Mem:* Fel Soc Photog Scientists & Engrs; Am Chem Soc; Sigma Xi. *Res:* Dyes and spectral sensitization of silver halide using various spectroscopic techniques, including sub nanosecond fluorescence lifetime measurements. *Mailing Add:* Nine Park Pl Rochester NY 14625-2163

MUENTER, JOHN STUART, b Cleveland, Ohio, May 10, 38; m 68; c 1. PHYSICAL CHEMISTRY, SPECTROSCOPY. *Educ:* Kenyon Col, BA, 60; Stanford Univ, PhD(chem). 65. *Prof Exp:* Fel Stanford Univ, 65-66; NIH fel, Harvard Univ, 66-68; asst prof, 69-75, assoc prof, 75-81, PROF PHYS CHEM, UNIV ROCHESTER, 81- *Mem:* Am Inst Physics; Sigma Xi; fel Am Phys Soc. *Res:* Spectroscopic studies of the electronic structure of small molecules and molecular clusters utilizing microwave and molecular beam electric resonance and laser spectroscopy. *Mailing Add:* Dept of Chem Univ of Rochester Rochester NY 14627

MUENZENBERGER, THOMAS BOURQUE, b New Orleans, La, Aug 13, 43; m 67. TOPOLOGY. *Educ:* Univ Fla, BS, 65, MS, 67; Univ Wyo, PhD(math), 72. *Prof Exp:* Asst prof, 73-80, ASSOC PROF MATH, KANS STATE UNIV, 80- *Concurrent Pos:* Consult. *Mem:* Am Math Soc; Math Asn Am; Sigma Xi. *Res:* Mathematics; general topology; fixed point theory; partially ordered spaces. *Mailing Add:* Dept of Math Kans State Univ Manhattan KS 66506

MUESSIG, PAUL HENRY, b Philadelphia, Pa, June 22, 49; m 72. AQUATIC ECOLOGY, FISH PHYSIOLOGY. *Educ:* La Salle Col, BA, 71; Fla State Univ, MS, 74. *Prof Exp:* Group leader & scientist, Ecol Serv, Tex Instruments Inc, 74-77; PROG MGR & SCIENTIST AQUATIC TOXICOL IMPACT ASSESSMENT, THERMAL ECOL, ACID PRECIPITATION, HYDRO POWER ASSESSMENT, ECOL ANALYSTS INC, 77- *Mem:* Ecol Soc Am; Estuarine Res Fedn; Can Soc Zoologists. *Res:* Aquatic Toxicology; industry environmental impact assessment and mitigation; thermal and stress physiology of fish; aquaculture. *Mailing Add:* Six Campfire Rd Wallkill NY 12589

MUESSIG, SIEGFRIED JOSEPH, b Freiburg, Germany, Jan 19, 22; nat US; m 49; c 2. ECONOMIC GEOLOGY. *Educ:* Ohio State Univ, BSc, 47, PhD(geol), 51. *Prof Exp:* Instr field geol, Ohio State Univ, 50; geologist, Mineral Deposits Br, US Geol Surv, 51-59; chief geologist, US Borax & Chem Corp, 59-66; minerals explor mgr, Tidewater Oil Co, Calif, 66-67; pres, Getty Oil Develop Co, Ltd, 73-84, vpres, Getty Mining Co, 84-85, minerals explor mgr, Getty Oil Co, 67-84; dir, Texaco Oil Developo Co, 85-87; PRES, CRYSTAL EXPLOR INC, 88-; DIR, CALLAHAN MINING CORP, 89- *Concurrent Pos:* Consult, Borax Consol, Ltd, 55; regional dir, Nat Defense Exec Reserve, 71-; dir region 9, Emergency Minerals Admin, Dept Interior, 71-, Nat Acad Sci Comt Geol, 78-82; minerals consult, 85-; distinguished mem, Soc Mining Engrs, 86. *Honors & Awards:* Ralph Marsden Award, Soc Econ Geologists, 90. *Mem:* AAAS; Am Inst Mining, Metall & Petrol Eng; Geol Soc Am; Soc Econ Geologists (vpres, 73-74, pres, 78-79); Am Asn Petrol Geologists. *Res:* Stratigraphy and structure; geology of saline deposits; geology of northeastern Washington; economic geology; geology of northwest Argentina; geology of and exploration for diamonds. *Mailing Add:* 1097 Charles St Pasadena CA 91103

MUETHER, HERBERT ROBERT, b Winfield, NY, Sept 27, 21; m 51; c 6. NUCLEAR PHYSICS. *Educ:* Queens Col, NY, BS, 42; Princeton Univ, AM, 47, PhD(physics), 51. *Prof Exp:* Instr physics, Princeton Univ, 49-50; lectr, Queens Col, NY, 50-52, instr, 52-55, asst prof, 55-59; assoc prof, 59-61, assoc chmn dept, 68-75, PROF PHYSICS, STATE UNIV NY STONY BROOK, 61-, DIR UNDERGRAD PROG PHYSICS, 75- *Concurrent Pos:* Res collabr, Brookhaven Nat Lab, 51-; consult, Frankford Arsenal, Pa, 58- *Mem:* AAAS; Am Phys Soc; Am Asn Physics Teachers; Sigma Xi. *Res:* Neutron physics. *Mailing Add:* Dept Physics State Univ NY Stony Brook NY 11794-3800

MUETING, ANN MARIE, b Norfolk, Nebr, Jan 11, 56. SYNTHETIC INORGANIC & ORGANOMETALLIC CHEMISTRY. *Educ:* Creighton Univ, BS, 78, MS, 80; Univ Minn, PhD(chem), 85. *Prof Exp:* Instr phys sci, Cathedral High Sch, Omaha, 79-80; chemist, US Army Corps Eng, 79-80 & Valmont Industs, 80-81; asst prof inorg chem, Drake Univ, 85-87; PATENT AGENT, 87- *Concurrent Pos:* Postdoctoral fel chem, Univ Minn, 89. *Mem:* Am Chem Soc. *Res:* Synthetic inorganic and organometallic chemistry; catalysis and the activation of hydrogen sulfide by transition metal complexes including the reactivity of sulfur and sulfhydryl ligands. *Mailing Add:* 1300 W Medicine Lake Dr Apt 118 Plymouth MN 55441-4854

MUFFLER, LEROY JOHN PATRICK, b Alhambra, Calif, Sept 19, 37; m 66; c 1. GEOPHYSICS, MINERALOGY-PETROLOGY. *Educ:* Pomona Col, BA, 58; Princeton Univ, MA, 61, PhD(geol), 62. *Prof Exp:* Geologist, Southwest States Br, 62, Alaska Geol Br, 62-64, Field Geochem & Petrol Br, 64-83, GEOLOGIST IGNEOUS & GEOTHERMAL PROCESSES BR, 83- *Concurrent Pos:* Vis scientist, NZ Dept Sci & Indust Res Geophys Div, 70, Ital Elec Agency, 76-77; mem, US Continental Sci Drilling Comt, Nat Res Coun, 81- *Honors & Awards:* Distinguished Serv Award, US Dept Interior, 90. *Mem:* Geol Soc Am; Am Geophys Union; Geothermal Resources Coun; Soc Econ Geologists; Int Geothermal Asn. *Res:* Geothermal resources; hydrothermal alteration; hot springs; Cenozoic volcanic rocks. *Mailing Add:* 961 Ilima Way Palo Alto CA 94306

MUFFLEY, HARRY CHILTON, b Urbana, Ill, Dec 2, 21; m 50. BIOLOGY, ORGANIC CHEMISTRY. *Educ:* Milliken Univ, BS, 49; Penn State, PhD, 79. *Prof Exp:* Res chemist, Fine Chem Div, Glidden Co, 50-52; res chemist, Rock Island Arsenal, 52-63, res phys scientist, US Army Weapons Command, 63-70; res phys scientist, Gen Thomas J Rodman Lab, 70-77; head hydraul fluids & lubricants div, US Army Armament Res & Develop Command, 77-80, dir, Valley Lab, 80-85; RETIRED. *Honors & Awards:* Wilbur Deutsch Mem Award, Am Soc Lubrication Eng, 67. *Mem:* AAAS; Am Chem Soc (pres, 61); Am Soc Lubrication Engrs (pres, 65); Am Defense Preparedness Asn. *Res:* Biomechanics; study of the unique or unusual characteristics of animal leading to new concepts in weapons or weapons systems; hydraulic fluids; corrosion preventives. *Mailing Add:* 3610 35th Ave Rock Island IL 61201

MUFSON, DANIEL, b Bronx, NY, Dec 24, 42; m 64; c 2. PHARMACEUTICAL CHEMISTRY. *Educ:* Columbia Univ, BS, 63, MS, 65; Univ Mich, PhD(pharmaceut chem), 68. *Prof Exp:* Res pharmacist, Parke Davis & Co, 68-71; sr scientist pharmaceut chem, Smith Kline & French Labs, 71-73, group leader, 73-74; mgr biopharmaceut & drug metab, 74-76; dir, Pharmaceut Res & Develop Dept, USV Pharmaceut Corp, Div Relvon, 76-80, dir, Pharmaceut Res & Develop Dept, Relvon Health Care Res & Develop, 80-85; vpres res & develop, Liposome Tech, Inc, 85-86, from vpres corp develop to sr vpres corp develop, 86-90; PRES, APOTHERX, 90- *Mem:* Am Pharmaceut Asn; Acad Pharmaceut Sci; Am Soc Hosp Pharmacists; Sigma Xi; Am Asn Pharmaceut Sci; Controlled Release Soc; Parenteral Drug Asn; Licensing Execs Soc. *Res:* Solubilization and dissolution studies on pharmaceuticals and endogenous lipids; biopharmaceutics and drug metabolism; pharmaceutical development. *Mailing Add:* 1301 Groventres Fremont CA 94539

MUFSON, R ALLAN, b New York, NY, June 10, 46; m 84; c 2. HEMATOPOIESIS. *Educ:* Brooklyn Col, City Univ NY, BA, 68; Brown Univ, PhD(cell biol), 74. *Prof Exp:* Nat Cancer Inst Individual fel, McArdle Lab Cancer Res, Univ Wis-Madison, 74-76, Nat Res Coun fel, 76-77; staff assoc, Inst Cancer Res, Col Physicians & Surgeons, Columbia Univ, 77-80; asst prof biochem toxicol, Inst Environ Med, NY Univ Med Ctr, 80-83; sr scientist & head sect cell biol, Genetics Inst, Cambridge, Mass, 84-88; SR SCIENTIST CELL BIOL, HOLLAND LAB BIOMED RES, AM RED CROSS, ROCKVILLE, MD, 88- *Concurrent Pos:* Adj asst prof, Inst Environ Med, NY Univ Med Ctr, 83-; mem, Adv Comt AIDS & Hematopoietic Suppression, NIH, Bethesda, Md, 88; consult, Genetics Inst, Cambridge, Mass, 88-89; int collabr, study Int Standard M-CSF, WHO, Hertfordshire, Eng, 91. *Mem:* Am Asn Cancer Res; Am Soc Molecular Biol & Biochem; NY Acad Sci. *Res:* Regulation of human monocyte differentiation by cytokines and possible relationships to HIV-1 replication; general regulation of hematopoiesis and its relationship to leukemogenesis; mechanism of action of tumor promoters. *Mailing Add:* Holland Lab Biomed Sci Am Red Cross 15601 Crabbs Branch Way Rockville MD 20855

MUFSON, STUART LEE, b Philadelphia, Pa, May 16, 46; m 72; c 2. ASTROPHYSICS, SUPERNOVA REMNANTS. *Educ:* Univ Pa, BA & MS(physics), 68; Univ Chicago, MS(astron), 70, PhD(astron), 74. *Prof Exp:* Res assoc, Nat Radio Astron Observ, 73-75; Nat Res Coun assoc, Marshall Space Flight Ctr, NASA, 75-77; from asst prof to assoc prof, 77-88, PROF ASTRON, IND UNIV, 88- *Concurrent Pos:* Prin investr, NASA, 77- *Mem:* Am Astron Soc; Int Astron Union. *Res:* High energy astrophysics, neutrino astronomy, cosmic ray research; evolution of supernova remnants. *Mailing Add:* Astron Dept Ind Univ Bloomington IN 47401

MUFTI, IZHAR H, b Batala, India, June 15, 31; Can citizen; m 61; c 2. APPLIED MATHEMATICS. *Educ:* D J Col, Univ Karachi, Pakistan, BSc, 51, MSc, 53; Univ BC, PhD(appl math), 60. *Prof Exp:* Lectr math, DJ Col, Univ Karachi, 51-56; asst, Univ BC, 56-60; asst res officer, Nat Res Coun Can, Ottawa, 60-66, assoc res officer, 66-76, sr res officer systs lab, Div Mech Eng, 76-90; RETIRED. *Mem:* Soc Indust & Appl Math. *Res:* Multi-body dynamics; manipulator dynamics and control; production scheduling; stability; optimization. *Mailing Add:* Six Massey Lane Blouchester ON K1J 6C8 Can

MUGA, MARVIN LUIS, b Dallas, Tex, Mar 1, 32. NUCLEAR CHEMISTRY. *Educ:* Southern Methodist Univ, BS, 53, MS, 54; Univ Tex, PhD, 57. *Prof Exp:* Res nuclear chemist, Lawrence Radiation Lab, Univ Calif, 57-60; from asst prof to assoc prof nuclear chem, 60-77, PROF CHEM, UNIV FLA, 77- *Concurrent Pos:* Fulbright lectr, 60. *Mem:* Am Chem Soc; Am Phys Soc. *Res:* Thin film scintillator detectors for dE dx measurements of energetic heavy ions; fission decay phenomena. *Mailing Add:* 406 Nuclear Sci Ctr Univ of Fla Gainesville FL 32611-2055

MUGGENBURG, BRUCE AL, b St Paul, Minn, May 2. 37; m 60; c 3. VETERINARY PHYSIOLOGY. *Educ:* Univ Minn, BS, 59, DVM, 61; Univ Wis, Madison, MS, 64, PhD(vet sci), 66. *Prof Exp:* From instr to asst prof vet sci, Univ Wis, Madison, 64-69; VET PHYSIOLOGIST, INHALATION TOXICOL RES INST, LOVELACE FOUND, 69- *Mem:* Am Thoracic Soc; Am Vet Med Asn; Am Physiol Soc; Health Physics Soc; Radiation Res Soc. *Res:* Therapy of radiation induced disease; deposition and clearance of particles from the lung; toxicity of inhaled radionuclides; pulmonary immune responses. *Mailing Add:* Inhalation Toxicol Res Inst PO Box 5890 Albuquerque NM 87185

MUGGERIDGE, DEREK BRIAN, b Godalming, Surrey, Eng, Oct 10, 43; m 65; c 2. OCEAN ENGINEERING, WAVE & ICE INTERACTION WITH OFFSHORE STRUCTURES. *Educ:* Calif State Polytech Univ, BS, 65; Univ Toronto, MASc, 66, PhD(aerospace eng), 70. *Prof Exp:* Indust postdoctoral fel res & develop, Fleet Mfg Co Ltd, 70-72; from asst prof to assoc prof struct anal, 72-82, PROF & DIR, OCEAN ENG, OCEAN ENG RES CENTRE, MEM UNIV NFLD, 82-, UNIV RES PROF, 90- *Concurrent Pos:* Spec lectr, Univ Toronto, 71; vis prof, Norweg Inst Technol, 76, Nat Res Coun Can, 79 & Univ Victoria, 88-89; pres res & develop, Offshore Design Assocs Ltd, 80- & Nfld Ocean Consults Ltd, 81-; partner, L N F Joint Venture, 84-90. *Mem:* Marine Technol Soc; Soc Naval Architects & Marine Engrs. *Res:* Wave and ice interaction with nearshore and offshore structures; ice mechanics; offshore evacuation systems; wave tank testing of marine and offshore structures. *Mailing Add:* Ocean Eng Res Ctr Mem Univ Nfld St John's NF A1B 3X5 Can

MUGGLI, ROBERT ZENO, b Richardton, NDak, Dec 6, 29; m 54; c 8. ORGANIC CHEMISTRY, CHEMICAL MICROSCOPY & MICRO INFRARED SPECTROSCOPY. *Educ:* St John's Univ, Minn, BA, 51; NDak State Univ, MS, 56; Kans State Univ, PhD(org chem). 60. *Prof Exp:* Paint chemist, Western Paint & Varnish, Minn, 53-54; anal chemist, Standard Oil Co, NDak, 54-55; res chemist, Sinclair Res Inc, Ill, 60-65; sr res assoc McCrone Assocs, Inc, Westmont, 65-91; RETIRED. *Mem:* Am Chem Soc; Sigma Xi. *Res:* Paint, petroleum, biological and organic chemistry; electron microscopy; x-ray diffraction; non-routine microanalytical chemistry; optical microscopy; infrared spectroscopy. *Mailing Add:* 1709 W Bonita Payson AZ 85541

MUGHABGHAB, SAID F, b Beirut, Lebanon, July 4, 34; m 63. NUCLEAR PHYSICS. *Educ:* Am Univ Beirut, BSc, 56, MSc, 59; Univ Pa, PhD(nuclear physics), 63. *Prof Exp:* Res assoc physics, 63-65, from asst physicist, to physicist 65-79, SR PHYSICIST, BROOKHAVEN NAT LAB, 80- *Concurrent Pos:* Adj assoc prof, Dowling Col, 70-73. *Mem:* Am Phys Soc. *Res:* Evaluation of nuclear data; compilation of nuclear data in the neutron field; photonuclear reactions; neutron total cross section measurements, study of nuclear structure and reaction mechanism with n,Gamma reactions; neutron physics and reactor calculations and design. *Mailing Add:* Dept Nuclear Energy 701 Brookhaven Nat Lab Upton NY 11973

MUGLER, DALE H, b Denver, Colo, Nov 8, 48; m 72; c 2. BIOMATHEMATICS. *Educ:* Univ Colo, BA, 70; Northwestern Univ, MA, 71, PhD(math), 74. *Prof Exp:* Asst prof math, Syracuse Univ, 74-75; from asst prof to assoc prof math, Univ Santa Clara, 75-89; ASSOC PROF MATH, UNIV AKRON, 89- *Concurrent Pos:* Alexander von Humboldt fel, 85. *Mem:* Inst Elec & Electronic Engrs; Math Asn Am; Soc Indust & Appl Math. *Res:* Complex function theory and related parts of applicable mathematics, especially differential and integral equations; signal processing applications and biomathematics. *Mailing Add:* Dept Math Sci Univ Akron Akron OH 44325-4002

MUGNAINI, ENRICO, b Siena, Italy, Dec 10, 37; m 61; c 2. NEUROANATOMY. *Educ:* Univ Pisa, MD, 62. *Prof Exp:* Trainee neuroanat, Univ Oslo, 63; asst prof anat, Med Sch, Univ Bergen, 64-66; assoc prof, Med Sch, Univ Oslo, 67-69; PROF BIOBEHAV SCI & HEAD LAB NEUROMORPHOL, UNIV CONN, 69- *Concurrent Pos:* Vis prof, Harvard Med Sch, 70; Sen Jacob Javits Neurosci Res Investr Award, NIH, 85-92. *Honors & Awards:* Decennial Camillo Golgi Award, Accademia Nazionale dei Lincei, Rome, 81; mem Norweg Acad Sci & Letters, Oslo, 84. *Mem:* AAAS; Soc Neurosci; Am Asn Anat; Am Asn Cell Biol; NY Acad Sci; Int Soc Develop Neurosci. *Res:* Developmental neurobiology with special reference to cerebellum and laminated gray matters; circuitry of cerebellum and precerebellar nuclei; vestibular system; acoustic system; myelin and glial cells; immunocytochemistry. *Mailing Add:* Dept Psychol Univ Conn U-154 Storrs CT 06269-4154

MUGWIRA, LUKE MAKORE, b Selukwe, Rhodesia, Mar 21, 40; m 65; c 2. SOIL CHEMISTRY. *Educ:* Lewis & Clark Col, BS, 65; Mich State Univ, MS, 67, PhD(soil chem), 70. *Prof Exp:* Teaching asst soil chem, Mich State Univ, 65-67, res asst, 67-70, fel, 70-71; assoc prof soil chem, Ala A&M Univ, 71-81, prof soil sci, 81-; AT DEPT LAND MGT, UNIV ZIMBABWE, SALISBURG. *Concurrent Pos:* Consult, Biochem Labs, 74 & Hayes Int Corp, City Investing Co, 75- *Mem:* Am Soc Agron; Soil Sci Soc Am. *Res:* Triticale adaptation to acid soils, its mineral nutrition and physiological characteristics related to fertilizer efficiency utilization; organic waste evaluation for crop production and role in environmental pollution. *Mailing Add:* Land Mgt Univ Zimbabwe PO Box MP167 Mt Pleasant Harre Zimbabwe

MUHLBAUER, KARLHEINZ CHRISTOPH, b Heidelberg, Ger, Dec 29, 30; US citizen; div; c 2. ENGINEERING MECHANICS. *Educ:* Mo Sch Mines, BS, 56, MS, 58; Vanderbilt Univ, PhD(struct mech), 68. *Prof Exp:* From instr to assoc prof eng mech, Univ Mo-Rolla, 56-73, prof, 73-81; mem staff, 81-85, mgr, Design Eval Sect, 85-90, PROF ENGR, AEROSPACE CORP, 90- *Concurrent Pos:* Consult, Naval Weapons Ctr, Calif, 69-72 & McDonnell Douglas Aircraft Co; Undergrad Educ Improv grant, Univ Mo-Rolla, 70-71. *Mem:* Am Soc Eng Educ; Am Soc Civil Engrs. *Res:* Stress analysis, folded plates, composite systems; individualized instruction techniques for undergraduate engineering education. *Mailing Add:* 2106 DuFour Ave Redondo Beach CA 90278

MUHLEMAN, DUANE OWEN, b Maumee, Ohio, Mar 7, 31; m 55; c 2. RADIO ASTRONOMY, PLANETARY SCIENCES. *Educ:* Univ Toledo, BS, 53; Harvard Univ, PhD(astron), 64. *Prof Exp:* Aeronaut res engr aerodyn, Nat Adv Comt Aeronaut, 53-55; commun res engr radar, Jet Propulsion Lab, Calif Inst Technol, 55-66; vis prof radar astron, Cornell Univ, 66-67; PROF PLANETARY SCI RADIO ASTRON, CALIF INST TECHNOL, 67- *Concurrent Pos:* Consult, Jet Propulsion Lab, 66-; staff mem, Owens Valley Radio Observ, Calif Inst Technol, 67-; NASA & NSF grants, 67-; US deleg gen assembly, Int Union Radio Sci, 76, 77 & 78. *Mem:* Am Astron Soc; Int Radio Sci Union; AAAS. *Res:* Planetary radio astronomy; galactic radio astronomy; experimental general relativity; celestial mechanics; radio propagation in electron plasmas; atmospheric science, microwave spectroscopy. *Mailing Add:* Dept Planetary Sci Calif Inst Technol Pasadena CA 91125

MUHLENBRUCH, CARL W(ILLIAM), b Decatur, Ill, Nov 21, 15; m 39; c 2. MATERIALS SCIENCE ENGINEERING. *Educ:* Univ Ill, BS, 37, CE, 45; Carnegie Inst Technol, MS, 43. *Prof Exp:* Res engr, Aluminum Co Am, Pa, 37-39; from instr to assoc prof civil eng, Carnegie Inst Technol, 39-48; assoc prof, Northwestern Technol Inst, 48-54; pres, 54-67, CHMN, TECSEARCH, INC, 67- *Honors & Awards:* Thompson Award, Am Soc Testing & Mat, 45. *Mem:* Am Soc Eng Educ; Am Soc Civil Engrs; Nat Soc Prof Engrs; Am Econ Develop Coun. *Res:* Mechanics and properties of engineering materials; engineering and industrial management; plant location; transportation engineering; city planning; regional studies of economic potential. *Mailing Add:* 4071 Fairway Dr Wilmette IL 60091

MUHLER, JOSEPH CHARLES, b Ft Wayne, Ind, Dec 22, 23; m 49; c 2. BIOCHEMISTRY. *Educ:* Ind Univ, BS, 45, DDS, 48, PhD(chem), 51. *Prof Exp:* From asst prof to prof chem, Ind Univ, Indianapolis, 51-61, res prof basic sci, Sch Dent, 61-72; RES PROF DENT SCI & DIR PREV DENT RES INST, SCH DENT, IND UNIV, FT WAYNE, 72- *Concurrent Pos:* Consult, Procter & Gamble Co, 49-; chmn biochem sect, Am Asn Dent Schs, 58; chmn dept prev dent, Sch Dent, Ind Univ, Indianapolis, 58-72, dir prev dent res inst, 68-72; consult, US Air Force Sch Aviation Med, 59-61, Off Surgeon Gen, US Army, 61-, Ft Knox, 62-, Mead Johnson Co, 61-, Gen Foods Corp, 64-69, Bur Med & Surg, US Navy, 64-, White Labs, 66-, & Dentsply Corp, 70- *Honors & Awards:* Award, Int Asn Dent Res, 68. *Mem:* Fel AAAS; fel Am Col Dent; fel Am Inst Chem; Am Chem Soc; Am Dent Asn. *Res:* Essentiality of trace elements; lipid metabolism. *Mailing Add:* PO Box 36 Howe IN 46746

MUHS, MERRILL ARTHUR, b San Francisco, Calif, May 9, 26; m 52; c 3. ORGANIC CHEMISTRY, ANALYTICAL CHEMISTRY. *Educ:* Univ Calif, BS, 49; Univ Wash, PhD(chem), 54. *Prof Exp:* Res chemist, Shell Develop Co, 54-73, supvr sr staff, 73; RETIRED. *Mem:* Am Chem Soc. *Res:* Gas chromatography; applied spectroscopy; characterization of odors; analysis of air and water pollutants; liquid chromatography; polymer analysis. *Mailing Add:* 942 Ashford Parkway Houston TX 77077-2402

MUIR, ARTHUR H, JR, b San Antonio, Tex, Aug 26, 31; m 88; c 2. MATERIALS SCIENCE. *Educ:* Williams Col, BA, 53; Calif Inst Technol, MS, 55, PhD(physics), 60. *Prof Exp:* Sr physicist, Atomics Int Div, NAm Rockwell Corp, 60-62, specialist Mossbauer effect, 62-63, mem tech staff, Sci Ctr, 63- 69, mem tech advan staff, 69-70, mgr int technol prog, 70-74, dir physics & chem dept, 74-76, mgr Com & Univ Progs, 76-80, proj mgr, Space Shuttle Prog, 80-82, PROG MGR, MAT SCI, ROCKWELL INT SCI CTR, 83- *Mem:* Am Phys Soc; Sigma Xi. *Res:* Applications of Mossbauer effect to solid state physics; nuclear spectroscopy; reactor physics; low energy nuclear physics; solar energy; research administration; minority engineering and science programs; aerospace materials science. *Mailing Add:* 1404 Cheswick Pl Westlake Village CA 91361

MUIR, BARRY SINCLAIR, b Belleville, Ont, July 21, 32; m 61; c 2. ZOOLOGY. *Educ:* Univ Toronto, BA, 56, MA, 58, PhD(zool), 61. *Prof Exp:* From asst prof to assoc prof zool, Univ Hawaii, 61-67; sr res scientist, Hydronautics, Inc, Md, 67-68; scientist, 68-71, asst dir res, Fisheries Res Bd Can, Marine Ecol Lab, Bedford Inst, 71-76; dir, Resource Br, Fisheries Mgt, Maritimes Region, Halifax, 76-79; DIR GEN, FISHERIES RES DIRECTORATE, DEPT FISHERIES & OCEANS, OTTAWA, 79- *Concurrent Pos:* Prin investr, NSF grant, 63-68. *Mem:* Sigma Xi; Am Fisheries Soc; Am Soc Ichthyologists & Herpteologists; Can Sos Zoologists. *Res:* Dynamics of fish populations; environmental influence on energy metabolism of fish, especially processes of growth and reproduction. *Mailing Add:* 3365 Clearwater Crescent South Keys Ottawa ON K1V 7S4 Can

MUIR, DEREK CHARLES G, b Montreal, Que, Oct 13, 49. TOXICOLOGY, ANALYTICAL CHEMISTRY. *Educ:* McGill Univ, BSc, 70, MSc, 73, PhD(agr chem), 77. *Prof Exp:* RES SCIENTIST ORG CHEMICALS TOXICOL, DEPT FISHERIES & OCEANS CAN, 77- *Concurrent Pos:* Adj prof soil sci dept, Univ Man, 78- *Mem:* Am Chem Soc; Chem Inst Can; Soc Environ Toxicol & Chem. *Res:* Persistence and degradation of xenobiotics in aquatic ecosystems; field and lab experiments; methodology development; uptake and metabolism of trace organics by aquatic animals. *Mailing Add:* Freshwater Inst 501 University Crescent Winnipeg MB R3T 0C7 Can

MUIR, DONALD RIDLEY, b Toronto, Ont, Jan 3, 29; m 54; c 2. PHYSICAL CHEMISTRY, RESEARCH ADMINISTRATION. *Educ:* Univ Toronto, BA, 51, MA, 52, PhD(phys chem), 54. *Prof Exp:* Mem staff, Res & Develop, Johnson & Johnson Ltd, 54-62; dir res & develop div, Columbia Cellulose Co, Ltd, BC, 62-69; dir res & develop, Oxford Paper Co, 69-73; pres, Sulphur Develop Inst Can, 73-87; RETIRED. *Mem:* Am Chem Soc; Tech Asn Pulp & Paper Indust; Can Pulp & Paper Asn; fel Chem Inst Can. *Res:* Research and development of new uses for sulphur. *Mailing Add:* 2816 Lionel Cres SW Calgary AB T3E 6B2 Can

MUIR, DOUGLAS WILLIAM, b Kalamazoo, Mich, May 18, 40; m 67; c 2. APPLIED NUCLEAR DATA, PARTICLE TRANSPORT CALCULATIONS. *Educ:* Southern Ill Univ, BA, 62; NMex State Univ, MS, 65, PhD(physics), 68. *Prof Exp:* Mem staff, Los Alamos Sci Lab, 68-78; nuclear data specialist, Int Atomic Energy Agency, 78-80; MEM STAFF, LOS ALAMOS NAT LAB, 80- *Mem:* Am Nuclear Soc; Am Phys Soc. *Res:* Calculation, measurement, compilation, processing and testing of nuclear data for calculations of neutron-photon transport and radiation effects. *Mailing Add:* Los Alamos Nat Lab Group T-2 MS B243 Los Alamos NM 87545

MUIR, JAMES ALEXANDER, b Jamestown, NY, Oct 29, 38; m 62; c 2. PHYSICS. *Educ:* Univ Rochester, BA, 60; Northwestern Univ, PhD(physics), 66. *Prof Exp:* From asst prof to assoc prof, 66-78, dir acad serv, Comput Ctr, 80-83, PROF PHYSICS, UNIV PR, RIO PIEDRAS, 78-, DIR, COMPUT CTR, 83- *Concurrent Pos:* Assoc scientist, PR Nuclear Ctr, 66-67; sr vis scholar, Univ Cambridge, 73-74; vis prof, Mich State Univ, 75-77; energy consult, PR Ctr Energy Environ Res, 76-79; user servs coordr, Comput Ctr, Univ PR, 78-80; spec asst to vchancellor comput, Cornell Univ, 86-87. *Mem:* Am Phys Soc; Am Crystallog Asn; Asn Comput Mach. *Res:* Crystallography; optical properties of solar selective surfaces; management of computer centers. *Mailing Add:* Dept Physics Univ PR Rio Piedras PR 00931

MUIR, MARIEL MEENTS, b Sioux Falls, SD, Nov 18, 39; m 62; c 2. CHEMISTRY. *Educ:* Grinnell Col, BA, 61; Northwestern Univ, PhD(chem), 65. *Prof Exp:* Instr chem, Univ Ill, Chicago Circle, 65-66; from asst prof to assoc prof, 66-77, dir, Dept Chem, 80-86, PROF CHEM, UNIV PR, RIO PIEDRAS, 77- *Concurrent Pos:* Assoc scientist, PR Nuclear Ctr, Rio Piedras, 66-68; sr vis scholar, Univ Cambridge, Eng, 73-74; grad prog coordr, Chem Dept, Univ PR, Rio Piedras, 78-80. *Mem:* Am Chem Soc; Sigma Xi. *Res:* Photochemistry and thermal reaction rates of transition metal complexes and interactions of complexes with molecules of biological importance. *Mailing Add:* Dept Chem Univ PR Box AW Univ Sta Rio Piedras PR 00931

MUIR, MELVIN K, b Johannesburg, SAfrica, Jan 25, 32; US citizen; m 57; c 2. SOIL CHEMISTRY, PLANT NUTRITION. *Educ:* Brigham Young Univ, BSc, 61; Pa State Univ, MSc, 63, PhD(agron), 66. *Prof Exp:* Mine officer, Johannesburg City Deep Mines, SAfrica, 52-57; asst soils, Pa State Univ, 61-66; asst prof, Mont State Univ, Univ, 57-59; environ assoc scientist, Kennecott Minerals Co, 69-83; ENVIRON ENGR, COUNTY HEALTH, SALT LAKE CITY, 84- *Mem:* Am Soc Agron; Soil Sci Soc Am. *Res:* Environmental research; amounts of hydrogen fluorine extractable ammonium in Pennsylvania soils; evaluation of slags as soil liming materials; minor element availability for plant growth; correlations of plant growth factors; influence of pollutants on air, water, soil and vegetation; methods development on analytical techniques for environment and biological samples. *Mailing Add:* 3176 S 2900 E Salt Lake City UT 84109

MUIR, PATRICK FRED, b Latrobe, Pa, Aug 18, 60; m 90. ROBOT FORCE CONTROL, MOBILE ROBOT CONTROL. *Educ:* Carnegie-Mellon Univ, BS, 82, MS, 84, PhD(elec & computer eng), 88. *Prof Exp:* Researcher, Int Bus Mach, T J Watson Res Ctr, 81, 82 & 83; SR MEM TECH STAFF, SANDIA NAT LABS, 88- *Concurrent Pos:* Mem, Tech Prog Comt Int Conf Robotics & Automation, Inst Elec & Electronic Engrs, 90. *Mem:* Inst Elec & Electronic Engrs. *Res:* Robot control, especially in-contact operations and mobile robot control; experimentally motivated robot modeling and parameter identification; model-based control; integration of sensors and the incorporation of uncertainty in robot controls; sensor design for control. *Mailing Add:* 4401 San Pedro NE No 301 Albuquerque NM 87109

MUIR, ROBERT MATHEW, b Laramie, Wyo, Oct 15, 17; m 47; c 3. PLANT PHYSIOLOGY. *Educ:* Univ Wyo, BA, 38; Univ Mich, MA, 41, PhD(bot), 46. *Prof Exp:* Asst prof bot, Pomona Col, 46-48; from asst prof to prof, 48-88, EMER PROF BOT, UNIV IOWA, 88- *Mem:* Am Soc Plant Physiol. *Res:* Relation of chemical constitution to growth regulator action; role of hormones in fruit development; vernalization; abscission; gibberellin and auxin physiology. *Mailing Add:* Dept Bot Univ Iowa Iowa City IA 52242

MUIR, THOMAS GUSTAVE, JR, b San Antonio, Tex, Aug 3, 38; div; c 2. ACOUSTICS. *Educ:* Univ Tex, Austin, BS, 61, MA, 65, PhD(mech eng), 71. *Prof Exp:* Res scientist, Appl Res Labs, Univ Tex, Austin, 61-86, Undersea Res Ctr, NATO, La Spezia, Italy, 86-89; COORDR BASIC RES & EDUC PROGS, APPL RES LABS, UNIV TEX, AUSTIN, 89- *Honors & Awards:* Jeffreys Award, 90. *Mem:* Fel Acoust Soc Am; Soc Explor Geophysicists; US Naval Inst; fel Brit Inst Acoust. *Res:* Underwater and airborne sonar and nonlinear acoustics; biomedical ultrasonics; sound propagation and scattering; seismo acoustics. *Mailing Add:* Appl Res Labs PO Box 8029 Austin TX 78713

MUIR, WILLIAM A, b Pittsburgh, Pa, Dec 8, 37; m 62; c 1. HUMAN GENETICS, ANATOMY. *Educ:* George Washington Univ, BS, 60, MS, 62; Univ Rochester, PhD(anat), 66. *Prof Exp:* Instr anat & human genetics, Sch Med, Univ Rochester, 65-66; fel human genetics, 66-74, ASST PROF MED, CASE WESTERN RESERVE UNIV, 74-, DIR HUMAN GENETICS, UNIV HOSPS, 76- *Mem:* Am Soc Human Genetics; NY Acad Sci. *Res:* Genetic and biochemical investigation of the thalassemias; hematology; biochemistry. *Mailing Add:* Prospect Hill Fredricksburg VA 22401

MUIR, WILLIAM ERNEST, b Portage la Prairie, Man, Sept 6, 40; m 68; c 2. AGRICULTURAL ENGINEERING. *Educ:* Univ Sask, BE, 62; Univ Ill, Urbana, MS, 64, PhD (agr eng), 67. *Prof Exp:* Jr res off food technol, Nat Res Coun Can, 62-63; from asst prof to assoc prof, 67-77, PROF AGR ENG, UNIV MAN, 77- *Concurrent Pos:* Res grants, Nat Res Coun Can, 68-69 & 73-78, Ctr Transp Studies, 68-71, Can Dept Agr, 68-83, Nat Sci Eng Res Coun, 79-93; vis scientist, Scottish Inst Agr Eng, 74-75; ed, Can Agr Eng, 81-84; consult, Food Technol Inst, Brasil, 82 & Food & Agr Orgn UN, India, 86; consult, Agr Can, Poland, 90. *Honors & Awards:* John Clark Award, Can Soc Agr Eng. *Mem:* Can Soc Agr Eng; Am Soc Agr Eng; Am Soc Eng Educ. *Res:* Storage and drying of cereal grains and oil seeds. *Mailing Add:* 713 Patricia Winnipeg MB R3T 3A8 Can

MUIR, WILLIAM W, III, b Bay City, Mich, July 8, 46. ANESTHESIOLOGY, ELECTROPHYSIOLOGY. *Educ:* Mich State Univ, BS, 68, DVM, 70; Ohio State Univ, MS, 72, PhD(physiol), 74; Am Col Vet Anesthesiol, dipl, 77; Am Col Vet Physiol & Pharm, dipl, 80. *Prof Exp:* Res assoc physiol, 70-74, from asst prof to assoc prof, 74-82, PROF PHYSIOL & PHARM, OHIO STATE UNIV, 82- *Concurrent Pos:* Ed, Am Vet Med Asn J, 78-, Am J Vet Res, 78- *Mem:* Am Chem Soc; Am Col Cardiol Vet; Am Heart Asn. *Res:* Electrophysiologic and hemodynamic effects of drugs, which effect the cardiovascular system. *Mailing Add:* Vet Clin Sci Ohio State Univ 1010 Vet Hosp Columbus OH 43210

MUIR, WILSON BURNETT, b Montreal, Que, Can, July 20, 32; m 55; c 3. SOLID STATE PHYSICS. *Educ:* McGill Univ, BSc, 53; Univ Western Ont, MSc, 55; Ottawa Univ, PhD(solid state physics), 62. *Prof Exp:* Sci officer, Radioactivity Div, Dept Mines & Tech Surv, Can, 55-57; physicist, Franklin Inst Labs, 61-64 & Noranda Res Centre, 64-66; physicist, 66-68, from asst prof to assoc prof, 68-83, PROF PHYSICS, MCGILL UNIV, 83- *Mem:* Am Phys Soc; Can Asn Physicists. *Res:* Electron transport; magnetization and Mossbauer effect in crystalline and amorphous metals and alloys, semiconductors and minerals. *Mailing Add:* Rutherford Bldg Dept Physics McGill Univ c/o W B Muir Box 607 Richford VT 05476

MUIRHEAD, E ERIC, b Recife, Brazil, Sept 13, 16; nat US; m 42; c 5. PATHOLOGY, MEDICINE. *Educ:* Baylor Univ, BA & MD, 39; Am Bd Path, dipl anat & clin path, 47, dipl blood banking, 74, dipl hemat, 75. *Prof Exp:* Intern, Baylor Univ Hosp, 39-40, resident path, 40-41, asst dir labs & instr clin path, Col Med, 41-43, asst prof, Dent Col, 46-48; med officer, USN, 43-46; from instr to asst prof clin path, Southwestern Med Sch, Univ Tex, 46-48, from assoc prof to prof path, 48-59, chmn dept, 50-56, chief, Div Hemat, Dept Internal Med, 51-59; prof clin path, Col Med, Wayne State Univ, 59-65; prof path & clin prof med, Ctr Health Sci, Univ Tenn, 65-80, prof path & chmn dept, Col Med, 80-86, prof med, 80-88, EMER PROF PATH & MED, COL MED, UNIV TENN, MEMPHIS, 86- *Concurrent Pos:* Consult, Brooke Gen Hosp, US Army, Ft Sam Houston, 48-67, emer consult, 67-; asst dir, William Buchanan Blood, Plasma & Serum Ctr, Dallas, 46-48; consult, Vet Admin Hosps, Dallas, 48-59, Vet Admin Path & Allied Sci Serv, Washington, DC, 63-64 & US Army Biol Labs, Ft Detrick, 66-71; dir, Labs & Blood Bank, Woman's Hosp, Detroit, 59-65 & Dept Path & Blood Bank, Baptist Mem Hosp, Memphis, 65-88; chmn, Awards Comt, Am Asn Blood Banks, 63-82, Comt Educ, 67-74, Coun Immunohemat, Am Soc Clin Pathologists, 65-67, Coun High Blood Pressure Res, Am Heart Asn, 82-84 & Coun Affairs Comt, 86- 88; mem, Coun Circulation, Am Heart Asn, 80-, bd dirs, 82-84 & 86-88, Steering Comt Med & Community Prog, 86-88; adj fac mem, Dept Biol, Memphis State Univ, 86- *Honors & Awards:* Emily Cooley Award, Am Asn Blood Banks, 63, John Elliott Mem Award, 68, Distinguished Serv Award, 82; Arthur C Corcoran Lectr, Coun High Blood Pressure Res, 79; Arthur Grollman Lectr, Univ Tex, Dallas, 81; Merck, Sharpe & Dohme Award, Int Soc Hypertension, 82; Ward Burdick Award, Am Soc Clin Pathologists, 86; Award of Merit, Am Heart Asn, 89. *Mem:* Am Physiol Soc; Am Soc Clin Invest; Am Asn Pathologists; Soc Exp Biol & Med; AAAS; Am Soc Clin Pathologists; AMA; Am Asn Blood Banks (pres, 56-57); fel Col Am Pathologists; fel Am Heart Asn. *Res:* Author of more than 400 technical publications. *Mailing Add:* Dept Path & Med Univ Tenn 899 Madison Ave Memphis TN 38146

MUIRHEAD, ROBB JOHN, b Adelaide, S Australia, July 7, 46; m 70; c 2. MATHEMATICAL STATISTICS. *Educ:* Univ Adelaide, BSc, 68, PhD(statist), 70. *Prof Exp:* From asst prof to assoc prof statist, Yale Univ, 70-78; assoc prof, 78-81, PROF STATIST, UNIV MICH, 81- *Mem:* Am Statist Asn; fel, Inst Math Statist; Royal Statist Soc. *Res:* Multivariate analysis and distribution theory; asymptotic methods; linear models. *Mailing Add:* Dept Statist Univ Mich Main Campus Ann Arbor MI 48109

MUIRHEAD, VINCENT URIEL, b Dresden, Kans, Feb 6, 19; m 43; c 3. AEROSPACE ENGINEERING. *Educ:* US Naval Acad, BS, 41; US Naval Postgrad Sch, BSAE, 48; Calif Inst Technol, AE, 48. *Prof Exp:* Ensign to commander, US Navy, 41-61; from asst prof to assoc prof aerodyn, 61-76, chmn dept, 76-88, prof aerospace eng, 76-89, EMER PROF AEROSPACE ENG, UNIV KANS, 89- *Concurrent Pos:* Consult, Black & Veatch, Consult Engrs, Mo, 64- *Mem:* Assoc fel Am Inst Aeronaut & Astronaut; Am Soc Eng Educ; Am Acad Mech. *Res:* Subsonic and supersonic aerodynamics; vortices; tornado damage to buildings and wind loadings on buildings from severe winds; shock tubes; dynamic ground effect. *Mailing Add:* Dept Aerospace Eng Univ Kans Lawrence KS 66045

MUJUMDAR, ARUN SADASHIV, b Karwar, India, Jan 14, 45; Can citizen. CHEMICAL ENGINEERING. *Educ:* Univ Bombay, BChemEng, 65; McGill Univ, MEng, 68, PhD(chem eng), 71. *Prof Exp:* Mech engr, Res Div, Carrier Corp, NY, 69-71; res assoc & univ fel, 71-74, asst prof, 74-78, assoc prof, 78-86, PROF, MCGILL UNIV & PULP & PAPER RES INST, CAN, 86- *Concurrent Pos:* Sabbatical leave, India, Japan, USA; ed, Advan Drying, 77-, Advan Tranport Processes, 78- & Int J Drying Technol, 89- *Honors & Awards:* Hemisphere Award Innovation Drying, 86; Josef Janus Medal, Czech Acad Sci, 90. *Mem:* Am Inst Chem Engrs; Am Soc Mech Engrs; Can Pulp & Paper Asn; Can Soc Chem Engrs; Ind Soc Heat Mass Tranfer; Tech Asn Pulp & Paper Indust; Sigma Xi. *Res:* Turbulence; heat and mass transfer; drying; fluidization. *Mailing Add:* Dept Chem Eng McGill Univ 3480 University St Montreal PQ H3A 2A7 Can

MUKA, ARTHUR ALLEN, b Adams, Mass, Oct 23, 24; m 52; c 5. ENTOMOLOGY. *Educ:* Univ Mass, BS, 50; Cornell Univ, MS, 52, PhD(econ entom), 54. *Prof Exp:* Asst entom, Cornell Univ, 50-54; assoc entomologist, Va Agr Exp Sta, 54-56; from asst prof to assoc prof, 56-65, PROF ENTOM, CORNELL UNIV, 65- *Concurrent Pos:* Rockefeller Found entomologist, Int Rice Res Inst, 65-66. *Mem:* Entom Soc Can. Am; Entom Soc Can. *Res:* Field and forage crop insect pests; vegetable insect pests; pest management; extension entomology; international agricultural development. *Mailing Add:* Dept Entom Cornell Univ Ithaca NY 14853-0999

MUKAI, CROMWELL DAISAKU, b Bostonia, Calif, Apr 13, 17; m 44; c 4. CHEMISTRY. *Educ:* Univ Calif, BS, 43; NY Univ, MS, 49, PhD(org chem), 55. *Prof Exp:* Res chemist, Gelatin Prods Corp Mich, 44-46; res chem chemist, Boyle-Midway Div, Am Home Prods Corp, 46-67, res assoc, 67-75; mgr, Anal Lab, Polychrome Corp, Yonkers, 75-80; RES MGR, DELEET MERCHANDISING CORP, NEWARK, 80- *Mem:* AAAS; Am Chem Soc. *Res:* Synthesis of amino acids; emulsion technology; petroleum additives; plastics; mechanisms of Grignard reactions; instrumental analysis; characterization and synthesis of resins; lithographic ink vehicles; graphic arts chemicals; aerosol technology; cleaners; detergents; emulsion polymerization. *Mailing Add:* 26 Brook St Berkeley Heights NJ 07922

MUKERJEE, BARID, b Suri, India, Oct 27, 28; m 59. GENETICS. *Educ:* Univ Calcutta, BSc, 51; Brigham Young Univ, MS, 56; Univ Utah, PhD(genetics), 58. *Prof Exp:* Demonstr biol, Vidyasagar Col, India, 51-55; asst prof, Westminster Col, 58-59; asst res prof, Univ Utah, 60-61; res assoc genetics, Columbia Univ, 61-63; from asst prof to assoc prof, 63-70, prof genetics, 70-80, PROF BIOL, MCGILL UNIV, 80- *Res:* Relationship of embryonic differentiation and malignancy; mechanism of chromosome differentiation. *Mailing Add:* Dept Biol McGill Univ 853 Sherbrooke St W Montreal PQ H3A 2M5 Can

MUKERJEE, PASUPATI, b Calcutta, India, Feb 13, 32; m 64. SURFACE CHEMISTRY, COLLOID CHEMISTRY. *Educ:* Univ Calcutta, BSc, 49, MSc, 51; Univ Southern Calif, PhD(colloid chem), 57. *Prof Exp:* Lab asst chem, Univ Southern Calif, 52-54, res fel, 54-56, lectr & res fel, 56-57; res assoc, Brookhaven Nat Lab, 57-59; reader phys chem, Indian Asn Cultivation Sci, 59-64; guest scientist chem, Van't Hoff Lab, Univ Utrecht, 64; sr scientist, Univ Southern Calif, 64-66; vis assoc prof, 66-67, PROF, SCH PHARM, UNIV WIS-MADISON, 67- *Concurrent Pos:* Vis asst prof, Univ Southern Calif, 57; hon lectr, Univ Calcutta, 61-64; vis prof, Indian Inst Technol, Kharagpur, India, 71-72; assoc mem, Comn Colloid & Surface Chem, Int Union Pure & Appl Chem, 75-83. *Mem:* Fel AAAS; Am Chem Soc; Int Asn Colloid & Interface Scientists; fel Acad Pharmaceut Sci; Sigma Xi; fel Am Inst Chemists. *Res:* Structure and properties of micelles; molecular microenvironment at interfaces; physical chemistry of dyes, drugs, bile salts, bilirubin and lipopolysaccharides; solubilization and chemical reactions in micellar systems; adsorption phenomena; fluorocarbon-hydrocarbon interactions. *Mailing Add:* Sch Pharm Univ Wis 425 N Charter St Madison WI 53706

MUKERJI, MUKUL KUMAR, b India, Jan 6, 38; Can citizen. ENTOMOLOGY, POPULATION ECOLOGY. *Educ:* Univ Calcutta, BSc, 57, MSc, 59; McGill Univ, PhD(entom), 65. *Prof Exp:* Res scientist insect pop, 67-80, SR RES SCIENTIST, RES BR AGR CAN, 80- *Concurrent Pos:* Nat Res Coun Can fel, 65-66; assoc ed, Can Entom, Entom Soc Can, 77- *Mem:* Entom Soc Can; Japanese Soc Pop Ecol; Entom Soc Am. *Res:* Insect population ecology; biocontrol; pest management; modelling; bioenergetics; insect morphology; survey. *Mailing Add:* Res Sta Res Br 107 Science Crescent Saskatoon SK S7N 0X2 Can

MUKHEDKAR, DINKAR, b Hyderabad, India, Feb 2, 36; Can citizen; m 64; c 2. ELECTRICAL ENGINEERING. *Educ:* Osmania Univ, India, BS, 57; Univ Nancy, DSc(power), 62. *Hon Degrees:* DSc, Univ Nancy, 62. *Prof Exp:* Design engr, Bhilai Steel Works, India, 57-59; Govt of France grant, Inst Radium, Paris, 62-63; asst prof elec eng, Indian Inst Technol, Bombay, 63-64; sr engr, Surveyer, Nenineger & Chenevert, 64-68; from asst prof to assoc prof, 68-74, PROF ELEC ENG, POLYTECH SCH, UNIV MONTREAL, 74- *Concurrent Pos:* Consult, Surveyer, Nenineger & Chenevert, 68- *Mem:* Fel Inst Elec & Electronics Engrs; Fr Soc Electricians; fel Eng Inst Can; Can Elec Asn; fel Inst Engrs, India; fel Inst Elec Eng, UK; fel Peruvian Soc Elec Eng. *Res:* Power systems; simulation techniques. *Mailing Add:* Ecole Polytech PB 6079 Sta A Montreal PQ H3C 3A7 Can

MUKHERJEA, ARUNAVA, b Calcutta, India, Aug 14, 41; m 71; c 1. MATHEMATICS. *Educ:* Univ Calcutta, MSc, 61; Wayne State Univ, PhD(math), 67. *Prof Exp:* Asst prof math, Eastern Mich Univ, 67-69; from asst prof to assoc prof, 69-75, PROF MATH, UNIV SOUTH FLA, 75- *Concurrent Pos:* Ed, J Theoret Probability, Plenum, NY & London. *Mem:* Inst Math Statist; Am Math Soc. *Res:* Probability; measure theory; analysis on semigroups. *Mailing Add:* Dept Math Univ South Fla Tampa FL 33620-5700

MUKHERJEE, AMAL, b Titagarh, West Bengal, India, Apr 27, 44; US citizen; m 75; c 1. HORMONE RECEPTORS, ENZYMOLOGY. *Educ:* Calcutta Univ, BS, 65, MS, 67, PhD(physiol), 72. *Prof Exp:* Asst prof med, Ischemic Heart Ctr, Univ Tex Health Sci Ctr, Dallas, 80-; AT DIAG SYSTS LABS, WEBSTER, TX. *Concurrent Pos:* Consult, Int Immunoassay Lab, 79-; prin investr, Am Heart Asn, 81- *Mem:* Am Physiol Soc; Int Soc Heart Res. *Res:* Involvement of autonomic receptors in cardiac excitation and contraction; development of radioimmunoassay of enzymes used as markers for detecting heart diseases. *Mailing Add:* Immuno Diag Ctr Inc 607 S Friendswood Dr No 27 Friendswood TX 77546

MUKHERJEE, AMIYA K, b Purnea, India, June 1, 36; m 62; c 2. MATERIALS SCIENCE, METALLURGY. *Educ:* Univ Calcutta, BSc, 54; Univ Sheffield, MSc, 59; Oxford Univ, DPhil(metall), 62. *Prof Exp:* Mgt trainee, Indian Iron & Steel Co, 54-56 & Stewart & Lloyds Ltd, Eng, 56-57; res metallurgist, Lawrence Radiation Lab, 62-65; sr scientist, Battelle Mem Inst, 65-66; from asst prof to assoc prof eng, 66-69, PROF ENG, UNIV CALIF, DAVIS, 69- *Concurrent Pos:* Fel, Univ Calif, 62-65; consult, Lawrence Radiation Lab, 68- & IBM Corp. *Mem:* Am Inst Mining, Metall & Petrol Engrs; Am Soc Metals; Brit Inst Metals; Brit Inst Metall; fel Am Inst Chemists. *Res:* Deformation mechanisms of crystalline materials as a function of temperature strain-rate and crystal structure; elevated temperature strength properties of alloys; fracture characteristics of engineering materials; environmental behavior of structural materials. *Mailing Add:* Dept of Mech Eng Col of Eng Univ of Calif Davis CA 95616

MUKHERJEE, ASIT B, b Suri, WBengal, India; m; c 1. MOLECULAR CYTOGENETICS, GENETICS OF AGING. *Educ:* Univ Utah, BS, 65, MS, 66, PhD(genetics), 68. *Prof Exp:* Teaching asst biol, Univ Utah, 65-67, res fel, 67-68; res assoc human genetics, State Univ NY Upstate Med Ctr, 68- 69; fel genetics, Med Ctr, Columbia Univ, 69-70; instr, Albert Einstein Col Med, 70-72; from asst prof to assoc prof, 72-83, PROF BIOL, FORDHAM UNIV, 83- *Concurrent Pos:* Vis prof, McGill Univ, 78-79; NIH fel, Columbia Univ; vis investr, Mem Sloan-Kettering Cancer Ctr, 90- *Mem:* AAAS; Tissue Cult Asn; Am Soc Cell Biol; Am Soc Human Genetics; Gerontol Soc Am. *Res:* Genetics of aging; molecular cytogenetics; human genetics. *Mailing Add:* Dept Biol Fordham Univ Bronx NY 10458

MUKHERJEE, DEBI PRASAD, b Krisnanagar, India, Oct 26, 39. CHEMICAL ENGINEERING, POLYMER SCIENCE. *Educ:* Jadavpur Univ, India, BChE, 61; Mass Inst Technol, SM, 65, ScD(chem eng), 69, Univ Conn, MBA, 79. *Prof Exp:* Asst chem eng, Mass Inst Technol, 65-68; sr res engr, Basic Polymer Res Sect, Goodyear Tire & Rubber Co, 69-74; sr res engr & group leader, Tech Specialist Biomat, Davis & Geck, Am Cyanamid Co, 74-87; res prog mgr, Dow Corning Wright, 87-90; RES SCIENTIST, UNION CARBIDE, 90- *Mem:* Am Chem Soc; Rheology Soc; Am Inst Chem Engrs. *Res:* Viscoelastic properties of elastin; molecular structure and physical properties of elastin; biopolymers; structure and property of polymers; fiber spinning; polymer melt rheology; sutures, molded medical devices and medical products development; orthopedics product development; biomaterials. *Mailing Add:* Union Carbide Bldg 98 Rm 214 PO Box 670 Bound Brook NJ 08805

MUKHERJEE, KALINATH, b Calcutta, India, Feb 19, 32; US citizen; m 59; c 3. SOLID STATE PHYSICS. *Educ:* Univ Calcutta, BE, 56; Univ Ill, Urbana, MS, 59, PhD(metall), 63. *Prof Exp:* Engr, Indian Iron & Steel Co, 56-57; res asst, Univ Ill, Urbana, 57-63, res assoc, 63-64; asst prof mat sci, State Univ NY Stony Brook, 64-67; assoc prof metall, Polytech Inst NY, 67-72, prof, 72-80, head, Dept Physics & Eng Metall, 74-80; PROF, COL ENG, MICH STATE UNIV, 80-, CHMN, DEPT METALL, MECHS & MAT SCI, 85- *Concurrent Pos:* Indian Inst Metals Awards, 56; NSF grant, 66-67; NSF fac res fel, Grumman Aerospace, 76; guest prof, Dept Metal Kunde, Cath Univ, Luven, Belg, 82; vis fac, Tokyo Inst Technol, Japan, 85. *Mem:* Fel Am Soc Metals; fel AAAS; Am Inst Mining, Metall & Petrol Engrs; Am Phys Soc. *Res:* Thermodynamics of point defects; diffusion-less phase transformations; solid state reactions kinetics; laser interactions with metals and alloys; lattice vibrations. *Mailing Add:* Col Eng Mich State Univ East Lansing MI 48824

MUKHERJEE, MUKUNDA DEV, nutrition, pediatrics, for more information see previous edition

MUKHERJEE, PRITISH, b Saugor, Madhya Pradesh, India, Sept 17, 55; m 87; c 1. ULTRAFAST LASER SPECTROSCOPY, NONLINEAR OPTICS. *Educ:* Univ Delhi, India, BS, 76, MS, 78; State Univ NY, Buffalo, MA, 82, PhD(elec eng), 86. *Prof Exp:* Postdoctoral fel, Los Alamos Nat Lab, 87-88; ASST PROF PHYSICS, UNIV SFLA, 88- *Honors & Awards:* Nat R&D 100 Award, Res & Develop Mag, 90. *Mem:* Am Phys Soc; Optical Soc Am; Inst Elec & Electronic Engrs; AAAS; Planetary Soc. *Res:* Ultrafast laser spectroscopy; study of non-linear laser-semiconductor interactions; utilization of laser spectroscopic techniques for analytical applications. *Mailing Add:* Dept Physics Univ SFla Tampa FL 33620

MUKHERJEE, TAPAN KUMAR, b Gorakhpur, India, Jan 5, 29; m 57; c 2. ORGANIC CHEMISTRY, SOLID STATE SCIENCE. *Educ:* Patna Univ, BS, 48, MS, 50, DSc(chem), 74; Wayne State Univ, PhD(chem), 56. *Prof Exp:* Lectr chem, Patna Univ, 50-52; asst, Wayne State Univ, 52-55, res fel, 55-56; lectr, Univ Bihar, 57-58; res chemist, Mass Inst Technol, 58-60, Gen Aniline & Film Corp, 60-61, Retina Found, 61-62 & Air Force Cambridge Res Lab, 62-74; prog mgr, Solar Energy Conversion, 74-75, prog mgr, Advan Energy & Resources, 75-78, prog mgr div appl res,78-81, dir Minerals & Primary Mat Processing Prog, 81-84, dir mat eng & processing prog, 85-86, dir environ eng prog, 86-87, DIR ENG RES CTRS PROG, NSF, 87- *Mem:* Am Chem Soc; Am Inst Mining, Metall & Petrol Engrs; Indian Chem Soc. *Res:* Syntheses of organic compounds; mechanism of reactions; stable free radicals; organic semiconductors; photoconductors; high temperature laser window materials, photovoltaic materials and devices. *Mailing Add:* Eng Res Ctrs Prog NSF 1800 G St NW Washington DC 20550

MUKHERJI, KALYAN KUMAR, b Calcutta, India, May 30, 39; m 65. GEOLOGY, GEOPHYSICS. *Educ:* Univ Calcutta, BSc, 59, MSc, 61; Univ Leeds, Dipl, 63; Univ Western Ont, PhD(geol), 68. *Prof Exp:* Demonstr geol, Univ Western Ont, 63-67; asst prof, 68-73, PROF GEOL, CONCORDIA UNIV, LOYOLA, MONTREAL, 73- *Concurrent Pos:* Fel, Carleton Univ, Ont, 68; Nat Res Coun Can grant, 69-; Geol Surv Can res grant, 69-70. *Mem:* Can Soc Petrol Geologists; Int Asn Sedimentologists; Indian Soc Earth Sci; Soc Econ Paleont & Mineral; Geol Asn Can. *Res:* Ordovician stratigraphy and sedimentation in southwest Ontario; carbonate sedimentation and petrology of Sicker Group, Vancouver Islands; thermo luminescence study of Middle Ordovician limestones and recent sediments; exploration geophysics; carbonate trace element geochemistry; oceanography; offshore mud bank and related sediment. *Mailing Add:* Dept of Geol Concordia Univ Loyola Montreal PQ H4B 1R6 Can

MUKHOPADHYAY, NIMAI CHAND, b Maharampur, West Bengal, Jan 17, 42; m 78; c 2. THEORETICAL PHYSICS. *Educ:* Univ Calcutta, BSc, 63, MSc, 65; Univ Chicago, SM, 70, PhD(physics), 72. *Prof Exp:* Vis mem res nuclear physics, Tata Inst Fundamental Res, Bombay, India, 66-68; resident assoc theoret physics, Argonne Nat Lab, 68-72; fel, Univ Md, 72-74; vis scientist, Europ Orgn Nuclear Res, Geneva, Switz, 74-75; physicist, Swiss Inst Nuclear Res, Villigen, 75-80, Swiss Fed Inst Reactor Res, Wurenlingen, 80-81; assoc prof, 81-86, PROF TEACHING & RES PHYSICS, RENSSELAER POLYTECH INST, 86- *Concurrent Pos:* Vis prof, Univ de Louvain, Belg, 75 & Int Sch Nuclear Physics, Erice, Italy, 76; vis scientist, Inst Nuclear Res, Holland, 77; vis prof, Joint Inst Nuclear Res, Dubna, USSR, 78, Phys Res Lab, Ahmedabad, India, 78 & Technion, Haifa, Israel, 79-80; invited lectr, Univ Fribourg, Switz, 81; vis prof, Paul Scherrer Inst, Switz, 87-88, Univ Va, 88-89. *Mem:* Sigma Xi; Am Phys Soc. *Res:* Theoretical medium energy physics as studied in lepton and meson factories; manifestations of fundamental interactions in atomic, nuclear and astrophysical phenomena. *Mailing Add:* Dept Physics Rensselaer Polytech Inst Troy NY 12180-3590

MUKHOPADHYAY, NITIS, b West Bengal, Dec 22, 50; m 78; c 1. PATTERN RECOGNITION, MULTIVARIATE ANALYSIS. *Educ:* Univ Calcutta, BSc, 70; Indian Statist Inst, MStat, 72, PhD(statist), 76. *Prof Exp:* Tutor, Monash Univ, Australia, 76-77, vis asst prof statist, Univ Minn, Minneapolis, 77-78, Univ Mo, Columbia, 78-79; asst prof, 79-81, ASSOC PROF STATIST, OKLA STATE UNIV, 81- *Mem:* Am Statist Asn; Inst Math Statist; Inst Math Statist; NY Acad Sci; Sigma Xi. *Res:* Applied and theoretical statistics using sequential analysis for reliability or clinical trials. *Mailing Add:* Dept Statist U-120 196 Auditorium Univ Conn Storrs CT 06268

MUKHTAR, HASAN, b Lucknow, India, Jan 1, 47; m 73; c 3. BIOCHEMISTRY, ENVIRONMENTAL SCIENCES. *Educ:* Lucknow Univ, India, BS, 65, MS, 67; Kanpur Univ, India, PhD(biochem), 71. *Prof Exp:* Fel biochem, Med Col Ga, Augusta, 74-76; vis scientist pharmacol, Nat Inst Environ Health Sci, NC, 76-79; asst prof, 80-84, ASSOC PROF DERMAT & ENVIRON HEALTH SCI, CASE WESTERN RESERVE UNIV, 85-; DIR DERMAT RES, VET ADMIN MED CTR, CLEVELAND, 85- *Concurrent Pos:* Vis scientist, Univ Leiden, Neth, 80. *Mem:* Am Soc Pharmacol & Exp Therapeut; Am Asn Cancer Res; Am Soc Photobiol; Soc Investigative Dermat; Int Soc Study Xenobiotics; AAAS; Soc Toxicol. *Res:* Mechanism of skin cancer formation by physical and chemical agents; industrial toxicology and occupational health; environmental health sciences and toxicology. *Mailing Add:* Vet Admin Med Ctr 10701 E Blvd Cleveland OH 44106

MUKI, ROKURO, b Morioka, Japan, Apr 27, 28; m 56; c 2. SOLID MECHANICS. *Educ:* Keio Univ, Japan, BS, 51, PhD(eng), 59. *Prof Exp:* Lectr math, Keio Univ, Japan, 56-58 & 60-61, assoc prof eng, 61-65; res assoc, Brown Univ, 58-60; vis assoc prof, Calif Inst Technol, 65-66, sr res fel, 66-67; assoc prof, 67-69, PROF APPL MECH, UNIV CALIF, LOS ANGELES, 69- *Mem:* Am Soc Mech Engrs; Sigma Xi. *Res:* Three dimensional theory of elasticity; thermal stress problems; linear viscoelasticity; elastic load diffusion problems; micromechanics of composite materials. *Mailing Add:* Sch of Eng & Appl Sci Univ Calif 4532 Boelter Hall Los Angeles CA 90024

MUKKADA, ANTONY JOB, b Kerala, India. MICROBIOLOGY, BIOCHEMISTRY. *Educ:* Univ Kerala, BSc, 57; Univ Delhi, MSc, 59, PhD(bot), 62. *Prof Exp:* Fel, Nat Inst Sci, India, 62-63; fel, St Thomas Inst Advan Studies, Ohio, 64-67; vis asst prof bact, 67-68, res assoc, 69-71, asst prof, 71-75, assoc prof biol, 75-79, PROF BIOL & MICROBIOL, UNIV CINCINNATI, 79- *Mem:* Am Soc Microbiol; Am Soc Parasitologists; AAAS; Int Soc Plant Morphologists; Sigma Xi. *Res:* Microbial physiology and metabolic regulation; biochemistry of intracellular parasitism with special reference to membrane transport and enzyme regulation in haemoflagellates. *Mailing Add:* Brodie Sci Complex Biol Univ Cincinnati Cincinnati OH 45221

MUKUNNEMKERIL, GEORGE MATHEW, b Kerala, India, Jan 10, 39; m 73; c 2. DEVELOPMENTAL BIOLOGY. *Educ:* Univ Kerala, BSc, 60; DePaul Univ, MS, 69; Univ Notre Dame, PhD(biol), 73. *Prof Exp:* Demonstr zool, St Berchmans' Col, 60-61; teacher biol, St Josephs High Sch, 61-65; NIH grant, DePaul Univ, 66-68; NSF fel, Univ Notre Dame, 71-73, res assoc radiation lab, 73-76; asst prof biol, NCent Col, 76-81 & Univ Ill, Chicago Circle, 81-85; asst prof biol, Univ Ill, Chicago Circle, 81-85; INSTR, GUILFORD TECH COMMUNITY COL, JAMESTOWN, 85- *Mem:* AAAS; Am Soc Zoologists; Sigma Xi. *Res:* Cytochemical and ultrastructural aspects of gametogenesis in insects with emphasis on mosquitoes; ontogenesis of hormonal control during vertebrate embryogenesis, particularly chick embryos; ultrastructure of hemopoiesis. *Mailing Add:* 3318 Warwick Dr Jamestown NC 27282

MULAIK, STANLEY B, b Sept 30, 02; m; c 1. ENTOMOLOGY, TERRESTRIAL CRUSTACEANS. *Educ:* Slippery Rock State Col, BS, 28; Cornell Univ, MS, 31; Univ Utah, PhD(entom), 54. *Prof Exp:* Teacher biol & physics, Ridgeway High Sch, Pa, 28-30 & sci & math, Edinburg Jr High Sch, Tex, 31-39; from instr to assoc prof, 42-70, EMER PROF BIOL, UNIV UTAH, 70- *Concurrent Pos:* Prof, Westminster Col Utah, 43-45. *Honors & Awards:* Connie Award, Nat Wildlife Fedn, 80. *Mem:* Fel AAAS; Am Nature Study Soc; Nat Wildlife Fedn; Conserv Educ Asn; Am Biol Teachers Soc. *Res:* Spiders; Mexican isopods; reptiles; acarina; conservation. *Mailing Add:* 1144 E 300 S Salt Lake City UT 84102-2506

MULAR, A(NDREW) L(OUIS), b Beulah, NDak, Dec 10, 30; m 57; c 4. MINERAL PROCESS ENGINEERING. *Educ:* Mont Col Mineral Sci & Technol, BSc, 57, MSc, 58. *Prof Exp:* Res engr metall, Mass Inst Technol, 58-60, Mich Technol Univ, 60-62 & Univ Calif, Berkeley, 62-63; from assoc prof to prof metall eng, Queen's Univ, Ont, 63-72; PROF MINERAL PROCESS ENG, UNIV BC, 72-, DEPT HEAD, 87- *Concurrent Pos:* Consult, 80- *Mem:* Am Inst Mining, Metall & Petrol Engrs; Can Inst Mining & Metall; Asn Prof Eng BC. *Res:* Comminution; flotation; process simulation; optimization; control; unit operations. *Mailing Add:* Univ BC 2075 West Brook Pl Univ BC 6350 Stores Rd Vancouver BC V6T 1W5 Can

MULARIE, WILLIAM MACK, b Duluth, Minn, Dec 4, 38; m 86; c 3. SOLID STATE PHYSICS. *Educ:* Univ Minn, Duluth, BA, 61; Univ Minn, Minneapolis, MSEE, 66, PhD(surface physics), 71. *Prof Exp:* Physicist, Minn Mining & Mfg Co, 61-63; res scientist solid state physics, Res Inst Advan Studies, 71-74; sr res engr, 3M Co, 74-77, res specialist, 77-79, supvr, Cent Res Lab, 79-80; corp res dir, Am Hoist & Derrick, 80-81; vpres & gen mgr, Interbond Technol, Inc, 81-82; exec vpres, Vac-Tec Systs, Inc, 82-84; gen mgr, Perkin-Elmer Inc, 85-86; lab mgr, 86-89, DIR, PGM DEVELOP, 3M CO, 90- *Mem:* Sigma Xi. *Res:* Physics of semiconductor surfaces; auger electron spectroscopy; glass corrosion; infrared detectors; optics; materials science engineering; magnetic materials. *Mailing Add:* 1940 Melody Hill Circle Excelsior MN 55331

MULARZ, EDWARD JULIUS, b Lakewood, Ohio, Nov 24, 43; m 67; c 2. COMBUSTION, AEROSPACE ENGINEERING. *Educ:* Univ Detroit, BS, 66; Northwestern Univ, PhD(mech eng), 71. *Prof Exp:* Engr combustion, 71-75, sr engr, 75-78, prog mgr combustion, 78-81, head, combustion section, propulsion lab, US Army Res & Tech Labs, NASA, 81-84, CHIEF, MODELING & VERIFICATION BRANCH, NASA-LEWIS RES CTR, 84- *Mem:* Am Inst Aeronaut & Astronaut; Sigma Xi; Am Soc Mech Engrs; Combustion Inst. *Res:* Heat transfer; thermo-dynamics; aircraft propulsion. *Mailing Add:* 4066 Brewster Dr Cleveland OH 44145

MULAS, PABLO MARCELO, b Atlixco, Mex, Apr 26, 39. PHYSICAL CHEMISTRY, NUCLEAR ENGINEERING. *Educ:* Univ Ottawa, BS, 60; Princeton Univ, PhD(chem eng), 65. *Prof Exp:* Resident res assoc, Jet Propulsion Lab, Nat Acad Sci, 67-68; prof nuclear sci, Nat Polytech Inst, Mex, 67-71, head dept nuclear eng, 69-71; dir, Reactor Lab, Nat Inst Nuclear Energy, Mex, 71-73; prof phys chem, CIEA Res, Nat Polytech Inst, Mex, 73-76; DIR, ENERGY RESOURCES DIV, MEX ELEC RES INST, MEX, 76- *Concurrent Pos:* Mem, Nat Prog Basic Sci Comt, Nat Coun Sci & Technol, Mex, 75-82 & Nat Prog Use Non-Renewable Resources Comt, 83-; vis prof, grad sect, Sch Eng, Univ Calif, Los Angeles, 68. *Mem:* Am Phys Soc; Am Nuclear Soc; Am Chem Soc; Acad Sci Res Mex; Mex Phys Soc; Sigma Xi. *Res:* Waste heat and low enthalpy fluid utilization; use in boilers of heavy fuel oils. *Mailing Add:* Aptdo Postal 100-C Col Laspalmas Cuernavaca Morelos 62050 Mexico

MULAY, LAXMAN NILAKANTHA, b Rahuri, India, Mar 5, 23; m 45. PHYSICAL & INORGANIC CHEMISTRY, SOLID STATE SCIENCE. *Educ:* Univ Bombay, MS, 46, PhD(phys chem), 50. *Prof Exp:* Daxina Merit fel, Karnatak Col, Bombay, 43-45, demonstr & lectr chem, 46-48, lectr, Inst Sci, 48-53 & 57-58; res assoc, Northwestern Univ, 53-55; res fel, Harvard Univ, 55-57; asst prof, Univ Cincinnati, 58-63; assoc prof, 63-67, chmn, Solid State Sci Prog, 67-77, PROF SOLID STATE SCI, PA STATE UNIV, 67- *Mem:* Am Chem Soc; Am Phys Soc; assoc Royal Inst Chem; fel Royal Soc Chem; sr mem Inst Elec & Electronic Engrs. *Res:* Magneto-chemistry applied to inorganic polymeric systems; nuclear and electron magnetic resonance and Mossbauer studies on metallocenes and biological systems; adsorption and catalysis; coordination compounds; superparamagnetic systems; high temperature superconductivity; semiconductor to metal transitions in tioxides, samarium sulfide; diffusion in solids. *Mailing Add:* 1011 Saxton Dr State College PA 16801-4899

MULAY, SHREE, b Patna, India, May 9, 41; Can citizen; c 2. FETAL MATURATION, DIABETES. *Educ:* Univ Delhi, India, BSc, 62; Univ McGill, Montreal, MSc, 66, PhD(agr chem), 69. *Prof Exp:* Postdoctoral fel endocrinol, McGill Univ, 69-72, lectr med, 73-75, asst prof, 75-81, ASSOC PROF MED, MCGILL UNIV, 81-, ASSOC MEM PHYSIOL, 86- *Concurrent Pos:* Supvr, endocrine Lab, Royal Victoria Hosp, 73-87, asst dir, 87- *Mem:* Endocrine Soc; Can Biochem Soc; NY Acad Sci; AAAS. *Res:* Mechanisms responsible for maldevelopment of the fetus in diabetic pregnancy; purifying heparin-binding proteins from fetal lungs. *Mailing Add:* Endocrine Lab Royal Victoria Hosp 687 Pine Ave W L2-05 Montreal PQ H3A 1A1

MULCAHEY, THOMAS P, b Gary, Ind, Sept 14, 31; div; c 2. CHEMICAL & NUCLEAR ENGINEERING. *Educ:* Purdue Univ, BSChE, 54, MSE, 59, PhD(nuclear eng), 63. *Prof Exp:* Mgr instrumentation & control, 70-85, group leader, Nat Battery Test Lab, 85-86, mgr anal & diag lab, 86-87, ENGR, ARGONNE NAT LAB, GROUP MEM, PYRO CHEM PROCESS DEVELOP, 87- *Mem:* Inst Elec & Electronics Engrs. *Res:* Coal conversion process, nuclear reactor instrumentation and control; battery testing and plutonium residue recovery process chemistry. *Mailing Add:* Argonne Nat Lab 9700 S Cass Ave Argonne IL 60439

MULCAHY, DAVID LOUIS, b Manchester, NH, Oct 16, 37; m 63; c 2. ECOLOGY, EVOLUTION. *Educ:* Dartmouth Col, BA, 59; Vanderbilt Univ, PhD(bot), 63. *Prof Exp:* Asst prof bot, Univ Ga, 63-66; vis scientist, Brookhaven Nat Lab, 66-68; asst prof, 68-71, assoc prof, 71-81, PROF BOT, UNIV MASS, AMHERST, 81- *Mem:* Ecol Soc Am; Genetics Soc Am; Am Genetic Asn; Soc Study Evolution; Bot Soc Am. *Res:* Gametophytic competition; population structure of trees; evolution of heterostyly; pollination systems. *Mailing Add:* Dept of Bot Univ of Mass Amherst MA 01003

MULCAHY, GABRIEL MICHAEL, b Jersey City, NJ, Feb 16, 29; m 58; c 7. PATHOLOGY. *Educ:* St Peter's Col, NJ, AB, 50; Georgetown Univ, MD, 54. *Prof Exp:* USPHS med officer, Navajo Indian Reservation, 55-57; resident path, USPHS Hosp, Seattle, Wash, 57-59, USPHS Hosp, Staten Island, NY, 59-61; chief path, USPHS Hosp, Detroit, Mich, 61-62; from instr to assoc prof path, Sch of Med, Creighton Univ, 62-69, actg chmn dept, 67; DIR PATH, JERSEY CITY MED CTR, 69-; PROF GEN & ORAL PATH, NJ DENT SCH & ASSOC CLIN PROF PATH, NJ MED SCH, COL MED & DENT NJ, 71- *Mem:* Am Soc Clin Pathologists; Int Acad Path; Am Soc Human Genetics; AAAS; AMA. *Res:* Pathology of familial tumors; cytogenetics. *Mailing Add:* Col Med & Dent NJ 112 Lembeck Ave Jersey City NJ 07305

MULCAHY, JOHN JOSEPH, b New York, NY, Jan 7, 41; m 70; c 4. UROLOGY. *Educ:* Georgetown Univ, MD, 66; Univ Minn, MS, 74; Univ Mich, PhD(physiol), 72. *Prof Exp:* Asst prof surg, Univ Ky, 74-78; assoc prof, 78-90 PROF UROL, IND UNIV MED CTR, 90-; CHIEF UROL, WISHARD MEM HOSP, 78- *Mem:* Am Urol Asn; Am Fertil Soc; Soc Univ Urol; AMA. *Res:* Male sexual dysfunction; urinary incontinence; prostatic obstruction. *Mailing Add:* Dept Urol Ind Univ Purdue Univ Med Ctr 1100 W Michigan St Indianapolis IN 46223

MULCARE, DONALD J, b New York, NY, July 27, 38; m 68; c 4. ZOOLOGY, DEVELOPMENTAL BIOLOGY. *Educ:* St Procopius Col, BS, 62; Univ Notre Dame, PhD(biol), 68. *Prof Exp:* Teaching asst biol, Univ Notre Dame, 62-66; lectr, St Mary's Col, Ind, 66-67; Nat Cancer Inst fel zool, Univ Mich, Ann Arbor, 68-69; from asst prof to assoc prof, 69-88, PROF BIOL, SOUTHEASTERN MASS UNIV, 88- *Mem:* Sigma Xi; Am Soc Zoologists; Soc Develop Biol; AAAS. *Res:* Oncology in amphibians; transmission of the Lucke renal adenocarcinoma in Rana pipiens; Rana palustris and their hybrids. *Mailing Add:* Dept of Biol Southeastern Mass Univ North Dartmouth MA 02747

MULCHI, CHARLES LEE, b Warren County, NC, Dec 2, 41; m 62; c 2. AGRONOMY, PLANT PHYSIOLOGY. *Educ:* NC State Univ, BS(crop sci) & BS(soil sci), 64, MS, 67, PhD(soil sci & plant physiol), 70. *Prof Exp:* Instr soil-plant rels & biochem, NC State Univ, 66-70; asst prof, 70-75, ASSOC PROF ENVIRON SCI & CROP PHYSIOL, UNIV MD, COLLEGE PARK, 75- *Concurrent Pos:* Consult power plant impact on crops and soils, air and soil pollution; mem, Tobacco Chemists Res Conf; mem Tobacco Workers Conf. *Mem:* Am Soc Agron; Crop Sci Soc Am. *Res:* Soil-plant relations; plant physiology with special interest in photosynthesis and photo-respiration; ozone air pollution effects on crops; aerosol salt effects on crops or cooling tower salt drift effects on vegetation; particulates and metals deposition on crops near power plants; heavy metal soil pollution resulting from municipal waste disposal in crop lands. *Mailing Add:* Dept Agron Univ Md College Park MD 20740

MULCRONE, THOMAS FRANCIS, b Chicago, Ill, Aug 5, 12. MATHEMATICS. *Educ:* Spring Hill Col, BS, 39; Catholic Univ, MS, 42; St Louis Univ, STL, 47. *Prof Exp:* Instr math, Spring Hill Col, 40-41 & 42-43; spec lectr, St Louis Univ, 43-47; asst prof, Spring Hill Col, 48-54; asst prof Loyola Univ, La, 54-60; assoc prof, Spring Hill Col, 61-75; EMER PROF MATH, CITY COL, LOYOLA UNIV, LA, 75- *Mem:* Math Asn Am; Nat Coun Teachers Math. *Res:* Modern geometry and algebra; semigroups; history of mathematics; statistics. *Mailing Add:* Dept Math City Col Loyola Univ 6363 St Charles Ave New Orleans LA 70118

MULDAWER, LEONARD, b Philadelphia, Pa, Aug 6, 20; m 50, 83; c 3. SOLID STATE PHYSICS, MEDICAL PHYSICS. *Educ:* Temple Univ, AB, 42, AM, 44; Mass Inst Technol, PhD(physics), 48. *Prof Exp:* Instr eng, Sci & Mgt War Training, 42-44, from asst prof to prof, 48-90, EMER PROF PHYSICS, TEMPLE UNIV, 90- *Concurrent Pos:* Consult, Labs, Meret Co; staff, Diag Radiol Res Lab, Temple Univ Med Sch, 72-; field ctr dir, Short Course Prog, NSF, 80- *Mem:* Am Phys Soc; Am Soc Metals; Am Crystallog Asn; Am Asn Physics Teachers; Sigma Xi. *Res:* X-ray diffraction studies of order, particle size and strain; electron diffraction studies of oxides; physiological acoustics; optical and transport properties of alloys; phase transformations; science education; radiological physics. *Mailing Add:* Dept Physics Temple Univ Philadelphia PA 19122

MULDER, CAREL, b Arnhem, Neth, Mar 19, 28; div; c 4. VIROLOGY & MICROBIOLOGY, BIOCHEMISTRY. *Educ:* Univ Leiden, BS, 51, Drs, 55; Oxford Univ, DPhil(microbiol), 63. *Prof Exp:* Instr molecular biol, Sch Med, Leiden Univ, 60-65; res assoc, Dept Chem, Harvard Univ, 65-67; NIH spec fel tumor virol, Sch Med, St Louis Univ, 67-68 & Salk Inst Biol Studies, 68-70; sr staff investr, Cold Spring Harbor Lab, 70-75; assoc prof microbiol & pharmacol, 75-80, PROF PHARMACOL, MOLECULAR GENETICS & MICROBIOL, SCH MED, UNIV MASS, 80- *Concurrent Pos:* Harvard jr fel, 65-67; vis prof, Sch Med, Univ Leiden, 75; res assoc prof microbiol & molecular genetics, Harvard Med Sch, 78-81; vis prof, London Univ, 86-87; Haddow fel, Inst Cancer Res, London, 86-88. *Mem:* Am Soc Microbiol; Soc Gen Microbiol; Neth Soc Biochem; AAAS. *Res:* Molecular biology of DNA tumor viruses and herpesvirus latency; HIV-acquired immune deficiency syndrome; Simian immunodeficiency viruses. *Mailing Add:* Dept Pharmacol Univ Mass Sch Med Worcester MA 01655

MULDER, DONALD WILLIAM, b Rehoboth, NMex, June 30, 17; m 43. CLINICAL NEUROLOGY. *Educ:* Calvin Col, AB, 40; Marquette Univ, MD, 43; Univ Mich, MS, 46. *Prof Exp:* Asst prof neurol, Univ Colo, 49-50; from instr to assoc prof, Mayo Grad Sch Med, Univ Minn, 50-64; chmn sect neurol, 66-71, pres staff, 70-71, sr consult, Mayo Clin, 50-82; prof neurol, Mayo Med Sch, Univ Minn, 64-82; RETIRED. *Mem:* Am Neurol Asn; Am Psychiat Asn; fel Am Acad Neurol; Sigma Xi. *Res:* Amyotropic lateral sclerosis; epilepsy; neuromuscular disease. *Mailing Add:* Mayo Clin Rochester MN 55901

MULDER, JOHN BASTIAN, b Hawarden, Iowa, Jan 20, 32; m 50; c 2. LABORATORY ANIMAL MEDICINE. *Educ:* Iowa State Univ, DVM, 56; Univ Mo, MS, 72, MEd, 73; Am Col Lab Animal Med, dipl, 72. *Prof Exp:* Captain, Vet Corps, US Army, 56-58; pvt pract vet med, Eagle Grove, Iowa, 58-69; dir animal care, Univ Mo, 69-73, Mich State Univ, 73-76 & Univ Kans, 76-85; DIR, UNIV ANIMAL RESOURCES, MED CTR, UNIV KANS, 85- *Concurrent Pos:* Consult, Menninger Found, 83-85, Am Asn Accreditation of Lab Animal Care, 83- & Kans City Vet Admin Med Ctr, 84- *Mem:* Sigma Xi; Am Vet Med Asn; Am Asn Lab Animal Sci (pres, 80-81); Am Soc Lab Animal Practitioners(pres, 76-77). *Res:* Animal behavior; anesthesiology; medical devices; wildlife rehabilitation; educational methodology; animal facility management. *Mailing Add:* 5744 N Via Umbrosa Tucson AZ 85715

MULDER, ROBERT UDO, b Sao Paulo, Brasil, Jan 23, 52; m 73; c 2. NEUTRON ACTIVATION ANALYSIS, MONITOR PROGRAMS. *Educ:* Univ Brasil, BS, 73; Univ Va, ME, 76, PhD, 81. *Prof Exp:* RES ASST PROF, UNIV VA, 84- *Concurrent Pos:* Dir, Reactor Fac, Univ Va, 84- *Mem:* Am Nuclear Soc. *Res:* Mixed radiation field dosimetry; research reactor applications. *Mailing Add:* 832 Harris Rd Charlottesville VA 22901-6467

MULDOON, MARTIN E, b County Mayo, Ireland, Feb 28, 39. SPECIAL FUNCTIONS. *Educ:* Nat Univ Ireland, BSc, 59, MSc, 60; Univ Alta, PhD(math), 66. *Prof Exp:* Asst prof, 66-71, assoc prof, 71-78, PROF MATH, YORK UNIV, 78- *Res:* Classical analysis with emphasis on special functions. *Mailing Add:* Dept Math & Statist York Univ North York ON M3J 1P3 CAN

MULDOON, THOMAS GEORGE, b Brooklyn, NY, May 13, 38. BIOCHEMISTRY. *Educ:* Queens Col, NY, BS, 60; Univ Louisville, PhD(biochem), 67. *Prof Exp:* Fel biochem, Med Ctr, Univ Kans, 67-69; from asst prof to assoc prof, 69-78, PROF ENDOCRINOL, MED COL GA, 78- *Concurrent Pos:* Mem biochem endocrinol study sect, NIH. *Mem:* Am Soc Biol Chemists; Endocrine Soc; NY Acad Sci; Soc Study Reprod; Soc Exp Biol & Med; Am Fertil Soc. *Res:* Tissue-specific interactions of steroid hormones and proteins; regulation of hormone receptor activity. *Mailing Add:* Dept Endocrinol Med Col Ga Augusta GA 30912

MULDOWNEY, JAMES, b Kilkenny, Ireland, Dec 25, 39. MATHEMATICS. *Educ:* Nat Univ Ireland, BSc, 61, MSc, 62; Univ Alta, PhD(math), 65. *Prof Exp:* PROF MATH, UNIV ALTA, 68- *Mem:* Am Math Soc; Math Asn Am; Soc Indust & Appl Math. *Res:* Qualitative theory of differential equations. *Mailing Add:* Math Dept Univ Alta Edmonton AB T6G 2G1 Can

MULDREW, DONALD BOYD, b Winnipeg, Man, Oct 17, 34; m 60; c 3. IONOSPHERIC PHYSICS, WAVE PROPAGATION. *Educ:* Univ Man, BSc, 57. *Prof Exp:* Sci officer, Defence Res Bd, Defence Res Telecommun Estab, 60-69; RES SCIENTIST, COMMUN RES CTR, DEPT COMMUN, 69- *Mem:* Am Geophys Union; Can Asn Physicists. *Res:* Electromagnetic and electrostatic wave propagation in plasmas; ionospheric-magnetospheric field-aligned irregularities; F-layer trough; ball lightning; plasma resonances and echoes; hot-plasma dispersion curves; HF-enhanced plasma lines; long-delay echoes. *Mailing Add:* Commun Res Ctr Box 11490 Sta H Ottawa ON K2H 8S2 Can

MULDROW, CHARLES NORMENT, JR, b Washington, DC, 1930; m 58; c 3. POLYMER CHEMISTRY. *Educ:* Col Charleston, BS, 50; Univ NC, MA, 54; Univ Va, PhD(phys chem), 58. *Prof Exp:* Instr chem, Univ NC, 51-52; res chemist, Shell Develop Co, 58-59; from res chemist to sr res chemist, Am Enka Corp, 59-65, head polyester develop sect, 65-70, head polymer develop, 70-76; proj leader, NL Industs, 76-82; vpres prod develop, Shamrock Chem, 82-84; assoc, Princeton Polymer Labs, 85-86, assoc tech dir, 88-89; SR CONSULT, ARTHUR D LITTLE, 89- *Mem:* Am Chem Soc; Soc Plastics Eng; Am Inst Chem; fel Nat Sci Found; Soc Advan Mat Process Eng. *Res:* Synthesis of dielectric materials; magnetochemistry; stabilizers for PVC; organic and inorganic solution thermodynamics; process development of plastics and synthetic rubber; reaction kinetics; synthetic fibers; synthetic inorganic and organometallic chemistry; coatings and inks; composite technology. *Mailing Add:* Five Knollwood Dr East Windsor NJ 08520

MULE, SALVATORE JOSEPH, b Trenton, NJ, Apr 7, 32; m 56; c 4. PHARMACOLOGY, BIOCHEMISTRY. *Educ:* Col Wooster, BA, 54; Rutgers Univ, MS, 55; Univ Mich, PhD(pharmacol), 61; Am Bd Forensic Toxicol, dipl. *Prof Exp:* Fel biochem & pharmacol, Univ Wis, 61-63; res pharmacologist, Addiction Res Ctr, NIMH, Ky, 63-68; dir, Off Substance Abuse Serv, Testing & Res Lab, 68-89; DIR TOXICOL, METWEST CLIN LABS, 90- *Concurrent Pos:* Consult to govt & indust, WHO. *Mem:* AAAS; Am Chem Soc; Am Soc Pharmacol & Exp Therapeut; Am Acad Forensic Sci; fel Am Inst Chem. *Res:* Drug metabolism as related to biochemical mechanisms associated with the action of narcotic analgesics; development of analytical techniques to detect drugs in biological materials. *Mailing Add:* Metwest Clin Labs 18408 Oxnard St Tarzana CA 91356

MULFORD, DWIGHT JAMES, b Greenville, Ill, Feb 9, 11; m 37. BIOCHEMISTRY. *Educ:* Greenville Col, BS, 33; St Louis Univ, PhD(biochem), 42. *Prof Exp:* Res assoc phys chem, Harvard Univ, 42-47, assoc, 47-49; from asst prof to assoc prof biochem, Univ Kans, Lawrence, 49-56; actg chmn dept, Univ Kans Med Ctr, Kansas City, 57-58, asst dean, 67-71, prof biochem, 56-77, dean, 71-77; RETIRED. *Concurrent Pos:* Tutor, Harvard Univ, 42-46; chief blood processing lab, Mass Dept Pub Health, 45-47, asst dir, 47-49; asst prof, Boston Univ, 46-47. *Mem:* Am Soc Biol Chem; Soc Exp Biol & Med. *Res:* Choline metabolism; plasma fractionation; stability of proteins; blood bank. *Mailing Add:* 1400 Lillac Ln Apt 303 Lawrence KS 66044

MULFORD, ROBERT ALAN, b Camden, NJ, Dec 13, 47; m 69; c 1. METALLURGY, MATERIALS SCIENCE. *Educ:* St Joseph's Col, BS, 69; Univ Pa, MSE, 72, PhD(metall), 74. *Prof Exp:* Asst metallurgist, Argonne Nat Lab, 74-79; metallurgist, Gen Elec Res & Develop Ctr, 79-81, METALLURGIST, KNOLLS ATOMIC POWER LAB, GEN ELEC CO, 81- *Mem:* Am Inst Mining, Metall & Petrol Engrs; AAAS. *Res:* Effects of microstructure and trace impurity segregation on fracture, corrosion and stress corrosion; auger spectroscopy; x-ray photoelectron spectroscopy. *Mailing Add:* Gen Elec Co Knolls Atomic Power Lab Box 1072 Schenectady NY 12301

MULFORD, ROBERT NEAL RAMSAY, b US, Oct 2, 22; m 51; c 1. PHYSICAL CHEMISTRY, PHYSICAL METALLURGY. *Educ:* Hofstra Col, BA, 47; Brown Univ, PhD(chem), 50. *Prof Exp:* Asst, Brown Univ, 46-49; staff chemist, 50-69, alt group leader, 69-74, GROUP LEADER, LOS ALAMOS NAT LAB, 74- *Mem:* Am Chem Soc. *Res:* Hydride chemistry; gas-metal equilibria; high temperature chemistry; plutonium chemistry. *Mailing Add:* 1235 46th St Los Alamos NM 87544-1943

MULHAUSEN, HEDY ANN, b Cleveland, Ohio, Dec 5, 40. BIOCHEMISTRY, INFORMATION SCIENCE. *Educ:* Ursuline Col, Ohio, AB, 62; Ohio State Univ, MS, 65, PhD(biochem), 67. *Prof Exp:* Res assoc biochem, Ohio State Univ, 68; assoc, Univ Ga, 68-69; BIOCHEM ED ANALYST, CHEM ABSTR SERV, OHIO STATE UNIV, 69-, TECH SERV REP, 80- *Mem:* Am Chem Soc. *Res:* Mechanisms of control in mammalian carbohydrate metabolism. *Mailing Add:* 488 E Dunedin Columbus OH 43214

MULHAUSEN, ROBERT OSCAR, b Chicago, Ill, June 7, 30; m 54; c 4. INTERNAL MEDICINE, MEDICAL ADMINISTRATION. *Educ:* Univ Ill, BS, 51, MD, 55; Univ Minn, MS, 64. *Prof Exp:* Chief med, St Paul Ramsey Hosp, 73-88; fel internal med, 56-59, from instr to assoc prof internal med, 59-73, from asst dean to assoc dean med sch, 67-73, PROF INTERNAL MED, UNIV MINN, MINNEAPOLIS, 73-, PROF ENVIRON HEALTH, 81-; ASSOC CHIEF STAFF, VET ADMIN MED CTR, MINNEAPOLIS, 88- *Concurrent Pos:* Fulbright res award grant, Rigshospitalet, Copenhagen, 65-66; dir, Midwest Ctr Occup Safety & Health, 81-90. *Mem:* AAAS; fel Am Col Physicians; Am Fedn Clin Res; Am Soc Nephrol. *Res:* Fluid, electrolyte and acid-base physiology; renal physiology; health services delivery. *Mailing Add:* Vet Admin Med Ctr One Veterans Dr Minneapolis MN 55417

MULHERN, JOHN E, JR, b Chicago, Ill, Mar 26, 26; m 50; c 3. PHYSICS. *Educ:* Okla State Univ, BS, 48; Boston Univ, MS, 49, PhD, 54. *Prof Exp:* Semiconductor physicist, Gen Elec Co, 51-54; from asst prof to assoc prof, 54-66, PROF PHYSICS, UNIV NH, 66- *Concurrent Pos:* Vis asst prof, Brandeis Univ, 57-58; consult, NASA Cambridge Res Ctr, 65-69; NIH spec res fel, 69-70; sr vis fel, Cavendish Lab, Eng, 69-70. *Mem:* Am Phys Soc. *Res:* Solid state physics; use of an electron microprobe to analyse the elemental content of biological tissue. *Mailing Add:* Dept of Physics Univ of NH Durham NH 03824

MULHOLLAND, GEORGE, b Philadelphia, Pa, July 12, 38; m 62; c 2. MECHANICAL ENGINEERING. *Educ:* NMex State Univ, BS, 61, MS, 62; Okla State Univ, PhD(mech eng), 67. *Prof Exp:* Sr engr, Jet Propulsion Labs, 65-66; asst prof, 66-76, assoc prof, 76-80, PROF MECH ENG, NMEX STATE UNIV, 80- *Mem:* Am Soc Mech Engrs; Sigma Xi. *Res:* Heat transfer and fluid mechanics. *Mailing Add:* 2900 Karen Dr Las Cruces NM 88001

MULHOLLAND, JOHN DERRAL, b Muncie, Ind, Sept 28, 34; m 57, 85; c 2. ASTRONOMY, ASTRONAUTICS. *Educ:* Purdue Univ, BSAE, 57; Univ Cincinnati, MS, 61, PhD, 65; Yale Univ, MS, 64. *Prof Exp:* Asst proj engr, Kett Tech Ctr, US Indust Inc, 57-58; staff engr, Ketco, Inc, 58-59; res asst & lectr aerodynamics, Univ Cincinnati, 59-60, instr astronaut, 60-64, res assoc astron, 64-65, asst prof, 65-66; mem tech staff, Jet Propulsion Lab, Calif Inst Technol, 66-71; res scientist, McDonald Observ, Univ Tex, Austin, 71-85; RES SCIENTIST, SPACE ASTRON LAB, UNIV FLA, GAINESVILLE, 85- *Concurrent Pos:* Consult, CTL Div, Studebaker-Packard Corp, 61-62; consult, Aerospace Res Lab, Wright-Patterson AFB, 65; consult mem working group on ephemerides for space res, Int Astron Union, 67-70; mem lunar laser ranging panel, Comt Space Res, Int Coun Sci Unions, chmn panel 1D, 75-79; res assoc, Ctr Astron Geodynamics Res, Grasse, France, 73-74, 76-77, 79-82, 84-85; consult, Encycl Britannica, 74-; sci counr, Ctr Res in Geodyn & Astron, Grasse, France; sr fel, NATO, 79-81; vis prof astrophysics, Univ Nice, France, 84-85. *Honors & Awards:* Bronze Medal, Nat Ctr Space Studies, France, 74. *Mem:* AAAS; Am Geophys Union; Am Astron Soc; Int Astron Union. *Res:* Celestial mechanics; theory of differential correction processes; orbital and rotational dynamics of the earth-moon system; astrometry and dynamics of natural satellites; geophysical application of laser ranging; planetary astrophysics. *Mailing Add:* Space Astron Lab 1810 NW 6th St Gainesville FL 32609

MULHOLLAND, ROBERT J(OSEPH), b St Louis, Mo, Jan 18, 40; m 72, 81. ELECTRICAL ENGINEERING, SYSTEMS THEORY. *Educ:* Washington Univ, BS, 61, MSc, 64, DSc(elec eng), 68. *Prof Exp:* Radar equip engr, Westinghouse Elec Corp, 61-62; instr elec eng, Washington Univ, 64-66; NSF fel, Univ Calif, Los Angeles, 68-69; from asst prof to assoc prof, 69-77, Okla State Univ, prof elec eng, 77-; AT DEPT ELEC ENG, UNIV OKLA. *Mem:* AAAS; Inst Elec & Electronics Engrs; Soc Indust & Appl Math; Ecol Soc Am; Sigma Xi. *Res:* Linear and nonlinear system theory; systems ecology; problems of energy and the environment. *Mailing Add:* Dept Elec Engr Univ Okla Norman OK 73019

MULICK, JAMES ANTON, b Passaic, NJ, June 17, 48; m 70; c 1. PEDIATRIC PSYCHOLOGY, ABNORMAL BEHAVIORAL DEVELOPMENT. *Educ:* Rutgers Col, AB, 70; Univ Vt, MA, 73, PhD(psychol), 75. *Prof Exp:* Fel clin child psychol, Child Develop Inst, Univ NC, 75-76; prog dir self-injurious behav proj, Univ Murdock Ctr, 76-77; dir psychol, Child Develop Ctr, RI Hosp, 78-83; clin asst prof pediat, Brown Univ, 78-83; adj assoc prof psychol, Univ RI, 79-83; ASSOC PROF PEDIAT & PSYCHOL, OHIO STATE UNIV, 83- *Concurrent Pos:* Chief psychologist, Eunice Kennedy Shriver Ctr Mental Retardation, Inc, 77-78; mem RI Gov's Comt Mental Retardation, 78-81; expert witness, Spec Litigation, Civil Rights Div, US Dept Justice, 80-82; co-prin investr, Nat Inst Mental Health grant, Ohio State Univ, 86-89; prin investr, Nat Inst Child Health & Human Develop grant subcontract, Ohio State Univ, 89- *Mem:* AAAS; fel Am Psychol Soc; fel Am Psychol Asn; Asn Behav Anal; Int Soc Res Aggression; Soc Behav Pediat. *Res:* Behavior analysis of biological, social and ecological variables in child development; treatment of severe developmental disabilities and psychopharmacology. *Mailing Add:* Dept Pediat Div Psychol Ohio State Univ 700 Childrens Dr Columbus OH 43205

MULIERI, BERTHANN SCUBON, b Menticle, Pa, May 4, 37; m 60. PHYSIOLOGY. *Educ:* Pa State Univ, BS, 58; Univ Vt, PhD(physiol, biophys), 68; Goddard Col, MA, 86. *Prof Exp:* Asst prof biol sci, Hunter Col, 67-70; USPHS fel, pharmacol inst, Univ Lund, 71-72; from instr to asst prof biol, Trinity Col, Vt, 72-74; res assoc, 74-77, res asst prof physiol & biophys, Univ Vt, 77-79; CONSULT, 79- *Concurrent Pos:* Health Educ & Welfare res fel physiol & biophys, Univ Vt, 74-75; mem bd, Creamery Educ Found. *Mem:* Biophys Soc; NY Acad Sci; Sigma Xi. *Res:* Mechanics of muscle contraction; desensitization at neuromuscular junction; electrophysiology of hypertrophied myocardium. *Mailing Add:* Rd Two Box 369 Lake Irroqquis Rd Hinesburg VT 05461

MULIERI, LOUIS A, b Brooklyn, NY, Dec 3, 35. MUSCLE ENERGETICS. *Educ:* Univ Vt, PhD(physiol & biophys), 68. *Prof Exp:* Res asst prof, 79-86, RES ASSOC PROF PHYSIOL & BIOPHYS, UNIV VT, 86- *Mem:* Am Physiol Soc; Biophys Soc; Soc Gen Physiologists. *Mailing Add:* Dept Physiol & Biophys Given Bldg Univ Vt 85 S Prospect St Burlington VT 05405

MULINOS, MICHAEL GEORGE, b Cairo, Egypt, Nov 24, 97; nat US; m 27; c 2. CLINICAL PHARMACOLOGY. *Educ:* Columbia Univ, AB, 21, AM, 22, MD, 24, PhD(physiol), 29. *Prof Exp:* Intern, St Vincent's Hosp, Pa, 24-25; instr pediat, Univ Minn, 25-26; vis asst, Univ Chicago, 26-27; from instr to assoc prof pharmacol, Col Physicians & Surgeons, Columbia Univ, 27-44; assoc prof, NY Med Col, 44-45; dir med res, Interchem Corp, NJ, 45-47; med dir, Com Solvents Corp, 53-63; consult & med dir, McCann-Erickson, Inc, 66-69; med dir, Erwin Wasey, Inc, 69-73, Res Consults Inc, NJ, 75-87; med dir & dir res, Unimed Corp, Somerville, NJ, 77-87; RETIRED. *Concurrent Pos:* Asst, Inst Child Guid, Minneapolis, Minn, 26; asst med dir, Life Exten

Inst New York. *Mem:* AAAS; Am Soc Pharmacol & Exp Therapeut; Asn Med Dirs; Soc Exp Biol & Med; Harvey Soc; Sigma Xi. *Res:* Pharmacology of gastrointestinal tract; toxicology of pharmaceuticals and irritation; physiology of autonomic nervous system; clinical testing of drugs; pharmacology of conditioned reflexes. *Mailing Add:* 42 Marian Terrace Easton MD 21601-3830

MULKS, MARTHA HUARD, b Waterville, Maine, June 9, 50; m 70; c 1. GENETIC CONTROL OF BACTERIAL VIRULENCE, IMMUNOBIOLOGY OF VACCINE PRODUCTION. *Educ:* Cornell Univ, Ithaca, BA, 71; Rensselaer Polytech Inst, Troy, PhD(microbiol), 77. *Prof Exp:* Bacteriologist, NY State Dept Health, 71-73; teaching asst biol, Rensselaer Polytech Inst, 73-77; res assoc, Tufts-New Eng Med Ctr, 77-80, res asst prof med, Tufts Univ Sch Med, 80-83; asst prof, 83-88, ASSOC PROF MICROBIOL, MICH STATE UNIV, 88- *Mem:* Am Soc Microbiol; AAAS; Sigma Xi. *Res:* Interactions between pathogenic bacteria and their hosts; genetic regulation of virulence factors of Neisseria gonorrhoeae; immunobiology of respiratory tract infections in food animals, as well as vaccine production. *Mailing Add:* Dept Microbiol Mich State Univ East Lansing MI 48824

MULLA, MIR SUBHAN, c 4. VECTOR CONTROL, AQUATIC ECOLOGY. *Educ:* Cornell Univ, BS, 52; Univ Calif, Berkeley, PhD(entom), 55. *Prof Exp:* Asst prof, 63-68, PROF ENTOM, UNIV CALIF, RIVERSIDE, 69- *Mem:* Soc Vector Ecol(vpres, 82, pres, 83); Entom Soc Am; Soc Invertebrate Path; Am Mosquito Control Asn. *Res:* Research and teaching in the area of medical entomology, vector control and ecology of aquatic ecosystems; study, evaluate and develop microbiol control agents against vectors of disease. *Mailing Add:* Dept Entom Univ Calif Riverside CA 92521

MULLAN, DERMOTT JOSEPH, b Omagh, Northern Ireland, Jan 10, 44; m 70; c 8. ASTROPHYSICS. *Educ:* Queen's Univ, Belfast, BS, 64, BS, 65; Univ Md, PhD(astron), 69. *Prof Exp:* Astronr, Armagh Observ, Northern Ireland, 69-72; presidential intern fel astron, 72-73, from asst prof to assoc prof, 73-82, PROF ASTROPHYS, BARTOL RES INST, UNIV DEL, 82- *Mem:* Am Astron Soc; Int Astron Union. *Res:* Physics of stellar flares; formation of sunspots and starspots; structure of magnetic fields in late-type stars; coronal propagation of cosmic rays; mass loss; magnetic convection; cosmic ray composition; non-thermal processes in stellar atmospheres. *Mailing Add:* Bartol Res Inst Univ Del Newark DE 19716

MULLAN, JOHN F, b County Derry, NIreland, May 17, 25; US citizen; m 59; c 3. NEUROSURGERY. *Educ:* St Columbus Col, Ireland, BAO, 42; Queen's Univ, Belfast, MB & BCh, 47; FRCS, 51; Am Bd Neurol Surg, dipl, 57. *Hon Degrees:* DSc, Queen's Univ, Belfast, 76. *Prof Exp:* Resident surg, Royal Victoria Hosp, Belfast, Ireland, 50-51; resident neurol surg, Montreal Neurol Inst, 53-55; from asst prof to assoc prof, 55-63, PROF NEUROL SURG, UNIV CHICAGO, 63-, CHMN DEPT, 67- *Concurrent Pos:* Fel, Middlesex Hosp, Univ London, 49-50 & Guys Hosp, 51. *Mem:* Fel Am Col Surg; Am Asn Neurol Surg; Am Acad Neurol Surg. *Res:* Head injury, intracranial aneurysm and pain; epilepsy. *Mailing Add:* Div Neurol Surg Univ Chicago Box 405 5812 Ellis Ave Chicago IL 60637

MULLANE, JOHN F, medicine, pharmacology, for more information see previous edition

MULLANEY, PAUL F, b New York, NY, Jan 26, 38; m 63; c 2. PHYSICS, BIOPHYSICS. *Educ:* Iona Col, BS, 59; Univ Del, MS, 63, PhD(physics), 65. *Prof Exp:* Asst prof physics, St Bonaventure Univ, 65-66; staff mem, Bio-Med Res Group, Los Alamos Sci Lab, 66-72, sect leader biophys, 72-73, group leader, biophys & instrumentation group H-10, 73-80; MGR BIOPHYS, BIOL DIV, OAK RIDGE NAT LAB, OAK RIDGE, TENN, 80- *Concurrent Pos:* Consult, Photoconductor Devices Div, Sylvania Elec Prod, Inc, Pa, 65-66; Nat Cancer Inst, Comt Cytol Automation, 73-75; vis scientist, Battelle Inst, Frankfurt, WGer, 78-79. *Honors & Awards:* Alexander von Humboldt Sr Am Scientist Prize, Bonn, WGer, 78. *Mem:* AAAS: Biophys Soc; Am Asn Physics Teachers (pres, 80-82). *Res:* Cellular biophysics; light scattering and fluorescence of cells; high speed cell analysis by flow methods; biomedical engineering. *Mailing Add:* 116 Claymor Lane Oak Ridge TN 37831

MULLEN, ANTHONY J, b Jermyn, Pa, Sept 2, 27. ELECTROMAGNETISM. *Educ:* Villanova Univ, BS, 50; Cath Univ Am, MS, 54; Bryn Mawr Col, PhD, 68. *Prof Exp:* Teacher, Archbishop Carroll High Sch, 54-59; prof elec eng, Villanova Univ, 59-71, chmn dept, 59-67; sr engr flying qual, Vertol Div, 72, ENG SUPVR ELECTROMAGNETIC PULSE ANALYTICAL ENG, BOEING AEROSPACE CO, 72- *Concurrent Pos:* Consult, Reentry Systs Div, Gen Elec Corp, 66-67; res assoc, Geophys Inst, Univ Alaska, 70-71. *Mem:* Am Phys Soc; Sigma Xi. *Res:* Electromagnetic radiation in plasmas; geomagnetic micropulsations. *Mailing Add:* 17327 158th Ave Se Renton WA 98055

MULLEN, BARBARA J, NUTRITION. *Prof Exp:* Staff res assoc, Univ Calif, Davis, 79-81; res asst, Univ Ga, 81-82, teaching asst, 82-83, res technician, 83-85, asst, 85-86, USDA fel, 86-88, POSTDOCTORAL ASSOC, DEPT FOODS & NUTRIT, UNIV GA, 88- *Concurrent Pos:* NIH res grant, 92-95. *Mem:* Am Inst Nutrit; Am Diabetes Asn; Soc Ingestive Behav; Soc Neurosci; NAm Asn Study Obesity. *Res:* Author of more than 20 publications. *Mailing Add:* Dept Foods & Nutrit Univ Ga Dawson Hall Athens GA 30602

MULLEN, GARY LEE, b Meadville, Pa, Apr 9, 47. NUMBER THEORY, COMBINATORICS. *Educ:* Allegheny Col, BS, 69; Pa State Univ, MA, 70, PhD(math), 74. *Prof Exp:* Grad asst, Pa State Univ, 69-74, from instr to assoc prof math, 74-88, asst dean, Col Sci, 82-89, PROF MATH, PA STATE UNIV, 88-, ASSOC DEAN, COL SCI, 89- *Mem:* Am Math Soc; Math Asn Am. *Res:* Finite field theory with applications in combinatorics, coding theory and cryptology. *Mailing Add:* Dept Math Pa State Univ University Park PA 16802

MULLEN, GARY RICHARD, b Ogdensburg, NY, Nov 16, 45; m 69; c 2. MEDICAL-VETERINARY ENTOMOLOGY, ECOLOGY. *Educ:* Northeastern Univ, BA, 68; Cornell Univ, MS, 70, PhD(entom), 74. *Prof Exp:* Med entomologist & adminr, Allegheny County Health Dept, Vector Control Prog, Pittsburgh, 74-75; from asst prof to assoc prof, Dept Zool & Entom, 75-89, PROF, DEPT ENTOM, AUBURN UNIV, 89- *Concurrent Pos:* Consult, Ala Poison Control Ctr, 81- & Environ Pest Control, 85-90; subj ed, J Entom Sci, 86-90; chair med-vet, Sect D, Entom Soc Am, 91-92. *Mem:* Entom Soc Am; Am Mosquito Control Asn; Acarological Soc Am, (pres, 84); Am Arachnological Soc; Soc Vector Ecol. *Res:* Ecology of insects and mites of medical and veterinary importance; biology and role of ceratopogonid midges as vectors of disease agents, Lyme disease; general acarology with emphasis on the parasitengona; acarine parasites of mosquitoes and other nematocerous flies. *Mailing Add:* Dept Entom & Agr Exp Sta Auburn Univ Auburn AL 36849-5413

MULLEN, JAMES A, b Malden, Mass, May 28, 28; m 61; c 4. APPLIED MATHEMATICS. *Educ:* Providence Col, BS, 50; Harvard Univ, MA, 51, PhD, 55. *Prof Exp:* Mem res staff, 55-65, prin scientist, 65-69, CONSULT SCIENTIST, RES DIV, RAYTHEON CO, 69- *Mem:* Soc Indust & Appl Math; Fel Inst Elec & Electronics Engrs; Am Asn Artificial Intel. *Res:* Statistical communication theory; noise in non-linear circuits and oscillators; artificial intelligence neural networks. *Mailing Add:* 337 S Main St Cohasset MA 02025-2028

MULLEN, JAMES G, b St Louis, Mo, Sept 17, 33; m 83; c 7. SOLID STATE PHYSICS. *Educ:* Univ Mo-Rolla, BS, 55; Univ Ill, MS, 57, PhD(physics), 60. *Prof Exp:* From asst physicist to assoc physicist, Argonne Nat Lab, 60-64; from asst prof to assoc prof, 64-75, PROF PHYSICS, PURDUE UNIV, LAFAYETTE, 75- *Concurrent Pos:* Consult, Argonne Nat Lab, 64-65; mem, Ad Hoc Panel Mossbauer Data; pres, World Technologies Inc, 81-85; fel, Univ Grotingen, Netherlands, 79-80; vis prof, Univ Mo, 86-87. *Mem:* Am Phys Soc; Am Asn Physics Teachers; Sigma Xi. *Res:* Studies of solid state diffusion in ionic and metallic crystals; Mossbauer studies of properties of solids; Mössbauer lineshape studies; Mössbauer diffraction. *Mailing Add:* Dept Physics Purdue Univ W Lafayette IN 47907

MULLEN, JAMES L, b Norristown, Pa, Jan 8, 42; m 71; c 2. NUTRITION. *Educ:* Harvard Univ, BA, 63; Univ Pa, MD, 67. *Prof Exp:* Surg resident, 67-73, asst prof, 75-78, ASSOC PROF SURG, SCH MED, UNIV PA, 78- *Concurrent Pos:* Chief surg, Philadelphia Vet Admin Med Ctr, 78-85. *Mem:* Am Soc Parental & Enteral Nutrit; Am Col Surgeons; Soc Univ Surgeons; Soc Surg Alimentary Tract. *Res:* Clinical nutrition; gastrointestinal surgery; nutrition and cancer; anorexia nervosa. *Mailing Add:* Dept Surg Sch Med Univ Penn 3400 Spruce St Four Silverstein Philadelphia PA 19104

MULLEN, JOSEPH DAVID, b Green Isle, Minn, Jan 6, 34; m 56; c 4. BIOCHEMISTRY, FOOD SCIENCE. *Educ:* Col St Thomas, BS, 56; Univ Minn, PhD(biochem), 62. *Prof Exp:* Group head, 61-72, dept head, 72-80, SR PRIN SCIENTIST, GEN MILLS, INC, 80- *Mem:* AAAS; Am Chem Soc; Am Asn Cereal Chem; Inst Food Technologists. *Res:* Protein chemistry; relation of protein structures to function in foods; new food product development including extruded protein and gum systems, edible protein films, gel technology, low calorie foods and nutrition; relation of food to dental health; satiety and appetite; seafood and health; cholesterol reducing foods. *Mailing Add:* Gen Mills Inc Bell Res Ctr 9000 Plymouth Ave N Minneapolis MN 55427

MULLEN, JOSEPH MATTHEW, b Washington, DC, June 1, 44; m 65. CHEMICAL PHYSICS, ASTROPHYSICS. *Educ:* Old Dom Univ, BS, 66; Univ Fla, PhD(physics), 72. *Prof Exp:* Asst prof physics, Valdosta State Col, 71-72; fel, Univ Fla, 72-74; prin staff oper anal, Oper Res Inc, Silver Spring, Md, 74-76; MEM STAFF, JEM ASSOCS, RESTON, VA, 76- *Mem:* AAAS; Sigma Xi; Am Phys Soc. *Res:* Determination of intermolecular potentials; development of instrumentation for ion-molecule reaction studies; theory of mass transfer in close binary star systems. *Mailing Add:* 8607 Forest St Annandale VA 22003

MULLEN, KENNETH, b London, Eng, Feb 28, 39; US citizen; m 61; c 3. STATISTICS, MATHEMATICS. *Educ:* Western Reserve Univ, BA, 61; Va Polytech Inst & State Univ, PhD(statist), 66. *Prof Exp:* Asst prof math, Radford Col, 64-65; asst prof biometry, Med Col Va, 65-67; sr statistician, Ciba, Ltd, 67-69; from asst prof to assoc prof, 69-81, PROF STATIST, UNIV GUELPH, 81- *Concurrent Pos:* Consult, Am Tobacco Co, 65-67 & Albemarle Paper Co, 66-67. *Mem:* Am Statist Asn. *Res:* Estimation problems associated with censored and truncated data; development of non-parametric procedures. *Mailing Add:* Dept Math & Statist Univ Guelph Guelph ON N1G 2W1 Can

MULLEN, LEO VINCENT, JR, rubber chemistry, for more information see previous edition

MULLEN, PATRICIA ANN, b Flushing, NY, July 10, 35. COSMETIC CHEMISTRY. *Educ:* Seton Hill Col, BA, 57; Mt Holyoke Col, MA, 61. *Prof Exp:* Res chemist, Charles Bruning Co, 57-59; res asst spectros, Mt Holyoke Col, 59-61; res chemist, Am Cyanamid Co, 63-74, group leader, 74-77; mgr, 77-80, mgr household prod line, 81-83, dir Appl Fragrance Technol, 83-85, DIR, FRAGRANCE APPLN LAB, NAARDEN INT USA, INC, 80-; PRES, INT PROD CREATIONS, LTD, 89- *Concurrent Pos:* Gen mgr, Exotherm Inc, 85-87; pres, Inatex, Ltd, 87-89. *Mem:* Am Chem Soc; Soc Cosmetic Chemists. *Res:* Ultraviolet and vacuum spectroscopy; photochemistry; spectropolarimetry; cosmetic chemistry; fragrances; polymers. *Mailing Add:* 41 Cutter Pl Babylon NY 11704-8301

MULLEN, RICHARD JOSEPH, b Leominster, Mass, Aug 15, 41; m 64; c 3. DEVELOPMENTAL GENETICS, NEUROSCIENCES. *Educ:* Fitchburg State Col, BS, 63; Univ NH, MS, 69, PhD(genetics), 71. *Prof Exp:* USPHS fel develop genetics, Harvard Med Sch, 71-74; res assoc neurosci, Children's

Hosp Med Ctr, 73-79; assoc prof, 79-83, PROF ANAT, UNIV UTAH SCH MED, 83-, ACTG CHMN ANAT, 87- *Concurrent Pos:* From instr to asst prof neuropath, Harvard Med Sch, 73-79; mem, NIH Study Sect, 85-88. *Mem:* Soc Neuroscience; Soc Develop Biol. *Res:* Mammalian developmental genetics; culture and manipulation of preimplantation embryos; use of experimental chimeric mice in studies of brain development. *Mailing Add:* 4739 Fairfield Rd Salt Lake City UT 84124-5612

MULLEN, ROBERT TERRENCE, b Chicago, Ill, Sept 25, 35; m 55; c 4. PHYSICAL CHEMISTRY. *Educ:* Univ Ill, BS, 57; Univ Calif, PhD(phys chem), 61. *Prof Exp:* Res assoc chem, Brookhaven Nat Lab, 61-63; sr chemist, Res Lab, Merck & Co, Inc, 63-65, sect leader, 65-67, res fel, 67-68, supvr automation & control dept, 68-70, mgr automation & control lab automation, 70-75, mgr lab & gen automation, 75-78, dir, 78-79, EXEC DIR AUTOMATION & CONTROL SYSTS, MERCK & CO, INC, 79- *Mem:* AAAS; Am Chem Soc; Soc Appl Spectros; NY Acad Sci. *Res:* Hot atom chemistry; photochemistry; Mossbauer spectroscopy; radioisotope tracer applications; laboratory and process automation. *Mailing Add:* Automation & Control Dept Merck & Co Inc Rahway NJ 07065

MULLENAX, CHARLES HOWARD, ecology, range science & management, for more information see previous edition

MULLENDORE, A(RTHUR) W(AYNE), b River Falls, Wis, Aug 5, 28; m 60. INORGANIC & ORGANOMETALLIC CHEMISTRY. *Educ:* Univ Wis, BS, 50; Mass Inst Technol, SM, 53, ScD(metall), 60. *Prof Exp:* Asst prof metall, Mass Inst Technol, 60-64; staff mem, Sandia Nat Labs, 64-87, div supvr, 69-78; RETIRED. *Concurrent Pos:* Distinguished mem tech staff, Sandia Nat Labs, 87-88. *Mem:* Am Inst Mining, Metall & Petrol Eng; Electron Micros Soc Am; Am Vaccum Soc. *Res:* High temperature metallurgy; electron microscopy; carbon materials; refractory materials, chemical vapor deposition; materials science engineering. *Mailing Add:* PO Box 11 Sandia Park NM 87047

MULLENDORE, JAMES ALAN, b Greenwood, Wis, Dec 6, 32; m 62; c 4. METALLURGICAL ENGINEERING. *Educ:* Univ Wis, BS, 54, MS, 55; Univ Ill, PhD(metall eng), 61. *Prof Exp:* Sr scientist, Avco Res & Develop Lab, 59-60, lead scientist, 60-61; asst prof metall eng, Univ Wis-Madison, 61-66; engr specialist, Sylvania Elec Prod, Inc, 66-71, head metals res, 71-80, SR ENGR SPECIALIST & HEAD ORDNANCE RES & DEVELOP RES, CHEM & METAL DIV, GTE SYLVANIA, 83- *Concurrent Pos:* Mem, Am Powder Metall Inst. *Mem:* Am Inst Mining, Metall & Petrol Engrs; Am Soc Metals; Am Defense Preparedness Asn; Am Powder Metall Inst. *Res:* Physical and process metallurgy of refractory metals and alloys. *Mailing Add:* Chem & Metals Div GTE Sylvania Towanda PA 18848

MULLENDORE, JAMES MYERS, speech pathology, audiology; deceased, see previous edition for last biography

MULLER, BURTON HARLOW, b New York, NY, May 11, 24; m 52; c 2. PHYSICS. *Educ:* Wesleyan Univ, BA, 44; Yale Univ, MS, 45; Univ Ill, PhD(physics), 54. *Prof Exp:* Asst physics, SAM Labs, Columbia Univ, 45; from asst prof to prof, 53-89, EMER PROF PHYSICS, UNIV WYO, 89- *Concurrent Pos:* NSF fac fel, Univ BC, 59-60 & Univ Nottingham, 65-66; vis prof, Univ Kent, 71-72; cong sci fel, Am Phys Soc, 81. *Mem:* Am Phys Soc; fel AAAS; Fedn Am Scientist. *Res:* Nuclear magnetic resonance; molecular motion. *Mailing Add:* Dept Physics Univ Wyo Laramie WY 82071

MULLER, CORNELIUS HERMAN, b Collinsville, Ill, July 22, 09; m 39; c 1. PLANT ECOLOGY, ALLELOPATHY. *Educ:* Univ Tex, BA, 32, MA, 33; Univ Ill, PhD(plant ecol), 38. *Prof Exp:* Asst bot, Univ Tex, 29-33; asst, Univ Ill, 34-38; ecologist, Ill State Natural Hist Surv, 38; asst botanist, Div Plant Explor & Introd, Bur Plant Indust, USDA, 38-42, rubber plant invests, 42, assoc botanist, Spec Guayule Res Proj, 42-45; from asst prof to prof, 45-76, EMER PROF BOT, UNIV CALIF, SANTA BARBARA, 76- *Concurrent Pos:* Instr grad sch, USDA, 41-42, collabr, 45-46; res assoc, Inst Tech & Plant Indust, Southern Methodist Univ, 45 & Santa Barbara Bot Garden, 48-76; fac res lectr, Univ Calif, Santa Barbara, 57, actg dean grad div, 61-62; adj prof bot, Univ Tex, Austin, 74- *Mem:* Eminent ecologist, Ecol Soc Am; Int Soc Chem Ecol. *Res:* Vegetation of the southwestern United States, Mexico and Central America; basic nature of the biotic community; biochemical inhibition among higher plants; plant competition and community interactions; taxonomy and evolution of American Quercus. *Mailing Add:* Dept Biol Sci Univ Calif Santa Barbara CA 93106

MULLER, DAVID EUGENE, b Austin, Tex, Nov 2, 24; m 44; c 2. APPLIED MATHEMATICS. *Educ:* Calif Inst Technol, BS, 47, PhD(physics), 51. *Prof Exp:* Res fel physics, Calif Inst Technol, 51-52; fel electronic digital comput, 52-53, res asst prof appl math, 53-56, res assoc prof, 56-60, res prof, 60-64, PROF MATH, UNIV ILL, URBANA, 64- *Concurrent Pos:* Consult, IBM Corp, 59-; Fulbright res scholar, Univ Tokyo, 61-62. *Mem:* AAAS; Am Phys Soc; Am Math Soc. *Res:* Switching and automata theory; error correcting codes. *Mailing Add:* Dept Math 273 Altgeld Hall Univ Ill Urbana IL 61801

MULLER, DIETRICH, b Leipzig, Ger, Sept 14, 36; m 68; c 3. PHYSICS, ASTROPHYSICS. *Educ:* Univ Bonn, Dipl physics, 61, PhD(physics), 64. *Prof Exp:* Res assoc physics, Univ Bonn, 64-68; res assoc physics, 68-70, from asst prof to assoc prof, 70-84, PROF, ENRICO FERMI INST & DEPT PHYSICS, UNIV CHICAGO, 85-, DIR, ENRICO FERMI INST, 86- *Concurrent Pos:* Prin investr, NASA Space Exp; mem, Nat Comt & Workshops on Space Sci & Astrophys. *Mem:* Fel Am Phys Soc; Am Astron Soc. *Res:* Experimental physics; mass spectroscopy; high energy astrophysics; cosmic ray research. *Mailing Add:* Enrico Fermi Inst Univ of Chicago 933 E 56th St Chicago IL 60637

MULLER, ERIC RENE, b Morija, Lesotho, Nov 5, 38; Can citizen; m 65; c 2. MATHEMATICS. *Educ:* Univ Natal, BSc, 60, MSc, 62; Univ Sheffield, PhD(theoret physics), 67. *Prof Exp:* Lectr math, Rhodes Univ, SAfrica, 61-64; lectr col appl arts & technol, Univ Sheffield, 64-67; asst prof math & physics, 67-71, assoc prof math, 71-80, PROF MATH, BROCK UNIV, 80- *Concurrent Pos:* Consult, Steltner Develop, 70-73 & Transp Develop Agency, Can Ministry Transport, 73-78. *Mem:* Inst Mgt Sci; Opers Res Soc Am; Math Asn Am; Soc Indust & Appl Math. *Res:* Mathematical models and analyses of transportation systems; theoretical solid state physics. *Mailing Add:* Dept of Math Brock Univ St Catharines ON L2S 3A1 Can

MULLER, ERNEST HATHAWAY, b Tabriz, Iran, Mar 4, 23; US citizen; m 51; c 3. GEOLOGY. *Educ:* Col Wooster, AB, 47; Univ Ill, MS, 49, PhD(geol), 52. *Prof Exp:* Geologist mil geol br, US Geol Surv, Alaska, 48, proj head, Bristol Bay Area, 49-54; asst prof geol, Cornell Univ, 54-59; from assoc prof to prof geol, 59-89, interim chmn, 70-71 & 80-81, EMER PROF GEOL, SYRACUSE UNIV, 89- *Concurrent Pos:* Geologist, NY State Sci Serv, 56-; mem SChile exped, Am Geog Soc, 59; mem exped, Katmai Nat Monument, Alaska, 63-64; res assoc, Mus Natural Hist, Reykjavik, Iceland, 68-69 & Churchill Falls Power Proj, Labrador, 70; Erskine fel, Univ Canterbury, Christchurch, NZ, 73-74; vis prof, Alaska Pac Univ, 79. *Mem:* AAAS; Geol Soc Am; Nat Asn Geol Teachers; Am Asn Univ Professors; Glaciol Soc. *Res:* Geomorphology; glacial, engineering and environmental geology; permafrost; denudation; drumlin origins; glacial geology of New York, Southwestern Alaska, Iceland, and South Island of New Zealand. *Mailing Add:* Dept Geol 204 Heroy Geol Lab Syracuse Univ Syracuse NY 13244-1070

MULLER, GEORGE HEINZ, b Ger, June 6, 19; US citizen; m 49. VETERINARY DERMATOLOGY. *Educ:* Tex A&M Univ, DVM, 43; Am Col Vet Internal Med, dipl, 74; Am Col Vet Dermat, dipl, 74. *Prof Exp:* Dir, Pittsburg Vet Hosp, 46-56; dir, Muller Vet Hosp, 56-87, dir, Vet Dermat Clin, 79-87; CLIN PROF DERMAT, SCH MED, STANFORD UNIV, 58- *Concurrent Pos:* Ed dermat, Current Vet Ther, 66-75; pres, Dermat Specialty Group, Am Col Vet Internal Med, 74-76. *Honors & Awards:* McCoy Mem Award, Wash State Univ, 69; Merit Award Dermat, Am Animal Hosp Asn, 70, Gaines Award, 84. *Mem:* Am Acad Vet Dermat (pres); affil Am Acad Dermat; Am Animal Hosp Asn; Am Vet Med Asn. *Res:* Small animal veterinary dermatology; comparative dermatology; canine and human demodicosis. *Mailing Add:* 2539 Oak Rd Walnut Creek CA 94596

MULLER, KARL FREDERICK, b Glen Ridge, NJ, Sept 5, 35; m 61; c 2. APPLIED MATHEMATICS, COMPUTER SCIENCE. *Educ:* Lafayette Col, BSME, 57; Syracuse Univ, MSEE, 65, PhD(appl math), 70. *Prof Exp:* Field engr, Leeds & Northrup Co, Philadelphia, 57-63; res engr, Syracuse Univ Res Corp, 64-66, group leader appl math, 66-69; mem tech staff, Mitre Corp, Bedford, Mass, 69-76; prin res engr, Avco Everett Res Labs, 76-79; mem staff, C S Draper Lab Inc, Cambridge, Mass, 79-82; SR PRIN ENGR, ELECTRO-OPTICS DIV, HONEYWELL, WILMINGTON, MA, 82- *Concurrent Pos:* Mem weather '85 study group, US Air Force Systs Command, 71-72; consult, PHI Comput Serv, Inc, Arlington, Mass, 72; sr engr, Raytheon Co, Wayland, Ma. *Res:* Detailed mathematical analysis, modeling and computer simulation of physical and probabalistic phenomena associated with seismic, and acoustic propagation, weather predictability, optimal control systems, high-sensitivity electro-optical sensors and detection, especially identification algorithms; fault tolerant digital architecture. *Mailing Add:* 9 San Jose Terr Stoneham MA 02180

MULLER, KENNETH JOSEPH, b Altadena, Calif, July 29, 45; m 75; c 2. NEUROBIOLOGY, NEURAL DEVELOPMENT. *Educ:* Univ Chicago, SB, 66; Mass Inst Technol, PhD(biol), 71. *Prof Exp:* Fel neurobiol anat, Harvard Med Sch, 71-75; staff mem neurobiol, Carnegie Inst Wash, 75-83; assoc prof, 77-81, assoc prof, dept biophys, Johns Hopkins Univ, 81-83; PROF PHYSIOL & BIOPHYS, SCH MED, UNIV MIAMI, FLA, 83- *Mem:* AAAS; Am Physiol Soc; Soc Neurosci; Soc Gen Physiologists. *Res:* Developmental neurobiology; synapse regeneration; neuronal signaling and integration; synaptic transmission. *Mailing Add:* Dept Physiol & Biophys Sch Med Univ Miami PO Box 016430 Biscayne Annex Miami FL 33101

MULLER, LAWRENCE DEAN, b Peoria, Ill, Nov 26, 41; m 65; c 3. DAIRY HUSBANDRY. *Educ:* Univ Ill, Urbana, BS, 64, MS, 66; Purdue Univ, Lafayette, PhD(animal sci), 69. *Prof Exp:* Asst prof animal sci, Purdue Univ, 69-71; assoc prof dairy sci, SDak State Univ, 71-76; assoc prof, 76-80, PROF DAIRY SCI, PA STATE UNIV, 80- *Mem:* Am Dairy Sci Asn; Am Soc Animal Sci; Am Inst Nutrit. *Res:* Animal production with emphasis on interrelationships between nutrition, physiology and management on animal productivity. *Mailing Add:* 160 Grove Circle Lemont PA 16851

MULLER, MARCEL WETTSTEIN, b Vienna, Austria, Nov 1, 22; nat US; m 47; c 3. SOLID STATE PHYSICS, ELECTROMAGNETISM. *Educ:* Columbia Univ, BS, 49, AM, 52; Stanford Univ, PhD(physics), 57. *Prof Exp:* Sr scientist, Varian Assocs, Calif, 52-66; PROF ELEC ENG, WASHINGTON UNIV, 66- *Concurrent Pos:* Lectr, Univ Zurich, 62-63; vis prof, Univ Colo, 68; Humboldt award, 76; vis scientist, Max Planck Inst Metals Res, Stuttgart, 76-77. *Mem:* Fel Am Phys Soc; fel Inst Elec & Electronics Engrs. *Res:* Microwave electronics; quantum electronics; solid state physics; applied magnetism. *Mailing Add:* Dept Elec Eng Washington Univ One Brookings Dr St Louis MO 63130

MULLER, MERVIN EDGAR, b Hollywood, Calif, June 1, 28; m 63; c 3. COMPUTER SCIENCE, STATISTICS. *Educ:* Univ Calif, Los Angeles, PhD(math), 54. *Prof Exp:* Instr math, Cornell Univ, 54-56; res assoc, Princeton Univ, 56-59; mem staff control planning, Data Processing Div, Int Bus Mach Corp, 59-60, mgr proj weld, 60-64; prof comput sci, Univ Wis-Madison, 64-70, prof comput sci & statist, 70-71, dir comput ctr, 64-70; dir dept comput activ, World Bank, 71-81, sr adv financial planning, 81-83, sr adv systs, 83-85; PROF, CHMN & ROBERT M CRITCHFIELD PROF COMPUT & INFO SCI, OHIO STATE UNIV, 85- *Concurrent Pos:* Mem bd dir, Am Fedn Info Processing, 71-73, chmn finance comt, 71-75; chmn comt statist comput, Int Statist Inst, 75-77; pres, Int Asn Statist Comput, 77-79; mem steering comt, Nat Res Coun panel sci computing, Nat Bur Standards, 80-82; prof info & technol, George Mason Inst, 85; mem, Comt Data for Sci & Technol, 85-93; mem coun, Int Statist Inst, 85-89; mem exec comt, Bd Trustees of the Adv Info Technol Res Ctr, 86-91; chmn sci & tech info bd, Nat Res Coun/Nat Acad Sci, 90-91. *Mem:* Am Math Soc; Int Statist Inst; Math Asn Am; fel Am Statist Asn; Asn Comput Mach. *Res:* Monte Carlo procedures and simulation; statistical design of experiments; use of computers in statistics, data processing and statistical control procedures; computer information systems and languages; management information systems; data base management systems. *Mailing Add:* Dept Comput & Info Sci 230 Civil & Aeronaut Eng Building 2036 Neil Ave Mall Columbus OH 43210-1277

MULLER, MIKLÓS, b Budapest, Hungary, Nov 24, 30; m 73; c 2. BIOLOGICAL CHEMISTRY, PARASITOLOGY. *Educ:* Med Univ Budapest, MD, 55. *Prof Exp:* Instr biol & histol, Med Univ Budapest, 50-55, asst prof, 55-64; res assoc biochem cytol, Rockefeller Inst, 64-65; Rask-Orsted fel & guest investr cell biol, Dept Physiol, Carlsberg Lab, Copenhagen, 65-66; from asst prof to assoc prof biochem cytol, 66-88, ASSOC PROF BIOCHEM PARASITOL, ROCKEFELLER UNIV, 88- *Concurrent Pos:* Ed, Molecular & Biochem Parasitol, 80-87; mem, steering comt sci working group, Leishmaniasis WHO Spec Prog Trop Dis, 77-82; mem, Biomed Sci Study Group, NIH, 84-87; mem, Int Comn Protozool, Int Union Biol Socs, 85-87; mem, sci coun, Int Inst Cellular & Molecular Biol, Brussels, Belg, 85-90; assoc ed, J Protozool, 88-90. *Honors & Awards:* S H Hutner Prize, Soc Protozool, 77. *Mem:* Soc Protozool (pres, 82-83); Am Soc Parasitol; Med Soc Study Venereal Dis; Am Microbiol Soc; Am Venereal Dis Asn; Am Soc Biol Chemists. *Res:* Physiology, biochemistry and evolution of parasitic and free-living protists; hydrogenosomes; action of antiprotozoal drugs. *Mailing Add:* Rockefeller Univ 1230 York Ave New York NY 10021

MULLER, NORBERT, b Hamburg, Ger, Jan 25, 29; nat US; m 58; c 11. PHYSICAL CHEMISTRY. *Educ:* Univ Calif, BS, 49; Harvard Univ, MA, 51, PhD(chem physics), 53. *Prof Exp:* From instr to assoc prof, 53-68, PROF CHEM, PURDUE UNIV, LAFAYETTE, 68- *Mem:* Am Chem Soc; Royal Soc Chem. *Res:* Molecular structure and spectra; nuclear magnetic resonance; surfactant chemistry; organofluorine chemistry. *Mailing Add:* Dept Chem Purdue Univ Lafayette IN 47907

MULLER, OLAF, b Tallinn, Estonia, Jan 14, 38; US citizen. INORGANIC & SOLID STATE CHEMISTRY, THIN FILMS. *Educ:* Western Reserve Univ, BA, 60, MS, 61; Pa State Univ, PhD(solid state sci), 68. *Prof Exp:* Res assoc solid state sci, Pa State Univ, 68-72; inorg chemist, Corp Res & Develop Ctr, Gen Elec Co, 72-75; SCIENTIST SOLID STATE SCI, WEBSTER RES CTR, XEROX CORP, 75- *Mem:* Am Chem Soc. *Res:* Synthesis and crystal chemistry of inorganic materials; property-composition relationships; magnetic materials; thin films. *Mailing Add:* 1169 Appian Dr Webster NY 14580

MULLER, OTTO HELMUTH, b Omaha, Nebr, Aug 7, 46; m 70; c 2. GRAVITY. *Educ:* Univ Rochester, BA, 70, MS, 72, PhD(geol), 74. *Prof Exp:* Instr geol, Univ Rochester, 71-74; asst prof geol, State Univ NY, Stony Brook, 74-76; asst prof geol, Colgate Univ, 76-82; geophysicist, US Geol Surv, 78-82; PROF GEOL, ALFRED UNIV, 82- *Mem:* Am Geophys Union. *Res:* Use of geometry and orientation of sheet intrusions to determine stress conditions and fracture propagation directions during emplacement; persistent zones of weakness in North America suggested by patterns of geophysical anomalies. *Mailing Add:* Geol Dept Alfred Univ Main St Alfred NY 14802

MULLER, RICHARD A, b New York, NY, Jan 6, 44; m 66; c 2. EXPERIMENTAL PHYSICS. *Educ:* Columbia Univ, AB, 64; Univ Calif, Berkeley, PhD(physics), 69. *Hon Degrees:* LHD, Am Univ, Switz, 89. *Prof Exp:* Asst res physicist, Space Sci Lab, 69-75, assoc res physicist, 75-78, assoc prof physics, 78-80, PROF PHYSICS, UNIV CALIF, BERKELEY, 80-, FAC SR SCIENTIST, LAWRENCE BERKELEY LAB, 78- *Concurrent Pos:* Lectr physics, Univ Calif, Berkeley, 72-74; consult to US Govt, 73-; MacArthur Found fel, 82-87. *Honors & Awards:* Tex Instruments Found Founders Prize, 77; Alan T Waterman Award, NSF, 78. *Mem:* Fel Am Phys Soc; Am Astron Soc; fel AAAS; Sigma Xi; Int Astron Union. *Res:* Cosmic microwave anisotropy; radioisotope dating using accelerators; adaptive optics; elementary particles; automated supernova search; instrumentation; mass extinctions; nemesis hypothesis; geomagnetic reversals. *Mailing Add:* 2831 Garber St 50/232 Univ Calif Berkeley CA 94705

MULLER, RICHARD S, b Weehawken, NJ, May 5, 33; m 57; c 2. SOLID-STATE DEVICE PHYSICS, SENSOR DEVICES. *Educ:* Stevens Inst Technol, MechE, 55; Calif Inst Technol, MSEE, 57, PhD(elec eng/physics), 62. *Prof Exp:* Res engr, Hughes Aircraft Co, 57-60,; instr, Univ Southern Calif, 59-62; from asst prof to assoc prof, 62-73, vchem admin, Dept Elec Eng & Comput Sci, 73-75, PROF ELEC ENG, UNIV CALIF, BERKELEY SENSOR & ACTUATOR CTR, 86-, DIR, 86- *Concurrent Pos:* Consult, Pac Semiconductors Inc, 59-62, Stanford Res Inst & Lawrence Berkeley Lab, 62-, Hughes Micro-electronics Labs, 63-64, Signetics Corp, 71-89 & Xerox Corp, 77-; NATO fel, Munich Tech Univ, 68-69; Fulbright res prof, Tech Univ Munich, Germany, 82-83; chmn sensors adv bd, Inst Elec & Electronics Engrs, chmn tranducers, Int Conf Sensors & Actuators, 91. *Mem:* Fel Inst Elec & Electronics Engrs; Electron Devices Soc. *Res:* Solid-state device research; integrated sensing devices. *Mailing Add:* Dir Berkeley Sensor & Actuator Ctr Dept Elec Eng & Comput Sci Cory Hall Univ Calif Berkeley CA 94720

MULLER, ROBERT ALBERT, b Passaic, NJ, Dec 5, 28; m 50; c 2. CLIMATOLOGY, PHYSICAL GEOGRAPHY. *Educ:* Rutgers Univ, BA, 58; Syracuse Univ, MA, 59, PhD(geog), 62. *Prof Exp:* Phys geogr, Pac Southwest Forest & Range Exp Sta, Calif, 62-64; lectr climat, Univ Calif, Berkeley, 64; from asst prof to assoc prof geog, Rutgers Univ, 64-68; assoc prof, 69-72, chmn dept geog & anthrop, 78-81, PROF GEOG, LA STATE UNIV, 72-, STATE CLIMATOLOGIST, 78-; DIR, SOUTHERN REGIONAL CLIMATE CTR, 90- *Mem:* Asn Am Geog; Am Geophys Union; Am Meteorol Soc; Am Asn State Climatologists. *Res:* Water balance and synoptic climatology evaluations of evapotranspiration loss, water yield, and river basin regimen including flooding. *Mailing Add:* Dept Geog La State Univ Baton Rouge LA 70803

MULLER, ROBERT E, b Brooklyn, NY, June 27, 21; m 49; c 4. CHEMICAL ENGINEERING. *Educ:* Polytech Inst Brooklyn, BChemE, 41, MChemE, 42, DSc(chem eng), 47. *Prof Exp:* Res engr plastics & atomic physics, Bakelite Co Div, Union Carbide & Carbon Corp, 42-46; res fel adsorption, Am Sugar Ref Co, 46-47; dir res & prod, Luxene, Inc, 47-54; from asst res dir to div res mgr, US Gypsum Co, 54-63; dir res & develop, MacAndrews & Forbes Co, 63-65; corp dir res & develop, Nat Can Corp, 65-67; PRES, ASTRON DENT CORP, 67- *Mem:* Am Chem Soc; Am Inst Chem Engrs. *Res:* High polymer and atomic research; paper and wood; research management. *Mailing Add:* Dir Eng Progs 131 Arts & Tech Bldg Univ Md Eastern Shore Princess Anne MD 21853

MULLER, ROBERT NEIL, b Santa Barbara, Calif, Aug 29, 46; m 70; c 2. PLANT ECOLOGY. *Educ:* Univ Calif, Riverside, BA, 69; Yale Univ, MFS, 72, PhD(plant ecol), 75. *Prof Exp:* Post-doctoral appointee, Argonne Nat Lab, 74-76, asst ecologist, 76-78; asst prof, 78-83, ASSOC PROF, DEPT FORESTRY, UNIV KY, 83- *Mem:* Ecol Soc Am. *Res:* Adaptations of species to their environment; factors affecting distributions of species within plant communities; structure and function of terrestrial ecosystems. *Mailing Add:* Dept of Forestry Univ of Ky Lexington KY 40506

MULLER, ROLF HUGO, b Aarau, Switz, Aug 6, 29; nat US; m 62; c 2. ELECTROCHEMISTRY. *Educ:* Swiss Fed Inst Technol, Dipl sc nat, 53, Dr sc nat(phys chem), 57. *Prof Exp:* Asst, Inst Phys Chem & Electrochem, Swiss Fed Inst Technol, 55-56; res chemist polychem, E I du Pont de Nemours & Co, 57-60; res assoc electrochem eng, Univ Calif, Berkeley, 61-62, staff scientist, Lawrence Radiation Lab, 62-66, asst head inorg mat res div, Lawrence Berkeley Lab, 70-75, asst head mat & molecular res div, 75-87, assoc head Mat & Chem Sci Div, 87-88, PRIN INVESTR PHYSICS & CHEM OF PHASE BOUNDARIES, LAWRENCE BERKELEY LAB, UNIV CALIF, 66-, LECTR CHEM ENG, 66-, STAFF SR SCIENTIST, 78-, ASSOC DIV DIR, 88- *Concurrent Pos:* Div co-chmn, exp methods, Int Soc Electrochem, 73-77; counr, 76-77; chmn, Technol Award, 77-81; secy-treas, Phys Electrochem Div, Electrochem Soc, 85-87, vchmn, 87-89, chmn, 89-91; guest lectr, Chongging Univ, China, 88. *Mem:* AAAS; Am Chem Soc; Optical Soc Am; Int Soc Electrochem; Electrochem Soc; Swiss Chem Soc. *Res:* Optical methods for the study of surfaces, thin film and boundary layers in liquid media; optical models for interpretation of measurements; nucleation and growth of electrochemical surface layers, structure, composition and microtopography; electrolytic deposition and dissolution of metals, effect of adsorbed materials; surface layers on electrodes for rechargeable galvanic cells; electrochemical processes at high current densities, enhancement of electrolytic mass transfer. *Mailing Add:* Mat Sci Div Lawrence Berkeley Lab 62-203 Berkeley CA 94720

MULLER, UWE RICHARD, MOLECULAR GENETICS, GENETIC ENGINEERING. *Educ:* Kans State Univ, PhD(microbiol), 76. *Prof Exp:* ASSOC PROF MICROBIOL, SCH MED, E CAROLINA UNIV, GREENVILLE, NC, 79- *Mailing Add:* Biotechnol Div Amoco Technol Co PO Box 3011 Greenville NC 60566

MULLER, WILLIAM A, b New York, NY, Aug 16, 42; c 2. MARINE BIOLOGY, MARINE ECOLOGY. *Educ:* Queens Col, NY, BS, 64, MS, 68; City Univ New York, PhD(biol), 72. *Prof Exp:* Res asst marine biol, Am Mus Natural Hist, 60-69; from asst prof to assoc prof, 72-75, chmn dept, 76-78, assoc dir natural sci ctr, 78-85, PROF LIFE SCI, NY INST TECHNOL, 87- *Concurrent Pos:* Pres, New York Sportfishing Coun, 77-78; consult, NY Dept Environ Conserv Shellfish Leasing, 77-; chmn squid-butterfish subpanel & mem bluefish subpanel, Mid Atlantic Fisheries Mgt Coun, 78-81; adv, Long Island State Parks & Recreation Comn, 79-; res assoc marine biol, City Univ New York, 69-78; mem NY State Marine Resources Adv Coun, 89- *Mem:* Sigma Xi; Soc Protozool; Am Soc Limnol & Oceanog; Am Fisheries Soc. *Res:* Salt marsh ecology, productivity and fisheries management; niche theory environmental stress. *Mailing Add:* Dept Life Sci NY Inst Technol 268 Wheatley Rd Old Westbury NY 11568

MULLER, WILLIAM HENRY, JR, b Dillon, SC, Aug 19, 19; m 46; c 3. SURGERY. *Educ:* The Citadel, BS, 40; Duke Univ, MD, 43; Am Bd Surg, dipl; Am Bd Thoracic Surg, dipl. *Prof Exp:* Intern, Johns Hopkins Hosp, 44, asst resident & asst surg, 44-46, instr surg & resident gen surg, 48-49, resident cardiovasc surg, 49; from asst prof to assoc prof surg, Sch Med, Univ Calif, Los Angeles, 49-54; S Hurt Watts prof surg & chmn dept, Univ Va, 54-76, surgeon- in-chief, Hosp, 54-76, vpres health affairs, Med Ctr, 76-89, S Hurt Watts prof surg, 82-90, VPRES & EMER PROF SURG, UNIV VA, 90- *Concurrent Pos:* Attend specialist, Wadsworth Vet Admin Hosp, Los Angeles, chief sect cardiovasc surg, Los Angeles County Gen Hosp, Torrence, consult, St John's Hosp, Santa Monica & Santa Monica Hosp, 49-54; mem exam bd, Am Bd Surg; chmn surg study sect, NIH; mem, President's Panel on Heart Dis, 72; mem Nat Joint Practice Comn of Med & Nursing, 72; mem bd trustees, Duke Univ & Duke Univ Med Ctr; chmn bd regents, Am Col Surgeons, 76-78, pres, 80. *Mem:* Soc Vascular Surg (pres, 66-67 & 68); Soc Clin Surg; Am Asn Thoracic Surg; AMA; fel Am Col Surg; Am Surg Asn (pres, 75). *Res:* Surgery of cardiovascular deformities; pulmonary hypertension; enzymatic debridement of wounds. *Mailing Add:* Dept Surg Box 474 Univ Va Med Ctr 3009 Charlottesville VA 22903

MULLER-EBERHARD, HANS JOACHIM, b Magdeburg, Ger, May 5, 27; m 77, 85; c 2. COMPLEMENT. *Educ:* Univ Gottingen, MD, 53; Univ Uppsala, DMSc, 61. *Hon Degrees:* MD Honoris Causa, Ruhr Univ, Bochum, Ger, 82. *Prof Exp:* Asst physician, Dept Med, Univ Gottingen, 53-54; asst & asst physician, Rockefeller Inst, 54-57; fel, Swedish Med Res Coun, Dept Clin

Chem, Univ Uppsala, 57-59; from asst prof to assoc prof biochem & immunol, Rockefeller Univ, 59-63; mem, Dept Exp Path, Scripps Clin & Res Found, 63-74, Cecil H & Ida M Green investr med res, 72-86, Chmn Dept Molecular Immunol, 74-82, assoc dir, Res Inst, 78-86, chmn Dept Immunol, 82-85, head, Div Molecular Immunol, 85-87; DIR, BERNHARD NOCHT INST TROP MED, HAMBURG, GER, 88- *Concurrent Pos:* Assoc physician, Rockefeller Univ, 59-63; mem Allergy & Immunol A Study Sct, NIH, 65-69; adj prof, Univ Calif, San Diego, 68-; Harvey lect, 70; prof, Univ Hamburg, 90-; vis prof internal med, Univ Tex, Houston, 90- *Honors & Awards:* Parke Davis Meritorious Award, Am Soc Exp Path, 66; Squibb Award, Infectious Dis Soc Am, 70; T Duckett Jones Mem Award, Helen Hay Whitney Found, 71; Harvey lectr, 70; Karl Landsteiner Mem Award, Am Asn Blood Banks, 74; Robert Koch Medal Gold, 87,; Philip Levine Award, Am Soc Clin Pathologists, 88; Rous Whipple Award, Am Asn Pathologists, 88; Modern Med Distinguished Achievement Award, 74; Ann Internist Award, Gairdner Found, Can, 74; Mayo H Soley Award, Western Soc Clin Res, 75. *Mem:* Nat Acad Sci; Am Soc Clin Invest; Am Asn Immunol; Am Soc Biol Chemists; Asn Am Physicians; Sigma Xi; Am Asn Pathologists. *Res:* Molecular biology and biochemistry of complement; cellular cytotoxicity; modulation of cellular functions by effector molecules. *Mailing Add:* Bernhard Nocht Inst Trop Med Bernhard Nocht Str 74 Hamburg 36 Germany

MULLER-EBERHARD, URSULA, b Gottingen, WGer, June 14, 28; US citizen; c 2. HEMATOLOGY, BIOCHEMISTRY. *Educ:* Univ Gottingen & Univ Freiburg, MD, 53. *Prof Exp:* Intern, Wyckoff Heights Hosp, Brooklyn, 54-55, asst res pediat, 55-56; asst res pediat, Univ Hosp, Bellevule Med Ctr, 56; Swedish Med Res Coun fel pediat, Univ Uppsala, Sweden, 57-59; sr asst res pediat, Bellevue Med Ctr, 59-60; asst pediat, Cornell Univ, 60-62, instr pediat & head div pediat hemat, Bellevue Med Ctr, 62-63; from assoc to assoc mem dept biochem, 63-75, mem, dept biochem, Scripps Clin & Res Found, 75-; PROF PEDIAT, PHARMACOL & BIOCHEM, CORNELL UNIV MED CTR. *Concurrent Pos:* Fel pediat hemat, Med Ctr, Cornell Univ, 60-62; Health Res Coun NY career scientist award, 62; USPHS res career develop award, 63-71; Eleanor Roosevelt Int Cancer fel, Am Cancer Soc, 79. *Honors & Awards:* Alexander von Humboldt Award, 79. *Mem:* Harvey Soc; Am Soc Hemat; Am Soc Exp Path; Am Soc Clin Invest; Am Asn Study Liver Dis; Am Soc Biol Chemists. *Res:* Biochemical hematology; developmental biochemistry. *Mailing Add:* Dept Pediat Div Pediat-Hematol Cornell Univ Med Ctr 1300 York Ave New York NY 10021

MULLER-PARKER, GISÈLE THÉRÈSE, b New York, NY. SYMBIOSIS, BIOLOGICAL. *Educ:* State Univ NY, BS, 75; Univ Del, MS, 78; Univ Calif, Los Angeles, PhD(biol), 84. *Prof Exp:* Res technologist, Artificial Upwelling Proj, Columbia Univ, 75-76; res asst, Sea Grant, Univ Del, 76-78; teaching asst biol, Univ Calif, Los Angeles, 78-84; res assoc, Univ Nebr, 84-86; sr res assoc, Ches Biol Lab, Univ Md, 86-89; RES SCIENTIST, SHANNON POINT MARINE CTR, WESTERN WASH UNIV, 89-, ASST PROF BIOL, 90- *Mem:* Am Soc Limnol & Oceanog; Am Soc Zoologists; Phycol Soc Am. *Res:* Coral reefs; symbiosis between algae and animals; nutrients and water quality. *Mailing Add:* Shannon Point Marine Ctr Western Wash Univ 1900 Shannon Point Rd Anacortes WA 98221

MULLER-SCHWARZE, DIETLAND, b Ger; m 65; c 2. ANIMAL BEHAVIOR, CHEMICAL ECOLOGY. *Educ:* Univ Freiburg, PhD(zool), 63. *Prof Exp:* Asst prof zool, Univ Freiburg, 63-65; asst prof, San Francisco State Univ, 65-68; assoc prof, Utah State Univ, 68-73; assoc prof, 73-78, PROF, COL ENVIRON SCI & FORESTRY, STATE UNIV NY, 78- *Mem:* Animal Behav Soc; Ecol Soc Am; Europ Chemoreception Res Orgn; Am Soc Mammalogists. *Res:* Vertebrate pheromones; chemical ecology; vertebrate behavior. *Mailing Add:* State Univ of NY Col of Environ Sci & Forestry Syracuse NY 13210

MÜLLER-SIEBURG, CHRISTA E, IMMUNOLOGY, MICROBIOLOGY. *Educ:* Univ Cologne, Dr rer nat, 83. *Prof Exp:* Postdoctoral fel, Med Ctr, Stanford Univ, 83-86, res assoc, 86-89; ASST MEM, MED BIOL INST, LA JOLLA, CALIF, 89- *Concurrent Pos:* Fel, Deutsche Forschungsgemeinschaft, 83-85; scholar, Am Leukemia Soc, 89-; chairperson, Workshop Stem Cells & Lymphocyte Lineages, 7th Int Cong Immunol, 89; lectr, Stanford Med Sch, 89, Calif Inst Technol & Univ Calif, San Diego, 90, Childrens Hosp, Los Angeles & Nat Cancer Inst, 91; mem, Spec Rev Comt, Nat Cancer Inst. *Mem:* Fedn Am Socs Exp Biol. *Res:* Regulation of early hemato-lymphoid differentiation in the mouse; isolation of precursors and stem cells at distinct differentiation stages; analysis of precursor and stem cell regulation by cytokines and stromal cells; characterization of stromal cell precursors; author of 19 technical publications and one book. *Mailing Add:* Med Biol Inst 11077 N Torrey Pines Rd La Jolla CA 92037

MULLHAUPT, JOSEPH TIMOTHY, b St Mary's, Pa, Feb 25, 32; m 57; c 5. PHYSICAL CHEMISTRY. *Educ:* Univ Rochester, BS, 54; Brown Univ, PhD(phys chem), 58. *Prof Exp:* From res chemist to sr res chemist, 58-67, res supvr phys chem, 67-69, SR RES SCIENTIST PHYS CHEM, LINDE DIV, UNION CARBIDE CORP, 69- *Mem:* AAAS; Am Phys Soc; Sigma Xi; Am Chem Soc. *Res:* Solid state chemistry; adsorption and surface chemistry; thermodynamics of phase equilibria. *Mailing Add:* Union Carbide Corp 61 E Park Dr Tonawanda NY 14150

MULLIGAN, BENJAMIN EDWARD, b Greensboro, NC, May 17, 36; m 63; c 2. SENSORY PSYCHOLOGY. *Educ:* Univ Ga, BA, 58; Univ Miss, MA, 61, PhD(sensory psychol), 64. *Prof Exp:* Asst prof, 64-69, ASSOC PROF SENSORY PSYCHOL, UNIV GA, 69- *Concurrent Pos:* Nat Inst Neurol Dis & Stroke grant, 65-70; partic, Int Cong Physiol Sci, 65. *Mem:* Acoust Soc Am; Optical Soc Am; Am Soc Cybernet; Soc Neurosci. *Res:* Sensory processes; psychophysics; mathematical models; noise pollution; communication. *Mailing Add:* Dept of Psychol Univ of Ga Athens GA 30602

MULLIGAN, BERNARD, b Montgomery, Ala, Aug 31, 34; m 64. THEORETICAL PHYSICS, NUCLEAR PHYSICS. *Educ:* Univ Ala, BS, 56; Mass Inst Technol, PhD(theoret physics), 62. *Prof Exp:* Vis asst prof, 61-63, from asst prof to assoc prof, 63-77, PROF PHYSICS, OHIO STATE UNIV, 77- *Mem:* Am Phys Soc. *Res:* Theory of nuclear phenomena; mathematical physics. *Mailing Add:* Dept of Physics Ohio State Univ 174 W18th St Columbus OH 43210

MULLIGAN, GEOFFREY C, b Palo Alto, Calif, May, 1958. NETWORKING & TRANSMISSION CONTROL PROTOCOL-INTENET PROTOCOL, UNIX OPERATING SYSTEM. *Educ:* USAF Acad, BS, 79, Univ Denver, MS, 88. *Prof Exp:* Syst analyst, Air Force Data Serv Ctr, Pentagon, Washington, DC, 79-81, software engr, First Info Syst Group, 81-84; instr computer sci, USAF Acad, 84-90; PRIN SOFTWARE ENGR, NETWORK SYST LAB, DIGITAL EQUIP CORP, 90- *Concurrent Pos:* Instr, Univ Denver, 88-90, Chapman Col, 89-90; consult, Telos Corp, 89, Sci Appln Int Corp, 89- *Res:* Research into data communication and networking and other advancing technologies to reduce the complexity of interconnection and use of computers, workstations and peripherals. *Mailing Add:* 3330 Whimbrel Ct Fremont CA 94555

MULLIGAN, JAMES ANTHONY, b Denver, Colo, Aug 31, 24; m 85; c 1. ANIMAL BEHAVIOR. *Educ:* St Louis Univ, AB, 47, STL, 57; Univ Calif, Berkeley, PhD(zool), 63. *Prof Exp:* From instr to assoc prof biol, St Louis Univ, 63-86; CONSULT, 86- *Concurrent Pos:* Frank M Chapman Mem Fund res grant, 64; NSF res grant, 65-71; fel, Woodrow Wilson Int Ctr Scholars, 71-72. *Mem:* Animal Behav Soc; Ecol Soc Am; Am Soc Zoologists; Am Ornith Union. *Res:* Social behavior and communication in animals by means of sound; field study and physical analysis of avian vocalizations; ontogeny and genetic analysis of bird vocalizations; vertebrate ecology; conservation of natural areas; ethics of the environmental crisis. *Mailing Add:* Trinity Episcopal Church Fishkill NY 12524-0484

MULLIGAN, JAMES H(ENRY), JR, b Jersey City, NJ, Oct 29, 20; m 47; c 2. ELECTRICAL ENGINEERING, COMPUTERIZED DESIGNS. *Educ:* Cooper Union, BEE, 43, EE, 47; Stevens Inst Technol, MS, 45; Columbia Univ, PhD, 48. *Prof Exp:* Mem tech staff, Bell Tel Labs, Inc, 41-44; sr engr, A B du Mont Labs, Inc, 45-47, chief engr, Transmitter Div, 48-49; from asst prof to prof elec eng, NY Univ, 49-68, chmn dept, 52-68; secy & exec officer, Nat Acad Eng, 68-74; dean, Sch Eng, 74-77, PROF ELEC ENG, UNIV CALIF, IRVINE, 74- *Concurrent Pos:* Consult, Res & Develop Bd, Off Secy Defense, Bell Tel Labs & Sprague Elec Co; mem comn human resources, Nat Res Coun, 79- *Honors & Awards:* Haraden Pratt Award, Inst Elec & Electronics Engrs. *Mem:* Nat Acad Eng (secy, 68-78); fel Inst Elec & Electronics Engrs (vpres, 70, pres, 71); fel Brit Inst Elec Engrs; Am Phys Soc; Nat Soc Prof Engrs. *Res:* Network theory; feedback systems; solid state electronics; applied mathematics in electrical engineering. *Mailing Add:* Dept Elec & Computer Eng Rm 444 Eng Bldg Univ Calif Irvine CA 92717

MULLIGAN, JOSEPH FRANCIS, b New York, NY, Dec 12, 20; m 68, 84. HISTORY OF PHYSICS. *Educ:* Boston Col, AB, 45, MA, 46; Cath Univ, PhD(physics), 51. *Prof Exp:* Instr physics, St Peter's Col, 46-47; instr, Fordham Univ, 55-57, from asst prof to assoc prof, 57-68, chmn dept, 57-64, dean grad sch arts & sci & dean fac, 64-67; dean grad studies & res, Baltimore County, 68-82, prof physics, 68-89, EMER PROF PHYSICS, UNIV MD, BALTIMORE COUNTY, 89- *Concurrent Pos:* Mem adv comt grad fels, Nat Defense Educ Act, 59-63, mem adv comt grad educ, NY State, 63-68; NSF fac fel, Univ Calif, San Diego, 61-62. *Mem:* AAAS; Am Phys Soc; Am Asn Physics Teachers; Sigma Xi; Hist Sci Soc. *Res:* History of physics in Germany at the end of the nineteenth century; author of 3 books and articles on physics; lives and contributions to physics of Hermann von Helmholtz, and Heinrich Hertz. *Mailing Add:* Dept of Physics Univ of Md Baltimore County Baltimore MD 21228

MULLIGAN, TIMOTHY JAMES, b Lowell, Mass, May 15, 55; m 87. FISH BIOLOGY, FISHERIES ECOLOGY. *Educ:* Univ Vt, BS, 77; Univ Cent Fla, MS, 81; Univ Md, PhD(marine sci), 87. *Prof Exp:* Postdoctoral fel fisheries, Univ Wash, Fisheries Res Inst, 87-89; ASST PROF FISHERIES, HUMBOLDT STATE UNIV, 89- *Mem:* Am Soc Ichthyologists & Herpetologists; Am Fisheries Soc; Sigma Xi; Western Soc Naturalists. *Res:* Ecology of estuarine and marine fishes; early life history of fishes; identification of walleye pollock stocks in the Bering Sea; identification of yellowtail rockfish stock along the Pacific Coast; ecology of bat rays in Northern California. *Mailing Add:* Dept Fisheries Humboldt State Univ Arcata CA 95521

MULLIKIN, H(ARWOOD) F(RANKLIN), b Baltimore, Md, June 27, 08; m; c 2. MECHANICAL & NUCLEAR ENGINEERING. *Educ:* Johns Hopkins Univ, BS, 30; Yale Univ, MS, 31, ME, 32, PhD(mech eng), 34. *Prof Exp:* Test engr, Gen Elec Co, 35; anal engr, Babcock & Wilcox Co, NY, 36-42; instr, City Univ New York, 42-44; design engr, Ebasco Servs, Inc, 44-46; assoc prof, Ill Inst Technol, 46-47; head dept mech eng, Mont State Univ, 47-67; chmn dept mech eng, Memphis State Univ, 67-75, consult, Ctr Nuclear Studies, 75-84; prof consult, Am Tech Inst, Brunswick, Tenn, 86; CONSULT, 86- *Mem:* Am Soc Mech Engrs. *Res:* Heat transfer; thermodynamics; dimensional analysis; metrication conversion; powerplant design; nuclear power. *Mailing Add:* 137 Wallace Rd Memphis TN 38117

MULLIKIN, RICHARD V(ICKERS), b Wilmington, Del, Apr 13, 23; m 49; c 2. CHEMICAL ENGINEERING. *Educ:* Mass Inst Technol, BS, 44. *Prof Exp:* Indust engr, E I du Pont de Nemours & Co, 44-48, plant assistance engr, 48-51, prod supvr, 51-55; res engr, Monsanto Co, 56-60, res specialist, 60-78; sr design engr, Cities Serv Co, 78-80; RETIRED. *Mem:* Am Inst Chem Engrs; Am Chem Soc. *Res:* Polymer process research. *Mailing Add:* 407 Karen Lane San Antonio TX 78209

MULLIKIN, THOMAS WILSON, b Tenn, Jan 9, 28; m 52; c 3. MATHEMATICS. *Educ:* Univ Tenn, AB, 50; Harvard Univ, MA, 54, PhD(math), 58. *Prof Exp:* Asst math, Oak Ridge Nat Lab, 47-48; US Navy, 50-52; mathematician, Rand Corp, 57-64; PROF MATH, PURDUE UNIV, 64-, ACTG VPRES & DEAN GRAD SCH, 91- *Mem:* AAAS; Am Math Soc; Soc Indust & Appl Math. *Res:* Differential and integral equations. *Mailing Add:* Dept Math Purdue Univ West Lafayette IN 47907

MULLIN, BETH CONWAY, b Philadelphia, Pa, Oct 28, 45. PLANT PHYSIOLOGY, CELL BIOLOGY. *Educ:* Earlham Col, BA, 67; NC State Univ, PhD(plant physiol), 72. *Prof Exp:* Investr RNA tumor viruses, Oak Ridge Nat Lab, 73-75, Nat Cancer Inst fel, 75-76; asst prof biochem, Wilmington Col, 76-77; from asst prof to assoc prof, 77-89, PROF PLANT PHYSIOL, UNIV TENN, 89- *Concurrent Pos:* NIH biomed res grant, 77-78; Lilly Found fel, Univ Tenn, 78-79; consult molecular carcinogenesis, Oak Ridge Nat Lab, 77- *Mem:* Am Soc Plant Physiologists; Sigma Xi; Plant Molecular Biol Asn. *Res:* Gene expression; molecular biology of symbiotic nitrogen fixation; actinorhizal symbiosis. *Mailing Add:* Hesler Biol Bldg Dept of Bot Univ of Tenn Knoxville TN 37916

MULLIN, BRIAN ROBERT, b Columbus, Ga, June 16, 45. MULTIPLE SCLEROSIS RESEARCH, MEDICAL STUDENT TEACHING. *Educ:* Univ Scranton, BS, 66; State Univ NY, Upstate Med Ctr, MD, 70; Am Bd Path, dipl, 73. *Prof Exp:* Researcher endocrinol, Sloan-Kettering Inst Cancer Res, 69-70; internship, Cleveland Metrop Gen Hosp, 70-71; residency path, Nat Cancer Inst, NIH, 71-73, fel biochem, Nat Inst Arthritis, Metabolism & Digestive Dis, 74-77; asst prof, Sch Med, Case Western Reserve Univ, 77-81; ASSOC PROF PATH, UNIFORMED SERV UNIV HEALTH SCI, 81- *Mem:* Paleopath Asn. *Res:* Gangliosides as autoantigens in multiple sclerosis. *Mailing Add:* 3900 Conn Ave NW Washington DC 20008

MULLIN, JAMES MICHAEL, b Philadelphia, Pa, Oct 8, 54. CELL CULTURE, RENAL PHYSIOLOGY. *Educ:* St Joseph's Col, BS, 76; Univ Pa, PhD(physiol), 80. *Prof Exp:* NIH fel, dept physiol, Univ Pa, 81-82; NSF fel, dept human genetics, Yale Univ, 82-84; res investr, Wistar Inst, 84-86; res assoc, 86-90, ASSOC INVESTR, LANKENAU MED RES CTR, 90- *Mem:* Am Physiol Soc. *Res:* Membrane transport and intermediary metabolism of cultured epithelial cells, genetic manipulation thereof and effects of chemical carcinogens thereon. *Mailing Add:* Med Res Ctr 100 Lancaster Ave Wynnewood PA 19096

MULLIN, MICHAEL MAHLON, b Galveston, Tex, Nov 17, 37; m 64; c 3. BIOLOGICAL OCEANOGRAPHY, ECOLOGY. *Educ:* Shimer Col, AB, 57; Harvard Univ, AB, 59, MA, 60, PhD(biol), 64. *Prof Exp:* NSF fel, 64; instr oceanog & res biologist, 64-65, asst prof & asst res biologist, 65-71, assoc prof oceanog & assoc res biologist, Inst Marine Resources, 71-77, chmn grad dept, 77-80, assoc dir, 80-87, PROF OCEANOG, SCRIPPS INST OCEANOG & RES BIOLOGIST, INST MARINE RESOURCES, UNIV CALIF, SAN DIEGO, 77-, DIR, MARINE LIFE RES GROUP, 87- *Concurrent Pos:* Mem, Ocean Sci Bd, Nat Res Coun, 79-81; sr Queen's fel marine sci, Australia, 81-82. *Mem:* Am Soc Limnol & Oceanog. *Res:* Ecology of marine plankton, especially energetics and population dynamics of zooplankton. *Mailing Add:* Marine Life Res Group Univ Calif San Diego A-018 La Jolla CA 92093-0218

MULLIN, ROBERT SPENCER, b Tazewell, Va, May 19, 12; m 38; c 2. PLANT PATHOLOGY. *Educ:* Hampden-Sydney Col, BS, 34; Va Polytech Inst, MS, 37; Univ Minn, PhD, 50. *Prof Exp:* Agt directing grain rust control, USDA, Va, 36-41, state leader, 41-44; assoc pathologist, Va Truck Exp Sta, Univ, Va, 45-46, plant pathologist & head dept, 48-58; assoc prof plant path, physiol & bot, Va Polytech Inst, 46-48; prof & plant pathologist, 58-77, EMER PROF & PLANT PATHOLOGIST, COOP EXTEN SERV, UNIV FLA, 77- *Concurrent Pos:* Consult plant path. *Mem:* Am Phytopath Soc; Sigma Xi. *Res:* Control of plant diseases, especially vegetable, ornamental and fruit crops. *Mailing Add:* 1132 SW 11 Ave Gainesville FL 32601

MULLIN, RONALD CLEVELAND, b Guelph, Ont, Aug 15, 36; m 71; c 2. MATHEMATICS. *Educ:* Univ Western Ont, BA, 59; Univ Waterloo, MA, 60, PhD(math), 64. *Prof Exp:* Lectr math, Univ Waterloo, 60-64, from asst prof to assoc prof, 64-68; prof, Fla Atlantic Univ, 68-69; assoc dean grad studies, 71-75, chmn dept combinatorics & optimization, 75-78, PROF MATH, UNIV WATERLOO, 69- *Mem:* Math Asn Am; Am Math Soc. *Res:* Cryptography; design theory. *Mailing Add:* Dept Combinatorics Optimization Univ Waterloo Waterloo ON N2L 3G1 Can

MULLIN, WILLIAM JESSE, b Brentwood, Mo, Dec 8, 34; m 61; c 3. THEORETICAL SOLID STATE PHYSICS. *Educ:* St Louis Univ, BS, 56; Washington Univ, PhD(theoret solid state physics), 65. *Prof Exp:* Res physicist, Aerospace Res Labs, Wright-Patterson AFB, Ohio, 64-65; res assoc physics, Univ Minn, Minneapolis, 65-67; from asst prof to assoc prof, 67-79, PROF THEORET SOLID STATE PHYSICS, UNIV MASS, AMHERST, 79- *Concurrent Pos:* Sci res coun fel, Univ Sussex, Eng, 73-74, 81-82; assoc prof, Ecole Normale Superirure, 88. *Mem:* Am Phys Soc. *Res:* Many-body theory; analysis of properties of quantum solids and liquids at low temperatures; polarized quantum fluids; author of one book. *Mailing Add:* Dept of Physics Hasbrouck Lab Univ of Mass Amherst MA 01003

MULLINAX, PERRY FRANKLIN, b Quebec, Que, June 7, 31; US citizen; m 57; c 2. RHEUMATOLOGY, IMMUNOLOGY. *Educ:* Duke Univ, BA, 51; Med Col Va, MD, 55; Am Bd Internal Med, dipl; Subspecialty Bd Rheumatol, dipl; Subspecialty Bd Allergy & Immunol, dipl; Subspecialty Bd Diag Lab Immunol, dipl. *Prof Exp:* Clin & res fel med, Mass Gen Hosp, 59-61; Helen Hay Whitney res fel, Sch Med, Washington Univ, 61-62; Helen Hay Whitney res fel, Mass Inst Technol, 62-63; asst prof, 63-67, asst dir, Clin Res Ctr, 70-75, assoc prof, 67-76, PROF MED, MED COL VA, 76- *Concurrent Pos:* Fel med, Harvard Med Sch, 59-61; res fel, Arthritis Found, 59-61. *Mem:* AAAS; Am Fedn Clin Res; Am Rheumatism Asn; fel Am Col Physicians. *Res:* Clinical immunology; immunochemistry; rheumatic diseases. *Mailing Add:* Dept Med Box 263 MCV Sta Med Col Va-Va Commonwealth Univ Richmond VA 23298

MULLINEAUX, DONAL RAY, b Weed, Calif, Feb 16, 25; m 51; c 3. GEOLOGY, VOLCANIC HAZARDS. *Educ:* Univ Wash, Seattle, BS, 47 & 49, MS, 50, PhD(geol), 61. *Prof Exp:* Field asst, 50, geologist, 50-88, EMER SCIENTIST, US GEOL SURV, 89- *Honors & Awards:* E B Burwell Award, Geol Soc Am. *Mem:* Geol Soc Am. *Res:* Geology of Puget Sound Basin; engineering geology; volcanic hazards in western United States and Hawaii; eruptive histories and products of Cascade Range volcanoes; Mount St Helens eruptions and effects. *Mailing Add:* US Geol Surv Mail Stop 903 Box 25046 Denver Fed Ctr Denver CO 80225

MULLINEAUX, RICHARD DENISON, b Portland, Ore, Feb 23, 23; m 47; c 2. ORGANIC CHEMISTRY. *Educ:* Univ Wash, Seattle, BS, 48; Univ Wis, PhD(org chem), 51. *Prof Exp:* Chemist, Shell Develop Co, 51-59, res supvr, 60-63, spec technologist, Wilmington Refinery, Shell Oil Co, 63-64, mgr, Aromatics Dept, Wood River Refinery, 65-66, asst mgr head off tech dept, 66-67, dir, Gen Sci Div, Shell Develop Co, Calif, 67-69, gen mgr mfg, transport & mkt, Shell Oil Co, 69-74, gen mgr mfg, transport & mkt-chem res & develop, 74, gen mgr res & develop prod, 75-78, gen mgr health safety & environ support, 78-83, ENVIRON CONSULT, SHELL OIL CO, 83- *Mem:* Am Chem Soc; NY Acad Sci; Am Asn Advan Sci. *Res:* Hydrocarbon chemistry; organo metallics; engine lubricants; petroleum refining; organic chemical products and processes. *Mailing Add:* 2 Hunters Br Houston TX 77024-4514

MULLINIX, KATHLEEN PATRICIA, b Boston, Mass, Mar 19, 44; m 66; c 3. BIOCHEMISTRY, MOLECULAR BIOLOGY. *Educ:* Trinity Col, DC, AB, 65; Columbia Univ, PhD(chem biol), 69. *Prof Exp:* NIH fel, Harvard Univ, 69-71, res assoc biol, 71-72; staff fel, Nat Inst Arthritis, Metab & Digestive Dis, NIH, 72-73, sr staff fel, 73-75, res chemist, Nat Cancer Inst, 75-79, asst dir intramural planning, 79-81; dep provost health sci, 81-82, vprovost, Columbia Univ, 82-87; SR VPRES NEUROGENETIC CORP, 87- *Mem:* AAAS; Am Chem Soc; Am Soc Microbiol; Am Soc Biol Chemists; Am Soc Cell Biol. *Res:* Regulation of gene expression; neurobiology; effects of steroid hormones on gene expression. *Mailing Add:* Neurogenetic Corp 215 Col Rd Paramus NJ 07652-1410

MULLINS, DAIL W, JR, b St Louis, Mo, Feb 9, 44; m 67; c 1. SCIENCE EDUCATION. *Educ:* Rhodes College, BA, 66; Memphis State Univ, MS, 69; Univ Ala, Birmingham, PhD(biochem), 75. *Prof Exp:* Lectr genetics, Dept Biol, Univ Ala, Birmingham, 74-75; res assoc biochem, Dept Biochem, Med Sch, Georgetown Univ, Washington, DC, 75-77; from res assoc to sr assoc, molecular biol, 77-81, sr res assoc biochem, 82-85, ASSOC PROF, SCH EDUC, UNIV ALA, BIRMINGHAM, 85- *Concurrent Pos:* Lectr genetics, Dept Biol, Univ Ala, Birmingham, 77-78, prin investr, Nat Cancer Inst grant, Dept Biochem, 81-83. *Mem:* Int Soc Study Orgin Life; Planetary Soc. *Res:* Orgin of life biochemistry; eukaryotic gene regulation. *Mailing Add:* 1300 28th S Birmingham AL 35294

MULLINS, DEBORRA E, ANGIOGENESIS, CELL BIOLOGY. *Educ:* Univ Fla, PhD(biochem & molecular biol), 79. *Prof Exp:* RES ASST PROF CELL BIOL, NY UNIV MED CTR, 84- *Res:* Mechanisms of cellular invasions. *Mailing Add:* Schering-Plough Corp B1-2 60 Orange St Bloomfield NJ 07003

MULLINS, DONALD EUGENE, b La Junta, Colo, Nov 2, 44; m 68; c 2. INSECT PHYSIOLOGY. *Educ:* Univ Colo, BA, 66; Colo State Univ, MS, 68; Va Polytech Inst & State Univ, PhD(entom), 71. *Prof Exp:* Lectr zool, Univ Western Ont, 71-73; from instr to assoc prof, 73-91, PROF ENTOM, VA POLYTECH INST & STATE UNIV, 91- *Mem:* Sigma Xi; AAAS; Entom Soc Am. *Res:* Physiology and biochemistry of nitrogen metabolism in insects as it relates to osmoregulation and excretion, particularly the role of stored urates; pesticide disposal using biological methods. *Mailing Add:* Dept Entom Va Polytech Inst & State Univ Blacksburg VA 24061

MULLINS, HENRY THOMAS, b Ghent, NY, Dec 14, 51; m 77. CARBONATE SEDIMENT, SEISMIC STATIGRAPHY. *Educ:* State Univ NY, Oneonta, BS, 73; Duke Univ, MS, 75; Univ NC, Chapel Hill, PhD(oceanog), 78. *Prof Exp:* Asst prof oceanog, Moss Landing Marine Labs, 78-; ASSOC PROF, SYRACUSE UNIV, NY. *Mem:* Am Asn Petrol Geol; AAAS; Geolog Soc Am; Sigma Xi; Soc Econ Paleontologists & Mineralogists. *Res:* Marine geology and geophysics; carbonate sedimentology; seismic stratigraphy; glacial later sedimentation. *Mailing Add:* Dept Geol Syracuse Univ Syracuse NY 13244

MULLINS, JAMES MICHAEL, b San Diego, Calif, Sept 22, 45; m 70; c 1. CELL BIOLOGY. *Educ:* Grinnell Col, BA, 68; Univ Tex, Austin, MA, 72, PhD(zool), 75. *Prof Exp:* Fel cell biol, Univ Colo, Boulder, 75-78; from asst prof to assoc prof, 78-89, PROF BIOL, CATH UNIV AM, 89- *Concurrent Pos:* NIH fel molecular, cellular & develop biol, Univ Colo, Boulder, 75-77. *Mem:* Am Soc Cell Biol; Bioelectromagnetics Soc. *Res:* Cell division; cellular motility; biological effects of electromagnetic fields. *Mailing Add:* Dept of Biol Cath Univ of Am Washington DC 20064

MULLINS, JEANETTE SOMERVILLE, b Salem, Ohio, Aug 1, 32; m; c 2. BOTANY-PHYTOPATHOLOGY. *Educ:* Wayne State Univ, BA, 55, MS, 62; NDak State Univ, PhD(bot), 75. *Prof Exp:* Sr bacteriologist, Henry Ford Hosp, Detroit, 55-66; bacteriologist, Providence Hosp, Southfield, Mich, 68; plant physiologist & res asst, Metab & Radiation Res Lab, Agr Res Serv, USDA, Fargo, NDak, 69-74; asst prof biol sci, 75-80, ASSOC PROF BIOL, CALIFORNIA STATE COL, PA, 80- *Concurrent Pos:* Cur, John Franklin Lewis Herbarium. *Mem:* AAAS; Am Inst Biol Sci; Am Soc Plant Taxonomists; Sigma Xi; Bot Soc Am. *Res:* Application of learning theory in botanical teaching: biosystematics of liatris scanosa complex; effects of surface applied herbicides on phyllpplane bacteria. *Mailing Add:* Dept Biol Sci Calif Univ Pa California PA 15419-0653

MULLINS, JOHN A, b Philadelphia, Pa, Feb 16, 31. CHEMISTRY. *Educ:* Univ Pa, BS, 58, PhD(chem), 64. *Prof Exp:* Fel photochem, Brandeis Univ, 64-66; lectr chem, Bucknell Univ, 66-67; asst prof, 67-74, assoc prof, 74-80, PROF NATURAL SCI, MICH STATE UNIV, 80- *Res:* Biophysical chemistry; history and philosophy of science. *Mailing Add:* Dept Natural Sci Mich State Univ 100 N Kedzie East Lansing MI 48824

MULLINS, JOHN THOMAS, b Richmond, Va, Nov 18, 32; m 55; c 3. BOTANY. *Educ:* Univ Richmond, BS, 55, MS, 57; Univ NC, Chapel Hill, PhD(bot), 60. *Prof Exp:* Asst prof bot & biol sci, Univ Fla, 59-64; NIH spec res fel, Harvard Univ, 64-65; assoc prof, 65-73, assoc chmn dept, 75-80, PROF BOT, UNIV FLA, 73- *Honors & Awards:* Nat Res Award, NIH, 88. *Mem:* AAAS; Bot Soc Am; Mycol Soc Am; Sigma Xi; Am Soc Plant Physiologist. *Res:* Regulatory mechanisms in hormonal control of sexual morphogenesis in fungi. *Mailing Add:* Dept Bot Univ Fla Gainesville FL 32611-2009

MULLINS, JOSEPH CHESTER, b Thomaston, Ga, Dec 5, 31; m 58; c 3. CHEMICAL ENGINEERING. *Educ:* Ga Inst Technol, BS, 54, MS, 60, PhD(chem eng), 65. *Prof Exp:* Res asst micromeritics, Eng Exp Sta, Ga Inst Technol, 58-62, asst res engr, 62-64; from asst prof to assoc prof, 65-77, PROF CHEM ENG, CLEMSON UNIV, 77- *Mem:* Am Inst Chem Engrs. *Res:* Thermodynamics and cryogenic engineering. *Mailing Add:* Dept Chem Eng Clemson Univ Clemson SC 29634-0909

MULLINS, LAWRENCE J, b New York, NY, Nov 7, 21; m 46; c 4. PHYSICAL CHEMISTRY. *Educ:* Queen's Col, NY, BS, 43; Univ NMex, PhD(chem), 57. *Prof Exp:* Staff mem, Los Alamos Sci Lab, 46-83; EXEC DIR, TRU ENG CO INC, 84- *Concurrent Pos:* Pvt consult, 84- *Honors & Awards:* Glen T Seaborg Actinide Separation Award, 88. *Mem:* Am Chem Soc; Am Nuclear Soc. *Res:* Extractive metallurgy of plutonium, plutonium chemistry and metallurgy; electrochemistry and electrorefining of plutonium metals; thermodynamic properties of nuclear materials; plutonium fuel cycles; fused salt chemistry; high temperature chemistry of plutonium, uranium, americum and rare earths; plutonium 238 heat sources. *Mailing Add:* 505 Rover Blvd Los Alamos NM 87544

MULLINS, LORIN JOHN, b San Francisco, Calif, Sept 23, 17; m 46; c 2. BIOPHYSICS. *Educ:* Univ Calif, BS, 37, PhD(biophys), 40. *Prof Exp:* Asst, Univ Calif, 38-40; asst physiol, Sch Med & Dent, Univ Rochester, 40-41, instr, 41-43; res assoc, Med Sch, Wayne State Univ, 46; Am-Scand Found fel, Inst Theoret Physics, Copenhagen, 47-48; Nat Res Coun Merck fel, Zool Sta, Naples & Johnson Res Found, Sch Med, Univ Pa, 48-49; Nat Res Coun Merck fel biophys, Johns Hopkins Univ, 49-50; assoc prof biol sci, Purdue Univ, 50-58; PROF BIOPHYS & HEAD DEPT, SCH MED, UNIV MD, BALTIMORE CITY, 59- *Concurrent Pos:* Mem Corp Bermuda Biol Sta, 51-; mem, Marine Biol Lab, Woods Hole, Mass, 56-; USPHS fel, Zoophysiol Lab, Univ Copenhagen, 56-57; mem bd sci counr, Nat Inst Neurol Dis & Stroke, 69-73; chmn, J Neurosci Res, 75-; ed, Ann Rev Biophys & Bioeng, 72-82. *Mem:* Am Physiol Soc; Biophys Soc; Am Chem Soc; Soc Gen Physiol; Sigma Xi. *Res:* Permeability of cells to ions; applications of radioisotopes to biological problems; active transport of ions; modes of anesthetic action. *Mailing Add:* Dept Biophysics Univ Md Baltimore MD 21201

MULLINS, RICHARD JAMES, MICROCIRCULATION, PLASMA VOLUME CONTROL. *Educ:* Tufts Univ, MD, 74. *Prof Exp:* ASST PROF SURG, SCH MED, UNIV LOUISVILLE, 83- *Res:* Microvascular permeability. *Mailing Add:* Sch Med Univ Louisville 530 S Jackson St Sch Med Univ Louisville Louisville KY 40292

MULLINS, ROBERT EMMET, b New York, NY, Sept 24, 37; m 64; c 2. MATHEMATICAL ANALYSIS. *Educ:* Iona Col, BS, 58; Univ Notre Dame, MS, 60; Northwestern Univ, PhD(function algebras), 65. *Prof Exp:* Instr, 64-65, asst prof, 65-70, ASSOC PROF MATH, MARQUETTE UNIV, 70- *Mem:* Am Math Soc. *Res:* Algebras of functions. *Mailing Add:* Dept Math Marquette Univ 1515 W Wisconsin Ave Milwaukee WI 53233

MULLINS, WILLIAM WILSON, b Boonville, Ind, Mar 5, 27; m 48; c 4. PHYSICS, METALLURGY. *Educ:* Univ Chicago, MS, 51, PhD(physics), 55. *Prof Exp:* Res physicist, Res Labs, Westinghouse Elec Corp, 55-59, adv physicist, 59-60; assoc prof metall eng, Carnegie Inst Technol, 60-63; prof & head dept, 63-66, dean, 66-70; dir, Ctr Joining Mat, 81-85, UNIV PROF APPL SCI, CARNEGIE-MELLON UNIV, 70- *Honors & Awards:* Mathewson Gold Medal, Am Inst Mining, Metall & Petrol Engrs, 63; Philip M McKenna Mem Award, 81. *Mem:* Nat Acad Sci; Am Phys Soc; Am Inst Mining, Metall & Petrol Engrs. *Res:* Metallic surfaces and interfaces; physical metallurgy; statistical mechanics of alloys; morphology of solid state transformations; diffusion; defect structures in crystalline lattices; particle flow and soil mechanics. *Mailing Add:* Dept Metall Eng & Metal Sci Carnegie-Mellon Univ Pittsburgh PA 15213

MULLISON, WENDELL ROXBY, b Philadelphia, Pa, Sept 24, 13. PLANT PHYSIOLOGY. *Educ:* Univ NMex, BA, 34; Univ Chicago, PhD(plant physiol), 38. *Prof Exp:* Instr biol, Purdue Univ, 40-44; plant physiologist olericulturist, Curacaosche Petrol Indust Maatschappij, Neth WIndies, 44-46; plant physiologist, Dow Chem Co, 45-50, asst tech dir in chg agr chem, Dow Chem Int, Ltd, 50-59, prod mgr, 59-62, dir info serv, Bioprod Dept, 62-65, mgr govt contract res & develop, 66-72, registr specialist, 72-79; HERBISIST CONSULT, 79- *Concurrent Pos:* Consult agr chem (herbicide specialist), 79- *Mem:* Am Soc Plant Physiol; Bot Soc Am; Weed Sci Soc Am; Soc Tropicology; Am Chem Soc; Soc Environ Tropicology & Chem. *Res:* Plant nutrition and hormones; herbicides. *Mailing Add:* 1412 N Parkway Midland MI 48640

MULLONEY, BRIAN, b Pittsfield, Mass, Feb 21, 42. ZOOLOGY. *Educ:* McGill Univ, BSc, 63; Univ Calif, MA, 66, PhD(invert zool), 69. *Prof Exp:* Postdoctoral fel zool, Univ Oxford, UK, 69-70; NIH postdoctoral fel biol, Univ Calif, San Diego, 70-72, asst res biologist, 73-74; from asst prof to assoc prof zool, 74-82, dir, Cellular Neurobiol Training Prog, 86-91, PROF ZOOL, UNIV CALIF, DAVIS, 82- *Concurrent Pos:* NIH & NSF res grants, 75-94; A P Sloan Found res fel, 75-77; fel, Neurosci Res Prog Intensive Study Prog, 77; Alexander von Humboldt res fel, Europ Molecular Biol Lab, Heidelberg, 80-81; mem, Neurobiol Rev Panel, NSF & Neurobiol Study Sect, USPHS-NIH, 83, Neurobiol Ad Hoc Panel, 90. *Mem:* Fel AAAS; Soc Exp Biol; Soc Neurosci; Am Physiol Soc; Am Soc Zoologists; Int Soc Neuroethology. *Res:* Cellular neurobiology. *Mailing Add:* Dept Zool Univ Calif Davis CA 95616

MULLOOLY, JOHN P, b Manhattan, NY, July 8, 37; m 69. BIOSTATISTICS, BIOMATHEMATICS. *Educ:* St Francis Col, BS, 59; Mich State Univ, MS, 61; Cath Univ Am, PhD(math statist), 66. *Prof Exp:* Math statistician, NIH, Md, 66-68; prof statist, Ore State Univ, 68-73; SR BIOSTATISTICIAN, KAISER FOUND HOSP, HEALTH SERV RES CTR, PORTLAND, ORE, 73- *Mem:* Biomet Soc; Am Statist Asn; Am Pub Health Asn. *Res:* Epidemiology, medical care and health services research; statistical methods in epidemiology; statistical inference; applied stochastic processes. *Mailing Add:* 4402 SE Flavel Portland OR 97215

MULRENNAN, CECILIA AGNES, b Everett, Mass, Aug 4, 25. BIOLOGY. *Educ:* Regis Col, Mass, AB, 46; Fordham Univ, MA, 57, PhD(genetics), 59. *Prof Exp:* Instr biol, 59-63, PROF BIOL DEPT, REGIS COL, MASS, 63- *Concurrent Pos:* NIH res grant, 60-61; Grass Found res grant, 68-71; extra mural assoc prog, NIH, 87. *Mem:* Sigma Xi. *Res:* Genetics of Drosophila; philosophy of science. *Mailing Add:* Dept Biol Regis Col 235 Wellesley St Weston MA 02193

MULRENNAN, JOHN ANDREW, JR, b Tallahassee, Fla, Mar 2, 34; m 59; c 3. MEDICAL ENTOMOLOGY, PARASITOLOGY. *Educ:* Univ Fla, BS, 57, MS, 59; Okla State Univ, PhD(entom & parasitol), 68. *Prof Exp:* US Navy, 59-, officer-in-chg, Dis Vector Ecol & Control Ctr, Alameda, Calif, 73-76, head vector control sect, Bur Med & Surg, 76-79, MED ENTOMOLOGIST, US NAVY, 59-; DIR OFF ENTOM, FLA DEPT HEALTH & REHAB SERV, 79- *Concurrent Pos:* Chmn, Armed Forces Pest Control Bd, 77-78. *Mem:* Am Mosquito Control Asn. *Res:* Mosquito control; cockroach control; toxicology of pesticides. *Mailing Add:* 8523 Goldeneye Ln Jacksonville FL 32217

MULROW, PATRICK J, b New York, NY, Dec 16, 26; m 53; c 4. MEDICINE. *Educ:* Colgate Univ, AB, 47; Cornell Univ, MD, 51; Am Bd Internal Med, dipl, 58, recert, 74; Endocrinol & Metab Spec Bd, dipl, 77. *Hon Degrees:* MSc, Yale Univ, 69. *Prof Exp:* Instr physiol, Med Col, Cornell Univ, 54-55; from instr to assoc prof med, Sch Med, Yale Univ, 57-69, chief, Endocrine Sect, 66-75, prof internal med, 69-75; PROF MED & CHMN DEPT, MED COL OHIO, 75- *Concurrent Pos:* USPHS res fel, 54-56, res grant, 57-66; Arthritis Res Found res fel, 56-57; clin investr, Vet Admin Hosp, West Haven, Conn, 58-61; attend, Yale-New Haven Hosp, 68-75; mem study sect, NIH, 70-74. *Mem:* Am Soc Clin Invest; Am Physiol Soc; Endocrine Soc; Am Fedn Clin Res; Asn Am Physicians. *Res:* Hypertension and endocrinology. *Mailing Add:* Dept of Med C S 10008 Toledo OH 43699

MULROY, JULIANA CATHERINE, b Pomona, Calif, June 12, 48. PLANT ECOLOGY, BOTANY. *Educ:* Pomona Col, BA, 70; Duke Univ, AM, 72, PhD(plant ecol), 79. *Prof Exp:* Res asst plant ecol, Duke Univ, 70-74, teaching asst bacteriol, 75-76, technician photobiol, 76-77; instr, 77-79, ASST PROF BIOL, DENISON UNIV, 79- *Mem:* Am Inst Biol Sci; Ecol Soc Am; Brit Ecol Soc; Am Asn Plant Taxonomists; Sigma Xi. *Mailing Add:* Dept of Biol Denison Univ Granville OH 43023

MULROY, MICHAEL JOSEPH, b Wyandotte, Mich, July 26, 31; m 65; c 2. ANATOMY. *Educ:* Our Lady of the Forest Sem, AB, 57; DePaul Univ, MS, 60; Univ Calif, San Francisco, PhD(anat), 68. *Prof Exp:* NIH fel auditory physiol, Harvard Med Sch-Mass Inst Technol, 68-70; instr anat, Harvard Med Sch, 72-74; asst prof, 74-77, assoc prof anat, Med Ctr, Univ Mass, 77-80; res assoc otolaryngol, Mass Eye & Ear Infirmary, Eaton Peabody Lab, 71-80; assoc prof, 80-86, PROF ANAT, MED COL GA, AUGUSTA, 86- *Concurrent Pos:* Teaching fel gross anat, Harvard Med Sch, 71- *Mem:* Am Asn Anatomists; Soc Neurosci; Am Soc Cell Biol; AAAS. *Res:* Structure and function of the inner ear; noise-induced deafness; congenital deafness; cardiac arrhythmias. *Mailing Add:* Sch Med Med Col Ga 1120 15th St Augusta GA 30912

MULROY, THOMAS WILKINSON, b Pomona, Calif, Sept 1, 46. PLANT ECOLOGY & TAXONOMY. *Educ:* Pomona Col, BA, 68; Univ Ariz, MS, 71; Univ Calif, Irvine, PhD(ecol & evolutionary biol), 76. *Prof Exp:* Instr bot, Pomona Col, 73-76, asst prof, 76-77; mgr natural sci & sr ecologist, sci div, Henningson, Durham & Richardson, 77-; AT URS CO. *Mem:* Ecol Soc Am; Am Soc Naturalists; Am Inst Biol Sci; AAAS; Sigma Xi. *Res:* Plant biogeography, evolution and systematics; desert ecology; ecology and evolution of succulent plants; systematics and ecology of Dudleya; adaptive morphology of plants; environmental impact analysis in terrestrial environments. *Mailing Add:* URS Co 1421 Chapala St Santa Barbara CA 93101-3014

MULSON, JOSEPH F, b Milwaukee, Wis, Feb 6, 29; m 49; c 2. ELECTRON PHYSICS. *Educ:* Rollins Col, BS, 56; Pa State Univ, MS, 61, PhD(physics), 63. *Prof Exp:* From asst prof to assoc prof physics, 63-73, chmn sci div, 70-72, PROF PHYSICS, ROLLINS COL, 73- *Concurrent Pos:* Cotrell grant, 64-65; mem, Nat Sci Stud Res Partic, 65-66. *Res:* Holography and laser applications. *Mailing Add:* 2635 N Seneca Pt Crystal River FL 32629

MULTER, H GRAY, b Syracuse, NY, July 7, 26; m 50; c 2. GEOLOGY. *Educ:* Syracuse Univ, AB, 49, MS, 51; Ohio State Univ, PhD(geol), 55. *Prof Exp:* Petrol geologist, Tex Co, Calif, 51-53; prof geol, Col Wooster, 55-69; dir, West Indies Lab, 69-75, chmn dept earth sci, 75-85, prof marine geol, 80-85, EMER PROF MARINE GEOL, FAIRLEIGH DICKINSON UNIV, 85- *Concurrent Pos:* Consult geologist, 80-; Fulbright prof, Marburg, WGer, 85-86. *Mem:* Geol Soc Am; Nat Asn Geol Teachers; Soc Econ Geologists & Mineralogists; Am Asn Prof Geologists. *Res:* Sedimentation; marine geology; enrivonmental geology. *Mailing Add:* 9855 Canaseraqa Rd Arkport NY 14807

MULUKUTLA, SARMA SREERAMA, b Andhra Pradesh, India, Nov 26, 38; m 56; c 3. NUMERICAL METHODS, COMPUTER-AIDED ANALYSIS & DESIGN. *Educ:* Banaras Hindu Univ, India, BSc, 58, MSc, 59; Univ Colo, Boulder, PhD(elec eng), 68. *Prof Exp:* Pool-officer, Indian Inst Technol, Kharagpur, India, 61-62, lectr elec eng, Madras, India, 62-64; res assoc elec eng, Univ Colo, Boulder, 64-68; asst prof, WVa Univ, Morgantown, 69-71; prof, Banaras Hindu Univ, India, 71-73; PROF ELEC ENG & DIR POWER-SYSTS ENG, NORTHEASTERN UNIV, BOSTON, 74- *Concurrent Pos:* Consult engr, Gen Elec Co, 66-81, McGraw Edison Co & Allegheny Power Serv Corp, 70-71, Marathon Elec Power Co, 76-78, Cambion Corp, 76-79, Ferrofluidics Corp, 82-84, New Eng Elec, 85-86 & Magnetic Res Int Corp, 86-; vis asst prof elec eng, Univ Colo, Boulder, 68-69; vis prof, Univ Iowa, Iowa City, 73-74; mem, Rotating Mach Comt & Power Eng Educ Comt, Inst Elec & Electronics Engrs. *Honors & Awards:* Meritorious Serv Award, Inst Elec & Electronics Engrs, Soc Computer Simulation, 85. *Mem:* Sr mem Inst Elec & Electronics Engrs. *Res:* Computer-aided analysis of three dimensional nonlinear electromagnetic field problems as applied to the design of electrical machinery; control systems; electromagnetic fields; energy conversion; power systems; author of numerous technical publications. *Mailing Add:* 12 Gannon Terr Framingham MA 01701

MULVANEY, JAMES EDWARD, b Brooklyn, NY, Aug 4, 29; m 52; c 4. ORGANIC CHEMISTRY. *Educ:* Polytech Inst Brooklyn, BS, 51, PhD(chem), 59. *Prof Exp:* Res chemist gen chem div, Allied Chem & Dye Corp, 51-53; asst, Polytech Inst Brooklyn, 55-59; res assoc chem, Univ Ill, 59-61; from asst prof to assoc prof, 61-71, PROF CHEM, UNIV ARIZ, 71- *Concurrent Pos:* Consult, Gen Motors, Arco, Am Optical, Allied Signal, IMI-Tech, Kodak & several law firms. *Mem:* AAAS; Am Chem Soc; Royal Soc Chem. *Res:* Organic synthesis; synthesis and mechanism of high polymer formation. *Mailing Add:* Dept Chem Univ Ariz Tucson AZ 85721

MULVANEY, THOMAS RICHARD, b Bellevue, Mich, July 4, 33; m 54; c 4. FOOD SCIENCE. *Educ:* Mich State Univ, BS, 56, MS, 59, PhD(food sci), 62. *Prof Exp:* Asst food eng, Mich State Univ, 56-59, asst food sci, 59-61, asst instr, 61-62, asst, 62; res engr, Alcoa Res Labs, Aluminum Co Am, Pa, 62-67, sr res scientist, 67-68; assoc prof food sci & technol, Univ Mass, Amherst, 68-71; sr scientist, 71-72, chief, Food Processing Sect, 72-84, spec asst to div dir food preserv process technol, 84-86, SCI ADV PROCESSING, FOOD & DRUG ADMIN, 86- *Mem:* AAAS; Am Chem Soc; Inst Food Technologists; fel Am Inst Chemists. *Res:* Thermal processing of foods and beverages; effect of sequestering agents on metals; chemical and physical changes in foods induced by packaging materials; nature of metallic flavors; consumer protection. *Mailing Add:* 8307 Forrester Blvd Springfield VA 22152

MULVEY, DENNIS MICHAEL, b Lockport, NY, Nov 17, 38; m 83; c 2. ORGANIC CHEMISTRY. *Educ:* Univ Pa, AB, 60; State Univ NY Buffalo, PhD(org chem), 65. *Prof Exp:* Res assoc, Columbia Univ, 64-65; sr res chemist, Merck Sharp & Dohme Res Labs, Rahway, 65-73, res fel, 73-77; group leader, Ortho Pharmaceut Corp, Raritan, 77-84; consult, 84-85; polymer design specialist, Pernwalt Corp, King of Prussia, Pa, 85-88; gen mgr, fire chem, Vega Biotech, Tucson, Ariz, 88-89; MGR, PEPTIDE MFG, PROCYTE CORP, KIRKLAND, WASH, 89- *Mem:* Am Chem Soc; Am Peptide Soc. *Res:* Heterocyclic compounds; reaction mechanisms; synthesis of nonclassical aromatic systems; photochemistry; chemistry of natural products; peptide synthesis/production; molecular modeling. *Mailing Add:* 18932 194th Ave NE Woodinville WA 98072

MULVEY, JAMES PATRICK, b New York, NY, June 23, 47; m; c 1. EXPERT SYSTEMS, KNOWLEDGE ENGINEERING. *Educ:* Western New Eng Col, BS, 75; Univ Ky, MS, 84. *Prof Exp:* Br chief supvr GS-13 TELIS, Dept Army, Blue Grass Army Depot, 75-81; grad asst fortran scientist & engr, Univ Ky, 81-83; tech staff CIM, Advan Mfg Technol, Hughes Aircraft, 84-86; CHIEF ENGR RES & DEVELOP (AT), GEN INFERENCE, 86- *Concurrent Pos:* Consult TMDE, USA Cent Test Equip Act, Dept Army, 77-81; rep, Dept Defense, Joint Logistics Comdr, ATE, 80-81 & US Govt, Indust Data Exchange Prog, 80-81; bd mem, Pantano Rotary Club, Tucson, Ariz, 89-90. *Res:* Artificial intelligence; expert systems; natural language; machine learning; neural networks; complete and incomplete knowledge inference techniques and representation; knowledge engineering and expert system implementation. *Mailing Add:* Gen Inference 4000 S Silverbeech Ave Tucson AZ 85730

MULVEY, MARGARET, b Waterbury, Conn, Apr 12, 52; c 2. POPULATION BIOLOGY. *Educ:* Univ Conn, BA, 74, MS, 77; Rutgers Univ, PhD(zool), 81. *Prof Exp:* VIS RESEARCHER, SAVANNAH RIVER ECOL LAB, 81-; ADJ ASST PROF, WAKE FOREST UNIV, 86- *Concurrent Pos:* Instr, Univ SC, Aiken. *Mem:* Am Genetic Soc; Genetics Soc Am; Soc Study Evolution; Am Malacological Union. *Res:* Population genetics; evolutionary aspects of host-parasite relationships. *Mailing Add:* Drawer E Savannah River Ecol Lab Aiken SC 29801

MULVEY, PHILIP FRANCIS, JR, b Worcester, Mass, Dec 22, 31; m 55; c 4. RADIOBIOLOGY, PHYSIOLOGY. *Educ:* Clark Univ, AB, 53; Bowling Green State Univ, MA, 55; Univ Buffalo, PhD(biol), 59; Suffolk Univ Sch Law, JD, 74. *Prof Exp:* Asst biol, Bowling Green State Univ, 54-55; res biochemist radioisotope serv, Vet Admin Hosp, 58-65; res physiologist, Electronics Res Ctr, NASA, Mass, 66; res physiologist, US Army Natick Labs, 66-68; lectr, 59-68, prof biol, Suffolk Inst, 68-80; PVT LAW PRACT, 77- *Concurrent Pos:* Asst, Sch Med, Boston Univ, 59-66. *Mem:* AAAS; Am Soc Zoologists; Am Physiol Soc. *Res:* Use of radioisotopes and antithyroid agents to study thyroid gland physiology; environmental law and management; use of activation analysis to determine trace element concentrations in biological systems. *Mailing Add:* 65 Paul Revere Rd Needham MA 02194

MULVIHILL, JOHN JOSEPH, b Washington, DC, Aug 20, 43; m 66; c 3. GENETICS, PEDIATRICS. *Educ:* Col of the Holy Cross, BS, 65; Dartmouth Med Sch, BMS, 67; Univ Wash, MD, 69. *Prof Exp:* Staff assoc epidemiol, 70-74, CHIEF CLIN GENETICS SECT, CLIN EPIDEMIOL BR, NAT CANCER INST, 74- *Concurrent Pos:* Mem, Comt Biol Effects Ionizing Radiation, Nat Acad Sci, 70-73; fel pediat, Sch Med, Johns Hopkins Univ, 72-74. *Mem:* Teratology Soc; Soc Pediat Res; Am Acad Pediat; Am Soc Human Genetics. *Res:* Epidemiology of cancer and genetic disorders defects in man, especially cancer families; animal models of congenital and genetic disease; medical genetics. *Mailing Add:* Univ Pittsburgh Pittsburgh PA 15261

MULVIHILL, MARY LOU JOLIE, b Chicago, Ill, Sept 28, 28. HUMAN PHYSIOLOGY. *Educ:* St Xavier Col, Ill, BA, 60; Purdue Univ, PhD(physiol), 67. *Prof Exp:* From asst prof to assoc prof biol, St Xavier Col, Ill, 67-72, chmn div natural sci, 69-72, vpres student affairs, 70-72; from assoc prof to prof biol, Harper Col, 72-88; RETIRED. *Concurrent Pos:* Consult physiol, W C Brown, Harper & Row & Scott Foresman, 73-87. *Mem:* Am Inst Biol Sci. *Res:* Effect of a vitamin A deficiency on the ultrastructure of the mosquito eye; human anatomy and disease. *Mailing Add:* 2121 S Meacham Rd Palatine IL 60067

MULY, EMIL CHRISTOPHER, JR, b Baltimore, Md, Mar 24, 34; m 60; c 3. TECHNICAL MANAGEMENT, RESEARCH ADMINISTRATION. *Educ:* Johns Hopkins Univ, BES, 56; Northwestern Univ, MS, 58, PhD(elec eng), 62. *Prof Exp:* Scientist electronics div, Martin Co, Md, 62-64; assoc prof elec eng, George Washington Univ, 64-65; asst dir res, Nat Res Corp, 65-72; prog mgr, Leeds & Northrup Co, 72-83; vpres oper, Intec Corp, 83-85; DIR, CONTROL ENG CTR, UNIV TENN, 85- *Mem:* Inst Elec & Electronics Engrs; Am Vacuum Soc; Soc Mfg Engrs; Soc Appl Spectros; Instrument Soc Am. *Res:* Research focuses on the theoretical and applied developments in the measurement and control fields; electron optics. *Mailing Add:* Control Eng Ctr Univ Tenn 101 Perkins Hall Knoxville TN 37996-2000

MUMA, MARTIN HAMMOND, b Topeka, Kans, July 24, 16; m 40; c 6. ARACHNOLOGY, DESERT ECOLOGY. *Educ:* Univ Md, BS, 39, MS, 40, PhD(entom), 43. *Prof Exp:* Lab asst, USDA, Md, 37-38; asst, Univ Md, 41-43, instr entom, 43-44, asst entomologist, 44-45; assoc exten entomologist, Univ Nebr, 45-48, assoc prof entom & assoc entomologist, 48-51; from assoc prof entom & assoc entomologist to prof & entomologist, 51-71, EMER PROF ENTOM & EMER ENTOMOLOGIST, CITRUS EXP STA, UNIV FLA, 71-; RES ASSOC, DIV PLANT INDUST, FLA DEPT AGR & CONSUMER SERV & WESTERN NMEX UNIV, 71- *Mem:* Am Arachnol Soc. *Res:* Taxonomy, systematics, biology, behavior and ecology of arid-land arachnids. *Mailing Add:* PO Box 135 San Simeon AZ 85632

MUMA, NANCY A, b Baltimore, Md, June 21, 58. NEUROPATHOLOGY. *Educ:* Western Md Col, BA, 79; Univ Louisville, MA, 81, PhD, 85. *Prof Exp:* Polysomnograph technician, Baltimore Regional Sleep Dis Ctr, Baltimore City Hosps & Johns Hopkins Univ, 79-80; instr, Dept Psychol, Univ Louisville, 82-83, res fel, 85; res fel, Dept Environ Health Sci, Sch Hyg & Pub Health, Johns Hopkins Univ, 85-86 & Dept Path, Neuropath Lab, Sch Med, 86-88, instr, 89-90, ASST PROF, DEPT PATH, NEUROPATH LAB, SCH MED & DEPT ENVIRON HEALTH SCI, SCH HYG & PUB HEALTH, JOHNS HOPKINS UNVI, 90- *Concurrent Pos:* Nat Res Serv Award, 86-88; Nat Inst Aging scholar Alzheimer's Dis, 89- *Mem:* Am Soc Pharmacol & Exp Therapeut; Soc Neurosci. *Res:* Author of numerous technical publications. *Mailing Add:* Neuropath Lab Sch Med Johns Hopkins Univ 509 Path Bldg 600 N Wolfe St Baltimore MD 21205-2181

MUMBACH, NORBERT R, b Buffalo, NY, June 6, 20; m 48; c 5. APPLIED CHEMISTRY. *Educ:* Canisius Col, BS, 42. *Prof Exp:* Instr phys chem, Canisius Col, 42-43; instr chem oper, Lake Ont Ord Works, 43; analytical chemist, Manhattan Proj, Linde Air Prods Co, 43-46, res chemist, 46-52, mass spectrometrist, Linde Div, Union Carbide Corp, 52-56, spectroscopist, 59-61, res chemist, Linde Div, 61-77; RETIRED. *Mem:* Am Chem Soc; Sigma Xi. *Res:* Synthetic gemstones; hydrothermal methods; new silicate phases created at high pressures and temperatures. *Mailing Add:* Genessee Rd East Concord NY 14055

MUMFORD, DAVID BRYANT, b Worth, Eng, June 11, 37; m 59; c 4. ALGEBRAIC GEOMETRY. *Educ:* Harvard Univ, BA, 57, PhD, 61. *Hon Degrees:* DSc, Univ Warwick, 83. *Prof Exp:* Jr fel, 58-61, from asst prof to prof, 62-77, chmn dept, 81-84, HIGGINS PROF MATH, HARVARD UNIV, 77- *Concurrent Pos:* Vis prof, Tata Inst Fundamental Res, 67-68, 78-79, Inst Higher Sci studies, 76-77; Nuffield prof, Univ Warwick, 70-71; Mac Arthur fel, 87-92; vpres, Int Math Union, 90-94. *Honors & Awards:* Fields Medal, Int Cong Mathematicians, 74. *Mem:* Nat Acad Sci; Am Math Soc; Inst Elec & Electronics Engrs. *Res:* Geometric invariant theory; author of several journals and books. *Mailing Add:* Dept Math Harvard Univ Cambridge MA 02138

MUMFORD, DAVID LOUIS, b Salt Lake City, Utah, May 2, 32; m 55; c 5. PLANT PATHOLOGY. *Educ:* Brigham Young Univ, BS, 56, MS, 58; Univ Minn, PhD(plant path), 62. *Prof Exp:* Plant pathologist, USDA, Mich State Univ, 63-67; PLANT PATHOLOGIST, USDA, CROPS RES LAB, UTAH STATE UNIV, 67- *Mem:* Am Phytopath Soc. *Res:* Virus diseases of sugar beets; disease resistance in sugar beets. *Mailing Add:* 1285 E 1500 N Logan UT 84321

MUMFORD, GEORGE, prosthodontics; deceased, see previous edition for last biography

MUMFORD, GEORGE SALTONSTALL, b Milton, Mass, Nov 13, 28; m 49; c 4. ASTRONOMY EDUCATION. *Educ:* Harvard Univ, AB, 50; Ind Univ, MA, 52; Univ Va, PhD(astron), 55. *Prof Exp:* Instr math, Randolph-Macon Woman's Col, 52-53; instr, 55-56, from asst prof to assoc prof, 56-68, dean, Col Liberal Arts, 69-79, dean, Grad Sch Arts & Sci, 80-86, PROF MATH, TUFTS UNIV, 68- *Concurrent Pos:* Vis astronr, Kitt Peak Nat Observ, 62-79;

actg dir, Wright Prog Innovation Sci Educ. *Mem:* AAAS; Am Astron Soc; Am Asn Variable Star Observers; Am Phys Soc; Int Astron Union. *Res:* Cataclysmic variable stars; close binaries; history of astronomy. *Mailing Add:* Dept Physics & Astron Tufts Univ Medford MA 02155

MUMFORD, RUSSELL EUGENE, b Casey, Ill, May 26, 22; m 47; c 3. ANIMAL ECOLOGY. *Educ:* Purdue Univ, BS, 48, MS, 52, PhD, 61. *Prof Exp:* Res biologist, State Dept Conserv, Ind, 48-50; teacher natural hist, Fla Audubon Soc, 50-51; res biologist, State Dept Conserv, Ind, 52-55; asst mus zool, Univ Mich, 55-57; PROF VERT NATURAL HIST, PURDUE UNIV, LAFAYETTE, 58- *Mem:* Am Soc Mammal; Wilson Ornith Soc; Am Ornith Union. *Res:* Life history and distribution of birds and mammals; bat banding. *Mailing Add:* 2244 Huron Rd West Lafayette IN 47907

MUMFORD, WILLARD R, b McMinnville, Ore, Aug 1, 33; m 55; c 3. CURRICULUM DEVELOPMENT & SYSTEMS ENGINEERING TECHNOLOGIES PROGRAM, ARTICULATION AGREEMENTS. *Educ:* Univ Md, BS, 56; Southern Methodist Univ, BSME, 65; Tex A&M Univ, MSME, 69. *Prof Exp:* Sect chief, Ballistic Missile Flight Anal Sect, Ogden Air Logistics Ctr, Utah, 70-72, br chief, Reliability & Value Eng Br, Serv Eng Div, 72-73, div chief, Logistics Systs Mgt Div, 73-76; prof aerospace studies & detachment comdr, Angelo State Univ, San Angelo, Tex, 76-79; dep comdr, Electromagnetic Compatibility Anal Ctr, Annapolis, Md, 79-81; CHMN, ENG & TECHNOLOGIES DIV, ANNE ARUNDEL COMMUNITY COL, 81- *Concurrent Pos:* Consult, NJ Dept Higher Educ, 86-87 & NSF, 91; founder & chair, Md Coun Eng & Technologies, 89-; mem, Task Force High Tech Careers, Md Higher Educ Comn, 90-91. *Honors & Awards:* George Washington Honor Medal, Freedoms Found, 78. *Mem:* Am Soc Eng Educ. *Res:* Microstructure of titanium alloys; curriculum development in engineering technologies using the Dacum process; developing a model for tech prep education. *Mailing Add:* Eng & Technologies Div Anne Arundel Community Col 101 College Pkwy Arnold MD 21012

MUMM, ROBERT FRANKLIN, b Urbana, Ill, Oct 16, 35. BIOMETRICS, GENETICS. *Educ:* Univ Ill, BS, 57, MS, 58; Univ Nebr, PhD(quant genetics), 67. *Prof Exp:* Assoc dir res, Crow's Hybrid Corn Co, Ill, 61-65; instr biomet, 65-67, from asst prof to assoc prof agron, 67-75, PROF AGRON & CONSULT BIOMET CTR, UNIV NEBR, LINCOLN, 75- *Mem:* Am Soc Agron; Crop Sci Soc Am; Am Statist Asn; Sigma Xi. *Res:* Genetic variance components; computer simulation studies of their distribution and effect of choice of mating deisgn on their estimation. *Mailing Add:* 3145 Calvert Lincoln NE 68502-5216

MUMMA, ALBERT G, b Findlay, Ohio, June 2, 06; m 27; c 3. ENGINEERING. *Educ:* Newark Col Eng, DEng, 70. *Prof Exp:* Chief bur ships & coordr shipbuilding, conversion & repair, Dept Defense, 55-59; vpres & group exec, Worthington Corp, 59-64, exec vpres & dir-in-chg all domestic opers, 64-67, pres & chief operating officer, 67, chmn bd, 67-71; chmn, Am Shipbuilding Comn, 71-73; CONSULT MGT, 73- *Concurrent Pos:* Past mem res coun, Nat Acad Sci, past chmn numerous comts; past mem maritime transp res bd, Eng & Indust Res Comt, Nat Res Coun; trustee, Drew Univ & St Barnabas Med Ctr, Livingston, NJ. *Honors & Awards:* Jerry Land Gold Medal. *Mem:* Nat Acad Eng; hon mem & fel Soc Naval Architects & Naval Engrs (past pres); hon mem Am Soc Naval Engrs (past pres). *Mailing Add:* Bald Peak Colony Club PO Box 405 Melvin Village NH 03850

MUMMA, MARTIN DALE, b Gideon, Mo, Jan 21, 36; m 56; c 3. GEOLOGY. *Educ:* Univ Mo, AB, 58, MA, 60; La State Univ, PhD(geol), 65. *Prof Exp:* Field geologist, Magnolia Petrol Co, 58; geologist, Esso Prod Res Co, Standard Oil Co, NJ, 65-66; sr geologist, Humble Oil & Refining Co, 66-68; assoc prof geol, Eastern Ky Univ, 68-69; from asst prof to prof geol, Eastern Wash State Col, 69-82; PRES, PROF TRAINING RESOURCES, INT, 82- *Mem:* Soc Econ Paleont & Mineral; Am Asn Petrol Geologists; Paleont Soc. *Res:* Sedimentary petrology and environments; biostratigraphy; micropaleontology. *Mailing Add:* Prof Training Resources Int 15915 E 131 S Broken Arrow OK 74011

MUMMA, MICHAEL JON, b Lancaster, Pa, Dec 3, 41; m 66; c 2. ATOMIC & MOLECULAR PHYSICS, PLANETARY & COMETARY SPECTROSCOPY. *Educ:* Franklin & Marshall Col, AB, 63; Univ Pittsburgh, PhD(physics), 70. *Prof Exp:* Space scientist, Goddard Space Flight Ctr, NASA, 70-76, head Infrared & Radio Astron Br, 76-84, acting chief, Lab Extraterrestrial Physics, 84-85, head planetary systs br, 85-90, CHIEF SCIENTIST, PLANETARY & ASTROPHYS SCI, LAB EXTRATERRESTRIAL PHYSICS, GODDARD SPACE FLIGHT CTR, NASA, 90-; PROF, PHYSICS DEPT, PA STATE UNIV, 82- *Concurrent Pos:* Univ lectr, 70-; mem working/study groups, Dept Defense, NASA, Nat Bur Standards, Nat Acad Sci, 74-; prin investr, planetary & cometary astron, molecular astrophys, infrared imaging & spectros, 74-; adj sr res assoc, Physics Dept, Pa State Univ, 79-82; expert witness, 80- *Honors & Awards:* Except Sci Achievement Medal, NASA, 88. *Mem:* Am Geophys Union; fel Am Phys Soc; Am Inst Physics; AAAS; Am Astron Soc; Int Astron Union. *Res:* Composition, dynamics and structure of planetary, solar stellar and terrestrial atmospheres, cometary composition and structure; natural infrared lasers on Jupiter, Venus and Mars; fundamental atomic and molecular cross-sections and structure, remote sensing at ultraviolet through radio wavelengths, laser physics, instrument development at ultraviolet through sub-millimeter wavelengths; author or co-author of over 120 publications. *Mailing Add:* Code 690 Lab Extraterrestrial Physics Goddard Space Flight Ctr NASA Greenbelt MD 20771

MUMMA, RALPH O, b Carlisle, Pa, June 20, 34; m 58; c 1. BIOCHEMISTRY, ENTOMOLOGY. *Educ:* Juniata Col, BS, 56; Pa State Univ, PhD(chem), 60. *Prof Exp:* Fel biochem, Pa State Univ, 60-61, asst prof, 61-66, from asst prof to prof chem pesticides, 66-90, DISTINGUISHED PROF ENVIRON QUAL, PA STATE UNIV, 90- *Mem:* Am Chem Soc; Entom Soc Am; AAAS; Am Inst Biol Sci. *Res:* Environmental quality, pesticie metalbolism, methods and residues; chemical ecology. *Mailing Add:* Pesticide Res Lab Dept Entom Pa State Univ University Park PA 16802

MUMME, JUDITH E, b Los Angeles, Calif, Sept 23, 43. MATHEMATICS. *Educ:* Calif State Univ, BA, 69; Calif Lutheran Univ, MA, 79. *Prof Exp:* Math specialist, County Sch Off, 83-85; PROF MATH, UNIV CALIF, SANTA BARBARA, 85- *Mem:* Nat Coun Supvrs Math; Am Asn Curric Develop; Nat Coun Teachers Math. *Mailing Add:* 640 Saphire Ave Ventura CA 93004

MUMME, KENNETH IRVING, US citizen. CHEMICAL ENGINEERING. *Educ:* Lawrence Col, BS, 54; Univ Maine, Orono, MS, 66, PhD(chem eng), 70. *Prof Exp:* Instrumentation engr, Kimberly-Clark Corp, Wis, 56-58, physicist, 58-63; lectr mach comput, 63-70, ASSOC PROF CHEM ENG, UNIV MAINE, ORONO, 70- *Concurrent Pos:* Royal Norweg Coun Sci & Indust Res fel, Norweg Inst Technol, 71-72. *Mem:* Am Inst Chem Engrs. *Res:* Chemical process control and optimization; river pollution control systems. *Mailing Add:* Dept of Chem Eng Univ of Maine Orono ME 04469

MUMPTON, FREDERICK ALBERT, b Rome, NY, Dec 14, 32; m 54; c 5. MINERALOGY. *Educ:* St Lawrence Univ, BS, 54; Pa State Univ, MS, 56, PhD(geochem), 58. *Prof Exp:* Res chemist, Linde Div, Union Carbide Corp, 58-60, res geochemist, Nuclear Div, 60-65; mineral group leader, Mining & Metals Div, NY, 65-69; assoc prof, 69-74, PROF, DEPT EARTH SCI, STATE UNIV NY COL BROCKPORT, 74- *Concurrent Pos:* Chmn, Int Comt Natural Zeolites, 76-; ed-in-chief, Clays & Clay Minerals, 78-; vis scientist, Los Alamos Nat Lab, 84. *Mem:* Geochem Soc; fel Mineral Soc Am; Clay Minerals Soc; Sigma Xi; Int Zeolite Asn. *Res:* Silicate chemistry; synthetic and clay mineralogy; mineralogy and utilization of zeolite minerals; mineralogy of asbestos and serpentinites; mineral resources; environmental mineral science. *Mailing Add:* Dept Earth Sci State Univ NY Brockport NY 14420

MUMTAZ, MOHAMMAD MOIZUDDIN, b Hyderabad, India, Aug 25, 49; US citizen; m 78; c 1. TOXICOLOGY, METABOLISM. *Educ:* Osmania Univ, India, BSc, 71, MSc, 73; Ore State Univ, MS, 77; Univ Md, PhD(toxicol & entom), 84. *Prof Exp:* Teaching asst, dept chem, Ore State Univ, 73-76, res asst, dept entom, 76-77; res fel psychiat, Ill State Psychiat Inst, Chicago, 78-79; res instr psychopharmacol, dept psychiat, Med Col Va, 79-81; res asst entom, Univ Md, 81-84; fel path, 84-85, RES ASSOC NUTRIT, PREV MED COMMUNITY HEALTH, UNIV TEX MED BR, GALVESTON, 86- *Mem:* Am Soc Mass Spectrometry; Am Chem Soc. *Res:* Relationship between toxicity (specific) of chemicals and their biotransformation including qualitative and quantitative analyses of endogeneous compounds, xenobiotics and their metabolites in biological matrices. *Mailing Add:* ECAD Off Res & Develop US Environ Protection Agency 26 W King St Cincinnati OH 45268

MUN, ALTON M, b Honolulu, Hawaii, Apr 1, 23; m 55; c 5. DEVELOPMENTAL BIOLOGY. *Educ:* Univ Southern Calif, BA, 49; Univ Ill, MS, 51; Ind Univ, PhD(zool), 56. *Prof Exp:* Res assoc avian embryol, Wash State Univ, 56-59; asst invest exp embryol, Carnegie Inst, 59-61; from assoc prof to prof, 61-86, EMER PROF ZOOL, UNIV MAINE, 86- *Mem:* Am Soc Zool; Sigma Xi; Soc Develop Biol; fel AAAS. *Res:* Experimental embryology; zoology; effects of antisera on chick embryos; enhancement of growth of chick host spleens by homologous adult organ fragments; homograft reaction in the chick embryo; parthenogenetic development in unfertilized turkey eggs; teratological effects of trypan blue in the chick embryo. *Mailing Add:* RR 1 Box 1590 Bar Harbor ME 04609-9716

MUNA, NADEEM MITRI, b Jerusalem, Palestine, Apr 8, 28; US citizen; m 69; c 2. IMMUNOLOGY, MICROBIOLOGY. *Educ:* Univ Dubuque, BSc, 51; Univ SDak, MA, 54; Univ Utah, PhD(microbiol), 68. *Prof Exp:* Clin Sacred Heart Hosp, Yankton, SDak, 54-56, Thomas D Dee Mem Hosp, Ogden, Utah, 58-65; pres & dir res mfg, Microbiol Res Corp, 68-78; PRES & DIR IMMUNOL PROD, IMMUNO-DIAG PROD INC, 78- *Mem:* Am Soc Microbiol; Sigma Xi. *Res:* Cancer immunology by immunofluorescence; identification and localization of cancer material for isolation and possible use for immunization as a means of prevention and treatment. *Mailing Add:* 69 W 3600 S Bountiful UT 84010

MUNAN, LOUIS, b New York, NY, Feb 10, 21; div; c 4. EPIDEMIOLOGY. *Educ:* City Col New York, BA, 48; George Washington Univ, AB, 48, MSc, 50. *Prof Exp:* Res analyst, Nat Acad Sci-Nat Res Coun, 49-51; res analyst, Prev Med Div, Off Surgeon Gen, Dept Army, 51-55, chief res & develop sect, Med Info & Intel Div, 55-56; Fulbright exchange prof, Schs Med, Lima & Guayaquil, 57-58; statistician, Pan-Am Sanit Bur, WHO, 58-61, res scientist, Off Res Coord, Pan-Am Health Orgn, 61-68; chmn dept, 68-75, assoc prof epidemiol, Fac Med, 68-81, HEAD EPIDEMIOL LAB, UNIV SHERBROOKE, 82- *Concurrent Pos:* Assoc, Sch Med, George Washington Univ, 50-58; vis prof, Malaria Eradication Training Ctr, Kingston, Jamaica, 62 & 63, Univ Nancy, 77 & 82, Univ Toronto, 79, Univ McGill, 80 & Univ Yaoundí, Cameroon, 89; consult, Univ Montreal, Alta Cancer Bd, Quebec Dept Health & Health Unit Asn Alta. *Mem:* Fel Am Pub Health Asn; fel Am Col Epidemiol; Soc Epidemiol Res; Int Epidemiol Asn. *Res:* Epidemiologic studies of biochemical and hematology characteristics of populations. *Mailing Add:* Epidemiol Lab Box 11422 Cent PO Edmonton AB T5J 3K6 Can

MUNASINGHE, MOHAN P, b Colombo, Sri Lanka, July 25, 45; m 70; c 2. ENVIRONMENTAL-ENERGY MODELS, POWER SYSTEMS PLANNING. *Educ:* Cambridge Univ, Eng, BA, 67, MA, 68; Mass Inst Tech, SM, 69, EE, 69; McGill Univ, Can, PhD(elec eng), 73; Concordia Univ, Can, MA, 75. *Prof Exp:* Asst dir, IIQE, Concordia Univ, Can, 73-75; pres, Sri Lanka Energy Mgr Asn, 83-85; exec chmn, Computer & Info Tech Coun, Sri Lanka, 83-86; sr energy adv to pres, Off Pres, Sri Lanka, 82-87; DIV CHIEF ENVIRON POLICY & RES, WORLD BANK, WASHINGTON, DC, 75- *Concurrent Pos:* Vis lectr, Univ Sri Lanka, Katnbedde, 69-70; vis prof, Am Univ, Washington, DC, 77-82, ITPD, State Univ NY, Stony Brook, NY, 83-88; sr res fel, CIDCM, Univ Md, 87-; vis prof energy & resources, Energy Ctr, Univ Pa, 87- *Honors & Awards:* Int Award Outstanding Achievement, Latin Am & Caribbean Coun Energy, 88; Prize Except Contrib, Int Asn Energy Econ, 87; Gold Medal Outstanding Scientist Sri Lanka, Lions Int

Orgn, 85. *Mem:* Inst Elec & Electronics Engrs; Am Phys Soc; Sigma Xi; Am Econ Asn; Third World Acad Sci. *Res:* Author of 34 books and over 150 technical papers in refereed leading international journals of physics, electrical engineering, energy, water resources, economics and informatics. *Mailing Add:* 4201 E West Hwy Chevy Chase MD 20815

MUNCH, G, b San Cristobal, Mex, June 9, 21; m 47; c 4. INTERSTELLAR MATTER. *Educ:* Nat Univ Mex, BS, 38, MS, 44; Univ Chicago, PhD(astron, astrophys), 47. *Prof Exp:* From instr to asst prof astrophys, Yerkes Observ, Univ Chicago, 47-51; from asst prof to prof, Calif Inst Technol, 51-77; dir, 78-89, EMER DIR, MAX PLANCK INST ASTRON, 89- *Honors & Awards:* NASA Medal for Except Sci Achievement, 74. *Mem:* Nat Acad Sci; fel Am Acad Arts & Sci. *Res:* Stellar atmospheres; interstellar matter; planetary atmospheres. *Mailing Add:* Torre Marina III - 13 E Calle Ciruelo s/n Aguadulce - Almeria 04720 Spain

MUNCH, JESPER, b Denmark, Sept 6, 45. EXPERIMENTAL PHYSICS. *Educ:* Mass Inst Technol, BS, 68; Univ Chicago, MS, 70, PhD(physics), 74. *Prof Exp:* Res asst electron optics & holography, Univ Chicago, 68-74; SR SCIENTIST, LASERS & COHERENT OPTICS, TRW SYSTS, 75- *Concurrent Pos:* Lectr, Sch Eng, Univ Calif, Los Angeles, 78- *Res:* Experimental research in lasers; coherent optics; frequency stability; electron holography; free electron lasers; fusion diagnostics. *Mailing Add:* 7215 Rindge Ave Playa Del Ray CA 90291

MUNCH, JOHN HOWARD, b St Louis, Mo, Feb 9, 38; m 65; c 2. SYNTHESIS OF SURFACTANTS, PROCESS IMPROVEMENT. *Educ:* Swarthmore Col, BA, 60; Univ Wis, Madison, PhD(org chem), 66. *Prof Exp:* Asst prof chem, Dickinson Col, 65-69; RES CHEMIST, PETROLITE CORP, ST LOUIS, 69- *Mem:* AAAS; Am Chem Soc; Royal Soc Chem. *Res:* Organic synthesis and reaction mechanisms; surface-active organic compounds; aromatic elcetrophilic substitution reactions; phenol-formaldehyde chemistry; sulfonation. *Mailing Add:* 9 Douglass Lane Kirkwood MO 63122

MUNCH, RALPH HOWARD, b Lafayette, Ind, May 5, 11; m 35; c 5. PHYSICAL CHEMISTRY. *Educ:* Univ NC, BS, 31, MS, 32; Northwestern Univ, PhD(phys chem), 35. *Prof Exp:* Res asst, Rockefeller Found, Chicago, 35-37; res chemist org div, Res Dept, Monsanto Chem Co, 37-41, group leader, 41-54, sect leader, 54-56, asst dir res, 56-60, res assoc, 60-64, sr res specialist, Monsanto Co, 64-67, advan scientist, 67-70, sr sci fel, 70-74, distinguished sci fel, Monsanto Indust Chem Co, 74-76; CONSULT, 76- *Concurrent Pos:* Vchmn, Gordon Res Conf Instrumentation, 53, chmn, 54. *Mem:* Fel AAAS; Am Chem Soc; fel Instrument Soc Am (vpres, 46). *Res:* Spectroscopy; process control instrumentation; gas chromatography; dielectrics. *Mailing Add:* 303 Planthurst Rd Webster Groves MO 63119

MUNCH, THEODORE, bacteriology, science education; deceased, see previous edition for last biography

MUNCHAUSEN, LINDA LOU, b New Orleans, La, Aug 30, 46. ORGANIC CHEMISTRY. *Educ:* Southeastern La Univ, BS, 68; Univ Ark, PhD(chem), 73. *Prof Exp:* Fel biol, Oak Ridge Nat Lab, 73-75; fel chem, La State Univ, 75; INSTR ORG CHEM, DEPT CHEM, SOUTHEASTERN LA UNIV, 75- *Mem:* Am Chem Soc; Royal Soc Chem; Am Soc Photobiol. *Mailing Add:* Dept Chem Box 372 Southeastern La Univ 100 W Dakota Hammond LA 70401

MUNCHMEYER, FREDERICK CLARKE, b Washington, DC, Mar 26, 22; m 64; c 4. NAVAL ARCHITECTURE, MARINE ENGINEERING. *Educ:* US Coast Guard Acad, BS, 42; Mass Inst Technol, MS, 48; Univ Mich, PhD(naval archit & marine eng), 78. *Prof Exp:* Comn officer, US Coast Guard, 42-63; prof mech eng, Univ Hawaii, 63-82; prof naval archit & marine eng & chmn dept, Univ New Orleans, 82-87, emer prof, 87-; RETIRED. *Concurrent Pos:* Sr engr, Westinghouse Ocean Res & Eng Ctr, 68-69; vis prof, Tech Univ, Berlin, 78-79; prin investr, Univ Hawaii, 75-82; consult prof, Northwestern Polytech Univ, Xian, China, 83- *Mem:* Sigma Xi; Soc Naval Archit & Marine Engrs. *Res:* Computer aided design of doubly curved surfaces; applications of differential geometry to the design and analysis of surfaces. *Mailing Add:* 226 Rue St Peter Metairie LA 70005

MUNCK, ALLAN ULF, b Buenos Aires, Arg, July 4, 25; US citizen; m 57; c 3. ENDOCRINOLOGY. *Educ:* Mass Inst Technol, BS, 48, MS, 49, PhD(biophys), 56. *Prof Exp:* Nat Cancer Inst fel steroid biochem, Worcester Found Exp Biol, 57-58; from asst prof to assoc prof, 59-67, PROF PHYSIOL, DARTMOUTH MED SCH, 67- *Concurrent Pos:* Res career develop award, NIH, 63-72; assoc ed, J Steroid Biochem, 67-; ed bd, J Biol Chem, 82-87. *Mem:* Biophys Soc; Endocrine Soc; Physiol Soc. *Res:* Physiological and molecular mechanisms of action of glucocorticoids. *Mailing Add:* Dept Physiol Dartmouth Med Sch Hanover NH 03756

MUNCY, ROBERT JESS, b Narrows, Va, Apr 13, 29; m 54; c 1. ZOOLOGY. *Educ:* Va Polytech Inst & State Univ, BS, 50, MS, 54; Iowa State Col, PhD(fishery biol), 57. *Prof Exp:* Asst zool, Iowa State Col, 54-57; fishery biologist, Chesapeake Biol Lab, Md, 57-59; asst prof wildlife mgt, La State Univ, 59-65; asst prof, Colo State Univ, 65-66; prof zool & entom & unit leader, Iowa Coop Fishery Unit, Fish & Wildlife Serv, Iowa State Univ, 66-79; unit leader & prof, Miss Coop Fish & Wildlife Res Inst, Miss State Univ, 79-89; RETIRED. *Concurrent Pos:* Vis lectr, Univ Md, 58-59 & Mt Lake Sta, Va, 64; fishery biologist, Food & Agr Orgn, UN, Zambia, Africa, 72-73. *Mem:* Am Fisheries Soc; Sigma Xi. *Res:* Fishery biology and limnology. *Mailing Add:* Rte 3 Box 299 Moneta VA 24121

MUNCZEU, HERMAN J, b Buenos Aires, Arg, June 9, 27; m 54; c 2. PHYSICS. *Educ:* Univ Buenos Aires, Lic, 54, PhD(physics), 58. *Prof Exp:* Res investr, AEC, Arg, 54-60; from asst prof to assoc prof physics, Univ Buenos Aires, 61-66; vis assoc prof, Northwestern Univ, 66-69; assoc prof, 69-71, PROF PHYSICS & ASTRON, UNIV KANS, 71- *Concurrent Pos:* Univ Buenos Aires res fel elem particle physics, Univ Rome, 58-60. *Mem:* Am Phys Soc. *Res:* Theory of elementary particles. *Mailing Add:* Dept Physics Univ Kans Lawrence KS 66049

MUNDAY, J(OHN) C(LINGMAN), chemistry; deceased, see previous edition for last biography

MUNDAY, JOHN CLINGMAN, JR, b Plainfield, NJ, June 10, 40; m 65; c 3. REMOTE SENSING, WATER QUALITY. *Educ:* Cornell Univ, AB, 62; Univ Ill, PhD(biophys), 68. *Prof Exp:* Res asst photosynthesis, Univ Ill, 65-68, res assoc, 68; physicist, Air Force Missile Develop Ctr, Holloman Air Force Base, NMex, 68-69; assoc marine scientist, Va Inst Marine Sci, 69-71; asst prof geog, Univ Toronto, 71-75; assoc marine scientist, Va Inst Marine Sci, 75-83; ASSOC DEAN & PROF NATURAL SCI, SCH PUB POLICY, REGENT UNIV, 83- *Concurrent Pos:* Nat Res Coun res associateship, 68-69; asst prof, Univ Va & Col William & Mary, 69-71; assoc prof marine sci, Col William & Mary, 75-81, prof 81-83. *Mem:* AAAS; Asn Am Geographers; Int Asn Energy Econ; Am Soc Photogram & Remote Sensing; Am Sci Affil. *Res:* Spectroscopy, photosynthesis and membrane physiology of algae; missile reentry spectral photography; estuarine oil pollution; remote sensing of coastal water quality and circulation; land satellite data processing; science and technology public policy. *Mailing Add:* Regent Univ Virginia Beach VA 23463

MUNDAY, THEODORE F, b Baton Rouge, La, July 7, 37; m 63; c 3. INORGANIC CHEMISTRY. *Educ:* Cornell Univ, AB, 59; Iowa State Univ, PhD(inorg chem), 64. *Prof Exp:* Res chemist, 64-68, sr res chemist, 68-78, res assoc, 78-88, SR RES ASSOC, FMC CORP, 88- *Mem:* Iron & Steel Soc Am Inst Mining Engrs. *Res:* Process development; phase equilibria; high temperature reactions; molten salts; sol stability; distillation; phosphorus and phosphates; hydrogen peroxide; peroxide monopropellant catalysts; aluminosilicates. *Mailing Add:* 16 Steven Rd Kendall Park NJ 08824

MUNDEL, AUGUST B(AER), b New York, NY, Dec 21, 11. QUALITY CONTROL, BATTERIES. *Educ:* Cooper Union Inst Technol, BS, 32; Univ Mich, MS, 34. *Prof Exp:* Res engr, Sonotone Corp, 35-50, dir qual control, 50-55, dir eng & qual assurance, 55-60, vpres, 60-69; dir, Westchester Community Health Plan, 75-83; CONSULT, 69- *Concurrent Pos:* Chmn, US Tech Adv Group, Int Orgn Standardization, 74-; ed, Qual Eng Quart, 86- *Honors & Awards:* E R Ott Award, Am Soc Qual Control, 82. *Mem:* Am Soc Qual Control; Syst Safety Soc; Inst Elec & Electronics Engrs; Am Soc Testing & Mat. *Mailing Add:* 34 Sammis Lane White Plains NY 10605-4726

MUNDELL, PERCY MELDRUM, b Vancouver, BC, Dec 14, 21; m 50; c 3. ORGANIC CHEMISTRY. *Educ:* Univ BC, BA, 43, MA, 45; Ohio State Univ, PhD(chem), 53. *Prof Exp:* Asst chem, Univ BC, 42-45, instr, 45-46; asst, Ohio State Univ, 47-51; from asst prof to prof chem, Miami Univ, 52-87; RETIRED. *Mem:* Am Chem Soc. *Res:* Structural studies of natural products. *Mailing Add:* PO Box 288 Oxford OH 45056-0288

MUNDELL, ROBERT DAVID, b Greensburg, Pa, Aug 30, 36; m 62; c 2. ANATOMY, CELL BIOLOGY. *Educ:* Waynesburg Col, BS, 57; Univ Pittsburgh, PhD(anat, cell biol), 65. *Prof Exp:* Instr anat & histol, Sch Med, Tufts Univ, 64-66; asst prof histol, 66-68, head dept, 68-70, assoc prof histol & anat, 68-72, PROF HISTOL & ANAT, UNIV PITTSBURGH, 72-, HEAD DEPT ANAT, 70-, ASSOC DEAN, SCH DENT MED, 88- *Concurrent Pos:* NSF res grant, Tufts Univ & Univ Pittsburgh, 66-68. *Mem:* Am Asn Anatomists; AAAS; Sigma Xi; Am Asn Dent Schs. *Res:* Craniofacial development. *Mailing Add:* 630 Salk Hall Sch Dent Med Univ Pittsburgh Pittsburgh PA 15213

MUNDIE, LLOYD GEORGE, b Udney, Ont, Dec 15, 16; nat US; m 42; c 2. OPTICS. *Educ:* Univ Sask, BSc, 35, MSc, 37; Purdue Univ, PhD(physics), 43. *Prof Exp:* Instr physics, Purdue Univ, 39-47; infrared physicist, Naval Ord Lab, 47-51; physicist, Nat Bur Standards, 51-54; physicist, Univ Mich, 54-57; head infrared & optics dept, Bendix Systs Div, 57-61; dir basic res lab, Lockheed-Calif Co, 61-65; PHYSICIST, RAND CORP, 65- *Mem:* Fel AAAS; fel Optical Soc Am. *Res:* Spectroscopy; infrared. *Mailing Add:* 9322 Bianca Ave Northridge CA 91325

MUNDKUR, BALAJI, b Mangalore, India, Dec 27, 24; nat US; m 46. CYTOLOGY. *Educ:* Univ Bombay, India, BSc, 45; Washington Univ, PhD(bot, genetics), 51. *Prof Exp:* Assoc mycol, Indian Agr Res Inst, 47; res assoc, Univ Southern Ill, 50-53; sr sci officer, Indian Cancer Res Ctr, 53-55; assoc bacteriologist, Univ PR, 55-58; spec res fel anat, USPHS, Chicago, 58-60; assoc prof biol sci, Univ Conn, 60-89; RETIRED. *Res:* Electron microscopy; application of cytological freeze-drying techniques in cytochemistry and electron microscopy; lecture anthropology, religion and religious arts. *Mailing Add:* 97 Dunhampound Rd Storrs CT 06268

MUNDORFF, SHEILA ANN, b Rochester, NY, Dec 14, 45; m 88. CARIOLOGY, DENTAL RESEARCH. *Educ:* Nazareth Col Rochester, BS, 67; Univ Rochester, MS, 84. *Prof Exp:* Lab tech, cariogenic potential of foods, Eastman Dental Ctr, 67-69, res asst, 69-71, RES ASSOC, EASTMAN DENTAL CTR, 71- *Concurrent Pos:* Mem, animal resource group, Am Dental Asn Health Found, 81- 83, invited participant animal caries models working group, sci consensus conf, 81-85, ed referee, Caries Res, 87-88, lectr, Basil G Bibby Scientific Seminar, Eastman Dental Ctr. *Honors & Awards:* Basil G Bibby Lectr, Eastman Dental Ctr, 84. *Mem:* Int Asn Dental Res; Am Asn Dental Res; Am Chem Soc. *Res:* Cariogenic potential of foods and caries-protective food components; use of flouridated sucrose in the control of caries; microbiology of caries. *Mailing Add:* 22 Belmeade Rd Rochester NY 14617-3602

MUNDT, PHILIP A, b Sioux Falls, SDak, Oct 2, 27; m 51; c 3. PETROLEUM GEOLOGY. *Educ:* SDak Sch Mines & Technol, BS, 51; Washington Univ, MA, 53; Stanford Univ, PhD(geol), 55. *Prof Exp:* Geologist, Mobil Producing Co, 55-58, subsurface supvr, Mobil Mediterranean Inc, 58-63, staff geologist, Mobil Oil Corp, 63-65, explor supvr, Libya, 65-67, geol supvr, 67-69; explor mgr, US Nat Resources, Inc, 69-72; mgr geol & geochem res, Mobil Res Develop Corp, Mobil Prod Nigeria, 72-77, explor mgr, 77-79, explor mgr, Mobil Oil Indonesia, 79-82, mgr regional geol, Mobil Oil, 82-86; CONSULT, 86- *Mem:* Am Asn Petrol Geol; Sigma Xi. *Res:* Petroleum exploration. *Mailing Add:* 21503 Lakefront Dr Lago Vista TX 78645

MUNDY, BRADFORD PHILIP, b Warrensburg, NY, Nov 9, 38; m 63; c 3. CHEMISTRY. *Educ:* State Univ NY Albany, BS, 61; Univ Vt, PhD(chem), 65. *Prof Exp:* Res assoc chem, Univ Calif, Berkeley, 65-66; NIH fel, 66-67; asst prof, 67-71, assoc prof, 71-75, PROF CHEM, MONT STATE UNIV, 75- *Mem:* Am Chem Soc. *Res:* Synthesis of natural products; heterocyclic chemistry; biosynthesis. *Mailing Add:* Dept Chem Mont State Univ Bozeman MT 59715

MUNDY, GREGORY ROBERT, b Melbourne, Australia, June 16, 42; US citizen; m 66; c 3. ENDOCRINOLOGY, BONE CELL BIOLOGY. *Educ:* Univ Melbourne, BM, 66; Univ Tasmania, MD, 73. *Prof Exp:* Resident, Royal Hobart Hosp, Tasmania, 67-68, prof med registrar, 69-70, lectr med, 70-72; res assoc, dept pharmacol, Univ Rochester, 72-74; from asst prof to assoc prof med, Univ Conn, 74-80; PROF MED, UNIV TEX HEALTH SCI CTR, SAN ANTONIO, 80- *Concurrent Pos:* Am Cancer Soc fac res award, 76-81; prog dir, Gen Clin Res Unit, Univ Tex Health Sci Ctr & Vet Admin Hosp, 80- *Honors & Awards:* Fuller Albright Award, Am Soc Bone & Mineral Res, 82. *Mem:* Am Soc Clin Invest; Am Fedn Clin Res; AAAS; Endocrine Soc; Am Soc Bone & Mineral Res (secy-treas, 85); Asn Am Physicans. *Res:* Mechanisms of bone destruction and bone formation in malignant disease; hormonal control of calcium homeostasis. *Mailing Add:* Dept Med-Endocrinol Univ Tex Health Sci Ctr 7703 Floyd Curl Dr San Antonio TX 78284

MUNDY, ROY LEE, b Charlottesville, Va, Mar 4, 22; m 41; c 3. PHARMACOLOGY, TOXICOLOGY. *Educ:* Howard Col, BS, 48; Univ Ala, MS, 50; Univ Va, PhD(pharmacol), 57. *Prof Exp:* Asst prof pharmacol, Howard Col, 50-51; chief pharmacol dept, Walter Reed Army Inst Res, 55-66; from assoc prof to prof, 66-83, EMER PROF PHARMACOL, MED CTR, UNIV ALA, BIRMINGHAM, 83- *Mem:* Soc Toxicol; Soc Exp Biol & Med; Am Soc Pharmacol & Exp Therapeut. *Res:* Pharmacology of sulfhydryl radio-protectant chemicals; removal of radiation agents from the animal body; autonomic pharmacology. *Mailing Add:* 211 Snake Hill Rd Trussville AL 35173

MUNETA, PAUL, b Harlowton, Mont, Apr 21, 31; m 56; c 2. FOOD SCIENCE. *Educ:* Mont State Col, BS, 53; Cornell Univ, PhD(veg crops), 59. *Prof Exp:* Asst agr chem, 59-68, asst prof, 61-68, ASSOC PROF FOOD SCI & ASSOC FOOD SCIENTIST, UNIV IDAHO, 68- *Mem:* Am Chem Soc; Inst Food Technol; Int Asn Milk, Food & Environ Sanit. *Res:* Changes in nitrate and nitrite in cured meats; factors affecting microbiol conversion of nitrate to nitrite in foods; determination of chemical and physical methods to control nitrate reduction; color in fresh and processed fruits and vegetables (enzymatic blackening); carbohydrate metabolism in potato callus tissue. *Mailing Add:* Food Res Ctr Univ Idaho Moscow ID 83843-4199

MUNGALL, ALLAN GEORGE, b Vancouver, BC, Mar 12, 28; m 50; c 3. EXPERIMENTAL ATOMIC PHYSICS. *Educ:* Univ BC, BASc, 49, MASc, 50; McGill Univ, PhD(physics), 54. *Prof Exp:* Geophysicist, Calif Stand Co, Alta, 50; jr res officer physics, Nat Res Coun Can, 50-52; asst, McGill Univ, 54; from asst res officer to prin res officer, Nat Res Coun Can, 54-83; RETIRED. *Concurrent Pos:* Part-time prin res officer, Nat Res Coun Can, 83-86. *Mem:* Asn Prof Engr Ont; fel Inst Elec & Electronics Engrs. *Res:* Atomic frequency and time standards. *Mailing Add:* 33 Woodview Cres Gloucester ON K1B 3B1 Can

MUNGALL, WILLIAM STEWART, b Buffalo, NY, July 24, 45; m 67; c 2. SYNTHETIC ORGANIC CHEMISTRY. *Educ:* State Univ NY Buffalo, BA, 67; Northwestern Univ, PhD(org chem), 70. *Prof Exp:* Asst prof chem, 71-74, assoc prof, 74-80, PROF CHEM, HOPE COL, 80- *Concurrent Pos:* Consult, Hexcel Inc & Parke-Davis/Warner Lambert; vis prof, MIT, 86-87. *Mem:* Am Chem Soc. *Res:* Synthesis of biologically active organic compounds; polymer synthesis. *Mailing Add:* Dept of Chem Hope Col Holland MI 49423-3698

MUNGAN, NECMETTIN, b Mardin, Turkey, Mar 1, 34; m 62; c 4. PETROLEUM ENGINEERING, MATHEMATICS. *Educ:* Univ Tex, BS, 56, BA, 57, MS, 58, PhD(petrol eng), 62. *Prof Exp:* Scientist, Sinclair Res, Inc, 61-63, res scientist, 63-65 & Sinclair Oil & Gas Co, 65-66, sr res scientist, 66; chief res officer, Petrol Recovery Inst, Univ Calgary, 66-78; consult, Mungan Petrol Consults Ltd, 78-86; CHIEF TECH ADV, AEC OIL & GAS CO, 86- *Honors & Awards:* Cedric K Ferguson Award, Soc Petrol Engrs, 66, Lester C Uren Award, 90; Presentation Award, Am Inst Chem Engrs, 66; Distinguished Serv Award, Petrol Soc, 89. *Mem:* Soc Petrol Engrs; Can Inst Mining & Metall; Am Chem Soc; Am Inst Chem Engrs. *Res:* Application of surface chemistry to flow of fluids in petroleum reservoirs; formation damage due to clays; reservoir engineering and well logging; mathematical modeling of fluid flow in reservoirs; enhanced recovery of oil and gas from underground deposits; gas storage; heavy oil recovery. *Mailing Add:* AEC Oil & Gas Co 2400, 639 5th Ave SW Calgary AB T2P 0M9 Can

MUNGER, BRYCE LEON, b Everett, Wash, May 20, 33; m 57; c 4. HUMAN ANATOMY, CELL NEUROBIOLOGY. *Educ:* Washington Univ, MD, 58. *Prof Exp:* Intern path, Johns Hopkins Hosp, 58-59; asst prof anat, Washington Univ, 61-65; assoc prof, Univ Chicago, 65-66; chmn dept, 66-87, PROF ANAT, HERSHEY MED CTR, PA STATE UNIV, 66- *Honors & Awards:* Purkinje Medal, 87; Roche Award, 56; Borden Award, 58. *Mem:* AAAS; Am Asn Anat; Am Soc Cell Biol. *Res:* Comparative ultrastructure of sensory nerve endings, particularly mechanoreceptors and nerve development. *Mailing Add:* Dept Anat Hershey Med Ctr Pa State Univ Hershey PA 17033

MUNGER, CHARLES GALLOWAY, b Los Angeles, Calif, Dec 29, 12; m 36; c 2. CORROSION TECHNOLOGY, INORGANIC COATINGS. *Educ:* Pomona Col, BA, 35. *Prof Exp:* Dir res, Amercoat Corp, dir res & mfg, vpres, res & mfg, pres, 48-70; vpres int, Ameron Inc, 70-73; CONSULT COATINGS & CORROSION, GOVT & INDUST, 73- *Honors & Awards:* Frank Newman Speller Award, Nat Asn Corrosion Engrs, 68; John D Keane Award of Merit, Steel Struct Painting Coun, 86. *Mem:* Nat Asn Corrosion Engrs (vpres, 62, pres, 63); Am Chem Soc; fel Am Inst Chemists; Steel Struct Painting Coun. *Res:* Organic and inorganic coatings; high performance coatings for prevention of corrosion; marine corrosion; tanker corrosion; atomic energy coatings; sewer corrosion; materials failure due to corrosion; surface preparation for coatings and linings. *Mailing Add:* 3210 Sage Rd Fallbrook CA 92028-9373

MUNGER, PAUL R, b Hannibal, Mo, Jan 14, 32; m 54; c 4. FLUID MECHANICS, HYDRAULICS. *Educ:* Mo Sch Mines & Metall, BS, 58, MS, 61; Univ Ark, PhD(eng sci), 72. *Prof Exp:* From instr to asst prof civil eng, Mo Sch Mines & Metall, 58-64; assoc prof, 65-72, PROF CIVIL ENG, UNIV MO, ROLLA, 73-, DIR RES, INST RIVER STUDIES, 76-, EXEC DIR RES, INT INST RIVER & LAKE SYSTS, 85- *Concurrent Pos:* Prin investr, base line study Mo River, US Army Corp Engrs, 73-74, Lower Miss Valley Div potamology study, 75-76. *Mem:* Am Soc Civil Engrs; Nat Coun Eng Examrs (pres, 83-84); Nat Soc Prof Engrs; Am Soc Eng Educr; Sigma Xi. *Res:* Rivers, river mechanics and potamology; maintenance of navigation, including training works and erosion problems. *Mailing Add:* PO Box 682 Rolla MO 65401

MUNGER, STANLEY H(IRAM), b Detroit, Mich, Apr 29, 20; m 55; c 3. RESEARCH MANAGEMENT, CHEMICAL ENGINEERING. *Educ:* Purdue Univ, BSChE, 42. *Prof Exp:* Res engr, 42-44 & 46-52, res supvr, 52-58, res mgr, E I du Ponte Nemours & Co, Inc, 58-85; CONSULT, 85- *Mem:* AAAS; Am Chem Soc. *Res:* Polymer processing; photopolymerization; photographic films. *Mailing Add:* 33 Wardell Ave Rumson NJ 07760

MUNI, INDU A, b Amreli, India, Oct 24, 42; m 69; c 2. TOXICOLOGY. *Educ:* Univ Nagpur, BS, 64; NDak State Univ, MS, 66; Univ Miss, PhD(biochem, pharmacol), 68. *Prof Exp:* NIMH res fel, Univ Miss, 68-69; clin biochemist dept path, St Joseph's Hosp, Milwaukee, 69-74; sr res scientist, Miles, Inc, 74-76; mgr regulatory & clin affairs, J T Baker Diag, 77-78; dir spec serv, Ind Bio-Test Labs, 78-81; dir toxicol, Bioassay Systs Corp, 81-87; PRES, DYNAGEN, INC, 87- *Mem:* Am Asn Clin Chemists; Am Chem Soc; Regulatory Affairs Prof Soc; Soc Toxicol. *Res:* Clinical chemistry; immunochemical and biochemical toxicology. *Mailing Add:* DynaGen Inc 99 Erie St Cambridge MA 02139

MUNIAPPAN, RANGASWAMY NAICKER, b Coimbatore, India, June 1, 41; US citizen; m 72; c 1. BIOLOGICAL CONTROL, ECONOMIC ENTOMOLOGY. *Educ:* Univ Madras, BSc,63, MSc, 65; Okla State Univ, PhD(entom), 69. *Prof Exp:* Chief plant indust, Guam Dept Agr, 71-75; assoc prof entom, 75-76, assoc dean, Agr Exp Sta, 76-83, PROF ENTOM & DIR AGR EXP STA, UNIV GUAM, 83- *Concurrent Pos:* Fulbright Res Scholar, US Info Serv, 84-85; consult, Interam Develop Bank, Univ Guyana, 89. *Mem:* Pac Sci Asn; Entom Soc Am; AAAS; Am Inst Biol Sci; Am Registry Prof Entomologists; Sigma Xi. *Res:* Biological control of the giant African snail, tropical insect pests and weeds. *Mailing Add:* Agr Exp Sta Univ Guam UOG Sta Mangilao GU 96923

MUNIES, ROBERT, industrial pharmacy, research administration, for more information see previous edition

MUNIGLE, JO ANNE, b Los Angeles, Calif, Nov 14, 34. TOXICOLOGY. *Educ:* Conn Col, BA, 57; Cornell Univ, PhD(anat), 67. *Prof Exp:* Fel anat, McGill Univ, 66-68; mgr reprod & mutagenesis, 68-78, dir Toxicol, Safety, Health & Ecol Dept, 78-87, VPRES TOXICOL, REGULATORY AUDITING & COMPLIANCE, CIBA-GEIGY CORP, 88- *Mem:* AAAS; Genetics Soc Am; Sigma Xi. *Res:* Developmental biology and genetics. *Mailing Add:* 500 E 77th St New York NY 10162

MUNIR, ZUHAIR A, b Iraq, July 7, 34; US citizen; m 63; c 2. COMBUSTION SYNTHESIS OF HIGH-TEMPERATURE MATERIALS. *Educ:* Univ Calif, Berkeley, BS, 56, MS, 58, PhD(mat sci), 63. *Prof Exp:* Chem engr, Mt Copper Co, 56-57; from asst prof to prof mat sci, San Jose State Col, 62-73; PROF MECH ENG, UNIV CALIF, DAVIS, 73-, ASSOC DEAN GRAD STUDIES, 80- *Concurrent Pos:* Consult, Molectro Corp, 63-64, Lawrence Livermore Lab, 73-, Rockwell-Int, 75 & State Calif Energy Comn, 77; engr, IBM Corp, 64, consult, 65-; engr, Gen Elec Co, 65; vis mem fac, Sch Eng Sci, Fla State Univ, 67-68; vis prof, Univ Calif, Davis, 72-73; Von Humboldt award sr US scientists, 90. *Mem:* Fel Am Ceramics Soc; fel Am Soc Metals; Electrochem Soc; Am Inst Mech Engrs. *Res:* High temperature chemistry and thermodynamic properties; thermodynamics and kinetics of solid-gas reactions; surface phenomena and corrosion; combustion synthesis of high temperature materials. *Mailing Add:* Dept Mech Aero & Mat Eng Univ of Calif Davis CA 95616-5294

MUNK, BENEDIKT AAGE, b Fredericia, Denmark, Dec 3, 29. ELECTRICAL ENGINEERING. *Educ:* Tech Univ Denmark, MScEE, 54; Ohio State Univ, PhD, 68. *Prof Exp:* Res & develop engr, Rhode & Schwarz, Ger, 57-58; engr, A/S Nordisk Antenne Fabrik, Denmark, 59-60; res & develop engr, Andrew Corp, 61-63; sr res engr, NAm Aviation, Inc, 63-64; res assoc, ElectroSci Lab, Ohio State Univ, 64-68, asst supvr, 68-69, from asst prof to assoc prof elec eng, 71-86, ASSOC SUPVR ELECTROSCI LAB, OHIO STATE UNIV, 69-, PROF ELEC ENG, 86- *Mem:* Fel Inst Elec & Electronic Engrs. *Res:* Electromagnetic theory; antennas; scattering; arrays; radomes; absorbers; radar; camouflage; periodic surfaces. *Mailing Add:* ElectroSci Lab Ohio State Univ 1320 Kinnear Rd Columbus OH 43212

MUNK, MINER NELSON, b Napa, Calif, Nov 17, 34; m 55; c 2. APPLIED PHYSICS. *Educ:* Univ Calif, Berkeley, AB, 57, MA, 59, PhD(physics), 67. *Prof Exp:* Physicist, Aerojet Gen Corp, Calif, 59-62; sr physicist, Varian Aerograph, 67-75; physicist, Milton Roy Co, 75-78; CHIEF SCIENTIST, LDC/MILTON ROY, 78- *Mem:* Optical Soc Am; Am Chem Soc. *Res:* Physical methods of analysis; mass spectroscopy; liquid chromatography. *Mailing Add:* LDC/Milton Roy Co PO Box 10235 Riviera Beach FL 33419

MUNK, PETR, b Praha, Czech, Oct 31, 32; m 61; c 2. POLYMER THERMODYNAMICS, POLYMER CHARACTERIZATION. *Educ:* Inst Chem Technol, Prague, Czech, MS, 56; Inst Macromolecular Chem, Czech Acad Sci, PhD(phys chem macromolecules), 60, DSc(phys chem macromolecules), 67. *Prof Exp:* Head dept molecular hydrodyn, Inst Macromolecular Chem, Czech Acad Sci, 56-67; res scientist, Res Triangle Inst, NC, 68; vis assoc prof chem, 69-71, from asst prof to assoc prof, 71-88, PROF CHEM, UNIV TEX, AUSTIN, 88- *Concurrent Pos:* Vis prof, Univ Mainz, WGermany, 79-80; vis scientist, Bausch & Lomb, Rochester, NY, 81-82. *Mem:* Am Chem Soc. *Res:* Thermodynamics of macromolecular solutions and blends; sedimentation analysis; inverse gas chromatography; viscometry; streaming birefringence; light scattering. *Mailing Add:* Dept Chem Univ Tex Austin TX 78712

MUNK, VLADIMIR, b Pardubice, Czech, Feb 27, 25; m 50; c 2. MICROBIOLOGY, BIOCHEMISTRY. *Educ:* Prague Tech Univ, MS, 50, PhD(biochem), 55. *Prof Exp:* Head anal dept, Cent Res Inst Food Indust, 53-57, head microbiol dept, 58-63; sr microbiologist, Inst Microbiol, Czech Acad Sci, 63-69; PROF BIOL, STATE UNIV NY COL PLATTSBURGH, 69- *Concurrent Pos:* Univ fel & grant-in-aid, 70-71, Chase Chem Co fel, 71-72; exchange scholar, State Univ NY, 74. *Honors & Awards:* State Prize, Govt Czech Socialistic Repub, 68. *Mem:* AAAS; Am Soc Microbiol. *Res:* Industrial use of microorganisms; microbial production of vitamins and enzymes; fermentation of hydrocarbons; continuous cultivation of microorganisms. *Mailing Add:* Fac of Sci State Univ of NY Col Plattsburgh NY 12901

MUNK, WALTER HEINRICH, b Vienna, Austria, Oct 19, 17; nat US; m 53; c 2. PHYSICAL OCEANOGRAPHY. *Educ:* Calif Inst Technol, BS, 39, MS, 40; Univ Calif, PhD(oceanog), 47. *Hon Degrees:* Dr, Univ Bergen, Norway, 75 & Univ Cambridge, 86. *Prof Exp:* From asst prof to assoc prof geophys, 47-54, assoc dir, Inst Geophys & Planetary Physics, 59-82, PROF GEOPHYS, UNIV CALIF, SAN DIEGO, 54- *Concurrent Pos:* Guggenheim Found fel, Univ Oslo, 48, Cambridge Univ, 55 & 62; Fulbright fel, 81-82. *Honors & Awards:* Arthur L Day Medal, Am Geol Soc, 65; Sverdrup Gold Medal, Am Meteorol Soc, 66; Gold Medal, Royal Astron Soc, 68; Josiah Willard Gibbs Lectr, Am Math Soc, 70; Maurice Ewing Medal, Am Geophys Union & USN, 76; Agassiz Medal, Nat Acad Sci, 76; Capt Robert Dexter Conrad Award, Dept Navy, 78; Nat Medal Sci, 85; William Bowie Medal, Am Geophys Union, 89. *Mem:* Nat Acad Sci; Am Acad Arts & Sci; Am Philos Soc; foreign mem Royal Soc London; Leopoldina Ger Acad Res Natural Sci; fel Am Geophys Union; fel Am Meteorol Soc; fel Acoust Soc Am; fel Marine Technol Soc. *Res:* Ocean waves; tides; wind stress and ocean currents; rotation of the earth; ocean acoustics. *Mailing Add:* Inst Geophys & Planetary Sci Univ Calif San Diego La Jolla CA 92093

MUNKACSI, ISTVAN, b Budapest, Hungary, Apr 15, 27; m 54; c 2. ANATOMY. *Educ:* Med Univ Budapest, Dr med, 53; Univ Khartoum, Dr philos, 64; Univ Saskatchewan, Dr med ad lundem, 71. *Prof Exp:* Lectr anat, Med Univ Budapest, 53-57, sr lectr, 57-60; sr lectr, Univ Khartoum, 60-64; sr lectr, Med Univ Budapest, 64-66; vis asst prof, 66-68, assoc prof, 68-74, PROF ANAT, UNIV SASK, 74- *Mem:* Am Asn Anatomists; Can Asn Anat; Am Asn Clin Anatomists; Pan Am Asn Anat (secy gen, 72-75). *Res:* Comparative morphology; innervation of the kidney and distribution of the type of nephrons investigated in desert and laboratory mammals; blood vessels and arterio-venous connections of the kidney; dynamics of synapse formation in the sympathetic nervous tissue in tissue culture; clinical anatomy in orthopedic surgery and neurosurgery. *Mailing Add:* Dept of Anat Univ of Sask Saskatoon SK S7N 0W0 Can

MUNKRES, JAMES RAYMOND, b Omaha, Nebr, Aug 18, 30; m 64; c 2. MATHEMATICS. *Educ:* Nebr Wesleyan Univ, AB, 51; Univ Mich, AM, 52, PhD(math), 56. *Hon Degrees:* DSc, Nebr Wesleyan Univ, 90. *Prof Exp:* Instr math, Univ Mich, 55-57; instr, Princeton Univ, 57-58, Fine instr, 58-60; from asst prof to assoc prof, 60-66, PROF MATH, MASS INST TECHNOL, 66- *Concurrent Pos:* Sloan res fel, 65-67. *Mem:* Am Math Soc; Math Asn Am. *Res:* Differential and combinatorial topology. *Mailing Add:* Dept Math Mass Inst Technol 77 Massachusetts Ave Cambridge MA 02139

MUNN, DAVID ALAN, b Elyria, Ohio, Mar 27, 47; m 69; c 2. SOIL CHEMISTRY, AGRONOMY. *Educ:* Ohio State Univ, BS, 70, MS, 72, PhD(agron), 74. *Prof Exp:* Teaching & res assoc, Soils Agron Dept, Ohio State Univ, 70-74; vis asst prof & res assoc, Soil Sci Dept, NC State Univ, 74-76; ASSOC PROF SOILS & AGRON, AGR TECH INST, OHIO STATE UNIV, 76- *Mem:* Soil Sci Soc Am; Am Soc Agron; Sigma Xi; AAAS. *Res:* Plant nutrition, including phosphorus, potassium, nitrogen, lime and acidity. *Mailing Add:* Agr Tech Inst 1328 Dover Rd Wooster OH 44691

MUNN, GEORGE EDWARD, b Lawrence Co, Pa, Nov 29, 24. ORGANIC CHEMISTRY. *Educ:* Westminster Col, Pa, BS, 45; Univ Ill, PhD(org chem), 48. *Prof Exp:* Asst, Univ Ill, 45-47; Du Pont fel, Mass Inst Technol, 48-49; CHEMIST, E I DU PONT DE NEMOURS & CO, INC, 49- *Mem:* Am Chem Soc. *Res:* Synthetic organic chemistry; polymer chemistry. *Mailing Add:* 105 Allmond Ave Wilmington DE 19803-4901

MUNN, JOHN IRVIN, b Pittsburgh, Pa, Oct 28, 22; m 48; c 3. ENVIRONMENTAL HEALTH, PHARMACOLOGY. *Educ:* Ind Univ Pa, BS, 48; George Washington Univ, MS, 52; Georgetown Univ, PhD(biochem), 57. *Prof Exp:* Res biochemist, US Naval Med Res Inst, 48-52; res hematologist, Walter Reed Army Inst Res, 52-58; biochemist, Food & Drug Admin, 58-61; scientist adminr-pharmacologist, NIH, Md, 61-71; sr scientist, WHO, 71-76; asst to sci coordr environ cancer, Nat Cancer Inst, NIH, 76-80; CONSULT, 80- *Honors & Awards:* Superior Performance Award, Walter Reed Army Inst Res, 57. *Mem:* Soc Toxicol; Am Soc Pharmacol & Exp Therapeut. *Res:* Metabolism of gallium; biochemical methodology; hemoglobin; fatty acids and surface areas of erythrocytes; toxicity of emulsifiers in food; food standardization and contaminant monitoring; international environmental health policies; identification, assessment and control of carcinogenic environmental agents. *Mailing Add:* 11022 Marcliff Rd Rockville MD 20852

MUNN, ROBERT EDWARD, b Winnipeg, Man, July 26, 19; m 44; c 4. METEOROLOGY, AIR POLLUTION. *Educ:* McMaster Univ, BA, 41; Univ Toronto, MA, 45; Univ Mich, PhD(meteorol), 62. *Prof Exp:* Meteorologist, Meteorol Serv Can, 41-77; prof, Inst Environ Studies, Univ Toronto, 77-84; head, environ prog, Int Inst Appl Systs Anal, Austria, 85-89; INSTR, ENVIRON STUDIES, UNIV TORONTO, 89- *Concurrent Pos:* Vis prof, Univ Stockholm, 70; ed, Scope Monographs on Environ, Int Coun Sci Unions & J Boundary Layer Meteorol. *Honors & Awards:* Appl Meteorol Award, Am Meteorol Soc, 74, Paterson Medal, 75; Frank A Chambers Res Award, Air Pollution Control Asn, 84. *Mem:* Fel Am Meteorol Soc; fel Royal Meteorol Soc; Can Meterol & Oceanog Soc; fel AAAS; fel Royal Soc Can. *Res:* Micrometeorology; regional air pollution; environmental monitoring; environmental impact assessment. *Mailing Add:* Inst Environ Studies Univ Toronto Toronto ON M5S 1A4 Can

MUNN, ROBERT JAMES, b Southampton, Eng, Jan 31, 37; m 62; c 2. CHEMICAL PHYSICS. *Educ:* Bristol Univ, BSc, 57, PhD(chem), 61. *Prof Exp:* Res fel chem, Bristol Univ, 61-63; Harkness fel of Commonwealth Fund, Univ Md, 63-64; lectr chem, Bristol Univ, 64-65; lectr math, Queen's Univ, Belfast, 65-66; from asst prof to assoc prof chem physics, 66-72, PROF CHEM, UNIV MD, COLLEGE PARK, 72-, DIR, 70-, ASST DEAN, 75- *Concurrent Pos:* Consult, Lockheed Electronics, 70-71 & NSF, 70- *Mem:* AAAS; Am Phys Soc. *Res:* Scattering; educational technology; computer aided instruction. *Mailing Add:* Dept Chem Univ Md College Park MD 20740

MUNNECKE, DONALD EDWIN, b St Paul, Minn, May 30, 20; m 42; c 4. PHYTOPATHOLOGY. *Educ:* Univ Minn, BA, 42, MS, 49, PhD(plant path), 50. *Prof Exp:* Instr & jr plant pathologist, Univ Calif, Los Angeles, 51-53, from asst prof & asst plant pathologist to assoc prof & assoc plant pathologist, 53-61; from assoc prof & assoc plant pathologist to prof & plant pathologist, 61-85, EMER PROF PATH, UNIV CALIF, RIVERSIDE, 85 - *Concurrent Pos:* Guggenheim fel & Fulbrigh res scholar, Univ Gottingen, 65-66. *Mem:* Fel Am Phytopath Soc. *Res:* Ornamental plant diseases; chemical soil treatments for disease control; fungicide action in soils; ecological relations in control of Armillaria mellea. *Mailing Add:* Dept Plant Path Univ Calif Riverside CA 92521-0122

MUNNELL, EQUINN W, b Sayville, NY, June 28, 13; m 37; c 3. MEDICINE. *Educ:* Amherst Col, BA, 35; Cornell Univ, MD, 39; Am Bd Obstet & Gynec, dipl, 48. *Prof Exp:* Asst instr obstet & gynec, Col Med, NY Univ, 45-47; from asst prof to assoc prof, 50-59, PROF OBSTET & GYNEC, COL PHYSICIANS & SURGEONS, COLUMBIA UNIV, 69- *Concurrent Pos:* Asst gynecologist, Mem Hosp, New York, 45-47; asst obstetrician & gynecologist, Presby Hosp, New York, 47-57, assoc obstetrician & gynecologist, 57-69, attend obstetrician & gynecologist, 69-; assoc attend gynecologist, Francis Delafield Hosp, 53-62, attend gynecologist, 62- *Mem:* Fel Am Col Surg; fel Am Col Obstet & Gynec; Am Asn Obstet & Gynec; Am Cancer Soc; Soc Gynec Theologists. *Res:* Obstetrics and gynecology; gynecologic cancer. *Mailing Add:* 1011 Lexington Ave New York NY 10032

MUNNO, FRANK J, b New Castle, Pa, Jan 5, 36; m 58; c 3. NUCLEAR ENGINEERING. *Educ:* Waynesburg Col, BS, 57; Univ Fla, MS, 62, PhD(nuclear eng), 64. *Prof Exp:* From asst prof to assoc prof nuclear eng, 65-72, dir nuclear eng prog, 72-90, PROF NUCLEAR ENG, UNIV MD, COLLEGE PARK, 72- *Concurrent Pos:* Consult, Gen Physics Corp, 66-69 & Harry Diamond Lab, 70-78, Armed Forces Radiobiol Res Inst, 79- & Baltimore Gas & Elec Co, 82-; fel, Pub Health, 64, DuPont Fac, 65. *Mem:* Am Nuclear Soc; Am Inst Chem Engrs. *Res:* Heat transfer in two phase flow; nuclear radiation shielding and neutral particle transport; nuclear reactor dynamics. *Mailing Add:* Dept Mat & Nuclear Eng Univ of Md Col Eng College Park MD 20740

MUNNS, DONALD NEVILLE, b Sydney, Australia, Sept 6, 31; m 60; c 3. PLANT NUTRITION, SOIL SCIENCE. *Educ:* Univ Sydney, BScAgr, 54; Univ Calif, Berkeley, PhD(soil sci), 61. *Prof Exp:* Asst chemist, NSW Dept Agr, Australia, 54-57; res asst, Univ Calif, Berkeley, 57-60; from res officer to sr res officer, Commonwealth Sci & Indust Res Orgn, Canberra, 60-66; asst chemist, 66-68, assoc prof soils & plant nutrit, 68-76, PROF SOIL SCI, UNIV CALIF, DAVIS, 76- *Concurrent Pos:* Vis soil scientist, Univ Hawaii, 73-74, Agr Res Inst, Wagga, New S Wales, 86 & Univ Sydney, 87. *Mem:* Soil Sci Soc Am; Am Soc Agron. *Res:* Plant soil relationships; soil acidity; salinity; genetic variations in response of plants and microorganisma to substrate factors; plant microbe synbioses. *Mailing Add:* Dept Land Air & Water Resources Univ Calif Davis CA 95616

MUNNS, THEODORE WILLARD, b Peoria, Ill, June 11, 41. BIOCHEMISTRY. *Educ:* Bradley Univ, BS, 63; St Louis Univ, PhD(biochem), 70. *Prof Exp:* Res assoc biochem, 70-74, instr, 74-76, ASST RES PROF, BIOCHEM, ST LOUIS UNIV, 76-; ASST RES PROF, WASHINGTON UNIV, 80- *Concurrent Pos:* NSF fel, St Louis Univ, 70-71, NIH fel, 72- *Mem:* Am Chem Soc; Sigma Xi; NY Acad Sci; Am Soc Biol Chemists. *Res:* Nucleic acid and protein metabolism in neoplastic systems; immunochemistry. *Mailing Add:* Box 3 Wright City MO 63390

MUNNS, W(ILLIAM) O, b Tuxford, Sask, Dec 16, 26; m 53; c 5. CHEMICAL ENGINEERING. *Educ:* Univ Sask, BSc, 49; Univ Toronto, MBA, 76. *Prof Exp:* Res engr, Res & Develop Labs, 49-56, group leader chem develop, 56-57, group leader food & pharmaceut, 57-65, new prod mgr, 65-71, chief chemist, 71-80, nat qual assurance mgr, Can Packers Ltd, 80-86; CONSULT, 86- *Concurrent Pos:* Chmn res policy comt, Meat Packers Coun Can, 65-; chmn & chief exec off, Innovation Can Inc, 77-; consult, Int Develop Res Ctr & Can Int Develop Agency; chmn nat conf, Can Inst Food Sci & Technol, 85, chmn & chief exec off, Innocan Inovation Ctr Inc, 81-; consult, chmn & chief exec off, Windows On The World Inc, 88-, Housing & Int Devel, 88- *Mem:* Am Inst Chem Engrs; fel Chem Inst Can; Can Inst Food Sci & Technol; Inst Food Technol; sr mem Am Soc Qual Control. *Res:* Food processing; pharmaceutical processing; product planning and new product development; quality assurance; chemical laboratories, microbiological laboratories and quality control; innovation management; entrepreneurship; product and service development in the public and private sectors; housing; international development. *Mailing Add:* Two Woodark Rd Weston ON M9P 1M1 Can

MUNOZ, JAMES LOOMIS, b East Orange, NJ, Oct 31, 39. GEOCHEMISTRY. *Educ:* Princeton Univ, AB, 61; Johns Hopkins Univ, PhD(geol), 66. *Prof Exp:* Fel, Carnegie Inst Geophys Lab, 66-68; asst prof geol, 68-74, assoc prof, 74-87, PROF GEOL, UNIV COLO, BOULDER, 87- *Concurrent Pos:* Ed, Am Mineralogist, 86-89. *Mem:* AAAS; Geol Soc Am; Mineral Soc Am; Geochem Soc. *Res:* Application of thermodynamics to petrology; geochemistry of fluorine in igneous, metamorphic, and ore-forming processes. *Mailing Add:* Dept Geol Sci Univ Colo Box 250 Boulder CO 80309-0250

MUNOZ, JOHN JOAQUIN, b Guatemala, Dec 23, 18; nat US; m 47; c 4. IMMUNOLOGY, MICROBIOLOGY. *Educ:* La State Univ, BS, 42; Univ Ky, MS, 45; Univ Wis, PhD(med bact), 47; Am Bd Microbiol, dipl. *Prof Exp:* Jr & sr technician bact, Univ Ky, 42-44; asst microbiol, Univ Wis, 44-47; asst prof med bact, Sch Med, Univ Ill, 47-51; res assoc, Merck Sharp & Dohme Res Labs, 51-57; prof microbiol & pub health, chmn dept & dir Stella Duncan Mem Labs, Univ Mont, 57-61; res microbiologist, 61-88, EMER SCIENTIST, ROCKY MOUNTAIN LAB, NAT INST ALLERGY & INFECTIOUS DIS, 89- *Concurrent Pos:* Spec assignment, Pasteur Inst, Paris, 66-67; spec assignment, Walter & Eliza Hall Inst Med Res, Melbourne, Australia, 82-83. *Honors & Awards:* Director's Award, NIH, 79. *Mem:* Fel Am Acad Microbiol; Am Soc Microbiol; Am Asn Immunol; Soc Exp Biol & Med; Int Endotoxin Soc. *Res:* Biologically active substances from Bordetella pertussis; mechanisms of action of B pertussis active substances; mechanism of anaphylaxis and hypersensitivity; purification of bacterial antigens; serological techniques. *Mailing Add:* Rocky Mountain Lab Nat Inst Allergy & Infect Dis Hamilton MT 59840

MUÑOZ, MARIA DE LOURDES, b Mex, Mar 14, 52; m 78; c 2. PARASITOLOGY, MOLECULAR BIOLOGY. *Educ:* Escuela Nacional Ciencias Biologicas, BSc, 74; Ctr Invest & Advan Studies, MSc, 78, PhD(cell biol), 81. *Prof Exp:* Postdoctoral exp path, Ctr Invest & Advan Studies, 82-83; headline prof technol & sci invest, Nat Sch Prof Studies, Iztacala, Mex, 86-88; headline prof biostatist & computation, 89-91, HEADLINE PROF CELL BIOL & LAB TECH, DEPT GENETICS & MOLECULAR BIOL, CINESTAV IPN, 86- *Concurrent Pos:* Headline prof, Univ Autonoma Juarez Tabasco, Cinestav-IPN, 78, Univ Sonora, 80, Univ Autonoma Baja Calif & Univ Autonoma Tlaxcala, 81 & Univ Autonoma Aguascalientes, 89; mem, Appraiser Comt, Dept Spec Supports, Consejo Nacional Ciencia & Tecnologi, 89, acad appraiser, 90. *Mem:* Am Soc Trop Med & Hyg; Am Soc Cell Biol. *Res:* Mechanisms of pathogenicity in the parasite Entamoeba histolytica and diagnosis using monoclonal and polyclonal antibodies anti-electron-dense granules; electron-dense granules secretion induced by collagen type I; role of electron-dense granules-associated collagenolytic activity during tissue invasion; regulation of electron-dense granules by calmodulin. *Mailing Add:* Dept Genetics & Molecular Biol Cinestav IPN Apartado Postal 14-740 Mexico DF 07000 Mexico

MUNRO, DONALD W, JR, b Philadelphia, Pa, Dec 27, 37; m 61; c 2. PHYSIOLOGY, GENETICS. *Educ:* Wheaton Col, BS, 59; Pa State Univ, MS, 63, PhD(zool), 66. *Prof Exp:* Assoc prof zool, 66-71, PROF BIOL & HEAD DEPT, HOUGHTON COL, 71- *Mem:* Am Soc Zool; Am Physiol Soc; Am Inst Biol Sci; AAAS. *Res:* Effects of cold exposure and hibernation on the respiration rates and oxidative phosphorylation of various tissues of carp, trout, frogs, rats, hamsters and chipmunks. *Mailing Add:* Dept Biol Houghton Col Houghton NY 14744

MUNRO, HAMISH N, b Edinburgh, Scotland, July 3, 15; wid; c 4. BIOCHEMISTRY, NUTRITION. *Educ:* Glasgow Univ, BSc, 36, MB, 39, DSc(biochem), 56, MD, 83. *Hon Degrees:* Dr, Nancy, France, 82. *Prof Exp:* Clin tutor med, Victoria Infirmary, Glasgow, Scotland, 40-45, asst dir path, 42-45; lectr physiol, Glasgow Univ, 45-47, sr lectr biochem, 47-56, reader nutrit biochem, 56-64, prof biochem, 64-66; prof physiol chem, Mass Inst Technol, 66-79; prof med & nutrit & dir, 79-83, SR SCIENTIST, USDA HUMAN NUTRIT RES CTR, TUFTS UNIV, 83- *Concurrent Pos:* Rockefeller traveling fel, 48; Fleck lectr, Glasgow Univ, 60; mem protein requirements comt, WHO-Food & Agr Orgn, 63, 71, 81; chmn food & nutrit bd, US Nat Res Coun, 75-80; chmn, Dietary Allowances Comt; mem, Comm Life Sci, 85; adj prof, Mass Inst Technol, 79-90. *Honors & Awards:* Osborne & Mendel Award, Am Inst Nutrit, 68; Borden Award; Bristol-Myers Award; Bolton Medal, 88. *Mem:* Nat Acad Sci; Am Inst Nutrit; fel Royal Soc Edinburgh; Brit Inst Biol; Brit Biochem Soc. *Res:* Mammalian protein metabolism; protein synthesis control mechanisms and actions of hormones on protein and RNA metabolism; tissue analysis of mammals; nutrition and aging; proteins of iron storage; ferritin. *Mailing Add:* Human Nutrit Res Ctr Aging USDA Tufts Univ 711 Washington St Boston MA 02111

MUNRO, JAMES IAN, b Newcastle-upon-Tyne, Eng, July 10, 47; Can citizen; m 79; c 2. COMPUTER SCIENCE. *Educ:* Univ New Brunswick, BS, 68; Univ Brit Columbia, MS, 69; Univ Toronto, PhD, 71. *Prof Exp:* From asst prof to assoc prof, 71-81, PROF COMPUT SCI, UNIV WATERLOO, 81- *Concurrent Pos:* Dir, Inst Comput Res, Univ Waterloo; mem tech staff comput sci, AT& T Bell labs, Murray Hill, 84-85. *Mem:* Asn Comput Mach. *Res:* Efficiency of computation, particularly data structures; specific interest include complexity of comparison based problems; file structures; text dominated databases; computational complexity of algebraic and numeric problems. *Mailing Add:* Dept Comput Sci Univ Waterloo Waterloo ON N2L 3G1 Can

MUNRO, MICHAEL BRIAN, b Calgary, Alta, Oct 12, 48; m 71. MECHANICAL ENGINEERING. *Educ:* Univ Waterloo, BASc, 71, PhD(mech eng), 77; Mass Inst Technol, SM, 73. *Prof Exp:* ASSOC PROF MECH ENG, UNIV OTTAWA, 78- *Concurrent Pos:* Nat Res Coun Can scholar, 73-76; Nat Res Coun Can fel, Univ Cambridge, 77. *Mem:* Assoc mem Am Soc Mech Engrs; Can Med & Biol Eng Soc; Am Soc Metals; Soc Adv Mat & Process Eng; Am Soc Testing & Mat. *Res:* Engineering application of composite materials; automated manufacturing of composite materials; filament winding. *Mailing Add:* Dept of Mech Eng Univ of Ottawa Ottawa ON K1N 6N5 Can

MUNRO, RICHARD HARDING, b Pasadena, Calif, Jan 28, 43; m 67; c 2. SOLAR PHYSICS, ASTRONOMY. *Educ:* Harvey Mudd Col, BS, 64; Harvard Univ, MA, 69, PhD(astron), 73. *Prof Exp:* Scientist solar physics, High Altitude Observ, Nat Ctr Atmospheric Res, 72-85; software mgr, Martin Marietta Denver Aerospace, 85-87; PROG MGR, BALL AEROSPACE SYSTS GROUP, 87- *Concurrent Pos:* Prin investr, Joint Lyman Alpha/White Light Spartan 2 Exp, 79-85; lectr, Univ Colo, 81. *Mem:* Am Astron Soc; Am Geophys Union; Int Astron Union; Sigma Xi. *Res:* Investigation of the physical properties, the dissipation and transport of non-radiative energy and the initial acceleration of the solar wind in the transition region and corona of the solar atmosphere. *Mailing Add:* 2378 Dennison Lane Boulder CO 80303-5713

MUNRO, RONALD GORDON, US citizen. MODELING, DATABASES. *Educ:* Univ Mich, BS, 69, MS, 70; Univ Ore, PhD(physics), 76. *Prof Exp:* Res assoc fel physics & solid state mat, Nat Bur Standards/Nat Res Coun, 76-78, res physicist, physis at high pressure, Nat Bur Standards, 78-84, from res physicist to supvr physicist tribology, 84-87, PHYSICIST CERAMICS DIV, NAT INST STANDARDS & TECHNOL, 87- *Concurrent Pos:* Proj leader, struct ceramics database, Nat Inst Standards & Technol, 88- *Mem:* Am Phys Soc; Am Ceramics Soc. *Res:* Materials research; materials property data for advanced ceramics; tribology of ceramics. *Mailing Add:* A256 Bldg 223 Nat Inst Standards & Technol Gaithersburg MD 20899

MUNRO, WILLIAM DELMAR, b Cedaredge, Colo, Nov 22, 16; m 51; c 3. NUMERICAL ANALYSIS. *Educ:* Univ Colo, BA, 38; Univ Minn, MA, 40, PhD(math), 47. *Prof Exp:* Asst math, Univ Minn, Mineapolis, 38-41, instr, 41-43, asst prof math & mech, 45-49, from assoc prof to prof math, 49-69, from prof to emer prof comput, info & control sci & assoc head adept, 69-87; RETIRED. *Concurrent Pos:* Proj engr, Honeywell Inc, 43-; consult, Radio Corp Am, 45-, Maico Corp, 58-, Viron Div, GCA Corp, 65- & North Star Res Corp, 70-; vis res mathematician, Univ Calif, Los Angeles, 57-58; vis prof, Johns Hopkins Univ, 59-60. *Mem:* Am Math Soc; Math Asn Am; Soc Indust & Appl Math; Asn Comput Mach. *Res:* Theory of approximation; computers; numerical methods and analysis research; navigation computer for aircraft; orthogonal trigonometric sums with auxiliary conditions; high order precision by exact arithmetic. *Mailing Add:* Dept Comput Sci Univ Minn Minneapolis MN 55455

MUNROE, EUGENE GORDON, b Detroit, Mich, Sept 8, 19; nat Can; m 44; c 4. ENTOMOLOGY, ECOLOGY. *Educ:* McGill Univ, BSc, 40, MSc, 41; Cornell Univ, PhD(entom), 48. *Prof Exp:* Lectr & res asst, Inst Parasitol, Macdonald Col, McGill Univ, 46-50; agr res officer, Can Dept Agr, 50-65; sci adv, Sci Secretariat, Off Privy Counr, 65-67, head studies, 67-68; res scientist, Biosyst Res Inst, Can Dept Agr, 68-79; RES ASSOC, LYMAN ENTOM MUS, MCGILL UNIV, 81-; BIOSYSTEM RES CTR, AGRIC, OTTAWA, CAN, 88- *Concurrent Pos:* Vis lectr, Univ Calif, Berkeley, 60-61; mem, Steering Comt, Biol Coun Can, 65-66; res assoc, Entom Res Inst, Can Dept Agr, 65-68 & 80-81; Bishop Mus, Honolulu, 86- *Honors & Awards:* Queens Silver Jubille Medal, 78. *Mem:* Royal Soc Can; hon mem Lepidop Soc (pres, 59-60); Entom Soc Can (pres, 63-64); Entom Soc Am. *Res:* Science policy; research planning and management; management and conservation of renewable resources; ecology; biogeography; taxonomy; systematics. *Mailing Add:* 3093 Barlow Ct RR 1 Dunrobin ON K0A 1T0 Can

MUNROE, MARSHALL EVANS, b Gainesville, Ga, 18; m 47; c 1. MATHEMATICS. *Educ:* Univ Tex, BA, 40; Brown Univ, ScM, 41, PhD(math), 45. *Prof Exp:* Instr math, Brown Univ, 43-45; from instr to assoc prof, Univ Ill, 45-58; prof math & chmn dept, 59-81, EMER PROF MATH, UNIV NH, 81- *Concurrent Pos:* Vis prof, Cairo Univ, 65-66; vis prof, Univ Col Galway, 75-76. *Mem:* Am Math Soc; Math Asn Am. *Res:* Abstract integration theory; measure theory; modernization of calculus. *Mailing Add:* PO Box 304 Brooksville ME 04617

MUNROE, STEPHEN HORNER, b Baltimore, Md, June 3, 46; m 78; c 2. MOLECULAR BIOLOGY, BIOCHEMISTRY. *Educ:* Haverford Col, BA, 68; Ind Univ, PhD(biol chem), 74. *Prof Exp:* Res fel genetics, Children's Hosp Med Ctr, Harvard Med Sch, 74-77; res assoc cell biol, Worcester Found Exp Biol, 77-78; asst prof, 78-85, ASSOC PROF BIOL, MARQUETTE UNIV, 85- *Concurrent Pos:* Vis scientist, Cold Spring Harbor Lab, 87-88. *Mem:* Am Chem Soc; Am Soc Cell Biol; AAAS; Sigma Xi; Am Soc Microbiol. *Res:* Eukaryotic gene expression in RNA processing. *Mailing Add:* Dept Biol Marquette Univ Wehr Life Sci Bldg Milwaukee WI 53233

MUNSE, WILLIAM H(ERMAN), b Chicago, Ill, July 10, 19; m 42; c 2. CIVIL ENGINEERING. *Educ:* Univ Ill, BS, 42, MS, 44. *Prof Exp:* Engr, Champaign, Ill, 41; struct draftsman, Am Bridge Co, Ind, 42-43; spec asst civil eng, Univ Ill, 43-44; res engr, Lehigh Univ, 46-47; spec res asst prof, 47-52, res assoc prof, 52-55, prof, 55-81, EMER PROF CIVIL ENG, UNIV ILL, URBANA, 81- *Honors & Awards:* Adams Mem Award, Am Welding Soc, 61; W L Huber Res Prize, Am Soc Civil Engrs, 62. *Mem:* Am Soc Testing & Mat; Am Welding Soc; Am Rwy Eng Asn; Am Concrete Inst; Am Soc Eng Educ; Sigma Xi; Hon Mem Am Soc Civil Engrs, 83. *Res:* Design of structures of concrete and steel; metal structure; fatigue; riveted, bolted and welded structures; brittle fracture. *Mailing Add:* 2129 Civil Eng Bldg Univ Ill 205 N Mathews Urbana IL 61801-2397

MUNSEE, JACK HOWARD, b Niagara Falls, NY, Sept 27, 34; m 62; c 4. PHYSICS. *Educ:* Col Wooster, BA, 56; Case Western Reserve Univ, MS, 62, PhD(physics), 68. *Prof Exp:* Jr engr, Aeroprod Opers, 56-57; asst physics, Case Western Reserve Univ, 58-62, part-time instr, 64-68; instr physics & math, Col Wooster, 62-64; asst prof, 68-71, assoc prof, 71-85, PROF PHYSICS, CALIF STATE UNIV, LONG BEACH, 85- *Mem:* Am Phys Soc; Am Asn Physics Teachers. *Res:* Low energy nuclear physics; neutrinos; physics education; hyperfine structure of atoms. *Mailing Add:* Dept Physics & Astron Calif State Univ 1250 Bellflower Blvd Long Beach CA 90840

MUNSELL, MONROE WALLWORK, b New London, Conn, Jan 8, 25; m 54; c 2. LUBE OIL FORMULATION TECHNOLOGY. *Educ:* Carnegie-Mellon Univ, BS, 47, MS, 50, PhD, 55. *Prof Exp:* Res chemist, Esso Res & Eng Co, 55-64, sr chemist & proj leader, 64-68, res assoc, Esso Kagaku kk, Japan, 68-72, res assoc, Exxon Chem Co, 72-86; RETIRED. *Mem:* Am Chem Soc; Sigma Xi; Soc Automotive Eng. *Res:* Lubrication oil and fuel additives; technical service and product application in field of detergents and viscosity improvers; conduct lubricant technology training seminars for government oil companies in Asia and Latin America. *Mailing Add:* 180 Sutton Dr Berkeley Heights NJ 07922

MUNSICK, ROBERT ALLIOT, b Glen Ridge, NJ, Oct 21, 28; m 53, 65; c 3. OBSTETRICS & GYNECOLOGY. *Educ:* Cornell Univ, AB, 50; Columbia Univ, MD, 54, PhD(pharmacol), 62; Am Bd Obstet & Gynec, dipl, 66. *Prof Exp:* Intern med, Roosevelt Hosp, New York, 54-55; resident obstet & gynec, Sloane Hosp Women, 59-62; asst prof, Sch Med, Univ Colo, 62-65; prof, Sch Med, Univ NMex, 65-74, chmn dept, 64-73; PROF OBSTET & GYNEC, SCH MED, IND UNIV, INDIANAPOLIS, 74- *Concurrent Pos:* Josiah Macy, Jr Found fel, 57-62; NIH res grant, 62- *Mem:* Soc Gynec Invest; fel Am Col Obstet & Gynec; Asn Planned Parenthood Physicians; Am Gynec & Obstet Soc; Asn Profs Obstet & Gynec. *Res:* Education. *Mailing Add:* Dept Obstet & Gynec F-508 1001 W Tenth St Indianapolis IN 46202

MUNSON, ALBERT ENOCH, b Williamsport, Pa, Mar 12, 34; m; c 2. PHARMACOLOGY, TOXICOLOGY. *Educ:* Univ Buffalo, BS, 60; State Univ NY, Buffalo, MS, 68; Va Commonwealth Univ, PhD(pharmacol), 70. *Prof Exp:* Scientist, Mem Inst Buffalo, NY, 56-68; from asst prof to assoc prof microbiol & immunol, 70-83, from asst prof to assoc prof pharmacol & toxicol, 72-83, PROF MICROBIOL & IMMUNOL & PROF PHARMACOL & TOXICOL, MED COL, VA COMMONWEALTH UNIV, 83- *Concurrent Pos:* NIH res grants, 79-; pres, Immunotoxiocol Specialty Subsect, Soc Toxicol, 91-92; counr, Am Col Toxicol, 91- & Int Soc Immunopharmacol; mem, Immunotoxicol Comt, Nat Acad Sci, Toxicol Study Sect, Soc Toxicol, Nat Inst Environ Health Sci, Environ Protection Agency, NIH & Nat Inst Drug Abuse, Comt Environ Restoration Dept Energy Sites, 90; consult, WHO. *Mem:* Am Asn Cancer Res; Am Asn Microbiol; Am Asn Immunologists; Am Cancer Soc; Am Col Toxicol; Am Soc Pharmacol & Exp Therapeut; Int Soc Immunopharmacol; Reticuloendothelial Soc; Sigma Xi; Soc Toxicol. *Res:* Author of numerous technical publications. *Mailing Add:* Dept Pharmacol & Toxicol Med Col Va Commonwealth Univ Box 613 MCV Sta Richmond VA 23298

MUNSON, ALBERT G, b Baton Rouge, La, Jan 10, 31; m 64. AERONAUTICAL ENGINEERING, FLUID MECHANICS. *Educ:* La State Univ, BS, 51; Calif Inst Technol, MS, 52, AeroEng, 56; Stanford Univ, PhD(aeronaut & astronaut eng), 64. *Prof Exp:* Aerodyn engr, Hughes Aircraft Co, Calif, 56-58; aeronaut res scientist, Ames Res Ctr, Moffet Field, 58-62; mem tech staff, Nat Eng Sci Co, Calif, 64-65, mem sr staff, 66; prin staff engr, Martin Co, Fla, 66-67, dept staff engr, 67-69, sr staff engr, Denver Div, 69-73; prin scientist, Douglas Aircraft Co, McDonnell Douglas Corp, 73-77, sect chief acoustic res, 77-78, sect chief exterior acoustic design, 78-85; RES ENGR, NORTHROP CORP, ESD, 85- *Mem:* Assoc fel Am Inst Aeronaut & Astronaut; Acoust Soc Am; Sigma Xi. *Res:* Hypersonic aerodynamics; fluid mechanics. *Mailing Add:* Northrop Corp, ESD 2301 W 120th St Hawthorne CA 90251-5032

MUNSON, ARVID W, b Paterson, NJ, Aug 22, 33; m 54; c 2. AGRICULTURAL STATISTICS. *Educ:* Iowa State Univ, BS, 55, MS, 57; Okla State Univ, PhD(animal breeding), 66. *Prof Exp:* Anal statistician, Biomet Serv Staff, Agr Res Serv, USDA, 65-66, biometrician, 66-67, asst dir, 67-68, actg dir data processing, Data Syst Appln Div, 68; dir, Res Serv Div, 68-78, PRES, RALTECH SCI SERV DIV, RALSTON PURINA CO, 78- *Mem:* AAAS; Am Soc Animal Sic; Am Statist Asn; Am Mgt Asn; Inst Food Technologists; Sigma Xi. *Res:* Statistical methodology and computer techniques in animal science research and improvement. *Mailing Add:* 10226 Forist Lake Dr Great Falls VA 22066

MUNSON, BENJAMIN RAY, b Tonawanda, NY, Sept 19, 37; m 63; c 2. BIOCHEMISTRY, ONCOLOGY. *Educ:* Houghton Col, BA, 60; State Univ NY Buffalo, PhD(biochem, pharmacol), 68. *Prof Exp:* Teacher, NY High Schs, 60-62; sr cancer res scientist, Springville Labs, Roswell Park Mem Inst, 70-76, cancer res scientist IV, Exp Biol Dept, 76-85. *Concurrent Pos:* NIH res fel biochem, Springville Labs, Roswell Park Mem Inst, 68-70, fel physiol, Inst, 70-, Nat Cancer Inst res grant, 71-; CONSULT, 88- *Mem:* AAAS; Am Chem Soc; Int Soc Biochem Pharmacol; NY Acad Sci; Am Soc Microbiol. *Res:* Bacteriology; molecular biology; biochemistry; molecular genetics. *Mailing Add:* Dept Exp Biol Roswell Park Mem Inst 666 Elm St Buffalo NY 14263

MUNSON, BURNABY, b Wharton, Tex, Mar 20, 33. PHYSICAL CHEMISTRY, ANALYTICAL CHEMISTRY. *Educ:* Univ Tex, BA, 54, MA, 56, PhD(phys chem), 59. *Prof Exp:* Res chemist, Humble Oil & Refining Co, 59-62, sr res chemist, 62-64, sr res chemist, Esso Res & Eng Co, 64-66, res specialist, Tex, 66-67; assoc prof chem, 67-72, dir, Honors Prog, 86-89, PROF CHEM, UNIV DEL, 72- *Mem:* Fel AAAS; Am Chem Soc; Am Inst Chem; Am Soc Mass Spectrom; Sigma Xi. *Res:* Kinetics; mass spectrometry; reactions of gaseous ions and excited species. *Mailing Add:* Dept of Chem Univ of Del Newark DE 19716

MUNSON, DARRELL E(UGENE), b Rapid City, SDak, Jan 18, 33; m 69; c 2. METALLURGY. *Educ:* SDak Sch Mines & Technol, BS, 54; Stanford Univ, MS, 56, PhD(metall eng), 60. *Prof Exp:* Asst, Stanford Univ, 56-57, actg instr, 57-58; asst prof metall, Wash State Univ, 59-61; staff mem technol, 61-67, DIV SUPVR, SANDIA CORP, 67- *Concurrent Pos:* Jr metallurgist, Stanford Res Inst, 56-57. *Mem:* Am Soc Metals; Am Soc Eng Educ; Am Inst Mining, Metall & Petrol Engrs. *Res:* Structural metallurgy and deformation, particularly high temperature kinetics and mechanisms of deformation and shock wave propagation; wave propagation and fracture in composites. *Mailing Add:* 23 Cedar Hill Rd NE Albuquerque NM 87122

MUNSON, DONALD ALBERT, b New York, NY, May 13, 41; m 79; c 2. ZOOLOGY, PARASITOLOGY. *Educ:* Colgate Univ, AB, 63; Adelphi Univ, MS, 66; Univ NH, PhD(zool), 70. *Prof Exp:* Teaching asst zool, Univ NH, 66-68, teaching fel, 68-70; NIH fel parasitol, Med Sch, Tulane Univ, 70-72; asst prof biol, Hood Col, 72-76; asst prof biol, 76-83, assoc prof, 83-89, PROF BIOL, WASHINGTON COL, 89-, CHMN, 84- *Concurrent Pos:* Joseph H McLain prof & curator, McLain Prog Environ Studies, 90- *Mem:* Am Soc Parasitol; Am Soc Trop Med & Hyg; Wildlife Dis Asn; Am Microscop Soc; Soc Protozoologists; Am Soc Zool. *Res:* Ecology of free-living amoebae; host-parasite relations of fish parasites; ecology of estuarine and marine protozoa. *Mailing Add:* Dept of Biol Washington Col Chestertown MD 21620

MUNSON, EDWIN STERLING, b Akron, Ohio, Dec 29, 33; c 5. ANESTHESIOLOGY. *Educ:* Univ Tenn, MD, 57. *Prof Exp:* Intern, State Univ Iowa, 57-58; NIH trainee anesthesia, 63-64, from clin instr to asst clin prof anesthesia, Univ Calif, San Francisco, 63-65; asst prof, Univ Va, Charlottesville, 65-67; asst prof, Univ Calif, Davis, 67-71, assoc prof anesthesia & guest investr, Nat Ctr Primate Biol, 67-71; dir res training, 71-78, prof anesthesiol, Col Med, Univ Fla, 71-84; chief, anesthesiol serv, Vet Admin Med Ctr, Gainesville, 78-84; prof & chmn, dept anesthesiol, Univ Ky, 84-86, staff physician VA Med Ctr, Lexington, Ky, 84-86; prof, dept anesthesiol, Med Col Ohio, Toledo, 86-87; STAFF PHYSICIAN, GOOD SAMARITAN HOSP, LEXINGTON, KY, 87- *Concurrent Pos:* NIH trainee anesthesia, Univ Calif, San Francisco, 64. *Mem:* Am Soc Anesthetists; Int Anesthesia Res Soc; Am Soc Regional Anesthesia; Soc Ambulatory Anesthesia. *Res:* Pharmacology of anesthetic drugs. *Mailing Add:* Univ Ky Med Ctr Lexington KY 46536

MUNSON, H RANDALL, JR, b Washington, DC, Sept 2, 34. ORGANIC CHEMISTRY. *Educ:* Univ Md, College Park, BS, 58; Georgetown Univ, PhD(org chem), 69. *Prof Exp:* Res asst biochem pharmacol, Med Sch, Georgetown Univ, 62-64; Walter Reed Army Inst Res fel, Univ Va, 68-69; sr res chemist, 69-74, res assoc, 74-79, GROUP MGR, GASTROINTESTINAL CHEM GROUP & ASSOC DIR CHEM RES DEPT, A H ROBINS CO, 79- *Mem:* Am Chem Soc; Sigma Xi. *Res:* Medicinal chemistry; synthesis of antimalarials, antivirals and nucleosides; quantitative structure-activity relationships; synthesis of gastrointestinal drugs. *Mailing Add:* A H Robins Co Chem Res Dept 12712 Fontana St Leawood KS 66209-2319

MUNSON, J(OHN) C(HRISTIAN), b Clinton, Iowa, Oct 9, 26; m 50; c 2. ELECTRONICS ENGINEERING, SYSTEMS DESIGN & SYSTEMS SCIENCE. *Educ:* Iowa State Col, BS, 49; Univ Md, MS, 52, PhD, 62. *Prof Exp:* Electronic scientist, Naval Ord Lab, 49-68; supt acoust div, Naval Res Lab, 68-85; VPRES, ENG & SCI, INC, 83- *Concurrent Pos:* Asst prof exten sch, Univ Md, 64-67; ed, J Underwater Acoust; consult, Underwater Acoust, 83- *Mem:* Fel Inst Elec & Electronics Engrs; fel Acoust Soc Am; Sigma Xi; Acoust, Speech & Signal Processing Soc. *Res:* Application of information theory and computer techniques to underwater acoustic system developments; understanding of sound propagation characteristics of the ocean, especially its effect on acoustic systems; systems design and systems science. *Mailing Add:* 119 Marine Terr Silver Spring MD 20904

MUNSON, JAMES WILLIAM, b Perrysburg, Ohio, Aug 13, 43; m 66; c 3. ANALYTICAL CHEMISTRY. *Educ:* Ohio State Univ, BS, 67; Univ Wis, MS, 69, PhD(pharmaceut), 71. *Prof Exp:* Asst prof pharm, Univ Conn, 71-73; from asst prof to assoc prof pharm, Univ Ky, 73-77; res scientist, 78-80, res head, 80-85, DIR, UPJOHN CO, 85- *Mem:* Am Pharmaceut Asn; Acad Pharmaceut Sci. *Res:* Analysis of drugs in biological fluids and dosage forms. *Mailing Add:* Qual Control Upjohn Co Portage Rd Kalamazoo MI 49001

MUNSON, JOHN BACON, b Clifton Springs, NY, Nov 15, 32; m 59; c 3. NEUROSCIENCES. *Educ:* Union Col, NY, AB, 57; Univ Rochester, PhD(neurobiol), 65. *Prof Exp:* Fel neurophysiol, Inst Physiol, Univ Pisa, 65-66; res assoc physiol, 66-68, from instr to asst prof physiol & psychol, 69-73, asst prof neurosci, 71-73, ASSOC PROF NEUROSCI, PHYSIOL, NEUROSURG & PSYCHOL, COL MED, UNIV FLA, 73- *Concurrent Pos:* Vis scientist, Duke Med Ctr, 73 & Neurol Sci Inst, Portland, Ore, 75. *Mem:* AAAS; Am Physiol Soc; Soc Neurosci; Int Brain Res Orgn; Asn Res Vision & Ophthal; Sigma Xi. *Res:* Disfunction and recovery of function in damaged spinal motoneurons; spinal cord damage; effects and recovery; anatomy and physiology of the epilepsies; central correlates of eye movements. *Mailing Add:* Dept Neurosci J2 44 JHM HC Univ Fla Col Med Gainesville FL 32610

MUNSON, PAUL LEWIS, b Washta, Iowa, Aug 21, 10; m 31, 48, 87; c 3. PHARMACOLOGY, ENDOCRINOLOGY. *Educ:* Antioch Col, BA, 33; Univ Wis, MA, 37; Univ Chicago, PhD(biochem), 42. *Hon Degrees:* MA, Harvard Univ, 55. *Prof Exp:* Asst biochem, Univ Chicago, 39-42; res biochemist, Wm S Merrell Co, Ohio, 42-43; res biochemist & head endocrinol res, Armour & Co, Ill, 43-48; from res asst to res assoc pharmacol, Yale Univ, 48-50; from asst prof to prof, Sch Dent Med, Harvard Univ, 50-65; prof pharmacol, 65-70, chmn, Dept Pharmacol, 65-77, Sarah Graham Kenan Prof, 70-81, EMER SARAH GRAHAM KENAN PROF PHARMACOL & ENDOCRINOL, SCH MED, UNIV NC, CHAPEL HILL, 81- *Concurrent Pos:* Mem, Corticotropin Assay Study Panel, US Pharmacopoeia, 51-55; tutor, Harvard Univ, 55-58, lectr, 65-66; Claude Bernard vis prof, Univ Montreal, 64; mem, Gen Med B Study Sect, USPHS, 66-70, chmn, 69-70; sr adv comt, Laurentian Hormone Conf, 66-80; mem pharmacol test comt, Nat Bd Med Exam, 67-71; ed, Vitamins & Hormones, 68-81, Pharmacol Reviews, 77-81; mem pharmacol-toxicol prog comt, Nat Inst Gen Med Sci, 72-76; chairperson, Task Force Bone & Mineral, NIH Eval. *Honors & Awards:* Fred Conrad Koch Award, Endocrine Soc, 76; William F Neuman Award, Am Soc Bone Mineral Res, 82. *Mem:* Am Soc Pharmacol & Exp Therapeut (secy-treas, 71-72); Am Chem Soc; Endocrine Soc; Am Soc Biol Chemists; Asn Med Sch Pharmacol (secy, 72-73, pres, 74-76); Am Soc Bone Mineral Res. *Res:* Isolation, bioassay and mechanism of action of hormones, especially hypothalamic, pituitary, androgenic, parathyroid and calcitonin; steroid metabolism; calcium metabolism; mechanism of stimulation of adrenocorticotropic hormone secretion. *Mailing Add:* Dept Pharmacol CB# 7365 FLO Bldg Sch Med Univ NC Chapel Hill NC 27599-7365

MUNSON, ROBERT DEAN, b Stockport, Iowa, Mar 14, 27; m 50; c 3. SOIL FERTILITY, PLANT NUTRITION. *Educ:* Univ Minn, BS, 51; Iowa State Univ, MS, 54, PhD(soil fertil, agr econ), 57. *Prof Exp:* Instr high sch, Minn, 51-52; agr economist, Tenn Valley Authority, 57-58; agronomist, Am Postash Inst, 58-67, midwest dir, 67-81, northcentral dir, 81-87; PRES, ROBERT D MUNSON & ASSOC, 87-; proj assoc, Ctr Int Food & Agr Policy, 89-90, ADJ PROF SOIL SCI, UNIV MINN, 88- *Concurrent Pos:* Assoc ed, J Agron Educ, 73; mem panel fertilizer use res needs, NSF, 75; assoc ed, Soil Sci Soc Am J; study, Potassium, Calcium and Magnesium in Tropical & Subtropical Soils, Int Fertilizer Develop Ctr, 79; lectr, Soils & Fertilizer Inst, Acad Agr Sci, Hangzhou, People's Repub of China, 85. *Honors & Awards:* Agronomic Serv Award, Am Soc Agron, 76, Outstanding Award for Diagnosing Yield Limiting Factors, 90. *Mem:* Fel AAAS; fel Am Soc Agron; fel Soil Sci Soc Am; Int Soil Sci Soc; fel Crop Sci Soc Am; Sigma Xi; Am Soc Agr Consults; Am Soc Agron. *Res:* Factors influencing the availability of potassium and phosphorus; interaction of plant nutrient and management proactices in crop production; potassium availability, movement in soils and physiological function in plants and animals; cation balance and interrelationship of plant nutrients; diagnostic techniques in crop production; soil and plant analysis; availability of soil and fertilizer nitrogen as affected by residues; economics of fertilizer use and research methodology. *Mailing Add:* Robert D Munson & Assocs 2147 Doswell Ave St Paul MN 55108

MUNSON, RONALD ALFRED, b Lancaster, Pa, Aug 12, 33; m 67; c 4. PHYSICAL CHEMISTRY. *Educ:* Franklin & Marshall Col, BS, 55; Northwestern Univ, PhD(phys chem), 59. *Prof Exp:* NSF fel, Max Planck Inst Phys Chem, Gottingen, WGermany, 58-59; phys chemist, Gen Elec Co Res & Develop Ctr, 60-67; res chemist, 67-72, staff chemist, 72-82, CHIEF, OFF MINERAL INSTS, US BUR MINES, 82- *Mem:* Am Inst Mining, Metall & Petrol Eng; Am Chem Soc. *Res:* Chemical kinetics; electrochemistry; ultra high pressure synthesis; zeolites. *Mailing Add:* Off Mineral Insts US Bur Mines-MS1020 2401 E St NW Washington DC 20241-0001

MUNTEAN, RICHARD AUGUST, b Warren, Ohio, Sept 6, 49. CHEMISTRY, NUCLEAR CHEMISTRY. *Educ:* Youngstown State Univ, BS, 72, MS, 74; Univ Tenn, PhD(chem), 79. *Prof Exp:* Anal chemist, Tenn Valley Authority, 79-80; crystal growth scientist, Harshaw Chem Co, 80-82; res scientist, Battelle Columbus Labs, 82-83; PRES, MUNCON ASSOCS, INC, 84- *Concurrent Pos:* Sr consult, Radiation Measurement Systs, Inc, 83-84; Oak Ridge Assoc Univs fel. *Mem:* Am Chem Soc; Sigma Xi. *Res:* Dating and characterization of extraterrestrial matter; environmental effects of radiation generated by nuclear power plants; crystal growth technology as an outgrowth of the Bridgeman-Stockbarger method; isotopic calibration techniques developed for post-accident radiation monitors. *Mailing Add:* 748 E Lucius Ave Youngstown OH 44502-2436

MUNTZ, ERIC PHILLIP, b Hamilton, Ont, May 18, 34; m 64; c 2. GAS DYNAMICS, SPACE SCIENCE & TECHNOLOGY. *Educ:* Univ Toronto, BASc, 56, MASc, 57, PhD(aerophys), 61. *Prof Exp:* Physicist, Gen Elec Space Sci Lab, 61-65, group leader, 65-69; assoc prof aerospace eng, 69-71, assoc dir environ eng, 71-74, PROF AEROSPACE ENG, UNIV SOUTHERN CALIF, 71-, PROF RADIOL, 74- *Concurrent Pos:* Consult numerous corp, 69-; mem, Plasma & Fluids Panel, Rev Physics, NAS, 83-85; chmn, dept aerospace eng, Univ Southern Calif, 87- *Honors & Awards:* Contrib to Soc Award, Am Inst Aeronaut & Astronaut, 89. *Mem:* fel Am Inst Aeronaut & Astronaut; Am Phys Soc. *Res:* Statistical nature of gas flows; development of electron beam fluorescence technique for the study of rarefied gas flows; gaseous separation phenomena; high altitude exhaust plumes; gas dynamic lasers; radiological imaging techniques; radiation physics. *Mailing Add:* 1560 E California Blvd Pasadena CA 91106

MUNTZ, KATHRYN HOWE, ADRENERGIC RECEPTORS, CARDIOVASCULAR SYSTEM. *Educ:* Univ Tex, PhD(cell biol), 79. *Prof Exp:* Fel, dept path, 80-84, ASST PROF GROSS ANAT, HEALTH SCI CTR, UNIV TEX, DALLAS, 85- *Mailing Add:* Univ Tex Health Sci Ctr 5323 Harry Hines Blvd Dallas TX 75235

MUNTZ, RICHARD ROBERT, b Jersey City, NJ, Mar 6, 41; m 64. COMPUTER SCIENCE. *Educ:* Pratt Inst, BEE, 63; NY Univ, MEE, 66; Princeton Univ, PhD(elec eng), 69. *Prof Exp:* Elec engr commun facilities, Indust Mgr Off, US Navy, 63-64; mem tech staff comput, Bell Tel Labs, Inc, 64-66; asst prof, 69-76, ASSOC PROF COMPUT SCI, UNIV CALIF, LOS ANGELES, 76- *Mem:* Asn Comput Mach; Inst Elec & Electronics Engrs. *Res:* Computer operating systems; computer system modeling and analysis. *Mailing Add:* Univ Calif Los Angeles CA 90024

MUNTZ, RONALD LEE, b Bonaparte, Iowa, Sept 19, 45; m 66; c 3. ORGANOMETALLIC CHEMISTRY, ORGANIC CHEMISTRY. *Educ:* Iowa State Univ, BS, 68; Univ Ill, PhD(org chem), 72. *Prof Exp:* Res chemist, 72-78, supvr org res, 78-80, TECH MGR SWS SILICONES, DIV TECH CTR, STAUFFER CHEM CO, 80- *Mem:* Am Chem Soc. *Res:* Flame retardant synthesis; organic intermediates and organometallics; silicones. *Mailing Add:* 1360 Trenton Rd Adrian MI 49221-1359

MUNUSHIAN, JACK, b Binghamton, NY, Sept 6, 23. ELECTRICAL ENGINEERING. *Educ:* Univ Rochester, BS, 47; Univ Calif, Berkeley, PhD(elec eng), 54. *Prof Exp:* Head appl physics dept, Component Div, Hughes Aircraft Co, 54-61; head solid state electronics dept, Labs Div, Aerospace Corp, 61-67; PROF ELEC ENG, UNIV SOUTHERN CALIF, 67-, DIR GRAD CTR ENG SCI, 76- *Concurrent Pos:* Lectr, Univ Southern Calif, 54-62, chmn comput sci prog, 69-76; dir, Instr Television Network, 72- *Mem:* AAAS: Inst Elec & Electronics Engrs; Am Phys Soc. *Res:* Solid state electronics; educational technology. *Mailing Add:* Sch Eng Univ Southern Calif Univ Park Los Angeles CA 90089

MUNVES, ELIZABETH DOUGLASS, OBESITY, CLINICAL TEACHING. *Educ:* New York Univ, PhD(nutrit), 53. *Prof Exp:* CLIN PROF MED & NUTRIT, NJ MED SCH, 68- *Mailing Add:* UMDNJ NJ Med Sch 37 Washington Sq W No Five B New York NY 10011

MUNYER, EDWARD ARNOLD, b Chicago, Ill, May 8, 36; m 60, 81; c 3. MUSEUM ADMINISTRATION, VERTEBRATE ZOOLOGY. *Educ:* Ill State Univ, BSEd, 58, MS, 62. *Prof Exp:* Teacher pub schs, Ill, 58-59 & 60-63; instr, Ill State Univ, 63-64; cur zool, Ill State Mus, 64-67; assoc prof sci, Vincennes Univ, 67-70; asst cur educ,Fla State Mus, Univ Fla, 70-75, assoc cur educ 75-81; ASST DIR, ILL STATE MUS, 81- *Concurrent Pos:* Pres, Midwest Mus Conf, 90-92; regional counr bd dirs, Am Asn Mus, 90-92. *Mem:* Am Asn Mus; Wilson Ornith Soc; Sigma Xi. *Res:* Museum education; vertebrate natural history; raptor ecology; comparative anatomy; perceptual learning; science education. *Mailing Add:* Ill State Mus Spring & Edwards St Springfield IL 62706

MURA, TOSHIO, b Kanazawa, Japan, Dec 7, 25; US citizen; m 54; c 2. MECHANICS. *Educ:* Univ Tokyo, BS, 49, PhD(mech), 59. *Prof Exp:* Asst prof eng, Meiji Univ, Japan, 54-58; Air Force Off Sci Res fel & res assoc mat sci, 58-61, from asst prof to assoc prof, 61-66, WALTER P MURPHY PROF CIVIL ENG, NORTHWESTERN UNIV, 66- *Mem:* Nat Acad Eng; fel Am Soc Mech Engrs; Soc Eng Sci; fel Am Acad Mech; Am Phys Soc. *Res:* Micromechanics; dislocations; disclinations; plasticity; variational principle; thermostresses. *Mailing Add:* Dept of Civil Eng Northwestern Univ Evanston IL 60208

MURACA, RALPH JOHN, b Dunmore, Pa, June 5, 35. FLUID MECHANICS, DYNAMICS. *Educ:* Drexel Inst Technol, BS, 59; Univ Va, MS, 70; Va Polytech Inst & State Univ, PhD(aerospace eng), 72. *Prof Exp:* Engr prod support, Convair Astronautics, 59-60; engr struct dynamics, Martin Marietta Co, 60-62; aerospace technician, Aerospace & Dynamics, 62-76, mgr laminar flow control proj, 76-79, mgr shuttle test procedure specification life assessment, 79-81, asst chief syst eng div, 81-83, dep mgr, Larc Space Sta Off, 83-85, CHIEF, SYST ENG DIV, LANGLEY RES CTR, NASA, 85- *Concurrent Pos:* Instr, George Washington Univ-Tidewater Campus, 73-75 & Hampton Inst, Va, 75-76. *Honors & Awards:* Spec Achievement Award, NASA Langley Res Ctr, 72. *Mem:* Am Inst Astronaut & Aeronaut. *Res:* Boundary layer flows; laminar flow control for commercial transport aircraft. *Mailing Add:* 113 Browns Neck Rd Poquoson VA 23662

MURAD, EDMOND, b Bagdad, Iraq, Nov 29, 34; US citizen; m 80. PHYSICAL CHEMISTRY. *Educ:* NY Univ, BA, 55; Univ Rochester, PhD(phys chem), 59. *Prof Exp:* Res assoc phys chem, Nat Bur Standards, DC, 59-60; res assoc, Univ Wis, 60-61; res assoc chem phys, Univ Chicago, 61-63; res assoc phys chem, Cornell Univ, 63-64; res scientist, Aeronutronic Div, Ford Motor Co, 64-66; RES CHEMIST, AIR FORCE GEOPHYSICS LAB, 66- *Honors & Awards:* Loeser Award, 82. *Mem:* Am Chem Soc; Am Geophys Union; Am Phys Soc; Am Inst Aeronaut & Astronaut. *Res:* Mass spectrometry; high temperature thermodynamics; ion-neutral collision phenomena; aeronomy; space physics. *Mailing Add:* Geophysics Lab/PHK Hanscom AFB Bedford MA 01731

MURAD, EMIL MOISE, b Detroit, Mich, May 10, 26. OPERATIONS RESEARCH. *Educ:* Univ Southern Calif, AB, 49, MS, 51. *Prof Exp:* Res chemist, Standard Coil Prod, Inc, 51-53; res chemist, Hydroaire Div, Crane Co, 53-55; chief chemist, Marvelco Electronics Div, Nat Aircraft Corp, 55-57; prin scientist, Stromberg-Carlson Div, Gen Dynamics Corp, 57-59; res specialist, Autonetics Div, NAm Aviation, Inc, 59; dir res, Orbitec Corp, 60-62; sr tech specialist space div, Rockwell Corp, 62-68; PRES, QUANTADYNE ASSOCS, INC, 66- *Mem:* AAAS; Am Phys Soc; Electrochem Soc; Inst Elec & Electronics Eng; Opers Res Soc Am. *Res:* Development of hardware for producing quantum gravity effects based on models incorporating gluon/anti-gluon pairs as the mediators of gravity and not the gravition of the standard model. *Mailing Add:* PO Box 1475 Huntington Beach CA 92647-1475

MURAD, FERID, b Whiting, Ind, Sept 14, 36; m 58; c 5. CLINICAL PHARMACOLOGY. *Educ:* DePauw Univ, BA, 58; Western Reserve Univ, MD & PhD(pharmacol), 65. *Prof Exp:* From intern to resident med, Mass Gen Hosp, 65-67; sr asst surg, Nat Heart & Lung Inst, 67-69, sr staff fel res, 69-70; assoc prof pharmacol & internal med, Sch Med, Univ Va, 70-75, prof, 75-81, dir, Clin Res Ctr, 71-81, dir, Div Clin Pharmacol, 73-81; prof med & pharmacol, Stanford Univ, 81-88; chief med, Vet Admin Med Ctr, 81-88; vpres discovery, 88-89, VPRES RES & DEVELOP, ABBOTT LABS, 90- *Concurrent Pos:* Nat Inst Arthritis & Metab Dis grant, Sch Med, Univ Va, 71-; USPHS res career develop award, 72-; Nat Heart & Lung Inst res grant, 75- *Honors & Awards:* CIBA Award, 88. *Mem:* Am Soc Biol Chem; Asn Am Physicians; Endocrine Soc; Am Soc Pharmacol & Exp Therapeut; Am Soc Clin Invest. *Res:* Cyclic adenosine monophosphate and cyclic guanosine monophosphate metabolism; endocrinology; cardiovascular; clinical pharmacology. *Mailing Add:* Dept Pharmaceut Prods R&D Abbott Labs D-473 Bldg AP10 1 Abbott Park Rd Abbott Park IL 60064

MURAD, JOHN LOUIS, b Tyler, Tex, Dec 15, 32; m 58; c 4. ZOOLOGY. *Educ:* Austin Col, BA, 56; NTex State Univ, MA, 58; Tex A&M Univ, PhD(microbiol), 65. *Prof Exp:* Asst microbiol, N Tex State Univ, 56-58; teaching res med br, Univ Tex, 58-59; instr biol, Stephen F Austin State Col, 59-61; instr, Tex A&M Univ, 61-65; asst prof zool, 65-71, PROF ZOOL & DIR RES, COL LIFE SCI, LA TECH UNIV, 71- *Concurrent Pos:* Res partic, NSF-NAtlantic Treaty Orgn, Ger, 72, Eng, 74, Mex, 82-88, Italy, 77-80 & 83-85. *Mem:* AAAS; Soc Nematol; Am Soc Microbiol; Am Soc Testing & Mat; Am Inst Biol Sci. *Res:* Parasitology of wild and domestic animals, especially helminthic parasites; nematodes of soil, water and sewage; medical parasitology. *Mailing Add:* Col Life Sci La Tech Univ Box 10198 Ruston LA 71272

MURAD, SOHAIL, b May 4, 53; US citizen; m 79; c 2. CHEMICAL ENGINEERING THERMODYNAMICS. *Educ:* Univ Eng, Pakistan, BS, 74; Univ Fla, Gainesville, MS, 76; Cornell Univ, PhD(chem eng), 79. *Prof Exp:* Sr engr, Exxon Res & Eng Co, 81-82; asst prof, 79-86, ASSOC PROF CHEM ENG, UNIV ILL, CHICAGO, 86- *Concurrent Pos:* Res assoc, Ballistic Res Lab, Maryland, 85. *Mem:* Am Inst Chem Engrs; Am Chem Soc. *Res:* Computer simulations of liquids and liquid mixtures; engineering correlations for thermodynamic and transport properties. *Mailing Add:* Chem Eng Dept M/C 110 Univ Ill PO Box 4348 Chicago IL 60680-4348

MURAD, TURHON ALLEN, b Hammond, Ind, July 27, 44; m 68; c 2. BIOLOGICAL ANTHROPOLOGY. *Educ:* Ind Univ, Bloomington, AB, 68, MA, 71, PhD(bio-anthrop), 75; Am Bd Forensic Anthropology, Dipl. *Prof Exp:* PROF PHYS ANTHROP, CALIF STATE UNIV, CHICO, 72- *Concurrent Pos:* Forensic osteologist. *Mem:* Am Acad Forensic Sci; Am Asn Phys Anthropologists; Sigma Xi; Am Anthrop Asn. *Res:* North Alaskan Eskimo intrapopulation variation for palmar dermatoglyphics; skeletal biology; forensic medicine; general physical anthropology. *Mailing Add:* Dept Anthrop Calif State Univ Chico CA 95929-0400

MURAI, KOTARO, b San Francisco, Calif, Jan 10, 25; m 54; c 1. ORGANIC CHEMISTRY. *Educ:* Univ Nebr, BSc, 44, MSc, 45; Univ Minn, PhD(org chem), 49. *Prof Exp:* Asst, Univ Nebr, 44-45; asst, Univ Minn, 45-48; mem res staff, Pfizer Inc, 49-73, sr res investr, Pfizer Cent Res, Pfizer Med Res Labs, 73-88; RETIRED. *Mem:* AAAS; Am Chem Soc; NY Acad Sci; Am Inst Chem. *Res:* Physico-organic approaches to kinetics, antibiotics, alkaloids and steroids. *Mailing Add:* 106 Smith St Groton CT 06340

MURAI, MIYEKO MARY, b San Francisco, Calif, Jan 16, 13. NUTRITION. *Educ:* Univ Calif, BA, 34, MS, 50, MPH, 60, DrPH, 64. *Prof Exp:* Staff dietitian, St Luke's Int Med Ctr, Tokyo, Japan, 35-39; chief dietitian, Kaukini Hosp, Honolulu, Hawaii, 41-47; lab technician, Univ Calif, 50; technician res proj, US Dept Navy, 51; asst prof home econ, Univ Hawaii, 53-60; fel pub health nutrit, Sch Pub Health, Univ Calif, Berkeley, Children's Br, 59-64, lectr, Sch Pub health, 64-66, from asst prof to assoc prof, 66-80, EMER PROF PUB HEALTH, UNIV CALIF, 80- *Concurrent Pos:* Nat Res Coun fel, Pac Sci Bd, 51; mem, Food & Agr Orgn-WHO, 55; consult, Off Surg Gen, US Army. *Mem:* AAAS; Am Pub Health Asn; Am Dietetic Asn; NY Acad Sci. *Res:* Nutrition survey and food habits of Micronesia; food values of South Pacific foods; training public health nutritionists; development and improvement of the criteria for selection of public health nutrition students. *Mailing Add:* 2523 Rose Walk Berkeley CA 94708

MURAKAMI, MASANORI, b Kyoto, Japan, Nov 28, 43; m 71; c 2. PHYSICAL METALLURGY, DIFFUSION. *Educ:* Kyoto Univ, BS, 66, MS, 68, PhD, 71. *Prof Exp:* Fel, Univ Calif, Los Angeles, 71-75; MEM RES STAFF, THOMAS J WATSON RES CTR, IBM CORP, 75- *Mem:* Am Vacuum Soc; Japan Inst Metals; Sigma Xi. *Res:* Physical properties in thin films; phase transformations in metals; phase diagrams; x-ray diffraction techniques; low temperature physics; thin films. *Mailing Add:* Thomas J Watson Res Ctr PO Box 218 Yorktown Heights NY 10598

MURAKAMI, MASANORI, b Ashiya, Hyogo, Japan, May 16, 40; m 68; c 3. PLASMA CONFINEMENT EXPERIMENT, FUSION ENERGY. *Educ:* Nagoya Inst Technol, BS, 63; Kyoto Univ, MS, 65; Mass Inst Technol, PhD(nuclear eng), 69. *Prof Exp:* Teaching asst plasma physics, Mass Inst Technol, 66-67, res asst fusion, 68-69; res staff, 69-77, SR RES STAFF FUSION ENERGY, OAK RIDGE NAT LAB, 78- *Concurrent Pos:* Task force leader, Tokamak Fusion Test Reactor, Princeton Plasma Physics Lab, Princeton Univ, 84-85; vis scientist, Plasma Physics Lab, Kyoto Univ, 86; mem exec comt, Am Phys Soc, 88-89. *Mem:* Fel Am Phys Soc; Sigma Xi. *Res:* Plasma physics application to magnetic fusion energy; experimental plasma physics; plasma confinement in tokamaks and stellarators; plasma transport and modeling; laser scattering. *Mailing Add:* Fusion Energy Div B9201-2 PO Box 2009 Oak Ridge TN 37831-8072

MURAKAMI, TAKIO, b Kanazawa, Japan, Mar 17, 21; m 49; c 2. METEOROLOGY. *Educ:* Meteorol Col, dipl, 43 & 49; Univ Tokyo, ScD(meteorol), 60. *Prof Exp:* Chief gen circulation, Meteorol Res Inst, Tokyo, 53-67; res meteorologist, Meteorol Satellite Lab, Washington, DC, 67-69; PROF METEOROL, UNIV HAWAII, 69- *Concurrent Pos:* Fel, Sch Advan Study, Mass Inst Technol, 60-62; res meteorologist, Inst Trop Meteorol, Poona, India, 66-67. *Honors & Awards:* Meteorol Soc Japan Award, 54. *Mem:* Am Meteorol Soc; Meteorol Soc Japan. *Res:* Synoptic and theoretical tropical meteorology. *Mailing Add:* Dept of Meteorol Univ Hawaii-Manoa 2500 Campus Rd Honolulu HI 96822

MURAKISHI, HARRY HARUO, b San Francisco, Calif, Oct 21, 17; m 48; c 3. PLANT PATHOLOGY. *Educ:* Univ Calif, BS, 40; Univ NC, MS, 47; Univ Minn, PhD(plant path), 48. *Prof Exp:* Asst plant path, Univ NC, 44-45; asst, Univ Minn, 46-48; from asst plant pathologist to plant pathologist, Univ Hawaii, 48-56, head dept plant path, 52-56; assoc prof, Mich State Univ, 56-63, prof bot & plant path, 63-88; RETIRED. *Concurrent Pos:* Agent, USDA, 44; vis prof, Univ Calif, 55-56; Guggenheim Mem Found fel, 55-56. *Honors & Awards:* Ruth Allen Award, Am Phytopathological Soc, 80. *Mem:* Fel Am Phytopathological Soc; Tissue Culture Asn. *Res:* Plant virology; virus diseases of vegetables and orchids; plant tissue culture. *Mailing Add:* 724 12th St No 203 Wilmette IL 60091

MURALIDHARAN, V B, lipid metabolism, for more information see previous edition

MURAMOTO, HIROSHI, b Hilo, Hawaii, June 6, 22; m 56; c 2. PLANT BREEDING, GENETICS. *Educ:* NMex Col Agr & Mech Arts, BS, 55; Univ Ariz, PhD(agron), 58. *Prof Exp:* Asst plant breeder, 58-64, assoc prof, 64-75, assoc plant breeder, 64-80, ASSOC PROF PLANT SCI & ASSOC RES SCIENTIST PLANT BREEDING, AGR EXP STA, UNIV ARIZ, 80- *Res:* Cotton breeding and genetics. *Mailing Add:* 4880 Kay T Dr Tucson AZ 85745

MURANO, GENESIO, b Cairano, Italy, Oct 23, 41; m; c 2. PHYSIOLOGY, BIOCHEMISTRY. *Educ:* Univ Mass, BA, 64; Wayne State Univ, MS, 66, PhD(physiol), 68. *Prof Exp:* Res assoc, 68-70, from asst prof to assoc prof physiol, Wayne State Univ, 71-77; sr staff fel, Off Biologics, Food & Drug Admin, 77-78; assoc prof, 78-85, PROF PHYSIOL, UNIFORMED SERV UNIV HEALTH SCI, 85-; PHYSIOLOGIST, CTR BIOLOGICS, FOOD & DRUG ADMIN, 78-, DEP DIR, DIV HEMATOL, CTR BIOLOGICS EVAL & RES, 90- *Concurrent Pos:* NIH fel, Karolinska Inst, Sweden, 70-71. *Honors & Awards:* Dept Commerence Inventors Award, 82. *Mem:* NY Acad Sci; Int Soc Thrombosis & Haemostasis; AAAS; Am Heart Asn. *Res:* Biochemical interactions of clotting factors; fibrinogen structure; fibrinolytic agents; antithrombin. *Mailing Add:* Ctr Biologics Eval & Res 8800 Rockville Pike Bldg 29 Bethesda MD 20894

MURANY, ERNEST ELMER, b Avella, Pa, Mar 28, 23; m 57; c 2. EXPLORATION GEOLOGY. *Educ:* Kent State Univ, BS, 50; Univ Utah, PhD(geol), 63. *Prof Exp:* Stratigr, US Geol Surv, 52-53; sect chief, Mene Grande Oil Co, Gulf Oil Corp, 53-60; explor mgr & dist geologist, Sinclair Venezuelan Oil Co, Sinclair Oil Co, 63-68; vpres, Collman Indust, 69-71; consult geologist, Dallas, Tex, 70-71, staff consult, Venezuelan Petrol Corp, 71-73; sr staff geologist, Belco Petrol Corp, 73-77; sr explor adv, Ministry Petrol, Sultanate Oman, 77-80; district mgr, Bois D'Arc Corp, 81-82; explor adv, Superior Oil Int, 82-84; PRES, MUREX INT, INC, 84- *Concurrent Pos:* Sr assoc, Cambrian Assocs, 85. *Mem:* AAAS; Am Asn Petrol Geologists; Asn Venezuelan Mineralogists & Petrologists; Am Geophys Union. *Res:* Structural, geochemical, sedimentology, plate tectonics and mountain building as related to generation, migration and accumulation of oil in the Middle East, South America and the United States. *Mailing Add:* 1111 Verret Lane Houston TX 77090-1242

MURASHIGE, TOSHIO, b Kapoho, Hawaii, May 26, 30; m 53; c 5. PLANT CELL & ORGAN CULTURE. *Educ:* Univ Hawaii, BS, 52; Ohio State Univ, MS, 54; Univ Wis, PhD(plant physiol), 58. *Hon Degrees:* DSc, State Univ NY, 84. *Prof Exp:* Res assoc, Univ Wis, 58-59; asst prof, Univ Hawaii, 59-64; from asst plant physiologist to assoc plant physiologist, 64-72, assoc prof plant sci, 67-72, PROF HORT SCI & PLANT PHYSIOLOGIST, UNIV CALIF, RIVERSIDE, 72- *Concurrent Pos:* Exec coun, Tissue Culture Asn, 74-78; exec bd, In Vitro Cell Biol & Biotech Prog, State Univ NY at Plattsburgh, 85-; consult, Rockefeller Found, 85- *Honors & Awards:* CAN Res Award, 73; Norman Coleman Res Award of AAN, 76-; Gold Medal Award, Mass Hort Soc, 77; Alex Laurie Award for Educ & Res, SAF, 78; Charles Valentine Riley Award, 87. *Mem:* AAAS; Am Soc Plant Physiol; Bot Soc Am; Am Soc Hort Sci; Tissue Culture Asn; Sigma Xi. *Res:* Plant tissue culture; experimental morphogenesis; growth regulators; agricultural applications of plant tissue culture. *Mailing Add:* Dept of Bot & Plant Sci Univ of Calif Riverside CA 92521

MURASKIN, MURRAY, b Brooklyn, Ny, Aug 7, 35; m; c 3. PHYSICS. *Educ:* Mass Inst Technol, BS, 57; Univ Ill, MS, 59, PhD(physics), 61. *Prof Exp:* Res assoc physics, Univ Minn, 61-63; asst prof, Univ Nebr, Lincoln, 63-69; assoc prof, 69-74, PROF PHYSICS, UNIV NDAK, 74- *Mem:* Am Phys Soc. *Res:* Elementary particle physics. *Mailing Add:* Dept Physics Univ NDak Box 8135 Univ Sta Grand Forks ND 58202

MURASKO, DONNA MARIE, b New Brunswick, NJ, May 9, 50; m 82; c 2. INTERFERON, GERONTOLOGY. *Educ:* Pa State Univ, PhD(microbiol), 75. *Prof Exp:* Asst prof microbiol, 77-81, assoc prof microbiol & immunol, 81-86, assoc prof med, 84-86, PROF MED, MICROBIOL & IMMUNOL, MED COL PA, 87- *Mem:* Am Asn Immunol; Am Soc Microbiol; Int Soc Inferon Res; Am Asn Cancer Res. *Res:* Evaluation of regulation of immune responses by interferon with particular emphasis in the role of immune response in antitumor effect of interferon; investigation of mechanism of decreased lymphoproliferative responses of elderly humans and animals. *Mailing Add:* Dept Microbiol & Immunol Med Col Pa 3300 Henry Ave Philadelphia PA 19129

MURASUGI, KUNIO, b Tokyo, Japan, Mar 25, 29; m 55; c 3. MATHEMATICS. *Educ:* Tokyo Univ Educ, BSc, 52, DSc(math), 61; Univ Toronto, MA, 61. *Prof Exp:* Lectr math, Hosei Univ, 55-59, asst prof, 59-61; res asst, Univ Toronto, 61-62; res assoc, Princeton Univ, 62-64; from asst prof to assoc prof, 64-69, PROF MATH, UNIV TORONTO, 69- *Concurrent Pos:* Res grants, Nat Res Coun Can, 61-62, NSF, 62-64 & Can Coun grant; ed, Can J Math, 69-71; vis scientist, Princeton Univ, 71-72; vis prof, Univ Southwestern La, 78, Tsukuba Univ, Japan, 79 & Univ Geneva, Switz, 85. *Mem:* Am Math Soc; Can Math Soc; Math Soc Japan; fel Royal Soc Can. *Res:* Knot theory in combinatorial topology; infinite group theory. *Mailing Add:* 611 Cummer Ave Willowdale ON M2K 2M5 Can

MURATA, TADAO, b Takayama, Japan, June 26, 38; m 64; c 2. COMPUTER SYSTEMS, PETRI NET THEORY. *Educ:* Tokai Univ, Tokyo, BS, 62; Univ Ill, Urbana, MS, 64, PhD(elec eng), 66. *Prof Exp:* Res asst, Coordinated Sci Lab, Univ Ill-Urbana, 62-66; asst prof elec eng, Univ Ill-Chicago, 66-68; assoc prof commun eng, Tokai Univ, Tokyo, 68-70; asst prof, 70-72, assoc prof, 72-78, PROF ELEC ENG & COMPUT SCI, UNIV ILL-CHICAGO, 78- *Concurrent Pos:* Vis assoc prof, Univ Calif, Berkeley, 76-77; prin investr, NSF, 78-, US-Spain Joint Comt Sci & Technol Coop, 85-88; guest researcher, Ges fur Math und Datenuerarbeitung, Bonn, WGer, 79; vis scientist, Centre

Nat de la Recherche Sci, France, 81; panel mem, Nat Res Coun, Nat Acad Sci, 81-82 & 83-85; ed, Inst Elec & Electronic Engr, Trans Software Eng, 86; gen chmn, Int Workshop Petri Nets & Performance Models, 87. *Honors & Awards:* Donald G Fink Prize Award, Inst Elect & Electronic Engrs, 91. *Mem:* Fel Inst Elec & Electronics Engrs; Asn Comput Mach; Info Processing Soc Japan; Europ Asn Theoret Corp Sci; Am Asn Artificial Intel. *Res:* Petri nets and related computation models; concurrent-distributed computer systems; artificial intelligence applications; very-large-scale integration systems and computations; applied graph theory; circuit and system theory. *Mailing Add:* Dept Elec Eng & Computer Sci (M/C 154) Univ Ill Box 4348 Chicago IL 60680

MURAY, JULIUS J, physics; deceased, see previous edition for last biography

MURAYAMA, MAKIO, b San Francisco, Calif, Aug 10, 12; m 45. BIOCHEMISTRY. *Educ:* Univ Calif, BA, 39, MA, 40, PhD(immunochem), 53. *Prof Exp:* Asst biochem, Univ Calif, 39-42; res chemist, Bellevue Hosp, NY, 43-45; res biochemist, Univ Hosp, Univ Mich, 45-48; res biochemist res div, Harper Hosp, 50-54; res fel chem, Calif Inst Technol, 54-56; res assoc biochem, Univ Pa, 55-58; Nat Cancer Inst spec res fel, Cavendish Lab, Cambridge Univ, 58; BIOCHEMIST, NIH, 58- *Mem:* AAAS; Am Soc Biol Chem. *Res:* Protein chemistry; chemistry and structure of hemoglobin, especially electron microscopic studies of human sickle cell hemoglobin cable; molecular mechanism of human red cell sickling with hemoglobin S; etiology of acute mountain sickness, a vascular occlusive disease; molecular mode of action of 2-oxo-1 pyrrolidine acetamide. *Mailing Add:* Nat Inst of Health Bethesda MD 20892

MURAYAMA, TAKAYUKI, polymer science; deceased, see previous edition for last biography

MURBACH, EARL WESLEY, b Almira, Wash, Oct 10, 22; m 48; c 2. INORGANIC CHEMISTRY. *Educ:* Gonzaga Univ, BS, 43; Wash State Univ, MS, 49, PhD(chem), 52. *Prof Exp:* Anal chemist, Kaiser Aluminum Co, 46-47; chem engr, Div Indust Res, Wash State Univ, 49-50; chemist, Calif Res & Develop Corp Div, Standard Oil Co, Calif, 52-53; chemist, Nat Carbon Co, 53-54; sr chemist, Phillips Petrol Co, Idaho, 54-56; sr res engr, Atomics Int Div, NAm Aviation, Inc, 56-57, supvr chem develop, 57-70; mgr nuclear process develop, Allied Gulf Nuclear Serv, 70-83; ENG, LAWRENCE LIVERMORE NAT LAB, 84- *Mem:* Am Chem Soc; Am Nuclear Soc. *Res:* High temperature methods for reprocessing reactor fuel; chemical reaction kinetics at high temperatures; all aspects of nuclear fuel cycle. *Mailing Add:* 1091 Madison Livermore CA 94550

MURCH, LAURENCE EVERETT, b Beverly, Mass, Nov 18, 42; m 68; c 3. MICRO-COMPUTERS. *Educ:* Northeastern Univ, BSME, 65; Clarkson Col Technol, MSME, 68; Univ Mass, PhD(mech eng), 72. *Prof Exp:* Methods engr, United Shoe Mach Corp, 65; instr, Clarkson Col Technol, 67-68; res asst, 68-70, instr, 70-72, asst prof, 72-76, ASSOC PROF MECH ENG, UNIV MASS, 76- *Concurrent Pos:* Vis prof, Cranfield Inst Technol, Eng, 78-79. *Mem:* Soc Mfg Engrs; Am Soc Eng Educ. *Res:* Automatic assembly; programmable assembly. *Mailing Add:* Mech Eng Dept Eng Lab Bldg Univ Mass Amherst MA 01003

MURCH, ROBERT MATTHEWS, b Lackawanna, NY, June 27, 24; m 51; c 4. HYDROPHILIC POLYMERS, CONTROLLED RELEASE POLYMERS. *Educ:* Univ Mich, BS, 48; Pa State Univ, MS, 50; Univ Wis, PhD(inorg chem), 66. *Prof Exp:* Chem engr, Dow Corning Corp, 50-54, res chemist, 54-58, proj leader, 58-62; sr chemist, 65-67, res supvr, 67-71, RES ASSOC, WASH RES CTR, W R GRACE & CO, 71- *Mem:* Am Chem Soc; Sigma Xi. *Res:* Organosilicon; organophosphorus; organofluorine; inorganic polymers; fire retardant urethanes and polyester resins; flammability and smoke testing research; biocompatible polymers and controlled release systems. *Mailing Add:* 141 Brinkwood PO Box 28 Brinklow MD 20862

MURCH, S(TANLEY) ALLAN, b Grand Mere, Que, July 29, 29; US citizen; div; c 2. ENGINEERING MECHANICS. *Educ:* Wayne State Univ, BS, 52; Univ Mich, MS, 54, PhD(eng mech), 58. *Prof Exp:* Res asst, Eng Res Inst, Univ Mich, 54-58; asst res engr, Inst Eng Res, Univ Calif, Berkeley, 58-61; STRUCT ENG GROUP HEAD, CHEM SYSTS DIV, UNITED TECHNOLOGIES CORP, 61- *Mem:* AAAS; Am Soc Mech Engrs; Soc Eng Sci. *Res:* Mechanical behavior of solid propellants; viscoelasticity; failure analysis; plasticity; continuous media; fracture; shell theory. *Mailing Add:* PO Box 9234 San Jose CA 95157

MURCHISON, CRAIG BRIAN, b Buffalo, NY, Jan 14, 43; m 70; c 2. CATALYSIS, PROCESS DEVELOPMENT. *Educ:* Univ Rochester, BS, 65; Univ Minn, PhD(phys chem), 70. *Prof Exp:* Researcher, 70-81, assoc scientist, 81-85, sr assoc scientist, 85-89, RES SCIENTIST, DOW CHEM CO, 89- *Concurrent Pos:* Secy & mem exec comt, 11th N Am Meeting of the Catalysis Soc. *Honors & Awards:* Giuseppe Parravano Award, Mich Catalysis Soc, 88. *Mem:* Am Chem Soc; Am Inst Chem Engrs; fel Am Inst Chemists; Sigma Xi. *Res:* Industrial heterogeneous catalysis; synthesis of carbon-2-carbon-4 hydrocarbons, alcohols; catalyzed oxidations and oxydehydrogenations; Fischer-Topsch and organic carbon oxide chemistry. *Mailing Add:* 606 W Meadowbrook Dr Midland MI 48640

MURCHISON, PAMELA W, b Pittsburgh, Pa, June 14, 43; m 70; c 2. PHYSICAL CHEMISTRY. *Educ:* Carnegie Inst Technol, BS, 65; Univ Minn, PhD(phys chem), 69. *Prof Exp:* Vis lectr chem, Hamline Univ, 69-70; ASST PROF PHYS CHEM, CENT MICH UNIV, 70- *Mem:* Am Chem Soc. *Res:* Effective undergraduate chemical education. *Mailing Add:* 606 W Meadowbrook Midland MI 48640

MURCHISON, THOMAS EDGAR, b Kingsville, Tex, Aug 7, 32; div; c 2. VETERINARY PATHOLOGY. *Educ:* Tex A&M Univ, DVM, 55; Ohio State Univ, MSc, 57, PhD, 59. *Prof Exp:* Instr vet path, Ohio State Univ, 55-59; head path dept, Fla Dept Agr, 59-62; head exp path, Orange Mem Hosp, 62-65; head, Dawson Res Inst, 60-65, pres, Dawson Res Corp, 65-91, SCI DIR, DAWSON RES CONSULT CORP, 91- *Concurrent Pos:* Pres, Temson Co, 65- & Wossie Corp, 65- *Mem:* Soc Toxicol; Am Col Vet Path; Soc Toxicol Path; Am Bd Toxicol. *Res:* Pharmaceutical toxicology; veterinary pathology; comparative pathology; research administration; directed a non-clinical laboratory that has provided a high quality toxicology service to the chemical, cosmetic, pharmaceutical and food industries since 1962; experienced in conducting carcinogenicity, acute, subacute, and chronic toxicity studies, consultation in drug development and study design, on site monitoring and auditing of toxicology studies. *Mailing Add:* PO Box 621417 Orlando FL 32862-1417

MURCRAY, DAVID GUY, b Leadville, Colo, Jan 19, 24; m 45; c 2. ATMOSPHERIC PHYSICS. *Educ:* Univ Denver, BS, 48, PhD, 63; Okla State Univ, MS, 50. *Prof Exp:* Fel, Univ Kans, 50-51; res mathematician, Phillips Petrol Co, 51-52; res physicist res inst, 52-63, asst prof, 63, assoc prof, 66-69, PROF PHYSICS, UNIV DENVER, 69-, SR RES PHYSICIST, 63- *Concurrent Pos:* Assoc chmn, physics dept, Univ Denver, 77-, dir, Acad Res Ctr, 86- *Mem:* AAAS; fel Optical Soc Am; Am Geophys Union; Royal Meteorol Soc; Sigma Xi. *Res:* Upper atmospheric physics; infrared transmission in the upper atmosphere; radiation balance of the atmosphere; infrared spectroscopy; operations analysis. *Mailing Add:* Dept Physics Univ Denver Denver CO 80210

MURCRAY, FRANK JAMES, b Stillwater, Okla, Mar 12, 50; m 86; c 3. ATMOSPHERIC & PLANETARY PHYSICS. *Educ:* Univ Denver, BS, 72; Harvard Univ, AM, 73, PhD(physics), 78. *Prof Exp:* Res physicist physics, 78-80, asst res prof, 80-84, assoc res prof, 84-87, RES PROF, UNIV DENVER, 87-, ASSOC DIR, CHAMBERLIN OBSERVATORY. *Concurrent Pos:* NSF grad fel. *Honors & Awards:* Antarctic Serv Medal. *Mem:* Optical Soc Am; Sigma Xi. *Res:* Infrared measurements; stratospheric composition and radiation budget. *Mailing Add:* Dept of Physics University Park Campus Denver CO 80208-0202

MURDAUGH, HERSCHEL VICTOR, JR, b Columbia, SC, Mar 4, 28; 48; c 3. INTERNAL MEDICINE. *Educ:* Duke Univ, MD, 50. *Prof Exp:* Intern med, Grady Mem Hosp, Atlanta, Ga, 50-51; from asst resident to sr asst resident, Duke Univ Hosp, 53-56, instr, 56-57, assoc, 57-58; from asst prof to assoc prof med, Med Col Ala, 58-65, dir renal & electrolyte div, 58-61; assoc prof med & dir renal div, Sch Med, Univ Pittsburgh, 65-77; PROF & CHMN MED, UNIV SC, VET ADMIN HOSP, 77- *Concurrent Pos:* Res fel, Duke Univ & USPHS Hosp, 54-55; chief res, Vet Admin Hosp, Durham, NC, 56-57, clin investr, 57-58; trustee, Mt Desert Island Biol Lab, 62- *Mem:* Am Physiol Soc; Am Fedn Clin Res; Am Soc Clin Invest; fel Am Col Physicians; Soc Exp Biol & Med. *Res:* Renal physiology and disease; physiology of aquatic mammals. *Mailing Add:* Dept Internal Med Univ SC Columbia SC 29204

MURDAY, JAMES STANLEY, b Trenton, NJ, Sept 16, 42; m 67; c 3. SURFACE PHYSICS, SURFACE CHEMISTRY. *Educ:* Case Inst Technol, BS, 64; Cornell Univ, PhD(physics), 70. *Prof Exp:* Res physicist, 69-75, sect surface analysis, 75-81, head surface chem br, 81-87, SUPT, CHEM DIV, NAVAL RES LAB, 88- *Concurrent Pos:* Consult, Off Naval Res, 73- *Honors & Awards:* Chem Div Publ Award, NRL. *Mem:* Mat Res Soc; Am Chem Soc; Am Phys Soc; Am Vacuum Soc. *Res:* Interaction of energy forms with surfaces, surface chemical analysis, surface reactions; nanometer scale science/technology. *Mailing Add:* Code 6100 Naval Res Lab Washington DC 20375

MURDEN, W(ILLIAM) P(AUL), JR, b Newport News, Va, Mar 13, 24; m 47; c 3. MECHANICS, OPERATIONS ANALYSIS. *Educ:* Va Polytech Inst, BS, 49, MS, 51. *Prof Exp:* Instr appl mech, Va Polytech Inst, 49-51; sr analyst, Southwest Res Inst, 51-54; combat opers res group, Tech Opers, Inc, 55-56, dep dir, 56-58, mgr, Monterey Res Off, 58-60, dep dir proj Omega, 60-61, dir, Washington Res Off, 61-62, dep dir, Washington Res Ctr, 62-63; mgr opers anal, 64-71, dir opers & reliability anal, 71-76, DIR ENG TECHNOL, McDONNELL AIRCRAFT CO, 76- *Mem:* Sigma Xi; Am Inst Aeronaut & Astronaut. *Res:* Matrix stress analysis, rigid frames; rocket ballistics; six degree of freedom elastic impact; electromechanical transducers; operational simulation and experiments; war gaming; military operations analysis; weapon systems and cost-effectiveness analysis. *Mailing Add:* 627 Sarawood Lane St Louis MO 63141

MURDEN, WILLIAM ROLAND, JR, b Beaufort, NC, Aug 19, 22. HYDRAULICS, DREDGING OPERATIONS. *Educ:* Elizabethton Col, BS, 73; Heed Univ, MBA, 79. *Prof Exp:* Captain, US Army Air Corps, 43-45; engr aid, US Army Corps Engrs, Norfolk dist, Ft Belvoir, 45-51, engr aid, Oper Div, 51-52, construct engr, 52-53, chief oper & maintenance sect, 53-55, asst chief dams & reservoirs br, 55-56, construct mgt engr, Off Chief Engrs, Washington, DC, 56-58, construct mgt engr, Planning Div, 58-60, gen engr, 60-63, chief plant & supply br & supv gen engr, Construct-Oper Div, 63-79, chief dredging div, Water Resources Support Ctr, 79-88; RETIRED. *Mem:* Nat Acad Eng; Am Soc Mech Engrs; Soc Am Mil Engrs; Soc Naval Architects & Marine Engrs. *Res:* Author of books and articles in the fields of naval architecture and marine engineering. *Mailing Add:* 7604 Ridgecrest Dr Alexandria VA 22308

MURDESHWAR, MANGESH GANESH, b Bombay, India, Mar 25, 33; m 61; c 3. TOPOLOGY. *Educ:* Univ Bombay, BA, 54, MA, 56; Univ Alta, PhD(math), 64. *Prof Exp:* Lectr math, Wilson Col, Univ Bombay, 56-57, Khalsa Col, 58-59 & Parle Col, 59-61; asst prof, 64-77, ASSOC PROF MATH, UNIV ALTA, 78- *Concurrent Pos:* Nat Res Coun Can overseas fel, Univ Bombay, 66-68. *Mem:* Can Math Soc; Bombay Math Colloquium. *Res:* Point-set topology. *Mailing Add:* Dept of Math Univ of Alta Edmonton AB T6G 2G1 Can

MURDICK, PHILIP W, b Akron, Ohio, Nov 13, 28; m 52; c 4. VETERINARY MEDICINE. *Educ:* Ohio State Univ, DVM, 52, MS, 58, PhD(physiol), 64. *Prof Exp:* Instr vet med, 56-64, from asst prof to assoc prof, 64-69, PROF VET MED, OHIO STATE UNIV, 69-, CHMN DEPT VET CLIN SCI, 71- *Mem:* Am Vet Med Asn. *Res:* Veterinary obstetrics and diseases of the genitalia; development of methods for the detection of drugs illegally administered to race horses. *Mailing Add:* Col Vet Med Ohio State Univ 1935 Coffey Rd Columbus OH 43210

MURDOCH, ARTHUR, b DuBois, Nebr, Aug 25, 34; m 57; c 2. COMPUTER SCIENCE. *Educ:* Westmar Col, BA, 56; Yale Univ, MS, 58, PhD(org chem), 64. *Prof Exp:* From asst prof to assoc prof chem, Morningside Col, 62-68; chmn chem dept, 68-84, assoc prof, 68-80, PROF CHEM, MT UNION COL, 80- *Mem:* Fed Am Scientists; Am Asn Univ Professors; Am Chem Soc. *Res:* Organic reduction-oxidation polymers. *Mailing Add:* Dept Chem Mt Union Col Alliance OH 44601

MURDOCH, BRUCE THOMAS, b Prague, Okla, Mar 15, 40; m 69; c 2. NUCLEAR WELL LOGGING. *Educ:* Carleton Col, BA, 62; Rice Univ, MA, 66; Utah State Univ, PhD(physics), 75. *Prof Exp:* Develop engr, Goodyear Aerospace Corp, 67-70; fel nuclear physics, Univ Man, 74-76, prof assoc, 76-77; proj develop engr, Schlumberger Well Serv, 78-82; mgr nuclear systs develop, NL McCullough, 82-88; sr res scientist, Atlas Wireline Serv, 88-89, mgr nuclear systs, 89-90; MGR, NUCLEAR SENSOR DEVELOP, WESTERN ATLAS MWD, 90- *Mem:* Am Phys Soc; Soc Petrol Engrs; Soc Prof Well Log Analysts. *Res:* Nuclear techniques in oil well logging. *Mailing Add:* 3642 Fir Forest Spring TX 77388

MURDOCH, CHARLES LORAINE, b Atkins, Ark, Aug 23, 32; m 66; c 2. HORTICULTURE. *Educ:* Univ Ark, BS, 59, MS, 60; Univ Ill, PhD(agron), 66. *Prof Exp:* Res asst agron, Southwest Br Exp Sta, Univ Ark, 60-62 & Univ Ill, 62-66; res assoc, Univ Ark, 66-70; from asst prof to assoc prof, 70-78, PROF HORT, UNIV HAWAII, HONOLULU, 78- *Mem:* Am Soc Agron; Hort Sci Soc Am. *Res:* Turfgrass management; ecological and physiological aspects of turfgrass growth and development. *Mailing Add:* Dept of Hort Univ of Hawaii Honolulu HI 96822

MURDOCH, DAVID CARRUTHERS, b Tunbridge Wells, Eng, Mar 31, 12; m 50; c 3. MATHEMATICS. *Educ:* Univ BC, BA, 31, MA, 33; Univ Toronto, PhD(math), 37. *Prof Exp:* Sterling fel math, Yale Univ, 37-38, instr, 38-40; instr, Univ Sask, 40-42, asst prof, 42-44; from assoc prof to prof, 44-77, EMER PROF MATH, UNIV BC, 77- *Concurrent Pos:* Vis prof, Ford Found, Mass Inst Technol Proj, Birla Inst Math & Sci, Pilani, India, 66-68. *Mem:* Am Math Soc; Math Asn Am; Can Math Soc. *Res:* Non-commutative ideal theory and theory of rings; abstract algebra. *Mailing Add:* Dept of Math Univ of BC Vancouver BC V6T 1W5 Can

MURDOCH, JOSEPH B(ERT), b Cleveland, Ohio, Jan 31, 27; m 51; c 5. ELECTRICAL ENGINEERING. *Educ:* Case Inst Technol, BS, 50, PhD, 62; Univ NH, MS, 55. *Prof Exp:* Engr, Gen Elec Co, 50-52; from instr to assoc prof, 52-65, chmn dept, 66-77, PROF ELEC ENG, UNIV NH, 65- *Concurrent Pos:* Prof lectr, Case Inst Technol, 59-60; Phillips fel admin, Ore State Univ, 65-66. *Mem:* Inst Elec & Electronics Engrs; Am Soc Eng Educ. *Res:* Network synthesis and analysis; illumination and circuits. *Mailing Add:* Dept Elec Eng Univ NH Durham NH 03824

MURDOCH, JOSEPH RICHARD, synthesis-structure-function theory, for more information see previous edition

MURDOCH, WILLIAM W, b Glassford, Scotland, Jan 28, 39; m 63; c 1. POPULATION BIOLOGY, ECOLOGY. *Educ:* Univ Glasgow, BSc, 60; Oxford Univ, DPhil(ecol), 63. *Prof Exp:* Res assoc & instr ecol, Univ Mich, 63-65; from asst prof to assoc prof, 65-75, PROF BIOL SCI, UNIV CALIF, SANTA BARBARA, 75- *Concurrent Pos:* Vis lectr, Univ BC, 65. *Mem:* Ecol Soc Am; Brit Ecol Soc; Japanese Soc Pop Biol. *Res:* Population and community dynamics of organisms. *Mailing Add:* Dept Biol Sci Univ Calif Santa Barbara CA 93106

MURDOCK, ARCHIE LEE, b Arcola, Mo, Nov 5, 33; m 53; c 3. BIOCHEMISTRY. *Educ:* Southwest Mo State Univ, BS, 57; Univ Mo, MS, 60, PhD(biochem), 63. *Prof Exp:* Res assoc fel biochem, Brookhaven Nat Lab, 63-65; asst prof, 65-82, ASSOC PROF BIOCHEM, MED CTR, UNIV KANS, 82- *Mem:* Am Chem Soc; Sigma Xi. *Res:* Protein chemistry; molecular mechanisms of thermophily in bacteria; methodology. *Mailing Add:* 9725 Mission Rd Leawood KS 66206

MURDOCK, FENOI R, b Blackfoot, Idaho, Feb 2, 17; m 42; c 3. ANIMAL NUTRITION. *Educ:* Univ Idaho, BS, 38; Pa State Univ, PhD(agr), 43. *Prof Exp:* Res chemist, Borden Co, 43-49; asst prof dairy sci, Western Wash Exp Sta, 49-52, assoc prof dairy sci & assoc dairy scientist, 52-60, prof dairy sci & dairy scientist, 60-82, EMER SCIENTIST, WESTERN WASH RES & EXTEN CTR, WASH STATE UNIV, 82- *Mem:* Am Dairy Sci Asn; Sigma Xi. *Res:* Dairy cattle nutrition, utilization of forages and by-products as feeds, animal waste management to conserve nutrients and prevent air and water pollution. *Mailing Add:* 16614 N 63rd St Scottsdale AZ 85254-5652

MURDOCK, GORDON ALFRED, b Minneapolis, Minn, Jan 4, 23; m 50; c 5. PHYSICAL CHEMISTRY. *Educ:* Willamette Univ, BS, 50; Univ Ore, MA, 51, PhD, 54. *Prof Exp:* Asst, Univ Ore, 49-53; res chemist, 53-57, prof leader, 57-60, supvr printing grades develop, 60-68, supvr reprographic res, 68-69, TECH MGR, REPROGRAPHIC PROD, CROWN ZELLERBACH CORP, 69- *Mem:* Am Chem Soc; Tech Asn Pulp & Paper Indust; Soc Photog Sci & Eng. *Res:* Electrochemistry; corrosion studies; polarography; paper products; graphic arts; electrophotography and reproduction papers. *Mailing Add:* 1906 NW Couch Camas WA 98607

MURDOCK, GORDON ROBERT, b Redlands, Calif, Jan 4, 43; m 68; c 2. INVERTEBRATE ZOOLOGY, FUNCTIONAL MORPHOLOGY. *Educ:* Reed Col, AB, 65; Duke Univ, PhD(zool), 72. *Prof Exp:* Asst prof zool, Ariz State Univ, 70-75; fel, Duke Univ, 75-77; vis assoc prof zool, Clemson Univ, 77-78; dir, NC Marine Res Ctr, Ft Fisher, 78-81; CUR PUB EDUC, BELL MUS NATURAL HIST, UNIV MINN, 81- *Concurrent Pos:* Fel, Univ Manchester, 74-75; vis fac, W Indies Lab, Farleigh Dickinson Univ, 76; mem steering comt, Minn Asn Mus, 83-86, chair, 85-86. *Mem:* AAAS; Sigma Xi; Am Asn Mus. *Res:* Prey capture and utilization by sessile invertebrates, biological fluid mechanics, biomechanics, and functional morphology, especially of invertebrates. *Mailing Add:* Ten Church St SE Minneapolis MN 55455

MURDOCK, HAROLD RUSSELL, b Orange, NJ, Oct 15, 19; m 50; c 3. TOXICOLOGY. *Educ:* Davidson Col, BS, 43; Univ Rochester, MS, 49; Univ Buffalo, PhD(pharmacol), 51. *Prof Exp:* Pharmacologist, Food & Drug Admin, 70-79; head, regulatory opers, Burroughs Wellcome Co, 79-83; RETIRED. *Mem:* Sigma Xi; AAAS; Am Chem Soc; Am Asn Clin Pharmacol. *Mailing Add:* 8232 Morrow Mill Rd Chapel Hill NC 27516-9708

MURDOCK, JOHN THOMAS, b Lynn Grove, Ky, Nov 21, 27; m 49; c 3. SOIL SCIENCE. *Educ:* Univ Ky, BS, 52, MS, 53; Univ Wis, PhD(soils), 56. *Hon Degrees:* Dr, Inst Pertanian Bogor, 85. *Prof Exp:* Assoc prof, 55-72, asst dir int agr progs, 68-70, PROF SOILS, UNIV WIS-MADISON, 72-, ASSOC DIR INT AGR PROGS, 74- *Concurrent Pos:* Soils specialist, Univ Wis Contract, Univ Rio Grande do Sul, Brazil, 64-68; proj coordr, Midwestern Univs Consortium Int Activities Higher Agr Educ Proj, Bogor, Indonesia, 70-72; sr res adv spec prog agr res, Brasilia, Brazil, 72-73; dir, Int Agr Progs, 73-78; exec dir, Midwest Univs Consortium for Int Activities, 75-80; dir, Grad Educ Proj, Bogor, Indonesia, 80-85; prof, Int Agr, 85-88. *Mem:* Soil Sci Soc Am. *Res:* Soil fertility and management. *Mailing Add:* 240 Agr Hall Univ Wis Madison WI 53706-1562

MURDOCK, KEITH CHADWICK, b Garfield, Utah, Feb 5, 28; m 53; c 2. MEDICINAL & SYNTHETIC ORGANIC CHEMISTRY. *Educ:* Univ Utah, BA, 48, MA, 50; Univ Ill, PhD(chem), 53. *Prof Exp:* Asst pharmacol & chem, Univ Utah, 48; asst chem, Univ Ill, 50-53; chemist, Res Div, Am Cyanamid Co, NJ, 53-54 & Army Chem Ctr, Edgewood, Md, 55-56; res chemist, 56-63, PRIN RES CHEMIST, LEDERLE DIV, AM CYANAMID CO, 63- *Mem:* Am Chem Soc. *Res:* Mannich reaction; copolymerization; pharmaceutical and medicinal chemistry; heterocyclics; cancer and antiviral chemotherapy; nucleic acids; DNA intercalating agents. *Mailing Add:* 15 Birch St Pearl River NY 10965

MURDY, WILLIAM HENRY, b New Bedford, Mass, Dec 25, 28; m 52; c 2. BOTANY. *Educ:* Univ Mass, BS, 56; Washington Univ, PhD(bot), 59. *Prof Exp:* From instr to assoc prof, 59-71, chmn, Biol Dept, 71-74, 83-87, PROF BIOL, EMORY UNIV, 71-, CHARLES HOWARD CANDLER PROF BIOL, OXFORD COL, 87- *Concurrent Pos:* Scholar, Harvard Univ, 67-68; dean & CEO, Oxford Col, Emory Univ, 87- *Mem:* Bot Soc Am; Soc Study Evolution. *Res:* Systematics of plant species of granite outcrop communities in the Southeastern Piedmont; gene silencing in the genetic diplorization of polyploids in the plant genus Talinum; microgametophytic competition. *Mailing Add:* Dept of Biol Oxford Col Emory Univ Oxford GA 30267

MUREIKA, ROMAN A, b Lithuania, Aug 9, 44; Can citizen; m 66; c 2. PROBABILITY. *Educ:* Cath Univ Am, BA, 64, MA, 67, PhD(math), 69. *Prof Exp:* Asst prof math, Univ Alta, 68-76; from asst prof to assoc prof, 76-89, PROF STATIST, UNIV NB, 89- *Mem:* Inst Math Statist; Statist Soc Can. *Res:* Probability theory; empirical characteristic functions; stochastic processes; sampling theory; entropy. *Mailing Add:* Dept Math Univ NB Fredericton NB E3B 5A3 Can

MURER, ERIK HOMANN, b Norway, Aug 13, 31; c 2. BLOOD PLATELETS, DIABETES RESEARCH. *Educ:* Univ Oslo, Norway, PhD(biochem), 72. *Prof Exp:* Res assoc, Enzyme Inst, Univ Wis, 61-65; res asst prof, Temple Univ Med Sch, 72-81; BIOSCIENTIST, GRAD HOSP, PHILADELPHIA, 83-; ADJ ASSOC PROF, TEMPLE UNIV MED SCH, 90- *Concurrent Pos:* Sr res fel, NRSA, 83-85; res fel, Univ Oslo, 65-72. *Mem:* AAP; Biophys Soc; Int Soc Thrombos Haemostas. *Res:* Blood coagulation; platelet response to fluoride; platelet calcium secretion and metabolism; isolation and characterization of serine protease inhibitors from leeches and effect on blood coagulation; study of glucose transporters. *Mailing Add:* Dept Med Grad Hosp One Graduate Plaza Philadelphia PA 19146

MURGIE, SAMUEL A, b McKees Rocks, Pa, Nov 8, 57; m 82; c 2. SUPERCOMPUTING, FINITE ELEMENT ANALYSIS. *Educ:* Washington & Jefferson Col, BA, 79; Univ Pittsburgh, MA, 84. *Prof Exp:* Sci programmer, Westinghouse Bettis Power Lab, 80-84; systs analyst, 84-85, SUPVR, SWANSON ANALYSIS SYSTS, INC, 85- *Mem:* Inst Elec & Electronic Engrs Computer Soc. *Res:* Responsible for porting major finite element program (ANSYS) to mainframe and super computer systems including IBM, Convex, Cray, etc. *Mailing Add:* Swanson Analysis Sys Johnson Rd PO Box 65 Houston PA 15342

MURGITA, ROBERT ANTHONY, b Auburn, Maine, Nov 4, 42; m 75; c 1. IMMUNOREGULATION, BIOTECHNOLOGY. *Educ:* Univ Maine, BSc, 65; Univ Vt, MSc, 67; McGill Univ, PhD(immunol), 71. *Prof Exp:* Res asst/instr immunol, Dept Med, State Univ NY, Buffalo, 71-73; vis scientist, Dept Immunol, Univ Uppsala, Sweden, 75-78; from asst prof to assoc prof, 78-82, PROF IMMUNOL, DEPT MICROBIOL & IMMUNOL, MCGILL UNIV, 83-. CAN PACIFIC PROF BIOTECHNOL, 89- *Concurrent Pos:* Chmn, Dept Microbiol & Immunol, McGill Univ, 84-, managing dir, Sheldon Biotechnol Ctr, 89- *Mem:* Am Asn Immunologists; Can Soc Immunol; NY Acad Sci. *Res:* Identification and characterization of immunoregulatory mechanisms; determination of how soluble and cellular immunoregulating elements mediate their selective vs non-selective activities on limbs of the immune system; detailed structure/function analysis of immunoregulatory oncofetal molecules. *Mailing Add:* Dept Microbiol & Immunol Suite 511 McGill Univ 3775 University St Montreal PQ H3A 2B4 Can

MURGOLA, EMANUEL J, b Brooklyn, NY, Aug 24, 37; m; c 2. MOLECULAR GENETICS OF TRANSLATION. *Educ:* Yale Univ, PhD(molecular genetics), 70. *Prof Exp:* Fel Am Cancer Soc, Stanford Univ, 70-72; career investr fel, Am Heart Asn, Stanford Univ, 72-73; from asst prof to assoc prof, 73-85, PROF MOLECULAR GENETICS, UNIV TEX, M D ANDERSON CANCER CTR, 85- *Concurrent Pos:* prin investr, NIH, 74-, career develop award, 75-80; prin investr, Am Cancer soc, 74-90; prin investr, Robert A Welch Found, 83-86; Lectr, Int Union Biochem, Sweden, France & Ireland, 84, Swed Natural Sci Res coun, 84, organizer & lectr, Europ Molecular Biol Orgn Uppsala, Sweden,87, Annecy, France, 91; co-prin investr, NATO Collab Res Grant, 85-87; vis scientist, NIH & Med Res, France, 88-89. *Mem:* AAAS; Am Soc Biochem & Molecular biol; Genetics Soc Am; Am Soc Microbiol. *Res:* Using in vivo and in vitro mutational analyses to study the structure, function and interactions of transfer RNA and ribosomal RNA, particularly in the accurate translation of the genetic code. *Mailing Add:* Molecular Genetics Dept Box 11 M D Anderson Cancer Ctr 1515 Holcombe Blvd Houston TX 77030

MURIE, JAN O, b Okanogan, Wash, July 24, 39; m 61; c 2. BEHAVIORAL ECOLOGY. *Educ:* Colo State Univ, BS, 59; Univ Mont, MA, 63; Pa State Univ, PhD(zool), 67. *Prof Exp:* Fel & lectr, 67-69, from asst prof to assoc prof zool,75-84, PROF ZOOL, UNIV ALTA, 84- *Mem:* Ecol Soc Am; Animal Behav Soc; Am Soc Zoologists; Sigma Xi; Am Soc Naturalists; Wildlife Soc. *Res:* Ecology and behavior of small mammals, especially social behavior of sciurid rodents. *Mailing Add:* Dept Zool Univ Alta Edmonton AB T6G 2E9 Can

MURIE, RICHARD A, b Mt Pleasant, Ohio, Oct 3, 23; m 50; c 2. ANALYTICAL CHEMISTRY. *Educ:* Ohio Univ, BS, 50; Iowa State Univ, MS, 52, PhD(analytical chem), 55. *Prof Exp:* Assoc prof chem, Drake Univ, 51-52; res chemist, Monsanto Chem Co, 55-60; sr res chemist, Res Ctr, Diamond Alkali Chem Co, 60, group leader, 60-63; prin scientist, Allison Div, 63-68, supvr appl mat res, Tech Ctr, 68-70, SR RES CHEMIST, GEN MOTORS RES LABS, 70- *Concurrent Pos:* Instr, Eve Div, WVa State Col, 55-60; vis prof, Univ Guadalajara, 72-88; instr, Eve Div, Lawrence Inst Technol, Southfield, Mich, 72- *Mem:* Am Chem Soc; Instrument Soc Am; Soc Appl Spectros. *Res:* Instrumental methods of analysis, especially gas chromatography, differential thermal analysis and ultraviolet and infrared spectroscopy; development of high temperature battery systems and materials to serve as insulators in such systems; preparation of platinum and rhodium complexes for catalyst applications. *Mailing Add:* Gen Motors Res Lab Dept 14 12 Mile & Mound Rd Warren MI 48090

MURINO, CLIFFORD JOHN, b Yonkers, NY, Feb 10, 29; m 54; c 3. METEOROLOGY. *Educ:* St Louis Univ, BS, 50, MS, 54, PhD(geophys), 57. *Prof Exp:* From assoc prof to prof geophysics, Inst Technol, St Louis Univ, 67-77, vpres finance & res, 71-75; dir Atmospheric Technol Div, Nat Ctr Atmospheric Res, 75-80; PRES, DESERT RES INST, UNIV NEV SYST, 80- *Concurrent Pos:* Vpres res, St Louis Univ, 69-71; sci consult, Div Environ Sci, NSF, Washington, DC, 67-69. *Mem:* Am Meteorol Soc. *Res:* Satellite meteorology; radiation physics; atmospheric energetics; severe storms. *Mailing Add:* 4790 Warren Way Reno NV 89509

MURINO, VINCENT S, b New York, NY, July 30, 24; m 50; c 1. METEOROLOGY, RESEARCH ADMINISTRATION. *Educ:* St Louis Univ, BS, 51, MS, 52; Am Univ, MA, 67. *Prof Exp:* Meteorologist, 52-64, meteorol syst analyst, 64-73, exec officer systs develop off, 73-76, EXEC OFFICER, NAT WEATHER SERV, NAT OCEANIC & ATMOSPHERIC ADMIN, 76- *Concurrent Pos:* Environ Sci Serv Admin fel, 66-67. *Mem:* Am Meteorol Soc; Am Geophys Union. *Res:* Administration of national weather service programs. *Mailing Add:* 3430 Pine Tree Terr Lake Barcroft EGT Falls Church VA 22041

MURISON, GERALD LEONARD, b SAfrica, May 16, 39; US citizen; m 69; c 2. CELL BIOLOGY, DEVELOPMENTAL BIOLOGY. *Educ:* Univ Witwatersrand, BSc, 61, MSc, 63; Johns Hopkins Univ, PhD(biol), 69. *Prof Exp:* Jr lectr biochem, Univ Witwatersrand, 61-64; Pa Plan scholar & instr histol, Med Sch, Univ Pa, 69-70; asst prof biol, Univ Miami, 70-73; PROF BIOL, FLA INT UNIV, 73- *Mem:* AAAS; Soc Develop Biol; Am Soc Cell Biologists. *Res:* Biochemistry of development with emphasis on protein synthesis in tissues before and after birth; regulation of liver metabolism; protein biosynthesis; chemical carcinogenesis in mammalian cells in vitro. *Mailing Add:* Dept Biol Sci Fla Int Univ Miami FL 33199

MURLI, HEMALATHA, m 62; c 2. GENERAL SCIENCES. *Educ:* Univ Madras, India, BSc, 59; Johns Hopkins Univ, PhD(cytogenetics), 62. *Prof Exp:* Instrumentation specialist, Patna Univ, India, 75-77, sci info analyst, self-employed, 78-80; postdoctoral, Litton Bionetics Inc, 81-84, scientist I, 84- 85; scientist, Hazleton Biotechnol, 85-86; SR SCIENTIST, HAZLETON LABS AMERICA INC, 86- *Mem:* Environ Mutagenesis Soc. *Res:* Chinese hamster ovary cells and human leucocyte cultures; studying DNA damage induced by mutagens by sister chrometid exchange studies; clastogenicity by chromosomal aberration studies. *Mailing Add:* Hazleton Labs 5516 Nicholson Lane Kensington MD 20895

MURMAN, EARLL MORTON, b Berkeley, Calif, May 12, 42. FLUID MECHANICS, AERODYNAMICS. *Educ:* Princeton Univ, BSE, 63, MA, 65, PhD(aerospace eng), 67. *Prof Exp:* Aerospace engr aerodyn, Boeing Co, 67-71 & NASA-Ames Res Ctr, 71-74; sr res scientist fluid mech, Flow Industs Inc, 74-76, vpres, 76-80; PROF & HEAD, DEPT AERONAUT & ASTRONAUT, MASS INST TECHNOL, 80- *Concurrent Pos:* Lectr, Stanford Univ, 73. *Mem:* Nat Acad Eng; Am Inst Aeronaut & Astronaut. *Res:* Transonic flow, electrostatic precipitators; computational fluid dynamics. *Mailing Add:* Dept Aeronaut Mass Inst Technol 77 Massachusetts Ave Cambridge MA 02139

MURMANN, ROBERT KENT, b Chicago, Ill, Oct 7, 27; m 55; c 2. INORGANIC CHEMISTRY. *Educ:* Monmouth Col, BS, 49; Northwestern Univ, MS, 51, PhD(chem), 53. *Prof Exp:* Res assoc, Univ Chicago, 53-54; from instr to asst prof chem, Univ Conn, 54-57; assoc prof, 58-60, PROF CHEM, UNIV MO-COLUMBIA, 60- *Mem:* AAAS; Am Chem Soc. *Res:* Coordination compounds; rhenium chemistry; 0-18 isotopic exchange kinetics. *Mailing Add:* Dept Chem 102 Schlundt Hall Univ Mo Columbia MO 65211

MURNANE, THOMAS GEORGE, b Dallas, Tex, May 5, 26; m 53; c 5. VETERINARY MEDICINE, RESOURCE MANAGEMENT. *Educ:* Agr & Mech Col Tex, DVM, 47. *Prof Exp:* Pvt pract, Tex, 47-48; vet, Foot & Mouth Campaign, Joint US-Mex Comn, Mex, 48-49; US Army, 49-, vet lab officer, Area Med Labs, 49-56, vet adv, US Mil Mission, Repub Panama, 56-59, dir defense subsistence testing lab, Ill, 59-63, dep dir div vet med, Walter Reed Army Inst Res, 63-66, chief vet dept, Ninth Med Field Lab, 66-67, chief vet res div, US Army Med Res & Develop Command, 67-72, sr vet corps staff officer, Off Surgeon Gen, 72-74, chief vet corps career activities off, Army Med Dept Personnel Support Agency, Off Surgeon Gen, 74-76; brigadier gen & chief, US Army Vet Corps, 76-80; animal health specialist, Interam Inst Agr Sci, Mex, 80; RETIRED. *Concurrent Pos:* Consult vet pub health, Surgeon Gen, Dept Army, Washington, DC, 74-76. *Mem:* Am Vet Med Asn; US Animal Health Asn; Asn Mil Surgeons US; Am Asn Vet Lab Diagnosticians; Conf Res Workers Animal Dis. *Res:* Food hygiene; rabies; leptospirosis; encephalomyocarditis virus; foot and mouth disease; military participation in emergency animal disease programs; career management of health professionals. *Mailing Add:* 6804 Benito Ct Ft Worth TX 76126

MURNANE, THOMAS WILLIAM, b Cambridge, Mass, July 18, 36; m 65; c 3. ORAL SURGERY, ANATOMY. *Educ:* Tufts Univ, BS, 58, DMD, 62, PhD(anat), 68; Am Bd Oral & Maxillofacial Surg, dipl. *Prof Exp:* Fel anesthesia, Tufts Univ-Boston City Hosp, 63-64; fel, Queen Victoria Hosp, Eng, 65; NIH fel, Med Col Va, 65-66; sr instr anat, Sch Med, 67-68, asst prof oral surg, Sch Dent Med, 67-71, actg dean, 71-72, assoc dean, Sch Dent Med & chmn steering comt, 72-76, asst to pres, 76-78, vpres health progs develop, 78-79, vpres univ develop, 79-85, clin prof oral surg, Sch Dent Med, 71-88, LECTR ANAT, SCH MED, TUFTS UNIV, 71-, SR VPRES, 85- *Concurrent Pos:* Vis assoc oral surgeon, Boston City Hosp & Tufts New Eng Med Ctr, 68; consult, Coun Dent Educ, Am Dent Asn; mem coun oral surg, Pan-Am Med Asn. *Mem:* Am Dent Asn. *Res:* Joints; synovial membrane; connective tissue; electron and light microscopy; experimental pathology. *Mailing Add:* Sr VPres Tufts Univ Medford MA 02155

MURNICK, DANIEL E, b New York, NY, May 5, 41; m 69; c 2. LASER SCIENCE, PARTICLE BEAM PHYSICS. *Educ:* Hofstra Univ, BA, 62; Mass Inst Technol, PhD(physics), 66. *Prof Exp:* Instr physics, Mass Inst Technol, 66-67; tech staff, AT&T Bell Labs, 67-88; PROF PHYSICS & CHMN DEPT, RUTGERS UNIV, 88- *Concurrent Pos:* Grad fac mem, Rutgers Univ, 67-; assoc prof, Mass Inst Technol, 75-76; guest scientist, Brookhaven Nat Lab, 76-82. *Honors & Awards:* Estabrook Award, Hofstra Univ, 76; Humboldt Award, Fed Repub Ger, 83. *Mem:* Fel Am Phys Soc; Inst Elec & Electronics Engrs; AAAS; Sigma Xi. *Res:* Nuclear, atomic, solid state physics and quantum electronics; laser spectroscopy of glow discharges and studies of high energy beam excited short wavelength lasers. *Mailing Add:* Dept Physics Rutgers Univ Newark NJ 07102

MURNIK, MARY RENGO, b Manistee, Mich, Aug 30, 42; m 70; c 1. GENETICS. *Educ:* Mich State Univ, BS, 64, PhD(zool, genetics), 69. *Prof Exp:* Asst prof biol & genetics, Fitchburg State Col, 68-70; from asst prof to prof, Western Ill Univ, 70-80; PROF BIOL SCI & HEAD DEPT, FERRIS STATE UNIV, 80- *Mem:* AAAS; Genetics Soc Am; Sigma Xi; Behav Genetics Asn. *Res:* Mutagen assays, reproductive behavior and meiotic drive in Drosophila. *Mailing Add:* Dept Biol Sci Ferris State Univ Big Rapids MI 49307

MUROGA, SABURO, b Numazu, Japan, Mar 15, 25; m 56; c 4. COMPUTER SCIENCE. *Educ:* Univ Tokyo, BE, 47, PhD(info theory), 58. *Prof Exp:* Mem res staff, Nat Railway Pub Corp, Japan, 47-49; mem eng staff, Govt Radio Regulatory Comn, Japan, 50-51; mem res staff, Elec Commun Lab, Nippon Tel & Tel Pub Corp, Japan, 51-60 & IBM Res Ctr, Yorktown Heights, NY, 60-64; PROF COMPUT SCI, UNIV ILL, URBANA, 64- *Honors & Awards:* Inada Award, Inst Electronics & Commun Engrs, Japan. *Mem:* Fel Inst Elec & Electronic Engrs; Asn Comput Mach; Info Processing Soc Japan; Inst Electronics & Commun Engrs Japan. *Res:* Logical design; switching theory; integrated circuits; mathematical programming; integrated programming. *Mailing Add:* Dept Comput Sci Digital Sci Lab 1304 W Springfield Ave Urbana IL 61801

MUROV, STEVEN LEE, b Los Angeles, Calif, Oct 16, 40; m 89; c 3. PHOTOCHEMISTRY. *Educ:* Harvey Mudd Col, BS, 62; Univ Chicago, PhD(photochem), 67. *Prof Exp:* NIH fel photochem, Calif Inst Technol, 67-68; asst prof org chem, State Univ NY Stony Brook, 68-73; assoc prof phys sci, Sangamon State Univ, 73-78; vis assoc prof chem, Moorhead State Univ, 77-78; lectr chem, Calif State Col, Bakersfield, 78-79; INSTR CHEM, MODESTO JR COL, 79- *Concurrent Pos:* Proj dir, Eisenhower Sci grant, 90-93. *Mem:* Am Chem Soc; Sigma Xi; Nat Sci Teachers Asn. *Res:* Organic photochemistry; methods in science education at the elementary school level. *Mailing Add:* Dept Chem Modesto Jr Col Modesto CA 93350

MURPHEY, BYRON FREEZE, b Great Falls, Mont, Aug 12, 18; m 41; c 2. PHYSICS. *Educ:* Univ Mont, BA, 39; Univ Minn, MA, 41, PhD(physics), 48. *Prof Exp:* Physicist underwater ord, Naval Ord Lab, 41-45; physicist, Minn Mining & Mfg Co, 48-49, physics sect leader, 53-58; div supvr weapons effects, Sandia Corp, NMex, 49-53, div supvr underground explosions, 58-61, div supvr appl phys sci, 61-62, dept mgr nuclear burst physics, 62-67, dir underground exp, 67-71; dir appl res, Sandia Labs, 71-83; RETIRED. *Mem:* Fel Am Phys Soc; Am Asn Physics Teachers. *Res:* Magnetism; solid state physics; effects of nuclear weapons. *Mailing Add:* 756 Indian Hills Dr Moscow ID 83843-9308

MURPHEY, ROBERT STAFFORD, b Littleton, NC, Oct 29, 21; m 46; c 2. MEDICINAL CHEMISTRY. *Educ:* Univ Richmond, BS, 42; Univ Va, MS, 47, PhD(org chem), 49. *Prof Exp:* Res chemist medicinal chem, 48-53, dir chem res, 53-55, asst dir res, 55-57, dir res, 57-60, dir int res, 60-66, dir sci develop, 66-80, asst vpres, 67-73, vpres, 73-81, vpres sci affairs & corp develop, 82-83, SR VPRES SCI AFFAIRS & CORP DEVELOP, A H ROBINS & CO, INC, 83- *Mem:* AAAS; Am Chem Soc. *Res:* Organic chemistry. *Mailing Add:* 2300 Chancellor Rd Bon Air VA 23235

MURPHEY, RODNEY KEITH, b Minneapolis, Minn, May 6, 42. NEUROBIOLOGY, ENTOMOLOGY. *Educ:* Univ Minn, Minneapolis, BA, 65, MS, 67; Univ Ore, PhD(biol), 70. *Prof Exp:* NIH fel, Univ Calif, Berkeley, 70-71; asst prof zool, Univ Iowa, 71-74; vis asst prof biol, Univ Ore, Eugene, 74-75; res assoc biol, State Univ NY, Albany, 75-77, from assoc prof to prof biol, 77-87, dir, Neurobiol Res Ctr, 85-87; prog dir, NSF, Washington, DC, 87-89; DIR, NEUROSCI & BEHAV PROG, UNIV MASS, AMHERST, 89- *Concurrent Pos:* Javits Neurosci Investr, 86- *Mem:* AAAS; Brit Soc Exp Biol; Soc Neurosci; Am Soc Zoologists. *Res:* Developmental neurobiology; neural mechanisms of animal behavior; neurophysiology of invertebrates; synaptogenesis. *Mailing Add:* Neurosci & Behav Prog Morrill Sci Ctr Zool Univ Mass Amherst MA 01003-0027

MURPHEY, WAYNE K, b Glenolden, Pa, Sept 5, 27; m 52; c 5. WOOD TECHNOLOGY. *Educ:* Pa State Univ, BS, 52, MF, 53; Univ Mich, PhD, 61. *Prof Exp:* Engr res, Koppers Co, Inc, 53-55; instr forest prod, Ohio Agr Exp Sta, 55-60; asst prof wood utilization, 60-67, head dept wood sci & technol & in chg forestry res lab, 67-68, actg asst dean resident instr, Col Agr, 68-70, prof wood technol & asst dir sch forest resources, Pa State Univ, University Park, 70-78; head & prof forest sci, Tex A&M Univ, College Station, 78-81; FOREST PROD TECHNOLOGIST, SCI & EDUC, COOP STATE RES SERV, USDA, WASHINGTON, DC, 81- *Concurrent Pos:* Consult wood surg, Philippines, Morocco, Brazil & Dominican Repub. *Mem:* Sigma Xi. *Res:* Physical and mechanical properties of wood; adhesives; wood preservation and seasoning; biomass utilization; energy; effects of environment on wood and fiber properties. *Mailing Add:* 1666 Winchester Rd Annapolis MD 21401

MURPHREE, A LINN, b Houston, Miss, June 6, 45. PEDIATRIC OPHTHALMOLOGY. *Educ:* Univ Miss, BS, 67; Baylor Col Med, MD, 72. *Prof Exp:* Resident ophthal, affil hosps, Baylor Col Med, 73-76, chief resident, 75-76; head fel ophthalmic genetics & pediat, Wilmer Inst, John Hopkins Hosp, 76-77; asst prof ophthal & pediat, Univ Southern Calif, 78-83; HEAD, DIV OPHTHAL, CHILDRENS HOSP, LOS ANGELES, CALIF, 78-, DIR, CLAYTON FOUND CTR OCULAR ONCOL, 78-; ASSOC PROF OPHTHAL & PEDIAT, SCH MED, UNIV SOUTHERN CALIF, LOS ANGELES, 83-, DIR, PEDIAT & DEVELOP OPHTHAL, SCH MED-MED CTR, 83- *Concurrent Pos:* Mem, prof adv comt, Blind Childrens Ctr, Los Angeles; med adv bd, Nat Asn Visually Handicapped, 80- *Mem:* Am Acad Ophthal; Am Asn Pediat Ophthal & Stabismus; Am Orthoptic Coun; Asn Res Vision Ophthal; Ophthal Genetics Study Club; Int Soc Genetic Eye Dis. *Res:* Retinoblastoma: cloning the gene for retinoblastoma; determining function of the RB gene; esterase D linkage studies; tumor chromosomal analysis; role of oncogenes in the etiology of RB; tissue culture mantenance; chemotherapy effectiveness; pre and postnatal diagnosis of retinoblastoma gene carriers; genetic counseling; genetic eye disease: clinical description-definition of genetic eye disease; improved genetic counseling for parents of children with genetic desease; recombinant DNA technology and inherited eye diseases. *Mailing Add:* Div Ophthal Childrens Hosp 4650 Sunset Blvd Los Angeles CA 90027

MURPHREE, HENRY BERNARD SCOTT, b Decatur, Ala, Aug 11, 27; m 53; c 3. PHARMACOLOGY. *Educ:* Yale Univ, BA, 50; Emory Univ, MD, 59. *Prof Exp:* Instr pharmacol, Emory Univ, 59-61, intern med, Grady Mem Hosp, 59-61; asst chief, Pharmacol Sect, Bur Res Neurol & Psychiat, NJ Neuropsychiat Inst, 61-68, mem staff, Inst, 62-68; assoc prof psychiat, Univ Med & Dent NJ Robert Wood Johnson Med Sch, 68-71, mem prof staff, Rutgers Ctr Alcohol Studies, 68-72, assoc, Grad Fac Psychol, 69-72, prof pharmacol, 71-81, dir liaison psychiat, 72-77, actg chmn, 77-78, assoc dean acad affairs, 77-81, chmn, Dept Psychiat, 78-91, PROF PSYCHIAT, UNIV MED & DENT NJ ROBERT WOOD JOHNSON MED SCH, 71- *Concurrent Pos:* Consult, Princeton Hosp, NJ, 64-75, Comt Impaired Physicians, Med Soc NJ, 77-83, hon consult staff, Carrier Found, 79-81; lectr, Hahnemann Med Col, 65-73; chief psychiat, Raritan Valley Hosp, 72-77, consult staff psychiat, 77-81; vis prof, Ctr Alcohol Studies, Rutgers Univ, 75-85; mem, State NJ Sci Adv Comt, 81-, dir, Corp J Alcohol Studies, 81-83. *Mem:* AAAS; Am Soc Pharmacol & Exp Therapeut; NY Acad Sci; Soc Biol Psychiat; Am Col Neuropsychopharmacol; Am Psychiat Asn; Asn Acad Psychiat. *Res:* Human psychopharmacology, psychophysiology, psychometrics, electronics, computer techniques, as all these come together in the understanding of the biological correlates and determinants of behavior. *Mailing Add:* Dept Psychiat Univ Med & Dent NJ Robert Wood Johnson Med Sch 675 Hoes Lane Piscataway NJ 08854-5635

MURPHY, ALEXANDER JAMES, b New York, NY, May 19, 39; m 60; c 1. BIOCHEMISTRY. *Educ:* Brooklyn Col, BS, 62; Yale Univ, PhD(biochem), 67. *Prof Exp:* Am Heart Asn spec fel, Univ Calif, San Francisco, 67-70; asst mem dept contractile proteins, Inst Muscle Dis, 70-72; asst prof physiol, 72-74, asst prof, 74-77, assoc prof & chairperson dept biochem, 77-83, PROF & CHAIRPERSON, SCH DENT, UNIV OF THE PAC, 83- *Concurrent Pos:* NIH career develop award, 72-77. *Mem:* Biophys Soc; NY Acad Sci; Am Soc for Biochem & Molecular Biol. *Res:* Protein structure; active sites and mechanism of action of ion transport membrane proteins. *Mailing Add:* Dept Biochem Sch Dent Univ the Pac 2155 Webster San Francisco CA 94115

MURPHY, ALFRED HENRY, b Stockton, Calif, Apr 14, 18; m 48; c 2. CHAPARRAL BRUSH MANAGEMENT, WEED CONTROL. *Educ:* Ore State Univ, BS 41. *Prof Exp:* Specialist, agron dept, Hopland Field Sta, Univ Calif, 46- 51, supt & specialist range mgt, Agr Field Sta Admin, 51-86; RETIRED. *Mem:* Soc Range Mgt; Am Forage & Grassland Coun. *Res:* Management of plants and animals on grazing land; seeding of grasses and legumes; weed control; burning brush; management of oak hardwood range lands. *Mailing Add:* PO Box 1818 Healdsburg CA 95448

MURPHY, ALLAN HUNT, b Cambridge, Mass, Oct 29, 31; m 60; c 4. STATISTICAL METEOROLOGY, APPLIED METEOROLOGY & CLIMATOLOGY. *Educ:* Mass Inst Technol, SB, 54; Univ Mich, MS, 60, AM, 63, PhD(atmos & oceanic sci), 74. *Prof Exp:* Res asst, Mass Inst Technol, 49-54; meteorologist, Northeast Weather Serv, Mass, 54-55; phys sci asst, US Army Electronic Proving Ground, Ariz, 55-58; res asst, Univ Mich, 58-59; res assoc, Travelers Res Ctr, Conn, 59-61, consult, 61-63; res meteorologist, Univ Mich, 61-67, lectr, 64-67; res scientist, Travelers Res Corp, 67-69; res assoc & lectr, Univ Mich, 69-71; mem staff, Nat Ctr Atmospheric Res, 71-79; assoc prof, 79-81, PROF, ORE STATE UNIV, 81-, DIR, STATIST & CLIMATIC IMPACTS LAB, 80-; CLIMATOLOGIST, STATE ORE, 81- *Concurrent Pos:* Fac mem, Univ Hartford, 67-69; vis scholar, Inst Behav Sci, Univ Colo, 71-75; res scholar, Int Inst Appl Syst Anal, Austria, 74-75, consult, 75-76; adj prof, Grad Sch Bus Admin, Univ Colo, 75-79; vis scientist, Europ Ctr Medium Range Weather Forecasts, 80 & 81, Royal Neth Meteorol Inst, 81. *Mem:* AAAS; fel Am Meteorol Soc; Nat Weather Asn; Am Statist Asn; Opers Res Soc Am; Sigma Xi. *Res:* Application of methodology of statistics and operations research in atmospheric sciences, especially statistical and probabilistic forecasting, forecast evaluation and value and use of weather and climate information. *Mailing Add:* Dept of Atmospheric Sci Ore State Univ Corvallis OR 97331

MURPHY, ALLEN EMERSON, b Barnesville, Ohio, Aug 9, 21; m 43; c 3. GEOLOGY. *Educ:* Mt Union Col, AB, 43; WVa Univ, MS, 48; Syracuse Univ, PhD, 55. *Prof Exp:* Topog engr, US Coast & Geod Surv, 43; geol engr, Guy B Panero, 48; from asst prof to assoc prof geol, 48-61, prof geol & geog, 61-85, head dept, 48-85, EMER PROF, WVA UNIV, 85- *Mem:* Geol Soc Am. *Res:* Physical, historical geomorphology and general geology. *Mailing Add:* 156 A St Keyser WV 26726

MURPHY, ARTHUR THOMAS, b Hartford, Conn, Feb 15, 29; m 52; c 7. SYSTEMS ENGINEERING, COMPUTER PACKAGING. *Educ:* Syracuse Univ, BEE, 51; Carnegie-Mellon Univ, MS, 52, PhD(elec eng), 57. *Prof Exp:* Instr elec eng, Carnegie-Mellon Univ, 52-56; asst & assoc prof & head elec eng, Wichita State Univ, 56-61; vis assoc prof mech eng, Mass Inst Technol, 61-62; prof & dean eng, Widener Univ, 62-71, vpres & dean, 71-75; Brown prof mech eng & head dept, Carnegie-Mellon Univ, 75-80; mgr comput & automated systs, DuPont Electronics, 80-81, res fel, 81-85, sr res fel, 85-87, RES FEL, E I DUPONT DE NEMOUS & CO INC, 87- *Concurrent Pos:* Consult, Boeing Co, 57-67, Educ Career Systs, 71-75 & Cyclops Corp, E G Smith Div, 76; ed eng systs, Pergamon Press, 66-; vis prof control eng, Univ Manchester Inst Sci & Technol, Eng, 68-69; mem bd dir, Rumford Press, 75-; prof-in-indust, E I du Pont de Nemours & Co, Inc, 79-80; grad prog lectr, Pa State Univ, 83-87; vis scientist, Sony Res Ctr, Yokohama, Japan, 91- *Honors & Awards:* Western Elec Fund Award, Am Soc Eng Educ, 65. *Mem:* Am Soc Mech Engrs; Sigma Xi; fel Inst Elec & Electronic Engrs; Am Soc Eng Educ; Int Electronic Packaging Soc; fel AAAS. *Res:* Electronic interconnections and effects on computer performance, electromagnetic interference; control systems; computer aided design and manufacturing. *Mailing Add:* Exp Sta Box 80336 DuPont Electronics Wilmington DE 19880-0336

MURPHY, BERNARD T, b Hull, Eng, May 30, 32; m 59; c 3. PHYSICS. *Educ:* Univ Leeds, BSc, 53, PhD(physics), 59. *Prof Exp:* Physicist-engr, Mullard Res Labs, Eng, 56-59; supvry engr, Westinghouse Elec Co, Pa, 59-62; dir develop, Siliconix, Inc, Calif, 62-63; lab dir, Bell Tell Labs Inc, 63-89; CONSULT, 89- *Mem:* Am Phys Soc; fel Inst Elec & Electronics Engrs. *Res:* Medical physics; electron beam studies; integrated circuit structures. *Mailing Add:* Rd Two Box 114 Mertztown PA 19539

MURPHY, BRIAN BORU, b Detroit, Mich, Dec 17, 47; m; c 4. MATHEMATICS, ACTUARIAL MATHEMATICS. *Educ:* Wayne State Univ, BS, 68, MA, 69, PhD(math), 74. *Prof Exp:* NDEA Title IV fel, Wayne State Univ, 69-72; exchange fel, Univ Munich, 73-74; instr math, Wayne State Univ, 74-75; sr res analyst, Blue Cross & Blue Shield, Mich, 76-77, proj leader, 77-78, res assoc, 78-79; ACTUARY, GABRIEL, ROEDER, SMITH & CO, 79- *Mem:* Acad Actuaries; Soc Actuaries. *Res:* Matrix theory; applied statistics; actuarial mathematics. *Mailing Add:* Gabriel Roeder Smith & Co 407 E Fort Detroit MI 48226

MURPHY, BRIAN DONAL, b Dublin, Ireland, May 31, 39; US citizen; m 67; c 2. PHYSICS, COMPUTATIONAL PHYSICS. *Educ:* Nat Univ Ireland, BSc, 61, MSc, 63; Univ Va, PhD(physics), 73. *Prof Exp:* Res officer, Agr Inst, Dublin, 63-65; res assoc radiation physics, Med Col Va, 65-66; res asst physics, Univ Va, 68-72; res assoc, Univ Wis-Madison, 72-74; analyst comput appln, 74-77, sect head, 77-81, DEPT HEAD PHYSICS, TECH APPLN COMPUT SCI, MARTIN MARIETTA ENERGY SYSTS, OAK RIDGE NAT LAB, 81- *Concurrent Pos:* Comput & Telecommun Strategic Planning, Martin Marietta Energy Systs, 90. *Mem:* Am Phys Soc; AAAS; Sigma Xi. *Res:* Applications of computers in physics; particle beam physics; long range atmospheric transport of pollutants; nuclear and radiation physics. *Mailing Add:* Oak Ridge Nat Lab PO Box 2009 MS 8227 Oak Ridge TN 37831-8227

MURPHY, BRIAN LOGAN, b Hartford, Conn, Apr 24, 39; m 61; c 6. AIR POLLUTION, SCIENCE POLICY. *Educ:* Brown Univ, ScB, 61; Yale Univ, MS, 63, PhD(physics), 66. *Prof Exp:* Physicist, Mt Auburn Res Assocs, Inc, 65-75; chief scientist & dep mgr, Air Qual Studies Div, Environ Res & Technol, Inc, 75-80, gen mgr phys sci, 80-82; PRES, GRADIENT CORP, 85- *Concurrent Pos:* Vis lectr, Envrion Risk Assessment, Harvard Sch Pub Health; consult, 82-85. *Mem:* Air Pollution Control Asn; Soc Risk Analysis. *Res:* Risk analysis; chemical transport and fate; exposure analysis; hazerdous waste and hazerdous materials. *Mailing Add:* 101 Avalon Rd Waban MA 02168

MURPHY, BRUCE DANIEL, b Denver, Colo, Mar 16, 41; m 67. REPRODUCTIVE PHYSIOLOGY. *Educ:* Colo State Univ, BSc, 65, MSc, 69; Univ Sask, PhD(physiol), 73. *Prof Exp:* Asst prof zool, Univ Idaho, 72-73; from asst prof to prof biol, 73-87, PROF, DEPT OBSTET/GYNEC & DIR, REPRODUCTION BIOL RES UNIT, COL MED, UNIV SASK, 87- *Concurrent Pos:* Res consult, Ctr Nat Sci Invest, Cuba, 73-77; assoc mem vet physiol sci, Univ Sask; tech officer, Int Atom Energy Agency, United Nations, 88; mem, Agr Acad Agr & Vet Sci, 88. *Honors & Awards:* Res Assoc Award, Int Develop Res Ctr, 79. *Mem:* Soc Study Reprod; Can Physiol Soc; Can Soc Endocrinol & Metab; Endocrine Soc; Soc Study Fertil. *Res:* Reproductive physiology of ovarian function; implantation in mammals. *Mailing Add:* Dept Obstet/Gynec Royal Univ Hosp Saskatoon SK S7N 0X0 Can

MURPHY, C(HARLES) H(ENRY), JR, b Chicago, Ill, Sept 1, 27; m 52; c 6. AERONAUTICAL ENGINEERING. *Educ:* Georgetown Univ, BS, 47; Johns Hopkins Univ, MA, 48, MS, 52, PhD(aeronaut), 57. *Prof Exp:* Instr, Univ Hawaii, 49-50; aeronaut res engr, Ballistic Res Labs, 50-70, chief exterior ballistic lab, 70-76, CHIEF, LAUNCH & FLIGHT DIV, BALLISTIC RES LAB, ABERDEEN PROVING GROUND, 76- *Concurrent Pos:* Vis prof, Univ Ill, 60, Univ Va, 69, East China Inst Technol, 85. *Honors & Awards:* Mech & Control Flight Award, Am Inst Aeronaut & Astronaut, 76; Res & Develop Achievement Award, US Army, 79 & 86. *Mem:* Fel Inst Aeronaut & Astronaut. *Res:* Motion of symmetric configurations acted on by nonlinear aerodynamic forces and moments; effect of liquid payloads on projectile stability; use of vertical fire guns for upper atmosphere research. *Mailing Add:* PO Box 269 Upper Falls MD 21156

MURPHY, CATHERINE MARY, b Cambridge, Mass, Oct 16, 40; m 71; c 2. DISCRETE MATHEMATICAL STRUCTURES. *Educ:* Regis Col, AB, 62; Catholic Univ Am, AM, 65, PhD(algebra), 68. *Prof Exp:* Asst prof, 68-76, ASSOC PROF MATH, PURDUE UNIV, 76-, HEAD DEPT MATH, COMPUTER SCI & STATIST, 85- *Concurrent Pos:* Chairperson, Ind Sect, Mat Asn Am, 85-86. *Mem:* Math Asn Am; Am Math Soc; Sigma Xi. *Res:* Application of permutation groups and combinatorics to the characterization of specified sets of linear rankings. *Mailing Add:* Dept Math/Computer Sci & Statist Purdue Univ Calumet Hammond IN 46323

MURPHY, CHARLES FRANKLIN, b Des Moines, Iowa, Dec 13, 33; m 61; c 1. PLANT BREEDING. *Educ:* Iowa State Univ, BS, 56, PhD(crop breeding), 61; Purdue Univ, MS, 57. *Prof Exp:* Asst prof, 60-67, assoc prof, 67-78, PROF CROP SCI, NC STATE UNIV, 78- *Concurrent Pos:* Prog coordr germplasm, USDA, 81. *Mem:* AAAS; Am Soc Agron. *Res:* Effects of diverse polygenic systems on yield; yield components and other quantitative characters in oats; small grain breeding. *Mailing Add:* Dept Crop Sci NC State Univ Raleigh NC 27650

MURPHY, CHARLES FRANKLIN, b Ithaca, NY, June 9, 40; m 63; c 1. MEDICINAL CHEMISTRY. *Educ:* Rochester Inst Technol, BS, 63; Iowa State Univ, PhD(org chem), 66. *Prof Exp:* NSF fel org chem, Inst Chem, Strasbourg, France, 66-67; sr chemist, 71-77, asst dir res, 77-79, head, Org Chem Dept, Surrey, Eng, 79-80, dir, microbiol & fermentation prof res, 80-84, DIR, PHARMACEUT PROJS MGT, ELI LILLY & CO, 84- *Mem:* Am Chem Soc. *Res:* Natural products chemistry; medicinal chemistry. *Mailing Add:* Dept MC 815 Indianapolis IN 46285-0002

MURPHY, CHARLES THORNTON, b Boston, Mass, May 20, 38; m 69; c 4. ELEMENTARY PARTICLE PHYSICS. *Educ:* Princeton Univ, AB, 59; Univ Wis, MA, 61, PhD(physics), 63. *Prof Exp:* Res assoc physics, Univ Wis, 63-64; asst prof, Univ Mich, Ann Arbor, 64-68; from asst prof to assoc prof, Carnegie-Mellon Univ, 68-73; PHYSICIST, FERMI NAT ACCELERATOR LAB, 73- *Mem:* Am Phys Soc. *Res:* Experimental high energy physics; weak interactions; bubble chamber and particle beam technology. *Mailing Add:* Fermilab MS 219 Box 500 Batavia IL 60510

MURPHY, CLARENCE JOHN, b Manchester, NH, Apr 20, 34; m 60; c 3. ORGANIC CHEMISTRY, POLYMER CHEMISTRY. *Educ:* Univ NH, BS, 55, MS, 57; Univ Buffalo, PhD(organometallic chem), 62. *Prof Exp:* Res assoc chem, Mass Inst Technol, 60-61; from asst prof to assoc prof & chmn dept, Ithaca Col, 61-69; chmn dept, 69-83, PROF CHEM, E STROUDSBURG UNIV, 69- *Concurrent Pos:* NSF vis res prof, Cornell Univ, 67-69; res scientist, Lehigh Univ, 81-; prof, St Anselm Col, 86-87. *Mem:* AAAS; Am Chem Soc; Soc Appl Spectros; Sigma Xi; Coblenz Soc; Hist Sci Soc; Soc Plastics Engrs. *Res:* Infrared spectroscopy; polymer syntheses using monomers derived from natural products; polymer chemistry; synthesis and characterization of polymers; polymer phase relationships and transitions. *Mailing Add:* Dept Chem E Stroudsburg Univ East Stroudsburg PA 18301-2999

MURPHY, CLIFFORD ELYMAN, biology; deceased, see previous edition for last biography

MURPHY, COLLIN GRISSEAU, b Dayton, Ohio, Oct 25, 40; m 62; c 2. CELL BIOLOGY, DEVELOPMENTAL BIOLOGY. *Educ:* Ohio State Univ, BS, 62; Univ Calif, Berkeley, MA, 6S, PhD(zool), 66. *Prof Exp:* Res asst zool, 66-69, asst res zoologist, 69-70, NIH trainee genetics, 70-71, NIH spec fel, 71-73, asst res geneticist, Univ Calif, Berkeley, 73-76; res assoc electron micros, Dept Ophthal, 77-80, RES CELL BIOLOGIST, UNIV CALIF, SAN FRANCISCO, 81- *Concurrent Pos:* Lectr biol, San Francisco State Univ, 74-77. *Mem:* AAAS; Am Soc Cell Biol; Electron Micros Soc Am; Soc Develop Biol; Sigma Xi; Asn Res Vision & Ophthal. *Res:* Ultrastructure and cell biology of the vertebrate eye, pathogenesis of the glaucomas; developmental genetics. *Mailing Add:* Dept Ophthal Univ Calif Box 0730 San Francisco CA 94143

MURPHY, CORNELIUS BERNARD, b Worcester, Mass, Dec 10, 18; m 43; c 4. ANALYTICAL & PHYSICAL CHEMISTRY, THERMODYNAMICS & MATERIAL PROPERTIES. *Educ:* Col of the Holy Cross, BS, 41, MS, 42; Clark Univ, PhD(chem), 52. *Prof Exp:* From instr to asst prof chem, Col of the Holy Cross, 45-52; res chemist, Stamford Labs, Am Cyanamid Co, 52-55; develop chemist, Gen Eng Lab, Gen Elec Co, 55-57, mgr analytical chem, 57-58, mgr analytical & phys chem, 58-63, proj engr, 63-65; mgr mat analytical, Xerox Corp, 65-70, prog mgr, 70-72, mgr chem eng, 72-73, toner processing, 73-75, prin scientist, 76-77, mat coordr, 76-78, contrib scientist, 78-82; RETIRED. *Concurrent Pos:* Mem bd dirs, Delta Labs, NY. *Honors & Awards:* Kurnakov Medal, USSR Acad Sci, 85. *Mem:* Am Chem Soc; Electrochem Soc; NY Acad Sci; Int Confedn Thermal Analysis (pres, 68-71); NAm Thermal Anal Soc. *Res:* Chelation; phase equilibria; differential thermal analysis. *Mailing Add:* PO Box 631 Fairport NY 14450

MURPHY, DANIEL BARKER, b Richmond Hill, NY, Apr 7, 28; m 51; c 4. ORGANIC CHEMISTRY, CARBON & GRAPHITE. *Educ:* Fordham Univ, BS, 47, MS, 49; Pa State Univ, PhD(fuel sci), 58. *Prof Exp:* Asst chem, Fordham Univ, 47-49; instr, Univ Scranton, 49-51; res chemist, Picatinny Arsenal, US Dept Army, 51-54; asst fuel technol, Pa State Univ, 54-57; from instr to prof, 57-91, EMER PROF CHEM, LEHMAN COL, 91- *Mem:* Am Chem Soc; Royal Soc Chem; Am Carbon Soc; Sigma Xi. *Res:* Organic synthesis; nitrogen heterocyclics; propellants and explosives; carbon and graphite; kinetics and mechanism of the formation of solid carbon by pyrolysis of hydrocarbons in a flow system. *Mailing Add:* Dept Chem Herbert H Lehman Col Bronx NY 10468

MURPHY, DANIEL JOHN, b New York, NY, July 16, 12; m 35; c 2. PHYSICAL METALLURGY. *Educ:* US Mil Acad, BS, 35; Mass Inst Technol, MS, 39; Columbia Univ, PhD(phys metall), 52. *Prof Exp:* Head Pitman-Dunn labs, Frankford Arsenal, US Dept Army, 50-52, chief ord indust planning opers, Far East Command, 52-54; res metallurgist, Los Alamos Sci Lab, 54-57; prof, 57-80, EMER PROF, UNIV ARIZ, 80- *Concurrent Pos:* Consult, Los Alamos Sci Lab, 57- *Mem:* Sigma Xi; Am Soc Metals. *Res:* Crystal imperfections by precision density measurements; effects of alloy additions and transformation characteristics in uranium; hot laboratory metallurgy; high temperature properties of metals. *Mailing Add:* 2625 E Southern Ave Apt 29 Tempe AZ 85282

MURPHY, DANIEL JOHN, b Fall River, Mass, Dec 23, 35; m 61; c 3. ELECTRICAL ENGINEERING. *Educ:* Southeastern Mass Univ, BS, 60; Northeastern Univ, MS, 66, PhD(elec eng), 69. *Prof Exp:* Assoc engr, Appl Physics Lab, Johns Hopkins Univ, 60-62; instr elec eng, New Bedford Inst Technol, 62-65; from asst prof to assoc prof, 65-73, chmn dept, 70-74, PROF ELEC ENG, SOUTHEASTERN MASS UNIV, 73-, CHMN DEPT, 81- *Concurrent Pos:* Consult, Bristol Electronics, 63-65 & Naval Underwater Systs Ctr, Newport, 67-; chmn bd, Mitili, Inc, 81- *Mem:* Inst Elec & Electronics Engrs; Am Soc Eng Educ. *Res:* Linear systems; applications of optimal control and estimation theory; engineering education. *Mailing Add:* Dept Elec Eng Southeastern Mass Univ North Dartmouth MA 02748

MURPHY, DANIEL L, b N Tarrytown, NY, Oct 3, 29; m 34; c 2. PETROLEUM EXPLORATION DEVELOPMENT& PRODUCTION. *Educ:* Lehigh Univ, BA, 51; Univ MO, MA, 55; Univ Mich, PhD(econ geol), 61. *Prof Exp:* Regional mgr metal resources, Amax Inc, 63-68; chief geologist non metal resources, JLS Div, Occidental Petroleum, 68-71; vpres int mineral, Geomatrics Int Resources Develop, 71-74; pres int energy, Anschutz Corp, 74-80; PRES, INT AM GLOBAL RESOURCES & OIL MINERAL RESOURCES, 80- *Mem:* Geol Soc Am Fel; Asn In Petrol Negotiators; Am Asn Petroleum Geologist; Am Inst Minning & Metall; Petroleum Engrs Soc Econ Geologist. *Mailing Add:* 1501 Winrock No 10220 Houston TX 77057

MURPHY, DENNIS L, b Milwaukee, Wis, Dec 30, 36; m; c 3. CLINICAL SCIENCES. *Educ:* Marquette Univ, BS, 58, MD, 63, ScM, 63; Am Bd Psychiat & Neurol, dipl, 71. *Prof Exp:* Res asst & res fel, Dept Physiol, Marquette Univ, 58-62; resident, Dept Psychiat & Behav Sci, Johns Hopkins Hosp & Johns Hopkins Univ Student Health Serv, Baltimore, 64-66; clin assoc, Sect Psychosom Med, Adult Psychiat Br, NIH, Bethesda, 66-68, chief, Clin Res Unit, Sect Psychiat, Lab Clin Sci, 68-73, chief, Sect Clin Neuropharm, Lab Clin Sci, 73-76, chief, Clin Neuropharm Br, 77-85, ASSOC DIR CLIN RES, DIV INTRAMURAL RES PROG, NIH, BETHESDA, 83-, CHIEF, LAB CLIN SCI & CHIEF, SECT CLIN NEUROPHARM, 85- *Concurrent Pos:* Psychopharm res award, Am Psychol Asn, 70; A E Bennett award clin res, Soc Biol Psychiat, 70; mem, Prog Comt, Am Psychiat Asn, 75-77. *Honors & Awards:* Int Anna-Monika Found Award, 71; Hofheimer Prize for Res, Am Psychiat Asn, 71; Meritorious Serv Award, Alcohol, Drug Abuse & Ment Health Admin, 77. *Mem:* Am Col Neuropsychopharm; Psychiat Res Soc; Soc Neurosci; Collegium Int Neuropsychopharmacologium. *Res:* Serotonin; obsessive-compulsive disorder; monoamine oxidase and MAO inhibitors; memory; cholinergic function; aging; Alzheimer's disease; affective disorder; antidepressant and antimanic drugs; catecholamines. *Mailing Add:* NIH NIMH Lab Clin Sci Bldg 10 Rm 3D41 Bethesda MD 20892

MURPHY, DONAL B, b New Haven, Conn, Aug 21, 44. IMMUNOLOGY, GENETICS. *Educ:* Bowdoin Col, AB, 66; Univ Mich, Ann Arbor, PhD(immunogenetics), 74. *Prof Exp:* Fel path, Stanford Univ, 74-77; asst prof, 77-83, ASSOC PROF PATH, SCH MED, YALE UNIV, 83- *Mem:* AAAS; Am Asn Immunologists. *Mailing Add:* Wadsworth Ctr NY State Dept Health Empire State Plaza Albany NY 12201

MURPHY, DONALD G, b New York, NY, July 14, 34; m 82; c 2. BIOMEDICAL RESEARCH, RESEARCH ADMINISTRATION. *Educ:* Ore State Univ, BS, 56, PhD(nematol), 61. *Prof Exp:* NSF grant & asst prof nematol, Ore State Univ, 61-62; NIH fel & spec fel, Univ Hamburg, 62-65; res nematologist, Agr Res Serv, USDA, 65-67; grants assoc, NIH, 67; health scientist adminr, Nat Inst Child Health & Human Develop, NIH, 68-74, Nat Inst Aging, 75-77, chief basic aging prog, Nat Inst Aging, 77-81, dep dir, Div

Digestive Dis & Nutrit, Nat Inst Diabetes & Digestive & Kidney Dis, 81-90, DIR NIH OFF EXTRAMURAL STAFF TRAINING, NIH, 91- *Concurrent Pos:* Fel, Dept Path, Johns Hopkins Univ, 71-73. *Mem:* AAAS; fel Geront Soc; Tissue Cult Asn. *Res:* Marine biology; cellular aging; nematode phylogeny and bionomics; research administration. *Mailing Add:* NIH Extramural Staff Training Off Bldg 31 Rm 5B35 NIH Bethesda MD 20892

MURPHY, DOUGLAS BLAKENEY, b Jan 25, 45; US citizen; m 67; c 2. CELL & DEVELOPMENTAL BIOLOGY. *Educ:* Univ Rochester, AB, 67; Syracuse Univ, MS, 69; Univ Pa, PhD(biol), 73. *Prof Exp:* NIH fel molecular biol, Univ Wis, 73-76; asst prof biol, Kansas State Univ, 76-78; from asst prof to assoc prof, 78-88, PROF CELL BIOL & ANAT, JOHNS HOPKINS SCH MED, 88- *Concurrent Pos:* NIH Res Career Dev Award, 80-85; NAS exchange scientist, USSR, 84 & 87; Fulbright sr scholar, Moscow State Univ, USSR, 85- *Mem:* Am Soc Cell Biol. *Res:* Cell biology of microtubules and cytoplasmic filaments; molecular mechanisms of cytoplasmic motility and intracellular transport. *Mailing Add:* Dept of Cell Biol & Anat 725 N Wolfe St Baltimore MD 21205

MURPHY, EDWARD G, b Sheffield, Eng, Dec 6, 21; Can citizen; m 47; c 1. MEDICINE, PEDIATRICS. *Educ:* Univ London, MB & BS, 45; Royal Col Physicians & Surgeons, dipl, 50; Royal Col Physicians & Surgeons Can, cert, 53; FRCPS(C), 72. *Prof Exp:* Intern med & surg, Guy's Univ, County Hosp, Pembury, Eng, 45; intern surg, St Mary's Hosp, Roehampton, 46; intern med, Edgware Gen Hosp, 49; intern, Evelina Children's Hosp, 49-50; intern med & pediat, 52-53, ELECTROENCEPHALOGRAPHER, HOSP SICK CHILDREN, TORONTO, ONT, 55-, CONSULT, 56-; ASSOC PROF MED & PEDIAT, UNIV TORONTO, 74- *Concurrent Pos:* Fel neurol serv, Hosp Sick Children, Toronto, Ont, 53-55; neurol consult, Ont Crippled Children's Ctr, 63-; assoc, Univ Toronto, 67-71, asst prof, 71-74. *Mem:* Can Med Asn; Can Pediat Soc; fel Can Soc Electroencephalog; assoc Can Neurol Soc; Brit Med Asn. *Res:* Neuromuscular disorders. *Mailing Add:* 1379 Clarkson N Mississauga ON L5J 2W6 Can

MURPHY, EDWARD JOSEPH, biophysics, for more information see previous edition

MURPHY, ELIZABETH WILCOX, nutrition, food chemistry, for more information see previous edition

MURPHY, EUGENE F(RANCIS), b Syracuse, NY, May 31, 13; m 55; c 2. MATERIALS ENGINEERING, COMMUNICATIONS. *Educ:* Cornell Univ, ME, 35; Syracuse Univ, MME, 37; Ill Inst Technol, PhD, 48. *Prof Exp:* Asst mech eng, Syracuse Univ, 35-36; mech engr, Ingersoll-Rand Co, 37-39; instr heat power, Ill Inst Technol, 39-41; from instr to asst prof mech eng, Univ Calif, 41-48; staff engr comt artificial limbs, Nat Acad Sci, Wash, 45-48; asst dir, Prosthetic & Sensory Aids Serv Res, 48-54, chief, Res & Develop Div, 54-73, dir, Res Ctr Prosthetics, 73-78, dir, Off Technol Transfer, 78-83; RETIRED. *Concurrent Pos:* Adv fel, Mellon Inst, 47-48; Fulbright lectr, Soc & Home for Cripples, Denmark, 57-58; actg dep dir, Rehabilitative Eng Res & Develop Serv, Vet Admin, Wash, 75-77; ed, Bulletin Prosthetics Res, 77-83; consult, 83- *Honors & Awards:* Meritorious Serv Award, Vet Adminr, 71; Biomed Eng Leadership Award, Alliance Eng Med & Biol, 83. *Mem:* Nat Acad Eng; fel Am Soc Mech Engrs; fel AAAS; assoc fel NY Acad Med; fel Int Soc Prosthetics & Orthotics; fel Rehab Eng Soc North Am; Acoust Soc Am; Am Soc Eng Educ; Am Soc Heating, Refrig & Air Conditioning Engrs; Am Soc Testing & Mat; Asn Educ & Rehab of Blind & Visually Impaired; Biomed Eng Soc; emer mem NY Acad Sci; emer mem Optical Soc Am; sr mem Soc Biomat; emer mem Int Soc Optical Eng. *Res:* Rehabilitative engineering, including prostheses, orthoses, sensory aids, mobility aids, and surgical implants; mechanical engineering; author of 15 chapters in 12 books. *Mailing Add:* 511 E 20th St MB New York NY 10010

MURPHY, FREDERICK A, b New York, NY, June 14, 34; m 60; c 4. VIROLOGY, IMMUNOLOGY. *Educ:* Cornell Univ, BS, 57, DVM, 59; Univ Calif, Davis, PhD(comp path), 64. *Hon Degrees:* MD, Univ Turku, Finland, 86. *Prof Exp:* Chief viral path br, Ctr Dis Control, HEW, 64-78; assoc dean, Col Vet Med & Biomed Sci, Colo State Univ, Ft Collins, 78-88; DIR, CTR INFECTIOUS DIS, CTR DIS CONTROL, 87- *Concurrent Pos:* Pres, Int Comt Taxon of Viruses; chmn, Am Comt Arthropod-borne Viruses; hon fel, John Curtin Sch Med Res, Australian Nat Univ, 70-71; chmn virol div, Int Union Microbiol Socs, 81-84; chmn prog comt, Fifth Int Cong Virol, 78-81; hon fel, John Curtin Sch Med Res, Australian Nat Univ, 71; James Law lectr, Cornell Univ, 83; mem, USSR Acad Med Sci, 88 & Ger Acad Natural Sci, Leopoldina, 85. *Honors & Awards:* K F Meyer Gold Headed Cane Award Epidemiol, Am Vet Epidemiol Soc & Am Vet Med Asn, 86; Snowden Lectr, Australian Soc Microbiol & Australian Nat Animal Health Lab of the Commonwealth Sci & Indust Res Orgn, 91. *Mem:* Am Soc Virol; Am Asn Immunol; Soc Exp Biol & Med; Am Soc Microbiol; Infectious Dis Soc Am. *Res:* Pathogenesis of viral diseases and encephalitis; electron microscopy; veterinary medicine; viral ultrastructure. *Mailing Add:* Ctr Infectious Dis Bldg 1 Rm 6013 Ctr Dis Control Atlanta GA 30333

MURPHY, FREDERICK VERNON, b Washington, DC, Mar 26, 38; m 65; c 2. EXPERIMENTAL PHYSICS. *Educ:* Georgetown Univ, BS, 59; Princeton Univ, MA, 61, PhD(physics), 67. *Prof Exp:* Instr physics, Princeton Univ, 66-67; asst res physicist, Univ Calif, Santa Barbara, 67-75; mem staff, Varian Assoc Inc, Palo Alto, 75-80, mgr res & develop, Res X-ray Imaging Div, 77-80; res scientist, Telesensory Systs, Inc, 80-83; prod eng mgr, thin film technol div, 84-88, ELEC ENG MGR, MED EQUIP, VARIAN ASSOC INC, PALO ALTO, 88- *Mem:* Am Vacuum Soc; Am Asn Physicists Med. *Res:* Semiconductor metalization; technology for the handicapped; tactile displays; elementary particle physics; photon cross sections; counters; spark and streamer chambers; x-ray computerized tomography; secondary particle beam design; electron accelerators for cancer radiotherapy. *Mailing Add:* 430 Sherwood Way Menlo Park CA 94025

MURPHY, GEORGE EARL, b Portland, Ore, Oct 17, 22; m; c 2. SUICIDE. *Educ:* Ore State Univ, BS, 49; Wash Univ, MD, 52. *Prof Exp:* From intern to asst resident med, Highland-Alameda County Hosp, Oakland, Calif, 52-54; fel psychosom med, Sch Med, Wash Univ, 54-55; clin & res fel psychiat, Mass Gen Hosp, 55-56; asst resident, Renard Hosp, Barnes Hosp Group, St Louis, Mo, 56-57; from instr to prof psychiat, 57-90, dir Psychiat Outpatient Dept, 76-90, EMER PROF PSYCHIAT, SCH MED, WASH UNIV, 90- *Concurrent Pos:* Consult forensic psychiat, St Louis County Med Exam Off, 74-79, forensic psychiatrist, 79- *Mem:* Fel Am Psychiat Asn; Psychiat Res Soc; Am Psychopath Asn; Sigma Xi. *Res:* Clinical and epidemiologic studies in suicide, alcoholism, drug addiction, affective disorder and life stress; cognitive therapy of depression. *Mailing Add:* 6211 McPherson Ave St Louis MO 63130

MURPHY, GEORGE GRAHAM, b Clarksville, Tenn, Aug 31, 43; div; c 1. HERPETOLOGY. *Educ:* Austin Peay State Univ, BS, 65; Miss State Univ, MS, 67, PhD(zool), 70. *Prof Exp:* Assoc prof, 69-80, PROF BIOL & CHMN DEPT, MID TENN STATE UNIV, 80- *Mem:* Sigma Xi; Soc Study Amphibians & Reptiles; Herpetologist's League; Am Soc Ichthyologists & Herpetologists. *Res:* Behavior of turtles; reproductive biology of turtles. *Mailing Add:* Box 60 Mid Tenn State Univ Murfreesboro TN 37130

MURPHY, GEORGE WASHINGTON, b Hot Springs, Ark, Jan 2, 19; m 45, 67; c 4. PHYSICAL CHEMISTRY. *Educ:* Univ Ark, AB, 40; Univ NC, PhD(phys chem), 46. *Prof Exp:* Asst chem, Univ NC, 40-42; res chemist, US Naval Res Lab, 42-45; from instr to asst prof chem, Univ Wis, 46-51; assoc chemist, Argonne Nat Lab, 51-53; prof chem & chmn dept, State Univ NY Col Teachers, Albany, 53-56; chmn dept, 60-68, from assoc prof to prof, 56-86, EMER PROG CHEM, UNIV OKLA, 86- *Mem:* Am Chem Soc. *Res:* Theory of solutions; thermodynamics; photochemical conversion of solar energy; electrochemistry. *Mailing Add:* 2328 Ashwood Norman OK 73071

MURPHY, GORDON J, b Milwaukee, Wis, Feb 16, 27; m 48; c 2. ELECTRONIC SYSTEMS, SIGNAL PROCESSING. *Educ:* Milwaukee Sch Eng, BS, 49; Univ Wis, MS, 52; Univ Minn, PhD, 56. *Prof Exp:* Asst prof elec eng, Milwaukee Sch Eng, 49-51; syst engr automatic control & inertial guid, AC Spark Plug Div, Gen Motors Corp, 51-52; from instr to asst prof elec eng, Univ Minn, 52-57; assoc prof, 57-60, chmn dept, 60-70, PROF ELEC ENG & COMPUT SCI, NORTHWESTERN UNIV, 60-, DIR, LAB DESIGN ELECTRONIC SYSTS, 87- *Concurrent Pos:* Consult, Aeronaut Div, Minneapolis-Honeywell Regulator Co, 54-57, AC Spark Plug Div, Gen Motors Corp, 59-62 & many others. *Mem:* Fel Inst Elec & Electronic Engrs; Sigma Xi. *Res:* Microprocessor-based control systems for industrial and consumer applications; speech processing and electronic systems for real-time data processing. *Mailing Add:* Dept Elec Eng Northwestern Univ Evanston IL 60208

MURPHY, GRATTAN PATRICK, b Parsons, Kans, Sept 15, 35; m 61; c 3. MATHEMATICS. *Educ:* Rockhurst Col, BS, 57; St Louis Univ, MS, 62, PhD(math), 66. *Prof Exp:* Tech analyst comput prog & data reduction, McDonnell Aircraft Co, 59-61; instr math, St Louis Univ, 62-65; from asst prof to assoc prof, 65-81, PROF MATH, UNIV MAINE, ORONO, 81- *Concurrent Pos:* Guest prof, Univ Freiburg, 71-72; vis prof, Univ Wash, 85; vis prof, Western Wash Univ, 86; chmn, Math Dept, Univ Maine, 86- *Mem:* Am Math Soc; Math Asn Am. *Res:* Geometry of generalized metric spaces. *Mailing Add:* Dept of Math Univ of Maine Orono ME 04469

MURPHY, HENRY D, b Hartshorne, Okla, Mar 21, 29. HISTOLOGY, ANATOMY. *Educ:* Univ Calif, Berkeley, AB, 58, MA, 60; Univ Calif, San Francisco, PhD(anat), 65. *Prof Exp:* From asst prof to assoc prof anat & physiol, 65-77, PROF BIOL, SAN JOSE STATE UNIV, 77- *Res:* Endocrine research on role of follicle stimulating hormone on the testes of rats; comparative histological study of marine mammals, seals, sea-lions, porpoises and various whales. *Mailing Add:* Dept of Biol San Jose State Univ Wash Sq San Jose CA 95192

MURPHY, JAMES A, b Philadelphia, Pa, July 28, 35; m 63; c 4. PHYSICAL CHEMISTRY, SURFACE CHEMISTRY. *Educ:* St Joseph's Col, BS, 57; Iowa State Univ, PhD(phys chem), 63. *Prof Exp:* Sr chemist, 63-66, res chemist, 66-69, sr res chemist, 69-72, proj leader, 72-76, RES ASSOC, CORNING GLASS WORKS, 76- *Concurrent Pos:* Ed, Newsletter, Corning Sect, Am Chem Soc, 74-80. *Mem:* Am Chem Soc. *Res:* Materials research, especially thin films; surface chemistry and the interaction of solid, liquids and gases with solids; optical wave guide materials and process research. *Mailing Add:* 262 Delevan Ave Corning NY 14830-3357

MURPHY, JAMES CLAIR, b Salt Lake City, Utah, July 29, 31. PATHOLOGY. *Educ:* Utah State Univ, BS, 57; Wash State Univ, DVM, 61; Colo State Univ, PhD(path), 66. *Prof Exp:* Fel path, Col Vet Med, Colo State Univ, 62-66; res fel, Harvard Med Sch, 66-67; pathologist, Hazleton Labs, Va, 67-68, supvr teratol sect, 67-68; instr, Sch Med, Tufts Univ, 68-74, asst prof surg, 74-75; VET PATHOLOGIST MED DEPT & DIR RES ANIMAL LAB, DIV LAB ANIMAL MED, MASS INST TECHNOL, 75- *Concurrent Pos:* Mem spec sci staff, New Eng Med Ctr Hosps, 68-75, vet & dir res animal lab, 70-75. *Mem:* Am Vet Med Asn; Am Col Vet Path; Int Acad Path. *Res:* Pathogenesis of infectious diseases. *Mailing Add:* Div Lab Animal Med Mass Inst Technol Cambridge MA 02139

MURPHY, JAMES FRANCIS, physical chemistry, for more information see previous edition

MURPHY, JAMES GILBERT, b Brooklyn, NY, July 25, 19; m 47; c 8. ORGANIC CHEMISTRY. *Educ:* St Francis Col, NY, BS, 47; Polytech Inst Brooklyn, MS, 50; Georgetown Univ, PhD(chem), 59. *Prof Exp:* Asst chemist, Nat Oil Prod Co, NJ, 40-42; chemist, Evans Res & Develop Corp, NY, 45-51; res chemist, NIH, 52-71; CONSULT CHEMIST, 72- *Mem:* Am Chem Soc. *Res:* Organic sulfur compounds; medicinal chemistry; biological substrates. *Mailing Add:* 829 Robert St Venice FL 34285

MURPHY, JAMES JOHN, b Scranton, Pa, Apr 1, 39; m 69; c 2. TOXICOLOGY. *Educ:* Lafayette Col, AB, 61; St John's Univ, MA, 66; Temple Univ, PhD(physiol/biophys), 79; Am Bd Toxicol, dipl, 80. *Prof Exp:* Asst prof psychol, Bloomsburg State Col, 67-70; fel, Temple Univ Health Sci Ctr, 70-74, lectr anat & physiol, 75; toxicologist, Stanford Res Inst, 76-78; TOXICOLOGIST, OFF TOXIC SUBSTANCE, US ENVIRON PROTECTION AGENCY, 79- *Concurrent Pos:* Lab instr physiol & histol, Pa Col Podiat Med, 73-75; consult, indust hyg & toxicol, 78. *Mem:* Soc Toxicol; Soc Risk Analysis. *Res:* Hazard identification and risk assessment; industrial toxicology; hydrogen sulfide as a geothermal and petrochemical air pollutant; gastrointestinal models of adrenergic pharmacology, cell cycle in tumors. *Mailing Add:* US Environ Protection Agency 401 M St SW TS-796 Washington DC 20460

MURPHY, JAMES JOSEPH, b New York, NY, Apr 29, 38; m 61; c 3. EXPERIMENTAL SOLID STATE PHYSICS. *Educ:* St Joseph's Col, Pa, BS, 59; Fordham Univ, MS, 61, PhD(physics), 71. *Prof Exp:* From instr to assoc prof, 61-76, chmn dept, 66-75, dir, Acad Develop Sci & Technol, 82-85, dir sci & technol, 85-87, PROF PHYSICS, IONA COL, 76-, CHMN DEPT, 87- *Concurrent Pos:* Mem adj fac, Bergen Community Col, 69-76. *Mem:* AAAS; Am Asn Physics Teachers; Soc Col Sci Teachers; Nat Sci Teachers Asn; Humanities & Technol Asn; Soc Hist of Technol; Fedn Am Scientists. *Res:* Magnetism in transition metals; scientific and technological literacy curriculum development; radon in indoor environments. *Mailing Add:* Dept of Physics Iona Col New Rochelle NY 10801

MURPHY, JAMES LEE, b Grand Ledge, Mich, Aug 29, 40; m 65; c 4. MATHEMATICS. *Educ:* Univ Detroit, BA, 64; Mich State Univ, MS, 66, PhD(math), 70. *Prof Exp:* Admin asst math, Mich State Univ, 66-69, instr, 69-70; asst prof math, Calif State Col, San Bernardino, 70-80; ASST PROF MATH, CALIF STATE UNIV, CHICO, 80- *Mem:* Math Asn Am. *Res:* Piecewise linear topology in Euclidean four-space. *Mailing Add:* Dept Math Calif State Univ Chico CA 95929

MURPHY, JAMES SLATER, b New York, NY, June 2, 21; m 48, 64; c 6. MICROBIOLOGY. *Educ:* Johns Hopkins Univ, MD, 45. *Prof Exp:* Intern med, Johns Hopkins Univ, 45-46; USPHS fel, 48-50, Am Cancer Soc fel, 50-51; from asst to asst prof, 51-60, ASSOC PROF VIROL & MED, ROCKEFELLER UNIV, 60- *Mem:* AAAS; Soc Exp Biol & Med; Am Soc Microbiol; Harvey Soc; Am Asn Immunologists; Sigma Xi. *Res:* Virology; influenza; virus development cycle; genetics; mammalian cell cycle kinetics; bacteriophage; aging and nutrition of Crustacea; effect of cytokines on human cells in tissue culture. *Mailing Add:* Virol & Med Dept Rockefeller Univ 1230 York Ave New York NY 10021

MURPHY, JOHN CORNELIUS, b Wilmington, Del, Feb 28, 36; m 58; c 7. THERMAL IMAGING & SPECTROSCOPY. *Educ:* Cath Univ Am, BA, 57, PhD(physics), 71; Univ Notre Dame, MS, 59. *Prof Exp:* Physicist, appl physics lab, Johns Hopkins Univ, 59-82, prin res assoc, dept mat, sci & eng, 80, Dunning prof, 83-84, res prof mat sci dept, 84-89, PRIN STAFF PHYSICIST, APPL PHYSICS LAB, JOHNS HOPKINS UNIV, 84-, PROF, MAT SCI DEPT & BIOMED ENG DEPT, 89- *Mem:* Am Phys Soc; Optical Soc Am; Mat Res Soc. *Res:* Microwave-optical double resonance experiments on excitation migration in solids; photocatalysis; elastic wave propagation in soils using electro-optical methods; photo acoustic spectroscopy; electron spin resonance of electro generated radical ions in solution, including double resonance in chemiluminescence; photothermal imaging; corrosion including remote sensing; thermal and thermoacoustic imaging and spectroscopy; microwave and magnetic properties of solids; metrology including laser sensing and interferometry; non-destructive evaluation of materials; near field imaging (tunneling, force) of organic conductors and films. *Mailing Add:* Johns Hopkins Appl Physics Lab Johns Hopkins Rd Laurel MD 20707

MURPHY, JOHN FRANCIS, b Cranston, RI, Aug 27, 22; m 45; c 2. GEOLOGY. *Educ:* Dartmouth Col, AB, 47, AM, 49. *Prof Exp:* Geologist, 51-65, chief org fuels br, Colo, 65-66, geologist, Heavy Metals Br, 67-68, dep assoc chief geologist, 68-72, dep chief off energy resources, US Geol Surv, Potomac Md, 72-77; MEM STAFF, OFF ENERGY RESOURCES, US GEOL SURV, DENVER, 77- *Mem:* AAAS; fel Geol Soc Am; Am Asn Petrol Geologists; Soc Econ Geologists. *Res:* Structural geology; stratigraphy; petrology. *Mailing Add:* 2011 Mt Zion Dr Golden CO 80401

MURPHY, JOHN JOSEPH, b Tucson, Ariz, July 28, 40; m 64; c 4. BIOCHEMISTRY, PESTICIDE METABOLISM. *Educ:* Univ Ariz, BS, 62, MS, 63; Purdue Univ, Lafayette, PhD(biochem), 66. *Prof Exp:* NSF fel, King's Col, Univ London, 66-67; res biochemist, Agr Res Ctr, Stauffer Chem Co, Calif, 67-74; sr res biochemist, 74-75, group leader biochem, 75-90, MGR METAB/RESIDUE, METHODOLOGY, AGR DIV, MOBAY CORP, 90- *Mem:* Am Chem Soc; Am Inst Biol Sci; Am Soc Plant Physiologists; Am Soc Agron; Weed Sci Soc Am. *Res:* Metabolism and mechanisms of action of pesticides in plants, animals and soil. *Mailing Add:* Mobay Corp 17745 S Metcalf Stilwell KS 66085

MURPHY, JOHN JOSEPH, b Scranton, Pa, Oct 2, 20; m 44; c 6. UROLOGY, SURGERY. *Educ:* Univ Scranton, BS, 42; Univ Pa, MD, 45. *Prof Exp:* Asst instr surg, Harrison Dept Surg Res, Sch Med, Univ Pa, 48-52; sr instr surg, Dept Urol, Sch Med, Univ Mich, 52-53; assoc urol, Hosp, 53-56, instr, 56-58, from asst prof to assoc prof, 56-64, PROF UROL, SCH MED, UNIV PA, 64- *Concurrent Pos:* Fel, Harrison Dept Surg Res, Sch Med, Univ Pa, 48-51, Am Cancer Soc fel, 51-52; Harrison fel urol surg & Am Cancer Soc fel, Dept Urol, Sch Med, Univ Mich, 52-53; Ravidin traveling fel, 52-53; consult urologist, Vet Admin Hosp, 53- & Children's Seashore House, Atlantic City, NJ; consult, Univ Hosp, Pa & sr consult, Mercy Catholic Med Ctr. *Mem:* Fel, Am Cancer Soc; Am Soc Exp Path; Am Surg Asn; Am Urol Asn; Am Col Surg; Am Asn Genito-Urinary Surg. *Res:* Lymphatic system of the kidney; hydrodynamics of the urinary tract; hypertension as related to the kidney; renal healing; pyelonephritis; cineradiography in urology. *Mailing Add:* Mercy Fitzgerald Med Off Bldg 1501 Lansdowne Ave Suite 304 Darby PA 19023

MURPHY, JOHN MICHAEL, b Madison, Wis, May 30, 35; m 59; c 4. MECHANICAL & AERONAUTICAL ENGINEERING. *Educ:* Purdue Univ, BS, 57, MS, 59, PhD(mech eng), 64. *Prof Exp:* Asst, Jet Propulsion Ctr, Purdue Univ, 55-56, asst aeronaut eng, 57-59, asst & instr mech eng, 60-64; tech opers officer, Missile & Space Off, US Army Eng Res & Develop Labs, Ft Belvoir, Va, 59-60; sr engr & prin engr, Thiokol Chem Corp, Ala, 64; chief tech group, air augmented rocket task team, 64-66; sr res scientist, air augmented rocket studies, Martin Co, 66-69; sr res scientist, Propulsion Systs Res & Develop, Martin Marietta Corp, 69-73, mgr external tank propulsion, 73-75, mgr amine fuels, 75-78, mgr propulsion eng, 78-79, mgr space shuttle reaction control syst propellent tanks, 79-81, mgr propellant systs, 81-84, mgr propulsion eng, 84-85, dir, Tech opers, Int, 85-86, proj dir, Shuttle-Return-To-Flight, 86-89, MGR SATELLITE SERVICERY, MARTIN MARIETTA CORP, 89- *Concurrent Pos:* Asst prof, Univ Ala, 64-66; lectr, Univ Colo, 67-71; adj prof, Colo State Univ; mem, Joint Army-Navy-NASA-Air Force Air Augmented Performance Comt & Liquid Rocket Tech Comt, Am Inst Aeroneut & Astronaut; consult, McGrawHill Info Systs 69-80, Baumgartner Co, 74-78, Battelle Labs, 81-86 & Delphic Assocs, Inc, 90- *Honors & Awards:* Thomas Jefferson Award, Martin Marietta Corp, 68; Wyld Propulsion Award, Am Inst Aeronaut & Astronaut, 90. *Mem:* Assoc fel Am Inst Aeronaut & Astronaut; Combustion Inst; Am Inst Chem Engrs. *Res:* Liaison and design of large space environment simulator; means of reclaiming water in space; radial flow turbines; chemical laser mixing dynamics; rocket combustion; air augmented rocket technology; liquid and solid rocket fuel combustion; liquid propellant feed systems; satellite servicing systems. *Mailing Add:* 6185 W Summit Dr Littleton CO 80123

MURPHY, JOHN N, b Pittsburgh, Pa, July 14, 39; m 62; c 1. MINING ENGINEERING, ELECTRICAL ENGINEERING. *Educ:* Univ Pittsburgh, BS, 61; Duquesne Univ, MBA, 67. *Prof Exp:* Res supvr indust hazards & commun, 61-78, RES DIR, PITTSBURGH RES CTR, US BUR MINES, 78- *Honors & Awards:* Distinguished Serv Award, Dept Interior. *Mem:* Sr mem Inst Elec & Electronics Engrs; Am Inst Mining, Metall & Petrol Engrs. *Res:* Research on mine safety and health; improved mine technology and the environmental problems relating to current and past mining activities. *Mailing Add:* US Bur Mines Cochrans Mill Rd PO Box 18070 Pittsburgh PA 15236

MURPHY, JOHN R, b Bridgeport, Conn, Oct 7, 41. BIOCHEMISTRY, MICROBIOLOGY. *Educ:* Univ Conn, BA, 65, MS, 69, PhD(microbiol), 72. *Prof Exp:* Tutor biol, Harvard Univ, 73-74, from asst prof to assoc prof, Dept Microbiol & Molecular Genetics, 74-83; Lawrence J Henderson assoc prof health sci & technol, Harvard-Mass Inst Technol, 79-83; PROF MED, DEPT MED, RES PROF BIOCHEM, SCH MED & CHIEF, SECT BIOMOLECULAR MED, EVANS DEPT CLIN MED, BOSTON UNIV, 83-, RES PROF MICROBIOL, 87- *Concurrent Pos:* Res fel biol, Harvard Univ, 72-74, mem bd tutors biochem sci, 79-83; res fel, Med Found, Inc, 74-76, mem res adv comt, 80-88; lectr microbiol, Med Sch, Harvard Univ, 74-81 & mechanisms microbial pathogenesis, Harvard-Mass Inst Technol, 75-82; USPHS & Nat Inst Allergy & Infectious Dis career develop award, 76-81; mem, US-Japan Coop Med Sci Prog, Cholera Panel, Nat Inst Allergy & Infectious Dis, 76-82 & 84-87 & Res Adv Comt Betty Lea Stone fel, Am Cancer Soc, 86-; fel, Japan Soc Prom Sci, 78; vis prof, Res Inst Microbial Dis, Osaka Univ, 78; consult, WHO, Geneva, Switz, 79-80. *Honors & Awards:* Pierce Immunotoxin Award, 90. *Mem:* Am Soc Microbiol; AAAS; Infectious Dis Soc Am; Am Soc Biol Chemists; Protein Soc. *Res:* Regulation, secretion and mechanism of diphtheria toxin action; development of hybrid toxin genes for the construction of eukaryotic cell receptor specific chimeric toxins; author of more than 100 technical publications. *Mailing Add:* Dept Med Sect Biomolecular Med Univ Hosp 88 E Newton St Boston MA 02118-2393

MURPHY, JOHN RIFFE, b Hooker, Okla, Apr 12, 42; m 62; c 4. EXPERIMENTAL STATISTICS. *Educ:* Panhandle State Col, BS, 64; Okla State Univ, MS, 67, PhD(statist), 73. *Prof Exp:* Math statistician, Control Systs Div, Environ Protection Agency, Research Triangle Park, NC, 71-72; SR STATISTICIAN, ELI LILLY & CO, 74- *Concurrent Pos:* Lectr statist, Butler Univ, 75. *Mem:* Am Statist Asn; Biomet Soc; Am Soc Qual Control. *Res:* Development and study of statistical procedures for obtaining an objective grouping in a set of observed means, especially multiple decision procedures; statistical quality control. *Mailing Add:* Lilly Corp Ctr Eli Lilly & Co Indianapolis IN 46285

MURPHY, JOHN THOMAS, b Yonkers, NY, Mar 14, 38; c 4. MEDICAL PHYSIOLOGY, NEUROLOGY. *Educ:* Columbia Univ, MD, 63; McGill Univ, PhD(neurol & neurosurg), 68. *Prof Exp:* Intern med, Columbia Univ, 63-64, resident surg, 64-65; fel electroencephalography & clin neurophysiol, Montreal Neurol Inst, McGill Univ, 65-66, res fel neurol & neurosurg, 65-68, lectr physiol, 68; asst prof, State Univ NY, 68-70; assoc prof, 70-73, PROF PHYSIOL, UNIV TORONTO, 73-, CHMN DEPT, 75- *Concurrent Pos:* Invited res lectr, Int Conf Nat Ctr Sci Res, Aix-Marseille, 74. *Mem:* Am Physiol Soc; Can Physiol Soc; Soc Neurosci; Int Brain Res Orgn; Can Soc Electroencephalographers, Electromyographers & Clin Neurophysiologists. *Res:* Brain mechanisms in control of voluntary movement. *Mailing Add:* 190 Cundles Rd E Barrie ON L4M 4S5 Can

MURPHY, JOSEPH, b Montreal, Que, Nov 6, 32; m 58; c 2. SOLID STATE PHYSICS. *Educ:* McGill Univ, BSc, 56, MSc, 58, PhD(physics), 63. *Prof Exp:* Sr physicist, US Naval Ord Lab, Calif, 60-63; sr physicist, Solid State Sci Dept, Westinghouse Res & Develop Ctr, 63-70, sr physicist, 70-; RETIRED. *Mem:* Am Phys Soc. *Res:* Theoretical nuclear physics; microwave and optical properties of solids; theory of interaction of localized defects with each other and with lattice vibrations. *Mailing Add:* 6361 Bahia Del Mar Blvd St Petersburg FL 33715

MURPHY, JOSEPH ROBISON, b Salt Lake City, Utah, June 14, 25; m 46; c 5. RAPTOR ECOLOGY. *Educ:* Brigham Young Univ, AB, 50, MA, 51; Univ Nebr, PhD(zool), 57. *Prof Exp:* From instr to asst prof zool, Univ Nebr, 51-60; from asst prof to prof, 60-85, chmn dept, 68-74, EMER PROF ZOOL,

BRIGHAM YOUNG UNIV, 85- *Concurrent Pos:* Pres, Raptor Res Found, 74-78. *Mem:* Sigma Xi; Cooper Ornith Soc; Am Ornithologists Union. *Res:* Ecology of predatory birds, especially American eagles. *Mailing Add:* Bean Life Sci Mus Brigham Young Univ Provo UT 84602

MURPHY, JUNEANN WADSWORTH, b Chickasha, Okla, Mar 13, 37; m 67; c 2. MEDICAL MYCOLOGY, IMMUNOLOGY. *Educ:* Univ Okla, BS, 59, MSS, 61, MS, 65, PhD(microbiol), 69. *Prof Exp:* Res asst, Sch Med, Tulane Univ, 59; med technologist, Cent State Hosp, 61, instr, 63-64; res asst med mycol, 62-63, vis asst prof, 69-70, asst prof, 70-81, ASSOC PROF MICROBIOL, UNIV OKLA, 81-, DIR MED TECHNOL, 78- *Concurrent Pos:* Vis asst prof clin immunol, Univ Colo Health Sci Ctr, 80; NIH, res grants, 83- *Mem:* Am Soc Microbiol; Med Mycol Soc Am; Am Asn Immunologists; Int Soc Human & Animal Mycoses. *Res:* Host-parasite relationships in systemic mycotic diseases, with a primary interest in host defense mechanisms and their regulation in Cryptococcosis. *Mailing Add:* Dept of Bot-Microbiol 770 Van Vleet Oval Univ of Okla 660 Parrington Oval Norman OK 73019

MURPHY, KEITH LAWSON, b Toronto, Ont, June 12, 32; m 53; c 4. CIVIL ENGINEERING. *Educ:* Univ Toronto, BASc, 54; Univ Wis, MSc, 59, PhD(civil eng), 61. *Prof Exp:* Jr design engr, City of Hamilton, Can, 54-55; asst proj engr, Greater Winnipeg Water & Sanit Dist, 55-57; instr civil eng, Univ Wis, 58-59, proj assoc, 59-60; from asst prof to prof civil eng, McMaster Univ, 61-76; prin, Beak Consults Ltd, 76-88; IMPERIAL OIL/WTC PROF ENVIRON SYSTS ENG, MCMASTER UNIV, 88-, PROF CIVIL ENG, 88- *Concurrent Pos:* Vis prof, Univ Newcastle, 67-68. *Mem:* Am Water Works Asn; Water Pollution Control Fedn; Can Soc Chem Engrs; Can Asn Water Pollution Res & Control; Am Soc Civil Engrs; Asn Environ Engr Prof. *Res:* Process of water and waste water treatment involving phase separation and biological treatments; effects of flow patterns. *Mailing Add:* Dept Civil Eng McMaster Univ 1280 Main St W Hamilton ON L8S 4L7 Can

MURPHY, KENNETH ROBERT, b Oneonta, NY, Oct 13, 40; m 80; c 2. ENVIRONMENTAL ENGINEERING, ATMOSPHERIC SCIENCE. *Educ:* Syracuse Univ, AB, 62; Rensselaer Polytech Inst, MS, 64, PhD(environ eng), 69. *Prof Exp:* Res engr, Dow Chem Co, 68-69; air pollution control engr, Real Estate Div, Gen Elec Co, 69-72, sr proj engr, Serv Eng Div, 72-77, sr environ engr, Energy Systs Div, 77-81, mgr, IDS Prog, Gen Elec Environs Systs, 81-91, LEAD ENGR, ENVIRON CHEM & MASS SPECTROMETRY, GEN ELEC CO, SCHENECTADY, 91- *Concurrent Pos:* Lectr hazardous waste treatment, Pa State Univ, Middletown. *Mem:* Air Pollution Control Asn. *Res:* Advanced systems for measurement and control of atmospheric emissions from energy utilization processes. *Mailing Add:* 129 Acorn Dr Schenectady NY 12302

MURPHY, LARRY S, b Greenfield, Mo, Dec 15, 37; m 59; c 2. AGRONOMY, PLANT PHYSIOLOGY. *Educ:* Univ Mo, BS, 59, MS, 60, PhD(agron), 65. *Prof Exp:* Instr soils, Univ Mo, 60-65; from asst prof to prof agron, Kans State Univ, 65-78; GREAT PLAINS DIR, POTASH & PHOSPHATE INST, 78- *Concurrent Pos:* Res assoc, Mich State Univ, 71-72. *Honors & Awards:* Geigy Award, Am Soc Agron, 73. *Mem:* Hon mem Nat Fertilizer Solutions Asn; fel Am Soc Agron; fel Soil Sci Soc Am. *Res:* Nitrate accumulation in forage crops and water supplies; wheat, corn, grain sorghum, forage production and quality; micronutrient nutrition of plants; evaluation of P fertilizers; water pollution; animal waste disposal. *Mailing Add:* 1015 N Juliette Manhattan KS 66502

MURPHY, LEA FRANCES, b Hutchinson, Kans, Dec 11, 54. POPULATION DYNAMICS, CONTINUUM MECHANICS. *Educ:* Temple Univ, BA, 76; Carnegie-Mellon Univ, MS & PhD(math), 80. *Prof Exp:* Res asst, Math Dept, Carnegie-Mellon Univ, 76-80; ASST PROF MATH, ORE STATE UNIV, 80- *Concurrent Pos:* Res asst, Mech Dept, Johns Hopkins Unvi, 79; mathematician, Naval Undersea Warfare Eng Sta, 81. *Mem:* Am Math Soc; Soc Indust & Appl Math; Resource Modeller's Asn. *Res:* Continuous mathematical modeling of biological phenomenon, including harvesting of age structured populations and variation of individual growth rates within age structured populations. *Mailing Add:* Dept Math Ore State Univ Corvallis OR 97331

MURPHY, MARJORY BETH, b Page, Nebr, July 21, 25. CELL PHYSIOLOGY, BIOCHEMISTRY. *Educ:* Nebr Wesleyan Univ, BA, 47; Univ Colo, MA, 53; Univ Ill, PhD(cell physiol), 61. *Prof Exp:* PROF CHEM, PHILLIPS UNIV, 53- *Mem:* AAAS; Sigma Xi. *Res:* Enzymes involved in membrane transport; active sites of enzymes. *Mailing Add:* 122 N Adolpha Enid OK 73703

MURPHY, MARTIN JOSEPH, JR, b Colorado Springs, Colo, Dec 29, 42; m 65; c 5. EXPERIMENTAL HEMATOLOGY. *Educ:* Regis Col, Colo, BS, 64; NY Univ, MS, 67, PhD(physiol, hemat), 69. *Prof Exp:* Reader & lect asst, Grad Sch Arts & Sci, NY Univ, 65-68; instr biol, Nassau Community Col, 68-69; asst mem, St Jude Children's Res Hosp, Memphis, 73-75; assoc, Sloan-Kettering Inst Cancer Res, 75-79; assoc prof, 79-83, PROF MED, SCH MED, WRIGHT STATE UNIV, DAYTON, OHIO, 83- *Concurrent Pos:* Damon Runyon res fel, Inst Cellular Path, Hopital Bicetre, Kremlin-Bicetre, France, 69-70; NIH res fel, Paterson Labs, Christie Hosp & Holt Radium Inst, Manchester, Eng, 70-71; Leukemia Soc Am spec fel immunol, John Curtin Sch Med Res, Australian Nat Univ, 71-73; dir hemat training prog, Sloan-Kettering Inst Cancer Res & Bob Hipple Lab Cancer Res, Mem Sloan-Kettering Cancer Ctr, 77-79; dir, Hipple Cancer Res Ctr, Dayton, Ohio, 79- *Mem:* NY Acad Sci; Electron Micros Soc Am; Soc Exp Biol & Med; Am Asn Cancer Res; Sigma Xi. *Res:* Physiology of blood cell production in health and disease. *Mailing Add:* Hipple Cancer Res Ctr 4100 S Kettering Blvd Dayton OH 45439-2092

MURPHY, MARY EILEEN, b Baltimore, Md, Nov 14, 54. PHYSIOLOGICAL ECOLOGY, NUTRITIONAL ECOLOGY. *Educ:* Towson State Univ, BS, 76; Wash State Univ, PhD(zoophysiol), 84. *Prof Exp:* ASST PROF BIOL & ZOOL, WASH STATE UNIV, 87- *Concurrent Pos:* Mem bd dirs, Cooper Ornith Soc, 89-92. *Mem:* Am Inst Nutrit; Cooper Ornith Soc; Am Soc Zoologists; Am Ornith Union. *Res:* Amino acid and protein metabolism in wild birds; how changes in protein metabolism through the course of the annual cycle or in response to nutritional challenges influences the birds energy budget, food choice and water intake; keratin synthesis and molt. *Mailing Add:* Dept Zool Wash State Univ Pullman WA 99164-4236

MURPHY, MARY ELLEN, b Hartford, Conn, Sept 10, 28. GEOCHEMISTRY, ANALYTICAL CHEMISTRY. *Educ:* St Joseph Col, Conn, BS, 50; Wesleyan Univ, MA, 56; Fordham Univ, PhD(chem), 65. *Hon Degrees:* DSc, Univ Hartford, 75. *Prof Exp:* Chemist, Naugatuck Chem Div, US Rubber Co, 50-51 & Travelers Ins Co, 51-54; res chemist, Monsanto Chem Co, 56-58; from asst prof to assoc prof, 65-73, PROF CHEM, ST JOSEPH COL, CONN, 73- *Concurrent Pos:* NASA fel, Univ Glasgow, 66-67, Oil & Gas Br, US Geol Surv, Denver, Colo, 80 & Inst Chimie, Univ Louis Pasteur, Strasbourg, France, 81. *Mem:* AAAS; Am Chem Soc; Geochem Soc. *Res:* Geochemistry of organic matter in rocks and meteorites; oil shales; correlation of petroleum and source rocks. *Mailing Add:* 605 Stevens Ave Portland ME 04103

MURPHY, MARY LOIS, b Nebr, Oct 16, 16. MEDICINE. *Educ:* Univ Nebr Col Arts & Sci, BA, 39; Univ Nebr Col Med, MD, 44. *Hon Degrees:* DSc, Univ Nebr, 74. *Prof Exp:* Pediat resident, St Christophers Hosp Children, Philadelphia, 45-46; path resident, Childrens Hosp, Washington DC, 46-47; pediat resident, Childrens Hosp Philadelphia, 47-48, asst chief resident, 48-49; chief resident, Camden Municipal Hosp Contagious Dis, 49-50; spec fel med, Mem Hosp Cancer & Allied Dis, NY, 51-53, clin asst pediatrician, 53-54, asst attend pediatrician, 54-58, assoc attend pediatrician, 58-66, actg chmn, dept pediat, 65-66, chmn, 66-76; instr med, 52-53, res assoc med, 53-54, asst prof med, 54-57, from asst prof to assoc prof pediat, 57-70, PROF PEDIAT, CORNELL UNIV MED COL, 70-; MEM, SLOAN-KETTERING INST CANCER RES, 70- *Concurrent Pos:* Res fel, Children's Hosp Philadelphia, Univ Pa, 49-51; res fel, Sloan-Kettering Inst Cancer Res, NY, 51-54, asst mem, 54-60, assoc mem, 60-70; vis investr, Jackson Lab, Bar Harbor, Maine, 60-64; attend pediatrician, Mem Hosp Cancer & Allied Dis, 66- *Mem:* Soc Pediat Res; Am Soc Clin Oncol; Am Asn Cancer Res; Am Fedn Clin Res; Am Pediat Soc; Am Soc Pediat Hemat & Oncol; Teratology Soc. *Res:* Pediatrics; leukemia; cancer; teratogenesis. *Mailing Add:* Dept Pediat Mem Sloan-Kettering Cancer Ctr 1275 York Ave New York NY 10021

MURPHY, MARY NADINE, b Waucoma, Iowa, June 22, 33. MICROBIOLOGY. *Educ:* Clarke Col, AB, 54; Purdue Univ, NSF fel, 61-64, MS, 64, PhD(biol sci), 65. *Prof Exp:* Instr, 59-61 & 65-66, from asst prof to assoc prof, 66-78, chmn dept, 66-82, PROF BIOL, MUNDELEIN COL, 78-, VPRES ACAD AFFAIRS, 85- *Concurrent Pos:* Shell Merit fel, Stanford Univ, 70; evaluator, NCent Asn Cols & Schs, 73-; Am Coun Educ fel, 74-75. *Mem:* AAAS; Am Soc Microbiol. *Res:* Cellular differentiation; physiology of aquatic fungi. *Mailing Add:* Dept Biol Mundelein Col 6363 N Sheridan Rd Chicago IL 60660

MURPHY, MICHAEL A, b Spokane, Wash, Mar 11, 25; m 47; c 2. GEOLOGY, STRATIGRAPHY. *Educ:* Univ Calif, Los Angeles, PhD, 54. *Prof Exp:* Subsurface geologist, Shell Oil Co, 53-54; from asst prof to assoc prof geol, 54-67, PROF GEOL, UNIV CALIF, RIVERSIDE, 67- *Concurrent Pos:* NSF grant, 58, 61, 67 & 69; prof, Univ Cenap, Brazil, 59-60. *Mem:* Paleont Soc; fel Geol Soc Am. *Res:* Biostratigraphy of silurian-deronian conodonts, graptolites and lower cretaceous ammonites. *Mailing Add:* 2324 Oakenshield Rd Davis CA 95616

MURPHY, MICHAEL JOHN, b Milwaukee, Wis. COMBUSTION ENGINEERING, ALTERNATIVE VEHICLE FUELS. *Educ:* Marquette Univ, BS, 70; Univ Wis, MS(chem), 72, MS(eng), 74, PhD(mech eng), 78. *Prof Exp:* RES SCIENTIST, BATTELLE COLUMBUS LABS, 78- *Mem:* Am Chem Soc; Combustion Inst; Air & Waste Mgt Asn; Am Soc Advan Sci. *Res:* Alternative fuels; indoor air pollution; combustion systems. *Mailing Add:* Battelle Columbus Labs 505 King Ave Columbus OH 43201

MURPHY, MICHAEL JOSEPH, b Butte, Mont, Feb 12, 23. PHYSICAL GEOLOGY. *Educ:* Univ Notre Dame, AB, 45, BS, 51; Univ Calif, Berkeley, MS, 53. *Prof Exp:* From instr to asst prof, 53-65, asst chmn dept, 66-70, ASSOC PROF GEOL, UNIV NOTRE DAME, 65-, CHMN DEPT, 70- *Concurrent Pos:* NSF fel, Columbia Univ, 60-61. *Mem:* AAAS; fel Geol Soc Am; Mineral Soc Am; Nat Asn Geol Teachers; Sigma Xi. *Res:* Isomorphic mineral systems. *Mailing Add:* Dept Earth Sci Univ Notre Dame Notre Dame IN 46556-1020

MURPHY, MICHAEL JOSEPH, b Albany, NY, Feb 2, 53. FETAL ENDOCRINOLOGY, COMPARATIVE PHYSIOLOGY. *Educ:* Siena Col, Loudonville, BS, 75; Col St Rose, Albany, MS, 80; State Univ NY, PhD(biol), 82. *Prof Exp:* Teaching asst embryol & histol, Siena Col, 75; res asst embryol, State Univ NY, Albany, 75-80; instr anat & histol, State Univ NY, Cobleskill, 80; instr embryol, Col St Rose, 81-82; asst prof histol, Siena Col, Loudonville, 82; PROF BIOL, STATE UNIV NY, COBLESKILL, 84- *Concurrent Pos:* Researcher, State Univ NY, Albany, 82. *Mem:* AAAS; Am Soc Zoologists. *Res:* Hormonal control of osmoregulatior during avian development; effects of arginine vasotocin and prolactin on embyonic fluids and renal function; embryoculture techniques. *Mailing Add:* Dept Biol State Univ NY Cobleskill Cobleskill NY 12043

MURPHY, MICHAEL ROSS, b San Diego, Calif, July 7, 53; m 77; c 3. METABOLISM, COMPUTER MODELING. *Educ:* Calif State Polytech Univ, BS, 74; Univ Calif, Davis, MS, 75, PhD(nutrit), 80. *Prof Exp:* Asst prof nutrit, dept dairy sci, 80-85, asst prof, div nutrit sci, 81-85, ASSOC PROF

NUTRIT, DEPT ANIMAL SCI & DIV NUTRIT SCI, UNIV ILL, URBANA-CHAMPAIGN, 85- *Mem:* Am Dairy Sci Asn; Am Soc Animal Sci; Am Inst Nutrit. *Res:* Ruminant nutrition, metabolism and mathematical modeling of physiological systems; the nutrition of dairy cattle in early lactation is currenty being emphasized. *Mailing Add:* 315 Animal Sci Lab 1207 W Gregory Dr Urbana IL 61801

MURPHY, PATRICIA A, b Eureka, Calif, July 27, 51. FOOD PROTEINS, FOOD TOXICOLOGY. *Educ:* Univ Calif, Davis, BS, 73, MS, 75; Mich State Univ, PhD(food sci), 79. *Prof Exp:* Sea grant trainee food sci, Inst Marine Resources, Univ Calif, Davis, 73-75; asst, Dept Food Sci & Human Nutrit, Mich State Univ, 75-78, NIH asst, 78-79; ASST PROF FOOD TECHNOL, IOWA STATE UNIV, 79- *Mem:* Inst Food Technologists; Am Chem Soc; AAAS; Am Oil Chemists Soc; Am Inst Nutrit. *Res:* Maximizing use of soy protein for human consumption by understanding basic interactions of soy protein with non-protein constituents, off-flavors, phytoestrogens, complex protein food systems through basic research on individual soy proteins. *Mailing Add:* Dept Food Technol Iowa State Univ Ames IA 50011

MURPHY, PATRICK AIDAN, b Liverpool, Eng, June 4, 37; m 64, 84; c 6. MICROBIOLOGY, MEDICINE. *Educ:* Univ Liverpool, BSc, 57, MB & ChB, 60; Oxford Univ, DPhil, 66; FRCP. *Prof Exp:* UK Med Res Coun fel, Oxford Univ, 64-67; fel microbiol, 67-68, from instr to assoc prof 68-85, PROF MICROBIOL & MED, SCH MED, JOHNS HOPKINS UNIV, 85- *Mem:* Fel Infectious Dis Soc Am; Am Asn Immunologists; Am Fedn Clin Res; fel Royal Col Physicians. *Res:* Pathogenesis of fever; infectious diseases. *Mailing Add:* Dept Med Blalock 1141 Johns Hopkins Hosp 600 N Wolfe St Baltimore MD 21205

MURPHY, PATRICK JOSEPH, b Chicago, Ill, June 11, 40; m 66; c 4. BIOCHEMISTRY. *Educ:* Loyola Univ Chicago, BS, 62; San Diego State Col, MS, 64; Univ Calif, Los Angeles, PhD(biochem), 67. *Prof Exp:* Sr scientist, Eli Lilly & Co, 67-73, res scientist, 73-77, res assoc, 77, head, Drug Disposition, 77-79, head, 79-83, dir, Cardiovasc Pharmacol & Drug Disposition, 83-86, dir, Toxicol & Drug Disposition, 86-89, DIR, DRUG METAB & CLIN ANAL SERV, ELI LILLY & CO, 89- *Concurrent Pos:* Vchmn, Gordon Conf Drug Metabolism, 80, chmn, 81. *Mem:* AAAS; Am Chem Soc; Am Soc Pharmacol & Exp Therapeut; Sigma Xi. *Res:* Drug metabolism; biochemical pharmacology; studies of enzymes involved in metabolism of endogenous and exogenous compounds. *Mailing Add:* Eli Lilly & Co Wishard Mem Hosp 1001 W Tenth St Indianapolis IN 46202-6700

MURPHY, PAUL HENRY, b Boston, Mass, July 7, 42; m 65; c 3. NUCLEAR MEDICINE, MEDICAL PHYSICS. *Educ:* Univ Kans, MS, 68, PhD(radiation biophys), 70. *Prof Exp:* asst prof, 71-79, ASSOC PROF NUCLEAR MED & PHYSICS, BAYLOR COL MED, 79- *Concurrent Pos:* Advan sr fel med physics, Univ Tex M D Anderson Hosp & Tumor Inst Houston, 70-71. *Mem:* Soc Nuclear Med; Health Physics Soc; Am Asn Physicists in Med; Soc Magnetic Resonance Med; Am Assoc Univ Prof. *Res:* Nuclear medicine imaging; nuclear magnetic resonance imaging; computer applications in nuclear medicine. *Mailing Add:* Dept Radiol Baylor Col Med Houston TX 77030

MURPHY, PETER GEORGE, b New York, NY, Feb 23, 42; m 67; c 3. ECOLOGY, BOTANY. *Educ:* Syracuse Univ, BS, 63, MS, 68; Univ NC, Chapel Hill, PhD(plant ecol), 70. *Prof Exp:* Res assoc trop ecol, PR Nuclear Ctr, Rio Piedras, 63-66; asst prof, 70-75, assoc prof, 75-81, PROF BOT & ECOL, MICH STATE UNIV, 81- *Concurrent Pos:* AID ecol adv, Indonesia, 73-74; mem directorate, US Man & Biosphere Prog, Trop Forest Panel, 75-; ed, Trop Ecol, 81-; prin investr, Man & Biosphere Prog, Dept Energy, 80- & NSF, 81- *Mem:* AAAS; Ecol Soc Am; Am Inst Biol Sci; Asn Trop Biol; Int Soc Trop Ecol. *Res:* Structure and function of tropical and temperate forest ecosystems; primary productivity; sand dune ecosystems; radiation ecology. *Mailing Add:* Dept Bot & Plant Path 166 Plant Biol Bldg Mich State Univ East Lansing MI 48824-1312

MURPHY, PETER JOHN, b Brooklyn, NY, July 29, 39. HYDRAULIC ENGINEERING, FLUID MECHANICS. *Educ:* Webb Inst Naval Archit, BS, 62; Johns Hopkins Univ, PhD(fluid mech), 68. *Prof Exp:* Vis prof, Univ Valle, Colombia, 68-74; asst prof, Cornell Univ, 74-78; asst prof civil engr, Univ Mass, 78-84; CONSULT, 84- *Mem:* Int Asn Hydraul Res. *Res:* Sediment transport; environmental turbulence; measurement techniques. *Mailing Add:* 53 Rolling Ridge Rd Amherst MA 01002

MURPHY, RANDALL BERTRAND, b Pasadena, Calif, May 6, 54; m. BIOPHYSICAL CHEMISTRY, NEUROSCIENCES. *Educ:* Univ Southern Calif, BS, 70; Univ Calif, Los Angeles, PhD(phys chem), 75. *Prof Exp:* Res chemist dept chem, Univ Calif, 74-75; Marion & Eugene Bailey fel biochem, Bone Res Lab, Univ Calif, Los Angeles, 75-76; res chemist, US Nat Bur Standards, 76-77; asst prof, 77-84, ASSOC PROF CHEM, NY UNIV, 85- *Concurrent Pos:* NIH res fel, Lab Biophys Chem, 76-77; med res coun, Neurochem Pharmacol Unit, Cambridge, 83; adj assoc prof biochem in psychiat, Cornell Univ Med Col, 85. *Mem:* Biophys Soc; AAAS; NY Acad Sci; Soc Neurosci. *Res:* Molecular biochemistry of receptors; olfaction; cholecystokinin receptors; sigma receptors. *Mailing Add:* Dept Chem NY Univ Washington Sq New York NY 10003

MURPHY, RAY BRADFORD, b USA, June 7, 22; m 54; c 5. MATHEMATICAL STATISTICS, APPLIED STATISTICS. *Educ:* Princeton Univ, AB, 43, MA, 48, PhD(math), 51. *Prof Exp:* From instr to asst prof math, Carnegie Inst Technol, 49-52; mem tech staff, 52-58, dept head qual theory, 58-67, dept head appl statist, 67-77, STATIST CONSULT, BELL LABS, HOLMDEL, 77- *Concurrent Pos:* Numerical Data Adv Bd, NRC. *Honors & Awards:* Fel, Am Soc Testing & Mat, 87. *Mem:* Am Math Soc; Am Soc Testing & Mat; Am Statist Asn; Inst Math Statist; Sigma Xi. *Mailing Add:* 436 Little Silver Point Rd Little Silver NJ 07739

MURPHY, RICHARD ALAN, b Twin Falls, Idaho, July 4, 38; m 61; c 2. PHYSIOLOGY. *Educ:* Harvard Univ, AB, 60; Columbia Univ, PhD(physiol), 64. *Prof Exp:* NIH fel physiol, Max Planck Inst Med Res, Heidelberg, 64-66; res assoc, Univ Mich, 66-68; from asst prof to assoc prof, 68-77, PROF PHYSIOL, SCH MED, UNIV VA, 77- *Concurrent Pos:* NIH career develop award, 71; assoc ed, Am J Physiol 84-90; spec topics ed, Ann Rev Physiol, 86-89. *Mem:* AAAS; Am Physiol Soc; Biophys Soc; Soc Gen Physiologists; Am Soc Biochem & Molecular Biol; Int Soc Heart Res. *Res:* Biochemistry of the contractile and regulatory proteins of vascular smooth muscle; contractile properties of arterial smooth muscle. *Mailing Add:* Dept Physiol Univ Va Sch Med Box 449 Charlottesville VA 22908

MURPHY, RICHARD ALLAN, b Evergreen Park, Ill, Feb 23, 41; c 3. MEDICAL MICROBIOLOGY, DENTAL MICROBIOLOGY. *Educ:* Loyola Univ, BS, 63; Univ Ill, MS, 66, PhD(microbiol), 71. *Prof Exp:* Teaching asst, 64-70, asst prof, 70-76, ASSOC PROF ORAL MED & DIAG SCI, UNIV ILL MED CTR, 76- *Concurrent Pos:* Mem bd dirs, Int Found Microbiol. *Mem:* Am Soc Microbiol; Int Found Microbiol (exec secy); Asn Practrs Infection Control; Nat Registry Microbiologists-Am Acad Microbiol. *Res:* Role of bacterial enzymes and toxins in pathogenesis; microbiology of periodontal diseases; microbiology of endodontic pathologies; dental microbiology; caries microbiology; dental plaque formation. *Mailing Add:* Dept Oral Med & Diag Sci Univ Ill 801 S Paulina MC 838 Chicago IL 60612

MURPHY, RICHARD ARTHUR, NERVE GROWTH FACTOR, EPIDERMAL GROWTH FACTOR. *Educ:* Rutgers Univ, PhD(anat), 72. *Prof Exp:* ASSOC PROF ANAT, HARVARD MED SCH, 76- *Mailing Add:* Dept Anat Harvard Med Sch 25 Shattuck St Boston MA 02115

MURPHY, ROBERT CARL, b Wheeler, Pa, Dec 18, 19; m 45; c 3. ANATOMY. *Educ:* Geneva Col, BS, 49; Univ Wis, MS, 52, PhD, 55. *Prof Exp:* Asst zool, Univ Wis, 49-52, asst anat, 52-54, instr, 54-55; asst prof, Univ Iowa, 55-57; from asst prof to prof, 57-85, EMER PROF ANAT, TERRE HAUTE CTR MED EDUC, SCH MED, IND UNIV, 85- *Res:* Cells and tissues of the lymphoid system; immunology. *Mailing Add:* 6225 Sara Myers Dr West Terre Haute IN 47885-9199

MURPHY, ROBERT CARL, b Seymour, Ind, Dec 15, 44; m 65; c 2. ORGANIC CHEMISTRY, PHARMACOLOGY. *Educ:* Mt Union Col, BS, 66; Mass Inst Technol, PhD(org chem), 70. *Prof Exp:* NIH trainee & Harvard Univ fel, Mass Inst Technol & Harvard Univ, 70-71; from asst prof to assoc prof, 71-80, PROF PHARMACOL, UNIV COLO MED CTR, DENVER, 80- *Concurrent Pos:* Assoc ed, Org Mass Spectrometry, 74-76; career develop award, NIH, 76-81; vis scientist, Karolinska Inst, Stockholm, Sweden, 78-79. *Honors & Awards:* Merit Award, NIH, 89-99. *Mem:* Am Chem Soc; Am Soc Mass Spectrometry (pres, 90-92); Am Soc Pharmacol & Exp Therapeut; Am Soc Biochem & Molecular Biol. *Res:* Application of stable isotopes and mass spectrometry to biomedical research; structure determination of complex lipids and pharmacologically active molecules by mass spectrometry; pharmacologic studies of leukotrienes and lipoxygenase products; pharmacologic studies of platelet activating factor. *Mailing Add:* Dept Pediat Nat Jewish Ctr Immunol & Respiratory Med Denver CO 80206

MURPHY, ROBERT EMMETT, meat sciences, for more information see previous edition

MURPHY, ROBERT FRANCIS, b Brooklyn, NY, Aug 25, 53; div; c 2. ENDOCYTOSIS, FLOW CYTOMETRY. *Educ:* Calif Inst Technol, PhD(biochem), 79. *Prof Exp:* Asst prof, 83-89, ASSOC PROF BIOL SCI, CARNEGIE MELLON UNIV, 89- *Concurrent Pos:* Consult, Becton-Dickinson, Immunocytometry Systs, 82-; consult, Becton-Dickinson Immunocytometry Systs, 82-; assoc mem, Pittsburgh Cancer Inst, 86- *Honors & Awards:* Presidential Young Investr Award, NSF, 84. *Mem:* Am Soc Cell Biol; Soc Anal Cytol; AAAS; NY Acad Sci; Sigma Xi. *Res:* Cell biology, endocytosis mutants; developed use of flow cytometry for the study of endocytosis; developed biphasic acidification model for control of ligand processing; active in application of computers in biology. *Mailing Add:* Dept Biol Sci Carnegie Mellon Univ Pittsburgh PA 15213

MURPHY, ROBERT PATRICK, b Iowa, Aug 11, 43; m 86; c 1. RETINAL VASCULAR & CHOROIDAL DISEASE. *Educ:* St Louis Univ, BS, 65; Northwestern Univ, MD, 69. *Prof Exp:* Intern, Milwaukee County Gen Hosp, 69-70; resident internal med, Univ Calif, 72-75; resident ophthalmol, Stanford Univ Med Ctr, 75-78; res fel retinal, Wilmer Ophthalmol Inst, 79-80; ASSOC DIR, CTR RETINAL VASCULAR RES, JOHNS HOPKINS MED INST, 80-, ASSOC PROF OPHTHALMOL, 84- *Mem:* Am Acad Ophthalmol; Asn Res Vision & Ophthalmol; Macular Soc; Retina Soc. *Res:* Retinal and choridal neovascularization; early treatment diabetic retinopathy and macular photocoagulation. *Mailing Add:* Johns Hopkins Hosp Wilmer Inst Maumenee 215 600 N Wolfe St Baltimore MD 21205

MURPHY, ROBERT T, pesticide chemistry, for more information see previous edition

MURPHY, ROY EMERSON, b Indianapolis, Ind, Sept 30, 26; m 51; c 2. PARALLEL COMPUTER SYSTEMS DESIGN, ADAPTIVE PROCESS CONTROLS. *Educ:* Purdue Univ, BSEE, 50; Univ Conn, MSEE, 56; Stanford Univ, PhD(mgt sci), 62. *Prof Exp:* Dir, Dynamics Anal Lab, Gen Dynamics Corp, 50-58; vpres res & develop, Quantum Sci Corp, 62-69; pres, Info Telecommun Corp, 69-74; proj mgr, Unisys, Inc, 74-83; dir res & develop, 83-88, PRES AZURAY, INC, 88- *Concurrent Pos:* Assoc prof, Univ Conn, 56-58 & Stanford Univ, 62-65; varian chair econ, Stanford Univ, 62-66; consult, Varian Assocs, Inc, 65-69 & Computer Usage Co, 70-73. *Mem:* Inst Elec & Electronic Engrs; Asn Comput Mach; Sigma Xi. *Res:* Adaptive processes; information theory; computer science; numerical analysis; computer circuit design; software engineering; parallel computer systems programming. *Mailing Add:* 1140 Larkin Valley Rd Watsonville CA 95076

MURPHY, ROYSE PEAK, b Norton, Kans, May 2, 14; m 41; c 3. PLANT BREEDING. *Educ:* Kans State Univ, BS, 36; Univ Minn, MS, 38, PhD(plant breeding, genetics), 41. *Prof Exp:* Asst, Div Agron & Plant Genetics, Univ Minn, 36-37, from instr to asst prof, 37-42; assoc prof, Mont State Univ, 42-46; assoc prof plant breeding, 46-48, prof, 48-79, head dept, 53-64, dean univ fac, 64-67, EMER PROF PLANT BREEDING, CORNELL UNIV, 79- *Mem:* AAAS; Am Soc Agron; Genetics Soc Am; Am Inst Biol Sci; Crop Sci Soc Am. *Res:* Plant genetics and breeding with perennial forage legumes and grasses. *Mailing Add:* Dept Plant Breeding & Biometry 607 Bradfield Hall Cornell Univ Ithaca NY 14853

MURPHY, RUTH ANN, b Ft Worth, Tex; m 66; c 2. PHYSICAL CHEMISTRY, EDUCATION ADMINISTRATION. *Educ:* Univ Tex, Austin, BS, 61, PhD(chem), 67. *Prof Exp:* Lectr chem, Southwestern Univ, Georgetown, Tex, 66; vis prof chem, Univ NMex, 70-81; asst prof chem & math, Univ Albuquerque, 72-78, chem coordr, 78-84; CHMN, DIV SCI & MATH, HOWARD PAYNE UNIV, 84- *Mem:* Nat Sci Teachers Asn; Am Chem Soc. *Res:* Hydrogen bonding studies by proton magnetic resonance using Fortran to generate and evaluate equilibrium constants at various temperatures, leading to calculation of enthalpies of hydrogen bonding. *Mailing Add:* Dept Phys Sci Howard Payne Univ Brownwood TX 76801-2794

MURPHY, SHELDON DOUGLAS, toxicology, pharmacology; deceased, see previous edition for last biography

MURPHY, STANLEY REED, b Guthrie, Okla, Nov 3, 24; m 57; c 1. PHYSICS. *Educ:* Fresno State Col, BA, 48; Univ Wash, PhD(physics), 59. *Prof Exp:* Res engr, Boeing Airplane Co, 50-52; assoc physicist, Appl Physics Lab, 52-54, physicist, 54-57, sr physicist, 57-64, asst dir, 64-68, prof oceanog, Col Arts & Sci, Prof mech & ocean eng, Col Eng & Dir div, Marine Resources, Univ Wash, 68-; AT APPL PHYSICS LAB, UNIV WASH, SEATTLE. *Concurrent Pos:* Mem, Wash Comn Oceanog, 69-, vchmn, 70-71; adj prof, Inst Marine Study, Univ Wash, 73- *Mem:* Am Phys Soc. *Res:* Acoustics; oceanography; instrumentation. *Mailing Add:* 201 NW 48th Ave Seattle WA 98107

MURPHY, TED DANIEL, b Stanley, NC, Apr 21, 36; m 57; c 1. BIOLOGY, ZOOLOGY. *Educ:* Duke Univ, AB, 58, MA, 60, PhD(zool), 63. *Prof Exp:* Asst prof biol, State Univ NY Binghamton, 63-70; assoc prof, Siena Col, 70-72; PROF BIOL, CALIF STATE COL, BAKERSFIELD, 72-, DIR ENVIRON STUDIES AREA & FACIL ANIMAL CARE & TREATMENT, 80- *Mem:* AAAS; Am Soc Ichthyologists & Herpetologists; Ecol Soc Am; Sigma Xi. *Res:* Ecology of amphibians and reptiles; vaptor biology. *Mailing Add:* Dept Biol Calif State Col 9001 Stockdale Hwy Bakersfield CA 93311-1099

MURPHY, TERENCE MARTIN, b Seattle, Wash, July 1, 42; m 69; c 1. PLANT BIOCHEMISTRY, PHYSIOLOGY. *Educ:* Calif Inst Technol, BS, 64; Univ Calif, San Diego, PhD(cell biol), 68. *Prof Exp:* USPHS fel, Univ Wash, 69-70; from asst prof to assoc prof, 71-82, PROF BOT, UNIV CALIF, DAVIS, 82 - *Concurrent Pos:* Chair, Dept Bot, Univ Calif, 86-90. *Mem:* Am Soc Photobiol; Am Soc Plant Physiologists,; Scand Soc Plant Physiol; AAAS. *Res:* Photochemistry and photobiology of RNA and DNA and plant cell membranes; effects of stress (ultraviolet, pathogenic organisms) on plant cells. *Mailing Add:* Dept Bot Univ Calif Davis CA 95616

MURPHY, THOMAS A, b Kewanee, Ill, Aug 26, 37; m 62; c 3. BIOCHEMISTRY, ANIMAL PHYSIOLOGY. *Educ:* Knox Col, BA, 59; Yale Univ, MS, 62, PhD(biol), 64. *Prof Exp:* Biochemist, US Army, Edgewood Arsenal, 64-65; res assoc biochem, Cornell Univ, 66-67; biologist, US Fed Water Pollution Control Admin, 67-70; prog analyst, Washington DC, 70-72, dir nonpoint res div, 72-75, deputy asst adminr, 75-79, LAB DIR, US ENVIRON PROTECTION AGENCY, CORVALLIS, ORE, 79- *Concurrent Pos:* Staff mem Natural Resources & Environ Pres Reorgn Proj, Washington DC, 76-77. *Res:* Effects of anthropogenic stresses on ecological systems and on methods for assessing the condition of ecological systems relative to environmental stresses. *Mailing Add:* Corvallis Environ Res Lab US EPA 200 SW 35th St Corvallis OR 97333

MURPHY, THOMAS DANIEL, b Franklin, Pa, Apr 30, 34; m 58; c 3. STATISTICAL TRAINING, QUALITY CONTROL. *Educ:* Univ Md, BS, 57; Rutgers Univ, MS, 64. *Prof Exp:* Res statistician, Am Cyanamid Co, 67-69, group leader process automation & statist, 71-78, sr res statistician, 78-80, group leader statist & comput, 80-84, supvr Clin Testing Lab, 84-87, PRIN RES STATISTICIAN, AM CYANAMID CO, 87- *Concurrent Pos:* Lectr, Rutgers Univ, 67-71, Ctr Prof Advan, 73-; chair, Chem Div, Am Soc Qual Control, 82; chair subcomt E11-03, Am Soc Testing & Mat, 87- *Mem:* Fel Am Soc Qual Control; Am Soc Testing & Mat; Am Inst Chem Engrs; Am Statist Asn. *Mailing Add:* 33 Junard Dr Morristown NJ 07960

MURPHY, THOMAS JAMES, b Brooklyn, NY, Feb 17, 42; m 68; c 2. STATISTICAL MECHANICS. *Educ:* Fordham Univ, BS, 63; Rockefeller Univ, PhD(physics), 68. *Prof Exp:* Res staff physicist, Yale Univ, 68-69; asst prof, 69-75, ASSOC PROF CHEM, UNIV MD, COLLEGE PARK, 75- *Res:* Equilibrium and non-equilibrium statistical mechanics of Coulomb systems; Brownian motion of interacting particles; density dependence of transport coefficients of gases. *Mailing Add:* Dept Chem Univ Md College Park MD 20742-2021

MURPHY, THOMAS JOSEPH, b Pittsburgh, Pa, Oct 4, 41; m. ATMOSPHERIC TRANSPORT, WATER CHEMISTRY. *Educ:* Univ Notre Dame, BS, 63; Iowa State Univ, PhD(photochem), 67. *Prof Exp:* NIH fel org chem, Ohio State Univ, 67-68; from asst prof org chem to assoc prof chem, 68-81, PROF CHEM, DEPAUL UNIV, 81- *Mem:* Am Chem Soc; Int Asn Great Lakes Res; Int Soc Limnol; AAAS. *Res:* Collection and measurement of atmospheric deposition and its effect on bodies of water. *Mailing Add:* Chem Dept De Paul Univ 25 E Jackson Blvd Chicago IL 60604-2218

MURPHY, TIMOTHY F, b Chicago, IL, Aug 21, 50; m 79; c 2. INFECTIOUS DISEASES, MICROBIAL PATHOGENESIS. *Educ:* NY Univ, BS, 72; Tufts Univ, MD, 76. *Prof Exp:* Intern med, NY Hosp Cornell Med Ctr, 76-77, resident, 77-79; fel infectious dis, Tufts Univ Sch Med, 79-81; asst prof, 81-87, ASSOC PROF MED, STATE UNIV NY, BUFFALO, 87-, ASSOC PROF MICROBIOL, 87- *Concurrent Pos:* Prin investr, Outer Membrane Proteins of H influenza, NIH, 85- & Pathogenesis of Branhamella catarrhalis, 89-; prog dir, Microbial Pathogenesis Grad Group, State Univ NY, Buffalo, 88-; mem, Bact & Mycol 2, NIH Study Sect, 89-; chief infectious dis, Buffalo Vet Admin Med Ctr, 90- *Mem:* Fel Am Col Physicians; fel Infectious Dis Soc Am; Am Fedn Clin Res; Am Soc Microbiol. *Res:* Molecular and antigenic studies of bacterial outer membrane proteins; mechanisms of pathogenesis and to develop vaccines to prevent infection. *Mailing Add:* Buffalo Vet Admin Med Ctr Med Res 151 3495 Bailey Ave Buffalo NY 14215

MURPHY, WALTER THOMAS, b Medford, Mass, Oct 5, 28; m 55; c 5. ORGANIC CHEMISTRY, POLYMER CHEMISTRY. *Educ:* Boston Col, BS, 50, MS, 52. *Prof Exp:* Lab supvr, Main Plant, 52-53, jr res chemist, Res Ctr, 53-57, res chemist, 57-64, sr res chemist, 64-78, RES ASSOC, RES CTR, B F GOODRICH CO, 78- *Mem:* Am Chem Soc. *Res:* Polyurethane polymers; catalysis; structure versus mechanical properties; spandex fiber; adhesives; coatings, poromeric films; formulation and processing of reactive liquid polymers of epoxy, butadiene and acrylonitrile. *Mailing Add:* 1091 Taft Ave Cuyahoga Falls OH 44223

MURPHY, WILLIAM FREDERICK, b Dunkirk, NY, June 10, 39; m 61; c 3. MOLECULAR SPECTROSCOPY. *Educ:* Case Inst Technol, BS, 61; Univ Wis, PhD(phys chem), 66. *Prof Exp:* Fel, 66-68, asst res officer, 68-72, assoc res officer, 72-81, SR RES OFFICER, NAT RES COUN, CAN, 81- *Concurrent Pos:* Invited investr, Inst Struct of Matter, CSIC, Madrid, Spain, 84-85. *Mem:* Am Phys Soc; Optical Soc Am; Soc Appl Spectros; Coblentz Soc; Sigma Xi. *Res:* Raman spectroscopy; Raman gas phase intensities and band contours; instrumentation for Raman spectroscopy. *Mailing Add:* Steacie Inst for Molecular Sci 100 Sussex Dr Ottawa ON K1A 0R6 Can

MURPHY, WILLIAM G(ROVE), b Pittsburgh, Pa, July 19, 21; m 42; c 2. CIVIL ENGINEERING. *Educ:* Univ Ill, BS, 43, MS, 48. *Hon Degrees:* LHD, Nashotah House, 88. *Prof Exp:* Instr eng drawing, Univ Ill, 46, instr theoret & appl mech, 46-49; from asst prof to assoc prof, Marquette Univ, 49-67, asst chmn dept, 56-57, chmn, Dept Theoret & Appl Mech, 57-66, chmn, Dept Civil Eng, 72-83, prof, 67-87, EMER PROF, MARQUETTE UNIV, 87- *Concurrent Pos:* Trustee, Nashtoh House, Wis, 79-88, hon trustee, 89- *Mem:* Am Soc Civil Engrs; Am Soc Eng Educ; Am Rwy Eng Asn; Am Concrete Inst; Int Soc Soil Mech & Found Engrs; Sigma Xi. *Res:* Soil mechanics; foundation engineering; structural mechanics; material behavior. *Mailing Add:* 7303 Maple Terr Wauwatosa WI 53213

MURPHY, WILLIAM HENRY, JR, microbiology, for more information see previous edition

MURPHY, WILLIAM J(AMES), b Lansing, Mich, Dec 21, 27; m 76; c 3. METALLURGICAL ENGINEERING, RESEARCH MANAGEMENT. *Educ:* Wayne State Univ, BS, 49; Lehigh Univ, MS, 51, PhD(metall eng), 55. *Prof Exp:* Metall engr, Aircraft Gas Turbine Div, Gen Elec Co, 51-52; asst prof metall eng, Lehigh Univ, 53-57; div chief, US Steel Corp, 59-72, mgr steel prod develop, 72- 81, dir contract res & tech serv, Tech Ctr, 81-87; VPRES OPERS, R J LEE GROUP INC, 87- *Mem:* Fel Am Soc Metals; Am Inst Mining, Metall & Petrol Engrs; Am Welding Soc; Iron & Steel Inst. *Res:* Physical metallurgy, structure and properties of steel, steel processing. *Mailing Add:* R J Lee Group 350 Hochberg Rd Monroeville PA 15146

MURPHY, WILLIAM MICHAEL, b West DePere, Wis, Oct 17, 41; m 66; c 2. CROP PRODUCTION, SOIL FERTILITY. *Educ:* Univ Wis, BS, 65, MS, 69, PhD(crop sci), 72. *Prof Exp:* Res asst forages & soils, Univ Wis, 67-70; res fel, Ford Found, 70-72; forage specialist, Southern Ill Univ, 72-74; res agronomist forages, Ore State Univ, 74-79; ASSOC PROF PLANT SOIL SCI, UNIV VT, 79- *Concurrent Pos:* Prin investr, Symbiotic Nitrogen Fixation Prog, Chile, 77- *Mem:* Am Soc Agron; Soil Sci Soc Am; Sigma Xi. *Res:* Legume inoculation and nodulation, soil fertility, pasture renovation and management. *Mailing Add:* Plant Soil Sci Dept Univ Vt Burlington VT 05401

MURPHY, WILLIAM PARRY, JR, b Boston, Mass, Nov 11, 23; c 3. MEDICINE. *Educ:* Univ Ill, MD, 47. *Prof Exp:* Instr med, Harvard Med Sch, 49-51, res assoc, 53-55; dir res, Dade Reagents, 55-57; pres, 57-77, BD CHMN, CORDIS CORP, 77- *Concurrent Pos:* Res fel med, Peter Bent Brigham Hosp, Boston, 49-51, asst med, 53-55; chief engr, Fenwal Labs, Mass, 49 & 54; with lab biol control, NIH, 51-53; res assoc, Miami Heart Inst, 56-68; res assoc prof biophys & chmn div, Med Sch, Univ Miami, 58-70; pres, Cordis Dow Corp, 70-75, bd chmn, 75-80. *Honors & Awards:* Award, Am Roentgen Ray Soc, 48. *Mem:* AAAS; AMA; Inst Elec & Electronics Engrs. *Res:* Artificial internal organs; transfusion; biology instrumentation; heart disease; biophysics. *Mailing Add:* 11901 Old Cutler Rd Miami FL 33170

MURPHY, WILLIAM R, b Blue Earth, Minn, Jan 13, 28; m 60; c 1. PHARMACEUTICAL CHEMISTRY, ANALYTICAL CHEMISTRY. *Educ:* Col St Thomas, BS, 51. *Prof Exp:* Mgr anal chem, Toiletries Div, Gillette Co, 61-75; tech dir, Barcolene Co, 75-76; dir qual control, 76-78, opers mgr, 78-80, DIR PRODS & PROCESS DEVELOP, PERRIGO CO, 80- *Mem:* Soc Cosmetic Chemists; Am Chem Soc. *Res:* Develop products and processes for private label health and beauty aids and OTC drugs. *Mailing Add:* 984 Laketown Dr Holland MI 49423

MURR, BROWN L, JR, b Atlanta, Ga, Feb 23, 31; m 53, 73; c 2. PHYSICAL ORGANIC CHEMISTRY. *Educ:* Emory Univ, AB, 52, MS, 53; Ind Univ, PhD(chem), 61. *Prof Exp:* NSF fel org chem, Mass Inst Technol, 61-62, res assoc, 62; from asst prof to assoc prof, 62-71, chmn dept, 73-76, PROF ORG CHEM, JOHNS HOPKINS UNIV, 71- *Concurrent Pos:* Vis scientist, Dept

Embryol, Carnegie Inst, Washington, DC, 76- *Mem:* Am Chem Soc; Royal Soc Chem. *Res:* Reaction mechanisms; kinetics; stereochemistry; kinetic isotope effects. *Mailing Add:* Dept of Chem Johns Hopkins Univ Baltimore MD 21218

MURR, LAWRENCE EUGENE, b Lancaster, Pa, Apr 7, 39; m 58; c 2. METALLURGY, MATERIALS SCIENCE. *Educ:* Pa State Univ, BS, 62, MS, 64, PhD(solid state sci), 67; Albright Col, BSc, 63. *Prof Exp:* Instr eng mech, Pa State Univ, 62-66, res asst solid state sci, Mat Res Lab, 66-67; asst prof mat sci & elec eng, Univ Southern Calif, 67-72; prof metall & mat eng, John D Sullivan Ctr For In-Situ Mining Res, NMex Inst Mining & Technol, 72-81, head dept & dir, 72-80; vpres, Acad Affairs & Res & Prof Mat Sci, 81-85, dir off acad & res progs, Ore Grad Ctr, 85-89; prof mat sci eng & founder, Monolithic Superconductors, Inc, 87-90; MURCHISON PROF & CHMN, DEPT METALL & MAT ENG, UNIV TEX, EL PASO, 89- *Concurrent Pos:* Consult, Res Div, US Naval Weapons Lab, 66-67; vpres res, NMex Tech Res Found, 73-80, pres, 80-81; consult metall, Los Alamos Sci Labs, 75-; vchmn, NMex Joint Ctr for Mat Sci, 76-79, chmn, 79-80. *Honors & Awards:* Phys Sci Award, Electromicros Soc Am, 72. *Mem:* Inst Elec & Electronic Engrs; Electron Micros Soc Am; fel Am Soc Metals; Am Inst Metal Engrs Metall Soc; Int Metallograph Soc. *Res:* Biophysics-effects of electric fields on plants; metal physics; transmission electron microscopy in the study of structure and properties of solids; studies of solid interfacial free energy; hydrometallurgy, environmental metallurgy; information delivery, system/telecommunication networking; superconductor technology. *Mailing Add:* Dept Metall & Mat Eng Univ Tex El Paso TX 79968-0520

MURRAY, BERTRAM GEORGE, JR, b Elizabeth, NJ, Sept 24, 33; m 73. ORNITHOLOGY. *Educ:* Rutgers Univ, AB, 61; Univ Mich, MS, 63, PhD(zool), 67. *Prof Exp:* Lectr biol, Cornell Univ, 67-68; asst prof natural sci, Mich State Univ, 68-71; from asst prof to assoc prof, 71-81, PROF BIOL, RUTGERS UNIV, NEW BRUNSWICK, 81- *Mem:* Am Ornithologists Union; Wilson Ornithol Soc; Cooper Ornithol Soc. *Res:* Ecology, behavior and evolution of birds; migration; orientation; territoriality; paleontology. *Mailing Add:* Dept Biol Sci Rutgers Univ New Brunswick NJ 08903

MURRAY, BRUCE C, b New York, NY, Nov 30, 31; m 54, 71; c 5. ASTRONOMY, GEOLOGY. *Educ:* Mass Inst Technol, SB, 53, SM, 54, PhD(geol), 55. *Prof Exp:* Explor & exploitation geologist, Calif Co, La, 55-58; geophysicist, Geophys Res Directorate, L G Hanscom Field, Mass, 58-60; res fel, 60-63, assoc prof, 63-68, dir jet propulsion lab, 76-82, PROF PLANETARY SCI, CALIF INST TECHNOL, 68- *Concurrent Pos:* Guest observer, Mt Wilson & Palomar Observ, 60-65, staff assoc, 65-69; consult, Rand Corp, 61-75; co-investr, TV Exp, Mariner 4, 65, Mariner 6 & 7, 69, Mariner 9, 71, leader imaging team, Mariner Venus Mercury 73 Mission; Guggenheim fel, 75-76; mem sci team, Phobos 88, 86-89, Mars 94, 89- *Honors & Awards:* Except Sci Achievement Award, NASA, 69. *Mem:* AAAS; Am Astron Soc; Am Geophys Union. *Res:* Planetary exploration; geology and geophysics of the surfaces of the moon and planets; techniques of space photography. *Mailing Add:* Div of Geol & Planetary Sci Calif Inst of Technol 1201 E Calif Blvd Pasadena CA 91125

MURRAY, CALVIN CLYDE, b Oakboro, NC, Aug 5, 07; m 34; c 1. AGRONOMY. *Educ:* NC State Col, BS, 32; Univ Ga, MS, 38; Cornell Univ, PhD(plant breeding), 45. *Prof Exp:* High sch teacher, NC, 32-35; asst agronomist, Soil Conserv Serv, USDA, NC, 35; from asst prof to prof agron, Univ Ga, 36-46; prof & agronomist, La State Univ, 46-48; dir exp sta, 48-50, dean col agr, 50-68, regents prof int educ, dir inter-instnl progs, int affairs & exec dir southern consortium int educ, 68-74, EMER DEAN COL AGR & REGENTS PROF INT EDUC, UNIV GA & EMER DIR INT AFFAIRS, UNIV GA SYST, 74- *Concurrent Pos:* Consult, AID, World Bank, 74-85 & Inter Am Develop Bank, 75-; mem, Int Exec Serv Corps, 75- *Res:* International education; administration; institutional development overseas. *Mailing Add:* 236 West View Dr Athens GA 30606

MURRAY, CHRISTOPHER BROCK, b Meadville, Pa, Jan 16, 37; m 61; c 2. MATHEMATICS. *Educ:* Rice Univ, BA, 58; Univ Tex, PhD(math), 64. *Prof Exp:* Spec instr math, Univ Tex, 61-63; engr scientist, Tracor, Inc, 64-66; ASST PROF MATH, UNIV HOUSTON, 66- *Mem:* Am Math Soc; Math Asn Am. *Res:* Mathematical analysis. *Mailing Add:* Dept of Math Univ of Houston 4800 Calhoun Rd Houston TX 77004

MURRAY, DAVID WILLIAM, b Calgary, Alta, July 19, 30; m 56; c 4. CIVIL ENGINEERING, MECHANICS. *Educ:* Univ Alta, BSc, 52; Univ London, MSc, 54, Univ Calif, PhD(civil eng), 67. *Prof Exp:* Design engr, T Lamb, R N McManus & Assoc, 54-56, Montreal Eng Co Ltd, 56-57 & Green, Blankstien, Russel & Assoc, 57-60; asst prof struct, Univ Man, 57-60; from asst prof to assoc prof, 60-70, PROF CIVIL ENG, UNIV ALTA, 70-, chmn dept, 82-87. *Honors & Awards:* Gold Medal in Civil Eng, Asn Prof Engrs Alta, 52. *Mem:* Am Soc Civil Engrs; Can Soc Civil Engrs. *Res:* Finite elements; inelastic buckling; soil consolidation; permafrost thermal analysis; inelastic analysis of nuclear containments; concrete constitutive theory; structures. *Mailing Add:* Dept Civil Eng Univ Alta Edmonton AB T6G 2M7 Can

MURRAY, DONALD SHIPLEY, b Philadelphia, Pa, June 30, 16; wid; c 3. ACADEMIC ADMINISTRATION, APPLIED STATISTICS. *Educ:* Univ Pa, BS, 37, AM, 40, PhD(statist), 44. *Prof Exp:* Instr statist & acct, 37-44, from asst prof to assoc prof statist, 45-57, prof, 57-80, chmn dept, 64, EMER PROF STATIST, UNIV PA, 80- *Concurrent Pos:* Treas, Am Inst Indian Studies, 64-73, pres asst for Indian opers, 73-80, treas & asst secy, 80-, trustee, 87-; dir, Nat Conf Admin Res, 66-72; mem, Nat Adv Coun Arthritis & Metab Dis, 69-72; mem, Grants Admin Adv Comt, Dept Health, Educ & Welfare, 69-72, chmn, 70-72; asst chmn, Dept S Asia Regional Studies, Univ Pa, 77-79, chmn, 79-80; secy-treas, La Napoule Found, La Napoule, France & Philadelphia, 78-81. *Mem:* Am Statist Asn; Asn Asian Studies. *Res:* Development of models for financial management of private institutions of higher education. *Mailing Add:* C-111 Spring House Estates Spring House PA 19477

MURRAY, EDWARD CONLEY, b Mullen, Nebr, Sept 25, 31; m 53; c 2. INORGANIC CHEMISTRY. *Educ:* Nebr State Teachers Col, Kearney, AB, 52; Univ Colo, Boulder, MS, 63, PhD(inorg chem), 69. *Prof Exp:* Jr chemist, Ames Lab, AEC, Iowa, 52-55; proj engr, Aeronaut Res Lab, Wright-Patterson AFB, Ohio, 59-60; sr res chemist, Am Potash & Chem Corp, 69-70; sr res chemist, Kerr-McGee Corp, 70-78; PRES, MURLIN CHEM INC, 78- *Mem:* Am Chem Soc. *Res:* Preparation of high purity inorganic chemicals; surface chemistry of titanium dioxide pigments; calcium phosphate compounds. *Mailing Add:* Murlin Chem Inc Balligomingo Rd West Conshohocken PA 19428

MURRAY, FINNIE ARDREY, JR, b Burgaw, NC, May 30, 43; m 64; c 3. REPRODUCTIVE IMMUNOLOGY. *Educ:* NC State Univ, BS, 66, MS, 68; Univ Fla, PhD(reproductive physiol), 71. *Prof Exp:* Res assoc, Dept Molecular, Cellular & Develop Biol, Univ Colo, 70-71; instr, Dept Zool, Univ Tenn, 71-72, asst prof, 72-74; from asst prof to assoc prof, dept animal sci, Ohio Agr Res & Develop Ctr, Wooster, 74-84; ASSOC PROF, DEPT ZOOL & BIOMED SCI, OHIO UNIV, 85- *Concurrent Pos:* Assoc prof, Dept Animal Sci, Ohio State Univ, 77-84; chmn, Interdisciplinary Doctoral Prog, Molecular & Cell Biol, Ohio Univ, 85-89; chmn, Dept Zool & Biomed Sci, Ohio Univ, 89- *Mem:* Am Soc Animal Sci; AAAS; Soc Study Reproduction; Am Asn Immunol; Am Soc Microbiol; Am Zool Soc. *Res:* Immunology of maternal-fetal relationship; regulatory immunology of reproduction and lactation; function of the uterus in embryonic development; biochemistry of uterine secretions; endocrine control of uterine and ovarian function. *Mailing Add:* Dept Zool & Biomed Sci Ohio Univ Irvine Hall Athens OH 45701-2979

MURRAY, FRANCIS E, b Grande Prairie, Alta, Nov 17, 18; m 39; c 4. PHYSICAL CHEMISTRY. *Educ:* Univ Alta, BSc, 50; McGill Univ, PhD(phys chem), 53. *Prof Exp:* Asst prof phys chem, Univ Man, 53-55; asst prof, Can Serv Col, Royal Roads, 55-56; res chemist, BC Res Coun, 56-61; res coord, Consol Paper Corp, 61-62; head, Div Chem, BC Res Coun, 62-68; assoc prof chem eng, Fac Appl sci, Univ BC, 68-70, prof chem eng & head dept, 70-83; RETIRED. *Concurrent Pos:* Consult, Air Pollution Br, USPHS. *Mem:* Can Pulp & Paper Asn; Tech Asn Pulp & Paper Indust; Air Pollution Control Asn. *Res:* Molecular association forces; critical temperature phenomena; Kraft pulping chemical recovery and pulp purification. *Mailing Add:* 6282 Thomson Terr RR 5 Ducan BC V9L 4T6 Can

MURRAY, FRANCIS JOSEPH, b New York, NY, Feb 3, 11; m 35; c 6. MATHEMATICS. *Educ:* Columbia Univ, AB, 32, MA, 33, PhD(math), 35. *Prof Exp:* From instr to prof math, Columbia Univ, 36-60; prof math, 60-80, dir spec res numerical anal, 60-70, dir undergrad studies, 76-80, EMER PROF MATH, DUKE UNIV, 80- *Concurrent Pos:* Consult ed, Math Tables & Other Aids to Comput, Div Math, Nat Res Coun, 53-57. *Mem:* Am Math Soc; Asn Comput Mach. *Res:* Partial differential equations; linear spaces; rings of operators; Hilbert space; mathematical machines; aids to computation. *Mailing Add:* 1012 Norwood Ave Durham NC 27707

MURRAY, FRANCIS JOSEPH, b Jersey City, NJ, Oct 16, 20; m 47; c 5. MEDICAL SCIENCE, HEALTH SCIENCES. *Educ:* St Peter's Col, BS, 42; Purdue Univ, PhD(bact), 48. *Prof Exp:* Asst bact, Lederle Labs, Am Cyanamid Co, NY, 44; asst serol, Purdue Univ, 44-46; asst chief bacteriologist, Wm S Merrell Co, 48-51, head dept microbiol, 51-56, exec asst to dir res, 56-59, dir sci rels, 60-67, dir sci & com develop & cent sci serv, 67-69, vpres, Richardson-Merrell Inc, 69-81, vpres, Merrell-Dow Pharmaceut Inc, 81-82; RETIRED. *Mem:* AAAS; Am Soc Microbiol; Am Asn Immunol; Am Soc Clin Pharmacol & Therapeut; Lic Exec Soc. *Res:* Antibiotics; immunology; virology; chemotherapy. *Mailing Add:* 50 Ravenwood Dr Weston CT 06883

MURRAY, FREDERICK NELSON, b Tulsa, Okla, Apr 21, 35; m 67; c 2. STRUCTURAL GEOLOGY, STRATIGRAPHY. *Educ:* Univ Tulsa, BGE, 57; Univ Wash, BS, 62; Univ Colo, MS, 62, PhD(geol), 66. *Prof Exp:* Jr geologist, Pan Am Petrol Corp, 57-58; asst geologist, Ill State Geol Surv, 65-67; asst prof geol, Allegheny Col, 68-71; geologist, US Geol Surv, 71-72; chief geologist, 72-74, mgr explor & land, 74-83, CONSULT GEOLOGIST, MAPCO INC, 84- *Mem:* Am Asn Petrol Geologists; Am Geophys Union; Geol Soc Am; Sigma Xi; Am Inst Prof Geologists. *Res:* Computer application in geology; geologic mapping; coal geology. *Mailing Add:* 3734 E 81st Pl Tulsa OK 74137

MURRAY, GARY JOSEPH, b Toronto, Ont, Aug 26, 50. GLYCOPROTEINS, RECEPTORS. *Educ:* Univ Waterloo, BSc, 72, PhD(chem), 77. *Prof Exp:* Lectr, Univ Waterloo, 77; vis fel, NIMH, Bethesda, 77-80; vis assoc, Develop & Metab Neurol Br, Nat Inst Neurol & Commun Disorders & Stroke, 80-85; RES ASSOC & CONSULT, NAT GAUCHER FOUND, 85- *Mem:* Am Chem Soc; NY Acad Sci. *Res:* Modification of proteins with carbohydrate or oligosaccharide chains; various lectin receptors on cell membranes; biochemistry and biosynthesis of lysosomal glycoprotein enzymes, including both cell biologic and enzyme kinetics approaches. *Mailing Add:* Bldg 10 Rm D04 NINDS NIH Bethesda MD 20892

MURRAY, GEORGE CLOYD, b Minneapolis, Minn, May 20, 34; m 57; c 2. NEUROPHYSIOLOGY, PHYSICS. *Educ:* George Washington Univ, BS, 59; Univ Colo, MS, 62; Johns Hopkins Univ, PhD(biophys), 68. *Prof Exp:* Assoc staff engr, Appl Physics Lab, Johns Hopkins Univ, 54-59; instr & teaching assoc physics, Univ Colo, 59-62; asst prof, George Washington Univ, 62; instr & res assoc biophys, Johns Hopkins Univ, 62-68; staff fel neurophysiol, Lab Neurophysiol, Nat Inst Neurol Dis & Blindness, 68-72, head, Communicative Disorders Sect & Biomed Eng Sect, C & FR, Nat Inst Neurol Dis & Stroke, 72-74, spec asst to dir & dept dir div blood dis & resources, Nat Heart, Lung & Blood Inst, 74-76, DIR, OPPE, NAT INST NEUROL & COMMUN DIS & STROKE, NIH, 76- *Res:* Biophysical study of the mechanisms of excitable biological membrane systems. *Mailing Add:* 1706 Mark Lane Rockville MD 20852

MURRAY, GEORGE R(AYMOND), JR, decision analysis, operations research, for more information see previous edition

MURRAY, GEORGE T(HOMAS), b Waynesburg, Ky, Feb 6, 27; m 58; c 2. METALLURGY. *Educ:* Univ Ky, BS, 49; Univ Tenn, MS, 51; Columbia Univ, ScD(metall), 58. *Prof Exp:* Metallurgist, Oak Ridge Nat Lab, 49-52, res lab, Bendix Aviation Co, 52-54 & Bridgeport Brass Co, 54-55; vpres, Mat Res Corp, 57-78; PROF METALL ENG & HEAD DEPT, CALIF POLYTECH STATE UNIV, 78- *Concurrent Pos:* Res assoc, Columbia Univ, 54-57; lectr, Cooper Union, 57- *Mem:* Am Soc Metals; Am Inst Mining, Metall & Petrol Engrs. *Res:* Solid state diffusion and reactions; irradiation effects in metals; mechanical properties; hydrogen embrittlement. *Mailing Add:* Dept of Metall Eng Calif Polytech State Univ San Luis Obispo CA 93401

MURRAY, GLEN A, b Sidney, Mont, Mar 1, 39; m 61; c 2. CROP PHYSIOLOGY, AGRONOMY. *Educ:* Mont State Univ, BS, 62, MS, 64; Univ Ariz, PhD(agron), 67. *Prof Exp:* Assoc prof & assoc crop physiologist, 67-77, PROF PLANT SCI & CROP PHYSIOLOGIST, UNIV IDAHO, 77- *Mem:* Am Soc Agron; Crop Sci Soc Am. *Res:* Cold hardiness of grass seedlings; photoperiodic studies on legumes; associating changes from vegetative to reproductive states; regulation of nitrogen distribution in wheat plants; protein and yield studies on Austrian winter peas, growth regulators; alternative crops for Idaho, especially sunflower, safflower and meadowfoam; row spacing, date of seeding and photoperiod studies. *Mailing Add:* Dept Plant & Soil Sci Univ Idaho 410 S Polk Moscow ID 83843

MURRAY, GROVER ELMER, b Maiden, NC, Oct 26, 16; m 41; c 2. GEOLOGY. *Educ:* Univ NC, BS, 37; La State Univ, MSc, 39, PhD(geol), 42. *Prof Exp:* Res geologist, State Geol Surv, La, 38-41; geologist, Magnolia Petrol Co, 41-48; prof stratig geol, La State Univ, 48-55, chmn dept, 50-53, Boyd prof geol, 55-66, vpres & dean, 63-65, vpres acad affairs, La State Univ Syst, 65-66; pres, 66-76, pres, Sch Med, 69-76, UNIV PROF & PROF GEOSCI, SCH MED, TEX TECH UNIV, LUBBOCK, 76- *Concurrent Pos:* With Ark Fuel Oil Corp, 51-60; dir, Orgn Trop Studies, Inc, 64-65 & Gulf Univs Res Corp, 64-66, pres, 65-66; consult, 49-; mem, Am Comn Stratig Nomenclature, 51-54 & 57-63; ed, J Paleont, 52-54; mem, Int Comn Stratig Nomenclature, 54-; ed, Bull Am Asn Petrol Geologists, 59-63; mem, US Nat Comt Geol, 65-69, chmn, 64-68; deleg, House Soc Reps, Am Geol Inst, 58-63 & 65-68; mem bd gov, ICASALS, Inc, 67-76; mem, Nat Sci Bd, 68-80; mem bd dirs, Tex Partners of Americas, 71-; mem, Nat Adv Comt Oceans & Atmosphere, 75-77; dir, Ashland Oil, Inc, 76-; pres & dir, Global Explor, Inc, 78- *Mem:* Geol Soc Am; hon mem Soc Econ Paleontologists & Mineralogists (pres, 62-63); Paleont Soc; hon mem Am Asn Petrol Geologists (pres, 64-65); Am Geophys Union; 1st hon mem Am Inst Prof Geologists (pres, 78); Am Geol Inst (pres, 79-80). *Res:* Structural and field geology; geomorphology; geophysics; micropaleontology; stratigraphy of the Gulf coast and southern Appalachians; petroleum geology of coastal plain. *Mailing Add:* 4609 Tenth St Lubbock TX 79416

MURRAY, HAROLD DIXON, b Neodesha, Kans, May 25, 31; m 554; c 1. MALACOLOGY. *Educ:* Ottawa Univ, BA, 52; Kans State Col Pittsburg, MSc, 53; Univ Kans, PhD(zool), 60. *Prof Exp:* Asst biol, zool & parasitol, Univ Kans, 55-60, instr limnol & invert zool, 60-61; from asst prof to assoc prof, 61-74, chmn, dept, 75-84, PROF BIOL, TRINITY UNIV, TEX, 74- *Concurrent Pos:* Actg dean, Trinity Univ, Tex, 87-88. *Mem:* Am Malacol Union (pres, 74); Sigma Xi; Am Soc Parasitol; Western Soc Malacologists. *Res:* Geographical distribution of Unionidae and Unionicolidae; biology and ecology of Thiaridae. *Mailing Add:* Dept Biol Trinity Univ San Antonio TX 78212

MURRAY, HAYDN HERBERT, b Kewanee, Ill, Aug 31, 24; m 44; c 3. CLAY MINERAOLOGY, ECONOMIC GEOLOGY. *Educ:* Univ Ill, BS, 48, MS, 50, PhD(geol), 51. *Prof Exp:* From asst prof to assoc prof geol, Ind Univ, 51-57; dir appl res, Ga Kaolin Co, 57-59, dir res & mfg, 59-63, vpres, 63-64, exec vpres, 64-73; chmn dept geol, 73-84, PROF GEOL, IND UNIV, 84- *Concurrent Pos:* Clay mineralogist, State Geol Surv, Ind, 51-57; mem exec comt, Working Comt Genesis & Age of Kaolins, UNESCO, 73-83. *Honors & Awards:* Hal Williams Hardinge Award, Am Inst Mining, Metall & Petrol Engr, 76. *Mem:* Fel Am Ceramic Soc (vpres, 74-75); Am Chem Soc; Mineral Soc Am; distinguished mem Am Inst Mining, Metall & Petrol Engr; distinguished mem Clay Minerals Soc (pres, 65-66); Soc Mining Engrs (pres, 88); fel Geol Soc Am; fel Tech Asn Pulp & Paper Industs. *Res:* Geology, economic uses and chemistry of clay minerals; beneficiation of metallic, non-metallic, and coals using high intensity magnetic separation. *Mailing Add:* Dept Geol Sci Ind Univ Bloomington IN 47405

MURRAY, JAMES GORDON, b Flint, Mich, July 8, 27; m 47; c 4. ORGANIC POLYMER CHEMISTRY. *Educ:* Univ Mich, BS, 50; Dartmouth Univ, MA, 52; Duke Univ, PhD(chem), 55. *Prof Exp:* Chemist, Monsanto Chem Co, 55-59 & Gen Elec Co, 59-67; group leader polyolefins, Mobil Chem Co, 67-74, res assoc, 74-88; RETIRED. *Mem:* Am Chem Soc. *Res:* Polymers; organometallics; catalysis; petroleum chemistry. *Mailing Add:* RD No 1 Box 89A Whiting VT 05778-9801

MURRAY, JAMES W, b Berwyn, Alta, Sept 11, 33; m 60. GEOLOGY. *Educ:* Univ Alta, BSc, 56; Princeton Univ, MA, 63, PhD(geol), 64. *Prof Exp:* Geologist, Texaco Explor Co, Alta, 57-61; fel, Inst Oceanog, 64-65, asst prof geol, 65-69, assoc prof geol & oceanog, 69-74, actg head, Dept Geol, 71-72, PROF GEOL SCI, UNIV BC, 74-, DIR UNIV & INDUST LIAISON, 90- *Concurrent Pos:* Mem subcomt stratig, paleont & fossil fuels res, Nat Adv Comt Can, 65-68; co-chmn, Marine Geol Prog Int Geol Cong, Montreal, 72. *Mem:* Fel Geol Soc Am; Am Asn Petrol Geologists; Soc Econ Paleontologists & Mineralogists. *Res:* Origin and distribution of recent marine sediments in fjords and delta; origin of continental shelves and slopes. *Mailing Add:* Dept Geol Sci Univ BC 2075 Westbrook Pl Vancouver BC V6T 1W5 Can

MURRAY, JAY CLARENCE, b Lapoint, Utah, June 27, 29; m 49; c 2. GENETICS, AGRONOMY. *Educ:* Utah State Univ, BS, 51; Colo Agr & Mech Col, MS, 55; Cornell Univ, PhD, 59. *Prof Exp:* Asst agron, Colo Agr & Mech Col, 53-55; asst genetics, Cornell Univ, 55; from asst prof to assoc prof, 59-67, PROF AGRON, OKLA STATE UNIV, 67-, ASSOC DIR AGR EXP STA, 68- *Res:* Physiological genetics of neurospora and photoperiodism in Gossypium; geneome complementation in the tetraploid species of Gossypium; genetics of fiber differentiation in cotton; quantitative genetic studies of fiber properties and earliness of cotton; genetics, cytogenetics, and breeding of forage grasses and legumes. *Mailing Add:* Dept Agron Okla State Univ Stillwater OK 74078

MURRAY, JEANNE MORRIS, b Fresno, Calif, July 6, 25. BIOCHEMISTRY-HUMAN VACCINES FOR VIRUSES, BEHAVIOR ETHOLOGY- PREDICTING BEHAVIOR UNDER STRESS. *Educ:* Morris Harvey Col, BS, 57; Ga Inst Technol, MS, 66; Am Univ, PhD(pub admin) & PhD(mgt sci), 81. *Prof Exp:* Res scientist statist anal & computer sci, Ga Inst Technol, 59-68; computer scientist, Defense Intel Agency, 68-69; info & computer scientist systs res, Delex Systs Inc, 69-70; mgt analyst pub bldgs serv, Gen Ser Admin, 71-74; mgt consult & assoc prof computer sci & commun, Northern Va Community Col, 75-76; PRES, SEQUOIA ASSOCS, INC, 81- *Concurrent Pos:* Ed printout, Rich Electronic Computer Ctr, Ga Tech, 63-67, prin investr, 67-68; adj prof computer sci, Am Univ, 68-73; adv computer sci, Maylasia, appl by US Dept State, 70; bd dir, Am Soc Cybernetics, 70-71; lectr, Dept Continuing Educ, Univ Va, Falls Church, 73-76; consult, Pres Carter's Transition Team, 76-77; organizor & facilitator, Panel Int Mkt High Technol Presence Dept Defense Technol Controls, Inst Elec & Electronics Engrs, 84; prin investr, stress study, Fed Emergency Mgt Agency, 84-87, Human Vaccine Res Proj, Sequoia Assocs Inc, 85-; mem, Nat Comt Technol Transfer, Inst Elec & Electronics Engrs, 85-86. *Mem:* NY Acad Sci; sr mem Inst Elec & Electronics Engrs; Int Soc Systs Sci; Inst Noctic Sci. *Res:* Neurophysiology-processing of visual information by the human brain; biochemistry-research on a human vaccine for certain viruses; psychology-prediction of human responses to stress using personality assessment system; policy sciences-research and development of a methodology for policy design and strategic planning; forecasting studies. *Mailing Add:* 2915 N 27th St Arlington VA 22207

MURRAY, JOAN BAIRD, b Rochester, NY, Nov 20, 26; m 52; c 2. VERTEBRATE ZOOLOGY, URBAN ECOLOGY. *Educ:* Alfred Univ, BA, 48; Syracuse Univ, MA, 50, PhD, 53. *Prof Exp:* Asst prof biol, Western Col, 60-68; teacher, 68-86, prof, 86-90, EMER PROF BIOL, DEKALB COL, 91- *Concurrent Pos:* Grammar sch teacher & demonstr, Univ Ibadan, 65-66. *Mem:* Am Soc Zoologists; Am Inst Biol Sci; EAfrican Wildlife Soc. *Res:* Teaching a heterogeneous student population; biological writing; interactive approaches; drought effects on Lake Lanier, Georgia; bird calls and inanimate alarms. *Mailing Add:* Sci Div DeKalb Col Clarkston GA 30021-2396

MURRAY, JOHN FREDERIC, b Mineola, NY, June 8, 27; m 49; c 3. INTERNAL MEDICINE. *Educ:* Stanford Univ, AB, 49, MD, 53. *Hon Degrees:* DSc, Univ Paris, 83. *Prof Exp:* From intern to asst resident med, San Francisco Hosp, 52-54; from resident to sr resident, Kings County Hosp, New York, 54-56; res fel, Am Col Physicians, Post-Grad Sch Med, Univ London, 56-57; from instr to assoc prof med & physiol, Univ Calif, Los Angeles, 57-59; chief chest serv, San Francisco Hosp, 66-89; assoc prof, 66-69, PROF MED, UNIV CALIF, SAN FRANCISCO, 69-, STAFF MEM CARDIOVASC RES INST, SCH MED, 66- *Concurrent Pos:* Attend specialist, Vet Admin, 59-; chmn pulmonary training comt, Nat Heart & Lung Inst, 70-72; ed, Am Rev Respiratory Dis; chmn pulmonary acad award comt, Nat Heart, Lung & Blood Inst, 74-75, mem pulmonary dis adv comt, 75-79, mem adv coun, 85- *Honors & Awards:* Col Medalist, Am Col Chest Phys, 85. *Mem:* Asn Am Physicians; Am Soc Clin Invest; Am Fedn Clin Res; Am Physiol Soc; fel Royal Col Physicians. *Res:* Cardiopulmonary physiological techniques in clinical medicine. *Mailing Add:* Chest Serv San Francisco Hosp San Francisco CA 94110

MURRAY, JOHN JOSEPH, b New York, NY, Oct 16, 37; m 60; c 3. ORGANIC CHEMISTRY. *Educ:* Manhattan Col, BS, 59; Fordham Univ, PhD(org chem), 64. *Prof Exp:* Res chemist, Gen Chem Div, Allied Chem Corp, NJ, 64-70; from asst prof to assoc prof, 70-76, PROF CHEM, MIDDLESEX COUNTY COL, 76- *Mem:* Am Chem Soc. *Res:* Synthesis of new fluorinated compounds; chemistry of carbenes and their diazo precursors. *Mailing Add:* 190 Skyline Dr Millington NJ 07946

MURRAY, JOHN RANDOLPH, b PEI, Aug 18, 16; m 42; c 2. PHARMACOLOGY. *Educ:* Univ Alta, BSc, 40, MSc, 50; Ohio State Univ, PhD(pharmacol), 55. *Prof Exp:* Asst dispenser, Dunford Drug Co, Ltd, Alta, 40-41; asst dept mgr, Parke, Davis & Co, Ont, 41-42; from lectr pharm to prof, Sch Pharm, Univ Alta, 46-59; dir, Sch Pharm, Univ Man, 59-70, dean fac pharm, 70-81; RETIRED. *Mem:* AAAS; Pharmacol Soc Can; Can Pharmaceut Asn; Asn Fac Pharmaceut Can; Can Found Adv Pharm. *Res:* Hypertension and the antihypertensive drugs. *Mailing Add:* 1731 Ridgewood Ave St Paul MN 55110

MURRAY, JOHN ROBERTS, b Campwhite, Ore, Aug 8, 43; m 75; c 2. LASERS, NONLINEAR OPTICS. *Educ:* Mass Inst Technol, SB, 65, PhD(physics), 70. *Prof Exp:* PHYSICIST, LAWRENCE LIVERMORE NAT LAB, UNIV CALIF, 72- *Concurrent Pos:* Topical ed, J Optical Soc Am, 84-91; div ed, Appl Optics, 91- *Mem:* Optical Soc Am. *Res:* Physics and technology of high peak and average power lasers; optical propagation; nonlinear optics. *Mailing Add:* Livermore Nat Lab Univ Calif Box 5508 L-490 Livermore CA 94551

MURRAY, JOHN WOLCOTT, b Flushing, NY, Jan 9, 09; m 38; c 2. CHEMISTRY. *Educ:* Colgate Univ, AB, 30; Johns Hopkins Univ, PhD(chem), 33. *Prof Exp:* Asst chem, Johns Hopkins Univ, 32-33, instr, 33-34; asst, Rockefeller Inst, 34-39; chief chemist, Thomasville Stone & Lime Co, 39-42; from asst prof to prof, 42-71, EMER PROF CHEM, VA POLYTECH INST & STATE UNIV, 71- *Mem:* Am Chem Soc. *Res:* Raman spectra; molecular models; dissociation of phenols; movement of water in cell models; factors controlling nature of cave deposits. *Mailing Add:* 6 Dogwood Circle Blacksburg VA 24060-6227

MURRAY, JOSEPH BUFORD, b Birmingham, Ala, July 29, 33; m 55; c 2. GEOLOGY. *Educ:* Univ Chattanooga, BS, 55; Univ Tenn, Knoxville, MS, 60; Case Western Reserve Univ, PhD(geol), 71. *Prof Exp:* Asst prof geol & geog, Grove City Col, 60-66; CHIEF GEOLOGIST, DEPT NATURAL RESOURCES, DIV GEOL & WATER RESOURCES, GEOL SURV, 69- *Concurrent Pos:* Mgr, Hydrogeol Eng-Sci, Inc, Houston, Tex. *Mem:* Geol Soc Am; Soc Econ Paleontologists & Mineralogists; Am Asn Petrol Geol. *Res:* Environmental applications of groundwater research. *Mailing Add:* 12463 Deep Spring Lane Houston TX 77077-2925

MURRAY, JOSEPH E, b Milford, Mass, Apr 1, 19. ORGAN TRANSPLANTATION. *Educ:* Col of the Holy Cross, AB, 40; Harvard Med Sch, 43, Am Bd Surg, cert, 52; Am Bd Plastic Surg, cert, 54. *Hon Degrees:* MRACP, 85, FRCS(L), 87, FRCS(I), 88. *Prof Exp:* Dir, Lab Surg Res, Harvard Med Sch & Peter Bent Brigham Hosp, 55-66; EMER PROF SURG, HARVARD MED SCH & EMER CHIEF PLASTIC SURG, BRIGHAM & WOMEN'S HOSP & CHILDREN'S HOSP MED CTR, BOSTON, MA. *Concurrent Pos:* Mem, Comt Tissue Transplantation, Nat Res Coun, 56-69, Surg Studies Sect, NIH, 62-66, Immunobiol Study Sect, NIH, 67-71, ad hoc comt, Harvard Med Sch; chmn First & Second Human Kidney Transplant Conf, Nat Res Coun, 63 & 65; co-ed, Transplantation J, 63-71. *Honors & Awards:* Nobel Prize in Med, 90; Francis Amory Prize, AAAS, 62; Gold Medal, Int Soc Surgeons, 63; Hon Award, Am Asn Plastic Surgeons, 69; George B Kunkel MD Award, 77; Nat Kidney Found Gift of Life, 79; Olof AF Acred Medal, Swedish Soc Med, 90; Lifetime Achievement Award, Nat Kidney Found Inc, 90; Minnie Rosen Award High Achievement in Service to Mankind, Ross Univ, 91; Am Surg Asn Medal, 91. *Mem:* Nat Acad Sci; Am Asn Plastic Surgeons (pres, 64-65); fel Am Col Surgeons (first vpres, 83-84); hon mem Am Soc Transplant Surgeons; Am Surg Asn (first vpres, 78); Soc Univ Surgeons; Soc Head & Neck Surgeons; Transplantation Soc. *Res:* Transplantation procedures; first human kidney transplant, 54; drugs to suppress the immune system thus allowing use of kidneys from relatives and cadavers; plastic surgery & skin grafts; repair of inborn facial defects in children; first successful human renal cadaveric transplant, 62. *Mailing Add:* 108 Abbott Rd Wellesley Hills MA 02181

MURRAY, JOSEPH JAMES, JR, b Lexington, Va, Mar 13, 30; m 57; c 3. ZOOLOGY. *Educ:* Davidson Col, BS, 51; Oxford Univ, BA, 54, MA, 57, DPhil(zool), 62. *Prof Exp:* Instr biol, Washington & Lee Univ, 56-58; from asst prof to assoc prof, 62-73, co-dir, Mountain Lake Biol Sta, 64-91, PROF BIOL, UNIV VA, 73- *Mem:* AAAS; Soc Study Evolution; Am Soc Naturalists; Am Soc Ichthyologists & Herpetologists; Genetics Soc Am. *Res:* Genetics of populations of gastropods. *Mailing Add:* Dept of Biol Univ of Va Charlottesville VA 22901

MURRAY, KENNETH MALCOLM, JR, nuclear physics, for more information see previous edition

MURRAY, LAWRENCE P(ATTERSON), JR, b Hollins, Va, Nov 30, 30; m 52; c 3. CHEMICAL ENGINEERING. *Educ:* Va Polytech Inst, BSc, 52, MSc, 54, PhD(chem eng), 56. *Prof Exp:* Sr engr tech develop, E I du Pont de Nemours & Co, Inc, 55-61, supvr, 61-66, sr res supvr, Spunbonded Div, 65-76, sr res engr, 76-82; RETIRED. *Mem:* Am Inst Chem Engrs. *Res:* High polymer development; synthetic fibers and elastomers; textile operations; solvent recovery; spunbonded technology. *Mailing Add:* 190 Cherokee Rd Hendersonville TN 37075

MURRAY, LEO THOMAS, b New York, NY, May 15, 37; m 60; c 5. INORGANIC CHEMISTRY, ORGANIC CHEMISTRY. *Educ:* Manhattan Col, BS, 58; Purdue Univ, PhD(inorg chem), 63. *Prof Exp:* Res chemist, 62-63, sr res chemist, 63-66, sect head oral prod, 66-70, sect head laundry res, 70-72, tech coordr mkt, 72-73, dir contract purchasing, 73-74, mgr household specialties res, 74-78, assoc dir, 78-80, DIR DEVELOP, COLGATE-PALMOLIVE CO, 80- *Mem:* Am Chem Soc; Asn Res Dir; Indust Res Inst. *Res:* Chemistry organoboranes, specifically amine-boranes; chemistry of oxidizing agents, specifically N-kalo and peroxide types; development of household and toiletries consumer products. *Mailing Add:* Colgate-Palmolive Co 909 River Rd Piscataway NJ 08854-5503

MURRAY, M(URRAY) JOHN, b Palmerston North, NZ, Oct 30, 22; m 46; c 4. MEDICINE. *Educ:* Univ NZ, MB & ChB, 46, MD, 53; Univ Otago, NZ, DSc, 70; Royal Australasian Col Physicians, dipl, 67; Royal Col Physicians Edinburgh, dipl, 69. *Prof Exp:* Registr, St Stephens Hosp, London, Eng, 50-51; asst physician, Wellington Hosp, NZ, 52-55; from instr to assoc prof, 56-69, PROF MED, UNIV MINN, MINNEAPOLIS, 69-, DIR MED SERV, UNIV HOSP, 70- *Concurrent Pos:* Hartford Found res grant, 68-69. *Mem:* Fel Am Col Cardiol; Am Gastroenterol Asn; Royal Col Physicians; fel Royal Soc Med. *Res:* Coronary artery disease; portal hypertension; iron absorption. *Mailing Add:* Dept Med PO Box 385 Mayo Univ Minn Minneapolis MN 55455

MURRAY, MARION, b Evanston, Ill, Feb 27, 37. NEUROBIOLOGY. *Educ:* McGill Univ, BSc, 59; Harvard Univ, MA, 61; Univ Wis-Madison, PhD(physiol), 64. *Prof Exp:* Res asst neurobiol, Rockefeller Univ, 67-69; asst prof anat, Pritzker Sch Med, Univ Chicago, 69-77; MEM STAFF, DEPT ANAT, MED COL PA, PRITZKER SCH MED, UNIV CHICAGO, 77- *Concurrent Pos:* NIH fel anat, McGill Univ, 64-67. *Mem:* Am Asn Anat; Soc Neurosci. *Res:* Structure and function of nerve cells with special emphasis on synthesis and transport of protein. *Mailing Add:* Med Col of Pa Dept of Anat 3300 Henry Ave Philadelphia PA 19129

MURRAY, MARVIN, b Milwaukee, Wis, June 10, 27; m 59; c 2. CLINICAL PATHOLOGY. *Educ:* Marquette Univ, BS, 48; Mich State Univ, MS, 50, PhD(physiol), 56; Wayne State Univ, MD, 55. *Prof Exp:* Res path resident path, Univ Wis, 56-59; from asst prof to assoc prof, 59-71, dir clin path, 59-78, PROF PATH, SCH MED, UNIV LOUISVILLE, 71-, ADJ ASSOC PROF CHEM, 65- *Concurrent Pos:* Nat Cancer Inst trainee, Univ Wis, 56-59, dir clin labs, Louisville Gen Hosp, 59-78. *Honors & Awards:* Lederle Med Fac Award, 64. *Mem:* Sigma Xi. *Res:* Blood coagulation; protein chemistry; clinical analytical chemistry. *Mailing Add:* Med Sch Univ Louisville Box 35260 Louisville KY 40292

MURRAY, MARY AILEEN, b Washington, DC, Nov 11, 14. BOTANY. *Educ:* Univ Ariz, BA, 36, MS, 38; Univ Chicago, PhD(bot), 45. *Prof Exp:* Pub sch teacher, Ariz, 37-43; asst bot, Univ Chicago, 43-45, res assoc & instr, 45-48; from asst prof to assoc prof bot, DePaul Univ, 48-82; RETIRED. *Mem:* AAAS; Am Inst Biol Sci; Sigma Xi. *Res:* Plant morphology and anatomy. *Mailing Add:* 1223 Granville Ave Chicago IL 60660

MURRAY, MARY PATRICIA, kinesiology, medical research, for more information see previous edition

MURRAY, PATRICK ROBERT, b Los Angeles, Calif, Jan 15, 48; m 70; c 3. MEDICINE, PATHOLOGY. *Educ:* St Mary's Col, Calif, BS, 69; Univ Calif, Los Angeles, MS, 72, PhD(microbiol), 74. *Prof Exp:* Res fel clin microbiol, Mayo Clin & Mayo Found, 74-76; asst prof med, Wash Univ Sch Med, 76-82; assoc dir, 76, DIR CLIN MICROBIOL, BARNES HOSP, ST LOUIS, 77-, DIR POSTDOCTORAL TRAINING PROG, 82-; ASSOC PROF MED, WASH UNIV SCH MED, 83-; CONSULT, ST LUKE'S HOSP, 85- *Concurrent Pos:* Mem, Nat Comt Clin Lab Standards, 80-; chmn, exam comt, Am Bd Med Microbiol, 82- & mem, joint standards & exam comt, 85-; mem, bd dirs, Southwestern Asn Clin Microbiol, 83-; chmn, Clin Microbiol Div, Am Soc Microbiol, 84- *Mem:* Fel Am Acad Med Microbiol; fel Infectious Dis Soc; Sigma Xi; Am Asn Pathologists; Am Soc Microbiol; Ny Acad Sci; Med Mycol Soc Am; Am Fedn Clin Res. *Res:* New diagnostic and therapeutic tests for clinical microbiology. *Mailing Add:* Wash Univ Sch Med 660 S Euclid Ave Box 8118 St Louis MO 63110

MURRAY, PETER, b Rotherham, Yorks, UK, Mar 13, 20; m 47; c 3. METALLURGY, CERAMICS. *Educ:* Univ Sheffield, BSc, 41, PhD(metall, ceramics), 49. *Prof Exp:* Res chemist, Steetley Co, Ltd, 41-46; sr sci off ceramics sect, Atomic Energy Res Estab, 49-51, prin sci off, 51-55, sr prin scientist, 55-60, head metall div, 60-64, asst dir-mem bd mgt, 64-67; sr consult, Westinghouse Electric Corp, 67-68, tech dir, 68-69, mgr fuels, mats & sodium technol, 69-70, mgr technol, Advan Reactors Div, 70-74, dir, Res Lab, 74-75, mgr technol, Advan Reactors Div, 75-76, mgr bus develop, Advan Nuclear Systs Div, 76-77, chief scientist, Advan Power Systs Div, 77-81, DIR NUCLEAR PROGS, GOVT BUS DEVELOP, WESTINGHOUSE ELEC CORP, 81- *Mem:* Nat Acad Eng; Am Nuclear Soc; Am Ceramic Soc; fel Royal Inst Chem; Brit Ceramic Soc (pres, 65). *Res:* Nuclear reactor technology; materials science and technology. *Mailing Add:* Westinghouse Elec Corp 1801 K St NW Washington DC 20006

MURRAY, RAYMOND CARL, b Fitchburg, Mass, July 2, 29; m 55; c 2. SEDIMENTARY PETROLOGY. *Educ:* Tufts Col, BS, 51; Univ Wis, MS, 52, PhD, 55. *Prof Exp:* Asst, Tufts Col, 50-51 & Univ Wis, 52-55; res geologist, Shell Develop Co Div, Shell Oil Co, 55-62, mgr prod geol res, 62-66; assoc prof geol, Univ NMex, 66-67; prof geol & chmn dept, Rutgers Univ, 67-77; ASSOC VPRES RES & DEAN GRAD SCH, UNIV MONT, MISSOULA, 77- *Concurrent Pos:* Sr field geologist, State Geol Surv, Wis, 52, 54 & Buchans Mining Co, Ltd, 53; instr, Edgewood Col, 53 & 55; vis lectr, Univ Calif, 62; vis scientist, Rijswijk, Neth, 63-64; vis prof, Univ Mont, 75; auth & expert witness forensic geol, 73- *Mem:* Fel Geol Soc Am; Soc Econ Paleontologists & Mineralogists; Am Asn Petrol Geologists. *Res:* Petrology of recent sediments and carbonate and evaporite rocks; forensic geology. *Mailing Add:* Univ of Mont Missoula MT 59801

MURRAY, RAYMOND GORBOLD, b Tokyo, Japan, May 12, 16; m 38, 56, 75; c 6. HISTOLOGY. *Educ:* Monmouth Col, SB, 37; Univ Chicago, PhD(histol), 42. *Prof Exp:* Asst histol, Univ Chicago, 39-43, res assoc, 43-46; instr, Med Sch, Tufts Col, 46-48; asst prof path, Dent Sch, Northwestern Univ, 48-49; assoc prof, 49-65, chmn dept anat & physiol, 73-76, PROF ANAT, MED SCH, IND UNIV, BLOOMINGTON, 65- *Mem:* AAAS; Electron Micros Soc Am; Am Soc Cell Biol; Am Asn Anat; Soc Neurosci. *Res:* Tissue culture of thymus; parenteral nutrition; histopathology of x-rays; histopathology and distribution of radioisotopes; morphology and function of lymphocytes; fine structure of taste buds. *Mailing Add:* 2329 Winding Brook Circle Bloomington IN 47401

MURRAY, RAYMOND HAROLD, b Cambridge, Mass, Aug 17, 25; div; c 8. CARDIOLOGY. *Educ:* Univ Notre Dame, BS, 45; Harvard Univ, MD, 48; Am Bd Internal Med, dipl, 55, recert, 77; Am Bd Cardiovasc Dis, dipl, 60. *Prof Exp:* Intern med, Peter Bent Brigham Hosp, Boston, Mass, 48-49; resident, Roosevelt Hosp, New York, 49-50; res assoc cardiol, Nat Heart Inst, 50-53; instr med, Univ Mich, 53-54; from assoc prof to prof med, Sch Med, Ind Univ Indianapolis, 67-77, chmn dept community health sci, 72-77; chmn dept med, 77-89, PROF DEPT MED, COL HUMAN MED, MICH STATE UNIV, EAST LANSING, 77- *Concurrent Pos:* Fel coun clin cardiol, Am Heart Asn, 64. *Mem:* Cent Soc Clin Res; Am Fedn Clin Res; Am Col Physicians; Am Physiol Soc. *Res:* Health care delivery; cardiovascular physiology. *Mailing Add:* Dept of Med Col of Human Med East Lansing MI 48824

MURRAY, RAYMOND L(EROY), b Lincoln, Nebr, Feb 14, 20; m 41, 67, 79; c 7. PHYSICS, NUCLEAR ENGINEERING. *Educ:* Univ Nebr, BS, 40, MA, 41; Univ Tenn, PhD(physics), 50. *Prof Exp:* Asst physics, Univ Nebr, 40-41 & Univ Calif, 41-43, res physicist, Radiation Lab, 42-43; asst dept supt, Tenn Eastman Corp, 43-47; res physicist, Carbide & Carbon Chem Co, 47-50; prof, 50-57, head dept physics, 60-63, head dept nuclear eng, 63-74, Burlington prof, 57-80, EMER PROF NUCLEAR ENG, NC STATE UNIV, 80- *Concurrent Pos:* Consult, Oak Ridge Nat Lab, 50-68, AMF Atomics, Conn, 55-64, Alco Prod, Inc, 56-62, Lockheed Aircraft Corp, Ga, 58-62 & Westinghouse Elec Corp, Pa, 58-62; mem, Gov Sci Adv Comt, NC, 61-62; consult, Int Atomic Energy Agency, 63 & 80; exec ed for US, J Nuclear Energy, 63-73; consult, Atomic Power Develop Assocs, Mich, 70-72, Duke Power Co, NC, 71-85, US Arms Control & Disarmament Agency, 77-78, Dept Energy, 80-81, Bechtel Power Corp, 83-90 & Los Alamos Nat Lab, 88-; mem, chmn, NC Radiation Protection Comn, 79-87; mem adv coun, Inst Nuclear Power Opers, 85-87 & 89-; mem, vchmn & chmn, NC Low-Level Radioactive Waste Mgt Authority, 87- *Honors & Awards:* Oliver Max

Gardner Award, Univ NC, 65; Arthur Holly Compton Award, Am Nuclear Soc, 70; Glenn Murphy Award, Am Soc Eng Educ, 76; Donald Fink Award, Int Elec & Electronics Engrs, 88. *Mem:* Fel Am Phys Soc; fel Am Nuclear Soc; Am Soc Eng Educ. *Res:* Nuclear reactor theory and design analysis; radioactive waste management; nuclear criticality safety. *Mailing Add:* Dept Nuclear Eng NC State Univ Raleigh NC 27695

MURRAY, RICHARD BENNETT, b Marietta, Ga, Dec 5, 28; m 56; c 2. SOLID STATE PHYSICS. *Educ:* Emory Univ, AB, 47; Ohio State Univ, MS, 50; Univ Tenn, PhD(physics), 55. *Prof Exp:* Asst physics, Oak Ridge Gaseous Diffusion Plant, 47-48; physicist, Oak Ridge Nat Lab, 55-66; assoc prof physics, Univ Del, 66-68, actg chmn dept, 75-76, Univ coordr grad studies, 79-85, assoc provost grad studies, 86-88, PROF PHYSICS, UNIV DEL, 69-, ACTG PROVOST VPRES, ACAD AFFAIRS, 88- *Concurrent Pos:* Vis assoc prof, Univ Del, 62-63; lectr physics, Univ Tenn, 63-66; vchmn, Coun Oak Ridge Assoc Univs, 83-85, mem bd dirs, 83-, chmn, 85-87, pres, 88; prin investr, Dept Energy & NSF res grant; mem bd trustees, Southeastern Univs Res Asn, 88-, bd dirs, Del Inst Med Educ & Res, 89- *Mem:* AAAS; Am Asn Physics Teachers; Sigma Xi; fel Am Phys Soc. *Res:* Luminescence and scintillation phenomena in solids; color centers; channeling; ion penetration and radiation damage; graduate studies administration. *Mailing Add:* Dept of Physics Univ Del Newark DE 19716

MURRAY, ROBERT FULTON, JR, b Newburgh, NY, Oct 19, 31; m 56; c 4. MEDICAL GENETICS. *Educ:* Union Col, NY, BS, 53; Univ Rochester, MD, 58; Am Bd Internal Med, dipl, 66; Univ Wash, MS, 68. *Prof Exp:* Resident med, Colo Gen Hosp, 59-62; sr surgeon, USPHS, NIH, 62-65; sr fel med genetics, Sch Med, Univ Wash, 65-67; from asst prof to assoc prof pediat & med, 67-75, chief, Med Genetics Unit, Dept Pediat & Child Health, 68-75, PROF PEDIAT & MED, COL MED, HOWARD UNIV, 75-, CHIEF, DIV MED GENETICS, DEPT PEDIAT & CHILD HEALTH, 75- *Concurrent Pos:* Mem nat adv coun, Nat Inst Gen Med Sci, 71-75; chmn ad hoc comt sickle cell trait, Armed Forces, 72; mem comt inborn errors metab, Nat Res Coun, Nat Acad Sci, 72-75; prog dir, MS & PhD Prog Genetics & Human Genetics, Grad Sch Arts & Sci, Howard Univ, 74-, chmn dept, 76-, grad prof, 76-, prof oncol, 77-; mem bd dirs, Inst Soc Ethics & Life Sci; mem coun, Inst Med-Nat Acad Sci, 83-85, co-chmn, Panel Opportunities for Res on Prev & Treatment Alcohol Probs, 87-88. *Mem:* Inst Med-Nat Acad Sci; Am Soc Human Genetics; fel AAAS; fel Am Col Physicians; fel Inst Soc Ethics & Life Sci; Sigma Xi; Genetics Soc Am; Soc Social Biol. *Res:* Studies of factors influencing genetic counseling; genetic and developmental variations in isoenzymes; bioethics; inherited susceptibility to disease; disease prevention-health promotion; science education. *Mailing Add:* Div Med Genetics Box 75 Howard Univ Col Med Washington DC 20059

MURRAY, ROBERT GEORGE EVERITT, b Ruislip, Eng, May 19, 19; m 44, 85; c 3. BACTERIOLOGY. *Educ:* Cambridge Univ, BA, 41, MA, 45; McGill Univ, MD, CM, 43. *Hon Degrees:* DSc, Univ Western Ont, 85; Univ Guelph, 88. *Prof Exp:* Lectr, 45-47, from asst prof to prof, 47-84, head dept, 49-74, EMER PROF BACT & IMMUNOL, UNIV WESTERN ONT, 84- *Concurrent Pos:* Ed, Can J Microbiol, 54-60 & Bact Rev, 69-79; hon consult, St Joseph's Hosp, London, Ont, 60-84; mem, Bergey's Manual Trust, 64-91, chmn, 76-90; mem, Int Comt Bact Nomenclature, 66-, chmn, 82-90; gov bd, Biol Coun Can, 66-72; assoc ed, Int J Syst Bacteriol, 82-; mem exec bd, Int Union Microbiol Socs, 82-86. *Honors & Awards:* Coronation Medal, 53; Harrison Prize, Royal Soc Can, 57, Award, 60-61 & Flavelle Medal, 84; Prize, Can Soc Microbiol, 63; Centennial Medal, Govt Can, 67; Jubilee Medal, 78; Porter Award, US Fedn Cult Collections, 87. *Mem:* Fel Am Acad Microbiol; hon mem Am Soc Microbiol (vpres, 71-72, pres, 72-73); Am Soc Cell Biologists; fel Royal Soc Can; hon mem Can Soc Microbiol (pres, 51-52). *Res:* Bacterial cytology and physiology; ultrastructure of bacteria and relation of structure to function, with emphasis on the cell wall and macromolecular arrangement. *Mailing Add:* Fac Med Dept Microbiol & Immunol Univ Western Ont London ON N6A 5C1 Can

MURRAY, ROBERT KINCAID, b Glasgow Scotland, Dec 18, 32; m 59; c 4. BIOCHEMISTRY. *Educ:* Glasgow Univ, MB, ChB, 56; Univ Mich, MS, 58; Univ Toronto, PhD(biochem), 61. *Prof Exp:* From asst prof to assoc prof, 61-73, PROF BIOCHEM, UNIV TORONTO, 73- *Mem:* Am Soc Biochem & Molecular Biol; Am Asn Pathologists; Am Soc Cell Biol; Am Asn Cancer Res. *Res:* Biochemistry of glycosphingolipids; biochemistry of cancer. *Mailing Add:* Dept Biochem Univ Toronto Med Sci Bldg Toronto ON M5S 1A8 Can

MURRAY, ROBERT WALLACE, b Brockton, Mass, June 20, 28; m 51; c 7. PHYSICAL ORGANIC CHEMISTRY. *Educ:* Brown Univ, AB, 51; Wesleyan Univ, MA, 56; Yale Univ, PhD(chem), 60. *Prof Exp:* Asst chem, Wesleyan Univ, 54-56 & Yale Univ, 56-57; res chemist polymer chem, Olin-Mathieson Chem Corp, 56-57; mem tech staff chem, Bell Tel Labs, Inc, 59-63, res supvr, 63-68; prof chem, 68-80, chmn dept, 75-80, CURATORS' PROF CHEM, UNIV MO-ST LOUIS, 81- *Concurrent Pos:* Mem, Nat Adv Comt Air Pollution Res Grants, 70-73; consult, Panel Vapor Phase Org Air Pollutants from Hydrocarbons, Nat Acad Sci & Nat Inst Environ Health Sci; vis prof, Univ Karlsruhe, WGer, 82, Univ Col Cork, Ireland, 89. *Honors & Awards:* Am Chem Soc Award, 74, Midwest Award, 89. *Mem:* AAAS; Am Chem Soc; Am Inst Chemists; Am Soc Photobiol; NY Acad Sci. *Res:* Oxidation of organic compounds; singlet oxygen and ozone chemistry; air pollution chemistry; chemistry of aging; carbene chemistry; reaction mechanisms; enzyme modelling; dioxirane chemistry. *Mailing Add:* 1810 Walnutway Dr Creve Coeur MO 63146

MURRAY, RODNEY BRENT, b Philadelphia, Pa, July 12, 49; m. PHARMACOLOGY. *Educ:* Temple Univ, BA, 71, PhD(pharmacol), 79. *Prof Exp:* From adj asst prof to adj assoc prof, Biomed Eng & Sci Inst, Drexel Univ, 77-87; ASST PROF, DEPT PHARMACOL, MED COL & DIR, OFF ACAD COMPUT, THOMAS JEFFERSON UNIV, 87- *Concurrent Pos:* Lectr, Dept Pharmacol, Pa Col Podiatric Med, 76-80 & Sch Med, Temple Univ, 80-81; Nat Inst Drug Abuse postdoctoral fel, 79-82; consult, Dept Biobehav Sci, Univ Conn & Addiction Res Ctr, Nat Inst Drug Abuse, 82; dir computer serv & sr pharmacologist, Biosearch, Inc, 82-86, vpres sci affairs, 86-87; chmn, Computer-Assisted Instr Task Force, Health Sci Libr Consortium, 87- *Mem:* Am Asn Med Systs & Informatics; Asn Develop Computer-Based Instrnl Systs; Soc Neurosci; fel Am Soc Pharmacol & Exp Therapeut. *Res:* Pharmacologic calculations with computer programs; author of 19 technical publications. *Mailing Add:* Thomas Jefferson Univ 11th & Walnut Sts Philadelphia PA 19107

MURRAY, ROGER KENNETH, JR, b Buffalo, NY, July 9, 42; m 80; c 1. SYNTHESIS, PHOTOCHEMISTRY. *Educ:* Cornell Univ, AB, 64; Mich State Univ, MS, 66, PhD(chem), 69. *Prof Exp:* Fel chem, Princeton Univ, 69-70, instr, 70-71; asst prof, 71-76, ASSOC PROF CHEM, UNIV DEL, 76- *Concurrent Pos:* Teacher & scholar, Camille & Henry Dreyfus Found, 76-81; vis prof, Fulbright-Hayes Found, Univ De Reims, France, 77-78; Lank exchange prof, Universite De Montreal, 81. *Mem:* Am Chem Soc; Sigma Xi. *Res:* Synthesis and chemistry of cage compounds related to adamantane and its derivatives; chemistry of compounds containing strained rings; organic photochemistry; conformational influences on mass spectrometric behavior. *Mailing Add:* Dept Chem Univ Del Newark DE 19716

MURRAY, ROYCE WILTON, b Birmingham, Ala, Jan 9, 37; m 57, 82; c 5. ANALYTICAL CHEMISTRY. *Educ:* Birmingham-Southern Col, BS, 57; Northwestern Univ, PhD(analytical chem), 60. *Prof Exp:* From instr to assoc prof, 60-69, actg chmn dept, 70-71, prof, 70-79, chmn dept, 80-85, KENAN PROF CHEM, UNIV NC, CHAPEL HILL, 79-, CHAIR, NATURAL SCI DIV, 87- *Concurrent Pos:* Alfred P Sloan res fel, 69-72; prog dir chem analysis, NSF, 71-72; Guggenheim fel, 80-81; Japan Soc Promotion Sci fel, 79-80; ed, J Analytical Chem, Am Chem Soc, 91. *Honors & Awards:* Carl Wagner Mem Award, Electrochem Soc, 87; Charles N Reilly Award, Soc Electroanal Chem, 88; Electrochem Group Medal, Royal Soc Chem, 89; Electrochem Award, Div Analytical Chem, Am Chem Soc, 90, Analytical Chem Award, 91. *Mem:* Nat Acad Sci; Soc Electroanalytical Chem (pres-elect); Electrochem Soc; Am Chem Soc. *Res:* Electroanalytical chemistry; molecular design of electrode surfaces; electrochemically reactive polymers; transport and electron transfer dynamics at interfaces and in films; electrocatalysis; solid state chemistry; electrochemistry at superconducting electrodes. *Mailing Add:* Dept Chem Univ NC Chapel Hill NC 27599-3290

MURRAY, STEPHEN PATRICK, b New York, NY, Oct 4, 38; m 62; c 3. PHYSICAL OCEANOGRAPHY. *Educ:* Rutgers Univ, AB, 60; La State Univ, MS, 63; Univ Chicago, PhD(geophys), 66. *Prof Exp:* NSF fel, 66-67; from asst prof to assoc prof, 67-77, asst dir, 75-85, PROF MARINE SCI, LA STATE UNIV, BATON ROUGE, 77-, DIR, COASTAL STUDIES INST, 85- *Concurrent Pos:* Consult, Bangladesh, 80. *Mem:* Am Geophys Union; Am Meteorol Soc; Sigma Xi; Estuarine Res Fedn. *Res:* Coastal oceanography, including generation and trajectories of coastal currents; land-sea interaction and the turbulent diffusion of solid particles under shoaling waves; dynamics of straits; dynamical oceanography of coastal currents and currents in straits in middle east environments such as Egypt, Saudi Arabia and Sudan. *Mailing Add:* Coastal Studies Inst La State Univ Howe-Russell Bldg W302 Baton Rouge LA 70803

MURRAY, STEPHEN S, b New York, NY, Aug 28, 44; m 65; c 2. X-RAY ASTRONOMY, COSMOLOGY. *Educ:* Columbia Univ, BS, 65; Calif Inst Technol, PhD(physics), 71. *Prof Exp:* Staff scientist x-ray astron, Am Sci & Eng, Inc, 71-73; ASTROPHYSICIST, CTR ASTROPHYS, SMITHSONIAN ASTROPHYS OBSERV, 73- *Concurrent Pos:* Assoc, Harvard Col Observ, 74- *Mem:* Am Astron Soc; Int Astron Union; Soc Photo-Optical Instrumentation Engr. *Res:* Observational x-ray astronomy, particularly extragalactic objects; development and use of high sensitivity, high resolution x-ray imaging detectors for extragalactic observations; advanced X-ray astrophysics; high resolution camera instruments. *Mailing Add:* Ctr Astrophys 60 Garden St Cambridge MA 02138

MURRAY, STEVEN NELSEN, b Los Angeles, Calif, Sept 7, 44; m 66; c 2. MARINE ECOLOGY, PHYCOLOGY. *Educ:* Univ Calif, Santa Barbara, BA, 66, MA, 68; Univ Calif, Irvine, PhD(phycol), 71. *Prof Exp:* From asst prof biol to assoc prof, 71-77, PROF BIOL, CALIF STATE UNIV, FULLERTON, 78- *Concurrent Pos:* Calif State Univ grants, 72, 73, 75, 87, 88, 89 & 90. *Mem:* Phycol Soc Am; Int Phycol Soc; Am Soc Limnol & Oceanog; Ecol Soc Am; Brit Phycol Soc; Western Soc Naturalists. *Res:* Marine algal ecology, including studies of seaweed distributions and production; successional events in intertidal communities; physiological ecology of seaweeds; marine pollution. *Mailing Add:* Dept Biol Sci Calif State Univ Fullerton CA 92634

MURRAY, T J, b Halifax, NS, May 30, 38; m 60; c 4. NEUROLOGY, MEDICAL EDUCATION. *Educ:* Dalhousie Univ, MD, 63; FRCP(C), 69; FACP, 80. *Hon Degrees:* LLD, St Francis Xavier Univ, 89. *Prof Exp:* prof neurol, 74-77, res assoc psychol, 77-78, assoc prof health prof & family med, 78-83, prog dir neurol, 78-85, PROF PHYSIOTHER, DALHOUSIE UNIV, 83-, DIR MULTIPLE SCLEROSIS RES UNIT, 80-, DEAN MED, 85- *Concurrent Pos:* Chief med,Camp Hill Hosp, Halifax, NS, 74-80; hon consult, London Hosp, London, Eng, 85-; Commonwealth travelling fel, 85; Commonwealth Scholar, 87-88. *Honors & Awards:* Officer Order Can, 91. *Mem:* Am Acad Neurol (vpres, 81-83); Can Cong Neurol Sci (pres, 82); Can Neurol Soc (pres, 82-84); Royal Soc Med, Eng; fel Am Col Physicians; Am Col Physicians; Soc Hist Med (pres). *Res:* Multiple sclerosis and other neurological disorders; medical history. *Mailing Add:* Off Dean Dalhousie Med Sch Dalhousie Univ Halifax NS B3H 4H7 Can

MURRAY, THOMAS FRANCIS, EPILEPSY, NEURORECEPTORS. *Educ:* Univ Wash, PhD(pharmacol), 79. *Prof Exp:* Asst prof, 83-86, ASSOC PROF PHARMACOL, ORE STATE UNIV, 86- *Mailing Add:* Col Pharm Ore State Univ Corvallis OR 97331

MURRAY, THOMAS HENRY, b Philadelphia, Pa, July 30, 46; m 78; c 4. BIOETHICS. *Educ:* Temple Univ, BA, 68; Princeton Univ, PhD(social psychol), 76. *Prof Exp:* Instr, New Col, Sarasota, Fla, 71-75; from asst prof to assoc prof interdisciplinary studies, Western Col Miami Univ, 75-80; assoc social & behav studies, Hastings Ctr, 80-84; from assoc prof to prof ethics & pub policy, Inst Med Humanities, Med Br, Univ Tex, 84-87; DIR & PROF BIOETHICS, CTR BIOMED ETHICS, SCH MED, CASE WESTERN RESERVE UNIV, 87- *Concurrent Pos:* Bd mem, Asn Integrative Studies, 80-87; Adv ed, Social Sci & Med, An Int J, 82-; founder & ed, Med Humanities Rev, 85-; mem, Comt Substance Abuse Res & Educ, US Olympic Comt, 85- & Ethics Working Group, Prog Adv Comt, Human Genome Initiative, NIH, 89-; chair, Fac Asn, Soc Health & Human Values, 89-90; chair, Human Genome Initiative, Task Force on Genetic Testing & Ins, 91-; consult, US Cong Off Technol Assessment; reviewer, NSF. *Mem:* Soc Health & Human Values; Soc Values Higher Educ; Am Pub Health Asn; Asn Integrative Studies (pres, 83); fel Hastings Ctr; fel Environ Health Inst; fel Aspen Inst. *Res:* Genetics; aging; children and health policy; author of 100 publications. *Mailing Add:* Ctr Biomed Ethics Sch Med Case Western Reserve Univ Cleveland OH 44106-4976

MURRAY, THOMAS PINKNEY, b Charleston, SC, Oct 8, 42; m 65; c 2. ORGANIC CHEMISTRY. *Educ:* Western Carolina Univ, BS, 64; Appalachian State Univ, MA, 66; Va Polytech Inst & State Univ, PhD(chem), 69. *Prof Exp:* Res assoc chem, Univ Alta, 69-71, Vanderbilt Univ, 71-72; from asst prof to assoc prof, 72-82, PROF CHEM, UNIV N ALA, 82- *Concurrent Pos:* Consult, Tenn Valley Authority, 76-; consult org nitrogen fertilizers, Int Fertilizer Develop. *Mem:* Am Chem Soc; Sigma Xi. *Res:* Biosynthesis of phenolic plant metabolites, isolation and characterization of new metabolites; organic synthesis; formaldehyde resin chemistry; characterization of industrially important compounds and materials. *Mailing Add:* Dept of Chem Univ of N Ala Florence AL 35630

MURRAY, WALLACE JASPER, b Quantico, Va, July 13, 40; m 64; c 3. MEDICINAL CHEMISTRY. *Educ:* San Diego State Univ, BS, 64; Univ Calif, San Francisco, PhD(pharmaceut chem), 74. *Prof Exp:* Instr chem, Mass Col Pharm, 74-75; asst prof med chem, 75-78, assoc prof biomed chem, 78-85, ASSOC PROF PHARM SCI, MED CTR, UNIV NEBR-OMAHA, 85- *Mem:* Sigma Xi; Am Chem Soc; Am Asn Pharmaceut Sci. *Res:* Topological indices in structure-activity relationships; 31-P-NMR metabolic studies of diabetic hearts; toxicity of halogenated hydrocarbons. *Mailing Add:* Dept Pharmacol Sci Col Pharm Univ Nebr Med Ctr Omaha NE 68105

MURRAY, WILLIAM DOUGLAS, b Guelph, Ont, Sept 14, 50. BIOTECHNOLOGY, MICROBIOLOGY. *Educ:* Univ Waterloo, BSc, 74, MSc, 76, PhD(microbiol), 79. *Prof Exp:* Res assoc, 79-81, asst res officer, 81-85, ASSOC RES OFFICER, NAT RES COUN CAN, 85- *Honors & Awards:* W B Pearson Medal, 79. *Mem:* Can Soc Microbiol; Am Soc Microbiol; Soc Indust Microbiol. *Res:* Microbiol production of flavors and fragrances; biological production of fuels; mechanisms that regulate cellulose fermentation. *Mailing Add:* Library Parliment Sci & Technol Div Ottawa ON K1A 0R6 Can

MURRAY, WILLIAM J, b Janesville, Wis, July 20, 33; c 3. ANESTHESIOLOGY, CLINICAL PHARMACOLOGY. *Educ:* Univ Wis, BS, 55, PhD(pharmacol & toxicol), 59; Univ NC, MD, 62. *Prof Exp:* Instr pharmacol, Univ NC, 59-62, asst prof surg, 62-65; asst dir, Drug Availability, HEW Dept, Food & Drug Admin, 68-69; assoc prof pharmacol, anesthesiol & clin pharmacol, Univ Mich, 69-72; assoc prof anesthesiol, 72-81, PROF ANESTHESIOL, DUKE UNIV MED CTR, 81- *Mem:* Am Med Asn; Am Soc Clin Pharmacol & Therapeut; Am Pharmaceut Asn; Am Soc Pharmacol & Exp Therapeut; Int Anesthesia Res Soc; Am Soc Anesthesiologists. *Res:* Basic and clinical pharmacologic studies of analgesics and anesthetics; clinical drug bioavailability; particulate matter in gas and IV lines; coring of vial stoppers; toxicology; operating room satellite pharmacy and lab development; establish surgical outpatient database to evaluate lab study over-prescription; computer program to develop medical history and recommended lab studies. *Mailing Add:* Dept of Anesthesiol Duke Univ Med Ctr Box 3094 Durham NC 27710

MURRAY, WILLIAM R, b Ottawa, Ont, Dec 4, 24; US citizen; c 2. ORTHOPEDIC SURGERY. *Educ:* St Patrick's Col, Ottawa, BSc, 47; McGill Univ, MD & CM, 52. *Prof Exp:* From instr to assoc prof, 58-72, chief orthop out patient surg clin, 58-73, vchmn orthop surg, 76-78, PROF ORTHOP SURG, SCH MED, UNIV CALIF, SAN FRANCISCO, 72-, CHMN DEPT, 78- *Concurrent Pos:* Adv, Bur Hearings & Appeals, Soc Security Admin, Dept Health, Educ & Welfare, 64-72; mem, Arthritis Found. *Mem:* AMA; Am Acad Orthop Surg; Am Orthop Asn; Am Rheumatism Asn; dipl mem Pan Am Med Asn. *Res:* Total hip joint replacement arthroplasty; rheumatoid arthritis. *Mailing Add:* Dept Orthop Surg Univ Calif San Francisco CA 94143

MURRAY, WILLIAM SPARROW, b Wilkes Barre, Pa, July 15, 26; m 52; c 2. SCIENCE ADMINISTRATION. *Educ:* Juniata Col, BS, 50; Univ Md, MS, 52, PhD(entom), 63. *Prof Exp:* Agt entomologist, USDA, 51; dist entomologist, US Army Corps Engrs, 52-55; entomologist, Norfolk Dist, US Dept Navy, 55-58, dist entomologist, River Commands, Washington, DC, 58-62, entomologist, Dept Navy, 63-64; consult, Nat Pesticide Prob, House of Rep, US Cong, 64-65; asst exec secy, Fed Comt Pest Control, 65-69; exec secy, Working Group Pesticides, President's Cabinet Comt Environ, 69-71; staff dir hazardous mat adv comt, 71-72, phys sci adminr, Off Pesticide Progs, 73-78, assoc dep asst adminr, 78-79, dir toxics & pesticides, Off Health & Ecol Effects, Off Res & Develop, Environ Protection Agency, 79-82; EXEC NUCLEAR MED & ULTRASOUND IMAGING & LITHOTRIPSY, NAT MFR ELEC ASN, 82- *Concurrent Pos:* Consult, Nat Plant & Animal Dis & Quarantine Probs, House Appropriations Comt, US Cong, 64-65 & 67. *Honors & Awards:* Qual Increase Award, Off of Pesticide Progs, Environ Protection Agency, 74. *Mem:* NY Acad Sci; Sigma Xi; Entom Soc Am. *Res:* National pesticide problem; incidence and effects of pesticides and other pollutants on human health and the environment. *Mailing Add:* 1281 Bartonshire Way Potomac Woods Rockville MD 20854

MURRELL, HUGH JERRY, b Tyler, Tex, Oct 20, 37; m 63; c 3. ONCOLOGY. *Educ:* Univ Tex, Galveston, MD, 62. *Prof Exp:* DIR RADIATION ONCOL, BOONE HOSP CTR, COLUMBIA, MO, 84- & UNIV MO HOSP, 85- *Mem:* Fel Am Col Radiol; Am Soc Therapeut Radiol & Oncol; Am Soc Clin Oncol. *Mailing Add:* 1600 Broadway Suite 102 Columbia MO 65201

MURRELL, JAMES THOMAS, JR, b Dickson, Tenn, Mar 17, 42; m 60; c 2. SYSTEMATIC BOTANY. *Educ:* Austin Peay State Col, BS, 64; Vanderbilt Univ, PhD(syst bot), 69. *Prof Exp:* NIH trainee, Univ Miami, 68-69; asst prof biol, George Peabody Col, 69-75; asst prof, Miss Univ Women, 75-78, assoc prof biol & dean, grad sch, 78-89, dean Arts & Sci, 80-89, vpres acad affairs, 82-89; VPRES & ACAD DEAN, HIWASSEE COL, 89- *Res:* Chemotaxonomy; pollination biology. *Mailing Add:* Office VPres & Acad Dean Hiwassee Col Madisonville TN 37354

MURRELL, KENNETH DARWIN, b Burley, Idaho, Jan 19, 40; m 65; c 2. PARASITOLOGY, IMMUNOLOGY. *Educ:* Chico State Col, AB, 62; Univ NC, Chapel Hill, MSPH, 63, PhD(parasitol), 69. *Prof Exp:* NIH trainee microbiol, Univ Chicago, 69-71; res zoologist, Naval Med Res Inst, 71-78; chief, Helminthic Dis Lab, Animal Parasitol Inst, 78-87, AREA DIR, AGR RES SERV, USDA, 88- *Mem:* Am Asn Parasitol (pres, 85); Am Soc Parasitol (vpres, 89); Am Asn Immunol. *Res:* Fundamental mechanisms of immunity to animal parasites; immunochemistry of parasite antigens; epidemiology and diagnoses of food-borne; parasitic zoonoses. *Mailing Add:* Midwest Area USDA Agr Res Serv 1815 University Peoria IL 61604

MURRELL, LEONARD RICHARD, b Stamford Centre, Ont, June 17, 33; m 68. ANATOMY. *Educ:* McMaster Univ, BSc, 57, MSc, 58; Univ Minn, Minneapolis, PhD(anat), 64. *Prof Exp:* Asst biol, McMaster Univ, 57-58; from instr to asst prof anat, Univ Minn, Minneapolis, 64-67; from asst prof to assoc prof, 67-74, PROF ANAT, UNIV TENN, MEMPHIS, 74- *Concurrent Pos:* Am Diabetes Asn res fel, Univ Minn, Minneapolis, 64-66; ed-in-chief, J Tissue Cult Methods, 80-85. *Mem:* Am Asn Anat; Am Diabetes Asn; Brit Soc Cell Biol; Tissue Cult Asn; Can Asn Anat. *Res:* Human anatomy; experimental diabetes; functional cytodifferentiation; organ culture. *Mailing Add:* Dept Anat & Neurobiology Univ Tenn Memphis Memphis TN 38163-9997

MURRILL, EVELYN A, b Sheffield, Ala, April 14, 30. ORGANIC CHEMISTRY. *Educ:* Fontbonne Col, BS, 50; Univ Minn, PhD(organic chem), 66. *Prof Exp:* Instr chem, Avila Col, 65-68; assoc, Chem Dept, Midwest Res Inst, 68-70; res fel, Univ Kans, 70-71; sr prin & sr adv, Chem Sci Dept, 71-89, ASST DIR LIFE SCI & PRIN ADV, CHEM DEPT, MIDWEST RES INST, 89- *Concurrent Pos:* Nat Sci Fel, Univ Minn, 64; Nat Sci Res Fel, Ill Inst Technol, 67-68. *Mem:* Am Chem Soc; Sigma Xi. *Res:* Development of methods for analysis of organic molecules in biological motives or complex mixtures. *Mailing Add:* Midwest Res Inst 410 Bolker Blvd Kansas City MO 64110

MURRILL, PAUL W(HITFIELD), b St Louis, Mo, July 10, 34; m 59; c 3. CHEMICAL ENGINEERING. *Educ:* Univ Miss, BS, 56; La State Univ, MS, 62, PhD(chem eng), 63. *Prof Exp:* Engr, Pittsburgh Plate Glass Co, 59-60 & Ethyl Corp, 62: lectr process control, 62-63, from asst prof to assoc prof chem eng & mech, indust & aerospace eng, La State Univ, Baton Rouge, 63-67, assoc prof chem eng, 67-68, head dept, 67-69, dean acad affairs, 69-70, vchancellor, 69-74, provost, 70-74, prof chem eng, 68-80, chancellor, 74-80; sr vpres & dir, Ethyl Corp, 81-82; chmn & chief exec officer, 82-87, SPEC ADV, GULF STATES UTILITIES, 88- *Concurrent Pos:* Consult ed, Intext Educ Publ, NY, 65-72; proj mgr & prin investr, US Dept Defense Proj Themis Study in Digital Automata, 67-70; dir, Nuclear Systs, Inc, 68-71, 74-77; dir, United Way, 69-79, pres, 77-79; dir, STL Electronics, Inc, 70-72; trustee, Gulf South Res Inst, 70-85; dir, Boy Scouts Am, 71-77, Foxboro Co, 74-; mem, Comn Future Blacks in Higher Educ & Black Cols & Univs, HEW & Air Univ Bd Visitors, US Air Force; indust consult numerous companies; dir, First Miss Corp, 67, First Chem Corp, 87-, Tidewater, Inc, 81-, Foxboro Co, 74- *Honors & Awards:* Nat Donald Eckman Award, Instrument Soc Am, 76. *Mem:* Am Inst Chem Engrs; Instrument Soc Am; Sigma Xi. *Res:* Process control and dynamics; formulation of mathematical models and simulation techniques; dynamic aspects of unit operations in chemical process industries; automatic control theory; systems engineering; computer applications; digital control. *Mailing Add:* 206 Sunset Blvd Baton Rouge LA 70808

MURRIN, LEONARD CHARLES, b Iowa City, Iowa, Oct 9, 43; m 68; c 3. PHARMACOLOGY. *Educ:* St John's Col, Calif, BA, 65; Yale Univ, PhD(pharmacol), 75. *Prof Exp:* Fel, dept pharmacol & exp ther, Johns Hopkins Univ Sch Med, 75-78; from asst prof to assoc prof, 78-88, PROF PHARMACOL, UNIV NEBR MED CTR, OMAHA, 88- *Mem:* AAAS; Soc Neurosci; Am Soc Pharmacol & Exp Therapeut; Am Soc Neurochem; Int Soc Neurochem; Int Soc Develop Neurosci; Int Brain Res Orgn. *Res:* Neurotransmitter systems in the central nervous system and their development; factors involved in control of neurotransmitter synthesis, release and catabolism. *Mailing Add:* Dept Pharmacol Univ Nebr Med Ctr 600 S 42nd St Omaha NE 68198-6260

MURRIN, THOMAS J, b New York, NY, Apr 30, 29; m 51; c 8. DEFENSE & AEROSPACE SYSTEMS. *Educ:* Fordham Univ, BS, 51. *Hon Degrees:* DSc, Duquesne Univ, 89. *Prof Exp:* Mat engr, Transformer Div, Westinghouse Elec Corp, Sharon, Pa, 52-55, supt factory planning, Distrib Transformer Plant, Athens, Ga, 55-59, Europ mfg rep, Geneva, Switz, 59-61, gen mgr Motor & Gearing Div, Buffalo, NY, 61-65, corp vpres mfg, 65-67, group vpres defense, 67-71, exec vpres Defense & Pub Systs Group, 71-74, sr exec vpres, 74, pres, energy & advan technol group, Pub Systs Co, 83-87; dep secy, US Dept Com, 87-89; DEAN, SCH BUS & ADMIN, DUQUESNE UNIV, 91- *Concurrent Pos:* US deleg, NATO Indust Adv Group; mem, Defense Policy Adv Comt on Trade, Dept Defense, chmn, Subcomt Trade Relations with Japan; mem, President's Comn Indust Competitiveness, 84; distinguished serv prof, Carnegie-Mellon Univ. *Honors & Awards:* James

Forrestal Mem Award, Nat Security Indust Asn, 83, Excellence in Mfg Award, 89; Mfg Mgt Award, Soc Mfg Engrs, 86. *Mem:* Fel Nat Acad Eng; Aerospace Industs Asn. *Mailing Add:* Sch Bus & Admin Duquesne Univ Rm 406 Rockwell Hall Pittsburgh PA 15282

MURRISH, DAVID EARL, b Glasgow, Mont, Jan 28, 37; m 65; c 1. COMPARATIVE PHYSIOLOGY. *Educ:* Calif State Univ, Los Angeles, BA, 53, MA, 65; Univ Mont, PhD(comp physiol), 68. *Prof Exp:* Res assoc comp physiol, Duke Univ, 68-70; asst prof biol, Case Western Reserve Univ, 70-77; ASSOC PROF BIOL, STATE UNIV NY BINGHAMTON, 77- *Mem:* Am Physiol Soc; Am Soc Zoologists. *Res:* Respiratory and metabolic physiology of birds and mammals. *Mailing Add:* Dept Biol Sci State Univ NY Binghamton NY 13901

MURRMANN, RICHARD P, b South Bend, Ind, Aug 3, 40; m 61; c 2. PHYSICAL CHEMISTRY, SOIL SCIENCE. *Educ:* Purdue Univ, BS, 62; Cornell Univ, MS, 63, PhD(soil sci chem), 66. *Prof Exp:* Res assoc soil chem, Cornell Univ, 66; res chemist, US Army Cold Regions Res & Eng Lab, NH, 66-74; assoc area dir, Agr Res Serv, NSDA, 74-87; DIR, USDA APPALACHIAN SOIL/WATER CONSERV LAB, 87- *Mem:* Am Soc Agron; Soil Sci Soc Am; Sigma Xi. *Res:* Inorganic phosphate in soil; electrical conductivity and diffusivity of ions; adsorption of heavy metal ions by minerals; chemistry of trace components in atmosphere, soil and water; land treatment wastewater. *Mailing Add:* Appalachian Soil & Water Conserv Res Lab PO Box 867 Airport Rd Beckley WV 25802-0867

MURSKY, GREGORY, b Ukraine, Feb 13, 29; US citizen; m 52; c 2. GEOLOGY. *Educ:* Univ BC, BSc, 56; Stanford Univ, MS, 60, PhD(geol), 63. *Prof Exp:* Geologist, Eldorado Nuclear Ltd, Ont, 56-57, chief geologist, 57-59; Nat Res Coun Can fel, 63-64; from asst prof to assoc prof geol, 64-68, chmn dept geol sci, 66-68, PROF GEOL, UNIV WIS-MILWAUKEE, 68- *Mem:* AAAS; Soc Econ Geol; Mineral Soc Am; Mineral Asn Can; Can Inst Mining & Metall. *Res:* Mineralogy; economic geology; petrology; geochemistry. *Mailing Add:* Dept of Geol Sci Univ of Wis Milwaukee WI 53201

MURTAGH, FREDERICK REED, b Philadelphia, Pa, Nov 20, 44; m 68; c 2. NEURORADIOLOGY. *Educ:* MD, Sch Med, Temple Univ. *Prof Exp:* PROF RADIOL, UNIV SFLA, TAMPA, 79-, DIR NEURO-RADIOL, 78. *Concurrent Pos:* Fel neuroradiol, Jackson Mem Hosp & Univ Miami, 78; pres, Southeastern Neuroradiol Soc, 89-90. *Mem:* Am Med Asn; Radiol Soc NAm; Am Soc Neuroradiol. *Res:* Computed tomography, hydrocephalus, aneurysms and cerebral circulation, spinal stenosis, magnetic resonance imaging. *Mailing Add:* Dept Radiol Univ SFla Tampa FL 33612

MURTHA, JOSEPH P, b South Connellsville, Pa, July 18, 31; m 54; c 5. CIVIL ENGINEERING. *Educ:* Carnegie Inst Technol, BS, 53, MS, 55; Univ Ill, PhD(civil eng), 61. *Prof Exp:* Instr civil eng, Carnegie Inst Technol, 54-55; res assoc, Univ Ill, 58-61, from asst prof to prof, 61-66, dir water resources ctr, 63-66; dir, Amphibious & Harbor Div, US Naval Civil Eng Lab, Calif, 66-67; sr staff mem, Nat Eng Sci Co, 67-68; mgr ocean eng, Western Offshore Drilling & Explor Co, 68-69; PROF STRUCT & HYDRAUL ENG, UNIV ILL, URBANA, 69-; DIR, ADVAN CONSTRUCT TECHNOL CTR, UNIV ILL, 86- *Concurrent Pos:* Fulbright-Hays sr res fel, US-UK Ed Comn, 76; vis prof eng, Heriot-Watt Univ, 76-77. *Mem:* Am Soc Civil Engrs; Am Geophys Union; Seismol Soc Am. *Res:* Construction engineering; structural dynamics; ocean engineering; water resources. *Mailing Add:* Newmark Lab 205 N Mathews St Urbana IL 61801

MURTHY, A S KRISHNA, b Bangalore, India, Feb 8, 32; m 65; c 1. HISTOCHEMISTRY, EXPERIMENTAL PATHOLOGY. *Educ:* Univ Mysore, BSc, 50; Univ Bombay, MSc, 55, PhD(biochem), 61. *Prof Exp:* Res asst histochem & exp path, Indian Cancer Res Ctr, Bombay, 52-57, asst res officer endocrinol, 57-61; res assoc morphol, Chicago Med Sch, 61-63; res assoc histochem, Children's Cancer Found, Boston, 63-65; res officer, Indian Coun Med Res, New Delhi, 65-67; res assoc histochem, exp path & biochem, Children's Cancer Res Found, Boston, 67-73; SR SCIENTIST, TSI MASON RES INST, WORCESTER, 73- *Concurrent Pos:* Ill Rheumatism & Arthritis Found fel, Chicago Med Sch, 61-62, Chicago Heart Asn fel, 62; dipl, Am Bd Toxicol. *Honors & Awards:* Dr Khanolkar Prize, Indian Asn Path & Bact, 63. *Mem:* AAAS; Am Soc Zoologists; Am Asn Cancer Res; Endocrine Soc. *Res:* Functional endocrine tumors and their induction in animals; endocrine interrelationships; carcinogens; chemical analysis of tumors. *Mailing Add:* TSI Mason Res Inst 57 Union St Worcester MA 01608-1182

MURTHY, ANDIAPPAN KUMARESA SUNDARA, b Sivakasi, India; m; c 2. CHEMICAL ENGINEERING, COMPUTER SCIENCE. *Educ:* Indian Inst Technol, BTech, 66; Columbia Univ, MS, 68, EngScD, 74. *Prof Exp:* Engr res, Digvijay Cement Co, Jamnagar, India, 66-67; res engr inorganic, 68-69, sr engr process, 69-76, engr assoc simulation, 76-78, mgr process, 78-80, mgr fuels res, 80-84, TECH OFFICER SCI & TECHNOL, ALLIED CHEM INC, 84- *Concurrent Pos:* Adj prof, Columbia Univ & NJ Inst Tech. *Honors & Awards:* PC Ray Award, Indian Inst Chem Engr, 66. *Mem:* Am Inst Chem Engrs. *Res:* Mathematical modeling of processes; computer simulation; chemical reaction engineering; separation processes; chemical thermodynamics; numerical methods; fuels and synfuels res. *Mailing Add:* 8 Pilgrim Ct Convent Station NJ 07961

MURTHY, GOPALA KRISHNA, dairying, for more information see previous edition

MURTHY, KRISHNA K, b Bangalore; US citizen. IMMUNOLOGY, MOLECULAR BIOLOGY. *Educ:* Univ Bangalore, India, PUC, 65; Univ Agr Sci, DVM, 71, MS, 74; Cornell Univ, PhD(immunol), 79. *Prof Exp:* Fel immunol, Cornell Univ, 75-79; res assoc, Univ Ga, 79-81 & 82-84; asst prof, Univ Basel, Switz, 81-82; ASSOC PROF PEDIAT & DIR IMMUNOL LAB, DEPT PEDIAT, LA STATE UNIV MED CTR, 84- *Concurrent Pos:* Vis mem, Basel Inst Immunol, 81-82; prin investr, NIH, 82-85; vis prof, Univ Gifu, 83; consult, Nat Heart Lung & Blood Inst, 85- *Mem:* NY Acad Sci; Sigma Xi; Int Soc Develop & Comp Immunol; Am Asn Immunologists. *Res:* Basic and clinical immunological studies to delineate the role of T lymphocytes and their products on the regulation of immune responses and in mediating autoimmune disorders; monoclonal antibodies. *Mailing Add:* Southwest Found Biomed Res 7620 NW Loop 41D San Antonio TX 78227

MURTHY, MAHADI RAGHAVANDRARAO VEN, b Bangalore, India, Nov 3, 29. NUCLEIC ACID STRUCTURE, HEREDITARY DISEASES. *Educ:* Univ Mysore, BSc, 49; Indian Inst Sci, Bangalore, PhD(biochem), 55. *Prof Exp:* Res assoc biochem, Med Br, Univ Tex, 61-63; from asst prof to assoc prof, 64-69, PROF BIOCHEM, FAC MED, LAVAL UNIV, 69-, DIR MOLECULAR NEUROBIOL LAB, 69- *Concurrent Pos:* Fel biochem, Texas A&M Univ, 55-58; fel, Sch Med, Yale Univ, 58-59; res assoc, Univ Tex Med Ctr, Galveston, 59-63; vis prof, France-Québec Exchange Prog, 77-, Australian Nat Univ, 78, Ind Inst Sci & Max Planck Inst, Gottingen, 79, Univ Clermont Ferrand, France, 85-86 & Mahidol Univ, Thailand, 89 & 90; consult, Coun Sci Ind Res, India, 90, Nat Res Coun, Morocco, 90; mem deleg ind biochem, People's Rep China, 90. *Honors & Awards:* M Sreenivasaya Award, Coun Indian Inst Sci, 53. *Mem:* AAAS; Can Biochem Soc; Int Soc Neurochem; NY Acad Sci; Am Soc Biochem & Molecular Biol. *Res:* Regulation of protein and nucleic acid synthesis in tissues during growth; nucleic acid and protein structure; development of clinical diagnostic methods for familial and genetic diseases; aging of brain and Alzheimer's disease. *Mailing Add:* Dept Biochem Fac Med Laval Univ Quebec PQ G1K 7P4 Can

MURTHY, RAMAN CHITTARAM, b Bangalore, India, US citizen; c 1. BIOLOGY, ENVIRONMENTAL HEALTH. *Educ:* Univ Mysore, BSc, 57, MSc, 61; Univ Notre Dame, MS, 66; Univ Cincinnati, PhD(biol), 72. *Prof Exp:* Fel toxicol, Dept Environ Health, Med Ctr, Univ Cincinnati, 70-74, res assoc, 74-76; ASST PROF BIOL, CENT STATE UNIV, 74-; ADJ ASST PROF ENVIRON HEALTH, MED CTR, UNIV CINCINNATI, 76- *Concurrent Pos:* Asst prof zool, Univ Mysore, 61-64; teaching asst biol, Univ Notre Dame, 64-66 & Univ Cincinnati, 68-70; res assoc biol, Univ Chicago, 66-68. *Mem:* Environ Mutagen Soc; Am Genetic Asn; Genetics Soc Am; Sigma Xi; Acad Kettering Fels. *Res:* Toxicology, genetic toxicology and teratogenic studies; physiological and biochemical responses to environmental pollutante; effects of drugs on the hemotological changes and membrane changes in red blood cells; effects as seen by light and electron microscope. *Mailing Add:* Dept Biol Cent State Univ Wilberforce OH 45384

MURTHY, SRINIVASA K R, b Bangalore, India, June 12, 49; US citizen. BUSINESS & STRATEGIC MANAGEMENT, VIDEOPHONY. *Educ:* Bangalore Univ, India, BS, 67, MS, 69; Mysore Univ, India, MS, 71. *Prof Exp:* Res engr, Bharat Electronics Ltd, India, 70-71; sr sci officer, Indian Space Res Orgn, 71-79; proj engr, Systs & Appl Sci Corp, Anaheim, Calif, 80-83; div dir, IMR Systs Corp, Roslyn, Va, 83-84; systs eng mgr, Gen Elec, Portsmouth, Va, 84-85; SERV & BUS MGR, AT&T BELL LABS, 85- *Concurrent Pos:* Vis prof electronics & computer eng, Calif State Univ, Pomona & Fullerton, 79-82; mem, gov bd, Inst Elec & Electronics Engrs Eng Mgt Soc, 86-, tech activ bd, Inst Elec & Electronics Engrs Computer Soc, 87- & prog comt, Inst Elec & Electronics Engrs Lect Tours, Australia, Singapore & India, 89; founder, Inst Elec & Electronics Engrs Network J, 87; overseas chmn & prog co-chair, numerous conferences, Inst Elec & Electronics Engrs Eng Mgt Soc; chmn, Int Adv comt, Inst Elec & Electronics Engrs Tencon Conf, India, 89-91 & Melbourne, Australia, 90-92. *Honors & Awards:* Distinguished Achievement Award, Dept Space, Govt India, 75; Outstanding Contrib Award, Inst Elec & Electronics Engrs Eng Mgt Soc, 90. *Mem:* Inst Elec & Electronics Engrs Eng Mgt Soc; Inst Elec & Electronics Engrs Systs Man & Cybernet Soc; Inst Elec & Electronics Engrs Computer Soc. *Res:* Strategic managment; international competition; videophony, integrated services digital network competitive analysis. *Mailing Add:* Five Polo Club Dr Tinton Falls NJ 07724

MURTHY, VADIRAJA VENKATESA, b Bombay, India, Mar 27, 40; US citizen; m 69; c 2. ENZYMOLOGY, DIAGNOSTIC ENZYMOLOGY IN CLINICAL MEDICINE. *Educ:* Univ Bombay, BSc Hons, 59, MSc, 61; Univ Md, PhD(biol chem), 68. *Prof Exp:* Res asst biochem, Purdue Univ, 61-63; res fel, Med Sch, Univ Md, 63-68; sci off pharmacol, St John's Med Col, Bangalore, India, 68-69; sr res biochemist & asst group leader, USV Pharmaceut Corp, 70-71; res assoc toxicol, Toxicol Ctr, Pharmacol Dept, Univ Iowa, 71-72; vis scientist environ toxicol, Nat Inst Environ Health, 72-74; sr res assoc, Pharmacol Dept, Emory Univ, 74-75; adj asst prof biochem, Chem Dept, Atlanta Univ, 75-76; assoc prof, Biol Dept, 76-79, prof biol & co-dir, Minority Biomed Support Prog, Talladega Col, 79-83; ASST PROF, DEPT LAB MED, ALBERT EINSTEIN COL MED, 83-; DIR CHEM, SPEC CHEM LABS, BRONX MUNIC HOSP, 89- *Concurrent Pos:* Guest lectr, Seminar Dermatol, Med Sch, Harvard Univ, 68; minority biomed support consult analytical chem, Stillman Col, 80. *Mem:* Am Asn Cancer Res; Am Chem Soc; NY Acad Sci; Am Fedn Clin Res; Sigma Xi; Am Asn Clin Chem. *Res:* Enzymes in the diagnostic medicine; immunodiagnostics development and research; new nonradioactive immunoassays. *Mailing Add:* 100 Lindbergh Blvd Teaneck NJ 07666-5347

MURTHY, VARANASI RAMA, b Visakhapatnam, India, July 2, 33; m 59; c 2. GEOCHEMISTRY. *Educ:* Andhra Univ, India, BSc, 51; Yale Univ, MS, 55, PhD, 57. *Prof Exp:* Res fel geol, Calif Inst Technol, 57-59; res asst geochem, Univ Calif, San Diego, 59-62, asst prof, 62-65; assoc prof, 65-69, chmn dept 71-83, actg dean, Inst Tech, 84, PROF GEOCHEM, UNIV MINN, MINNEAPOLIS, 69-, VPRES ACAD AFFAIRS & PROVOST, 85- *Mem:* Am Geophys Union; Geochem Soc; Geol Soc Am. *Res:* Petrology; cosmochemistry and lunar investigation; early crustal and mantle evolution in the earth. *Mailing Add:* Dept Earth Sci 106A Pittsburg Hall Univ Minn Minneapolis MN 55455

MURTHY, VEERARAGHAVAN KRISHNA, b Pudukottah, India, Feb 27, 34; c 3. BIOCHEMISTRY, PHYSIOLOGY. *Educ:* Univ Madras, BS, 53; Univ Bombay, MS, 60, PhD(biochem), 64. *Prof Exp:* Res asst biochem, Vallabhbhai Patel Chest Inst, Univ Delhi, 55-57; sci officer, Indian Cancer Res Ctr, AEC, Govt India, 57-64; asst prof med & biochem, 74-84, ASSOC PROF INTERNAL MED & BIOCHEM, UNIV NEBR MED CTR, OMAHA, 84- *Concurrent Pos:* Res fel, Univ Fla, 64-68 & Univ Toronto, 68-74. *Mem:* Fel Royal Inst Chem; Can Biochem Soc; Am Diabetes Asn; Am Physiol Soc. *Res:* Diabetes and lipid metabolism; hormones and lipids; cardiac muscle metabolism; muscle contraction; drug metabolism in cancer. *Mailing Add:* Dept Biochem Univ Nebr Med Ctr 42 & Dewey Ave Omaha NE 68105

MURTHY, VISHNUBHAKTA SHRINIVAS, b Kanker, India, Jan 1, 42; m 68; c 2. PHARMACOLOGY. *Educ:* Univ Indore, India, BS & MB, 65, MD, 68; Univ Manitoba, PhD(pharmacol), 72. *Prof Exp:* Demonstr & lectr pharmacol, Mahatma Gandhi Mem Med Col, Indore, India, 66-69; sr scientist, Warner-Lambert Res Inst, 72-74; sr res investr, Squibb Inst Med Res, 74-; ASST PROF PHARMACOL & DIR CLIN PHARMACOL, MT SINAI MED CTR. *Concurrent Pos:* Adj asst prof physiol, Rutgers Med Sch, Col Med & Dent NJ, 74- *Mem:* NY Acad Sci. *Res:* Physiology of cardiovascular homeostasis and its modification in cardiovascular diseases; pharmacological modulation of experimental myocardial ischemia, infarction and oxygen transport to tissue. *Mailing Add:* Off Dir Clin Pharmacol Mt Sinai Med Ctr 950 N 12th St PO Box 343 Milwaukee WI 53233

MURTI, KURUGANTI GOPALAKRISHNA, b Masulipatam, India, Feb 16, 43; m 72; c 3. VIROLOGY, MOLECULAR BIOLOGY. *Educ:* Andhra Univ, India, BSc, 61, MSc, 63; Univ Colo, Boulder, PhD(biol), 72. *Prof Exp:* Res fel zool, Andhra Univ, India, 63-65; res officer, Zool Surv India, Calcutta, 65-67; teaching asst biol, Univ Colo, Boulder, 68-72, res assoc cell biol, 72-75; res assoc virol, 75-77, asst mem, 77-81, assoc mem, 81-85, MEM & DIR, ELECTRON MICROS, ST JUDE CHILDREN'S RES HOSP, MEMPHIS, 88- *Concurrent Pos:* Asst prof, Univ Tenn, Memphis, 81-85; prin investr, Am Cancer Soc, 85- *Mem:* Am Soc Cell Biol; Electron Microscope Soc Am; Am Inst Biol Sci. *Res:* Studies of virus-infected mammalian cells to understand the molecular basis of the mechanism of virus-infection. *Mailing Add:* Virol & Molecular Biol St Jude Children's Res Hosp PO Box 318 Memphis TN 38101-0318

MURTY, DANGETY SATYANARAYANA, b Visakhapatnam, India, Dec 28, 27; m 52; c 4. PHYSICS. *Educ:* Govt Arts Col, India, BSc, 48; Presidency Col, MA, 50; Andhra Univ, MSc, 51, DSc(ionosphere physics), 56. *Prof Exp:* Lectr appl physics, Andhra Univ, India, 52-57; Colombo Plan res scholar, 57-58; lectr appl physics, Andhra Univ, 58-60; assoc prof physics & actg head dept, Tex Southern Univ, 60-63; from assoc prof to prof, 63-85, chmn dept, 63-72, CHMN DEPT PHYSICS, ST MARY'S UNIV, NS, 85- *Concurrent Pos:* Vis scholar, Univ Calif, Berkeley, 77-78. *Mem:* Can Asn Physicists; Inst Elec & Electronics Engrs; Am Asn Physics Teachers; fel Brit Inst Elec Engrs; fel Brit Inst Electronics & Radio Engrs. *Res:* Mossbauer effect; radio astronomy; low energy nuclear physics. *Mailing Add:* Saint Marys Univ Robie St Halifax NS B3H 3C3 Can

MURTY, DASIKA RADHA KRISHNA, b Guntur, India, Dec 13, 31; m 49; c 4. ORGANIC CHEMISTRY. *Educ:* Andhra Univ, India, BSc, 51, MSc, 52; Fla State Univ, PhD(org chem), 60. *Prof Exp:* Asst org chem, Fla State Univ, 55-60; fel, Wayne State Univ, 60; sr chemist, Tracerlab Div, Lab for Electronics, Inc, 61-63; sr res scientist, 63-68, supvr, Radiomed Synthesis Sect, 68-70, head, Radiopharmaceut Res Sect, 70-72, head, In-Vitro Diag Sect, 72-79, ASST DIR CLIN ASSAY RES & DEVELOP, E R SQUIBB & SONS, 79- *Mem:* AAAS; Am Inst Clin Chemists; Am Chem Soc; Soc Nuclear Med. *Res:* Heterocyclics synthesis, reaction mechanisms; synthesis of radiochemicals and radiopharmaceuticals; clinical radioassay research and development. *Mailing Add:* 755 Hoover Dr North Brunswick NJ 08902

MURTY, KATTA GOPALAKRISHNA, b Pandillapalli, India, Sept 9, 36; m 64; c 2. OPERATIONS RESEARCH. *Educ:* Madras Univ, BSc, 55; Univ Calif, Berkeley, MS, 66, PhD(opers res), 68. *Prof Exp:* Consult asst prof statist & opers res, Indian Statist Inst, Calcutta, 58-65; Univ Calif, Berkeley, 65-68; assoc prof, 68-80, PROF OPERS RES, UNIV MICH, ANN ARBOR, 80- *Concurrent Pos:* Fulbright travel grant, 61-62. *Mem:* Opers Res Soc Am; Math Prog Soc; Sigma Xi. *Res:* Mathematical programming; branch and bound algorithms; complementarity problem; network flows; convex polyhedra. *Mailing Add:* Dept Indust & Opers Eng Univ Mich Ann Arbor MI 48109-2117

MURTY, RAMA CHANDRA, b Vizianagaram, India, July 1, 28; Can citizen; m 62; c 1. PHYSICS. *Educ:* Andhra Univ, India, BSc, 47; Univ Bombay, MSc, 50, dipl librarianship, 51; Univ Western Ont, MSc, 58, PhD(physics), 62. *Prof Exp:* Demonstr physics, Wilson Col, Bombay, 47-52; librn, Express Newspapers Ltd, India, 52; sr master physics, Harrison Col, Barbados, West Indies, 52-57; demonstr, 57-58, sr demonstr, 58-61, res assoc geophys, 59-61, lectr physics, 61-63, asst prof, 63-68, ASSOC PROF PHYSICS, UNIV WESTERN ONT, 68- *Concurrent Pos:* Scanner, Tata Inst Fundamental Res, India, 50; Nat Res Coun Can grant, 65-; contract, Meteorol Br, Can Dept Transportation, 65-72; chmn comn VII, Can Div, Int Sci Radio Union. *Mem:* NY Acad Sci; Am Asn Physics Teachers; Am Meteorol Soc; Can Asn Physicists; Brit Inst Physics. *Res:* Physics of lightning; atmospheric electricity; sferics and meteorology. *Mailing Add:* Dept Physics Univ Western Ont London ON N6A 5B9 Can

MURTY, TADEPALLI SATYANARAYANA, b Rambhotlapalem, India, Aug 5, 38; m 73. PHYSICAL OCEANOGRAPHY. *Educ:* Andhra Univ, India, BSc, 55, MSc, 59; Univ Chicago, MS, 62, PhD(geophys), 67. *Prof Exp:* Lectr physics, Osmania Col, India, 59-60; res asst geophys, Univ Chicago, 60-67; res scientist I, 67-69, res scientist II, Can Dept Environ, 69-85, SR RES SCIENTIST, FISHERIES & OCEANS, INST OCEAN SCI, 85- *Concurrent Pos:* Can mem int tsunami comt, Int Union Geod & Geophys, 71- *Honors & Awards:* Distinguished Res Medal, Univ Chicago, 67; Appl Oceanog Prize, Can Meteorol & Oceanog Soc, 83. *Mem:* Am Geophys Union; Am Meteorol Soc; Am Soc Limnol & Oceanog; Seismol Soc Am; Am Math Soc. *Res:* Theoretical research in physical oceanography using numerical integration techniques. *Mailing Add:* Inst Ocean Sci PO Box 6000 9860 W Saanic Sidney BC V8L 4B2 Can

MURVOSH, CHAD M, b Toronto, Ohio, Aug 10, 31; m 65; c 3. STREAM ECOLOGY, AQUATIC INSECTS. *Educ:* Kent State Univ, BS, 53; Ohio State Univ, MS, 58, PhD(zool, entom), 60. *Prof Exp:* Instr entom, Ohio State Univ, 60-61; instr zool, Ohio Wesleyan Univ, 61; med entomologist, Entom Res Div, USDA, 62-64; from asst prof to assoc prof zool, 64-78, PROF BIOL, UNIV NEV, LAS VEGAS, 78- *Concurrent Pos:* Grant, Desert Res Inst, Univ Nev, 65-66; Regents grant, Univ Nev, Las Vegas, 78-79; Barrick res grant, 84-85. *Mem:* Entom Soc Am; Ecol Soc Am; Soc Study Evolution; Soc Syst Zool; NAm Benthological Soc. *Res:* Aquatic insect ecology. *Mailing Add:* Dept of Biol Sci Univ of Nev 4505 S Maryland Las Vegas NV 89154

MUSA, JOHN D, b Amityville, NY, June 11, 33; m. SOFTWARE RELIABILITY ENGINEERING. *Educ:* Dartmouth Col, BA, 54, MS, 55. *Prof Exp:* Mem, Guidance Group, AT&T Bell Labs, 58-61, Discrimination Group, 61-63, Nike X Syst Group, 63, supvr, Guidance Simulation Group, 63-66, Sprint Guidance & Interceptor Response Group, 66-68, Guidance Group, 68, BMDC Syst Design Group, 68-69, Mgt Control Group, 69-72, Human Factors Test Group, 72-74, Batch Graphics & Data Mgt Group, 74-75, Computer Graphics, 75-80, Computer Measurements & Security, 80-83, Computer Measurements & Capacity Planning, 83-85, SUPVR, SOFTWARE QUALITY, AT&T BELL LABS, 85- *Concurrent Pos:* Chmn, Tech Comt Software Eng, Inst Elec & Electronic Engrs, 82-84, mem gov bd, 84-85, vpres publ, 84-85, vpres tech activities, 86, 2nd vpres, 86; chmn, steering comt, Int Conf Software Eng, 82-84; distinguished lectr, Inst Elec & Electronic Engrs Computer Soc, 80-83; co-dir, NATO Adv Study Inst, 85; mem, US sci deleg software eng, People's Repub China. *Honors & Awards:* Meritorious Serv Awards, Inst Elec & Electronic Engrs Computer Soc. *Mem:* Sr mem Inst Elec & Electronic Engrs; Inst Elec & Electronic Engrs Computer Soc. *Res:* Software reliability engineering; author of several books and articles. *Mailing Add:* AT&T Bell Labs 600 Mountain Ave Rm 2D248 Murray Hill NJ 07974

MUSA, MAHMOUD NIMIR, b Arraba, Palestine, Mar 22, 43; US citizen. PSYCHOPHARMACOLOGY. *Educ:* Am Univ Beirut, BS, 64; Univ Wis, MS, 66, PhD(pharm), 72; Med Col Wis, MD(med), 79. *Prof Exp:* Teaching asst pharm, Univ Wis, 64-71, res assoc, 72-75; asst prof pharm, Idaho State Univ, 75-76; resident phsician, Ill State Psychiat Inst, 79-83; PVT PRACT MED & PSYCHIAT, 83-; ASSOC PROF PSYCHIATRY, CHICAGO MED SCH, 87- *Concurrent Pos:* Lectr, Ill State Psychiat Inst, 82-83; dir, Chicago Inst Ment Sci, 84-87. *Honors & Awards:* Sci Res Award, Ill Psychiat Soc, 82. *Mem:* Am Psychiat Asn; Am Pub Health Asn; AAAS; Am Col Clin Pharmacol. *Res:* Clinical pharmacokinetics and determinants of therapeutic response to psychoactive drugs. *Mailing Add:* 1115 S Plymouth Apt 102 Chicago IL 60605

MUSA, SAMUEL A, US citizen. RESEARCH & DEVELOPMENT MANAGEMENT, STRATEGIC TECHNOLOGY PLANNING. *Educ:* Rutgers Univ, BS & BA, 61; Harvard Univ, MS, 62, PhD(appl physics), 65. *Prof Exp:* Res scientist, Gen Precision Aerospace Res Ctr, 65-66; asst prof elec eng, Univ Pa, 66-71; proj leader, Inst Defense Analysis, 71-78; dep dir commun, Command & Control Policy, Off Undersecy Defense, 78-79, staff specialist electronic warfare, 79-82, dep dir, Mil Systs Technol, 82-83; VPRES RES & ADVAN TECHNOL, E-SYSTS, 83- *Concurrent Pos:* Vis lectr math, Stevens Inst Technol, 65-66; consult, Computer Command & Control Co, 67-71; tech ed, Inst Elec & Electronic Engrs Trans Geosci & Remote Sensing, 75-80; prof lectr eng, George Washington Univ, 78-83; mem, Air Force Sci Adv Bd, 86- *Mem:* Fel Inst Elec & Electronic Engrs; Sigma Xi; Asn Old Crows; Nat Security Indust Asn; Am Inst Aeronaut & Astronaut. *Res:* Strategic technology development. *Mailing Add:* E-Systs Inc PO Box 660248 Dallas TX 75266-0248

MUSACCHIA, X J, b Brooklyn, NY, Feb 11, 23; m 50; c 4. RESEARCH ADMINISTRATION, EDUCATIONAL ADMINISTRATION. *Educ:* St Francis Col, NY, BS, 44; Fordham Univ, MS, 47, PhD(biol), 49. *Prof Exp:* Instr biol, Marymount Col, NY, 47-49; instr comp physiol, St Louis Univ, 49-51, from asst prof to prof, 51-65; prof physiol, Univ Mo-Columbia, 65-78, assoc dean, grad sch & assoc dir res, 72-78, dir, Dalton Res Ctr, 74-78; prof physiol & biophys, dean grad sch & assoc univ provost res, Univ Louisville, 78-90; RETIRED. *Concurrent Pos:* Co-dir, Arctic Res Prog, St Louis Univ, 49-52, actg dir, Biol Labs, 52-53; vis scientist, Am Physiol Soc, 63-65; sr investr, Dalton Res Ctr, 65-74. *Mem:* Fel AAAS; Am Physiol Soc; Am Soc Zoologists; Sigma Xi; Soc Exp Biol Med. *Res:* Environmental physiology; biochemistry of hibernation in reptiles and mammals; radiation biology and comparative physiology of intestinal absorption; physiology of depressed metabolism, hypothermia and hibernation; gravitational physiology. *Mailing Add:* Dept Physiol & Biophysics Univ of Louisville Louisville KY 40292

MUSAL, HENRY M(ICHAEL), JR, b Chicago, Ill, Aug 18, 31; m 52; c 4. ELECTRICAL ENGINEERING, ENGINEERING PHYSICS. *Educ:* Ill Inst Technol, BS, 54, MS, 57, PhD(elec eng), 65. *Prof Exp:* Jr engr, Sinclair Res Labs, 53-54; instr elec eng, Ill Inst Technol, 54-56; sr engr, Cook Res Labs, 56-58 & Bendix Systs Div, 58-61; staff scientist, GM Defense Res Labs, Calif, 61-67; CONSULT SCIENTIST, LOCKHEED PALO ALTO RES LAB, 67- *Mem:* Inst Elec & Electronics Engrs; Am Phys Soc. *Res:* Laboratory generation of plasma for basic investigations; plasma diagnostics; gas lasers; radar systems and electromagnetic wave scattering; interaction of electromagnetic fields and laser radiation with plasmas and materials. *Mailing Add:* Lockheed Palo Alto Res Lab Dept 91-60 Bldg 256 3251 Hanover St Palo Alto CA 94304

MUSCARI, JOSEPH A, b Chicago, Ill, May 13, 35; m 60; c 6. PHYSICS. *Educ:* Beloit Col, BS, 57; Johns Hopkins Univ, MS, 63; Wash State Univ, PhD(physics), 66. *Prof Exp:* Teacher high sch, 59-63; chief, Optical Physics Sect, 66-70, PROG MGR, MARTIN MARIETTA CORP, 70- *Concurrent Pos:* Prin investr, Skylab Prog Exp, Martin Marietta Corp. *Mem:* Optical Soc Am; Am Vacuum Soc. *Res:* Nuclear physics; gamma ray spectroscopy; optical physics; vacuum ultraviolet spectroscopy; space contamination. *Mailing Add:* 2413 W Costilla Littleton CO 80120

MUSCATELLO, ANTHONY CURTIS, b Princeton, WVa, Sept 25, 50; m 85; c 3. SEPARATIONS CHEMISTRY, RADIOCHEMISTRY. *Educ:* Concord Col, BS, 72; Fla State Univ, PhD(inorg chem), 79. *Prof Exp:* Res assoc, Argonne Nat Lab, 79-81; sr res chemist, Rocky Flats Plant, Rockwell Int, 81-85, res specialist I, 85-87, res specialist II, 87-88; STAFF MEM, LOS ALAMOS NAT LAB, 88- *Mem:* Am Chem Soc; Planetary Soc. *Res:* Separations chemistry of actinides and lanthandes-solvent extraction using bifunctional organophosphorus compounds, ion-exchange chromatography; formation and dissociation kinetics of transplutonium element chelates; radiochemical techniques. *Mailing Add:* 9921 Perry Ct Westminster CO 80030-2637

MUSCATINE, LEONARD, b Trenton, NJ, Sept 7, 32; m 57; c 4. INVERTEBRATE ZOOLOGY, BIOLOGICAL OCEANOGRAPHY. *Educ:* Lafayette Col, BA, 54; Univ Calif, Berkeley, MA, 56, PhD(zool), 61. *Prof Exp:* Fel biochem, Howard Hughes Med Inst, 61-62; fel plant biochem, Scripps Inst Oceanog, 62-63, res biologist, 63-64; assoc prof, 64-74, chmn dept, 76-79, PROF BIOL, UNIV CALIF, LOS ANGELES, 74- *Concurrent Pos:* NIH fel, 61-63; NSF res grant, 63, pres; Guggenheim fel, Oxford Univ, 70-71. *Mem:* Am Soc Zoologists (pres, 87); Marine Biol Asn UK; Am Soc Limnol & Oceanog; Int Soc Reef Studies. *Res:* Coelenterate physiology; symbiosis of invertebrates and unicellular algae; biology of corals and coral reefs. *Mailing Add:* Dept Biol Univ Calif Los Angeles CA 90024

MUSCH, DAVID C(HARLES), b Havre de Grace, Md, Aug 11, 54. CLINICAL RESEARCH ON OPHTHALMIC DISEASE, CLINICAL TRIALS. *Educ:* Calvin Col, BS, 76; Univ Mich Sch Public Health, MPH, 78; Univ Mich Rachham Grad Sch, PhD(epidemiol), 81. *Prof Exp:* Res investr clin res, Dept Ophthalmol, 81-86, asst res scientist, 86-89, ASSOC RES SCIENTIST CLIN RES, UNIV MICH, 89- *Concurrent Pos:* Ed bd mem, Ophthalmol, off J Am Acad Ophthalmol, 87-; prin investr, Sponsored Res Grants in Clin Vision Res. *Mem:* Am Acad Ophthalmol; AAAS; Am Col Epidemiol; Asn Res in Vision & Ophthalmol; Soc Epidemiol Res. *Res:* The identification of factors that place an individual at higher risk of developing ophthalmic disorders; corneal transportation. *Mailing Add:* W K Kellogg Eye Ctr Univ Mich 1000 Wall St Ann Arbor MI 48105-1994

MUSCHEK, LAWRENCE DAVID, b Philadelphia, Pa, Apr 28, 43; m 64; c 3. BIOCHEMICAL PHARMACOLOGY. *Educ:* Philadelphia Col Pharm & Sci, BSc, 65; Mich State Univ, PhD(biochem), 70. *Prof Exp:* Mich Heart Asn fel cardiovasc pharmacol, Mich State Univ, 70-72; res scientist, 72-73, sr scientist, 73-76, group leader cardiovasc biochem, 76-78, head biochem mechanisms sect, 78-79, DIR, DEPT BIOL RES, MCNEIL LABS, INC, 76- *Mem:* Am Chem Soc; AAAS. *Res:* Discovery and development of new agents effective in the treatment and/or prevention of thrombosis; mechanisms responsible for myocardial ischemia and infarction. *Mailing Add:* 700 E Brigantine Ave Brigantine NJ 08203

MUSCHEL, LOUIS HENRY, b New York, NY, July 4, 16; m 46; c 1. IMMUNOLOGY. *Educ:* NY Univ, BS, 36; Columbia Univ, AM, 38; Yale Univ, MS, 51, PhD(microbiol), 53. *Prof Exp:* Asst supvr, Serum Diag Dept, Div Labs & Res, NY State Dept Health, 39-41 & 46; chief, Dept Spec Serol & exec officer, Fourth Area Lab, Brooke Med Ctr, US Army Med Serv Corps, Tex, 46-47, chief lab serv, 20th Sta Hosp, Clark Field, PI, 47-48, Depts Serol & Chem, Second Area Lab, Ft Meade, Md, 48-50, 406th Med Gen Lab, Far East Command, 53-56, Exp Immunol Sect, Dept Appl Immunol, Walter Reed Army Inst Res, Washington, DC, 56-58, Dept Serol, 58-62; from assoc prof to prof microbiol, Med Sch, Univ Minn, Minneapolis, 62-70; mem staff, Res Dept, Am Cancer Soc, 70-88; SCI ADV, ISRAEL CANCER RES FUND, 88- *Concurrent Pos:* Abstractor, Biol Abstr, 47-48 & Chem Abstr, Am Chem Soc, 55-62; mem, Bact & Mycol Study Sect, Div Res Grants, NIH, 59- & Grants Rev Comt, Minn Chap, Arthritis & Rheumatism Found, 65-70; consult, Walter Reed Army Inst Res, Washington, DC, 63- & Vet Admin Hosp, Minneapolis, 65-; adj prof microbiol, Columbia Univ, 77-84; adj prof path, NY Univ, 84- *Mem:* Am Soc Microbiol; Soc Exp Biol & Med; Am Asn Immunol; Am Asn Cancer Res; NY Acad Sci; Sigma Xi. *Res:* Immunochemistry; natural resistance mediated by the immune bactericidal reaction; serology of syphilis; immunohematology. *Mailing Add:* 3333 Henry Hudson Pkwy Apt 8A Bronx NY 10463-3233

MUSCHIK, GARY MATHEW, b Rice Lake, Wis, July 22, 44; m 70; c 2. ORGANIC & ANALYTICAL CHEMISTRY. *Educ:* Wis State Univ, River Falls, BS, 66; Kans State Univ, PhD(chem), 72. *Prof Exp:* Chemist, Agr Res Serv, USDA, 72-73; scientist, 73-77, HEAD CHEM SYNTHETICS & ANALYSIS, FREDERICK CANCER RES CTR, NAT CANCER INST, 77- *Mem:* Am Chem Soc; Am Soc Mass Spectrometry; AAAS. *Res:* Polycyclic aromatic hydrocarbon synthesis, separation, identification, metabolism and carcinogenicity studies; synthesis and development of liquid crystal GLC liquid phases for novel separations; development of new methodologies in analytical chemistry and organic syntheses. *Mailing Add:* Frederick Cancer Res Ctr PR I PO Box B Bldg 467 Frederick MD 21701

MUSCHIO, HENRY M, JR, b New York, NY, Apr 25, 31; m 57; c 4. HUMAN GENETICS, SCIENCE ADMINISTRATION. *Educ:* Syracuse Univ, AB, 52; Fordham Univ, MS, 57, PhD(biol), 63. *Prof Exp:* Instr biol sci, Fairleigh Dickinson Univ, 58-62; asst prof, Montclair State Col, 62-66; assoc prof, 66-68, PROF BIOL SCI, DUTCHESS COMMUNITY COL, 68-, HEAD DEPT, 66 - *Concurrent Pos:* Dir & lectr, NSF Inserv Inst Modern Biol, Montclair State Col, 65-66; dir, NSF grant, Norrie Point Proj, Dutchess Community Col, 79-82; mem bd dirs, Rehab Progs Inc, Poughkeepsie & Anderson Sch, Staatsburg, NY; dir, HCOP Grant Proj-Allied Health Technol, Dutchess Community Col, 85-88. *Mem:* AAAS; NY Acad Sci; Nat Sci Teachers Asn. *Res:* Human cytogenetics and cytological research related to the effects of chemical agents and their effects on the human karyotype and various human cell lines in vitro, with consideration of ethical and moral issues and values. *Mailing Add:* Dept Allied Health & Biol Sci Dutchess Community Col Poughkeepsie NY 12601-1595

MUSCHLITZ, EARLE EUGENE, JR, b Palmerton, Pa, Apr 23, 21; m 53; c 2. PHYSICAL CHEMISTRY, CHEMICAL PHYSICS. *Educ:* Pa State Univ, BS, 41, MS, 42, PhD(phys chem), 47. *Prof Exp:* Asst, Pa State Univ, 43-46; instr phys chem, Cornell Univ, 47-51; from asst res prof to assoc prof, 51-58, chmn dept, 73-77, prof, 58-86, EMER PROF CHEM, UNIV FLA, 86- *Concurrent Pos:* NSF sr fel, 63-64; vis fel, Joint Inst Lab Astrophys, Boulder, Colo, 68; Alexander von Humboldt sr scientist award, Gottingen, WGer, 78. *Mem:* Sigma Xi; Am Chem Soc; fel Am Phys Soc; Am Soc Mass Spectrometry. *Res:* Ion, electron and excited atom scattering in gases; negative ions; molecular beams; mass spectrometry; molecular structure; upper atmosphere phenomena. *Mailing Add:* 4850 NW 20th Pl Gainesville FL 32605

MUSCOPLAT, CHARLES CRAIG, b St Paul, Minn, Aug 13, 48; m 69; c 2. IMMUNOLOGY, VETERINARY MEDICINE. *Educ:* Univ Minn, BA, 70, PhD(vet microbiol), 75. *Prof Exp:* Instr, Cornell Med Sch, 75-76; assoc, Sloan-Kettering Inst, NY, 75-76; res assoc immunol, Univ Minn, 76-78, assoc prof, 78-81; PRES & CHIEF OPERATING OFFICER MOLECULAR GENETICS, INC, MINNETONKA, MINN, 81- *Concurrent Pos:* adj prof, Col Vet Med, Univ Minn, 83-; mem bd Agr, NAS, 84-87. *Mem:* Am Asn Immunologists; Am Soc Microbiol; Conf Res Workers Animal Dis; Sigma Xi. *Res:* Immunology of respiratory disease in cattle; immunology of cancer and immune regulation. *Mailing Add:* New Business Molecular Genetic Inc 10320 Bren Rd E Minnetonka MN 55343

MUSE, JOEL, JR, b Williamston, NC, July 11, 41; div; c 2. INDUSTRIAL ORGANIC CHEMISTRY. *Educ:* Univ NC, Chapel Hill, AB, 63; Univ Md, College Park, PhD(chem), 68. *Prof Exp:* Sr res chemist, 68-77, sect head, 77-80, mgr, 80-82, DEPT MGR, GOODYEAR TIRE & RUBBER CO, AKRON, 82- *Mem:* Am Chem Soc; Sigma Xi; Tech Asn Pulp & Paper Indust. *Res:* Process and product development; rubber chemicals; hydroquinone and derivates; petrochemicals; resins and speciality polymers. *Mailing Add:* 1089 Kevin Dr Kent OH 44240

MUSER, MARC, b Baltimore, Md, Feb 18, 54; m 82; c 2. INORGANIC CHEMISTRY. *Educ:* Wash Col, BS, 75; Univ Md, College Park, MS, 80. *Prof Exp:* Jr chemist, Alcolac, 75-76; chemist, M G Burdette Gas Prod, 81-82; quality control mgr, Kanasco Ltd, 82-83; TECH DIR, COURTNEY INDUSTS, 83- *Mem:* Am Chem Soc; Tech Asn Pulp & Paper Indust. *Res:* Inorganic polymers used in water treatment. *Mailing Add:* 1920 Benhill Ave Baltimore MD 21226

MUSGRAVE, ALBERT WAYNE, b Eads, Colo, Jan 22, 23; m 43; c 4. GEOPHYSICS, ENGINEERING. *Educ:* Colo Sch Mines, ScD(geophys eng), 52. *Prof Exp:* From geophys trainee to interpreter, Seismic Surv, Magnolia Petrol Co, Socony Mobil Oil Co, Inc, 47-49, seismologist seismic interpretation, Mobil Oil Corp, 50, seismic party chief, Seismic Surv, 52-53, res geophysicist geophys explor, 54-60, supt spec probs, 60-65, sr geophys scientist, Geophys Serv Ctr, 65-72, sr scientist, Seismic Res, Mobil Res & Develop, 72-80; RETIRED. *Honors & Awards:* Van Diest Gold Medal, Colo Sch Mines, 61. *Mem:* Soc Explor Geophys; Sigma Xi; Am Asn Petrol Geologists. *Res:* Geophysical engineering; physics; geology; mathematics; electronics; seismology; gravity; magnetism; well logging. *Mailing Add:* 6404 Lavendale Dallas TX 75230

MUSGRAVE, F STORY, b Boston, Mass, Aug 19, 35; m; c 6. PHYSIOLOGY, SURGERY. *Educ:* Syracuse Univ, BS, 58; Univ Calif, Los Angeles, MBA, 59; Marietta Col, BA, 60; Columbia Univ, MD, 64; Univ Ky, MS, 66; Univ Houston, MA, 87 & 90. *Prof Exp:* Intern surg, Med Ctr, Univ Ky, 64-65; SCIENTIST-ASTRONAUT, JOHNSON SPACECRAFT CTR, NASA, 67- *Concurrent Pos:* US Air Force fel aerospace physiol & med & Nat Heart Inst fel, Univ Ky, 65-67; instr physiol & biophys, Med Ctr, Univ Ky, 69-; fel surg, Denver Gen Hosp, 69-; astronaut, spaceflight missions STS-6, Spacelab No 2, STS-33 & STS-44. *Honors & Awards:* Except Serv Medal, NASA, 74, 85 & 87; Spaceflight Medal, 83, 85 & 89. *Mem:* AAAS; Aerospace Med Asn; Am Inst Aeronaut & Astronaut; AMA; Civil Aviation Med Asn. *Res:* Design and development of Space Shuttle extravehicular activity equipment and procedures; temperature regulation; physical fitness. *Mailing Add:* NASA Code CB Houston TX 77058

MUSGRAVE, MARY ELIZABETH, b Ithaca, NY, June 4, 54. PLANT STRESS PHYSIOLOGY, SPACE BIOLOGY. *Educ:* Cornell Univ, AB, 77; Duke Univ, PhD(bot & cell & molecular biol), 86. *Prof Exp:* Postdoctoral fel space biol, NASA, 86-87; ASST PROF PLANT STRESS PHYSIOL, DEPT PLANT PATH & CROP PHYSIOL, LA STATE UNIV, 87- *Mem:* Am Soc Plant Physiologists; Crop Sci Soc Am; Am Soc Gravitational & Space Biol. *Mailing Add:* Dept Plant Path & Crop Physiol La State Univ Baton Rouge LA 70803-1720

MUSGRAVE, STANLEY DEAN, b Hutsonville, Ill, Jan 26, 19; m 44; c 2. ANIMAL BREEDING, ANIMAL NUTRITION. *Educ:* Univ Ill, BS, 47, MS, 48; Cornell Univ, PhD(animal breeding), 51. *Prof Exp:* Asst animal husb, Cornell Univ, 47-50; asst prof dairy prod, Univ Ill, 50-51; from asst prof to prof dairying & head dept, Okla State Univ, 51-68; chmn dept animal & vet sci, 68-73 & 80-84, prof, 68-85, EMER PROF ANIMAL & VET SCI, UNIV MAINE, ORONO, 85- *Concurrent Pos:* Prog chmn, Am Dairy Sci Asn, 66; consult, Mossoro Advan Sch Agr, Brazil, 74-, Univ Mosul, Iraq, 73-79, Livestock Prod, 85- & Voca, Usaid, 89-; interim assoc dir, Maine Agr Exp Sta, 83. *Honors & Awards:* Distinguished Serv Award, Am Soc Animal Sci, 86.

Mem: AAAS; fel Am Soc Animal Sci; Am Genetics Asn; Am Registry Prof Animal Scientists; Am Dairy Sci Asn. *Res:* Milk component analysis; dairy cattle nutrient, health requirements; scanning electron microscopic studies of feed and age effect on gastrointestinal epithelium; dairy and livestock management systems development and analysis; animal production. *Mailing Add:* Dept Animal, Vet & Aquatic Sci Univ Maine 15 Kelly Rd Orono ME 04473-1312

MUSHAK, PAUL, b Dunmore, Pa, Dec 9, 35. BIOCHEMISTRY, CHEMISTRY. *Educ:* Univ Scranton, BS, 61; Univ Fla, PhD(chem), 70. *Prof Exp:* Res asst clin biochem & toxicol, Clin Res Labs, Sch Med, Univ Fla, 67-69; NIH res assoc metalloenzym, Dept Molecular Biophys & Biochem, Yale Univ, 69-71; ASST PROF METAL BIOCHEM & PATH, UNIV NC, CHAPEL HILL, 71- *Concurrent Pos:* Sr mem, Nat Inst Environ Health Sci proj prog heavy metal path, Univ NC, Chapel Hill, 71-, Environ Protection Agency & Inter-univ Consortium Environ Studies grants, 71-; consult, Nat Inst Environ Health Studies, 71-; mem & consult, Inter-univ Consortium Environ Studies, 71- *Mem:* AAAS; Am Chem Soc; Sigma Xi. *Res:* Metalloenzymology; trace metal analysis; metabolism and biochemical effects of metal chelating agents; heavy metal toxicology; organometallic chemistry of the nickel triad metals. *Mailing Add:* 811 Onslow St Durham NC 27705

MUSHER, DANIEL MICHAEL, b New York, NY, Feb 27, 38; m 67; c 3. MEDICINE, INFECTIOUS DISEASES. *Educ:* Harvard Univ, BA, 59; Columbia Univ, MD, 63. *Prof Exp:* Intern resident med, Columbia Div, Bellevue Hosp, New York, 63-65; chief internal med, USAF Hosp, Laredo AFB, Tex, 65-67; resident NIH trainee infectious dis, Tufts-New Eng Med Ctr, 67-71; from asst prof to assoc prof, 71-76, PROF MED MICROBIOL & IMMUNOL, BAYLOR COL MED, 76- *Concurrent Pos:* Chief infectious dis, Houston VA Med Ctr, 71-; assoc ed, J Infect Dis, 84-88. *Mem:* Am Asn Immunologists; Am Fedn Clin Res; Am Soc Clin Invest; Am Soc Microbiol; Soc Exp Biol & Med. *Res:* Infectious diseases and host response; immunologic aspects of syphilis; infections due to pneumococci, haemophilus, staph aureus, osteomyelitis. *Mailing Add:* Infectious Dis Sect Vet Admin Hosp Houston TX 77030

MUSHETT, CHARLES WILBUR, b Elizabeth, NJ, Apr 1, 14; m 39. PATHOLOGY. *Educ:* NY Univ, AB, 39, MS, 41, PhD(vert morphol), 44. *Prof Exp:* Technician, Merck Inst Therapeut Res, Merck & Co Inc, 33-35, lab asst, 35-37, sr worker, Bact & Path Dept, 37-40, assoc hemat & path, 40-43, from assoc head to head dept path, 43-56, from asst dir to dir sci rels, Merck Sharp & Dohme Res Labs, 57-66, dir int sci rels, 66-70, dir sci indust liaison, Merck Sharp & Dohme Res Labs, 70-79; CONSULT, 79- *Concurrent Pos:* Merck foreign fel, Denmark & Ger, 52-53. *Mem:* AAAS; Endocrine Soc; Am Soc Exp Path; fel NY Acad Sci; fel Int Soc Hemat. *Res:* Experimental animal pathology and hematology in relation to nutrition, infection and toxicology of drugs; blood coagulation and anticoagulants. *Mailing Add:* 82 Parkway Dr Clark NJ 07066

MUSHINSKI, J FREDERIC, b New Brighton, Pa, Mar 18, 38; m 71. MOLECULAR GENETICS, CANCER. *Educ:* Yale Univ, BA, 59; Harvard Med Sch, MD, 63. *Prof Exp:* Intern med, Med Ctr, Duke Univ, 63-64; res assoc biochem, 65-70, sr investr, Lab Cell Biol, 70-88, CHIEF, ONCOGENE SECT, LAB GENETICS, NAT CANCER INST, 88- *Concurrent Pos:* USPHS fel, Res Training Prog, Med Ctr, Duke Univ, 64-65; William O Moseley traveling fel from Harvard Univ, Max Planck Inst Exp Med, 69-70; assoc ed, J Immunol, 87- *Mem:* Am Asn Cancer Res; AAAS; Am Soc Biol Chemists; Am Asn Immunologists. *Res:* Molecular biology of cancer; immunology; messenger RNA; genomic DNA; oncogenes. *Mailing Add:* Lab of Genetics Nat Cancer Inst Bldg 37 Rm 2B26 NIH Bethesda MD 20892

MUSHOTZKY, RICHARD FRED, b New York, NY, June 18, 47; m 81. ASTROPHYSICS. *Educ:* Mass Inst Technol, BS, 68; Univ Calif, San Diego, MS, 70, PhD(physics), 76. *Prof Exp:* Res assoc physics, Univ Calif, San Diego, 69-76, res fel, 76-77; Nat Res Coun assoc, 77-79, ASTROPHYSICIST, GODDARD SPACE FLIGHT CTR, NASA, 79- *Concurrent Pos:* Sci working group interdisciplinary scientist, Advan X-Ray Astrophys Facil, 85- *Honors & Awards:* NASA Medal Except Sci Achievement, 84; John C Lindsay Mem Award, 85. *Mem:* Am Astron Soc. *Res:* X-ray astronomy concentrating on x-ray observations; clusters of galaxies, seyfert galaxies, BL Lac objects. *Mailing Add:* Goddard Space Flight Ctr 666 Greenbelt MD 20771

MUSIC, JOHN FARRIS, b Childress, Tex, Oct 5, 21; m 42; c 2. PHYSICAL CHEMISTRY, PHYSICS. *Educ:* Univ Tex, BA, 46, PhD(phys chem), 51. *Prof Exp:* Res scientist, Gen Elec Co, 51-52, supvr graphite & mat develop, 52-54, process tech, 54-56, mgr, 56-60, proj analyst, 60-61, consult analyst, 61-65, mgr, Div Analysis & Planning, 65-68, mgr aerospace analysis & planning, Aerospace Group, Valley Forge Space Technol Ctr, 68-69, mgr, Group Planning Oper, Info Systs Group, 69-71; PRES, MACRO OPERS, 71- *Concurrent Pos:* Pres, Performance Consult Group, Ltd. *Mem:* Am Chem Soc; Am Phys Soc; Inst Mgt Sci. *Res:* Properties of matter; chemicals, materials, nuclear energy, aerospace, computers, economics; management of research and development; coupling of science and technology to business; leadership and management of the enterprise; mathematics. *Mailing Add:* 239 E King St Lancaster PA 17602

MUSICK, GERALD JOE, b Ponca City, Okla, May 24, 40; m 62; c 2. ENTOMOLOGY, INSECT PEST MANAGEMENT. *Educ:* Okla State Univ, BS, 62; Iowa State Univ, MS, 64; Univ Mo-Columbia, PhD(entom), 69. *Prof Exp:* Asst entom, Iowa State Univ, 62-64; instr, Univ Mo-Columbia, 64-69; from asst prof to assoc prof entom, Ohio Agr Res & Develop Ctr, 69-76; assoc prof & head dept entom & fisheries, Univ Ga, Coastal Plain Exp Sta, Tifton, 76-79; prof & head dept entom, 79-87, DEAN & DIR, COL AGR & HOME ECON, ARK EXP STA, UNIV ARK, FAYETTEVILLE, 87- *Mem:* Sigma Xi; Entom Soc Am. *Res:* Biology and control of insect pests of corn. *Mailing Add:* 2229 Golden Oaks Dr Fayetteville AR 72703

MUSICK, JACK T(HOMPSON), b Cleveland, Va, Oct 6, 27; m 55; c 3. AGRICULTURAL ENGINEERING. *Educ:* Va Polytech Inst, BS, 53; Okla State Univ, MS, 55. *Prof Exp:* Res agr engr, 55-65, res agr engr & dir, 66-67, agr engr, Southwestern Great Plains Res Ctr, 68-80, AGR ENGR & RES LDR, CONSERV & PROD RES LAB, AGR RES SERV, USDA, 80- *Mem:* Am Soc Agr Engrs; Am Soc Agron; Soil Conserv Soc Am. *Res:* Irrigation water management research in the southwestern Great Plains. *Mailing Add:* 5502 Floyd Ave Amarillo TX 79106

MUSICK, JAMES R, b Mendota, Ill, Mar 24, 46; div. NEUROPHYSIOLOGY, NEUROCHEMISTRY. *Educ:* Northwestern Univ, BA, 68, PhD(biol), 75. *Prof Exp:* Res assoc, 73-74, res instr, 74-75, instr, Dept Physiol, 75-80, ASST PROF PHYSIOL, UNIV UTAH COL MED, 80- *Concurrent Pos:* Fel, Muscular Dystrophy Asn, 75-76; investr, Marine Biol Lab, Woods Hole, Mass, 76. *Mem:* AAAS; Soc Neurosci; NY Acad Sci; Sigma Xi. *Res:* Mechanisms of short and long term regulation of synaptic transmission; neurotrophism; mechanisms of action of toxins. *Mailing Add:* 908 S 11th No 3 Salt Lake City UT 84105

MUSICK, JOHN A, b Jan 12, 41; div; c 2. ICHTHYOLOGY, ECOLOGY. *Educ:* Rutgers Univ, AB, 62; Harvard Univ, MA, 64, PhD, 69. *Prof Exp:* Fisheries biologist, US Fish & Wildlife Serv, 62; teaching fel comp anat & gen biol, Harvard Univ, 62-63, anthrop, 63-64, ichthyol, 65 & 67; assoc prof, Univ Va, 67-81; from asst prof to assoc prof, 67-81, PROF MARINE SCI, COL WILLIAM & MARY, UNIV VA, 81- *Concurrent Pos:* Assoc marine scientist, Va Inst Marine Sci, 67-80, sr marine scientist, 80-83, head, Dept Ichthyol, 80-82 & Vert Ecol & Syst Sect, 82-; sci collabr, Capes Hatteras & Lookout Nat Seashores, US Park Serv; mem, Adv Comt Vertebrates, Smithsonian Inst Oceanog Sorting Ctr, 77-, Rev Panel Biol Oceanog, Nat Sci Found & Rev Panel Oceanog Facil Support, 77; chair, Sci & Statist Comt for Summer Flounder, Atlantic States Marine Fisheries Comn, 74-84; mem, Comt Shark Mgt, Mid-Atlantic Fishery Mgt Coun, 86-; mem, Sea Turtle Working Group, Int Union Conserv of Nature, 90- *Mem:* Fel AAAS; Am Soc Ichthyologists & Herpetologists; Ecol Soc Am; Am Fisheries Soc; Soc Cons Biol; fel Explorers Club. *Res:* Community ecology of demersal fishes; structure and function of deep-sea ecosystems; systematics and zoogeography of fishes and reptiles; sea turtle ecology and conservation; elasmobranch biology. *Mailing Add:* Va Inst Marine Sci Gloucester Point VA 23062

MUSIEK, FRANK EDWARD, b Union City, Pa, July 4, 47; m 72; c 1. AUDITORY NEUROPHYSIOLOGY, AUDITORY PSYCHOPHYSICS. *Educ:* Edinborough Col, BS, 68; Kent State Univ, MA, 71; Case Western Reserve Univ, PhD(audiol), 75. *Prof Exp:* ASSOC PROF AUDIOL, DARTMOUTH-HITCHCOCK MED CTR, DARTMOUTH COL, 75- *Concurrent Pos:* Adj asst prof psychol, Dartmouth Col, 77- *Mem:* Acoust Soc Am; Am Speech, Language & Hearing Asn; Soc Neurosci; Deafness Res Found. *Res:* Auditory neurophysiology and its clinical application to neuroaudiology; electrophysiological measures and psychophysical measures to evaluate dysfunction of the higher auditory system; vestibular physiology, particularly electronystagmography; development of various test procedures to monitor higher auditory function. *Mailing Add:* Dept Otolaryngol-Audiol Dartmouth-Hitchcock Med Ctr Hanover NH 03755

MUSINSKI, DONALD LOUIS, b Winsted, Conn, Mar 29, 46; m 71; c 1. PHYSICS, ENGINEERING. *Educ:* Trinity Col, BS, 68; Univ Rochester, MA, 70, PhD(physics), 73. *Prof Exp:* Fel, Dept Appl Physics, Cornell Univ, 73-75; sr res scientist physics & eng, 75-79, mgr, 79-80, SR MGR TARGET & CRYOG TECHNOL DEVELOP, DEPT MATS & TARGET TECHNOL, KMS FUSION INC, 80- *Mem:* Am Phys Soc; Am Vacuum Soc; Prod Develop & Mgt Asn. *Res:* Target technology for inertial confinement fusion experiments. *Mailing Add:* Dept Mat & Target Technol KMS Fusion Inc 3621 S State Rd Park Dr Boc 1778 Ann Arbor MI 48106

MUSKA, CARL FRANK, b Milwaukee, Wis, Feb 3, 48; m 70. AQUATIC TOXICOLOGY, PHYSIOLOGICAL ECOLOGY. *Educ:* Univ Tex, Austin, BA, 70; Tex A&M Univ, MS, 73; Ore State Univ, PhD(fishery biol), 77. *Prof Exp:* Staff res asst toxicol, Ore State Univ, 76-77; RES TOXICOLOGIST, HASKELL LAB TOXICOL & INDUST MED, E I DU PONT DE NEMOURS & CO, 77- *Concurrent Pos:* Mem comt environ, US Chamber Com, 80- *Mem:* Am Fisheries Soc; Am Soc Testing & Mat. *Res:* Interactive effects of toxicants and environmental parameters on biological systems; growth and bioenergetics of aquatic organisms; comparative vertebrate pharmacology; use of aquatic organisms as animal models for carcinogenicity research. *Mailing Add:* Savannah River Plant Bldg 773-11A Aiken SC 29808

MUSKAT, JOSEPH BARUCH, b Marietta, Ohio, Sept 20, 35; m 59; c 3. NUMBER THEORY, COMPUTER SCIENCE. *Educ:* Yale Univ, AB, 55; Mass Inst Technol, SM, 56, PhD(math), 61. *Prof Exp:* From asst prof to assoc prof math, Univ Pittsburgh, 61-69; vis assoc prof, 69-70, ASSOC PROF MATH, BAR-ILAN UNIV, ISRAEL, 70- *Concurrent Pos:* res assoc comput, Univ Pittsburgh, 61-69; NSF fel, 63-69; chmn dept math, Bar-Ilan Univ, 71-74; vis assoc prof, Univ Ill, 76-77, Univ Pittsburgh, 83-84. *Mem:* Am Math Soc; Math Asn Am; Asn Comput Mach. *Res:* Reciprocity laws; cyclotomy; use of computers in number theory. *Mailing Add:* Dept Math & Comp Sci Bar-Ilan Univ Ramat-Gan 52900 Israel

MUSKAT, MORRIS, b Riga, Latvia, Apr 21, 06; US citizen; wid; c 3. PETROLEUM. *Educ:* Ohio State Univ, BA & MA, 26, Calif Inst Technol, PhD(physics), 29. *Prof Exp:* Instr physics & chem, Bowling Green Univ, 26-27; physicist & dir physics, Gulf Res, Gulf Oil Corp, 29-50, tech coordr, 50-61, tech adv, 61-71; RETIRED. *Concurrent Pos:* Chmn, Comt Petrol Reserves, Am Petrol Inst, 55-71. *Honors & Awards:* Lucas Medal, Am Inst Mining, Metall & Petrol Engrs, 53; L C Uren Award, Soc Petrol Engrs, 69. *Mem:* Nat Acad Eng; AAAS; Am Phys Soc; hon mem Am Inst Mining, Metall & Petrol Engrs; fel NY Acad Sci; Soc Petrol Engrs. *Res:* Reservoir engineering; geophysics; lubrication; author and co-author of more than 85 papers. *Mailing Add:* 1000 San Pasqual Apt 20 Pasadena CA 91106

MUSKATT, HERMAN S, b New York City, NY, May 2, 31. PHYSICAL STATIGRAPHY. *Educ:* City Col NY, BA, 55, Syracuse Univ, MS, 63, PhD(geol), 69. *Prof Exp:* Instr geol, City College NY, 55-60; PROF GEOL, SYRACUSE UNIV, 64- *Concurrent Pos:* Consult geol, 55- *Honors & Awards:* Ward Medal, City Col NY, 55. *Mem:* Fel Geol Soc Am; Soc Econ Paleontologist & Minerologist; Am Asn Petrol Geologist. *Mailing Add:* Dept Geol Utica Col Syracuse Univ Utica NY 13502

MUSKER, WARREN KENNETH, b Chicago, Ill, Apr 17, 34; c 2. INORGANIC CHEMISTRY. *Educ:* Bradley Univ, BS, 55; Univ Ill, PhD(org chem), 59. *Prof Exp:* Asst boron hydrides, Univ Mich, 61-62; from asst prof to assoc prof, 62-70, PROF INORG CHEM, UNIV CALIF, DAVIS, 75- *Concurrent Pos:* Alexander von Humboldt fel, 70-71. *Mem:* Am Chem Soc. *Res:* Influence of ligand structure on the stereochemistry and reactivity of metal complexes; copper II oxidations; medium ring complexes of transition metals; thioether cation radicals and dications; organosulfur chemistry. *Mailing Add:* Dept of Chem Univ of Calif Davis CA 95616

MUSS, DANIEL R, b Birmingham, Ala, Apr 5, 28; m 65; c 2. PHYSICS. *Educ:* Mass Inst Technol, BS, 48; Calif Inst Technol, MS, 53; Univ Pittsburgh, PhD(physics), 61. *Prof Exp:* Sr physicist, Westinghouse Res Labs, 48-64, mgr silicon device develop, 64-69, dir solid state res, 72-75, res dir, Pub Systs Co, 75-77, mgr solid state res & develop div, 77-81, mgr appl sci div, 81-88; RETIRED. *Concurrent Pos:* Assoc ed, Inst Elec & Electronics Engrs Transactions on Electron Devices. *Mem:* Am Phys Soc; sr mem Inst Elec & Electronics Engrs. *Res:* Defect structure of metals; semiconductors; semiconductor devices. *Mailing Add:* 5619 Marlborgh Rd Pittsburgh PA 15235

MUSSELL, HARRY W, b Paterson, NJ, Nov 10, 41; m 64. PLANT PATHOLOGY, PLANT BIOCHEMISTRY. *Educ:* Drew Univ, AB, 65; Duke Univ, MF, 65; Purdue Univ, PhD(bot), 68. *Prof Exp:* PLANT PATHOLOGIST, BOYCE THOMPSON INST PLANT RES, INC, 68- *Honors & Awards:* Ciba Sci Res Award, 62. *Mem:* AAAS; Am Phytopath Soc; Am Soc Plant Physiol; Bot Soc Am; Am Inst Biol Sci; Sigma Xi. *Res:* Physiology of parasitism, particularly enzymology of pathogenesis in plants; disease tolerance. *Mailing Add:* Boyce Thompson Inst for Plant Res Tower Rd Cornell Univ Ithaca NY 14853

MUSSELMAN, NELSON PAGE, b Luray, Va, Mar 20, 17. TOXICOLOGY, ANALYTICAL CHEMISTRY. *Educ:* Western Md Col, AB, 38. *Prof Exp:* Chemist, Armco Steel Corp, 51-54; res chemist, US Army Edgewood Arsenal, 54-73; environ health asst, Air Mgt Admin, State of Md, 75-87; RETIRED. *Mem:* Fel AAAS; Am Chem Soc. *Res:* Vapor and carbon monoxide toxicity; chemical analysis and methods; evaluation of chemical protective devices; air pollution. *Mailing Add:* 1101 N Calvert St Baltimore MD 21202

MUSSEN, ERIC CARNES, b Schenectady, NY, May 12, 44; m 69; c 2. APICULTURE, INSECT PATHOLOGY. *Educ:* Univ Mass, BS, 66; Univ Minn, MS, 69, PhD(entomol), 75. *Prof Exp:* EXTEN APICULTURIST, UNIV CALIF, 76- *Mem:* Entomol Soc Am; Sigma Xi. *Res:* Honey bee diseases; crop pollination; judicious use of bee-toxic insecticides. *Mailing Add:* Entomol Exten Univ Calif Davis CA 95616

MUSSER, DAVID MUSSELMAN, b Bowmansville, Pa, Apr 30, 09; m 38; c 2. ORGANIC CHEMISTRY. *Educ:* Pa State Col, BS, 31; Ga Inst Technol, MS, 33; Univ Wis, PhD(org chem), 37. *Prof Exp:* Indust fel, Mellon Inst, 37-42; res chemist, Pac Mills, NJ, 42-46; sr chemist, Deering Milliken Res Trust, Conn, 46-47; head textile res & develop, Onyx Oil & Chem Co, 47-52; dir res, Refined Prods Corp, 52-62; dir res, Raytex Chem Corp, 62-85; CONSULT, 85- *Mem:* Am Chem Soc; Am Asn Textile Chemists & Colorists. *Res:* Cellulose; textile finishing agents. *Mailing Add:* 821 Lawrence Dr Emmaus PA 18049

MUSSER, DAVID REA, b Sherman, Tex, Aug 24, 44; m 66; c 2. PROGRAMMING METHODOLOGY, SOFTWARE SPECIFICATION & VERIFICATION. *Educ:* Austin Col, BA, 66; Univ Wis, MA, 68, PhD(comput sci), 71. *Prof Exp:* Asst prof comput sci, Univ Tex, Austin, 70-73; vis asst prof comput sci & res staff mem, Math Res Ctr, Univ Wis, Madison, 73-74; res staff mem comput sci, Info Sci Inst, Univ Southern Calif, 74-79; comput scientist, Gen Elec Res & Develop Ctr, Schenectady, NY, 79-87; RES PROF COMPUT SCI, RENSSELAER POLYTECH INST, 88- *Mem:* Asn Comput Mach. *Res:* Generic software libraries and program verification; formal specification of programs; automatic theorem proving; computer algebra; analysis of algorithms. *Mailing Add:* Comput Sci Dept Rensselaer Polytech Inst Amos Eaton Hall Troy NY 12180

MUSSER, JOHN H, b Sharon, Pa, Feb 2, 49; m 76; c 2. ETHICAL DRUG DISCOVERY & DEVELOPMENT, BIOTECHNOLOGY ENTREPRENEURAL ENVIRONMENT. *Educ:* Fla State Univ, BA, 72; Univ Calif, Santa Cruz, PhD(chem), 76. *Prof Exp:* Chemist res, Upjohn Co, 76-78; group leader res, Rorer Rhome-Poulenc, Revlon Health Care, 78-82; mgr res, Wyeth, 82-88, dir res, Wyeth-Ayerst, 88-91; CHIEF SCIENTIST RES, GLYCOMED, 91- *Concurrent Pos:* Ed, Med Chem Res, Birkhauser, 91- *Mem:* Am Chem Soc; NY Acad Sci; AAAS. *Res:* Design, synthesis, submission of patents, analysis, pharmacological evaluation, GMP production and formulation of new drug candidates; therapeutic areas include inflammation, CV, CNS and viral. *Mailing Add:* Glycomed Inc 860 Atlantic Ave Alameda CA 94501

MUSSER, MICHAEL TUTTLE, b Williamsport, Pa, Jan 31, 42; m 66; c 3. ORGANIC CHEMISTRY. *Educ:* Purdue Univ, BS, 63, PhD(org chem), 68. *Prof Exp:* Res chemist, Intermediates Div, Plastics Dept, Exp Sta, E I Du Pont De Nemours & Co, 67-72 & Nylon Intermediates Div, Polymer Intermediates Dept, 72-74, sr res chemist, 74-80, RES ASSOC, SABINE RIVER WORKS, E I DU PONT DE NEMOURS & CO, INC, TEX, 80- *Mem:* Am Chem Soc; Sigma Xi. *Res:* Homogeneous catalysis as a route to organic intermediates. *Mailing Add:* 1529 Lindenwood Dr Orange TX 77630

MUSSINAN, CYNTHIA JUNE, b Elizabeth, NJ, Dec 23, 46. ANALYTICAL CHEMISTRY. *Educ:* Georgian Court Col, BA, 68; Rutgers Univ, MS, 75. *Prof Exp:* Sr chemist, 68-76, res chemist, 76-83, proj chemist, 83-86, DIR, INT FLAVORS & FRAGRANCES, INC, 87- *Mem:* Am Chem Soc; Analytical Lab Mgr Asn. *Res:* Isolation, identification and synthesis of the volatile and nonvolatile flavor constituents of foods. *Mailing Add:* Int Flavors & Fragrances 1515 Hwy 36 Union Beach NJ 07735

MUSSON, ALFRED LYMAN, animal science; deceased, see previous edition for last biography

MUSSON, ROBERT A, INFLAMMATION, MACROPHAGES. *Educ:* Univ Conn, PhD(biomed sci), 76. *Prof Exp:* Personnel mgr, Environ Response, Inc, 85-86; dir prod develop, Response Technol Inc, 86-90; SR SCI ADV, OFF PROG PLANNING & EVAL, NAT HEART, LUNG & BLOOD INST, 90- *Res:* Cell biology. *Mailing Add:* Off Prog Planning & Eval Nat Heart Lung & Blood Inst Bldg 31 Rm 5A03 Bethesda MD 20892

MUSTACCHI, HENRY, b Alexandria, Egypt, Oct 2, 30; US citizen; m 54; c 2. DIAZONIUM SALTS, DIAZOTYPE MATERIALS. *Educ:* Strasbourg Univ, BSc, 53, PhD(phys chem), 58. *Prof Exp:* Chief chemist appl chem, Hall Harding Ltd, 61-66; mgr res & develop appl chem, GAF Ltd, 66-77; dir res & develop appl chem, 77-86, VPRES, ANDREWS PAPER & CHEM CO, 86- *Concurrent Pos:* Chmn Drawing Off Mat Mfrs Tech Comt, 74-76; mem, Asn Reprographic Mats Tech Comt, 78- *Mem:* Am Chem Soc. *Res:* Design and synthesize new light sensitive molecules and other chemicals using in the diazotype reprographic field; develop new reprographic processes. *Mailing Add:* Andrews Paper Chem Co Inc PO Box 509 One Channel Dr Port Washington NY 11050

MUSTACCHI, PIERO OMAR DE ZAMALEK, b Cairo, Egypt, May 29, 20; nat US; m 48; c 2. MEDICINE. *Educ:* Italian Lyceum, Cairo, BS, 37; Fuad First Univ, Cairo, MB, ChB, 44. *Hon Degrees:* Dr, Univ Marseille, France. *Prof Exp:* Asst resident path, Med Sch, Univ Calif, 49-51; clin instr med, Univ Calif, San Francisco, 53-58, from clin asst prof to clin prof med & prev med, 58-90, consult, Hemat Clin, 54-58, asst dir continuing educ med & health sci, 64-69, assoc dir, 69-74, vchmn, Dept Prev Med, 65-66, assoc dir, Spec Serv Exten Prog Med Educ, 74-75, CLIN PROF MED & EPIDEMIOL, MED SCH, UNIV CALIF, SAN FRANCISCO, 90- *Concurrent Pos:* Res fel, Am Cancer Soc, 49-52; fel, Sloan-Kettering Inst, 51-53; vis instr, Fac Med, Cairo Univ, 50; resident, Mem Hosp Cancer & Allied Dis, New York, 51-53; physician-in-chg, Hemat & Lymphoma Clin, St Mary's Hosp, San Francisco, 54-56, consult, 68; mem consult tumor bd, Children's Hosp, San Francisco, 55-56, consult, 58-, head off epidemiol & biomet, 60-63; consult, staff, Franklin Hosp, 70-85; physician, Ital Consulate, San Francisco, hon vconsul Italy, 71-90; chmn, Comt Continuing Educ, Children's Hosp, 72-78; med consult, Work Clin, Univ San Francisco Med Ctr, 74-75; head, Occup Epidemiol, 80-90 & Int Health Educ, Dept Epidemiol Univ Calif, San Francisco, 86-90. *Mem:* AAAS; Am Soc Clin Invest; AMA; fel Am Col Physicians; Am Soc Environ & Occup Health. *Res:* Epidemiology of cancer; general ecology and education. *Mailing Add:* 3838 California San Francisco CA 94118

MUSTAFA, MOHAMMED G, b Bangladesh, Nov 3, 41; US citizen; m 68; c 3. NUCLEAR PHYSICS. *Educ:* Univ Dacca, BSc, 62; Yale Univ, MS, 67, PhD(physics), 70. *Prof Exp:* Res assoc physics, Oak Ridge Nat Lab, 70-73, Univ Md & Goddard Space Flight Ctr, 73-77; PHYSICIST, LAWRENCE LIVERMORE LAB, 77- *Concurrent Pos:* Nat Res coun fel, 75-77. *Mem:* Am Phys Soc; fel Int Atomic Energy Agency. *Res:* Theoretical studies of the physics and chemistry of nuclear fission and heavy-ion collisions. *Mailing Add:* Dir Env & Occup Health Sci Sch Pub Health Univ Calif Los Angeles CA 90024

MUSTAFA, SHAMS, b Karachi, Pakistan, Oct 8, 52; US citizen; m 82; c 2. FOOD CHEMISTRY, PETROLEUM & PETROCHEMICALS. *Educ:* Punjab Univ, BS, 69, MS, 71. *Prof Exp:* Sr chemist-in-chg, Assoc Consult Engrs, Ltd, 72-77; chemist, Am Standards Testing Bur, New York, 77-78; sr chemist-in-chg grain & veg oil lab, 79-88, CHIEF CHEMIST, AGR LAB, CALEB BRETT INC, 88-; MGR, AGR, EW SAYBOLT & CO INC, 89- *Honors & Awards:* Smalley Award, Am Oil Chemists Soc, 87 & 88, Doughtie Award. *Mem:* Am Oil Chemists Soc; Am Chem Soc; fel Am Inst Chemists; NY Acad Sci; AAAS; Asn Off Analytical Chemists; Int Union Pure & Apl Chem; Am Mgt Asn; Am Mkt Asn; Royal Soc of Chem. *Res:* Saline corrosion of mild steel and its prevention. *Mailing Add:* 1417 Meeker Loop LaPlace LA 70068

MUSTAFA, SYED JAMAL, b Lucknow, India, July 10, 46; m 73; c 3. MEDICAL RESEARCH. *Educ:* Lucknow Univ, BS, 62, MS, 65, PhD(biochem), 70. *Prof Exp:* Fel, Indust Toxicol Res Ctr, Lucknow Res Ctr, Lucknow, India, 70-71, Dept Physiol, Univ Va, 71-74; from asst prof to assoc prof pharmacol, Col Med, Univ S ALA, 74-80; assoc prof, 80-83, PROF, DEPT PHARMACOL, MED SCH, E CAROLINA UNIV, GREENVILLE, NC, 83- *Concurrent Pos:* Fel, Coun Sci & Indust Res & Indian Coun Med Res, New Delhi, India, 70-71, NIH, 71-74. *Mem:* Am Physiol Soc; Sigma Xi; NY Acad Sci; Am Heart Asn; Int Soc Heart Res; Am Soc Pharmacol & Exp Ther; fel Am Col Clin Pharmacol. *Res:* Field of cardiology and pharmacology; the study of the relationship between vasoactive agents and blood flow. *Mailing Add:* Dept Pharmacol Med Sch E Carolina Univ Greenville NC 27834

MUSTARD, JAMES FRASER, b Toronto, Ont, Oct 16, 27; m 52; c 6. PATHOLOGY. *Educ:* Univ Toronto, MD, 53; Cambridge Univ, PhD, 56; FRCP, 65. *Hon Degrees:* DSc, Univ Western Ont, 83, McGill, 86, Toronto, 88, McMaster, 90, BC, 90; LLD, Calgary, 87; DU, Ottawa, 87. *Prof Exp:* Can Heart Found sr res assoc med, Univ Toronto, 58-63, from asst prof to assoc prof path, 63-66, assoc med, 63-66, asst prof med, 65; chmn dept path, fac med, McMaster Univ, 66-72, prof path, 66-88, dean fac health sci, 75-80, vpres health sci, 80-82, EMER PROF PATH, MCMASTER UNIV, 88-; PRES & SR FEL, CAN INST ADVAN RES, 82- *Concurrent Pos:* Fel, Coun

Arteriosclerosis, Am Heart Asn, 60-65, Coun Thrombosis, 71-; mem, Int Comt Haemostasis & Thrombosis, Am Heart Asn, 66-, sr adv comt, 71-, Int Soc Thrombosis & Haemostasis, 69-, coun mem, 69-75, pres, 79-81; chmn, Health Res Comt, Ont Coun Health, 66-73, chmn, Task Force Health Planning, Ont Coun Health, 73-74; chmn, Med Adv Comt, Can Heart Found, 72-76, dir, 71-; mem, Expert Adv Panel Cardiovasc Dis, WHO, 72; chmn, Ont Adv Coun Occup Health & Occup Safety, 77-83; mem Royal comn matters health & safety arising from use of asbestos in Ont, 80-83; chmn task force med manpower, Coun Ont Univs, 82-83 & task force rev primary care, Ont Ministry Health & Govt Ont, 82-83; mem, Med Adv Comt, Alcoholic Beverage Med Res Fedn, 82-91, bd trustees, 89-; mem bd dir, Strategic Planning Comt, 85-, Steel Co Can Inc, Toronto, 90; mem, Exec Mgt & Compensation Comt, 91-, Atomic Energy Can, Ltd, Ottawa, Premier's Econ Coun, Ont, 86-, Premier's on Health Strategy, 88-, Prime Minister's Adv Bd, Sci & Technol, Ottawa, 87-, vchmn, 88-; chmn bd dir, Ont Workers Compensation Inst, Toronto, 90- *Honors & Awards:* Gairdner Found Int Award, 67; Izaak Walton Killam Prize, 87; Robert P Grant Award, Cong Thrombosis & Haemostasis, 87; Distinguished Career Award, Int Soc Thrombosis & Haemostasis. *Mem:* Fel Royal Soc Can; Am Soc Hemat (secy, 64-67, pres, 70); fel Am Asn Physicians; Can Soc Clin Invest (pres, 65-66); Am Soc Exp Path. *Res:* Blood and vascular disease. *Mailing Add:* Can Inst Advan Res 179 John St Suite 701 Toronto ON M5T 1X4 Can

MUSTARD, MARGARET JEAN, b Bayfield, Ont, Feb 18, 20; nat US. BOTANY, HORTICULTURE. *Educ:* Univ Miami, BS, 42, MS, 50; Ohio State Univ, PhD(hort), 58. *Prof Exp:* Lab technician, Sub-trop Exp Sta, Univ Fla, 42-45; instr hort, 45-50, from asst prof to assoc prof, 50-68, prof, 68-85, EMER PROF TROP BOT, UNIV MIAMI, 85- *Mem:* AAAS; Am Soc Hort Sci; Bot Soc Am; Sigma Xi. *Res:* Anatomical and morphological aspects of botany; fruit setting in horticultural plants. *Mailing Add:* 1204 Placetas Coral Gables FL 33146

MUSTER, DOUGLAS FREDERICK, b Milwaukee, Wis, Nov 2, 18; m 44; c 5. APPLIED MECHANICS. *Educ:* Marquette Univ, BS, 40; Ill Inst Technol, MS, 49, PhD(appl mech), 55. *Prof Exp:* Asst mech, Ill Inst Technol, 46-48, instr, 48-50, asst prof mech eng, 50-53; vibrations engr, Gen Eng Lab, Gen Elec Co, 53-60, mech systs engr, 60-61; chmn dept mech eng, Univ Houston, 62-72, dir, Off Eng Pract Progs, 77-79, prof mech eng, 61-89, Brown & Root prof eng, 67-89, EMER PROF, UNIV HOUSTON, 89- *Concurrent Pos:* Pres tech comt 108, mech vibration & shock, Int Standards Orgn, 67-83; consult forensic eng. *Mem:* Am Soc Eng Educ; fel Am Soc Mech Engrs; fel Inst Mech Engrs; fel Acoust Soc Am; fel Inst Acoust; fel AAAS; fel Nat Acad Forensic Engrs; fel Inst Engrs Australia. *Res:* Vibration; structure borne sound; rotor dynamics; underwater acoustics; design methods. *Mailing Add:* Dept Mech Eng Univ Houston Houston TX 77004

MUT, STUART CREIGHTON, b Dallas, Tex, July 27, 24; m 47; c 5. GEOPHYSICS. *Educ:* Rice Inst, BS, 47, MS, 48. *Prof Exp:* Asst physicist, Atlantic Refining Co, 48-49, admin asst, Res Admin, 49-51, sr physicist, 51-56, supvry physicist, 56-59, dir res & develop, Explor Sect, 59-61, mgr, Eng Div, Producing Dept, 61-63, Eastern Dist, 63-66, vpres eastern region, NAm Producing Div, 66-81, SR VPRES RES & ENG, ATLANTIC RICHFIELD CO, 81-; SR VPRES, ARCO RES TECHNOL. *Mem:* Soc Explor Geophysicists; Soc Petrol Engrs; Inst Elec & Electronics Engrs. *Res:* Petroleum exploration geophysics; petroleum engineering; geology. *Mailing Add:* 4818 Heather Brook Dr Dallas TX 75243

MUTCH, GEORGE WILLIAM, b Ann Arbor, Mich, June 22, 43; m 66. CHEMICAL KINETICS, PHYSICAL CHEMISTRY. *Educ:* Andrews Univ, BA, 66; Univ Calif, Davis, PhD(chem), 73. *Prof Exp:* Teaching asst chem, Univ Calif, Davis, 67-70, res asst, 70-73; from asst prof to assoc prof, 73-88, PROF CHEM, ANDREWS UNIV, 88- *Mem:* Am Chem Soc; Am Phys Soc; Sigma Xi. *Res:* Chemical dynamics of high energy unimolecular decomposition processes. *Mailing Add:* 10789 Garr Rd Berrien Springs MI 49104

MUTCH, PATRICIA BLACK, b Alexandria, La, Sept 5, 43; m 66. PUBLIC HEALTH & EPIDEMIOLOGY, RESEARCH. *Educ:* Andrews Univ, BS, 65; Loma Linda Univ, dietetic internship cert, 66; Univ Calif, Davis, PhD(nutrit), 72. *Prof Exp:* Therapeut dietitian, Hinsdale Sanitarium & Hosp, 66-67; res asst nutrit, Univ Calif, Davis, 67-72; asst prof home econ, Andrews Univ, 72-76, coordr health maj, 75-77, dir dietetic educ, 73-84, dir off res, 87-90, PROF NUTRIT, ANDREWS UNIV, 76-, DIR, INST ALCOHOLISM/DRUG DEPENDENCE, 84- *Concurrent Pos:* Mem nutrit adv coun, Seventh-day Adventist Church, 75-; comnr, State Mich Nutrit Comn, 77-80. *Mem:* Am Home Econ Asn; Am Dietetic Asn; Seventh-day Adventist Dietetic Asn (pres, 77-78); Soc Nutrit Educ; Am Pub Health Asn. *Res:* Interaction of ethanol and caffeine; methods of dietetic education and practice; epidemiology drug use in church organ; program evaluation; drug prevention methodology. *Mailing Add:* 10789 Garr Rd Berrien Springs MI 49103

MUTCHLER, CALVIN KENDAL, b Oceola, Ohio, Jan 25, 26; m 51; c 6. SOIL EROSION, SEDIMENT YIELD. *Educ:* Ohio State Univ, BAgrEng, 51, MS, 52; Univ Minn, PhD, 70. *Prof Exp:* Instr soil & water struct, Univ Conn, 52-53; tech man, B F Goodrich Co, Ohio, 53-54; agent, Miss, 54-56, agr engr, 56-58, agr engr, Minn, 58-72, HYDRAUL ENGR, SEDIMENTATION LAB, AGR RES SERV, USDA, MISS, 72- *Mem:* Am Soc Agr Engrs; Soil Conserv Soc Am. *Res:* Mechanics of erosion, especially raindrop splash erosion; soil and water conservation; applied research using erosion plots; sediment yield research. *Mailing Add:* 216 Carol Lane Oxford MS 38655

MUTCHLER, GORDON SINCLAIR, b Iowa City, Iowa, Mar 18, 38; m 63; c 1. NUCLEAR PHYSICS, ELEMENTARY PARTICLE PHYSICS. *Educ:* Mass Inst Technol, BS, 60, PhD(physics), 66. *Prof Exp:* Res assoc nuclear physics, Los Alamos Sci Lab, Univ Calif, 66-68; from res assoc to sr res assoc, 68-73, asst prof, 73-76, assoc prof, 76-82, PROF PHYSICS, T W BONNER NUCLEAR LABS, RICE UNIV, 82- *Concurrent Pos:* Sabbatical leave, Sweizerisches Inst Nuclear Forschung, Switzerland, 81-82. *Mem:* Am Phys Soc; Sigma Xi. *Res:* Investigation of nuclear structure and nucleon-nucleon interactions using intermediate energy particles. *Mailing Add:* T W Bonner Nuclear Labs Rice Univ PO Box 1892 Houston TX 77251

MUTCHMOR, JOHN A, b Ft William, Ont, Aug 21, 29; m 55; c 1. INSECT PHYSIOLOGY. *Educ:* Univ Alta, BSc, 50; Univ Minn, MS, 55, PhD(entom), 61. *Prof Exp:* Tech officer entom, Sci Serv Lab, Can Dept Agr, 50-51 & Chatham Entom Lab, 56-61; from asst prof to assoc prof zool & entom, 62-70, PROF ZOOL & ENTOM, IOWA STATE UNIV, 70- *Mem:* Entom Soc Am; Sigma Xi. *Res:* Low temperature adaption of poikilotherms and the influence of temperature and thermal adaptation on their dispersion and physiology. *Mailing Add:* Zool Dept Iowa State Univ Ames IA 50011

MUTEL, ROBERT LUCIEN, b St Albans, NY, June 22, 46; m 70; c 3. RADIO ASTRONOMY. *Educ:* Cornell Univ, AB, 68; Univ Colo, PhD(astrogeophys), 75. *Prof Exp:* Physicist, Environ Sci Serv Admin, Dept of Com, 68-70; from asst prof to assoc prof, 75-85, PROF ASTRON, UNIV IOWA, 85- *Concurrent Pos:* Exec officer, US Very Long Baseline Interferometer Consortium, 86-90; fac scholar, Univ Iowa, 83-85. *Honors & Awards:* Antarctic Serv Metal US, NSF, 73. *Mem:* Am Astron Soc; Royal Astron Soc; AAAS; Int Astron Union. *Res:* Radio astronomy, including very long baseline interferometry, stellar masers, radio binary stars and scintillations. *Mailing Add:* Dept Physics & Astron Univ Iowa Iowa City IA 52242

MUTH, CHESTER WILLIAM, b Antioch, Ohio, May 23, 22; m 49; c 5. ORGANIC CHEMISTRY. *Educ:* Ohio Univ, BS, 43; Ohio State Univ, PhD(chem), 49. *Prof Exp:* Synthetic org chemist, Eastman Kodak Co, 43-44, jr chemist, Tenn Eastman Corp, 44-45; asst, Ohio State Univ, 45-49; from asst prof to prof chem, 49-87, assoc chmn dept, 82-87, EMER PROF CHEM, WVA UNIV, 87- *Mem:* Am Chem Soc. *Res:* Synthetic organic chemistry with emphasis on tertiary amine- N-oxides and redox reactions of nitrobenzyl alcohols. *Mailing Add:* Dept Chem WVa Univ Morgantown WV 26506-6045

MUTH, EGINHARD JOERG, b Beuthen, Germany, Sept 12, 28; US citizen; div; c 4. MICROCOMPUTERS, ROBOTICS. *Educ:* Karlsruhe Tech Univ, Dipl Ing, 51; Polytech Inst Brooklyn, MS, 65, PhD(syst sci), 67. *Prof Exp:* Test engr, Oerlikon Eng Co, Switz, 51-55; develop engr, Allen Bradley Co, 55-56; design engr, Gen Elec Co, 56-63, sr analyst comput, 63-65, consult engr, 65-67, mgr process control, 67-69; assoc prof indust & systs eng, 69-73, PROF INDUST & SYSTS ENG, UNIV FLA, 73- *Concurrent Pos:* Adj lectr, Univ Fla, 65-67 & Polytech Inst Brooklyn, 68; Fulbright scholar, WGermany, 76-77; fac assoc, IBM, 83-84. *Mem:* Inst Elec & Electronics Engrs; Opers Res Soc Am; Inst Indust Engrs; Soc Mfg Engrs. *Res:* Stochastic processes; reliability modeling; industrial applications of microcomputers; robotics. *Mailing Add:* Dept Indust & Systs Eng Univ Fla Gainesville FL 32611

MUTH, ERIC ANTHONY, b Bethesda, Md, July 23, 48; m 74; c 2. PSYCHOPHARMACOLOGY, NEUROCHEMISTRY. *Educ:* Cornell Univ, BA, 73; George Washington Univ, PhD(pharmacol), 81. *Prof Exp:* NIMH biologist, 74-80; supvr neurochem, Wyeth Labs, 81-84, mgr neuropsychopharmacol, 84-87, assoc dir neuropsychopharmacol, Wyeth-Ayerst Res, 87-89, DIR, CNS PHARMACOL, WYETH-AYERST RES, WYETH LABS, 89- *Mem:* AAAS; Soc Neurosci; Am Chem Soc; NY Acad Sci. *Res:* Neurochemical mechanism of action of psychoactive drugs, especially neuroleptics and antidepressants; central nervous system receptor activity and effects on neurotransmitter activity of psychoactive drugs; novel psychotherapeutic drug discovery research. *Mailing Add:* Wyeth Labs PO Box 8299 Philadelphia PA 19101

MUTH, GILBERT JEROME, b Modesto, Calif, Mar 1, 38; m 59; c 2. BOTANY, FLORISTICS. *Educ:* Pac Union Col, BA, 61, MA, 67; Univ Calif, Davis, PhD(bot), 76. *Prof Exp:* Teacher biol & math, Napa Jr Acad, 61-66; from instr to assoc prof, 66-80, PROF BIOL, PAC UNION COL, 80- *Concurrent Pos:* Sci fac fel, NSF, 71-72. *Mem:* Sigma Xi. *Res:* Computerizing herbarium label data for interactive instant retrieval to a terminal or line printer of raw or summarized data with the primary use being toward rare and endangered plants. *Mailing Add:* 305 Sky Oaks Dr Angwin CA 94508

MUTH, WAYNE ALLEN, b Denver, Colo, Mar 25, 32; m 54; c 3. ENGINEERING, COMPUTER SCIENCE. *Educ:* Univ Colo, BS, 54; Iowa State Univ, MS, 60, PhD(mech eng & math), 63. *Prof Exp:* Asst prof naval sci, Iowa State Univ, 57-59, asst prof mech eng, 59-63; res scientist, Aerospace Div, Martin-Marietta Corp, 63-65; asst prof info processing sci, Southern Ill Univ, Carbondale, 65-69; PROF COMPUT SCI & DIR COMPUT CTR, W VA UNIV, 69- *Mailing Add:* Comput Ctr WVa Univ Morgantown WV 26506

MUTHARASAN, RAJAKKANNU, b India, Jan 1, 47; m 74; c 2. CHEMICAL ENGINEERING, BIOCHEMICAL ENGINEERING. *Educ:* Indian Inst Technol, Madras, BS, 69; Drexel Univ, MS, 71, PhD(chem eng), 73. *Prof Exp:* Fel, Univ Toronto, 73-74; from asst prof to assoc prof, 74-84, PROF CHEM ENG, DREXEL UNIV, 84- *Concurrent Pos:* Consult, Metals & Biochem Indust, 80- *Mem:* Am Inst Chem Engrs; Am Chem Soc; Am Inst Metall Engrs. *Res:* Control of bio-reactors; biochemical process engineering; bio-instrumentation; removal of micron-sized particles from molten metal systems. *Mailing Add:* Dept Chem Eng Drexel Univ Philadelphia PA 19104-2875

MUTHUKRISHNAN, SUBBARATNAM, b India, Dec 27, 42; m 77; c 2. BIOCHEMISTRY. *Educ:* Univ Madras, BS, 63, MS, 65; Indian Inst Sci, Bangalore, PhD(biochem), 70. *Prof Exp:* Res assoc, Univ Chicago, 71-73; res fel, Roche Inst Molecular Biol, 73-76; vis scientist, NIH, 76-80; asst prof, 80-85, ASSOC PROF BIOCHEM, KANS STATE UNIV, 85- *Res:* Hormonal control of gene expression in plants; messenger RNA structure and function. *Mailing Add:* Dept Biochem Kans State Univ Willard Hall Manhattan KS 66506

MUTHUKUMAR, MURUGAPPAN, Indian citizen. POLYMER DYNAMICS. *Educ:* Univ Madras, India, BSc, 70, MSc, 72; Univ Chicago, PhD(chem), 79. *Prof Exp:* Fel physics, Cavendish Lab, Univ Cambridge, 79-81; asst prof chem, Ill Inst Technol, 80-; AT POLYMER RES INST, UNIV MASS, AMHERST, MASS. *Res:* Polymerdynamics and hydrodynamics of solutions; viscoelasticity of polymers; flow through porous media; diffusion controlled chemical reactions; energy transport in solution. *Mailing Add:* Polymer Res Inst Univ Mass Amherst MA 01003-0004

MUTIS-DUPLAT, EMILIO, b Cucuta, Colombia, Sept 6, 32 ; US citizen; m 58; c 3. GEOLOGY, PETROLOGY. *Educ:* Nat Univ Colombia, geologist, 60; Tex A&M Univ, MS, 69; Univ Tex, Austin, PhD(geol), 72. *Prof Exp:* From asst prof to assoc prof geol, Nat Univ Colombia, 61-67, dean fac, 64-65, chmn dept, 65-67; explor geologist, Tex Land & Trading Co, Austin, 73-74; assoc prof earth sci, 74-81, PROF GEOL, UNIV TEX, PERMIAN BASIN, 81-, CHMN DEPT, 82- *Concurrent Pos:* Fel, Univ Tex, Austin, 72-73. *Mem:* Fel Geol Soc Am; fel Geol Soc London; Geochem Soc; Mineral Soc Am; Mineral Soc London; Soc Explor Geophys. *Res:* Origin of migmatites; origin of augen gneiss; amphibolite facies in regional metamorphism; marbles in regional metamorphic rocks, geology Llano region, Texas. *Mailing Add:* Dept Geol Univ of Tex Permian Basin Odessa TX 79762-8301

MUTMANSKY, JAN MARTIN, b New Rochelle, NY, Apr 26, 41; m 65; c 4. MINE DUST CONTROL, MINE VENTILATION. *Educ:* PA State Univ, BS, 64, MS, 66, PhD(mining eng), 68. *Prof Exp:* Trainee mining eng, Ingersoll-Rand Co, 60-62 & US Bur Mines, 63-64; systs analyst, Kennecott Copper Corp, 68-69; asst prof, Univ Utah, 69-73; assoc prof, WVa Univ, 73-77; assoc prof mining eng, coordr, Mining Ctr, Fayette Campus, 77-81, assoc prof mining eng, Univ Park Campus, 81-88, PROF MINING ENG, UNIV PARK CAMPUS, PA STATE UNIV, 88- *Concurrent Pos:* US Bur Mines fel, 70-71; mem bd dirs, Soc Mining Metall & Explor Inc, 83-89. *Mem:* Soc Mining Engrs. *Res:* Mine dust control, mine ventilation, mine operations research. *Mailing Add:* Dept Mining Eng Pa State Univ University Park PA 16802

MUTO, PETER, b Chicago, Ill, Apr 23, 24; m 45; c 5. ENVIRONMENTAL SCIENCES,. *Educ:* Wis State Univ, Stevens Point, BS, 48; Phillips Univ, MEd, 51. *Prof Exp:* Teacher high sch, Okla, 49-51 & Iowa, 51-54; asst prof chem, 54-61, ASSOC PROF PHYS SCI, UNIV WIS-RIVER FALLS, 62- *Concurrent Pos:* Retired Lt Col, US Army. *Mem:* Nat Sci Teachers Asn. *Res:* Improvement of instruction in science for non-scientists at the college level. *Mailing Add:* Dept of Chem Univ of Wis River Falls WI 54022

MUTSCH, EDWARD L, b Madelia, Minn, Mar 15, 39; m 61. ORGANIC CHEMISTRY. *Educ:* St Olaf Col, BS, 61; Univ Minn, PhD(org chem), 65. *Prof Exp:* Sr res investr, G D Searle & Co, Ill, 65-67; sr res chemist, Biochem Res Cent Res Lab, Minn Mining & Mfg Co, 67-74; mgr med chem, 74-76, mgr chem res & develop, 76-78, LAB MGR, NEW MOLECULE RES, RIKER LABS, INC, 78- *Mem:* Am Chem Soc. *Res:* Synthesis of nucleoside antimetabolites. *Mailing Add:* 5772 Deer Trail St Paul MN 55125-9755

MUTSCHLECNER, JOSEPH PAUL, b Ft Wayne, Ind, July 6, 30; m 55; c 3. ASTROPHYSICS, HYDRODYNAMICS. *Educ:* Ind Univ, AB, 52, MA, 54; Univ Mich, PhD(astron), 63. *Prof Exp:* Scientist, US Naval Ord Test Sta, 55-57 & 61-63; staff mem scientist, Los Alamos Sci Lab, Univ Calif, 63-67; ASSOC PROF ASTRON, IND UNIV, BLOOMINGTON, 67- *Mem:* Am Astron Soc; Int Astron Union. *Res:* Solar and stellar atmosphere and abundances. *Mailing Add:* 1610 S Sage Los Alamos NM 87544

MUTTALIB, KHANDKER ABDUL, b Rajshahi, Bangladesh, Dec 1, 52; m 77; c 2. CONDENSED MATTER. *Educ:* Dhaka Univ, Bangladesh, BSc, 75, MSc, 76; Princeton Univ, PhD(physics), 82. *Prof Exp:* Res assoc, Univ Chicago, 82-84; res scientist, Brookhaven Nat Lab, 84-86; lectr & res scientist physics, Yale Univ, 86-87; ASST PROF PHYSICS, UNIV FLA, 87- *Mem:* Am Phys Soc. *Res:* Electronic transport in disordered quantum conductors; mesoscopic systems; strongly correlated fermi-systems; superconductivity; diffusion of atoms adsorbed on surfaces. *Mailing Add:* Physics Dept 215 Williamson Hall Gainesville FL 32611

MUTTER, WALTER EDWARD, b New York, NY, Nov 13, 21; m 63; c 2. PHYSICS. *Educ:* Polytech Inst Brooklyn, BS, 42; Mass Inst Technol, PhD(physics), 49. *Prof Exp:* Engr, Radio Corp Am, 42-46; res assoc physics, Mass Inst Technol, 46-49; engr, 49-59, SR ENGR, IBM CORP, 59- *Honors & Awards:* Sr Medal, Am Inst Chemists, 42. *Mem:* Am Phys Soc; Am Chem Soc; Electrochem Soc; Inst Elec & Electronics Engrs. *Res:* Electronic processes in crystals and semiconductors; semiconductor devices; vacuum tubes. *Mailing Add:* Eight Bobrick Rd Poughkeepsie NY 12601

MUTTER, WILLIAM HUGH, b Orange, NJ, July 28, 34. MECHANICAL ENGINEERING. *Educ:* US Merchant Marine Acad, BS, 56; Columbia Univ, MS, 57. *Prof Exp:* Mem tech staff power systs lab, Bell Tel Labs, 62-66; sr proj engr vending mach systs, Trodyne Corp, 66-69; proj mgr auto typewriter systs, Quindar Electronics, 69-73; mech eng group leader pneumatic tube systs, Airmatic Systs Div, Mosler Safe Co, 73-75; MGR SYSTS DESIGN & MFR, PRINTING PROD DIV, HOECHST CELANESE CORP, 75- *Res:* Machine design; associated control circuitry. *Mailing Add:* Printing Prod Div Hoechst Celanese Corp PO Box 3700 Somerville NJ 08876-1258

MUTTON, DONALD BARRETT, b New Toronto, Ont, Oct 29, 27; m 53; c 2. CHEMISTRY. *Educ:* Univ Toronto, BASc, 49, MASc, 51, PhD(cellulose chem), 53. *Prof Exp:* Asst res chemist, Int Cellulose Res, Ltd, 52-55, asst in-chg, Pioneering Res Div, 55-58, asst mgr, Basic Res Div, 58-60, mgr, 60-62, dir basic res & spec serv, 62-70, dir sci, 70-71, DIR RES, CIP RES LTD, 71-, VPRES & DIR RES, 72- *Mem:* Can Pulp & Paper Asn; Chem Inst Can (treas, 62-64); Tech Asn Pulp & Paper Indust; Asn Advan Sci Can. *Res:* Pulp and paper technology; wood and cellulose chemistry. *Mailing Add:* 535 Thorne St Hawkesbury ON K6A 2N7 Can

MUUL, ILLAR, b Tallinn, Estonia, Feb 18, 38; US citizen; m 61, 74; c 2. ECOLOGY, ANIMAL BEHAVIOR. *Educ:* Univ Mass, BS, 60; Univ Mich, MS, 62, PhD(zool), 65. *Prof Exp:* Researcher ecol virus transmission, Walter Reed Army Inst Res, Washington, DC, 65-68, chief, Dept Ecol, Army Med Res Unit, Inst Med Res, Malaysia, 68-74, chief, Environ Res Requirements Br, US Army Med Bioeng Lab, 74-77, exec officer & res prog mgr, Armed Forces Res Inst Med Sci, Bangkok, Thailand, 77-80, coordr int progs, US Army Med Res & Develop Command, Ft Detrick, Md, 80-87; conserv consult, Smithsonian Inst, 87-88; PRES, INTERGRATED CONSERV RES, HARPERS FERRY, WVA, 88- *Concurrent Pos:* Lectr, Eastern Mich Univ, 65; coordr, Res & Conserv Prog, China & Malaysia, 88- *Honors & Awards:* A Brazier Howell Award, Am Soc Mammal. *Mem:* Am Soc Mammal; AAAS; Malaysian Soc Parasitol & Trop Med. *Res:* Environmental physiology; ethology; systematics of flying squirrels of the world; ecological factors involved in disease transmission; tropical ecology; population dynamics of mammals; zoogeography; environmental quality; coordination of international research programs; conservation of tropical rain forests. *Mailing Add:* PO Box 920 Harpers Ferry WV 25425

MUUS, JYTTE MARIE, biochemistry; deceased, see previous edition for last biography

MUVDI, BICHARA B, b Barranquilla, Colombia, Aug 16, 27; nat US; m 52; c 3. ENGINEERING MECHANICS. *Educ:* Syracuse Univ, BME, 52, MME, 54; Univ Ill, PhD, 61. *Prof Exp:* Instr & res assoc, Syracuse Univ, 52-56; sr engr, Martin Co, 56-58; assoc prof eng mech, Mich Technol Univ, 58-63; chmn, Dept Civil Eng & Eng Mech, 64-81, actg dean Col Eng & Technol, 70-72, PROF CIVIL ENG & CONSTRUCT, BRADLEY UNIV, 81- *Mem:* Am Soc Eng Educ; Am Soc Civil Engrs. *Res:* Influence of environment on materials behavior and structural mechanics. *Mailing Add:* Col Eng & Technol Bradley Univ Peoria IL 61606

MUZIK, THOMAS J, b Lorain, Ohio, Dec 21, 19; m 45; c 3. PLANT PHYSIOLOGY. *Educ:* Univ Mich, AB, MS, 42, PhD(bot), 50. *Prof Exp:* Res botanist, Firestone Plantations, Liberia, 42-47; plant physiologist, Fed Exp Sta, PR, 49-56; assoc prof agron, 56-62, PROF AGRON & AGRONOMIST, WASH STATE UNIV, 62- *Concurrent Pos:* Vis prof, Univ Madrid, 70-71; distinguished vis prof, US Air Force Acad, 76-77. *Mem:* AAAS; Bot Soc Am; Weed Sci Soc Am; Am Soc Agron. *Res:* Growth and development; environmental relationships; growth-regulators and herbicides. *Mailing Add:* Dept of Agron & Soils Wash State Univ Pullman WA 99164

MUZYCZKO, THADDEUS MARION, b Chicago, Ill, Jan 14, 36; m 65; c 2. POLYMER CHEMISTRY. *Educ:* Mich State Univ, BA, 59; Roosevelt Univ, MS, 68. *Prof Exp:* Proj engr, Chicago Pump Div, FMC Corp, 60-61; chemist & supvr chem, Richardson Co, 61-69, mgr & tech dir graphic arts, Res & Develop Div, Melrose Park, Ill, 70-78; VPRES RES & DEVELOP, SAMUEL BINGHAM CO, CHICAGO, 78- *Concurrent Pos:* Lectr polymer chem, Roosevelt Univ, 69-; instr polymer technol, Col DuPage, 71-, coordr plastics technol prog, 75-76; mem, Plastics Technol Adv Comt, Col Dupage, 77- *Mem:* Am Chem Soc; Soc Plastics Engrs; Fedn Socs Coatings & Technol. *Res:* Polymer morphology; photopolymers; polymer characterizations; surface coatings; polymer design and engineering. *Mailing Add:* 530 W 36th St Downers Grove IL 60515-1637

MUZYKA, DONALD RICHARD, b Northampton, Mass, Aug 23, 38; m 61; c 3. METALLURGY, MATERIALS SCIENCE. *Educ:* Univ Mass, BS, 60; Rensselaer Polytech Inst, MS, 66; Dartmouth Col, PhD(mat sci), 67. *Prof Exp:* Metallurgist, Pratt & Whitney Aircraft, United Technol Corp, 60-63; supvr high temperature alloy res, Carpenter Technol Corp, 66-73, mgr alloy res & develop, 73-75, mgr high temperature alloy metall, 75-76, gen mgr res & develop lab, 76-77, gen mgr distrib, 77-79, dir vpres-tech, 79-81; dir technol & opers planning, Engr Prod Group, Cabot Corp, 82-84, gen mgr, Cabot Refractory Metals, 85-86, gen mgr, Cabot Electronic Mat & Refractory Metals, 87, chmn bd, TANCO, 88, pres, Cabot Ceramics, 88, vpres & gen mgr, Cabot & Electronic Mat & Refractory Metals, 89, vpres corp res & develop, Cabot Corp, 89; PRES & CHIEF OPERATING OFFICER, SPEC METALS CORP, NY, 90- *Concurrent Pos:* Steel Indust Adv Panel, Off Technol Assessment, US Cong. *Mem:* Fel Am Soc Metals; Am Inst Mining, Metall & Petrol Engrs; Am Ceramic Soc. *Res:* High temperature alloys; superalloys; powder and process metallurgy. *Mailing Add:* Spec Metals Corp Middle Settlement Rd New Hartford NY 13413

MYCEK, MARY J, b Shelton, Conn, Dec 19, 26. BIOCHEMISTRY, PHARMACOLOGY. *Educ:* Brown Univ, BA, 48; Yale Univ, PhD(biochem), 55. *Prof Exp:* Instr biochem, Yale Univ, 54-55; sr res biochemist, NY State Psychiat Inst, 57-61; res assoc biochem, Col Physicians & Surgeons, Columbia Univ, 59-61; res assoc pharmacol, Univ Med & Dent, NJ Med Sch, 61-63, from asst prof to prof, 63-90; RETIRED. *Concurrent Pos:* USPHS fel, Rockefeller Inst, 55-57; USPHS grants, 61-66, 67-71 & 72-84; mem pharmacol study sect, NIH, 74-78, Pharmacol Sci Study Sect, Nat Inst Gen Med Sci, 80-84. *Mem:* AAAS; Am Chem Soc; Am Soc Pharmacol & Exp Therapeut; fel NY Acad Sci; Sigma Xi. *Res:* mechanism of barbiturate tolerance in CNS; drug metabolism. *Mailing Add:* 34 Laurel Ave Derby CT 06418

MYCHAJLONKA, MYRON, b Ansbach, WGermany, July 11, 47; US citizen; m 70; c 2. MICROBIAL PHYSIOLOGY, CELL SURFACES. *Educ:* Syracuse Univ, BS, 69, PhD(microbiol), 75. *Prof Exp:* NIH res fel microbiol, Temple Univ Med Sch, 75-77; assoc dir, Res Microbiol Unit, Upstate Med Ctr, State Univ NY, 79-82; PROF, UNIV MICH, DEARBORN, 82- *Concurrent Pos:* Res assoc, Temple Univ Med Ctr, 74-75; vis lectr, Southeastern Mass Univ, 77-78; lectr, Upstate Med Ctr, State Univ NY, 78-79; consult, Am Sterilizer Co, 80-81; prin investr, Am Heart Asn, Mich, 83-84; pres, Am Soc Micro-Mich Branch, 87-88. *Mem:* Am Soc Microbiol; NY Acad Sci; Sigma Xi. *Res:* Component parts, assembly, cell envelope enlargement and division of bacterial cell surfaces; regulation of bacterial cell surface activities and integration with other cellular events for usage in chemotherapeutics and biotechnology. *Mailing Add:* Dept Natural Sci Univ Mich 4901 Evergreen Rd Dearborn MI 48128-1491

MYCIELSKI, JAN, b Wisniowa, Poland, Feb 7, 32; m 59. MATHEMATICS. *Educ:* Wroclaw Univ, MA, 55, PhD(math), 57. *Prof Exp:* Full researcher, Nat Ctr Sci Res, Paris, 57-58; adj, Inst Math, Polish Acad Sci, 58-63, docent, 63-68, prof, 68-69; PROF MATH, UNIV COLO, BOULDER, 69- *Concurrent Pos:* Vis prof, Univ Calif, Berkeley, 61-62 & 70, Case Western Reserve Univ, 67 & Univ Colo, Boulder, 67. *Honors & Awards:* Polish Math Soc Award, 56, Stefan Banach Prize, 66. *Mem:* Am Math Soc; Math Asn Am; Polish Math Soc; Asn Symbolic Logic. *Res:* Logic and foundations; artificial intelligence; theory of games; measure theory; topological algebra. *Mailing Add:* Dept of Math Univ of Colo Boulder CO 80309-0426

MYER, CAROLE WENDY, b New York, NY, Mar 14, 44; m 71. DIAGNOSTIC RADIOLOGY. *Educ:* Cornell Univ, DVM, 67; Ohio State Univ, MS, 74. *Prof Exp:* Vet, Dueland Vet Clin, 67-70; resident vet radiol, 70-73, clin instr, 73-74, asst prof, 74-83, ASSOC PROF VET RADIOL, VET COL, OHIO STATE UNIV, 83- *Mem:* Am Col Vet Radiol (secy, 75-78 & 80-83, pres elect, 78-79, pres, 79-80); Am Vet Med Asn; Int Vet Radiol Soc. *Res:* Diagnostic veterinary radiology; ultrasound. *Mailing Add:* Vet Teaching Hosp Ohio State Univ 1935 Coffey Rd Columbus OH 43210

MYER, DONAL GENE, b Toledo, Ohio, May 4, 30; m 51; c 3. FISH PARASITES, PARASITE LIFE-HISTORIES. *Educ:* Ohio State Univ, BSc, 51, MSc, 53, PhD(zool), 58. *Prof Exp:* Asst instr zool, Ohio State Univ, 57-58; from asst prof to assoc prof, 58-70, chmn dept diol sci, 74-77, PROF ZOOL, SOUTHERN ILL UNIV, 70- *Concurrent Pos:* Asst dean, Grad Sch, Southern Ill Univ, 64-70; Am Coun Educ acad admin intern, Fla State Univ, 67-68; fel trop med, Med Sch, La State Univ, 70; prin investr, Ill Water Res Ctr grant, Univ Ill, Urbana Champaign, 79-81; dean, Sch Sci, Southern Ill Univ, 84- *Mem:* Am Soc Parasitol; Am Micros Soc; Sigma Xi; AAAS; NAm Benth Soc. *Res:* Taxonomy, distribution, life history and ecology of parasites of fish. *Mailing Add:* Dept Biol Sci Southern Ill Univ Edwardsville IL 62025

MYER, GEORGE HENRY, b Bronx, NY, Dec 25, 37; m 76; c 1. GEOLOGY, MINERALOGY. *Educ:* Univ Calif, Santa Barbara, BA, 59; Yale Univ, PhD(geol), 65. *Prof Exp:* Asst prof geol, Univ Maine, 65-70; asst prof, 70-77, ASSOC PROF GEOL, TEMPLE UNIV, 77- *Concurrent Pos:* Consult, 79- *Mem:* Am Crystallog Asn; Mineral Soc Am; Geol Soc Am; Sigma Xi. *Res:* Mineralogy and metamorphic petrogenesis; claysources in the Isthmus of Ierapetra, Eastern Crete and mineralogical ceramic analysis of Bronze Age pottery. *Mailing Add:* Dept Geol Temple Univ Philadelphia PA 19122

MYER, GLENN EVANS, b Kingston, NY, Sept 16, 41; m 61. OCEANOGRAPHY, METEOROLOGY. *Educ:* State Univ NY Plattsburgh, BS, 65; State Univ NY Albany, MS, 69, PhD(atmospheric physics), 71. *Prof Exp:* Teacher physics, Plattsburgh High Sch, NY, 65-66; asst prof meteorol, 71-76, ASSOC PROF EARTH SCI, STATE UNIV NY PLATTSBURGH, 76-, DIR NORTH COUNTRY PLANETARIUM, 71-, DIR & CHMN LAKES & RIVERS RES LAB, 72- *Mem:* Am Phys Soc; Am Asn Physics Teachers; Am Soc Limnol & Oceanog; Int Asn Great Lakes Res; Royal Astron Soc Can. *Res:* Computer modeling and field measurements related to turbulent transport processes; applications of fluid dynamics to problems of physical limnology, atmospheric transport and modeling of planetary atmospheres. *Mailing Add:* RD 2 Box 293 West Chazy NY 12992

MYER, JON HAROLD, b Heilbronn, Ger, Sept 29, 22; nat US; m 48; c 4. EXPERIMENTAL PHYSICS. *Educ:* Hebrew Tech Col, BEE, 41. *Prof Exp:* Instrument maker, Anglo Iranian Oil Co, Iran, 42-44; instrument designer, Hebrew Tech Col, 44-46, eng consult, 46-47; instrumentologist, Dept Chem, Univ Southern Calif, 47-53; sub-lab head, Semiconductor Div, Hughes Aircraft Co, 53-60, mgr laser metall, 60-66, sr staff engr, Theoret Studies Dept, Hughes Res Labs, Malibu, Calif, 66-70, Chem Physics Dept, 70-74, Optical Electronics Dept, 74-77, Optical Physics Dept, 77-80, Optical Circuits Dept, Hughes Res Labs, 80-86, SR SCIENTIST, ENG DIV, RADAR SYSTS GROUP, HUGHES AIRCRAFT CO, 86- *Concurrent Pos:* Lectr, Calif State Lutheran Col, 73-78; instr, advan tech educ prog, Hughes Aircraft Co, 83. *Mem:* Sigma Xi; Am Phys Soc; sr mem Inst Elec & Electronics Engrs; Optical Soc Am. *Res:* Physical instrumentation and apparatus design; semiconductor devices; laser applications; bubble domains; forensic science and technology; fiber optics and optical circuits; ophthalmic surgery instrumentation; head up displays. *Mailing Add:* 22931 Gershwin Dr Woodland Hills CA 91364-3827

MYER, YASH PAUL, b Jullundur City, India, May 5, 32; m 59; c 2. PHYSICAL BIOCHEMISTRY. *Educ:* Punjab Univ, BSc, 53, MSc, 55; Univ Ore, PhD(chem), 61. *Prof Exp:* Lectr chem, SD Col, Punjab, India, 53-55; jr sci officer, Coun Sci & Indust Res, Punjab Univ, 55-57; res assoc biochem, Sch Med, Yale Univ, 61-66; from asst prof to assoc prof, 66-74, PROF CHEM, STATE UNIV NY ALBANY, 74- *Mem:* Am Chem Soc; Am Soc Biol Chem; Biophys Soc. *Res:* Macromolecular conformation and structure. *Mailing Add:* Dept of Chem State Univ of NY Albany NY 12222

MYERHOLTZ, RALPH W, JR, b Bucyrus, Ohio, July 29, 26; m 51; c 2. POLYMER CHEMISTRY. *Educ:* Purdue Univ, BS, 50; Northwestern Univ, PhD(org chem), 54. *Prof Exp:* Asst proj chemist, Standard Oil Co, Ind, 54-55, proj chemist, 55-58, group leader high polymers, 58-60; group leader, Amoco Chem Corp, 60-66, res assoc, 66-69, dir, Polymer Physics Div, 69-86; RETIRED. *Mem:* Am Chem Soc; Soc Plastics Engrs. *Res:* Structure-property relationships of high polymers; rheology, dynamic mechanical properties, stability and crystallization of high polymers; anionic polymerization processes; catalysis and hydrocarbon isomerization. *Mailing Add:* 1125 Cricket Reel Greenfield IN 46140-2805

MYERS, ALAN LOUIS, b Cincinnati, Ohio, Sept 26, 32; m 57; c 2. CHEMICAL ENGINEERING. *Educ:* Univ Cincinnati, BS, 60; Univ Calif, Berkeley, PhD(chem eng), 64. *Prof Exp:* Chem engr, Andrew Jergens Co, Ohio, 49-51, 55-59; from asst prof to assoc prof, 64-71, chmn chem & biochem eng, 77-80, PROF CHEM ENG, SCH CHEM ENG, UNIV PA, 71- *Concurrent Pos:* Vis prof, Univ Graz, Austria, 75-76, Univ Tokyo, 86; fel, Japan Soc Promotion Sci, 86. *Mem:* Am Inst Chem Engrs; Am Chem Soc. *Res:* Adsorption and surface science; thermodynamics; statistical mechanics. *Mailing Add:* Dept Chem & Biochem Eng Univ Pa 220 S 33rd St Philadelphia PA 19174

MYERS, ANNE BOONE, b New Haven, Conn, May 9, 58. CHEMICAL PHYSICS. *Educ:* Univ Calif, Riverside, BS, 80; Univ Calif, Berkeley, PhD(chem), 84. *Prof Exp:* Postdoctoral fel, Univ Pa, 85-86; asst prof, 87-90, ASSOC PROF CHEM, UNIV ROCHESTER, 90- *Mem:* Am Chem Soc; Am Phys Soc; AAAS. *Res:* Chemical physics: resonance raman, four-wave mixing, and ultrafast spectroscopic studies of electronic and vibrational relaxation and photochemical reaction dynamics, especially in liquids. *Mailing Add:* Dept Chem Univ Rochester Rochester NY 14627-0216

MYERS, ARTHUR JOHN, b South Haven, Mich, Aug 27, 18. GEOMORPHOLOGY. *Educ:* Kalamazoo Col, BA, 41; Mich Col Mining, BS & MS, 49; Univ Mich, PhD(geol), 57. *Prof Exp:* Asst prof, 51-61, assoc prof, 61-73, PROF GEOL, UNIV OKLA, 73- *Mem:* AAAS; Geol Soc Am; Am Asn Petrol Geol. *Res:* Permian and Pleistocene fluviatile deposits of northwestern Oklahoma; geomorphology as it reflects rock types, structure, climate and time; geologic mapping using aerial photographs and field observations; photogrammetry. *Mailing Add:* 830 Van Vleet Oval Rm 163 Norman OK 73019

MYERS, BASIL R, engineering, mathematics; deceased, see previous edition for last biography

MYERS, BENJAMIN FRANKLIN, JR, b Steelton, Pa, Sept 3, 26; m 56; c 2. PHYSICAL CHEMISTRY. *Educ:* Pa State Univ, BS, 50; Northwestern Univ, PhD(chem), 55. *Prof Exp:* Res chemist, Union Oil Co, Calif, 55-59; res assoc & instr phys chem, Princeton Univ, 59-62; staff scientist, Gen Dynamics/Convair, 62-69 & Sci Applications, Inc, 69-74; STAFF SCIENTIST, GA TECHNOL, INC, 74- *Concurrent Pos:* Consult, Sci Applications, Inc, 74-78. *Mem:* AAAS; Am Phys Soc; Combustion Inst. *Res:* Shock phenomena; chemical kinetics; energy exchange processes; atmospheric chemistry; nuclear reactor fission product transport. *Mailing Add:* Ga Technol Inc 10955 John Jay Hopkins Dr San Diego CA 92121

MYERS, BETTY JUNE, b Ashland, Ohio, Apr 18, 28. PARASITOLOGY. *Educ:* Ashland Col, BA & BSc, 49; Univ Nebr, MA, 51; McGill Univ, PhD(parasitol), 59. *Prof Exp:* Lab asst biol, Ashland Col, 47-49, instr, 51-52; asst biol, Univ Nebr, 49-51; instr parasitol, McGill Univ, 52-53, res assoc, 53-59, asst prof parasitol, 59-64; parasitologist, Southwest Found Res & Educ, 64-69, asst found scientist, 69-73; health sci adminr, Div Res Grants, NIH, 73-90; RETIRED. *Concurrent Pos:* Fisheries Res Bd Can res grant, 54-58; consult, Food & Agr Orgn, 57-; Mem, Can Comt Freshwater Fisheries Res & Sci Adv Bd, Ashland Col, 59-; Nat Res Coun Can fel, 59-60; consult, NIH, 61; parasitologist, Arctic Unit, Fisheries Res Bd Can, 62-63; coun mem-at-large, Am Soc Parasitol, 74-75, vpres, 75, assoc ed newsletter, 73- *Mem:* AAAS; Am Soc Mammal; Am Soc Parasitol; Am Soc Trop Med & Hyg; Am Micros Soc (pres, 74). *Res:* Parasites of marine mammals, fish and primates; anisakiasis; schistosomiasis haematobium; primate ecological relationships of host and parasites; host-phylogenetic relationships; zoogeography; systematic experimental studies; helminthology. *Mailing Add:* 131 High St Ashland OH 44805

MYERS, BLAKE, b Seattle, Wash, July 15, 23; m 44; c 2. MECHANICAL ENGINEERING. *Educ:* Univ Wash, Seattle, BSME, 47; Mass Inst Technol, SMME, 48. *Prof Exp:* Instr mech eng, Univ Wash, Seattle, 46; assoc, Mass Inst Technol, 47-48; design engr, Richmond Refinery, Standard Oil Co, Calif, 48-50; mech engr, Calif Res & Develop Co, 50-53; leader, Apparatus Eng Div, 56-58 & Propulsion Eng Div, 58-67, MECH ENGR, LAWRENCE LIVERMORE LAB, UNIV CALIF, 53-, HEAD RES PROG ENG DIV, 67- *Concurrent Pos:* Consult, Kaiser Eng, 67-; mem exec comn tech group controlled fusion, Am Nuclear Soc, 71- *Mem:* Sigma Xi. *Res:* Design and analysis of power, propulsion, accelerator and nuclear systems. *Mailing Add:* 4650 Almond Circle Livermore CA 94550

MYERS, BRYAN D, b Cape Town, SAfrica, Nov 15, 36; US citizen. NEPHROLOGY. *Educ:* Univ Cape Town, MB & ChB, 59. *Prof Exp:* Intern & med registrar, Dept Med, Univ Cape Town & Groote Schuur Hosp, SAfrica, 60-65; med registrar, Dept Med, Cent Middlesex Hosp, Univ London, Eng, 66-67 & UK Med Res Coun, 67-68; sr consult nephrologist, Beilinson Med Ctr, Univ Tel Aviv, Israel, 69-71, chief nephrol, Meir Hosp, 71-76; from asst prof to assoc prof, 76-86, PROF MED, SCH MED, STANFORD UNIV, 86-, CHIEF, DIV NEPHROLOGY, 89- *Concurrent Pos:* Vis prof, numerous US & foreign univs, 78-90; dir, Hemodialysis Ctr, Stanford Univ, 76-80, dir clin nephrology, 81-, actg chief, Div Nephrology, 84-; mem, Spec Study Sect & site vis, NIH, 81-87, Abstract Rev Comt, Am Soc Nephrology, 81-85, Res Fel & Grants Comt, Nat Kidney Found, 84-86, Kidney & Urol Adv Bd, Dept Health & Human Serv, 86- , Med Sci Rev Comt, Juv Diabetes Found, 88-; assoc ed, Am J Kidney Dis, 86- *Mem:* Royal Col Physicians; AAAS; Am Soc Nephrology; Int Soc Nephrology; Am Fedn Clin Res; Am Physiol Soc; Am Col Physicians. *Res:* Author of more than 100 technical publications. *Mailing Add:* Div Nephrology Sch Med Stanford Univ S215 300 Pasteur Dr Stanford CA 94305-5114

MYERS, CARROL BRUCE, b Asheville, NC, Sept 6, 43; m 65; c 2. ALGEBRA. *Educ:* Berea Col, BA, 65; Univ Ky, MA, 67, PhD(math), 70. *Prof Exp:* Asst prof, 70-72, assoc prof math, 72-80, PROF MATH & COMPUT SCI, AUSTIN PEAY STATE UNIV, 80- *Mem:* Math Asn Am; Nat Coun Teachers Math; Asn Comput Mach. *Res:* Ring theory; module theory. *Mailing Add:* Dept of Math & Comput Sci Austin Peay State Univ College St PO Box 8346 Clarksville TN 37040

MYERS, CHARLES, b Wilkes-Barre, Pa, June 24, 43; m 66; c 2. CANCER DRUG PHARMACOLOGY. *Educ:* Wesleyan Univ, BA, 65; Univ Pa Sch Med, MD, 69, MD(internal med), 71; Nat Cancer Inst, MD(oncol), 73. *Prof Exp:* Sr investr, 74-78, sect head, 78-81, chief clin chmn, 81-88, CHIEF, MED, NAT CANCER INST, 88- *Mem:* Am Soc Clin Invest; Am Asn Cancer Res; Am Soc Cancer Res; AAAS; Am Soc Clin Pharmacol & Therapeut. *Res:* Development of new agents active against human cancer. *Mailing Add:* 1111 Balston Rd Rockville MD 20852

MYERS, CHARLES CHRISTOPHER, b Richwood, WVa, June 12, 34; m 60; c 3. FORESTRY, BIOMETRY. *Educ:* WVa Univ, BS, 60; Syracuse Univ, MS, 62; Purdue Univ, PhD(forestry), 66. *Prof Exp:* From instr to asst prof forestry, Purdue Univ, 62-67; technician, US AID, 67-69; asst prof, Univ Vt, 69-73; ASSOC PROF FORESTRY, SOUTHERN ILL UNIV, CARBONDALE, 73- *Concurrent Pos:* NSF grant, 71. *Mem:* Soc Am Foresters. *Res:* Applications of statistics and computers to forest inventory. *Mailing Add:* Dept Forestry Agr Bldg 0186D Southern Ill Univ Carbondale IL 62901-4411

MYERS, CHARLES EDWIN, b Philadelphia, Pa, Dec 15, 40; m 63; c 3. ENVIRONMENTAL ENGINEERING, SCIENCE POLICY. *Educ:* Pa State Univ, BS, 62; Univ Md, MS, 65. *Prof Exp:* Res chemist, Dept Interior, Bur Mines, 65-67; res assoc water transfer & membranes, AeroChem Res Labs, Inc, 67-68; res engr ion exchange & membrane technol, Permutit Co, 68-69; chemist water pollution control, Fed Water Pollution Control Admin, 69-71; chem engr water qual sanit eng, US Environ Protection Agency, 71-74; exec secy interagency arctic res coord comt, 74-80, INTERAGENCY/INT COORDR, NSF, 80- *Concurrent Pos:* Partner & consult engr, Myers, Tobin, Trax & Assocs, 73-74; adj prof, Montgomery Col, Takoma Park, Md, 84. *Mem:* Am Inst Chem Engrs. *Res:* Membranes and ion exchange; water quality; sanitary engineering; pollution control; solid waste management; integrated utility systems; arctic; cold regions engineering; antarctic; environmental impact analysis; conservation; pollution control regulation; international science policy. *Mailing Add:* NSF 1800 G St NW Washington DC 20550

MYERS, CHARLES R, b Milwaukee, Wis, June 30, 56; m 79. BIOGEOCHEMISTRY. *Educ:* Carroll Col, BS, 78; Univ Wis, MS, 85, PhD, 87. *Prof Exp:* FEL MICROBIOL, UNIV WIS, MILWAUKEE CTR GREAT LAKES STUDIES, 87- *Mem:* Am Soc Microbiol; AAAS; Sigma Xi. *Res:* Microbial cycling of iron and manganese in aquatic environments; biochemistry and physiology of microbial manganese and iron reduction. *Mailing Add:* Ctr Great Lakes Studies Univ Wis 600 E Greenfield Ave Milwaukee WI 53204

MYERS, CHARLES WILLIAM, b St Louis, Mo, Mar 4, 36. HERPETOLOGY. *Educ:* Univ Fla, BS, 60; Southern Ill Univ, MA, 62; Univ Kans, PhD(zool), 70. *Prof Exp:* Vis scientist herpet, Gorgas Mem Lab, Panama, 64-67; asst cur, 68-73, assoc cur, 73-78, chmn dept, 80-87, CUR HERPET, AM MUS NATURAL HIST, 78- *Mem:* Am Soc Ichthyologists & Herpetologists; Soc Study Amphibians & Reptiles. *Res:* Systematics of neotropical amphibians and reptiles. *Mailing Add:* Dept Herpet Am Mus Natural Hist New York NY 10024

MYERS, CLIFFORD EARL, b Jefferson City, Tenn, June 1, 29; m 53; c 2. PHYSICAL & HIGH TEMPERATURE CHEMISTRY. *Educ:* Carson-Newman Col, BS & BA, 51; Purdue Univ, MS, 53, PhD(inorg chem), 56. *Prof Exp:* Asst, Purdue Univ, 51-54; grad res chemist, Inst Eng Res, Univ Calif, 54-55; res assoc chem eng, Univ Ill, 55-56; from asst prof to assoc prof chem, Lynchburg Col, 56-58; asst prof, State Univ NY Col Ceramics, Alfred Univ, 58-63; assoc prof, 63-78, PROF CHEM, STATE UNIV NY BINGHAMTON, 78- *Concurrent Pos:* Vis staff mem, Los Alamos Sci Lab, 64-65; vis scientist, Univ Brussels, Belg, 69-70 & Lawrence Berkeley Lab, 84; vis prof, Ames Lab, Iowa State Univ, 77 & 78, Univ Vienna, Austria, 90-91; collabr, Los Alamos Nat Lab, 83-87; vis scientist, Upsala Univ, Sweden, 91. *Mem:* Am Chem Soc; Sigma Xi; Electrochem Soc. *Res:* High temperature vaporization processes; thermodynamic stabilities, structure and bonding in refractory substances and high temperature molecules. *Mailing Add:* Dept Chem State Univ NY Binghamton NY 13901

MYERS, DALE DEHAVEN, b Kansas City, Mo, Jan 8, 22; m 43; c 2. AERONAUTICAL ENGINEERING, SPACE AERONAUTICS. *Educ:* Univ Wash, BS, 43. *Hon Degrees:* PhD, Whitworth Col, 70. *Prof Exp:* Chief engr, Missile Develop Div, NAm Aviation, 46-57, vpres & weapons systs mgr, 57-63, asst div dir advan systs, Rockwell Int Corp, El Segundo, Calif, 63-64, vpres & gen mgr CSM progs, 64-69, vpres & mgr space shuttle prog, 69-70; assoc adminr manned space flight, NASA, 70-74; pres, Rockwell Int Corp & corp vpres, Rockwell Int, 74-77; under-secy, Dept Energy, 77-79; consult, Dale Myers & Assocs, 84-86; dep adminr, NASA, 86-89; PRES, DALE MYERS & ASSOC, INC, 89- *Concurrent Pos:* Distinguished Serv Medal, Apollo Mission, NASA, 71 & 74. *Honors & Awards:* Distinguished Serv Medal, Dept Energy, 79. *Mem:* Nat Acad Eng; fel Am Inst Aeronaut & Astronaut; Am Astronaut Soc. *Mailing Add:* Dale Myers & Assoc Inc PO Box 2518 Leucadia CA 92024

MYERS, DAVID, b Philadelphia, Pa, Sept 18, 06; m 30; c 2. MEDICINE. *Educ:* Univ Pa, 27; Temple Univ, MD, 30; Am Bd Otolaryngol, dipl, 35. *Prof Exp:* Intern, Temple Univ, 30-32, preceptor, Temple Univ & trainee, Temple Univ Hosp, 32-40, mem staff dept otorhinol, Temple Univ Med Sch, 32-55, prof & chmn dept, 55-62; prof otorhinolaryngol, 62-71, chmn dept, 64-71, prof otolaryngol, 71-76, EMER PROF OTOLARYNGOL & HUMAN COMMUN, SCH MED, UNIV PA, 76- *Concurrent Pos:* Dir inst otol, Presby Hosp, Philadelphia, 62- *Mem:* Fel Am Otol Soc; Am Laryngol, Rhinol & Otol Soc; fel Am Col Surg; fel Am Acad Ophthal & Otolaryngol. *Res:* Otolaryngology. *Mailing Add:* Philadelphian 2401 Pennsylvania Ave Philadelphia PA 19130

MYERS, DAVID DANIEL, b Morris, Minn, Dec 19, 32; m 59; c 1. LABORATORY ANIMAL SCIENCE, ANIMAL PATHOLOGY. *Educ:* Univ Minn, BS, 55, DVM, 57; Univ Ill, MS, 62, PhD(vet path & microbiol), 65, Am Col Lab Animal Sci, dipl, 84. *Prof Exp:* Poultry vet, Fla Livestock Bd, 57-58; animal pathologist, Ill Dept Agr, 58-59; instr vet path, Univ Ill, 59-61; assoc staff scientist, 65-69, staff scientist, 69-75, sr staff scientist, Jackson Lab, 75-82; DIR, RES ANIMAL RESOURCE CTR, SLOAN-KETTERING INST, CORNELL UNIV MED COL, 82- *Concurrent Pos:* Assoc ed, Lab Animal Sci; actg assoc dir lab animal resources, Wide Ctr, Cornell Univ, 80-81. *Mem:* AAAS; Am Vet Med Asn; Am Asn Lab Animal Sci; Asn Gnotobiol; Am Soc Lab Animal Practrs; NY Acad Sci. *Res:* Histopathology; pathogenesis of infectious diseases. *Mailing Add:* PO Box 201 Mt Desert Island ME 04660

MYERS, DAVID RICHARD, b Harvey, Ill, Dec 16, 48; m 75. SOLID-STATE PHYSICS, ELECTRICAL ENGINEERING. *Educ:* Univ of Ill, Chicago Circle, BSE, 71; Univ Ill, Urbana-Champaign, MS, 73, PhD(elec eng), 77. *Prof Exp:* Res asst semiconductor phys, Coord Sci Lab, Univ Ill, Urbana-Champaign, 72-77; Nat Bur Standards-Nat Res Coun res assoc, Electronic Technol Div, Nat Bur Standards, 77-78, physicist, Electronic Devices Div, 78-81; MEM TECH STAFF, SANDIA NAT LABS, ALBUQUERQUE, NMEX, 81- *Mem:* Inst Elec & Electronics Engrs; Am Phys Soc; Sigma Xi. *Res:* Ion implantation; ion range particularly theory and measurement; radiation damage and annealing of semiconductors; deep levels in semiconductors; growth and characterization of compound semiconductors; lasers; superconductivity; compound semiconductor device physics; process development for compound semiconductors. *Mailing Add:* 5705 Carruthers NE Albuquerque NM 87111

MYERS, DIRCK V, b New York, NY, Aug 24, 35; m 59; c 3. BIOCHEMISTRY. *Educ:* Dartmouth Col, AB, 57; Univ Wash, PhD(biochem), 62. *Prof Exp:* Res fel biol, Harvard Univ, 62-64; sr res scientist, Squibb Inst Med Res, NJ, 64-68; sr scientist, Coca-Cola Co, 68-77, prin investr, 77-81, mgr chem sci, 81-91. *Mem:* Am Chem Soc; Sigma Xi. *Res:* Protein chemistry; peptide synthesis; enzymology. *Mailing Add:* Coca Cola Co PO Drawer 1734 Atlanta GA 30301

MYERS, DONALD ALBIN, b Denver, Colo, May 17, 36; m 58; c 2. SPACE SYSTEMS. *Educ:* Colo Sch Mines, PE, 58; Univ Colo, PhD(physiol), 73. *Prof Exp:* Pipeline engr, US Army Corps Engrs, 58-60; design engr, Martin Marietta Corp, 60-65; develop engr, AiResearch Mfg Co, Garrett Corp, 65-67; staff engr, Martin Marietta Corp, 67-69; consult, US Govt, Washington, DC, 73-87; LAB SCIENTIST, HUGHES AIRCRAFT CO, 87- *Mem:* AAAS; Sigma Xi. *Res:* Requirements analysis to develop advanced government space and communications systems. *Mailing Add:* 492 Alabama Dr Herndon VA 22070

MYERS, DONALD ARTHUR, b Seattle, Wash, May 30, 21. BIO-STRATIGRAPHY. *Educ:* Stanford Univ, BA, 43. *Prof Exp:* Field geologist, US Geol Survey, NMex, 48-84; RETIRED. *Mem:* Fel Geol Soc Am; Paleont Soc. *Res:* Fusulintd and Foraminzfera. *Mailing Add:* 1240 Cody St Lakewood CO 80215

MYERS, DONALD EARL, b Chanute, Kans, Dec 29, 31; m 54; c 2. MATHEMATICS. *Educ:* Kans State Univ, BS, 53, MS, 55; Univ Ill, PhD, 60. *Prof Exp:* Asst math, Kans State Univ, 53-55 & Univ Ill, 55-58; assoc prof, Millikin Univ, 58-60; from asst prof to assoc prof, 60-68, PROF MATH, UNIV ARIZ, 68- *Concurrent Pos:* Co-dir, Sec Sci Training Prog, NSF, Univ Ariz, 61, dir, 63, mem, Adv Panel Judge Proposals, 63-66 & writing team, Minn Math & Sci Teaching Proj, Univ Minn, 63-66; vis lectr, Teachers Col, 66; chmn fac, Univ Ariz, 77-79; consult, India Prog, NSF-AID, 67, US Geol Surv, 82-85; pres, Math Geologists US, 85-87; vis prof, Univ Paris-XII, 86. *Mem:* AAAS; Math Asn Am; Am Math Soc; Inst Math Statist; Int Asn Math Geologists. *Res:* Analysis; theory of distributions; geostatistics; statistics applied to problems in mining; hydrology; soil physics; geological engineering and remote sense. *Mailing Add:* Dept Math Univ Ariz Tucson AZ 85721

MYERS, DONALD ROYAL, b Cleveland, Ohio, Dec 18, 13; m 40; c 1. ORGANIC CHEMISTRY. *Educ:* Ohio State Univ, AB, 35, PhD(chem), 40. *Prof Exp:* Asst anal chem, Ohio State Univ, 35-40; res engr, Battelle Mem Inst, Ohio, 40-42; fel, Carnegie Inst Technol, 42-44; res chemist, Upjohn Co, 44-54, sect head, Chem Dept, 54-60, mgr chem res prep dept, 60-79; RETIRED. *Mem:* Sigma Xi. *Res:* Steroids; medicinal chemicals. *Mailing Add:* 8908 East G Ave Kalamazoo MI 49004

MYERS, DREWFUS YOUNG, JR, organic polymers, colloids and surfaces, for more information see previous edition

MYERS, EARL A(BRAHAM), b York Co, Pa, Jan 15, 29; m 56; c 2. AGRICULTURAL ENGINEERING. *Educ:* Pa State Univ, BS, 50, MS, 52; Mich State Univ, PhD, 60. *Prof Exp:* From instr to prof, 51-77, EMER PROF AGR ENG, PA STATE UNIV, 77- *Concurrent Pos:* Consult land application of wastewater, 62-77; wastewater irrigation specialist, consult, Williams & Works, Grand Rapids, Mich, 77- *Honors & Awards:* Karl M Mason Award. *Mem:* Am Soc Agr Engrs; Water Pollution Control Fedn; Irrigation Asn. *Res:* Soil and water area; small watershed hydrology; terraces; tile drainage; wastewater irrigation systems design; development of heads for distributing waste water year-around, including subzero weather. *Mailing Add:* 164 W Hamilton Ave State College PA 16801

MYERS, EARL EUGENE, b Ruffsdale, Pa, Nov 5, 24; m 54; c 4. PETROLEUM CHEMISTRY. *Educ:* Thiel Col, BS, 47; Western Reserve Univ, MS, 49, PhD(org chem), 51. *Prof Exp:* Res chemist, Esso Res & Eng Co, 51-52; res chemist, Gulf Res & Develop Co, 53-62, sr res chemist, 62-85; RETIRED. *Concurrent Pos:* Consult, 86-90. *Mem:* AAAS; Am Chem Soc; Sigma Xi. *Res:* Preparation of chemicals for use as petroleum additives; relationships of structure to activity; oil-soluble polymers; synthetic lubricants. *Mailing Add:* 921 Treasure Lake Dubois PA 15801-9021

MYERS, ELLIOT H, b Cleveland, Ohio. LUBRICATOR, TRIBOLOGY. *Educ:* Case Inst Technol, BS, 59. *Prof Exp:* Chemist, Brookpark Inc, 59-61; chemist, Bryerlyte Corp, 61-63, chief chemist, 63-66; tech officer, Addex Mfg co, 66-67; chemist, Penreco, Penzoil Co Div, 67-73, tech sales rep, 73-80; lab mgr, 80-84, TECH DIR, BROOKS TECHNOL, DIV PREMIER INDUST CORP, 84- *Concurrent Pos:* Chmn grease res comt, Am Soc Testing & Mats. *Mem:* Am Chem Soc; Asn Iron & Steel Engrs; Soc Tribologists & Lubrication Engrs; Nat Lubricating Grease Inst; Am Soc Testing & Mats. *Mailing Add:* 3304 E 87th St Cleveland OH 44127-1849

MYERS, EUGENE NICHOLAS, b Philadelphia, Pa, Nov 27, 33; m 56; c 2. OTOLARYNGOLOGY. *Educ:* Univ Pa, BS, 54; Temple Univ, MD, 60. *Prof Exp:* PROF OTOLARYNGOL & CHMN DEPT, SCH MED, UNIV PITTSBURGH, 72- *Concurrent Pos:* Consult, Children's Hosp & Vet Admin Hosp, Pittsburgh, 72-; chief dept otolaryngol, Eye & Ear Hosp, Pittsburgh, 72-; bd gov, Am Col Surg, 81- *Mem:* Am Acad Facial, Plastic & Reconstruct Surg; Am Acad Otolaryngol Head & Neck Surg; Am Soc Head & Neck Surg; Am Otol Soc; Asn Res Otolaryngol. *Res:* Clinical trials head and neck surgery. *Mailing Add:* Dept Otolaryngol Univ Pittsburgh 230 Lothrop St Pittsburgh PA 15213

MYERS, GARDINER HUBBARD, b Washington, DC, Jan 16, 39; m 63; c 2. PHYSICAL CHEMISTRY. *Educ:* Princeton Univ, AB, 59; Univ Calif, Berkeley, PhD(chem), 65. *Prof Exp:* Asst prof, 65-72, ASSOC PROF CHEM, UNIV FLA, 72- *Mem:* Am Chem Soc. *Res:* Gas kinetics; singlet oxygen; chemical education. *Mailing Add:* Dept of Chem Univ of Fla Gainesville FL 32611

MYERS, GEORGE E, b Detroit, Mich, Aug 9, 26; m 53; c 6. PHYSICAL CHEMISTRY, POLYMER CHEMISTRY. *Educ:* Univ Southern Calif, BS, 48, MS, 49; Harvard Univ, PhD, 52. *Prof Exp:* Res chemist, Oak Ridge Nat Lab, 52-56; proj scientist, Plastics Div, Union Carbide Corp, 56-63; sr tech specialist & sect chief, Lockheed Propulsion Co, Calif, 63-75; RES CHEMIST, FOREST PROD LAB, 75- *Mem:* Am Chem Soc; Forest Prods Res Soc; Sigma Xi. *Res:* Physical chemistry of proteins, clays, ion-exchange resins and polymers; solid propellants; wood adhesives. *Mailing Add:* USDA-FS Div One Grifford Pinchot Dr Madison WI 53705-2398

MYERS, GEORGE HENRY, b New York, NY, Feb 21, 30; div; c 3. ELECTRICAL ENGINEERING, BIOMEDICAL ENGINEERING. *Educ:* Mass Inst Technol, SB & SM, 52; Columbia Univ EngScD, 59. *Prof Exp:* Mem tech staff, Bell Tel Labs, Inc, 52-59, supvr guid & control, 59-65; from assoc prof to prof elec eng, NY Univ, 65-69; mgr biomed eng lab, Riverside Res Inst, New York, 69-74; tech dir pacemaker ctr, Newark Beth Israel Med Ctr, 74-78; treas, Sonometrics Systs, Inc, 78-83; PRES, MEDSYS, INC, 83- *Mem:* Sr mem Inst Elec & Electronics Eng; Am Soc Artificial Internal Organs; Biomed Eng Soc; Sigma Xi. *Res:* Ultrasonics; pacemakers and control of respiration; digital and analog computers. *Mailing Add:* 340 Sea Isle Key Harmon Cove Secaucus NJ 07094-2211

MYERS, GEORGE SCOTT, JR, b Monte Vista, Colo, Mar 21, 34; m 61, 86; c 8. ANIMAL NUTRITION, DAIRY CONSULTANT. *Educ:* Colo State Univ, BS, 56; Univ Conn, MS, 58; Cornell Univ, PhD(animal nutrit), 66. *Prof Exp:* Asst animal nutrit, Univ Conn, 56-58 & Cornell Univ, 60-63; sr nutritionist, Ciba Res Farm, NJ, 63-69; sr res nutritionist, Squibb Agr Res Ctr, 69-70; PRES, MYERS ANIMAL SCI CO (MASCO), 70- *Concurrent Pos:* Consult, sales & mkt, livestock mgt, farming. *Res:* Carotene; vitamins A and E requirements of cattle, sheep and swine; rumen metabolism; ration formulation; animal husbandry management; animal health research; feeding experiments. *Mailing Add:* Myers Animal Sci Co 604 W Florinda St Hanford CA 93230-3629

MYERS, GERALD ANDY, b Boelus, Nebr, Sept 23, 28; m 53, 86; c 6. PLANT ANATOMY. *Educ:* Kearney State Col, AB, 51; Colo State Col, AM, 57; SDak State Col, PhD(plant sci), 63. *Prof Exp:* Teacher elem sch, Nebr, 51-52; teacher, jr high sch, Idaho, 52-55, high sch, 55-56; asst biol, Colo State Col, 56-57; elem teacher & prin, Ill, 57-58; instr bot, 58-64, asst prof, 64-68, assoc prof, 68-72, head dept, 72-81, PROF BIOL, SDAK STATE UNIV, 72- *Concurrent Pos:* NSF instr, Ind Univ, 59, SDak State Col, 60, Univ Wash, 61, Fla State Univ, 67 & Pa State Univ, 68; consult, Biol Sci Curric Study, SDak, 63-79, chmn, Testing Comt, 66-67; consult sci process approach, AAAS, 68-79; consult, Intermediate Sci Curric Study, 69-79; Danforth assoc; fel, Ohio State Univ, 72 & Univ Munchen, W Ger, 88-89. *Mem:* Inst Soc, Ethics & Life Sci. *Res:* Plant morphogenesis and developmental plant anatomy. *Mailing Add:* Dept Biol SDak State Univ Brookings SD 57007

MYERS, GLEN E(VERETT), b Los Angeles, Calif, Mar 6, 34; m 63; c 2. MECHANICAL ENGINEERING. *Educ:* Rensselaer Polytech Inst, BME, 56; Stanford Univ, MS, 57, PhD(mech eng), 62. *Prof Exp:* From asst prof to assoc prof, 62-71, PROF MECH ENG, UNIV WIS-MADISON, 71- *Concurrent Pos:* NSF sci fac fel, Stanford Univ, 69-70. *Mem:* Am Soc Mech Engrs; Am Soc Eng Educ. *Res:* Finite-difference and finite-element methods for conduction heat transfer. *Mailing Add:* Dept Mech Eng Univ Wis Madison WI 53706

MYERS, GLENN ALEXANDER, b Saginaw, Mich, Oct 18, 49; m 82; c 2. NONLINEAR DYNAMICS, BLOOD PRESSURE & HEART RATE. *Educ:* Mich State Univ, BS, 71; Univ Mich, MS, 74; Univ Calif, Berkeley, PhD(elec eng & computer sci), 85. *Prof Exp:* Engr, Gen Motors Corp, Saginaw, Mich, 67-70; proj engr, Tech Commun Int, Palo Alto, Calif, 74-79; dir computer systs, Cardiol Sect, Northwestern Univ Med Sch, 83-86; PRES & CHIEF EXEC OFFICER, MICROMEASUREMENTS INC, IOWA CITY, IOWA, 79-; ASST PROF BIOMED ENG, UNIV IOWA, IOWA CITY, 86- *Concurrent Pos:* Asst prof elec eng & computer sci, Northwestern Univ, 84-86; asst prof elec eng & computer sci, Univ Iowa, 86-, dir, Instruct Comput Lab, Dept Biomed Eng, 88-; chmn, Inst Elec & Electronics Engrs, Comt on Pub Policy on Computer Security & Privacy, 89- & Int Pupil Soc, 89-; vis scientist, Mass Inst Technol, 91. *Honors & Awards:* Distinguished Serv Award, Inst Elec & Electronics Engrs Computer Soc, 89. *Mem:* AAAS; sr mem Inst Elec & Electronics Engrs; sr mem Inst Elec & Electronic Engrs Computer Soc; Eng Med & Biol Soc; Am Soc Eng Educ; Asn Comput Mach. *Res:* Developing smart instruments using model reference adaptive signal processing to estimate noninvasively the integrity of the autonomic nervous system and organs it innervates, eg the heart and the eye. *Mailing Add:* 1218 Eng Bldg Univ Iowa Iowa City IA 52242

MYERS, GRANT G, b Avoca, Iowa, Dec 6, 30; m 52; c 2. ELECTRICAL ENGINEERING. *Educ:* Univ Iowa, BSEE, 57, MSEE, 58, PhD(elec engr), 65. *Prof Exp:* Elec engr develop, Collins Radio Co, 58-62, res, 62-63; asst prof elec eng, Univ Nebr, Lincoln, 65-67; elec eng in internal med & physiol, Sch Med, Univ Nebr, Omaha, 67-69, assoc prof physiol, Sch Med & elec eng, Sch Eng, 69-74, asst dir cardiovasc eng, 65-70; MEM FAC, UNIV NEBR, LINCOLN, 74- *Concurrent Pos:* Nat Insts Health res grant, 67; Fulbright award, Uruguay, 67; NSF grant, 79. *Mem:* Inst Elec & Electronics Engrs; Sigma Xi. *Res:* Solid state devices and integrated circuits; bioengineering; control systems. *Mailing Add:* Dept Elec Eng Univ Nebr 209 N WSEC Lincoln NE 68505-0511

MYERS, HARVEY NATHANIEL, b Tampa, Fla, Aug 26, 46; m 70; c 2. ANALYTICAL CHEMISTRY. *Educ:* Morehouse Col, BS, 69; Univ Ill, Urbana, MS, 71, PhD(chem), 74. *Prof Exp:* Fel comput based teaching, Univ Ill, Urbana, 74; asst prof chem, Chicago State Univ, 74-76, coordr comput assisted instr, 75-76; res scientist, Upjohn Co, 76-83; GROUP MGR, ANALYTICAL SERV, 83- *Mem:* AAAS; Am Chem Soc; Sigma Xi. *Res:* Curriculum development; bio-organic chemistry for computer assisted instruction. *Mailing Add:* The Upjohn Co 7000 Portage Rd Kalamazoo MI 49001

MYERS, HOWARD, b New York, NY, Jan 27, 28; m 76; c 3. CHEMICAL PHYSICS. *Educ:* Univ Chicago, PhB, 48, BS, 51, MS, 58. *Prof Exp:* Mem tech staff, Hughes Res Labs, 54-56; res specialist, Douglas Aircraft Co, Inc, 56-61; mem tech staff, Aerospace Corp, 61-63, mgr reentry physics, 63-66; mem tech staff, TRW Systs Group, 66-68; mgr plasma physics, McDonnell Douglas Astronaut, 69-70, mgr planetary atmospheric physics, 70-73, proj scientist, Planetary Progs, 73-78, instrumentation mgr, Fusion Energy Progs, 78-79, sr tech specialist, 79-80; SR SYSTS ENGR, SYSTS INTEGRATION DIV, GEN ELEC CO, 80- *Concurrent Pos:* Pres & tech dir, CPRL, Inc, 68- *Mem:* Am Astron Soc; Am Phys Soc; Am Geophys Union; fel Am Inst Chemists; Am Chem Soc; NY Acad Sci. *Res:* Thermodynamics and reaction kinetics of solids, gases and plasmas; applications to reentry physics, communications satellites and space exploration. *Mailing Add:* 699 N Valley Rd Paoli PA 19301

MYERS, HOWARD M, b Brooklyn, NY, Dec 12, 23; m 72; c 3. PHARMACOLOGY. *Educ:* Western Reserve Univ, DDS, 49; Univ Calif, MS, 53; Univ Rochester, PhD(pharmacol), 58; San Francisco State Col, MA, 64. *Hon Degrees:* MA, Univ Pa, 74. *Prof Exp:* Asst dent med, Sch Dent, Univ Calif, San Francisco, 4951, from instr to asst prof, 51-59, assoc prof dent med & biochem, 59-65, prof oral biol & lectr biochem, 65-71, vchmn dept biochem, Sch Med, 67-71, prof biochem & biophys, 71-72; prof biochem & chmn dept, Sch Dent, Univ Pac, 71-74; dir, Ctr Oral Health Res, 74-78, PROF PHARMACOL, SCH DENT MED, UNIV PA, 74- *Concurrent Pos:* Nat Inst Dent Res spec res fel, Dept Med Physics, Karolinska Inst, Sweden, 64-65; trainee, Advan Seminar Res Educ, Am Col Dentists, 63, mentor, 64; consult, Stanford Res Inst, 63-74; mem, Dent Training Comt, Nat Inst Dent Res, 65-69; dent res consult, Vet Admin Hosp, San Francisco, 66-74; mem, Dent Study Comt, Div Res Grants, NIH, 69-73; ed, Monographs Oral Sci, 70-; consult, Cooper Labs, 74-79; invited prof, Univ Geneva, Switz, 80-81; Fogarty sr int res fel, 80-81. *Honors & Awards:* Zyma Found Award, 81. *Mem:* AAAS; Am Asn Dent Res (pres, 74); Int Asn Dent Res. *Res:* Mineral metabolism; composition of saliva; surface properties of tooth and bone mineral. *Mailing Add:* 171 Ridge Ave Philadelphia PA 19104

MYERS, IRA LEE, b Madison Co, Ala, Feb 9, 24; m 43; c 4. PREVENTIVE MEDICINE, PUBLIC HEALTH. *Educ:* Howard Col, BS, 45; Univ Ala, MD, 49; Harvard Univ, MPH, 53; Am Bd Prev Med, dipl pub health, 67. *Prof Exp:* Chief epidemic intel serv officer & asst to chief epidemiol br, Commun Dis Ctr, USPHS, 49-55; admin officer & asst state health officer, 55-62, state health officer, Dept Pub Health, Ala, 63-86; asst clin prof prev med, Med Col, Ala, 57-86; RETIRED. *Concurrent Pos:* Secy, Ala Bd Med Examrs, 62-73; chmn, Ala Water Improv Comn, 63-82. *Honors & Awards:* McCormack Award, Asn State & Territorial Health Offices. *Mem:* Am Thoracic Soc; AMA; Med Asn. *Res:* Epidemiology of acute and chronic disease; problems of the aged, including medical care and nursing. *Mailing Add:* 925 Green Forest Dr Montgomery AL 36109

MYERS, JACK DUANE, b New Brighton, Pa, May 24, 13; m 46; c 4. CLINICAL MEDICINE. *Educ:* Stanford Univ, AB, 33, MD, 37. *Prof Exp:* From intern to asst resident, Stanford Univ Hosps, 36-38; from asst resident to resident, Peter Bent Brigham Hosp, Boston, 39-42; assoc, Emory Univ, 46-47; assoc prof med, Duke Univ, 47-55; prof & chmn dept, 55-70, univ prof, 70-85, EMER UNIV PROF MED, UNIV PITTSBURGH, 85- *Concurrent Pos:* Chmn, Am Bd Internal Med, 67-70 & Nat Bd Med Examrs, 71-75; mem, Nat Adv Coun Arthritis & Metab Dis, 70-74. *Mem:* Inst Med-Nat Acad Sci; Am Soc Clin Invest (secy-treas, 55-57); master Am Col Physicians (pres, 76); Am Physiol Soc; Asn Am Physicians. *Res:* Clinical investigation of circulatory system of man, particularly the hepatic blood flow; diagnostic use of computers in clinical medicine. *Mailing Add:* 220 N Dithridge St Dithridge House No 900 Pittsburgh PA 15213

MYERS, JACK EDGAR, b Boyds Mills, Pa, July 10, 13; m 37; c 4. PHOTOBIOLOGY, PHOTOSYNTHESIS. *Educ:* Juniata Col, BS, 34; Mont State Col, MS, 36; Univ Minn, PhD(bot), 39. *Hon Degrees:* DSc, Juniata Col, 66. *Prof Exp:* Nat Res Coun fel, Smithsonian Inst, 39-41; asst prof physiol, 41-46, from assoc prof to prof zool, 46-55, prof, 55-83, EMER PROF BOT & ZOOL, UNIV TEX, AUSTIN, 83- *Concurrent Pos:* Guggenheim fel, 60;

sci ed, Highlights for Children, 61. *Honors & Awards:* Kettering Award, Am Soc Plant Physiologists, 74. *Mem:* Nat Acad Sci; Am Soc Plant Physiologists; Phycol Soc Am; Am Soc Photobiol (pres, 75). *Res:* Photosynthesis; plant pigments; biological effects of radiation; physiology of algae. *Mailing Add:* Dept Zool Univ Tex Austin TX 78712

MYERS, JACOB MARTIN, b Mercersburg, Pa, Aug 16, 19; m 45; c 2. PSYCHIATRY. *Educ:* Princeton Univ, AB, 40; Johns Hopkins Univ, MD, 43; Am Bd Psychiat & Neurol, dipl, 49. *Prof Exp:* Exec med officer, 51-62, med dir, 62-70, psychiatrist-in-chief, 70-81, psychiatrist-in-residence, 81-85, EMER PSYCHIATRIST-IN-CHIEF, PA HOSP, 85-; EMER PROF PSYCHIAT, SCH MED, UNIV PA, 85-; CLIN PROF PSYCHIAT, JEFFERSON MED COL, 85- *Concurrent Pos:* From asst prof to prof psychiat, Sch Med, Univ Pa, 54-85, consult, Vet Admin Hosp, Coatesville, 55-65; US Naval Hosp, Philadelphia, 57-75; mem, Accreditation Coun, Psychiat Facil, Joint Comn Accreditation Hosps, 76-80; HCFA consult, 85- *Honors & Awards:* Bowis Award, Am Col Psychiatrists, 75. *Mem:* AAAS; Am Psychopath Asn; AMA; fel Am Psychiat Asn; Am Col Psychiat (pres, 71-72). *Res:* Clinical evaluation of treatment of hospitalized psychiatric patients. *Mailing Add:* 141 Seminary Ave Gettysburg PA 17325

MYERS, JAMES HURLEY, b Memphis, Tenn, Sept 28, 40; m 63; c 2. PHYSIOLOGY. *Educ:* Memphis State Univ, BS, 63; Univ Tenn, Memphis, PhD(physiol), 69. *Prof Exp:* Instr physiol, Memphis State Univ, 68-69; asst prof, 71-77, ASSOC PROF PHYSIOL, SOUTHERN ILL UNIV, CARBONDALE, 77- SCH MED, 71-, CURRIC COORDR, SCH & REP TO AM ASN MED COLS GROUP STUDENT AFFAIRS, 71- *Concurrent Pos:* USPHS fel biol, Brookhaven Nat Lab, 69-71. *Mem:* Assoc Am Physiol Soc; Geront Soc. *Res:* Physiology of circulation; radiation injury in primates; radioisotope techniques; physiology of aging. *Mailing Add:* Dept Physiol Sch Med Southern Ill Univ Carbondale IL 62901

MYERS, JAMES R(USSELL), b Middletown, Ohio, June 17, 33. METALLURGY. *Educ:* Univ Cincinnati, BS, 56; Univ Wis, MS, 57; Ohio State Univ, PhD(metall eng), 64. *Prof Exp:* Proj engr, Wright Air Develop Div, Wright-Patterson AFB, 59-60, mat engr, Aeronaut Systs Div, 60-62; assoc prof metall, Sch Civil Eng, US Air Force Inst Technol, 62-71, prof, 71-79; DIR, JRM ASSOCS, 79- *Concurrent Pos:* Corrosion consult, US Air Force Civil Eng, worldwide. *Mem:* Nat Asn Corrosion Engrs; Am Soc Metals; Am Inst Mining, Metall & Petrol Engrs; fel Brit Inst Corrosion Sci & Technol. *Res:* Thermodynamic activity measurements; stress-corrosion cracking; oxidation; anodic polarization behavior; corrosion in potable waters; protective coating; phase diagram determinations; cathodic protection. *Mailing Add:* 719 Charles St Middletown OH 45042

MYERS, JAMES ROBERT, b Columbus, Ind, June 26, 54; m 79; c 2. PLANT BREEDING & GENETICS, SOMATIC CELL GENETICS. *Educ:* Kans State Univ, BS, 78; Univ Wis, Madison, MS, 81, PhD(plant breeding & genetics), 84. *Prof Exp:* Fel, Dept Agronomy, Univ Ky, Lexington, 84-85, res specialist, 85-87; ASST PROF PLANT SOILS & ENTOM SCI, UNIV IDAHO RES & EXT CTR, 87- *Mem:* AAAS; Crop Sci Soc Am; Am Soc Agronomy; Sigma Xi. *Res:* Genetics of dry bean, phaseolus vulgaris, at the whole-plant and cellular levels with development and release of improved bean cultivars. *Mailing Add:* Res & Extension Ctr Univ Idaho Kimberly ID 83341

MYERS, JEFFREY, b Philadelphia, Pa, Feb 8, 32; div; c 1. PATHOLOGY, INFORMATION SCIENCE. *Educ:* Univ Pa, AB, 52; Temple Univ, MD, 57, MSc, 62; McGill Univ, PhD(path), 65. *Prof Exp:* Pathologist, Allentown Hosp, 65-66; pathologist, Philadelphia Gen Hosp, 66-69, chief surg path, 69-77; DIR LAB & CHIEF PATH, BURLINGTON COUNTY MEM HOSP, 77- *Concurrent Pos:* Asst prof, Sch Med, Temple Univ, 65-66; consult, Wyeth Labs, 66-77; asst prof path, Sch Med, Univ Pa, 66-78. *Mem:* AAAS; fel Col Am Path; NY Acad Sci. *Res:* Use of computers to analyze pathology data; sources of errors in medical information systems; validity of computerized information. *Mailing Add:* 175 Madison Ave Mt Holly NJ 08060

MYERS, JOHN ADAMS, b Elk Point, SDak, June 12, 32; m 58; c 5. HEAT AND MASS TRANSFER, APPLIED MATHEMATICS. *Educ:* Univ Kans, BS, 58, MS, 60, PhD(heat transfer), 64. *Prof Exp:* Instr, Univ Kans, 59-63; from asst prof to assoc prof, 63-82, PROF CHEM ENG, VILLANOVA UNIV, 82- *Concurrent Pos:* Consult to chem indust, 64- *Mem:* Am Inst Chem Engrs; Am Soc Eng Educ; Sigma Xi. *Res:* Heat transfer accompanied by phase change; optimal design. *Mailing Add:* Dept Chem Eng Villanova Univ Villanova PA 19085

MYERS, JOHN ALBERT, b Sandusky, Ohio, Mar 13, 43; m 73; c 5. ORGANIC CHEMISTRY. *Educ:* Carson-Newman Col, BS & BA, 65; Univ Fla, PhD(chem), 70. *Prof Exp:* Res grant, Mich State Univ, 70-71; assoc prof, 71-80, PROF CHEM, NC CENT UNIV, 80- *Mem:* Am Chem Soc. *Res:* Heterocyclic synthesis; radiation sensitizers; dipolar cycloaddition reactions. *Mailing Add:* Dept Chem NC Cent Univ Box 19791 Durham NC 27707-0099

MYERS, JOHN E(ARLE), b Swalwell, Alta, Sept 14, 23; nat US; m 46; c 2. CHEMICAL ENGINEERING. *Educ:* Univ Alta, BSc, 44; Univ Toronto, MASc, 46; Univ Mich, PhD(chem eng), 52. *Prof Exp:* Lectr chem eng, Univ Toronto, 46-47; instr, Univ Mich, 47-50; from asst prof to prof, Purdue Univ, 50-66; chmn dept chem & nuclear eng, 66-76, PROF CHEM ENG, UNIV CALIF, SANTA BARBARA, 66-, DEAN COL ENG, 76- *Concurrent Pos:* Fulbright lectr, Univ Leeds, 56-57 & Univ Toulouse, 63-64. *Mem:* Am Chem Soc; Am Inst Chem Engrs; Sigma Xi. *Res:* Heat transfer; fluid flow. *Mailing Add:* Dept of Chem & Nuclear Eng Univ of Calif Santa Barbara CA 93106

MYERS, JOHN MARTIN, b Portland, Ore, June 8, 35; m 59; c 3. APPLIED PHYSICS, RELATIVITY. *Educ:* Calif Inst Technol, BS, 56; Harvard Univ, MS, 57, PhD(appl physics), 62. *Prof Exp:* Jr engr, Raytheon Co, 56-57, engr, 57-60, res scientist, 60-62, sr res scientist, 62-65, prin res scientist, 65-67; opers res analyst, Off Asst Secy Defense, 67-68; asst adminr, Model City Admin, Boston, 68-70; CONSULT, 70- *Concurrent Pos:* Consult, timing & synchronization, 70- *Res:* General relativity and synchronization; boundary value problems; systems research; concurrency, choice; electromagnetic theory; principles for biological structure. *Mailing Add:* 102 Chestnut St Boston MA 02108

MYERS, LAWRENCE STANLEY, JR, b Memphis, Tenn, Apr 29, 19; m 42; c 3. RADIATION BIOPHYSICS, ENVIRONMENTAL SCIENCES. *Educ:* Univ Chicago, 41, PhD(phys chem), 49. *Prof Exp:* Asst chem, Metall Lab, Manhattan Eng Dist, Chicago, 42-44; assoc chemist, Clinton Lab, Tenn, 44-46; asst, Inst Nuclear Studies, Univ Chicago, 46-48, chemist, Univ, 48-49; assoc chemist, Argonne Nat Lab, 49-52; asst prof biophys, nuclear med & radiol, Sch Med, Univ Calif, Los Angeles, 53-70, res radiobiologist, Lab Nuclear Med & Radiation Biol, 53-76, lectr radiol, Sch Med, 70-76, adj prof radiol sci, 76-82, adj prof radiation oncol, 80-82; sci adv, Armed Forces Radiobiol Res Inst, 82-87; RETIRED. *Concurrent Pos:* Biophysicist, Biol Br, Div Biol & Med, AEC, 72-74; assoc ed, Radiation Res, 74-77; vis scientist, Armed Forces Radiobiol Res Inst, 87- *Mem:* Am Soc Photobiol; AAAS; Radiation Res Soc; Biophys Soc; Sigma Xi; Health Physics Soc; NY Acad Sci. *Res:* Effects of ionizing radiation and environmental contaminants on nucleic acids, nucleoproteins and simple biological systems. *Mailing Add:* 11810 Coldstream Dr Potomac MD 20854-3612

MYERS, LLOYD E(LDRIDGE), land & water resources, for more information see previous edition

MYERS, LYLE LESLIE, b Salem, Ore, June 11, 38; m 60; c 3. BIOCHEMISTRY, IMMUNOCHEMISTRY. *Educ:* Ore State Univ, BS, 60; Mont State Univ, MS, 62; Purdue Univ, PhD(biochem), 66. *Prof Exp:* From asst prof to assoc prof, 66-78, PROF VET BIOCHEM, VET RES LAB, MONT STATE UNIV, 79- *Mem:* Conf Res Workers Animal Dis; Am Soc Microbiol. *Res:* Immunological and biochemical characteristics of bacteria that cause neonatal enteritis of young livestock. *Mailing Add:* Vet Res Lab Mont State Univ Bozeman MT 59717

MYERS, MARCUS NORVILLE, b Boise, Idaho, May 30, 28; m 50; c 3. ANALYTICAL CHEMISTRY. *Educ:* Brigham Young Univ, BS, 50, MS, 52; Univ Utah, PhD(phys chem), 65. *Prof Exp:* Engr, Hanford Works, Gen Elec Co, 51-57, chemist, Idaho, 57-61, Vallecitos Atomic Lab, 61-62; from res asst to res assoc, 62-67, asst res prof, 67-68, ASSOC RES PROF PHYS CHEM, UNIV UTAH, 78- *Mem:* Am Chem Soc. *Res:* Field flow fractionation; theory of all forms of chromatography; high pressure gas chromatography. *Mailing Add:* Dept Chem Univ Utah Salt Lake City UT 84112

MYERS, MARK B, b Winchester, Ind, Oct 14, 38; m 59; c 4. MATERIALS SCIENCE, CERAMICS. *Educ:* Earlham Col, AB, 60; Pa State Univ, PhD(solid state technol), 64. *Prof Exp:* Mem sci staff, Xerox Res Labs, NY, 64-68, mgr mat sci br, 68-71, mgr mat res lab, 71-75, mgr, Res Ctr Can Ltd, 75-78, VPRES, MAT ENG DEPT, XEROX RES LABS, NY, 78- *Concurrent Pos:* Assoc prof, Univ Rochester, 70-75. *Mem:* Am Phys Soc; Am Ceramic Soc. *Res:* Thermodynamics and kinetics of glass formation; phase transitions; glass transition phenomena; two phase glass ceramics; chalcogenide materials. *Mailing Add:* 2204 Five Mile Line Rd Pensfield NY 14526

MYERS, MAUREEN, b Providence, RI, Aug 28, 43. MEDICAL RESEARCH. *Educ:* Brown Univ, BA, 64; Georgetown Univ, PhD(microbiol), 75. *Prof Exp:* Lab technician, Virus Diag Lab, Wash Hosp Ctr, 65-66; med technician, Virus Diag Lab, Clin Ctr, NIH, 66-67, biologist, Nat Cancer Inst, 67-71, staff fel, Lab Exp Path, Nat Inst Arthritis, Metab & Digestive Dis, 75-78, sr staff fel, 78-79, grants assoc, Div Res Grants, 79-80, antiviral substances prog officer, Develop & Appl Br, Microbiol & Infectious Dis Prog, Nat Inst Allergy & Infectious Dis, 80-86, chief, Treatment Res Br, AIDS Prog, 86-89, asst dir, Div AIDS, 89-90, asst dir prev, Off of Dir, 90-91; SR ASSOC DIR, VIROL GROUP, CLIN RES, BOEHRINGER INGELHEIM PHARMACEUT INC, 91- *Concurrent Pos:* Chmn, Interferon Nomenclature Comt, 81-85; exec secy, Interferon Info Exchange Group, 81-85; mem, Drug Selection Comt, Nat Inst Allergy & Infectious Dis/Nat Cancer Inst, 85-86; chmn, Data & Safety Monitoring Bd, Burroughs Wellcome, 86; mem, AIDS Preclin Decision Network Comt, 87, AIDS Clin Drug Develop Comt, 87-90. *Mem:* Infectious Dis Soc Am; Am Soc Microbiol; AAAS; Am Soc Virol; Sigma Xi; Inter-Am Soc Chemother; Int AIDS Soc; Int Soc Interferon Res; Int Soc Antiviral Res; Am Diabetes Asn. *Res:* Therapies for HIV infection and its sequelae. *Mailing Add:* Boehringer Ingeheim Pharmaceut Inc 90 E Ridge Box 368 Ridgefield CT 06877

MYERS, MAX H, b Lynchburg, Va, July 2, 36; m 59; c 2. BIOMETRICS, STATISTICS. *Educ:* Bridgewater Col, BA, 58; Va Polytech Inst, MS, 60; Univ Minn, PhD(biomet), 71. *Prof Exp:* USPHS officer, End Results Sect, Biomet Br, 60-62, math statistician, 62-73, HEAD END RESULT SECT, BIOMET BR, FIELD STUDIES & STATIST, DIV CANCER CAUSE & PREV, NAT CANCER INST, 73- *Mem:* Am Statist Asn; Biomet Soc. *Res:* Epidemiology of cancer patient survival including detailed study of factors related to prognosis; statistical methodology for evaluating multifactor relationships to survival. *Mailing Add:* 24916 Woodfield School Rd Gaithersburg MD 20882

MYERS, MELVIL BERTRAND, JR, b New Orleans, La, Sept 12, 28; m 54; c 3. MEDICINE. *Educ:* Tulane Univ, MD, 51; Am Bd Surg, dipl, 58. *Prof Exp:* Clin instr, 56-71, assoc prof, 71-75, PROF SURG, SCH MED, LA STATE UNIV MED CTR, NEW ORLEANS, 75-; PRIN INVESTR SURG RES, VET ADMIN HOSP, NEW ORLEANS, 63-, SR SURGEON, TOURO INFIRMARY, 65- *Concurrent Pos:* Grants, Am Cancer Soc, Southeast Surg Cong, Ethicon, Inc, Warren-Teed, Inc & John A Hartford Found; sr surgeon, Charity Hosp, New Orleans, 61-; attend physician, US Vet Admin Hosp, 66-71, staff physician, 71- *Mem:* Fel Am Col Surgeons; Am Heart Asn; Plastic Surg Res Coun; Southern Surg Asn; Am Soc Plastic &

Reconstruct Surgeons. *Res:* Wound healing and revascularization; cause of tissue necrosis; tissue changes following devascularization and revascularization; mechanism of ventricular fibrillation following ischemia; various clinical surgical problems. *Mailing Add:* Dept Surg Vet Admin Hosp Vet Admin Med Ctr 1601 Perdido New Orleans LA 70146

MYERS, MICHAEL KENNETH, b Portland, Ore, May 18, 39; m 62, 79; c 2. FLUID MECHANICS, ACOUSTICS. *Educ:* Willamette Univ, BA, 62; Columbia Univ, BS, 62, MS, 63, PhD(eng mech), 66. *Prof Exp:* From asst prof to assoc prof civil eng & eng mech, Columbia Univ, 66-73; assoc prof, 73-78, PROF, JOINT INST ADVAN FLIGHT SCI & TECH DIR AEROACOUSTICS PROG, GEORGE WASHINGTON UNIV, 78- *Concurrent Pos:* Mem, Eng Mech Div Elasticity Comt, Am Soc Civil Engrs, 76-84; mem, Aeroacoustics Tech Comt, Am Inst Aeronaut & Astronaut, 79-82 & 88, chmn, Aeroacoust Tech Comt, 91-; gen chmn, 12th Aeroacoustics Conf, Am Inst Aeronaut & Astronaut, 89; assoc ed, J Am Inst Aeronaut & Astronaut, 91- *Mem:* Assoc fel Am Inst Aeronaut & Astronaut; Soc Eng Sci; Acoust Soc Am. *Res:* Wave propagation; propagation of sonic booms; asymptotic solution of hyperbolic partial differential equations; linear and nonlinear acoustics; duct acoustics; atmospheric sound propagation. *Mailing Add:* NASA Langley Res Ctr JIAFS Mail Stop 269 Hampton VA 23665-5225

MYERS, OVAL, JR, b Roachdale, Ind, July 28, 33; m 59; c 2. PLANT GENETICS, PLANT BREEDING. *Educ:* Wabash Col, BA, 58; Dartmouth Col, MA, 60; Cornell Univ, PhD(genetics), 63. *Prof Exp:* From instr to asst prof bot & bact, Univ Ark, Fayetteville, 63-68; assoc prof bot & plant indust, 68-75, PROF PLANT & SOIL SCI, SOUTHERN ILL UNIV, CARBONDALE, 75- *Concurrent Pos:* Educ specialist, Southern Ill Univ & Food & Agr Orgn, Brazil, 72-74; maize breeding consult, Southern Ill Univ & USAID, Zambia, Africa, 83-85. *Mem:* Fel AAAS; Sigma Xi; Am Soc Agron. *Res:* Genetics and plant breeding of Zea mays and Glycine max; developmental plant morphology. *Mailing Add:* Dept of Plant & Soil Sci Southern Ill Univ Carbondale IL 62901

MYERS, PAUL WALTER, b Schenectady, NY, Jan 15, 23; m 44; c 4. NEUROSURGERY. *Educ:* Albany Med Col, MD, 46; Am Bd Neurol Surg, dipl, 60. *Hon Degrees:* DSc, Albany Med Col. *Prof Exp:* Intern med, Ellis Hosp, Schenectady, NY, 46-47; US Air Force, 51-, resident surg, Ellis Hosp, 52-53, resident neurol surg, Albany Med Ctr, 53-56, chief neurol surg, 58-71, commander, Wilford Hall Air Force Med Ctr, 71, surgeon gen, US Air Force, 78-82; PROF NEUROSURG, UNIFORMED SERV UNIV, EBERT SCH MED, 82- *Concurrent Pos:* Hon flight surgeon designation, Govt Chile, 68; bd trustees, Falcon Found, USAF Acad; clin prof neurosurg, Univ Tex Health Sci Ctr, San Antonio. *Mem:* AMA; Cong Neurol Surgeons; Am Asn Neurol Surg; Am Col Surgeons; Soc Neurol Surgeons. *Res:* Cervical injuries; neuroanatomy; combat induced penetrating cerebral wounds. *Mailing Add:* Lt Gen USAF RET 1119 Homeric Dr San Antonio TX 78213

MYERS, PETER BRIGGS, b Washington, DC, Apr 24, 26; m 48, 73; c 2. NUCLEAR PHYSICS, SYSTEMS ANALYSIS. *Educ:* Worcester Polytech Inst, BSEE, 46; Oxford Univ, PhD(physics), 50. *Hon Degrees:* DHumLitt, Col Idaho, 73. *Prof Exp:* Rhodes scholar, 47-50; mem tech staff, Switching Res Dept, Bell Tel Labs, Inc, 50-59; staff scientist, Motorola Inc, 59-62; mgr res & advan technol, Martin Co, 62-64; mgr res & advan technol, Bunker-Romo Corp, 64-65, vpres res & develop, 65-66; mgr advan technol, Magnavox Res Labs, Calif, 66-74, dir, Advan Systs Anal Off, Magnavox Co, Md, 74-78, dir fed liason, 78-79; STAFF DIR, BD RADIOACTIVE WASTE MGT, NAT RES COUN, NAT ACAD SCI, 79- *Mem:* Fel AAAS; Inst Mgt Sci; Opers Res Soc Am; Am Phys Soc; fel Inst Elec & Electronics Engrs; Sigma Xi. *Res:* Solid state device physics; solid state circuits; magnetic logic and memory; communications navigation and position location systems analysis and technology; solid state integrated circuits. *Mailing Add:* 2101 Const Ave NW Washington DC 20418

MYERS, PHILIP, b Baltimore, Md, June 10, 47; m 69; c 1. MAMMALOGY, SYSTEMATICS. *Educ:* Swarthmore Col, BA, 69; Univ Calif, Berkeley, PhD(zool), 75. *Prof Exp:* Asst prof, 75-82, ASSOC PROF ZOOL, UNIV MICH, ANN ARBOR, 82- *Mem:* Assoc Sigma Xi; Am Soc Mammalogists; Asn Syst Collections; Soc Syst Zool. *Res:* Ecology and evolution of mammals; population biology; biosystematics of mammals. *Mailing Add:* Mus of Zool Univ of Mich Ann Arbor MI 48109

MYERS, PHILIP CHERDAK, b Elizabeth, NJ, Nov 18, 44; m 72; c 3. RADIO ASTRONOMY. *Educ:* Columbia Univ, AB, 66; Mass Inst Technol, PhD(physics), 72. *Prof Exp:* Staff scientist radio physics & astron, Res Lab Electronics, Mass Inst Technol, 72-75, from asst prof to assoc prof physics, 75-82; ASTROPHYSICIST, SMITHSONIAN ASTROPHYS OBSERV & LECTR ASTRON, HARVARD UNIV, 84- *Concurrent Pos:* Sr res assoc, Goddard Inst Space Studies, NASA, 79-80 & 82-84; vis scientist, Ctr Astrophys, 82-84. *Mem:* Am Astron Soc; Int Union Radio Sci; Int Astron Union; AAAS. *Res:* Molecular clouds, star formation, magnetic fields. *Mailing Add:* Ctr For Astrophys MS 42 Harvard Univ Cambridge MA 02138

MYERS, PHILLIP S(AMUEL), b Webber, Kans, May 8, 16; m 43; c 5. MECHANICAL ENGINEERING. *Educ:* McPherson Col, BS, 40; Kans State Col, BS, 42; Univ Wis, PhD(mech eng), 47. *Prof Exp:* From instr to prof, 42-86, EMER PROF MECH ENG, UNIV WIS-MADISON, 86- *Concurrent Pos:* Consult, US Army, Echlin, Deere, Nelson Industs, NAS Comts. *Honors & Awards:* Horning Mem Award, Soc Automotive Engrs; Medal of Honor Award, Soc Automotive Engrs; Diesel & Gas Power Awards, Am Soc Mech Engrs; Sci Medal, Japan Soc Automotive Engrs; Dugald Clerk Award, Inst Mech Engrs. *Mem:* Nat Acad Eng; AAAS; fel Am Soc Mech Engrs; fel Soc Automotive Engrs (pres, 69). *Res:* Combustion in internal combustion engines; diesel combustion; thermodynamics; heat power; flame-temperature measurements in internal combustion engines; pyrolysis of propane; fuel sprays and vaporization; welding heat transfer. *Mailing Add:* Dept Mech Eng Univ Wis 1513 University Ave Madison WI 53706

MYERS, R(ALPH) THOMAS, b Maidsville, WVa, Mar 28, 21; div; c 3. PHYSICAL INORGANIC CHEMISTRY. *Educ:* WVa Univ, AB, 41, PhD(phys org chem), 49. *Prof Exp:* Asst, Manhattan Proj, Columbia Univ, 44-46; assoc prof chem & head dept, Waynesburg Col, 48-51; asst prof phys chem & consult, Res Found, Colo Sch Mines, 51-56; from asst prof to assoc prof chem, 56-77, PROF CHEM, KENT STATE UNIV, 77- *Mem:* AAAS; Am Chem Soc; Sigma Xi. *Res:* Nonaqueous solvents; dielectric constant; forces between molecules in liquids. *Mailing Add:* Dept Chem Kent State Univ Kent OH 44242-0002

MYERS, RAYMOND HAROLD, b Charleston, WVa, Oct 13, 37; m 59; c 2. MATHEMATICAL STATISTICS. *Educ:* Va Polytech Inst, BSc, 59, MSc, 61, PhD(statist), 64. *Prof Exp:* From asst prof to assoc prof, 63-71, PROF STATIST, VA POLYTECH INST & STATE UNIV, 71- *Mem:* Am Statist Asn. *Res:* Experimental design and analysis; response surface techniques. *Mailing Add:* Dept of Statist Va Polytech Inst & State Univ Blacksburg VA 24061

MYERS, RAYMOND REEVER, b New Oxford, Pa, Jan 23, 20; m 43; c 3. CHEMISTRY. *Educ:* Lehigh Univ, AB, 41, PhD(chem), 52; Univ Tenn, MS, 42. *Prof Exp:* Res chemist, Cent Res Labs, Monsanto Co, 42-46 & Jefferson Chem Co, 46-50; asst chem, Lehigh Univ, 50-52, res assoc, 52-53, from res asst prof to res prof, 53-65; prof chem, Kent State Univ, 65-85, chmn dept, 65-77; RETIRED. *Concurrent Pos:* Res dir, Paint Res Inst, 64-83; former consult, Nat Bur Standards, R T Vanderbilt Co & Air Reduction Co; ed, J Rheol. *Honors & Awards:* Borden Award, Am Chem Soc, 71; Morrison Award, NY Acad Sci, 58; Heckel Award, Fedn Soc Coatings Technol, 73, Matiello lectr, 75; Matiello Lectr, Fedn Socs Coatings Technol, 75; First Distinguished Award, Soc Rheology, 85. *Mem:* Am Chem Soc; fel Am Inst Chemists; Soc Rheol; Brit Soc Rheol; fel NY Acad Sci. *Res:* Rheology of coatings; adhesion; application of spectra in catalysis; structure of matter; research administration. *Mailing Add:* 43 Wall St Bethlehem PA 18018

MYERS, RICHARD F, b Hammond, Ind, Feb 1, 31; m 51; c 4. VERTEBRATE ZOOLOGY, ACADEMIC ADMINISTRATION. *Educ:* Earlham Col, AB, 52; Cornell Univ, MS, 54; Univ Mo, PhD, 64. *Prof Exp:* From asst prof to assoc prof zool, Cent Mo State Univ, 59-67; assoc prof zool & ecol, Univ Mo-Kansas City, 67-72; dir, Nat Weather Serv Training Ctr, 72-91; RETIRED. *Concurrent Pos:* Inst Int Educ-US AID consult, Bangladesh, 69-70; US consult training & educ, Iranian Meteorol Orgn, 78. *Honors & Awards:* Silver Medal, US Dept Com, 81. *Mem:* Am Soc Mammals; Sigma Xi; Am Meteorol Soc. *Res:* Mammalogy; wildlife biology; movement and migration patterns; population ecology, especially bats. *Mailing Add:* 13404 Woodland Kansas City MO 64146

MYERS, RICHARD HEPWORTH, b Austin, Tex, Nov 1, 47; m 78; c 2. BEHAVIOR GENETICS, CYTOGENETICS. *Educ:* Univ Kans, BA, 69; Ga State Univ, MEd, 73, MA, 76, PhD(psychol), 79. *Prof Exp:* Teacher spec educ, 69-74; psychol tech, Genetics Lab, Ga Mental Health Inst, 75-80; clin asst prof, Dept Psychiat, Emory Med Sch, 80; asst prof, 80-87, ASSOC PROF, DEPT NEUROL, BOSTON UNIV MED CTR, 87-, NEUROPSYCHOLOGIST,UNIV HOSP, 80- *Concurrent Pos:* Consult genetics, Dept Neurol, Mass Gen Hosp, 87-; lectr, Dept Neurol, Harvard Med Sch, 80- *Mem:* Behav Genetics Asn; AAAS; Am Soc Human Genetics; Am Psychol Asn. *Res:* Midlife onset inherited neurological disorders and factors related to age of onset and rate of progression; genetic testing by DNA polymorphism linkage analysis. *Mailing Add:* Dept Neurol Boston Univ Sch Med 720 Harrison Ave Suite 1105 Boston MA 02118

MYERS, RICHARD LEE, b Doylestown, Pa, Oct 26, 44; m 66. ANALYTICAL CHEMISTRY. *Educ:* Calif Inst Technol, BS, 66; Univ Wis-Madison, PhD(analytical chem), 71. *Prof Exp:* Mem tech staff, NAm Rockwell Sci Ctr, 71-73, prog mgr, Air Monitoring Ctr, Rockwell Int Corp, 73-75, cent region dir, Environ Monitoring & Serv Ctr, 75-80; mem staff, 80-81, PRES, MEAD COMPUCHEM, MEAD TECH LABS, 81- *Mem:* Am Chem Soc; Electrochem Soc. *Res:* Application of gas chromatography-mass spectrometry and minicomputers to chemical and physical measurements; ambient air and water pollution measurement techniques and instrumentation; environmental systems and technology. *Mailing Add:* 4831 Saxlin Dr No 219 New Smyrna Beach FL 32169-4419

MYERS, RICHARD SHOWSE, b Jackson, Miss, Oct 26, 42; m 65; c 2. PHYSICAL CHEMISTRY. *Educ:* Miss Col, BS, 64; La State Univ, MS, 66; Emory Univ, PhD(chem), 68. *Prof Exp:* From asst prof to assoc prof, 68-77, PROF CHEM, DELTA STATE UNIV, 77-, DIR, INSTNL RES & PLANNING, 89- *Mem:* Am Chem Soc; Sigma Xi. *Res:* Formulation and properties of microemulsions; surface thermodynamics and surface tension of nonelectrolyte solutions; kinetics of adsorption and desorption on soil. *Mailing Add:* 202 Kethley Delta State Univ Cleveland MS 38733

MYERS, ROBERT ANTHONY, b Brooklyn, NY, Feb 22, 37; m 67. SOLID STATE PHYSICS. *Educ:* Harvard Univ, AB, 58, AM, 59, PhD(appl physics), 64. *Prof Exp:* Physicist, IBM Watson Res Ctr, NY, 63-68, mem corp tech comt staff, IBM Corp, 68-72, mgr terminal technol, 73-83, mgr plans & oper systs, IBM Watson Res Ctr, 84-85; dir technol, IBM, Japan Yamato Lab, 86-87, tech asst lab dir, 87-89; PROG MGR, TECHNOL REV BD, IBM ENTERPRISE SYSTS, 89- *Concurrent Pos:* Secy, Sci Adv Comt, IBM Corp, 71-72; chair, indust adv comt, Inst imaging Sci, Polytech Inst, NY, 83-84; mem adv comt, Dept Mech & Aerospace Engr, Princeton Univ, 85-; mem adv comt, Inst Syst Sci, Nat Univ, Singapore, 86-87. *Mem:* Am Phys Soc; Inst Elec & Electronics Engrs. *Res:* Application of solid state technology to computer input/output devices. *Mailing Add:* IBM Corp Rte 100 Box 100 Somers NY 10587

MYERS, ROBERT DURANT, b Philadelphia, Pa, Oct 25, 31; m 53; c 4. PSYCHOBIOLOGY, NEUROBIOLOGY. *Educ:* Ursinus Col, BS, 53; Purdue Univ, MS, 54, PhD, 56. *Hon Degrees:* Dr, Univ Granada, Spain, 84. *Prof Exp:* Asst psychol, Purdue Univ, 54-55; PROF, PHARMACOL &

PSYCHIAT MED, SCH MED, EAST CAROLINA UNIV; fel, Neurol Sci Group, Sch Med, Johns Hopkins Univ, 60-61; prof psychol, Purdue Univ, 65-72, prof psychol & biol sci, 72-78, coordr neurobiol training prog, 70-78, dir psychobiol prog, 73-78; prof psychiat & pharmacol, Univ NC Sch Med, 78-87; dir, Bowles Biomed Res Lab, 82-87; CONSULT, 87- *Concurrent Pos:* Res psychologist, Rome Air Develop Ctr, Griffiss AFB, 57-58; consult, NSF, 60-, US Vet Admin, 72-, Med Res Coun Can & Australian Sci Res Coun, 75-, Schick Shadel Labs, 77-79, Nat Found March of Dimes, 74-; vis scientist, Nat Inst Res Eng, 63-65 & 69-70; chmn adv comt, psychol sci, Purdue Univ, 68-72, mem comt, 72-78; adv panel, NSF, Psychobiol Prog, 71-72, Neurobiol Prog, 72-74; Sigma Xi res award, Purdue Univ, 71; sci adv bd, Mary Cullen Res Trust, 77-83; vis prof, La Trobe Univ, Australia, 75 & 82; co-chmn, NY Acad Sci Conf Neurochem Anal, 85. *Honors & Awards:* Medal of Belgrade Univ, Yugoslavia, 85. *Mem:* Fel AAAS; Int Brain Res Orgn; NY Acad Sci; fel Am Col Neuropharmacol; Am Physiol Soc; Am Soc Pharmacol & Exp Therapeut; Int Soc Biomed Res Alcoholism; Soc Neurosci; Sigma Xi. *Res:* Neural mechanisms controlling feeding, drinking, emotional behavior and thermoregulation; transmitter synthesis, turnover and release in brain stem; physiology and pharmacology of hypothalamus; role of amines and ions; alcohol and drug administration. *Mailing Add:* Dept Pharmacol & Psychiat Med Sch Med E Carolina Univ Greenville NC 27858

MYERS, ROBERT FREDERICK, b Trenton, NJ, Feb 23, 16; m 40; c 3. METEOROLOGY, CHEMICAL ENGINEERING. *Educ:* Va Polytech Inst, BS, 36. *Prof Exp:* Observer, US Weather Bur, 39-41, jr instrument engr, 41-42; sr inspector eng, Bur Ord, US Navy, 42-43; meteorologist, US Weather Bur, 45-57, liaison officer, Ga, 47-48, meteorologist, Tenn, 48-54, meteorologist in charge res, 54-55; chief data handling br, US Air Force Cambridge Res Lab, 55-60, sr res engr, 60-81, CONSULT, AIR FORCE GEOPHYS LAB, 81- *Mem:* Am Meteorol Soc. *Res:* Meteorological instrumentation including data acquisition, communication and display; micrometeorological research; ozone research; meteorological satellite ground stations with integral interactive computer. *Mailing Add:* 80 Willow Rd Nahant MA 01908

MYERS, ROLLIE JOHN, JR, b Nebr, July 15, 24; m 50; c 2. PHYSICAL CHEMISTRY. *Educ:* Calif Inst Technol, BS, 47, MS, 48; Univ Calif, PhD(chem), 51. *Prof Exp:* From instr to assoc prof, 51-62, PROF CHEM, UNIV CALIF, BERKELEY, 62-, ASST DEAN COL CHEM, 73-, PRIN INVESTR, INORG MAT RES DIV, 73- *Concurrent Pos:* Guggenheim fel, 57-58; int fac award, Am Chem Soc-Petrol Res Fund, 65-66. *Mem:* Am Chem Soc; Am Phys Soc. *Res:* Spectroscopy; magnetic resonance; microwave and molecular structure. *Mailing Add:* Dept Chem Univ Calif 2120 Oxford St Berkeley CA 94720

MYERS, RONALD BERL, b Battle Creek, Mich, Jan 31, 44; m 67; c 2. BACTERIAL STRAIN DEVELOPMENT. *Educ:* Mich State Univ, BS, 67, MS, 68, PhD(bot), 74. *Prof Exp:* Teaching fel bact physiol, Mich State Univ, 74-79; res scientist, 79-84, mgr develop, 84-85, DIR PHENYLALANINE RES, RES & DEVELOP, NUTRASWEET CO, ILL, 85-, DIR FERMENTATION. *Mem:* Am Soc Microbiol; Soc Indust Microbiol. *Res:* Development of mutant bacterial strains to produce commercial quantities of L-phenylalanine; improvement of microbial transformation of sterols into steroids; fermentation development. *Mailing Add:* 813 Pine Forest Lane Prospect Heights IL 60070

MYERS, RONALD ELWOOD, b Chicago Heights, Ill, Sept 24, 29; m 57, 81; c 5. NEUROPATHOLOGY, PERINATOLOGY. *Educ:* Univ Chicago, AB, 50, PhD(neuroanat), 55, MD, 56. *Prof Exp:* Intern, Univ Chicago Clins, 56-57; res officer, Walter Reed Army Inst Res, 57-60; dir, Lab Neurol Sci, Spring Grove State Hosp, Baltimore, Md, 63-64; chief Lab Perinatal Physiol, Nat Inst Neurol Dis & Stroke, 64-80; ASSOC CHIEF STAFF FOR RES, CINCINNATI VET ADMIN MED CTR, 80- *Concurrent Pos:* Spec fel physiol & neurol med, Sch Med, Johns Hopkins Univ, 60-63; guest lectr, First Mem Ignatz Semmelweis Seminar, 75, Am Psychol Asn, 76, Int Neuropsychol Cong, 78 & Can Investr Reproduction, 80; distinguished lectr, Women & Infants' Hosp RI, 75; spec lectr, Swiss Asn Neuropathologists, 78; med legal consult, 83- *Honors & Awards:* Purkinje lectr, Slovak Acad Sci, 63; 14th Crittenden Mem lectr, Kansas City, 74; Hershenson lectr, Boston Hosp Women, 78; Albert von Bezold Medal, Soc Path & Clin Physiol, Ger Dem Rep, 87. *Mem:* Am Soc Neuropath; Am Physiol Soc; Am Acad Neurol; Soc Gynec Invest; Pavlovian Soc NAm. *Res:* Physiological pyschology; fiber connections of the brain; experimental neuropathology; perinatalogy. *Mailing Add:* Cincinnati VA Med Ctr 3200 Vine St Cincinnati OH 45220

MYERS, RONALD EUGENE, b Hanover, Pa, Aug 12, 47; m 72; c 2. INORGANIC MODIFICATION OF POLYMERIC MATERIALS. *Educ:* Gettysburg Col, BA, 69; Purdue Univ, PhD(inorg chem), 77. *Prof Exp:* Advan res & develop chemist, 77-80, sr res & develop chemist, 80-83, res & develop assoc, 83-88, SR RES & DEVELOP ASSOC, B F GOODRICH CO, 88- *Concurrent Pos:* Instr chem, Baldwin Wallace Col, 78; Mem, Am Ceramic Soc Presidential Comt Pre Col Educ, 90-; Vis Scholars Prog, Ohio Acad Sci, 90- *Mem:* Am Chem Soc; Am Ceramic Soc; AAAS; fel Am Inst Chemists; Sigma Xi. *Res:* Electrically conducting polymers; flame retardants for polymers; polyphosphazenes; specialty glasses and ceramics; inorganic modification of polymers; high temperature composites. *Mailing Add:* B F Goodrich Res & Develop Ctr 9921 Brecksville Rd Brecksville OH 44141

MYERS, RONALD FENNER, b East Haven, Conn, July 22, 30; m 59; c 3. NEMATOLOGY. *Educ:* Univ Conn, BS, 57, MS, 59; Univ Md, PhD(plant path, nematol), 64. *Prof Exp:* Nematologist, USDA, Md, 59-61; asst bot, Univ Md, 61-63; asst prof, Univ Conn, 64-65; from asst prof to assoc prof nematode physiol, 65-75, dir, Rhiz Res Group, 79-81, PROF PLANT PATH, RUTGERS UNIV, NEW BRUNSWICK, 75- *Concurrent Pos:* Res prof, Mem Univ Nfld, 73-74. *Mem:* Soc Nematol; Soc Europ Nematol. *Res:* Physiology and biochemistry of nematodes; culture and nutrition of nematodes; nematode detection, chemical control, and control recommendations. *Mailing Add:* Dept Plant Path Cook Col Rutgers Univ New Brunswick NJ 08903-0231

MYERS, RONALD G, b Eldon, Mo, Aug 12, 33; m 55; c 7. ELECTRICAL ENGINEERING. *Educ:* Univ Mo, BS, 56, MS, 57. *Prof Exp:* Res asst elec eng, Univ Mo, 56-57; asst, Argonne Nat Labs, 57-59; radar engr, Emerson Elec Co, 59-65; res specialist, Monsanto Co, Mo, 65-66, eng group leader digital prod, 66-69, eng mgr, Boulder Tech Ctr, 69-70; VPRES ENG & MEM BD DIRS, TECNETICS, INC, 70- *Concurrent Pos:* Consult digital instrument design & develop. *Mem:* Sr mem Inst Elec & Electronics Engrs. *Res:* Electromagnetic testing of metals; radar display and video processing; transient testing of systems; digital voltmeters, counter-timers and instruments; power supplies. *Mailing Add:* Dept Eng Motorola GEG 8201 E McDowell MS 2240 Scottsdale AZ 85252

MYERS, ROY MAURICE, b Scottdale, Pa, Sept 24, 11; m 39; c 4. BOTANY. *Educ:* Ohio State Univ, BSc, 34, MA, 37, PhD(plant physiol), 39. *Prof Exp:* Teacher high sch, Ohio, 34-35; asst bot, Ohio State Univ, 35-38 & Northwestern Univ, 38-40; instr, Boise Jr Col, 40-42; instr biol, Denison Univ, 42-45; from asst prof to prof biol, 45-78, cur, Herbarium, 69-78, EMER PROF BIOL, WESTERN ILL UNIV, 78- *Concurrent Pos:* Chmn, Dept Biol Sci, Western Ill Univ, 53-69; mem dirs & consult, UN Asn, 82- *Mem:* AAAS; Am Inst Biol Sci; Soc Econ Bot; Sigma Xi; UN Asn. *Res:* Economic botany; plant taxonomy. *Mailing Add:* 23463 Weschester Blvd Port Charlotte FL 33980

MYERS, SAMUEL MAXWELL, JR, b Florence, SC, Jan 30, 43. SOLID STATE PHYSICS. *Educ:* Duke Univ, BS, 65, PhD(physics), 70. *Prof Exp:* Sandia Corp fel, 70-72, MEM STAFF, SANDIA NAT LABS, 72- *Mem:* Am Phys Soc; Am Inst Metall Engrs Metall Soc. *Res:* Ion implantation and ion beam analysis are employed in fundamental studies of metals; hydrogen effects, diffension and tapping of solutes; phase diagrams, metastalle phases and corrosion. *Mailing Add:* Org 1112 Sandia Nat Labs Albuquerque NM 87185

MYERS, STEVEN RICHARD, b Russell, Ky, July 17, 56. CHEMICAL CARCINOGENESIS, HEMOGLOBIN-XENOBIOTIC ADDUCTS. *Educ:* Univ Ky, Lexington, BS, 79, PhD(pharmacol), 86; Marshall Univ, Huntington, WVa, MS, 82. *Prof Exp:* Res asst, Dept Pharmacol, Univ Ky Col Med, 83-86, asst res prof, 88-91; postdoctoral, Div Biol & Med Res, Argonne Nat Labs, Univ Chicago, 86-87; ASST PROF, DEPT PHARMACOL & TOXICOL, UNIV LOUISVILLE SCH MED, 91- *Mem:* Sigma Xi; Am Asn Cancer Res; Soc Toxicol; Am Chem Soc. *Res:* Chemical and biochemical characterization of protein-carcinogen adducts; use of hemoglobin as a biomarker in exposure to xenobiotics, drug metabolism, analytical techniques in drug metabolism and DNA-protein adducts. *Mailing Add:* Dept Pharmacol & Toxicol Univ Louisville Sch Med Louisville KY 40292

MYERS, THOMAS DEWITT, b Wilmington, Del, Apr 8, 38; m 61; c 3. OCEANOGRAPHY. *Educ:* Bridgewater Col, BA, 59; Univ NC, MA, 65; Duke Univ, PhD(zool-oceanog), 68. *Prof Exp:* Fishery biologist, Biol Lab, US Bur Commercial Fisheries, 61-62; asst prof biol sci, Univ Del, 67-70, asst prof marine studies, 70-74; mgr, Life Systs Dept, Roy F Weston, Inc, 74-77; res dir, Booz-Allen & Hamilton, Inc, 77-82; pres, Thomas D Myers & Assoc, 82-89; VPRES, APPLIED ENVIRON INC, 89- *Concurrent Pos:* Adj asst prof marine sci, Univ Del, 74-78. *Mem:* AAAS; Am Chem Soc; Water Pollution Control Fedn. *Res:* Environmental management audits; environmental impact of advanced energy technology; marine pollution and ocean waste disposal. *Mailing Add:* Applied Environ Inc 11800 Sunrise Valley Dr Suite 1200 Reston VA 22091

MYERS, THOMAS WILMER, b Lamar, Mo, Nov 25, 39; m 61; c 3. MECHANICAL ENGINEERING. *Educ:* Univ Mo, BS, 61, MS, 63, PhD(mech eng), 65. *Prof Exp:* Instr thermodyn heat transfer, Univ Mo, 64-65; vis res assoc, Aerospace Res Labs, Wright-Patterson AFB, 65-66; res scientist, 66-76, GROUP LEADER, TECH CTR, DEERE & CO, 76- *Mem:* Am Soc Mech Engrs; Sigma Xi. *Res:* Fluid mechanics; hydraulic and electrohydraulic controls and control systems. *Mailing Add:* 757 E Valley Dr Rte 1 Bettendorf IA 52722

MYERS, VERNE STEELE, b Hillsdale, Mich, Apr 11, 07; m 32; c 2. STATISTICS. *Educ:* Hillsdale Col, BS, 30; Columbia Univ, BS, 32, MS, 35. *Prof Exp:* Develop engr, Tex Corp, 32-34; develop engr, M W Kellogg Co, 34; econ statistician, Tidewater Oil Co, 35-41; supvr mgt planning, Lockheed Aircraft Corp, 41-71; MGT ENG CONSULT, 71- *Concurrent Pos:* Lectr, Pasadena City Col, 54-56; lectr, Univ Southern Calif, 56-68. *Mem:* Economet Soc; Am Inst Indust Engrs. *Res:* Statistical relationships existing in economic factors in the general economy; the petroleum industry; the aircraft industry; specific corporations; statistical engineering applied to wholistic health practice. *Mailing Add:* 4610 Commonwealth La Canada CA 91011

MYERS, VERNON W, b New Castle, Pa, Feb 16, 19; m 47; c 5. PHYSICS. *Educ:* Geneva Col, BS, 40; Syracuse Univ, MA, 42; Yale Univ, PhD(physics), 47. *Prof Exp:* Instr physics, Yale Univ, 43-44; physicist, Naval Res Lab, DC, 44 & Argonne Nat Lab, 47-48; from asst prof to assoc prof physics, Pa State Univ, 48-63; physicist, Nat Bur Standards, 63-81; CONSULT, 81- *Concurrent Pos:* Guest scientist, Brookhaven Nat Lab, 52-53, 60-61 & 63-66; Fulbright prof, Univ Philippines, 61-62. *Mem:* Am Phys Soc. *Res:* Molecular quantum mechanics; nuclear physics. *Mailing Add:* 7704 Hacker Dr Minocqua WI 54548

MYERS, VICTOR (IRA), b Casa Blanca, NMex, June 8, 21; m 40; c 3. AGRICULTURAL ENGINEERING. *Educ:* Univ Idaho, BS, 49, MS, 55. *Hon Degrees:* PhD, Univ Idaho, 82. *Prof Exp:* Civil engr, Soil Conserv Serv, 49-51, civil engr hydraul eng, Ore, 51-52; assoc prof agr eng, Univ Idaho, 52-56; proj supvr irrig & drainage res, Agr Res Serv, USDA, Nev, 56-60, res invests leader, Soil & Water Lab, Tex, 60-69; dir, Remote Sensing Inst, S Dak State Univ, 69-86; RETIRED. *Concurrent Pos:* Mem comt remote sensing for agr purposes, Nat Acad Sci, 66-69, rev bd, NASA; consult, Egypt remote sensing proj, 72, 75, 78, 80 & 81, Bangladesh remote sensing training prog, 78, Syrian land use & soils classification, 79-82, Senegal Nat Plan for land use

& develop, 79 & 82 & UN desertification prog. *Mem:* Am Soc Agr Engrs; Sigma Xi; Am Soc Civil Engrs; Am Soc Photogram. *Res:* Irrigation; water resources management; drainage; hydrology; remote sensing. *Mailing Add:* 440 Dakota Ave Brookings SD 57006

MYERS, WALTER LOY, b Joliet, Ill, Mar 13, 33; m 59. MOLECULAR BIOLOGY. *Educ:* Univ Ill, BS, 55, DVM, 57, MS, 59; Univ Wis, PhD(vet sci), 62. *Prof Exp:* From asst prof to prof vet path & hyg, Univ Ill, Urbana, 61- 73; PROF IMMUNOL, SCH MED, SOUTHERN ILL UNIV, SPRINGFIELD, 73-, CHMN, DEPT MED MICRO & IMMUNOL, 79- *Mem:* AAAS; Am Asn Cancer Res; Am Asn Immunologists; Am Soc Microbiol. *Res:* Inhibition of retrovirus replication by antisense oligodeoxynucleotides; use of tumor associated anti-idiotypic antibodies as biological response modifiers. *Mailing Add:* Southern Ill Univ Sch Med PO Box 19230 Springfield IL 62794

MYERS, WARREN POWERS LAIRD, b Philadelphia, Pa, May 2, 21; m 44; c 4. INTERNAL MEDICINE, ONCOLOGY. *Educ:* Yale Univ, BS, 43; Columbia Univ, MD, 45; Univ Minn, MS, 52; Am Bd Internal Med, dipl, 53 & 77. *Prof Exp:* Intern, Philadelphia Gen Hosp, 45-46; intern med, Maimonides Hosp, NY, 48-49; from asst prof to assoc prof, 54-68, PROF MED, MED COL CORNELL UNIV, 68- *Concurrent Pos:* Fel, Mem Hosp, NY, 48; Eleanor Roosevelt Found fel, Cambridge, Eng, 62-63; clin asst, Mem Hosp, NY, 52-54, from asst attend physician to attend physician, 54-59, assoc chmn dept med, 64-67, chmn dept med, 68-77, vpres for educ affairs, 77-81, chmn clin educ affairs, 81-; asst, Sloan-Kettering Inst Cancer Res, 52-56, assoc, 56-60, assoc mem, 60-69, mem, 69-; head metab & renal studies sect, Div Clin Invest, 57-66, head calcium metab lab, 67-78; asst attend physician, NY Hosp, 59-68, attend physician, 68-; vis physician, Bellevue Hosp, 60-68; consult, Grasslands Hosp, Valhalla, 66-68; mem clin cancer training comt, Nat Cancer Inst, 70-73, chmn, 71-73, mem & chmn clin cancer educ comt, 75-78; Eugene Kettering prof, Mem Sloan Kettering Cancer Ctr, 79- *Mem:* AAAS; Endocrine Soc; Harvey Soc; AMA; Am Asn Cancer Res. *Res:* Medical oncology; calcium metabolism; clinical endocrinology. *Mailing Add:* RFD 2 Box 376 Cornell Univ Med Col 1300 York Ave White River Junction VT 05001

MYERS, WAYNE LAWRENCE, b Adrian, Mich, Sept 17, 42; m 62; c 3. FORESTRY, BIOMETRY. *Educ:* Univ Mich, Ann Arbor, BS, 64, MF, 65, PhD(forestry), 67. *Prof Exp:* Res scientist, Forest Res Lab, Can Dept Fisheries & Forestry, Ont, 66-69; from asst prof to assoc prof forestry, Mich State Univ, 69-78; ASSOC PROF FOREST BIOMETRICS, PA STATE UNIV, 78- *Mem:* Soc Am Foresters; Am Soc Photogram; Am Statist Asn. *Res:* Forest biometry; quantitative ecology; remote sensing. *Mailing Add:* New Delhi - ID Dept State Washington DC 20520-9000

MYERS, WILLARD GLAZIER, JR, b Albany, NY, Aug 9, 35; m 62; c 4. ENVIRONMENTAL SCIENCES. *Educ:* Rensselaer Polytech Inst, BChE, 58, MS, 65, PhD(environ eng), 69. *Prof Exp:* Design engr, Boeing Co, Wash, 58-64; RES ASSOC ENVIRON SCI, RENSSELAER POLYTECH INST, 69- *Mem:* Am Chem Soc; Am Nuclear Soc; Am Soc Mass Spectrometry; Sigma Xi. *Res:* Application of mass spectrometry to environmental science. *Mailing Add:* 5646 Wells Ct San Jose CA 95123

MYERS, WILLIAM GRAYDON, CYCLOTRON-GENERATED SHORT-LIVED RADIO ISOTOPES FOR MEDICINE. *Educ:* Ohio State Univ, PhD(theoret chem), 39, MD, 41. *Prof Exp:* Prof radiol, Ohio State Univ, 33-79; PROF RADIOL, GRAD SCH MED SCI, CORNELL UNIV, NEW YORK, 79- *Mailing Add:* 2724 Wexford Rd Columbus OH 43221

MYERS, WILLIAM HOWARD, b Dodge Co, Nebr, Nov 17, 08; m 32, 75; c 3. MATHEMATICS. *Educ:* Stanford Univ, AB, 34, PhD(math), 39; Univ Calif, MA, 35. *Prof Exp:* Asst math, Stanford Univ, 35-36, instr, 36-39; instr, Univ Utah, 39-40; from asst prof to prof, 40-74, EMER PROF MATH, SAN JOSE STATE UNIV, 74- *Mem:* Math Asn Am. *Res:* Linear groups; algebra; analysis. *Mailing Add:* 2352 Sunny Vista Dr San Jose CA 95128

MYERS, WILLIAM HOWARD, b Oak Ridge, Tenn, Jan 26, 46; m 67; c 2. INORGANIC CHEMISTRY. *Educ:* Houston Baptist Col, BA, 67; Univ Fla, PhD(chem), 72. *Prof Exp:* Instr chem, Univ Fla, 68-69; fel, Ohio State Univ, 72-73; asst prof, 73-82, ASSOC PROF CHEM, UNIV RICHMOND, 82- *Concurrent Pos:* Res chemist, Ethyl Corp, 81-82; vis res assoc prof, Chem Dept, Univ Va, 90-91. *Mem:* Am Chem Soc; Sigma Xi. *Res:* Halogenation reactions of amine-boranes and investigations of steric effects on reactivity in such systems; catalysis, particularly involving olefin metathesis. *Mailing Add:* Dept Chem Univ Richmond Richmond VA 23173

MYERSON, ALBERT LEON, b New York, NY, Nov 14, 19; m 53; c 3. COMBUSTION CHEMISTRY, PHYSICS OF FLUIDS. *Educ:* Pa State Univ, BS, 41; Univ Wis, PhD(phys chem), 48. *Prof Exp:* Asst org chem, Off Sci Res & Develop & Nat Defense Res Comt, Columbia Univ, 41-42, org chem & phys chem, Manhattan Proj, SAM labs, 42-45; asst chem, Univ Wis, 46-48; mem staff, Franklin Inst, 48-56; mgr phys chem, Missile & Space Vehicle Dept, Gen Elec Co, 56-60; prin res phys chemist, Cornell Aeronaut Lab, Inc, 60-69; res assoc, Exxon Res & Eng Res Labs, 69-79; sr staff scientist & head, Phys Chem Div, Mote Marine Lab, 79-85; ASSOC, PRINCETON SCI ENTERPRISES, INC, 85- *Concurrent Pos:* Self-employed consult, 85- *Mem:* Am Chem Soc; Am Phys Soc; Combustion Inst. *Res:* Reduction of nitric oxide as a pollutant; airborne marine pollution; reactions and properties of gaseous uranium hexafluoride; combustion mechanisms; shock-tube chemical kinetics; recombination of atoms on surfaces; chemistry of missile re-entry. *Mailing Add:* 4147 Rosas Ave Sarasota FL 34233-1614

MYERSON, GERALD, b New York, NY, July 10, 51; m 85; c 1. NUMBER THEORY, COMBINATORICS & FINITE MATHEMATICS. *Educ:* Harvard Univ, AB, 72; Stanford Univ, MS, 75; Mich Univ, PhD(math), 77. *Prof Exp:* Asst prof math, State Univ NY, Buffalo, 77-85 & Southwest Tex State Univ, 86; SR LECTR MATH, MACQUARIE UNIV, 87- *Concurrent Pos:* Vis asst prof math, Univ BC, 82-83 & Univ Tex, Austin, 85-86; vis assoc prof math, Brigham Young Univ, 90-91. *Mem:* Am Math Soc; Math Asn Am. *Res:* Various branches of number theory, including uniform distribution, finite fields, Dedekind sums and measures of polynomials. *Mailing Add:* Math Macquarie Univ NSW 2122 Australia

MYERSON, RALPH M, b New Britain, Conn, July 21, 18; m 43; c 2. INTERNAL MEDICINE. *Educ:* Tufts Univ, BS, 38, MD, 42. *Prof Exp:* Intern, Boston City Hosp, 42-43, resident med, 46-48; ward physician, Vet Admin Hosp, Wilmington, Del, 48-52, asst chief med serv, Vet Admin Hosp, Philadelphia, 53-67, chief med serv, 67-72, chief staff, 72-75; assoc dir, Clin Serv Dept, Smith, Kline & French, 75-77, group dir med affairs, 77-83, sr consult gastroenterol, 83-85,; CLIN PROF MED, MED COL PA, 75-, ACTG ASSOC DEAN, GRAD MED EDUC, 90- *Concurrent Pos:* Instr, Sch Med, Tufts Univ, 47-48; clin prof, Med Col Pa, 77-66, prof med, 66-75. *Mem:* Fel Am Col Physicians; Am Fedn Clin Res; Int Soc Internal Med; fel Am Col Gastroenterol; Am Gastroenterol Asn; AMA. *Res:* Alcoholism; liver disease; application of new diagnostic techniques and evaluation in clinical medicine; gastroenterology; hepatology. *Mailing Add:* 310 Maplewood Ave Merion Station PA 19066

MYHRE, BYRON ARNOLD, b Fargo, NDak, Oct 22, 28; m 53; c 2. PATHOLOGY, IMMUNOHEMATOLOGY. *Educ:* Univ Ill, BS, 50; Northwestern Univ, MS, 52, MD, 53; Univ Wis, PhD, 62. *Prof Exp:* Resident path, Univ Wis, 56-60; asst prof med microbiol & immunol, Sch Med, Marquette Univ, 62-64, asst prof path, 64-66; assoc clin prof, Univ Southern Calif, 66-69, assoc prof path, 69-72; PROF PATH, UNIV CALIF, LOS ANGELES, 72-; DIR, BLOOD BANK, HARBOR-UCLA MED CTR, 72-, CHIEF CLIN PSYCHOL, 88- *Concurrent Pos:* Nat Inst Arthritis & Metab Dis fel, Univ Wis, 60-62; res grants, Ortho Found, 63-64, NIH, 65-66 & 72-73, Am Nat Red Cross, 67-68, Nat Heart Inst, 73-75; assoc med dir, Milwaukee Blood Ctr, Wis, 62-65; sci dir, Los Angeles-Orange Counties Res Cross Blood Ctr, 66-72. *Honors & Awards:* Distinguished Serv Award, Am Soc Clin Path. *Mem:* Am Soc Clin Path; Am Soc Exp Path; Am Asn Blood Banks; Col Am Pathologists. *Res:* Blood banking; immunopathology; histochemistry; cryobiology. *Mailing Add:* Dept Path Harbor Med Ctr Univ Calif Los Angeles Torrance CA 90509

MYHRE, DAVID V, b Lloydminster, Sask, Jan 4, 32; m 57; c 2. INFORMATION MANAGER, CARBOHYDRATE CHEMISTRY. *Educ:* Concordia Col, Moorhead, Minn, BA, 54; NDak State Univ, MS, 55; Univ Minn, PhD(biochem), 62. *Prof Exp:* res chemist, 62-87, INFO MGR, MIAMI VALLEY LABS, PROCTER & GAMBLE CO, 87- *Mem:* Am Chem Soc. *Res:* Flavor and food chemistry; carbohydrate chemistry; synthesis of glycoconjugates. *Mailing Add:* Miami Valley Labs Procter & Gamble Co Cincinnati OH 45239-8707

MYHRE, PHILIP C, b Tacoma, Wash, Mar 13, 33; m 85; c 1. ORGANIC CHEMISTRY. *Educ:* Pac Lutheran Univ, BA, 54; Univ Wash, PhD(chem), 58. *Prof Exp:* NSF fel, Nobel Inst Chem, Stockholm, 58-60; from asst prof to assoc prof, 60-69, PROF CHEM, HARVEY MUDD COL, 69- *Concurrent Pos:* Vis assoc, Calif Inst Technol, 67-68; guest prof, Swiss Fed Inst Technol, 71-72; chmn dept chem, Harvey Mudd Col, 74-75, 81-; vis scientist, IBM Res Labs, 84-86. *Mem:* AAAS; Am Chem Soc; Coun Undergrad Res. *Res:* Mechanisms of organic reactions; nuclear magnetic resonance spectroscopy. *Mailing Add:* Dept of Chem Harvey Mudd Col Claremont CA 91711

MYHRE-HOLLERMAN, JANET M, b Tacoma, Wash, Sept 24, 32; m 54, 88; c 1. OPERATIONS RESEARCH. *Educ:* Pac Lutheran Univ, BA, 54; Univ Wash, MA, 56; Univ Stockholm, Fil Lic, 68. *Prof Exp:* Res engr, Boeing Co, 56-58; lectr math, Harvey Mudd Col, 61-62; PROF MATH, CLAREMONT MCKENNA COL, 62- *Concurrent Pos:* Consult, US Navy, 68-; assoc ed, Technometrics, 70-75; guest prof math, Univ Stockholm & Eidgenossische Tech Hochschule, Zurich, 71-72; pres, Math Anal Res Corp, 74-; dir, Inst Decision Sci, 75- *Mem:* Am Statist Asn; Inst Math Statist; Am Soc Qual Control. *Res:* Reliability theory. *Mailing Add:* Dept Math Claremont McKenna Col Claremont CA 91711

MYKKANEN, DONALD L, b Bovey, Minn, Jan 7, 32; m 56; c 5. MATERIALS SCIENCE, MECHANICAL ENGINEERING. *Educ:* Univ Calif, Los Angeles, BS, 54; Univ Ill, MS, 55, PhD(mech eng), 61. *Prof Exp:* Instr mech eng, Univ Ill, 56-61; res & develop specialist, Missiles & Space Systs Div, Douglas Aircraft Co, Inc, 61-63, sect chief Saturn prod design, 63-65, sect chief nonmetall res proj, 64-68, prog mgr nuclear effects, McDonnell Douglas Astronaut Co, 68-75; PRES, ETA CORP, 75- *Concurrent Pos:* Lectr, Systs Analysis & Mgt, Univ Southern Calif; consult, Nuclear Radiation Effects. *Mem:* Am Soc Metals; AAAS; Am Inst Aeronaut & Astronaut; Inst Elec & Electronic Engrs. *Res:* Research and development and engineering analysis of high technology, military systems; survivability and vulnerability of military electronic systems; missile and space systems development; fragment weapon development. *Mailing Add:* 4523 La Canada Fallbrook CA 92028

MYKLES, DONALD LEE, b Stockton, Calif, Oct 23, 50; m 78; c 1. PROTEIN METABOLISM, MUSCLE BIOCHEMISTRY. *Educ:* Univ Calif, Berkeley, PhD(zool), 79. *Prof Exp:* asst prof, 85-89, ASSOC PROF BIOL, COLO STATE UNIV, 89- *Concurrent Pos:* Fel biol div, Oak Ridge Nat Lab, 79-85; Fulbright scholar, Univ Heidelberg, 90. *Honors & Awards:* NSF Presidential Young Investr, 89. *Mem:* AAAS; Am Soc Zool; Am Soc Cell Biol; Sigma Xi; Crustacean Soc. *Res:* Protein synthesis and degradation in crustacean muscles; regulation of calcium dependent proteolysis in muscle atrophy; neuronal control of gene expression during fiber transformation. *Mailing Add:* Dept Biol Colo State Univ Ft Collins CO 80523

MYLES, CHARLES WESLEY, b Bethesda, Md, Nov 21, 47; m 69; c 2. THEORETICAL SOLID STATE PHYSICS, SEMICONDUCTOR PHYSICS. *Educ:* Univ Mo-Rolla, BS, 69; Wash Univ, MO, 71, PhD(physics), 73. *Prof Exp:* Fel solid state theory, Battelle Mem Inst, 73-75; res assoc &

instr, Swiss Fed Inst Technol, Lausanne, Switz, 75-77; res asst prof, Physics Dept, Univ Ill, Urbana-Champaign, 77-78; from asst prof to assoc prof, 78-87, PROF SOLID STATE THEORY, DEPT PHYSICS, TEX TECH UNIV, 87-, CHMN, 91- *Concurrent Pos:* Consult, Mat Res Lab & Dept Physics, Univ Ill, Urbana-Champaign, 78-83 & Dept Elec Eng, Univ Southern Calif, 81-; vis scientist, Sandia Nat Labs, 85; vis prof, Wash Univ, St Louis, 87-88. *Mem:* Am Phys Soc; AAAS; Sigma Xi. *Res:* Deep impurity levels in semiconductors; electronic and structural properties of semiconductor alloys; properties of magnetic materials; magnetic resonance; quantum size effects in metal particles; solid molecular hydrogen. *Mailing Add:* Dept Physics & Eng Physics Tex Tech Univ Lubbock TX 79409-1051

MYLES, DIANA GOLD, b Topeka, Kans, May 8, 44. MEMBRANE BIOLOGY, REPRODUCTIVE BIOLOGY. *Educ:* Univ Calif, Berkeley, BA, 67, PhD(bot), 73. *Prof Exp:* Fel, Dept Biol Sci, Stanford Univ, 73-77, res assoc, Depts Physiol & Anat, Lab Human Reprod & Reprod Biol, 77-81; asst prof, 81-87, ASSOC PROF PHYSIOL, HEALTH CTR, UNIV CONN, 87- *Mem:* Sigma Xi; Am Soc Cell Biol; Soc Study Reproduction. *Res:* Topographical organization and dynamics of the mammalian sperm cell surface studies; role of the sperm cell surface in fertilization. *Mailing Add:* Dept Physiol Health Ctr Univ Conn Farmington CT 06030

MYLES, KEVIN MICHAEL, b Chicago, Ill, July 18, 34; m 56; c 3. FUEL CELLS, COAL COMBUSTION & CONVERSION. *Educ:* Univ Ill, Urbana, BS, 56, MS, 57, PhD(phys metall), 63. *Prof Exp:* Phys metallurgist, chem eng div, 57-78, asst mgr, Nuclear Fuel Reprocessing Prog, 78-80, dep dir fossil energy progs, 82-85, sect head, coal combustion & conversion, 80-87, MGR, FUEL CELLS, ARGONNE NAT LAB, 87- *Concurrent Pos:* Prof mat sci, Midwest Col Eng, 70-82. *Mem:* Am Soc Metals; Am Inst Mining, Metall & Petrol Engrs; Sigma Xi. *Res:* Nuclear fuel reprocessing; high temperature lithium-metal sulfide batteries; sodium chemistry; physical metallurgy of nuclear materials; thermodynamics of transition elements; fluidized-bed combustion coal; combustion emissions control technologies; thermal destruction of municipal solid waste; molten carbonate and solid oxide fuel cells. *Mailing Add:* 1231 60th Pl Downers Grove IL 60516

MYLONAS, CONSTANTINE, b Athens, Greece, June 24, 16; m 55; c 2. ENGINEERING. *Educ:* Athens Tech Univ, ScB, 39; London Univ, PhD(eng), 49. *Prof Exp:* Construct engr, Ergon Construct Co, Greece, 39-41; design engr struct, Off Struct Design, Greek Ministry Pub Works, 41-44; res engr stress analysis, Aero-Res, Ltd, Eng, 48-51; prof & dir lab testing mat, Athens Tech Univ, 51-53; assoc prof, 53-60, PROF ENG, BROWN UNIV, 60- *Concurrent Pos:* Lectr, Athens Tech Univ, 41-44; Guggenheim fel, 59-60; expert, Am del, Int Inst Welding, 62- *Mem:* AAAS; Am Soc Mech Engrs; Soc Exp Stress Anal; Am Soc Testing & Mat; Adams Memorial mem Am Welding Soc; Sigma Xi. *Res:* Stress analysis; photoelasticity; plasticity; structures; mechanical properties of materials under static and dynamic locating; mechanics of fracture; radiation damage in solids; strength of cemented structures. *Mailing Add:* N Kountourioti 5 Kifissia 14563 Greece

MYLREA, KENNETH C, b Merlin, Ont, Sept 29, 35; US citizen; m 62; c 2. CLINICAL ENGINEERING. *Educ:* Univ Mich, BSEE, 63, MS, 65, PhD(bioeng), 68. *Prof Exp:* Head, Human Performance Ctr, Univ Mich, 63-65, asst res engr, psychol dept, 65-68, res assoc bioeng, 68-69; assoc scientist, instrumentation, Univ Wis-Madison, 69-75; PROF ELEC & COMPUT ENG, UNIV ARIZ, 75-, DIR CLIN ENG, 75- *Concurrent Pos:* Mem, Technol Transfer Comt, Nat Acad Engr, 70-73, chmn, Remote Diagnosis Comt, 72-73; head biomed instrumentation & assoc dir, Adv Ctr, Univ Wis, 71-75; consult clin eng, WHO, Singapore, 75; sect ed, Annals Biomed Eng, 80-; adj assoc prof anesthesiol, Univ Ariz, 80- *Mem:* Sr mem Inst Elec & Electronics Engrs; Engrs in Med & Biol Soc; sr mem Biomed Eng Soc; Asn Advan Med Instrumentation. *Res:* Operating room monitoring and instrumentation; instrumentation for medical care; anesthetic delivery; collection, integration and display of patient data in the operating room. *Mailing Add:* Elec & Comput Eng Dept Univ Ariz Tucson AZ 85721

MYLROIE, JOHN EGLINTON, b Philadelphia, Pa, June 13, 49; m 70. GEOMORPHOLOGY, PALEOECOLOGY. *Educ:* Syracuse Univ, BS, 71; Rensselaer Polytech Inst, PhD(geol), 77. *Prof Exp:* Tech specialist biol, State Univ NY Albany, 72-74; asst prof geol, Murray State Univ, 77-; AT DEPT GEOL & GEOG, MISS STATE UNIV. *Concurrent Pos:* Ed, Western Ky Speleol Surv, 77-; chmn, Nat Speleol Soc Comt Cave Ownership & Mgt, 78-81. *Mem:* Fel Nat Speleol Soc; Cave Res Found; Am Geol Inst; Sigma Xi. *Res:* Karst geomorphology and hydrology, evolution of karst systems, effects of glaciation on karst systems; origin of life, interaction of paleoecosystem parameters. *Mailing Add:* Dept Geol & Geog Miss State Univ Mississippi MS 39762

MYLROIE, VICTOR L, b Ogden, Utah, Feb 7, 37; m 62; c 5. CATALYTIC HYDROGENATION DYE INTERMEDIATES, ORGANIC SYNTHESIS. *Educ:* Brigham Young Univ, BS, 65, MA, 68. *Prof Exp:* Mgr egg prod, Woodward Bros Poultry, 65-67; res chemist high pressure catalysis, Eastman Kodak Co, 68-70, supvr, 71-72, res chemist spec org synthesis, 72-74, proj engr copier catalysis, 74-78, SUPVR HIGH PRESSURE LAP, EASTMAN KODAK CO, 78- *Mem:* NY Acad Sci; Org Reactions Catalysis Soc; Catalysis Soc NAm; Photographic Scientists & Engrs. *Res:* Catalysis of organic reaction by heterogenous catalysis and related pressure reactions; catalytic hydrogenation to produce dye intermediates, dye couplers and organic specialty chemicals. *Mailing Add:* 19 Hanford Way Fairport NY 14450

MYLROIE, WILLA W, b Seattle, Wash, May 30, 17; m 40, 66; c 2. TRANSPORTATION ENGINEERING, REGIONAL PLANNING. *Educ:* Univ Wash, BS, 40, MS, 52. *Prof Exp:* Checker, Planning Div, State Hwy Dept Wash, 40-41; jr & asst engr, mil & civil construct, US Army Engrs, 41-45; assoc static mech courses, Univ Wash, 48-50, res engr hwys, 51-55; assoc prof civil eng, Purdue Univ, 56-58; res engr, Wash State Dept Transp, 58-69, head, Res & Spec Assignments Div, 69-81; CONSULT, 82- *Concurrent Pos:* Planning assoc, Wash State Bd Dirs, 70-74; assoc mem, Transp Res Bd, Nat Acad Sci-Nat Res Coun, coun mem opers & maintenance, 73-76; int bd dirs, Inst Transp Engrs, 73-76; mem, Eng Col Adv Bd, Wash State Univ, 77-85 & Eng Col vis Comt, Univ Wash, 78-86; invitational res partic, Nat Conf Future Nat Safety Prog, 77; affil prof, Univ Wash, 81-84; res consult, King County Design Comn, 81-88. *Honors & Awards:* Edmund Friedman Award, Am Soc Civil Engrs, 78; Tech Coun Award, Inst Transp Engrs, 82. *Mem:* Inst Transp Engrs; Am Soc Civil Engrs; Sigma Xi. *Res:* Comprehensive land use planning; transportation research; resource management; transportation and land use. *Mailing Add:* 7501 Boston Harbor Rd NE Olympia WA 98506

MYODA, TOSHIO TIMOTHY, b Mukden, Manchuria, Mar 17, 29; m 63; c 2. MICROBIOLOGY, MOLECULAR BIOLOGY. *Educ:* Hokkaido Univ, BS, 49, MS, 52; Iowa State Univ, PhD(bact), 59. *Prof Exp:* Asst microbiol, Hokkaido Univ, 52-54, instr, 54-59; asst bact, Iowa State Univ, 56-59; res fel, Nat Res Coun Can, 59-60; res fel microbiol, Western Reserve Univ, 60-64; chief microbiol, Inst Microbial Chem, Japan, 64-66; res assoc & instr, La Rabida-Univ Chicago Inst, 66-67; assoc chief microbiol & immunochemist, Alfred I Du Pont Inst, 67-73, dir burn res, 77-78, affil med staff, 72-86, chief microbiol dept, 73-84, dir microbiol genetics, 78-86, sr res scientist, 85-86; PRES, UNITECH ASN, USA, INC, 86-; PROF, TOKAI UNIV MED SCH, 89- *Concurrent Pos:* Vis prof, Valparaiso Univ, 67; ed asst, Int Union Microbiol Soc, 77-82, asst vpres, 82-84, asst pres, 84-89; adj prof, Sch Life & Health Sci, Univ Del, 86- *Honors & Awards:* Distinguished Serv Medal, Int Union Microbiol Soc, 82. *Mem:* AAAS; Am Soc MIcrobiol; The Biochem Soc; Am Chem Soc; Am Acad Microbiol; US Fed Cult Collection. *Res:* Applied microbiology; molecular genetics, recombinant DNA, cloning, mapping and expression of genes; culture collections. *Mailing Add:* PO Box 354 Rockland DE 19732-0354

MYRBERG, ARTHUR AUGUST, JR, b Chicago Heights, Ill, June 28, 33; c 2. BEHAVIOR ETHOLOGY, MARINE BIOLOGY. *Educ:* Ripon Col, BA, 54; Univ Ill, MS, 58; Univ Calif, Los Angeles, PhD(zool), 61. *Prof Exp:* Asst, Ill Natural Hist Surv, 57; asst zool, Univ Ill, 57-58 & Univ Calif, Los Angeles, 58-61; fel, Max Planck Inst Behav Physiol, Seewiesen, Ger, 61-64; from asst prof to assoc prof, 64-71, PROF MARINE SCI, UNIV MIAMI, 71- *Mem:* Am Soc Zoologists; Am Soc Ichthyol & Herpet; fel Animal Behav Soc; Am Inst Biol Sci; fel Am Inst Fishery Res Biologists; Am Elasmobranch Soc; Int Assoc Fish Ethologists. *Res:* Ichthyology; comparative behavior of fishes, particularly those of tropical waters with emphasis on the families Cichlidae, Pomacentridae and Acanthuridae; underwater acoustics and its biological significance; bioacoustics; sensory biology and behavior of sharks; bioacoustics. *Mailing Add:* Rosenstiel Sch Marine Atmospheric Sci Univ Miami 4600 Rickenbacker Causeway Miami FL 33149-1098

MYRES, MILES TIMOTHY, b London, Eng, May 16, 31; Brit & Can citizen; c 1. ORNITHOLOGY, CONSERVATION BIOLOGY. *Educ:* Univ Cambridge, BA, 53, MA, 58; Univ BC, MA, 57, PhD(zool), 60. *Prof Exp:* Res officer, Edward Grey Inst Field Ornith, Oxford Univ, 59-61; asst prof biol, Lakehead Col, 62-63; from asst prof zool to assoc prof zool, Univ Calgary, 63-87; RETIRED. *Concurrent Pos:* Nat dir, Can Nature Fedn, 72-74; mem coun, Pac Seabird Group, 73-74; chmn, Can Nat Sect, Int Coun Bird Preserv, 73-74 & 75-76. *Mem:* Elective mem Am Ornithologists Union; Brit Ornithologists Union; Soc Conserv Biol. *Res:* Bird migration; ecology of birds; North Pacific seabirds; man-land-fauna interactions; cultural aspects of environmental conservation at different times in history; relationship between amateur naturalists and professional biologists. *Mailing Add:* Gouray Cottage Gorey Village Grouville Jersey Channen Island JE3 9EP England

MYRIANTHOPOULOS, NTINOS, b Cyprus, July 24, 21; m 55; c 3. HUMAN GENETICS. *Educ:* George Washington Univ, BS, 52; Univ Minn, MS, 54, PhD(genetics), 57. *Prof Exp:* Res assoc neurol, Univ Ill, 55-57; res geneticist, 57-63, proj consult, 55-57, HEAD SECT EPIDEMIOL & GENETICS, DEVELOP NEUROL BR, NAT INST NEUROL & COMMUN DIS & STROKE, 63- *Concurrent Pos:* Assoc prof neurol, George Washington Univ, 58-; instr grad prog, NIH, 58-; dir, Genetic Counseling Ctr, 58- *Mem:* Am Soc Human Genetics; NY Acad Sci; Soc Study Social Biol; Teratology Soc. *Res:* Genetics and epidemiology of neurological disease; congenital malformations, prevalence, incidence and mutation rates; metabolic etiology; clinical and behavioral genetics. *Mailing Add:* 7913 Orchid St NW Bethesda MD 20892

MYRICK, ALBERT CHARLES, JR, b Visalia, Calif, July 8, 39. MARINE MAMMAL AGE DETERMINATION, MARINE MAMMAL STRESS PHYSIOLOGY. *Educ:* George Wash Univ, BA, 71; Univ Calif, Los Angeles, MA, 74, PhD(biol), 79. *Prof Exp:* Mus specialist, Dept Paleobiol, Smithsonian Inst, Washington, DC, 64-72; teaching fel vertebrate paleo mammal evolution comp anat, Dept Biol, Univ Calif, Los Angeles, 72-75; Smithsonian fel, Dept Paleobiol, Smithsonian Inst, 75; curatorial asst earth sci, Los Angeles County Mus Natural Hist, 74-76; WILDLIFE BIOLOGIST, NAT MARINE FISHERIES SERV, DEPT COMMERCE, LAJOLLA, CALIF, 77- *Mem:* AAAS; Soc Marine Mammal; Am Soc Mammalogists. *Res:* Age determination and age-related biological parameters of marine mammals especially small species of odontocetes and serenians; time calibration of layers and etiology of layered tissue in teeth and bones of mammals; effects of chronic stress on marine mammal populations. *Mailing Add:* 5644 Adams Ave San Diego CA 92115

MYRICK, HENRY NUGENT, b Cisco, Tex, Apr 30, 35; m. ENVIRONMENTAL SCIENCES. *Educ:* Lamar Univ, BS(chem) & BS(biol), 57; Rice Univ, MS, 59; Washington Univ, St Louis, ScD, 62. *Prof Exp:* Instr environ sci, Rice Univ, 62-63; instr, Harvard Univ, 63-65, fel, 63-64; from asst prof to assoc prof, Cullen Col Eng, Univ Houston, 65-74; PRES, PROCESS CO INC. *Honors & Awards:* H P Eddy Award, Water Pollution Control Fedn, 61. *Mem:* Am Inst Chem Engrs; Am Inst Chemists; Am Water Works Asn; Water Pollution Control Fedn; Sigma Xi; Air & Waste Mgt Asn. *Res:* Environmental engineering; process kinetics of processes for air, water and solid waste conversion or treatment; analytical chemical and biological methods for waste characterization. *Mailing Add:* Process Co Inc 2123 Winnock Blvd Houston TX 77057

MYRICK, MICHAEL LENN, b Greensboro, NC, Dec 5, 61; m 87. MOLECULAR PHOTOPHYSICS. *Educ:* NC State Univ, BS, 85; NMex State Univ, PhD(phys chem), 88. *Prof Exp:* Postdoctoral asst, Lawrence Livermore Nat Lab, 89-90, staff chemist, 90-91; ASST PROF PHYS CHEM, UNIV SC, COLUMBIA, 91- *Concurrent Pos:* Consult, Optical Sensor Consults, 89-91; co-owner, Laser Raman Analysis, Inc, 91- *Res:* Molecular photophysics and electronic phenomena, investigated with optical spectroscopy and scanning tunneling microscopy; author of over 35 publications. *Mailing Add:* Dept Chem Univ SC Columbia SC 29208

MYROLD, DAVID DOUGLAS, b Milwaukee, Wis, Oct 16, 55; m 79; c 3. SOIL MICROBIOLOGY, FOREST SOILS. *Educ:* Mich Technol Univ, BS, 77; Wash State Univ, MS, 79; Mich State Univ, PhD(microbiol), 84. *Prof Exp:* Asst prof, 84-89, ASSOC PROF SOIL MICROBIOL, ORE STATE UNIV. *Concurrent Pos:* NSF presidential young investr award, 86. *Mem:* Soil Sci Soc Am; Am Soc Microbiol; Ecol Soc Am; AAAS. *Res:* Nitrogen cycling in forest and agricultural ecosystems, especially 15 nitrogen tracer studies; autecology of Frankia in soil, especially use of gene probes. *Mailing Add:* Dept Crop & Soil Sci Strand Agr Hall Ore State Univ Corvallis OR 97331

MYRON, DUANE R, b Birmingham, Ala, Jan 3, 43; m 62; c 2. BIOCHEMISTRY, NUTRITION. *Educ:* Univ NDak, BS, 65, PhD(biochem), 70. *Prof Exp:* Fel biochem, St Jude Children's Res Hosp, 70-72 & Mem Univ Nfld, 72-73; fel biochem & nutrit, 73-74, RES ASSOC BIOCHEM & NUTRIT, HUMAN NUTRIT LAB, AGR RES SERV, UNIV NDAK, USDA, 74- *Mem:* Sigma Xi; AAAS; Am Chem Soc. *Res:* Biochemical and metabolic role of trace mineral nutrients in laboratory animals and in humans. *Mailing Add:* PO Box 51A Thomson ND 58278

MYRON, HAROLD WILLIAM, b New York, NY, Apr 28, 47; m; c 2. THEORETICAL SOLID STATE PHYSICS. *Educ:* City Univ New York, BA, 67; Iowa State Univ, PhD(physics), 72. *Prof Exp:* Res asst physics, Ames Lab, US AEC, 69-72; res assoc, Dept Physics, Magnetic Theory Group, Northwestern Univ, 72-74, vis asst prof, 74-75; scientist physics, High Field Magnet Lab, Inst Metal Physics, Univ Nijmegen, Neth, 75-85; PROG LEADER, ARGONNE NAT LAB, 85- *Mem:* Am Phys Soc. *Res:* Ab initio calculations of electron and phonon properties in metals including dielectric functions of transition metal compounds. *Mailing Add:* Argonne Nat Lab Argonne IL 60439

MYRON, THOMAS L(EO), b Pittsburgh, Pa, June 12, 23; m 49; c 11. CHEMICAL ENGINEERING. *Educ:* Univ Pittsburgh, BS, 48, MS, 49. *Prof Exp:* Asst lab instr chem eng, Univ Pittsburgh, 48-49; asst chief raw mat eng div, Res Lab, 59-61, chief, Raw Mat Div, 61-76, chief, Chem Div, 76-84, MGR, CHEM & POLY DIV, RES LAB, US STEEL CORP, UNIVERSAL, 84- *Mem:* Am Inst Chem Engrs; Am Inst Mining, Metall & Petrol Engrs. *Res:* Ore agglomeration and reduction; pollution abatement; coal and coke; chemicals. *Mailing Add:* 650 Broughton Rd Bethel Park PA 15102

MYRONUK, DONALD JOSEPH, b Kapuskasing, Ont; m 66; c 3. MECHANICAL ENGINEERING, FORENSIC ENGINEERING. *Educ:* Queen's Univ, Ont, BSc, 61, MSc, 65; Univ Ill, Urbana, PhD(mech eng), 69. *Prof Exp:* Design engr, Chalk River Proj, Atomic Energy Can Ltd, 61-62; res asst, Univ Ill, Urbana, 64-69; from asst prof to assoc prof, 69-78, assoc dean eng, 81-86, PROF MECH ENG, SAN JOSE STATE UNIV, 78- *Concurrent Pos:* Res engr fire & earthquake induced fire damage, Sci Serv Inc. *Mem:* Soc Automotive Eng; Air Pollution Control Asn; Am Soc Mech Eng; Am Soc Eng Educ; Soc Forensic Engrs & Scientists. *Res:* Clean air cars; compressible gas dynamics; biomedical engineering, especially devices for the handicapped; fire research, flammability of fuels, hydraulic fluids and lubricants; fire resistant materials; disaster mitigation; off-highway construction vehicle fires. *Mailing Add:* Dept Mech Eng San Jose State Univ Washington Sq San Jose CA 95192

MYRVIK, QUENTIN N, b Minneota, Minn, Nov 9, 21; m; c 1. MICROBIOLOGY, IMMUNOLOGY. *Educ:* Univ Wash, Seattle, BS, 48, MS, 50, PhD(microbiol), 52. *Prof Exp:* From asst prof to assoc prof, Dept Microbiol, Sch Med, Univ Va, 52-63; prof microbiol & immunol, 63-90, chmn dept, 63-81, EMER PROF, BOWMAN GRAY SCH MED, WAKE FOREST UNIV, 90-; VPRES & SR SCIENTIST, MUSCULOSKELETAL SCI RES INST, 90- *Concurrent Pos:* Mem, Bact & Mycol Study Sect, NIH, 67-72, chmn, 71-72, mem, Allergy & Immunol Study Sect, 77-81, Med Sci Adv Panel NASA, Am Inst Biol Sci, 80-83, Space Med Adv Bd, 84- & Sci Rev Comt, Nat Inst Environ Health, 87-, chmn, 90-; Am Soc Microbiol vis prof, Dept Parasitol & Microbiol, Sch Med, Univ Antioquia, Colombia, 70 & Univ Uruguay, 73; ed, J Immunol, 72-75 & J Leukocyte Biol, 82-; ed-in-chief, J Reticuloendothelial Soc, 74-81; mem, Space Biol & Med Task Group, Space Sci Bd, NASA, 84-89. *Mem:* Am Acad Microbiol; Am Soc Microbiol; Am Asn Immunologists; Am Thoracic Soc; Reticuloendothelial Soc (vpres, 69, pres, 70); Sigma Xi. *Res:* Author of numerous technical publications. *Mailing Add:* Musculoskeletal Sci Res Inst 2190 Fox Mill Rd Herndon VA 22071

MYSAK, LAWRENCE ALEXANDER, b Saskatoon, Sask, Jan 22, 40; m 74; c 2. PHYSICAL OCEANOGRAPHY, CLIMATE MODELING. *Educ:* Univ Alta, BSc, 61; Univ Adelaide, MSc, 63; Harvard Univ, AM, 64, PhD(appl math), 66. *Prof Exp:* US Navy fel, Harvard Univ, 66-67; from asst prof to prof math & oceanog, Univ BC, 67-86; Atmospheric Environ Serv/Nat Sci & Eng Res Coun of Can, sr indust res prof, dir climate res group, 86-90, DIR CTR CLIMATE & GLOBAL CHANGE RES, MCGILL UNIV, 90- *Concurrent Pos:* Vis res assoc, Ore State Univ, 68; sr visitor, Cambridge Univ, 71-72; travel fel, Nat Res Coun Can, 71-72 & 76-77; vis lectr, Soc Indust & Appl Math, 75-76 & 78-79; consult scientist & guest lectr, Inst Ocean Sci, Patricia Bay, 76; vis scientist, Nat Ctr Atmospheric Res, 77; vis prof, Naval Postgrad Sch, Monterey, 81, Fed Inst Technol, Zurich, 82-83; fel, Killiam Mem, 82-83; mem, IOC/Spec Comt on Oceanog Res on climatic changes and the oceans. *Honors & Awards:* Pres Prize, Can Meteorol & Oceanog Soc, 80. *Mem:* Can Appl Math Soc; Can Meteorol & Oceanog Soc; Am Meteorol Soc; The Oceanog Soc. *Res:* Ocean and climate dynamics; ocean circulation modelling; air-ice-sea interactions; interannual and decadal variability of the climate; climate modelling; influence of climate on fisheries. *Mailing Add:* Dept Meteorology McGill Univ Montreal PQ H3A 2K6 Can

MYSELS, ESTELLA KATZENELLENBOGEN, b Berlin, Ger, Jan 12, 21; nat US; m 53. CHEMISTRY. *Educ:* Univ Calif, BS, 42, PhD(chem), 46. *Prof Exp:* Chemist, Richfield Oil Co, 42-44; instr chem, Univ Calif, 46-47 & Univ Southern Calif, 47-50; asst, Sloan-Kettering Inst Cancer Res, 50-53; res assoc chem, Univ Southern Calif, 54-66; lectr chem, Salem Col, 67, prof, 67-70; lectr chem, Univ Calif, San Diego, 75; consult, 80-86; RETIRED. *Concurrent Pos:* Chmn dept chem, Salem Col, 68-70. *Mem:* Am Chem Soc. *Res:* Infrared and ultraviolet spectroscopy; electrophoresis; surfactant solutions; organic and physical chemistry. *Mailing Add:* 8327 La Jolla Scenic Dr La Jolla CA 92037

MYSELS, KAROL JOSEPH, b Krakow, Poland, Apr 14, 14; nat US; m 53. COLLOID CHEMISTRY, SURFACE CHEMISTRY. *Educ:* Univ Lyon, Lic-es-sc, Ing chim, 37; Harvard Univ, PhD(inorg chem), 41. *Prof Exp:* Res asst chem, Stanford Univ, 41-42, res assoc, 43-45; instr, NY Univ, 45-47; from asst prof to prof, Univ Southern Calif, 47-66; assoc dir res, R J Reynolds Industs, 66-70; sr res adv, Gen Atomic Co, 70-79; PRIN, RES CONSULT, 79- *Concurrent Pos:* Mem staff, Shell Develop Co, 40-42; NSF fac fel, 57-58, sr fel, Strasbourg Ctr Macromolecule Res, France, 62-63; Guggenheim fel, 65-66; Rennebohm lectr, Univ Wis, 64, Pharm Alumni lectr, 67; John Watson Mem Lectr, Va Polytech Inst, 68; assoc mem, Comn Colloid & Surface Chem, Int Union Pure & Appl Chem, 65-69, titular mem, 69-73, chmn, 73-79, mem, Div Comt Phys Chem Div, 75-79; Am Chem Soc tour lectr, 71, 74, 77, 78, 80-82, 87 & 88; res chemist, Dept Med, Univ Calif, San Diego, 80-, Dept Chem, 85-; invited lectr, Col de France, Paris, 87, Japanese Colloid Symp, Kyoto, 87; vis prof, Sci Univ Tokyo, Japan, 87. *Honors & Awards:* Kendall Award, Am Chem Soc, 64; Phi Lambda Upsillon Lectr, Univ Okla, 74. *Mem:* AAAS; Am Chem Soc; Am Inst Chemists. *Res:* Surfactant solutions; surface tension; soap films; evaporation control; osmosis; conductivity; electrophoresis; diffusion; rheology; intermolecular forces; thermochemical water splitting; gas-cooled nuclear reactor design and applications; membranology; research tactics and strategy. *Mailing Add:* 8327 La Jolla Scenic Dr La Jolla CA 92037

MYSEN, BJORN OLAV, b Oslo, Norway, Dec 20, 47; m 75; c 2. PETROLOGY, GEOCHEMISTRY. *Educ:* Univ Oslo, BSc, 69, MA, 71; Pa State Univ, PhD(geochem), 74. *Prof Exp:* Res assoc geochem, Mineral Mus Univ Oslo, 70-71; fel exp petrol, Geophys Lab, Carnegie Inst Wash, 74-77; lectr petrol, Johns Hopkins Univ, 74-76; STAFF MEM GEOCHEMIST GEOCHEM & PETROL, GEOPHYS LAB, CARNEGIE INST WASHINGTON, 77- *Honors & Awards:* F W Clarke Award, Geochem Soc, 77; Reusch Medal, Geol Soc Norway, 79. *Mem:* Am Geophys Union; fel Mineral Soc Am; Geochem Soc; Royal Norwegian Acad Sci & Lett. *Res:* Experimental petrology relevant to igneous processes in the earth and terrestrial planets with emphasis on the role of volatiles, trace element partitioning, phase equilibria and physical properties and structure of silicate melts and glasses at high pressure and temperature. *Mailing Add:* Geophys Lab 5251 Broad Branch Rd NW Washington DC 20015-1305

MYSER, WILLARD C, b Cuyahoga Falls, Ohio, Apr 22, 23; m 43; c 4. ZOOLOGY. *Educ:* Kent State Univ, BS, 44; Ohio State Univ, MS, 47, PhD, 52. *Prof Exp:* Asst zool, 45-47, asst instr, 47-48, from instr to assoc prof, 48-61, asst chmn dept, 61-68, PROF ZOOL, OHIO STATE UNIV, 61- *Concurrent Pos:* Res assoc, Argonne Nat Lab, 56. *Mem:* Radiation Res Soc. *Res:* Radiation biology; cytology. *Mailing Add:* Bot Zool Bldg Ohio State Univ Columbus OH 43210

MYSLINSKI, NORBERT RAYMOND, b Buffalo, NY, Apr 14, 47; wid. ORAL BIOLOGY, NEUROSCIENCE. *Educ:* Canisius Col, BS, 69; Univ Ill, PhD(pharmacol), 73. *Prof Exp:* Res assoc pharmacol & biochem, Tufts Univ, 73-75; asst prof pharmacol, 75-80, co-dir, Myo-Facial Pain Prog, 81-84, ASSOC PROF PHYSIOL & DIR GRAD PROG, UNIV MD, BALTIMORE, 80-,. *Concurrent Pos:* Consult, publishers, journals & granting orgns, 77; ed, Md Soc Med Res, 77-82; lectr, Community Col Baltimore, 80-82; res fel, Univ Bristol, Eng, 84-85. *Mem:* Sigma Xi; Int Asn Dental Res; Am Physiol Soc; Soc Neurosci; Int Brain Res Orgn. *Res:* Physiology and pharmacology of the nervous system; drugs and classical electrophysical techniques regarding neurotransmitters involved in both motor and sensory systems; oral-facial neuroscience; facial pain, mastication and jaw reflexes; author of over 60 publications. *Mailing Add:* Dept Physiol Sch Dent Univ Md 666 W Baltimore St Baltimore MD 21201

MYSLOBODSKY, M S, b Vilna, Poland, Jan 2, 37; m 60. PSYCHOLOGY. *Educ:* Charkow State Univ, MD, 60; Inst Higher Nerv Activ, PhD(neuroradiology), 65. *Hon Degrees:* DSc, Coun Higher Educ USSR, 71. *Prof Exp:* MEM STAFF, PSYCHOPATH LAB, NIMH. *Mailing Add:* Psychol Psychopath Lab NIMH 9000 Rockville Pike Bldg 10 Rm 4C110 Bethesda MD 20892

MYTELKA, ALAN IRA, b Somerville, NJ, Jan 30, 35; m 56; c 2. ENVIRONMENTAL SCIENCES, CHEMICAL ENGINEERING. *Educ:* Polytech Inst Brooklyn, BChE, 58; Newark Col Eng, MSChE, 61; Rutgers Univ, PhD(environ sci), 67. *Prof Exp:* Instr chem, Newark Col Eng, 58-64; environ scientist, Aero Chem Res Labs, Inc, Ritter-Pfaudler Corp, Princeton, 66-70; ASST INVESTR & ASST CHIEF ENGR, INTERSTATE SANITATION COMN, 70- *Mem:* Am Inst Chemists; Air Pollution Control Asn; Am Inst Chem Engrs; Water Pollution Control Asn. *Res:* Oxidation in aqueous systems, especially ionizing-radiation treatment of industrial wastes. *Mailing Add:* Nine Redwood Rd Martinsville NJ 08836

MYTTON, JAMES W, b Kansas City, Mo, Feb 18, 27; m 69; c 1. GEOLOGY. *Educ:* Dartmouth Col, AB, 49; Univ Wyoming, MA, 51. *Prof Exp:* Geologist, US Geol Surv, 51-58; photogeologist, Knox, Bergman & Shearer, 61-63; geologist, US Geol Surv, 63-86; RETIRED. *Mem:* Fel Geol Soc Am; Am Asn Petrol Geologists. *Res:* Investigation of fissionable materials in sedimentary rocks; coal resource studies; regional stratigraphic studies; photogeologic interpretation; evaluation of mineral resources on a regional scale; investigation of phosphate and related commodities; waste disposal studies; studies related to underground nuclear testing; geologic studies related to geothermal energy. *Mailing Add:* 2720 Simms St Lakewood CO 80215

N

NA, GEORGE CHAO, b Liao-Ning, China, March 23, 47; US citizen; m 73. PHYSICAL BIOCHEMISTRY. *Educ:* Tunghai Univ, BS, 69; Boston Univ, PhD(chem), 76. *Prof Exp:* Fel biochem res, Grad Dept Biochem, Brandeis Univ, 75-81; RES CHEMIST, EASTERN REGIONAL RES CTR, USDA, 81- *Mem:* Am Chem Soc; Biophys Soc. *Res:* Protein physical biochemistry. *Mailing Add:* Sterling Drug Co 25 Great Valley Pkwy Malvern PA 19355-1314

NAAE, DOUGLAS GENE, b Graettinger, Iowa, Dec 24, 46; m 67; c 2. ORGANIC CHEMISTRY, POLYMER CHEMISTRY. *Educ:* Univ Iowa, BS, 69, MS, 71, PhD(org chem), 72. *Prof Exp:* Res assoc solid state org chem, Univ Minn, Minneapolis, 73-74; from asst prof org chem to prof, Univ Ky, 74-80; RES CHEMIST, GULF OIL CHEMICALS CO, 80- *Mem:* Am Chem Soc. *Res:* Polymer chemistry; Ziegler-Natt catalysis; properties and synthesis of polyolefins; transition metal chemistry; organometallic chemistry; organic chemistry; spectroscopy. *Mailing Add:* 3901 Briarpark Houston TX 77042-5301

NAAKE, HANS JOACHIM, b Leipzig, Ger, Jan 2, 25; nat; m 56; c 3. ACOUSTICS, PHYSICS. *Educ:* Univ Gottingen, dipl, 51, PhD(physics), 53. *Prof Exp:* Sci co-worker physics, Physics Inst, Univ Gottingen, 53-57; physicist, 57-69, mgr appl physics, Major Appliance labs, 69-79, SR SCIENTIST APPL PHYSICS, ENG SCI LAB, GEN ELEC CO, 80- *Concurrent Pos:* Adj prof, Univ Louisville, 69- *Mem:* Acoust Soc Am; Ger Phys Soc. *Res:* General acoustics; vibrations; sound propagation in liquids and solids; solid state and semiconductor physics; thermoelectricity, especially thermoelectric refrigeration. *Mailing Add:* 8603 Cool Brook Ct Louisville KY 40291

NAAR, JACQUES, b Salonica, Greece, Aug 15, 30; nat US; m 53; c 3. CIVIL ENGINEERING. *Educ:* Free Univ Brussels, Ing Civ, 54, PhD, 60; Mass Inst Technol, MS & CE, 58. *Prof Exp:* Asst, Free Univ Brussels, 54-56 & Mass Inst Technol, 56-58; proj engr, Gulf Res & Develop Co, 58-61; res assoc, Olin Mathieson Chem Corp, 61-62; SR STAFF SCIENTIST, SRI INT, 62- *Concurrent Pos:* McCarty-Little chair gaming & res tech, US Naval War Col, 72-74. *Mem:* Am Soc Civil Engrs. *Res:* Systems evaluation and analysis. *Mailing Add:* Res & Analysis Div SRI Int 333 Ravenswood Menlo Park CA 94025

NAAR, RAYMOND ZACHARIAS, b Salonica, Greece; US citizen. POLYMERS, MECHANICAL ENGINEERING. *Educ:* Advan Sch Textiles, Verviers, Belg, Ingenieur, 55; Mass Inst Technol, SM, 57, ScD(mat sci), 74. *Prof Exp:* Engr polymers, Dewey & Almy, 58-60; res mgr, Cabot Corp, 60-68; res fel polymers & mat, Mass Inst Technol, 68-72; asst prof chem eng, Tufts Univ, 72-77; mgr new prod res, Huyck Res Ctr, Huyck Corp, 77-79; MGR COMM DEVELOP, PLASTICS TECHNOL DEPT, GENERAL ELECTRIC CO, 79- *Concurrent Pos:* Tech consult, 68-72; mkt consult, 73-77; chmn, bd trustees, Plastics Inst Am. *Mem:* Soc Plastics Engrs; Am Chem Soc; Am Inst Chem Engrs. *Res:* Polymers; biocompatible materials; fibers and fibrous structures. *Mailing Add:* 43 Marlboro Rd Delmar NY 12054-1316

NABELEK, ANNA K, b Warszawa, Poland, Mar 1, 34; US citizen; m 66; c 2. SPEECH COMMUNICATION, ROOM ACOUSTICS. *Educ:* Warsaw Politechnic, BS, 55, MS, 59; Polish Acad Sci, PhD, 66. *Prof Exp:* Res assoc, Central Inst Deaf, St Louis, 68-73; res assoc, Gallaudet Col, 71-73; RES PROF, UNIV TENN, 73- *Mem:* Fel Acoust Soc Am; Am Speech Lang & Hearing Asn. *Res:* Communication abilities of listeners with perceptual deficits; patterns of hearing aid use by the elderly. *Mailing Add:* 701 Chateugay Rd Knoxville TN 37923

NABER, EDWARD CARL, b Mayville, Wis, Sept 12, 26; m 53; c 2. POULTRY NUTRITION. *Educ:* Univ Wis, BS, 50, MS, 52, PhD(biochem & poultry sci), 54. *Prof Exp:* Asst poultry husb, Univ Wis, 53-54; asst poultry nutritionist, Clemson Col, 54-56; from asst prof to assoc prof, 56-63, prof poultry sci, 63-88, chmn dept, 69-86, PROF FOOD SCI & NUTRIT, OHIO STATE UNIV, 76- *Concurrent Pos:* Vis prof, Univ Wis, 64-65; mem, Animal Nutrit Comt, Nat Res Coun, 72-77; mem gov bd, Am Inst Biol Sci, 77-83; mem bd dirs, Coun Agr Sci & Technol, 83-86; vis scientist, Inst Kleintierzucht Celle, WGer, 86. *Honors & Awards:* Fel, Poulty Sci Asn. *Mem:* Am Chem Soc; Worlds Poultry Sci Asn; Poultry Sci Asn (pres, 76-77); Am Inst Nutrit; Am Inst Biol Sci. *Res:* Nutrition and metabolism in the avian species; vitamin metabolism and protein formation in the avian embryo; energy utilization in the chick; amino acid and lipid metabolism in the laying hen. *Mailing Add:* Dept Poultry Sci Ohio State Univ 674 W Lane Ave Columbus OH 43210

NABESHIMA, TOSHITAKA, b Nagoya, Japan, Apt 9, 43; m 70; c 2. NEUROPSYCHOPHARMACOLOGY, BEHAVIORAL PHARMACOLOGY. *Educ:* Gifu Pharmaceut Univ, BS, 68; Osaka Univ, MS, 70; Tohoku Univ, PhD(pharmacol), 77. *Prof Exp:* Instr pharmacol, Fac Pharmacol Sci, Meijo Univ, 73-82; from asst prof to assoc prof, 82-89, PROF & DIR PHARMACOL & HOSP PHARM, SCH MED, NAGOYA UNIV, 90- *Concurrent Pos:* Vis asst prof, Univ Miss Med Ctr, 78-81, res consult, 83; adv bd, Conf Med Biol Brain, 90- & Aichi-Soc Hosp Pharm, 90-; secy, Japanese Soc Pharmacists, 90-; vpres, Aichi-Soc Hosp Pharm, 91- *Honors & Awards:* Incentive Award, Pharmaceut Soc Japan, 86; Miyata Senjii Award, Miyata Sci Res Prom, 87. *Mem:* Am Soc Neurosci; Int Brain Res Orgn; Am Soc Pharmacol & Exp Therapeut; Fedn Am Soc Exp Biol; Col Int Neuro-psychopharmacol. *Res:* Nueropsychopharmacology of drugs and behavior. *Mailing Add:* Dept Hosp Pharm Sci Med Nagoya Japan

NABIGHIAN, MISAC N, b Bucharest, Rumania, Dec 5, 31; US citizen; m 66; c 2. GEOPHYSICS. *Educ:* Mining Inst Bucharest, BS, 54; Columbia Univ, PhD(geophys), 67. *Prof Exp:* Geophysicist, Geol Comt, Bucharest, 55-57; res scientist, Geophys Inst, Rumanian Acad Sci, 57-62; res asst, Lamont Geol Observ, NY, 63-67; geophysicist, 67-80, sr res geophysicist, 80-86, CONSULT GEOPHYSICIST, NEWMONT EXPLOR LTD, 86- *Concurrent Pos:* Adj prof geophysics, Columbia Univ, 67-76 & Univ Ariz, 76- *Mem:* Soc Explor Geophys; Europ Asn Explor Geophys; Am Geophys Union; Royal Astron Soc; hon mem SEG, 87. *Res:* Theoretical research for the development of new exploration and interpretive techniques in geophysical prospecting. *Mailing Add:* 1700 Lincoln St 26th Floor Denver CO 80203

NABOR, GEORGE W(ILLIAM), b Chippewa Falls, Wis, Dec 17, 29; m 52; c 3. PETROLEUM ENGINEERING. *Educ:* Univ Wis, BS, 50; Purdue Univ, PhD(chem eng), 53. *Prof Exp:* Sr res technologist, Mobil Res & Develop Corp, 52-63, res assoc, 63-64, sect supvr, 64-70, mgr, Reservoir Mech Res, Dallas Res Lab, 70-90; RETIRED. *Mem:* Soc Petrol Engrs. *Res:* Fluid flow in porous media; reservoir engineering; digital computation; applied mathematics and physics. *Mailing Add:* 4138 Allencrest Dallas TX 75244-7305

NABORS, CHARLES J, JR, cytology, anatomy; deceased, see previous edition for last biography

NABORS, MURRAY WAYNE, b Carlisle, Pa, Oct 4, 43; m 66. PLANT PHYSIOLOGY, PLANT BREEDING. *Educ:* Yale Univ, BS, 65; Mich State Univ, PhD(bot), 70. *Prof Exp:* NSF grant & vis asst prof biol, Univ Ore, 70; asst prof, Univ Santa Clara, 70-72; ASSOC PROF BIOL, COLO STATE UNIV, 72- *Mem:* AAAS; Am Soc Plant Physiol. *Res:* Phytochrome; water relations; tissue culture. *Mailing Add:* Dept of Bot & Plant Path Colo State Univ Ft Collins CO 80523

NABOURS, ROBERT EUGENE, b Tucson, Ariz, Nov 27, 34; m 54; c 3. ELECTRICAL ENGINEERING. *Educ:* Univ Ariz, BSEE, 57, PhD(elec eng), 65; Stanford Univ, MS, 59. *Prof Exp:* Engr, Lenkurt Elec Co, Calif, 57-58; student instr elec eng, Stanford Univ, 58-59; instr, Univ Ariz, 59-62, engr, Appl Res Lab, 62-63; res engr, Tucson Res Lab, Bell Aerosysts Co, 63-65; sr engr, Burr-Brown Res Corp, 65-66, chief engr & mgr eng dept, 66-68; vpres, Kinnison & Nabours Consult Engrs, Inc, 68-71; prin, Robert E Nabours Consult Elec Engr, 71-78; sr vpres, Johannessen & Grand Consult Engrs, 78-80; dir elec eng, Finical & Dombrowski, Architects & Engrs, 80-82; PRES, CONSULT ELEC ENGRS, INC, TUCSON, ARIZ, 82- *Mem:* Inst Elec & Electronic Engrs; Nat Soc Prof Engrs; Instrument Soc Am; Illum Eng Soc; Nat Acad Forensic Engrs. *Res:* Adaptive communication systems; synthesis of optimum systems; analog/hybrid computer systems for random process simulation; construction electrical engineering; forensic engineering. *Mailing Add:* Consult Elec Engrs Inc 5201 Salida Del Sol Tucson AZ 85718

NABRIT, SAMUEL MILTON, b Macon, Ga, Feb 21, 05. EMBRYOLOGY. *Educ:* Morehouse Col, BS, 25; Brown Univ, MS, 28, PhD(biol), 32. *Hon Degrees:* Numerous hon degrees from US Univ. *Prof Exp:* Instr zool, Morehouse Col, 25-27, prof, 28-31; prof, Atlanta Univ, 32-55; pres, Tex Southern Univ, 55-66; comnr, US AEC, 66-67; exec dir, Southern Fels Fund, 67-81; RETIRED. *Concurrent Pos:* Exchange prof, Atlanta Univ, 30, dean grad sch; Gen Educ Bd fel, Columbia Univ, 43; res fel, Univ Brussels, 50; coordr, Carnegie Exp Grant-in-Aid Res Prog; mem sci bd, NSF, 56-60; mem corp, Marine Biol Lab, Woods Hole; mem, Marine Biol Labs; Atomic Energy Comn, 66-67; exec dir, Nat Fel Fund, 67-81. *Honors & Awards:* Distinguished Grad Sch Achievement Award, Brown Univ, 82. *Mem:* Inst Med-Nat Acad Sci; Soc Develop Biol; Nat Asn Res Sci Teaching; Nat Inst Sci (pres, 45); fel AAAS; Am Soc Zool. *Res:* Neuroembryology; role of fin rays in regeneration of tail-fins of fishes. *Mailing Add:* 686 Beckwith St SW Atlanta GA 30314

NACAMU, ROBERT LARRY, b New York, NY, Oct 17, 44; m 66; c 3. CERAMIC SCIENCE. *Educ:* Rutgers Univ, BS, 66, PhD(ceramics), 69. *Prof Exp:* Res scientist, Lavino Div, Int Minerals & Chem, 69-74; staff res engr, Kaiser Refractories Div, Kaiser Aluminum & Chem Corp, 74-84; MGR, PROCESS & QUAL ENG, NAT REFRACTORIES & MINERALS CORP, 84- *Mem:* Am Ceramic Soc; Nat Inst Ceramic Eng; Am Inst Mining, Metall & Petrol Eng. *Res:* Basic refractory brick and monolithics. *Mailing Add:* 6538 Lansing Ct Pleasanton CA 94566

NACE, DONALD MILLER, b Hanover, Pa, Nov 28, 24; m 45; c 2. PHYSICAL CHEMISTRY. *Educ:* Lehigh Univ, BS, 47, MS, 49; Pa State Univ, PhD(chem), 56. *Prof Exp:* Asst, Nat Printing Ink Res Inst, 47-49; res chemist, Mobil Res & Develop Corp, 49-58, sr res chemist, 58-74, res assoc, 74-87; RETIRED. *Mem:* Am Chem Soc. *Res:* Catalysis in petroleum processing. *Mailing Add:* Four Oak Rd Woodbury NJ 08096

NACE, HAROLD RUSS, b Collingswood, NJ, July 5, 21; m 44. ORGANIC CHEMISTRY. *Educ:* Lehigh Univ, BS, 43; Mass Inst Technol, PhD(org chem), 48; Brown Univ, MS, 57. *Prof Exp:* Res chemist, Merck & Co, Inc, 44-45; from instr to assoc prof, 48-59, PROF CHEM, BROWN UNIV, 59- *Concurrent Pos:* Res chemist, Jackson Lab, E I du Pont de Nemours & Co, 56-57, consult, 57-60; consult, William S Merrell Co div, Richardson-Merrell Inc, 60-68, Wyeth Labs, 68- & Bd Rev, US Pharmacopeia, 80- *Mem:* AAAS; Am Chem Soc; fel Am Inst Chemists; fel NY Acad Sci. *Res:* Stereochemistry and partial synthesis of steroids; dipole moment studies of organic compounds; prostaglandin syntheses. *Mailing Add:* 28 Chapin Rd Barrington RI 02806-4407

NACE, PAUL FOLEY, b Brooklyn, NY, May 6, 17. DIABETES RESEARCH. *Educ:* NY Univ, PhD(histol), 51. *Prof Exp:* Prof biol, Col Staten Island, 71-83; RETIRED. *Mem:* Am Zool Soc; Histochem Soc Am; Am Physiol Soc. *Mailing Add:* City Univ NY 815 Ocean Terr Staten Island NY 10301

NACHAMKIN, JACK, b New York, NY, Mar 23, 40; div; c 2. MATHEMATICAL PHYSICS. *Educ:* Polytech Inst Brooklyn, BS, 60; Rensselaer Polytech Inst, PhD(physics), 64. *Prof Exp:* Physicist fel appl nuclear, Univ Kans, 64-66, Chalk River Lab, Can, 66-68; physicist appl nuclear, Los Alamos Nat Lab, 68-80; mem staff, Exxon Enterprises, 80-84; MEM STAFF, LOCKHEED, 84- *Mem:* Am Phys Soc; Inst Elec & Electronics Engrs. *Res:* Nuclear structure; hydrodynamics; radiation transport; space charge electronics; electric vehicle research. *Mailing Add:* 13326 Racquet Ct Poway CA 32064-4903

NACHBAR, MARTIN STEPHEN, b New York, NY, July 17, 37; m 62; c 1. MEDICINE, MICROBIOLOGY. *Educ:* Union Col, BS, 58; NY Univ, MD, 62. *Prof Exp:* From intern to resident, Bellevue Hosp, New York, 62-66; USPHS med scientist training fel, 66-69, from instr to asst prof med, 69-72, asst prof med & microbiol, 72-74, ASSOC PROF MED & MICROBIOL, MED CTR, NY UNIV, 75- *Concurrent Pos:* NY Heart Asn sr investr, 69. *Honors & Awards:* Career Scientist Award, Irma T Hirschl Trust, 74. *Res:* Dietary effects of lectins. *Mailing Add:* Dept Med & Microbiol NY Univ Med Ctr 550 First Ave New York NY 10016

NACHBAR, WILLIAM, b Brooklyn, NY, Apr 25, 23; m 52; c 1. APPLIED MATHEMATICS. *Educ:* Cornell Univ, BME, 44; NY Univ, MS, 48; Brown Univ, PhD(appl math), 51. *Prof Exp:* Res assoc appl math, Brown Univ, 49-51; staff mem, Math Servs Unit, Boeing Airplane Co, 51-55; sect head mech, Appl Math Dept, Missiles & Space Div, Lockheed Aircraft Corp, 55-60, staff scientist, 60-63; assoc prof, Dept Aeronaut & Astronaut, Stanford, 63-65; PROF APPL MECH & ENG SCI, UNIV CALIF, SAN DIEGO, 65- *Concurrent Pos:* Res assoc, Stanford Univ, 61-63. *Mem:* Assoc fel, Am Inst Aeronaut & Astronaut; Am Soc Mech Eng; Combustion Inst. *Res:* Solid mechanics and structural analysis; stability theory; energy policy and technology. *Mailing Add:* Dept Appl Mech & Eng Sci B-010 Univ Calif San Diego Box 109 La Jolla CA 92093

NACHBIN, LEOPOLDO, b Recife, Brazil, Jan 7, 22; m 56; c 3. MATHEMATICS. *Educ:* Univ Brazil, MS, 43, PhD(math), 47. *Hon Degrees:* Dr, Univ Pernambuco, 66. *Prof Exp:* Prof math, Univ Brazil, 50-61; vis prof, Univ Paris, 61-63; prof, 63-67, GEORGE EASTMAN PROF MATH, UNIV ROCHESTER, 67- *Concurrent Pos:* Fels, Guggenheim Found, 49-50, 57-58 & Rockefeller Found, 56-57; head div math res, Nat Res Coun Brazil, 55-56, mem gen bd, 60-61; mem, Inter-Am Comt Math Educ, 61-75; hon prof, Univ de Campinas, 89. *Honors & Awards:* Moinho Santista Found Prize, 62; Houssay Sci Prize, Orgn Am States, 82. *Mem:* Brazilian Acad Sci; Lisbon Acad Sci; Acad Sci Latin Am; Acad Sci Spain; Acad Sci Arg. *Res:* Approximation theory; holomorphy; functional analysis. *Mailing Add:* Dept Math Univ Rochester Rochester NY 14627

NACHLINGER, R RAY, b Taylor, Tex, Dec 4, 44; m 65; c 1. ENGINEERING MECHANICS. *Educ:* Univ Tex, BEngS, 66, MS, 67, PhD(eng mech), 68. *Prof Exp:* Res engr, Univ Tex, 66-68; from asst prof to assoc prof mech eng, Univ Houston, 68-80; PRES, ULTRAMARINE, INC, 80- *Mem:* Am Soc Civil Engrs; Soc Eng Sci; Soc Naval Architects & Marine Engrs. *Res:* Theoretical and applied solid and continuum mechanics. *Mailing Add:* 10910 Wickersham Houston TX 77042

NACHMAN, ARJE, b Salzburg, Austria, Sept 12, 46; US citizen; m 71. APPLIED MATHEMATICS, PHYSICAL MATHEMATICS. *Educ:* Wash Univ, BS, 68; NY Univ, PhD(math), 73. *Prof Exp:* Asst prof math, Tex A&M Univ, 73-78; assoc prof math, Old Dominion Univ, 79-81; sr res scientist, Southwest Res Inst, 81-84; prof math, Hampton Univ, 84-85; PROG MGR MATH, AIR FORCE OFF SCI RES, 85- *Mem:* Soc Indust & Appl Math. *Res:* Theoretical fluid mechanics; fully nonlinear solid mechanics; theoretical combustion; neoelastic fracture mechanics. *Mailing Add:* Air Force Off Sci Res-NM Bolling AFB Washington DC 20332

NACHMAN, JOSEPH F(RANK), b Toledo, Ohio, Jan 22, 18; m 43; c 2. PHYSICAL METALLURGY. *Educ:* Univ Toledo, BS, 40; Ohio State Univ, MSc, 47. *Prof Exp:* Metallurgist phys metall, US Naval Ord Lab, 48-56; mgr alloy develop, Denver Res Inst, 56-63; specialist, Atomics Int Div, Rockwell Int, Inc, 63-66; chief appl sci, advan methods & mat, Solar Div, Int Harvester Co, 66-81, res staff specialist, Solar Turbines Int, 77-81; pres, Metall Consult Serv, Inc, 81-86; RETIRED. *Honors & Awards:* Letter Commendation, Chief Navy Bur Ord, 45; Civilian Meritorious Award, US Dept Navy, 53. *Mem:* AAAS; Am Soc Metals; Nat Asn Corrosion Engrs. *Res:* Nonstrategic aluminum-iron base alloys; research on magnetic alloys; dispersion hardening of metals; turbine superalloys; high-damping alloys; rare earth metals; nuclear materials and cladding alloys; metal hydrides. *Mailing Add:* 7060 Murray Park Dr San Diego CA 92119

NACHMAN, RALPH LOUIS, b Bayonne, NJ, June 29, 31; m 58; c 2. MEDICINE. *Educ:* Vanderbilt Univ, AB, 53, MD, 56. *Prof Exp:* From instr to asst prof, 63-68, assoc prof, 68-80, PROF MED, MED COL, CORNELL UNIV, 80-, CHIEF DIV HEMAT, 70-; ASSOC ATTEND PHYSICIAN, NY HOSP, 70- *Concurrent Pos:* Res fel med, New York Hosp-Cornell Med Ctr, 62-63; dir labs clin path, NY Hosp, 63-70. *Mem:* AAAS; Am Fedn Clin Res; Am Soc Hemat; Am Physiol Soc. *Res:* Immunological aspects of hematologic diseases, particularly platelet abnormalities. *Mailing Add:* Cornell Univ Med Col 525 E 68th St New York NY 10021

NACHMAN, RONALD JAMES, b Takoma Park, Md, Feb 1, 54; m 76, 89. NEUROPEPTIDE & ANALOG SYNTHESIS & CONFORMATION. *Educ:* Revelle Col, Univ Calif, San Diego, BS, 76; Stanford Univ, PhD(org chem), 81. *Prof Exp:* Res asst, Scripps Inst Oceanog, 74-76; res chemist, Western Regional Res Ctr, 81-89, RES CHEMIST VET TOXICOL & ENTOM RES LAB, USDA, 89- *Concurrent Pos:* Vis scientist, Labs Neuroendocrinol, Salk Inst, 85, Dept Molecular Biol, Res Inst Scripps Clin, 88-89. *Honors & Awards:* Curric Competitive Grant Award, 85; Cert of Merit, USDA, 88 & 91. *Mem:* NY Acad Sci; AAAS; Am Chem Soc; Sigma Xi; Entom Soc Am; Agr Chem Soc Japan. *Res:* Insect neuropeptide chemistry, identification, synthesis, conformation, molecular biology and structure activity relationships; nucleoside chemistry; constituents of poisonous plants and tetrodotoxin and analogs. *Mailing Add:* Vet Toxicol Entom Res Lab Rte 5 Box 810 College Station TX 77840

NACHMIAS, VIVIANNE T, ANATOMY, CELL BIOLOGY. *Educ:* Swarthmore Col, BA, 52; Radcliffe Col, MA 53; Univ Rochester, MD, 57. *Prof Exp:* NIH postdoctoral fel, Inst Neurol Sci, Univ Pa, 57-59, res assoc, Dept Anat, Sch Med, 61-65; res assoc, Dept Biol, Haverford Col, 65-68, asst prof, 69-71, res assoc & instr, 71-73; assoc prof, 73-82, PROF ANAT, SCH MED, UNIV PA, 82- *Concurrent Pos:* Vis scientist, MRC Unit Molecular Biol, Cambridge, Eng, 68-69, Dept Biol, Brandeis Univ, 78-79 & Dept Cell Biol, Stanford Univ, 86; Sloan fel, Haverford Col, 69-71; spec reviewer, Cell Biol Study Sect, NIH, 73-74, mem, 76-80; session chairperson, Am Soc Cell Biol, 74, 77 & 78, mem coun, 78-81, Legis Alert Comt, 82 & Pub Policy Comt, 83-87; lectr, NIH, 86 & 87, Gladstone Found, 86 & Univ Pa, 91; prin investr, Pa Muscle Inst, 88-91; mem, Physiol Processes Rev Comt, NSF, 91. *Mem:* Sigma Xi; Am Asn Anatomists; AAAS; Am Soc Cell Biologists; Electron Micros Soc Am; Biophys Soc; Soc Exp Biol & Med; Am Heart Asn; Am Soc Hemat. *Res:* Actomyosin systems and cell motility; author of numerous technical publications. *Mailing Add:* Dept Anat Sch Med Univ Pa 138C Anat-Chem Bldg Philadelphia PA 19104

NACHREINER, RAYMOND F, b Richland Center, Wis, Apr 29, 42; m 68; c 3. VETERINARY PHYSIOLOGY, ENDOCRINOLOGY. *Educ:* Iowa State Univ, DVM, 66; Univ Wis, PhD(endocrinol), 72. *Prof Exp:* Res asst endocrinol, Univ Wis, 68-72; asst prof physiol, Auburn Univ, 72-77; PROF, ENDOCRINE DIAG SECT, ANIMAL HEALTH DIAG LAB, MICH STATE UNIV, 77- *Concurrent Pos:* Sabbatical, Int Atomic Energy Agency, Vienna, Austria, 85-86. *Honors & Awards:* Burr Beach Award, Univ Wis, 72. *Mem:* Soc Study Fertil; Soc Study Reprod; Am Vet Med Asn; Endocrine Soc. *Res:* Population control in animals; endocrine diagnostic procedures in domestic animals; endocrine pharmacokinetics; bovine cystic ovarian disease. *Mailing Add:* Large Animal Clin Sci A13 Vet Clin Ctr Mich State Univ East Lansing MI 48824

NACHT, SERGIO, b Buenos Aires, Argentina, Apr 13, 34; m 58; c 4. PRODUCT PERFORMANCE EVALUATION, DERMATOLOGICAL RESEARCH ADMINISTRATION. *Educ:* Univ Buenos Aires, BA, 58, MS, 60, PhD(biochem), 64. *Prof Exp:* Asst prof biochem, Univ Buenos Aires, 60-65; asst prof med, Univ Utah, 65-70; res scientist, Alza Corp, Palo Alto, Calif, 70-76; sr investr, 73-76, asst dir, 76-79, dir, 74-83, dir biomed res, Richardson Vicks, Inc, 83-87; SR VPRES, RES & DEVELOP, ADVANCED POLYMER SYSTEMS, INC, REDWOOD CITY, CALIF, 87- *Concurrent Pos:* Lectr, dermat dept, State Univ NY, Dowstate Med Ctr, 77-87. *Mem:* Soc Investigative Dermat; Soc Cosmetic Chemists; Am Physiol Soc; Dermat Found; Am Acad Dermat. *Res:* Biochemistry and biophysics of human skin and hair; new and improved topical delivery systems for drugs and cosmetic ingredients; new methodologies for the measurement of biophysical properties of skin and hair. *Mailing Add:* 289 Quinnhill Ave Los Altos CA 94024

NACHTIGALL, GUENTER WILLI, b Hamburg, Ger, Jan 1, 29; nat US; m 50; c 1. PHYSICAL ORGANIC CHEMISTRY. *Educ:* Columbia Univ, BS, 62; Univ Colo, PhD(org chem), 68. *Prof Exp:* Chemist, Indust Chem Div, 61, chemist cent res div, 62-63, res chemist, 64-74, sr res chemist, Chem Res Div, Am Cyanamid Co, 74-86; RETIRED. *Mem:* Am Chem Soc; Sigma Xi. *Res:* Carbonium ion reactions, particularly those of bridged polycyclic compounds; asymmetric organic pharmaceuticals; industrial organic process research and development. *Mailing Add:* PO Box 846 Crested Butte CO 81224-0846

NACHTMAN, ELLIOT SIMON, b Cleveland, Ohio, June 8, 23; m 45; c 3. MECHANICAL METALLURGY, TRIBOLOGY. *Educ:* Wooster Col, BA, 45; Ill Inst Technol, MS, 50, PhD(metall eng), 69; cert mgt, Univ Chicago, 56. *Prof Exp:* Jr chemist, Harshaw Chem Co, Ohio, 45-46; res chemist, Manhattan Proj, Univ Chicago, 46-48; instr metall, Ill Inst Technol, 48-50; prod engr, LaSalle Steel Co, Ind, 50-52, mgr prod eng, 52-56, dir res & develop, 56-66, vpres res & develop, 66-71; MGR & PRIN, TOWER OIL & TECHNOL CO, 71- *Concurrent Pos:* Mem adv panel metall, NSF, 70-; adj prof, Univ Ill, Chicago Circle, 75- *Mem:* Fel Am Soc Metals; Metall Soc; Am Inst Mining, Metall & Petrol Engrs; Soc Lubrication Engrs. *Res:* Plastic deformation of metals; surface reactions; ferrous metallurgy; machinability of metals; lubrication. *Mailing Add:* Tower Oil & Technol Co 205 W Randolph St Chicago IL 60606

NACHTRIEB, NORMAN HARRY, b Chicago, Ill, Mar 4, 16; m 41, 53; c 1. PHYSICAL CHEMISTRY. *Educ:* Univ Chicago, BS, 36, PhD(chem), 41. *Prof Exp:* Analytical chemist, State Geol Surv, Ill, 37-38; head, Analytical Sect, Columbia Chem Div, Pittsburgh Plate Glass Co, Ohio, 41-43; res chemist, Manhattan Dist, Metall Lab, Chicago, 43-44; alternate group leader anal group, Los Alamos Sci Lab, NMex, 44-46; from asst prof to assoc prof, 46-53, prof chem, James Franck Inst, 46-86, chmn dept, 62-71, master, Phys Sci Col Div & assoc dean, Div Phys Sci, 73-81, prof, 53-86, EMER PROF CHEM, UNIV CHICAGO, 85- *Concurrent Pos:* Adv ed, Encyclopaedia Britannica, 55-; sr fel NSF, 59-60, mem adv panel, Sci Educ Div, 65-; sr fel, Nat Sci Found, 59-60; vis prof chem, Univ Ill, Chicago, 86-89. *Mem:* Am Chem Soc; fel Am Phys Soc. *Res:* Spectrochemical analysis of solutions; fused salts; metals; extraction of metal halides by organic solvents; electrode potentials; diffusion in crystalline solids and liquids; high pressure chemistry; nuclear magnetic resonance; solid state chemistry; magnetic, electrical and optical properties of metal-molten salt solutions; author of two books on chemistry. *Mailing Add:* PO Box 61 Palos Park Chicago IL 60464

NACHTSHEIM, PHILIP ROBERT, b New York, NY, Dec 26, 28; m 56; c 4. AERONAUTICAL ENGINEERING. *Educ:* Ohio State Univ, BS, 52; Case Inst Technol, MS, 58, PhD(aeronaut eng), 63. *Prof Exp:* Instr mech eng, Case Inst Technol, 55-61; res engr, Lewis Res Ctr, NASA, Ohio, 61-65, asst chief, Thermal Protection Br, 65-80, ASST CHIEF, ENTRY TECHOL BR,

AMES RES CTR, NASA, 80- *Mem:* Am Inst Aeronaut & Astronaut. *Res:* Planetary entry technology; interaction of intense radiation with material; patterned ablation effects; hydrodynamic stability. *Mailing Add:* 563 Dublin Way Sunnyvale CA 94087

NACHTWEY, DAVID STUART, b Seattle, Wash, Aug 9, 29; m; c 3. RADIOBIOLOGY, RADIOLOGICAL HEALTH. *Educ:* Univ Wash, BA, 51; Univ Tex, MA, 56; Stanford Univ, PhD(biol sci), 61. *Prof Exp:* Nat Cancer Inst fel, Biol Inst, Carlsberg Found, Denmark, 61-62; res biologist, Cellular Radiobiol Br, US Naval Radiol Defense Lab, 62-68; assoc prof radiation biol, Ore State Univ, 68-74; ed, Climatic Impact Assessment Prog Monograph 5, 74-75; bioenviron effects off, 75-81, chief biomed appl br, 81-84, chief biomed res br, 84-86, mgr, med Sci Space Sta Off, 86-87, RADIOL HEALTH OFF, NASA JOHNSON SPACE CTR, NASA, 81-, CHIEF SCIENTIST, MED SERVS, 87- *Concurrent Pos:* Mem comt photobiol, Nat Res Coun; mem, High Let Panel of Comt on Interagency Radiation Res & Policy Coordr; consult, Nat Coun on Radiation Protection & Measurements. *Honors & Awards:* Medal Exceptional Sci Achievement, NASA, 81. *Mem:* Sigma Xi; Radiation Res Soc; Am Soc Photobiol. *Res:* Cell division processes in protozoa, algae, and mammalian cells; effects of ultraviolet and ionizing radiation on cells; mutagenesis; recovery from radiation damage; ecosystem responses to solar ultraviolet radiation; skin cancer incidence modeling; space radiation carcinogenesis. *Mailing Add:* Med Sci Div Mail Code SD12 NASA Johnson Space Ctr Houston TX 77058-3696

NACOZY, PAUL E, b Los Angeles, Calif, Apr 15, 42. ASTRONOMY, CELESTIAL MECHANICS. *Educ:* San Diego State Col, BA, 64; Yale Univ, MS, 66, PhD(astron), 68. *Prof Exp:* Asst prof, 68-74, ASSOC PROF AEROSPACE ENG, UNIV TEX, AUSTIN, 74- *Mem:* Am Astron Soc; Int Astron Soc; Am Inst Aeronaut & Astronaut. *Res:* Celestial mechanics and aerospace mechanics, especially series-solutions of the motions of space vehicles, asteroids, and comets; earth-moon-space vehicle dynamical system. *Mailing Add:* Dept of Aerospace Eng Univ of Tex Austin TX 78712

NACY, CAROL ANNE, b Tokyo, Japan, 1948; m 81; c 5. MACROPHAGE ACTIVATION, PARASITE IMMUNOLOGY. *Educ:* Cath Univ Am, PhD(microbiol), 78. *Prof Exp:* SR RES SCIENTIST, WALTER REED ARMY INST RES, 80- *Concurrent Pos:* Adj assoc prof, Dept Biol, Catholic Univ Am, 80-; outstanding young investr award, Soc Leukocyte Biol, 84; ed, J Immunol, 86-91, Infection & Immunity, 90-95; chmn, Immunol Div, Am Soc Microbiol, 86-88, div group rep, 88-90. *Mem:* Am Acad Microbiol; Am Soc Microbiol; Soc Leukocyte Biol (secy, 86-90, pres-elect, 91); Am Asn Immunologists; Am Soc Trop Med & Hyg. *Mailing Add:* Dept Cellular Immunol Walter Reed Army Inst Res 9620 Medical Center Dr Suite 200 Rockville MD 20850

NADAL-GINARD, BERNARDO, b Arta, Baleares, Spain, March 14, 42; c 2. CARDIOLOGY. *Educ:* Univ Barcelona, MD, 65; Yale Univ, PhD(biol), 75. *Prof Exp:* Intern, Sch Cardioangiol, Univ Barcelona, 65-66, resident, 66-68; resident, Nat Heart Inst, 68-70, chief resident, 70-72; NIH Fogarty Int fel, Yale Univ, 72-73, Pop Coun fel & asst clin prof pediat, 73-75; from asst prof to assoc prof cell biol, Albert Einstein Col Med, 75-81; CHMN DEPT CARDIOL, CHILDREN'S HOSP MED CTR, BOSTON, 82-; PROF PEDIAT, HARVARD MED SCH, 82-, PROF CELLULAR & MOLECULAR PHYSIOL, 87- *Concurrent Pos:* Mem, Molecular Cytol Study Sect, NIH, 83-; bd mem, Coun Basic Sci, Am Heart Asn, 85-; dir PhD prog, Harvard Univ-Mass Inst Technol, 85-, prof, Health Sci & Technol Prog, 86-; investr, Howard Hughes Med Inst, 86- *Mem:* Am Heart Asn; Am Col Cardiol; AAAS; Am Soc Cell Biol; Am Soc Microbiol. *Res:* Genetic regulation of all differentiation with particular emphasis on myogenesis; cardiac development and hypertrophy; post-transcriptional regulation of contractile protein genes; generation of protein isoform complexity through promoter selection and alternate splicing. *Mailing Add:* Dept Cardiol Children's Hosp 300 Longwood Ave Boston MA 02115

NADAS, ALEXANDER SANDOR, b Budapest, Hungary, Nov 12, 13; US citizen; m 41; c 3. PEDIATRIC CARDIOLOGY. *Educ:* Med Univ Budapest, MD, 37; Wayne State Univ, MD, 45; Am Bd Pediat, dipl & cert pediat cardiol. *Prof Exp:* Intern, Fairview Park Hosp, Cleveland, Ohio, 39-40 & Wilmington Gen Hosp, Del, 40-41; resident pediat, Mass Mem Hosps, Boston, 41-42; vol asst med serv, Children's Hosp, Boston, 42-43, asst resident, 43; chief resident, Children's Hosp, Mich, 43-45; pvt pract, Greenfield, Mass, 45-49; instr pediat, 50-52, clin assoc, 52-55, from asst clin prof to clin prof, 55-69, prof, pediat, 69-82, chief cardiol dept, 69-82, EMER PROF PEDIAT, HARVARD MED SCH, 82-; EMER CHIEF CARDIOL DEPT, CHILDREN'S HOSP, 82- *Concurrent Pos:* Res fel pediat, Harvard Med Sch, 49; Guggenheim fel, 70; instr, Wayne State Univ, 43-45; asst physician & assoc chief cardiol div, Children's Hosp Med Ctr, 49-50, from assoc physician to physician, Sharon Sanatorium, 50-51; assoc physician & assoc cardiologist, Sharon Cardiovasc Unit, 51-52, cardiologist, 52-66, sr assoc med & Good Samaritan Div, 62-, chief cardiol div, 66-69; Fulbright prof, State Univ Groningen, 56-57; consult, Brigham & Women's Hosp, Mass Gen Hosp, Boston & Newton Wellesley Hosp. *Mem:* Am Heart Asn; Am Acad Pediat; Soc Pediat Res; Am Pediat Soc. *Res:* Applied cardiovascular physiology and physiology of congenital heart disease; natural history of congenital heart disease with special emphasis on clinical-physiologic correlations; multicenter clinical trials in congenital heart disease. *Mailing Add:* Children's Hosp 300 Longwood Ave Boston MA 02115

NADASEN, ARUNA, b Durban, SAfrica; US citizen. ELASTIC SCATTERING, KNOCKOUT REACTIONS. *Educ:* Rhodes Univ, SAfrica, MSc, 67; Ind Univ, MS, 71, PhD(nuclear physics), 77. *Prof Exp:* Teaching asst physics, Ind Univ, 70-71, assoc instr, 72-73, res asst, 73-77; res assoc, Univ Md, 77-81, lectr, 81-82; asst prof, 82-87, ASSOC PROF PHYSICS, UNIV MICH, DEARBORN, 87- *Concurrent Pos:* Lectr, Univ Durban, Westville, SAfrica, 71-72; prin investr, Exp Nuclear Physics, Univ Mich, Dearborn, 83- *Mem:* Am Phys Soc; Am Asn Physics Teachers; Nat Sci Teachers Asn. *Res:* Experimental nuclear physics; interaction between few nucleon systems; light ion elastic scattering; investigation of clustering in nuclei via knockout reactions. *Mailing Add:* Dept Natural Sci Univ Mich 4901 Evergreen Rd Dearborn MI 48128

NADDOR, ELIEZER, operations research, software systems, for more information see previous edition

NADDY, BADIE IHRAHIM, b Haifa, Palestine, Dec 31, 33; US citizen; m 63; c 3. PHYSICAL CHEMISTRY, SOIL CHEMISTRY. *Educ:* Am Univ Beirut, BS, 57; Kans State Univ, PhD(soil & phys chem), 63. *Prof Exp:* Chmn div sci, Henderson State Col, 63-65; dir labs, Jordan Govt, 65-67; CHMN DIV MATH & SCI, COLUMBIA STATE COL, 68- *Mem:* Am Chem Soc; Sigma Xi. *Res:* Effect of cation exchange on the dielectric constants of minerals; x-ray diffraction of minerals saturated with different cations. *Mailing Add:* Dent Chem Columbia State Col PO Box 1315 Columbia TN 38401

NADEAU, BETTY KELLETT, b Topeka, Kans, Apr 23, 65. MICRO PALEONTOLOGY. *Educ:* Univ Kans, AB, 27. *Prof Exp:* Prof micro paleont, NY Univ, 60-65; RETIRED. *Honors & Awards:* Howard Erasmus Award, Univ Kans, 60. *Mem:* Am Soc Petrol Geologist fel; fel Geol Soc Am. *Mailing Add:* Westview Lane South Norwalk CT 06854

NADEAU, HERBERT GERARD, b Cranston, RI, Aug 1, 28; m 51; c 4. PHYSICAL CHEMISTRY, COMBUSTION TECHNOLOGY. *Educ:* Providence Col, BS, 51. *Prof Exp:* Anal res chemist, Geigy Chem Corp, 51-55; chief, Sect Analytical Chem, Olin Mathieson Chem Corp, 55-65; head analytical res serv, Upjohn Co, 65, mgr cellular plastics, 66-68, mgr analytical & phys res, 68-74, mgr phys & flammability res, 74-85; assoc dir fire res, Dow Chem, 85-86; RETIRED. *Concurrent Pos:* Mem, Prods Res Comt, 75-80. *Honors & Awards:* Distinguished Serv Award, Citation & Cert Appreciation, Soc Plastics Indust, 75. *Mem:* Am Chem Soc; Int Isocyanate Inst; Soc Plastics Indust. *Res:* Organic analysis; infrared; vapor phase chromatography; high vacuum techniques; boron compounds; commercial products; urethane chemistry; polymers; flammability testing and research. *Mailing Add:* RR 2 Meredith NH 03253

NADEAU, REGINALD ANTOINE, b St Leonard, NB, Dec 18, 32; m 57; c 2. CARDIOVASCULAR PHYSIOLOGY, CARDIOLOGY. *Educ:* Loyola Col, Can, BA, 52; Univ Montreal, MD, 57; FRCP(C), 62. *Prof Exp:* From asst prof to assoc prof, 64-70, prof physiol, 72-75, PROF MED, FAC MED, UNIV MONTREAL, 75- *Concurrent Pos:* Career investigator, Med Res Coun Can, 65; dir res, Cardiol Hosp Sacre Coeur, Montreal. *Mem:* Can Physiol Soc; Can Cardiovasc Soc; Am Col Cardiol. *Res:* Clinical cardiology. *Mailing Add:* Dept Med Fac Med Univ Montreal Montreal PQ H3C 3J7 Can

NADEL, ETHAN RICHARD, b Washington, DC, Sept 3, 41. PHYSIOLOGY. *Educ:* Williams Col, BA, 63; Univ Calif, Santa Barbara, MA, 66, PhD(biol), 69. *Prof Exp:* Asst prof, 70-76, ASSOC PROF, DEPTS EPIDEMIOL & PUB HEALTH & PHYSIOL, SCH MED, YALE UNIV, 76- *Concurrent Pos:* NIH fel, Sch Med, Yale Univ, 69-70, USPHS grant, 70-; from asst fel to assoc fel, John B Pierce Found Lab, 70-; partic, US-Japan Prog Human Adaptability, 72-73; mem, Environ Physiol Comn, Int Union Physiol Sci, 77-; Hall Mem lectr, Univ Louisville, 79. *Mem:* AAAS; Am Physiol Soc; fel Am Col Sports Med. *Res:* Physiological regulations against hyperthermia. *Mailing Add:* John B Pierce Found Lab 290 Congress Ave New Haven CT 06519

NADEL, JAY A, b Philadelphia, Pa, Jan 21, 29; m 60; c 3. PULMONARY PHYSIOLOGY, PHARMACOLOGY. *Educ:* Temple Univ, AB, 49; Jefferson Med Col, MD, 53. *Prof Exp:* Trainee heart & lung res, 58-62, clin instr med, 61-62, asst clin prof, 62-64, from asst prof to assoc prof, 64-70, PROF MED & PHYSIOL, CARDIOVASC RES INST, MED CTR, UNIV CALIF, SAN FRANCISCO, 70-, MEM STAFF, 64- *Concurrent Pos:* Mem, Cardiovasc B Study Sect, Nat Heart & Lung Inst, 71-75; adv, Comn State-Wide Air Pollution Res, Univ Calif; mem, Comn Surv Prof Manpower Pulmonary Dis; dir, Multidisciplinary Res Training Prog in Pulmonary Dis, Nat Heart, Lung & Blood Inst, 74-, prog dir, Neuro Humoral Control of Lungs & Airway, 78-, mem, Pulmonary Dis Adv Comt, 81-85. *Mem:* Am Physiol Soc; Asn Am Physicians; Thoracic Soc (pres, 73-); Am Soc Clin Invest. *Res:* Cell biology of airways, mucous secretion, respiration, smooth muscle and automic nerves; sensory neuropeptides. *Mailing Add:* Physiol Univ Calif San Francisco CA 94143

NADEL, NORMAN A, b New York, NY, Apr 10, 27; m 52; c 2. UNDERGROUND CONSTRUCTION. *Educ:* City Col New York, BCE, 49. *Prof Exp:* Engr, 59-63, proj mgr, 63-66, vpres, 66-70, PRES, MACLEAN GROVE & CO, INC, 70- *Honors & Awards:* Construct Mgt Award, Am Soc Civil Engrs, 86. *Mem:* Nat Acad Eng; fel Am Soc Civil Engrs. *Mailing Add:* 14 Reynwood Manor Greenwich CT 06831

NADELHAFT, IRVING, b New York, NY, Nov 4, 28; m 56; c 2. NEUROBIOLOGY. *Educ:* City Col New York, BS, 49; Syracuse Univ, PhD, 56. *Prof Exp:* Res assoc, Carnegie Inst Technol, 58-61, asst prof physics, 61-67; asst prof neurol surg, 74-, ASSOC PROF NEUROL SURG & PHARMACOL, MED SCH, UNIV PITTSBURGH; RES PHYSICIST, VET ADMIN HOSP, PITTSBURGH, 69- *Concurrent Pos:* Ford Found fel, Europ Orgn Nuclear Res, 56-57; NIH spec fel, Dept Anat & Cell Biol, Sch Med, Univ Pittsburgh & Dept Biol, Mass Inst Technol, 67-69; guest res physicist, Brookhaven Nat Lab, 62-63; adj asst prof pharmacol, Med Sch, Univ Pittsburgh, 69- *Mem:* AAAS; Biophys Soc; Soc Neurosci; Am Phys Soc. *Res:* High energy physics; weak interactions; particle physics; axoplasmic flow; tracer techniques; computer simulation. *Mailing Add:* 6929 Rosewood St Pittsburgh PA 15208

NADER, ALLAN E, b Chicago, Ill, Dec 24, 37; m 65; c 1. ORGANIC CHEMISTRY. *Educ:* Ill Inst Technol, BS, 60; Western Mich Univ, MA, 63; Purdue Univ, PhD(org chem), 67. *Prof Exp:* Res chemist, 66-71, sr res chemist, Petrochem Dept, 71-80, sr res chemist, Cent Res Dept, 80-84, res assoc, 85-89, SR RES ASSOC, DU PONT CHEMICALS, E I DU PONT DE NEMOURS & CO, INC, 89- *Mem:* Am Chem Soc; Sigma Xi. *Res:* Reaction mechanism studies; hydrocarbon nitration and oxidation; metal catalysis; chemical process research; chemical synthesis; photosensitive polymers. *Mailing Add:* 328 Snuff Mill Rd Wilmington DE 19807

NADER, BASSAM SALIM, b Aug 5, 52; m 77; c 4. SKIN-CARE CHEMICALS, ADVANCED AERONAUTICAL & SPACE LUBRICANTS. *Educ:* Am Univ, Beirut, BS, 74, MS, 76; Fordham Univ, PhD(chem), 80. *Prof Exp:* Postdoctoral chem-org synthesis, Univ Wis-Madison, 79-82; sr res chemist, 82-86, proj leader, 86-91, RES LEADER, CENT RES, DOW CHEM CO, 91- *Mem:* Am Chem Soc. *Res:* New synthetic emollients and skin-care chemicals; novel monomers for advanced polymeric systems; advanced lubricants for aeronautical and space applications; new methodologies in organofluorine chemistry and their industrial applications. *Mailing Add:* Dow Chem Co 1707 Bldg Midland MI 48674

NADER, JOHN S(HAHEEN), b Farrell, Pa, Nov 16, 21; m 54; c 6. AIR POLLUTION MEASUREMENT, INSTRUMENT DESIGN. *Educ:* Univ Cincinnati, EE, 44, MS, 53. *Prof Exp:* Proj engr instrumentation, Sperti, Inc, 45; develop engr, Russell R Gannon Co, 45-46; instr physics, Univ Cincinnati, 46-49; physicist measurements & instrumentation, Robert A Taft Sanit Eng Ctr, 49-57, chief instrumentation air pollution eng res, 57-62, chief phys res develop, Lab Eng & Phys Sci, 62-64, chief phys measurements, Div Air Pollution, 64-67; chief, Nat Ctr Air Pollution Control, Ohio, 67-70; chief stationary source emissions measurement methods, Optics & Radiation Sect, Nat Environ Res Ctr, Environ Protection Agency, 70-74, chief, Stationary Source Emissions Res Br, Environ Sci Res Lab, 75-79; CONSULT AIR QUAL MEASUREMENTS, 79- *Mem:* Int Standards Orgn. *Res:* Nuclear physics; air pollution; automation and control instrumentation; aerosol physics. *Mailing Add:* 2336 New Bern Ave Raleigh NC 27610

NADKARNI, RAMACHANDRA ANAND, b Karwar, India, Oct 28, 38; US citizen; m 70; c 3. ANALYTICAL CHEMISTRY, RADIOCHEMISTRY. *Educ:* Univ Bombay, BSc, 59, MSc, 61, PhD(anal chem), 65. *Prof Exp:* Res assoc radiochem, Univ Ky, 67-70; res officer, Univ Bombay, 70-72; res mgr analytical chem, Cornell Univ, 72-78; sr staff chemist analytical chem, Exxon Res & Eng Co, 78-86, ANALYSIS PROCESS LEADER, EXXON CHEM CO, 86- *Honors & Awards:* Fulbright Award, US Educ Found in India, 67. *Mem:* Am Chem Soc; Am Soc Testing & Mat; Sigma Xi. *Res:* Geochemistry; trace elements in bio-environmental processes; environmental chemistry; coal and shale analysis; petroleum analysis. *Mailing Add:* Exxon Chem Co PO Box 536 Linden NJ 07036

NADLER, CHARLES FENGER, b Chicago, Ill, Nov 8, 29; m 53; c 3. INTERNAL MEDICINE, ZOOLOGY. *Educ:* Dartmouth Col, AB, 51; Northwestern Univ, MD, 55. *Prof Exp:* From intern to asst resident med, Barnes Hosp, St Louis, Mo, 55-57 & 59-60; res assoc, Div Mammals, Field Mus Natural Hist, 65-85; from instr to assoc, 61-67, asst prof, 67-71, ASSOC PROF MED, MED SCH, NORTHWESTERN UNIV, 71- *Concurrent Pos:* Fel hemat, Med Ctr, Univ Colo, 60-61; attend physician, Passavant Mem Hosp, Chicago, Ill, 61-72; attend physician, Northwestern Mem Hosp, Chicago, 73-; assoc mammal, Mus Natural Hist, Univ Kans, 73-87; activity leader, US-USSR Environ Agreement, 77- *Mem:* Am Fedn Clin Res; Soc Exp Biol & Med; Am Soc Mammal; Soc Syst Zool; Am Col Physicians. *Res:* application of cytogenetics and comparative biochemistry of proteins to the evolution of Asian and North American mammals. *Mailing Add:* 707 N Fairbanks Ct Chicago IL 60611

NADLER, GERALD, b Cincinnati, Ohio, Mar 12, 24; m 47; c 3. INDUSTRIAL ENGINEERING. *Educ:* Purdue Univ, BSc, 45, MSc, 46, PhD(indust eng), 49. *Prof Exp:* Asst indust eng, Purdue Univ, 46-48; plant indust engr, Cent Wis Canneries, 48; instr indust eng, Purdue Univ, 48-49; from asst prof to prof, Wash Univ, 49-64, head dept, 55-64; chmn dept, Univ Wis-Madison, 64-67, 71-75, prof indust eng, 64-83; PROF & CHMN, DEPT INDUST & SYSTEMS ENER, UNIV SOUTHERN CALIF, 83-, IBM PROF ENG MGMT, 86- *Concurrent Pos:* Consult indust engr, 49-; vis prof, Univ Birmingham, Eng, 59, Waseda Univ, Japan, 63-64, Ind Univ, 64, Univ Louvain, Belg, 75 & Technion-Israel Inst Technol, 75-76; vpres gen opers, Artcraft Mfg Co, 56-57; mem bd dirs, Intertherm Inc, 69-85; vis lectr numerous US & foreign univs; pres, Breakthrough Thinking Co, 89- *Honors & Awards:* Gilbreth Medal, 61; Ed Award, Hosp Mgt Mag, 66. *Mem:* Nat Acad Eng; fel AAAS; fel Am Inst Indust Engrs; Am Soc Eng Educ; Inst Mgt Sci; Sigma Xi; Am Asn Univ Prof; Inst Indust Engrs (pres, 89-90); fel Inst Advan Eng. *Res:* System design strategies; health care and hospital systems; planning largescale complex systems; engineering concepts for nonengineers; planning and design methods; author or coauthor of numerous publications. *Mailing Add:* Indust Systs Eng Dept Univ Southern Calif University Park Los Angeles CA 90089-0193

NADLER, HENRY LOUIS, b New York, NY, Apr 15, 36; m 57; c 4. PEDIATRICS, HUMAN GENETICS. *Educ:* Colgate Univ, AB, 57; Northwestern Univ, MD, 61; Univ Wis, MS, 65. *Prof Exp:* From intern to resident pediat, Med Ctr, NY Univ, 61-63, chief resident & inst pediat, 63-64; instr, Sch Med, Univ Wis, 64-65; assoc, 65-66, from asst prof to assoc prof, 67-70, PROF PEDIAT, MED SCH & GRAD SCH & CHMN, DEPT PEDIAT, MED SCH, NORTHWESTERN UNIV, 70-; PRES, MICHAEL REESE HOSP & MED CTR, 88- *Concurrent Pos:* Res fel pediat, Children's Mem Hosp, Chicago, 64-65, head, Div Genetics, 69-, chief staff pediat, 70-81; Irene Heinz & John La Porte Given res prof pediat, Children's Mem Hosp, Chicago; prof & dean pediat, Med Sch, Wayne State Univ, 81-88. *Honors & Awards:* E Mead Johnson Award, Am Acad Pediat, 73; Meyer O Cantor Award, Int col Surgeons, 87. *Mem:* Am Soc Clin Invest; Am Soc Human Genetics; Am Pediat Soc; Soc Pediat Res; Soc Exp Biol & Med. *Res:* Human biochemical genetics; chromosomal disorders and inborn errors of metabolism; prenatal detection of genetic diseases. *Mailing Add:* Michael Reese Hosp & Med Ctr-Off Pres Lake Shore Dr 31st St Chicago IL 60616

NADLER, J VICTOR, NEUROPHARMACOLOGY, EPILEPSY. *Educ:* Yale Univ, PhD(pharmacol), 72. *Prof Exp:* ASSOC PROF PHARMACOL, DUKE UNIV, 83- *Mailing Add:* Dept Pharmacol Med Ctr Duke Univ PO Box 3813 Durham NC 27710

NADLER, KENNETH DAVID, b Bronx, NY, Sept 18, 42; m 67; c 2. PLANT PHYSIOLOGY, BIOCHEMISTRY. *Educ:* Rensselaer Polytech Inst, BSc, 63; Rockefeller Univ, PhD(life sci), 68. *Prof Exp:* NIH res fel biol, Revelle Col, Univ Calif, San Diego, 68-70; from asst prof to assoc prof, 70-83, PROF BOT, MICH STATE UNIV, 83- *Concurrent Pos:* Lectr, Univ Calif, San Diego, 68; sabbatical leave, Dept Genetics, John Innes Inst, UK, 79-80; vis prof, gene manipulation prog, McGill Univ, 83, Ben Gurion Univ, Israel, 84. *Mem:* AAAS; Am Soc Microbiol; Am Soc Plant Physiologists. *Res:* Biochemical mechanisms with which subcellular activities are integrated into cellular metabolism; heme formation in legume root nodules and root nodule bacteria; biochemistry of chlorophylls and hemes; control of differentiation of plastids, mitochondria and cells in relation to porphyrin biosynthesis; genetics of the Rhizobium-legume root nodule symbiosis. *Mailing Add:* 166 Plant Biol Bldg Mich State Univ East Lansing MI 48823

NADLER, MELVIN PHILIP, b Malden, Mass, May 20, 40; m 71, 86; c 2. FT-IR SPECTROSCOPY. *Educ:* Northeastern Univ, AB, 63; Cornell Univ, PhD(phys chem), 69. *Prof Exp:* NSF fel under R Weiner, Northeastern Univ, 61-63; res assoc, State Univ NY Binghamton, 68-70; supvr air & water pollution control, Remington Rand Div, Sperry Rand Corp, 71-72; RES CHEMIST, NAVAL WEAPONS CTR, 72- *Mem:* Am Chem Soc. *Res:* FT-IR spectroscopy surface reactions; polymer structure; combustion of solid propellants. *Mailing Add:* 525 S Sanders Ridgecrest CA 93555

NADLER, NORMAN JACOB, b Montreal, Que, Dec 24, 27; m 53. ENDOCRINOLOGY, ANATOMY. *Educ:* McGill Univ, BSc, 47, MD, CM, 51, PhD(thyroid), 55. *Prof Exp:* Lectr, 57-59, asst prof, 59-64, ASSOC PROF ANAT, McGILL UNIV, 65- *Concurrent Pos:* Consult med, Jewish Gen Hosp, Montreal, 59- *Mem:* Endocrine Soc; Am Asn Anatomists; Can Asn Anat. *Res:* Biophysical approach to morphological physiological aspects of biological structure. *Mailing Add:* PO Box 560 Westmount Postal Sta Westmount PQ H3Z 2T6 Can

NADLER, RONALD D, b Newark, NJ, Jan 19, 36; m 79; c 3. ANIMAL BEHAVIOR, PRIMATOLOGY. *Educ:* Univ Calif, Los Angeles, BA, 60, MA, 63, PhD(physiol psychol), 65. *Prof Exp:* USPHS res fels, Oxford Univ, 65-66 & Univ Wash, 66-67; asst prof psychiat, State Univ NY Downstate Med Ctr, 67-71; develop biologist primatol, 71-80, assoc res prof behav biol, 80-82, RES PROF REPROD BIOL, YERKES PRIMATE RES CTR, EMORY UNIV, 82- *Concurrent Pos:* State Univ NY Res Found grant, State Univ NY Downstate Med Ctr, 68-70, NSF grant, 69-71, USPHS grants, 69-70, 71-72 & 84-87; NSF grants, Yerkes Primate Res Ctr, Emory Univ, 71-73 & 71-74; NSF res grant, 75-79 & 80-83; adj prof, Dept Psychol, Emory Univ, 84-, Ga State Univ, 84-; vis res prof, Karisoke Res Ctr, 81, Int Ctr Med Res de Franceville, 82-84. *Mem:* Int Primatol Soc; AAAS; Int Acad Sex Res; Am Soc Primatologists. *Res:* Comparative, developmental research on socio-sexual behavior of the great apes, chimpanzee, gorilla and orang-utan; emphasis is placed on physiological correlates of behavior, especially hormonal. *Mailing Add:* Yerkes Regional Primate Res Ctr Emory Univ Atlanta GA 30322

NADOL, BRONISLAW JOSEPH, JR, b Cambridge, Mass, Oct 2, 43; m 70; c 2. OTOLARYNGOLOGY. *Educ:* Harvard Col, BA, 66; Johns Hopkins Univ, MD, 70; Am Bd Otolaryngol, dipl, 75. *Prof Exp:* Intern surg, Beth Israel Hosp, 70-71, residency surg, 71-72, residency otolaryngol, 72-75; from instr to prof, 76-80, ASSOC PROF OTOLARYNGOL, HARVARD MED SCH, 80-, CHMN DEPT. *Concurrent Pos:* Asst surgeon otolaryngol, Mass Eye & Ear Infirmary, 75-80, assoc surgeon, 80-, chief dept otolaryngol. *Mem:* Am Acad Ophthal & Otolaryngol. *Res:* Electron microscopy of the human inner ear; clinical electrocochleography. *Mailing Add:* 243 Charles St Boston MA 02114

NADOLNEY, CARLTON H, b New York, NY, Apr 15, 33. CARCINOGENICITY & MUTAGENICITY,. *Educ:* Brooklyn Col, BS, 56; NY Univ, MS, 61, PhD(cell biol), 65. *Prof Exp:* Fel, Biol Dept, Brookhaven Nat Lab, Upton, NY, 70-73; cancer specialist, Int Cancer Res Data Bank, 78-79; TOXICOLOGIST, US ENVIRON PROTECTION AGENCY, 79- *Concurrent Pos:* Ed, Tissue Culture Asn TCA Report, 81-86. *Mem:* AAAS; Am Soc Cell Biol; Soc Toxicol; NY Acad Sci; Sigma Xi; Tissue Cult Asn. *Res:* Assuring the quality and integrity of laboratory data submitted to the Environmental Protection Agency under the Toxic Substances Control Act; drug metabolism. *Mailing Add:* Office Toxic Substances TS-778 US Environ Protection Agency 401 M St SW Washington DC 20460

NAEGELE, EDWARD WISTER, JR, b Philadelphia, Pa, Sept 30, 23; m 50; c 1. ORGANIC CHEMISTRY. *Educ:* Temple Univ, AB, 48, MA, 50, PhD(chem), 55. *Prof Exp:* Technologist, E I du Pont de Nemours & Co, 55-57; supvr basic res, Acheson Dispersed Pigments Co, 57-58; assoc prof chem, 58-62, PROF CHEM & CHMN DEPT, GROVE CITY COL, 62- *Mem:* AAAS; Am Chem Soc. *Res:* Heterocyclic compounds. *Mailing Add:* 319 State St Grove City PA 16127-1145

NAEGER, LEONARD L, b St Louis, Mo, July 19, 41; m 66; c 2. PHARMACOLOGY. *Educ:* St Louis Col Pharm, BS, 63, MS, 65; Univ Fla, PhD(pharmacol), 70. *Prof Exp:* Asst prof, Sch Dent Med, Wash Univ, 72-77; ASST PROF PHARMACOL, ST LOUIS COL PHARM, 70- *Concurrent Pos:* Lectr, Med Ctr, St Louis Univ, 71- *Res:* Ethanol metabolism. *Mailing Add:* 1537 Louisville Ave St Louis MO 63139

NAESER, CHARLES RUDOLPH, b Mineral Point, Wis, Nov 13, 10; m 36; c 2. INORGANIC CHEMISTRY. *Educ:* Univ Wis, BS, 31; Univ Ill, MS, 33, PhD(inorg chem), 35. *Prof Exp:* Asst gen chem, Univ Ill, 32-35; from instr to asst prof inorg chem, 35-42, from assoc prof to prof chem, 45-76, chmn dept, 48-50, 51-53 & 56-76, EMER PROF CHEM, GEORGE WASHINGTON UNIV, 76- *Concurrent Pos:* Chief, Chem Group, US Geol Surv, 53-56, consult, 56-75; consult, Off Saline Water, 62-72. *Honors & Awards:* Am Inst Chem Honor Award, 62. *Mem:* AAAS; Am Inst Chem; Am Chem Soc; Geochem Soc. *Res:* Inorganic chemistry of rare earths, beryllium, uranium, rhenium and selenium; electro reduction of less common metals; fluoroplatinates; geochemistry; radioactive waste disposal; desalinization. *Mailing Add:* 6654 Van Winkle Dr Falls Church VA 22044

NAESER, CHARLES WILBUR, b Washington, DC, July 2, 40; m 82; c 2. GEOLOGY. *Educ:* Dartmouth Col, AB, 62, MA, 64; Southern Methodist Univ, PhD(geol), 67. *Prof Exp:* GEOLOGIST, US GEOL SURV, 67- *Concurrent Pos:* Vis prof, Univ Alaska, 72, Univ Wash, 75 & Univ Ariz, 78; adj prof, Dartmouth Col, 79- & Univ Wyo, 84- *Mem:* Fel Geol Soc Am; Am Geophys Union. *Res:* Geochronology, specifically the use of fission track dating of minerals. *Mailing Add:* US Geol Surv Box 25046 Denver Fed Ctr Stp 424 Denver CO 80225

NAESER, NANCY DEARIEN, b Morgantown, WVa, Apr 15, 44; m 82. FISSION-TRACK DATING. *Educ:* Univ Ariz, BS, 66; Victoria Univ, PhD(geol), 73. *Prof Exp:* Geol field asst & phys sci technician, US Geol Surv, Flagstaff, Ariz, 66; sci ed, New Zealand Dept Sci & Indust Res, Wellington, 74-76; res assoc, Univ Toronto, 76-79; res assoc, 79-81, GEOLOGIST, US GEOL SURV, DENVER, COLO, 81- *Concurrent Pos:* Asst Ed, NZ J Geol & Geophys, 74-76; adj prof, Dartmouth Col, 85-, Univ Wyo, 84- *Mem:* Fel Geol Soc Am; Am Asn Petrol Geologists; Am Quaternary Asn; Soc Econ Paleontologists & Mineralogists; Geol Soc NZ. *Res:* Fission-track dating applied to geologic studies, including tephrochronology and thermal history of sedimentary basins. *Mailing Add:* Mail Stop 424 US Geol Surv Fed Ctr Denver CO 80225

NAEYE, RICHARD L, b Rochester, NY, Nov 27, 29; m 55; c 3. PATHOLOGY. *Educ:* Colgate Univ, AB, 51; Columbia Univ, MD, 55; Am Bd Path, cert, anat path & clin path. *Prof Exp:* Instr path, Col Physicians & Surgeons, Columbia Univ, 57-58; trainee, Med Col, Univ Vt, 58-60, from asst prof to prof, 60-67; PROF PATH & CHMN DEPT, HERSHEY MED CTR, PA STATE UNIV, 67- *Concurrent Pos:* Marckle scholar, 60-65; asst attend pathologist, Mary Fletcher Hosp, Burlington, Vt, 60-63, assoc attend pathologist, 63-67; mem path A study sect, NIH, 68-72; mem adv bd, Armed Forces Inst Path, 72-76; mem comt epidemiol & vet follow-up, Nat Acad Sci, 72-75; mem, Coun Cardiopulmonary Dis, Am Heart Asn, Inc, 72-; consult, World Health Orgn. *Mem:* Col Am Path; Am Soc Exp Path; Am Soc Clin Path; Int Acad Path. *Res:* Prenatal and neonatal disorders; fetal growth and development; pulmonary vascular disease; sudden infant death syndrome. *Mailing Add:* Dept Path Pa State Univ Hershey Med Ctr Hershey PA 17033

NAFE, JOHN ELLIOTT, b Seattle, Wash, July 22, 14; m 41; c 2. PHYSICS, GEOPHYSICS. *Educ:* Univ Mich, BS, 38; Wash Univ, MS, 40; Columbia Univ, PhD(physics), 48. *Prof Exp:* Asst physics, Wash Univ, 38-39; asst, Columbia Univ, 40-41, instr, 46-49; asst prof, Univ Minn, 49-51; dir res, Hudson Lab, Columbia Univ, 51-53, res assoc, Lamont-Doherty Geol Observ, 53-55, adj assoc prof geophys, 55-58, prof, 58-80, chmn dept geol, 62-65, emer prof geophys, 80-81; RETIRED. *Concurrent Pos:* Vis fel, Cambridge Univ, 71-72; hon prof, Dept Geophys & Astron, Univ BC, Can, 80- *Mem:* Fel AAAS; fel Am Geophys Union; fel Am Phys Soc; fel Geol Soc Am. *Res:* Atomic beams; hyperfine structure of deuterium and hydrogen; seismology; marine geophysics; underwater sound. *Mailing Add:* 5775 Toronto Rd Suite 909 Vancouver BC V6T 1X4 Can

NAFF, JOHN DAVIS, b Atlanta, Ga, Nov 12, 18; m 42; c 5. GEOLOGY. *Educ:* Univ Ala, AB, 39, MA, 40; Univ Kans, PhD, 60. *Prof Exp:* From assoc prof to prof geol, Okla State Univ, 50-85; RETIRED. *Concurrent Pos:* Staff geologist, Juneau Icefield Res Prog, NSF lect series; ed, Geol Sect, Okla Acad Sci, 68- *Mem:* Soc Econ Paleont & Mineral; Am Asn Petrol Geologists; Nat Asn Geol Teachers (secy). *Res:* Invertebrate paleontology; stratigraphy; mountaineering; photography; electronics; miniaturization and packaging of equipment. *Mailing Add:* 1711 W Fifth Ave Stillwater OK 74074

NAFF, MARION BENTON, b Lexington, Ky, Mar 23, 18; m 46. ORGANIC CHEMISTRY. *Educ:* Univ Ky, BS, 41, MS, 46; Ore State Col, PhD(org chem), 50. *Prof Exp:* Asst prof chem, Western Carolina State Col, 50-51 & Bowling Green State Univ, 51-55; assoc prof, Loyola Univ, La, 55-58 & Dickinson Col, 58-66; CHEMIST, DRUG SYNTHESIS & CHEM NAT CANCER INST, 66- *Concurrent Pos:* Vis assoc prof chem, Brown Univ, 64-65. *Mem:* Fel AAAS; Am Chem Soc. *Res:* Synthesis of heterocyclics; acetylenic chemistry; chemical kinetics; medicinal chemistry; organic synthesis and analysis; instrumentation. *Mailing Add:* 5352 Pooks Hill Rd Bethesda MD 20814-2005

NAFIE, LAURENCE ALLEN, b Detroit, Mich, Aug 9, 45; m 68; c 2. PHYSICAL CHEMISTRY. *Educ:* Univ Minn, Minneapolis, BChem, 67; Univ Ore, MS, 69, PhD(chem), 73. *Prof Exp:* Sci & eng asst nuclear physics, Nuclear Effects Lab, Edgewood Arsenal, Md, 69-71; res assoc infrared circular dichroism, Univ Southern Calif, 73-75; from asst prof to assoc prof, 75-82, PROF PHYS CHEM, SYRACUSE UNIV, 82-, CHMN DEPT, 84- *Concurrent Pos:* Alfred P Sloan fel, 78-82. *Honors & Awards:* Coblentz Soc Award, 81. *Mem:* Am Chem Soc; Optical Soc Am; Am Phys Soc; Soc Appl Spectros; Coblentz Soc. *Res:* Vibrational optical activity, including experimental and theoretical research in vibrational circular dichroism and Raman circular intensity differential scattering; resonance Raman spectroscopy and the theory of Raman line shapes; fourier transform infrared spectroscopy. *Mailing Add:* Dept Chem Syracuse Univ Syracuse NY 13244-1200

NAFISSI-VARCHEI, MOHAMMAD MEHDI, b Arak, Iran, Sept 23, 36; m 69; c 1. ORGANIC CHEMISTRY. *Educ:* Tehran Univ, Iran, Licentiate, 60; Miami Univ, Ohio, MSc, 66; Mass Inst Technol, PhD(org chem), 69. *Prof Exp:* Instr chem, Tehran Univ, 60-65; res assoc spectros, Mass Inst Technol, 69-70; sr scientist med chem, 71-74, PRIN SCIENTIST, SCHERING CORP, 74- *Mem:* Am Chem Soc; Chem Soc Eng. *Res:* Medicinal chemistry; anti-infective agents. *Mailing Add:* PO Box 1623 West Caldwell NJ 07006

NAFOOSI, A AZIZ, mathematics, for more information see previous edition

NAFPAKTITIS, BASIL G, b Athens, Greece, Dec 23, 29; m 64; c 3. BIOLOGY. *Educ:* Am Univ Beirut, BSc, 62, MS, 63; Harvard Univ, PhD(biol), 66. *Prof Exp:* From asst prof to assoc prof, 67-79, PROF BIOL, UNIV SOUTHERN CALIF, 79- *Mem:* Am Soc Ichthyologists & Herpetologists; Am Soc Zoologists; Soc Syst Zool. *Res:* Ichthyology, particularly distribution, systematics and ecology of deep-sea fishes; bioluminescence. *Mailing Add:* Dept Biol Sci Univ Southern Calif Univ Park 3616 Trousdale AHF103 Los Angeles CA 90089-0371

NAFTOLIN, FREDERICK, b New York, NY, Apr 7, 36; m 87; c 2. OBSTETRICS & GYNECOLOGY, NEUROENDOCRINOLOGY & REPRODUCTION ENDOCRINOLOGY. *Educ:* Univ Calif, Berkeley, BA, 58, San Francisco, MD, 61; Oxford Univ, Eng, DPhil(neuroendocrinol), 70. *Hon Degrees:* MA(hon) McGill Univ & Yale Univ. *Prof Exp:* Asst chief gynec serv, USPHS Hosp, Seattle, Wash, 66-68; from asst prof to assoc prof obstet & gynecol, Univ Calif, San Diego, 70-73; assoc prof, Harvard Med Sch, 73-75; prof & chmn, Fac Med, McGill Univ, 75-78; PROF & CHMN OBSTET & GYNECOL, SCH MED YALE UNIV, 78-, PROF BIOL, 83-, DIR, CTR RES REPRODUCTIVE BIOL, 86- *Concurrent Pos:* Res fel, Univ Wash, 66-68, NIH, Oxford Univ, 68-70; fel, Guggeheim & Fogarty, 82-83; Berlex, fel, 91-92. *Mem:* Endocrine Soc; Soc Gynecol Invest (pres, 91-92); Soc Psychoneuroendocrinol; Int Soc Neuroendocrinol. *Res:* Investigation of the relationship of hormones to reproductive function; brain/pituitary-sex steroid interaction; molecular biology; cell biology; infertility. *Mailing Add:* Dept Obstet & Gynecol Sch Med Yale Univ 333 Cedar St Box 3333 New Haven CT 06510-8063

NAFZIGER, RALPH HAMILTON, b Minneapolis, Minn, Aug 9, 37. CHEMICAL METALLURGY. *Educ:* Univ Wis, BS, 60; Pa State Univ, PhD(geochem), 66. *Prof Exp:* Geol asst, US Geol Surv, 63; res assoc geochem, Pa State Univ, 66-67; res chemist, 67-79, RES SUPVR, US BUR MINES, 79- *Concurrent Pos:* Mem electroslag & plasma arc melting, Nat Mat Adv Bd, 74-75; abstractor, Mineral Abstracts, 77-; mem, Process Fundamentals Comt, Minerals, Metals & Mat Soc, 80-, vchmn & chmn, Copper, Nickel & Cobalt Comt, 86-, Extractive & Processing Div, Exec Prog Publ Comts, 89-, Awards Comt, 90-; mem Fuels Comt, Am Foundrymen's Soc, 80-87, charge Mat Comt, 84-87, Cupola Comt, 87-, mem rev bd, Met Trans, 80- *Honors & Awards:* Meritorious Serv Award, US Dept Interior, 90. *Mem:* Mineral Soc Am; Am Inst Mining & Metall Engrs; Am Soc Metals; Sigma Xi (vpres, 72, pres, 73). *Res:* Thermochemistry of metal-nonmetal systems; electroslag melting of metals; electric furnace reduction and smelting of lower-grade titaniferous, chromite, manganese and ferrous ores. *Mailing Add:* US Bur Mines 1450 Queen Ave SW Albany OR 97321-2198

NAG, ASISH CHANDRA, b Jamalpur, Bengal, India, Jan 1, 32; m 61; c 2. CELL BIOLOGY. *Educ:* Calcutta Univ, India, BS, 53; Univ Hawaii, MS, 66; Univ Alta, Can, PhD(zool), 70. *Prof Exp:* Res assoc biol, Univ Pa, 71-72; res assoc med & biochem, Univ Chicago, 73-75; from asst prof to assoc prof biol, 75-86, PROF BIOL, OAKLAND UNIV, 87- *Concurrent Pos:* Prin investr, Am Heart Asn Mich grant, 77-, NIH grant, 81-84, NSF grant, 87-90, Gen Motors biomed res grant, 85-86; ed, Newslett, speaker, Cardiac Morphogenesis Workshop, Nat Heart, Blood & Lung Inst, Bethesda, Md, 84- *Mem:* Am Soc Cell Biol; Electron Micros Soc Am; Sigma Xi; AAAS. *Res:* Control mechanisms of cell proliferation and differentiation in embryonic and adult cardiac muscle cells; the expression of myosin isoforms. *Mailing Add:* Dept Biol Sci Oakland Univ University Dr & Squirrel Rochester MI 48309-4401

NAG, SUBIR, therapeutic radiology, brachytherapy, for more information see previous edition

NAGAI, JIRO, b Nagano, Japan, Sept 26, 27; Can citizen; m 59; c 2. ANIMAL BREEDING. *Educ:* Univ Tokyo, BA, 52, DAgr(animal breeding), 61. *Prof Exp:* Instr animal breeding, Univ Tokyo, 55-65; RES SCIENTIST ANIMAL GENETICS, AGR CAN, 65- *Concurrent Pos:* Vis prof, NC State Univ, 74-75. *Mem:* Genetics Soc Can; Can Asn Lab Animal Sci; Japan Exp Animal Res Asn; Am Soc Animal Sci. *Res:* Animal breeding through experiments using mice and cattle; selection for long-term performance in mice, long-term performance of crosses from the selected lines of mice, and micromanipulation of mouse and bovine embryos. *Mailing Add:* 28 Kinnear St Ottawa ON K1A 3R6 Can

NAGAMATSU, HENRY T, b Garden Grove, Calif, Feb 13, 16; m 42; c 2. FLUID PHYSICS, PLASMA PHYSICS. *Educ:* Calif Inst Technol, BS, 38 & 39, MS, 40, PhD(aeronaut), 49. *Prof Exp:* Asst, Calif Inst Technol, 38-41; theoret aerodynamicist, Douglas Aircraft Co, 41-42; theoret aerodynamicist, Curtiss-Wright Aircraft Corp, 42-43, head aeronaut res, Res Lab, 43-46; asst sect head, Jet Propulsion Lab, Calif Inst Technol, 46-49, sr res fel & dir hypersonic res, Aeronaut Dept, 49-55; res assoc, Res & Develop Ctr, Gen Elec Co, 55-78; res prof, 78-80, PROF AERONAUT ENG, RENSSELAER POLYTECH INST, 80- *Concurrent Pos:* Consult, US Naval Ord Test Sta, 49-56, Rand Corp, 50-59, Atlas Proj, Gen Dynamics/Convair, 50-53, Midwest Res Inst, 52-55, Off Sci Res, US Dept Air Force, 57- & Aeronaut Lab, Wright-Patterson AFB, 60-; adj prof, Rensselaer Polytech Inst, 56-64; mem, Eng Noise Subcomt, Nat Acad Sci, 68-71. *Mem:* AAAS; fel Am Phys Soc; fel Am Inst Aeronaut & Astronaut; NY Acad Sci. *Res:* High temperature gas dynamics associated with intercontinental ballistic missiles, satellites and space vehicles; applied mathematics; magnetohydrodynamics; high temperature physics; physical chemistry; fluid mechanics; rockets and missiles; jet noise and acoustics; transonic flows; arc physics. *Mailing Add:* 1046 Cornelius Ave Schenectady NY 12309

NAGARKATTI, MITZI, b Karnataka, India, March 10, 52; m 78; c 1. CANCER IMMUNOLOGY, IMMUNOTOXICOLOGY. *Educ:* Bangalore Univ, BSc, 70; Karnatak Univ, MS, 74; Defense R & D Estab, PhD(immunol), 81. *Prof Exp:* Fel immunol, McMaster Univ Med Ctr, 81-83; res assoc microbiol, Univ Ky Med Ctr, 83-86; ASST PROF BIOL, VA POLYTECH INST & STATE UNIV, 86- *Concurrent Pos:* Prin investr, NIH, 86-87 &

87-91. *Mem:* Am Asn Immunologists. *Res:* Studying the mechanism by which anticancer drugs such as BCNU (nitrosoureas) bring about tumor rejection; the role of the immune cells (lymphocytes) in causing tumor rejection following the drug injection; the mechanism by which the environmental pollutants suppress the immune response. *Mailing Add:* Dept Biol Div Microbiol & Immunol Va Polytech Inst & State Univ Blacksburg VA 24061

NAGARKATTI, PRAKASH S, b Dharwad, Karnataka, July 5, 52; m 78; c 1. IMMUNOLOGY, CANCER IMMUNOLOGY. *Educ:* Karnatak Univ, BSc, 71 & MS, 74; Defence Res & Develop Estab, PhD(immunol), 81. *Prof Exp:* Fel immunol, McMaster Univ Med Ctr, 81-83; res assoc, Univ Ky Med Ctr, 83-85, res asst prof, 85-86; ASST PROF BIOL, VA POLYTECH INST & STATE UNIV, 86- *Concurrent Pos:* Prin investr, NIH grant, 86-; young scientist award, Nat Acad Sci, 81. *Honors & Awards:* Young Scientist Award, Nat Acad Sci, 81. *Mem:* Am Asn Immunologists. *Res:* Study cause of autoimmune diseases, particularly the mechanism by which lymphocytes react with self-antigens and bring about destruction of self-tissues; mechanism by which environmental pollutants alter the immune responses. *Mailing Add:* Dept Biol Div Microbiol VPI & State Univ 5029A Derring Hall Blacksburg VA 24060

NAGASAWA, HERBERT TSUKASA, b Hilo, Hawaii, May 31, 27; m 51; c 2. MEDICINAL CHEMISTRY. *Educ:* Western Reserve Univ, BS, 50; Univ Minn, PhD(org chem), 55. *Prof Exp:* Fel biochem, Univ Minn, 55-57; sr chemist, Radioisotope Serv, 57-61, sr scientist, Lab Cancer Res, 61, PRIN SCIENTIST, MED RES LABS, VET ADMIN HOSP, MINNEAPOLIS, 61- *Concurrent Pos:* From asst prof to assoc prof, Col Pharm, Univ Minn, 59-72, prof, 73-; assoc ed, J Med Chem, 72-, actg ed, 73, sr ed, 85; Vet Admin res career scientist, 78. *Mem:* Am Chem Soc; Am Asn Cancer Res; NY Acad Sci; Soc Toxicol; Am Soc Pharmacol & Exp Therapeut; Res Soc Alcoholism. *Res:* Design and synthesis of trapping agents for the detoxication of xenobiotic substances (including ethanol) that are activated to toxic metabolites in vivo and latentiated (prodrug) forms of biologically active substances. *Mailing Add:* Vet Admin Med Ctr One Veterans Dr Minneapolis MN 55417-2300

NAGATA, KAZUHIRO, b Shiga, Japan, May 12, 47; m 72; c 2. MOLECULAR BIOLOGY. *Educ:* Kyoto Univ, PhD(DSc), 79. *Prof Exp:* Assoc prof, 79-86, PROF, CHEST DIS RES INST, KYOTO UNIV, 86- *Mem:* Japan Soc Cell Biol; Am Soc Cell Biol; Japan Soc Cancer Res; Japan Soc Biochem; Japan Soc Molecular Biol; Japan Soc Hyperthermic Oncol. *Res:* Strss proteins; cell differentiation and cytoskeletons; cytokines and their receptors. *Mailing Add:* Dept Cell Biol Chest Dis Res Inst Kyoto Univ Sakyo-Ku Kyoto 606 Japan

NAGEL, ALEXANDER, b New York, NY, Sept 13, 45. MATHEMATICS. *Educ:* Harvard Univ BA, 66; Columbia Univ, PhD(math), 71. *Prof Exp:* PRO MATH, WIS UNIV, 77- *Concurrent Pos:* Guggenheim fel, Guggenheim Found, 87. *Mailing Add:* 829 Sauk-Ridge Trail Madison WI 53717-1186

NAGEL, CHARLES WILLIAM, b St Helena, Calif, Dec 8, 26; m 51; c 5. FOOD SCIENCE. *Educ:* Univ Calif, BA, 50, PhD(microbiol), 60. *Prof Exp:* Bacteriologist, US Dept War, Dugway Proving Grounds, Utah, 51-52; lab technician food technol, Univ Calif, 52-54; sr lab technician & coop agt food technol & preserv of refrig poultry, USDA & Univ Calif, 54-60; from asst prof to prof fruit & veg processing, Wash State Univ, 60-71; res dir, United Vintners Inc, Calif, 71-73; actg chmn dept, 78-79, PROF FOOD SCI & TECHNOL, WASH STATE UNIV, 73- *Concurrent Pos:* Chmn food sci exec comt, Wash State Univ, 64-68. *Mem:* Fel Inst Food Technologists; Am Soc Enol & Viticult (secy-treas, 88-); Sigma Xi (secy-treas, 75-77). *Res:* Sanitation; fruit and vegetable products processing; pectic enzymes of bacteria; food fermentations; enology; phenolic compounds. *Mailing Add:* Dept Food Sci & Human Nutrit Wash State Univ Pullman WA 99164-6376

NAGEL, DONALD LEWIS, b Blue Island, Ill, May 24, 41; m 63; c 1. CHEMICAL CARCINOGENESIS, ORGANIC CHEMISTRY. *Educ:* Knox Col, BA, 67; Univ Nebr-Lincoln, PhD(org chem), 71. *Prof Exp:* Chemist, Libby, McNeill & Libby, 63-67; asst org chem, Univ Nebr-Lincoln, 67-71; from instr to assoc prof cancer, 71-86, PROF CANCER & ASSOC DIR, EPPLEY INST RES CANCER, UNIV NEBR MED CTR, 87- *Mem:* Am Chem Soc; Sigma Xi. *Res:* Chemical carcinogenesis with special emphasis on problems relating to mechanisms of action, organic chemical applications, nuclear magnetic resonance. *Mailing Add:* 600 S 42nd St Omaha NE 68198-6085

NAGEL, EDGAR HERBERT, b San Diego, Calif, Mar 24, 38; div; c 2. ANALYTICAL CHEMISTRY. *Educ:* Valparaiso Univ, BS, 60; Northwestern Univ, PhD(chem), 65. *Prof Exp:* From instr to asoc prof, 63-76, 63-76, PROF CHEM, VALPARAISO UNIV, 76- *Concurrent Pos:* Consult, Argonne Nat Lab, 65-79. *Mem:* Am Chem Soc. *Res:* Gas chromatography; electrochemistry; computer applications in teaching. *Mailing Add:* Dept Chem Valparaiso Univ Valparaiso IN 46383

NAGEL, EUGENE L, b Quincy, Ill, Aug 12, 24; c 3. ANESTHESIOLOGY. *Educ:* Cornell Univ, BEE, 49; Wash Univ, MD, 59. *Prof Exp:* Intern, St Luke's Hosp, St Louis, Mo, 59-60; resident anesthesiol, Presby Hosp, New York, 60-62; from asst prof to assoc prof, Sch Med, Univ Miami, 65-73; prof anesthesiol, Univ Calif, Los Angeles, 74-77; PROF ANESTHESIOL & CHMN DEPT, JOHNS HOPKINS UNIV SCH MED, 77- *Concurrent Pos:* Attend physician, Jackson Mem Hosp, Miami, Fla, 63-74, clin dir anesthesiol, 66-74; consult, Am Heart Asn, 69, Nat Registry of Emergency Med Technicians, 71-, Am Col Surgeons, Robert Wood Johnson Found & Bur Med Serv, Dept Health, Educ & Welfare, 72-; chmn sect clin care, Am Soc Anesthesiol, 73-74; chmn comn emergency med serv, AMA, 74- *Mem:* AMA; Am Physiol Soc; Am Soc Anesthesiol; Am Col Cardiol; Soc Critical Care Med. *Res:* Emergency care; telemetry; sudden cardiac death. *Mailing Add:* 567 Ave KSE Winter Haven FL 33880

NAGEL, FRITZ JOHN, b Ger, Oct 20, 19; nat US; m 53; c 3. ORGANIC CHEMISTRY. *Educ:* Univ Notre Dame, BSChE, 41, MS, 42. *Prof Exp:* Engr mat, Gen Elec Co, 50-52; tech dir, Capac Plastics, Inc, 52-53; dir res, Congoleum-Nairn, Inc, 53-56; vpres, Polymer Processes, Inc, 56-68; res & develop mgr, Signal Oil & Gas Co, 68-70; vpres, Chapman Chem Co, 70-83; RETIRED. *Concurrent Pos:* Res engr, Res Labs, Westinghouse Elec Corp, 43-48. *Mem:* Am Chem Soc. *Res:* High polymers; laminates; insulating varnishes; adhesives; protective coatings; plastics; electrical insulation; fungicides. *Mailing Add:* 1264 E Massey Rd Memphis TN 38120-3232

NAGEL, G(EORGE) W(OOD), b Pittsburgh, Pa, Jan 28, 15; m 40; c 5. ELECTRONICS. *Educ:* Carnegie Inst Technol, BS, 36; Univ Pittsburgh, MS, 41. *Prof Exp:* Res engr electronics, Res Labs, Westinghouse Elec Corp, 36-38, design engr, Radio Div, 39-46, res engr electronics, 46-52, adv engr, 52-55, new prod, 55-62 & appl physics, 62-65, adv engr solid state applns, 65-84; RETIRED. *Concurrent Pos:* Consult, 84- *Res:* Television; pulse and frequency modulation radar; industrial electronics; military vehicle guidance; document scanners. *Mailing Add:* 234 Cascade Rd Pittsburgh PA 15221

NAGEL, GLENN M, b Blue Island, Ill, Apr 16, 44; m 66; c 2. BIOCHEMISTRY, MOLECULAR BIOLOGY. *Educ:* Knox Col, BA, 66; Univ Ill, PhD(biol chem), 71. *Prof Exp:* Scholar molecular biol & biochem, Univ Calif, Berkeley, 70-72; asst prof chem & molecular biol, 72-76, ASSOC PROF CHEM & MOLECULAR BIOL, CALIF STATE UNIV, FULLERTON, 76- *Concurrent Pos:* NIH res fel, 70-72. *Honors & Awards:* Nat Res Serv Award, NIH, 90; Alfred P Sloan Found Award, 90. *Mem:* Am Chem Soc. *Res:* Structure and function of oligomeric enzymes; metabolic control; specific interactions between proteins and nucleic acids. *Mailing Add:* Dept Chem & Biochem Calif State Univ Fullerton CA 92634

NAGEL, HAROLD GEORGE, b Natoma, Kans, May 17, 40; m 69. ECOLOGY, ENTOMOLOGY. *Educ:* Ft Hays Kans State Col, BS, 62, MS, 64, Kans State Univ, PhD(entom), 69. *Prof Exp:* Instr biol, Ft Hays Kans State Col, 64-65; res asst entom, Kans State Univ, 66-69; prof, 69-84, CHMN BIOL, KEARNEY STATE COL, 84- *Mem:* Central States Entom Soc; Mosquito Control Asn. *Res:* Prairie plant insect interrelationships; soil invertebrate ecology; butterfly distribution. *Mailing Add:* Dept of Biol Kearney State Col Kearney NE 68847

NAGEL, HARRY LESLIE, b Brooklyn, NY, Jan 29, 43; m 68; c 7. OPERATIONS RESEARCH. *Educ:* Brooklyn Col, BS, 64; NY Univ, MS, 66, PhD(opers res), 73. *Prof Exp:* Analyst, Stauffer Chem Co, 67-69; PROF QUANT ANALYSIS, ST JOHN'S UNIV, 69- *Concurrent Pos:* Adj prof, Fordham Univ, 73-75; consult, Mobil Oil, Nat Industs, Gem Stores, Plainview Tire, MLR Mgt Co, FWF Mgt & A Link & Co, 73- *Mem:* Am Statist Asn. *Res:* Regression analysis with outliers; personal computer education using overhead equipment; author on basic programming. *Mailing Add:* St John's Univ Jamaica NY 11439

NAGEL, RONALD LAFUENTE, b Santiago, Chile, Jan 18, 36; US citizen; m 60; c 3. HEMATOLOGY. *Educ:* Univ Chile, MD, 60. *Prof Exp:* Asst resident, Hosp Salvador, Sch Med, Univ Chile, 60-63; int fel, NIH, 63-64; res fel, Albert Einstein Col Med, 64-67, assoc med, 67-69, from asst prof to assoc prof, 69-78, head, Div Hemat, 82-84, PROF MED, ALBERT EINSTEIN COL MED, 78-, DIR, UNIFIED DIV HEMAT, 84- *Concurrent Pos:* Mem, Exec Comt, Hemolytic Anemia Study Group, 74- *Honors & Awards:* Award in Black, Found Res & Educ Sickle Cell Dis, 73. *Mem:* Am Soc Clin Invest; Am Soc Biol Chemists; Am Soc Hemat; Int Soc Hemat; Am Asn Physicians. *Res:* The molecular, cellular and clinical aspects of sickle cell anemia and other hemoglobinopathies; the structural and functional relationships in hemoglobin; malaria. *Mailing Add:* Dept of Med 1300 Morris Park Ave Albert Einstein Col of Med Bronx NY 10461

NAGEL, SIDNEY ROBERT, b New York, NY, Sept 28, 48; m 89. SOLID STATE PHYSICS. *Educ:* Columbia Univ, BA, 69; Princeton Univ, MA, 71, PhD(physics), 74. *Prof Exp:* Res assoc solid state physics, div eng, Brown Univ, 74-76; from asst prof to assoc prof physics, 76-84, dir, Mat Res Lab, 87-91, PROF JAMES FRANCK INST & DEPT PHYSICS, UNIV CHICAGO, 84- *Concurrent Pos:* Alfred P Sloan res fel, 79-81; mem, sci tech adv comt, Argonne Nat Lab, 85-91. *Mem:* Fel Am Phys Soc; AAAS. *Res:* Electronic structure of metals and alloys; transport in metals; amorphous materials; glass transition; relaxation in solids. *Mailing Add:* James Franck Inst Univ Chicago Chicago IL 60637

NAGEL, TERRY MARVIN, b Rochester, Minn, Mar 25, 43. INORGANIC CHEMISTRY, PHYSICAL CHEMISTRY. *Educ:* Macalester Col, BA, 65; Univ Minn, Minneapolis, PhD(chem), 70. *Prof Exp:* Asst prof chem, Monmouth Col, 70-74; asst prof, Kalamazoo Col, 74-75; asst prof chem, Lakeland Col, 75-80; MEM FAC, CHEM DEPT, COL IDAHO, 80- *Mem:* Am Chem Soc. *Res:* Mechanisms of electron-transfer reactions. *Mailing Add:* Chem Dept Col Idaho 2112 Cleveland Blvd Caldwell ID 83605-4494

NAGELL, RAYMOND H, b Rochester, NY, Apr 10, 27; m 49; c 1. GEOLOGY. *Educ:* Univ Rochester, BS, 51, MA, 52; Stanford Univ, PhD(geol), 58. *Prof Exp:* Geologist, Cerro Corp, 52-55 & Cia Minera Cuprum, 56-57; chief geologist, Industria e Comercio de Minerios SAm, 57-61; geologist, Shenon & Full, 61-62 & US Geol Surv, 63-70; geologist, Bethlehem Steel, 70-76, geologist coal invests, 76-83; CONSULT, 83- *Concurrent Pos:* Tech consult, Industria e Comercio de Minerios, SAm, 72-76. *Mem:* Soc Econ Geol; Geol Soc Am; Am Inst Mining, Metall & Petrol Engrs; Brazilian Geol Soc. *Res:* Economic geology; ore deposits. *Mailing Add:* Geol Dept Bethlehem Steel Corp Bethlehem PA 18016

NAGER, GEORGE THEODORE, b Zurich, Switz, Dec 1, 17; m 50; c 2. OTOLARYNGOLOGY. *Educ:* Zurich Univ, MD, 47; Swiss Bd Otolaryngol, dipl, 54; Am Bd Otolaryngol, dipl, 59. *Prof Exp:* PROF LARYNGOL & OTOL, SCH MED, JOHNS HOPKINS UNIV, 68-, CHMN DEPT, 70-, CHIEF OTOL, JOHNS HOPKINS HOSP, 70- *Concurrent Pos:* Consult

Baltimore City Hosps, Greater Baltimore Med Ctr, Good Samaritan Hosp & USPHS Hosp, Baltimore. *Honors & Awards:* Sir Win Wilde Medal, 80. *Mem:* Am Acad Ophthal & Otolaryngol; Am Col Surg; Am Laryngol, Rhinol & Otol Soc; Am Otol Soc; hon mem Egyptian & Irish Otol Soc. *Res:* Monographs, chapters and numerous articles concerned with the clinical, radiological and pathological aspects of diseases, tumors, malformations and trauma involving the ear. *Mailing Add:* Carnegie 600 N Wolfe St Baltimore MD 21205

NAGER, URS FELIX, b Zurich, Switz, May 15, 22; nat US; m 51; c 3. ORGANIC CHEMISTRY, BIOCHEMISTRY. *Educ:* Swiss Fed Inst Technol, Chem Eng, 45, PhD(org chem), 49. *Prof Exp:* Res biochemist, Com Solvents Corp, Ind, 59-52; sr res chemist, Burke Res Co, Mich, 52-58; res assoc, Squibb Inst Med Res, 58-60, res supvr, 60-69, asst dir, 69-82, dir, Squibb Pharmaceut, 82-89; RETIRED. *Mem:* Am Chem Soc; NY Acad Sci; Swiss Chem Soc. *Res:* Antibiotics; enzymes; steroids; designed experimentation. *Mailing Add:* Bunkerhill Rd RD1 Princeton NJ 08540-9801

NAGERA, HUMBERTO, b Havana, Cuba, May 23, 27; Brit citizen; m 52; c 3. PSYCHIATRY, PSYCHOANALYSIS. *Educ:* Maristas Col, BS, 45; Havana Univ, MD, 52. *Prof Exp:* prof psychiat, Univ Mich, 68-73, prof child psychiat & chief youth serv, 73-79, dir child psychoanal study prog, 68-79, clin prof psychiat, 79-87, dir, Long Term Psychother Clin, 82-87, EMER PROF, UNIV MICH, 87-; prof psychiat, 68-73, prof child psychiat & chief youth serv, 73-79, dir child psychoanal study prog, 68-79, clin prof psychiat, 79-87, dir, Long Term Psychother Clinic, 82-87, EMER PROF, UNIV MICH, 87-; PROF PSYCHIAT & DIR, CHILDREN & ADOLESCENT INPATIENT UNIT, UNIV SFLA, 87- *Concurrent Pos:* Pvt pract, 79-; training psychoanalyst, Mich Psychoanal Inst, 68-87. *Mem:* Brit Psychoanal Soc; Am Psychoanal Asn; Int Psychoanal Asn; Asn Prof Child Psychiatrists; Cuban Med Asn in Exile. *Res:* Child development, psychopathology and clinical psychiatry. *Mailing Add:* 5202 Dwire Ct Tampa FL 33647-1016

NAGHDI, P(AUL) M(ANSOUR), b Iran, Mar 29, 24; nat US; wid; c 3. FLUIDS, MECHANICS. *Educ:* Cornell Univ, BS, 46; Univ Mich, MS, 48, PhD(eng mech), 51. *Hon Degrees:* DSc, Nat Univ Ireland, 87. *Prof Exp:* From instr to prof eng mech, Univ Mich, 49-58; res prof, Miller Inst, 63-64 & 71-72, chmn div appl mech, 64-69, PROF ENG SCI, UNIV CALIF, BERKELEY, 58- *Concurrent Pos:* Mem adv comt, Sch Math Study Group, NSF, 58-64; Guggenheim fel, 58; ed, Proc, US Nat Cong Appl Mech, 54; mem, US Nat Comt Theoret & Appl Mech, 72-, chmn, 78-80. *Honors & Awards:* George Westinghouse Award, Am Soc Eng Educ, 62; Timoshenko Medal, Am Soc Mech Engrs, 80; A C Eringen Medal, Soc Engr Sci, 86. *Mem:* Nat Acad Eng; fel & hon mem Am Soc Mech Engrs; fel Acoust Soc Am; Sigma Xi; Soc Rheol; fel Soc Eng Sci. *Res:* Continuum mechanics; theory of elasticity; elastic shells and plates; theory of plasticity. *Mailing Add:* 530 Vistamont Ave Berkeley CA 94708

NAGI, ANTERDHYAN SINGH, b Lahore, India, Apr 1, 33; Can citizen; m 64; c 2. THEORETICAL SUPERCONDUCTIVITY. *Educ:* Punjab Univ, India, BSc Hons, 56, MSc, 57; Univ Delhi, PhD(physics), 61. *Prof Exp:* From asst prof to assoc prof, 67-78, PROF PHYSICS, UNIV WATERLOO, 78- *Concurrent Pos:* Alexander Von Humboldt fel, Inst Theoret Physics, Univ Frankfurt, 70; Nat Res Coun Can grant, 67-79; Nat Sci & Eng Res Coun Can grant, 79-; mem policy comt, J Low Temperature Physics, 86- *Mem:* Can Asn Physicists; Am Phys Soc. *Res:* Theory of antiferromagnetic superconductors; magnetic impurities in superconductors; heavy fermion superconductors; theory of high temperature superconductors. *Mailing Add:* Physics Dept Univ of Waterloo Waterloo ON N2L 3G1 Can

NAGLE, BARBARA TOMASSONE, b Philadelphia, Pa, Dec 8, 47; m 72. PHARMACOLOGY. *Educ:* Drexel Univ, BS, 70; Hahnemann Med Sch, MS, 72, PhD(pharmacol), 75. *Prof Exp:* Res technician pharmacol clin lab, Hahnemann Med Col, 70-72; res, Wills Eye Res Inst, 75-76; ASST PROF PHARMACOL & PHYSICOL PHILADELPHIA COL OSTEOP MED, 76-; CLIN STUDIES ASSOC, SMITH KLINE CORP. *Concurrent Pos:* Investr, Fight for Sight Inc grant, Wills Eye Res Inst, 77-78. *Mem:* NY Acad Sci; Am Physiol Soc; Sigma Xi; AAAS. *Res:* Pharmacokinetics of systemically administered drugs within special tissues of the body such as aqueous humor within the eye and parotid salivary secretions. *Mailing Add:* 850 Mill Rd Bryn Mawr PA 19010

NAGLE, DARRAGH (EDMUND), b New York, NY, Feb 25, 19; m 49; c 3. PHYSICS. *Educ:* Calif Inst Technol, BS, 40; Columbia Univ, AM, 42; Mass Inst Technol, PhD(physics), 47. *Prof Exp:* Lectr, Columbia Univ, 41-42; res assoc, Metall Lab, Univ Chicago, 43; group leader, Argonne Nat Lab, 43-44; asst group leader, Los Alamos Sci Lab, 44-45; res assoc, Mass Inst Technol, 45-47, instr, 47-48; Fulbright fel, Cambridge Univ, 48-49; asst prof physics, Univ Chicago, 49-55; mem staff, 55-62, group leader, 62-65, assoc div leader, 65-68, alt div leader medium energy physics, 68-85, SR FEL, LOS ALAMOS NAT LAB, 85- *Concurrent Pos:* Guggenheim fel, 53; mem, Stanford Linear Accelerator Policy Bd, 70-73, Agronne Nat Lab Physics Rev Bd, 77-80, Bates Linac Prog Comt, Mass Inst Technol, 73-76 & Subpanel on Non-Accelerator Physics, 89; trustee, Santa Fe Inst, 86- *Mem:* Fel Am Phys Soc; Am Astron Soc. *Res:* Physics of particles; nuclear physics; astrophysics; accelerator physics. *Mailing Add:* Los Alamos Nat Lab PO Box 1663 Los Alamos NM 87545

NAGLE, DENNIS CHARLES, b Dolgeville, NY, Jan 13, 45; m 69; c 5. CERAMIC COMPOSITES, CERAMIC COATING TECHNOLOGY. *Educ:* Alfred Univ, BS, 67; Pa State Univ, PhD(mats sci), 72. *Prof Exp:* Sr engr, Pratt & Whitney Aircraft, 72-76; PRIN ENGR, MARTIN MARIETTA CORP, 76- *Mem:* Am Ceramic Soc. *Res:* Processing of carbon/carbon composites; development of carbon and graphite materials for fuel cell and alumunium reduction cell applications; development of ceramic materials for stealth technology and high temperature applications; development of ceramic-ceramic composites and metal matrix composites; granted 26 United States patents. *Mailing Add:* 10148 Tanfield Ct Ellicott City MD 21043-5808

NAGLE, EDWARD JOHN, b Chicago, Ill, May 29, 41; m 70; c 5. MANUFACTURING GENERAL, BUILDING CODE COMPLIANCE. *Educ:* Southern Ill Univ, BA, 67, MS, 68. *Prof Exp:* Prof technol, 76-77, DIR AUDIO-VISUAL & TECHNOL DIV, TRI-STATE UNIV, 77- *Concurrent Pos:* Consult, 75- *Mem:* Soc Mfg Engrs. *Res:* Integration of computer aided design into manufacturing and into metrology. *Mailing Add:* 601 Westview Dr Angola IN 46703

NAGLE, FRANCIS J, b Lynn, Mass, July 1, 24; m 60; c 11. EXERCISE PHYSIOLOGY, CARDIOVASCULAR PHYSIOLOGY. *Educ:* Univ Nebr, BS, 51, MA, 53; Boston Univ, EdD(health, phys educ), 59; Univ Okla, PhD(physiol), 66. *Prof Exp:* Asst prof phys educ, Univ Fla, 56-62; sect chief biodynamics br, Civil Aeromed Res Inst, Fed Aviation Agency, 62-64; from asst prof to assoc prof, 66-75, PROF PHYSIOL & PHYS EDUC, BIODYNAMICS LAB, UNIV WIS-MADISON, 75-, DIR LAB, 73- *Mem:* AAAS; Am Asn Health Phys Educ & Recreation; Am Asn Univ Prof; Am Physiol Soc. *Res:* Cardiovascular physiology and metabolism in stress. *Mailing Add:* 2237 Rugby Row Madison WI 53705

NAGLE, FREDERICK, JR, b Queens, NY, Jan 30, 37; m 57; c 2. PETROLOGY, MINERALOGY. *Educ:* Lafayette Col, BA, 58; Princeton Univ, MA, 61, PhD(geol), 66. *Prof Exp:* Asst prof geol, Juniata Col, 64-68; from asst prof to assoc prof geol, 68-78, PROF GEOL, UNIV MIAMI, 78- *Concurrent Pos:* NSF col teacher res fel, 67-68. *Mem:* AAAS; Mineral Soc Am; Geol Soc Am; Am Geophys Union. *Res:* Igneous and metamorphic petrology; K-Ar dating; Caribbean Island arc geology; ore deposits. *Mailing Add:* Dept Geol Univ Miami Univ Sta Coral Gables FL 33124

NAGLE, H TROY, b Booneville, Miss, Aug 31, 42; m 89. MEDICAL SENSORS, MEDICAL INSTRUMENTATION. *Educ:* Univ Ala, BSEE, 64, MSEE, 66; Auburn Univ, PhD(elec Eng), 68; Univ Miami, MD, 81. *Prof Exp:* From asst prof to prof elec eng, Auburn Univ, 70-84; PROF COMPUTER ENG, NC STATE UNIV, 84- *Concurrent Pos:* Vis prof, Swiss Fed Inst Technol, 78; res prof biomed eng, Univ NC, Chapel Hill, 88- *Honors & Awards:* New Technol Award, NASA, 68; Centennial Medal, Inst Elec & Electronics Engrs, 84. *Mem:* Fel Inst Elec & Electronics Engrs; Inst Elec & Electronics Engrs Eng in Med & Biol Soc; Inst Elec & Electronics Engrs Indust Electronics Soc (pres, 84-85). *Res:* Medical electronics and systems; fault-tolerant, testable systems design; microelectrode arrays; biomedical instrumentation; biomedical signal processing. *Mailing Add:* Elec & Computer Eng NC State Univ Box 7911 Raleigh NC 27695

NAGLE, JAMES JOHN, b Wilkes-Barre, Pa, Nov 10, 37; m 84. EVOLUTION, SCIENCE EDUCATION. *Educ:* Bloomsburg State Col, BS, 62; NC State Univ, MS, 65, PhD(genetics), 67. *Prof Exp:* Instr genetics, NC State Univ, 66-67; from asst prof to assoc prof zool & bot, 67-78, PROF BIOL, DREW UNIV, 78- *Concurrent Pos:* Adj prof, Hunter Col, City Univ New York, 71-74. *Res:* Experimental evolution; biological education, especially biology for non-majors; social implications of biology. *Mailing Add:* Dept Biol Drew Univ 36 Madison Ave Madison NJ 07940

NAGLE, JOHN F, b Easton, Pa, Sept 29, 39; m 80; c 2. PHYSICS, BIOPHYSICS. *Educ:* Yale Univ, BA, 60, MS, 62, PhD(physics), 65. *Prof Exp:* NATO fel physics & statist mech, King's Col, London, 65-66; res assoc statist mech, Cornell Univ, 66-67; asst prof physics, 67-72, assoc prof physics & biol sci, 72-78, PROF PHYSICS & BIOL SCI, CARNEGIE-MELLON UNIV, 78- *Concurrent Pos:* A P Sloan Found fel, 69-71; Guggenheim fel, 79-80. *Mem:* Fel Am Phys Soc; Biophys Soc; Sigma Xi; NY Acad Sci. *Res:* Phase transitions; condensed matter physics and chemistry; membrane biophysics. *Mailing Add:* Dept Physics Biol Carnegie-Mellon Univ 5000 Forbes Ave Pittsburgh PA 15213

NAGLE, RAY BURDELL, b Jefferson, Wis, Feb 10, 39; m 63; c 2. IMMUNOPATHOLOGY, RENAL PATHOLOGY. *Educ:* Wash State Univ, BS, 60; Univ Wash, MD, 64, PhD(path), 77. *Prof Exp:* Asst exp path, Walter Reed Army Inst Res, 70-72; from asst prof to assoc prof path, Univ Md, 72-75; assoc prof path, 76-81, PROF PATH, UNIV ARIZ, 81- *Concurrent Pos:* Consult, Vet Admn Med Ctr, 76- *Mem:* Am Soc Cell Biol; Int Acad Path; Am Soc Clin Path; Am Asn Pathologist; Am Soc Nephrol; NY Acad Sci. *Res:* Immunopathology; Characterization & applications of monoclonal antibodies; basic biology of intermediate filamicuts. *Mailing Add:* Dept Path Ariz Hlth Sci Ctr Rm 5230 Basic Sci Tucson AZ 85724

NAGLE, RICHARD KENT, b Detroit, Mich, Feb 19, 47; m 69; c 2. NONLINEAR DIFFERENTIAL EQUATIONS. *Educ:* Univ Mich BS, 68, MA, 69, PhD(math), 75. *Prof Exp:* Asst prof, Univ Mich, Dearborn, 75-76; from asst prof to assoc dor, ctr math serv, 70-84, ASSOC PROF MATH, UNIV SOUTH FLA, 80-, DIR, CTR MATH SERV, 85- *Mem:* Am Math Soc; Soc Indust & Appl Math; Math Asn Am; Sigma Xi. *Res:* Nonlinear partial differential equations; nonlinear ordinary differential equations; nonlinear functional analysis. *Mailing Add:* Dept Math Phy 114 Univ SFla Tampa FL 33620-5700

NAGLE, WILLIAM ARTHUR, b W Reading, Pa, May 16, 43; m 70, 83; c 2. RADIATION BIOPHYSICS, MOLECULAR BIOLOGY. *Educ:* Albright Col, BS, 65; Univ Okla, Norman, MS, 66; Univ Tex Southwestern Med Sch, PhD(radiation biol), 72. *Prof Exp:* Fel, Harvard Med Sch, 72-74; asst prof, 74-79, ASSOC PROF RADIOL & PHYSIOL-BIOPHYS, UNIV ARK MED SCI, 79- *Concurrent Pos:* Student travel grant, Radiation Res Soc, 71; radiation biologist, Med Res Serv, Vet Admin Hosp, Little Rock, 74-; consult, Nat Cancer Inst, 77, consult radiation study sect AHR, 78. *Mem:* Biophys Soc; Radiation Res Soc. *Res:* Molecular radiobiology of mammalian cells; DNA damage and repair; influence of cell energy metabolism on recovery from radiation and hyperthermia injury. *Mailing Add:* Dept of Radiol Physiol Univ Ark Med Sch 4301 W Markham St Little Rock AR 72205

NAGLER, ARNOLD LEON, b New York, NY, Aug 18, 32; m 61; c 2. PATHOLOGY, PHYSIOLOGY. *Educ:* City Col New York, BS, 53; NY Univ, MD, 58, PhD(path), 60. *Prof Exp:* Res assoc, Col Med, NY Univ, 58-60 & Mt Sanai Hosp, 60-61; PRIN INVESTR, NIH GRANT ALBERT EINSTEIN COL MED, 70- *Concurrent Pos:* NIH fel, NY Univ, 60-61. *Mem:* NY Acad Sci; Fedn Am Socs Exp Biol; Sigma Xi. *Res:* Pathophysiology of shock; determination of the mechanisms involved in the pathogenesis of shock and the mortality therefrom; methods of circumventing lethality; choline and nutritional deficiencies and the pathology resulting therefrom. *Mailing Add:* 72 Hazlewood Dr Jericho NY 11753

NAGLER, CHARLES ARTHUR, b Whitsett, Pa, May 10, 16; m 47; c 2. PHYSICAL METALLURGY. *Educ:* Univ Mich, BSE, 38, MSE, 39; Univ Minn, PhD(phys metall), 45. *Prof Exp:* Metallurgist, Eng Res Inst, Univ Mich, 38-39; asst phys metall, Univ Minn, 39-41, from instr to asst prof, 41-46; assoc prof metall eng, Wayne State Univ, 46-80; RETIRED. *Concurrent Pos:* Chief metallurgist, Twin City Cartage Corp, 42-45; pres, Metall Consult, Inc, 46-; mem, Hwy Res Bd, Nat Acad Sci-Nat Res Coun. *Mem:* Am Soc Metals; Soc Automotive Engrs; Am Soc Testing & Mat; fel Am Inst Chem; NY Acad Sci. *Res:* Physical and extractive metallurgy; failure of materials of engineering in service. *Mailing Add:* 7272 McCandlish Rd Grand Blanc MI 48439

NAGLER, ROBERT CARLTON, b Iowa City, Iowa, July 4, 23; m 47; c 5. ORGANIC CHEMISTRY. *Educ:* William Penn Col, BS, 47; Univ Mo, MA, 49; Univ Iowa, PhD(chem), 53. *Prof Exp:* Asst chem, Univ Mo, 47-49; asst, Univ Iowa, 49-51, instr, 52-53; asst prof, Purdue Univ, 53-56; from asst prof to prof chem, 56-86, asst chmn dept, 68-78, chmn, 78-86, EMER PROF CHEM, WESTERN MICH UNIV, 86- *Concurrent Pos:* USAID sci adv, Nigeria, 62-64. *Mem:* AAAS; Sigma Xi; Am Chem Soc. *Res:* Nitrogen-magnesium-halide reagents; Grignard reactions; synthesis of anti-tumor agents; fluoride analysis. *Mailing Add:* Dept of Chem Western Mich Univ 3332 Mc Cracken Kalamazoo MI 49008

NAGLER-ANDERSON, CATHRYN, b Brooklyn, NY, May 7, 57. IMMUNOLOGY. *Educ:* Columbia Univ, BA, 79; NY Univ, MS, 83, PhD(immunol), 86. *Prof Exp:* Res asst immunol, Sch Med, NY Univ, 79-81; NIH training fel, Ctr Cancer Res, Mass Inst Technol, 86-89, postdoctoral assoc, 89-90; ASST PROF PEDIAT, MED SCH, HARVARD UNIV, 90-; ASSOC IMMUNOLOGIST, CHILDREN'S SERV, MASS GEN HOSP, 90- *Concurrent Pos:* Nat Inst Allergy & Infectious Dis nat serv res award, 88; tutor biol, Harvard Univ, 89, Roche II award res autoimmunity, 90; Nat Found Ileitis & Colitis career develop award, 90. *Mem:* Fedn Am Socs Exp Biol. *Res:* Author of various publications in medical journals. *Mailing Add:* Mass Gen Hosp E Rm 3308 Bldg 149 13th St Charlestown MA 02129

NAGLIERI, ANTHONY N, b New York, NY, Apr 15, 30; m 55; c 1. INDUSTRIAL ORGANIC CHEMISTRY. *Educ:* Fordham Univ, BS, 51, MS, 53; Columbia Univ, PhD(chem), 59. *Prof Exp:* Proj leader, 59-64, res assoc, 64-68, sect head res & develop, 68-74, asst dir res, 74-76, DIR EXPLOR RES, HALCON RES, MONTVALE, NJ, 76- *Mem:* Am Chem Soc. *Res:* Liquid and vapor phase oxidations; catalysis; carbonyletim; free radical chemistry. *Mailing Add:* Ten Johnson St Monmouth Beach NJ 07750-1409

NAGODAWITHANA, TILAK WALTER, b Colombo, Sri Lanka; US citizen; m 69; c 2. YEAST TECHNOLOGY, EXTRACTS & AUTOLYSATES. *Educ:* Univ Ceylon, BSc, 63; Univ Philippines, MS, 69; Cornell Univ, PhD(microbiol), 74. *Prof Exp:* Chemist, Sri Lanka Sugar Corp, 63-65, gen mgr, 65-70; sr res scientist, Jos Schlitz Brewing Co, 74-77; res mgr, Anheuser Busch Inc, St Louis, 77-81; RES MGR, UNIVERSAL FOODS CORP, MILWAUKEE, 81- *Mem:* Am Soc Microbiol; Inst Food Technologists; Soc Indust Microbiol; Am Chem Soc. *Res:* Yeast research; yeast physiology and genetics; extracts, autolysates and nutritional tests in the biotechnology processes. *Mailing Add:* Universal Foods Corp Technical Center 6143 N 60th St Milwaukee WI 53218-1696

NAGODE, LARRY ALLEN, b New Deal, Mont, Nov 15, 38; m 63; c 2. VETERINARY PATHOLOGY. *Educ:* Colo State Univ, DVM, 63; Ohio State Univ, MSc, 65, PhD(vet path), 68. *Prof Exp:* Morris Animal Found res fel vet path, Ohio State Univ, 63-65, Nat Cancer Inst res fel, 65-68; Nat Cancer Inst res fel biochem, Med Sch, Univ Pa, 69-70; asst prof vet path, 70-77, ASSOC PROF VET PATH, COL VET MED, OHIO STATE UNIV, 77- *Concurrent Pos:* Sabbatical, Path Inst, Univ Bern, Switz. *Res:* Metabolism, mechanism of action and therapy of vitamin D metabolites in normal and diseased states; mechanism of action of enterotoxins of escherichia coli. *Mailing Add:* Dept Vet Path Ohio State Univ Col Vet Med 1925 Coffey Rd Columbus OH 43210

NAGY, ANDREW F, b Budapest, Hungary, May 2, 32; m 65. AERONOMY, ATMOSPHERIC PHYSICS. *Educ:* Univ New South Wales, BE, 57; Univ Nebr, MSc, 59; Univ Mich, MSE, 60, PhD(elec eng), 63. *Prof Exp:* Design engr, Elec Control & Eng Co, 56-57; instr elec eng, Univ Nebr, 57-59; asst res engr, 60-63, from asst prof to assoc prof elec eng, 63-71, PROF ATMOSPHERIC SCI & PROF ELEC ENG, UNIV MICH, ANN ARBOR, 71- *Concurrent Pos:* Ed, Rev Geophys Space Physics; mem comt solar space physics, Nat Res Coun. *Mem:* Inst Elec & Electronics Eng; Am Geophys Union; AAAS. *Res:* Theoretical and experimental studies of the chemistry and physics of the terrestrial and planetary atmospheres. *Mailing Add:* Dept Elec Eng Univ Mich Ann Arbor MI 48109

NAGY, BARTHOLOMEW STEPHEN, b Budapest, Hungary, May 11, 27; nat US; m 52, 67; c 2. ORGANIC GEOCHEMISTRY. *Educ:* Pazmany Peter Univ, Hungary, BA, 48; Columbia Univ, MA, 50; Pa State Univ, PhD(mineral), 53. *Prof Exp:* Asst, Pa State Univ, 49-53; res engr, Pan Am Oil Co, 53-55; res assoc & supvr geophys res, Cities Serv Res & Develop Co, Okla, 55-57; from asst prof to assoc prof chem, Fordham Univ, 57-65; assoc res geochemist, Univ Calif, San Diego, 65-68; PROF GEOSCI, UNIV ARIZ, 68- *Concurrent Pos:* Vis assoc prof, Univ Calif, San Diego, 63-65; mem adv comt, Lunar Sci Inst, 72; managing ed, Precambrian Res, Elsevier Pub Co, Amsterdam, Neth, 72-, ed-in-chief, 84-; mem sample allocation comt, Vatican Meteorite Collection, 85- *Mem:* Geochem Soc; Am Chem Soc; Int Soc Study Origin Life. *Res:* Origin of life on earth; amino acids; hydrocarbons; meteorites; clay mineralogy; x-ray crystallography; petroleum geochemistry and exploration; organic matter in ore deposit evolution. *Mailing Add:* 520 Gould-Simpson Dept Geosci Univ Ariz Tucson AZ 85721

NAGY, BELA FERENC, b Nagybanhegyes, Hungary, May 15, 26; US citizen; m 58; c 2. BIOCHEMISTRY. *Educ:* Eotvos Lorand Univ, Budapest, dipl biol & chem, 53; Brandeis Univ, PhD biochem, 64. *Prof Exp:* Asst prof biochem, Eotvos Lorand Univ, 53-56; Nat Acad Sci res fel, Rockefeller Inst, 57, res assoc, 57-59; res assoc, NY Univ, 59-60; Muscular Dystrophy Asn Am spec fel, Inst Muscle Dis, 60; NIH spec fel, Brandeis Univ, 61-64; res assoc, Retina Found, 64-70, staff scientist, Boston Biomed Res Inst, 70-78; ASSOC PROF, DEPT NEUROL & DEPT PHARMACOL & CELL BIOPHYS, UNIV CINCINNATI, COL MED, 78- *Concurrent Pos:* NIH res grant, 67-69 & career develop award, 67-72; res assoc neuropath, Harvard Univ, 68-69; prin assoc neurol (biochem), Harvard Univ, 69-78. *Mem:* AAAS; Am Chem Soc; Brit Biochem Soc; Biophys Soc; Am Soc Biol Chem. *Res:* Chemistry and physiology of muscle contraction; chemistry and physical chemistry of proteins. *Mailing Add:* Dept Neurol & Pharmacol Unv Cincinnati Med Ctr Cincinnati OH 45267

NAGY, DENNIS J, b Perth Amboy, NJ, Oct 8, 50; m 74; c 1. POLYMER CHEMISTRY. *Educ:* Lebanon Valley Col, BS, 72, MA, 75; Lehigh Univ, PhD(chem), 79. *Prof Exp:* Sr res asst, E R Squibb & Sons, Inc, 74-76; PRIN RES CHEMIST, AIR PROD & CHEM, INC, 79- *Mem:* Am Chem Soc. *Res:* Polymer characterization and analysis; correlation of polymer structure with properties and performance. *Mailing Add:* Air Prod & Chem Inc 6406 Tupelo Rd Allentown PA 18104-9578

NAGY, GEORGE, b Budapest, Hungary, July 7, 37; Can citizen; m 63; c 2. DOCUMENT-IMAGE PROCESSING, TERRAIN VISIBILITY. *Educ:* McGill Univ, BEng, 59, MEng, 60; Cornell Univ, PhD(elec eng), 62. *Prof Exp:* Res assoc cognitive systs, Cornell Univ, 62-63; staff mem comput sci, Watson Res Ctr, IBM Corp, 63-72; prof & chmn, Univ Nebr, 72-81; PROF, RENSSELAER POLYTECH INST, 85- *Concurrent Pos:* Res assoc, Cornell Univ, 66; vis lectr, Univ Nebr, 67; vis prof, Univ Montreal, 68-69; vis scientist, IBM Watson Res Ctr & IBM San Jose Res Lab, 76, Bell Labs, 81, CNR Italy, 81, Univ Genova, 86. *Mem:* Sr mem Inst Elec & Electronics Engrs; Asn Comput Mach. *Res:* Pattern recognition; optical character recognition; speech; remote sensing; image processing; man-computer interface; geographic data processing; computational geometry; computer vision. *Mailing Add:* ECSE Rensselaer Polytech Inst Troy NY 12180-3590

NAGY, JULIUS G, b Balatonboglar, Hungary, Aug 7, 25; US citizen; m 49; c 2. NUTRITION, BACTERIOLOGY. *Educ:* Wayne State Univ, BS, 60; Colo State Univ, MS, 63, PhD(wildlife nutrit), 66. *Prof Exp:* Animal scientist, 63-65, from instr to assoc prof, 65-77, PROF WILDLIFE BIOL, COLO STATE UNIV, 77- *Mem:* Wildlife Soc. *Res:* Wildlife nutrition and physiology, especially the rumen microbiological digestion of wild ruminants. *Mailing Add:* Dept of Fishery Biol Colo State Univ Ft Collins CO 80523

NAGY, KENNETH ALEX, b Santa Monica, Calif, July 1, 43; m 67; c 2. ENVIRONMENTAL PHYSIOLOGY. *Educ:* Univ Calif, Riverside, AB, 67, PhD(biol), 71. *Prof Exp:* Actg asst prof zool, Univ Calif, Los Angeles, 71-72, adj asst prof biol, Dept Biol & asst res zoologist environ biol, Nuclear Med & Radiation Biol Lab, 72-77, assoc prof, Dept Biol & assoc res zoologist environ biol, 77-83, PROF, DEPT BIOL & BIOMED ENVIRON SCI LAB, UNIV CALIF, LOS ANGELES, 83- *Concurrent Pos:* Univ Calif Regents Grad Intern fel, 67-71; Fulbright Fel, 86-87. *Mem:* Sigma Xi; AAAS; Am Soc Ichthyologists & Herpetologists; Ecol Soc Am; Am Soc Zool; Am Soc Mammal. *Res:* Physiology and behavior of desert vertebrates as these relate to the animal's survival in nature, including water, mineral, nitrogen and energy balance in field animals measured with isotopically-labeled water. *Mailing Add:* Lab Biomed & Environ Sci Univ Calif Los Angeles CA 90024-1786

NAGY, STEPHEN MEARS, JR, b Yonkers, NY, Apr 1, 39; c 2. ALLERGY, IMMUNOLOGY. *Educ:* Princeton Univ, AB, 60; Tufts Univ, MD, 64; Am Bd Internal Med, dipl, 71; Am Bd Allergy & Immunol, dipl, 74. *Prof Exp:* Intern med, Baltimore City Hosps, 64-65, resident, 65-66, 68-69; fel allergy & immunol, Johns Hopkins Hosp, 69-71; from asst prof to assoc prof, 73-84, PROF CLIN MED, SCH MED, UNIV CALIF, DAVIS, 84- *Concurrent Pos:* Fel med, Sch Med, Johns Hopkins Univ, 64-66, 68-71. *Mem:* Am Acad Allergy & Clin Immunol; AMA. *Res:* Evaluation and management of allergic and asthmatic diseases of adults and children. *Mailing Add:* 4801 J St Ste A Sacramento CA 95819

NAGY, STEVEN, b Fords, NJ, Apr 7, 36; m 80; c 3. CITRUS CHEMISTRY. *Educ:* La State Univ, BS, 60; Rutgers Univ, MS, 62, PhD(physiol & biochem), 65; Univ SFla, ME, 77. *Prof Exp:* Anal chemist, US Pub Health Serv, 62-65; res assoc, Lever Brothers, 65-67; res chemist, USDA, 68-78; RES SCIENTIST, FLA DEPT CITRUS, 79- *Concurrent Pos:* Adj prof, Univ Fla, 79-; vchmn, Agr & Food Div, Am Chem Soc, 81-82, chmn elect, 82-83. *Honors & Awards:* Distinguished Serv Award, Am Chem Soc, 87. *Mem:* Fel Am Chem Soc; Phytochem Soc NAm; Am Soc Hort Sci; Inst Food Technologists. *Res:* Methods development to monitor storage abuse of product; nutrient stability; off-flavor development; can corrosion of citrus products; lipid chemistry of mycorrhizal infections; citrus chemosystematics and leaf proteins; nonenzymic browning reactions. *Mailing Add:* 103 Arietta Shores Dr Auburndale FL 33823

NAGY, THERESA ANN, b Wheeling, WVa, July 16, 46. ASTROPHYSICS. *Educ:* West Liberty State Col, BS, 68; Tex A&M Univ, MS, 70; Univ Pa, PhD(astron), 74. *Hon Degrees:* LHD, West Liberty State Col, 90. *Prof Exp:* Analyst & programmer, Anal Inc, 72-73; staff scientist, Comput Sci Corp, Nasa Goddard Space Flight Ctr, 73-83; chair, Dept Physics, Ind State Univ, 83-88; vis sr scientist, NASA, Washington DC, 88-90; DIR ACAD AFFAIRS, PA STATE UNIV, FAYETTE, 90- *Mem:* Am Astron Soc; Sigma Xi. *Res:* Data reduction and retrieval of space satellite data; computation of synthetic light curves for contact binary star systems; computer generation and manipulation for stellar and nonstellar astronomical data; image processing; cosmic ray muon intensity research. *Mailing Add:* Pa State Univ Fayette Campus PO Box 519 Uniontown PA 15401

NAGY, ZOLTAN, b Budapest, Hungary, Aug 16, 33; US citizen; m 56; c 1. ELECTROCHEMISTRY. *Educ:* Univ Veszprem, Hungary, dipl chem eng, 56; Univ Akron, MS, 62; Univ Pa, PhD(phys chem), 72. *Prof Exp:* Res chemist, Chem Div, PPG, Inc, 57-63; sr res engr, Philadelphia Div, Honeywell, Inc, 63-65; res group leader, Betz Lab, Inc, 65-67; res fel, Univ Pa, 67-71; res assoc, Diamond Shamrock Corp, 72-76; STAFF CHEMIST, ARGONNE NAT LAB, 76- *Concurrent Pos:* Mem bd dirs, Electrochem Soc, 88-90. *Mem:* Am Chem Soc; Electrochem Soc; Int Soc Electrochem; AAAS. *Res:* Electrode phenomena; electrode kinetics; investigation of electrode surfaces with synchrotrom x-ray techniques; application of computers to electrode kinetic investigations; electrochemical aspects of corrosion; electrochemical technology; thermodynamics of ionic solutions. *Mailing Add:* Argonne Nat Lab 9700 S Cass Ave Argonne IL 60439-4837

NAGYLAKI, THOMAS ANDREW, b Budapest, Hungary, Jan 29, 44; US citizen; m 69. THEORETICAL POPULATION GENETICS. *Educ:* McGill Univ, BS, 64; Calif Inst Technol, PhD(physics), 69. *Prof Exp:* Res assoc physics, Univ Colo, 69-71; vis asst prof, Ore State Univ, 71-72; proj assoc med genetics, Univ Wis-Madison, 72-74, asst scientist med genetics & math res ctr, 74-75; from asst prof to prof biophys & theoret biol, 75-83, prof molelcular genetics & cell biol, 84-89, PROF ECOLOGY & EVOLUTION, UNIV CHICAGO, 89- *Res:* Theoretical population genetics; geographical structure of populations; stochastic processes in population genetics; linkage and selection; gene conversion. *Mailing Add:* Dept Ecology & Evolution Univ of Chicago Chicago IL 60637

NAGYVARY, JOSEPH, b Szeged, Hungary, Apr 18, 34; m 63; c 3. ORGANIC CHEMISTRY. *Educ:* Univ Zurich, PhD(org chem), 62. *Prof Exp:* Res asst peptide synthesis, CIBA, Ltd, Switz, 62; fel nucleotides, Univ Cambridge, 62-64; res asst prof nucleotides synthesis, Univ Conn, 64-65; asst prof nucleic acids, Sch Med, Creighton Univ, 65-68; assoc prof biochem, 68-74, PROF BIOCHEM, COL AGR, TEX A&M UNIV, 74- *Concurrent Pos:* Swiss Regional Scholar fel, 62-63; NIH grant, 64-66; consult, CIBA, Ltd, 64. *Mem:* Am Chem Soc; Am Soc Biol Chem. *Res:* Nucleic acid chemistry; synthesis of nucleotide di-and triesters; nutrient-fiber interactions; Italian violin varnish. *Mailing Add:* Dept Biochem & Biophys Tex A&M Univ College Station TX 77843-1248

NAHABEDIAN, KEVORK VARTAN, b Boston, Mass, Oct 31, 28; m 57; c 1. CHEMISTRY. *Educ:* Mass Inst Technol, SB, 52; Univ Vt, MS, 54; Univ NH, PhD(chem), 59. *Prof Exp:* Asst chem, Univ Vt, 52-54, instr, 62-63; instr, Lafayette Col, 54-55; asst, Univ NH, 55-57, fel, 57-59; res assoc, Brown Univ, 59-61; res chemist, Qm Res & Eng Ctr, 61-62; asst prof chem, Union Col, NY, 63-68; PROF CHEM & CHMN DEPT, STATE UNIV NY COL GENESEO, 68- *Mem:* Am Chem Soc. *Res:* Physical organic chemistry; electrophilic and nucleophilic aromatic substitution; acid and base catalysis; donor functions for basic media. *Mailing Add:* Ten Melody Lane Geneseo NY 14454

NAHAS, GABRIEL GEORGES, b Alexandria, Egypt, Mar 4, 20; nat US; m 54; c 3. PHARMACOLOGY. *Educ:* Univ Toulouse, BA, 37, MD, 44; Univ Rochester, MS, 49; Univ Minn, PhD(physiol), 53. *Hon Degrees:* DSc, Univ Uppsala, 88. *Prof Exp:* Chief lab exp surg, Marie Lannelongue Hosp, Paris, 53-55; asst prof physiol, Univ Minn, 55-57; chief respiratory sect, Walter Reed Army Inst Res, 57-59; assoc prof anesthesiol, 59-62, RES PROF ANESTHESIOL, COL PHYSICIANS & SURGEONS, COLUMBIA UNIV, 62- *Concurrent Pos:* Mem med adv bd, Coun Circulation & Basic Sci, Am Heart Asn; mem comt trauma, Nat Res Coun, 64; consult, Oceanog Inst, Monaco; adj prof anesthesiol res, Univ Paris, 68. *Honors & Awards:* Mem, Order Brit Empire; Officer, Order Orange Nassau; Presidential Medal Freedom with Gold Palm; Comdr, Legion of Honor, Laureate Fr Acad Med. *Mem:* Am Physiol Soc; Am Soc Pharmacol & Exp Therapeut; Brit Pharmacol Soc; Harvey Soc. *Res:* Acid-base regulation, catecholamines and metabolism; mechanisms of action of marijuana and cocaine; drug dependence. *Mailing Add:* Place Below Columbia Univ Columbia Univ 630 W 168th St New York NY 10032

NAHATA, MILAP CHAND, b Sardar Shahr, India, Oct 20, 50; m 78; c 1. PEDIATRIC PHARMACOLOGY. *Educ:* Univ Jodhpur, BS, 70; Univ Bombay, BS, 73; Duquesne Univ, MS, 75, Pharm D, 77. *Prof Exp:* From asst prof to assoc prof, 77-88, PROF PHARM & PEDIAT, OHIO STATE UNIV, 88-; DIR, INFECTIOUS DIS LAB, CHILDREN'S HOSP, 88- *Concurrent Pos:* Consult, Univ Toledo & Univ Fla, 87-88; mem, Bd of External Examiners, Univ Calcutta, 83-; vis prof, Univ Utah, 88- *Honors & Awards:* George Hopkins Award, Parenteral Drug Asn, 84; Pfizer Award, 87; Award for Sustained Contributions to the Literature of Hosp Pharm, Am Soc Hosp Pharmacists, 87; Educ Award, Am Col Clin Pharm, 90. *Mem:* AAAS; Am Col Clin Pharm; Am Soc Clin Pharmacol & Therapeut; Am Soc Hosp Pharmacists; Am Asn Cols Pharm; Soc Pediat Res. *Res:* Develop safe and effective dosage regimens of drugs for the treatment of diseases affecting infants and children; antimicrobial therapy. *Mailing Add:* Col Pharm Ohio State Univ 500 W 12th Ave Columbus OH 43210

NAHHAS, FUAD MICHAEL, b Sidon, Lebanon, Jan 29, 27; m 53; c 3. PARASITOLOGY, MEDICAL MICROBIOLOGY. *Educ:* Univ of the Pac, AB, 58, MA, 60; Purdue Univ, PhD(biol sci), 63. *Prof Exp:* Res fel parasitol, Fla State Univ, 63-64; from asst prof to assoc prof, 64-71, PROF BIOL, UNIV OF THE PAC, 71- *Concurrent Pos:* Dir microbiol, Dameron Hosp, Stockton, Calif. *Mem:* Am Soc Parasitol; Am Soc Microbiol. *Res:* Parasites of vertebrates; taxonomy and life history studies; geographic distribution of parasites; the antibiogram as an aid in the identification of bacteria. *Mailing Add:* Dept Biol Sci Univ Pac Stockton CA 95211

NAHM, MOON H, b Seoul, Korea, Mar 23, 48; US citizen; m 83; c 2. LABORATORY MEDICINE, IMMUNOLOGY. *Educ:* Wash Univ, AB, 70, MD, 74. *Prof Exp:* Intern-resident internal med, Jewish Hosp, St Louis, 74-76; resident lab med, Barnes Hosp, St Louis, 77-80; ASST PROF PATH & MED, WASH UNIV, 80- *Mem:* Am Asn Immunol; Am Asn Clin Chem; Acad Clin Lab Physicians & Scientists; Clin Immunol Soc; Soc Analytical Cytol. *Res:* Development of human B cells following antigenic stimuli. *Mailing Add:* Wash Univ Sch Med 660 S Euclid Ave St Louis MO 63110

NAHMAN, NORRIS S(TANLEY), b San Francisco, Calif, Nov 9, 25; m 53; c 4. ELECTRICAL ENGINEERING. *Educ:* Calif State Polytech Col, BS, 51; Stanford Univ, MS, 52; Univ Kans, PhD(elec eng), 61. *Prof Exp:* Res engr, Stanford Univ & electronics engr, San Francisco Naval Shipyard, 52; electronic scientist, Nat Security Agency, 52-55; instr elec eng, Univ Kans, 55-61, from assoc prof to prof, 61-66, dir electronics res lab, 58-66, prin investr, Proj Jayhawk, 56-64; sci consult, Radio Stand Lab, Nat Bur Standards, 66-70, chief, Pulse & Time Domain Sect, Electromagnetics Div, 70-73; prof elec eng & chmn dept, Univ Toledo, 73-75; chief, Time Domain Metrol Sect, Electromagnetics Div, Nat Bur Standards, 75-76, group leader, Picosecond Transition Phenomena, 76-77, chief, Time Domain Metrol Sect, Electromagnetic Technol Div, 78-79, sr scientist, electromagnetic waveform metrol, 80-83, group leader, electromagnetic fields characterization, 84-85; vpres, Picosecond Pulse Labs, 86-90; CONSULT ELEC ENGR, 90- *Concurrent Pos:* Tech adv, Dept Defense, 58-61; consult, Space Tech Labs, Calif, 62, Martin Co, Md, 61-62 & Wilcox Elec Co, Mo, 63-66; adj prof, Univ Colo, Boulder, 66-73 & 75-; prin prof & distinguished lectr, pulses & time domain tech & measurements, Ctr Nat Study of Telecommunications, France, 78; Harbin Inst Technol, 82; mem fac, NATO Advan Study Inst, Italy, 83. *Honors & Awards:* Inst Elec & Electronic Engrs Instrumentation & Measurement Soc Award Tech Accomplishment & Leadership, 88. *Mem:* Int Radio Sci Union; fel Inst Elec & Electronics Engrs; Sigma Xi. *Res:* Transient response of distributed networks; generation, transmission and measurement of electrical and optical picosecond pulses; interaction of electromagnetic waves and materials; dielectric dispersion; low temperatures; normal and super conductors. *Mailing Add:* 375 Erie Dr Boulder CO 80303

NAHMIAS, ANDRE JOSEPH, b Alexandria, Egypt, Nov 20, 30; US citizen; m 56; c 3. VIROLOGY, PEDIATRICS. *Educ:* Univ Tex, Austin, BA, 50, MA, 52; Univ Mich, MPH, 53; George Washington Univ, MD, 57. *Prof Exp:* Intern, USPHS Hosp, Staten Island, NY, 57-58; resident pediat, Boston City Hosp, Mass, 60-62, clin assoc, 62-64; from asst prof to assoc prof pediat & prev med, 64-70, PROF PEDIAT, SCH MED, EMORY UNIV, 70-, PROF PATH, 77-, ASSOC PROF DENT, 83- *Concurrent Pos:* NIH spec res fel virol, Mass Mem Hosp & res assoc microbiol, Sch Med, Boston Univ, 62-64; var study grants, USPHS, Am Cancer Soc, US Army & Nat Found & lectr var univs, US, Europe & Australia, 64-; NIH res career develop award, 66-71; consult, Am Red Cross, 68-72; mem study sect, Grants Rev Bd, Nat Commun Dis Ctr, 69-70 & Nat Cancer Inst, 83-; chief div infectous dis, immounol, Emory Univ, 64- *Honors & Awards:* Mead Johnson Award Pediat Res, 74; P R Edwards Award, Am Soc Microbiol, 77; Macy Fac Award, Univ Western Australia,76. *Mem:* AAAS; fel Am Pub Health Asn; fel Am Acad Pediat; Soc Pediat Res. *Res:* Bacterial infections, particularly staphylococcus, pertussis and leptospirosis; viral infections, particularly herpes viruses and relation to cancer; serological techniques, particularly immunofluorescence; immune mechanisms, particularly cellular immunity; clinical and epidemiological aspects of infectious diseases; fetal and neonatal diseases and neurological diseases. *Mailing Add:* Dept Pediat & Infect Dis Emory Univ 69 Butler St SE Atlanta GA 30303

NAHMIAS, STEVEN, b New York, NY, June 19, 45. OPERATIONS RESEARCH, STATISTICS. *Educ:* Queens Col, BA, 68; Columbia Univ, BS, 68; Northwestern Univ, MS, 71, PhD(opers res), 72. *Prof Exp:* From asst prof to assoc prof indust eng, Univ Pittsburgh, 72-78; PROF DECISION SCI, DEPT DECISION & INFO SCI, UNIV SANTA CLARA, 80- *Concurrent Pos:* Programmer & Analyst, IBM Corp, 67-68; consult, Ingalls Shipbuilding Div, Litton Indust, Xerox Res Ctr & Lex Automotive, 74-75; vis assoc prof opers res & indust eng, Stanford Univ, 78-79, vis prof indust eng, 81. *Mem:* Opers Res Soc Am; Inst Mgt Sci. *Res:* Stochastic inventory models with emphasis on perishable inventories and reparable item systems; fuzzy systems theory. *Mailing Add:* Dept Decision & Info Sci Univ Santa Clara Santa Clara CA 95053

NAHORY, ROBERT EDWARD, b McKeesport, Pa, Mar 1, 38; m 60; c 3. PHYSICS, SOLID STATE PHYSICS. *Educ:* Carnegie-Mellon Univ, BS, 60; Purdue Univ, MS, 62, PhD(physics), 67. *Prof Exp:* Mem tech staff physics, Bel Tel Labs, 67-82, supvr, 82-83, DIST MGR, BELL COMMUN RES, 84- *Concurrent Pos:* Vis scientist, Nat Ctr Telecommuns Studies, Paris, 80. *Mem:* Fel Am Phys Soc; Inst Elec & Electronics Engrs. *Res:* Semiconductor physics, especially optical properties; characteristics of new semiconductor materials; behavior of semiconductors under high excitation; lasers; optoelectronic devices; superlattices; oscillatory photoconductivity. *Mailing Add:* Bell Commun Res Rm 3Z-167 Red Bank NJ 07701-7040

NAHRWOLD, DAVID LANGE, b St Louis, Mo, Dec 21, 35; m 58; c 4. SURGERY, GASTROENTEROLOGY. *Educ:* Ind Univ, AB, 57, MD, 60; Am Bd Surg, dipl, 68; Am Bd Thoracic Surg, dipl, 69. *Prof Exp:* Intern surg, Med Ctr, Ind Univ, Indianapolis, 60-61, resident, 61-65, asst prof, Sch Med, 68-70; assoc prof surg, 70-73, assoc dean patient care, 78-80, prof surg, vchmn dept & chief div gen surg, Col Med, Pa State Univ, 73-82, assoc provost & dean health affairs, 81-82; PROF SURG & CHMN, DEPT SURG, NORTHWESTERN UNIV MED SCH, 82-; SURGEON-IN-CHIEF, NORTHWESTERN MEM HOSP, 82- *Concurrent Pos:* Scholar, Univ Calif, Los Angeles, 65-66; mem, Surg & Bioeng Study Sect, NIH, 78-; mem, Nat Digestive Dis Adv Bd, 85-90; Am Col Surgeons rep, Coun Med Specialty Soc, 88-89; dir, Am Bd Surg, 90-; chmn bd dirs, Soc Clin Surg, 91. *Mem:* Am Surg Asn; Soc Clin Surg (secy, 84-86, pres, 90); fel Am Col Surgeons; Soc Surg Alimentary Tract (secy, 85-89); Soc Univ Surgeons; Am Fedn Clin Res; Am Gastroenterol Asn; AMA; assoc mem Am Physiol Soc; Asn Acad Surg; Int Biliary Asn; Int Surg Group; Sigma Xi; Int Soc Surg; Soc Am Gastrointestinal Endoscopic Surgeons. *Res:* Gastrointestinal physiology. *Mailing Add:* Dept Surg Northwestern Univ Med Sch 250 E Superior St Suite 201 Chicago IL 60611

NAHRWOLD, MICHAEL L, b St Louis, Mo, Nov 23, 43; m 71; c 3. METABOLIC & TOXIC EFFECTS OF ANESTHETICS. *Educ:* Ind Univ, MD, 69. *Prof Exp:* PROF & CHMN DEPT ANESTHESIOL, MED CTR, UNIV NEBR, 86- *Mailing Add:* Anesthesia-Fesler Hall 204 Ind Univ Sch Med 1120 South Dr Indianapolis IN 46223

NAIB, ZUHER M, b Aleppo, Syria, Dec 10, 27; US citizen; m 58; c 3. PATHOLOGY. *Educ:* Univ Geneva, BS, 49, MD, 52; Am Bd Path, dipl, 60. *Prof Exp:* From instr to asst prof cytopath, Univ Md, 58-63; assoc prof, 63-67, PROF PATH & PROF GYNEC & OBSTET, EMORY UNIV, 67- *Concurrent Pos:* Fel path, Univ Va, 54-58; mem staff, Cytopath Div, Grady Mem Hosp, Atlanta, Ga. *Mem:* Am Cytol Soc; fel Am Soc Clin Pathologists. *Res:* Exfoliative cytopathology. *Mailing Add:* Dept Path Sch Med Emory Univ 69 Butler St Atlanta GA 30303

NAIBERT, ZANE ELVIN, b Cedar Rapids, Iowa, Nov 19, 31; m 58. SCIENCE EDUCATION, CHEMISTRY. *Educ:* Coe Col, BA, 54; Univ Iowa, MS, 57, PhD(sci educ, chem), 64. *Prof Exp:* Asst anal chem, La State Univ, 54-55; asst, Univ Iowa, 55-57, asst & teacher, Lab Schs, 59-61; teacher, Algona Community Schs, 58-59; asst prof chem, State Univ NY Col Cortland, 61-67; assoc prof & chmn, Div Sci & Math, 67-71, PROF CHEM, MONTGOMERY COL, 71- *Mem:* AAAS; Am Chem Soc. *Mailing Add:* Dept Chem Montgomery Col Takoma Park MD 20912

NAIDE, MEYER, b Russia, Mar 13, 07; nat US; m 33; c 2. MEDICINE. *Educ:* Univ Pa, AB, 29, MD, 32. *Prof Exp:* In chg vascular clin, Woman's Med Col Pa, 52-58, assoc prof med, 58-67; dir peripheral vascular div, Grad Hosp, 64-77, asst prof med, Univ, 67-77, EMER DIR PERIPHERAL VASCULAR DIV, GRAD HOSP, UNIV PA, 77- *Concurrent Pos:* In chg vascular clin, Einstein Med Ctr, 36-72; assoc, Univ Pa, 46-; pvt practr. *Mem:* AAAS; AMA; Am Col Physicians; Am Col Angiol; Int Cardiovasc Soc. *Res:* Peripheral vascular disease. *Mailing Add:* 2034 Spruce St Philadelphia PA 19103

NAIDER, FRED R, b New York, NY, Jan 31, 45. PEPTIDE CHEMISTRY. *Educ:* Polytech Inst Brooklyn, PhD(chem), 71. *Prof Exp:* PROF CHEM, COL STATEN ISLAND, 73- *Mem:* Am Chem Soc; Am Soc Microbiologists; Am Soc Biol Chem; AAAS. *Mailing Add:* Dept Chem City Univ NY Col Staten Island 50 Bay St Staten Island NY 10301

NAIDU, ANGI SATYANARAYAN, b Sundargarh, India, Oct 21, 36; m 69; c 3. SEDIMENTOLOGY, MARINE GEOLOGY. *Educ:* Andhra Univ, India, BSc(hons), 59, MSc, 60, PhD(geol), 68. *Prof Exp:* Demonstr geol, Andhra Univ, India, 60-61, Univ Grants Comn jr res fel, 66-69; asst prof marine sci, Univ Alaska, 69-71, marine geochemist, Inst Marine Sci, 71-76, assoc prof, 76-86, PROF MARINE SCI, UNIV ALASKA, 86- *Concurrent Pos:* Geol consult Arctic environ. *Mem:* Soc Econ Paleont & Mineral; Geochem Soc; Clay Minerals Soc; Int Asn Study Clays. *Res:* Marine geochemistry; lithological and chemical facies changes in recent sediments of arctic and tropical deltas and in subarctic fjordal environment, their present and paleoenvironmental implications; Cenozoic sedimentary history of Arctic Ocean; clay mineralogy; environmental impact of oil spills. *Mailing Add:* Inst Marine Sci Univ Alaska 335 Irving II Fairbanks AK 99775-1080

NAIDU, JANAKIRAM RAMASWAMY, b Bangalore, India, Nov 15, 31; US citizen; m 64. RADIOECOLOGY, ENVIRONMENTAL POLLUTION. *Educ:* Univ Bombay, BS, 55; Univ Washington, MS, 63; Ore State Univ, PhD(oceanog radioecol), 74. *Prof Exp:* Jr res scholar marine wood borers, Forest Res Inst, Dehra Dun, India, 56-59; sr scientist radioecol, Bhabha Atomic Res Ctr, Bombay, India, 59-68; proj ecologist, Dames & Moore, Los Angeles, 74-75; ECOLOGIST ENVIRON MONITORING & DOSES ASSESSMENT, BROOKHAVEN NAT LAB, 75- *Concurrent Pos:* Res scholar, US Atomic Energy Comn, 61-63 & 69-74; adj prof, Marine Sci Res Ctr, State Univ NY, Stony Brook, 76- *Mem:* Marine Biol Asn UK; plenary mem Health Physics Soc. *Res:* Effect on wood treated by preservatives against the attack of marine wood borers; distribution of radioactivity in the marine environment; effects of nuclear fallout on human beings. *Mailing Add:* Four Thornwood Circle Setauket NY 11733

NAIDU, SEETALA V, b Akiveedu, India, Jan 23, 57; m 87; c 1. POSITRON ANNIHILATION STUDIES, RADIATION EFFECTS & DEFECTS IN SOLIDS. *Educ:* Andhra Univ, India, BSc, 79; Mysore Univ, India, MSc, 80; Inst Physics, Bhubanesway, India, predoctoral (advan dipl) physics, 81; Calcutta Univ, India, PhD(physics), 88. *Prof Exp:* Postdoctoral res assoc & lectr, Dept Physics, Univ Tex, Arlington, 86-88; ASST PROF, PHYSICS DEPT, GRAMBLING STATE UNIV, LA, 88- *Mem:* Am Phys Soc. *Res:* Material science; positron annihilation studies; resistivity and hall effect measurements; electron microscopy; high-Tc superconductors; conducting polymers; semiconductors; radiation effects; ion-implantation; surface modification; radiation damage and defects in solids. *Mailing Add:* Dept Physics Grambling State Univ Grambling LA 71245

NAIDUS, HAROLD, b New York, NY, Apr 11, 21; m 43; c 2. CHEMISTRY. *Educ:* Univ Ill, AB, 41, MS, 42; Polytech Inst Brooklyn, PhD (polymer chem), 44. *Prof Exp:* Sr res chemist, Publicker Industs, Pa, 44-48; res dir, Am Polymer Corp Div, Borden Co, 48-55, tech dir & vpres, Polyvinyl Chem, Inc, 55-62; assoc prof chem, 62-69, asst dean, 69-76, dir lib arts progs, 76-79, assoc dean, Northeastern Univ, 76-83, dir, sci & health progs, 79-83; RETIRED. *Concurrent Pos:* Consult. *Mem:* Am Chem Soc; AAAS; NY Acad Sci; Sigma Xi. *Res:* Kinetics of polymerization; physical properties and preparation of polymers; monomer synthesis; organic synthesis. *Mailing Add:* Seven Warwick Terr Marblehead MA 01945

NAIK, DATTA VITTAL, b Goa, India, Mar 5, 47; m 71; c 2. ANALYTICAL CHEMISTRY, INORGANIC CHEMISTRY. *Educ:* Univ Bombay, BSc, 67; Univ Notre Dame, PhD(chem), 72. *Prof Exp:* Asst prof pharmaceut analysis, Col Pharm, Univ Fla, 73-75; asst prof analytical chem, Manhattanville Col, 75-77; from asst prof to assoc prof analytical chem, 77-90, CHMN DEPT CHEM & PHYSICS, MONMOUTH COL, 82-, PROF ANAL CHEM, 90- *Concurrent Pos:* Assoc chem, Univ Fla, 72-73; fac res fel, US Air Force, Tyndall AFB, Fla, 84; NASA, 86, 87. *Mem:* Am Chem Soc; Sigma Xi. *Res:* Development of new and sensitive analytical methods using spectroscopic techniques; process gas analysis by mass spectrometry; indoor-outdoor air quality. *Mailing Add:* Dept Chem & Physics Monmouth Col West Long Branch NJ 07764-1898

NAIK, TARUN RATILAL, b Ahmedabad, India, April 22, 40; US citizen; m 66; c 2. STRUCTURAL ENGINEERING, GEOTECHNICAL ENGINEERING. *Educ:* Gujarat Univ, India, BE, 62; Univ Wis, Madison, MS, 64, PhD(struct eng), 72. *Prof Exp:* Jr engr, Consult Eng Co, Chicago & Madison, 64-67; researcher & lectr struct eng & mech, Univ Wis, Madison, 67-72; exec vpres, Soils & Eng Serv, Inc, 72-75; ASSOC PROF STRUCT ENG & GEOTECH, UNIV WIS, MILWAUKEE, 75- *Concurrent Pos:* Consult var spec proj, NAm, 72-; lectr, Univ Wis, Exten, 67- & Mex Cement & Concrete Inst, Mexico City, 77- *Mem:* Am Soc Civil Engrs; Am Concrete Inst; Am Soc Testing & Mat; Nat Soc Prof Engrs; Am Soc Eng Educ; Soc Exp Mech. *Res:* Properties of concrete (eg use of by products for making concrete nondestructive testing of concrete and rapid testing of plastic concrete); timber engineering; machinery foundations; deep foundations; structures subjected to dynamic loads. *Mailing Add:* Dept Civil Eng Univ Wis PO Box 784 Milwaukee WI 53201

NAIL, BILLY RAY, b Roby, Tex, Jan 19, 33; m 52; c 3. ALGEBRA. *Educ:* Hardin-Simmons Univ, BA, 56; Univ Ill, Urbana, MA, 62, PhD(math), 67. *Prof Exp:* Teacher math, High Sch, Tex, 57-61; instr, Wayland Baptist Col, 62-64; from assoc prof to prof, Morehead State Univ, 67-72, chmn div math sci, 67-72; dean, 72-86, PROF MATH, CLAYTON STATE COL, 72- *Mem:* Math Asn Am; Sigma Xi. *Res:* Lie algebras. *Mailing Add:* Clayton State Col Morrow GA 30260

NAIMAN, ROBERT JOSEPH, b Pasadena, Calif, July 31, 47. AQUATIC ECOLOGY. *Prof Exp:* Nat Res Coun Can fel estuarine ecol, Pac Biol Sta, Fisheries & Oceans, Can, 74-76, 78; res assoc stream ecol, Dept Fisheries & Wildlife, Ore State Univ, 76-77; asst cur estuarine & fish ecol, Acad Natural Sci Philadelphia, 77-78; assoc scientist, Woods Hole Oceanog Inst, 78-85; DIR, CTR WATER & ENVIRON, UNIV MINN, 85- *Concurrent Pos:* Dir, Matamek Res Sta, Que, Can; Nat Sci Res Ctr, Toulouse, France, 84-88. *Mem:* Am Fisheries Soc; Am Soc Limnol & Oceanog; Ecol Soc Am; Desert Fishes Coun. *Res:* Aquatic ecology with emphasis on lotic ecosystems and landscape dynamics; terrestrial-aquatic interactions mediated by beavers, periphyton and macrophytal primary production; detritus dynamics; carbon and nutrient expont from watersheds; decompositon dynamics of wood and leves; biology of Atlantic salmon and brook trout; stream ecology. *Mailing Add:* Dir Ctr Streamside Univ Wash Seattle WA 98195

NAIMARK, GEORGE MODELL, b New York, NY, Feb 5, 25; m 46; c 3. BIOCHEMISTRY. *Educ:* Bucknell Univ, BS, 47, MS, 48; Univ Del, PhD(biochem), 51. *Prof Exp:* Res fel ultrasonics & biochem, Biochem Res Found, 48-51; res biochemist, Clevite-Brush Develop Co, 51; asst dir biochem labs, Strong, Cobb & Co, Inc, 51-53, asst dir control, 53-54, dir, 54; asst to dir med res dept, White Labs, Inc, 54-58, dir sci serv, 58-60; dir, Burdick & Becker, Inc, 60-61; vpres & dir res & develop, Dean L Burdick Assocs, Inc, New York, 61-66; PRES, NAIMARK & BARBA, INC, NEW YORK, 66- *Mem:* AAAS; Am Chem Soc; Am Inst Chem; NY Acad Sci. *Res:* Product development; promotional utilization of scientific and medical information. *Mailing Add:* 87 Canoe Brook Pkwy Summit NJ 07901

NAIMI, SHAPUR, ELECTROPHYSIOLOGY, COMPUTERS IN CARDIOLOGY. *Educ:* Birmingham Univ, Eng, MD, 53. *Prof Exp:* DIR, INTENSIVE CARDIAC CARE UNIT, TUFTS NEW ENG MED CTR, 70- *Mailing Add:* Dept Med Tufts Univ Med Sch 171 Harrison Ave Boston MA 02111

NAIMPALLY, SOMASHEKHAR AMRITH, b Bombay, India, Aug 31, 31; m 55; c 3. TOPOLOGY. *Educ:* Univ Bombay, BSc, 52, MSc, 54 & 58; Mich State Univ, PhD(math), 64. *Prof Exp:* Lectr math, Ruparel Col, India, 52-58; prof, Kirti Col, India, 59-61; teaching asst, Mich State Univ, 61-64; asst prof, Iowa State Univ, 64-65; assoc prof, Univ Alta, 65-69; prof, Indian Inst Technol, 69-71; vis prof, Lakehead Univ, 71-74, prof math, 74-90; PROF MATH, CARLETON UNIV, 90- *Concurrent Pos:* Fel, Inst Sci India, 52-53; Fulbright scholar, 61; vis prof math, Kuwait Univ, 88-90. *Mem:* Am Math Soc; Math Asn Am; Indian Math Soc; Can Math Cong. *Res:* General topology; proximity and uniform spaces; function spaces; semi-metric, developable spaces; compactifications; convexity; proximity. *Mailing Add:* Dept Math Carleton Univ Ottawa ON K1S 5B6 Can

NAINI, MAJID M, COMPUTER SCIENCE, ENGINEERING. *Educ:* Univ Tehran, Iran, BS, 76; City Univ New York, MS, 79; Univ Pa, PhD(comput sci & eng), 85. *Prof Exp:* Res asst, Dept Comput Sci, City Univ New York, 77-79; fel, Dept Comput & Info Sci, Univ Pa, 79-85; ASST PROF COMPUT

SCI, FLA ATLANTIC UNIV, BOCA RATON, 87- *Concurrent Pos:* Consult & co-dir, Hardware Dept, Daisy Computers Inc, Philadelphia, 81-82; speaker, Honeywell Corp, Minneapolis, IBM Watson Res Ctr, Yorktown Heights, Univ Ill, Chicago & Univ Hawaii Manoa, Honolulu, 85; guest speaker, Portland State Univ, Univ Central Fla & Univ Nevada, Las Vegas, 86; asst prof, Univ Hawaii, Honolulu, 86; speaker, Int Asn Mini & Micro Comput Int Conf, Miami Beach, 88. *Mem:* Inst Elec & Electronics Engrs. *Res:* Computer architecture and organization; logic design; parallel and dedicated architecture; data flow systems; data base machines; artificial intelligence; semantics and compiler design; design and development of a parser machine; design and development of a bottom-up evaluator which accepts the semantic equations with conditional rules; the dedicated architectures to simulate the nerve's signal to the brain for different parts of the human body; author of articles and textbooks on computers and computer architecture. *Mailing Add:* Dept Computer Eng Col Eng Fla Atlantic Col Boca Raton FL 33431

NAIPAWER, RICHARD EDWARD, b Passaic, NJ, Jan 28, 45; m 67; c 2. ORGANIC CHEMISTRY, AROMA CHEMISTRY. *Educ:* Rutgers Univ, BA, 66, MS, 68, PhD(org chem), 71. *Prof Exp:* Teaching asst chem, Rutgers Univ, 66-67; res chemist, Givaudan Corp, 67-71, sr res chemist, 71-85, group leader chem res, 85-88, DIR RES SERV, GIVAUDAN CORP, 88- *Concurrent Pos:* NASA fel, 66-67. *Mem:* Am Chem Soc; Am Inst Chemists; NY Acad Sci; Soc Cosmetics Chemists; Am Soc Photobiol. *Res:* Isolation and synthesis of monoterpenoids and sesquiterpenoids; stereochemistry of heterogeneous catalytic hydrogenations; synthesis of aroma-fragrance and flavor chemicals; structure-odor relationships. *Mailing Add:* Givaudan Corp 125 Delawanna Ave Clifton NJ 07015-5034

NAIR, C(HELLAPPAN) RAJAGOPALAN, GASTROENTEROLOGY, NUTRITION. *Educ:* Univ Allahabad, India, PhD(med, nutrit & biochem), 70. *Prof Exp:* MEM RES FAC & LAB DIR, MEHARRY MED COL, 83- *Res:* Nutritional biochemistry; chemical carcinogenesis. *Mailing Add:* Dept Med Meharry Med Col Hubbard Hosp Nashville TN 37208

NAIR, CHANDRA KUNJU, b Trichur, India, May 20, 44; US citizen; m 76; c 3. CORONARY & VALVULAR HEART DISEASE. *Educ:* Bombay Univ, BS, 64, MD, 72; Armed Forces Med Col, MBBS, 68. *Prof Exp:* Registr cardiol, Bombay Univ, 70-73; clin instr, 77-78, from asst prof to assoc prof, 78-90, PROF INTERNAL MED, CREIGHTON UNIV, OMAHA, 90- *Mem:* Am Heart Asn; Am Soc Internal Med; fel Am Col Physicians; fel Am Col Cardiol; Am Fedn Clin Res; Am Inst Ultrasound in Med. *Res:* Arrhythmias; exercise testing; echocardiography; computerized tomography. *Mailing Add:* Cardiac Ctr Sch Med Creighton Univ 601 N 30th St Suite 3810 Omaha NE 68131

NAIR, GANGADHARAN V M, b Madras, India, Jan 26, 30; m 71. PLANT PATHOLOGY, MYCOLOGY. *Educ:* Univ Madras, BSc, 51; Aligarh Muslim Univ, MSc, 53; Univ Wis-Madison, PhD(plant path, mycol), 64. *Prof Exp:* Mycologist, Indian Agr Res Inst, New Delhi, 55-59; res scientist, Univ Wis-Madison, 64-68; asst prof environ sci, Univ Wis-Green Bay, 68-69; UN expert, Food & Agr Orgn, UN Develop Prog, Italy, 69-71; assoc prof environ control, 71-79, PROF PLANT & FOREST PATH & MOCOL & INT CONSERV NATURAL RESOURCES, UNIV WIS-GREEN BAY, 79-, DIR INT PROGS, 74- *Concurrent Pos:* NSF fel plant path, Univ Wis-Madison, 64-66, NSF sr fel, 66-68; external exam doctoral thesis, Aligarh Muslim Univ, 69-; UN expert, Food & Agr Orgn, UN Italy, 69-71. *Mem:* Fel Nat Acad Sci India; Am Inst Biol Sci; Indian Phytopath Soc; Forestry Asn Nigeria; Int Union Forest Res Orgn; Am Phytopath Soc. *Res:* International control programs of plant-forest tree diseases; weedicide-Sylvicide applications in the establishment of exotic tree species in developing countries; host parasite interactions of vascular wilt pathogens; electron microscopy and chemotherapy. *Mailing Add:* Dept Sci Environ Change 2420 Nicolet Dr Green Bay WI 54301

NAIR, K AIYAPPAN, b Trivandrum, India, Jan 7, 36; m 66; c 1. MATHEMATICAL STATISTICS, COMPUTER SCIENCE. *Educ:* Univ Kerala, BSc, 56, MSc, 58; State Univ NY Buffalo, PhD(statist), 70. *Prof Exp:* Statist asst, Damodar Valley Corp, India, 59; res off statist, Univ Kerala, 59-63, lectr, 63-66; PROF MATH, EDINBORO, UNIV PA, 70- *Mem:* Am Statist Asn; Asn Comput Mach. *Mailing Add:* Dept Math & Comp Sci Edinboro State Col Edinboro PA 16412

NAIR, MADHAVAN G, b Kerala, India, May 15, 40; US citizen; m 71; c 2. ANTIFOLATES, EXPERIMENTAL THERAPEUTICS. *Educ:* Univ Kerala, BSc, 59, MSc, 62; Fla State Univ, PhD(org chem), 69. *Prof Exp:* Assoc prof, 76-79, PROF BIOCHEM, UNIV S ALA, 80-, ACTG CHMN, BIOCHEM DEPT, 90- *Concurrent Pos:* Sci consult, Nat Cancer Inst, NIH & Nat Inst Allergy & Infectious Dis, 73-90; prin investr cancer chemother, NIH, 79-; consult, Am Radiolabeled Chem, Inc, 82- & Burroughs Wellcome Co, 87- *Mem:* Am Asn Cancer Res; Am Soc Biochem & Molecular Biol; Am Soc Pharmacol & Exp Therapeut; Am Chem Soc. *Res:* Medicinal chemistry; anticancer drug development; metabolism of antifolates; drug delivery; drug targeting; radiolabeling; chemical synthesis. *Mailing Add:* Biochem Dept Univ S Ala Mobile AL 36688

NAIR, PADMANABHAN, b Singapore, Nov, 9, 31; m 59; c 3. BIOCHEMISTRY. *Educ:* Univ Travancore, India, BSc, 51; Univ Bombay, MSc, 55, PhD(biochem), 57. *Prof Exp:* Res officer chem path, All India Inst Med Sci, 58-60; res assoc biol, McCollum-Pratt Inst, Johns Hopkins Univ, 60-62; res assoc biochem, 62-64, dir biochem res div, Sinai Hosp, Baltimore, 64-83; RES CHEMIST USDA, ARS, 83- *Concurrent Pos:* Fel, Indian Coun Med Res, 57-58; Fulbright res grant, 60-63; instr, Sch Med, Johns Hopkins Univ, 64-69, lectr, 72-80, asst prof, 80-88, assoc prof, Sch Hyg & Pub Health, 91-; res chemist, Lipid Nutrit Lab, Beltsville Human Nutrit Res Ctr. *Mem:* AAAS; Am Inst Nutrit; Am Oil Chem Soc; NY Acad Sci; Am Soc Biochem & Molecular Biol; Am Chem Soc. *Res:* Biochemistry of fat soluble vitamins; fatty acids, sterols and bile acids; biochemistry and nutrition in cancer. *Mailing Add:* Beltsville Human Nutrit Res Ctr Rm 105 Bldg 308 BARC-e Beltsville MD 20705

NAIR, PANKAJAM K, b Kerala, India; c 2. MECHANISM OF CHEMORECEPTION. *Educ:* Univ Madras, BSc, 51, MA, 53, Univ Iowa, PhD(physiol bot), 64. *Prof Exp:* Res assoc, Univ Iowa, 64-67; res scientist, St Vincent Charity Hosp, 67-80; dir cardiovasc res, Cleveland Res Inst, 80-83; sr res scientist, 83-85, SR BIOMED SCI, CARDIOVASC RES, LA TECH UNIV, 85- *Concurrent Pos:* Prin investr, 70-86, comt mem, NIH, 84. *Honors & Awards:* Fyson Prize, Univ Madras, 53. *Mem:* Am Physiol Soc; Soc Exper Biol & Med; Int Soc Oxygen transp; Sigma Xi. *Res:* Co developer of the oxygen microelectrode; study oxygen transport to tissue, mechanism of chemoreception, brain ischemia, oxygen consumption; oxygen metabolism in brain. *Mailing Add:* Dept Biomed Eng La Tech Univ PO Box 10348 Tech Sta Ruston LA 71272

NAIR, RAMACHANDRAN MUKUNDALAYAM SIVARAMA, b North Parur, India, Nov 15, 38; m 73. NATURAL PRODUCT CHEMISTRY. *Educ:* Kerala Univ, India, BSc, 57, MSc, 59; Poona Univ, PhD(chem), 64. *Prof Exp:* Sr res fel org synthesis & natural prod, Nat Chem Lab, Poona, India, 64-66; res fel fungal metabolites, NY Bot Garden, 66-69; sr researcher org synthesis, Univ Paris, France, 69-70; sci pool officer org synthesis, Indian Inst Technol, Bombay, 70-71; res fel, NY Bot Garden, NY, 71-73, res assoc fungal prod, 73-80; CHEMIST BIO-ORG, MARINE NATURAL PROD, NY AQUARIUM, 81- *Concurrent Pos:* Jr res fel, Coun Sci & Indust Res, India, 61-64, sr res fel, 64-66; adj prof, Lehman Col, Long Island Univ, Bronx, NY, 85- *Mem:* Am Chem Soc; Am Inst Chemists; Am Soc Pharmacog. *Res:* Study of the structure, chemistry, biogenesis and biological activity of secondary metabolites of fungi and of certain higher plants; synthesis of natural products. *Mailing Add:* 159-34 Riverside Dr W New York NY 10032

NAIR, SHANKAR P, b Calicut, India, Apr 16, 26; m 62; c 2. POPULATION GENETICS, CYTOGENETICS. *Educ:* Univ Madras, BSc, 48, MSc, 61; Wash Univ, PhD(genetics), 66. *Prof Exp:* Asst prof zool, Madras Univ, 61-68; res assoc genetics, Univ Tex, Austin, 68-70; ASSOC PROF GENETICS, SOUTHERN ILL UNIV, EDWARDSVILLE, 70-, CHMN DEPT BIOL, 80- *Concurrent Pos:* Vis prof, Wash Univ, 72-73 & Univ Hawaii, 73-74. *Mem:* AAAS; Genetics Soc Am; Soc Study Evolution; Scientists' Inst Pub Info. *Res:* Cytogenetic effect of environmental agents on cell cultures. *Mailing Add:* 13 Glen Hollow Edwardsville IL 62025

NAIR, SREEDHAR, b Trivandrum, India, July 28, 28; nat US; m 54; c 3. PULMONARY PHYSIOLOGY, PHARMACOLOGY. *Educ:* Univ Travancore, India, ISc, 45; Univ Madras, MB, BS, 51; Am Bd Internal Med & Am Bd Pulmonary Dis, dipl. *Prof Exp:* Res physician med & pulmonary dis, Metrop Hosp & NY Med Col, 54-57, res assoc med, 57-58; asst prof physiol & pharmacol, NY Med Col, 58-64, asst clin prof med, 64-77; ASSOC CLIN PROF MED, SCH MED, YALE UNIV, 77-; PROF RESPIRATORY TECHNOL, UNIV BRIDGEPORT, 78-; CLIN PROF MED, SCH MED, YALE UNIV, 81- *Concurrent Pos:* Sr attend physician, Norwalk Hosp, Conn, 68-, dir, Dept Chest Dis, 70- *Mem:* Fel Am Col Physicians; Am Thoracic Soc; fel Am Col Chest Physicians; AMA. *Res:* Pulmonary diseases; physiology of respiration. *Mailing Add:* 15 Stevens Norwalk CT 06850

NAIR, SREEKANTAN S, statistics, operations research, for more information see previous edition

NAIR, VASAVAN N P, b Trivandrum, India, April 25, 34; Can citizen; m 58; c 3. PSYCHONEUROENDOCRINOLOGY, PSYCHOPHARMACOLOGY. *Educ:* Univ Kerale, MD, 59; Univ Mysore, DPM, 63; Glasgow Col, MRCP, 65; FRCP(C), 72; FRCPsych, 81; FRCPRCPS, 81. *Prof Exp:* Dir res & med educ psychiat, Saskatchewan Hosp, Weyburn, 68-69, NBattleford, 69-71; coordr res & lectr, 72, asst prof, 72-76, dir res, 76-79, assoc prof, 81-85, PROF, DOUGLAS HOSP RES CTR, MCGILL UNIV, 85-, DIR PSYCHIAT, 80- *Concurrent Pos:* Chmn, Prog Comt, Can Col Neuropsychopharmacol, 79-; temp adv, WHO, 81- *Mem:* AAAS; Can Col Neuropsychopharmacol; Collegium Internationale Neuropsychopharmacologicum; Am Psychiat Asn; NY Acad Sci. *Res:* Clinical psychopharmacology; neuroendocrine effects of psychotropic drugs in schizophrenia and tardive dyskinesia; new psychotropic drugs. *Mailing Add:* Dir Sci Douglas Hosp Res Ctr 6875 LaSalle Blvd Verdun PQ H4H 1R3 Can

NAIR, VASU, b Suva, Fiji Islands, Jan 7, 39; m 78; c 1. NUCLEOSIDE & NUCLEOTIDE CHEMISTRY, ANTIVIRAL COMPOUNDS. *Educ:* Univ Otago, NZ, BS, 63; Univ Adelaide, PhD(org chem), 66. *Prof Exp:* Res assoc chem, Univ Ill, Urbana, 67-68; res fel, Harvard Univ, 68-69; from asst prof to assoc prof, 69-79, PROF CHEM, UNIV IOWA, 80- *Concurrent Pos:* Res grants & awards, NSF, US Army Med Res, Am Diabetes Asn, Am Cancer Soc & NIH; distinguished scholar award, Univ Adelaide, Australia, 87. *Mem:* Am Chem Soc; AAAS; Int Union Pure & Appl Chem; Int Soc Antiviral Res. *Res:* Synthetic organic chemistry; nucleosides and nucleotides; antiviral compounds. *Mailing Add:* Dept Chem Univ Iowa Iowa City IA 52242

NAIR, VELAYUDHAN, b India, Dec 29, 28; US citizen; m 57; c 3. PHARMACOLOGY. *Educ:* Benares Univ, BPharm, MS, 48; Univ London, PhD(med), 56, DSc, 76. *Prof Exp:* Res assoc pharmacol, Col Med, Univ Ill, Chicago, 56-58; asst prof, Univ Chicago, 58-63; assoc prof, 63-66, PROF PHARMACOL & THERAPEUT, CHICAGO MED SCH, 66-, VCHMN DEPT, 71-, DEAN SCH GRAD & POSTDOCTORAL STUDIES, 76- *Concurrent Pos:* Dir lab neuropharmacol & biochem, Psychiat Inst, Michael Reese Hosp, 63-66, dir therapeut res, 66-70. *Honors & Awards:* Morris Parker Award, 72. *Mem:* Radiation Res Soc; Soc Toxicol; Int Soc Chronobiol; Int Brain Res Orgn; fel NY Acad Sci; Am Soc Pharmacol & Exp Therapeut; fel Am Col Clin Pharmacol. *Res:* Blood-brain barrier; effects of environmental toxicants in pregnancy on biochemical and functional development in the offspring; radiation effects on the nervous system; radiation pharmacology; circadian rhythms in drug action. *Mailing Add:* Sch Grad & Postdoctoral Studies Chicago Med Sch Univ Health Sci North Chicago IL 60064

NAIR, VIJAY, b Konny, India, Oct 5, 41; m 75; c 2. ORGANIC SYNTHESIS. *Educ:* Univ Kerala, BSc, 60; Banaras Hindu Univ, MSc, 62, PhD(org chem), 67; Univ BC, PhD, 69. *Prof Exp:* Fel org synthesis, Univ Chicago, 69-71, Univ Toronto, 71-72 & Columbia Univ, 72-74; SR RES CHEMIST, LEDERLE LABS, DIV AM CYANAMID CO, 74- *Concurrent Pos:* Vis scholar, Columbia Univ, 78-79. *Mem:* Am Chem Soc; Royal Soc Chem. *Res:* Total synthesis of modified steroids; steroidal alkaloids; macrolides; methods for thiocarbamylation; synthesis of sulfur and nitrogen heterocycles; glycosides; polyanionic immunomodulators; circular dichroism studies; enzymatic reactions of organic compounds. *Mailing Add:* Regional Res Lab CSIR Trivandrum 695019 Kerala State India

NAIRN, ALAN EBEN MACKENZIE, b Newcastle on Tyne, Eng, Sept 9, 27; m; c 4. GEOLOGY. *Educ:* Univ Durham, BSc, 51; Glasgow Univ, PhD(geol), 54. *Prof Exp:* Asst geophys, Cambridge Univ, 54-55; asst, King's Col, Univ Durham, 56-58, Turner & Newall fel, 58-62, lectr, 62-65; vis prof, 63-64, from assoc prof to prof geol & geophys, Case Western Reserve Univ, 65-73; PROF GEOL, UNIV SC, 73- *Concurrent Pos:* Guest prof, Univ Bonn, 65-66, NICE, 75, Catana, 86, Naples, 90. *Honors & Awards:* Brit Geol Soc Lyell Fund, 63. *Mem:* Geol Soc London; Am Geol Soc. *Res:* Paleomagnetism of Mediterranean rocks for interpretation of continental drift, megatectonic deformation; paleoclimates and origin of glaciations; development and changes of climatic belts; regional geology; paleogeographic syntheses. *Mailing Add:* Earth Sci & Resources Inst Univ SC Columbia SC 29208

NAIRN, JOHN GRAHAM, b Toronto, Ont, Aug 23, 28; m 54; c 4. PHARMACEUTICS, PHYSICAL PHARMACY. *Educ:* Univ Toronto, BScPharm, 52; State Univ NY Buffalo, PhD(chem), 59. *Prof Exp:* Retail pharmacist, 52-54; from asst prof to assoc prof, 58-73, PROF PHARM, UNIV TORONTO, 73- *Concurrent Pos:* Grants, Can Found Adv Pharm, 61-65, Nat Res Coun Can, 63-66, Nu Chapter, Rho Phi Fraternity, 62 & 65, Univ Toronto, 64-65, Med Res Coun Can, 67, 83-, Ont Ministry Natural Resources, 80-83 & Ont Ministry Health, 81-83; mem, Pharm Exam Bd, Can. *Mem:* Asn Faculties Pharm Can; Ont Col Pharmacists. *Res:* Ion exchange resins in pharmacy; kinetics of drug decomposition and stabilization; microencapsulation; prolonged release drugs. *Mailing Add:* Fac of Pharm Univ of Toronto 19 Russell St Toronto ON M5S 1A1 Can

NAIRN, RODERICK, b Dumbarton, Scotland, Mar 25, 51; US citizen; m 71; c 1. MOLECULAR IMMUNOLOGY, CELL SURFACE BIOCHEMISTRY. *Educ:* Univ Strathclyde, Scotland, BSc, 73; Univ London, Eng, PhD(biochem), 76. *Prof Exp:* Scholar biochem, Med Res Coun, Nat Inst Med Res, London, Eng, 73-76; res fel microbiol & immunol, Albert Einstein Col Med, New York, 76-81; asst prof, 81-86, DIR, STUDENT BIOMED RES PROGS, 89- *Mem:* Biochem Soc Eng; Am Chem Soc; Am Asn Immunologists; AAAS; Am Soc Microbiol; Am Soc Cell Biol. *Res:* Structure and function of immunologically relevant proteins and genes. *Mailing Add:* Dept Microbiol & Immunol M6749/0620 Univ Mich Sch Med 1301 E Catherine Ann Arbor MI 48109-0620

NAISMITH, JAMES POMEROY, b Dalllas, Tex, Aug 4, 36; m 57; c 4. SURFACE WATER TREATMENT, WATER TRANSMISSION. *Educ:* Cornell Univ, BCE, 58, MS, 59. *Prof Exp:* Fac instr hydrol engr, Cornell Univ, 59-60; asst engr, State Calif Water Pollution Control, 60-61; from asst engr to pres, Naismith Engrs, Inc, 61-89; MGR & DIST ENGR, SAN PATRICIO MUNIC WATER DIST, 89- *Concurrent Pos:* Adj prof, Tex A&M Univ, 74-88; chair comts, Am Soc Civil Engrs, 80-, dir, 83-85. *Mem:* Am Soc Civil Engrs; Am Water Works Asn. *Res:* Multiple reports in technical and planning areas involving wasterwater treatment, water treatment, regionalized facilities, surface water flow and other civil engineering topics. *Mailing Add:* 11153 Jackson Terr Corpus Christi TX 78410

NAISTAT, SAMUEL SOLOMON, b Worcester, Mass, Mar 6, 17; m 42; c 3. PHYSICAL CHEMISTRY. *Educ:* Worcester Polytech Inst, BS, 37, MS, 39; Univ Wis, PhD(chem), 44. *Prof Exp:* Asst physics, Worcester Polytech Inst, 37-39; asst chem, Univ Wis, 39-41, instr physics, 43-44; res engr, Westinghouse Elec Corp, NJ, 44-45; res chemist, Congoleum-Nairn, Inc, 45-50; res chemist, Buffalo Electro-Chem Co, Inc, 50-55; group leader & asst to mgr, Becco Chem Div, FMC Corp, 55-57, sect mgr, Inorg Res & Develop Dept, 57-61, proj evaluator & tech consult, 61-63; supvr propulsion tech, Rocketdyne Solid Rocket Div, NAm Aviation, Inc, Tex, 63, res specialist, 64-65; from asst prof to assoc prof, 65-74, PROF CHEM, STEPHEN F AUSTIN STATE UNIV, 74- *Mem:* Am Chem Soc. *Res:* Physical chemistry of macromolecules; manufacture and applications of peroxygen chemicals; chemistry and energetics of propellants; energy conversion processes. *Mailing Add:* 1819 Heather Nacogdoches TX 75961

NAITO, HERBERT K, b Honolulu, Hawaii, Nov 25, 42; m 88. CLINICAL BIOCHEMISTRY, PHYSIOLOGY. *Educ:* Iowa State Univ, PhD(physiol), 71; Lake Erie Col, MBA, 88. *Prof Exp:* HEAD, SECT LIPIDS, NUTRIT & METAB DIS, CLEVELAND CLIN FOUND, 71- *Concurrent Pos:* Clin consult, dept cardiol, Cleveland State Univ, clin prof, dept chem; chmn, Lab Standardization Panel, Nat Cholesterol Educ Prog, NIH; bd dir, Am Asn Clin Chem, Am Col Nutrit, Nat Acad Clin Biochemists. *Honors & Awards:* George Grannis Award, Nat Acad Clin Biochem, 81. *Mem:* Fel Nat Acad Clin Biochem; fel Am Col Nutrit; Am Asn Clin Chem; Soc Exp Biol & Med; Am Inst Nutrit; Endocrine Soc; fel Asn Clin Scientists. *Res:* Nutrition; cholesterol; heart disease. *Mailing Add:* Chief Clin Chem Vet Admin Med Ctr 10701 East Blvd Cleveland OH 44106

NAITOH, PAUL YOSHIMASA, b Japan, Feb 1, 31; US citizen; m 59; c 2. PSYCHOPHYSIOLOGY. *Educ:* Yamaguchi Univ, BA, 53; Univ Minn, MA, 56, PhD(psychol), 64. *Prof Exp:* Asst psychol, Univ Minn, 53-56, psychiat, 57-58 & vet radiol, 58-61; head psychophysiol lab, Neuropsychiat Inst, Univ Calif, Los Angeles, 65-67, staff psychologist, Inst & asst prof med psychol, Univ, 66-67; head, Behav Res Br, Psychophysiol Div, US Navy Med Neuropsychiat Res Unit, 67-74, head, Behav Res Br, Psychophysiol Div, Naval Health Res Ctr, 74-79, SUPVRY RES PSYCHOLOGIST, 79- *Concurrent Pos:* Trainee, Nat Inst Ment Health interdisciplinary res training prog, 64-66; res assoc, Nat Ctr Sci Res, France, 74-75; liaison psychologist, Off Naval Res, Tokyo, 80-81; adj prof, US Int Univ, 81-83. *Mem:* Am Electroencephalographic Soc; Biomet Soc; Asn Psychophysiol Study Sleep. *Res:* Psychophysiological analyses of electroencephalography; psychophysiological correlates of alcoholism; psychophysiology of sleep and sleep loss. *Mailing Add:* Naval Health Res Ctr PO Box 85122 San Diego CA 92138

NAITOVE, ARTHUR, b New York, NY, Mar 25, 26; m 46; c 5. SURGERY, PHYSIOLOGY. *Educ:* Dartmouth Col, BA, 45; NY Univ, MD, 48; Am Bd Surg, dipl, 58. *Prof Exp:* Asst chief surg, White River Junction Vet Admin Hosp, Vt, 58; from instr to assoc prof, 59-81, PROF SURG, DARTMOUTH MED SCH, 81-, DIR SURG LABS, 71- *Concurrent Pos:* USPHS fel physiol, Dartmouth Med Sch, 59-61; attend, White River Junction Vet Admin Hosp, 58-, consult surg, 70-; prin investr, USPHS res grant, Dartmouth Med Sch, 61-76; consult, Hitchcock Clin, Hanover, NH, 70-; assoc dean, Acad & Student Affairs, Dartmouth Med Sch, Hanover, NH, 80. *Mem:* NY Acad Sci; Am Gastroenterol Asn; fel Am Col Surg; New Eng Surg Soc. *Res:* Abdominal surgery; gastrointestinal tract and physiology of gastrointestinal tract, particularly relating to hemodynamic events, their control and their influence on secretory processes. *Mailing Add:* 20 Rip Rd Hanover NH 03755-1614

NAJAFI, AHMAD, b Isfahan, Iran, Jan 25, 50; m 81. NUCLEAR MEDICINE. *Educ:* Nat Univ Iran, BS, 73; Univ Manchester, England, MS, 75; Univ Bath, England, PhD (chem), 78. *Prof Exp:* ASST PROF RADIOCHEM, UNIV TEX MED BR, 83-; ASST PROF RADIO CHEM, UNIV SOUTHERN CALIF, 88. *Mem:* Am Chem Soc; Soc Nuclear Med. *Res:* Development of new radiopharmaceuticals for use in clinical imaging of human organs. *Mailing Add:* 16826 Tulsa St Granada Hills CA 91344-5038

NAJAR, RUDOLPH MICHAEL, b San Fernando, Calif, June 11, 31; m 70; c 4. MATHEMATICS, PHYSICS. *Educ:* St Mary's Col Calif, BS, 54; Univ Calif, Berkeley, MA, 61; Univ Notre Dame, MS, 62, PhD(math), 69. *Prof Exp:* From instr to asst prof math, St Mary's Col Calif, 67-70, chmn dept, 69-70; from asst prof to assoc prof, Univ Wis-Whitewater, 70-81, prof math, 81-90, assoc dean, Col Lett & Sci, 79-90; PROF MATH, CALIF STATE UNIV, FRESNO, 90- *Mem:* Am Math Soc; Soc Advan Chicanos & Native Am Sci. *Res:* Homological algebra and category theory; elementary number theory. *Mailing Add:* Dept Math Calif State Univ Fresno Fresno CA 93740-0108

NAJARIAN, HAIG HAGOP, b Nashua, NH, Jan 5, 25; m 57; c 3. PARASITOLOGY. *Educ:* Univ Mass, BS, 48; Boston Univ, MA, 49; Univ Mich, PhD(zool), 53. *Prof Exp:* Asst biol, Boston Univ, 49; asst zool, Univ Mich, 49-51; asst prof biol, Northeastern Univ, 53-55; assoc res parasitologist, Parke, Davis & Co, 55-57; scientist, Bilharziasis Control Proj, WHO, Iraq, 58-59; USPHS trainee trop med & parasitol, Med Br, Univ Tex, 59-60, asst prof microbiol, 60-66; assoc prof, 66-68, chmn div sci & math, 67-71, chmn dept biol, 71-75, PROF BIOL, UNIV SOUTHERN MAINE, 68- *Mem:* AAAS; Am Soc Parasitol; Am Micros Soc; Am Soc Trop Med & Hyg. *Res:* Life histories of digenetic trematodes; morphology of aspidogastrid trematodes and parasitic copepods; experimental chemotherapy of schistosomiasis, malaria, amebiasis, paragonimiasis and intestinal helminths; biharziasis control; ecology of bulinid snails; filariasis; haemobartonellae; haemoflagellates; experimental amebiasis; author of books Patterns of Medical Parasitology and Sex Lives of Animals Without Backbones. *Mailing Add:* Dept of Biol Univ of Southern Maine Portland ME 04103

NAJARIAN, JOHN SARKIS, b Oakland, Calif, Dec 22, 27;; c 4. SURGERY. *Educ:* Univ Calif, AB, 48, MD, 52; Am Bd Surg, dipl. *Prof Exp:* From intern to resident surg, Sch Med, Univ Calif, 52-60; prof surg & vchmn dept, dir surg res labs & chief transplantation serv, Sch Med, Univ Calif, San Francisco, 63-67; PROF SURG & CHMN DEPT, COL MED SCI, UNIV MINN, MINNEAPOLIS, 67- *Concurrent Pos:* NIH spec res fel immunopath, Sch Med, Univ Pittsburgh, 60-61; NIH assoc & sr fel tissue transplantation immunol, Scripps Clin & Res Found, La Jolla, Calif, 61-63; Markle Award, 64-69; NIH spec consult clin res training comt, Nat Inst Gen Med Sci, 65-69. *Mem:* AAAS; fel Am Col Surgeons; Soc Exp Biol & Med; Am Soc Exp Path; Am Fedn Clin Res. *Mailing Add:* Univ Minn Hosps 516 Del St SE Box 195 Minneapolis MN 55455

NAJJAR, TALIB A, b Baghdad, Iraq, July 1, 38; US citizen; m 65; c 4. DENTISTRY, MAXILLOFACIAL SURGERY. *Educ:* Univ Baghdad, BDS, 60; Ala Univ, DMD, 72; NY Univ, MSc, 65; McGill Univ, PhD(path), 65; Am Bd Oral & Maxillofacial Surg & Am Bd Oral Path, dipl. *Prof Exp:* Instr oral surg, Univ Baghdad, 60-62; asst teacher, Yale Univ, 63-65; asst teacher oral path, McGill Univ, 65-67; asst prof, Montreal Univ, 67-70, Ala Univ, 70-73; PROF MAXILLOFACIAL SURG, COL MED & DENT NJ, 76- *Concurrent Pos:* Intern, Yale New Haven Hosp, 63-64, residency, 64-65; Med Res Coun Can fel, 65-67. *Mem:* Int Asn Dent Res; fel Int Asn Maxillofacial Surgeons; Int Asn Oral Pathologists; Am Asn Maxillofacial Surgeons; fel Am Acad Oral Path. *Res:* Experimental osteomycitis; electron microscopy of oral and maxillofacial pathologic lesion bone healing and remodeling both clinical and experimental. *Mailing Add:* Univ Med & Dent NJ Med Sch 150 Bergen St Newark NJ 07103-2406

NAJJAR, VICTOR ASSAD, b Zalka, Lebanon, Apr 15, 14; nat US; div; c 3. PEDIATRICS, BIOCHEMISTRY. *Educ:* Am Univ, Beirut, MD, 35; Am Bd Nutrit, dipl. *Prof Exp:* From instr to assoc prof pediat, Harriet Lane Home, Johns Hopkins Univ, 39-57; prof microbiol & head dept, Sch Med, Vanderbilt Univ, 57-68; Am Cancer Soc prof molecular biol, prof pediat & chmn div protein chem, Sch Med, 68-84, EMER PROF PEDIAT & MOLECULAR BIOL, TUFTS UNIV, 84- *Concurrent Pos:* Nat Res Coun fels, Sch Med, Washington Univ, 46-48 & Sch Biochem, Cambridge Univ, 48-49; Irving McQuarrie lectr, 57; ed-in-chief, Molecular & Cellular Biochem, 72-82; Rosie & Max Varon vis prof, Weizman Inst Sci, 88. *Honors & Awards:* Mead Johnson Award, 51; Fulbright-Hays Award, 76. *Mem:* NY Acad Sci; Am Soc

Biol Chemists; Am Pediat Soc; Am Soc Clin Invest; Pediat Res Soc. *Res:* Vitamin metabolism and human requirement; mammalian and bacterial enzymology; mechanism of enzyme action; immunochemistry. *Mailing Add:* Div Protein Chem Tufts Univ Sch Med Boston MA 02111

NAKA, F(UMIO) ROBERT, b San Francisco, Calif, July 18, 23; m 49; c 4. ELECTRONICS, SPACE SYSTEMS. *Educ:* Univ Mo, BS, 45; Univ Minn, MS, 47; Harvard Univ, ScD(elec eng, electron optics), 51; Harvard Grad Sch Bus Admin, cert, 67. *Prof Exp:* Asst math, Univ Mo, 44-45; instr elec eng, Univ Minn, 45-47; staff mem radar develop, Lincoln Lab, Mass Inst Technol, 51-54, assoc group leader, 54-56, group leader, 56-59; dept head, Mitre Corp, Mass, 59-60, assoc tech dir, Appl Sci Labs, 60-62, tech dir, 62-68, chief scientist, 68-69; dep under secy Air Force for space systs, US Air Force, 69-72; dir, Detection & Instrumentation Systs, Raytheon Co, 72-75; chief scientist, US Air Force, 75-78; corp vpres, Sci Applications, Inc, 78-82; dir, Simmonds Precision Prods, Inc, 78-84; vpres eng & planning, GTE Govt Systs Corp, 82-88; PRES, CERA, INC, 88-; DIR, AEROSPACE CORP, 88- *Concurrent Pos:* Mem, Space Prog Adv Coun, NASA, 70-77, Air Force Studies Bd, Nat Acad Sci, 72-75, 79-82 & Air Force Sci Adv Bd, 75-78, 84-; dir, Hercules Aerospace Co, 84-87; dir, CAE Link Corp, 88-; vis scholar, Ctr Electromagnetics Res, Northeastern Univ, 88- *Mem:* Sr mem Inst Elec & Electronic Engrs; Sigma Xi; fel Explorers Club; AAAS; NY Acad Sci. *Res:* Radar; space systems; electron optics; electromagnetic wave propagation; radar techniques. *Mailing Add:* Ctr Electromagnetics Res Inc Six Edmonds Rd Concord MA 01742-2649

NAKADA, HENRY ISAO, b Los Angeles, Calif, Oct 12, 22; m 45; c 3. BIOCHEMISTRY. *Educ:* Temple Univ, BA, 48, PhD, 53. *Prof Exp:* Res assoc, Inst Cancer Res, Philadelphia, 50-54; mem, Scripps Clin & Res Found, 54-62; from assoc prof to prof, 62-78, EMER PROF BIOCHEM, UNIV CALIF, SANTA BARBARA, 62- *Mem:* Am Soc Biol Chemists; Soc Exp Biol & Med; Am Chem Soc. *Res:* Mucopolysaccharide, carbohydrate and amino acid metabolism; commercial fishing; research on octopus; oil spills; public relations of fishermen and Alaska's natives. *Mailing Add:* PO Box 908 Homer AK 99603

NAKADA, MINORU PAUL, b Los Angeles, Calif, Jan 15, 21; m 53; c 3. PHYSICS. *Educ:* Univ Calif, AB, 47, PhD(cosmic rays, physics), 52. *Prof Exp:* Physicist, Lawrence Radiation Lab, Univ Calif, 52-61 & Jet Propulsion Lab, Calif Inst Technol, 61-62; PHYSICIST, GODDARD SPACE FLIGHT CTR, NASA, GREENBELT, 62- *Mem:* Am Phys Soc; Am Geophys Union. *Res:* Space plasma physics. *Mailing Add:* 17928 Pond Rd Ashton MD 20861

NAKADA, YOSHINAO, b Los Angeles, Calif, Mar 14, 18; m 44; c 2. PHYSICS. *Educ:* Calif Inst Technol, BS, 40, MS, 41. *Prof Exp:* Chief chemist, Nobell Res Found, 46-48; res physicist, West Precipitation Corp, 48-55, Magnavox Co, 55-56, Hughes Aircraft Co, 56-59, A C Spark Plug Div, Gen Motors Corp, 59-61 & Hughes Aircraft Co, 61-78; RETIRED. *Mem:* Am Phys Soc. *Res:* Real time data processing; electrical discharge; infrared. *Mailing Add:* 4227 Don Mariano Dr Los Angeles CA 90008

NAKADOMARI, HISAMITSU, b Japan, Sept 15, 35; US citizen; m 67; c 2. ELECTROCHEMISTRY. *Educ:* Kans State Col of Pittsburgh, BA, 61; Ft Hays Kans State Col, MS, 65; Colo State Univ, PhD(chem), 74. *Prof Exp:* Asst prof chem, St Gregory's Col, 65-68; sr res assoc bioelectrochem, Brookhaven Nat Lab, 74-76; asst prof chem, George Mason Univ, 76-81; vpres res, Safety Devices, Inc, 81-85; CONSULT, 85- *Concurrent Pos:* Fel, Colo State Univ, 73-74. *Mem:* Sigma Xi; Am Chem Soc; AAAS. *Res:* Ion transport phenomena in lipid bilayer membranes; studies of the electrode-solution interfaces; thermodynamic studies of the solvent effects on the properties of electrolyte solutions; the investigation of new ion selective electrodes; electrochemical studies of corrosion inhibition. *Mailing Add:* 8704 Stonewall Rd Manassas VA 22110

NAKAGAWA, SHIZUTOSHI, b Toyonaka, Osaka, Japan, July 7, 47; m; c 4. BIOLOGY, CULTURE SYSTEMS. *Educ:* Nikon Univ, BA, 71, MA, 73; Hiroshima Univ, PhD(microbiol), 79. *Prof Exp:* Jr scientist, 73-76, sr scientist, 76-85, DIR, TOXICOL, PHARMACOL & CELL BIOL, WAKUNAGA PHARMACEUT CO, LTD, 85- *Concurrent Pos:* Guest scientist, Nat Inst Arthritis, Diabetes, Digestive & Kidney Dis, NIH, 80-82. *Mem:* Am Soc Cell Biol; Am Tissue Cult Asn. *Res:* Bioactive substances in garlic; demonstrated various biological activities on biosynthetic human-EGF using cultured cells and neonatal animals. *Mailing Add:* 4-4-21 Hattore-Minami-Machi Toyonaka Osaka 560 Japan

NAKAGAWA, YASUSHI, ENZYME MECHANISMS, BIOORGANIC CHEMISTRY. *Educ:* Univ Calif, PhD(biochem), 72. *Prof Exp:* DIR, KIDNEY STONE LAB, UNIV CHICAGO, 79- *Res:* Protein chemistry. *Mailing Add:* Nephrol Dept Box 453 Univ Chicago Med Sch 950 E 59th St Chicago IL 60637

NAKAHARA, SHOHEI, b Hiroshima. Japan, Jan 3, 42; m 78; c 2. MATERIALS SCIENCE. *Educ:* Hiroshima Univ, BE, 65, ME, 67; Stevens Inst Technol, PhD(metall), 73. *Prof Exp:* MEM TECH STAFF, BELL LABS, 73- *Mem:* Electrochem Soc; Sigma Xi. *Res:* Electron microscopy; electrodeposition; thin film phenomena. *Mailing Add:* 2C-144 AT&T Bell Labs 600 Mountain Ave Murray Hill NJ 07974

NAKAI, SHURYO, b Kanazawa, Japan, Dec 13, 26; m 52; c 3. FOOD CHEMISTRY, PROTEIN CHEMISTRY. *Educ:* Univ Tokyo, BSc, 50, PhD(dairy tech), 62. *Prof Exp:* Res chemist, Okayama Plant, Meiji Milk Prod Co Ltd, 50-51, sect head dairy chem, Cent Res Lab, Meiji Milk Prod Co, 52-62; res assoc, Univ Ill, 62-66; asst prof dairying, 66-70, assoc prof food chem, 70-75, PROF FOOD CHEM, UNIV BC, 75- *Concurrent Pos:* Expert comt dairy prods, Can Comt on Food, 78-; grant selecting comt, Natural Sci Eng Coun, Can, 85-88. *Honors & Awards:* W J Eva Award, Can Inst Food Sci Technol, 82; Res Award, Am Egg Bd, 82; Killam Res Prize, 86. *Mem:* Am Dairy Sci Asn; fel Can Inst Food Sci & Technol; Inst Food Technol; Am Chem Soc; Soc Agr Chem Japan. *Res:* Chemistry of food proteins; chemical studies on food products. *Mailing Add:* Dept Food Sci Univ BC 6650 NW Marine Dr Vancouver BC V6T 2A2 Can

NAKAJIMA, MOTOWO, b Tokyo, Japan, Sept 16, 51; m 76; c 2. TUMOR BIOLOGY, TUMOR BIOCHEMISTRY. *Educ:* Univ Tokyo, BS, 75, MS, 78, PhD, 81. *Prof Exp:* Proj investr tumor biol, 81-82, res assoc & res instr, 82-86, ASST PROF TUMOR BIOL, UNIV TEX M D ANDERSON CANCER CTR, 86- *Concurrent Pos:* Asst prof, Univ Tex Grad Sch Biomed Sci, 86- *Mem:* Am Asn Cancer Res; Am Soc Cell Biol; Japanese Biochem Soc; Tissue Cult Asn; NY Acad Sci; AAAS. *Res:* Mechanisms of tumor metastasis; tumor degradative enzymes; tumor-normal cell interactions; roles of extracellular sulfated glycans in tumor invasion and metastasis. *Mailing Add:* Dept Tumor Biol Box 108 Univ Texas M D Anderson Cancer Ctr 1515 Holcombe Blvd Houston TX 77030

NAKAJIMA, NOBUYUKI, b Tokyo, Japan, Nov 3, 23; c 3. POLYMER CHEMISTRY, POLYMER PHYSICS. *Educ:* Univ Tokyo, BS, 45; Polytech Inst Brooklyn, MS, 55; Case Inst Technol, PhD(phys chem), 58. *Prof Exp:* Asst polymer & phys chem, Naval Air Force Res Ctr, Japan, 44-45; prod engr, Chem Div, Osaka Gas Co, 45-51; from res asst to res assoc, Case Inst Technol, 55-60; res chemist, W R Grace & Co, 60-63, sect leader, Polymer Chem Div, 63-66, asst to vpres chem, Res Div, 65-66; tech supvr plastics div, Allied Chem Corp, NJ, 66-67, mgr polymer physics & anal plastics div, 67-71; res fel, Tech Ctr, BF Goodrich Chem Co, 71-84; AT INST POLYMER ENG, COL POLYMER SCI & ENG, 84- *Concurrent Pos:* Assoc ed, Rubber Chem & Technol. *Mem:* Am Chem Soc; Soc Rheol; Am Phys Soc; Sigma Xi. *Res:* Elastomer rheology and processing; polymer solution thermodynamics; rheology of polymer melts and solution; molecular weight distribution of polymers; polymer morphology, processing and structural analysis. *Mailing Add:* Inst Polymer Eng Univ Akron Col Polymer Sci & Eng Akron OH 44325

NAKAJIMA, SHIGEHIRO, b Kobe, Japan, July 3, 31; m 57; c 2. PHYSIOLOGY, NEUROSCIENCE. *Educ:* Univ Tokyo, MD, 55, PhD(physiol), 61. *Prof Exp:* Res fel neurophysiol, Dept Neurol, Col P & S, Columbia Univ, 62-64; asst res zoologist neurophysiol, Brain Res Inst, Univ Calif Los Angeles, 64-65; assoc prof physiol, Juntendo Univ Sch Med, Tokyo, 65-67; res fel physiol, Univ Cambridge, 67-69; from assoc prof to prof biol sci, Purdue Univ, 69-88; PROF PHARMACOL, UNIV ILL COL MED, 88- *Concurrent Pos:* Fulbright travel grant, 62-64; ad hoc mem, Physiol Study Sect, NIH, 81-83; mem sci prog adv comt, NIH & Nat Inst Neurol & Commun Dis & Stroke, 85-86. *Mem:* Physiol Soc; Am Physiol Soc; Soc Neurosci; Biophys Soc; Soc Gen Physiologists. *Res:* Physiology of muscle; physiology and pharmacology of brain neurons; neuroscience. *Mailing Add:* Pharmacol Dept Univ Ill Col Med 835 S Wolcott Ave MC 868 Chicago IL 60612

NAKAJIMA, YASUKO, b Osaka, Japan, Jan 8, 32; m 57; c 2. NEUROBIOLOGY, CELL BIOLOGY. *Educ:* Univ Tokyo, MD, 55, PhD(anat), 62. *Prof Exp:* Instr anat, Univ Tokyo, 62-67; vis res fel zool, Univ Cambridge, 67-69; from assoc prof to prof neurobiol, PURDUE UNIV, WEST LAFAYETTE, 76-88; PROF ANAT & CELL BIOL, UNIV ILL COL MED, CHICAGO, 88- *Concurrent Pos:* Vis res fel anat, Col Physicians & Surgeons, Columbia Univ, 62-64; asst res anat, Med Sch, Univ Calif, Los Angeles, 64-65; vis res, Univ Col, London, 78-79; mem, Marine Biol Lab, Woods Hole. *Mem:* Am Asn Anat; Am Soc Cell Biol; Soc Neurosci; Biophys Soc. *Res:* Neurobiology at the cellular level; electron microscopy, electrophysiology, and tissue culture. *Mailing Add:* Dept Anat & Cell Biol (m/c 512) Univ Ill Col Med 808 S Wood St Chicago IL 60612

NAKAMOTO, KAZUO, b Kobe, Japan, Mar 1, 22; m 50; c 3. PHYSICAL CHEMISTRY. *Educ:* Osaka Univ, BS, 45, DSc, 53. *Prof Exp:* Res asst chem, Osaka Univ, 45-46, res assoc, 46-51, lectr, 51-57, assoc prof, 57; res fel, Clark Univ, 57-58, asst prof, 58-61; from assoc prof to prof, Ill Inst Technol, 61-69; WEHR PROF CHEM, MARQUETTE UNIV, 69- *Concurrent Pos:* Res fel, Iowa State Univ, 53-55; Fulbright exchange scholar, 53. *Honors & Awards:* Alexander Von Humboldt Award, 74; Soc Promotion Sci Award, Japan, 77 & 79; Distinguished Sci Res Award, Sigma Xi, 88. *Mem:* Am Chem Soc; Chem Soc Japan; hon mem Soc Appl Spectros; Sigma Xi. *Res:* Electronic, vibrational and nuclear magnetic resonance spectra. *Mailing Add:* Dept Chem Marquette Univ Milwaukee WI 53233

NAKAMOTO, TETSUO, b Kure-city, Japan, Dec 20, 39; m 80; c 2. NEONATAL NUTRITION. *Educ:* Nihon Univ, Tokyo, DDS, 64; Univ Mich, MS, 66 & MS, 71; Univ NDak, Grand Forks, MS, 69; Mass Inst Technol, PhD(nutrit biochem & metab), 78. *Prof Exp:* Teaching asst physiol, Univ NDak, Grand Forks, 67-68, res asst, 68-69; NIH fel, Univ Mich, 69-72, teaching fel, 71-72; NIH fel, nutrit biochem & metab, Mass Inst Technol, 72-78; asst prof, 78-84, ASSOC PROF PHYSIOL, LA STATE MED CTR, NEW ORLEANS, 84- *Concurrent Pos:* vis prof, Nihon Univ, Tokyo, Japan, 87. *Mem:* Int Asn Dent Res; Am Asn Dent Res; Soc Exp Biol & Med; Am Inst Nutrit; NY Acad Sci; Am Physiol Soc. *Res:* Neonatal nutrition. *Mailing Add:* La State Univ Med Ctr 1100 Fla Ave New Orleans LA 70119-2799

NAKAMOTO, TOKUMASA, b Kohala, Hawaii, July 8, 28; m 50; c 3. BIOCHEMISTRY. *Educ:* Univ Chicago, BA, 56, PhD(biochem), 59. *Prof Exp:* Res assoc biochem, Univ Chicago, 59-62; res assoc, Rockefeller Inst, 62-64, asst prof, 64-65; asst prof, 65-67, ASSOC PROF BIOCHEM, SCH MED, UNIV CHICAGO, 67- *Concurrent Pos:* Mem, Franklin McLean Mem Res Inst. *Res:* Biosynthesis of proteins. *Mailing Add:* Dept Biochem Box 420 Univ Chicago 5801 Ellis Ave Chicago IL 60637

NAKAMURA, EUGENE LEROY, b San Diego, Calif, June 8, 26; m 61; c 1. BIOLOGICAL OCEANOGRAPHY, FISH BIOLOGY. *Educ:* Univ Ill, BS, 50, MS, 51. *Prof Exp:* Res asst zool, Univ Hawaii, 51-56; fishery biologist, Biol Lab, US Bur Com Fisheries, Hawaii, 56-70; DIR, PANAMA CITY LAB, NAT MARINE FISHERIES SERV, 70- *Mem:* AAAS; Am Fisheries Soc; Am Soc Ichthyologists & Herpetologists; Am Soc Limnol & Oceanog; Am Inst Fishery Res Biol (pres, 79-80); Sigma Xi; Am Inst Biol Sci. *Res:* Biology, ecology and fishery of marine fishes. *Mailing Add:* Nat Marine Fish Serv 3500 Delwood Beach Rd Panama City FL 32408-7403

NAKAMURA, KAZUO, microbial genetics, phycology, for more information see previous edition

NAKAMURA, MITSURU J, b Los Angeles, Calif, Dec 17, 26; m 51; c 3. MICROBIOLOGY. *Educ:* Univ Calif, Los Angeles, AB, 49; Univ Southern Calif, MS, 50, PhD(med sci), 56; Am Bd Med Microbiol, dipl, 62. *Prof Exp:* Asst, Sch Med, Univ Calif, 50-52; from asst prof to assoc prof, Northeastern Univ, 52-56; assoc prof, 57-63, prof & chmn dept microbiol, 63-86, EMER PROF MICROBIOL, UNIV MONT, 86- *Concurrent Pos:* Res assoc, Sch Med, Boston Univ, 55-56; responsible investr, Comn Enteric Infections, Armed Forces Epidemiol Bd, Off Surgeon Gen, US Dept Army, 57-62; fel, Sch Med, La State Univ, 59 & 60; fel, Univ Costa Rica, 63; Am Acad Microbiol fel, 67; Nat Acad Sci interacad awards to visit Poland, 76, Yugoslavia, 78 & Hungary, 82, 84, Bulgaria, 85, Czechoslovakia, 87; vis prof, Univ Med Sch, Hungary, 87-89, Yang Ming Med Col, Taipei, Taiwan, 88, Fac Med Sci, Univ West Indies, Trinidad & Tobago. *Mem:* Fel AAAS; fel Am Pub Health Asn; Am Soc Microbiol; Am Soc Trop Med & Hyg; Soc Exp Biol & Med. *Res:* Physiology of Shigella; water pollution; effects of ultraviolet irradiation on bacteria; cultivation of Protozoa; Clostridium perfringens food poisoning. *Mailing Add:* Dept Microbiol Univ Mont Missoula MT 59812

NAKAMURA, ROBERT MASAO, b Los Angeles, Calif, Sept 18, 35; m 66; c 3. REPRODUCTIVE BIOLOGY, BIOPHYSICS. *Educ:* Occidental Col, BA, 59; Univ Southern Calif, MS, 64, PhD(biochem), 68. *Prof Exp:* Res asst biochem, Univ Southern Calif, 59-61, res assoc biophys, Allan Hancock Found, 66-67; asst res biochemist, Dept Obstet & Gynec, Harbor Gen Hosp, Torrance, Calif, 67-68, res biochemist, 68-69; asst prof reprod biol, 69-76, ASSOC PROF OBSTET & GYNEC & PATH & PHYSIOL, SCH MED, UNIV SOUTHERN CALIF, 76- *Concurrent Pos:* Consult, Ford Found, 69-; mem task force human reproduction, WHO; managing ed, Contraception, 69-71. *Mem:* AAAS; Royal Soc Chem; Am Chem Soc; NY Acad Sci; Endocrine Soc. *Res:* Hormonal profiles in normal menstruating women; effects of contraceptive steroids on the serum hormone levels; biophysical properties of cervical mucus; properties of protein hormones; investigating methods to prevent the spread of AIDS. *Mailing Add:* 446 S Boyle Ave Los Angeles CA 90033

NAKAMURA, ROBERT MOTOHARU, b Montebello, Calif, June 10, 27; m 57; c 2. PATHOLOGY, IMMUNOPATHOLOGY. *Educ:* Whittier Col, AB, 49; Temple Univ, MD, 54. *Prof Exp:* Chief clin path, Long Beach Vet Admin Hosp, 59-60; pathologist, Atomic Bomb Casualty Comn, Japan, 60-61; instr path, Sch Med, Univ Calif, Los Angeles, 61-62, asst prof, 62-65; pathologist, St Joseph Hosp, Orange, Calif, 68-69; from assoc prof to prof path & dir clin labs, Orange County Med Ctr, Univ Calif, Irvine, 69-74; HEAD DEPT PATH, HOSP SCRIPPS CLIN, 74-, ADJ PROF PATH, UNIV CALIF, SAN DIEGO, 75- *Concurrent Pos:* USPHS spec fel exp path, Scripps Clin & Res Found, Univ Calif, 65-68; pathologist, Los Angeles County Harbor Gen Hosp, Calif, 61-65; consult dept path, Orange County Gen Hosp, 62-65; prof path, Univ Calif, Irvine, 72-74. *Mem:* AAAS; AMA; fel Am Col Path; fel Am Soc Clin Path; Am Soc Exp Path. *Res:* Clinical pathology; general area of cell proliferation; immunopathology, specifically autoimmune diseases and immunological tolerance. *Mailing Add:* Dept Path Scripps Clin & Res FNON N Torrey Pine Rd La Jolla CA 92037

NAKAMURA, SHOICHIRO, b Osaka, Japan, Oct 20, 35; m 64; c 2. COMPUTATIONAL FLUID DYNAMICS, APPLIED MATHEMATICS. *Educ:* Kyoto Univ, BS, 58, PhD(nuclear eng), 67. *Prof Exp:* Vis res fel nuclear eng, Univ Calif, Berkeley, 66-70; engr nuclear reactor design, Hitachi Ltd, Tokyo, 58-69; fel nuclear eng, 69-70, from asst prof to assoc prof, 70-78, PROF MECH ENG, OHIO STATE UNIV, 78- *Mem:* Am Inst Aeronaut & Astronaut; Soc Indust & Appl Math. *Res:* Nuclear engineering; numerical methods; computational fluid dynamics. *Mailing Add:* Dept Mech Eng Ohio State Univ 206 W 18th Ave Columbus OH 43210

NAKANE, PAUL K, b Yokahama, Japan, Oct 20, 35; m 59; c 3. HISTOCHEMISTRY, ELECTRON MICROSCOPY. *Educ:* Huntingdon Col, BA, 58; Brown Univ, MS, 61, PhD(cytol), 63. *Prof Exp:* Res assoc histochem, Sch Med, Stanford Univ, 63-65; instr path, Univ Mich, Ann Arbor, 65-67, asst prof cell biol in path, 67-68; asst prof, 68-69, assoc prof, 69-72, prof path, Sch Med, Med Ctr, Univ Colo, Denver, 73-; PROF, DEPT CELL BIOL, TOKAI UNIV SCH MED. *Concurrent Pos:* Cancer Inst trainee, 59-63. *Mem:* Am Asn Cell Biol; Histochem Soc (pres, 78-79); Fedn Am Soc Exp Biol; AAAS. *Res:* Ultrastructural localization of antigens by enzyme-labeled antibody. *Mailing Add:* Dept of Cell Biol Tokai Univ Sch Med Isehara Kanagawa Japan

NAKANISHI, KEITH KOJI, b Honolulu, Hawaii, Mar 20, 47; m 71; c 1. SEISMOLOGY. *Educ:* Occidental Col, BA, 69; Univ Calif, Los Angeles, MS, 71, PhD(geophysics & space physics), 78. *Prof Exp:* SEISMOLOGIST, EARTH SCI DEPT, LAWRENCE LIVERMORE NAT LAB, UNIV CALIF, 77- *Mem:* Seismol Soc Am; Am Geophys Union. *Res:* Seismic monitoring for verifying nuclear test ban treaties. *Mailing Add:* L-205 Lawrence Livermore Nat Lab PO Box 808 Livermore CA 94550

NAKANISHI, KOJI, b Hong Kong, May 11, 25; m 47; c 2. ORGANIC CHEMISTRY. *Educ:* Nagoya Univ, BSc, 47, PhD(chem), 54. *Hon Degrees:* DSc, Williams Col, 87. *Prof Exp:* Garioa fel, Harvard Univ, 50-52; asst prof chem, Nagoya Univ, 55-58; prof, Tokyo Kyoiku Univ, 58-63 & Tohoku Univ, Japan, 63-69; PROF CHEM, COLUMBIA UNIV, 69- *Concurrent Pos:* Consult, Syntex/Zoecon Corp, 65- & Lederle Labs, 69-; dir res, Int Centre Insect Physiol & Ecol, Kenya, 69-77; dir, Suntory Inst Bioorg Res, Osaka, 79- *Honors & Awards:* Chem Soc Japan Award, 54 & 79; Pure Chem Award, Chem Soc Japan, 54; Cult Award, Asahi Press, Japan, 68; E Guenther Award, Am Chem Soc, 78; Centenary Medal, British Chem Soc, 79; E E Smissman Medal, Univ Kansas, 79; Remsen Award Am Chem Soc, Maryland Sect, 81; Res Achievement Award, Am Soc Pharmacol; Alcon Award, Opthalmology, 86; Paul Karrer Gold Medal, Univ Zurich, 86. *Mem:* Am Chem Soc; Chem Soc France; Royal Soc Chem; Chem Soc Japan; fel Am Acad Arts & Sci; Swiss Chem Soc. *Res:* Isolation and structural studies of physiologically active natural products; applications of spectroscopy to structure determination; visual pigments. *Mailing Add:* Dept Chem Columbia Univ New York NY 10027

NAKANO, JAMES HIROTO, microbiology, experimental pathology; deceased, see previous edition for last biography

NAKASHIMA, TADAYOSHI, b Yokkaichi, Japan, Dec 1, 22; m 47; c 1. BIOCHEMISTRY, EXOBIOLOGY. *Educ:* Nagoya Pharmaceut Col, BP, 43; Taihoku Imp Univ, BS, 46; Kyushu Univ, PhD(biochem), 61. *Prof Exp:* Fel biochem, Univ Hawaii, 62-64; res scientist biochem, Inst Molecular Evolution, 65-73, from res asst prof to res assoc prof, 73-81, RES PROF BIOCHEM, INST MOLECULAR & CELLULAR EVOLUTION, UNIV MIAMI, 81- *Concurrent Pos:* Vis res scientist, Inst Animal Physiol, Univ Bonn, Ger, 66-69. *Mem:* Int Soc Study Origin Life; Am Chem Soc; Sigma Xi. *Res:* Nucleic acid; amino acid interaction, genetic code, peptide synthesis on the model ribosomes, including prebiological chemistry. *Mailing Add:* 7400 SW 159th Terr Miami FL 33157

NAKASONE, HENRY YOSHIKI, b Kauai, Hawaii, July 6, 20; m 48; c 2. HORTICULTURE. *Educ:* Univ Hawaii, BA, 43, MS, 52, PhD(genetics), 60. *Prof Exp:* Asst hort, Agr Exp Sta, Univ Hawaii, 48-52, instr plant propagation, Col Agr & jr horticulturist, Agr Exp Sta, 52-58, asst prof plant propagation & trop pomol & asst horticulturist, 58-60, assoc prof & assoc horticulturist, 60-69, chmn dept, 75-80, prof hort & horticulturist, 69-82, EMER PROF HORT, COL AGR & EMER HORTICULTURIST, AGR EXP STA, UNIV HAWAII, 82-; CONSULT HORTICULTURIST, 82- *Concurrent Pos:* Consult, Heinz Alimentos, Mex, 66-72, Comn Fruit Cult, Mex Govt & DaCosta Bros, Jamaica, 72-, Bank of Mex FIRA Prog, Nat Comn Fruit Crops, Mex, 73-78, pvt indust, Japan, 74, Japanese Prefectural Govt, 82, UN/FAO India prog, 81, SPAC proj, 84, USAID, Port, 83, Peru, 84 & Jamaica, 85, UN/FAO SPAC, 86, USAID Belize, E Caribbean Islands, 87. *Honors & Awards:* Res Excellence Award, Am Soc Hort Sci, 77. *Mem:* Fel AAAS; Am Soc Hort Sci. *Res:* Plant breeding and culture of tropical crops; genetics of tropical crops; development of disease resistance in papaya. *Mailing Add:* Univ Hawaii 4207 B Huanui St Honolulu HI 96816

NAKATA, HERBERT MINORU, b Pasadena, Calif, Mar 10, 30; m 60; c 3. BACTERIOLOGY. *Educ:* Univ Ill, BS, 52, MS, 56, PhD, 59. *Prof Exp:* From instr to assoc prof, 59-71, PROF BACT, WASH STATE UNIV, 71-, CHMN DEPT MICROBIOL, 68- *Mem:* AAAS; Am Soc Microbiol; Sigma Xi. *Res:* Bacterial physiology, particularly the biochemical processes associated with sporulation of aerobic bacilli. *Mailing Add:* Dept of Microbiol Wash State Univ Pullman WA 99164-4233

NAKATANI, ROY E, b Seattle, Wash, June 8, 18; m 55; c 5. FISH BIOLOGY. *Educ:* Univ Wash, BS, 47, PhD(fisheries), 60. *Prof Exp:* Fishery biologist, Fisheries Res Inst, 47-48; res assoc blood chem salmonoids, Univ Wash, 52-58, anal past Alaskan fisheries data, 58-59; biol scientist, Hanford Labs, Gen Elec Co, 59-62, mgr aquatic biol, 62-66; mgr ecol, Pac Northwest Labs, Battelle Mem Inst, 66-70; from assoc prof to prof, 70-88, assoc dir fisheries res inst & prog dir, Div Marine Rescources, 70-88, EMER PROF FISHERIES, UNIV WASH, 88- *Concurrent Pos:* Consult, Wash Pub Power Supply Syst, ARCO & Puget Power & Light; mem panel radioactivity in marine environ & biol effects of ionizing radiation subcomt environ effects, Nat Acad Sci-Nat Res Coun. *Mem:* AAAS; Am Inst Fishery Res Biol (nat secy); Sigma Xi; Int Acad Fishery Sci; Am Fisheries Soc. *Res:* Radiation biology of aquatic organisms; water pollution; nuclear power plant siting; oil pollution. *Mailing Add:* Fisheries Res Inst Univ Wash Seattle WA 98195

NAKATO, TATSUAKI, b Okayama, Japan, Jan 17, 42; m 79; c 1. HYDRAULIC MODELING, SEDIMENT TRANSPORT. *Educ:* Nagoya Univ, Japan, BS, 66, MS, 68; Univ Iowa, PhD(mech & hydraul), 74. *Prof Exp:* Res asst, 71-74, asst res scientist, 75-78, RES SCIENTIST, INST HYDRAUL RES, UNIV IOWA, 78- ; ADJ ASST PROF CIVIL & ENVIRON ENG, 77- *Concurrent Pos:* Adj asst prof hydraul, Inst Hydraul Res, Univ Iowa, 77- *Mem:* Japan Soc Civil Eng; Am Soc Civil Engrs; Sigma Xi. *Res:* Hydraulic models, including pump-intake structures and thermal-discharge schemes; analyzing mathematical models of various types of sediment-transport phenomena. *Mailing Add:* Inst Hydraul Res Univ Iowa Iowa City IA 52242

NAKATSU, KANJI, b Greenwood, BC, June 19, 45; m 71; c 2. PHARMACOLOGY. *Educ:* Univ Alta, BSc, 64, MSc, 68; Univ BC, PhD(pharmacol), 71. *Prof Exp:* Student pharmacol, Univ Alta, 66-68, Univ BC, 68-71; fel, Stanford Univ, 71-73; from asst prof to assoc prof, 73-86, PROF PHARMACOL, QUEEN'S UNIV, 86- *Concurrent Pos:* Consult, Dept Med, Div Respirology, Queen's Univ, 76-; vis scientist, Baylor Col Med, 79-80, Univ of BC, 87-88. *Mem:* Can Pharmacol Soc; Am Soc Pharmacol & Exp Therapeut; Soc Toxicol Can. *Res:* Mechanism of action and disposition of organonitrate vasodolators; drug distribution system to improve patient compliance. *Mailing Add:* Dept Pharmacol & Toxicol Queen's Univ Kingston ON K7L 3N6 Can

NAKATSUGAWA, TSUTOMU, b Kochi-Ken, Japan, Apr 17, 33; m 65; c 3. ENVIRONMENTAL & INSECTICIDE TOXICOLOGY. *Educ:* Univ Tokyo, BAgr, 57; Iowa State Univ, MS, 61, PhD(insect toxicol), 64. *Prof Exp:* Asst entomologist, Nat Inst Agr Sci, Tokyo, 57-60; res assoc insect toxicol, Iowa State Univ, 64-68; from asst to assoc prof, 68-76, PROF INSECTICIDE TOXICOL, COL ENVIRON SCI & FORESTRY, STATE UNIV NY, 76- *Mem:* AAAS; Am Chem Soc; Entom Soc Am; Soc Toxicol; Soc Environ Toxicol & Chem; Am Asn Study Liver Dis. *Res:* Detoxication and health effects of insecticides and environmental toxicants with emphasis on liver toxicology. *Mailing Add:* Fac Environ & Forest Biol Col Environ Sci & Forestry Syracuse NY 13210

NAKAYAMA, FRANCIS SHIGERU, b Honolulu, Hawaii, July 1, 30; m; c 1. SOIL CHEMISTRY. *Educ:* Univ Hawaii, BS, 52; Iowa State Univ, MS, 55, PhD(soil fertil), 58. *Prof Exp:* Asst, Iowa State Univ, 53-58; CHEMIST, US WATER CONSERV LAB, USDA, 58- *Concurrent Pos:* vis prof, Univ Calif Davis. *Mem:* Fel AAAS; fel Am Inst Chem; fel Am Soc Agron; Am Chem Soc; Asn Adv Indust Crops. *Res:* Arid zone crops (guayule rubber production); interrelation between water quality, trickle irrigation operation and soil water movement; neutron soil moisture equipment. *Mailing Add:* US Water Conserv Lab 4331 E Broadway Phoenix AZ 85040

NAKAYAMA, ROY MINORU, plant breeding, horticulture; deceased, see previous edition for last biography

NAKAYAMA, TAKAO, b Sacramento, Calif, Sept 19, 13; m 41; c 2. PHYTOPATHOLOGY, HORTICULTURE. *Educ:* Univ Calif, BS, 37. *Prof Exp:* Plant pathologist, Ohara Inst Agr Res, 40-46; agriculturist & dir, Chofu Hydroponic Farm, Japan, 46-61; horticulturist, US Army Procurement Agency Japan, 62-74; horticulturist, Pac Air Force Procurement Ctr, Japan, 74-81; RETIRED. *Concurrent Pos:* Agr consult, Land Auth, Govt PR, 59, Repub Korea, 64-65 & Repub China, Thailand & SVietnam, 66-68. *Mem:* Am Phytopath Soc; Phytopath Soc Japan. *Res:* Wheat scab and vegetable diseases; vegetable production; hydroponic vegetable production; technical and economical implications surrounding production of fresh fruits and vegetables in Japan and far eastern countries. *Mailing Add:* 1-12-8 Nishishiba Kanazawa-Ku Yokohama 236 Japan

NAKAYAMA, TOMMY, b Ballico, Calif, Mar 15, 28; m 57; c 4. FOOD SCIENCE. *Educ:* Univ Calif, Berkeley, BS, 51, MS, 52, PhD(agr chem), 57. *Prof Exp:* Jr specialist, Univ Calif, Berkeley, 54-57, asst specialist, 57-58; asst prof food sci & technol, Univ Calif, Davis, 59-63; res supvr, Miller Brewing Co, 63-66; assoc prof food sci, Univ Ga, 66-70; prof, Univ Hawaii, 70-77; PROF FOOD SCI & HEAD DEPT, UNIV GA, 77- *Mem:* Fel Inst Food Technologists; Am Chem Soc; Am Soc Advan Sci; Sigma Xi. *Res:* Carotenoids and polyphenolic compounds in foods; agricultural waste utilization; small scale processing of foods. *Mailing Add:* Dept Food Sci & Technol Ga Sta Experiment GA 30223-1797

NAKHASI, HIRA LAL, b Srinagar, Kashmir, India, July 1, 50; US citizen; m 76; c 2. MOLECULAR VIROLOGY, VIRAL-CELL INTERACTION. *Educ:* J&K Univ, India, BS, 68, MS Univ, India, MS, 71, PhD(biochem), 76. *Prof Exp:* Res assoc biochem, Ohio State Univ, Columbus, 74; lectr biochem, MS Univ, Baroda, India, 75-77; vis fel molecular biol, Nat Cancer Inst, NIH, Bethesda, Md, 77-79; staff res assoc, Columbia Univ, NY, 79-82; staff fel, Nat Cancer Inst, NIH, 82-84; sr staff fel molecular biol, 84-89, RES CHEMIST, CTR BIOL EVE RES, FOOD & DRUG ADMIN, BETHESDA, MD, 89- *Concurrent Pos:* Invited lectr, Biotechnol Ctr, MS Univ, Baroda, India, 87, Nat Virol Inst, Pune, India, 88, Nat Inst Immunol, New Delhi, India, 88. *Mem:* Am Soc Biol Chemists; Soc Exp Biol; AAAS; Am Soc Microbiol. *Res:* Viral cell interactions; gene regulation at the transcription and translational level in a simple model system such as rubella virus; the regulation of milk protein gene expression in malignant mammary glands; developmental biology; molecular parasitology. *Mailing Add:* Ctr Biol Eve Res FDA 8800 Rockville Pike Bethesda MD 20892

NAKON, ROBERT STEVEN, b Brooklyn, NY, May 1, 44. INORGANIC CHEMISTRY. *Educ:* DePaul Univ, BS, 65; Tex A&M Univ, PhD(inorg chem), 71. *Prof Exp:* Fel inorg chem, Iowa State Univ, 71-73 & Memphis State Univ, 73-74; from asst prof to assoc prof, 74-84, PROF INORG CHEM, WVA UNIV, 85- *Mem:* Am Chem Soc; AAAS. *Res:* Metal complexes as catalysts for biomimetic reactions; metal ion interactions with "Good's" buffers and other biologically important substances; correlation of reactivity with the structures of metal chelates as intermediates. *Mailing Add:* Dept Chem WVa Univ PO Box 6045 Morgantown WV 26506-6045

NALBANDIAN, JOHN, b Providence, RI, Nov 26, 32; m 70; c 1. ORAL PATHOLOGY, ELECTRON MICROSCOPY. *Educ:* Brown Univ, AB, 54; Harvard Univ, DMD, 58. *Prof Exp:* Res assoc periodont, Sch Dent Med, Harvard Univ, 62-63, assoc, 63-64, asst prof oper dent, 64-69; prof gen dent, Sch Dent Med, Hosp, Univ Conn, Hartford, 69-74, head dept, 69-71; PROF PERIODONT, HEALTH CTR, UNIV CONN, FARMINGTON, 74- *Concurrent Pos:* Fel periodont, Sch Dent Med, Harvard Univ, 58-61; USPHS spec fel, Nat Inst Dent Res, 61-62; actg head dept periodont, Health Ctr, Univ Conn, 74-75; Fogarty Sr Int fel, 77-78. *Mem:* AAAS; Int Asn Dent Res. *Res:* Dental aspects of aging; dental embryology; ultrastructure of oral tissues and of bone; bone resorption; dental caries; dental plaque; periodontal disease. *Mailing Add:* Dept Periodont Univ Conn Health Ctr 263 Farmington Ave Farmington CT 06032

NALDRETT, ANTHONY JAMES, b London, Eng, June 23, 33; m 60; c 3. GEOLOGY. *Educ:* Cambridge Univ, BA, 56, MA, 62; Queen's Univ, MSc, 61, PhD(geol), 64. *Prof Exp:* Geologist, Falconbridge Nickel Mines Ltd, 57-59; fel geochem, Geophys Lab, Carnegie Inst Wash, 64-67; from asst prof to assoc prof, 67-72, PROF GEOL, UNIV TORONTO, 72-, UNIV PROF, 84- *Concurrent Pos:* Chmn, Comn Exp Petrol, Int Union Geol Sci; ed, J Petrol, 73-82; mem, Earth Sci Grant Selection Comt, Nat Res Coun, 75-78, chmn, 77-78; mem, Panel Geochem & Petrol, NSF, 80-82; chmn, comt appl mineral, Lab Mineral Asn, 84. *Honors & Awards:* Barlow Medal, Can Inst Mining & Metall, 74; Dun Derry Medal, Geol Asn Can, 80; Soc Medal, Soc Econ Geologists, 82. *Mem:* Geol Soc Am; Can Inst Mining & Metall; Geol Asn Can; Soc Econ Geol; Mineral Asn Can; Mineral Soc Am. *Res:* Geology and geochemistry of nickel deposits; geology of the Sudbury area; petrology of mafic and ultramafic rocks; experimental geochemistry of sulfide, sulfide-oxide and sulfide-silicate systems. *Mailing Add:* Dept Geol Univ Toronto St George Campus Toronto ON M5S 1A1 Can

NALEWAJA, JOHN DENNIS, b Browerville, Minn, Oct 7, 30; m 59; c 4. WEED SCIENCE. *Educ:* Univ Minn, BS, 53, MS, 59, PhD(agron), 62. *Prof Exp:* PROF WEED SCI, NDAK STATE UNIV, 62- *Mem:* Fel Weed Sci Soc Am (pres, 84); N Central Weed Control Confer (pres, 79). *Res:* Basic and applied aspects of weed science. *Mailing Add:* Crop & Weed Sci Dept NDak State Univ State Univ Sta Fargo ND 58105

NALL, BARRY T, b Bakersfield, Calif, Nov 23, 48. PHYSICAL BIOCHEMISTRY, PROTEIN FOLDING. *Educ:* Stanford Univ, PhD(biochem), 76. *Prof Exp:* ASST PROF BIOCHEM & MOLECULAR BIOL, SCH MED, UNIV TEX, 79- *Mem:* Am Soc Cell Biol; Am Heart Asn; Biophysical Soc. *Mailing Add:* Dept Biochem & Molecular Biol Univ Tex Health 7703 Floyd Curl Dr San Antonio TX 78284-7760

NALL, RAY(MOND) W(ILLETT), b Flaherty, Ky, Nov 21, 39; c 3. LIMNOLOGY, BOTANY. *Educ:* Western Ky Univ, BS, 61; Univ Louisville, PhD(bot), 65. *Prof Exp:* Staff biologist, 65, RESOURCE PROJS MGR, TENN VALLEY AUTHORITY, 73- *Concurrent Pos:* Asst prof, Murray State Univ, 70-72, adj prof, 70-73. *Mem:* Am Soc Limnol & Oceanog; Sigma Xi; Nat Management Asn. *Res:* Water pollution ecology; life histories of deer and wild turkey; improved techniques for management of natural resources. *Mailing Add:* Forestry Bldg Tenn Valley Authority Norris TN 37828

NALLEY, DONALD WOODROW, b Easley, SC, Aug 14, 32; c 2. ELECTRICAL ENGINEERING, COMPUTER SCIENCE. *Educ:* Clemson Univ, BSEE, 61, PhD(elec eng), 71; Univ Ark, MS, 65. *Prof Exp:* Engr, Naval Ord Lab, Md, 61-65 & Union Bleachery, SC, 65; instr elec eng, Clemson Univ, 66-68; consult, Comput Sci Corp, Ala, 69; asst prof elec eng & dir Burton Comput Ctr, McNeese State Univ, 70-76; SUBSYSTS MGR, UNITED SPACE BOOSTERS, INC, 76- *Concurrent Pos:* Burton Found grant, Burton Comput Ctr, McNeese State Univ, 71-74. *Mem:* Inst Elec & Electronics Engrs; Simulation Coun. *Res:* Fault isolation; feature extraction; x-ray diagnostics. *Mailing Add:* US Army Missile Command Bldg 5452 AMcpm-Am-E Redstone Arsenal AL 35898

NALLEY, SAMUEL JOSEPH, b Benton, Ark, May 5, 43; m 62; c 2. ATOMIC PHYSICS, MOLECULAR PHYSICS. *Educ:* State Col Ark, BS, 65; Univ Tenn, Knoxville, MS, 67, PhD(physics), 71. *Prof Exp:* From asst prof to assoc prof, 71-85, PROF PHYSICS, CHATTANOOGA STATE TECH COMMUNITY COL, 85- *Concurrent Pos:* mem, Laser Safety Training. *Mem:* Am Phys Soc; Am Asn Physics Teachers; Laser Inst Am. *Res:* Atomic and molecular collision. *Mailing Add:* Dept Physics Chattanooga State Tech Com Col Chattanooga TN 37406

NALOS, ERVIN JOSEPH, b Prague, Czech, Sept 10, 24; nat US; m 47; c 3. PHYSICS. *Educ:* Univ BC, BASc, 46, MASc, 47; Stanford Univ, PhD(elec eng), 51. *Prof Exp:* Res assoc microwave physics, 50-54; group leader microwave tube develop & res, Microwave Lab, Gen Elec Co, Calif, 54-59, sci rep to Europe, Res Lab, 59-62; staff engr, Off Vpres Res & Develop, 62-69, mgr appl technol, Military Airplane Systs Div, 69-71, supvr civil & com systs, 71-78, SUPVR ELECTROMAGNETICS & ENG TECHNOL, BOEING AEROSPACE CO, 78- *Honors & Awards:* Baker Award, Inst Radio Eng, 59. *Mem:* Sr mem Inst Elec & Electronics Eng; Sigma Xi. *Res:* Electron physics and electronics; high power microwave devices. *Mailing Add:* Boeing Aerospace PO Box 3999 MS 84-06 Seattle WA 98124

NAM, SANG BOO, b Kyung Nam, Korea; US citizen; m 68; c 2. CONDENSED MATTER PHYSICS, SUPERCONDUCTIVITY. *Educ:* Seoul Nat Univ, BS, 58; Univ Ill, MS, 61, PhD(physics), 66. *Prof Exp:* Res assoc physics, Univ Ill, 66; res fel, Rutgers State Univ, 66-68; asst prof, Univ Va, 68-71; vis prof, Belfer Grad Sch, Yeshiva Univ, 71-74; Nat Res Coun-Nat Acad Sci sr res fel, 74-76, 80-86; res prof, Univ Dayton, 76-80; PROF PHYSICS, UNIV RES CTR, WRIGHT STATE UNIV, 80- *Concurrent Pos:* Vis prof physics, Seoul Nat Univ, 70. *Mem:* Fel Am Phys Soc; Sigma Xi; AAAS. *Res:* Superconductivity; semiconductor physics, laser and non-linear physics and phase transition and magnetism. *Mailing Add:* 7735 Peters Pike Dayton OH 45414-1713

NAM, SEHYUN, b Seoul, Korea, Feb 11, 49; m 75; c 2. RHEOLOGY, POLYMER PROCESSING & MATERIALS SCIENCE. *Educ:* Seoul Nat Univ, BSE, 71; Ore State Univ, MSE, 75; Univ Mich, MS, 78; Univ Tenn, PhD(polymer eng), 82. *Prof Exp:* Sr engr, 82-86, RES SPECIALIST, 3M CO, 86-, ADJ PROF RHEOLOGY, 90- *Mem:* Soc Rheology; Soc Plastic Engrs; Soc Inc Polymer Processing. *Res:* Polymer rheology; relationship between structure, properties and processing. *Mailing Add:* 201-1W-28 3M Ctr St Paul MN 55194

NAMBA, RYOJI, b Honolulu, Hawaii, Jan 31, 22; m 48; c 4. ENTOMOLOGY. *Educ:* Mich State Col, BS, 48, MS, 50; Univ Minn, PhD(entom), 53. *Prof Exp:* From asst entomologist to assoc entomologist, 53-68, ENTOMOLOGIST, AGR EXP STA, UNIV HAWAII, 68-, PROF ENTOM, 76- *Mem:* Entom Soc Am. *Res:* Leafhoppers; insect transmission of plant pathogens. *Mailing Add:* Dept Entom Fil Univ Hawaii 3050 Meile Rm 310 Honolulu HI 96822

NAMBA, TATSUJI, b Changchun, China, Jan 29, 27. NEUROLOGY, PHARMACOLOGY. *Educ:* Okayama Univ, MD, 50, PhD(med), 56. *Prof Exp:* Asst med, Med Sch, Okayama Univ, 56-57, lectr, 57-62; res assoc, 62-64, from asst attend physician to assoc attend physician, 64-71, dir neuromuscular dis labs, 66-71, ATTEND PHYSICIAN, MAIMONIDES MED CTR, 71-, HEAD ELECTROMYOGRAPHY CLIN, 66-, DIR NEUROMUSCULAR DIS DIV, 71- *Concurrent Pos:* Res fel, Maimonides Med Ctr, 59-62; Fulbright fel, 59-62; consult, Fukuyama Defense Force Hosp, Japan, 57-59; from instr to assoc prof, State Univ NY Downstate Med Ctr, 59-76; from asst vis physician to assoc vis physician, Kings County Hosp Ctr, 65-73, vis physician, 73-; attend physician, State Univ Hosp, 66-; mem, Med Adv Bd, Myasthenia Gravis Found; prof med, State Univ NY Health Sci Ctr, Brooklyn, 76- *Mem:* Am Col Physicians; Am Acad Neurol; Am Soc Pharmacol & Exp Therapeut; Am Soc Clin Pharmacol & Chemother; AMA; Am Asn Electrodiag Med. *Res:* Basic and clinical research of skeletal muscle and neuromuscular diseases; clinical pharmacology of neuromuscular agents. *Mailing Add:* Maimonides Med Ctr 4802 Tenth Ave Brooklyn NY 11219

NAMBI, PONNAL, b Nagercoil, India, Sept 28, 45; US citizen; m 67; c 2. MOLECULAR MECHANISMS OF SIGNAL TRANSDUCTION, HORMONE & DRUG RECEPTOR INTERACTIONS. *Educ:* Scott Christian Col, Madras, India, BSc, 66; Delhi Univ, India, BEd, 73; Univ Tenn, MS, 75; Univ Calgary, PhD(biochem), 79. *Prof Exp:* Instr biochem, Univ Tenn, Memphis, 79-81; res assoc, Howard Hughes Med Inst, Duke Univ, 81-84; assoc sr investr, Smith Kline & French Labs, 84-86, sr investr, 86-88, ASST DIR PHARMACOL, SMITHKLINE BEECHAM PHARMACEUT, 88- *Concurrent Pos:* Demonstr, S T Hindu Col, Nagercoil, India; DTEA,

Higher Secondary Sch, Delhi, India, higher secondary sch sci teacher; grad teaching asst, Univ Calgary, Can. *Mem:* Am Soc Pharmacol & Exp Therapeut; Am Soc Hypertension. *Res:* Hormone and drug receptor interactions; inter-reception regulation; receptor structure and function in normal and disease states; biochemical and molecular mechanisms of receptor regulation; second messenger roles and identification of nonpeptide novel antagonists for peptide hormone receptors as potential therapeutic drugs; numerous publications. *Mailing Add:* Dept Pharmacol L-521 Smithkline Beecham 709 Swedland Rd Swedland PA 19479

NAMBIAR, KRISHNAN P, b Sivapuram, Kerala, India, Oct 8, 51. PROTEIN ENGINEERING, SYNTHETIC ORGANIC. *Educ:* Univ Calicut, Kerala, India, BSc, 71, MSc, 73; Univ New Brunswick, PhD(organic chem), 78. *Prof Exp:* Fel org chem, Harvard Univ, Cambridge, 78-81, RES ASSOC BIOORG CHEM, 81-; ASSOC PROF, DEPT CHEM, UNIV CALIF. *Mem:* Sigma Xi; Am Chem Soc. *Res:* structural enzymology; understanding structural aspects of enzymes in detail. *Mailing Add:* Dept Chem Univ Calif Davis CA 95616

NAMBOODIRI, MADASSERY NEELAKANTAN, b Kothamangalam, India, Oct 18, 35; m 63; c 1. NUCLEAR CHEMISTRY. *Educ:* Kerala Univ, India, BSc, 57; State Univ NY Stony Brook, PhD(chem), 72. *Prof Exp:* Sci officer radiochem, Bhabha Atomic Res Ctr, Bombay, India, 58-67; res scientist nuclear chem, Cyclotron Inst, Tex A&M Univ, 72-83; CHEMIST, LAWRENCE LIVERMORE NAT LAB, 83- *Concurrent Pos:* Res affil chem, Argonne Nat Lab, 60-61. *Mem:* Am Phys Soc; Am Chem Soc. *Res:* Nuclear reactions, especially heavy ion induced reactions, with emphasis on fission, heavy ion fusion, deep inelastic reactions and intermediate energy reactions. *Mailing Add:* Nuclear Chem Div L234 Lawrence Livermore Nat Lab Livermore CA 94550

NAMBU, YOICHIRO, b Tokyo, Japan, Jan 18, 21; US citizen; m 45; c 2. ELEMENTARY PARTICLE PHYSICS. *Educ:* Univ Tokyo, BS, 42, DSc, 52. *Hon Degrees:* DSc, Northwestern Univ, 87. *Prof Exp:* Asst, Univ Tokyo, 45-49; from asst prof to prof, Osaka City Univ, 49-56; res assoc, Univ Chicago, 54-56, from assoc prof to prof physics, 56-71, distinguished serv prof, 71-90, chmn dept, 74-77, EMER PROF, DEPT PHYSICS, UNIV CHICAGO, 91- *Concurrent Pos:* Mem, Inst Advan Study, 52-54; Harry Pratt Judson distinguished serv prof, Univ Chicago, 77. *Honors & Awards:* Dannie Heineman Prize Math Physics, Am Phys Soc, 70; J Robert Oppenheimer Prize, 76; Order of Cult, Govt Japan, 78; US Nat Medal Sci, 82; Max Planck Medal, 85; P A M Dirac Medal, Int Ctr Theoret Physics, 86. *Mem:* Nat Acad Sci; Am Phys Soc; Am Acad Arts & Sci; Japan Acad. *Res:* Field theory; theory of elementary particles; theory of superconductivity. *Mailing Add:* 5530 S University Ave Chicago IL 60637

NAMER, IZAK, b Istanbul, Turkey, Nov 14, 52; US citizen; m 75; c 2. COMBUSTION, FLUID MECHANICS. *Educ:* City Col New York, BE, 75; Univ Calif, Berkeley, MS, 76, PhD(mech eng), 80. *Prof Exp:* Res asst, Univ Calif, Berkeley, 75-80; ASST PROF MECH ENG, DREXEL UNIV, 80- *Concurrent Pos:* Consult, Lotepro Corp, New York, 81, Gen Eng Assoc, Philadelphia, 81; prin investr res proj, Nat Sci Found, NASA & Mick A Navlin Found, 83-85; tech comt propellants & combustion, Am Inst Aeronaut & Astronaut, 85-86. *Mem:* Am Inst Aeronaut & Astronaut; Am Soc Mech Engrs; Combustion Inst; Am Soc Eng Educ; Sigma Xi; Air Pollution Control Asn. *Res:* Studies on turbulent jets, turbulent flames, spray combustion, free convection flows, and incineration of hazardous waste; study of combustion characteristics of liquid fuel sprays; inceration of toxic organic wastes; study of fundamental turbulent flows; turbulent jets. *Mailing Add:* Reman Dev Corp 134 W Tenth St New York NY 10014

NAMEROFF, MARK A, b Philadelphia, Pa, May 16, 39. DEVELOPMENTAL BIOLOGY. *Educ:* Univ Pa, BA, 60, MD, 65, PhD(anat embryol), 66. *Prof Exp:* Instr anat, Sch Med, Univ Pa, 66-67; staff mem, Armed Forces Inst Path, Washington, DC, 67-70; asst prof, 70-75, ASSOC PROF BIOL STRUCT, UNIV WASH, 75- *Res:* Cell differentiation; embryonic muscle cells. *Mailing Add:* Dept of Biol Struct Univ Wash SM-20 Seattle WA 98195

NAMIAS, JEROME, b Bridgeport, Conn, Mar 19, 10; m 38; c 1. METEOROLOGY. *Educ:* Mass Inst Technol, MS, 41. *Hon Degrees:* ScD, Univ RI, 72, Clark Univ, 84. *Prof Exp:* Asst aerology, Blue Hill Meteorol Observ, Harvard Univ, 33-36; res assoc, Mass Inst Technol, 35-40; chief extended forecast div, US Weather Bur, Nat Oceanic & Atmospheric Agency, 41-71; RES METEOROLOGIST, SCRIPPS INST OCEANOG, UNIV CALIF, SAN DIEGO, 72- *Concurrent Pos:* Meteorologist, Trans World Airlines, Inc, 34; assoc, Woods Hole Oceanog Inst, 54-, mem sci vis comt, 64-; res meteorologist, Scripps Inst Oceanog, 68-71; vis scholar, Rockefeller Study & Conf Ctr, Bellagio, Italy, 77. *Honors & Awards:* Meisinger Award, 38; Award Extraordinary Sci Accomplishment, Am Meteorol Soc, 55; Sverdrup Gold Medal, Am Meteorol Soc, 81; Lect, Univ Stockholm, 50; Distinguished Lectr, Pa State Univ, 62. *Mem:* Nat Acad Sci; fel Am Meteorol Soc; fel AAAS; fel Am Geophys Union; Royal Meterol Soc; Am Acad Arts & Sci; fel Explorers Club. *Res:* Long range weather forecasting and general circulation of atmosphere; aerology; large scale air-sea interaction. *Mailing Add:* Scripps Inst of Oceanog Univ of Calif at San Diego La Jolla CA 92093

NAMKOONG, GENE, b New York, NY, Jan 25, 34; m 56; c 3. FOREST GENETICS, EVOLUTION. *Educ:* State Univ NY, BS, 56, MS, 58; NC State Univ, PhD(genetics), 63. *Prof Exp:* Res forester, 58-60, plant geneticist, 63-71, pioneering res scientist pop genetics, Forest Serv, USDA, 71-; AT GENETICS DEPT, NC STATE UNIV. *Concurrent Pos:* From asst prof to assoc prof genetics & forestry, NC State Univ, 63-71, prof, 71-; vis prof, Univ Chicago, 68; consult, BC Forest Serv, 72- & Repub Korea Forest Serv, 74-; adj prof, Shaw Univ, 75-; group leader genetics, Int Union Forest Res Orgn, 76-; sci policy adv, Dept Admin, NC, 77-78; chmn genetics working group, Soc Am Foresters, 78-80; vis prof, Oxford Univ, 79 & Univ Goettingen, 83. *Honors & Awards:* Sci Achievement Award, Int Union Forest Res Orgns, 71. *Mem:* Biomet Soc; Genetics Soc Am; Am Soc Nat; Soc Study Evolution; Sigma Xi. *Res:* Mathematical, population genetics, particularly with respect to forest tree species. *Mailing Add:* Genetics Dept Box 7614 NC State Univ Raleigh NC 27695-7614

NAMM, DONALD H, b Hamden, Conn, Feb 10, 40; m 63; c 2. PHARMACOLOGY. *Educ:* Rensselaer Polytech Inst, BS, 61; Albany Med Col, PhD(pharmacol), 65. *Prof Exp:* Instr pharmacol, Emory Univ, 67-68; asst prof, Sch Med, Univ Okla, 68-69; sr res pharmacologist, 69-75, GROUP LEADER, RES LABS, BURROUGHS-WELLCOME CO, 75- *Concurrent Pos:* Fel pharmacol, Emory Univ, 65-67. *Mem:* Am Soc Pharmacol & Exp Therapeut. *Res:* Heart metabolism; blood vessel metabolism; drug-enzyme interactions. *Mailing Add:* Burroughs Wellcome Co 3030 Cornwallis Rd Research Triangle Park NC 27709

NAMMINGA, HAROLD EUGENE, b Scotland, SDak, July 26, 45; m 73; c 1. FISHERIES ECOLOGY, LIMNOLOGY. *Educ:* Univ SDak at Springfield, BS, 67; Univ SDak at Vermillion, MA, 69; Okla State Univ, PhD(zool), 75. *Prof Exp:* Instr biol, Kearney State Col, 69-71; environ scientist-ecol, Technol Res & Develop, Inc Div, Benham-Blair & Affil, 75-78; RES COORDR FED AID, OKLA DEPT WILDLIFE CONSERV, 78- *Mem:* Am Fisheries Soc. *Res:* Fish commmunity structure, predator-prey relationships and effects of fishing mortality on populations of sportfish; secondary interests include heavy metals in aquatic ecosystems and the effects of metals on aquatic community structure. *Mailing Add:* 2720 Julies Trail Edmond OK 73034-1513

NAMY, JEROME NICHOLAS, b Cleveland, Ohio, Aug 11, 38; m 63; c 3. PETROLOGY, STRATIGRAPHY. *Educ:* Western Reserve Univ, BA, 60; Univ Tex, Austin, PhD(geol), 69. *Prof Exp:* Explor geologist, Pan Am Petrol Corp, Standard Oil Co, Ind, 67-70; from asst prof to assoc prof geol, Baylor Univ, 70-78; mem staff, 78-85, VPRES, TEXLAND PETROL, 85- *Mem:* Geol Soc Am; Am Asn Petrol Geologists; Sigma Xi; Soc Econ Paleontologists & Mineralogists; Soc Explor Geophysicists; Soc Prof Well Log Analysts. *Res:* The stratigraphy and petrology of carbonate and sedimentary rocks; petroleum geology. *Mailing Add:* Texland Petrol 500 Throckmorton Suite 3402 Ft Worth TX 76102

NANCARROW, WARREN GEORGE, b Texarkana, Tex, Aug 10, 23; m 47; c 4. PETROLEUM ENGINEERING. *Educ:* Tex A&M Univ, BS, 47. *Prof Exp:* Petrol engr, Standard Oil Co, Ind, 47-54; petrol engr, De Golyer & Macnaughton, Inc, 54-66, vpres, 66-68, sr vpres petrol eng, 68-78, CHMN BD, DE GOLYER & MACNAUGHTON, INC, 78- *Mem:* Am Petrol Inst; Am Inst Mining, Metall & Petrol Engrs. *Res:* Petroleum drilling, production and reservoir study. *Mailing Add:* 4625 Greenville Suite 305 Dallas TX 75206

NANCE, DWIGHT MAURICE, b Bartlesville, Okla, Jan 31, 43. NEUROSCIENCE, PSYCHOBIOLOGY. *Educ:* Okla State Univ, BS, 65, MS, 67, PhD(psychol), 69. *Prof Exp:* Fel physiological psychol, Univ Houston, 69-70; fel reproductive physiol, Worcester Found Exp Biol, 70-72; res anatomist neuroendocrinol, Univ Calif, Los Angeles, 72-77; asst prof anat, Univ SFla, 77-79; ASSOC PROF ANAT, DALHOUSIE UNIV, 79- *Concurrent Pos:* NIMH sr fel, Worcester Found Exp Biol, 70-71; trainee reproductive physiol, 71-72; co-prin investr, Nat Inst Arthritis & Metab Dis, NIH grant, Univ Calif, Los Angeles, 72-77; co-investr, Nat Inst Child Health & Human Develop, NIH grant, Univ SFla, 77-80; prin investr, Can Med Res Coun, Dalhousie Univ, 81- *Mem:* Soc Neurosci; Am Asn Anatomists. *Mailing Add:* Dept Path Path Bldg Rm P/220/218 Univ Man 770 Eannatyne Ave Winnepeg MB R3E 0W3 Can

NANCE, FRANCIS CARTER, b Manila, Philippines, Jan 1, 32; US citizen; m 59; c 4. SURGERY, PHYSIOLOGY. *Educ:* Univ Tenn, MD & MS, 59. *Prof Exp:* Instr surg, 65-67, from asst prof to assoc prof surg & physiol, 67-73, PROF SURG & PHYSIOL, LA STATE UNIV SCH MED, NEW ORLEANS, 73- *Concurrent Pos:* Am Cancer Soc fel, Univ Pa, 63-64. *Mem:* AAAS; Am Gastroenterol Soc; Am Col Surgeons; Am Asn Surg Trauma; Am Surg Asn. *Res:* Gastrointestinal physiology; effects of microbial flora on various gastrointestinal functions and diseases; burns. *Mailing Add:* Dept Surg-Physiol Med Sch La State Univ 1542 Tulane Ave New Orleans LA 70112

NANCE, RICHARD DAMIAN, b St Ives, Cornwall, UK, Oct 25, 51; m 82; c 2. TECTONICS, GEODYNAMICS. *Educ:* Leicester Univ, UK, BS Hons, 72; Cambridge Univ, PhD(geol), 78. *Prof Exp:* Asst prof geol, St Francis Xavier Univ, NS, 76-80; sr res geologist, Exxon Prod Res Co, 82-83; from asst prof to assoc prof, 80-90, PROF GEOL, OHIO UNIV, 91- *Concurrent Pos:* Consult, Inst Environ Studies, La State Univ, 77-81, Cominco Am Inc, 84-85 & Argonne Nat Lab, 84-88; mem, Ohio Univ geol curric comt, 80-84, grad comt, 80-86, field camp dir, 82-88, chmn curric comt, 85-87, mem faculty develop comt, 86-89, mem faculty senate, 86-87, chmn grad comt, Ohio Univ, 87- *Honors & Awards:* Numerous Res Grants, 73-90. *Mem:* Geol Soc Am; Geol Asn Can; Am Asn Petrol Geologists; Am Geophys Union; Sigma Xi. *Res:* Tectonostratigraphic evolution of the Avalonian-Cadomian belt in the northern Appalachian and western Europe; kinematic and technolthermal history of the Avalon terrane in Southern New Brunswick and Nova Scotia, Canada; empirical analysis of supercontinent periodicity in plate tectonics and cycles in tectono-bio-geochemical history of earth; dextral transpression and syntectonic, late Paleozoic sedimentation in southern New Brunswick; author of over 60 articles. *Mailing Add:* Dept of Geol Sci Clippinger Labs Ohio Univ Athens OH 45701

NANCE, RICHARD E, b Raleigh, NC, July 22, 40; m 62; c 2. COMPUTER SCIENCE, OPERATIONS RESEARCH. *Educ:* NC State Univ, BS, 62, MS, 66; Purdue Univ, PhD(opers res), 68. *Prof Exp:* From asst prof to assoc prof comput sci & opers res, Southern Methodist Univ, 68-73; head, Dept Comput Sci, Va Polytech Inst & State Univ, 73-79; comput scientist, Naval Surface Weapons Ctr, 79-80; PROF, DEPT COMPUT SCI, VA POLYTECH INST

& STATE UNIV, 80-, DIR, SYSTS RES CTR, 84- *Concurrent Pos:* Area ed comput struct & tech, Opers Res, 78-83 & simulation, gaming & info systs, Inst Indust Engrs Trans, 74-82; sr res assoc, Imp Col, UK, 80; ed, Opers Res Soc Am J Comput, 87- *Mem:* Asn Comput Mach; Opers Res Soc Am; Inst Mgt Sci; Am Soc Info Sci; Inst Indust Engrs. *Res:* Digital simulation theory; mathematical models of information networks; computer systems modeling and performance evaluation; software engineering. *Mailing Add:* Dept Comput Sci Va Polytech Inst & State Univ Blacksburg VA 24061

NANCE, WALTER ELMORE, b Manila, Philippines, Mar 25, 33; US citizen; m 57; c 2. HUMAN GENETICS, INTERNAL MEDICINE. *Educ:* Univ of South, SB, 54; Harvard Univ, MD, 58; Univ Wis, PhD(med genetics), 68. *Prof Exp:* From intern to resident med, Sch Med, Vanderbilt Univ, 58-61, asst prof, Sch Med, Vanderbilt Univ, 64-69; prof genetics & med, Sch Med, Ind Univ, Indianapolis, 69-75; PROF HUMAN GENETICS & CHMN DEPT, MED COL VA, 75-, PROF MED & PEDIAT, 76- *Concurrent Pos:* Mem, Genetics Training Comt, Nat Inst Gen Med Sci, 71-74; prin investr, Ind Univ Human Genetics Ctr, 74-75; consult, Nat Inst Neurol Dis & Stroke, Annual Surv Hearing Impaired Children & Youth & Genetics Sect, WHO; mem, Epidemiol & Dis Control Study Sect, NIH, 75-79. *Mem:* AAAS; Math Asn Am; Am Soc Human Genetics (secy, 71-74); Int Soc Twin Studies; fel Am Col Physicians; Sigma Xi. *Res:* Medical genetics; hereditary deafness; human twin studies; population genetics; analysis of human genetic polymorphisms; genetically determined disorders of metabolism. *Mailing Add:* Dept Human Genetics Med Col Va 1101 E Marshall St Richmond VA 23292

NANCOLLAS, GEORGE H, b Wales, Brit, Sept 24, 28; m 54; c 2. PHYSICAL CHEMISTRY, INORGANIC CHEMISTRY. *Educ:* Univ Wales, BSc, 48, PhD(phys chem), 51; Glasgow Univ, DSc, 63. *Prof Exp:* Res assoc, Univ Manchester, 51-53; lectr, Glasgow Univ, 53-65; provost fac natural sci & math, 70-75, PROF CHEM, STATE UNIV NY BUFFALO, 65- *Concurrent Pos:* Fel, Univ Wales, 51-52; vis scientist, Brookhaven Nat Lab, 63-64; adj prof chem, Sch Dent, State Univ NY, Buffalo, 70-; chmn anal chem comn on equilibria, Int Union Pure & Appl Chem. *Honors & Awards:* Schoellkopf Medal, Am Chem Soc, 77. *Mem:* Fel AAAS; Faraday Soc; Am Chem Soc; Royal Soc Chem; fel Royal Inst Chem. *Res:* Formation of metal complexes and ion-pairs; inorganic ion exchangers; kinetics of crystal growth and dissolution; the electrical double layer. *Mailing Add:* Dept Chem NY State Univ Buffalo NY 14214

NANDA, DAVE KUMAR, plant breeding, genetics, for more information see previous edition

NANDA, JAGDISH L, b Punjab, India, Feb 1, 33; m 63; c 2. MATHEMATICS. *Educ:* Univ Delhi, BA, 53, MA, 55; Ind Univ, PhD(geom), 61. *Prof Exp:* Asst math, Govt India, New Delhi, 55-57; res assoc, Wright-Patterson AFB, Ohio, 60-62; asst prof, Univ Dayton, 61-63; assoc prof, Villanova Univ, 63, Univ Delhi, 63-64; PROF MATH, EASTERN ILL UNIV, 64- *Mem:* Math Asn Am; Am Math Soc. *Res:* Geometry; relativity. *Mailing Add:* Dept Math Eastern Ill Univ Charleston IL 61920

NANDA, RAVINDER, b Bombay, India, Sept 12, 36; m 62; c 2. INDUSTRIAL ENGINEERING. *Educ:* Banaras Hindu Univ, BSc, 60; Univ Ill, MS, 60, PhD(indust eng), 62. *Prof Exp:* Asst indust eng, Univ Ill, 59-62; asst prof, Univ Miami, Fla, 62-67; from asst prof to assoc prof, NY Univ, 67-73; dir, Indust Eng Prog, Polytech Inst NY, 75-84, ASSOC PROF, POLYTECH UNIV, 73- *Concurrent Pos:* Consult, Case & Co Ohio, 62, Trans World Airlines, 69-70, Port Authority NY & NJ, 69-83, Nashua Corp, 77 & Am Mgt Asn, 80; NSF Inst res grants, 65-67 & initiation res grant, 65-67; mem, Vol Int Tech Assistance; Danforth Assoc. *Mem:* Sr mem Am Inst Indust Engrs; Inst Mgt Sci. *Res:* System analysis and design; operational planning and control; work design productivity improvement; human resource development. *Mailing Add:* Polytech Univ 333 Jay St Brooklyn NY 11201

NANDA, RAVINDRA, b Layallpur, India, Feb 19, 43; c 2. ORTHODONTICS, TERATOLOGY. *Educ:* Univ Lucknow, BDS, 64, MDS, 66; Roman Cath Univ, Nijmegen, PhD(med), 69. *Prof Exp:* Res assoc orthod, Roman Cath Univ, Nijmegen, 67-70; asst prof, Col Dent Surg, Loyola Univ, Chicago, 70-73; assoc prof orthod, 73-79, PROF ORTHOD, UNIV CONN HEALTH CTR, FARMINGTON, 79- *Mem:* Teratol Soc; Int Asn Dent Res; Europ Orthod Soc; Indian Dent Asn; Am Asn Orthod. *Res:* Clinical orthodontics; growth and development of face; tooth development; radiotracer studies; craniofacial orthopedics. *Mailing Add:* 185 Deercliff Rd Avon CT 06001

NANDAKUMAR, KRISHNASWAMY, b Srirangam, India, Oct 24, 51; Can citizen; m 79. COMPUTATIONAL FLUID DYNAMICS, MULTICOMPONENT DISTILLATION. *Educ:* Madras Univ, BTech, 73; Univ Sask, MSc, 75; Princeton Univ, PhD(chem eng), 79. *Prof Exp:* Fel chem eng, Univ Alta, 79-81; res engr chem eng, Gulf Can Ltd, 81-83; assoc prof, 83-87, PROF CHEM ENG, UNIV ALTA, 87- *Concurrent Pos:* Res fel, Univ Erlangen, Nurnberg, Ger, 89-90; Alexander von Humboldt Stiftung fel, Ger, 89-90. *Honors & Awards:* Albright & Wilson Americas Award, Can Soc Chem Eng, 91. *Mem:* Am Inst Chem Engrs; Can Soc Chem Eng; fel Chem Inst Can. *Res:* Computational and experimental study of bifurcation phenomena in fluid mechanics and heat transfer; flow through porous media and multicomponent distillation. *Mailing Add:* Dept Chem Eng Univ Alta Edmonton AB T6G 2G6 Can

NANDEDKAR, ARVINDKUMAR NARHARI, b Nagpur, India, Apr 8, 37; m 64; c 2. BIOCHEMISTRY, CLINICAL CHEMISTRY. *Educ:* Univ Nagpur, BSc, 59, MSc, 61; Univ Delhi, PhD(med biochem), 66. *Prof Exp:* Asst res officer, Indian Coun Med Res, V Patel Chest Inst, Univ Delhi, 66; NIH res assoc biochem, Georgetown Univ, 66-68; from instr to assoc prof, 68-84, asst dir acad reinforcement prog, 73-81, actg chmn dept biochem, 78-80, PROF BIOCHEM, HOWARD UNIV, 84- *Concurrent Pos:* AEC fel, Howard Univ, 68-71; vis assoc prof, Cornell Univ Med Ctr, 75-; consult clin biochem, New York Hosp, 75-; consult, Path Lab, Doctors Hosp, Lanham, Md, 76- *Mem:* Am Asn Clin Chemists; fel Am Acad Clin Toxicol; AAAS; Am Chem Soc; fel Am Inst Chem; Am Soc Biol Chem; Soc Toxicol. *Res:* Lipid biochemistry, including clinical application to tuberculosis and anaphylaxis; epilepsy and drug metabolism; protein chemistry; binding and carrier of metals in body fluids. *Mailing Add:* 3302 Enterprise Rd Mitcheville MD 20716

NANDI, JYOTIRMOY, b Calcutta, India, May 12, 49. GASTROENTEROLOGY. *Educ:* Univ Calcutta, PhD(biochem), 75. *Prof Exp:* Res assoc, 80-85, RES SUPVR, DEPT SURG, STATE UNIV NY UPSTATE MED CTR, 85- *Mem:* Am Physiol Soc; Biophys Soc; NY Acad Sci; AAAS. *Mailing Add:* Dept Med State Univ NY Upstate Med Ctr 750 E Adams St Rm 6800 Syracuse NY 13210

NANDI, SATYABRATA, b North Lakhimpur, India, Dec 1, 31; nat US; m 57. ZOOLOGY, ENDOCRINOLOGY. *Educ:* Univ Calcutta, BSc, 49, MSc, 51; Univ Calif, PhD(zool), 58. *Prof Exp:* Demonstr zool, Bethune Col, India, 49-53; lectr, City Col Calcutta, 53; asst biophys, Saha Inst Nuclear Physics, Calcutta, 53-54; asst zoologist, 54-56, res zoologist, 56-57, jr res zoologist & lectr, 58-59, asst res endocrinologist, 59-61, actg asst prof zool, 61-62, from asst prof to prof, 62-70, Miller prof, 70-71, chmn Dept Zool, 71-73, dir, Cancer Res Lab, 74-85, RES ENDOCRINOLOGIST CANCER RES LAB, UNIV CALIF, BERKELEY, 68-, PROF ZOOL, 71- *Concurrent Pos:* Guggenheim fel, Netherlands Cancer Inst, Amsterdam, 67-68; vis scientist, Virus Res Inst, Kyoto Univ, 65; bd gov, Int Asn for Breast Cancer Res. *Mem:* Am Soc Zoologists; Endocrine Soc; Am Asn Cancer Res. *Res:* Tumor biology, including endocrinology, growth regulation, virology, genetics and hormone receptor. *Mailing Add:* Dept Cell & Molecular Biol Univ Calif 4079 LSB Berkeley CA 94720

NANDI, SATYENDRA PROSAD, b Aug 1, 27; US citizen; m 59; c 2. FUEL SCIENCE & TECHNOLOGY, PHYSICAL CHEMISTRY. *Educ:* Dacca Univ, BSc, 49, MSc, 50; Pa State Univ, PhD(fuel technol), 63. *Prof Exp:* Scientist phys chem, Central Fuel Res Inst, 52-60; res assoc fuel sci, Pa State Univ, 63-65, sr res assoc, 68-74; chemist fuel chem, Argonne Nat Lab, 74-76; CHEMIST FUEL PROCESS RES, INST GAS TECHNOL, 76- *Concurrent Pos:* Consult, St Regis Paper Co, 69-71. *Mem:* Am Chem Soc; Am Carbon Soc. *Res:* Surface chemistry of coals; preparation and characterization of active carbons and molecular line carbons; kinetics of gas solid reactions; Fischer Tropsch synthesis. *Mailing Add:* 5717 Washington St Downers Grove IL 60516-1324

NANDY, KALIDAS, b Calcutta, India, Oct 1, 30; m 61; c 2. GERIATRICS, NEUROANATOMY. *Educ:* Univ Calcutta, MD, 53; Univ Lucknow, MSurg, 60; Emory Univ, PhD(anat), 63. *Prof Exp:* Lectr anat, Univ Calcutta, 54-57, asst prof, 57-60, reader, 60-61; from asst prof to prof anat, Emory Univ, 63-75; prof, 75-81, RES PROF ANAT & NEUROL, BOSTON UNIV, 81-; DEP DIR & DIR RES, GERIATRIC RES EDUC & CLIN CTR, BEDFORD VET ADMIN HOSP, 75- *Concurrent Pos:* Indian Coun Med Res-Rockefeller Found fel, 59-60; Tull fel, 62-63. *Mem:* Am Asn Anat; Geront Soc; Soc Neurosci. *Res:* Neurobiology of aging and senile dementia; gero-pharmacological agents. *Mailing Add:* Dept Anat Neurol Boston Univ Sch Med 80 E Concord St Boston MA 02118

NANES, ROGER, b Brooklyn, NY, Jan 25, 44; m 68; c 3. MOLECULAR SPECTROSCOPY. *Educ:* Harpur Col, BA, 65; Johns Hopkins Univ, PhD(phys chem), 70. *Prof Exp:* Nat Res Coun Can fel spectros, Div Pure Physics, Nat Res Coun Can, Ont, 70-71 & Univ Western Ont, 71-72; asst prof, 72-75, assoc prof, 75-80, PROF PHYSICS, CALIF STATE UNIV, FULLERTON, 80- *Mem:* Am Chem Soc; Am Phys Soc; Soc Appl Spectros; Optical Soc Am. *Res:* Spectroscopy and molecular structure; vapor phase electronic spectroscopy of polyatomic molecules; electric and magnetic field effects on spectra; air pollution. *Mailing Add:* Dept of Physics Calif State Univ Fullerton CA 92634

NANEVICZ, JOSEPH E, b Buckley, Wash, Sept 11, 25; m 55; c 4. ELECTRICAL ENGINEERING. *Educ:* Univ Wash, BS, 51, MS, 53; Stanford Univ, PhD(elec eng), 58. *Prof Exp:* Instr elec eng, Univ Wash, 52-53; engr, Boeing Co, 53-54; res engr, SRI Int, 54-67, prog mgr, 67-69, dep dir, Electromagnetic Sci Lab, 79-88, DIR, APPL ELECTROMAGNETICS OPTICS LAB, SRI INT, 89- *Concurrent Pos:* NAm ed, J Electrostatics. *Mem:* Sr mem Inst Elec & Electronic Engrs. *Res:* Electromagnetics; noise; plasmas; electrostatics; spacecraft charging; lightning. *Mailing Add:* SRI Int 333 Ravenswood Menlo Park CA 94025

NANKERVIS, GEORGE ARTHUR, b Meriden, Conn, Apr 1, 30; m 54; c 2. PEDIATRICS, MICROBIOLOGY. *Educ:* Princeton Univ, AB, 52; Univ Rochester, PhD(microbiol), 59, MD, 62. *Prof Exp:* From asst prof to assoc prof pediat, Case Western Reserve Univ, 67-76, prof, 76-79; prof & chmn, dept pediat, Med Col Ohio, 79-85; PROF & CHMN PEDIAT, CHILDRENS HOSP, MED CTR OHIO, AKRON, 85- *Concurrent Pos:* Teaching fel pediat, Harvard Univ, 64-65; res fel infectious dis, Case Western Reserve Univ, 65-67. *Mem:* Sigma Xi. *Res:* Diagnostic virology and vaccine evaluation; study of pediatric populations with respect to immune status to infectious diseases; congenital viral infections. *Mailing Add:* Childrens Hosp Med Ctr Ohio 281 Locust St Akron OH 44308

NANKIVELL, JOHN (ELBERT), physics, for more information see previous edition

NANN, HERMANN, b Cologne, WGer, Oct 8, 40; div; c 3. MESON PRODUCTION NEAR THRESHOLD. *Educ:* Univ Frankfurt, WGer, dipl, 65, PhD(physics), 67, Dr habil, 74. *Prof Exp:* Asst res, Inst Kernphysik, Univ Frankfurt, 68-72, dozent, Dept Physics, 72-74; vis asst prof, Physics Dept, Mich State Univ, 74-76, vis assoc prof, 76-77; sr res assoc, Physics Dept, Northwestern Univ, 77-79; assoc prof, 79-84, PROF PHYSICS DEPT, IND UNIV, 84- *Concurrent Pos:* Res visitor, KFA-Juelich, W Ger, 87-88. *Mem:*

Am Phys Soc; Sigma Xi. *Res:* One, two and three nucleon transfer; electron induced knockout reactions; inelastic proton scattering; proton induced pion production; meson production near threshold. *Mailing Add:* Physics Dept Ind Univ Bloomington IN 47405

NANNELLI, PIERO, b Montelupo, Italy, Sept 29, 35; m 61; c 2. POLYMER CHEMISTRY. *Educ:* Univ Florence, DSc(org chem), 61. *Prof Exp:* Res assoc inorg polymers, Univ Ill, 61-63; asst prof coord chem, Univ Florence, 63-65; sr res chemist, 65-70, proj leader, 70-72, group leader, 72-83, RES MGR, PENNWALT CORP, 83- *Mem:* Am Chem Soc; Mat Res Soc; Am Ceramic Soc. *Res:* Inorganic polymers; coordination compounds; high performance structural materials and coatings; electroactive polymers; lubricant additives; plastic additives. *Mailing Add:* 557 General Armstrong Rd King of Prussia PA 19406-1613

NANNEY, DAVID LEDBETTER, b Abingdon, Va, Oct 10, 25; m 51; c 2. GENETICS. *Educ:* Okla Baptist Univ, AB, 46; Ind Univ, PhD(zool), 51. *Prof Exp:* From asst prof to assoc prof zool, Univ Mich, 51-58; fel, Calif Inst Technol, 58-59; prof zool, 59-76, prof genetics & develop, 76-86, PROF ECOL ETHOLOGY & EVOLUTION, UNIV ILL, URBANA, 87- *Honors & Awards:* Humboldt Prize, WGer Govt, 84. *Mem:* Am Genetics Soc (pres, 82-83); Genetics Soc Am; Soc Protozool; Am Soc Zool; fel Am Acad Arts & Sci. *Res:* Formal genetics, cytogenetics, developmental genetics and evolutionary genetics of ciliated protozoa. *Mailing Add:* Dept Ecol Ethology & Evolution Univ Ill 505 S Gregory Ave Urbana IL 61801

NANNEY, LILLIAN BRADLEY, b Clarksville, Tenn, Jan 21, 52; m 73; c 2. BIOLOGICAL SCIENCES. *Educ:* Vanderbilt Univ, BA, 73; Austin Peay State Univ, MS, 76; La State Univ Med Ctr, PhD(anat), 80. *Prof Exp:* Asst prof anat & plastic surg, 80-88, ASSOC PROF ANAT & CELL BIOL, VANDERBILT UNIV, 88- *Mem:* Am Asn Cell Biologists; Soc Invest Dermat; Am Burn Asn; AAAS. *Res:* Involved in skin related research; study growth factors in hyperproliferative skin diseases and wound healing; in-situ hybridization; electron microscopy; immunohistochemistry. *Mailing Add:* S-2221 MCN Vanderbilt Med Ctr Nashville TN 37232

NANNEY, THOMAS RAY, b Concord, NC, Apr 21, 31; m 54; c 2. PHYSICAL CHEMISTRY, COMPUTER SCIENCE. *Educ:* Univ NC, BS, 53; Univ SC, PhD(phys chem), 62. *Prof Exp:* Chemist, E I du Pont de Nemours & Co, Inc, 53-54; asst prof chem, 60-66, assoc prof chem, 66-77, dir comput ctr, 67-73, assoc prof, 70-72, chmn dept, 73-86, PROF COMPUT SCI, FURMAN UNIV, 72-, HERMAN N HIPP PROF COMPUT SCI, 87- *Concurrent Pos:* Vis asst prof & USPHS & Dept Defense Advan Res Projs Agency fel, 64-65; consult. *Mem:* AAAS; Asn Comput Mach; Sigma Xi. *Res:* Programming languages; artificial intelligence. *Mailing Add:* Comput Sci Dept Furman Univ Greenville SC 29613

NANZ, ROBERT AUGUSTUS ROLLINS, b Baltimore, Md, Apr 3, 15; m 39, 86. FOOD SCIENCE, NUTRITION. *Educ:* Rutgers Univ, BS, 37; Columbia Univ, MS, 39. *Prof Exp:* Food chemist, Quaker Maid Co, Inc, Great Atlantic & Pac Tea Co, NY, 37-38; biochemist, Watchung Labs, NJ, 38-39; nutrit specialist, Walker-Gordon Lab Co, Inc, 39-43; asst to coordr res, Spec Prod Div, Borden Co, NY, 46-47; dir food tech sect, Foster D Snell, Inc, 47-50; res chemist, Fla Citrus Canners Coop, 50-51; tech rep, Crown Can Co Div, Crown Cork & Seal Co, Fla, 51-53; pres, Fla Chemists & Engrs, Inc, 53-60; pres, Sci Assocs, Inc, 60-62; aerospace technologist, Food & Nutrit Group, Biomed Res Off, Manned Spacecraft Ctr, NASA, Tex, 62-67; asst dir, Nat Ctr Fish Protein Concentrate, US Dept Com, DC, 67-68; proj dir tech develop, Aquatic Sci Inc, Fla, 69-70; CONSULT FOODS & NUTRIT, 70- *Concurrent Pos:* Instr, Col Boca Raton, 71-74; instr, Palm Beach Jr Col, 78; lectr, Nutrit & Food Serv, 85-; consult, Hospitality Mgt, 85-88. *Mem:* Inst Food Technologists. *Res:* Food processing, formulations; marketing; nutritional factors in food. *Mailing Add:* Apt 701 300 NE 20th St Boca Raton FL 33431

NANZ, ROBERT HAMILTON, JR, b Shelbyville, Ky, Sept 14, 23; c 2. GEOLOGY. *Educ:* Miami Univ, AB, 44; Univ Chicago, PhD(geol), 52. *Prof Exp:* Res geologist, Shell Develop Co, 47-58, mgr geol dept, 58-59, dir explor res, 59-64, explor mgr, 64-67, vpres explor & prod res div, 67-70, vpres explor, Shell Oil Co, 70-75, VPRES WESTERN E&P REGION, SHELL OIL CO, 75- *Mem:* Geol Soc Am; Am Asn Petrol Geol. *Res:* Petroleum geology. *Mailing Add:* 18118 Longcliffe Dr Houston TX 77084

NANZETTA, PHILIP NEWCOMB, b Wilmington, NC, June 4, 40; m 62; c 2. RESEARCH ADMINISTRATION, TECHNOLOGY TRANSFER. *Educ:* NC State Univ, BS, 62; Univ Ill, MS, 63, PhD(math), 66. *Prof Exp:* Res assoc math, Case Western Reserve Univ, 66-67; asst prof, Univ Fla, 67-70; assoc prof, St Mary's Col Md, 70-74; dean, Fac Natural Sci & Math, 74-79, dir, Ctr Environ Res, 78-79, vpres acad affairs, 79-82; PROG MGR, AUTOMATED MFG RES FACIL, STOCKTON STATE COL, 82-, MGR, NIST MTC PROG, 88-, DIR LIFE TECHNOL CTR PROG, 89- *Concurrent Pos:* Mem, NJ Natural Resource Coun, 78-80, NJ Pinelands Comn, 80-82 & gov sci adv panel, 81-82. *Honors & Awards:* DOC Gold Medal, 88. *Mem:* AAAS; Am Math Soc. *Res:* Topology; structure spaces; lattice theory; genetics; automated manufacturing. *Mailing Add:* Automated Mfg Res Facil NIST Gaithersburg MD 20899

NAPADENSKY, HYLA S, m 56; c 2. EXPLOSIVE & PROPELLANT SAFETY. *Educ:* Univ Chicago, BS & MS. *Prof Exp:* Design anal engr, Int Harvester, 52-57; instr, mech dept, Ill Inst Technol, 64-66, sr eng adv, 57-81, dir res, Explosion Sci & Eng Dept, Res Inst, 81-88, BD OVERSEAS, ARMOUR COL ENG, ILL INST TECHNOL, 88-; VPRES NAPADENSKY ENERGETICS INC, 88- *Mem:* Nat Acad Eng; Combustion I; Sigma Xi. *Res:* Accidental fires and explosions during manufacture, transport and storage of explosives, propellants and pyrotechnics; explosive and initiation mechanisms; facility siting; systems safety and risk analysis. *Mailing Add:* Napadensky Energetics Inc 650 Judson Ave Evanston IL 60202-2551

NAPHTALI, LEONARD MATHIAS, b Brooklyn, NY, Aug 6, 27; m 58; c 2. HAZARDOUS WASTE, SYNTHETIC FUELS. *Educ:* Cooper Union, BChE, 49; Univ Mich, MSE, 50, PhD(chem eng), 54. *Prof Exp:* Res engr process metall, Sci Lab, Ford Motor Co, 54-56; prof, Polytech Inst Brooklyn, 56-66; mgr comput appln comput serv, Realtime Systs Inc, 66-68; dir petrochem, Sci Resources Corp, 68-70; dep city adminr urban admin, Off of Mayor, New York, NY, 71-72; chief process engr chem processes, Heyward-Robinson Co, 72-74; mgr process anal, Chem Construc Corp, 74-75; prog mgr coal gasification, Off Asst Secy Fossil Energy, US Dept Energy, 76-85; environ engr, US Environ Protection Agency, 85-87; VPRES, RESOURCE MOBILIZATION INC, 83-; ADJ PROF, NY UNIV, 85- *Concurrent Pos:* Adj prof, Wayne State Univ, 55-56, NY Univ, 54-76 & 85-, Columbia Univ, 75-77; consult, var clients, 56-; vis prof, Israel Inst Technol, 65-66; Fulbright vis prof, 65-66; atomic energy fel, 51-53. *Mem:* Am Inst Chem Engrs; Am Chem Soc; AAAS. *Res:* Fossil fuel processes; coal gasification; process synthesis; process design; separation processes; distillation; chemical reactor design; computer aided design. *Mailing Add:* 575 West End Ave New York NY 10024

NAPIER, DOUGLAS HERBERT, b Sidcup, Kent, Eng, Aug 23, 23; Can citizen; m 51; c 1. LOSS PREVENTION, HAZARD CONTROL. *Educ:* Univ London, Eng, BSc, 44, MSc, 47, PhD(chem eng), 51. *Prof Exp:* Res chemist, Philips Mitcham Works Ltd, 42-45, United Anodizing, 45-46; works control chemist, Borax Consol, 46-47; asst lectr chem, Univ Aberdeen, 47-48; scientist in charge, combustion sect, Brit Coal Utilization Res Asn, 51-57; head, chem projs sect, Vickers Res Ltd, 57-65; sr lectr indust safety, Imp Col Sci & Technol, London, 65-80; prof, 80-89, EMER PROF INDUST HAZARD CONTROL ENG, UNIV TORONTO, 88-; TECH DIR HAZARD & RISK, CONCORD ENVIRON CORP, 88- *Concurrent Pos:* Mem, Standing Comt Hazardous Mat, Nat Fire Code, Nat Res Coun, 85. *Mem:* Fel Royal Soc Chem; fel Inst Energy; Soc Chem & Indust; Brit Occup Hyg Soc; Am Inst Chem Eng. *Res:* Development of risk assessment and of models for hazard analysis as applied to the chemical and process industries; elucidation of the mechanism of ignition systems and of propagation of surface combustion. *Mailing Add:* Dept Chem Eng & Appl Chem Univ Toronto Toronto ON M5S 1A4 Can

NAPIER, ROGER PAUL, b Rochester, NY, Apr 1, 38; m 65. ORGANIC CHEMISTRY. *Educ:* St John Fisher Col, BS, 59; Univ Rochester, PhD(chem), 63. *Prof Exp:* Fel organophosphorus chem, Rutgers Univ, 63-65; PROJ LEADER, MOBIL CHEM CO, 65- *Mem:* Sr mem Am Chem Soc. *Res:* Natural products; nitrogen heterocycles; organophosphorus chemistry; pesticide chemistry. *Mailing Add:* Mobil Chem Co Chem Prods Div Box 250 Edison NJ 08818-0250

NAPIER, T(AVYE) CELESTE, ELECTROPHYSIOLOGY, BASAL GANGLIA. *Educ:* Tex Tech Univ, PhD(pharmacol), 81. *Prof Exp:* Res assoc, 85-86, ASST PROF, BIOL SCI RES CTR, NC SCH MED, 86- *Mailing Add:* 2160 S First Ave Maywood IL 60153

NAPKE, EDWARD, b Zahle, Lebanon, Jan 21, 24; Can citizen; m 61; c 3. MEDICINE, PHYSIOLOGY. *Educ:* Univ NB, Fredericton, BSc, 45; Univ Toronto, MD, 51, dipl pub health, 68. *Prof Exp:* Res officer aviation physiol, Defence Med Res Labs, 57-58; med officer, Dept Health & Welfare, Food & Drug Directorate, Health Protection Br, Govt Can, 63-65, med officer, Drug Adverse Reaction & Poison Control Progs, 65-90; CONSULT, 90- *Concurrent Pos:* Res fels, Karolinska Inst, Sweden, 58-61, C H Best Inst, 61-63 & Univ Toronto; consult, Drug Monitoring Prog, WHO, 68-90, Copenhagen Europ Off, 90-; mem Child Resistant Packaging Comt, Can Standards Asn, 70. *Honors & Awards:* Queens Jubilee Award; Award of Merit, Can Asn Poison Control Ctrs. *Mem:* AAAS; Can Asn Poison Control Ctrs; Int Soc Biometeorol. *Res:* G stress and unconsciousness; drug reaction and interreaction detection by epidemiological methods; poison control and prevention; human toxicology; hypnotherapy; fatty tissue studies; biometeorological studies; blood clotting; cerebrospinal fluid and central nervous system studies. *Mailing Add:* 124 Amberwood Crescent Nepean ON K2E 7H8 Can

NAPLES, FELIX JOHN, b Quadrelle, Italy, July 7, 12; nat US; m 41; c 3. ORGANIC CHEMISTRY. *Educ:* Youngstown State Univ, AB, 33; Univ Vt, MS, 34; Ind Univ, PhD(org chem), 36. *Prof Exp:* Asst, Ind Univ, 34-36; instr chem & physics, Youngstown State Univ, 36-37, assoc prof, 40-43; head dept chem, Springfield Jr Col, 37-40; res chemist, Goodyear Tire & Rubber Co, 43-45, sr res chemist, 45-77; RETIRED. *Concurrent Pos:* Lectr, Univ Akron, 78-89. *Mem:* Am Chem Soc. *Res:* Polymerization; solubility of amino acids; chemical equilibrium; organic synthesis; synthetic rubber; chemical derivatives of diene rubbers. *Mailing Add:* 466 Roslyn Ave Akron OH 44320-1241

NAPLES, JOHN OTTO, b Long Beach, Calif, Apr 2, 47; m 69; c 2. ORGANIC CHEMISTRY, POLYMER CHEMISTRY. *Educ:* Stanford Univ, BS, 69; Univ Calif, Los Angeles, PhD(org chem), 74. *Prof Exp:* RES FEL, ROHM & HAAS CO, 74- *Mem:* Am Chem Soc. *Res:* Synthesis of reactive polymers for use as ion exchangers, adsorbents, chromatographic media or membranes, for use in selective separations on analytical, preparative or industrial scale; synthesis of pour point depressants, viscosity index improvers or sludge dispersants for use in engine, hydraulic or gear oil. *Mailing Add:* Rohm & Haas Co Norristown & McKean Rds Spring House PA 19477

NAPLES, VIRGINIA L, b Worcester, Mass; m. FUNCTIONAL MORPHOLOGY, VERTEBRATE PALEONTOLOGY. *Educ:* Univ Mass, BS, 72, MS, 75, PhD(zool), 80. *Prof Exp:* Biol fac biol comp & vert anat, Dept Biol, Mt Holyoke Col, 78-79; fel gross anat, Dept Oral Anat, Univ Ill Med Ctr, 80-83; ASSOC PROF, DEPT BIOL SCI, NORTHERN ILL UNIV, DE KALB, 83- *Concurrent Pos:* Prin investr, NSF grants, 81- *Mem:* Paleont Soc; Am Soc Zoologists; Sigma Xi; Soc Study Evolution; Am Soc Mammalogists. *Res:* Functional morphology, paleontoloy, evolution and systematics of the Xenarthra. *Mailing Add:* Dept Biol Sci Northern Ill Univ De Kalb IL 60115-2861

NAPOLI, JOSEPH LEONARD, b Poughkeepsie, NY, Aug 13, 48; m 87. LIPID METABOLISM, HORMONAL CONTROL OF PROTEIN EXPRESSION. *Educ:* Siena Col, BS, 70; Univ Mich, MS, 72, PhD(med chem), 75. *Prof Exp:* Res fel biochem, Univ Wisc, 75-79; asst prof biochem, South Western Med School, Dallas, 79-85; ASSOC PROF BIOCHEM, STATE UNIV NY, BUFFALO, 85- *Concurrent Pos:* Consult, Fed Am Socs Exp Biol, 85 & Pub Health Serv, 85. *Mem:* Am Soc Biochem & Molecular Biol; Am Soc Bone Mineral Res; Am Inst Nutrit. *Res:* The metabolism and function of retinoids; determining what cells and enzymes synthesize active retinoids and how active retinoids direct differentiation, embryogenesis, and morphogenesis in target cells. *Mailing Add:* Dept Biochem Sch Med & Biomed Sci State Univ NY Buffalo Buffalo NY 14214

NAPOLITANO, LEONARD MICHAEL, b Oakland, Calif, Jan 8, 30; m 55; c 3. ANATOMY. *Educ:* Univ Santa Clara, BS, 51; St Louis Univ, MS, 54, PhD(anat), 56. *Prof Exp:* Instr anat, Med Col, Cornell Univ, 56-58; from instr to asst prof, Sch Med, Univ Pittsburgh, 58-64; assoc prof, 64-70, PROF ANAT, SCH MED, UNIV NMEX, 70-, DEAN SCH MED, 72-, DIR MED CTR, 77- *Mem:* Am Asn Anat; Am Asn Cell Biol; Electron Micros Soc Am. *Res:* Autonomic nervous system; fine structure of adipose tissue, heart and myelin. *Mailing Add:* Dept Anat Univ of NMex Sch of Med Albuquerque NM 87131

NAPOLITANO, LEONARD MICHAEL, JR, b New York, NY, Nov 7, 56; m 79. COMPUTER ARCHITECTURE. *Educ:* Mass Inst Technol, BS(am studies), 78, BS(archit), 78, BS(civil eng), 78, MS, 79; Stanford Univ, PhD(comput archit), 86- *Prof Exp:* SR MEM TECH STAFF, SANDIA NAT LAB, 79- *Mem:* Sigma Xi; Inst Elec & Electronics Engrs; Asn Comput Mach. *Res:* Theory, design and construction of high performance, special-purpose computers; parallel processor communication networks; use of parallel and pipelined computers. *Mailing Add:* 825 El Quanito Dr Danville CA 94526

NAPOLITANO, RAYMOND L, b New York, NY, Feb 7, 47; m 70; c 1. MICROBIOLOGY, BIOCHEMISTRY. *Educ:* Manhattan Col, BS, 68; St John's Univ, MS, 70, PhD(biochem), 73. *Prof Exp:* Fel elec microbiol, Brooklyn Col, City Univ New York, 73-74, asst prof biol, New York City Col, 74-78; asst prof biol, Manhattanville Col, 78-80; MEM STAFF, ANIMAL HOSP, NEW YORK ZOOL SOC, 80- *Concurrent Pos:* Adj asst prof biol, C W Post Col, Long Island Univ, 74-; res assoc, NY Zool Soc, 75- *Mem:* Sigma Xi; Am Soc Microbiol; Am Soc Protozoologists; Am Soc Zoologists; NY Acad Sci. *Res:* Isolation and axenic cultivation of free living ciliates from various sources and parasitic amoebae with special emphasis on the importance of lipids in the physiological role. *Mailing Add:* 29 Jacobs Rd Thiells NY 10984

NAPORA, THEODORE ALEXANDER, b Ridgewood, NJ, Sept 14, 27. BIOLOGICAL OCEANOGRAPHY. *Educ:* Columbia Univ, BS, 51; Univ RI, MS, 53; Yale Univ, PhD(biol), 64. *Prof Exp:* ASSOC PROF OCEANOG, UNIV RI, 64-, ASST DEAN GRAD SCH OCEANOG, 71-, DEAN STUDENTS, 77- *Mem:* AAAS; Am Soc Limnol & Oceanog. *Res:* Plankton ecology; composition and distribution of oceanic zooplankton; physiology of deep-sea organisms. *Mailing Add:* Dept Oceanog Univ of RI Kingston RI 02881

NAPPI, ANTHONY JOSEPH, b New Britain, Conn, Oct 21, 37; m; c 3. INSECT PHYSIOLOGY, PATHOLOGY. *Educ:* Cent Conn State Col, BS, 59, MS, 64; Univ Conn, PhD(entom, zool), 68. *Prof Exp:* Instr biol, Cent Conn State Col, 64-65; res asst entom, Univ Conn, 65-67; univ res grant, State Univ NY Col Oswego, 68-69, NSF res grant, 71-72, asst prof biol, 68-70, assoc prof, 71-76, prof, 76-81, PROF BIOL & CHMN DEPT, LOYOLA UNIV, CHICAGO, 81- *Concurrent Pos:* NIH res grant, 74; Am Cancer Soc scholarship, 75. *Mem:* AAAS; Am Inst Biol Sci; Am Soc Zool; Soc Invertebrate Path; Am Soc Parasitologists. *Res:* Cellular immune mechanisms of insects against metazoan parasites; insect pathology; parasitology. *Mailing Add:* Dept Biol Loyola Univ Lakeshore Campus 6525 N Sheridan Rd Chicago IL 60626

NAPTON, LEWIS KYLE, b Bozeman, Mont, Nov 15, 33; m 60; c 2. ARCHAEOLOGY, PHYSICAL ANTHROPOLOGY. *Educ:* Mont State Univ, BS, 59; Univ Mont, MA, 65; Univ Calif, Berkeley, PhD(anthrop), 70. *Prof Exp:* Wenner-Gren fel, Univ Calif, Berkeley, 68-69, asst prof anthrop, 70-71; assoc prof, 72-74, prof anthrop, 74-81, PROF ARCHAEOL, CALIF STATE COL, STANISLAUS, 81- *Concurrent Pos:* NSF fel, Univ Calif, Berkeley, 70-; NSF res assoc, Cent Australian Exped, 73-74. *Res:* North American archaeology; paleoanthropology; environmental archaeology, prehistoric man in arid environments. *Mailing Add:* Dept Anthrop Geog Calif State Col Stanislaus 801 W Monte Vista Ave Turlock CA 95380

NAQVI, IQBAL MEHDI, b New Delhi, India, Jan 6, 39; m 64; c 2. DEVICE PHYSICS, SOLID STATE ELECTRONICS. *Educ:* Univ Panjab, WPakistan, BS, 58; Youngstown Univ, BEng, 60; Univ Pa, MS, 61; Cornell Univ, PhD(electrophys), 69. *Prof Exp:* Sr engr electronic data processing div, Honeywell, Inc, 61-66; res asst elec eng, Cornell Univ, 66-69; asst prof elec eng, Univ Hawaii, 70-73; device physicist, Rockwell Int, 76-78; head device characterization lab, Hughes Aircraft Co, 78-; SR CONSULT ENG, BASIC FOUR INFO SYSTS, TUSTIN, CALIF. *Concurrent Pos:* Vis lectr, Calif State Univ, Fullerton, 77, Calif State Polytech Univ, Pomona, 78 & Univ Calif, Irvine, 82- *Mem:* AAAS; Inst Elec & Electronics Engrs. *Res:* Microwave solid state devices; noise in semiconductor devices; semiconductor device processing; integrated circuit technology; semiconductor memories; electronic circuit design; computer aided circuit design. *Mailing Add:* 14771 Doncaster Rd Irvine CA 92714

NAQVI, S REHAN HASAN, b Amroha, India, Apr 28, 33; m 74. REPRODUCTIVE ENDOCRINOLOGY, CONTRACEPTIVE STUDIES. *Educ:* Lucknow Univ, India, BS, 51; Aligarh Univ, India, MS, 59; Purdue Univ, PhD(reprod physiol), 67. *Prof Exp:* Sci teacher biol phys chem, I M Intermediate Col, Amroha, India, 51-57; asst prof zool & fish physiol, Aligarh Univ, India, 63; teaching & res asst biol, Purdue Univ, 64-67; res fel, dept obstet & gynec, Kans Univ Med Ctr, 67-69, res assoc, 70-71; res instr reproductive biol, dept obstet & gynec, Sch Med, Wash Univ, 71-72; sect head reproductive physiol, Mason Res Inst, Worcester, 72-78, DIR REPROD PHYSIOL, TSI/MASON INST, 79-, VPRES, 87- *Concurrent Pos:* NIH fel steroid biochem, Worcester Found Exp Biol, Mass, 69-70. *Mem:* Endocrine Soc; Soc Study Reproduction. *Res:* Biological and biochemical studies in the areas of reproductive endocrinology and chemical contraception using animal models; development of radioimmunoassays and pharamacokinetic studies. *Mailing Add:* TSI/Mason Res Inst 57 Union St Worcester MA 01608

NAQVI, SAIYID ISHRAT HUSAIN, b Saharanpur, India, June 29, 31; m 79; c 3. NUCLEAR PHYSICS, ASTRONOMY. *Educ:* Univ Lucknow, BSc, 51, MSc, 53; Univ Man, MSc, 56, PhD(nuclear physics), 61. *Prof Exp:* Lectr physics, Univ Man, 56-61; asst prof, St Paul's Col, Man, 61-65 & Univ Man, 65-66; asst prof, 66-68, assoc prof physics, 68-77, PROF PHYSICS, UNIV REGINA, 77-, HEAD, DEPT PHYSICS, 87- *Concurrent Pos:* Vis prof, Atomic Energy Can, Ltd, Chalk River, 62, Mass Inst Technol, 63, Copenhagen Univ Observ, 72-73 & TRIUMF, Vancouver, 83, 85-86. *Mem:* Int Astron Union; Royal Astron Soc Can; Can Asn Univ Teachers. *Res:* Beta and gamma ray spectroscopy; low and medium energy nuclear physics; photoelectric photometry; low-mass pseudo-scalar boson; intermediate energy nuclear physics. *Mailing Add:* Dept Physics & Astron Univ Regina Regina SK S4S 0A2 Can

NARA, HARRY R(AYMOND), b New York Mills, Minn, Sept 19, 21; m 46; c 3. ENGINEERING MECHANICS. *Educ:* Case Inst Technol, BS, 46, MS, 48, PhD(eng mech), 51. *Prof Exp:* From instr to assoc prof eng mech, Case Western Reserve Univ, 46-57, prof struct & mech & head dept, 57-61, assoc head, Eng Div, 61-64, vprovost, 64-67, assoc dean, Sch Eng, 68-73, chmn, Comput Eng & Sci Dept, 85-86, prof, 61-86, EMER PROF ENG, CASE WESTERN RESERVE UNIV, 86- *Concurrent Pos:* Pres, Dicar Corp, 72-76. *Mem:* Am Soc Civil Engrs; Am Soc Mech Engrs; Soc Exp Stress Anal; Am Soc Eng Educ; Sigma Xi. *Res:* Structures; soil mechanics; dynamics; reinforced plastics; adhesives. *Mailing Add:* 7174 Kinsman Rd Novelty OH 44072

NARAHARA, HIROMICHI TSUDA, b Tokyo, Japan, Oct 24, 23; US citizen; m 54; c 4. BIOCHEMISTRY, METABOLISM. *Educ:* Columbia Univ, BA, 43, MD, 47; Am Bd Internal Med, dipl, 55. *Prof Exp:* USPHS res fel med, Univ Wash, 53-56, res instr, 56-58; USPHS spec res fel biol chem, Wash Univ, 58-60, from asst prof to assoc prof, 60-70; MEM STAFF, DIV LABS & RES, NY STATE DEPT HEALTH, 70-; ASSOC PROF BIOCHEM, ALBANY MED COL, 70- *Mem:* Am Diabetes Asn; Endocrine Soc; Am Soc Biol Chem. *Res:* Intermediary metabolism of carbohydrates; muscle physiology; effect of hormones and muscle contraction on carbohydrate metabolism; cell membranes; toxicology. *Mailing Add:* Div Environ Health Assessment NY State Dept Health Albany NY 12203-3310

NARAHASHI, TOSHIO, b Fukuoka, Japan, Jan 30, 27; m 56; c 2. NEUROTOXICOLOGY, NEUROPHARMACOLOGY. *Educ:* Univ Tokyo, BS, 48, PhD(insect neurotoxicol), 60. *Prof Exp:* Res assoc physiol, Univ Chicago, 61-62, asst prof, 62; asst prof, 62-63, from asst prof to prof, 65-69, vchmn dept, 73-75, prof physiol & pharmacol, Med Ctr, Duke Univ, 69-77; PROF & CHMN, PHARMACOL DEPT, MED SCH, NORTHWESTERN UNIV, CHICAGO, 77-, A N RICHARDS PROF, 82-, J EVANS PROF, 86- *Concurrent Pos:* Res assoc, Fac Agr, Univ Tokyo, 51-65; Javits neurosci investr award, NIH, 86- *Honors & Awards:* Japanese Soc Appl Entom & Zool Prize, 55; Cole Award, Biophys Soc, 81; K P DuBois Award, Soc Toxicol, 88, Toxicol Merit Award, 91; Burdick & Jackson Int Award, Am Chem Soc, 89. *Mem:* Soc Neurosci; Am Soc Pharmacol & Exp Therapeut; Am Physiol Soc; Soc Toxicol; Biophys Soc; Sigma Xi; fel AAAS. *Res:* Electrophysiology and pharmacology of nerve and muscle membrane ion channels and synaptic junctions in general; basic insect neurophysiology; neurotoxicology of insecticides; neurotoxicology of environmental toxicants. *Mailing Add:* Dept Pharmacol Northwestern Univ Med Sch 303 E Chicago Ave Chicago IL 60611-3008

NARAIN, AMITABH, b India, July 30, 55; m 85. FLUID MECHANICS, CONTINUUM MECHANICS. *Educ:* Indian Instit of Technol, BTech, 78, Univ Minn, MS, 80, PhD(mech), 83. *Prof Exp:* ASST PROF, ENG MECH, MICH TECHNOL UNIV, 83- *Concurrent Pos:* Consult, math, Res Ctr, Univ Wis, 84; prin investr, NAG 3-711, NASA-Lewis, 86-; vis prof, Univ Minn, 88- *Mem:* Am Soc Mech Eng; Soc Rheology; Soc Natural Philos; Am Acad Mech; Soc Eng Sci. *Res:* Shock propogation in viscoelastic flows; constitutive modeling of viscoelastic liquids; flow and stability analysis for pure vapors condensing over cold surfaces. *Mailing Add:* Dept Mech Eng Mich Technol Univ Houghton MI 49931

NARANG, SARAN A, b Agra, India, Sept 10, 30; c 1. ORGANIC CHEMISTRY, MOLECULAR BIOLOGY. *Educ:* Panjab Univ, India, BSc, 51, MSc, 53; Univ Calcutta, PhD(org chem), 60. *Prof Exp:* Sr res fel chem, Indian Asn Cultivation Sci, Calcutta, 59-62; res assoc, Johns Hopkins Univ, 62-63; proj assoc molecular biol, Inst Enzyme Res, Univ Wis-Madison, 63-66; asst res officer, Div Pure Chem, 66-67, assoc res officer, Div Biochem & Molecular Biol, 67-73, SR RES OFFICER, DIV BIOL SCI, NAT RES COUN CAN, 73- *Concurrent Pos:* Adj prof, Carleton Univ, 73- *Honors & Awards:* Coochbihar Professorship Mem Award, Indian Asn Cultivation Sci, Calcutta, 74. *Res:* Chemico-enzymatic synthesis of DNA and RNA and their biological roles; DNA-protein recognition and studies viroids. *Mailing Add:* Div Biol Sci Nat Res Coun Can 100 Sussex Dr Ottawa ON K1A 0R6 Can

NARASIMHACHARI, NEDATHUR, b Nellore, India, Apr 15, 26; m 44; c 3. PHARMACOLOGY, ANALYTICAL BIOCHEMISTRY. *Educ:* Andhra Univ, India, BSc Hons, 46, MSc, 47; Delhi Univ, India, PhD(org chem), 51. *Prof Exp:* Lectr chem, Delhi Univ, 49-53; chief org chemist, Antibiotics Res Ctr, Hindustan Antibiotics, Pimpri, Poona, India, 53-68; admin res scientist psychiat, Thudichum Psychiat Res Lab, Galesburg, Ill, 68-75; admin res scientist psychopharmacol, Ill State Psychiat Inst, Chicago, 75-78; PROF PSYCHIAT & PHARMACOL, MED COL VA, VA COMMONWEALTH UNIV, RICHMOND, 78- *Concurrent Pos:* WHO fel, Geneva, 53; Nat Res Coun fel, 60-62. *Mem:* Soc Biol Chemists; Soc Pharmacol & Exp Therapeut; Soc Neurosci; Am Soc Neurochem; Am Soc Mass Spectrometry; Am Chem Soc. *Res:* Chemistry of natural products; microbial metabolites; biochemistry and behavior; biochemical correlates of mental disorders; drug metabolism; mode of action of neuroleptic drugs. *Mailing Add:* Dept Psych & Pharmacol Med Col Va Box 710 MCV Sta Richmond VA 23298

NARASIMHAN, MANDAYAM A, b Tarikere, India, July 5, 49; m 77. ELECTRICAL ENGINEERING, COMPUTER SCIENCE. *Educ:* Bangalore Univ, BE, 69; Indian Inst Sci, Bangalore, ME, 71; Univ Tex, Arlington, PhD(elec eng), 75. *Prof Exp:* Lectr elec eng, Bangalore Univ, 72-73; mem tech staff, elec eng, 74-79, MGR, RES & DEVELOP, TEX INSTRUMENTS, 79-, SR MEM TECH STAFF, 81- *Mem:* Sigma Xi; Inst Elec & Electronics Engrs. *Res:* Digital image processing; pattern recognition; artificial intelligence; robotics and industrial automation; intelligent systems. *Mailing Add:* Box 655621 Ms354 Tex Instruments Inc Dallas TX 75265

NARASIMHAN, MYSORE N L, b Mysore City, India, July 7, 28; US citizen; m 49; c 3. ENGINEERING SCIENCE, CONTINUUM MECHANICS. *Educ:* Univ Mysore, MSc, 51; Indian Inst Technol, Kharagpur, PhD(math), 58. *Prof Exp:* Lectr math, Lingaraj Col, India, 51-55; asst lectr, Indian Inst Technol, Kharagpur, 55-58; lectr, Indian Inst Technol, Bombay, 58-61, asst prof, 61-62, assoc prof, 64-65; res prof, Math Res Ctr, Univ Wis, 62-64; assoc prof, Univ Calgary, 65-66; PROF MATH, ORE STATE UNIV, 66- *Concurrent Pos:* Vis prof, Princeton Univ, 72-73; guest scientist, Polish Acad Sci, 79-80; res scientist, US Army Res Off, 84-85. *Honors & Awards:* Iyengar Mem Prize, Mysore, 51. *Mem:* Am Math Soc; Soc Indust & Appl Math; US Soc Eng Sci. *Res:* Non-Newtonian fluid flows; flow through elastic tubes; porous channel and magnetohydrodynamic flows; stability of fluid flows; microcontinuum theory; liquid crystal theory; thermodynamics; micro and nonlocal continuum mechanics; fracture mechanics; turbulent flows; lubrication flows. *Mailing Add:* Dept Math Ore State Univ Corvallis OR 97331-4605

NARASIMHAN, THIRUPPUDAIMARUDHUR N, b Madras City, India, Oct 6, 35; m 62; c 1. HYDROGEOLOGY, CIVIL ENGINEERING. *Educ:* Univ Madras, BSc, 56; Univ Calif, Berkeley, MS, 71, PhD(eng sci), 75. *Prof Exp:* Geol asst, Geol Surv India, 56-57, asst geologist, 57-64, geologist, 64-70; res engr, Lawrence Berkeley Lab, 75-76, staff scientist, 76-80, staff sr scientist, 80-90; PROF MINERAL ENG & SOIL SCI, UNIV CALIF, BERKELEY, 90- *Concurrent Pos:* Res hydrologist, Nat Coun Appl Econ Res, New Delhi, India, 68, prof-in-residence, 82-; lectr, Dept Mat Sci & Mineral Eng, Univ Calif, Berkeley, 77-82; vis prof, Lamont Doherty Geol Observ, Columbia Univ, New York, 85-86. *Honors & Awards:* Oscar E Meinzer Award, Geol Soc Am, 86. *Mem:* Fel Geol Soc India; Am Geophys Union; fel Geol Soc Am. *Res:* Flow through porous media; mathematical modeling of groundwater systems; geothermal reservoir engineering; well testing; flow in fractured rocks; chemical transport in groundwater systems; water in geologic processes. *Mailing Add:* Dept Mats Sci & Mineral Eng Univ Calif Berkeley CA 94720

NARATH, ALBERT, b Berlin, Ger, Mar 5, 33; US citizen; m 58, 76; c 4. ELECTRONICS ENGINEERING. *Educ:* Univ Cincinnati, BS, 55; Univ Calif, Berkeley, PhD(phys chem, molecular spectros), 59. *Prof Exp:* Mem tech staff & dept mgr solid state res, Sandia Nat Labs, 59-68, dir solid state sci res, 68-71, managing dir phys sci, 71-73, vpres res & systs develop, Sandia Nat Labs, 73-82, exec vpres, 82-84; VPRES, AT&T BELL LABS, 84- *Mem:* Fel Am Phys Soc; AAAS; Nat Acad Eng. *Res:* Nuclear magnetic resonance in nonmetallic magnetic crystals and in transition metals and intermetallic compounds; properties of ferromagnets and antiferromagnets. *Mailing Add:* Sandia Nat Labs PO Box 5800 Albuquerque NM 87185

NARAYAN, JAGDISH, b Kanpur, India, Oct 15, 48; US citizen; m 73; c 1. MATERIALS SCIENCE, SOLID STATE PHYSICS. *Educ:* Indian Inst Technol, BTech, 69; Univ Calif, Berkeley, MS, 70, PhD(mat sci), 71. *Prof Exp:* Res metallurgist mat sci; Lawrence Berkeley Lab & Univ Calif, Berkeley, 71-72; mem res staff mat sci & solid state physics, Oak Ridge Nat Lab, 72-84; DISTINGUISHED PROF & DIR, NC STATE UNIV, 84-; DIR DIV MAT RES, NSF, 90- *Honors & Awards:* India President's Gold Medal, Indian Inst Technol, Kanpur & Govt India, 69; Sustained Basic Res Award, Dept Energy, 81; IR-100 Award, 79, 81 & 82. *Mem:* Fel Nat Acad Sci; fel Am Phys Soc; Am Inst Mining, Metall & Petrol Engrs; fel Am Phys Soc; fel AAAS; fel Am Soc Metals Int. *Res:* Defects in oxides; radiation damage in metals; ion implantation and defect physics of semiconductors; laser-solid interactions; electric microscopy. *Mailing Add:* Dept Mat Eng NC State Univ Raleigh NC 27695

NARAYAN, KRISHNAMURTHI ANANTH, b Secunderabad, India, Oct 1, 30; m 61; c 2. NUTRITIONAL BIOCHEMISTRY, FOOD SCIENCE. *Educ:* Madras Univ, BS, 49; Osmania Univ, India, MS, 51; Univ Ill, Urbana, PhD(food technol), 57. *Prof Exp:* Res assoc phys chem, Wash State Univ, 57-60, res assoc agr chem, 61-62; sci officer, Nutrit Res Lab, 60-61; from asst prof to assoc prof food chem, Univ Ill, Urbana, 62-71; RES BIOCHEMIST, SSD/BIOCHEM, US ARMY NATICK RES & DEVELOP CTR, 71- *Concurrent Pos:* Nat Cancer Inst career develop award, 66- *Mem:* Am Oil Chem Soc; Am Inst Nutrit; NY Acad Sci; Inst Food Technol. *Res:* Lipids in cancer; lipoprotein metabolism; disc electrophoresis and lipoproteins; wheat and serum proteins; oxidized lipid-protein complexes; lipoproteins in cancer; liver plasma membranes; essential fatty acid deficiency; absorption, transport and utilization of lipids; nutrient degradation; bioavailability of proteins and fats; glycerol metabolism. *Mailing Add:* SSD/Biochem US Army Natick RD&E Ctr Natick MA 01760

NARAYAN, LATHA, b Coimbatore, India, Oct 5, 45; m 68; c 2. MICROBIOLOGY. *Educ:* Madras Univ, India, BSc, 66; Univ Detroit, MS, 72. *Prof Exp:* Microbiologist, Difco Labs, 78-84, mgr prod eval, 84-86, asst dir qual control, 86-88, DIR QUAL CONTROL, DIFCO LABS, 88- *Mem:* Am Soc Microbiol; Am Soc Qual Control. *Res:* Current technologies in microbiological testing and best methods to implement quality management applications. *Mailing Add:* Difco Labs 920 Henry St Detroit MI 48231

NARAYAN, OPENDRA, b Essequibo, Guyana, Nov 28, 36; Can citizen; m 63; c 3. VIROLOGY. *Educ:* Univ Toronto, DVM, 63; Univ Guelph, PhD(virol), 70. *Prof Exp:* Veterinarian, pvt pract, Man, 63-65; Can Med Res Coun fel, 70-72; asst prof neurol, 72-74, ASSOC PROF COMP MED, SCH MED, JOHNS HOPKINS UNIV, 74- *Mem:* AAAS; Am Asn Microbiol; Am Asn Neuropath. *Res:* Slow virus infections; mechanisms of virus infections of the brain; viral teratology. *Mailing Add:* Dept Comparitive Med John Hopkins Univ 720 Rutland Baltimore MD 21205

NARAYAN, TV LAKSHMI, b Udamalpet, India, June 5, 37; m 68; c 2. ORGANIC CHEMISTRY, POLYMER CHEMISTRY. *Educ:* Univ Madras, BSc, 58; Annamalai Univ, Madras, MSc, 61; Univ 21 Pa, PhD(chem), 65. *Prof Exp:* Fel polymer chem, Univ Ariz, 65-66; res chemist, Am Cyanamid Co, Conn, 66-69; sr res chemist, 69-80, res assoc, 80-87, SR RES ASSOC, BASF CORP, 87- *Mem:* Am Chem Soc; Royal Soc Chem. *Res:* Organophosphorous and sulfur chemistry; thermostable polymers; engineering thermoplastics; flame retardant polymers; isocyanate chemistry and production; polyurethane chemistry. *Mailing Add:* Corp Res & Develop BASF Corp Wyandotte MI 48192

NARAYANA, TADEPALLI VENKATA, mathematics, for more information see previous edition

NARAYANAMURTI, VENKATESH, b Bangalore, India, Sept 9, 39; m 61; c 3. EXPERIMENTAL SOLID STATE PHYSICS. *Educ:* Univ Delhi, BS, 58, MS, 60; Cornell Univ, PhD(physics), 65. *Prof Exp:* Res assoc physics, Cornell Univ, 64-65, instr, 67-68; asst prof, Indian Inst Technol, Bombay, 65-66; mem tech staff physics, Bell Labs, 68-76, head, Semiconductor Electronics Res Dept, 76-81, dir, Solid State Electronics Res Lab, 81-87; VPRES, RES SANDIA NAT LAB, 87- *Mem:* Fel Am Phys Soc; fel Acad Sci India; AAAS; fel Inst Elec & Electronics Engrs; Royal Swedish Acad Eng Sci. *Res:* Phonons in solids and liquid helium; second sound and sound propagation in matter; superconductivity; metal-insulator transitions under pressure; energy transport in semiconductors. *Mailing Add:* Org 1000 Sandia Nat Lab Albuquerque NM 87185

NARAYANAN, A SAMPATH, b Tamilnadu, India, Jan 14, 41; m 75; c 2. CELL GROWTH & MATRIX SYNTHESIS, INFLAMMATION WOUND REPAIR. *Educ:* Univ Madras, India, PhD(biochem), 67. *Prof Exp:* Res officer, Univ Madras, India, 65-67; res assoc, 71-75, from asst prof to assoc prof, 75-86, PROF RES, UNIV WASH, SEATTLE, 86- *Concurrent Pos:* US vis fel, NIH, 69-71; chairperson, Gordon Res Conf Periodont Dis, 81; prin investr, NIH grants, 88-, peer rev; lectr, Univ Wash; assoc ed, J Perrodont Res; consult, Regeneron Corp. *Mem:* Am Soc Biol Chemists; Am Soc Cell Biol; Int Asn Dent Res; Can Biomed Soc. *Res:* Mechanisms of alterations in connective tissues such as skin, bone, heart disease and inflammatory diseases. *Mailing Add:* Dept Path SM-30 Sch Med Univ Wash Seattle WA 98195

NARAYANASWAMY, ONBATHIVELI S, b Madras, India, May 13, 36; m 67. ENGINEERING MECHANICS. *Educ:* Univ Madras, BE, 58; Univ Sask, MS, 62; Case Western Reserve Univ, PhD(eng mech), 65. *Prof Exp:* Jr sci officer, Atomic Energy Estab, India, 59-60; lectr, Indian Inst Technol, 60-61; sr res scientist, 65-69, sr res engr, 69-72, assoc prin res engr, 72-81, staff scientist, Ford Motor Co, 81-86; ENG CONSULT, 87- *Honors & Awards:* Ross Coffin Purdy Award, Am Ceramic Soc, 73. *Mem:* Fel Am Ceramic Soc; Soc Exp Stress Anal. *Res:* Viscoelasticity applied to glass fabrication problems; glass science; theoretical and experimental stress analysis. *Mailing Add:* Comput Simulations 26610 Hass Ave Dearborn Heights MI 48127

NARCOWICH, FRANCIS JOSEPH, b Gary, Ind, Jan 23, 46; c 1. PARTIAL DIFFERENTIAL EQUATIONS, OPERATOR THEORY. *Educ:* De Paul Univ, BS, 68; Princeton Univ, MA, 70, PhD(math physics), 72. *Prof Exp:* NSF fel, Princeton Univ, 68-71, fel, 71-72; from asst to assoc prof, 72-87, PROF MATH, TEX A&M UNIV, 87- *Concurrent Pos:* Appl mathematician, Zenith Radio Corp, 69, 70 & 71; consult; vis mem, Courant Instit, 85-86. *Mem:* Am Math Soc. *Res:* Relationship between classical and quantum mechanics. *Mailing Add:* Dept Math Tex A&M Univ College Station TX 77843-3368

NARDELLA, FRANCIS ANTHONY, b Clarksburg, WVa, Aug, 5, 42; m 70; c 2. RHEUMATOLOGY, INTERNAL MEDICINE. *Educ:* WVa Univ, AB, 64, MD, 68. *Prof Exp:* Intern, Hennepin County Gen Hosp, Minneapolis, 68-69; gen med officer, US Navy, 69-72; resident med, 72-74, fel rheumatology, 74-77, instr, 77-79, asst prof med, 77-84, ASSOC PROF MED, UNIV WASH, SEATTLE, 84- *Concurrent Pos:* Consult, IMRE Corp, Seattle, 83- *Mem:* Am Rheumatism Asn; Am Asn Immunologists; Am Fedn Clin Res. *Res:* Role bacteria and viruses in the production of rheumatoid factors in rheumatoid arthritis. *Mailing Add:* Dept Med RG-28 Univ Wash Seattle WA 98195

NARDI, JAMES BENJAMIN, b Clinton, Ind, Oct 9, 48. DEVELOPMENTAL BIOLOGY. *Educ:* Purdue Univ, BS, 70; Harvard Univ, PhD(biol), 75. *Prof Exp:* Fel biol, Med Res Coun, Lab Molecular Biol, 75-76; res assoc biol, Dept Genetics & Develop, 76-83, RES SCIENTIST, UNIV ILL, 83. *Concurrent Pos:* NATO fel, NSF, 75-76; vis asst prof, Univ Ill, 78-79; NSF grants, 81, 83 & 88. *Mem:* Soc Exp Biol; AAAS; Soc Develop Biol. *Res:* Cell interactions during development. *Mailing Add:* Dept of Entom Univ of Ill Urbana IL 61801

NARDI, JOHN CHRISTOPHER, b Cleveland, Ohio, June 10, 46; m 69; c 2. ELECTROCHEMISTRY. *Educ:* Bowling Green State Univ, BS, 68, MA, 69. *Prof Exp:* Prof chem, US Air Force Acad, 74-76, res scientist molten salts, Seiler Res Lab, 76-77; STAFF ELECTROCHEMIST, EVEREADY BATTERY CO, 77- *Mem:* Electrochem Soc. *Res:* Fundamental research of aqueous based batteries (primary) utilizing manganese dioxide; fundamental mechanisms of lithium nonaqueous and molten salt electrochemistry. *Mailing Add:* 3398 Tyler Dr Brunswick OH 44212-3726

NARDI, VITTORIO, b Ravenna, Italy, Oct 9, 30; m 56; c 4. STATISTICAL PHYSICS, PLASMA PHYSICS. *Educ:* Univ Rome, Italy, PhD(physics & field theory), 54; Italian Ministry Educ, Rome, Libero Docente, 67. *Prof Exp:* Fel res scientist, Univ Rome, 54-55, Univ Amsterdam, 55-56, Univ Padua, 56-59; lectr & prof statist mech & thermodynamics, Univ Padua & Ferrara, Italy, 59-63; vis res scientist, NY Univ, 62-63 & Courant Inst Math Sci, NY Univ, 64-67; vis res prof electrodynamics, waves propagation & gen physics, 67-79, RES PROF PHYSICS, STEVENS INST TECHNOL, 79- *Concurrent Pos:* Consult, Nat Inst Nuclear Physics, Rome, 57-64, Nat Inst Electrotech, Turin, 72-78 & Lawrence Livermore Lab, 76-78; Fulbright Scholar, 62-63; prin investr particle beam physics, plasma focus & particle beam generators res, Physics Dept, Stevens Inst Technol, 72-; ed, Energy, Storage, Compression & Switching Conf Proceedings, 74- *Honors & Awards:* Davis Res Award, Stevens Inst Technol, 73. *Res:* Dynamics and structure of particle beams; propagation of relativistic electron beams of high density; plasma focus and pulsed power systems. *Mailing Add:* Dept Physics & Eng Physics Stevens Inst Technol Hoboken NJ 07030

NARDO, SEBASTIAN V(INCENT), b Brooklyn, NY, Dec 25, 17; m 42; c 4. AEROSPACE ENGINEERING. *Educ:* Polytech Inst Brooklyn, BME, 40, MAE, 42, PhD(appl mech), 49. *Prof Exp:* Aeronaut engr, Chance Vought Aircraft, Tex, 42-45; from instr to assoc prof aeronaut eng, 46-60, PROF AEROSPACE ENG, POLYTECH INST NEW YORK, FARMINGDALE, 60- *Mem:* Soc Exp Stress Analysis; Am Inst Aeronaut & Astronaut; Am Soc Eng Educ; NY Acad Sci; Am Soc Mech Engrs. *Res:* Heat conduction; stability and stresses in plate and shell structures; dynamics; solar energy. *Mailing Add:* 270 Foster Ave Malverne NY 11565

NARDONE, JOHN, materials engineering, mechanical engineering, for more information see previous edition

NARDONE, ROLAND MARIO, b Brooklyn, NY, Mar 29, 28; m 51; c 4. PHYSIOLOGY. *Educ:* Fordham Univ, BS, 47, MS, 49, PhD(biol), 51. *Prof Exp:* Instr, St Francis Col, 48-51 & St Louis Univ, 51-52; from asst prof to assoc prof, 52-63, PROF BIOL, CATH UNIV AM, 63- *Concurrent Pos:* Dir, Ctr Advan Training Cell & Molecular Biol, 82-; pres, R & M Nardone Assocs, Inc. *Mem:* Tissue Cult Asn (secy, 72-76); AAAS; Am Soc Cell Biol. *Res:* Cell division; physiology of cells in culture; cytotoxicity in vitro; development of an in vitro model for the evaluation of the potential neurotoxicity of chemicals; microwave effects in vitro. *Mailing Add:* Dept Biol Cath Univ Am Washington DC 20064

NARDUCCI, LORENZO M, b Torino, Italy, May 25, 42; m 65; c 3. QUANTUM OPTICS. *Educ:* Univ Milan, PhD(physics), 64. *Prof Exp:* Asst prof quantum electronics, Univ Milan, 65-66; from asst prof physics, to assoc prof Worcester Polytech Inst, 66-76; from assoc prof to prof. 76-88, FRANCIS K DAVIS PROF PHYSICS, DREXEL UNIV, 88- *Concurrent Pos:* ed J Optics Commun; assoc ed J Physical Rev A. *Honors & Awards:* Drexel Univ Res Award, 80. *Mem:* Fel Am Phys Soc; fel Optical Soc Am; Sigma Xi; AAAS. *Res:* Laser theory; interaction of radiation and matter; quantum statistics; light scattering and phase transitions; bistability; nonlinear optics. *Mailing Add:* Dept Physics & Atmospheric Sci Drexel Univ Philadelphia PA 19104

NARENDRA, KUMPATI S, b Madras, India, Apr 14, 33; m 61. CONTROL SYSTEMS. *Educ:* Univ Madras, BE, 54; Harvard Univ, SM, 55, PhD(appl physics), 59; Yale Univ, MA, 68. *Prof Exp:* Lectr appl physics, Harvard Univ, 59-61, asst prof, 61-65; assoc prof appl sci, 65-68, chmn elec eng dept, 84-87, PROF APPL SCI, YALE UNIV, 68-, DIR, CTR SYSTS SCI, 81-, CHMN ELEC ENG DEPT, 84- *Concurrent Pos:* Consult, Boston Div, Minneapolis Honeywell Regulator Co, Mass, 59-61, Sperry Rand Res Ctr, 61-64, Dynamics Res Corp, 65-67, Bell Aerosysts Co, NY, 66-67, Sikorsky Aircraft, 67-73, Long Lines Div, Am Tel & Tel Co, 75-81 & Borg-Warner Corp, 79-81, Gen Motors Res Labs, 82-85; vis assoc prof, Indian Inst Sci, Bangalore, 64-65; ed, J Cybernetics & Info Sci; assoc ed, Inst Elec & Electronics Engrs-Control Systs Soc & Trans Automatic Control; assoc ed, Circuits Systs & Signal Processing & Int J Adaptive Control & Signal Processing. *Honors & Awards:* Franklin V Taylor Award, Inst Elec & Electronics Engrs-SMC Soc, 72. *Mem:* Fel Inst Elec & Electronics Engrs; fel Inst Elec Engrs UK; fel Am Asn Advan Sci. *Res:* Stability theory; large scale systems; adaptive control learning automata; neural networks. *Mailing Add:* 35 Old Mill Rd Woodbridge CT 06525

NARIBOLI, GUNDO A, b Dharwar, India, Sept 2, 25; m 47, 84; c 4. APPLIED MATHEMATICS. *Educ:* Univ Bombay, BSc, 47, MSc, 52; Karnatak Univ, India, MSc, 54; Indian Inst Technol, Kharagpur, PhD(appl math), 59. *Prof Exp:* Lectr math, Col Eng & Technol, Hubli, India, 52-55 & Indian Inst Technol, Kharagpur, 56-59; reader, Univ Bombay, 59-62; assoc prof, Iowa State Univ, 62-64; reader, Univ Bombay, 64-66; assoc prof math & eng mech, 66-69, PROF ENG MECH, IOWA STATE UNIV, 69- *Concurrent Pos:* Reviewer, Math Rev. *Mem:* Soc Indust & Appl Math; Math Asn Am; Am Math Soc; Sigma Xi; Am Acad Mech; Indian Soc Theoret & Appl Mech. *Res:* Linear and nonlinear waves; group-invariant solutions; Baeklund transformations; perturbation methods; method of perturbation for waves in bounded media; traffic flow theory. *Mailing Add:* Eng Sci & Mech Dept Iowa State Univ Ames IA 50010

NARICI, LAWRENCE ROBERT, b Brooklyn, NY, Nov 15, 41. MATHEMATICAL ANALYSIS. *Educ:* Polytech Inst Brooklyn, BS, 62, MS, 63, PhD(math), 66. *Prof Exp:* From instr to asst prof math, Polytech Inst Brooklyn, 65-67; assoc prof, 67-72, PROF MATH, ST JOHN'S UNIV, NY, 72- *Concurrent Pos:* Vis prof, Univ Fed Rio de Janeiro, 87. *Mem:* Am Math Soc; Math Asn Am; Mex Math Soc; Math Soc France; Israel Math Union. *Res:* Non-Archimedean Banach spaces and algebras; topological algebras; functional analysis; valuation theory. *Mailing Add:* Dept of Math St John's Univ Jamaica NY 11439

NARO, PAUL ANTHONY, b Scranton, Pa, Aug 17, 34; m 57; c 2. ORGANIC CHEMISTRY, TOXICOLOGY. *Educ:* Temple Univ, AB, 56; Pa State Univ, PhD(chem), 60. *Prof Exp:* Res chemist, Socony Mobil Oil Co, Inc, 59-61, sr res chemist, Mobil Oil Corp, 61-72, asst supvr, 64-66, admin mgr, Mobil Res & Develop Corp, 72-78, mgr toxicol opers, 78-84, MGR, QUAL ASSURANCE & INFO SCI, MOBIL OIL CORP, 84- *Mem:* Am Chem Soc; Am Col Toxicol; Soc Qual Assurance. *Res:* Hydrocarbon synthesis; organic sulfur compounds; polymer chemistry; heterogeneous catalysis; computer applications. *Mailing Add:* 11 Forrest Lane Trenton NJ 08628

NARSKE, RICHARD MARTIN, b Berwyn, Ill, July 4, 42; m 65; c 2. ORGANIC CHEMISTRY, INSTRUMENTAL ANALYSIS. *Educ:* Augustana Col, BA, 64; Univ Iowa, MS, 66, PhD(chem), 68. *Prof Exp:* Asst prof chem, Univ Tampa, 68-70, assoc prof, 70-78; head, Anal & Microbiol Lab, Ophthal Res & Develop, Milton Roy Co, 78-80; PROF CHEM AUGUSTANA COL, 80- *Concurrent Pos:* Chief chem consult, Intersci Inc, 70-78; consult, Martin-Marrieta, 82. *Mem:* Am Chem Soc (treas-secy, 69-71); Am Inst Chemists; Sigma Xi. *Res:* Gas chromatographic analysis of pesticide residues; forensic chemistry; organic metallic complexes; instrumental methods of analysis. *Mailing Add:* Chem Dept Augustana Col Augustana Col Rock Island IL 61201

NARULA, SUBHASH CHANDER, b Bannu, India, Jan 20, 44; US citizen. APPLIED STATISTICS, OPERATIONS RESEARCH. *Educ:* Univ Delhi, BE, 65; Univ Iowa, MS, 69, PhD(indust & mgt eng), 71. *Prof Exp:* Supvr prod, Hindustan Mach Tools Ltd, 65-68; asst prof indust eng, State Univ NY, Buffalo, 71-77; assoc prof, Sch Mgt, Rensselaer Polytech Inst, 77-83; PROF, DEPT DECISION SCI & BUS LAW, SCH BUS, VA COMMONWEALTH UNIV, 83- *Mem:* Fel Am Statist Asn; Royal Statist Soc; Opers Res Soc Am; Inst Mgt Sci; Math Prog Soc; Int Statist Inst; Am Soc Qual Control. *Res:* Regression analysis; sensitivity analysis; location-allocation analysis; multi-criteria decision making. *Mailing Add:* Dept Decision Sci & Bus Law Va Commonwealth Univ 1015 Floyd Ave Richmond VA 23284-4000

NARVAEZ, RICHARD, b New York, NY, May 4, 30; m 63; c 7. COMPLEX MATERIALS ANALYSIS, NEW METHODS DEVELOPMENT. *Educ:* Col City New York, BS, 51; NY Univ, PhD(phys chem), 63. *Prof Exp:* Anal chemist, Explosives Dept, Du Pont, 51-55, chemist, 57-58, res chemist, 63-65, res & sr res chemist, Textile Fibers Dept, 65-85, sr res chemist, Polymer Prod Dept, 85-90; CONSULT, 90- *Mem:* Am Chem Soc. *Res:* Design efficient methods for analysis of complex or trace materials; depolymerization and derivatization techniques for polymer analysis; chromatographic, spectroscopic techniques and their interpretation. *Mailing Add:* 708 Burnley Rd Wilmington DE 19803-1729

NASAR, SYED ABU, b Gorakhpur, India, Dec 25, 32; m 61; c 2. ELECTRICAL ENGINEERING. *Educ:* Agra Univ, BSc, 51; Univ Dacca, BScEE, 55; Tex A&M Univ, MS, 57; Univ Calif, Berkeley, PhD(elec eng), 63. *Prof Exp:* Lectr elec eng, Ahsanullah Eng Col, Dacca Univ, Pakistan, 55-63; from asst prof to assoc prof, EPakistan Univ Eng & Technol, 63-66; assoc prof, Gonzaga Univ, 66-68; assoc prof, 68-70, dir grad studies, 80-87, PROF ELEC ENG, UNIV KY, 70-, CHAIR, 89- *Concurrent Pos:* Brit Coun visitor, Imp Col, Univ London, 64; Sigma Xi-Sci Res Soc Am res award, 64-65; Sigma Xi res awards, 66-67 & 69-70; NSF res grants, 67-68, 69-70, 70-72 & 74-88; Ford Motor Co res grant, 69; consult numerous cos, 76- *Honors & Awards:* Aurel Vlaicu Award, Romanian Nat Acad, 78. *Mem:* Fel Inst Elec & Electronics Engrs; fel Brit Inst Elec Engrs. *Res:* Linear electric machines; novel electric machines. *Mailing Add:* Dept of Elec Eng Univ of Ky Lexington KY 40506

NASATIR, MAIMON, b Chicago, Ill, Apr 16, 29; div; c 3. CELL BIOLOGY. *Educ:* Univ Chicago, PhB, 50; Univ Pa, PhD(bot), 58. *Prof Exp:* Instr biol, Univ Pa, 58-59; lectr, Haverford Col, 59; USPHS fel, Univ Brussels, 59-60 & Univ Ill, 60-61; asst prof bot, Brown Univ, 61-66, asst to dean, Pembroke Col, 63-65; chmn dept biol, 66-70, PROF BIOL, UNIV TOLEDO, 66- *Concurrent Pos:* Lalor fel, 62 & 63; adj prof dept physiol, Med Col Ohio, 68-71; vis scientist, Brandeis Univ, 76, 87. *Mem:* AAAS; NY Acad Sci; Bot Soc Am; Am Soc Cell Biol. *Res:* Biochemical cytology; plant physiology; cellular biology. *Mailing Add:* Dept Biol Univ Toledo Toledo OH 43606

NASCI, ROGER STANLEY, b Pittsburgh, Pa, Jan 30, 52; m 80. BEHAVIORAL ECOLOGY, MEDICAL ENTOMOLOGY. *Educ:* Ohio Univ, BS, 74, MS, 76; Univ Mass, PhD(entom), 80. *Prof Exp:* Fel, Vector Biol Lab, Univ Notre Dame, 79-80, NIH fel, 80-; AT DEPT BIOL & ENVIRON SCI, MCNEESE STATE UNIV, LAKE CHARLES,. *Mem:* AAAS; Am Soc Trop Med & Hyg; Mosquito Control Asn; Entom Soc Am; Sigma Xi. *Res:* Mosquito behavior under natural conditions with emphasis on host-mosquito interaction; variations in behavior that may influence the dynamics of mosquito-borne disease system. *Mailing Add:* Dept Biol & Environ Sci McNeese State Univ Lake Charles LA 70609

NASH, CARROLL BLUE, b Louisville, Ky, Jan 29, 14; m 41. BIOLOGY, PARAPSYCHOLOGY. *Educ:* George Washington Univ, BS, 34; Univ Md, MS, 36, PhD, 39. *Prof Exp:* Instr zool, Univ Ariz, 39-41; assoc prof biol, Pa Mil Col, 41-44; asst prof, Am Univ, 44-45; prof, Wash Col, Md, 45-48; prof biol, 48-80, dir Parapsychol Lab, 56-86, EMER PROF BIOL, ST JOSEPH UNIV, PA, 80- *Honors & Awards:* William McDougall Award, Parapsychol Lab, Duke Univ, 60. *Mem:* AAAS; Parapsychol Asn; Sigma Xi. *Res:* Extrasensory perception; precognition; psychokinesis. *Mailing Add:* 16493 Harado Ct San Diego CA 92128

NASH, CHARLES DUDLEY, JR, b New York, NY, May 28, 26; m 49; c 4. MECHANICAL ENGINEERING. *Educ:* Yale Univ, BE, 49; Ohio State Univ, MS, 51, PhD(mech eng), 59. *Prof Exp:* Mech engr, Rocket Sect, Armament Lab, Air Develop Ctr, Wright-Patterson AFB, 49-50, aircraft armament engr, 50-51, rocket design engr, 51-52, ord engr, 52-53, physicist, Ballistics & Terminal Effects Sect, 53-54; from instr to asst prof mech eng, Ohio State Univ, 55-62; assoc prof, Univ Maine, 62-64; actg chmn dept mech eng & appl mech, 66-67, mem, pres selection comt, 67-68, dir, univ honors colloquium, 69-70, PROF MECH ENG, UNIV RI, 64- *Concurrent Pos:* Consult, US Naval Underwater Systs Ctr-Newport Lab, 65- *Mem:* Am Soc Eng Educ; Am Soc Mech Engrs; Am Inst Aeronaut & Astronaut; Soc Eng Sci; Am Math Soc; Sigma Xi. *Res:* Socio-technological and socio-economic problems; fatigue failure; reliability; vibrations; materials science; thermodynamics; applied mechanics and mathematics; systems analysis. *Mailing Add:* 2420 Kingstown Rd Kingston RI 02881

NASH, CHARLES PRESLEY, b Sacramento, Calif, Mar 15, 32; m 55; c 3. PHYSICAL CHEMISTRY. *Educ:* Univ Calif, BS, 52; Univ Calif, Los Angeles, PhD(chem), 58. *Prof Exp:* Actg instr chem, Univ Calif, Los Angeles, 56; from instr to assoc prof, 57-70, PROF CHEM, UNIV CALIF, DAVIS, 70- *Concurrent Pos:* Consult, Lawrence Livermore Lab, Univ Calif, 57-68; vis sr lectr, Imp Col, Univ London, 68; distinguished vis prof, US Air Force Acad, 79; chmn, Acad Senate, 87-91. *Mem:* Sigma Xi; Am Chem Soc. *Res:* Solution chemistry; vibrational spectroscopy; amino acids. *Mailing Add:* Dept Chem Univ Calif Davis CA 95616

NASH, CLINTON BROOKS, b Gunnison, Miss, Jan 3, 18; m 46; c 1. PHARMACOLOGY. *Educ:* Univ Tenn, BS, 50, MS, 52, PhD(pharmacol), 55. *Prof Exp:* Sr pharmacologist, Res Labs, Mead Johnson & Co, 54-57, group leader pharmacol, 57-58; from asst prof to assoc prof, 58-65, actg chmn, 75-77 & 79-81, PROF PHARMACOL, UNIV TENN CTR HEALTH SCI, MEMPHIS, 65- *Mem:* Am Soc Pharmacol & Exp Therapeut; Am Heart Asn; Soc Exp Biol & Med; Soc Toxicol. *Res:* Cardiovascular effects of anesthetic agents; intraocular pressures; peripheral vasodilators; catecholamine content of various tissues; coronary blood flow; antiarrhythmic agents; cardiovascular actions of vasopressin, reserpine and digitalis. *Mailing Add:* 5344 Timmons Ave Memphis TN 38119

NASH, COLIN EDWARD, resource management, for more information see previous edition

NASH, DAVID, b London, Eng, Sept 10, 37. BIOCHEMICAL GENETICS, CYTOGENETICS. *Educ:* Univ London, BSc, 60; Univ Cambridge, PhD(genetics), 63. *Prof Exp:* Wis Alumni Res Found fel zool, Univ Wis, 63-64; asst prof genetics, 65-70, assoc prof genetics, 70-75, assoc chmn, 78-79, PROF GENETICS, UNIV ALTA, 75-, CHMN DEPT, 83- *Mem:* Genetics Soc Am; Genetics Soc Can; Can Soc Cell Biol. *Res:* Nucleotide metabolism in Drosophila; studies on mutnats of Drosophilia; defective in purine biosynthesis. *Mailing Add:* Dept Genetics Univ Alta Edmonton AB T6G 2M7 Can

NASH, DAVID ALLEN, b Aug 24, 42; m 66; c 2. PEDIATRIC DENTISTRY. *Educ:* Milligan Col, BA, 64; Univ Ky, DMD, 68; Univ Iowa, MS, 70; WVa Univ, EdD, 84. *Prof Exp:* Asst prof, Sch Dent, La State Univ, 70-73; from assoc prof to prof & chair pediatric dent, WVa Univ Sch Dent, 73-87; DEAN & PROF DENT, UNIV KY COL DENT, 87- *Concurrent Pos:* Harry W Bruce Jr Legis fel, Am Asn Dent Sch, 85; Fogarty Sr Int fel, NIH, 80, res fel, 69-70; vis prof, Cath Univ, Neth, 80; prin investr, Pew Mem Trust-Nat Dent Educ Strategic Planning Prog, WVa Univ, 86; consult, Coun Dent Educ, Am Dent Asn, 80-; reviewer, J Am Dent Asn, 85- *Mem:* Col Dipl Am Bd Pediat Dent (pres, 88); Am Asn Dent Sch; fel Am Acad Pediat Dent; Am Asn Dent Res; Am Dent Asn. *Res:* higher education administration; pediatric dentistry. *Mailing Add:* Col Dent D-136 Chandler Med Ctr Univ Ky Lexington KY 40536-0084

NASH, DAVID BYER, b Cambridge, Mass, Jan 21, 49. QUATERNAY AGE DATING, THERMAL REMOTE SENSING. *Educ:* Colo Col, BA, 71; Univ Mich, MS, 74, PhD(geol), 77. *Prof Exp:* ASST PROF GEOMORPHOL, UNIV CINCINNATI, 77- *Concurrent Pos:* Nat Res Coun res fel, Jet Propulsion Lab, NASA, 81- *Mem:* Geol Soc Am; Sigma Xi; Am Soc Photogram. *Res:* Morphologic dating and computer modeling of fault scarps and other hillslopes; thermal remote sensing of buried bedrock faults; stabilization of landslide prone areas by tree roots; catastrophism in fluvial systems. *Mailing Add:* Dept Geol Univ Cincinnati 13 223 OT Cincinnati OH 45221

NASH, DAVID HENRY GEORGE, b Ash Vale, Eng, June 19, 43. APPLIED MATHEMATICS, SOFTWARE DEVELOPMENT. *Educ:* Univ Calif, Riverside, BA, 65; Univ Calif, Berkeley, MA, 67, PhD(math), 70. *Prof Exp:* Actg asst prof math, Univ Hawaii, 69-70; Woodrow Wilson intern, Va State Col, 70-71; lectr, Univ Calif, Berkeley, 71; assoc sr res mathematician, Res Labs, Gen Motors Corp, 72-77, sr staff analyst, Corp Prod Planning Group, 77-81; PRES, SCI MKT CORP, 81- *Concurrent Pos:* Assoc prof math, Drexel Univ, 82-86; pres, Sci Software Assoc, Inc, 82-89; vpres res, Autofacts, Inc, 86-88, sr vpres, 88-, managing dir, Int & Powertrains, 90- *Mem:* Am Math Soc; Math Asn Am; Int Inst Forecasters. *Res:* Mathematics of forecasting. *Mailing Add:* PO Box 461 Bala Cynwyd PA 19004

NASH, DONALD JOSEPH, b New York, NY, Dec 20, 30; m 54; c 3. GENETICS, ZOOLOGY. *Educ:* Univ Mich, BS, 51; Univ Kans, MA, 57; Iowa State Univ, PhD(genetics), 60. *Prof Exp:* Asst prof genetics, Pa State Univ, 60-62; asst prof zool, Rutgers Univ, 62-65; assoc prof radiation biol & zool, 65-66, assoc prof zool, 66-71, PROF ZOOL, COLO STATE UNIV, 71- *Mem:* AAAS; Genetics Soc Am; Am Genetic Asn; Am Soc Mammal; Behav Genetics Asn; Sigma Xi. *Res:* Physiological and quantitative genetics; radiation biology; behavioral genetics; reproductive genetics. *Mailing Add:* Dept of Biol Colo State Univ Ft Collins CO 80523

NASH, DONALD ROBERT, b Pittsfield, Mass, Nov 15, 38; m 63; c 1. IMMUNOBIOLOGY, MONOCLONAL ANTIBODY. *Educ:* Am Int Col, BA, 61; Boston Col, MS, 63; Univ NC, Chapel Hill, PhD(bact, immunol), 67. *Prof Exp:* Asst prof immunol, Univ Hawaii, 69-70; head immunobiol res, 72-78, RES ASSOC PROF, IMMUNOL/MICROBIOL DEPT, HEALTH SCI CTR, UNIV TEX, TYLER, 78-, DIR, HYBRIDOMA CORE, 84- *Concurrent Pos:* Res fel immunol, Univ NC, Chapel Hill, 67-68; Belg Am Educ Fund res fel, Cath Univ Louvain, 68-69; sr res fel immunol, Ref & Training Ctr, WHO, Switz, 70-72; consult, M D Anderson Hosp & Tumor Inst, 75-78; adj prof med microbiol, Univ Tex, Tyler, 78- *Honors & Awards:* Belg Am Educ Found Award, 68-69. *Mem:* Am Thoracic Soc; Am Asn Immunologists; NY Acad Sci. *Res:* Humoral and cellular immunity; drug resistance in bacteria; monoclonal antibodies. *Mailing Add:* Univ Tex Health Sci Ctr Box 2003 Tyler TX 75701

NASH, DOUGLAS B, b Elgin, Ill, Dec 2, 32; m 64. GEOLOGY, SPACE PHYSICS. *Educ:* Univ Calif, Berkeley, AB, 60, MA, 62. *Prof Exp:* From assoc scientist to sr scientist, Jet Propulsion Lab, Calif Inst Technol, 62-68, res group supvr, 68-70, prin investr, Lunar Sample Anal, 69-74, consult, 74-76, res scientist, 76-81, mrg, Planetology & Oceanog Sect, 81-83, mgr, Planetary Sci Res Prog, 83-88, PROJ SCIENTIST, LUNAR OBSERVER, JET PROPULSION LAB, 88-; FOUNDER & PRES, SAN JUAN CAPISTRANO RES INST, 88- *Concurrent Pos:* Mayor & city councilman, San Luan Capistrano, Calif, 74-78; dir & pres, San Juan Capistrano Hist Soc, 79-83. *Honors & Awards:* NASA Except Sci Achievement Award, 89. *Mem:* Am Geophys Union; Am Astron Soc; AAAS. *Res:* Lunar luminescence; lunar surface optical properties; proton irradiation effects on rocks; x-ray diffraction analysis of rock glass; instrument development for lunar and planetary geological analysis; surface properties of Galilean satellites; spectroscopy of planetary materials; planetary sciences; physics and chemistry of sulphur and related compounds; ion bombardment and sputtering studies; reflectance and emission spectroscopy of rock materials. *Mailing Add:* 32906 Ave Descanso Jet Propulsion Lab San Juan Capistrano CA 92675

NASH, EDMUND GARRETT, b Manitowoc, Wis, Nov 19, 36; m 61; c 2. ORGANIC CHEMISTRY. *Educ:* Lawrence Col, BS, 59; Univ Colo, PhD(chem), 65. *Prof Exp:* Sr res chemist, Gen Mills, Inc. 65-66; res assoc, Johns Hopkins Univ, 66-67; from asst prof to assoc vpres, 84-87, PROF CHEM, FERRIS STATE UNIV, 77-, VPRES ACAD AFFAIRS, 87- *Mem:* AAAS; Am Chem Soc; Sigma Xi. *Res:* Chemistry of organic nitrogen compounds; nuclear magnetic resonance of systems with restricted rotation; organic polymer chemistry; higher education management. *Mailing Add:* Acad Affairs Off Ferris State Univ Big Rapids MI 49307

NASH, EDWARD THOMAS, b New York, NY, July 31, 43; m 70. EXPERIMENTAL HIGH ENERGY PHYSICS. *Educ:* Princeton Univ, AB, 65; Columbia Univ, MA, 67, PhD(physics), 70. *Prof Exp:* Res assoc physics, Nevis Cyclotron Lab, Columbia Univ, 70 & Lab Nuclear Sci, Mass Inst Technol, 70-71; proj mgr, Tagged Photon Beam Facil, Proton Lab, Fermi Nat Accelerator Lab, 72-75, head, Internal Target Lab, 76-77, proj mgr, Tagged Photon Spectrometer, Proton Lab, 77-79, dep chmn, Physics Dept, 79-83, head, Advan Computer Prog, 83-89, STAFF PHYSICIST, FERMI NAT ACCELERATOR LAB, BATAVIA, ILL, 72-, HEAD, COMPUT DIV, 89- *Concurrent Pos:* Mem, Res Briefing Panel Computer Archit, Nat Acad Sci, 84 & Comput/Networking Rev Panel, Ill Technol Challenge Grant Prog, Gov Sci Adv Comt, 90; int adv comt, Comput High Energy Physics Conf, Calif, 87, int sci adv comt, NMex, 90, Conf Comp HEP, Eng, 89; sci adv comts, Italy, 88 & Japan, 91; specialist ed, Computer Physics Commun, 86-; ed consult, Encycl Appl Physics, Am Inst Physics, 88-; Carnegie sci fel, Ctr Int Security & Arms Control, Stanford Univ, 88-89. *Mem:* Fel Am Phys Soc; assoc Sigma Xi. *Res:* Fundamental forces and symmetries; searches for and studies of the properties of new particles; study of the interaction of photons with matter at very high energy; study of charmed particle dynamics; computer science, parallel processors. *Mailing Add:* Fermi Nat Accelerator Lab PO Box 500 Batavia IL 60510

NASH, FRANKLIN RICHARD, b Brooklyn, NY, July 23, 34. PHYSICS. *Educ:* Polytech Inst New York, BS, 55; Columbia Univ, PhD(physics), 62. *Prof Exp:* MEM TECH STAFF PHYSICS RES, BELL LABS, 63- *Mem:* Am Phys Soc; Inst Elec & Electronics Engrs. *Res:* Semiconductor lasers. *Mailing Add:* AT&T Bell Labs 2A-440 Murray Hill NJ 07974

NASH, HAROLD ANTHONY, b Corvallis, Ore, Sept 28, 18; m 46; c 2. BIOCHEMISTRY. *Educ:* Ore State Col, BS, 40; Purdue Univ, PhD(biochem), 47. *Prof Exp:* Asst agr chem, Purdue Univ, 40-44, asst chemist, 42-44 & 46-47; res chemist, Pitman-Moore Co, 47-55, dir chem res, 55-60, dir pharmaceut res, 60-61, asst to tech dir, 61-63; dir res biosci, NStar Res & Develop Inst, 63-64, dir biosci div, 64-70; staff assoc, 70-71, ASSOC DIR, CTR BIOMED RES, POP COUN, 72- *Mem:* Am Chem Soc; AAAS. *Res:* Chemistry of natural products; sustained release systems; fertility control; contraceptive development. *Mailing Add:* 27 Kohring Circle S Harrington Park NJ 07640-1917

NASH, HAROLD EARL, b Lindsay, Calif, July 14, 14; m 40. PHYSICS. *Educ:* Univ Calif, Berkeley, BA, 38. *Prof Exp:* Radio engr, US Signal Corps, McClellan Field, Calif, 42-44; res assoc, Underwater Sound Lab, Harvard Univ, 44-45; sect leader, US Naval Underwater Sound Lab, 45-50, div head, 50-60, from assoc tech dir to tech dir systs develop, 60-70, tech dir, US Naval Underwater Systs Ctr, Conn, 70-75; consult staff, Sonalysts, Inc, 78-83; CONSULT, UNDERWATER ACOUSTICS, 75- *Mem:* Fel Acoust Soc Am; fel Inst Elec & Electronics Engrs. *Res:* Sonar systems; laboratory administration. *Mailing Add:* PO Box 314 Quaker Hill CT 06375-0314

NASH, HARRY CHARLES, b Cleveland, Ohio, Mar 24, 27; m 51; c 12. SOLID STATE PHYSICS, OPTICS. *Educ:* John Carroll Univ, BS, 50, MS, 51; Case Inst Technol, PhD(physics), 58. *Prof Exp:* From instr to assoc prof, 51-64, PROF PHYSICS, JOHN CARROLL UNIV, 64-, CHMN DEPT, 71- *Mem:* Optical Soc Am; Am Asn Physics Teachers. *Res:* Optical properties of absorbing thin films; elastic constants of single crystals; spectroscopy; emission spectroscopy. *Mailing Add:* Dept Physics John Carroll Univ Cleveland OH 44118

NASH, HOWARD ALLEN, b New York, NY, Nov 5, 37; m 63; c 2. GENETIC RECOMBINATION, GENERAL ANESTHESIA. *Educ:* Tufts Univ, BS, 57; Univ Chicago, MD, 61, PhD(physiol), 63. *Prof Exp:* Intern, Univ Chicago, 63-64; res assoc, Lab Neurochem, USPHS, 64-68, res med officer, Lab Neurochem, 68-84, CHIEF, SECT MOLECULAR GENETICS, LAB MOLECULAR BIOL, NIMH, 84- *Concurrent Pos:* Assoc ed, Cell, 85-; consult, E I Du Pont de Nemours & Co, 86-87; co-chmn, Gordon Res Conf on Nucleic Acids, 88-; co-org, Woods Hole Workshop on Site-Specific Recombination, 90. *Mem:* Nat Acad Sci; Am Acad Arts & Sci; Am Soc Biochem & Molecular Biol; Am Soc Microbiol. *Res:* Mechanism of genetic recombination, in particular, the set of site-specific DNA rearrangements that integrate and excise the genome of a bacterial virus. *Mailing Add:* Lab Molecular Biol Bldg 36 Rm 1B-08 NIMH 9000 Rockville Pike Bethesda MD 20892

NASH, J(OHN) THOMAS, b Glen Cove, NY, July 30, 41; m 66; c 1. GEOLOGY, GEOCHEMISTRY. *Educ:* Amherst Col, BA, 63; Columbia Univ, MA, 65, PhD(geol), 67. *Prof Exp:* GEOLOGIST, US GEOL SURV, 67- *Mem:* Geol Soc Am; Am Inst Mining, Metall & Petrol Engrs; Mineral Soc Am; Mineral Asn Can. *Res:* Geochemistry of mineral deposits; fluid inclusions; geology of uranium deposits; clay mineralogy; exploration geochemistry. *Mailing Add:* US Geol Surv Fed Ctr Box 25046 Mail Stop 973 Denver CO 80225

NASH, JAMES LEWIS, JR, b Drakesboro, Ky, Sept 24, 26; m 51; c 2. POLYMER CHEMISTRY. *Educ:* Western Ky State Col, BS, 48; Univ Fla, MS, 50, PhD(chem), 53. *Prof Exp:* Asst chem, Univ Fla, 48-49, 51-53; sr chemist, E I du Pont de Nemours & Co Inc, 53-56, group supvr, 56-60, sr res chemist, 60-69, tech serv specialist, 69-79, tech mkt assoc, Textile Fibers Dept, 79-85; RETIRED. *Res:* Textile chemistry. *Mailing Add:* 1375 Candle Ct Charlotte NC 28234

NASH, JAMES RICHARD, b Trenton, NJ, Nov 9, 31. PHYSICAL CHEMISTRY, ENGINEERING. *Educ:* Mt St Mary's Col, BS, 53; Univ Notre Dame, PhD(diffusion kinetics), 58. *Prof Exp:* Res asst reaction kinetics, Univ Notre Dame, 53-57, res assoc radiation, 59-61; guest scientist for Frankford Arsenal, Brookhaven Nat Lab, 58-59; supvr subsyst eng, Space Div, NAm Rockwell Corp, 61-67, syst engr, 67-69; syst engr, Systs Group, TRW, Inc, 69-74; mgr, Shuttle Power Systs, Space Div, Rockwell Int, 74-78; SR PROJ ENGR, TRW SYSTS, 78- *Mem:* Am Chem Soc; Radiation Res Soc; Am Inst Aeronaut & Astronaut; Sigma Xi. *Res:* Space studies; Apollo spacecraft fuel cell system; radiation chemistry; photochemistry; kinetics. *Mailing Add:* 700 Esplanade Condo 11 Redondo Beach CA 90277

NASH, JOHN CHRISTOPHER, b Tunbridge Wells, Eng, Sept 9, 47; Can citizen. COMPUTATIONAL STATISTICS, NUMERICAL ANALYSIS. *Educ:* Univ Calgary, BSc, 68; Oxford Univ, DAM, 69, PhD(math), 72. *Prof Exp:* Res fel, Univ Alta, 72-73; economist, Agr Can, 73-80; sr consult, Hickling Partners, 80-81; ASSOC PROF STATIST, FAC ADMIN, UNIV OTTAWA, 81- *Concurrent Pos:* Treas, Amnesty Int, Can, 79-82; pres, Nash Info Serv Inc, 80- *Mem:* Asn Comput Mach; Soc Indust & Appl Math; Statist Soc Can; Am Statist Asn. *Res:* Numerical linear algebra; function minimization; nonlinear parameter estimation; risk analysis, forecasting; library management software; microcomputer system design and integration. *Mailing Add:* 1975 Bel Air Dr Ottawa ON K2C 0X1 Can

NASH, JONATHON MICHAEL, b Little Rock, Ark, Aug 10, 42; m 72; c 2. ENERGY SYSTEMS. *Educ:* Univ Miss, BSME, 66, MS, 70, PhD(mech eng), 73. *Prof Exp:* Eng officer, US Army Corps Engrs, 68-70; fel, Univ Miss, 70-73, sr assoc engr, 73-74, staff engr, 75-77, proj develop engr, 77-80, adv engr, 80-81, mgr tech planning, Fed Syst Div, 81-83, proj segment mgr, 84-88, PROG MGR, AIR TRAFFIC CONTROL PROGS, INT BUS MACH CORP, 88-, SR ENGR, 83- *Concurrent Pos:* Res & develop reserve officer, US Army Mobility Equip Res & Develop Command, 71-88. *Honors & Awards:* Tudor Medal, Soc Am Military Engr, 78; New Technol Award, NASA, 79. *Mem:* Fel Am Soc Mech Engrs; Am Soc Heating, Refrig & Air Conditioning Engrs; assoc fel Am Inst Aeronaut & Astronaut; Am Inst Chem Engrs; Soc Am Mil Engrs; Sigma Xi. *Res:* Analysis and evaluation of energy conversion processes and their applications including synthetic fuel processes and solar thermal systems; high temperature cryogenic refrigeration. *Mailing Add:* 300 Rockwell Terrace Frederick MD 21701

NASH, KENNETH LAVERNE, b Joliet, Ill, July 6, 50; m 86; c 3. RADIOCHEMISTRY,. *Educ:* Lewis Univ, Lockport, Ill, BA, 72; Fla State Univ, Tallahassee, MS, 75, PhD(inorg chem), 78. *Prof Exp:* Res chemist, Dow Chem Co, 75; res assoc, Argonne Nat Lab, 79-81; res chemist, US Geol Surv, 81-86; CHEMIST, ARGONNE NAT LAB, 86- *Mem:* Am Chem Soc; AAAS; Am Nuclear Soc; Sigma Xi. *Res:* Investigations of the thermodynamics and kinetics of metal ion complexation and oxidation; reduction reactions in aqueous and non aqueous solutions; solvent extraction separations chemistry with primary focus on the actinide elements; radioactive waste disposal. *Mailing Add:* Argonne Nat Lab 9700 S Cass Ave Argonne IL 60439-4831

NASH, LEONARD KOLLENDER, b New York, NY, Oct 27, 18; m 45; c 2. PHYSICAL CHEMISTRY. *Educ:* Harvard Univ, BS, 39, MA, 41, PhD(anal chem), 44. *Prof Exp:* Asst chem, Harvard Univ, 43-44; res assoc, Columbia Univ, 44-45; instr, Univ Ill, 45-46; from instr to prof chem, Harvard Univ, 46-81, chmn dept, 71-74, William R Kenan prof, 81-86; RETIRED. *Res:* Thermodynamics; statistical mechanics. *Mailing Add:* 11 Field Rd Lexington MA 02173

NASH, MURRAY L, physical chemistry, chemical engineering; deceased, see previous edition for last biography

NASH, PETER, b St Paul, Minn, July 11, 45; m 68; c 2. MEDICAL MICROBIOLOGY, IMMUNOLOGY. *Educ:* Lawrence Univ, BA, 67; Univ Hawaii, MS, 69; Colo State Univ, PhD(microbiol), 72, Carnegie Mellon Univ, 79. *Prof Exp:* Res asst microbiol, Colo State Univ, 69-72; environ control & safety officer, Ball State Univ, 77, assoc prof biol, 72-77; assoc prof biol, Mankato State Univ, 77-84; DIR RES, MICROBIAL DIAG, CAMAS DIAG CO, 84- *Concurrent Pos:* Researcher fac res grants, Ball State Univ, 73-77, Mankato State Univ, 77-, CORE res grant, 76-77; consult, Marsh Supermkts, Inc, 74-77, Process Supplies, Inc, 76-77 & Muncie Clin, Inc, 75-77; consult, biocompatible implants development. *Honors & Awards:* Gordon Rosene Cancer Award, Am Cancer Soc, Ball State Univ, 76; Nat Registry Microbiol Award, Am Soc Microbiol, 77. *Mem:* Sigma Xi; Am Soc Microbiol; Am Soc Testing & Mat; Wildlife Dis Asn; Nat Environ Health Asn. *Res:* Medical microbiology in vaccine development; tumor immunology and the effect of nutrition on tumor development and immune response repression; microbial detection assay systems; three patents on microbial diagnostic. *Mailing Add:* 18811 Mapleleaf Dr Eden Prairie MN 55346

NASH, PETER HOWARD, b Sidcup, Eng, Feb 20, 17; Can citizen; m 51; c 4. RESEARCH ADMINISTRATION. *Educ:* Cambridge Univ, BA, 38, MB, BCh, 41, MA, 45, MD, 50; Univ London, DPH, 47, DIH, 53. *Prof Exp:* Intern surg, Middlesex Hosp, London, 41-42; resident internal med, Metrop Hosp, London, 49-50; asst dir, Slough Indust Health Serv, 50-53; regional med dir occup health, Bell Tel Co Can, 54-57; med dir, Abbott Labs, Ltd, 57-64, dir sci affairs, 64-82; CONSULT PHARMACEUT MED & OCCUP HEALTH, 82- *Concurrent Pos:* Rockefeller fel prev med, 46-48; res fel indust toxicol, Harvard Med Sch, 48; lectr indust health, London Sch Hyg & Trop Med, 50-53; asst physician, Royal Victoria Hosp, Montreal, Can, 57-82. *Mem:* NY Acad Sci; Can Med Asn; Brit Med Asn; Pharmacol Soc Can. *Res:* Pharmaceutical, medical and research development; occupational health. *Mailing Add:* 1375 Regent Rd Montreal PQ H3P 2K9 Can

NASH, RALPH GLEN, b Del Norte, Colo, July 26, 30; m 57; c 3. SOIL SCIENCE, CHEMISTRY. *Educ:* Colo State Univ, BS, 58, MS, 61, PhD(soil sci), 63. *Prof Exp:* Soil scientist, Agr Environ Qual Inst, Admin-Agr Res, USDA, 65-89; STUDY DIR, EPL BIO-ANALYSIS SERV, INC, DECATUR, IL, 90- *Mem:* Am Soc Agron; Soil Sci Soc Am; Weed Sci Soc Am; Am Chem Soc. *Res:* Toxic and residual interactions which result from a combination of two or more pesticides added to soils; plant absorption of pesticides; pesticide degradation; persistence and movement in soil and plants; design and development of an agricultural terrestrial microcosm for following fate of pesticides in plant, soil, water and air; soil and plant analytical pesticide methods; comparative volatization of pesticides from soils; modeling of pesticide disipation; ground water contamination. *Mailing Add:* 1320 Meadowview Dr Decatur IL 62526-9204

NASH, RALPH ROBERT, b South Bend, Ind, May 17, 16; m 42; c 2. PHYSICAL METALLURGY. *Educ:* Purdue Univ, BS, 42; Rensselaer Polytech Inst, MS, 48, PhD(metall), 55. *Prof Exp:* Metallurgist, Aluminum Co Am, 42-44; from instr to assoc prof phys metall, Rensselaer Polytech Inst, 48-57; phys metallurgist, Div Res, Atomic Energy Comn, Washington, DC, 57-60; sci liaison, Off Naval Res, London & Washington, DC, 60-63; mgr solid state mat res prog, NASA, 63-67, chief mat sci br, Div Res, 67-71, exec secy, Off Advan Res & Technol Res Coun, 71-77, MGR INDEPENDENT RES & DEVELOP, NASA, 77- *Concurrent Pos:* Consult, Rensselaer Polytech Inst, 50-58 & Dow Chem Co, 55-58. *Mem:* Am Phys Soc; Am Soc Metals; Am Inst Mining, Metall & Petrol Engrs. *Res:* Plastic deformation of solids; solid state physics and chemistry of solids; surface phenomena; interaction of radiation with matter. *Mailing Add:* Eight George St Gaithersburg MD 20877

NASH, REGINALD GEORGE, b LaValle, Wis, Nov 20, 22; m 52; c 3. PARASITOLOGY. *Educ:* William Penn Col, BA, 48; Univ Iowa, MS, 52; Mich State Univ, PhD(zool), 64. *Prof Exp:* Asst prof biol, Northern Ill Univ, 52-54; instr natural sci, Mich State Univ, 54-58; PROF BIOL, UNIV WIS-WHITEWATER, 58- *Res:* Immunologic studies involving infections with roundworms Trichinella spiralis, Ascaris lumbricoides and Toxocara canis. *Mailing Add:* Dept Biol Univ Wis Whitewater 800 W Main St Whitewater WI 53190

NASH, ROBERT ARNOLD, b Brooklyn, NY, July 6, 30; m 52; c 3. PHARMACEUTICS, INDUSTRIAL CHEMISTRY. *Educ:* Brooklyn Col Pharm, BS, 52; Rutgers Univ, MS, 54; Univ Conn, PhD, 58. *Prof Exp:* Asst, Rutgers Univ, 53-54 & Univ Conn, 54-57; res assoc, Merck, Sharp & Dohme Res Labs, 57-60; proj leader, mgr pharmaceut prod develop, Lederle Labs Div, Am Cyanamid Co, 60-76; dir pharmaceut develop, Purdue Frederick Co, 76-81; ASSOC PROF INDUST PHARM, ST JOHN'S UNIV, 81- *Concurrent Pos:* Consult pharmaceut & indust pharm. *Honors & Awards:* Richardson Award, 57. *Mem:* Acad Pharmaceut Sci; Am Chem Soc; Int Soc Pharmaceut Engrs; Am Asn Pharmaceut Scientists. *Res:* Application of the principles of physical chemistry and chemical engineering to industrial pharmaceutical technology; process validation. *Mailing Add:* St John's Univ Jamaica NY 11439

NASH, ROBERT JOSEPH, b Coventry, Eng, Sept 12, 39; m 63; c 2. SURFACE CHEMISTRY. *Educ:* Univ Wales, BSc, 62; Bristol Univ, PhD(phys chem), 66. *Prof Exp:* Res assoc surface chem, Amherst Col, 65-66, asst prof chem, 66-67; res assoc surface chem, Case Western Reserve Univ, 67-69; scientist, 70-73, SR SCIENTIST, XEROX CORP, 73- *Concurrent Pos:* Ed, CHEMunications, 78-83. *Mem:* Am Chem Soc; Sigma Xi. *Res:* Palladium-hydrogen system; hysteresis in absorption processes; surface chemistry of metals; uses of gas chromatography in surface chemistry; surface chemistry of pigments and polymers; surface potentials; xerographic materials; mechanochemical degradation of polymers. *Mailing Add:* 1200 Severn Ridge Webster NY 14580

NASH, ROBERT T, b Columbus, Ohio, Sept 20, 29; m 61; c 3. ELECTRICAL ENGINEERING. *Educ:* Ohio State Univ, BSc, 52, MSc, 55, PhD(elec eng), 61. *Prof Exp:* From instr to assoc prof elec eng, Ohio State Univ, 57-66; ASSOC PROF ELEC ENG, VANDERBILT UNIV, 66- *Mem:* Am Astron Soc; Inst Elec & Electronics Engrs. *Res:* Radio astronomy; decision theory. *Mailing Add:* Dept Elec Engr Mgmt Tech Vanderbilt Univ Box 1553 Sta B Nashville TN 37235

NASH, VICTOR E, b Frankfort, Ky, Sept 27, 28; m 56; c 4. SOIL CHEMISTRY, MINERALOGY. *Educ:* Univ Ky, BS, 51, MS, 52; Univ Mo, PhD(soils), 55. *Prof Exp:* Res geochemist, Cities Serv Res & Develop Co, 56-59; from asst prof to assoc prof soils, 59-68, assoc agronomist, 59-68, PROF SOILS & AGRONOMIST, MISS STATE UNIV, 68- *Mem:* Am Soc Agron; Clay Minerals Soc; Soil Sci Soc Am; Int Soil Sci Soc. *Res:* Cation exchange of soil colloids and interaction of soil colloids; soil micromorphology and mineralogy; non-crystalline minerals in soils. *Mailing Add:* Dept Agron Miss State Univ PO Box 5248 Mississippi State MS 39762

NASH, WILLIAM A(RTHUR), b Chicago, Ill, Sept 5, 22; m 53; c 2. MECHANICS. *Educ:* Ill Inst Technol, BS, 44, MS, 46; Univ Mich, PhD(eng mech), 49. *Prof Exp:* Asst engr, Armour Res Found, Ill Inst Technol, 44-45, instr mech inst, 45-46; instr, Univ Mich, 47-49; asst prof, Univ Notre Dame, 49-50; head, Plates & Shells Sect, David Taylor Model Basin, US Dept Navy, 50-54; from assoc prof to prof mech, Univ Fla, 54-67, chmn eng sci & mech, 64-67; PROF CIVIL ENG, UNIV MASS, AMHERST, 67- *Concurrent Pos:* Ed, Int J Nonlinear Mech, 65-; hon prof, Shanghai Univ Technol, 85. *Honors & Awards:* Award, Am Soc Eng Educ, 58 & 63; Humboldt Sr US Scientist Prize, 87. *Mem:* Fel Am Soc Mech Engrs; Am Soc Eng Educ; Am Inst Aeronaut & Astronaut; Int Asn Shell & Spatial Struct. *Res:* Mathematical investigations in the theory of elasticity and the theory of plates and shells; theoretical and applied mechanics; earthquake engineering. *Mailing Add:* Dept of Civil Eng Univ of Mass Amherst MA 01002

NASH, WILLIAM DONALD, b Shreveport, La, Jan 17, 47; m 69; c 2. SYNTHETIC ORGANIC CHEMISTRY, PETROLEUM PROCESSING. *Educ:* McNeese State Univ, La, BS, 70; Tex A&M Univ, PhD(org chem), 74. *Prof Exp:* res chemist, El Paso Prod Co, 74-83; PENNZOIL PROD CO, 84- *Mem:* Am Chem Soc. *Mailing Add:* 12220 Rock Oak The Woodlands TX 77380

NASH, WILLIAM HART, b Oct 8, 25; US citizen; c 3. ELECTRICAL ENGINEERING, PLASMA PHYSICS. *Educ:* Univ Wis-Madison, BS, 49, MS, 56; Univ Chicago, MBA, 74. *Prof Exp:* Trainee, McGraw Edison Power Systs, 49-50, jr engr, 50-51, asst engr, 51-52, engr, 52-60, chief engr, 60-67, mgr new prod, 67-68, mgr advan develop, 68-71, mgr res & develop, Power Systs Div, 71-76, DIR, THOMAS A EDISON TECH CTR, MCGRAW EDISON CO, 76-, STAFF CONSULT- *Mem:* Sr mem Inst Elec & Electronics Engrs; fel Brit Inst Elec Engrs. *Res:* Plasma physics applied to power vacuum interrupters; solid state physics applied to high voltage lightning arresters. *Mailing Add:* 1824 Drexel Blvd South Milwaukee WI 53172

NASH, WILLIAM PURCELL, b Boston, Mass, Mar 20, 44; m 66; c 2. PETROLOGY, GEOCHEMISTRY. *Educ:* Univ Calif, Berkeley, BA, 65, PhD(geol), 71. *Prof Exp:* From asst prof to assoc prof, 70-78, PROF GEOL, UNIV UTAH, 78-, CHMN, DEPT GEOL & GEOPHYS, 80- *Mem:* Am Geophys Union; Mineral Soc Am; Geochem Soc. *Res:* Field, chemical and thermodynamic methods applied to the origin, evolution and crystallization of igneous rocks. *Mailing Add:* Dept Geol & Geophys Univ Utah 717 Browning Sci Bldg Salt Lake City UT 84112

NASHED, MOHAMMED ZUHAIR ZAKI, b Aleppo, Syria, May 14, 36; m 59; c 4. MATHEMATICS. *Educ:* Mass Inst Technol, SB, 57, SM, 58; Univ Mich, MS & PhD(math), 63. *Prof Exp:* From asst prof to prof math, Ga Inst Technol, 63-76; vis prof, Univ Mich, 76-77; PROF MATH, UNIV DEL, 77- *Concurrent Pos:* Assoc prof, Am Univ Beirut, 67-69; vis prof, Math Res Ctr, Univ Wis-Madison, 67, 70-72; ed-in-chief, J Numerical Functional & Optimization & J Integral Equations. *Honors & Awards:* Lester Ford Award, Math Asn Am, 67. *Mem:* AAAS; Am Math Soc; Math Asn Am; Soc Indust & Appl Math; Opers Res Soc Am. *Res:* Nonlinear functional analysis; iterative methods for operator equations; numerical analysis; optimization; mathematical programming; integral equations; generalized inverses; ill-posed problems; random operators. *Mailing Add:* Dept Math & Elec Eng Univ Del Newark DE 19716

NASHED, WILSON, b Damanhour, Egypt, Feb 16, 19; nat US; m 54; c 1. PHARMACY. *Educ:* Univ Cairo, BS, 39; Purdue Univ, MS, 51, PhD(pharm), 54. *Prof Exp:* Hosp pharmacist, Egyptian Govt, 40-41; tech dir, Delta Labs, Egypt, 41-50; sr res chemist, 54-55, res group leader, 56-62, asst dir prod coord, 62-65, mgr regulatory affairs, 65-69, dir sci info, 69-74, assoc dir res, 74-75, dir tech serv & res facil, 75-78, dir sci affairs, 78-79, dir new technol, 79-84, CONSULT, JOHNSON & JOHNSON, 84- *Mem:* Am Pharmaceut Asn. *Res:* Pharmaceutical research and product development. *Mailing Add:* Johnson/Johnson Res Lab New Brunswick NJ 08901

NASHMAN, ALVIN E, b Dec 16, 26; m; c 3. ELECTRICAL ENGINEERING. *Educ:* Col City NY, BS, 48; NY Univ, MS, 51. *Hon Degrees:* ScD, Pacific Univ, 68; George Washington Univ, 86. *Prof Exp:* Dir, Guidance Telecommunications Lab & Oper, Intelcom Corp, 52-65; PRES, SYSTEMS GROUP, COMPUT SCI CORP, 65-, VPRES & MEM, BD DIRS, 65- *Concurrent Pos:* Bd dirs, Armed Forces Commun & Electronics Asn Int, 84- *Honors & Awards:* Albert Meyer Award, Armed Forces Commun & Electronics Asn, 80. *Mem:* Fel Inst Elect & Electronic Engrs; Am Inst Aeronaut & Astronaut; Nat Space Club; Nat Security Indust Asn. *Res:* Author of five books on electronics. *Mailing Add:* Computer Sciences Corp 3160 Fairview Park Dr Falls Church VA 22042

NASHOLD, BLAINE S, b Lennox, SDak, Nov 12, 23; m 48; c 4. NEUROSURGERY, NEUROPHYSIOLOGY. *Educ:* Ind Univ, AB, 43; Ohio State Univ, MSc, 44; Univ Louisville, MD, 49; McGill Univ, MSc, 54. *Prof Exp:* Instr neuroanat, McGill Univ, 53; asst neurosurg, Bowman Gray Sch Med, 56-57; from asst prof to assoc prof, 57-75, PROF NEUROSURG, SCH MED, DUKE UNIV, 75- *Concurrent Pos:* Chief neurosurg sect, Vet Admin Hosp, Durham, NC, 57-59; chmn, Coop Studies Intervertebral Disc Dis & Parkinsonism, Vet Admin, 60-; mem, Cong French Speaking Neurosurgeons, 64. *Mem:* Am Asn Neurol Surg; Am Acad Neurol; Asn Res Nerv & Ment Dis; Am Acad Cerebral Palsy. *Res:* Stereotactic neurosurgical problems in relation to extrapyramidal diseases and problems of central pain; neurochemistry of brain function. *Mailing Add:* Surg Med Ctr Duke Univ Box 3807 Durham NC 27710

NASIM, ANWAR, genetics, for more information see previous edition

NASJLETI, CARLOS EDUARDO, periodontics, for more information see previous edition

NASJLETTI, ALBERTO, BASOACTIVE HORMONES, LIPID METABOLISM. *Educ:* Cuyo Univ, Mendoza, Argentina, 65. *Prof Exp:* PROF PHARMACOL, UNIV TENN, MEMPHIS, 75- *Mailing Add:* Dept Pharmacol NY Med Col Valhalla NY 10595

NASKALI, RICHARD JOHN, b Jefferson, Ohio, Dec 11, 35. BOTANY. *Educ:* Ohio State Univ, BSc, 57, MSc, 61, PhD(bot), 69. *Prof Exp:* Instr bot, Ohio State Univ, 60-67; ASST PROF BOT, UNIV IDAHO, 67- *Mem:* AAAS; Bot Soc Am; Sigma Xi. *Res:* Developmental plant anatomy, particularly of flowering and plant chimeras; internode elongation in monocotyledons; aquatic macrophytes of Pacific Northwest. *Mailing Add:* Arboretum Dir Univ Idaho 205 CEB Moscow ID 83843

NASON, HOWARD KING, b Kansas City, Mo, July 12, 13; m 34. CHEMISTRY. *Educ:* Univ Kans, AB, 34; Harvard, AMP, 50. *Prof Exp:* Chief chemist, Anderson-Stolz Corp, 35-36; res chemist, Org Div, Monsanto Chem Co, 36-39, asst dir res plastics div, 39-44, dir develop cent res dept, 44-46, assoc dir, 46-48, dir, 48-50, asst to vpres, 50-51, dir res org div, 51-56, vpres & gen mgr res & eng div, 56-60, pres, Monsanto Res Corp, 60-76; pres, IRI Res Corp, 76-82; CONSULT, 82- *Concurrent Pos:* Mem adv comt isotopes & radiation develop, AEC, 64-68, labor-mgt adv comt, 65; mem, President's Comn Patent Syst, 65-68; mem patent adv comt, US Patent Off, 68; trustee-at-large, Univs Res Asn, Inc, 71-76; vpres & mem exec comt, Atomic Indust Forum, Inc, 71-73; chmn, Aerospace Safety Adv Panel, 72-77, trustee, Charles F Kettering Found, 73-84; mem, Nat Mat Adv Bd, Nat Acad Eng, 73-77; consult, 76-; trustee, Acad Sci, St Louis, 84- *Honors & Awards:* Chevron de l'ordre du Merite Agricole, France. *Mem:* AAAS; Am Chem Soc; Am Soc Testing & Mat; Am Inst Chem Engrs; Am Inst Chemists. *Res:* Plastics; plasticizers; industrial microbiology; physical testing; water treatment; industrial application of chemicals; protective coatings; management of innovation. *Mailing Add:* Howard K Nason Assocs 230 S Brentwood Blvd Box F St Louis MO 63105

NASON, ROBERT DOHRMANN, b San Francisco, Calif, Dec 9, 39. SEISMOLOGY. *Educ:* Calif Inst Technol, BS, 61; Univ Calif, San Diego, PhD, 71. *Prof Exp:* Seismologist, Earthquake Mechanism Lab, Nat Oceanic & Atmospheric Admin, Calif, 66-73; EARTHQUAKE GEOPHYSICIST, US GEOL SURV, 73- *Mem:* Am Geophys Union; Geol Soc Am; Seismol Soc Am; Earthquake Eng Res Inst. *Res:* Heat-flow and marine tectonics; earthquakes and earthquake tectonics; movement on the San Andreas fault; fault creep; earthquake damage and seismic intensity. *Mailing Add:* 744 24th Ave San Francisco CA 94121

NASRALLAH, HENRY A, b Apr 30, 47; US citizen; m 72; c 2. NEUROPSYCHOBIOLOGY, NEUROPSYCHOPHARMACOLOGY. *Educ:* Am Univ Beirut, BS, 68, MD, 72. *Prof Exp:* Asst prof psychiat, Univ Calif, San Diego, 77-79; from assoc prof to prof, Univ Iowa, 79-85; PROF & CHMN, DEPT PSYCHIAT, OHIO STATE UNIV, 85- *Concurrent Pos:* Res fel psychopharmacol, Nat Inst Ment Health, 77; chief psychiat, Vet Admin Med Ctr, Iowa City, 79-85; prin investr, Pharmacol Subtype Tardive Dyskinesia, 78-84, Brain Imaging in Schizophrenia, 80-85, Structure & Function of the Brain in Schizophrenia, 86-89; vis prof, var Med Schls, 79-; ed-in-chief, J Schizophrenia Res, 88- *Mem:* Am Col Neuropsychopharmacol; Soc Biol Psychiat; Am Psychiat Asn; Soc Neurosci; Am Acad Clin Psychiatrists. *Res:* Neurobiology of severe psychotic disorders, especially schizophrenia; the use of rain imaging techniques as well as neuropsychopharmacological methods are employed to generate new knowlede and understanding of the etiology and treatment of schizophrenia. *Mailing Add:* Dept Psychiat Ohio State Univ Col Med 473 W 12th Ave Columbus OH 43210

NASRALLAH, MIKHAIL ELIA, b Kafarmishky, Lebanon, Feb 1, 39; c 3. BIOLOGY, GENETICS. *Educ:* Am Univ Beirut, BSc, 60; Univ Vt, MS, 62; Cornell Univ, PhD, 65. *Prof Exp:* Res assoc physiol genetics, Cornell Univ, 65-66; asst prof plant breeding & biol, Cornell Univ & State Univ NY Col Cortland, 66-67; from asst prof to assoc prof biol, 67-74, PROF BIOL, STATE UNIV NY COL CORTLAND, 74- *Res:* Physiological genetics of self-incompatible plants; characterization of self-incompatibility antigens; cytochemical, enzymatic and immunogenetic studies with pollen and stigmatic proteins involved in cell-cell recognition. *Mailing Add:* Dept Biol Col Cortland PO Box 2000 Cortland NY 13045

NASS, MARGIT M K, b Stuttgart, Ger, Sept 24, 31; US citizen; m 69. DNA PROTEIN INTERACTION, MITOCHONDRIAL DNA FUNCTION. *Educ:* Columbia Univ, BS, 54, MA, 55, PhD(molecular biol & biophys), 61. *Hon Degrees:* MA, Univ Pa, 71. *Prof Exp:* Fel molecular biol, Univ Stockholm, 61-64; asst prof molecular biol, 64-69, assoc prof molecular biol & microbiol, 70-79, PROF RADIATION THER & ONCOL, UNIV PA SCH MED, 80-, PROF PHARMACOL & TOXICOL, SCH VET MED, 85-

Concurrent Pos: Postdoctoral fel awards, NSF & Am Cancer Soc, 61-64; actg chmn, dept therapeut res, Univ Pa Sch Med, 71-79; consult, Molecular Biol Study Sect, NIH, 79-82; adv ed, J Subcellular Biochem, 72-84; vis prof & lectr, Univ Tel-Aviv, Israel, 71; prin investr, Univ Pa Sch Med, 73-; consult reviewer, Int Cancer Res Data Bank, Nat Cancer Inst, 76-; 5 year career develop award, NIH. *Mem:* Am Soc Biol Chemists; NY Acad Sci; Am Soc Cell Biologists; AAAS; Genetic Toxicol Asn; Am Soc Microbiol; Soc Gen Physiologists. *Res:* The molecular mechanisms that control DNA replication, DNA methylation & gene expression in cancer cells; the selective use of anticancer drugs to inhibit cancer over normal cells; mitochondrial pathology. *Mailing Add:* Dept Radiation Oncol Univ Pa Sch Med B26 Anat Chem Bldg 6058 Philadelphia PA 19104

NASS, ROGER DONALD, b Merrill, Wis, Nov 9, 32; m 54; c 2. WILDLIFE MANAGEMENT. *Educ:* Univ Wis, BS, 60; Univ Mo, MS, 63. *Prof Exp:* Biologist animal damage control res, Tex, 63-65, biologist forest animal res, Colo, 65-66, biologist rodent res, Hawaii, 66-72, biologist predator res, Idaho, 72-74, proj leader predator res, US Fish & Wildlife Serv, Idaho, 74-86, PRED RES, USDA/APHIS/ADC, 86- *Mem:* Wildlife Soc. *Res:* Predator-prey relationships; animal damage control methodology. *Mailing Add:* 3643 N 3100 E Twin Falls ID 83301

NASSAU, KURT, b Stockerau, Austria, Aug 25, 27; US citizen; m 49. SOLID STATE CHEMISTRY, PHYSICS. *Educ:* Bristol Univ, BSc, 48; Univ Pittsburgh, PhD(phys chem), 59. *Prof Exp:* Res chemist, Glyco Prod Co, Inc, Pa, 49-54; chemist, Dept Metab, Walter Reed Army Med Ctr, Washington, DC, 54-56; mem tech staff, Bell Labs, Murray Hill, 59-89; CONSULT, 90- *Mem:* Am Chem Soc; Optic Soc Am; Am Asn Crystal Growth; fel Am Mineral Asn; fel Am Ceramic Soc. *Res:* Preparation of crystals, glasses and their physical and chemical properties; solid state and crystal chemistry and physics; laser, magnetic, piezoelectric, ferroelectric and vitreous materials. *Mailing Add:* 154A Guinea Hollow Rd Lebanon NJ 08833

NASSER, DELILL, b Terre Haute, Ind, July 17, 29. MICROBIOLOGY, MICROBIAL GENETICS. *Educ:* Ind State Col, BS, 50; Purdue Univ, MS, 55, PhD(microbiol), 63. *Prof Exp:* Res asst, Purdue Univ, 55-56; res microbiologist, Eli Lilly & Co, 56-58; instr, Purdue Univ, 58-60, NIH res trainee microbiol, 63-64; NIH res trainee, Univ Wash, 64-67; asst prof bact, Univ Fla, 67-70, assoc prof microbiol, 70-72; assoc res biochemist, Univ Calif, San Francisco, 72-78; assoc prog dir, Genetic Biol, 79-84, PROG DIR, EUCARYOTIC GENETICS, NSF, 84- *Mem:* Genetics Soc Am; AAAS. *Res:* Glucose metabolism of Fusarium; bioconversion of steroids; biosynthesis of flagellin; transformation of the property of flagellation in Bacillus subtilis; genetic and enzymatic studies of the synthesis of aromatic amino acids in Bacillus subtilis; eucaryotic chromosome structure and regulation. *Mailing Add:* NSF Eucaryotic Genetics Prog 1800 G St NW Rm 325 Washington DC 20550

NASSER, KARIM WADE, b Shweir, Lebanon, Dec 9, 26; Can citizen; c 5. CIVIL ENGINEERING. *Educ:* Am Univ Beirut, BA, 48, BSc, 49; Univ Kans, MSc, 52; Univ Sask, PhD(civil eng), 65. *Prof Exp:* Engr, Trans-Arabian Pipe Line Co, Lebanon, 49-51, design engr, 51-53, supvr engr, Pump Sta, 53-54, maintenance supvr, 54-56; res assoc prestressed concrete, Lehigh Univ, 56-59; gen mgr, J M Wright Ltd, Can, 59-62; lectr civil eng, 62-64, from asst prof to assoc prof, 64-71, PROF CIVIL ENG, UNIV SASK, 71- *Concurrent Pos:* Pres, Victory Construct Ltd, 64; comnr, Crown Investment Review Comn, Sask, 82; pres, NHF Eng, 76-; pres, Int Construct Co Ltd, 81-85; mem, Sci Coun Can, 84-90. *Honors & Awards:* Wason Medal for Res, Am Concrete Inst, 71; Eng Achievement Award, Asn Prof Engrs Sask, 84. *Mem:* Am Concrete Inst; Am Soc Civil Engrs; Eng Inst Can; Can Standards Asn; Am Soc Testing & Mat; Sci Coun Can. *Res:* Concrete and structures; creep of concrete at elevated temperatures; behavior of prestressed concrete members; behavior and design criterion for beams with large openings; K-slump tester, K-5 strength tester and K-situ air tester and fluid level controller; K-tester for removal of concrete forms; mini air meter. *Mailing Add:* Dept Civil Eng Univ Sask Saskatoon SK S7N 0W0 Can

NASSER, TOURAI, b Tehran, Iran, Feb 28, 44; Can citizen; m 75; c 2. TOTAL QUALITY MANAGEMENT, SAFETY & RISK MANAGEMENT. *Educ:* Manchester Univ, UK, Bsc, 67; Imp Col, UK, MSc, 69; London Univ, UK, PhD(struct eng), 74. *Prof Exp:* Mgr Offshore Tech Prog, Det Norske Veritas Ltd, Calgary, Can, 82-84, vpres, 84-88, pres, 88-89; PRES, CENTRE FRONTIER ENG RES, 89- *Concurrent Pos:* Mem, Can Standards Asn Steering Comt, Offshore Struct Prog, 84-, tech comt design steel & concrete offshore struct, 84-, Prov Alta, Premier's Coun Sci & Technol, subcomt nat strategies, 90-, Polartech Conf Int Comt, 91. *Mem:* Asn Prof Engrs, Geologists & Geophysicists Alta; Inst Civil Engrs London UK. *Res:* Application of reliability engineering to design and operation of structural systems; structural mechanics, particularly non-linear and dynamic behaviour of structures; design and operation of offshore exploration and production systems; effective integration of engineering research and practice. *Mailing Add:* Ctr Frontier Eng Res 200 Karl Clark Rd Edmonton AB T6N 1E2 Can

NASSERSHARIF, BAHRAM, b Apr 6, 60; US citizen; m 81; c 3. SUPERCOMPUTING, ARTIFICIAL INTELLIGENCE. *Educ:* Ore State Univ, BS, 80, PhD(nuclear eng), 82. *Prof Exp:* Staff mem, Los Alamos Nat Lab, 83-86; ASST PROF NUCLEAR ENG & DIR, TEX ENG EXP STA, TEX A&M UNIV, 86-, DIR, COMPUTER CTR, 89-, ASST PROF COMPUTER SCI, 90-; SR SCIENTIST, SCI APPLICATIONS INT CORP, 88- *Concurrent Pos:* Asst prof nuclear eng, Ore State Univ, 83; NSF presidential young investr, 86; consult, Scientech, Inc, 86-87 & Mgt Anal Co, 87-88; mem fac, Northwest Col & Univ Asn Sci, 87-88. *Mem:* Fel AAAS; Am Nuclear Soc; Am Asn Artificial Intel; Inst Elec & Electronic Engrs Computer Soc; Soc Indust & Appl Math; Asn Comput Mach. *Res:* Computational methods for design, analysis, and simulation of engineering systems; space nuclear power systems. *Mailing Add:* Supercomputer Ctr Texas A&M Univ College Station TX 77845-3363

NASSI, ISAAC ROBERT, b New York City, NY, Feb 24, 49; m 70; c 2. SOFTWARE SYSTEMS. *Educ:* State Univ NY, Stony Brook, BS, 70, MS, 72, PhD(comput sci), 74. *Prof Exp:* Sr software engr, Softech, Inc, 74-76; prin software engr, Digital Equip Corp, 76-78, consult software engr, 78-82; VPRES SYST ENG, ONTEL CORP, 82-; DIR E COAST RES, APPLE COMPUTER. *Concurrent Pos:* Consult, US Army, 78-79; distinguished rev, Defense Advan Res Proj Agency, 79-; assoc ed, J Comput Languages, 79-; adj instr, Boston Univ, 74-79; prog comt, Boston Sicplan, 76-82. *Mem:* Asn Comput Mach; Inst Elec & Electronics Engrs. *Res:* Design and implementation of programming languages; distributed systems; software engineering; personal computing. *Mailing Add:* 41 Carter Dr Framingham MA 01701

NASSOS, PATRICIA SAIMA, b San Francisco, Calif, Sept 10, 51; m 81. FOOD MICROBIOLOGY, CLINICAL MICROBIOLOGY. *Educ:* Univ Calif, Berkeley, AB, 73, PhD(microbiol), 81. *Prof Exp:* Med technologist, Ralph K Davies Med Ctr, San Francisco, 73-76; RES MICROBIOLOGIST, WESTERN REGIONAL RES LAB, USDA, 81- *Mem:* Am Soc Microbiol; Am Soc Clin Pathologists. *Res:* Effect of dietary fiber on ureolytic gut microflora; incipient bacterial spoilage in ground beef products. *Mailing Add:* 574 Sahara St Santa Rosa CA 95403

NATALE, NICHOLAS ROBERT, b Philadelphia, Pa, Oct 30, 53; m 74, 85; c 2. ASYMMETRIC SYNTHESIS, LANTHANIDE CHEMISTRY. *Educ:* Drexel Univ, BS, 76, PhD(org chem), 79. *Prof Exp:* Res fel, Colo State Univ, 79-81; asst prof, 81-86, ASSOC PROF ORG CHEM, UNIV IDAHO, 87- *Mem:* Am Chem Soc; Sigma Xi. *Res:* Use of heterocycles as masked annulating agents; synthesis of spirocyclic terpenes; selective reductions. *Mailing Add:* Dept Chem Univ Idaho Moscow ID 83843

NATALINI, JOHN JOSEPH, b Norristown, Pa, Apr 27, 44. BIOLOGICAL RHYTHMS, VERTEBRATE BIOLOGY. *Educ:* Villanova Univ, BS, 66; Northwestern Univ, MS & PhD(biol), 71. *Prof Exp:* From asst prof to assoc prof, 71-81, PROF BIOL, QUINCY COL, 81-, CHMN DIV, 86- *Concurrent Pos:* Instr anat & physiol, Blessing Hosp Sch Nursing, Quincy, Ill, 74-82. *Mem:* Sigma Xi; Int Soc Chroniobiol; AAAS; Am Inst Biol Sci. *Res:* Phase response curves to light and the means of entrainment of biological rhythms to various zeitgebers; circannual rhythms of gerbils. *Mailing Add:* Div Sci & Math Quincy Col Quincy IL 62301

NATANI, KIRMACH, b Milwaukee, Wis, June 5, 35; div; c 1. CEREBRAL ARCHITECTURE & BEHAVIOR. *Educ:* Oklahoma Univ, MSc, 71; Oklahoma Univ Col Med, PhD(bio psychol), 77. *Prof Exp:* Particle physics rec technician, anti-proton studies, Berkeley Lawrence Lab, Univ Calif, Berkeley, 58-63; Peace Corp Vol, Thailand, 63-65; res asst sleep psychophysiol & res assoc altered states, Okla Med Res Found, 66-75; res psychologist, cognitive processes, Okla City Vet Admin Hosp, 75-77; nat res coun res assoc, neuro psychol & crew technol, US Air Force Sch AErospace Med, San Antonio, Tex, 77-79; engr human factors, McDonnell Douglas Corp, 79-82, sr engr, 82-86, lead engr, 86-88, PRIN SPEC ENGR, SYSTS SAFETY MCDONNELL DOUGLAS CORP, 88- *Concurrent Pos:* Consult, Bio Dynamics Inc, Cambridge, Mass, 70, dept psychiat, LA Childrens Hosp, Calif, 70-75, dept psychiat, Mass Gen Hosp, Boston, 71; psychol instr, div social sci, Oscar Rose Jr Col, Okla, 75-76; adj asst prof, dept psychiat & behav sci, Okla Univ Col Med, 77-79; mem, Nat Acad Sci adv comt, USSR & E Europe, 77-80, exchange vis to USSR, Nat Acad Sci, 72; neurosci reviewer, N Atlantic Treaty Orgn, Nat Res Coun & Ford Found doctoral fel progs, Nat Acad Sci, Washington, DC, 78-87; partic, Tomahawk, 83-; adj instr, Human Factor Resource Methods, Lindenwood Col, 89- *Honors & Awards:* Roche Award, 73. *Mem:* Am Psychol Asn; Human Factors Soc; Soc Psycho Physiol Res. *Res:* Small group behavior with emphasis to space; sleep psychophysiology and altered states of consciousness; cerebral architecture and non-verbal measures of intelligence; man-machine interface problems; 31 publications. *Mailing Add:* 2842 Gainsboro Ct Bel Nor MO 63121-4717

NATANSOHN, SAMUEL, b Rzeszow, Poland, June 18, 29; US citizen; m 51; c 4. CERAMICS, MATERIALS SCIENCE. *Educ:* Brooklyn Col, BA, 55, MA, 59. *Prof Exp:* SR STAFF SCIENTIST, GEN TEL & ELEC LAB INC, 55- *Honors & Awards:* Leslie H Warner Award. *Mem:* Am Chem Soc; Sigma Xi; Am Ceramic Soc; Metall Soc; Am Soc Testing & Mat. *Res:* Complexation reactions in solutions; magnetic materials; luminescent phenomena in solids; crystal growth; inorganic synthesis; hydrometallurgy; structural ceramics. *Mailing Add:* GTE Labs 40 Sylvan Rd Waltham MA 02254

NATAPOFF, MARSHALL, b New York, NY, May 5, 25; m 78; c 2. PHYSICS, ELECTRONICS. *Educ:* Cornell Univ, BA, 48; NY Univ, MS, 56; Stevens Inst Technol, PhD(physics), 68. *Prof Exp:* Tech writer electronics, Warner Inc, 59-60; instr physics, City Univ New York, 57-60 & NJ Inst Technol, 60-63; engr electronics, Radio Corp Am, 57-63; ASSOC PROF PHYSICS, NJ INST TECHNOL, 63- *Mem:* Am Phys Soc; Am Asn Physics Teachers. *Res:* Solid state physics; properties of dilute metallic alloys; determination of activation energies theoretically; calculation of atomic radii; influence of time on quantum mechanical systems; cancellation in pseudo potential theory. *Mailing Add:* Dept Physics NJ Inst Technol 323 High St Newark NJ 07102

NATARAJAN, KOTTAYAM VISWANATHAN, b Cochin, India, Apr 21, 33; c 4. OCEANOGRAPHY, MICROBIOLOGY. *Educ:* Univ Travancore, India, BS, 52; Banaras Hindu Univ, MS, 55; Univ Alaska, PhD(marine sci), 65. *Prof Exp:* Demonstr bot, NSS Col, Kerala, 52-53 & Vivekananda Col, Madras, 55-56; lectr, Mar Ivanios Col, Kerala, 56-57; res asst, Indian Agr Res Inst, 57-60; scientist, Kaiser Found Res Inst, Calif, 60-61; res asst bot, Univ Calif, Berkeley, 61-62; sr res asst marine sci, Univ Alaska, 62-65, asst prof, 65-70; PROF SCI, GREATER HARTFORD COMMUNITY COL, 70- *Concurrent Pos:* Mem sci fac fel panel, NSF, 74 & mem Comprehensive Assistance Undergrad Sci, 76; mem, Bd Educ, Rocky Hill, Conn, 81-87. *Mem:* AAAS. *Res:* Nitrogen fixation by blue-green algae; general physiology of algae; vitamins of the sea. *Mailing Add:* Dept of Sci Greater Hartford Community Col Hartford CT 06105-2354

NATARAJAN, VISWANATHAN, b Chidambaram, India, Aug 15, 48; m 77; c 1. NEUROCHEMISTRY, LIPID METABOLISM. *Educ:* Univ Bombay, India, BSc, 68, MSc, 70; Indian Inst Sci, Bangalore, PhD(biochem), 75. *Prof Exp:* Res fel, 75-78, res assoc, 78-81, ASST PROF BIOCHEM, HORMEL INST, UNIV MINN, 81- *Mem:* Am Soc Neurochem. *Res:* Structure, function and metabolism of phospholipids in biological membranes; lipid metabolism in ischemia; diabetes and peripheral neuropathy; drug-induced phospholipidoses; metabolism and function of inositol lipids. *Mailing Add:* Dept Pulmonary Med Wishard Hosp 1001 W Tenth St OPW 425 Indianapolis IN 46202

NATELSON, BENJAMIN HENRY, BEHAVIORAL MEDICINE, NEURO-CARDIOLOGY. *Educ:* Univ Pa, MD, 67. *Prof Exp:* Prof neurosci, Univ Med & Dent NJ; CHIEF, PRIMATE NEURO-BEHAV UNIT, VET ADMIN MED CTR. *Mailing Add:* Primate-Neuro Behav Unit Vet Admin Med Ctr East Orange NJ 07019

NATELSON, SAMUEL, b New York, NY, Feb 28, 09; m 37; c 4. CLINICAL CHEMISTRY. *Educ:* NY Univ, ScM, 30, PhD(chem), 31. *Prof Exp:* Instr chem, NY Univ, 28-31; res chemist in-chg, NY Testing Lab, 31-32; res biochemist, Jewish Hosp Brooklyn, 33-49; chmn dept biochem, Rockford Mem Hosp, 49-57, St Vincent's Hosp, New York, 57-58 & Roosevelt Hosp, New York, 58-65; chmn, Dept Biochem, Michael Reese Hosp, 65-79; ADJ PROF ENVIRON PRACT, COL VET MED, UNIV TENN, KNOXVILLE, 79- *Concurrent Pos:* Lectr, Grad Sch, Brooklyn Col, 47-49 & 57-65, New York Polyclin Med Sch & Hosp, 62-65 & Ill Inst Technol, 71-79. *Honors & Awards:* Van Slyke Award Clin Chem, 61; Ames Award, Am Asn Clin Chemists, 65; Chicago Clin Chem Award, 72; Sci Award, Ill Clin Lab Assoc, 71. *Mem:* AAAS; Am Microchem Soc; Harvey Soc; Soc Appl Spectros; Am Chem Soc; Am Asn Biol Chemists. *Res:* Citric acid metabolism in humans; infant feeding; radiopaques; surface tension; vapor pressure; sterols; vitamin D; resins; alkaloids; synthetic organic chemistry; organic analysis; microanalysis; instrumentation; nitrogen metabolism; epilepsy, neuropeptides, guanidino compounds. *Mailing Add:* 925 Southgate Rd Knoxville TN 37919

NATH, AMAR, b Agra, India, Nov 28, 29; m 57; c 1. PHYSICAL CHEMISTRY, SOLID STATE CHEMISTRY. *Educ:* Agra Univ, MSc, 50, 70; Moscow State Univ, PhD(chem), 61. *Prof Exp:* Sci officer, Bhabha Atomic Res Ctr, 51-66; sr chemist, dept chem, Univ Calif, Los Angeles, 66-67 & 69-70; res chemist, Lawrence Radiation Lab, Univ Calif, Berkeley, 67-69; PROF CHEM, DREXEL UNIV, 70- *Concurrent Pos:* Vis scientist, KFA, J06lich, WGer, 81-82. *Honors & Awards:* Vis fel, USSR, 58-61; DSc, Agra Univ, 70. *Mem:* AAAS; Am Chem Soc; Am Phys Inst; Sigma Xi. *Res:* Hot-atom chemistry of solids; isotopic exchange in solid cobalt chelates; high temperature superconductivity; detoxification of asbestos; Mossbauer studies of after-effects of Auger events; Mossbauer spectroscopy of vitamin B12 and hemoglobin. *Mailing Add:* Dept Chem Drexel Univ Philadelphia PA 19104

NATH, DILIP K, b Calcutta, India, Dec 4, 33; US citizen; m 66; c 3. MATERIALS SCIENCE. *Educ:* Univ Calcutta, PhD(ceramics), 64. *Prof Exp:* Sr res scientist mat sci, Gen Elec Co, 68-80; MGR OPTICAL FIBER RES & DEVELOP, ITT CORP, 80- *Concurrent Pos:* Fel Zementforshung, Dusseldorf, WGer, 64-65 & Pa State Univ, 65-68. *Mem:* Am Ceramic Soc; Electrochem Soc. *Res:* To discover and understand the inorganic luminescent materials with regards to phosphor technology; high quality quartz for lamps and other applications. *Mailing Add:* 132 Old Canal Way Weatogue CT 06089

NATH, JAYASREE, b Oct 22, 39; m; c 1. HEMATOLOGY, PHYSIOLOGY. *Educ:* Calcutta Univ, India, BSc Hons, 59, MSc, 61, PhD(biochem), 67. *Prof Exp:* Sr sci asst, Indian Inst Exp Med, Calcutta, 65-67; Brit Med Res Coun fel, Dept Biochem, Univ Birmingham, UK, 67-70; res assoc, Dept Biol, Univ Va, Charlottesville, 71-76; expert scientist, Lab Clin Invest, Nat Inst Allergy & Infectious Dis, NIH, 81-85; scientist, Dept Biol, Univ Va, Charlottesville, 85-88; PHYSIOLOGIST, DEPT HEMAT, WALTER REED ARMY INST RES, 88- *Concurrent Pos:* Vis assoc, Dept Biol, Univ Va, Charlottesville, 76-81. *Mem:* Am Soc Cell Biol; Am Fedn Clin Res; Fedn Am Socs Exp Biol; NY Acad Sci; AAAS. *Res:* Human neutrophil physiology; role of the cytoskeleton in cell motility and inflammatory responses; cell biology of acute inflammatory processes; signal-transduction mechanisms in cellular activation; basic mechanisms of cellular toxicity: studies of microbial toxins of military importance; cell biology of growth and differentiation; author of numerous technical publications. *Mailing Add:* Dept Hematol Div Med Walter Reed Army Inst Res Rm 1056 Bldg 40 Washington DC 20307-5100

NATH, JOGINDER, b Joginder nagar, India, May 12, 32; m 69; c 2. BIOCHEMICAL GENETICS, CYTOGENETICS. *Educ:* Panjab Univ, India, BS, 53, MS, 55; Univ Wis, PhD(agron), 60. *Prof Exp:* Res assoc cryobiol, Am Found Biol Res, Madison, Wis, 60-63; asst prof physiol, Southern Ill Univ, 64-66; assoc prof genetics,66-72, PROF GENETICS & REPROD PHYSIOL, WVA UNIV, 72-, CHMN, GENETICS & DEVELOP BIOL PROG, 75. *Concurrent Pos:* NSF res grant, WVa Univ, 67- *Mem:* Electron Micros Soc Am; Soc Cryobiol; Indian Soc Genetics & Plant Breeding; Environ Mutagenesis Soc; Sigma XI. *Res:* Cytology of genus Selaginella; cytogenetic studies of some species of Paniceae and Phleum; cytogenetics and origin of wheat; cryobiological studies on semen, blood cells, plasma, pollen and bacteria; electron microscopy of frog oocytes; environmental mutagensis with special reference to air-pollutants, carcinogenesis; antimutagenesis; mutagenesis and cryobiology. *Mailing Add:* Genetics & Develop Biol/Plant & Soil Sci WVa Univ Morgantown WV 26506-6108

NATH, K RAJINDER, b Ferozepur, India, May 26, 37; m 73; c 1. DAIRY MICROBIOLOGY, FOOD SCIENCE. *Educ:* Delhi Univ, BSc Hons, 59; Agra Univ, MSc, 61; Cornell Univ, PhD(food sci), 69. *Prof Exp:* Res assoc dairy technol, Ohio State Univ, 69-70; res assoc food sci, Cornell Univ, 70-75; GROUP LEADER, KRAFT INC, 75- *Concurrent Pos:* Proj leader & secy, Qual Assurance Consumer Foods Proj, NE-83, 72-75. *Mem:* Am Dairy Sci Asn; Inst Food Technologists. *Res:* Development of microbial culture for food use; microbial interaction in foods; stimulation and inhibition of lactic acid bacteria; protein hydrolysis and protein modification; cheese ripening and cheese flavor. *Mailing Add:* Kraft Inc Res & Dev Kraft Ct Glenview IL 60025

NATH, NRAPENDRA, b Meerut, India, Oct 13, 40; US citizen; m 74; c 2. MICROBIOLOGY, VETERINARY SCIENCE. *Educ:* Agra Univ, BVSc, 63, MVSc, 65; All-India Inst Med Sci, PhD(microbiol), 70. *Prof Exp:* Instr vet virol, Univ Guelph, 70-71; res fel coagulation, McMaster Univ, 71-72; res instr, Temple Univ Med Ctr, 72-74; res scientist hepatitis, 74-81, SR RES SCIENTIST, HEPATITIS & TISSUE CULTURE, AM RED CROSS BLOOD SERV LAB, 81-, RES MGR, DANDEX. *Concurrent Pos:* Res fel, Dir Gen Health Serv, 66-68 & Indian Coun Med Health, 68-70. *Mem:* Am Soc Microbiol; AAAS. *Res:* Studies in the natural history of hepatitis B virus; test development for human T-cell lymphotropic virus type 3, hepatitis B virus and syphilis. *Mailing Add:* 900 Saybrook Lane Buffalo Grove IL 60089

NATH, PRAN, b Panjab, Pakistan, Sept 9, 39; m 69; c 2. HIGH ENERGY PHYSICS, STRING THEORY. *Educ:* Univ Delhi, BSc, 58 & MSc, 60; Stanford Univ, PhD(physics), 64. *Prof Exp:* Res physicist, Univ Calif Riverside, 64-65; Mellon fel physics, Univ Pittsburgh, 65-66; from asst prof to assoc prof, 66-75, PROF PHYSICS, NORTHEASTERN UNIV, 75- *Concurrent Pos:* Vis scientist, Europ Orgn Nuclear Res, Geneva, Switz, 73-74, sci assoc, 79-80; vis scholar, 79-80 & 86-87, affil dept, Harvard Univ, 87- *Honors & Awards:* Robert D Kline Univ Lectr, 83. *Mem:* fel Am Phys Soc; Europ Phys Soc; AAAS; NY Acad Sci. *Res:* Implications of the unification of gravity along with the electro-weak and strong interactions which are explored using supersymmetry and superstring theory; exploration of physics beyond the standard electro-weak model of Glashow, Weinberg & Salam. *Mailing Add:* Dept Physics Northeastern Univ Boston MA 02115

NATH, RAVINDER KATYAL, b Jullundar, India, Apr 9, 42; m 71; c 2. RADIOLOGICAL PHYSICS. *Educ:* Univ Delhi, BS, 63, MS, 65; Yale Univ, PhD(physics), 71. *Prof Exp:* Res staff physicist, 71-73, res assoc, 73-76, from asst prof to assoc prof, 76-85, PROF THERAPEUT RADIOL, YALE UNIV, 85. *Honors & Awards:* Med Physics Award, Am Asn Physicists Med, 75. *Mem:* Am Phys Soc; Am Asn Physicists Med; Health Physics Soc; Radiation Res Soc; AAAS. *Res:* Radiological physics and radiobiology related to radiation therapy. *Mailing Add:* Dept Therapeut Radiol PO Box 3333 New Haven CT 06510

NATHAN, ALAN MARC, b Rumford, Maine, Sept 17, 46; m 70. EXPERIMENTAL NUCLEAR PHYSICS. *Educ:* Univ Md, BS, 68; Princeton Univ, MA, 72, PhD(physics), 75. *Prof Exp:* Res assoc physics, Brookhaven Nat Lab, 75-77; From asst prof to assoc prof, 77-85, asst dean, Col Eng, 80-81, PROF PHYSICS, UNIV ILL, URBANA, 85- *Mem:* Am Phys Soc. *Res:* Nuclear spectroscopy and fundamental interactions of nuclei and nucleons; photonuclear reactions; intermediate energy nuclear physics. *Mailing Add:* Dept Physics Univ Ill 1110 W Green St Urbana IL 61801

NATHAN, CHARLES C(ARB), b Ft Worth, Tex, Oct 15, 19; m 48; c 4. PHYSICAL CHEMISTRY, CHEMICAL ENGINEERING. *Educ:* Rice Univ, BSChE, 40; Univ Pittsburgh, MS, 42, PhD(chem), 49. *Prof Exp:* Asst chem, Univ Pittsburgh, 40-42; chemist & chem engr, Monsanto Chem Co, 42-44; asst, Purdue Univ, 44; group leader & res assoc petrol prod res, Texaco, Inc, 48-63; mgr coating res & eng, AMF Tuboscope, 63-64; consult, Proj Mohole, 65; process specialist petrol prod & utilization, Monsanto Co, 65-67; sr res assoc, Betz Labs, Inc, 67-68, sect head, 68-71, sr corrosion engr, 71-77; dir, NMex Energy Inst, 77-81, PROF PETROL ENG, NMEX INST MINING & TECHNOL, SOCORRO, 81- *Concurrent Pos:* Observer & lectr, Bikini Atom Bomb Tests, 46; mem, Fossil Energy Adv Comt, US Dept Energy, 78-81. *Mem:* Fel Am Inst Chem Engrs; Nat Asn Corrosion Engrs; Am Soc Qual Control; NY Acad Sci; Nat Soc Prof Engrs. *Res:* Corrosion of ferrous metals; water treatment; industrial quality control; thermodynamics of boron hydrides and hydrocarbons; atomic energy; nuclear physics; statistics; energy research and applications; corrosion of metals. *Mailing Add:* NMex Inst of Mining & Technol Campus Sta Socorro NM 87801

NATHAN, DAVID G, b Cambridge, Mass, May 25, 29. PEDIATRICS. *Educ:* Harvard Univ, BA, 51, MD, 55; Am Bd Internal Med, cert, 62; Am Bd Pediat, cert, 74. *Prof Exp:* House officer med, Peter Bent Brigham Hosp, Boston, Mass, 55-56, sr resident, 58-59, sr assoc med, 67; clin assoc, Nat Cancer Inst, NIH, 56-58; res assoc med, Harvard Med Sch, 59-63, assoc med, 63-66, from asst prof to prof, 66-77, ROBERT A STRANAHAN PROF PEDIAT, HARVARD MED SCH, BOSTON, MASS, 77-; PHYSICIAN-IN-CHIEF, DEPT MED, CHILDREN'S HOSP, BOSTON, MASS, 85- *Concurrent Pos:* Fel, Med Found, Boston, Mass, 59-61; assoc med, Children's Hosp Med Ctr, 63-67, res assoc hemat, 65-66, assoc 66-68, chief, Div Hemat, 68-73, sr assoc med, 68-85, chief, Div Hemat & Oncol, 74-84; chief, Div Hemat & Oncol, Dana Farber Cancer Inst, Boston, Mass, 74-84, pediatrician-in-chief, 74-85; numerous vis professorships & hon lectrs. *Honors & Awards:* Nat Medal Sci, 90; Distinguished Career Award, Am Soc Pediat Hemat/Oncol, 91. *Mem:* Inst Med-Nat Acad Sci; fel AAAS; AAAS; Am Asn Cancer Res; Am Fedn Clin Res; Soc Pediat Res; Am Pediat Soc; Am Acad Pediat; Am Soc Clin Invest; Am Soc Clin Oncol. *Res:* Author or co-author of 5 books and over 230 publications. *Mailing Add:* Children's Hosp 300 Longwood Ave Boston MA 02115

NATHAN, HENRY C, biology; deceased, see previous edition for last biography

NATHAN, KURT, b Essen, Ger, June 27, 20; nat US; m 48; c 1. CIVIL ENGINEERING, HYDROLOGY. *Educ:* Cornell Univ, BS, 46, MS, 48; Rutgers Univ, BSAE, 55. *Prof Exp:* Asst agr eng & mod lang, Cornell Univ, 46-48; asst prof agr eng, Nat Agr Col, 48-51; res assoc, 51-55, from asst prof to prof, 55-82, EMER PROF AGR ENG, RUTGERS UNIV, 82- *Concurrent Pos:* Pres & prin engr, Conserv Eng, PA, 81- *Mem:* Nat Soc Prof Engrs; Am

Soc Agr Engrs; Am Water Resources Asn; Am Soc Civil Engrs. *Res:* Drainage; engineering aspects of soil and water conservation; site development engineering; surface water hydrology; water resources. *Mailing Add:* 144 Dayton Ave Somerset NJ 08873

NATHAN, LAWRENCE CHARLES, b Corning, Calif, Nov 26, 44; m 88; c 4. COORDINATION CHEMISTRY. *Educ:* Linfield Col, BA, 66; Univ Utah, PhD(inorg chem), 71. *Prof Exp:* PROF CHEM, SANTA CLARA UNIV, 70- *Concurrent Pos:* Petrol Res Fund res grant, 71. *Mem:* Sigma Xi; Am Chem Soc. *Res:* Preparation and characterization of new transition metal coordination complexes. *Mailing Add:* Dept of Chem Santa Clara Univ Santa Clara CA 95053

NATHAN, MARC A, b Great Falls, Mont, Sept 14, 37; m 57; c 4. NEUROPHYSIOLOGY, CARDIOVASCULAR PHYSIOLOGY. *Educ:* Wash State Univ, BS, 60; Univ Wash, MS, 62, PhD(psychol), 67. *Prof Exp:* Res assoc physiol psychol, Sch Med, Univ Wash, 67-68, res psychologist radiobiol, Sch Aerospace Med, 68-71; instr, Med Col, Cornell Univ, 72-73, asst prof neurol, 73-78; ASSOC PROF PHARMACOL, UNIV TEX HEALTH SCI CTR, SAN ANTONIO, 78- *Concurrent Pos:* Dir field neurobiol & behav, Med Col, Cornell Univ, 73-77; res career develop award, 76-81. *Mem:* AAAS; Am Physiol Soc; Inst Elec & Electronics Engrs; Soc Neurosci; Am Heart Asn. *Res:* Neural control of the cardiovascular system; neurogenic hypertension; emotional behavior. *Mailing Add:* Dept Pharmacol CUNY Med Sch 138th St & Convent Ave New York NY 10031

NATHAN, MARSHALL I, b Lakewood, NJ, Jan 22, 33; m 55, 71; c 2. PHYSICS. *Educ:* Mass Inst Technol, BS, 54; Harvard Univ, PhD(physics), 58. *Prof Exp:* Staff mem, IBM Corp, 58-71, mgr coop phenomena group, 71-74, consult to dir res, 74-75, mgr optical solid state technol group, 75-77, mgr, Semiconductor Physics & Device Dept, 77-79, mgr, Semiconductor Miscrostruct Device Physics Group, IBM Corp, 79-87; PROF ELEC ENG, UNIV MINN, 87- *Honors & Awards:* David Sarnoff Prize, Inst Elec & Electronics Engrs, 80. *Mem:* Fel Am Phys Soc; fel Inst Elec & Electronics Engrs. *Res:* Solid state physics; semiconductor devices; optics. *Mailing Add:* Dept Elec Eng Univ Minn 123 Church St SE Minneapolis MN 55455

NATHAN, PAUL, b Chicago, Ill, June 18, 24; m 53; c 4. PHYSIOLOGY. *Educ:* Univ Chicago, PhB, 46, PhD(physiol), 53. *Prof Exp:* Biochemist, Galesburg State Res Hosp, Ill, 53-55; from instr to assoc prof, 55-77, PROF PHYSIOL, COL MED, UNIV CINCINNATI, 77- ASST PROF EXP SURG, 66-, DIR DEPT CELL BIOL & IMMUNOL, SHRINERS BURNS INST, 66- *Concurrent Pos:* Advan res fel, Am Heart Asn, 59-61; res assoc, May Inst Med Res, Cincinnati Jewish Hosp, 55-65; estab investr, Am Heart Asn, 61-66; res collabr, Brookhaven Nat Lab, 64-71. *Mem:* AAAS; Am Physiol Soc; Int Soc Burn Injuries. *Res:* Transplantation; physiology of digestion; immunology; burn injury. *Mailing Add:* Enquay Pharm Assoc 411 Oak St Suite 305 Cincinnati OH 45221

NATHAN, RICHARD ARNOLD, b New York, NY, Sept 25, 44; m 66; c 2. ORGANIC CHEMISTRY. *Educ:* Mass Inst Technol, BS, 65; Polytech Inst Brooklyn, PhD(org chem), 69. *Prof Exp:* Chemist, Rohm and Haas Co, 65 & Polaroid Corp, 69; proj leader, Org Chem Div, 70-74, assoc mgr, Org & Struct Chem Sect, 74-76, mgr, Org, Anal & Environ Chem Sect, 76-79, ASSOC DIR, CORP TECH DEVELOP, COLUMBUS LABS, BATTELLE MEM INST, 79- *Mem:* Am Mgt Asn; Am Defense Preparedness Asn; AAAS; Am Nuclear Soc; Atomic Indust Forum; Sigma Xi. *Res:* Photochemistry and bioorganic chemistry. *Mailing Add:* 4621 Nugent Dr Columbus OH 43220

NATHAN, RICHARD D, CARDIAC ELECTROPHYSIOLOGY, MEMBRANE BIOPHYSICS. *Educ:* Univ Fla, PhD(physics), 71. *Prof Exp:* ASSOC PROF PHYSIOL, TEX TECH UNIV, 82- *Mailing Add:* Health Sci Ctr Tex Tech Univ Sch Med Lubbock TX 79430

NATHAN, RONALD GENE, b Paterson, NJ, Feb 22, 51; m 74; c 2. MEDICAL EDUCATION, STRESS. *Educ:* Cornell Univ, BA, 73; Univ Houston, MA, 75, PhD(clin psych), 78. *Prof Exp:* Clin asst prof, dept psychol, Univ NC, Chapel Hill, 78-79; from asst prof to assoc prof, depts psychiat & family med & dir med psychology, La State Univ Sch Med, 78-87; ASSOC PROF FAMILY PRAT & PSYCHIAT, ALBANY MED COL, 87-, DIR EDUC DEVELOP, DEPT FAMILY PRACT, 87- *Concurrent Pos:* Dir biofeedback clinic, Sch Med, La State Univ, 83-87; mem, adv bd, Col Nursing, Northwestern State Univ La, 83-87; consult, Libbey Glass, 83-84, Merrill Lynch, 83, AT&T, 85, Blue Cross/Blue Shield, 85, Univ Mo-Kansas City Med Sch, 85, Mansfield Community Hosp, Dallas, 85, ERA Real Estate, 86 & Sch Med, La State Univ, 87, Fleet-Norstar Bank, Albany, 88. *Mem:* Am Psychol Asn; Soc Behav Med; Psychologists Family Med & Primary Care; fel Am Inst Stress; Soc Teachers Family Med. *Res:* Stress management training; assessment of depression; medical interviewing. *Mailing Add:* Dept Family Pract (A-46) Albany Med Col Albany NY 12208

NATHANIEL, EDWARD J H, b Guntur, India, Apr 21, 28; m 53; c 3. ANATOMY. *Educ:* Univ Madras, MB, BS, 52; Univ Calif, Los Angeles, MS, 58, PhD(anat), 62. *Prof Exp:* Demonstr anat, Christian Med Col, Vellore, 52-54; civil asst surgeon & actg chief med officer, Leprosy Treatment & Study Ctr, Turukoilur, 55-56; res anatomist, Sch Med, Univ Calif, Los Angeles, 57-62; chief electron micros lab path, Cedars of Labanon Hosp, 62-64; assoc prof anat & path, Med Col Ga, 64-66; asst prof anat, McGill Univ, 66-68; ASSOC PROF ANAT, SCH MED, UNIV MAN, 68- *Concurrent Pos:* Med Res Coun Can fel, Sch Med, Univ Man, 60-72, Nat Cancer Inst Can fel, 69-72; USPHS fel, Cedars of Lebanon Hosp, Los Angeles, 63-66; USPHS fel, Med Col Ga, 64-66. *Mem:* Am Asn Anatomists; Can Asn Anat; Electron Micros Soc Am; Am Soc Cell Biol; Am Soc Exp Path. *Res:* Electron microscopic research in experimental neurology with emphasis on demyelination and remyelination, postnatal development of nervous system, platelet morphology in artificially induced thrombi in experimental situations and chemotherapy of experimental mouse tumors. *Mailing Add:* Dept Anat Univ Man Sch Med 750 Williams Ave Winnipeg MB R3E 0W2 Can

NATHANS, DANIEL, b Wilmington, Del, Oct 30, 28; m 56; c 3. MOLECULAR BIOLOGY, GENETICS. *Educ:* Univ Del, BS, 50; Washington Univ, MD, 54. *Hon Degrees:* ScD, Univ Del, Wash Univ, Rockefeller Univ, Yale Univ. *Prof Exp:* Resident, Columbia-Presby Med Ctr, 57-59; from asst prof to prof microbiol, 62-76, PROF MOLECULAR BIOL & GENETICS, JOHNS HOPKINS UNIV, 80-; SR INVESTR, HOWARD HUGHES MED INST, 82- *Honors & Awards:* Nobel Prize Cowinner in Med, 78. *Mem:* Nat Acad Sci; Am Acad Arts & Sci; Am Philos Soc. *Res:* Molecular genetics. *Mailing Add:* Howard Hughes Med Inst Johns Hopkins Univ 725 N Wolfe St Baltimore MD 21205

NATHANSON, BENJAMIN, b New York, NY, Jan 9, 29. HAZARDOUS WASTES ANALYSIS, ANALYTICAL INSTRUMENTATION. *Educ:* City Col New York, BS, 47; Columbia Univ, MA, 49; NY Univ, PhD(chem), 65. *Prof Exp:* Grad asst chem, NY Univ, 50-54; lectr, Chem Dept, Hunter Col, 54-55; instr, Finch Col, 66-67; lectr, Chem Dept, Pace Col, 67-68; ASSOC CHEMIST, GRADE II, DEPT ENVIRON PROTECTION, CITY OF NEW YORK, 68- *Concurrent Pos:* Adj asst prof, Chem Dept, Pratt Inst, 67-68, Cooper Union, 76-77, 85. *Mem:* Am Chem Soc; Am Soc Testing & Mat. *Res:* Instrumental analysis of air and water pollutants and toxic wastes utilizing atomic absorption; x-ray fluorescence; gas chromatography; high-pressure liquid chromatography; gas chromatography-mass spectroscopy; ion chromatography; analysis of airborne particulates for trace metals; analyzing gasoline for lead; trace gas analysis; fuel oil contaminants. *Mailing Add:* 470 W 24th St #7G New York NY 10011

NATHANSON, FRED E(LIOT), b Baltimore, Md, Jan 12, 33; m 56; c 2. ELECTRICAL ENGINEERING. *Educ:* Johns Hopkins Univ, BS, 55; Columbia Univ, MS, 56. *Prof Exp:* Assoc engr, Appl Physics Lab, Johns Hopkins Univ, 56-60, sr staff engr, 60-63, prin staff engr & asst supvr advan radar technol, 63-70; mgr, Washington opers, 70-86, DIR CORP PLANNING, TECHNOL SERV CORP, 86- *Concurrent Pos:* Assoc mem, Adv Group Electron Devices. *Mem:* Fel Inst Elec & Electronics Engrs; Int Union Radio Sci. *Res:* Radar design; radar signal processing; optical electronic techniques. *Mailing Add:* Ga Tech Res Inst 11400 Dorchester Lane Rockville MD 20852

NATHANSON, H(ARVEY) C(HARLES), b Pittsburgh, Pa, Oct 22, 36; m 63; c 2. SOLID STATE ELECTRONICS. *Educ:* Carnegie Inst Technol, BSEE, 58, MSEE, 59, PhD(elec eng), 62. *Prof Exp:* Instr elec eng, Carnegie Inst Technol, 59-60; sr engr, Westinghouse Elec Corp, 62-66, fel eng, 66-67, mgr silicon jct physics, 67-72, mgr microelectronic devices, 72-90, CHIEF SCIENTIST, ELECTRONIC SCI DIV RES LABS, WESTINGHOUSE ELEC CORP, 90- *Mem:* Fel Inst Elec & Electronic Engrs; Electron Device Soc (pres, 78-80). *Res:* Solid state devices; hyper-abrupt pn semiconductor junctions; surface controlled pn junction breakdown; silicon-silicon dioxide surface stability; mechanically resonant metal-insulator-semiconductor systems; microwave devices; imaging devices. *Mailing Add:* 5635 Marlborough Rd Pittsburgh PA 15235

NATHANSON, MELVYN BERNARD, b Philadelphia, Pa, Oct 10, 44; m 78; c 2. HISTORY & PHILOSOPHY OF SCIENCE. *Educ:* Univ Pa, BA, 65; Univ Rochester, MA, 68, PhD(math), 72. *Prof Exp:* Prof math, Southern Ill Univ, Carbondale, 71-81; dean grad sch & prof math, Rutgers Univ, Newark, 81-86; PROVOST, VPRES ACAD AFFAIRS & PROF MATH & COMPUT SCI, LEHMAN COL, CITY UNIV NY, 86- *Concurrent Pos:* Int Res & Exchanges Bd fel, fac mech & math, Moscow State Univ, 72-73; asst to Andre Weil, Inst Advan Study, 74-75 & 90-91; assoc prof math, Brooklyn Col, 75-76; guest, Rockefeller Univ, 75-76; hon res fel math, Harvard Univ, 77-78; vpres & trustee, asn mems Inst Adv Study, Princeton, NJ. *Mem:* Am Math Soc; Math Asn Am; AAAS; Arms Control Asn; fel NY Acad Sci. *Res:* Number theory; algebra; combinatorial theory. *Mailing Add:* Provost & Vpres Acad Affairs Lehman Col, CUNY Bronx NY 10468

NATHANSON, NEAL, b Boston, Mass, Sept 1, 27; m 54; c 3. VIROLOGY, EPIDEMIOLOGY. *Educ:* Harvard Univ, AB, 49, MD, 53. *Prof Exp:* Chief poliomyelitis surveillance unit, USPHS Commun Dis Ctr, 55-57; res assoc anat & asst prof, Sch Med, Johns Hopkins Univ, 57-63, from assoc prof to prof epidemiol, 63-79; CHMN, DEPT MICROBIOL, SCH MED, UNIV PA, 79- *Concurrent Pos:* Ed-in-chief, Am J Epidemiol, 64-79. *Mem:* Am Epidemiol Soc; Am Asn Immunol; Am Soc Microbiol; Am Soc Trop Med & Hyg; Am Pub Health Asn. *Res:* Neurotropic viruses; neuropathology; AIDS. *Mailing Add:* Dept Microbiol Sch Med Univ Pa Philadelphia PA 19104

NATHANSON, NEIL MARC, b Philadelphia, Pa, Dec 11, 48. MUSCARINIC ACETYLCHOLINE RECEPTORS, SIGNAL TRANSDUCTION. *Educ:* Univ Pa, BA, 70; Brandeis Univ, PhD(biochem), 75. *Prof Exp:* Muscular Dystrophy Asn, Inc Fel, Lab Biochem Genetics, Nat Heart & Lung Inst, 75-76; postdoctoral Fel, Univ Calif, San Francisco, 76-79; from asst prof to assoc prof, 79-90, PROF DEPT PHARMACOL, UNIV WASH, 90- *Concurrent Pos:* Estab investr, Am Heart Asn, 85-90. *Mem:* AAAS; Soc Neurosci; Am Soc Cell Biol; Am Soc Biochem & Molecular Biol. *Res:* Cell and molecular biology of neurotransmitter action and signal transduction; regulation, function and molecular cloning of muscarinic receptors and G-proteins. *Mailing Add:* Dept Pharmacology, SJ-30 Univ Washington Seattle WA 98195

NATHANSON, WESTON IRWIN, b Detroit, Mich, May 2, 38; m 58; c 2. SOFTWARE SYSTEMS DESIGN & ENGINEERING, DIGITAL FILTERING. *Educ:* Univ Calif, Los Angeles, BA, 61, MA, 63, PhD(math), 70. *Prof Exp:* Aeronaut engr, Douglas Aircraft Co, 61-66; prof math, Calif State Univ, Northridge, 66-81; adv engr, Fed Syst Div, IBM Corp, Calif, 81-87; sr scientist, Intercom Systs Corp, Woodland Hills, Calif, 87-89; PROF MATH, CALIF STATE UNIV, NORTHRIDGE, 89- *Concurrent Pos:* Eng analyst, Litton Aero Prod, 78-81. *Mem:* Math Prog Soc; Math Asn Am; Soc Indust & Appl Math. *Res:* Discrete optimization; linear programming. *Mailing Add:* Calif State Univ Northridge Dept Math 18111 Nordhoff Northridge CA 91330

NATHENSON, MANUEL, b Charleroi, Pa, Feb 17, 44; m 68; c 1. ENGINEERING, GEOPHYSICS. *Educ:* Carnegie-Mellon Univ, BS, 65; Stanford Univ, MS, 67, PhD(aeronaut eng), 71. *Prof Exp:* MECH ENGR GEOTHERMAL ENERGY, US GEOL SURV, 72- *Concurrent Pos:* Mem, Geothermal Resources Coun. *Mem:* Am Phys Soc; Am Geophys Union; AAAS. *Res:* Geothermal energy. *Mailing Add:* US Geol Surv 345 Middlefield Rd MS 910 Menlo Park CA 94025

NATHENSON, STANLEY G, b Denver, Colo, Aug 1, 33; m 59; c 2. MOLECULAR IMMUNOLOGY, IMMUNOBIOLOGY. *Educ:* Reed Col, BA, 55; Wash Univ, MD, 59. *Prof Exp:* From asst prof to assoc prof microbiol & immunol, 66-73, PROF CELL BIOL, MICROBIOL & IMMUNOL, ALBERT EINSTEIN COL MED, 73- *Concurrent Pos:* Nat Found Fel pharmacol, Wash Univ, 60-62; Helen Hay Whitney Found fel, Queen Victoria Hosp, Sussex, Eng, 64-67. *Mem:* Nat Acad Sci; Transplantation Soc; Am Asn Immunol. *Res:* Molecular genetic studies of major histocompatibility genes, and structural analysis of their products; immunochemistry and biochemistry of mammalian cell membranes; studies in specific interactions between MHC products, antigenic peptides and T cell receptors. *Mailing Add:* Dept Microbiol & Immunol Albert Einstein Col Med Bronx NY 10461

NATHER, ROY EDWARD, b Helena, Mont, Sept 23, 26; m 62; c 3. ASTRONOMY, PHYSICS. *Educ:* Whitman Col, BA, 49; Univ Cape Town, PhD(astron), 72. *Prof Exp:* Physicist, Hanford Works, Gen Elec Co, 47-51, Calif Res & Develop Corp, 51-53, Tracerlab, Inc, 53-56 & Gen Atomics Div, Gen Dynamics Corp, 56-60; programmer comput div, Royal McBee, Inc, 60-61 & Packard Bell Comput Co, 61-62; tech dir, Sharp Lab, Beckman Inst, 62-67; spec res assoc, 67-73, assoc prof, 73-80, PROF ASTRON, UNIV TEX, AUSTIN, 80- *Concurrent Pos:* Rex G Baker, Jr & McDonald Observ Centennial Res Prof Astronom, 87. *Res:* Application of high speed electronic techniques to study of short timescale astronomical phenomena, in particular, the evolution of interacting binary stars and variable white dwarfs. *Mailing Add:* Dept of Astron Univ of Tex Austin TX 78712

NATHWANI, BHARAT N, b Bombay, India, Jan 20, 45; US citizen; m 70. HEMATOPATHOLOGY. *Educ:* Bombay Univ, MBBS, 69, MD, 72; Am Bd Path, dipl, 77. *Prof Exp:* Pathologist, City of Hope Med Ctr, 77-84; PROF PATH & CHIEF HEMATOPATH, SCH MED, UNIV SOUTHERN CALIF, 84- *Concurrent Pos:* NIH grant on comput-aided diagnoses of lymph node dis, 86- *Mem:* AAAS; Am Soc Hemat; Am Soc Oncol; Am Soc Clin Path; Int Acad Path. *Res:* Building "experts systems" and highly interactive videodisc systems for learning, teaching and diagnosis in the field of surgical pathology. *Mailing Add:* HMR 204 2025 Zonal Ave Los Angeles CA 90033

NATION, JAMES EDWARD, b Springfield, Ill, Aug 22, 33. NEUROLINQUISTICS. *Educ:* Ill State Univ, BS, 59; Univ Wis, MS, 60, PhD(speech path), 64. *Prof Exp:* Asst prof speech path, Univ Ga, 64-66; from asst prof to assoc prof, 66-80, prof speech path, Dept Commun Sci, Case Western Reserve Univ, 80-86, SPEECH-LANGUAGE PATHOLOGIST, TUCSON UNIFIED SCH DISTRICT, 85- *Concurrent Pos:* Assoc dir org dis, Cleveland Hearing & Speech Ctr, 66-67, dir, Dept Speech Path, 70-74; assoc adj prof, Dept Speech Path, Cleveland State Univ, 70-74 & Kent State Univ, 76-77; consult speech path, Dept Pediat, Cleveland Metrop Gen Hosp, 70-; speech path craniofacial team, Univ Hosps, Cleveland, 78-84; proj dir, Rehab Serv Admin Training grant, 77-84; sr clin instr, Dept Pediat, Case Western Reserve Univ, 79-84; mem, Nat Coun Grad Prog Speech-Language Path & Audiol, 78-84, secy & treas, 80-84. *Honors & Awards:* Fel Am Speech-Language-Hearing Asn, 79; Distinguished Serv Award, Nat Coun Grad Progs, 82. *Mem:* Am Speech & Hearing Asn; Am Cleft Palate Asn. *Res:* Speech and language disorders of children and adults with emphasis on the underlying physical and biological processes that account for normal and disordered behavior. *Mailing Add:* 2600 Skyline Dr #19 Tucson AZ 85718

NATION, JAMES LAMAR, b Webster Co, Miss, Mar 3, 36; m 59; c 3. INSECT PHYSIOLOGY. *Educ:* Miss State Univ, BS, 57; Cornell Univ, PhD(entom), 60. *Prof Exp:* From asst prof to assoc prof biol sci, 60-70, assoc prof entom & nematol, 70-72, PROF ENTOM & NEMATOL, UNIV FLA, 72- *Concurrent Pos:* NSF grants, 61-65 & 88-91; mem staff, Univ Guelph, 69-70; USDA Coop grant, 70-72 & 80-83, USDA Competitive grant, 85-87 & 87-91; Environ Protection Agency grant, 80-83; vis scientist, Swiss Fed Res Sta, Wadenswil, Switzerland, 83. *Mem:* AAAS; Entom Soc Am; Int Soc Chem Ecol. *Res:* Sex pheromones in insects; insect-plant interactions. *Mailing Add:* Dept of Entom & Nematol Univ of Fla Gainesville FL 32611-0740

NATION, JOHN, b Bridgwater, Eng, Aug 8, 35; m 61; c 2. RELATIVISTIC BEAMS, ACCELERATORS. *Educ:* Univ London, BSc & ARCS, 57, DIC & PhD(plasma physics), 60. *Prof Exp:* Consult plasma physics, Nat Comt for Nuclear Energy, Rome, Italy, 60-62; staff scientist, Cen Elec Generating Bd, Eng, 62-65; from asst prof to assoc prof, 65-77, from asst dir to assoc dir, Lab Plasma Studies, 75-84, PROF ELEC ENG, CORNELL UNIV, 78-, DIR SCH ELEC ENG, 84- *Concurrent Pos:* Lectr, Chelsea Col Sci & Technol, Eng, 64-65; sr vis fel, Sci Res Coun, London, Eng, 73-74; consult phys dynamics, La Jolla Inst, La Jolla, Calif, Jaycor, Del Mar, Calif, Los Alamos Nat Lab, NMex & Schaeffer Assocs, Arlington, Va. *Honors & Awards:* Centennial Medal, IEEE, 86. *Mem:* Sr mem Inst Elec & Electronics Engrs; fel Am Phys Soc. *Res:* Investigation of the physics of intense electron and ion beams and their applications to accelerators; relativistic electron beams; collective accelerators; high power microwave generation. *Mailing Add:* Dept Elec Eng Cornell Univ Phillips Hall Ithaca NY 14853

NATIONS, CLAUDE, b Marlow, Okla, July 9, 29; m 53; c 4. CELL BIOLOGY. *Educ:* Univ Okla, BS, 53; Okla State Univ, MS, 58, PhD(bot), 67. *Prof Exp:* Teacher, Okla City Bd Educ, Okla, 55-57; teacher, Wichita Bd Educ, Kans, 58-62; consult sci educ, Mo State Dept Educ, 63-64; instr biol, bot & plant physiol, Okla State Univ, 64-66; asst prof bot, Univ Tex, Arlington, 66-68; asst prof, 68-76, ASSOC PROF BIOL, SOUTHERN METHODIST UNIV, 76- *Concurrent Pos:* Fel, Lab Cancer Res, Univ Wis, 72-73. *Mem:* Am Soc Cell Biol. *Res:* Gene regulation by nuclear proteins. *Mailing Add:* Dept of Biol Southern Methodist Univ Dallas TX 75275

NATIONS, JACK DALE, b Prairie Grove, Ark, Oct 18, 34; m 57; c 4. PALEOECOLOGY. *Prof Exp:* Geologist, Stand Oil Co Tex, 61-64, Pan Am Petrol Corp, 64, Bur Land Mgt, 79-80; assoc prof, 69-80,bur reclamation, 86-87, PROF GEOL, NORTHERN ARIZ UNIV, 80-, 86-87, REGENTS PROF, 88- *Concurrent Pos:* Grant paleoecol & biostratig of Cenozoic Basins, 75-76, Cretaceous biostratig & peleocol of Ariz, 76-81; chmn, Ariz Oil & Gas Conserv Comn, 76-91; off, Ariz Rep Interstate Oil Compact Comn, 79-; pres, Ariz & Nev Acad Sci, 87-88. *Mem:* Am Asn Petrol Geol; Paleont Soc; Soc Econ Paleont & Mineral; Int Paleont Union; Am Inst Prof Geologists; Geol Soc Am. *Res:* Paleoecology and biostratigraphy Cenozoic Basins; systematics, phylogeny and paleobiogeography of fossil crabs; Cretaceous paleoecology and biostratigraphy. *Mailing Add:* Dept Geol Northern Ariz Univ PO Box 6030 Flagstaff AZ 86011

NATKE, ERNEST, JR, CELL VOLUME REGULATION. *Educ:* State Univ NY Upstate Med Ctr, PhD(pharmacol), 78. *Prof Exp:* RENAL PHYSIOLOGIST, DIV NEPHROLOGY, DEPT MED, WINTHROP UNIV HOSP, 82- *Res:* Hormonal influence of an ion and water transport. *Mailing Add:* Winthrop Univ Hosp 259 First Ave Mineola NY 11501

NATOLI, JOHN, b Clearfield, Pa, July 5, 50. SURFACE & COLLOID SCIENCE. *Educ:* Univ Pittsburgh, BS, 72; Carnegie-Mellon Univ, MS, 77, PhD(chem eng), 80. *Prof Exp:* Anal biochemist, Biodecision Labs, Div Mylan Pharmaceut, 73-74; asst prof chem eng, Pa State Univ, 79-81; SR SCIENTIST, ROHM & HAAS CO, 81- *Mem:* Am Inst Chem Engrs; Am Chem Soc; AAAS. *Res:* Water soluble polymers; emulsion polymerization; phase behavior of surfactant solutions; membranes and membrane bound transport systems. *Mailing Add:* 504 Welsh Rd Ambler PA 19002

NATOWITZ, JOSEPH BERNARD, b Saranac Lake, NY, Dec 24, 36; m 61; c 2. NUCLEAR CHEMISTRY. *Educ:* Univ Fla, BS, 58; Univ Pittsburgh, PhD(chem), 65. *Prof Exp:* Asst chem, Univ Pittsburgh, 61-65; res assoc, State Univ NY, Stony Brook, 65-67; from asst prof to assoc prof, 67-76, head dept, 81-85, PROF CHEM, TEX A&M UNIV, 76- *Concurrent Pos:* Res collabr, Brookhaven Nat Lab, 65-67; vis scientist, Max Planck Inst; vis prof, Tokyo Univ, 79, Caen Univ, 85-86 & Univ Catholique de Louvain, 87; chmn, Gordon Res Conf, 84; res award, Tex A&M Univ, 88. *Honors & Awards:* Alexander Von Humboldt Sr Scientist Award, 78. *Mem:* Am Chem Soc; Chem Inst Can; fel Am Phys Soc. *Res:* Nuclear reaction studies; fission, light fragment emission, angular momentum and excitation energy limits to nuclear stability; properties of highly excited nuclei. *Mailing Add:* Dept Chem Tex A&M Univ College Station TX 77843

NATOWSKY, SHELDON, b Brooklyn, NY, May 6, 44; m 68; c 2. INDUSTRIAL ORGANIC CHEMISTRY. *Educ:* Brooklyn Col, BS, 66; Cornell Univ, PhD(phys org chem), 73. *Prof Exp:* Res specialist surfactants, GAF Corp, 73-75; dir res & develop labs, Carson Chem Inc, Div Quad Chem Corp, 75-78; prod develop chemist, Clorox Co, 78-79; mgr oil field chem, prod develop & anal serv, Chemlink Petrol, Inc, 79-86; mkt mgr, Well Stimulation Additives, 86-88, MGR, BUS DEVELOP, 86- *Mem:* Am Chem Soc; Am Oil Chemists Soc; Soc Cosmetic Chemists; Nat Asn Corrosion Engrs; Soc Petrol Engrs; Chem Mkt Res Asn; Planning Forum; Com Develop Asn. *Res:* Management of synthesis, formulation, analysis and governmental regulations relating to oil field production, chemicals manufacture, sales and applications; surfactant formulation cosmetics; consumer products, marketing and business development. *Mailing Add:* Kerr-McGee Corp Kerr-McGee Ctr PO Box 25861 Oklahoma City OK 73125-0861

NATRELLA, JOSEPH, b Brooklyn, NY, Feb 7, 19. COMPUTER SCIENCE, STATISTICS. *Educ:* New York City Col, BA, 43; Am Univ, MA, 57. *Prof Exp:* Staff math & statist, Fed Govt; RETIRED. *Mailing Add:* 25 N Floyd St Alexandria VA 22304

NATTA, CLAYTON LYLE, b San Fernando, Trinidad, WI, Nov 11, 32; US citizen. SICKLE CELL ANEMIA. *Educ:* McMaster Univ, Toronto, BA, 57; Univ Toronto, MD, 61; FRCP(C), 72; FRCPath, 90. *Prof Exp:* From instr to asst prof med, 70-81, ASSOC PROF CLIN MED, COLUMBIA UNIV COL PHYSICIANS & SURGEONS, 81- *Concurrent Pos:* Chief hemat, Harlem Hosp Ctr, 70-, assoc attend physician, 80-; assoc attend physician, Columbia Presby Med Ctr, 85- *Mem:* Am Soc Hemat; Int Soc Hemat; Am Fedn Clin Res; Am Soc Clin Nutrit; Am Inst Nutrit; Am Soc Internal Med. *Res:* Study of factors that affect the clinical severity of sickle cell anemia; study of anti sickling agents; study of nutritional factors that affect sickle cell anemia. *Mailing Add:* Columbia Univ Col Physicians & Surgeons 630 W 168th St New York NY 10032

NATTIE, EUGENE EDWARD, b Alexandria, Va, June 15, 44; m 70; c 2. PULMONARY PHYSIOLOGY, CONTROL OF BREATHING. *Educ:* Dartmouth Col, BA, 66, BMS, 68; Harvard Univ, MD, 71. *Prof Exp:* Intern med, Peter Bent Brigham Hosp, 71-72; fel, 72-75, from asst prof to assoc prof, 75-85, PROF PHYSIOL, SCH MED, DARTMOUTH COL, 85- *Concurrent Pos:* Adv dean students, Dartmouth Col. *Res:* Control of breathing; central chemoreceptor function; control of blood pressure. *Mailing Add:* Dept of Physiol Dartmouth Med Sch Hanover NH 03755

NATTRESS, JOHN ANDREW, b Lansdowne, Pa, June 16, 20; m 43, 75; c 5. INDUSTRIAL ENGINEERING. *Educ:* Drexel Inst Tech, BS, 43; Ga Inst Technol, MS, 50. *Hon Degrees:* DEng, Embry-Riddle Aeronaut Univ, 69. *Prof Exp:* Head indust tech dept, Southern Tech Inst, Ga, 49-53; chief indust eng, Norwood Mfg Co, 53-54; dir, Charlotte Tech Inst, 54-55; asst prof indust eng, NC State Col, 56-57; assoc prof, 57-65, actg dean col eng, 66-68, assoc dean, 68-78, exec vpres, 78-85, PROF INDUST & SYST ENG, UNIV FLA, 65- *Mem:* Am Soc Eng Educ; Am Inst Indust Engrs. *Res:* Critical path scheduling and schedule evaluation techniques; work measurement and simplification; engineering economics. *Mailing Add:* 5960 SW 35th Way Gainesville FL 32608

NATUK, ROBERT JAMES, b Doylestown, Pa, Aug 11, 56; m 86. VACCINE DEVELOPMENT, VIRAL IMMUNOLOGY. *Educ:* Trenton State Col, BA, 78; Rutgers Univ, MS, 81, PhD(microbiol), 84. *Prof Exp:* Fel, Dept Path, Univ Mass Med Ctr, 84-87; sr molecular biologist, 87-89, RES SCIENTIST, WYETH-AYERST RES, 89- *Mem:* Am Asn Immunologists. *Res:* Chemotaxis and accumulation of natural killer cells at site of virus replication; immunogenicity studies of recombinant adenovirus - HIV vaccines. *Mailing Add:* Wyeth-Ayerst Res PO Box 8299 Philadelphia PA 19101

NATZKE, ROGER PAUL, b Greenleaf, Wis, June 15, 39; m 64; c 2. DAIRY SCIENCE, PHYSIOLOGY. *Educ:* Univ Wis, BS, 62, MS, 63, PhD(dairy sci), 66. *Prof Exp:* From asst prof to prof dairy sci, Cornell Univ, 66-81; CHMN & PROF, DAIRY SCI, UNIV FLA, 81- *Concurrent Pos:* Mem, Nat Mastitis Coun. *Honors & Awards:* Milking Mastitis Award, Am Dairy Sci Asn. *Mem:* Am Dairy Sci Asn; Am Soc Animal Sci. *Res:* Factors affecting screening tests; effect of sanitary practices on mastitis; free stall management factors. *Mailing Add:* Dept Dairy Sci 701 IFAS Gainesville FL 32611-0701

NAU, CARL AUGUST, preventive medicine; deceased, see previous edition for last biography

NAU, DANA S, b Urbana, Ill, Dec 29, 51. ARTIFICIAL INTELLIGENCE. *Educ:* Univ Mo-Rolla, BS, 74; Duke Univ, AM, 78, PhD(comput sci), 79. *Prof Exp:* From asst prof, 79-84, TO ASSOC PROF, COMPUT SCI, UNIV MD, 84- *Concurrent Pos:* Comput scientist, IBM Res, 76-77; consult, IBM Res, 76-77; comput scientist, Nat Bur Standards, 81; rev panel, Nat Sci Found, Pres Investr Awards Prog, 84; consult, TRW, 85; vis prof, Univ Rochester, 85; comput scientist, General Motors Res Labs, 86; prog comt, Conf Expert Sys Govt, 86; comput scientist, Nat Bur Standards, 86; rev panelist, Nat Sci Found, Res Initiation Grants Prog, 88. *Honors & Awards:* Nat Sci Found, Pres Young Investr Award, 84-89. *Mem:* Asn Comput Mach; Am Asn Artificial Intelligence; IEEF Comput Soc. *Res:* Artificial intelligence research, especially searching and problem-solving techniques, knowledge-based computer systems, diagnostic problem solving geometric reasoning, planning, and applications to automated manufacturing. *Mailing Add:* Comput Sci Dept Univ Md College Park MD 20742

NAU, RICHARD WILLIAM, b Lakefield, Minn, May 28, 41; m 63; c 2. MATHEMATICS. *Educ:* SDak Sch Mines & Technol, BS, 63, MS, 65; Univ Va, PhD(appl math), 70. *Prof Exp:* Res assoc, Boeing Co, 65-66; asst prof math, Clarkson Col, 66-67; actg asst prof comput sci, Univ Va, 70; assoc prof, 70-80, PROF MATH & COMPUT SCI, CARLETON COL, 80- *Concurrent Pos:* Fulbright lectr, 80-81. *Mem:* Math Asn Am; Asn Comput Mach; Soc Indust & Appl Math. *Res:* Asymptotic methods; optimization methods; computer graphics. *Mailing Add:* Dept of Math/Comp Sci Carleton Col One N Col St Northfield MN 55057

NAU, ROBERT H(ENRY), b Burlington, Iowa, Apr 21, 13; m 50; c 3. ELECTRICAL ENGINEERING. *Educ:* Iowa State Col, BS, 35, EE, 41; Agr & Mech Col, Tex, MS, 37. *Prof Exp:* Instr physics, Agr & Mech Col, Tex, 35-36, instr elec eng, 36-37; design & develop engr, Westinghouse Elec Corp, Pa, 37-42, design & develop engr heating & vent, Buffalo, NY, 46-47; design & develop engr & consult, Allis-Chalmers Mfg Co, Mass, 46-47; asst prof elec eng, Univ Ill, 47-52; prof in charge power & circuit courses, Dept Elec Eng, Univ Santa Clara, 52-55; head prof elec eng, Ohio Northern Univ, 55-57; PROF ELEC ENG, UNIV MO-ROLLA, 57- *Concurrent Pos:* Instr, Amateur Trapshooting Asn, 81-91; hon mem, Blue Key, 83; hon prof military sci, Univ Mo-Rolla, 83 & mem, Order Golden Shillelaugh, 83. *Honors & Awards:* Educ Award, Inst Elec & Electronic Engrs, 82 & Award Honor, 84; Ranking Lectr, Univ Mo-Rolla; Scholar Fund named in honor, Robert H Nau Perpetual Endowed Scholar Fund, Univ Mo, 83. *Mem:* Am Soc Eng Educ; Nat Soc Prof Engrs; fel Inst Elec & Electronics Engrs; Sigma Xi; Am Asn Eng Socs. *Res:* Magnetic fluxes in three-phase circuits; arc interruption by magnetic means in free air, by compressed air and by oil flow; mechanical vibration of electric power transmission lines; electric and magnetic circuits; educational methods; circuits; network analysis and synthesis; solid state physics; servomechanisms; five textbooks, seventy-two technical publications. *Mailing Add:* Dept of Elec Eng Univ of Mo Rolla MO 65401

NAUDASCHER, EDUARD, b Sofia, BG, Aug 3, 29; c 4. HYDRAULIC ENGINEERING. *Educ:* Univ Karlsruhe, Dr CEng. *Prof Exp:* Res asst, Inst Hydromech T H Karlsruhe, Ger, 54-59; res engr, Iowa Inst Hydraul Res, Univ Iowa, from asst prof to assoc prof, Dept Mech & hydraul, 60-68; PROF HYDRAUL ENG & HEAD, INST HYDROMECH, UNIV KARLSRUHE, GER, 68- *Concurrent Pos:* Guest scientist, Max Planck Inst Aerodyn, Goettingen, 66-67; res assoc, St Anthony Falls Hydraul Lab, Univ Minn, Minneapolis. *Honors & Awards:* Huber Civil Eng Res Prize, Am Soc Civil Engrs, 68; Hydraul Struct Medal, Am Soc Civil Engrs, 87. *Mem:* Am Soc Civil Engrs; Am Soc Eng Educ; Int Asn Hydraul Res. *Res:* Hydraulics, turbulence and flow-induced oscillations; series of 150 scientific publications. *Mailing Add:* Hauffstrasse 19 Karlsruhe 51 D-7500 Germany

NAUENBERG, MICHAEL, b Berlin, Ger, Dec 19, 34; US citizen; m 69; c 2. ELEMENTARY PARTICLE AND CONDENSED MATTER PHYSICS. *Educ:* Mass Inst Technol, BS, 55; Cornell Univ, PhD(physics), 59. *Prof Exp:* Asst prof physics, Columbia Univ, 61-64; vis res prof, Stanford Univ, 64-66; chmn dept, 83-85, PROF PHYSICS, UNIV CALIF, SANTA CRUZ, 66-, DIR, INST NONLINEAR SERV, 87- *Concurrent Pos:* Consult, Brookhaven Nat Lab, 62-64 & NSF; Guggenheim fel, 63-64; A P Sloan fel, 64-66; NSF grant, 66-; dep dir, Inst Theoret Physics, Univ Calif, Santa Barbara, 81-83; Alexander Von Humboldt Award, 89-92. *Mem:* Am Phys Soc. *Res:* Elementary particles and their interactions; condensed matter physics; astrophysics. *Mailing Add:* Dept Physics Univ Calif Santa Cruz CA 95064

NAUENBERG, URIEL, b Berlin, Ger, Dec 16, 38; m 59; c 2. COAL CONVERSION, SYNTHETIC FUELS TECHNOLOGY. *Educ:* Columbia Univ, BA, 59, PhD(high energy physics), 63. *Prof Exp:* From instr to asst prof, Princeton Univ, 63-69; assoc prof, 69-72, PROF, DEPT PHYSICS, UNIV COLO, 72- *Mem:* Am Phys Soc; AAAS; Fedn Am Scientists. *Res:* Testing the predictions of the electroweak theory mass energies. *Mailing Add:* Dept Physics Univ Colo Campus Box 390 Boulder CO 80309-0390

NAUGHTEN, JOHN CHARLES, b Chicago, Ill, Jan 29, 42; m 66; c 1. HISTOLOGY, GENETICS. *Educ:* Univ Chicago, AB, 64; Univ Iowa, MS, 68, PhD(zool), 71. *Prof Exp:* Res assoc neuroembryol, Univ Iowa, 71-72; asst prof biol, Univ Wis-Eau Claire, 72-77; asst prof biol, 77-78, asst prof anat, Univ SDak, 78-81; asst prof, 81-85, ASSOC PROF BIOL, NORTHERN STATE COL, ABERDEEN, SDAK, 86- *Concurrent Pos:* Vis asst prof zool, Univ Iowa, 74, 75 & 76; vis scientist, NIH, NCI, 87. *Mem:* AAAS; Am Soc Zool; Soc Develop Biol; Sigma Xi; Int Soc Differentiation; Int Pigment Cell Soc. *Res:* Vertebrate embryology, tissue interactions and differentiation, biology of pigment cell pattern, metamorphosis. *Mailing Add:* Northern State Univ 12th & Jay Sts Box 672 Aberdeen SD 57401

NAUGHTON, JOHN, SCIENTIFIC ADMINISTRATION. *Prof Exp:* FAC MEM, BUFFALO HEALTH SCI CTR, STATE UNIV NY. *Mailing Add:* SUNY at Buffalo Health Scis Ctr Sch Med 3435 Main St Buffalo NY 14214

NAUGHTON, MICHAEL A, b UK, June 23, 26; US citizen; c 4. BIOPHYSICS, OBSTETRICS & GYNECOLOGY. *Educ:* Univ St Andrews, BSc, 52; Cambridge Univ, PhD, 59. *Prof Exp:* Res asst, Cambridge Univ, 54-56; res assoc, Mass Inst Technol, 59-62; assoc prof biophys, Johns Hopkins Univ, 62-67; sr prin res scientist, Div Animal Genetics, Commonwealth Sci, Indust & Res Orgn, 67-70; PROF BIOPHYS, OBSTET & GYNEC, SCH MED, UNIV COLO, DENVER, 70- *Concurrent Pos:* NIH career develop award, 64. *Mem:* Brit Biochem Soc; Am Soc Biol Chemists. *Res:* Molecular biology and immunology of proteins. *Mailing Add:* Dept Obstet & Gynec Univ Colo Med Ctr Denver CO 80262

NAUGLE, DONALD, b Wetumpka, Okla, Apr 23, 36; m 58; c 2. LOW TEMPERATURE, SOLID STATE PHYSICS. *Educ:* Rice Univ, BA, 58; Tex A&M Univ, PhD(physics), 65. *Prof Exp:* Res fel physics, Tex A&M Univ, 65-66; res assoc, Univ Md, College Park, 67-69; asst prof, 69-75, assoc prof, 75-81, PROF PHYSICS, TEX A&M UNIV, 81- *Concurrent Pos:* NATO fel, Univ Gottingen, 66-67. *Mem:* Am Phys Soc; Am Vacuum Soc; Mat Res Soc; Am Soc Metals. *Res:* Ultrasonics; fluid transport properties; superconductivity; amorphous metals. *Mailing Add:* 1113 Westover College Station TX 77840

NAUGLE, NORMAN WAKEFIELD, b Saginaw, Tex, Jan 9, 31; wid; c 2. NUMERICAL ANALYSIS, APPLIED MATHEMATICS. *Educ:* Tex A&M Univ, AB, 53, MS, 58, PhD(physics), 65. *Prof Exp:* Asst math, Tex A&M Univ, 55-57 from instr to asst prof, 57-64; mathematician, Manned Spacecraft Ctr, NASA, 64-68; assoc prof, 68-88, PROF MATH, TEX A&M UNIV, 88- *Concurrent Pos:* Teacher, Allen Mil Acad, 57-59 & Alvin Jr Col, 65-67; consult, Appl Res Corp, Tex, 68-72. *Mem:* Math Asn Am; Soc Indust & Appl Math. *Res:* Numerical analysis; digital picture data processing; photoclinometry; molecular structure calculations for vibrational and rotational analysis. *Mailing Add:* Dept of Math Tex A&M Univ College Station TX 77843

NAUMAN, CHARLES HARTLEY, b Philadelphia, Pa, June 6, 37; m 68. ENVIRONMENTAL TOXICOLOGY. *Educ:* Davis & Elkins Col, BS, 62; Univ Ark, Fayetteville, MS, 65; Northwestern Univ, Evanston, PhD(biol), 72; Columbia Univ, MPH, 79. *Prof Exp:* Assoc radiobiol, Brookhaven Nat Lab, 65-68, tech collabr, Brookhaven Nat Lab-Columbia Univ, 71-72, asst biologist, Brookhaven Nat Lab, 73-74, assoc biologist, 75-79; environ toxicologist, 79-86, sci adv, 87-88, MATRIX MGR, US ENVIRON PROTECTION AGENCY, 89- *Mem:* AAAS; Environ Mutagen Soc; Radiation Res Soc; Am Chem Soc; USPHS Prof Asn. *Res:* Health and environmental exposure assessment and risk analysis; biomarkers of exposure. *Mailing Add:* 4201 Quadrel St Las Vegas NV 89129

NAUMAN, EDWARD BRUCE, b Kansas City, Mo, Oct 3, 37; m 59; c 2. POLYMERS, REACTION ENGINEERING. *Educ:* Kans State Univ, BS, 59; Univ Tenn, MS, 61; Univ Leeds, PhD(chem eng), 63. *Prof Exp:* Demonstr chem eng, Univ Leeds, 62-63; engr, Union Carbide Corp, Bound Brook, 63-65, proj scientist, 66-67, res scientist, 67-69, group leader & technol mgr, 69-73, bus strategy mgr, New York, 73-75, prod mgr, 75-77; dir res & develop, Xerox Corp, Rochester, NY, 77-81; prof & chmn, Dept Chem Eng, 81-84, PROF CHEM ENG & DIR INDUST LIAISON, RENSSELAER POLYTECH INST, 84- *Mem:* AAAS; fel Am Inst Chem Engrs; Sigma Xi; Am Chem Soc; Asn Consult Chemists & Chem Engrs. *Res:* Polymer separations and reactions; numerical analysis, simulation and process control; polymer processing; chemical reaction engineering with emphasis on mixing phenomena. *Mailing Add:* Dept Chem Eng Rensselaer Polytech Inst 110 Eighth St Troy NY 12180

NAUMAN, EDWARD FRANKLIN, chemistry, materials engineering; deceased, see previous edition for last biography

NAUMAN, ROBERT KARL, b Allentown, Pa, Feb 26, 41; m 68; c 2. MICROBIOLOGY. *Educ:* Pa State Univ, BS, 63; Univ Mass, MS, 65, PhD(microbiol), 68. *Prof Exp:* Res asst microbiol, Univ Mass, 63-68; res assoc, Med Ctr, WVa Univ, 68-69, instr, 69-70; asst prof, 70-78, ASSOC PROF MICROBIOL, SCH DENT, UNIV MD, BALTIMORE, 78- *Mem:* Am Soc Microbiol; fel Am Acad Microbiol. *Res:* General microbiology; electron microscopy, microbial cytology, oral microbiology and periodontal diseases; motility, ultrastructure and function; spirochetes. *Mailing Add:* Dept of Microbiol Univ of Md Sch of Dent 666 W Baltimore St Baltimore MD 21201

NAUMAN, ROBERT VINCENT, b East Stroudsburg, Pa, Dec 6, 23; m 55; c 4. PHYSICAL CHEMISTRY. *Educ:* Duke Univ, BS, 44; Univ Calif, PhD(chem), 47. *Prof Exp:* Res assoc chem, Cornell Univ, 47-52; asst prof, Univ Ark, 52-53; from asst prof to assoc prof, 53-63, dir, grad studies, 81-84, prof, 63-90, EMER PROF CHEM, LA STATE UNIV, BATON ROUGE, 90- *Concurrent Pos:* Fulbright-Hays lect award, Santa Maria Tech Univ, Valparaiso, Chile, 66-67; vis prof, Fac Sci, Univ Chile, 71-72; Fulbright res grant, Univ Chile, 85; distinguished prof, Santa Maria Tech Univ, Valparaiso,

Chile, 86- *Mem:* AAAS; Am Chem Soc; Am Phys Soc. *Res:* Molecular spectra; internal energy conversion; photochemistry; light scattering; sodium silicates; structure, size and shape of molecules; detergents; photoacoustic spectroscopy. *Mailing Add:* Dept Chem La State Univ Baton Rouge LA 70803

NAUMANN, ALFRED WAYNE, b Farmington, Iowa, May 8, 28; m 52; c 4. PHYSICAL CHEMISTRY. *Educ:* Grinnell Col, BA, 51; Iowa State Univ, PhD, 56. *Prof Exp:* Asst, Iowa State Univ, 51-56; assoc chemist, 56-71, SR RES SCIENTIST, UNION CARBIDE CORP, 71- *Mem:* AAAS; Am Chem Soc. *Res:* Hydrometallurgical separation processes; surface and colloid chemistry; preparation and properties of high performance ceramics; heterogeneous catalysis. *Mailing Add:* 1587 Virginia St E Charleston WV 25311

NAUMANN, HUGH DONALD, b Newport, Ark, Oct 26, 23; m 45; c 2. FOOD MICROBIOLOGY. *Educ:* Univ Mo, BS, 49, MS, 50, PhD(meat technol), 56. *Prof Exp:* From asst instr to instr animal husb, Univ Mo, 50-53; asst prof, Cornell Univ, 53-55; from asst prof to assoc prof animal husb, 55-68, chmn dept, 75-81, PROF FOOD SCI & NUTRIT, UNIV MO-COLUMBIA, 68- *Concurrent Pos:* Fulbright sr res scholar, Commonwealth Sci & Indust Orgn, 65-66, Fulbright sr lectr, Gida Fermantasyon Teknolojisi Kurusu Ege Univ, Turkey, 74-75; res fel, Animal & Dairy Sci Res Inst, Irene, SAfrica, 82 & 87. *Honors & Awards:* Distinguished Serv Award, Am Meat Sci Asn, 90. *Mem:* Inst Food Technologists; Food Distrib Res Soc; Am Soc Animal Sci. *Res:* Meat technology; evaluation of quality attributes of meat, formulation of meat food products and processing meats, particularly the effects of sanitation, temperature environment, gas environment, and packaging upon the stability of fresh, frozen and cured meats. *Mailing Add:* Food Sci & Nutrit Univ Mo 122 Eckles Hall Columbia MO 65211

NAUMANN, ROBERT ALEXANDER, b Dresden, Ger, June 7, 29; nat US; m 61; c 2. PHYSICAL CHEMISTRY. *Educ:* Univ Calif, BS, 49; Princeton Univ, MA, 51, PhD(phys chem), 53. *Prof Exp:* From instr to assoc prof, 52-73, PROF CHEM & PHYSICS, PRINCETON UNIV, 73- *Concurrent Pos:* Procter & Gamble fac fel, 59-60; Sr US scientist award, Alexander von Humboldt Found, 78 & 83; vis prof physics, Tech Univ Munich, 88. *Honors & Awards:* Sr US Scientist Award, Alexander von HumboldtFound, 78 & 83. *Mem:* Fel AAAS; Am Chem Soc; fel Am Phys Soc. *Res:* Radioactivity; inorganic chemistry; nuclear physics. *Mailing Add:* Dept Physics Princeton Univ PO Box 708 Jadwin Hall Princeton NJ 08540

NAUMANN, ROBERT JORDAN, b Gillespie, Ill, July 1, 35; m 59; c 3. SOLID STATE PHYSICS, MATERIALS SCIENCES. *Educ:* Univ Ala, BS, 57, MS, 62, PhD(physics), 70. *Prof Exp:* Physicist, Army Ballistic Missile Agency, 57-60; physicist, Marshall Space Flight Ctr, NASA, 60-64, br chief, 64-70, div chief instrumentation sci, 70-77, chief Low Gravity Sci Div, Space Sci Lab, 77-90, chief scientist, Mat Processes Space, 79-90, proj scientist, Space Sta, 85-90; ASSOC DEAN RES & PROF MAT SCI, UNIV ALA HUNTSVILLE, 90-; CHIEF SCIENTIST, SPACE INDUSTS INT, INC, 90- *Concurrent Pos:* Instr, Athens State Col, 64-; prof physics, Univ Ala, Huntsville, 70- *Honors & Awards:* Medal Except Sci Achievement, NASA, Medal Except Serv; Lloyd V Berkner Award, Am Astron Soc. *Mem:* Sigma Xi; Am Inst Aeronaut & Astronaut. *Res:* Crystal growth and characterization; solidification phenomena; fluid-chemical processes; bio and chemical physics; cell and protein separation. *Mailing Add:* Off Dean Col Sci Univ Ala Huntsville AL 35899

NAUNTON, RALPH FREDERICK, b London, Eng, Sept 26, 21; m; c 2. OTOLARYNGOLOGY. *Educ:* Univ London, MB, BS, 45. *Prof Exp:* Resident surg & dir audiol, Univ Col Hosp, Univ London, 47-50; Med Res Coun res fel, Cent Inst for Deaf, Mo, 51; sci officer, Med Res Coun Eng, 51-52; res asst, Cent Inst for Deaf, Mo, 52-53; sci officer, Med Res Coun Eng, 53-54; instr surg, Sch Med, Univ Chicago, 54-57, from asst prof to prof otolaryngol, 57-80, chmn dept, 68-80; DIR COMMUN DISORDERS PROG, NAT INST NEUROL & COMMUN DISORDERS & STROKE, NIH, 80- *Concurrent Pos:* Asst surgeon, St John's Hosp, London, 47-50; Univ London fel, Holland, 49; clin asst, Royal Nat Throat, Nose & Ear Hosp, 51-54. *Mem:* Am Speech & Hearing Asn; fel Royal Soc Med. *Res:* Otology. *Mailing Add:* Fed Bldg Ic-11 7550 Wisconsin Ave Bethesda MD 20892

NAUS, JOSEPH IRWIN, b New York, NY, Mar 14, 38; m 61; c 4. APPLIED STATISTICS, SURVEY SAMPLING. *Educ:* City Col New York, BBA, 59; Harvard Univ, MS, 61, PhD(statist), 64. *Prof Exp:* Analyst opers res, Appl Sci Div, Mass Inst Technol, 62-63 & Inst Naval Studies, Franklin Inst, 63-64; assoc prof statist, City Col New York, 67-68; asst prof, 64-66, assoc prof, 67-74, PROF STATIST, RUTGERS UNIV, 74- *Concurrent Pos:* Dir statist, Rutgers Univ, 73-77, 81-82 & 83-86. *Mem:* Am Statist Asn; Inst Math Statists. *Res:* Applied probability and statistics; distributional theoretic problems, particularly with the unusual clustering of points in time or space; approaches for screening and editing data for errors. *Mailing Add:* Dept Statist Rutgers Univ-State Univ NJ New Brunswick NJ 08903

NAUSEEF, WILLIAM MICHAEL, NEUTROPHIL FUNCTION, INFECTIOUS DISEASE. *Educ:* State Univ NY, MD, 84. *Prof Exp:* ASST PROF MED, UNIV IOWA HOSPS & CLINS, 83- *Res:* Lysosomal enzyme synthesis; membrane protein cytoskeleton. *Mailing Add:* Dept Internal Med Univ Iowa Iowa City IA 52242

NAUSS, KATHLEEN MINIHAN, b Cambridge, Mass, Apr 9, 40; m 65; c 2. TOXICOLOGY. *Educ:* Regis Col, AB, 61; Univ Pa, PhD(biochem), 65. *Prof Exp:* Fel, Retina Found, 65-67; res fel, Univ Hohenbeim, Stuttgart, 67-70; res fel, Children's Can Res Found, 71-73; lectr, biochem, Regis Col, 73-76; res scientist, Mass Inst Technol, 76-86; DIR SCI REV & EVAL, HEALTH EFFECTS INST, CAMBRIDGE, 87- *Mem:* AAAS; Am Inst Nutrit; Soc Toxicol. *Res:* Effects of dietary factors on cancer and immune functions; developing and managing research programs dealing with the health effects of automotive emissions. *Mailing Add:* Health Effects Inst 141 Portland St Cambridge MA 02139

NAUTA, WALLE J H, b Medan, Indonesia, June 8, 16; US citizen; m 42; c 3. ANATOMY, NEUROANATOMY. *Educ:* State Univ Utrecht, MD, 42, PhD(anat), 45. *Hon Degrees:* Dr, Univ Rochester, 75, Univ Zürich, 83, Univ Autonoma, Madrid, 84, Northeastern Ohio Univ Col Med, 89. *Prof Exp:* Lectr anat, State Univ Utrecht, 42-46; assoc prof, State Univ Leiden, 46-47 & Univ Zurich, 47-51; neurophysiologist, Walter Reed Army Inst Res, 51-64; prof, 64-75, Inst Prof, 73-86, EMER PROF, NEUROANAT, MASS INST TECHNOL, 86- *Concurrent Pos:* Prof, Univ Md, 55-64; mem bd, Found Fund Res Psychiat, 59-62; mem res career develop awards comt, NIMH, 66-; mem Biol Stain Comn, 57-; pres, Soc Neurosci, 73. *Honors & Awards:* Karl Spencer Lashley Award, Am Philos Soc, 64; Henry Gray Award, Am Asn Anat, 73; Ralph Gerard Award, Soc Neurosci, 83, Bristol-Myers Award Neurosci, 88. *Mem:* Nat Acad Sci; Am Philos Soc; Am Asn Anat; Am Acad Arts & Sci; Swiss Asn Anat. *Res:* Neuroanatomy; neurophysiology. *Mailing Add:* Mass Inst Technol Rm E25-618 Cambridge MA 02139

NAVAB, FARHAD, b Tehran, Iran, Sept 12, 38; m; c 3. GASTROENTEROLOGY. *Educ:* Cambridge Univ, BA, 60, PhD, 76; Univ London, MA, MB & BChir, 63; Am Bd Internal Med, dipl, 85, dipl gastroenterol, 87; FRCP, 88. *Prof Exp:* From asst prof to assoc prof med, Univ Tehran, Iran, 72-79; assoc prof, Div Gastroenterol, Univ Ark Med Sci, 80-87, assoc prof physiol & biophys, 85-87; CHIEF, GASTROENTEROL DIV, BAYSTATE MED CTR, 87-; PROF MED, SCH MED, TUFTS UNIV, 91- *Concurrent Pos:* Consult physician, Brook Gen Hosp, London, 79-80, St Vincent Infirmary, Little Rock, Ark, 85-87, Doctors Hosp, Little Rock, 86-87 & Mercy Hosp, Springfield, Mass, 89-; attend physician, Vet Admin Med Ctr, Little Rock, Univ Ark Med Sci, 81-87 & Baystate Med Ctr, 87-; staff physician, Div Gastroenterol, John L McClellan Mem Vet Admin Hosp, 86-87; assoc staff physician, Cooley Dickinson Hosp, 90- *Mem:* Royal Col Physicians; Brit Soc Gastroenterol; Am Gastroenterol Asn; fel Am Col Physicians; sr mem Am Fedn Clin Res; Am Soc Gastrointestinal Endoscopy; Am Physiol Soc; fel Am Col Gastroenterol; Fedn Am Socs Exp Biol; Am Asn Study Liver Dis. *Res:* Author of more than 50 technical publications. *Mailing Add:* Baystate Med Ctr 759 Chestnut St Springfield MA 01199

NAVALKAR, RAM G, b Bombay, India, May 7, 24; m 66; c 2. MICROBIOLOGY, IMMUNOLOGY. *Educ:* St Xavier's Col, India, BSc, 46; Univ Bombay, PhD(microbiol), 56. *Prof Exp:* Res officer leprosy, Acworth Leprosy Hosp, Bombay, 58-60; proj assoc tuberc, Sch Med, Univ Wis, 60-63; res assoc, Gothenburg Univ, 64-65; proj assoc tuberc & leprosy, Sch Med, Univ Wis, 66-67; from asst prof to prof med microbiol, Meharry Med Col, 67-80; PROF & CHMN, DEPT MICROBIOL & IMMUNOL, MOREHOUSE SCH MED, ATLANTA, GA, 80- *Concurrent Pos:* Fel microbiol, Sch Med, Stanford Univ, 56-58; NIH grant, 67- & Fogarty Int Ctr travel grant, India, 84; consult, Stanford Res Inst, 57-58, MARC Prog, Tenn State Univ, Nashville, 81, Univ Degli, Studi, Bologna, 88; mem, site vis comt, NIH, 75, ad hoc rev comt, 78 & 79, proj site vis comt, Altanta, 78; prin coordr, Immunol Leprosy Proj, Al-Azhar Univ, Cairo, Egypt, 77; mem, Int Leprosy Asn Workshop, Mex, 78, Biosafety Comt, Atlanta Univ Ctr, 82; vis prof, Ky State Univ, 88, Div Biol, Tenn State Univ, Nashville, 89 & 90, Div Natural Sci, Selma Univ, Ala, 90; prin investr, Minority Student Sci Career Support Prog, Am Soc Microbiol, 89-; reviewer & chmn, Spec Study Sect, Nat Inst Allergy & Infectious Dis, NIH, 90. *Honors & Awards:* B K Dhurandhar Gold Medal, 56; Major Gen Sahib Singh Sokhey Outstanding Researcher Award, 76. *Mem:* Fel Am Acad Microbiol; Soc Exp Biol Med; AAAS; Am Soc Microbiol; Int Leprosy Asn. *Res:* Antigenic studies on mycobacteria; study of Mycobacterium leprae by comparative analysis, using various immunochemical techniques; specific antigens of Mycobacterium leprae in relation to their biological activity; immune mechanisms in Mycobacterial infections. *Mailing Add:* 2564 Creek Indian Trail Jonesboro GA 30236

NAVANGUL, HIMANSHOO VISHNU, b Wai, India, Dec, 22, 40; US citizen; m 64; c 2. CHEMISTRY. *Educ:* Univ Poona India, BS, 61, MS, 63, PhD(chem), 67. *Prof Exp:* Postdoctoral chem, Nat Chem Lab, Poona, India, 66-67, Univ Zurich, Switz, 67-69, Univ Wyoming, Laramie, 69-72; vis asst prof chem, Univ Mo, Kansas City, 72-75, NE Mo State Univ, Kirksville, 75-76; assoc prof chem, Al Fateh Univ, Tripoli, Libya, 76-79; vis fac chem, Clemson Univ, SC, 79-80; CHMN & PROF CHEM, NC WESLEYAN COL, ROCKY MT, 80- *Concurrent Pos:* Fac res fel, Caltech JPL/NASA, Pasadena, Calif, 85, 86; fac res fel, NASA/USAF, Hanscom AFB, 88; vis prof, Univ NC Chapel Hill, 89-90. *Mem:* Am Chem Soc; Sigma Xi; Am Inst Chemists. *Res:* Science education and related research; molecular systems involved in vision and their spectroscopic investigation; molecular modelling. *Mailing Add:* Dept Chem NC Wesleyan Col 3400 N Wesleyan Blvd Rocky Mt NC 27804

NAVAR, LUIS GABRIEL, b El Paso, Tex, Mar 24, 41; m 65; c 4. PHYSIOLOGY, BIOPHYSICS. *Educ:* Tex A&M Univ BS, 62; Univ Miss, PhD(biophys, physiol), 66. *Prof Exp:* Instr physiol & biophys, Sch Med, Univ Miss, 66-67, from asst prof to assoc prof, 67-74; assoc prof, 75-76, PROF PHYSIOL & BIOPHYS, SCH MED, UNIV ALA, BIRMINGHAM, 76-, PROF MED, 83- *Concurrent Pos:* Nat Inst Arthritis & Metab Dis fel physiol & biophys, Med Ctr, Univ Miss, 66-69 & spec fel med, Duke Univ, 72; Nat Heart & Lung Inst res career develop award, 74; mem, Coun Kidney & Cardiovasc Dis, Am Heart Asn; assoc ed, Am J Physiol-Renal. *Mem:* Am Soc Nephrology; Am Heart Asn; Am Physiol Soc; NY Acad Sci; Sigma Xi; Microcirculation Soc. *Res:* Control of renal hemodynamics; regulation of sodium excretion; pathophysiology of high blood pressure; regulation of extracellular fluid volume. *Mailing Add:* Dept Physiol Tulane Univ Med Sch 1430 Tulane Ave New Orleans LA 70112

NAVARRA, JOHN GABRIEL, b Bayonne, NJ, July 3, 27; m 47; c 2. EARTH SCIENCES, SCIENCE EDUCATION. *Educ:* Columbia Univ, AB, 49, MA, 50, EdD(sci educ), 54. *Prof Exp:* Assoc prof chem, physics & sci educ, ECarolina Univ, 54-58; prof sci & chmn dept, 58-68, PROF GEOSCI & GEOG, JERSEY CITY STATE COL, 68-; DIR, LEARNING RESOURCES LABS, 67- *Concurrent Pos:* Consult, NC State Bd Educ Sci Curriculum Study, 58, State of Calif NDEA Workshops Strengthening Elem

Sch Sci, 60-61 & State Adv Comt on Sci, 60-62; partic, White House Conf Children & Youth Golden Anniversary, 60; teacher educ study, Nat Asn State Dir Teacher Educ & Cert, 60-61; Am Inst Physics coordr vis scientists prog physics for high schs, NJ, 61-64; sci ed, Arabian-Am Oil Co, Saudi Arabia, 63. *Mem:* Am Geol Soc; AAAS. *Res:* Applications of chromatography; development and refinement of broad areas of science in elementary, junior high school and college curriculum. *Mailing Add:* Dept Geosci & Geog Jersey City State Col 2039 Kennedy Blvd Jersey City NJ 07305

NAVARRO, JOSEPH ANTHONY, b New Britain, Conn, July 6, 27; m 51; c 3. MATHEMATICAL STATISTICS. *Educ:* Cent Conn State Univ, BS, 50; Purdue Univ, MS, 52, PhD(math statist), 55. *Prof Exp:* Asst math, Purdue Univ, 51, math statist, 51-55; consult statistician, Gen Elec Co, 55-59; mem res staff, IBM Corp, 59-64; mem res staff, Inst Defense Anal, 64-70, asst dir systs eval div, 70-72; exec vpres, 72-86, pres, Syst Planning Corp, 86-87; dep undersecy defense testing & eval, US Dept Defense, 86-87; pres, Wackenhut Appl Technol Ctr, 89-90; PRES, JAN ASSOC INC, 87- *Mem:* Sigma Xi; Opers Res Soc Am; Am Statist Asn. *Res:* Use of operations research, probability and mathematical statistics in analysis weapon system acquisition; systems analysis; test and evaluation. *Mailing Add:* 7825 Fulbright Ct Bethesda MD 20817

NAVARRO-BERMUDEZ, FRANCISCO JOSE, b San Jose, Costa Rica, Aug 4, 35. MATHEMATICS. *Educ:* Mass Inst Technol, BS, 59; Harvard Univ, AM, 60; Bryn Mawr Col, PhD(math), 77. *Prof Exp:* Prof math, Univ Costa Rica, 61-63; ASSOC PROF MATH, WIDENER UNIV, 64- *Mem:* Am Math Soc; Math Asn Am; Sigma Xi. *Res:* Real variables; measure theory. *Mailing Add:* Dept Math Widener Univ Chester PA 19013

NAVE, CARL R, b Newport, Ark, July 21, 39; m 62; c 2. PHYSICS. *Educ:* Ga Inst Technol, BEE, 61, MS, 64, PhD(physics), 66. *Prof Exp:* Fel physics, Univ Col NWales, 66-67; asst prof, 68-71, ASSOC PROF PHYSICS, GA STATE UNIV, 71- *Concurrent Pos:* Auth. *Mem:* Am Phys Soc; AAAS; Audio Eng Soc; Am Asn Physics Teachers. *Res:* Determination of molecular structure and study of intramolecular interactions by microwave spectroscopy; electron spin resonance studies of radiation damage in organic crystals; sound analysis and musical acoustics. *Mailing Add:* Dept Physics & Astron Ga State Univ University Plaza Atlanta GA 30303

NAVE, FLOYD ROGER, b Moline, Ill, Oct 7, 25; m 49; c 4. GEOMORPHOLOGY, PALEONTOLOGY. *Educ:* Augustana Col, AB, 49; Univ Iowa, MS, 52; Ohio State Univ, PhD, 68. *Prof Exp:* Instr geol, Augustana Col, 51-52; geologist, Gen Petrol Corp, Calif, 52-53; from asst prof to prof geol, 53-88, chmn dept, 64-74, EMER PROF GEOL, WITTENBURG UNIV, 89- *Concurrent Pos:* Mem staff, US Geol Surv, 56-57; NSF sci fac fel, 58-59. *Mem:* Geol Soc Am; Nat Asn Geol Teachers; Sigma Xi; Am Quaternary Asn. *Res:* Environmental geology. *Mailing Add:* Dept Geol Wittenberg Univ Springfield OH 45501

NAVE, PAUL MICHAEL, b Lancaster, Pa, June 3, 43; m 65; c 2. ORGANIC CHEMISTRY. *Educ:* Memphis State Univ, BS, 65; Iowa State Univ, PhD(org chem), 69. *Prof Exp:* Asst prof, 69-74, ASSOC PROF CHEM, ARK STATE UNIV, 74- *Mem:* Am Chem Soc. *Res:* Metal ion oxidation of organic compounds; ligand transfer oxidation of free radicals. *Mailing Add:* Ark State Univ PO Box 111 State University AR 72467

NAVIA, JUAN MARCELO, b Havana, Cuba, Jan 16, 27; US citizen; m 50; c 4. NUTRITION. *Educ:* Mass Inst Technol, BS & MS, 50, PhD, 65; FRSH, 72. *Prof Exp:* Tech dir, Cuba Indust & Com Co, 50-52; assoc prof nutrit & food sci, Univ Villanueva, Cuba, 55-61; res assoc, Mass Inst Technol, 61-65, assoc prof nutrit biochem, 66-69; sr scientist, Inst Dent Res & prof biochem, comp med, oral biol & nutrit sci, 69-85, dir res training, Sch Dent, 73-85, DIR, JOHN J SPARKMAN CTR INT PUB HEALTH EDUC, UNIV ALA, 81-, DEAN, SCH PUB HEALTH, 89- *Concurrent Pos:* Dir, FIM Nutrit Lab, 52-55; asst dir, Cuban Inst Tech Invest, 55-61; adv to dir, Nat Inst Dent Res, 70-72; mem, select comt nutrit & human needs, US Senate, 73; sr Int Fogarty fel, 79; mem adv bd, Fogarty Int Ctr, NIH, Washington, DC, 85-88. *Honors & Awards:* San Esteban Conde de Canongo Award, Cuban Acad Sci, 54; H Trendley Dean Mem Award, Int Asn for Dental Res, 90. *Mem:* Int Asn Dent Res; fel AAAS; Am Inst Nutrit; Inst Food Technologists; Am Inst Chemists; Sigma Xi; Am Asn Pub Health; Am Chem Soc; NY Acad Sci. *Res:* Nutrition; international public health; oral biology; fluoride metabolism; dental caries. *Mailing Add:* Sch Pub Health Univ Ala Birmingham AL 35294

NAVLAKHA, JAINENDRA K, b Indore, India, Aug 26, 50; m 79; c 2. SOFTWARE METRICS, PROGRAM VERIFICATION. *Educ:* Birla Inst Technol & Sci, Pilani, India, BE Hons 72; Indian Inst Technol, Kanpur, India, MS, 74; Case Western Reserve Univ, PhD(computer sci), 78. *Prof Exp:* From asst prof to assoc prof, 78-87, PROF COMPUTER SCI, FLA INT UNIV, 87-, DIR, 89- *Concurrent Pos:* Distinguished lectr, Inst Elec & Electronics Engrs, 84, distinguished visitor, Computer Soc, 85-86; consult, IBM, Arg, 85; prin investr, NCR Corp grant, 85-86; invited lectr numerous places. *Mem:* Asn Comput Mach; Inst Elec & Electronics Soc. *Res:* Prediction of errors and effort required in the maintenance phase of software systems using software complexity metrics; software productivity metrics; expert systems for varied applications. *Mailing Add:* Sch Computer Sci Fla Int Univ Miami FL 33199

NAVON, DAVID H, b New York, NY, Oct 28, 24; m 47; c 2. ELECTRONIC PHYSICS, MICROELECTRONICS. *Educ:* City Col New York, BEE, 47; NY Univ, MS, 50; Purdue Univ, PhD(physics), 53. *Prof Exp:* Instr physics, Mohawk Col, 47 & Queen's Univ, NY, 47-50; res assoc, Purdue Univ, 53-54; asst dir res, Transition Electronic Corp, 54-60, dir semiconductor res, 60-65; vis assoc prof elec eng, Mass Inst Technol, 65-68; PROF ELEC ENG, UNIV MASS, AMHERST, 68- *Concurrent Pos:* Lectr, Hebrew Univ, Jerusalem, 74, Nanjing Inst Technol & Futan Univ, Shanghi, 82; fac fel, USGAO, 78. *Mem:* Inst Elec & Electronics Engrs. *Res:* Semiconductor electronics; microelectronics; numerical, computer-aided analysis of semiconductor devices. *Mailing Add:* Dept of Elec Eng Univ of Mass Amherst MA 01003

NAVRATIL, GERALD ANTON, b Troy, NY, Sept 5, 51; m 76; c 3. PLASMA PHYSICS. *Educ:* Calif Inst Technol, BS, 73; Univ Wis-Madison, MS, 74, PhD(plasma physics), 76. *Prof Exp:* Proj assoc physics, Univ Wis-Madison, 76-77; asst prof eng sci, Mech & Nuclear Eng Dept, 77-78, asst prof, 78-83, assoc prof appl physics, 83-88, PROF APPL PHYSICS & CHMN, DEPT APPL PHYSICS & NUCLEAR ENG, COLUMBIA UNIV, 88- *Concurrent Pos:* Alfred P Sloan res fel, 84; vis fel, Princeton Univ, 85-86; exec comt, Univ Fusion Asn, 86-88; assoc ed, Physics of Fluids, 87-89. *Mem:* Fel Am Phys Soc; Univ Fusion Asn (secy & treas, 86-89, vpres, 90, pres, 91). *Res:* Trapped particle instabilities in plasma; high pressure limits of plasma confinement in tokamaks; controlled fusion research; high temperature plasma diagnostic development. *Mailing Add:* 215 SW Mudd Bldg Columbia Univ New York NY 10027

NAVRATIL, JAMES DALE, b Denver, Colo, Jan 20, 41; m 67; c 4. INDUSTRIAL CHEMISTRY, ANALYTICAL CHEMISTRY. *Educ:* Univ Colo, Boulder, BA, 70, MSc, 72, PhD(analytical chem), 75. *Prof Exp:* Analytical lab technician analytical chem, Dow Chem USA, 61-66, chem res & develop master technician, 66-68, sr chemist, 70-73, res chemist, 73-75; sr res chemist res & develop, Rockwell Int, 75-76, res specialist, 76-77, group leader I, 77-78; first officer, Int Atomic Energy Agency, 78-81; head dept mineral processing & extractive metall, Univ New South Wales, 87-89; mgr chem res, 81-87, sr proj mgr, Chem Waste Mgt, 89-91, SR SCIENTIST, ROCKWELL INT, 91- *Concurrent Pos:* Res assoc, Univ Colo, Boulder, 75-76, instr, 76-78; adj prof, Colo Sch Mines, Golden, Colo, 84-87. *Honors & Awards:* Rockwell Int Engr of Year, 77; IR-100 Award, 83 & 85. *Mem:* Fel AAAS; Am Chem Soc; Am Nuclear Soc; Am Inst Chem Engrs. *Res:* Chemical separations methods; liquid chromatography; solvent extraction research; chemical synthesis and characterization; actinide chemistry; environmental science; extractive metallurgy. *Mailing Add:* Rockwell Int Mail Stop T006 6633 Canoga Ave Canoga Park CA 91303

NAVROTSKY, ALEXANDRA, b New York, NY, June 20, 43. CHEMISTRY. *Educ:* Univ Chicago, BS, 63, MS, 64, PhD(chem), 67. *Prof Exp:* Res assoc theoret metall, Clausthal Tech Univ, 67-68; res assoc geochem, Pa State Univ, 68-69; from asst prof to assoc prof, 69-77, prof chem, 77-81, PROF CHEM & GEOL, ARIZ STATE UNIV, 81- *Concurrent Pos:* Alfred P Sloan Found fel, 73. *Mem:* Am Geophys Union; Mineral Soc Am; Am Ceramic Soc. *Res:* Thermodynamics; phase equilibria and high temperature calorimetry; oxides and oxide solid solutions; order-disorder; geochemistry; geothermal fluids. *Mailing Add:* Dept Chem-Geol Princeton Univ Princeton NJ 08544

NAWAR, TEWFIK, b Cairo, Egypt, Sept 6, 39; Can citizen. NEPHROLOGY. *Educ:* Einshams Univ, Cairo, MB, BCh, 63; McGill Univ, MSc, 72; Col Physicians & Surgeons Can, FRCP(C), 72; Am Bd Internal Med, dipl nephrol, 78. *Prof Exp:* Med Res Coun fel, Montreal Clin Res Inst, 70-72; from asst prof to assoc prof, 72-84, PROF MED, MED SCH, UNIV SHERBROOKE, 84-, CHMN DEPT, 85. *Concurrent Pos:* Med Res Coun fel, Renal Div, Med Sch, Wash Univ, 71-72. *Mem:* Can Soc Nephrol; NY Acad Sci; Am Soc Nephrol; Int Soc Nephrol; Can Soc Clin Invest; Royal Col Physicians & Surgeons, Can. *Res:* Hypertension; clinical nephrology. *Mailing Add:* Univ Hosp Ctr Univ of Sherbrooke Sherbrooke PQ J1K 2R1 Can

NAWAR, WASSEF W, b Cairo, Egypt, May 17, 26; m 53; c 1. FOOD CHEMISTRY. *Educ:* Univ Cairo, BSc, 47, MS, 50; Univ Ill, PhD(food sci), 59. *Prof Exp:* Asst dairy tech, Univ Ill, 50-52 & 57-59; from asst prof to assoc prof, 59-70, PROF FOOD SCI, UNIV MASS, AMHERST, 70- *Concurrent Pos:* Res grants, Sigma Xi, 61, USPHS, 65-69 & AEC, 65-; mem comt food irradiation, Nat Acad Sci-Nat Res Coun, 74- *Mem:* Am Chem Soc; Am Oil Chem Soc; AAAS; Inst Food Technol. *Res:* Flavor chemistry; thermal decomposition of fats; effects of ionizing radiation on fats. *Mailing Add:* Dept of Food Sci & Nutrit Univ of Mass Amherst MA 01003

NAWORSKI, JOSEPH SYLVESTER, JR, b Glassmere, Pa, May 3, 37; m 65; c 3. CHEMICAL ENGINEERING. *Educ:* Univ Pittsburgh, BS, 59, MS, 62; Cornell Univ, PhD(chem eng), 66. *Prof Exp:* Res engr, Res Labs, Aluminum Co Am, 59-62 & Res & Develop Div, Sun Oil Co, 66-68; asst prof chem eng, Va Polytech Inst & State Univ, 68-72; supvr process develop, Stauffer Chem Co, 72-77, mgr process develop, 77- *Honors & Awards:* Kirkpatrick Award. *Res:* Chemical reaction engineering; oxychlorination of ethylene; chemical process development. *Mailing Add:* 430 Ferguson Dr Bldg 3 Mountain View CA 94043

NAWROCKY, ROMAN JAROSLAW, b Przemysl, Poland, Apr 30, 32; US citizen; m 66; c 1. ELECTRICAL ENGINEERING. *Educ:* Univ Man, BSEE, 56; Polytech Inst Brooklyn, MSEE, 63; Polytech Inst New York, PhD(elec eng), 75. *Prof Exp:* Develop engr elec eng, Can Gen Elec Co, 56-58, Nat Co, 58-59, Mass Inst Technol, 59-61 & Bendix Corp, 63-64; SR RES ENGR, BROOKHAVEN NAT LAB, 64- *Mem:* Sigma Xi; Inst Elec & Electronics Engrs. *Res:* Automatic control; communications; instrumentation. *Mailing Add:* 105 Quaker Path Box 154 Stony Brook NY 11790

NAWY, EDWARD GEORGE, b Baghdad, Iraq, Dec 21, 26; US citizen; m 49; c 2. CIVIL & STRUCTURAL ENGINEERING. *Educ:* Univ Baghdad, Dipl, 48; Imp Col, Univ London, Dipl, 51; Mass Inst Technol, CE, 59; Univ Pisa, DrEng(concrete), 67. *Prof Exp:* Dep head struct div, Israel Water Planning Authority, Tel Aviv, 51-57; res engr, Mass Inst Technol, 57-59; PROF CIVIL ENG, RUTGERS UNIV, NEW BRUNSWICK, 59-, CHMN, DEPT CIVIL ENVIRON ENG, 80- *Concurrent Pos:* Chmn, Nat Comt on Cracking, Am Concrete Inst, 66-; adv, Fed Aviation Admin, Washington, DC, 69- *Honors & Awards:* H L Kennedy Award of Excellence, Am Concrete Inst. *Mem:* Fel Am Soc Civil Engrs; Nat Soc Prof Engrs; NY Acad Sci; fel Brit Inst Civil Eng. *Res:* Concrete structural systems, particularly cracking in two-way slabs; ultimate load of large diameter concrete pipes; plastic hinge rotation in reinforced and prestressed concrete systems; off-shore airports; fiber glass reinforcement in concrete. *Mailing Add:* Dept of Civil Eng Rutgers Univ New Brunswick NJ 08903

NAYAK, DEBI PROSAD, b W Bengal, India, Apr 1, 37; m 65; c 2. VIROLOGY, GENETIC ENGINEERING. *Educ:* Univ Calcutta, BVSc, 57; Univ Nebr, Lincoln, MS, 63, PhD(virol), 64. *Prof Exp:* Actg asst prof virol, 64-66, asst res virologist, 66-68, from asst prof to assoc prof, 68-76, PROF VIROL, UNIV CALIF, LOS ANGELES, 77- *Concurrent Pos:* Cancer Res Coord Comt & Calif Inst Cancer Res fels, Univ Calif, Los Angeles, 69, Calif Div Am Cancer Soc Sr Dernham fel, 69-74, Am Cancer Soc res grant, 72-74, Nat Cancer Inst res grant, 74-78, Nat Inst Allergy & Infectious Dis res grant, 75- & NSF, 79-81; NIH Study Sect, 85-89, 90-93. *Mem:* AAAS; Am Soc Microbiol. *Res:* Influenza virus genome and its translation, transcription, and replication; structure and genesis of defective interfering influenza viral genome; mechanism of interference; cloning, sequencing and expression of viral genes in both bacterial and eukaryotic systems; manipulation of cloned genes for producing medically important products; transport, sorting, assembly of viral proteins. *Mailing Add:* Dept Microbiol 143-239 Chs Univ Calif 405 Hilgard Ave Los Angeles CA 90024-1747

NAYAK, RAMESH KADBET, b Udipi, India, Sept 6, 34; US citizen; m 58; c 2. REPRODUCTIVE PHYSIOLOGY, CELL BIOLOGY. *Educ:* Univ Madras, BS, 54; Univ Bombay, MS, 56; Univ RI, MS, 64; Ore State Univ, PhD(physiol), 70. *Prof Exp:* Lectr zool, Inst Sci, Bombay, 56-61; res asst biol, Childrens Cancer Res Found, 64-65; fel, Univ Nebr, 70-72; res assoc anat, George Washington Univ, 72-75; assoc prof zool, Kuwait Univ, 75-78; HEALTH SCIENTIST ADMINR, NIH, BETHESDA, 78- *Concurrent Pos:* Scientist biol, Smithsonian Sci Info Exchange, 72-75. *Mem:* Electron Micros Soc Am; Am Soc Cell Biol; Am Soc Animal Sci; Am Inst Biol Sci; Am Anat Soc. *Res:* Electron microscopic studies of mammalian oviduct; effect of contraceptive steroids on the cardiovascular system. *Mailing Add:* 5333 Westbard Ave Westwood Bldg Rm 233 Bethesda MD 20892

NAYAK, TAPAN KUMAR, b Kalera, India, 1957; m; c 1. STATISTICAL INFERENCE, SOFTWARE RELIABILITY. *Educ:* Univ Calcutta, India, BSc, 76; Indian Statist Inst, MStatist, 79; Univ Pittsburgh, PhD(statist), 83. *Prof Exp:* Lectr statist, Ramakrishna Mission Residential Col, India, 79-80; asst prof, 83-89, ASSOC PROF STATIST, GEORGE WASHINGTON UNIV, WASH, DC, 89- *Concurrent Pos:* Vis scientist, Indian Statist Inst, Calcutta, India, 90. *Mem:* Am Statist Asn. *Res:* Diversity analysis; reliability analysis, especially software reliability; statistical inference; income inequality; inference using Pitman nearness criterion; randomized response design; estimation of population size. *Mailing Add:* Dept Statist George Washington Univ Washington DC 20052

NAYAR, JAI KRISHEN, b Kisumu, East Africa, Jan 3, 33; US citizen; m 64; c 3. INSECT PHYSIOLOGY, MEDICAL ENTOMOLOGY. *Educ:* Univ Delhi, BSc, 54, MSc, 56; Univ Ill, Urbana, PhD(entom), 62. *Prof Exp:* Sr res asst entom, Indian Agr Res Inst, Delhi, 56-58; med entomologist, Div Health, Fla Med Entom Lab, Vero Beach, Fla, 63-79; med entomologist, 79-82, PROF, FLA MED ENTOM LAB, UNIV FLA, VERO BEACH, 82- *Concurrent Pos:* Nat Res Coun Can award, Univ Man, 62-63; adj assoc prof entom & nematol, Univ Fla, Gainesville, 75-, affil assoc prof preventive med, Col Vet Med, 81-86. *Mem:* Entom Soc Am; Am Soc Trop Med & Hyg. *Res:* Biology and physiology of mosquitoes of medical and veterinary importance. *Mailing Add:* Fla Med Entom Lab IFAS Univ Fla 200 Ninth St SE Vero Beach FL 32962

NAYFEH, ALI HASAN, b Shuweikah, Jordan, Dec 21, 33; m 65; c 4. MECHANICS, APPLIED MATHEMATICS. *Educ:* Stanford Univ, BS, 62, MS, 63, PhD(aeronaut, astronaut), 64. *Prof Exp:* Prin res scientist, Heliodyne Corp, Calif, 64-68; mgr math physics dept, Aerotherm Corp, 68-71; prof eng mech, Va Polytech Inst & State Univ, 71-74; dean eng & vpres Eng Affairs, Yarmouk Univ, Jordan, 80-84; UNIV DISTINGUISHED PROF ENG MECH, VA POLYTECH INST & UNIV, 75- *Honors & Awards:* Physics Award 81; Res Award, Yarmouk Univ, 82. *Mem:* Soc Eng Sci; Am Phys Soc; Am Soc Mech Engrs; Am Inst Aeronaut & Astronaut. *Res:* Perturbation methods; nonlinear dynamics and chaos; structural dynamics; ship motion; aeroacoustics; structural acoustics; hydrodynamic stability; aerodynamic; nonlinear waves; flight mechanics. *Mailing Add:* Dept of Eng Mech Va Polytech Inst & State Univ Blacksburg VA 24061

NAYFEH, MUNIR HASAN, b Shuweikah-Tulkarem, Jordan, Dec 13, 45; m 73; c 4. ATOMIC PHYSICS. *Educ:* Am Univ Beirut, BSc, 68, MSc, 70; Stanford Univ, PhD(physics), 74. *Prof Exp:* Fel physics, Oak Ridge Nat Lab, 74-76, res physicist, 76-77; lectr, Yale Univ, 77-78; asst prof, 78-80, ASSOC PROF PHYSICS, UNIV ILL, URBANA-CHAMPAIGN, 80- *Honors & Awards:* Indust Res-100 Award. *Res:* High resolution laser spectroscopy; atomic collisions; coherence and quantum optics; multiphoton ionization of atoms and molecules. *Mailing Add:* Dept Physics Univ Ill Urbana-Champaign 110 W Green St Urbana IL 61801

NAYFEH, SHIHADEH NASRI, b Merj 'Youn, Lebanon. BIOCHEMISTRY, ENDOCRINOLOGY. *Educ:* Am Univ Beirut, BS & teaching dipl, 59, MS, 61; Univ NC, Chapel Hill, PhD(biochem), 64. *Prof Exp:* Fel, Harvard Univ, 64-65; investr biochem, Lebanese Agr Res Inst, 65-67 & Univ Pa, 67-68; asst prof biochem & endocrinol & dir endocrinol lab, NC Mem Hosp, 68-72; assoc prof, 72-82, PROF BIOCHEM, NUTRIT & PEDIAT, SCH MED, UNIV NC, CHAPEL HILL, 82- *Mem:* AAAS; Soc Study Reproduction; Endocrine Soc; NY Acad Sci; Am Soc Biol Chemists. *Res:* Mechanisms of action of polypeptide hormones in normal and tumour cells. *Mailing Add:* Dept Biochem & Nutrit Univ NC Sch Med Chapel Hill NC 27514

NAYLOR, ALFRED F, b South River, NJ, Oct 17, 27; m 50; c 3. GENETICS. *Educ:* Univ Chicago, AB, 50, PhD(zool), 57. *Prof Exp:* Asst, Univ Chicago, 51-54 & 55-57; asst prof zool, Univ Okla, 57-60; asst prof genetics, McGill Univ, 60-64; GENETICIST, NAT INST NEUROL DIS & STROKE, 64- *Mem:* Am Soc Human Genetics. *Res:* Population and human genetics; biometry; population ecology. *Mailing Add:* 6303 Bannockburn Dr Bethesda MD 20014

NAYLOR, AUBREY WILLARD, b Union City, Tenn, Feb 5, 15; m 40; c 2. PLANT PHYSIOLOGY. *Educ:* Univ Chicago, BS, 37, MS, 38, PhD(bot), 40. *Prof Exp:* Rockefeller asst, Univ Chicago, 38; mem staff, Bur Plant Indust, USDA, 38-40; instr bot, Univ Chicago, 40-44, naval radio, 42-44; instr bot, Northwestern Univ, 44-45; Nat Res Coun fel biol sci, Boyce Thompson Inst, 45-46; asst prof plant physiol, Univ Wash, 46-47 & Yale Univ, 47-52; from assoc prof to prof, 52-72, James B Duke prof, 72-85, JAMES B DUKE EMER PROF PLANT PHYSIOL, DUKE UNIV, 85- *Concurrent Pos:* Res partic & consult, Oak Ridge Inst Nuclear Studies, 54-64; consult, Oak Ridge Nat Lab, 57-58, NSF, 60-65, Res Triangle Inst, 68-, Biol Div, Tenn Valley Authority, 69-75 & Educ Testing Serv, 72-85, Schaper & Brummer Pharmaceut Co, 86-, AKZO Salt Inc, 90-; NSF sr fel & vis prof, Univ Bristol, 58-59, prog dir metab biol, NSF, 61-62; mem, Comn Undergrad Educ in Biol Sci, chmn panel interdisciplinary activ; chmn comt examr, Grad Rec Exam Biol, 66-72; mem bd, Southeastern Plant Environ Labs, 68-80. *Honors & Awards:* Charles Reid Barnes Life Mem Award, Am Soc Plant Physiol, 81; Distinguished Serv Award, Southern Sect Am Soc Plant Physiologists, 81. *Mem:* Am Soc Plant Physiol (secy, 53-55, vpres, 56, pres, 61, archivist, 87-); Japanese Soc Plant Physiologists; Australian Soc Plant Physiologists; Biochem Soc; Sigma Xi; Scandinavian Soc Plant Physiol; fel AAAS; Soc Exp Biol. *Res:* Photophysiology; growth regulation; enzymes; amino acid metabolism. *Mailing Add:* Dept Bot Duke Univ Durham NC 27706

NAYLOR, BENJAMIN FRANKLIN, b Gilroy, Calif, Nov 15, 17; m 46; c 3. PHYSICAL CHEMISTRY. *Educ:* San Jose State Univ, AB, 40; Stanford Univ, MA & PhD(phys chem), 43. *Prof Exp:* Phys chemist, US Bur Mines, Calif, 43-45; chemist, Stand Oil Co Calif, 45; from instr to assoc prof chem, San Jose State Univ, 45-51, head dept, 51-61, prof chem & coordr gen chem, 51-80, adj prof chem, 80-84; RETIRED. *Concurrent Pos:* Res, Lawrence Livermore Lab, 69-71. *Mem:* Am Chem Soc. *Res:* Chemical thermodynamics; high-temperature heat contents of titanium carbide and titanium nitride; specific heats of metals at high temperatures by AC-DC method. *Mailing Add:* 258 Washington Ct Sebastopol CA 95472-3140

NAYLOR, BRUCE GORDON, b Midale, Sask, Aug 19, 50. VERTEBRATE PALEONTOLOGY, HERPETOLOGY. *Educ:* Univ Sask, BSc, 72; Univ Alta, PhD(vert paleont), 78. *Prof Exp:* Fel vert paleont, Univ Toronto, 78-80; lectr, Univ Calif, Berkeley, 79; asst prof, 80-82, ADJ PROF GEOL ZOOL, UNIV ALBERTA, 82-; CUR, VERT PALEONT, TYRRELL MUS, 82-, ASST DIR. *Concurrent Pos:* Nat Res Coun Can fel, 78-80. *Mem:* Soc Vert Paleont; Soc Study Evolution; fel Geol Asn Can; Paleont Soc. *Res:* Phylogenetic relationships and functional morphology of fossil and recent salamanders. *Mailing Add:* Tyrrell Mus Paleont Box 7500 Drumheller AB T0J 0Y0 Can

NAYLOR, CARTER GRAHAM, b Denver, Colo, May 22, 42; m 64; c 3. PETROLEUM CHEMISTRY, SURFACTANTS. *Educ:* Calif Inst Technol, BS, 64; Univ Colo, PhD(org chem), 69. *Prof Exp:* From res chemist to sr res chemist, Jefferson Chem Co, 69-80; sr res chemist, Texaco Chem Co, 80-81, proj chemist, 81-86, sr proj chemist, 86-90, PROJ LEADER, TEXACO CHEM CO, 90- *Mem:* Am Chem Soc; Am Oil Chemists Soc. *Res:* Petrochemicals, especially surfactants; enhanced oil recovery; detergents. *Mailing Add:* Texaco Chem Co Box 15730 Austin TX 78761

NAYLOR, DAVID L, b Chesterfield, England, UK, Feb 15, 59; m 87; c 1. INTEGRATED OPTICS, MICROFABRICATION TECHNOLOGY. *Educ:* Oxford Univ, BA, 80; Univ Southern Calif, PhD(elec eng), 88. *Prof Exp:* Res assoc, uk Atomic Energy Authority Culham Lab, Dept Physics, Oxford Univ, 80-81; res asst, Elec Eng Dept, Univ Southern Calif, 81-88; ASST PROF ELEC ENG, ELEC ENG & COMPUTER SCI DEPT, UNIV ILLINOIS, CHICAGO, 88-, ASSOC DIR, MICROFABRICATION APPLN LAB, 90- *Mem:* Inst Elec & Electronic Engrs; Inst Elec & Electronic Engrs-Lasers & Electrooptics Soc; Optical Soc Am; Int Soc Optical Eng. *Res:* Integrated optics; optical interconnects; polymer, ferroelectric, electrooptic and dielectric optical waveguide materials and devices; optical fiber/waveguide/semiconductor integration using VLSI and micro fabrication procedures. *Mailing Add:* Elec Eng & Computer Sci Dept MC154 Univ Ill PO Box 4348 Chicago IL 60680

NAYLOR, DENNY VE, b Twin Falls, Idaho, Oct 26, 37; m 59; c 2. SOIL CHEMISTRY. *Educ:* Univ Idaho, BS, 59, MS, 61; Univ Calif, Berkeley, PhD(soil sci), 66. *Prof Exp:* Assoc prof, 66-77, PROF SOILS, UNIV IDAHO, 77- *Mem:* AAAS; Am Soc Agron; Soil Sci Soc Am; Sigma Xi. *Res:* Nutrients in soil-water systems; water quality and agricultural practices; soil organic matter chemistry. *Mailing Add:* Dept of Plant & Soil Sci Univ of Idaho Moscow ID 83843

NAYLOR, DEREK, b Eng, Nov 9, 29; m 60; c 3. APPLIED ANALYSIS. *Educ:* Univ London, BSc, 51, PhD(aerodyn), 53. *Prof Exp:* Res assoc, Brown Univ, 53-54; aerodynamicist, A V Roe & Co, 54-56; asst prof appl math, Univ Toronto, 56-62; sr lectr, Royal Col Sci, Glasgow, 62-63; assoc prof, 63-65, PROF APPL MATH, UNIV WESTERN ONT, 65- *Res:* Integral transforms; asymptotic analysis. *Mailing Add:* Dept Appl Math Univ Western Ont London ON N6A 5B8 Can

NAYLOR, GERALD WAYNE, b Keener, Ala, Feb 15, 22; m 44; c 3. AGRONOMY, PLANT PHYSIOLOGY. *Educ:* Auburn Univ, BS, 47, MS, 49; NC State Univ, PhD(agron), 53; Southeastern Baptist Theol Sem, BD, 56. *Prof Exp:* Minister & hosp chaplain, NC Baptist Hosp, 53-60, res fel, 60-61; assoc prof, 61-71, PROF BIOL, CARSON-NEWMAN COL, 71- *Concurrent Pos:* Partic, NSF res prog, Univ Tex, 64-66 & acad year exten, 64-66, 67-69; proj dir, NSF Student Sci Training Prog, 72-79. *Mem:* AAAS; Am Soc Plant Physiol. *Res:* Water pollution studies of Lake Cherokee and the resultant serious problem of fish kill. *Mailing Add:* Dept Biol Carson-Newman Col Russell Ave Jefferson City TN 37760

NAYLOR, HARRY BROOKS, b Minn, Mar 30, 14; m 40; c 3. BACTERIOLOGY. *Educ:* Univ Minn, BS, 38; Cornell Univ, PhD(bact), 43. *Prof Exp:* Dairy chemist, Sheffield Farms Co, 46-47; prof dairy indust, 47-50, prof bact, 50-77, EMER PROF MICROBIOL, CORNELL UNIV, 77- *Concurrent Pos:* Fulbright-Hays lectureship, Univ Alexandria, 66-67; Orgn Am States lectureship, Univ Campinas, Brazil, 72, 73, sabbatical res leave, Univ Campinas, 75; spec assignment at Fed Univ Rio de Janeiro, 78; mem staff, Pasco Lab, Inc, 78-85. *Mem:* Am Soc Microbiol; Am Acad Microbiol. *Res:* Bacterial physiology; virology. *Mailing Add:* 5390 Belfern Dr Bellingham WA 98226

NAYLOR, MARCUS A, JR, b Oberlin, Ohio, Apr 27, 20; m 43; c 2. CHEMICAL ENGINEERING. *Educ:* Col Wooster, BA, 42; Johns Hopkins Univ, MA, 43, PhD(org chem), 45. *Prof Exp:* Lab instr chem, Johns Hopkins Univ, 42-44; res chemist, Plastics Dept, E I Du Pont de Nemours & Co, Inc, 44-50, supvr, Res Div, Polychem Dept, 50-55, com investr, Planning Div, 55-56, sect mgr, Res Div, 56-58, dir gen prod res & develop, 58-59, asst dir, Res Div, Indust & Biochem Dept, 59-67, asst dir, Indust Chem Sales Div, 67-72, dir lab, Chem Dyes & Pigments Dept, 73-80; RETIRED. *Mem:* Am Chem Soc; Sigma Xi; AAAS. *Res:* Research administration; organic synthesis; reaction mechanisms; chemistry of high polymers. *Mailing Add:* 4706 Washington St Ext Wilmington DE 19809

NAYLOR, PAUL HENRY, b Easton, Md, Jan 11, 48; m 80; c 3. BIOCHEMICAL ENDOCRINOLOGY. *Educ:* Washington Col, Md, Bs, 70; Johns Hopkins Univ, MA, 72; Univ Tex, Galveston, PhD(biochem), 77. *Prof Exp:* Instr chem, Univ Md, Baltimore County, 72-74; res affil biochem endocrinol, Roswell Park Mem Inst, 78-80; res fel, Harvard Med Sch, 80-82, res assoc biol chem, 82-84; res asst prof, 84-88, RES ASSOC PROF, GEORGE WASHINGTON UNIV, 89- *Concurrent Pos:* McLaughland fel, Univ Tex Med Br, 77-78; training fel, Nat Cancer Inst, 78-80. *Mem:* Endocrine Soc; AAAS; NY Acad Sci. *Res:* Immunological and endocrinological roles of the thymus; thymic peptides, monoclonal antibodies, immunochemistry; development of AIDS vaccines and diagnostics; mechanism of neoplastic transformation by hormones and chemical carcinogens. *Mailing Add:* Dept Biochem George Washington Univ 2121 Eye St NW Washington DC 20052

NAYLOR, RICHARD STEVENS, b Lakeland, Fla, July 15, 39. GEOLOGY, GEOCHEMISTRY. *Educ:* Mass Inst Technol, BS, 61; Calif Inst Technol, PhD(geol), 67. *Prof Exp:* Asst prof geol, Mass Inst Technol, 67-74; ASSOC PROF EARTH SCI & CHMN DEPT, NORTHEASTERN UNIV, 74- *Mem:* Geol Soc Am; Am Geophys Union; Geochem Soc; Sigma Xi. *Res:* Geology and geochronology of northern Appalachian Mountain system; geology and geochronology of mantled gneiss domes. *Mailing Add:* Dept Earth Sci 14HO Northeastern Univ Boston MA 02115

NAYLOR, ROBERT ERNEST, JR, b Nashville, Ark, July 14, 32; m 63; c 1. CHEMICAL PHYSICS. *Educ:* Univ SC, BS, 51; Harvard Univ MA, 54, PhD(chem physics), 56. *Prof Exp:* Res chemist, Film Dept, E I du Pont de Nemours & Co, Inc, 56-62, res supvr, 62-64, res mgr, 64-66, tech supt, 66-67, lab dir, 67-71, tech mgr, 71-74, prod & tech dir, Film Dept, 74-76, tech dir, Atomic Energy Div, 76-79, dir, Res & Develop Planning, Corp Plans Dept, 79-80, dir res, Cent Res & Develop Dept, 81; VPRES & CORP DIR RES, ROHM & HAAS CO, 82- *Mem:* Am Chem Soc. *Res:* Polymer science; nuclear energy. *Mailing Add:* Rohm & Haas Co Res Lab 727 Norristown Rd Spring House PA 19477

NAYMARK, SHERMAN, b Duluth, Minn, May 12, 20; m 42; c 2. NUCLEAR ENGINEERING. *Educ:* US Naval Acad, BS, 41; Mass Inst Technol, MS, 46. *Prof Exp:* Asst chief engr, USS Saratoga, 41-44; repair supt, Norfolk Naval Shipyard, 46-48; sr scientist, Argonne Nat Labs, 48-52; dir Naval Reactors Schenectady Opers, Off AEC, 52-54; proj engr & mgr submarine intermediate reactor, Knolls Atomic Power Lab, Gen Elec Co, Schenectady, 54-56, mgr reactor design, Atomic Power Equip Dept, San Jose, 56-60, mgr fuel develop, 60-65, mgr nuclear mat & propulsion opers, Cincinnati, 65-67, mgr & engr turnkey opers, Atomic Power Equip Dept, San Jose, 67-69; pres, 70-85, CHMN, QUADREX CORP, 85- *Mem:* Fel Am Nuclear Soc; AAAS; assoc Am Public Power Asn. *Res:* Nuclear power. *Mailing Add:* 21 Forrest St Los Gatos CA 95032

NAYMIK, DANIEL ALLAN, b Lorain, Ohio, Mar 8, 22; m 44. MATHEMATICS, QUANTUM PHYSICS. *Educ:* Univ Mich, BS, 47, MS, 48, PhD(physics), 58. *Prof Exp:* Mem tech staff semiconductor adv develop, Bell Tel Labs, NJ, 58-64; mem sr staff physics of thin films, Gen Dynamics/ Electronics, 64-66, prin engr, Comput Sci Dept, 66-69; sr staff res scientist, Comput Res Dept, Amoco Prod Co, 69-86; RETIRED. *Mem:* Am Phys Soc. *Res:* Computer and management sciences; thin film physics; high energy and mathematical physics; quantum mechanics; electron scattering; semiconductor device development; computer graphics systems. *Mailing Add:* 6435 S Knoxville Ave Tulsa OK 74136

NAYUDU, Y RAMMOHANROY, b Masulipatam-Andhr, India, Jan 13, 22; US citizen; m 43; c 3. MARINE GEOLOGY, PETROLOGY. *Educ:* Univ Bombay, BS, 45, MSc, 47; Univ Wash, PhD(geol), 59. *Prof Exp:* Lectr geol, Univ Rangoon, 51-53; geologist, Burma Geol Dept, Ministry Mines, 54-55; res instr geol oceanog, Univ Wash, 59-61; res asst geologist, NSF fel, Scripps Inst, Univ Calif, 61-63; res asst prof marine geol, Univ Wash, 63-65, res assoc prof, 65-68; dep dir, Inst Marine Sci, Alaska, 68-69; prof marine geol, Univ Alaska, 69-71; dir div marine & coastal zone mgt, Alaska Dept Environ Conserv, 71-74; SCI ADV TO GOV, ALASKA, 70-; DIR DIV FISHERIES & NATURAL RESOURCES, CENT COUN TLINGIT & HAIDA INDIANS OF ALASKA, 77- *Concurrent Pos:* NSF grants, 59-63, 65 & 66-67; dir, Ore Coastal Zone Mgt Asn, 76-77. *Mem:* Am Asn Petrol Geologists; Soc Econ Paleontologists & Mineralogists; Geol Soc Am; fel Brit Geol Asn; Int Asn Sedimentol. *Res:* Deep sea sediments and submarine volcanics. *Mailing Add:* Alaska Pac Univ 4101 University Dr Anchorage AK 99508

NAYYAR, RAJINDER, b Khanna, India, June 14, 36; US citizen; m 69; c 2. NEUROLOGY, MICROSCOPIC ANATOMY. *Educ:* Panjab Univ, India, BSc, 57, MSc, 59; Univ Delhi, PhD(histochem & cytogenetics), 64. *Prof Exp:* Asst lectr zool, Univ Delhi, 63-64, chmn dept zool, H R Col, 64-65; asst prof neurol & anat, Med Sch, Northwestern Univ, Chicago, 67-77; dir, Electron Microscope, Histol & Histochem Labs, Neurol Serv, Vet Admin Lakeside Hosp, 67-77 & Electron Microscope, Immunofluorescence, Cytogenetics & Histol Labs, Dept Path, W Surburban Hosp, 77-84; asst prof, 77-81, ASSOC PROF ANAT, STRITCH SCH MED, LOYOLA UNIV, 81- *Concurrent Pos:* Joseph P Kennedy fel anat, Univ Western Ont, 65-66, Med Res Coun fel, 66-67; co-investr neonatal endotoxemia, dept pediat, Loyola Univ Med Ctr, 84-; consult, Edward Hines Vet Admin Hosp, 84- *Mem:* Electron Micros Soc Am; Am Asn Anatomists; Am Soc Cell Biol; Histochem Soc; Int Soc Nephrology. *Res:* Fish chromosome studies; histochemistry of fish oocytes; sex-chromatin studies; histochemistry of the diabetic retina; electron microscopy and histochemistry; the effect of antibiotics on the brain; cytogenesis of lysosomes; electromagnetic autoradiography; chemically induced myelopathy; preputial gland secretion; effect of castration; experimental pathology at light and electron microscope level involving immunologic and endotoxic shocks in mice and rats. *Mailing Add:* Dept Anat Loyola Univ Stritch Sch Med 2160 S First Maywood IL 60153

NAZARIAN, GIRAIR MIHRAN, physical chemistry; deceased, see previous edition for last biography

NAZAROFF, GEORGE VASILY, b San Francisco, Calif, Apr 12, 38; m 63; c 3. THEORETICAL CHEMISTRY. *Educ:* Univ Calif, Berkeley, BS, 59; Univ Wis-Madison, PhD(chem), 65. *Prof Exp:* NSF fel, 65-66; asst prof chem, Col Natural Sci, Mich State Univ, 66-72; ASSOC PROF CHEM, IND UNIV, SOUTH BEND, 72- *Concurrent Pos:* Res Corp starter grant, 66- *Mem:* Am Chem Soc. *Res:* Perturbation theory; generalized Hartree-Fock formalisms; natural spin orbitals; resonant scattering; electron-diatomic collision theory. *Mailing Add:* Dept Chem Ind Univ 1700 Mishawaka Ave PO Box 7111 South Bend IN 46634

NAZAROFF, WILLIAM W, b Lynwood, Calif, Sept 3, 55; m 78; c 4. AIR QUALITY ENGINEERING. *Educ:* Univ Calif, Berkeley, AB, 78, MEng, 80; Calif Inst Technol, PhD(environ eng sci), 89. *Prof Exp:* Staff scientist, Lawrence Berkeley Lab, 80-88, ASST PROF ENVIRON ENG, UNIV CALIF, BERKELEY, 88- *Concurrent Pos:* Assoc ed, Health Physics J, 87-90; fac assoc, Lawrence Berkeley Lab, 88-; NSF presidential young investr, 90. *Mem:* Am Asn Aerosol Res; Health Physics Soc; Air & Waste Mgt Asn; Sigma Xi; Asn Environ Eng Professors. *Res:* Physical and chemical processes that govern air pollutant concentrations and fates with an emphasis on indoor environments. *Mailing Add:* Dept Civil Eng Univ Calif Berkeley CA 94720

NAZEM, FARAMARZ FRANZ, b Rasht, Iran, Jan 22, 43; US citizen; m 69; c 2. RHEOLOGY, CHEMICAL ENGINEERING. *Educ:* Ohio State Univ, BSChE, 68; Washington Univ, MS, 71, DSc(chem eng), 73. *Prof Exp:* Proj assoc rheology, Univ Wis-Madison, 73-74, asst scientist, 74-75; staff res scientist, Union Carbide Corp, 75-80, res scientist, 80-81, group leader, 81-82, sr group leader, 82-83, res mgr, 83, DEVELOP MGR, UNION CARBIDE CORP, 83- *Mem:* Am Inst Chem Engrs; Soc Rheol; Soc Plastics Engrs. *Res:* Flow and deformation of viscoelastic materials; fundamentals of melt spinning; elongational viscosity of composite materials; rheology of liquid crystals; graphite electrode manufacturing. *Mailing Add:* Union Carbide Corp 777 Old Sawmill River Rd Tarrytown NY 10591

NAZERIAN, KEYVAN, b Tehran, Iran, Dec 21, 34; m 59; c 3. VIROLOGY, ELECTRON MICROSCOPY. *Educ:* Univ Tehran, DVM, 58; Mich State Univ, MS, 60, PhD(virol), 65. *Prof Exp:* Asst virol, Mich State Univ, 59-60; vis scientist, Pub Health Inst, Padua, 60-62; asst virol, Mich State Univ, 63-65; head electron micros lab, South Jersey Med Res Found, 65-66; MICROBIOLOGIST, REGIONAL POULTRY RES LAB, USDA, 66- *Mem:* AAAS; Am Soc Microbiol; Electron Micros Soc Am; NY Acad Sci. *Res:* Biochemical, biophysical and morphological studies of animal viruses, particularly oncogenic viruses and their interaction with susceptible hosts. *Mailing Add:* 5955 Eagles Way Haslett MI 48804

NAZRI, GHOLAM-ABBAS, b Malayer, Iran, Apr, 8, 51. SPECTROELECTROCHEMISTRY, ENERGY CONVERSION DEVICES. *Educ:* Tehran Univ, BS & MS, 75; Case Western Reserve Univ, PhD(chem), 81. *Prof Exp:* Res assoc corrosion, Inst Nuclear Sci & Technol, 74-78; res assoc electrochem, Case Western Reserve Univ, 81-82 & Mat & Molecular Div, Lawrence Berkeley Lab, Univ Calif, 82-84; SR RES SCIENTIST ELECTROCHEM, RES LAB, GEN MOTORS CORP, 84- *Mem:* Am Chem Soc; Electrochem Soc; Am Physics Soc; Mat Res Soc. *Res:* Electrochemistry; advanced energy conversion devices; electrocatalysis; corrosion and passivation; design and development of new techniques for nontraditional electrochemistry and material science; spectroelectrochemistry; solid state chemistry; solid state devices; spectroscopy. *Mailing Add:* 4700 Haddington Lane Bloomfield Hills MI 48304

NAZY, JOHN ROBERT, b Alamosa, Colo, July 12, 33; m 56; c 5. INDUSTRIAL ORGANIC CHEMISTRY. *Educ:* Regis Col, Colo, BS, 54; Northwestern Univ, PhD(org chem), 59. *Prof Exp:* Chemist, Union Carbide Chem Co, 58-60; proj leader, 60-66, tech supvr, 66-69, sect leader, Tech Serv Dept, 69-73, tech dir, 73-76, marketing mgr petroleum, 76-77, dir indust mgrs, 77-79, dir prod mgrs, 79-82, VPRES, FINANCE/ADM, FUNCTIONAL PROD GROUP, HENKEL CORP, 82- *Mem:* Am Mgt Asn; Am Chem Soc; Am Soc Personnel Admin. *Res:* Organoboron and fatty acid chemistry; polymer synthesis; guar and natural gums. *Mailing Add:* 1625 Wheatgrass Dr Reno NV 89509

NEADERHOUSER PURDY, CARLA CECILIA, b Rome, NY, June 15, 45; m 68, 82; c 2. LIMIT THEOREMS, STATISTICAL MECHANICS. *Educ:* Cornell Univ, BA, 67; Univ Ill, Urbana, MA, 69, PhD(math), 75. *Prof Exp:* ASST PROF MATH, TEX A&M UNIV, 75- *Concurrent Pos:* Prin investr NSF grants, 76-77. *Mem:* Am Math Soc; Math Asn Am. *Res:* Limit behavior of ramdom fields with application to statistical mechanics; asymptotic expansions and clustering behavior, along with related computational problems. *Mailing Add:* Comput Sci Univ Cincinnati Mail Location 008 Cincinnati OH 45221

NEAGLE, LYLE H, b Mutual, Okla, Nov 6, 31; m 64; c 2. ANIMAL NUTRITION, BIOCHEMISTRY. *Educ:* Okla State Univ, BS, 53; Iowa State Univ, PhD(animal nutrit), 60. *Prof Exp:* Asst dir animal nutrit & res, Supersweet Feeds Div, Int Milling Co, Minn, 60-67; mgr res, Allied Mills, Inc, 67-72, dir res, 72-81, VPRES RES & DEVELOP, CONTINENTAL GRAIN CO, 81- *Mem:* Am Soc Animal Sci. *Mailing Add:* 824 Sandstone Dr Libertyville IL 60048

NEAL, C LEON, b Carolean, NC, Sept 13, 38; m 60; c 2. ENERGY ENGINEERING, AIR CONDITIONING ENGINEERING. *Educ:* NC State Univ, BS, 60; Purdue Univ, MS, 62. *Prof Exp:* Flight test engr VTOC, Patuxent Naval Air Test Ctr, Patuxent River, md, 60; teaching asst statist & strength mat, Purdue Univ, West Lafayette, Ind, 60-62; asst aeronaut eng, Cornell Aeronaut Lab, Buffalo, NY, 62-66, assoc & proj engr, 66-69; applications eng mech, NC Sci & Technol Res Ctr, Research Triangle Park, 69-81; actg dir, NC Energy Inst, Research Triangle Park, 81-82; SR PROJ ENGR, NC ALTERNATIVE ENERGY CORP, RESEARCH TRIANGLE PARK, 83- *Concurrent Pos:* Consult, Building Resources Mgt Corp, San Francisco, Calif, 88-89, Proctor Eng, San Francisco, Calif, 90-91, Ctr Energy & Urban Environ, Minneapolis, Minn, 91- *Honors & Awards:* Nat Awards Prog for Energy Innovation, Dept Energy, 88. *Mem:* Am Solar Energy Soc; Am Soc Heating, Refrig & Air Conditioning Engrs. *Res:* Actual energy efficiency of residential and small commercial air conditioning; training of air conditioning, heat pump technicians; field test of air conditioning performance; residential energy efficiency. *Mailing Add:* 3506 Carriage Dr Raleigh NC 27612

NEAL, DONALD WADE, b Hopewell, Va, June 23, 51. BASIN ANALYSIS, CARBONATE PETROLOGY. *Educ:* Col William & Mary, BS, 73; Eastern Ky Univ, MS, 75; WVa Univ, PhD(geol), 79. *Prof Exp:* Res assoc stratig, WVa Geol & Encon Surv, 77-79; asst prof, 79-87, ASSOC PROF GEOL, EAST CAROLINA UNIV, 87- *Mem:* Geol Soc Am; Soc Econ Paleontologists & Mineralogists; Paleont Soc. *Res:* Basin analysis of upper paleozoic sediments in the Central Appalachians-includes stratigraphy, petrology, geochemistry and paleontology. *Mailing Add:* Dept Geol East Carolina Univ Greenville NC 27858-4353

NEAL, HOMER ALFRED, b Franklin, Ky, June 13, 42; m 62; c 2. EXPERIMENTAL HIGH ENERGY PHYSICS. *Educ:* Ind Univ, BS, 61; Univ Mich, MS, 63, PhD(physics), 66. *Hon Degrees:* DSc, Ind Univ, 84. *Prof Exp:* NSF fel, Europ Orgn Nuclear Res, 66-67; from asst prof to assoc prof physics, Ind Univ, Bloomington, 67-72, prof, 72-81, dean res & grad develop, 76-81; provost, State Univ NY, Stony Brook, 81-86; PROF & CHAIR, DEPT PHYSICS, UNIV MICH, 87- *Concurrent Pos:* Alfred P Sloan Found fel, Ind Univ, Bloomington, 68-; chmn zero gradient synchrotron accelerator users orgn, mem zero gradient synchrotron prog comt, 70-72; mem bd trustees, Argonne Univs Asn, 71-74 & 77-80, Univ Res Asn, 83-; J S Guggenheim fel, Stanford Univ, 80-81; mem, Nat Sci Bd, 80-86; bd dirs, NY Sea Grant Inst, 82-, State Univ NY Res Found, 83-84, Ogden Corp, 85-; mem bd regents, Smithsonian Inst, Washington, DC, 89-; mem, Int Affairs Comt, Strategic & Int Studies, Washington, DC, 89-; mem bd overseers, Superconducting Super Collider, 89- *Mem:* Fel Am Phys Soc; Sigma Xi; AAAS. *Res:* Experimental studies of elementary particle interactions. *Mailing Add:* Dept Physics Univ Mich Ann Arbor MI 48109-1120

NEAL, J(AMES) P(RESTON), III, electrical engineering; deceased, see previous edition for last biography

NEAL, JAMES THOMAS, b Detroit, Mich, Feb 9, 36; m 60; c 4. GEOLOGY. *Educ:* Mich State Univ, BS, 57, MS, 59. *Prof Exp:* Geologist, Can Cliffs Ltd, 57 & Albanel Minerals Ltd, 59; US Air Force, 60-79, proj officer geol res, Air Force Cambridge Res Labs, 60-63, proj scientist, 63-66, chief geotech br, 66-68, from instr to assoc prof geog, US Air Force Acad, 68-73, chief ground shock & cratering, Civil Eng Res Div, US Air Force Weapons Lab, 73-78, staff scientist, Air Force Systs Command Hq, 78-79; MEM TECH STAFF, SANDIA NAT LAB, 80- *Honors & Awards:* Outstanding Res & Develop Award, US Air Force, 66. *Mem:* Sigma Xi; fel Geol Soc Am. *Res:* Engineering and military geology; geology of playas; remote sensing; site selection and evaluation methodology; geology of salt domes. *Mailing Add:* 903 Martingale Lane SE Albuquerque NM 87123-4303

NEAL, JOHN ALEXANDER, b Aliquippa, Pa, Aug 7, 40; m 64; c 2. INORGANIC CHEMISTRY. *Educ:* Eastern Wash State Col, BA, 66; Univ Wash, PhD(inorg chem), 70. *Prof Exp:* Res assoc chem, Wash State Univ, 71-72; res chemist, 72-75, group leader, 75-76, RES DIR, GA-PAC CORP, 76- *Mem:* Am Chem Soc. *Res:* Factors which influence formation and stability of transition metal complexes in polydentate systems; sterochemistry of polydentate complexes; oxidative degradation of metal complexes & colloidal properties of thermodynamically stable sols. *Mailing Add:* Bellingham Div Ga-Pac Corp 1754 Thorne Rd Tacoma WA 98421-3207

NEAL, JOHN ALVA, b Omaha, Nebr, Mar 6, 38; m 61; c 3. CIVIL ENGINEERING, MATERIALS SCIENCE. *Educ:* Ga Inst Technol, BME, 61; Univ Ill, Urbana, MS, 62, PhD, 65. *Prof Exp:* Asst prof civil eng, 65-70, dir construct & rehab, Off Facil Planning, 70-73, asst vpres facil planning, 73-78, vpres facil planning, 78-82, ASSOC PROF CIVIL ENG, STATE UNIV NY BUFFALO, 70-; TEST ENGR, NAT CTR EARTHQUAKE ENG RES, 87- *Mem:* Am Concrete Inst; assoc Am Soc Civil Engrs; Am Soc Eng Educ. *Res:* Fatigue of plain concrete. *Mailing Add:* Dept Civil Eng 221 R8 Engr W State Univ NY Buffalo N Campus Buffalo NY 14260

NEAL, JOHN LLOYD, JR, b Concordia, Kans, Oct 18, 37; m 62; c 2. MICROBIOLOGY, SOIL SCIENCE. *Educ:* Ore State Univ, BSc, 60, MSc, 63, PhD(soil microbiol), 68. *Prof Exp:* Res asst microbiol, Ore State Univ, 60-67; res scientist, Can Agr Res Sta, 67-77; asst prof, 77-83, ASSOC PROF MICROBIOL, VA POLYTECH STATE UNIV, 83- *Mem:* Am Soc Microbiol; Sigma Xi; AAAS; Soil Sci Soc Am. *Res:* Asymbiotic and symbiotic nitrogen fixation; interrelationship between plant roots and soil microorganisms; soil biochemistry; microbial transformations in soil. *Mailing Add:* Dept Biol VA Polytech Inst & State Univ Blacksburg VA 24061-0406

NEAL, JOHN WILLIAM, JR, b St Louis, Mo, Nov 17, 37; m 73. ENTOMOLOGY. *Educ:* Univ Mo-Columbia, BS, 61, MS, 64; Univ Md, College Park, PhD(entom), 70. *Prof Exp:* Mus aid & entomologist, Dept Mammals (Iran), US Nat Mus, 63-65; fac res asst, Univ Md, College Park, 66-68, instr biol control, 68-70; RES ENTOMOLOGIST, FLORIST & NURSERY CROPS LAB, PLANT SCI INST, AGR RES SERV, USDA, 70- *Mem:* Entom Soc Am; Sigma Xi. *Res:* Biological and chemical control pests of ornamental plants; screening germplasm for insect-host resistance; insect pheromones. *Mailing Add:* Agr Res Serv USDA 10300 Baltimore Blvd Beltsville MD 20705-2350

NEAL, MARCUS PINSON, JR, b Columbia, Mo, Apr 22, 27; m 61; c 3. RADIOLOGY. *Educ:* Univ Mo, AB, 49, BS, 51; Univ Tenn, MD, 53; Am Bd Radiol, cert radiol, 58 & radiol in nuclear med, 59. *Prof Exp:* Intern, Med Col Va Hosp, 53-54; res assoc path, Sch Med, Univ Mo, 54; resident radiol, Univ Wis Hosps, 54-57, from instr to asst prof, Sch Med, Univ Wis, 57-63; assoc prof radiol, Va Commonwealth Univ, 63-66, chmn, Div Diag Radiol, 65-71, asst dean, Sch Med & dir, Regional Med Progs, 68-71, dir, continuing educ med & grad educ med, 69-71, interim dean, Sch Med, 71, asst vpres health sci, 71-73, provost health sci, 73-78, assoc dean, Sch Med, Med Educ Qual Assurance, 78-79, PROF RADIOL, MED COL VA, VA COMMONWEALTH UNIV, 66-, DIR RADIOL, GRAD MED EDUC, 79- *Concurrent Pos:* Radiologist, Cent Wis Colony, Madison, 59-63 & Vet Admin Hosp, 61-63; consult, Wis Diag Ctr, 61-63, US Air Force Hosp, Truax Field, 63 & Vet Admin Hosp, Richmond, Va, 63-; pres, Va Coun Health & Med Care, 70-74 & 78-80. *Mem:* Fel Am Col Radiol; AMA; Radiol Soc NAm; Brit Inst Radiol. *Res:* Diagnostic radiology and medical education research. *Mailing Add:* Med Col Va Sta Box 295 Va Commonwealth Univ Richmond VA 23298-0295

NEAL, MICHAEL WILLIAM, b Rochester, NY, July 11, 46; m 79. REGULATORY TOXICOLOGY, ENVIRONMENTAL RISK ASSESSMENT. *Educ:* Hartwick Col, BA, 69; Syracuse Univ, PhD(biochem), 75. *Prof Exp:* Res assoc chem carcinogenesis, Upstate Med ctr, State Univ NY, 74-78; SR RES FEL REGULATORY TOXICOLOGIST, SYRACUSE RES CORP, 78- *Mem:* Am Chem Soc; Environ Mutagen Soc. *Res:* Develop documentation on the toxic properties of chemicals for the federal government to support regulatory action and determine the adequacy of available toxicologic data and the need for additional testing. *Mailing Add:* Syracuse Res Corp Merrill Lane Syracuse NY 13210

NEAL, RALPH BENNETT, engineering, for more information see previous edition

NEAL, RICHARD ALLAN, b Waverly, Iowa, July 27, 39; m 62; c 3. FISHERIES. *Educ:* Iowa State Univ, BS, 61, MS, 62; Univ Wash, PhD(invert fishery biol), 67. *Prof Exp:* Supvry fishery biologist, Nat Marine Fisheries Serv, 66-77; aquacult adv, 77-80, dep dir gen, Int Ctr Living Aquatic Resources Mgt, 80-83, dir gen, 83-85, SR FISHERIES SPECIALIST, AGENCY, INT DEVELOP, 85- *Mem:* Am Fisheries Soc; World Aquacult Soc; Asian Fisheries Soc. *Res:* Freshwater fishery biology; ecological studies of paralytic shellfish poisoning; fishery population dynamics; penaeid shrimp culture; aquaculture research; international fisheries development. *Mailing Add:* 2120 Bonita La Dena Way El Cajon CA 92019-2324

NEAL, RICHARD B, b Lawrenceburg, Tenn, Sept 5, 17; m 44; c 2. MICROWAVE PHYSICS. *Educ:* US Naval Acad, BS, 39; Stanford Univ, PhD(physics), 53. *Prof Exp:* Field serv engr, Sperry Gyroscope Co, 41-42, fire control serv supt, 42-46, res engr, 46-47; res assoc, Stanford Univ, 50-62, assoc dir tech div, Linear Accelerator Ctr, 62-82, prof physics res, 80-85; RETIRED. *Mem:* Nat Acad Eng; Sigma Xi; Am Phys Soc. *Res:* Microwave and accelerator physics; high energy linear electron accelerators. *Mailing Add:* 735 Camino Santa Barbara Solana Beach CA 92075

NEAL, ROBERT A, b Casper, Wyo, Apr 21, 28; m 58; c 3. TOXICOLOGY, BIOCHEMISTRY. *Educ:* Univ Denver, BS, 49; Vanderbilt Univ, PhD(biochem), 63. *Prof Exp:* From asst prof to assoc prof, biochem, Sch Med, Vanderbilt Univ, 64-75, dir, ctr environ toxicol, 73-81, prof, 75-81; PRES, CHEM INDUST INST TOXICOL, 81- *Concurrent Pos:* NIH res fel toxicol, Univ Chicago, 63-64; mem, Food Protection Comt & Toxciol Subcomt, Nat Acad Sci; mem, Toxicol Study Sect, Prog Comt Multiple Factors Causation Environ Induced Dis & bd Toxicol & environ Health Hazards, NIH. *Honors & Awards:* George H Scott Mem Award, 85. *Mem:* AAAS; Am Soc Pharmacol & Exp Therapeut; Soc Toxicol; Am Inst Nutrit; Am Asn Biol Chemists. *Res:* Natural product chemistry; isolation and identification of natural products; detoxification mechanisms. *Mailing Add:* 2700 Toxey S Dr Raleigh NC 27609

NEAL, SCOTTY RAY, b Redlands, Calif, July 12, 37; m 58; c 2. APPLIED MATHEMATICS, TELECOMMUNICATIONS. *Educ:* Univ Calif, Riverside, BA, 61, MA, 63, PhD(math), 65. *Prof Exp:* Res mathematician, US Naval Weapons Ctr, China Lake, Calif, 64-67; mem tech staff, 67-73, SUPVR, TRAFFIC RES GROUP, BELL TEL LABS, 73- *Mem:* Am Math Soc; Oper Res Soc. *Res:* Optimal design strategy for stochastic networks. *Mailing Add:* Bell Tel Labs 185 Monmouth Pkwy Rm 2B- 115 West Longbranch NJ 07764

NEAL, THOMAS EDWARD, b Royal Oak, Mich, May 2, 42; m 67; c 3. ANALYTICAL CHEMISTRY, TEXTILE CHEMISTRY. *Educ:* Univ Mich, BS, 64; Univ NC, PhD(analytical chem), 70. *Prof Exp:* Res asst inorg chem, Univ Mich, 64-65; res chemist, 70-73, sr res chemist, Textile Fibers Dept, 73-78, mkt develop rep, 78-80, TECH SERV SUPVR, 80-, MGR, KEVLAR MKT, E I DU PONT DE NEMOURS & CO, INC, 88- *Concurrent Pos:* Qual mgr. *Res:* Electroanalytical chemistry in non-aqueous systems; end use research of synthetic polymer products in textile applications; high modulus fiber applications in rigid structures. *Mailing Add:* DuPont Fibers Chestnut Run Plaza Laurel Run Wilmington DE 19880-0705

NEAL, VICTOR THOMAS, b Dell Rapids, SDak, Nov 1, 24; m 48; c 2. PHYSICAL OCEANOGRAPHY. *Educ:* Univ Notre Dame, BS, 48; Univ NDak, MEd, 54; Ore State Univ, PhD(phys oceanog), 65. *Prof Exp:* Geophysicist, Carter Oil Co, 48-49; teacher, Various Sec Schs & Jr Cols, 50-62; instr phys oceanog, Ore State Univ, 64; asst prof, US Naval Postgrad Sch, 64-66; from asst prof to assoc prof oceanog, Ore State Univ, 66-89, dir, Latin Am Oceanog Prog & Marine Resource Mgt Prog, 70-89; RETIRED. *Mem:* Am Geophys Union; Coastal Soc; Sigma Xi. *Res:* Estuarine, coastal and arctic oceanography. *Mailing Add:* 6902 Fawn Ridge Dr NW Albany OR 97321

NEAL, WILLIAM JOSEPH, b Princeton, Ind, Nov 19, 39; m 59; c 3. SEDIMENTARY PETROLOGY, COASTAL GEOLOGY. *Educ:* Univ Notre Dame, BS, 61; Univ Mo, MA, 64, PhD(geol), 68. *Prof Exp:* Fel geol, McMaster Univ, 67-68; asst prof, Ga Southern Col, 69-71; from asst prof to assoc prof, 71-79, CHMN, DEPT GEOL, GRAND VALLEY STATE UNIV, 75-79, 88-, PROF, 79- *Concurrent Pos:* Adj prof, Skidaway Inst Oceanog, Ga, 69-71; fel, Duke Univ 76-77, vis scientist, 80-81, 91; Luso-Am fel, Geol Surv Portugal, 88. *Mem:* Am Asn Petrol Geol; Soc Econ Paleont & Mineral; Int Asn Sedimentologists; Nat Asn Geol Teachers; Sigma Xi. *Res:* Heavy minerals; recent and Pleistocene deep-sea sediments; coastal hazards; carbonate petrology; Pennsylvanian cyclothems; ancient turbidites; barrier islands. *Mailing Add:* Dept Geol Grand Valley State Univ Allendale MI 49401

NEALE, ELAINE ANNE, b Philadelphia, Pa, May 20, 44; m 67; c 2. ELECTRON MICROSCOPY, NEUROCYTOLOGY. *Educ:* Rosemont Col, AB, 65; Georgetown Univ, PhD(biol), 69. *Prof Exp:* Asst res neuromorphologist, Ment Health Res Inst, Univ Mich, Ann Arbor, 70-75; NEUROCYTOLOGIST, BEHAV BIOL BR, NEUROBIOL SECT, NAT INST CHILD HEALTH & HUMAN DEVELOP, NIH, 75- *Concurrent Pos:* NIH staff fel, Nat Cancer Inst, 69-70 & Nat Inst Child Health & Human Develop, 73-; NIH fel, Univ Mich, Ann Arbor, 70-73. *Mem:* Am Soc Cell Biol; Am Asn Women Sci; Sigma Xi. *Res:* Structure-function relationships in the nervous system; techniques for the ultrastructural localization of specific macromolecules; ultrastructural anatomy. *Mailing Add:* NIH Bldg 36 Rm 2A-21 Bethesda MD 20892

NEALE, ERNEST RICHARD WARD, b Montreal, Que, July 3, 23; m 50; c 2. GEOLOGY. *Educ:* McGill Univ, BSc, 49; Yale Univ, MS, 50, PhD(geol), 52. *Hon Degrees:* LLD, Univ Calgary, 77, DSc, Mem Univ, 89. *Prof Exp:* Asst, Yale Univ, 51-52; asst prof, Rochester Univ, 52-54; geologist, Geol Surv Can, 54-60 & 65-67, head Pre-Cambrian shield sect, 67-68; prof geol & head dept, Mem Univ Nfld, 68-76; head geol info, Inst Sedimentary & Petrol Geol, 76-81; acad vpres, Mem Univ Nfld, 82-87, CONSULT, 88- *Concurrent Pos:* Field geologist, Que Dept Mines, 47-53; actg head, Appalachian Sect, Geol Surv Can 59-60, head, 60-62; Brit Commonwealth Geol Liaison Off, Eng, 63-65; ed, Can J of Earth Sci, 74-; vis prof, Univ BC, 74-75; adj prof, Univ Calagary, 76-81; chmn Nat Bd, Sci publ, 82-87; nat lectr, Sigma Xi, 76-77; chmn Nat Bd, Sci Rev Comm Biol & Chem Defence, 90-; off, Order of Can, 90. *Honors & Awards:* Bancroft Award, Royal Soc Can, 75; Queen's Jubilee Medal, 77; Ambrose Medal, Geol Asn Can, 86. *Mem:* Am Geol Soc; Royal Soc Can; Geol Asn Can (pres, 72); Mineral Asn Can; Can Geosci Coun, (pres, 75-76); Asn Earth Sci Ed. *Res:* Appalachian geology and mineral resources; Canadian science policy; editorial policies and practices. *Mailing Add:* 5108 Carney Rd NW Calgary AB T2L 1G2 Can

NEALE, ROBERT S, b Abington, Pa, Mar 19, 36; m 57; c 3. ORGANIC CHEMISTRY. *Educ:* Amherst Col, AB, 57; Univ Ill, PhD(org chem), 61. *Prof Exp:* Org chemist, Union Carbide Res Inst, 60-67, RES SCIENTIST, UNION CARBIDE CHEM & PLASTICS, 67- *Mem:* Am Chem Soc. *Res:* Chemistry of nitrogen free radicals, especially those derived from N-Halo compounds; chemistry of hydroperoxide oxidations; synthesis of organosilicon compounds; organosilicon chemistry. *Mailing Add:* 2499 Carrington Ct Wilmington NC 28409

NEALE, WILLIAM MCC(ORMICK), JR, system analysis, management information systems, for more information see previous edition

NEALEY, RICHARD H, b Lawrence, Mass, May 30, 36; m 60; c 4. ORGANIC CHEMISTRY. *Educ:* Merrimack Col, BSc, 57; Univ Conn, MSc, 59; Brown Univ, PhD(chem), 63. *Prof Exp:* Res chemist, Ethyl Corp, Mich, 62-63; sr res chemist, Monsanto Res Corp, 63-68; mgr org chem res, Tech Opers Inc, 68-69; TECH MFG SPEC, ADVAN MFG & ENGþN AM MFG DIV, XEROX CORP, 69- *Res:* Organometallic chemistry; heterocyclic synthesis; photo-sensitizing dyes; solvent recovery; manufacturing research; synthesis of novel imaging materials. *Mailing Add:* 59 Coachman Dr Rochester NY 14526

NEALON, THOMAS F, JR, b Jessup, Pa, Feb 24, 20; m 46; c 4. SURGERY, THORACIC SURGERY. *Educ:* Scranton Univ, BS, 41; Jefferson Med Col, MD, 44. *Prof Exp:* Am Cancer Soc fel surg, Jefferson Med Col, 51-53; from instr to prof, 55-68; PROF SURG, NY UNIV, 68-; DIR SURG, ST VINCENT'S HOSP & MED CTR, NEW YORK, 68- *Concurrent Pos:* Consult, Greenwich Hosp, Conn, Holy Name Hosp, Teaneck, NJ, St Agnes Hosp, White Plains, NY & St Vincent's Med Ctr of Richmond, Staten Island. *Mem:* Am Col Chest Physicians; Am Surg Asn; Am Asn Thoracic Surg; Am Col Surg; Am Soc Artificial Internal Organs. *Res:* Cancer; cardiorespiratory physiology during operations; gastrointestinal surgery. *Mailing Add:* St Vincents Hosp Med Ctr 170 W 11th St New York NY 10011

NEALSON, KENNETH HENRY, b Iowa City, Iowa, Oct 8, 43. MARINE MICROBIOLOGY. *Educ:* Univ Chicago, BS, 65, MS, 66, PhD(microbiol), 69. *Prof Exp:* NIH fel, Harvard Univ, 69-71; asst prof biol, Univ Mass, Boston, 71-73; ASST PROF MARINE BIOL, SCRIPPS INST OCEANOG, 73- *Mem:* AAAS; Am Soc Microbiol. *Res:* Physiology, biochemistry and genetics of luminous bacteria; study of the symbiotic relationship between luminous bacteria and marine luminous fishes; physiology and biochemistry of manganese oxidizing bacteria. *Mailing Add:* Dept Biol Sci Univ Wis PO Box 413 Milwaukee WI 53201

NEALY, CARSON LOUIS, b Natchitoches, La, Dec 24, 38;; m 66; c 1. ANALYTICAL CHEMISTRY. *Educ:* Northwestern State Col, La, BS, 60; Fla State Univ, MS, 63, PhD(nuclear & inorg chem), 65. *Prof Exp:* Nuclear chemist, Shell Develop Co, Houston, 65-70, anal chemist, Shell Oil Co, 70-72; mgr anal chem energy syst group, Rockwell Int Corp, Rocketdyne, 72-84, sr scientist 84-87; tech asst to the dir, 87-88, DEPUTY DIR, NEW BRUNSWICK LAB, US DEPT ENERGY, CHICAGO OPERS OFF, 88- *Mem:* Am Chem Soc; Sigma Xi. *Res:* Nuclear reaction spectroscopy in nuclear structure studies; neutron activation analysis; gas chromatography; instrument development; nuclear fuel analysis; nuclear methods in analytical chemistry; combustion analysis; environmental chemistry; geochemistry. *Mailing Add:* 220 Walker Ave Clarendon Hills IL 60514

NEALY, DAVID LEWIS, b Monticello, NY, June 29, 36; m 62; c 2. ORGANIC CHEMISTRY. *Educ:* Duke Univ, BS, 58; Cornell Univ, PhD(org chem), 63. *Prof Exp:* NSF fel org chem, Mass Inst Technol, 63-64; from res chemist to sr res chemist, Tenn Eastman Co, 64-69, res assoc, 70, head phys & analytical chem div, 71-73, supt, Fiber Develop Div, 74-76, mkt staff, 76-77, asst supt, Organic Chem Div, 78-79, dir, chem res div, 79-87, DIR, SPECIALTY CHEM BUSINESS TEAMS, EASTMAN CHEM DIV, EASTMAN KODAK CO, 88- *Mem:* Am Chem Soc. *Res:* Organic polymer chemistry; synthetic Organic chemistry; chemical kinetics. *Mailing Add:* 102 Crown Colony Kingsport TN 37660-9765

NEARY, JOSEPH THOMAS, b Carbondale, Pa, Oct 14, 43; m 67; c 2. PROTEIN PHOSPHORYLATION, CALCIUM HOMEOSTASIS. *Educ:* Univ Scranton, BS, 65; Univ Pittsburgh, PhD(biochem), 69. *Prof Exp:* Postdoctoral fel, Univ Ill, 69-71; res fel med, Harvard Med Sch & Mass Gen Hosp, 71-74, asst biochem, 74-78; biochemist, Marine Biol Lab, 78-85; res asst prof, 85-87, RES ASSOC PROF, VET ADMIN MED CTR, UNIV MIAMI, 87- *Mem:* Am Soc Biochem & Molecular Biol; Soc Neurosci; Am Soc Neurochem; Int Soc Neurochem; AAAS; Sigma Xi. *Res:* Signal transduction mechanisms in the brain; molecular basis of neurological disorders. *Mailing Add:* Res Serv 151 Vet Admin Med Ctr 1021 NW 16th St Miami FL 33125

NEARY, MICHAEL PAUL, analytical chemistry, for more information see previous edition

NEAS, ROBERT EDWIN, b Sheldon, Mo, May 7, 35; m 57; c 4. ANALYTICAL CHEMISTRY. *Educ:* Cent Methodist Col, AB, 57; Southern Ill Univ, Carbondale, MS, 65; Univ Mo-Columbia, PhD(chem), 70. *Prof Exp:* Analyst chem, Mallinckrodt Chem Works, 57-58, chemist, 58-59, supvr, 59-61, asst to dir qual control, 61-62, res asst chem, 62-63; instr, Univ Mo-Mo Exp Sta Lab, 65-66; from asst prof to assoc prof chem, 69-85. *Concurrent Pos:* Univ res coun grant, Western Ill Univ, 70-71; Environ Impact Study Group, Argonne Nat Lab, 76-78; Lake Study Group, Ill Environ Protection Agency, 79; Drinking Water Study, Ill Energy Res Corp, 85; vis assoc prof chem, Inst Environ Studies, Univ Ill, 75. *Mem:* Am Chem Soc. *Res:* Envionmental analytical chemistry; trace analysis of waters; spectrophotometry of complex ions; ion selective electrodes; analytical chemistry studies of natural waters and drinking waters being carried out in attempts to correlate seasonal variations in mutagenicity with chemical composition. *Mailing Add:* Western Ill Univ 519A Currens Hall Macomb IL 61455-1396

NEASE, ROBERT F, b Walters, Okla, June 4, 31. ELECTRICAL ENGINEERING. *Educ:* Tex Tech Col, BS, 51; Mass Inst Technol, SM, 53, ScD, 57. *Prof Exp:* Asst, Mass Inst Technol, 51-52 & 54-55, instr, 56; sr res engr & supvr autonetics, NAm Aviation, Inc, 57-60; mgr systs integration dept, aeronutronic, Ford Motor Co, 60-63; SR TECH STAFF & CHIEF SCIENTIST, AUTONETICS DIV, ROCKWELL INT, 63- *Res:* Autonatic control; avionics and space systems analysis. *Mailing Add:* 701 Lamark Dr Anaheim CA 92802

NEATHERY, MILTON WHITE, b Chapel Hill, Tenn, Apr 15, 28; m 50; c 1. ANIMAL NUTRITION. *Educ:* Univ Tenn, BS, 50, MS, 55; Univ Ga, PhD(animal nutrit), 73. *Prof Exp:* Farm mgr dairy, Minglewood Farm, Tenn, 50-51; fieldman dairy, Nashville Pure Milk Co, 53-54; asst dairy, Univ Tenn, 55-56; asst animal husbandman animal nutrit, Ga Mountain Exp Sta, 56-59; from asst prof to assoc prof, 59-86, PROF DAIRY SCI & ANIMAL NUTRIT, UNIV GA, 86- *Honors & Awards:* Gustav Bohstedt Award, 80- *Mem:* Sigma Xi; Am Dairy Sci Asn; fel Am Soc Animal Sci. *Res:* Mineral metabolism in animals; primarily trace mineral metabolism in ruminants using radioisotopes. *Mailing Add:* Animal & Dairy Sci Dept Univ Ga Livestock-Poultry Bldg Athens GA 30602

NEATHERY, RAYMOND FRANKLIN, b Conroe, Tex, Aug 31, 39; m 62; c 3. BIOMECHANICS,ENGINEERING TECHNOLOGY. *Educ:* John Brown Univ, BS, 61; NMex State Univ, MSME, 64; Univ Ark, Fayetteville, PhD(eng sci), 70. *Prof Exp:* Asst prof mech eng, John Brown Univ, 64-67; assoc prof, LeTourneau Col, 67-71; sr res engr, Biomed Sci Dept, Gen Motors Res Labs, 71-76; assoc prof & head, 76-80, chmn dept, 80-85, PROF MECH DESIGN TECHNOL, OKLA STATE UNIV, 80- *Concurrent Pos:* NASA-Am Soc Eng Educ Systs Design Inst fel, NASA Manned Spacecraft Ctr, 70, consult mach design, biomech, accident reconstruct & safety. *Honors & Awards:* NSF Fac Fel. *Mem:* Am Soc Mech Engrs; Am Soc Eng Educ; Soc Automotive Engrs. *Res:* Solid mechanics; mechanical modeling of humans for crash testing; machine design. *Mailing Add:* Mech Design Technol Okla State Univ Stillwater OK 74078

NEATHERY, THORNTON LEE, b Atlanta, Ga, Mar 12, 31; m 56; c 3. GEOLOGY. *Educ:* Univ Ala, BS, 56, MS, 64. *Prof Exp:* Geologist, Reynolds Metals Co, 56-62; asst vpres, Textile Rubber & Chem Co, 63; geologist, Geol Surv Ala, 64-73, chief geologist, 73-76, asst state geologist & dir budget & res devel, 76-83, sr geologist, 83-87; PRIN GEOLOGIST/OWNER, NEATHERY & ASSOCS, 86- *Concurrent Pos:* Consult geologist. *Mem:* Sigma Xi; fel Geol Soc Am; Soc Econ Geologist; Mineral Soc Am; Soc Mining Engrs. *Res:* Regional geologic mapping in southern Piedmont and folded Appalachians with emphasis on sedimentation, metamorphism, and structural evolution as applied to distribution of ore deposits. *Mailing Add:* 3033 Firethorn Dr Tuscaloosa AL 35405

NEAVEL, RICHARD CHARLES, b Philadelphia, Pa, Oct 21, 31; m 58; c 3. FUEL SCIENCE. *Educ:* Temple Univ, BA, 54; Pa State Univ, MS, 57, PhD(geol), 66. *Prof Exp:* Coal petrologist, Ind Geol Surv, 57-61; res assoc, Dept Geol, Pa State Univ, 61-66; staff geologist, Humble Oil & Refining Co, 67; geologist, Synthetic Fuel Lab, 68-69, group leader, 70, res assoc, Gasification Lab, 71-75, sr res assoc, 75-79, SCI ADV, COAL RES LAB, EXXON RES & ENG CO, 79- *Concurrent Pos:* Consult, Inst Gas Technol, 59-63; chmn, fuels sci, Gordon Res Conf, 75 & Fuel Chem Div, Am Chem Soc, 81. *Honors & Awards:* R A Glenn Award, Am Chem Soc, 79 & Storch Award, 80. *Mem:* Am Chem Soc; AAAS; Sigma Xi; Am Soc Testing & Mat. *Res:* Characterization of coals and relationships between coal properties and utilization. *Mailing Add:* Exxon Res & Eng Co PO Box 4255 Baytown TX 77522

NEAVES, WILLIAM BARLOW, b Spur, Tex, Dec 25, 43; m 65; c 2. ANDROLOGY, CELL BIOLOGY. *Educ:* Harvard Univ, AB, 66, PhD(anat), 69. *Prof Exp:* Lectr vet anat, Univ Nairobi, 70-71; lectr anat, Med Sch, Harvard Univ, 72; asst prof, 72-74, assoc prof & dir anat, 74-77, assoc dean grad sch, 77-80, PROF CELL BIOL, UNIV TEX HEALTH SCI CTR DALLAS, 77-, DEAN, GRAD SCH, 80- *Concurrent Pos:* Rockefeller Found fel anat, Harvard Univ & Univ Nairobi, 70-71; res assoc, Los Angeles County Mus, 70-73; consult, Ford Found, 73-74; assoc ed, Anat Record, 75-; vis prof, Univ Nairobi, 78. *Honors & Awards:* Young Andologist Award, Am Soc Andrology, 83. *Mem:* AAAS; Am Asn Anat; Soc Study Reproduction; Am Soc Andrology. *Res:* Reproductive biology; spermatogenesis; testicular endocrinology; contraception. *Mailing Add:* Exec Vpres Academic Affairs Univ Tex SW Med Ctr 5323 Harry Hines Blvd Dallas TX 75235

NEBEKER, ALAN V, b Salt Lake City, Utah, Apr 8, 38; m 60; c 3. ENTOMOLOGY, AQUATIC ECOLOGY. *Educ:* Univ Utah, BS, 61, MS, 63, PhD(zool), 66. *Prof Exp:* Res aquatic biologist entom, Nat Water Qual Lab, 66-71, RES AQUATIC BIOLOGIST, ENVIRON PROTECTION AGENCY, ENTOM & FISHERIES, WESTERN FISH TOXICOL LAB, 71- *Mem:* Entom Soc Am; Am Entom Soc; Am Fisheries Soc. *Res:* Water pollution toxicology; systematics of aquatic insects; water quality criteria for protection of aquatic life; bioassay analysis. *Mailing Add:* 2922 NW Angelica Pl Corvallis OR 97330

NEBEKER, THOMAS EVAN, b Richfield, Utah, May 10, 45; m 64; c 3. FOREST ENTOMOLOGY. *Educ:* Col Southern Utah, BS, 67; Utah State Univ, MS, 70; Ore State Univ, PhD(entom), 74. *Prof Exp:* Teaching asst zool & entom, Utah State Univ, 67-70; NSF trainee pest pop ecol, Ore State Univ, 70-73; fel pop ecol, Utah State Univ, 73-74; from asst prof to assoc prof, 74-83, PROF FOREST ENTOM, MISS STATE UNIV, 83- *Mem:* Entom Soc Am; Sigma Xi; Can Entom Soc. *Res:* Population biology of forest insects with emphasis on the dynamics of southern pine beetle populations; parasite and predator efficiency studies utlizing behavior patterns; host/pest/microorganisms interactions. *Mailing Add:* Dept of Entom Miss State Univ PO Box 5328 Mississippi State MS 39762

NEBEL, CARL WALTER, b Dover, NJ, July 25, 37; m 60; c 1. ORGANIC CHEMISTRY, ENVIRONMENTAL SCIENCE. *Educ:* Tusculum Col, BS, 58; Cornell Univ, MS, 61; Univ Del, PhD(org chem), 65. *Prof Exp:* Von Humboldt fel, Govt Ger, Univ Karlsruhe, 68; asst prof chem, Univ Del, 68-70; asst gen mgr, Welsbach Corp, 70-76; vpres & chief operating officer ozone technol, PCI Ozone Corp, 76-88. *Concurrent Pos:* Postdoctoral fel, Univ Karlsruhe, 68. *Mem:* Am Chem Soc; Water Pollution Control Fedn. *Res:* Organic ozone reactions; application of ozone to air and water purification. *Mailing Add:* 45 Dogwood Terr Millington NJ 07946

NEBENZAHL, LINDA LEVINE, b Duluth, Minn, Oct 4, 49; m 71; c 2. PHYSICAL ORGANIC CHEMISTRY, SURFACE CHEMISTRY. *Educ:* Univ Minn, Minneapolis, BA, 71; Univ Calif, Berkeley, PhD(org chem), 75. *Prof Exp:* STAFF CHEMIST, 75-, SR ENGR/PROJ MGR, THIN FILM HEAD MFG, IBM CORP, 87; RETIRED. *Concurrent Pos:* Adv engr, IBM Corp, 81. *Mem:* Am Chem Soc; Sigma Xi; Inst Elec & Electronics Engrs. *Res:* Light scattering of surfactants and polymer solutions; fluid flow through porous media. *Mailing Add:* 1196 Stafford Dr Cupertino CA 95014

NEBERT, DANIEL WALTER, b Portland, Ore, Sept 26, 38; m 60, 81; c 6. PEDIATRICS. *Educ:* Univ Ore, BA, 61, MS & MD, 64. *Prof Exp:* From intern to resident pediat, Ctr Health Sci, Univ Calif, Los Angeles, 64-66; res assoc biochem, Nat Cancer Inst, 66-68; res investr pharmacol, Sect Develop Enzym, Nat Inst Child Health & Human Develop, NIH, 68-70, head sect develop pharmacol, Lab Biomed Sci, 70-74, chief Neonatal & Pediat Med Br, 74-75, chief, Lab Develop Pharmacol, 75-89 PROF & HEAD, MOLECULAR TOXICOL LAB, DEPT ENVIRON HEALTH & PROF PEDIAT, SCH MED, UNIV CINCINNATI, OHIO, 89- *Concurrent Pos:* Adj prof, US Uniformed Health Sci Med Sch, Bethesda, 80-89; Wellcome vis prof, Biochem & Molecular Bio Div, Fed Am Soc Exp Biol, 90. *Honors & Awards:* Frank Ayrey Fel Clin Pharmacol, Eng, 84; Bernard B Brodie Award, 86; Ernst A Sommer Mem Lectr, Ore Health Sci Univ, Portland, 88. *Mem:* AAAS; Am Soc Pharmacol & Exp Therapeut; Am Soc Biol Chemists; Sigma Xi; Am Soc Clin Invest; Am Soc Pediat Res; Am Soc Microbiol; Genetics Soc Am; Endocrine Soc; Soc Toxicol. *Res:* Application of recombinant DNA technology, mammalian cell culture, and inbred animal strains to molecular genetics, clinical pharmacology, toxicology, cancer research and pharmacogenetic disorders; teratology; developmental biology. *Mailing Add:* Dept Environ Health Univ Cincinnati Med Ctr Cincinnati OH 45267-0056

NEBGEN, JOHN WILLIAM, b Independence, Mo, May 20, 3S; div; c 2. PHYSICAL CHEMISTRY, INORGANIC CHEMISTRY. *Educ:* Washington Univ, AB, 56; Univ Pa, PhD(chem), 60. *Prof Exp:* Assoc chemist, Midwest Res Inst 60-65, sr chemist, 65-68, prin chemist, 68-78, sr adv chem, 78-82; vis lectr, Ohio Wesleyan Univ, 82-83; asst prof, 83-88, ASSOC PROF CHEM, EUREKA COL, 88- *Mem:* Am Chem Soc; Water Pollution Control Fedn. *Res:* Molecular structure; infrared, nuclear magnetic resonance and electron spin resonance spectroscopy; inorganic synthesis; portland cement manufacture; wastewater treatment; water pollution abatement; environmental systems analysis; water desalting and reuse; waste utilization; corrosion chemistry. *Mailing Add:* Eureka Col PO Box 91 Eureka IL 61530

NEBIKER, JOHN HERBERT, b Eastport, Maine, May 26, 36; m 62; c 2. ENVIRONMENTAL ENGINEERING. *Educ:* Mass Inst Technol, SB, 58;. *Hon Degrees:* DSc, Swiss Fed Inst Techol, 66. *Prof Exp:* Asst prof sanit & water resources eng, Vanderbilt Univ, 65-67; asst prof civil eng, Univ Mass, 67-69; vpres, Curran Assocs, Inc, 69-72; sanit engr, WHO, 72-75; mgr environ planning, Malcolm Pirnie, Inc, 75-79; SANIT ENGR, WORLD BANK, 79- *Mem:* Am Soc Civil Engrs; Am Water Works Asn; NY Acad Sci. *Res:* Physical processes in sanitary engineering; public health engineering. *Mailing Add:* World Bank Group 66 Avenue D'Lena Paris 75116 France

NEBOLSINE, PETER EUGENE, b Annapolis, Md, Apr 6, 45; m 67; c 2. PHYSICS, OPTICS. *Educ:* Lafayette Col, BS, 67; Univ Rochester, PhD(optics), 72. *Prof Exp:* Sr scientist physics, Avco Everett Res Lab, 72-73; prin scientist physics, 74-80, MGR EXP RES, PHYS SCI INC, 80- *Mem:* Optical Soc Am. *Res:* Laser applications especially laser propulsion, material interaction and sound generation; hypervelocity impact experimentalist. *Mailing Add:* Phys Sci Inc Dascomb Res Park PO Box 3100 Andover MA 01810

NECE, RONALD ELLIOTT, b Seattle, Wash, May 25, 27; m 51; c 2. CIVIL ENGINEERING. *Educ:* Univ Wash, BS, 49; Lehigh Univ, MS, 51; Mass Inst Technol, ScD(civil eng), 58. *Prof Exp:* Instr civil eng, Rutgers Univ, 51-52; instr fluid mech & civil eng, Mass Inst Technol, 52-56, asst prof hydraul & civil eng, 56-59; from asst prof to assoc prof civil eng, 59-67, PROF CIVIL ENG, UNIV WASH, 67- *Mem:* Am Soc Civil Engrs; Am Soc Mech Engrs; Int Asn Hydraul Res. *Res:* Fluid mechanics; hydraulic engineering; hydrodynamics in natural bodies of water; coastal engineering. *Mailing Add:* Dept Civil Eng Univ Wash Seattle WA 98195

NECHAY, BOHDAN ROMAN, b Prague, Czech; nat US; m 88; c 2. PHARMACOLOGY, THERAPEUTICS. *Educ:* Univ Minn, DVM, 53. *Prof Exp:* Pvt practr, Minn, 53-56; asst prof pharmacol, Col Med, Univ Fla, 61-66; asst prof pharmacol & urol, Med Ctr, Duke Univ, 66-68; assoc prof, Univ Tex Med Br Galveston, 68-78, prof pharmacol & toxicol, 78-90; RETIRED. *Concurrent Pos:* Fel pharmacol, Col Med, Univ Fla, 56-60; Am Heart Asn fel, 58-60; NIH fel pharmacol, Univ Uppsala, 60-61; vis mem grad fac, Tex A&M Univ, 81-; pres, metal's specialty Sect, Soc Toxicol, 85-86; Fac fel, NASA, 88 & 89. *Mem:* Soc Toxicol; Am Soc Nephrology; Am Soc Pharmacol & Exp Therapeut. *Res:* Electrolyte physiology and pharmacology; environmental occupational toxicology; kidney and cardiovascular system. *Mailing Add:* Dept of Pharmacol & Toxicol Univ of Tex Med Br 231 Galveston TX 77550

NECKERS, DOUGLAS, b Corry, Pa, Aug 15, 38; m 60; c 2. ORGANIC CHEMISTRY, PHOTOCHEMISTRY. *Educ:* Hope Col, AB, 60; Univ Kans, PhD(org chem), 63. *Prof Exp:* From asst prof to assoc prof chem, Hope Col, 64-71; assoc prof, Univ NMex, 71-73; PROF CHEM & CHMN DEPT, 73-, DISTINGUISHED RES PROF, BOWLING GREEN STATE UNIV, 86- *Concurrent Pos:* Fel, Harvard Univ, 63-64; vis lectr, Ohio State Univ, 65 & Univ Ill, 70; vis prof, State Univ Groningen, 68-69 & Roman Cath Univ Nijmegen, 75; Alfred P Sloan Found fel, 71; exec dir, Ctr Photochem Sci, 86- *Honors & Awards:* Leo Friend Award; Paul & Ruth Olscamp Res Award, 87; Mead Imaging Pres Award, 87; Paul Block, Jr Award, 87. *Mem:* Sigma Xi (vpres, 76-77, pres, 77-78); Am Chem Soc; Am Asn Univ Prof. *Res:* Polymer chemistry; polymer based reagents in synthesis; photopolymerization. *Mailing Add:* Dept Chem Bowling Green State Univ Bowling Green OH 43404

NEDDENRIEP, RICHARD JOE, b Leipsic, Ohio, June 3, 30; m 57; c 2. PHYSICAL CHEMISTRY. *Educ:* Miami Univ, BA, 53; Univ Wis, PhD(phys chem), 58. *Prof Exp:* Res chemist, Linde Div, Union Carbide Corp, NY, 57-65, group leader, 65-69; group leader, Betz Labs, Inc, 69-72, mgr prod res, 72-73, asst dir res, 73-74, asst vpres, 74-85; CONSULT, 85- *Mem:* Am Chem Soc. *Res:* Reaction kinetics, particularly free radical reactions and the radiolysis of organic materials; adsorption; catalysis, particularly with molecular sieves; water and air purification; corrosion and scale inhibition. *Mailing Add:* 39 Willow Brook Dr Doylestown PA 18901

NEDDERMAN, HOWARD CHARLES, physics; deceased, see previous edition for last biography

NEDELSKY, LEO, b Russia, Oct 28, 03; nat US; m 41; c 3. THEORETICAL PHYSICS. *Educ:* Univ Wash, Seattle, BS, 28; Univ Calif, MS, 31, PhD(theoret physics), 32. *Prof Exp:* Instr physics, Univ Calif, 32-35 & Hunter Col, 35-40; fel Gen Educ Bd, 40-41, dir res basic nursing educ, 54-55, prof, 41-75, EMER PROF PHYS SCI & EXAM, UNIV CHICAGO, 75- *Concurrent Pos:* Lectr, Baker & Ottawa Univs, 48, Univ Pa, 52-53, Univ Wash, Seattle, 56, Univ Mo, 58, Northern Mich Col, 58-59, Univs Colombia, El Salvador, Guatemala, Mex, Puerto Rico, Venezuela, Cuba & Brazil, 66; consult, Univ Wash, Seattle, 54-57, NY State Univ, 57, Am Bd Radiol, 58, CBS-TV, 58, Univ Ill, 59-60, Michael Reese Hosp, 60; res assoc dent educ, Univ Ill, 65-71; consult, WHO, 66-68. *Mem:* AAAS; Am Phys Soc; Am Asn Physics Teachers. *Res:* Physical sciences in general education. *Mailing Add:* 5807 Dorchester Ave, Apt 9E Chicago IL 60637

NEDICH, RONALD LEE, b Chicago, Ill, 41; c 2. PHYSICAL PHARMACY, INDUSTRIAL PHARMACY. *Educ:* St Louis Col Pharm, BS, 65; Purdue Univ, MS, 68, PhD(phys pharm), 70. *Prof Exp:* Sr res pharmacist, Baxter-Travenol Labs, Inc, 70-73, mgr pharm develop, 73-76, sr mgr pharm develop, 76-78, dir pharm develop, 78-83, dir tech develop, 83-85; dir, Pilot Plant Div, Wyeth Labs, Inc, 85-87; SR VPRES OPERS, LEMMON CO, 87- *Mem:* Acad Pharmaceut Sci; Am Pharm Asn; Am Chem Soc; Am Asn Pharmaceut Sci. *Res:* Dosage form design; physical chemical principles associated with pharmaceutical dosage forms; pharmacokinetics; process scale up and validation. *Mailing Add:* 1508 Morgan Lane Wayne PA 19087

NEDOLUHA, ALFRED K, b Vienna, Austria, Sept 13, 28; m 57; c 1. THEORETICAL SOLID STATE PHYSICS. *Educ:* Univ Vienna, PhD(physics), 51. *Prof Exp:* Staff mem, Felten & Guilleaume, A G, Austria, 51-56, head high voltage lab, 56-57; physicist, White Sands Missile Range, NMex, 57-59; Naval Ord Lab, Corona, 59-62, res physicist, Naval Electronics Lab Ctr, 62-70, res physicist, San Diego, 70-75; chief, Electronics Br, Europ Res Off, US Army, 75-79; res physicist, 79-82, head, 82-88, EMER HEAD, ELECTRONIC MAT SCI DIV, NAVAL OCEAN SYSTS CTR, SAN DIEGO, 88- *Mem:* Am Phys Soc. *Res:* Solid state theory. *Mailing Add:* Naval Ocean Syst Ctr Code T55 271 Catalina Blvd San Diego CA 92152-5000

NEDWICK, JOHN JOSEPH, b Ranshaw, Pa, Jan 11, 22; m 61; c 3. INDUSTRIAL ORGANIC CHEMISTRY, CHEMICAL ENGINEERING. *Educ:* Univ Louisville, AB, 47; Univ Pa, MS, 53. *Prof Exp:* Chemist, Rohm and Haas Co, 48-56, group leader high pressure chem, 56-63, group leader polymer chem, 63-85, prof leader, Chem Process Eng Dept, 73-85; RETIRED. *Mem:* Am Chem Soc; Am Inst Chem Engrs. *Res:* High pressure research and process development; acetylene reactions; continuous bench scale pilot plants; polymer chemistry; plastics; coatings; process development of agricultural chemicals and health products. *Mailing Add:* 133 Forrest Ave Southampton PA 18966

NEE, M COLEMAN, b Taylor, Pa, Nov 14, 17. MATHEMATICS. *Educ:* Marywood Col, AB, 39, MA, 43; Univ Notre Dame, MS, 59. *Prof Exp:* Teacher math & Latin, Marywood Sem, 43-55; asst prof math, 59-70, pres, Marywood Col, 70-88; RETIRED. *Res:* Group theory of algebra. *Mailing Add:* Marywood Col Marywood Col 2300 Adams Ave Scranton PA 18509-1598

NEE, MICHAEL WEI-KUO, b Berkeley, Calif, Apr 21, 55. METALLOPORPHYRINS. *Educ:* Univ Santa Clara, BS, 77; Calif Inst Technol, PhD(chem), 81. *Prof Exp:* NIH postdoctoral fel, Univ Calif, Santa Barbara, 81-83; ASSOC PROF ORG CHEM, OBERLIN COL, 83- *Concurrent Pos:* Vis scientist, Univ Calif, Berkeley, 90-91. *Mem:* Am Chem Soc; AAAS. *Res:* Metalloporphyrins; metal alkoxides and aryloxides; metal oxos. *Mailing Add:* Dept Chem Oberlin Col Oberlin OH 44074

NEE, VICTOR W, b Soochow, China, Apr 8, 35; m 61; c 4. FLUID MECHANICS. *Educ:* Univ Taiwan, BS, 57; Johns Hopkins Univ, PhD(fluid mech), 67. *Prof Exp:* Asst civil eng, Univ Taiwan, 57-60; asst fluid mech, Johns Hopkins Univ, 60-65; from asst prof to assoc prof, 65-73, PROF FLUID MECH, UNIV NOTRE DAME, 73- *Concurrent Pos:* Mem adv comt, Environ Protection Agency, USDA, 74-; oversea fel, Acad Sinica, Taiwan. *Mem:* Am Phys Soc. *Res:* Fluid mechanics; turbulence; turbulent boundary layers; laminar flows; air pollution; heat and mass transfer; inventor and legal expert. *Mailing Add:* Dept Aerospace & Mech Eng Univ Notre Dame Notre Dame IN 46556

NEECE, GEORGE A, b Pine Bluff, Ark, Sept 18, 39; m 62. PHYSICAL CHEMISTRY. *Educ:* Rice Univ, AB, 61; Duke Univ, PhD(phys chem), 64. *Prof Exp:* Chemist, US Army Res Off-Durham, 64-67; res assoc chem, Cornell Univ, 67-68; asst prof, Univ Ga, 68-71; dir chem prog, Off Naval Res, 72-85, chemist, 71-72; DIR, RES & TECHNOL, ARMY RES OFF, 85- *Mem:* Sigma Xi; Am Chem Soc. *Res:* Statistical mechanics. *Mailing Add:* 3102 Camelot Ct Durham NC 27705-5405

NEEDELS, THEODORE S, b Cleveland, Ohio, Apr 26, 22; m 54. NUCLEAR PHYSICS. *Educ:* Ohio State Univ, BS, 43, PhD(physics), 50. *Prof Exp:* Physicist, Los Alamos Scientific Lab, 50-55; oper anal, Res Anal Corp, 55-63; sr scientist, Booz-Allen Corp, 63-67; open anal, Mitre Corp, 67-72; scientist, EC-Systs, 72-73; PHYSICAL SCIENTIST, DEPT ENERGY, WASHINGTON, DC, 73- *Mem:* Am Phys Soc. *Res:* Transportation and environmental specialist in radioactive materials. *Mailing Add:* 4304 47th St NW Washington DC 20016

NEEDHAM, CHARLES D, b Chicago, Ill, Sept 17, 37. PHYSICAL CHEMISTRY. *Educ:* Carnegie Inst Technol, BS, 59; Univ Minn, PhD(chem), 65. *Prof Exp:* Staff fel phys biol, NIH, 65-67; ADV CHEMIST, GEN TECHNOL DIV, IBM CORP, 67- *Mem:* Am Chem Soc. *Res:* Raman spectroscopy and optical analysis techniques. *Mailing Add:* IBM Corp Zip 56A Rte 52 Hopewell Jct NY 12533-6531

NEEDHAM, GERALD MORTON, b Caldwell, Idaho, Aug 4, 17; m 42; c 2. BACTERIOLOGY. *Educ:* Col Idaho, BS, 40; Univ Minn, PhD(bact), 47; Am Bd Med Microbiol, dipl. *Prof Exp:* Bacteriologist, State Dept Health, Minn, 41-42; instr bact, Med Sch, Univ Minn, 41-46; med bacteriologist, Mayo Clin, 46-68; assoc dir div educ, Mayo Sch Health Rel Sci, Mayo Found, 68-80, assoc dean student affairs, Mayo Med Sch, 71-77, asst dean, 77-82; Consult, 82- *Concurrent Pos:* Consult bacteriologist, Econ Labs, St Paul, 41-46. *Mem:* AAAS; Am Soc Microbiol; fel Am Acad Microbiol; fel Am Pub Health Asn. *Res:* Medical bacteriology; antibiotics; tuberculosis; action of a few antibacterial substances on resting cells. *Mailing Add:* 806 14th Ave SW Rochester MN 55902

NEEDHAM, GLEN RAY, b Lamar, Colo, Dec 25, 51; m 83; c 3. ACARINE PHYSIOLOGY, ENTOMOLOGY. *Educ:* Southwestern Okla State Univ, BS, 73; Okla State Univ, MS, 75, PhD(entom), 78. *Prof Exp:* Asst prof, 78-84, ASSOC PROF ENTOM, OHIO STATE UNIV, 84- *Concurrent Pos:* consult, Nat Inst Health. *Mem:* Entom Soc Am; Acarol Soc Am; AAAS; Sigma Xi; Soc Vector Ecologists. *Res:* Disease transmission by ticks; off-host physiology of ticks; coldhardiness by ticks; host-parasite interactions and host immunity to ticks; biology & control of honey-bee mites. *Mailing Add:* Dept Entomol Acarol Lab 484 W 12th Ave Columbus OH 43210-1292

NEEDHAM, THOMAS E, JR, b Newton, NJ, Apr 12, 42. PHARMACEUTICS. *Educ:* Univ RI, BS, 65, MS, 67, PhD(pharmaceut sci), 70. *Prof Exp:* Pharmacist, Galen Drug, Inc, 65-68; instr pharm, Univ RI, 67-69; pharmacist, Pinault Drug, Inc, 69-70; from asst prof to assoc prof, Sch Pharm, Univ Ga, 70-79; assoc dir pharmaceut develop, Travenol Labs, 79-82, dir drug delivery, 82-86 & new prod develop, 86-87; sect head, Drug Delivery & Line Exten, G D Searle & Co. *Mem:* Am Pharmaceut Asn; Acad Pharmaceut Sci; AAAS; Am Asn Pharmaceut Scientists. *Res:* Development and evaluation of new drug delivery products; effects of selected variables on the bioavailability of drugs in these systems. *Mailing Add:* 175 Pine Hill Rd Wakefield RI 02879

NEEDLEMAN, ALAN, b Philadelphia, Pa, Sept 2, 44; m 70; c 2. COMPUTATIONAL MECHANICS, FRACTURE MECHANICS. *Educ:* Univ Pa, BS, 66; Harvard Univ, MS, 67, PhD(eng), 70. *Prof Exp:* Instr appl math, Mass Inst Technol, 70-72, asst prof, 72-75; from asst prof to assoc prof eng, 75-81, dean eng, 88-91, PROF ENG, BROWN UNIV, 81- *Concurrent Pos:* mem, NSF Adv Comt Mech & Struct Systs, 90- *Mem:* Fel John Simon Guggenheim Found; fel Am Soc Mech Engrs. *Res:* Computational micromechanics of plastic flow and fracture in metals and metal-matrix composites; ductile fracture mechanics; finite element methods; plastic instability phenomena. *Mailing Add:* Div Eng Brown Univ Providence RI 02912

NEEDLEMAN, HERBERT L, PSYCHIATRY, PEDIATRICS. *Educ:* Univ Pa, MD, 52; Am Bd Pediat, dipl; Am Bd Psychiat & Neurol, dipl. *Prof Exp:* Assoc prof pediat, Harvard Med Sch; dir, Behav Sci Div, Children's Hosp, Pittsburgh; DIR, LEAD STUDIES, SCH MED, UNIV PITTSBURGH & PROF PSYCHIAT & PEDIAT. *Concurrent Pos:* Consult various US govt agencies & more than fifty invited lectrs at int confs. *Honors & Awards:* Sarah Poiley Medal, NY Acad Sci; Charles A Dana Award. *Mem:* Inst Med-Nat Acad Sci; Am Acad Pediat; Am Acad Child & Adolescent Psychiat; Am Pediat Soc; AAAS; Sigma Xi. *Res:* Effects of lead on higher order social behavior; author of over 80 publications. *Mailing Add:* Sch Med Univ Pittsburgh Pittsburgh PA 15260-0001

NEEDLEMAN, PHILIP, b Brooklyn, NY, Feb 10, 39. PROSTAGLANDINS, CARDIOVASCULAR PHARMACOLOGY. *Educ:* Philadelphia Col Pharm & Sci, BSc, 60, MSc, 62; Univ Md Med School, PhD(pharmacol), 64. *Prof Exp:* Fel, 65-67, from asst prof to prof, 67-75, PROF, WASH UNIV, SCH MED, ST LOUIS, 75-, DEPT PHARMACOL, 76-; CORP VPRES RES & DEVELOP & CHIEF SCIENTIST, 91- *Concurrent Pos:* Investr, Am Heart Asn, 68-73; Study Sect & Task Force, Nat Inst Health, 78. *Honors & Awards:* John Jacob Abel Award, Am Pharmacol Soc, 74; Res Career Develop Award, Nat Inst Health, 74, 76; Wellcome Creesy Award in Clin Pharmacol, 77, 78, 80, 87; Cochems Thrombosis Res Prize, 80. *Mem:* Nat Acad Sci; Am Physiol Soc; Am Pharmacol Soc; Brit Pharmacol Soc; Am Biol Chem Soc. *Res:* Author numerous articles in various journals. *Mailing Add:* 326 New Salem Creve Coeur MO 63141

NEEDLEMAN, SAUL BEN, b Chicago, Ill, Sept 25, 27; m 54; c 4. BIOCHEMISTRY, NEUROCHEMISTRY. *Educ:* Ill Inst Technol, BS, 50, MS, 55; Northwestern Univ, PhD(med & biochem), 57. *Prof Exp:* Res assoc biochem, Col Med, Univ Ill, 52-53; asst chem, Ill Inst Technol, 53-55; asst chem, Northwestern Univ, 55-57, from instr to asst prof neurobiochem, 57-72; prof biochem & head dept, Roosevelt Univ, 72-74; clin diag specialist, Abbott Labs, 74-76, coordr clin res, 76-79; dir clin affairs, Consumers Prod Div, Schering-Plough, 79-81; DIR MED AFFAIRS, HOLLISTER INC, 81- *Concurrent Pos:* Group leader, Helene Curtis Industs, Inc, 57-58; sr res chemist, Nalco Chem Co, 58-62; sr res biochemist, Abbott Labs, 62-66; asst chief radioisotope serv, Vet Res Hosp, Chicago, 66-71. *Mem:* Am Chem Soc; Sigma Xi; Am Soc Biol Chemists; Am Soc Cell Biol; Biophys Soc. *Res:* Protein sequence determination; mechanism of enzyme action; lysosomal enzyme diseases; Wilson's disease; organic synthesis of biochemically active substances; Collagen disease metabolism. *Mailing Add:* Dir Med Affairs Hollister Inc PO Box 250 2000 Hollister Dr Libertyville IL 60048

NEEDLER, GEORGE TREGLOHAN, b Sommerside, PEI, Feb 2, 35; m 59, 84; c 5. DYNAMICAL OCEANOGRAPHY, DEEP SEA POLLUTION. *Educ:* Univ BC, BSc, 58, MSc, 59; McGill Univ, PhD(theoret physics), 63. *Prof Exp:* Res scientist phys oceanog, 62-79, dir, Atlantic Oceanog Lab, Bedford Inst Oceanog, Dartmouth, NS, Can, 80-85; DIR & CHIEF SCIENTIST, WORLD OCEAN CIRCULATION EXP, 85- *Concurrent Pos:* Res assoc phys oceanog, Dalhousie Univ, 67- *Mem:* Can Meteorol & Oceanog Soc; Int Union Geodesy & Geophysics; fel Royal Soc Can. *Res:* Large-scale dynamical oceanography including thermocline theory, climate research, and the interpretation of geochemical data; deep ocean pollution, especially as related to the disposal of radio-active wastes. *Mailing Add:* Deacon Lab Inst Oceanogr Sci Wormley Godalming GU8 5UB England

NEEDLES, HOWARD LEE, b Bloomington, Calif, June 26, 37; m 83; c 4. FIBER SCIENCE, TEXTILE CHEMISTRY. *Educ:* Univ Calif, Riverside, AB, 59; Univ Mo, PhD(org chem), 63. *Prof Exp:* Res chemist, Western Regional Res Lab, USDA, 63-69; asst prof textile sci & asst textile chemist, Agr Exp Sta, 69-74, assoc prof, 74-77, PROF TEXTILE & MAT SCI, UNIV CALIF, DAVIS, 77- *Concurrent Pos:* Vis prof, NC State Univ, 75-76, Univ Leeds, 82-83, Univ Otago, 90, 91; prog chair, Cellulose, Paper, Textile Div, Am Chem Soc, 85-88, chmn, 89. *Mem:* Fiber Soc; Am Asn Textile Chem &

Colorists; Am Chem Soc; Soc Dyers & Colourists; Textile Inst; Soc Forensic Eng & Sci. *Res:* Modification and properties of fibers and polymers; textile flammability; color relationships in dyed fibers and films. *Mailing Add:* Div of Textiles & Clothing Univ of Calif Davis CA 95616

NEEFE, JOHN R, b Philadelphia, Pa, July 17, 43; m 68; c 2. INTERFERON, CELLULAR CYTOTOXICITY. *Educ:* Harvard Univ, BA, 65; Univ Pa, MD, 69. *Prof Exp:* Res, Baltimore City Hosp, 69-71; clin assoc, Nat Cancer Inst, 71-76, investr, 76-78; from asst prof to assoc prof, Georgetown Univ, 78-86; assoc prof, 86-89, PROF, UNIV KY, 89- *Mem:* Am Asn Immunologists; AAAS; Am Asn Cancer Res; Am Soc Clin Oncol; Am Col Physicians. *Mailing Add:* Dept Med Univ Ky Markey Cancer Ctr Lexington KY 40536

NEEL, JAMES VAN GUNDIA, b Middletown, Ohio, Mar 22, 15; m 43; c 3. GENETICS. *Educ:* Col Wooster, AB, 35; Univ Rochester, PhD(genetics), 39, MD, 44. *Hon Degrees:* DSc, Col Wooster, 59, Univ Rochester, 74, Med Col Ohio, 81. *Prof Exp:* Asst, Univ Rochester, 35-39; instr zool, Dartmouth Col, 39-41; from intern to asst resident med, Strong Mem Hosp & Rochester Munic Hosp, NY, 44-46; dir field studies, Atomic Bomb Casualty Comn, Nat Res Coun, 47-48; assoc geneticist, Inst Human Biol, Sch Med, Univ Mich, 48-56, from asst prof to assoc prof internal med, 49-56, prof, 56-66, chmn, Dept Human Genetics, 56-81, Lee R Dice Prof, 66-85, EMER PROF HUMAN GENETICS & INTERNAL MED, SCH MED, UNIV MICH, ANN ARBOR, 85- *Concurrent Pos:* Mem, comt res probs sex, Nat Res Coun, 49-56, panel genetics, comt growth, 51-56, comt atomic casualties, 51-54, adv comt, Atomic Bomb Casualty Comn, 57-70, comt epidemiol & vet follow-up studies, 65-68, 75-78; mem comt genetic effects of atomic radiation, Nat Acad Sci, 55-58, comt int biol prog, 65-70, selection comt sr res fels, NIH, 56-60 & gen res training grant comt, 57-58, chmn genetic training grant comt, 58-63, mem comt int ctrs for med res & training, 61-65, comt int res, 65-69, expert adv panel radiation, WHO, 57-61, expert adv panel human genetics, 61, mem coun, Nat Acad Sci, 70-73; US deleg, US-Japan Coop Med Sci Prog, 71-78; pres, Sixth Int Cong Human Genetics, 81; coun, Nat Inst Aging, 84-87, health & environ res adv comt, Dept Energy, 84-85; radiation adv comt, Science Adv Bd, Environ Protection Agency, 85-89; vets adv comt, Environ Hazards, Vets Admin, 85- *Honors & Awards:* Lasker Award, Am Pub Health Asn, 60; Allen Award, Am Soc Human Genetics, 65; Nat Medal of Sci, 74; Medal, Smithsonian Inst, 81. *Mem:* Nat Acad Sci; Inst Med; Asn Am Physicians; Am Philos Soc; Am Soc Naturalists; Am Acad Arts & Sci. *Res:* Genetics of man. *Mailing Add:* Dept Human Genetics Univ Mich Ann Arbor MI 48109-0618

NEEL, JAMES WILLIAM, b Turlock, Calif, July 20, 25; m 56. SOIL SCIENCE, BOTANY. *Educ:* Univ Calif, Berkeley, BS, 49, Univ Calif, Los Angeles, PhD(bot sci), 64. *Prof Exp:* From lab technician to prin lab technician, Atomic Energy Proj, Univ Calif, Los Angeles, 49-59, asst plant physiol, Dept Irrig & Soil Sci, 59-63; from asst prof to assoc prof, 63-69, chmn dept, 69-71, PROF BIOL, SAN DIEGO STATE UNIV, 69-, ASSOC DEAN, COL SCI, 76- *Concurrent Pos:* Consult, Rand Corp, Calif, 63-71. *Mem:* AAAS; Am Soc Agron; Soil Sci Soc Am; NY Acad Sci; Sigma Xi. *Res:* Fate of soluble and insoluble forms of radionuclides in soils and availability to plants; effects of heavy metals on plant systems. *Mailing Add:* Dept Biol San Diego State Univ 5300 Campanile Dr San Diego CA 92182

NEEL, JOE KENDALL, SR, limnology; deceased, see previous edition for last biography

NEEL, LOUIS EUGENE FELIX, b Lyon, France, Nov 22, 04. MAGNETISM. *Educ:* Strasburg, France, DSc, 32. *Prof Exp:* Pres, Polytech Nat Inst, Grenoble, 71-76; pres, Coun Higher Nuclear Safety, 81-86; RETIRED. *Honors & Awards:* Nobel Prize in Physics, 70; Gold Medal, Nat Ctr Sci Res, 65. *Mem:* French Acad Sci; USSR Acad Sci; Romanian Acad Sci; Am Acad Sci; Royal Soc Neth Acad; Royal Soc London. *Mailing Add:* 15 Rue Marcel Allegot Meudon-Bellevue 92190 France

NEEL, THOMAS H, b Oxnard, Calif, Mar 7, 41; c 2. EXPLORATION OPERATIONS. *Educ:* Stanford Univ, BS, 62, MS, 63. *Prof Exp:* Geologist & geophysicist, Mobil Oil Libya, Ltd, 64-67; geologist, Atlantic Richfield Co, Long Beach, 67-68, planning coord, Planning Dept, Long Beach, 68-70, sr geologist, S La Dist, 70-71, staff geologist, Dallas ,71-73, lease sales magr, STex/Offshore Dist, Houston, 73-75, explor mgr, S Alaska Dist, 75-79, vpres, dist mgr, Dallas, 79-81, vpres, explor opers eastern US, Atlantic Richfield Co, 81-84; sr vpres, explor & prod & minerals explor, Kerr McGee Corp, Okla City, 84-85; pres, Depco, Inc & DeKalb Petrol Corp, 85-88, PRES & CHIEF OPER OFF, DEKALB ENERGY CO, DENVER, 88- *Mailing Add:* 4211 S Bellaire Englewood CO 80110

NEEL, WILLIAM WALLACE, b Thomasville, Ga, Feb 4, 18; m 52; c 2. ENTOMOLOGY. *Educ:* Emory Univ, BA, 40; Univ Fla, MS, 49; Tex A&M Univ, Tex, PhD(entom), 54. *Prof Exp:* Field supvr, Disease Control, USPHS, 45-47; asst entomologist, Inter-Am Inst of Agr Sci, 48-49, 52-53; asst prof entom & asst entomologist, Agr Exp Sta, Miss State Univ, 54-62; field res, Chemagro Corp, 62-63; asst prof entom & asst entomologist, Agr Exp Sta, Univ WVa, 63-66; assoc prof & assoc entomologist, Agr Exp Sta, Miss State Univ, 66-77, prof entom & entomologist, 77-85; RETIRED. *Mem:* Entom Soc Am; Sigma Xi. *Res:* Forest and pecan entomology with special interest in seed and cone insects; toxicant research on pecan and seed and cone insects; pheromone isolation of pecan weevil. *Mailing Add:* 2146 Tanglewood Rd Decatur GA 30033

NEELEY, CHARLES MACK, b Pine Bluff, Ark, Mar 7, 42; m 68. PHYSICAL CHEMISTRY, POLYMER CHEMISTRY. *Educ:* Univ Ark, Fayetteville, BS, 65, PhD(chem), 69. *Prof Exp:* Lab technician, Houston Chem Corp, 62-63; asst phys chem, Univ Ark, Fayetteville, 65-69; fel, Univ Fla, 69-70; res chemist polymer phys chem, Plastics Lab, 70-75, sr chemist, Polymer Develop Polyethylene & Polypropylene Divs, Polymer Develop Div, 75-77, mgt staff asst, 77-79, sr chemist, Appln Lab, 79-86, RES ASSOC, POLYETHYLENE- POLYPROPYLENE RES, 86- *Mem:* AAAS; Am Chem Soc. *Res:* Pyrolysis gas chromatography; rheology; gas phase photochemistry; gas phase kinetics. *Mailing Add:* NF 57 Lake Cherokee Longview TX 75603

NEELIN, JAMES MICHAEL, b London, Ont, Dec 4, 30; m 53; c 4. MOLECULAR EVOLUTION, CELL DIFFERENTIATION. *Educ:* Univ Toronto, BA, 53, PhD(biochem), 58. *Prof Exp:* Jr scientist, Atlantic Tech Sta, Fisheries Res Bd, Can, 53- 54, asst scientist, 54-55; Coun fel, Div Appl Biol, Nat Res Coun Can, 58-59, asst res officer, 59-62; res assoc chicken histones, Stanford Univ, 62-63; assoc res officer, Div Biosci, Nat Res Coun Can, 64-70, sr res officer, 71; chmn dept biol, 71-74 & 75-78, dean sci, 84-86, PROF BIOL & BIOCHEM, CARLETON UNIV, 71-, DIR, BIOCHEM INST, 79-83 & 89- *Concurrent Pos:* Vis scientist, IBMC du CNRS, Strasbourgh, 74-75; vis prof, Université Laval, 79-80 & 89; Co-ed, Can J Biochem, 81-83 & Can J Biochem Cell Biol, 84. *Mem:* Can Biochem Soc; Can Soc Cell Molecular Biol; Can Fedn Biol Sci; Can Hist Sci Technol Asn. *Res:* Chemistry, evolution and metabolism of linker histones; history of biology and biochemistry in Canada. *Mailing Add:* Dept Biol Carleton Univ Ottawa ON K1S 5B6 Can

NEELY, BROCK WESLEY, b London, Ont, Apr 28, 26; nat US; m 53; c 3. BIOCHEMISTRY, PHYSICAL CHEMISTRY. *Educ:* Univ Toronto, BS, 48; Mich State Univ, PhD(biochem), 52. *Prof Exp:* Res Found fel chem, Ohio State Univ, 52-53; Rockefeller fel, Univ Birmingham, 53-54; res assoc, G D Searle & Co, Ill, 54-55; RES ASSOC BIOCHEM, DOW CHEM CO, 57-, RES ASSOC ENVIRON SCI, 73- *Mem:* Am Chem Soc; Sigma Xi; NY Acad Sci; Soc Environ Toxicol & Chem. *Res:* Structure- activity relationships; environment research; math modeling. *Mailing Add:* 4302 Cruz Dr Midland MI 48640

NEELY, CHARLES LEA, JR, b Memphis, Tenn, Aug 3, 27; m 57; c 2. MEDICINE, HEMATOLOGY. *Educ:* Princeton Univ, AB, 50; Washington Univ, MD, 54. *Prof Exp:* Intern, Bellevue Hosp, New York, 54-55; resident, Barnes Hosp, St Louis, 55-57, fel med & NIH trainee chemother, 57-58; from instr to assoc prof, 58-71, chief, Dept Oncol, 76 & Dept Hemat, 81, PROF MED, COL MED, UNIV TENN, MEMPHIS, 71- *Concurrent Pos:* Actg dir, Memphis Regional Cancer Ctr, 79-87 & Cancer Prog, Univ Tenn, 79-87. *Mem:* Am Soc Hemat; fel Am Col Physicians; Am Fedn Clin Res; Am Soc Clin Oncol; AMA; Sigma Xi. *Res:* Hemolytic anemias and oncology. *Mailing Add:* 440 Goodwyn St Memphis TN 38111-3312

NEELY, PETER MUNRO, b Los Angeles, Calif, Dec 31, 27; m 65; c 2. SYSTEMATICS, ECOLOGY. *Educ:* Univ Calif, Los Angeles, BA, 52, PhD(bot), 60. *Prof Exp:* Asst bot, Univ Calif, Los Angeles, 52-63; asst prof bot & res assoc, Biol Sci Comput Ctr, Chicago, 63-68; assoc prof statist biol, Dept Bot, 68-71, assoc dir comput ctr, 69-77, ASSOC PROF SYSTS & ECOL, UNIV KANS, 71- *Concurrent Pos:* USPHS fel, Sch Pub Health, Univ Calif, Los Angeles, 61-62. *Res:* Application of computer processing and statistical techniques to biological problems; grouping algorithms and development of general theory of classifications; development of interactive pedagogic programs; bioethics and ecological scarcity. *Mailing Add:* Dept Biol Univ Kans Lawrence KS 66045

NEELY, ROBERT DAN, b Senath, Mo, Oct 6, 28; m 53; c 2. PLANT PATHOLOGY. *Educ:* Univ Mo, BS, 50, PhD(bot), 57. *Prof Exp:* Plant pathologist, Ill Natural Hist Surv, 51-90; RETIRED. *Concurrent Pos:* Adj prof plant path, Univ Ill, Urbana-Champaign, 72- *Honors & Awards:* Author's Citation, Int Soc Arboriculture, 74. *Mem:* Mycol Soc Am; Am Phytopath Soc. *Res:* Diseases of shade and forest trees and woody ornamentals. *Mailing Add:* Rte 1 Box 1060-6 Scott City MO 63780

NEELY, STANLEY CARRELL, b Abilene, Tex, Sept 11, 37; m 59; c 4. PHYSICAL CHEMISTRY. *Educ:* Southern Methodist Univ, BS, 60; Yale Univ, PhD(phys chem), 65. *Prof Exp:* Asst prof, 65-69, ASSOC PROF CHEM, UNIV OKLA, 69-, ASST CHMN CHEM, 81- *Concurrent Pos:* Fulbright fel, Karachi Univ, Pakistan, 89-90. *Mem:* Am Chem Soc. *Res:* Molecular and solid state spectroscopy; molecular interactions. *Mailing Add:* Chem Dh-208 Univ Okla 660 Parrington Oval Norman OK 73019

NEELY, WILLIAM CHARLES, b Cave City, Ark, Nov 22, 31; m 57; c 2. PHYSICAL CHEMISTRY. *Educ:* Miss State Col, BS, 53; La State Univ, MS, 60, PhD(chem), 62. *Prof Exp:* Chemist, Nylon Div, Chemstrand Corp, 53-54, res chemist, Chemstrand Res Ctr, Inc, 62-66; from asst prof to assoc prof, 66-89, PROF CHEM, AUBURN UNIV, 89- *Mem:* Am Chem Soc. *Res:* Molecular spectroscopy and photochemistry of electronic excitation energy transfer processes applied to biological systems and excited states of oxygen; aqueous ozone reactions; biomass conversion processes; chemical reactions in electrical discharge plasmas. *Mailing Add:* Dept Chem Auburn Univ Auburn AL 36830

NEEMAN, MOSHE, b Latvia, Apr 1, 19; nat US; m 60; c 4. ORGANIC CHEMISTRY, BIOCHEMISTRY. *Educ:* Univ London, BSc, 43; Hebrew Univ, MSc, 45, PhD, 47. *Prof Exp:* Asst, D Sieff Res Inst, Israel, 36-44; with indust, 45-46; indust safety & hyg expert, Govt Israel, 48-52; dep dir, Israel Res Coun & dir, Lab Appl Org Chem, 52-56; sr lectr, Israel Inst Technol, 56; vis res assoc chem, Univ Wis, 56-59; assoc res prof chem, 64-72, co-chmn chem prog, 69-73, chmn, PhD Prog Sch Grad & Prof Educ, SUNY, 72-82, head, Carcinogenesis Lab, 73-82, ASSOC CANCER RES SCIENTIST, ROSWELL PARK MEM INST, 59-; VPRES RES ORTHO KENETICS RES FOUND, AMHERST, NY, 82- *Concurrent Pos:* Chmn standards comt, Israel Standards Inst, 48-50; dir, Inst Indust Hyg, 50-52; res prof biol, Niagara Univ, 69-; res prof biol, Canisuis Col, 69-80. *Honors & Awards:* Szold Award, Israel, 57. *Mem:* Am Chem Soc; fel Royal Soc Health; fel Am Inst Chemists; fel Royal Soc Chem. *Res:* Heterocyclic syntheses; diazoalkanes; medicinal chemistry; mechanism of carcinogenesis and tumor promotion; chemistry and reaction mechanisms of steroids; steroidal alkaloids; steroid hormone metabolism. *Mailing Add:* Five Sedgmoor Ct Williamsville NY 14221

NEENAN, JOHN PATRICK, b Detroit, Mich, Nov 3, 43; US citizen. NUCLEIC ACID CHEMISTRY, ENZYMOLOGY. *Educ:* Wayne State Univ, BS, 69; Univ Calif, Santa Barbara, PhD(chem), 73. *Prof Exp:* Fel pharmacol, Sch Med, Yale Univ, 73; fel med chem, Sch Med, Univ Pa, 74; sr res chemist, Asn Stevens Inc, Detroit, 74-75; res assoc chem, Mich Cancer

Found, 75-77; asst prof biochem, Ill Benedictine Col, 77-79; res assoc, Univ Ariz, 79-81; RES ASSOC BIOCHEM, BOWLING GREEN STATE UNIV, 81- *Concurrent Pos:* Adj lectr, Univ Mich, Dearborn, 76-79; mem fac, Argonne Nat Lab, 78; prin investr, US Army contract, Ill Benedictine Col, 78-79; prin investr, US Army contract, Univ Ariz, 80-81. *Mem:* Am Chem Soc. *Res:* Design of nucleoside and nucleotide analogs as enzyme inhibitors and antiviral, antitumor and antiparasite agents. *Mailing Add:* Dept of Chem Rochester Inst of Tech One Lomb Memorial Dr Rochester NY 14623

NEEPER, DONALD ANDREW, b New York, NY, Aug 9, 37; m 58; c 4. SOLAR ENERGY, ENVIRONMENTAL REMEDIATION. *Educ:* Pomona Col, BA, 58; Univ Wis, MS, 60, PhD(physics), 64. *Prof Exp:* Res assoc, James Franck Inst, Univ Chicago, 66-68; staff mem, 68-79, group leader, 79-83, STAFF MEM, LOS ALAMOS NAT LAB, 83- *Concurrent Pos:* Vis assoc prof, Northern Ariz Univ, 83-84. *Mem:* Asn Ground Water Scientists & Engrs; Int Solar Energy Soc; AAAS. *Res:* Thermal design and energy conservation for small buildings; evaluation of advanced concepts in thermal transport, thermal storage and glazings; soil venting. *Mailing Add:* 2708 Walnut St Los Alamos NM 87544-2050

NEEPER, RALPH ARNOLD, b Toledo, Ohio, Sept 29, 40; m 73; c 2. INFORMATION SYSTEMS SECURITY, PHOTOGRAMMETRY. *Educ:* Purdue Univ, BS, 63, MS, 72. *Prof Exp:* Cartogr & mathematician, Defense Mapping Agency, Aerospace Ctr, 68-77, computer specialist, Defense Mapping Sch, 77-81; computer systs accreditor, Off Asst Chief Staff, Intel, US Army, 81-84, computer specialist, Mgt Systs Support Agency, 84-85; systs adminr, Orgn Joint Chiefs, 85-86; security prog mgr, US Intel & Security Command, 86-90; SYSTS ANALYST, US MIL COMMUNITY ACT IV, BERLIN, 90- *Mem:* AAAS. *Res:* Developing techniques, procedures and policies for the protection of information held and processed in computers. *Mailing Add:* US Mil Community Activ Berlin DCSI Security Dale City NY 09742

NEER, EVA JULIA, b Warsaw, Poland; US citizen. SIGNAL TRANSDUCTION, HORMONE ACTION. *Educ:* Barnard Col, BA, 59; Col Physicians & Surgeons, Columbia Univ, MD, 63. *Prof Exp:* Intern med, Georgetown Univ Hosp, DC, 63-64; fel biol, Yale Univ, Conn, 65-66; fel, 66-68, res assoc, 67-76, from asst prof to assoc prof, 76-91, PROF MED, HARVARD MED SCH, BOSTON, 91- *Concurrent Pos:* Estab investr, Am Heart Asn, 71-76; tutor biochem sci, Harvard Univ, Cambridge, 72-; ed, J Biol Chem, 78-83 & 85-91, J Selond Messengers & Phosphoproteins, 81-, Endocrine Rev, 91-; mem, NIH Study Sect, 84-88. *Mem:* Am Soc Biol Chemists; Endocrine Soc; Soc Neurosci. *Res:* Chemistry of the early events in the action of hormones which act through guanine nucleotide binding proteins; defining the structure and subunit interactions of guanine nucleotide binding proteins. *Mailing Add:* Cardiovasc Div Brigham & Womens Hosp 75 Francis St Boston MA 02115

NEER, KEITH LOWELL, b Springfield, Ohio, Feb 18, 49; m 71. MEAT SCIENCE. *Educ:* Ohio State Univ, BSc, 72, MSc, 73; Univ Nebr, PhD(animal sci), 75. *Prof Exp:* Technician meats, Ohio State Univ, 72-73; instr animal sci, Univ Nebr, Lincoln, 75-76; asst prof animal sci, Va Polytech Inst & State Univ, 76-77; MEM STAFF RES & DEVELOP, KROGER CO, 77- *Mem:* Am Meat Sci Asn; Am Soc Animal Sci; Inst Food Sci; Sigma Xi. *Res:* Investigating various feeding regimes and their effect on beef palatability; evaluating mechanical tenderization, pressing, power cleaving and cooking methods on beef palatability. *Mailing Add:* 2090 Natchez Trace Batavia OH 45103

NEER, ROBERT M, b Celina, Ohio, Nov 12, 35; m 58; c 2. ENDOCRINOLOGY. *Educ:* Harvard Col, BA, 57; Columbia Univ, MD, 61. *Prof Exp:* From instr to asst prof, 68-77, ASSOC PROF MED, HARVARD MED SCH, 77- *Concurrent Pos:* Asst physician, Mass Gen Hosp, 71-74, assoc physician, 74-81, dir, Gen Clin Res Ctr, 77-90, PHYSICIAN, MASS GEN HOSP, 81-; consult, US Food & Drug Admin, 79-85, Illum Eng Res Soc, 78-82. *Mem:* Endocrine Soc; Am Soc Bone & Mineral Res. *Res:* Osteoporosis; metabolic bone disease; calcium-regulating hormones. *Mailing Add:* Mass Gen Hosp Boston MA 02114

NEESBY, TORBEN EMIL, b Copenhagen, Denmark, Apr 21, 09; nat US; m 39; c 3. CLINICAL CHEMISTRY. *Educ:* Tech Univ Denmark, MSc, 32; Columbia Pac Univ, PhD, 79. *Prof Exp:* Consult chemist, 32-40; tech dir, Norsk Sulfo, Norway, 40-43; private bus, Denmark, 43-48; tech dir, Am Sulfo, Inc, NY, 48-51; head org chem, Carroll Dunham Smith Pharmacal Co, 51-56; head lab, E F Drew & Co, 56-58; asst dir tech servs, Bristol Myers, Inc, 58; sr scientist, Schieffelin & Co, 58-61; sr biochemist surg res, Harbor Gen Hosp, Univ Calif, 61-68; CLIN BIOCHEMIST, VALLEY MED CTR, 68-; AT T E NEESBY, INC. *Mem:* AAAS; Am Chem Soc; Soc Clin Ecol; Am Acad Environ Med; Inst Food Technologists; Am Pub Health Asn. *Res:* Structure-activity relationship; isotopes; investigation of mediator compounds; research on exogenous and endogenous polypeptide factors with effect on growth of various organs; ecology and allergy. *Mailing Add:* T E Neesby Inc 2227 N Pleasant Ave Fresno CA 93705-3663

NEESON, JOHN FRANCIS, b Buffalo, NY, Dec 9, 36; m 63; c 1. NUCLEAR PHYSICS. *Educ:* Canisius Col, BS, 58; Univ Mich, MS, 60; State Univ NY Buffalo, PhD(physics), 65. *Prof Exp:* From asst prof to assoc prof physics, St Bonaventure Univ, 65-68; assoc prof, State Univ NY Col Brockport, 68-69; assoc prof, 69-74, assoc dean col arts & sci, 69-74, dir instnl res, 74-77, PROF PHYSICS, ST BONAVENTURE UNIV, 74-, CHMN DEPT, 77- *Concurrent Pos:* Consult, Clarke Bros Co Div, Dresser Industs, 65-67 & Dresser-Clark Div, 67-; sr scientist, Western NY Nuclear Res Ctr, 69- *Mem:* Am Phys Soc; Am Nuclear Soc; Am Asn Physics Teachers; Am Asn Higher Educ. *Res:* Low energy nuclear physics, especially nuclear structure of spherical nuclei; ultrasonics, light sound wave interaction. *Mailing Add:* Dept of Physics St Bonaventure Univ St Bonaventure NY 14778

NEET, KENNETH EDWARD, b St Petersburg, Fla, Sept 24, 36; m 60; c 4. ENZYMOLOGY, PROTEINS. *Educ:* Univ Fla, BSCh, 58, MS, 60, PhD(biochem), 65. *Prof Exp:* Fel biochem, Univ Calif, Berkeley, 65-67; from asst prof to prof biochem, Sch Med, Case Western Reverse Univ, 67-90; prof gen med sci, 87-90, PROF & CHMN BIOL CHEM, CHICAGO MED SCH, 90- *Concurrent Pos:* Macy fac scholar vis prof, Stanford Univ, 80-81. *Mem:* AAAS; Am Chem Soc; Am Soc Biochem & Molecular Biol; NY Acad Sci; Soc Neurosci. *Res:* Protein chemistry; enzyme regulation and mechanisms; subunit interactions; neurobiology. *Mailing Add:* Dept Biol Chem UHS Chicago Med Sch 3333 Green Bay North Chicago IL 60064

NEFF, ALVEN WILLIAM, b Lafayette, Ind, Sept 13, 23; m 48; c 4. BIOCHEMISTRY. *Educ:* Kans State Col, BS, 47, MS, 48; Purdue Univ, PhD, 52. *Prof Exp:* Biochemist, Up John Co, 52-84; RETIRED. *Mem:* AAAS; Am Chem Soc. *Res:* Drug and pesticide metabolism; drug residue in animals; pesticide residues in plants. *Mailing Add:* 474 Fineview Coopersville MI 49404

NEFF, BRUCE LYLE, b Philadelphia, Pa, Apr 15, 50; m 76; c 2. POLYMER PHYSICAL CHEMISTRY. *Educ:* Tulane Univ, BS, 72; Mass Inst Technol, PhD(phys chem), 77. *Prof Exp:* Res chemist polymer spectros, Tenn Eastman Co, Eastman Kodak Co, 77-79, res chemist polymer phys chem, Eastman Chem Div, 79-84; RES CHEMIST, POLYMER CHARACTERIZATION, DUPONT CO, 84- *Mem:* Am Phys Soc; Am Chem Soc. *Res:* Investigations of polymer structure and branching by size exclusion chromatography; light scattering; solid state nuclear magnetic resonance spectroscopy; thermal and rheological analysis of coatings and polymers. *Mailing Add:* 2325 Kennwynn Rd Wilmington DE 19810

NEFF, CARROLL FORSYTH, b Pigeon, Mich, Jan 10, 08; m 30, 57; c 2. NUTRITION. *Educ:* Washington Univ, AB, 29; Univ Calif, MA, 30. *Prof Exp:* With res lab, Ralston Purina Co, Mo, 26-28 & 30-31; head biol lab, Anheuser-Busch, Inc, 31-38; mgr plant & labs, Sterling Drug, Inc, Ga, 38-50, plant mgr, SAfrica, 50-51, consult chemist, Ga, 51-56; with res dept, New Prod Liaison, Foremost Dairies, Inc, 57-62; TECH CONSULT FOOD, DRUGS & COSMETICS, 62- *Concurrent Pos:* Scholar, Food Law Inst, Emory Univ, 54; western rep, Food & Drug Res Labs, Inc, NY, 68-72; lectr-consult, Nat Agr Col Chapingo, Mex, 74-76 & Calsec Consults, 80- *Mem:* Am Chem Soc; Am Pharmaceut Asn; Am Mkt Asn; Inst Food Technologists. *Res:* Chemotherapy of poultry coccidiosis; yeast vitamins; diet and immunity; milk derivatives; special dietary use foods; food and drug regulations. *Mailing Add:* 1201 Liberty St No Three El Cerrito CA 94530

NEFF, HERBERT PRESTON, JR, b Knoxville, Tenn, Jan 1, 30; m 53; c 4. ELECTRICAL ENGINEERING. *Educ:* Univ Tenn, Knoxville, BS, 53, MS, 56; Auburn Univ, PhD(elec eng), 67. *Prof Exp:* Jr engr, Univ Tenn, Knoxville, 52-56, asst prof elec eng, 56-65; asst, Auburn Univ, 65-66; ASSOC PROF ELEC ENG, UNIV TENN, KNOXVILLE, 66- *Concurrent Pos:* Consult, Auburn Univ, 66-68, Oak Ridge Nat Labs, 69- & Goddard Space Flight Ctr, NASA, 72- *Mem:* Inst Elec & Electronics Engrs; Sigma Xi. *Res:* Electromagnetic fields; antennas and propagation. *Mailing Add:* 2234 Keller Bend Rd Knoxville TN 37922

NEFF, JOHN DAVID, b Cedar Rapids, Iowa, July 30, 26; m 52. MATHEMATICS. *Educ:* Marquette Univ, BNS, 46; Coe Col, BA, 49; Kans State Univ, MS, 51; Univ Fla, PhD(math), 56. *Prof Exp:* Mem tech staff, Bell Tel Labs, NY, 52-53; instr math, Univ Fla, 53-55; from instr to asst prof, Case Inst Technol, 56-61; from asst prof to assoc prof, 61-72, actg dir math, 70-72, dir, 72-78, PROF MATH, GA INST TECHNOL, 72- *Mem:* Am Math Soc; Math Asn Am; Inst Math Statist; Soc Indust & Appl Math; Nat Coun Teachers Math. *Res:* Differential equations; probability. *Mailing Add:* Sch Math Atlanta GA 30332-0160

NEFF, JOHN S, b Milwaukee, Wis, Nov 24, 34; m 60; c 2. ASTRONOMY, ASTROPHYSICS. *Educ:* Univ Wis, BS, 57, MS, 58, PhD(astron), 61. *Prof Exp:* Res assoc, Yerkes Observ, Chicago, 61-64; asst prof, 64-68, ASSOC PROF ASTRON, UNIV IOWA, 68- *Concurrent Pos:* NSF res grant, 65-67. *Mem:* Am Astron Soc; fel Royal Astron Soc; Int Astron Union. *Res:* Stellar photometry and spectrophotometry; planetary spectrophotometry; design of astronomical instruments; stellar classification and investigation of galactic structure. *Mailing Add:* 2305 Macbride Dr Iowa City IA 52246

NEFF, LAURENCE D, b Santa Ana, Calif, Jan 11, 38; m 60; c 2. PHYSICAL CHEMISTRY. *Educ:* John Brown Univ, BA, 59; Univ Ark, MS, 62, PhD(phys chem), 64. *Prof Exp:* Sr chemist, NAm Aviation, Inc, 64; res chemist, Beckman Instruments, Inc, 64-66; res specialist mat sci, NAm Rockwell Corp, 66-68; from asst prof to assoc prof chem, ETex State Univ, 68-78, prof, 78-81; proj chemist, 81-88, TECH COORDR, TEXACO, USA, 88- *Mem:* AAAS; Am Chem Soc. *Res:* Heterogeneous catalysis; application of infrared spectroscopy to gas-solid phase interactions. *Mailing Add:* 4601 Lake Shore Dr Port Arthur TX 77640

NEFF, LOREN LEE, b Seattle, Wash, Sept 14, 18; m 40; c 1. CHEMISTRY, CORROSION. *Educ:* Univ Wash, Seattle, BS, 39, PhD(chem), 43. *Prof Exp:* Asst chem, Univ Wash, Seattle, 39-43; supvr res dept, Union Oil Co Calif, 43-81; RETIRED. *Mem:* Am Chem Soc. *Res:* Lubricating oil additives; lubricating oils; activity of colloidal electrolytes; determination of the thermodynamic activity of 1-dodecane sulfonic acid in aqueous solutions at forty degrees Centigrade by electromotive force measurements; petroleum industry corrosion; chemicals. *Mailing Add:* 425 W Brookdale Pl Fullerton CA 92632

NEFF, MARY MUSKOFF, b Jacksonville, Fla, Jan 20, 30; m 52. COMBINATORICS & FINITE MATHEMATICS. *Educ:* Purdue Univ, BS, 51, MS, 52; Univ Fla, PhD(math), 56. *Prof Exp:* Tech staff, Bell Labs, 52-53; intrst math, Univ Fla, 55-56; from instr to asst prof, John Carroll Univ, 56-61; asst prof, 61-68, ASSOC PROF MATH, EMORY UNIV, 68- *Concurrent Pos:* Vis lectr, Math Asn Am, 76- *Mem:* Math Asn Am; Am Math Soc. *Res:* The structure of near-rings; hyperidentities and varieties of clones and semigroups. *Mailing Add:* Dept of Math Emory Univ Atlanta GA 30322

NEFF, RAYMOND KENNETH, b New York, NY, May 1, 42; m 69. BIOMETRICS. *Educ:* Dartmouth Col, AB, 64; Harvard Univ, SM, 67. *Hon Degrees:* DSc, Harvard Univ, 77. *Prof Exp:* Lectr biostatist, Harvard Univ, 74-78 & 81-82, asst prof, 78-81; ASSOC PROF BIOSTATIST, DARTMOUTH COL, 82- *Concurrent Pos:* Pres, New Eng Regional Comput, 73-76; dir sci comput, Sidney Farber Cancer Inst, Boston, 79-82, Health Sci Comput Facil, Harvard Univ, 71-82; consult, Nat Toxicol Prog, NIH, 80-81; dir acad comput, Dartmouth Col, 82- *Mem:* Am Public Health Asn; Asn Comput Mach; Biomet Soc; AAAS; Am Statist Asn. *Res:* Multivariate analysis of epidemiologic data; statistical and interactive computing. *Mailing Add:* Case Western Reserve Univ 2040 Adelbert Rd Rm 30 Cleveland OH 44106

NEFF, RICHARD D, health physics; deceased, see previous edition for last biography

NEFF, RICHMOND C(LARK), b DeKalb, Ill, July 30, 23; m 44; c 4. ENGINEERING MECHANICS. *Educ:* Purdue Univ, BS, 43, MS, 48, PhD(eng mech), 54. *Prof Exp:* Instr eng mech, Purdue Univ, 47-52; from asst prof to prof civil eng, Univ Ariz, 52-79; RETIRED. *Mem:* Soc Exp Stress Analysis; Am Soc Eng Educ; Sigma Xi. *Res:* Photoelastic analysis of structural and machine components; numerical solutions of elasticity and plasticity equations. *Mailing Add:* 4214 E Blanton Tucson AZ 85712

NEFF, ROBERT JACK, b Kansas City, Mo, Jan 22, 21; m 48; c 3. BIOLOGY, PHYSIOLOGY. *Educ:* Univ Mo, AB, 42, MA, 48, PhD(zool), 51. *Prof Exp:* Instr anat, Sch Med, Johns Hopkins Univ, 51-52; from asst prof to assoc prof biol, 52-64, ASSOC PROF MOLECULAR BIOL, VANDERBILT UNIV, 64- *Concurrent Pos:* Vis assoc prof, Univ Calif, 57-58; NIH spec fel, Biol Inst, Carlsberg Found, Copenhagen, 65-66. *Mem:* AAAS; Soc Protozool; Am Soc Cell Biol; NY Acad Sci. *Res:* Cellular physiology; nuclear-cytoplasmic control mechanisms; cellular osmoregulation; macromolecular organization of protoplasm; cytodifferentiation; encystment; cell growth-division cycle. *Mailing Add:* 2116 Westwood Ave Nashville TN 37212

NEFF, STUART EDMUND, b Louisville, Ky, Oct 3, 26; m 48; c 3. BIOLOGY. *Educ:* Univ Louisville, BS, 54; Cornell Univ, PhD(limnol), 60. *Prof Exp:* Asst limnol, Cornell Univ, 54-59, instr, 60; from asst prof to assoc prof, Va Polytech Inst, 60-68; assoc res prof, 68-72, res prof biol, Water Resources Lab, Univ Louisville, 72-85; PROF BIOL, TEMPLE UNIV, 85- *Mem:* Soc Syst Zool; Entom Soc Am; Royal Entom Soc London. *Res:* Hydrobiology; immature stages of aquatic insects; taxonomy and biology of acalyptrate Diptera; biology of Chironomidae. *Mailing Add:* Dept Biol Temple Univ Philadelphia PA 19122

NEFF, THOMAS O'NEIL, b Jacksonville, Fla, Nov 19, 09; m 39; c 3. ENGINEERING MECHANICS. *Educ:* Univ Fla, BSEE, 32, MSE, 52, CE, 56. *Prof Exp:* Inspector, State Rd Dept, Fla, 32-38; party chief, State Hwy Dept Ga, 38-39; dist engr, Works Projs Admin, Fla, 39-41; sect head, US Corps Engrs, 41-43, assoc engr, Dist Off, 45-46; resident engr, G S Broadway, Consult Engrs, 43-44; prof eng mech, Univ Fla, 46-67, prof mech eng, 67-76, emer assoc prof mech eng, 76-87; pres, Neff & Assoccs Inc, 80-87; RETIRED. *Mem:* Fel Am Soc Civil Engrs. *Res:* Structural design and analysis. *Mailing Add:* 1708 SW 35th Pl Gainesville FL 32608

NEFF, THOMAS RODNEY, geology; deceased, see previous edition for last biography

NEFF, VERNON DUANE, b Rochester, NY, Sept 16, 32; m 55; c 4. PHYSICAL CHEMISTRY. *Educ:* Syracuse Univ, BS, 53, PhD(phys chem), 60. *Prof Exp:* Spectroscopist, Gen Tire & Rubber Co, 59-61; from asst prof to assoc prof, 61-81, PROF PHYS CHEM, KENT STATE UNIV, 81- *Concurrent Pos:* Consult, Gen Tire & Rubber Co, 61- *Mem:* Am Chem Soc; Am Phys Soc. *Res:* Infrared spectroscopy; quantum chemistry. *Mailing Add:* Dept of Chem Kent State Univ Kent OH 44242

NEFF, WILLIAM DAVID, b Portland, Ore, Mar 25, 45; div; c 2. ATMOSPHERIC PHYSICS. *Educ:* Lewis & Clark Col, BA, 67; Univ Wash, MS, 68; Univ Colo, PhD, 80. *Prof Exp:* Commissioned officer, 68-73, physics scientist, 73-80, SUPVRY PHYSICIST, ATMOSPHERIC STUDIES, NAT OCEANIC & ATMOSPHERIC ADMIN, BOULDER, 81- *Mem:* Am Meteorol Soc. *Res:* Applications of acoustic remote sensing techniques to boundary layer meteorology, particularly relating to air quality meteorology in complex terrain. *Mailing Add:* NOA ERL Atmospheric Studies Prog RE WP7 325 Broadway Boulder CO 80303-3328

NEFF, WILLIAM DUWAYNE, b Lomax, Ill, Oct 27, 12; m 37; c 2. NEUROSCIENCES. *Educ:* Univ Ill, AB, 36; Univ Rochester, PhD(psychol), 40. *Prof Exp:* Res assoc psychol, Swarthmore Col, 40-42; res scientist, Div War Res, Columbia Univ & Univ Calif, 41-46; from asst prof to prof, Univ Chicago, 46-59, psychol & physiol, 59-61; dir, Lab Physiol Psychol, Bolt, Beranek, Newman, Inc, Mass, 61-63; prof psychol, Ind Univ, Bloomington, 63-64, dir, Ctr Neural Sci, 65-68, res prof, 64-82, EMER RES PROF, IND UNIV, BLOOMINGTON, 82- *Concurrent Pos:* Consult, Nat Acad Sci, Nat Res Coun, NSF, NIH, & NASA; mem, Otolaryngol Res Group, Int Brain Res Orgn & sci liaison off, London Br, Off Naval Res, 53-54. *Honors & Awards:* Hearing Res Award, Beltone Inst. *Mem:* Nat Acad Sci; fel AAAS; fel Am Physiol Soc; fel Acoust Soc Am; Soc Exp Psychol (secy-treas, 52-59). *Res:* Brain functions; neural mechanisms of sensory discrimination; physiological acoustics. *Mailing Add:* 3505 Bradley Bloomington IN 47405

NEFF, WILLIAM H, b May 13, 31; US citizen; m 50; c 6. ZOOLOGY, PHYSIOLOGY. *Educ:* Pa State Univ, BS, 56, MS, 59, PhD(zool), 66. *Prof Exp:* Instr anat & physiol, 59-65, asst prof zool, 66-75, ASSOC PROF BIOL, PA STATE UNIV, 75-, COORDR BIOL SCI, COMMONWEALTH CAMPUSES, 66- *Res:* Adaptation to environmental stress; gross metabolic and electrolyte response during cold acclimation; histological changes in reproductive tract of season breeding mammals. *Mailing Add:* Dept of Biol Pa State Univ University Park PA 16802

NEFF, WILLIAM MEDINA, b San Francisco, Calif, Oct 27, 29; m 52; c 4. EMBRYOLOGY. *Educ:* Stanford Univ, AB, 51, PhD(biol sci, statist), 58. *Prof Exp:* From instr to assoc prof biol, Knox Col, 56-68; assoc prof, Chico State Col, 68-70; assoc prof embryol, 70-81, ASSOC PROF BIOL SCI & OCEANOG, CITY COL SAN FRANCISCO, 81- *Mem:* Western Soc Naturalists; Nat Asn Underwater Instrs; Sigma Xi. *Res:* Development of the skin; cells of the dermis. *Mailing Add:* Dept of Biol & Oceanog City Col of San Francisco 50 Phelan Ave San Francisco CA 94112

NEFF-DAVIS, CAROL ANN, PHARMACOKINETICS, DRUG METABOLISM. *Educ:* Ohio State Univ, MS, 72. *Prof Exp:* RES SPECIALIST, COL VET MED, UNIV ILL, 79- *Mailing Add:* Res Spec Basic Sci Univ Ill 3615 Vet Med Urbana IL 61801

NEFKENS, BERNARD MARIE, b Heerlen, Neth, July 27, 34; US citizen; m 61; c 3. ELEMENTARY PARTICLE PHYSICS. *Educ:* Univ Utrecht, doctorandus, 59; Univ Nijmegem, Dr(physics), 67. *Prof Exp:* Res assoc physics, Purdue Univ, 59-62; res asst prof, Univ Ill, 62-66; from asst prof to assoc prof, 66-74, PROF PHYSICS, UNIV CALIF, LOS ANGELES, 74- *Concurrent Pos:* Visitor, Europ Orgn Nuclear Res, 58-59 & 71-73; vis scientist, Saclay, 78-79 & 88-89; US Dept Energy res grant, 67- *Mem:* Fel Am Phys Soc. *Res:* Fundamental properties of elementary particles; test of basic symmetries such as time reversal invariance, charge symmetry, cp invariance; standard model of quarks and leptons. *Mailing Add:* Physics Dept Univ Calif Los Angeles CA 90024-1547

NEFSKE, DONALD JOSEPH, b Detroit, Mich, Dec 18, 38; m 83. ACOUSTICS, STRUCTURAL-DYNAMICS. *Educ:* Univ Detroit, BS, 62; Univ Mich, MS, 64, PhD(eng mech), 69. *Prof Exp:* Res engr, 69-70, sr res engr, 70-85, SR STAFF ENGR, GEN MOTORS RES LABS, 85- *Mem:* Am Soc Mech Engrs; Acoust Soc Am; Soc Automotive Engrs; Am Inst Aeronaut & Astronaut; Soc Eng Sci; Sigma Xi. *Res:* Acoustics; structural-dynamics, noise and vibration; structural-acoustic interaction. *Mailing Add:* Eng Mech Dept Gen Motors Res Lab Warren MI 48090-9055

NEFT, NIVAED, b Shooks, Minn, Apr 1, 22. BIOCHEMISTRY, MEDICAL TECHNOLOGY. *Educ:* Col St Benedict, BS, 46; St Cloud Univ, MT, 52; Utah State Univ, PhD(biochem), 72. *Prof Exp:* Sci instr, Cathedral High Sch, St Cloud, 48-51; med technologist, St Cloud Hosp, 52-56 & Clinica Font Martelo, Humacao, PR, 56-58; chief med technologist & teaching supvr, St Benedict's Hosp, Ogden, 59-64; sci instr anat, physiol & chem, St Benedict's Sch Nursing, 64-68; assoc prof integrated sci core, Sch Allied Health, Weber State Col, 73-77; ASSOC PROF ORG & BIOCHEM, COL ST BENEDICT, 77- *Mem:* AAAS; affil mem Am Soc Clin Pathologists; Am Chem Soc. *Res:* Chemistry of antimycin A and antimicrobial agents; integrated science core curriculum. *Mailing Add:* 511 Ninth Ave St Cloud MN 56303

NEGELE, JOHN WILLIAM, b Cleveland, Ohio, Apr 18, 44; m 67; c 2. THEORETICAL PHYSICS. *Educ:* Purdue Univ, Lafayette, BS, 65; Cornell Univ, PhD(theoret physics), 69. *Prof Exp:* NATO fel & vis physicist, Niels Bohr Inst, Copenhagen, Denmark, 69-70; from vis asst prof to assoc prof, 70-79, PROF PHYSICS, MASS INST TECHNOL, 79-, DIR, CTR THEORET PHYSICS, 89- *Concurrent Pos:* Alfred P Sloan Foun res fel, 72; consult, Los Alamos Sci Lab, 73, Argonne Nat Lab, 77 & Lawrence Livermore Lab, 77; mem, Brookhaven Nat Lab, Tandem Prog Adv Comt, 73-77, co-ed, Advan Nuclear Physics, 77-; fel, Japan Soc Prom Sci, 81; John Simon Guggenheim fel, 82; head, theoret Div Dept Physics, Mass Inst Technol, 88-; NATO post doctoral fel, 69-70. *Mem:* Fel Am Phys Soc; fel AAAS; Am Asn Prom Sci. *Res:* Theoretical physics, including nuclear condensed matter, and particle physics; many-body theory; microscopic theory of nuclear structure; hadronic structure; lattice gauge theory; computational physics. *Mailing Add:* Rm 6-308 Dept Physics Mass Inst Technol 77 Mass Ave Cambridge MA 02139

NEGGERS, JOSEPH, b Amsterdam, Netherlands, Jan 10, 40; US citizen; m 65; c 2. MATHEMATICS. *Educ:* Fla State Univ, BS, 59, MS, 60, PhD(math), 63. *Prof Exp:* Asst prof math, Fla State Univ, 63-64; sci asst math, Univ Amsterdam, 64-65; lectr pure math, King's Col, London, 65-66; vis assoc prof, Univ PR, 66-67; from asst prof to assoc prof, 67-79, PROF MATH, UNIV ALA, 79- *Mem:* Am Math Soc; Math Asn Am. *Res:* Algebra; derivations and automorphisms on local-rings, associated structures; partially ordered sets. *Mailing Add:* Box 870350 Univ of Ala Tuscaloosa AL 35487-0350

NEGIN, MICHAEL, b Tampa, Fla, Dec 19, 42; div; c 3. ELECTRICAL ENGINEERING, BIOMEDICAL ENGINEERING. *Educ:* Univ Fla, BEE, 64, MSE, 65, PhD(elec eng), 68; Temple Univ, MS, 75. *Prof Exp:* From asst prof to prof Elec Eng, Drexel Univ, 71-80; prof, dept comput & info sci, Temple Univ, 80-82; vpres res & develop, Kulicke & Soffa, 83; PRES, MNEMONICS, INC, 84- *Concurrent Pos:* Nat Inst Gen Med Sci spec fel, 72-74. *Mem:* Sigma Xi; Inst Elec & Electronics Engrs; Soc Mfg Engrs; Mach Vision Assoc. *Res:* Early detection of breast cancer by computerized thermography and radiography; sleep electroencephalographic analysis by computer; digital simulation of biosystems; electromyographic and electroencephalographic studies and the relationship between these bioelectric events; dental radiography by digital computer; machine vision and image processing; industrial and manufacturing automation. *Mailing Add:* Mnemonics, Inc PO Box 352 Moorestown NJ 08057-0352

NEGISHI, EI-ICHI, b Shinkyo, Repub China, July 14, 35; Japanese citizen; m 60; c 2. ORGANIC CHEMISTRY. *Educ:* Univ Tokyo, BE, 58; Univ Pa, PhD(chem), 63. *Prof Exp:* Res chemist, Teijin Ltd, 58-66; res assoc, Purdue Univ, 66-72; from asst prof to assoc prof, Syracuse Univ, 72-79; PROF, PURDUE UNIV, 79- *Honors & Awards:* J Simon Guggenheim Mem Found Fel, 87. *Mem:* Am Chem Soc; Sigma Xi; Japan Chem Soc. *Res:* Organic and organometallic chemistry; development of new selective synthetic methods and applications to the synthesis of natural products of biological and medicinal interest. *Mailing Add:* Dept Chem Purdue Univ West Lafayette IN 47907

NEGISHI, MASAHIKO, MOLECULAR BIOLOGY. *Educ:* Osaka Univ, Japan, PhD(biochem), 72. *Prof Exp:* VIS SCIENTIST, NAT INST ENVIRON HEALTH SCI, NIH, 83- *Res:* Steroid hydroxylases in liver microsomes; sex-dependent gene regulation. *Mailing Add:* Lab Pharmacol Bldg 19 Rm 1910 Nat Inst Environ Health Sci PO Box 12233 Research Triangle Park NC 27709

NEGRO-VILAR, ANDRES F, b Buenos Aires, Argentina, Jan 1, 40; m; c 3. ENDOCRINOLOGY, MEDICAL RESEARCH. *Educ:* Univ Buenos Aires, MD, 63; Univ Sao Paulo, PhD(physiol), 69. *Prof Exp:* Head, Lab Protein & Polypeptide Hormones, Inst Neurobiol, Buenos Aires, Arg, 74-77; from assoc prof to prof physiol, Univ Tex Health Sci Ctr, Dallas, Tex, 77-83; HEAD, SECT REPRODUCTIVE NEUROENDOCRINOL, NAT INST ENVIRON HEALTH SCI, NIH, 83-, CHIEF LAB MOLECULAR & INTEGRATIVE NEUROSCI, 87- *Concurrent Pos:* Mem, adv comt Sci Prog, Endocrine Soc, 82-90; Adv, Human Reproduction Prog, WHO, 84-; preceptor, Pharmacol Res Assoc Prog, Nat Inst Gen Med Sci, NIH, 84-; chmn, Toxicol Panel, Spec Prog Human Reproduction, WHO, Geneva, Switz, 86; mem, Biochem Endocrinol Study Sect, 89-92; ed-in-chief, Endocrine Rev, 91-95. *Mem:* Endocrine Soc; Am Physiol Soc; Soc Neurosci; Soc Study Reproduction; AAAS. *Res:* Neuroendocrinology, neuropharmacology and neurochemistry; molecular mechanisms of hormone action; molecular biology of brain peptides; reproductive endocrinology. *Mailing Add:* 639 Kensington Dr Chapel Hill NC 27514

NEGUS, NORMAN CURTISS, b Portland, Ore, Sept 20, 26; m 48; c 4. ZOOLOGY. *Educ:* Miami Univ, BA, 48, MA, 50; Ohio State Univ, PhD(zool), 56. *Prof Exp:* Asst zool, Miami Univ, 48-50; res fel, Ohio State Univ, 51-55; from instr to prof, Tulane Univ, 55-70; prof zool, 70-76, PROF BIOL, UNIV UTAH, 76- *Concurrent Pos:* Vis investr, SEATO Med Res Lab, Thailand, 65-66. *Mem:* Am Soc Mammalogists; Ecol Soc Am; Am Soc Zoologists; Am Asn Anatomists; Wildlife Soc. *Res:* Mammalian population dynamics; reproductive physiology; molting in mammals; orientation and movements of mammals. *Mailing Add:* Dept Biol Univ Utah 201 S Biol Bldg Salt Lake City UT 84112

NEHER, CLARENCE M, b Twin Falls, Idaho, May 14, 16; m 39; c 4. ORGANIC CHEMISTRY. *Educ:* Manchester Col, AB, 37; Purdue Univ, MS, 39, PhD(chem), 41. *Prof Exp:* Asst, Purdue Univ, 37-40; res chemist, Ethyl Corp, 41-45, res supvr, 45-51, asst dir res & develop labs, 51-54, proj mgr, 54-57, dir com develop, 57-63, spec assignment, 63-64, vpres & gen mgr, Plastics Div, 64-69, sr vpres, 69-81, mem bd dirs, 70-81; RETIRED. *Mem:* Mfg Chem Asn; Am Chem Soc; Chem Mkt Res Asn; Com Chem Develop Asn; Soc Plastics Indust. *Res:* Chlorination; polymers; chlorination of aliphatics and aromatics; cracking of chlorocarbons; chlorination methane and ethane. *Mailing Add:* 861 Delgado Dr Baton Rouge LA 70808

NEHER, DAVID DANIEL, b McCune, Kans, July 12, 23; m 50; c 3. SOILS. *Educ:* Kans State Univ, BS, 46, MS, 48; Utah State Univ, PhD(soil sci), 59. *Prof Exp:* Instr soils, Kans State Univ, 48-49; from asst prof to assoc prof, 49-65, PROF SOILS, TEX A&I UNIV, 65- *Concurrent Pos:* Interim dean agr, Tex A&I Univ. *Mem:* Soil Sci Soc Am; Am Soc Agron; Crop Sci Soc Am; Soil Conserv Soc Am. *Res:* Causes for soil salinity problems developing in Kleberg County, Texas. *Mailing Add:* Col Agr PO Box 156 Tex A&I Univ Kingsville TX 78363

NEHER, DEAN ROYCE, b Enterprise, Kans, Feb 10, 29; m 53; c 4. PHYSICS, COMPUTER SCIENCE. *Educ:* McPherson Col, BS, 54; Univ Kans, MS, 59, PhD(physics), 64. *Prof Exp:* From asst prof to assoc prof, 61-68, prof physics, 68-82, dir, compt ctr, 70-79, PROF COMPT SCI & PHYSICS, BRIDGEWATER COL, 82- *Mem:* Am Asn Physics Teachers. *Res:* Nuclear physics; use of computers in teaching. *Mailing Add:* Bc Box 24 Bridgewater VA 22812

NEHER, ERWIN, b Landberg, Ger, March 20, 44; m 78; c 5. CELL CHANNELS. *Educ:* Tech Univ Munich, Vordipl, 65, PhD, 70, Univ Wisconsin-Madison, MS, 67. *Prof Exp:* Res assoc, Max Planck Inst Psychiat, 70-72; res assoc, 72-83, RES DIR, MAX PLANCK INST BIOPHYS CHEM, 83- *Concurrent Pos:* Res assoc, Yale Univ, 75-76; Fairchild scholar, Calif Inst Tech, 88-89. *Honors & Awards:* Nobel Prize Physiol/Med, 91; Louisa Gross-Horwitz Prize, Columbia Univ, 86. *Mailing Add:* Max Planck Inst Biophys Chem Am Fassberg, Postfach 2841 D-3400 Gottingen Germany

NEHER, GEORGE MARTIN, b Chicago, Ill, June 4, 21; m 45. VETERINARY SCIENCE. *Educ:* Purdue Univ, BS, 47, MS, 50, PhD(endocrinol), 53, DVM, 67. *Prof Exp:* From asst prof to assoc prof vet sci, 54-66, PROF VET PHYSIOL & PHARMACOL, SCH VET SCI & MED, PURDUE UNIV, WEST LAFAYETTE, 66- *Mem:* Assoc Am Vet Radiol Soc; assoc Am Rheumatism Asn; Conf Res Workers Animal Dis; Am Vet Med Asn; Soc Study Reproduction. *Res:* Histochemistry and pathogenesis of the arthritides of animals; experimental hypopituitarism in swine; physiologic effects of noise. *Mailing Add:* 7025 S 100 Lafayette IN 47805

NEHER, LELAND K, b Porterville, Calif, Dec 2, 20; m 43; c 7. NUCLEAR PHYSICS. *Educ:* Pomona Col, BA, 43; Univ Calif, PhD(physics), 53. *Prof Exp:* MEM STAFF, LOS ALAMOS NAT LAB, 53- *Mem:* AAAS; Am Phys Soc. *Res:* Nuclear science. *Mailing Add:* 205 Rio Bravo Los Alamos NM 87544

NEHER, MAYNARD BRUCE, b Greenville, Ohio, Apr 2, 23; m 44; c 3. CHEMISTRY. *Educ:* Manchester Col, AB, 44; Univ Calif, San Diego, AB, 86; Purdue Univ, PhD(chem), 47. *Prof Exp:* Asst chem, Purdue Univ, 44-47; asst prof, Univ Ohio, 47-51; res chemist, chem dept, Columbus div, Battelle Mem Inst, 51-56, asst div chief, 56-71, res chemist, 71-78, qual assurance coordr, 78-84; SOFTWARE ENGR, CUBIC CORP, SAN DIEGO, 86. *Mem:* Am Chem Soc. *Res:* Nitroparaffin derivatives as insecticidal compounds; aldol condensation of fluoroform with acetone; preparation of selected diaryl nitroalkanes and their derivatives; application of computer techniques in organic analysis; structural organic chemistry; mass spectroscopy; analytical quality control and quality assurance; gas-liquid chromatography. *Mailing Add:* 12537 Caminito de la Gallarda San Diego CA 92128

NEHER, ROBERT TROSTLE, b Mt Morris, Ill, Nov 1, 30; m 54; c 3. PLANT TAXONOMY, ENVIRONMENTAL BIOLOGY. *Educ:* Manchester Col, BS, 53; Univ Ind, MAT, 55, PhD(bot), 66; Bethany Sem, MRE, 57. *Prof Exp:* From asst prof to assoc prof, 57-67, chmn dept, 67-78, PROF BIOL, UNIV LA VERNE, 67-, CHMN NATURAL SCI DIV, 72- *Concurrent Pos:* NSF fac fel, 62-63; mem environ qual comn, City of La Verne, 72-76, mem city coun, 76-; aquaculture consult, Am/China, 81; Los Angeles Co Watershed Comnr, 76-90; Los Angeles Co Mosquito Abatement Comnr, 90- *Mem:* AAAS; Sigma Xi; Am Soc Plant Taxon. *Res:* Systematic studies in Tagetes, stressing chemotaxonomy and cytogenetics; study and development of environmental control models; development of multi-stage aquaculture systems. *Mailing Add:* Dept Biol Univ La Verne 1950 Third St La Verne CA 91750

NEHLS, JAMES WARWICK, b Memphis, Tenn, June 30, 26; m 53; c 3. INORGANIC CHEMISTRY. *Educ:* Univ Tenn, BS, 48, MS, 49, PhD(chem), 52. *Prof Exp:* Asst chem, Univ Tenn, 49-52; chemist, E I du Pont de Nemours & Co Tenn, 52, Ind, 52-53, & SC, 53-60; chemist, Oak Ridge Opers, US Atomic Energy Comn, 60-65, asst to dir, Res & Develop Div, 65-73; chemist, Res & Tech Support Div, 73-78, mgr, Isotope Prod & Distribution, 81-88, MGR, DATA SYSTS & DEVELOP, US DEPT ENERGY, 88- *Mem:* AAAS; Am Chem Soc. *Res:* Transplutonium elements; radioisotope production; exchange reactions; radiochemistry. *Mailing Add:* 121 Balboa Circle Oak Ridge TN 37830

NEHORAI, ARYE, b Haifa, Israel, Sept 10, 51; US citizen; m 79; c 2. SIGNAL PROCESSING, SYSTEM IDENTIFICATION. *Educ:* Technion, BSc, 76, MSc, 79; Stanford, PhD(elec eng), 83. *Prof Exp:* Res/teaching asst, Technion, Israel, 76-79; res asst, Stanford Univ, 79-83, res assoc, 83-84; res engr, Systs Control Technol, 84-85; asst prof, 85-89, ASSOC PROF, YALE UNIV, 89- *Concurrent Pos:* Prin investr, USAF, 87-93, US Navy, 91-93; assoc ed, Trans Acoust, Speech & Signal Processing, Inst Elec & Electronics Engrs, 87-89, Circuits, Systs & Signal Processing, 90; vis prof, Uppsala Univ, Sweden, 89 & Technion, Israel, 89-90; consult, Defense Sci Orgn, Singapore, 90-91. *Mem:* Inst Elec & Electronics Engrs; Sigma Xi. *Res:* Signal processing - adaptive filtering, sensor array processing (source localization, sonar, radar), system identification; biomedical engineering; communications; algorithm development and performance analysis. *Mailing Add:* Dept Elec Eng Yale Univ PO Box 2157 New Haven CT 06520

NEHRKORN, THOMAS, b Fulda, Ger, Jan 8, 57. NUMERICAL WEATHER PREDICTION. *Educ:* Univ Hamburg, Ger, Vordiplom, 78; Colo State Univ, Ft Collins, MS, 81; Mass Inst Technol, PhD(meteorol), 85. *Prof Exp:* Res asst, Colo State Univ, 79-81 & Mass Inst Technol, 81-85; STAFF SCIENTIST, ATMOSPHERIC & ENVIRON RES, INC, 85- *Mem:* Am Meteorol Soc. *Res:* Use of optimal interpolation and adjoint technique for data assimilation with atmospheric models; formulation and optimization of spectral models; prediction and diagnosis of atmospheric humidity, clouds, and precipitation. *Mailing Add:* AER Inc 840 Memorial Dr Cambridge MA 02139

NEI, MASATOSHI, b Miyazaki, Japan, Jan 2, 31; m 63; c 2. POPULATION GENETICS, EVOLUTION. *Educ:* Miyazaki Univ, BS, 53; Kyoto Univ, MS, 55, PhD(genetics), 59. *Hon Degrees:* DSc, Kyushu Univ, 77. *Prof Exp:* Instr, Kyoto Univ, 58-62; geneticist, Nat Inst Radiol Sci, Japan, 62-64, chief geneticist, 65, head lab, 65-69; from assoc prof to prof, Brown Univ, 69-72; actg dir, Univ Tex, Houston, 78-80 & 86-87, prof pop genetics, Ctr Demog & Pop Genetics, 72-90; DISTINGUISHED PROF BIOL & DIR, INST MOLECULAR EVOLUTIONARY GENETICS, PENN STATE UNIV, UNIVERSITY PARK, PA, 90- *Concurrent Pos:* Assoc ed, Genetics Soc Am, 77-85; managing ed, Molecular Biol & Evol, 83- *Honors & Awards:* Japan Soc Human Genetics Award, 77; Kihara Prize, Genetics Soc Japan, 90. *Mem:* AAAS; fel Am Acad Arts & Sci; Am Soc Naturalists; Am Soc Human Genetics; Soc Study Evolution; Genetics Soc Am. *Res:* Molecular evolution; population genetics; human evolution. *Mailing Add:* Dept Biol Penn State Univ 328 Mueller Lab University Park PA 16802

NEIBLING, WILLIAM HOWARD, b Highland, Kans, Mar 6, 52; m 77; c 2. EROSION MECHANICS, SEDIMENT TRANSPORT. *Educ:* Kans State Univ, BS, 74, MS, 76; Purdue Univ, PhD(agr eng), 83. *Prof Exp:* AGR ENGR, ARG RES SERV, USDA, 76- *Mem:* Am Soc Agr Engrs; Soil Conservation Soc Am; Sigma Xi. *Res:* Erosion mechanics; sediment transport; identify variables important in the erosion/deposition process and quantify relationships relating these variables to prediction of the detachment of soil by raindrop impact and overland flow and subsequent transport or deposition of sediment. *Mailing Add:* Dept Agr Eng Univ Wyo Univ Sta PO Box 3295 Laramie WY 82071

NEIBURGER, MORRIS, meteorology, cloud physics; deceased, see previous edition for last biography

NEIDELL, NORMAN SAMSON, b New York, NY, Mar 11, 39; m 63; c 5. GEOPHYSICS. *Educ:* NY Univ, BA, 59; Univ London, DIC, 61; Cambridge Univ, PhD(geod, geophys), 64. *Prof Exp:* Res geophysicist, Gulf Res & Develop Co, 64-68; geophys researcher, Seismic Comput Corp, 68-72; assoc prof, Univ Houston, 80-89; CHMN & PRES, N S NEIDELL & ASSOCS. *Concurrent Pos:* Consult & dir, Geoquest Int, Inc, 73-80; assoc ed marine opers, Geophysics, 76-80; exec vpres, founder & dir, Zenith Explor Co, Inc, 77-; managing founder & partner, Delphian Signals Ltd, 80- *Honors & Awards:* Best Presentation Award, Soc Explor Geophysicists, 74. *Mem:* Soc Explor Geophysicists; Europ Asn Explor Geophysicists; Soc Photo-Optical Instrument Engrs. *Res:* Signal analysis and computer processing in the earth sciences; digital computer applications in geology and geophysics; oil and gas explorations. *Mailing Add:* 315 Vanderpool Lane Houston TX 77024

NEIDERHISER, DEWEY HAROLD, b Masontown, Pa, Jan 22, 35. BIOCHEMISTRY. *Educ:* Duquesne Univ, BS, 56; Univ Pittsburgh, MS, 59; Univ Wis, PhD(biochem), 63. *Prof Exp:* ASST PROF BIOCHEM & RES CHEMIST, VET ADMIN MED CTR, SCH MED, CASE WESTERN RESERVE UNIV, 62- *Mem:* Res Soc Alcoholism; AAAS; Am Gastroenterol Asn; Am Oil Chemist Soc. *Res:* Biochemistry of human cholesterol gallstone formation; alcoholism. *Mailing Add:* 258 North St Cleveland OH 44106

NEIDERS, MIRDZA ERIKA, b Riga, Latvia, Aug 21, 33; US citizen; div; c 2. ORAL PATHOLOGY, IMMUNOLOGY. *Educ:* Univ Mich, DDS, 58; Univ Chicago, MS, 61; State Univ NY Buffalo, cert periodont, 74. *Prof Exp:* Instr dent, Zoller Dent Clin, Univ Chicago, 61-62; from asst prof to assoc prof oral path, 62-70, PROF ORAL PATH & ORAL BIOL, SCH DENT, STATE UNIV NY, BUFFALO, 70- *Concurrent Pos:* Consult oral path, Vet Admin Hosp, 65-73; res assoc, Roswell Park Mem Inst, 63-72; counr, Am Acad Oral Path, 73-76; guest scientist immunol, NIH, 74-77, mem OBMS sect & BBS sect, NIH grant review bd, 77-81. *Mem:* Am Acad Oral Path; Am Acad Periodont; Am Dent Asn; Int Asn Dent Res; Sigma Xi. *Res:* Immunoglobulin synthesis; cell detachment; in vitro and in vivo studies on bone resorption; periodontal disease; cellular immunology; B gingivalis collagluase and other enzymes. *Mailing Add:* 310 Foster Hall Sch of Dent State Univ NY Buffalo NY 14214

NEIDHARDT, FREDERICK CARL, b Philadelphia, Pa, May 12, 31; m 56, 77; c 3. MICROBIOLOGY, BIOCHEMISTRY. *Educ:* Kenyon Col, BA, 52; Harvard Univ, PhD(bact), 56. *Hon Degrees:* DSc, Kenyon Col; DSc, Purdue Univ. *Prof Exp:* Am Cancer Soc res fel, Inst Pasteur, Paris, 56-57; Harold C Ernst res fel, Harvard Med Sch, 57-58, instr bact & immunol, 58-59, assoc, 59-61; from assoc prof biol sci to prof & assoc head dept, Purdue Univ, 61-70; prof microbiol & chmn dept, 70-82, DISTINGUISHED UNIV PROF MICROBIOL & IMMUNOL, MED SCH, UNIV MICH, ANN ARBOR, 82- *Concurrent Pos:* Mem microbial chem study sect, NIH, 65-69, 89-; NSF sr fels, Univ Inst Microbiol, Copenhagen, 68-69; mem comn scholars, Ill Bd Higher Educ, 73-80; mem microbiol comt, Nat Bd Med Examr, 75-83; Alexander Von Humboldt Sr US Scientist Award, 79. *Honors & Awards:* Award Bact & Immunol, Eli Lilly & Co, 66. *Mem:* AAAS; Am Soc Microbiol (pres, 81-82); Am Soc Biol Chem & Molecular Biol; Soc Gen Physiol; Genetics Soc Am. *Res:* Regulation of gene expression in bacteria; regulation of bacterial metabolism; regulation of macromolecule synthesis in bacteria; genetics; molecular biology. *Mailing Add:* Dept Microbiol 6643 Med Sci II Univ of Mich Med Sch Ann Arbor MI 48109-0620

NEIDHARDT, WALTER JIM, b Paterson, NJ, June 19, 34; m 62; c 2. LOW TEMPERATURE PHYSICS, QUANTUM PHYSICS. *Educ:* Stevens Inst Technol, ME, 56, MS, 58, PhD(physics), 62. *Prof Exp:* Instr, Newark Col Eng, 62-63, asst prof, 63-67, ASSOC PROF PHYSICS, NJ INST TECHNOL, 67- *Concurrent Pos:* Consult ed, J Am Sci Affil, 68- *Mem:* Am Phys Soc; Am Inst Physics Teachers; fel Am Sci Affil; Sigma Xi; NY Acad Sci. *Res:* Application of quantum physics to low temperature phenomena; examination of the nature of science, its proper domains and limits; emphasis on those integrative concepts that are common to science, philosophy and religion. *Mailing Add:* 146 Park Ave Randolph NJ 07869

NEIDIG, DONALD FOSTER, b Harrisburg, Pa, Aug 6, 44; m 79. ASTRONOMY. *Educ:* Dickinson Col, BS, 66; Pa State Univ, MS, 68, PhD(astron), 76. *Prof Exp:* Fac mem sci, East Pennsboro High Sch, Pa, 68-69; instr math & physics, Alliance Col, 71-73; ASTROPHYSICIST SOLAR PHYSICS, AIR FORCE PHILLIPS LAB, GEOPHYS DIRECTORATE, 76- *Concurrent Pos:* Partic, Skylab Solar Flare Workshops, NASA & NSF, 76-77; partic, Solar Maximum Mission Flare Workshops, NASA, 83-84, Solar-Terrestrial Predictions Workshops, NOAA, 79, co-chair, 84; co-chair, Workshop XV & XXVII Comt on Space Res meeting, 88; mem, NASA Mgt Oper Working Group, Solar Physics, 86, NASA Working Group Solar Cycle Predictions, 88. *Mem:* Am Astron Soc; Sigma Xi; Int Astron Union. *Res:* Physics of solar flares; solar activity forecasting. *Mailing Add:* Phillips Lab Geophys Directorate Sacramento Peak Observ Sunspot NM 88349

NEIDIG, HOWARD ANTHONY, b Lemoyne, Pa, Jan 25, 23; m 46, 72; c 4. PHYSICAL CHEMISTRY, ORGANIC CHEMISTRY. *Educ:* Lebanon Valley Col, BS, 43; Univ Del, MS, 46, PhD(chem), 48. *Prof Exp:* Instr chem, Univ Del, 46-48; from asst prof to assoc prof, 48-59, HEAD DEPT, LEBANON VALLEY COL, 51-, PROF CHEM, 59- *Concurrent Pos:* Prog ed, Modular Lab Prog Chem, 70- *Mem:* AAAS; Am Chem Soc; Am Inst Chemists; Nat Sci Teachers Asn; Sigma Xi. *Res:* Reaction mechanism; molecular rearrangements; mechanism of oxidation and reduction reactions in organic chemistry. *Mailing Add:* Chem Educ Resources PO Box 357 Palmyra PA 17078

NEIDINGER, RICHARD DEAN, b Topeka, Kans, Feb 8, 55; m 78. BANACH SPACES, NUMERICAL ANALYSIS. *Educ:* Trinity Univ, BA, 77; Univ Tex, Austin, MA, 80, PhD(math), 84. *Prof Exp:* ASSOC PROF MATH, DAVIDSON COL, NC, 84- *Honors & Awards:* George Pólya Award, Math Asn Am. *Mem:* Am Math Soc; Math Asn Am; Asn Comput Mach. *Res:* The isomorphic theory of Banach Spaces; characterization of operators which map closed bounded convex sets to closed sets; automatic differentiation and APL; arbitrary order, multivariable, automatic differentiation. *Mailing Add:* Dept Math Davidson Col Davidson NC 28036-1719

NEIDLE, ENID ANNE, b New York, NY, Apr 6, 24; m 49; c 2. PHARMACOLOGY, PHYSIOLOGY. *Educ:* Vassar Col, AB, 44; Columbia Univ, PhD(physiol), 49. *Hon Degrees:* DSc, Georgetown Univ, 86. *Prof Exp:* Assoc pharmacol, Jefferson Med Col, 49-50; instr biol, Brooklyn Col, 50-54; from instr to prof physiol & pharmacol, 55-78, PROF PHARMACOGY & CHMN, COL DENT & GRAD FAC ARTS & SCI, NY UNIV, 78-, DIR RES, 80- *Concurrent Pos:* USPHS grant, 60; asst exec dir, Sci Affairs, Am Dental Asn. *Mem:* AAAS; Am Asn Dent Schs; Sigma Xi; Am Physiol Soc; Harvey Soc; hom mem Am Dental Asn; hon fel Int Col Dentists; Int Asn Dent Res. *Res:* Vasomotor innervation of orofacial structures; circulation in the dental pulp; contribution of the mandibular nerve to growth, development and vasomotion in orofacial structures. *Mailing Add:* Am Dental Assoc 211 E Chicago Ave Chicago IL 60611

NEIDLEMAN, SAUL L, b New York, NY, Oct 3, 29; m 56; c 4. BIOCHEMISTRY, ENZYMOLOGY. *Educ:* Mass Inst Technol, MS, 52; Univ Ariz, PhD(biochem), 59. *Prof Exp:* Biophysicist, Peter Bent Brigham Hosp, Boston, 53-54; res assoc agr biochem, Univ Ariz, 58-59; sr res microbiologist, Squibb Inst Med Res Div, Olin Corp, 59-67, res assoc, 67-69, sect head microbial biochem, 69-72; sr scientist, Cetus Corp, 73-80, dept dir, 80-82, distinguished res fel, New Ventures Res, 82-90, assoc dir, chem dept, 85-90; SR DIR, PROJ ACQUISITION & PLANNING, BIOSOURCE GENETICS CORP, 90- *Mem:* NY Acad Sci; Am Soc Microbiol; Soc Indust Microbiol; Am Chem Soc; Am Oil Chem Soc. *Res:* Antibiotic biosynthesis; microbial metabolism; enzyme reactions; oil microbiology. *Mailing Add:* Biosource Genetics Corp 3333 Vaca Valley Pkwy Vacaville CA 95688

NEIDLINGER, HERMANN H, b Bolanden, Ger, Feb 13, 44; US citizen; m 73; c 3. STRUCTURE & PROPERTY RELATIONSHIPS OF POLYMERS, POLYMERS FOR SOLAR ENERGY APPLICATIONS. *Educ:* Johannes Gutenberg Univ, Mainz, Cand Nat, 68, Dipl Chem, 71, DrRerNat, 75. *Prof Exp:* NATO res fel, Stanford Univ, 75-76; from asst prof to assoc prof, Dept Polymer Sci, Univ Southern Miss, 76-82; sr scientist, Solar Energy Res Inst, 82-89; DIR POLYMER RES & DEVELOP, SOLA/BARNES-HIND, PILKINGTON VISION CARE, 89- *Concurrent Pos:* Res chemist, Badische Anilin-und Sodafabrik AG, 70-71, consult, 79; vis res fel, Stanford Univ, 72; prin investr, US Dept Energy, 77-83; consult, Solar Energy Res Inst, 81-82, Cities Servs, 81- & Alcon Labs, 87-88. *Mem:* Am Chem Soc; Soc Plastics Engrs. *Res:* Macromolecular design and structure/property relationships for polymers in ophthalmic, solar and mobility control applications; environmental stabilization of polymers; photon addressable polymer materials. *Mailing Add:* 810 Kifer Rd Sunnyvale CA 94086

NEIE, VAN E(LROY), b Clifton, Tex, Nov 1, 38; m 63; c 2. PHYSICS EDUCATION. *Educ:* McMurry Col, BA, 61; NTex State Univ, MS, 66; Fla State Univ, PhD(sci educ), 70. *Prof Exp:* Instr physics, NTex State Univ, 64-67; teacher, Fla High Sch, Tallahassee, 68-69; asst prof, 70-76, ASSOC PROF PHYSICS & EDUC, PURDUE UNIV, WEST LAFAYETTE, 76- *Mem:* Nat Asn Res Sci Teaching; Nat Sci Teachers Asn; Asn Educ Teachers Sci; Am Asn Physics Teachers; Soc Col Sci Teachers. *Res:* Information theory and its application to lexical communication in science; problem solving in science. *Mailing Add:* Dept Physics & Educ Purdue Univ West Lafayette IN 47907

NEIHEISEL, JAMES, b Cincinnati, Ohio, June 3, 27; m 53; c 2. ENGINEERING GEOLOGY, GEOCHEMISTRY. *Educ:* Ohio State Univ, BS, 50; Univ SC, MS, 58; Ga Inst Technol, PhD(geophys sci), 73. *Prof Exp:* Chemist, Va Carolina Chem Corp, SC, 55-57; chief geol & petrog sect, South Atlantic Div Lab, Corps Eng, US Army, 58-77; GEOLOGIST, ENVIRON PROTECTION AGENCY, 77- *Concurrent Pos:* Spec lectr, Ga Inst Technol, 65-71; instr, Ga State Univ, 66-77, George Mason Univ, 81-82, Kennesaw State Col, 88-90; mem comt interagency radiation res & policy coord. *Mem:* Fel Geol Soc Am; affil Am Soc Civil Engrs. *Res:* Standard development for radon and low level radioactive waste; ocean disposal regulations and investigations of radioactive waste; remedial measures for radium contaminated soils at superfund sites. *Mailing Add:* Environ Protection Agency Off Radiation Prog ANR 461 Washington DC 20460

NEIHOF, REX A, b Ponca City, Okla, Oct 31, 21; m 49; c 3. HYDROCARBON FUEL CONTAMINATION. *Educ:* Tex Technol Col, BS, 43; Univ Minn, PhD(biochem), 50. *Prof Exp:* Chem engr, Gates Rubber Co, Colo, 43-45; asst, Univ Minn, 46-48, instr physiol chem, 48-49; phys chemist, NIH, 49-55, 57-58; Nat Heart Inst spec fel, Physiol Inst, Univ Uppsala, 55-57; RES CHEMIST, US NAVAL RES LAB, 58- *Concurrent Pos:* Vis scientist, Univ Calif, San Diego, 70-71. *Mem:* Am Soc Microbiol; Am Chem Soc; Sigma Xi; Soc Indust Microbiol. *Res:* Physical chemistry and electrochemistry of membrane processes; permselective membranes; biophysical studies on biological membranes; microbial cell walls; marine fouling; microbial deterioration of hydrocarbons; microbial ecology; low temperature properties of hydrocarbons. *Mailing Add:* Chem Div US Naval Res Lab Washington DC 20375-5000

NEIKIRK, DEAN P, b Oklahoma City, Okla, Oct 31, 57; m; c 3. ADVANCED FABRICATION TECHNIQUES. *Educ:* Okla State Univ, BS, 79; Calif Inst Technol, MS, 81, PhD(appl physics), 84. *Prof Exp:* Asst prof, 84-88, ASSOC PROF ELEC ENG, UNIV TEX, AUSTIN, 88- *Concurrent Pos:* Eng Found fac award, Univ Tex, Austin, 84-85. *Mem:* Am Phys Soc; sr mem Inst Elec & Electronic Engrs; Am Vacuum Soc; Soc Photo-Optical Instrumentation Engrs. *Res:* Advanced fabrication techniques, such as molecular beam epitaxy, for millimeter and submillimeter-wave device development; novel structures for high frequency generation, detection, wave guiding, and radiation for monolithic integration. *Mailing Add:* Dept Elec Eng Univ Tex Austin TX 78712

NEIL, GARY LAWRENCE, b Regina, Sask, June 13, 40; m 62; c 2. PHARMACEUTICAL RESEARCH & DEVELOPMENT. *Educ:* Queen's Univ, Ont, BSc, 62; Calif Inst Technol, PhD(chem), 66. *Prof Exp:* Res assoc cancer res, 66-74, res head cancer res, 74-79, res mgr, Exp Biol Res, 79-81; vpres, Discover Res, 85-87; dir cancer res, Upjohn Co, Kalamazoo, 81-82, exec dir therapeutics, 82-85, vpres, Biotech & Basic Res, 88-89; sr vpres res & develop, 89-90, EXEC VPRES, WYETH-AYERT RES, RADNOR, PA, 90- *Mem:* Am Asn Cancer Res; Am Soc Pharmacol & Exp Therapeut; Am Chem Soc. *Res:* Enzyme kinetics; mode of action of antitumor agents; drug metabolism and pharmacokinetics. *Mailing Add:* Wyeth-Ayert Res PO Box 8299 Philadelphia PA 19101-1245

NEIL, GEORGE RANDALL, b Springfield, Mo, Apr 11, 48; m 78. ELECTRON LASERS. *Educ:* Univ Va, BS, 70; Univ Wis, MS, 73, PhD(nuclear eng), 77. *Prof Exp:* SCIENTIST, TRW CORP, 77- *Concurrent Pos:* Prin investr, Free Electron Laser Prog, TRW Corp, 80- *Mem:* Am Phys Soc; Sigma Xi; Soc Photo-Optical Instrumentation Engrs. *Res:* Free electron lasers; lasers and plasma physics research including controlled fusion and isotope separation, neutron and x-ray sources. *Mailing Add:* 203 Queens Dr W Williamsburg VA 23185

NEIL, THOMAS C, b Tacoma, Wash, Dec 21, 34; m 63; c 5. ORGANIC CHEMISTRY, ANALYTICAL CHEMISTRY. *Educ:* Earlham Col, AB, 56; Pa State Univ, MS, 60, PhD(org chem), 64. *Prof Exp:* Asst prof chem, Baldwin-Wallace Col, 64-66; assoc prof, 66-80, PROF CHEM, KEENE STATE COl, 80- *Concurrent Pos:* Mem, Bd Dir, Harsyd Chem, Inc, 82-83; pres, Avani Syst, 80- *Res:* Photochemistry; reactions of carbenes, waxes and macromolecules. *Mailing Add:* Dept Sci Keene State Col 229 Main St Keene NH 03431

NEILAND, BONITA J, b Eugene, Ore, June 5, 28; m 55. PLANT ECOLOGY, RESOURCE MANAGEMENT. *Educ:* Univ Ore, BS, 49; Ore State Univ, BA, 51; Univ Wis, PhD(bot), 54. *Prof Exp:* Instr biol, Univ Ore, 54-55; asst prof, Gen Exten Div, Ore State Syst Higher Educ, 55-60; from asst prof to prof bot & land resources, Univ Alaska, 61-87, head, Dept Land Resources & Agr Sci, 71-73, dir instr & pub serv, Sch Agr & Land Resources Mgt, 75-87, EMER PROF BOT & LAND RESOURCES, 87-; RETIRED. *Concurrent Pos:* NSF grant, 55-57 & 62-70; mem contract group, Proj Chariot, AEC, Alaska, 61-62; McIntire-Stennis Fund grant, 70-82; Bur Land Mgt grant, 71-78; US Forest Serv contracts, 74-84; co-prin investr proj, Arctic Willow Reestab, Alyeska Pipeline Serv Co, 78-81; instr community educ, Cent Oregon Community Col, 91- *Mem:* Fel AAAS; Ecol Soc Am; Arctic Inst NAm; Brit Ecol Soc; Soil Conserv Soc Am; Sigma Xi. *Res:* Comparisons of burned and unburned forest lands; analysis of forest and bog communities; vegetation, topography and ground ice correlations in the Fairbanks area; revegetation of denuded lands. *Mailing Add:* 69715 Holmes Rd Sisters OR 97759

NEILAND, KENNETH ALFRED, b Portland, Ore, Feb 18, 29; m 55; c 1. INVERTEBRATE ZOOLOGY. *Educ:* Reed Col, BA, 50; Univ Ore, MA, 53. *Prof Exp:* Asst parasitol, Reed Col, 50-51; asst physiol, Univ Ore, 51-53; asst zool, physiol & parasitol, Univ Calif, Los Angeles, 53-54; res fel physiol, Univ Ore, 55-56; instr biol sci, Ore Col Educ, 57-59; res biologist parasitol & comp physiol, Alaska Dept Fish & Game, 59-81; RETIRED. *Mem:* Am Soc Parasitologists; Sigma Xi. *Res:* Comparative parasitology; biology of helminth parasites; comparative physiology of molting in Crustacea; wildlife diseases and parasites. *Mailing Add:* 69715 Holmes Rd Sisters OR 97759

NEILANDS, JOHN BRIAN, b Glen Valley, BC, Sept 11, 21; nat US. BIOCHEMISTRY, MICROBIOLOGY. *Educ:* Univ Toronto, BS, 44; Dalhousie Univ, MSc, 46; Univ Wis, PhD(biochem), 49. *Prof Exp:* Nat Res Coun chemist, SAM Med Inst, Stockholm, 49-51; instr biochem, Univ Wis, 51-52; from asst prof to assoc prof, 52-61, PROF BIOCHEM, UNIV CALIF, BERKELEY, 61- *Concurrent Pos:* Guggenheim Found fel, 58-59. *Mem:* Am Chem Soc; Am Soc Biol Chem; Biochem Soc; Bertrand Russell Soc. *Res:* Bioinorganic chemistry; chemistry and biochemistry of iron compounds; microbial iron transport; membranes; cell surface receptors. *Mailing Add:* Dept Biochem Univ Calif Berkeley CA 94720

NEILD, A(LTON) BAYNE, engineering, mechanical engineering, for more information see previous edition

NEILD, RALPH E, b Georgetown, Ill, Apr 14, 24; m 49; c 4. HORTICULTURE, AGRICULTURAL CLIMATOLOGY. *Educ:* Univ Ill, Urbana, BS, 49; Iowa State Univ, MS, 51; Kans State Univ, PhD, 70. *Prof Exp:* Agr & opers researcher, Libby McNeill & Libby, 51-64; assoc prof, 64-74, PROF HORT, UNIV NEBR-LINCOLN, 74- *Concurrent Pos:* Consult, Libby McNeil & Libby, Imp Govt, Iran & Hashemite Kingdom, Jordan. *Mem:* Am Soc Hort Sci. *Res:* Crop ecology; operations research; crop geography. *Mailing Add:* 1530 Skyline Dr Lincoln NE 68506

NEILER, JOHN HENRY, b Mt Oliver, Pa, Dec 21, 22; m 47; c 3. NUCLEAR PHYSICS. *Educ:* Univ Pittsburgh, BS, 47, MS, 50, PhD(physics), 53. *Prof Exp:* Instr physics, Univ Pittsburgh, 47-51; physicist, Oak Ridge Nat Lab, 53-62, vpres-tech dir, Oak Ridge Tech Enterprises Corp, 62-67, VPRES-TECH DIR, ORTEC INC, 67- *Concurrent Pos:* Lectr, Univ Tenn, 57-; vis scientist, Am Inst Physics-Am Asn Physics Teachers Prog, 57-63. *Mem:* Am Asn Physics Teachers; Sigma Xi. *Res:* Neutron and gamma ray spectrometry; nanosecond pulsing and timing techniques; neutron cross section measurements; nuclear spectrometry with semiconductor diode detectors; fission fragment energy correlations. *Mailing Add:* 853 W Outer Dr Oak Ridge TN 37830

NEILL, ALEXANDER BOLD, b Jersey City, NJ, Sept 27, 19; m 47; c 4. ORGANIC CHEMISTRY. *Educ:* Lehigh Univ, BS, 41, MS, 47, PhD(chem), 49. *Prof Exp:* Chemist, Hercules Powder Co, 41, shift supvr smokeless powder, 42-44, shift supvr rocket powder, 44-45; develop chemist, Carwin Co, 49-50; sr res chemist, Norwich-Eaton Pharmaceut, 51-58, admin asst to dir res, 58-62, admin asst to vpres res, 62-65, chief scheduling & control, 65-67, chief document, 67, mgr info serv, 67-74, dir regulatory affairs, 74-77, dir res oper, 77-79, dir prod regist, 80-84; RETIRED. *Mem:* Am Chem Soc; Sigma Xi. *Mailing Add:* 17 Ridgeland Rd Norwich NY 13815

NEILL, JIMMY DYKE, b Merkel, Tex, Mar 6, 39; m 60; c 2. PHYSIOLOGY, ENDOCRINOLOGY. *Educ:* Tex Tech Col, BS, 61; Univ Mo, MS, 63, PhD(physiol), 65. *Prof Exp:* Nat Inst Child Health & Human Develop res fel physiol, Sch Med, Univ Pittsburgh, 65-67, instr, 67-69; from asst prof to assoc prof, Sch Med, Emory Univ, 69-76, William Patterson Timmie prof, Div Basic Health Sci, 76-79; PROF & CHMN DEPT PHYSIOL & BIOPHYSICS, UNIV ALA, BIRMINGHAM, 79- *Concurrent Pos:* Nat Inst Child Health & Human Develop career develop award, 70-75. *Mem:* Soc Study Reproduction; Am Physiol Soc; Endocrine Soc. *Res:* Neuroendocrinology; Molecular Recognition. *Mailing Add:* Dept Physiol & Biophysics Univ Ala Univ Sta Birmingham AL 35294

NEILL, ROBERT LEE, b Sedan, Kans, July 25, 41; m 63; c 1. BOTANY, ORNITHOLOGY. *Educ:* Kans State Teachers Col, BSE, 63, MS, 68; Univ Okla, PhD(bot), 70. *Prof Exp:* ASSOC PROF BIOL, UNIV TEX, ARLINGTON, 70- *Concurrent Pos:* Res partic, NSF Res Partic Prog, 66. *Mem:* Sigma Xi; Cooper Ornith Soc; Wilson Ornith Soc; Ecol Soc Am; Nat Audobon Soc; Nat Asn Biol Teachers. *Res:* Plant allelopathic studies; plant bird interactions and investigation of causes of high incidence of bill abnormalities in birds; shorebird migration. *Mailing Add:* Dept of Biol Univ of Tex Arlington TX 76019

NEILL, WILLIAM ALEXANDER, b Nashville, Tenn. MEDICINE, CARDIOLOGY. *Educ:* Amherst Col, BA, 51; Cornell Univ, MD, 55. *Prof Exp:* NIH fel, Peter Bent Brigham Hosp, Boston, 59-61; instr med, Mass Mem Hosp, 61-63; assoc prof, Sch Med, Univ Ore, 63-76; prof med, Sch Med, Tufts Univ, 76-84; chief cardiol, Boston Vet Admin Med Ctr, 76-82; PROF MED, RUSH MED COL, 84-; DIR CARDIOL, MACNEAL HOSP, BERWYN, ILL, 84- *Concurrent Pos:* Mem staff, USPHS Commun Dis Ctr, 56-68; fel, Physiol Inst, Dusseldorf, Ger, 69-70; fel coun clin cardiol, Am Heart Asn. *Mem:* Am Fedn Clin Res; Am Physiol Soc; Am Col Cardiol. *Res:* Coronary circulation; muscle metabolism; tissue oxygen supply; cardiac rehabilitation. *Mailing Add:* Dept Cardiol 1725 W Harrison Chicago IL 60612

NEILL, WILLIAM HAROLD, b Wynne, Ark, Oct 21, 43; m 64; c 1. FISH BIOLOGY. *Educ:* Univ Ark, BS, 65, MS, 67; Univ Wis, PhD(zool), 71. *Prof Exp:* Res fishery biologist, Nat Marine Fisheries Serv, Nat Oceanic & Atmospheric Admin, US Dept Commerce, 71-74; assoc prof, 75-83, PROF FISHERIES, TEX A&M UNIV, 83- *Concurrent Pos:* Affil prof zool, Univ Hawaii, 73-75; mem tech comt, USDA Southern Regional Aquacult Ctr, 87-90. *Mem:* Am Fisheries Soc; Am Inst Fishery Res Biologists; AAAS; Sigma Xi. *Res:* Behavioral and physiological ecology of fishes, with emphasis on behavioral regulation of environment and intra-habitat distribution, aquaculture; modeling processes involved in fish ecology. *Mailing Add:* Dept Wildlife & Fisheries Sci Tex A&M Univ College Station TX 77843-2258

NEILSEN, GERALD HENRY, b Kingston, Ont, Jan 10, 48; m 72; c 2. SOIL FERTILITY, PLANT NUTRITION. *Educ:* Queen's Univ, BSc, 70, MSc, 72; McGill Univ, PhD(soil sci), 77. *Prof Exp:* Lectr soil sci, Fac Agr, McGill Univ, 72-73; res scientist, Soil Res Inst Agr Can, 77-78; RES SCIENTIST SOIL SCI, AGR CAN RES STA, SUMMERLAND, BC, 78- *Concurrent Pos:* Res scientist, Macaulay Soil Res Inst, Aberdeen, Scotland, 86-87. *Mem:* Can Soil Sci Soc; Int Soil Sci Soc; Soil Sci Soc Am; Am Soc Hort Sci; Int Soc Hort Sci. *Res:* Soil fertility; nutrient leaching; runoff water quality; soil chemistry; plant nutrient content; fruit tree calcium, nitrogen, phosphorus, magnesium, potassium, zinc nutrition, fruit quality; fertigation. *Mailing Add:* 482 Scott Ave Penticton BC V2A 2J8 Can

NEILSEN, IVAN ROBERT, b Rulison, Colo, Aug 12, 15; m 37; c 2. PHYSICS. *Educ:* Pac Union Col, AB, 36; Stanford Univ, MS, 48, PhD(physics), 52. *Prof Exp:* Instr physics, Glendale Union Acad, 36-38, San Diego Union Acad, 38-40 & Modesto Union Acad, 40-43; from instr to asst prof, Pac Union Col, 43-48; res assoc, Microwave Lab, Stanford Univ, 48-51; assoc prof, Pac Union Col, 51-52, prof & chmn phys sci div, 52-64, dir data processing lab, 58-64; prof physiol & biophys, 64-69, coordr, Sci Comput Facil, 65-76, prof biomath & chmn dept, 69-81, PROF RADIOL & PHYSIOL, SCH MED, LOMA LINDA, 81- *Concurrent Pos:* Res consult, Hansen Labs, Stanford Univ, 52-; consult, Calif State Dept Educ, 64. *Mem:* AAAS; Am Phys Soc; Am Asn Physics Teachers; Asn Comput Mach; Inst Elec & Electronics Engrs; Radiation Res Soc. *Res:* Applied electromagnetic field theory; high power pulsed klystrons; linear electron accelerators; chemical and biological reactions induced by ionizing radiation; computer models of living systems. *Mailing Add:* 1434 Bella Vista Crest Redlands CA 92373

NEILSON, GEORGE CROYDON, b Vancouver, BC, Apr 4, 28; c 3. NUCLEAR PHYSICS. *Educ:* Univ BC, BA, 50, MA, 52, PhD(physics), 55. *Prof Exp:* Physicist, Radiation Sect, Defence Res Bd, 55-58, head radiation sect, 58-59; from asst prof to assoc prof, 59-66, PROF PHYSICS, UNIV ALTA, 66- *Mem:* Am Phys Soc; Can Asn Physicists. *Res:* Measurement of the energy angular distribution, gamma ray correlation and polarization of neutrons and protons produced by deuteron bombardment of light nuclei; angular correlation of cascade gamma rays. *Mailing Add:* 5150 Westminister Ave Ladner BC V4K 2J2 Can

NEILSON, GEORGE FRANCIS, JR, b Portland, Ore, Jan 19, 30; m 55; c 3. PHYSICAL CHEMISTRY. *Educ:* Ore State Univ, BS, 51, MS, 53; Ohio State Univ, PhD, 62. *Prof Exp:* Res chemist, Cent Res Dept, E I du Pont de Nemours & Co, 58-62; res scientist, 62-67, sr scientist, Owens-Ill, Inc, 67-78; mem tech staff, Jet Propulsion Lab, 78-88; RES PROF, MAT SCI ENG DEPT, UNIV ARIZ, 88- *Mem:* Am Chem Soc; Am Phys Soc; Am Crystallog Asn; Am Ceramic Soc. *Res:* Small-angle x-ray scattering; kinetics and mechanisms of nucleation and crystallization; phase transformation processes in glass systems; microstructure of amorphous and polycrystalline materials; space processing of glass. *Mailing Add:* 1541 E Placita Lupita Tucson AZ 85718

NEILSON, JAMES MAXWELL, geology; deceased, see previous edition for last biography

NEILSON, JOHN WARRINGTON, b Saskatoon, Sask, Feb 13, 18; m 47; c 3. DENTISTRY. *Educ:* Univ Sask, BA, 39; Univ Alta, DDS, 41; Univ Mich, MSc, 46; Am Bd Periodont, dipl. *Prof Exp:* From asst prof to assoc prof periodont & oral path, Univ Alta, 46-52; assoc prof periodont, Univ Wash, 52-57; dean fac dent, Univ Man, 57-77, prof oral biol, 57-83; RETIRED.

Concurrent Pos: Consult, USPHS Hosp, Seattle, Wash, 54-57, Royal Can Dent Corps, 60-75, Winnipeg Gen & Children's Hosps, 60-83; examr oral med, Nat Dent Exam Bd Can, 58-60; mem assoc comt dent res, Nat Res Coun Can, 59-60; mem dent adv comt, Nat Health & Welfare, 65-67; mem coun higher learning, Prov Man, 65-67; pres, Asn Can Fac Dentistry, 66-68. *Mem:* Fel Am Col Dent; Am Acad Oral Path; Am Acad Periodont; Can Dent Asn; Can Acad Periodont (pres, 61-62); fel Int Col Dent; fel Royal Col Dent Can. *Res:* Oral pathology and medicine; periodontology; effect of irritation on supporting tissues of the dentition. *Mailing Add:* 8503 Kal view Vernon BC V1B 1W9 Can

NEILSON, ROBERT HUGH, b Pittsburgh, Pa, Jan 24, 48. INORGANIC CHEMISTRY. *Educ:* Carnegie-Mellon Univ, BS, 69; Duke Univ, PhD(chem), 73. *Prof Exp:* Res assoc inorg chem, Univ Tex, Austin, 74-75; asst prof, Duke Univ, 75-78; asst prof inorg chem, 78-81, ASSOC PROF CHEM, TEX CHRISTIAN UNIV, 81- *Concurrent Pos:* US Army Res Off grant, 77-84, Off Naval Res grant, 79-84. *Mem:* Am Chem Soc; Sigma Xi. *Res:* Preparative chemistry of the main group elements including inorganic polymers. *Mailing Add:* 6513 Lawndale Dr Ft Worth TX 76134

NEILSON, RONALD PRICE, b Portland, Ore, Feb 21, 49; m 74; c 2. GLOBAL CLIMATE CHANGE & ECOLOGICAL EFFECTS OF GLOBAL CLIMATE CHANGE. *Educ:* Univ Ore, BA, 71; Portland State Univ, MS, 75; Univ Utah, PhD(biol), 81. *Prof Exp:* Res assoc biol, NMex State Univ, 82-84; res assoc biol, Ariz-Sonora Desert Mus, 84-85; sr scientist environ sci, Univ Utah Rest Inst, 85-87; asst prof, 87-90, ASSOC PROF ENVIRON SCI, ORE STATE UNIV, 90- *Concurrent Pos:* Res assoc, Univ Ariz, 84-85; proj leader, US Environ Protection Agency, 87-90. *Honors & Awards:* William S Cooper Award, Ecol Soc Am, 87. *Mem:* AAAS; Am Inst Biol Sci; Am Geophys Union; Am Quaternary Asn; Am Soc Naturalists; Ecol Soc Am; Int Asn Landscape Ecol; Sigma Xi. *Res:* The causal relations between global scale weather patterns and plant distributions from plant physiology in the microhabitat to global jetstream patterns; global scale biosphere modeling. *Mailing Add:* US Environ Protection Agency 200 SW 35 St Corvallis OR 97333

NEIMAN, GARY SCOTT, b Chicago, Ill, Oct 2, 47; m 72; c 2. CRANIOFACIAL ANOMALIES, VOICE DISORDERS. *Educ:* Univ Ill, Urbana, BS, 69, MA, 71, PhD(speech sci), 73. *Prof Exp:* Speech pathologist, Facial Deformity Team, Carle Found Hosp, 73-77; asst prof speech path, Kans State Univ, 73-77, clin dir, Speech & Hearing Clin, 74-77; ASSOC PROF SPEECH PATH, KENT STATE UNIV, 77-, DIR, SCH SPEECH PATH & AUDIOL, 79 - *Concurrent Pos:* Mem med adv bd, Kent Vis Nurse Assoc, 79 -; speech pathologist & co-dir, Akron Craniofacial Clin, 77- *Mem:* Am Speech & Hearing Asn; Am Cleft Palate Asn; Int Asn Logopedics & Phoniatrics. *Res:* Normal and abnormal aspects of speech and voice production; psychosocial aspects of esophageal speech and cleft palate speech; pre-linguistic variables affecting the acquisition of language in children with orofacial anomalies. *Mailing Add:* Mus Speech Ctr Kent State Univ Speech A104 Kent OH 44242

NEIMARK, HAROLD CARL, b Detroit, Mich, July 25, 32; m 69; c 1. MICROBIOLOGY, IMMUNOLOGY. *Educ:* Univ Calif, Los Angeles, BA, 54, PhD(microbiol), 60. *Prof Exp:* Res assoc, Inst Microbiol, 59-60; from instr to asst prof, 60-71, ASSOC PROF MICROBIOL & IMMUNOL, STATE UNIV NY DOWNSTATE MED CTR, 72- *Concurrent Pos:* NIH grants, 67-77; collabr, USDA; Fogerty Sr Int fel, 78. *Mem:* AAAS; Am Soc Microbiol; Brit Soc Gen Microbiol; Int Orgn Mycoplasmology. *Res:* Genetics and physiology of microorganisms; mycoplasmas; infectious diseases; bacterial evolution. *Mailing Add:* Dept MIC Sta Univ NY 450 Clarkson Ave Box 44 Brooklyn NY 11203

NEIMS, ALLEN HOWARD, b Chicago, Ill, Oct 24, 38; m 61; c 3. BIOCHEMISTRY, PEDIATRICS. *Educ:* Univ Chicago, BA & BS, 57; Johns Hopkins Univ, MD, 61, PhD(physiol chem), 66. *Prof Exp:* NIH fel, Lab Neurochem, Nat Inst Neurol Dis & Stroke, 68-70; asst prof pediat, Johns Hopkins Univ, 70, asst prof physiol chem, 70-72; from asst prof to prof pharmacol & assoc prof to prof pediat, McGill Univ, 72-78; from assoc prof to prof, pharmacol & pediat & chmn pharmacol, 74-89, DEAN, COL MED, UNIV FLA, 89- *Concurrent Pos:* Physician, Johns Hopkins Hosp & J F Kennedy Inst, Baltimore, 70-72. *Honors & Awards:* Henry Strong Denison Award, 61. *Mem:* Asn Med Sch Pharmacol; Am Soc Clin Pharmacol Therapy; Am Soc Pharmacol & Exp Therapeut; Am Pediat Soc. *Res:* Developmental pharmacology and therapeutics; developmental biology; clinical pharmacology; pharmacology. *Mailing Add:* Dean Col of Med Univ Fla Gainesville FL 32610

NEISENDORFER, JOSEPH, b Chicago, IL, Apr 22, 45. ALGEBRAIC TOPOLOGY. *Educ:* Princeton Univ, PhD(math), 72. *Prof Exp:* Inst Adv Study, 80-81; PROF MATH, UNIV ROCHESTER, 85- *Mailing Add:* Univ Rochester Rochester NY 14627

NEISH, GORDON ARTHUR, b Saskatoon, Sask, May 1, 49; m 78; c 2. MYCOLOGY, BOTANY. *Educ:* Acadia Univ, BSc, 70; Univ BC, PhD(bot), 77. *Prof Exp:* Asst prof bot, Univ RI, 77-78; RES SCIENTIST MYCOL, CAN DEPT AGR, 78-, HEAD, MYCOL SECT, BIOSYSTEMATICS RES INST, 83- *Concurrent Pos:* Sci secy, Nat Sci & Eng Res Coun Can, 84- *Mem:* Mycol Soc Am; Brit Mycol Soc; Sigma Xi; Can Phytopath Soc. *Res:* Hyphomycete systematics; mycotoxins. *Mailing Add:* Agr Can No 61 Res Sta PO Box 3000-Main Lethbridge AB T1J 4B1 Can

NEISWENDER, DAVID DANIEL, b Palmdale, Pa, Oct 6, 30; m 55; c 2. PETROLEUM CHEMISTRY. *Educ:* Lebanon Valley Col, BS, 53; Pa State Univ, MS, 55, PhD(chem), 57. *Prof Exp:* Res chemist, Cent Res Div, Mobil Oil Corp, 57-60, sr res chemist, 60-62, asst supvr, 62-64, res assoc, Paulsboro Lab, 64-80, admin mgr, Prod Res Div, Mobile Res & Develop Corp, 80-90; RETIRED. *Mem:* Am Chem Soc. *Res:* Development and testing of automotive engine, transmission and gear oils; chemistry of electrical discharges; synthesis of petrochemicals; preparation and reactions of organoboron compounds; hydrocarbon oxidation; design and testing of synthetic lubricants; lubricant contributions to fuel economy. *Mailing Add:* Three Colonial Lane Cherry Hill NJ 08003

NEITHAMER, RICHARD WALTER, b Wesleyville, Pa, Aug 3, 29; m 58; c 1. INORGANIC CHEMISTRY. *Educ:* Allegheny Col, BS, 51; Univ Ind, PhD(inorg chem), 57. *Prof Exp:* Asst prof chem, Lebanon Valley Col, 55-59, East Tex State Univ, 59-61, Rose Polytech Inst, 61-64; assoc prof & coord chem, 64-67, PROF CHEM, ECKERD COL, 67-, CHMN, COLLEGIUM NATURAL SCI, 72- *Concurrent Pos:* Vis scientist, Ind Acad Sci, 63-64 & Fla Acad Sci, 65-66; consult, US Naval Weapons Lab, 65-66, contract res, 66-69. *Mem:* AAAS; Am Chem Soc; fel Am Inst Chem; Sigma Xi. *Res:* Coordination and metal chelate compounds; polarography. *Mailing Add:* Dept of Chem Eckerd Col St Petersburg FL 33733

NEITZEL, GEORGE PAUL, b Atlanta, Ga, Nov 28, 47; m 74; c 4. HYDRODYNAMIC STABILITY. *Educ:* Rollins Col, BS, 69; Johns Hopkins Univ, MS, 74, PhD(fluid mech), 79. *Prof Exp:* Mathematician & aerospace engr, US Army Ballistic Res Lab, 69-79; from asst prof to prof Mech & Aerospace Eng, Ariz State Univ, 79-90; PROF MECH ENG, GA INST TECHNOL, 90- *Concurrent Pos:* Presidential young investr, NSF, 84-89; res fel, Alexander von Humboldt Found, 85-86. *Mem:* Am Phys Soc; Am Inst Aeronaut & Astronaut; Soc Indust & Appl Math; Sigma Xi; Am Acad Mech. *Res:* Theoretical and experimental research in fluid mechanics emphasizing stability of laminar flows, particularly unsteady and rotating flows; computation of fluid flows and fluid mechanics of materials processing. *Mailing Add:* George W Woodruff Sch Mech Eng Ga Inst Technol Atlanta GA 30332-0405

NEL, LOUIS DANIEL, b Barkly West, SAfrica, June 5, 34; m 56; c 4. FUNCTIONAL ANALYSIS & EXPONENTIAL LAWS. *Educ:* Univ Stellenbosch, BSc, 54, MSc, 58; Cambridge Univ, PhD(math), 62. *Prof Exp:* Lectr math, Univ Stellenbosch, 56-62; sr lectr, Univ Cape Town, 62-65; prof, Port Elizabeth Univ, 66-68; assoc prof, 68-76, PROF MATH, CARLETON UNIV, 76- *Mem:* SAfrican Math Soc (secy, 63-67); Am Math Soc; Can Math Soc. *Res:* Functional analysis and exponential laws; categories in functional analysis; topology and topological algebra which uphold an exponential laws; application of the intrinsic functorial calculus of such categories to solve various problems. *Mailing Add:* Comp Sci Dept Univ of Waterloo Waterloo ON N2I 3G1 Can

NELB, GARY WILLIAM, b Waterbury, Conn, Sept 8, 52. POLYMER PHYSICAL CHEMISTRY. *Educ:* Dartmouth Col, AB, 74; Univ Wis, PhD(chem), 78. *Prof Exp:* RES CHEMIST, E I DU PONT DE NEMOURS & CO INC, 78-; ACCT MGR. *Mem:* Am Chem Soc; Sigma Xi. *Res:* Polymer chemistry especially fibers and fibrous materials. *Mailing Add:* Fiber Dept Du Pont Co Chestnut Run Plaza Willmington DE 19880-0705

NELB, ROBERT GILMAN, organic chemistry; deceased, see previous edition for last biography

NELKIN, MARK, b New York, NY, May 12, 31; m 52; c 2. THEORETICAL PHYSICS. *Educ:* Mass Inst Technol, SB, 51; Cornell Univ, PhD, 55. *Prof Exp:* Res assoc, Knolls Atomic Power Lab, Gen Elec Co, 55-57; mem res staff, Gen Atomic Div, Gen Dynamics Corp, 57-62; assoc prof eng physics, 62-67, PROF APPL PHYSICS, CORNELL UNIV, 67- *Concurrent Pos:* Vis res assoc, State Univ Utrecht, 60-61; Guggenheim fel, Orsay, Paris, 68-69; vis prof, Col France, Paris, 76; vis scientist, Courant Inst, NY Univ, 83-84; vis prof, Univ de Paris VI, France, 81. *Mem:* Fel Am Phys Soc. *Res:* Statistical physics; turbulent fluid flow. *Mailing Add:* Sch of Appl & Eng Physics Cornell Univ Ithaca NY 14853

NELLES, JOHN SUMNER, b Regina, Sask, Dec 2, 20; m 43; c 1. MECHANICAL ENGINEERING. *Educ:* Univ Sask, BSc, 49. *Prof Exp:* Design engr, Dom Eng Works, Que, 49-51, prod engr, 51-53; work shops engr, Atomic Energy Can Ltd, 54-55, develop engr, 55-75, head indust liaison, 75-78; indust contracts officer, Atomic Energy Can Res Co, 78-79, com opers officer, Chalk River Nuclear Lab, 79-82; RETIRED. *Mem:* Eng Inst Can. *Res:* Development of nuclear fuels. *Mailing Add:* Box 802 Deep River ON K0J 1P0 Can

NELLIGAN, WILLIAM BRYON, b Northampton, Mass, Jan 19, 20; m 42. PHYSICS. *Educ:* Rensselaer Polytech Inst, BS, 50, MS, 51. *Prof Exp:* Designer elec power, Gen Elec Co, Mass, 42-44; asst physics, Rensselaer Polytech Inst, 50-51; physicist, Res Lab, Schlumberger Well Surv Corp, 51-53, sr res physicist, 53-65, res proj physicist, 65-87, CONSULT, SCHLUMBERGER-DOLL RES, 87- *Mem:* AAAS; Am Phys Soc; Am Nuclear Soc; Soc Petrol Engrs; Sigma Xi; NY Acad Sci. *Res:* Applied nuclear physics; electronic instrumentation; mathematics; structural chemistry. *Mailing Add:* 00 Candlewood Vista Danbury CT 06811-3750

NELLIS, LOIS FONDA, b Dayton, Ohio, Nov 30, 26. MICROBIOLOGY. *Educ:* Hobart & William Smith Cols, BA, 46; Smith Col, MA, 48; Purdue Univ, PhD(bact), 62. *Prof Exp:* From instr to assoc prof, 48-68, chmn dept, 69-74, PROF BIOL, HOBART & WILLIAM SMITH COLS, 68- *Concurrent Pos:* Res mem, Bergey's Manual Comn, 53-55 & 61; Geneva City bacteriologist, 50-52; United Health Found of western NY grant, 68-70; NIH co-proj dir, Dept Pharmacol, Med Sch, State Univ NY Buffalo, 70-76; vis prof med, Univ Rochester, 75-76; vis prof, dept food sci & technol, NY State Agr Exp Sta, Cornell Univ, Geneva, 79-80. *Mem:* AAAS; Am Soc Microbiology. *Res:* Myxobacteria; R factors and tetracycline resistance in Escherichia coli; fermentation of foods. *Mailing Add:* Dept Biol Hobart Col Geneva NY 14456

NELLIS, STEPHEN H, b Arkansas City, Kans, Jun 22, 42; m 65; c 2. BIOMEDICAL ENGINEERING. *Educ:* Univ Kans, BS, 65; Univ Mo, MS, 67; Univ Va, PhD(biomed eng), 72. *Prof Exp:* Prod eng, Western Elec, Lee's Summit, Mo, 65-66; fel, Univ Calif, San Diego, 72-74; from asst prof to assoc prof med, Milton S Hershey Med Ctr, Pa State Univ, Hershey, Pa, 74-83; assoc prof med & dir cardiovascular res lab, 83-88, PROF MED & DIR CARDIOVASCULAR RES LAB, UNIV WIS, MADISON, WI, 88- *Concurrent Pos:* Postdoctoral NIH res fel, 72-74; grants from var asns; NIH Site Rev Comt, 83 & 85. *Mem:* Am Fedn Clin Res; Biomed Eng Soc; Am Heart Asn; Microcirculatory Soc; Cent Soc Clin Res; fel Am Heart Asn Coun Circulation. *Res:* Coronary microcirculation of the intact beating heart; measurement of the transport of materials in the microcirculation; investigation into myocardial metabolism. *Mailing Add:* Dept Med Div Cardiol Univ Wis 600 Highland Ave Madison WI 53792

NELLIS, WILLIAM J, b Chicago, Ill; c 3. CONDENSED MATTER PHYSICS. *Educ:* Loyola Univ Chicago, BS, 63; Iowa State Univ, MS, 65, PhD(physics), 68. *Prof Exp:* Post doctorate, Mat Sci Div, Argonne Nat Lab, 68-70; asst prof physics, Monmouth Col, 70-73; PHYSICIST, LAWRENCE LIVERMORE NAT LAB, 73-; ASSOC DIV LEADER & HEAD PHYSICS DEPT, HIGH PRESSURE SCI CTR, UC INST GEOPHYSICS & PLANETARY PHYSICS, 84- *Mem:* Am Physical Soc fel (chmn, Topical Group Shock Compression of Condensed Matter, 87; Am Geophysical Union; Mat Res Soc; Am Assoc Adv Sci. *Res:* Properties of condensed matter at high shock pressures and temperatures and on the shock compaction and synthesis of novel materials. *Mailing Add:* Lawrence Livermore Nat Lab PO Box 808 MS L-299 Livermore CA 94550

NELLOR, JOHN ERNEST, b Omaha, Nebr, Oct 31, 22; m 46; c 3. PHYSIOLOGY, ENDOCRINOLOGY. *Educ:* Univ Calif, BS, 50, PhD(comp physiol), 55. *Prof Exp:* From instr to prof physiol, Col Human Med & Natural Sci, Mich State Univ, 55-69, dir, Endocrine Res Unit, 64-69, asst to assoc vpres res develop & dir, Ctr Environ Qual, 71-76; DEAN, GRAD SCH & COORDR RES, UNIV NEV, RENO, 76- *Concurrent Pos:* Mem staff, NSF, 66-67, prog dir metab biol, 67-68; mem, US Nat Comn for UNESCO Man & Biosphere Prog, 75- *Mem:* Am Physiol Soc; Soc Study Reproduction; Sigma Xi. *Res:* Comparative reproductive physiology; hormones and body defense; adrenal-pituitary hormones and aging. *Mailing Add:* 3247 San Simeon Ct Reno NV 89509

NELLUMS, ROBERT (OVERMAN), b Nashville, Tenn, Sept 19, 21; m 47; c 4. CHEMICAL ENGINEERING. *Educ:* Vanderbilt Univ, BSChE, 42. *Prof Exp:* Control analyst, Org Div, Monsanto Co, 42-43, asst prod supvr, 43-44, proj engr, 46-48, prod supvr, 48-51, group leader pilot plant, 51-55, sect leader chem eng, 55-57, asst dir res, 58-59, asst dir eng, 59-62, mgr purchasing, 62-66, dir purchasing admin & control, 66-71, dir intermediates, Monsanto Textiles Co, 71-76, dir textile intermediates, Monsanto Intermediates Co, 76-80, prod dir textile intermediates, Monsanto Int Div-Europe, Monsanto Co, 80-83, bus dir intermediates prod, Monsanto Fibers & Intermediates Co, 83-85; dir, process develop, McNeil Speciality Products Co, 86-87; CONSULT, NORAMCO INC, 88- *Mem:* Am Chem Soc; Sigma Xi; Am Mgt Asn; Am Inst Chem Engrs; Soc Chem Indust. *Res:* Process development; scale-up; plant design. *Mailing Add:* 853 Elm Tree Lane 800 N Lindbergh Blvd Kirkwood MO 63122

NELMS, GEORGE E, b Ark, Feb 6, 27; m 50; c 4. ANIMAL BREEDING. *Educ:* Ark State Col, BS, 51; Ore State Col, MS, 54, PhD(genetics), 56. *Prof Exp:* Instr animal husb, Ore State Col, 55-56; asst prof animal sci, Univ Ariz, 56-59; PROF ANIMAL BREEDING, UNIV WYO, 59- *Mem:* Am Soc Animal Sci; Am Genetic Asn; Sigma Xi; Coun Agr Sci & Tech. *Res:* Reproductive and environmental physiology; genetics of beef cattle. *Mailing Add:* 1719 Ord Laramie WY 82071

NELP, WILL B, b Pittsburgh, Pa, July 30, 29; m 52, 69; c 4. INTERNAL MEDICINE, NUCLEAR MEDICINE. *Educ:* Franklin Col, BA, 51; Johns Hopkins Univ, MD, 55. *Hon Degrees:* DSc, Franklin Col, 67. *Prof Exp:* NIH fel med & radiol, Johns Hopkins Univ, 60-62; from asst prof to assoc prof, 62-71, PROF MED & RADIOL, UNIV WASH, 71-, CHIEF DIV NUCLEAR MED & HEAD CLIN NUCLEAR MED, UNIV HOSP, 62- *Concurrent Pos:* Instr, Johns Hopkins Univ, 61-62; Nat Inst Arthritis & Metab Dis training grant, 63-69; consult, Providence Hosp, Seattle, Wash, 64-68, Nat Heart Inst, 67-68 & Nat Heart & Lung Inst, 68-; consult adv radiopharmaceut, Food & Drug Admin, 70; consult, Children's Orthop Hosp, Seattle Vet Admin Hosp, Harborview Med Ctr & USPHS Hosp, Seattle. *Mem:* Am Fedn Clin Res; Soc Nuclear Med (vpres, 69-70, pres, 73-74); fel Am Col Physicians; fel Am Col Nuclear Physicians. *Res:* Physiologic and clinical investigations in nuclear medicine. *Mailing Add:* Dept Nuclear Med Univ Wash 1959 NE Pacific Seattle WA 98195

NELSEN, ROGER BAIN, b Chicago, Ill, Dec 20, 42; div. BIVARIATE DISTRIBUTION THEORY. *Educ:* DePauw Univ, BA, 64; Duke Univ, PhD(math), 69. *Prof Exp:* PROF MATH & CHMN DEPT, LEWIS & CLARK COL, 69- *Concurrent Pos:* Sabbatical lectr, Univ Mass, 83-84; vis prof, Mt Holyoke Col, 86, 88 & 90-91. *Mem:* Math Asn Am; Amer Math Soc; Sigma Xi. *Res:* Statistics, probability, combinatorics. *Mailing Add:* Dept Math Lewis & Clark Col Box LC110 Portland OR 97219

NELSEN, STEPHEN FLANDERS, b Chicago, Ill, Apr 17, 40; m 62; c 1. ORGANIC CHEMISTRY. *Educ:* Univ Mich, BS, 62; Harvard Univ, PhD(chem), 65. *Prof Exp:* From asst prof to assoc prof, 65-75, PROF ORG CHEM, UNIV WIS-MADISON, 75- *Concurrent Pos:* Vis scientist, Hahn-Meitner Inst, Berlin, 79, 81, 82, 84 & 85. *Honors & Awards:* Humboldt-Stifting, 88. *Mem:* Am Chem Soc. *Res:* Physical organic chemistry; physical and chemical properties of free radicals; electrochemistry; conformational analysis; electron transfer reactions; preparation of organic molecules. *Mailing Add:* 1101 W University Ave Madison WI 53706

NELSEN, THOMAS SLOAN, b Tacoma, Wash, Aug 4, 26; m 45; c 2. SURGERY. *Educ:* Univ Wash, BS, 47, MD, 51; Am Bd Surg, dipl, 59. *Prof Exp:* From instr to asst prof, Univ Chicago, 57-60; from asst prof to assoc prof, 60-71, PROF SURG, SCH MED, STANFORD UNIV, 71- *Mem:* Am Col Surgeons; Inst Elec & Electronics Engrs; Sigma Xi. *Res:* Surgery of neoplasms; gastrointestinal physiology and surgery. *Mailing Add:* Dept of Surg Stanford Univ Sch of Med Stanford CA 94305

NELSESTUEN, GARY LEE, b Galesville, Wis, Sept 10, 44; m 67; c 2. BIOCHEMISTRY. *Educ:* Univ Wis-Madison, BS, 66; Univ Minn, St Paul, PhD(biochem), 70. *Prof Exp:* NIH fel biochem, Univ Wis, 70-72; asst prof biochem, Univ Minn, St Paul, 72-76, assoc prof, 76-80. *Concurrent Pos:* Estab investr, Am Heart Asn, 75-80. *Mem:* Am Chem Soc; AAAS. *Res:* Protein-membrane associations including protein Kihase C; membrane attack complex and blood coagulation. *Mailing Add:* 1479 Gortner Ave St Paul MN 55108

NELSON, A CARL, JR, b West Chester, Pa, Jan 2, 26; m 50; c 4. MATHEMATICAL STATISTICS, MATHEMATICS. *Educ:* Mass Inst Technol, SB, 46; Univ Del, MS, 48. *Prof Exp:* Instr math, Univ Del, 48-50, 51-52 & 53-56; scientist, Bettis Atomic Power Div, Westinghouse Elec Corp, Pa, 56-60, fel scientist, 60; statistician, Res Triangle Inst, 60-63, sr statistician, 63-73; statist consult, 73-75; sr prof scientist, Pedco Environ Inc, 75-82; at Dept Math, Univ NC, Wilmington, 82-88; RETIRED. *Mem:* Statist Asn; Biomet Soc; Am Soc Qual Control. *Res:* Applied research in application of statistics to physical sciences, particularly the fields of environmental analysis, occupational and highway safety, quality assurance, systems analysis, and the statistical design of experiments for developing mathematical models. *Mailing Add:* 3408 Regency Dr Wilmington NC 28412

NELSON, A GENE, b Galesburg, Ill, Sept 9, 42; m 64; c 3. AGRICULTURAL ECONOMICS. *Educ:* Western Ill Univ, BS, 64; Purdue Univ, MS, 67, PhD(agr econ), 69. *Prof Exp:* Asst prof, 69-74, assoc prof, 74-79, PROF AGR ECON, ORE STATE UNIV, 79- *Mem:* Am Agr Econ Asn; Am Soc Farm Mgrs & Rural Appraisers. *Res:* Systems analysis of beef and forage production; decision making under risk and uncertainty. *Mailing Add:* Dept Agr & Resource Econ Ore State Univ Corvallis OR 97331

NELSON, ALAN R, b Logan, Utah, June 11, 33; m 59; c 3. ENDOCRINOLOGY. *Educ:* Northwestern Univ, BS, 55, MD, 58; Am Bd Internal Med, dipl, 66, cert endocrinol & metab, 79. *Prof Exp:* Pres, Utah Prof Rev Orgn, 71-75; ASSOC, MEMORIAL MED CTR, 64-; CLIN PROF, DEPT INTERNAL MED, COL MED, UNIV UTAH. *Concurrent Pos:* Mem, Nat Prof Stand Rev Coun, 73-77; mem, Adv Panel Health Ins, Comt Ways & Means, US House Rep, 75, var comts, Nat Acad Sci, 85- & Secy's Adv Comt, Food & Drug Admin, 90-91; chmn, Nat Adv Bd Correctional Health Care, 82 & Nat Comn Vaccine Injury Compensation, 83-85; chmn bd, AMA, 86-88; pvt pract internal med & endocrinol. *Mem:* Inst Med-Nat Acad Sci; AMA (pres elect, 88-89, pres, 89-90); World Med Asn (pres elect, 90-); fel Am Col Physicians; Am Soc Internal Med; Endocrine Soc. *Res:* Medical utilization review and quality assessment; author of numerous scientific publications. *Mailing Add:* Mem Med Ctr 2000 S 900 E Salt Lake City UT 84105

NELSON, ALBERT WENDELL, b Boston, Mass, June 2, 35; m 59; c 3. CARDIOVASCULAR DISEASES. *Educ:* Cornell Univ, DVM, 59; Colo State Univ, MS, 62, PhD(path), 65; Am Col Vet Surgeons, dipl, 74. *Prof Exp:* Vet, private practice, 59-60; from asst prof to assoc prof surg, 65-75, PROF CLIN SCI, COLO STATE UNIV, 75-, DIR, VET TEACHING HOSP, 90- *Concurrent Pos:* NIH res grants, Colo State Univ, 65-68, 71-73; Colo Heart Asn res grant, 71-73; Nat Heart & Lung Inst Contract, 73-77. *Mem:* Am Vet Med Asn; Am Col Vet Surgeons. *Res:* Cardiovascular pathology and physiology, primarily in relation to the microcirculation; reconstructive surgery relative to animal and human problems. *Mailing Add:* Vet Teaching Hosp Colo State Univ Ft Collins CO 80521

NELSON, ALLEN CHARLES, b Plum City, Wis, July 13, 32; m 54; c 4. MYCOLOGY, MICROBIOLOGY. *Educ:* Wis State Univ, River Falls, BS, 54; Univ SDak, MA, 61; Univ Wis, PhD(bot), 64. *Prof Exp:* From asst prof to assoc prof, 64-68, chmn dept biol, 67-73, PROF, UNIV WIS-LA CROSSE, 68- *Mem:* Mycol Soc Am; Sigma Xi; Am Soc Microbiol. *Res:* Ascomycetes; morphological and cytological studies. *Mailing Add:* Dept of Biol Univ of Wis-LaCrosse La Crosse WI 54601

NELSON, ARNOLD BERNARD, b Valley Springs, SDak, Aug 26, 22; m 43; c 4. ANIMAL NUTRITION. *Educ:* SDak State Col, BS, 43, MS, 48; Cornell Univ, PhD(animal husb), 50. *Prof Exp:* Asst animal husb, SDak State Col, 46-47, asst animal husbandman, 47-48; asst, Cornell Univ, 48-50; from asst prof to assoc prof, Okla State Univ, 50-62; prof, 63-71, PROF ANIMAL RANGE SCI & HEAD DEPT, NMEX STATE UNIV, 71- *Mem:* AAAS; Am Soc Animal Sci; Soc Range Mgt; Am Dairy Sci Asn. *Res:* Ruminant nutrition; applied cattle nutrition. *Mailing Add:* Dept Animal Husbandry NMex State Univ University Park NM 88070

NELSON, ARTHUR ALEXANDER, JR, b New Roads, La, June 12, 46; m 78; c 2. PHARMACY. *Educ:* Northeast La Univ, BS, 69, MS, 71; Univ Iowa, PhD(pharm), 73. *Prof Exp:* Asst prof pharm, Med Ctr, Univ Ill, 73-76; assoc prof pharm, Univ SC, 76-84; dean Col Pharm, Univ Nebr, 84-87; DEAN COL PHARM, IDAHO STATE UNIV, 87- *Concurrent Pos:* Proj dir, Ill Dept Ment Health & Deviation Disabilities, 75-76; consult, Ill State Pharmaceut Asn, 74-76 & Col Pharm, Univ Nebr, 75-76. *Mem:* Am Soc Hosp Pharmacists; Acad Pharmaceut Sci. *Res:* Behavioral and administrative practices of pharmacists; economics of health care delivery with particular interest in the pharmaceutical component. *Mailing Add:* Dean Col Pharm Idaho State Univ Pocatello ID 83209

NELSON, ARTHUR EDWARD, b Orange, NJ; m 52; c 3. STRUCTURAL GEOLOGY, METAMORPHIC GEOLOGY. *Educ:* Upsala Col, BS, 49; Univ Tenn, MS, 54. *Prof Exp:* GEOLOGIST, US GEOL SURV, 50- *Mem:* Geol Soc Am. *Res:* Volcanic geology and stratigraphy of Puerto Rico; stratigraphy of medium to high grade metamorphic rocks; tectonics and structural geology of Appalachian orogenic belt. *Mailing Add:* 926 National Ctr US Geol Surv Nat Ctr Reston VA 22092

NELSON, ARTHUR KENDALL, b Washburn, Wis, Aug 28, 32; m 61; c 1. CHEMISTRY. *Educ:* Univ Wis, BS, 54; Univ Minn, PhD(chem), 59. *Prof Exp:* Asst prof, Macalester Col, 59-60; res chemist, Stauffer Chem Co, 60-64; sr tech specialist, Nalco Chem Co, 64-68; mem staff, Rauland Div, Zenith Radio Corp, 68-76; mgr bus develop, Carus Chem Co, 77-81, VPRES & GEN MGR, LA SALLE TRANSPORT CO, SUBSID CARUS CORP, 81- *Mem:* Am Chem Soc. *Mailing Add:* 1267 W Highland Ave Elgin IL 60123-5101

NELSON, ARTHUR L(EE), b Dallas, Tex, Jan 29, 15; m 35; c 2. ENGINEERING. *Educ:* Univ Ark, BSEE, 38. *Prof Exp:* Engr, RCA Mfg Co, Inc, 35-37; proj engr, Farnsworth TV & Radio Corp, 38-41; chief engr & prod mgr, Aircraft Accessories Corp, 41-43; pres, Nelson Elec Corp, 43-49; engr, Fla Power Corp, 49-51; tech adv, US Air Force, 51-53; consult engr, 53-54; consult, RCA Serv Co, Inc, 54-56; sr engr, Scripps Inst, Univ Calif, 56-59; supvry engr, US Navy Electronics Lab, 59-63; founder & pres, Electro Oceanics, Inc, 63-65; founder & chmn bd dirs, Hotsplicer Corp, 68-74; RETIRED. *Concurrent Pos:* Consult ocean engr; founder & owner, Arthur L Nelson & Co, 65-88. *Mem:* Marine Technol Soc. *Res:* Oceanography; high-voltage engineering; development of methods and equipment for splicing and terminating high-voltage underground cables. *Mailing Add:* 1508 Circa del Largo 307B Lake San Marcos CA 92069

NELSON, BERNARD CLINTON, medical entomology; deceased, see previous edition for last biography

NELSON, BERNARD W, b San Diego, Calif, Sept 15, 35; m; c 4. MEDICAL ADMINISTRATION. *Educ:* Stanford Univ, BA, 57, MD, 61. *Prof Exp:* Asst prof & asst dean med, Stanford Univ, 65-67, assoc dean, 67-71; assoc prof & assoc dean med, Univ Wis, Madison, 74-77; exec vpres, Kaiser Family Found, 79-86; CHANCELLOR, UNIV COLO HEALTH SCI CTR, DENVER, 86-, PROF PREV MED, 87- *Concurrent Pos:* Mem bd dirs, Nat Med Fels, Inc, 69-77, vpres, 72-75, pres, 75-77; nat vchmn, Group Student Affairs, AAMC, 72-73, Comt Health Manpower, 73, Data Develop Liaison Comt, 73-74 & 75-77, chmn, Task Force Student Financing, 76-; mem adv coun, Environ Health, Sch Hyg & Pub Health, Johns Hopkins Univ, 81; mem bd trustees, Morehouse Sch Med, Atlanta, Ga, 81-83. *Mem:* Inst Med-Nat Acad Sci; Sigma Xi. *Mailing Add:* Univ Colo Health Sci Ctr 4200 E Ninth Ave Box A09S Denver CO 80262

NELSON, BRUCE WARREN, b Cleveland, Ohio, Mar 17, 29; m 56; c 1. SEDIMENTOLOGY, CLAY MINERALOGY. *Educ:* Harvard Col, AB, 51; Pa State Univ, MS, 54; Univ Ill, PhD(geol), 55. *Prof Exp:* From assoc prof to prof geol, Va Polytech Inst, 55-63; prof & head dept, Univ SC, 63-74, dean col arts & sci, 66-72, dean grad sch & vprovost advan studies & res, 72-74; asst provost & dean, Sch Continuing Educ, 74-77, assoc provost, 77-81, PROF ENVIRON SCI, UNIV VA, 74- *Concurrent Pos:* Geologist, US Geol Surv, 51-55; geologist, Ohio Geol Surv, 52-54; vis scientist, Am Geol Inst, 64-69; vis prof, Univ Va, 70-71; Fulbright lectr, Dept Geol, Univ Malaya, Kuala Lumpur, Malaysia, 82-83, Mahatura Gandi Inst, Moka, Mauritius, 88-89. *Mem:* Fel AAAS; fel Mineral Soc Am; fel Geol Soc Am; Soc Econ Paleontologists & Mineralogists. *Res:* Sedimentary mineralogy, geochemistry, and petrology; recent sedimentary processes; diagenesis; chemistry of natural waters; estuarine environment. *Mailing Add:* 36 University Circle Univ Va Charlottesville VA 22903

NELSON, BURT, b Milwaukee, Wis, Mar 10, 22; m 47. ASTRONOMY. *Educ:* Univ Wis, BS, 51, MS, 52, PhD(philos), 59. *Prof Exp:* Asst prof astron & phys sci, 57-61, assoc prof astron, 61-66, PROF ASTRON, SAN DIEGO STATE COL, 66- *Mem:* Am Astron Soc. *Res:* Astronomical photoelectric photometry. *Mailing Add:* Dept Astron San Diego State Univ 5300 Campanile Dr San Diego CA 92182

NELSON, CARLTON HANS, b Wabasha, Minn, Dec 16, 37; div; c 2. GEOLOGY. *Educ:* Carleton Col, BA, 59; Univ Minn, MS, 62; Ore State Univ, PhD(oceanog), 68. *Prof Exp:* Ranger naturalist, Nat Park Serv, 59-61 & 63; teaching asst, Lehigh Univ, 61-62; field asst, US Geol Surv, 62; instr phys sci, Portland State Col, 62-63; res asst, Ore State Univ, 63-67; GEOLOGIST, US GEOL SURV, 66- *Concurrent Pos:* Vis asst prof, Chapman Col, 66, San Jose State Col, 68-69 & Calif State Col, Hayward, 70-71; actg asst prof, Stanford Univ, 73; vis prof, Univ Barcelona, Spain & Univ Utrecht, The Netherlands, 81; lectr, Norweg Petrol Directorate, 81 & Soc Econ Paleontologists & Mineralogists, US & Europe, 84- *Mem:* Fel Geol Soc Am; AAAS; Soc Econ Paleont & Mineral; Int Asn Sedimentol; Sigma Xi; Am Asn Petrol Geologists. *Res:* Geological limnology; Pleistocene geology; sedimentology; geological oceanography; epicontinental shelf and deep-sea fan sedimentation; placer and trace metal dispersal in marine sediments; marine geology; bottom feeding processes of mammals. *Mailing Add:* US Geol Surv 345 Middlefield Rd Menlo Park CA 94025

NELSON, CHARLES A, b Buffalo, NY, June 26, 36; m 64; c 2. BIOCHEMISTRY. *Educ:* Cornell Col, BS, 57; Univ Iowa, MS, 60, PhD(biochem), 62. *Prof Exp:* Res assoc, Duke Univ, 61-66; asst prof, 66-79, ASSOC PROF BIOCHEM, MED CTR, UNIV ARK, LITTLE ROCK, 79- *Concurrent Pos:* NIH fel, 62-64. *Mem:* Am Soc Biol Chemists. *Res:* Subunit structure of serum lipoproteins; hybridoma-monoclonal antibodies. *Mailing Add:* Univ Ark Med Sci 4301 W Markham Little Rock AR 72205

NELSON, CHARLES ARNOLD, b Chadron, Nebr, Oct 11, 43; m 71, 88; c 1. HIGH ENERGY & THEORETICAL PHYSICS. *Educ:* Univ Colo, BS, 65; Univ Md, PhD(theoret physics), 68. *Prof Exp:* Res assoc high energy theoret physics, City Col New York, 68-70 & La State Univ, Baton Rouge, 70-72; Nat Res Coun-Nat Bur Standards fel, Nat Bur Standards, Washington DC, 72-73; from asst prof to assoc prof, 73-78, PROF PHYSICS, STATE UNIV NY, BINGHAMTON, 84- *Concurrent Pos:* Consult, Ctr Particle Theory, Univ Tex, Austin, 70-72 & Ctr Theoret Physics, Univ Md, 72-73; sabbatical leave, Fermilab & RIFP, Kyoto Univ & AERI, Nihon Univ, Tokyo, 80-81. *Mem:* Am Phys Soc. *Res:* Particles and fields in theoretical high energy physics; mathematical physics. *Mailing Add:* Dept Physics State Univ NY Binghamton NY 13902-6000

NELSON, CHARLES G(ARTHE), b Northport, Mich, Mar 4, 33. ELECTRICAL ENGINEERING. *Educ:* Mich State Univ, BS, 55; Stanford Univ, MS, 59, PhD(elec eng), 62. *Prof Exp:* Proj engr, Microwave Sect, Zenith Radio Res Corp, 62-64; asst prof elec eng, 65-71, PROF ELEC ENG, CALIF STATE UNIV, SACRAMENTO, 71-, CHMN DEPT. *Mem:* Inst Elec & Electronics Engrs; Inst Noise Control Eng. *Res:* High efficiency microwave tubes; acoustics and noise pollution measurements; instrumentation. *Mailing Add:* Dept of Eng 6000 J St Calif State Univ Sacramento CA 95819

NELSON, CHARLES HENRY, b Boston, Mass, July 28, 41; m 66; c 2. ENTOMOLOGY. *Educ:* Univ Mass, BS, 63, MS, 67, PhD(entom), 69. *Prof Exp:* From asst prof to assoc prof, 69-78, PROF BIOL, UNIV TENN, CHATTANOOGA, 78-, DEPT HEAD, 81- *Mem:* Entom Soc Am; Am Entom Soc; Soc Syst Zool; Willi Hennig Soc. *Res:* Systematics and morphology of the Plecoptera; systematic entomology. *Mailing Add:* Dept Biol Univ Tenn Chattanooga TN 37403

NELSON, CLARENCE NORMAN, b Starbuck, Minn, June 6, 09; m 35; c 2. OPTICS. *Educ:* St Olaf Col, BA, 31; Ohio State Univ, MA, 33. *Prof Exp:* Physicist, Res Labs, Eastman Kodak Co, 33-53, res assoc, 53-74; RETIRED. *Concurrent Pos:* Lectr image sci, Rochester Inst Technol, 81-84. *Mem:* Optical Soc Am; Soc Photog Scientists & Engrs. *Res:* Optics; physics of the photographic process; sensitometry; vision; tone reproduction; modulation transfer; communication theory; image science; American standards on image evaluation; theory of the photographic process; information storage capacity; detective quantum efficiency. *Mailing Add:* 235 Danbury Circle S Rochester NY 14618-2752

NELSON, CLIFFORD MELVIN, JR, b Chicago, Ill, Nov 8, 37; m 86. HISTORY OF GEOLOGY, PALEONTOLOGY. *Educ:* Univ Ill, Urbana, BS, 60; Mich State Univ, MS, 63; Univ Calif, Berkeley, PhD(paleontol), 74. *Prof Exp:* Teaching asst geol, Mich State Univ, 61-63; teaching asst paleontol, Univ Calif, Berkeley, 63-66; instr geol, Cabrillo Col, 67-69 & 75-76; lectr geol, Calif State Univ, Hayward, 70-71; geologist & assoc historian, 76-80, STAFF GEOLOGIST, US GEOL SURV, 80-, CHIEF HIST PROJ OFF SCI PUBL, 87- *Concurrent Pos:* Res fel, Smithsonian Inst, 74-75, res collabr, 90-; res assoc, Mus Paleontol, Univ Calif, Berkeley, 75-; mem, educ comt, Am Geol Inst, 82-85; chmn, US Comt Hist Geol, 85-90. *Mem:* Sigma Xi; AAAS; fel Geol Soc Am; Paleontol Soc; Hist Sci Soc; fel Linnean Soc London; corresp mem Int Comn Hist Geol Sci; Orgn Am Historians. *Res:* History of ideas and institutions in the earth sciences, especially United States Geological Survey and predecessor agencies from 1867; evolution and distribution of Cenozoic marine molluscs especially Neptunidae. *Mailing Add:* US Geol Surv 904 National Ctr Reston VA 22092

NELSON, CLIFFORD VINCENT, b Boston, Mass, Sept 23, 15; m 41; c 2. CARDIOVASCULAR PHYSIOLOGY. *Educ:* Mass Inst Technol, BS, 42; Univ London, PhD(eng electrocardiol), 53. *Prof Exp:* Asst biol eng, Mass Inst Technol, 40; engr, Submarine Signal Co, Mass, 42-47; res engr, Sanborn Co, 48; researcher, EEG Lab, Mass Gen Hosp, 49; asst res prof med & biophys, Col Med, Univ Utah, 54-56; res assoc cardiol & res, Maine Med Ctr, 56-83; RETIRED. *Concurrent Pos:* Am Heart Asn fel, 53-55, estab investr, 56-61; Nat Heart Inst res career award, 62-83; adj assoc res prof, Boston Univ, 66-70; hon res fel, Baker Med Res Inst, Royal Melbourne Hosp, Australia, 69-70; spec res fel, Nat Cancer Inst, 41. *Mem:* Fel Am Col Cardiologists; Biophys Soc; Am Physiol Soc; Inst Elec & Electronic Engrs; Biomed Eng Soc. *Res:* Vector-cardiology; electrophysiology. *Mailing Add:* 187 Flaggy Meadow Rd Gorham ME 04038-9211

NELSON, CRAIG EUGENE, b Mankato, Kans, May 21, 40; m 62; c 2. COMMUNITY ECOLOGY, EVOLUTIONARY BIOLOGY. *Educ:* Univ Kans, AB, 62; Univ Tex, MA, 64, PhD(zool), 66. *Prof Exp:* Asst prof zool, 66-71, dir environ studies, 71-87, assoc prof, 77-87, PROF BIOL, PUB & ENVIRON AFFAIRS, IND UNIV, BLOOMINGTON, 87- *Mem:* Soc Study Evolution; Ecol Soc Am; Am Soc Naturalists; Soc Study of Amphibians & Reptiles; Sigma Xi. *Res:* Ecological and evolutionary theory; community structure; ecological and evolutionary processes in amphibia and reptiles; sex-determination in reptiles; microhylid frogs; teaching of evolution and critical thinking. *Mailing Add:* Dept Biol Ind Univ Bloomington IN 47405

NELSON, CURTIS JEROME, b Mitchell Co, Iowa, Mar 25, 40; m 60; c 2. CROP PHYSIOLOGY. *Educ:* Univ Minn, St Paul, BS, 61, MS, 63; Univ Wis-Madison, PhD(agron), 66. *Prof Exp:* Res assoc, Cornell Univ, 66-67; from asst prof to assoc prof, 67-75, PROF FORAGE PHYSIOL, UNIV MO-COLUMBIA, 75- *Concurrent Pos:* Vis scientist, Welsh Plant Breeding Sta, Aberystwyth Wales, UK, 73-74; NSF fel, NATO, 73-74; assoc ed, Crop Science, 75-78; vis scientist, Zurich, Switz, 80-81; assoc ed, Plant Physiol, 87-89; Curator's prof, Univ Mo, 89. *Mem:* Am Forage & Grassland Coun; fel Am Soc Agron; fel Crop Sci Soc Am (pres, 88); Am Soc Plant Physiol; Brit Grassland Soc. *Res:* Crop physiology and biochemistry; genetic control of photosynthesis; carbon metabolism; yield expression of forage grasses; management of forage legumes and grasses; fructans and other water-soluble carbohydrates. *Mailing Add:* Dept Agron Univ Mo Columbia MO 65211

NELSON, CURTIS NORMAN, b Rochester, NY, Jan 31, 41; m 61; c 3. NEUROSURGERY, BRAIN RESEARCH. *Educ:* Princeton Univ, BSE, 63; Univ Rochester, PhD(physiol), 70, MD, 72. *Prof Exp:* Intern surg, Mary Hitchcock Mem Hosp, 72-73; resident neurosurg, Mass Gen Hosp, 73-78; ASSOC PROF NEUROSURG, UNIV ROCHESTER, 78- *Res:* Cerebral blood flow and stroke. *Mailing Add:* 300 White Spruce Blvd Rochester NY 14623

NELSON, D KENT, b Ft Collins, Colo, Mar 8, 39; m 60; c 4. ANIMAL NUTRITION. *Educ:* Colo State Univ, BS, 61; Mich State Univ, MS, 64; Iowa State Univ, PhD(animal nutrit), 68. *Prof Exp:* Asst prof dairy sci, Wash State Univ, 68-69; assoc prof dairy sci, Iowa State Univ, 69-77; PRES, NELSON FARM CONSULT, INC, 77- *Concurrent Pos:* Guest lectr, Am Soybean Asn, Japan & Korea, 76. *Mem:* Am Soc Animal Sci; Am Dairy Sci Asn. *Res:* Calf nutrition; nonprotein nitrogen utilization; dairy cattle nutrition. *Mailing Add:* Nelson Dairy Consult Inc RR 2 Box 100 Decorah IA 52101

NELSON, DARRELL WAYNE, b Aledo, Ill, Nov 28, 39; m 61; c 2. SOIL CHEMISTRY, SOIL MICROBIOLOGY. *Educ:* Univ Ill, Urbana, BS, 61, MS, 63; Iowa State Univ, PhD(soil chem), 67. *Prof Exp:* Asst & assoc prof, 68-77, prof soil biochem & microbiol, Purdue Univ, West Lafayette, 77-84; prof & chmn, dept agron, 84-88, DEAN AGR RES, UNIV NEBR, 88- *Concurrent Pos:* Dir, Water Resources Res Ctr, Purdue Univ, 82-84; mem, Coun Agr Sci & Technol. *Honors & Awards:* Agronomy Award, Ciba-Geigy Inc, 75; Agron Achievement Award, 83; Environ Qual Res Award, 85. *Mem:* Fel Am Soc Agron; fel Soil Sci Soc Am; Int Soil Sci Soc; fel AAAS; Coun Agr Sci & Technol. *Res:* Chemistry of nitrogen in soils and sediments; effect of fertilizer use on the environment; nature and properties of soil organic matter. *Mailing Add:* Agr Res Div Univ Nebr Lincoln NE 68583-0704

NELSON, DARREN MELVIN, b Lincoln, Nebr, Aug 15, 25; m 53; c 4. ANIMAL PHYSIOLOGY, ENDOCRINOLOGY. *Educ:* Univ Nebr, BS, 54; Univ Ill, PhD(animal physiol), 65. *Prof Exp:* Res asst animal sci, Univ Nebr, 54; asst prof animal husb, Calif State Polytech Col, 54-58; instr animal sci, Purdue Univ, 58-60; res asst, Univ Ill, 60-65; NIH fel as trainee in endocrinol, Sch Med, Univ Kans, 65-66; assoc res prof gynec & obstet, Med Ctr, Univ Okla, 66-67; assoc prof biol, Univ Redlands, 67-68; PROF ANIMAL SCI, CALIF STATE UNIV, FRESNO, 68- *Concurrent Pos:* Vis prof, Univ Mo, Columbia, 75. *Mem:* AAAS; Soc Exp Biol & Med; Am Fertil Soc; Am Soc Animal Sci; Poultry Sci Asn; NY Acad Sci; Am Physiol Soc; Soc Study Reproduction. *Res:* Neuroendocrine regulation of reproductive processes in mammals of both sexes; early neonatal differentiation of the central nervous system as influenced by steroid administration in mammals and avian species. *Mailing Add:* Dept Animal Sci Calif State Univ Fresno CA 93710-0001

NELSON, DAVID, b Cape Girardeau, Mo, Jan 2, 18. MATHEMATICAL LOGIC. *Educ:* Univ Wis, BA, 39, MA, 40, PhD, 46. *Prof Exp:* Asst prof math, Amherst Col, 42-46; from asst prof to prof, 46-86, chmn dept, 56-68, EMER PROF MATH, GEORGE WASHINGTON UNIV, 86- *Concurrent Pos:* Consult, Nat Res Coun, 60-63. *Mem:* Am Math Soc; Math Asn Am; Asn Symbolic Logic. *Res:* Theory of recursive functions; intuitionistic mathematics. *Mailing Add:* Dept of Math George Washington Univ Washington DC 20006

NELSON, DAVID A, b Moscow, Idaho, Jan 29, 61; m 82; c 1. NEURAL NETWORKS, REAL TIME EXPERT SYSTEMS. *Educ:* Univ Idaho, BS, 83, MS, 84; Univ Mass, PhD(chem eng), 89. *Prof Exp:* Engr, Gen Foods, 84; PROCESS ENGR, CHEVRON RES & TECHNOL, 89- *Mem:* Am Inst Chem Engrs; Instrument Soc Am; Int Neural Network Soc. *Res:* Neural networks; expert systems; artificial intelligence technologies. *Mailing Add:* 5237 James Ave Oakland CA 94618

NELSON, DAVID ALAN, b Melrose, Mass, June 13, 31; m 56; c 5. ANALYTICAL CHEMISTRY. *Educ:* Mass Inst Technol, BS, 53; Univ RI, MS, 55; Univ NH, PhD(chem), 60. *Prof Exp:* Res assoc chem, Univ Ore, 60-62; asst prof, 62-67, assoc prof, 67-81, PROF CHEM, UNIV WYO, 81- *Concurrent Pos:* Petrol Res Fund grant, 62-63; USPHS grants, 65-68 & 69-71; NSF grant, 74-78; Dept Energy grant, 79-82. *Mem:* Am Chem Soc. *Res:* High performance liquid chromatography; synthesis of surface modified silica gels; analysis of petroleum products; water tracing compounds; separation of peptides and proteins by high-performance liquid chromatography. *Mailing Add:* Dept Chem Univ Wyo Laramie WY 82070

NELSON, DAVID BRIAN, b Lincoln, Nebr, Oct 23, 40; m 64; c 2. PLASMA PHYSICS. *Educ:* Harvard Univ, AB, 62; NY Univ, MA, 65, PhD(math), 67. *Prof Exp:* Elec engr, Guy B Panero Engrs, 63-64; res asst, Courant Inst, NY Univ, 64-66; res staff mem eng physics, Oak Ridge Nat Lab, 66-71, res staff mem plasma physics, 71-79; chief, Fusion Theory & Comput Serv Br, 79-84, dir Appl Plasma Physics Div, Off Fusion Energy, 84-87, dir Sci Comput, 87, EXEC DIR OFF ENERGY RES, US DEPT ENERGY, 87- *Concurrent Pos:* Adv comt civil defense, Nat Acad Sci, 69-72; vis mem plasma physics, Courant Inst, NY Univ, 75-76. *Mem:* Am Phys Soc. *Res:* Theoretical plasma physics and applications to contolled nuclear fusion. *Mailing Add:* Off of Energy Res ER-Two US Dept Energy Washington DC 20585

NELSON, DAVID ELMER, b Harvey, Ill, Sept 14, 33; m 60; c 4. SUBMARINE ACOUSTIC REFLECTIVITY, DESIGN DIGITAL SYSTEMS. *Educ:* Purdue Univ, BSEE, 55, MSEE, 57; Univ Rochester, PhD(commun), 75. *Prof Exp:* Engr, Gen Dynamics Electronics Div, 57-70; sr engr, Marine Resources, Inc, 70-71; DIR SPEC PROJ, HYDROACOUST INC, 71- *Mem:* Inst Elec & Electronic Engrs; Acoust Soc Am; Soc Explor Geophys. *Res:* Design and conduct measurements of submarine acoustic reflectivity for US Navy; design digital signals for use in seismic marine vibrator oil exploration. *Mailing Add:* Hydroacoustics Inc PO Box 23447 Rochester NY 14692

NELSON, DAVID HERMAN, b Houston, Tex, Mar 28, 43; m 65; c 2. VERTEBRATE ECOLOGY, AQUATIC ECOLOGY. *Educ:* Baylor Univ, BA, 66, MA, 68; Mich State Univ, PhD(zool), 74. *Prof Exp:* Asst prof biol, Adrian Col, 73-75; res assoc ecol, Savannah River Ecol Lab, Univ Ga, 75-77; asst prof, 77-80, ASSOC PROF BIOL, UNIV S ALA, 80- *Honors & Awards:* Roosevelt Mem Award, Am Mus Natural Hist, 69; Pres Award, Am Soc Ichthyologists & Herpetologists, 70. *Mem:* Ecol Soc Am; Am Inst Biol Sci; Am Soc Ichthyologists & Herpetologists; Soc Study Amphibians & Reptiles; Sigma Xi; Herpetologist's League. *Res:* Thermal ecology of aquatic organisms; biological effects of heated reactor effluents; temperature tolerances, temperature prefences and thermal stress; ecology, movements and activity patterns of amphibians and reptiles; herpetology; natural history. *Mailing Add:* Dept Biol Univ S Ala Mobile AL 36688

NELSON, DAVID L, b Topeka, Kans, 1944. PEDIATRICS, ALLERGY IMMUNOLOGY. *Educ:* Univ Kans, MD, 70. *Prof Exp:* Sr invest, 75-87, CHIEF IMMUNOL SECT, METAB BR, NAT CANCER INST, NIH, 87- *Concurrent Pos:* CAPT, USPHS, 72- *Mem:* Am Fedn Clin Res; Am Soc Clin Invest; Am Asn Immunologists; Soc Pediat Res; Clin Immunol Soc. *Mailing Add:* Nat Cancer Inst Bldg 10 Rm 4N112 NIH Bethesda MD 20892

NELSON, DAVID L, b Council, Idaho, Mar 18, 48. NEUROPHARMACOLOGY. *Educ:* Univ Colo, PhD(pharmacol), 76. *Prof Exp:* Asst prof, 79-85, ASSOC PROF, COL PHARMACOL, UNIV ARIZ, 85- *Mem:* Soc Neurosci; AAAS; Am Soc Pharmacol & Exp Therapeut; Sigma Xi (pres, 85-86). *Mailing Add:* Dept MC907 Lilly Corp Ctr Indianapolis IN 46285

NELSON, DAVID LEE, b Fairmont, Minn, June 19, 42; m 88; c 1. BIOCHEMISTRY. *Educ:* St Olaf Col, Northfield, Minn, BS, 64; Stanford Univ, PhD(biochem), 69. *Prof Exp:* Fel biol chem, Harvard Med Sch, 69-71; ASST PROF BIOCHEM, UNIV WIS-MADISON, 71- *Mem:* Am Soc Biochem & Molecular Biol. *Res:* Biochemistry of sensory transductions in eukaryotic microbes; biochemistry of secretion; proteins kinases. *Mailing Add:* Dept Biochem Univ Wis 420 Henry Mall Madison WI 53706

NELSON, DAVID LLOYD, b Council, Idaho, Mar 18, 48; m 70; c 3. NEUROPHARMACOLOGY. *Educ:* Idaho State Univ, BS, 71; Univ Colo, PhD(pharmacol), 76. *Prof Exp:* Postdoctoral fel neuropharmacol, Col France, 76-79; from asst prof to assoc prof pharmacol, Univ Ariz, 79-90; RES SCIENTIST, ELI LILLY & CO, 90- *Concurrent Pos:* Prin investr, NIH grant, 81-90; NIH res career develop award, 85-90; mem, NLS-2 Study Sect, NIH, 85-89; adj assoc prof, Univ Ariz, 90- *Mem:* Soc Neurosci; Am soc Pharmacol & Exp Therapeut; AAAS. *Res:* Structure activity relationships at serotonergic receptors; ligand receptor interactions; serotonin and second messenger systems. *Mailing Add:* Lilly Res Labs Mail Drop 0815 Lilly Corp Ctr Indianapolis IN 46285

NELSON, DAVID LYNN, b Sacramento, Calif, Dec 6, 42. PHYSICAL CHEMISTRY. *Educ:* Augustana Col, BA, 65; Univ Waterloo, PhD(phys chem), 69. *Prof Exp:* Vis asst prof phys chem, Univ Windsor, 70-72 & Rensselaer Polytech Inst, 72-75; sci officer phys chem, Off Naval Res, 75-89; DIR, SOLID STATE CHEM & POLYMERS PROG, DIV MAT RES, NSF, 89- *Mem:* Am Chem Soc; Sigma Xi. *Res:* Spectroscopy and instrumentation; electrochemistry; surface chemistry and photochemistry. *Mailing Add:* Div Mat Res NSF 1800 G St NW Washington DC 20550

NELSON, DAVID MICHAEL, b Madison, Wis, Nov 21, 46; m 69; c 2. BIOLOGICAL OCEANOGRAPHY, CHEMICAL OCEANOGRAPHY. *Educ:* Dartmouth Col, AB, 69; Univ Alaska, PhD(oceanog), 75. *Prof Exp:* Scholar biol, Woods Hole Oceanog Inst, 75-76, investr, 76-77; from asst prof to assoc prof, 77-86, PROF OCEANOG, ORE STATE UNIV, 86- *Concurrent Pos:* Mem, Ocean Sci Adv Panel, NSF, 85-87; mem, Sci Adv Comt Provasoli-Guillard Ctr Marine Phytoplankton, 89-91, Sci Oversight Comt US Res Icebreaker, Nathaniel B Palmer, 90-92. *Mem:* Am Soc Limnol & Oceanog; Phycol Soc Am; Sigma Xi. *Res:* Nutrient dynamics of the near surface ocean; marine silicon cycle; silicon metabolism in marine diatoms; chemistry and biology of the southern ocean. *Mailing Add:* Col Oceanog Ore State Univ Corvallis OR 97331

NELSON, DAVID ROBERT, b Stuttgart, Ger, May 9, 51; m 75; c 2. THEORETICAL PHYSICS, PHYSICAL CHEMISTRY. *Educ:* Cornell Univ, AB, 72, MS, 74, PhD(physics), 75. *Hon Degrees:* MA, Harvard Univ, 80. *Prof Exp:* Res assoc chem dept, Cornell Univ, 75; jr fel physics, Harvard Soc, 75-78; assoc prof, 78-80, PROF PHYSICS, HARVARD UNIV, 80- *Concurrent Pos:* A P Sloan fel, 79-81; MacArthur fel, 84-89. *Honors & Awards:* Initiatives in Res Award, Nat Acad Sci, 86. *Mem:* Am Phys Soc; Am Acad Arts & Sci. *Res:* Static and dynamic critical phenomena, superfluidity and melting; turbulence; properties of glasses; polymer physics; high Tc superconductors. *Mailing Add:* Dept Physics Harvard Univ Cambridge MA 02138

NELSON, DAVID TORRISON, b Decorah, Iowa, May 16, 27; m 57; c 4. OPTICS. *Educ:* Luther Col, Iowa, BA, 49; Univ Rochester, MA, 55; Iowa State Univ, PhD(physics), 60. *Prof Exp:* Asst physics, Univ Rochester, 49-53; instr, Luther Col, Iowa, 54-57; asst, Iowa State Univ, 58-60; from asst prof to assoc prof, 60-67, chmn dept, 72-86, PROF PHYSICS, LUTHER COL, IOWA, 67- *Concurrent Pos:* NSF sci fac fel, Stanford Univ, 67-68; vis prof eng, Ariz State Univ, 74, 82; mem, Gov Sci Adv Coun, 79- & Iowa Energy Policy Coun, 84-86. *Mem:* Am Phys Soc; Am Asn Physics Teachers; Optical Soc Am; Acoustical Soc Am; Int Solar Energy Soc. *Res:* Solar energy. *Mailing Add:* Dept of Physics Luther Col Decorah IA 52101

NELSON, DENNIS RAYMOND, b New Rockford, NDak, Feb 7, 36; m 61; c 3. BIOCHEMISTRY. *Educ:* NDak State Univ, BS, 58, MS, 59; Univ NDak, PhD(biochem, chem, physiol), 64. *Prof Exp:* Res chemist, 64-71, RES LEADER METAB & RADIATION RES LAB, AGR RES, USDA, 71- *Concurrent Pos:* Assoc prof biochem, NDak State Univ, 72- *Mem:* Am Chem Soc; Sigma Xi; Am Soc Biol Chemists; AAAS. *Res:* Structure, biosynthesis and hormonal control of insect cuticular hydrocarbons; mass spectra of insect methylalkanes; biochemistry of photoperiodic induction of dispause; mode of action of insect hormones. *Mailing Add:* USDA/ARS Biosci Res Lab Fargo ND 58105

NELSON, DIANE RODDY, b Knoxville, Tenn, July 10, 44; m 66. INVERTEBRATE ZOOLOGY. *Educ:* Univ Tenn, Knoxville, BS, 66, MS, 68, PhD(invert zool), 73. *Prof Exp:* Instr biol, 68-69, from instr to assoc prof gen sci, 69-78, assoc prof biol sci, 78-82, PROF BIOL SCI, EAST TENN STATE UNIV, 82- *Mem:* Am Inst Biol Sci; Am Micros Soc; Am Soc Zoologists; Soc Syst Zool; Int Soc Meiobenthologists. *Res:* Systematics and ecology of tardigrades or water bears, Phylum: Tardigrada. *Mailing Add:* Box 23590A E Tenn State Univ Johnson City TN 37614

NELSON, DON B, organic chemistry, for more information see previous edition

NELSON, DON HARRY, b Salt Lake City, Utah, Nov 28, 25; m 49; c 3. MEDICINE. *Educ:* Univ Utah, BA, 45, MD, 47. *Prof Exp:* Res instr biochem, Univ Utah, 50, asst res prof, 52; res assoc med, Harvard Med Sch, 55-56, instr, 57, assoc, 58-59; dir metab ward, Peter Bent Brigham Hosp, Boston, 57-59; from assoc prof to prof med, Sch Med, Univ Southern Calif, 59-66; PROF MED, SCH MED, UNIV UTAH, 66-, CHIEF, DIV ENDOCRINOL, 80- *Mem:* Endocrine Soc (vpres, 69); Am Soc Clin Invest; Am Physiol Soc; Am Asn Physicians. *Res:* Endocrinology; control of adrenal secretion; mechanism of action of adrenal steroids. *Mailing Add:* Dept Internal Med Rm 4C216 Univ Utah Med Ctr 50 N Medical Dr Salt Lake City UT 84112

NELSON, DON JEROME, b Pilger, Nebr, Aug 17, 30. ELECTRICAL ENGINEERING. *Educ:* Univ Nebr, BSc, 53, MSc, 58; Stanford Univ, PhD(elec eng), 62. *Prof Exp:* Mem tech staff, Bell Tel Labs, Inc, 53-55; instr elec eng, 55-58, from asst prof to assoc prof, 60-63, dir Comput Ctr, 63-71, PROF ELEC ENG & COMPUT SCI, UNIV NEBR, LINCOLN, 67- *Concurrent Pos:* Developer, Law Info Retrieval Syst, Nebr Univ, 64-, Remote Operating Syst, 67-, Bill Drafting Syst, Nebr Legis, 71-84, Load Proj Syst, Nebr Pub Power Dist, 76-, Power Prod Modeling Syst, 74-, Packet Switching Commun Simulator, 84-; circuit switching commun simulator, 88-, optical fiber optics database, 90- *Mem:* Inst Elec & Electronics Engrs; Asn Comput Mach. *Res:* System simulation and modeling. *Mailing Add:* Elec Eng Univ Nebr Lincoln NE 68588-0511

NELSON, DONALD CARL, b Minneapolis, Minn, June 28, 31; m 53; c 3. HORTICULTURE, PLANT PHYSIOLOGY. *Educ:* Univ Minn, BS, 53, PhD(hort), 61. *Prof Exp:* Asst veg crops, Univ Minn, 53-55; from asst prof to assoc prof, 61-73, PROF HORT, NDAK STATE UNIV, 73- *Mem:* Europ Asn Potato Res; Potato Asn Am (pres, 79-80); Weed Soc Am. *Res:* Physiology and culture of potatoes. *Mailing Add:* Box 428 Custer SD 57730

NELSON, DONALD FREDERICK, b East Grand Rapids, Mich, July 4, 30; m 54; c 2. PHYSICS. *Educ:* Univ Mich, BS, 52, MS, 53, PhD(physics), 59. *Prof Exp:* Fel physics, Univ Mich, 58-59; mem tech staff, Bell Tel Labs, Inc, 59-67; prof physics, Univ Southern Calif, 67-68; mem tech staff, Res Div, AT&T Bell Labs, 68-87; PROF PHYSICS, WORCESTER POLYTECH INST, 87- *Concurrent Pos:* Vis lectr, Princeton Univ, 76 & Beijing, Shanghai & Jinan, Chinese Acad Sci, 85; mem, NJ Gov Sci Adv Comt, 83-87, Am Nat Standards Inst Comm Safe Use Lasers, 79-, laser safety standards comm, Optical Soc Am, 86-90. *Mem:* Fel Am Phys Soc; Acoust Soc Am; Optical Soc Am. *Res:* Scattering of polarized electrons; laser properties; diode lasers and electroluminescence; electro-optic diode light modulators; optical waveguides; semiconductor and dielectric luminescence; nonradiative recombination; acousto-optic interactions; Brillouin scattering; electrodynamics of elastic dielectrics; nonlinear electroacoustics; continuum mechanics; drift velocities in semi-conductors; lattice dynamics; optical properties of superlattices. *Mailing Add:* Dept Physics Worcester Polytech Inst Worcester MA 01609-2280

NELSON, DONALD J, b Harvey, Ill, Feb 2, 38; m 60; c 2. BIOCHEMICAL PHARMACOLOGY. *Educ:* Oberlin Col, BA, 60; Yale Univ, PhD(pharmacol), 65. *Prof Exp:* Fel pharmacol, Case Western Reserve Univ, 65-67, instr, 67-69; sr res biochemist, 69-80, GROUP LEADER, BURROUGHS WELLCOME CO, 80- *Concurrent Pos:* Adj prof pharmacol, Univ NC, 76- *Mem:* NY Acad Sci; Am Soc Biol Chemists. *Res:* Synthesis of pyrimidine antimetabolites; purification and kinetics of thymidylate kinase from tumors and Escherichia coli; metabolism of thiopurines; control mechanisms in purine and pyrimidine biosynthesis; metabolic effects of allopurinol; adenosine deaminase inhibitors; anti-parasitic drugs and Leishmania; metabolism of antiviral nucleoside analogs. *Mailing Add:* Dept Exp Ther Burroughs Wellcome Co 3030 Cornwallis Rd Research Triangle Park NC 27709

NELSON, DONALD JOHN, b Perth Amboy, NJ, July 24, 45; m 67; c 2. BIOCHEMISTRY. *Educ:* Rutgers Univ, BS, 67; Univ NC, Chapel Hill, PhD(biochem), 72. *Prof Exp:* Fel biochem, Dept Pharmacol, Stanford Univ, 72-74 & Dept Chem, Univ Va, 74-75; PROF CHEM, CLARK UNIV, 75- *Concurrent Pos:* NIH fel gen med sci, 74-75. *Mem:* Am Chem Soc. *Res:* Metal ion and small molecule binding to proteins by nuclear magnetic resonance and fluorescence spectroscopy; protein evolution and polymorphism; intermolecular associations in nucleoside-drug complexes. *Mailing Add:* Dept Chem Clark Univ Worcester MA 01610

NELSON, DOUGLAS A, b Windom, Minn, Jan 20, 27; m 56; c 4. CLINICAL PATHOLOGY, HEMATOLOGY. *Educ:* Univ Minn, BA, 50, BS & MD, 54. *Prof Exp:* Intern, Philadelphia Gen Hosp, 54-55; resident path anat, Mallory Inst Path, Boston City Hosp, 55-58; med fel specialist & res clin path, Univ Minn, 58-60, instr lab med, 60-63, asst prof & asst dir clin labs, 63-64; assoc prof path & assoc dir clin path, 64-69, PROF PATH & ASSOC DIR CLIN PATH, DIV CLIN PATH, STATE UNIV NY UPSTATE MED CTR, 69- *Concurrent Pos:* Sr teaching fel path, Sch Med, Boston Univ, 57-58 & Harvard Med Sch, 57-58; hon consult hemat, Royal Postgrad Med Sch, Hammersmith Hosp, London, Eng, 70-71; trustee, Am Bd Path, 79-90. *Mem:* Am Soc Clin Path; Am Soc Hemat; NY Acad Sci; AMA; Acad Clin Lab Physicians & Scientists; Int Soc Hemat; Am Asn Pathologists. *Res:* Cellular pathology and cytochemistry of hematopoietic system. *Mailing Add:* Div of Clin Path State Univ Hosp Syracuse NY 13210

NELSON, DOUGLAS O, AUTONOMIC NEUROPHARMACOLOGY. *Educ:* Univ Ill, PhD(physiol), 80. *Prof Exp:* Asst prof, 81-86, ASSOC PROF PHYSIOL, NORTHWESTERN UNIV MED SCH, 86- *Mailing Add:* 303 E Chicago Ave Northwestern Univ Med Sch Chicago IL 60611

NELSON, EARL EDWARD, b New Richmond, Wis, Jan 11, 35; m 60; c 1. PLANT PATHOLOGY. *Educ:* Ore State Univ, BS, 57, PhD(plant path), 62. *Prof Exp:* Res forester, Pac Northwest Forest Exp Sta, US Forest Serv, 57-59, plant pathologist, 59-63; ASST PROF PLANT PATH, ORE STATE UNIV, 70-, US FOREST SERV PLANT PATHOLOGIST, FORESTRY SCI LAB, 63-, PROJ LEADER, 74- *Mem:* Mycol Soc Am; Am Phytopath Soc; Sigma Xi. *Res:* Forest disease; root diseases of northwest conifers; ecology of root pathogens emphasizing antagonism by soil fungi; dwarf mistletoes of northwest conifers. *Mailing Add:* 23617 Henderson Rd Corvallis OR 97333

NELSON, EDWARD BLAKE, b Altoona, Pa, Dec 12, 43; m 64; c 2. MEDICINE, PHARMACOLOGY. *Educ:* Pa State Univ, BS, 65; Mich State Univ, PhD(biochem), 70; Univ Tex Med Br, MD, 74. *Prof Exp:* Nat Heart & Lung Inst fel biochem, Univ Tex Southwestern Med Sch, Dallas, 70-71; sr scientist endocrinol, Univ Tex Med Br, Galveston, 71-76; asst prof internal med, pharmacol & therapeut, State Univ NY Buffalo, 77-79; asst prof internal med, Baylor Col Med, 79-89; DIR CLIN RES, CARDIOVASC DIV, MERCK, SHARPE & DOHME RES LABS, 89- *Mem:* Am Col Physicians; Am Soc Clin Pharm. *Res:* Biochemical pharmacology; clinical pharmacology. *Mailing Add:* Cardiovasc Div Merck Sharpe & Dohme Res Labs West Point PA 19486

NELSON, EDWARD BRYANT, b McHenry, Ky, July 26, 16; m 41; c 2. PHYSICS. *Educ:* Western Ky State Col, BS, 37; Vanderbilt Univ, MS, 38; Columbia Univ, PhD, 49. *Prof Exp:* Asst physics, Columbia Univ, 38-41, lectr, 41-43; asst prof, Western Ky State Col, 43-44; res physicist, Manhattan Proj, 44-46; lectr physics, Columbia Univ, 46-49; from asst prof to assoc prof, 49-63, PROF PHYSICS & ASSOC HEAD DEPT, UNIV IOWA, 63-, PROF ASTRON, 77- *Concurrent Pos:* NSF sr fel, Cambridge Univ, 56-57; vis lectr, Univ Exeter, 61-62. *Mem:* Am Phys Soc; Am Asn Physics Teachers. *Res:* Nuclear physics; reactions in light nuclei; nuclear models. *Mailing Add:* 1719 NW 23rd Ave Gainesville FL 32605-3049

NELSON, ELDON CARL, b Dunkirk, Ohio, Dec 13, 35; m 57; c 2. BIOCHEMISTRY, NUTRITION. *Educ:* Ohio State Univ, BSc, 57, MSc, 60, PhD(nutrit), 63. *Prof Exp:* From instr to assoc prof, 63-75, PROF BIOCHEM, OKLA STATE UNIV, 75-, PROF & HEAD, DEPT FOOD, NUTURIT & INST ADMIN, 91- *Concurrent Pos:* Vis assoc res biochemist, Univ Calif, Davis, 75. *Mem:* Am Inst Nutrit; Am Chem Soc; Am Soc Biol Chemists. *Res:* Metabolism and metabolic function of the vitamins A; nutrition; lipid metabolism. *Mailing Add:* Dept Biochem Okla State Univ Stillwater OK 74078

NELSON, ELDON LANE, JR, b Morehead City, NC, May 10, 42; c 1. MEDICAL EDUCATION, PHYSIOLOGY. *Educ:* E Carolina Univ, Greenville, BS & BA, 64, MA, 71; Univ Fla, PhD(physiol), 75. *Prof Exp:* Res technician appl toxicol, Hazelton Labs, 67-68; res asst, Dept Biol, E Carolina Univ, 68-70; res asst, Univ Fla, 71-74, res assoc, Dept Physiol, 74-75; from asst prof to assoc prof physiol, Col Osteop Med, Okla State Univ, Tulsa, 75-84; vpres, Prof Learning Asn, 84-90; ASSOC DIR MED EDUC, MICH HEALTH CTR, DETROIT, 87- *Concurrent Pos:* NIH & Pub Health Serv biomed res develop grant, 80; Health, Educ & Welfare, Pub Health Serv & NIH res grant, 82; vis prof, Northeastern State Univ, 84-86; assoc dir, Educ Prog Develop, Mich Osteop Med Ctr, 86-87; res dir, Nat Osteop Found, 87-90; Am Podiatry Med Asn res grant, 87 & 88; Dept Health & Human Serv training grant, 88. *Mem:* Am Physiol Soc; AAAS; Sigma Xi; Am Asn Univ Prof. *Res:* Physiologic mechanisms for salt and water balance; renal-endocrine interaction; response to environmental stress; hypothalmic regulation of pituitary function. *Mailing Add:* Dept Med Educ Mich Health Ctr 2700 Martin Luther King Jr Blvd Detroit MI 48208-2596

NELSON, ELDRED (CARLYLE), b Starbuck, Minn, Aug 14, 17; m 46, 63; c 1. PHYSICS, COMPUTER SCIENCES. *Educ:* St Olaf Col, AB, 38; Univ Calif, PhD(physics), 42. *Prof Exp:* Instr physics & res assoc, Radiation Lab, Univ Calif, 42-43; group leader theoret physics div, Los Alamos Sci Lab, NMex, 43-46; asst prof physics, Univ Chicago, 46-47; partner & consult math physics, Frankel & Nelson, 47-48; head comput systs dept, Res & Develop Labs, Hughes Aircraft Co, 48-54, head adv electron lab & assoc dir res & develop labs, 54; assoc dir comput systs div, Ramo Wooldridge Corp, 54-58, head Army data processing systs proj, 58-60, dir intellectronic systs lab, 60-61, prog & appl math lab, TRW Systs, 62, dir comput & data reduction ctr, 63-69, dir technol planning & res, TRW Systs, 69-83; RETIRED. *Concurrent Pos:* Res assoc, Calif Inst Technol, 46; lectr, Univ Calif, 47-48; prof, Univ Southern Calif, 52-53; consult comput technol, 83- *Mem:* Am Phys Soc; Inst Mgt Sci; NY Acad Sci; Asn Comput Mach; Sigma Xi. *Res:* Nuclear physics; quantum field theory; electronic digital computer; computer software; research on computer software technology, including the theoretical basis for software reliability, program structure, data structures and computer security. *Mailing Add:* 1808 Melhill Way Los Angeles CA 90049

NELSON, ELTON GLEN, b Elgin Ore, Sept 15, 10; wid; c 2. TEXTILES. *Educ:* Ore State Col, BS, 37, MS, 46; Univ Minn, PhD, 61. *Prof Exp:* Shipping point inspector, State Dept Agr, Ore, 36; coop fiber flax, USDA & Ore State, 37-48; consular attache, US Dept State, India, 48-49; proj leader field crops res br, Agr Res Serv, USDA, 49-57, head, Cordage Fibers Sect, 57-60; asst chief agron & soils br, Off Food, Agency Int Develop, 60-61, asst sci adv, Latin Am, 61-63; asst dir bus serv & analysis div, Off Textiles, US Dept Commerce, 63-70; fiber specialist, Plant Genetics & Germplasm Inst, Agr Res Serv, USDA, 70-84; consult, Cordage Inst, 72-81; ADV LONG VEGETABLE & MISCELLANEOUS FIBERS, NAT AGR LIBR, USDA, 84- *Mem:* Soc Econ Bot. *Res:* Long vegetable fibers, hard and soft fibers; cordage and industrial textiles. *Mailing Add:* 7813 Old Chester Rd Bethesda MD 20817

NELSON, ERIC ALAN, b San Bernardino, Calif, May 28, 49; m 71; c 2. PLANT PHYSIOLOGY, FORESTRY. *Educ:* Occidental Col, BA, 71; Ore State Univ, MS, 74, PhD(forest sci), 78. *Prof Exp:* Lab instr, Ore State Univ, 72, res asst, 72; forestry technician res, Forest Serv, USDA, 72; res asst, Ore State Uni Univ, 72-77; plant physiologist res, US Forest Serv, USDA, 78-79; res assoc, NDak State Univ, 79-80; res physiologist, 80-87, SR RES PHYSIOLOGIST, WESTVACO FOREST RES, 87- *Mem:* Am Soc Plant Physiologists; Sigma Xi; Soc Am Foresters; Plant Growth Regulator Soc Am. *Res:* Seedling physiology; physiology during dormancy; storage compounds; endogenous plant hormones; water relations; nursery production of seedlings for reforestation; photoperiodic responses. *Mailing Add:* Forest Sci Lab Westvaco Corp PO Box 1950 Summerville SC 29484

NELSON, ERIC LOREN, b Los Angeles, Calif, June 29, 24; m 48; c 3. MOLECULAR PHARMACOLOGY. *Educ:* Univ Calif, Los Angeles, BA, 47, PhD(microbiol), 51. *Prof Exp:* Asst bact, Univ Calif, Los Angeles, 48-51, assoc infectious dis, Med Sch, 51-52, instr, 52-55, asst prof, 55-58, asst prof bact, 58-60; sci dir, Allergan Pharmaceut Corp, Calif, 61-63, vpres, 63-72; pres, Nelson Res & Develop Co, 72-86; CONSULT, 86- *Concurrent Pos:* Res assoc, Univ Chicago, 55-57. *Mem:* Am Soc Microbiol; Soc Exp Biol & Med; Int Soc Chemotherapy. *Res:* Speciation and brucellaphage of brucella; hemoglobin particles; radiation infection; cellular immunity; ophthalmology; bioreceptors and molecular design. *Mailing Add:* 14 Trafalgar Newport Beach CA 92660

NELSON, ERIC V, b Green Bay, Wis, Jan 9, 40; m 61; c 4. ENTOMOLOGY, APICULTURE. *Educ:* Univ Wis, BS, 61, MS, 63; Univ Man, PhD(entom), 66. *Prof Exp:* Res entomologist, Entom Res Apicult Br, Agr Res Serv, USDA, 66-67; from asst prof to assoc prof, 67-79, PROF BIOL, OHIO NORTHERN UNIV, 80- *Concurrent Pos:* Vis assoc prof entom, Univ Man, 73-74. *Mem:* Entom Soc Am; Sigma Xi. *Res:* Physiology of bee diseases; diets for bees. *Mailing Add:* Dept of Biol Ohio Northern Univ Ada OH 45810

NELSON, ERLAND, b Blair, Nebr, June 4, 28. NEUROLOGY, NEUROPATHOLOGY. *Educ:* Carthage Col, BA, 47; Columbia Univ, MD, 51; Univ Minn, PhD(neurol, path), 61. *Hon Degrees:* DSc, Carthage Col, 73. *Prof Exp:* Armed Forces Inst Path fel, Max Planck Inst, Munich, Ger, 55-57, NIH fel, 59-60; assoc prof neurol & neuropath, Univ Minn, 61-64; prof neurol & head dept, Sch Med, Univ Md, Baltimore City, 64-81; prof & vchmn Affil Hosp, Rush Med Col, 81-84; chmn, Dept Neurol, Lovelace Med Ctr, Albuquerque, NMex, 82-90; CONSULT, 90- *Concurrent Pos:* Chmn, Dept Neurol, Mount Sinai Hosp Med Ctr, 81- *Mem:* Asn Univ Professors Neurol (secy-treas, 68-72); Am Acad Neurol; Am Neurol Asn; Am Asn Neuropath. *Res:* Clinical neurology; electron microscopy of central nervous system infections, neoplasms and leukoencephaopathy; ultrastructure of intracranial arteries; atherosclerosis; cerebrovascular disease. *Mailing Add:* Ocean Reef Club PO Box 196 Key Largo FL 33037

NELSON, EVELYN MERLE, algebra, theoretical computer science; deceased, see previous edition for last biography

NELSON, FRANK EUGENE, b Harlan, Iowa, Dec 5, 09; m 40; c 3. FOOD MICROBIOLOGY. *Educ:* Univ Minn, BS, 32, MS, 34; Iowa State Col, PhD(dairy bact), 36. *Prof Exp:* Lab technician, Univ Minn, 32-33, instr dairy bact, 36-37; from asst prof to assoc prof bact, Kans State Col, 37-43; prof dairy bact & res prof, Iowa State Univ, 43-60; food scientist, Agr Exp State & prof food sci, 60-77, prof microbiol & med technol, 69-77, EMER PROF NUTRIT, FOOD SCI & MICROBIOL, UNIV ARIZ, 77- *Concurrent Pos:* Ed, Am Dairy Sci Asn J, 47-52. *Honors & Awards:* Borden Award, Am Dairy Sci Asn, 53, Award of Honor, 71. *Mem:* Am Soc Microbiol; Am Dairy Sci Asn (vpres, 64-65, pres, 65-66). *Res:* Lipolytic and proteolytic activities of bacteria; factors affecting resistance to heat; psychrophilic bacteria. *Mailing Add:* 3960 E Ina Rd Tucson AZ 85718

NELSON, FREDERICK CARL, b Braintree, Mass, Aug 8, 32; m 55; c 4. ACOUSTICS. *Educ:* Tufts Univ, BS, 54; Harvard Univ, MS, 55, PhD(appl mech), 61. *Prof Exp:* From instr to assoc prof, 55-71, chmn dept, 69-80, PROF MECH ENG, TUFTS UNIV, 71-, DEAN ENG, 80- *Concurrent Pos:* NSF res grants, Tufts Univ, 63-64, 66-68; vis res fel, Inst Sound & Vibration Res, Univ Southampton, 67; vis prof, Nat Inst Appl Sci, Lyon France, 77, 82. *Honors & Awards:* Centennial Medal, Am Soc Mech Engrs; Seijong Medal, Korea Advan Inst Sci & Technol; Anniversary Medal, Nat Inst Appl Sci, de Lyon. *Mem:* AAAS; Am Soc Mech Engrs; Acoust Soc Am. *Res:* Structural dynamics and damping; mechanical acoustics and noise control. *Mailing Add:* Dean Eng Tufts Univ Medford MA 02155

NELSON, GARY JOE, b Oakland, Calif, Sept 27, 33; m 59; c 1. BIOPHYSICS, BIOCHEMISTRY. *Educ:* Univ Calif, Berkeley, BS, 55, PhD(biophys), 60. *Prof Exp:* Res fel heart dis, Donner Lab, Univ Calif, Berkeley, 60-63; sr staff scientist, Lawrence Livermore Lab, Univ Calif, 63-73; assoc res scientist, NY State Inst Basic Res in Ment Retardation, 73-74; grants assoc, NIH, 74-75; health scientist adminr, Nat Heart, Lung & Blood Inst, 75-83; RES CHEMIST, WESTERN HUMAN NUTRIT RES CTR, US DEPT AGR, 83- *Concurrent Pos:* Nat Heart Inst fel, 60-62; estab investr, Am Heart Asn, 62-63; ed, Coun Arteriosclerosis Newsletter, 78-84; assoc ed, Lipids, 89- *Mem:* AAAS; Am Soc Biol Chemists; Am Oil Chem Soc. *Res:* Human nutrition and lipid metabolism; membrane structure and function; lipid biochemistry; analytical instrumentation; chromatography; platelet function; heart disease and atherosclerosis; disorders of lipid metabolism, nutrition, blood clotting. *Mailing Add:* Biochem Unit Western Human Nutrit Res Ctr PO Box 29997 Presidio San Francisco CA 94129

NELSON, GAYLE HERBERT, b Dayton, Wash, Mar 17, 26; m 46; c 2. ANATOMY. *Educ:* Walla Walla Col, BA, 47; Univ Md, MS, 53; Univ Mich, PhD(anat), 57. *Prof Exp:* Asst biol, Walla Walla Col, 47-49; instr, Washington Missionary Col, 49-51, instr & actg head dept, 51-53; from instr to assoc prof anat, Loma Linda Univ, 57-66; from assoc prof to prof anat, Kansas City Col Osteop Med, 66-79, chmn dept, 71-79; asst dean basic sci, 80-85, PROF ANAT, COL OSTEOP MED PAC, 79- *Concurrent Pos:* Consult anat, Nat Bd Examinr for Osteop Physicians & Surgeons, Inc, 75-87; consult pre-clinical area, Comt Cols, Am Osteop Asn, 84- *Res:* Lymphatic system; biology and taxonomy of Coleoptera. *Mailing Add:* Col Osteop Med Pacific Col Plaza Pomona CA 91766-1889

NELSON, GEORGE, b Sterling, Ill, May 28, 42. MATHEMATICAL LOGIC. *Educ:* Oberlin Col, BA, 64; Inst Technol, PhD(math), 68. *Prof Exp:* PROF MATH, IOWA UNIV, 68- *Mem:* Asn Symbolic Logic; Am Math Soc. *Mailing Add:* Dept Math Univ Iowa Iowa City IA 52242

NELSON, GEORGE DRIVER, b Charles City, Iowa, July 13, 50; m 71; c 2. ASTRONOMY, SOLAR PHYSICS. *Educ:* Harvey Mudd Col, BS, 72; Univ Wash, MS, 74, PhD(astron), 78. *Prof Exp:* Researcher solar physics, Sacramento Peak Observ, 75-76; researcher astrophys, Astron Inst Netherlands, Utrecht, 76-77; fel, Joint Inst Lab Astrophys, Univ Colo, 78; ASTRONAUT, JOHNSON SPACE CTR, NASA, 78- *Mem:* AAAS; Am Astron Soc. *Res:* Convection in stellar atmospheres, radiation driven hydrodynamics. *Mailing Add:* One Lawrence Rd St Louis MO 63141

NELSON, GEORGE HUMPHRY, b Charleston, SC, Nov 24, 30; m 56; c 4. BIOCHEMISTRY, OBSTETRICS & GYNECOLOGY RESEARCH. *Educ:* Col Charleston, AB, 51; Med Col SC, MS, 53, PhD(biochem), 55; WVa Univ, MD, 62. *Prof Exp:* USPHS res fel, 55-56; asst prof biochem, Univ SDak, 56-58; instr, WVa Univ, 58-62; asst prof biochem, 62-68, from instr to asst prof obstet & gynec, 62-68, assoc prof biochem & obstet & gynec, 68-74, PROF OBSTET & GYNEC, MED COL GA, 74- *Concurrent Pos:* Intern, Eugene Talmadge Mem Hosp, 64. *Honors & Awards:* Foundation Prize Award, Am Asn Obstet Gynec, 74. *Mem:* AMA; Soc Gynec Invest; Sigma Xi; NY Acad Sci. *Res:* Lipid metabolism in normal and abnormal pregnancy; fetal maturity evaluation; Prostagladins and reproduction; Fetal pulmonary maturation. *Mailing Add:* Dept Obstet & Gynec Med Col Ga Augusta GA 30902

NELSON, GEORGE LEONARD, b Marshall, Minn, Dec 8, 43; m 64. ORGANIC CHEMISTRY, PROSTAGLANDIN OLIGOMERS. *Educ:* St John's Univ, Minn, BS, 65; Univ Wis-Madison, PhD(chem), 69. *Prof Exp:* NIH fel, Columbia Univ, 69-70; asst prof, 70-76, assoc prof, 76-80, PROF ORGANIC CHEM, ST JOSEPH'S COL, 80- *Concurrent Pos:* Cottrell Res Corp grant, 70-71; vis prof, Hahneman Med Col, 75-, Univ Wis, 75 & Univ Pa, 81-82. *Mem:* Am Chem Soc; NY Acad Sci; Sigma Xi. *Res:* Mechanistic organic chemistry; thermal rearrangements; synthetic organic chemistry; development of new synthetic techniques; heart and brain chemistry; preparation of biologically active compounds; prostaglandin oligomers. *Mailing Add:* Dept Chem St Joseph's Univ Philadelphia PA 19131

NELSON, GEORGE WILLIAM, b Mansfield, Ohio, Apr 27, 38; m 61; c 4. GAMMA RAY SPECTRUM ANALYSIS. *Educ:* Case Inst Technol, BS, 60, MS, 65, PhD(nuclear eng), 66. *Prof Exp:* Engr, Los Alamos Sci Lab, 61-62; from asst prof to assoc prof, 66-87, PROF NUCLEAR ENG, NUCLEAR REACTOR LAB, ARIZ, 87- *Concurrent Pos:* Dir, Nuclear Reactor Lab, Ariz, 73-; first off tech serv, Dept Safeguards, Int Atomic Energy Agency, 78-80. *Mem:* Am Soc Nondestructive Testing; Inst Elec & Electronics Engrs; Am Nuclear Soc; Sigma Xi. *Res:* Cavity nuclear reactors; pulsed neutron experimentation; neutron thermalization; nuclear materials safeguards; neutron activation analysis; subcritical multiplication analysis. *Mailing Add:* 201 E Hyde St Tucson AZ 85704

NELSON, GERALD CLIFFORD, b Benson, Minn, Aug 21, 40; m 64. NUCLEAR PHYSICS. *Educ:* St Olaf Col, BA, 62; Iowa State Univ, PhD(physics), 67. *Prof Exp:* Res assoc, Iowa State Univ, 67-68; res assoc, Lawrence Radiation Lab, Univ Calif, Berkeley, 68-70; RES ASSOC, SANDIA LABS, 70- *Mem:* AAAS; Am Phys Soc. *Res:* Properties of x-rays from high atomic number elements; gamma ray energies, intensities and internal conversion of coefficients. *Mailing Add:* 11509 Nassau Dr NE Albuquerque NM 87111

NELSON, GERALD DUANE, b Detroit Lakes, Minn, July 9, 33; m 57; c 3. ELECTRICAL ENGINEERING. *Educ:* Univ Minn, BEE, 59; Univ Iowa, MSEE, 62, PhD(elec eng), 68. *Prof Exp:* Engr, Collins Radio Co, 59-60; teaching asst elec eng, Univ Iowa, 60-62; res engr, Honeywell, Inc, 62-66; teaching asst elec eng, Univ Iowa, 67; prin res engr, Honeywell, Inc, 68-69; assoc prof elec eng, SDak State Univ, 69-78, pattern recognition specialist, Remote Sensing Inst, 69-74; STAFF SCIENTIST, SPERRY UNIVAC, 78- *Mem:* Am Asn Artificial Intel; Inst Elec & Electronics Engrs; Sigma Xi. *Res:* Pattern recognition systems and theory; expert systems development; statistical communication theory; information processing; dynamic and integer programming approaches to the selection of pattern features. *Mailing Add:* 12740 Portland Circle Burnsville MN 55337

NELSON, GILBERT HARRY, b Manhattan, NY, Sept 20, 27; m 55; c 3. CLINICAL CHEMISTRY. *Educ:* Wagner Lutheran Col, BS, 52; Purdue Univ, MS, 56, PhD(biochem), 58. *Prof Exp:* Biochemist, Christian Hansen's Lab, Inc, Wis, 57-58 & New Castle State Hosp, 58-64; clin chemist, Miami Valley Hosp, 64-90; RETIRED. *Concurrent Pos:* Assoc prof clin chem, Univ Dayton, 69-79; pres, Clin Chem Consults, Inc. *Mem:* AAAS; Am Chem Soc; Am Asn Clin Chemists; NY Acad Sci; Sigma Xi. *Mailing Add:* 293 E Spring Valley Pike Centerville OH 45459

NELSON, GORDON ALBERT, b Bentley, Alta, Nov 29, 25; m 54; c 3. PLANT PATHOLOGY. *Educ:* Univ Alta, BSc, 49, MSc, 51; SDak State Univ, PhD(plant path), 61. *Prof Exp:* Res officer bact, Defense Res Bd, Alta, 54-55; lab scientist, Prov Dept Agr, 55-57; res asst plant path, SDak State Univ, 57-61; res scientist, Can Dept Agr, Nfld, 61-66, res scientist, Res Sta, Alta, 66-90; RETIRED. *Concurrent Pos:* Nat Res Coun Can res fel, 61; vis lectr, Mem Univ, 63-65. *Mem:* Am Phytopath Soc; Can Phytopath Soc; Potato Asn Am. *Res:* Dairy bacteriology; bacterial plant diseases; soil-borne plant diseases; potato viruses; survival and symptomology of the potato ring rot pathogens. *Mailing Add:* 1411 Henderson Lake Blvd Res Sta Lethbridge AB T1K 3B9 Can

NELSON, GORDON L(EON), b Maynard, Minn, Dec 28, 19; m 42; c 5. AGRICULTURAL ENGINEERING. *Educ:* Univ Minn, BAgrE, 42; US Naval Acad, cert, 45; Okla State Univ, MS, 51; Iowa State Univ, PhD(theoret & appl mech, agr eng), 57. *Prof Exp:* Sr agr engr, Portland Cement Asn, 46-47; from assoc prof to prof agr eng, Okla State Univ, 47-69; chmn dept, 69-81, PROF AGR ENG, OHIO STATE UNIV, 69- *Concurrent Pos:* NSF sr fel, Univ Calif, Berkeley & Davis, 64-66; dir, Ford Found proj through Ohio State Univ, Punjab Agr Univ, India, 69-73; consult, Okla Agr Exp Sta; mem bd dirs,

Coun Agr Sci & Technol, 74-82, Am Soc Agr Engrs, 69-71 & 79-81. *Honors & Awards:* Award, Am Soc Agr Eng-Metal Bldg Mfrs Asn, 60; Massey Ferguson Award, Am Soc Agr Eng, 87. *Mem:* Fel Am Soc Agr Engrs; Am Soc Eng Educ. *Res:* Light structures engineering; environmental control engineering for livestock; wind force effects on low profile structures; similitude and theory of models; process engineering for agricultural products; development of agricultural engineering educational and research programs; mechanical engineering; author of over 115 technical papers. *Mailing Add:* 590 Woody Hayes Dr Columbus OH 43210-1273

NELSON, GORDON LEIGH, b Palo Alto, Calif, May 27, 43. POLYMER FLAMMABILITY, COMBUSTION PRODUCT TOXICITY. *Educ:* Univ Nev, Reno, BS, 65; Yale Univ, MS, 67, PhD(chem), 70. *Hon Degrees:* DSc, William Carey Col, 88,. *Prof Exp:* Res chemist, Corp Res & Develop, Gen Elec Co, 70-74, mgr, Combustibility Technol, Plastics Bus Div, 74-79, mgr, Environ Protection Oper, Plastics Bus Oper, 79-82; vpres mats sci & technol, Springborn Labs, 82-83; PROF POLYMER SCI & CHMN DEPT, UNIV SOUTHERN MISS, 83- *Concurrent Pos:* Consult to indust, 83- *Honors & Awards:* Henry Hill Award, Am Chem Soc, 86. *Mem:* Am Chem Soc; Am Soc Testing & Mat; Comput & Bus Equip Mfrs Asn; Soc Plastics Indust; Nat Fire Protection Asn; Soc Plastics Engrs. *Res:* Polymer flammability; combustion product toxicity; smoke formation; polymer degradation; aging and weathering; C-13 nuclear magnetic resonance; organic coating adhesion to plastic substrates; polymer processing fumes. *Mailing Add:* Col Sci & Liberal Arts Fla Inst Technol Melbourne FL 32901-6988

NELSON, GREGORY A, b Minneapolis, Minn, Mar 1, 52. BIOTECHNOLOGY. *Educ:* Cal Inst Technol, BS, 74, Harvard Univ, MA, 76, PhD(cell develop), 79. *Prof Exp:* Grad student, Harvard Univ, div of Med Sci, 74-79; res fel Path, Dept Path, Harvard Med Sch , Boston, MA, 80-82; MEM TECH STAFF & GROUP LEADER, BIOTECHNOL GROUP, 88- *Concurrent Pos:* Res asst, div Biol Caltech, 73-74. *Mem:* AAAS; Am Soc Cell Biol; Genetics Soc Am. *Mailing Add:* Cal Inst Technol, Jet Propulsion Lab 4800 Oak Grove Dr Pasadena CA 91109

NELSON, GREGORY VICTOR, b Minneapolis, Minn, Nov 16, 43; m 67; c 2. MOLECULAR MODELING, EXPERT SYSTEMS. *Educ:* St Olaf Col, BA, 65; Univ Calif, Berkeley, PhD(chem), 68. *Prof Exp:* Asst prof chem, Drew Univ, 68-73, assoc prof, 73-79; MEM STAFF, CELANESE RES CO, 79- *Concurrent Pos:* Vis prof, Inst Inorg Chem, Univ Fribourg, Switz, 74-75; consult, Chem Solve, Inc, Morristown, NJ, 78. *Mem:* AAAS; Am Chem Soc; Sigma Xi. *Res:* Organometallic complexes of transition metal carbonyls, metalolefin complexes; stereoscopic computer drawing; aromatic radical anions; crown complexes; electron spin resonance; molecular modelling; conformational analysis; expert systems; parallel computing. *Mailing Add:* Hoechst Celanese Res Div 86 Morris Ave Summit NJ 07901

NELSON, HARLAN FREDERICK, b Otranto Twp, Iowa, Aug 6, 38; m 61; c 3. AERONAUTICAL ENGINEERING. *Educ:* Iowa State Univ, BS, 61; Purdue Univ, MS, 64, PhD(aeronaut eng), 68. *Prof Exp:* PROF AEROSPACE ENG, UNIV MO-ROLLA, 68- *Concurrent Pos:* Consult. *Mem:* Am Inst Aeronaut & Astronaut; Combustion Inst; Sigma Xi; Am Soc Mech Eng. *Res:* Thermophysics; radiation transfer. *Mailing Add:* Dept of Mech & Aerospace Eng Univ of Mo Rolla MO 65401

NELSON, HAROLD STANLEY, b New Britain, Conn, Jan 17, 30; m 53; c 3. MEDICINE. *Educ:* Harvard Univ, AB, 51; Univ Mich, MS, 69; Emory Univ, MD, 55. *Prof Exp:* Chmn dept med, US Army Hosp, Ft Rucker, 62-64 & 5th Gen Hosp, 64-67; chmn allergy-immunol serv, Fitzsimons Army Med Ctr, 69-86; SR STAFF PHYSICIAN, DEPT MED, NAT JEWISH CTR, IMMUNOL & RESPIRATORY MED, 86- *Concurrent Pos:* Fel allergy-immunol, Univ Mich, 67-69. *Mem:* Fel Am Acad Allergy & Immunol; fel Am Col Allergy & Immunol; Am Thoracic Soc. *Res:* Beta-adrenergic bronchodilators, efficacy and subsensitivity; diagnostic tests in allergy. *Mailing Add:* 4970 S Fulton St Greenwood Village CO 80111

NELSON, HARRY ERNEST, b Rockford, Ill, Sept 21, 13; m 41; c 3. MATHEMATICS, ASTRONOMY. *Educ:* Augustana Col, AB, 35; Univ Wis, PhM, 40; Univ Iowa, PhD(math), 50. *Prof Exp:* Teacher high sch, Ill, 35-37; teacher math, Luther Col, 37-42; teacher, Gustavus Adolphus Col, 42-46; prof math, 46-81, DIR, JOHN DEERE PLANETARIUM, AUGUSTANA COL, ILL, 67- *Concurrent Pos:* Mem Univ Ky team, Int Coop Admin, Indonesia, Bandung, 58-60. *Mem:* Math Soc Am; Int Planetarium Soc; Planetary Soc. *Res:* Meteors. *Mailing Add:* 2439 17th Ave Rock Island IL 61201

NELSON, HARRY GLADSTONE, b Chicago, Ill, Feb 4, 22; m 44, 86; c 2. ENTOMOLOGY, ECOLOGY. *Educ:* Univ Chicago, BS, 45. *Prof Exp:* Asst zool, Univ Chicago, 45-47; instr biol, Gary Col, 47-48; teacher zool, Herzl Br, Chicago City Jr Col, 48-51; from lectr to assoc prof, Roosevelt Univ, 51-69, actg chmn dept, 65-66, chmn, 66-72, prof, 69-89, EMER PROF BIOL, ROOSEVELT UNIV, 89- *Concurrent Pos:* Teacher, Herzl Br, Chicago City Jr Col, 54; assoc cur, Field Mus Natural Hist, 58- *Mem:* Coleopterist Soc; Am Entomol Soc; AAAS; NAm Benthol Soc. *Res:* Aquatic biology; invertebrate zoology and entomology; evolution, ecology, distribution, taxonomy and morphology of the Dryopoidea in North America and Southern Asia. *Mailing Add:* Dept Insects Field Mus Natural Hist Roosevelt Rd & Lake Shore Dr Chicago IL 60605

NELSON, HARVARD G, b Logan, Utah, Aug 29, 19; m 47; c 1. DAIRY SCIENCE. *Educ:* Utah State Univ, BS, 41; Ohio State Univ, MS, 42; Univ Minn, PhD(dairy indust), 52. *Prof Exp:* Explosives chemist, Pantex Ord Plant, 42-43; PROF FOOD SCI, SOUTHEASTERN LA UNIV, 50- *Concurrent Pos:* Tech dir, Hammond Milk Corp, 51-58; consult, Brown's Velvet Dairy Prod, Inc, 55-; lab dir, Goldhill Foods Corp, 61-64. *Mem:* Am Dairy Sci Asn; Sigma Xi; Inst Food Technol. *Res:* Quality control work in seafood, vegetable and dairy products. *Mailing Add:* 1419 Ellis Dr Hammond LA 70401

NELSON, HERBERT LEROY, b Eddyville, Iowa, June 15, 22; m 43; c 4. PSYCHIATRY, ADMINSTRATIVE PSYCHIATRY. *Educ:* Univ Iowa, BA, 43, MD, 46; Am Bd Neurol & Psychiat, dipl psychiat, 53; cert admin psychiat, 58. *Prof Exp:* Intern, Univ Hosps, Iowa City, Iowa, 46-47; psychiat physician, US Vet Admin Hosp, Knoxville, Iowa, 49-51; chief acute male treatment serv, Ore State Hosp, Salem, 51-55, clin dir, 55-63, asst supt, 58-63; dir, Iowa Ment Health Authority, 68-82; from asst prof to assoc prof, 63-73, prof, 73-84, EMER PROF PSYCHIAT, UNIV IOWA, 84- *Concurrent Pos:* Proj dir, Iowa Comprehensive Ment Health Planning, 63-66; asst dir, State Psychiat Hosp, Iowa City, 66-74. *Mem:* AMA; fel Am Psychiat Asn; Am Asn Psychiat Adminr; fel Am Col Ment Health Admin. *Res:* Mental health administration; therapeutic community organization; community psychiatry. *Mailing Add:* Psychiat Hosp 500 Newton Rd Iowa City IA 52242

NELSON, HOMER MARK, b Malad, Idaho, Dec 11, 32; m 69; c 4. SOLID STATE PHYSICS. *Educ:* Brigham Young Univ, BS, 53, MS, 54; Harvard Univ, PhD(physics), 60. *Prof Exp:* Assoc prof, 59-71, PROF PHYSICS, BRIGHAM YOUNG UNIV, 71- *Mem:* Am Phys Soc. *Res:* Magnetic resonance under high pressure; molecular beams; plasma physics. *Mailing Add:* Dept Physics 178 Eyring Sci Ctr Brigham Young Univ Provo UT 84602

NELSON, HOWARD GUSTAVE, b St Paul, Minn, Jan 16, 38; m 70; c 2. MATERIALS SCIENCE. *Educ:* Wash State Univ, BS, 60, MS, 63; Univ Calif, Los Angeles, PhD(mat sci), 70. *Prof Exp:* Process metallurgist, NAm Aviation, Rockwell Int Corp, 60-62; RES SCIENTIST & MGR, AMES RES CTR, NASA, 63- *Concurrent Pos:* Chief Test Eng & Anal Branch NASA Ames Res Ctr. *Honors & Awards:* Except Sci Achivement Medal, NASA, 75, Except Engr Achievement Medal, 88. *Mem:* Am Soc Metals; Am Inst Mining, Metall & Petrol Engrs. *Res:* Understanding the influences of the chemical environment on the fracture behavior of materials. *Mailing Add:* 213-3 NASA Ames Res Ctr Moffett Field CA 94035

NELSON, IRAL CLAIR, b Eugene, Ore, Apr 18, 27; m 55; c 3. HEALTH PHYSICS. *Educ:* Univ Ore, BS, 51, MA, 55; Am Bd Health Physics, dipl, 62, recert, 81, 85. *Prof Exp:* Engr, Hanford Atomic Prod Opers, Gen Elec Co, 55-64, mgr external dosimetry, 64-65; sr res scientist, 65-67, res assoc, 67-72, mgr radial health res sect, 72-76, mgr environ technol, 76-80, PAC NORTHWEST LABS, BATTELLE MEM INST, 80- *Concurrent Pos:* Staff scientist, Nat Environ Protection Agency. *Mem:* Health Physics Soc; Soc Risk Analysis. *Res:* Radiation dosimetry; methods for determining the fate of radionuclides deposited in the body from accidental intake; environmental consequences of construction, operation and decommissioning of nuclear power production and commercial and national defense waste management facilities. *Mailing Add:* 2105 Putnam Richland WA 99352

NELSON, IVORY VANCE, b Curtiss, La, June 11, 34; m 60; c 4. ANALYTICAL CHEMISTRY. *Educ:* Grambling Col, BS, 59; Univ Kans, PhD(chem), 63. *Prof Exp:* From assoc prof to prof chem, Southern Univ, Baton Rouge, 63-67, chmn, Div Natural Sci, Shreveport, 67-68; asst dean, Prairie View A&M Univ, 68-71, vpres res & spec progs, 71-82, actg pres, 82-83; exec asst to chancellor, Tex A&M Univ, 83-86; CHANCELLOR, ALAMO COMMUNITY COL DIST, 86- *Concurrent Pos:* Vis prof, Loyola Univ, 67-68; consult, Oak Ridge Assoc Univs, 69-70. *Mem:* AAAS; Am Chem Soc; NY Acad Sci. *Res:* Higher education; administration; behavior of metal ions in nonaqueous solvent. *Mailing Add:* Alamo Community Col Dist 811 W Houston PO Box 3800 San Antonio TX 77284

NELSON, J(OHN) BYRON, b Buffalo Ctr, Iowa, Feb 7, 37; m 63, 81; c 3. ERGONOMICS, ENGINEERING ECONOMICS. *Educ:* Iowa State Univ, BSIE, 59; Purdue Univ, MSIE, 65, PhD(indust eng), 68. *Prof Exp:* Qual control engr, US Gypsum Co, 59-60, personnel supvr, 60-61; indust engr, Sylvania Electronics Prod, Inc, 61-63; indust engr, Radio Corp Am, 66; instr indust eng, Purdue Univ, 63-68; asst prof indust eng & opers res, Va Polytech Inst & State Univ, 68-73; prof eng mgt, Univ Mo-Rolla, 73-82, exten consult, 76-82; CHAIR & PROF INDUST ENG, WESTERN NEW ENG COL, 82- *Honors & Awards:* Res Award, Inst Indust Engrs, 66. *Mem:* Inst Indust Engrs; Am Soc Eng Educ. *Res:* Dynamic visual perception; occupational safety and health; life cycle costing and inflation of engineering projects; engineering economics. *Mailing Add:* Dept Indust Eng Western New Eng Col 1215 Wilbraham Rd Springfield MA 01119

NELSON, JACK RAYMOND, b Fargo, NDak, July 25, 34; m 55; c 5. RANGE ECOLOGY, WILDLIFE ECOLOGY. *Educ:* NDak State Univ, BS, 60, MS, 61; Univ Idaho, PhD(range ecol), 69. *Prof Exp:* Instr range sci, 64-69, asst prof range & wildlife ecol, 69-74, assoc prof, 74-80, PROF WILDLIFE HABITAT MGT, WASH STATE UNIV, 80- *Concurrent Pos:* Consult, Spokane Indian Tribe, Bur Indian Affairs, 69-70, Key Chem Inc, Wash, 70-71, US Forest Serv, 76, US Army Corp Engrs, 79 & Nova Corp, 86. *Mem:* Wildlife Soc; Soc Range Mgt. *Res:* Big game range ecology; big game livestock competition; multiple land-use management; habitat management. rehabilitation. *Mailing Add:* Dept Natural Resource Sci Wash State Univ Pullman WA 99164

NELSON, JAMES ARLY, b Livingston, Tex, Feb 8, 43; c 3. PHARMACOLOGY. *Educ:* Univ Houston, BS, 65, MS, 67; Univ Tex Med Br Galveston, PhD(pharmacol), 70. *Prof Exp:* Res assoc pharmacol, Brown Univ, 70-72; sr biochemist, Southern Res Inst, 72-76; asst prof pharmacol & toxicol, Univ Tex Med Br, Galveston, 76-79; assoc prof pediat pharmacol, PROF, UNIV TEX M D ANDERSON CANCER CTR, 90- *Concurrent Pos:* Assoc prof pharmacol, Med Sch, Univ Tex, Houston, 79- *Mem:* Am Asn Cancer Res; Am Soc Pharmacol & Exp Therapeut. *Res:* Biochemical pharmacology; cancer chemotherapy; ion and drug transport; environmental toxicology. *Mailing Add:* Dept Exp Pediat Univ Tex M D Anderson Cancer Ctr Houston TX 77030

NELSON, JAMES DONALD, b Paducah, Ky, Apr 30, 43. COMPUTER SCIENCE. *Educ:* Univ Ky, BS, 65, MS, 67, PhD(math), 70. *Prof Exp:* Assoc prof math, 77-84, ASSOC PROF & CHMN DEPT COMPUT SCI, WESTERN MICH UNIV, 84- *Mem:* Asn Comput Mach; Math Asn Am. *Res:* Analysis. *Mailing Add:* Comp Sci Dept Western Mich Univ Kalamazoo MI 49008

NELSON, JAMES H, JR, b Marietta, Ohio, June 28, 26; c 2. OBSTETRICS & GYNECOLOGY. *Educ:* Marietta Col, BS, 49; NY Univ, MD, 54. *Prof Exp:* Spec fel, Gynec Tumor Serv, State Univ NY Downstate Med Ctr, 57-58, from instr to prof obstet & gynec, 61-76, chmn dept, 69-76, dir gynec tumor serv, 64-76; JOE V MEIGS PROF GYNEC, HARVARD MED SCH, 76-; CHIEF GYNEC & VINCENT MEM HOSP, MASS GEN HOSP, 76- *Concurrent Pos:* Am Cancer Soc adv clin fel, 61-64; from asst attend physician to assoc attend, Kings County Hosp, 61-66, vis attend, 66-76; fel med, Cancer Chemother Div, Mem Ctr Cancer & Allied Dis, 62-63; res fel, Sloan-Kettering Res Inst, 62-63, vis investr chemother, 63-68; consult, US Naval Hosps, St Albans, NY, 63-65, Bethesda, Md, 66 & consult & lectr, Philadelphia, Pa, 67; mem div gynec oncol, Am Bd Obstet & Gynec. *Mem:* AAAS; Am Cancer Soc; Am Col Surgeons; Am Asn Obstet & Gynec; Am Gynec Soc. *Res:* Gynecologic malignancy; thymo-lymphatic system. *Mailing Add:* 401 Sabra Dr Wilmington NC 28405

NELSON, JAMES HAROLD, b Gosnell, Ark, Apr 26, 36; m 74; c 2. HEALTH SCIENCE ADMINISTRATION, MEDICAL DEVICES. *Educ:* Ark State Univ, BS, 61, MS, 69; Okla State Univ, PhD(med Entom), 72. *Prof Exp:* Grad res asst, Okla State Univ, 69-72; postdoctoral, US Dept Agr, 72; br chief, US Army Environ Hyg Agency, 72-76; res area mgr, US Army Biomed Res & Develop Lab, 76-79, res coordr, 79-83, dep div chief, 83-85, actg div chief, 85, DIV CHIEF, RES & DEVELOP MED MAT, US ARMY BIOMED RES & DEVELOP LAB, 85- *Concurrent Pos:* Consult, Dir Engrs, US Army, Ft Detrick, 76- & Atlantic Dist US Army Corps Engrs, 80-84; mem, Res Panel Fed Working Group Pest Mgt, 77-81; chair, Equip Comn, Armed Forces Pest Mgt Bd, 79-83; assoc ed, J Am Mosquito Control Asn, 82-88; guest lectr, Acad Health Sci, US Army, 86-88. *Honors & Awards:* Recognition Award, Sigma Xi, 88. *Mem:* AAAS; NY Acad Sci; Am Pub Health Asn; Am Mosquito Control Asn; Asn Mil Surgeons US; Sigma Xi. *Res:* research and development of conventional and chemical defense medical material for field Army use; suppression and control of disease vectoring arthropods. *Mailing Add:* 2419 Tabor Dr Middletown MD 21769-9006

NELSON, JAMES S, b St Louis, Mo, Mar 19, 33; m 56; c 2. NEUROPATHOLOGY. *Educ:* St Louis Univ, MD, 57. *Prof Exp:* Asst path, St Louis Univ, 57-59; instr neuropath, Columbia Univ, 59-60; asst path, St Louis Univ, 60-61; Nat Inst Neurol Dis & Blindness spec fel neurochem, 61-63; instr path, Washington Univ, 63-64; state neuropathologist, Div Ment Dis, Mo, 64-65; from asst prof to assoc prof path, Sch Med, St Louis Univ, 65-73; assoc prof path & assoc prof path in pediat, 73-75, PROF PATH, PROF PATH IN PEDIAT & DIR DIV NEUROPATH, DEPT PATH, SCH MED, WASHINGTON UNIV, 75- *Concurrent Pos:* Clin clerk, Nat Hosp Neurol Dis, Queen Square, London, Eng, 61; consult, St Louis State Hosp, 65-66, St Louis City Hosp, St Johns Mercy Hosp & St Lukes Hosp, 68-79; neuropathologist, St Louis Univ Hosps & Glennon Mem Hosp Children, 65-73; from asst pathologist to assoc pathologist, Barnes & Allied Hosps & St Louis Childrens Hosp, 73-82, pathologist, 82-; vis scientist, Inst Neuropath, Frei Univ Berlin, 79-80; US Sr Scientist Award, Alexander von Humboldt Found, West Germany, 79-80; mem, Test Comt Neuropath, Am Bd Path, 85- *Mem:* AAAS; Am Asn Neuropath; Int Soc Neurochem; Am Inst Nutrit; Am Asn Pathologists; AMA. *Res:* Diagnosis and treatment of brain tumors; pathogenesis and treatment of organophosphate intoxication; effects of vitamin E deficiency on the mammalian nervous system. *Mailing Add:* Dept Neuropathol Armed Forces Inst Pathol Washington DC 20306-6000

NELSON, JERRY ALLEN, b Durango, Colo, Feb 22, 23; m 44; c 3. ORGANIC CHEMISTRY. *Educ:* Univ Wash, BS, 46, PhD(chem), 50. *Prof Exp:* Res chemist, E I du Pont de Nemours & Co, Inc, Wilmington, 50-53, res supvr, 53-56, head, Textile Chem & Intermediate Res Div, 56-60, head textile & indust chem, Res & Develop Div, 60-69, head, New Prod Div, Org Chem Dept, Res & Develop Div, 69-76, head, Dyes Prod Develop Div, 76-78, res & develop staff, Chem & Pigment Dept, 78-82; RETIRED. *Mem:* AAAS; Am Chem Soc. *Res:* Polymer, surface and fluorine chemistry. *Mailing Add:* 1012 Baylor Dr Newark DE 19711-3130

NELSON, JERRY EARL, b Glendale, Calif, Jan 15, 44; m 65; c 2. ASTROPHYSICS, EXPERIMENTAL HIGH ENERGY PHYSICS. *Educ:* Calif Inst Technol, BS, 65; Univ Calif, Berkeley, PhD(physics), 72. *Prof Exp:* Fel particle physics, 72-75, DIV FEL ASTROPHYS, LAWRENCE BERKELEY LAB, UNIV CALIF, 75- *Mem:* Am Phys Soc; Am Astron Soc. *Res:* Pulsars, x-ray sources and black holes; optical pulsars; electron-positron colliding beam physics; astronomical optics. *Mailing Add:* Astron Dept Univ Calif Berkeley CA 94720

NELSON, JERRY REES, b Payson, Utah, Apr 22, 47; m 67; c 2. MICROBIOLOGY, IMMUNOLOGY. *Educ:* Univ Utah, BS, 69, MS, 73, PhD(microbiol & immunol), 76. *Prof Exp:* Cult cur microbiol, Univ Utah, 69-74; technician, Deseret Pharmaceutical, 74-75; lab dir qual control, Microbiol Develop & Control Inc, 75-85; PRES, NELSON LABS, 85- *Mem:* Am Soc Microbiol; Am Pub Health Asn; Am Asn Lab Animal Sci; AAAS. *Res:* Biomaterials toxicology; tissue culture toxicology; bacterial aerosols, enumeration and sizing; barrier properties of nonwoven materials, endotoxin detection. *Mailing Add:* Nelson Labs 4535 S 2300 E PO Box 17557 Salt Lake City UT 84117

NELSON, JOHN ARCHIBALD, b Can, Nov 25, 16; m 46; c 3. ORGANIC CHEMISTRY. *Educ:* Univ Alta, MSc, 39; McGill Univ, PhD(org chem), 45. *Prof Exp:* Biochemist, Animal Dis Res Inst, Can, 45-46; chemist, Ciba Co, Ltd, Can, Ciba-Geigy Corp, 46-49, Ciba Pharmaceut Prod, Inc, 49-69, mgr process res, Ciba-Beigy Corp, 69-72, asst dir clin prep process res, 72-77, asst vpres develop & control, 77-79, dir tech doc, 80-81; RETIRED. *Mem:* Am Chem Soc; NY Acad Sci; Sigma Xi. *Res:* Process research in pharmaceuticals. *Mailing Add:* 14 Coursen Way Morris Plains NJ 07950

NELSON, JOHN ARTHUR, b Sturgeon Bay, Wis, Jan 16, 38; m 62; c 2. CHEMISTRY, CHEMICAL ENGINEERING. *Educ:* Univ Wis-Madison, BS, 60; Univ Ariz, PhD(chem), 66. *Prof Exp:* Metallurgist, Res Dept, Inland Steel Co, Ind, 60-62, res engr, 66-68, sr res engr, 68-69; sr phys chemist, Res Dept, Whirlpool Corp, 69-70, sr res chemist, 71-80; MEM STAFF, CIBA PHARMACEUT INC, 80- *Mem:* Am Chem Soc; Electrochem Soc; Nat Asn Corrosion Engrs; Sigma Xi. *Res:* Corrosion of metals; treatment and processing of metals; physics and chemistry of processing systems. *Mailing Add:* 3983 Highland Ave Benton Harbor MI 49022

NELSON, JOHN D, b Duluth, Minn, Sept 16, 30. PEDIATRICS. *Educ:* Univ Minn, BS, 52, MS, 54. *Prof Exp:* Fel infectious dis, Univ Tex Southwestern Med Sch, 59-60; from instr to assoc prof, 60-70, PROF PEDIAT, UNIV TEX HEALTH SCI CTR, DALLAS, 70- *Concurrent Pos:* Nat Inst Allergy & Infectious Dis res fel, 60-62 & res career develop award, 63-73; vis prof, Med Ctr, Univ Colo, Denver, 62; mem antibiotics panel drug efficacy study, Nat Acad Sci-Nat Res Coun, 66-68; consult staff, John Peter Smith Hosp, Ft Worth, Tex, 74-; pres, Int Cong Chemother, 79; adv panel pediat, US Pharmcopeial Conv, 80. *Mem:* Am Soc Microbiol; Infectious Dis Soc Am; Soc Pediat Res; Am Pediat Soc. *Res:* Pediatric infectious diseases. *Mailing Add:* Dept of Pediat Univ of Tex Health Sci Ctr 5323 Harry Hines Blvd Dallas TX 75235

NELSON, JOHN DANIEL, b Duluth, Minn, July 4, 50; m 73; c 3. OPTHALMOLOGY, CORREAL & EXTERNAL DISEASE. *Educ:* Univ Minn, BS, 72, MD, 75. *Prof Exp:* Asst prof, 82-88, ASSOC PROF OPHTHAL, UNIV MINN, 88-; CHIEF OPHTHAL, ST PAUL-RAMSEY MED CTR, 84- *Concurrent Pos:* Chief ophthal, Ramsey Clin, 84-; clin dir, Dry Eye & Tear Res Ctr, 85-; med dir, St Paul-Ramsey Sch Med Technologists, 86- & Nat Sjogren's Syndrome Asn, 90-; dir bd, Ramsey Health Care, Inc, 87-, Ramsey Clin, 88- & Perry Assurance, 88-; dir, Minn Asn Pub Teaching Hosps, 90-; assoc med dir, Minn Lions Eye Bank, 90- *Mem:* Asn Res Vision & Ophthal; Am Acad Ophthal; Am Col Surgeons; Contact Lens Asn Ophthalmologists; Pan-Am Asn Ophthal. *Res:* On ocular surface - how it heals and how cells communicate with each other; tears, their composition and properties as they relate to conditions of excess or deficiency. *Mailing Add:* Dept Ophthal Dry Eye & Tear Res Ctr St Paul-Ramsey Med Ctr 640 Jackson St St Paul MN 55101

NELSON, JOHN FRANKLIN, b Twin Falls, Idaho, Sept 27, 34; m 58; c 2. ORAL PATHOLOGY, ORAL MEDICINE. *Educ:* Univ Minn, BS, 57, DDS, 59; Univ Pa, cert oral med, 68; Armed Forces Inst Path, cert oral path, 71; George Washington Univ, MAEd, 78. *Prof Exp:* Officer, Dent Corp, US Army, 59-62 & 63-79; prof oral path & med, Col Dent, Univ Iowa, 79-84; prof & chmn, Dept Oral Diag & Radiol, 84-90, ASST DEAN CLIN AFFAIRS, BAYLOR COL DENT, 87- *Concurrent Pos:* Consult oral med & path, numerous orgns including Repub of Vietnam, Univ Saigon, and various US Army installations. *Mem:* Am Dent Asn; fel Am Acad Oral Path; Int Asn Dent Res; Asn Dent Schs; Asn Military Surgeons; Sigma Xi; fel Am Col Dentists; fel Int Col Dentists; Am Acad Oral Med. *Res:* Case reports epidemiological analysis, clinical features and applications of oral lesions and conditions; healing; clinic administration; higher dental education. *Mailing Add:* Off Clin Affairs Baylor Col Dent 3302 Gaston Ave Dallas TX 75246

NELSON, JOHN HENRY, b Ogden, Utah, Mar 25, 40; m 64; c 5. INORGANIC CHEMISTRY, ORGANOMETALLIC CHEMISTRY. *Educ:* Weber State Col, AS, 61; Univ Utah, BS, 64, PhD(inorg chem), 68. *Prof Exp:* Esso fel inorg chem, Tulane Univ, La, 68-70; from asst prof to assoc prof, 70-79, PROF INORG CHEM, UNIV NEV, RENO, 79- *Concurrent Pos:* NIH fel, 65-68; Univ Nev Res Adv Bd grants, Reno, 71-80 & 89; Petrol Res Fund grants, 71-74, 75-77, 84-87 & 89-91; vis scientist chem, China Lake Naval Weapons Ctr, 72 & fel, Am Asn Eng Educ, 80; Res Corp grants, 73-76; NSF grant, 77-78; US Off Surface Mining grants, 79-83; Exxon Corp grant, 81-82; Fulbright fel chem, Univ Louis Pasteur de Strasbourg, Univ de Rennes, France, 81-82; US Bur Mines grants, 79-91; Echo Bay grant, 90-91; Hexel Corp grant, 90-91. *Mem:* AAAS; Am Chem Soc; NY Acad Sci; Sigma Xi; fel Royal Soc Chem. *Res:* Synthesis, physical properties, structure, reactions and catalytic properties of coordination and organometallic compounds; environmental chemistry related to precious metals refining. *Mailing Add:* Dept Chem Univ Nev Reno NV 89557-0020

NELSON, JOHN HOWARD, b Chicago, Ill, May 29, 30; m 52; c 3. TECHNICAL MANAGEMENT. *Educ:* Purdue Univ, BS, 52, MS, 53; Univ Minn, PhD(biochem), 61. *Prof Exp:* Res biochemist, Gen Mills, Inc, Minn, 60-61, supvr chem, microbiol & phys testing, 61-64, dept head refrig foods res, James Ford Bell Res Ctr, 65-66, head frozen food res, 66-68; dir corp res, 68, vpres res & develop, Peavey Co, Minn, 68-76, dir int ventures res div, 71-76; partner, Johnson, Powell & Co & Sci prog advisor, Charles F Kettering Res Lab, 76-77; vpres res, Am Maize Prod Co, 77-80, vpres mkt & prod develop, 80-82; vpres corp develop, Roman Meal Co, 82-84, vpres & chief op officer, 84-86; corp dir res & develop, 86-88, vpres res & develop, 88, VPRES SCI & TECHNOL, MCCORMICK & CO, 88- *Concurrent Pos:* Mem gen comt on DOD Food Prog & chmn comt on cereal & gen prods, Adv Bd Mil Personnel Supplies, Nat Acad Sci, 73-82; adv bd, Univ Minn, 74-56, Univ Ill, 80-85; fel, League for Int Food Educ, 84; chmn I/A bd, Towson State Univ, 89-; mem chancellor's Adv Comt Univ MD, 89- *Honors & Awards:* William F Geddes Award, Am Asn Cereal Chemists, 79. *Mem:* Am Asn Cereal Chemists (past pres, 73-74); Am Chem Soc; Inst Food Technologists; Am Oil Chemists Soc; AAAS. *Res:* Lipids of cereal grain; new instrumental methods of analysis for food ingredients and products; new food product research and development; market and business planning. *Mailing Add:* 5K Alexander Chase Sparks MD 21152

NELSON, JOHN HOWARD, b Bozeman, Mont, Feb 5, 26; m 52; c 3. FOOD SCIENCE. *Educ:* Mont State Col, BS, 50; Univ Wis, MS, 51, PhD(dairy & food indust), 53. *Prof Exp:* Res fel, Univ Wis, 53-54; dir res, Dairyland Food Labs Inc, 54-61, vpres res & develop, 61-72, vpres corp develop, 72-75, vpres sci affairs, 75-77; mgr, regulatory compliance, Kraft Inc, 77-78, corp dir, 79-

81, vpres qual assurance & regulatory compliance, 81-87; SR RES PROG MGR, UNIV WIS, 87- *Concurrent Pos:* Mem food stand adv comt, State of Wis, 67-73, 77- *Mem:* Asn Food & Drug Off, Am Chem Soc; Am Dairy Sci Asn; Inst Food Technol. *Res:* Industrial enzymes for food and dairy field; enzyme modified food ingredients; flavors; food law; enzymes and fermentations in food processing; food safety. *Mailing Add:* 306 Ozark Trail Madison WI 53705-2535

NELSON, JOHN KEITH, b Oldham, UK, July 3, 43; m 68; c 2. ELECTRICAL POWER ENGINEERING, DIELECTRICS. *Educ:* Univ London, BSc, 65, PhD(elec power eng), 69. *Prof Exp:* Lectr elec & electronic eng, Queen mary Col, London, 69-78; reader, Univ London, 78-79; res mgr, Gen Elec Corp Res & Develop, 79-82; prof elec power eng, 82-87, DEPT HEAD, ELEC POWER ENG, RENSSELAER POLYTECH INST, 87- *Concurrent Pos:* External examr, Univ Sri Lanka, 71-83; prin investr, numerous res contracts, 72-; consult, numerous indust co, 82- & US Foreign Rels Coord Unit, 85-87; chmn, Deis Educ Comt, Inst Elec & Electronics Engrs, 83-90, tech vpres, 90- & mem, Subcomt Digital Measurements, 87-; mem, Joint Working Group Electrification, Int Conf Large High Voltage Elec Systs, 90- *Honors & Awards:* Snell Premium & J R Beard Award, Inst Elec Engrs, 72. *Mem:* Fel Inst Elec & Electronics Engrs; fel Inst Elec Engrs; Inst Elec & Electronics Engrs Dielectrics & Elec Insulation Soc (vpres, 91-). *Res:* Physical phenomena limiting the application of power system electrical plant with special emphasis on electrostatic and dielectric processes. *Mailing Add:* Dept Elec Power Eng Rensselaer Polytech Inst Troy NY 12180-3590

NELSON, JOHN MARVIN, JR, b Richmond, Mo, May 19, 33; m 62; c 4. ENTOMOLOGY. *Educ:* North Cent Bible Col, BA, 56; Evangel Col, BS, 64; Southern Ill Univ, Carbondale, MS, 66, PhD(zool), 70. *Prof Exp:* Asst prof gen zool, Glenville State Col, 69-71; from asst prof to assoc prof, 71-83, PROF BIOL, ORAL ROBERTS UNIV, 83-, BIOL CHMN, 86- *Mem:* Entom Soc Am; Am Arachnol Soc; Am Inst Biol Sci; Lepidopterists Soc. *Res:* Nest ecology and nest symbionts of Polistes wasps; morphology of Polistes larvae; spider distribution and ecology; lepidoptera distribution in Oklahoma. *Mailing Add:* Dept Biol Oral Roberts Univ Tulsa OK 74171

NELSON, JOHN WILLIAM, b St Louis, Mo, July 18, 26; m 53; c 3. ENVIRONMENTAL PHYSICS. *Educ:* Univ Calif, Los Angeles, BA, 47; Wash Univ, BA, 49; Univ Tex, MA, 52, PhD(physics), 59. *Prof Exp:* Flight forecaster meteorol, US Weather Bur, La, 50-51; asst prof, Kans State Univ, 62-66; res assoc physics, 59-62, from asst prof to assoc prof, 66-77, PROF PHYSICS, FLA STATE UNIV, 77- *Mem:* Am Phys Soc; Inst Elec & Electronics Engrs; Am Asn Physics Teachers; Sigma Xi. *Res:* Low energy nuclear physics; application of nuclear techniques to environmental problems. *Mailing Add:* Dept Physics Fla State Univ Tallahassee FL 32306

NELSON, JOSEPH EDWARD, b Decatur, Ga, May 4, 32; m 52. MATHEMATICS. *Educ:* Univ Chicago, AM, 53, PhD(math), 55. *Prof Exp:* Mem, Inst Advan Study, 56-59; from asst prof to assoc prof, 59-70, PROF MATH, PRINCETON UNIV, 70- *Mem:* Am Math Soc. *Res:* Functional analysis. *Mailing Add:* Dept Math Princeton Univ Princeton NJ 08544

NELSON, JOSEPH SCHIESER, b San Francisco, Calif, Apr 12, 37; Can citizen; m 63; c 4. ICHTHYOLOGY. *Educ:* Univ BC, BSc, 60, PhD(zool), 65; Univ Alta, MSc, 62. *Prof Exp:* Res assoc zool, Indiana Univ, 65-67, asst dir univ biol stas, 67-68; from asst prof to assoc prof, 68-78, assoc chmn dept, 76-81, PROF ZOOL, UNIV ALTA, 78- *Concurrent Pos:* Mem, Fisheries & Oceans Res Adv Coun, Gov Can, 81-85; gov, Am Soc Ichthyol & Herpet & chair, Comt Names of Fishes, Am Fisheries Soc, Am Soc Ichthyol & Herpet. *Mem:* Soc Study Evolution; Am Soc Ichthyol & Herpet; Am Fisheries Soc; Can Soc Zool; Can Soc Environ Biol (pres, 72-74); Europ Ichthyol Union. *Res:* Hybridization in cyprinid and catostomid fishes; systematics of gasterosteid, psychrolutid, creediid, and percophid fishes; classification of world fishes; pelvic skeleton absence in sticklebacks. *Mailing Add:* Dept Zool Univ Alta Edmonton AB T6G 2E9 Can

NELSON, JUDD OWEN, b Stoughton, Wis, Sept 10, 47; m 68; c 2. ENVIRONMENTAL TOXICOLOGY, PESTICIDE CHEMISTRY. *Educ:* Univ Wis-Madison, BS, 69, MS, 72, PhD(entom), 74. *Prof Exp:* Proj asst, Dept Entom, Univ Wis, 68-69, res asst, 69-74; res assoc, 74-76, asst prof, 76-81, ASSOC PROF, DEPT ENTOM, UNIV MD, 81- *Mem:* Am Chem Soc; Entom Soc Am; AAAS; Sigma Xi. *Res:* Chemistry, metabolism and toxicology of pesticides and environmental contaminants; insect pheromone perception; insecticide resistance; chemical mutagenesis and carcinogenesis. *Mailing Add:* 4000 Tennyson Rd Hyattsville MD 20782

NELSON, KAREN ANN, b Spokane, Wash, Feb 16, 48. IMMUNE REGULATION, TUMOR IMMUNOLOGY. *Educ:* Univ Wash, BS, 70, PhD(path), 75. *Prof Exp:* Fel, Wallenberg Lab, Univ Lund, Sweden, 75-77; res assoc tumor immunol, Fred Hutchinson Cancer Res Ctr, 78-84; scientist, HLA, Genetic Syst Corp, 84-88; STAFF, DEPT PATHOL, MED SCH UNIV WASH, 83-; LAB DIR, DEPT HISTOCOMPATABILITY, PUGET SOUND BLOOD CTR, 88- *Mem:* AAAS; Asn Women Sci. *Res:* Identification of the cells, molecules and genes involved in regulation of the immune response to neoplastic cells and manipulation of this response towards rejection of tumors. *Mailing Add:* 921 Terry Ave Seattle WA 98121

NELSON, KARIN BECKER, b Chicago, Ill; c 4. CHILD NEUROLOGY. *Educ:* Univ Chicago, MD, 57. *Prof Exp:* Assoc neurologist, Children's Hosp, Washington, DC, 67-71; asst prof, 70-72, ASSOC CLIN PROF NEUROL, GEORGE WASHINGTON UNIV, 72-; MEM, NEUROEPIDEMIOL BR, NAT INST NEUROL DIS & STROKE, NIH, 73- *Concurrent Pos:* Child Neurol Soc liaison, Nat Inst Neurol & Commun Dis & Stroke, 75-, mem clin res panel, 77-; Nat Inst Neurol & Commun Dis & Stroke rep, Interagency Group Metal Retardation, 75- & Interagency Collab Group Hyperactiv, 76-79; consult, Nat Inst Child Health & Human Develop, 76-79; mem, Food & Drug Admin Peripheral & Cent Nerv Syst Drug Adv Comn, 78-80, chmn, 81-82; mem exec bd, Am Acad Neurol, 89-91; mem adv bd, Int Sch Neurosci, Venice. *Honors & Awards:* United Cerebral Palsy Award, 90; Spec Achievement Award, US Pub Health Serv, 81. *Mem:* Child Neurol Soc; Am Acad Neurol; Am Acad Cerebral Palsy; Am Epilepsy Soc; Am Neurol Asn. *Res:* Epidemiologic studies of the etiology of childhood neurologic disorders. *Mailing Add:* NIH 7550 Wisconsin Ave Rm 700 Bethesda MD 20892

NELSON, KARL MARCUS, SEPSIS, TRAUMA. *Educ:* Loyloa Univ, PhD(physiol), 79. *Prof Exp:* RES SCIENTIST, BAPTIST MED CTR, 84- *Mailing Add:* Baptist Med Ctr 701 Princeton Ave Birmingham AL 35211

NELSON, KAY LEROI, b Richmond, Utah, Apr 4, 26; m 47; c 6. PHYSICAL ORGANIC CHEMISTRY. *Educ:* Utah State Univ, BS, 48; Purdue Univ, PhD(chem), 52. *Prof Exp:* Asst chem, Purdue Univ, 48-50, instr, 53-54; fel org chem, Off Naval Res, Univ Calif, Los Angeles, 52-53; asst prof, Wayne State Univ, 54-56; from assoc prof to prof, 56-90, chmn dept, 68-71, EMER PROF ORG CHEM, BRIGHAM YOUNG UNIV, 90- *Concurrent Pos:* Vis prof, Ore State Univ, 71-72; vis scientist, Texas A & M, 85. *Mem:* Sigma Xi; Am Chem Soc. *Res:* Classification of organic reactions by mechanistic steps; vapor pressure correlations by functional type; kinetics and mechanisms of organic reactions; natural products by selective extraction from bark and berries. *Mailing Add:* Dept Chem Brigham Young Univ Provo UT 84602

NELSON, KEITH, b San Francisco, Calif, Nov 25, 34; m 82; c 2. ZOOLOGY, MARINE BIOLOGY. *Educ:* Univ Calif, Berkeley, AB, 56 & 59, MA, 61, PhD(zool), 63. *Prof Exp:* NIMH res fel zool, State Univ Leiden, 63-65; asst prof, Univ Md, College Park, 65-68; assoc prof biol, San Francisco State Col, 68-69; dir, Int Dirtyupsquird Found, 69-73; RES ASSOC, UNIV CALIF, DAVIS & BODEGA MARINE LAB, 73- *Concurrent Pos:* Vis assoc prof, Tel-Aviv Univ, 71; adj prof, Sonoma State Univ, 72-78. *Mem:* World Maricult Soc; AAAS; Am Soc Naturalists. *Res:* Temporal patterning of behavior; motivation and memory; genesis of biological patterning; population genetics and evolution; ecological theory; marine biology; growth inhibitor substances and pheromones; crustacean growth and reproduction; biogeography; aquacultural and fisheries genetics. *Mailing Add:* Bodega Marine Lab PO Box 247 Bodega Bay CA 94923

NELSON, KEITH A, b New York, NY, Dec 8, 53; m 81; c 2. CONDENSED MATTER DYNAMICS. *Educ:* Stanford Univ, BS, 76, PhD(phys chem), 81. *Prof Exp:* Instr phys chem, Stanford Univ, 80; teaching fel, Univ Calif, Los Angeles, 81-82; asst prof, 82-87, ASSOC PROF CHEM, MASS INST TECHNOL, 87- *Concurrent Pos:* A P Sloan fel, 87. *Honors & Awards:* Presidential Young Investr Award, 85; Coblentz Prize, 88. *Mem:* Optical Soc Am; Am Phys Soc; Am Chem Soc. *Res:* Condensed matter rearrangements and relaxational phenomena; ultrafast time-resolved spectroscopy; dynamics of phase transitions, chemical reactions, simple and glass-forming liquids; picosecond-femtosecond spectroscopy. *Mailing Add:* Dept Chem Rm 6-231 Mass Inst Technol Cambridge MA 02139

NELSON, KENNETH FRED, b Council Grove, Kans, Sept 18, 42; m 65; c 1. MEDICINAL CHEMISTRY. *Educ:* Univ Kans, BSc, 65; Univ Wash, PhD(med chem), 70. *Prof Exp:* ASST PROF PHARM, UNIV WYO, 70- *Mem:* AAAS; Am Chem Soc. *Res:* Medicinal chemistry, especially in the field of analgetics. *Mailing Add:* Dept Pharm Univ Wyo Box 3375 Laramie WY 82071

NELSON, KENNETH WILLIAM, b Superior, Wis, Sept 27, 17; m 42; c 2. INDUSTRIAL HYGIENE. *Educ:* Superior State Col, Wis, BEd, 38; Univ Utah, MS, 57. *Prof Exp:* Teacher high sch, Wis, 38-39; lab asst toxicol, US Food & Drug Admin, Washington, DC, 40-41, jr chemist, 41-42; indust hygienist, 46-50, chief hygienist, 50-57, dir dept hyg, 58-66, dir dept environ sci, 66-74, vpres environ affairs, Asarco Inc, 74-82; RETIRED. *Honors & Awards:* Cummings Mem Award, Am Indust Hyg Asn, 81. *Mem:* AAAS; Am Chem Soc; Am Indust Hyg Asn (pres, 58); Am Acad Indust Hyg (pres, 75). *Mailing Add:* 1894 Millcreek Way Salt Lake City UT 84106

NELSON, KLAYTON EDWARD, b Vivian, SDak, May 15, 17; m 42; c 2. PLANT PATHOLOGY. *Educ:* SDak State Col, BS, 39; Univ Calif, PhD(plant path), 49. *Prof Exp:* Teacher high sch, SDak, 39-41; asst plant path, 47-49, from lectr to assoc prof viticult, 50-64, from jr viticulturist to assoc viticulturist, 50-64, PROF VITICULT & VITICULTURIST, UNIV CALIF, DAVIS, 64- *Concurrent Pos:* Agent-plant pathologist, Bur Plant Indust, USDA, 49- *Mem:* Am Phytopath Soc; Am Soc Hort Sci; Am Soc Enol. *Res:* Post-harvest pathological and physiological problems of table grapes. *Mailing Add:* Div Viticulture Univ Calif Davis CA 95616

NELSON, KURT HERBERT, b Sweden, Dec 8, 24; US citizen; m 49; c 3. ANALYTICAL CHEMISTRY. *Educ:* Reed Col, BA, 48; Univ Wash, PhD(analytical chem), 53. *Prof Exp:* Analytical res chemist, Phillips Petrol Co, 53-60; res chemist, Tektronix, Inc, 60-62; res specialist, Autonetics Div, NAm Aviation, Inc, 62-64, mem tech staff, Rocketdyne Div, NAm Rockwell Corp, Calif, 64-70; sr res chemist, Burlington Industs, 70-72; Sr res assoc, 72-81, SUPVR, INT PAPER CO, 81- *Mem:* Am Chem Soc; Am Inst Chemists; Tech Asn Pulp & Paper Indust; Can Pulp & Paper Asn. *Res:* Analytical methods development; high pressure liquid chromatography; nondispersive x-ray analysis; instrument research; paper and textile analysis; environmental chemistry and pollution; ultratrace analysis; gas chromatography; spectrophotometry; data correlation; quality control. *Mailing Add:* 14 Virginia Ave Monroe NY 10950-2216

NELSON, KYLER FISCHER, b Litchfield, Minn, Sept 16, 38; m 61; c 3. ELECTROOPTICS. *Educ:* Hamline Univ, BS, 60; Purdue Univ, MS, 62; Univ Utah, PhD(physics), 68. *Prof Exp:* Fel, Univ Utah, 68; scientist, Webster Res Ctr, Xerox Corp, 69-80; SR RES SPECIALIST, 3M CO, 80- *Mem:* Am Phys Soc; Sigma Xi; Soc Info Display. *Res:* Charge transport in insulating media; liquid crystals; photoelectrophoresis; photoconductivity. *Mailing Add:* 48 Pleasant Lk Rd E St Paul MN 55127

NELSON, LARRY DEAN, b Newton, Kans, Aug 5, 37; m 72. APPLIED MATHEMATICS, COMPUTER SCIENCE. *Educ:* Phillips Univ, BA, 59; Kans State Univ, MS, 62; Ohio State Univ, PhD(math), 65. *Prof Exp:* Mathematician, Battelle Mem Inst, Ohio, 62-65; res assoc comput sci, Ohio State Univ Res Found, 65; mem tech staff, Bellcomm, Inc, DC, 65-68, supvr numerical methods & systs studies group, Appl Math Dept, 68-70, supvr data applns group, Data Systs Develop Dept, 70-72; supvr, Mgt Info Systs Dept, Bell Tel Labs, 72-77; Supvr rate & tariff planning div, Tariffs & Cost Dept, Am Tel & Tel, 77-79; dep adminr, Res & Spec Progs Admin, US Dept Transp, 79-81; PRES, MCS, INC, 81-; supvr govt commun systs, AT&T Bell Labs, 85-89, mgr, govt mkt team, AT&T-NS, 89-90, SUPVR C3I SYSTS ENG, AT&T BELL LABS, 90- *Mem:* Asn Comput Mach; Math Prog Soc; Am Math Soc; Inst Elec & Electronics Eng; NY Acad Sci; Computer Soc. *Res:* Information systems; computers; telecommunications. *Mailing Add:* 440 New Jersey Ave SE Washington DC 20003

NELSON, LAWRENCE BARCLAY, b New York, NY, Jan 9, 31; m 55; c 2. CHEMISTRY. *Educ:* NY Univ, BA, 51, PhD(chem), 55. *Prof Exp:* Res investr, NJ Zinc Co, Pa, 55-56; res assoc, Socony Mobil Oil Co, Inc, 56-60; asst to vpres res & mfg, 60-62, Witco Chem Corp, 60-62, corp res dir, 62-76, vpres tech, Sonnenborn Dir, 76-80, corp vpres & gem mgr, 81-84, group vpres petrol, Witco Corp, 84-90, GROUP VPRES CORP TECH, WITCO CORP, 90- *Mem:* Am Chem Soc; Am Petrol Inst. *Res:* Industrial and petroleum chemistry; isotope tracer techniques; fused salt electrolysis; crystal growth; technical management. *Mailing Add:* Witco Corp 520 Madison Ave New York NY 10022

NELSON, LENIS ALTON, b Walnut Grove, Minn, Sept 22, 40; m 65; c 2. AGRONOMY. *Educ:* SDak State Univ, BS, 62; NDak State Univ, MS, 68, PhD(agron), 70. *Prof Exp:* Voc agr instr, Pub Sch, Minn, 62-64; asst supt agron, Southeast SDak Exp Farm, SDak State Univ, 64-66; res asst, NDak State Univ, 66-68; PROF AGRON, UNIV NEBR, LINCOLN, 70- *Mem:* Am Soc Agron; Crop Sci Soc Am. *Res:* Alternate Crops; variety testing of grain crops such as corn, oats, wheat, barley, sorghum and soybeans. *Mailing Add:* 342 Keim Hall Univ of Nebr Lincoln NE 68583-0915

NELSON, LEONARD, b Philadelphia, Pa, Oct 29, 20; m 43; c 1. PHYSIOLOGY, REPRODUCTIVE PHYSIOLOGY. *Educ:* Univ Pa, AB, 42; Univ Minn, MA, 50, PhD, 53. *Prof Exp:* Asst zool, Wash Univ, 46-47; asst zool, Univ Minn, 47-48; from instr to asst prof physiol, Univ Nebr, 48-56; res assoc & asst prof anat, Univ Chicago, 56-58; Lalor Found fel, 58-59; assoc prof physiol, Emory Univ, 59-66; prof physiol & chmn dept, Med Col Ohio, 67-88; RETIRED. *Concurrent Pos:* Mem corp, Marine Biol Lab, Woods Hole, Mass, 54; Pop Coun, Inc fel, 56-58; USPHS sr res fel, Emory Univ, 59-60; USPHS career develop award, 60-66; Commonwealth Found fel, Cambridge Univ, 63-64; vis prof, Physiol Labs, Cambridge Univ, 63-64; prog dir develop biol, NSF, 67; Josiah Macy, Jr Found fac scholar award, 75-76; fac Med Dijon, France & Univ Geneva, Switz, 83-84; res prof, Zool Sta, NSF, Naples, Italy, 85- *Mem:* Fel AAAS; Soc Gen Physiol; Soc Study Fertil; Am Physiol Soc; Soc Study Reproduction. *Res:* Physiology of reproduction; gamete transport and fertilization; regulatory control of sperm cell function by neurotransmitters, modulators, hormones; effects of pesticides. *Mailing Add:* Dept Physiol Med Col Ohio Toledo OH 43699-0008

NELSON, LEONARD C(ARL), b Albia, Iowa, Aug 26, 20; m 46; c 4. MECHANICAL ENGINEERING. *Educ:* Iowa State Col, BS, 43; Mo Sch Mines, MS, 49; Northwestern Univ, PhD(eng), 54. *Prof Exp:* Asst prof mech eng, Mo Sch Mines, 47-50; lectr, Northwestern Univ, 53-54; assoc prof mech eng, NC State Col, 54-56; prof eng & dean, 56-61, PRES, WVA INST TECHNOL, 61- *Mem:* Am Soc Mech Engrs; Am Soc Eng Educ; Sigma Xi. *Res:* Thermodynamics and heat transfer. *Mailing Add:* Off Pres WVa Inst Technol Montgomery WV 25136

NELSON, LLOYD STEADMAN, b Norwich, Conn, Mar 29, 22; m 47; c 3. MATHEMATICS. *Educ:* Univ NC, BS, 43; Univ Conn, PhD(chem), 50. *Prof Exp:* Asst chem, Univ NC, 42-43; instr physics, Univ Conn, 43-44, asst chem, 46-48; instr org chem, Ill Inst Technol, 49-51; chemist, Silicone Prod Dept, Gen Elec Co, 51-53, res assoc, Res Lab, 53-56, consult statistician, Lamp Div, 56-58, mgr appl math lab, Major Appliance Div, 68-80; DIR STATIST METHODS, NASHUA CORP, 80- *Concurrent Pos:* Spec lectr, Case Western Reserve Univ, 63-65; ed, Indust Qual Control, 65-67 & J Qual Technol, 68-70, Am Soc Qual Control. *Honors & Awards:* Shewhart Medal, 78; Deming Medal, 85. *Mem:* Fel AAAS; fel Am Soc Qual Control; fel Am Statist Asn; Sigma Xi. *Res:* Application of mathematics to scientific and business problems. *Mailing Add:* 17 Jefferson Dr Londonderry NH 03053-3647

NELSON, LOUISE MARY, b Kirkland Lake, Ont, May 13, 51; m 80; c 2. SOIL MICORBIOLOGY. *Educ:* Univ Western Ont, BSc, 72; Univ Calgary, PhD(microbiol ecol), 76. *Prof Exp:* Fel soil microbiol, Dept Microbiol, McGill Univ, 76-78; Rhodes vis fel soil microbiol, Dept Agr Sci, Univ Oxford, Eng, 78-79; res assoc, Nat Res Coun Can, Sask, 79-80, asst res officer, Prairie Regional Lab, 80-82, assoc res officer, 83-90, SR RES OFFICER, PLANT BIOTECHNOL INST, NAT RES COUN CAN, SASK, 90- *Concurrent Pos:* Assoc ed, Can J Microbiol, 82-83; sect ed, Microbial Ecology, Can J Microbiol, 83-88; adj prof, Soil Sci Dept, Univ Sask, 85-; 1st vpres, Can Soc Microbiol, 87-88, pres, 88-89; indust liaison officer, Plant Biotechnol Inst, 87-89. *Mem:* Can Soc Microbiologists; Am Soc Microbiol. *Res:* Plant-microbe interactions including symbiotic dinitrogen fixation and plant-growth promoting rhizobacteria in legume & non-legume crops; plant biotechnology; plant transformation. *Mailing Add:* Plant Biotechnol Inst Nat Res Coun Can Saskatoon SK S7N 0W9 Can

NELSON, LYLE ENGNAR, b Donnybrook, NDak, Jan 6, 21. SOILS, PLANT NUTRITION. *Educ:* NDak State Univ, BS, 48, Cornell Univ, MS, 50, PhD(soils), 52. *Prof Exp:* Soil scientist, USDA, 48; asst, Cornell Univ, 49-52; asst prof agron & asst agronomist, Agr Exp Sta, Miss State Univ, 52-55; vis assoc prof soils, Univ Philippines, 55-57; from assoc prof & assoc agronomist to prof & agronomist, 57- 86, EMER PROF AGRON, AGR EXP STA, MISS STATE UNIV, 86- *Concurrent Pos:* Vis prof, NC State Univ, 65; vis scientist, Rothamsted Exp Sta, Eng, 71-72; vis scientist, Int Rice Res Inst & vis prof, Univ Philippines-Los Banos, 78-79; consult, Winrock Int, India, 90- *Mem:* AAAS; Soil Sci Soc Am; Am Soc Agron; Soil Conserv Soc Am; Int Soc Soil Sci; Am Inst Biol Sci. *Res:* Soil fertility; soil reaction and plant growth; nutrition and nutrient cycling in forest stands. *Mailing Add:* Dept Agron Box 5248 Mississippi State MS 39762

NELSON, MARGARET CHRISTINA, b Louisville, Ky, Nov 13, 43. NEUROPHYSIOLOGY, ETHOLOGY. *Educ:* Swarthmore Col, BA, 65; Univ Pa, MA, 67, PhD(physiol psychol), 70. *Prof Exp:* Nat Inst Ment Health fel biol, Tufts Univ, 70-72; asst prof psychol, Brandeis Univ, 72-75; res fel neurobiol, Harvard Med Sch, 75-80; mem staff, Langmuir Lab, 80-, AT DEPT NEUROBIOL & BEHAV, CORNELL UNIV. *Mem:* AAAS; Am Soc Zool; Animal Behav Soc; Soc Neurosci. *Res:* Behavior and neurophysiological correlates of behavior in insects. *Mailing Add:* 124 1/2 Judd Falls Rd Ithaca NY 14850

NELSON, MARITA LEE, b Torrance, Calif. HUMAN ANATOMY, ENDOCRINOLOGY. *Educ:* Univ Calif, Los Angeles, BS, 57, MS, 59; Univ Calif, Berkeley, PhD(anat), 68. *Prof Exp:* Assoc phys educ, Univ Calif, Los Angeles, 59-60; instr, Ill State Univ, 60-64; actg asst prof anat, Univ Calif, Berkeley, 68-69; asst prof, Schs Med & Dent, Georgetown Univ, 69-72; actg asst prof, Univ Calif, Berkeley, 72-74; asst prof, Sch Dent, Univ Pac, 73-74; assoc prof anat, 74-82, PROF ANAT & REPROD BIOL, UNIV HAWAII, 82- *Mem:* Am Asn Anatomists; Sigma Xi; Endocrine Soc; Soc Study Reprod; Asn Women Sci; Am Asn Clin Anatomists. *Res:* Hormonal regulation of puberty; effects of high altitude on growth and pituitary function; effects of stress on reproductive maturation; plastination techniques for biologic specimens; medical education. *Mailing Add:* Dept Anat & Reprod Biol Univ Hawaii 1960 E W Rd Honolulu HI 96822

NELSON, MARK ADAMS, b Payson, Utah, Sept 28, 49. DESIGN ENGINEERING AND MANAGEMENT, MANUFACTURING MANAGEMENT. *Educ:* Brigham Young Univ, BS, 73. *Prof Exp:* Advan Develop engr, Tex Instruments, 73-74; mgr mfg, Galigher Co, 77-83 & Becton Dickinson, Salt Lake City, 83-86; mgr eng, Wicat Corp, 86-88; DIR OPERS, OHMEDA, SALT LAKE CITY, 88- *Honors & Awards:* Cert, Am Prod & Inventory Control Soc, 86. *Mem:* Am Prod & Inventory Control Soc; Computer-Aided Soc Am. *Res:* Manufacturing and engineering systems; project planning and production execution; product introduction of a Raman scattering analyzer; design and manufacturing of medical instrumentation and medical disposables. *Mailing Add:* 385 Troy Way Salt Lake City UT 84107

NELSON, MARK JAMES, b Lynwood, Calif, May 16, 52; m 84; c 2. ENZYMOLOGY, INORGANIC BIOCHEMISTRY. *Educ:* Calif Inst Technol, BS, 74; Stanford Univ, PhD(chem), 80. *Prof Exp:* Postdoctoral res fel, Mass Inst Technol, 80-84; PRIN INVESTR, E I DU PONT DE NEMOURS & CO, 84- *Mem:* Am Chem Soc; Am Soc Biochem & Molecular Biol. *Res:* Chemical mechanisms employed by metalloenzymes and use of that knowledge for rational inhibitor design; non-heme iron enzymes; application of electron paramagnetic resonance in biochemistry. *Mailing Add:* Cent Res Dept Exp Sta 328-250B E I du Pont de Nemours & Co Wilmington DE 19880-0328

NELSON, MARTIN EMANUEL, physics, for more information see previous edition

NELSON, MARY ANNE, b Denver, Colo, Oct 1, 52. MOLECULAR GENETICS. *Educ:* Univ Colo, BA, 75, PhD(biol), 82. *Prof Exp:* Res asst, Univ Colo, 73-76 & 79-82; res biologist, Int Inst Genetics & Biophysics, 77-79; fel, Univ Rome, Italy, 83-85, vis prof biol, 85-86, res biologist, 86-87; fel, Univ Wis, 88-90; ASST PROF BIOL & GENETICS, UNIV NMEX, 91- *Res:* Molecular control of sexual development in Neurospora; control of gene expression in the filamentous fungi; genetic analysis of fungal sexual differentiation. *Mailing Add:* Dept Biol Univ NMex Albuquerque NM 87131

NELSON, MARY LOCKETT, b New Orleans, La, July 24, 14; m 68. CELLULOSE CHEMISTRY. *Educ:* Newcomb Col, BA, 34; Tulane Univ, MS, 36, PhD(phys chem), 57. *Prof Exp:* Plant physiologist, Forest Serv, La, USDA, 36-40, seed technologist, Agr Mkt Serv, Washington, DC, 40-42, chemist, Southern Regional Res Lab, 42-64, sr res chemist, Southern Regional Res Ctr, 64-80; RETIRED. *Mem:* Fel AAAS; fel Am Inst Chem; Fiber Soc; Am Chem Soc; Sigma Xi. *Res:* Cellulose fine structure; crystallinity; accessibility; IR spectra; swelling; crosslinking; changes in cotton fiber during growth; heat damage to cotton; heats of combustion and solution; storage of pine seed. *Mailing Add:* 6848 Louisville St New Orleans LA 70124

NELSON, MERRITT RICHARD, b New Richmond, Wis, Oct 11, 32; m 56; c 3. PLANT PATHOLOGY. *Educ:* Univ Calif, Berkeley, BS, 55; Univ Wis, PhD, 58. *Prof Exp:* Asst, Univ Wis, 55-57; asst plant pathologist, 58-61, assoc prof, 61-67, PROF PLANT PATH, AGR EXP STA, UNIV ARIZ, 67-, HEAD DEPT, 76- *Mem:* Am Phytopath Soc; Soc Gen Microbiol; Am Soc Virol; Asn Appl Biologists. *Res:* Biology, pathology and epidemiology of plant viruses. *Mailing Add:* Dept Plant Pathology Univ Ariz Tucson AZ 85721

NELSON, NATHAN, b Avihail, Israel, Jan 1, 38; m 64; c 3. PHOTOSYNTHESIS, MOLECULAR BIOLOGY. *Educ:* Tel-Aviv Univ, BSc, 65, MSc, 66 & PhD(plant biochem), 70. *Prof Exp:* Post doc fel, Cornell Univ, Ithaca, 70-72; sr lectr, Technion, Haifa, Israel, 72-77, from assoc prof to prof membrane biochem, 77-85; mem, 85-86, LAB HEAD MEMBRANE BIOCHEM, ROCHE INST MOLECULAR BIOL, 86- *Concurrent Pos:* Vis prof, Bioctr, Univ Basel, Switz, 78-79, Cornell Univ, Ithaca, 83-84. *Mem:* Am Soc Biochem & Molecular Biol; Int Soc Neurochem. *Res:* Photosynthesis, bioenergetics, biological membranes, channels and ion pumps, secretary vesicles; biogenesis, assembly and molecular biology of protein complexes in membranes. *Mailing Add:* Roche Inst Molecular Biol Nutley NJ 07110

NELSON, NEAL STANLEY, b Chicago, Ill, Jan 1, 34; m 66. PHARMACOLOGY, RADIOBIOLOGY. *Educ:* Univ Ill, BS, 55, DVM, 57; Univ Chicago, PhD(pharm), 64. *Prof Exp:* Res assoc pharmacol, Univ Chicago, 63-65; Nat Cancer Inst spec fel, Milan, 65-66; res scientist, Div Radiol Health, Robert A Taft Sanit Eng Ctr, Ohio, 66-67; dep chief toxicol studies sect, Div Biol Effects, Bur Radiation Health, Hazelton Lab, USPHS, Va, 67-71, res scientist & dep chief toxicol studies sect, Off Res & Monitoring, Twinbrook Res Lab, 71-73, RADIOBIOLOGIST, OFF RADIATION PROGS, ENVIRON PROTECTION AGENCY, 73- *Concurrent Pos:* Mem comt guide rev, Inst Lab Animal Resources, Nat Acad Sci-Nat Res Coun, 65 & mem comt stand for cats, 72-76. *Mem:* AAAS; Am Vet Med Asn; Am Soc Lab Animal Practitioners; NY Acad Sci; Sigma Xi. *Res:* Metabolism of radioisotopes; pathology of alpha-emitting isotopes; effects of ionizing radiation; high resolution autoradiography and intracellular localization of labeled compounds. *Mailing Add:* 8102 Ashtonbirch Dr Springfield VA 22152

NELSON, NEIL DOUGLAS, b Yankton, SDak, Sept 22, 44; c 3. FORESTRY, BIOTECHNOLOGY. *Educ:* Iowa State Univ, BS, 66; Univ Wis-Madison, MS, 68, PhD(plant physiol), 73. *Prof Exp:* Wood scientist plant physiol, Forest Prod Lab, US Forest Serv, Madison, 71-77; res plant physiologist tree physiol, US Forest Serv, Rhinelander, Wis, 77-81, proj leader, 82, biotech prog leader, Forestry Sci Lab, 83-85; PRES, FORGENE INC, 86- *Concurrent Pos:* Fulbright Scholar, 68-69; Fulbright fel, CSIRO Forest Prod Lab, Melbourne, Australia, 75-76. *Honors & Awards:* Wood Award, Forest Prod Res Soc, 73. *Mem:* Soc Am Foresters; Sigma Xi. *Res:* Biotechnology; genetic engineering; tissue culture; photosynthesis; yield physiology; plant hormones; wood formation; secondary metabolism in woody plants; tree genetics & clonal propagation. *Mailing Add:* Forgene Inc PO Box 1370 Rhinelander WI 54501

NELSON, NELS M, b Baker, NDak, May 30, 19; m 57; c 1. MICROBIOLOGY. *Educ:* Jamestown Col, BS, 41; Univ Wash, MS, 53, PhD(microbiol), 55. *Prof Exp:* From asst prof to prof, 55-80, EMER PROF MICROBIOL, MONT STATE UNIV, 80- *Mem:* AAAS; Am Soc Microbiol. *Res:* Microbial physiology. *Mailing Add:* 330 Lindley Pl Bozeman MT 59715

NELSON, NILS KEITH, b Leadwood, Mo, Nov 10, 26. ORGANIC CHEMISTRY. *Educ:* Mo Sch Mines, BS, 46; Univ Ill, MS, 47, PhD(chem), 49. *Prof Exp:* Asst gen & org chem, Univ Ill, 46-49; instr, Univ Maine, 49-51; res chemist, Shell Oil Co, 51-62; asst prof, 62-66, ASSOC PROF ORG CHEM, PURDUE UNIV, CALUMET, 66- *Mem:* Am Chem Soc. *Res:* Stereochemistry of substituted aryl amines; Darzens reaction. *Mailing Add:* 2219 169th St No 39 Hammond IN 46323-2019

NELSON, NORMAN ALLAN, b Edmonton, Alta, July 26, 27; nat US; m 55; c 2. PHARMACEUTICAL CHEMISTRY. *Educ:* Univ Alta, BSc, 49; Univ Wis, PhD(chem), 52. *Prof Exp:* Res assoc, Mass Inst Technol, 52-53, from instr to asst prof chem, 53-59; res assoc, 59-81, RES HEAD CARDIOVASC DIS RES, UPJOHN CO, 81-, ASSOC DIR, 85- *Mem:* Am Chem Soc. *Res:* Steroidal hormone analogs; organic synthesis; prostaglandin chemistry; ionophores; renin inhibitors. *Mailing Add:* 1643 S 36th St Galesburg MI 49053-9643

NELSON, NORMAN CROOKS, b Hibbing, Minn, July 24, 29; m 55; c 3. SURGERY, ENDOCRINOLOGY. *Educ:* Tulane Univ, BS, 51, MD, 54. *Prof Exp:* Clin & res fel surg, Harvard Med Sch & Mass Gen Hosp, Boston, 62-63; from instr to prof surg, Sch Med, Univ New Orleans, 63-73, from assoc dean to dean sch med, 69-73; PROF SURG, MED CTR, UNIV MISS, 73-, VCHANCELLOR HEALTH AFFAIRS & DEAN, SCH MED, 73- *Honors & Awards:* Arthur M Shipley Award, Southern Surg Asn, 69. *Mem:* Endocrine Soc; Int Soc Surg; Am Surg Asn; Soc Univ Surgeons; Soc Head & Neck Surgeons. *Res:* Mineral and carbohydrate metabolism. *Mailing Add:* Sch Med Univ Miss Med Ctr Jackson MS 39216

NELSON, NORMAN NEIBUHR, mathematics education; deceased, see previous edition for last biography

NELSON, NORTON, environmental medicine, biochemistry; deceased, see previous edition for last biography

NELSON, NORVELL JOHN, US citizen; c 2. MATERIALS SCIENCE. *Educ:* Ill Inst Technol, BS, 64; Stanford Univ, PhD(chem), 67. *Prof Exp:* Res assoc & instr inorg chem, Northwestern Univ, 67-69; res assoc organometallic chem, Stanford Univ, 69-70; sr res chemist org photochem, Eastman Kodak Co, 70-74; sr engr solid state chem, Varian Assocs, Inc, 74-78; sr res chemist, Catalytica Assocs, Inc, 78-80; vpres res, AFS, 80-82; vpres res & co-founder, PSI Star, 82-90; VPRES, STAR ETCH CORP, 90- *Mem:* Am Chem Soc; Electrochem Soc; AAAS; Mat Res Soc. *Res:* Industrial process chemistry; development of new analytical techniques; electronic materials preparation; process chemistry development for the electronics industry. *Mailing Add:* 3445 Greer Rd Palo Alto CA 94303

NELSON, OLIVER EVANS, JR, b Seattle, Wash, Aug 16, 20; m 63. PLANT BIOCHEMICAL GENETICS. *Educ:* Colgate Univ, AB, 41; Yale Univ, MS, 43, PhD, 47. *Hon Degrees:* Dr Agr, Purdue Univ. *Prof Exp:* Assoc geneticist, Purdue Univ, 47-54, geneticist, 54-69; prof genetics, Univ Wis-Madison, 69-81, Brink prof, 81-91, chmn, Lab Genetics, 86-89, EMER PROF GENETICS, UNIV WIS-MADISON, 91- *Concurrent Pos:* Vis investr, Nat Forest Res Inst & Biochem Inst, Univ Stockholm, 54-55; NSF sr fel biol, Calif Inst Technol, 61-62. *Honors & Awards:* John Scott Medal; Hoblitzelle Award; Browning Award; Donald F Jones Medal. *Mem:* Nat Acad Sci; Am Acad Arts & Sci; Am Soc Plant Physiologists; Crop Sci Soc Am; Am Genetics Soc; AAAS. *Res:* Effect of transposable elements on gene function; mutations affecting starch synthesis in cereal endosperms. *Mailing Add:* Genetics Lab Univ Wis Madison WI 53706

NELSON, PAUL EDWARD, b Franklin Twp, Wis, May 26, 27; m 50; c 3. PLANT PATHOLOGY. *Educ:* Univ Calif, BS, 51, PhD(plant path), 55. *Prof Exp:* From asst prof to assoc prof plant path, Cornell Univ, 55-65; assoc prof, 65-67, PROF PLANT PATH, PA STATE UNIV, 67- *Concurrent Pos:* Fel, Am Phytopath Soc. *Honors & Awards:* Merit Award, Am Phytopath Soc. *Mem:* Am Phytopath Soc; Mycol Soc Am. *Res:* Biology, taxonomy and mycotoxicology of fusarium species; pathological anatomy of diseased plants. *Mailing Add:* 211 Buckhout Lab Dept Plant Path Pa State Univ University Park PA 16802

NELSON, PAUL VICTOR, b Somerville, Mass, May 4, 39; m 64. PLANT NUTRITION. *Educ:* Univ Mass, BS, 60; Pa State Univ, MS, 61; Cornell Univ, PhD(floricult), 64. *Prof Exp:* Staff res specialist, Geigy Chem Corp, NY, 64-65; from asst prof to assoc prof, 65-73, PROF HORT SCI, NC STATE UNIV, 73- *Concurrent Pos:* NSF travel grant, Int Hort Cong, Israel, 70; leave of absence, Lab Plant Physiol Res, Agr Univ, Wageningen, Holland, 71-72; Dept Hort, Univ Hawaii, Honolulu, 87-88. *Honors & Awards:* Res Award, Am Carnation Soc, 67; Gunlogson Res Med, Am Hort Soc, 86. *Mem:* Am Soc Hort Sci; Soc Am Florists; Int Soc Hort Sci; Am Hort Soc. *Res:* Foliar analysis and fertilization of floricultural crops; efficiency of uptake and utilization of nutrients in plants; plant root media. *Mailing Add:* Dept Hort Sci NC State Univ Box 7609 Raleigh NC 27695

NELSON, PETER REID, b Norwich, Conn, Aug 11, 49; m 81. MATHEMATICAL STATISTICS. *Educ:* Case Inst Technol, BS, 71; Case Western Reserve Univ, MS, 73, PhD(math & statist), 75. *Prof Exp:* Vis asst prof statist, Ohio State Univ, 75-76, asst prof, 76-81; SR STATISTICIAN, G D SEARLE & CO, 81- *Concurrent Pos:* Prin investr res libr type Markov chains, NSF grant, 78-79. *Mem:* Am Statist Asn; Am Soc Qual Control. *Res:* The analysis of means. *Mailing Add:* 32 Ichabod Rd Simsbury CT 06070

NELSON, PHILIP EDWIN, b Shelbyville, Ind, Nov 12, 34; m 55; c 3. FOOD SCIENCE, AGRICULTURE. *Educ:* Purdue Univ, BS, 56, PhD(food sci), 67. *Prof Exp:* Plant mgr food processing, Blue River Packing Co, 56-60; from instr to asst prof food sci, 60-70, dir, Food Sci Inst, 75-86, PROF FOOD SCI, PURDUE UNIV, 74-, HEAD DEPT, 84- *Concurrent Pos:* Guest specialist, US Info Agency, 75-76; chmn subcomt world food & nutrit study, Nat Acad Sci, 76-77, food additive surv, 80-; consultant, 80- *Honors & Awards:* Indust Achievement Award, Inst Food Technologists, 76. *Mem:* Sigma Xi; fel Inst Food Technologists; AAAS; Coun Agr Sci & Technol. *Res:* Fruit and vegetable preservation including aseptic processing and prevention of food losses. *Mailing Add:* Dept of Food Sci Smith Hall Purdue Univ West Lafayette IN 47907

NELSON, PHILLIP GILLARD, b Albert Lea, Minn, Dec 3, 31; m 55; c 4. NEUROSCIENCES, CELL BIOLOGY. *Educ:* Univ Chicago, MD, 56, PhD(physiol), 57. *Prof Exp:* Intern, Philadelphia Gen Hosp, 57-58; from sr surgeon to surgeon, USPHS, 58-67; actg chief sect spinal cord, Nat Inst Neurol Dis & Blindness, 64-69; chief behav biol br, 69-75, CHIEF, LAB DEVELOP NEUROBIOL, NAT INST CHILD HEALTH & HUMAN DEVELOP, 76- *Concurrent Pos:* Hon res asst, Dept Biophys, Univ Col, Univ London, 62-63; lectr physiol, George Washington Univ, 64- *Mem:* AAAS; Am Physiol Soc; Int Soc Develop Neurosci; Am Soc Cell Biol; Soc Neurosci. *Res:* Single unit and integrative activity in the central nervous system; tissue culture of nervous tissue. *Mailing Add:* NIH Lab Develop Neurobiol Bldg 36 Rm 2A21 Bethesda MD 20892

NELSON, R WILLIAM, b Logan, Utah, May 23, 31; m 52; c 9. HYDROLOGY, EARTH SCIENCE. *Educ:* Univ Idaho, BS, 54; Colo State Univ, MS, 59. *Prof Exp:* Drainage engr, Agr Res Serv, USDA, Idaho, 54-56, engr, Colo, 56-60; engr, Hanford Atomic Prod, Gen Elec Co, 60-63, sr engr, 63-65; sr res scientist, Pac Northwest Lab, Battelle Mem Inst, Wash, 65-66, res assoc, 66-68; sr staff specialist, Sci Systs Dept, Comput Sci Corp, 68-75 & Boeing Comput Serv Richland Inc, 75-78; staff scientist, Geosci Sect, Pac Northwest Labs, Battelle Mem Inst, 78-80, staff scientist, Hydrologic Systs Sect, 80-; AT INTERA ENVIRON CONSULT. *Concurrent Pos:* Deleg, Int Conf Ground Disposal Radioactive Wastes, Chalk River, 61; US rep, Int Radioisotopes Hydrol Conf, Int Atomic Energy Agency, Tokyo, 63; mem int adv comt, Inventory Groundwater Models, Int Asn Sci Union, 75-77. *Honors & Awards:* Meinzer Award, Geol Soc Am, 78. *Mem:* Am Geophys Union; Soil Sci Soc Am; Sigma Xi. *Res:* Research and development in application of mathematics to flow liquids in porous media including contaminant transport methods for application in groundwater quality management; groundwater quality control standards and assessment methods for evaluating the enviromental consequences. *Mailing Add:* Intera Environ Consult Las Vegas NV 99352

NELSON, RALPH A, b Minneapolis, Minn, June 19, 27; m 54; c 5. NUTRITION, PHYSIOLOGY. *Educ:* Univ Minn, BA, 50, MD, 53, PhD(physiol), 61; Am Bd Internal Med, dipl, 81. *Prof Exp:* Intern med, Cook County Hosp, 53-54; fel path, Hosps Univ, Minn, 54-55, res assoc neurophysiol, 55-56, fel physiol, Mayo Found, 57-60; asst prof nutrit, Cornell Univ, 61-62; assoc physiol, Med Sch, Case Western Reserve Univ, 62-67; asst prof physiol, Mayo Grad Sch, 67-73; asst prof physiol, Mayo Med Sch, 73-78, assoc prof nutrit, 74-78; chmn sect nutrit, Mayo Clin, 67-78; assoc prof internal med & chmn nutrit sect, Sch Med, Univ SDak, 78-79; PROF NUTRIT, COL MED, UNIV ILL, URBANA-CHAMPAIGN, 79- *Concurrent Pos:* Dir med res, George H Scott Res Lab, Fairview Park Hosp, Cleveland, 62-67; consult internal med, Danville Vet Admin Hosp, 79-; dir res & head, Nutrit Support Serv, Carle Found Hosp, Urbana, 80- *Mem:* Am Physiol Soc; Am Inst Nutrit; Am Soc Clin Nutrit; Am Gastroenterol Soc. *Res:* Gastroenterology, membrane transport, physiology and structural aspects of vitamin function; applied nutrition of obesity and chronic renal disease. *Mailing Add:* Dept Med Col Med Carle Found 611 W Park St Urbana IL 61801

NELSON, RALPH FRANCIS, b Hartford, Conn, Sept 20, 45; m 72; c 2. NEUROPHYSIOLOGY, NEUROANATOMY. *Educ:* Amherst Col, BA, 67; Johns Hopkins Univ, PhD(biophysics), 72. *Prof Exp:* Fel, 72-74, staff fel, 74-78, res physiologist, Nat Eye Inst, 78-83, RES PHYSIOLOGIST, NAT INST NEUROL DISORDERS & STROKE, DEPT HEALTH & HUMAN SERV, NIH, 83- *Concurrent Pos:* Fel, Max Planck Inst Physcol & Clin Res, Bad Nauheim, Fed Repub Ger, 81. *Mem:* Asn Res Vision & Opthamol; Soc Gen Physiologists; Soc Neurosci. *Res:* Electrophysiology and neuroanatomy of retinal neurons; neural circuitry and neuronal interactions, structure and function. *Mailing Add:* Nat Inst Health Bldg 36 Rm 2C02 Bethesda MD 20892

NELSON, RANDALL BRUCE, b Swissvale, Pa, Feb 15, 48; div; c 2. ORGANIC & NATURAL PRODUCTS CHEMISTRY. *Educ:* Washington & Jefferson Col, BA, 70; Dartmouth Col, PhD(org chem), 75. *Prof Exp:* NIH fel dept chem, Columbia Univ, 75-76; res chemist, Arapahoe Chem Div, Syntex Corp, 76-78; res chemist chem, 78-80, res group leader, Olympic Res Div, 80-85, tech mkt rep, 85-87, MGR INDUST SALES, ITT RAYONIER CHEM PROD, 88- *Concurrent Pos:* Consult, chem, 84- *Honors & Awards:* Sr Award in Chem, Pittsburgh Sect, Am Chem Soc, 70. *Mem:* Sigma Xi; Am Chem Soc. *Res:* New synthetic methods in organic and heterocyclic chemistry; process development and modern synthetic reagents in production scale equipment; chemistry of metal hydrides; pulping and bleaching processes; paper and dissolving pulp chemistry; lignin chemistry; vanillin process chemistry. *Mailing Add:* 18000 Pac Hwy S Suite 900 Seattle WA 98188

NELSON, RAYMOND ADOLPH, b Spokane, Wash, Apr 24, 26; m 52, 78; c 5. RADIOPHYSICS. *Educ:* Wash State Univ, BS, 50, MS, 52; Stanford Univ, PhD(physics), 61. *Prof Exp:* Mem tech staff, Bell Tel Labs, Inc, 52-55; res engr, 55-61, physicist, 61-70, SR RES PHYSICIST, RADIO PHYSICS LAB, SRI INT, 70- *Mem:* Am Geophys Union. *Res:* Foundations of statistical mechanics; ionospheric physics. *Mailing Add:* Radio Physics Lab SRI Int 333 Ravenswood Menlo Park CA 94025

NELSON, RAYMOND JOHN, b Chicago, Ill, Oct 8, 17; m 42; c 3. MATHEMATICS. *Educ:* Grinnell Col, AB, 41; Univ Chicago, PhD(philos), 49. *Prof Exp:* Prof philos, Univ Akron, 46-52; mathematician, Int Bus Mach Corp, 52-55; staff engr, Link Aviation, Inc, 55-56; prof math, 56-65, dir comput ctr, 56-65, PROF PHILOS, CASE WESTERN RESERVE UNIV, 65- *Concurrent Pos:* Mem bd dirs, CHI Corp, Ohio; consult, Rockefeller Found, 74- *Mem:* Am Math Soc; Asn Symbolic Logic; Asn Comput Mach; Am Philos Asn. *Res:* Mathematical logic; automata theory; philosophy of science; application of mathematical models to psychological attitudes such as belief and perception. *Mailing Add:* 2400 Demington Dr Cleveland Heights OH 44106

NELSON, REX ROLAND, b Greenville, Mich, Sept 14, 24; m 54; c 2. ACOUSTICS. *Educ:* Kenyon Col, BA, 49; Univ Calif, Los Angeles, MS, 51; Pa State Univ, PhD(high pressure physics), 59. *Prof Exp:* Prof physics, Occidental Col, 59-87; RETIRED. *Mem:* Am Phys Soc; Am Asn Physics Teachers; Acoust Soc Am; Sigma Xi. *Res:* Polymorphic transitions and phase transitions at high pressure; acoustics; acoustics of musical instruments. *Mailing Add:* 5033 Caesena Way Oceanside CA 92056

NELSON, RICHARD BARTEL, b Weiser, Idaho, Sept 13, 40. CIVIL ENGINEERING, ENGINEERING MECHANICS. *Educ:* Willamette Univ, BA, 63; Columbia Univ, BS, 63, MS, 64, DEngSc(civil eng), 68. *Prof Exp:* Asst prof, 68-76, ASSOC PROF ENG, UNIV CALIF, LOS ANGELES, 76- *Mem:* Am Inst Aeronaut & Astronaut. *Res:* Structural analysis; optimum structural design; wave propagation in elastic media; structural stability. *Mailing Add:* 3555 Grandview Blvd Los Angeles CA 90066

NELSON, RICHARD BURTON, b Powell, Wyo, Dec 10, 11; c 1. ELECTRONICS ENGINEERING. *Educ:* Calif Inst Technol, BS, 35; Mass Inst Technol, PhD(physics), 38. *Prof Exp:* Mem electron optics lab, Radio Corp Am Mfg Co, 38-41; asst res physicist, Nat Res Coun Can, 41-42; res assoc, Gen Elec Co, 42-50; sr engr, Litton Industs, 50-51; sr engr, Varian Assocs, 51-57, mgr klystron develop, 57-60, mgr tube res & develop, 60-63, chief engr, Tube Div, 63-74, CONSULT, VARIAN ASSOCS, 74- *Mem:* Fel Inst Radio Engrs. *Res:* Vacuum tubes; high-power magnetrons and klystrons. *Mailing Add:* 27040 Dezahara Way Los Altos Hills CA 94022

NELSON, RICHARD CARL, b Stillwater, Minn, May 1, 15; m 43; c 2. PHYSICS, BIOPHYSICS. *Educ:* Univ Minn, AB, 35, PhD(plant physiol), 38. *Prof Exp:* Agent tung res, USDA, 39; res fel plant physiol, Univ Minn, 40-42; spectroscopist, Armour & Co, 42-43; chief chemist, Citrus Concentrates, Inc, Fla, mgr, Pectin Prods Div, 45-46; res assoc, Northwestern Univ, 46-49; from assoc prof to prof, 49-78, EMER PROF PHYSICS, OHIO STATE UNIV, 78- *Concurrent Pos:* Caleb-Dorr fel, Univ Minn, 36-37; mem comt heat attenuation in clothing systs, Qm Res & Develop Adv Bd, 59-62; mem, Qm Res & Develop Adv Bd, 59-62; consult, Minn Mining & Mfg Co, 60-70; HSPHS res career prog award, 64-74; vis prof, Phys Chem Inst, Univ Marburg, 65. *Mem:* Fel Am Phys Soc. *Res:* Photoconductivity; sensitization by dyes; electronic processes in organic solids; large molecules adsorbed on ionic crystals. *Mailing Add:* Dept Physics Ohio State Univ Columbus OH 43210

NELSON, RICHARD D(OUGLAS), b Modesto, Calif, Apr 17, 41; m 61; c 3. ENTOMOLOGY, PLANT PATHOLOGY. *Educ:* Univ Calif, Davis, BS, 63, MS, 66, PhD(entom), 68. *Prof Exp:* Asst res biologist, Biol Res Div, Stauffer Chem Co, 63-64; entomologist, plant pathologist & dir, Res Div, Driscoll Strawberry Assoc, Inc, 68-; PRES, PLANT SCI, INC. *Mem:* AAAS; Entom Soc Am; Entom Soc Can; Acaralogical Soc Am; Int Soc Hort Sci. *Res:* Biology, population dynamics, and integrated pest management strategies used for control of insects, mites, diseases, and weeds in strawberries; physiological studies with strawberries, including nutrition and tissue culturing. *Mailing Add:* Plant Sci Inc 342 Green Valley Rd Watsonville CA 95076

NELSON, RICHARD L(LOYD), b Mansfield, Ohio, Nov 18, 34; m 53; c 2. ELECTRICAL ENGINEERING, ECONOMICS. *Educ:* Univ Cincinnati, EE, 57; Ohio State Univ, MSc, 64. *Prof Exp:* Sr res engr, Battelle Mem Inst, 60-63; mgr prod & mkt, Ray Data Corp, 63-64; adminstr mgt, Battelle Develop Corp, Ohio, 64-67; dir res, Gilbarco Inc, 68-69; mgr new bus develop group, Jersey Enterprises, Standard Oil Co NJ, 69-72; founder & pres, QWIP Systs Div, 72-78, chief exec, Advan Projs Div, 78-81, vpres, Exxon Enterprises, Inc, 78-81; PRES, QUEST RES, INC, 81- *Concurrent Pos:* Consult, US Navy, 60-61 & Battelle Mem Inst, 63-64. *Mem:* Fel AAAS; sr mem Inst Elec & Electronics Engrs; Am Soc Naval Engrs. *Res:* Cost-effectiveness analysis; development of techniques for application of digital systems to management reporting; control systems; management of research and new technology business-venture development. *Mailing Add:* Bodines Rd Bodines PA 17722

NELSON, RICHARD ROBERT, plant pathology; deceased, see previous edition for last biography

NELSON, ROBERT A, b Tracy, Minn, Mar 26, 35; m 60; c 4. VETERINARY MEDICINE, TOXICOLOGY. *Educ:* Univ Minn, BS, 58, DVM, 60. *Prof Exp:* Vet, Hanover Animal Hosp, Forest Park, Ill, 60-62 & Albrecht Animal Hosp, Denver, 62; instr vet surg, Col Vet Med, Univ Minn, 62-63; sr vet, Biochem Res Lab, 3M Co, 63-68, supvr, Animal Lab, 68-71, supvr, Animal Lab & toxicol, Riker Res Lab, 71-72, mgr, Animal Lab & toxicol, 72-75, mgr safety eval, 75-78, lab mgr safety eval, 78-79, mgr, Develop Labs, 79-81, DIR REGULATORY AFFAIRS & QUAL ASSURANCE, RIDER RES LAB, 3M CO, 81- *Concurrent Pos:* Guest lectr, Col Med, Univ Minn, 68- *Mem:* Drug Info Asn; Soc Toxicol. *Res:* Laboratory animal care; biopolymer surgery and toxicology; testing of potential drugs, agrichemicals, biopolymers and industrial chemicals. *Mailing Add:* 3M Pharmaceut 3M Ctr Bldg 270-3A St Paul MN 55144

NELSON, ROBERT ANDREW, b Detroit, Mich, Apr 16, 43; m 68. POLYMER CHEMISTRY, CELLULOSE CHEMISTRY. *Educ:* Wayne State Univ, BS, 67; Univ Mich, MS, 70, PhD(polymer chem), 72. *Prof Exp:* Group leader alloy analysis, Detroit Testing Lab, 67; assoc scientist polymer chem, 72-74, scientist cellulose chem, 75-76, scientist polymer chem, 76-80, SR SCIENTIST POLYMER CHEM, XEROX CORP, 80- *Mem:* Am Chem Soc; AAAS; Sigma Xi. *Res:* Basic physicochemical properties of cellulose and paper including cellulose/water interactions, electrical conduction mechanisms and mechanical properties; polymer synthesis; kinetics, structure-property relationships. *Mailing Add:* 453 Maplewood Lane Webster NY 14580-2824

NELSON, ROBERT B, b Casper, Wyo, Apr 2, 35; m 58; c 2. PHARMACOLOGY. *Educ:* Univ Wyo, BS, 57; Univ Calif, San Francisco, MS, 63, PhD(pharmacol), 65. *Prof Exp:* Asst prof, Idaho State Univ, 65-70; assoc prof, 70-77, PROF PHARMACOL, SCH PHARM, UNIV WYO, 77- *Mem:* Am Col Pharm. *Res:* Actions of narcotic analgesics on respiration and other general functions. *Mailing Add:* Dept Pharmacol Univ Wyo Box 3375 Laramie WY 82071

NELSON, ROBERT B, b July 13, 29; US citizen; m 56; c 2. STRUCTURAL GEOLOGY. *Educ:* Ore State Univ, BS, 51; Univ Wash, MS, 56, PhD(structure, petrog), 59. *Prof Exp:* From instr to assoc prof, 60-74, PROF GEOL, UNIV NEBR-LINCOLN, 74- *Mem:* Fel Geol Soc Am; Am Geophys Union; fel Geol Soc London. *Res:* Deformations in sedimentary and metamorphic strata. *Mailing Add:* 1330 N 79th St Lincoln NE 68505

NELSON, ROBERT D, US citizen. PATHOLOGY, IMMUNOLOGY. *Educ:* Univ Minn, PhD(genetics), 69. *Prof Exp:* From instr to assoc prof, 73-88, PROF INFECTIOUS DIS, DEPR DERMAT, UNIV MINN, 85- *Mem:* Am Asn Immunol; Am Asn Path; Surg Infectious Dis; Am Burn Asn. *Res:* Mechanisms of acquired loss of neutrophil chemotactic function associated with thermal injury, trauma, infection; mechanisms of energy associated with chronic condidiasis. *Mailing Add:* Dept Derm & Microbiol Univ Minn Med Sch Box 124 UMHC Minneapolis MN 55455

NELSON, ROBERT LEON, b Billings, Mont, Aug 17, 45; m 70; c 2. MEDICAL ONCOLOGY, CLINICAL PHARMACOLOGY. *Educ:* Ore State Univ, BS, 68, MS & MD, 71; Am Bd Internal Med, dipl, 75; Am Bd Med Oncol, dipl, 77. *Prof Exp:* Intern internal med, Albert Einstein Col Med, 71-72, resident, 72-73; fel med oncol, Nat Cancer Inst, 73-75, fel clin pharmacol, 74-75; CLIN PHARMACOLOGIST ONCOL, LILLY RES LABS, 75- *Concurrent Pos:* Assoc prof med & consult oncol, Sch Med, Ind Univ, 75-, assoc prof pharmacol, 78- *Mem:* Am Col Physicians; Am Fedn Clin Res. *Res:* Clinical pharmacology and initial clinical trials of new anticancer drugs; pharmacokinetics and optimization of clinical dosage regimens. *Mailing Add:* Lilly Res Lab 1001 W Tenth St Indianapolis IN 46202

NELSON, ROBERT M, b Los Angeles, Calif. PLANETARY SATELLITES, SOLID STATE SPECTROSCOPY. *Educ:* City Univ NY, BS, 66; Wesleyan Univ, MA, 69; Univ Pittsburgh, PhD(earth & planetary sci), 77. *Prof Exp:* Nat Acad Sci res assoc, NASA, 78-80; prof, dept physics, Calif State Univ, Los Angeles, 81-82; SR SCIENTIST, JET PROPULSION LAB, CALIF INST TECHNOL, 80- *Concurrent Pos:* Mem comt, Div Planetary Sci, Am Astron Soc. *Mem:* Am Astron Soc; Am Geophys Union; Am Inst Physics; AAAS. *Res:* Telescope and spacecraft observations of planetary satellites, particularly those of Jupiter, Saturn and Uranus; laboratory spectroscopy of solid state materials of planetary interest. *Mailing Add:* 183-501 Jet Propulsion Lab 4800 Oak Grove Dr Pasadena CA 91109

NELSON, ROBERT MELLINGER, b Burlington, Iowa, May 17, 18. ORTHODONTICS. *Educ:* Univ Iowa, BA, 41, DDS, 50, MA, 51. *Prof Exp:* Practicing orthodontist, Chicago, 51-52; assoc prof orthod, 53-65, PROF ORTHOD & HEAD DEPT, SCH DENT, UNIV NC, CHAPEL HILL, 65- *Res:* Facial growth and development. *Mailing Add:* 903 Coker Dr Chapel Hill NC 27514

NELSON, ROBERT NORTON, b Cincinnati, Ohio, Nov 1, 41; m 65, 89; c 2. PHYSICAL CHEMISTRY. *Educ:* Brown Univ, ScBChem, 63; Mass Inst Technol, PhD(phys chem), 69. *Prof Exp:* Fel molecular collisions, Dept Chem, Univ Fla, 68-70, interim asst prof chem, 69-70; asst prof, 70-82, ASSOC PROF CHEM, GA SOUTHERN COL, 82- *Concurrent Pos:* Vis asst prof chem, Colgate Univ, 77-78; vis lectr, Univ Ga, 81; vis sci, NASA Goddard Space Flight Ctr, 85- *Mem:* Am Chem Soc; Am Phys Soc; Sigma Xi; Soc Appl Spectros. *Res:* Molecular collisions at thermal and near thermal energies; effusive flow of gases; laser spectroscopy; formation of cosmic granules. *Mailing Add:* Dept Chem 8064 Ga Southern Univ Statesboro GA 30460-8064

NELSON, ROBERT S, medicine; deceased, see previous edition for last biography

NELSON, ROGER EDWIN, b New York, NY, Feb 1, 40; m 64; c 2. TECHNICAL MANAGEMENT, APPLIED RESEARCH. *Educ:* Rutgers Univ, BS, 62; Seton Hall Univ, MS, 66, PhD(phys chem), 69. *Prof Exp:* Develop scientist prod develop, Lever Bros Co, 64-75; tech mgr, PQ Corp, 75-79; TECH DIR, REHEIS CHEM CO, 79- *Mem:* Am Chem Soc; Soc Cosmetic Chemists; Am Pharmaceut Soc. *Res:* Management of inorganic chemical research on aluminum and other light metal compounds for applications in the cosmetic, pharmaceutical and other specialty chemicals industries. *Mailing Add:* 1078 Colfax Ave Pompton Lakes NJ 07442-2186

NELSON, ROGER PETER, b Bridgeport, Conn, Dec 15, 42; m 64; c 3. ORGANIC CHEMISTRY. *Educ:* Fairfield Univ, BS, 64; Univ Mich, PhD(org chem), 67. *Prof Exp:* Sr res scientist, chem res & develop, 72-74, proj leader food chem res & develop, 74-78, proj leader process res & develop, 78-82, proj leader tech info, 83-85, MGR TECH INFO, PFIZER, INC, 86- *Mem:* Am Chem Soc. *Res:* Construction and conformational properties of bridged bicyclic systems; chemistry and mode of action of antihypertensives; agents affecting gastrointestinal function; food protein research; immobilized enzymes; chemical bioprocess development. *Mailing Add:* Pfizer Inc Eastern Point Rd Groton CT 06340-5196

NELSON, RONALD HARVEY, b Union Grove, Wis, Aug 10, 18; m 40; c 4. ANIMAL HUSBANDRY. *Educ:* Univ Wis, BA, 39; Okla Agr & Mech Col, MS, 41; Iowa State Col, PhD(animal breeding), 44. *Prof Exp:* From asst prof to assoc prof, Mich State Univ, 46-49, head dept, 50-84, prof animal husb, 49-85; RETIRED. *Concurrent Pos:* Chief of party, Agr Proj, Balcarce, Argentina, 66-68. *Honors & Awards:* Animal Sci Award in Int Animal Agr, 78; Animal Sci Indust Award, 84. *Mem:* AAAS; fel Am Soc Animal Sci. *Res:* Sheep and beef cattle breeding; lamb mortality and factors affecting it; effect of inbreeding on a herd of Holstein-Friesian cattle. *Mailing Add:* 1545 N Harrison East Lansing MI 48823

NELSON, RUSSELL ANDREW, b Grand Forks, NDak, May 13, 13; m 39. INTERNAL MEDICINE. *Educ:* Univ Minn, AB, 33; Johns Hopkins Univ, MD, 37. *Hon Degrees:* DSc, Univ Miami, 72. *Prof Exp:* Intern, asst resident & resident med, Johns Hopkins Hosp, 37-44, dir med clins, 45-50, dir admin, 52-63, pres, 63-72; CONSULT MED EDUC, 72- *Concurrent Pos:* Mem fac, Sch Med, Johns Hopkins Univ, 40-52, asst dir, Hosp, 45-47, assoc admin, 47-48, adj prof pub health admin, Sch Hyg & Pub Health, 52-55, lectr, 55-57; chmn adv comt hosp facil & serv, USPHS, 56-; mem, President's Comn Health Manpower, 61; Nuffield Trust fel, London, 73. *Mem:* Inst Med-Nat Acad Sci; master Am Col Physicians; AAAS; AMA; Am Asn Med Cols. *Mailing Add:* 605 Ocean Dr Apt 8L Key Biscayne FL 33149

NELSON, RUSSELL C, b Hackensack, NJ, Nov 3, 25; m 52; c 2. METALLURGICAL ENGINEERING. *Educ:* Lehigh Univ, BS, 48; Colo Sch Mines, MS, 49, DSc(metall), 51. *Prof Exp:* Asst metall, Colo Sch Mines, 49-51; mineral engr, Oak Ridge Nat Lab, 51-53; sr engr, Chem & Metall Div, Sylvania Elec Prod, Inc, 53-54, engr in chg chem develop lab, 54-55, metall testing lab, 55-58, head metall res lab, 58-61; assoc prof, 61-65, PROF MECH ENG, UNIV NEBR, LINCOLN, 65-, MEM GRAD FAC, 62-, ASSOC DEAN GRAD STUDIES & APPL DENT, 78-, ORTHOPEDICS, 83-, PROF, MECH ENG & SPEC ASST DEAN ENG, 87- *Concurrent Pos:* Consult metall eng. *Mem:* Am Inst Mining, Metall & Petrol Engrs; Am Soc Metals; Am Soc Eng Educ; Soc Biomat; Sigma Xi; Am Powder Metall Group. *Res:* Powder metallurgy, mechanical behavior of materials, metallurgical aspects of fracture; biomaterials. *Mailing Add:* 900 Moraine Dr Lincoln NE 68510

NELSON, RUSSELL MARION, b Salt Lake City, Utah, Sept 9, 24; m 45; c 10. SURGERY. *Educ:* Univ Utah, BA, 45, MD, 47; Univ Minn, PhD(surg), 54; Am Bd Surg, dipl; Am Bd Thoracic Surg, dipl. *Hon Degrees:* ScD, Brigham Young Univ, 70. *Prof Exp:* From intern to sr resident, Univ Minn Hosps, 47-55; from asst prof to assoc clin prof surg, Col Med, 55-69, res prof, 70, DIR TRAINING PROG CARDIOVASC & THORACIC SURG, UNIV UTAH AFFIL HOSPS, 67- *Concurrent Pos:* Nat Heart Inst res fel, 49-50; first asst resident, Mass Gen Hosp, 53-54; Nat Cancer Inst trainee, 53-55; Markle scholar, 57-59; mem, White House Conf Youth & Children, 60; chmn div thoracic & cardiovasc surg, Latter-day Saints Hosp, 66-72; dir med serv, Utah Biomed Test Lab, 70-73; mem bd dirs, Am Bd Thoracic Surg, 72- & Int Cardiol Found; pvt pract, 59- *Mem:* Am Surg Asn; Soc Univ Surgeons; Am Asn Thoracic Surg; AMA; fel Am Col Surgeons; Sigma Xi. *Res:* Development of artificial heart-lung machine for open heart surgery; cardiovascular surgery; physiology of shock; physiological mechanisms involved in the etiology and treatment of ventricular fibrillation and other cardiac arrhythmias. *Mailing Add:* 324 Tenth Ave Salt Lake City UT 84103

NELSON, S(TUART) O(WEN), b Pilger, Nebr, Jan 23, 27; m 53, 79; c 2. ELECTRICAL ENGINEERING, PHYSICS. *Educ:* Univ Nebr, BSc, 50, MSc, 52, MA, 54; Iowa State Univ, PhD, 72. *Hon Degrees:* DSc, Univ Nebr, 89. *Prof Exp:* Asst physics, Univ Nebr, 52-54; proj leader, Farm Electrification Res Br, Agr Res Serv, 54-59, invest leader, 59-72, res leader, Lincoln, Nebr, 72-76, RES AGR ENGR, AGR RES SERV, USDA, ATHENS, GA, 76- *Concurrent Pos:* Res assoc, Univ Nebr-Lincoln, 54-59, from assoc prof to prof agr eng, 60-76; adj prof, Univ Ga, 76-; assoc ed, Int Microwave Power Inst, 75-85; mem coun, Agr Sci & Technol. *Honors & Awards:* Tech Paper Award, Am Soc Agr Engrs, 65; Decade Award Sci Contrib to Jour Microwave Power, Int Microwave Power Inst, 81; Fed Engr of the Yr, Nat Soc Prof Engrs, 85; Super Serv Award, USDA, 86, Prof Yr, Orgn Prof Employees, 87. *Mem:* AAAS; fel Am Soc Agr Engrs; fel Int Microwave Power Inst; Inst Elec & Electronics Engrs; Agr Sci & Technol; Nat Soc Prof Engrs. *Res:* Electromagnetic energy for insect control and seed treatment to improve germination; electrical properties of agricultural products; radiofrequency dielectric properties of insects, grain and seeds; frequency, moisture and density dependence of dielectric properties; methods and techniques for measuring dielectric properties; agricultural microwave power applications. *Mailing Add:* Russell Res Ctr USDA Agr Res Serv PO Box 5677 Athens GA 30613

NELSON, SAMUEL JAMES, b Vancouver, BC, June 2, 25; m 53; c 2. STRATIGRAPHY, PALEONTOLOGY. *Educ:* Univ BC, BASc, 48, MASc, 50; McGill Univ, PhD(stratig, paleont), 52. *Prof Exp:* Asst prof geol, Univ NB, 52-54; from asst prof to prof geol, Univ Calgary, 54-91; RETIRED. *Concurrent Pos:* Consult to var oil industs. *Mem:* Fel Geol Asn Can; hon mem Can Soc Petrol Geologists. *Res:* Ordovician and Permocarboniferous stratigraphy and paleontology. *Mailing Add:* Box 51 Satsuma BC V0N 2Y0 Can

NELSON, SHELDON DOUGLAS, b Idaho Falls, Idaho, Aug 12, 43; m 65; c 5. SOIL CONSERVATION, SOIL MORPHOLOGY. *Educ:* Brigham Young Univ, BS, 67; Univ Calif, Riverside, PhD(soil sci), 71. *Prof Exp:* Soil scientist, Agr Res Serv, USDA, 67-72; assoc prof, 72-90, PROF AGRON, BRIGHAM YOUNG UNIV, 90- *Concurrent Pos:* Environ consult, Eryring Res Inst, 75-76; vis sciientist, Univ Tarapaca Chile, 84; coop scientist, USDA, 83, 88 & 91. *Mem:* Am Soc Agron; Soil Sci Soc Am; Int Soil Soc; Coun Agr Sci & Technol; Sigma Xi. *Res:* Subsurface and drip irrigation practices; water quality; water movement in soils; soil physics; soil salinity; iron nutrition in plants; soil stabilization with polymers. *Mailing Add:* Dept Agron Widb 253 Brigham Young Univ Provo UT 84602

NELSON, SIDNEY D, b Seattle, Wash, Aug 18, 45. MEDICINAL CHEMISTRY. *Educ:* Univ Calif, San Francisco, PhD(chem), 74. *Prof Exp:* PROF MED CHEM, UNIV WASH, 80- *Honors & Awards:* John J Abel Pharmacol Award. *Mem:* Am Chem Soc; Am Soc Mass Spectrometry; Am Soc Pharmacol & Exp Therapeut; fel AAAS; Am Asn Pharmaceut Sci. *Res:* Mechanisms of reactive metabolic formation. *Mailing Add:* Dept Med Chem BG-20 Univ Wash Seattle WA 98195

NELSON, SIGURD OSCAR, JR, b Marquette, Mich, Jan 5, 37; m 60. ARACHNOLOGY. *Educ:* Northern Mich Univ, BS, 64; Mich State Univ, MS, 66, PhD(zool), 71. *Prof Exp:* Asst prof biol, Adrian Col, Mich, 71-72; chmn dept, 72-90, PROF BIOL, STATE UNIV NY COL OSWEGO, 72- *Mem:* Sigma Xi; Am Arachnological Soc; Am Micros Soc. *Res:* Systematics and ecology of pseudoscorpions. *Mailing Add:* Dept of Biol State Univ of NY Oswego NY 13126-3599

NELSON, STANLEY REID, b Kidder, SDak, Dec 20, 28; m 58; c 3. PHARMACOLOGY, NEUROSURGERY. *Educ:* Univ SDak, BA, 49, MS, 57; Tulane Univ, MD, 59. *Prof Exp:* Intern surg, Univ NC, 59-60; resident neurosurg, Univ Miss, 60-64; fel neurochem, Washington Univ, 64-66; from instr to assoc prof pharmacol & neurosurg, 66-73, prof pharmacol & assoc prof neurosurg, 73-81, prof & chmn, Dept Anat, 81-87, PROF ANAT, UNIV KANS MED CTR, KANSAS CITY, 87- *Mem:* Soc Neurosci; Am Soc Pharmacol & Exp Therapeut; Am Asn Anat. *Res:* Effect of drugs on brain glycolysis; neurochemistry of head injury; effect of anticholinesterase; agents on regional brain activity; role of peptides in neuronal function. *Mailing Add:* Dept Anat Univ Kans Med Ctr 39th & Rainbow Blvd Kansas City KS 66103

NELSON, STEPHEN GLEN, b Frederick, Okla, July 22, 47. CORAL REEF ECOLOGY, AQUACULTURE. *Educ:* San Diego State Univ, BS, 70, MS, 75; Univ Calif, Davis, PhD(ecol), 76. *Prof Exp:* Res assoc water sci & eng, Univ Calif, Davis, 76-77; asst prof, 77-80, dir, 82-85, ASSOC PROF BIOL, MARINE LAB, UNIV GUAM, 80- *Mem:* Ecol Soc Am; Am Soc Limnol & Oceanog; AAAS; World Maricult Soc. *Res:* Coral reef ecology; nitrogen metabolism of aquatic species. *Mailing Add:* Marine Lab Univ Guam UOG Sta Mangilao GU 96923

NELSON, TALMADGE SEAB, b Booneville, Ark, Jan 25, 28. ANIMAL NUTRITION. *Educ:* Univ Ark, BSA, 52; Univ Ill, MS, 52; Cornell Univ, PhD(animal nutrit), 59. *Prof Exp:* Res farm mgr, Western Condensing Co, Wis, 51-52; field supvr feed & fertilizer inspection, State Plant Bd, Ark, 52-54; res assoc poultry husb, Cornell Univ, 54-58, asst, 58-59; sr res biochemist, Int Minerals & Chem Corp, 59-63, supvr animal nutrit res, 63-68; UNIV PROF ANIMAL & POULTRY SCI, UNIV ARK, FAYETTEVILLE, 68- *Honors & Awards:* Nutrit Res Award, Am Feed Indust Asn. *Mem:* Am Soc Animal Sci; Poultry Sci Asn; Am Inst Nutrit; Animal Nutrit Res Coun. *Res:* Animal nutrition, especially requirements and functions; minerals. *Mailing Add:* Dept Animal & Poultry Sci Univ Ark Fayetteville AR 72701

NELSON, TERENCE JOHN, b Sioux City, Iowa, May 12, 39; m 63; c 2. GENERAL PHYSICS. *Educ:* Iowa State Univ, BS, 61, PhD(physics), 67; NY Univ, MEE, 63. *Prof Exp:* Mem tech staff, Bell Tel Labs, 61-63; AEC fel physics, Ames Lab, Iowa State Univ, 67-68 & Lawrence Radiation Lab, Univ Calif, 68-69; instr, Iowa State Univ, 69-70; mem tech staff, Bell Tel Labs, 70-83, DISTINGUISHED MEM PROF STAFF, BELL COMMUN RES, 84- *Mem:* Am Phys Soc; Sigma Xi. *Res:* Lasers; magnetic domain research and device development; mathematical physics in particle and group theory; display device research and development. *Mailing Add:* Bell Commun Res 20219 Murray Hill NJ 07974

NELSON, THEODORA S, b Phillips, Nebr, Dec 18, 13. MATHEMATICS. *Educ:* Nebr State Teachers Col, Kearney, BS, 42; Univ Ill, Urbana, BS, 46; Univ Nebr, EdD, 59. *Prof Exp:* Teacher pub schs, Nebr, 32-41; teacher high sch, Nebr, 42-45; asst math, Univ Ill, Urbana, 45-46; from asst prof to prof, 46-79, EMER PROF MATH, KEARNEY STATE COL, 79- *Concurrent Pos:* Part-time asst, Univ Ill, 45-46 & Univ Nebr, 57-58. *Mem:* Math Asn Am. *Res:* Mathematics seminar and research; projective geometry; history of mathematics. *Mailing Add:* 622 W 25th Kearney NE 68849

NELSON, THOMAS CHARLES, forestry; deceased, see previous edition for last biography

NELSON, THOMAS CLIFFORD, b Columbus, Ohio, July 24, 25. MICROBIOLOGY. *Educ:* Queens Col, NY, BS, 46; Columbia Univ, MA, 46, PhD, 51. *Prof Exp:* Asst prof, Vanderbilt Univ, 51-52; proj assoc genetics, Univ Wis-Madison, 52-53, USPHS res fel, 53-54, proj assoc bot, 55-57; USPHS res fel microbiol, Rutgers Univ, 54-55; sr microbiologist, Eli Lilly & Co, 57-65; asst prof microbiol, 65-70, ASSOC PROF BOT & ZOOL, UNIV WIS-MILWAUKEE, 70- *Res:* Microbial genetics; industrial microbiology; algal biochemistry. *Mailing Add:* Dept of Biol Sci Univ Wis Milwaukee PO Box 413 Milwaukee WI 53201

NELSON, THOMAS EDWARD, SKELETAL MUSCLES, ANESTHETIC EFFECTS. *Educ:* Okla State Univ, PhD(nutrit), 70. *Prof Exp:* PROF ANESTHESIOL & DIR RES ANESTHESIOL, MED BR, UNIV TEX, 76- *Res:* Malignant Hyperthermia. *Mailing Add:* Dept Anesthesiol Health Sci Ctr Univ Tex 6431 Fannin St MSB 5-020 Houston TX 77030

NELSON, THOMAS EUSTIS, JR, b Sharon, Mass, May 3, 22; m 47; c 3. PHARMACOLOGY, PHYSIOLOGY. *Educ:* Antioch Col, BA, 47; Univ Southern Calif, MS, 51, PhD(med physiol), 56. *Prof Exp:* Asst physiol, Sch Med, Univ Southern Calif, 50-56, instr, 56-57; from asst prof to assoc prof pharmacol, Sch Med & Dent, Univ PR, 57-61, actg head dept, 59-60; asst prof, Med Ctr, Univ Colo, 61-68; assoc prof, Univ Tex Dent Br Houston, 68-70; chmn dept pharmacol, 70-76, PROF PHARMACOL & CHMN DEPT BIOMED SCI, SCH DENT MED, SOUTHERN ILL UNIV, EDWARDSVILLE, 76-, DIR RES ADMIN, 80- *Concurrent Pos:* Consult, Am Dent Asn Comn & Coun Dent Educ, 80- *Mem:* Am Soc Pharmacol & Exp Therapeut; Int Asn Dent Res; Microcirculatory Soc; Am Physiol Soc; Sigma Xi. *Res:* Physiology and pharmacology of the cardiovascular and nervous systems. *Mailing Add:* Dept of Biomed Sci Southern Ill Univ Dent 800 College Ave Edwardsville IL 62026

NELSON, THOMAS EVAR, b Chicago, Ill, Oct 13, 34; m 59. BIOCHEMISTRY. *Educ:* Univ Ill, Urbana, BS, 57, MS, 59; Univ Minn, St Paul, PhD(biochem), 65. *Prof Exp:* Asst, Univ Ill, Urbana, 58-59; asst, Univ Minn, St Paul, 59-64, instr, 63; instr, Univ Minn, Minneapolis, 66-67; ASST PROF BIOCHEM, BAYLOR COL MED, 68- *Concurrent Pos:* Res fel, Univ Minn, Minneapolis, 65-68; USPHS fel, 66-68. *Mem:* Sigma Xi; AAAS; Am Chem Soc; fel Am Inst Chemists; Am Soc Biol Chemists. *Res:* Action pattern, specificity, mode to attack, transglucosylation and cleavage mechanisms of carbohydrases; fine structure of polysaccharides; enzyme purification techniques; carbohydrate chemistry; enzyme interrelationships in glycogen storage diseases; glycogen debranching enzyme; control of glycogen metabolism; specificity and mechanism of hormone action; cystic fibrosis factor. *Mailing Add:* 3614 Montrose Blvd No 906 Baylor Col Med One Baylor Plaza Houston TX 77006

NELSON, THOMAS LOTHIAN, b Baranquilla, Colombia, Jan 17, 22; US citizen; m 55; c 2. PEDIATRICS, ALLERGY. *Educ:* Univ Calif, Berkeley, AB, 43; Univ Calif, San Francisco, MD, 46. *Prof Exp:* Resident pediat, Med Ctr, Univ Calif, San Francisco, 49-51, instr, 51-56, asst clin prof pediat & lectr psychiat, Sch Med, 56-61; from assoc prof to prof pediat, Col Med, Univ Ky, 61-64; chmn dept, 63-64, assoc dean, 78-90, PROF PEDIAT, IRVINE-CALIF COL MED, UNIV CALIF, 64- *Concurrent Pos:* Pediatrician, Sonoma State Hosp, Eldridge, 51-52, chief physician, 52-54, asst supt med serv, 54-56, supt & med dir, 56-61; chief physician pediat & contagious dis serv, Los Angeles County Gen Hosp Unit II, 64-68; prog consult, Nat Inst Child Health & Human Develop, 64-68; mem, Fed Hosp Coun, 64-67; dir pediat prog, Orange County Med Ctr, 68-78. *Res:* Clinical immunology; mental retardation. *Mailing Add:* Dept Pediat Med Ctr Univ Calif Irvine Orange CA 92668

NELSON, VERNON A, b Norwood, Mass, Apr 17, 39; m 61; c 2. ENTOMOLOGY. *Educ:* Univ Mass, BS, 63, MS, 64; Pa State Univ, PhD(entom), 68. *Prof Exp:* Instr entom, Pa State Univ, 64-68; asst prof, 68-77, ASSOC PROF BIOL, SOUTHERN CONN STATE COL, 77- *Mem:* Entom Soc Am. *Res:* Arthropods of public health importance; taxonomy and biology; aquatic entomology. *Mailing Add:* Dept Biol Southern Conn State Col 501 Crescent St New Haven CT 06515

NELSON, VERNON RONALD, b Webster, SDak, Jan 20, 21; m 45; c 1. ELECTRONICS. *Educ:* Augustana Col, BA, 44; Univ Colo, MA, 51, DSc(sci educ), 53. *Prof Exp:* Assoc prof, 46-54, PROF PHYSICS, AUGUSTANA COL, SDAK, 54- *Mem:* Am Asn Physics Teachers; Nat Sci Teachers Asn; Inst Elec & Electronics Engrs. *Res:* Wave analysis; electronics in music; square wave generator. *Mailing Add:* Dept Physics Augustana Col Sioux Falls SD 57197

NELSON, VICTOR EUGENE, b Denver, Colo, Jan 8, 36; m 58; c 3. ZOOLOGY. *Educ:* Augustana Col, Ill, BA, 59; Univ Colo, PhD(zool, physiol), 64. *Prof Exp:* Asst biol, Univ Colo, 60-63, asst acarine physiol, 61-63; res assoc, Univ Kans, 64-65, asst prof & res assoc acarine physiol & entom, 65-66, asst prof biol, entom & insect biochem, 66-71; assoc prof, 71-76, PROF BIOL, BAKER UNIV, 76-, CHMN DEPT, 72- *Concurrent Pos:* Res grant, Univ Kans, 66-67. *Mem:* AAAS; Am Inst Biol Sci; Am Soc Zoologists; Sigma Xi. *Res:* Physiological ecology of terrestrial arthropods, particularly water relations. *Mailing Add:* Dept of Biol Baker Univ Baldwin City KS 66006

NELSON, W(INSTON) L(OWELL), b Logan, Utah, Apr 21, 27; m 49; c 2. ELECTRICAL ENGINEERING. *Educ:* Univ Utah, BS, 50; Columbia Univ, MS, 53, PhD(elec eng), 59. *Prof Exp:* Asst, Columbia Univ, 50-51 & 52-54, staff engr electronics res labs, 54-57, from instr to asst prof elec eng, univ, 57-60; mem tech staff, Bell Tel Labs, Inc, 60-62, supvr commun theory & control studies group, 62-66, supvr systs anal group, Radar Res Dept, 66-67, supvr control systs res group, 67-70, supvr systs anal group, Ocean Systs Anal Dept, 70-77, MEM TECH STAFF, BELL TEL LABS, INC, 77- *Concurrent Pos:* Chmn appln comt, Am Automatic Control Coun. *Mem:* Inst Elec & Electronics Engrs; Instrument Soc Am. *Res:* Control system theory; computer-aided system design; optimal control techniques; system performance evaluation; cybernetics; computer data analysis; estimation technique. *Mailing Add:* AT&T-Bell Lab Rm 2T-406 600 Mountain Ave Murray Hill NJ 07974

NELSON, WALTER GARNET, b London, Eng, Oct 26, 50; m 75. MARINE ECOLOGY. *Educ:* Duke Univ, AB, 72, PhD(zool), 78. *Prof Exp:* Fel marine ecol, Harbor Br Inst, 78-79, oil pollution, Royal Norwegian Coun Indust & Sci Res, 79-80; res assoc math model of population, Nat Res Coun, US Environ Protection Agency, 80-81; asst prof, 81-84, ASSOC PROF BIOOCEANOG, DEPT OCEANOG & OCEAN ENG, FLA INST TECHNOL 84- *Concurrent Pos:* Dir, Indian River Marine Sci Res Ctr, 86- *Honors & Awards:* Fulbright Fel, 88. *Mem:* AAAS; Ecol Soc Am; Crustacean Soc; Sigma Xi. *Res:* Ecosystem dynamics of marine benthic cummunities; benthic and pelagic community relationships in estuarine and near-shore ecosystems; biological effects of beach nourishment. *Mailing Add:* Dept Oceanog & Ocean Eng Fla Inst Technol Melbourne FL 32901

NELSON, WALTER RALPH, b St Paul, Minn, Mar 24, 37; m 60; c 2. HEALTH PHYSICS. *Educ:* Univ Calif, Berkeley, AB, 63; Univ Wash, MS, 64; Stanford Univ, PhD(health physics & dosimetry), 73. *Prof Exp:* STAFF MEM RADIATION PHYSICS, STANFORD LINEAR ACCELERATOR CTR, 64- *Concurrent Pos:* Consult, Varian Assocs, Calif, 67-75 & McCall Assocs, 69-; lectr, Sch Radiation Protection & Dosimetry, Italy, 75; dir, Sch Comput Shielding & Dosimetry, Italy, 78; sci assoc, Europ Orgn Nuclear Res, 78-79; Adj prof, Calif State Univ, San Jose, 86-; Consult, Varian Assocs, Calif, 88- *Honors & Awards:* Distinguished Sci Achievement Award, Health Physics Soc. *Mem:* Fel, Health Physics Soc. *Res:* Electromagnetic cascade calculations; muon production and transport; radiation dosimetry; electron accelerator shielding; medical physics calculations. *Mailing Add:* Stanford Linear Accelerator Ctr PO Box 4349 Stanford CA 94309

NELSON, WAYNE BRYCE, b Chicago, Ill, Aug 17, 36; m 58; c 3. STATISTICS, RELIABILITY. *Educ:* Calif Inst Technol, BS, 58; Univ Ill, Urbana, MS, 59, PhD(statist), 65. *Prof Exp:* Statistician, Corp Res & Dev, Gen Elec Co, 65-89; PVT CONSULT & TRAINER INDUST, 89- *Concurrent Pos:* Adj prof, Union Col, 66- & Rensselaer Polytech Inst, 69-; Am Statist Asn fel, 73; pvt consult & instr, 63-; sr res fel, Nat Inst Standards & Technol, 90-91. *Honors & Awards:* Brumbaugh Award, Am Soc Qual Control, 69, Youden Prize, 70, Wilcoxon Prize, 72; Brumbaugh Award & Publ Award, GE Corp, 81. *Mem:* fel Am Statist Asn; Biomet Soc; fel Inst Elec & Electronic Engrs; fel Am Soc Qual Control. *Res:* Accelerated testing; reliability and life data analysis; quality control; statistical computing and prediction methods. *Mailing Add:* 739 Huntingdon Dr Schenectady NY 12309

NELSON, WAYNE FRANKLIN, b Altona, Ill, Jan 16, 20; m 45; c 2. INORGANIC CHEMISTRY, PHYSICAL CHEMISTRY. *Educ:* Augustana Col, Ill, AB & BS, 41. *Prof Exp:* Asst chemist, Am Container Corp, Ill, 40-41, plant chemist, NY, 41-43, from asst plant mgr to plant mgr, 43-49, dir res, WVa, 49-57; dir res & develop, A Schulman, Inc, 57-84; RETIRED. *Mem:* Am Chem Soc; Soc Plastics Engrs; fel Am Inst Chemists. *Res:* Soft and hard rubber compounds; styrene plastics; colors and blends; polyolefins; polypropylene-natural; special compounds. *Mailing Add:* 1240 Walton Dr Akron OH 44313

NELSON, WENDEL LANE, b Mason City, Nebr, Apr 7, 39; m 68; c 2. MEDICINAL CHEMISTRY. *Educ:* Idaho State Univ, BS, 62; Univ Kans, PhD(pharmaceut chem), 65. *Prof Exp:* From asst prof to assoc prof, 65-76, prof pharmaceut chem, 76-80, PROF MED CHEM, UNIV WASH, 80- & CHMN DEPT, 84- *Mem:* Am Chem Soc; NY Acad Sci; Am Asn Cols Pharm. *Res:* Mechanisms of drug action; stereochemistry and conformational analysis; drug metabolism. *Mailing Add:* Sch Pharm BG-20 Univ Wash Seattle WA 98195

NELSON, WERNER LIND, b Sheffield, Ill, Oct 17, 14; m 40; c 2. AGRONOMY, SOIL. *Educ:* Univ Ill, BS, 37, MS, 38; Ohio State Univ, PhD(soil physics), 40. *Prof Exp:* Instr, Univ Idaho, 40-41; asst agronomist, Agr Exp Sta, NC State Univ, 41-44, assoc agronomist, 44-47, prof agron, 47-54, in charge soil fertil res, 51-54; midwest dir, Am Potash Inst, 55-67, regional dir, 62-67; sr vpres, Potash Inst, 67-77, sr vpres, Potash & Phosphate Inst, 77-85; CONSULT, 85- *Concurrent Pos:* Dir soil testing div, State Dept Agr, NC, 49-52; chmn soil test work group, Int Soil Fertil Cong, Ireland; mem fertilizer surv team, Food & Agr Orgn, UN, Asia & Far East Region, 59, fertilizer indust adv panel, 60-75; bd of dirs, Coun Agr Sci Technol, 72-81; adj prof agron, Purdue Univ, 73-85. *Mem:* AAAS; fel & hon mem Am Soc Agron (pres, 68-69); fel Soil Sci Soc Am (pres, 60-61); hon mem, Nat Fertilizer Solutions Asn. *Res:* Plant nutrition; soil fertility evaluation and management; fertilizers; limiting factors in crop production; maximum yield; top profit yield; intensive crop management diagnostic approach, interactions. *Mailing Add:* 1800 Happy Hollow Rd West Lafayette IN 47906

NELSON, WILBUR C, aerodynamics; deceased, see previous edition for last biography

NELSON, WILFRED H, b Evanston, Ill, May 23, 36; m 74; c 4. INORGANIC CHEMISTRY. *Educ:* Univ Chicago, BSc & MSc, 59; Univ Minn, PhD(inorg chem), 63. *Prof Exp:* Res assoc inorg chem, Univ Ill, Urbana, 63-64; from asst prof to assoc prof, 64-77, PROF CHEM, UNIV RI, 77- *Concurrent Pos:* Res Corp grants, 65-66 & 70-71; grant, Sydney Univ, 70-71; NSF grant, 66-72, 78-80 & 81-82; Dept Interior grants, 74-83; ARO grants, 83-91; mem, Nat Res Coun, 79; grant, Stanford Univ, 80. *Mem:* Am Chem Soc; Soc Appl Spectros; Sigma Xi; Biophys Soc; Soc Appl Bact. *Res:* Synthesis and structure of organo tin complexes; light scattering; ultraviolet-resonance Raman and fluorescene study of microorganisms and taxonomic markers; rapid microbiological identification and detection. *Mailing Add:* Dept of Chem Univ of RI Kingston RI 02881

NELSON, WILLIAM ARNOLD, b Lethbridge, Alta, June 24, 18; m 48; c 3. ENTOMOLOGY. *Educ:* Univ Alta, BSc, 44; McGill Univ, MSc, 48, PhD(med entom), 57. *Prof Exp:* Asst wheat stem sawfly, Field Crop Entom Sect, Res Sta, Can Dept Agr, 43-47, res scientist, Vet Med Sect, 48-82; RETIRED. *Res:* Physiology of host-parasite relationships; humoral factors. *Mailing Add:* 1020 Fern Crescent Lethbridge AB T1K 2W3 Can

NELSON, WILLIAM FRANK, b Cleveland, Ohio, May 4, 24. PHYSICS. *Educ:* Univ Akron, BS, 48, MS, 49; Wash State Univ, PhD(physics), 56. *Prof Exp:* Head, Mat Res Group, Long Beach Div, Douglas Aircraft Co, 53-60; sr res scientist & head solid state physics group fundamental res, Owens-Ill, Inc, 60-64, dir fundamental res, Tech Ctr, 64-69, ASSOC DIR RES & DIR ADVAN TECHNOL LAB, GTE LABS, INC, 69- *Concurrent Pos:* Adj prof, Univ Toledo, 63-69. *Mem:* AAAS; fel Am Phys Soc; Sigma Xi. *Res:* Physics of solids. *Mailing Add:* 548 Parkside Dr Akron OH 44313

NELSON, WILLIAM HENRY, b Huntsville, Ala, Nov 24, 43. MAGNETIC RESONANCE. *Educ:* Auburn Univ, BS, 66; Duke Univ, PhD(physics), 70. *Prof Exp:* Asst prof physics, Hollins Col, Va, 70-73; instr, Duke Univ, 73-74; from asst prof to assoc prof, 74-88, PROF PHYSICS, GA STATE UNIV, 88- *Concurrent Pos:* Res assoc, Microwave Lab, Duke Univ, 73-74. *Mem:* Am Phys Soc; Am Asn Physics Teachers; Sigma Xi; Radiation Res Soc. *Res:* Use of magnetic resonance, electron spin resonance and electron-nuclear double resonance, for study of radiation damage and molecular structure. *Mailing Add:* Dept of Physics Ga State Univ Atlanta GA 30303

NELSON, WILLIAM PIERREPONT, III, b New Orleans, La, Jan 9, 20; m 44; c 3. MEDICINE, INTERNAL MEDICINE. *Educ:* Wesleyan Univ, BA, 41; Cornell Univ, MD, 44; Am Bd Internal Med, dipl, 52. *Prof Exp:* Life Ins Med Res Fund fel metab, Sch Med, Yale Univ, 50-51; instr med, Albany Med Col, 51-53; asst prof, 53-56, asst dean col, 56-66, PROF MED & PROF POSTGRAD MED, ALBANY MED COL, 56- *Concurrent Pos:* Chief metab & endocrine sect, Vet Admin Hosp, 52-54, chief med serv, 54-56, chief ambulatory care, 74-77, assoc chief staff educ, 77-86; consult, NY State Bd Health, 86-89; pres, NY State Soc Internal Med, 76-77. *Mem:* Fel Am Col Physicians; Am Soc Int Med; Sigma Xi. *Res:* Metabolism and endocrinology. *Mailing Add:* 319 Highgate Dr Slingerlands NY 12159-9531

NELSON-REES, WALTER ANTHONY, b Havana, Cuba, Jan 11, 29; US citizen. GENETICS, CYTOLOGY. *Educ:* Emory Univ, AB, 51, MA, 52; Univ Calif, Berkeley, PhD(genetics), 60. *Prof Exp:* NIH training grant & res assoc genetics, Univ Calif, Berkeley, 60-61, from asst res geneticist to assoc res geneticist, Naval Biosci Lab, Sch Pub Health, 61-73, CONSULT, NAVAL BIOSCI LAB, SCH PUB HEALTH, UNIV CALIF, BERKELEY, 63-, ASSOC CHIEF CELL CULT DIV, 69-, LECTR CYTOL & CYTOGENETICS, 71- *Concurrent Pos:* Fulbright res scholar cytogenetics, Max Planck Inst Marine Biol, Ger, 61-62; consult, Breast Cancer Task Force, Nat Cancer Inst, NIH, 75-; co-prin investr, Nat Cancer Inst-NIH-Univ Calif, Berkeley, 76-; adj fac mem, W A Jones Cell Sci Ctr, 77. *Mem:* AAAS; Tissue Cult Asn. *Res:* Induction of chromosome aberrations in Tradescantia; factors influencing sex determination in coccid insects; heterochromatin and fertility factors in male mealy bugs; cinemicrography of animal cells; chromosome banding and other methods for cell line identification and detection of cellular cross contamination. *Mailing Add:* 6000 Contra Costa Rd Oakland CA 94618

NEMAT-NASSER, SIAVOUCHE, b Tehran, Iran, Apr 14, 36; m 65; c 3. APPLIED & STRUCTURAL MECHANICS, GEOMECHANICS. *Educ:* Sacramento State Col, BS, 60; Univ Calif, Berkeley, MS, 61, PhD(struct mech), 64. *Prof Exp:* Asst prof eng, Sacramento State Col, 61-62; fel struct mech, Northwestern Univ, 64-65, sr res fel, 65-66; asst prof appl mech, Univ Calif, San Diego, 66-69, assoc prof, 69-70; prof civil eng & appl math, Northwestern Univ, 70-85; PROF APPL MECH & ENG SCI, UNIV CALIF, SAN DIEGO, 85- *Concurrent Pos:* Mem, Ctr Study Democratic Insts; vis prof dept solid mech, Tech Univ Denmark, 72-73; ed, Mech Today; chief ed int jour, Mech Mat; Alburz Educ Found, 75. *Mem:* Am Soc Mech Engrs; Am Soc Civil Engrs; Soc Eng Sci (vpres-pres, 78-80); fel Am Acad Mech; Earthquake Eng Res Inst; Am Asn Univ Profs; NZ Nat Soc Earthquake Eng; Int Soc Soil Mech & Found Eng; Am Geophys Union; Soc Exp Mech. *Res:* Vibration and stability; continuum mechanics; elasticity; plasticity; waves in composites; flow, fracture and general constitutive behavior of geophysical, geotechnical and technological materials (experiment and theory); earthquake and geothermal energy research. *Mailing Add:* Dept Eng Sci Univ Calif La Jolla CA 92093-0310

NEMATOLLAHI, JAY, b Astara, Azerbaijan, Dec 21, 25; US citizen; m 52; c 3. MEDICINAL CHEMISTRY, MICROBIOLOGY. *Educ:* Univ Tehran, PharmD, 48; Univ Calif, MA, 54, PharmD, 58, PhD(pharmaceut chem), 63. *Prof Exp:* Fel, Univ Calif, 63; asst prof, prof pharmaceut chem, Univ RI, 63-64; assoc prof, Tex Southern Univ, 64-65; asst prof, 67-71, assoc prof pharmaceut chem, Col Pharm, Univ Tex, Austin, 71-; AT DEPT MED CHEM, COL PHARM, UNIV HOUSTON. *Concurrent Pos:* Adj grad fac, Univ Houston; Fulbright fel, USSR, 80. *Mem:* Am Chem Soc; AAAS. *Res:* Synthesis of organic medicinals; spectroscopy. *Mailing Add:* PO Box 2335 Rohnert Park CA 94928-1679

NEMEC, JOSEF, b Ostresany, Czech, Sept 7, 29; m 75; c 1. SYNTHETIC ORGANIC CHEMISTRY, NATURAL PRODUCT CHEMISTRY. *Educ:* Inst Chem Technol, MS, 54; Czech Acad Sci, PhD(org chem), 58. *Prof Exp:* Scientist org chem, Czech Acad Sci, 54-61; sr res chemist, Inst Chem Technol, Czech Acad Sci, 61-69; res fel org chem, Wayne State Univ, 69-70; sr res scientist, Squibb Inst Med Res, 70-75; mem, St Jude Children's Res Hosp, 75-84; ADJ PROF, DEPT MED CHEM, UNIV TENN, MEMPHIS, 79-; SR SCIENTIST & HEAD MED CHEM SECT, FREDERICK CANCER RES & DEVELOP CTR, 84- *Mem:* Am Chem Soc; Royal Soc Chem; AAAS. *Res:* Isolation, synthesis and modification of natural products and metabolities including glycosides, nucleosides, conjugated monoclonal antibodies and antineoplastic agents; micro and semimicro experimental techniques. *Mailing Add:* Frederick Cancer Res & Develop Ctr PO Box B Bldg 325 Frederick MD 21702-1201

NEMEC, JOSEPH WILLIAM, b Philadelphia, Pa, Mar 24, 22; m 48; c 3. ORGANIC CHEMISTRY. *Educ:* Temple Univ, AB, 43; Ind Univ, MA, 44; Pa State Univ, PhD(chem), 49. *Prof Exp:* Sr org chemist, Rohm & Haas Co, 49-61, lab head, 61-69, res supvr, 69-73, mgr process res, 73-85; RETIRED. *Mem:* Am Chem Soc. *Res:* Process research and development; catalysis; acrylate monomer technology. *Mailing Add:* 931 Washington Lane Rydal Jenkintown PA 19046

NEMEC, STANLEY, b St Louis, Mo, Feb 3, 35; m 60; c 2. PLANT PATHOLOGY. *Educ:* Auburn Univ, BS, 60; Okla State Univ, MS, 64; Ore State Univ, PhD(plant path), 67. *Prof Exp:* Landscape technician, Harland Bartholomew & Assoc, City Planners, 60-61; plant pathologist, Ore Dept Agr, 63-66; res plant pathologist, Southern Ill Univ, Carbondale, 66-72, RES PLANT PATHOLOGIST, USDA, FLA, 72- *Concurrent Pos:* Fla Inst Phosphate Res grant, 85. *Mem:* Am Phytopath Soc; Soc Nematol. *Res:* Botany; horticulture; biochemistry. *Mailing Add:* Agr Res Serv USDA US Hort Field Lab 2120 Camden Rd Orlando FL 32803

NEMECEK, GEORGINA MARIE, b Mineola, NY, Aug 27, 46. ATHEROGENESIS, METABOLIC REGULATION. *Educ:* Mount Holyoke Col, AB, 68; Univ Pa, PhD(pharmacol), 72. *Prof Exp:* Res assoc biochem, Sch Med, Univ Mass, 72-74, from asst prof to assoc prof, 74-83; sr scientist, 83-85, FEL SR SCI STAFF, SANDOZ RES INST, 86-, GROUP LEADER MOLECULAR BIOL, 87- *Concurrent Pos:* Am Heart Asn fel, 74; young investr, Nat Heart, Lung & Blood Inst, 77-81; NATO vis scientist, Free Univ, Brussels, Belgium, 79; Guest Scientist, Sea Pharm, Inc, 85; vis res molecular biologist, Dept Molecular Biol, Princeton Univ, 87; job rotation, Dept Biotechnol, Sandoz AG, Basel, Switz, 88. *Mem:* Am Soc Pharmacol & Exp Therapeut; Tissue Cult Asn; NY Acad Sci; Sigma Xi. *Res:* Regulation of gene expression; role of peptide growth factors in atherogenesis; regulation of vascular cell function; cholesterol biosynthesis. *Mailing Add:* Sandoz Res Inst 59 Rte 10 East Hanover NJ 07936-1080

NEMENZO, FRANCISCO, zoology; deceased, see previous edition for last biography

NEMER, MARTIN JOSEPH, b Philadelphia, Pa, Nov 26, 29. BIOCHEMISTRY. *Educ:* Kenyon Col, BA, 52; Harvard Univ, MS, 55, PhD(biochem), 58. *Hon Degrees:* DSc, Kenyon Col, 77. *Prof Exp:* Fel, Univ Brussels, 58-59; res fel, Stazione Zool, Naples, Italy, 59-60; res assoc biochem, 60-63, from asst mem to mem, 63-86, SR MEM, INST CANCER RES, 86- *Mem:* AAAS; Am Soc Biol Chemists; Soc Develop Biol. *Res:* Chemical embryology; regulation of protein and nucleic acid synthesis. *Mailing Add:* Inst Cancer Res Fox Chase Cancer Ctr Philadelphia PA 19111

NEMERE, ILKA M, b Los Angeles, Calif, Jan 1, 53; m 78. CALCIUM TRANSPORT, MEMBRANE & MICROTUBULE BIOLOGY. *Educ:* Univ Calif, San Diego, BA, 74, Los Angeles, PhD(molecular biol), 80. *Prof Exp:* Fel biochem, Riverside, 80-83, fel cell biol, San Diego, 83-84, ASST RES BIOCHEM, UNIV CALIF, RIVERSIDE, 84- *Concurrent Pos:* Vis asst prof, Univ Calif, Los Angeles, 85, lectr, Riverside, 87 & 88. *Mem:* Am Soc Biochem & Molecular Biol; Am Soc Cell Biol; Endocrine Soc; Sigma Xi; Am Soc Bone & Mineral Res; AAAS. *Res:* Role of vesicular flow & microtubules in 1, 25-dihydroxyvitamin D3-mediated calucum transport; subcellular localization & function of calbindin-D28K; mechanism of nongenomic actions of 1, 25-dihydroxyvitamin D3. *Mailing Add:* Dept Biochem Univ Calif Riverside CA 92521

NEMEROFF, CHARLES BARNET, b Bronx, NY, Sept 7, 49; m 80; c 3. PSYCHOPHARMACOLOGY, NEUROBIOLOGY. *Educ:* City Col NY, BS, 70, Northeastern Univ, 73, Univ NC, Chapel Hill, PhD(neurobiol), 76, MD, 81. *Prof Exp:* Res fel, Sch Med, Univ NC, 77-83, clin instr psychiat, 82-83; from asst prof pharmacol to assoc prof pharmacol, 83-89, asst prof physchiat, 85-86, PROF PSYCHIAT & PHARMACOL, MED CTR, DUKE UNIV, 89- *Concurrent Pos:* Mem, neurosci res rev comt, NIMH, 83-87, prin investr, 83- & grant Alzheimer's dis, 85- *Honors & Awards:* A E Bennett Award, Soc Biol Psychiat, 79; Curt P Richter Award, Int Soc Psychoneuroendocrinol, 85; Anna Monika Prize, Res Depression, 87; Efron Award, Am Col Neuropsychopharm, 87; J Folch-PI Award, Am Soc Neurochem, 87; Kempf Award Psychobiol, Am Psychiat Asn, 89. *Mem:* Am Col Neuropsychopharmacol; Endocrine Soc; Soc Biol Psychiat; Am Soc Neurochem; Soc Neurosci; Am Psychiat Asn; Am Col Psychiatrists; Psychiat Res Soc. *Res:* Elucidation of the biological bases of Alzheimer's disease, schizophrenia and severe depression using neurochemical methods. *Mailing Add:* 2008 Jo Mac Rd Chapel Hill NC 27516

NEMEROW, NELSON L(EONARD), b Syracuse, NY, Apr 16, 23; m 47; c 3. SANITARY ENGINEERING. *Educ:* Syracuse Univ, BSChE, 44; Rutgers Univ, MS, 49, PhD(sanit eng), 51. *Prof Exp:* Res engr, Johns Manville Corp, 44 & 46; from asst prof to assoc prof civil eng, NC State Col, 51-57; prof, Syracuse Univ, 58-76; RES PROF, UNIV MIAMI, 76- *Concurrent Pos:* Lectr, Stanford Univ, 65-66; consult, Environ Protection Agency, 67-78, UN Indust Develop Orgn, Vienna, Austria, 74-78, Nat Water Comn, 75-76 &

WHO, 80-; adj prof, Fla Atlantic Univ, 89- *Mem:* Am Soc Civil Engrs; Am Soc Sanit Engrs; Am Acad Environ Engrs. *Res:* Water; sewage; industrial wastes; stream pollution; solid wastes disposal; economics of water resources; oil pollution of beaches; aquaculture; industrial complexes; author of 180 technical publications & 8 textbooks. *Mailing Add:* 227 SE Eighth St Dania FL 33004

NEMERSON, YALE, b New York, NY, Dec 15, 31; m 58; c 3. HEMATOLOGY, BIOCHEMISTRY. *Educ:* Bard Col, BA, 53; NY Univ, MD, 60. *Prof Exp:* Intern, Lenox Hill Hosp, 60-61; resident, Bronx Vet Admin Hosp, 61-62; fel hemat, Montefiore Hosp, 62-64; from instr to prof, Sch Med, Yale Univ, 64-75; prof med, State Univ NY Stony Brook, 75-77; PHILLIP J & HARRIET L GOODHART PROF MED, MT SINAI SCH MED, 77- *Concurrent Pos:* Leukemia Soc fel, 63-64; NIH res grant, 64-76; investr, Am Heart Asn, 67-72, res grant, 67-73. *Mem:* AAAS; Am Soc Hemat; Am Soc Exp Path; Am Soc Clin Invest. *Res:* Lipid-protein interactions in blood coagulation. *Mailing Add:* Mt Sinai Med Ctr One Gustav Levy Pl Box 1269 New York NY 10029

NEMETH, ABRAHAM, b New York, NY, Oct 16, 18; m 44. MATHEMATICS, COMPUTER SCIENCE. *Educ:* Brooklyn Col, BA, 40; Columbia Univ, MA, 42; Wayne State Univ, PhD(math), 64. *Prof Exp:* Instr math, Brooklyn Col, 46, Manhattan Col, 53-54 & Manhattanville Col, 54-55; from instr to prof math, Univ Detroit, 55-85; RETIRED. *Mem:* Math Asn Am; Asn Comput Mach. *Res:* Computer science; Nemeth braille code of mathematics and scientific notation. *Mailing Add:* 20764 Knob Woods Dr No 201 Southfield MI 48076

NEMETH, ANDREW MARTIN, b Philadelphia, Pa, Sept 13, 26; m 61; c 4. ANATOMY. *Educ:* Johns Hopkins Univ, AB, 49, MD, 53. *Prof Exp:* Intern pediat, Johns Hopkins Hosp, 53-54; NSF fel biochem, Col Physicians & Surgeons, Columbia Univ, 54-56; assoc anat, 56-59, from asst prof to assoc prof, 59-68, PROF ANAT, SCH MED, UNIV PA, 68- *Mem:* Am Asn Anatomists. *Res:* Regulation of enzyme formation in developing tissues; induction of enzymes; fetal development. *Mailing Add:* Univ Pa 36th & Hamilton Walk Philadelphia PA 19104

NEMETH, EDWARD JOSEPH, b Glassport, Pa, Oct 1, 38; m 68; c 1. COAL CONVERSION. *Educ:* Univ Pittsburgh, BS, 60, PhD(chem eng), 67; Univ Ill, MS, 62. *Prof Exp:* Engr, Textile Fibers Dept, E I du Pont de Nemours & Co, Inc, 63-64; res engr raw mat, Chem Div, US Steel, 67-69, sr res engr, 69-76, supvr chem mat, 76-81, mgr facil planning, 81-82, dir res, 82-86; DIR RES, ARISTECH CHEM CORP, 86- *Mem:* Am Inst Chem Engrs. *Res:* Polypropylene; unsaturated polyster resins; acrylics; coal chemicals; industrial, intermediate and specialty chemicals. *Mailing Add:* Aristech Chem Corp 1000 Tech Ctr Dr Monroeville PA 15146

NEMETH, EVI, b Cooperstown, NY, June 7, 40; m 63; c 1. COMBINATORICS, COMPUTER SCIENCE. *Educ:* Pa State Univ, BA, 61; Univ Waterloo, PhD(math), 71. *Prof Exp:* Gen mgr oceanog, Estuarine & Oceanic Technol Corp, 72-75; asst prof math & comput sci, Fla Atlantic Univ, 75-76; asst prof math & comput sci, State Univ NY Col Technol, 76-80; ASSOC PROF COMPUT SCI, UNIV COLO, 80- *Concurrent Pos:* Mathematician math & comput sci, US Dept Com, 76-84; HEW grant, 78; consult, Pattern Anal & Recognition Corp, 77-80. *Mem:* Math Asn Am; Asn Comput Mach. *Res:* Combinatorics; existence theorems for designs; combinatorial algorithms; information retrieval; encryption, coding theory, networks. *Mailing Add:* Comput Sci Dept Univ Colo Box 430 Boulder CO 80309-0430

NEMETH, JOSEPH, b Ithaca, NY, Dec 29, 40; m 76; c 3. MATERIALS ENGINEERING. *Educ:* Alfred Univ, BS, 63, MS, 66; Univ Windsor, PhD(mat eng), 71. *Prof Exp:* Res assoc mech properties, Air Force Lab, State Univ NY Col Ceramics, Alfred Univ, 63-66; res assoc crystal growth, Univ Windsor, 66-70; res engr, Champion Spark Plug Co, 70-76, staff res engr, 70-76, sr res engr, 76-81, sect head, Process Control, 81-84, sect head develop, 84-87; mgr ceramic appln, Form Physics Corp, 87-88; tech dir, EDO Can, Calgary, 88-89; vpres ceramics, O Hommel Corp, Pittsburg, 89-90; VPRES OPERS, QUEST TECH CORP, 90- *Concurrent Pos:* Lectr, Univ Windsor, 68-70. *Mem:* Am Ceramic Soc; Nat Inst Ceramic Engrs; Am Soc Metals. *Res:* Mechanical and electrical properties of insulating and semiconducting materials; microstructure and fracture behavior of oxide materials. *Mailing Add:* Quest Tech Corp 6750 Nancy Ridge Way San Diego CA 92121

NEMETH, LASZLO K, b Szeged, Hungary, June 7, 32; m 56; c 4. PETROLEUM ENGINEERING. *Educ:* Rensselaer Polytech Inst, BS, 55; Pa State Univ, MS, 57; Tex A&M Univ, PhD(petrol & natural gas eng), 66. *Prof Exp:* Reservoir & sec recovery engr, Texaco, Inc, 57-63; consult, Nat Oil Co Arg, 66-70; sr res engr, Tenneco Oil Co, 70-71; consult engr, Butler, Miller & Lents Ltd, 71-76; exec vpres, 76-80, PRES, J R BUTLER & CO, 80- *Concurrent Pos:* Univ Ind lectr, US, Arg & Chile; sr vpres, GeoQuest Int, Inc. *Mem:* Soc Petrol Engrs; Am Inst Mining, Metall & Petrol Engrs; Int Oil Scouts Asn. *Res:* Reservoir engineering and simulation; secondary recovery; improved recovery methods; miscible flooding; gas and condensate recovery. *Mailing Add:* 5542 Dumfries Houston TX 77096

NEMETH, MARGARET ANN, b Cleveland, Ohio. STATISTICS. *Educ:* Cleveland State Univ, BA, 70, MS, 71; Va Polytech Inst & State Univ, PhD(statist), 78. *Prof Exp:* Instr math, Va State Col, 72-73 & Richard Bland Col, 73-74; asst prof statist, Tex A&M Univ, 78-84; math statistician, USDA, New Orleans, 84-85; RES SPECIALIST, MONSANTO, ST LOUIS, MO, 85- *Concurrent Pos:* Consult statistician, US Dept Energy grant, 78-81. *Mem:* Am Statist Asn; Inst Math Statist; Biomet Soc; Asn Women Sci; Royal Statist Soc. *Res:* Multivariate statistical inference when the underlying distribution belongs to the class of elliptically symmetric distributions; invariant test procedures; path analysis; graphical techniques for multivariate distributions; experimental design. *Mailing Add:* Monsanto Co 800 N Lindbergh Blvd St Louis MO 63167

NEMETH, RONALD LOUIS, b Endicott, NY, Mar 4, 41; m 66. COLLOID AND SURFACE SCIENCE. *Educ:* Clarkson Col Technol, BS, 64, MS, 66, PhD(phys chem), 69. *Prof Exp:* Res chemist, 68-80, SR TECHNOL SPECIALIST, MONSANTO CO, 80- *Mem:* Am Chem Soc. *Res:* Stability of aqueous dispersions; molecular structure and mechanical property relationships of polymers. *Mailing Add:* 384 Col Hwy Southampton MA 01073-9375

NEMETH, ZOLTAN ANTHONY, b Sopron, Hungary, Apr 27, 31; US citizen; m 54; c 2. CIVIL & TRANSPORTATION ENGINEERING. *Educ:* Budapest Tech Univ, Dipl, 54; Ohio State Univ, MS, 63, PhD(transp eng), 68. *Prof Exp:* PROF CIVIL ENG, OHIO STATE UNIV, 68-; DIR, OHIO TRANS TECHNOL TRANSFER CTR, 88- *Concurrent Pos:* Mem comt, Transp Res Bd, Nat Acad Sci-Nat Res Coun, 70- *Mem:* Inst Transp Engrs; Am Soc Civil Engrs; Am Soc Testing & Mat. *Res:* Traffic control; transportation safety. *Mailing Add:* Dept Civil Eng Ohio State Univ Columbus OH 43210-1275

NÉMETHY, GEORGE, b Budapest, Hungary, Oct 11, 34; US citizen; m 75; c 2. BIOPHYSICAL CHEMISTRY. *Educ:* Lincoln Univ, Pa, BA, 56; Cornell Univ, PhD(phys chem), 62. *Prof Exp:* Phys chemist, Gen Elec Res Lab, NY, 62-63; asst prof phys chem, Rockefeller Univ, 63-72; vis prof, Dept Biochem, Univ Paris, Orsay, 72-74; vis assoc prof, Dept Chem, State Univ NY Binghamton, 74-75; sr res assoc, Dept Chem, Cornell Univ, 75-89; PROF, DEPT BIOMATH SCI, MT SINAI SCH MED, 89- *Concurrent Pos:* NATO fel & vis lectr, Instituto Superiore di Sanita, Rome, 70; lectr, Med Ctr, NY Univ, 71-72; mem biophys sci training comt, Nat Inst Biomed Sci, 71-72; vis prof, Dept Chem, Univ Napoli, Italy, 84, 86; Europ Molecular Biol Orgn sr scientist fel, 73-74. *Honors & Awards:* Pius XI Gold Medal, Pontifical Acad Sci, 72. *Mem:* Am Soc Biochem & Molecular Biol; NY Acad Sci; Am Chem Soc. *Res:* Statistical thermodynamics, liquid structure; thermodynamic properties of aqueous solutions; structure of proteins, conformations, thermodynamics, structure and enzymatic activity; protein conformation; collagen structure and assembly. *Mailing Add:* Dept Biomath Sci Box 1023 Mt Sinai Sch Med One Gustave Levy Pl New York NY 10029

NEMHAUSER, GEORGE L, b New York, NY, July 27, 37; m 59; c 2. OPERATIONS RESEARCH. *Educ:* City Col New York, BChE, 58; Northwestern Univ, MS, 59, PhD(opers res), 61. *Prof Exp:* From asst prof to assoc prof opers res, Johns Hopkins Univ, 61-69; prof opers res, Cornell Univ, 69-85, dir, Sch Opers Res & Indust Eng, 77-83; CHANDLER PROF, SCH INDUST & SYSTS ENG, GA INST TECHNOL, 85- *Concurrent Pos:* Vis lectr, Univ Leeds, 63-64; vis res prof, Ctr Opers Res & Econometrics, Cath Univ Louvain, 69-70 & 83-84; NSF fac fel, 69-70; dir res, Ctr Opers Res & Economet, Cath Univ Louvain, 75-77; ed-in-chief, J Opers Res Soc Am, 75-78; found ed, Oper Res Letters, 81- *Honors & Awards:* Lanchester Prize, Opers Res Soc Am, 77, 90 & Kimball Medal, 88. *Mem:* Nat Acad Eng; Inst Mgt Sci; Soc Indust & Appl Math; Math Prog Soc (pres, 88-91); Opers Res Soc Am (vpres, 80-81, pres, 81-82); Am Inst Indust Engrs. *Res:* Theory and computational aspects of mathematical programming; mathematical modelling of complex systems. *Mailing Add:* Sch Indust & Systs Eng Ga Inst Technol Atlanta GA 30332

NEMIR, PAUL, JR, b Navasota, Tex, Aug 30, 20; m 49; c 3. SURGERY. *Educ:* Univ Tex, AB, 40, MD, 44. *Prof Exp:* From instr to assoc prof surg, 48-69, dean grad sch med, 59-64, dir div grad med, 64-69, PROF SURG, SCH MED, UNIV PA, 69- *Concurrent Pos:* Surgeon-in-chief, Grad Hosp & consult cardiovasc surg, US Naval Hosp, 48- *Mem:* AMA; Soc Univ Surgeons; Soc Vascular Surg; Am Asn Thoracic Surg; Soc Surg Alimentary Tract; Am Surg Asn. *Res:* Pulmonary function and embolism; diseases of the esophagus and esophageal motor function; mechanism of toxicity of hemoglobin derivatives; studies on vascular prosthetics. *Mailing Add:* Grad Hosp 19th & Lombard Sts Philadelphia PA 19146

NEMIR, ROSA LEE, b Waco, Tex, July 16, 05; m 34; c 3. PEDIATRICS. *Educ:* Univ Tex, BA, 26; Johns Hopkins Univ, MD, 30. *Hon Degrees:* DSc, Colgate Univ, 74. *Prof Exp:* From instr to assoc prof, Med Col, 33-50, assoc prof, Postgrad Med Sch, 50-53, PROF PEDIAT, SCH MED, NY UNIV, 53- *Concurrent Pos:* Lectr, Sch Nursing, Bellevue Hosp, 34-49, from asst vis pediatrician to vis pediatrician, 37-; attend pediatrician, Univ Hosp, 50-, vis physician & in chg chest unit, Children's Med Serv, 60, dir children's chest clin, 61-; attend pediatrician, Gouvernour Hosp, 50-58, consult, 58; consult, NY Infirmary, 54-82, dir pediat educ & res, 66-73; vis prof, Col Physicians & Surgeons, Columbia Univ, 58-59; dir cont med educ, Am Med Womens Asn, 77-81. *Honors & Awards:* Off, Medal of Cedars of Lebanon, Repub Lebanon; Elizabeth Blackwell Award, Am Med Women's Asn, 70. *Mem:* Soc Pediat Res; Am Pediat Soc; Am Col Chest Physicians; Am Thoracic Soc; Soc Adolescent Med; Am Med Women's Asn (pres, 64); Med Women's Int Asn (vpres, 70-74). *Res:* Pneumonia in children; nutrition; tuberculosis; virology. *Mailing Add:* NY Univ Sch of Med 550 First Ave New York NY 10016

NEMITZ, WILLIAM CHARLES, b Memphis, Tenn, July 27, 28. PURE MATHEMATICS. *Educ:* Southwestern at Memphis, BS, 50; Ohio State Univ, MS, 56, PhD(math), 59. *Prof Exp:* Instr math, Ohio State Univ, 59-60; asst prof, Univ Kans, 60-61; from asst prof to prof math, Southwestern at Memphis, 61-85; RETIRED. *Concurrent Pos:* NSF grant, 63-64, 71-73. *Mem:* Am Math Soc. *Res:* Algebraic structures related to intuitionistic logics, particularly implicative semi-lattices. *Mailing Add:* 31247 Morlock Apt 104 Livonia MI 48152-1638

NEMOTO, TAKUMA, US citizen; m 60; c 4. BREAST CANCER. *Educ:* School Med, Keio Univ, Tokyo, MD, 54. *Prof Exp:* sr cancer res & assoc chief breast surg, Rouswell Park Mem Inst, 63-88,; RES ASSOC PROF, STATE UNIV NY, 75- *Mem:* Am Col Surgeons; Am Soc Clin Oncol; Am Inst Cancer Res; Soc Surgical Oncol. *Mailing Add:* Breast Care Ctr 2121 Main Buffalo NY 14214

NEMPHOS, SPEROS PETER, b New York, NY, July 8, 30; m 55; c 3. POLYMER CHEMISTRY. *Educ:* Ursinus Col, BS, 52; Univ Del, MS, 55, PhD(phys chem), 57. *Prof Exp:* Res chemist, Monsanto Co, 56-63, specialist, 63-65, group leader, 65-74, mgr res, 74-85; MGR RES & DEVELOP, NOVACOR CHEMICALS, LEOMINSTER, MASS, 86- *Mem:* Am Chem Soc. *Res:* Polymer synthesis; kinetics and characterization of vinyl polymers; polymer degradations and stabilization; plastic foams; emulsion and suspension polymerizations; graft polymer systems; barrier resins; plastics fire safety; membrane science; separational technology; condensation polymers; degradable plastics. *Mailing Add:* PO Box 1031 Leominster MA 01453-1031

NENNO, ROBERT PETER, b Buffalo, NY, Mar 3, 22; c 4. PSYCHIATRY. *Educ:* Univ Notre Dame, BS, 43; Loyola Univ, MD, 47; Am Bd Psychiat & Neurol, dipl. *Prof Exp:* Intern, E J Meyer Mem Hosp, Buffalo, NY, 47-48; resident psychiat, Vet Admin Hosp, Minneapolis, 48-50; resident, Vet Admin Hosp, Downey, Ill, 50-51; asst prof psychiat & asst dir dept, Sch Med, Georgetown Univ, 53-56, assoc prof, 56-58; prof psychiat & chmn dept, Seton Hall Col Med, 58-63; med dir & chief exec officer, NJ State Hosp, Marlboro, 63-68; interim med dir, Raritan Bay Ment Health Ctr, 70; prof psychiat, NJ Med Sch, Col Med & Dent NJ, Newark, 73-76; attend, Martland Hosp, 76-77; dir dept psychiat, Jersey City, Med Ctr, 73-76, actg med dir, 75-76; clin prof psychiat, NJ Med Sch, 76-77; med dir, Pitt County Mental Health Ctr, 77-83; CLIN PROF PSYCHIAT, SCH MED, E CAROLINA UNIV, 77- *Concurrent Pos:* Asst dir psychiat div, Georgetown Univ Hosp, 53-56; consult, DC Gen Hosp, Mt Alto Vet Admin Hosp & Cent Intel Agency, 53-58; asst examr, Am Bd Psychiat & Neurol, 56-; clin prof psychiat, Sch Med, Rutgers Univ, 66-73; pvt pract psychiat, 68-; mem staff, Overlook Hosp, Summit, NJ, 68-77; consult, Pitt County Mem Hosp, Greenville NC, 77- *Mem:* Fel Am Psychiat Asn; Am Pub Health Asn; Am Asn Social Psychiat; Acad Psychoanal; fel, World Health Org. *Res:* Vocational and social rehabilitation of the mentally ill; nicotinic acid in the treatment of schizophrenia; use of librium analogs in psychiatry. *Mailing Add:* PO Box 354 Grimesland NC 27837

NEPTUNE, JOHN ADDISON, b Barnesville, Ohio, Nov 27, 19; m 47; c 1. INORGANIC CHEMISTRY. *Educ:* Muskingum Col, BS, 42; Univ Wis, MS, 49, PhD(chem), 52. *Prof Exp:* Instr chem, Muskingum Col, 43-44 & 45-48; shift foreman, Chem Refining Div, Tenn Eastman Corp, 44-45; asst prof chem, Bowling Green State Univ, 49-50; instr pharmaceut chem, Univ Wis, 52-55; from asst prof to assoc prof, 55-61, chmn dept, 73-86, PROF CHEM, SAN JOSE STATE UNIV, 61- *Mem:* Am Chem Soc. *Res:* Removal of gaseous pollutants from air; reactions in non-aqueous solvents. *Mailing Add:* 50 Cherokee Lane San Jose CA 95127

NEPTUNE, WILLIAM EVERETT, b Lawton, Okla, Apr 24, 28; m 50; c 2. PHYSICAL CHEMISTRY. *Educ:* Okla Baptist Univ, BS, 50; Univ Okla, MS, 52, PhD(chem), 54. *Prof Exp:* From asst prof to prof chem, Okla Baptist Univ, 54-80, dean lib arts, 61-73, vpres acad affairs, 73-78, actg pres, 76-77, provost, 78-80; NAT CONSULT, HOME MISSION BD, 81- *Concurrent Pos:* Consult, Saline Waters Proj, US Dept Interior, 60-61, dir, 61-65; consult, NCent Asn Cols & Sec Schs, 62-82, mem, Comn Insts Higher Educ, 69-76, exec bd comm, 71-76 & Coun Res & Serv, 78-80. *Mem:* Am Chem Soc; Royal Soc Chem; NY Acad Sci; fel Am Inst Chemists. *Res:* Electrochemistry; chemistry of carbon; history and philosophy of science; chemical education. *Mailing Add:* Rte 2 Box 30 Eufaula OK 74432

NERBUN, ROBERT CHARLES, JR, b Waukegan, Ill, Apr 19, 46; m 72; c 7. NUCLEAR STRUCTURE. *Educ:* Univ Wis-River Falls, BS, 68; Case Western Reserve Univ, MS, 71, PhD(exp nuclear physics), 73. *Prof Exp:* Lectr, Col Physics, Case Western Reserve Univ, 69-70; PROF PHYSICS, UNIV SC, SUMTER, 73-, CHMN SCI, MATH, & ENG, 76- *Concurrent Pos:* Vis prof physics & astron, Univ Wis-River Falls, 75; bd trustees, SC Gov's Sch Sci & Math. *Mem:* Am Asn Physics Teachers; Am Acad Natural Family Planning. *Mailing Add:* Div Sci Math & Eng Univ SC 200 Miller Rd Sumter SC 29150-2498

NEREM, ROBERT MICHAEL, b Chicago, Ill, July 20, 37; m 58, 78; c 4. CELLULAR BIOMECHANICS, FLUID MECHANICS. *Educ:* Univ Okla, BS, 59; Ohio State Univ, MS, 61, PhD(aeronaut eng), 64. *Hon Degrees:* Dr, Univ Paris. *Prof Exp:* From instr to assoc prof, Ohio State Univ, 61-72, prof aeronaut & astronaut eng, 72-79, assoc dean, Grad Sch, 75-79; prof & chmn mech eng, Univ Houston, 79-86; PARKER H PETIT DISTINGUISHED CHAIR ENG IN MED, GA TECH, 86- *Concurrent Pos:* Consult, Goodyear Aerospace Corp, Ohio, 62-74, Aro, Inc, Tenn, 64-65 & Space & Info Systs Div, NAm Aviation, Inc, Calif, 65; Sci Res Coun sr vis fel, Imp Col, Univ London, 70; consult, Gen Elec Co, 71-72 & Technol Inc, 80; chmn, US Nat Comt Biomechanics, 88-91; pres, Int Fedn Med & Biol Eng, 88-91. *Honors & Awards:* Lissner Award, Am Soc Mech Engrs, 89; Alza Distinguished Lectr, 91. *Mem:* Nat Acad Eng; Fel Am Phys Soc; fel AAAS; fel Am Soc Mech Engrs; Am Physiol Soc; fel Am Heart Asn. *Res:* Biological fluid mechanics; cellular engineering; hemodynamics and atherogenesis; heart disease; cell culture engineering; bioprocessing And biotechnology. *Mailing Add:* Sch Mech Eng Ga Inst Technol Atlanta GA 30332-0405

NERENBERG, MORTON ABRAHAM, b Montreal, Que, Mar 17, 36; m 57; c 3. JOSEPHSON JUNCTIONS, NON-LINEAR SYSTEMS. *Educ:* McGill Univ, BSc, 57, PhD(theoret physics), 65. *Prof Exp:* From asst prof to assoc prof math, 62-80, PROF APPL MATH, UNIV WESTERN ONT, 80- *Concurrent Pos:* Vis scientist, Univ Paris, 70-71 & Univ Tel Aviv, 83. *Mem:* Can Appl Math Soc; Soc Indust & Appl Math. *Res:* Mathematical study of chaotic and coherent behavior in nonlinear systems, in particular superconducting Josephson Junctions; Green's function methods applied to study of impurities in ionic crystals; reconstruction of dynamics from time senes of data. *Mailing Add:* Dept Appl Math Univ Western Ont London ON N6A 5B9 Can

NERESON, NORRIS (GEORGE), b Gaylord, Minn, Nov 4, 18; c 3. LASERS. *Educ:* Concordia Col, Minn, BA, 39; Univ Denver, MS, 41; Cornell Univ, PhD(nuclear physics), 43. *Prof Exp:* Asst physics, Cornell Univ, 42-43; mem staff, Physics Div, Los Alamos Sci Lab, 43-45; asst prof physics, Univ NMex, 46-47; mem staff, Laser Div, Los Alamos Sci Lab, 48-82; RETIRED. *Concurrent Pos:* Int Atomic Energy Agency vis prof, Brazil, 64-65. *Res:* Solid state physics and lasers; application of molecular spectroscopy to isotope separation employing diode lasers. *Mailing Add:* PO Box 366 Los Alamos NM 87544

NERI, ANTHONY, b Italy, Jan 15, 48; US citizen; m 76; c 2. TISSUE CULTURE. *Educ:* Loyola Univ, BS, 71; Calif State Univ, Northridge, MS, 75; Univ Calif, Irvine, PhD(cell biol), 81. *Prof Exp:* Res assoc cancer res, Sch Med, Univ Calif, Los Angeles, 71-75; asst instr, Univ Calif, Irvine, 75-80; RES FEL, UNIV SOUTHERN CALIF CANCER CTR, 81- *Mem:* Sigma Xi. *Res:* Cellular and biochemical events associated with the process of cancer metastasis and detection. *Mailing Add:* 539 Prospect St Nutley NJ 07110

NERI, RUDOLPH ORAZIO, b Barre, Mass, Sept 11, 28; m 55; c 4. ENDOCRINOLOGY. *Educ:* Col Holy Cross, BS, 50; NY Univ, MS, 58, PhD(biol), 63. *Prof Exp:* Asst biologist, Worcester Found Exp Biol, 51-55; from asst scientist to prin scientist, 55-70, res fel, 70-79, ASSOC DIR, CLIN RES, SCHERING CORP, 79- *Concurrent Pos:* Adj prof physiol, Farleigh Dickinson Univ Dental Sch, 72- *Mem:* AAAS; Am Physiol Soc; Endocrine Soc; Am Soc Dermat. *Res:* Effects of anti-estrogens, anti-androgens and anti-progesterones on reproductive processes in laboratory animals; rat skin homotransplantation studies, immune and reticuloendothelial response. *Mailing Add:* 15 Annette Ave Hawthorne NJ 07506

NERI, UMBERTO, b Rimini, Italy, Sept 7, 39; m 64; c 1. MATHEMATICS. *Educ:* Univ Chicago, BS, 61, MS, 62, PhD(math), 66. *Prof Exp:* From asst prof to assoc prof, 66-81, PROF MATH, UNIV MD, COLLEGE PARK, 81- *Concurrent Pos:* NSF fel, Univ Md, College Park, 68-69; Nat Res Coun Italy fel, Univ Genoa, 69-70, Univ Pisa, 73-74, Univ Bari, 75, Univ Cagliari, 77 & 79, Univ Minn, 80, Ind Univ & Univ Nice, 82, Scuola Normale Superiore, Pisa, 84, Univ Cagliari, 82, 84 & 85. *Mem:* Am Math Soc; Ital Math Asn. *Res:* Singular integral operators; partial differential equations; harmonic analysis. *Mailing Add:* Dept of Math Univ of Md College Park MD 20742

NERING, EVAR DARE, b Gary, Ind, July 18, 21; m 42; c 2. MATHEMATICS. *Educ:* Ind Univ, AB, 42, AM, 43; Princeton Univ, AM, 47, PhD(math), 48. *Prof Exp:* Asst, Ind Univ, 42-44; jr physicist, Appl Physics Lab, Johns Hopkins Univ, 44-45; instr math, Princeton Univ, 45-46, asst, 46-47; instr, Rutgers Univ, 47-48; from asst prof to assoc prof, 60-74, chmn dept, 62-70, PROF MATH, ARIZ STATE UNIV, 74-, DIR DEPT, 78- *Concurrent Pos:* Technician, Manhattan Proj, Ind Univ, 43-44; mathematician, Goodyear Aircraft Corp, 53-54. *Mem:* Am Math Soc; Math Asn Am. *Res:* Algebraic function and game. *Mailing Add:* Dept Math Ariz State Univ Tempe AZ 85281

NERKEN, ALBERT, b New York, NY, Aug 21, 12. CHEMISTRY. *Educ:* Cooper Union, BS, 33. *Prof Exp:* Chemist, Sinclair Ref Co, Pa, 33-35; jr physicist, Nat Adv Comt Aeronaut, Langley Field, Va, 35-37; chief chemist, Polin Labs, NY, 37-43; res chemist, Kellex Corp, 43-45; partner & engr, 45-61, chmn bd & treas, 61-71, vchmn bd, 71-75, CHMN BD, VEECO INSTRUMENTS INC, 75- *Mem:* Am Chem Soc; Am Vacuum Soc. *Res:* Airplane dynamics; plastics from agricultural products; high vacuum techniques; gas analysis. *Mailing Add:* Veeco Instruments Terminal Dr Plainview NY 11803-2304

NERLAND, DONALD EUGENE, b Webster City, Iowa, Apr 29, 46; m 83; c 1. PHARMACOLOGY, ORGANIC CHEMISTRY. *Educ:* Univ Iowa, BS, 69; Univ Kans, PhD(med chem), 74. *Prof Exp:* Teaching assoc pharmacol, Univ Minn, 74-77; asst prof, 77-83, ASSOC PROF PHARMACOL, UNIV LOUISVILLE, KY, 83- *Mem:* Am Soc Pharmacol & Exp Therapeut; AAAS; Am Chem Soc. *Res:* Drug and xenobiotic metabolism involving microsomal and soluble enzyme systems; metabolism of inhaled organic compounds. *Mailing Add:* Dept Pharmacol Univ Louisville Louisville KY 40292

NERLICH, WILLIAM EDWARD, b Los Angeles, Calif, May 6, 23; m 57; c 4. MEDICINE. *Educ:* Univ Southern Calif, MD, 47; Am Bd Internal Med, dipl, 55. *Prof Exp:* From head physician to chief physician, 52-66, DIR INTERN TRAINING, LOS ANGELES COUNTY HOSP, 66-; DIR OFF EDUC & ASST MED DIR, LOS ANGELES COUNTY-UNIV SOUTHERN CALIF MED CTR, 69-; PROF MED, SCH MED, UNIV SOUTHERN CALIF, 71- *Concurrent Pos:* From asst prof med & asst dean student affairs to assoc prof med & asst dean student affairs & assoc dean admis, Univ Southern Calif, 58-81; mem, Physician's Asst Exam Comt, Calif State Bd Med Examr, 71-79. *Mem:* Asn Am Med Cols; Am Col Physicians. *Res:* Medical education; physical diagnosis. *Mailing Add:* 1200 N State St Los Angeles CA 90033

NERO, ANTHONY V, JR, b Salisbury, Md, Apr 11, 42. ENVIRONMENTAL RISK ASSESSMENT, INDOOR AIR QUALITY. *Educ:* Fordham Univ, BS, 64; Stanford Univ, PhD(physics), 71. *Prof Exp:* Res fel nuclear physics, Kellogg Radiation Lab, Calif Inst Technol, 70-72; asst prof physics, Princeton Univ, 72-75; SR SCIENTIST & PRIN INVESTR, APPL SCI DIV, LAWRENCE BERKELEY LAB, 75- *Concurrent Pos:* Phys sci officer, US Arms Control & Disarmament Agency, 78; lectr, Univ Calif, Berkeley, 79; consult, Elec Power Res Inst, Nat Acad Sci, Orgn Econ Coop & Develop, Environ Protection Agency, Nat Comt Radiation Protection & Measurements, World Health Orgn. *Mem:* Fel Am Phys Soc; AAAS; Sigma Xi; Air Pollution Control Asn; Health Physics Soc. *Res:* Health and safety impacts, as well as general environmental effects of energy technologies, with emphasis on indoor and outdoor air pollution; instrumentation for environmental radiation monitoring; alternative nuclear reactor types. *Mailing Add:* 2738 Benvenue Ave Berkeley CA 94705

NERODE, ANIL, b Los Angeles, Calif, June 4, 32; m 70; c 3. MATHEMATICAL LOGIC, APPLIED MATHEMATICS. *Educ:* Univ Chicago, BA, 49, BS, 52, MS, 53, PhD, 56. *Prof Exp:* Sr mathematician, Inst Syst Res, Univ Chicago, 54-56, group leader, 56-57; mem, Inst Adv Study, 57-58, 62-63; vis asst prof math, Univ Calif, 58-59; from asst prof to assoc prof, Cornell Univ, 59-65, actg dir, Ctr Appl Math, 65-66, chmn dept, 82-87, PROF MATH, CORNELL UNIV, 65-, DIR MATH, SCI INST, 87- *Concurrent Pos:* NSF fel, Cornell Univ, 57-58; mem consult bur, Math Asn Am, 61-; prin investr, NSF grants, 61- 90; mem comt appln math, Nat Res Coun, 67-70; ed, J Symbolic Logic, 68-82, J Pure & Appl Logic, 82-, Future Generation Comput Systs, 84-, Math & Artificial Intel, 90- & Logical Methods in Computer Sci, 90-; vis prof, Monash Univ, Australia, 70, 74 & 78, Univ Chicago, 76; mem math sci adv comt, NSF, 72-75; prin investr, Environ Protection Agency grants, 78-; vis prof, Mass Inst Technol, 80 & Univ Calif, San Diego, 81, 84 & 85. *Mem:* Am Math Soc; Math Asn Am; Soc Indust & Appl Math; Asn Symbolic Logic; Asn Computer Mach; Inst Elec & Electronic Engrs Computer Soc. *Res:* Mathematical logic; recursive functions; automata; intelligent systems; computation theory. *Mailing Add:* Dept Math Cornell Univ Ithaca NY 14850

NERSASIAN, ARTHUR, b Salem, Mass, July 14, 24; m 46; c 3. ORGANIC CHEMISTRY. *Educ:* Mass Inst Technol, BS, 49; Univ Mich, MS, 51, PhD(org chem), 54. *Prof Exp:* Asst, Dow Chem Co for Univ Mich, 50-51, res fel & instr chem, 52-53; res chemist, Org Chem Dept, Elastomers Div, E I du Pont de Nemours & Co, Inc, 54-57, Elastomer Chem Dept, 57-79, Polymer Prod Dept, 80-81, sr tech consult, 82-85; RETIRED. *Mem:* Am Chem Soc. *Res:* Rubber chemicals; Hypalon synthetic rubbers; polymer structural analysis; resilient foams; thermoplastic polyurethanes; Viton fluorohydrocarbon rubbers; effect of sour fuels and alcohol containing fuels on rubber properties; rubbers in automotive applications; oil additives-rubber compatibility; fluids interactions and solubility parameters of rubber. *Mailing Add:* 5225 S Covewood Terr Floral City FL 32636-2162

NES, WILLIAM DAVID, b Bethesda, Md, Aug 16, 53; m 76; c 3. PLANT LIPID BIOCHEMISTRY. *Educ:* Gettysburg Col, BA, 75; Drexel Univ, MS, 77; Univ Md, PhD(plant biochem), 79. *Prof Exp:* Fel, Dept Nutrit, Univ Calif Berkeley, 79-80, Bi-Nat Agri Res Develop, US/Israel, 80-81; res chemist & proj leader, USDA Berkeley 82-86, PROJ LEADER, USDA ATHENS, 87; ADJ ASSOC PROF CHEM, AUBURN UNIV, 88- *Concurrent Pos:* Instr, Berkeley Exten Sch, Univ Calif, 80-81; assoc ed Lipids, 86-; plenary lectr, chmn & co-org, Nat & Int Sterol & Plant Lipid meetings; ed J Am Oil Chem Soc, 88-, Exp Mycol, Inform, 89- *Mem:* NY Acad Sci; Sigma Xi; Soc Plant Physiol; Phytochem Soc NAm; Am Oil Chemists Soc; Am Chem Soc; Am Soc Biochem & Molecular Biol. *Res:* Chemistry, biological functions and phylogenetic significance of steroids and other lipids in plants; author of various books, memorial issues in Lipids, J Chromatography and steriods. *Mailing Add:* USDA ARS Richard B Russell Ctr PO Box 5677 Athens GA 30613

NES, WILLIAM ROBERT, biochemistry; deceased, see previous edition for last biography

NESBEDA, PAUL, b Trieste, Italy, June 20, 21; nat US; m 49; c 5. SYSTEM ENGINEERING, COMMUNICATION SYSTEMS. *Educ:* Univ Pisa, PhD(math), 43. *Prof Exp:* Asst prof math analysis, Univ Trieste, 44-46; fel, Univ Paris, 46-47; mem staff, Inst Adv Study, 47-48; instr math anal, Catholic Univ, 48-52; mathematician & engr, RCA Corp, 52-58, leader tech staff, 58-64, staff eng scientist, Aerospace Div, 65-74; MEM TECH STAFF, THE MITRE CORP, 74- *Concurrent Pos:* Lectr, Boston Col, 59-60. *Mem:* Am Inst of Aeronaut & Astronaut; Sigma Xi; Armed Forces Commun & Electronics Asn. *Res:* Functional and combinatorial analysis; probability; information processings and decision theory; digital communications and imaging systems; electro-optical systems; bioengineering; system analysis; defense electronics systems. *Mailing Add:* 10 Blodgett Rd Lexington MA 02173

NESBET, ROBERT KENYON, b Cleveland, Ohio, Mar 10, 30; m 58; c 3. THEORETICAL PHYSICS. *Educ:* Harvard Univ, AB, 51; Cambridge Univ, PhD(physics), 54. *Prof Exp:* Mem staff, Lincoln Lab, Mass Inst Technol, 54-56; asst prof physics, Boston Univ, 56-62; STAFF MEM, RES LAB, IBM CORP, 62- *Concurrent Pos:* Nat Cancer Inst spec res fel, Inst Pasteur, France, 60-61; exchange prof, Univ Paris, 73; assoc ed, J Chem Physics, 71-73 & J Computational Physics, 70-74; vis prof, Univ Kaiserslautern, WGermany, 79-80. *Mem:* Fel Am Phys Soc; AAAS; Sigma Xi. *Res:* Theoretical atomic and molecular physics; quantum theory of finite many-particle systems; computational physics. *Mailing Add:* IBM ARC K31/802 650 Harry Rd San Jose CA 95120-6099

NESBIT, RICHARD ALLISON, b Whitter, Calif, Jan 17, 35; m 57; c 1. BIORESEARCH INSTRUMENTS, CLINICAL INSTRUMENTS. *Educ:* Univ Calif, Los Angeles, BS, 58, MS, 60, PhD(eng), 63. *Prof Exp:* Mem tech staff, Aerospace Corp, El Segundo, CA, 58-63; sr scientist, Lear Siegler Inc, Santa Monica, 63-64; eng mgr, 64-66, res & develop mgr, 66-75, dir res, 75-80, VPRES RES & DEVELOP, BECKMAN INSTRUMENTS INC, FULLERTON, CA, 80- *Concurrent Pos:* Lectr eng, Univ Calif, Los Angeles, 62-66; mgt consult, 66; lectr mgt sci, Redlands, Calif, 69. *Mem:* Sigma Xi; Soc Indust Appl Math. *Res:* Bioanalytical instrumentation and clinical diagnostic reagents and instruments. *Mailing Add:* 1849 Avenida Del Norte Fullerton CA 92633

NESBITT, BRUCE EDWARD, b Crawfordsville, Ind, Aug 28, 51; m 84; c 2. ECONOMIC GEOLOGY. *Educ:* Carleton Col, BA, 73; Univ Mich, MS, 76, PhD(econ geol), 79. *Prof Exp:* Teaching fel geol, Univ Mich, 73-78, lectr econ geol, 78-79; assoc res, Pa State Univ, 79-80; from asst prof to assoc prof, 80-89, PROF ECON GEOL, UNIV ALTA, 89- *Mem:* Soc Econ Geol; Mineral Soc Am; Sigma Xi; Am Geophys Union; Can Mineral Asn. *Res:* Metamorphosed ore deposits and the genesis and geochemistry of precious metal deposits. *Mailing Add:* Dept Geol Univ Alberta Edmonton AB T6G 2E3 Can

NESBITT, CARL C, b Yerington, Nev, Nov 20, 58; m 87; c 3. ENVIRONMENTAL PROCESSES, MINERAL PROCESSING. *Educ:* Univ Nev, Reno, BS, 80, MS, 85, PhD(metall eng), 90; Univ Mich, MSE, 89. *Prof Exp:* Metall engr, Anadonda Minerals Co, 80-83; ASST PROF MINERALS PROCESSING, DEPT METALL & MAT ENG, MICH TECHNOL UNIV, 90- *Concurrent Pos:* Mem, Educ & Pub Affairs Comt, Metall Soc, 90- *Mem:* Soc Mining Engrs; Metall Soc. *Res:* Novel ways of removing heavy metals from waste waters of mineral processing and other industries; precious metal hydrometallurgy; mineral processing. *Mailing Add:* Dept Metall & Mat Eng Mich Technol Univ 1400 Townsend Dr Houghton MI 49931

NESBITT, CECIL JAMES, b Ft William, Ont, Oct 10, 12; nat US; m 38. ACTUARIAL MATHEMATICS. *Educ:* Univ Toronto, BA, 34, MA, 35, PhD(math), 37. *Prof Exp:* Mem, Inst Advan Study, 37-38; from instr to prof, 38-80, assoc chmn dept, 62-67, chmn, 70-71, EMER PROF MATH, UNIV MICH, ANN ARBOR, 80- *Concurrent Pos:* Coun mem, Conf Bd Math Sci, 68-75; co-chmn data registry comt, Cystic Fibrosis Found, 73-78; res dir, Actuarial Educ & Res Fund, 80-86. *Mem:* Am Math Soc; Inst Math Statist; Soc Actuaries (vpres, 85-87). *Res:* Actuarial theory. *Mailing Add:* Dept Math Univ Mich Ann Arbor MI 48109-1003

NESBITT, HERBERT HUGH JOHN, b Ottawa, Ont, Feb 7, 13; m 44; c 4. ENTOMOLOGY. *Educ:* Queen's Univ, Ont, BA, 37; Univ Toronto, MA, 39, PhD(invert zool), 44; Univ Leiden, DSc, 51. *Hon Degrees:* DSc, Carleton Univ. *Prof Exp:* Agr scientist, Div Entom, Dept Agr, Ont, 39-48; from asst prof to prof, Carleton Univ, 48-78, dir, Div Sci, 60-63, dean fac sci, 63-74, clerk senate, 75-81, EMER PROF BIOL, CARLETON UNIV, 78- *Concurrent Pos:* Chmn Bd Govs, Algonquin Col, 81-86, Carleton Univ, 72-76. *Mem:* fel Royal Entom Soc London; sci fel Zool Soc London; fel Linnean Soc London; fel Entom Soc Can. *Res:* Nervous system of insects; comparative morphological and taxonomic work on Acari. *Mailing Add:* 50 Littleton Gardens Rockcliffe Park ON K1L 5B6 Can

NESBITT, JOHN B, b State College, Pa, Apr 18, 24; div; c 3. SANITARY ENGINEERING. *Educ:* Pa State Univ, BS, 48; Mass Inst Technol, SM, 49, ScD, 52. *Prof Exp:* Asst sanit eng, Mass Inst Technol, 49-52; instr hydraul, McCoy Col, Johns Hopkins Univ, 52-53; from asst prof to assoc prof civil eng, 53-65, PROF CIVIL ENG, PA STATE UNIV, 65- *Concurrent Pos:* Sanit eng designer, Whitman, Requardt & Assocs, Md, 52-53; mem USA nat comt, Int Asn Water Pollution Res, 69-78. *Mem:* Am Soc Civil Engrs; Water Pollution Control Fedn; Asn Environ Eng Prof; Am Water Works Asn; Sigma Xi. *Res:* Liquid and solid municipal and industrial wastes; removal of radioactive contaminants from water supplies; treatment of industrial wastes containing the cyanide ion; water filtration; removal of phosphorus from municipal wastewaters. *Mailing Add:* 4465 Hillcrest Ave State College PA 16803

NESBITT, ROBERT EDWARD LEE, JR, b Albany, Ga, Aug 21, 24; m 47; c 2. OBSTETRICS & GYNECOLOGY. *Educ:* Vanderbilt Univ, BA, 44, MD, 47; Am Bd Obstet & Gynec, dipl, 56. *Prof Exp:* Asst instr, Johns Hopkins Hosp, 48-52; obstetrician & gynecologist in chief, US Army Hosp, Ger, 52-54; asst prof, Sch Med, Johns Hopkins Univ, 54-56; prof & chmn dept, Albany Med Col, 58-61; PROF OBSTET & GYNEC & CHMN DEPT, STATE UNIV NY UPSTATE MED CTR, 61- *Concurrent Pos:* Fel, Am Asn Maternal & Child Health Dirs; obstetrician & gynecologist in chief, Albany Hosp, 56-61 & Crouse-Irving Hosp, 63-70; mem, Pub Health Coun State NY, 62-67; chief obstet & gynec, State Univ Hosp, 64-; assoc examr, Am Bd Obstet & Gynec. *Mem:* Soc Gynec Invest; Pan-Am Med Asn; fel Am Col Obstet & Gynec; fel Am Col Surg; NY Acad Sci. *Res:* Cytologic, cytochemical and histochemical study and diagnosis of early cervical cancer; prenatal and placental pathology; cytohormonal diagnosis; experimental production of abruptio placenta; reproductive endocrinology; animal experimentation; hormonal influence on placentation; fetal anoxia; immunoglobulin patterns in normal and toxemic pregnancy; in vitro placental perfusion studies; infertility. *Mailing Add:* 750 E Adams St Syracuse NY 13210

NESBITT, STUART STONER, b Aledo, Ill, Jan 29, 21; m 51; c 2. CHEMISTRY. *Educ:* Monmouth Col, BS, 43; Univ Tex, MA, 44, PhD(chem), 49. *Prof Exp:* Lab asst, Monmouth Col, 42-43; lab instr, Univ Tex, 43-44, asst, 46-49; res chemist, Mid-Continent Petrol Corp, 49-55; supvr prod res, Sunray D-X Oil Co, 55-69, mgr fuels & indust prod res, DX Div, 69-70, chief prod develop, Appl Res Dept, Tulsa Lab, 70-71, mgr, 71-72, chief prod serv, 72-79, mgr, Tulsa Refinery Lab, Sun Oil Co, 79-86; RETIRED. *Mem:* Am Chem Soc. *Res:* Alkoxyacetaldehydes; allylic chlorides; petroleum. *Mailing Add:* 8203 S Jamestown Tulsa OK 74137-1618

NESBITT, WILLIAM BELTON, horticulture, plant breeding; deceased, see previous edition for last biography

NESENBERGS, MARTIN, b Jelgava, Latvia, Nov 8, 28; m 54; c 1. COMMUNICATIONS THEORY, COMPUTER NETWORKS. *Educ:* Univ Denver, BS, 52; NY Univ, MS, 58; Univ Colo, PhD(appl math), 67. *Prof Exp:* Mem tech staff commun res, Bell Tel Labs, 53-60; MATHEMATICIAN COMMUN RES, INST TELECOMMUN SCI, DEPT COMMERCE, 60- *Concurrent Pos:* Mem, US Delegation, Int Telecommun Union/Int Radio Consult Comt, Geneva, 67-72, Comn C, US Nat Int Union Radio Sci, 81-; ed, Commun Mag, Inst Elec & Electron Engrs, 74-75; vis prof, elec eng & telecommun, Univ Colo, 80-; mem, Armed Forces Commun & Electronics Asn. *Mem:* Sr mem Inst Elec & Electronics Engrs. *Res:* Communications theory subareas such as random processes, error-coding, applied mathematics, traffic theory and topology synthesis have been applied to various communication networks. *Mailing Add:* Nat Telecommun & Info Admin Inst Telecommun Sci-4 325 Broadway Boulder CO 80303

NESHEIM, MALDEN CHARLES, b Rochelle, Ill, Dec 19, 31; m 77; c 3. NUTRITION. *Educ:* Univ Ill, BS, 53, MS, 54; Cornell Univ, PhD, 59. *Prof Exp:* Asst animal sci, Univ Ill, 53-54; asst poultry husbandry, 56-59, from asst prof to assoc prof animal nutrit, 59-74, prof & dir, Div Nutrit Sci, 74-87, vpres, planning & budgeting, 87-89, PROVOST CORNELL UNIV, 89- *Concurrent Pos:* NIH spec fel, Cambridge Univ, 72-73, overseas fel, Churchill Col, 72-73; vis fel, Sidney Sussex Col, 83-84. *Mem:* Am Inst Nutrit; Brit Nutrit Soc. *Res:* Amino acid metabolism; amino acid and protein requirements; gastrointestinal physiology; nutrition and parasitic infection. *Mailing Add:* Day Hall Cornell Univ Ithaca NY 14853

NESHEIM, MICHAEL ERNEST, b San Francisco, Calif, Feb 5, 45; m 66, 85; c 4. FEEDBACK CONTROL SYSTEMS IN BIOLOGY. *Educ:* Univ Minn, BA, 67, PhD(biochem), 78. *Prof Exp:* Fel hemat res, Mayo Clin, 76-79, res assoc, 79-81, asst prof biochem & hemat res, Mayo Clin Found, 81-84; PROF MED & BIOCHEM, QUEENS UNIV, 84- *Concurrent Pos:* Technician endocrine res, Mayo Clin, 68-72. *Honors & Awards:* Louis N Katz Award, Am Heart Asn, 81. *Mem:* Am Soc Molecular Biol & Biochem. *Res:* Enzymology and protein chemistry of blood coagulation and fibrinolysis; concepts of control mechanisms in biological systems; computer simulation of coagulation and fibrinolysis. *Mailing Add:* Dept Biochem Queens Univ Kingston ON Can

NESHEIM, ROBERT OLAF, b Monroe Center, Ill, Sept 13, 21; m 81; c 3. NUTRITION. *Educ:* Univ Ill, BS, 43, MS, 50, PhD(animal nutrit), 51. *Prof Exp:* Swine res specialist, Gen Mills, Inc, 51-52; head swine feed res, Quaker Oats Co, 52-59, mgr livestock feed res, 59-64; prof animal sci & head dept, Univ Ill, 64-67; assoc dir, Quaker Oats Co, 67-69, dir, 69, vpres res & develop, 69-77, vpres sci & technol, 77-83; PRES, ADVAN HEALTHCARE, 84- *Concurrent Pos:* Chmn, Animal Nutrit Res Coun, 65-66; mem, US Nat Comt, Int Union Nutrit Sci, 68-74; mem, food industs adv comt & chmn, food & nutrit liason comt, Nutrit Found, mem bd trustees, 78-80; mem adv comt, US Meat Animal Res Ctr, USDA, 71-75; mem, food & nutrit bd, Nat Acad Sci, 72-78, mem exec comt, 72-76; mem, Comt Food Sci & Technol; mem food indust liaison comt, Am Med Asn; mem,food adv comt, Off Technol Assessment, US Cong, 75-80; mem bd trustees, Food Safety Coun & mem exec comt, 76-79; mem bd trustees, BioSci Info Serv, 78-84, vchmn, 80, chmn, 82-83; chmn,comt military nitrit res, Food & Nutrit Bd, 82-; fel, AAAS & Am Inst Nutrit. *Mem:* AAAS; Am Inst Nutrit; Fedn Am Soc Exp Biol (treas, 74-78); Soc Nutrit Educ; Am Pub Health Asn. *Res:* Amino acid nutrition; energy utilization; obesity treatment. *Mailing Add:* Advan Health Care 2801 Salinas Hwy Bldy F Monterey CA 93940

NESHEIM, STANLEY, b Chicago, Ill, Apr 24, 30; m 58; c 4. ANALYTICAL CHEMISTRY. *Educ:* Brooklyn Col, BS, 56; George Washington Univ, MS, 62. *Prof Exp:* Chemist, US Bur Mines, 56-57; RES CHEMIST, US FOOD & DRUG ADMIN, 57- *Mem:* AAAS; Am Chem Soc; Am Oil Chem Soc; Sigma Xi. *Res:* Edible fats and oil characterization and chemical analysis; food contaminants; mycotoxin chemistry and analysis. *Mailing Add:* 3008 Tennyson St NW Washington DC 20015

NESHYBA, STEVE, b Jourdanton, Tex, Oct 8, 27; m 50; c 6. PHYSICAL OCEANOGRAPHY. *Educ:* Univ Tex, BSEE, 49, MS, 54; Tex A&M Univ, PhD(phys oceanog), 65. *Prof Exp:* Res engr, Elec Eng Res Lab, Univ Tex, 53-54; aerophys engr, Gen Dynamics/Convair, 54-57, sr aerophys engr, 57-60; asst prof elec eng, Arlington State Col, 60-62; fel oceanog, Tex A&M Univ, 62-64; assoc prof, 65-76, PROF PHYS OCEANOG, ORE STATE UNIV, 76- *Concurrent Pos:* Planning dir, Prof Sch Appl Sci & Mgt Marine Resources, 86; auth; coodr, Int Sea Grant Proj, Ore State, 87-89. *Res:* Deep sea hydrography; analyses of time-dependent motions of water bodies; polar ocean studies; microscale mixing processes. *Mailing Add:* Dept of Oceanog Ore State Univ Corvallis OR 97331

NESLINE, FREDERICK WILLIAM, JR, b Baltimore, Md, Dec 31, 26; m 57; c 3. ELECTRICAL ENGINEERING. *Educ:* Univ Md, BSEE, 51; Yale Univ, MSEE, 52, PhD(elec eng), 56; Northeastern Univ, MBA, 75. *Prof Exp:* From instr to asst prof elec eng, Mass Inst Technol, 55-58; asst mgr systs anal dept, Missile Systs Div, Raytheon Co, 58-61, performance dept mgr, 61-63, mgr systs performanc dept, Space & Info Systs Div, 63-64, mgr systs eng ctr, 64-66; pres, Appl Analysis, Inc, 66-68; consult aerospace systs eng, 68-70; CONSULT SCIENTIST, MISSILE SYSTS DIV, RAYTHEON CO, 70- *Honors & Awards:* Excellence in Technol Award, Raytheon Co. *Mem:* fel Am Inst Aeronaut & Astronaut; sr mem Inst Elec & Electronic Engrs. *Res:* Digital and analog smoothing, filtering and prediction using noisy measurement data; automatic feedback control system design, analysis and synthesis; homing missile guidance, navigation and autopilot control; computer optimization of large scale systems. *Mailing Add:* 53 Baskin Rd Lexington MA 02173

NESNOW, STEPHEN CHARLES, b New York, NY, Dec 19, 41. ONCOLOGY, BIO-ORGANIC CHEMISTRY. *Educ:* Bucknell Univ, BS, 63; NY Univ, MS, 66, PhD(org chem), 68. *Prof Exp:* Fel cancer res, Sloan-Kettering Inst Cancer Res, 68-70; res assoc, McArdle Inst Cancer Res, 70-74; asst scientist human oncol, Univ Wis, 74-76; chief metab effects sect, 77-79, CHIEF, CARCINOGENESIS & METAB BR, US ENVIRON PROTECTION AGENCY, 79- *Concurrent Pos:* Am Cancer Soc grant, 74-75; mem, Cancer Res Ctr, Univ NC, 77-; asst prof path, Univ NC, 77-78, adj assoc prof, 83-; adj mem, W Alton Jones Cell Sci Ctr, 78-79. *Honors & Awards:* Bronze Medal, US Environ Protection Agency, 80. *Mem:* Am Chem Soc; Am Asn Cancer Res; Tissue Culture Asn. *Res:* Chemical carcinogenesis; biochemical oncology; environmental carcinogenesis; oncogenic transformation of cells in culture; carcinogen metabolism. *Mailing Add:* Health Effects Res Lab US Environ Protection Agency Research Triangle Park NC 27711

NESS, GENE CHARLES, b Bemidji, Minn, Dec 17, 44; m 66, 89; c 2. BIOCHEMISTRY, MOLECULAR BIOLOGY. *Educ:* Bemidji State Univ BS, 66; Univ NDak, PhD(biochem), 71. *Prof Exp:* Fel lipid metabolism, Dept Physiol Chem, Vet Admin Hosp, 71-73; sr scientist clin chem, Div Am Hosp Supply, Dade, 74; from asst prof to assoc prof, 74-86, PROF BIOCHEM & MOLECULAR BIOL, UNIV SFLA, 86- *Concurrent Pos:* Prin investr, NIH, 75-; consult, Metabolism Study Sect, NIH, 82; mem, Physiol Chem Study Sect, NIH, 84. *Mem:* Am Soc Biochem & Molecular Biol. *Res:* Regulation of hydroxymethylglutaryl-coenzyme A reductase; effects of hormones on RNA levels, immunoreactive protein and posttranslation modification; identification of DNA sequences required for hormonal regulation; regulation of cholesterol 72 hydroxylase. *Mailing Add:* Dept Biochem & Molecular Biol Col Med Univ SFla 12901 N Bruce B Downs St Tampa FL 33612

NESS, LINDA ANN, b Albert Lea, Minn, Oct 29, 47. MATHEMATICS. *Educ:* St Olaf Col, BA, 69; Harvard Univ, MA & PhD(math), 75. *Prof Exp:* Asst prof math, Univ Wash, 75-80; mem fac math, Harvard Univ, 80-; AT DEPT MATH, CARLETON COL. *Mem:* Sigma Xi. *Res:* Algebraic and differential geometry. *Mailing Add:* Dept Math Carleton Col One N College St Northfield MN 55057

NESS, NATHAN, b New York, NY, Jan 18, 16; m 53; c 1. FLUID MECHANICS. *Educ:* Polytech Inst Brooklyn, BAeroEng, 43, MS, 49, PhD(appl mech), 52. *Prof Exp:* Stress analyst, Lockheed Aircraft Corp, Calif, 44-45, flight test analyst, 46-47; res asst prof fluid mech, Polytech Inst Brooklyn, 52-56; res engr, Gen Elec Co, Pa, 56-59 & Radio Corp Am, NJ, 59-64; from prof to emer prof aerospace eng, WVa Univ, 64-88; RETIRED. *Concurrent Pos:* Consult, Gen Elec Co, 55-56. *Res:* Mass addition effects on skin friction and heat transfer; wake studies on hypersonic reentry missiles; circulation control of bluff-ended bodies; theory of bubble formation in gas-particulate fluidized beds. *Mailing Add:* 417 Pocahontas Ave Morgantown WV 26505

NESS, NORMAN FREDERICK, b Springfield, Mass, Apr 15, 33; m 56; c 2. SPACE PHYSICS. *Educ:* Mass Inst Technol, BS, 55, PhD(geophys), 59. *Prof Exp:* Res geophysicist inst geophys, Univ Calif, Los Angeles, 59-60, asst prof geophys, 60-61; res physicist, Goddard Space Flight Ctr, NASA, 61-66, staff scientist, 66-68, head extraterrestrial physics br, 68-69, chief lab extraterrestrial physics, 69-86; PRES & PROF PHYSICS, BARTOL RES INST, UNIV DEL, NEWARK, 87- *Concurrent Pos:* Nat Acad Sci-Nat Res Coun res assoc, 60-61; consult, 57-; vis assoc prof, Univ Md, 65-68; liaison scientist, ONR-London. *Honors & Awards:* Exceptional Sci Achievement Medal, NASA, 66, 81 & 86; Flemming Medal, US Govt, 68; Space Sci Award, Am Inst Aeronaut & Astronaut, 72; John Adam Fleming Medal, Am Geophys Union, 65. *Mem:* Nat Acad Sci; fel Am Geophys Union; Sigma Xi. *Res:* Experimental investigation of magnetic fields in the magnetosphere and interplanetary space; satellite and space probe studies; measurement of planetary magnetic fields. *Mailing Add:* Bartol Res Inst Univ of Delaware Newark DE 19716

NESS, ROBERT KIRACOFE, b York, Pa, Apr 29, 22; m 44; c 3. CARBOHYDRATE CHEMISTRY. *Educ:* Lebanon Valley Col, BS, 43; Ohio State Univ, MS, 45, PhD(org chem), 48. *Prof Exp:* Asst, Ohio State Univ, 43-45; assoc prof, Lebanon Valley Col, 47-48; fel, NIH, 48-50, sr asst scientist, 50-54, from scientist to sr scientist, 54-63, scientist dir, 63-77; RETIRED. *Concurrent Pos:* USPHS, 50-77. *Mem:* Am Chem Soc; emer mem NY Acad Sci. *Res:* Carbohydrates; reaction mechanisms; sugar benzoates; chemistry of ribose and deoxyribose; synthesis of deoxynucleosides; glycals; vinyl glycosides; amino sugars. *Mailing Add:* 4216 Dresden St Kensington MD 20895

NESSEL, ROBERT J, b Bronx, NY, Dec 28, 36; m 60; c 2. PHARMACY, PHARMACEUTICAL CHEMISTRY. *Educ:* Columbia Univ, BS & MS, 60; Purdue Univ, PhD(indust pharm), 63. *Prof Exp:* Pharmacist, Mt Sinai Hosp, NY, 58-59; asst pharm, Purdue Univ, 60-61; sr res pharmacist, Squibb Inst Med Res, 63-66; head med prod tech serv, Merck & Co, 66-69, head animal formulations res & develop, 69-74, assoc dir regulatory affairs, Int, 74-75, dir regulatory affairs, 75-81, SR DIR REGULATORY AFFAIRS, MERCK, SHARP & DOHME RES LABS, 81- *Mem:* Am Pharmaceut Asn; Acad Pharmaceut Sci; Am Asn Pharm Scientists. *Res:* Industrial pharmacy. *Mailing Add:* Merck Sharp & Dohme Res Labs Rahway NJ 07065

NESSELSON, EUGENE J(OSEPH), b Omaha, Nebr, May 30, 28; m 54; c 2. ENVIRONMENTAL & CHEMICAL ENGINEERING. *Educ:* Univ Omaha, BS, 49; Univ NC, MS, 51; Univ Wis, PhD(sanit eng), 53. *Prof Exp:* Chem engr waste disposal, Standard Oil Co Ind, 54-55; chief process develop, Taft Sanit Eng Ctr, USPHS/EPA, 55-57; sr chem engr, Gen Elec Co, 57-61; prin chem engr, Am-Standard Corp, 61-64; dir air & water mgt, Velsicol Chem Corp, 64-68; CONSULT ENVIRON MGT, 68- *Mem:* Am Chem Soc; Am Water Works Asn; Am Inst Chem Engrs; Water Pollution Control Fedn; Air & Waste Mgt. *Res:* Water treatment; industrial wastes; air and water pollution abatement; industrial hygiene; chemical process engineering; solid/hazardous waste disposal; industrial environmental affairs. *Mailing Add:* 2620 W Fitch Ave Chicago IL 60645-3102

NESSMITH, JOSH T(HOMAS), JR, b Bulloch Co, Ga, June 18, 23; m 48; c 4. ELECTRICAL ENGINEERING. *Educ:* Ga Sch Technol, BEE, 47; Univ Pa, MSEE, 57, PhD(elec eng), 65. *Prof Exp:* Engr, Civil Aeronaut Admin, 47-52; design engr, RCA Corp, 52-53, proj engr, 53-59, syst & proj leader, 59-60, RCA Tradex-Press prog mgr, 60-64, mgr syst eng, Missile & Surface Radar, 65-83; PRIN RES ENGR, RSAL GTRI, GA INST TECHNOL, 83- *Concurrent Pos:* Eng consult, gout & indust. *Mem:* Fel Inst Elec & Electronic Engrs; assoc Am Inst Aeronaut & Astronaut. *Res:* Advanced radars; scatterers; phased arrays; wide band radar operation; weapon system engineering. *Mailing Add:* 4340 Cornwallis Ct Marietta GA 30068

NESTE, SHERMAN LESTER, b Decorah, Iowa, Sept 23, 43; m 72; c 1. METEORITICS. *Educ:* Luther Col, BA, 65; Mich State Univ, MS, 67; Drexel Univ, PhD(physics), 75. *Prof Exp:* RES PHYSICIST METEORITICS, SPACE SCI LAB, GEN ELEC CO, 67- *Mem:* AAAS; Am Geophys Union. *Res:* Establishing an experimental model of the asteroid/meteoroid environment in the region of space between the orbits of Earth and Jupiter. *Mailing Add:* 187 Belmont Rd King of Prussia PA 19406

NESTELL, MERLYND KEITH, b Fletcher, NC, Oct 27, 37; m 82; c 2. MATHEMATICS, GEOLOGY. *Educ:* Andrews Univ, BA, 57; Univ Wis, MA, 59; Ore State Univ, PhD(math), 66, Princeton Univ, MA, 77. *Prof Exp:* Instr math, Southern Col, 59-61; instr, Ore State Univ, 63-65; sr res scientist, Pac Northwest Lab, Battelle Mem Inst, 65-69; ASSOC PROF MATH & GEOL, UNIV TEX, ARLINGTON, 69- *Mem:* Paleont Soc Japan; Paleont Soc; Soc Econ Paleont & Mineral. *Res:* Functional analysis; integral and integro-differential equations, radiative transfer theory; paleontology; Upper Paleozoic biostratigraphy. *Mailing Add:* Dept of Math/Geol Univ of Tex Box 18088 Uta Station Arlington TX 76019

NESTER, EUGENE WILLIAM, b Johnson City, NY, Sept 15, 30; m 59; c 2. MICROBIOLOGY. *Educ:* Cornell Univ, BS, 52; Western Reserve Univ, PhD, 59. *Prof Exp:* Am Cancer Soc res fel genetics, Sch Med, Stanford Univ, 59-62; instr microbiol, 62-63, from asst prof to assoc prof microbiol & genetics, 63-72, PROF MICROBIOL & IMMUNOL, UNIV WASH, 72- *Mem:* Am Soc Microbiol. *Res:* Genetics and biochemistry of enzyme regulation; bacterial-plant relationships. *Mailing Add:* Dept of Microbiol/Immunol Univ of Wash Seattle WA 98195

NESTICO, PASQUALE FRANCESCO, b Catanzaro, Italy, May 6, 45; nat US. MEDICINE. *Educ:* Villanova Univ, BSEE, 72; Temple Med Sch, MD, 80. *Prof Exp:* med intern, Hahnemann Univ Hosp, 80-81, med resident, 81-83; chief med resident, St Agnes Med Ctr, 82-83; cardiol fel, 83-85, chief fel, 84-85, ASST PROF CARDIOL, SUDDEN CARDIAC DEATH, LIKOFF CARDIOVASC INST, HAHNEMANN UNIV HOSP, 85- *Concurrent Pos:* Mem staff, Westinghouse Co, 69-72; affil, John F Kennedy Hosp, 81-83, Nazareth Hosp, 81-83 & Albert Einstein Hosp, 84; dir student prog, St Agnes Med Ctr, 86; res scientist, Grad Hosp Res Prog, 87; bd dirs, Am Heart Asn, 87- *Mem:* AMA; Am Col Physicians; Am Col Cardiol. *Res:* Dose ranging study of intravenous Nicarpidine in mildly hypertensive post-operative patients; cardiac arrhythmia treatment study; comparison of Quinidine with Pronestyl for the treatment of premature ventricular complexes; controlled open-label use of eneainide hydrochloride in patients with ventricular tachycardia and heart disease. *Mailing Add:* 3115 S 19th St Philadelphia PA 19145

NESTOR, C WILLIAM, JR, b Nashville, Tenn, Mar 12, 31; m 52; c 5. COMPUTATIONAL PHYSICS. *Educ:* Vanderbilt, BA, 52, MS, 54, PhD(physics), 69. *Prof Exp:* Physicist, 55-62, COMPUT CONSULT, OAK RIDGE NAT LAB, 62- *Mem:* AAAS; Sigma Xi; Soc Indust Appl Math. *Res:* Shock wave propagation in water-moderated nuclear reactors under severe accident conditions. *Mailing Add:* Bldg 9104-2 M-S 8058 PO Box 2009 Oak Ridge TN 37831

NESTOR, JOHN JOSEPH, JR, b Miami, Fla, Jan 21, 45; m 74; c 2. POLYPEPTIDE CHEMISTRY. *Educ:* Polytech Inst NY, BS, 66; Univ Ariz, PhD(org chem), 71. *Prof Exp:* Assoc fel, Cornell Univ, 71-74; staff researcher, 74-78, sr staff researcher, 78-82, head dept peptide res, 82-85, asst dir, 85-87, VPRES & DIR, INST BIO-ORG CHEM, SYNTEX RES, 87- *Honors & Awards:* Sci Award, Syntex, 85. *Mem:* Am Chem Soc. *Res:* Agonists and antagonist of oxytocin and vasopressin; immunostimulatory glycopeptide adjuvant analogs; agonists and antagonists of luteinizing hormone-releasing hormone; unnatural amino acid synthesis; asymmetric synthesis of amino acids; immunostimulatory peptides; bradykinin antagonists. *Mailing Add:* Inst Bio-Org Chem Syntex Res 3401 Hillview Ave Palo Alto CA 94304

NESTOR, KARL ELWOOD, b Kasson, WVa, Dec 17, 37; m 58; c 2. GENETICS. *Educ:* Univ WVa, BS, 59, MS, 61; Ohio State Univ, PhD(genetics), 64. *Prof Exp:* Asst genetics, Univ WVa, 59-61; asst, Ohio Agr Res & Develop Ctr, 61-64, asst instr, 65, from asst to assoc prof, 65-78, PROF GENETICS, OHIO AGR RES & DEVELOP CTR, 78- *Honors & Awards:* Nat Turkey Found Res Award, Poultry Sci Asn, 76. *Mem:* Poultry Sci Asn; Worlds Poultry Sci Asn. *Res:* Physiological genetics of chickens and turkeys; poultry physiology. *Mailing Add:* Dept Poultry Sci Ohio Agr Res & Develop Ctr Wooster OH 44691

NESTOR, ONTARIO HORIA, b Youngstown, Ohio, Sept 20, 22; m 43; c 2. CRYSTAL GROWTH, CRYSTAL CHARACTERIZATION. *Educ:* Marietta Col, AB, 43; Univ Minn, MS, 49; Univ Buffalo, PhD(physics), 60. *Prof Exp:* Instr physics, Marietta Col, 42-43; res asst, SAM Labs, Columbia Univ, 43-44; res asst, Univ Minn, 47-48; res assoc, Linde Div, Union Carbide Corp, 49-68, sr res assoc, 68-71; dir technol, Crystal Optics Res, Inc, 71-74; mgr crystal develop, Harshaw-Filtrol Partnership, 74-87; CONSULT, 88- *Mem:* Am Phys Soc. *Res:* Growth of crystals for nuclear radiation detection, laser and optical applications; alkali halides, alkaline earth fluorides and oxides. *Mailing Add:* 402 E Fourth St Lakeside OH 43440

NESTRICK, TERRY JOHN, US citizen; m; c 1. CHROMATOGRAPHIC SEPARATIONS. *Educ:* Oakland Univ, BS, 68, MS, 72; Univ Umeå Sweden, PhD, 88. *Prof Exp:* ASSOC SCIENTIST, MICH DIV ANALYTICAL SCI, DOW CHEM CO, 72- *Mem:* Am Chem Soc. *Res:* Environmental and commercial product trace level organic analyses, specializing in chlorinated dibenzo-p-dioxins and related compounds. *Mailing Add:* 4520 Washington St Midland MI 48640

NESTVOLD, ELWOOD OLAF, b Minot, NDak, Mar 19, 32; wid; c 2. INFORMATION SCIENCE, GEOPHYSICS. *Educ:* Augsburg Col, BA, 52; Univ Minn, MS, 59, PhD(physics), 62. *Prof Exp:* Asst physics, Univ Minn, 56-59, instr, 59-61; physicist, Shell Develop Co, 62-63, res physicist, 63-65, res assoc & sect head, 65-68, mgr geophys dept, 68-71, sr staff geophysicist, 71-72, mgr geophysics, Western Region, 72-74, Int Region, 74-75, sr staff geophysicist, Pecten Cameroon, 76; chief geophysicist & dep mgr explor, Woodside Petrol Develop Pty, Ltd, Perth, Australia, 77-78; mgr, Explor & Prod Processing Ctr, Koninklijke/Shell Explor & Prod Lab, Rijswijk, Neth, 79-81, chief geophysicist, 81-86, DIR, GEOPHYS & TOPOG, ROYAL DUTCH/SHELL GROUP, THE HAGUE, NETH, 86- *Concurrent Pos:* Consult, Lighting & Transients Res Inst, Minn, 57-62. *Mem:* Inst Elec & Electronic Engrs; Soc Explor Geophys; Europ Asn Explor Geophysicists; NY Acad Sci; Sigma Xi; Europ Asn Petrol Geoscientists. *Res:* Prediction and filter theory; digital-computer techniques for acoustic and seismic signal processing; application of probability theory to hydrocarbon detection; performance evaluation of computer systems for geophysical processing; 3D seismic technology. *Mailing Add:* SIPM PO Box 162 The Hague 2501 An Netherlands

NETA, PEDATSUR, b Tripoli, Libya, Jan 1, 38; m 59; c 2. PHYSICAL CHEMISTRY, ORGANIC CHEMISTRY. *Educ:* Hebrew Univ, Jerusalem, MSc, 60; Weizmann Inst Sci, PhD(phys chem), 65. *Prof Exp:* Res assoc radiation chem, Soreq Nuclear Res Ctr, Israel, 60-66; Nat Res Coun Can fel, Univ Toronto, 66-67; AEC fel, Ohio State Univ, 67-68; Nat Acad Sci-Nat Res Coun fel, US Army Natick Labs, 68-69; res fel, Radiation Res Labs, Mellon Inst Sci, Carnegie-Mellon Univ, 69-74, sr res chemist, 74-76; assoc prof specialist, Radiation Lab, Univ Notre Dame, 76-80, assoc fac fel, 80-83; RES CHEMIST, NAT INST STANDARDS & TECHNOL, GAITHERSBURG, MD, 83- *Concurrent Pos:* Assoc ed, Radiation Res, 75-79; regional ed, Radiation Physics & Chem, 87- *Honors & Awards:* Silver Medal, Dept Com, 87. *Mem:* Am Chem Soc; Radiation Res Soc. *Res:* Physical organic chemistry; free radical reactions; radiation chemistry; electron transfer processes; metalloporphyrins. *Mailing Add:* Chem Kinetics Div Nat Inst Standards & Technol Gaithersburg MD 20899

NETA, RUTH, b Lodz, Poland; US citizen; m 59; c 2. IMMUNOLOGY. *Educ:* Tel-Aviv Univ, MS, 67; Univ Notre Dame, PhD(immunol), 79. *Prof Exp:* Res assoc immunol, Ohio State Univ, 67-68; res assoc, Sch Med, Univ Pittsburgh, 69-77; res assoc immunol, Univ Notre Dame, 79-82; SR INVESTR, ARMED FORCES RADIOBIOLOGY RES INST, 82- *Concurrent Pos:* Nat res serv award, NIH, 79-82; assoc prof, Dept Microbiol, George Washington Univ Med Sch. *Mem:* Am Asn Immunologists; Am Acad Microbiol. *Res:* Cell-mediated immunity; delayed hypersensitivity; immune regulation; cytokines; radioprotection. *Mailing Add:* Dept Exp Hemat Armed Forces Radiobiol Res Inst Bethesda MD 20814

NETER, JOHN, b Ger, Feb 8, 23; US citizen; m 51; c 2. APPLIED STATISTICS. *Educ:* Univ Buffalo, BA, 43; Univ Pa, MBA, 47; Columbia Univ, PhD(bus statist), 52. *Prof Exp:* Asst prof bus statist, Syracuse Univ, 49-55; prof quant anal, Univ Minn, Minneapolis, 55-75; prof mat sci & statist, 75-85, C Herman & Mary Virginia Terry prof mgt serv & statist, 86-89, C HERMAN & MARY VIRGINIA TERRY EMER PROF MGT SCI & STATIST, COL BUS ADMIN, UNIV GA, 90- *Concurrent Pos:* Ford Found fac res fel, Univ Minn, 57-58; supvry math statistician, US Bur Census, 59-60, consult, 61-65; ed, The Am Statistician, 76-80; chmn, panel qual control fed assistance progs, NAS, 86-88; assoc ed, Decision Sci, 73-74; chair, Sect Statist, AAAS, 91. *Honors & Awards:* Distinguished Serv Award, Decision Sci Inst, 81; Creative Res Award, Univ Ga, 84. *Mem:* Fel AAAS; fel Am Statist Asn (pres, 85); Inst Mgt Sci; Inst Math Statist; fel Decision Sci Inst (pres, 78-79); Int Statist Inst. *Res:* Uses of statistics in accounting; measurement errors; linear models. *Mailing Add:* Col Bus Admin Univ Ga Athens GA 30602

NETHAWAY, DAVID ROBERT, b San Diego, Calif, Aug 6, 29; m 64; c 3. NUCLEAR CHEMISTRY. *Educ:* Univ Calif, BS, 51, MS, 57; Wash Univ, PhD(chem), 59. *Prof Exp:* Chemist, Gen Elec Co, Wash, 51-53; chemist, Calif Res & Develop Co, 53; CHEMIST, LAWRENCE LIVERMORE NAT LAB, UNIV CALIF, 53- *Mem:* Am Chem Soc; Am Phys Soc. *Res:* Nuclear charge distribution in fission; low-energy nuclear reactions; decay scheme studies. *Mailing Add:* Lawrence Livermore Nat Lab L234 Livermore CA 94550

NETHERCOT, ARTHUR HOBART, JR, b Evanston, Ill, June 16, 23; m 44; c 4. PHYSICS. *Educ:* Northwestern Univ, BA, 44, MS, 46; Univ Mich, PhD, 50. *Prof Exp:* Union Carbide & Carbon Corp fel, Columbia Univ, 51-52, res assoc physics, 52-57; PHYSICIST RES CTR, IBM CORP, 57- *Mem:* Fel Am Phys Soc. *Res:* Microwave physics; solid state physics; materials science. *Mailing Add:* 107 Mt Hope Blvd Hastings NY 10706

NETHERCUT, PHILIP EDWIN, b Indianapolis, Ind, Apr 3, 21; m 49; c 3. ORGANIC CHEMISTRY. *Educ:* Beloit Col, BS, 42; Lawrence Col, MS, 44, PhD(org chem), 49. *Prof Exp:* Res chemist, Watervliet Paper Co, Mich, 49-50; process control engr, Scott Paper Co, Pa, 51, res group leader, 52-54, res mgr, 55-56; tech secy, Tech Asn Pulp & Paper Indust, 57-58, exec secy, 59-74, treas, 63-74, exec dir, 75-82, bd vchmn, 82-86; RETIRED. *Concurrent Pos:* Pres, Coun Eng & Sci Soc Execs, 69. *Mem:* Fel Tech Asn Pulp & Paper Indust; Am Soc Asn Execs. *Res:* Pulp and paper technology; industrial research administration. *Mailing Add:* 9240 Huntcliff Trace Atlanta GA 30350-1603

NETHERTON, LOWELL EDWIN, b Fairfield, Ill, Feb 9, 22; m 46; c 4. INORGANIC CHEMISTRY, PHYSICAL CHEMISTRY. *Educ:* Western Ill Univ, BS, 44; Univ Wis, PhD(chem), 50. *Prof Exp:* Anal chemist, Sinclair Refining Co, 44; asst, Univ Wis, 48-50; res chemist, Victor Chem Co, 50-57, dir res, Victor Chem Div, Stauffer Chem Co, 57-63, mgr prod, 63-65; dir res, Mich, 65-78, DIR SCI & TECH RELS, BASF WYANDOTTE CORP, 78- *Mem:* Am Chem Soc; Soc Plastics Indust; Indust Res Inst; Am Inst Chem Engrs; Am Inst Chemists. *Res:* Phosphorus compounds; electrochemistry of rare elements; chemistry of rhenium, tungsten and tantalum; chlor-alkali electrochemistry; urethanes, polyethers, isocyanates, alkyleneoxides and auxiliaries; foamed plastics; textile and paper specialties; surfactants and detergents. *Mailing Add:* 1574 Country Squire Ct Decatur GA 30033-1812

NETI, RADHAKRISHNA MURTY, b Nandigama, India, June 20, 33; m 60; c 2. PHYSICAL CHEMISTRY. *Educ:* Hindu Col, Masulipatam, India, BS, 53; Banaras Hindu Univ, MS, 55, dipl mod Europ lang, 56, PhD(chem), 60. *Prof Exp:* Lectr chem, Banaras Hindu Univ, 60; res assoc radiation chem, Univ Notre Dame, 60-62; res assoc bot & biophys, Univ Ill, 62-66; RES SCIENTIST, BECKMAN INDUST, INC, 66- *Honors & Awards:* John C Vaaler Award, 70. *Mem:* AAAS; Am Chem Soc. *Res:* Photo, radiation and electro-analytical chemistry; air and water quality instrumentation. *Mailing Add:* Dept 371 Beckman Instruments Inc 200 S Kramer Blvd Brea CA 92621

NETRAVALI, ARUN NARAYAN, b Bombay, India, May 26, 46; m 74. ELECTRICAL ENGINEERING. *Educ:* Indian Inst Technol, Bombay, BS, 67; Rice Univ, MS, 69, PhD(elec eng), 70. *Prof Exp:* Asst elec eng, Rice Univ, 67-70; asst dir elec eng, Optimal Data Corp, 70-72; mem tech staff picture processing, 72-78, HEAD, VISUAL COMMUN RES DEPT, AT&T BELL LABS, 78- *Concurrent Pos:* Adj prof, Dept Elec Eng, Rutgers Univ, 79-; ed, Signal Processsing & Commun Electronics, Inst Elec & Electronics Engrs Trans Communs. *Mem:* Sr mem Inst Elec & Electronics Engrs. *Res:* Processing of television and graphical images for the purpose of efficient transmission, enhancement and restoration; digital signal processing. *Mailing Add:* AT&T Bell Labs 600 Mountain Ave Rm 3d-406 Murray Hill NJ 07974

NETSKY, MARTIN GEORGE, b Philadelphia, Pa, May 15, 17; m 46. NEUROLOGY, NEUROPATHOLOGY. *Educ:* Univ Pa, AB, 38, MS, 40, MD, 43. *Prof Exp:* From intern to resident neurol, Hosp Univ Pa, 43-44; Weil fel neuropath, Montefiore Hosp, New York, 46-47; from asst neuropathologist to assoc neuropathologist, Montefiore Hosp, 47-54; from assoc prof to prof neurol, Bowman Gray Sch Med, 55-61, prof neuropath, 55-61, dir sect neurol, 57-59, chmn dept, 59-61; vis prof path, Univ Med Sci, Thailand, 61; prof neuropath, Sch Med, Univ Va, 62-75; PROF PATH, SCH MED, VANDERBILT UNIV, 74- *Concurrent Pos:* Lectr, US Naval Hosp, St Albans, NY, 48, consult pathologist, 50-54; from adj attend physician to assoc attend physician, Montefiore Hosp, 49-54; assoc, Col Physicians & Surgeons, Columbia Univ, 52-54; mem sci adv bd, Armed Forces Inst Path. *Mem:* Am Neurol Asn; Am Asn Neuropath (pres, 63); Asn Res Nerv & Ment Dis; Am Acad Neurol; Am Soc Exp Path. *Res:* Permeability of living membranes; autonomic nervous system; clinicopathologic aspects of human brain tumors; congenital and degenerative neurologic disorders; medical education; learning. *Mailing Add:* Dept of Path Vanderbilt Univ Sch of Med Nashville TN 37232

NETT, T M, REPRODUCTIVE ENDOCRINOLOGY, RADIOIMMUNOASSAY RECEPTOR ASSAY. *Educ:* Mont State Univ, BS, 68; Wash State Univ, PhD(reprod physiol), 72. *Prof Exp:* Fel, 72-74, from asst prof to assoc prof, 74-84, PROF REPROD PHYSIOL, COLO STATE UNIV, 84- *Concurrent Pos:* Consult, Int Atomic Energy Agency, 76-82, Sandoz Pharmaceut, Inc, 86-; ad hoc reviewer, Contraceptive Develop Br, NIH, 77- *Mem:* Soc Study Reprod; Am Soc Animal Sci; Endocrine Soc. *Res:* Regulation of the synthesis and secretion of gonadotropins in domestic and laboratory animals, particularly as it relates to reproductive quiescence and abnormalities. *Mailing Add:* Dept Physiol Colo State Univ Ft Collins CO 80523

NETTEL, STEPHEN J E, b Prague, Czech, Aug 12, 32; US citizen. SOLID STATE PHYSICS. *Educ:* McGill Univ, BEng, 54; Mass Inst Technol, PhD(physics), 60. *Prof Exp:* Jr res physicst, Univ Calif, San Diego, 60-61; staff mem, Int Bus Mach Corp, Switz, 61-65; res fel appl physics, Harvard Univ, 65-66; asst prof physics, 66-69, ASSOC PROF PHYSICS, RENSSELAER POLYTECH INST, 69- *Res:* Theoretical solid state physics. *Mailing Add:* Dept of Physics Rensselaer Polytech Inst Troy NY 12181

NETTESHEIM, PAUL, b Cologne, Ger, Sept 11, 33; US citizen; m; c 3. PULMONARY PATHOBIOLOGY. *Educ:* Med Sch, Munich, WGer, MD & DMS, 59. *Prof Exp:* Res biologist, Biol Div, Oak Ridge Nat Lab, 63-69, group leader respiratory carcinogenesis, Biol Div, 69-77; CHIEF, LAB PULMONARY PATHOBIOL, NAT INST ENVIRON HEALTH SCI, 77- *Concurrent Pos:* Adj prof, Dept Path, Med Sch, Univ NC, 77-, Duke Univ, 78- *Mem:* Am Asn Cancer Res; Am Thoracic Soc; Am Asn Pathologists; AAAS; Int Asn Study Lung Cancer; Soc Toxicol Pathologists; Am Soc Cell Biol; Europ Asn Cancer Res. *Res:* Cancer; biology. *Mailing Add:* Lab Pulmonary Path Nat Inst Environ Health Sci Bldg 101 Rm D240 PO Box 12233 Research Triangle Park NC 27709

NETTLES, JOHN BARNWELL, b Dover, NC, May 19, 22; m 56; c 3. OBSTETRICS & GYNECOLOGY. *Educ:* Univ SC, BS, 41; Med Col SC, MD, 44. *Prof Exp:* Intern gen med, Garfield Mem Hosp, DC, 44-45; res fel path, Med Col Ga, 46-47; res staff obstet & gynec, Univ Ill Hosps, 47-51, from instr to asst prof, Univ, 51-57; asst prof, Sch Med, Univ Ark, 57-60, assoc prof & med educ nat defense coordr, 60-67, prof obstet & gynec, 67-69; prof obstet & gynec, Univ Okla, Oklahoma City, 69-75; chmn dept, 75-81, PROF OBSTET & GYNEC, MED COL, UNIV OKLA-TULSA, 75-, DIR STUDENT EDUC, 81- *Concurrent Pos:* Examr, Am Bd Obstet & Gynec, 64-; mem, surg forum comt, Am Col Surgeons, 67-84; dir, Tulsa Residency Training Prog Obstet & Gynec & Tulsa Obstet & Gynec Educ Found, 69-81. *Mem:* AAAS; AMA; Asn Mil Surgeons US; Am Col Obstet & Gynec; Int Soc Advan Humanistic Studies Gynec; fel Royal Soc Med; fel Royal Soc Health. *Res:* Renal function and structure; kidney biopsy; newborn and fetal morbidity and mortality; genital malignancy; obstetric anesthesia and analgesia; physiology and toxemia of pregnancy. *Mailing Add:* Univ Okla-Col Med-Tulsa 2808 S Sheridan Tulsa OK 74129

NETTLES, WILLIAM CARL, JR, b Anderson, SC, Dec 21, 34; m 67. ENTOMOLOGY. *Educ:* Clemson Univ, BS, 55, MS, 59; Rutgers Univ, PhD(entom), 62. *Prof Exp:* Res asst agr chem, Clemson Univ, 57-59; res entomologist, Agr Res Serv, 62-88, RES ENTOMOLOGIST, SUBTROP AGR RES LAB, BIO CONTROL PEST RES UNIT, USDA, 88- *Concurrent Pos:* Secy, Sect B (Physiol, Biochem & Toxicol), Entom Soc Am, 74, vchmn, 75, chmn, 76, mem exec comt, Southeastern Br, 80. *Mem:* AAAS; Entom Soc Am; Int Soc Chem Ecol. *Res:* Physiology and biochemistry of insects; biological control. *Mailing Add:* Subtrop Agr Res Lab Biocontrol Pest Res Lab USDA 2413 E Hwy 83 Weslaco TX 78596

NETTLETON, DONALD EDWARD, JR, b New Haven, Conn, Mar 16, 30; m 57; c 4. ORGANIC CHEMISTRY, BIOCHEMISTRY. *Educ:* Yale Univ, BS, 52; Rice Inst, PhD, 56. *Prof Exp:* Res assoc biochem, Med Col, Cornell Univ, 56-58; sr chemist, dept biochem, Bristol Labs, Inc div, Bristol-Myers Co, 58-86; ADJ PROF, DEPT FOREST CHEM, SCH ENVIRON SCI & FORESTRY, STATE UNIV NY, SYRACUSE, 85- *Mem:* Am Chem Soc; Am Soc Pharmacog. *Res:* Isolation of antitumor and antibiotic agents from fermentation liquors and higher plants; syntheses of derivatives and analogs of physiologically active natural products. *Mailing Add:* RFD 1 Box 524 Jordan NY 13080

NETTLETON, G(ARY) STEPHEN, b Albert Lea, Minn, May 7, 46; m 68; c 2. ANATOMY, HISTOCHEMISTRY. *Educ:* McPherson Col, BS, 68; Univ Minn, PhD(anat), 76. *Prof Exp:* From instr to assoc prof, 76-90, PROF ANAT SCI & NEUROBIOL, SCH MED, UNIV LOUISVILLE, 90-, ASSOC DEAN, STUDENT AFFAIRS. *Concurrent Pos:* Ed asst, Stain Technol, 77-80, asst ed, 80-87, ed, 88-90; ed, Biotech & Histochem, 91- *Mem:* Histochem Soc; Biol Stain Comn; Sigma Xi; Am Chem Soc; Am Asn Anat; Soc Biol Eds; Soc Clin Anat. *Res:* Chemical reaction mechanisms of biological stains; histochemical approaches to the study of tissues. *Mailing Add:* Dept Anat Sci & Neurobiol Health Sci Ctr Univ Louisville Louisville KY 40292

NETTLETON, WILEY DENNIS, b Noble, Ill, June 8, 32; m 56; c 4. SOIL SCIENCE, GEOLOGY. *Educ:* Univ Ill, BS, 57, MS, 58; NC State Univ, PhD(soil classification & genesis), 66. *Prof Exp:* Trainee, Western Soil Surv Invest Unit, Soil Conserv Serv, 56, soil scientist, 57 & 58-65, res soil scientist, Soil Surv Lab, 65-72, supvry res soil scientist, 72-76, RES SOIL SCIENTIST, NAT SOIL SURV LAB, USDA, 77- *Concurrent Pos:* Res asst, NC State Univ, 60-65; lectr, Univ Calif, Riverside, 70. *Mem:* Soil Sci Soc Am; Am Soc Agron; Sigma Xi. *Res:* Use of mineralogy and micromorphology in the laboratory and geomorphology in the field as tools for studying genesis and classification of soils from the Midwest, Southeast and Western United States. *Mailing Add:* Natl Soil Survey Lab Fed Bldg Rm 345 Lincoln NE 68508

NETZEL, DANIEL ANTHONY, b Chicago, Ill, Feb 16, 34; m 58; c 3. PHYSICAL CHEMISTRY, NUCLEAR MAGNETIC RESONANCE. *Educ:* Univ Ill, BS, 57; Univ Mo-Kansas City, MS, 61; Northwestern Univ, PhD(phys & analytical chem), 75. *Prof Exp:* Assoc res chemist, Midwest Res Inst, 58-67; res chemist, DeSoto Inc, 67-69; supvr spectros, Northwestern Univ, 69-74; res assoc chem & supvr sect, Laramie Energy Tech Ctr, 75-82; SR RES SCIENTIST, WESTERN RES INST, 82- *Concurrent Pos:* Adj prof chem, Univ Wyo, 78-88. *Mem:* Am Chem Soc; Soc Appl Spectros; Int Soc Magnetic Resonance. *Res:* Identification, structural characterization and analytical applications of nuclear magnetic resonance to fossil fuels. *Mailing Add:* Western Res Inst PO Box 3395 Univ Sta Laramie WY 82071

NETZEL, RICHARD G, b Shawano, Wis, May 13, 28; m 51; c 3. PHYSICS, ACADEMIC ADMINISTRATION. *Educ:* Univ Wis, BS, 50, MS, 56, PhD(physics), 60. *Prof Exp:* From asst prof to prof physics, Univ Wis-Oshkosh, 60-68, asst vpres prog develop & staffing, 68-70, asst acad vpres, 70-71; actg dep dir educ, AAAS, 71-72; vpres acad affairs, Metrop State Col, 72-78, actg pres, 78-79; assoc vpres, Consortium State Col, Colo, 79-83; PROF PHYSICS, METROP STATE COL, COLO, 83- *Mem:* AAAS; Am Asn Higher Educ. *Res:* Low temperature physics; superconducting transition temperatures of titanium; science education. *Mailing Add:* Dept Biol Metrop State Col 1006 11th St Denver CO 80204

NETZEL, THOMAS LEONARD, b Wausau, Wis, Dec 5, 46; m 68; c 3. CHEMICAL PHYSICS, BIOPHYSICS. *Educ:* Univ Wis-Madison, BS, 68; Yale Univ, MPhil, 70, PhD(chem), 72. *Prof Exp:* Mem tech staff chem physics & appl econ, Bell Labs, 72-77; assoc chemist, 77-80, chemist, Brookhaven Nat Lab, 80-85; STAFF RES CHEMIST, AMOCO CORP, 85- *Mem:* Inter Am Photochem Soc; Am Chem Soc. *Res:* Development and application of picosecond spectroscopic techniques to probe the dynamics of inorganic, organometallic and biological systems; observation and modeling of electron and energy transfer processes; construction of molecular, photochemical, transducers for DNA-probes and catalysis initiators; photochemistry; inorganic chemistry; catalysis; biotechnology. *Mailing Add:* Dept Chem Ga State Univ University Plaza Atlanta GA 30303-3044

NETZER, DAVID WILLIS, b Washington, DC, Feb 25, 39; m 61; c 1. AERONAUTICAL & MECHANICAL ENGINEERING. *Educ:* Va Polytech Inst, BSME, 60; Purdue Univ, MSME, 62, PhD, 68. *Prof Exp:* Engr, Aerojet Gen Corp, 62-64; assoc prof, 68-80, PROF AERONAUT, NAVAL POSTGRAD SCH, 80- *Mem:* Combustion Inst; Am Inst Aeronaut & Astronaut. *Res:* Solid propellant combustion-acceleration sensitivity; particulate behavior; hybrid rocket and solid fuel ramjet internal ballistics; numerical analysis of reacting flows; pollution from turbojet test cells. *Mailing Add:* 9367 Bur Oak Pl Salinas CA 93901

NEU, ERNEST LUDWIG, b Frankfurt, Ger, Aug 19, 15; nat US; m 46; c 2. CHEMISTRY. *Educ:* Univ Nancy, BS, 37; Univ Caen, MS, 38. *Prof Exp:* Res chemist, Celotex Corp, 38-42; chief chemist dicalite & perlite, Great Lakes Carbon Corp, 52-59, asst tech dir, Mining & Mineral Prod Div, Grefco Inc, 59-62, tech dir, 62-66, managing dir, Dicalite Europe Nord, 59-73, gen mgr, Int Div, 66-73, vpres, 73-81, managing dir, Permalite Europe, SA, 73-81; RETIRED. *Mem:* Am Chem Soc; Am Inst Chem Eng; NY Acad Sci. *Res:* Filteraids; diatomite; perlite; acoustical and thermal insulation; fungi diseases. *Mailing Add:* 205 Camino de Las Colinas Redondo Beach CA 90277-2598

NEU, HAROLD CONRAD, b Omaha, Nebr, Aug 19, 34; m 62; c 3. MICROBIOLOGY, PHARMACOLOGY. *Educ:* Creighton Univ, AB, 56; Johns Hopkins Univ, MD, 60. *Prof Exp:* From intern to resident med, Columbia-Presby Med Ctr, 60-62; res assoc biochem, Nat Inst Arthritis & Metab Dis, 62-64; chief resident med, Columbia-Presby Med Ctr, 64-65; assoc, 65-66, from asst prof to assoc prof med, 66-75, PROF MED & PHARMACOL, COL PHYSICIANS & SURGEONS, COLUMBIA UNIV, 74- *Concurrent Pos:* Career scientist, New York Health Res Coun, 65-71; assoc attend, Columbia-Presby Med Ctr, 70-75, hosp epidemiologist, 71-, attend physician, 75-; distinguished investr, Am Col Clin Pharm, 90. *Honors & Awards:* Borden Award, Borden Found, 60; Hoechst-Russell Award, Am Soc Microbiol, 83; Squibb Lectr, Rutgers Univ, 84. *Mem:* Am Soc Microbiol; Am Col Clin Pharm; Am Soc Clin Invest; fel Am Col Physicians; Infectious Dis Soc Am. *Res:* Resistance of bacteria to antibiotics; bacterial surface enzymes; new antimicrobial agents; pharmacology of antibiotics; infectious diseases; medicinal chemistry. *Mailing Add:* Col Physicians & Surgeons Columbia Univ 630 W 168th St New York NY 10032

NEU, JOHN TERNAY, b Commerce, Tex, Apr 23, 20; div; c 2. AEROSPACE RESEARCH. *Educ:* Agr & Mech Col, Tex, BS, 42; Univ Calif, Berkeley, PhD(phys chem), 49. *Prof Exp:* Instr chem, Univ Calif, Berkeley, 49; res chemist, Calif Res Corp, 49-56; sr staff scientist & chief optics technol, Gen Dynamics & Convair Aerospace, 56-75; Prin physicist, IRT Corp, 75-81; PRES, SURFACE OPTICS CORP, 81- *Res:* Space physics; optical properties of surfaces; space optical instrumentation; vehicle signatures; vehicle survivability. *Mailing Add:* Surface Optics Corp 9929 Hibert St Sweet C San Diego CA 92131

NEUBAUER, BENEDICT FRANCIS, b Bird Island, Minn, Mar 14, 38; m 61; c 3. PLANT ANATOMY. *Educ:* St John's Univ, Minn, BA, 60; Iowa State Univ, PhD(plant anat), 65. *Prof Exp:* Asst prof bot & plant anat, 65-70, ASSOC PROF BOT, UNIV MAINE, ORONO, 70- *Mem:* Sigma Xi; Bot Soc Am. *Res:* Anatomy of vascular plants. *Mailing Add:* Dept Bot Unv Maine Deering Hall Orono ME 04469

NEUBAUER, L(OREN) W(ENZEL), b St James, Minn, June 23, 04; m 46; c 1. AGRICULTURAL & CIVIL ENGINEERING. *Educ:* Univ Minn, BS, 26, MS, 32, PhD(civil eng), 48. *Prof Exp:* Instr math, Col Eng, Univ Minn, 26-28, instr agr eng, 29-39; from asst prof to prof, 40-71, EMER PROF AGR ENG, UNIV CALIF, DAVIS, 71- *Honors & Awards:* L J Markwardt Wood Eng Res Award, Forest Products Res Soc, 74; MBMA Award Agr Eng, Am Soc Agr Engrs, 84. *Mem:* Am Soc Agr Engrs; Am Soc Testing & Mat; Am Geophys Union; Forest Prod Res Soc. *Res:* Farm structures; adobe construction; wood durability; strength of wood, wood beams and wood columns, and concrete; solar heating and cooling. *Mailing Add:* 3422 Monte Vista Ave Davis CA 95616

NEUBAUER, RUSSELL HOWARD, b Detroit, Mich, Feb 15, 44; m 69; c 4. LYMPHOCYTE FUNCTIONS, MONOCLONAL ANTIBODIES. *Educ:* Mich State Univ, BS, 66, MS, 68; Univ Md, PhD(microbiol), 72. *Prof Exp:* Virologist, Litton Bionetics, Inc, 72-76; immunologist, Frederick Cancer Res Facil, Nat Cancer Inst, 76-83; RES SUPVR, E I DU PONT DE NEMOURS & CO, 83- *Concurrent Pos:* Mem, Herpesvirus Adv Team-1, Simian Herpeviruses, World Health Orgn, Int Agency Res Cancer, 77-83. *Mem:* Am Asn Cancer Res; Am Asn Immunologists; Am Soc Microbiol; Int Asn Comp Res Leukemia & Related Dis; Tissue Cult Asn. *Res:* Use of monoclonal antibodies to identify norman and malignant cell populations; the effect of growth factors on normal and transformed cells; the effect of lymphoid disease on normal immunological functions. *Mailing Add:* 1184 Pynchon Rd Westchester PA 19036

NEUBAUER, WERNER GEORGE, b White Plains, NY, Apr 18, 30; m 54; c 2. PHYSICS. *Educ:* Roanoke Col, BS, 52; Cath Univ Am, PhD(acoust), 68. *Prof Exp:* Physicist, Electronics Br, Sound Div, Naval Res Lab, 53-57, physicist, Propagation Br, 57-58, sect head microacoust sect, Acoust Res Br, 58-69, actg br head, Phys Acoust Br, Acoust Div, 69-70, sect head micro acoust sect, Phys Acoust Br, 70-79, spec asst target characteristics, Acoust Div, 79-82. *Concurrent Pos:* Consult, 82- *Honors & Awards:* Applied Res Award, Sigma Xi, 81. *Mem:* Fel Acoust Soc Am; Sigma Xi. *Res:* Radiation, reflection and diffraction of waves; properties of elastic media; optical visualization of acoustic waves; underwater acoustic radiation, reflection and scattering. *Mailing Add:* 4603 Quarter Charge Dr Annandale VA 22003

NEUBECK, CLIFFORD EDWARD, b Erie, Pa, Nov 6, 17; m 43; c 8. CHEMISTRY. *Educ:* Univ Pittsburgh, BS, 39, PhD(biochem), 43. *Prof Exp:* Asst chem, Univ Pittsburgh, 39-41, asst biochem, 41-43; biochemist, Rohm and Haas Co, 43-50, sr chemist, 50-59, sr scientist, 59-71, sr biochemist, 71-76, head, Enzyme Lab, 76-82; RETIRED. *Concurrent Pos:* Consult, 82-90. *Mem:* Am Chem Soc; Soc Indust Microbiol; Am Soc Enologists. *Res:* Commercial production of enzymes; utilization of enzymes; microbial fermentations. *Mailing Add:* 45 Horsham Rd Hatboro PA 19040

NEUBECKER, ROBERT DUANE, b Lackawanna, NY, Oct 10, 25; m 50; c 5. PATHOLOGY. *Educ:* Univ Rochester, AB, 46, MD, 49; Am Bd Path, dipl, 56. *Prof Exp:* Intern med, Vanderbilt Univ Hosp, 49-50; intern path, Strong Mem Hosp, Rochester, NY, 50; asst, Sch Med & Dent, Univ Rochester, 52-54, instr, 54-55; pathologist, Armed Forces Inst Path, 55-57, chief obstet, gynec & breast path br, 57-61; pathologist & dir res & develop, St Joseph's Hosp, Marshfield, Wis, 61-66; assoc dir labs, Mercy Hosp, 66-86; RETIRED. *Mem:* Am Asn Path & Bact; Int Acad Path. *Res:* Pathologic anatomy with clinical pathologic correlations in the field of female genital diseases. *Mailing Add:* Dept of Lab Med Mercy Hosp Oshkosh WI 54901

NEUBERGER, DAN, b Zagreb, Yugoslavia, Feb 19, 29; nat US; m 56; c 2. PHOTOGRAPHIC CHEMISTRY. *Educ:* Columbia Univ, BA, 50; Univ Rochester, PhD(phys chem), 53. *Prof Exp:* Asst, Univ Rochester, 50-52; sr res chemist, Eastman Kodak Co, 53-63, res assoc, 63-86; RETIRED. *Res:* Photographic science. *Mailing Add:* 95 Wendover Rd Rochester NY 14610

NEUBERGER, HANS HERMANN, b Mannheim, Ger, Feb 17, 10; nat US; m 39, 57; c 2. METEOROLOGY. *Educ:* Univ Hamburg, DSc(meteorol), 36. *Prof Exp:* Res asst to Dr Christian Jensen, Univ Hamburg, 36-37; instr geophys, 37-41, asst prof, 41-43, from assoc prof to prof, 43-70, chief div, 45-54, head dept, 54-61 & 65-67, EMER PROF METEOROL, PA STATE UNIV, 70- *Concurrent Pos:* Tech ed, Weatherwise, 48-53; consult, US Weather Bur, Turkey, 54-55; assoc, Army-Navy Vision Comt & mem subcomt visibility & atmospheric optics, Nat Res Coun; Am Meteorol Soc rep div earth sci, Nat Acad Sci-Nat Res Coun, 57-60; UN tech expert, Pakistan, 61; vis prof, Univ SFla, 71-74, emer prof geog, 74- *Mem:* Fel Am Meteorol Soc; Am Geophys Union; Sigma Xi. *Res:* Design of mine safety device and meteorological equipment; atmospheric optics and pollution. *Mailing Add:* 1805 Burlington Circle Sun City Center FL 33573-5219

NEUBERGER, JACOB, b Ger, June 8, 27; nat US; c 7. SOLID STATE PHYSICS. *Educ:* Johns Hopkins Univ, BS, 50; NY Univ, MS, 53, PhD(physics), 58. *Prof Exp:* Res scientist, NY Univ, 54-57; asst prof physics, Rutgers Univ, 57-60; from asst prof to assoc prof, 60-82, PROF PHYSICS, QUEENS COL, NY, 82- *Concurrent Pos:* Lectr, City Col New York, 54-61; guest, Brookhaven Nat Lab, 67-68. *Mem:* AAAS; Am Phys Soc; Am Asn Physics Teachers; Sigma Xi. *Res:* Surface states of solids; defects and diffusion in solids. *Mailing Add:* Dept of Physics Queens Col 6530 Kissena Blvd Flushing NY 11367

NEUBERGER, JOHN STEPHEN, b New York, NY, June 29, 38; m 80; c 1. ENVIRONMENTAL EPIDEMIOLOGY. *Educ:* Cornell Univ, BME, 61; Columbia Univ, MBA, 67; Johns Hopkins Univ, MPH, 74, DrPH, 77. *Prof Exp:* Consult systs analyst, Rand Corp, 68-69; researcher, Riverside Res Inst, 69-71; consult analyst, Inst Space Studies, NASA, 71-75; environ epidemiologist & environ health specialist, Environ Sci Lab, 76-77, asst prof, 78-84, ASSOC PROF EPIDEMIOL & ENVIRON HEALTH, MED SCH, UNIV KANS, 84- *Concurrent Pos:* Consult, Md Dept Health & Ment Hyg, 73-74, Reg Planning Coun, 75-76 & Reg VII, US Environ Protection Agency, 79-81. *Mem:* Am Pub Health Asn; NY Acad Sci; Soc Epidemiol Res; Soc Occup & Environ Health; Sigma Xi. *Res:* Environmental and occupational carcinogenesis; health effects of toxic chemicals; epidemiology of brain and lung cancer. *Mailing Add:* Dept Prev Med Univ Kans Med Ctr Rainbow Blvd at 39th Kansas City KS 66103

NEUBERGER, JOHN WILLIAM, b Ventura, Iowa, Aug 14, 34; m 59; c 2. MATHEMATICS. *Educ:* Univ Tex, BA, 54, PhD(math), 57. *Prof Exp:* Spec instr math, Univ Tex, 56-57; instr, Ill Inst Technol, 57-59; asst prof, Univ Tenn, 59-63; from assoc prof to prof, Emory Univ, 63-77; prof math, 77-87, REGENTS PROF, UNIV N TEX, 87- *Concurrent Pos:* Mem, Inst Air Weapons Res, Univ Chicago, 59-65, consult, Oak Ridge Nat Lab, 59, 65 & 79-89; Alfred P Sloan res fel, 67-69; mem, Inst Advan Study, 68; NSF res grant, 70-72, 74 & 77-79; mem, Inst Defense Analysis, 73, stanby mem, Commun Res Div, 73; vis prof, Univ Ky, 73; ed, Houston J Math, 75-90, ed, Int J Math, 81-; consult, Inst Defense Anal, Sci & Technol Div, 86-, Formal Syst Design & Develop, 86- *Mem:* Am Math Soc; Soc Indust & Appl Math; Edinburgh Math Soc; London Math Soc. *Res:* Partial differential equations; numerical analysis; functional analysis; computational fluid dynamics; real variables. *Mailing Add:* Dept Math Univ N Tex Denton TX 76203

NEUBERT, JEROME ARTHUR, b Mankato, Minn, Dec 1, 38; m 66. PHYSICS. *Educ:* Univ Kans, BS, 62; Calif State Col, Los Angeles, MS, 66; Pa State Univ, PhD(eng acoust), 70. *Prof Exp:* SR RES PHYSICIST, NAVAL OCEAN SYSTS CTR, 63- *Mem:* Acoust Soc Am; Sigma Xi. *Res:* Engineering acoustics; applied mathematics and physics; sound propagation in stochastic media; underwater acoustics. *Mailing Add:* 321 Murray Dr El Cajon CA 92020

NEUBERT, RALPH LEWIS, chemical engineering, for more information see previous edition

NEUBERT, THEODORE JOHN, b Rochester, NY, Jan 10, 17. PHYSICAL CHEMISTRY. *Educ:* Univ Rochester, BS, 39; Brown Univ, PhD(phys chem), 42. *Prof Exp:* Res assoc, US Naval Res Lab, 42; chemist, Manhattan Proj, Chicago, 42-45; scientist, Argonne Nat Lab, 45-47, fel inst nuclear studies, 47-49; from asst prof to prof, 49-88, EMER PROF CHEM, ILL INST TECHNOL, 88- *Concurrent Pos:* Consult, Argonne Nat Lab, 49-60. *Mem:* AAAS; Am Chem Soc; Am Phys Soc. *Res:* Physics and chemistry of solid state. *Mailing Add:* 6578 Plastermill Rd Box 282 Victor NY 14564-0282

NEUBERT, VERNON H, b Cabot, Pa, Dec 23, 27; m 50; c 5. STRUCTURAL DYNAMICS, VIBRATION CONTROL. *Educ:* Carnegie Inst Technol, BS, 48, MS, 49. *Hon Degrees:* DENG, Yale Univ, 57. *Prof Exp:* Design engr, struct, Pittsburgh Indust Eng Corp, 49-50; res engr, Res Labs, Aluminum Co Am, 52-54; part-time instr appl mech, Yale Univ, 54-57; supvr appl mech, Elec Boat Div, Gen Dynamics Corp, 57-62; assoc prof eng mech, 62-66, PROF ENG MECH, PA STATE UNIV, 66- *Concurrent Pos:* Lectr eve sch, New London Exten, Univ Conn, 57-62. *Mem:* Am Soc Civil Engrs; Am Soc Mech Engrs; Acoust Soc Am. *Res:* Applied and solid mechanics; stress and vibration analysis; optimum vibration control. *Mailing Add:* Rd 1 Box 254 Centre Hall PA 16828

NEUBIG, RICHARD ROBERT, b Fountain Hill, Pa, Apr 10, 53; m 79; c 2. PHARMACOLOGY, BIOCHEMISTRY. *Educ:* Univ Mich, BS, 75; Harvard Univ, MD, PhD, 81. *Prof Exp:* House off, 81-84, from instr to asst prof pharmacol, 83-89, asst prof internal med, 84-90, ASSOC PROF PHARMACOL, UNIV MICH, ANN ARBOR, 89-, ASSOC PROF INTERNAL MED, 90- *Mem:* AAAS; Am Soc Pharmacol & Exp Ther; Biophys Soc. *Res:* Biophys of adrenergic receptor mechanism; G-Proteins. *Mailing Add:* M6322 Med Sci Bldg One Ann Arbor MI 48109-0626

NEUBORT, SHIMON, b New York, NY, Aug 21, 42; m 67; c 5. CANCER BIOLOGY, RADIOBIOLOGY. *Educ:* Yeshiva Univ, BA, 64; Albert Einstein Col Med, PhD(biochem), 71. *Prof Exp:* Res fel biochem, Albert Einstein Col Med, 71-72, res assoc radiobiol, 72-74, instr, 74-76, asst prof, 76-80; sr biochemist cancer biol, New York Univ, 80-84; PROJ DIR, NEUROL RES CTR, HELEN HAYES HOSP, 84- *Concurrent Pos:* Chief chemist, Crown-Kesser Wine Cellars, 81- *Mem:* Am Asn Cancer Res; Radiation Res Soc. *Res:* Response of mammalian cells and their nucleic acid metabolism to ionizing radiation and chemical agents; radiosensitizers; hyperthermia; chemical carcinogens; chemical analysis of cell components and metabolites. *Mailing Add:* Neurol Res Ctr NY State Health Dept Rte 9W West Haverstraw NY 10993

NEUCERE, JOSEPH NAVIN, b Hessmer, La, Feb 21, 32; m 71; c 1. BIO-ORGANIC CHEMISTRY, IMMUNOCHEMISTRY. *Educ:* La State Univ, Baton Rouge, BS, 60. *Prof Exp:* Chemist, 61-68, res chemist plant biochem, Southern Regional Res Ctr, USDA, 68-91. *Mem:* Am Chem Soc; AAAS. *Res:*

General characterization of proteins and enzymes in relation to species differentiation, biological value, pest resistance and other factors leading to improved quality of plant foods. *Mailing Add:* Southern Regional Res Ctr USDA PO Box 19687 New Orleans LA 70179

NEUDECK, GEROLD W(ALTER), b Beach, NDak, Sept 25, 36; m 62; c 2. ELECTRICAL ENGINEERING, SOLID STATE ELECTRONICS. *Educ:* Univ NDak, BSEE, 59, MSEE, 60; Purdue Univ, PhD(elec eng), 69. *Prof Exp:* Asst prof elec eng, Univ NDak, 60-64; from asst prof to assoc prof, 68-77, asst dean, 88-90, PROF ELEC ENG, PURDUE UNIV, 77- *Concurrent Pos:* Consult, CTS Corp, 66-; lectr, Western Elec Co, Inc, 69-71, Delco Electronics, 80, Bell Telephone Labs, 81, Int Electrodynamics, 81, D & M Corp, Twain Sage Inc 80-, Monolithic Sensors, 90. *Honors & Awards:* D D Ewing Award; A A Potter Award. *Mem:* Fel Inst Elec & Electronic Engrs; Am Vacuum Soc. *Res:* Integrated circuits and devices, germanium-silicon heterojunctions, high frequency noise in diodes, transistor and intergrated circuit fabrication, amorphous silicon thin films, 3 dimensional integrated devices; silicon selective epitaxial films. *Mailing Add:* Sch of Elec Eng Purdue Univ West Lafayette IN 47907

NEUDECK, LOWELL DONALD, b Billings, Mont, Aug 28, 37; m 66; c 2. PHYSIOLOGY, ENDOCRINOLOGY. *Educ:* Univ Mont, BA, 64, MA, 68; Univ Conn, PhD(zool), 77. *Prof Exp:* From asst prof to assoc prof biol, 76-86, PROF BIOL, NORTHERN MICH UNIV, 86- *Mem:* Sigma Xi; Am Soc Zool; Am Soc Mammalogists. *Res:* Mammalian, renal physiology; mammalian kidneys, physiology, morphology changes during growth and aging including hormonal effects on these changes; sexual dimorphism in mammalian kidneys. *Mailing Add:* Dept Biol Northern Mich Univ Marquette MI 49855

NEUE, UWE DIETER, b Neunkirchen, Ger, July 16, 48; c 2. HIGH PERFORMANCE LIQUID CHROMATOGRAPHY. *Educ:* Univ Saarbrn06cken, MS, 73, PhD(chem), 76. *Prof Exp:* Sr res chemist, Waters Assoc, Millipore Corp, 76-80, mgr core res, 80-84, mkt mgr, GmbH, Ger, 84-85, DIR PROD DEVELOP, WATERS, MILLIPORE CORP, 84- *Honors & Awards:* IR 100 Award, Indust Res & Develop, 80. *Mem:* AAAS. *Res:* High performance liquid chromatography and related techniques, from instrument design to the design of stationary phases and packed beds; applications of HPLC; surface chemistry; hydrodynamics; chemical engineering of packed beds. *Mailing Add:* Waters Chromatography Div Millipore Corp Milford MA 01757c

NEUENDORFFER, JOSEPH ALFRED, b New York, NY, Feb 28, 18; m 42; c 3. PHYSICS, OPERATIONS ANALYSIS. *Educ:* Mass Inst Technol, SB & SM, 41; Johns Hopkins Univ, PhD(physics), 51. *Prof Exp:* Staff mem, Acoustics Lab, Mass Inst Technol, 41-42; opers analyst, Columbia Res Group M, 42-43; opers analyst naval opers, Off Sci Res & Develop, 43-46, opers analyst, Opers Eval Group, 48-51, dep dir, 51-56; opers analyst, Staff Comdr in Chief Naval Opers, 58-61; staff comdr, Submarine Force, Atlantic, 61-62; dir naval objectives group, Ctr Naval Analyses, 63-64; analyst staff comdr, US Sixth Fleet, 65-66, sr scientist, Ctr Naval Analysis, 67-68, staff comdr in chief, Atlantic Fleet, 69-70, sr scientist, Ctr Naval Analysis, 71-74, staff comdr, Surface Forces, Atlantic, 75-76, sr scientist, Ctr Naval Analysis, Univ Rochester, 77-80; consult, Ctr Naval Analysis & Ketron, Inc, 81-85; RETIRED. *Mem:* Oper Res Soc Am; Am Phys Soc. *Res:* Development comparison and testing of naval tactics and equipment; sonar research, war gaming, simulation of military operations and search theory; angular distribution of the products of nuclear reactions. *Mailing Add:* 911 Allison St Alexandria VA 22302

NEUFELD, ABRAM HERMAN, b Russia, Apr 26, 07; nat Can; m 37; c 2. CLINICAL PATHOLOGY. *Educ:* Univ Man, BSc, 34, MSc, 35, PhD(med biochem), 37; McGill Univ, MD & CM, 50. *Prof Exp:* Instr biochem, Univ Man, 35-36; lectr, McGill Univ, 36-41, asst prof endocrinol, 41-43; med biochemist, Queen Mary Vet Hosp, Montreal, 46-55, chief serv biochem & radioisotopes, 55-6O; prof, 60-72, EMER PROF CLIN PATH, FAC MED, UNIV WESTERN ONT, 72- *Concurrent Pos:* Hon lectr, McGill Univ, 58-60; sr consult, Can Dept Vet Affairs, 50-60; ed, Med Serv J, Can, 47-59. *Honors & Awards:* Physiol Res Medal, 36. *Mem:* Can Soc Clin Invest; Am Soc Hemat; Can Med Asn; Chem Inst Can; Can Physiol Soc; Can Asn Path. *Res:* Metabolism of normal and pathological lipids and proteins, especially in arteriosclerosis, myelomatosis and lipidoses; hemoglobins; radioactive isotopes in clincal investigation; chemical pathology. *Mailing Add:* 1071 Colborne St London ON N6A 4B4 Can

NEUFELD, BERNEY ROY, b Sask, Aug 3, 41; US citizen; m 63; c 2. MOLECULAR BIOLOGY. *Educ:* Columbia Union Col, BA, 63; Loma Linda Univ, MA, 65; Ind Univ, Bloomington, PhD, 68. *Prof Exp:* Asst prof biol, Loma Linda Univ, 68-78; prof biol, Southwestern Adventist Col, 78-85; VPRES, DEVELOP & ALUMNI RELATIONS, PAC UNION COL, 91- *Concurrent Pos:* Mem extended day fac, Riverside City Col, 70-71; lectr, Univ Calif Exten, 71; NIH spec res fel, Calif Inst Technol, 71-73, vis fel, 73-74; assoc prof biol & head dept, Mid East Col, Beirut, Lebanon, 75-76. *Mem:* Genetics Soc Am; Am Soc Microbiol. *Res:* Evolution of restriction enzyme genes in salmonella. *Mailing Add:* Pac Union Col Howell Mountain Rd Angwin CA 94508

NEUFELD, DANIEL ARTHUR, b Fresno, Calif, Aug 26, 45; div; c 3. EXPERIMENTAL MORPHOLOGY, HISTOLOGY. *Educ:* Univ Calif, Los Angeles, BA, 68; Calif State Univ, Long Beach, MA, 72; Tulane Univ, PhD(human anat), 75. *Prof Exp:* Asst prof human anat, George Washington Univ Med Ctr, 75-77; asst prof 77-80, ASSOC PROF ANAT, SCH MED, UNIV SDAK, 80- *Mem:* Am Asn Anat; Am Soc Cell Biol; Soc Develop Biol. *Res:* Attempted induction of extremity regeneration in mammals. *Mailing Add:* Dept of Anat Univ SDak Sch Med 414 E Clark St Vermillion SD 57069

NEUFELD, ELIZABETH FONDAL, b Paris, France, Sept 27, 28; nat US; m 51; c 2. ENZYMOLOGY, HUMAN GENETICS. *Educ:* Queen's Col NY, BS, 48; Univ Calif, Berkeley, PhD(comp biochem), 56. *Hon Degrees:* Dr, Univ Rene Descartes, Paris, 78, Hahnemann Univ, 84; DSc, Russell Sage Col, 80. *Prof Exp:* Res biochemist, Nat Inst Arthritis, Metab & Digestive Dis, NIH, 63-73, chief, Sect Human Biochem Genetics, 73-79, chief, Genetics & Biochem Br, Nat Inst Arthritis, Diabetes & Digestive & Kidney Dis, 79-84, dep dir, Div Intramural Res, 81-83; PROF BIOL CHEM & CHMN DEPT, SCH MED, UNIV CALIF, LOS ANGELES, 84- *Concurrent Pos:* USPHS fel, Univ Calif, Berkeley, 56-57, asst res biochemist, 57-63; coun mem, Am Soc Biol Chemists, 77-80; mem bd sci counr, Nat Inst Mental Health, 77-81; mem sci adv coun, Tay-Sachs & Allied Dis Asn, 83-; mem adv coun, Nat Inst Diabetes Digestive & Kidney Dis, 87-91; coun mem, Am Soc Biochem & Molecular Biol, 88-91. *Honors & Awards:* Dickson Prize, 74; G Burroughs Mider Lectr, NIH, 74; Hillebrand Award, 75; Harvey Lectr, 79; Gairdner Found Int Award, 81; William Allan Award & Lectr, Am Soc Human Genetics, 82; Albert Lasker Clin Med Res Award, 82; J Henry Wilkinson Mem Award, Int Soc Clin Enzym, 83; Elliott Cresson Medal, Franklin Inst, 84; Wolf Prize Med, 88. *Mem:* Nat Acad Sci; Inst Med-Nat Acad Sci; Am Soc Human Genetics; Am Chem Soc; Am Soc Biol Chemists; Am Soc Cell Biol; Tissue Cult Asn; Am Soc Biochem & Molecular Biol (pres-elect, 91-); Am Acad Arts & Sci; fel AAAS. *Res:* Human biochemical genetics; mucopolysaccharidoses; tay-sachs disease; synthesis and transport of lysosomal enzymes; inherited disorders of lysosomal functions. *Mailing Add:* Dept Biol Chem Sch Med Univ Calif Los Angeles CA 90024-1737

NEUFELD, GAYLEN JAY, b Beaver Co, Okla, Feb 25, 39; m 61; c 3. CELL PHYSIOLOGY. *Educ:* Tabor Col, BA, 61; Kans State Univ, MS, 63; Univ Tex, Austin, PhD(protein chem), 66. *Prof Exp:* Damon Runyon Cancer Fund fel dept genetics, Univ Melbourne, 66-67; from asst prof to assoc prof, 67-78, PROF BIOL, EMPORIA STATE UNIV, 78- *Concurrent Pos:* Res physiologist, NIH, Res Triangle Park, NC, 77-78; res biochemist, NIH, Whitney Marine Lab, Fla, 83-84. *Mem:* AAAS; Sigma Xi; Am Soc Zoologists; Am Inst Biol Sci; Nat Asn Biol Teachers. *Res:* Structural characteristics of phycocyanin and phycoerythrin; effects of molting hormone on protein and ribonucleic acid synthesis in the third instar larvae of Calliphora; role of sodium potassium-atpase in osmoegulation in the blue crab; Anion-Sensitive ATPase in plasma membranes. *Mailing Add:* Div Biol Sci Emporia State Univ Emporia KS 66801

NEUFELD, GORDON R, b Mar 10, 37; c 2. ANESTHESIOLOGY. *Educ:* Univ Alta, BS, 59, MD, 62. *Hon Degrees:* MA, Univ Pa, 71. *Prof Exp:* ASSOC PROF ANESTHESIA, SCH MED, UNIV PA, 75- *Mem:* AAAS; Am Soc Anesthesiologists. *Res:* Transport of respiratory gases ventilation and perfusion in lungs. *Mailing Add:* Univ Pa Hosp 3400 Spruce St Philadelphia PA 19104

NEUFELD, HAROLD ALEX, b Paterson, NJ, Mar 23, 24; m 50; c 4. BIOCHEMISTRY. *Educ:* Rutgers Univ, BS, 49; Univ Rochester, PhD(biochem), 53. *Prof Exp:* Res fel biochem, Sch Med, Univ Rochester, 53-54, instr, 54-55; biochemist crops div, US Dept Army, 55-57, biochemist chem br, 57-70; teacher, Frederick County Sch Syst, 70-72; biochemist, Phys Sci Div, US Army Inst Infectious Dis, Ft Detrick, 72-85, prin investr, 80-85; RETIRED. *Concurrent Pos:* Lectr, Frederick Community Col, 57-65, adj prof chem, 85-; lectr biochem, Hood Col, Frederick, Md, 74-79; adj prof chem, Fredeum Community Col, 85- *Mem:* Am Chem Soc; Am Soc Biol Chem; Sigma Xi; Soc Exp Biol Med. *Res:* Purification and kinetics of enzymes concerned with biological oxidation; properties of enzymes in proliferating tissue; relationship between hormones and enzymes; respiratory enzymes in bacteria, fungi and animal; chemiluminescence and bioluminescence; effect of T-2 Mycotoxin on Mammalian metabolism. *Mailing Add:* 117 W 14th St Frederick MD 21701

NEUFELD, JERRY DON, b Isabella, Okla, Jan 20, 40; m 60; c 2. ANALYTICAL CHEMISTRY, FORENSIC CHEMISTRY. *Educ:* Tabor Col, BA, 61; Univ Hawaii, PhD(chem), 72. *Prof Exp:* Teacher math & chem, Hillsboro High Sch, Kans, 61-63 & Kodaikanal Sch, S India, 63-67; fel & res assoc chem, Univ NC, 72-73; prof & chmn dept, Anderson Col, 73-80; LECTR, CALIF STATE UNIV, FRESNO, 80- *Mem:* Am Chem Soc; Soc Appl Spectros: Nat Sci Teachers Asn; Am Sci Affil. *Res:* Infrared optic properties of polyatomic substances; development of low cost instrumental analysis experiments for a chemistry laboratory; trace analysis in forensic chemistry. *Mailing Add:* 6814 Ave 384 Dinuba CA 93618

NEUFELD, RONALD DAVID, b Brooklyn, NY, Feb 10, 47; m 68; c 3. ENVIRONMENTAL ENGINEERING. *Educ:* Cooper Union, BEChE, 67; Northwestern Univ, MS, 68, PhD(environ health eng), 73; Am Acad Environ Engrs, dipl, 80. *Prof Exp:* Sanit engr environ eng, Hydrotechnic Corp, 67; USPHS trainee, Northwestern Univ, 67-68; chem engr, Rohm & Haas Corp, 68-70; NSF trainee civil & environ eng, Northwestern Univ, 70-73; from asst prof to assoc prof, 77-82, PROF CIVIL ENG, UNIV PITTSBURGH, 82- *Concurrent Pos:* NSF res grant, 74-76; chem engr, US Dept Energy, 76-78, res contract, 77-; Am Iron & Steel Inst res grant, 77-; Environ Protection Agency res grant, 78-; consult design firms, A-C Valley Corps, Neville Chem Co, CH2M-Hill Co; sr Fulbright fel, Casali inst Applied Chem, Jerusalem, 83-84; expert witness, Cercia litigation, 90-91. *Mem:* Am Soc Civil Engrs; Am Inst Chem Engrs; Water Pollution Control Fedn; Int Asn Water Pollution; Asn Environ Eng Profs; Sigma Xi; Am Acad Environ Engrs; Am Water Works Asn. *Res:* Industrial and advanced waste treatment; environmental process control technology; biological and physical chemical processes; toxic and hazardous sludge and solid waste disposal; coal gasification and liquefaction waste treatment; biological nitrification; thiocyanate bio-degradation; heavy metal wastes; coal refuse piles; polychlorinated byphenyl biodegradation; electroplating wastes; cyanides. *Mailing Add:* 6558 Bartlett St Pittsburgh PA 15217

NEUFFER, MYRON GERALD, b Preston, Idaho, Mar 4, 22; m 43; c 7. GENETICS. *Educ:* Univ Idaho, BS, 47; Univ Mo, MA, 48, PhD(genetics), 52. *Prof Exp:* Asst field crops, 47-51, from asst prof to prof genetics, 51-70, chmn dept, 67-69, prof biol sci, 70-75, PROF AGRON, UNIV MO, COLUMBIA, 76- *Mem:* AAAS; Genetics Soc Am; Am Genetic Asn. *Res:* Gene mutation in maize. *Mailing Add:* Dept Agron 202 Curtis Hall Univ of Mo Columbia MO 65211

NEUGARTEN, BERNICE L, AGING HEALTH. *Educ:* Univ Chicago, BA, 36, MA, 37, PhD, 43. *Hon Degrees:* DSc, Univ S Calif, 80; PhD, Univ Nijmegen, Neth, 88. *Prof Exp:* From asst prof to prof, Univ Chicago, 69-73, dir, Grad Training Prog, 59-80, prof, Sch Social Serv Admin, 78-80; prof, Sch Educ & Dept Sociol, Northwestern Univ, 80-88; ROTHSCHILD DISTINGUISHED SCHOLAR, UNIV CHICAGO, 88-, FAC ASSOC, CHAPIN HALL CTR CHILDREN, 89- *Concurrent Pos:* Assoc ed, J Geront, 58-61; mem, Fed Coun Aging, 78-80; dep chair, 1981 White House Conf Aging, 79-81; mem, Comt Pub Policy Studies, Univ Chicago, 79-80; consult ed, J Appl Develop Psychol, 79-82; mem, Proj Health Prom Older People, NIMH, 83-84; served on numerous adv panels & couns of nat & int sci asns & pub bodies. *Honors & Awards:* Sandoz Int Prize, Int Asn Geront, 87. *Mem:* Sr mem Inst Med-Nat Acad Sci; Geront Soc Am (pres, 69); fel Am Acad Arts & Sci. *Mailing Add:* Ctr Aging Health & Soc Univ Chicago Chicago IL 60637

NEUGEBAUER, CHRISTOPH JOHANNES, b Dessau, Ger, Apr 21, 27; nat US; m 58; c 3. MATHEMATICS. *Educ:* Univ Dayton, BS, 50; Ohio State Univ, MS, 52, PhD(math), 54. *Prof Exp:* From instr to assoc prof, 54-62, PROF MATH, PURDUE UNIV, WEST LAFAYETTE, 62- *Mem:* Am Math Soc. *Res:* Analysis. *Mailing Add:* Dept Math Purdue Univ West Lafayette IN 47907

NEUGEBAUER, CONSTANTINE ALOYSIUS, b Dessau, Ger, Apr 20, 30; nat US; m 58; c 4. PHYSICAL CHEMISTRY, SEMICONDUCTOR PACKAGING. *Educ:* Union Col, NY, BS, 53; Univ Wis, PhD(chem), 57. *Prof Exp:* Fel phys chem, Univ Wis, 57; res assoc, Info Sci Lab, Res & Develop Ctr, 57-76, mgr, Power Module & Hybrid Unit, 76-84, Semiconductor Packaging Prog, 84-87, ADV, ELEC ASSEMBLIES, CORP RES & DEVELOP, GEN ELEC CO, 88- *Mem:* Inst Elec & Electronics Engrs; Am Phys Soc; Am Vacuum Soc. *Res:* Calorimetry; thermodynamics; structure and properties of thin films; large scale integration; hybrids; power semiconductor packaging; very large scale integration packaging. *Mailing Add:* Corp Res & Develop Gen Elec Co PO Box 8 KWC-1603 Schenectady NY 12301

NEUGEBAUER, GERRY, b Gottingen, WGermany, Sept 3, 32; US citizen; m 56; c 2. ASTROPHYSICS. *Educ:* Cornell Univ, AB, 54; Calif Inst Technol, PhD, 60. *Prof Exp:* US Army, Sta Jet Propulsion Lab, 60-62; from asst prof to prof, 62-85, HOWARD HUGHES PROF PHYSICS, CALIF INST TECHNOL, 85-, CHMN, DIV PHYSICS, MATH & ASTRON, 88- *Concurrent Pos:* Mem, NASA Astron Missions Bd, 69-70; prin investr infrared radiometer, Mariner, 69 & 71; dir, Asn Univ Res Astron Bd, 70-73; co-investr, Mariner 2 Infrared Radiometer, Pioneer F & G Infrared Radiometer, Mariner Venus-Mercury, Viking Infrared Thermal Mapper & Shuttle Infared Telescope Facility-Multiband Imaging Photometer, 73; staff mem, Hale Observ, 70-80; KPNO time allocation comt, 73-76; mem, Intermediate Range Task Force, Infrared Telescope Design Comt, NASA, 74-79; chmn, Study on Infrared & Submillimeter Astron from Space, Nat Acad Sci, 75; US prin scientist, Infrared Astron Satellite; vis comt, Asn Univ Res Astron, 77-80; exec mem, Interdisciplinary Sci Comn E Res Astrophysics from Space, Comt Space Res, 79-; mem infrared subcomt ultraviolet, optical & infrared panel, Astron Surv Comt, 79; actg dir, Palomar Observ, 80-81, dir, 81-; Comt on Space Res, Int Coun of Sci Unions. *Honors & Awards:* Except Sci Achievement Medal, NASA, 72 & 84; Richtmyer Lectr Award, 85; Space Sci Award, Am Inst Aeronaut & Astronaut, 85; George Darwin Lectr, Royal Astron Soc, 86; Rumford Premium, Am Acad Arts & Sci, 86. *Mem:* Nat Acad Sci; Am Acad Arts & Sci; Am Astron Soc; Royal Astron Soc; Int Astron Union; Sigma Xi; Am Philos Soc. *Mailing Add:* Down Lab Physics 320-47 Calif Inst Technol Pasadena CA 91125

NEUGEBAUER, OTTO E, astronomy; deceased, see previous edition for last biography

NEUGEBAUER, RICHARD, b New York, NY, Aug 18, 44; m 74. EPIDEMIOLOGY. *Educ:* Univ Chicago, BA, 65; Columbia Univ, MA, 68, PhD(hist), 76, MPH, 79. *Prof Exp:* Instr hist, Alfred Univ, 68-70; SERGIEVSKY SCHOLAR NEURO-EPIDEMIOL, GERTRUDE H SERGIEVSKY CTR, FAC MED, COLUMBIA UNIV, 79- *Concurrent Pos:* NIMH res training fel hist, Columbia Univ, 76-78 & NIMH res training fel epidemiol, 78-79; prin investr, NIH grants, 78-79, 79-80 & 81-84; res scientist, NY State Psychiat Inst; instr, Prog Health & Soc, Barnard Col. *Mem:* Am Pub Health Asn; Soc Epidemiol Res; AAAS. *Res:* Psychosocial and biological factors in organic and functional psychiatric disorders. *Mailing Add:* Gertrude H Sergievsky Ctr 630 W 168th St New York NY 10027

NEUGROSCHL, DANIEL, b Akron, Ohio, Nov 20, 66. METAL-POLYMER ADHESION. *Educ:* Yeshiva Univ, NY, BA, 87; Columbia Univ, MS, 89. *Prof Exp:* Res engr, Int Bus Mach, T J Watson Res Ctr, 89-90; res asst, 87-89, TEACHING ASST MAT SCI, COLUMBIA UNIV, 90- *Mem:* Soc Mining Engrs. *Res:* Interconnection and metallization issues related to high performance packaging: metal-polyimid adhesion, solder-metal reaction kinetics, chip joining and removal; production of superconductor bulk samples with enhanced critical current density capabilities; study of blood complement adsorption onto titanium surfaces to help understand biocompatibility. *Mailing Add:* 9515 Midwood Rd Silver Spring MD 20910

NEUHALFEN, ANDREW J, b Henry, Ill, Oct 5, 61; m 87. METALORGANIC CHEMICAL VAPOR DEPOSITION, III-V COMPOUND SEMICONDUCTOR. *Educ:* Univ Ill, BS, 83; Northwestern Univ, PhD(mat sci), 92. *Prof Exp:* Develop engr, Motorola, Inc, 83-88; RES FEL, NORTHWESTERN UNIV, 88- *Concurrent Pos:* Newport res award, Optical Soc Am, 89-90 & 90-91. *Mem:* Inst Elec & Electronic Engrs; Optical Soc Am; Soc Appl Spectros; Int Soc Hybrid Microelectronics. *Res:* Metalorganic chemical vapor deposition epitaxial layer growth and defect level structure characterization of rare-earth doped III-V compound semiconductors and their alloys; primary emphasis on InP and InGaP. *Mailing Add:* 308 S Main St Algonquin IL 60102

NEUHAUS, FRANCIS CLEMENS, b Huntington, WVa, May 5, 32; m 55; c 4. BIOCHEMISTRY. *Educ:* Duke Univ, BS, 54, PhD(biochem), 58. *Prof Exp:* NSF fel, Univ Newcastle-on-Tyne, Eng, 58-59 & Univ Ill, 59-60, instr biochem div, 59-60, asst prof chem, 60-61; from asst prof to prof chem, 61-74, PROF BIOCHEM & MOLECULAR BIOL & CELL BIOLOGY, NORTHWESTERN UNIV, 74- *Concurrent Pos:* USPHS res career develop award, 66-71. *Mem:* Am Chem Soc; Am Soc Microbiol; Am Soc Biol Chem. *Res:* Biosynthesis of bacterial cell wall components; mechanism of antibiotic action; membrane reactions. *Mailing Add:* Dept Biochem & Molecular & Cell Biol Northwestern Univ 2153 Sheridan Rd Evanston IL 60208

NEUHAUS, OTTO WILHELM, b Ger, Nov 18, 22; nat US; m 47; c 3. BIOCHEMISTRY. *Educ:* Univ Wis, BS, 44; Univ Mich, MS, 47, PhD(biochem), 53. *Prof Exp:* Res chemist, Merck & Co, 44-46; asst biochem, Univ Mich, 46-49; res chemist, Huron Milling Co, 51-54; instr physiol chem, Col Med, Wayne State Univ, 54-58, res assoc anat, 54-58, asst prof physiol chem, 58-65, assoc prof biochem, 65-66; chmn dept biochem, Sch Med, Univ SDak, 66-75 & 82-86, chmn div biochem, physiol & pharmacol, 76-82, prof biochem, 66-88, EMER PROF BIOCHEM, SCH MED, UNIV SDAK, 88- *Mem:* AAAS; Am Chem Soc; Soc Exp Biol & Med; Am Soc Biol Chemists. *Res:* Proteins of biological fluids; control of plasma protein biosynthesis; transport of amino acids; renal reabsorption and disposal of proteins. *Mailing Add:* Dept Biochem & Molecular Biol Univ SDak Sch Med Vermillion SD 57069

NEUHAUSER, DUNCAN B, b Philadelphia, Pa, June 20, 39; m 65; c 2. CLINICAL DECISION ANALYSIS. *Educ:* Harvard Univ, BA, 61; Univ Mich, MHA, 63; Univ Chicago, MBA, 66, PhD(bus admin), 71. *Prof Exp:* Res assoc health mgt, Ctr Health Admin Studies, Univ Chicago, 65-70; from asst prof to assoc prof, Dept Health Servs Admin, Harvard Sch Public Health, 70-79; assoc dir, Health Systs Mgt Ctr, 79-85, PROF EPIDEMIOL & COMMUNITY HEALTH, MED SCH, CASE WESTERN RESERVE UNIV, 79-, PROF MED, MED SCH, 81-, KECK FOUND SR RESEARCHER, 82- *Concurrent Pos:* Consult med, Mass Gen Hosp, 75-80, Cleveland Metrop Gen Hosp, 81-; fac ed, Health Matrix, 82-; ed, Med Care, 83-; adj mem med staff, Cleveland Clin Found, 84-; consult, res comt, Ohio Permanent Med Group, 84-; co-dir, Health Syst Mgt Ctr, 85-; adj prof organ behav, Weatherhead Sch Mgt, 79- *Honors & Awards:* Co-winner James A Hamilton Bk Award, Am Col Hosp Adminr, 74. *Mem:* Inst Med-Nat Acad Sci; Am Pub Health Asn; Int Soc Technol Assessment Healthcare; Soc Med Decision Making; Am Hosp Asn. *Res:* Organization of medical care delivery in general and in hospitals; health management, cost effectiveness & clinical decision analysis. *Mailing Add:* Dept Epidemiol & Biostatist Sch Med Case Western Reserve Univ Cleveland OH 44106

NEUHOLD, JOHN MATHEW, b Milwaukee, Wis, May 18, 28; m 52; c 1. AQUATIC ECOTOXICOLOGY. *Educ:* Utah State Univ, BS, 52, MS, 54, PhD(fish biol), 59. *Prof Exp:* Biologist, State Dept Fish & Game, Utah, 52-55, asst fed aide coord, 55-56; asst fish toxicologist, Utah State Univ, 56-58, from asst prof to prof fish biol, 58-89, actg dir, Ecol Ctr, 66-68, dir, 68-78, asst dean, Col Natural Resources, 61-89, EMER PROF FISH BIOL, UTAH STATE UNIV, 89- *Concurrent Pos:* Prog dir ecosyst anal, NSF, 71-72; trustee, Inst Ecol, 73-74, dir, 74-76; sci adv bd, Environ Protection Agency, 74-, distinguished vis scientist, 84-85. *Mem:* Soc Environ Chem & Toxicol; Am Fisheries Soc; Ecol Soc Am. *Res:* Fish toxicology, production in aquatic habitat, population dynamics; pollution biology. *Mailing Add:* 1254 Island Dr Logan UT 84321-4339

NEUHOUSER, DAVID LEE, b Leo, Ind, Mar 28, 33; m 54; c 4. MATHEMATICS. *Educ:* Manchester Col, BS, 55; Univ Ill, MS, 59; Fla State Univ, PhD(math educ), 64. *Prof Exp:* High sch teacher, Iowa, 55-57 & Ind, 57-58; from instr to assoc prof math & head dept, Manchester Col, 59-71; PROF MATH & HEAD DEPT, TAYLOR UNIV, 71- *Concurrent Pos:* Consult to sch systs, Northern Ind, 64- *Mem:* Math Asn Am. *Res:* Methods of teaching mathematics, particularly the discovery method. *Mailing Add:* Dept Math Taylor Univ Upland IN 46989

NEUMAN, CHARLES HERBERT, b Los Angeles, Calif, Feb 8, 37; m 58; c 2. PHYSICS. *Educ:* Calif Inst Technol, BS, 58; Univ Ill, MA, 60, PhD(physics), 63. *Prof Exp:* Res assoc physics, Univ Calif, Riverside, 63-65; res physicist, Chevron Oil Field Res Co, 65-76, sr res physicist, 76-80, Sr res assoc, 80-88, FORMATION EVAL SPECIALIST, CHEVRON, USA, 88- *Mem:* Am Phys Soc; Soc Petrol Engrs; Soc Prof Well Log Analysts. *Res:* Defects in solids; effects of pressure on transition metal oxides; nuclear spin echo in liquids; x-ray diffraction of minerals; radioactive measurement of liquid content in porous media. *Mailing Add:* 35080 Garden Dr Slidell LA 70460

NEUMAN, CHARLES P(AUL), b Pittsburgh, Pa, July 26, 40; m 67. APPLIED MATHEMATICS, ELECTRICAL ENGINEERING. *Educ:* Carnegie-Mellon Univ, BS, 62; Harvard Univ, SM, 63, PhD(appl math), 68. *Prof Exp:* Mem tech staff, Bell Tel Labs, Inc, 67-69; from asst prof to assoc prof elec eng, 69-78, prof elec eng, 78-83, PROF ELEC & COMPUT ENG, CARNEGIE-MELLON UNIV, 83- *Mem:* AAAS; Inst Elec & Electronics Engrs; Soc Indust & Appl Math; Inst Mgt Sci; Sigma Xi. *Res:* Robotics; control engineering and adaptive control; microcomputer control; control and systems science; Lyapunov stability theory; sensitivity analysis of dynamic systems; signal processing. *Mailing Add:* Dept of Elec & Comput Eng Carnegie-Mellon Univ Pittsburgh PA 15213-3890

NEUMAN, MICHAEL R, b Milwaukee, Wis, Nov, 25, 38; m 73; c 1. BIOENGINEERING & BIOMEDICAL ENGINEERING. *Educ:* Case Inst Technol, BS, 61, MS, 63, PhD, 66; Case Western Reserve Univ, MD, 74. *Prof Exp:* Asst prof, 66-70, ASSOC PROF, CASE WESTERN RESERVE UNIV, 70- *Concurrent Pos:* Res assoc, Dept Obstet & Gynecol, Cleveland Metrop Gen Hosp, 66-, Univ Hosp Cleveland, 74-; guest prof, Universitat Frauenklinik, Zurich, 80. *Mem:* Am Vacuum Soc; Inst Elec & Electronics Engrs; AAAS; Soc Gynec Invest; Biomed Eng Soc; Inter Soc on Biotalimetry. *Res:* Design, development and application of biomedical electronic instrumentation and transducers for clinical and basic science measurements and basic patient monitoring; blood gas measurement; perinatology; obstetrics; gynecology; pediatrics. *Mailing Add:* Electronics Design Ctr Case Western Reserve Univ Cleveland OH 44106

NEUMAN, RICHARD STEPHEN, b San Francisco, Calif, June 12, 45; Can citizen; m 69; c 2. PHARMACOLOGY. *Educ:* San Francisco State Col, BA, 68; Univ Alta, PhD(pharmacol), 74. *Prof Exp:* From asst prof to assoc prof, 74-88, PROF PHARMACOL, MEM UNIV NFLD, 88- *Concurrent Pos:* Asst prof res grants, Res Drug Abuse, 77-78, Med Res Coun Can, 78- *Mem:* Pharmacol Soc Can; Neurosci Soc; Inst Elec & Electronic Engrs Computer Soc; Asn Comput Mach. *Res:* Synaptic transmission in the mammalian central nervous system; monoamines and epilepsy; excitatory amino acid pharmacology. *Mailing Add:* Fac Med Mem Univ Nfld St John's NF A1B 3V6 Can

NEUMAN, ROBERT BALLIN, b Washington, DC, Feb 28, 20; m 49; c 2. STRUCTURAL GEOLOGY, STRATIGRAPHY-SEDIMENTATION. *Educ:* Univ NC, BS, 41; Johns Hopkins Univ, PhD(geol), 49. *Prof Exp:* Geologist, US Geol Surv, 49-80; RES ASSOC, DEPT PALEOBIOL, MUS NATURAL HIST, SMITHSONIAN INST, 80- *Concurrent Pos:* Lectr, Johns Hopkins Univ, 56-57 & Univ Oslo, 70-71; chmn US working group, Appalachian-Caledonide Orogen Proj int working group, Int Geol Correlation Prog, 74-85; vis prof, Va Tech, 80, Univ Mich, 88; distinguished prof, Univ Del, 82; guest res worker, Univ Bergen, 84-85. *Mem:* AAAS; Paleont Soc; Geol Soc Am; Soc Econ Paleont & Mineral. *Res:* Ordovician paleontology and paleogeography of the region bordering the North Atlantic Ocean. *Mailing Add:* E-308 Mus Natural Hist Smithsonian Inst Washington DC 20560

NEUMAN, ROBERT C, JR, b Chicago, Ill, Aug 21, 38; m 81; c 2. PHYSICAL ORGANIC CHEMISTRY. *Educ:* Univ Calif, Los Angeles, BS, 59; Calif Inst Technol, PhD(org chem), 63. *Prof Exp:* NSF res fel, Columbia Univ, 62-63; from asst prof to assoc prof, 63-72, PROF CHEM, UNIV CALIF, RIVERSIDE, 72- *Concurrent Pos:* Vis lectr dept chem, Princeton Univ, 71-72; NIH spec res fel, 71-72. *Mem:* Am Chem Soc; Sigma Xi. *Res:* Effects of high pressure on chemical systems; free radical chemistry. *Mailing Add:* Dept Chem Univ Calif 900 Univ Ave Riverside CA 92521-0403

NEUMAN, ROSALIND JOYCE, b Detroit, Mich, July 19, 38; m 60; c 4. COMPUTER SCIENCE, ALGEBRA. *Educ:* Wash Univ, St Louis, Mo, AB, 60, AM, 74, PhD(math), 81. *Prof Exp:* Sr assoc data processing & comput sci, 81-84, lectr, 81-87, DIR OF INSTR, MATH & COMPUT SCI, WASH UNIV, 84-, FEL, 87- *Mem:* Asn Comput Mach; Math Asn Am; Am Math Soc; Asn Women Math. *Mailing Add:* 848 S Meramec St Louis MO 63105

NEUMAN, SHLOMO PETER, b Zilina, Czech, Oct 26, 38; US citizen; m 65; c 3. SUBSURFACE HYDROLOGY, HYDROGEOLOGY. *Educ:* Hebrew Univ, Jerusalem, BSc, 63; Univ Calif, Berkeley, MS, 66, PhD(eng sci), 68. *Prof Exp:* Asst res engr, Univ Calif, Berkeley, 68-70, actg asst prof civil eng, 70; res scientist hydrol, Agr Res Orgn, Israel, 70-74; vis assoc prof civil eng, Univ Calif, Berkeley, 74-75; prof, 75-88, REGENTS PROF HYDROL, UNIV ARIZ, 88- *Concurrent Pos:* Res consult, Israel Inst Technol, 70-74, Lawrence Berkeley Lab, 75-80 & other gov & pvt consult. *Honors & Awards:* Robert E Horton Award, Am Geophys Union, 69; O E Meinzer Award, Geol Soc Am, 76; Birdsall Distinguished Lectr, Geol Soc Am, 86-87; Sci Award, Nat Water Well Asn, 88. *Mem:* Am Inst Mining, Metall & Petrol Engrs; Nat Water Well Asn; fel Am Geophys Union; fel Geol Soc Am; Soc Petrol Engr; Int Asn Hydrogeologists. *Res:* Subsurface flow and contaminant transport modeling; finite element techniques; well hydraulics; parameter estimation; stochastic subsurface hydrology; hydrology of fractured rocks; spatial variability of hydraulic and transport parameters; subsurface nuclear waste disposal. *Mailing Add:* Dept Hydrol & Water Res Univ Ariz Tucson AZ 85721

NEUMANN, A CONRAD, b Oak Bluffs, Mass, Dec 21, 33; m 62; c 3. OCEANOGRAPHY. *Educ:* Brooklyn Col, BS, 55; Tex A&M Univ, MS, 58; Lehigh Univ, PhD(geol), 63. *Prof Exp:* Res assoc sedimentary geol, Woods Hole Oceanog Inst, 58-60; asst prof marine geol, Lehigh Univ, 63-65; asst prof marine sci, Univ Miami, 65-69; prog dir marine geol & geophys, NSF, 69-70; assoc prof marine sci, Univ Miami, 69-72; actg dir marine sci curric, 73-74, Bowman & Gordon Gray Prof, 86-89, PROF MARINE SCI, MARINE SCI PROG, UNIV NC, 72- *Concurrent Pos:* Trustee, Bermuda Biol Sta Res, Inc, 72-76. *Mem:* AAAS; Soc Econ Paleont & Mineral; fel Geol Soc Am; Sigma Xi; Am Asn Petrol Geologists. *Res:* Sedimentology; recent carbonate sediments of the Bermudas, Bahamas and Florida; biological erosion of limestone; marine geology by research submersibles; Quaternary history of sea level; deep flanks of carbonate platforms. *Mailing Add:* Curric Marine Sci Univ NC Chapel Hill NC 27599-3300

NEUMANN, CALVIN LEE, b Coldwater, Mich, Sept 13, 38; m 61; c 1. STATISTICALLY DESIGNED CONSUMER RESEARCH. *Educ:* Wayne State Univ, BS, 62, PhD(org chem), 66. *Prof Exp:* Technician res labs, Ethyl Corp, Mich, 58-59; sr chemist, Res Dept, R J Reynolds Indust, Inc, 66-70, group leader, 70-74, sr res & develop chemist, 74-82, sr staff scientist, 82-91; INSTR, WINSTON SALEM STATE UNIV, 90- *Mem:* Am Chem Soc. *Res:* Conformational analysis; tobacco and health relationships; research and development of tobacco and tobacco related products; multivariate optimization of products through consumer research; process and product optimization analysis. *Mailing Add:* 1224 Huntingdon Rd Winston Salem NC 27104-1717

NEUMANN, FRED WILLIAM, b Chicago, Ill, Sept 28, 18; m 44, 89; c 3. ORGANIC CHEMISTRY, ANALYTICAL CHEMISTRY. *Educ:* Univ Ill, BS, 40; Ind Univ, PhD(org chem), 44. *Prof Exp:* Jr chemist, Merck & Co, NJ, 40-41; asst chem, Ind Univ, 41-43; chemist, Gen Aniline & Film Corp, 44-55; group leader, Dow Chem, USA, 55-68, analysis specialist, 68-72, res specialist, 72-81; RETIRED. *Mem:* Am Chem Soc; Sigma Xi. *Res:* Phenol and bisphenol A processes; synthetic organic chemistry of diazotypes, amidines and phenols; identification of dyes, diazotypes and organic commercial products; organic air-borne pollutants; iodine and bromine processes. *Mailing Add:* 2501 Lambros Dr Midland MI 48642

NEUMANN, GERHARD, b Frankfurt, Ger, Oct 8, 17; US citizen; m 46; c 3. AIRCRAFT ENGINEERING. *Educ:* Mittweida Saxony, Ger, Mech Eng. *Hon Degrees:* DSc, Armed Forces Col, Salem State Col; DEng, Milwaukee Col Eng. *Prof Exp:* Gen mgr, Flight Propulsion Div, Gen Elec Co, 61-68, vpres & group exec, 68-80; RETIRED. *Honors & Awards:* Collier Trophy, 58; Goddard Gold Medal, 69; French Legion Honor, 78; Guggeheim Medal, 79; Ann Award, Am Soc Mech Engrs, 91. *Mem:* Nat Acad Eng; hon fel Am Inst Aeronaut & Astronaut. *Mailing Add:* 53 Ocean View Rd Swampscott MA 01907

NEUMANN, HELMUT CARL, b Berlin, Germany, Dec 24, 16; nat US; m 46; c 4. ORGANIC CHEMISTRY. *Educ:* Polytech Inst Brooklyn, BS, 38; Fed Inst Technol, Zurich, DSc(chem), 49. *Prof Exp:* Analytical & develop chemist, Fritzsche Bros Inc, 38-42, asst lab head, 45-46; res fel, Univ Pa, 49-50; res chemist, White Labs, Inc, 50-51; res assoc, Sterling-Winthrop Res Inst, 51-59, sr res chemist, 59-82; RETIRED. *Concurrent Pos:* Ed, Eastern New York Chemist, 65-68. *Mem:* AAAS; Am Chem Soc. *Res:* Terpenes; steroids; antifertility agents; research on steroids and related compounds in search for medicinal applications. *Mailing Add:* 1508 Yale Dr Winchester VA 22601

NEUMANN, HENRY MATTHEW, b Minneapolis, Minn, July 15, 24. INORGANIC CHEMISTRY, NUCLEAR CHEMISTRY. *Educ:* Col St Thomas, BS, 47; Univ Calif, PhD(chem), 50. *Prof Exp:* Instr chem, Northwestern Univ, 50-54, asst prof, 54-56; assoc prof, 56-60, PROF CHEM, GA INST TECHNOL, 60- *Mem:* Am Chem Soc; Sigma Xi. *Res:* Solvation phenomena; kinetics and reaction mechanisms of complex ions; chemical education. *Mailing Add:* Sch Chem Ga Inst Technol Atlanta GA 30332

NEUMANN, HERSCHEL, b San Bernardino, Calif, Feb 3, 30; m 51; c 2. ATOMIC & MOLECULAR PHYSICS. *Educ:* Univ Calif, Berkeley, BA, 51; Univ Ore, MA, 59; Univ Nebr, Lincoln, PhD(theoret phys), 65. *Prof Exp:* Physicist, Hanford Labs, Gen Elec Co, Wash, 51-57; instr phys, Univ Nebr, Lincoln, 65; PROF & CHAIR, DEPT PHYSICS, UNIV DENVER, 85- *Mem:* Am Asn Physics Teachers; Am Phys Soc. *Res:* Atomic and molecular collision theory; ion and atom scattering from solids; physics education. *Mailing Add:* Dept of Physics Univ of Denver Denver CO 80208-0202

NEUMANN, JOACHIM PETER, b Berlin, Ger, June 15, 31; US citizen; m 62; c 2. METALLURGY & PHYSICAL METALLURGICAL ENGINEERING. *Educ:* Tech Univ, Berlin, Dipl Ing, 56; Univ Calif, Berkeley, PhD(metall), 65. *Prof Exp:* Metall engr, Sherritt Gordon Mines, Ltd, 56-57 & Aluminum Co Can, 57-58; asst prof metall, Univ Calif, Los Angeles, 63-69; tech adv, UNESCO, Spain & Venezuela, 69-74; sr scientist, Univ Wis-Milwaukee, 74-78; metallurgist, Bur Mines, Albany, 78-80; assoc prof metall eng, Univ Ala, Tuscaloosa, 80-85; PROF METALL ENG, UNIV WIS-MILWAUKEE, 85- *Concurrent Pos:* Consult, various indust firms. *Mem:* Am Soc Metals Int; Am Inst Mining, Metall & Petrol Engrs; Am Soc Eng Educ. *Res:* Chemical metallurgy; physical metallurgy; relationship between structure and properties of materials. *Mailing Add:* Mat Dept Univ Wis Milwaukee PO Box 784 Milwaukee WI 53201

NEUMANN, MARGUERITE, b West Bend, Wis, May 7, 14. ORGANIC CHEMISTRY, MEDICINAL CHEMISTRY. *Educ:* Mundelein Col, BS, 42; Univ Iowa, MS, 43; St Louis Univ, PhD(chem), 54. *Prof Exp:* Asst prof chem, Mundelein Col, 43-50 & 54-57; chmn dept, 57-70, prof chem, 57-85, EMER PROF CHEM & DIR INSTITUTIONAL RES, CLARKE COL, 85- *Mem:* Am Chem Soc; Sigma Xi; Am Asn Univ Women. *Res:* Biochemistry. *Mailing Add:* 1993 1/2 Asbury Rd Dubuque IA 52001

NEUMANN, NORBERT PAUL, b Chicago, Ill, Oct 13, 31; m 56; c 3. BIOCHEMISTRY. *Educ:* St Peters Col, BS, 53; Okla State Univ, MS, 55; Univ Wis, PhD(biochem), 58. *Prof Exp:* Res assoc biochem, Rockefeller Inst, 58-61; from instr to asst prof microbiol, Rutgers Univ, 61-67, asst prof sch med, 67-70; dir biochem, 70-74, asst dir res, 74-81, dir immunol res, Ortho Res Inst Med Sci, 81-; AT PURDUE FREDERICK RES CTR BIOL. *Mem:* AAAS; Am Chem Soc; Am Soc Human Genetics; Soc Complex Carbohydrates; NY Acad Sci. *Res:* Relationship between structure and function in enzymes; methods of protein isolation and characterization; diagnostic chemistry and immunology; lymphokines and immunoregulation. *Mailing Add:* 25 Hunt Ct Jericho NY 11753

NEUMANN, PAUL GERHARD, b Insterburg, Ger, June 26, 11; nat US; m 48; c 4. OCEANOGRAPHY, METEOROLOGY. *Educ:* Univ Berlin, Dr rer nat(geophys), 39. *Prof Exp:* Res scientist, Ger Marine Observ, 39-45 & Ger Hydrographic Off, 45-47; docent, Univ Hamburg, 47-51; from assoc prof to prof oceanog, NY Univ, 51-73; prof earth & planetary sci, 73-80, EMER PROF & RESIDENT PROF, CITY COL NEW YORK, 80- *Concurrent Pos:* Consult, NSF, 66, 67 & 72. *Mem:* Am Meteorol Soc; Am Geophys Union; Ger Geophys Soc. *Res:* Air-sea interaction; ocean waves and currents; internal waves; physical oceanography; oceanography of the tropical Atlantic Ocean. *Mailing Add:* 11 Goodwin St Hastings-on-Hudson NY 10706

NEUMANN, PETER G, New York, NY, Sept 21, 32. COMPUTER SECURITY, SOFTWARE ENGINEERING. *Educ:* Hawaii Univ, AB, 54, SM, 55 & PhD(appl math), 60. *Hon Degrees:* Dr rer nat, Darmstadt Tech High Sch, 60. *Prof Exp:* Mem tech staff comput sci, Bell Telephone Labs,

60-70; vis MacKay lect electrical eng & comput sci, Univ Calif Berkeley, 70-71; FROM RES ENGR TO STAFF SCIENTIST, COMPUT SCI LAB, SRI INT, 71- *Concurrent Pos:* Fulbright grantee, 58-60; nat lectr, Asn Comput Mach, 69-70; chmn comt comput & pub policy, Asn Comput Mach, 85-; Nat Acad Sci comts, 81-82, 89-90. *Mem:* Asn Comput Mach; Inst Elec & Electronics Engrs; Sigma Xi; Am Asn Advan Sci. *Res:* Development of computer systems with critical requirements such as security, reliability and human safety. *Mailing Add:* SRI EL-243 Menlo Park CA 94025-3493

NEUMANN, RONALD D, b Watertown, Wis, Oct 10, 47; m. NUCLEAR MEDICINE. *Educ:* Carroll Col, BS, 70; Yale Univ, MD, 74; Am Bd Nuclear Med, cert. 79. *Prof Exp:* Asst prof diag radiol, Sch Med, Yale Univ, 79-83, assoc prof diag radiol & path, 83-86; ASSOC CLIN PROF DIAG RADIOL, SCH MED, GEORGE WASHINGTON UNIV, 86-; PROG DIR, NUCLEAR MED RESIDENCY TRAINING PROG CLIN CTR, NIH, 86-, SR INVESTR, NAT CANCER INST, 86-, CHIEF, DEPT NUCLEAR MED, 88- *Concurrent Pos:* Actg chief, Nuclear Med Serv, Vet Admin Med Ctr, 79-85; attend physician, Yale New Haven Hosp, 79-85; dep chief, Dept Nuclear Med, NIH, 85-88. *Mem:* Soc Nuclear Med; Int Acad Path; AAAS; Asn Univ Radiologists; Am Col Nuclear Physicians; Am Col Chest Physicians; Europ Asn Nuclear Med; Metastasis Res Soc. *Res:* Cancer; nuclear medicine. *Mailing Add:* NIH Warren Grant Magnuson Clin Ctr Nuclear Med Dept Bldg 10 Rm 1C401 Bethesda MD 20892

NEUMANN, STEPHEN MICHAEL, b La Porte, Ind, Aug 7, 52; m 74. ORGANOMETALLIC CHEMISTRY. *Educ:* Ind Univ, BS, 74; Univ Wis, PhD(org chem), 78. *Prof Exp:* RES CHEMIST, EASTMAN KODAK CO, 78- *Mem:* Am Chem Soc. *Res:* Organometallic chemistry, specifically transition metal formyl complexes and their properties; properties of dyes for photographic image transfer. *Mailing Add:* Color Hard Copy Media RL Bldg 65 Floor 1 Eastman Kodak Co Rochester NY 14650-1824

NEUMARK, GERTRUDE FANNY, b Nueremberg, Ger, Apr 29, 27; nat US; m 50. SEMICONDUCTOR PHYSICS, MATERIALS SCIENCE. *Educ:* Columbia Univ, BA, 48, PhD(chem), 51; Radcliffe Col, MA, 49. *Prof Exp:* Asst chem, Columbia Univ, 48, asst, Barnard Col, 50-51; adv res physicist, Sylvania Elec Prod, Inc, Gen Tel & Electronics Corp, 52-60; staff physicist, Philips Labs, NAm Philips Co, Inc, 60-85; PROF MAT SCI, COLUMBIA UNIV, 85- *Concurrent Pos:* Anderson fel, Am Asn Univ Women, 51-52; adj assoc prof, Fairleigh Dickinson Univ, 73-74; vis prof, Columbia Univ, 82-83, adj prof, 83-85. *Mem:* Am Chem Soc; fel Am Phys Soc; NY Acad Sci; Sigma Xi; Electrochem Soc; Soc Women Engrs; Mat Res Soc. *Res:* Luminescence and electroluminescence; non-radiative recombination; conductivity in wide band-gap semiconductors; influence of dielectric screening on semiconductor transport and recombination properties; photovoltaic effect. *Mailing Add:* Div Metall & Mat Sci 1145 SW Mudd Bldg Columbia Univ New York NY 10027

NEUMEIER, LEANDER ANTHONY, b St Louis, Mo, Feb 15, 33; m 63; c 1. METALLURGICAL ENGINEERING, PHYSICAL METALLURGY. *Educ:* Univ Mo-Rolla, BS, 59, MS, 60. *Prof Exp:* Technician metall, Uranium Div, Mallinckrodt Chem Works, 58; metallurgist, 60-74, actg res dir, 79, SUPVRY METALLURGIST, US BUR MINES, 74- *Mem:* Sigma Xi; Am Soc Metals; Am Inst Mining, Metall & Petrol Engrs. *Res:* Physical and process metallurgy; ferrous and nonferrous; cast irons; magnesium alloys; secondary materials; powder metallurgy; coating and soldering; mineral treatment and extraction. *Mailing Add:* 301 Lariat Lane Rolla MO 65401

NEUMEYER, JOHN L, b Munich, Ger, July 19, 30; US citizen; m 56; c 3. MEDICINAL & ORGANIC CHEMISTRY. *Educ:* Columbia Univ, BS, 52; Univ Wis, PhD(med chem), 61. *Prof Exp:* Pharmaceut chemist, Ethicon, Inc Div, Johnson & Johnson, 52-53, group leader pharmaceut res, 55-57; sr res chemist, Cent Res Labs, Niagara Chem Div, FMC Corp, NJ, 61-63; sr staff chemist, Arthur D Little, Inc, 63-69; dir, Grad Sch Pharm & Allied Health Prof, 76-85, PROF MED CHEM, NORTHEASTERN UNIV, 69-, PROF CHEM, 75-; DIR RES & CHMN BD, RES BIOCHEM, INC, 82- *Concurrent Pos:* Adj prof, Mass Col Pharm, 64-69; mem adv panel secy comn on pesticides & their relationship to environ health, HEW, 69-70; mem comt rev, US Pharmacopeia, 69-85; consult, Arthur D Little, Inc, 69- & Environ Protection Agency, 70-72; mem, Mass State Pesticide Bd, 73-75; bk rev ed, 74-87, bd publ, 91-, J Med Chem, 74-87; vis prof chem, Univ Konstanz, WGer, 75-76; Sr Hays-Fulbright fel, 75-76; mem study sect, NIH, 81-; vis scientist, McLean Hosp, Harvard Med Sch, 85-; distinguished univ prof, Northeastern Univ, 80- *Honors & Awards:* Lunsford-Richardson Award, 61; Res Achievement Award, Med Chem, Acad Pharmaceut Sci, 82. *Mem:* Am Chem Soc; Am Pharmaceut Asn; AAAS; Fel Acad Pharmaceut Sci; Soc Neurosci; Fel Am Asn Pharm Sci; Sigma Xi; Am Asn Pharmacol & Exp Therapeut. *Res:* Chemistry of biologically-active compounds of natural and synthetic origin; aporphines; isoquinolines; chemistry of heterocyclics, antimalarials; central nervous system active compounds and dopamine receptors. *Mailing Add:* Dept Med Chem Northeastern Univ Boston MA 02115

NEUMILLER, HARRY JACOB, JR, b Peoria, Ill, Dec 25, 29; m 57; c 4. ORGANIC CHEMISTRY. *Educ:* Knox Col, BA, 51; Univ Ill, MS, 52, PhD(org chem), 56. *Prof Exp:* Res chemist, Eastman Kodak Co, NY, 55-59; from asst prof to assox prof, 59-85, PROF CHEM, KNOX COL, ILL, 85- *Mem:* Am Chem Soc; Sigma Xi. *Res:* Orientations in additions to quinones and quinone-like compounds. *Mailing Add:* 1225 N Cherry St Galesburg IL 61401

NEUNZIG, HERBERT HENRY, b Richmond Hill, NY, May 11, 27; m 55; c 2. ENTOMOLOGY. *Educ:* Cornell Univ, MS, 55, PhD(entom), 57. *Prof Exp:* From asst prof to assoc res prof entom, 57-68, PROF ENTOM, NC STATE UNIV, 68- *Mem:* Entom Soc Am; Lepidopterists' Soc; Sigma Xi. *Res:* Taxonomy of lepidotera and megaloptera. *Mailing Add:* Dept of Entom NC State Univ Raleigh NC 27695-7613

NEUPERT, WERNER MARTIN, b Worcester, Mass, Dec 19, 31; m 59; c 2. SPECTROSCOPY, SOLAR PHYSICS. *Educ:* Worcester Polytech Inst, BS, 54; Cornell Univ, PhD(physics), 60. *Prof Exp:* Vis asst prof physics, Univ Calif, Santa Barbara, 59-60; head, Solar Plasmas Br, 72-84, PHYSICIST, GODDARD SPACE FLIGHT CTR, NASA, 60-, PROJ SCIENTIST, SOUNDING ROCKETS, 89- *Mem:* Int Astron Union; Am Phys Soc; Am Astron Soc. *Res:* Studies of the solar extreme ultraviolet and x-ray spectrum using rocket and satellite-borne instrumentation. *Mailing Add:* Code 680 Goddard Space Flight Ctr NASA Greenbelt MD 20771

NEURATH, ALEXANDER ROBERT, b Bratislava, Czech, May 8, 33; US citizen. VIROLOGY. *Educ:* Inst Tech, Bratislava, Czech, DiplIng, 57; Vienna Tech Univ, ScD(microbiol), 68. *Prof Exp:* Res scientist, Plant Physiol Inst, Czech Acad Sci, 57-59; dept head vet virol, Bioveta, Nitra, Czech, 59-61; res scientist, Inst Virol, Czech Acad Sci, 61-64; res fel virol, Wistar Inst, 64-65; sr virologist, Wyeth Labs, Inc, 65-72; SR INVESTR, L KIMBALL RES INST, NEW YORK BLOOD CTR, 72- *Res:* Analysis of trace elements in plants; analytical biochemistry; biochemistry of viruses including, myxoviruses, adenoviruses, rabies and pseudorabies; hepatitis B; human immuno-deficiency virus; synthetic peptide immunogens. *Mailing Add:* New York Blood Ctr 310 E 67th St New York NY 10021

NEURATH, HANS, b Vienna, Austria, Oct 29, 09; nat US; m 36, 60; c 1. BIOCHEMISTRY. *Educ:* Univ Vienna, PhD(colloid chem), 33. *Hon Degrees:* DSc, Univ Geneva, 70; Dr, Univ Tokushima, 77. *Prof Exp:* Res fel, Univ Col, Univ London, 34-35; res fel biochem, Univ Minn, 35-36; instr & Baker res fel chem, Cornell Univ, 36-38; from asst prof to assoc prof biochem sch med, Duke Univ, 38-46, prof phys biochem, 46-50; chmn dept biochem, 50-75, prof biochem, 50-80, sci dir, Ger Cancer Res Ctr, 80-81, EMER PROF BIOCHEM, UNIV WASH, 82- *Concurrent Pos:* Mem comt biol chem, Div Chem & Chem Eng, Nat Res Coun, 54-58, mem exec comt, Off Biochem Nomenclature, 64-; consult, NIH, 54-; Guggenheim fel, 55-56; Phillips visitor, Haverford Col, 59; mem, US Nat Comt Biochem, 62-63 & US-Japan Coop Sci Prog Adv Panel on Med Sci, 64-65; comn ed, Int Union Biochem, 65-; mem comt res life sci, Nat Acad Sci, 66-70; mem, Nat Bd Grad Educ, 71-74; guest prof, Alexander von Humboldt Found, WGermany & hon prof, Univ Heidelberg, 75; 1st distinguished fac lectr, Univ Wash, 76; vis prof, Japan Soc Prom Sci, 77 & NSF vis prof, Univ Hawaii, 78; foreign sci mem, Max Planck Soc, WGermany, 82-; ed, Biochem, 61-91; hon prof, Univ Heidelberg, 81- *Honors & Awards:* Darling Lectr, Allegheny Col, 60; Awardee, Alexander von Humboldt Found, WGermany & Univ Heidelberg, 75; Kelly Lectr, Purdue Univ, 77; Smith, Kline & French Distinguished Lectr, Univ Mich, 78; Stein & Moore Award, Protein Soc, 89. *Mem:* Nat Acad Sci; fel AAAS; fel Am Acad Arts & Sci; hon mem Biochem Soc Japan; fel NY Acad Sci. *Res:* Protein structure and function; enzymes; proteolytic enzymes; evolution; physical biochemistry of macromolecules. *Mailing Add:* Dept Biochem Univ Wash Seattle WA 98195

NEUREITER, NORMAN PAUL, b Macomb, Ill, Jan 24, 32; m 59; c 4. SCIENCE POLICY. *Educ:* Univ Rochester, AB, 52; Northwestern Univ, PhD, 57. *Prof Exp:* Instr basic sci, Northwestern Univ, 56-57; res chemist, Humble Oil & Refining Co, Tex, 57-63; asst prog dir off int sci activ, NSF, Washington, DC, 63-64; actg prog dir, US-Japan Coop Sci Prog, 64-65; dep sci attache, US Dept State, Am Embassy, Bonn, 65-67, sci attache, Am Embassy, Warsaw, 67-69, tech asst off sci & technol, Exec Off of the President, 69-73; dir, East-West Bus Develop, Tex Instruments Inc, 73-74, mgr int bus develop, 74-77, pres & mgr Europe Div, 77-80, vpres, Corp Staff, 82-89, MGR, CORP RELATIONS, TEXAS INSTRUMENTS INC, 80-; MGR, CORP RELATIONS/ASIA & PRES, TEX INSTRUMENTS ASIA LTD. *Concurrent Pos:* Guide, Am Nat Exhib, Moscow, 59; mem bd int orgn & prog, Nat Acad Sci, Off Secy, Washington, DC, 73-77; mem, Adv Comt Int Progs, NSF, 78-83. *Mem:* AAAS; Am Chem Soc. *Res:* International cooperation in science; relationships of science to foreign policy; role of technology in international trade; East-West trade. *Mailing Add:* PO Box 655474 M/S 236 Dallas TX 75265

NEURINGER, JOSEPH LOUIS, b Brooklyn, NY, Jan 16, 22; m 46; c 4. MATHEMATICAL PHYSICS. *Educ:* Brooklyn Col, BA, 43; Columbia Univ, MA, 48; NY Univ, PhD(physics), 51. *Prof Exp:* Meteorologist, US Army Air Force, 44-46; asst jet propulsion, NY Univ, 47-51; eng specialist, Repub Aviation Corp, 51-55; prin scientist res & adv develop div, Avco Corp, 55-56; chief scientist plasma propulsion & power lab, Repub Aviation Corp, 56-62; sr consult scientist res & tech labs, Avco Space Systs Div, 62-67, prin staff scientist, 67-70; prof math, Univ Lowell, 70-85; RETIRED. *Concurrent Pos:* Consult, Avco space Syst Div & Div Support Command, US Army, 77-85. *Mem:* AAAS; Am Phys Soc; assoc fel Am Inst Aeronaut & Astronaut; NY Acad Sci; Sigma Xi. *Res:* Magnetohydrodynamics; ferrohydrodynamics; application of plasma physics and aerophysics to propulsion; generation of electricity; space flight; differential equations. *Mailing Add:* 9121 F SW 20th St Boca Raton FL 33428-7741

NEURINGER, LEO J, b New York, NY, Nov 20, 28; m 54; c 3. HIGH FIELD NUCLEAR MAGNETIC RESONANCE. *Educ:* Rensselaer Polytech Inst, BS, 51; Univ Pa, PhD(physics), 57. *Prof Exp:* Physicist optics, Nat Bur Standards, 51-52; staff scientist infrared detectors, Res Div, Raytheon Co, 56-63; staff mem semiconductors, 63-67, SR SCIENTIST HIGH FIELD SUPERCONDUCTORS, FRANCIS BITTER NAT MAGNET LAB, MASS INST TECHNOL, 67-, GROUP LEADER MOLECULAR BIOPHYS, 73-, DIR, HIGH FIELD NUCLEAR MAGNETIC RESONANCE FACIL, 76-, PROG DIR, MAGNETIC RESONANCE IMAGING FACIL, 82- *Concurrent Pos:* Instr physics, Tufts Univ, 57-58, vis prof, 69-70; vis prof, Hebrew Univ, Jerusalem, 71, Univ Chile, 73. *Mem:* AAAS; fel Am Phys Soc; Sigma Xi; Biophys Soc. *Res:* High field nuclear magnetic resonance of proteins and tissue; magnetic resonance imaging; high field superconductivity. *Mailing Add:* 61 Glenn Rd Wellesley Hills MA 02181

NEUSCHEL, SHERMAN K, b Buffalo, NY, Dec 21, 13. GEOLOGY MAPPING & ADMINISTRATION. *Educ:* Denison Univ, BA, 37. *Prof Exp:* Res scientist, US Geol Surv, Washington, DC, 42-74; RETIRED. *Mem:* Fel Geol Soc Am; Sigma Xi; Am Asn Petrol Geologists. *Res:* Mapping with airborne magnetic and airborne radioactivity. *Mailing Add:* 7501 Democracy Blvd B324 Bethesda MD 20817

NEUSE, EBERHARD WILHELM, b Berlin, Ger, Mar 7, 25; m 63; c 2. POLYMER CHEMISTRY, ORGANOMETALLIC CHEMISTRY. *Educ:* Hanover Tech Univ, BS, 48, MS, 50, PhD(org chem), 53; Univ Witwatersrand, DSc(polymer chem), 76. *Prof Exp:* Res asst chem heterocyclics, Hanover Tech Univ, 51-53; head appln lab, O Neynaber & Co, AG, Ger, 54-57; res assoc plastics lab, Princeton Univ, 57-59; head polymer lab missile & space systs div, Douglas Aircraft Co, 60-70, chief plastics & elastomers develop sect, 67-70; sr lectr chem, 71-73, READER & PROF MACROMOLECULAR CHEM, UNIV WITWATERSRAND, 73- *Concurrent Pos:* Mem, Nat Adv Comt Plastics Educ; Mem comt, Macromolecular Div, Int Union Pure & Appl Chem, 75-89. *Mem:* Am Chem Soc; NY Acad Sci; SAfrican Chem Inst. *Res:* Organic and organometallic chemistry of monomeric and polymeric compounds; development of polymeric materials for high temperature applications; organo-sulfur chemistry; charge-transfer complexes; carbohydrate modification; polymers for biomedical applications. *Mailing Add:* Dept Chem Univ the Witwatersrand WITS 2050 Johannesburg South Africa

NEUSHUL, MICHAEL, JR, b Shanghai, China, Dec 27, 33; nat US. BOTANY. *Educ:* Univ Calif, Los Angeles, BA, 55, PhD(bot), 59. *Prof Exp:* Asst bot, Univ Calif, Los Angeles, 55-56; res biologist, Scripps Inst, 56-58; asst bot, Univ Calif, Los Angeles, 58-59; NSF fel, Univ London, 59-60; from instr to asst prof bot, Univ Wash, 60-63; from asst prof to assoc prof, 63-73, PROF BOT, UNIV CALIF, SANTA BARBARA, 73-, PROF MARINE PHYCOL, 77- *Concurrent Pos:* Botanist, Arg Antarctic Exped, 57-58; vis fel, Swiss Fed Inst Technol, Zurich, 69-70; partic, Scripps Mex Oceanog Cruises. *Mem:* Am Phycol Soc; Bot Soc Am; Brit Phycol Soc; Int Soc Plant Morphol; Arg Antarctic Asn. *Res:* Marine algology; ultrastructure; development; sublittoral ecology; antarctic marine algae; algae development. *Mailing Add:* Dept of Biol Sci Univ of Calif Santa Barbara CA 93106

NEUSTADT, BERNARD RAY, b Washington, DC, May 7, 43; m 64; c 7. MEDICINAL CHEMISTRY. *Educ:* Columbia Univ, AB, 64; Brandeis Univ, PhD(org chem), 69. *Prof Exp:* Jr chemist, Arthur D Little, Inc, 66; sr chemist, 69-74, prin scientist, 74-78, RES SECT LEADER, SCHERING CORP, 78- *Mem:* Am Chem Soc. *Res:* Medicinal organic chemistry. *Mailing Add:* Schering Corp Bloomfield NJ 07003

NEUSTADTER, SIEGFRIED FRIEDRICH, b Ger, July 5, 23; nat US. MATHEMATICS, OPERATIONS RESEARCH. *Educ:* Univ Calif, PhD(math), 48. *Prof Exp:* Lectr math, Univ Calif, 48-49; Peirce instr, Harvard Univ, 49-52; mem staff, Lincoln Lab, Mass Inst Technol, 52-58; PROF MATH, SAN FRANCISCO STATE UNIV, 58- *Concurrent Pos:* Mem staff res lab, Sylvania Elec Prod, Inc, Lexington, Mass, 62-65. *Mem:* Am Math Soc; Sigma Xi. *Res:* Applied mathematics; mathematical analysis; mathematical biology. *Mailing Add:* Dept Math 1600 Holloway Ave San Francisco CA 94132

NEUTRA, MARIAN R, b Chicago, Ill, Aug 31, 38; m 64, 77; c 3. INTESTINAL PHYSIOLOGY, ELECTRON MICROSCOPY. *Educ:* Univ Mich, BA, 60; McGill Univ, PhD(cell biol & anat), 66. *Prof Exp:* Vis asst prof cell biol, Univ Del Valle, Colombia, 71-73; asst prof, 73-80, ASSOC PROF ANAT, HARVARD MED SCH, 80- *Concurrent Pos:* Prin investr, Cystic Fibrosis Found, 74-77 & 81- & NIH res grant, 78-; res comt, Cystic Fibrosis Found, 79-; assoc ed, Anatomical Rec, 82- *Mem:* Am Soc Cell Biol; Am Asn Anatomists; Electron Microscope Soc Am; Fedn Am Scientists. *Res:* Cell biology of the intestinal epithelium; glycoprotein secretion, macromolecular transport, bacterial adherance, intercellular junctions and membrane specializations in humans and other mammals. *Mailing Add:* Dept Pediat Childrens Hosp 300 Longwood Ave Boston MA 02115

NEUTS, MARCEL FERNAND, b Ostend, Belg, Feb 21, 35; m 59; c 4. STATISTICS, OPERATIONS RESEARCH. *Educ:* Univ Louvain, Lic math, 56; Stanford Univ, MSc, 59, PhD(statist), 61. *Prof Exp:* Instr math, Univ Lovanium, Leopoldville, 56-57; from asst prof to prof math & statist, Purdue Univ, 62-76; Unidel Chair prof Statist & comput sci, Univ Del, 76-78; AT DEPT SYSTS INDUST ENG, UNIV ARIZ, 78- *Concurrent Pos:* Consult, Gen Motors Res Labs, 64-66 & Bell Labs, 80; vis prof, Cornell Univ, 68-69; dept ed appl stochastic model, Mgt Sci, 74-80; ed, J Appl Probability, 79-; fac vis, IBM Res Ctr, 81; founding ed, Stochastic Models, 84- *Honors & Awards:* Lester R Ford Award, Math Asn Am. *Mem:* Fel Inst Math Statist; Math Asn Am; Opers Res Soc Am; Inst Elec & Electronic Engrs; Sigma Xi. *Res:* Probability theory; numerical methods in probability; queueing theory; Markov chains; general stochastic processes; communication engineering. *Mailing Add:* Dept Systs Indust Eng Univ Ariz Tucson AZ 85721

NEUVAR, ERWIN W, b Hallettsville, Tex, Mar 13, 30; m 56; c 2. INORGANIC CHEMISTRY. *Educ:* Tex A&M Univ, BS, 52, PhD(inorg chem), 62. *Prof Exp:* Chemist, Hanford Atomic Prod Oper, Gen Elec Co, 55-58; SR CHEMIST, MINN MINING & MFG CO, 62- *Res:* Ion exchange membrane technology; radiochemical analysis; microwave spectroscopy; fluorine and polymer chemistry; gas chromatography. *Mailing Add:* 2214 Beech St St Paul MN 55119

NEUWIRTH, JEROME H, b Brooklyn, NY, Mar 7, 31; m 57; c 2. MATHEMATICS. *Educ:* City Col New York, BS, 52; Univ Ill, MS, 54; Mass Inst Technol, PhD(math), 59. *Prof Exp:* Asst prof math, Rutgers Univ, 59-63; mathematician, NASA, 63-65; assoc prof math, Hunter Col, 65-67; assoc prof, 67-73, PROF MATH, UNIV CONN, 73- *Concurrent Pos:* Vis prof, TU Delft, Netherlands, 90. *Mem:* Am Math Soc. *Res:* Harmonic analysis. *Mailing Add:* Dept of Math Univ of Conn Storrs CT 06268

NEUWIRTH, MARIA, b Zagreb, Yugoslavia, Nov 27, 44; Can citizen. ELECTRON MICROSCOPY SERVICES. *Educ:* Univ Toronto, BSc, 67, MSc, 71, PhD(insect physiol), 75. *Prof Exp:* Lectr histol, Univ Toronto, 75-76; NIH fel, 77-80 & Nat Res Coun, 80-82; fac lectr, McGill Univ, 82-85; SECT HEAD, ALTA ENVIRON CTR, 85- *Mem:* Can Micros Soc. *Res:* Supervise electron microscopy service; develop new techniques for client departments. *Mailing Add:* Alta Environ Ctr Bag 4000 Vegreville AB T0B 4L0 Can

NEUWIRTH, ROBERT SAMUEL, b New York, NY, July 11, 33; m 57; c 4. OBSTETRICS & GYNECOLOGY. *Educ:* Yale Univ, BS, 55, MD, 58. *Prof Exp:* Intern surg, Columbia-Presby Med Ctr, 58-59, resident obstet & gynec, 59-64; from asst prof to assoc prof, Columbia Univ, 64-71; prof, Albert Einstein Col Med, 71-72; PROF OBSTET & GYNEC, COLUMBIA UNIV, 72-; DIR OBSTET & GYNEC, WOMAN'S HOSP, ST LUKE'S MED CTR, 74-; AT DEPT OBSTERICS, COLUMBIA UNIV. *Concurrent Pos:* Am Cancer Soc grant, Columbia Univ, 63-64; dir obstet & gynec, Bronx-Lebanon Hosp Ctr, 67-72; consult, Nat Inst Child Health & Human Develop, 71- & Fertil Control, WHO, 72-; consult, Wausau Ins Co. *Mem:* Soc Gynec Invest; Am Col Obstet & Gynec. *Res:* Gynecologic endoscopy; fertility control; methods of female sterilization; infertility and reproductive failure. *Mailing Add:* 1111 Amsterdam Ave New York NY 10025

NEUZIL, EDWARD F, b Chicago, Ill, Oct 12, 30; m 55; c 2. NUCLEAR CHEMISTRY. *Educ:* NDak State Col, BS, 52; Purdue Univ, MS, 54; Univ Wash, PhD(nuclear chem), 59. *Prof Exp:* From asst prof to assoc prof, 59-66, PROF CHEM, WESTERN WASH UNIV, 66- *Mem:* Am Chem Soc; Fedn Am Scientists. *Res:* Nuclear fission of lighter elements; geochemistry; chemical kinetics; radiation damage. *Mailing Add:* Dept Chem Western Wash Univ 516 High St Bellingham WA 98225

NEUZIL, JOHN PAUL, b Decorah, Iowa, Aug 8, 42; m 66; c 1. MATHEMATICS. *Educ:* Univ Iowa, BA, 64, MS, 66, PhD(math), 69. *Prof Exp:* Asst prof math, 69-77, ASSOC PROF MATH, KENT STATE UNIV, 77- *Mem:* Am Math Soc. *Res:* Geometric topology. *Mailing Add:* Dept Math Kent State Univ 301A Merrill Hall Kent OH 44242

NEUZIL, RICHARD WILLIAM, b Chicago, Ill, Sept 4, 24; wid; c 2. CHEMISTRY. *Educ:* Roosevelt Univ, BS, 50. *Prof Exp:* Spectroscopist, Universal Oil Prod Co, 50-55, supvr, 55-58, proj leader radiation chem, 58-59, proj leader air pollution control, 59-63, assoc res coordr catalyst eval, 63-66, res coordr, Catalyst Eval, 66-76, res mgr separations res, 76-84; RETIRED. *Mem:* AAAS; Am Chem Soc. *Res:* Determination of physical and thermochemical constants; thermodynamic properties; continuous process instrumentation; methods for separation and purification of chemical compounds for use on an industrial scale. *Mailing Add:* 527 Eldon Pl Downers Grove IL 60516

NEVA, FRANKLIN ALLEN, b Cloquet, Minn, June 8, 22; m 47; c 3. MICROBIOLOGY, INTERNAL MEDICINE. *Educ:* Univ Minn, MD, 46; Am Bd Internal Med, dipl, 54. *Prof Exp:* House officer med, Boston City Hosp, 46-47, asst resident internal med, Harvard Med Servs, 49-50; Nat Res Coun fel poliomyelitis, dept microbiol, Sch Pub Health, Harvard Univ, 50-51, Nat Found Infantile Paralysis fel virol, Res Div, Infectious Dis, Children's Hosp, Harvard Med Sch, 51-53; asst prof res bact & instr res med, Sch Med, Univ Pittsburgh, 53-55; from asst prof to assoc prof trop pub health, Sch Pub Health, Harvard Univ, 55-64, John Laporte Given prof, 64-69; CHIEF LAB PARASITIC DIS, NAT INST ALLERGY & INFECTIOUS DIS, 69- *Concurrent Pos:* Area consult, Vet Admin, 57-64; mem, Comn Parasitic Dis & assoc mem, Comn Virus Infections, Armed Forces Epidemiol Bd, 60-68 & 65-68; mem, Latin Am Sci Bd, Nat Acad Sci-Nat Res Coun, 63-68; mem, bd sci counr, Nat Inst Allergy & Infectious Dis, 66-69, Study Sect Virus & Rickettsial Dis, 68-70; mem med adv bd, Leonard Wood Mem Found, 68-70. *Honors & Awards:* Bailey K Ashford Award, 65; Smadel Award, Infectious Dis Soc Am, 85; Presidential Meritorious Exec Rank Award, 85. *Mem:* Soc Exp Biol & Med; Asn Am Physicians; Infectious Dis Soc Am; Am Soc Trop Med & Hyg. *Res:* Virus, rickettsial and parasitic diseases; clinical infectious diseases. *Mailing Add:* Bldg 4 Rm 126 NIH Bethesda MD 20892

NEVAI, PAUL, b Budapest, Hungary, July 20, 48; US citizen; m 82. ORTHOGONAL POLYNOMIALS, APPROXIMATION THEORY. *Educ:* Leningrad State Univ, MSc, 71; Hungarian Acad Scis, PhD(math), 73. *Prof Exp:* PROF MATH, OHIO STATE UNIV, 77- *Concurrent Pos:* Prin investr, NSF, 77-; sr res Fulbright fel, Hungarian Acad Sci, 85-86; vis prof, Univ Paris, Orsay, 85, Catholic Univ Louvain la Neuve, Belgium, 86; distinguished vis prof, Univ Alta, 86; Carolina res prof, Univ SC, 87-88; dir, NATO Advan Study Inst on Orthogonal Polynomials, 88-89. *Honors & Awards:* K Renyi Prize, Univ Budapest, 71; G Grunwald Prize, Bolyai Math Soc, Budapest, 73. *Mem:* Am Math Soc; Soc Indust Appl Math. *Res:* Contemporary theory of general orthogonal polynomials. *Mailing Add:* Dept Math Ohio State Univ 231 W 18th Ave Columbus OH 43210

NEVALAINEN, DAVID ERIC, b Moose Lake, Minn, June 30, 44; m 66; c 1. LABORATORY MEDICINE, DIAGNOSTICS. *Educ:* Univ Minn, Minneapolis, BS, 66, PhD(path), 72. *Prof Exp:* Instr lab sci, Hibbing Area Tech Inst, Minn, 69-72; res assoc lab med, Hibbing Gen Hosp, Minn, 69-72; lab supvr, Fairbanks Mem Hosp, Alaska, 72-73; asst prof med technol, Mich Technol Univ, 73-76, assoc prof, 76-79, assoc dir dept, 73-79; assoc prof med technol, Univ Wis-Milwaukee, 79-81; sr hematologist, 81-82, clin proj mgr, 82-85, sr tech mgr, Venture Technol, 85-86, MGR CLIN AFFAIRS, ABBOTT DIAGNOSTICS, ABBOTT LABS, 86- *Concurrent Pos:* Consult, Diag Div, Abbott Labs, Dallas, 76-79; assoc ed, Am J Med Technol. *Honors & Awards:* Distinguished Serv Award, Am Soc Clin Pathologists, 88. *Mem:* Am Soc Clin Pathologists; Am Asn Blood Banks; AAAS. *Res:* Development, manufacture and testing of diagnostic products for use in the clinical laboratory. *Mailing Add:* Dept 49C Abbott Labs Abbott Park IL 60064

NEVE, RICHARD ANTHONY, b Los Angeles, Calif, Nov 3, 23; m 53; c 2. BIOCHEMISTRY, MARINE CHEMISTRY. *Educ:* Loyola Univ, BS, 48; Univ San Francisco, MS, 51; Univ Ore, PhD(biochem), 56. *Prof Exp:* Biologist, US Naval Radiol Defense Lab, Calif, 49-51; asst biochem, Med Sch, Univ Ore, 51-56, res assoc, 56-58; USPHS fel, Univ Calif, 58-60; dir chem lab, Providence Hosp, 60-62; assoc prof biol & chmn dept, Seattle Univ, 62-66; prof biochem & dean grad sch, Cent Wash State Col, 66-70; prof marine path & toxicol, Univ Alaska, 70-90; CHIEF SCIENTIST, LAND & WATER RESOURCES, 90- *Mem:* Am Chem Soc; Am Fisheries Soc. *Res:* Porphyrin metabolism; hemoglobin synthesis; iron metabolism; application of enzymology; biochemical evolution of hemoglobins; marine carotenoids; biochemical behavior patterns; paralytic shellfish poisoning; aquaculture; stress and disease in aquatic fish and invertebrates; arctic resources. *Mailing Add:* 17108 SE 29th Ct Bellevue WA 98008-5671

NEVELS, ROBERT DUDLEY, b Hopkinsville, Ky, Apr 14, 46; m 87; c 2. ANTENNA THEORY, MILLIMETER WAVE DEVICES. *Educ:* Univ Ky, BS, 69; Ga Inst Technol, MS, 73; Univ Miss, PhD(elec eng), 79. *Prof Exp:* Asst prof electromagnetics, Univ Miss, 77-78; ASST PROF ELECTROMAGNETICS, TEX A&M UNIV, 78- *Concurrent Pos:* Assoc ed, Inst Elec & Electronic Engrs Trans Antennas & Propagation, 86-, Microwave & Optical Technol Letters, 87-; mem, Union Radio Sci Int Comm B. *Mem:* Inst Elec & Electronic Engrs; Sigma Xi. *Res:* Electromagnetic scattering by integral equation techniques. *Mailing Add:* Dept Elec Eng Tex A&M Univ College Station TX 77843

NEVES, RICHARD JOSEPH, b New Bedford, Mass. UNIONIDAE, ALOSID FISHES. *Educ:* Univ RI, BS, 68; Univ Maine, MS, 73; Univ Mass, PhD(fisheries), 77. *Prof Exp:* Asst leader, Va Coop Fishery Res Unit, 78-83, actg leader, 83-85, UNIT LEADER, VA COOP FISH & WILDLIFE RES UNIT, VA TECH, 85-, ADJ PROF, 78- *Mem:* Am Fisheries Soc; Am Malacological Union; Am Inst Fishery Res Biologists; NAm Benthological Soc. *Res:* Biology of freshwater mussels, endangered and threatened species, stream ecology, and anadromous fishes. *Mailing Add:* Dept Fisheries & Wildlife Sci Va Tech Blacksburg VA 24061-0321

NEVEU, DARWIN D, b Green Bay, Wis, Feb 26, 33. ANALYTICAL CHEMISTRY. *Educ:* Wis State Univ, Oshkosh, BS, 58. *Prof Exp:* Chemist, Freeman Chem Co, 63-65; asst chief chemist, WVa Pulp & Paper Co, 65-66; chemist, Newport Army Ammunition Ctr, FMC Corp, 66-67 & Celanese Plastics, 67-69; mgr org lab, Crobaugh Labs, 69-70; LAB DIR CHEM ANAL & CONSULT, NALIN-NEVEU LABS, 70- *Concurrent Pos:* Fac adv, Wooster Agr Tech Inst, 73-79; guest lectr, Columbus Tech Inst, 75-84. *Mem:* Am Chem Soc; Nat Asn Corrosion Engrs; Soc Plastics Engrs; Am Soc Metals; Soc Appl Spectros. *Mailing Add:* Nalin-Neveu Memorial Lab 2641 Cleveland Ave Columbus OH 43211

NEVEU, MAURICE C, b Nashua, NH, Feb 3, 29; m 55; c 5. FUELS & EXPLOSIVES, BIOCHEMISTRY. *Educ:* Univ NH, BS, 52, MS, 55; Ill Inst Technol, PhD(chem), 59. *Prof Exp:* Asst, Univ NH, 53-54; asst, Ill Inst Technol, 54-58; instr, Ohio Northern Univ, 58-59, asst prof chem, 59-60; asst prof, Longwood Col, 60-64; ASSOC PROF CHEM, STATE UNIV NY COL FREDONIA, 64- *Concurrent Pos:* Fuel res, NASA; explosives res, US Air Force. *Mem:* Am Chem Soc; Sigma Xi; Am Asn Univ Prof. *Res:* Mechanisms of organic reactions; chemical kinetics; catalysis; isotope effects; mechanism of ester hydrolysis; enzymology; metabolic pathways by carbon-14 tracers; differential scanning calorimetry; phase diagrams of binary and ternary hydrocarbon systems from DSC data; mechanisms of explosives detonations; synthesis of insensitive explosives; fuel degradation mechanisms. *Mailing Add:* 18 Hanover St Silver Creek NY 14136

NEVID, JEFFREY STEVEN, b Brooklyn, NY. CLINICAL PSYCHOLOGY, BEHAVIOR THERAPY. *Educ:* State Univ NY at Binghamton, BA, 72; State Univ NY at Albany, PhD(clin psychol), 76. *Prof Exp:* Asst prof psychol, Hofstra Univ, 77-81; ASSOC PROF & DIR CLIN PSYCHOL, ST JOHN'S UNIV, 81- *Mem:* Am Psychol Asn. *Res:* Writings on behavior therapy, abnormal psychology, and psychological adjustment; behavior therapy for anxiety; smoking reduction and obesity. *Mailing Add:* Dept Psychol St Johns Univ Grand Central & Utopia Jamaica NY 11439

NEVILL, GALE E(RWIN), JR, b Houston, Tex, Nov 17, 33; m 54; c 2. AERONAUTICAL & ASTRONAUTICAL ENGINEERING. *Educ:* Rice Univ, BA, 54, BSME, 55, MSME, 57; Stanford Univ, PhD(eng mech), 61. *Prof Exp:* Res engr, McEvoy Co, 58; res asst mech, Stanford Univ, 59-60; sr res engr, Southwest Res Inst, 60-64; from assoc prof to prof mech, 68, chmn dept, 68-73, PROF ENG SCI, UNIV FLA, 68- *Concurrent Pos:* Consult, Lawrence Livermore Nat Lab, 76-88. *Mem:* Am Soc Mech Engrs; Am Soc Eng Educ; Am Inst Aeronaut & Astronaut; Am Asn Artificial Intel. *Res:* Design theory and methodology; automated design; robotics. *Mailing Add:* Dept Eng Sci Col Eng Univ Fla Gainesville FL 32611

NEVILL, WILLIAM ALBERT, b Indianapolis, Ind, Jan 1, 29; m 79; c 5. ORGANIC CHEMISTRY. *Educ:* Butler Univ, BS, 51; Calif Inst Technol, PhD(chem biol), 54. *Prof Exp:* Res chemist, Procter & Gamble Co, 54; from asst prof to assoc prof chem, Grinnell Col, 56-67, chmn dept, 64-67; prof, Purdue Univ & chmn dept chem, Indianapolis Campus, 67-70; asst dean acad affairs, 70-71, actg dean, 38th St Campus, 71-72, actg dean, Sch Eng & Technol, 72-74, dean, Sch Sci, 72-79, dir grad studies, Ind Univ-Purdue Univ, Indianapolis, 79-83; acad chancellor, 83-85, Prof, LA STATE UNIV, 83- *Concurrent Pos:* Consult, Eli Lilly, 71-81 & Lilly Indust Coatings, 80-83, Foreign Sci, 86-, Kendall/Hunt Publ Co, 87- *Mem:* Am Chem Soc. *Res:* Mechanisms of organic reactions relating to small ring compounds, synthesis of nitrogen heterocycles and industrial coating polymers; biotechnology applications. *Mailing Add:* Dept Chem 8515 Youree Dr Shreveport LA 71115

NEVILLE, DAVID MICHAEL, JR, MEMBRANE BIOLOGY, TOXIN TRANSPORT. *Educ:* Univ Rochester, MD, 59. *Prof Exp:* SECT CHIEF, NIMH, NIH, 60- *Res:* Immuno-toxins. *Mailing Add:* 9624 Parkwood Dr Bethesda MD 20814

NEVILLE, DONALD EDWARD, b Los Angeles, Calif, Apr 5, 36; m 77; c 1. THEORETICAL HIGH ENERGY PHYSICS. *Educ:* Loyola Univ, Los Angeles, BS, 57; Univ Chicago, SM, 60, PhD(physics), 62. *Prof Exp:* Fel physics, Univ Calif, Berkeley, 62-64, Europ Orgn Nuclear Res, Geneva, 64-65 & Lawrence Radiation Labs, Livermore, 65-67; assoc prof, 67-80, PROF PHYSICS, TEMPLE UNIV, 80- *Mem:* Am Phys Soc; AAAS. *Res:* Impact of particle physics on cosmology and astrophysics; quantum gravity. *Mailing Add:* Dept of Physics Temple Univ Philadelphia PA 19122

NEVILLE, JAMES RYAN, b San Jose, Calif, May 13, 25; m 49; c 2. MEDICAL PHYSIOLOGY, SCIENCE ADMINISTRATION. *Educ:* Stanford Univ, AB, 49, MA, 51, PhD(physiol), 55. *Prof Exp:* Teaching asst physiol, Stanford Univ, 52-55; aviation physiologist, US Air Force Sch Aerospace Med, 55-57, from asst prof to assoc prof biophys, 57-63, chief biophys sect, 63-70; clin res physiologist, 70-79, asst dir res contract, Letterman Army Inst Res, 79-84; PRES, CHEMTRONICS INC, 85- *Mem:* AAAS; Biophys Soc; Instrument Soc Am; Aerospace Med Asn; Am Physiol Soc; Int Soc Oxygen Transp Tissues; Sigma Xi. *Res:* Respiratory physiology; gas transport; enzymes; methodology and bioinstrumentation; polarography. *Mailing Add:* PO Box 9106 La Jolla CA 92038-9106

NEVILLE, JANICE NELSON, b Schenectady, NY, Dec 1, 30; m 53; c 2. NUTRITION, PUBLIC HEALTH & EPIDEMIOLOGY. *Educ:* Carnegie Inst Technol, BS, 52; Univ Ala, MS, 53; Univ Pittsburgh, MPH, 62, DSc(hyg), 64. *Prof Exp:* Instr diet ther & clin dietitian, Sch Nursing, Hillman Clins, Univ Ala, 54; res dietitian, Grad Sch Pub Health, Univ Pittsburgh, 56-61, res asst nutrit, 61-64, asst res prof, 64-65; from asst prof to assoc prof, 65-74, chmn dept, 74-82, PROF NUTRIT, CASE WESTERN RESERVE UNIV, 74- *Concurrent Pos:* Food & Drug Admin Adv Comt, 77-78; bd trustees, Am Heart Asn, NE Ohio Affil, 75-92, mem, Coun Epidemiol, chmn, Food Sci Prod Rev Comt, Am Heart Asn, Nat Ctr, mem, Heart Guide Pub Interest & Oversight Panel; mem, NCent Regional Res Comt, Am Heart Asn, 84-86, Res Task Force, 84, pres 88-89, coordr, Speaker House Deleg, 83-84, pres & chmn bd dirs, 87-88; bd dirs, Am Dietetic Asn, 82-84 & 86-88. *Mem:* AAAS; Am Dietetic Asn; NY Acad Sci; Am Pub Health Asn; Soc Nutrit Educ; Am Col Nutrit; Am Heart Asn. *Res:* Diet therapy; community health, obesity and weight control, serum lipids, factors affecting food choice and response to counseling. *Mailing Add:* Dept Nutrit Case Western Reserve Univ Cleveland OH 44106-2839

NEVILLE, MARGARET COBB, b Greenville, SC, Nov 4, 34; m 57; c 2. CELL BIOLOGY. *Educ:* Pomona Col, BA, 56; Univ Pa, PhD(physiol), 62. *Prof Exp:* Res assoc cell physiol, Dept Molecular Biol, Pa Hosp, Philadelphia, 64-68; from instr to assoc prof, 68-82, PROF PHYSIOL, MED CTR, UNIV COLO, DENVER, 82- *Concurrent Pos:* Prin investr, NIH grant, 74-; res career develop award, NIH, 75-80. *Mem:* Am Physiol Soc; Soc Cell Biol. *Res:* Mammary gland biology; lactation; epithelial transport and secretion. *Mailing Add:* Dept of Physiol C240 Univ of Colo Med Ctr Denver CO 80262

NEVILLE, WALTER EDWARD, JR, b Rabun Gap, Ga, May 5, 24; m 48; c 2. ANIMAL SCIENCE. *Educ:* Univ Ga, BS, 47; Univ Mo, MS, 50; Univ Wis, PhD(genetics, animal husb), 57. *Prof Exp:* Asst county agr agent agr exten, Univ Ga, 47-49; asst animal breeding, Coastal Plain Exp Sta, Univ Ga, 50-51, assoc, 51-77, prof animal breeding, 77-89, emer prof, 89-; RETIRED. *Honors & Awards:* Sigma Xi Res Award, 75, 86. *Mem:* AAAS; Am Soc Animal Sci; Coun Agr Sci & Technol; Sigma Xi. *Res:* Crossbreeding of beef cattle and heritability estimates among various economic traits in beef cattle; physiology of reproduction in farm animals; nutritive requirements of beef cattle and their calves. *Mailing Add:* 2429 Madison Dr Tifton GA 31794-2570

NEVILLE, WILLIAM E, b Fairbury, Nebr, Apr 13, 19; m 58; c 5. SURGERY. *Educ:* Univ Nebr, BS & MD, 43. *Prof Exp:* Assoc prof surg, Univ Ill Col Med, 62-71; PROF SURG & DIR CARDIOTHORACIC SURG, COL MED & DENT NJ, 71- *Mem:* Fel Am Col Surg; Soc Thoracic Surg; Am Asn Thoracic Surg; Int Cardiovasc Soc; Am Surg Asn. *Res:* Cardiothoracic surgery. *Mailing Add:* Dept Surg Univ Med & Dent NJ 100 Bergen St Newark NJ 07103

NEVIN, ROBERT STEPHEN, b New York, NY, Oct 20, 33; m 55. ORGANIC CHEMISTRY, POLYMER CHEMISTRY. *Educ:* Queens Col, NY, BS, 55; St John's Univ, NY, MS, 57; State Univ NY Col Forestry, Syracuse, PhD(chem), 61. *Prof Exp:* Res chemist, Esso Res & Eng Co, 61-63; sr res chemist, J T Baker Chem Co, 63-67; sr res chemist, 67, RES SCIENTIST, LILLY RES LAB, ELI LILLY & CO, 67- *Mem:* Am Chem Soc. *Res:* Organic chemistry of polymers; biocompatibility of polymers; monomer synthesis; chemical modification of polymers. *Mailing Add:* Lilly Res Lab MC727 Eli Lilly & Co Indianapolis IN 46285-0002

NEVINS, ARTHUR JAMES, b New York, NY, Sept 22, 37; m 60; c 2. COMPUTER SCIENCE. *Educ:* Mass Inst Technol, BS, 59; Univ Rochester, MA, 62, PhD(econ), 65. *Prof Exp:* Fel, Carnegie Inst Technol, 64-65; sr scientist logistics res proj, George Washington Univ, 65-69, sr staff scientist, Inst Mgt Sci, 69-72; res scientist, Artificial Intel Lab, Mass Inst Technol, 72-74; PROF COMPUT INFO SYSTS, GA STATE UNIV, 82-, MEM FAC, 74- *Mem:* Asn Comput Mach. *Res:* Artifical intelligence, especially computer programs for proving theorems, balancing assembly lines, decomposing shapes, and learning concepts; knowledge-based decision support systems. *Mailing Add:* Dept Comput Info Systs Ga State Univ Atlanta GA 30302-4015

NEVINS, DONALD JAMES, b San Luis Obispo, Calif, July 6, 37; m 62; c 2. PLANT PHYSIOLOGY. *Educ:* Calif State Polytech Col, BS, 59; Univ Calif, Davis, MS, 61, PhD(plant physiol), 65. *Prof Exp:* NIH res assoc chem, Univ Colo, 65-67; from asst prof to prof bot, Iowa State Univ, 67-84; DEPT HEAD VEG CROPS, UNIV CALIF, 84- *Concurrent Pos:* Vis prof, Osaka City Univ, Japan Soc Prom Sci, 74-75. *Mem:* AAAS; Am Soc Plant Physiol; Am Chem Soc; Japanese Soc Plant Physiologists. *Res:* Physiology of cell walls, growth and development. *Mailing Add:* Dept Veg Crops Univ Calif Davis CA 95616

NEVINS, WILLIAM MCCAY, b Pasedena, Calif, Aug 28, 48; m 79; c 1. NON-INDUCTIVE CURRENT DRIVE, WAVES & INSTABILITIES. *Educ:* Univ Calif, Berkeley, AB, 70, PhD(physics), 79. *Prof Exp:* Res asst, Univ Calif, Berkeley, 74-78; res assoc, Princeton Plasma Physics Lab, 78-79; physicist, 79-91, INT THERMONUCLEAR EXP REACTOR PROG LEADER, LAWRENCE LIVERMORE NAT LAB, 91- *Concurrent Pos:* Div & heating phys task leader, Int Exp Thermonuclear Reactor Study, 88- *Mem:* Fel Am Phys Soc. *Res:* Design of the International Thermonuclear Experimental Reactor; engineering test results for the purpose of developing magnetics confinement fusion as a viable source of commercial electric power. *Mailing Add:* Lawrence Livermore Nat Lab PO Box 5511 L-644 Livermore CA 94550

NEVIS, ARNOLD HASTINGS, neurology, biophysics; deceased, see previous edition for last biography

NEVISON, CHRISTOPHER H, b Philadelphia, Pa, Nov 24, 45; m 72; c 2. PARALLEL ALGORITHMS, DISCRETE EVENT SIMULATION. *Educ:* Dartmouth Col, BA, 67; Oxford Univ, BA, 69; Stanford Univ, MSc, 70, PhD(math), 74. *Prof Exp:* Lectr math, Univ Calif, Davis, 73-74; from asst prof to assoc prof math, Colgate Univ, 74-83, assoc prof computer sci, 83-84, PROF COMPUTER SCI, COLGATE UNIV, 84-, CHMN DEPT, 86- *Concurrent Pos:* Mem, Comt Teaching Undergrad Math, Math Asn Am, 82-86; prin investr, NSF, 89-92; vis lectr, Univ Wales, Col Cardiff, 91. *Mem:* Math Asn Am; Opers Res Soc Am; Asn Comput Mach. *Res:* Methods of discrete event simulation using parallel computers; implementation and analysis of parallel algorithms. *Mailing Add:* Computer Sci Dept Colgate Univ Hamilton NY 13346

NEVITT, MICHAEL VOGT, b Lexington, Ky, Sept 7, 23; m 46; c 4. METAL PHYSICS. *Educ:* Univ Ill, BS, 44, PhD(metall eng), 54; Va Polytech Inst, MS, 51. *Prof Exp:* Res metallurgist, Olin Mathieson Chem Corp, 46-48; from asst prof to assoc prof metall eng, Va Polytech Inst, 48-55 & head dept, 53-55; assoc phys metallurgist, Metall Div, Argonne Nat Lab, 55-64, group leader, 55-66, sr phys metallurgist, 64, dir metall div, 66-69, dep lab dir, 69-81, sr scientist, 81-90; ADJ PROF, DEPT PHYSICS & ASTRON, CLEMSON UNIV, 90- *Concurrent Pos:* Mem adv comt, Polysci Corp, 54; vis prof, Univ Sheffield, 65-66; mem univ adv comts, Lehigh Univ, 76-79 & Univ Ky, 78-79; consult ed, J Contemp Physics, 77-88 & Res Mechanica Lett, 80-82; mem adv comt mat res, NSF, 80-82; accreditation team, Accreditation Bd Eng Technol, 83- *Honors & Awards:* Achievement Award, Dept Energy, 90. *Mem:* Fel Am Soc Metals; Sigma Xi; Metall Soc; AAAS; Mat Resource Soc. *Res:* Physical and engineering metallurgy of nuclear materials; alloy theory of transition and actinide elements; magnetic properties; advanced ceramics. *Mailing Add:* One Boatswain Keowee Key Salem SC 29676

NEVITT, THOMAS D, b Kewanee, Ill, July 22, 25; m 49; c 3. ORGANIC CHEMISTRY, PETROLEUM CHEMISTRY. *Educ:* Bradley Univ, BS, 47; Iowa State Col, MS, 50, PhD(phys org chem), 53. *Prof Exp:* PROJ MGR, AMOCO OIL CO, 54- *Mem:* Am Chem Soc; Royal Soc Chem; Sigma Xi. *Res:* Catalysis; chemical kinetics; reaction engineering. *Mailing Add:* 1515 Lacabra Dr Albuquerque NM 87123

NEVIUS, TIMOTHY ALFRED, b Springfield, Ohio, Nov 27, 52; m 83; c 1. LIQUID CHROMATOGRAPHY DETECTORS, PROCESS INSTRUMENTATION. *Educ:* Wright State Univ, BSc, 78; Purdue Univ, PhD(chem), 84. *Prof Exp:* Res asst chem, Purdue Univ, 78-84; sr chemist, Dow Chem Co, 84-86; RES DIR, ANSPEC CO, INC, 86- *Mem:* Am Chem Soc; Asn Analytical Chemists; Am Soc Testing & Mat. *Res:* High performance analytical instruments, including thermal, spectroscopic, process and chromatographic fields; development of instruments for biotechnological, pharmaceuticals and analytical laboratories. *Mailing Add:* 622 Springbrook Ct Saline MI 48176

NEVLING, LORIN IVES, JR, b St Louis, Mo, Sept 23, 30; m 57; c 5. PLANT TAXONOMY. *Educ:* St Mary's Col, BS, 52; Wash Univ, AM, 57, PhD(bot), 59. *Hon Degrees:* DH, Lewis Univ, 85. *Prof Exp:* Researcher, Mo Bot Garden, 59; from asst cur to assoc cur, Arnold Arboretum, Harvard Univ, 59-69, cur, 69-73, supvr, Gray Herbarium, 63-72, assoc cur, 63-69, coordr bot syst collections, 72-73; cur & chmn dept bot, Field Mus Natural Hist, 73-77, asst dir sci & educ, 78-80, dir, 80-85, PRES MUS MGT CONSULT, 86-; CHIEF, ILL NATURAL HIST SERV, 87- *Concurrent Pos:* lectr biol, Harvard Univ, 68-69 & 71-73; mem, Conserv Comn, City Boston, 73. *Honors & Awards:* George R Cooley Prize, 70. *Mem:* Bot Soc Am; Am Soc Plant Taxon (secy, 66-71, pres, 77); Int Asn Plant Taxon; Am Inst Biol Sci; Sigma Xi. *Res:* Taxonomy of the Thymelaeaceae; flora of Veracruz, Mexico. *Mailing Add:* RR I Box 195 607 E Peabody Dr Monticello IL 61856

NEVO, EVIATAR, b Tel-Aviv, Israel, Feb 2, 29; div; c 1. MOLECULAR BIOLOGY, ZOOLOGY. *Educ:* Hebrew Univ, MSc, 58, PhD(biol), 64. *Prof Exp:* Vis prof zool, Univ Tex, 64-65; fel biol, Harvard Univ, 65-66; res assoc genetics, Hebrew Univ, 67-68, lectr, 68-70, sr lectr, 70-71; sr res fel biol, Univ Chicago, 72-73; res assoc, Mus Vert Zool, Univ Calif, Berkeley, 72-73; assoc prof, 73-75, PROF BIOL, UNIV HAIFA, 75-, DIR, INST EVOLUTION, 77-, INCUMBENT CHAIR EVOLUTIONARY BIOL, 84- *Concurrent Pos:* Res grants, US-Israel Binational Sci Found, 74-82, US-Israel Agr & Develop Fund, 80-84 & Wolfson Found, 86-89; fel, Guggenheim Found, 78-80. *Mem:* Soc Study Evolution (vpres, 78); Am Soc Naturalists; fel AAAS; Genetics Soc Am; Genetical Soc Israel; Zool Soc Israel; fel Explorers Club; fel AAAS; NY Acad Sci; fel Linnean Soc London. *Res:* Genetic variation in natural populations of plants and animals and its ecological correlates; origin and evolution of new species (speciation); quality of the marine environment; wild genetic resources for crop improvement; structure and evolution of agression in animals; 330 papers in evolutionary biology. *Mailing Add:* Inst Evolution Univ Haifa Mt Carmel 31999 Israel

NEW, JOHN COY, JR, b Little Rock, Ark, Jan 14, 48; m 75; c 2. VETERINARY MEDICINE, EPIDEMIOLOGY. *Educ:* Tex A&M Univ, BS, 70, DVM, 71; Univ Minn, MPH, 77. *Prof Exp:* Res assoc vet econ, Univ Minn, 75-77; ASSOC PROF PUB HEALTH, UNIV TENN, KNOXVILLE, 77- *Mem:* Am Vet Med Asn; Am Col Vet Prev Med; Conf Pub Health Veterinarians; Asn Teachers Vet Pub Health & Prev Med. *Res:* Zoonotic diseases; economics of animal diseases; diseases of wildlife and pet birds; food hygiene. *Mailing Add:* Dept Environ Pract PO Box 1071 Knoxville TN 37901

NEW, MARIA IANDOLO, b New York, NY, Dec 11, 28; m 49; c 3. PEDIATRICS. *Educ:* Cornell Univ, AB, 50; Univ Pa, MD, 54; Am Bd Pediat, dipl, 60. *Prof Exp:* Intern med, Bellevue Hosp, New York, 54-55; asst resident pediat, 55-57, asst pediatrician, Clin Res Ctr, 57-59, pediatrician, Outpatient Dept, 59-63, res investr diabetic study group, Comprehensive Care & Teaching Prog, 58-61, instr pediat, 58-63, asst prof & asst attend pediatrician, 63-68, assoc prof & assoc attend pediatrician, 68-71, PROF PEDIAT & ATTEND PEDIATRICIAN, NEW YORK HOSP-CORNELL MED CTR, 71-, DIR PEDIAT METAB & ENDOCRINE CLIN & DIV HEAD PEDIAT ENDOCRINOL, 64-, ASSOC DIR PEDIAT CLIN RES CTR, 66- *Concurrent Pos:* Fel pediat metab & renal dis, New York Hosp-Cornell Med Ctr, 57-58, res fel med, 62-64, Harold & Percy Uris prof pediat endocrinol & metab, 78; vis physician, Rockefeller Univ, 71-, consult, Albert Einstein Col Med, 74- & United Hosp, Port Chester, NY, 77-; vchmn Dept Pediat, 74-80, chmn, Pediat, NY Hosp-Cornell Med Ctr, 1980- *Mem:* AAAS; Am Acad Pediat; Am Pediat Soc; Soc Pediat Res; Am Fedn Clin Res; Endocrine Soc. *Res:* Pediatric endocrinology and renal diseases; juvenile hypertension; pediatric pharmacology; growth and development from the biochemical viewpoint. *Mailing Add:* 525 E 68th St New York NY 10021

NEWBALL, HAROLD HARCOURT, b Columbia, SC, 42. RESPIRATORY PHYSIOLOGY & TOXICOLOGY. *Educ:* Loma Linda Univ, MD, 68. *Prof Exp:* CHIEF PHYSIOL DIV, US BIOMED LAB, 82-; AT JOHN HOPKINS. *Res:* Chemical mediators. *Mailing Add:* NIH Bldg 29 Rm 307 Bethesda MD 20892

NEWBERGER, BARRY STEPHEN, b Huntington, WVa, June 19, 45; m 75. LINEAR STABILITY THEORY, ACCELERATOR PHYSICS. *Educ:* Carnegie Inst Technol, BS, 67; Princeton Univ, PhD(astrophys sci), 76. *Prof Exp:* Staff mem plasma physics, Laser Theory Group, Los Almos Sci Lab, 72-77, staff mem plasma physics, Intense Particle Beam Theory Group, 77-84; scientist, Mission Res Corp, Albuquerque, NMex, 84-87; RES SCIENTIST, INST FUSION STUDIES, UNIV TEX-AUSTIN, 87- *Concurrent Pos:* Vis res scientist, Inst Fusion Studies, Univ Tex-Austin, 82-; guest scientist, Los Alamos Nat Lab, 84-; consult, Mission Res Corp, 87-; vis scientist, Nat Lab High Energy Physics, 90. *Mem:* Am Phys Soc; Sigma Xi. *Res:* Accelerator physics especially as applied to high energy physics experiments; theoretical plasma physics with emphasis on the linear stability theory, physics of charged particle beam interactions with plasma, advanced accelerator concepts and low frequency instabilities in magnetic fusion plasmas; mathematical and computational physics as applied in these areas. *Mailing Add:* Inst Fusion Studies, RLM 11-222 Univ Texas, 26th & Speedway Austin TX 78712-1060

NEWBERGER, EDWARD, b New York, NY, Feb 15, 40. MATHEMATICS. *Educ:* City Col New York, BS, 61; Ind Univ, PhD(partial differential equations), 69. *Prof Exp:* Asst prof, 70-74, ASSOC PROF MATH, STATE UNIV NY COL BUFFALO, 74- *Mem:* Am Math Soc; Am Math Asn; NY Acad Sci; Math Soc France. *Res:* Pseudo-differential operators; Gevrey classes; asymptotic Gevrey classes; partial differential equations. *Mailing Add:* Dept of Math State Univ NY 1300 Elmwood Ave Buffalo NY 14222

NEWBERGER, MARK, b Brooklyn, NY, Mar 26, 42. CHEMICAL ENGINEERING. *Educ:* Cooper Union, BChE, 62; Univ Mich, MS, 63, PhD, 67. *Prof Exp:* From develop engr to sr dev engr, 63-69; sr process engr, Eng Div, 69-74, process supvr, 75-79, MGR PROCESS ANALYSIS, AM CYANAMID CO, WAYNE, NJ, 79- *Concurrent Pos:* Adj instr, Jersey City State Col, 68-71, consult, 70-71; sr lectr, Stevens Inst Technol, 75-79. *Mem:* Am Inst Chem Engrs. *Res:* Process control; dynamics; optimization and optimal control; large system simulation; mathematical modeling; chemical reactor engineering, distillation design and analysis; computer applications. *Mailing Add:* Am Cyanamid Co Chem Group Eng Cyanamid Plaza Wayne NJ 07470

NEWBERGER, STUART MARSHALL, b New York, NY, Oct 4, 38; m 64; c 1. MATHEMATICS. *Educ:* City Col New York, BEE, 60; Mass Inst Technol, PhD(math), 64. *Prof Exp:* From instr to asst prof math, Univ Calif, Berkeley, 64-69; ASSOC PROF MATH, ORE STATE UNIV, 69- *Mem:* Am Math Soc; Math Asn Am. *Res:* Functional analysis, nonlinear analysis; operator theory, including unbounded operators in Hilbert Space and partial differential operators; generalized function theory; mathematical physics. *Mailing Add:* 1845 NW Garfield Corvallis OR 97330

NEWBERNE, JAMES WILSON, b Adel, Ga, Dec 1, 23; m 49; c 3. VETERINARY PATHOLOGY. *Educ:* Ala Polytech Inst, DVM, 50, MS, 54. *Prof Exp:* Instr path, Ala Polytech Inst, 52-55; pathologist, Pitman-Moore Co, 55-57, asst dir path dept, 57-61; head dept path & toxicol, Wm S Merrell Co, 62-69, dir drug safety & metab, Richardson-Merrell Inc, 69-76, vpres drug safety assessment, Merrell Nat labs div, Richardson-Merrell Inc, 76-80, VPRES & DIR, DRUG SAFETY ASSESSMENT, MERRELL DOW RES CTR, MERRELL DOW PHARMACEUT INC, 81-; ASSOC CLIN PROF PATH, COL MED, UNIV CINCINNATI, 63-, CLIN PROF LAB MED, MED CTR, 74- *Concurrent Pos:* Consult, Toxicol Protocol Comt, Nat Acad Sci-Nat Res Coun, 75-; mem health ministry, France, 76- *Mem:* Am Vet Med Asn; Am Col Vet Path; Am Asn Lab Animal Sci; NY Acad Sci; Soc Toxicol; Sigma Xi. *Res:* Toxicopathology; experimental pathology. *Mailing Add:* Merrell Dow Res Inst 2110 E Galbrath Rd Cincinnati OH 45215

NEWBERNE, PAUL M, b Adel, Ga, Nov 4, 20; m 45; c 2. NUTRITION, PATHOLOGY & TOXICOLOGY. *Educ:* Auburn Univ, DVM, 50, MSc, 51; Univ Mo, PhD(biochem, nutrit), 58. *Prof Exp:* Instr path, Auburn Univ, 50-51; instr microbiol, Univ Mo, 54-56, instr agr chem, 56-57, fel, NIH, 57-58; animal pathologist, Auburn Univ, 58-62; prof, 62-86, EMER PROF NUTRIT PATH, MASS INST TECHNOL & BOSTON UNIV SCH MED, 86- *Concurrent Pos:* Fel, Am Inst Nutrit, 89. *Honors & Awards:* Borden Award, Am Inst Nutrit, 88. *Mem:* AAAS; Am Inst Nutrit; Am Col Vet Path; Am Vet Med Asn; Am Asn Pathologists; hon mem Am Soc Toxicol Pathologists. *Res:* Nutritionally-induced experimental cancer; cardiovascular disease; nutritionally-induced teratology; nutritional pathology and toxicology; interaction of nutrition and toxicology. *Mailing Add:* Prof Dept Path Boston Univ Sch Med 80 E Concord St Boston MA 02118

NEWBERRY, ANDREW TODD, b Orange, NJ, Aug 30, 35; m 58; c 2. INVERTEBRATE ZOOLOGY, BIOSYSTEMATICS. *Educ:* Princeton Univ, AB, 57; Stanford Univ, PhD(biol), 65. *Prof Exp:* Nat Acad Sci-Nat Res Coun fel, 64-65; from asst prof to assoc prof, 65-81, PROF BIOL, UNIV CALIF, SANTA CRUZ, 81- *Concurrent Pos:* Adj res assoc, Cal Acad Sci, 84- *Mem:* AAAS; Western Soc Nat. *Res:* Invertebrate reproduction and development; evolution of development; biology of ascidian tunicates; biosystematics. *Mailing Add:* Dept Biol Univ of Calif Santa Cruz CA 95064

NEWBERRY, J(AMES) R(AYMOND), chemical engineering, waste treatment, for more information see previous edition

NEWBERRY, WILLIAM MARCUS, b Columbus, Ga, Nov 13, 38; m 65; c 3. INTERNAL MEDICINE. *Educ:* Northwestern Univ, BA, 60; Emory Univ, MD, 64. *Prof Exp:* Asst prof int med, Southwestern Med Sch, Univ Tex, 70-71; assoc prof, 71-75, dean col med & prof int med, 75-83, VPRES ACAD AFFAIRS, MED UNIV SC, 83- *Concurrent Pos:* Actg dean col med, Med Univ SC, 74-75; interim pres, Med Univ SC, 82, Winthrop Col, 86. *Mem:* Am Fedn Clin Res; Reticuloendothelial Soc. *Res:* Infectious diseases, specifically the epidemiology of the systemic mycoses, and the cellular immune responses to the granulomatous infections. *Mailing Add:* Med Univ SC 171 Ashley Ave Charleston SC 29425

NEWBERY, A CHRIS, b Broxbourne, Eng, July 12, 23; Can citizen; m 54; c 2. MATHEMATICS. *Educ:* Cambridge Univ, BA, 48; Univ London, BA, 53, PhD(math), 62; Univ BC, MA, 58. *Prof Exp:* Lectr math, Univ BC, 56-62; mathematician, Boeing Co, 62-63; asst prof, Univ Alta, Calgary, 63-65; assoc prof math & comput, Univ Ky, 65-67; mathematician, Boeing Co, Wash, 67-69; assoc prof math, 69-72, PROF COMPUT SCI, UNIV KY, 72- *Mem:* Soc Indust & Appl Math; Can Math Cong. *Res:* Numerical analysis; application of computers to problems, particularly in curve-fitting; linear algebra and polynomial problems. *Mailing Add:* Dept Math Univ Ky Lexington KY 40506

NEWBOLT, WILLIAM BARLOW, b Berea, Ky, Sept 29, 34; m 62; c 1. PHYSICS. *Educ:* Berea Col, BA, 56; Vanderbilt Univ, MS, 59, PhD(physics), 63. *Prof Exp:* From instr to assoc prof, 62-73, PROF PHYSICS, WASHINGTON & LEE UNIV, 73- *Mem:* Am Phys Soc; Am Asn Physics Teachers; Health Physics Soc. *Res:* Nuclear spectroscopy; Mossbauer effect. *Mailing Add:* Dept Physics Washington & Lee Univ Lexington VA 24450

NEWBORG, MICHAEL FOXX, b Philadelphia, Pa, Mar 23, 48; m 72; c 2. IMMUNOPHARMACOLOGY, CELLULAR IMMUNOBIOLOGY. *Educ:* Univ Md, BS, 70, PhD(zool), 77. *Prof Exp:* Postdoctoral assoc, Trudeau Inst, 77-80; from asst prof to assoc prof, Univ New Eng, 80-85; res scientist, 85-87, sr res scientist, 87-89, sr res investr, 89-90, MGR IMMUNOPHARMACOL, PFIZER CENT RES, 90- *Mem:* Am Asn Immunologists. *Res:* Discovery of novel immunotherapeutants; immuno modulations; cell immo growth factors. *Mailing Add:* Pfizer Central Res Eastern Point Rd Groton CT 06340

NEWBORN, MONROE M, b Cleveland, Ohio, May 21, 38; m 75; c 2. THEOREM PROVING BY COMPUTERS, GAME PLAYING BY COMPUTERS. *Educ:* Rensselaer Polytech Inst, BEE, 60; Ohio State Univ, MS, 62, PhD(elec eng), 67. *Prof Exp:* From asst prof to assoc prof computer sci, Columbia Univ, New York, 67-75; assoc prof, 75-81, dir, 76-84, PROF COMPUTER SCI, MCGILL UNIV, MONTREAL, 81- *Concurrent Pos:* Vis prof, Technion, Haita, Israel, 73-74 & 83-84; JSPS fel, Kyoto Univ & Kyoto Japan Univ, 84; consult, Centre Res Info Montreal, 88- *Honors & Awards:* Outstanding Contrib Award, Asn Comput Mach, 89. *Mem:* Asn Comput Mach; Int Computer Chess Asn (pres, 83-86). *Res:* Automated theorem proving; game playing by computers. *Mailing Add:* Sch Computer Sci McGill Univ 3480 University St Montreal PQ H3Y 1Y1

NEWBOUND, KENNETH BATEMAN, b Winnipeg, Man, Mar 12, 29; m 47; c 4. PHYSICS. *Educ:* Univ Man, BSc, 40, MSc, 41; Mass Inst Technol, PhD(physics), 48. *Prof Exp:* Physicist naval res, Nat Res Coun Can, 41-43; from assoc prof to prof, 48-84, assoc dean sci, 64-76, dean sci, 76-81, EMER PROF PHYSICS, UNIV ALTA, 84- *Mem:* Can Asn Physicists. *Res:* Atomic spectroscopy; spectrographic analysis; precision wave-length measurements; underwater acoustics. *Mailing Add:* 8910 Windsor Rd Edmonton AB T6G 2A2 Can

NEWBRUN, ERNEST, b Vienna, Austria, Dec 1, 32; US citizen; m 56; c 3. ORAL BIOLOGY & PERIODONTOLOGY, BIOCHEMISTRY. *Educ:* Univ Sydney, BDS, 54; Univ Rochester, MS, 57; Univ Ala, DMD, 59; Univ Calif, San Francisco, PhD(biochem), 65. *Hon Degrees:* D Odontol, Univ Lund. *Prof Exp:* Res assoc, Eastman Dent Dispensary, Rochester, NY, 56-57 & Med Ctr, Univ Ala, 57-59; lectr biochem, Univ Calif, 65, assoc res dentist & assoc prof oral biol, 65-70, chmn sect biol sci, 72-77, PROF ORAL BIOL, SCH DENT, UNIV CALIF, SAN FRANCISCO, 70- *Concurrent Pos:* Nat Health & Med Res Coun dent res fel, 60-61; res teacher trainee, Med Ctr, Univ Calif, San Francisco, 61-63, fel, 63-64; biochem consult oral calculus study, Sect Epidemiol, USPHS, San Francisco, 64-65; Am Col Dent trainee, Phys Biol Sect, Inst Advan Educ Dent Res, 65; USPHS res career develop award, 65-70; vis scientist, Sch Dent, Univ Lund, 67-68; mem, Nat Caries Prog Adv Comt, Nat Inst Dent Res, 72-75 & Dent Drug Prod Adv Comt, Food & Drug Admin, 74-78. *Honors & Awards:* Int Res Year Award, Acad Int Dent Studies, 85. *Mem:* Am Inst Oral Biol; Int Asn Dent Res; Europ Orgn Caries Res; Fel AAAS; Am Soc Microbiol; Am Dent Asn. *Res:* Dental caries; microradiography and microhardness of enamel; chemistry; mucoprotein chemistry and biosynthesis; bacterial polysaccharides, chemistry and synthesis; dental plaque, chemistry and microbiology. *Mailing Add:* Dept Stomatology Univ Calif Sch Dent San Francisco CA 94143-0512

NEWBURG, EDWARD A, b Indianapolis, Ind, Dec 22, 29; m 59. APPLIED MATHEMATICS. *Educ:* Purdue Univ, BS, 52, MS, 53; Univ Ill, PhD(math), 58. *Prof Exp:* Mathematician, Nuclear Div, Combustion Eng, Inc, 58-61; res scientist, Travelers Res Ctr, Inc, 61-66; assoc prof math, Worcester Polytech Inst, 66-70; assoc prof, Va Commonwealth Univ, 70-74; PROF MATH & HEAD DEPT, ROCHESTER INST TECHNOL, 74- *Concurrent Pos:* From adj asst prof to adj assoc prof, Hartford Grad Ctr, Rensselaer Polytech Inst, 60-70. *Mem:* Soc Indust & Appl Math. *Mailing Add:* Dept Math Rochester Inst Technol 1 Lomb Memorial Dr Rochester NY 14623

NEWBURGER, JEROLD, pharmocokinetics; deceased, see previous edition for last biography

NEWBURGH, ROBERT WARREN, b Sioux City, Iowa, Mar 22, 22; m 47, 79; c 4. BIOCHEMISTRY. *Educ:* Univ Iowa, BS, 49; Univ Wis, MS, 51, PhD(biochem), 53. *Prof Exp:* Asst biochem, Univ Wis, 49-53; res assoc, Sci Res Inst, Ore State Univ, 53-54, from asst prof to assoc prof, 54-61, asst dir, Sci Res Inst, 62-72, chmn, Dept Biochem & Biophys, 68-76, dean, Grad Sch, 76-80, prof biochem, 61-80; sect head, molecular genetics biosci, NSF, 79-82; HEAD, BIOL SCI DIV, OFFICE NAVAL RES, 82- *Concurrent Pos:* Am Cancer Soc grant & assoc prof, Univ Conn, 60-61; vis prof, Univ Calif, San Diego, 70-71; consult, NIH, 66-74. *Mem:* Am Chem Soc; Am Soc Biol Chem; Am Soc Neurochem; Toxicol Soc; Am Soc Microbiol. *Res:* Neural development; insect biochemistry. *Mailing Add:* 9600 Overlea Dr Rockville MD 20850

NEWBURGH, RONALD GERALD, b Boston, Mass, Feb 21, 26; m 57, 70; c 2. ELECTROMAGNETISM. *Educ:* Harvard Univ, AB, 45; Mass Inst Technol, PhD(physics), 59. *Prof Exp:* Sr chemist, Electronics Corp Am, 49-53, consult, 53-55 & 57-59; res physicist, Comstock & Wescott, Inc, 59-61; res physicist, Air Force Cambridge Res Labs, 61-76; sect chief, Rome Air Develop Ctr, Hanscom AFB, 76-87; TEACHER, SCI DEPT, WORCESTER ACAD, 87- *Mem:* Am Phys Soc; Sigma Xi. *Res:* Physics of rotating systems; special and general relativity. *Mailing Add:* Ten Hillside Terr Belmont MA 02178

NEWBURN, RAY LEON, JR, b Rock Island, Ill, Jan 9, 33; m 68; c 2. ASTRONOMY. *Educ:* Calif Inst Technol, BS, 54, MS, 55. *Prof Exp:* Res engr, 56-60, sr scientist lunar & planetary sci sect, 60-62, sci specialist, 62-65, sci group supvr, 65-72, mem tech staff, 73-85, SCI TEAM LEADER, JET PROPULSION LAB, CALIF INST TECHNOL, 80-, LEADER, INT HALLEY WATCH, WESTERN HEMISPHERE OFF, 81-91, SR MEM TECH STAFF, 85- *Concurrent Pos:* Comt mem, Div Planetary Sci, Am Astron Soc, 82-85. *Honors & Awards:* Except Serv Medal, NASA, 87. *Mem:* AAAS; Am Astron Soc; Astron Soc Pac; Int Astron Union. *Res:* Ground based and space probe research in astronomy of the solar system, especially photometry and physical modeling of comets; asteroid #2955 named Newburn. *Mailing Add:* 3226 Emerald Isle Dr Glendale CA 91206

NEWBURY, DALE ELWOOD, b Danville, Pa, May 15, 47; m 69; c 2. MATERIALS SCIENCE, ANALYTICAL CHEMISTRY. *Educ:* Lehigh Univ, BS, 69; Oxford Univ, PhD(metall), 72. *Prof Exp:* res staff metall, 72-79, GROUP LEADER, MICROANALYSIS RES GROUP, NAT BUR STANDARDS, 79- *Concurrent Pos:* Nat Res Coun fel & res metallurgist, Nat Bur Standards, 72-73; tech chmn, Microbeam Analysis Soc, 78-79 & Nat Coun, Microbeam Analysis Soc, 79-81. *Honors & Awards:* Hardy Gold Medal, Am Inst Metall Engrs, 73; Corning Award, Microbeam Analysis Soc, 75; US Dept Com Bronze Medal, 80, Silver Medal, 81 & Gold Medal, 86; Arthur S Flemming Award, 86; Macres Award, Microbeam Analysis Soc, 88; IR-100 Award, 87. *Mem:* Microbeam Analysis Soc (pres, 85); Am Soc Testing & Mat; Sigma Xi. *Res:* Development of techniques of microanalysis by electron and ion microbeams; calculation of the properties of electron interactions in solids by Monte Carlo simulation techniques; studies of contrast mechanisms in scanning electron microscopy; analytical electron microscopy. *Mailing Add:* Chem Bldg 222/A113 Nat Inst Standards & Technol Gaithersburg MD 20899

NEWBURY, ROBERT W(ILLIAM), b Winnipeg, Man, June 16, 39; c 2. SCIENCE EDUCATION. *Educ:* Univ Man, BSc, 62, MSc, 64; Johns Hopkins Univ, PhD(eng sci), 68. *Prof Exp:* Prof eng & earth sci, Univ Man, 68-75; res scientist, Can Dept Fisheries & Environ, 75-87; CONSULT, 87- *Concurrent Pos:* Mem, Int Hydrol Decade; vis prof, Univ Man, 75- & Univ Calgary, 87- *Res:* Physical hydrology; geomorphology; water resources engineering; permafrost hydrology; reservoir erosion and sedimentation; stream rehabilitation design and training. *Mailing Add:* Newbury Hydraul Ltd PO Box 1173 Gibsons BC V0N 1V0 Can

NEWBY, FRANK ARMON, JR, b Columbus, Kans, Dec 4, 32; div; c 4. PHYSICAL CHEMISTRY. *Educ:* Univ Kans, BS, 54, PhD(chem), 64. *Prof Exp:* From asst prof to assoc prof phys chem, 59-70, PROF PHYS CHEM, E TENN STATE UNIV, 70- *Mem:* AAAS; Am Chem Soc; Sigma Xi. *Res:* Chemical information storage, retrieval and use by individuals. *Mailing Add:* Dept Chem E Tenn State Univ Box 23350A Johnson City TN 37614

NEWBY, JOHN R, b Kansas City, Mo, Nov 17, 23; m 50; c 3. STEEL SHEET PRODUCTS, ELECTROPLATING. *Educ:* Univ Kansas City, BA, 47; Colo Sch Mines, MetE, 49; Univ Cincinnati, MS, 63. *Prof Exp:* Chemist, Bar Rusto Plating Corp, 49-50; supvr, United Chromium Corp, 50-52; prin res metallurgist, Armco Inc, 52-85; PRIN, JOHN NEWBY CONSULT, 85- *Honors & Awards:* Award of Merit, Am Soc Metals Int, 73 & Am Soc Testing & Mat, 86. *Mem:* Am Soc Metals Int; Am Soc Testing & Mat; Soc Automotive Engrs; NAm Deep Drawing Res Group (secy, 63-84). *Res:* Formability and ductility testing of metallic materials; writer of national (USA) and international (ISO) standards for test methods; inventor of ductile, high strength steel. *Mailing Add:* 100 Marymount Ct PO Box 584 Middletown OH 45042-0584

NEWBY, NEAL D(OW), b Alden, Kans, Mar 7, 99; m 24; c 1. ELECTRICAL ENGINEERING. *Educ:* Univ Kans, BS, 22, EE, 48. *Prof Exp:* Engr, Develop & Res Dept, Am Tel & Tel Co, New York, 22-34; mem tech staff, Bell Tel Labs, 34-63; RES CONSULT, 63- *Concurrent Pos:* Mem, Am Inst Physics. *Mem:* Asn Comput Mach; sr mem Inst Elec & Electronics Engrs; Am Phys Soc. *Res:* High speed machine switching and signaling systems; electronic circuitory and data storage; electronic computers and data processing systems; microwave logic; radar; magnetic devices and circuitry. *Mailing Add:* 640 Alta Vista St Apt No 312 Santa Fe NM 87501-4149

NEWBY, NEAL DOW, JR, b New York, NY, Mar 18, 26; m 64; c 2. THEORETICAL PHYSICS. *Educ:* Columbia Univ, BS, 49; Harvard Univ, MA, 51; Ind Univ, PhD(physics), 59. *Prof Exp:* Instr math, Univ Ohio, 52-53; res assoc physics, Univ Calif, Berkeley, 59-61; asst prof, Univ Southern Calif, 61-63; physicist, Autonetics, 64-67; lectr, Calif State Col, 69; prof physics, Edinboro State Col, 69-80; INSTR, NORTHERN NMEX COMMUNITY COL, 82- *Concurrent Pos:* Tutor, San Juan-Santa Clara Indian Pueblos, 81-82. *Mem:* Am Phys Soc. *Res:* Classical mechanics; relativity. *Mailing Add:* Box 1072 San Juan Pueblo NM 87566

NEWBY, WILLIAM EDWARD, b Kansas City, Mo, Nov 17, 23; m 49; c 3. PHYSICAL CHEMISTRY, ORGANIC CHEMISTRY. *Educ:* Univ Kans City, BA, 47; Northwestern Univ, PhD(chem), 50. *Prof Exp:* Res chemist, Jackson Lab, E I Du Pont De Nemours & Co Inc, 50-52, res supvr, 53-58, res supvr petrol lab, 58-59, res supvr plant tech sect, 59-60, res supvr, Jackson Lab, 60-68, supvr mkt res, 68-70, mkt planning mgr, Org Chem Dept, Dyes & Chem Div, 70-78, mkt res assoc, chem & pigments dept, 78-85; RETIRED. *Mem:* Am Chem Soc; Chem Mkt Res Asn. *Res:* Catalytic hydrogenation; fuel deposit and combustion phenomena; synthesis and application of dyes for synthetic fibers. *Mailing Add:* 37 Paxon Dr Penarth Wilmington DE 19803

NEWCOMB, ELDON HENRY, b Columbia, Mo, Jan 19, 19; m 49; c 3. PLANT CELL BIOLOGY. *Educ:* Univ Mo, AB, 40, AM, 42; Univ Wis, PhD(bot), 49. *Prof Exp:* From asst prof to assoc prof, Univ Wis-Madison, 49-58, chmn dept, 82-88, prof bot, 58-90, EMER PROF BOT, UNIV WIS MADISON, 90- *Concurrent Pos:* Guggenheim fel, Univ Calif, 51-52; NSF sci fac fel, Harvard Univ, 63-64; consult, Shell Develop Co, 54-59; US managing ed, Protoplasma, 69-73; mem photorespiration exped of R/V Alpha Helix to Great Barrier Reef, 73; Fulbright sr res scholar, Australian Nat Univ, 76; mem, Univ Nations Oceanog Lab Syst R/V Alpha Helix Rev Comt, 78-79; dir, Inst Plant Develop, Univ Wis-Madison, 79-84; Folke Skoog prof bot. *Mem:* Nat Acad Sci; Bot Soc Am; Am Soc Cell Biol; Am Soc Plant Physiologists. *Res:* Plant cell fine structure in relation to function; plant peroxisomes; cellular specialization in legume root nodules. *Mailing Add:* Dept Bot Univ Wis Madison WI 53706

NEWCOMB, HARVEY RUSSELL, b Bismarck, NDak, Oct 6, 16; m 53; c 2. MICROBIOLOGY. *Educ:* Denison Univ, AB, 39; Syracuse Univ, MS, 52, PhD(microbiol), 54. *Prof Exp:* Res bacteriologist, Borden Co, NY, 53-55; res asst prof microbiol, Dept Bact & Bot & Biol Res Labs, Syracuse Univ, 55-64; lab dir, Raritan Bay & Hudson-Champlain Water Pollution Control Projs, USPHS, NJ, 64-65; prof microbiol, Col Osteop Med & Surg, 65-82, chmn dept, 65-77; RETIRED. *Mem:* AAAS; Am Soc Microbiol; Am Pub Health Asn; NY Acad Sci. *Res:* Radiation effects on microorganisms; bacterial spores; preservation and wholesomeness of irradiated foods; physiology and industrial production of lactic acid bacteria; physiology of wood-rotting basidiomycetes; gnotobiotic technology; metabolic functions of serotonin. *Mailing Add:* 3104 Mary Lynn Dr Urbandale IA 50322

NEWCOMB, MARTIN, b Mishawaka, Ind, Nov 17, 46; m 67; c 1. CHEMISTRY. *Educ:* Wabash Col, BA, 69; Univ Ill, PhD(chem), 73. *Prof Exp:* Fel chem, Univ Calif, Los Angeles, 73-75; from asst prof to assoc prof, 75-85, PROF CHEM, TEX A&M UNIV, 85- *Concurrent Pos:* Camille & Henry Dreyfus teacher-scholar, Dreyfus Found, 80-85. *Mem:* Am Chem Soc; Royal Chem Soc; AAAS; Sigma Xi. *Res:* Synthetic and mechanistic organic chemistry; electron transfer reactions and host-guest chemistry; free radical chemistry. *Mailing Add:* Dept Chem Tex A&M Univ College Sta TX 77843-3255

NEWCOMB, ROBERT LEWIS, b Oceanside, Calif, Aug 2, 32; m 58; c 3. STATISTICS, SURVEY SAMPLE DESIGN. *Educ:* Univ Redlands, BA, 59; Univ Calif, Santa Barbara, PhD(math), 67. *Prof Exp:* Res assoc statist, Univ Calif, Santa Barbara, 68-69; SR LECTR SOC SCI, UNIV CALIF, IRVINE, 69- *Mem:* Math Asn Am; Am Statist Asn; Am Eval Asn. *Res:* Computer-graphics in statistics education. *Mailing Add:* Sch Soc Sci Univ Calif Irvine CA 92717

NEWCOMB, ROBERT WAYNE, b Glendale, Calif, June 27, 33; m 54; c 2. MICROSYSTEMS, SYSTEMS THEORY. *Educ:* Purdue Univ, BS, EE, 55; Stanford Univ, MSc, 57; Univ Calif, Berkeley, PhD(elec eng), 60. *Prof Exp:* Res intern, SRI, Stanford Univ, 55-57, assoc prof elec eng, 63-70; PROF ELEC ENG, UNIV MD, COL PARK, 70- *Concurrent Pos:* Fulbright fel, Australia, 63-64; invited prof, Cath Univ Louvain, 67-68; Fulbright fel, Malaysia, 76; hon prof, Univ Politenica de Madrid, Spain, 85. *Mem:* Fel Inst Elec & Electronics Eng; Soc Indust & Appl Math; Math Asn Am; Australian Inst Radio & Electronics Engrs. *Res:* Network theory; microsystems and systems theory; nonlinear systems via the semistate, design of hysteresis; new number bases for computers, especially the fibonacci computer; adaptive hearing aids, surface acoustic wave theory; neural-type microsystems; systems theory as applied to university administration; robotics; Kemp echo hearing theory. *Mailing Add:* Dept of Elec Eng Univ of Md College Park MD 20742

NEWCOMB, THOMAS F, b Buffalo, NY, June 22, 27; m 51; c 6. INTERNAL MEDICINE, HEMATOLOGY. *Educ:* Univ Pittsburgh, BS, 49, MD, 51; Am Bd Internal Med, dipl. *Prof Exp:* From intern med to resident hemat, Univ Pa, 51-53; resident med, Vet Admin Hosp, Seattle, Wash, 53-54; res fel hemat, Univ Wash, 54-55; Fulbright scholar med, Rikshospitalet Coagulation Lab, Oslo, Norway, 55-56; sr asst resident, Peter Bent Brigham Hosp, Boston; jr assoc med, Hosp & invstr, Howard Hughes Med Inst, 57-59; assoc dir, Richard C Curtis Hemat Lab, 58-59. *Concurrent Pos:* Instr, Harvard Med Sch, 58-59; consult, Vet Admin Hosp, Lake City, Fla, 62-68, assoc chief of staff res & educ & chief research serv, Gainesville, 68-72; prof med, Univ Tex Med Sch, 78-85; assoc prof med, Duke Univ, 85- *Mem:* Fel Am Col Physicians; Am Soc Hemat; Am Fedn Clin Res; Int Soc Hemat. *Res:* Hemostasis. *Mailing Add:* Vet Admin Med Ctr 508 Fulton St Durham NC 27705

NEWCOMB, WILLIAM A, b San Jose, Calif, Sept 4, 27; m 64; c 2. MAGNETOHYDRODYNAMICS. *Educ:* Cornell Univ, BA, 48, PhD(theoret physics), 52. *Prof Exp:* Physicist, Proj Matterhorn, Forrestal Res Ctr, Princeton Univ, 52-55; PHYSICIST, LAWRENCE LIVERMORE LAB, UNIV CALIF, 55- *Concurrent Pos:* Adj prof, Univ Calif, Davis. *Mem:* Am Phys Soc; Math Soc Am. *Res:* Magneto-hydrodynamics and plasma physics. *Mailing Add:* Lawrence Livermore Nat Lab Univ Calif L630 Livermore CA 94550

NEWCOMBE, DAVID S, b Boston, Mass, June 28, 29; m 65; c 3. ENVIRONMENTAL HEALTH SCIENCES, ARTHRITIS. *Educ:* Amherst Col, BA, 52; McGill Univ, MD, CM, 56. *Prof Exp:* Intern med, Boston City Hosp, 56-57; resident, Med Ctr, Duke Univ, 59-61; New Eng Rheumatism Soc res fel, Boston Univ, 61-62, asst, 62-63; res fel biochem, Harvard Univ, 63-65; asst prof, Sch Med, Univ Va, 65-67; assoc prof med, Col Med, Univ Vt, 67-77, dir rheumatology unit, 70-77; assoc prof, 77-82, PROF ENVIRON HEALTH SCI & MED, JOHNS HOPKINS UNIV, 82- *Concurrent Pos:* NIH spec fel, 62-63; Am Cancer Soc res fel, Harvard Med Sch, 63-65. *Mem:* Am Rheumatism Asn; Sigma Xi; Am Soc Biochem & Molecular Biol; Soc Toxicol. *Res:* Biochemical and biological mechanisms that regulate inflammatory and immune reactions; biology and regulation of lipid mediators; immunotoxicology; toxicology; occupational health. *Mailing Add:* Dept Environ Health Sci Johns Hopkins Univ 615 N Wolfe St Baltimore MD 21205

NEWCOMBE, HOWARD BORDEN, b Kentville, NS, Sept 19, 14; m 42; c 3. RADIATION GENETICS, HUMAN GENETICS. *Educ:* Acadia Univ, BSc, 35; McGill Univ, PhD(genetics), 39. *Hon Degrees:* DSc, McGill Univ, 66, Acadia Univ, 70. *Prof Exp:* 1851 sci res scholar, John Innes Hort Inst, Eng, 39-40; sci officer, Brit Ministry Supply, 40-41; res assoc genetics, Carnegie Inst Wash, 46-47; head, Biol Br, Atomic Energy Can, Ltd, 47-70, head, Pop Res Br, 70-79; CONSULT, 79- *Concurrent Pos:* Sci adv, Sci Comt Atomic Radiation, UN, 55-66; mem expert adv panel human genetics, WHO, 61-88, mem, Int Comn Radiol Protection, 65-77; vis prof, Ind Univ, 63. *Mem:* Am Soc Human Genetics (pres, 65); Genetics Soc Am (secy, 56-58); Radiation Res Soc; Genetics Soc Can (pres, 64-65); Royal Soc Can. *Res:* Genetics; epidemiology; public health. *Mailing Add:* 67 Hillcrest Ave PO Box 135 Deep River ON K0J 1P0 Can

NEWCOME, MARSHALL MILLAR, b Chicago, Ill, Nov 22, 26; m 49; c 2. ANALYTICAL CHEMISTRY. *Educ:* Ill Inst Technol, BS, 49; Univ Wash, Seattle, PhD(chem), 54. *Prof Exp:* Res chemist, 54-74, SUPVR ANALYTICAL CHEM, MORTON CHEM CO, 74- *Mem:* Am Chem Soc. *Res:* Electronic instrumentation in chemical analysis. *Mailing Add:* 546 W Kimball Ave Woodstock IL 60098

NEWCOMER, WILBUR STANLEY, b Turbotville, Pa, Nov 25, 19; m 46; c 4. PHYSIOLOGY. *Educ:* Pa State Univ, BS, 41; Cornell Univ, MS, 42, PhD(zool), 48. *Prof Exp:* Asst chem, Lycoming Col, 38-39; asst zool, Cornell Univ, 42-44 & 46-47; from instr to asst prof biol, Hamilton Col, NY, 47-50; from asst prof to assoc prof, 50-58, PROF PHYSIOL, OKLA STATE UNIV, 58- *Concurrent Pos:* NIH spec res fel, 62. *Honors & Awards:* Hutyra Medal, Univ Vet Sci, Hungary. *Mem:* Soc Exp Biol & Med; Am Physiol Soc. *Res:* Physiology of thyroid and adrenal glands, especially in birds. *Mailing Add:* Prof Eng Dept Physiol Okla State Univ Stillwater OK 74075

NEWELL, ALLEN, b San Francisco, Calif, Mar 19, 27; m 47; c 1. COMPUTER SCIENCE, PSYCHOLOGY. *Educ:* Stanford Univ, BS, 49; Carnegie Inst Technol, PhD(indust admin), 57. *Prof Exp:* Res scientist, Rand Corp, 50-61; res scientist inst prof systs & commun sci, 61-67, UNIV PROF COMPUT SCI, CARNEGIE-MELLON UNIV, 67- *Concurrent Pos:* Lectr, Carnegie-Mellon Univ, 57-61; consult, Rand Corp, 61-; mem adv comt comput, Stanford Univ, 66-70; mem comput sci study sect, NIH, 67-71; chmn panel comput, Comt Res Life Sci, Nat Acad Sci, 67-69; consult, Xerox Res Lab, Calif, 71- *Honors & Awards:* Harry Goode Mem Award, Am Fedn Info Processing Socs, 71; John Danz Lectr, Univ Wash, 72; A M Turing Award, Asn Comput Mach, 75. *Mem:* Nat Acad Sci; AAAS; Asn Comput Mach; Am Psychol Asn; Inst Elec & Electronics Eng; Am Asn Artificial Intel. *Res:* Computer programs that exhibit intelligence; information processing psychology; programming systems and computer structures. *Mailing Add:* Dept Computer Sci Carnegie-Mellon Univ Pittsburgh PA 15213

NEWELL, DARRELL E, b Audubon, Iowa, Sept 24, 26; m 47; c 2. ELECTRICAL ENGINEERING. *Educ:* Iowa State Univ, BS, 52; Univ Iowa, MS, 56, PhD(elec eng), 58. *Prof Exp:* Engr, Collins Radio, 52-54 & 58-59; from instr to assoc prof elec eng, Univ Iowa, 54-61; eng mgr, Bendix Corp, 61-65; pres, Newell Labs, Inc, 65-66, dir eng, CTS Knights Corp, 66-68; assoc prof indust & technol, 69-77, PROF INDUST & TECHNOL, NORTHERN ILL UNIV, 77- *Concurrent Pos:* Mem bd dirs, CTS Knights Corp, 69- & Bodelle Corp, 71- *Mem:* Inst Elec & Electronics Engrs (treas). *Res:* Originated temperature compensated crystal oscillators. *Mailing Add:* Dept of Indust & Technol Northern Ill Univ W Lincoln Hwy DeKalb IL 60115

NEWELL, FRANK WILLIAM, b St Paul, Minn, Jan 14, 16; m 42; c 4. OPHTHALMOLOGY. *Educ:* Loyola Univ, Ill, MD, 39; Univ Minn, MSc, 42; Am Bd Ophthal, dipl. *Prof Exp:* Fel, Univ Minn, 40-42; res fel, Northwestern Univ, 46-47, instr, 47-50, assoc, 50-53; assoc prof, 53-55, chmn, Dept Ophthal, 53-81, PROF OPHTHAL, SCH MED, UNIV CHICAGO, 55- *Concurrent Pos:* Ed-in-chief, Am J Ophthal, 65-91. *Mem:* Am Ophthal Soc (pres, 86-87); Pan-Am Asn Ophthal (pres, 81-83); Am Acad Ophthal & Otolaryngol (pres, 75). *Res:* Pharmacology and physiology of the eye. *Mailing Add:* Univ Chicago Visual Sci Ctr 939 E 57th St Chicago IL 60637

NEWELL, GORDON FRANK, b Dayton, Ohio, Jan 26, 25; m 49; c 2. APPLIED MATHEMATICS. *Educ:* Union Col, BS, 45; Univ Ill, PhD(physics), 50. *Prof Exp:* Lectr physics & math, Union Col, 45-46; asst physics, Univ Ill, 46-49, fel, 50; fel appl math, Univ Md, 50-51, res assoc, 51-52; res assoc, Brown Univ, 53-54, from asst prof to prof, 54-66; PROF TRANSP ENG, UNIV CALIF, BERKELEY, 66- *Concurrent Pos:* Sloan res fel, 56-59; Fulbright fel, 63-; vis prof, Univ Calif, Berkeley, 65-66. *Mem:* Soc Indust & Appl Math; Opers Res Soc Am. *Res:* Operations research; transportation and traffic engineering. *Mailing Add:* Dept Civil Eng Univ of Calif 2120 Oxford St Berkeley CA 94720

NEWELL, GORDON WILFRED, b Madison, Wis, Aug 27, 21; m 48; c 4. ENVIRONMENTAL TOXICOLOGY, LAB ANIMAL CARE. *Educ:* Univ Wis, BA, 43, MS, 44, PhD(biochem), 48; Am Bd Indust Hyg, dipl. *Prof Exp:* Novadel-Agene fel, Univ Wis, 48-49; res biochemist, Wallace & Tiernan Co, Inc, 49-50; sr biochemist, Stanford Res Inst, 50-66, dir div indust biol, 66-68, dir, Dept Toxicol, 68-78, assoc dir, Bd Toxicol & Environ Health Hazards, Nat Res Coun, Nat Acad Sci, 78-82; sr prog mgr, health studies, Elec Power Res Inst, 82-88; CONSULT ENVIRON TOXICOL, 88- *Concurrent Pos:* Consult adv comn animal resources, NIH, 65-68; coun mem, Am Asn Accreditation Animal Care, 67-78; assoc ed, J Lab Animal Care, 64-; bd mem, 81-85, chmn, 83-85, bd mem, Prof Standards Evaluation Bd, Acad Toxicol Sci, 87-90; bd dir, Toxicol Lab Accreditation Bd, 78-84; consult, EPA, FDA, NIH, 62-88. *Honors & Awards:* Dipl, Acad Toxicol Sci, Am Acad Industrial Hygiene. *Mem:* Soc Toxicol; Am Col Toxicol (pres-elect, 86, pres, 87); Environ Mutagen Soc (counr, 77-79, vpres, 81, pres, 82); Soc Risk Anal (treas, 80-81, organizing comt mem, 80-81); Int Soc Regulatory Toxicol & Pharmacol. *Res:* Animal nutrition and metabolism; toxicological studies on chemicals and food products; biochemical toxicology of food additives, environmental, industrial, and military chemicals, drugs, and pesticides; toxicity of pollutants to fish and wildlife; nutrition and metabolism; mutagenesis, carcinogenesis, teratology, and inhalation toxicology. *Mailing Add:* 4163 Hubbartt Dr Palo Alto CA 94306

NEWELL, JON ALBERT, b St Louis, Mo, Aug 5, 41; m 84; c 3. BIOCHEMISTRY. *Educ:* Okla State Univ, BS, 63, PhD(biochem), 67. *Prof Exp:* Technician analytical chem, Enid Bd Trade Lab, 59-62; res asst toxicol, Okla State Univ, 60-63, asst flavor chem, 63-67; sr res chemist food & fermentation res, 67-75, res mgr biochem, 75-80, res mgr process optimization, 80-83, RES MGR, ANALYTICAL CTR, CORP RES & DEVELOP, ANHEUSER-BUSCH CO, INC, 83- *Concurrent Pos:* Nestle fel, Inst Food Technologists, Univ Chicago, 66. *Mem:* Am Chem Soc; Inst Food Technologists. *Res:* Use of unconventional protein sources in foods; identification of flavorful constituents of foods; process control using microprocessors; estimation of shelf life of foods. *Mailing Add:* Anheuser Busch Co, Inc Corp R&D 1101 Wyoming St Bldg 156-2 St Louis MO 63118-2627

NEWELL, JONATHAN CLARK, b Worcester, Mass, Oct 13, 43; m; c 2. PHYSIOLOGY, BIOMEDICAL ENGINEERING. *Educ:* Rensselaer Polytech Inst, BEE, 65, MEngr, 68; Albany Med Col, PhD(physiol), 74. *Prof Exp:* Supv engr, 70-72, from asst prof to assoc prof, 74-85, PROF BIOMED ENG, RENSSELAER POLYTECH INST, 85-; PROF PHYSIOL, ALBANY MED COL, 79- *Concurrent Pos:* Consult, Trauma Res Unit, Albany Med Col, 74- *Mem:* Am Physiol Soc; Inst Elec & Electronics Engrs; NY Acad Sci; Biomed Eng Soc. *Res:* Electrical impedance imaging; modelling of physiological systems; pulmonary hemodynamics and mechanics. *Mailing Add:* Biomed Eng Dept Rensselaer Polytech Inst Troy NY 12181

NEWELL, KATHLEEN, b Stafford, Kans, Sept 1, 22. NUTRITION. *Educ:* Kans State Univ, BS, 44; Univ Wis, MS, 51; Univ Tenn, PhD(nutrit), 73. *Prof Exp:* Admin ther & teaching dietitian, Butterworth Hosp, Grand Rapids, Mich, 45-47; ther & teaching dietitian, Univ Colo Med Ctr, 47-49, educ dir dietetic intern, 52-56; asst prof dietetics, Loretto Heights Col & Glockner Penrose Hosp, Colo Springs, 56-58; asst prof home econ, Univ Wyo, 58-62; asst prof food & nutrit, Kans State Univ, 62-69, from asst prof to prof, 72-88; RETIRED. *Mem:* Am Dietetic Asn; Sigma Xi; Geront Soc; Soc Nutrit Educ; Nutrit Today Soc; Am Inst Nutrit. *Res:* Nutrition education; human nutrition. *Mailing Add:* 3109 Heritage Ct Manhattan KS 66506

NEWELL, MARJORIE PAULINE, b Holden, Alta; nat US. ORGANIC CHEMISTRY. *Educ:* Univ Fla, BS, 53, PhD(biochem), 58. *Prof Exp:* Asst biochem, Univ Fla, 55-58; res chemist org chem, res dept, R J Reynolds Tobacco Co, 58-59, sr scientist, 59-83. *Mem:* AAAS; Am Chem Soc. *Res:* Physical and chemical properties of tobacco and smoke; tracer techniques with radioisotopes. *Mailing Add:* 3901 Guinevere Lane Winston-Salem NC 27104

NEWELL, NANETTE, b Pensacola, Fla, Aug 20, 51. RECOMBINANT DNA TECHNOLOGY. *Educ:* Lewis & Clark Col, BS, 73; Johns Hopkins Sch Med, PhD(biochem), 78. *Prof Exp:* Fel, Univ Wis, 78-80; asst prof genetics, Reed Col, 80-81; analyst, off Technol Assessment, 81-84; dir, 84-85, BIOTECHNOL CONSULT, CALGENE INC, 85- *Mem:* AAAS; Asn Women Sci; Am Chem Soc. *Res:* Commercial development of biotechnology. *Mailing Add:* PO Box 13454 Research Triangle Park NC 27709

NEWELL, NORMAN DENNIS, b Chicago, Ill, Jan 27, 09; m 73. GEOLOGY, EARTH SCIENCES. *Educ:* Univ Kans, BS, 29, AM, 31; Yale Univ, PhD(geol), 33. *Prof Exp:* Asst geologist, Geol Surv, Kans, 29-33; from instr to asst prof geol, Univ Kans, 34-37; assoc prof, Univ Wis, 37-45; prof geol, Columbia Univ, 45-78; cur, 45-77, EMER CUR, AM MUS NATURAL HIST, NY, 78-; EMER PROF GEOL, COLUMBIA UNIV, 78- *Concurrent Pos:* Sterling fel, Yale Univ, 33-34; geologist, Geol Surv, Kans, 35-37; invited lectr, many insts world-wide; US Dept State deleg, Int Geol Cong, Moscow, 37; co-ed, Paleont, 39-42; consult, Govt Peru, 42-43; leader, Exped Geol & Petrol Resources Lake Titicaca, Peru-Bolivia, 43-44, Am Mus Natural Hist-Columbia Univ Exped, Peru, 47 & Andros Island, Bahamas, 50 & 51; leader invests limestone reefs, WTex, 49-52; leader, Pac Sci Bds, Nat Res Coun Exped Coral Atoll, Raroia, Tuamotu Group, SPac, 52; mem, Smithsonian Coun, 66 -; mem, Scripps Inst Oceanog Exped Carmarsel, Micronesia, 67; exchange scholar, Nat Acad Sci, USSR, 78; partic, numerous expeds throughout the world. *Honors & Awards:* Mary Clark Thompson Medal, Nat Acad Sci, 60; Medal, Univ Hiroshima, 64; Hayden Mem Award, Acad Natural Sci, Philadelphia, 65; Verrill Medal, Yale Univ, 66; Gold Medal, Am Mus Natural Hist, 78; Paleont Soc Medal, 79; Raymond C Moore Medal, Soc Econ Paleontologists & Mineralogists, 80; Sci Freedom & Responsibility Award, AAAS, 87; Penrose Medal, Geol Soc Am, 90. *Mem:* Nat Acad Sci; fel Geol Soc Am; Soc Econ Paleontologists & Mineralogists; fel Paleont Soc (vpres, 48, pres, 61); Soc Study Evolution (pres, 49); Am Acad Arts & Sci. *Res:* Invertebrate paleontology; micropaleontology; stratigraphy; petroleum geology; coral reefs; geology of South America. *Mailing Add:* Am Mus Natural Hist Central Park W & 79 St New York NY 10024

NEWELL, REGINALD EDWARD, b Peterborough, Eng, Apr 9, 31; US citizen; m 54; c 4. METEOROLOGY, CLIMATE. *Educ:* Univ Birmingham, BSc, 54; Mass Inst Technol, SM, 56, ScD(meteorol), 60. *Prof Exp:* Res asst meteorol, 54-60, mem staff, Div Sponsored Res, 60-61, FROM ASST PROF TO PROF METEOROL, MASS INST TECHNOL, 61- *Concurrent Pos:* Pres, Int Comn Climat, Int Asn Meteorol & Atmospheric Physics 77-83; mem, Int Comn Atmospheric Chem & Global Pollution, Int Asn Meteorol & Atmospheric Physics, 71-83. *Mem:* Am Meteorol Soc; Am Geophys Union; Sigma Xi; fel Royal Meteorol Soc. *Res:* Climate variations; atmospheric trace constituents; planetary circulations; atmospheric general circulation; global ozone cycle, including recent trends; role of carbon dioxide in controlling atmospheric temperature; global sea surface temperature patterns and changes in past 100 years; global carbon monoxide budget from aircraft and space shuttle measurements; energy budget of atmosphere and ocean. *Mailing Add:* Dept Earth Atmospheric & Planetary Sci 54-1824 Mass Inst Technol Cambridge MA 02139

NEWELL-MORRIS, LAURA, b Whitehall, NY, May 16, 33. PHYSICAL ANTHROPOLOGY. *Educ:* Univ NMex, BA, 54; Northwestern Univ, MA, 57; Univ Wash, PhD(anthrop), 66. *Prof Exp:* Asst cur archaeol, State Mus NY, 54-55; res asst phys anthrop, Fels Res Inst, 59-60; from instr to assoc prof, 64-81, PROF PHYS ANTHROP, UNIV WASH, 81- *Concurrent Pos:* Assoc ed, Am J Phys Anthrop & NIH res award, 75; consult ed, Am J Primatol. *Mem:* Am Anthrop Asn; Am Asn Phys Anthrop; Brit Soc Study Human Biol; Int Primatol Soc; Am Soc Primatol; Human Biol Coun. *Res:* Growth, especially of non-human primates and population as a unit of study. *Mailing Add:* Dept of Anthrop Univ of Wash Seattle WA 98195

NEWEY, HERBERT ALFRED, b Logan, Utah, June 11, 16; m 41; c 3. ORGANIC POLYMER CHEMISTRY. *Educ:* Utah State Col, BS, 38; Mass Inst Technol, PhD(org chem), 41. *Prof Exp:* Res chemist, Am Cyanamid Co, 41-46; chemist, Shell Develop Co Div, Shell Oil Co, 46-62, supvr res, 62-73; consult chemist, 73-84; RETIRED. *Mem:* Am Chem Soc. *Res:* Epoxy and polyester resins for surface coatings, laminates and adhesives. *Mailing Add:* 730 Los Palos Dr Lafayette CA 94549-5358

NEWHART, M(ARY) JOAN, b Oak Park, Ill. COMPUTER SCIENCE. *Educ:* Mundelein Col, Chicago, BS, 49; St Louis Univ, MS, 65. *Prof Exp:* Instr chem, Sch Sisters of Charity, BVM, 52-65 & 70-81, Univ Javeriana Bogota, Colombia, 65-70; student computer sci, Univ Wis-Milwaukee, 82-84; instr computer sci, Carmel High Sch, Mundelein, Ill, 84-89; dir, Computer Ctr, Mundelein Col, Chicago, 89-91; COORDR MICROCOMPUTER LAB, LOYOLA UNIV, CHICAGO, 91- *Concurrent Pos:* Instr chem & biomed sci, Meharry Med Sch, 74-81. *Mailing Add:* 589 Glenview Ave Highland Park IL 60035

NEWHOUSE, ALBERT, b Cambrai, France, May 31, 14; nat US; wid; c 2. MATHEMATICS, COMPUTER SCIENCE. *Educ:* Univ Chicago, PhD(math), 40. *Prof Exp:* Asst Math, Tulane Univ, 39-41; instr, Univ Ala, 41-42, Univ Nebr, 42-44 & Rice Inst, 44-46; from instr to prof math, 46-70, prof, 70-78, EMER PROF COMPUT SCI, UNIV HOUSTON, 78- *Concurrent Pos:* Consult, Res Lab, Humble Oil & Ref Co, 58, Camco, Inc, 61-70 & Symbiotics Int, 70-74; chmn, dept math, Univ Houston, 52-54, asst dir comput & data processing ctr, 56-60. *Mem:* AAAS; Asn Comput Mach; Am Math Soc; Soc Indust & Appl Math; Math Asn Am. *Res:* Modern algebra; theory of groups; numerical methods; formal languages. *Mailing Add:* 7907 Braesview Lane Houston TX 77071

NEWHOUSE, JOSEPH PAUL, b Waterloo, Iowa, Feb 24, 42; m 68; c 2. ECONOMICS OF HEALTH & MEDICAL CARE. *Educ:* Harvard Univ, BA, 63, PhD(econ), 69. *Prof Exp:* SR CORP FEL, RAND CORP, 68-; MACARTHUR PROF HEALTH POLICY & MGT, HARVARD UNIV, 88- *Concurrent Pos:* Ed, J Health Econ, 80- *Honors & Awards:* David N

Kershaw Award, Asn Pub Policy & Mgt, 83; Baxter Am Found Prize, 88. *Mem:* Inst Med-Nat Acad Sci; Am Econ Asn; Asn Pub Policy & Mgt; Asn Health Sci Res; Economet Soc. *Res:* Economics of health and medical care, particularly the demand for medical care, reimbursement of health care providers and medical malpractice. *Mailing Add:* Div Health Policy Res & Educ Harvard Univ 25 Shattuck St Parcel B-First Floor Boston MA 02115

NEWHOUSE, KEITH N, b Chambers, Nebr, Mar 11, 24; m 47; c 3. MECHANICAL ENGINEERING. *Educ:* Univ Nebr, BS, 48, MS, 49. *Prof Exp:* Instr mech eng, Univ Nebr, 48-51, asst prof mech eng & asst utilities engr, 51-56, assoc prof & power plant engr, 56-60; res engr, Jet Propulsion Lab, Calif Inst Technol, 63-65; prof mech enf, Univ Nebr, Lincoln, 65-87; RETIRED. *Honors & Awards:* Western Elec Fund Award, 65. *Mem:* Am Soc Mech Engrs; Nat Soc Prof Engrs; Am Soc Heating, Refrig, Air Conditioning Engrs. *Res:* Heat transfer. *Mailing Add:* 500 Mulder Dr Lincoln NE 68510

NEWHOUSE, RUSSELL C(ONWELL), b Clyde, Ohio, Dec 17, 06; wid; c 1. ELECTRICAL ENGINEERING. *Educ:* Ohio State Univ, BEE, 29, MS, 30. *Prof Exp:* Dir Kwajalein Field Sta, Bell Tel Labs, Inc, 30-67, dir, Nike-X Radar Systs Lab, 67-71; consult engr & mgr radar dept, 71-74, CONSULT ENGR, TELEDYNE BROWN ENG, 74- *Concurrent Pos:* Consult, Air Navig Develop Bd, 54; mem President's Task Force air traffic control, Proj Beacon, 61; off electronics, Mutual Weapons Develop Prog, Off Asst Secy Defense. *Honors & Awards:* Lawrence Sperry Award, Inst Aeronaut Sci, 38; Pioneer Award, Aerospace & Electronics Group, Inst Elec & Electronics Engrs, 67. *Mem:* Fel Inst Elec & Electronics Engrs; Sigma Xi; AAAS. *Res:* Radio altimeters; radar; electronic computers; air navigation and traffic control. *Mailing Add:* 31 Francisco Ave West Caldwell NJ 07006

NEWHOUSE, VERNE FREDERIC, b Tulsa, Okla, May 7, 30; m 53; c 5. MEDICAL ENTOMOLOGY. *Educ:* Pac Lutheran Col, AB, 53; Wash State Univ, MS, 55, PhD(entom), 60. *Prof Exp:* Asst zool, Wash State Univ, 54-55 & 56-57; med entomologist, Rocky Mountain Lab, USPHS, 57-63, res entomologist, Arbovirus Vector Lab, 64-75, res entomologist, Leprosy & Rickettsia Br, Virol Div, 75-80, viral rickettsial zoonoses br, Ctr Dis Control, 81-87; RETIRED. *Mem:* Sigma Xi. *Res:* Ecology of arthropod-borne rickettsial and viral agents and their vectors. *Mailing Add:* 423 Stonewood Dr Stone Mountain GA 30087

NEWHOUSE, VERNON LEOPOLD, b Mannheim, Ger, Jan 30, 28; US citizen; m 50; c 4. BIOMEDICAL ENGINEERING. *Educ:* Univ Leeds, BSc, 49, PhD(physics), 52. *Prof Exp:* Mem staff, Ferranti Comput Lab, 51-54; mem staff, Res Lab, Radio Corp Am, 54, proj engr, Comput Dept, 54-57; physicist, Gen Elec Res & Develop Ctr, NY, 57-67; prof elec eng, Sch Elec Eng, Purdue Univ, 67-73, prof med eng & coordr, 73-81; DISTINGUISHED PROF ELEC & COMPUT ENG, DREXEL UNIV, 82- *Concurrent Pos:* Consult, NSF, 74. *Mem:* Am Phys Soc; fel Inst Elec & Electronics Engrs; Am Inst Ultrasound Med. *Res:* Acoustic flow measurement and imaging for medical and industrial applications. *Mailing Add:* Biomed Eng & Sci Inst Drexel Univ Philadelphia PA 19104

NEWHOUSE, W JAN, phycology, science education, for more information see previous edition

NEWIRTH, TERRY L, b Morristown, NJ, Sept 22, 45; m 74; c 2. MECHANISTIC ORGANIC & BIOORGANIC CHEMISTRY. *Educ:* Bryn Mawr Col, AB, 67; Mass Inst Technol, PhD(org chem), 71. *Prof Exp:* Postdoctoral biochem, Univ Pa Med Sch, 71-73 & Path Dept, Temple Univ Med Sch, 73-75; asst prof, 75-88, ASSOC PROF CHEM, HAVERFORD COL, 88- *Mem:* Am Chem Soc; AAAS. *Res:* Mechanistic organic chemistry, especially reactions related to bioorganic chemistry. *Mailing Add:* Dept Chem Haverford Col Haverford PA 19041

NEWITT, EDWARD JAMES, b Scranton, Pa, Mar 18, 27; m 48; c 5. PHYSICAL ORGANIC CHEMISTRY. *Educ:* Imp Col, Univ London, BSc, 53, PhD, 57. *Prof Exp:* Works chemist indust chem, Boots Pure Drug Co, Eng, 48-49; asst chem, Brit Coal Utilization Res Asn, 49-53; lectr inorg & phys chem, Imp Col, Univ London, 53-57; res chemist, Plastics Dept, Exp Sta, 57-59, res supvr, 59-61, sr res chemist, Wash Works, 61-62, res supvr, Electrochem Dept, Exp Sta, 62-69, prod mgr mkt, 69-70, res supvr, 70-71, develop supvr polymer prod mkt, Plastics Dept, 72-73, new prod specialist, 73-74, consult, Energy & Math Dept, 74-76, specialist, feedstocks, Central Res & Develop Dept, 76-77, SPECIALIST, PATENT DIV, E I DU PONT DE NEMOURS & CO, INC, 77- *Mem:* Am Chem Soc; Royal Inst Chemists. *Res:* Chemical kinetics; high temperature chemistry and polymer chemistry; alternate raw materials and energy sources. *Mailing Add:* 12 Ringfield Chadds Ford PA 19317

NEWKIRK, DAVID ROYAL, organic chemistry, biochemistry, for more information see previous edition

NEWKIRK, GARY FRANCIS, b Paterson, NJ, June 25, 46; m 68; c 3. AQUACULTURAL BREEDING. *Educ:* Rutgers Univ, BSc, 68; Duke Univ, PhD(zool), 74. *Prof Exp:* Fel ecol genetics, Dalhousie Univ, 73-75, res assoc ecol genetics, 75-77, asst prof, 77-84, ASSOC PROF ECOL GENETICS, DALHOUSIE UNIV, 84- *Mem:* World Maricult Soc; Nat Shellfish Asn; Aquacult Asn Can. *Res:* Aquaculture genetics of marine species; development of tropical bivalve culture. *Mailing Add:* Dept Biol Dalhousie Univ Halifax NS B3H 4J1 Can

NEWKIRK, GORDON ALLEN, JR, astrophysics; deceased, see previous edition for last biography

NEWKIRK, HERBERT WILLIAM, b Jersey City, NJ, Nov 23, 28; m 52; c 2. INORGANIC CHEMISTRY, PHYSICAL CHEMISTRY. *Educ:* Polytech Inst Brooklyn, BS, 51; Ohio State Univ, PhD(chem), 56. *Prof Exp:* Res chemist, Allied Chem & Dye Corp, 51-52; res engr, Gen Elec Co, 56-59; mat scientist, Radio Corp Am, 59-60 & Lawrence Livermore Lab, Univ Calif, 60-69; guest prof, Philips Res Lab, Aachen, WGer, 69-71; prog mgr, 71-78, MAT SCIENTIST, LAWRENCE LIVERMORE LAB, UNIV CALIF, 78- *Res:* High energy density and high temperature chemistry; irradiation chemistry of nuclear fuels and structural materials; electronic properties of solids; inorganic synthesis, characterization and single crystal growth of materials; technology transfer processes, materials characterization. *Mailing Add:* 1141 Madison Ave Livermore CA 94550

NEWKIRK, JOHN BURT, b Minneapolis, Minn, Mar 24, 20; m 51; c 3. BIOENGINEERING, PHYSICAL METALLURGY. *Educ:* Rensselaer Polytech Inst, BMetE, 41; Carnegie Inst Technol, MS, 47, DSc(phys metall), 50. *Prof Exp:* Fulbright fel crystallog, Cavendish Lab, Cambridge Univ, 50-51; res metallurgist, Res Lab, Gen Elec Co, 51-59; prof phys metall, Cornell Univ, 59-64; Phillipson prof phys metall, 64-75, prof phys chem, 75-85, EMER PHILLIPSON PROF PHYS CHEM, UNIV DENVER, 85- *Concurrent Pos:* Consult, Gen Elec Co, 60-65, Atomics Int, Inc, 63-65, Jet Propulsion Lab, 64-65, 3M Co, & Denver Med Specialties, Inc, 75-77, Codman & Shurtlff, 77-; pres, Denver Biomat Inc, 68-77, pres, Colo Biomed Inc, 69-; lectr, Univ Denver, 75. *Honors & Awards:* V C Huffsmith Res Award, Denver Res Inst, 73. *Mem:* Fel Am Soc Metals. *Res:* Characterization of solids, with special emphasis on heat resisting alloys; biomaterials and implantable biodevices for flow control of abnormal body liquids; surgical devices. *Mailing Add:* Colo Biomed Inc 6851 Hwy 73 Evergreen CO 80439

NEWKIRK, LESTER LEROY, b Kansas City, Kans, June 2, 20; m 44; c 3. SPACE PHYSICS, NUCLEAR PHYSICS. *Educ:* Kans State Col, BS, 43, MS, 48; Iowa State Univ, PhD(physics), 51. *Prof Exp:* Physicist microwave lab, Res & Develop Dept, Hughes Aircraft Co, 51-53; mem staff, Lawrence Radiation Lab, Univ Calif, 53-58 & Los Alamos Sci Lab, 58-61; Staff scientist, Lockheed Missiles & Space Co, Palo Alto, 61-84; RETIRED. *Mem:* Am Phys Soc; Am Geophys Union. *Res:* Radiation belt theory; nuclear weapons design; solar x-ray physics; atmospheric neutron theory. *Mailing Add:* 240 Silvia Ct Los Altos CA 94022

NEWKOME, GEORGE RICHARD, b Akron, Ohio, Nov 26, 38; m 86; c 1. ORGANIC CHEMISTRY. *Educ:* Kent State Univ, BS, 61, PhD(org chem), 66. *Prof Exp:* Res chemist, Firestone Tire & Rubber Co, 61-62; asst instr, Kent State Univ, 64-65; res assoc, Princeton Univ, 66-68; from asst prof to prof, 68-86, assoc exec dir, res & develop, Ctr Energy Studies, 82-83, exec dir, 83-86, EMER PROF CHEM, LA STATE UNIV, 86-; prof chem, dean, Grad Sch & vprovost, Res & Grad Studies, 86-88, VPRES RES, UNIV SFLA, 88- *Concurrent Pos:* NATO sr fac fel sci, 76-77; vis prof, Stanford Univ, 77, Univ Bonn, 77 & Nat Univ Mex, 82; adv, John Wiley & Sons Publ, 83-, grants prog, State Bd Regents, 83- & NSF Exp Prog, 85-; consult, Standard Oil Ohio, 84-, Teltech Resource Network, 85- & Dow Chem, 87-; adj prof, Emory Univ, 84-87. *Mem:* AAAS; Am Chem Soc; NY Acad Sci; Swiss Chem Soc; Brit Chem Soc. *Res:* Synthesis and structure elucidation of enzyme models and natural products; macromolecular molecules; inorganic complexation and organometallics; mechanistic aspects of organic reactions; x-ray analyses of conformationally mobile systems; chemical and biochemical aspects of anit-tumor reagents; homogeneous industrial catalysts; transport of neutral molecules in membranes. *Mailing Add:* Dept Chem Univ S Fla Tampa FL 33620

NEWLAND, GORDON CLAY, b Kingsport, Tenn, Feb 26, 27; m 53; c 2. POLYMER CHEMISTRY, PHOTOCHEMISTRY. *Educ:* ETenn State Univ, BS, 49. *Prof Exp:* Res chemist, 50-57, SR RES CHEMIST, TENN EASTMAN RES LAB DIV, EASTMAN KODAK, 57- *Mem:* Am Chem Soc; AAAS. *Res:* Photochemistry of polymer degradation and stabilization mechanisms; photoinitiated polymerization. *Mailing Add:* 1917 E Sevier Ave Kingsport TN 37664

NEWLAND, HERMAN WILLIAM, animal husbandry; deceased, see previous edition for last biography

NEWLAND, LEO WINBURNE, b Nocona, Tex, Sept 15, 40. SOIL CHEMISTRY, WATER CHEMISTRY. *Educ:* Tex A&M Univ, BS, 64; Univ Wis, MS, 66, PhD(soils), 69. *Prof Exp:* Agr res specialist, Frito Lay Inc, 64-66; res asst soils & water, Univ Wis, 66-68; fel biol, 68-69, ASST RES SCIENTIST & PROF GEOL & BIOL, TEX CHRISTIAN UNIV, 69-, DIR ENVIRON SCI PROG, 71- *Concurrent Pos:* Sabbatical Environ Chem, Univ Amsterdam, 78, Univ Bayreuth, Ger, 84; ed, J Toxicol & Environ Chem. *Mem:* Am Chem Soc; Soil Sci Soc Am; Water Pollution Control Fedn; Am Soc Agron. *Res:* Chemical pollution of surficial waters and soils, specifically pesticidal pollution and toxic metals. *Mailing Add:* Dept Biol & Geol Tex Christian Univ Ft Worth TX 76129

NEWLAND, ROBERT JOE, b Lansing, Mich, Jan 30, 46; m 67; c 2. ORGANIC CHEMISTRY. *Educ:* Kalamazoo Col, BA, 68; Wayne State Univ, PhD(chem), 74. *Prof Exp:* Chemist, Res & Develop Dept, Dupont, 73-75; vis asst prof, Univ Ill, 75-77; asst prof, Lafayette Col, 77-83; asst prof, 83-88, ASSOC PROF, GLASSBORO STATE COL, 88- *Mem:* Am Chem Soc. *Res:* Synthetic organic chemistry; medicinal chemistry; biochemistry. *Mailing Add:* Dept Phys Sci Glassboro State Col Glassboro NJ 08028

NEWLANDS, MICHAEL JOHN, b London, Eng, Mar 10, 31; m 54; c 2. INORGANIC CHEMISTRY, STRUCTURAL CHEMISTRY. *Educ:* Cambridge Univ, BA, 53, PhD(chem), 57. *Prof Exp:* Fel, Inst Sci & Technol, Manchester, 58-60, lectr, 60-67; assoc prof, 67-72, PROF CHEM, MEM UNIV NFLD, 72- *Concurrent Pos:* Vis prof, Latrobe Univ, 82. *Mem:* Sigma Xi; Chem Inst Can. *Res:* Organometallic chemistry of main group elements, analytical chemistry of environmental pollutants, x-ray crystallography. *Mailing Add:* Dept Chem Mem Univ Nfld St John's NF A1B 3X7 Can

NEWLIN, CHARLES W(ILLIAM), b Terre Haute, Ind, Feb 3, 24; m 47; c 2. CIVIL ENGINEERING. *Educ:* Rose Polytech Inst, BS, 47; Harvard Univ, MS, 49; Northwestern Univ, PhD, 65. *Prof Exp:* Asst prof civil eng, Swarthmore Col, 49-61; from assoc prof to prof, Ariz State Univ, 61-76, chmn dept, 68-76; civil engr, Dames & Moore, 74-83; CONSULT CIVIL ENG, 83- *Concurrent Pos:* Civil engr, Soil Conserv Serv, 58-61; fac fel, NSF, 65- *Mem:* Am Soc Civil Engrs; Am Soc Eng Educ; Nat Soc Prof Engrs. *Res:* Soil mechanics; earth dams; shear and consolidation of soils. *Mailing Add:* 7223 N 15 Place S Phoenix AZ 85020

NEWLIN, OWEN JAY, b Des Moines, Iowa, Feb 6, 28; m 52; c 4. AGRONOMY. *Educ:* Iowa State Univ, BS, 51, MS, 53; Univ Minn, PhD, 55. *Prof Exp:* Asst agron, Iowa State Univ, 51-53; asst agron & plant genetics, Univ Minn, 53-55; asst prod res, Pioneer, Cent Div, 55-56, dir prod res, 56-60, asst prod mgr & dir prod res, 60-64, prod mgr, 64-67, pres, Cent Div, 67-78, VPRES, PIONEER HI-BRED INT, INC, 78-, MEM BD DIRS, 63- *Mem:* Am Soc Agron. *Res:* Corn production and breeding. *Mailing Add:* 3315 48th Pl Des Moines IA 50310

NEWLIN, PHILIP BLAINE, b Apr 26, 23; US citizen; c 2. CIVIL ENGINEERING. *Educ:* Univ Ariz, BS, 46; Mo Sch Mines, MS, 49. *Prof Exp:* From instr to prof, 46-86, EMER PROF CIVIL ENG, UNIV ARIZ, 86- *Concurrent Pos:* Hwy res engr, US Forest Serv; consult, US Army Signal Corps & Pima County Hwy Dept; consult, earth resource observational systs prog, earth resources survs, US Geol Surv. *Mem:* Am Soc Civil Engrs; Am Soc Photogram; Am Cong Surv & Mapping. *Res:* Photogrammetry and highway engineering; legal aspects of land surveying and boundaries. *Mailing Add:* Dept Civil Eng-Eng Mech Univ Ariz Bldg 72 Rm 206 Tucson AZ 85721

NEWMAN, A KIEFER, electrical engineering, computer science, for more information see previous edition

NEWMAN, B(ARRY) G(EORGE), b Manchester, Eng, May 23, 26; Can citizen; m 55; c 3. AERODYNAMICS, FLUID MECHANICS. *Educ:* Cambridge Univ, BA, 47; Univ Sydney, PhD(aerodyn), 51. *Prof Exp:* Sci officer flight res, Royal Australian Air Force, 51-53; res officer, Nat Aeronaut Estab, Nat Res Coun Can, 53-55; lectr aeronaut, Cambridge Univ, 55-58; chmn dept mech eng, 69-72, PROF AERODYN, MCGILL UNIV, 59-, CANADAIR CHAIR, 59- *Concurrent Pos:* Vis prof, Laval Univ, 58-59; consult, Canadair Ltd, 58-70; Defence Res Bd Can grants, 59-74; mem, assoc comt aerodyn, Nat Res Coun Can, 61-64, 66-69 & 78-; mem, Can nat comt, Int Union Theoret & Appl Mech, 65-68; consult, Can Pratt & Whitney, 73- & Pulp & Paper Res Inst Can, 74-; invited lectr, Cancam, 81; vchmn, Acad Sci, Royal Soc Can, 85-86. *Honors & Awards:* Busk Mem Prize, Royal Aeronaut Soc, 60; Turnbull lectr, Can Aeronaut & Space Inst, 69. *Mem:* Fel Can Aeronaut & Space Inst; fel Royal Aeronaut Soc; fel Royal Soc Can; Am Soc Mech Engrs. *Res:* Turbulent boundary layer separation; separation control by geometric and aerodynamic means; Coanda effect; aerodynamics of air-cushion vehicles; jets and wakes in streaming flow; insect flight; flow past flexible structures. *Mailing Add:* Dept Mech Eng McGill Univ 817 Sherbrooke St W Montreal PQ H3A 2K6 Can

NEWMAN, BERNARD, b New York, NY, Sept 17, 13; m 46; c 2. CHEMISTRY. *Educ:* City Col New York, BS, 35, MS, 43; NY Univ, PhD, 55. *Prof Exp:* Lab technician, Bronx County Anal Labs, NY, 35-37; chief lab technician, Morrisania City Hosp, 37-38; biochemist, Hosp Daughters Jacob, 38-42, dir chem labs, 46; chief biochemist, Rystan Co, Inc, NY, 46-47 & Vet Admin Hosp. NY, 46-47; biochemist, S Shore Res Lab, 48-58; dir, Newing Labs, Inc, 58-78; sr res scientist, Res Div, NY Univ, 56-74; RETIRED. *Concurrent Pos:* Dir, Police Lab, Police Dept, Suffolk County, 60-74; adj assoc prof, Suffolk Community Col, 62-67; prof, Grad Dept Marine Sci, Long Island Univ, 67-78; consult, Kings Park State Hosp, NY; vis prof, chem dept, State Univ NY, Stony Brook, 80-81. *Mem:* AAAS; Am Soc Microbiol; Am Chem Soc; fel Am Pub Health Asn; fel Am Acad Forensic Sci. *Res:* Criminalistics; toxicology. *Mailing Add:* NCR 66 Box 4 Sandisfield MA 01255-9701

NEWMAN, BERTHA L, b Caldwell, Idaho, Mar 29, 26. ANATOMY. *Educ:* Col Idaho, BS, 48; Univ Ore, MA, 50; Univ Iowa, PhD(anat), 58. *Prof Exp:* Instr zool, Idaho State Col, 49-53; asst prof neuroanat, Univ NDak, 58; res fel neurophysiol, Univ Wash, 58-60; asst prof, 60-65, ASSOC PROF NEUROANAT, SCH MED, UNIV SOUTHERN CALIF, 65- *Mem:* Soc Neurosci; Am Asn Anatomists; Am Physiol Soc. *Res:* Electrophysiological changes associated with electrical self-stimulation; ultrastructure of the median eminence of neonatal and adult rats; subcellular changes of blood vessels in hypertension; electronmicroscopy of subcellular fractions of hypertensive rat brains. *Mailing Add:* Dept of Anat & Cell Biol Univ Southern Calif 2025 Zonal Ave Los Angeles CA 90033

NEWMAN, CHARLES MICHAEL, b Chicago, Ill, Mar 1, 46; m 70; c 2. STATISTICAL MECHANICS, STOCHASTIC PROCESSES. *Educ:* Mass Inst Technol, BS(math) & BS(physics), 66; Princeton Univ, MA, 68, PhD(physics), 71. *Prof Exp:* Asst prof math, New York Univ, 71-73; asst prof, Ind Univ, 73-75, assoc prof, 75-79; prof math, Univ Ariz, 79-90, regents prof, 90-91; PROF MATH, COURANT INST, NEW YORK UNIV, 89- *Concurrent Pos:* Consult, Bell Telephone Labs, 75; vis assoc prof, Technion Israel Inst Technol, 75-76, Univ Ariz, 79; assoc ed, J Statist Physics, 82-85, J Math Physics, 83-85, Annls of Probability, 85-; Sloan fel, 78-81, Guggenheim Mem fel, 84-85. *Mem:* Am Math Soc; fel Inst Math Statist. *Res:* Mathematical physics and probability theory with an emphasis on statistical mechanics, stochastic processes and mathematical biology. *Mailing Add:* Courant Inst Math 251 Mercer St New York NY 10012

NEWMAN, CLARENCE WALTER, b Lake Providence, La, Aug 3, 32; m 54; c 2. ANIMAL NUTRITION. *Educ:* La State Univ, BS, 54; Tex A&M Univ, MS, 58; La State Univ, PhD(animal sci), 65. *Prof Exp:* Instr animal sci, La Agr Exp Sta, 58-60; spec lectr animal nutrit, La State Univ, 61-62; assoc prof, 64-75, PROF ANIMAL SCI, MONT STATE UNIV, 75- *Concurrent Pos:* Consult. *Mem:* Am Inst Nutrit; Am Soc Animal Sci; Sigma Xi; Am Asn Cereal Chemists. *Res:* Nutritive value of barley cultures; covered and naked isotypes; high-lysine and high-low betaglucan barleys. *Mailing Add:* Dept Animal & Range Sci Mont State Univ Bozeman MT 59717-0920

NEWMAN, DARRELL FRANCIS, b Fort Knox, Kentucky, Mar 22, 40; m 66. NUCLEAR WASTE MANAGEMENT, NUCLEAR FUEL CYCLE. *Educ:* Kansas State Univ, BS, 63; Univ Wash, MS, 70. *Prof Exp:* Engr, radiation dosimetry, Hanford Labs, Gen Elec, 63; 1st lieutenant, Ordinance Corps, US Army, 63-65; sr res engr, reactor phys, Pac Northwest Lab, Battelle Mem Inst, 65-77, tech leader res reactor oper, Reactor Physics Dept, 70-73 & fuels refabrication, 77-79, prin engr, advan reactor design, 79-80, prog mgr, extended fuel burnup, 80-82 & spent fuel mgt, 82-85, prog mgr, Nuclear Waste Systs, 85-86, proj mgr, eng technol, 86-89, PROJ MGR, HEAVY WATER REACTOR DESIGN, PAC NORTHWEST LAB, BATTELLE MEM INST, 89- *Concurrent Pos:* Consult, naval fuels div, United Nuclear Corp, 74-75 & light water reactors div, US Dept Energy, 80-82; consult, Naval Fuels Div, United Nuclear Corp, 74-75; tech leader, Res Reactor Oper, Reactor Phys Dept, Pacific Northwest Lab/Battelle Meml Inst; lic prof eng, State Wash, 71- *Mem:* NY Acad Sci; Am Nuclear Soc; Nat Soc Prof Eng; AAAS. *Res:* Research and development activites in storage, handling, packaging, transportation and disposal of high level radioactive waste; predict behavior of nuclear fuel both during operation in a reactor and following its discharge; application of high-temperature superconductors to electric power equipment; reactor core design; heavy water reactor technology; awarded one US patent. *Mailing Add:* 1100 McMurray Richland WA 99352

NEWMAN, DAVID BRUCE, JR, b Boston, Mass, Aug 15, 48; m 83. STATISTICAL COMMUNICATIONS, CODING THEORY. *Educ:* Univ RI, BS, 70; Pa State Univ, MS, 71, PhD(elec eng), 74; Am Univ, JD, 83. *Prof Exp:* Analyst, Cent Intel Agency, 74-83; patent atty, Finnegan, Henderson, Farabow, Garrett & Dunner, 83-85; ASSOC PROF ENG & APPL SCI, GEORGE WASHINGTON UNIV, 85-, PROF LECTR LAW, 90- *Concurrent Pos:* Prof lectr eng, George Washington Univ, 83-85; patent atty, 85-; mem, Comt Commun & Info Policy, Inst Elec & Electronic Engrs, 86-, Comt on Defense Res & Develop, 86-88; chmn, Info Security sub-comt, Comt on Commun & Info Policy, Inst Elec & Electronic Engrs, 87-89, Radio/Space Issues sub-comt, 86-87; ed, Inst Elec & Electronic Engrs J on Selected Areas in Commun, 89 & 92. *Mem:* Sr mem Inst Elec & Electronic Engrs; Int Union Radio Sci. *Res:* Data communications; encryptions, computer communications networks; computer modelling and simulation; author of book, author and co-author of numerous publications; patent. *Mailing Add:* Centennial Sq PO Box 2728 La Plato MD 20646

NEWMAN, DAVID EDWARD, b Lilybrook, WVa, Jan 4, 47. PHYSICS. *Educ:* Mass Inst Technol, BS & MS, 70; Princeton Univ, PhD(physics), 74. *Prof Exp:* Fel Univ Michigan, 74-77, asst res scientist physics, 77-82; sr staff scientist, Gen Atomic Co, 82-89; PRES, PHYSICS SOLUTIONS, 89- *Mem:* Am Phys Soc; AAAS. *Res:* Particle physics; nuclear physics; precision measurements; plasma physics; accelerators,; superconductivity and instrumentation; magnets. *Mailing Add:* Physics Solutions PO Box 3333 Suite 244 Encinitas CA 92024

NEWMAN, DAVID S, b New York, NY, Sept 18, 36; m 59; c 4. PHYSICAL CHEMISTRY. *Educ:* Earlham Col, AB, 57; NY Univ, MS, 60; Univ Pa, PhD(chem), 65. *Prof Exp:* Teacher, Newtown High Sch, 59-60; instr chem & physics, Bronx Community Col, 60; teacher, Roosevelt High Sch, 60; res assoc phys chem, Princeton Univ, 64-65; from asst prof to assoc prof phys chem, 65-74, PROF PHYS CHEM, BOWLING GREEN STATE UNIV, 74- *Concurrent Pos:* Prof adv continuing educ appointee, Argonne Nat Lab, 67, consult, Chem Div; Cotrell res grant, Res Corp, 67-68; sr Fulbright fel, 74-75; sr Fulbright lectr, US Govt, 74-75; NSF fac resident, Argonne Nat Lab, 79, fossil fuels res fel, 80, vis scientist, 81; Fulbright prof, NIH, Trondheim, Norway, 84-85. *Mem:* Am Chem Soc; Electrochem Soc. *Res:* Chemistry of fused salts; structure of electrolyte solutions; electrochemistry; chemistry of solid electrolytes. *Mailing Add:* Dept Chemistry Bowling Green State Univ Bowling Green OH 43403-0002

NEWMAN, DAVID WILLIAM, b Pleasant Grove, Utah, Oct 26, 33; m 56. PLANT PHYSIOLOGY, SCIENCE EDITING & WRITING. *Educ:* Univ Utah, BS, 55, MS, 57, PhD, 60. *Prof Exp:* Asst bot, Univ Utah, 54-55; from asst prof to prof, 60-87, EMER PROF BOT, MIAMI UNIV, 87- *Mem:* Am Soc Plant Physiol; Am Chem Soc. *Res:* Lipid and protein metabolism of plants. *Mailing Add:* 1710 Destry Lane Cottonwood AZ 86326

NEWMAN, ERIC ALLAN, b New York, NY, May 9, 50; m 75; c 2. VISION. *Educ:* Mass Inst Technol, BS, 71, MS, 72, PhD(biol), 77. *Prof Exp:* Fac assoc biol, Hampshire Col, 72-73; from asst scientist to assoc scientist, Eye Res Inst, 80-86, sr scientist, 87-90; ASSOC PROF DEPT PHYSIOL, UNIV MINNEAPOLIS, 90- *Concurrent Pos:* Prin investr, Membrane Physiol & Function of Glial Cells Proj, 82- *Mem:* Soc Neurosci; Asn Res Vision & Ophthal. *Res:* Physiological and biophysical properties of glial cells in the vertebrate central nervous system; generation of the electroetinogram; function of glial cells. *Mailing Add:* Dept Physiol Univ Minneapolis 6-255 Millard Hall 435 Delaware St SE Minneapolis MN 55455

NEWMAN, EUGENE, b New York, NY, Sept 14, 30; m 52; c 2. NUCLEAR PHYSICS. *Educ:* Polytech Inst Brooklyn, BS, 52; Yale Univ, MS, 57, PhD(physics), 60. *Prof Exp:* Physicist, 52-54 & 60-74, sect head isotopes, 74-84, SECT HEAD TECH RESOURCES, OAK RIDGE NAT LAB, 84- *Mem:* Am Phys Soc; Am Nuclear Soc. *Res:* Use of low to medium energy cyclotron produced nucleons to investigate nuclear spectroscopy and reaction mechanisms; methods of isotopic enrichment and production of radioisotopes and enriched stable isotopes. *Mailing Add:* Opers Div Bldg 3047 Oak Ridge Nat Lab PO Box 2008 Oak Ridge TN 37831-6021

NEWMAN, EZRA, b New York, NY, Oct 17, 29; m 58; c 2. THEORETICAL PHYSICS. *Educ:* NY Univ, BA, 51; Syracuse Univ, MA, 55, PhD, 56. *Prof Exp:* Asst physics, Syracuse Univ, 52-56; from instr to assoc prof, 57-66, PROF PHYSICS, UNIV PITTSBURGH, 66- *Concurrent Pos:* Vis lectr, Syracuse Univ, 60-61; vis prof, King's Col, Univ London, 64-65 & 68-69; mem, Comt Int Soc Gen Relativity & Gravitation, pres, 86-89. *Mem:* Fel Am Phys Soc. *Res:* General theory of relativity with emphasis on gravitational radiation and the theory of twistors. *Mailing Add:* Dept of Physics Univ of Pittsburgh Pittsburgh PA 15260

NEWMAN, FRANKLIN SCOTT, b Rozel, Kans, July 31, 31; m 54; c 4. MICROBIOLOGY, VIROLOGY. *Educ:* Southwestern Col Kans, AB, 53; Kans State Univ, MS, 57, PhD(bact), 62. *Prof Exp:* Asst prof microbiol, Mont State Univ, 61-65; asst prof, Med Br, Univ Tex, 65-68; assoc prof microbiol, 68-75, dir, Wami Med Prog, 74-84, PROF VET VIROL, VET RES LAB, MONT STATE UNIV, 75- *Concurrent Pos:* Asst dean, Univ Wash Sch Med, 75-84. *Mem:* AAAS; Am Soc Microbiol; Sigma Xi. *Res:* Infectious diseases of domestic animals; bacterial virology; phage-host relationships in pathogenic bacteria. *Mailing Add:* 515 S Eighth Bozeman MT 59715

NEWMAN, GEORGE ALLEN, b Las Cruces, NMex, Mar 15, 41; m 64; c 2. ECOLOGY, ORNITHOLOGY. *Educ:* Baylor Univ, BSc, 64, MSc, 66; Tex A&M Univ, PhD(wildlife sci), 75. *Prof Exp:* From asst prof to assoc prof, 75-79, Cullen prof, 80-81, PROF BIOL, HARDIN-SIMMONS UNIV, 79- *Concurrent Pos:* Collabr, Nat Park Serv, 72-; mem citizens adv coun, Tex Air Control Bd, 72-74; consult, Ecol Audits Inc, 73. *Mem:* Fel Welder Wildlife Found; Ecol Soc Am; Am Ornithologists Union; Wilson Ornith Soc; Cooper Ornith Soc. *Res:* Avian population studies of Guadalupe Mountain Range, Texas and West Texas. *Mailing Add:* Dept Biol Hardin-Simmons Univ 2200 Hickory Abilene TX 79698

NEWMAN, HOWARD ABRAHAM IRA, b Chicago, Ill, July 5, 29; m 55; c 4. BIOCHEMISTRY, PHYSIOLOGY. *Educ:* Univ Ill, BS, 51, MS, 56, PhD(food technol), 58; Am Bd Clin Chem, dipl. *Prof Exp:* Asst, Univ Ill, 54-58, res assoc, 58; res assoc physiol, Univ Tenn, 58-59, asst prof, 59-65; asst prof biochem, Case Western Reserve Univ, 66-68; assoc prof path, 68-79, PROF PATH, OHIO STATE UNIV, 79-, ASSOC PROF PHYSIOL CHEM, 68- *Concurrent Pos:* Mem coun arteriosclerosis, Am Heart Asn. *Mem:* AAAS; Am Physiol Soc; NY Acad Sci; Am Inst Chem; Am Asn Clin Chem. *Res:* Cholesterol, triglyceride and phospholipid metabolism in the intact animal, especially in the atherosclerotic intima; cholesterol esterases; analytical lipid techniques; lipoprotein phenotyping in clinical chemistry; mycobacterial phage lipids; hypolipoproteinemic drugs; cell membrane changes in carcinogenesis; childhood autism. *Mailing Add:* Dept Path Ohio State Univ Col Med 320 W Tenth Ave Columbus OH 43210

NEWMAN, J NICHOLAS, b New Haven, Conn, Mar 10, 35. OCEAN ENGINEERING. *Educ:* Mass Inst Technol, SB, 56, SM, 57, ScD, 60. *Prof Exp:* Res student, Cambridge Univ, 58-59; res naval architect, David Taylor Res Ctr, 59-67; assoc prof 67-70, PROF NAVAL ARCHIT, DEPT OCEAN ENG, MASS INST TECHNOL, 70- *Concurrent Pos:* Adj lectr, Am Univ, 62-67; vis lectr, Univ Calif, Berkeley, 63; vis prof, Univ New S Wales, Australia, 73, Univ Adelaide, 74, Univ Trondheim, Norway, 81-82; fel, Guggenheim Mem Found, 73-74, Australian-Am Educ Found, 73-74, Norwegian Tech & Sci Coun, 81-82; actg head, Dept Ocean Eng, Mass Inst Technol, 79-80; consult, USN Dept, Justice Dept & var pvt firms. *Honors & Awards:* Bronze Medal, Royal Inst Naval Architects, 76; Davidson Medal, Soc Naval Architects & Marine Engrs, 88; George Weinblum Mem Lectr, 88 & 89. *Mem:* Nat Acad Eng; fel Soc Naval Architects & Marine Engrs; AAAS. *Res:* Marine hydrodynamics; author or co-author of over 90 publications. *Mailing Add:* Dept Ocean Eng Mass Inst Technol Rm 5-324 Cambridge MA 02139

NEWMAN, JACK HUFF, b Roanoke, Va, Aug 15, 29; m 56; c 3. BACTERIOLOGY, BIOCHEMISTRY. *Educ:* Va Polytech Inst, BS, 56, MS, 59. *Prof Exp:* Jr biochemist, Smith Kline & French Labs, 58-62; biochemist, 62-77, GROUP MGR, A H ROBINS CO, INC, 77- *Concurrent Pos:* Biochemist, US Army Reserve. *Mem:* Am Chem Soc; Health Physics Soc. *Res:* Radioisotopes; drug metabolism; quantitatively administer radioactive experimental drugs to different animal species by different routes, collect biological samples, process and use appropriate radioactive analyses procedures; calculate raw data. *Mailing Add:* 8106 Diane Lane Richmond VA 23227-1709

NEWMAN, JAMES CHARLES, JR, b Memphis, Tenn, Oct 12, 42; m 64; c 4. ENGINEERING MECHANICS, MATERIALS SCIENCE. *Educ:* Univ Miss, BS, 64; Va Polytech Inst & State Univ, MS, 69, PhD(eng mech), 74. *Prof Exp:* RES ENGR FATIGUE & FRACTURE, NASA, 64- *Concurrent Pos:* Lectr fracture mech, Fed Avaiation Admin. *Honors & Awards:* Sci Achievement Award, NASA, 77; George Rankin Irwin Award, Am Soc Testing & Mat, 81. *Mem:* Am Soc Testing & Mat. *Res:* Fatigue, crack propagation and fracture of materials under service loading and environmental conditions. *Mailing Add:* Langley Res Ctr NASA Mail Stop 188E Hampton VA 23665-5225

NEWMAN, JAMES EDWARD, b Brown Co, Ohio, Dec 22, 20; m 49; c 3. BIOMETEOROLOGY. *Educ:* Ohio State Univ, BS, 47, MS, 49. *Prof Exp:* Asst, Ohio Agr Exp Sta, 47-49; from asst prof to assoc prof agron & climat, 49-69, prof, 69-87, EMER PROF AGRON, PURDUE UNIV, WEST LAFAYETTE, 87- *Concurrent Pos:* Partic comn agr meteorol, World Meteorol Orgn, Ont, Can, 62; New World ed, Jour Agr Meteorol, 63; vis prof, Univ Calif, Riverside, 65-66; vis scientist, Inst Agr Sci, Univ Alaska, 70; ed-in-chief, Int J Agr Meteorol, 74-76; mem select comt, Nat Defense Univ, 77-78; adv bd, Int Soc Biometeorol, 79-84; mem comt atmospheric sci, Nat Res Coun-Nat Acad Sci, 79-81. *Honors & Awards:* Soils & Crops Award, Am Soc Agron, 65. *Mem:* Fel AAAS; fel Am Soc Agron; Am Meteorol Soc; Ecol Soc Am; Int Soc Biometeorol (pres, 87-90); Sigma Xi. *Res:* Radiant energy flux and plant responses in both natural and mono culture phyto-environments; adaptation of cereal grains; crop modeling; climatic variability/changes on world food production; author of approximately 200 scientific and semi-technical papers. *Mailing Add:* Dept Agron Purdue Univ West Lafayette IN 47907

NEWMAN, JOHN ALEXANDER, b Lethbridge, Alta, Jan 23, 32; m 59; c 4. ANIMAL BREEDING, ANIMAL GENETICS. *Educ:* Univ Alta, BSc, 55; Univ Edinburgh, dipl animal genetics, 58, PhD(animal genetics), 60. *Prof Exp:* RES SCIENTIST ANIMAL BREEDING, AGR CAN RES STA, 55- *Concurrent Pos:* Chmn, Can Beefcattle Record Performance Tech Comt, 75-81; mem, Can Agr Res Coun, 75-79; assoc ed, Can J Animal Sci, 77-80. *Mem:* Agr Inst Can (vpres, 74-75); Genetics Soc Can; Am Soc Animal Sci; Can Soc Animal Sci (pres, 72-73). *Res:* Beef cattle breeding and genetics. *Mailing Add:* Box 1237 Lacombe AB T0C 1S0 Can

NEWMAN, JOHN B(ULLEN), b Okmulgee, Okla, Sept 27, 38; m 73; c 5. APPLIED MECHANICS, ENGINEERING MECHANICS. *Educ:* Stanford Univ, BS, 61, MS, 62, PhD(eng mech), 65. *Prof Exp:* Assoc engr, Lockheed Missiles & Space Co, 62-63; sr engr, 65-73, fel engr, 73-80, ADV ENGR, BETTIS ATOMIC POWER LAB, WESTINGHOUSE ELEC CORP, 80- *Mem:* Am Soc Mech Engrs; Am Inst Aeronaut & Astronaut; Am Acad Mech; Am Nuclear Soc; Sigma Xi. *Res:* Elastic-plastic instabilities; approximation methods in solid mechanics; constitutive relations in continuum mechanics; modelling of behavior of nuclear reactor fuel rods; fluid-solid interactions; non-linear, time dependent, structural analysis methods and computational techniques for engineering application. *Mailing Add:* 4830 McAnulty Rd Pittsburgh PA 15236

NEWMAN, JOHN HUGHES, b Baltimore, Md, Oct 1, 45; m 78; c 2. PULMONARY MEDICINE, SCIENCE EDUCATION. *Educ:* Harvard Univ, BA, 67; Columbia Univ, MD, 71. *Prof Exp:* Intern, Columbia-Presby Med Ctr, 71-72, resident, 72-73; sr resident, Johns Hopkins Hosp, 73-74; major, US Army Med Corps, 74-76; fel pulmonary med, Med Ctr, Univ Colo, 76-79; asst prof, 79-83, prof pulmonary med, sch med, 91, ELSA S HANIGAN CHAIR, VANDERBILT UNIV, 86-; CHIEF PULMONARY MED SERV, ST THOMAS HOSP, 84- *Concurrent Pos:* Rap Study Sect, NIH, NLLBI. *Honors & Awards:* Clin Investr Award, NIH, 81. *Mem:* Am Physiol Soc; Am Thoracic Soc. *Res:* Lung injury and respiratory failure; pulmonary hypertension; oxygen toxicity. *Mailing Add:* 4730 Clendenin Rd Nashville TN 37220

NEWMAN, JOHN JOSEPH, b Wolf Point, Mont, Jan 15, 36; m 64; c 4. ELECTRICAL ENGINEERING, MAGNETISM. *Educ:* Mont State Col, BSEE, 58; Univ NMex, MSEE, 61; Univ Santa Clara, PhD(elec eng), 68. *Prof Exp:* Staff mem res & develop, Sandia Corp, 58-61; sr elec engr reliability, Lockheed Missiles & Space Co, 61-64; teaching asst, Univ Santa Clara, 64-67; SR STAFF MEM ELEC ENG, MEMOREX CORP, 67-, PRIN STAFF SCIENTIST. *Mem:* Inst Elec & Electronics Engrs. *Res:* Magnetic recording, especially theory, media and processes; magnetic measurements; instrumentation development and design. *Mailing Add:* 5212 Ralston Dr San Jose CA 95124

NEWMAN, JOHN SCOTT, b Richmond, Va, Nov 17, 38; c 2. ELECTROCHEMISTRY, CHEMICAL ENGINEERING. *Educ:* Northwestern Univ, BS, 60; Univ Calif, Berkeley, MS, 62, PhD(chem eng), 63. *Prof Exp:* From asst prof to assoc prof chem eng, 63-70, PROF CHEM ENG, UNIV CALIF, BERKELEY, 70-, PRIN INVESTR MAT RES DIV, 63- *Honors & Awards:* Young Author's Prize, Electrochem Soc, 66 & 69, David C Grahame Award, 85, Henry B Linford Award, 90. *Mem:* Am Inst Chem; Electrochem Soc. *Res:* Design and analysis of electrochemical systems; transport properties of concentrated electrolytic solutions; mass transfer. *Mailing Add:* Dept Chem Eng Univ Calif Berkeley CA 94720-0001

NEWMAN, JOSEPH HERBERT, b Brooklyn, NY, Feb 2, 25; m 50; c 2. CHEMICAL ENGINEERING. *Educ:* Polytech Inst New York, BChE, 45, MChE, 47. *Prof Exp:* Res engr, Flintkote Co, 45-51; sect head, M W Kellogg Co, 51-53; asst mgr aeronaut div, Curtiss Wright Co, Woodridge, NJ, 53-59; gen mgr, Tishman Res Corp, 59-65, vpres, 66-72, sr vpres, 73-76, exec vpres, 77-79, pres, 80-88; RETIRED. *Concurrent Pos:* Mem adv panel bldg res sect, Nat Bur Standards, 65-68; mem comt urban technol, Nat Res Coun, 67-69; vpres, Tishman Realty & Construct Co, Inc, 67-73, 1st vpres, 74-76, sr vpres, 77-79, exec vpres & dir, 80-86, dir emer, 86-; mem panel housing technol, US Dept Com, 68-70; chmn bldg res adv bd, Nat Acad Sci, 72-73, mem comn sociotech systs, 74-77; chmn comt tech transfer, Nat Acad Eng, 73-74; dir, Nat Inst Building Sci, 76-83; sr vpres, Tishman Construct & Res Co, Inc, 77-78; chmn, Nat Inst Building Sci, 80-82; dir, Polytech Univ, 85-87, Shared Technol, Inc, 88- *Mem:* Nat Acad Eng; Am Inst Chem Engrs; Nat Inst Bldg Sci; Am Chem Soc. *Res:* Building science and construction technology innovation. *Mailing Add:* Llewellyn Park West Orange NJ 07052

NEWMAN, KARL ROBERT, b Mt Pleasant, Mich, Feb 26, 31; m 61; c 3. GEOLOGY. *Educ:* Univ Mich, BS, 53, MS, 54; Univ Colo, PhD(geol), 61. *Prof Exp:* Geologist, Magnolia Petrol Co, 54 & Palynological Res Lab, 59-61; sr res geologist, Pan-Am Petrol Corp, 61-66; asst prof geol, Mont Col Mineral Sci & Technol, 66-67; assoc prof, Cent Wash State Col, 67-71, chmn dept, 69-71; assoc prof, Colo Sch Mines, 71-80, prof geol, 80-90. *Concurrent Pos:* Consult geologist, Geol Explor Assoc, Ltd, 66- *Mem:* Soc Econ Paleontologists & Mineralogists; Am Asn Petrol Geologist; Geol Soc Am; Am Inst Prof Geologists. *Res:* Stratigraphy; geology and palynology of Rocky Mountain basins; field geology; upper Cretaceous and lower Tertiary palynomorphs; geology of coal and oil shale. *Mailing Add:* Box 3789 Evergreen CO 80439

NEWMAN, KENNETH WILFRED, b Lincoln, Nebr. ALGEBRA. *Educ:* City Univ New York, BS, 65; Cornell Univ, PhD(math), 70. *Prof Exp:* Fel math, McGill Univ, 69-71; asst prof math, Univ Ill, Chicago Circle, 71-78; ASSOC PROF MATH, UNIV MASS, BOSTON, 78- *Mem:* Am Math Soc. *Res:* Theory of Hopf algebras. *Mailing Add:* Dept Math & Computer Sci Univ Mass Harbor Campus Boston MA 02125

NEWMAN, LEONARD, b New York, NY, Jan 15, 31; m 53; c 2. ATMOSPHERIC & ANALYTICAL CHEMISTRY. *Educ:* Polytech Inst Brooklyn, BS, 52; Mass Inst Technol, PhD(chem), 56. *Prof Exp:* Asst, Mass Inst Technol, 52-55; scientist, Nat Lead Co, 56-57; scientist, 58-63, sr scientist, 63-80, assoc head, Atmospheric Sci Div, 77, HEAD, ENVIRON CHEM DIV, BROOKHAVEN NAT LAB, 78- *Concurrent Pos:* Vis scientist, Royal Inst Technol, Sweden, 62-63; mem, Comt on Nuclear Methods for Investigating Air Pollution, Nat Acad Sci, 68; consult, Gen Pub Utilities

Corp, 74, Public Serv Comn, Wis, 75, Environ Criteria Assessment Off, Environ Protection Agency, 80, Empire State Elec Energy Res Corp & State Univ Res Found, NY, 81 & Antimony Oxide Indust Asn, 84-; adj prof, NY Univ, 84-, Ga Inst Tech, 91-; mem rev comt, Chem Technol Div, Univ Chicago, 86-, Chem Res & Develop Ctr, US Army, 87- *Mem:* AAAS; Am Chem Soc; Am Soc Testing & Mat; NY Acad Sci; Air & Waste Mgt Asn. *Res:* Atmospheric chemistry; analytic chemistry of air pollutants; complex ion equilibria; actinide chemistry; nuclear reactor chemistry and fuel processing; kinetic mechanisms; electrochemistry; impacts of air pollution; global climate change chemistry. *Mailing Add:* Dept Appl Sci Bldg 426 Brookhaven Nat Lab Upton NY 11973

NEWMAN, LESTER JOSEPH, b St Louis, Mo, June 15, 33. CYTOGYENETICS. *Educ:* Wash Univ, BA, 55; Univ Mich, MA, 60; Wash Univ, PhD(zool), 63. *Prof Exp:* NIH trainee, Wash Univ, 60-63; asst prof zool, Ore State Univ, 63-64; actg head dept, 65-66, from asst prof to assoc prof, 64-78, PROF BIOL, PORTLAND STATE UNIV, 78- *Mem:* AAAS; Genetics Soc Am. *Res:* Cytogenetics of diptera. *Mailing Add:* Dept Biol Portland State Univ Portland OR 97207

NEWMAN, LOUIS BENJAMIN, b New York, NY; m. REHABILITATION MEDICINE. *Educ:* Ill Inst Technol, ME, 21; Rush Med Col, MD, 33; Am Bd Phys Med & Rehab, dipl, 47. *Prof Exp:* Attend physician, Cook County Hosp, Ill, 33-42; PROF REHAB MED, MED SCH, NORTHWESTERN UNIV, CHICAGO, 46- *Concurrent Pos:* Chief rehab med serv, Vet Admin Hosps, Hines, 46-53 & Vet Res Hosp, Chicago, 53-67; consult rehab med, Vet Admin Med Centers & several community hosps, Chicago Area, 67-; mem med adv bd, Vis Nurse Asn, Nat Found, Arthritis Found, Rehab Comt of Inst Med, United Parkinson Found & others; lectr rehab med, Col Med, Univ Ill & Stritch Sch Med, Loyola Univ Chicago; lectr, Chicago Med Sch, Univ Health Sci. *Honors & Awards:* Davis Award, Asn Phys & Ment Rehab, 56. *Mem:* AMA; Am Cong Rehab Med (vpres, 60); Am Acad Phys Med & Rehab (pres, 59); Am Asn Electromyog & Electrodiag; Int Soc Rehab Disabled. *Mailing Add:* 400 E Randolph St 3219 Chicago IL 60601-7308

NEWMAN, M(ORRIS) M, b Poland, Sept 7, 09; nat US; m 38; c 3. ELECTRICAL ENGINEERING. *Educ:* Univ Minn, BSEE, 31, MSEE, 37. *Prof Exp:* Asst prof elec eng, Univ Minn, 44-46; res dir, Lightning & Transients Res Inst, Minneapolis, 46-71; res prof & dir, Lightning Res Oceanic Lab, 72-76, Lightning & Transients Res Inst, 76-80 & Lightning & Oceanics Inst, 80-; RES DIR, COOP PROG, LIGHTING & TRANSIEN, ST PAUL; PROF, MATH DEPT, UNIV CALIF, SANTA BARBARA. *Concurrent Pos:* Assoc res prof, Univ Fla, 60-63; vis prof, Univ Miami, 63-72 & Friends World Col, 66- *Mem:* AAAS; Am Meteorol Soc; fel Inst Elec & Electronics Engrs; Am Geophys Union. *Res:* High voltage engineering; lightning studies and radio interference. *Mailing Add:* Math Dept Univ Calif Santa Barbara CA 93106

NEWMAN, MELVIN MICKLIN, b Chicago, Ill, Dec 20, 21; m 49; c 2. SURGERY, PULMONARY PHYSIOLOGY. *Educ:* Univ Chicago, BS, 41, MD, 44. *Prof Exp:* From asst resident to instr surg, Univ Chicago, 46-52; from asst prof to assoc prof, State Univ NY Downstate Med Ctr, 54-59; chief, Nat Jewish Hosp, 59-68; assoc prof surg, Med Ctr, Univ Colo, Denver, 61-84; City of Hope Med Ctr, Duarte, 84-87; CLIN ASSOC PROF SURG, CIGNA HEALTHPLANS OF SOUTHERN CALIF, UNIV SOUTHERN CALIF, 87- *Concurrent Pos:* Nat Res Coun fel, Univ Chicago, 52; NIH grants, Nat Jewish Hosp, Denver, Colo, 62-68; NIH grant, Univ Colo, Denver, 70-72. *Mem:* Am Soc Artificial Internal Organs (pres, 60); Soc Univ Surgeons; Am Asn Thoracic Surg; Soc Thoracic Surg; Am Thoracic Soc. *Res:* Shock; microcirculation; pulmonary ventilation and circulation; vascular prostheses. *Mailing Add:* 1750 E Mountain Street Pasadena CA 91104

NEWMAN, MELVIN SPENCER, b New York, NY, Mar 10, 08; m 33; c 4. ORGANIC SYNTHESIS. *Educ:* Yale Univ, BS, 29, PhD(chem), 32. *Hon Degrees:* DSc, Univ New Orleans, 75, Bowling Green State Univ, 78, Ohio State Univ, 79. *Prof Exp:* Nat Tuberc Asn fel, Yale Univ, 32-33; Nat Res Coun fel chem, Col Physicians & Surgeons, Columbia Univ, 33-34 & Harvard Univ, 34-36; instr chem, Ohio State Univ, 36-39, Elizabeth Clay Howald scholar, 39-40, from asst prof to prof, 40-65, regents prof, 65-78; RETIRED. *Concurrent Pos:* Guggenheim fel, 49 & 51; Fulbright lectr, Glasgow Univ, 57 & 67; ed, J, Am Chem Soc. *Honors & Awards:* Award, Am Chem Soc, 61, Roger Adams Award, 79; Wilbur Cross Medal, Yale Univ, 70. *Mem:* Nat Acad Sci; AAAS; Am Chem Soc; The Chem Soc. *Res:* Synthetic and theoretical organic chemistry. *Mailing Add:* 2239 Onandaga Dr Columbus OH 43221

NEWMAN, MICHAEL CHARLES, b Bridgeport, Conn, Feb 21, 51; m 80; c 2. ECOTOXICOLOGY. *Educ:* Univ Conn, BA, 74, MS, 78; Rutgers Univ, MS, 80, PhD(environ sci), 81. *Prof Exp:* Teaching environ biol, Univ Calif, San Diego, 82-83; res assoc environ sci, 81-82, asst, 83-90, ASSOC ECOLOGIST ECOTOXICOL, SAVANNAH RIVER ECOL LAB, UNIV GA, 90- *Concurrent Pos:* Vis fac, Savannah River Ecol Lab, Univ Ga, 83; lectr, Univ Calif, San Diego, 83, Univ SC, Aiken, 84-88. *Mem:* AAAS; Am Soc Limnol & Oceanog; Am Chem Soc; Am Malacol Union; Soc Environ Toxicol & Chem. *Res:* Geochemical factors determining the bioavailability of trace elements; kinetics of trace element accumulation in aquatic biota; environmental toxicology; chemical limnology; aquatic chemistry of trace elements; invertebrate pathology. *Mailing Add:* Savannah River Ecol Lab PO Drawer E Aiken SC 29801

NEWMAN, MICHAEL J(OHN), b Oak Park, Ill, Apr 12, 48; div. APPLIED THEORETICAL PHYSICS, ASTROPHYSICS. *Educ:* Rice Univ, BA, 70, MS, 73, PhD(astrophysics), 75; La State Univ, Baton Rouge, MS, 71. *Prof Exp:* Res assoc astrophysics, Rice Univ, 75; res fel physics, W K Kellogg Radiation Lab, Calif Inst Technol, 75-77; staff mem, Max Planck Inst Physics & Astrophysics, Munich, WGer, 77-78; STAFF MEM, LOS ALAMOS NAT LAB, 78- *Concurrent Pos:* Asst to pres, Am Phys Soc, 76-77; spec sci adv to the asst to the secy defense, atomic energy, 90-; Woodrow Wilson fel. *Mem:* Am Astron Soc; fel Royal Astron Soc; Astron Soc Pac; fel Am Phys Soc; AAAS. *Res:* Applied theoretical physics; program management; nuclear astrophysics; theoretical astrophysics; stellar structure and evolution; star formation; nucleosynthesis; interactions with the interstellar medium; astronomical influences on terrestrial climate. *Mailing Add:* Los Alamos Nat Lab ADNWT MS A105 PO Box 1663 Los Alamos NM 87545

NEWMAN, MORRIS, b New York, NY, Feb 25, 24; m 48; c 2. MATHEMATICS. *Educ:* NY Univ, BA, 45; Columbia Univ, MA, 46; Univ Pa, PhD(math), 52. *Prof Exp:* Lectr math, Columbia Univ, 45; Instr, Univ Del, 48-51; res mathematician, 51-63, chief numerical anal sect, 63-70, sr res mathematician, Nat Bur Standards, 70-76; PROF MATH, UNIV CALIF, SANTA BARBARA, 76- *Concurrent Pos:* Ed, Math Comput, 75-86; assoc ed, J Linear & Multilinear Algebra, 73- & Letters in Linear Algebra, 79- *Honors & Awards:* Dept Commerce Gold Medal Award, 66. *Mem:* Am Math Soc; Math Asn Am; London Math Soc. *Res:* Number theory; group theory; matrix theory; structure of matrix groups over rings; automorphic and modular functions. *Mailing Add:* Dept Math Univ Calif Santa Barbara CA 93106

NEWMAN, MURRAY ARTHUR, b Chicago, Ill, Mar 6, 24; m 52; c 1. ICHTHYOLOGY. *Educ:* Univ Chicago, BS, 49; Univ Calif, MA, 51; Univ BC, PhD(zool), 60. *Prof Exp:* Cur fishes, Univ Calif, Los Angeles, 51-53; cur inst fishes, Univ BC, 53-56; DIR, VANCOUVER PUB AQUARIUM, 56- *Concurrent Pos:* Field expeds to Can Arctic, Amazon & Micronesia. *Mem:* Can Mus Asn; Int Union Zool Park Dirs; Am Asn Zool Parks & Aquariums. *Res:* Behavior of fishes; marine ecology; systematic ichthyology. *Mailing Add:* Vancouver Pub Aquarium Box 3232 Vancouver BC V6B 3X8 Can

NEWMAN, NORMAN, b Brooklyn, NY, Mar 13, 39; m 62; c 2. ORGANIC CHEMISTRY. *Educ:* Brooklyn Col, BS, 59; Univ Minn, PhD(org chem), 64. *Prof Exp:* Res specialist, 63-80, DIV SCIENTIST, MINN MINING & MFG CO, 80- *Mem:* Am Chem Soc; Soc Photog Sci & Eng. *Res:* Photographic science and chemistry; photochemical systems; reaction mechanisms. *Mailing Add:* Minn Mining & Mfg Co 3M Bldg 209-2C-08 St Paul MN 55144-1000

NEWMAN, PAUL HAROLD, b Washington, DC, Apr 25, 33; m 58; c 3. INFORMATION SCIENCE. *Educ:* Antioch Col, BSc, 56. *Prof Exp:* Sr res assoc human eng, Am Insts Res, 56-60; res engr, Boeing Co, 60-61; systs specialist info processing systs design, Syst Develop Corp, 61-72, mgr control staff, 73; systs design mgr, Dept Social & Health Serv, State Wash, 73-75, asst chief, Off Info Systs, 75-80, mgr, Systs Eng, 80-85, actg chief, Off Info Systs, 85-88, mgr software technol, 88-90, MGR OPER SUPPORT SERV, DEPT SOCIAL & HEALTH SERV, STATE WASH, 90- *Mem:* AAAS; Soc Eng Psychol; Human Factors Soc; Asn Comput Mach. *Res:* information processing systems design and control; educational requirements; systems analyses; human factors analyses. *Mailing Add:* Dept Social & Health Serv Mail Stop PJ-13 Olympia WA 98504-0095

NEWMAN, PAULINE, b New York, NY, June 20, 27. INDUSTRIAL CHEMISTRY, LAW. *Educ:* Vassar Col, AB, 47; Columbia Univ, AM, 48; Yale Univ, PhD(chem), 52; NY Univ, LLB, 58. *Prof Exp:* Lab instr, Columbia Univ, 48 & Yale Univ, 48-50; res chemist, Am Cyanamid Co, 51-54; patent atty, FMC Corp, 54-69, dir patent & licensing dept, 69-84; CONSULT, 84- *Concurrent Pos:* Specialist natural sci, UNESCO, 61-62; mem patent adv comt, Res Corp, 72-84; mem nat bd, Med Col Pa, 76-84; mem adv comt, Indust Innovation, Domestic Policy Rev, 79; mem adv comt, Int Intellectual Property, State Dept, 74-84; Judge, US Court Appeals, 84- *Mem:* AAAS; Am Chem Soc; Am Inst Chem; Soc Chem Indust; fel NY Acad Sci. *Res:* Chemistry of high polymers; oxidation-reduction reactions; patent law; physical organic chemistry. *Mailing Add:* 2700 Virginia Ave NW Apt 114 Washington DC 20037-1908

NEWMAN, R(OBERT) W(EIDENTHAL), b Cleveland, Ohio, May 14, 14; m 81; c 3. OPERATIONS RESEARCH. *Educ:* Mass Inst Technol, BS, 36. *Prof Exp:* Works engr glass tech, Pitney Glass Works, Gen Elec Co, Ohio, 36-41, works engr electron tube mfg, Buffalo Tube Works, 41-47, mfg engr, Schenectady Tube Works, 47-55, consult oper res & synthesis, Mgt Consult, Serv, NY, 55-62; econ decision models, Acct Serv, 62-69; mgr corp planning serv, Gen Elec Co, 69-76; RETIRED. *Concurrent Pos:* Assoc prof, NY Univ, 69-76; consult corp planning. *Honors & Awards:* US Navy Develop Award, 46. *Mem:* AAAS; Am Chem Soc; fel Royal Numis Soc; Inst Elec & Electronic Engrs; Archeol Inst Am. *Res:* Electron tube technology; cathode materials and processes; operations research and management science; management and decision making process; corporate planning. *Mailing Add:* 381 Eden Ave Springdale CT 06907

NEWMAN, RICHARD HOLT, b Mebane, NC, Aug 12, 32; m 55; c 3. RADIOCHEMISTRY. *Educ:* Elon Col, BA, 54; Univ SDak, MA, 60. *Prof Exp:* Chemist I water analysis, NC State Bd Health, 54-55, Chemist II pollution control, 57-58; assoc scientist analytical res, 60-67, res prof radiochem, 68-74, PROJ LEADER SMOKE MECHANISM, PHILIP MORRIS, INC, 74- *Mem:* Am Chem Soc; Sigma Xi. *Res:* Study of precursor product relationship between tobacco and smoke and elucidating mechanisms for smoke formation utilizing both radioactive and stable isotopes as tracers. *Mailing Add:* Philip Morris Inc PO Box 26583 Richmond VA 23261

NEWMAN, ROBERT ALWIN, b Winchester, Mass, July 11, 48. BIOCHEMISTRY, PHARMACOLOGY. *Educ:* Univ RI, BS, 70; Univ Conn, MS, 73, PhD(pharmacol), 75. *Prof Exp:* Res fel cell & molecular biol, Med Col Ga, 75-76; res assoc biochem, Univ Vt, 76-77, asst prof pharmacol, Dept Pharmacol, Col Med & staff mem, Vt Reg Cancer Ctr, 77-84; CHIEF, SECT PHARMACOL, DEPT MED ONCOL, M D ANDERSON HOSP & TUMOR INST, HOUSTON, TEX, 84- *Mem:* Am Asn Cancer Res; Am Soc Clin Oncologists; Am Cancer Soc; Sigma Xi; Am Soc Pharmacol & Exp Therapeut. *Res:* Experimental cancer chemotherapy (basic and clinical); cancer biology. *Mailing Add:* Dept Med Oncol M D Anderson Hosp Box 52 1515 Holcombe Blvd Houston TX 77030

NEWMAN, ROBERT WEIDENTHAL, b Cleveland, Ohio, May 14, 14; m 82; c 3. RESOURCE ALLOCATION UNDER UNCERTAINTY, MANAGEMENT AT LARGE INSTITUTIONS. *Educ:* Mass Inst Technol, BS, 36. *Prof Exp:* Works engr glass technol, Pilnzy Glass Works, Gen Elec, Mela Park, 36-41; works engr & mgr, Electronic Tubes, Buffalo Tube Works, Gen Elec, 42-47; mfg eng & mgr, Electronic Tubes, Schnectady Tube Works, 47- 53; consult, Opnatims Res, Gen Elec Co Hq, NY, 54-62; mgr, Corp Planning, Gen Elec Co Hq, NY & Fairfield, Conn, 63-76; CONSULT, 76- *Concurrent Pos:* Adj prof mgt, Grad Sch Bus, NY Univ, 60-74; vchmn, Nat Planning Asn, 63-65. *Mem:* Am Soc Qual Control; Am Chem Soc; Inst Elec & Electronic Engrs; Nat Planning Asn; fel Royal Numismatic Soc; Archaeological Inst Am. *Res:* Structure of glass, to understand its strength and etching; glass furnace design; decision theory; resource allocation for long term corporate profitability. *Mailing Add:* 381 Eden Rd Stamford CT 06907

NEWMAN, ROGER, b New York, NY, Aug 16, 25; m 46; c 3. CHEMISTRY. *Educ:* Columbia Univ, AB, 44, MA, 46, PhD(phys chem), 49. *Prof Exp:* Du Pont fel, Harvard Univ, 49-50; fel chem, Calif Inst Technol, 50-52; res assoc semiconductors, Res Labs, Gen Elec Co, 52-59; head dept physics, Mat Res Lab, Hughes Prod Div, Hughes Aircraft Co, 59-60, mgr semiconductor mat dept, Hughes Res Labs, 60-61; mgr solid state sci dept, Sperry Rand Res Ctr, 61-78; dept mgr, Rockwell Int, 78-80; dept head, 80-89, PRIN SCIENTIST, AEROSPACE CORP, 89- *Mem:* Fel Am Phys Soc; fel Inst Elec & Electronics Eng; Electrochem Soc. *Res:* Physics and technology of solid state electronic devices and circuits. *Mailing Add:* Aerospace Corp M2/238 P O Box 92957 Los Angeles CA 90009

NEWMAN, ROGERS J, b Ramar, Ala, Dec 22, 26; m 51; c 3. MATHEMATICS. *Educ:* Morehouse Col, AB, 48; Atlanta Univ, MA, 49; Univ Mich, PhD(math), 61. *Prof Exp:* Instr physics & math, Bishop Col, 49-50; instr math, Grambling Col, 50-51; instr math & physics, Jackson State Col, 51-53; instr, Southern Univ, 53-55; jr instr math, Univ Mich, 59-60; chmn dept math, 61-74, PROF MATH, SOUTHERN UNIV, BATON ROUGE, 60- *Concurrent Pos:* NSF fac fel, Imp Col, Univ London, 70-71. *Mem:* Am Math Soc; Math Asn Am; Nat Inst Sci (vpres, 64). *Res:* Complex variables. *Mailing Add:* 7860 Emile St Baton Rouge LA 70807

NEWMAN, SARAH WINANS, b Hannibal, Mo, Apr 23, 41; m 81. NEUROANATOMY, NEUROPSYCHOLOGY. *Educ:* Cornell Univ, BA, 63, PhD(anat), 69. *Prof Exp:* Instr anat, State Univ NY Downstate Med Ctr, 68-70; from asst prof to assoc prof, 70-83, PROF ANAT, UNIV MICH, ANN ARBOR, 83- *Mem:* AAAS; Am Asn Anatomists; Soc Neurosci; Sigma Xi. *Res:* Structure and function of the olfactory and vomeronasal pathways in the central nervous system. *Mailing Add:* Dept Anat Univ of Michigan Ann Arbor MI 48109-0010

NEWMAN, SEYMOUR, b New York, NY, July 9, 22; m 43; c 2. POLYMER SCIENCE, PLASTICS. *Educ:* City Col New York, BS, 42; Columbia Univ, MA, 47; Polytech Inst Brooklyn, PhD, 49. *Prof Exp:* Res assoc, Southern Regional Res Labs, USDA, 49-51, Cornell Univ, 51-52 & Allegany Ballistics Lab, Hercules Powder Co, 52-55; group leader & scientist, Monsanto Co, 56-67; staff scientist sci res staff, Plastics Develop Ctr, Ford Motor Co, Detroit, 67-69, mgr, Polymer Sci Dept, 69-73, sr staff scientist, 73-86, mgr adv functional components, 77-86; RETIRED. *Concurrent Pos:* Adj assoc prof, Univ Mass, Amherst, 67-; mem technol assessment panel, Engrs Joint Coun, 75-; adj prof, Univ Detroit, 80-81 & Wayne State Univ, 90-; mem bd dir, Engr Prop Div Specialities, 80-81, Ad Hoc Panel, Polymer Sci & Engr, Nat Res Coun, 80-81; consult, Off Technol Assessment, US Cong, 84-85; pvt consult, 86-88; instr, Lawrence Inst Technol, 88. *Mem:* Am Chem Soc; Am Phys Soc; Soc Plastic Engrs; Soc Automotive Engrs. *Res:* Physical chemistry of high polymers; dynamic mechanical behavior; crystallinity; dilute solution behavior; fiber and film properties; strength properties; rheology; coatings; processing; composites; plastic materials; design; automotive component development; author or co-author of 60 technical publications and holder of 21 US patents. *Mailing Add:* 6340 Celeste Ct West Bloomfield MI 48322-1313

NEWMAN, SIMON LOUIS, b Jacksonville, Fla, Oct 31, 47; m 72; c 1. MACROPHAGE BIOLOGY. *Educ:* Emory Univ, BA, 69; Univ Ala, MS, 71, PhD(immunol), 78. *Prof Exp:* Res asst, Univ Ala Med Ctr, 71-72; asst path, Univ Fla Col Med, 72-73; fel, Nat Jewish Hosp & Res Ctr, 78-81; fel, Univ NC, Chapel Hill, 81-82 res asst prof immunol, 82-86. *Concurrent Pos:* Young Investr Award, Am Asn Clin Immunol & Allergy, 79; Investr Award, Nat Arthritis Found, 82. *Mem:* Am Asn Immunologists; Am Soc Microbiol; AAAS. *Res:* Expression and function of complement and Fc receptors on differentiating monocytes, the physical and biochemical mechanisms by which these receptors trigger phagocytosis, and their role in mediating the phagocytosis and killing of microorganisms. *Mailing Add:* 313 Carol St Carrboro NC 27510

NEWMAN, SIMON M(EIER), information science, communication; deceased, see previous edition for last biography

NEWMAN, STEPHEN ALEXANDER, b Auburn, NY, Apr 12, 38; m 64; c 2. PROCESS SIMULATION, SYNFUELS. *Educ:* Rensselaer Polytech Inst, BChE, 60; Mass Inst Technol, SM, 62; Rutgers Univ, PhD(phys chem), 76. *Prof Exp:* Res engr, M W Kellogg Co, 62-67; supvr process design, Foster Wheeler Corp, 67-70, tech data coordr, 70-74, tech data mgr, 74-79, process technol mgr, 79-81, TECHNOL MGR, FOSTER WHEELER CORP, 81- *Concurrent Pos:* Consult, Nat Bur Standards, 79; mem, Eval Panel, Nat Res Coun, 80; ed, Am Chem Soc text, 80, Am Inst Chem Engr text, 81, Ann Arbor Sci/Butterworth Group text, 82, 83 & Gulf Publ, 85. *Honors & Awards:* Organizing Design Award, Inst Phys Properties, Am Inst Chem Engrs. *Mem:* NY Acad Sci; Am Petrol Inst; Am Inst Chem Engrs; Sigma Xi. *Res:* Thermodynamic and physical property data and correlations; computer simulation of process plants; conceptual design of synthetic fuel plants. *Mailing Add:* 941 Douglas Terrace Union NJ 07083-6523

NEWMAN, STEVEN BARRY, b New York, NY, May 19, 52; m 73; c 1. CLOUD PHYSICS, ATMOSPHERIC SCIENCE. *Educ:* City Col New York, BS, 73; State Univ NY Albany, MS, 75, PhD(atmospheric sci), 78. *Prof Exp:* Teaching asst atmospheric sci, State Univ NY Albany, 73-77, instr meteorol, Col Oneonta, 77-78; ASST PROF METEOROL & EARTH SCI, CENT CONN STATE COL, 78- *Concurrent Pos:* Fac res fel, Air Force Geophys Lab, Hanscom AFB, Mass, 81. *Mem:* Am Meteorol Soc. *Res:* Hail and hail dynamics; cloud-precipitation physics and weather modification. *Mailing Add:* Physics/Earth Sci Dept Cent Connecticut State Univ New Britain CT 06050

NEWMAN, STUART ALAN, b New York, NY, Apr 4, 45; m 68; c 2. DEVELOPMENTAL BIOLOGY. *Educ:* Columbia Univ, AB, 65; Univ Chicago, PhD(chem physics), 70. *Prof Exp:* Fel theoret biol, Univ Chicago, 70-72; vis fel biol sci, Univ Sussex, Eng, 72-73; instr anat, Univ Pa, 73-75; asst prof biol sci, State Univ NY Albany, 75-79; assoc prof, 79-84, PROF ANAT, NY MED COL, 84-, PROF CELL BIOL & MED, 88- *Concurrent Pos:* Consult biotechnology, Friends of The Earth & Nat Coun Churches; prin investr, NSF, 76-79, & 86-, NIH, 79-; mem exec coun, Coun Responsible Genetics; Fogarty Sr Int fel, Australia, 89. *Mem:* Am Soc Cell Biol; AAAS; NY Acad Sci; Soc Develop Biol; Teratology Soc; Soc Math Biol. *Res:* Molecular and biophysical mechanics of cellular pattern and tissue morphogenesis; protein structure; mechanisms of evolution; dynamics of biochemical networks. *Mailing Add:* Dept Cell Biol & Anat Basic Sci Bldg NY Med Col Valhalla NY 10595

NEWMAN, WALTER HAYES, b Birmingham, Ala, Mar 15, 38; m 64; c 3. PHARMACOLOGY. *Educ:* Auburn Univ, BS, 62, MS, 63; Med Col SC, PhD(pharmacol), 67. *Prof Exp:* From instr to assoc prof, 66-78, PROF PHARMACOL, MED UNIV SC, 78- *Concurrent Pos:* Vis prof, Univ Munich, Fed Repub Ger, 86-87. *Honors & Awards:* Sr Fulbright Fel. *Mem:* Am Soc Pharmacol & Exp Therapeut; Sigma Xi. *Res:* Cardiac hypertrophy & heart failure. *Mailing Add:* Dept Pharmacol Med Univ SC Charleston SC 29425

NEWMAN, WILEY CLIFFORD, JR, b Europa, Miss, Apr 15, 31; m 58; c 1. PHYSIOLOGY. *Educ:* Vanderbilt Univ, AB, 53; Univ Tenn, PhD(physiol), 65. *Prof Exp:* Instr, Univ Tenn, Memphis, 66-68; asst prof, 68-76, ASSOC PROF PHYSIOL, SCH MED, TULANE UNIV, 76- *Mem:* AAAS. *Res:* Endocrine physiology; experimental mammary cancer induction. *Mailing Add:* 301 Gordon River Ridge LA 70112

NEWMAN, WILLIAM ALEXANDER, b Colebrook, NH, Nov 14, 34; m 60; c 2. HYDROLOGY & WATER RESOURCES. *Educ:* Boston Univ, AB, 57, AM, 59; Syracuse Univ, PhD(geol), 71. *Prof Exp:* From instr to assoc prof, 68-86, PROF GEOL, NORTHEASTERN UNIV, 86- *Concurrent Pos:* Consult geologist, Maine Geol Surv, 78-82, Nuclear Regulatory Comn, 79, Tech Adv Serv Attys, 82-86, Dames & Moore, 90-; Marion & Jasper Found Acad Fel, 83; Weeks fund grant, Univ Wis-Madison, 83; Hydrogeologist, Mass Hazardous Waste Facil Site Safety Coun, 84-88; provost grant, Northeastern Univ, Boston, Mass, 85. *Mem:* Asn Eng Geologists; Sigma Xi; Int Glaciol Soc. *Res:* Pleistocene geology of New England; groundwater aquifers associated with Pleistocene sediments, Pleistocene stratigraphy of Boston Basin; mechanics of drumlin formation. *Mailing Add:* Dept Geol Northeastern Univ 14 Holmes Hall Boston MA 02115

NEWMAN, WILLIAM ANDERSON, b San Francisco, Calif, Nov 13, 27; c 5. MARINE BIOLOGY. *Educ:* Univ Calif, Berkeley, AB, 53, MA, 54, PhD(zool), 62. *Prof Exp:* Actg instr zool, Univ Calif, Berkeley, 60-61, asst prof oceanog, Univ Calif, San Diego, 62-63; asst prof marine biol, Harvard Univ, 63-65; asst prof biol oceanog, 65-69, assoc prof oceanog, 69-75, PROF OCEANOG, SCRIPPS INST OCEANOG, UNIV CALIF, SAN DIEGO, 75- *Concurrent Pos:* NSF fel, 62; mem adv comt arthropods, Smithsonian Oceanog Sorting Ctr, 64-67; mem comt ecol of interoceanic canal, Nat Acad Sci, 69-70; mem biol sci comt, World Book Encyclop, 71-83; joint adv bd, Hawaii Inst Marine Biol, 82-; trustee, San Diego Nat Hist Mus, 88- *Mem:* fel AAAS; Crustacean Soc; Sigma Xi; Western Soc Nat. *Res:* Systematics and biogeography of the Crustacea, especially the Cirripedia; biology and near surface geology of oceanic islands; oceanography. *Mailing Add:* Scripps Inst Oceanog Univ of Calif San Diego La Jolla CA 92093-0202

NEWMAN, WILLIAM L, b Rockford, Ill, July 14, 20; m 46; c 2. GEOLOGY, APPLIED ECONOMIC & ENGINEERING. *Educ:* Beloit Col, BS, 42. *Prof Exp:* Geologist, ore distr, Anaconda Copper Mining Co, Butte, 46-48; grad instr, Dept Geol, Univ Mont, Missoula, 48-50; proj geologist, uranium studies, US Geol Surv, 50-57, staff asst, geol admin, Dir's Office 57-59, proj chief , metallogen maps, 60-62, staff geol, 62-67, unit chief, general int publ, 67-74, geol reports, Office Sci Publ, 67-74, deputy chief, Office Sci Publ, 74-80; CONSULT GEOLOGIST, 80- *Concurrent Pos:* mem interdept Comt Sci Exhibits, 70-80, publications comt, Am Geol Inst, 74-80; ed, The Cross Section, 70-80; Secy-Treas, Exec Comt, Am Geol Inst, 71-73. *Mem:* Fel Geol Soc Am; Fel Am Assoc Adv Sci; sr fel Soc Econ Geol; Am Asn Petrol Geologists; Assoc of Earth Sci Ed. *Res:* Exploration and geochemical studies of Colorado Plateau uranium deposits; military geology, preparation of general interest articles and scripts describing a broad spectrum of research activities for United States Geological Survey publication and for print media, films, radio, TV, and classroom use. *Mailing Add:* 5624 E Wethersfield Rd Scottsdale AZ 85254

NEWMARK, HAROLD LEON, b New York, NY, July 21, 18; m 49, 87; c 2. ORGANIC CHEMISTRY, BIOCHEMISTRY. *Educ:* City Col New York, BS, 39; Polytech Inst Brooklyn, MS, 50. *Prof Exp:* Res dir, Vitarine Co, Inc, NY, 50-59; res chemist & group leader appl res, Hoffman-LaRoche, Inc, Nutley, NJ, 59-66, asst dir prod develop, 66-81, consult, 81; sr scientist, Ludwig Inst Cancer Res, Toronto, 81-84; SCIENTIST, MEM SLOAN KETTERING CANCER CTR, NY, 84- *Concurrent Pos:* Mem, XV revision, US Pharmacopoeia; adj prof chem biol & pharmacog, Col Pharm, Rutgers Univ, NJ, 87- *Mem:* AAAS; Am Chem Soc; Am Pharmaceut Asn; NY Acad

Sci; Am Pharm Asn; Am Asn Cancer Res. *Res:* Biochemistry, nutrition, vitamins, carotenoids, parenterals, animal health drugs and prevention of carcinogenesis; applications to foods and pharmaceuticals. *Mailing Add:* 11 Claremont Dr Maplewood NJ 07040

NEWMARK, JONATHAN, b New York, NY,. MEDICINE, NEUROSCIENCES. *Educ:* Harvard Col, AB, 74; Columbia Univ, MD, 78. *Prof Exp:* Intern internal med, Roosevelt Hosp, 78-79; resident path, State Univ NY, Brooklyn, 79-80; NRSA fel neurochem, Nat Inst Neurol Disorders & Stroke, Bethesda, 80-82; resident neurol, Boston City Hosp, Boston, 82-84; fel occup neurol, Mass Gen Hosp, Boston, 84-85; asst prof neurol, Sch Med, Univ Louisville, 85-90; fel neuromuscular dis, Hosp Univ Pa, Philadelphia, 90-91. *Concurrent Pos:* Clin instr, Sch Med, Boston Univ, 82-84 & Harvard Med Sch, Boston, 84-85; clin consult, Int Conf Symphony & Opera Musicians, 85-; assoc anat, Sch Med, Univ Louisville, 87-90. *Mem:* Am Acad Neurol; Performing Arts Med Asn. *Res:* Focal demyelination in rat fimbria induced by focal stereotactic injection of lysophosphatidyl choline a new model for focal demyelinazing disease; medical/neurological problems of performing artists. *Mailing Add:* 236 Edgemont Ave Ardmore PA 19003-2705

NEWMARK, MARJORIE ZEIGER, biochemistry, for more information see previous edition

NEWMARK, RICHARD ALAN, b Urbana, Ill, Nov 11, 40; m 65; c 2. ANALYTICAL CHEMISTRY, SPECTROSCOPY. *Educ:* Harvard Univ, AB, 61; Univ Calif, Berkeley, PhD(chem), 65. *Prof Exp:* NSF fel, Mass Inst Technol, 64-66; asst prof chem, Univ Colo, Boulder, 66-69; res chemist, 69-72, res specialist, 72-76, sr res specialist, 76-81, STAFF SCIENTIST, 3M CO, 81- *Mem:* Am Chem Soc; Soc Appl Spectros; Sigma Xi. *Res:* Nuclear magnetic resonance studies; conformational analyses of polymers. *Mailing Add:* 3M Bldg 201-BS-05 St Paul MN 55144-1000

NEWMARK, STEPHEN, b San Antonio, Tex, Feb 24, 43. ENDOCRINOLOGY, NUTRITION & METABOLISM. *Educ:* Univ Pa, MD, 69. *Prof Exp:* Assoc prof, 80-85, PROF ENDOCRINOL, UNIV OKLA, 85-; DIR, SW METAB & DIABETES CTR, 80- *Mem:* Am Col Physicians; Endocrine Soc; Am Diabetes Soc; Am Inst Nutrit; Am Col Nutrit; Am Fedn Clin Res. *Mailing Add:* 700 Shadow Lane Las Vegas NV 89106

NEWNAN, DONALD G(LENN), industrial engineering, for more information see previous edition

NEWNHAM, ROBERT EVEREST, b Amsterdam, NY, Mar 28, 29. PHYSICS. *Educ:* Hartwick Col, BS, 50; Colo State Univ, MS, 52; Pa State Univ, PhD(physics), 56; Cambridge Univ, PhD(crystallog), 60. *Prof Exp:* Assoc prof elec eng, Mass Inst Technol, 59-66; assoc prof solid state sci, 66-71, PROF SOLID STATE SCI, PA STATE UNIV, UNIVERSITY PARK, 71-, SECT HEAD, 77- *Mem:* Am Phys Soc; Am Crystallog Asn; Am Ceramic Soc; Mineral Soc Am. *Res:* Crystal and solid state physics; x-ray crystallography. *Mailing Add:* Dept Geosci Pa State Univ Main Campus University Park PA 16802

NEWNHAM, ROBERT MONTAGUE, b Bromley, Eng, Aug 11, 34; m 59; c 3. FOREST MANAGEMENT, FOREST MENSURATION. *Educ:* Univ Wales, BS, 56; Univ BC, MF, 58, PhD(forestry), 64. *Prof Exp:* Asst exp officer forest ecol, Nature Conserv, Grange over Sands, Eng, 60-62; res scientist forest mgt & forest mensuration, Forest Mgt Inst, Ottawa, 64-79, dir, Petawawa Nat Forestry Inst, 79-83, RES SCIENTIST FOREST MGT MODELLING, PETAWAWA NAT FORESTRY INST, CAN FORESTRY SERV, CHALK RIVER, ONT, 83- *Honors & Awards:* Elwood Wilson Award, Can Pulp & Paper Asn, 76. *Mem:* Can Inst Forestry; Commonwealth Forestry Asn; Int Union Forest Res Orgn. *Res:* Planning logging operations; systems analysis; applications of computers to forest research; development of forest management models; forest mensuration. *Mailing Add:* Petawawa Nat Forestry Inst Forestry Can Box 2000 Chalk River ON K0J 1J0 Can

NEWROCK, KENNETH MATTHEW, b Brooklyn, NY, Oct 11, 44; m 81; c 2. DEVELOPMENTAL BIOCHEMISTRY, MOLECULAR NEUROLOGY. *Educ:* Franklin & Marshall Col, BA, 67; Ind Univ, Bloomington, PhD(biol), 74. *Prof Exp:* USPHS genetics trainee, Ind Univ, 67-74; NIH res fel, Fox Chase Cancer Ctr, 74-76, res assoc biochem, 76-78; asst prof biol develop, McGill Univ, 78-84; EXEC SECY MOLECULAR NEUROL, NEUROL C STUDY SECT, NIH, 84- *Concurrent Pos:* NSF undergrad res partic, Bryn Mawr Col, 67; res partic, marine embroyol, NSF, Bermuda Biol Sta, 69; vis scientist, div cancer treat, develop therapeut prog, Lab Molecular Pharmacol, NIH. *Mem:* Am Soc Cell Biol; Am Soc Develop Biol; Fedn Am Socs Exp Biol; Soc Neurosci. *Res:* Developmental biochemistry, chromatin chemistry, and gene regulation; importance of egg structure to early development; histone and histone genes, their regulation and their importance to development chromatin structure; chromatin structure. *Mailing Add:* Neurol C Study Sect NIH-DRG WW232 5333 Westbard Ave Bethesda MD 20892

NEWROCK, RICHARD SANDOR, b New York, NY, Aug 7, 42; div; c 1. SOLID STATE PHYSICS. *Educ:* Rensselaer Polytech Inst, BS, 64; Rutgers Univ, MS, 66, PhD(physics), 70. *Prof Exp:* Res fel physics, Cornell Univ, 70-72, adj asst prof, 72-73; from asst to assoc prof, 73-82, PROF PHYSICS, UNIV CINCINNATI, 82-, DEPT HEAD, 82- *Mem:* Am Phys Soc; Sigma Xi. *Res:* Metals and alloys; electrical and thermal magnetoresistivity; magnetic susceptibility; superconductivity; thin films, cermets and inhomogeneous materials; electrical and thermal properties; superconducting arrays. *Mailing Add:* Dept Physics ML-11 Univ Cincinnati Cincinnati OH 45221

NEWROTH, PETER RUSSELL, b Sheffield, Eng, Oct 12, 45; Can citizen; m 84. MARINE BOTANY, WATER RESOURCE MANAGEMENT. *Educ:* Univ NB, BS, 66, PhD(marine biol), 70. *Prof Exp:* Fel bot, Univ BC, 70-72; biologist, Water Invest Br, Water Resources Serv, BC, 72-80; BIOLOGIST & MGR, AQUATIC STUDIES BR, MINISTRY ENVIRON, 80- *Mem:* Aquatic Plant Mgt Soc; NAm Lake Mgt Soc; Int Asn Aquatic Plant Biologists. *Res:* Life histories, taxonomy and morphology of marine Rhodophyta; management and ecology of freshwater macrophytes. *Mailing Add:* Sect Water Qual Br Ministry Environ Victoria BC V8V 1X5 Can

NEWSAM, JOHN M, b Edinburgh, Scotland, UK; m 84; c 2. DIFFRACTION METHODS, ZEOLITES. *Educ:* Oxford Univ, Eng, BA, 76, MA & PhD(chem), 80. *Prof Exp:* Res chemist, Exxon Res & Engr Co, 82-85, sr chemist, 85-87, staff chemist, 87-89, sr staff chemist, 89-90; TECH DIR, BIOSYM TECHNOLOGIES, INC, 90- *Concurrent Pos:* Res fel physics, Tohoku Univ & Japan Spallation Neutron Source, 80-82; consult, Struct Comn Int Zeolite Asn, 83- *Honors & Awards:* Corday-Morgan Medal, 89. *Mem:* Royal Soc Chem; Am Chem Soc; Am Crystallog Asn; Am Inst Physics. *Res:* Structural characterization of complex crystalline materals by diffraction and related techniques; neutron scattering; computer modelling methods; x-ray and neutron diffraction; synchrotron x-ray diffraction. *Mailing Add:* Biosym Technologies Inc 10065A Barnes Canyon Rd San Diego CA 92121

NEWSOM, BERNARD DEAN, b Oakland, Calif, Feb 8, 24; m 45; c 2. ENVIRONMENTAL PHYSIOLOGY. *Educ:* Univ Calif, AB, 49, PhD(physiol), 60; Univ San Francisco, MS, 54. *Prof Exp:* Investr & physiologist, US Naval Radiol Defense Lab, 49-61; sr staff scientist, Life Sci Lab, Gen Dynamics/Convair, 61-68; res analyst, Med Res & Opers Directorate, Manned Spacecraft Ctr, 68-72, proj mgr biomed res, Ames Res Ctr, NASA, 72-77; CONSULT, RES PLANNING & MGT, 77- *Mem:* Sigma Xi; Am Physiol Soc; Radiation Res Soc; Aerospace Med Asn; NY Acad Sci. *Res:* Physiological and performance changes of man in a rotational environment; adaptation and tolerance to prolonged exposures to angular velocities and perturbations; biological interpretation of complex space stresses of vibration, acceleration, null gravity and radiation; effect of abnormal environments on radiation sequela in terms of stress tolerance, performance and longevity; Apollo; Skylab; environmental impact of proposed space power satellite; biological effects of power transmission by microwaves; NASA technology transfer; space utilization. *Mailing Add:* 26645 Altamont Rd Los Altos Hills CA 94022

NEWSOM, DONALD WILSON, b Shongaloo, La, Nov 14, 18; m 44; c 2. HORTICULTURE. *Educ:* La State Univ, BS, 47, MS, 48; Mich State Univ, PhD(hort), 52. *Prof Exp:* Asst prof agr, Tex Col Arts & Indust, 50; assoc horticulturist, Clemson Col, 51-54; horticulturist, USDA, 54-57; head dept, 66-85, PROF HORT, LA STATE UNIV, BATON ROUGE, 57-, EMER HEAD DEPT, 85- *Concurrent Pos:* Prof liaison rep, AVI Pub Co, 85-86; hort consult, 85- *Honors & Awards:* Outstanding Admin Award, Southern Region, Am Soc Hort Sci, 85. *Mem:* Fel AAAS; fel Am Soc Hort Sci (vpres, 80); Am Soc Plant Physiol; Am Forestry Asn; Sigma Xi; NY Acad Sci. *Res:* Post harvest physiology and chemical composition of fruits and vegetables, especially flavor components. *Mailing Add:* Dept of Hort La State Univ Baton Rouge LA 70803

NEWSOM, GERALD HIGLEY, b Albuquerque, NMex, Feb 11, 39; m 72; c 2. ASTRONOMY, ATOMIC & MOLECULAR PHYSICS. *Educ:* Univ Mich, Ann Arbor, BA, 61; Harvard Univ, MA, 63, PhD(astron), 68. *Prof Exp:* Res asst, Imperial Col, Univ London, 68-69; from asst prof to assoc prof, 69-82, PROF ASTRON, OHIO STATE UNIV, 82- *Concurrent Pos:* Acting asst dean, Ohio State Univ, 65-66; res asst, Physikalisches Inst Univ Bonn, W Ger, 78. *Mem:* Am Astron Soc; Int Astron Union. *Res:* Observation and interpretation of the spectrum of SS 433; classification of energy levels in neutral atoms; measurement of oscillator strengths for neutral and singly ionized atomic spectral lines. *Mailing Add:* Dept Astron Ohio State Univ 174 W 18th Ave Columbus OH 43210-1106

NEWSOM, HERBERT CHARLES, b Whittier, Calif, Oct 25, 31; m 55, 83; c 3. ORGANIC CHEMISTRY, PESTICIDE CHEMISTRY. *Educ:* Whittier Col, BA, 53; Univ Southern Calif, PhD(org chem), 59. *Prof Exp:* Asst chem, Univ Southern Calif, 55-59; res chemist, 59-65, sr res chemist, 65-80, MGR CHEM ECON, US BORAX RES CORP, 80- *Mem:* AAAS; Am Chem Soc; Nat Asn Corrosion Engrs. *Res:* Process and product economics; market research; organoboron and free radical chemistry; kinetics; herbicide residue analysis; photolysis; herbicide degradation; process development; EPA pesticide registration studies; corrosion testing; engineering economics; analytical chemistry. *Mailing Add:* 220 S Calle Grande Orange CA 92669-4412

NEWSOM, LEO DALE, entomology; deceased, see previous edition for last biography

NEWSOM, RAYMOND A, b Tarrant Co, Tex, Jan 8, 31; m 50; c 2. ORGANIC CHEMISTRY. *Educ:* Ariz State Univ, BS, 53; Univ Ariz, MS, 57; Univ Iowa, PhD(org chem), 60. *Prof Exp:* Res chemist, Plastics Div Res, 60-61 & Hydrocarbons Div, 62-65, process chemist, Process Technol Dept, 65-75, SR PROCESS SPECIALIST, PROCESS TECHNOL DEPT, MONSANTO CO, TEXAS CITY, TEX, 75- *Mem:* Am Chem Soc; Sigma Xi. *Res:* Organic syntheses and reaction mechanisms; product and process development for monomers. *Mailing Add:* 1209 Plantation Dr Dickinson TX 77539

NEWSOM, WILLIAM S, JR, b Wynne, Ark, Dec 31, 18; m 46; c 3. AGRICULTURAL CHEMISTRY. *Educ:* Univ Ark, BS, 48. *Prof Exp:* Chemist, Lion Oil Co, 48-55 & Monsanto Chem Co, 55-57; sr res chemist, Int Minerals & Chem Corp, Ill, 57-67; sr res engr, Ga Inst Technol, 67-68; SECT HEAD, OCCIDENTAL CHEM CO, 68- *Mem:* Am Chem Soc; Brit Fertilizer Soc. *Res:* Fertilizer technology; pesticides; plant growth regulators; animal feed technology; high temperature defluorination; fluorine technology. *Mailing Add:* 194 Shelby Dr Lake City FL 32055

NEWSOME, DAVID ANTHONY, b Winston-Salem, NC, Apr 16, 42. OPHTHALMOLOGY. *Educ:* Duke Univ, AB, 64; Col Physicians & Surgeons, Columbia Univ, MD, 68. *Prof Exp:* Resident ophthal, Harvard Univ, 73-76; lectr physiol, Simmons Col, Boston, 75-77; fel, Bascom Palmer Eye Inst, Univ Miami, 77; sr staff ophthalmologist, Nat Eye Inst, 77-79, chief, Sect Retinal & Ocular Connective Tissue Dis, 79-82; dir lab retinal degenerations, Wilmer Inst, Johns Hopkins Univ, 82-85; dir, Retinal Degeneration Res, Eye Ctr, La State Univ, 86-88; DIR, SENSORY & ELECTROPHYSIOLOGY RES UNIT, TOURO INFIRMARY, NEW ORLEANS, 88-; CLIN PROF OPHTHAL, TULANE UNIV SCH MED, NEW ORLEANS, 88- *Concurrent Pos:* Res fel med, Mass Gen Hosp, 73-76; consult, Pan Am Health Orgn, 79 & Nat Geog Soc, 81-85; lectr macular dis, Wilmer Ophthal Inst, 80-82; comt chmn res sect, Asn Res Vision & Ophthal, 80-83; vis sci study sect, NIH, 86-; retinal specialist, Eye Surg Ctr La, New Orleans, 88- *Mem:* AAAS; Asn Res Vision & Ophthal; Am Acad Ophthal; Am Soc Cell Biol. *Res:* Clinical and laboratory investigations into the cellular and tissue mechanisms of serious eye diseases affecting retina, vitreous, choroid and other ocular tissues. *Mailing Add:* 1401 Foucher St New Orleans LA 70115

NEWSOME, RICHARD DUANE, b Kalamazoo, Mich, Aug 19, 31; m 54; c 2. PLANT ECOLOGY, BOTANY. *Educ:* Western Mich Univ, BS, 54; Univ Sask, MS, 63, PhD(plant ecol), 65. *Prof Exp:* From instr to assoc prof, 65-84, PROF BOT & ECOL, BELOIT COL, 84- *Concurrent Pos:* Vchmn, Wis Nat Areas Preserv Coun. *Mem:* Ecol Soc Am; Am Inst Biol Scientists; Nat Sci Teachers Asn; Bot Soc Am; Nature Conservancy. *Res:* Responses of grasslands across a moisture gradient to controlled burning; landscape ecology and use; grassland ecology; environmental education. *Mailing Add:* Dept Biol Beloit Col Beloit WI 53511

NEWSOME, ROSS WHITTED, b Lynchburg, Va, Nov 6, 35. PHYSICS. *Educ:* Mass Inst Technol, SB, 57; Univ Mich, Ann Arbor, MS, 58, PhD(physics), 63. *Prof Exp:* Univ Mich res asst high energy physics, Lawrence Radiation Lab, Univ Calif, Berkeley, 59-60; from res asst to res assoc exp nuclear physics, Univ Mich, Ann Arbor, 60-64; mem staff, Los Alamos Sci Lab, Univ Calif, 64-66; mem tech staff systs eng, Bellcomm, Inc, Washington, DC, 66-72; MEM TECH STAFF CUSTOMER SYSTS ENG CTR, BELL LABS, 72- *Mem:* AAAS; Am Phys Soc; Sigma Xi; NY Acad Sci. *Res:* Transmission of infrared radiation through the atmosphere; capabilities of thermal infrared instruments for remote sensing of surface targets from satellites; measurements and analyses of telephone traffic for small business customers; utilization of hierarchical data bases to store and extract large quantities of information. *Mailing Add:* PO Box 375 Oakhurst NJ 07755

NEWSON, HAROLD DON, b Salt Lake City, Utah, July 11, 24; m 48; c 4. MEDICAL ENTOMOLOGY. *Educ:* Univ Utah, BA, 49, MS, 50; Univ Md, PhD(entom), 59; Am Registry Prof Entomologists. *Prof Exp:* From res entomologist to med entom consult to Surgeon Gen, US Army, 51-70; assoc prof, Mich State Univ, 70-76, prof entom, microbiol & pub health, 76-89; RETIRED. *Concurrent Pos:* Mem, US Armed Forces Pest Control Bd, 62-70, chmn, 63-67; mem res subcomt, Fed Comt Pest Control, 64-66; mem study group, Off Environ Sci, Smithsonian Inst, 72-74; chmn region V, USPHS Vector Control Group, 73-76; res consult, US Army Res & Develop Command, 79-84. *Mem:* Am Soc Trop Med & Hyg; Am Mosquito Control Asn; Entom Soc Am. *Res:* Ecology; transmission and control of arthropod-borne diseases of medical and veterinary importance. *Mailing Add:* 8946 Sunset W Traverse City MI 49684-1516

NEWSTEAD, JAMES DUNCAN MACINNES, b Camberley, Eng, Oct 11, 30; m 62; c 1. CELL BIOLOGY, HISTOLOGY. *Educ:* Univ BC, BA, 54, MA, 56; Ore State Univ, PhD(cell biol), 62. *Prof Exp:* From instr to asst prof zool, Ore State Univ, 60-63; USPHS fel fine structure, Univ Wash, 63-65; from asst prof to assoc prof anat, 65-74, PROF ANAT, UNIV SASK, 74- *Mem:* AAAS; Am Soc Zoologists; Am Soc Cell Biol. *Res:* Circulation and ion transport in gills of fish; fine structure of cell division in protozoa. *Mailing Add:* Dept of Anat Univ of Sask Saskatoon SK S7N 0W0 Can

NEWSTEIN, HERMAN, b Philadelphia, Pa, May 4, 18; m 53; c 1. METEOROLOGY. *Educ:* Temple Univ, BS, 48, MEd, 51; NY Univ, MS, 53, PhD(meteorol), 57. *Prof Exp:* Meteorologist, US Weather Bur, 41-53, res meteorologist, 53-62; PROF PHYSICS & ATMOSPHERIC SCI, DREXEL UNIV, 62- *Concurrent Pos:* Sci consult, Nat Acad Sci, 58; adj assoc prof, NY Univ, 58-62. *Mem:* AAAS; Am Meteorol Soc; Am Geophys Union; Air Pollution Control Asn; Am Asn Physics Teachers; Sigma Xi. *Res:* Experimental and theoretical atmospheric physics. *Mailing Add:* 6027 4th Ave N St Petersburg FL 33710

NEWSTEIN, MAURICE, b Philadelphia, Pa, Feb 13, 26; m 57; c 2. THEORETICAL PHYSICS. *Educ:* Temple Univ, AB, 49; Mass Inst Technol, PhD(physics), 54. *Prof Exp:* Asst univ observ, Harvard Univ, 54-55; physicist, Tech Res Group, Inc, 55-67; res scientist electrophys, 67-70, PROF ELECTROPHYS, GRAD CTR, POLYTECH INST NY, 70- *Res:* Atomic scattering problems; plasma physics; applications of microwave and optical spectroscopy; quantum electronics. *Mailing Add:* Polytech Univ Rte 110 Farmington NY 11735

NEWTON, AMOS SYLVESTER, b Shingletown, Calif, July 26, 16; m 42; c 2. CHEMISTRY. *Educ:* Univ Calif, BS, 38; Univ Mich, MS, 39, PhD(phys chem), 41. *Prof Exp:* Chemist, Eastman Kodak Co, NY, 41-42, Manhattan Proj, Iowa State Col, 42-46 & Eastman Kodak Co, NY, 46-62; chemist, 62-85, EMER, LAWRENCE BERKELEY LAB, UNIV CALIF, 85- *Concurrent Pos:* Consult, Lawrence Berkeley Lab, Univ Calif, 46-62. *Mem:* AAAS; Am Soc Mass Spectros; Am Chem Soc. *Res:* Use of radioisotopes as tracer; chemistry of heavy elements; radiochemistry of fission products; radiation chemistry; mass spectrometry; molecular beam studies; environmental chemistry; marine chemistry; fuel science. *Mailing Add:* 1665 Thousand Oaks Blvd Berkeley CA 94707-1541

NEWTON, ARTHUR R, b Melbourne, Australia, July 1, 51. COMPUTER DESIGN OF ELECTRICAL SYSTEMS. *Educ:* Univ Melbourne, BEE, 72, MEE, 75; Univ Calif, Berkeley, PhD, 78. *Prof Exp:* PROF ELEC ENG, UNIV CALIF, BERKELEY, 79- *Mem:* Fel Inst Elec & Electronics Engrs. *Mailing Add:* Dept EEC Univ Calif Berkeley, Cory Hall, Rm 512 Berkeley CA 94720

NEWTON, (WILLIAM) AUSTIN, US citizen. DEVELOPMENTAL GENETICS. *Educ:* Univ Calif, PhD(biochem), 64. *Prof Exp:* Asst prof, 66-72, ASSOC PROF BIOL, PRINCETON UNIV, 72- *Res:* Biochemistry and genetics of gene expression and development in microorganisms. *Mailing Add:* Dept Biol Princeton Univ Princeton NJ 08544

NEWTON, CHESTER WHITTIER, b Los Angeles, Calif, Aug 17, 20; m 48; c 4. METEOROLOGY. *Educ:* Univ Chicago, SB, 46, SM, 47, PhD(meteorol), 51. *Prof Exp:* Weather observer, US Weather Bur, 39-41, meteorologist, 48; asst meteorol, Univ Chicago, 47-48, synoptic analyst, 48-51; synoptic analyst, Univ Stockholm, 51-53 & Woods Hole Oceanog Inst, 53; res assoc meteorol, Univ Chicago, 53-56, asst prof, 56-61; chief scientist, Nat Severe Storms Proj, US Weather Bur, 61-63; sr scientist, Nat Ctr Atmospheric Res, 63-86, vis scientist, 86-90; RETIRED. *Concurrent Pos:* Affil prof, Pa State Univ, 65-67; mem steering comt, Earth Sci Curric Proj, 65-68; ed, Monthly Weather Rev, 73-77; assoc ed meteorol, Rev Geophys, 84-86. *Honors & Awards:* C F Brooks Award, Am Meteorol Soc, 83. *Mem:* Fel AAAS; fel Am Meteorol Soc (pres, 79); Am Geophys Union; foreign mem Royal Meteorol Soc. *Res:* Synoptic meteorology; atmospheric general circulation and global energy balance; fronts and jet streams; structure of and physical processes in cyclone formation; thunderstorms and severe local storms. *Mailing Add:* Nat Ctr for Atmospheric Res Box 3000 Boulder CO 80307

NEWTON, DAVID C, b Middletown, Conn, Apr 27, 39; m 58; c 2. APICULTURE, ANIMAL BEHAVIOR. *Educ:* Cent Conn State Univ, BS, 61; Wesleyan Univ, MALS, 65; Univ Ill, PhD(entom), 67. *Prof Exp:* Teacher pub schs, 61 & high sch, 61-64; ASSOC PROF BIOL, CENT CONN STATE COL, 67- *Concurrent Pos:* Conn Res Comn grant, 68-71; USDA study grant, 73-76. *Mem:* AAAS; Animal Behav Soc; Bee Res Asn; Am Inst Biol Sci. *Res:* Behavior studies of honey bees; behavior studies of honey bees relating to nest cleaning and disease resistance. *Mailing Add:* 3 Deborah Lane Farmington CT 06032

NEWTON, ELISABETH G, b Bennettsville, SC, Aug 5, 33. GOVERNMENT RELATIONS. *Educ:* Univ SC, BA, 56. *Prof Exp:* Geologist, US Dept Interior, 56-87; PRES, E G NEWTON & ASSOC, 88- *Mem:* Fel Geol Soc Am; Am Petrol geologists; Am Inst Geol. *Res:* Geocivics. *Mailing Add:* 8370 Greensboro Dr Suite 4-814 McLean VA 22102

NEWTON, GEORGE LARRY, b Lincolnton, NC, Mar 29, 45; m 68; c 1. ANIMAL NUTRITION, WASTE MANAGEMENT. *Educ:* NC State Univ, BS, 67, PhD(animal sci), 72; Va Polytech Inst & State Univ, MS, 70. *Prof Exp:* Res assoc animal sci, Univ Ky, 72-73; asst prof, 73-80, ASSOC PROF ANIMAL SCI, UNIV GA, 80- *Mem:* Am Soc Animal Sci; Am Registry Cert Animal Scientists; Sigma Xi. *Res:* Ruminant and non-ruminant nutrition; environmental and nutritional interactions; livestock waste management. *Mailing Add:* Dept Animal Sci Coastal Plain Sta Box 748 Tifton GA 31793

NEWTON, HOWARD JOSEPH, b Oneida, NY, June 16, 49; m 87; c 2. STATISTICAL ANALYSIS. *Educ:* Niagara Univ, BS, 71; State Univ NY Buffalo, MA, 73, PhD(statist sci), 75. *Prof Exp:* res asst prof statist sci & tech specialist surg, statist lab, Res Found State NY, State Univ NY Buffalo, 75-78; from asst prof to assoc prof, 78-88, PROF STATIST, TEX A&M UNIV, 88-, DEPT HEAD, 90- *Mem:* Am Statist Asn; Inst Math Statist. *Res:* Statistical computing with particular emphasis on the analysis of multiple time series having rational spectra. *Mailing Add:* Dept Statist Texas A&M Univ College Station TX 77843

NEWTON, JACK W, PHOTOSYNTHESIS, NITROGEN FIXATION. *Educ:* Univ Wis-Madison, PhD(biochem), 54. *Prof Exp:* res biochemist, USDA, 60-89; RETIRED. *Mailing Add:* 1032 Birch Creek Dr Wilmington NC 28403

NEWTON, JOHN CHESTER, analytical chemistry, for more information see previous edition

NEWTON, JOHN MARSHALL, b Popejoy, Iowa, May 20, 13; m 41; c 3. CARBOHYDRATE CHEMISTRY. *Educ:* Iowa State Col, BS, 36, PhD(chem), 41. *Prof Exp:* Res chemist, Clinton Indust, 41-42, from asst supvr to res supvr, 42-49; dir tech sales serv, Clinton Corn Processing Co, 49-63, tech asst to vpres sales, 63-78; RETIRED. *Mem:* AAAS; Am Asn Cereal Chem; Am Chem Soc; Tech Asn Pulp & Paper Indust; Am Asn Textile Chem & Colorists. *Res:* Carbohydrate and enzyme chemistry; concentration, characterization and properties of soybean amylase; chemistry of starch, lactic acid, oil and plant proteins; fermentation of lactic acid. *Mailing Add:* 1425 Seventh St NW Clinton IA 52732

NEWTON, JOHN S, b Oneonta, NY, Sept 21, 08; m 30, 46; c 6. ELECTRICAL & MECHANICAL ENGINEERING. *Educ:* Ore State Col, BS, 30. *Prof Exp:* Design engr, East Pittsburgh Works, Westinghouse Elec Corp, 30-39, mgr appl eng, Steam Div, 39-44, asst mgr eng, 44-48; vpres eng, Baldwin-Lima-Hamilton Corp, 48-51, locomotive div, 51-54, testing mach div, 54-55; vpres eng, Goodman Mfg Co, 55-62, exec vpres, 62-63, pres, 63-65, vpres & gen mgr, Goodman Div, Westinghouse Air Brake Co, 65-66; pres & chief exec off, Poor & Co, Ill, 66-69; pres, chief exec off & dir, Portec, Inc, 69-74; PRES, NEWTON ENG, 76- *Mem:* Am Soc Mech Engrs. *Res:* Removal of hard cemented materials from the earth, especially mining of ores and tunneling; sources of energy for the future-photovoltaic, wind; alternate sources of energy for transportation vehicles and their design; design, prototype manufacturing and testing of 4-wheel drive hybrid passenger automobiles; emphasis on diesel power-electric powered passenger cars with fuel consumption in the 75 to 100 mpg range. *Mailing Add:* Newton Eng Co 22 W 450 Ahlstrand Dr Glen Ellyn IL 60137

NEWTON, JOSEPH EMORY O'NEAL, b Orlando, Fla, Apr 5, 27; m 58; c 2. PSYCHIATRY, PHYSIOLOGICAL PSYCHOLOGY. *Educ:* Emory Univ, BS, 52, MD, 55. *Prof Exp:* Fel psychiat, Pavlovian Lab, Phipps Psychiat Clin, Sch Med, Johns Hopkins Univ, 56-57, USPHS fel, 58-61; instr psychiat, Pavlovian Lab, Phipps Psychiat Clin, Sch Med, Johns Hopkins Univ, 62-66, asst prof, 66-68; res physiologist, 68-74, PHYSIOLOGIST, NEUROPSYCHIAT RES LAB, VET ADMIN HOSP, NORTH LITTLE ROCK, ARK, 74- *Concurrent Pos:* Investr psychophysiol res lab, Vet Admin Hosp, Perry Point, Md, 58-61. *Mem:* AAAS; Am Physiol Soc; Soc Psychophysiol Res; Pavlovian Soc NAm; Sigma Xi. *Res:* Conditional reflex studies in dogs and opossums; cardiovascular conditioning; effects of cerebral cortical ablations on acquired emotional reactions. *Mailing Add:* Neuropsychiat Res Lab Vet Admin Hosp North Little Rock AR 72114

NEWTON, MARSHALL DICKINSON, b Boston, Mass, July 15, 40; m 63; c 2. THEORETICAL CHEMISTRY. *Educ:* Dartmouth Col, BA, 61, MA, 63; Harvard Univ, PhD(chem), 66. *Prof Exp:* NSF fel chem, Oxford Univ, 66-67; NIH fel, Carnegie-Mellon Univ, 67-68, res assoc, 68-69; assoc chemist, 69-73, chemist, 73-81, SR CHEMIST, BROOKHAVEN NAT LAB, 81- *Concurrent Pos:* Carnegie-Mellon Univ & Mellon Inst res fel, 68; adj prof, Chem Dept, Stonybrook Univ, 86- *Mem:* Am Chem Soc. *Res:* Calculation of molecular potential energy surfaces; analysis of molecular bonding in terms of electronic structure; theory of solvation phenomena and charge transfer kinetics. *Mailing Add:* Chem Dept Brookhaven Nat Lab Upton NY 11973

NEWTON, MELVIN GARY, b Millen, Ga, Feb 18, 39. ORGANIC CHEMISTRY. *Educ:* Ga Inst Technol, BS, 61, PhD(chem), 66. *Prof Exp:* Asst prof, 67-77, ASSOC PROF CHEM, UNIV GA, 77- *Concurrent Pos:* Fel, Univ Ill, 65-67. *Mem:* Am Chem Soc. *Res:* Structural chemistry, single crystal x-ray diffraction; structure and mechanism in organophosphorus chemistry; organometallic structural chemistry. *Mailing Add:* 196 Hidden Hills Lane Athens GA 30605-4203

NEWTON, MICHAEL, b Hartford, Conn, Oct 24, 32; m 54; c 3. FOREST ECOLOGY, WEED SCIENCE. *Educ:* Univ Vt, BS, 54; Ore State Univ, BS, 59, MS, 60, PhD(bot), 64. *Prof Exp:* Res asst forest herbicides, 59-60, instr forest mgt, 60-64, asst prof forest sci, 64-68, assoc prof forest mgt, 68-75, PROF FOREST ECOL, ORE STATE UNIV, 75- *Concurrent Pos:* Consult to numerous corps, 62-; fel, Univ Tenn, 69-70; consult comt effects of herbicides in Vietnam, Nat Acad Sci, 72-73, mem study of pest problems, 73-74. *Mem:* AAAS; Soc Am Foresters; Ecol Soc Am; Weed Sci Soc Am. *Res:* Quantitative forest ecology; usage of herbicides to manipulate components of forest ecosystems; development of theory and practice in forest manipulation. *Mailing Add:* Dept Forest Sci Ore State Univ 154 Peavy Hall Corvallis OR 97331

NEWTON, PAUL EDWARD, b Terre Haute, Ind, Oct 19, 45; m 70; c 3. INHALATION TOXICOLOGY. *Educ:* Rose Polytech Inst, BS, 67, MS, 69; Med Col Wis, PhD(physiol), 79; Am Bd Toxicol, dipl, 84. *Prof Exp:* Sr biomed engr, dept environ med, Med Col Wis, 70-79; head, respiratory toxicol dept, Dept Community & Environ Med, Univ Calif, Irvine, 79-85; dir inhalation toxicol, Am Biogenics Corp, 85-86; DIR INHALATION TOXICOL, BIO/DYNAMICS, 86- *Mem:* Soc Toxicol; Am Physiol Soc; Am Thoracic Soc; Am Col Toxicol. *Res:* Acute, subchronic and chronic inhalation testing. *Mailing Add:* Bio Dynamics Inc Box 2360 East Millstone NJ 08875

NEWTON, R(OBERT) E(UGENE), mechanical engineering, for more information see previous edition

NEWTON, RICHARD WAYNE, b Baytown, Tex, Aug 26, 48; m 68; c 2. ELECTRICAL ENGINEERING. *Educ:* Tex A&M Univ, BS, 70, MS, 71, PhD(elec eng), 77. *Prof Exp:* Engr & scientist, Lockheed Electronics Co, Inc, 71-73; prog mgr, Remote Sensing Ctr, 73-77; asst prof, 77-81, assoc dir, Remote Sensing Ctr, 77-80, ASSOC PROF ELEC ENG, TEX A&M UNIV, 81-, DIR, REMOTE SENSING CTR, TEX ENG EXP STA, 80- *Concurrent Pos:* Prin, Innovative Develop Eng Assocs, 80-; pres, Aerial Surveys, Inc, 77-; consult, Zesco, Inc, 80-; Univ Space Res Asn representative, Landsat D Tech Users Comt, NASA, 81-; mem, various workshops, comts & working groups, NASA, 74-; mem, Automated In-Situ Water Quality Workshop, Enivron Protection Agency, 77. *Mem:* Am Geophys Union; Inst Elec & Electronics Engrs; Sigma Xi. *Res:* Techniques of extracting agricultural, hydrologic and oceanographic information from remote sensing measurements ranging from microwave to visible wavelengths; automated sensor systems for remote sensing applications. *Mailing Add:* 6107 Lansford Lane Colleyville TX 76034-5233

NEWTON, ROBERT ANDREW, b Oakville, Wash, Sept 23, 22; m 46; c 4. PHYSICAL ORGANIC CHEMISTRY. *Educ:* Univ Wash, BS, 47, PhD(chem), 53. *Prof Exp:* Mem staff res, E I du Pont de Nemours & Co, 53-56; chemist, Dow Chem Co, 56-57, from res chemist to sr res chemist, 57-68, res specialist, 68-74, sr res specialist, 74-80, assoc scientist, 80-85; RETIRED. *Mem:* Am Chem Soc; Sigma Xi; AAAS. *Res:* Aklylene oxide chemistry and polymerization products; chromatography; chemical kinetics; isolation and identification of trace components; process studies, especially related to productivity and quality enhancement. *Mailing Add:* 53 Pin Oak Ct Lake Jackson TX 77566

NEWTON, ROBERT CHAFFER, b Bellingham, Wash, June 11, 33; m 67; c 2. PETROLOGY. *Educ:* Univ Calif, Los Angeles, AB, 56, MA, 58, PhD(geol), 63. *Prof Exp:* From asst prof to assoc prof, 63-71, PROF GEOL, DEPT GEOPHYS SCI, UNIV CHICAGO, 71- *Concurrent Pos:* Ed, J Geol, 84-; counr, Am Mineral Soc, 85-87. *Honors & Awards:* Volcanology, Geochem & Petrol Award, Am Geophys Union, 85. *Mem:* Am Geophys Union; Am Mineral Soc; Geol Soc Am; Geol Soc India; Mineral Soc Can. *Res:* Experimental investigation of the high-temperature, high-pressure stabilities of minerals. *Mailing Add:* 5406 S University Chicago IL 60615

NEWTON, ROBERT MORGAN, b Salem, Mass, Jan 3, 48; m 69; c 1. GLACIAL GEOLOGY. *Educ:* Univ NH, BA, 70; State Univ NY Binghamton, MA, 72; Univ Mass, PhD(geol), 78. *Prof Exp:* Vis prof geol, Brock Univ, 76-78; ASST PROF GEOL, SMITH COL, 78- *Mem:* Geol Soc Am; Int Glaciological Soc; Clay Minerals Soc; Sigma Xi. *Res:* Geomorphology; groundwater geology; low temperature geochemistry and clay petrology. *Mailing Add:* Dept Geol Smith Col Northampton MA 01060

NEWTON, ROGER GERHARD, b Germany, Nov 30, 24; nat US; m 53; c 3. THEORETICAL PHYSICS. *Educ:* Harvard Univ, AB, 49, AM, 50, PhD(physics), 53. *Prof Exp:* Mem, Inst Advan Study, 53-55, from asst prof to prof physics, 55-78, chmn dept, 73-80, DISTINGUISHED PROF PHYSICS, IND UNIV, BLOOMINGTON, 78- *Concurrent Pos:* Jewett fel, 53-55; NSF sr fel, Univ Rome, 62-63, Univ Montpellier, 71-72 & Inst Advan Study, Princeton Univ, 79; dir, Inst Advan Study, Ind Univ, 82-86; coun deleg, AAAS, 86-89. *Mem:* Fel AAAS; fel Am Phys Soc; Sigma Xi; Fed Am Scientists; NY Acad Sci. *Res:* Field and scattering theories; nuclear and high energy physics; elementary particles; quantum mechanics. *Mailing Add:* Dept Physics Ind Univ Bloomington IN 47405

NEWTON, SHEILA A, b Baton Rouge, La, Nov 3, 54. BIOCHEMISTRY, CELL BIOLOGY. *Educ:* Univ Mass, PhD(biochem), 83. *Prof Exp:* Res assoc, Cancer Ctr, 82-85, ASST PROF ONCOL, HOWARD UNIV, 86-, STAFF INVESTR, CANCER CTR, 85- *Mailing Add:* 9207 Bulls Run Pkwy Bethesda MD 20817

NEWTON, STEPHEN BRUINGTON, b Freeport, Ill, Dec 24, 34; m 58; c 3. TECHNICAL WRITING, TECHNICAL EDITING. *Educ:* Univ Mo, BA, 56; Univ Ill, MS, 61, PhD(bact), 65. *Prof Exp:* Prod develop scientist, Pillsbury Co, 64-71; sr group leader, Quaker Oats Co, 71-76; assoc ed, Food Prod Develop, 76-79; sr food technologist, Sara Lee, 79-81; TECH WRITER, 81- *Mem:* Inst Food Technologists. *Res:* Production of polysaccharides and extracellular enzymes by microbes; food product development. *Mailing Add:* 862 Coventry Lane Crystal Lake IL 60014

NEWTON, THOMAS ALLEN, b Buffalo, NY, May 30, 43. ORGANIC CHEMISTRY. *Educ:* Hobart Col, BS, 65; Bucknell Univ, MS, 68; Univ Del, PhD(org chem), 73. *Prof Exp:* Res assoc, Rensselaer Polytech Inst, 73-76, asst prof, Williams Col, 76-78; ASST PROF COLBY COL, 78- *Mem:* Am Chem Soc. *Res:* The synthesis of nucleoside analogues as potential antitumor/antiviral agents; the photochemistry of enaminonitriles; orbital symmetry controlled reactions; sigmatropic rearrangements. *Mailing Add:* Dept 96 Univ S Maine Falmouth St Portland ME 04103

NEWTON, THOMAS HANS, b Berlin, Ger, May 9, 25; US citizen; c 2. RADIOLOGY. *Educ:* Univ Calif, Berkeley, BA, 49; Univ Calif, San Francisco, MD, 52. *Prof Exp:* From asst prof to assoc prof, 59-68, PROF RADIOL, MED CTR, UNIV CALIF, SAN FRANCISCO, 68-, CHIEF SECT NEURORADIOL, 77- *Concurrent Pos:* Consult, Ft Miley Vet Admin Hosp & Letterman Gen Hosp, San Francisco, Martinez Vet Admin Hosp & Oaknoll Naval Hosp, Oakland. *Mem:* Asn Univ Radiol; Am Soc Neuroradiol; Neurosurg Soc Am. *Res:* Neuroradiology. *Mailing Add:* Dept Radiol & Neurosurg Univ Calif Med Ctr 513 Parnassus Ave San Francisco CA 94143

NEWTON, THOMAS WILLIAM, b Berkeley, Calif, June 26, 23; m 48; c 3. INORGANIC CHEMISTRY, PHYSICAL CHEMISTRY. *Educ:* Univ Calif, Berkeley, BS, 43, PhD(chem), 49. *Prof Exp:* Chemist, Manhattan Proj, Univ Calif & Tenn Eastman, 44-46; chemist staff mem, Los Alamos Sci Lab, 49-85; RETIRED. *Concurrent Pos:* Vis prof, State Univ NY Stonybrook, 67. *Mem:* AAAS; Am Chem Soc. *Res:* Actinide chemistry; equilibrium and kinetics of reactions in aqueous solutions. *Mailing Add:* 4589 Trinity Dr Los Alamos NM 87544

NEWTON, TYRE ALEXANDER, b Morris, Okla, Dec 28, 21; m 46; c 2. MATHEMATICAL ANALYSIS. *Educ:* Colo State Univ, BS, 49; Univ Ga, MA, 51, PhD(math), 52. *Prof Exp:* Instr math, Univ Nebr, 52-55; from asst prof to assoc prof, Colo State Univ, 55-58; from asst prof to prof, 58-86, EMER PROF MATH, WASH STATE UNN, 86- *Mem:* Am Math Soc; Soc Indust & Appl Math; Math Asn Am; Am Soc Eng Educ; Sigma Xi. *Res:* Infinite series; finite differences; functional analysis; qualitative theory of ordinary differential equations; using the analog computer to illustrate mathematical concepts. *Mailing Add:* Dept Math Wash State Univ Pullman WA 99163

NEWTON, VICTOR JOSEPH, b Boston, Mass, Apr 9, 37. THEORETICAL NUCLEAR PHYSICS. *Educ:* Spring Hill Col, BS, 61, MA, 62; Mass Inst Technol, PhD(physics), 66. *Prof Exp:* Lectr physics, Loyola Col Md, 68-69; asst prof, 69-73, chmn dept, 76-85, ASSOC PROF PHYSICS, FAIRFIELD UNIV, 73- *Mem:* AAAS; Am Phys Soc; Am Asn Physics Teachers; Fed Am Scientists; Sigma Xi. *Res:* Many channel scattering theory; three and four nucleon systems as applied to nuclear systems, including weak interactions. *Mailing Add:* 4042 Congress St Fairfield CT 06430

NEWTON, WILLIAM ALLEN, JR, b Traverse City, Mich, May 19, 23; m 45; c 4. PEDIATRICS, PATHOLOGY. *Educ:* Alma Col, Mich, BSc, 43; Univ Mich, MD, 46. *Prof Exp:* Intern, Wayne County Gen Hosp, Eloise, Mich, 47; fel pediat path, Children's Hosp Mich, 48, fel pediat hemat, 49-50; resident pediat, Children's Hosp Philadelphia, 50; from instr to asst prof path, 52-59, assoc prof path & pediat, 59-66, PROF PATH & PEDIAT, OHIO STATE UNIV, 66-; DIR LABS, CHILDREN'S HOSP, 52- *Concurrent Pos:* Sci res grant. *Mem:* Soc Pediat Res; Am Asn Cancer Res; Soc Pedag Path; Am Soc Clin Oncol; Int Soc Pedag Oncol; Col Am Pathologists; Sigma Xi. *Res:* Cancer chemotherapy in children, particularly leukemia and brain tumors; red cell enzyme deficiency of glucose 6-phosphate dehydrogenase; causes and types and drug treatment of childhood cancer; chronic hemolytic anemia in man. *Mailing Add:* Pediat 700 Children's Dr The Ohio State Univ Col of Med Columbus OH 43210

NEWTON, WILLIAM EDWARD, b London, Eng, Nov 10, 38; m 77; c 4. NITROGEN FIXATION, BIOINORGANIC. *Educ:* Nottingham Univ, Eng, BSc, 61; Royal Inst Chem, grad, 65; Univ London, PhD(chem), 68. *Prof Exp:* Anal chemist, Rayner & Co, Ltd, London, 62-66; teaching asst chem, Northern Polytechnic, London, 66-68; res fel chem, Harvard Univ, 68-69; staff scientist, Charles F Kettering Res Lab, 69-71, sect head, 71-73, investr, 73-77, res mgr & sr investr, Nitrogen Fixation Mission, 77-84; RES LEADER, PLANT DEVELOP-PRODUCTIVITY, USDA-AGR RES SERV, WESTERN REGIONAL RES CTR, ALBANY, CALIF, 84- *Concurrent Pos:* Adj prof agron & range sci, Univ Calif, Davis. *Mem:* Royal Soc Chem; Am Chem Soc; Am Soc Biochem & Molecular Biol; AAAS. *Res:* Biosynthesis and structure of nitrogenase iron-molybdenum cofactor; mechanism of biological nitrogen fixation; bioinorganic chemistry to probe role of molybdenum in enzymes; heavy metal sequestration. *Mailing Add:* USDA-ARS Western Regional Res Ctr 800 Buchanan St Albany CA 94710

NEY, EDWARD PURDY, b Minneapolis, Minn, Oct 28, 20; m 42; c 4. PHYSICS, ASTROPHYSICS. *Educ:* Univ Minn, BS, 42; Univ Va, PhD(physics), 46. *Prof Exp:* Asst physics, Univ Minn, 40-42; res assoc, Univ Va, 43-46, from asst prof to assoc prof, 46-47; from asst prof to prof physics, 47-74, chmn, astron dept, 74-78, REGENT'S PROF PHYSICS & ASTRON, UNIV MINN, MINNEAPOLIS, 74- *Concurrent Pos:* Consult, Naval Res Lab, DC, 43-44 & Gen Dynamics & Convair. *Honors & Awards:* Except Achievement Medal, NASA. *Mem:* Nat Acad Sci; AAAS; fel Am Phys Soc; Am Astron Soc; Am Geophys Union; Am Acad Arts & Sci. *Res:* Mass spectroscopy; cosmic rays; atmospheric physics; astrophysics; infrared astronomy. *Mailing Add:* Dept of Astron Univ of Minn Minneapolis MN 55455

NEY, PETER E, b Brno, Czech, July 6, 30; US citizen; m 55; c 2. MATHEMATICS. *Prof Exp:* Instr math, Cornell Univ, 58-60, asst prof indust eng, 60-63; vis asst prof statist, Stanford Univ, 63-64; assoc prof indust eng, Cornell Univ, 64-65; assoc prof math, 65-69, chmn dept, 74-77, PROF MATH, UNIV WIS-MADISON, 69- *Concurrent Pos:* Grants, Off Naval Res & NSF, 58-65 & 70-; prin investr, NIH grant, 65-70; Guggenheim fel, 71-72; vis prof, Israel Inst Technol & Weizmann Inst Sci, 71-72 & Univ Helsinki, 84; Fulbright sr res scholar, 84, ed, Annels Probability, 88- *Mem:* Am Math Soc; fel Inst Math Statist. *Res:* Probability; stochastic processes; branching processes; Markov chains. *Mailing Add:* Dept of Math Univ of Wis Madison WI 53706

NEY, RONALD E, JR, ENVIRONMENTAL CHEMISTRY. *Educ:* James Madison Univ, BS, 59; Pac Western Univ, PhD, 89. *Prof Exp:* Supvr chemist & sect chief, Hazard Eval Div, Environ Fate Br, Environ Protection Agency, 78-80, chemist & dir staff, Hazardous Eval Div, Off Pesticides & Toxic Substances, 80-81, environ scientist, solid waste & emergency waste, Land Disposal Br, 81-86; RETIRED. *Concurrent Pos:* Consult, 86- *Mem:* Nat Registry Environ Professionals. *Mailing Add:* 3818 Charles Stewart Dr Fairfax VA 22033-2418

NEY, WILBERT ROGER, b Rockford, Ill, Nov 28, 29; m 58; c 2. PHYSICS. *Educ:* Yale Univ, BS, 57; George Washington Univ, JD, 64. *Prof Exp:* Physicist, Nat Bur Stand, 58-60, sci asst, 60-64; EXEC DIR, NAT COUN RADIATION PROTECTION & MEASUREMENTS, 64- *Concurrent Pos:* Secy, Nat Comt Radiation Protection & Measurements, 61-64; tech secy, Int Comn Radiation Units & Measurements, 61-79, exec secy, 79- *Mem:* Radiol Soc NAm; Health Physics Soc; Radiation Res Soc. *Res:* Law; radiation, protection, quantities, units and effects. *Mailing Add:* Nat Coun Rad Prot & Meas 7910 Woodmont Ave Bethesda MD 20014

NEYNABER, ROY H(AROLD), b Highland Park, Mich, July 4, 26; m 51; c 4. ATOMIC PHYSICS. *Educ:* Univ Wis, BS, 49, MS, 51, PhD(physics), 55. *Prof Exp:* Asst physics, Univ Wis, 51-55; sr staff scientist, Gen Dynamics/Convair, 55-69; mgr atomic physics br, Gulf Energy & Environ Systs Co, 69-73; mgr, Atomic Physics Dept, IRT Corp, 73-77, sci adv, 78-80; STAFF SCIENTIST, LA JOLLA INST, 80- *Concurrent Pos:* Physicist, Los Alamos Sci Lab, 51, Liberty Powder Co, 53; adj prof physics, Univ Calif, San Diego, 80- *Mem:* Sigma Xi; fel Am Phys Soc. *Res:* Gaseous electronics; atomic beams; atomic scattering experiments; particle-surface interactions; small angle scattering of x-rays; lattice imperfections in solids. *Mailing Add:* 4471 Braeburn Rd San Diego CA 92116

NEZRICK, FRANK ALBERT, b Mansfield, Ohio, Apr 1, 37; m 60; c 3. ELEMENTARY PARTICLE PHYSICS. *Educ:* Case Inst, BS, 59, MS, 62, PhD(physics), 65. *Prof Exp:* Vis scientist grant elem particle physics, Europ Orgn Nuclear Res, 65-68; PHYSICIST, NAT ACCELERATOR LAB, 68- *Concurrent Pos:* Mem, high energy physics staff, US Dept Energy, 82-83, high energy physics adv panel, 85-86. *Mem:* Am Phys Soc. *Res:* Searches for source of dark matter; precision optical polarization experiments; magnetic monopole search; development of particle focussing systems; weak interaction physics; astrophysics experiments. *Mailing Add:* MS-306 Fermilab PO Box 506 Batavia IL 60510

NG, AH-KAU, PATHOLOGY. *Educ:* Temple Univ, PhD(immunol & microbiol), 76. *Prof Exp:* assoc prof path, Columbia Univ, 84-88. *Res:* Tumor immunology; transportation immunology. *Mailing Add:* 72 Bowdon St Portland ME 04102

NG, BARTHOLOMEW SUNG-HONG, b Canton, China, Sept 10, 46; m 73. APPLIED MATHEMATICS, FLUID DYNAMICS. *Educ:* St Josephs Col, Ind, BS, 68; Univ Chicago, MS, 70, PhD(appl math), 73. *Prof Exp:* Syst engr, Int Bus Mach Corp, 68; res asst math, Univ Chicago, 68-69, teaching asst & lectr, 69-73; fel, Univ Toronto, 73-75; res assoc, Indianapolis Ctr Advan Res, 75-76; from asst prof to assoc prof, 75-83, PROF MATH, IND UNIV-PURDUE UNIV, INDIANAPOLIS, 83- *Concurrent Pos:* Vis asst prof math, Rensselaer Polytech Inst, 79 & vis assoc prof, 80; assoc prof math, Old Dominion Univ, 79-80. *Mem:* Am Math Soc; Soc Indust & Appl Math; Sigma Xi. *Res:* Hydrodynamic stability; asymptotic and numerical techniques in applied mathematics. *Mailing Add:* Dept Math Sci Ind Univ-Purdue Univ PO Box 647 Indianapolis IN 46202

NG, EDWARD WAI-KWOK, US citizen; m; c 3. APPLIED MATHEMATICS, COMPUTER SCIENCE APPLICATIONS. *Educ:* Univ Minn, BA, 62; Columbia Univ, MA, 65, PhD, 67. *Prof Exp:* Sr engr, 67-68, sr scientist, 68-70, TECH MGR SPACE SCI DATA SYSTS, JET PROPULSION LAB, CALIF INST TECHNOL, 70- *Concurrent Pos:* Adj assoc prof, Univ Southern Calif, 74-77. *Mem:* Asn Comput Mach; Soc Indust & Appl Math; fel AAAS. *Res:* Mathematical software; mathematics of computation; numerical and symbolic computation; computer science applications; space science data systems. *Mailing Add:* Jet Propulsion Lab Calif Inst Technol Pasadena CA 91109

NG, KAM WING, b Hong Kong, Oct 26, 51; US citizen; m 74; c 2. UNDERWATER ACOUSTICS. *Educ:* Cooper Union, BE, 73; Rensselaer Polytech Inst, MS, 75; Univ RI, PhD, 88. *Prof Exp:* Acoust engr, Pratt & Whitney, United Technol, 73-78; proj engr noise control, ITT Grinnel Corp, 78-81; MECH ENGR, NAVAL UNDERWATER SYSTS CTR, US NAVY, 82- *Concurrent Pos:* Tech paper reviewer, Am Soc Mech Engrs, 81-; consult acoust, Navy Technol Transfer Prog, 83-, adj fac, 89- *Mem:* Am Soc Mech Engrs; Acoust Soc Am; Inst Noise Control Eng; Instrument Soc Am. *Res:* Acoustics and noise control engineering; jet noise, valve noise, pump noise and flow induced noise and vibration; quiet valve; active noise control; transient signal analysis; launcher noise. *Mailing Add:* Ten Candleberry Rd Barrington RI 02806

NG, LAWRENCE CHEN-YIM, b Canton, China, Dec 21, 46; US citizen; m 74; c 3. SIGNAL PROCESSING, TARGET TRACKING. *Educ:* Mass Inst Technol, BS, 70, MS, 73; Univ Conn, PhD(elec eng), 83. *Prof Exp:* Sr engr control syst, Elec Boat Div, Gen Dynamics, Groton, Conn, 74-78; proj leader signal processing, Naval Underwater Syst Ctrs, New London, Conn, 78-85; GROUP LEADER SIGNAL & IMAGE PROCESSING, ENG RES DIV, LAWRENCE LIVERMORE NAT LAB, CALIF, 85- *Concurrent Pos:* Adj fac, Northwestern Polytechnic Univ, Fremont, Calif, 86-89. *Mem:* Inst Elec & Electronic Engrs; Am Inst Aeronaut & Astronaut; Sigma Xi. *Res:* Space based defense systems; vehicle guidance and control; target tracking; signal and image processing. *Mailing Add:* Lawrence Livermore Nat Lab L-156 PO Box 808 Livermore CA 94550

NG, SIMON S F, Can citizen; m; c 2. CIVIL ENGINEERING. *Educ:* Univ BC, BS, 62; Univ Windsor, MS, 64, PhD(civil eng), 67. *Prof Exp:* Engr, Hydroelec Design Div, Int Power & Eng Consult, 62-63; scientist, IBM Corp, 66-67; from asst prof to assoc prof, 67-75, chmn dept, 79-82, PROF CIVIL ENG, UNIV OTTAWA, 75- *Mem:* Am Acad Mech; Can Soc Civil Eng. *Res:* Static and dynamic behavior of plates and shells; experimental stress analysis; sandwich construction; material behavior at abnormal temperature; vibration of bridge structures. *Mailing Add:* Dept Civil Eng Univ Ottawa Ottawa ON K1N 6N5 Can

NG, TAI-KAI, b Hong Kong, Apr 29, 59. CONDENSED MATTER PHYSICS, MATHEMATICAL PHYSICS. *Educ:* Univ Hong Kong, BSc, 81; Northwestern Univ, PhD(physics), 87. *Prof Exp:* Postdoctoral assoc physics, Mass Inst Technol, 87-89; POSTDOCTORAL ASSOC CONDENSED MATTER PHYSICS, AT&T BELL LABS, 89- *Mem:* Am Phys Soc. *Res:* Theoretical studies high-Tc superconductors; low-dimensional quantum spin systems; strongly correlated metals; mesoscopic systems, quantum transport; disordered electronic systems; density-functional theory. *Mailing Add:* AT&T Bell Labs Rm 1D-361 600 Mountain Ave Murray Hill NJ 07974-2070

NG, TIMOTHY J, b Oakland, Calif, Apr 15, 50; m 76. PLANT BREEDING, HORTICULTURE. *Educ:* Univ Calif, Berkeley, BS, 69; Purdue Univ, MSc, 72, PhD(plant breeding), 76. *Prof Exp:* Lab technician chem, Eng Sci Inc, 69-70; res asst hort, Purdue Univ, 70-76; from asst prof to assoc prof, 77-90, PROF HORT, UNIV MD, COLLEGE PARK, 90- *Concurrent Pos:* Instr, Ore State Univ, 74-75. *Honors & Awards:* Meadows Award, Am Soc Hort Sci, 78. *Mem:* AAAS; Am Soc Hort Sci; Sigma Xi. *Res:* Vegetable breeding and improvement; breeding for disease resistance, improved horticultural characteristics and improved nutritional quality; investigations in postharvest physiology, ripening and storage life. *Mailing Add:* Hort Dept Univ Maryland College Park MD 20742-5611

NG, WING-FAI, b Hong Kong, Sept 17, 56; Brit citizen; m 84; c 2. PROPULSION, TURBOMACHINERY. *Educ:* Northeastern Univ, Boston, BSc, 79; Mass Inst Technol, MS, 80, PhD, 84. *Prof Exp:* Co-op eng, Gen Elec Co, 76-78; res asst, Mass Inst Technol, 79-83; PROF MECH ENG, VA POLYTECH INST & STATE UNIV, 84- *Concurrent Pos:* Res engr, Langley Res Ctr, NASA, 84, Lewis Res Ctr, 85; prin investr, Langley Res Ctr, 84- & Air Force Off Sci Res, 85- *Mem:* Am Inst Aeronaut & Astronaut; Am Soc Eng Educ; Am Soc Mech Engrs; Sigma Xi. *Res:* Internal flows; propulsion; fluid dynamics; thermodynamics of turbomachinery; science education. *Mailing Add:* Dept Mech Eng Va Polytech Inst & State Univ Blacksburg VA 24061-0238

NGAI, KIA LING, b Canton, China, May 20, 40; US citizen; m 67; c 3. CONDENSED MATTER PHYSICS, POLYMER PHYSICS. *Educ:* Univ Hong Kong, BSc, 62; Univ Southern Calif, MS, 64; Univ Chicago, PhD(physics), 69. *Prof Exp:* Staff mem, Lincoln Lab, Mass Inst Technol, 69-71; SUPVRY RES PHYSICIST, NAVAL RES LAB, 71- *Mem:* Am Phys Soc; Am Chem Soc. *Res:* Relaxations in complex correlated systems. *Mailing Add:* Code 6807 Naval Res Lab Washington DC 20375-5000

NGAI, SHIH HSUN, b China, Sept 15, 20; nat US; m 48; c 3. ANESTHESIOLOGY. *Educ:* Nat Cent Univ, China, MD, 44. *Prof Exp:* From instr to assoc prof, 49-65, PROF ANESTHESIOL, COL PHYSICIANS & SURGEONS, COLUMBIA UNIV, 65-, PROF PHARMACOL, 74- *Concurrent Pos:* Asst anesthesiologist, Presby Hosp, New York, 49-54, from asst attend anesthesiologist to attend anesthesiologist, 57-87. *Mem:* Am Soc Anesthesiol; Am Physiol Soc; Am Soc Pharmacol & Exp Therapeut; Asn Univ Anesthetists. *Res:* Neural control of respiration and circulation; pharmacology of anesthetics and agents affecting respiration and circulation. *Mailing Add:* 622 W 168th St New York NY 10032

NGO, THAT TJIEN, b Indonesia, Mar 10, 44; m 77. IMMUNOCHEMISTRY. *Educ:* Univ Sask, BSc, 70, PhD(biochem), 74. *Prof Exp:* Sr investr, Clin Res, Inst Montreal, 76-78; sr res scientist, Miles Lab, 78-79; biochemist, Univ Calif, Irvine, 79-81; dir, Immutron, Inc, 81-86; pres, Ceo, Bioprobe Int, Inc, 83-86. *Mem:* Am Chem Soc. *Res:* Enzymology; analytical chemistry; clinical chemistry; biotechnology. *Mailing Add:* 14272 Franklin Ave Suite 106 Tustin CA 92680

NGUYEN, CAROLINE PHUONGDUNG, b Saigon, Vietnam; US citizen. ENERGY STORAGE SYSTEMS, THIN FILM GAS SENSORS. *Educ:* Univ Calif, Berkeley, BS, 85; Univ Wash, Seattle, MS, 87. *Prof Exp:* Asst engr water hardening, Water Technol Ctr, Richmon, Calif, 85; teaching asst chem kinetics, Univ Wash, Seattle, 86, res asst gas sensors, 85-87, res asst mass sensors, 87; equip engr NiH2 battery for Hubble Space Telescope, 88-89, RESPONSIBLE EQUIPMENT ENGR RES/SPECIALIST ENGR, SPACECRAFT ENERGY STORAGE SYSTS, LOCKHEED MISSILES & SPACE CO, SSD, 88- *Mem:* Soc Women Engrs. *Res:* Investigating and evaluating different energy storage systems for near and far terms space applications; sodium sulfur battery technology; developing concepts to increase the electrical power subsystem efficiency while reducing the system cost. *Mailing Add:* 4540 Romano Dr Stockton CA 95207

NGUYEN, CHARLES CUONG, b Danang, Vietnam, Jan 1, 56; US citizen; m 89. ROBOTICS & CONTROL OF MANIPULATORS, ROBOT VISION NEURAL NETWORKS. *Educ:* Konstanz Univ, WGer, Dipl Ing, 79; George Washington Univ, MS, 80, DSc, 82. *Prof Exp:* Engr, Siemens Corp, Ger, 78-79; asst prof, 82-87, ASSOC PROF ELEC ENG, CATHOLIC UNIV AM, 87- *Concurrent Pos:* Prin investr, NASA grants, 86-; dir, Ctr A I & Robotics, Cath Univ, 87-89 & Robotics & Control Lab, 89-; consult, NASA, Questech & Meridian Corp, 87-; guest ed, Computers & Elec Eng, Int J, 90-91. *Mem:* Sr mem Inst Elec & Electronic Engrs; sr mem Soc Mfg Engrs; Sigma Xi. *Res:* Control of robot manipulators; time-varying multivariable systems; robot vision; neural network controller; control of large space structures. *Mailing Add:* Dept Elec Eng Cath Univ Am Washington DC 20064

NGUYEN, DONG HUU, nuclear science & engineering, for more information see previous edition

NGUYEN, HENRY THIEN, b Vinh Long, Vietnam, Jan 7, 54; US citizen; m 77; c 2. MOLECULAR GENETICS. *Educ:* Pa State Univ, BS, 77, MS, 78; Univ Mo, Columbia, PhD(plant genetics), 82. *Prof Exp:* Asst prof plant genetics, Okla State Univ, 82-84; asst prof, 84-89, ASSOC PROF PLANT GENETICS, TEX TECH UNIV, 89- *Concurrent Pos:* NSF presidential young investr award, 86; dir, Inst Biotechnol, Tex Tech Univ, 89-; Crop Sci Soc Am young crop scientist award, 90. *Mem:* Am Soc Agron; Crop Sci Soc Am; Int Soc Plant Molecular Biol; Genetics Soc Am; Am Soc Plant Physiologists; AAAS. *Res:* Molecular genetics and regulation of gene expression in cereal crop plants grown under drought and heat stress conditions; molecular genetic strategies for crop improvement. *Mailing Add:* Dept Agron Mail Stop 2122 Tex Tech Univ Lubbock TX 79409

NGUYEN, HIEN VU, b Saigon, Vietnam, Dec 15, 43; m 72; c 3. CHEMICAL ENGINEERING, POLYMER CHEMISTRY. *Educ:* Mont State Univ, BS, 66; Univ Wis-Madison, MS, 68, PhD(chem eng), 70. *Prof Exp:* Fel chem eng, McGill Univ, Montreal, 71-72; res assoc polymer chem, Univ Montreal, 72-74; sr develop engr non woven, Res & Develop Div, Johnson & Johnson, Montreal, 74-76; res assoc specialty paper, Polyfibron Div, W R Grace & Co, Cambridge, Mass, 76-78; prin scientist, Chicopee Res, Div Johnson & Johnson, 78- *Mem:* Am Chem Soc; Am Inst Chem Eng; Sigma Xi. *Res:* Mathematical and computer simulation of fluid transport; computer science general; applied mathematics; hydrogels; non-woven and paper technology; radiation chemistry. *Mailing Add:* Chicopee Res 2351 US Rte 130 Dayton NJ 08810

NGUYEN, LUU THANH, b Saigon, SVietnam, Oct 11, 54; US citizen; m 88. ELECTRONIC PACKAGING, PACKAGING RELIABILITY. *Educ:* Univ Southern Calif, BS, 76, MS, 77; Mass Inst Technol, MS, 79, PhD(mech eng), 84. *Prof Exp:* Staff engr, Jet Propulsion Lab, Calif Inst Technol, 75-77; mem res staff, Int Bus Mach Corp T J Watson Res Ctr, 84-88; sr scientist, Philips Res & Develop Ctr, 88-91; PRIN ENGR, NAT SEMICONDUCTOR CORP, 91- *Concurrent Pos:* AAAS & Indust Res Inst fel, 84; mem, Modern Plastics Mgt Adv Panel, 86-87, Moldflow Steering Comt, 86 & bd dirs, Soc Plastics Engrs, 87-88; tech prog chmn, Soc Plastics Engrs, 87-88. *Mem:* Am Soc Mech Engrs; Int Soc Hybrid Microelectronics; Soc Plastics Engrs. *Res:* Packaging assembly reliability issues; chemorheology of reactive polymer systems; reactive flow simulation; statistical process control and taguchi-related design of experiments; electrostatic discharge protection; very large scale integrated circuit processing and reliabilty; author of one book and numerous publications. *Mailing Add:* Nat Semiconductor Corp PO Box 58090 MS 29-100 Santa Clara CA 95052

NGUYEN, THUAN VAN, b Namdinh, Vietnam, Oct 7, 29; m 57; c 3. ELECTRICAL ENGINEERING, COMPUTER SCIENCE. *Educ:* Chu Van An Col, Hanoi, Baccalaureat Complet, 51; Univ Hanoi, Lic en Droit, 52; Univ NMex, MA, 60, PhD(elec eng), 69; Stanford Univ, MS, 62. *Prof Exp:* Instr, US Army Lang Sch, 55-59; asst prof elec eng, Univ Alta, 62-65; assoc prof, 69-75, actg head dept, 71-73, PROF ELEC ENG, UNIV OF THE PAC, 75- *Concurrent Pos:* Lectr, Univ Alta Community Col, 63-65; off res grant, Univ of the Pac, 70-72; Nat Sci Found Regional Comput Network grant, Stanford Univ, 71-73; biomed, res, Pacific Med Ctr, 73-74. *Mem:* Am Soc Eng Educ; Inst Elec & Electronics Engrs; Sigma Xi; Soc Adv Normal Ethics. *Res:* Network theory; signal processing; communications systems; computer simulation and applications; ocean energy generation. *Mailing Add:* 1454 W Euclid Ave Stockton CA 95204-2903

NGUYEN-DINH, PHUC, b Hanoi, Vietnam, Apr 12, 49; US citizen. CHEMOTHERAPY & IMMUNOLOGY OF MALARIA. *Educ:* Lycee J J Rousseau, Saigon, Vietnam, BS, 66; Univ Geneva, Switz, MD, 73; Harvard Univ, Cambridge, Mass, MPH, 77. *Prof Exp:* Res assoc parasitol, Rockefeller Univ, 77-79; vis scientist, 79-83, med officer, 83-87, CHIEF MALARIA IMMUNOL ACTIV, CTR DIS CONTROL, ATLANTA, 87- *Mem:* AAAS; Am Soc Trop Med & Hyg; Royal Soc Trop Med & Hyg. *Res:* Various aspects of malaria such as drug resistance, immunology and approaches for control, conducted both in the lab and in field locations. *Mailing Add:* Malaria Br Div Parasitic Dis Ctr Dis Control Atlanta GA 30333

NGUYEN-HUU, CHI M, molecular embryology,mammalian genetics, for more information see previous edition

NI, CHEN-CHOU, b Feb 1, 27; US citizen; m 64. OPTICS, FLUIDS. *Educ:* Nat Taiwan Univ, BS, 55; Univ Minn, MS, 60; Ill Inst Technol, PhD(physics), 70. *Prof Exp:* Eng officer civil eng, Bur Pub Works, Chinese Navy, China, 55-57; jr lectr, Nat Taiwan Univ, 57-58; civil engr, Ill State Hwy, 60-66; RES PHYSICIST, US NAVAL RES LAB, 70- *Mem:* Am Phys Soc; Sigma Xi. *Res:* Structural dynamics; fluid dynamics; fiber optics; mechanics. *Mailing Add:* 1530 Pathfinder Lane McLean VA 22101

NI, I-HSUN, b Jiangsu, China, Mar 24, 46; Can & Taiwan citizen; m 76; c 1. FISHERIES BIOLOGY & MANAGEMENT, MARINE MAMMALS. *Educ:* Nat Taiwan Univ, BSc, 69, MSc, 72; Univ BC, PhD(fisheries mgt), 78. *Prof Exp:* Lectr biol, Non-commissioned Officers Sch, Taiwan, 69-70; res scientist redfish, 78-85, RES SCIENTIST & HEAD, MARINE MAMMALS SECT, NW ATLANTIC FISHERIES CTR, CAN DEPT FISHERIES & OCEANS, ST JOHN'S, 85- *Mem:* Am Fisheries Soc; Can Soc Zoologists; Am Soc Ichthyologists & Herpetologists; Soc Marine Mammalogy. *Res:* Comparative fish population studies, fish population dynamics and biomass assessment; population ecology and demography of marine mammals, particularly the harp seal. *Mailing Add:* Dept Fisheries & Oceans PO Box 5667 St John's NF A1C 5X1 Can

NIBLACK, JOHN FRANKLIN, b Oklahoma City, Okla, Mar 5, 39; m 77; c 2. PHARMACOLOGY, BIOCHEMISTRY. *Educ:* Okla State Univ, BS, 60; Univ Ill, Urbana, MS, 65, PhD(biochem), 68. *Prof Exp:* Res biochemist, 68-69, res projs leader, 69-72, res mgr, 72-75, asst dir, Dept Pharmacol, 75-76, dir, Dept Pharmacol, Cent Res Div, 76-80, dir Med Prod Res, 81-84, vpres, Med Prod Res, 84-85, exec vpres, Cent Res Div, 86-90, PRES, CENT RES DIV, PFIZER INC, 90- *Mem:* Am Soc Microbiol; AAAS. *Res:* Biochemical mechanisms of drugs; drugs affecting immune responses and general lymphoreticular function; pharmacology. *Mailing Add:* Cent Res Div Pfizer Inc Groton CT 06340

NIBLER, JOSEPH WILLIAM, b Silverton, Ore, May 9, 41; m 64; c 2. PHYSICAL CHEMISTRY. *Educ:* Ore State Univ, BS, 63; Univ Calif, Berkeley, PhD(chem), 66. *Prof Exp:* NSF fel chem, Cambridge Univ, 66-67; from asst prof to assoc prof, 67-78, PROF CHEM, ORE STATE UNIV, 78- *Concurrent Pos:* Vis scientist, Nat Res Lab, Washington, DC, 75-76, Tech Univ, Munich, 82-83 & Univ Colo, 89-90; Alexander von Humboldt award, 82-83; fel, Joint Inst Lab Astrophys, 89-90. *Mem:* Am Optical Soc; Am Phys Soc; Am Chem Soc. *Res:* Infrared and Raman spectroscopy of matrix isolated molecules; nonlinear optical spectroscopy of gases; energy transfer in solids. *Mailing Add:* Dept of Chem Ore State Univ Corvallis OR 97331

NIBLETT, CHARLES LESLIE, b Wolfeboro, NH, Feb 15, 43; m 61; c 2. PLANT PATHOLOGY, VIROLOGY. *Educ:* Univ NH, BS, 65; Univ Calif, PhD(plant path), 69. *Prof Exp:* From asst prof to assoc prof, Kans State Univ, 69-80; prof & chmn, 80-86, PROF, DEPT PLANT PATH, UNIV FLA, 80- *Concurrent Pos:* Vis prof, Cornell Univ, 77-78 & Wash Univ, 87-88. *Mem:* Am Phytopath Soc. *Res:* Plant transformation for virus resistance; plant virology; viroids. *Mailing Add:* Plant Path Dept Univ Fla Gainesville FL 32611

NICASTRO, DAVID HARLAN, b Los Angeles, Calif. FORENSIC ENGINEERING, FAILURE INVESTIGATION. *Educ:* Pomona Col, BA, 83; Univ Tex, MS, 85. *Prof Exp:* Res asst eng, Univ Texas, Austin, 83-85; FORENSIC ENGR, LAW ENG, 85- *Mem:* Am Soc Chem Engrs; Am Soc Testing & Mat. *Res:* Development of taxonomy for construction failure classification; development of "matrix method" of objective determination of contributing causes to failures; development of curriculm for forensic science education. *Mailing Add:* Law Eng Inc 13831 NW Freeway Suite 500 Houston TX 77040

NICCOLAI, NILO ANTHONY, b Pittsburgh, Pa, May 21, 40; m 64; c 4. SOFTWARE DESIGN, SYSTEMS SOFTWARE. *Educ:* Carnegie-Mellon Univ, BS, 62, MS, 63, PhD(math), 68. *Prof Exp:* Mathematician, US Bur Mines, Pa, 67-68; math analyst, Off Chief Staff, US Army, 68-70; asst prof, 70-77, ASSOC PROF MATH, UNIV NC, CHARLOTTE, 77- *Concurrent Pos:* Consult. *Mem:* Comput Soc; Sigma Xi; Asn Comput Mach. *Res:* Programming languages; operating systems; data base; theorem-proving. *Mailing Add:* Dept Math Univ NC Uncc Sta Charlotte NC 28223

NICE, CHARLES MONROE, JR, b Parsons, Kans, Dec 21, 19; m 40; c 6. RADIOLOGY. *Educ:* Univ Kans, AB, 39, MD, 43; Univ Colo, MS, 48; Univ Minn, PhD(radiol), 56. *Prof Exp:* From instr to assoc prof, Univ Minn, 51-58; PROF RADIOL, SCH MED, TULANE UNIV, 58- *Mem:* AAAS; Radiol Soc NAm; Am Roentgen Ray Soc; AMA; Asn Am Med Cols. *Res:* Acquired radioresistance in mouse tumors; clinical diagnostic radiology. *Mailing Add:* Dept Radiol Sch Med Tulane Univ 1415 Tulane Ave New Orleans LA 70112

NICELY, KENNETH AUBREY, b Slab Fork, WVa, Feb 25, 38; m 64. BOTANY. *Educ:* WVa Univ, AB, 59, MS, 60; NC State Univ, PhD(bot), 63. *Prof Exp:* Asst prof bot, Va Polytech Inst, 63-64; from asst prof to assoc prof, 64-77, PROF BOT, WESTERN KY UNIV, 77- *Mem:* Am Soc Plant Taxon; Int Asn Plant Taxon. *Res:* Taxonomy of flowering plants. *Mailing Add:* Dept of Biol Western Ky Univ Bowling Green KY 42101

NICELY, VINCENT ALVIN, b Botetourt Co, Va, Feb 10, 43; m 66; c 1. PHYSICAL CHEMISTRY, ANALYTICAL CHEMISTRY. *Educ:* WVa Wesleyan Col, BS, 65; Mich State Univ, PhD(phys chem), 69. *Prof Exp:* Res assoc quantum chem, Mich State Univ, 69-70; res chemist, 70-72, sr res chemist, 72-78, res assoc, 78-86, SR RES ASSOC, TENN EASTMAN CO DIV, EASTMAN KODAK CO, 86- *Mem:* Am Chem Soc. *Res:* Spectroscopy; quantum chemistry; applications of computers in chemistry; polymer physical properties. *Mailing Add:* 2605 Suffolk St Kingsport TN 37660

NICHAMAN, MILTON Z, b Apr 19, 31; US citizen; m 58; c 3. PUBLIC HEALTH NUTRITION, CARDIOVASCULAR DISEASES. *Educ:* Brandeis Univ, AB, 53; Tufts Univ, MD, 57; Univ Pittsburgh, ScD, 64. *Prof Exp:* From intern to resident med, Boston City Hosp, 57-59; surgeon, USPHS Heart Dis & Stroke Control Prog, Med Col SC, 59-61, surgeon, Grad Sch Pub Health, Univ Pittsburgh, 61-64, sr surgeon & chief lab, Field & Training Sta, San Francisco, 64-67, sr surgeon & chief epidemiol, 67-70; dir nutrit prog, Ctr Dis Control, HEW, 71-82; PROF NUTRIT & EPIDEMIOL, SCH PUB HEALTH, UNIV TEX, 82- *Concurrent Pos:* Lab instr, Brandeis Univ, 53-54; res assoc, Univ Pittsburgh, 63-65; lectr, Univ Calif, 67; mem arteriosclerosis coun, Am Heart Asn; adj prof, Sch Pub Health, Univ Tex, 75- *Mem:* AMA; Am Heart Asn; Am Fedn Clin Res; Am Soc Clin Nutrit; Am Pub Health Asn. *Mailing Add:* Dept Nutrit & Epidemiol Univ Tex Health Sci Ctr PO Box 20036 Houston TX 77225

NICHOL, CHARLES ADAM, b Fergus, Ont, May 3, 22; nat US; m 47; c 3. PHARMACOLOGY, BIOCHEMISTRY. *Educ:* Univ Toronto, BS, 44; McGill Univ, MS, 46; Univ Wis, PhD(biochem), 49. *Prof Exp:* From instr to asst prof, Western Reserve Univ, 49-52; asst prof, Sch Med, Yale Univ, 53-56; res prof, State Univ NY Buffalo, 56-70; dir dept exp therapeut, Roswell Park Mem Inst, 56-70, dir res, Wellcome Res Labs, 69-72, head med biochem, 72-86; sr scientist, Glaxo Res Labs, 86-88, DIR SCI AFFAIRS, GLAXO INC, 88- *Concurrent Pos:* Am Cancer Soc scholar, 52-56; mem biochem-pharmacol panel, Cancer Chemother Nat Serv Ctr, NIH, 57-58, mem drug eval panel, 58-60, chmn exp therapeut comt, 59-60, consult grants & fels div, 59-62, mem cancer chemother study sect, 59-62, spec adv comt, 64-69; adj prof pharmacol, Duke Univ, 70- & Univ NC, 72-; mem, sci adv comt chemotherapy, Am Cancer Soc, 71-76; bd sci counrs, Nat Inst Exp Health Sci, 71-75; bd dirs, Am Asn Cancer Res, 69-72 & Leukemia Soc Am, 74-79. *Mem:* Am Soc Biol Chem; Am Chem Soc; Soc Pharmacol & Exp Therapeut; Soc Exp Biol & Med; Am Asn Cancer Res. *Res:* Cancer chemotherapy; mechanism of action of folic acid antagonists; metabolism of folic acid and vitamin B-12; drug resistance in bacterial and mammalian cells; nutrition and tumor growth; corticosteroids and transaminase enzymes; drug-metabolizing enzymes; lipophilic inhibitors of dihydrofolate reductase; tetrahydrobiopterin biosynthesis and function. *Mailing Add:* Glaxo Res Inst Glaxo Inc Five Moore Dr Research Triangle Park NC 27709

NICHOL, FRANCIS RICHARD, JR, b Baltimore, Md, Feb 27, 42; m 60; c 3. VIROLOGY. *Educ:* Pa State Univ, BA, 64, MS, 66, PhD(microbiol), 68. *Prof Exp:* Res scientist, Upjohn Co, 67-72, clin res assoc, 72-74, sr res scientist, 74-75; PRES, INST BIOL RES & DEVELOP, INC, 75- *Mem:* AAAS; assoc mem Am Soc Clin Pharmacol & Therapeut; Am Soc Microbiol. *Res:* Research and development of interferon stimulators; inhibitors of tumor virus enzymes; chemotherapeutic agents for virus infections; stimulators of host-defense mechanisms; cell mediated immunity; post-marketing surveillance; clinical trials, phase II & III. *Mailing Add:* 42 Braeburn Lane Newport Beach CA 92660

NICHOL, JAMES CHARLES, b Onoway, Alta, Apr 6, 22; nat US; m 48; c 2. PHYSICAL CHEMISTRY. *Educ:* Univ Alta, BSc, 43, MSc, 45; Univ Wis, PhD(phys chem), 48. *Prof Exp:* Instr Univ Alta, 44-46; proj assoc, Univ Wis, 48-49; assoc prof chem, Willamette Univ, 49-57; from asst prof to assoc prof, 57-62, PROF CHEM, UNIV MINN, DULUTH, 62- *Concurrent Pos:* Calif Res Corp fel, Yale Univ, 53-54; vis prof, Inst Enzyme Res, Univ Wis, 65-66. *Mem:* Am Chem Soc; The Chem Soc; Sigma Xi. *Res:* Transport properties of liquids. *Mailing Add:* Dept Chem Univ Minn Duluth MN 55812

NICHOLAIDES, JOHN J, III, b Statesville, NC, Aug 1, 44; m 67; c 1. SOIL FERTILITY. *Educ:* NC State Univ, BS, 66, MS, 69; Univ Fla, PhD(soil chem), 73. *Prof Exp:* Peace Corps vol agron, 69-70; vis asst prof soil fertil eval, Int Soil Fertil Eval & Improv Proj, NC State Univ, 73-75, vis asst prof soil fertil, 75-77, asst prof, 77-81, assoc prof & coordr, Trop Soils Res Prog, 81-; ASSOC DEAN INT AGR, UNIV ILL. *Concurrent Pos:* NDEA Title IV fel, US Govt/Univ Fla, 70-73; mem, Nat Soil Fertil Adv Comn, Govt Costa Rica, 74-75; mem, Nat Agr Adv Comn, Govt Nicaragua, 74-75; coordr consortium on soils of the tropics, Title XII Activ, 77-81; consult, World Bank, US & foreign govts, tech adv comt, private indust & Rockefeller Found, 76- *Mem:* Sigma Xi. *Res:* Developing economical agronomic systems for soils of the tropics; relationships and correlations of soil test values and plant tissue nutrient content with crop response to fertilizers. *Mailing Add:* Univ Ill 114 Mumford Hall Urbana IL 61801

NICHOLAS, HAROLD JOSEPH, b St Louis, Mo, Mar 1, 19; m 52; c 2. BIOCHEMISTRY. *Educ:* Univ Mo, BS, 41; St Louis Univ, PhD(biochem), 50. *Prof Exp:* Res chemist, Hercules Powder Co, 41-44 & Parke, Davis & Co, 44-45; asst prof biochem, Univ Kans, 50-53, assoc prof obstet, Med Ctr, 55-63; prof biochem, 63-85, DIR EXP MED, INST MED EDUC & RES, SCH MED, ST LOUIS UNIV, 63-, EMER PROF BIOCHEM, 85- *Mem:* Sigma Xi; Am Soc Biol Chemists; Am Acad Neurol; Int Soc Neurochem; Am Soc Plant Physiol. *Res:* Metabolism of plant and animal steroids; terpene biosynthesis; chemistry of the central nervous system. *Mailing Add:* c/o Carol F Shabel Inst Med Educ Res 295 Oak Path Ct Ballwin MO 63011-3826

NICHOLAS, JAMES A, b Portsmouth, Va, Apr 15, 21; m 52; c 3. SURGERY. *Educ:* NY Univ, BA, 42; Long Island Col Med, MD, 45; Am Bd Orthop Surg, dipl, 55. *Prof Exp:* From instr to asst prof surg & orthop, 53-70, PROF ORTHOP, MED COL, CORNELL UNIV, 70-; ADJ PROF PHYS EDUC, NY UNIV, 70- *Concurrent Pos:* Attend orthop surgeon, New York Hosp, 53-; attend, Hosp for Spec Surg, 53-, chief metab bone dis clin, 59-64; adj attend orthop surg, Lenox Hill Hosp, 53-, dir dept orthop surg, 70-, dir, Inst Sports Med & Athletic Trauma; secy bd trustees, Philip D Wilson Res Found; pvt pract; mem exec subcomt athletic injuries skeletal systs, Nat Acad Sci, Presidents' Coun on Phys Fitness. AEC res grant; NIH res grant. *Mem:* AAAS; Am Orthop Asn; Am Trauma Soc; Am Orthop Soc Sports Med (pres); Am Acad Orthop Surg; Orthop Res Educ Found; NY Acad Sci; Orthop Res Soc. *Res:* Disturbances of knee joint, especially prosthetics, instability and replacement; athletic injuries; muscle physiology; radioisotope study of metabolic bone disease; osteoporosis; metabolic response to injury; sports medicine research; relationship between the cardiovascular system and the musculoskeletal system, conceptualized as the "linkage system". *Mailing Add:* 130 E 77th St New York NY 10021

NICHOLAS, KENNETH M, b Jamaica, NY, July 20, 47; m 71; c 2. HOMOGENEOUS CATALYSIS. *Educ:* State Univ NY Stony Brook, BS, 69; Univ Tex, Austin, PhD(chem), 72. *Prof Exp:* Res fel org chem, Dept Chem, Brandeis Univ, 72-73; asst prof, Boston Col, 73-79, assoc prof org & inorg chem, dept chem, 79-84; PROF, DEPT CHEM, UNIV OKLA, 84- *Concurrent Pos:* Vis prof, Celanese Res Co, 81; A P Sloan fel, 80-84. *Mem:* Am Chem Soc. *Res:* Organometallic complexes as synthetic intermediates and reagents; activation of small molecules by coordination; homogeneous catalysis. *Mailing Add:* Dept Chem Univ Okla Norman OK 73019

NICHOLAS, LESLIE, b Philadelphia, Pa, Dec 22, 13. DERMATOLOGY, SYPHILOLOGY. *Educ:* Temple Univ, BS, 35, MD, 37. *Prof Exp:* From assoc prof to prof, 52-79, EMER PROF DERMAT, HAHNEMANN MED UNIV, 79- *Concurrent Pos:* Lectr, Div Grad Med, Univ Pa, 49-67; venereal dis prog specialist, Philadelphia Dept Pub Health, 65-72; tech consult, Int Union Against Venereal Dis & Treponematoses, 78-80. *Honors & Awards:* Lectureship named in honor, Leslie Nicholas Lectureship, 90. *Mem:* Am Venereal Dis Asn (pres, 77-78); Soc Invest Dermat; Am Acad Dermat; Acad Psychosom Med. *Res:* Necrobiosis lipoidica diabeticorum; demodex folliculorum; erysipeloid; syphilis. *Mailing Add:* 1521 Locust St Philadelphia PA 19102-3778

NICHOLAS, PAUL PETER, b Ohrid, Yugoslavia, Aug 6, 38; US citizen; m 62; c 1. ORGANIC CHEMISTRY. *Educ:* Univ Ill, Urbana, BS, 60; Cornell Univ, PhD(org chem), 64. *Prof Exp:* Sr res chemist org chem, 64-67, sect leader polymer chem res, 67-69, res assoc, 69-73, sr res assoc, 73-78, MGR NEW TECHNOL, CORP RES, B F GOODRICH RES CTR, 78- *Mem:* Sigma Xi; Am Chem Soc. *Res:* Synthesis and reaction mechanisms; chemistry of nitrenes; phase transfer catalysis; heterogeneous high temperature catalysis; halogenation; polymer crosslinking and stabilization. *Mailing Add:* 4775 Canterbury Lane Broadview Heights OH 44147

NICHOLES, PAUL SCOTT, b American Fork, Utah, Apr 28, 16; m 41; c 3. BACTERIOLOGY. *Educ:* Brigham Young Univ, AB, 41; Univ Cincinnati, PhD(bact), 46. *Prof Exp:* Asst bact, Brigham Young Univ, 40-41; from asst prof to prof bact, 47-76, PROF MICROBIOL & MED TECHNOL IN APPL PHARMACEUT SCI, UNIV UTAH, 76- *Concurrent Pos:* Consult, Amalgamated Sugar Co & Deseret Pharmaceut Co. *Mem:* AAAS; Am Soc Microbiol; NY Acad Sci. *Res:* Immunology; bacterial metabolism; antigenic structure of bacterium tularense; aerobiology; microbiology. *Mailing Add:* 3350 Monte Verde Dr Salt Lake City UT 84109

NICHOLLS, DORIS MCEWEN, b Bayfield, Ont, Jan 24, 27; m 52. BIOCHEMISTRY, ENZYMOLOGY. *Educ:* Univ Western Ont, BSc, 49, MSc, 51, PhD(biochem), 56, MD, 59. *Prof Exp:* Demonstr bot, Univ Western Ont, 47-51; assoc physiol, George Washington Univ, 59-60; dir clin invest unit lab, Westminster Hosp, London, Ont, 60-65; assoc prof biochem, 65-70, PROF BIOCHEM, YORK UNIV, 70- *Concurrent Pos:* Consult, Nat Heart Inst, 59-60; res assoc, Univ Western Ont, 60-62. *Mem:* Am Soc Biol Chemists; Can Biochem Soc; Brit Biochem Soc; Am Chem Soc; Can Soc Cell Biol; AAAS. *Res:* Protein synthesis in mammalian tissue; regulation of cell metabolism; heavy metal and aluminum toxicity; metal toxicity. *Mailing Add:* Dept Biol York Univ North York ON M3J 1P3 Can

NICHOLLS, GERALD P, b Orange, NJ, Nov 12, 43; m 65; c 2. HEALTH PHYSICS, RADIOECOLOGY. *Educ:* Trenton State Col, BA, 65, MA, 68; Temple Univ, MS, 73, PhD(radioecol), 79. *Prof Exp:* From instr to asst prof, 68-80, assoc prof physics, Trenton State Col, 80-84; res scientist health physics, 81-85, asst dir, 85-89, DEP DIR, DIV ENVIRON QUAL, BUR RADIATION PROTECTION, NJ, 89- *Honors & Awards:* Earth Day Award, EPA, 87. *Mem:* Health Physics Soc; AAAS; hon mem Am Asn Radon Scientists & Technologists, 89. *Res:* Gamma spectrometry applied to environmental monitoring for natural and man made radioactive materials; movement of radioactive materials in the food chain leading to man and the application of computers to these problems; health hazards of radon. *Mailing Add:* 361 W Burlington St Bordentown NJ 08505

NICHOLLS, J(AMES) A(RTHUR), b Detroit, Mich, Feb 12, 21; m 45; c 3. AERONAUTICAL ENGINEERING. *Educ:* Wayne State Univ, BS, 50; Univ Mich, MS, 51, PhD(aeronaut eng), 60. *Prof Exp:* From asst to res engr, Aeronaut Eng Labs, Univ Mich, Ann Arbor, 50-64, instr aeronaut & astronaut eng, Univ, 55-58, lectr, 58-60, from assoc prof to prof, 60-88, head, Gas Dynamics Labs, 61-85, EMER PROF AEROSPACE ENG, UNIV MICH, ANN ARBOR, 88- *Concurrent Pos:* Consult, 58- *Honors & Awards:* S S Attwood Award, 80. *Mem:* fel Am Inst Aeronaut & Astronaut; Combustion Inst. *Res:* Gas dynamics; heterogeneous and homogeneous combustion; engine generated pollutants. *Mailing Add:* Dept Aerospace Eng Univ Mich Ann Arbor MI 48109

NICHOLLS, PETER, b Southampton, Eng, July 13, 35; m 61. BIOENERGETICS, ELECTRON TRANSPORT. *Educ:* Cambridge Univ, UK, BA, 56, PhD(biochem), 59, ScD, 76. *Prof Exp:* Res assoc, Ore State Col, 59-60; res assoc, Univ Pa, 60-61; res fel, St Johns Col, Cambridge, UK, 61-63;

from asst prof to assoc prof biochem, State Univ NY, Buffalo, 63-69; fel, Bristol Univ, UK, 70; sr res fel med sci, Sidney Sussex Col, Cambridge, UK, 71-73; lectr biochem, Odense Univ, Denmark, 73-75; PROF BIOL SCI, BROCK UNIV, ST CATHARINES, ONT, 75- *Concurrent Pos:* Vis scientist, ARC Inst Animal Physiol, Babraham, UK, 71-73. *Mem:* Can Biochem Soc; Am Soc Biol Chemists; Hist Sci Soc; Biochem Soc, UK; Sci for Peace, Can; Biophys Soc Can (secy, 88-). *Res:* Bioenergetics of mitochondria and reconstituted vesicle systems; mechanism of action of heme enzymes, especially cytochrome c oxidas. *Mailing Add:* Dept Biol Sci Brock Univ St Catharines ON L2S 3A1 Can

NICHOLLS, PETER JOHN, b Kent, Gt Brit, Nov 29, 45; m 68; c 3. PURE MATHEMATICS. *Educ:* Imp Col, Univ London, BSc, 67; Cambridge Univ, PhD(math), 70. *Prof Exp:* Tutorial fel math, Univ Lancaster, 70-71; asst prof, 71-77, ASSOC PROF MATH, NORTHERN ILL UNIV, 77- *Mem:* Am Math Soc; London Math Soc; Sigma Xi. *Res:* Functions of a complex variable; discrete groups and Riemann surfaces. *Mailing Add:* Dept of Math Scis Northern Ill Univ De Kalb IL 60115

NICHOLLS, RALPH WILLIAM, b Richmond, Eng, May 3, 26; m 52. PHYSICS. *Educ:* Univ London, BSc, 46, PhD(physics), 51, DSc(spectros), 61. *Prof Exp:* Demonstr physics, Imp Col, Univ London, 45-46, sr demonstr astrophys, 46-48; instr physics, Univ Western Ont, 48-50, lectr, 50-52, from asst prof to prof, 52-62, sr prof, 62-65; chmn dept, 65-69, prof, 65-83, DISTINGUISHED PROF PHYSICS, YORK UNIV, 83-, DIR, CTR RES EXP SPACE SCI, 65- *Concurrent Pos:* Consult, Nat Bur Standards, 59-60; vis prof, Stanford Univ, 64 & 68. *Mem:* Int Astron Union; fel Am Phys Soc; fel Optical Soc Am; fel Royal Astron Soc; fel Royal Soc Can; fel Can Aeronaut & Space Inst. *Res:* Spectroscopy; astrophysics; aeronomy; chemical physics; atmospheric science. *Mailing Add:* Dept Physics York Univ 4700 Keele St North York ON M3J 1P3 Can

NICHOLLS, ROBERT LEE, b Lincoln, Nebr, June 11, 29; m 58; c 3. GEOTECHNICAL ENGINEERING. *Educ:* Univ Colo, BS, 51; Iowa State Univ, MS, 52, PhD(civil eng), 57. *Prof Exp:* Res asst construct mat, Iowa State Univ, 55-57; mat & geotech engr, Gannett, Gleming, Corddry & Carpenter, 57-59; from asst prof to assoc prof, 59-70, PROF CIVIL ENG, UNIV DEL, 70-; GEOTECH & CONSTRUCT MAT CONSULT, E I de NEMOURS & CO, 61- *Mem:* Am Soc Civil Engrs; Transp Res Bd; Am Concrete Inst. *Res:* Composite construction materials; civil engineering systems; geotechnical engineering. *Mailing Add:* Dept of Civil Eng Univ of Del Newark DE 19711

NICHOLS, ALEXANDER VLADIMIR, b San Francisco, Calif, Oct 9, 24; m 55; c 3. MEDICAL PHYSICS, BIOPHYSICS. *Educ:* Univ Calif, AB, 49, PhD(biophys), 55. *Prof Exp:* Res biophysicist, 55-59, lectr med physics & biophys, 59-61, from asst prof to prof, 61-77, vchmn div med physics, 67-71, chmn, 71-72, PROF BIOPHYSICS, DONNER LAB, UNIV CALIF, BERKELEY, 77- *Mem:* Sigma Xi. *Res:* Structure and function of lipoproteins; atherosclerosis; biophysics of lipid-protein structures. *Mailing Add:* Univ Calif Donner Lab Berkeley CA 94720

NICHOLS, AMBROSE REUBEN, JR, b Corvallis, Ore, June 21, 14; m 38; c 3. PHYSICAL CHEMISTRY. *Educ:* Univ Calif, BS, 35; Univ Wis, PhD(inorg chem), 39. *Hon Degrees:* LHD, Nat Univ, 76. *Prof Exp:* Asst chem, Univ Wis, 36-39; instr, San Diego State Col, 39-43; res chemist, Radiation Lab, Manhattan Proj, Univ Calif, 43-45; from asst prof to prof chem, San Diego State Col, 45-61; pres, 61-70, prof chem, 70-76, EMER PROF CHEM, SONOMA STATE COL, 76-, EMER PRES, 83- *Concurrent Pos:* Analyst, Procter & Gamble Co, Ohio, 37; asst physicist, Radio & Sound Lab, US Dept Navy, 42-43; prin chemist, Oak Ridge Nat Lab, 51-52; sr staff scientist chem & consult, Convair Sci Res Lab, 56-60; vis prof chem, Purdue Univ, 59; res collabr, Brookhaven Nat Lab, 70. *Mem:* Am Chem Soc; Sigma Xi. *Res:* Chemistry of manganese; solution chemistry of uranium; autoxidation of manganous hydroxide; high temperature electrochemistry. *Mailing Add:* Dept of Chem Sonoma State Univ Rohnert Park CA 94928

NICHOLS, BARBARA ANN, b Long Beach, Calif, Nov 16, 21; m 82; c 2. ANATOMY, BACTERIOLOGY. *Educ:* Univ Calif, Los Angeles, BA, 43; Univ Calif, Berkeley, PhD(zool), 68. *Prof Exp:* Teaching asst zool, Univ Calif, Berkeley, 63-66, instr, 66-67; res fel path, Sch Med, 68-72, from asst prof to assoc prof microbiol, 72-85, DIR ELECTRON MACROS LAB, PROCTOR RES FOUND OPHTHAL, SCH MED, UNIV CALIF, SAN FRANCISCO, 72-, ADJ PROF OPHTHAL, 85- *Concurrent Pos:* NIH training grant, Sch Med, Univ Calif, San Francisco, 68-69, NIH fel, 69-71, Nat Tuberc & Respiratory Dis Asn fel, 71-72. *Honors & Awards:* US Pub Health Serv Res Career Develop Award, Nat Inst of Allergy & Infectious Dis. *Mem:* Am Soc Cell Biol; Electron Micros Soc Am; Am Soc Microbiol; AAAS; Asn Res Vision & Ophthal; Soc Protozoologists; Am Soc Parasitol. *Res:* Intercellular parasitism, the development and function of monocytes and macrophages in normal and pathologic states; cell biology of Toxoplasma gondii; host-parasite interactions in toxoplasmosis; functions of the conjunctival epithelium in normal and pathologic states. *Mailing Add:* Proctor Found for Res in Ophthal Univ Calif Med Sch San Francisco CA 94143

NICHOLS, BENJAMIN, b Staten Island, NY, Sept 20, 20; m 42; c 2. ELECTRICAL ENGINEERING. *Educ:* Cornell Univ, BEE, 46, MEE, 49; Univ Alaska, PhD(geophys), 58. *Prof Exp:* From instr to assoc prof elec eng, 46-59, dir elem sci study, 64-65, PROF ELEC ENG, CORNELL UNIV, 59-, ASST DEAN ENG, 80- *Concurrent Pos:* Ford Found fel, 50-51; mem, US comt, Int Sci Radio Union, 57-; lectr, Cornell Aeronaut Lab, 59-60. *Mem:* AAAS; Sigma Xi; Am Soc Eng Educ. *Res:* Science education; science policy. *Mailing Add:* 109 Llenroc Ct Ithaca NY 14850

NICHOLS, BUFORD LEE, JR, b Ft Worth, Tex, Dec 21, 31; m; c 3. MEDICINE, NUTRITION. *Educ:* Baylor Univ, BA, 55, MS, 58; Yale Univ, MD, 60; Am Bd Pediat, dipl; Am Bd Nutrit, dipl. *Prof Exp:* Instr physiol, Col Med, Baylor Univ, 56-57; pediat internship, Yale-New Haven Med Ctr, 60-61; pediat resident, Johns Hopkins Hosp, 61-63; instr pediat, Sch Med, Yale Univ, 63-64; instr physiol & pediat, 64-66, from asst prof to assoc prof pediat, 66-67, instr physiol, 66-74, chief, Sect Nutrit & Gastroenterol, Dept Pediat, 70-85, ASSOC PROF COMMUNITY MED, BAYLOR COL MED, 75-, PROF PHYSIOL & PEDIAT, 77-, HEAD, SECT NUTRIENT PHYSIOL, 79-; DIR, USDA CHILDREN'S NUTRIT RES CTR, HOUSTON, 79- *Concurrent Pos:* Chief resident pediat, Yale-New Haven Med Ctr, 63-64; Bristol-Myers award, 84. *Mem:* Am Acad Pediat; Am Soc Clin Nutrit; Am Col Nutrit (vpres, 75-76, pres, 77-79). *Res:* Environmental effects upon growth and development in the infant, especially alterations in body composition and muscle physiology in malnutrition; chronic diarrhea and malnutrition. *Mailing Add:* 9917 Oboe Houston TX 77025

NICHOLS, CARL WILLIAM, b State Center, Iowa, June 1, 24; m 55; c 3. PLANT PATHOLOGY. *Educ:* Mich State Col, BS, 48; Univ Idaho, MS, 49; Univ Calif, Davis, PhD(plant path), 52. *Prof Exp:* Tech plant path, Univ Idaho, 48-49; asst, Univ Calif, 49-51, jr specialist, 51-53; from assoc plant pathologist to plant pathologist, Calif Dept Food & Agr, 53-62, prog supvr, 62-66, chief bur plant path, 66-71, chief spec serv & asst to div chief, Div Plant Indust, 71-72 & 76-78, asst dir plant indust, 72-76, chief lab serv, Div of Plant Indust, 78-85; regulatory consult, Calif Asn Nurserymen, 85-88; PRIN CONSULT, NICHOLS BIOL CONSULTS, 85- *Concurrent Pos:* Assoc, Exp Sta, Univ Calif, Davis, 67-; mem, Western Plant Bd, 72-84, Nat Plant Bd, 74-84, Interstate Pest Control Compact Tech Adv Comt, 75-84; collabr, USDA, 80-82. *Mem:* AAAS; Am Phytopath Soc; Am Inst Biol Sci; Int Asn Plant Path. *Res:* Regulatory plant pest control. *Mailing Add:* 6336 La Cienega Dr North Highlands CA 95660

NICHOLS, CHESTER ENCELL, b Boston, Mass, Dec 28, 35; m 62; c 2. GEOSTATISTICS, ENVIRONMENTAL SCIENCE. *Educ:* Cornell Univ, AB, 60; Univ Iowa, MS, 65; Univ Mo-Rolla, PhD, 77; Nat Cert Comn Chem, cert. *Prof Exp:* Teaching asst geol, Univ Iowa, 61-62; geologist, US Bur Mines, 62-63; teaching asst geol, Univ Mo-Rolla, 64-66; explor geologist, 68-73, sr mines geologist, 73-75, proj geologist-geochemist, 75-78, sr explor geologist, Grand Junction, Colo, 78-80, sr staff geologist, Union Carbide Corp, Reno, Nev, 81-82; MINERALS EXPLORATION CONSULT, RENO, NEV, 83- *Concurrent Pos:* Prin lectr, Bur Land Mgt Mineral Assessment Workshop, 83; consult, Lode Gold & Platinum Deposits, 83-; environ scientist mobilization augmentee, US Army Res Off, 84-88; chief geologist, Placer Mgt Group, 85-86; founding ed, publ & ed-in-chief, Explore, 87-90; sr staff geologist, Miramar Mining Corp, 90-91. *Mem:* Asn Explor Geochemists; Sigma Xi; AAAS; Geol Soc Am; fel Am Inst Chemists. *Res:* Gold and tungsten exploration in Great Basin; applied geochemistry; uranium exploration in Wyoming (Precambrian), Colorado Plateau (Jurassic) and Texas Gulf Coast (Tertiary); exploration geophysics; astrogeology; structural geology; environmental science; geostatistics. *Mailing Add:* 1192 La Via Way Sparks NV 89431-3166

NICHOLS, COURTLAND GEOFFREY, b Wilmington, Del, May 16, 34; m 59; c 3. PLANT BREEDING. *Educ:* Pa State Univ, BS, 56; Univ Wis, PhD(hort, plant path), 63. *Prof Exp:* Res assoc tomato breeding, Campbell Inst Agr Res, 63-68; PLANT BREEDER, FERRY MORSE SEED CO, 69- *Mem:* Am Phytopath Soc; Am Soc Hort Sci; Sigma Xi; Tomato Genetics Corp. *Res:* Processing and fresh market types for high quality disease resistance and stress tolerance; carrot breeding, especially hybrids. *Mailing Add:* Ferry Morse Seed Co PO Box 1010 San Juan Bautista CA 95045

NICHOLS, DAVID EARL, b Covington, Ky, Dec 23, 44; c 2. MEDICINAL CHEMISTRY. *Educ:* Univ Cincinnati, BS, 69; Univ Iowa, PhD(med chem), 73. *Prof Exp:* Fel pharmacol, Univ Iowa, 73-74; from asst prof to assoc prof, 74-84, PROF MED CHEM, SCH PHARM, PURDUE UNIV, WEST LAFAYETTE, 84- *Honors & Awards:* Fel, Acad Pharmaceut Sci. *Mem:* Am Chem Soc; Soc Neurosci; Am Asn Col Pharm; Am Pharm Asn; Am Soc Pharmacol & Exp Therapeut. *Res:* Synthesis and study of structure-activity relationships of centrally active drugs and neurotransmitter congeners; study of mode of action of hallucinogenic or psychedelic, and antipsychotic drugs; molecular pharmacology; development of drugs for psychiatry. *Mailing Add:* Dept Med Chem & Pharmacog Purdue Univ Sch Pharm & Pharmacal Sci West Lafayette IN 47907

NICHOLS, DAVIS BETZ, b Carlisle, Ky, Nov 19, 40; m 65; c 3. INFRARED SENSOR SYSTEMS, LASER OPTICS. *Educ:* Wheaton Col Ill, BS, 62; Univ Ky, PhD(nuclear physics), 66. *Prof Exp:* Res assoc nuclear physics, Van de Graaff Lab, Ohio State Univ, 66-67, fel, Van de Graaff Accelerator Lab, 67-68; res fel physics, Kellogg Radiation Lab, Calif Inst Technol, 68-70; MEM STAFF, BOEING AEROSPACE CO, 70- *Mem:* Am Phys Soc. *Res:* Infrared and laser instrumentation, high power laser beam profiling, laser damage of optical components, laser-induced gas breakdown, laser effects, pulsed chemical lasers; high power laser testing of infrared and ultraviolet optics; nuclear spectroscopy; inelastic neutron scattering. *Mailing Add:* Boeing Aerospace & Electronics MS 8H-37 PO Box 3999 Seattle WA 98124

NICHOLS, DONALD RAY, b Omaha, Nebr, Mar 26, 27; m 48; c 4. ENGINEERING GEOLOGY. *Educ:* Univ Nebr, BS, 50. *Prof Exp:* Geologist & geomorphologist, US Geol Surv, Washington, DC, 51-64, staff geologist, Eng Geol Subdiv, 64-67, res geologist, Menlo Park, Calif, 67-75, chief earth sci appln prog, Reston, Va, 75-79, hazards info coordr, 77-79, chief br eng geol, Denver, Colo, 79-83, res geologist, US Geol Surv, 83-89; CONSULT, 89- *Concurrent Pos:* Vchmn land use planning adv group, Joint Calif Legis Comt Seismic Safety, 70-74; mem, Earthquake Hazards Reduction Implementation Plan Working Group, 77-78. *Mem:* Geol Soc Am; Asn Eng Geol; Int Asn Eng Geol. *Res:* Engineering and glacial geology; permafrost; geomorphology; earth sciences applications to land-use planning. *Mailing Add:* 978 Coneflower Dr Golden CO 80401

NICHOLS, DONALD RICHARDSON, b Minneapolis, Minn, Feb 22, 11; wid; c 6. CLINICAL MEDICINE. *Educ:* Amherst Col, AB, 33; Univ Minn, MD, 38, MS, 42; Am Bd Internal Med, dipl. *Prof Exp:* From asst prof to prof clin med, Mayo Grad Sch Med, Univ Minn, 48-72, prof med, Mayo Med Sch, 72-81, sr consult med, Mayo Clin, 73-81; RETIRED. *Concurrent Pos:* Consult, Mayo Clin, 43-81, head sect infectious dis, 61-69, chmn div infectious dis & internal med, 70-73. *Mem:* AMA; Cent Soc Clin Res; fel Am Col Physicians; Infectious Dis Soc Am; Sigma Xi. *Res:* Clinical investigation of antibiotic agents and the treatment of infectious diseases. *Mailing Add:* 207 Fifth Ave SW Rochester MN 55902

NICHOLS, DOUGLAS JAMES, b Jamaica, NY, Feb 19, 42; m 64; c 3. PALEONTOLOGY, PALYNOLOGY. *Educ:* NY Univ, BA, 63, MS, 66; Pa State Univ, PhD(geol), 70. *Prof Exp:* Sci asst micropaleont, Am Mus Natural Hist, 63-65; lectr geol, City Col New York, 65-66; asst prof, Ariz State Univ, 70 & State Univ NY Col Geneseo, 70-74; palynologist, Chevron Oil Co, 74-78; GEOLOGIST, US GEOL SURV, 78- *Mem:* Soc Econ Paleontologists and Mineralogists; Palaeont Asn; Am Asn Stratig Palynologists (pres, 82-83); Int Fedn Palynological Socs. *Res:* Palynology; biostratigraphy; evolution; paleoecology. *Mailing Add:* US Geol Surv MS 919 Box 25046 Denver CO 80225-0046

NICHOLS, DUANE GUY, b Middlebourne, WVa, July 18, 37; m 63; c 2. CHEMICAL ENGINEERING. *Educ:* WVa Univ, BSChE, 59; Univ Del, MChE, 63, PhD(chem eng), 68. *Prof Exp:* Asst prof physics, Del State Col, 63-68; from asst prof to assoc prof chem eng, WVa Univ, 68-77; sr chem engr, Res Triangle Inst, 78-80; sr res engr, Conoco, Inc, 80-86; GROUP LEADER, CONSOLIDATION COAL CO, 87- *Concurrent Pos:* Vis lectr, Del Acad Sci, 66-68 & Am Inst Chem Engrs, 77-79; tech chmn, Am Inst Chem Engrs & Air Pollution Control Asn, 5th & 6th Nat Conf on Energy & Environ, 77-79; vis assoc prof, WVa Univ, 78. *Honors & Awards:* Whitehill Chem Award. *Mem:* Am Inst Chem Engrs; Am Chem Soc. *Res:* Mass transfer; digital and analog computer simulation; fluidized bed applications; coal conversion processes; environmental and biomedical engineering. *Mailing Add:* Consol Res & Develop 4000 Brownsville Rd Library PA 15129

NICHOLS, EUGENE DOUGLAS, b Rovno, Poland, Feb 6, 23; US citizen; m 51. MATHEMATICS. *Educ:* Univ Chicago, BS, 49; Univ Ill, MA, 53, MEd, 54, PhD(math educ), 56. *Prof Exp:* Instr math, Roberts Wesleyan Col, 50-51, Univ Ill, 51-53,& 54-56, URBANAHigh Sch, 53-54; PROF MATH EDUC, FLA STATE UNIV, 56- *Mem:* Am Math Soc; Math Asn Am; Asn Comput Mach; AAAS. *Res:* Foundations of mathematics, with emphasis on linguistics of mathematics. *Mailing Add:* 3386 W Lake Shore Dr Tallahassee FL 32312

NICHOLS, FREDERIC HONE, b Boston, Mass, Nov 28, 37; m 71; c 3. BIOLOGICAL OCEANOGRAPHY. *Educ:* Hamilton Col, AB, 60; Univ Wash, MS, 68, PhD(oceanog), 72. *Prof Exp:* OCEANOGR, US GEOL SURV, 71-, CHIEF, WATER RESOURCE DIV BR REGIONAL RES, 90- *Concurrent Pos:* Mem, comt syst resources invert zool, NSF, 74-77; vis scientist, Univ Kiel, Ger, 75-76; mem, Calif Water Resources Control Bd, San Francisco Bay & Estuary Adv Comt, 76-78; mem gov bd, Tiburon Ctr Environ Studies, San Francisco State Univ, 77-; mem, Panel on Longterm Ecol Measurements, NSF, 78; assoc chief, Br Pac Marine Geol, US Geol Surv, 79-82; mem, San Francisco Bay Tech Adv Comt, 82-86; workshop biol effects coastal ocean suspended sediment transport, BECOSST, NSF, 84-86; sci & tech adv comt, San Francisco Bay Conserv & Develop Comn, 86- & Environ Protection Agency, San Francisco Estuary Proj, 87-; organizing comt NSF workshop land-sea interface, Am Soc Limnol & Oceanog, 86-87; Nat Res Coun Chesapeake Bay Panel, comt systs assessment marine environ monitoring, 87-88; guest partic, NSF Ecol Panel DOI Minerals Mgt Serv OCS Environ Studies Prog, 88; NSF Panel, evaluate Land-Margin Ecosystem Res proposals, 88; Calif Acad Sci fels nominating comt, 88-; exec comt, Pac Div AAAS, 88-, pres, 89-90; Calif State Atty Gen Shell Oil Spill Tech Panel, 88-89; chmn, organizing comt, Estuarine Res Fedn, 91, biennial meeting, 88- *Honors & Awards:* Super Serv Award, Dept of Interior, 88. *Mem:* Am Soc Limnol & Oceanog; Ecol Soc Am; Estuarine Res Fedn; Sigma Xi; AAAS. *Res:* Marine and estuarine benthic ecology. *Mailing Add:* US Geol Surv 345 Middlefield Rd MS-496 Menlo Park CA 94025

NICHOLS, G(EORGE) STARR, b Argyle, Wis, Aug 11, 18; m 41; c 3. CHEMICAL ENGINEERING. *Educ:* Univ Wis, BS, 41, PhD(chem eng), 52. *Prof Exp:* Res engr, E I Du Pont De Nemours & Co, Inc, 41-89; STAFF ENGR, WESTINGHOUSE SAVANNAH RIVER CORP, 89- *Concurrent Pos:* Res engr, Argonne Nat Lab, 51-52; instr, Augusta Col. *Mem:* Am Chem Soc; Am Inst Chem Engrs. *Res:* Chemical kinetics; prevention of hazardous reaction. *Mailing Add:* 1224 Johns Rd Augusta GA 30904

NICHOLS, GEORGE, JR, biochemistry, oceanography, for more information see previous edition

NICHOLS, GEORGE MORRILL, physical chemistry, organic chemistry, for more information see previous edition

NICHOLS, HERBERT WAYNE, b Bessemer, Ala, Feb 24, 37. BIOLOGY, PHYCOLOGY. *Educ:* Univ Ala, BS, 59, MS, 60, PhD(phycol), 63. *Prof Exp:* Asst prof, 63-65, assoc prof bot & co-chmn dept biol, 65-76, ASSOC PROF BIOL, WASH UNIV, 76- *Concurrent Pos:* Instr, Marine Biol Lab, Woods Hole, 64-65; vis lectr, Univ Tex, 63; NSF grant, 65-67; Am Cancer Soc grant, 65; USPHS res grant, 67-; Off Naval Res res grant, 67-; Rockefeller grant, 70-73; Kellog Found grant. *Mem:* AAAS; Bot Soc Am; Am Micros Soc; Am Phycol Soc (treas, 70-72, vpres, 72-73); Int Soc Plant Morphologists. *Res:* Morphogenetic studies of algae; aging in photosynthetic algae; aquatic microbiology. *Mailing Add:* Dept of Biol Wash Univ Campus Box 1137 St Louis MO 63130

NICHOLS, J WYLIE, PHOSPHO-LIPID TRANSFER. *Educ:* Univ Calif, Davis, PhD(physiol), 79. *Prof Exp:* ASST PROF PHYSIOL, EMORY UNIV, 82- *Res:* Lipid metabolism. *Mailing Add:* Dept Physiol Woodruff McAdmiral Bldg Emory Univ Sch Med Atlanta GA 30322

NICHOLS, JACK LORAN, b Drumheller, Can, Dec 3, 39. BIOCHEMISTRY. *Educ:* Univ Alta, BS, 60, MS, 63, PhD(biochem), 67. *Prof Exp:* Res fel, Lab Molecular Biol, Cambridge, Eng, 68-70; assoc prof microbiol, Med Ctr, Duke Univ, 70-76; ASSOC PROF MICROBIOL, UNIV VICTORIA, BC, 76- *Mem:* Am Soc Microbiol; Am Soc Biol Chemists. *Res:* Structure and function of nucleic acids. *Mailing Add:* Dept Biochem & Microbiol 3473 W 39th Ave Victoria BC V6N 3A3 Can

NICHOLS, JAMES CARLILE, b Evanston, Ill, Aug 17, 50; m 71; c 1. COMPUTER MODELING, DATA ANALYSIS. *Educ:* Univ Fla, BS, 72, ME, 74. *Prof Exp:* Design engr, Bessent, Hammack & Ruckman, 72 -73; engr two, City of Tampa, 75-76; STAFF ENGR, WATER & AIR RES, INC, 76- *Mem:* Am Waterworks Asn; Nat Soc Prof Engrs; Soc Environ Toxicol & Chem; Water Pollution Control Fedn. *Res:* Stormwater management; hydrologic, hydraulic and water quality modeling; environmental impacts; toxicology, risk management and hazardous waste management; water supply treatment and distribution; wastewater treatment and disposal; data analysis; statistical analysis. *Mailing Add:* 2758 SW 14th Dr Gainesville FL 32608

NICHOLS, JAMES DALE, b Waynesboro, Va, May 24, 49; m 70; c 2. FISH & WILDLIFE SCIENCES. *Educ:* Wake Forest Univ, BS, 71; La State Univ, MS, 73; Mich State Univ PhD (wildlife ecol), 76. *Prof Exp:* Res biologist, 76-85, TEAM LEADER, VERTEBRATE POP DYNAMICS, BIOMETRICS, US FISH & WILDLIFE SERV, 85- *Concurrent Pos:* Instr, Latin Am Workshop Migratory Birds, US Fish & Wildlife Serv, 83-84; lectr, Wildlife Stat Workshop, Southeastern Coop Wildlife & Fisheries Proj, NC State Univ, 83-87; vis prof, Dept Zool, Univ Guelph, 85; adj assoc prof, Dept Wildlife & Range Sci Univ Fla, 87- *Honors & Awards:* Wildlife Pub Award, Wildlife Soc, 91. *Mem:* Ecol Soc Am; Wildlife Soc. *Res:* Dynamics of exploited animal populations; the influence of environmental and individual variables on survival and reproductive rates, and the estimation of parameters for free-ranging animal populations. *Mailing Add:* Br Migratory Bird Res US Fish & Wildlife Serv Patuxent Wildlife Res Ctr Laurel MD 20708

NICHOLS, JAMES RANDALL, b Wilmington, Del, Mar 10, 31; m 54; c 3. PHYSICAL CHEMISTRY. *Educ:* Univ Del, BS, 53; Pa State Univ, MS, 57, PhD(fuel technol), 61. *Prof Exp:* Sr chemist, Chem Res Dept, Atlas Chem Industs, Inc, 61-67, res chemist, 67-72, RES CHEMIST, CHEM RES DEPT, ICI AMERICAS, INC, 72- *Mem:* AAAS; Am Chem Soc. *Res:* Physical-chemical structure studies on activated carbon; chemical nature of carbon surface; characterization of supported catalysts; thermal analysis; x-ray diffraction and electron microscopy of polymer and pharmaceutical products MS, GC/MS of organic compounds. *Mailing Add:* 117 Alders Dr Blue Rock Manor Wilmington DE 19803

NICHOLS, JAMES ROBBS, b Jackson, Tenn, May 30, 26; m; c 1. ANIMAL REPRODUCTIVE PHYSIOLOGY, ANIMAL POPULATION-GENETICS. *Educ:* Univ Tenn, BS, 49; Univ Minn, MS, 55, PhD(dairy), 57. *Prof Exp:* Asst county agent, Carroll County, Tenn, 49-50; instr animal husb, Univ Tenn Jr Col, 50-54; assoc prof, Univ Tenn, Martin, 57-59; assoc prof, dairy sci, Pa State Univ, 59-64; exec vpres & gen mgr, Select Sires, Inc, 71-73; head & prof, Dept Dairy Sci, Va Polytech Inst & State Univ, 64-69, assoc dean & dir, 69-71 & 73-75, DEAN COL, VA POLYTECH INST & STATE UNIV, 75-, DIR, VA AGR EXP STA, 78- *Concurrent Pos:* Int travel consult & lectr, numerous foreign countries on grants from var agencies & foreign govt; chair, Coun Admin Heads Agr, 82-83 & Div Agr, Nat Asn State Univs & Land-Grant Cols, 83-84. *Mem:* Sigma Xi; Southern Asn Agr Scientists (pres, 82). *Res:* Research administration in the life sciences, food, and agriculture. *Mailing Add:* Va Agr Exp Sta Col Agr & Life Sci Va Polytech Inst & State Univ Blacksburg VA 24061-0402

NICHOLS, JAMES ROSS, b Kansas City, Mo, June 26, 44; m 65; c 3. ZOOLOGY. *Educ:* Abilene Christian Col, BS, 66; Univ Mich, Ann Arbor, MS, 68; Univ Mo-Columbia, PhD(zool), 73. *Prof Exp:* Asst prof, 72-75, assoc prof, 78-81, prof biol, Univ Cent Ark, 81-; AT DEPT BIOL, ABILENE CHRISTIAN UNIV. *Mem:* Am Soc Zoologists; Am Inst Biol Sci; AAAS; Nat Asn Biol Teachers. *Res:* Cellular and comparative physiology, especially ionic and water regulation; effects of antibiotics and endocrine control; temperature and light effects. *Mailing Add:* Dept of Biol Abilene Christian Univ Abilene TX 79699

NICHOLS, JAMES T, b Salina, Kans, Jan 5, 30; m 59; c 6. RANGE MANAGEMENT, AGRONOMY. *Educ:* Ft Hays Kans State Univ, BS, 60, MS, 61; Univ Wyo, PhD(range mgt), 64. *Prof Exp:* Asst prof range mgt, SDak State Univ, 64-69; assoc prof, 69-74, PROF AGRON, UNIV NEBR, 74- *Honors & Awards:* Range Mgt Serv Award, Soc Range Mgt, 76; Grassland Serv Award, Nebr Forage & Grassland Coun, 84; Merit Cert Award, Am Forage & Grassland Coun, 84; Outstanding Achievement Award, Soc Range Mgt, 89; Distinguished Serv Agr Award, Am Soc Farm Mgrs & Rural Appraisers, 90. *Mem:* Soc Range Mgt; Soc Agron; Am Forage & Grassland Coun. *Res:* Range improvement through grazing management and range renovation; meadow improvement and management; irrigated pasture production; pasture management; complementary forages. *Mailing Add:* W Cent Res Exten Ctr Box 46A Rt Four North Platte NE 69101

NICHOLS, JOE DEAN, b Skiatook, Okla, July 14, 31; m 55; c 1. SOIL SCIENCE. *Educ:* Okla A&M Col, BS, 55, MS, 56. *Prof Exp:* Party leader soil sci, Soil Conserv Serv, USDA, Pawhuska, Okla, 59-62, soil scientist specialist, Pauls Valley, 62-65, soil correlator, Stillwater, 65-69, state soil scientist, Denver Colo, 69-71, asst prin soil correlator, 71-76, head soil staff, 76-87, HEAD SOIL INTERPRETATIONS STAFF, SNTC, SOIL CONSERV SERV, USDA, 87- *Concurrent Pos:* Chmn, Div Five, Soil Sci Soc Am, 82. *Mem:* Soil Conserv Soc Am (pres, 85-86); Soil Sci Soc Am; Int Soc Soil Sci; World Asn Soil & Water Conserv. *Res:* Soil genesis, morphology and soil classification; vis twelve foreign countries for consulting trips for AID-Soil Conservation Service projects and for meetings. *Mailing Add:* PO Box 6567 Ft Worth TX 76115

NICHOLS, JOHN C, b Chicago, Ill, Feb 28, 39; m 61; c 2. MATHEMATICS. *Educ:* Blackburn Col, BA, 60; Southern Ill Univ, MS, 62; Univ Iowa, PhD(math), 66. *Prof Exp:* Asst prof math, Monmouth Col, Ill, 66-71; assoc prof 71-76, PROF MATH, THIEL COL, 76-, CHMN, DEPT MATH & COMPUT SCI, 85- *Mem:* Math Asn Am; Am Math Soc. *Res:* Homological algebra as related to the theory of local rings. *Mailing Add:* Dept Math Thiel Col Greenville PA 16125

NICHOLS, JOSEPH, b New York, NY, July 9, 17; m 51; c 2. ORGANIC CHEMISTRY. *Educ:* City Col New York, BS, 38; Univ Minn, PhD(org chem), 43. *Prof Exp:* Asst physiol chem, Univ Minn, 39-42, Nat Defense Res Coun res fel, 42-43; res chemist, Interchem Corp, NY, 43-49; chief, Dept Org Chem, Ethicon, Inc, 51-57, assoc dir res, 57-65, dir res, Collagen Prod, 65-68; pres, Princeton Biomedix Inc, 68-76; pres, Helitrex Inc, 76-85; vpres res, Am Biomat Corp, 85-86; PRES, PRODEX INC, 86- *Mem:* AAAS; Am Chem Soc; NY Acad Sci; Soc Biomaterials; Am Soc Artificial Internal Organs. *Res:* Quinones; fatty acids; biomaterials medical products. *Mailing Add:* Prodex Inc PO Box 7064 Princeton NJ 08543

NICHOLS, KATHLEEN MARY, b Amsterdam, NY, May 29, 48. CELL BIOLOGY, ANIMAL BEHAVIOR. *Educ:* State Univ NY, Albany, BS, 70, PhD(biophysics), 76; Col St Rose, MS, 72. *Prof Exp:* Res assoc fel biophys cell biol, State Univ NY, Albany, 76-79; res assoc fel toxicol, Albany Med Col, 79-80; ASST PROF BIOL, RUSSELL SAGE COL, NY, 80- *Concurrent Pos:* Res assoc, State Univ NY, Albany, 80, 81 & 82. *Mem:* Biophys Soc; Am Soc Cell Biol; AAAS. *Res:* Control of the mobility of cellular flagella. *Mailing Add:* Biol Dept Russell Sage Col Troy NY 12180

NICHOLS, KATHRYN MARION, b Santa Monica, Calif, Apr 30, 46. STRATIGRAPHY, SEDIMENTARY PETROLOGY. *Educ:* Univ Calif, Riverside, BS, 68; Stanford Univ, PhD(geol), 72. *Prof Exp:* NSF res asst, Stanford Univ, 70-72, NSF res assoc, 73-74, lectr geol, 75-76; GEOLOGIST, US GEOL SURV, 78- *Concurrent Pos:* Petrol Res Fund fel, 74-76; Nat Res Coun fel, 76-77; Gilbert fel, US Geol Surv, 86-87. *Mem:* Geol Soc Am; Soc Econ Paleontologists & Mineralogists. *Res:* Sedimentology of carbonate rocks; Mississippian stratigraphy of the northern rocky mountains; conodont biostratigraphy. *Mailing Add:* Fed Ctr US Geol Surv Denver CO 80225

NICHOLS, KENNETH DAVID, b Cleveland, Ohio, Nov 13, 07; m 32; c 2. ENGINEERING. *Educ:* US Mil Acad, BS, 29; Cornell Univ, CE, 32, MCE, 33; Univ Iowa, PhD(hydraul), 37. *Prof Exp:* Asst dir, US Waterways Exp Sta, US Army, Miss, 33-34 & 35-36, instr civil & mil eng, US Mil Acad, 37-41, prof mech, 47-48, area engr, Rome Air Depot, NY & Pa Ord Works, 41-42, dep dist engr, Manhattan Dist, 42-43, dist engr, 43-47, chief armed forces spec weapons proj & Army mem mil liaison comt to US Atomic Energy Comn, 48-50, dep dir guided missiles, Off Secy Defense, 50-53, chief res & develop, US Army, 52-53; gen mgr, US Atomic Energy Comn, 53-55; consult engr, 55-87; auth, 87; RETIRED. *Concurrent Pos:* Dir, Atomic Indust Forum, 56-59 & 64-70, Detroit Edison Co, 62-80 & Fruehauf Corp, 64-82. *Honors & Awards:* Collinwood Prize, Am Soc Civil Engrs, 38; Nicaraguan Medal of Merit; Comdr, Order of the Brit Empire. *Mem:* Nat Acad Engrs; fel Am Nuclear Soc; hon mem Am Soc Mech Engrs. *Res:* Hydraulic model; atomic energy; nuclear engineering management. *Mailing Add:* 8101 Connecticut Ave C707 Chevy Chase MD 20815-2810

NICHOLS, KENNETH E, b Brems, Ind, Dec 17, 20; m 41; c 4. PLANT PHYSIOLOGY. *Educ:* Valparaiso Univ, AB, 49; Univ Chicago, MS, 53, PhD(physiol), 61. *Prof Exp:* From assoc prof to prof biol, Valparaiso Univ, 53-88; RETIRED. *Mem:* Am Soc Plant Physiol; Phycol Soc Am; Int Phycol Soc. *Res:* Biosynthesis of plant pigments. *Mailing Add:* Dept Biol Valparaiso Univ Valparaiso IN 46383

NICHOLS, LEE L(OCHHEAD), JR, b Richmond, Va, June 5, 23; c 3. ELECTRICAL ENGINEERING, MICROWAVES. *Educ:* Va Mil Inst, BS, 47; Ohio State Univ, MS, 51; Va Polytech Inst, PhD, 70. *Prof Exp:* Dir eng & head dept elec eng, 68-87, PROF ELEC ENG, VA MIL INST, 47- *Concurrent Pos:* Vis fac mem, Nat Judicial Col, Reno, Nevada, 82, 84 & 85; prin consult, Fla Radar Comn, 80-81 & consult, police traffic radar, thirteen states & DC, Nat Highway Traffic Safety Admin, 80-81; chmn, Southeastern Asn Elec Eng Dept Heads, 83-84 & pres, Southeastern Ctr Elec Eng Educ, 84-85. *Mem:* Am Soc Eng Educ; Inst Elec & Electronics Engrs. *Res:* Instrumentation and measurement of physical quantities by electrical means and analog devices as applied to basic physical problems. *Mailing Add:* 902 Bowyer Lane Lexington VA 24450

NICHOLS, MICHAEL CHARLES, b Minneapolis, Minn, Jan 24, 51; m 81; c 2. HYDROBIOLOGY. *Educ:* Univ Mich, Flint, AB, 73; Univ Ga, MS, 78; Ga Tech, MSHP, 85. *Prof Exp:* Res asst, Inst Ecol, Univ Ga, 77-78; SR BIOLOGIST, CENT LAB, GA POWER CO, 78-, HEALTH PHYSICS & CHEM OPERS SUPVR, 85- *Mem:* Am Soc Limnol & Oceanog; AAAS; Soc Power Indust Biologists; Health Physics Soc. *Res:* Monitoring environmental effects of nuclear power plant operations; investigation of effects of hypolimnetic releases form hydroelectric impoundments on downstream water quality; personnel dusimetry. *Mailing Add:* 5131 Maner Rd Smyrna GA 30080

NICHOLS, NATHAN LANKFORD, b Jackson, Mich, Nov 16, 17; m 41; c 5. PHYSICS. *Educ:* Western Mich Univ, AB, 39; Univ Mich, MS, 45; Mich State Univ, PhD(physics), 53. *Prof Exp:* Teacher high sch, SDak, 39-40 & Mich, 40-43; instr physics, Ill Col, 43-44 & Univ Mich, 44-45; asst, Mich State Univ, 46-48; prof & head dept, Alma Col, 49-55; from assoc prof to prof physics, Western Mich Univ, 55-81; RETIRED. *Concurrent Pos:* Manhattan Proj, Univ Mich, 45. *Mem:* Optical Soc Am; Am Asn Physics Teachers. *Res:* Near infrared spectroscopy; optics. *Mailing Add:* 726 E Gull Lake Dr Augusta MI 49012-9712

NICHOLS, ROBERT LESLIE, b Boston, Mass, June 10, 04; m 35; c 1. GEOLOGY. *Educ:* Tufts Univ, BS, 26; Harvard Univ, MA, 30, PhD(geol), 40. *Hon Degrees:* DSc, Eastern Ky Univ, 74, Tufts Univ, 78. *Prof Exp:* Master, Montpelier Sem, 26-27 & Milton Acad, 27-28; asst geol, Harvard Univ, 28-30; instr, Tufts Univ, 29-36, from asst prof to prof, 36-45, actg head dept geol, 36-40, head dept, 40-69; distinguished prof, Eastern Ky Univ, 69-74; EMER PROF GEOL, TUFTS UNIV, 74- *Concurrent Pos:* Ranger naturalist, Nat Park Serv, 30-31; instr, Boston Adult Educ Ctr, 33-35; res assoc, Mus Northern Ariz, 40-41; from asst geologist to assoc geologist, US Geol Surv, 40-74; geologist, Ronne Antarctic Res Exped, 47-48; geologist, US Navy Arctic Task Force, 48; geologist, Am Geog Soc-Nat Hist Mus Arg Exped, Patagonia, 49; mem, Juneau, Alaska Ice Cap Exped, 50; leader, US Army Transp Corps Exped, Inglefieldland, Greenland, 53; geologist, Oper Deep Freeze, Antarctica, US Navy, 57-58 & Int Geophys Year, 58-59; leader, Tufts Col-NSF Antarctic Exped, 59-60 & 60-61, Northwest Greenland Exped, 63 & 65; field worker, NMex, Wash, Ore, New Eng, Alaska, Antarctica, Arctic & Patagonia. *Honors & Awards:* Bellingshausen Mem Medal, Acad Sci Soviet Union, 75. *Mem:* AAAS (secy, Geol & Geog Sect, 55-56); fel Geol Soc Am; fel Am Geog Soc; Nat Asn Geol Teachers (vpres, 59); fel Royal Geog Soc. *Res:* Geomorphology; vulcanology; high-alumina clay. *Mailing Add:* 10901 Johnson Blvd No 604 Seminole FL 33542

NICHOLS, ROBERT LORING, b Newton, NJ, Mar 3, 46; m 76; c 5. CROP PROTECTION. *Educ:* Yale Univ, BA, 68; Univ Conn, MS, 77, PhD(agron), 80. *Prof Exp:* Sr lang analyst, US Army Security Agency, 68-72; tech dir polymer sci, Tech Rubber, Inc, 72-73; teaching fel soil sci, Univ Conn, Storrs, 74-76; instr, Univ Conn, Torrington, 76; res assoc, Univ Conn, Storrs, 77-80; res agronomist Forage Weed Control, Agr Res Serv, Univ Ga, USDA, 80-84; sr biochem specialist, PPG Industr, 84-87, CO-CRD RPD WEST US MAAG 88, RDD MANALEN US MAAG 89-90. *Mem:* Am Soc Agron; Am Soc Hort Sci; Weed Sci Soc Am. *Res:* Crop protection. *Mailing Add:* MAAG Agr Chem 5690 58th Ave PO Box 6430 Vero Beach FL 32961-6430

NICHOLS, ROBERT TED, b Lewis, Iowa, Dec 30, 25; m 55; c 3. PHYSICS. *Educ:* Iowa State Univ, BS, 50, MS, 55, PhD(physics), 60. *Prof Exp:* Asst beta & gamma ray spectros, Ames Lab, AEC, 50-51; instr physics & math, Bethel Col, Minn, 54-56; asst prof physics, Gustavus Adolphus Col, 60-63; mem tech staff & physicist, Hughes Res Labs, Calif, 63-66; assoc prof physics, 66-74, chmn dept, 74-78, PROF PHYSICS, CALIF LUTHERAN COL, 74- *Concurrent Pos:* Consult mil prod group, Minneapolis-Honeywell Regulator Co, 62-63. *Mem:* Am Asn Physics Teachers; Sigma Xi. *Res:* Radiation physics; linear energy transfer for dosimetry; small angle beta scattering; beta and gamma ray spectroscopy; computer generation of random number distributions. *Mailing Add:* 3317 S Kings Ave Springfield MO 65807-5087

NICHOLS, ROGER LOYD, microbiology; deceased, see previous edition for last biography

NICHOLS, ROY ELWYN, b Leonardsville, NY, July 10, 09; m 32; c 2. VETERINARY PHYSIOLOGY. *Educ:* Univ Toronto, DVM, 33, DVSc, 43; Ohio State Univ, MSc, 34, PhD(vet hemat), 41. *Prof Exp:* Asst vet surg, Col Vet Med, Ohio State Univ, 34-35, instr, 35-41; asst prof col agr & assoc, Agr Exp Sta, Purdue Univ, 41-42 & 45-47; dean col vet med, State Col Wash, 47-50; lectr & res assoc, 50-51, prof, 51-72, EMER PROF VET SCI, UNIV WIS-MADISON, 72- *Honors & Awards:* Borden Award, Am Vet Med Asn. *Mem:* Am Soc Animal Sci; Am Vet Med Asn; Am Dairy Sci Asn; Conf Res Workers Animal Dis. *Res:* Veterinary hematology, physiology and surgery; instrument for photographic recording of erythrocyte sedimentation; ruminology. *Mailing Add:* 6209 Mineral Point Rd Madison WI 53705

NICHOLS, RUDOLPH HENRY, b Bellaire, Mich, Dec 6, 11; m 39; c 1. ACOUSTICS. *Educ:* Hope Col, AB, 32; Univ Mich, AM, 33, PhD(physics), 39. *Prof Exp:* Lab asst physics, Univ Mich, 35-36, asst acoust & physics, 36-38, res assoc, 39; physicist, Owens-Ill Glass Co, 40-41; spec res assoc, Cruft Lab, Harvard Univ, 41-44, from asst dir to assoc dir, Electro-Acoust Lab, 44-46; mem tech staff, Bell Labs, 46-76; consult acoust, Off Naval Res, 76-84; AT PHYSICS DEPT, NAVAL POSTGRAD SCH, 84- *Mem:* Fel Acoust Soc Am; Sigma Xi. *Res:* Audio communications; sound reduction in vehicles; vibration isolation; acoustical materials; hearing and hearing aids; underwater acoustics. *Mailing Add:* Physics Dept Code 61/NO Naval Postgrad Sch Monterey CA 93943

NICHOLS, WARREN WESLEY, b Collingswood, NJ, May 16, 29; m 53; c 3. CYTOGENETICS, PEDIATRICS. *Educ:* Rutgers Univ, BS, 50; Jefferson Med Col, MD, 54; Univ Lund, PhD(med cytogenetics), 64, DrPhil, 66. *Prof Exp:* Intern, Cooper Hosp, Camden, NJ, 54-55; fel, Children's Hosp Philadelphia, 55-56; chief pediat, Lake Charles AFB, 57-59; assoc, Sch Med & instr, Grad Sch Med, Univ Pa, 59-67, from asst prof to assoc prof, Sch Med, 64-73; asst dir, Inst Med Res, 63-81, vpres res, 81-83, S Emlen Stokes Prof Genetics, 75-84; prof human genetics & pediat, Sch Med, Univ Pa, 73-88. *Concurrent Pos:* Chief med staff, Camden Munic Hosp Contagious Dis, 59-61; assoc mem, S Jersey Med Res Found, 59-65; asst physician, Children's Hosp Philadelphia, 59-65, assoc physician & assoc hematologist, 65-73, sr physician, Div Metab & Genetics, 73-81, univ assoc, Div Human Genetics & Teratol, 81-; mem pediat staff, Cooper Hosp, Camden, NJ, 59-65, consult staff, 65-; consult staff, Our Lady of Lourdes Hosp, Camden, 59-66, assoc pediat, 66-83; NIH res career develop award, 63-72; assoc prof, Univ Lund, 66-; mem, Human Cytogenetics Study Group, 66-; consult secy comn pesticides & their relationship to environ health, Dept Health, Educ & Welfare, 69; mem human embryol & develop study sect, NIH, 73-78; mem panel, US-Japan Coop Med Sci Prog, 72-74; co-chmn USA-USSR Prog on Mammalian Somatic Cell Genetics Related to Neoplasia, 76-80; clin prof pediat, NJ Col Med & Dent, 79-; mem bd sci counsrs, Div Cancer Cause & Prev, Nat Cancer Inst, 79-82. *Mem:* AAAS; Environ Mutagen Soc (secy, 73-76); AMA; Am Soc Clin Invest; Genetics Soc Am. *Res:* Spontaneous and induced gene and chromosome mutations and their role in carcinogenesis, aging and hereditary disease; emphasis is on virus induced cellular genetic changes and high risk cancer individuals and families. *Mailing Add:* Genetic & Cell Toxicol Merck Sharp & Dohme Res Labs West Point PA 19486

NICHOLS, WILLIAM HERBERT, b Cleveland, Ohio, Mar 15, 28. STATISTICAL MECHANICS, FIBER OPTICS. *Educ:* West Baden Col, AB, 50; Mass Inst Technol, SB, 55, PhD(physics), 58; Weston Col, STL, 61. *Prof Exp:* Instr physics & algebra, Loyola Acad, 52-53; res asst physics, Inst Theoret Physics, Univ Vienna, 62-63; asst prof, Univ Detroit, 63-67; from asst prof to assoc prof, 67-74, PROF PHYSICS, JOHN CARROLL UNIV, 74- *Concurrent Pos:* Pres, Ohio Sect, Am Asn Physics Teachers, 84-85. *Mem:* Am Phys Soc; Inst Theol Encounter Sci & Technol; Am Asn Physics Teachers; Sigma Xi. *Res:* Theory of optical fibers as sound sensors. *Mailing Add:* Dept of Physics John Carroll Univ Cleveland OH 44118

NICHOLS, WILLIAM KENNETH, b Seattle, Wash, Sept 25, 43; m 73. PHARMACOLOGY. *Educ:* Univ Wash, BS, 66; Univ Minn, PhD(phamacol), 71. *Prof Exp:* asst prof, 71-78, ASSOC PROF PHARMACOL, DEPT BIOCHEM PHARMACOL & TOXICOL, COL PHARM & COL MED, UNIV UTAH, 78- *Concurrent Pos:* NIH career develop award, Nat Inst Arthritis, Metabol & Digestive Dis, 78-83. *Mem:* Am Asn Col Pharmacol; Am Soc Pharmacol Exp Therapeut. *Res:* Biochemical pharmacology; endocrinology; diabetes; immunology; leukocyte metabolism and function; cyclic nucleotides; allergy; hormonal and cyclic nucleotide modulation of cellular functions; alterations in cellular immune responses and vascular tissue in the diabetic state. *Mailing Add:* Dept Biochem Pharmacol & Toxicol Col Pharm Univ Utah Salt Lake City UT 84112

NICHOLS, WILMER WAYNE, b Booneville, Miss, Aug 12, 34; m 62; c 2. CARDIOVASCULAR PHYSIOLOGY. *Educ:* Delta State Univ, BS, 60; Univ Southern Miss, MS, 66; Univ Ala, PhD(physiol, biophys), 70. *Prof Exp:* Instr physiol, Med Sch, Univ Ala, 70-71; asst prof med physics, Inst Med Physics, Holland, 71-72; asst prof physiol, Sch Med, Johns Hopkins Univ, 72-74; PROF MED, COL MED, UNIV FLA, 74- *Concurrent Pos:* Consult, Millar Instruments Inc, Tex, 74-; mem, Cardiovasc Catheter Standards Subcomt, 74- *Mem:* Am Fedn Clin Res; Am Physiol Soc; Cardiol Soc Holland; Am Heart Asn; Asn Advan Med Instrumentation; Am Col Cardiol. *Res:* Pulsatile hemodynamics in man; coronary blood flow; myocardial ischemia; neutrophil function; endothelial function. *Mailing Add:* Dept Med Div Cardiol Univ Fla Med Col J Hillis Miller Health Ctr Gainsville FL 32610

NICHOLSON, ARNOLD EUGENE, b Jasper, Ind, Apr 9, 30. PHARMACEUTICAL CHEMISTRY. *Educ:* Purdue Univ, BS, 52, MS, 54, PhD(pharmaceut chem), 56. *Prof Exp:* Sr res chemist, Smith Kline & French Labs, 56-64; asst to tech dir pharmaceut develop, Stuart Div, Atlas Chem Industs, Inc, 64-72; PROJ ADMINR REGIONAL AFFAIRS, HYLAND LABS, DIV BAXTER LABS & MEM STAFF, TRAVENOL LAB DIV, BAXTER TRAVENOL LABS, INC, 72- *Mem:* Am Chem Soc; Am Pharmaceut Asn; Am Soc Hosp Pharmacists; Int Asn Biol Standardization. *Res:* Pharmaceutical research and development. *Mailing Add:* Baxter Healthcare Hyland Div 550 N Brand Blvd Glendale CA 91203-1900

NICHOLSON, BRUCE LEE, b Baltimore, Md, Feb 13, 43; m 65; c 1. MICROBIOLOGY. *Educ:* Univ Md, College Park, BS, 65, PhD(microbiol), 69. *Prof Exp:* Asst prof microbiol, 69-74, assoc prof, 74-79, PROF MICROBIOL & ZOOL & CHMN, DEPT MICROBIOL, UNIV MAINE, 79- *Mem:* Am Soc Microbiol; AAAS; Wildlife Dis Asn; Tissue Culture Asn; Am Fisheries Soc. *Res:* Biochemical and biophysical characteristics of fish viruses and their interactions with susceptible cells; serological methods for detection and identification of fish viruses. *Mailing Add:* Dept Biochem, Microbiol & Molecular Biol Univ Maine Orono ME 04469

NICHOLSON, D ALLAN, b Waterloo, Iowa, June 22, 39; m 62; c 2. ORGANOMETALLIC CHEMISTRY. *Educ:* Cornell Col, BA, 60; Northwestern Univ, PhD(inorg chem), 65. *Prof Exp:* Staff chemist, 65-70, sect head anal chem, 70-72, sect head toxicol, 72-74, assoc dir soap & toilet goods technol div, 74-79, ASSOC DIR PACKAGED SOAP & DETERGENTS DIV, PROCTER & GAMBLE CO, 79- *Mem:* Am Chem Soc. *Res:* Non-transition metals such as germanium, silicon; organophosphorus chemistry; analytical chemistry. *Mailing Add:* 472 Lakeridge Ave Cincinnati OH 45231

NICHOLSON, DANIEL ELBERT, b Waco, Tex, Sept 1, 26; m 51. PHYSICAL CHEMISTRY. *Educ:* Baylor Univ, BS, 46; Univ Tex, MA, 48, PhD(phys chem), 50. *Prof Exp:* Tutor, Univ Tex, 46-48; chemist, Oak Ridge Nat Lab, 50-51; from res chemist to sr res chemist, Humble Oil & Ref Co, 52-61; staff mem, Los Alamos Sci Lab, 62; res assoc, Richfield Oil Corp, 63-67; sect head optical spectros, Micro Data Opers, 67-68, sect mgr, Electro-Optical Systs, Inc, 68-76; tech coord, Witco Chem Co, 76-78; MEM STAFF, ENG DIV-ENFORCEMENT, SOUTH COAST AIR QUAL MGT DIST, 78- *Res:* Kinetics of hydrocarbon decomposition; precision calorimetry; infrared spectroscopy of absorbed species; electrical and magnetic properties of catalysts; Raman spectroscopy; thermodynamic properties; catalytic studies of hydrode-sulfurization; hydrogenation of aromatic hydrocarbons. *Mailing Add:* 2209 E California Blvd San Marino CA 91108

NICHOLSON, DAVID WILLIAM, mechanics, for more information see previous edition

NICHOLSON, DONALD PAUL, b Pershing, Iowa, Jan 3, 30; m 54; c 2. MICROBIOLOGY, VIROLOGY. *Educ:* Iowa State Col, BS, 51; Univ SDak, BSM, 59; Univ Iowa, MD, 61. *Prof Exp:* Asst radiation res, Univ Iowa, 56-57, clin intern med, 61-62, USPHS fel, 62-65, from instr to asst prof microbiol, 62-68, asst dir clin lab serv, 68-71, asst prof path & supvr clin microbiol, 71-76, investr res, develop, eval for clin microbiol, 76-81, ASST PROF PATH, UNIV IOWA, 81-, DIR, CLIN VIROL LAB, 84- *Mem:* AAAS; Am Soc Microbiol; Sigma Xi. *Res:* Areas related to clinical virology. *Mailing Add:* Dept Path Univ Iowa Rm 254B MRC Iowa City IA 52242

NICHOLSON, DOUGLAS GILLISON, inorganic chemistry; deceased, see previous edition for last biography

NICHOLSON, DWIGHT ROY, b Racine, Wis, Oct 3, 47; m 69. PLASMA PHYSICS. *Educ:* Univ Wis-Madison, BS, 69; Univ Calif, Berkeley, PhD(plasma physics), 75. *Prof Exp:* Res assoc plasma physics, Univ Colo, Boulder, 75-77, asst prof Astro-Geophys Dept, 77-78; from asst to assoc prof, 78-86, PROF PHYSICS, DEPT PHYSICS & ASTRO, UNIV IOWA, IOWA CITY, 86-, CHMN DEPT, 85- *Concurrent Pos:* Univ fac scholar, Univ Iowa, Iowa City, 83-85. *Mem:* AAAS; Union Radio Sci Int; fel Am Phys Soc; Am Geophys Union; Int Neural Network Soc. *Res:* Plasma theory; nonlinear waves in plasma; plasma turbulence; chaos; neural networks. *Mailing Add:* Dept Physics & Astron Univ Iowa Iowa City IA 52242-1410

NICHOLSON, EUGENE HAINES, b St Louis, Mo, Nov 10, 07. MATHEMATICS, PHYSICS. *Educ:* Wash Univ, BS, 31, MS, 37, PhD(math, physics), 41. *Prof Exp:* Elec engr, Union Elec Co, 41-60; CONSULT, 60- *Mem:* AAAS; Am Math Soc. *Res:* Applied mathematics; mathematical physics; energy conversion; systems engineering; nuclear power. *Mailing Add:* 5232 Lansdowne Ave St Louis MO 63109

NICHOLSON, HOWARD WHITE, JR, b Brooklyn, NY, Dec 18, 44; m 71. HIGH ENERGY PHYSICS. *Educ:* Hamilton Col, BA, 66; Mass Inst Technol, BA, 66; Calif Inst Technol, PhD(physics), 71. *Prof Exp:* From asst prof to assoc prof, 71-80, PROF PHYSICS, MT HOLYOKE COL, 80- *Concurrent Pos:* Res Corp grants, Mt Holyoke Col, 72, 75; Dept Energy Contract, 78- *Mem:* Am Phys Soc. *Res:* Experimental high energy particle physics. *Mailing Add:* Dept of Physics Shattuck Hall Mt Holyoke Col South Hadley MA 01075

NICHOLSON, ISADORE, organic chemistry, for more information see previous edition

NICHOLSON, J(OHN) CHARLES (GODFREY), b Great Malvern, Eng, Feb 8, 42. NEUROBIOLOGY. *Educ:* Univ Birmingham, BSc, 63; Univ Keele, PhD(commun), 68. *Prof Exp:* Sci officer math physics, UK Atomic Energy Authority, 63-65; vis investr neurobiol, Inst Biomed Res, Am Med Asn-Educ & Res Found, Ill, 67-69, asst mem, 69-70; from asst prof to assoc prof, Univ Iowa, 70-76; PROF PHYSIOL & BIOPHYS, MED CTR, NY UNIV, 76- *Mem:* NY Acad Sci; Soc Neurosci; Inst Elec & Electronics Engrs. *Res:* Physiology and biophysics of the brain cell microenvironment with emphasis on the diffusion and modulation of ions and neuroactive molecules in the cerebellum. *Mailing Add:* Dept Physiol & Biophys NY Univ Med Ctr 550 First Ave New York NY 10016

NICHOLSON, J W G, b Crapaud, PEI, Jan 19, 31; m 55; c 4. ANIMAL NUTRITION. *Educ:* McGill Univ, BS, 51; Cornell Univ, MS, 56, PhD(animal nutrit), 59. *Prof Exp:* RES SCIENTIST ANIMAL NUTRIT, CAN DEPT AGR, 51- *Honors & Awards:* Can Packers Medal, Can Soc Animal Sci. *Mem:* Can Soc Animal Sci; Agr Inst Can; Am Soc Animal Sci; Am Dairy Sci Asn. *Res:* Nutrition of ruminant animals. *Mailing Add:* Res Sta Agr Can PO Box 20280 Fredericton NB E3B 4Z7 Can

NICHOLSON, LARRY MICHAEL, b Nevada, Mo, Nov 22, 41; m 64; c 2. BIOCHEMISTRY. *Educ:* Kans State Univ, BS, 63, PhD(biochem), 68. *Prof Exp:* Asst prof chem, Univ Mo-Rolla, 67-75; from asst prof to assoc prof, 76-84, PROF CHEM, FT HAYS STATE UNIV, 84- *Mem:* Am Chem Soc; Sigma Xi. *Res:* Enzyme chemistry; peroxidase enzymes; biological halogenation; biosynthesis of thyroxine. *Mailing Add:* 1315 E 17th St Hays KS 67601-4099

NICHOLSON, MARGIE MAY, b San Antonio, Tex, June 10, 25; m 51. PHYSICAL CHEMISTRY. *Educ:* Univ Tex, BS, 46, MA, 48, PhD(phys chem), 50. *Prof Exp:* Phys chemist, US Naval Ord Lab, 51-52; res chemist, Humble Oil & Refining Co, 52-56, sr res chemist, 56-62; electrochemist, Stanford Res Inst, 62-64; res specialist, Rocketdyne Div, NAm Aviation, Inc, 64-65, sr tech specialist, 65, mem tech staff, Atomic Int Div, NAm Rockwell Corp, Canoga Park, 65-74, mem tech staff, Electronics Res Ctr, 74-77, mgr org electron devices, 77-79, mem tech staff, Autonetics Strategic Systs Div, 80-83, mem tech staff, 83-87, MEM TECH STAFF, THOUSAND OAKS SCI CTR, ROCKWELL INT, ANAHEIM, 87- *Mem:* Am Chem Soc; Electrochem Soc; Sigma Xi. *Res:* Electrochemistry; theory and applications of voltammetric techniques; electrochemical power sources; electrode kinetics; nonaqueous electrochemistry; semiconductor electrodes; electrochromic displays; organic solid state electrochemistry. *Mailing Add:* 2209 California Blvd San Marino CA 91108

NICHOLSON, MORRIS E(MMONS), JR, b Indianapolis, Ind, Feb 15, 16; m 43; c 3. PHYSICAL METALLURGY, CORROSION. *Educ:* Mass Inst Technol, SB, 39, ScD, 47. *Prof Exp:* Res asst, Mass Inst Technol, 39-41 & 46-47; res metallurgist, Standard Oil Co, Ind, 47-48, sect head, Phys Metall, 48-50; asst prof, Inst Study of Metals, Univ Chicago, 50-55; dept head, Univ Minn, Minneapolis, 56-62, prof metall, 56-85, dir cont educ eng & sci, 73-85; RETIRED. *Honors & Awards:* Distinguished Serv Award, Am Soc Eng Educ, 81. *Mem:* Am Soc Metals; Metall Soc; Corrosion Soc; Am Soc Eng Educ. *Res:* Ferrous metals and alloys; plastic flow and fracture; x-ray metallography; transformations in metals; textures and recrystallization mechanisms in low carbon steels; corrosion mechanisms in phosphate solutions; pitting corrosion. *Mailing Add:* 1776 N Pascal Ave Falcon Heights MN 55113

NICHOLSON, NICHOLAS, b New York, NY, June 9, 38; m 64. EXPERIMENTAL NUCLEAR PHYSICS. *Educ:* Polytech Inst Brooklyn, BS, 60; WVa Univ, MS, 62, PhD(physics), 65. *Prof Exp:* Half-time instr physics, WVa Univ, 63-64, asst, 64-65; physicist, Div Res, USAEC, Washington, DC, 65-67; PHYSICIST, LOS ALAMOS SCI LAB, 67- *Res:* Nuclear spectroscopy, alpha, beta, and gamma; angular correlations; x-ray physics; x-ray spectroscopy; nuclear safeguards; radiography. *Mailing Add:* 415 Rover Blvd Los Alamos NM 87544

NICHOLSON, PATRICK STEPHEN, b London, Eng, Oct 25, 36; Can citizen; m 65; c 3. SOLID STATE ELECTROLYSIS, ULTRASONIC NDE CERAMICS. *Educ:* Univ Leeds, Eng, BSc, 63; Univ Calif, MA, 65, PhD(ceramic eng), 67. *Prof Exp:* From asst to assoc prof, 69-78, PROF, CERAMIC ENG, MCMASTER UNIV, 78- *Concurrent Pos:* Vis asst prof, Univ Utah, 67-69; vis res scientist, refractories, Carborundum, US, 76-77; prin investr, Dept Nat Defense, Can, 80-; consult, Kenyon & Kenyon Law Firm, 84-, Domglacs, 82-; prin investr, Ont Hydro, 86-; pres ed bd Can Univ-Indust Coun Adv Ceramics, 86- *Mem:* Fel Am Ceramic Soc; fel Can Ceramic Soc; fel Can Univ-Indust Coun Adv Ceramics; Int Inst Sintering. *Res:* Non-destructive evaluation of model defects in ceramics; static fatigue of bioactive ceramics gas/solid reactions, ceramics for energy storage systems-development of new ceramic membranes for medium temperature steam electrolysis/fuel cells; improved strength of ceramics by flaw-free processing; improved ceramics for superconductors; ceramic/ceramic and ceramic/metal composites. *Mailing Add:* Dept Mat Sci & Eng McMaster Univ 1280 Main St Hamilton ON L8S 4L7 Can

NICHOLSON, RALPH LESTER, b Lynn, Mass, Aug 25, 42; m 74. PHYTOPATHOLOGY. *Educ:* Univ Vt, BA, 64; Univ Maine, MS, 67; Purdue Univ, PhD(plant path), 72. *Prof Exp:* from asst prof to assoc prof, 72-82, PROF PLANT PATH, PURDUE UNIV WEST LAFAYETTE, 82- *Concurrent Pos:* Purdue rep, NCent Region, Corn & Sorghum Dis Comt, USDA-Coop State Res Serv, 72- *Mem:* Am Phytopath Soc; Sigma Xi; Can Phytopath Soc; Am Chem Soc. *Res:* Study of host biochemical response to infection with emphasis on stress compounds; recognition of the pathogen by the host, and histopathology related to disease physiology and time of host response; phenolic compound metabolism and synthesis; integrated pest management. *Mailing Add:* Dept of Bot & Plant Path Purdue Univ West Lafayette IN 47907-1155

NICHOLSON, RICHARD BENJAMIN, b Tacoma, Wash, Sept 8, 28; m 52; c 4. NUCLEAR PHYSICS, COMPUTER SOFTWARE. *Educ:* Univ Puget Sound, BS, 50; Cornell Univ, MS, 60; Univ Mich, PhD(nuclear sci), 63. *Prof Exp:* Physicist mat lab, Puget Sound Naval Shipyard, 51-53; physicist, Atomic Power Develop Assocs, 54-58 & 59-61; assoc prof nuclear eng, Univ Wis, 63-66; assoc physicist, Argonne Nat Lab, 66-69, sr physicist, 69-72; prof nuclear eng, Ohio State Univ, 72-74; mgr licensing uranium enrichment, Exxon Nuclear Co Inc, 74-79; physicist, Laser Enrichment, 79-81; CONSULT PHYSICS, 81- *Concurrent Pos:* Consult, Atomic Power Develop Assocs, 61-; physicist, Lawrence Radiation Lab, Univ Calif, 65; consult adv comt reactor safeguards, AEC; vis scientist, Joint Res Ctr, EEC, 89-91. *Mem:* Fel Am Nuclear Soc; Am Phys Soc. *Res:* Physics and safety of fast nuclear reactors, especially theoretical problems in Doppler effect and accident analysis; theory of high temperature plasmas; uranium laser enrichment; computer simulation of nuclear reactor accidents. *Mailing Add:* Rte 1 Box 1439 Lopez Island WA 98261

NICHOLSON, RICHARD SELINDH, b Des Moines, Iowa, Apr 5, 38; m 58; c 2. CHEMISTRY. *Educ:* Iowa State Univ, BS, 60; Univ Wis-Madison, PhD(chem), 64. *Prof Exp:* Res assoc chem, Iowa State Univ, 59-60; NSF fel, Univ Wis, 60-61, res assoc, 63-64; from asst prof to assoc prof, Mich State Univ, 64-71; prog dir, 71-75, actg head Chem Synthesis & Analysis Sect & dept dir, Chem Div, 75-76; spec asst to dir, NSF, 76-77, dept asst dir math & phys sci, 80-85, asst dir math & phys sci, 85-89, DIR, CHEM DIV, NSF, 77-; EXEC OFFICER, AAAS, 89- *Concurrent Pos:* NSF fel, 64-71; consult, US Army Electronic Command, 67; guest worker, NIH, 71-; exec secy, President's Comt on Nat Medal of Sci, 77-, consult, Off Sci & Technol Policy, Exec Off Pres, 78-81; publ, J Science, 89- *Honors & Awards:* Eastman Kodak Award, Univ Wis, 64. *Mem:* Fel AAAS; Am Chem Soc; Brit Polarographic Soc; Int Electrochem Soc. *Res:* Electrochemistry; electrode kinetics; applications of computers in instrumentation; mass spectrometry, especially chemical ionization. *Mailing Add:* Am Asn Advan Sci 1333 H St NW Washington DC 20005

NICHOLSON, THOMAS DOMINIC, astronomy, navigation; deceased, see previous edition for last biography

NICHOLSON, VICTOR ALVIN, b Stafford, Kans, Dec 16, 41; m 70; c 4. MATHEMATICS. *Educ:* Okla State Univ, BS, 62, MS, 64; Univ Iowa, PhD(math), 68. *Prof Exp:* Asst prof math, Park Col, 64-65; asst prof, 68-76, ASSOC PROF MATH, KENT STATE UNIV, 76- *Concurrent Pos:* Software develop consult, 80-81. *Mem:* Am Math Soc; Math Asn Am; Sigma Xi. *Res:* Geometric topology; topology of manifolds. *Mailing Add:* Dept Math Kent State Univ Kent OH 44242

NICHOLSON, W(ILLIAM) J(OSEPH), b Tacoma, Wash, Aug 24, 38; m 64; c 4. CHEMICAL ENGINEERING. *Educ:* Mass Inst Technol, SB, 60, SM, 61; Cornell Univ, PhD(chem eng), 65; Pac Lutheran Univ, MBA, 69. *Prof Exp:* Sr develop engr, Hooker Chem Corp, 64-69 & Battelle Northwest, Battelle Mem Inst, 69-70; planning analyst, Potlatch Forests, Inc, corp enery coordr, 75-81, MGR, CORP ENERGY SERV, POTLATCH CORP, 81- *Concurrent Pos:* Chmn, Elec Comt, Am Paper Inst, 81- *Mem:* Tech Asn Pulp & Paper Indust; Am Chem Soc; Am Inst Chem Engrs; Sigma Xi; AAAS. *Res:* Corporate planning; application of planning to systematic business management; chemical process development; energy development, supply, use, and economics. *Mailing Add:* Potlatch Corp Suite 610 244 Calif St San Francisco CA 94111

NICHOLSON, WAYNE LOWELL, b Canton, NY, Mar 26, 58. SPORE PHOTOCHEMISTRY, MICROBIAL DEVELOPMENT. *Educ:* State Univ NY, Potsdam, BS(biol) & BS(chem), 80; Univ Wis- Madison, PhD(genetics), 87. *Prof Exp:* ASST PROF MICROBIOL, TEX COL OSTEOP MED, 90- *Concurrent Pos:* Adj asst prof, Biochem & Molecular Biol Dept, Tex Col Osteop Med & Univ NTex, 90- *Mem:* Am Soc Microbiol; AAAS. *Res:* Endospore formation in Bacillus subtilis; ultraviolet radiation resistance of bacterial spores. *Mailing Add:* Dept Microbiol & Immunol Tex Col Osteop Med 3500 Camp Bowie Blvd Ft Worth TX 76107

NICHOLSON, WESLEY LATHROP, b Andover, Mass, Jan 31, 29; m 52, 83; c 4. STATISTICS. *Educ:* Univ Ore, BA, 50, MA, 52; Univ Ill, PhD(math statist), 55. *Prof Exp:* Asst math, Univ Ore, 50-52 & Univ Ill, 52-55; instr, Princeton Univ, 55-56; statistician, Gen Elec Co, 56-64; sr res assoc, Pac Northwest Lab, 65-71, SR STAFF SCIENTIST, BATTELLE-NORTHWEST, BATTELLE MEM INST, 71- *Concurrent Pos:* Lectr, Univ Wash, Richland Campus, 56-71, chmn math prog, 71-79; adj prof math, Wash State Univ, 66- *Mem:* Fel Am Statist Asn; Inst Math Statist. *Res:* Derivation and application of statistical methodology to physical and engineering sciences; construction of stochastic models; analysis of large data sets; statistical graphics. *Mailing Add:* Battelle Pac Northwest Lab PO Box 999 Richland WA 99352

NICHOLSON, WILLIAM JAMIESON, b Seattle, Wash, Nov 21, 30; m 57; c 4. PHYSICS, ENVIRONMENTAL HEALTH. *Educ:* Mass Inst Technol, BS, 52; Univ Wash, PhD(physics), 60. *Prof Exp:* Instr, Univ Wash, 60; physicist, Watson Res Lab, Int Bus Mach Corp, NY, 60-68; asst prof, 69-73, ASSOC PROF COMMUNITY MED, MT SINAI SCH MED, 73- *Concurrent Pos:* Adj assoc prof, Fordham Univ, 64. *Mem:* AAAS; Am Phys Soc; NY Acad Sci. *Res:* Occupational and environmental health; analysis and effect of airborne micro-particulates. *Mailing Add:* Mt Sinai Sch Med One Gustave L Levy Pl Box 1057 New York NY 10029

NICHOLSON, WILLIAM ROBERT, b Camden, NJ, Jan 25, 25; m 50; c 4. FISHERIES. *Educ:* Rutgers Univ, BSc, 50; Univ Maine, MSc, 53. *Prof Exp:* Proj leader water fowl biol, Md Game & Inland Fish Comn, 51-55; fishery res biologist, Ctr Estuarine & Fisheries Res, Nat Marine Fisheries Serv, 56-84; RETIRED. *Mem:* Am Inst Biol Sci; Am Fisheries Soc; Am Inst Fishery Res Biologists. *Res:* Population dynamics of marine fishes. *Mailing Add:* Rte 3 Box 177 Beaufort NC 28516

NICHOLSON-GUTHRIE, CATHERINE SHIRLEY, b Jackson, Miss; m 61; c 1. GENETICS. *Educ:* Auburn Univ, BS, 57; Fla State Univ, MS, 60; Ind Univ, Bloomington, PhD(genetics), 72. *Prof Exp:* Instr physiol, Fla State Univ, 60; res asst molecular biol, Calif Inst Technol, 60-62; instr biol, Boston State Col, 63-64; NIH trainee, 67-71; vis asst prof biol, Univ Evansville, 72-73; mem adj fac, 74-76, MEM FAC & ADJ RES SCIENTIST, SCH MED, IND UNIV, INDIANAPOLIS, 76- *Concurrent Pos:* Consult, Mead Johnson & Co, 76; Sarah Berliner fel, Am Asn Univ Women, 78-79; mass media sci fel, AAAS, 79; prof staff mem comt sci & tech, US House Rep, 81; consult, comt environ & pub works, US Senate, 81. *Honors & Awards:* First Place Res Award, Ala Acad Sci, 56. *Mem:* Sigma Xi; AAAS; Genetics Soc Am. *Res:* Inheritance of chlorophyll and lamellae; GABA receptors; relationship between membrane structure and function. *Mailing Add:* 700 Drexel Dr Evansville IN 47712

NICKANDER, RODNEY CARL, b Aitkin, Minn, July 5, 38; m 57; c 2. PHARMACOLOGY. *Educ:* SDak State Univ, BS, 60; Purdue Univ, MS, 63, PhD(pharmacol), 64. *Prof Exp:* Pharmacol res assoc, 64-77, head, Dept Immunol & Connective Tissue Res, 77-80, DIR IMMUNOL, CONNECTIVE TISSUE & PULMONARY RES, LILLY RES LABS, ELI LILLY & CO, 80- *Res:* Detection and pharmacological evaluation of new anti-inflammatory agents and analgesics; pharmacological role of metabolites. *Mailing Add:* 6960 Olive Indianapolis IN 46285

NICKEL, ERNEST HENRY, mineralogy, for more information see previous edition

NICKEL, GEORGE H(ERMAN), b Brawley, Calif, Mar 10, 37; m 63; c 3. PHYSICS & CHAOS, NON LINEAR DYNAMICS & PULSE POWER. *Educ:* San Diego State Univ, AB, 58; Univ Ill, MS, 60; Univ Calif, Davis, PhD(appl sci & eng), 66. *Prof Exp:* Proj officer physics, Air Force Weapons Lab, 61-70; mil staff mem, Los Alamos Sci Lab, 70-72; from adj asst prof to assoc prof, 77-81, ADJ ASSOC PROF PHYSICS, AIR FORCE INST TECHNOL, WRIGHT-PATTERSON AFB, 81-; STAFF MEM, LOS ALAMOS NAT LAB, LOS ALAMOS, NMEX, 81- *Concurrent Pos:* Adj instr, Univ Albuquerque, 67-69; adj prof, Brevard Community Col, 73-77 & Fla Inst Technol, 73-77. *Res:* Plamsa physics; thermal physics; statistical mechanics. *Mailing Add:* Los Alamos National Lab MS-B259 Los Alamos NM 87545

NICKEL, JAMES ALVIN, b Grants Pass, Ore, Sept 27, 25; m 52; c 3. APPLIED MATHEMATICS. *Educ:* Willamette Univ, BA, 49; Ore State Col, MS, 51, PhD(appl math, anal), 57. *Prof Exp:* Asst math, Ore State Col, 49-50; asst appl math, Ind Univ, 51-53, asst math, 53; instr, Willamette Univ, 53-57, asst prof, 57-59; from assoc prof to prof & chmn dept, Oklahoma City Univ, 59-67; sr res mathematician, Dikewood Corp, 67-69, prin res mathematician, 69-70; mgr anal sect, Technol Inc, 70-71; math physicist, Lockheed Electronics Co, 71-72; PROF MATH, UNIV TEX PERMIAN BASIN, 72-, CHMN, DEPT MATH & COMPUT SCI, 79- *Concurrent Pos:* Statistician, State Hwy Dept, Ore, 56-59; consult, Systs Res Ctr, Okla. *Mem:* Am Math Soc; Math Asn Am; Am Statist Asn; Sigma Xi. *Res:* Systems design and simulation; applied mathematics and statistics; citizenship education. *Mailing Add:* 3942 Monclair Ave Odessa TX 79762

NICKEL, PHILLIP ARNOLD, b Deadwood, SDak, Oct 10, 37; m 59; c 2. PARASITOLOGY, ORNITHOLOGY. *Educ:* Ore State Univ, BS, 62; Kans State Univ, MS, 66, PhD(entom), 69. *Prof Exp:* Asst prof, Calif Lutheran Col, 69-74, prof biol sci, 74-86; VECTOR CONTROL BIOLOGIST, VENTURA CO, 88- *Concurrent Pos:* Dir, Med Technol Prog, Calif State Univ, Northridge, mem, Health Adv Comt & assoc prof, Dept Biol, 80-86. *Mem:* Am Inst Biol Sci; Nat Asn Adv Health Professions; Am Soc Parasitologist; AAAS. *Res:* Mites associated with insects; helminths of bats; medical technology; invertebrate zoology; vector biology. *Mailing Add:* Environ Health Dept 800 S Victoria St Ventura CA 93009

NICKEL, VERNON L, b Sask, Can, May 1, 18; US citizen; m 41; c 3. ORTHOPEDIC SURGERY. *Educ:* Loma Linda Univ, MD, 44; Univ Tenn, MSc, 49. *Prof Exp:* Fel orthop surg, Campbell Clin, Memphis, Tenn, 48-49; head orthopedist & chief surg serv, Rancho Los Amigos Hosp, Downey, Calif, 53-64, med dir, 64-69; dir dept orthop surg & rehab, 69-75, prof orthop surg & rehab, Loma Linda Univ, 69-78; adj prof orthop & med, George Washington Univ & dir, Rehab Eng Res & Develop Serv, Vets Admin, 78-80; adj prof surg & rehab, Loma Linda Univ, 69-80; MED DIR, SHARP REHAB CTR, 80-; PROF SURG, ORTHOP & REHAB, UNIV CALIF MED CTR, SAN DIEGO, 80-; AT DEPT OF PED SURG, LOMA LINDA UNIV. *Concurrent Pos:* Pvt pract, 50-78; Fulbright lectr, Cairo Univ, 61; clin prof, Univ Southern Calif, 65- & Univ Calif, Irvine-Calif Col Med, 66-78. *Mem:* Am Orthop Asn; Am Acad Orthop Surg; Am Soc Surg Hand; hon mem British Orthop Asn. *Mailing Add:* 1614 Caminito Blvd La Jolla CA 92037

NICKELL, CECIL D, b Rochester, Ind, Jan 9, 41; m 62; c 2. PLANT GENETICS, SOYBEAN GENETICS. *Educ:* Purdue Univ, BS, 63; Mich State Univ, MS, 65, PhD(biomet), 67. *Prof Exp:* Asst instr plant genetics, Mich State Univ, 63-67; from asst prof to assoc prof Plant breeding & genetics, Kans State Univ, 67-74; PROF PLANT BREEDING, UNIV ILL, URBANA, 79- *Mem:* Crop Sci Soc Am; Am Soc Agron; Sigma Xi. *Res:* Genetic studies concerned with changes in plant populations under stresses; application of selection procedures for the improvement of soybeans. *Mailing Add:* Dept Agron Univ Ill 1102 S Goodwin Urbana IL 61801

NICKELL, LOUIS G, b Little Rock, Ark, July 10, 21; m 42; c 3. PLANT PHYSIOLOGY. *Educ:* Yale Univ, BS, 42, MS, 47, PhD(physiol), 49. *Prof Exp:* Res assoc plant physiol, Brooklyn Bot Garden, 49-51; plant physiologist, Pfizer Co, 51-53, head phytochem lab, 53-61; head dept physiol & biochem, Exp Sta, Hawaiian Sugar Planters Asn, 61-65, asst dir, 65-75; vpres, res div, W R Grace & Co, 75-78; vpres, Res & Develop, Velsicol Chem Corp, 78-85; PRES, NICKELL RES INC, 85- *Concurrent Pos:* Responsible scientist, Cell Nutrit Exhibit, Worlds Fair, Brussels, 58; trustee, Hawaiian Bot Gardens Found, 62-65; chmn Hawaiian sect, Am Chem Soc, 67-68, comt Int Activ, 85-; mem, Gov Adv Comt Sci & Technol, Hawaii, 64-70, chmn, 70-75; mem, State Task Force Energy Policy, Hawaii, 73-75, vchmn, 74-75; bd trustees, Am Soc Plant Physiol, 84-87; mem, Bd Sci & Technol Int Develop, Nat Res Coun, 85-; mem bd dirs, Crop Genetics Int Corp, 81- *Honors & Awards:* Belknap Prize for Excellence Biol Studies, 42. *Mem:* Am Soc Plant Physiol (treas, 75-81); Bot Soc Am; Am Chem Soc; Plant Growth Regulator Soc Am (vpres, 80, pres, 81); Agron Soc; Crop Sci Soc Am. *Res:* Sugarcane physiology and biochemistry; tissue and cell culture; medicinal and economic botany; antibiotics; plant growth substances; pesticides; research administration; microbiology. *Mailing Add:* Nickell Res Inc 24 Sierra Dr Hot Springs Village AR 71909

NICKELL, WILLIAM EVERETT, b Hazel Green, Ky, July 29, 16; m 42. PHYSICS. *Educ:* Berea Col, BA, 40; Univ Iowa, MS, 43, PhD(physics), 54. *Prof Exp:* Researcher, Univ Iowa, 43-44, res assoc, 44-45, instr physics, 47-51, res assoc, 51-52; from asst prof to prof, SDak State Univ, 53-63; from assoc prof to prof physics & astron Southern Ill Univ, 63-86; RETIRED. *Mem:* Sigma Xi; Am Phys Soc; Am Asn Physics Teachers. *Res:* Nuclear physics; radio proximity fuze; missile vibrations. *Mailing Add:* RR 2 Box 386 Murphysboro IL 62966

NICKELSEN, RICHARD PETER, b Lynbrook, NY, Oct 1, 25; m 50; c 3. STRUCTURAL GEOLOGY. *Educ:* Dartmouth Col, BA, 49; Johns Hopkins Univ, MA, 51, PhD, 53. *Prof Exp:* Asst, Johns Hopkins Univ, 51-53; asst prof geol, Pa State Univ, 53-59; assoc prof, 59-63, chmn dept geol & geog, 59-76, PROF GEOL, BUCKNELL UNIV, 64- *Concurrent Pos:* NATO fel, Norway, 65-66, NSF res grant, 68-69; consult, Amoco Prod Co, 79-82. *Mem:* AAAS; Geol Soc Am; Am Geophys Union. *Res:* Genesis of joints and rock cleavage; Appalachian tectonics; regional joint patterns, regional strain variation and stages of deformation; Caledonide stratigraphy and tectonics (Norway); ambient environmental parameters of deforming rocks; pressure and temperature of deformation from fluid inclusion studies in syntectonic veins; Alleghany orogeny trend sequence; Appalachian foreland fault and cleavage duplexes cut by out- of-sequence faults; Southwest Utah deformation sequence. *Mailing Add:* 432 Pheasant Ridge Rd Lewisburg PA 17837

NICKELSON, ROBERT L(ELAND), b Livingston, Mont, Sept 13, 27; m 52; c 5. CHEMICAL ENGINEERING. *Educ:* Mont State Col, BS, 51, MS, 52; Univ Minn, PhD(chem eng), 57. *Prof Exp:* Asst, Univ Minn, 52-53; from asst prof to assoc prof chem eng, 56-64, PROF CHEM ENG, MONT STATE UNIV, 64- *Mem:* Am Soc Eng Educ; Am Inst Chem Engrs. *Res:* Reverse osmosis; mass transfer in fluid beds; reaction kinetics; application of mathematics to chemical engineering. *Mailing Add:* Dept Chem Eng Mont State Univ Bozeman MT 59715

NICKERSON, BRUCE GREENWOOD, b 1950; m 71; c 3. SUDDEN INFANT DEATH SYNDROME. *Educ:* Univ Calif, Los Angeles, MD, 76. *Prof Exp:* DIR, PULMONARY FUNCTION LAB, CHILDREN'S HOSP NOTHERN CALIF, 81- *Mem:* Am Physiol Soc; Am Thoracic Soc; Am Col Chest Physicians. *Res:* Infant pulmonary function test. *Mailing Add:* Dept Pulmonary Med Children's Hosp Northern Calif 51st & Grove Sts Oakland CA 94609

NICKERSON, HELEN KELSALL, mathematics; deceased, see previous edition for last biography

NICKERSON, JOHN CHARLES, III, physics, for more information see previous edition

NICKERSON, JOHN DAVID, b Halifax, NS, Feb 12, 27; m 52; c 2. PHYSICAL CHEMISTRY. *Educ:* Mt Allison Univ, BSc, 48; Dalhousie Univ, MSc, 50; Univ Toronto, PhD(chem), 54. *Prof Exp:* Chemist fatty acid hydrogenation, Fisheries Res Bd Can, 48-50; demonstr, Univ Toronto, 51-53; develop chemist carbon & graphite, Nat Carbon Co, 53-57; prin res chemist, Agr Prod, Int Minerals & Chem Corp, Fla, 57-62; mem staff, Southern Nitrogen Co, Ga, 62-64; mgr chem res, Armour Agr Chem Co, 64-68; dir res & develop, 68-75, DIR DEVELOP & TECH SERV, AGRI CHEM DIV, USS, 75- *Mem:* Am Chem Soc; Chem Inst Can. *Res:* Fatty acid hydrogenation; dielectric constants of low boiling liquids; oxidation of carbon and graphite; alkali resistance of carbon and graphite; fertilizer chemistry and animal feed supplements. *Mailing Add:* Stone Container Corp Tech Eng 2150 Park Lane Suite 400 Atlanta GA 30345

NICKERSON, JOHN LESTER, physiology; deceased, see previous edition for last biography

NICKERSON, JOHN MUNRO, GENETICS. *Educ:* Univ Tex, PhD(human genetics), 80. *Prof Exp:* SR STAFF FEL, NAT EYE INST, NIH, 80- *Mailing Add:* 18405 Shady View Lane Brookville MD 20833-2842

NICKERSON, KENNETH WARWICK, b Attleboro, Mass, Nov 19, 42; m 72; c 2. MICROBIAL BIOCHEMISTRY. *Educ:* Rutgers Univ, BS, 63; Univ Cincinnati, PhD(chem), 69. *Prof Exp:* USPHS fel biophys, Ore State Univ, 69-70, genetics, Univ Wis, 70-71; fel microbial insecticides, Northern Regional Res Lab, Sci & Educ Admin-Agr Res Serv, USDA, Nat Acad Sci/Nat Res Coun, 71-73; fel plant path, 73-75, assoc prof, 75-85, PROF BIOL SCI, UNIV NEBR-LINCOLN, 85- *Concurrent Pos:* NIH res career develop award, Univ Nebr-Lincoln, 79- *Mem:* Am Soc Microbiol; Entom Soc Am; Am Soc Biol Chemists; Soc Gen Microbiol; Soc Invert Pathol; Mycological Soc Am. *Res:* Bacillus thuringiensis; microbial insecticides; fungal dimorphism; calcium-calmodulin; bacterial detergent resistance. *Mailing Add:* Dept of Biol Sci Univ of Nebr Lincoln NE 68583

NICKERSON, MARK, b Montevideo, Minn, Oct 22, 16; Can citizen; m 42; c 3. PHARMACOLOGY, THERAPEUTICS. *Educ:* Linfield Col, AB, 39; Brown Univ, ScM, 41; Johns Hopkins Univ, PhD(embryol), 44; Univ Utah, MD, 50. *Hon Degrees:* DSc, Med Col Wis, 74. *Prof Exp:* Res biochemist, Nat Defense Res Comn, Johns Hopkins Univ, 43-44; instr pharmacol, Col Med, Univ Utah, 44-47, from asst prof to assoc prof, 47-51; assoc prof, Univ Mich, 51-54; prof pharmacol & med res, Fac Med, Univ Man, 54-57, prof pharmacol & therapeut & chmn dept, 57-67; chmn dept, 67-75, prof, 67-82, EMER PROF PHARMACOL & THERAPEUT, MCGILL UNIV, 82- *Concurrent Pos:* Norman Bethune prof, Peoples Repub China, 75. *Honors & Awards:* Abel Award, 49; Upjohn Award, 78. *Mem:* Am Soc Pharmacol & Exp Therapeut (pres); Soc Exp Biol & Med; Pharmacol Soc Can (pres); Can Physiol Soc; fel Royal Soc Can; fel German Acad Natural Res; Int Union Pharmacol (treas, 66-72); Am Physiol Soc; Can Soc Clin Invest. *Res:* Drugs blocking sympathetic nervous system; cardiovascular and autonomic nervous system physiology and pharmacology; shock; clinical pharmacology. *Mailing Add:* Dept Pharmacol & Therapeut McGill Univ 3655 Drummond Montreal PQ H3G 1Y6 Can

NICKERSON, NORTON HART, b Quincy, Mass, Apr 14, 26; m 54; c 3. PLANT MORPHOLOGY, PLANT ECOLOGY. *Educ:* Univ Mass, BS, 49; Univ Tex, MA, 51; Wash Univ, PhD(bot), 53. *Prof Exp:* Instr bot, Univ Mass, 53-56 & Cornell Univ, 56-58; from asst prof to assoc prof, Wash Univ, 58-63; assoc prof bot, 63-81, PROF ENVIRON STUDIES, TUFTS UNIV, 81-, DIR ENVIRON STUDIES PROG, 83- *Concurrent Pos:* Res fel, Calif Inst Technol, 54 & 55; NSF sci fac fel, 58; morphologist, Mo Bot Garden, 58-63; mem Mass Bd Environ Mgt, 77-80; mem, Mass Agr Lands Preserv Comt, 78- & chmn, Mass Hazardous Waste Facil Site Safety Coun, 80-85; dir, Phippen-Lacroix Herbarium, Tufts Univ, 85-; ed, Rhodora, 81- *Mem:* AAAS; Bot Soc Am; Ecol Soc Am; Sigma Xi. *Res:* ethnobotany; conservation; ecology of wetlands and coasts; mangroves. *Mailing Add:* Dept Biol Tufts Univ Medford MA 02155

NICKERSON, PETER AYERS, b Hyannis, Mass, Feb 19, 41. PATHOLOGY, ENDOCRINOLOGY. *Educ:* Brown Univ, AB, 63; Clark Univ, MA, 65, PhD(biol), 68. *Prof Exp:* Res instr, 67-69, from res asst prof to assoc prof, 69-80, PROF PATH, STATE UNIV NY BUFFALO, 80- *Mem:* Am Soc Cell Biol; Endocrine Soc; Am Soc Exp Path; Electron Micros Soc Am; Am Soc Zool; Am Asn Anatomists. *Res:* Structure and function of the adrenal cortex; adrenal ultrastructure; hypertension and the adrenal cortex; gerbil adrenal cortex; oxygen toxicity and lung; hypertension and role of calcium in vascular smooth muscle; lung cell biology; pulmonary vasculature. *Mailing Add:* Dept Path State Univ NY Buffalo NY 14214

NICKERSON, RICHARD G, b Harwich, Mass, Nov 20, 27; m 57; c 3. LATEX VINYL POLYMERS, PVC RESINS. *Educ:* Univ Mass, Amherst, BS, 50; Northwestern Univ, PhD(org chem & biochem), 55; Boston Univ, MBA, 83. *Prof Exp:* Res chemist, Cellophane Tech Sect, DuPont, Richmond, Va, 54-55 & W R Grace, Dewey & Almy, Cambridge, Mass, 57-60; postdoctoral, Polytech Inst Brooklyn, 55-57; vpres res & develop, Electronautics Corp, Maynard, Mass, 60-61, pres, 61-63; proj leader, Borden Chem, Leom, Mass, 63-65, group leader, 65-67, develop mgr, 67-81, lab mgr, 81-87; PRES & MANAGING DIR, BOSTON PROF INT, INC, 87- *Mem:* Am Chem Soc; Soc Plastics Engrs. *Res:* Correlation of polymeric criteria to manufacturing variables and application performance for all markets for emulsion latex polymers; latex vinyl polymers based on vinyl acetate, acrylics, styrene, butadiene, vinyl chloride, ethylene for all markets; PVC resins for all markets including injection molding, extrusion, calenderizing, transfer molding; awarded seven US patents and one German patent. *Mailing Add:* 16 Valleywood Rd Hopkinton MA 01748-1635

NICKERSON, ROBERT FLETCHER, b Stoneham, Mass, Mar 25, 30; m 60; c 1. COMPUTER SCIENCES, TECHNICAL MANAGEMENT. *Educ:* Tufts Univ, BS, 52, MS, 53; Univ Calif, PhD(chem), 58. *Prof Exp:* Asst chem, Tufts Univ, 52-53; asst, Univ Calif, 53-58, chemist, Lawrence Livermore Lab, 58-71; comput analyst, Calif Inst Technol, 72-81; mgr, Sierra Geophysics Inc, 81-84, dir systs develop, 84-88; pvt consult, 88-89; CONSULT, WINERY ADMIN, 89- *Res:* Small computer applications. *Mailing Add:* 6479 137th Ave N E Apt 374 Redmond WA 98052

NICKERSON, STEPHEN CLARK, b Melros, Mass, Mar 14, 50; m 78; c 3. MAMMARY GLAND HISTOLOGY, MASTITIS. *Educ:* Univ Maine, BS, 72; Va Polytech Inst & State Univ, MS, 76 & PhD(dairy sci), 80. *Prof Exp:* Fel res assoc, dept animal sci, Purdue Univ, 79-80; fel res scientist, Animal Sci Inst, US Dept Agr, 80-81; asst prof, La State Univ Agr Ctr, 81-84, assoc prof & lab dir, 84-90; PROF & LAB DIR, MASTITIS RES LAB, 91- *Concurrent Pos:* Prin investr, US Dept Agr, 83-; grad fac, La State Univ, 83-; consult, Holstein Asn, 86-, ed bd mem, J Dairy Sci, 87-90, consult, ELI LILLY, 87- *Mem:* Am Dairy Sci Asn; Nat Mastitis Coun; Soc Experimental Biol & Med. *Res:* Elicidating local intramammary immunities; prevention of mastitis; vaccination; immunostimulation; biotechnological means; writer of 60 referred journals and articles. *Mailing Add:* Mastitis Res Lab Hill Farm Res Sta Rt One Box Ten Homer LA 71040

NICKLAS, ROBERT BRUCE, b Lakewood, Ohio, May 29, 32; m 60; c 1. CELL BIOLOGY. *Educ:* Bowling Green State Univ, BA, 54; Columbia Univ, MA, 56, PhD(zool), 58. *Prof Exp:* From instr to asst prof zool, Yale Univ, 58-64, fel sci, 63-64, assoc prof zool, 64-65; assoc prof, 65-71, PROF ZOOL, DUKE UNIV, 71- *Concurrent Pos:* Fel J S Guggenheim Found, Max Planck Inst, Tübingen, Germany, 72-73; mem Am Cancer Soc Adv Comt Virol & Cell Biol, 75-78; mem exec subcomt, Adv Comt Physiol, Cellular & Molecular Biol, NSF, 79. *Mem:* Fel AAAS; Am Soc Cell Biol; Soc Gen Physiologists; Genetics Soc Am; Am Soc Naturalists. *Res:* Cell biology; chromosome movement in mitosis; evolution of chromosome cycles; chromosome movement studied by micromanipulation of living cells using computer-enhanced video microscopy and correlated electron microscopy. *Mailing Add:* Dept Zool Duke Univ Durham NC 27706

NICKLE, DAVID ALLAN, b Portland, Ore, May 18, 44; m 74; c 3. SYSTEMATIC ENTOMOLOGY. *Educ:* Temple Univ, BA, 70; Univ Fla, MS, 73, PhD(entom), 76. *Prof Exp:* Res scientist insect attractants, behav & basic biol, Res Lab, 76-79, RES ENTOMOLOGIST, SYST ENTOM LAB, USDA, 79- *Concurrent Pos:* Res scientist stored prod entom, Univ Fla, 76-79. *Honors & Awards:* Bailey Award, Am Peanut Res & Educ Asn, 79. *Mem:* Soc Syst Zool (treas, 80-85); Entom Soc Am; Sigma Xi; Orthopterists' Soc. *Res:* Systematics and acoustic behavior of Tettigoniidae; systematics of Orthoptera and Isoptera. *Mailing Add:* NHB 168 Syst Entom Lab USDA c/o Nat Mus Natural Hist Smithsonian Inst Washington DC 20560

NICKLE, WILLIAM R, b Bridgeport, Conn, July 20, 35; m 64. NEMATOLOGY. *Educ:* State Univ NY Col Forestry, Syracuse, BS, 56; Univ Idaho, MS, 58; Univ Calif, PhD(nematol), 63. *Prof Exp:* Res officer entomophilic nematodes, Res Inst, Can Dept Agr, 62-65; NEMATOLOGIST, NEMATOL INVEST, CROPS RES DIV, AGR RES SERV, USDA, 65- *Mem:* Soc Nematol; Soc Syst Zool; Entom Soc Can. *Res:* Taxonomy, morphology and biology of plant parasitic, insect parasitic and mycophagus nematodes. *Mailing Add:* Plant Nematol Inst 106 Biosci Bldg Beltsville MD 20705

NICKLES, ROBERT JEROME, b Madison, Wis, Mar 22, 40; m 63; c 1. NUCLEAR PHYSICS, MEDICAL PHYSICS. *Educ:* Univ Wis, BS, 62, PhD(nuclear physics), 68; Univ Sao Paulo, MS, 67. *Prof Exp:* Res assoc nuclear physics, Sch Med, Univ Wis-Madison, 68-69; James A Picker Found res fel, Niels Bohr Inst, Copenhagen Univ, 69-71; James A Picker Found res fel med physics, 71-73, asst prof radiol, 73-77, ASSOC PROF RADIOL, SCH MED, UNIV WIS-MADISON, 77- *Mem:* Am Phys Soc. *Res:* Study of heavy ion transfer reactions; short-lived isotope production utilizing the helium-jet technique; development of an intense neutron source for cancer therapy. *Mailing Add:* Radiol 1530a Med Sci Ctr Univ of Wis Sch of Med Madison WI 53706

NICKLOW, ROBERT MERLE, b St Petersburg, Fla, Oct 11, 36; m 58; c 3. SOLID STATE PHYSICS. *Educ:* Ga Inst Technol, BS, 58, MS, 60, PhD(x-ray diffraction), 64. *Prof Exp:* Asst res physicist, Eng Exp Sta, Ga Inst Technol, 63; PHYSICIST, OAK RIDGE NAT LAB, 63- *Honors & Awards:* Sidhu Award, 68. *Mem:* Fel Am Phys Soc; Am Crystallog Asn. *Res:* Crystal physics; lattice dynamics; neutron and x-ray diffraction; study of lattice dynamics and spin waves by means of coherent inelastic neutron scattering. *Mailing Add:* Solid State Div Bldg 3025 MS 6031 Oak Ridge Nat Lab PO Box 2008 Oak Ridge TN 37831

NICKOL, BRENT BONNER, b Agosta, Ohio, June 22, 40; m 64; c 2. PARASITOLOGY. *Educ:* Col Wooster, BA, 62; La State Univ, MS, 63, PhD(zool), 66. *Prof Exp:* From asst prof to assoc prof, 66-75, PROF ZOOL, UNIV NEBR, LINCOLN, 75- *Concurrent Pos:* Dir, Cedar pt Biol Sta, 75-79; vis cur asst, Houston Zool Gardens, 81; assoc ed, Invertebrates, Midland Naturalist, 85-; ed, Trans of Neb Acad Sci, 80-85, J Parasitol, 89- *Mem:* Wildlife Dis Asn; Am Soc Parasitol; Coun Biol Ed. *Res:* Epizootiology of the Acanthocephala. *Mailing Add:* Sch Biolog Sci Univ Nebr Lincoln NE 68588-0118

NICKOLLS, JOHN RICHARD, b Easton, Pa, Mar 6, 50; m 83; c 1. MASSIVELY PARALLEL COMPUTER SYSTEMS, SIGNAL PROCESSING. *Educ:* Univ Ill, BS, 72; Stanford Univ, MS, 74, PhD(elec eng), 77. *Prof Exp:* Mem res staff image processing, Ampex Corp, 77-81; dir software eng, Picture Element Ltd, 81-84; dir eng, Silicon Solutions, Zycad Corp, 84-88; VPRES, HARDWARE ENG MASSIVELY PARALLEL COMPUTER CORP, 88- *Mem:* Asn Comput Mach; Inst Elec & Electronics Engrs. *Res:* Massively parallel computer architecture and design; digital signal processing; image processing and compression; language and compiler design. *Mailing Add:* 390 Cherry Ave Los Altos CA 94022

NICKOLS, G ALLEN, b Springfield, Ill, Apr 9, 51; m 71; c 4. PHARMACOLOGY. *Educ:* Univ Mo, BA, 73, Sch Med, PhD(pharmacol), 77. *Prof Exp:* Fel, pharmacol, Univ Va Sch Med, 77-80; asst prof pharmacol, Southern Il Univ Sch Med, 80-86; assoc prof pharmacol, Univ Tex Med Br, 86-91; SR RES SPECIALIST, MONSANTO CORP RES, 91- *Concurrent Pos:* Fac develop fel, Pharmaceut Mfrs Asn Fedn. *Mem:* Sigma Xi; Am Soc Cell Biol; Am Fedn Clin Res; Tissue Cult Asn; Fedn Am Socs Exp Biol; Endocrine Soc; Am Soc Pharmacol & Exp Therapeut; Am Soc Bone & Mineral Res. *Res:* Regulation of cellular cyclic adenosine monophosphate metabolism, endothelial and vascular smooth muscle pharmacology, calcium metabolism and parathyroid hormone-vitamin D interrelationships; cardiovascular pharmacology. *Mailing Add:* Monsanto Res Corp 800 N Lindbergh Blvd Bldg T3P St Louis MO 63167

NICKOLS, NORRIS ALLAN, b Ellensburg, Wash, July 8, 28; m 58; c 2. NUCLEAR PHYSICS. *Educ:* Cent Wash Col Educ, BA, 52; Univ Calif, Berkeley, PhD(physics), 60. *Prof Exp:* Physicist, Lawrence Radiation Lab, Univ Calif, 59-61; res scientist, Lockheed Calif Co, 61-63; res specialist, Space & Info Systs Div, NAm Aviation, Inc, 63-67; STAFF MEM, LOS ALAMOS NAT LAB, 68- *Mem:* Am Phys Soc; Sigma Xi. *Res:* High energy physics strange particles; high energy muon scattering; hyperfragments; nuclear weapons. *Mailing Add:* 784 45th St Los Alamos NM 87544

NICKON, ALEX, b Poland, Oct 6, 27; nat US; m 50; c 3. ORGANIC CHEMISTRY. *Educ:* Univ Alta, BSc, 49; Harvard Univ, MA, 51, PhD(chem), 53. *Prof Exp:* Vis lectr org chem, Bryn Mawr Col, 53; Nat Res Coun Can fel, Birkbeck Col, London, 53-54 & Ottawa, Ont, 54-55; from asst prof to prof, 55-75, VERNON K KRIEBLE PROF CHEM, JOHNS HOPKINS UNIV, 75- *Concurrent Pos:* NSF sr fel, Imp Col, London, 63-64 & Univ Munich, 71-72; sr ed, J Org Chem, 65-71; Am exec ed, Tetrahedron Reports, 78- *Mem:* Am Chem Soc; The Chem Soc; Sigma Xi. *Res:* Carbanions and carbocations; stereochemistry; reaction mechanisms; carbene chemistry; biologically important reactions. *Mailing Add:* 1009 Painters Lane Cockeysville MD 21030

NICKS, ORAN WESLEY, b Eldorado, Tex, Feb 2, 25; m 55; c 4. RESEARCH ADMINISTRATION, AERONAUTICS. *Educ:* Univ Okla, BS, 48. *Prof Exp:* Aeronaut engr, NAm Aviation, Inc, Calif, 48-58; proj engr, Chance-Vought Aircraft, Inc, Tex, 58-60; head lunar flight systs, NASA, Washington, DC, 60-61, dir, Lunar & Planetary Progs, 61-67, dep assoc adminr space sci & appln, 67-70, actg assoc adminr advan res & technol, 70, dep dir, Langley Res Ctr, Hampton, VA, 70-80; RES ENGR, TEX A&M UNIV, 80-, DIR, SPACE RES CTR, 85- *Concurrent Pos:* Dir, Soaring Soc Am, 83-89, chmn, Tech Bd, 83-89; chmn, Ctr Space Power, 88- & Tex Space Grant Consortium, 89- *Mem:* Fel Am Astronaut Soc; assoc fel Am Inst Aeronaut & Astronaut. *Res:* Aerodynamic research on total energy sensors for aircraft; applications of laminar flow technologies to improve wing efficiencies; automated spacecraft for scientific exploration of planets; space station research and technology, commercial applications. *Mailing Add:* Dir Space Res Ctr Tex A&M Univ College Station TX 77843-3118

NICKUM, JOHN GERALD, b Rochester, Minn, Aug 7, 35; m 85; c 4. EDUCATION, CONSERVATION. *Educ:* Mankato State Col, BSc, 57; Univ SDak, MA, 61; Univ Southern Ill, PhD(zool), 66. *Prof Exp:* Teacher high sch, Minn, 57-59, jr high sch, 59-60; asst prof biol, Western Ky State Col, 65-66; from asst prof wildlife mgt to assoc prof wildlife & fisheries, 66-73; asst leader, NY Coop Fishery Res Unit, 73-77, leader, 77-80, asst prof natural resources, Cornell Univ, 73-80, dir aquaculture prog, 74-80; leader, Iowa Coop Fishery Res Unit & assoc prof animal ecol, Iowa State Univ, 80-85; fishery biologist, Fish Hatcheries, 85-90, NAT AQUACULTURE COORDR, US FISH & WILDLIFE SERV, WASHINGTON, DC, 90- *Concurrent Pos:* Pres, Educ Sect, Am Fisheries Soc; tech ed, Aquatic Resources Educ Curriculum; pres-elect, Fish Culture Sect, Am Fisheries Soc; Predoctoral Fel, NSF, 63-65. *Mem:* Am Fisheries Soc; World Aquaculture Soc. *Res:* Aquaculture; application of aquaculture in the solution of recreational and environmental problems; warm-water fish management; pond management; urban recreational fisheries; use of wastewater in aquaculture; culture of cool-water fishes; research administration and program management, fish health management; aquaculture policy. *Mailing Add:* US Fish & Wildlife Serv 1849 C St NW ARLSQ 820 Washington DC 20240

NICO, WILLIAM RAYMOND, b Aurora, Ill, Mar 23, 40; m 67; c 3. MATHEMATICS. *Educ:* Loyola Univ, Ill, BS, 62; Univ Calif, Berkeley, MA, 64, PhD(math), 66. *Prof Exp:* Res assoc eng mech, Stanford Univ, 66-67; from asst prof to prof math, Tulane Univ, La, 67-84; PROF MATH & COMPUTER SCI, CALIF STATE UNIV, HAYWARD, 83- *Mem:* Am Math Soc; Math Asn Am; Inst Elec & Electronic Engrs Computer Soc; Asn Comput Mach. *Res:* Computational complexity; secure communication protocals; programming languages; automata; semigroups; transformation monoids, extension theory of monoids and categories; homological algebra. *Mailing Add:* Dept Math & Comput Sci Calif State Univ Hayward CA 94542

NICODEMUS, DAVID BOWMAN, b Kobe, Japan, July 1, 16; US citizen; wid. PHYSICS. *Educ:* DePauw Univ, AB, 37; Stanford Univ, PhD(physics), 46. *Prof Exp:* Asst, Stanford Univ, 37-41, asst physicist, Off Sci Res & Develop proj, 42-43; physicist, Los Alamos Sci Lab, Calif, 43-46; instr physics, Stanford Univ, 46-49, actg asst prof, 49-50; from asst prof to prof physics, Ore State Univ, 50-86, asst dean sch sci, 62-65, actg dean, 65-66, dean, 66-86, EMER PROF PHYSICS & EMER DEAN, ORE STATE UNIV, 86- *Concurrent Pos:* Consult, Los Alamos Sci Lab, 56-57. *Res:* X-rays; nuclear physics. *Mailing Add:* 3525 S W Williamette Corvallis OR 97333

NICODEMUS, FRED(ERICK) E(DWIN), b Osaka, Japan, July 25, 11; US citizen; m 35. OPTICAL PHYSICS. *Educ:* Reed Col, AB, 34. *Prof Exp:* Radio engr & physicist, Air Force Cambridge Res Labs, 46-55; advan develop engr & eng specialist, Sylvania Electronic Defense Labs, Gen Tel & Electronics Corp, Calif, 55-69; physicist, Michelson Lab, US Naval Weapons Ctr, 69-74; physicist, Nat Bur Standards, 74-81; Nat Bur Standards prin investr, Catholic Univ Am, 81-85; CONSULT, GTE GOVT SYSTS CORP, 85- *Concurrent Pos:* Mem, Nat Acad Sci-Nat Acad Eng-Nat Res Coun adv panel to heat div, Nat Bur Standards & liaison to ad hoc panel on radiometry & photom, 70-74; consult, Int Tech Comt Photom & Radiometry, Int Comn Illum, 72- *Mem:* Fel Optical Soc Am; Sigma Xi; Int Comn Illum; fel Soc Photo-Optical Instrumentation Engrs. *Res:* Clarification of basic radiometric relations, definitions, and nomenclature. *Mailing Add:* 720 Brentwood Pl Los Altos CA 94024-5425

NICOL, CHARLES ALBERT, b Ft Worth, Tex, Apr 24, 25; m 56; c 1. MATHEMATICS. *Educ:* Univ Tex, PhD(math), 54. *Prof Exp:* Instr math, Univ Tex, 54-55; asst prof, Ill Inst Technol, 55-59 & Univ Okla, 59-60; asst head dept math & comp sci, 73-76, ASSOC PROF MATH, UNIV SC, 60- *Mem:* AAAS; Am Math Soc; Math Asn Am; Sigma Xi. *Res:* Number theory; algebra and combinatorial problems. *Mailing Add:* Dept of Math Univ of SC Columbia SC 29208

NICOL, DAVID, b Ottawa, Ont, Aug 16, 15; nat US; m 47; c 1. PALEONTOLOGY. *Educ:* Tex Christian Univ, BA, 37, MS, 39; Stanford Univ, MA, 43, PhD(paleont), 47. *Prof Exp:* Asst prof geol, Univ Houston, 47-48; assoc cur Mesozoic & Cenozoic inverts, US Nat Mus, 48-58; assoc prof geol, Southern Ill Univ, 58-64; from assoc prof to prof, Univ Fla, 65-85, emer prof geol, 82; RETIRED. *Mem:* Paleont Soc. *Res:* Pelecypods. *Mailing Add:* PO Box 14376 University Sta Gainesville FL 32604

NICOL, JAMES, b Dundee, Scotland, Aug 24, 21; nat US; m 48. ENGINEERING PHYSICS, LOW TEMPERATURE PHYSICS. *Educ:* St Andrews Univ, BSc, 46 & 48; Union Col, MS, 50; Ohio State Univ, PhD(physics), 52. *Prof Exp:* Instr physics, Ohio State Univ, 52-53; from asst prof to assoc prof, Amherst Col, 53-57; physicist, Arthur D Little, Inc, 57-62; dir res & vpres, Cryonetics Corp, 62-64; vpres, 77-86, PHYSICIST, ARTHUR D LITTLE, INC, 64- *Mem:* Am Phys Soc. *Res:* Phenomena below one degree absolute; thermal conductivity and electron tunneling in superconductors; electric power transmission; geomagnetism; underwater acoustic transmission. *Mailing Add:* HC 33 Box 171 Bald Head Rd Arrowsic ME 04530

NICOL, MALCOLM FOERTNER, b New York, NY, Sept 13, 39; c 3. CHEMICAL PHYSICS. *Educ:* Amherst Col, BA, 60; Univ Calif, Berkeley, PhD(chem), 63. *Prof Exp:* Res asst, 63-64, from actg asst prof to assoc prof, 65-75, PROF CHEM, UNIV CALIF, LOS ANGELES, 75- *Concurrent Pos:* Sloan Found fel, 73-77; guest prof, Univ G H Paderborn, WGer, 79; assoc ed, J Phys Chem, 80-90, sr ed, 91- *Honors & Awards:* Herbert Newby McCoy Award, 84. *Mem:* Am Chem Soc; Am Phys Soc; AAAS; Sigma Xi; Am Geophys Union. *Res:* Spectroscopy, structure and bonding in solids and macromolecules under extreme conditions of pressures; temperature and time. *Mailing Add:* Dept Chem & Biochem Univ of Calif Los Angeles CA 90024-1569

NICOL, SUSAN ELIZABETH, b New York, NY, Apr 30, 41. NEUROCHEMISTRY, HUMAN GENETICS. *Educ:* Mt Holyoke Col, BA, 62; Columbia Univ, MA, 66; Univ Minn, PhD(behav genetics), 72. *Prof Exp:* Asst res scientist med genetics, NY State Psychiat Inst, 64-67; fel pharmacol, Univ Minn, 72-74; Nat Inst Arthritis, Metab & Digestive Dis fel endocrinol extramural prog, 74-75; res fel pharmacol, Univ Minn, 75-76; asst prof psychiat, Univ Minn, St Paul-Ramset Med Ctr, 76-80; INTERN, CLIN PSYCHOL, HENNEPIN COUNTY MED CTR, IN, 81- *Concurrent Pos:* Prin investr NIMH grant, 77-80. *Res:* Biological psychiatry; human biochemical genetics; hormone action; cyclic nucleotides; clinical psychology. *Mailing Add:* Hennepin County Med Ctr 710 Park Ave Minneapolis MN 55415

NICOLAE, GEORGE G, b Focsani, Romania, Oct 2, 43; m 77; c 2. SYNTHETIC CHEMISTRY, ORGANIC CHEMISTRY. *Educ:* Univ Bucharest, MS, 66, PhD, 74. *Prof Exp:* Asst prof, Univ Bucharest, 67-73 & Polytech Inst Bucharest, 73-79; res chemist, Wallace A Erickson, Chicago, 80; RES DIR, CONFI-DENTAL CO, CHICAGO, 81- *Mem:* Int Asn Dental Res; Am Asn Dental Res; Am Chem Soc. *Res:* Heterocyclic chemistry; phenoxathiins derivatives; octahydro-xanthylium compounds; octahydroacridinium compounds; pyrylium salts; acrylic monomers; urethane derivatives. *Mailing Add:* 1920 N Clybourn Chicago IL 60614

NICOLAENKO, BASIL, b Paris, France, Mar 23, 42; US citizen. MATHEMATICS. *Educ:* Univ Paris, Lic es sci, 65; Univ Mich, PhD(math), 68. *Prof Exp:* Staff assoc appl sci, Brookhaven Nat Lab, 68-69; prof math, Courant Inst Math Sci, NY Univ, 69-74; staff scientist, 74-81, PROF MATH, LOS ALAMOS NAT LAB, 81- *Concurrent Pos:* Prof math, Univ Paris-Orsay, France, 77-78. *Mem:* Am Math Soc; Soc Math Francaise; Soc Indust & Appl Math. *Res:* Nonlinear functional analysis and applications. *Mailing Add:* Los Alamos Natl Lab UCD Group T-7 Math Anal Mail Stop B-284 PO Box 1663 Los Alamos NM 87545

NICOLAI, VAN OLIN, b Barrington, Ill, Jan 18, 24; m 55. OPTICAL PHYSICS, LASERS. *Educ:* Univ Ill, BS, 49, MS, 51, PhD(physics), 54. *Prof Exp:* Res assoc physics, Univ Ill, 54-55; asst prof, Univ NDak, 55-57; asst prof, Southern Ill Univ, 57-59; tech adv physics, Off Naval Res, 60-86; RETIRED. *Mem:* Am Phys Soc. *Res:* Optical properties of solids, laser applications and surface physics. *Mailing Add:* 206 Apple Blossom Ct Vienna VA 22181

NICOLAIDES, ERNEST D, organic chemistry, for more information see previous edition

NICOLAIDES, JOHN DUDLEY, b Washington, DC, Feb 13, 23; m 45; c 1. AERODYNAMICS. *Educ:* Lehigh Univ, BA, 46; Johns Hopkins Univ, MSE, 53; Cath Univ Am, PhD, 62. *Prof Exp:* Chief aerodynamicist, Proj Dragonfly, Gen Elec Co, 46-48; aeronaut res scientist, Ballistic Res Lab, US Army Ord Dept, 48-53; chief exterior ballistician, US Naval Bur Ord, 53-56, asst aerodyn, hydrodyn & ballistics, 56-58, sci adv astronaut, 58-59, tech dir naval astronaut, US Bur Naval Weapons, 59-61; dir prog rev & resources mgt, Off Space Sci & Applns, NASA, 61-62, spec asst to assoc adminr space sci & applns, 62-64; prof aerospace eng & chmn dept, Univ Notre Dame, 64-74; head, Dept Aeronaut Eng, 75-80, prof aeronaut & mech eng, Calif Polytech State Univ, 80-83; PRES, AERO, 69- *Concurrent Pos:* Lectr, Univ Md, 58-64 & Cath Univ Am, 59-64; aerospace consult, Govt & Indust, 64- *Mem:* Sigma Xi; assoc fel Am Inst Aeronaut & Astronaut. *Res:* Aerodynamics; space sciences; ballistics; hydrodynamics. *Mailing Add:* 2048 Skylark Lane San Luis Obispo CA 93401

NICOLAIDES, R A, b London Eng, May 12, 46. MATHEMATICS. *Educ:* Univ London, PhD(comput sci), 72. *Prof Exp:* PROF MATH, CARNEGIE-MELLON UNIV, 80- *Honors & Awards:* Nat Innovators Prize, NASA, 85. *Mailing Add:* Math Dept Carnegie-Mellon Univ 5000 Forbes Ave Pittsburgh PA 15213

NICOLAISEN, B(ERNARD) H(ENRY), b Cleveland, Ohio, Nov 12, 20; m 51; c 3. CHEMICAL ENGINEERING. *Educ:* Case Inst Technol, BS, 42, MS, 47. *Prof Exp:* Jr chem engr eng res, Monsanto Chem Co, 42-44; asst org chem, Case Inst Technol, 44-45, 46-47 & C F Prutton Assocs, 48; chem engr eng res, Mathieson Chem Corp, 48-50, asst mgr eng develop, 50-53, asst dir eng res, 53-54, asst dir res & develop, Chem Div, Olin Mathieson Chem Corp, 54-60, dir develop, 60-65, tech asst to vpres mfg, 65-66, dir process eng, 66-70, mgr automotive & org prod, 70-71, sr tech adv, 72-73, mgr process technol, 74-78, dir advan technol, Olin Corp, 78-85; RETIRED. *Mem:* Am Inst Chem Engrs. *Res:* Organic chemical research and chemical process development. *Mailing Add:* 16410 Shady Elms Dr Houston TX 77059

NICOLAU, GABRIELA, b Budapest, Hungary, June 11, 28; US citizen; m 51; c 2. DRUG METABOLISM, PHARMACOKINETICS. *Educ:* Univ Bucharest, Rumania, BS, 50, MS, 51, PhD(org chem), 63. *Prof Exp:* Asst prof org chem, Dept Chem, Univ Bucharest, 52-63, assoc prof org & biochem, 63-70; res assoc lipid metab, Pub Health Res Inst, New York, 72-75; sr res scientist, 75-78, group leader & prin res scientist drug metab, 78-88, DEPT HEAD, METAB RES, MED RES DIV, AM CYANAMID, 88- *Concurrent Pos:* Ed, Chem Annals J, Univ Bucharest, 61-64. *Honors & Awards:* NAt Chem Res Award, Dept Educ, Rumanian Govt, 64. *Mem:* Rumanian Chem Soc; Am Chem Soc; Am Asn Pharmaceut Scientists. *Res:* Optical activity; inclusion compounds (urea, thiourea, cyclodextrins); cholesterol and lipid metabolism; pharmacokinetics and drug metabolism of various drugs: antineoplastics, antiarthritics, synthetic prostaglandins; analytical methods development in biological fluids; in vitro metabolism; skin metabolism. *Mailing Add:* 300 Winston Dr Apt 2512 Cliffside Park NJ 07010

NICOLETTE, JOHN ANTHONY, b Chicago, Ill, Apr 2, 35; m 62; c 2. PHYSIOLOGY, ENDOCRINOLOGY. *Educ:* Dartmouth Col, AB, 56; Univ Ill, MS, 61, PhD(physiol), 63. *Prof Exp:* USPHS fel, Nat Cancer Inst, 63-66; asst prof biol sci, 66-71, asst dean grad col, 71-73, ASSOC PROF BIOL SCI, UNIV ILL, CHICAGO CIRCLE, 71-, ASSOC DEAN, COL LIBERAL ARTS & SCI, 87- *Concurrent Pos:* Asst dean, Col Liberal Arts & Sci, 80-87. *Res:* Biochemical mechanism of estrogen action. *Mailing Add:* Dept Biol Sci-M/C 228 Univ Ill Box 4348 Chicago IL 60680

NICOLL, CHARLES S, b Toronto, Ont, Apr 11, 37; US citizen; m 75; c 3. PHYSIOLOGY, ENDOCRINOLOGY. *Educ:* Mich State Univ, BS, 58, MS, 60, PhD(physiol), 62. *Prof Exp:* Res zoologist, Univ Calif, Berkeley, 62, Am Cancer Soc fel endocrine-tumor probs, 62-64; staff fel tumor-endocrinol, Nat Cancer Inst, 64-66; from asst prof to assoc prof physiol, 66-74, PROF PHYSIOL, UNIV CALIF, BERKELEY, 74- *Concurrent Pos:* Postdoctoral fel, Am Cancer Soc, 62-64; Miller prof, Miller Inst Basic Res Sci, Univ Calif, Berkeley, 74-75. *Honors & Awards:* Sigma Xi Award, Mich State Univ Chap, 61; Medal of City of Nice, France, Meritorious Res, 77; fel, AAAS, 80; Irving I Geschwind Mem Lectr, Div Comp Endocrinol, Am Soc Zoologists, 85; Grace Pickford Medal, Outstanding Res Comp Endocrinol, 89. *Mem:* Am Cancer Soc; Am Soc Zool; Endocrine Soc; Am Physiol Soc; fel AAAS. *Res:* Mammary and pituitary physiology; comparative aspects of prolactin physiology; growth regulation; cancer. *Mailing Add:* Dept Integrative Biol Univ Calif LSA 281 Berkeley CA 94720

NICOLL, JEFFREY FANCHER, b Washington, DC, Feb 25, 48; m 70; c 2. CONDENSED MATTER PHYSICS. *Educ:* Mass Inst Technol, BS(elec eng), BS(physics) & BS(math), 70, PhD(physics), 75. *Prof Exp:* Res asst physics, Ctr Mat Sci & Eng, Mass Inst Technol, 72-76, res assoc, Ctr Theoret Physics, 76-78; vis asst prof physics, Inst Phys Sci & technol, Univ Md, College Park, 78-82; dir res, Pangaro, Inc, 82-87; RES STAFF, INST DEFENSE ANALYSIS, 87-, ASST DIR, SCI & TECHNOL DIV, 89- *Mem:* Am Phys Soc. *Res:* Phase transitions; renormalization-group; cosmology. *Mailing Add:* Sci & Technol Div Inst Defence Analyses 1801 W Beauregard St Alexandria VA 22311

NICOLL, ROGER ANDREW, b Camden, NJ, Jan 15, 41; m 70. NEUROPHYSIOLOGY, NEUROPHARMACOLOGY. *Educ:* Lawrence Univ, BA, 63; Univ Rochester, MD, 68. *Prof Exp:* Intern med, Univ Chicago, 68-69; res assoc neurophysiol, NIMH, 69-73; assoc prof, 77-80, PROF PHARMACOL & PHYSIOL, UNIV CALIF, SAN FRANCISCO, 80- *Res:* Electrophysiology and pharmacology of neuromonal circuits and synapses in vertebrate central nervous system; mechanisms of synaptic plasticity; mechanisms of presynaptic and postsynaptic inhibition in the central nervous system. *Mailing Add:* Dept Pharmacol & Physiol Univ Calif San Francisco CA 94143

NICOLLE, FRANCOIS MARCEL ANDRE, b Nancy, France, Feb 25, 37; Can citizen; m 64; c 2. CHEMISTRY, POLLUTION CONTROL. *Educ:* Advan Nat Sch Agr & Food Indust, Paris, biochem engr, 61. *Prof Exp:* Indust engr, Pernod, France, 61-62; res engr, Int Cellulose Res Ltd, 64-73, sr res engr, CIP Res Ltd, 73-76, res assoc, 76-79, asst mgr process develop, CIP Res Ltd 79-86, dir environ, CIP, Inc, 86-89; DIR, ENVIRON CAN PAC FOREST PROD, LTD, 89- *Honors & Awards:* Douglas Jones Award, Can Pulp & Paper Asn, 74. *Mem:* Can Pulp & Paper Asn; Chem Inst Can; Can Soc Chem Eng; Asn Prof Engrs Ont. *Res:* Pulping; bleaching; by-products; pollution control involving biological treatment of newsprint and sulfite mills waste water; water and energy conservation by countercurrent washing during bleaching and water-reuse; air pollution. *Mailing Add:* 580 Allen St Hawkesbury ON K6A 2M2 Can

NICOLLS, KEN E, b Albuquerque, NMex, Nov 28, 35; m 64; c 3. ANATOMY, BIOSTATISTICS. *Educ:* Colo State Univ, BS, 59, MS, 61, PhD(anat), 69. *Prof Exp:* Res asst wildlife mgt, Colo Game & Fish Dept, 61-62; range conservationist, Worland Dist, Bur Land Mgt, 63-64; asst prof anat, Sch Med, Univ NDak, 69-76, mem fac anat, 69-76; fel, Cardiovascular Ctr, Col Med Univ Iowa, 76-78; asst prof, 78-80, ASSOC PROF ANAT, NORTHERN ARIZ UNIV, DEPT BIOL SCI, 80- *Concurrent Pos:* NSF fac res grant, Univ NDak, 70, 73-76, NIH instnl res grant, Sch Med, 70-71, 74-76. *Mem:* Am Asn Anatomists. *Res:* Mechanisms of action particularly axoplasmic flow, associated with interrelationships of light parameters; morphophysiology of the pituitary gland; structural defects of kidney tubules during genesis of cystic kidney disease. *Mailing Add:* Dept Biol Sci Box 5640 Flagstaff AZ 86011

NICOLOFF, DEMETRE M, b Lorain, Ohio, Aug 31, 33; c 3. CARDIOVASCULAR SURGERY, BIOENGINEERING. *Educ:* Ohio State Univ, BA, 54, MD, 57; Univ Minn, PhD(surg), 65, PhD(physiol), 67. *Prof Exp:* Instr surg, Univ Minn, 64-65; staff surgeon, Vet Admin Hosp, 65-69; asst prof surg, Univ Minn Hosps, Minneapolis, 69-71, assoc prof, 71-79. *Mem:* Am Med Asn; Soc Surg Alimentary Tract; Am Col Surg; Asn Acad Surg; Soc Thoracic Surg; Sigma Xi. *Mailing Add:* 1492 Hunter Dr Wayzata MN 55391

NICOLOSI, GREGORY RALPH, b Toledo, Ohio, Sept 8, 43; m 69. PHYSIOLOGY. *Educ:* Mich State Univ, BS, 65; Ohio State Univ, PhD(physiol), 71. *Prof Exp:* Instr anat & physiol, Ohio Northern Univ, 69-70; asst prof physiol, Col Med, Ohio State Univ, 71-72; asst prof, 72-77, ASSOC PROF PHYSIOL & ASSOC DEAN, COL MED, UNIV SOUTH FLA, 77- *Mem:* Am Physiol Soc; Biophys Soc. *Res:* Cardiovascular physiology; hemodynamics; spinal cord injury. *Mailing Add:* Dept Physiol Box 52-12901 Bruce B Downs Blvd Tampa FL 33612

NICOLOSI, JOSEPH ANTHONY, b New York, NY, Dec 29, 50; m 83; c 2. TECHNICAL MANAGEMENT. *Educ:* NY Inst Technol, BS, 72; Polytech Inst NY, MS & PhD(physics), 82. *Prof Exp:* Mem tech staff, Philips Res Labs, 72-82, SR SCIENTIST & DEVELOP MGR, PHILIPS ELECTRONIC INST, 82- *Mem:* Sigma Xi; Am Crystallog Asn; Soc Photo Optic Instrument Engrs; Mat Res Soc. *Res:* Research and development of methods and instrumentation for x-ray analysis by diffraction and fluorescence spectroscopy; development of instrumentation for microanalysis by energy dispersive spectroscopy. *Mailing Add:* 15 Joseph Lane Bardonia NY 10954

NICOLSON, MARGERY O'NEAL, biochemistry, for more information see previous edition

NICOLSON, PAUL CLEMENT, b Brooklyn, NY, June 3, 38; m 64; c 2. ANALYTICAL CHEMISTRY, SURFACTANTS POLYMERS. *Educ:* WVa Wesleyan Col, BS, 60; Ariz State Univ, MS, 65, PhD(phys-org chem), 66. *Prof Exp:* Proj leader org anal phys chem, Geigy Chem Corp, 65-69, group leader anal res, 69-71, mgr, 71-78, assoc dir cent res dept, 78-82, EXEC DIR RES, CIBA VISION CARE, CIBA-GEIGY CORP, 82- *Mem:* AAAS; Am Chem Soc; fel Am Inst Chemists; Asn Res Vision & Ophtal. *Res:* Physical organic chemistry; instrumentation; polymer chemistry. *Mailing Add:* Ciba Vision Care 2910 Amwiler Ct Atlanta GA 30360

NICOSIA, SANTO VALERIO, b Catania, Italy, Dec 12, 43; US citizen; m 69; c 3. REPRODUCTIVE PATHOBIOLOGY, CYTOPATHOLOGY. *Educ:* Catholic Univ, Sacred Heart Sch Med & Surg, Rome, Italy, MD, 67; Univ Ill, Chicago, MS, 71; Am Bd Path, dipl, 78. *Prof Exp:* Resident anat path, Michael Reese Hosp & Med Ctr, Chicago, Ill, 69-72; fel, Sch Med, Univ Pa, 72-73, asst prof obstet, gynec & path, 73-79, assoc prof, 79-; AT DEPT PATH, UNIV S FLA. *Concurrent Pos:* Dir, Electron Micros Unit & prin investr, Div Reproductive Biol, Sch Med, Univ Pa, 73-; prin investr, Marine Biol Lab, Woods Hole, Mass, 76, 78 & 79; Ad Hoc ed, Fertil & Steril, 73-78, Biol Reproductive, 76-, Scanning Elec Microsc, Annual Symposia, 76-79, Biol Bulletin, 80- & Am J Obstet Gynecol, 81-; consult, Kimberly-Clark Corp, 80-81. *Mem:* Am Tissue Cult Asn; Soc Study Reproduction; Int Acad Path; AAAS; Am Soc Cell Biol. *Res:* Regulation of mucus secretion in vertebrate and invertebrate cells; epithelialstromal interactions in developing and adult reproductive tissues; ultrastructure of mammalian fertilization; maturation and atresia of ovarian follicles; pathobiology of ovarian surface epithelium. *Mailing Add:* 13813 Shady Shores Dr Tampa FL 33613

NIDAY, JAMES BARKER, b Nashville, Tenn, May 21, 17; m 45; c 4. NUCLEAR CHEMISTRY. *Educ:* Univ Chicago, SB, 42, SM, 60. *Prof Exp:* Jr chemist, Tenn Valley Auth, 42-45; asst radiochem, Univ Chicago, 47-52; chemist, Calif Res & Develop Co, 52-53; chemist, Lawrence Livermore Lab, Univ Calif, 53-83. *Concurrent Pos:* Consult, 83- *Mem:* Am Phys Soc; Am Chem Soc; Sigma Xi. *Res:* Study of the fission process through radio-chemical study of fission yields and of the ranges of fission fragments in matter; computer acquisition and analysis of data in gamma ray spectroscopy. *Mailing Add:* 4440 Entrada Dr Pleasanton CA 94566

NIDEN, ALBERT H, b Philadelphia, Pa, Aug 17, 27; m 55; c 3. INTERNAL MEDICINE. *Educ:* Univ Pa, AB, 49, MD, 53; Am Bd Internal Med, dipl. *Prof Exp:* Instr pharmacol, Univ Pa, 54-55; from instr to assoc prof internal med & chest dis, Sch Med, Univ Chicago, 57-68; prof med & chief pulmonary dis, Temple Univ, 69-73; prof med & chief Pulmonary Dis Div, Charles R Drew Univ Med & Sci, 73-90; prof med, Sch Med, Univ Calif, Los Angeles, 79-90; PROF MED, UNIV SOUTHERN CALIF, SCH MED, LOS ANGELES, 73-79 & 90- *Concurrent Pos:* Hon res asst, Inst Path, Ger, 64; hon res assoc dept anat, Kyushu, 64-65; consult, Nat Heart & Lung Inst, 68-72; consult, Comt Med & Biol Effects of Environ Pollutants, Nat Res Coun; assoc dean fac affairs, Charles R Drew Univ Med & Sci, 80-87. *Mem:* Am Thoracic Soc; Am Fedn Clin Res; fel Am Col Chest Physicians; Am Physiol Soc. *Res:* Electron microscopy of lung; chest diseases; pulmonary physiology; physiology and pharmacology of the pulmonary circulation and ventilation; effects of air pollutants on lungs; lung metabolism; diffuse infiltrative lung disease; sarcoidosis. *Mailing Add:* Univ Southern Calif Sch Med 2025 Zonal Ave Los Angeles CA 90033

NIEBAUER, JOHN J, b San Francisco, Calif, July 7, 14; m 43; c 4. MEDICINE. *Educ:* Stanford Univ, BA, 37, MD, 42; Am Bd Orthop Surg, dipl, 53. *Prof Exp:* Res orthop surg, Stanford Univ Hosp, 43-44; Gibney fel, Hosp for Spec Surg, New York, 44-45; assoc clin prof, Sch Med, Stanford Univ, 46-69; CHIEF DEPT HAND SURG, PAC MED CTR, 69- *Concurrent Pos:* Adj clin asst prof surg, Sch Med, Stanford Univ, 69-; clin prof orthop surg, Univ Calif, San Francisco, 67-; dir orthop & hand surg, Res Proj, Inst Med Sci, Presby Med Ctr, San Francisco. *Mem:* AMA; Am Orthop Asn; Am Acad Orthop Surg; Am Col Surg; Am Soc Surg of Hand. *Mailing Add:* 101 Crown Rd Kentfield CA 94904

NIEBEL, B(ENJAMIN) W(ILLARD), b Hunan, China, May 7, 18; nat US; m 42; c 4. INDUSTRIAL ENGINEERING, MATERIALS SCIENCE. *Educ:* Pa State Univ, BS, 39, MS, 49, IE, 52. *Prof Exp:* Chief indust engr, Lord Mfg Co, Pa, 39-47; from instr to prof indust eng & head dept, Pa State Univ, 47-79; RETIRED. *Concurrent Pos:* Consult to various indust concerns; Int Coop Admin consult, Mex & Peru. *Honors & Awards:* Frank & Lillian Gilbreth Indust Eng Award, Am Inst Indust Engrs, 76. *Mem:* Fel Am Inst Indust Engrs (vpres educ, 71-73); Am Soc Eng Educ; Soc Mfg Engrs. *Res:* Product design of medical instruments, calculus disintegrator, animated intestinal tube, hemorrhoidal excisor, surgical strip stitch, mole remover; motion and time study; materials and processes and their influence on the design of products. *Mailing Add:* Dept Indust Eng University Park PA 16801

NIEBERGALL, PAUL J, b Newark, NJ, Sept 5, 32; m 57; c 4. PHARMACY. *Educ:* Rutgers Univ, BSc, 53; Univ Mich, MSc, 58, PhD(pharm), 62. *Prof Exp:* From asst prof to prof pharm, Phila Col Pharm & Sci, 61-75; DIR CORP PROD DEVELOP, MARION LABS, 75-; AT DEPT OF PHARM, MED UNIV OF SC. *Concurrent Pos:* Prin investr, Nat Inst Allergy & Infectious Dis res grant, 65-68. *Mem:* Am Pharmaceut Asn; Acad Pharmaceut Sci. *Res:* Physical pharmacy; dissolution rates; solubilization of drugs through amide fusion; metal ion-penicillin interactions. *Mailing Add:* Dept Pharm Med Univ SC 171 Ashley Ave Charleston SC 29425

NIEBERLEIN, VERNON ADOLPH, b Dayton, Ohio, June 28, 18; m 58; c 4. CHEMICAL ENGINEERING, MATERIALS SCIENCE. *Educ:* Univ Dayton, BChE, 39; Univ Ala, MS, 65, PhD(chem eng), 70. *Prof Exp:* Chemist, Allison Div, Gen Motors Corp, 40-43; res chemist, P R Mallory & Co, Inc, 43-48; plating engr electrochem, Mat Bearing Div, Am Brake Shoe Co, 48-53; chemist extraction metall, US Bur Mines, 53-59; chemist mat, US Army Missile Command, 59-82. *Mem:* Soc Advan Mat & Process Engrs. *Res:* High-temperature materials, fibers, transport phenomena. *Mailing Add:* 2419 Henry St Huntsville AL 35801

NIEBYL, JENNIFER ROBINSON, b Can, Dec 5, 42; US citizen; m 75; c 2. OBSTETRICS & GYNECOLOGY, MATERNAL-FETAL MEDICINE. *Educ:* McGill Univ, BSc, 63; Yale Univ, MD, 67. *Prof Exp:* Intern med, New York Hosp-Cornell Univ, 68, asst resident obstet & gynec, 70; asst & chief resident, 73, fel, 76-78, DIR DIV MATERNAL & FETAL MED, JOHNS HOPKINS HOSP, 82-; ASSOC PROF, DEPT OBSTET & GYNEC, SCH MED, JOHNS HOPKINS UNIV, 80-, ASSOC PROF PEDIAT, 85- *Concurrent Pos:* Assoc examr, Am Bd Obstet & Gynec, 84-; mem bd dirs, Soc Perinatal Obstetricians; co-ed-in-chief, Am J Perinatol. *Mem:* Soc Gynec Invest; Soc Perinatal Obstetricians. *Res:* Drug use in pregnancy; use of drug therapy for premature labor. *Mailing Add:* Univ Iowa Hosps Head Dept Gynecol & Obstet Iowa City IA 52242

NIEDBALSKI, JOSEPH S, RIBOSOMAL PROTEIN, MEMBRANE PROTEIN. *Educ:* Univ Okla, PhD(biochem), 83. *Prof Exp:* FEL, SAMUEL R NOBLE FOUND, 84- *Res:* Membrane protein. *Mailing Add:* Samuel R Noble Found 2510 Hwy 70 E PO Box 2180 Ardmore OK 73402

NIEDENFUHR, FRANCIS W(ILLIAM), b Chicago, Ill, Jan 23, 26; m 51; c 2. ENGINEERING MATHEMATICS. *Educ:* Univ Mich, BSc, 50, MSc, 51; Ohio State Univ, PhD(eng mech), 57. *Prof Exp:* Engr, NAm Aviation, Inc, 51-52; instr eng mech, Ohio State Univ, 52-57, from assoc prof to prof, 57-66; dir technol assessment, Defense Advan Res Projs Agency, 66-76; CONSULT ENGR, C3 DIV, MITRE CORP, 76- *Concurrent Pos:* Mem staff, Inst Defense Anal, 64-66; adj prof, Howard Univ, 67-74; lectr, Univ Md, 77- *Honors & Awards:* Distinguished Civilian Serv Award, Secy Defense, 76. *Mem:* Am Soc Eng Educ; Int Asn Shell Struct; Am Acad Mech; Am Soc Mech Engrs; Am Inst Aeronaut & Astronaut; Sigma Xi; Inst Elec & Electronic Engrs. *Res:* Operations analysis and engineering mechanics. *Mailing Add:* 3737 Fessenden St NW Washington DC 20016

NIEDENZU, KURT, b Fritzlar, Ger, Mar 12, 30; m 58; c 4. INORGANIC CHEMISTRY. *Educ:* Univ Heidelberg, Dipl, 55, PhD(chem), 56. *Prof Exp:* Sci asst, Univ Heidelberg, 55-57, instr inorg & anal chem, 57; chemist, Chem Sci Div, Off Ord Res, US Dept Army, Duke Univ, 58-62, off chief scientist, Army Res Off, 62-67; res adminr, Wintershall AG, Kassel, Ger, 67-68; assoc prof, 68-73, PROF CHEM, UNIV KY, 73- *Concurrent Pos:* Vis prof, Gmelin Inst, Max-Planck Soc, 74-75; Alexander von Humboldt Found US Sr Scientist Award, 74. *Mem:* Am Chem Soc; Royal Soc Chem; Ger Chem Soc. *Res:* Synthesis and structure of phosphorus, sulfur, boron and nitrogen compounds, especially isoelectronic systems; organometallic synthesis; spectroscopy. *Mailing Add:* 724 Haverhill Dr Lexington KY 40503

NIEDERHAUSER, WARREN DEXTER, b Akron, Ohio, Jan 2, 18; m 49; c 1. ORGANIC CHEMISTRY, POLYMERS. *Educ:* Oberlin Col, AB, 39; Univ Wis, PhD(org chem), 43. *Hon Degrees:* DSc, Oberlin Col, 84. *Prof Exp:* Res chemist, 43-51, head surfactant synthesis group, 51-55, head chem sect redstone div, 55-59, res supvr, 59-65, asst res dir, 66-73, dir pioneering res, 73-83, CONSULT, ROHM & HAAS CO, 83- *Concurrent Pos:* Mem limited war comt, Dept of Defense, 61-63, mem ord sci adv panel, US Dept Army, 61-62 & mem chem & biol warfare adv comt, 61-63. *Honors & Awards:* Henry Hill Award, Am Chem Soc, 85. *Mem:* Am Chem Soc (pres, 84); fel Am Inst Chem; AAAS. *Res:* Ion exchange; environmental control; petroleum chemicals; tough, clear acrylic plastics; dispersants and low foam surfactants;

high energy rocket propellants; developed first use and manufacture process for epoxidized soybean oil (PVC plasticizer-stabilizer); vinyl impact modifiers and processing aids. *Mailing Add:* Res Div Lab Rohm & Haas Co Spring House PA 19477

NIEDERJOHN, RUSSELL JAMES, b Schenectady, NY, June 13, 44; m 69; c 2. ELECTRICAL ENGINEERING, COMPUTER ENGINEERING. *Educ:* Univ Mass, BS, 67, MS, 68, PhD(elec eng), 71. *Prof Exp:* Comput programmer, IBM, 67; res asst elec eng, Univ Mass, 68-71; from asst prof to assoc prof, 71-80, PROF ELEC ENG & COMPUT ENG, MARQUETTE UNIV, 80-, CHMN DEPT, 87- *Concurrent Pos:* Prin investr, several res, educ & design grants, 72-; consult, Gen Elec Co, 76-78, Rome Air Develop Ctr, 77 & Eaton Corp, 78-85; dir NSF cause grant, 78-81. *Honors & Awards:* C Holmes Mac Donald Outstanding Elec Eng Prof, 78; Western Elec Fund Award, 81. *Mem:* Inst Elec & Electronic Engrs; Am Soc Eng Educ; Acoust Soc Am; Audio Eng Soc; Am Asn Univ Prof; Sigma Xi. *Res:* Speech and signal processing; speech in noise enhancement; graduate and undergraduate educational methods. *Mailing Add:* Dept Elec Eng Marquette Univ Milwaukee WI 53233

NIEDERKORN, JERRY YOUNG, b St Louis, Mo, Oct 31, 46; m 69; c 2. IMMUNOLOGY, PARASITOLOGY. *Educ:* Cent Methodist Col, BA, 68; Cent Mo State Univ, MS, 72; Univ Ark, PhD(zool), 77. *Prof Exp:* PROF OPHTHAL, SOUTHWESTERN MED SCH, UNIV TEX, 77- *Mem:* Am Soc Parasitologists; AAAS; Sigma Xi; Am Asn Immunologists; Am Soc Microbiol. *Res:* Immunoparasitology; ocular immunology; tumor immunology. *Mailing Add:* Dept Ophthal Southwestern Med Sch Univ Tex 5323 Harry Hines Blvd Dallas TX 75235

NIEDERLAND, WILLIAM G, b Schippenbeil, Ger, Aug 29, 04; US citizen; m 52; c 3. PSYCHIATRY, MEDICAL SCIENCES. *Educ:* Univ W rzburg, MD, 29; Univ Genoa, MD, 34; State Univ NY, MD, 41. *Hon Degrees:* Dr, Julius-Maximilian Univ, Bavaria. *Prof Exp:* Clin asst prof med, Univ Philippines, 39-40; assoc prof psychol, Univ Tampa, 45-47; clin prof, 55-77, EMER CLIN PROF PSYCHIAT, STATE UNIV NY DOWNSTATE MED CTR, 77- *Concurrent Pos:* Training psychoanalyst, State Univ NY Downstate Med Ctr, 58-74, supv psychoanalyst, 58-, emer training psychoanalyst, 75-; chief consult psychiatrist, Altro Health & Rehab Serv, New York, 58-76; ed consult, Am Imago, NY & Detroit, 63-; consult psychiatrist, Hackensack Gen Hosp, NJ, 72- *Mem:* Fel Am Psychiat Asn; Am Psychoanal Asn; Int Psychoanal Asn. *Res:* Artistic and cultural creativity; psychobiographical psychohistorical as well as psychogeographical studies; clinical research in the fields of depression and survival after social and natural catastrophes; author of more than 200 scientific articles and essays and four books. *Mailing Add:* 108 Glenwood Rd Englewood NJ 07631

NIEDERMAN, JAMES CORSON, b Hamilton, Ohio, Nov 27, 24; m 51; c 4. INTERNAL MEDICINE, EPIDEMIOLOGY. *Educ:* Johns Hopkins Univ, MD, 49; Kenyon Col, AB, 45. *Hon Degrees:* DSc, Kenyon Col, 81. *Prof Exp:* Intern med, Osler Med Serv, Johns Hopkins Hosp, 49-50; from asst resident to assoc resident, Yale-New Haven Med Ctr, 50-55; instr prev med, Sch Med, 55-58, asst prof epidemiol & prev med, 58-66, assoc clin prof, 66-76, CLIN PROF EPIDEMIOL & MED, SCH MED, YALE UNIV, 76- *Concurrent Pos:* Fel, Silliman Col, Yale Univ, 64-; mem bd counrs, Smith Col, Mass, 73-77; trustee, Kenyon Col, Ohio, 73-; mem nat coun, Johns Hopkins Med. *Mem:* Am Epidemiol Soc; Infectious Dis Soc Am; Sigma Xi. *Res:* Infectious mononucleosis; EB virus infections; clinical epidemiological studies of virus infections. *Mailing Add:* Dept of Epidemiol & Pub Health Yale Univ Sch of Med 333 Cedar St New Haven CT 06510

NIEDERMAN, RICHARD, PERIODONTOLOGY, CELL MOTILITY. *Educ:* Harvard Univ, DMD(dent), 76. *Prof Exp:* HEAD & ASSOC MEM STAFF, DEPT CELL BIOL, FORSYTH DENT CTR, 84- *Mailing Add:* 140 The Fenway Boston MA 02115

NIEDERMAN, ROBERT AARON, b Norwich, Conn, Jan 19, 37. MOLECULAR BIOLOGY, BIOCHEMISTRY. *Educ:* Univ Conn, BS, 59, MS, 61; Univ Ill, Urbana, DVM, 64, PhD(bact), 67. *Prof Exp:* Atomic Energy Comn fel biochem, Mich State Univ, 67-68; fel physiol chem, Roche Inst Molecular Biol, 68-70; from asst prof to assoc prof microbiol, 70-80, PROF BIOCHEM, RUTGERS UNIV, PISCATAWAY, 80- *Concurrent Pos:* Merck Co Found fac develop award, Rutgers Univ, New Brunswick, 71-; USPHS res grant, Nat Inst Gen Med Sci, 73 & 79; NSF res grants, 74, 76, 79, 82, 85; USPHS res career develop award, Nat Inst Gen Med Sci, 75-80; res fel biochem, Univ Bristol, Eng, 77-78. *Mem:* AAAS; Am Soc Biol Chem; Am Soc Microbiol. *Res:* Membrane biochemistry; mechanisms of bacterial membrane differentiation and assembly; membrane structure-function relationships. *Mailing Add:* Dept Molecular Biol & Biochem Rutgers Univ PO Box 1059 Piscataway NJ 08854

NIEDERMAYER, ALFRED O, b Munich, Ger, Aug 8, 21; US citizen; m 46; c 3. ANALYTICAL CHEMISTRY, PHARMACEUTICAL CHEMISTRY. *Educ:* Rutgers Univ, BA, 50. *Prof Exp:* SR RES SCIENTIST, E R SQUIBB & SONS, 50-, SECT HEAD, 80- *Mem:* Am Chem Soc; Sigma Xi. *Res:* Analytical separations, particularly gas chromatography; dissolution studies of pharmaceutical dosage forms. *Mailing Add:* 17 Overbrook Rd Piscataway NJ 08854

NIEDERMEIER, ROBERT PAUL, b Waukesha, Wis, Sept 8, 18; m 45; c 2. DAIRY HUSBANDRY. *Educ:* Univ Wis, BS, 40, MS, 42, PhD(dairy husb, biochem), 48. *Prof Exp:* From instr to assoc prof, 47-56, PROF DAIRY HUSB, UNIV WIS-MADISON, 57-, CHMN DEPT, 63- *Mem:* Am Soc Animal Sci; Am Dairy Sci Asn (vpres, 75-76, pres, 76-77); AAAS. *Res:* Forage preservation and utilization; mineral studies of parturient paresis in dairy cattle. *Mailing Add:* 5817 Driftwood Ave Madison WI 53706

NIEDERMEIER, WILLIAM, b Evansville, Ind, Apr 1, 23; m 45; c 5. IMMUNOCHEMISTRY, IMMUNOLOGY. *Educ:* Purdue Univ, BS, 46; Univ Ala, MS, 53, PhD(biochem), 60. *Prof Exp:* Res chemist, Mead Johnson & Co, 46-49; instr nutrit, Sch Med, Northwestern Univ, 49-50; from asst prof to assoc prof biochem, Univ Ala, Birmingham, 60-87, sr scientist, Comprehensive Cancer Ctr, 75-87, scientist, Multipurpose Arthritis Ctr, 79-87; RETIRED. *Concurrent Pos:* Chmn, Ala Sect, Am Chem Soc, 68, counr, 70-86; cong sci counr, Am Chem Soc, 74-85; nat tour speaker, Soc Appl Spectros, 73; treas, Ala sect, Am Chem Soc, 66-67, mem comt sect activity, 69-70. *Mem:* AAAS; Am Chem Soc; Soc Appl Spectros; Am Soc Biol Chem; Am Asn Immunol; Sigma Xi; Soc Complex Carbohydrates. *Res:* Carbohydrate composition of immunoglobins; significance of antibody to mouse mammary tumor virus in breast cancer; effect of adrenal corticosteroids on electrolyte metabolism; trace element metabolism in rheumatoid arthritis; biochemical and biophysical properties of hyaluronic acid; composition and role of carbohydrates in immunoglobulins; significance of antibody to mouse mammary tumor virus in women with breast cancer. *Mailing Add:* 2016 Crest Lane Birmingham AL 35226

NIEDERMEYER, ERNST F, b Schonberg, Ger, Jan 19, 20; US citizen; m 46; c 5. NEUROLOGY. *Educ:* Innsbruck Univ, MD, 47. *Prof Exp:* Resident neurol & psychiat, Innsbruck Univ Hosp, 48-50; French Govt fel & foreign asst neurol, Salpêtrière Hosp, Paris, 50-51; resident neurol, psychiat & EEG, Innsbruck Univ Hosp, 51-55, asst prof, 55-60; from asst prof to assoc prof EEG, Univ Iowa, 60-65; assoc prof, Johns Hopkins Univ & Electroencephalographer-in-chg, 65-87, prof neurol, 87-90, EMER PROF NEUROL, JOHNS HOPKINS UNIV, 90- *Honors & Awards:* Hans-Berger Prize, EEG Soc, Ger, 88. *Mem:* Am EEG Soc; Am Epilepsy Soc; Austrian EEG Soc; corresp mem Ger EEG Soc; Peruvian Neuropsychol Soc; Am EEG Soc (pres, 90-91). *Res:* Clinical electroencephalography with particular emphasis on epilepsy; clinical epileptology. *Mailing Add:* Dept Neurol Johns Hopkins Hosp 600 N Wolfe St Baltimore MD 21205

NIEDERPRUEM, DONALD J, b Buffalo, NY, Sept 3, 28; m 51; c 4. MICROBIOLOGY, BIOCHEMISTRY. *Educ:* Univ Buffalo, BA, 49, MA, 56, PhD(biol), 59. *Prof Exp:* USPHS fel bact, Univ Calif, Berkeley, 59-61; from asst prof to assoc prof microbiol, 61-68, PROF MICROBIOL, MED CTR, IND UNIV, INDIANAPOLIS, 68-, PROF IMMUNOL, 77- *Concurrent Pos:* Lederle med fac award, 62-65. *Mem:* Am Soc Microbiol; Bot Soc Am; Am Soc Plant Physiol; Am Soc Biol Chem; Soc Develop Biol. *Res:* Biochemical basis of cellular regulation of differentiation and morphogenesis. *Mailing Add:* Dept of Microbiol & Immunol Ind Univ Med Ctr 1100 W Michigan St Indianapolis IN 46223

NIEDRACH, LEONARD WILLIAM, b Weehawken, NJ, Sept 11, 21; m 50; c 3. CHEMISTRY. *Educ:* Univ Rochester, BS, 42; Harvard Univ, MA, 47, PhD(chem), 48. *Prof Exp:* Analytical chemist, Gen Elec Co, Mass, 43-44; jr chemist, Univ Chicago, 44-45 & Monsanto Chem Co, Ohio, 45; res assoc, 48-58, phys chemist, 58-71, mgr membrane & sensor projs, 71-77, phys chemists, 77-87, CONSULT PHYS CHEMIST, GEN ELEC CO, 87- *Mem:* AAAS; fel Am Inst Chem; Am Chem Soc; Electrochem Soc. *Res:* inorganic preparations; electrodeposition; inorganic separations; fuel cells; electrochemical devices for medical applications , electrochemical sensors; water chemistry at high temperatures. *Mailing Add:* Res & Develop Ctr Gen Elec Co PO Box 8 Schenectady NY 12301

NIEDZIELSKI, EDMUND LUKE, b Brooklyn, NY, Nov 14, 17; m 46; c 5. PETROLEUM CHEMISTRY, FUEL TECHNOLOGY & PETROLEUM ENGINEERING. *Educ:* St John's Univ, NY, BS, 38; Fordham Univ, MS, 40, PhD(org chem), 43. *Prof Exp:* Asst, Fordham Univ, 38-42; res chemist, E I Du Pont de Nenoros & Co, 46-48, res chemist, Petrol Lab, 48-62, res chemist, Jackson Lab, 62-66, sr res chemist, Petrol Lab, 66-86; SR CONSULT, CONDUX INC, 86- *Mem:* AAAS; fel Am Inst Chem; NY Acad Sci. *Res:* Petroleum additives; synthetic fluids and lubricants; redistribution reactions; gas separation membranes. *Mailing Add:* 1007 Piper Rd Wilmington DE 19803

NIEFORTH, KARL ALLEN, b Melrose, Mass, July 7, 36; m 58, 89; c 4. MEDICINAL CHEMISTRY, PSYCHOPHARMACOLOGY. *Educ:* Mass Col Pharm, BS, 57; Purdue Univ, MS, 59, PhD(med chem), 61. *Prof Exp:* From asst prof to assoc prof med chem, 61-75, asst dean, Sch Pharm, 67-76, assoc dean, Sch Pharm, 76-81, PROF MED CHEM, UNIV CONN, 75-, DEAN, SCH PHARM, 81- *Concurrent Pos:* Lectr, Yale Univ, 70-76. *Mem:* Am Chem Soc; Am Pharmaceut Asn; Am Asn Col Pharm. *Res:* Design, synthesis and biological testing of compounds in the area of psychotherapeutics; hypoglycemics or drug antagonists. *Mailing Add:* Box U-91 Univ of Conn Storrs CT 06269-2092

NIEHAUS, MERLE HINSON, b Enid, Okla, Mar 25, 33; m 54; c 2. INTERNATIONAL DEVELOPMENT. *Educ:* Okla State Univ, BS, 55, MS, 57; Purdue Univ, PhD(plant breeding), 64. *Prof Exp:* Instr agron, Imp Ethiopian Col Agr & Mech Arts, Harar, 59-61; instr, Purdue Univ, 63; asst dir res, Advan Seed Co, 64; from asst prof to prof agron, Ohio Agr Res & Develop Ctr, 64-78, assoc chmn dept, 75-78; PROF AGRON & HEAD DEPT, NMEX STATE UNIV, 78-; DEAN, COL AGR SCI, COLO STATE UNIV, FT COLLINS, 85- *Mem:* Am Soc Agron; fel Crop Sci Soc Am; Sigma Xi; fel AAAS. *Res:* Soybean breeding and variety development; research and education administration. *Mailing Add:* 1113 Springwood Dr Ft Collins CO 80525

NIEHAUS, WALTER G, JR, b Minneapolis, Minn, Dec 13, 37. BIOCHEMISTRY. *Educ:* Univ Minn, BS, 62, PhD(biochem), 64. *Prof Exp:* Nat Heart Inst fel biochem, Univ Ill, 65-66 & Karolinska Inst, Sweden, 66-67; res assoc, Univ Minn, 67-68; from asst prof to assoc prof, Pa State Univ, 68-75; ASSOC PROF BIOCHEM, VA POLYTECH & STATE UNIV, 75- *Mem:* Am Soc Microbiol; Am Soc Biol Chem; Am Chem Soc. *Res:* Regulation of fungal secondary metabolism; biosynthesis of aflatoxin; enzyme reaction mechanisms; inhibitor induced cooperativity of mannitol-l-phosphate dehydrogenase from aspergillus parasiticus. *Mailing Add:* Dept of Biochem & Nutrit Va Polytech Inst Blacksburg VA 24061

NIEKAMP, CARL WILLIAM, b Catskill, NY, Jan 12, 43; m 64; c 2. PHYSICAL CHEMISTRY, BIOCHEMISTRY. *Educ:* Hope Col, BA, 65; Purdue Univ, PhD(phys chem), 71. *Prof Exp:* Res assoc chem, Yale Univ, 71-73; asst prof, Hanover Col, 73-78; res chemist, A E Staley Mfg Co, 78-80, sr res chemist, 80-81, lab mgr, 81-82, sr lab mgr, 83-84, group mgr, 84-85, SCI MGR FOOD & INDUST PROD MGR, A E STALEY MFG CO, 85- *Concurrent Pos:* Humboldt Found res fel, 76-77. *Mem:* Am Chem Soc; fel Am Inst Chemists. *Res:* Carbohydrate chemistry of mono-, oligo-, and polysaccharides; biotechnology, enzyme-catalyzed reactions and fermentation; physical and chemical investigations of biomaterials; process development. *Mailing Add:* 2900 Crestwood Dr East Lansing MI 48823-2319

NIELAN, PAUL E, b Buffalo, NY, May 7, 57; m 89. MULTIBODY DYNAMICS. *Educ:* State Univ NY Buffalo, BS, 78; Stanford Univ, MS, 79, PhD(mech eng), 86. *Prof Exp:* MECH ENGR, APPL MECH DEPT, SANDIA NATIONAL LABS, 80- *Mem:* Am Soc Mech Engrs; Am Inst Aeronaut & Astronaut. *Res:* Application of computerized symbolic manipulation to problems in multibody dynamics including spacecraft, vehicles, robotics, and mechanisms; mesh generation and nonlinear finite elements. *Mailing Add:* 349 Merrilee Pl Danville CA 94526

NIELL, ARTHUR EDWIN, b Lubbock, Tex, Sept 15, 42. RADIO ASTRONOMY, GEODESY. *Educ:* Calif Inst Technol, BS, 65; Cornell Univ, PhD(radio astron), 71. *Prof Exp:* Res fel radio astron, Queen's Univ, Ont, 71-72; resident res assoc, Nat Res Coun-Nat Acad Sci, 72-74; sr engr radio astron & earth physics, Jet Propulsion Lab, 74-76, mem tech staff, 76-84; SR SCIENTIST, HAYSTACK OBSERV, MASS INST TECHNOL, 84- *Mem:* Am Astron Soc; Int Radio Sci Union; Int Astron Union; Am Geophys Union. *Res:* Geodetic measurements using radio astronomical observations; investigation of the structure and physics of extragalactic radio sources and galactic stars. *Mailing Add:* Haystack Observ Mass Inst Technol Westford MA 01886

NIELSEN, ALVIN HERBORG, b Menominee, Mich, May 30, 10; m 42; c 1. PHYSICS. *Educ:* Univ Mich, BA, 31, MSc, 32, PhD(physics), 34. *Prof Exp:* Asst physics, Univ Mich, 31-34; from instr to assoc prof physics, 35-46, head dept, 56-69, dean, Col Lib Arts, 63-77, prof physics, 46-80, EMER PROF, UNIV TENN KNOXVILLE, 80-, EMER DEAN, COL LIB ARTS, 77- *Concurrent Pos:* Fulbright scholar, Inst Astrophys, Belg, 51-52; hon fel, Ohio State Univ; consult, Union Carbide Corp, Oak Ridge Nat Lab, & Off Ord Res; mem vis sci prog, Am Inst Physics-Am Asn Physics Teachers, 64-71, Nat Acad Sci-Nat Res Coun comt basic res adv to US Army Res Off, 64-70 & div chem physics, Am Inst Physics. *Mem:* Fel AAAS; fel Am Phys Soc; fel Optical Soc Am; Am Asn Physics Teachers. *Res:* Infrared spectra of polyatomic molecules; infrared detectors. *Mailing Add:* Dept Physics 401 Univ Tenn Knoxville TN 37916

NIELSEN, ARNOLD THOR, b Seattle, Wash, Sept 2, 23; m 47; c 3. ORGANIC CHEMISTRY. *Educ:* Univ Wash, BSc, 44, PhD(org chem), 47. *Prof Exp:* Res chemist, Chas Pfizer & Co, 47-48; asst prof chem, Univ Idaho, 49-52; res assoc, Purdue Univ, 52-55; instr, Rutgers Univ, 55-57; asst prof, Univ Ky, 57-59; res chemist, 59-80, SR RES SCIENTIST, MICHELSON LAB, NAVAL WEAPONS CTR, 80- *Honors & Awards:* William B McLean Award, 90. *Mem:* Am Chem Soc; Sigma Xi; Int Soc Heterocyclic Chem. *Res:* Aldol condensation; nitro compounds; stereochemistry; nitrogen heterocyclics; explosives. *Mailing Add:* Michelson Lab Code 38503 Naval Weapons Ctr China Lake CA 93555

NIELSEN, CARL EBY, b Los Angeles, Calif, Jan 22, 15; m 38; c 3. PHYSICS, SOLAR ENERGY. *Educ:* Univ Calif, AB, 34, MA, 40, PhD(physics), 41. *Prof Exp:* Instr physics, Iowa State Univ, Calif, Berkeley, 41-45, lectr, 45-46; asst prof, Univ Denver, 46-47; from asst prof to prof physics & astron, 47-85, PROF EMER PHYSICS, OHIO STATE UNIV, 85- *Concurrent Pos:* Ford fel, Europ Orgn Nuclear Res, Geneva, 58-59; scientist, Midwestern Univs Res Asn, 60-61; consult thermonuclear div, Oak Ridge Nat Lab, 61-70; vis scientist, Culham Lab, UK Atomic Energy Authority, 66; consult, Los Alamos Sci Lab, 71-75 & Lawrence Livermore Lab, 74-77. *Mem:* Am Phys Soc; Int Solar Energy Soc; Am Soc Heating, Refrig & Air Conditioning Engrs; Am Solar Energy Soc. *Res:* Cloud chambers; collective phenomena in beams and plasmas; solar energy and solar ponds. *Mailing Add:* Dept Physics Ohio State Univ 174 W 18 Ave Columbus OH 43210

NIELSEN, DAVID GARY, b Longview, Wash, Nov 18, 43; m 64; c 2. ENTOMOLOGY. *Educ:* Willamette Univ, BA, 66; Cornell Univ, MS, 69, PhD(entom), 70. *Prof Exp:* From asst prof to assoc prof, 70-80, PROF ENTOM, OHIO AGR RES & DEVELOP CTR, OHIO STATE UNIV, 80- *Concurrent Pos:* Consult entom, insect host-plant relationships, urban forest pest mgt. *Honors & Awards:* Arboricultural Res Award, Int Soc Arboricult, 86. *Mem:* Entom Soc Am; Int Soc Arboricult; AAAS. *Res:* Behavioral ecology and suppression of insects which attack woody ornamental plants; urban forest pest management; insect host plant stress relationships. *Mailing Add:* Dept of Entom Ohio Agr Res & Develop Ctr Wooster OH 44691

NIELSEN, DONALD R, b Phoenix, Ariz, Oct 10, 31; m 53; c 5. SOIL PHYSICS. *Educ:* Univ Ariz, BS, 53, MS, 54; Iowa State Univ, PhD(soil physics), 58. *Hon Degrees:* DSc, Ghent State Univ, Belgium, 86. *Prof Exp:* Res assoc soil physics, Iowa State Univ, 54-58; from asst prof to assoc prof, Univ Calif, Davis, 58-68, dir, Kearney Found Soil Sci, 70-75, chmn dept land, air & water resources, 75-77, assoc dean, 71-80, exec assoc dean, 86-89, chmn agron & range sci, 89-91, PROF SOIL PHYSICS, UNIV CALIF, DAVIS, 68- *Concurrent Pos:* NSF sr fel, 65-66; consult, Int Atomic Energy Agency, Vienna, Austria, 74-75. *Mem:* Sigma Xi; Soil Sci Soc Am (pres, 84); Am Agron Soc; Am Geophys Union; Europ Geophys Soc; Int Soc Soil Sci. *Res:* Leaching and miscible displacement of inorganic and organic solutes in soils; soil water properties; pesticide behavior in soils; microbiological transformations during leaching; geo-statistical concepts in relation to developing technologies for sampling field soils. *Mailing Add:* LAWR-Veihmeye Hall Univ Calif Davis CA 95616

NIELSEN, DONALD R, b Oak Park, Ill, Oct 11, 30; m 54; c 3. ORGANIC CHEMISTRY. *Educ:* Knox Col, BA, 52; Univ Kans, PhD(org chem), 56. *Prof Exp:* Res chemist, Chem Div, Pittsburgh Plate Glass Co, 58-64, res supvr, 64-68, sr res assoc, Chem Div, PPG Industs, Inc, 68-; RETIRED. *Mem:* Am Chem Soc. *Res:* Organic synthesis. *Mailing Add:* 9400 River Styx Rd Wadsworth OH 44281

NIELSEN, FORREST HAROLD, b Junction City, Wis, Oct 26, 41; m 64; c 2. NUTRITION, BIOCHEMISTRY. *Educ:* Univ Wis-Madison, BS, 63, MS, 66, PhD(biochem), 67. *Prof Exp:* Res chemist, Beltsville, Md, 69-70, res chemist, 70-86, DIR, HUMAN NUTRIT RES CTR, AGR RES SERV, USDA, 86- *Concurrent Pos:* Adj prof biochem, Univ NDak, 71- *Honors & Awards:* Klaus Schwarz Commemorative Medal, Int Asn Bioinorg Scientists, 90. *Mem:* Am Inst Nutrit; Soc Environ Geochem & Health; Soc Exp Biol & Med; Int Asn Bioinorgan Scientists; Am Soc Magnesium Res; Int Soc Trace Element Res in Humans. *Res:* Trace element nutrition and metabolism; arsenic, boron, nickel, silicon, vandium and the newer essential trace elements. *Mailing Add:* Human Nutrit Res Ctr Agr Res Serv USDA Box 7166 Univ Sta Grand Forks ND 58202-7166

NIELSEN, GERALD ALAN, b Frederic, Wis, Nov 10, 34; m 55; c 4. SOIL SCIENCE, ECOLOGY. *Educ:* Univ Wis, BS, 58, MS, 60, PhD(soil sci), 63. *Prof Exp:* Asst prof soil sci, Univ Wyo team-US Agency Int Develop, Afghanistan, 63-67; assoc prof, 67-73, PROF SOIL SCI, MONT STATE UNIV, 73- *Mem:* AAAS; Soil Sci Soc Am; Am Soc Agron; Soil Conserv Soc Am; Int Soc Soil Sci; Can Soil Sci Soc. *Res:* Soil genesis, classification and ecology; interpretation of soil survey for land use planning; soil inventories and land potential evaluation; geographic information systems for agriculture. *Mailing Add:* Dept Plant & Soil Sci Mont State Univ Bozeman MT 59715

NIELSEN, HARALD CHRISTIAN, b Chicago, Ill, Apr 18, 30; m 53; c 3. PHYSICAL BIOCHEMISTRY. *Educ:* St Olaf Col, BA, 52; Mich State Univ, PhD(biochem), 57. *Prof Exp:* Chemist, Northern Regional Res Ctr, USDA, 57-87; RETIRED. *Mem:* AAAS; Am Chem Soc; Am Asn Cereal Chem. *Res:* Isolation and physical-chemical characterization of cereal grain proteins and related products. *Mailing Add:* 2318 N Gale Ave Peoria IL 61604

NIELSEN, HELMER L(OUIS), b Fredericia, Denmark, Oct 13, 21; US citizen; m 42. THERMODYNAMICS, AERONAUTICS. *Educ:* Univ Calif, Berkeley, BS, 50, MS, 52; Von Karman Inst Fluid Dynamics, Belg, Dipl Ing, 68. *Prof Exp:* Aeronaut scientist, Ames Res Ctr, NASA, 51-59; assoc prof aeronaut eng, Univ Wash, 59-61; prof mech eng, San Jose State Univ, 61-91, chmn dept, 80-85; RETIRED. *Concurrent Pos:* Consult, Lockheed Aircraft Corp, 62-63, Precision Data Inc, 78 & Elec Power Res Inst, 78-80; NASA-Am Soc Eng Educ fel, 69-70. *Mem:* Assoc fel Am Inst Aeronaut & Astronaut; Am Soc Mech Engrs; Am Soc Heating Refrig Air Conditioning Engrs. *Res:* Astronautical engineering; aerodynamics; boundary layer theory; energy; high speed convective heat transfer; thermodynamics; technical, social, political aspects. *Mailing Add:* Dept Mech Eng San Jose State Univ San Jose CA 95192

NIELSEN, JENS JUERGEN, plant pathology, for more information see previous edition

NIELSEN, JOHN MERLE, b Logan, Utah, Aug 31, 28; m 57; c 5. SILICONES, MATERIALS SCIENCE. *Educ:* Brigham Young Univ, BS, 52, Purdue Univ, MS, 54, PhD(org chem), 57. *Prof Exp:* Res & prod develop chemist silicon fluids, Silicon Prod Dept, Gen Elec Co, 56-71, mat engr plastics & chem, Corp Consult Serv, 72-75, mat engr, Plastics & Chem, Lubricants, Coatings, Health & Safety Mat, Corp Res & Develop, 75-85, mat engr, Corp Eng & Mfg, 85-90; EXEC CONSULT, MAT INFO SERV, GE CONSULT SERV INC, 90- *Concurrent Pos:* Ed, Mat Safety Data Sheet Collection, Gen Elec Co, 76-84; subj ed safety, health & envrion, Encyclopedia Mat Sci & Eng, 80-86. *Mem:* Am Chem Soc; Am Soc Testing & Mat. *Res:* Oxidative and thermal stability of silicones; synthesis and hydrolysis of chlorosilanes; synthesis of silicone fluids; specifications, materials information and health and safety information on industrial materials. *Mailing Add:* 243 Petrose Circle Orange CT 06477

NIELSEN, JOHN P(HILLIP), physical metallurgy, materials science; deceased, see previous edition for last biography

NIELSEN, JOHN PALMER, civil engineering, mathematics, for more information see previous edition

NIELSEN, KAJ LEO, b Nyker, Denmark, Dec 3, 14; nat US; m 43; c 2. MATHEMATICS. *Educ:* Univ Mich, AB, 36; Syracuse Univ, MA, 37; Univ Ill, PhD(math), 40. *Prof Exp:* Asst math, Syracuse Univ, 36-37; asst math, Univ Ill, 37-40, instr, 40-41; Carnegie fel, Brown Univ, 41, instr, 41-42; from instr to asst prof, La State Univ, 42-45; sr mathematician, US Naval Ord Plant, Ind, 45-48, head math div, 48-58; chief opers anal, Allison Div, Gen Motors Corp, 58-60; head anal staff, Defense Systs Div, Gen Motors Corp, 60-61; dir systs anal div, Battelle Mem Inst, Ohio, 63-71; lectr, 57-60, prof, 61-63, head dept, 71-85, EMER PROF MATH, BUTLER UNIV, 85- *Concurrent Pos:* Proj anal engr, Chance Vought Aircraft, Inc, Conn, 44-45; lectr, Purdue Univ, 55-58, Otterbein Col, 70; consult, 85- *Mem:* Am Math Soc; Math Asn Am; Am Comput Mach; Sigma Xi. *Res:* Partial differential equations of elliptic type; exterior ballistics; fire control; teaching of applied mathematics; numerical methods; operations and systems analysis; man-machine and management systems; computer science. *Mailing Add:* 2226 Del Mar Dr North Ft Myers FL 33903

NIELSEN, KENNETH FRED, soil fertility, plant physiology, for more information see previous edition

NIELSEN, KENT CHRISTOPHER, b Pocatello, Idaho, Sept 5, 45; m; c 2. STRUCTURAL GEOLOGY, TECTONOPHYSICS. *Educ:* Univ NC, BS, 68, MS, 72; Univ BC, PhD(geol), 78. *Prof Exp:* From instr to asst prof, 76-82, ASSOC PROF GEOL, UNIV TEX, DALLAS, 82- *Concurrent Pos:* NSF prin investr, 77-79, 82-84. *Mem:* Geol Soc Am; Am Geophys Union. *Res:* Field structural geology, fold analysis, cleavage development, metamorphic petrology; experimental rock deformation, high temperature creep, diffusion mechanisms, pressure solution. *Mailing Add:* Prog of Geosci Univ of Tex at Dallas PO Box 688 Richardson TX 75083-0688

NIELSEN, KLAUS H B, b Copenhagen, Denmark, Aug 18, 45; Can citizen; m 66; c 2. IMMUNOCHEMISTRY, IMMUNOASSAYS. *Educ:* Univ Guelph, BSc, 69, MSc, 71; Univ Glasgow, PhD(immunol), 74. *Prof Exp:* Asst prof immunol, Univ Guelph, 74-77; scientist, Animal Dis Res Inst, Agr Can, 77-80; assoc prof, Tex A&M Univ, 80-83; SCIENTIST, ANIMAL DIS RES INST, AGR CAN, 83- *Mem:* Am Asn Immunologists; Brit Soc Immunol; Am Asn Vet Immunologists; Can Soc Immunol; Sigma Xi. *Res:* Ruminant immune response to infectious agents; immunoglobulin structure and function; complement; monoclonal antibodies. *Mailing Add:* PO Box 11300 Sta H Nepean ON V2H 8P9 Can

NIELSEN, LAWRENCE ARTHUR, b Minneapolis, Minn, Aug 7, 34; m 62; c 2. ORGANIC CHEMISTRY, PATENTS. *Educ:* Univ Minn, BS, 56; Univ Nebr, MS, 59, PhD(org chem), 62. *Prof Exp:* Res chemist, Chemstrand Res Ctr, Inc, Monsanto Co, 62-70; head patent liaison, 70-85, HEAD PATENT DEPT, BURROUGHS WELLCOME CO, 85- *Mem:* Am Chem Soc; Sigma Xi. *Res:* Heterocyclic synthesis; medicinal chemistry. *Mailing Add:* 416 Ridgecrest Dr Chapel Hill NC 27514

NIELSEN, LAWRENCE ERNIE, b Pilot Rock, Ore, Dec 17, 17; m 42; c 1. POLYMER SCIENCE, COMPOSITE MATERIALS. *Educ:* Pac Univ, AB, 40; State Col Wash, MS, 42; Cornell Univ, PhD(phys chem), 45. *Prof Exp:* Asst phys chem, State Col Wash, 40-42; lab asst, Cornell Univ, 42-45; phys chemist, Monsanto Co, 45-55, sr scientist, 55-77; POLYMER CONSULT, 77- *Concurrent Pos:* Fel, Harvard Univ, 52; leader, Juneau Ice Field Proj, Alaska, 53; affil prof, Wash Univ, 65-76. *Honors & Awards:* Bingham Award, Soc Rheol, 76; Int Res Award, Soc Plastics Engrs, 81. *Mem:* Am Chem Soc; Am Phys Soc; Soc Rheol. *Res:* Molal volumes of electrolytes; Raman, infrared and ultraviolet spectroscopy; fractionation of proteins; molecular structure of high polymers and its relation to physical and mechanical properties; properties of composite materials; glaciology and flow properties of ice. *Mailing Add:* 3208 NW Lynch Way Redmond OR 97756

NIELSEN, LEWIS THOMAS, b Salt Lake City, Utah, Aug 6, 20; m 83. ENTOMOLOGY. *Educ:* Univ Utah, BA, 41, MA, 47, PhD, 55. *Prof Exp:* From instr to prof, 46-89, EMER PROF BIOL & ENTOM, UNIV UTAH, 89 - *Concurrent Pos:* prin investr, Mosquito Res, Scand, 78, 79 & 85 & Nfld, 81; ed, Mosquito Systematics (J Am Mosquito Control Asn). *Honors & Awards:* Medal of Honor, Am Mosquito Control Asn, 88. *Mem:* Am Mosquito Control Asn (vpres, 75-76, pres, 77). *Res:* Systematics, biology and distribution of mosquitoes of Holarctic Region; medical entomology; nematode parasites of mosquitoes. *Mailing Add:* Dept Biol Univ Utah Salt Lake City UT 84112

NIELSEN, MERLYN KEITH, b Omaha, Nebr, Oct 9, 48; m 70. ANIMAL BREEDING. *Educ:* Univ Nebr-Lincoln, BS, 70; Iowa State Univ, MS, 72, PhD(animal sci), 74. *Prof Exp:* Asst prof, 74-78, ASSOC PROF, UNIV NEBR-LINCOLN, 78-, MEM FAC ANIMAL SCI, 76- *Mem:* Am Soc Animal Sci; Biomet Soc; Sigma Xi. *Res:* Accurate identification of additive genetic differences in beef cattle; planning crossbreeding systems. *Mailing Add:* A218 Animal Sci Univ Nebr Lincoln NE 68583-0908

NIELSEN, MILO ALFRED, b Madelia, Minn, Aug, 20, 38; m 58; c 3. FOOD SCIENCE. *Educ:* Univ Minn, BS, 65, MS, 67, PhD(food sci), 71. *Prof Exp:* Sr food scientist, 71-72, mgr pet food res, 72-78, asst dir pet food res, 78-82, dir basic res, Carnation Res Lab, Van Nuys, Calif, 82-84; dir res SVC, Calreco Inc, Van Nuys, Calif, 85-88; GROUP LEADER, VITORECO LTD, KEMPTTAL, SWITZ, 88- *Mem:* Inst Food Technologists; Am Chem Soc. *Res:* Emulsion stability and oxidation. *Mailing Add:* 17520 Orna Dr Granada Hills CA 91344

NIELSEN, N NORBY, b Denmark, Mar 29, 28; US citizen; m 54; c 2. STRUCTURAL ENGINEERING, EARTHQUAKE DESIGN. *Educ:* Tech Univ Denmark, MS, 54; Calif Inst Technol, PhD(civil eng), 64; Univ Hawaii, MBA, 85. *Prof Exp:* Instr civil eng, Univ Southern Calif, 56-58, asst prof, 58-60; asst prof, Univ Ill, Urbana, 64-66, assoc prof, 66-70; chmn, 72-81, PROF CIVIL ENG, UNIV HAWAII, 70-, CHMN, 86- *Concurrent Pos:* UNESCO expert, Int Inst Seismol & Earthquake Eng, 67-68; Erskine fel, Univ Canterbury, 76. *Honors & Awards:* Moisseiff Award, Am Soc Civil Engrs, 72. *Mem:* Am Soc Civil Engrs; Am Soc Eng Educ; Am Concrete Inst; Am Acad Mech; Earthquake Eng Res Inst. *Res:* Earthquake engineering and design; behavior of reinforced concrete multistory buildings subjected to earthquakes. *Mailing Add:* Col Eng Univ Hawaii 2540 Dole St Honolulu HI 96822

NIELSEN, N OLE, b Edmonton, Alta, Mar 3, 30; m 55; c 3. VETERINARY PATHOLOGY. *Educ:* Univ Toronto, DVM, 56; Univ Minn, PhD(vet path), 63. *Prof Exp:* Private practice, 56-57; res assoc lab animal care & radiation res, Med Dept, Brookhaven Nat Lab, 60-61; from instr to prof vet path, Western Col Vet Med, Univ Sask, 57-82, head dept, 68-74, dean, 74-82; DEAN, ONT VET COL, 85- *Concurrent Pos:* Med Res Coun Can vis scientist, Int Escherichia Ctr, Copenhagen, 70-71; vis scholar, Dept Path, Univ Calif, San Diego, 83-84. *Mem:* AAAS; Am Col Vet Path; Am Vet Med Asn; Can Vet Med Asn (pres, 68-69). *Res:* Relationship of E coli to enteric disease; diseases of swine; enteric pathology; comparative pathology. *Mailing Add:* Ont Vet Col Univ Guelph Guelph ON N1G 2W1 Can

NIELSEN, NIELS CHRISTIAN, b Madison, Wis, July 24, 42; m 66; c 2. MOLECULAR BIOLOGY. *Educ:* Univ Wis-Madison, BS, 66; Vanderbilt Univ, PhD(molecular biol), 72. *Prof Exp:* Res assoc genetics, Inst Genetics, Univ Copenhagen, Denmark, 72-74 & assoc instr biochem, dept biochem & biophys, Univ Calif, Davis, 74-77; from adj asst prof to adj assoc prof, 78-85, ADJ PROF AGRON, DEPT AGRON, PURDUE UNIV, 82-; RES PLANT GENETICIST, AGR RES SER, USDA, 78- *Concurrent Pos:* Marshall fel, Am Scand Found, 72; vis scientist, Plant Breeding Inst, Trumpington, Cambridge, Eng, 84-85. *Honors & Awards:* Archer-Daniels-Midland Award, Am Oil Chem Soc, 86, 88. *Mem:* Am Soc Plant Physiol; Am Soc Biol Chemists; Am Soc Agron; Int Soc Plant Molecular Biol. *Res:* Structure and function of plant genes; molecular sources of genetic variation; site directed mutagenesis; crop improvement. *Mailing Add:* Dept Agron Purdue Univ USDA Agr Res Serv Lafayette IN 47907

NIELSEN, NORMAN RUSSELL, b Pittsburgh, Pa, Sept 8, 41; m 63; c 1. KNOWLEDGE-BASED SYSTEMS & EVALUATION. *Educ:* Pomona Col, BA, 63; Stanford Univ, MBA, 65, PhD(opers & systs anal), 67. *Prof Exp:* From asst prof to assoc prof opers & systs anal, Stanford Univ, 66-73; sr res engr, 73-74, mgr info systs group, Stanford Res Inst, 74-75; comput syst prog mgr, SRI Int, 75-79, prod dir & dir, Advan Comput Syst Dept, 79-86, dir Intelligent Systs Lab, 86-88, ASSOC DIR INFO TECHNOL CTR, SRI INT, 88-; DIR, BERGSTROM CAPITAL CORP, 76- *Concurrent Pos:* From asst dir to dept dir, Stanford Comput Ctr, 66:72. *Mem:* AAAS; Asn Comput Mach; Inst Mgt Sci; Am Asn Artificial Intelligence; Inst Elec & Electronic Engrs Comput Soc. *Res:* Knowledge-based systems, expert systems; computer resource allocation; computer modeling; system simulation; computer networks; management information systems; parallel-processing computer systems; distributed systems; computer performance measurement and evaluation; computer system design. *Mailing Add:* Info Technol Ctr 333 Ravenswood Ave Menlo Park CA 94025

NIELSEN, PAUL HERRON, b Berkeley, Calif, June 14, 43; m 69; c 2. SOLID STATE PHYSICS. *Educ:* Univ Chicago, BS, 64, MS, 65, PhD(physics), 70. *Prof Exp:* Assoc scientist, Xerox Corp, 69-71, scientist, 71-78, sr scientist, 78-79; scientist, Univ Del, 79-80, mgr electronic mat, 80-83; VPRES, NIELSEN-KELLERMAN CO, INC, 78- *Concurrent Pos:* Assoc prof, Bartel Res Found, 80-83; Consult, computers & physics, 80- *Mem:* Am Phys Soc; Am Phys Soc; Am Vacuum Soc; Int Solar Energy Soc. *Res:* Ultraviolet photoemission spectroscopy; energy levels and electronics structures of molecules, solids, and interfaces; photosensitization, photovoltaic conversion, amorphous silicon; design and development of electronic systems employing microcomputers as a major component. *Mailing Add:* 1817 Shipley Rd Wilmington DE 19803

NIELSEN, PETER ADAMS, b Evanston, Ill, Oct 12, 26; m 52; c 2. MICROBIOLOGY. *Educ:* Williams Col, BA, 50; Columbia Univ, MA, 56, PhD(bot), 60. *Prof Exp:* Biologist, Lederle Labs, Am Cyanamid Co, 51-55; asst bot, Barnard Col, Columbia Univ, 55-59; microbiologist, Lederle Labs, Am Cyanamid Co, NY, 59-67; head, Microbiol Sect, 67- 82, SR RES INVESTR, RICHARDSON-VICKS, INC, 82- *Mem:* Am Soc Microbiol; Soc Indust Microbiol. *Res:* Analytical microbiology; microbiological quality control; quality assurance of health and personal care products. *Mailing Add:* 27 Firehouse Rd Trumbull CT 06611

NIELSEN, PETER JAMES, b North Platte, Nebr, Feb 21, 38; m 61; c 3. CELL PHYSIOLOGY, PROTOZOOLOGY. *Educ:* Midland Col, BS, 60; Univ Nebr, MS, 65, PhD(zool physiol), 68. *Prof Exp:* Teaching asst physiol, Univ Nebr, 64-66; from asst prof to assoc prof, 68-82, PROF BIOL SCI, WESTERN ILL UNIV, 83- *Concurrent Pos:* Post doctoral fel, Ill Inst Technol Res Inst, 79. *Mem:* AAAS; Am Soc Zool; Soc Protozool; Tissue Cult Asn; Sigma Xi. *Res:* Effects on the phenomena of aging using the protozoan, Tetrahymena pyriformis, as a research tool; continuous culture of Tetrahymena as a tool for studying metabolism; tissue culture of chick embryo fibroblasts; studying effects of various toxins on in vitro aging. *Mailing Add:* Dept Biol Sci Western Ill Univ Macomb IL 61455

NIELSEN, PETER TRYON, b Durham, NC, Dec 24, 33; m 60; c 2. PLANT PHYSIOLOGY, BIOPHYSICS. *Educ:* Duke Univ, BS, 57; Univ NC, Chapel Hill, PhD(algal physiol), 65. *Prof Exp:* NIH fel biophys, East Anglia, 65-66; asst prof plant physiol & biophys, State Univ NY Col Plattsburgh, 66-70; ASSOC PROF BIOL, JAMES MADISON UNIV, 70- *Mem:* Am Soc Plant Physiol; Bot Soc Am. *Res:* Ion accumulation and transport; photosynthesis and role of light reactions in ion transport; effects of heavy metal ions and algal physiology. *Mailing Add:* Dept of Biol James Madison Univ Harrisonburg VA 22807

NIELSEN, PHILIP EDWARD, b Chicago, Ill, July 18, 44; m 71; c 3. TRADE-OFF ANALYSIS, ZERO-ORDER PHYSICS. *Educ:* Ill Inst Technol, BS, 66; Case Western Reserve Univ, MS, 68, PhD(physics), 70. *Prof Exp:* Physicist, Air Force Weapons Lab, US Air Force, 70-74; asst prof physics, Air Force Inst Technol, 74-77, assoc prof, 77-79, dep head dept, 78-79; DIR, DIRECTORATE AEROSPACE STUDIES, AIR FORCE SYSTS COMMAND, 80- *Mem:* AAAS; Am Phys Soc. *Res:* System effectiveness modeling; interaction of high-intensity lasers with matter; laser-plasma interactions; transport phenomena in metals and alloys. *Mailing Add:* 9138 Payne Farm Lane Spring Valley OH 45370

NIELSEN, ROBERT PETER, industrial chemistry, for more information see previous edition

NIELSEN, STUART DEE, b Green River, Wyo, Oct 26, 32; m 54; c 5. INDUSTRIAL HYGIENE, ANALYTICAL CHEMISTRY. *Educ:* Univ Wyo, BS, 54; Univ Wash, PhD(org chem), 62. *Prof Exp:* Sr chemist, Rohm and Haas Co, Pa, 62-66; sr res chemist, 66-68, res scientist, Gen Tire & Rubber Co, 68-78; mem staff, 78-79, sect leader, 79-84, STAFF, LOS ALAMOS NAT LAB, 84- *Mem:* Am Chem Soc; Am Soc Mass Spectros; Soc Appl Spectros. *Res:* Polymer synthesis and properties; kinetics and mechanisms of polymerization processes; pollution control methods and analysis. *Mailing Add:* 114 Sherwood Blvd White Rock NM 87544

NIELSEN, SURL L, b Long Beach, Calif, 36. NEUROLOGICAL PAHTOLOHY. *Educ:* Univ San Francisco, MD, 61. *Prof Exp:* Intern, King Co Hosp, Seattle, 61-62; res path, 65-68; fel neuropath, Mass Gen Hosp, 68-70; LAB DIR PATH, DIAG PATH MED GROUP. *Mailing Add:* Diag Path Group 2420 J St Sacramento CA 95816

NIELSEN, SUSAN THOMSON, b Astoria, Ore, Apr 26, 47; m 69. PHARMACOLOGY. *Educ:* Univ Chicago, BS, 69; Univ Rochester, PhD(pharmacol), 74. *Prof Exp:* Fel radiol biol & biophys, Univ Rochester, 75-78; fel pharmacol, ICI Americas Inc, 78-79, res pharmacologist, 80-81; SUPVR GI PHARMACOL, WYETH LABS, INC, 81- *Mem:* AAAS; Am Chem Soc; NY Acad Sci. *Res:* Pharmacology of receptors; mechanism of action of histamine, steroid and protein hormones; hormone receptors; cyclic nucleotides and hormone stimulated adenyl cyclase activity. *Mailing Add:* 1817 Shipley Rd Wilmington DE 19803

NIELSEN, SVEND WOGE, b Herning, Denmark, Apr 4, 26; nat US; m 52; c 3. VETERINARY PATHOLOGY. *Educ:* Herning Gym, Denmark, Artium, 45; Royal Vet Col, DVM, 51; Ohio State Univ, MSc, 57, PhD(path), 59. *Prof Exp:* Res path, Angell Mem Animal Hosp, Boston, 51-52; res asst, Ont Vet Col, Can, 52-53, lectr, 53-55, asst prof, 55; from instr to assoc prof vet path, Ohio State Univ, 55-60; PROF PATHOBIOL, UNIV CONN, 60- *Concurrent Pos:* NIH spec fel, Univ Cambridge, 67-68; dir, WHO Collab Lab Urogenital Tumors of Animals, 68-74; dir, Northeastern Res Ctr Wildlife Dis, 72-; vis prof, Cornell Univ, 74 & Univ Calif, Davis, 81-82; fel, Armed Forces Inst Path, Washington, DC, 88. *Mem:* Int Acad Path; Am Vet Med Asn; Am Col Vet Path (pres, 71-72); Wildlife Dis Asn. *Res:* Comparative oncology; pathology of vitamin A deficiency and toxicosis in animals; diseases of wildlife in the northeastern United States; classification of urogenital neoplasms of animals; pathology of lead and mercury poisoning; defining the problem of environmental pollutants in diseases of wildlife and particularly their effects on the immune system; investigating new, emerging anthropozoonoses and defining the role of free-living species in their transmission (Lyme disease, larval migrans, chlamydiosis, and canine dirofilariasis); studying neoplasia of free-living species with emphasis on viral and pollutant-related tumors. *Mailing Add:* 498 Gurleyville Rd Storrs CT 06268

NIELSON, CLAIR W, b Pocatello, Idaho, Dec 10, 35. PHYSICS. *Educ:* Mass Inst Technol, SB, 57, PhD(physics), 62. *Prof Exp:* From instr to asst prof physics, Swarthmore Col, 63-68; MEM STAFF, LOS ALAMOS NAT LAB, 68- *Mem:* Am Phys Soc. *Res:* Atomic and molecular theory; numerical simulation of plasma. *Mailing Add:* PO Box 456 Los Alamos NM 87544

NIELSON, DENNIS LON, b Urbana, Ill, Jan 13, 48; m 70. ECONOMIC GEOLOGY. *Educ:* Beloit Col, BA, 70; Dartmouth Col, MA, 72, PhD(geol), 74; Univ Utah, MBA, 86. *Prof Exp:* staff geologist, Anaconda Co, 74-78; geologist, Earth Sci Lab, 78-79, proj mgr, 79-85, sect mgr geol, 81-85, ASSOC DIR TECHNOL, UNIV UTAH RES INST, 85- *Concurrent Pos:* Instr, Calderas & Hydrothermal Systs, Yellowstone Inst, 79- *Mem:* Geol Soc Am; Am Geophys Union; Soc Econ Geologists; Geothermal Resources Coun. *Res:* Structural controls of geothermal systems; genesis of uranium concentrations in igneous and metamorphic systems; geothermal exploration technology. *Mailing Add:* Earth Sci Lab Univ Utah Res Inst 391-C Chipeta Way Salt Lake City UT 84108

NIELSON, ELDON DENZEL, b Salt Lake City, Utah, Dec 4, 20; m 42; c 3. BIOCHEMISTRY. *Educ:* Univ Utah, AB, 46; Univ Ill, PhD(biochem), 48. *Prof Exp:* Instr biochem, Univ Southern Calif, 48-49; sect head, Upjohn Co, 49-56; head biochem dept, Armour Pharmaceut Co, 56-58; head biochem, Sterling-Winthrop Res Inst, 58-62; mgr biol res div, RJ Reynolds Tobacco Co, 62-69; vpres phys sci, Mead Johnson Res Ctr, 69-75, DIR LICENSING, MEAD JOHNSON & CO, 75- *Mem:* AAAS; Soc Indust Microbiol; Am Chem Soc; NY Acad Sci. *Res:* Adrenal steroids; fermentation; natural products; tobacco; pulmonary physiology; pharmaceutical synthesis; chemical development. *Mailing Add:* 11790 S Nicklaus Rd Sandy UT 84092

NIELSON, GEORGE MARIUS, b Wadsworth, Ohio, May 17, 34. MATHEMATICS. *Educ:* Ohio Wesleyan Univ, BA, 56; Univ Wis, MS, 57, PhD(math), 63. *Prof Exp:* Instr math, Ohio Wesleyan Univ, 59-60; asst prof, 63-70, ASSOC PROF MATH, KALAMAZOO COL, 70- *Mem:* Am Math Soc; Math Asn Am. *Res:* Lie algebras. *Mailing Add:* Dept of Math Kalamazoo Col Kalamazoo MI 49007

NIELSON, HOWARD CURTIS, b Richfield, Utah, Sept 12, 24; m 48; c 7. STATISTICS, MATHEMATICS. *Educ:* Univ Utah, BS, 47; Univ Ore, MS, 49; Stanford Univ, MBA, 56, PhD(bus admin & statist), 58. *Prof Exp:* Actg instr math, Univ Utah, 46-47; asst, Univ Ore, 47-49; sr statistician, Calif & Hawaii Sugar Refining Corp, 49-51; res economist, Stanford Res Inst, 51-57; assoc prof econ, Brigham Young Univ, 57-60, assoc prof statist, 60-61, chmn dept, 60-63, prof, 61-76 & 78-82, dir, Ctr Bus & Econ Res, 71-72, EMER PROF STATIST, BRIGHAM YOUNG UNIV, 82- *Concurrent Pos:* Consult, Hercules Powder Co, 60-65; mgr & consult, C-E-I-R, Inc, 63-65; prin scientist, GCA Corp, 65-67 & Fairchild Semiconductor Corp, 67; consult, EG&G, Inc, 67; econ develop consult, Ford Found, Jordan, 70; mem, Gov Econ Resources Adv Coun, 67-69, Gov Sci Adv Comt, 72-75; Utah State Rep, 67-74, US Rep, 83-90; mem Asn Comn Higher Educ, 76-78. *Mem:* Am Statist Asn; Sigma Xi. *Res:* Economic forecasting; demographic studies and projections; statistical methods in industry; sampling survey methods; experimental design; probability; reliability; operations research. *Mailing Add:* Dept Statist Brigham Young Univ Provo UT 84601

NIELSON, JANE ELLEN, b Clinton, SC, Jan 23, 43; m 84; c 2. GEOLOGY. *Educ:* George Washington Univ, AB, 65; Univ Mich, MS, 68; Stanford Univ, PhD(geol), 74. *Prof Exp:* Res asst, Smithsonian Inst, 67; instr geol, Northern Ariz Univ, 68-69; GEOLOGIST, US GEOL SURV, 78- *Concurrent Pos:* Lectr earth sci, Calif State Univ, Hayward, 77-78. *Mem:* Fel Geol Soc Am; Geochem Soc; Mineral Soc Am; AAAS; Asn Women Geoscientists. *Res:* Field mapping, petrology and stratigraphic correlation of miocene rocks in extended terrains of the western US; petrology and process modeling of the upper mantle from petrologic and geochemical studies of ultramatic modules from basaltic lavas and of ultramatic rocks in alpine periodotite massifs. *Mailing Add:* US Geol Surv 345 Middlefield Rd MS-975 Menlo Park CA 94025

NIELSON, LYMAN J(ULIUS), sanitary engineering, personnel administration, for more information see previous edition

NIELSON, MERVIN WILLIAM, b Provo, Utah, Apr 7, 27; m 56; c 3. ENTOMOLOGY. *Educ:* Utah State Univ, BS, 49, MS, 50; Ore State Univ, PhD(entom), 55. *Prof Exp:* Asst entomologist, Univ Ariz, 55-56, assoc prof, 68-77; entomologist, 56-68, res leader sci & educ admin-agr res, USDA, 68-83; ADJ PROF BRIGHAM YOUNG UNIV, MONTE L BEAN LIFE SCI MUS, 83- *Concurrent Pos:* Asst, Utah State Univ, 49-50; asst & instr, Ore State Univ, 51-55; adj prof, Univ Ariz, 77-83. *Mem:* AAAS; Entom Soc Am; Soc Syst Zool. *Res:* Taxonomy of the Cicadellidae; biology of leafhoppers and aphids; insect transmission of plant viruses; development of crop resistance to insects. *Mailing Add:* Monte L Bean Life Sci Mus Brigham Young Univ Provo UT 84602

NIELSON, READ R, b Omaha, Nebr, Aug 4, 28; m 54; c 4. PHYSIOLOGY. *Educ:* Grinnell Col, BA, 50; Univ Iowa, MS, 52; Marquette Univ, PhD(physiol), 61. *Prof Exp:* Asst prof, 61-67, ASSOC PROF ZOOL, MIAMI UNIV, 67- *Concurrent Pos:* NIH res grant, 62-65. *Mem:* Am Physiol Soc. *Res:* Thyroid physiology and regulation of metabolism; effect of electrical stimulation on glycogen concentration in skeletal muscle during acute inanition; metabolic changes in the intact rat and excised tissues after thyroidectomy; changes in succinic dehydrogenase activity related to feeding. *Mailing Add:* Dept of Zool Miami Univ Oxford OH 45056

NIEM, ALAN RANDOLPH, b New York, NY, Mar 7, 44; m 67. GEOLOGY. *Educ:* Antioch Col, BS, 66; Univ Wis-Madison, MS, 68, PhD(geol), 71. *Prof Exp:* Asst geol, Univ Wis-Madison, 66-70; asst prof, 70-76, ASSOC PROF, ORE STATE UNIV, 76- *Mem:* Am Asn Petrol Geol; Soc Econ Paleont & Mineral; Geol Soc Am. *Res:* Sedimentation; sedimentary petrography; volcaniclastic sediments; stratigraphy of forearc and subduction zone rocks of the Pacific Northwest; hydrogeology. *Mailing Add:* Dept Geosci Ore State Univ Corvallis OR 97331-5506

NIEM, WENDY ADAMS, b New York, NY, Feb 11, 46; m 67. BASIN ANALYSIS, PALEOGEOGRAPHIC RECONSTRUCTIONS. *Educ:* Univ Wis-Madison, BS, 70; Ore State Univ, Corvallis, MS, 76. *Prof Exp:* Teaching asst, Dept Geog, Ore State Univ, 72-73, teaching asst, Dept Geol, 74, instr, 77, RES ASSOC GEOL, ORE STATE UNIV, 80-84 & 89-; CONSULT GEOLOGIST, 81- *Concurrent Pos:* Res asst geol, US Geol Surv, 76-77 & 78-81. *Res:* Basin analysis; paleogeographic reconstructions; sedimentary petrography including modern heavy mineral sands; sedimentation and distribution of volcaniclastic sedimentary rocks. *Mailing Add:* Dept Geol Ore State Univ Corvallis OR 97331-5506

NIEMAN, GARY FRANK, SURFACTANT PHYSIOLOGY. *Educ:* State Univ NY Col Geneseo, BS, 72. *Prof Exp:* RES ASST, DEPT RESPIRATORY THER, UPSTATE MED CTR, STATE UNIV NY, SYRACUSE, 81- *Mailing Add:* Upstate Med Ctr State Univ NY 750 E Adams St Syracuse NY 13210

NIEMAN, GEORGE CARROLL, b Dayton, Ohio, Dec 25, 38; m 60; c 3. PHYSICAL CHEMISTRY, SPECTROSCOPY. *Educ:* Carnegie Inst Technol, BS, 61; Calif Inst Technol, PhD(chem), 65. *Prof Exp:* Asst prof chem, Univ Rochester, 64-70; assoc prof, Muskingum Col, 70-76, prof chem, 76-79; ASSOC PROF, MONMOUTH COL, 79- *Mem:* Am Chem Soc. *Res:* Molecular crystals; nonradiative transitions; energy transfer; triplet states. *Mailing Add:* 1208 Lawnway Dr Monmouth IL 61462

NIEMAN, RICHARD HOVEY, b Pasadena, Calif, Nov 7, 22; m 46; c 2. PLANT PHYSIOLOGY. *Educ:* Univ Southern Calif, AB, 49, MS, 53; Univ Chicago, PhD(plant physiol), 55. *Prof Exp:* Asst histol & morphol, Univ Chicago, 51-54, plant physiol, 54, res assoc biochem, 55-57; plant physiologist, US Salinity Lab, 57-88; RETIRED. *Mem:* AAAS; Am Soc Plant Physiol; Am Inst Biol Sci; Sigma Xi; NY Acad Sci; physiological-biochemical basis of salt tolerance of plants. *Res:* Influence of salinity and drought on ion uptake, metabolism and bioenergetics of plant cells; physiological and biochemical basis of salt tolerance of plants. *Mailing Add:* 5755 North View Pl Riverside CA 92506

NIEMAN, TIMOTHY ALAN, b Cincinnati, Ohio, Dec 31, 48; m 70; c 2. CHEMILUMINESCENCE. *Educ:* Purdue Univ, BS, 71; Mich State Univ, PhD(anal chem), 75. *Prof Exp:* Asst prof, 75-81, ASSOC PROF ANAL CHEM, UNIV ILL, URBANA, 81- *Concurrent Pos:* Us ed, Mikrochimica Acta, 86- *Mem:* Am Chem Soc; Soc Appl Spectros; Soc Electroanal Chem. *Res:* Analytical chemistry involving techniques and instrumentation for trace analysis of species in solution, chemiluminescence detection in liquid chromatography, sensors; bipolar pulse conductance; coulostatic electroanalysis. *Mailing Add:* Dept Chem Univ Ill 1209 W Calif St Urbana IL 61801

NIEMANN, MARILYN ANNE, IMMUNOLOGY. *Educ:* City Univ NY, PhD(biol), 76. *Prof Exp:* ASST PROF, UNIV ALA, BIRMINGHAM, 83- *Mailing Add:* 2214 Cahaba Rd Birmingham AL 35223

NIEMANN, RALPH HENRY, b Farley, Mo, Mar 16, 22; m 48; c 3. MATHEMATICS. *Educ:* Park Col, BA, 47; Purdue Univ, MS, 49, PhD(math), 54. *Prof Exp:* Assoc prof math, Worcester Polytech Inst, 54-59; asst prof, 59-63, chmn math sect, Dept Math & Statist, 67-68, actg chmn dept, 68-69, PROF MATH, COLO STATE UNIV, 63- *Concurrent Pos:* Lectr, Math Asn High Sch Lectr Prog, 61, 62; consult, Summer Sci Inst Prog, India, 64-65. *Mem:* Math Asn Am. *Res:* Mathematical analysis. *Mailing Add:* Dept of Math Colo State Univ Ft Collins CO 80523

NIEMANN, THEODORE FRANK, b Burlington, Iowa, July 31, 39. POLYMER CHEMISTRY, COMPUTER SCIENCE. *Educ:* Univ Iowa, BS, 61; Univ Kans, PhD(chem), 67. *Prof Exp:* Res chemist, BF Goodrich Res Ctr, Ohio, 67-71; SCIENTIST, GLIDDEN CO, STRONGSVILLE, OHIO, 71- *Mem:* Am Chem Soc. *Res:* Computer applications in chemistry, laboratory automation and real-time systems. *Mailing Add:* 11442 Harbour Light Dr North Royalton OH 44133

NIEMCZYK, HARRY D, b Grand Rapids, Mich, July 17, 29; m 56; c 4. ENTOMOLOGY. *Educ:* Mich State Univ, BS, 57, MS, 58, PhD(entom), 61. *Prof Exp:* Asst instr entom, Mich State Univ, 60-61; entomologist, Can Dept Agr, 61-64; from asst prof to assoc prof, 64-71, PROF ENTOM, OHIO STATE UNIV & OHIO AGR RES & DEVELOP CTR, 71- *Concurrent Pos:* Consult entomologist, 74-; mem, educ staff, Golf Course Superintendants Asn, 76-; ed adv, Weeds, Trees & Turf, 81-; sem instr, Prof Lawn Care Asn, 84- *Mem:* Entom Soc Am; Int Turfgrass Soc; Am Soc Agron. *Res:* Biology, ecology and control of insects associated with agricultural crops and turfgrasses; the fate and movement of pesticides applied to turf. *Mailing Add:* 2935 E Smithville Western Wooster OH 44691

NIEMCZYK, THOMAS M, b Madison, Wis, Mar 17, 47; m 81; c 1. ATOMIC SPECTROSCOPY, CHEMICAL INSTRUMENTATION. *Educ:* Univ Wis, BS, 69; Mich State Univ, PhD(chem), 73. *Prof Exp:* From asst prof to assoc prof, 80-85, PROF CHEM, UNIV NMEX, 85- *Concurrent Pos:* Consult, Air Force Weapons Lab, Albuquerque, 75-77, Los Alamos Nat Lab, 81-, Deuel & Assoc, 81-; vis prof, Univ Wis, 79; Assoc Western Univs fel, Sandia Nat Lab, 87-88. *Mem:* Am Chem Soc; Am Optical Soc; Soc Appl Spectros. *Res:* Analytical spectroscopy and instrumentation; plasma diagnostics, the development of atomic vapor sources; energy transfer from excited state nitrogen molecules; the development of intelligent instrumentation systems. *Mailing Add:* Dept Chem Univ NMex Albuquerque NM 87131

NIEMEIER, RICHARD WILLIAM, b Akron, Ohio, May 16, 45; m 66; c 3. ENVIRONMENTAL HEALTH. *Educ:* Thomas More Col, AB, 67; Univ Cincinnati, MS, 69, PhD(environ health sci), 73. *Prof Exp:* NIH fel, 74-75, asst prof, Dept Environ Health, Col Med, Univ Cincinnati, 75-76; CHIEF, ACUTE & SUBCHRONIC TOXICOL SECT, EXP TOXICOL BR, DIV BIOMED & BEHAV SCI, NAT INST OCCUP SAFETY & HEALTH, HEW, 76- *Concurrent Pos:* Adj asst prof, Dept Environ Health, Col Med, Univ Cincinnati, 76-; mem, Task Force Environmental Cancer, Heart & Lung Disease, Dept Health & Human Serv & mem, Subcomt Environ Mutigens. *Mem:* Sigma Xi. *Res:* Pulmonary metabolism of carcinogens using the isolated perfused lung; industrial hygiene surveys and characterization of carcinogenic occupational environments; effects of inhaled toxicants on pulmonary alveolar macrophages; factors affecting carcinogenic response; teratology, mutagenesis; cutaneous toxicology. *Mailing Add:* Nat Inst Occup Safety/Health 4767 Columbia Pkwy Cincinnati OH 45226

NIEMEYER, KENNETH H, b St Louis, Mo, Sept 15, 28; m 53; c 1. VETERINARY MEDICINE. *Educ:* Univ Mo, BS & DVM, 55, MS, 62. *Prof Exp:* From instr to prof, 55-78, head small animal clins, 63-68, PROF & ASSOC CHMN VET MED & SURG, UNIV MO-COLUMBIA, 74-, ASSOC DEAN, COL VET MED, 76- *Mem:* Am Vet Med Asn. *Res:* Clinical veterinary medicine, especially as applied to small animals. *Mailing Add:* W203 Vet Med Univ Mo Columbia MO 65211

NIEMEYER, SIDNEY, b Grand Rapids, Mich, Oct 27, 51; m 72; c 3. PHYSICS, SPECTROSCOPY & SPECTROMETRY. *Educ:* Calvin Col, AB, 73; Univ Calif, Berkeley, PhD(physics), 78. *Prof Exp:* Asst res physicist, Univ Calif, Berkeley, 78, res chemist, 79-80, STAFF PHYSICIST, LAWRENCE LIVERMORE NAT LAB, UNIV CALIF, SAN DIEGO, 80- *Mem:* Meteoritical Soc; Am Geophys Union; Am Sci Affil. *Res:* Cosmochemistry; geochemical evolution of the earth's mantle and crust; goechemical tracing of environmental processes. *Mailing Add:* Lawrence Livermore Natl Lab PO Box 808 L-232 Livermore CA 94550

NIEMI, ALFRED OTTO, b Grand Marais, Mich, Aug 13, 15; m 43; c 1. CONSERVATION. *Educ:* Mich State Univ, BS, 48, MA, 52, EdD(forestry ed), 60. *Prof Exp:* Teacher pub sch, Mich, 48-50; asst agr, Mich State Univ, 50-51; teacher pub sch, Mich, 51-56; asst prof agr & conserv forestry, 56-68, assoc prof conserv, 68-72, consult, Vista Training Ctr, 65, prof, 72-80, EMER PROF CONSERV, NORTHERN MICH UNIV, 80- *Concurrent Pos:* Asst, Mich State Univ, 58-59; coordr, Upper Mich Jr Acad Sci, Arts & Letters, 59-71. *Mem:* Conserv Educ Asn; Soil Conserv Soc Am. *Res:* Forestry; agriculture; conservation education. *Mailing Add:* 1940 Neidhart Marquette MI 49855

NIEMITZ, JEFFREY WILLIAM, b Orange, NJ, July 14, 50; m 73; c 3. OCEANOGRAPHY, GEOCHEMISTRY. *Educ:* Williams Col, BA, 72; Univ Southern Calif, PhD(geochem), 78. *Prof Exp:* ASSOC PROF GEOL, DICKINSON COL, 77- *Concurrent Pos:* Sedimentologist, Deep Sea Drilling Proj, 78-79. *Mem:* Geochem Soc; Sigma Xi; Am Geophys Union; Nat Asn Geol Teachers. *Res:* Geochemistry, sedimentology and tectonics of young rifting ocean basins (Gulf of California); trace element geochemistry and paleoclimate studies of various marine sediments; trace element pollution in nearshore sediments. *Mailing Add:* Dept Geol Dickinson Col Carlisle PA 17013

NIENHOUSE, EVERETT J, b Oak Park, Ill, Oct 29, 36; m 65. ORGANIC CHEMISTRY. *Educ:* Hope Col, AB, 58; Northwestern Univ, MSc, 62; State Univ NY Buffalo, PhD(org chem), 66. *Prof Exp:* asst prof org chem, 66-77, MEM FAC, FERRIS STATE COL, 77- *Mem:* Am Chem Soc; Royal Soc Chem. *Res:* Chemistry of bridged bicyclic systems; abnormal Grignard reactions; new synthetic reagents in organic chemistry; olefinic cyclizations. *Mailing Add:* Phys Sci Ferris State Col Big Rapids MI 49307

NIENHUIS, ARTHUR WESLEY, b Hudsonville, Mich, Aug 9, 41; m 68; c 4. NEMATOLOGY. *Educ:* Univ Calif, MD, 68. *Prof Exp:* Asst resident med, Mass Gen Hosp, 69-70; clin assoc, Nat Heart, Lung & Blood Inst, NIH, 70-72; clin fel hematol, Med Ctr, Children's Hosp, 72-73; chief, Clin Serv, Sect Clin Hematol, 73-77, CHIEF, CLIN HEMATOL BR, NAT HEART, LUNG & BLOOD INST, NIH, 77- *Concurrent Pos:* Deputy dir clin, Nat Heart, Lung & Blood Inst, 77- *Mem:* Asn Am Physicians; Am Soc Clin Invest; Am Soc Hematol; Am Fedn Clin Res. *Res:* Regulation of hemoglobin synthesis; structure and function of globin genes, and the molecular biology of thalassemia; evaluation and treatment of red blood cell disorders particularly thalassemia; sickle cell anemia and aplastic anemia. *Mailing Add:* Clin Ctr HIH Bldg 10 Rm 7D-18 Bethesda MD 20892

NIENSTAEDT, HANS, b Copenhagen, Denmark, Dec 23, 22; m 49; c 5. FOREST GENETICS. *Educ:* Yale Univ, MF, 48, PhD(path), 51. *Prof Exp:* Asst geneticist, Agr Exp Sta, Univ Conn, 51-55; geneticist, Lake States Forest Exp Sta, 53-60, chief lab, N Cent Forest Exp Sta, Inst Forest Genetics, 60-76, CHIEF PLANT GENETICIST, FORESTRY SCI LAB, US FOREST SERV, 70- *Concurrent Pos:* US rep, Forest Biol Comt, Tech Asn Pulp & Paper Indust, 76; chmn, adv comt, Col Nat Res, Univ Wis, 75-81 & Lake States for Tree Improv Comt, 66-78; secy, Study Group Forest Tree Improv, NAm Forestry Comn, 76-78 & 80. *Mem:* AAAS; Soc Am Foresters; Int Union Forest Res Orgns; Tech Asn Pulp & Paper Indust; Sigma Xi. *Res:* Genetics of conifers of northern North America; breeding of spruces, and pines of northern North America. *Mailing Add:* Santa Maria 35H Morelia MICH 58090 Mexico

NIER, ALFRED OTTO CARL, b St Paul, Minn, May 28, 11; m 37, 69; c 2. PHYSICS. *Educ:* Univ Minn, BE, 31, MS, 33, PhD(physics), 36. *Hon Degrees:* DSc, Univ Minn, 80. *Prof Exp:* Nat Res Coun fels, Harvard Univ, 36-38; from asst prof to assoc prof physics, Univ Minn, 38-43; physicist, Kellex Corp, NY, 43-45; chmn dept, 53-65, prof, 45-80, EMER PROF PHYSICS, UNIV MINN, MINNEAPOLIS, 80- *Honors & Awards:* Day Medal, Geol Soc Am, 56; AEC Award, 71; Except Sci Achievement Award, NASA, 77; Goldschmidt Medal, Geochem Soc, 84; Field & Franklin Award, Am Chem Soc, 85; Thomson Medal, Int Mass Spectrometry Conf, 85. *Mem:* Nat Acad Sci; fel Am Phys Soc; Am Philos Soc; Geochem Soc; Am Geophys Union; AAAS; Max Planck Ges, W Germany; Royal Swed Acad Sci. *Res:* Mass spectrometry; aeronomy. *Mailing Add:* Sch Physics & Astron Univ Minn Minneapolis MN 55455

NIERENBERG, WILLIAM AARON, b New York, NY, Feb 13, 19; m 41; c 2. PHYSICS. *Educ:* City Col New York, BS, 39; Columbia Univ, MA, 42, PhD(physics), 47. *Hon Degrees:* DSc, Univ Md, 81, NJ Inst Technol, 85. *Prof Exp:* Tutor physics, City Col New York, 39-42, res scientist, Manhattan Proj, 42-45; instr physics, Columbia Univ, 46-48; asst prof, Univ Mich, 48-50; assoc prof, Univ Calif, Berkeley, 50-53, prof, 54-65; assoc prof, Univ Paris, 60-62; dir, Scripps Inst Oceanog, 65-86, vchancellor Marine Sci, 69-86, EMER DIR, UNIV CALIF, SAN DIEGO, 86- *Concurrent Pos:* Dir, Hudson Labs, Columbia Univ, 53-54; mem mine adv comt, Nat Res Coun, 54-, consult comt nuclear constants, 58-; prof, Miller Inst Basic Res Sci, 57-59; consult, Nat Security Agency, 58-60 & President's Spec Proj Comt, 58-; asst secy gen sci, NATO, 60-62; adv at large, Dept of State, 68-, NATO sr sci fel, 69; mem, Nat Sci Bd, 72-78 & 82-88; chmn nat adv comt oceans & atmosphere, 71-75; NATO sr sci fel, 69; mem, White House Task Force Oceanog, 69-70; mem oil spill panel, Off Sci & Technol, 69; chmn, NASA adv coun, 78-82; mem, Space Panel, Naval Studies Bd, Nat Res Coun, 78-84; consult, Nat Sci Bd, 88-; chmn, Nat Acad Sci, US Nat Comt, Pac Sci Asn, 88- *Honors & Awards:* Procter Prize, Sigma Xi, 77; Richtmyer Mem Lectr, Am Asn Physics Teachers, 79; Charles H Davis lectr, US Naval Postgrad Sch & Naval War Col, 81; Richtmyer Mem lectr, Am Asn Physics Teachers, 79; Charles H Davis lectr, US Naval Postgrad Sch & Naval War Col, 81; Delmer S Fahrney Medal, The Franklin Inst, 87. *Mem:* Nat Acad Sci; fel Am Phys Soc; Am Acad Arts & Sci; Am Geophys Union; Nat Acad Eng; Am Philos Soc; Sigma Xi. *Res:* Gas diffusion; molecular and atomic beams; physical oceanography; nuclear moments. *Mailing Add:* Scripps Inst Oceanog 0221 Univ Calif San Diego La Jolla CA 92093

NIERING, WILLIAM ALBERT, b Scotrun, Pa, Aug 28, 24; m 55; c 2. PLANT ECOLOGY. *Educ:* Pa State Univ, BS, 48, MS, 50; Rutgers Univ, PhD(plant ecol), 52. *Prof Exp:* From asst prof to assoc prof, 52-64, PROF BOT, CONN COL, 64- *Concurrent Pos:* Mem Kapingamarangi Exped, Caroline Islands, 54; consult, Recreation & Open Space Proj, Regional Plan Asn, Inc, 58; dir, Conn Arboretum, 65-; assoc dir environ biol prog, Nat Sci Found, 67-68. *Honors & Awards:* Mercer Award, Ecol Soc Am, 67. *Mem:* AAAS; Ecol Soc Am; Bot Soc Am; Am Inst Biol Sci. *Res:* Vegetation science; herbicides; wetland ecology; applied ecology. *Mailing Add:* Dept of Bot Conn Col Box 1511 New London CT 06320

NIERLICH, DONALD P, b Ft Lewis, Wash, Aug 10, 35; m 61; c 3. MICROBIAL PHYSIOLOGY, MOLECULAR BIOLOGY. *Educ:* Calif Inst Technol, BS, 57; Harvard Univ, PhD(bact), 62. *Prof Exp:* NSF fels, Mass Inst Technol, 63-64 & Inst Biol & Phys Chem, Paris, France, 64-65; from asst prof to assoc prof, 65-74, PROF MICROBIOL, UNIV CALIF, LOS ANGELES, 74- *Concurrent Pos:* Mem, Molecular Biol Inst, Univ Calif, Los Angeles, 74-; ed, J Bact, 77-83. *Mem:* Am Soc Microbiol; Am Soc Biol Chem. *Res:* Regulation of gene expession, particularly the synthesis of RNA and its control. *Mailing Add:* Dept Anmolecular Genetics Univ Calif Los Angeles CA 90024-1489

NIES, ALAN SHEFFER, b Orange, Calif, Sept 30, 37; m 61; c 2. CLINICAL PHARMACOLOGY. *Educ:* Stanford Univ, BS, 59; Harvard Med Sch, MD, 63. *Prof Exp:* Resident & intern internal med, Univ Wash Hosp, 63-66; NIH fel clin pharmacol, Univ Calif, San Francisco, 66-68; chief clin pharmacol, Walter Reed Army Inst Res, 68-70; from asst prof to prof med & pharmacol, Sch Med, Vanderbilt Univ, 70-77; PROF MED & PHARMACOL, UNIV COLO MED CTR, 77- *Concurrent Pos:* Mem pharmacol study sect, NIH, 81-85, chmn, 83-85. *Honors & Awards:* Rawls-Palmer Award, Am Soc Clin Pharmacol Therapeut, 85. *Mem:* AAAS; Am Fedn Clin Res; Am Soc Clin

Invest; Am Soc Pharmacol & Exp Therapeut; Asn Am Physicians; Am Soc Clin Pharmacol Therapeut. *Res:* Effects of disease on drug metabolism and disposition in man; prostaglandins and circulation; aging and the adrenergic nervous system. *Mailing Add:* Div Clin Pharmacol C237 Univ Colo Health Sci Ctr 4200 E Ninth Ave Denver CO 80262

NIESEN, THOMAS MARVIN, b San Diego, Calif, Apr 10, 44; m 67; c 1. MARINE ECOLOGY, POPULATION BIOLOGY. *Educ:* Univ Calif, Santa Barbara, BA, 66; San Diego State Univ, MS, 69; Univ Ore, PhD(biol), 73. *Prof Exp:* From instr to assoc prof, 73-84, PROF BIOL, SAN FRANCISCO STATE UNIV, 84- *Concurrent Pos:* Nat Oceanog & Atmospheric Admin Hydrolab, 81, 83, 84; proj dir, US Fish & Wildlife Serv, 78-85, Proj Ishtar, 88. *Mem:* Sigma Xi; AAAS; Ecol Soc Am; Am Zool Soc; Am Soc Naturalists. *Res:* Marine invertebrates; esturarine community ecology; intertidal and nearshore subtidal ecology. *Mailing Add:* Dept Biol 1600 Holloway Ave San Francisco CA 94132

NIESSE, JOHN EDGAR, b Indianapolis, Ind, Nov 30, 27; m 58; c 2. MATERIALS SCIENCE. *Educ:* US Naval Acad, BS, 50; Mass Inst Technol, SM, 56, ScD(metall), 58. *Prof Exp:* Supvry engr, Crane Co, 59-60; supvry engr, Carborundum Co, 60-61, mgr process develop dept, 61-63, mgr ceramics & metall dept, 63-64, mgr technol br, NY, 64-67; chief ceramics res & develop sect, Space Systs Div, Avco Corp, Mass, 67-72; sr res group leader steel wire prod, Monsanto Co, 72-76, eng supt, 76-80, prin eng specialist mat technol, 80-89, eng group consult, 89-90; INDEPENDENT CONSULT, 90- *Mem:* Am Ceramic Soc; Nat Asn Corrosion Engrs; Am Soc Metals; Sigma Xi. *Res:* Abrasive materials; high temperature and wear resistant materials; corrosion resistant materials; composition, processing and application. *Mailing Add:* 424 Glan Tai Dr Manchester MO 63011

NIETO, MICHAEL MARTIN, b Los Angeles, Calif, Mar 15, 40; m 73; c 2. THEORETICAL PHYSICS. *Educ:* Univ Calif, Riverside, BA, 61; Cornell Univ, PhD(physics), 66. *Prof Exp:* Res assoc physics, Inst Theoret Physics, State Univ NY Stony Brook, 66-68; vis physicist, Niels Bohr Inst, Copenhagen, Denmark, 68-70; lectr & asst prof physics, Univ Calif, Santa Barbara, 70-71; sr res assoc physics, Purdue Univ, Lafayette, 71-72; MEM STAFF, LOS ALAMOS NAT LAB, 72- *Mem:* Fel Am Phys Soc; Int Asn Math Phys. *Res:* Theoretical physics in fields of high energy, astrophysics and quantum mechanics; recent work on weak interactions, DKP meson and Bhabha arbitrary spin wave equations, quantum phase operators, coherent states, photon mass, law of planetary distances, new gravitational forces, and supersymmetry in physical systems. *Mailing Add:* Theor Div T-8 MS-B285 Los Alamos Nat Lab Los Alamos NM 87545

NIEVERGELT, JURG, b Lucerne, Switz, June 6, 38; US citizen; m 65; c 2. COMPUTER SCIENCE, SOFTWARE SYSTEMS. *Educ:* Swiss Fed Inst Technol, Dipl math, 62; Univ Ill, PhD(math), 65. *Prof Exp:* From asst prof to prof, dept comput sci, Univ Ill, Urbana-Champaign, 65-77; chmn & William Rand Kenan prof, dept comput sci, Univ NC, Chapel Hill, 85-89; PROF, SWISS FED INST TECHNOL, ZURICH, 75- *Concurrent Pos:* Vis assoc prof, Univ Calif & Jet Propulsion Labs, 69, IBM Res Ctr, NY, 70 & CERN, Geneva, 71; vis prof, proj MAC, Mass Inst Technol, 72, Univ Grenoble & IRIA, Paris, 73, Univ Stuttgart, Germany, 74, IBM Res Ctr, Calif, 78, Univ Melbourne, & Univ Wollongong, Australia, 79; consult, Univ Philippines, 81, Nanjing Inst Technol, China, 84, Keio Univ, Japan, 91. *Mem:* Fel AAAS; fel Inst Elec & Electronic Engrs; Asn Comput Mach. *Res:* Computer software; algorithms and data structures; interactive systems; computers in education. *Mailing Add:* Informatik ETH Zurich CH-8092 Switzerland

NIEVERGELT, YVES, b Lausanne, Switz, Apr 23, 54. COMPLEX ANALYSIS, NUMERICAL ANALYSIS. *Educ:* Fed Polytech Sch, Lausanne, dipl, 76; Univ Wash, Seattle, MA, 78, MS, 84, PhC & PhD(math), 84. *Prof Exp:* Teaching asst math, Univ Wash, Seattle, 77-84; instr, St Olaf Col, Northfield, Minn, 84-85; asst prof, Univ Wash, Seattle, 85; ASST PROF MATH, EASTERN WASH UNIV, CHENEY, 85- *Mem:* Am Math Soc; Math Asn Am; Soc Indust & Appl Math. *Res:* Complex integral geometry and numerical inversions of Radon transforms, with occasional applications to computed tomography and nuclear magnetic resonance; application of numerical analysis to microbiology and pharmacology. *Mailing Add:* Eastern Wash Univ Cheney WA 99004

NIEWENHUIS, ROBERT JAMES, b Corsica, SDak, Sept 21, 36; m 58; c 2. ANATOMY, ELECTRON MICROSCOPY. *Educ:* Calvin Col, BS, 59; Mich State Univ, MS, 61; Univ Cincinnati, PhD(anat), 70. *Prof Exp:* Histopathologist, Procter & Gamble Co, 70-72; asst prof, Sch Med, Univ Cincinnati, 72-79, assoc prof anat, 79-84; PROF, PHILADELPHIA COL OSTEOP MED, 84- *Mem:* Electron Micros Soc Am; Am Asn Anatomists; Soc Study Reproduction; Sigma Xi. *Res:* Effects of cryptorchidism vasectomy and toxins upon the ultrastructure of the testis and associated ducts; reproductive biology. *Mailing Add:* Dept Anat Philadelphia Col Osteop 4190 City Ave Philadelphia PA 19131

NIEWIAROWSKI, STEFAN, b Warsaw, Poland, Dec 4, 28. PHYSIOLOGICAL CHEMISTRY. *Educ:* Warsaw Univ, MD, 52, PhD(biochem), 60. *Prof Exp:* Res assoc clin biochem, Inst Hematol, Warsaw, 51-61; prof biochem, Med Sch, Brolystok, 61-68; vis prof med, Tufts Univ Med Sch, 68-70; assoc prof path, McMaster Univ Health Sci Ctr, 70-72; RES PROF MED, TEMPLE UNIV HEALTH SCI CTR, 72- & PROF PHYSIOL, 75- *Concurrent Pos:* Vis scientist, Ctr Nat Transfusion, Paris, 59 & Tufts Univ, 65; fel, Ont Heart Found, 70-71; mem, Int Comt Haemastasis Thrombosis, 68-73; consult, Res Rev Comt NIHL, 75-; NIH grants, 72-82 & Am Heart Asn grant, 78-81. *Mem:* Int Soc Hemat; Int Soc Thrombosis & Haemastasis; Am Physiol Soc; Am Soc Hemat. *Res:* Molecular biology of the platelet and its significance for hemostasis, thrombosis and atherosclerosis; platelet interaction with enzymes, drugs, and plasma proteins; platelet secretory proteins. *Mailing Add:* 416 S 45 Philadelphia PA 19104

NIFFENEGGER, DANIEL ARVID, b Grinnell, Iowa, Apr 7, 30; m 53; c 2. BOTANY, AGRONOMY. *Educ:* Iowa State Univ, BS, 52, PhD(bot, 67; Mont State Univ, MS, 57. *Prof Exp:* Seed analyst, Mont State Univ, 55-64, asst agron, 58-62, asst prof agron, 62-64; asst agron, Iowa State Univ, 64-67; biometrician, 67-69, asst dir, Biomet Serv Staff, 69-72, prog analyst, N Cent Region, 72-76, chief prog planning & rev, Sci & Educ Admin-Agr Res, Peoria, Ill, 76-81, ASST DEP ADMINR, NAT PROG STAFF, AGR RES SERV, USDA, BELTSVILLE, MD, 81- *Mem:* Am Soc Agron; Asn Off Seed Anal; Biomet Soc; Am Statist Asn. *Res:* Seed testing methodology; use of seed test results to determine field seeding rates; development of a seed homogeneity test; seed tolerances; seed research priorities; grain sampling; tobacco fumigation. *Mailing Add:* 1029 Turnberry Lane Lexington KY 40515

NIGAM, BISHAN PERKASH, b Delhi, India, July 14, 28; m 56; c 3. PARTICLE PHYSICS, NUCLEAR PHYSICS. *Educ:* Univ Delhi, BSc, 46, MSc, 48; Rochester Univ, PhD(theoret physics), 54. *Prof Exp:* Lectr physics, Univ Delhi, 50-52 & 55-56; asst, Rochester Univ, 52-54; res fel, Case Inst, 54-55; Nat Res Coun Can res fel, 56-59; res assoc, Rochester Univ, 59-60, asst part-time prof physics, 60-61; assoc prof, State Univ NY Buffalo, 61-64; PROF PHYSICS, ARIZ STATE UNIV, 64- *Concurrent Pos:* Prin scientist, Basic Sci Res Lab, Gen Dynamics/Electronics, 60-61; prof, Univ Wis-Milwaukee, 66-67. *Mem:* Fel Am Phys Soc. *Res:* Theoretical and elementary particle physics; field theory. *Mailing Add:* Dept of Physics Ariz State Univ Tempe AZ 85287

NIGAM, LAKSHMI NARAYAN, b Fatehpur, India, Sept 17, 34; m 58; c 3. FLUID MECHANICS, MATHEMATICS. *Educ:* Univ Allahabad, BSc, 55, MSc, 57; Indian Inst Technol, Kharagpur, PhD(fluid mech), 61. *Prof Exp:* Sr res asst appl math, Indian Inst Technol, Bombay, 60-61, from assoc lectr to lectr, 61-66, asst prof, 66-67; from asst prof to assoc prof, 67-73, chmn dept, 73-78, PROF MATH, QUINNIPIAC COL, 73- *Concurrent Pos:* Adj asst prof, New Haven Col, 68-69; vis fac, Yale Univ, 75-76; adj fac, Southern Conn State Univ, 82-; pres, Videorama Inc, 84- *Mem:* Indian Math Soc. *Res:* Structure and propagation of shock waves in interstellar gas; gas shear flow past cylinders and wings when the effects of compressibility and viscosity are negligible. *Mailing Add:* Dept Math Quinnipiac Col Mt Carmel Ave Hamden CT 06518

NIGG, HERBERT NICHOLAS, b Detroit, Mich, July 9, 41; m 64; c 2. TOXICOLOGY. *Educ:* Mich State Univ, BS, 67; Univ Ill, Urbana-Champaign, PhD(entom), 72. *Prof Exp:* Nat Res Coun-USDA fel insect endocrinol, Insect Physiol Lab, USDA, Md, 72-74; from asst prof to assoc prof, 74-81, PROF ENTOM, INST FOOD & AGR SCI, CITRUS RES & EDUC CTR, UNIV FLA, LAKE ALFRED, 85- *Concurrent Pos:* Environ toxicol consult & ed-in-chief, Bull Environ Contamination & Toxicol, 81- *Mem:* Soc Toxicol; Entom Soc Am; Am Chem Soc; Sigma Xi. *Res:* Natural product toxicology and the relationship between exposure and pesticide agricultural worker health. *Mailing Add:* Inst Food & Agr Sci Univ Fla 700 Exp Sta Rd Lake Alfred FL 33850

NIGH, EDWARD LEROY, JR, b Hagerstown, Md, Aug 25, 27; m 78; c 3. PLANT PATHOLOGY, NEMATOLOGY. *Educ:* Colo Agr & Mech Col, BS, 52; Colo State Univ, MS, 56; Ore State Univ, PhD(plant path), 62. *Prof Exp:* Tech dir cotton res, Algodonera del Valle, SA, Mex, 54-62; from asst prof to assoc prof plant path & nematol, 62-67, head dept plant path, 67-76, prof plant path, assoc dean agr & assoc dir, Coop Exten Serv, 76-78, RES SCIENTIST & EXTEN NEMATOLOGIST, UNIV ARIZ, 78- *Mem:* Am Phytopath Soc; Soc Nematol. *Res:* Entomology; nematology; plant pathology. *Mailing Add:* 1161 W Sahuaro Lane Yuma AZ 85365

NIGH, HAROLD EUGENE, b Parnell, Mo, May 20, 32; m 55; c 4. SOLID STATE PHYSICS. *Educ:* Northwest Mo State Col, BS, 58; Iowa State Univ, PhD, 63. *Prof Exp:* Fel physics, Inst Atomic Res, 63-64; mem tech staff, Bell Labs, Allentown, 64-68, supvr surface physics, 68-73, supvr memory technol, 73-83, eng mgr, mgt operating syst prod, AT&T Technologies, 83-85, MGR SANTA CRUZ PLANT, AT&T TECHNOLOGIES, 85- *Mem:* Am Phys Soc; Inst Elec & Electronics Engrs. *Res:* Magnetic properties of solids; semiconductor surface physics. *Mailing Add:* Bell Tel Labs 555 Union Blvd Allentown PA 18103

NIGHSWONGER, PAUL FLOYD, b Alva, Okla, Apr 14, 23; m 51; c 6. PLANT ECOLOGY. *Educ:* Northwestern State Col, Okla, BS, 49; Univ Okla, MS, 67, PhD(bot), 69. *Prof Exp:* asst prof, 69-80, ASSOC PROF BIOL, NORTHWESTERN STATE COL, OKLA, 80- *Mem:* Soc Range Mgt. *Mailing Add:* Dept of Biol Northwestern Okla State Univ Alva OK 73717

NIGHTINGALE, ARTHUR ESTEN, b Millville, NJ, Dec 25, 19; m 52; c 1. HORTICULTURE. *Educ:* NJ State Teachers Col, Glassboro, BS, 42; Rutgers Univ, New Brunswick, BS & MEd, 49; Tex A&M Univ, PhD(hort), 66. *Prof Exp:* Exten agent, Rutgers Univ & USDA, 45-47; dir adult educ, Atlantic County Voc Schs, NJ, 49-51; field rep horticulturist, Calif Spray Chem Co, 51-53; area horticulturist, 54-60; dir educ progs, State NJ, 60-63; asst, 63-66, prof, 66-87, EMER PROF HORT, TEX A&M UNIV, 87- *Mem:* Am Soc Hort Sci; Am Inst Biol Sci; Am Hort Soc; fel Royal Hort Soc. *Res:* Chemical and environmental influences on plant growth and production. *Mailing Add:* Dept Hort Tex A&M Univ College Station TX 77843

NIGHTINGALE, CHARLES HENRY, b New York, NY, June 19, 39; m 62; c 3. BIOPHARMACEUTICS, PHARMACOKINETICS. *Educ:* Fordham Univ, BSPharm, 61; St John's Univ, NY, MS, 66; State Univ NY Buffalo, PhD(pharmaceut), 70. *Prof Exp:* Coordr educ & res, Mercy Hosp, New York, 64-66; from asst prof to assoc prof pharm, Univ Conn, 69-76, dir clin pharm prog, Sch Pharm, 70-76; assoc prof & chmn dept, State Univ NY Buffalo, 76-78; assoc res prof, 78-82, dir pharm serv, 78-87, RES PROF PHARM, UNIV CONN, 82-, VPRES, HARTFORD HOSP, 87- *Concurrent Pos:* Vis prof pharmaceut, Shanghai First Med Col, People's Repub China, 85. *Mem:* AAAS; Am Pharmaceut Asn; Acad Pharmaceut Sci; fel Am Col Clin Pharmacol; Am Soc Clin Pharmacol & Therapeut; fel Infectious Dis Soc Am;

Sigma Xi; NY Acad Sci; Am Soc Microbiol; Am Soc Hosp Pharmacists; Am Asn Clin Pharmacists. *Res:* Factors affecting drug absorption, distribution, metabolism and excretion; pharmacokinetics of drug therapy; antibiotic transport in tissue sites, protein binding; antibiotic pharmacokinetics. *Mailing Add:* Hartford Hosp 80 Seymour St Hartford CT 06115

NIGHTINGALE, DOROTHY VIRGINIA, b Ft Collins, Colo, Feb 21, 02. ORGANIC CHEMISTRY. *Educ:* Univ Mo, AB, 22, AM, 23; Univ Chicago, PhD(org chem), 28. *Prof Exp:* From instr to prof, 23-58, EMER PROF CHEM, UNIV MO-COLUMBIA, 72- *Concurrent Pos:* Hon fel, Univ Minn, 38; res assoc, Univ Calif, Los Angeles, 46-47. *Honors & Awards:* Garvan Award, Am Chem Soc, 59. *Mem:* Am Chem Soc. *Res:* Chemiluminescence of organomagnesium halides; alkylations and acylations in the presence of aluminum chloride; action of nitrous acid on alicyclic amines; reactions of nitroparaffins with alicyclic ketones. *Mailing Add:* 350 Ponca Pl Boulder CO 80303

NIGHTINGALE, ELENA OTTOLENGHI, b Leghorn, Italy, Nov 1, 32; US citizen; m 65; c 2. MEDICAL GENETICS, MICROBIAL GENETICS. *Educ:* Barnard Col-Columbia Univ, AB, 54; Rockefeller Univ, PhD(microbial genetics), 61; NY Univ, MD, 64. *Prof Exp:* Genetics training grant, 61-62; Am Cancer Soc res scholar, 62-64; instr med, NY Univ, 64-65; asst prof microbiol, Med Col, Cornell Univ, 65-70; asst prof microbiol, Sch Med, Johns Hopkins Univ, 70-73; UAP clin genetics fel, Georgetown Univ Hosp, 73-74; Sloan fel health & sci policy, Inst Med, Nat Acad Sci, 74- 75, sr staff officer, 75-76, dir, Div Health Prom & Dis Prev, 76- 80, sr prog officer, 79-82; clin pediat, Georgetown Univ, Washington, DC, 75-80, asst prof, 80-83; SPEC ADV TO PRES, CARNEGIE CORP, NY, 83-, SR PROG OFFICER, 89- *Concurrent Pos:* Res fel, Am Cancer Soc, 62-64; guest lectr, Albert Einstein Col Med, 64-69, Mount Sinai Sch Med, 69, Georgetown Univ Sch Med, 73-85, Johns Hopkins Univ Evening Col, 77-81, Nonatol fel teaching prof, Columbia Hosp Women, Washington, DC, 79-85; liaison repr, NIH Recombinant DNA Molecule Prog Adv Comt, Nat Acad Sci, 75-77, mem, Comt to rev Human Studies, 76-80; mem, Coord Comt Digestive Dis, Dept Health Educ Welfare, 78- 80; liaison, Comt Lab Created Biohazards, Inst Med, Nat Acad Sci, 78-80, sr prog officer & dir, Div Health Prom & Dis Prev, 79-82, sr scholar-in-residence, 82-83, mem, Comt Re-eval Polio Vaccine Policies, 87-88, chair, Comt Health & Human Rights, 88-90, mem, 90-91; mem, Office Technol Assess Comt Select Cong fels, 79; genetics consult, Sch Med, Georgetown Univ, 75-, mem, Planning Comt Int Ctr Interdisciplinary Studies Immunol, 78, clin asst prof pediat, 80-83, adj prof, 83-; mem adv comt, Am Coun Life Ins & Health Ins Asn Am, 78-84; mem ed adv bd, Nat Inst Mental Health Sci Monographs & Reports, 79-82; mem & exec officer, Working Group Health Prom & Dis Prev, Div Health Policy Res & Educ, Harvard Univ, 80-82, lectr social med & health policy, 84-; mem soc issues comt, Am Soc Human Genetics, 81-85; consult, President's Comn Study Ethical Prob Med & Biomed & Behavioral Res, 81-82; mem adv bd, Ctr Educ in Sci, Technol & Soc, Univ Colo, 82-; mem, Sci Adv Bd, Fondazione Giovanni Lorenzini, USA, 82-; mem, Task Force on Communicating Sci & Technol Risk to the Pub, Twentieth Century Fund, 83-84; mem, Dist Colo Comt World Health Day, 84-85; mem bd dirs, Aesculapius Int Med, 84-85; mem bd, Am Asn World Health, 85-; mem bd trustees, Sci Serv Inc, 85-; mem sci adv bd, Nat Pub Radio, 86-; mem, Comt Re-eval Polio Vaccine Polices, Inst Med, Nat Acad Sci, 87-, Planning Group Alcohol & Drug Abuse Studies, Nat Res Coun & Inst Med, 88-, chmn, Comt Health & Human Rights, Inst Med, 88-; paticipant numerous conf & wkshps, 79- *Mem:* Inst Med-Nat Acad Sci; fel NY Acad Sci; Am Soc Microbiol; Genetics Soc Am; Am Soc Human Genetics; Am Asn World Health; Harvey Soc; Sigma Xi; fel AAAS. *Res:* Bacterial transformations; somatic cell genetics; infectious diseases; health and health science policy; human rights and the health professions; maternal and child health; adolescent health and development; public health and preventive medicine; international health; medical genetics and genetic counseling; infectious diseases; health education; molecular genetics; medical education; societal and ethical aspects of health and science policy; author and co-author of numerous publications. *Mailing Add:* Carnegie Corp NY 2400 N St NW 6th Floor Washington DC 20037-1153

NIGHTINGALE, RICHARD EDWIN, b Walla Walla, Wash, June 3, 26; m 44; c 3. PHYSICAL CHEMISTRY, NUCLEAR ENGINEERING. *Educ:* Whitman Col, BA, 49; Wash State Univ, PhD(chem), 53. *Prof Exp:* Fel, Univ Minn, 52-54; sr engr, Hanford Labs, Gen Elec Co, 54-57, supvr nonmetallic mat, 57-65; mgr mat res & serv sect, 65-68, mgr ceramics dept, 68-69, mgr metall & ceramics dept, 69-70, mgr chem technol dept, 70-83, PROG MGR, PAC NORTHWEST LABS, BATTELLE MEM INST, 83- *Mem:* Am Chem Soc; fel Am Nuclear Soc; Am Carbon Soc. *Res:* Infrared spectroscopy and molecular structure; radiation damage effects; graphite structure and properties; coal research, nuclear fuel reprocessing and waste disposal. *Mailing Add:* Pac Northwest Labs Battelle Mem Inst PO Box 999 Richland WA 99352

NIGRELLI, ROSS FRANCO, protozoology, parasitology; deceased, see previous edition for last biography

NIGRO, NICHOLAS J, b Chicago, Ill, Sept 24, 34; m 54; c 6. MECHANICS. *Educ:* Mich Technol Univ, BS, 56; Iowa State Univ, MS, 59; Univ Iowa, PhD(mech), 65. *Prof Exp:* Struct designer, Victor Chem Co, 56-57; instr eng, Southern Ill Univ, 59-62; from asst prof to assoc prof, 65-79, PROF MECH ENG, MARQUETTE UNIV, 79- *Mem:* Am Soc Eng Educ. *Res:* Vibrations; dynamics; systems; wave propagation; computer aided engineering. *Mailing Add:* Dept of Mech Engrs Marquette Univ 1515 W Wisconsin Ave Milwaukee WI 53233

NIGROVIC, VLADIMIR, b Sarajevo, Yugoslavia, Mar 3, 34; US citizen; m 63; c 3. PHARMACOLOGY OF RELAXANTS, GENERAL PHARMACOLOGY. *Educ:* Univ Heidelberg, Fed Repub Germany, MD, 62. *Prof Exp:* Res assoc radiation biol, Nuclear Res Ctr, Fed Repub Germany, 61-66 & Sch Pub Health, Univ Minn, 66-68; from asst prof to assoc prof pharmacol, 68-75, & 79-88, from asst prof to assoc prof anesthesiol, 79-89, PROF PHARMACOL & ANESTHESIOL, MED COL OHIO, 89- *Concurrent Pos:* Fel, German Academic Exchange Serv, Fed Repub Ger, 59-61. *Honors & Awards:* B B Sankey Anesthesia Advan Award, 86, 88. *Mem:* Am Soc Exp Pharmacol Therapeut; Am Soc Anesthesiologists; German Soc Anesthesiologists; Soc Exp Med & Biol; Assoc Univ Anesthetists; Sigma Xi. *Res:* Pharmacology and toxicology of skeletal muscle relaxants. *Mailing Add:* Dept Anesthesiol Med Col Ohio CS No 10008 Toledo OH 43699

NIJENHUIS, ALBERT, b Eindhoven, Neth, Nov 21, 26; US citizen; m 55; c 4. MATHEMATICS. *Educ:* Univ Amsterdam, PhD(math), 52. *Prof Exp:* Vis fel, Princeton Univ, 52-53; mem, Inst Advan Study, 53-55; instr & res assoc, Univ Chicago, 55-56; from asst prof to prof math, 56-63, AFFIL PROF, UNIV WASH, 88-; EMER PROF MATH, UNIV PA, 87- *Concurrent Pos:* Mem, Inst Advan Study & Guggenheim fel, 61-62; Fulbright lectr, Univ Amsterdam, 63-64; prof, Univ Pa, 63-87; vis prof, Univ Geneva, 67-68 & Dartmouth Col, 77-78; corresp, Royal Neth Acad Sci. *Mem:* Am Math Soc; Math Asn Am; Asn Comput Mach. *Res:* Local and global differential geometry; theory of deformations in algebra and geometry; combinatorial analysis; algorithms. *Mailing Add:* 13727 41st Ave NE Seattle WA 98125-3820

NIJHOUT, H FREDERIK, b Eindhoven, Neth, Nov 25, 47; c 1. INSECT PHYSIOLOGY. *Educ:* Univ Notre Dame, BS, 70; Harvard Univ, MA, 72, PhD(biol), 74. *Prof Exp:* NIH staff fel, 75-77; asst prof, 77-81, assoc prof, 81-87, PROF ZOOL, DUKE UNIV, 87- *Mem:* Am Soc Zoologists; Sigma Xi; AAAS. *Res:* Insect development and physiology, with particular emphasis on endocrinology, polymorphism and pattern formation. *Mailing Add:* Dept of Zool Duke Univ Durham NC 27706

NIKAIDO, HIROSHI, b Tokyo, Japan, Mar 26, 32; m 63; c 2. MICROBIOLOGY, BIOCHEMISTRY. *Educ:* Keio Univ, Japan, MD, 55, DMedSc(microbiol), 61. *Prof Exp:* Asst microbiol, Med Sch, Keio Univ, Japan, 56-60 & Inst Protein Res, Osaka Univ, 61; res fel biol chem, Harvard Med Sch, 62, assoc bact & immunol, 63-64, asst prof, 65-69; from assoc prof to prof microbiol, 69-90, PROF BIOCHEM & MOLECULAR BIOL, UNIV CALIF, BERKELEY, 90- *Concurrent Pos:* USPHS res grant, 63- & Am Cancer Soc res grant, 69-86; mem, NIH Study Sect, 78-82. *Honors & Awards:* Ehrlich Award, 69; Hoechst-Roussel Award, 84. *Mem:* Am Soc Biol Chem; Am Soc Microbiol. *Res:* Biochemistry of bacterial cell wall and cell membrane, especially structure and functions of the outer membrane of gram-negative bacteria. *Mailing Add:* Dept Molecular Cell Biol Univ Calif Stanley Hall Berkeley CA 94720

NIKELLY, JOHN G, b Evanston, Ill, Apr 16, 29; m 61. ANALYTICAL CHEMISTRY. *Educ:* Univ Ill, BS, 52; Cornell Univ, PhD(chem), 56. *Prof Exp:* Chemist, Exxon Res & Eng Co, 56-62; PROF CHEM, PHILADELPHIA COL PHARM, 62- *Concurrent Pos:* Sci adv, Philadelphia Dist, Food & Drug Admin, 71-75. *Mem:* AAAS; Am Chem Soc; Sigma Xi. *Res:* Gas chromatography; high pressure liquid chromatography. *Mailing Add:* Chem Dept Philadelphia Col Pharm Philadelphia PA 19104

NIKI, HIROMI, b Japan; US citizen. ATMOSPHERIC CHEMISTRY. *Educ:* Int Christian Univ, Tokyo, BA, 59; Carnegie Mellon Univ, MS, 61, PhD(phys chem), 63. *Prof Exp:* Fel chem, Harvard Univ, 63-65; res scientist, Ford Motor Co US, 65-87; RES INDUST CHAIR CHEM, NAT SCI & ENG RES COUN CAN- ATMOSPHERIC ENVIRON SERV, YORK UNIV, 87- *Concurrent Pos:* Guest prof, Wuppertal Univ, WGer, 80, 81 & 86; chair, Nat Oxidants Res Comt, 90; mem, Sci Adv Comt, Southern Oxidants Studies, Int Global Atmospheric Chem, Hydrocarbon Measurement Intercomparison Proj Coord Comt, Alternative Fluorocarbon Environ Assessment Comt, NASA Site 4 Comt; grant rev panel mem, NSF, NIH, Dept Energy & Environ Protection Agency; mem, Air Quality Model Develop & Eval Prog, Environ Protection Agency; peer rev mem, Nat Acid Precipitation Assessment Prog, Nat Acad Sci Global Tropospheric Chem Spec Comt; mem adv comt, Nat Bur Standards; mem, Toxic Substances Protocols, Nat Acad Sci, Chemical Kinetics Comt. *Honors & Awards:* Frank A Chambers Award, Am Phys Chem Asn. *Mem:* Fel Am Petrol Inst. *Res:* Global kinetics; photochemistry; chemical processes in urban, regional and global atmospheres; global biogeochemical cycle; photochemical modelling; development and application of analytical methodologies to atmospheric trace gas measurements. *Mailing Add:* Dept Chem Ctr Atmospheric Chem 4700 Keele St North York ON M3J 1P3 Can

NIKIFORUK, GORDON, b Redfield, Sask, Nov 2, 22; m 50; c 2. BIOCHEMISTRY, DENTISTRY. *Educ:* Univ Toronto, DDS, 47; Univ Ill, MS, 50. *Prof Exp:* Prof prev dent, Univ Toronto, 57-64, chmn div dent res, 54-64; prof pediat, Sch Med & prof pediat dent & head dept, Sch Dent, Univ Calif, Los Angeles, 64-66, prof oral biol & head dept, 66-69; dean fac dent, 70-77, PROF PREV DENT, UNIV TORONTO, 70- *Concurrent Pos:* Prof serv, Colgate-Palmolive, 90- *Honors & Awards:* Dentistry for Children Prize, Astra, 83. *Mem:* AAAS; fel Royal Col Dent; Can Dent Asn; Can Soc Dent for Children; Int Asn Dent Res; Royal Col Dent Surgeons Ont (pres, 87-89). *Res:* Biochemistry of teeth and saliva; pediatric dentistry; caries. *Mailing Add:* Fac Dent Univ Toronto Toronto ON M5G 1G6 Can

NIKIFORUK, P(ETER) N, b St Paul, Alta, Feb 11, 30; m 57; c 2. ENGINEERING PHYSICS. *Educ:* Queen's Univ, Ont, BSc, 52; Univ Manchester, PhD(electron eng), 55, DSc, 70. *Prof Exp:* Serv officer, Defence Res Bd, Can, 56-57; systs engr, Canadair, Ltd, 57-59; head dept mech eng, 65-73, PROF MECH & CONTROL ENG, UNIV SASK, 60-, DEAN ENG, 73- *Concurrent Pos:* Mem, Nat Res Coun, 73-78, Sask Res Coun, 77-; chmn Sask Sci Coun, 78-83. *Honors & Awards:* Centennial Medal, Inst Elec & Electronic Engrs, Kelvin Premium, Inst Elec & Electronic Engrs (UK). *Mem:* Fel Brit Inst Physics; fel Brit Phys Soc; fel Brit Inst Elec Engrs; fel Brit Royal Soc Arts; fel Can Soc Mech Engrs (vpres, 86-88); fel Eng Inst Can; fel Can Acad Eng. *Res:* Nonlinear self-adaptive control systems; very high pressure hydraulic servomechanisms; computer applications. *Mailing Add:* Col Eng Univ Sask Saskatoon SK S7N 0W0 Can

NIKITOVITCH-WINER, MIROSLAVA B, b Kraljevo, Yugoslavia, May 13, 29; US citizen; m 55; c 2. ANATOMY, ENDOCRINOLOGY. *Educ:* Univ Belgrade, BS, 45; Sorbonne, cert sci, 46; Radcliffe Col, MA, 54; Duke Univ, PhD(anat, physiol), 57. *Prof Exp:* Asst, Sch Med, Duke Univ, 57-58; USPHS fels, Univ Lund, 58, Nobel Med Inst, Karolinska Med Sch, 59 & Maudsley Hosp, Univ London, 59-60; res assoc, 60-61, from asst prof to assoc prof, 61-73, PROF ANAT, PHYSIOL & BIOPHYS, MED CTR, UNIV KY, 73-, CHMN DEPT ANAT, MED CTR, 79- *Mem:* AAAS; Am Asn Anat; Endocrine Soc; Am Physiol Soc; Int Brain Res Orgn. *Res:* Neuroendocrinology of reproduction; indirect and direct determination of neurohumoral controls of gonadotropic hormone secretion; identification of releasing versus hypophysiotropic effects of hypothalamic humoral factors concerned with hormone secretion; control of prolactin secretion. *Mailing Add:* Dept Anat & Neurobiol Univ of Ky Med Ctr Lexington KY 40536-0084

NIKKEL, HENRY, b Munsterberg, Russia, Oct 8, 22; US citizen; m 49; c 2. PHYSICS, METALLURGY. *Educ:* Univ Mich, BS, 45, MS, 51. *Prof Exp:* Res engr, Bethlehem Steel Corp, 47-65; sr res engr, Youngstown Sheet & Tube Co, 65-69, res assoc, 69-77; microprobe analyst, 77-78, lab develop analyst, 78-85, PROD MAT ENGR, FORD MOTOR CO, 85- *Mem:* Am Soc Metals; Int Metallog Soc; Sigma Xi; Soc Automotive Engrs; Am Soc Testing & Mat. *Res:* Emission spectroscopy involving analytical techniques; excitation studies; instrument development and investigations in the vacuum ultraviolet; physical metallurgy, particularly electron microprobe techniques; materials evaluation and failure analysis; approval of and recommendations for materials and processing of materials applied to internal combustion engions. *Mailing Add:* 55 W Woodland Ave Columbiana OH 44408-1548

NIKLAS, KARL JOSEPH, b New York, NY, Aug 23, 48. BIOMATHEMATICS, GEOCHEMISTRY. *Educ:* City Col New York, BS, 70; Univ Ill, Urbana-Champaign, MS, 71, PhD(bot, math), 74. *Prof Exp:* Fel bot, Univ London & Cambridge Univ, 74-75; cur palaeobot, NY Bot Garden, 74-78; asst prof, 78-81, ASSOC PROF BOT, CORNELL UNIV, 81- *Concurrent Pos:* Fulbright-Hays fel, Int Educ Comt, 74; adj prof, City Univ New York, 75. *Res:* Application of mathematical theory to selected problems in plant morphology and evolution; organic geochemistry of fossil materials and associated rock strata; paleobotanical and biological problems. *Mailing Add:* Dept of Plant Biol & Ecol Cornell Univ Ithaca NY 14853

NIKLOWITZ, WERNER JOHANNES, b Germany, Sept 14, 23; m 52; c 2. NEUROBIOLOGY, ELECTRON MICROSCOPY. *Educ:* Univ Jena, MS, 52, PhD(biol), 54. *Prof Exp:* Asst prof biol, Univ Jena, 52-54; res assoc microbiol, Acad Inst Microbiol, Jena, Germany, 54-60; res assoc path, Univ Freiburg, 61-62; res assoc, Max Planck Inst Brain Res, Frankfurt, 62-68; sr res assoc toxicol, Med Ctr, Univ Cincinnati, 68-69, from asst prof to assoc prof environ health, 69-74; res assoc neuropath, Med Ctr, Univ Ind, Indanapolis, 74-76; res assoc radiation, Univ San Francisco, 78-80, sr res assoc bone res, 80-86; RETIRED. *Concurrent Pos:* NASA res grant. *Mem:* AAAS; Electron Micros Soc Am; Soc Neurosci; NY Acad Sci. *Res:* Experimental epilepsy; effects of drugs, toxic metals and radiation on the ultrastructure of brain components, enzymes, trace metals and neurotransmitters; effects of immobilization on bone. *Mailing Add:* 4180 Monet Circle San Jose CA 95136

NIKODEM, ROBERT BRUCE, b Oak Park, Ill June 2, 39; m 64. PHYSICAL CHEMISTRY, INORGANIC CHEMISTRY. *Educ:* Elmhurst Col, BS, 62; Purdue Univ, Lafayette, MS, 65; Va Polytech Inst & State Univ, PhD(inorg chem), 69. *Prof Exp:* Res supvr glass films, Libbey-Owens-Ford Co, 69-77; prod mgr, Photon Power Inc, 77-81; TECH MGR FILMS, LIBBEY-OWENS-FORD CO, 81- *Mem:* Am Chem Soc. *Res:* Rare earth oxides; compound semiconductors; thin films, both oxide and metallic; solid state chemistry; photovoltaic solar energy cells. *Mailing Add:* Libbey Owens Ford Co 1701 E Broadway Toledo OH 43605

NIKOLAI, PAUL JOHN, b Minneapolis, Minn, Mar 2, 31; m 60; c 2. MATHEMATICS. *Educ:* Col St Thomas, BA, 53; Ohio State Univ, MSc, 55, PhD(math), 66. *Prof Exp:* Mathematician, Univac Div, Sperry Rand Corp, 55-57; mathematician, Digital Comput Br, Aeronaut Res Lab, 58-60, res mathematician, Appl Math Res Lab, Aerospace Res Labs, 60-75, MATHEMATICIAN, STRUCT DIV, FLIGHT DYNAMICS LAB, AIR FORCE WRIGHT AERONAUT LABS, US DEPT AIR FORCE, 75- *Concurrent Pos:* Asst math, Ohio State Univ, 53-57; vis scholar math, Univ Calif, Santa Barbara, 68-69; mem, Special Interest Group Numerical Math, Asn Comput Mach. *Mem:* Am Math Soc; Soc Indust & Appl Math. *Res:* Numerical linear algebra, matrix theory and combinatorial computing. *Mailing Add:* 7312 Caribou Trail Dayton OH 45459

NIKOLAI, ROBERT JOSEPH, b Rock Island, Ill, Apr 6, 37; m 61; c 5. ORTHODONTICS, BIOMECHANICS. *Educ:* Univ Ill, Urbana, BS, 59, MS, 61, PhD(theoret & appl mech), 64. *Prof Exp:* Teaching asst theoret & appl mech, Univ Ill, Urbana, 59-61, instr, 61-64, res assoc, 63-64; from asst prof eng & eng mech to assoc prof eng mech, 64-71, asst dean grad sch, 71-72, assoc prof biomech in orthod, 71-75, actg dean, 87-88, ASSOC DEAN, GRAD SCH, ST LOUIS UNIV, 72-87 & 88-, PROF BIOMECH IN ORTHOD, 75-, UNIV MARSHAL, 89- *Concurrent Pos:* Affil prof civil & mech eng, Washington Univ, 80-; consult, several Orthods suppliers; sabbatical grant, Orthodontic Educ & Res Found, 81; res grants, Nat Inst Dent Res, 83-87. *Mem:* Am Acad Mech; Int & Am Asn Dent Res; Orthod Educ & Res Found; Sigma Xi. *Res:* Force and structural analyses, and design of orthodontic appliances (braces); biomechanical interaction of the orthodontic appliance with the dentition and the facial complex. *Mailing Add:* St Louis Univ Grad Sch 221 N Grand Blvd St Louis MO 63103

NIKOLIC, NIKOLA M, b Belgrade, Yugoslavia, Sept 14, 27; US citizen; m 60; c 1. PHYSICS. *Educ:* Univ Belgrade, BS, 50; Columbia Univ, MA, 59, PhD(physics), 62. *Prof Exp:* Asst prof physics, US Naval Postgrad Sch, 62-64, La State Univ, 64-68 & Old Dom Univ, 68-69; asst prof, 69-74, PROF PHYSICS, MARY WASHINGTON COL, 74- *Mem:* Am Phys Soc; Sigma Xi. *Res:* Nuclear physics. *Mailing Add:* 12 Winston Pl Fredericksburg VA 22405

NIKORA, ALLEN P, b Stuttgart, Ger, July 17, 55; US citizen. SOFTWARE RELIABILITY MODELING. *Educ:* Calif Inst Technol, BS, 77. *Prof Exp:* Mem tech staff, McDonell-Douglas Astronaut Co, 77-78 & Jet Propulsion Lab, 78-86; sr analyst, Ashton-Tate, 86-87; MEM TECH STAFF, JET PROPULSION LAB, 87- *Concurrent Pos:* Mem, Space Based Observ Systs Comt Standards, Am Inst Aeronaut & Astronaut & Tech Subcomt Software Reliability Eng, Inst Elec & Electronic Engrs. *Mem:* Inst Elec & Electronic Engrs Computer Soc; Inst Elec & Electronic Engrs Reliability Soc; Am Inst Aeronaut & Astronaut; Inst Elec & Electronic Engrs. *Res:* Software reliability from process-product metrics during design; relating operational reliability to reliability observed during test; software reliability modeling. *Mailing Add:* Jet Propulsion Lab MS125-233 4800 Oak Grove Dr Pasadena CA 91103

NIKOUI, NIK, b Tehran, Oct 24, 38; US citizen; m 60; c 3. DISPERSION OF HYDROCARBON RESINS IN AQUOUS MEDIA, POLYMERIZATION OF VINYL ACETATE. *Educ:* Univ Calif Los Angeles, BS, 66. *Prof Exp:* Sr adhesive specialist, DAP Inc, 86-89; DIR RES, ROBERTS CONSOL INDUSTS, 90- *Mem:* Am Chem Soc; Adhesive & Sealant Coun; Tech Asn Pulp & Paper Indust. *Res:* Development of new generation of adhesives to meet Environmental Protection Agency's standards as well as local agencies. *Mailing Add:* Roberts Consol Industs 600 N Baldwin Park Blvd Industry CA 91749

NILAN, ROBERT ARTHUR, b Can, Dec 26, 23; nat US; m 48; c 3. GENETICS. *Educ:* Univ BC, BSA, 46, MSA, 48; Univ Wis, PhD(genetics), 51. *Prof Exp:* Asst, Univ BC, 46-48 & Univ Wis, 48-51; res assoc, 51-52, from asst prof & asst agronomist to prof agron & agronomist, 52-65, chmn prog genetics, 65-79, PROF GENETICS & AGRONOMIST, WASH STATE UNIV, 65-, DEAN DIV SCI, 79- *Concurrent Pos:* Fulbright teaching scholar, 59; Guggenheim fel, 59-60; USPHS spec fel, 67-68. *Mem:* Genetics Soc Am; Am Soc Agron; Environ Mutagen Soc. *Res:* Cytogenetical, mutagenetical and breeding studies in barley. *Mailing Add:* Program in Genetics Wash State Univ Pullman WA 99164

NILAN, THOMAS GEORGE, b White Plains, NY, Sept 4, 26; m 61; c 6. SURFACE FINISH MEASUREMENT & CONTROL, IMPERFECTIONS IN METALS. *Educ:* Columbia Univ, BS, 50; Univ Ill, MS, 56, PhD(physics), 61. *Prof Exp:* Physicist, Res Lab, US Steel Corp, 50-55, sr res physicist, 60-69, assoc res consult, 69-83; sr fel, Mellon Inst, 84-88; CONSULT, 88- *Concurrent Pos:* Mem B46 standards comt, Surface Qual, Am Soc Mech Engrs. *Mem:* Int Lead Zinc Res Orgn. *Res:* Ferromagnetic domains; radiation damage in metals; ultra-high pressure effects on phase equilibria, kinetics of transformation and micro-structure of metals; surface physics; imperfections in metals. *Mailing Add:* 8945 Eastwood Rd Pittsburgh PA 15235-1402

NILES, GEORGE ALVA, b Flagstaff, Ariz, Oct 4, 26; m 48; c 4. AGRONOMY, PLANT BREEDING. *Educ:* NMex State Univ, BS, 48; Okla State Univ, MS, 49; Tex A&M Univ, PhD(plant breeding), 59. *Prof Exp:* Instr agron, NMex State Univ, 49; res assoc, Okla Agr Exp Sta, 51-53; from instr to assoc prof, 53-74, PROF AGRON, TEX A&M UNIV, 74- *Mem:* AAAS; Am Soc Agron; Crop Sci Soc Am; Int Soc Biometeorol; Sigma Xi. *Res:* Cotton genetics and breeding; host plant resistance; crop-climate studies; cotton production systems; cotton physiology; insect pest management. *Mailing Add:* Dept Agron Tex A&M Univ College Station TX 77843

NILES, JAMES ALFRED, b Eureka, Calif, Apr 24, 45; m 67; c 2. AGRICULTURAL ECONOMICS. *Educ:* Univ Calif, Davis, BS, 67, MS, 68, PhD(agr econ), 72. *Prof Exp:* Asst prof agr econ, Food & Resource Econ Dept, Univ Fla, 73-78; ASSOC PROF, INST AGRBUS, GRAD SCH BUS, UNIV SANTA CLARA, 78- *Mem:* Am Asn Agr Economists. *Res:* Commodity futures markets; price forecasting of agricultural commodities and modelling; computer simulation of commodity systems. *Mailing Add:* 245 Johnson Ave Los Gatos CA 95032

NILES, NELSON ROBINSON, b Southampton, NY, May 27, 24; m 51; c 5. PATHOLOGY, ANATOMY. *Educ:* Cornell Univ, MD, 47. *Prof Exp:* From instr to assoc prof, 52-67, PROF PATH ANAT, MED SCH, UNIV ORE, 67- *Concurrent Pos:* Res fel path, Royal Col Surg, 62-63. *Mem:* Int Acad Path. *Res:* Cardiovascular pathology; coronary artery disease; congenital heart disease and cardiac surgery; histochemistry. *Mailing Add:* Dept Path Ore Health Sci Univ 3181 SW S Jackson Park Rd Portland OR 97201

NILES, PHILIP WILLIAM BENJAMIN, b Fairfield, Calif, Jan 4, 36; m 73; c 4. PASSIVE SOLAR DESIGN TOOLS, SOLAR ENERGY. *Educ:* Univ Calif, Berkeley, BS, 57, MS, 58. *Prof Exp:* Sr res engr res & develop, Rocketdyne, NAm Aviation Corp, 58-62 & 65; consult, Dept Geophys & Astron, Rand Corp, Santa Monica, 67-68; PROF, DEPT MECH ENG, CALIF POLYTECH STATE UNIV, 67- *Concurrent Pos:* Solar consult, Skytherm Processes & Eng, Los Angeles, 73-75; solar thermal analyst, HUD contract, Calif Polytech State Univ Found, 73-74, NSF grant, 74-75; system design engr, Energy Res Develop Admin contract, Calif Polytech State Univ Found, 76-78, proj dir, Calif Energy Comn, 78-80; prin investr, Calif State Energy Comn contract, Calif Polytech State Univ, 78-80; consult, USAID, Africa, Assocs Rural Develop, 84-85. *Mem:* Int Solar Energy Soc; Am Soc Heating Refrig & Air Conditioning Engrs; AAAS. *Res:* Application of solar energy to the environmental control of buildings, particularly passive methods. *Mailing Add:* Dept of Mech Eng Calif Polytech State Univ San Luis Obispo CA 93407

NILES, WESLEY E, b Taos, NMex, July 17, 32. BOTANY. *Educ:* NMex State Univ, BS, 59, MS, 61; Univ Ariz, PhD(bot), 68. *Prof Exp:* Assoc cur, New York Bot Garden, 66-68; asst prof, 68-71, ASSOC PROF BOT, UNIV NEV, LAS VEGAS, 71- *Mem:* Am Soc Plant Taxon; Int Asn Plant Taxon. *Res:* Angiosperm taxonomy. *Mailing Add:* Dept of Biol Univ of Nev 4505 S Maryland Pkwy Las Vegas NV 89154

NILGES, MARK J, b Berea, Ohio, Aug 7, 52. ELECTRON PARAMAGNETIC RESONANCE. *Educ:* Case Western Reserve Univ, BS, 74; Univ Ill, PhD(chem), 79. *Prof Exp:* Fel, Cornell Univ, 79-81; RES BIOPHYSICIST, UNIV ILL, 81- *Mem:* Am Chem Soc. *Res:* Electron paramagnetic resonance and nuclear quadrupole coupling in transition metal ion complexes; computer simulation of electron paramagnetic resonance spectra; ultrahigh vacuum studies of metal surfaces using magnetic resonance; electron paramagnetic resonance of melanin and tree radicals involved in cancer. *Mailing Add:* Med Sci Bldg Univ Ill 506 S Matthews Ave Urbana IL 61801

NILSEN, WALTER GRAHN, b Brooklyn, NY, Nov 13, 27; m 59; c 3. CHEMICAL PHYSICS. *Educ:* Columbia Univ, AB, 50, PhD(phys chem), 56; Cornell Univ, MS, 52; Seton Hall Sch Law, JD, 73. *Prof Exp:* Staff mem chem, Opers Eval Group, Mass Inst Technol, 56-59; mem tech staff, 59-69, PATENT ATTORNEY, BELL LABS, NJ, 69- *Mem:* Am Chem Soc; Am Phys Soc; Sigma Xi. *Res:* Magnetic resonance; relaxation mechanisms; masers; lasers; Raman spectroscopy. *Mailing Add:* Six Fern Way Berkeley Heights NJ 07922

NILSON, ARTHUR H(UMPHREY), b Wilkinsburg, Pa, Aug 27, 26; m 81; c 4. STRUCTURAL ENGINEERING. *Educ:* Stanford Univ, BS, 48; Cornell Univ, MS, 56; Univ Calif, Berkeley, PhD, 67. *Prof Exp:* Struct eng pract, 48-54; instr civil eng, 54-56, from asst prof to assoc prof struct eng, 56-69, chmn dept, 78-83, PROF STRUCT ENG, CORNELL UNIV, 69- *Concurrent Pos:* Lectr, Swiss Fed Tech Inst, Lausanne, 71; vis res fel, Politecnico di Milano, 75; vis scholar, Salford Univ, 75-77; consult, H H Robertson Co, NY, Geiger Berger Asn, NY & Thompson & Lichtner Co, Boston & US Corp Engrs. *Honors & Awards:* Wason Medal for Mat Res, Am Concrete Inst, 74, 84, 86. *Mem:* Fel Am Soc Civil Engrs; fel Am Concrete Inst; Prestressed Concrete Inst; Int Asn Bridge & Struct Engrs; Sigma Xi. *Res:* Teaching; behavior; analysis and design of structures, particularly reinforced and prestressed concrete; concrete materials, notably high-strength concrete. *Mailing Add:* Dept Civil & Environ Eng Cornell Univ 218 Hollister Hall Ithaca NY 14853

NILSON, EDWIN NORMAN, b Wethersfield, Conn, Feb 13, 17; m 41; c 3. APPROXIMATION OF ANALYTIC FUNCTIONS IN THE COMPLEX DOMAIN, THEORY & APPLICATION OF REAL & COMPLEX SPLINE FUNCTIONS. *Educ:* Trinity Col, BS, 37; Harvard Univ, MA, 38, PhD(math), 41. *Hon Degrees:* ScD, Trinity Col, 63. *Prof Exp:* Instr math, Univ Md, 41-42; asst prof math, Mt Holyoke Col, 42-47; instr math, US Naval Acad, Annapolis, 44-46; anal engr aerodyn, United Aircraft Res Labs, 46-48; from asst to assoc prof math, Trinity Col, 48-56; from proj engr to chief engr, Pratt & Whitney Aircraft, 56-82; pres, Edwin N Nilson Assocs, Ltd, 82-86; RETIRED. *Concurrent Pos:* Adj prof math, Hartford Grad Ctr, Rensselaer Polytechnic Inst, 58-62; adj prof, Trinity Col, 62-65. *Honors & Awards:* George J Mead Award for Outstanding Contrib to Eng, 82. *Mem:* Math Asn Am. *Mailing Add:* 82 Kenmore Rd Bloomfield CT 06002

NILSON, ERICK BOGSETH, b Aurora, Nebr, Feb 6, 27; m 55; c 2. AGRONOMY. *Educ:* Univ Nebr, BS, 50, MS, 55; Kans State Univ, PhD(agron), 63. *Prof Exp:* Soil scientist, Soil Conserv Serv, USDA, Nebr, 51-52 & Kans, 52-53; county exten agent, State Col Wash, 55-57; exten area agronomist, Iowa State Univ, 57-61; asst prof agr & biol sci, Eastern NMex Univ, 63-65; from asst prof to assoc prof, 65-76, PROF AGRON, KANS STATE UNIV, 76- *Mem:* Weed Sci Soc Am. *Res:* Crop physiology; cell length in wheat; temperature influence on DNA and RNA in grain sorghum seedlings. *Mailing Add:* Dept Agron Kans State Univ Manhattan KS 66502

NILSON, JOHN ANTHONY, b Regina, Sask, Nov 14, 36; m 59; c 2. LASERS. *Educ:* Univ Sask, BE, 59, MSc, 60; Univ London, PhD(elec eng) & dipl, Imp Col, 65. *Prof Exp:* Mem sci staff, Plasma & Space Physics Res Dept, RCA Ltd, Que, 65-71; staff scientist, 71-74, tech dir, 74-85, DIR ADV RES, LUMONICS INC, 85- *Mem:* Can Asn Physicists; Inst Elec & Electronics Engrs; Optical Soc Am. *Res:* Pulsed transverse excitation TEA carbon dioxide, HF/DF, eximer, lasers; glow discharge plasmas; laser applications. *Mailing Add:* Lumonics Inc 105 Schneider Rd Kanata ON K2K 1Y3 Can

NILSSON, NILS JOHN, b Saginaw, Mich, Feb 6, 33; m 58; c 2. COMPUTER SCIENCE, ELECTRICAL ENGINEERING. *Educ:* Stanford Univ, MS, 56, PhD(elec eng), 58. *Prof Exp:* Head artificial intel group, 61-67, sr res engr, 67-69, staff scientist, 69-80, DIR, ARTIFICIAL INTEL CTR, SRI INT, 80- *Mem:* Asn Comput Mach. *Res:* Learning machines; artificial intelligence; pattern recognition; radar signal processing. *Mailing Add:* 150 Coquito Way Menlo Park CA 94025

NILSSON, WILLIAM A, b New York, NY, Jan 16, 31; m. OCCUPATIONAL HEALTH. *Educ:* Univ Ill, Urbana, BS, 57; Univ Calif, Berkeley, PhD(org chem), 62. *Prof Exp:* Asst, Radiation Lab, Univ Calif, Berkeley, 60-61; asst prof chem, Western Wash State Col, 61-64; mgr prod develop, Purex Corp, Ltd, Calif, 65-70; teacher chem, Glendale Col, 71-73; PUB HEALTH CHEMIST, HAZARDOUS MAT, CALIF DEPT HEALTH, 74- *Mailing Add:* 2644 Doray Circle Monrovia CA 91016

NIMAN, JOHN, b Latakia, Syria, June 10, 38; US citizen. APPLIED MATHEMATICS. *Educ:* Polytech Inst NY, BS, 65; Univ Wis, MS, 68; Columbia Univ, PhD(math) & MA, 69. *Prof Exp:* Res asst geophys, Hudson Labs, Columbia Univ, 64-65; from asst prof to assoc prof, 67-77, PROF MATH & EDUC, HUNTER COL, 78- *Concurrent Pos:* George N Shuster fac fel, 70; consult math, New York Bd Educ, 70 & Educ Assoc Prog, La Guardia Community Col, 71; res award hist math, Fed Repub Ger, 75; math consult, Metrop Mus Art, 77-78; mem, New York Bd Educ, 78; Fulbright fel, Sr Scholars, Univ Freiberg; vis prof math, Vassar Col. *Mem:* AAAS; Math Asn Am; NY Acad Sci; Nat Coun Teachers Math. *Res:* Curriculum development; Hilbert; mathematics education in the elementary school. *Mailing Add:* Dept of Math & Educ Hunter Col New York NY 10021

NIMER, EDWARD LEE, b Denver, Colo, Jan 1, 23; m 46; c 5. PHYSICAL CHEMISTRY. *Educ:* Univ NMex, BS, 44; Univ Utah, MS, 49, PhD(chem), 52. *Prof Exp:* Sr res chemist, Explor Process Res, 52-55, petrochem process develop & eval, 55-62, SR RES CHEMIST, POLYMER PROCESS DEVELOP, CHEVRON RES CO, STANDARD OIL CO CALIF, 62- *Mem:* Am Chem Soc; Am Soc Mech Eng. *Res:* Flow, creep and failure of solid materials; exploratory process research in petrochemicals; polarography; polymer polymerization, finishing, compounding, conversions and spinning processes; petroleum cracking processes; sulfur and applications. *Mailing Add:* 8 Alasdair Ct San Rafael CA 94903

NIMITZ, WALTER W VON, US citizen. ELECTRICAL ENGINEERING. *Educ:* Tech Univ Munich, MS, 50; Int Col, PhD, 80. *Prof Exp:* Engr, Inst Electromed, Univ Munich, 49; mem staff electronic serv, Sears Roebuck & Co, 51-57; proj leader, Indust Physics Dept, 57-60, sr res physicist & sr proj leader, Dept Appl Physics, 60-66, sect mgr, 66-69, asst dir, Dept Appl Physics, 69-74, DIR INDUST APPLNS, DIV APPL PHYSICS, SOUTHWEST RES INST, 74- *Mem:* Am Soc Mech Engrs; sr mem Instrument Soc Am. *Res:* Engineer acoustics and mechanics, machinery and plant dynamics. *Mailing Add:* 214 Landblade Lane San Antonio TX 78213-3350

NIMMO, BRUCE GLEN, b Attleboro, Mass, Dec 14, 38; m 62; c 2. MECHANICAL ENGINEERING, SOLAR ENERGY. *Educ:* Clarkson Col Technol, BME, 60; Univ Fla, MS, 61; Stanford Univ, PhD(mech eng), 68. *Prof Exp:* Proj engr heat transfer, Brookhaven Nat Lab, 61-64; res asst, Stanford Univ, 64-68; asst prof mech eng, Clarkson Col Technol, 68-70; assoc prof, Univ Cent Fla, 70-76, prof, 76; prof mech eng, Univ Petrol & Minerals, Saudi Arabia, 76-83; simulation engr, Naval Training Systs Ctr, 84-90; PROG DIR, THERMAL ENERGY SYSTS, FLA SOLAR ENERGY CTR, 90- *Concurrent Pos:* Chmn mech eng, Univ Petrol & Minerals, 77-78 & head res inst solar prog, 78-83; fel, Atomic Energy Comn. *Mem:* Am Soc Mech Eng; Am Solar Energy Soc; Sigma Xi. *Res:* Experimental heat transfer; thermal analysis; solar thermal applications. *Mailing Add:* 7220 N US 1 No 204 Cocoa FL 32927-5063

NIMNI, MARCEL EFRAIM, b Buenos Aires, Arg, Feb 1, 31; US citizen; m 83; c 4. BIOCHEMISTRY, NUTRITION. *Educ:* Univ Buenos Aires, BS, 54, PhD(pharmacol), 60; Univ Southern Calif, MS, 57. *Prof Exp:* Biochemist, AEC, Arg, 58-60; head biol sect, Don Baxter, Inc, Calif, 62-63; asst prof biochem & nutrit, Sch Dent, 64-67, asst prof med, Sch Med, 67-68, assoc prof, 68-73, PROF MED & BIOCHEM, SCH MED, UNIV SOUTHERN CALIF, 73- *Concurrent Pos:* Mem, Pathobiochem Study Sect, NIH, 80-84; ed, Collagen, Connective Tissue Res. *Honors & Awards:* Founders Award, Soc Biomat, 86. *Mem:* Fel AAAS; Am Inst Nutrit; Am Soc Biol Chem; Am Rheumatism Asn; Soc Exp Biol & Med. *Res:* Biochemistry of collagen; mechanism of defects in molecular aggregation; nature and biosynthesis of the crosslinks in collagen from skin, bone and cartilage; collagen as a biomaterial; collagen modification for bioprosthesis. *Mailing Add:* Orthopedic Hosp of Los Angeles 2400 S Flower St Los Angeles CA 90007

NIMS, JOHN BUCHANAN, b Monmouth, Maine, Dec 1, 24; m 47; c 2. PHYSICS. *Educ:* Boston Univ, BA, 50, MA, 51. *Prof Exp:* Physicist reactor physics, Knolls Atomic Power Lab, Gen Elec Co, 51-54; physicist, Atomic Power Develop Assocs Inc, 54-59, sect head, 59-66, div head, 66-73; dir nuclear eng, Detroit Edison Co, 73-74, corp strategic planner, 74-76, mgr strategic planning, 76-86; RETIRED. *Mem:* Am Phys Soc; Am Nuclear Soc; World Future Soc. *Res:* Reactor physics, kinetics and safety. *Mailing Add:* 3245 Manistee Rd N Frederik MI 49733

NINE, HARMON D, b Detroit, Mich, July 8, 31; m 55; c 3. METAL PHYSICS. *Educ:* Univ Mich, BS, 53, MS, 54. *Prof Exp:* Res physicist, 58-61, sr res physicist, 61-80, staff res scientist, 80-86, SR STAFF RES SCIENTIST, GEN MOTORS RES LABS, 86- *Concurrent Pos:* Pres, N Am Deep Drawing Res Group, Am Soc Metals. *Honors & Awards:* McCuen Award; Kettering Award. *Mem:* Am Soc Metals. *Res:* Ultrasonics; ultrasonic attenuation in solids; fatigue of metals; spectroscopy; acoustic emission; friction in metal forming; sheet metal deformation. *Mailing Add:* Dept Physics Gen Motors Res Labs Warren MI 48090-9055

NING, ROBERT Y, b Shanghai, China, Mar 12, 39; m 66; c 2. ORGANIC CHEMISTRY, BIOTECHNOLOGY. *Educ:* Rochester Inst Technol, BS, 63; Univ Ill, PhD(bio-org chem), 66; Fairleigh Dickinson Univ, MBA, 79. *Prof Exp:* Res fel, Calif Inst Technol, 66-67; sr chemist, 67-75, res fel, 75-78, group leader, Hoffmann-La Roche Inc, 78-85; sci investr, 85-88, REGULATORY AFFAIRS, BIOPHARMACEUT DEVELOP, HYBRITECH, INC, 88- *Mem:* Am Chem Soc; Protein Soc. *Res:* Industrial chemical and biochemical production processes; synthetic organic chemistry; chemistry of heterocycles; drug design. *Mailing Add:* Hybritech Inc PO Box 269006 San Diego CA 92196-9006

NING, TAK HUNG, b Canton, China, Nov 14, 43; US citizen; m 75; c 2. VLSI TECHNOLOGY, SILICON DEVICE PHYSICS. *Educ:* Reed Col, BA, 67; Univ Ill, MS, 68, PhD(physics), 71. *Prof Exp:* Fel solid state electronics, Dept Elec Eng, Univ Ill, 71-72, res asst prof, 72-73; res staff mem, 73-78, MGR IBM RES CTR, 78- *Honors & Awards:* JJ Ebers Award, Inst Elec & Electronic Engrs, 89, Jack A Morton Award, 91. *Mem:* Am Phys Soc; fel Inst Elec & Electronic Engrs. *Res:* Semiconductor device physics and technology; silicon transistor device physics and technology. *Mailing Add:* IBM Res Ctr PO Box 218 Yorktown Heights NY 10598

NININGER, ROBERT D, b Brookings, SDak, Mar 28, 19; m 43; c 4. ECONOMIC GEOLOGY. *Educ:* Amherst Col, BA, 41; Harvard Univ, MA, 42. *Prof Exp:* Geologist, Strategic Minerals Invest, US Geol Surv, 42-43; dep asst dir raw mat, AEC, 47-54, asst dir, 55-71; asst dir raw mat, Div Nuclear Fuel Cycle & Prod, US Energy Res & Develop Admin, 72-76; dir, Uranium Resources, 77-79, consult nuclear raw mat, US Dept Energy, 79-84; CONSULT, 84- *Mem:* Fel AAAS; Soc Min Engrs; Am Nuclear Soc; Soc Econ Geol. *Res:* Geology of radioactive materials; uranium resources and supply. *Mailing Add:* 14627 Crossway Rd Rockville MD 20853-1934

NINKE, WILLIAM HERBERT, b Toledo, Ohio, Aug 25, 37; m 61; c 3. COMMUNICATIONS SCIENCE, COMPUTER SCIENCE. *Educ:* Case Inst Technol, BS, 59, MS, 61, PhD(digital systs), 64. *Prof Exp:* Mem tech staff, Bell Labs, Murray Hill, NJ, 63-69, head, Digital Systs Res Dept, Holmdel, NJ, 69-78, head, image processing & display res dept, Bell Labs, Holmdel, NJ, 78- 86; DIR, COMPF ROBT SYSTEMS RES LAB, BELL LABS, 86- *Mem:* Sr mem Inst Elec & Electronics Engrs; fel Soc Info Display. *Res:* Visual telecommunications services; computer-driven display consoles; image synthesis; image analysis; robotics; computer aided design. *Mailing Add:* AT&T Bell Labs Rm 4e502 Crawfords Corner Rd Holmdel NJ 07733

NINO, HIPOLITO V, b Bogota, Colombia, Sept 28, 24; US citizen; m 55; c 3. CLINICAL & MEDICAL DEVICE REGULATION. *Educ:* Nat Univ Colombia, degree microbiol, 50; Univ Wis-Madison, MS, 52, PhD(biochem), 58. *Prof Exp:* Asst prof biochem, Nat Univ Colombia, 54-56; assoc prof, Univ Valle, Colombia, 56-57; lab dir clin chem, Citizens Hosp, Barberton, Ohio, 57-60 & St Joseph's Hosp, Syracuse, 60-63; asst prof biochem med & path & dir, Lab Clin Chem, Hosp, Univ NC, 63-70; chief nutrit biochem, Ctr Dis Control, Atlanta, 70-75; dir lab clin chem, Christ Hosp, Cincinnati, 75-78; dir res & develop diag, Beckman Instruments, Inc, 78-80; dir, div clin lab devices, Bur Med Devices, 80-84, ASSOC DIR, DIV LIFE SCI, CTR DEVICES & RADIOL HEALTH, OFF SCI & TECHNOL, FOOD & DRUG ADMIN, 84- *Concurrent Pos:* Instr biochem, State Univ NY Med Ctr, Syracuse, 60-63; consult interdept comt nutrit nat defense, NIH, 61-62; Rockefeller fel. *Honors & Awards:* Int Fel Award, Am Asn Clin Chem, 83. *Mem:* Am Asn Clin Chem; Sigma Xi; Acad Clin Lab Physicians & Scientists. *Res:* Clinical chemistry; methodology and quality assurance; vitamins and trace elements; clinical, epidemiological and analytical methods; regulatory aspects of clinical laboratory devices; biotechnology applications to clinical labs. *Mailing Add:* 17007 Catalpa Ct Rockville MD 20855

NIP, WAI KIT, b Hong Kong, May 5, 41; m 70; c 2. FOOD SCIENCE AND TECHNOLOGY. *Educ:* Chung-Hsing Univ, Taiwan, BS, 62; Tex A&M Univ, MS, 65, PhD(food technol), 69. *Prof Exp:* Assoc prof food sci, Chung-Hsing Univ, Taiwan, 69-74; res assoc, Univ Wis, 74-76; from asst prof to assoc prof food sci, 76-90, PROF FOOD SCI, UNIV HAWAII, 90- *Concurrent Pos:* NIH fel, Univ Wis, 75-76. *Mem:* Inst Food Technologists; Am Asn Cereal Chemists; Am Soc Hort Sci; World Aquaculture Soc. *Res:* Food processing technology; control of spoilage; chemical constituents in foods. *Mailing Add:* Dept Food Sci & Human Nutrit Univ Hawaii Honolulu HI 96822

NIPPER, HENRY CARMACK, b Alexander City, Ala, Mar 31, 40; m 66; c 2. ANALYTICAL CHEMISTRY, PATHOLOGY. *Educ:* Emory Univ, AB, 60; Purdue Univ, MS, 66; Univ Md, PhD(anal chem), 71; Am Bd Clin Chem, cert, 74. *Prof Exp:* Analytical chemist, E I du Pont de Nemours & Co, 60-63; teaching asst chem, Purdue Univ, 63-65, instr, 65-66; teaching asst chem, Univ Md, 66-68, Gillette-Harris res fel, 69-70, fel clin chem, Univ Hosp & Sch Med Dept Path, 71-73, from instr to asst prof path, 73-83, assoc mem grad fac, 76-83; sci dir clin chem, Beth Israel Hosp, Boston, 83-86; ASSOC PROF PATH, SCH MED, CREIGHTON UNIV, 86-, DIR, FORENSIC CHEM LAB, 89- *Concurrent Pos:* Chief, Clin Chem & Emergency Lab Opers, Vet Admin Med Ctr, Baltimore, Md, 73-83, chief, Clin Chem Res Lab, Res Serv, 74-83; lectr, USPHS Hosp Med Technol Prog, 74-80, NIH, 75-76, Essex Community Col, 75-82, Med Sch, Johns Hopkins Univ, 75-83; mem clin adv bd, Beckman Instruments, 82-83; mem bd dirs, Am Bd Clin Chem, 82-; instr path, Harvard Med Sch, 83-86; sci dir clin chem, St Joseph Hosp, Omaha, 86-; consult, 84-; mem Adv Comt Clin Chem & Toxicol, CORH, Food & Drug Admin, 88- *Honors & Awards:* Joseph H Roe Award, Am Asn Clin Chem, 78. *Mem:* AAAS; Am Acad Forensic Sci; Sigma Xi; Am Asn Clin Chem; Am Chem Soc; fel Nat Acad Clin Biochem; Nat Registry Clin Chem (pres). *Res:* Clinical biochemistry of myocardia infarction; kinetic (reaction rate) methods in clinical chemistry and toxicology; electrochemistry and chromatography in analytical chemistry and toxicology. *Mailing Add:* Dept Path Creighton Univ 601 N 30th St Omaha NE 68131

NIPPES, ERNEST F(REDERICK), b New York, NY, Feb 1, 18; m 39, 69; c 5. MATERIALS SCIENCE ENGINEERING. *Educ:* Rensselaer Polytech Inst, BS, 38, MetE, 40, PhD(metall), 42. *Prof Exp:* Asst metall eng, 38-39, instr, 39-45, from asst prof to assoc prof, 45-54, dir welding res, 48-61, chmn dept mat eng, 61-65, dir res div, 65-71, dir off res & sponsored progs, 71-75, PROF METALL ENG, RENSSELAER POLYTECH INST, 54- *Concurrent Pos:* Nat Acad Sci exchange visit to USSR, 61, 77, 79 & 81; pres, Rensselaer Res Corp, 65-70. *Honors & Awards:* Award, Am Soc Metals, 56; Miller Mem Medal, Am Welding Soc, 59. *Mem:* Fel Am Soc Metals; Am Welding Soc (vpres, 65-68, pres, 68-69); Am Soc Testing & Mat; Am Soc Eng Educ; Am Inst Mining, Metall & Petrol Engrs. *Res:* Metallurgy and welding. *Mailing Add:* Dept Mat Eng Rensselaer Polytech Inst Troy NY 12180-3590

NIPPO, MURN MARCUS, b New York, NY, Feb 8, 44; m 74; c 2. ANIMAL BEHAVIOR, ANIMAL NUTRITION. *Educ:* Univ Maine, BS, 65, MS, 68; Univ RI, PhD(animal nutrit), 76. *Prof Exp:* From asst prof to assoc prof, 76-87, CHMN ANIMAL SCI, DEPT FISH, ANIMAL & VET SCI, UNIV RI, 88- *Concurrent Pos:* Teaching comt, 83-86, Environ & Animal Welfare Comt, Am Soc Animal Sci, 89- *Mem:* Am Asn Lab Animal Sci; Am Soc Animal Sci; Nat Asn Cols & Teachers Agr; NY Acad Sci. *Res:* Interrelationship of stress and animal behavior in farm animals. *Mailing Add:* Woodward Hall Univ RI Kingston RI 02881

NIRDOSH, INDERJIT, b Ferozepur, Panjab, India, Feb 8, 43; Can citizen; m 69; c 2. PROCESS ENGINEERING. *Educ:* Panjab Univ, India, BSc, 66, MSc, 68; Birmingham, Univ, UK, PhD(chem eng), 74. *Prof Exp:* Lectr chem eng, Panjab Univ, Chandigarh, India, 68-71; res student chem eng, Birmingham Univ, UK, 71-74; sr chem engr, Brilex Chemicals, Amritsar, India, 76-77; res fel chem eng, McMaster Univ, Hamilton, Ont, 75-76 & 77-81; from asst prof to assoc prof, 81-87, PROF CHEM ENG, LAKE HEAD UNIV, THUNDER BAY, ONT, 87- *Concurrent Pos:* Managing partner, Sundew Chemicals, Ghaziabad, India, 76-77; consult, Nat Acad Sci, Wash, 84-85. *Mem:* Can Soc Chem Engrs; Can Inst Mining & Metall; Chem Inst Can. *Res:* Hydrometallurgy; flotation; electrolysis; ion-exchange; solvent extraction; process flowsheet development. *Mailing Add:* 494 Ryerson Crescent Thunder Bay ON P7C 5R8 Can

NIRENBERG, LOUIS, b Hamilton, Ont, Feb 28, 25; nat US; m 48; c 2. MATHEMATICS. *Educ:* McGill Univ, BSc, 45; NY Univ, MS, 47, PhD, 49. *Hon Degrees:* DSc, McGill Univ, 86, Univ Pisa, 90 & Univ Paris, Dauphine, 90. *Prof Exp:* From res asst to res assoc, 45-54, from asst prof to assoc prof, 52-57, PROF MATH, NY UNIV, 57- *Concurrent Pos:* Nat Res Coun fel, NY Univ, 51-52; fel, Inst Advan Study, 58; Sloan Found fel, 58-60; Fulbright lectr, 65; Guggenheim fel, 66-67 & 75-76; dir, Courant Inst Math Sci, 70-72; hon prof, Nankai Univ, 87 & Zhejiang Univ, 88; foreign corresp, Acad Sci, France, 89. *Honors & Awards:* Bocher Prize, Am Math Soc, 59; Crafoord prize, Swedish Acad Sci, 82. *Mem:* Nat Acad Sci; Am Math Soc; Am Acad Arts & Sci; Acad Dei Lincei; Am Philos Soc. *Res:* Partial differential equations; differential geometry; complex analysis. *Mailing Add:* Courant Inst NY Univ 251 Mercer St New York NY 10012

NIRENBERG, MARSHALL WARREN, b New York, NY, Apr 10, 27; m 61. ZOOLOGY, CHEMISTRY. *Educ:* Univ Fla, BS, 48, MS, 52; Univ Mich, PhD(biochem), 57. *Hon Degrees:* DSc, Univ Mich, 65, Univ Chicago, 65, Yale Univ, 65, Univ Windsor, 66, George Washington Univ, 72 & Weizmann Inst, Israel, 78. *Prof Exp:* Asst zool, Univ Fla, 45-50, res assoc, Nutrit Lab, 50-52; Am Cancer Soc fel, Nat Inst Arthritis & Metab Dis, 57-59, USPHS fel, Sect Metab Enzymes, 59-60, res biochemist, 60-62; head sect biochem genetics, 62-66, RES BIOCHEMIST & CHIEF BIOCHEM GENETICS LAB, NAT HEART, LUNG & BLOOD INST, 66- *Honors & Awards:* Nobel Prize in Med, 68. *Mem:* Nat Acad Sci; Pontif Acad Sci; Am Chem Soc; Biophys Soc; Soc Develop Biol; AAAS. *Mailing Add:* Lab Biochem Genetics Nat Heart Lung & Blood Inst Bethesda MD 20892

NIRSCHL, JOSEPH PETER, b Boca Raton, Fla, Apr 27, 45; m 67; c 2. FOOD PROCESS DEVELOPMENT. *Educ:* Carnegie Inst Technol, BS, 67; Univ Wis, PhD(chem eng), 72. *Prof Exp:* Res engr, Miami Valley Labs, Procter & Gamble, 71-73, process develop group leader, Household Cleaning Prod Div, 74-76; mgr food process eng, 77-80, DIR PROCESS ENG, BRISTOL-MYERS SQUIBB PHARMACEUT & MEAD JOHNSON NUTRIT DIV, EVANSVILLE, 80-, VPRES PHARMACEUT OPERS, 90- *Mem:* Am Inst Chem Engrs; Pharmaceut Mfrs Asn; Parental Drug Asn; Am Chem Soc. *Res:* Pharmaceutical product development; food product development and sterilization; pharmaceutical process development; food process development. *Mailing Add:* 5633 Cliftmeere Newburgh IN 47630-1849

NISBET, ALEX RICHARD, b Plainview, Tex, Apr 14, 38; m 80; c 1. ELECTROANALYTICAL CHEMISTRY. *Educ:* Univ Tex, BS, 59, PhD, 63. *Prof Exp:* Assoc prof, 63-73, PROF CHEM, OUACHITA BAPTIST UNIV, 73- *Mem:* AAAS; Electrochem Soc; Am Chem Soc; Soc Electroanal Chem. *Res:* Chronopotentiometry of metals. *Mailing Add:* Ouachita Baptist Univ Box 3685 Arkadelphia AR 71923

NISBET, EUAN G, archaean geology, early history of life, for more information see previous edition

NISBET, JERRY J, b Palisade, Colo, Nov 8, 24; m 48; c 2. SCIENCE EDUCATION, PLANT ANATOMY. *Educ:* Colo State Col Educ, AB, 49, AM, 50; Purdue Univ, PhD, 58. *Prof Exp:* Instr, Ball State Teachers Col, 50-52, asst prof, 54-56; asst, Purdue Univ, 52-53, res fel, 53-54, instr, 56-58; from asst prof to prof biol, Univ Eval, 58-83, coodr, 74-77; dir, Inst Environ Studies, Ball State Univ, 74-78; RETIRED. *Concurrent Pos:* Dir, NSF Summer Inst Biol, 60, 62-64 & 66-68, dir, NSF Acad Yr Inst, 68-71; nat dir, Outstanding Biol Teacher Awards Prog, Nat Asn Biol Teachers, 65-67, dir, Region III, 66-68; mem, Sci Adv Comt, Ind Dept Pub Instr, 72-76 & Ind Lt Govr's Sci Adv Comt, 73-78; secy, Ind Acad Sci, 72-74, pres, 78; secy, Cent States Univ, Inc, 68-70, chmn, Coun, 72, pres, Bd, 76. *Mem:* Fel AAAS; Am Inst Biol Sci; Nat Asn Res Sci Teaching; Asn Midwest Col Biol Teachers; Nat Sci Teachers Asn. *Res:* Development and testing of technologically based systems of instruction; interpretation of science for the non-scientist; study of the ultrastructure of plant cell walls. *Mailing Add:* PO Box 340 Palisade CO 81526-0340

NISBET, JOHN S(TIRLING), b Darval, Scotland, Dec 10, 27; US citizen; m 53; c 2. ELECTRICAL ENGINEERING. *Educ:* Univ London, BS, 50; Pa State Univ, MS, 57, PhD(elec eng), 60. *Prof Exp:* Trainer engr res & develop, Nash & Thompson Ltd, Eng, 44-51; engr, Decca Radar Ltd, 51-53 & Can Westinghouse, 53-55; res assoc, 55-60, from asst prof to assoc prof ionospheric res, 60-67, PROF ELEC ENG, PA STATE UNIV, 67-, DIR, IONOSPHERIC RES LAB, 71- *Concurrent Pos:* Nat Sci Found fel, Belg Inst Spatial Aeronomy, 65-66; Fulbright fel, Kharkov Univ, USSR, 79; Nat Res Coun-Nat Acad Sci fel, Goddard Space Flight Ctr, 80. *Mem:* Am Geophys Union; sr mem Inst Elec & Electronics Engrs; Inst Sci Radio Union; Sigma Xi. *Res:* Physics of ionosphere. *Mailing Add:* Dept Elec Eng Pa State Univ University PA 16802

NISBET-BROWN, ERIC ROBERT, b Vancouver, BC, July 10, 56; Can citizen. IMMUNOREGULATION, ANTIGEN PROCESSING. *Educ:* Univ BC, BSc, 76; Univ Alta, PhD(immunol), 81, MD, 83. *Prof Exp:* Fel immunol, Univ Alta, 81-83, Hosp Sick Children, Toronto, 83-85; ASST PROF IMMUNOL, UNIV TORONTO, 86 -; TORONTO WEST HOSP. *Concurrent Pos:* Dep med dir, Can Red Cross Blood Transfusion Serv, Toronto, 86 - *Mem:* Am Asn Immunologists; Can Asn Immunologists. *Res:* Regulation of immune responses by T-lymphocytes; mechanism of antigen processing and presentation; immunogenetics. *Mailing Add:* Immunol Lab Toronto West Hosp 399 Bathurst St MC5-405 Toronto ON M5T 2S8 Can

NISENOFF, MARTIN, b New York, NY, Dec 25, 28; wid; c 3. LOW TEMPERATURE PHYSICS. *Educ:* Worcester Polytechnic Inst, Mass, BS, 50; Purdue Univ, MS, 52, PhD(physics), 60. *Prof Exp:* Res assoc physics, Purdue Univ, 60-61; res scientist physics, Sci Res Staff, Ford Motor Co, Dearborn, Mich, 61-70; physicist, Cryog Appln Group, Stanford Res Inst, Menlo Park, Calif, 70-72; res scientist physics, Cryog & Superconductivity Br, 72-79, res scientist physics, 79-88, CONSULT, MICROWAVE TECHNOL BR, NAVAL RES LAB, WASHINGTON, DC, 88- *Mem:* Am Phys Soc. *Res:* Low temperature physics; superconducting electronics; Josephson effects; superconducting magnetometry; applications of superconductive electronic devices and systems. *Mailing Add:* Naval Res Lab Code 6850-1 Washington DC 20375-5000

NISHI, YOSHIO, b Yokohama, Japan, Mar 1, 40; m 60; c 2. INTEGRATED CIRCUITS, SILICON PROCESS TECHNOLOGY. *Educ:* Waseda Univ, Japan, BS, 62; Univ Tokyo, Japan, PhD(elec eng), 73. *Prof Exp:* Mem tech staff, silicon process & mat, Toshiba R & D Ctr, 62-68; res assoc solid state physic, Stanford Electron Labs, 68-69; supvr device physics, Toshiba R & D Ctr, 70-75, sect mgr device technol, 75-77, dept mgr memory technol, Toshiba semiconductor-Device Eng Lab, 79-86; LAB DIR SILICON PROCESS R & D, HEWLETT-PACKARD LABS, 86- *Concurrent Pos:* Consult prof, elec eng dept, Stanford Univ, 86-, CIS mentor, Ctr Integrated Systs, Stanford Univ, 87, adv bd, Microelectron Lab, Santa Clara Univ, 86-, tech adv bd, Sematech, 88-, consult tech modeling assocs, 88- *Mem:* fel Inst Elec & Electronics Engrs; Japan Soc Appl Phys; Electrochemical Soc; Japan Inst Elec & Electronic Engrs. *Res:* Semiconductor process technology for ultra-large scale integrated circuits, device physics for field effect transistors including non-volatile memory and dynamic memory cells. *Mailing Add:* Bldg 25 U 3500 Deer Creek Rd Palo Alto CA 94303

NISHIBAYASHI, MASARU, b Los Angeles, Calif, May 6, 23; m 49; c 3. PHYSICAL CHEMISTRY. *Educ:* Univ Cincinnati, BS, 49, PhD(phys chem), 53. *Prof Exp:* Res chemist, Aerojet-Gen Corp, 53-60, sr res chemist, 60-63, sr chem specialist, Aerojet Ord Co, 63-86; RETIRED. *Mem:* Sigma Xi. *Res:* Physical and chemical properties of propellants and explosives; interior ballistics; gun propellants. *Mailing Add:* 1339 Beech Hill Ave Hacienda Heights CA 91745

NISHIDA, TOSHIRO, b Nagasaki, Japan, Jan 6, 26; m 60; c 3. FOOD CHEMISTRY. *Educ:* Kyoto Univ, MS, 52; Univ Ill, PhD(food chem), 56. *Prof Exp:* Res asst agr chem, Univ Osaka Prefecture, 47-49; res chemist, Osaka Soda Co, Japan, 52-53; from asst prof to assoc prof food chem, 58-70, PROF FOOD SCI, UNIV ILL, URBANA, 70- *Concurrent Pos:* Grants, NIH, 57- & Ill Heart Asn, 65- *Mem:* Am Chem Soc; Am Oil Chem Soc; Am Soc Biol Chemists; Am Inst Nutrit; Sigma Xi. *Res:* Chemistry and metabolism of lipids and lipoproteins. *Mailing Add:* Burnside Res Lab Univ Ill Urbana IL 61803

NISHIE, KEICA, pharmacology, toxicology, for more information see previous edition

NISHIKAWA, ALFRED HIROTOSHI, b San Francisco, Calif, Apr 23, 38; m 61; c 2. BIOCHEMISTRY. *Educ:* Univ Calif, Berkeley, AB, 60; Ore State Univ, PhD(enzym), 65. *Prof Exp:* Lab technician biochem, Univ Calif, Berkeley, 60-61; res assoc, Ore State Univ, 65-69; scientist, Xerox Res Labs, NY, 69-71; sr biochemist, 71-75, res fel, Chem Res Div, 76-80, asst dir, Biopolymer Res Dept, 81-83, dir, Biopolymer Res Dept, Hoffmann-La Roche Inc, 84-85; dir protein biochem, Smith Kline & French Labs, 85-88, GROUP DIR PROTEINS & MACROMOLECULES, SMITH KLINE BEECHAM, 89- *Concurrent Pos:* Chmn, Gordon Res Conf on Immobilized Species, 82. *Mem:* AAAS; Am Chem Soc; NY Acad Sci; Sigma Xi. *Res:* Enzyme isolation and purification; physical-chemical studies of chemically modified proteins; affinity chromatography and immobilized enzyme reactors; recombinant proteins pharmaceuticals. *Mailing Add:* 31 Caenarvon Lane Haverford PA 19041-1048

NISHIKAWARA, MARGARET T, b Vancouver, BC, Feb 3, 23. PHYSIOLOGY. *Educ:* Univ Toronto, BA, 47, MA, 48, PhD(physiol), 52. *Prof Exp:* Res assoc physiol, Univ Toronto, 52-54; from asst prof to assoc prof, 54-72, PROF PHYSIOL, OHIO STATE UNIV, 72- *Mem:* Sigma Xi; Am Physiol Soc; Endocrine Soc. *Res:* Physiologic and endocrine control of metabolic pathways. *Mailing Add:* Dept Physiol 4196 Graves Hall Ohio State Univ Main Campus Columbus OH 43210

NISHIMOTO, ROY KATSUTO, b Lihue, Hawaii, Oct 29, 44; m 70; c 1. WEED SCIENCE, VEGETABLE CROPS. *Educ:* Ore State Univ, BS, 66, MS, 67; Purdue Univ, PhD(weed sci), 70. *Prof Exp:* Asst prof, 70-74, assoc prof, 74-79, PROF WEED SCI, DEPT HORT, UNIV HAWAII, 79- *Mem:* Asian Pac Weed Sci Soc (treas, 73-); Weed Sci Soc Am; Am Soc Hort Sci. *Res:* Weed control in horticultural crops; Cyperus rotundus tuber dormancy; absorption and translocation of herbicides; phosphorus fertility in tropical soils and vegetable crops. *Mailing Add:* Dept Hort One St John 102 Univ Hawaii Monoa 2500 Campus Rd St John 102 Honolulu HI 96822

NISHIMURA, JONATHAN SEI, b Berkeley, Calif, Sept 30, 31; m 55; c 2. BIOCHEMISTRY. *Educ:* Univ Calif, AB, 56, PhD(biochem), 59. *Prof Exp:* USPHS fel biochem, Sch Med, Tufts Univ, 59-62, sr instr, 62-64, asst prof, 64-69; assoc prof, 69-71, PROF BIOCHEM, UNIV TEX HEALTH SCI CTR SAN ANTONIO, 71- *Concurrent Pos:* NIH Pathobiochem Study Sect, 82-86. *Mem:* Am Soc Biol Chem. *Res:* Mechanism of enzyme action. *Mailing Add:* Dept Biochem Univ Tex Med Sch 7703 Floyd Curl Dr San Antonio TX 78284

NISHIOKA, DAVID JITSUO, b Los Angeles, Calif, Aug 12, 45. CELLULAR BIOLOGY, DEVELOPMENTAL BIOLOGY. *Educ:* Calif State Univ, Fullerton, AB, 71, MA, 72; Univ Calif, Berkeley, PhD(zool), 76. *Prof Exp:* Scholar biol, Univ Calif, San Diego, 76-77 & Stanford Univ, 77-78; asst prof, 78-84, ASSOC PROF BIOL, GEORGETOWN UNIV, 84- *Concurrent Pos:* NSF grant, 80-83 & 86-88, NIH grant, 84-88. *Mem:* Am Soc Cell Biol; Soc Develop Biol; Am Soc Zool; Sigma Xi. *Res:* Spermatogenesis, fertilization, and activation of development; localization of sperm proteins during spermatogenesis and role of sperm proteins in fertilization and activation of development. *Mailing Add:* Dept Biol Georgetown Univ Washington DC 20057

NISHIOKA, RICHARD SEIJI, b Hilo, Hawaii, Mar 9, 33; m 61; c 3. NEUROENDOCRINOLOGY, IMMUNOCYTOCHEMISTRY. *Educ:* Univ Hawaii, BA, 56, MA, 59; Univ Tokyo, PhD, 87. *Prof Exp:* Res zoologist, 59-65, from asst specialist to specialist, 65-88, RES ENDOCRINOLOGIST, INTEGRATIVE BIOL, UNIV CALIF, BERKELEY, 88- *Mem:* Am Soc Zoologists; AAAS. *Res:* Comparative neuroendocrinology; ultrastructure and immunocytochemical aspects of neurosecretion formation, transport and release; innervation of endocrine organs; fish development; endocrinology of salmon smoltification; endocrinology and physiology of striped bass. *Mailing Add:* 974 Tulare Ave Albany CA 94707

NISHIOKA, YUTAKA, b Apr 1, 48. MECHANISMS OF SEX DETERMINATIONS, DEVELOPMENTAL BIOLOGY. *Educ:* Columbia Univ, PhD(microbiol), 78. *Prof Exp:* Asst prof, 80-86, ASSOC PROF BIOL, MCGILL UNIV, 86- *Mem:* Am Soc Microbiol; Am Soc Human Genetic; Genetic Soc Can. *Res:* Mouse embryology. *Mailing Add:* Dept Biol Univ McGill 1205 Dr Penfield Ave Montreal PQ H3A 1B1 Can

NISONOFF, ALFRED, b New York, NY, Jan 26, 23; div; c 2. IMMUNOCHEMISTRY, IMMUNOBIOLOGY. *Educ:* Rutgers Univ, BS, 42; Johns Hopkins Univ, PhD(chem), 51. *Prof Exp:* AEC fel, Med Sch, Johns Hopkins Univ, 51-52; sr res chemist, US Rubber Co, Conn, 52-54; sr cancer res scientist, Roswell Park Mem Inst, NY, 54-57, assoc cancer res scientist, 57-60; from assoc prof to prof microbiol, Univ Ill, Urbana, 60-66; prof microbiol, Univ Ill Col Med, 66-69, head dept biochem, 69-75; PROF BIOL, ROSENSTIEL RES CTR, BRANDEIS UNIV, 75- *Concurrent Pos:* NIH career res award, 62-69; mem study sects allergy & immunol, NIH & Nat Multiple Sclerosis Soc; foreign corresp, Belgian Royal Acad Med, 77-; mem coun, Am Asn Immunol, 85- *Honors & Awards:* Medal, Pasteur Inst, 71. *Mem:* Nat Acad Sci; Am Asn Immunol (pres, 90-91); fel Am Acad Arts & Sci. *Res:* Mechanism of biosynthesis of antibodies and their genetic control; mechanism of formulation of antibodies, including genetic control; properties of antibodies responsible for allergic reactions. *Mailing Add:* Dept Biol Rosenstiel Res Ctr Brandeis Univ Waltham MA 02254

NISS, HAMILTON FREDERICK, b Milwaukee, Wis, Apr 29, 23; m 49; c 2. BIOCHEMISTRY. *Educ:* Univ Wis, PhB, 45, MS, 47; Purdue Univ, PhD, 58. *Prof Exp:* Asst prof biol, Sam Houston State Col, 47-49; instr bact, Purdue Univ, 49-57; asst prof microbiol, Syracuse Univ, 57-61; sr microbiologist, Ely Lilly & Co, 61-88; RETIRED. *Mem:* AAAS; Am Soc Microbiol; Sigma Xi. *Res:* Antibiotic biosynthesis; chemical and microbiological assay development. *Mailing Add:* 3710 N Grant Ave Indianapolis IN 46218

NISSAN, ALFRED H(ESKEL), b Baghdad, Iraq, Feb 14, 14; US citizen; m 40; c 1. CHEMICAL ENGINEERING, PHYSICAL CHEMISTRY. *Educ:* Univ Birmingham, BSc, 37, PhD, 40, DSc, 43. *Prof Exp:* Lectr petrol prod, Univ Birmingham, 40-47; dir res, Bowater Pulp & Paper Co, Eng, 47-53; res prof textile eng, Univ Leeds, 53-57; chem eng, Rensselaer Polytech Inst, 57-62; corp res dir, Westvaco Corp, 62-79, vpres, 67-79; CONSULT, 79- *Concurrent Pos:* Consult, WVa Pulp & Paper Co, 57-62 & Flame Warfare Comt, Ministry of Supply, Eng, 42-57; adj prof, Rensselaer Polytech Inst, 62-69, State Univ NY Col Environ Sci & Forestry, 79-; hon vis prof, Univ Uppsala, Sweden, 74. *Honors & Awards:* Mitscherlich Medal, Zellcheming, Ger, 80; Gold Medal, Tech Asn Pulp & Paper Indust, 82. *Mem:* Fel Am Inst Chem; fel AAAS; fel Tech Asn Pulp & Paper Indust; fel Am Inst Chem Engrs; fel Int Acad Wood Sci. *Res:* Fluid mechanics and other transport phenomena; rheology; fibers and fibrous systems. *Mailing Add:* 6A Dickel Rd Scarsdale NY 10583

NISSELBAUM, JEROME SEYMOUR, b Hartford, Conn, Dec 21, 25; m 49; c 3. BIOCHEMISTRY, PHYSIOLOGY. *Educ:* Univ Conn, BA, 49; Tufts Univ, PhD(biochem), 53. *Prof Exp:* Asst biochem, Tufts Univ, 50-53, USPHS res fel, 53-55, instr, Sch Med, 54-57; asst, 57-60, ASSOC, SLOAN-KETTERING INST CANCER RES, 60-, ASSOC MEM, 67-; ASSOC PROF BIOCHEM, CORNELL UNIV, 68- *Concurrent Pos:* Res assoc, Sloan-Kettering Div, Cornell Univ, 57-68; asst biochemist, Mem Hosp, Mem Sloan-Kettering Cancer Ctr, 72-75, from asst to assoc attend biochemist, Mem Hosp, 75-81, attend biochemist & clin mem, 88. *Mem:* AAAS; Am Asn Clin Chemists; Am Chem Soc; Am Soc Biol Chem; NY Acad Sci; Harvey Soc; Clin Ligand Assay Soc. *Res:* Enzymology; clinical biochemistry; methods development; protein chemistry; immunochemistry. *Mailing Add:* 1275 York Ave New York NY 10021

NISSENSON, ROBERT A, b Chicago, Ill, Dec 17, 49. ENDOCRINOLOGY, MECHANISM OF HORMONE ACTION. *Educ:* Northwestern Univ, PhD(physiol), 76. *Prof Exp:* Asst prof, 78-84, ASSOC ADJ PROF MED & PHYSIOL, UNIV CALIF, SAN FRANCISCO, 84-; ASSOC RES CAREER SCIENTIST, VET ADMIN MED CTR, SAN FRANCISCO, CALIF, 86- *Mem:* Endocrine Soc; Am Physiol Soc; Am Soc Bone & Mineral Res. *Mailing Add:* Vet Admin Med Ctr 111 & 4150 Clement St San Francisco CA 94121

NISSIM-SABAT, CHARLES, b Sofia, Bulgaria, Feb 1, 38; US citizen; m 67; c 1. COSMOLOGY, HIGH ENERGY PHYSICS. *Educ:* Columbia Univ, AB, 59, MA, 60, PhD(physics), 65. *Prof Exp:* Res assoc high energy physics, Fermi Inst, Univ Chicago, 65-67; from asst prof to assoc prof, 67-74, PROF PHYSICS & CHMN DEPT, NORTHEASTERN ILL UNIV, CHICAGO, 74- *Mem:* AAAS. *Res:* General relativity and high energy physics, particularly experimental weak interactions and mu-mesic x-rays. *Mailing Add:* Dept Physics Northeastern Ill Univ 5500 N St Louis Ave Chicago IL 60625

NISSLEY, S PETER, ENDOCRINOLOGY, GROWTH FACTORS. *Educ:* Univ Pa, MD, 64. *Prof Exp:* SR INVESTR, NIH, 72- *Mailing Add:* NIH Bldg Ten Rm 3B38 Bethesda MD 20892

NISTERUK, CHESTER JOSEPH, b New York, NY, Sept 3, 28. STATISTICAL MECHANICS, SIGNAL PROCESSING. *Educ:* Polytech Inst Brooklyn, BEE, 49, MS, 54, PhD(physics), 67. *Prof Exp:* From instr to assoc prof elec eng, 51-79, head dept, 69-77, PROF ELEC ENG, MANHATTAN COL, 79-, CHMN, GRAD STUDIES ELEC & COMPUT ENG, 85- *Concurrent Pos:* Prof lectr, St John's Univ, NY, 56-57; lectr, Adelphi Univ, NY, 56-57. *Mem:* Inst Elec & Electronics Engrs; Am Phys Soc; Am Asn Physics Teachers; Am Soc Eng Educ. *Res:* Quantum statistical mechanics; quantum electronics; biological potentials. *Mailing Add:* Dept Elec Eng Manhattan Col Bronx NY 10471

NISULA, BRUCE CARL, b Hartford, Conn, Jan 15, 45; m 66; c 2. ENDOCRINOLOGY. *Educ:* Dartmouth Col, BA, 66; Dartmouth Med Sch, BMS, 67; Harvard Med Sch, MD, 69. *Prof Exp:* Intern, Peter Bent Brigham Hosp, Boston, 69-70; resident, Children's Hosp Med Ctr, Boston, 70-71 & Georgetown Univ Med Ctr, 74-76; fel, Nat Inst Child Health & Human Develop, NIH, 71-74, sr investr, 76-83, sect chief, 83-89, BR CHIEF, NAT INST CHILD HEALTH & HUMAN DEVELOP, NIH, 89- *Mem:* Am Soc Clin Invest; Am Thyroid Asn; Endocrine Soc; Am Fedn Clin Res. *Res:* Glycoprotein hormones; thyrotropin; follitropin; chorionic gonadotropin; luteinizing hormone. *Mailing Add:* NIH Rm 10/10N262 Bethesda MD 20854

NISWANDER, JERRY DAVID, dentistry, human genetics; deceased, see previous edition for last biography

NISWENDER, GORDON DEAN, b Gillette, Wyo, Apr 21, 40; m 64; c 2. REPRODUCTIVE ENDOCRINOLOGY. *Educ:* Univ Wyo, BS, 62; Univ Nebr, MS, 64; Univ Ill, PhD(animal sci), 67. *Prof Exp:* Res asst reprod physiol, Univ Nebr, 62-64 & Univ Ill, 64-65; NIH res fel, Univ Mich, Ann Arbor, 67-68, asst prof, 68-71; from asst prof to assoc prof, 71-75, PROF REPRODUCTION PHYSIOL, COLO STATE UNIV, 75- *Honors & Awards:* Animal Physiol & Endrocrinol Award, Am Asn Animal Scientists, 81; Ayerst Award, Endocrine Soc, 83. *Mem:* AAAS; Soc Study Reproduction (pres, 81); Am Soc Animal Sci; Endocrine Soc. *Res:* Reproductive biology, endocrinology, immunology. *Mailing Add:* Dean Physiol Colo State Univ Ft Collins CO 80523

NITECKI, DANUTE EMILIJA, b Lithuania, Apr 22, 27; US citizen; c 1. CHEMISTRY, BIOCHEMISTRY. *Educ:* Univ Chicago, MS, 56, PhD(chem), 61. *Prof Exp:* NIH fel, 61-63; asst res biochemist, Med Ctr, 63-69, assoc res biochemist, 69-77, res biochemist, Sch Med, Univ Calif, San Francisco, 77-82; SR SCIENTIST & DIR CHEM DEPT, CETUS CORP, EMERYVILLE, CO, 82- *Mem:* Am Chem Soc; Royal Soc Chem; Protein Soc. *Res:* Synthesis of peptides used as haptens in immunochemical studies; synthesis of boron containing aromatic compounds of physiological interest; syntheses of well defined small molecular weight antigens used to investigate the initiation and the progress of cellular and humoral immune response; synthesis of crosslinking agents and their use to crosslink various biomaterials. *Mailing Add:* 2296 Virginia St Berkeley CA 94709

NITECKI, MATTHEW H, b Poland, Apr 30, 25; nat US; m 64; c 2. EVOLUTIONARY BIOLOGY. *Educ:* Univ Chicago, MS, 58, PhD, 68. *Prof Exp:* Asst cur, 65-69, assoc cur, 69-75, CUR, FIELD MUS NATURAL HIST, 75-, CUR, UNIV CHICAGO, 55- *Concurrent Pos:* Vis investr, Inst Geol & Geophys, USSR, 78; guest scientist, Acad Sci USSR, 81, exchange scholar, Nat Acad Sci, USSR, 84, Fulbright-Hays res scholar, Norway, 85-86; lectr, Univ Chicago, 82- *Mem:* Geol Soc Am; Am Asn Petrol Geologists; Soc Econ Paleontologists & Mineralogists; Soc Study Evolution. *Res:* Paleobiology of paleozoic invertebrates, algae and problematic fossils; history and sociology of science; theoretical evolutionary biology with emphasis on common problems of zoology, botany and paleontology. *Mailing Add:* Field Mus Natural Hist Roosevelt Rd & Lake Shore Dr Chicago IL 60605

NITECKI, ZBIGNIEW, b Plymouth, Eng, May 12, 46. COMPUTER PROGRAMMER, SYSTEM ANALYSIS. *Educ:* Univ Chicago, BS, 65; Univ Calif, Berkeley, PhD(math), 69. *Prof Exp:* Prof math, Yale Univ, 69-71 & New York City Col, 71-72; PROF MATH, TUFTS UNIV, 72- *Concurrent Pos:* Prog dir, Geometric Div, Nat Sci Found, 82-83. *Mem:* Am Math Soc. *Mailing Add:* Dept Math Tufts Univ Medford MA 02155

NITHMAN, CHARLES JOSEPH, b Belleville, Ill, Jan 14, 37; m 59; c 2. PHARMACY. *Educ:* Okla State Univ, BS, 59; Univ Okla, BSPh, 62, Univ Okla, MSc, 70; Mercer Univ, PharmD, 74. *Prof Exp:* Clin instr pharm, Univ Okla, 68-70; asst dir pharm, St Anthony Hosp, Oklahoma City, 70-72; asst prof, 72-78, ASSOC PROF PHARM, SOUTHWESTERN OKLA STATE UNIV, 72-, HEAD, CLIN PHARM DEPT, 81- *Mem:* Sigma Xi; Am Soc Hosp Pharmacists; Am Pharmaceut Asn; Am Asn Col Pharm. *Res:* Drug distribution systems in hospitals. *Mailing Add:* 4520 NW 32nd Pl Oklahoma City OK 73122

NITOWSKY, HAROLD MARTIN, b Brooklyn, NY, Feb 12, 25; m 54; c 2. MEDICINE, GENETICS. *Educ:* NY Univ, AB, 44, MD, 47; Univ Colo, MSc, 51; Am Bd Pediat, dipl, 56; Am Bd Med, 82. *Prof Exp:* USPHS fel, Univ Colo, 50-51; from instr to assoc prof, Sch Med, Johns Hopkins Univ, 53-67; PROF PEDIAT & GENETICS, ALBERT EINSTEIN COL MED, 67- *Concurrent Pos:* Pediatrician, Johns Hopkins Hosp, 53-; res assoc & adj attend pediatrician, Sinai Hosp, 55-, dir pediat res, 58-; sr investr, Nat Asn Retarded Children, 60-65. *Mem:* AAAS; Soc Pediat Res; Sigma Xi; Am Fedn Clin Res; Am Inst Nutrit. *Res:* Somatic cell genetics; inborn errors of metabolism; genetic counseling. *Mailing Add:* Dept of Pediat Albert Einstein Col of Med Bronx NY 10461

NITSCHE, JOHANNES CARL CHRISTIAN, b Olbernhau, Ger. PARTIAL DIFFERENTIAL EQUATIONS. *Educ:* Univ Gottingen, dipl, 50; Univ Leipzig, PhD(math), 51. *Prof Exp:* Scientist, Max-Planck Inst, Gottingen, 50-52; prin asst, Tech Univ, Berlin-Charlottenburg, 52-55; Fulbright fel, Stanford Univ, 55-56; vis assoc prof, Univ Cincinnati, 56-57; assoc prof, 57-60, head, Sch Math, 71-78, PROF MATH, UNIV MINN, 60- *Concurrent Pos:* Prin investr, US Govt & NSF grants, 57-; vis prof, Univ Puerto Rico, 60-61, Univ Hamburg, 65, Tech Univ Vienna, 68, Univ Bonn, 71 & 75, Univ Heidelberg, 79; assoc ed, Contemporary Math, 80-; von Humboldt sr scientist award, 81. *Honors & Awards:* Lester R Ford Award, Math Asn Am, 75. *Mem:* AAAS; Am Math Soc; Math Asn Am; Soc Natural Philosophy. *Res:* Mathematical analysis (approximation theory, calculus of variations, differential geometry, minimal surface theory, partial differential equations) and its applications to related disciplines; minimal surfaces. *Mailing Add:* Dept Math 302 Vincent Hall Univ Minn 206 Church St SE Minneapolis MN 55455

NITSCHKE, J MICHAEL, b Berlin, Germany, Apr 27, 39. NUCLEAR PHYSICS. *Educ:* Brunswick Tech Univ, MS, 65, Dr rer nat(physics), 68. *Prof Exp:* PHYSICIST NUCLEAR CHEM & PHYSICS, LAWRENCE BERKELEY LAB, 71- *Mem:* Am Phys Soc; Am Chem Soc. *Res:* Physical, chemical and nuclear properties of the heaviest elements; nuclei far from stability; on-line isotope analysis; radioactive nuclear beams. *Mailing Add:* Lawrence Berkeley Lab 71-259 One Cyclotron Rd Berkeley CA 94720

NITSOS, RONALD EUGENE, b Sacramento, Calif, Aug 26, 37; m 66; c 2. HORTICULTURE. *Educ:* Sacramento State Col, BA, 63; Ore State Univ, MA, 66, PhD(plant physiol), 69. *Prof Exp:* ASSOC PROF BIOL, SOUTHERN ORE STATE COL, 69- *Concurrent Pos:* Res grant, Southern Ore Col, 69-71; proj dir, Cause Grant for Greenhouse, NSF, 79-81. *Mem:* Am Soc Plant Physiologists; NAm Mycol Asn; Native Plant Soc. *Res:* Plant physiology, especially roles of univalent cations in plants; needs and effects of univalent cations in enzyme activation. *Mailing Add:* Dept Biol Southern Ore State Col 1250 Siskiyou Blvd Ashland OR 97520

NITTLER, LEROY WALTER, b Shickley, Nebr, Jan 10, 21; wid; c 2. AGRICULTURE. *Educ:* Univ Nebr, BS, 49, MS, 50; Cornell Univ, PhD(plant breeding), 53. *Prof Exp:* Res asst, 50-53, from asst prof to prof, 53-80, head dept, 68-73, EMER PROF SEED INVESTS, COL AGR & LIFE SCI, CORNELL UNIV, 80- *Mem:* Am Soc Agron. *Res:* Varietal purity testing of grain, forage crop and turf grass seed; response of grain, grass and legume seedlings to photoperiod, light quality, light intensity and temperature; response of seedlings to chemicals and nutrient elements. *Mailing Add:* 6041 S Hill Rd Middlesex NY 14507

NITTROUER, CHARLES A, b Philadelphia, Pa, June 20, 50; m 76. GEOLOGICAL OCEANOGRAPHY. *Educ:* Lafayette Col, BA, 72; Univ Wash, MS, 74, PhD(oceanog), 78. *Prof Exp:* Res assoc oceanog, Univ Wash, 78; ASST PROF MARINE SCI, NC STATE UNIV, 78- *Mem:* Am Geophys Union; Geol Soc Am; Soc Econ Paleontologists & Mineralogists; Int Asn Sedimentologists. *Res:* Marine sedimentology, especially sediment transport and accumulation on continental shelves and adjacent environments. *Mailing Add:* Dept Marine Sci State Univ NY Stony Brook Main Stony Brook NY 11794

NITZ, DAVID F, b Shighaw, Mich, July 18, 50. PHYSICS, HIGH ENERGY PHYSICS. *Educ:* Univ Rochester, PhD(physics), 77. *Prof Exp:* Res assoc, 77-81, asst res scientist, 81-86, ASSOC RES SCIENTIST, UNIV MICH, 86- *Mem:* Am Phys Soc. *Mailing Add:* Randall Physics Lab Mich Univ Ann Arbor MI 48109

NITZ, OTTO WILLIAM JULIUS, b Sigourney, Iowa, June 25, 05; m 36. ORGANIC CHEMISTRY. *Educ:* Elmhurst Col, BS, 29; Univ Iowa, MS, 33, PhD(org chem), 36. *Prof Exp:* Instr, Elmhurst Col, 29-32; asst, Univ Iowa, 33-36; prof chem, Parsons Col, 36-40 & Northern Mont Col, 40-51; chief chemist, Ky Synthetic Rubber Corp, 51-52; prof chem, Univ Wis-Stout, 52-71, emer prof, 71-; RETIRED. *Mem:* Am Asn Retired Persons; Nat Retired Teachers Asn. *Res:* Vanillin derivatives; author of text books and manuals. *Mailing Add:* 1103 Third Ave Menomonie WI 54751

NIU, JOSEPH H Y, inorganic chemistry, organic chemistry, for more information see previous edition

NIU, MANN CHIANG, b Peking, China, Oct 31, 14; nat US; m 43; c 2. CYTOLOGY, BIOENGINEERING & BIOMEDICAL ENGINEERING. *Educ:* Peking Univ, AB, 36; Stanford Univ, PhD, 47. *Prof Exp:* Asst zool, Peking Univ, 36-40, lectr, 40-44; res assoc embryol, Stanford Univ, 44-52, res biologist, 52-55; asst prof gen physiol, Rockefeller Univ, 55-60; prof biol, Temple Univ, 60-82; RETIRED. *Concurrent Pos:* Guggenheim fel, 54 & 55; vis prof, Rockefeller Univ & Groteborg Univ Sweden, 70; hon prof, Inst Zool, Acad Sinica, Beijing, 73-, co-founder & sci adv, Inst Develop Biol, 78-; hon prof lab reproduction eng, Hunan Med Univ, 88. *Honors & Awards:* Lillie Award, 57. *Mem:* Am Soc Biochem & Molecular Biol; Am Tissue Cult Asn; Am Soc Cell Biol; Soc Exp Biol & Med; Int Soc Develop Biol; Am Soc Zoologists. *Res:* Origin of pigment cells; cell transformation and cancer; physiology of color changes; nucleocytoplasmic interactions; causal analysis of embryonic induction; DNA synthesis in enucleated eggs; mechanisms of mRNA mediated genetic change; genetic manipulation in higher organisms. *Mailing Add:* 7950 Montgomery Ave Elkins Park PA 19117

NIV, YEHUDA, nuclear structure, for more information see previous edition

NIVEN, DONALD FERRIES, b Aberdeen, Scotland, Mar 5, 47; Brit & Can citizen; m 65; c 2. MICROBIOLOGY, BIOCHEMISTRY. *Educ:* Univ Aberdeen, BSc, 70, PhD(biochem), 73. *Prof Exp:* Res asst biochem, Univ Aberdeen, 70-73; res fel biol, Univ Kent, 73-76; res assoc, 76-79, asst prof, 79-85, ASSOC PROF MICROBIOL, MACDONALD COL, MCGILL UNIV, 85- *Concurrent Pos:* assoc ed, Can J Microbiol, 85-87. *Mem:* Brit Soc Gen Microbiol; Can Soc Microbiologists; Am Soc Microbiol. *Res:* Microbiology physiology and biochemistry, iron acquisition; bioenergetics and pathogenicity. *Mailing Add:* Dept Microbiol Macdonald Col McGill Univ 21111 Lakeshore Rd Ste Anne de Bellevue PQ H9X 1C0 Can

NIX, JAMES RAYFORD, b Natchitoches, La, Feb 18, 38; m 61; c 2. THEORETICAL NUCLEAR PHYSICS. *Educ:* Carnegie Inst Technol, BS, 60; Univ Calif, Berkeley, PhD(physics), 64. *Prof Exp:* NATO fel, Niels Bohr Inst, Univ Copenhagen, 64-65; fel, Lawrence Radiation Lab, Univ Calif, Berkeley, 66-68; mem staff, 68-77, group leader nuclear theory, 77-89, MEM STAFF, LOS ALAMOS NAT LAB, 89- *Concurrent Pos:* Chmn, SuperHilac Prog Adv Comt, Lawrence Berkeley Lab, Univ Calif, 74-76, Gordon Res Conf Nuclear Chem, 76; mem, Physics Div Adv Comt, Oak Ridge Nat Lab, 75-77, chmn, 76; mem vis comt, Nuclear Sci Div, Lawrence Berkeley Lab, Univ Calif, 78-80, chmn, 79-80; mem prog adv comt, Holifield Heavy Ion Res Facil, Oak Ridge Nat Lab, 84-85, Superhilac prog adv comt, Lawrence Berkeley Lab, Univ Calif, 87-88, physics div comt, Oak Ridge Nat Lab, 91-; distinguished lectr nuclear sci, Univ Rochester, 71, guest prof, Heavy Ion Lab GSI, Darmstadt, Ger, 73, vis prof, Brazilian Ctr Pesquisas Physics, Rio de Janeiro, 74; mem exec comt div nuclear physics, Am Phys Soc, 73-75; consult, Calif Inst Technol, 76 & 79; Alexander von Humboldt sr US scientist award, 80-81. *Mem:* Fel Am Phys Soc; AAAS; Sigma Xi. *Res:* Theory of nuclear fission, very-heavy-ion reactions, superheavy nuclei, nuclear ground-state masses and deformations, large-amplitude collective nuclear motion, nuclear dissipation and relativistic nucleus-nucleus collisions. *Mailing Add:* Nuclear Theory T-2 MS B243 Los Alamos Nat Lab Los Alamos NM 87545

NIX, JOE FRANKLIN, b Malvern, Ark, Aug 28, 39. GEOCHEMISTRY, ANALYTICAL CHEMISTRY. *Educ:* Ouachita Baptist Col, BS, 61; Univ Ark, MS, 63, PhD(chem), 66. *Prof Exp:* Assoc prof chem, 66-74, PROF CHEM, OUACHITA BAPTIST UNIV, 74- *Mem:* AAAS; Am Chem Soc; Am Geophys Union; Am Soc Limnol & Oceanog; Sigma Xi. *Res:* Radioactive fallout; neutron activation analysis of meteorites; geochemistry of hot springs and geysers; geochemistry of impoundments. *Mailing Add:* Box 3780 Ouachita Baptist Univ Arkadelphia AR 71923

NIX, WILLIAM DALE, b King City, Calif, Oct 28, 36; m 58; c 3. MATERIALS SCIENCE. *Educ:* San Jose State Col, BS, 59; Stanford Univ, MS, 60, PhD(mat sci), 63. *Prof Exp:* From asst prof to assoc prof, Stanford Univ, 63-72, dir, Ctr Mat Res, 68-70, assoc chmn, Dept Mat Sci & Eng, 75-87, PROF MAT SCI, STANFORD UNIV, 72-, LEE OTTERSON PROF ENG, 89- *Concurrent Pos:* Participant, Ford's Residencies in Eng Pract Prog, WD, Manly Stellite Div, Union Carbide Corp, Ind, 66. *Honors & Awards:* Western Elec Award Eng Teaching, 64; Bradley Stoughton Award, Am Soc Metals, 70, Campbell Lectr, Am Soc Metals Int, 89; Mathewson Gold Medal, Am Inst Mining, Metall & Petrol Engrs, 79, Mehl Medal, 88. *Mem:* Nat Acad Eng; fel Am Soc Metals; fel Inst Mining, Metall & Petrol engrs; fel Am Inst Mining & Metall Engrs; Mat Res Soc. *Res:* Imperfections in crystals; dislocation theory and mechanical properties of crystals; mechanical properties of thin films; high temperature creep and fracture. *Mailing Add:* Dept Mat Sci & Eng Stanford Univ Stanford CA 94305

NIXON, ALAN CHARLES, b Workington, Eng, Oct 10, 08; US citizen; m 37, 67; c 2. PHYSICAL & ORGANIC CHEMISTRY, FUEL CHEMISTRY. *Educ:* Univ Saskatchewan, BSc, 29, MSc, 31; Univ Calif, Berkeley, PhD(chem), 34. *Prof Exp:* Instr chem, Univ Calif, Berkeley, 34-37; res chemist, Shell Develop Co, Emeryville, Calif, 37-54, res supvr, 54-70; secy, 79-84, PRES, CALSEC CONSULTS, INC BERKELEY, 84-; CONSULT CHEM, CHEM CONSULT SERV, 70- *Concurrent Pos:* Bd dirs, Am Chem Soc, 72-74; chair, Coun Sci Soc Presidents, 73-74. *Honors & Awards:* Asn Tech Professionals Award, 81; Henry Hill Award, Am Chem Soc, 84. *Mem:* Fel AAAS; Am Chem Soc (pres, 73); Coun Sci Soc Presidents; Combustion Soc; Am Inst Aeronaut & Astronaut; fel Am Inst Chemists; Am Aging Asn. *Res:* Rates of alcoholysis of trytyloacyl chlorides; treatment and oxidative stability of gasoline and jet fuel; effect of radiation on jet fuel; advanced fuels for aircraft engines using their vaporizing and endothermic properties for cooling; possibilities of improving aircraft safety by improved fuel properties. *Mailing Add:* 2727 Marin Ave Berkeley CA 94708

NIXON, CHARLES MELVILLE, b Newton, Mass, Jan 22, 35; m 60; c 2. WILDLIFE ECOLOGY. *Educ:* Northeastern Univ, BS, 57; Pa State Univ, MS, 59. *Prof Exp:* Res biologist, Ohio Div Wildlife, 59-70; RES BIOLOGIST WILDLIFE, ILL NATURAL HIST SURV, 70- *Honors & Awards:* Publ Award, Wildlife Soc, 75. *Mem:* Wildlife Soc; Am Soc Mammalogists; Soc Am Foresters. *Res:* Ecology and management of forest-dwelling wildlife. *Mailing Add:* Nat Resources Bldg Ill Nat Hist Surv 607 E Peabody Champaign IL 61820

NIXON, CHARLES WILLIAM, b Carlinville, Ill, Sept, 28, 25. GENETICS, HISTOLOGY. *Educ:* Univ Ill, BS, 46, MS, 48; Brown Univ, PhD(biol), 51. *Prof Exp:* Asst prof biol, Northeastern Univ, 51-57 & Simmons Col, 57-60; biologist, Bio-Res Inst, Inc, 60-70; pvt consult genetics, 70-73; res assoc, Mass Inst Technol, 73-79; CONSULT GENETICS, 79- *Concurrent Pos:* Ed, Sempervium Fanciers Asn Newsletter, 75- *Res:* Mammalian and plant genetics. *Mailing Add:* 37 Ox Bow Lane Randolph MA 02368

NIXON, CHARLES WILLIAM, b Wellsburg, WVa, Aug 15, 29; m 56; c 2. HEARING PROTECTION DEVICES, VOICE COMMUNICATIONS. *Educ:* Ohio State Univ, BS, 52, MS, 53, PhD(res audiol), 60. *Prof Exp:* Instr spec educ, Ohio & WVa Pub Schs, 54-56; res audiologist bioacoust, 56-67, SUPVRY RES AUDIOLOGIST BIOACOUST & BIOCOMMUN, AEROSPACE MED RES LAB, 67- *Concurrent Pos:* Mem, Nat Supersonic Transp Prog, 61-67; comt chair, Am Speech & Hearing Asn, 65-67; chair, Am Nat Standards Inst, 65-; US rep, Hearing Protectors Standard, 68-; USAF rep, Exec Coun, Comn Hearing, Bioacoust & Biomech, Nat Acad Sci, 76-; ad hoc fac, Biomed Eng, Wright State Univ, 82-85; chair, Joint Dir Labs-Robotics Panel, Dept Defense, 87-88; mem, Am Nat Standards Inst Working Groups, 88- *Honors & Awards:* Meritorious Serv Medal, Dept Defense, 86. *Mem:* Nat Acad Sci; Res Soc Am; Am Nat Standards Int; fel Acoust Soc Am; Int Standards Orgn. *Res:* Bioacoustics; biocommunications; perception of digital speech; long duration noise exposure effects; effects of infrasound on people; auto air bag inflation; effects of sonic booms on people; effects of whole body vibration of speech; helium speech intelligibility; positive pressure; noise effects on hearing. *Mailing Add:* 4316 Sillman Pl Kettering OH 45440

NIXON, DONALD MERWIN, b Topeka, Kans, Nov 11, 35; m 62; c 3. AGRICULTURAL ECONOMICS. *Educ:* Colo State Univ, BS, 65, MS, 66, PhD(agr mkt), 69. *Prof Exp:* Instr poultry sci & mkt, Colo State Univ, 65-69; assoc prof bus & agr mkt, 69-70, assoc prof, 70-81, PROF AGR ECON, TEX A&I UNIV, 81- *Concurrent Pos:* Res grants, Houston Livestock & Rodeo Asn, 70, 80 & 82-83, Rio Farms, 72 & Perry Found, 75. *Mem:* Am Mkt Asn; Am Agr Econ Asn; Inst Food Technologists; Poultry Sci Asn. *Res:* Agricultural marketing; economic analysis among beef cattle common to south Texas; taste panel evaluation; economics of crop insurance; use of grain sorghum and beef futures by south Texas producers; evaluate prickly pear as emergency forage; market prices of southern Texas beef; truck farming; financial situation of US Farms. *Mailing Add:* Col Agr Tex A&I Univ Kingsville TX 78363

NIXON, ELRAY S, b Escalante, Utah, Feb 5, 31; m 57; c 4. PLANT ECOLOGY, PLANT TAXONOMY. *Educ:* Brigham Young Univ, BS, 57, MS, 61; Univ Tex, PhD(bot), 63. *Prof Exp:* Asst prof biol, Chadron State Col, 63-65 & Southern Ore Col, 65-66; from asst prof to assoc prof, 66-72, PROF BIOL, STEPHEN F AUSTIN STATE UNIV, 72- *Res:* Floristic ecology; soil-plant relationships. *Mailing Add:* PO Box 13003 SFA Station Nacogdoches TX 75962

NIXON, EUGENE RAY, b Mt Pleasant, Mich, Apr 14, 19; m 45; c 2. PHYSICAL CHEMISTRY. *Educ:* Alma Col, ScB, 41; Brown Univ, PhD(chem), 47. *Prof Exp:* Res chemist, Manhattan Dist, Brown Univ, 42-46, res assoc, Off Naval Res, 46-47, instr chem, 47-49; from instr to assoc prof, 49-65, dir, mat res lab, 69-72, PROF CHEM, UNIV PA, 65- *Concurrent Pos:* Vdean grad sch, Univ Pa, 58-65. *Mem:* Am Chem Soc; Coblentz Soc; Am Phys Soc; Sigma Xi. *Res:* Molecular spectroscopy and structure; laser chemistry. *Mailing Add:* Dept Chem Univ Pa Philadelphia PA 19104

NIXON, JOHN V, b Sept 15, 40; c 2. CARDIOLOGY, INTERNAL MEDICINE. *Educ:* Univ Manchester, Eng, MB, 65, MD, 62. *Prof Exp:* PROF MED & DIR ECHOCARDIOGRAPHY, MED COL VA, 86- *Concurrent Pos:* Assoc dir, Heart Sta, Med Col Va, 86- *Mailing Add:* Box 128 MCV Station Richmond VA 23298-0128

NIXON, PAUL R(OBERT), b Kenya, EAfrica, June 23, 24; US citizen; m 50; c 1. AGRICULTURAL ENGINEERING. *Educ:* Iowa State Univ, BS, 52, MS, 55; Stanford Univ, MS, 66. *Prof Exp:* Asst soil conserv off, Kenya Dept Agr, EAfrica, 43-45; agr engr, Soil Conserv Serv, US Dept Agr, 52-54; assoc irrig res, Iowa State Univ, 55-56; prof leader hydroclimat res, Agr Res Serv, USDA, 56-65, res invest leader, 65-71, agr engr, Rio Grande Soil & Water Res Ctr, 71-82, agr engr, subtropical res lab, 80-88, res leader remote sensing, 82-88; RETIRED. *Concurrent Pos:* Collabr, water mgt studies, India & Yugoslavia. *Mem:* Am Soc Agr Engrs; Am Meteorol Soc; Am Soc Photogram & Remote Sensing. *Res:* Water supply and use; consumption of water as related to climate, vegetation and availability of soil water; multispectral remote sensing using satellite and aircraft data; developing narrowband video for remote sensing. *Mailing Add:* 126 Tanglewood Dr Fredericksburg TX 78624

NIXON, SCOTT WEST, b Philadelphia, Pa, Aug 24, 43; m 65; c 2. ECOLOGY, OCEANOGRAPHY. *Educ:* Univ Del, BA, 65; Univ NC, Chapel Hill, PhD(bot), 70. *Prof Exp:* Res assoc, 69-70, from asst prof to assoc prof, 70-80, PROF OCEANOG, UNIV RI, 80-, DIR SEA GRANT PROG, 83- *Concurrent Pos:* co-ed-in-chief, Estuaries. *Mem:* Am Soc Limnol & Oceanog; Estuarine Res Fedn. *Res:* Ecological systems; nutrient cycling; energetics; simulation. *Mailing Add:* Sch of Oceanog Univ of RI Kingston RI 02881

NIXON, WILFRID AUSTIN, b Harpenden, Eng, Sept 17, 59; m 80; c 4. ICE MECHANICS, COLD REGIONS ENGINEERING. *Educ:* Cambridge Univ, BA, 81, PhD(eng), 85. *Hon Degrees:* MA, Cambridge Univ, 85. *Prof Exp:* Res asst prof eng, Thayer Sch Eng, Dartmouth Col, 85-87; RES ENGR, IOWA INST HYDRAUL RES, UNIV IOWA, 87-, ASST PROF CIVIL ENG, DEPT CIVIL & ENVIRON ENG, 89- *Concurrent Pos:* Chmn, Mech Comt, Offshore Mech & Arctic Eng, Am Soc Mech Eng, 89- & Working Group Testing methods, Int Asn Hydraul Res, 90- *Mem:* Am Soc Civil Eng; assoc mem Am Soc Mech Eng; Int Glaciological Soc; Am Geophys Union. *Res:* Mechanical properties of ice and brittle materials; fracture and fatigue of ice; experimental and analytical studies of ice-structure interaction; methods of removing ice from roads; novel arctic construction methods. *Mailing Add:* Iowa Inst Hydraul Res Univ Iowa Iowa City IA 52242

NIYOGI, SALIL KUMAR, b Calcutta, India, Feb 1, 32; US citizen; m 64; c 2. GROWTH FACTORS PROTEIN ENGINEERING. *Educ:* Univ Calcutta, BS, 53, MS, 56; Northwestern Univ, PhD(biochem), 61. *Prof Exp:* Res assoc, Dept Chem, Stanford Univ, 61-62, Dept Biochem, Univ Maryland, 63-64 & Dept Biophys, Johns Hopkins Univ, 64-66; SR SCIENTIST BIOCHEM, BIOL DIV, OAK RIDGE NAT LAB, 66- *Concurrent Pos:* Prof, Oak Ridge Grad Sch Biomed Sci, Univ Tenn, 70-; expert, UN Develop Prog. *Mem:* Fel AAAS; Am Soc Biol Chemists; Am Soc Microbiol. *Res:* Human epidermal growth factor; structure-function analysis; site-directed mutagenesis; molecular biology of transcription. *Mailing Add:* Biol Div Oak Ridge Nat Lab PO Box 2009 Oak Ridge TN 37832-8077

NIZEL, ABRAHAM EDWARD, b Boston, Mass, July 27, 17; m 42; c 2. PREVENTIVE DENTISTRY, NUTRITION. *Educ:* Tufts Univ, DMD, 40, MSD, 52. *Prof Exp:* Intern oral surg, Worcester City Hosp, 40-41; instr periodont, Sch Dent Med, Tufts Univ, 52-60, asst clin prof, 61-66, assoc prof nutrit, 66-72, prof nutrit & prev dent, 73-; RETIRED. *Concurrent Pos:* Res assoc, Mass Inst Technol, 52-70, vis assoc prof nutrit & metab, 71-73, vis prof, 73-; guest lectr, Eastman Dent Ctr, 60- & Forsyth Dent Ctr, 69-; grants, Nutrit Found, 67-72 & Nat Dairy Coun, 68-72; consult, Am Dent Asn & Food & Nutrit Bd, Nat Acad Sci; consult ed, J Am Dent Asn & J Prev Dent; clin prof nutrit, Forsyth Sch Dent Hyg. *Honors & Awards:* Prev Dent Award, Am Dent Asn. *Mem:* AAAS; fel Am Col Dent; fel Am Asn Dent Sci; Int Asn

Dent Res; Am Dent Asn. *Res:* Oral health-nutrition interrelationships, particularly the effect of dietary phosphate supplements on inhibition of experimental caries; model nutrition teaching program for dental schools and schools of dental hygiene; relative plaque removal effectiveness of different toothbrushes. *Mailing Add:* Tufts Univ Sch of Dent Med One Kneeland St Rm 301-05 Boston MA 02111

NJUS, DAVID LARS, b Honolulu, Hawaii, Oct 17, 48; m 84; c 2. BIOPHYSICS. *Educ:* Mass Inst Technol, BS, 70; Harvard Univ, PhD(biophys), 75. *Prof Exp:* Fel biochem, Oxford Univ, 75-78; from asst prof to assoc prof, 78-86, PROF BIOL, WAYNE STATE UNIV, 86- *Concurrent Pos:* NATO fel, Oxford Univ, 75-76; NIH res, 76-78; investr, Am Heart Asn, 83-88. *Mem:* Fel AAAS; Biophys Soc; Am Soc Biochem & Molecular Biol; Am Heart Asn; Am Soc Neurochem. *Res:* Catecholamine metabolism in the adrenal medulla; membrane biophysics; energetics of organelle membranes. *Mailing Add:* Dept Biol Sci Wayne State Univ Detroit MI 48202

NOACK, MANFRED GERHARD, b Olbersdorf, Ger, Jan 25, 36; m 62; c 2. INORGANIC CHEMISTRY, INDUSTRIAL CHEMISTRY. *Educ:* Tech Univ, Munich Vordiplom, 59, Diplomchemiker, 62, Dr rer nat(chem), 64; Univ New Haven, MBA, 75. *Prof Exp:* Teaching asst chem, Munich Tech Univ, 62-64; res assoc, Univ Md, 64-67; res chemist, Olin Corp, 67-75, group leader res & develop, 75-80, consult scientist, 80-89, SR CONSULT SCIENTIST, OLIN RES CTR, 89- *Mem:* Am Chem Soc; Nat Asn Corrosion Engrs; Sigma Xi. *Res:* Chemistry of metal carbonyls; magnetic resonance phenomena in solutions; homogeneous catalysis; corrosion inhibition; applications of redox chemistry; corrosion science and prevention; industrial and municipal water treatment. *Mailing Add:* Olin Res Ctr 350 Knotter Dr Cheshire CT 06410-0586

NOAKES, DAVID LLOYD GEORGE, b Hensall, Can, Aug 3, 42; m 66; c 1. ETHOLOGY. *Educ:* Univ Western Ont, BS, 65, MS, 66; Univ Calif, Berkeley, PhD(zool), 71. *Prof Exp:* Lectr zool, Edinburgh Univ, 70-72; from asst prof to assoc prof, 72-86, PROF ZOOL, UNIV GUELPH, 86-, DIR INST ICHTHYOL, 90- *Mem:* Animal Behav Soc; Can Soc Zoologists; Sigma Xi; Am Soc Ichthyologists & Herpetologists. *Res:* Behavioral ontogeny, social behavior and social systems, feeding and reproductive ecology, physiological basis of behavior, evolution of behavior, especially of fishes. *Mailing Add:* Dept Zool Col Biol Sci Univ Guelph Guelph ON N1G 2W1 Can

NOAKES, JOHN EDWARD, b Windsor, Ont, May 21, 30; US citizen; m 61; c 2. GEOCHEMISTRY, OCEANOGRAPHY. *Educ:* Champlain Col Plattsburg, BS, 53; Tex A&M Univ, MS, 59, PhD(chem oceanog), 62. *Prof Exp:* Soils engr, NY State Soil Div, 53-55; res chemist, Clark Cleveland Pharmaceut Co, 55-57; res chemist, Tex A&M Univ, 58-59, res found, 59-61; asst prof chem oceanog, Univ Alaska, 61-62; res scientist, Oak Ridge Assoc Univs, 62-68; dir gen res servs, 70-81, PROF GEOL, UNIV GA, 68-, DIR, CTR APPL ISOTOPE STUDIES, 81- *Concurrent Pos:* Dir geochronology lab, Univ Ga, 68-70. *Mem:* Am Chem Soc. *Res:* Geochemistry of marine environment; tritium, radioactive carbon and uranium geochronology; development of nuclear radiation analytical techniques for measuring low levels of radiation. *Mailing Add:* Dept Geol & Anthrop Univ Ga Athens GA 30602

NOALL, MATTHEW WILCOX, b Salt Lake City, Utah, Mar 16, 24; m 50; c 2. BIOCHEMISTRY. *Educ:* Univ Utah, BA, 48, MA, 49, PhD(biochem), 52. *Prof Exp:* USPHS fel, Nat Inst Arthritis & Metab Dis, 52-54; Am Cancer Soc fel, Sch Med, Tufts Univ, 54-56; asst prof obstet & gynec, Sch Med, Wash Univ, 56-62; asst prof biochem, 62-66, ASSOC PROF PATH & BIOCHEM, BAYLOR COL MED, 66- *Mem:* Am Chem Soc; NY Acad Sci. *Res:* Toxicology; chemical carcinogenesis. *Mailing Add:* S430 Lymbar Dr Baylor Col Med Houston TX 77096

NOBACK, CHARLES ROBERT, b New York, NY, Feb 15, 16; m 38; c 4. ANATOMY, EMBRYOLOGY. *Educ:* Cornell Univ, BS, 36; NY Univ, MS, 38; Univ Minn, PhD(anat), 42. *Prof Exp:* Asst prof, Med Col Ga, 41-44; from asst prof to assoc prof, Long Island Col Med, 44-49; from asst prof to prof, 49-86, actg chmn dept, 74-75, EMER PROF ANAT & SPEC LECTR, COL PHYSICIANS & SURGEONS, COLUMBIA UNIV, 86- *Concurrent Pos:* NIH res grants, 53-81; mem nerv & sensory syst res eval comt, Vet Admin, 69-72. *Honors & Awards:* James Arthur lectr, 59. *Mem:* AAAS; Histochem Soc; Harvey Soc; Am Asn Anat; Soc Neurosci; NY Acad Sci (secy, 61-63). *Res:* Development of mammalian skeleton; reproduction in the primates; histochemistry of mammalian tissue; regeneration of neural tissues; comparative neuroanatomy; nutrition and nervous system; physical anthropology. *Mailing Add:* Dept Anat & Cell Biol Col Physics Columbia Univ New York NY 10032-3795

NOBACK, RICHARDSON K, b Richmond, Va, Nov 7, 23; m 47; c 3. INTERNAL MEDICINE. *Educ:* Cornell Univ, MD, 47; Am Bd Internal Med, dipl. *Prof Exp:* Nat Heart Inst res fel, 49-50; instr med, Col Med, Cornell Univ, 50-53; asst prof, Col Med, State Univ NY Upstate Med Ctr, 55-56; assoc prof, Med Ctr, Univ Ky, 56-63, asst dean, 56-58, dir univ health serv, 59-63; assoc dean, 64-69, dean, 69-78, PROF INTERNAL MED, MED SCH, UNIV MO-KANSAS CITY, 64-, HEAD, DIV GERIATRICS, 80- *Concurrent Pos:* Asst dir comprehensive care & teaching prog, Cornell Univ, 52-53; med dir, Syracuse Dispensary, 55-56; consult, Med Col, Univ Tenn & John Gaston Hosp, 55; exec med consult, Norfolk Area Med Ctr Authority, Va, 64-70; exec dir, Kansas City Gen Hosp & Med Ctr, 64-69; dir, Mo Geriat Educ Ctr, 85- *Mem:* AMA; Asn Am Med Cols; NY Acad Sci. *Res:* Medical education; health care programs; cardiovascular and renal diseases. *Mailing Add:* Dept Med Univ Mo Med Sch 2411 Holmes Kansas City MO 64108

NOBE, KEN, b Berkeley, Calif, Aug 26, 25; m 57; c 3. CHEMICAL ENGINEERING. *Educ:* Univ Calif, BS, 51, PhD(eng), 56. *Prof Exp:* Jr chem engr polymer res, Res Labs, Air Reduction Corp, 51-52; from asst prof to assoc prof, 57-68, chmn dept chem & nuclear eng, 78-83, chmn dept chem eng, 83-84, PROF ENG, UNIV CALIF, LOS ANGELES, 68- *Concurrent Pos:* Mem tech staff, Space Tech Labs, TRW, Inc, 58-59. *Mem:* Electrochem Soc; Nat Asn Corrosion Engrs; Am Chem Soc. *Res:* Electrochemistry; corrosion and catalysis; photoelectrochemistry. *Mailing Add:* Dept Chem Eng Univ Calif Los Angeles CA 90024

NOBEL, JOEL J, b Philadelphia, Pa, Dec 8, 34; c 3. MEDICINE, BIOMEDICAL ENGINEERING. *Educ:* Haverford Col, BA, 56; Univ Pa, MA, 58; Thomas Jefferson Univ, MD, 63. *Prof Exp:* Intern, Presbyterian Hosp, Philadelphia, 63-65; resident surg, Pa Hosp, Philadelphia, 64-65; submarine med officer, US Navy, 66-67, proj officer submarine rescue & survival, Mil Opers Br, Submarine Med Res Ctr, 67-68; PRES, EMERGENCY CARE RES INST, 68- *Concurrent Pos:* Consult, Foreign Policy Res Inst, 56-59; consult biomed eng, 64-68; prin investr, Pa Heart Asn grant, 65-68; fel anesthesiol, Thomas Jefferson Univ Hosp, 68-70; prin investr, HEW grant, 68-72; mem bd dir, Consumers Union, 76-79 & 80-; trustee, CITECH. *Mem:* Asn Advan Med Instrumentation; Biomed Eng Soc; Soc Critical Care Med; Soc Advan Med Systs; Am Hosp Asn. *Res:* National health policy; health care technology assessment; medical equipment evaluation; hospital risk management; accident investigation; biomedical engineering design and development; underwater physiology; emergency medical services. *Mailing Add:* 5200 Butler Pike Plymouth Meeting PA 19462

NOBEL, PARK S, b Chicago, Ill, Nov 4, 38; m 65; c 2. BIOPHYSICAL PLANT ECOLOGY. *Educ:* Cornell Univ, BEP, 61; Calif Inst Technol, MS, 63; Univ Calif, Berkeley, PhD(biophys), 65. *Prof Exp:* NSF fels chloroplasts, Tokyo Univ, 65-66 & King's Col, London, 66-67; from asst prof to assoc prof molecular biol, 67-75, PROF BIOL, UNIV CALIF, LOS ANGELES, 75-, ASSOC DIR, LAB BIOMED & ENVIRON SCI, 89- *Concurrent Pos:* Guggenheim fel, Australian Nat Univ, 73-74. *Mem:* Agron Soc Am; Am Soc Plant Physiologists; Ecol Soc Am; Bot Soc Am; Scand Soc Plant Physiologists. *Res:* Biophysical aspects of plant physiology, especially plant-environment interactions and ecology; emphasis on desert plants, including cacti and agaves; writer of books on plant biophysics and desert plants. *Mailing Add:* Dept Biol Univ Calif Los Angeles CA 90024-1606

NOBIS, JOHN FRANCIS, b Helena, Mont, Jan 23, 21; m 47; c 4. ORGANIC CHEMISTRY. *Educ:* Col St Thomas, BS, 42; Iowa State Univ, PhD(org chem), 48. *Prof Exp:* Instr org chem, Iowa State Univ, 46-48, asst prof, 48; asst prof, Xavier Univ, Ohio, 48-51; group leader, Sodium Div, Nat Distillers Chem Co, 51-56, res supvr, 56-59, asst mgr metals & sodium res, Nat Distillers & Chem Corp, US Indust Chem Co, 59; mgr, Prod Develop Dept, Armour Indust Chem Co, 59-60; dir com develop, Formica Corp, Am Cyanamid Co, 60-86, dir mkt serv, 63-65, asst to pres, 65-66, dir environ med & safety affairs, 78-86; DIR HEALTH, SAFETY & TRAINING, PEI ASSOCS, INC, 87- *Mem:* AAAS; Am Chem Soc. *Res:* Sodium and organosodium chemistry; synthetic organic, organosilicon chemistry; antimalarials; anti-tuberculars; heterocycles; organometallics; polyolefins; plastics. *Mailing Add:* 7698 Pineglen Dr Cincinnati OH 45224

NOBLE, ALLEN G, b Astoria, NY, Jan 28, 30; m 59; c 3. MATERIAL CULTURE. *Educ:* Syracuse Univ, BA, 51; Univ Md, College Park, MA, 53; Univ Ill, Urbana, PhD(geog), 57. *Prof Exp:* Planning consult, City of Rantoul, Ill, 55-57; officer, US Foreign Serv, US Dept State, 57-63; assoc prof geog, Calif State Col, Pa, 63-64; PROF GEOG & HEAD DEPT, UNIV AKRON, 64- *Concurrent Pos:* External examr, Univ Madras, Univ Rajasthan, Univ Calcutta, Aligarh Muslim Univ & Utkal Univ, India, 73-; exec dir, Pioneer Am Soc, 77-85; regional counr, Asn Am Geographers, 79-82. *Honors & Awards:* Fulbright lectr, Univ Peredeniya, Sri Lanka, 80; Honors Award, Asn Am Geogr, 89. *Mem:* Asn Am Geographers; Sigma Xi; Can Asn Geographers; Pioneer Am Soc; Nat Coun Geog Educ; Am Geog Soc. *Res:* Material culture; cultural studies of South Asia; noise pollution. *Mailing Add:* 414 Merriman Rd Akron OH 44303

NOBLE, ANN CURTIS, b Harlingen, Tex, Nov 6, 43. FOOD SCIENCE. *Educ:* Univ Mass, Amherst, BS, 66, PhD(food sci), 70. *Prof Exp:* Asst prof food sci, Univ Guelph, 70-73; asst prof, 74-86, PROF ENOL, UNIV CALIF, DAVIS, 86- *Mem:* Am Chem Soc; Inst Food Technol; Asn Chemoreception Sci; Am Soc Vitic Enol; Europ Chemo Receptor Orgn. *Res:* Flavor chemistry and sensory evaluation. *Mailing Add:* Dept Viticult & Enol Univ Calif Davis CA 95616

NOBLE, BERNICE, IMMUNOPATHOLOGY. *Educ:* State Univ NY, Buffalo, PhD(microbiol), 75. *Prof Exp:* ASSOC PROF MICROBIOL & IMMUNOPATH, SCH MED, STATE UNIV NY, 79- *Mailing Add:* Microbiol/203 Sherman State Univ NY Health Sci Ctr 3435 Main St Buffalo NY 14214

NOBLE, CHARLES CARMIN, b Syracuse, NY, May 18, 16; m 42; c 5. CIVIL ENGINEERING. *Educ:* US Mil Acad, West Point, NY, BS, 40; Mass Inst Technol, MS, 48; George Washington Univ, AM, 64. *Prof Exp:* 2nd Leitenant, US Army, 40, from Major to General, 40-69, exec officer, Manhattan Proj, Oak Ridge, 46-47, planner, Atomic Energy Comn, 47-48, Army Gen Staff, 48-51, Supreme Hq, Allied Power Europe, 51-54, dep dist engr, NY, 54-56, comdr, Eng Combat Group, Ft Benning, Ga, 57-58, dist engr, Louisville, 58-60, dir, Atlas & Minuteman Intercontinental Ballistic Missile Construct Prog, 60-63, chief engr, UN Command, US Forces Korea, 64-66, dir construction, Off Secy Defense, 66-67, dir civil works, Corps Engrs, 67-69, chief engr, US Forces, Europe, 69-70 & Vietnam, 70-71; pres, Miss River Comn, 71-74; proj mgr, C T Main Corp, 74-78, dir & exec vpres, 78-81, dir, pres & chief oper officer, 81-83; pres & chief exec officer, 84; RETIRED. *Concurrent Pos:* Defense rep, Mex-US Joint Comn Mutual Disaster Assistance, 68; eng agent, Atlantic-Pac Interoceanic Sea Level Canal Study Comn, 67-69; fed adv, Coun Regional Econ Develop, 67-69; pres, US Army Coastal Eng Res Bd, 67-69; def mem, Comt Multiple Uses Coastal Zone, 68-69; chmn & fel mem, Red River Compact Comn, 71-74; consult, UN Develop Orgn, 78; dir & pres, Rondaxe Lake Asn, 86-90, dir, Webb Property Owner Asn, 87-90. *Honors & Awards:* Wheeler Medal, Soc Am Mil Engrs,

62. *Mem:* Nat Acad Eng; US Comt Large Dams; Am Soc Civil Engrs; Soc Am Mil Engrs; Am Consult Engrs Coun; Nat Soc Prof Engrs. *Res:* Geomorphology and behavior of major alluvial rivers (Mississippi River, Paraguay River, etc); potomological studies to support engineering improvements for navigation and flood control. *Mailing Add:* 55 Fieldstone Dr Londonderry NH 03053

NOBLE, EDWIN AUSTIN, b Bethel, Vt, Dec 15, 22; m 48; c 3. GEOLOGY. *Educ:* Tufts Univ, BS, 46; Univ NMex, MS, 50; Univ Wyo, PhD(geol), 61. *Prof Exp:* Geologist, US Geol Surv, Boston, 52-54 & AEC, 54-62 & 63-65; adv nuclear raw mat, Int Atomic Energy Agency, Arg, 62-63; assoc prof geol, Univ NDak, 65-69, prof & chmn dept, 69-77; dep chief, Off Energy Resources, US Geol Surv, USAID, Pakistan, 77-84, chief of party, 85-90; RETIRED. *Concurrent Pos:* Asst state geologist, NDak State Geol Surv, 65-69, state geologist & dir, 69-77; supvr oil & gas, State NDak, 69-77; ed, Asn Am State Geologists, 71-77. *Mem:* Geol Soc Am; Am Asn Petrol Geologists; hon mem Asn Am State Geologists; Soc Econ Geologists. *Res:* Genesis of ore deposits in sedimentary rocks. *Mailing Add:* 11407 Great Meadow Dr Reston VA 22091

NOBLE, ELMER RAY, b Pyong Yang, Korea, Jan 16, 09; US citizen; m 32; c 4. PARASITOLOGY. *Educ:* Univ Calif, AB, 31, AM, 33, PhD, 36. *Prof Exp:* From instr to prof, Univ Calif, Santa Barbara, 74-87; RETIRED. *Concurrent Pos:* Chmn dept biol sci, Univ Calif, Santa Barbara, 47-51, dean div letters & sci, 51-59, actg provost, 56-58, vchancellor, 58-62; consult, Govt Indonesia, 60. *Mem:* Am Soc Parasitol (pres, 81); Soc Protozool (pres, 71-72); Am Micros Soc (vpres, 71). *Res:* Cytology pf parasitic protozoa; life history of myxosporidia, trypanosomes and amoebae; parasitism in deep-sea fishes. *Mailing Add:* 1250 Dover Lane Santa Barbara CA 93103

NOBLE, GLENN ARTHUR, b Pyong Yang, Korea, Jan 16, 09; US citizen; m 35; c 3. PARASITOLOGY. *Educ:* Univ Calif, AB, 31, MA, 33; Stanford Univ, PhD(parasitol), 40. *Prof Exp:* Asst zool, Col Pac, 33-35; instr biol, San Francisco City Col, 35-46, chmn dept, 39-46; consult, US Mil Govt, Korea, 46-47; vis prof parsitol, Med Sch, Seoul Nat Univ & Severance Union Med Col, 47; prof, 47-73, head dept, 49-71, EMER PROF BIOL, CALIF POLYTECH STATE UNIV, SAN LUIS OBISPO, 73- *Concurrent Pos:* Calif Fish & Game res grant, 43; Fulbright prof, Univ Philippines, 53-54 & Nat Taiwan Univ, 61-62; res grants, NSF, 56-57 & 63-66, NIH, 59-61 & Gorgas Mem Lab, Panama, 69. Soc Protozool. *Mem:* Am Soc Parasitologists; Soc Protozoologists; Am Soc Parasitol; Philippine Soc Advan Res. *Res:* Parasitology; protozoology. *Mailing Add:* Dept Biol Sci Calif Polytech State Univ San Luis Obispo CA 93407

NOBLE, GORDON ALBERT, b Joliet, Ill, June 20, 27. SOLID STATE & CHEMICAL PHYSICS, PHYSICS EDUCATION. *Educ:* Univ Chicago, PhB, 47, SB, 49, SM, 51, PhD(chem), 55. *Prof Exp:* Solid state physicist, Zenith Radio Corp, 54-60; res physicist, IIT Res Inst, 60-66, sr scientist, 66-68; from asst prof to assoc prof, 68-84, EMER PROF PHYSICS, N PARK COL, 84- *Mem:* AAAS; Am Phys Soc; Am Chem Soc; Sigma Xi. *Res:* Electron paramagnetic and nuclear magnetic resonance; color centers; solid state optical spectroscopy. *Mailing Add:* 906 N Larkin Ave Joliet IL 60435

NOBLE, JOHN DALE, b Glendale, Calif, Nov 21, 34; m 62; c 2. PHYSICS. *Educ:* Univ Wyo, BS, 56, MS, 59; Univ BC, PhD(physics), 65. *Prof Exp:* Elec engr, Missile Systs Div, Lockheed Aircraft Corp, 56-57; instr physics, Univ Wyo, 59-61; lectr, Univ BC, 64-65; from asst prof to assoc prof, 65-73, chmn dept, 77-84, PROF PHYSICS, WESTERN ILL UNIV, 73- *Concurrent Pos:* Mem fac, Univ Wyo, 71-72. *Mem:* AAAS; Am Asn Physics Teachers. *Res:* Nuclear magnetic resonance; liquid-gas critical points. *Mailing Add:* Dept of Physics Western Ill Univ Macomb IL 61455

NOBLE, JOHN F, pharmacology, for more information see previous edition

NOBLE, JULIAN VICTOR, b New York, NY, June 7, 40; m 60; c 3. THEORETICAL PHYSICS, MATHEMATICAL BIOPHYSICS. *Educ:* Calif Inst Technol, BS, 62; Princeton Univ, MA, 63, PhD(physics), 66. *Prof Exp:* Res assoc theoret physics, Univ Pa, 66-68, asst prof, 68-71; ASSOC PROF PHYSICS, UNIV VA, 71- *Concurrent Pos:* Sloan Found fel, 71-73; mem, Prog Comt, Space Radiation Effects Lab, Va, 74- *Mem:* AAAS; Sigma Xi. *Res:* Application of quantum-mechanical collision theory to nuclear reaction studies, both to determine nuclear properties and to develop methods of handling general strong-interaction problems; nuclear reaction studies at intermediate energies; dynamics of interacting populations; functional techniques in stochastic theories. *Mailing Add:* Dept of Physics Univ of Va Charlottesville VA 22901

NOBLE, NANCY LEE, b Chattanooga, Tenn, Mar 1, 22. EDUCATION ADMINISTRATION, BIOCHEMISTRY. *Educ:* Emory Univ, MS, 49, PhD(biochem), 53. *Prof Exp:* Asst, Org Res Lab, Chattanooga Med Co, Tenn, 43-48; lab asst biochem, Sch Med, Emory Univ, 49-50; from res instr to res asst prof, 53-57, from asst prof to assoc prof, 57-86, ASSOC DEAN FAC AFFAIRS, UNIV MIAMI, 81-, PROF BIOCHEM & MED, SCH MED, 86- *Concurrent Pos:* Dir biochem res lab, Miami Heart Inst, 53-56; investr labs cardiovasc res, Howard Hughes Med Inst, 56-70; mem coun arteriosclerosis & coun basic sci, Am Heart Asn. *Honors & Awards:* Ciba Found Awards, 57 & 58. *Mem:* Fel Geront Soc; Soc Exp Biol & Med; Biochem Soc; fel Am Inst Chemists; NY Acad Sci; Sigma Xi; Am Physiol Soc. *Res:* Connective tissue metabolism and disorders. *Mailing Add:* D2-6 Fac Affairs Box 016960 Miami FL 33101

NOBLE, PAUL, JR, b Ind, Oct 16, 22; m 46; c 3. PHYSICAL CHEMISTRY. *Educ:* Reed Col, AB, 43; Rochester Univ, PhD(org chem), 50. *Prof Exp:* Jr chemist, Shell Develop Co, 43-44; asst, Reed Col, 46-47; assoc res chemist, Calif Res Corp, 50-52; res chemist & head, Phys Org Div, Merrill Co, 52-54; res chemist, Phys Org Sect, Western Labs, Arthur D Little, Inc, 54-58; res scientist, Lockheed Missiles & Space Co Inc, 58-62, staff scientist & group leader, 62-63, sr staff scientist & sr mem res lab, 63-70, mgr chem lab, 70-72, prog engr, 72-89; RETIRED. *Mem:* Am Chem Soc; Royal Soc Chem. *Res:* Organic synthesis; kinetics and reaction mechanism; solid propellants; composite materials; aliphatic polynitro compounds. *Mailing Add:* 45 Arbuelo Way Los Altos CA 94022

NOBLE, REGINALD DUSTON, b Huntington, WVa, Nov 15, 35; c 3. PLANT PHYSIOLOGY. *Educ:* Marshall Univ, AB, 57, MA, 60; Ohio State Univ, PhD(bot), 69. *Prof Exp:* Teacher high sch, 57-59, dept chmn, 59-61; asst prof biol & phys sci, Marshall Univ, 62-66; teaching assoc bot, Ohio State Univ, 66-67; from asst prof to assoc prof, 68-78, PROF BIOL, BOWLING GREEN STATE UNIV, 78-, CHMN DEPT, 80- *Concurrent Pos:* US Proj Leader, US/USSR Environ Protection Agreement, 82- *Mem:* Am Soc Plant Physiologists; Am Inst Biol Sci; AAAS. *Res:* Photosynthetic studies; including effects of air pollutants, especially sulphuric dioxide and ozone, on photosynthesis. *Mailing Add:* Dept Biol Bowling Green Univ Bowling Green OH 43403

NOBLE, RICHARD DANIEL, b Newark, NJ, Oct 14, 46; m 79. CHEMICAL ENGINEERING. *Educ:* Stevens Inst Technol, BE, 68, ME, 69; Univ Calif, Davis, PhD(chem eng), 76. *Prof Exp:* Proj engr, Nat Starch & Chem Co, 68-71; asst prof chem eng, Univ Wyo, 76-81; chem engr, Nat Bur Standards, 81-87; RES PROF, UNIV COLO, 87- *Concurrent Pos:* Consult, Nat Bur Standards, 80-81; adj assoc prof, Univ Colo, 81-87. *Mem:* Am Inst Chem Eng; Am Soc Eng Educ. *Res:* Facilitated transport; chemical complexation; liquid-liquid interfacial kinetics; mathematical modeling. *Mailing Add:* Chem Eng Dept Univ Colo PO Box 424 Boulder CO 80309-0424

NOBLE, ROBERT HAMILTON, b Alton, Ill, June 16, 16; m 40; c 3. OPTICAL PHYSICS. *Educ:* Antioch Col, BS, 40; Ohio State Univ, PhD(physics), 46. *Prof Exp:* Asst engr, Globe Industs, Ohio, 40-42; asst physics, Ohio State Univ, 42-46, res assoc, Univ Res Found, 46-47; asst prof physics, Mich State Col, 47-53; res physicist, Leeds & Northrup Co, 53-55; engr, Perkin-Elmer Corp, 55-64; prof optical sci, Univ Ariz, 64-74; prof & head optics dept, Nat Inst Astrophys, Optics & Electronics, Mex, 74-77; res physicist, Inst Astron, Univ Nat Auton Mex, 77-90; RETIRED. *Honors & Awards:* First Ann Award, Mex Acad Optics, 88. *Mem:* Fel AAAS; Am Phys Soc; fel Optical Soc Am; Mex Phys Soc; fel Soc Photo-Optical Instrumentation Engrs. *Res:* Instrumentation for astronomy. *Mailing Add:* 5435 Vale Way San Diego CA 92115

NOBLE, ROBERT LAING, physiology; deceased, see previous edition for last biography

NOBLE, ROBERT LEE, b Hominy, Okla, July 16, 23; m 47; c 4. ANIMAL NUTRITION. *Educ:* Okla State Univ, BS, 48, MS, 52; Kans State Univ, PhD(animal nutrit), 60. *Prof Exp:* From instr to prof aminal sci & indust, Okla State Univ, 49-85; RETIRED. *Mem:* Am Soc Animal Sci. *Res:* Applied research with sheep. *Mailing Add:* 2138 W Admiral Rd Stillwater OK 74074

NOBLE, ROBERT VERNON, b Ithaca, NY, Jan 1, 23; m 48; c 4. ACADEMIC ADMINISTRATION. *Educ:* Cornell Univ, AB, 46; Univ Fla, MA, 50. *Prof Exp:* From instr to asst prof corresp, Univ Fla, 50-57, head dept corresp study, 57-62; assoc prof, Fla Inst Continuing Univ Studies, 62-64; educ & training officer, Div Nuclear Educ & Training, AEC, 64-67, educ & training specialist, 67-72, staff asst to dir, 72-74, asst to dir, Div Admin Serv, US ERDA, 74-76, Freedom Info & Privacy Act admin officer, US Dept Energy, 76-78; RETIRED. *Mem:* Am Nuclear Soc; Sigma Xi; NY Acad Sci. *Mailing Add:* Rte 2 Box 2077 Melrose FL 32666

NOBLE, ROBERT WARREN, JR, b Washington, DC, Feb 14, 37; m 62; c 3. BIOPHYSICAL CHEMISTRY. *Educ:* Mass Inst Technol, BA, 59, PhD(biophys), 64. *Prof Exp:* Fel biophys, Mass Inst Technol, 64; Nat Cancer Inst fel biochem, Univ Rome & Regina Elena Inst, 64-66; res assoc, Cornell Univ, 67-68; estab invest, Am Heart Asn, 73-78; asst prof, 68-72, assoc prof, 72-76, PROF MED & BIOCHEM, STATE UNIV NY BUFFALO, 76- *Mem:* Biophys Soc; Am Soc Biol Chem; Protein Soc. *Res:* Reactions of hemeproteins with ligands; interactions between subunits of allosteric proteins and the structural basis for allosteric effects; reactions of antibodies with protein antigens; kinetics of liganding reactions. *Mailing Add:* Biochem 1007 Veterans SUNY Health Sci Ctr 3935 Main St Buffalo NY 14214

NOBLE, VINCENT EDWARD, b Detroit, Mich, Nov 28, 33; div; c 1. PHYSICAL OCEANOGRAPHY. *Educ:* Wayne State Univ, AB, 55, MS, 57, PhD(physics), 60. *Prof Exp:* Jr engr electronics & math anal, Res Labs, Bendix Aviation Corp, 54-56; res assoc solid state physics, Wayne State Univ, 58-60; assoc res physicist, Great Lakes Res Div, Univ Mich, Ann Arbor, 60-68; res physicist, US Naval Oceanog Off, US Naval Res Lab, 68-72, spec asst Navy Environ Remote Sensing, 72-78, head, Space Sensing Applications Br, 79-90, DEP DIR, CTR FOR ADVAN SPACE SENSING, US NAVAL RES LAB, 90- *Mem:* Am Phys Soc; Am Geophys Union; Am Soc Limnol & Oceanog; Sigma Xi. *Res:* Physical limnology; air-sea interaction; polar and remote sensing oceanography. *Mailing Add:* 7615 Mendota Pl Springfield VA 22150

NOBLES, LAURENCE HEWIT, b Spokane, Wash, Sept 28, 27; m 48; c 2. GLACIOLOGY. *Educ:* Calif Inst Technol, BS & MS, 49; Harvard Univ, PhD(geol), 52. *Prof Exp:* From instr to assoc prof, 52-67, dean admin, 72-81, PROF, NORTHWESTERN UNIV, EVANSTON, 67-, VPRES ADMIN & FINANCIAL PLANNING, MED SCH, 80- *Concurrent Pos:* Consult/evaluator, NCent Asn Cols & Sec Schs, 67-75; from asst dean to assoc dean, Col Arts & Sci, Northwestern Univ, Evanston, 66-70, actg dean, 70-72; pres, Chicago Acad Sci, 73-79; trustee, Adler Planetarium, 81- *Mem:* AAAS; Geol Soc Am; Am Asn Petrol Geologists; Am Geophys Union; Glaciol Soc. *Res:* Geomorphology; glacial geology. *Mailing Add:* 3528 Lark San Diego CA 92103

NOBLES, WILLIAM LEWIS, b Meridian, Miss, Sept 11, 25; m 48; c 2. PHARMACEUTICAL CHEMISTRY. *Educ:* Univ Miss, BS, 48, MS, 49; Univ Kans, PhD(pharmaceut chem), 52. *Prof Exp:* From asst prof to prof pharm & pharmaceut chem, Univ Miss, 52-68; PRES, MISS COL, 68- *Concurrent Pos:* Pfeiffer mem res fel, 55-58 & 59-60; NSF fel, Univ Mich, 58-59; dean, Grad Sch, Univ Miss, 60-68. *Honors & Awards:* Found Award, Am Pharmaceut Asn 66; Nat Rho Chi Award in Montreal, Can, 69. *Mem:* Am Chem Soc; Am Pharmaceut Asn; NY Acad Sci; Royal Soc Chem; AAAS; Sigma Xi. *Res:* Medicinal chemistry; pharmaceutical product development. *Mailing Add:* Miss Col Box 4186 Clinton MS 39058

NOBLESSE, FRANCIS, b Souplicourt, France, May 2, 46; US citizen; m 72. HYDRODYNAMICS. *Educ:* Univ Toulouse, France, Engr, 69; Univ Iowa, MS, 71, PhD(mech & hydraul), 74. *Prof Exp:* Res engr, Inst Hydraul Res, Univ Iowa, 75; res assoc, Dept Aeronaut & Astronaut, Stanford Univ, 75-77; asst prof hydrodyn, Dept Ocean Eng, Mass Inst Technol, 77-81; sr res staff, ORI Inc, 81; RES NAVAL ARCHITECT, DAVID TAYLOR RES CTR, BETHESDA, 82- *Concurrent Pos:* Vis prof, Univ Nantes, France, 84, Univ Poitiers, France, 91. *Mem:* Soc Indust & Appl Math; Soc Naval Architects & Marine Engrs; Am Geophys Union; Sigma Xi. *Res:* Development of analytical-numerical methods for calculating free-surface flows about ships and offshore structures. *Mailing Add:* David Taylor Res Ctr Bethesda MD 20084-5000

NOBLET, RAYMOND, b Hiawassee, Ga, Aug 5, 43. INSECT PHYSIOLOGY, INVERTEBRATE PATHOLOGY. *Educ:* Univ Ga, BS, 65, MS, 67, PhD(entom), 70. *Prof Exp:* Asst prof, 70-75, assoc prof entom & econ zool, 75-77, PROF ENTOM, CLEMSON UNIV, 75- *Mem:* Entom Soc Am; Soc Invert Pathol. *Res:* Invertebrate immunity; malariology; parasitology; physiological and ecological investigations pf insect vectors of disease. *Mailing Add:* 217 Waldrop-Stone Rd Central SC 29630

NOBUSAWA, NOBUO, b Osaka, Japan, May 15, 30; m 61; c 2. ALGEBRA. *Educ:* Osaka Univ, BS, 53, MS, 55, PhD(math), 58. *Prof Exp:* Asst prof math, Univ Alta, 62-66; assoc prof, Univ RI, 66-67; assoc prof, 67-71, PROF MATH, UNIV HAWAII, 71- *Mem:* Am Math Soc; Can Math Cong; Math Soc Japan. *Res:* Ring theory and number theory. *Mailing Add:* Dept Math/PS6 320 Univ Hawaii at Manoa 2500 Campus Rd Honolulu HI 96822

NOCENTI, MERO RAYMOND, b Masontown, Pa, Sept 7, 28; m 55; c 3. PHYSIOLOGY. *Educ:* Univ WVa, AB, 51, MS, 52; Rutgers Univ, PhD(endocrinol), 55. *Prof Exp:* Waksman-Merck fel, Rutgers Univ, 55-56; from instr to prof physiol, Col Physicians & Surgeons, Columbia Univ, 56-90; RETIRED. *Concurrent Pos:* Managing ed, Proc, Soc Exp Biol & Med, 74, ed, 80-89. *Mem:* AAAS; Am Physiol Soc; Sigma Xi; Soc Exp Biol & Med (exec secy, 80-89). *Res:* Endocrine physiology; hormonal influences on electrolyte and water balance and on connective tissue. *Mailing Add:* 284 Wayfair Circle Franklin Lakes NJ 07417

NOCETI, RICHARD PAUL, b Pittsburgh, Pa, Jan 16, 47; m 70; c 4. ORGANIC CHEMISTRY. *Educ:* Duquesne Univ, BS, 68, MS, 74, PhD(org chem), 79. *Prof Exp:* Supvry res chemist, 75-85, CHIEF, EXPLOR CHEM BR, PROCESS SCI, PITTSBURGH ENERGY TECH CTR, US DEPT ENERGY, 85- *Mem:* Am Chem Soc. *Res:* Natural product chemistry; synthetic organic chemistry; mechanism of coal liquefaction and coal chemistry; methane activation and conversion; catalyst chemistry, electrochemistry; coal surface chemistry. *Mailing Add:* 5530 Baptist Rd Pittsburgh PA 15236

NOCKELS, CHERYL FERRIS, b Chicago, Ill, July 20, 35; m 57. ANIMAL NUTRITION. *Educ:* Colo State Univ, BS, 57, MS, 59; Univ Mo, PhD(animal nutrit), 65. *Prof Exp:* Res technician, 59-60, asst poultry scientist biochem, 64-70, assoc prof, 70-74, PROF ANIMAL SCI DEPT, COLO STATE UNIV, 74- *Mem:* Am Soc Animal Sci; Poultry Sci Asn; Am Inst Nutrit; NY Acad Sci; Sigma Xi. *Res:* Nutritional and biochemical studies involving both monogastric and ruminant animals. *Mailing Add:* 421 E Douglas Rd Ft Collins CO 80524

NODA, KAORU, b Hilo, Hawaii, Oct 16, 24; m 53; c 4. PARASITOLOGY. *Educ:* Grinnell Col, BA, 50; Univ Iowa, MS, 53, PhD(zool), 56. *Prof Exp:* Asst zool, Univ Iowa, 51-56; asst parasitologist, Univ Hawaii, Hilo, 57-59, asst prof sci, 59-63, assoc prof & dir, Hilo Campus, 62-68, prof biol, 69-85; RETIRED. *Concurrent Pos:* USPHS res grant, 60; scholar, Univ Calif, Los Angeles, 65-66; provost, Univ Hawaii, Hilo, 68-70, asst chancellor, 70-72; vis prof & researcher, Univ Tokyo, 72-73. *Mem:* AAAS; Am Soc Parasitol. *Res:* Trematodes, particularly the family Heterophyidae. *Mailing Add:* 276 Pohakulani St Hawaii Community Col 1175 Manono St Hilo HI 96720

NODA, LAFAYETTE HACHIRO, b Livingston, Calif, Mar 13, 16; m 47; c 2. BIOCHEMISTRY. *Educ:* Univ Calif, BS, 39, MA, 43; Stanford Univ, PhD, 50. *Prof Exp:* Res assoc, Stanford Univ, 50-51; asst prof, Univ Wis, 53-56; biochemist, US Naval Med Res Inst, Md, 56-57; from assoc prof to prof, 57-82, chmn, dept, 60-65, EMER PROF BIOCHEM, DARTMOUTH MED SCH, 82- *Concurrent Pos:* Res fel, Enzyme Inst, Univ Wis, 51-53; Guggenheim fel, 68-69; Japan Soc Promotion Sci fel, 80. *Mem:* Am Soc Biol Chem. *Res:* Purification, kinetics, structure and mechanism of action of enzymes. *Mailing Add:* Dept Biochem Dartmouth Med Sch Hanover NH 03755

NODAR, RICHARD (H)ENRY, b Schenectady, NY, June 15, 35. AUDIOLOGY. *Educ:* Col Geneseo, State Univ NY, BS, 62; Purdue Univ, MS, 64, PhD(audiol), 67. *Prof Exp:* Asst prof audiol, Purdue Univ, 66-67 & Mich State Univ, 67-68; assoc prof, Syracuse Univ, 68-75; HEAD, SECT COMMUN DISORDERS, DEPT OTOLARYNGOL & COMMUN DISORDERS, CLEVELAND CLIN, 75- *Mem:* Fel Am Speech-Lang-Hearing Asn; Am Acad Otolaryngol; Am Asn Ment Deficiency; Acad Rehabilitative Audiol; Am EEG Soc. *Res:* Application of brain stern auditory evoked potential testing to clinical populations; origin, nature, measurement and diagnostic significance of tinnitus aurium. *Mailing Add:* 9500 Euclid Ave One Clinic Ctr Cleveland OH 44195-5034

NODIFF, EDWARD ALBERT, b US, Nov 25, 26; m 50; c 2. ORGANIC CHEMISTRY. *Educ:* Temple Univ, BA, 48. *Prof Exp:* Res assoc chem, Res Inst, 49-55, proj dir, 55-60, dir org chem res, 60-80, PRIN SCIENTIST, FRANKLIN RES CTR, 80- *Mem:* Am Chem Soc. *Res:* Organic fluorine and medicinal chemistry. *Mailing Add:* 1600 Placid St Philadelphia PA 19152

NODULMAN, LAWRENCE JAY, b Chicago, Ill, May 6, 47. EXPERIMENTAL HIGH ENERGY PHYSICS. *Educ:* Univ Ill, Urbana-Champaign, BS, 69, MS, 70, PhD(physics), 73. *Prof Exp:* Res assoc physics, Univ Ill, Urbana-Champaign, 73-75; asst res physicist, Univ Calif, Los Angeles, 75-79; asst physicist, 79-83, PHYSICIST, ARGONNE NAT LAB, 83- *Mem:* AAAS; Am Phys Soc. *Res:* Particle production in electron-positron interactions; neutrino interactions; weak and electromagnetic interactions of hadrons. *Mailing Add:* Argonne Nat Lab HEP-362 Argonne IL 60439

NODVIK, JOHN S, b Canonsburg, Pa, July 2, 30; m 53; c 3. PHYSICS. *Educ:* Carnegie Inst Technol, BS(physics) & BS(math), 52, MS, 52; Univ Calif, Los Angeles, PhD(physics), 58. *Prof Exp:* From asst prof to assoc prof physics, 58-70, PROF PHYSICS, UNIV SOUTHERN CALIF, 70- *Mem:* Am Phys Soc. *Res:* Relativistic spin; nuclear optical model. *Mailing Add:* Dept of Physics Univ Southern Calif Univ Park Los Angeles CA 90089

NOE, BRYAN DALE, b Peoria, Ill, Mar 1, 43; m 65; c 3. BIOCHEMISTRY, CELL BIOLOGY. *Educ:* Goshen Col, BA, 65; WVa Univ, MA, 67; Univ Minn, PhD(anat), 71. *Prof Exp:* USPHS res fel, Univ Minn, 71-72; asst prof, 73-77, assoc prof anat, 77-83, DIR GRAD STUDIES, DEPT ANAT & CELL BIOL, EMORY UNIV, 77-, PROF ANAT & CELL BIOL, SCH MED, 83- *Mem:* Am Soc Cell Biologists; Am Asn Anatomists; Corp mem Marine Biol Lab; Am Diabetes Asn; Endocrine Soc. *Res:* Biosynthesis of glucagon and somatostatin; proteolytic processing of precursors; conversion enzyme characterization. *Mailing Add:* Dept Anat & Cell Biol Emory Univ Sch Med Atlanta GA 30322

NOE, ERIC ARDEN, b Bluffton, Ohio, Dec 24, 43; m. ORGANIC CHEMISTRY. *Educ:* Univ Cincinnati, BS, 65; Calif Inst Technol, PhD(chem), 71. *Prof Exp:* Fel chem, Univ Southern Calif, 70-71 & Univ Calif, San Francisco, 71-72; fel, Wayne State Univ, 72-75, res assoc, 75-77; from asst prof to assoc prof, 77-86, PROF CHEM, JACKSON STATE UNIV, 86- *Mem:* Am Chem Soc. *Res:* Conformational analysis using dynamic nuclear magnetic resonance spectroscopy; organic synthesis. *Mailing Add:* Dept of Chem Jackson State Univ 1400 John R Lynch St Jackson MS 39217

NOE, FRANCES ELSIE, b Beacon Falls, Conn, May 23, 23; m 56; c 2. PHYSIOLOGY. *Educ:* Middlebury Col, BA, 44; Yale Univ, MN, 47; Univ Vt, MD, 54. *Prof Exp:* Intern, Mary Hitchcock Mem Hosp, 54-55; Mich Heart Asn fel, Harper Hosp, Detroit, 55-56; resident pulmonary med, Henry Ford Hosp, 56-57; Rands fel med, Wayne State Univ, 57-58, res assoc anesthesiol, Col Med, 58-59, from instr to asst prof, 59-65; res assoc, 65-70, CHIEF PULMONARY PHYSIOL SECT, DIV RES, SINAI HOSP, DETROIT, 70-; ASST PROF, SCH MED, WAYNE STATE UNIV, 75- *Mem:* Sigma Xi; Int Anesthesia Res Soc; Am Soc Anesthesiol. *Res:* Anesthesiology; cardiopulmonary physiology, computer applications, especially as applied to anesthesiology. *Mailing Add:* Sinai Hosp Div Res 6767 W Outer Dr Detroit MI 48235

NOE, JERRE D(ONALD), b McCloud, Calif, Feb 1, 23; m 43, 83; c 3. COMPUTER SCIENCE, ELECTRICAL ENGINEERING. *Educ:* Univ Calif, BS, 43; Stanford Univ, PhD(elec eng), 48. *Prof Exp:* Res assoc, Radio Res Lab, Harvard Univ, 43-45; develop engr, Hewlett-Packard Co, 46-48; res engr, Stanford Res Inst, 48-54, asst dir div eng res, 54-60, dir eng, Sci Div, 61-64, exec dir eng sci & indust develop, 64-68; chmn dept computer sci, 68-76, PROF COMPUTER SCI, UNIV WASH, 68- *Concurrent Pos:* Lectr, Stanford Univ, 52-67; mem comt, Nat Joint Comput Conf & chmn, Prof Group Electronic Comput, 56-57, Army Sci Bd, 85-; vis prof, Vrije Univ, Amsterdam, 76-77; vis res scientist, GMD, Bonn, WGer, 86-87; bd mem, Int Comput Sci Inst, 91-; consult, Nat Res Coun, 91- *Mem:* Inst Elec & Electronic Engrs. *Res:* Computer system modeling; measurement and evaluation; computer networks. *Mailing Add:* Dept Comput Sci FC-35 Univ Wash Seattle WA 98195

NOE, LEWIS JOHN, b Cleveland, Ohio, Oct 26, 41; m 68; c 1. CHEMISTRY. *Educ:* Western Reserve Univ, AB, 63; Case Western Reserve Univ, PhD(chem), 67. *Prof Exp:* Fel chem, Univ Pa, 67-69; asst prof, 69-74, ASSOC PROF CHEM, UNIV WYO, 74- *Concurrent Pos:* NSF res grant, Univ Wyo, 71-73. *Mem:* Am Chem Soc; Sigma Xi. *Res:* Ultraviolet, visible and infrared spectroscopy of molecular crystals; applications of Stark and Zeeman effects to spectroscopy of molecular crystals. *Mailing Add:* Chem Phys Sci Bldg Univ of Wyo Laramie WY 82070

NOEHREN, THEODORE HENRY, b Buffalo, NY, Sept 6, 17; m 40; c 2. INTERNAL MEDICINE, PULMONARY DISEASES. *Educ:* Williams Col, BA, 38; Univ Rochester, MD, 42; Univ Minn, MS, 50. *Prof Exp:* Mayo Found fel, Univ Minn, 47-49; asst prof pub health & prev med, Univ Utah, 49-52; asst prof internal med, State Univ NY Buffalo, 52-69; assoc prof med, Univ Utah, 69-75; dir pulmonary dis div, Holy Cross Hosp, 74-77; pvt pract, 77-85; RETIRED. *Concurrent Pos:* Consult, USPHS, 50; Markle scholar, 53-58; Fulbright lectr, Univ Helsinki, 66; consult & sr cancer res internist, Roswell Park Mem Cancer Res Hosp. *Mem:* Am Thoracic Soc; AMA; Am Col Chest Physicians. *Res:* Pulmonary physiology; role of lungs as an excretory organ; action of intermittent positive pressure in pulmonary diseases; factors altering the rate of excretion of inert gases by the lungs. *Mailing Add:* 662 E 4025 South #G Salt Lake City UT 84107

NOEL, BRUCE WILLIAM, b York, Pa, Sept 13, 34; m 64. ELECTRONICS ENGINEERING. *Educ:* Drexel Inst Technol, BSEE, 64; Case Inst Technol, MSEE, 66; Univ NMex, PhD(elec eng), 71. *Prof Exp:* Staff mem, Sandia Labs, 66-68; res engr, Univ NMex, 71-72; staff mem, 72-87, group leader, 87-90, STAFF MEM, LOS ALAMOS NAT LAB, 90- *Concurrent Pos:* Consult,

Sandia Labs, 68-72. *Res:* Optoelectronics applications in remote high-temperature measurements, nuclear radiation imaging, and in uranium-enrichment centrifuge diagnostics; network and spectrum analysis applications in electronics systems analysis; electronic circuit design and instrumentation; radiation effects on solid state devices; characteristics and applications of image intensifers. *Mailing Add:* Los Alamos Nat Lab PO Box 1663 Los Alamos NM 87545

NOEL, DALE LEON, b Wichita, Kans, May 21, 36; m 62; c 2. ANALYTICAL CHEMISTRY. *Educ:* Friends Univ, BA, 58; Wichita State Univ, MS, 60; Kans State Univ, PhD(chem), 70. *Prof Exp:* Asst prof chem, Friends Univ, 61-65, Eastern Nazarene Col, 67-69; assoc prof, Kearney State Col, 70-74; PRIN SCIENTIST, INT PAPER CO, 74- *Mem:* Am Chem Soc; Tech Asn Pulp & Paper Indust. *Res:* Gas chromatography; mass spectrometry; environmental analysis; headspace analysis. *Mailing Add:* Corp Res Ctr Int Paper Co Long Meadow Rd Tuxedo NY 10987

NOEL, GERALD THOMAS, b West Chester, Pa, Oct 10, 34; m 62; c 3. SOLID STATE PHYSICS. *Educ:* Drexel Univ, 61; Temple Univ, MS, 66. *Prof Exp:* Mem tech staff, RCA Labs, 61-68, tech dir hybrid circuits facil, Astro Electron Div, 68-72; mem prof staff res, Univ Pa-Energy Ctr, 72-77; RES LEADER, BATTELLE COLUMBUS LABS, 77- *Mem:* Am Inst Physics; Inst Environ Sci; Am Asn Physics Teachers; Mat Res Soc; Nat Tech Asn; Am Vacuum Soc. *Res:* Energy technology, primarily solar; semiconductor technology; device development; photovoltaic devices; infrared detectors; photovoltaic systems design; solar heating and cooling systems; vacuum deposition techniques; thin film materials properties. *Mailing Add:* Battelle Columbus Labs 505 King Ave Columbus OH 43201

NOEL, JAMES A, b Williamstown, Pa, Aug 11, 22; m 45; c 3. GEOLOGY. *Educ:* Lehigh Univ, BA, 49; Dartmouth Col, MA, 51; Ind Univ, PhD(geol), 56. *Prof Exp:* Sr geologist, Creole Petrol Corp, 56-60; assoc prof geol math, Northwestern State Col, La, 60-64; res assoc, Res Ctr Union Oil, 64-66; from assoc prof to prof geol & chmn dept, Wright State Univ, 66-71, asst dir, Western Ohio Br Campus, 71-73, asst dean, 73-74, prof geol, 74-; AT DEPT GEOL, ASHLAND COL. *Mem:* Am Asn Petrol Geologists; Am Inst Prof Geologists; Sigma Xi; Nat Asn Geol Teachers; Soc Explor Geophys. *Res:* Computer systems for gravity and magnetic residual calculation and automatic mapping; combined geological and gravity interpretations using computer systems; sedimentary environments. *Mailing Add:* Dept Geol Ashland College Ashland OH 44805

NOEL, JAN CHRISTINA, b Portland, Ore, April 7, 49; m 73. INTERNATIONAL DEVELOPMENT, INTERNATIONALIZATION OF EDUCATION. *Educ:* Washington State Univ, DVM, 73. *Prof Exp:* Res asst vet path, Wash State Univ, 70-73; state vet, State Wash Racing Comn, 73-74; res scientist trop dis, Int Lab Res Animal Dis, 74-78; assoc dir, Int Develop & Animal Health, 78-89, ACTING DIR, INT PROG DEVELOP, WASH STATE UNIV, 90- *Concurrent Pos:* Consult, Consortium Int Develop, 79-, US Agency Int Develop, 79-, Int Develop Mgt Ctr, Univ Md, 81-; proj dir, US AID Projs, Wash State Univ, 79-, actg dir, Int Develop Coop; mem, Coun Dirs, Int Progs. *Mem:* Asn US Univ Dirs Int Agr Progs; Am Vet Med Asn. *Res:* Plan, manage and conduct research training on international agricultural and institutional development, primarily in developing countries and in US educational institutions; research and development of livestock and animal health, with emphasis on Africa; research and training on internationalization of US education and on development cooperation international. *Mailing Add:* PO Box 2684CS Pullman WA 99165

NOELKEN, MILTON EDWARD, b St Louis, Mo, Dec 5, 35; m 62; c 2. BIOCHEMISTRY. *Educ:* Wash Univ, BA, 57, PhD(phys chem), 62. *Prof Exp:* Researcher chem, Duke Univ, 62-64; assoc & res chemist, Eastern Regional Res Lab, USDA, 64-67; from asst prof to assoc prof, 67-81, PROF BIOCHEM, UNIV KANS MED CTR, 81- *Concurrent Pos:* Actg chmn dept biochem, Univ Kans Med Ctr, 73-74; vis prof, Fed Univ Minas Gerais, Brazil, 78. *Mem:* AAAS; Am Chem Soc; Am Soc Biol Chem; Sigma Xi; Biophys Soc. *Res:* Physical chemistry of proteins; structure of basement membranes. *Mailing Add:* 9310 W 82nd Terr Overland Park Kansas City KS 66204

NOELL, WERNER K, physiology, for more information see previous edition

NOER, RICHARD JUUL, b Madison, Wis, July 3, 37; m 67; c 2. SUPERCONDUCTIVITY, FIELD EMISSION. *Educ:* Amherst Col, BA, 58; Univ Calif, Berkeley, PhD(physics), 63. *Prof Exp:* Physicist, Atomic Energy Res Estab, Eng, 63-64; asst prof physics, Amherst Col, 64-66; from asst prof to assoc prof, 66-75, PROF PHYSICS, CARLETON COL, 75- *Concurrent Pos:* Vis physicist, Lab de Physique des Solides, Orsay, France, 72-73, Ames Lab, Iowa State Univ, 77-80, 82-84 & Univ Geneva, Switz, 80-81, 84-85 & Cornell Univ, 86 & 88-90. *Mem:* Am Phys Soc; Am Asn Physics Teachers; Sigma Xi. *Res:* Low temperature solid state physics(metals and superconducters, electron tunneling); surface physics (anomalous field emission). *Mailing Add:* Dept Physics Carleton Col Northfield MN 55057

NOER, RUDOLF JUUL, b Menominee, Mich, Apr 25, 04; m 33; c 2. SURGERY. *Educ:* Univ Wis, AB, 24; Univ Pa, MD, 27. *Prof Exp:* Asst anat, Univ Wis, 32-34, resident surg, 34-37, instr, 36-37; fel surg res, Col Med, Wayne State Univ, 37-38, from instr to assoc prof surg, 38-49, prof surg & appl anat, 49-52; prof surg & head dept, Sch Med, Univ Louisville, 52-70; prof, 70-75, EMER PROF SURG, COL MED, UNIV S FLA, 75- *Concurrent Pos:* Assoc, Detroit Receiving Hosp, Mich, 46-52; consult, Detroit Marine Hosp, USPHS, 47-48; sr consult, Vet Admin Hosp, Dearborn, Mich, 48-52, area consult, Vet Admin, 55-70, Tampa, 72-; dir surg, Louisville Gen Hosp, Ky, 52-70; ed, J Trauma, 61-68. *Mem:* Am Surg Asn (vpres, 63-64); Am Asn Anatomists; Am Asn Surg of Trauma (pres, 64); Int Surg Soc; fel Am Col Surg (vpres, 64-65). *Res:* Intestinal obstruction; intestinal circulation in man and animals; physiologic responses of intestine to distention; diverticular disease of the colon; adjuvant chemotherapy for breast cancer. *Mailing Add:* 1000 Cannon Valley Dr Northfield MN 55057

NOERDLINGER, PETER DAVID, b New York, NY, May 3, 35; wid; c 5. ASTROPHYSICS, PLASMA PHYSICS. *Educ:* Harvard Univ, AB, 56; Calif Inst Technol, PhD(physics), 60. *Prof Exp:* From instr to asst prof physics, Univ Chicago, 60-66; assoc prof, Univ Iowa, 66-68; from assoc prof to prof, NMex Inst Mining & Technol, 68-71; prof astron, Mich State Univ, 71-82; Los Alamos Nat Lab, 82-85; Jet Propulsion Lab, 85-87; MICROCOSM, INC, 87- *Concurrent Pos:* Sr resident res assoc, Nat Res Coun, NASA-Ames Res Ctr, 71, 74 & 75, NSF, 74 & 75; vis scientist, Smithsonian Astrophys Observ, Mass, 73-, High Altitude Observ, Nat Ctr Atmospheric Res, 77-78; vis prof, Univ Calif, Santa Cruz, 71; NSF fel, 77-78; vis sr res assoc, Univ Colo, 77-78; mem cosmology comn, Int Astron Union; actg chmn dept astron & astrophys, Mich State Univ, 74-75. *Mem:* Am Astron Soc; Am Asn Physics Teachers; fel Royal Astron Soc; Am Phys Soc; Fedn Am Scientists; Sigma Xi. *Res:* Quasi-stellar objects; cosmology; theoretical astrophysics; radiative transfer; comets; interstellar matter; properties of quasi-stellar objects; mass loss from stars and quasi-stellar objects; plasma processes in stars and quasars; theoretical cosmology; astronautical engineering. *Mailing Add:* 2517 Grand Summit Rd Torrance CA 90505

NOETHER, GOTTFRIED EMANUEL, mathematical statistics; deceased, see previous edition for last biography

NOETZEL, DAVID MARTIN, b Waseca, Minn, Feb 19, 29; m 50; c 6. ENTOMOLOGY, ZOOLOGY. *Educ:* Univ Minn, BA, 51, MS, 56. *Prof Exp:* From asst prof to assoc prof agr entom, NDak State Univ, 56-65; asst prof biol, Concordia Col, Moorhead, Minn, 65-70; from instr to assoc prof, 70-88, PROF ENTOM, EXTEN ENTOMOLOGIST, UNIV MINN, ST PAUL, 89- *Mem:* Entom Soc Am; Am Soc Mammal; Bee Res Asn. *Res:* Animal ecology; ornithology; economic zoology. *Mailing Add:* Entom Dept Univ of Minn St Paul MN 55108

NOF, SHIMON Y, b Haifa, Israel, 1946; Israeli & US citizen; m; c 2. COMPUTER-INTEGRATED MANUFACTURING, INDUSTRIAL ROBOTICS. *Educ:* Technion, Israel Inst Technol, BS, 69, MS, 72; Univ Mich, Ann Arbor, PhD(indust & opers eng), 76. *Prof Exp:* Programmer syst analyst, Prod & Logistic Syst, Dept Defense, Israel, 68-72; instr prod mgt, Shenkar Col, Ramat-Gan, Israel, 71-72; lectr opers mgt & quantitative methods, Sch Mgt, Univ Mich, Dearborn, 74-75, asst prof, 76; sr syst analyst, Res & Develop Div, Mfg Data Systs, Inc, Ann Arbor, Mich, 76-77; sr lectr indust eng, Fac Eng, Tel-Aviv Univ, Israel, 80-81; corp consult, CIM, CAD/CAM, Scitex Inc, Herzlia, Israel, 84-86; corp consult, programmable assembly computerized mfg, Tadriran Ltd, Tel-Aviv, Israel, 84-86; asst prof indust eng, 77-80, assoc prof, 82-84 & 86-88, PROF ROBOTICS, CIM & PROD SYSTS, SCH INDUST ENG, PURDUE UNIV, 88- *Concurrent Pos:* Consult, numerous orgn US, Can, Mex, PR, Japan, Europ community & Israel, 78-; prin investr, NSF, 78-93; vis sr lectr, Fac Indust Eng & Mgt, Technion, Israel Inst Technol, Haifa, 80-81, vis prof, 84-86; vis res scholar, Dept Mech Eng, Mass Inst Technol, Cambridge, 89. *Honors & Awards:* Significant Accomplishments in Mfg Systs Award, Inst Indust Engrs, 87. *Mem:* Asn Comput Mach; fel Inst Indust Engrs; Inst Mgt Sci. *Res:* Design and control of automated and robotic production systems; production and manufacturing information systems and computational models; applications of artificial intelligence in industrial operations and control; interactive robotic devices for disabled persons; intelligent scheduling systems; modeling of software development. *Mailing Add:* Sch Indust Eng Purdue Univ West Lafayette IN 47907-1287

NOFFSINGER, ELLA MAE, b Center, Colo, Mar 15, 34. NEMATOLOGY, PLANT PATHOLOGY. *Educ:* Colo State Univ, BS, 56, MS, 58. *Prof Exp:* Lab asst nematol, Sugar Beet Develop Found, Ft Collins, Colo, 56-58; lab asst nematol & plant path, Dept Plant Path, Univ Wis-Madison, 58-62; lab tech nematol, Univ Calif, Davis, 62-67; nematologist, Univ Calif Coop Prog, Univ Chile, 67-69; staff res assoc, 69-76, SR MUS SCIENTIST NEMATOL, UNIV CALIF, DAVIS, 76- *Concurrent Pos:* Assoc ed, Nematropica, Trop Am Nematologists, 75-; ed, Nematol Newsletter, Soc Nematologists, 77-80; mem exec bd, Soc Nematologists, 88-90. *Mem:* Soc Nematologists; Orgn Trop Nematologists; Sigma Xi; Sci Res Soc. *Res:* Taxonomy, systematics and distribution of the plant parasitic and free-living nematodes, especially fresh-water, marine and soil; taxonomic revisions of free-living marine and soil nematodes, from genus through superfamily. *Mailing Add:* Dept Nematol Univ Calif Hutchinson Hall 482 Davis CA 95616

NOFTLE, RONALD EDWARD, b Springfield, Mass, Mar 10, 39; m 64; c 1. INORGANIC CHEMISTRY, FLUORINE CHEMISTRY. *Educ:* Univ NH, BS, 61; Univ Wash, PhD(inorg chem), 66. *Prof Exp:* Res asst inorg chem, Univ Wash, 62-66, instr chem, 66; fel, Univ Idaho, 66-67; from asst prof to assoc prof, 67-79, chmn dept chem, 80-86, PROF CHEM, WAKE FOREST UNIV, 79- *Concurrent Pos:* Res chemist, Naval Res Lab, 75-76; secy-tres, Div fluorine chem, Am Chem Soc, 84-85, chmn-elect, 86, chmn, 87; sabbatical, Univ Southampton, Eng, 86. *Honors & Awards:* Bailey Prize, Am Inst Chemists. *Mem:* Am Chem Soc; fel Am Inst Chemists; Sigma Xi; Am Asn Univ Professors. *Res:* Synthesis and properties of compounds containing fluorine; chemistry of the halogens; electrochemistry; conducting polymers. *Mailing Add:* Dept Chem Wake Forest Univ Winston-Salem NC 27109

NOFZIGER, DAVID LYNN, b Wauseon, Ohio, June 13, 44; m 68; c 2. SOIL PHYSICS, FLUID TRANSPORT. *Educ:* Goshen Col, BA, 66; Purdue Univ, MS, 70, PhD(agron), 72. *Prof Exp:* Lectr & res fel soil physics, Ahmadu Bello Univ, Nigeria, 72-74; asst prof, 74-80, ASSOC PROF SOIL PHYSICS, OKLA STATE UNIV, 80- *Mem:* Am Soc Agron; Soil Sci Soc Am; Int Soil Sci Soc. *Res:* Water movement in saturated and unsaturated soils; utilization of small computers in data acquisition annd control. *Mailing Add:* Dept Agron Rm 265 Okla State Univ Stillwater OK 74078

NOGA, EDWARD JOSEPH, b Chicago, Ill, Sept 14, 53. AQUATIC ANIMAL MEDICINE, BIOTECHNOLOGY. *Educ:* Fla Atlantic Univ, BS, 74, MS, 77; Univ Fla, DVM, 82. *Prof Exp:* Res fel aquatic animal med, Cornell Univ, 78; fel comp path, Harvard Univ, 80; asst prof, 82-88, ASSOC PROF

AQUATIC ANIMAL MED, NC STATE UNIV, 88- *Concurrent Pos:* Lectr, Aquavet Prog, Woods Hole, Mass, 83-; prin investr, Sea Grant, Water Resources Res Inst, Bard, Environ Protection Agency, 83- *Honors & Awards:* Wilton Earle Award, Tissue Cult Asn, 77. *Mem:* AAAS; Int Asn Aquatic Animal Med; Am Fisheries Soc; Am Vet Med Asn; Tissue Cult Asn; World Aquacult Soc. *Res:* Major infectious disease problems affecting the commerical fisheries and aquaculture industries of North Carolina. *Mailing Add:* Dept CASS Col Vet Med NC State Univ 4700 Hillsborough St Raleigh NC 27606

NOGAMI, YUKIHISA, b Hamada, Japan, Oct 22, 29; m 58; c 3. THEORETICAL PHYSICS. *Educ:* Kyoto Univ, BSc, 52, DSc(physics), 61. *Prof Exp:* From asst to lectr physics, Univ Osaka Prefecture, 54-61; fel, Nat Res Coun Can, 61-63; res fel, Battersea Col Technol, 63-64; sr fel, 64-65, assoc prof, 65-69, PROF PHYSICS, MCMASTER UNIV, 69- *Mem:* Can Asn Physicists; Phys Soc Japan. *Res:* Theoretical physics, nuclear and subnuclear physics. *Mailing Add:* Dept Physics McMaster Univ Hamilton ON L8S 4M1 Can

NOGAR, NICHOLAS STEPHEN, b Chicago, Ill, Jan 19, 50; m 74; c 3. PHYSICAL CHEMISTRY, ATMOSPHERIC CHEMISTRY. *Educ:* Univ NMex, BS, 71; Univ Utah, PhD(phys chem), 76. *Prof Exp:* NSF fel chem, Univ Calif, Berkeley, 76-77; asst prof chem, Univ Nebr, 77-80; MEM STAFF, LOS ALAMOS NAT LAB, 80-, DEP GROUP LEADER, 84- *Mem:* Am Chem Soc; AAAS; Sigma Xi; Int Union Pure & Appl Chem. *Res:* Applications of lasers in physical, analytical and biological chemistry; electronic and vibrational photochemistry and kinetics; laser mass spectrometric measurement of trace components; CARS of biochemical systems; chemical kinetics and dynamics; application of lasers in physical and analytical chemistry. *Mailing Add:* CLS 2 MS G738 Los Alamos Nat Lab PO Box 1663 Los Alamos NM 87545

NOGES, ENDRIK, b Moisakula, Estonia, Apr 5, 27; nat US; m 51; c 3. ELECTRICAL ENGINEERING. *Educ:* Northwestern Univ, BS, 54, MS, 56, PhD(elec eng), 59. *Prof Exp:* From asst to instr, Northwestern Univ, 54-57; from asst prof to assoc prof, Univ Wash, 58-69, asst dean eng, 66-71, dir televised instr in eng, 83-86, assoc chmn dept, 86-88, actg chmn, 88-90, PROF ELEC ENG, UNIV WASH, 69-, ASSOC CHMN DEPT, 90- *Concurrent Pos:* Fulbright lectr, Finnish Inst Technol, 63-64; vis prof, Univ Karlsruhe, Ger, 72-73; consult, Boeing Co, 72-; fel, Deutsche Forschungsgemerscheft, Karlsruhe, Ger, 72-73. *Mem:* Am Soc Eng Educ; Inst Elec & Electronic Engrs; Sigma Xi. *Res:* Nonstationary and nonlinear feedback control systems; quantization in feedback control systems; pulse frequency modulated and other pulsed feedback systems; guidance and navigation systems. *Mailing Add:* Dept Elec Eng FT10 Univ Wash Seattle WA 98195

NOGGLE, JOSEPH HENRY, b Harrisburg, Pa, Mar 19, 36; m 60. PHYSICAL CHEMISTRY. *Educ:* Juniata Col, BS, 60; Harvard Univ, MS, 63, PhD(chem), 65. *Prof Exp:* Asst prof chem, Univ Wis-Madison, 65-71; assoc prof, 71-76, PROF CHEM, UNIV DEL, 76- *Mem:* AAAS; Am Chem Soc. *Res:* Nuclear magnetic resonance, including relaxation phenomena. *Mailing Add:* Dept Chem Univ Del Newark DE 19716

NOGRADY, THOMAS, b Budapest, Hungary, Oct 16, 25; Can citizen; m 50, 70; c 1. MEDICINAL CHEMISTRY, LIMNOLOGY. *Educ:* Eotvos Lorand Univ, Budapest, MSc, 48, PhD(org chem), 50. *Prof Exp:* Res chemist, Res Inst Pharmaceut Indust, Budapest, 50-56; res assoc org chem, Univ Vienna, 57 & Univ Montreal, 57-61; from asst prof to assoc prof, Loyola Col, Montreal, 61-70; prof biochem, Concordia Univ, 70-90; RETIRED. *Concurrent Pos:* Rockefeller scholar, Univ Vienna, 57; adj prof biol, Queen's Univ, Kingston, Ont, 84- *Mem:* Am Chem Soc; fel Chem Inst Can. *Res:* Molecular pharmacology; invertebrate physiology; limnology. *Mailing Add:* Dept Biol Queen's Univ Kingston ON K7L 3N6 Can

NOGUCHI, PHILIP D, b Sacramento, Calif, Jan 11, 49; m; c 2. MOLECULAR BIOLOGY. *Educ:* Univ Calif, Berkeley, AB, 70; George Washington Univ, MD, 74. *Prof Exp:* Lab instr gross anat, Med Sch, George Washington Univ, 71-72; sr res investr, Exp Biol Res Br, Div Path, Bur Biologics, Food & Drug Admin, 75-80, Div Biochem & Biophys, 80-81, chief, Cell Biol Br, 81-84, CHIEF, LAB CELLULAR & MOLECULAR BIOL, DIV BIOCHEM & BIOPHYS, CTR BIOLOGICS EVAL & RES, FOOD & DRUG ADMIN, 84- *Concurrent Pos:* Sr asst surgeon, USPHS, 75-76, surgeon, 76-83, sr surgeon, 83-; adj prof, Genetics Prog, Grad Sch Arts & Sci, George Washington Univ, 87- *Mem:* Tissue Cult Asn; Soc Analytical Cytol; Comn Officers Asn; AMA. *Res:* Invasion and metastases of human colon carcinoma cells; genetic control of tumor antigen expression; flow cytometry and cell sorting. *Mailing Add:* Div Biochem & Biophys Lab Cellular & Molecular Biol Ctr Biologics Eval & Res 800 Rockville Pike Bethesda MD 20892

NOHEL, JOHN ADOLPH, b Prague, Czech, Oct 24, 24; nat US; m 48; c 3. MATHEMATICS. *Educ:* George Washington Univ, BEE, 48; Mass Inst Technol, PhD(math), 53. *Prof Exp:* Asst math, George Washington Univ, 46-48; instr, Mass Inst Technol, 50-53; from asst prof to prof, Ga Inst Technol, 53-61; assoc prof, 61-64, chmn dept math, 68-70, PROF MATH, UNIV WIS-MADISON, 64-, DIR, MATH RES CTR, 79- *Concurrent Pos:* Mem, Math Res Ctr, 58-59, 77-; res sabbatical, Univ Paris, 65-66; vis prof, Ecole Polytech, Lausanne, 71-72; mem comt appl math, Nat Res Coun-Nat Acad Sci. *Mem:* fel AAAS; Am Math Soc; Soc Indust & Appl Math; Math Asn Am; Fedn Am Scientists. *Res:* Volterra functional differential equations, qualitative theory and applications. *Mailing Add:* Univ Wis 610 Walnut St Madison WI 53705

NOID, DONALD WILLIAM, b Marshalltown, Iowa, Feb 6, 49; m 77; c 2. PHYSICAL & THEORETICAL CHEMISTRY, CHEMICAL PHYSICS. *Educ:* Iowa State Univ, BS, 71; Univ Ill, MS, 73, PhD(chem), 76. *Prof Exp:* Res assoc chem, Univ Ill, 76; NSF fel, 76-77; CHEMIST, OAK RIDGE NAT LAB, 77- *Concurrent Pos:* Eugene P Wigner fel, Oak Ridge Nat Lab, 77-79; from adj asst prof to adj assoc prof, Univ Tenn, 81-90, adj prof, 90-; vis assoc prof, Univ Wis, 83-84; vis sr scientist, Inst Defense Analysis, Alexandria, Va, 85-86. *Mem:* Am Chem Soc; Am Phys Soc; Sigma Xi. *Res:* Chemical kinetics; molecular dynamics and laser chemistry; polymer chemistry. *Mailing Add:* Dept Chem PO Box X Oak Ridge TN 37830

NOISEUX, CLAUDE FRANCOIS, b Winchester, Mass, June 24, 53. APPLIED MATHEMATICS. *Educ:* Harvard Univ, AB, 75, MS, 77, PhD(appl math), 80. *Prof Exp:* FEL, HARVARD UNIV, 80- *Concurrent Pos:* Consult, Chase Inc, 80- *Mem:* Sigma Xi. *Res:* Study of wave propagation in harbor geometries and related Tsunami problems; modeling of turbulent boundary layer structures and the development of related approximate expansion procedures. *Mailing Add:* 31 Brook St Concordge MA 01742-2306

NOKES, RICHARD FRANCIS, b Deerfield, Mich, Mar 16, 34; m 56; c 6. VETERINARY MEDICINE, ANATOMY. *Educ:* Mich State Univ, BS, 56, DVM, 58. *Prof Exp:* Vet, pvt pract, 58-60 & Agr Res Serv, USDA, 60-61; asst prof, 62-74, ASSOC PROF BIOL, UNIV AKRON, 74- *Concurrent Pos:* Vet, Copley Rd Animal Hosp, Akron, 62-; consult, Akron Children's Zoo, 64- & Akron Gen Med Ctr, 65- *Mem:* Am Asn Zoo Vets. *Res:* Rheological studies of blood flow; developmental studies of sense organs in the dog; excretion pathways of friction reducing agents. *Mailing Add:* 1378 Copley Rd Akron OH 44320

NOLA, FRANK JOSEPH, b Miami, Fla, June 25, 30; m 57; c 3. ELECTRONICS ENGINEERING. *Educ:* Univ Miami, BS, 58. *Prof Exp:* Elec engr, Allis Chalmers, 58-59, Martin Marietta, 59-60 & Army Ballastic Missile Agen, Dept Defense, 60-62; ELEC ENGR, MARSHALL SPACE FLIGHT CTR, NASA, 62- *Honors & Awards:* IR 100 Award, Indust Res Mag, 79; Excalibur Award, US Congress, 79. *Res:* Design and develop electronic control systems for guidance and control of missiles, for controlling earth orbiting space craft and for controlling experiments. *Mailing Add:* 117 Westbury Dr SW Huntsville AL 35802

NOLAN, CHRIS, b Salt Lake City, Utah, July 3, 30; m 63; c 2. BIOCHEMISTRY. *Educ:* Univ Nev, BS, 52; Univ Utah, PhD(biochem), 61. *Prof Exp:* NIH fel, Univ Wash, 61-63; RES BIOCHEMIST, ABBOTT LABS, 63- *Mem:* Am Soc Biochem & Molecular Biol. *Res:* Tumor markers; hormone receptors. *Mailing Add:* 940 Greenleaf St Gurnee IL 60031

NOLAN, CLIFFORD N, b Harberson City, Fla, Oct 22, 25; m 54; c 3. CROPS, SOIL CONSERVATION & FERTILITY. *Educ:* Univ Fla, BSA, 51, PhD(soils), 60; Univ Ga, MSA, 57. *Prof Exp:* Soil technician soil testing, Col Agr, Univ Ga, 51-57; asst tech dir citrus prod, Soil Sci Found, Lakeland, Fla, 60-64; agronomist crop prod, Smith-Douglass Co, Norfolk, Va, 64-65; exten agronomist soils & corp prod, Col Agr, Clemson Univ, 65-87; RETIRED. *Mem:* Am Soc Agron; Soil Conserv Soc Am. *Res:* Educational programs in herbicides for field crops; production guidelines for corn, peanuts, small grain and forage crops; guidelines for conservation tillage and other soil and water conservation practices. *Mailing Add:* 214 Riffle View Dr Clemson SC 29631

NOLAN, EDWARD J, b Philadelphia, Pa; m 49; c 2. THERMODYNAMICS, SCIENCE ADMINISTRATION. *Educ:* Villanova Univ, BChE, 45; Univ Del, MChE, 52; Univ Pa, PhD(chem eng & biochem eng), 81. *Prof Exp:* Proj engr, Pennwalt Chem Co, 46-48, Day & Zimmermann, Inc, 48-50 & Selas Corp, 50-56; eng mgr, Gen Elec Co, 56-83. *Concurrent Pos:* Res assoc, Med Sch, Thomas Jefferson Univ, 72-78; lectr appl math, La Salle Univ, 52-56, chmn, 60- *Mem:* Am Soc Mech Engrs; Am Chem Soc; Am Inst Chem Engrs; Soc Indust & Appl Math; Am Math Soc; Soc Rheology. *Res:* Biochemical technology, including energy development and selected chemicals and fuel produced from biomass and renewable resources. *Mailing Add:* 565 Wanamaker Rd Jenkintown PA 19046

NOLAN, GEORGE JUNIOR, b Stilwell, Okla, Nov 3, 35; m 55; c 5. PHYSICAL CHEMISTRY. *Educ:* Northeastern State Col, BS, 58; Univ Ark, MS, 62, PhD(chem), 64. *Prof Exp:* Chemist, Phillips Petrol Co, 64-68; assoc prof chem, 68-73, PROF CHEM, NORTHEASTERN OKLA STATE COL, 73- *Mem:* Am Chem Soc. *Res:* Fundamental and development research in catalysis. *Mailing Add:* 281 Hickory Dr Northeastern Okla State Univ Tahlequah OK 74464

NOLAN, JAMES FRANCIS, b Scranton, Pa, Nov 16, 31; m 57; c 3. PHYSICS. *Educ:* Univ Scranton, BS, 54; Univ Pittsburgh, PhD(physics), 62. *Prof Exp:* Res physicist, Westinghouse Res Lab, 61-66; res physicist, Owens Ill, 66-72, mgr display technol, 72-77, mgr mat res, 77-84, tech dir, 84-87; mgr solar develop, Glasstech, Inc, 87-91; VPRES OPERS, SOLAR CELLS, INC, 91- *Mem:* Am Phys Soc. *Res:* Atomic collisions; medium energy ion-atom charge transfer and ionization; low energy electron-atom collisions; flat panel matrix displays; photovoltaic solar panels. *Mailing Add:* 4608 Vicksburg Dr Sylvania OH 43560

NOLAN, JAMES P, b Buffalo, NY, June 21, 29; m 56; c 4. MEDICINE. *Educ:* Yale Univ, BA, 51, MD, 55. *Prof Exp:* Instr, Yale Univ, 61-63; from asst prof to assoc prof, 63-69, PROF & ACTG CHMN MED, STATE UNIV NY BUFFALO, 69-; HEAD DEPT MED, BUFFALO GEN HOSP, 81- *Concurrent Pos:* Chief med, Buffalo Gen Hosp; actg chief, Erie Co Med Ctr. *Mem:* Am Col Physicians; Am Fedn Clin Res; Am Gastroenterol Soc; Am Asn Study Liver Dis; Reticuloendothelial Soc. *Res:* Role of bacterial endotoxins and the reticuloendothelial system in the initiation and perpetuation of liver disease. *Mailing Add:* Med/C221 E C Med Ctr 462 Grider St Buffalo NY 14215

NOLAN, JAMES ROBERT, b New York, NY, May 8, 23; m 46; c 3. BOTANY. *Educ:* Cornell Univ, BS, 61, PhD(plant anat & morphol), 67. *Prof Exp:* Instr biol, Antioch Col, 64-67; from asst prof to assoc prof biol, State Univ NY Plattsburgh, 67-85; RETIRED. *Mem:* Am Fern Soc; Bot Soc Am; Int Soc Plant Morphol; Int Soc Study Evolution. *Res:* Ontogeney and phylogeny of branching systems in plants. *Mailing Add:* Dept Biol State Univ NY Plattsburgh NY 12901

NOLAN, JANIECE SIMMONS, b Ft Worth, Tex; m; c 6. HOSPITAL ADMINISTRATION, PHYSIOLOGY. *Educ:* Univ Tex, BA, 61, MA, 63; Tulane Univ, PhD(biol), 68; Univ Calif, Berkeley, MPH, 75. *Prof Exp:* Res scientist aerospace med, Tex Nuclear Corp, 63-65; head cell biol, Gulf Southern Res Inst, 68-70; res physiologist, Vet Admin Hosp, 70-73, health care admin trainee, Vet Admin Hosp, Martine & Univ Calif,,Berkeley, 73-75; asst adminr ambulatory care serv, Univ Calif Med Ctr, San Francisco, 75-77; dir ambulatory care serv, 77-79, asst adminr outpatient & ancillary serv, 79-80, vpres, prof serv, 80-86, CHIEF OPER OFFICER, JOHN MUIR MED CTR, WALNUT CREEK, 86- *Concurrent Pos:* Post doctoral scholar, Dept Physiol-Anat, Univ Calif, Berkeley, 70-72. *Mem:* Am Col Hosp Adminr; Am Pub Health Asn; Geront Soc; Am Physiol Soc; Am Hosp Asn; Sigma Xi. *Res:* Delivery of outpatient health services; cell culture; cellular aging; plant senescence and growth regulators; ambulatory surgery. *Mailing Add:* John Muir Med Ctr Walnut Creek CA 94598

NOLAN, JOHN THOMAS, JR, b Boston, Mass, Apr 15, 30; m 55; c 5. PETROLEUM CHEMISTRY, RESEARCH DIRECTION. *Educ:* Cath Univ Am, AB, 51; Mass Inst Technol, PhD(org chem), 55. *Prof Exp:* Chemist, Texaco Inc, 55-58, group leader polymer res, 59-69, asst supvr ref res, 69-76, supvr coal res, 76-78, asst mgr sci planning, 78-82, asoc dir res, 82-87, DIR STRATEGIC RES, TEXACO, INC, 87- *Mem:* Am Chem Soc; NY Acad Sci; Sigma Xi; AAAS. *Res:* Organic syntheses; polymer syntheses, structures and testing; applied catalysis; refining processes; coal conversion. *Mailing Add:* 18 Relyea Terr Wappingers Falls NY 12590

NOLAN, MICHAEL FRANCIS, b Evergreen Park, Ill, July 28, 47; m 73. NEUROANATOMY. *Educ:* Marquette Univ, BS, 69; Med Col Wis, PhD(anat), 75. *Prof Exp:* Staff phys therapist, Kiwanis Children's Ctr, Curative Workshop, Milwaukee, 69; from instr to asst prof, 75-81, ASSOC PROF ANAT, COL MED, UNIV SOUTH FLA, 81- *Mem:* Am Phys Ther Asn; Am Pain Soc; Int Asn Study Pain. *Res:* Neuroanatomical aspects of pain transmission, perception and appreciation; neurocytological changes in neural dysfunction. *Mailing Add:* Dept Anat Col Med Univ SFla 12901 N 30th St Tampa FL 33612

NOLAN, RICHARD ARTHUR, b Omaha, Nebr, Nov 2, 37; m 67. MYCOLOGY, PHYSIOLOGY. *Educ:* Univ Nebr, Lincoln, BSc, 59, MSc, 62; Univ Calif, Berkeley, PhD(bot), 67. *Prof Exp:* NIH fel biochem, Univ Calif, Berkeley, 67-68, res biochemist, 68-69; asst prof biol, NMex State Univ, 69-70; univ fel, 70-71, from asst prof to assoc prof, 71-80, PROF BIOL, MEM UNIV NFLD, 80- *Mem:* Mycol Soc Am; Can Soc Microbiologists; Can Bot Asn; Soc Invert Path; Can Col Microbiol. *Res:* Comparative immunology and enzymology; nutritional requirements of and biochemical studies with fungal protoplasts; biological control of forest insect pests. *Mailing Add:* Dept Biol Mem Univ of Nfld St John's NF A1B 3X9 Can

NOLAN, RON SCOTT, marine biology, marine ecology, for more information see previous edition

NOLAN, STANTON PEELLE, b Washington, DC, May 29, 33; m 55; c 2. CARDIOVASCULAR SURGERY. *Educ:* Princeton Univ, AB, 55; Univ Va, Charlottesville, MD, 59, MS, 62. *Prof Exp:* From intern to resident surg, Med Ctr, Univ Va, Charlottesville, 59-65, Va Heart Asn res fel, Univ, 61-62, resident thoracic cardiovasc surg, Univ Va Hosp, 65-66; sr surgeon, Nat Heart Inst, 66-68; from asst prof to assoc prof, 68-74, surgeon-in-chg, 70-74, PROF SURG & SURGEON-IN-CHG, THORACIC CARDIOVASC SURG DIV, UNIV VA HOSP, 74-, CLAUDE A JESSUP PROF SURG, 81- *Concurrent Pos:* Am Cancer Soc clin fel, 63-64; estab investr, Am Heart Asn, 69-74, mem coun cardiovasc surg, 68, mem coun thrombosis, 70; attend staff, Med Ctr, Univ Va, 68-; consult cardiac surg, Vet Admin Hosp, Salem, Va, 68-; consult surg, Comm Va Children's Specialty Serv, 68- *Honors & Awards:* John Horsley Mem Prize Med Res, 62. *Mem:* Fel Am Col Surg; fel Am Col Cardiol; Am Heart Asn; Soc Thoracic Surgeons; Am Surg Asn. *Res:* Cardiovascular hemodynamics and pathophysiology. *Mailing Add:* Box 181 Univ Va Med Ctr Charlottesville VA 22908

NOLAN, THOMAS BRENNAN, b Greenfield, Mass, May 21, 01; m 27; c 1. GEOLOGY. *Educ:* Yale Univ, PhB, 21, PhD(geol), 24. *Hon Degrees:* LLD, St Andrews, 62. *Prof Exp:* Geologist, US Geol Surv, 24-44, from asst dir to dir, 44-65, res geologist, 65-; RETIRED. *Concurrent Pos:* Vpres, Int Union Geol Sci, 65-72. *Honors & Awards:* Spendiaroff Prize, Int Geol Cong, 33; Silver Medal, Tokyo Geog Soc, 65. *Mem:* Nat Acad Sci; fel Geol Soc Am (pres, 61); fel Soc Econ Geologists (pres, 50); fel Mineral Soc Am; fel Am Geophysics Union; hon fel Royal Soc Edinburgh; Am Philos Soc; hon mem Tokyo Geog Soc; AAAS; foreign mem Geol Soc London; Sigma Xi. *Res:* Geology and ore deposits of the Great Basin. *Mailing Add:* 2219 California St NW Washington DC 20008

NOLAN, VAL, JR, b Evansville, Ind, Apr 28, 20; m 46, 80; c 3. POPULATION BIOLOGY, EVOLUTIONARY ECOLOGY. *Educ:* Ind Univ, AB, 41, JD, 49. *Prof Exp:* Dep US Marshal, 41-42; agent, US Secret Serv, 42; from asst prof to prof law, 49-85, prof zool, 68-78, prof biol, 78-85, EMER PROF, IND UNIV, BLOOMINGTON, 85- *Concurrent Pos:* Guggenheim fel, 57; res scholar zool, Ind Univ, Bloomington, 57-68. *Honors & Awards:* Brewster Medal, Am Ornithologists Union, 86. *Mem:* Fel Am Ornith Union (vpres, 89-90); Ecol Soc Am; Animal Behav Soc; Ger Ornith Soc; Brit Ornith Union; Fel AAAS; Soc Study Reproduction. *Res:* Behavior and ecology of birds. *Mailing Add:* Dept Biol Ind Univ Bloomington IN 47405

NOLAND, JAMES STERLING, b Cape Girardeau, Mo, June 18, 33. ORGANIC POLYMER CHEMISTRY. *Educ:* Southeast Mo State Col, BS(chem), & BS(educ), 55; Univ Iowa, MS, 57, PhD(org chem), 60. *Prof Exp:* Res chemist polymer chem, 59-73, mgr, Aerospace Adhesives Res, 73-77, tech dir, Gen Prod Dept, 77-78, tech dir, Aerospace Prod Dept, Indust Chem Div, 78-80, MGR, STAMFORD RES, CHEM RES DIV, AM CYANAMID CO, 80- *Mem:* Am Chem Soc; Soc Advan Mat & Process Eng; Soc Advan Educ. *Res:* New polymer systems; resin systems for structural adhesives; composites; environmental resistant resins. *Mailing Add:* 121 Old Redding Rd North Redding CT 06875

NOLAND, JERRE LANCASTER, b Richmond, Ky, Feb 14, 21; m 50; c 3. BIOCHEMISTRY. *Educ:* Purdue Univ, BS, 42, MS, 44; Univ Wis, PhD(biochem, zool), 49. *Prof Exp:* Asst chem & biol, Purdue Univ, 43-44; asst biochem, Univ Wis, 48; Lalor fel, Marine Biol Lab, Woods Hole, 49; biochemist, Entom Br, Med Labs, US Army Chem Ctr, 49-54; chief biochemist, Res Lab, Wood Vet Admin Ctr, Wis, 54-58; chief med res lab, Vet Admin Hosp, Louisville, Ky, 58-73; assoc prof biochem, Univ Louisville, 59-73, res coordr, Dept Obstet & Gynec, Sch Med, 74-83; RETIRED. *Mem:* Am Chem Soc. *Res:* Endocrinology; sterol metabolism. *Mailing Add:* 4018 Brownlee Rd Louisville KY 40207-4532

NOLAND, PAUL ROBERT, b Chillicothe, Ill, Sept 28, 24; m 47; c 4. ANIMAL NUTRITION. *Educ:* Univ Ill, BS, 47, MS, 48; Cornell Univ, PhD, 51. *Prof Exp:* Asst, Univ Ill, 48 & Cornell Univ, 51; from asst prof to assoc prof animal nutrit & husb, 51-60, prof animal sci, 60-88, DEPT HEAD, ANIMAL SCI, UNIV ARK, FAYETTEVILLE, 88- *Concurrent Pos:* With Ark Agr Mission, Panama, 55-57. *Mem:* Am Soc Animal Sci; Am Inst Nutrit; Animal Nutrit Res Coun. *Res:* Swine nutrition and management; proteins, amino acids and mineral nutrition. *Mailing Add:* Dept Animal Sci Univ Ark Fayetteville AR 72701

NOLAND, WAYLAND EVAN, b Madison, Wis, Dec 8, 26. ORGANIC CHEMISTRY, HETEROCYCLIC CHEMISTRY. *Educ:* Univ Wis, BA, 48; Harvard Univ, MA, 50, PhD(phys org chem), 51. *Prof Exp:* Du Pont fel, 51-52, from asst prof to assoc prof, 52-62, PROF CHEM, UNIV MINN, MINNEAPOLIS, 62- *Concurrent Pos:* Vis instr chem, Univ BC, 56; consult, Sun Oil Co, 58-70; actg chief div org chem, Univ Minn, Minneapolis, 61-62, actg chmn dept chem, 67-69 & area coordr org chem, 72-74; secy, Org Syntheses, Inc, 69-79, vpres, 79-88, ed-in-chief, 79-88. *Mem:* Fel AAAS; Sigma Xi; Am Asn Univ Prof; Am Chem Soc; NY Acad Sci. *Res:* Heterocyclic nitrogen chemistry; synthesis and reactions of indoles and pyrroles; 1,3-cycloaddition reactions and ring expansions of isatogens; antimalarial compounds; rearrangements of nitronorbornenes; cycloaddition reactions of indenes; new rearrangements; reaction mechanisms; structure determination. *Mailing Add:* Sch Chem Univ Minn 207 Pleasant St SE Minneapolis MN 55455-0431

NOLASCO, JESUS BAUTISTA, b Manila, Philippines, Oct 5, 17; m 42; c 4. PHYSIOLOGY. *Educ:* Univ Philippines, MD, 40. *Prof Exp:* From instr to asst prof physiol, Univ Philippines, 40-52; prof & head dept, Far Eastern Univ, Manila, 52-64, sr consult hosp, 54-64; vis prof physiol, NJ Med Sch, Col Med & Dent, 64-66, assoc prof, 66-70, actg chmn dept, 74-75, prof physiol, 70-80; RETIRED. *Concurrent Pos:* Mem, Nat Res Coun Philippines, 40-; Rockefeller Found fel, Western Reserve Univ, 46-47; Williams-Waterman fel, Univ Berne, 62-63; mem comt nat med res prog, Nat Sci Develop Bd, Philippines, 62-64. *Mem:* Fel Am Col Clin Pharmacologists; Philippine Heart Asn (pres, 60-61); Am Physiol Soc. *Res:* Cardiovascular physiology, particularly electrophysiology of the heart. *Mailing Add:* 7546 Kestrel Dr Jacksonville FL 32222

NOLEN, GRANVILLE ABRAHAM, b Richmond, Ky, Apr 21, 26; m 50; c 5. TOXICOLOGY. *Educ:* Miami Univ, AB, 50. *Prof Exp:* Chief animal technician, Miami Valley Labs, Procter & Gamble Co, 54-64, staff res asst animal nutrit, Teratology & Physiol & Colony Mgt, 64-70, staff develop toxicologist, 70-87; CONSULT, 87- *Mem:* AAAS; Am Asn Lab Animal Sci; Behav Teratology Soc; Teratology Soc; Am Inst Nutrit; Soc Toxicology (pres, Reproductive & Develop Toxicol SpecialtySect, 85-86); Japanese Teratology Soc. *Res:* Animal nutrition, especially lipid and protein nutrition and metabolism; teratology, especially methods in relation to drug testing; behavioral effects of foods and chemicals; environmental effects on laboratory animals. *Mailing Add:* 316 Erickson Dr Oxford OH 45056

NOLEN, JERRY A, JR, b Washington, DC, Nov 17, 40; m 80; c 6. NUCLEAR PHYSICS, CHARGED PARTICLE BEAM OPTICS. *Educ:* Lehigh Univ, BS, 61; Princeton Univ, PhD(physics), 65. *Prof Exp:* Instr physics, Princeton Univ, 65-66; postdoctoralappointee nuclear physics, Argonne Nat Lab, 66-68; asst prof physics, Univ Md, 68-70; assoc prof, 70-76, PROF PHYSICS, MICH STATE UNIV, 76-, ASSOC DIR, NAT SUPERCONDUCTING CYCLOTRON LAB, 80- *Concurrent Pos:* Fel, Max Planck Inst Nuclear Physics, Heidelberg, Ger, 77. *Mem:* Fel Am Phys Soc. *Res:* Experimental nuclear physics; magnetic spectrometer design; superconducting magnets and cryogenics; accelerator physics and instrumentation development; heavy ion nuclear science. *Mailing Add:* Cyclotron Lab Mich State Univ East Lansing MI 48824

NOLF, LUTHER OWEN, b Solomon, Kans, July 16, 02. PARASITOLOGY. *Educ:* Kans State Col, BS, 26, MS, 29; Johns Hopkins Univ, ScD(med zool), 31. *Prof Exp:* Asst parasitol, Kans State Col, 27-29; asst helminth, Johns Hopkins Univ, 29-31; assoc zool, 31-37, from asst prof to prof zool, 57-70, EMER PROF ZOOL, UNIV IOWA, 70- *Concurrent Pos:* Consult, Vet Admin. *Mem:* Am Soc Trop Med & Hyg; Am Soc Parasitol; Am Micros Soc. *Res:* Helminthology; Trichinella spiralis; trematodes of fish; poultry parasites. *Mailing Add:* RR 7 Box 73 Iowa City IA 52240

NOLIN, JANET M, PROLACTIN, IMMUNOCYTOCHEMISTRY. *Educ:* Col St Rose, BS. *Prof Exp:* RES BIOLOGIST, UNIV RICHMOND. *Mailing Add:* 911 Spottswood Rd Richmond VA 23173

NOLL, CLIFFORD RAYMOND, JR, b Providence, RI, Dec 20, 22; m 49; c 3. HISTORY & PHILOSOPHY OF SCIENCE. *Educ:* Brown Univ, AB, 44; Univ Ill, MS, 50; Univ Wis, PhD(biochem), 52. *Prof Exp:* Instr biochem, Univ Mich, 52-56; biochemist, USDA, 56-58; asst prof chem, Goucher Col, 58-62; assoc prof, Wellesley Col, 62-65; res biochemist, Hartford Hosp, 65-67; fac mem, Franconia Col, 67-68; assoc prof 68-74, PROF SCI, GREATER HARTFORD COMMUNITY COL, 74- *Concurrent Pos:* Carnegie intern gen educ, Brown Univ, 55-56. *Mem:* AAAS. *Res:* History, philosophy and sociology of science. *Mailing Add:* Dept of Sci Greater Hartford Community Col 61 Woodland St Hartford CT 06105

NOLL, HANS, b Basel, Switz, June 14, 24; nat US; m 49; c 4. MOLECULAR BIOLOGY. *Educ:* Univ Basel, PhD(biochem), 50. *Prof Exp:* Fel biol standardization, State Serum Inst, Copenhagen, Denmark, 50-51; asst tuberc, Pub Health Res Inst, City of New York, Inc, 51-53, assoc, 54-56; asst res prof microbiol, Sch Med, Univ Pittsburgh, 56-58, from asst prof to assoc prof, 59-64; PROF BIOL SCI, NORTHWESTERN UNIV, EVANSTON, 64- *Concurrent Pos:* Sr res fel, USPHS, 59-63, career investr, 64, mem molecular biol study sect, NIH, 66-68; vis prof, Univ Hawaii, 69; pres, Molecular Instruments Co, 70; vis prof Basel Inst Immunol, 71; consult, Europ Molecular Biol Lab, Heidelberg, 74; vis res prof, Lab Molecular Embryol, Naples, 75 & Univ Palermo, 78. *Honors & Awards:* Lifetime Endowed Career Professorship, Am Cancer Soc, 66. *Mem:* AAAS; Am Chem Soc; Am Soc Biol Chem; NY Acad Sci. *Res:* Biochemistry of viruses; role of membrane proteins in sea urchin embryogenesis; molecular biology of nucleic acids and protein synthesis. *Mailing Add:* Dept Genetics Univ Hawaii Burns Med Sch 1960 East-West Rd Honolulu HI 96822

NOLL, JOHN STEPHEN, b Hungary, Oct 1, 44; Can citizen. CEREAL CHEMISTRY, BIOCHEMISTRY. *Educ:* Univ Winnipeg, BSc, 69; Univ Man, PhD(cereal biochem), 77. *Prof Exp:* Cereal chemist, Plant Breeding Inst, Univ Sydney, 77-78; CEREAL PHYSIOLOGIST, CAN DEPT AGR, 78- *Mem:* Am Asn Cereal Chemists. *Res:* Screening wheat varieties that are resistant to preharvest sprouting; screening two- and six-row barley lines, in the various barley breeding programs in Canada, for malting barley quality. *Mailing Add:* Can Dept Agr 195 Dafoe Rd Winnipeg MB R3T 2M9 Can

NOLL, KENNETH E(UGENE), b Brantwood, Wis, Aug 20, 36; m 59; c 1. ENVIRONMENTAL ENGINEERING. *Educ:* Mich Technol Univ, BSCE, 59; Univ Wash, MSCE, 66, PhD(air resources eng), 69. *Prof Exp:* Air sanit engr, Dept Pub Health, State of Calif, 63-68, sr air sanit engr, Air Resources Bd, 69-70; from assoc prof to prof air resources, Univ Tenn, Knoxville, 70-74; PROF ENVIRON ENG, ILL INST TECHNOL, 75-, CHMN, DEPT ENVIRON ENG, 90- *Concurrent Pos:* Chmn, Knox County Air Pollution Control Bd, 72-75; dir, Environ Protection Agency Regional Air Pollution Training Ctr, 78- *Mem:* Air Pollution Control Asn; Am Soc Civil Engrs. *Res:* Atmospheric aerosol technology; air monitoring; design of air pollution control devices. *Mailing Add:* Dept Environ Eng Ill Inst Technol Chicago IL 60616

NOLL, LEO ALBERT, b Colorado Springs, Colo, Aug 5, 32; m 75. CALORIMETRY, ADSORPTION. *Educ:* St Benedict's Col, BS, 55; Univ Colo, MS, 62, PhD(phys chem), 76. *Prof Exp:* Instr chem & math, Abbey Sch, Colo, 55-71; instr physics & sci, Nunawading High Sch, Australia, 73-75; fel, Univ Nebr, 76; res chemist, US Dept Energy, Bartlesville Energy Technol Ctr, 76-83; SR CHEMIST, NAT INST PETROL & ENERGY RES, 83- *Mem:* Soc Petrol Engrs; Sigma Xi; Am Chem Soc. *Res:* Thermodynamics of enhanced oil recovery systems, especially measuring the enthalpy of adsorption of components of micellar formulations on surfaces. *Mailing Add:* 2090 S Osage Bartlesville OK 74003

NOLL, WALTER, b Berlin, Ger, Jan 7, 25; nat US; m 79; c 2. MATHEMATICS, CONTINUUM MECHANICS. *Educ:* Univ Paris, Lic es Sci, 50; Tech Univ Berlin, diplom, 51; Ind Univ, PhD(math), 54. *Prof Exp:* Instr mech, Tech Univ, Berlin, 51-55; instr math, Univ Southern Calif, 55-56; assoc prof, 56-60, PROF MATH, CARNEGIE-MELLON UNIV, 60 - *Concurrent Pos:* Vis prof, Johns Hopkins Univ, 62-63, Oxford Univ & Univ of Pisa, 84-85. *Mem:* Am Math Soc; Math Asn Am; Soc Natural Philos. *Res:* Foundations of mechanics and thermodynamics; differential geometry; relativity. *Mailing Add:* 308 Field Club Ridge Rd Pittsburgh PA 15238

NOLLE, ALFRED WILSON, b Columbia, Mo, July 28, 19; m 46. PHYSICS. *Educ:* Southwest Tex State Teachers Col, BA, 38; Univ Tex, MA, 39; Mass Inst Technol, PhD(physics), 47. *Prof Exp:* Tutor physics, Univ Tex, 40-41; spec res assoc, Underwater Sound Lab, Harvard Univ, 41-45; asst prof eng res, Ord Res Lab, Pa State Col, 45; res assoc physics, Mass Inst Technol, 45-47, mem staff, Div Indust Coop, 47-48; from asst prof to assoc prof physics, 48-57, PROF PHYSICS, UNIV TEX, AUSTIN, 57- *Mem:* Fel Am Phys Soc; fel Acoustical Soc Am. *Res:* Solid state physics; magnetic resonance and relaxation; paramagnetic impurities and imperfections; physics of musical tone sources; ultrasonics. *Mailing Add:* Dept of Physics Univ of Tex Austin TX 78712

NOLLEN, PAUL MARION, b Lafayette, Ind, Feb 24, 34; m 66; c 2. PARASITOLOGY. *Educ:* Carroll Col, BS, 56; Univ Wis, Madison, MS, 57; Univ Purdue, PhD(parasitol), 67. *Prof Exp:* Teacher high sch, Wis, 60-62; instr biol, Univ Purdue, 62-65; from asst prof to assoc prof, 67-74, PROF ZOOL & PARASITOL, WESTERN ILL UNIV, 74- *Concurrent Pos:* Vis prof, Univ Iowa, 69-70; fac lectr, Western Ill Univ, 75. *Honors & Awards:* Herrick Award, 67; Res Award, Sigma Xi, 74. *Mem:* AAAS; Am Soc Parasitol; Am Inst Biol Sci. *Res:* Reproductive activities of digenetic trematodes; uptake and incorporation of nutrients by digenetic trematodes; egg-shell chemistry of trematodes; host-finding behavior of miracidia. *Mailing Add:* Dept Biol Sci Western Ill Univ Macomb IL 61455

NOLLER, DAVID CONRAD, b Elma, NY, Oct 1, 23; m 47; c 2. ORGANIC CHEMISTRY. *Educ:* Univ Buffalo, BA, 49, MA, 54. *Prof Exp:* Asst chem, Univ Buffalo, 49-52; res chemist, Lucidol Div, Pennwalt Corp, 52-57, head analyst & control lab, 57-60, supvr tech serv lab, 60-64, group leader patents, lit & safety, 64-66, group leader patents & lit, 66-69, patent agent, 69-70, tech info specialist, 70-82; RETIRED. *Concurrent Pos:* Vol, Creative Educ Found & Prob Solving Insts, Buffalo, NY & Fla & Sarasota County Schs. *Mem:* Am Chem Soc; Sigma Xi. *Res:* Organic peroxides with emphasis on safety and governmental regulation. *Mailing Add:* 1040 Sylvan Dr Sarasota FL 33580

NOLLER, HARRY FRANCIS, JR, b Oakland, Calif, June 10. 39; m 64; c 2. BIOCHEMISTRY. *Educ:* Univ Calif, Berkeley, AB, 60; Univ Ore, PhD(chem), 65. *Prof Exp:* NIH fel, Lab Molecular Biol, Med Res Coun, Cambridge, Eng, 65-66 & Inst Molecular Biol, Geneva, Switz, 66-68; from asst prof to prof biol, 68-87, ROBERT L SINSHEIMER PROF MOLECULAR BIOL, UNIV CALIF, SANTA CRUZ, 87- *Honors & Awards:* Harvey Lectr, 89. *Res:* Protein and nucleic acid chemistry; structure and function of ribosomes. *Mailing Add:* Sinsheimer Labs Univ Calif Santa Cruz CA 95064

NOLTE, KENNETH GEORGE, b E Carondelet, Ill, July 20, 41; m 63; c 3. ENGINEERING MECHANICS, STRUCTURAL ENGINEERING. *Educ:* Univ Ill, BSc, 64; Brown Univ, ScM, 66, PhD(eng), 68. *Prof Exp:* Res assoc solid mech, Brown Univ, 67-68; sr res engr Marine Opers, Amoco Prod Co, Amoco Corp, 68-78, res assoc hydraulic fracturing res, 78-84; ENG ADV, DOWELL-SCHLUMBERGER, 86- *Concurrent Pos:* Prin, Nolte-Smith Inc, 84-86. *Mem:* Am Soc Civil Engrs. *Res:* Vessel design; hydraulic fracturing of petroleum formation; numerical modeling. *Mailing Add:* 6726 S 69th East Ave Tulsa OK 74133-1736

NOLTE, LOREN W, b Napoleon, Ohio, Dec 23, 33; m 57; c 4. OCEAN ACOUSTICS SIGNAL PROCESSING. *Educ:* Northwestern Univ, BSEE, 56; Univ Mich, MSE, 60, PhD(elec eng), 65. *Prof Exp:* Coop student, Cook Res Labs, Ill, 53-56, jr engr, 56-57, engr, 57-58; res asst, Willow Run Labs, Univ Mich, 58-59, asst res engr, teaching fel & lectr, Dept Elec Eng, 59-65, assoc res engr, 65-66, res fel, Univ, 65-66; from asst prof to assoc prof, Duke Univ, 66-69; vis assoc prof, Univ Colo, Boulder, 69-70; assoc prof elec & biomed eng, 71-72, PROF ELEC & BIOMED ENG, DUKE UNIV, 72- *Mem:* Inst Elec & Electronics Engrs; Acoust Soc Am. *Res:* Extension of signal detection theory to adaptive receivers, design and performance; application of statistical communications to radar, sonar and communications problems. *Mailing Add:* Dept Elec Eng Duke Univ Durham NC 27706

NOLTIMIER, HALLAN COSTELLO, b Los Angeles, Calif, Mar 19, 37; m 61; c 2. GEOPHYSICS. *Educ:* Calif Inst Technol, BS, 58; Univ Newcastle, Eng, PhD(geophysics), 65. *Prof Exp:* Lectr physics & geophysics, Sch Physics, Univ Newcastle, Eng, 66-68; asst prof geol, Sch Geol & Geophys, Univ Okla, 68-71; assoc prof geol, Dept Geol, Univ Houston, 71-72; assoc prof, 72-78, PROF GEOPHYSICS, DEPT GEOL & MINERAL, OHIO STATE UNIV, 78- *Concurrent Pos:* Consult, Chevron Oil Feld Res, 70, Chevron Overseas Petrol, 71, Humble Res Lab, 72, Digital Resources Inc, 78-79 & Towner Petrol, 81- *Mem:* Fel Royal Astron Soc; Am Geophys Union; Sigma Xi; Soc Explor Geophysicists; AAAS. *Res:* Paleomagnetism of Mesozoic intrusives along the eastern margin of North America; paleomagnetism of lower Paleozoic limestones; paleomagnetism of coal; paleomagnetism of recent lake sediments with particular interest in short geomagnetic excursions. *Mailing Add:* 60 Walhalla Rd Columbus OH 43202

NOLTING, HENRY FREDERICK, b Indianapolis, Ind, Aug 15, 16; m 37; c 3. LUBRICANTS & ADDITIVES, ENERGY. *Educ:* Purdue Univ, BS, 38, MS, 39. *Prof Exp:* Tech asst plant opers, Standoil Ind, Amoco, 42-43, mgr, tech serv, 43-46, mgr, plant opers, 46-59, mgr supply planning, Am Oil, Amoco, 59-63, mgr & dir, Int Spec Opers, Amoco, UK, 63-66, mgr ref opers, Amoco, 67-69, mgr maintenance & coust, 69-77; PRES, NOLTING ASSOC INC, 77- *Concurrent Pos:* Chmn, Am Inst Chem Engrs Nat Meeting, Chicago, 56 & Am Inst Chem Engrs Mkt Energy Planning, 57; dir, Am Inst Chem Engrs, 58-60; co-chmn, Chem Eng Int Mkt, London, 64. *Mem:* Am Inst Chem Engrs. *Res:* Separation of crude oils into lubricant fractures; energy programs to determine next needed activities in development. *Mailing Add:* 221 Greyhound Pass Carmel IN 46032

NOLTMANN, ERNST AUGUST, biochemistry, enzymology, for more information see previous edition

NOMANI, M ZAFAR, b Hyderabad, India, June 14, 35. DIETARY FIBER, ENERGY METABOLISM. *Educ:* Osmania Univ, Bs, 55, MS, 60; Rutgers Univ, PhD(nutrit), 73. *Prof Exp:* PROF NUTRIT, WVA UNIV, 83- *Mem:* Am Inst Nutrit; Am Dietetic Asn; Inst Food Technologists; Am Home Econ Asn; Fel Int Col Nutrit. *Res:* Nutritional assessment; cholesterol; protein; Ramadan fasting. *Mailing Add:* Nutrit Sect 704 Allen Bldg Wva Univ Box 6122 Morgantown WV 26506

NOMURA, KAWORU CARL, b Deer Lodge, Mont, Apr 1, 22; m 47; c 4. SEMICONDUCTORS, INTEGRATED CIRCUITS. *Educ:* Univ Minn, BS, 48, MS, 49, PhD(elec eng), 53. *Prof Exp:* Res physicist, Honeywell, Inc, 53-60, vpres & gen mgr, 80-84, corp sr vpres, 84-86; RETIRED. *Mem:* Am Phys Soc. *Res:* Crystalline, galvanomagnetic and optical properties of semiconductors. *Mailing Add:* 363 Discovery Rd Port Townsend WA 98368

NOMURA, MASAYASU, b Hyogo-ken, Japan, Apr 27, 27; m 57; c 2. MOLECULAR BIOLOGY, MICROBIOLOGY. *Educ:* Univ Tokyo, BS, 51, PhD(microbiol), 57. *Prof Exp:* Res assoc microbiol, Univ Ill, 57-59; res assoc biol sci, Purdue Univ, 59-60; asst prof inst protein res, Osaka Univ, 60-63; from assoc prof to prof genetics, Inst Enzyme Res, Univ Wis-Madison, 63-70, prof genetics & biochem, 70-84; PROF BIOCHEM, UNIV CALIF, IRVINE, 84- *Concurrent Pos:* NIH grants, 61-63 & 64-; NSF grants, 63- *Honors & Awards:* US Steel Found Award Molecular Biol, 71; Japan Acad Award, 72. *Mem:* Nat Acad Sci; AAAS; Am Soc Microbiol; Genetics Soc Am; Am Soc Biol Chemists; Am Acad Arts & Sci; hon mem Royal Danish Acad Arts & Sci; hon mem Royal Netherland Acad Arts and Sci. *Res:* Structure, function and assembly of ribosomes; biosynthesis of nucleic acids and protein. *Mailing Add:* Dept Biol Chem Sch Med Univ Calif Irvine CA 92717

NOMURA, SHIGEKO, microbiology, molecular biology, for more information see previous edition

NOMURA, YASUMASA, b Kumamoto City, Japan, Sept 25, 21; c 2. AERODYNAMICS, ASTRODYNAMICS. *Educ:* Tokyo Inst Tech, DEng, 62. *Prof Exp:* Teacher math, Yatsushiro Jr High Sch, Japan, 46-47 & Kumamoto High Sch, 48-52; asst prof, Col Coast Guard, 52-57; asst prof, Nat Defense Acad, 57-62, head, Dept Aerodyn Sci, 71-73 & 81-83, prof aerodyn, 62-86; RETIRED. *Mem:* Am Inst Aeronaut & Astronaut; Japan Soc Aerodyn & Space Sci. *Res:* Aerodynamics, especially boundary layer problems; air pollution problems. *Mailing Add:* Kotubo 12443 Bushi 249 Kanagawa Japan

NONDAHL, THOMAS ARTHUR, b Monroe, Wis, Jan 11, 51; m 77. ELECTRICAL ENGINEERING. *Educ:* Univ Wis-Madison, BS, 73, MS, 74, PhD(elec eng), 77. *Prof Exp:* ELEC ENGR, GEN ELEC CO, 78- *Mem:* Inst Elec & Electronics Engrs; Sigma Xi. *Res:* Rotating and linear electric machines. *Mailing Add:* 2373 N 84 St Milwaukee WI 53226-1901

NONNECKE, IB LIBNER, b Copenhagen, Denmark, Oct 1, 22; nat Can; m 45; c 3. HORTICULTURE. *Educ:* Univ Alta, BSc, 45, MSc, 50; Ore State Univ, PhD, 58. *Prof Exp:* Head hort sect, Res Sta, Can Dept Agr, 46-63; pres, Asgrow Seed Co Can, Ltd, 63-66, coordr crop improv, Asgrow Seed Co, 66-68; assoc prof hort, 68-70, chmn dept hort sci, 74-85, PROF HORT & CHIEF VEG DIV, UNIV GUELPH, 70-, PROF VEG CROPS, 85- *Concurrent Pos:* Consult, Ib Nonnecke Consults, Inc. *Mem:* Am Soc Hort Sci; Can Soc Hort Sci; Sigma Xi; Agr Inst Can. *Res:* Plant breeding and genetics; statistics; field plot technique. *Mailing Add:* Nonnecke Assocs 160 Woolwich St Guelph ON N1H 3V3 Can

NONNENMANN, UWE, b Stuttgart, Ger, June 11, 61. AUTOMATIC PROGRAMMING, EXPERT SYSTEMS. *Educ:* Technische Univ Berlin, Dipl-Informatikes, 86. *Prof Exp:* Consult, Danet Gurblt, CGE, 84-86; MEM TECH STAFF, AT&T BELL LABS, 86- *Mem:* Am Asn Artificial Intel; Inst Elec & Electronics Engrs. *Res:* Software productivity tools based on artificial intelligence and software engineering methodologies; new software development processes. *Mailing Add:* 16 Beech Ave Berkeley Heights NJ 07922

NONOYAMA, MEIHAN, b Tokyo, Japan, Feb 25, 38; m 69; c 1. VIRAL ONCOLOGY, MOLECULAR VIROLOGY. *Educ:* Univ Tokyo, BS, 61, MS, 63, PhD(molecular biol), 66. *Prof Exp:* Vis scientist, microbiol, Wistar Inst, Philadelphia, 68-70; res asst prof, microbiol, Univ NC, 70-73; asst prof-lab chief to assoc prof, microbiol, Rush-Presby St Luke's Med Ctr, Univ Ill Med Ctr, Chicago, 73-76; dir, molecular virol, Life Sci Inc, 76-81; PRES, VIROL, SHOWA UNIV RES INST, 81-; PROF, SHOWA UNIV MED SCH, 81- *Concurrent Pos:* Fullbright fel, 66; Prin investr, NIH grant, 73-; prof, Univ S Fla Med Sch, 83- *Mem:* Am Asn Cancer Res; Am Soc Microbiol; Soc Exp Biol & Med; Intl Asn Res EB Virus. *Res:* Molecular biology & oncogenic herpesvirus; therapeutic agent for HIV. *Mailing Add:* Tampa Bay Res Inst 10900 Roosevelt Blvd St Petersburg FL 33716

NOODEN, LARRY DONALD, b Oak Park, Ill, June 10, 36; m 63; c 2. PLANT PHYSIOLOGY, BIOCHEMISTRY. *Educ:* Univ Ill, BSc, 58; Univ Wis, MSc, 59; Harvard Univ, PhD(biol), 63. *Prof Exp:* NIH res fel, Univ Edinburgh, 64-65; from asst prof to assoc prof, 65-76, PROF BOT, UNIV MICH, ANN ARBOR, 76- *Concurrent Pos:* NIH special res fel, Calif Inst Technol, 71-72; vis scientist, Boyce Thompson Inst, 79; vis fel, Res Sch Biol Sci, Australian Nat Univ, Canberra, 81, 83, 89, 90; Fulbright sr res fel, 83. *Mem:* AAAS; Am Soc Plant Physiol; Bot Soc Am; Crop Sci Soc Am; Japanese Soc Plant Physiol; Am Soc Agron; Scand Soc Plant Physiol; Am Chem Soc. *Res:* Biochemical regulation of plant development, plant senescence; regulation of pod development and senescence in soybean; biochemistry of senescence and hormones (especially cytokinin) regulating senescence. *Mailing Add:* 2148 E Delhi Rd Ann Arbor MI 48109

NOOKER, EUGENE L(EROY), b Cheyenne, Wyo, Oct 10, 22; m 49; c 1. ORDNANCE, MATERIALS SCIENCE. *Educ:* Univ Colo, BS, 47, MS, 49. *Prof Exp:* Assoc res engr, Eng Exp Sta, Univ Colo, 47-51; sr engr, Appl Physics Lab, Johns Hopkins Univ, 51-58, group supvr & prin prof staff, 58-67; vpres & dir develop eng div, G W Galloway Co, 67-73; CONSULT, 73- *Mem:* Am Soc Mech Engrs; Am Phys Soc; Newcomen Soc; Am Defense Preparedness Asn. *Res:* Warhead technology and explosive reactions; resistance welding processes; metal powder compaction. *Mailing Add:* PO Box 1383 San Luis Obispo CA 93406

NOOLANDI, JAAN, b Estonia, Aug 23, 42; Can citizen; m 72; c 3. CONDENSED MATTER PHYSICS. *Educ:* Univ Toronto, BSc, 65, PhD(physics), 70. *Prof Exp:* Nat Res Coun Can fel, Univ Oxford, 70-71 & Univ Calif, San Diego, 71-72; mem tech staff solid state physics, Bell Labs, NJ, 72-74; mem, sci staff solid state physics, Xerox Res Ctr Can Ltd, 74-81, sect mgr, 81-86; SR RES FEL, XEROX CORP, 86- *Concurrent Pos:* Vis asst prof, Univ Toronto, 75-76; visitor, Ctr Rech Macromolecules, Strasbourg, France, 81; vis scientist, Max Planck Inst, Mainz, Frg, 86, 88; vis prof, Politecnico Milano, Italy, 89. *Mem:* Fel Am Phys Soc; Can Asn Physicists; Europ Phys Soc. *Res:* Electrical conduction in amorphous materials; physics and chemistry of liquid-solid interfaces; structural and electronic properties of high temperature superconductors; exciton-phonon interaction in molecular solids; multicomponent polymer systems; polymer dynamics; pulsed field gel electrophoresis. *Mailing Add:* Xerox Res Ctr Can Ltd 2660 Speakman Dr Mississauga ON L5K 2L1 Can

NOONAN, CHARLES D, b San Francisco, Calif, July 16, 28; m; c 5. RADIOLOGY. *Educ:* Univ Calif, Berkeley, AB, 50; Univ Calif, San Francisco, MD, 53. *Prof Exp:* Resident, Cincinnati Gen Hosp, 56-59; from instr to assoc prof, 59-73, PROF RADIOL, SCH MED, UNIV CALIF, SAN FRANCISCO, 73- *Mailing Add:* Vet Admin Hosp 4150 Clement St San Francisco CA 94121

NOONAN, JACQUELINE ANNE, b Burlington, Vt, Oct 28, 28. PEDIATRIC CARDIOLOGY. *Educ:* Albertus Magnus Col, BA, 50; Univ Vt, MD, 54. *Prof Exp:* Asst prof pediat, Col Med & pediat cardiologist, Hosp, Univ Iowa, 59-61; assoc prof, 61-69, PROF PEDIAT, COL MED, UNIV KY, 69-, CHMN DEPT, 74-, PEDIAT CARDIOLOGIST, UNIV HOSP, 61- *Mem:* Am Pediat Soc; Am Acad Pediat; AMA; Am Col Cardiol; Soc Pediat Res. *Mailing Add:* Dept Pediat Univ Ky Col Med Lexington KY 40506

NOONAN, JAMES WARING, b Fall River, Mass, Dec 21, 44. DATA PROCESSING. *Educ:* Providence Col, AB, 66; Univ Md, College Park, MA, 69, PhD(math), 71. *Prof Exp:* Nat Res Coun res assoc, US Naval Res Lab, 70-71; asst prof math, Col of the Holy Cross, 71-78, assoc prof, 78-81, asst dean, 78-81; mgr consult, Data Gen Corp, 80-90; MGT INFO SYSTS, DELL COMPUTER, AUSTIN, TEX, 90- *Mem:* Sigma Xi. *Res:* Operations research; geometric function theory. *Mailing Add:* Dell Computer 11209 Metric Blvd Bldg H Austin TX 78758

NOONAN, JOHN ROBERT, b Springfield, Mo, Sept 26, 46; m 74. THIN FILM TECHNOLOGY, ELECTRICAL ENGINEERING. *Educ:* Wash Univ, BS, 68, MS, 70; Univ Ill, Urbana, PhD(elec eng), 74. *Prof Exp:* Engr comput design, Comput Sci Lab, Wash Univ, 68; NSF trainee, Univ Ill, 69-72, res asst semiconductor physics, 70-74; RES STAFF MEM SURFACE PHYSICS, SOLID STATE DIV, OAK RIDGE NAT LAB, 74- *Concurrent Pos:* Mem, Publ Comt & ed bd, J Vacuum Sci & Technol, Am Vacuum Soc, 85-88, Nat Symp Prog Comt, 88 & bd dirs, 88-90; mem, Am Inst Physics Publ Policy Comt, 84-87, Am Inst of Physics Subcomt on Jour, 86-91. *Mem:* Sigma Xi; Inst Elec & Electronic Engrs; Am Vacuum Soc; Am Phys Soc; AAAS. *Res:* Surface physics; polymer and ceramic thin films; optical emission from semiconductors; ultra high vacuum technology. *Mailing Add:* Solid State Div PO Box 2008 Oak Ridge TN 37831-6024

NOONAN, KENNETH DANIEL, b New York, NY, Jan 27, 48; m 72. BIOCHEMISTRY, CELL BIOLOGY. *Educ:* St Joseph's Col, Pa, BS, 69; Princeton Univ, PhD(biol), 72. *Prof Exp:* Jane Coffin Childs fel, Biocenter, Basel, Switz, 72-73; asst prof biochem, J Hillis Miller Health Ctr, Univ Fla, 78-81, 73-78, assoc prof, 78-81; PROF, UNIV MD, BALTIMORE CAMPUS, 81- *Concurrent Pos:* Dir mkt & prod develop, Bethesda Res Labs, Inc, 81-; dir, Immunol Res & Develop, BBL Microbiol Syst. *Mem:* AAAS; Am Soc Cell Biol. *Res:* Role of the plasma membrane in normal and transformed cell growth; mechanisms of nuclear envelope break down and resynthesis. *Mailing Add:* 2608 N Calvert St Baltimore MD 21030

NOONAN, NORINE ELIZABETH, b Philadelphia, Pa, Oct 5, 48. RESOURCE MANAGEMENT. *Educ:* Princeton Univ, PhD(cell biol), 76. *Prof Exp:* From asst prof to assoc prof, Univ Fla, 76-81, assoc prof, Georgetown Univ Sch Med, 81-83; prog analyst, Energy Sci Div, 83-87, BR CHIEF, SCI & SPACE PROGS, OFF MGT & BUDGET, 87- *Concurrent Pos:* Cong Sci fel, Am Chem Soc, 82-83. *Mem:* AAAS; Am Soc Cell Biol. *Res:* cell biology (nuclear and plasma membranes). *Mailing Add:* Energy Sci Div Off Mgt & Budget 725 17th St NW Washington DC 20503

NOONAN, THOMAS ROBERT, b 1912; m 40. BIOLOGICAL EFFECTS OF RADIATION. *Educ:* Univ Buffalo, MD, 39. *Prof Exp:* Sci asst to dir, Comp Animal Res Lab, 74-77; RETIRED. *Mailing Add:* 1030 W Outer Dr Oakridge TN 37830-8609

NOONAN, THOMAS WYATT, b Glendale, Calif, July 16, 33; m 66; c 3. ASTRONOMY, PHYSICS. *Educ:* Calif Inst Technol, BS, 55, PhD(physics), 61. *Prof Exp:* Physicist, Smithsonian Astrophys Observ, 61-62; vis asst prof physics, Univ NC, Chapel Hill, 62-65, asst prof, 65-68; assoc prof, 68-71, PROF PHYSICS, STATE UNIV NY COL BROCKPORT, 71- *Concurrent Pos:* State Univ NY Res Found fel, Hale Observ, 70. *Mem:* Am Astron Soc; Royal Astron Soc; Int Astron Union; Int Soc Gen Relativity & Gravitation. *Res:* Relativity; cosmology. *Mailing Add:* Dept of Physics State Univ of NY Col Brockport NY 14420

NOOR, AHMED KHAIRY, b Cairo, Egypt, Aug 11, 38; US citizen; m 66; c 1. STRUCTURAL MECHANICS & ENGINEERING. *Educ:* Cairo Univ, BS, 58; Univ Ill, Urbana-Champaign, MS, 61, PhD(struct mech), 63. *Prof Exp:* Asst struct mech, Cairo Univ, 58-59; asst prof aeronaut, Stanford Univ, 63-64; sr lectr struct mech, Cairo Univ, 64-67; vis sr lectr, Baghdad Univ, 67-68; sr lectr, Univ New South Wales, 68-71; sr res assoc, 71-72, prof Eng & Appl Sci, NASA Langley Res Ctr, George Washington Univ, 72-90; FURMAN W PERRY PROF AEROSPACE STRUCT & APPL MECH & DIR CTR COMPUTATIONAL STRUCT TECHNOL, NASA LANGLEY RES CTR, UNIV VA, HAMPTON, 90- *Concurrent Pos:* Mem comt large space systs, Nat Acad Sci, 78-, comput mech, 81-, aeronaut technol in year 2000, 83- *Mem:* Am Inst Aeronaut & Astronaut; fel Am Soc Mech Engrs; Am Soc Civil Engrs; fel Am Acad Mech; Int Asn Computational Mech; US Asn Computer Mech. *Res:* Computational mechanics, new computing systems; improved numerical techniques for nonlinear and dynamic problems; fibrous composite structures; shell structures; large space structures. *Mailing Add:* Mail Stop 210 NASA Langley Res Ctr Hampton VA 23665

NOORDERGRAAF, ABRAHAM, b Utrecht, Neth, Aug 7, 29; m 56; c 4. BIOPHYSICS, BIOENGINEERING. *Educ:* Univ Utrecht, BSc, 53, MSc, 55, PhD(biophysics), 56. *Hon Degrees:* MA, Univ Pa, 71. *Prof Exp:* Asst exp physics, Univ Utrecht, 52-53, res asst med physics, 53-55, res fel med physics, 56-58, sr res fel, 59-65; assoc prof elec eng, 64-70, chmn dept, 73-76, chmn, Grad Group Bioeng, 73-76, PROF VET MED, UNIV PA, 76-, PROF BIOMED ENG, 70-, PROF DUTCH CULT, 83-, PROF MED, ANESTHESIA, 90- *Concurrent Pos:* Vis fel ther res, Univ Pa, 57-58, NIH res grants, 66, 68, 70, 72, 74, 76, 78, 81, 84 & 88, Surgeon Gen, 74; mem spec study sect, NIH, 65-, mem cardiovasc study sect, 85-89; vis prof appl physics, Delft Univ Technol, 70-71; vis prof cardiol, Erasmus Univ Med Sch, Rotterdam, 70-71; mem coun circulation, Am Heart Asn, 72-; vis prof, Univ Miami, 70-79; consult, NATO Sci Affairs Div, 73- *Honors & Awards:* Herman C Burger Award, 78; S Reid Warren Award, 86; Christian & Mary Lindback Award, 88. *Mem:* Fel Inst Elec & Electronic Engrs; fel NY Acad Sci; fel AAAS; fel Explorers Club; fel Am Col Cardiol; fel Royal Soc Med, Eng; hon mem Cardiovasc Syst Dynamics Soc. *Res:* Mammalian cardiovascular system analysis; operation of the heart as a pump; dynamics of the microcirculation. *Mailing Add:* 620 Haydock Ln Haverford PA 19041

NORA, AUDREY HART, b Picayune, Miss, Dec 5, 36; m 68; c 2. PEDIATRICS, PUBLIC HEALTH. *Educ:* Univ Miss, BS, 58, MD, 61; Am Bd Pediat, cert, 68, cert hemat & oncol, 74; Univ Calif, MPH, 78. *Prof Exp:* Fel pediat hemat & oncol, Baylor Col Med, 64-66, instr pediat, 66-70, asst prof, 70-71; dir genetics & birth defects, Denver Children's Hosp, 71-78; asst clin prof, 77-78, ASSOC CLIN PROF PEDIAT, HEALTH SCI CTR, UNIV

COLO, DENVER, 78- *Concurrent Pos:* Assoc hematologist, Tex Children's Hosp, 66-71; consult, NIH Adv Comt Arteriosclerosis & Hypertension, 75-77, maternal & child health, Region VIII, Dept Health & Human Serv, 78-83; asst surg gen, Region VIII, Regional Health Admin, 83- *Honors & Awards:* Virginia Apgar Mem Award. *Mem:* Am Soc Human Genetics; Genetics Soc Am; Am Acad Pediat; Am Pub Health Asn. *Res:* Arteriosclerosis in the pediatric age group; teratology of medications, congenital malformations relating to medications taken during first trimester. *Mailing Add:* ASG & Regnal Health Adminr 1961 Stout St Denver CO 80294

NORA, JAMES JACKSON, b Chicago, Ill, June 26, 28; m 66; c 5. CARDIOLOGY, GENETICS. *Educ:* Harvard Univ, AB, 50; Yale Univ, MD, 54; Am Bd Pediat, dipl & cert; Univ Calif, Berkeley, MPH, 78; Am Bd Pediat & Cardiol, dipl & cert; Am Bd Med Genetics, cert, 82. *Prof Exp:* Am Heart Asn res fel cardiol, Univ Wis, 62-64, from instr to asst prof pediat, 62-65; from asst prof to assoc prof, Baylor Univ, 65-71, head div human genetics & dir birth defects ctr, 67-71; assoc prof pediat, 71-74, dir pediat cardiol, 71-78, PROF PEDIAT & DIR PREVENTIVE CARDIOL, UNIV COLO MED CTR, DENVER, 78-, PROF GENETICS & PREV MED, 79-, DIR GENETICS, ROSE MED CTR, 80- *Concurrent Pos:* NIH spec fel genetics, McGill Univ, 64-65; assoc dir cardiol, Tex Children's Hosp, 65-71, chief genetics serv, 67-71. *Honors & Awards:* Virginia Apgar Mem Award. *Mem:* Am Soc Human Genetics; fel Am Acad Pediat; Am Pediat Soc; Soc Pediat Res; Teratology Soc; Am Col Cardiol; Transplantation Soc. *Res:* Etiology of cardiovascular diseases. *Mailing Add:* Dept Biochem Biophys & Genetics A-007 Univ Colo Med Ctr Denver CO 80262

NORBACK, DIANE HAGEMEN, b Mar 22, 46; m; c 2. PATHOLOGY, HEMATOLOGY. *Educ:* Univ Wis, PhD(path), 73, MD, 74. *Prof Exp:* ASSOC PROF PATH & LAB MED, UNIV WIS, 81- *Res:* Hematopoietic skin cells. *Mailing Add:* Univ Wis Hosp 2551 Arboretum Dr Madison WI 53713-1007

NORBECK, EDWIN, b Seattle, Wash, June 10, 30; m 56; c 4. HEAVY-ION NUCLEAR PHYSICS. *Educ:* Reed Col, BA, 52; Univ Chicago, MS & PhD(physics), 56. *Prof Exp:* Res assoc physics, Univ Chicago, 56-57 & Univ Minn, 57-60; from asst prof to assoc prof, 60-67, PROF PHYSICS, UNIV IOWA, 67- *Concurrent Pos:* Fulbright res grant, France, 89. *Honors & Awards:* Comput Appl Nuclear & Plasma Physics Award, Inst Elec & Electronics Engrs, 87. *Mem:* Fel Am Phys Soc; Inst Elec & Electronic Engrs. *Res:* Experimental studies of nuclear reactions between heavy nuclei; use of MeV ion beams for analysis of thin films. *Mailing Add:* Dept of Physics & Astron Univ of Iowa Iowa City IA 52242

NORBERG, ARTHUR LAWRENCE, b Providence, RI, Apr 13, 38. TECHNOLOGY, RESEARCH ADMINISTRATION. *Educ:* Providence Col, BS, 59; Univ Vt, MS, 61; Univ Wis-Madison, PhD(hist sci), 74. *Prof Exp:* Instr physics, St Michael's Col, Vt, 61-63, asst prof, 64-68; assoc scientist, Westinghouse Elec Corp, 63-64; instr physics, Univ Wis, Whitewater, 68-71; res historian, Bancroft Libr, Univ Calif, Berkeley, 73-79; prog mgr ethics & values, NSF, 79-81; ASSOC PROF HIST TECHNOL, UNIV MINN, 81-, DIR, CHARLES BABBAGE INST, 81-; EXEC DIR, CHARLES BABBAGE FOUND, 83- *Concurrent Pos:* Mem adv coun, NASA, 87- *Mem:* Hist Sci Soc (treas, 75-80); Soc Hist Technol; AAAS; Sigma Xi. *Res:* Interplay between scientific development and technological innovation invarious settings, industry, government, academia, for the nineteenth and twentieth centuries; emphasizing topics in the later period-radio, television, computing and aerospace. *Mailing Add:* Charles Babbage Inst 103 Walter Libr Univ Minn 117 Pleasant St SE Minneapolis MN 55455

NORBERG, RICHARD EDWIN, b Newark, NJ, Dec 28, 22; m 47, 78; c 3. PHYSICS. *Educ:* DePauw Univ, BA, 43; Univ Ill, MA, 47, PhD(physics), 51. *Prof Exp:* Res assoc physics & control systs lab, Univ Ill, 51-53, asst prof, 53-54; vis lectr physics, 54-56, assoc prof, 56-58, PROF PHYSICS, WASH UNIV, 58-, CHMN DEPT, 62- *Concurrent Pos:* Sloan fel, 55-59. *Mem:* Fel Am Phys Soc; Mat Res Soc; Int Soc Magnetic Resonance. *Res:* Nuclear and electron spin resonance; solid state and low temperature physics; high pressures; hydrogen; rare gases; amorphous semiconductors and metallic alloys. *Mailing Add:* Dept of Physics Wash Univ St Louis MO 63130

NORBY, RODNEY DALE, b Carrington, NDak, Jan 4, 45; m 68; c 1. GEOLOGY. *Educ:* Univ NDak, BS, 67; Ariz State Univ, MS, 71; Univ Ill, PhD(geol), 76. *Prof Exp:* Res asst geol, 72-76, from asst geologist to assoc geologist, 76-85, GEOLOGIST, ILL STATE GEOL SURV, 86- *Concurrent Pos:* Fulbright fel, Australia, 67-68. *Mem:* Geol Soc Am; Soc Econ Paleont & Mineral; Paleont Soc; Paleont Asn. *Res:* Carboniferous stratigraphy; conodont biostratigraphy and taxonomy; coastal sedimentation and processes; micropaleontology. *Mailing Add:* Ill State Geol Surv 615 E Peabody Dr Champaign IL 61820-6964

NORCIA, LEONARD NICHOLAS, b Mountain Iron, Minn, Jan 1, 16; m 50; c 1. BIOCHEMISTRY, PHYSIOLOGY. *Educ:* Univ Minn, BChem, 46, PhD(physiol chem), 52. *Prof Exp:* Res fel biochem, Hormel Inst, Univ Minn, 52-55; from asst prof to assoc prof res biochem, Sch Med, Univ Okla, 56-60; asst prof, 60-69, assoc prof, 69-79, EMER ASSOC PROF BIOCHEM, SCH MED, TEMPLE UNIV, 79- *Concurrent Pos:* Biochemist, Okla Med Res Found, 55-60; mem coun arteriosclerosis, Am Heart Asn. *Mem:* Am Chem Soc; Am Oil Chemists Soc. *Res:* Metabolism of lipids. *Mailing Add:* Erin Arms 407 Fourth Ave Apt C Coralville IA 52241-2466

NORCROSS, BRUCE EDWARD, b Newport, Vt, Mar 14, 35; m 56; c 1. PHYSICAL ORGANIC CHEMISTRY. *Educ:* Univ Vt, BA, 56, MS, 57; Ohio State Univ, PhD(chem), 60. *Prof Exp:* Res assoc chem, Harvard Univ, 60-62; asst prof chem, State Univ NY Binghamton, 62-66, assoc dean, Harpur Col, 66-67, asst vpres acad affairs, 78-80, chmn dept chem, 84-90, ASSOC PROF CHEM, STATE UNIV NY BINGHAMTON, 66- *Mem:* AAAS; Am Chem Soc. *Res:* Organic reaction mechanisms and synthetic organic chemistry. *Mailing Add:* Dept Chem State Univ NY Binghamton NY 13901

NORCROSS, DAVID WARREN, b Cincinnati, Ohio, July 18, 41; m 67; c 2. ATOMIC PHYSICS. *Educ:* Harvard Col, AB, 63; Univ Ill, MS, 65; Univ Col, Univ London, PhD(physics), 70. *Prof Exp:* Res physicist, Sperry Rand Res Ctr, 65-67; res assoc physics, Univ Colo, 70-74; PHYSICIST, QUANTUM PHYSICS DIV, NAT INST STANDARDS & TECHNOL, 74-, CHIEF, 90- *Concurrent Pos:* Lectr, Dept Physics, Univ Colo, 74- *Honors & Awards:* Bronze Medal, Dept Com, 82. *Mem:* Fel Am Phys Soc; Fedn Am Scientists; AAAS. *Res:* Theory of electron-atom and electron-molecule interactions, atomic and molecular structure and radiative properties. *Mailing Add:* Joint Inst for Lab Astrophys Univ of Colo Boulder CO 80309-0440

NORCROSS, MARVIN AUGUSTUS, b Tansboro, NJ, Feb 8, 31; m 56; c 2. RESEARCH ADMINISTRATION, VETERINARY PATHOLOGY. *Educ:* Univ Pa, VMD, 59, PhD(path), 66. *Prof Exp:* Gen vet pract, Md, 59-60; vet, Animal Husb Res Div, Agr Res Serv, USDA, 60-62; res fel path, USPHS training grant cancer res, Sch Vet Med, Univ Pa, 62-66; asst vet pathologist, 66-69, assoc dir clin res, 69-72, sr dir animal sci res, Merck, Sharp & Dohme Res Lab, Rahway, NJ, 72-75; dir, Div Vet Med Res, Food & Drug Admin, Beltsville, 75-78, assoc dir res, 78-82, assoc dir Human Food Safety, Rockville, 82-84, dir, Off New Animal Drug Eval, Ctr Vet Med, 84-87; asst dep adminr, 87-88, DEP ADMINR, SCI & TECHNOL PROG, FOOD SAFETY & INSPECTION SERV, USDA, WASH, 88- *Concurrent Pos:* Mem animal health sci res adv bd, USDA, 78-84; adj prof, Va-Md Regional Col Vet Med, Blacksburg, Va, 80-85; consult, coun biologic & therapeutic agents, Am Vet Med Asn, 79-82; US Deleg Codex Comt, Residues Vet Drugs Foods, 88- *Honors & Awards:* Meritorous Presidential Rank Award, 89. *Mem:* AAAS; Am Asn Avian Path; Am Vet Med Asn; NY Acad Sci; Soc Toxicol Path; Sigma Xi. *Res:* Chemical and viral carcinogenesis; animal health drugs and biologics; public health and food safety. *Mailing Add:* 14304 Brickhowe Ct Germantown MD 20874

NORCROSS, NEIL LINWOOD, b Derry, NH, July 18, 28; m 53; c 2. IMMUNOLOGY. *Educ:* Univ Miami, AB, 50; Univ Mass, MS, 55, PhD(bact), 58. *Prof Exp:* Immunologist, Plum Island Animal & Dis Lab, Agr Res Serv, USDA, 57-60; from asst prof to assoc prof immunochem, 60-69, PROF IMMUNOCHEM, NY STATE UNIV COL VET MED, CORNELL UNIV, 69-, SECY OF COL, 73- *Concurrent Pos:* Mem, Nat Mastitis Coun. *Mem:* Am Acad Microbiol; Am Soc Microbiol; Sigma Xi. *Res:* Immunology of foot and mouth disease virus; production of foot and mouth vaccines; mastitis; streptococci of bovine origin; immunology of equine infectious anemia; autoimmune experimental myasthenia gravis. *Mailing Add:* NY State Col Vet Med Cornell Univ Ithaca NY 14850

NORD, GORDON LUDWIG, JR, b Cincinnati, Ohio, Apr 3, 42; div; c 1. GEOLOGY. *Educ:* Univ Wis, BS, 65; Univ Idaho, MS, 67; Univ Calif, Berkeley, PhD(geol), 73. *Prof Exp:* Res assoc geol, Case Western Reserve Univ, 71-74; GEOLOGIST, US GEOL SURV, 74- *Mem:* Geol Soc Am; Am Mineral Soc; Am Geophys Union. *Res:* The characterization and mechanisms of precipitation reactions, symmetry transitions and defect structures in natural and synthetic minerals and mineral aggregates as a function of geological history. *Mailing Add:* 959 Nat Ctr US Geol Surv Reston VA 22092

NORD, JOHN C, b Joliet, Ill, Mar 19, 38; m 68; c 3. FOREST ENTOMOLOGY. *Educ:* Univ Mich, BS, 60, MF, 62, PhD(forestry), 68. *Prof Exp:* PRIN RES ENTOMOLOGIST, SOUTHEAST FOREST EXP STA, US FOREST SERV, 64- *Mem:* Entom Soc Am; Georgia Entom Soc; Michigan Entom Soc. *Res:* Biology and ecology of forest insects; chemical control. *Mailing Add:* Forest Sci Lab Carlton St Athens GA 30602

NORDAN, HAROLD CECIL, b Vancouver, BC, Jan 21, 25; m 52; c 2. VERTEBRATE BIOLOGY. *Educ:* Univ BC, BA, 48, BSA, 50, MSA, 54; Ore State Univ, PhD(microbiol, biochem), 59. *Prof Exp:* Fel zool, Univ BC, 59-61, res assoc & sessional lectr, 62-65, from asst prof to assoc prof, 65-88; RETIRED. *Res:* Vertebrate physiology, especially bioenergetics and growth of wild species of ungulates. *Mailing Add:* 19240 59 A Ave Surrey BC V3F 7S8 Can

NORDBY, GENE M, b Anoka, Minn, May 7, 26; m 49, 75; c 3. CIVIL ENGINEERING. *Educ:* Ore State Univ, BS, 48; Univ Minn, MS, 49, PhD, 55. *Prof Exp:* Asst, St Anthony Falls Hydraul Res Lab, Minn, 48, asst civil eng, 49-50; struct designer, Pfieffer & Shultz, 50; asst prof civil eng, Univ Colo, 50-56; prog dir eng sci, NSF, 56-58; prof civil eng, head dept & dir, Transp & Traffic Inst, Univ Ariz, 59-62; prof aerospace, mech & civil eng & dean, Col Eng, Univ Okla, 62-70, vpres finance & admin, 70-77, prof civil, mech & aerospace eng, 76-77; prof civil eng & vpres bus & finance, Ga Inst Technol, 77-80; chancellor, Univ Colo, Denver, 80-85, prof civil eng, 85-86; head agr biosysts eng, 86-91, PROF AGR BIOSYSTS ENG, UNIV ARIZ, 91- *Concurrent Pos:* Assoc prof & res engr joint hwy res proj, Purdue Univ, 56-57; lectr, George Washington Univ, 56-58; consult ed, Macmillan Co, 62-70; mem eng educ & acreditation comt & chmn coord comt, Engrs Coun Prof Develop, 65-71, treas, 80-82, vpres, 83-84, pres-elect, 84-85, pres, 85-,; mem, Accreditation Bd Eng & Technol; pres, Tetracon Assocs, Inc, 69-; consult, NCent Asn Schls & Cols, 76-77 & 81-; mem bd dirs, Higher Educ & the Handicapped, Am Coun Educ, 80-83. *Honors & Awards:* Edmund Friedman Award, Am Soc Civil Engrs, 80. *Mem:* Fel Am Soc Civil Engrs; Am Soc Eng Educ; Nat Soc Prof Engrs. *Res:* Structural adhesives; composite materials; reinforced concrete; structural mechanics; higher education administration; accreditation of higher education programs; research administration and finance; facilities planning. *Mailing Add:* 4418 E Dianthus Pl Tucson AZ 85712

NORDBY, GORDON LEE, b Moscow, Idaho, Oct 3, 29; m 53; c 4. BIOLOGICAL SYSTEM SIMULATION. *Educ:* Stanford Univ, BS, 51, MS, 54, PhD(chem philos), 58. *Prof Exp:* Instr biochem, Stanford Univ, 58-59; res assoc, Harvard Med Sch, 60-63; asst prof biol chem & assoc res biophysicist, 63-68, ASSOC PROF BIOL CHEM, 68- *Concurrent Pos:* Am Cancer Soc res fel, 59-61. *Mem:* AAAS; Am Chem Soc. *Res:* Physical biochemistry; computer modeling; experimental design. *Mailing Add:* 2988 Burlington Ann Arbor MI 48105

NORDBY, HAROLD EDWIN, b New England, NDak, Nov 3, 31; m 58; c 2. ORGANIC CHEMISTRY, BIOCHEMISTRY. *Educ:* Concordia Col, Moorhead, Minn, BA, 53; Univ Ariz, MS, 59, PhD(biochem), 63. *Prof Exp:* Jr scientist lipids, Hormel Inst, Univ Minn, 53-56; res assoc dept agr biochem, Univ Ariz, 59-63; res chemist citrus, fruit & veg lab, 63-86, RES CHEMIST, CITRUS, HORT LAB, AGR RES SERV, USDA, 86- *Mem:* Am Oil Chem Soc; Sigma Xi. *Res:* Isolation and characterization of bitter products in grapefruit; synthesis and isolation of cycloporpenes; lipids and natural products, cold hardening, chilling injury. *Mailing Add:* 804 W Lake Jessie Dr NW Winter Haven FL 33880

NORDELL, WILLIAM JAMES, b Chicago, Ill, June 18, 30; m 55; c 2. STRUCTURAL ENGINEERING, ENGINEERING MECHANICS. *Educ:* Univ Ill, BS, 58, MS, 59, PhD(civil eng), 63. *Prof Exp:* Struct res engr, 63-69, DIR OCEAN STRUCTURES DIV, US NAVAL CIVIL ENG LAB, 69- *Mem:* Sigma Xi; Am Soc Civil Engrs. *Res:* Behavior of materials; structural dynamics; reinforced concrete; limit design; structural mechanics; ocean structures. *Mailing Add:* 373 Maryville Ave Ventura CA 93003

NORDEN, ALLAN JAMES, b Perkins, Mich, Nov 27, 24; m 46; c 6. PLANT BREEDING, GENETICS. *Educ:* Mich State Univ, BS, 49, MS, 50; Iowa State Univ, PhD(plant breeding), 58. *Prof Exp:* Asst farm crops, Mich State Univ, 49-50, agr exten agent, 50-52, res instr crops & soils, 52-55; asst farm crops, Iowa State Univ, 55-58; from asst prof to prof, 71-87, EMER PROF AGRON, UNIV FLA, 87- *Concurrent Pos:* Consult, Panama & Columbia, 67, Nicaragua, 68, Ghana, 69, Guyana, 69-71, Venezuela, 72, Senegal, 73, SAfrica, Zimbabwe, Malawi, 81, USAID Philippines, Thailand, Sengal, Niger, 89, Argentina, Nat Peanut Found, 90- *Honors & Awards:* Golden Peanut Res Award, Nat Peanut Coun, 73. *Mem:* Fel Am Soc Agron; Crop Sci Soc Am; Sigma Xi; Am Genetics Asn; fel Am Peanut Res & Educ Soc (pres, 78-79). *Res:* Plant breeding, genetics and physiological studies with peanuts. *Mailing Add:* Rte 2 Box 1651 High Springs FL 32643

NORDEN, CARROLL RAYMOND, b Escanaba, Mich, May 20, 23; m 51; c 4. ZOOLOGY. *Educ:* Northern Mich Univ, AB, 48; Univ Mich, MS, 51, PhD(zool), 58. *Prof Exp:* Asst prof biol, Univ Southwestern La, 57-63; from asst prof to assoc prof zool, 63-71, chmn dept, 69-74 & 76-78, PROF ZOOL, UNIV WIS-MILWAUKEE, 71- *Concurrent Pos:* Fishery biologist, State Dept Conserv, Mich, 53, 54 & 56; fishery biologist, US Fish & Wildlife Serv, 58 & 59. *Mem:* Am Fisheries Soc; Am Soc Ichthyologists & Herpetologists; Sigma Xi. *Res:* Ichthyology; taxonomy of fishes; fishery biology. *Mailing Add:* Dept of Zool Univ of Wis Milwaukee WI 53201

NORDEN, JEANETTE JEAN, b Ovid, Colo, Apr 15, 48. NEUROBIOLOGY. *Educ:* Univ Calif, Los Angeles, BA, 70; Vanderbilt Univ, PhD(psychol), 75. *Prof Exp:* NIH fel neuroanat, Duke Univ, 75-77; RES ASSOC, SCH MED, VANDERBILT UNIV, 77- *Concurrent Pos:* Vis res assoc histochem, Dent Res Ctr, Univ NC, Chapel Hill, 76. *Mem:* Soc Neurosci; Asn Res Vision & Ophthal. *Res:* Quantitative electron microscopy of synaptic connections of the retino-tectal projection in toads and goldfish; freeze fracture of developing retino-tectal synapses. *Mailing Add:* Dept of Cell Biol Vanderbilt Sch of Med Nashville TN 37232

NORDENG, STEPHAN C, b Chippewa Falls, Wis, May 24, 23; m 53; c 6. STRATIGRAPHY. *Educ:* Univ Wis, BS, 49, MS, 51, PhD(geol), 54. *Prof Exp:* Field geologist, Tex Petrol Co, 54-56; from assoc prof to prof geol, Mich Technol Univ, 57-88; RETIRED. *Concurrent Pos:* Consult, Copper Range Co, Calumet & Hecla. *Mem:* Geol Soc Am; Paleont Soc. *Res:* Applications of statistics and computers in geology. *Mailing Add:* 507 Dodge St Houghton MI 49931

NORDGREN, RONALD P, b Munisine, Mich, Apr 3, 36. ENGINEERING. *Educ:* Univ Mich, BSE, 57, MSE, 58; Univ Calif, Berkeley, PhD, 62. *Prof Exp:* Engr & res assoc, Shell Develop Corp, 63-90; DEPT CHMN & HERMAN GEORGE R BROWN PROF CIVIL ENG, RICE UNIV, 89- *Mem:* Nat Acad Eng; fel Am Soc Mech Engrs; Am Soc Civil Engrs. *Mailing Add:* 14935 Broadgreen Dr Houston TX 77090

NORDIN, ALBERT ANDREW, b McKeesport, Pa, Sept 7, 34; m 56; c 3. IMMUNOBIOLOGY. *Educ:* Univ Pittsburgh, BS, 56, MS, 59, PhD(microbiol), 62. *Prof Exp:* From instr to asst prof, Dept Microbiol, Univ Pittsburgh, 63-66; from asst to assoc prof microbiol, Univ Notre Dame, 66-72; RES CHEMIST, NAT INST AGING, NIH, 72- *Concurrent Pos:* Vis scientist, Swiss Inst Exp Cancer Res, 69-70; mem adv panel regulatory biol prog, NSF, 73-75; vis prof, Basel Inst Immunol, 75 & 80-81. *Mem:* Am Asn Immunol. *Res:* Transmembrane signalling mechanisms induced by mitogens/antigens and/or lymphokines that result in the activation, differentiation and proliferation of lymphoid cells and their involvement in age-associated immunodeficiencies. *Mailing Add:* Geront Res Ctr F S Key Med Ctr Baltimore MD 21224

NORDIN, GERALD LEROY, b Rockford, Ill, Aug 14, 44; m 64; c 2. INSECT PATHOLOGY. *Educ:* Univ Ill, Urbana, BS, 62, MS, 68, PhD(entom), 71. *Prof Exp:* Tech asst entom, Ill Natural Hist Surv, 64-66, res asst, 66-71; asst prof, 71-77, PROF ENTOM, UNIV KY, 77- *Mem:* Entom Soc Am; Soc Invert Path; Sigma Xi. *Res:* Forest entomology and insect pathology; microbial control of forest insects; studies on biology, ecology and control of forest and agronomic insect pests principally through the use of entomopathogens (fungi, microsporidia, viruses and nematodes). *Mailing Add:* S-225 Agr Sci Ctr-North Dept Entom Univ Ky Lexington KY 40506

NORDIN, IVAN CONRAD, b Lindsborg, Kans, May 25, 32; div; c 4. ORGANIC CHEMISTRY. *Educ:* Bethany Col, BS, 54; Univ Kans, PhD(org chem), 60. *Prof Exp:* From assoc res chemist to sr res chemist, Warner Lambert/Parke Davis, 60-71, res scientist org chem, 71-77, sect dir, chem cent nervous syst drugs, Pharm Res Div, 77-80, mgr, chem develop, 80-85; CHEMIST, KOCH CHEM, 88- *Concurrent Pos:* Instr, Schoolcraft Col, 71-73, & Hope Col, 85-87; mfrs consult. *Mem:* Am Chem Soc. *Res:* Claisen Rearr; heterocyclic chemistry; organic chemistry of nitrogen; medicinal chemistry of central nervous system; process development. *Mailing Add:* 17367 Lake Michigan Dr W Olive MI 49460

NORDIN, JOHN HOFFMAN, b Chicago, Ill, Oct 11, 34; m 56; c 3. BIOCHEMISTRY. *Educ:* Univ Ill, BS, 56; Mich State Univ, PhD(biochem), 61. *Prof Exp:* NIH res fel biochem, Univ Minn, 62-65; from asst prof to assoc prof, 65-77, PROF BIOCHEM & ADJ PROF ENTOM, UNIV MASS, 77- *Concurrent Pos:* NIH fel, Inst Biochem, Univ Lausanne, 71-72. *Mem:* Entom Soc Am; Soc Complex Carbohydrates; Am Soc Biochem & Molecular Biol. *Res:* Biochemistry of glycoconjugates and insect octellogemins. *Mailing Add:* Dept of Biochem Univ of Mass Amherst MA 01003

NORDIN, PAUL, b Kansas City, Mo, Feb 12, 29; div; c 2. APPLIED PHYSICS. *Educ:* Univ Calif, Berkeley, AB, 56, MA, 57, PhD(physics), 61. *Prof Exp:* Telegrapher-clerk, Southern Pac RR, 48-52; eng draftsman, Northrop Aircraft, Inc, 52-53, radio-radar mechanic, 53; asst, Univ Calif, Berkeley, 56-61; sr scientist, Aeronutronic Div, Ford Motor Co, 61-63; mem tech staff, Aerospace Corp, 63-69; sect head, Dept Missile Systs & Technol, TRW Systs Group, 69-78, mem tech staff, Vulnerability & Hardness Lab, 78-81, MEM TECH STAFF, SYSTS ENG, TRW ELECTRONICS & DEFENSE, SPACE & TECHNOL GROUP, REDONDO BEACH, 81- *Mem:* AAAS; Am Phys Soc; Am Inst Aeronaut & Astronaut; Sigma Xi; Inst Elec & Electronic Engrs. *Res:* Nuclear weapons effects; reentry vehicles; laser weapon effects; space system survivability; communication satellites. *Mailing Add:* PO Box 482 Bethpage NY 11714-0482

NORDIN, PHILIP, b Can, Mar 21, 22; nat US; m 47; c 2. BIOCHEMISTRY. *Educ:* Univ Sask, BSA, 49, MSc, 50; Iowa State Univ, PhD, 53. *Prof Exp:* Lectr biochem, Univ Toronto, 53-54; from asst prof to prof, 54-88, EMER PROF BIOCHEM & BIOCHEMIST, AGR EXP STA, KANS STATE UNIV, 88- *Mem:* Am Chem Soc; Am Soc Biol Chem. *Res:* Carbohydrate biosynthesis and analysis. *Mailing Add:* 1805 Virginia Dr Manhattan KS 66502

NORDIN, RICHARD NELS, b Rainy River, Ont, Can, June 21, 47; m 73; c 2. LIMNOLOGY, PHYCOLOGY. *Educ:* Univ NDak, BSc, 70, MSc, 71; Univ BC, PhD(biol), 74. *Prof Exp:* Biologist, BC Dept Lands, Forests & Water Resources, 74-75; water qual specialist, 75-80, coordr, Water Resource Studies, 80-82, SUPVR, LIMNOL STUDIES, BC MINISTRY ENVIRON, 82- *Mem:* Int Asn Theoret & Appl Limnol; Am Soc Limnol & Oceanog; Soc Can Limnologists; Am Fisheries Soc; NAm Lake Mgt Soc. *Res:* Protection and restoration of lakes, particularly the linkages between chemical and biological processes; derivation of quantitative standards for water quality. *Mailing Add:* BC Ministry Environ 765 Broughton St Victoria BC V8V 1X5 Can

NORDIN, VIDAR JOHN, b Ratansbyn, Sweden, June 28, 24; nat Can; m 47; c 2. FOREST PATHOLOGY & MENSURATION. *Educ:* Univ BC, BA, 46, BScF, 47; Univ Toronto, PhD(forest path), 51. *Prof Exp:* Asst forest path, Forest Path Lab, Can Dept Forestry, Toronto, 47-49, officer in chg, Fredericton, NB, 49-51, Calgary, Alta, 52-57, assoc dir forest biol, 58-64, res dir, 65-71; dean fac forestry & landscape archit, Univ Toronto, 71-84, prof forest biol, 71-86, exec dir, Peru Prog, 85-90, dir, Can Forestry Accreditation Proj, 87-88, sr consult, China/Can Univ Linkage Prog, 87-90, EMER PROF FOREST BIOL, UNIV TORONTO, 86- *Concurrent Pos:* Consult, 69-; chmn bd, Algonquin Forestry Authority Corp, 74-82; mem, Prov Parks Adv Coun, 75-77; dir, Int Soc Trop Foresters, 83-88, hon vpres, Can, 89- *Mem:* NAm Forestry Comn; Int Poplar Comn; Int Union Forest Res Orgns; Soc Am Foresters; Can Inst Forestry (vpres, 65-66, pres, 67); Int Soc Trop Foresters (treas, 83-87). *Res:* forestry education; research administration. *Mailing Add:* V J Nordin Assocs Inc 19 Qualicum St Ottawa ON K2H 7G9 Can

NORDINE, PAUL CLEMENS, b Grayling, Mich, Jan 7, 40; m 68; c 1. HIGH TEMPERATURE SCIENCE, HETEROGENOUS KINETICS. *Educ:* Mich State Univ, BS, 62; Univ Kans, PhD(chem), 70. *Prof Exp:* Res assoc, Yale Univ, 71-72, instr, chem eng, 72-74, from asst prof to assoc prof, 74-82; prin scientist, Midwest Res Inst, 82-87; consult, 87-90; VPRES & DIR RES, INTERSONICS, INC, 90- *Concurrent Pos:* Chmn, Gordon Res Conf High Temperature Chem, 88. *Mem:* Am Inst Chem Engrs; Am Chem Soc; AAAS; Sigma Xi; Metall Soc; Am Ceramic Soc. *Res:* High temperature science and materials science; chemical reaction engineering; heterogeneous kinetics; flourine corrosion; containerless processing; application of laser techniques to high temperature science; synthesis of ceramic fibers; technology transfer to industry. *Mailing Add:* 3453 Commercial Ave Northbrook IL 60062

NORDLANDER, JOHN ERIC, organic chemistry; deceased, see previous edition for last biography

NORDLANDER, PETER JAN ARNE, b Stockholm, Sweden, Nov 21, 55; m. SURFACE PHYSICS. *Educ:* Chalmers Univ, Gothenburg, Sweden, MSci, 80; PhD(theoret physics), 85. *Prof Exp:* Postdoctoral, T J Watson Res Ctr, IBM, 85-86; res asst prof physics, Vanderbilt Univ, Tenn, 87-88; sr postdoctoral physics, Rutgers Univ, NJ, 88-89; ASST PROF PHYSICS, RICE UNIV, TEX, 89- *Concurrent Pos:* Consult, AT&T Bell Labs, Murray Hill, NJ, 87-89; adj asst prof, Dept Physics, Vanderbilt Univ, Tenn, 88-; fel, Rice Quantum Inst, Houston, Tex, 89-; docent, Chalmers Univ, Sweden, 91- *Mem:* Am Phys Soc; Am Chem Soc. *Res:* Condensed matter theory with emphasis on electronic properties of surfaces; tunneling processes in chemical reactions. *Mailing Add:* Dept Physics Rice Univ Houston TX 77251

NORDLIE, BERT EDWARD, b Denver, Colo, July 21, 35; m 57; c 2. GEOCHEMISTRY, GEOLOGY. *Educ:* Univ Colo, BA, 60, MS, 65; Univ Chicago, PhD(geochem, petrol), 67. *Prof Exp:* Asst prof geol, Univ Ariz, 67-71, assoc prof geosci & chief scientist, 71-73; PROF GEOL & CHMN DEPT, IOWA STATE UNIV, 74- *Mem:* AAAS; Geochem Soc; Am Inst Chemists; Int Asn Volcanology & Chem Earth's Interior; Am Geophys Union; Sigma Xi. *Res:* Petrology; volcanology, especially magnetic gases. *Mailing Add:* Dept of Earth Sci Iowa State Univ Ames IA 50011

NORDLIE, FRANK GERALD, b Willmar, Minn, Jan 23, 32; m 60; c 2. ECOLOGY, ANIMAL PHYSIOLOGY. *Educ:* St Cloud State Col, BS, 54; Univ Minn, MA, 58, PhD(zool), 61. *Prof Exp:* From asst prof to assoc prof, 61-76, PROF ZOOL, UNIV FLA, 76- *Concurrent Pos:* Biol sci coordr, Univ Fla, 76-86, assoc chmn, 81-86, chmn, 86- *Mem:* Am Soc Limnol & Oceanog; Int Soc Pure & Appl Limnology; Ecol Soc Am; Estuarine Res Fedn; Am Soc Ichthyol & Herpet. *Res:* Adaptations for a euryhaline existence and energetics of natural aquatic communities. *Mailing Add:* Dept of Zool Univ of Fla Gainesville FL 32611

NORDLIE, ROBERT CONRAD, b Willmar, Minn, June 11, 30; m 59; c 3. BIOCHEMISTRY. *Educ:* St Cloud State Col, BS, 52; Univ NDak, MS, 57, PhD(biochem), 60. *Prof Exp:* Asst, Sch Med, Univ NDak, 55-58; Nat Cancer Inst fel, Inst Enzyme Res, Univ Wis, 60-62; Hill prof, 62-74, CHESTER FRITZ DISTINGUISHED PROF BIOCHEM, SCH MED, UNIV NDAK, 74-, WILLIAM EUGENE CORNATZER PROF & CHMN DEPT, 83- *Concurrent Pos:* Consult, Oak Ridge Assoc Univs; vis prof, Tokyo Biomed Inst; hon prof, San Marcos Univ, Lima, Peru, 82. *Mem:* AAAS; Am Soc Biol Chem; Am Chem Soc; Am Inst Nutrit; Soc Exp Biol & Med; Sigma Xi. *Res:* Enzymology; intermediary metabolism of carbohydrates; effects of hormones on enzymes; metabolic regulatory mechanisms; inorganic pyrophosphate metabolism; carbamyl phosphate metabolism; control of blood glucose levels. *Mailing Add:* Dept Biochem & Molecular Biol Univ of NDak Sch of Med Grand Forks ND 58202

NORDLUND, JAMES JOHN, b St Paul, Minn, Aug 11, 39; m 62; c 3. PIGMENTATION & CELL BIOLOGY, IMMUNOLOGY. *Educ:* St John's Univ, BA, 61; Univ Minn, BS, 63, MD, 65. *Hon Degrees:* MA, Yale Univ, 83. *Prof Exp:* From intern to resident, Duke Univ, 65-67; clin assoc, Pub Health Serv, Nat Cancer Inst, 67-69; resident internal med, Yale Univ, 69-70, resident dermat, 70-72, from asst prof to prof, 72-83; PROF & DIR DERMAT, UNIV CINCINNATI, 83- *Concurrent Pos:* Dermat Found fel, 72-74; physician, Yale Univ Health Serv, 72-79, chief dermat, 79-83; attend physician, Yale New Haven Hosp, 72-83; dermatologist, Fair Haven Community Free Health Clin, 75-79; bd dirs, Am Acad Dermat. *Mem:* Am Acad Dermat; Asn Professors Dermat; Int Fedn Pigment Cell Soc (sec/treas); Pan Am Soc Pigment Cell Res (pres); Soc Investigative Dermat. *Res:* Eliology of the disease vitiligo and its treatment by studying normal and vitiliginous melanocytes in culture; cloning genes causing depigmentation in mice; understanding the normal intercutaneons signals for melanocyte proliferation used in future therapy. *Mailing Add:* Dept Dermat Univ Cincinnati 231 Bethesda Ave Cincinnati OH 45267-0592

NORDLUND, R(AYMOND) L(OUIS), b Vancouver, BC, Feb 17, 29; m 54; c 3. CIVIL ENGINEERING. *Educ:* Univ BC, BASc, 51; Univ Tex, MS, 53; Univ Ill, PhD(civil eng), 56. *Prof Exp:* Sales engr, Finning Tractor Co, 51-52; asst soil mech, Univ Ill, 53-56; found engr & consult, Raymond Int Inc, 56-65; div mgr, 65-70, VPRES & CHIEF ENGR, FRANKI FOUND CO, 70- *Res:* Soil mechanics. *Mailing Add:* 4806 Bristow Dr Annandale VA 22003-5454

NORDMAN, CHRISTER ERIC, b Helsinki, Finland, Jan 23, 25; US citizen; m 52; c 4. PHYSICAL CHEMISTRY. *Educ:* Finnish Inst Technol, BS, 49; Univ Minn, PhD(phys chem), 53. *Prof Exp:* Asst phys chem, Univ Minn, 49-53; res assoc physics, Inst Cancer Res, 53-55; from instr to assoc prof, 55-64, PROF CHEM, UNIV MICH, ANN ARBOR, 64- *Concurrent Pos:* NIH spec fel, Oxford Univ, 71-72. *Mem:* AAAS; Am Phys Soc; Am Chem Soc; Am Crystallog Asn. *Res:* X-ray crystal structure analysis. *Mailing Add:* Dept Chem Univ Mich Ann Arbor MI 48109-1055

NORDMAN, JAMES EMERY, b Quinnesec, Mich, Apr 27, 34; m 59; c 6. ELECTRONIC DEVICES. *Educ:* Marquette Univ, BEE, 57; Univ Wis-Madison, MS, 59, PhD(elec eng), 62. *Prof Exp:* From instr to asst prof elec eng, Univ Wis-Madison, 59-67; mem tech staff, David Sarnoff Res Ctr, RCA Corp, NJ, 67-68; assoc prof, 68-74, PROF ELEC ENG, DEPT ELEC & COMPUT ENG, UNIV WIS-MADISON, 74- *Concurrent Pos:* Fulbright-Hayes travel grant, 72; consult, L'Air Liquide, Grenoble, France, 72-73. *Mem:* Inst Elec & Electronics Engrs; Am Vacuum Soc. *Res:* Solid state electronic devices; thin films; superconductors. *Mailing Add:* Dept Elec & Comput Eng Univ Wis Madison WI 53706

NORDMEYER, FRANCIS R, b Kankakee, Ill, Feb 1, 40; m 61; c 6. INORGANIC CHEMISTRY. *Educ:* Wabash Col, BA, 61; Wesleyan Univ, MA, 64; Stanford Univ, PhD(chem), 67. *Prof Exp:* Asst prof chem, Univ Rochester, 67-72; from asst prof to assoc prof, 72-82, PROF CHEM, BRIGHAM YOUNG UNIV, 82- *Mem:* AAAS; Am Chem Soc. *Res:* Mechanisms of inorganic reactions; ion chromatography. *Mailing Add:* Dept of Chem Brigham Young Univ Provo UT 84602-1022

NORDQUIST, EDWIN C(LYDE), b Salt Lake City, Utah, Aug 7, 21; m 43, 73; c 8. CIVIL ENGINEERING, GEOTECHNICAL ENGINEERING. *Educ:* Univ Utah, BSCE, 43; Iowa State Univ, MS, 51. *Prof Exp:* Engr, Douglas Aircraft Co, Calif, 43-44; engr, US Army, 44-46; from instr to assoc prof, 46-85, EMER ASSOC PROF CIVIL ENG, UNIV UTAH, 85- *Concurrent Pos:* Asst mat engr, Utah Dept Highways, 60-61, mat engr, 62, soils engr & consult, Pittsburgh Testing Lab, 63-85; assoc prof, Div Mat Sci & Eng, Univ Utah, 72-75; geotech consult, 85- *Mem:* Fel Am Soc Civil Engrs; Am Arbit Asn. *Res:* Soil mechanics and foundations; geotechnical. *Mailing Add:* 8321 S 1100 E Sandy UT 84094-0715

NORDQUIST, PAUL EDGARD RUDOLPH, JR, b Wash DC, Sept 7, 36. CRYSTAL GROWTH, ANALYTICAL CHEMISTRY. *Educ:* George Wash Univ, BS, 58; Univ Minn, PhD(inorg chem), 64. *Prof Exp:* Teaching asst inorg chem, Univ Minn, 58-60, 62-63; res chemist, Monsanto Co, 64-70; RES CHEMIST, US NAVAL RES LAB, 70- *Concurrent Pos:* Chemist, Nat Bur Standards, 58-59. *Mem:* Am Chem Soc; Sigma Xi; AAAS. *Res:* Materials preparation, purification, analysis and characterization; growth of III-V compounds; SIC and glasses; trace analysis. *Mailing Add:* Code 6872 US Naval Res Lab 4555 Overlook Ave SW Washington DC 20375-5000

NORDQUIST, ROBERT ERSEL, b Oklahoma City, Okla, Sept 10, 38; m 56; c 2. PATHOLOGY, VIROLOGY. *Educ:* Oklahoma City Univ, BA, 68; Univ Okla, PhD(med sci), 71. *Prof Exp:* Res assoc electron micros, Med Sch, Univ Okla, 62-66; res assoc, Scripps Clin & Res Found, 66-67; res assoc, 67-71, asst mem cancer, Okla Med Res found, 71-; asst prof path, Sch Med, Univ Okla, 71-; AT DEAN A MCGEE EYE INST. *Mem:* Electron Micros Soc Am; Tissue Cult Asn; Am Asn Cancer Res; Sigma Xi. *Res:* The search for human tumor viruses, utilizing the electron microscope coupled with immunologic and virologic techniques. *Mailing Add:* Dean A McGee Eye Inst 608 Stant L Young Blvd Oklahoma City OK 73104

NORDSCHOW, CARLETON DEANE, b Hampton, Iowa, June 7, 26; m 50; c 3. PATHOLOGY, BIOCHEMISTRY. *Educ:* Luther Col, Iowa, AB, 49; Univ Iowa, MD, 53, PhD(biochem), 64. *Prof Exp:* From instr to assoc prof path, Univ Iowa, 54-70, dir clin labs, 63-66; PROF PATH & CHMN DEPT CLIN PATH, SCH MED, IND UNIV, INDIANAPOLIS, 70- *Concurrent Pos:* Mem, Acad Clin Lab Physicians & Scientists; USPHS res grants, 63-65 & 66-68. *Mem:* AMA; Am Soc Clin Path; Am Chem Soc; Int Acad Path. *Res:* Comparative properties of polymers in health and disease. *Mailing Add:* Dept Path Ind Univ Med Ctr 1100 W Michigan UH-N-440 Indianapolis IN 46223

NORDSKOG, ARNE WILLIAM, b Two Harbors, Minn, Feb 21, 13; m 38; c 2. POULTRY SCIENCE. *Educ:* Univ Minn, BS, 37, MS, 40, PhD(animal breeding), 43. *Prof Exp:* Instr agr, Univ Alaska, 37-39; assoc prof animal indust, Mont State Col, 43-45; from assoc prof to prof, 45-83, EMER PROF ANIMAL SCI, IOWA STATE UNIV, 83- *Honors & Awards:* CPC-Int Res Award, Poultry Sci Asn, 72. *Mem:* Fel Poultry Sci Asn; hon mem Norwegian Poultry Asn; Genetics Soc Am; Poultry Sci Asn. *Res:* Poultry breeding and immunogenetics; factors affecting selection for growth rate in swine. *Mailing Add:* 201 Kildee Hall Iowa State Univ Ames IA 50010

NORDSTEDT, ROGER ARLO, b Wichita, Kans, Oct 16, 42; m 79; c 4. AGRICULTURAL ENGINEERING. *Educ:* Kans State Univ, BS, 64, MS, 66; Ohio State Univ, PhD(agr eng), 69. *Prof Exp:* Res asst feed technol, Kans State Univ, 64-65; res assoc agr eng, Ohio State Univ, 68-69; asst prof, 69-76, ASSOC PROF AGR ENG, UNIV FLA, 76- *Mem:* AAAS; Am Soc Agr Engrs. *Res:* Agriculture and quality of environment; agricultural pollution control; animal waste handling, treatment, utilization and/or disposal; lagoon treatment. *Mailing Add:* Dept of Agr Eng Rogers Hall Univ of Fla Gainesville FL 32611

NORDSTROM, BRIAN HOYT, b Oakland, Calif, Aug 5, 49; m 71; c 4. CHEMICAL EDUCATION. *Educ:* Univ Calif, Berkeley, AB, 71, MS, 75; Northern Ariz Univ, EdD, 89. *Prof Exp:* Instr chem, Calif State Univ, Chico, 75-78 & Calif Maritime Acad, 78-80; asst prof, Embry-Riddle Aero Univ, Prescott, 80-83, chmn dept, 85-86, assoc prof chem, 83-90, PROF CHEM, EMBRY-RIDDLE AERO UNIV, PRESCOTT, 90- *Concurrent Pos:* Instr, Yavapai Col, 81-84; consult, Oxycal Mfg Corp, 82-85; adj fac, Northern Ariz Univ, 90- *Mem:* Am Chem Soc. *Res:* Curriculum development; chemical kinetics; computational chemistry. *Mailing Add:* 3200 N Willow Creek Rd Embry-Riddle Aeronaut Univ Prescott AZ 86301-8662

NORDSTROM, DARRELL KIRK, b San Francisco, Calif, Nov 14, 46; m 73; c 2. GEOCHEMISTRY, WATER CHEMISTRY. *Educ:* Southern Ill Univ, BA, 69; Univ Colo, MS, 71; Stanford Univ, PhD(geochem), 77. *Prof Exp:* Res chemist, US Geol Surv, 74-76; asst prof environ geochem, dept environ sci, Univ Va, 76-80; GEOCHEMIST, US GEOL SURV, 80- *Concurrent Pos:* Consult, Calif State Water Qual Control Bd, US Geol Surv, 77-80, Swed Nuclear Fuel & Waste Mgt Agency, 81-, Atomic Energy Can, Ltd, US Environ Protection Agency, US Forest Serv; NSF fel. *Mem:* Sigma Xi; AAAS; Geol Soc Am; Geochem Soc; Mineral Soc Am; Int Asn Geochemists & Cosmochemists. *Res:* Modeling the deep groundwater chemistry in a granite; the geochemical processes of acid mine waters and the thermodynamic properties of rock-water interactions. *Mailing Add:* US Geol Surv MS-420 345 Middlefield Rd Menlo Park CA 94025-9998

NORDSTROM, J DAVID, b Minneapolis, Minn, Sept 30, 37; m 59; c 3. ORGANIC POLYMER CHEMISTRY. *Educ:* Gustavus Adolphus Col, BA, 59; Univ Iowa, PhD(chem), 63. *Prof Exp:* Res chemist polymers, Archer Daniels Midland Co, Minn, 63-68; sr res scientist, 68-70, MGR POLYMER RES & DEVELOP, PAINT PLANT, FORD MOTOR CO, 70-; AT E I DU PONT DE NEMOURS CO INC. *Mem:* Am Chem Soc. *Res:* Synthetic organic chemistry; thermoset polymer synthesis and evaluation; Development of polymers for automotive coatings. *Mailing Add:* E I du Pont de Nemours Co Inc 945 Stephenson Hwy PO Box 2802 Troy MI 48007

NORDTVEDT, KENNETH L, b Chicago, Ill, Apr 16, 39; c 3. THEORETICAL PHYSICS. *Educ:* Mass Inst Technol, BS, 60; Stanford Univ, MS, 62, PhD(physics), 65. *Prof Exp:* Staff physicist instrumentation lab, Mass Inst Technol, 63-65; FROM ASST PROF TO PROF PHYSICS, MONT STATE UNIV, 65- *Concurrent Pos:* NASA res grants, 65-73; Sloan fel, 71-73; NSF res grants, 85-; Nat sci bd, 87- *Mem:* Fel Am Phys Soc. *Res:* Gravitation; relativity and cosmology. *Mailing Add:* Dept of Physics Mont State Univ Bozeman MT 59715

NORDYKE, ELLIS LARRIMORE, b Houston, Tex, June 20, 42. BIOCHEMISTRY, NEUROCHEMISTRY. *Educ:* Univ Houston, BS, 68, MS, 70, PhD(biophys sci), 72. *Prof Exp:* NIMH fel neurochem, Tex Res Inst Ment Sci, 72-74; from asst prof to assoc prof, 74-83, dept head, 83-88, PROF BIOL, UNIV ST THOMAS, 83- *Mem:* AAAS. *Res:* Biochemistry of snakes. *Mailing Add:* Dept Biol Univ St Thomas 3812 Montrose Blvd Houston TX 77006

NOREIKA, ALEXANDER JOSEPH, b Philadelphia, Pa, Feb 24, 35; m 63; c 3. SOLID STATE PHYSICS, MOLECULAR BEAM EPITAXY. *Educ:* Drexel Inst, BS, 58; Univ Reading, PhD(physics), 66. *Prof Exp:* From jr to sr physicist, Philco Sci Labs, Ford Motor Co, Pa, 59-62; sr physicist, 66-81, FEL SCIENTIST, WESTINGHOUSE RES CTR, 81- *Mem:* Am Phys Soc;

Am Vacuum Soc. *Res:* Electron microscopy study of defect structures in single crystals; investigation of extended miscibility ranges in III-IV-V compounds; molecular beam epitaxy of III-V and II-VI compounds; study of infrared intrinsic photodetectors. *Mailing Add:* Westinghouse Res Ctr Beulah Rd 501 2C25 Pittsburgh PA 15235

NORELL, JOHN REYNOLDS, b Hutchinson, Kans, June 25, 37; m 59; c 2. CHEMISTRY. *Educ:* Bethany-Nazarene Col, BS, 59; Purdue Univ, PhD(org chem), 63. *Prof Exp:* Res chemist, Phillips Petrol Co, 64-66, group leader, 66-74, sect supvr, 74-75, sales mgr, 75-79, dir chem res, 79-83; PRES PROVESTA CORP, PHILLIPS PETROL CO, 83- *Concurrent Pos:* NIH fel, Inst Org Chem, Munich, Ger, 63-64. *Mem:* Am Chem Soc; Sigma Xi. *Res:* Biotechnology; organosulfur chemistry; olefins; fertilizer; petrochemicals; organic synthesis, mechanisms and spectroscopy; flame retardants; fertilizer chemistry and organic chemistry; fermentation technology and enzymes. *Mailing Add:* Provestar Corp 15 Phillips Bldg Bartlesville OK 74004

NOREN, GERRY KARL, b Minneapolis, Minn, June 22, 42; m 61; c 2. ORGANIC POLYMER CHEMISTRY. *Educ:* Univ Minn, BA, 66; Univ Iowa, PhD(org chem), 71. *Prof Exp:* Res chemist, Archer-Daniels-Midland Co, 66-67; res assoc chem, Univ Ariz, 71-72; res assoc chem, Calgon Corp, Merck Sharp & Dohme Res Labs, 72-76; supvr polymer chem, Desoto Inc, 76-79, tech mgr inves res, 79-83, res assoc, Polymer Develop, 83-87, sr res assoc, group res, 87-89; SR RES ASSOC, NEW VENTURES RES, DSM DESOTECH INC, 89- *Concurrent Pos:* Adj prof, Roosevelt Univ, 80 & 82, Elmhurst Col, 78, 83, 85 & 88, DePaul Univ, 85. *Honors & Awards:* Roon Award, 85. *Mem:* Am Chem Soc; Sigma Xi; Fedn Socs Coatings Technol. *Res:* Design and study of unique controlled release systems; synthesis and utilization of new monomers and polymeric systems; low energy crosslinking reactions; ion containing polymers; resins for high-solid and water-borne coatings; thermally stable polymers. *Mailing Add:* 1836 Sessions Walk Hoffman Estates IL 60195

NORFLEET, MORRIS L, b Nancy, Ky, Dec 15, 30; m 52; c 1. AGRICULTURE, BIOLOGY. *Educ:* Univ Ky, BS, 52; Purdue Univ, MS, 57, PhD(educ), 62. *Prof Exp:* Teacher, Spiceland Pub Schs, Ind, 52-58; pub relations asst, Ind Farm Bur Coop Asn, 58-60, market res anal, 60; instr educ, Purdue Univ, 60-62; assoc prof educ & dir student teaching, 62-65, prof educ & dir res & prog develop, 65-68, vpres res & develop, 68-76, actg pres, 76, PRES, MOREHEAD STATE UNIV, 77- *Concurrent Pos:* Dir Head Start Training prog, 66 & Proj Upward Bound, 66-68; coordr, Comput Assisted Instr Res Demonstration Proj, Ky, 67; interim dir, Appalachian Adult Basic Educ Res & Demonstration Ctr, 67; mem Steering Comt, Gov Efficiency Task Force, 66, Ky Sci & Technol Comn, 68-70 & Northeast Ky Crime Comn, 69-71; dir, Proj Newgate, 69; chmn, Gateway Comprehensive Health Planning Coun, 73, Comprehensive Planning Task Force, Ky Coun Higher Educ, 74-75, & chmn bd dir, Gateway Area Develop Dist, 77-; mem comn of gov relations of Am Coun Educ, 79- & adv comt eligibility of US Off of Educ, 78-; bd trustees, Campbellsville Col, 73-, bd dir, Area Health Educ Syst, 74-, Ky Coun Higher Educ, 77-, Gov Task Force Transp, 78- & Gov Appalachian Develop Coun, 78- consult, US Off Educ, US Off Econ Opportunity, Am Asn Jr Cols, NSF, Danforth Found, Danforth Assoc. *Mem:* Am Asn Teacher Educ; Am Asn Higher Educ; Am Educ Res Asn; Nat Col & Univ Res Adminr; Soc Col & Univ Planning. *Mailing Add:* 1190 Hwy 1664 Morehead State Univ Univ Blvd Nancy KY 42544

NORFORD, BRIAN SEELEY, b Gidea Park, Eng, Sept 15, 32; Can citizen; m 62. GEOLOGY, PALEONTOLOGY. *Educ:* Cambridge Univ, BA, 55, MA, 59, ScD, 77; Yale Univ, MSc, 56, PhD, 59. *Hon Degrees:* LLD, Univ Calgary, 89. *Prof Exp:* Paleontologist, Shell Oil Co, Can, 59-60; GEOLOGIST, GEOL SURV CAN, 60-, head western paleont sect, 67-72, head paleont subdiv, 72-77, 80-81, 86-88. *Concurrent Pos:* Chmn, Int Working Group Cambrian-Ordovician Boundary, 80-; mem bd dirs, Can Energy Res Inst, 81-85; chancellor, Univ Calgary, 82-86; mem bd dirs, Can Geosci Coun, 84-92, pres, 91; protocol comt, Calgary Olympic Games, 87-88. *Mem:* Can Soc Petrol Geologists; Am Paleont Soc; Brit Palaeont Asn; Geol Asn Can. *Res:* Lower Paleozoic stratigraphy; Ordovician and Silurian graptolites, brachiopods and trilobites. *Mailing Add:* Geol Surv of Can 3303 33rd St Calgary AB T2L 2A7 Can

NORGAARD, NICHOLAS J, b Aledo, Ill, May 28, 43; m 67; c 1. STATISTICS, COMPUTER SCIENCE. *Educ:* Univ Wis-Platteville, BS, 65, MS, 69; Univ Ga, MS, 72, PhD(statist), 75. *Prof Exp:* Asst math, Univ Wis-Platteville, 68-69; asst statist, Univ Ga, 69-70, NSF fel, 70-73; instr, 73-76, ASST PROT MATH & STATIST, WESTERN CAROLINA UNIV, 76- *Concurrent Pos:* Statist consult, Comput Ctr, Western Carolina Univ, 75- *Mem:* Am Statist Asn. *Res:* Estimation of parameters in continuous distributions using censored samples. *Mailing Add:* Dept Math & Sci Western Carolina Univ Cullowhee NC 28723

NORIN, ALLEN JOSEPH, b Chicago, Ill, July 30, 44; m 69; c 2. CELL BIOLOGY, TRANSPLANTATION IMMUNOLOGY. *Educ:* Roosevelt Univ, BS, 67; Univ Houston, MS, 70, PhD(biol), 72. *Prof Exp:* USPHS fel & res assoc microbiol, Univ Chicago, 72-75; Asst Prof Surg, Microbiol & Immunol, Montefiore Hosp & Med Ctr, Albert Einstein Col Med, Yeshiva Univ, 75-; assoc prof med, anat & cell biol, 85-90, DIR, PULMONARY IMMUNOL LAB, DIV PULMONARY MED, SUNY COL MED, BROOKLYN, 88- , PROF, MED ANAT & CELL BIOL, 90- *Concurrent Pos:* Immunologist, Manning Lab, NIH Prog, Proj Lung Transplantation, Montefiore Hosp, 75-; dir, Transplant Immunol & Immunogenetics, Suny Col Med, Brooklyn. *Mem:* Am Soc Microbiol; Am Inst Biol Sci; Fedn Am Scientists. *Res:* Role of cytolytic lymphocytes in rejection of tissue allografts; characterization of cells surface molecules in cytolytic lymphocyte-allograft and tumor interaction and lysis; mechanisms of immunologic tolerance. *Mailing Add:* SUNY Health Sci Ctr Brooklyn ABox 19 450 Clarkson Ave Brooklyn NY 11203-2098

NORING, JON EVERETT, b Pipestone, Minn, Nov 30, 54; m 78; c 1. CHEMICAL ENGINEERING. *Educ:* Univ Minn, BS, MS & PhD(mech eng). *Prof Exp:* MEM TECH STAFF, SANDIA NAT LAB, LIVERMORE, 81- *Mem:* Int Asn Hydrogen Energy; Solar Thermal Test Facil Users Asn. *Res:* Solar central receiver systems, predominately in the area of fuels and chemicals production; very high temperature thermal dissociation processes. *Mailing Add:* 754 Catalina Dr Livermore CA 94550

NORINS, ARTHUR LEONARD, b Chicago, Ill, Dec 2, 28; m 54; c 4. DERMATOLOGY. *Educ:* Northwestern Tech Inst, BS, 51; Northwestern Univ, MS, 53, MD, 55; Am Bd Dermat, dipl, 61, cert(dermatopath), 74. *Prof Exp:* Asst prof, Stanford Univ, 61-64; assoc prof, 64-69, PROF DERMAT & PATH, IND UNIV, INDIANAPOLIS, 69-, CHMN, DEPT DERMAT, 76- *Concurrent Pos:* Chief dermat, Riley Children's Hosp. *Mem:* Fel Am Col Physicians; Am Soc Dermatopath; Am Acad Dermat; Soc Invest Dermat; Soc Pediat Dermat; Am Dermat Asn. *Res:* Dermatopathology; photobiology. *Mailing Add:* Dermat Dept RG 524 1001 West Tenth St Indianapolis IN 46223

NORK, WILLIAM EDWARD, b Shenandoah, Pa, Aug 13. 34; m 60; c 4. HYDROGEOLOGY. *Educ:* Columbia Univ, AB, 60; Univ Buffalo, MA, 61. *Prof Exp:* Asst geol, Univ Buffalo, 60-61; from asst to instr geol, Univ Ariz, 61-64; scientist, Hazelton Nuclear Sci Corp, 64-66; dep proj mgr & asst mgr hydrogeol sect isotopes, Teledyne Inc, Calif, 66-70, mgr, Teledyne Isotopes, Nev, 70-71; res assoc, Desert Res Inst, Univ Nev, Las Vegas, 71-72; consult hydrologist, Hydro-Search, Inc, 72-77; CONSULT HYDROGEOLOGIST, WILLIAM E NORK, INC, 77- *Concurrent Pos:* Consult, 60-61. *Mem:* Nat Water Well Asn; Geol Soc Am; Am Water Resources Asn; Am Geophys Union. *Res:* Optimum use and development of water, especially ground water resources and water quality assurance for public use. *Mailing Add:* William E Nork Inc 1026 W First St Reno NV 89503

NORLING, PARRY MCWHINNIE, b Des Moines, Iowa, Apr 17, 39; m 65; c 2. POLYMER CHEMISTRY. *Educ:* Harvard Univ, AB, 61; Princeton Univ, PhD(polymer chem), 64. *Prof Exp:* Res assoc oxidation of polymers, Princeton Univ, 64-65; res chemist, Electrochem Dept, E I du Pont de Nemours & Co, Inc, 65-69, res supvr, 69-71, from tech supt to prod supt, Memphis Plant, 71-75, res mgr, Indust Chem Dept, Exp Sta, 75-78, dir safety & health, 78-80, lab dir, Dept Chem & Pigments, Exp Sta, 80-83, mgr res & develop, 83-84, tech res dir, Polymer Prod, 84- 90, PLANNING DIR RES & DEVELOP, E I DU PONT DE NEMOURS & CO INC, WILMINGTON, 91- *Mem:* Am Chem Soc. *Res:* Kinetics and mechanism of the oxidation of polymers; polymerization kinetics. *Mailing Add:* DuPont Res & Develop D9166 E I du Pont de Nemours & Co Inc Wilmington DE 19898

NORLYN, JACK DAVID, b Bellflower, Calif, Dec 10, 31; m 58; c 2. PLANT PHYSIOLOGY. *Educ:* Univ Calif, Davis, BS, 55, PhD(plant physiol), 82; Calif State Univ, Fresno, MS, 76. *Prof Exp:* Lectr hydroponics, 76, STAFF RES ASSOC SALT TOLERANCE PLANTS, UNIV CALIF, DAVIS, 71- *Concurrent Pos:* Consult biomass prod, 78- *Mem:* Sigma Xi. *Res:* Physiological genetics of salt tolerance in plants; development of a crop production system using seawater on sand; improving salt tolerance of crops; biomass production for energy. *Mailing Add:* Dept LAWR Univ Calif Davis CA 95616

NORMAN, ALEX, b New York, NY, Aug 7, 23; m 47; c 3. RADIOLOGY. *Educ:* NY Univ, BA, 44; Chicago Med Sch, MD, 48. *Prof Exp:* Nat Cancer Inst fel, Bellevue Hosp, 52; radiologist, Beth Israel Hosp, New York, 55-56; assoc attend radiol, Hosp Joint Dis, New York, 56-60, chief diag roentgenol, 60-66; asst prof clin radiol, Sch Med, NY Univ, 64-67; dir diag roentgenol, 66-70, DIR RADIOL, HOSP JOINT DIS & ORTHOP INST, 70-, CHMN DEPT, 85-; PROF RADIOL, NEW YORK UNIV SCH MED, 85- *Concurrent Pos:* NIH grant, Hosp Joint Dis, Bethesda, 62-64; prof radiol, Mt Sinai Sch Med, 67-85; hon mem chair radiol, Cent Univ, Venezuela, 71; examr, Am Bd Radiol, 74; distinguished lectr diag radiol, NY Roentgen Soc, 79; consult reviewer, J Bone & Joint Surg, 80; prof lectr, dept radiol, Mt Sinai Med Ctr, 86. *Honors & Awards:* Maimonides Award, 82. *Mem:* Fel NY Acad Med; fel Am Col Radiol; hon mem Venezuelan Radiol Soc. *Res:* Application of tomography and enlargement technique in the diagnosis of bone diseases. *Mailing Add:* NY Med Col Radiol Valhalla NY 10595

NORMAN, AMOS, b Vienna, Austria, Nov 25, 21; m 46; c 4. MEDICAL PHYSICS. *Educ:* Harvard Univ, AB, 43; Columbia Univ, MA, 47, PhD(biophys), 50. *Prof Exp:* AEC fel, Columbia Univ, 50-51; res biophysicist, 51-54, asst prof, 53-58, assoc prof, 58-63, prof radiol, 63-81, PROF RADIATION ONCOL & RADIOL SCI, UNIV CALIF, LOS ANGELES, 81- *Mem:* AAAS; Am Soc Photobiol; Am Soc Physicists in Med; Radiation Res Soc. *Res:* Cellular radiobiology; engineering in medicine. *Mailing Add:* Dept of Radiol Sci AR 259 Chs Univ of Calif 405 Hilgard Ave Los Angeles CA 90024

NORMAN, ANTHONY WESTCOTT, b Ames, Iowa, Jan 19, 38; m 75; c 3. BIOCHEMISTRY. *Educ:* Oberlin Col, BA, 59; Univ Wis, MS, 61; PhD(biochem), 63. *Prof Exp:* Res assoc biochem, Univ Wis, 59-63; from asst prof to assoc prof, 63-72, chmn dept, 76-81, PROF BIOCHEM, UNIV CALIF, RIVERSIDE, 72-, PROF & DIV DEAN, BIOMED SCI, 86- *Concurrent Pos:* NIH res grants, 64-, spec fel, 70-71 & career develop award, 71-76; Fulbright fel, 70-71. *Honors & Awards:* Ernst Oppenheimer Award, 77; Mead Johnson Award, 77; Merit Award, NIH, 86; Osborne-Mendel Award, 89; David Curnow Plenary Lectr, Australian Soc Clin Biochem, 89. *Mem:* Endocrinol Soc; Am Fedn Clin Res; AAAS; Am Chem Soc; Am Soc Biol Chemists; Am Soc Bone & Mineral Res. *Res:* Mechanism of action of vitamin D related to calcium metabolism; ion transport; mode of action of steroid hormones. *Mailing Add:* Div Biomed Sci & Dept Biochem Univ of Calif 1900 University Ave Riverside CA 92521-0121

NORMAN, ARLAN DALE, b Westhope, NDak, Mar 26, 40; m 66. INORGANIC CHEMISTRY. *Educ:* Univ NDak, BS, 62; Ind Univ, PhD(chem), 66. *Prof Exp:* Res assoc chem, Univ Calif, Berkeley, 65-66; from asst prof to assoc prof, 66-74, dept head, 80-84, PROF CHEM, UNIV COLO, BOULDER, 74- *Concurrent Pos:* Alfred P Sloan Found fel, 73-75. *Mem:* Am Chem Soc; Chem Soc London. *Res:* Chemistry of phosphous hydrides, phosphous-nitrogen rings, polymers and phosphine-metal complexes; photocatalysts. *Mailing Add:* Dept Chem Box 215 Univ Colo Boulder CO 80309-0215

NORMAN, BILLY RAY, b Luverne, Ala, Feb 10, 35. SCIENCE EDUCATION. *Educ:* Troy State Col, BS, 57; Univ Ga, EdD(sci), 65. *Prof Exp:* Asst prof sci, Campbell Col, 65-66; assoc prof, 66-77, PROF SCI EDUC, TROY STATE UNIV, 77- *Mem:* Nat Sci Teachers Asn. *Res:* Concepts of teaching science. *Mailing Add:* Dept of Sci Educ Troy State Univ Troy AL 36082

NORMAN, CARL EDGAR, b Cokato, Minn, Feb 1, 31; m 62; c 1. STRUCTURAL GEOLOGY, ROCK MECHANICS. *Educ:* Univ Minn, Minneapolis, BA, 57; Ohio State Univ, MS, 59, PhD(geol), 67. *Prof Exp:* Geologist, Humble Oil & Refining Co, 59-62; from instr to asst prof geol, 65-71, ASSOC PROF GEOL, UNIV HOUSTON, 71- *Concurrent Pos:* NSF instr sci equip prog grant, 68-69; univ fac res support prog grant, 69. *Mem:* Am Inst Prof Geologists; Geol Soc Am; Int Soc Rock Mech; Sigma Xi. *Res:* Mechanism of failure in rocks; behavior of rocks under varying conditions of load; active geologic faults in the Texas-Louisiana gulf coastal zone. *Mailing Add:* Dept Geosci Univ of Houston Houston TX 77204-5503

NORMAN, DAVID, b Philadelphia, Pa, Jan 29, 42. NEURORADIOLOGY. *Educ:* Ursinus Col, BS, 63; Univ Pa, Philadelphia. *Prof Exp:* Intern, med, Hosp Univ Pa, 67-68; Resident radiol, Columbia Presbyterian Hosp, NY, 68-71; chief radiol, US Army Med Corp, Danang, Vietnam, 71-72 & Fitzsimmons Gen Hosp, 72-73; NIH fel, 73-75, from asst prof to assoc prof, 75-83, PROF RADIOL, UNIV CALIF, SAN FRANCISCO, 83-, CHIEF NEURORADIOLOGY, 87- *Concurrent Pos:* Consult, VA Hosp, Ft Miley, San Francisco & Letterman Army Hosp, 76-, med adv bd, Gen Elec, 83-86, adv bd, Gen Elec Med Syst, 86- & educ adv bd, 88- *Mem:* Fel Am Col Radiol; Radiol Soc Am; Am Soc Neuroradiology; Am Roentgen Ray Soc; Asn Univ Radiologists; AMA. *Res:* Areas of magnetic resonance imaging, including use in testing spinal cord neoplasms and cryptic vascular malformations; research on brain ischemia includes magnetic resonance imaging of acute and chronic brain ischemia. *Mailing Add:* Dept Radiol Univ Calif San Francisco Med Ctr Box 0628 Rm L371 San Francisco CA 94143

NORMAN, DOUGLAS JAMES, b Seattle, Wash, Mar 6, 46; m 77; c 2. TRANSPLANTATION IMMUNOLOGY, IMMUNOGENETICS. *Educ:* Stanford Univ, BA, 68; Univ Wash Sch Med, MD, 72. *Prof Exp:* Fel res med, Harvard Med Sch, 76-79; fel nephrology, Peter Brent Brigham Hosp, 76-79, fel transplantation, 77-79; from asst prof to assoc prof, 79-89, PROF MED, ORE HEALTH SCI UNIV, 89- *Concurrent Pos:* Dir, Immunogenetics & Transplantation Lab, Ore Health Sci Univ, 79, dir, transplant med, 86- *Mem:* Transplantation Soc; Am Soc Transplant Physicians; Am Soc Nephrology; Int Soc Nephrology; Int Soc Heart Transplantation. *Res:* Transplantation immunobiology and immunogenetics. *Mailing Add:* 3181 SW Sam Jackson Park Rd Portland OR 97201

NORMAN, EDWARD, b New York, NY, Aug 7, 32; m 59; c 2. MATHEMATICS. *Educ:* City Col New York, BS, 54; Cornell Univ, PhD(math), 58. *Prof Exp:* Asst prof math, Mich State Univ, 58-61; res mathematician, Socony Mobil Oil Co, 61-64; asst prof math, Drexel Inst, 64-69; assoc prof math, Fla Technol Univ, 69-80; MEM FAC, DEPT MATH, UNIV CENT FLA, 80- *Mem:* Am Math Soc. *Res:* Analysis; stability of differential equations. *Mailing Add:* Dept Math Univ Cent Fla Box 25000 Orlando FL 32816

NORMAN, EDWARD COBB, b BC, Can, Oct 5, 13; US citizen; m 49; c 3. PUBLIC HEALTH EDUCATION, PSYCHIATRY. *Educ:* Univ Wash, BS, 35; Univ Pa, MD, 40; Tulane Univ, MPH, 65. *Prof Exp:* Psychiatrist, USPHS, 43-46; clin instr psychiat, Univ Ill Col Med, 49-53; from asst prof to assoc prof clin psychiat, 53-64, prof psychiat & prev med, Dept Trop Med & Pub Health, 64-79, dir ment health sect, Sch Pub Health & Trop Med, 67-69, EMER PROF PSYCHIAT & PREV MED, DEPT TROP MED & PUB HEALTH, TULANE UNIV, 79-; ASSOC DIR PAIN TREAT CTR, HOTEL DIEU HOSP, NEW ORLEANS, 78- *Concurrent Pos:* Pvt pract, Chicago, 49-53 & New Orleans, 53-; clin physician, Michael Reese Hosp, Chicago, 49-53 & Vet Admin Hosps, Gulfport, Miss & New Orleans, 53-64; consult, Southeast La Hosp, Mandeville & East La Hosp, Jackson, 53-60; on active staff, Sara Mayo Hosp, New Orleans, 58-64; sr vis physician, Charity Hosp La, New Orleans, 64-79, co-dir, Inter-Univ Forum Educr Community Psychiat, Duke Univ, 67-72; mem ad hoc grants rev comt, NIMH, 68-69; consult, New Orleans City Police Dept, 70-72 & Orleans Parish Sch Syst, 71-72; secy, Forum Improv Quality of Life, 74; mem, Am Psychiat Asn Task Force Eco-Psychiatry, 75; consult, La Rehab Inst, 76-87, Minister Health, S Vietnam, 70; courtesy staff, De Paul Hosp, 70-, Hotel Dieu Hosp, 78-; vpres, La Exec Serv Corps, 87- *Mem:* AAAS; Am Psychiat Asn; Am Acad Psychoanal; Am Pub Health Asn. *Res:* Evaluation of the education process; evaluation of mental health consultation. *Mailing Add:* 439 Pine St New Orleans LA 70118

NORMAN, ELIANE MEYER, b Lyon, France, Nov 15, 31; US citizen; m 58; c 2. BOTANY, TAXONOMY. *Educ:* Hunter Col, BA, 53; Wash Univ, MA, 55; Cornell Univ, PhD(plant taxon), 62. *Prof Exp:* Instr biol, Hobart Col, 55-56; instr natural sci, Mich State Univ, 59-60; asst prof bot, Rutgers Univ, 63-69; asst prof biol, 70-77, assoc prof, 77-84, PROF BIOL, STETSON UNIV, 84- *Concurrent Pos:* Vis scholar, Harvard Univ Herbaria, 87-88. *Mem:* Am Soc Plant Taxon; Asn Trop Biol; Int Asn Plant Taxon; Am Inst Biol Sci. *Res:* Taxonomic studies of Northwest Buddlejaceae; reproductive biology of Florida plants especially Annonaceae. *Mailing Add:* 1620 Druid Rd Maitland FL 32751

NORMAN, ERIC B, b Buffalo, NY, Jan 15, 51; m 73; c 2. NUCLEAR ASTROPHYSICS. *Educ:* Cornell Univ, AB, 72; Univ Chicago, MS, 74, PhD(physics), 78. *Prof Exp:* from res assoc to res asst prof physics, Univ Wash, 78-83; asst prof physics, Seattle Univ, 83-84; div fel Nuclear Sci, 84-89, SR PHYSICIST, LAWRENCE BERKELEY LAB, 89- *Concurrent Pos:* vis asst prof, physics Dept, Univ Calif, Berkeley, 88; distinguished lectr, Assoc Western Univs, 90-91. *Mem:* Am Phys Soc; Sigma Xi. *Res:* Experimental studies in nuclear astrophysics. *Mailing Add:* Bldg 88 Lawrence Berkeley Lab Berkeley CA 94720

NORMAN, FLOYD (ALVIN), b Hallettsville, Tex, July 31, 11; m 38; c 2. MEDICINE. *Educ:* Univ Tex, MD, 35; Am Bd Pediat, dipl, 40. *Prof Exp:* Clin prof pediat, Univ Tex Southwest Med Sch Dallas, 60-72; regional health adminr, Pub Health Serv, Region VI, Dept Health, Educ & Welfare, 72-78; RETIRED. *Mem:* AMA; Am Acad Pediat. *Mailing Add:* 11550 Wander Lane Dallas TX 75230

NORMAN, GARY L, b Boston, Ma, June 18, 50. CANCER RESEARCH, AIDS VIRUS. *Educ:* Univ Southern Calif, PhD(molecular biol), 80. *Prof Exp:* Res assoc, 82-86, ASST PROF CLIN PATH, UNIV SOUTHERN CALIF, 86- *Mem:* AAAS; Sigma Xi; Am Soc Cell Biol. *Mailing Add:* 2721 Sixth St Apt 206 Santa Monica CA 90405

NORMAN, HOWARD DUANE, b Liberty, Pa, Nov 4, 42; m 74; c 3. ANIMAL BREEDING. *Educ:* Pa State Univ, BS, 64, MS, 67; Cornell Univ, PhD(animal breeding), 70. *Prof Exp:* RES GENETICIST, ANIMAL SCI INST, BELTSVILLE AGR RES CTR, AGR RES SERV, USDA, 70- *Mem:* Am Dairy Sci Asn. *Res:* Developed procedures to improve the accuracy of estimated transmitting ability for milk yield and composition thereby increasing the genetic capability of the United States dairy population. *Mailing Add:* Bldg 263 BARC-E 10300 Baltimore Ave Beltsville MD 20705-2350

NORMAN, JACK C, b Taunton, Mass, June 16, 38; m 64; c 2. NUCLEAR CHEMISTRY, RADIOCHEMISTRY. *Educ:* Univ NH, BS, 60; Univ Wis-Madison, PhD(phys chem), 65. *Prof Exp:* Instr chem, Univ Wash, 65-66; asst prof, Univ Ky, 66-68; asst prof ecosysts anal, 68-71, assoc prof sci & environ change, 77-88, ASSOC PROF ECOSYSTS ANAL, UNIV WISGREEN BAY, 71-, PROF NATURAL & APPL SCI, 88- *Mem:* AAAS; Am Chem Soc; Am Phys Soc. *Res:* Pulp and paper chemicals recovery and clean-up; radionuclides in the environment. *Mailing Add:* Natural & Applied Scis Univ Wis-Green Bay Green Bay WI 54311-7001

NORMAN, JAMES EVERETT, JR, b Washington, DC, July 11, 39; m 75; c 1. EPIDEMIOLOGY, BIOSTATISTICS. *Educ:* Univ Ala, BS, 57, MS, 60; Va Polytech Inst, PhD(statist), 65. *Prof Exp:* Mathematician atmospheric res, US Army Ballistic Missile Agency, 60; physicist appl physics, Bendix Res Lab, 60-61; asst prof statist, Univ Mass, 65-67 & Univ Ga, 67-72; statistician epidemiol, Nat Acad Sci Med Follow Up Agency, 72-75; statistician epidemiol & statist dept, Radiation Effects Res Found, 75-77; statistician, Nat Res Coun-Nat Acad Sci Med Follow Up Agency, 77-85; RETIRED. *Mem:* Am Statist Asn. *Res:* Epidemiologic studies of chronic diseases; statistical methods and design in epidemiology; epidemiologic studies of multiple sclerosis; Hodkin's disease; liver cancer in relation to a US Army hypatitis eperdermia in World War II. *Mailing Add:* 9909 Montclair Ct Vienna VA 22181

NORMAN, JOE G, JR, b Brevard, NC, Aug 8, 47. TRANSITION METAL CHEMISTRY. *Educ:* Rice Univ, BA, 69; Mass Inst Technol, PhD(chem), 72. *Prof Exp:* From asst prof to assoc prof chem, 72-82, assoc dean, Grad Sch, 80-87, PROF CHEM, UNIV WASH, 82-, DEAN, COL ARTS & SCI, 87- *Mem:* Am Chem Soc; Chem Soc; Am Crystallog Asn. *Res:* Electronic structure of large molecules containing transition metals. *Mailing Add:* Col Arts & Sci GN-15 Univ Wash Seattle WA 98195

NORMAN, JOHN HARRIS, b Battle Creek, Mich, Apr 13, 29; m 50; c 3. PHYSICAL CHEMISTRY, NUCLEAR SCIENCE. *Educ:* Univ Mich, BS, 50; Univ Wis, PhD(chem), 54. *Prof Exp:* Chemist mass spectros, Olin Mathieson Chem Corp, 54-60; staff mem, 60-70, sr staff mem chem, 70-86, CONSULT, GEN ATOMIC CO, 86- *Concurrent Pos:* Vis scientist, J A Eri, Tokai, Japan, 87. *Mem:* Am Chem Soc; Am Nuclear Soc; Int Asn Hydrogen Energy. *Res:* Mass spectrometry; high temperature chemistry; transport phenomena; chemistry of fallout; nuclear chemistry; nuclear safety; thermochemical water splitting; thermoelectrics; solid state science; thermodynamics. *Mailing Add:* 8312 Sugarman Dr La Jolla CA 92037

NORMAN, JOHN MATTHEW, b Virginia, Minn, Nov 27, 42; m 67; c 2. AGRICULTURAL METEOROLOGY, ENVIRONMENTAL BIOPHYSICS. *Educ:* Univ Minn, BS, 64, MS, 67; Univ Wis-Madison, PhD(soil sci), 71. *Prof Exp:* From asst prof to assoc prof meteorol, Pa State Univ, University Park, 72-78; prof agron, Univ Neb, Lincoln, 78-88; PROF SOIL SCI, UNIV WIS-MADISON, 88- *Concurrent Pos:* Res fel, Dept Bot, Univ Aberdeen, Scotland, 71-72; vis prof, Commonwealth Sci & Indust Orgn, Div Forestry, Canberra, Australia, 83; vis prof, Dept Plant Sci, Univ Guelph, Ont, 84; ed, Agron Jour, 81-84; consult, San Diego Stat Univ, 85-87, LI-Cor, Inc, 82-, Mesomet Inc, 77. *Mem:* Sigma Xi; AAAS; Am Meteorol Soc; fel Am Soc Agron; fel Crop Sci Soc Am. *Res:* Studies of the interactions between plants and their environments including measurements of soil, plant and atmospheric characteristics and integrative modeling of the entire soil-plant-atmosphere system. *Mailing Add:* Dept Soil Sci Univ Wis-Madison 1525 Observatory Dr Madison WI 53706

NORMAN, L(EWIS) A(RTHUR), JR, economic geology, mining engineering; deceased, see previous edition for last biography

NORMAN, OSCAR LORIS, b Crawfordsville, Ind, Apr 28, 25; m 46; c 2. ORGANIC CHEMISTRY. *Educ:* Wabash Col, AB, 47; Northwestern Univ, MS, 49; Purdue Univ, PhD(chem), 53. *Prof Exp:* Sr res chemist, Int Mineral & Chem Corp, 52-58, mgr food prod res, 58-60, mgr biochem & chem res, 60-61; mgr prod develop, Sun Oil Co, 61-62, sect chief basic res, 62-64, mgr, Basic Res, 65-77, mgr, Prod Develop, 77-83; RETIRED. *Mem:* AAAS; Am Chem Soc; Sigma Xi. *Res:* Biochemistry. *Mailing Add:* 707 Bristol Rd Wilmington DE 19803

NORMAN, PHILIP SIDNEY, b Pittsburg, Kans, Aug 4, 24; m 55; c 3. IMMUNOLOGY, ALLERGY. *Educ:* Wash Univ, MD, 51. *Prof Exp:* Intern, Barnes Hosp, St Louis, Mo, 51-52; asst resident, Vanderbilt Univ Hosp, 52-54; USPHS fel, Rockefeller Inst, 54-56; from instr to assoc prof, 56-74, PROF MED, JOHNS HOPKINS UNIV, 75-, HEAD CLIN IMMUNOL DIV, 71-, PHYSICIAN, HOSP, 59- *Concurrent Pos:* Head allergy serv & physician, Good Samaritan Hosp, Baltimore, Md, 71-; physician, Francis Scott Key Med Ctr, 89- *Mem:* Am Acad Allergy (pres 75-76); Am Soc Clin Invest; Am Asn Immunol; NY Acad Sci; Am Clin Climat Soc; Am Asn Physics. *Res:* Antigens of ragweed; hay fever; asthma. *Mailing Add:* Johns Hopkins Asthma & Allergy Ctr 301 Bayview Blvd Baltimore MD 21224

NORMAN, REID LYNN, b Scott City, Kans, Feb 26, 44; m 67; c 2. NEUROENDOCRINOLOGY. *Educ:* Kans State Univ, BS, 66, MS, 68; Univ Kans, PhD(anat), 71. *Prof Exp:* Fel neuroendocrinol, Univ Calif, Los Angeles, 71-72; asst scientist & scientist reprod physiol, Ore Regional Primate Res Ctr, 72-83; from asst prof anat, Ore Health Sci Univ, 73-83; PROF & ASSOC CHMN, CELL BIOL & ANAT, HEALTH SCI CTR, TEX TECH UNIV, 83- *Concurrent Pos:* Prin investr, NIH grant, 74-; ed, Neuroendocrine Aspects Reproduction, 83 & Reproductive Biol Study Sect, 90-94. *Mem:* Am Asn Anatomists; Soc Study Reprod; Endocrine Soc; Am Physiol Soc; Soc Neurosci. *Res:* Anatomical and physiological regulation of anterior pituitary function by the central nervous system. *Mailing Add:* Cell Biol & Anat Health Sci Ctr Texas Tech Univ Lubbock TX 79430

NORMAN, RICHARD DAVIESS, b Franklin, Ind, Feb 7, 27; m 51; c 2. DENTISTRY, ANALYTICAL CHEMISTRY. *Educ:* Franklin Col Ind, AB, 50; Ind Univ, DDS, 58, MSD, 64. *Prof Exp:* Anal chemist, Eli Lilly & Co, 50-54; res asst chem, Ind Univ Sch Dent, 55-58, from instr to prof dent mat, 58-76; dir dent clin res, J & J Dent Prod Co, 76-79; prof prosthetics, Fairleigh-Dickinson Univ, 79-80; chmn dept, 80-85, PROF RESTORATIVE DENT, SOUTHERN ILL UNIV SCH DENT MED, 80-, DIR RES, 85- *Concurrent Pos:* Chmn, Working Group III, Am Nat Standard Comn, 73-; consult, 79-; mem, Oral Biol & Med Study Sect, NIH, 80-85 & Test Construct Comt, Am Dent Asn, 80-86; vis lectr restorative dent, Wash Univ, 81- *Mem:* Am Dent Asn; Int Asn Dent Res; Am Asn Dent Schs; Sigma Xi; Am Col Dentists. *Res:* Constituents of teeth; metals used in dentistry; investments; cements; intra-oral currents and pH; wear of materials; fluorides in dentistry. *Mailing Add:* Five Monterey Pl Alton IL 62002

NORMAN, ROBERT DANIEL, b New York, NY, Nov 3, 38; Can citizen; m 62; c 3. MATHEMATICS. *Educ:* Univ Toronto, BA, 60; Queens Univ, Ont, MA, 62; Univ London, PhD(math), 64. *Prof Exp:* Fel, 64-65, asst prof, 65-71, ASSOC PROF MATH, QUEENS UNIV, ONT, 71- *Mem:* Can Math Soc; Am Math Soc. *Res:* Dynamical systems. *Mailing Add:* Math and Stat Dept Queens Univ Kingston ON K7L 3N6 Can

NORMAN, ROBERT ZANE, b Chicago, Ill, Dec 16, 24; m 52; c 2. MATHEMATICS. *Educ:* Swarthmore Col, AB, 49; Univ Mich, AM, 50, PhD, 54. *Prof Exp:* Instr math, Princeton Univ, 54-56; assoc prof, 56-66, PROF MATH, DARTMOUTH COL, 66-, CHMN PROG IN MATH & SOC SCI, 71- *Mem:* Am Math Soc; Math Asn Am. *Res:* Theory of graphs; combinatorial analysis; mathematical models in the social sciences. *Mailing Add:* Dept Math & Soc Sci Dartmouth Col Hanover NH 03755

NORMAN, ROGER ATKINSON, JR, b Danville, Ky, Oct 16, 46; m 80. PHYSIOLOGY, CHEMICAL ENGINEERING. *Educ:* Univ Miss, BS, 68, MS, 71, PhD(biomed eng), 73. *Prof Exp:* Res assoc physiol, 72-73, from asst prof to assoc prof, Sch Med, 73-84, PROF, PHYSIOL & BIOPHYS, MED CTR, UNIV MISS, 84- *Concurrent Pos:* Coun High Blood Pressure Res, Am Heart Asn. *Mem:* Sigma Xi; AAAS; Am Phys Soc; Int Soc Hypertension; Am Heart Asn. *Res:* Cardiovascular physiology; hypertension. *Mailing Add:* Dept of Physiol & Biophys Univ of Miss Med Ctr 2500 N State St Jackson MS 39216

NORMAN, WESLEY P, b Marion, Ill, Aug 14, 28. DEVELOPMENTAL BIOLOGY, NEUROANATOMY. *Educ:* Southern Ill Univ, BA, 52, MA, 54; Univ Ill, PhD(anat), 67. *Prof Exp:* Asst histologist, Am Meat Inst Found, 62-64; instr anat, Col Med, Univ Ill, 66-67; asst prof, 67-72, ASSOC PROF ANAT, GEORGETOWN UNIV, 72- *Concurrent Pos:* Consult & lectr otolaryngol basic sci course, Armed Forces Inst Path, 67- & Walter Reed Inst Dent Res & Naval Dent Sch, Bethesda, Md, 68- *Mem:* Am Asn Anat; Am Soc Cell Biol; Am Soc Zool; Sigma Xi. *Res:* Regeneration of the forelimb of the adult newt, Diemictylus viridescens, specifically skeletal muscle regeneration; histochemistry; autoradiography of fibrillogenesis; electron microscopy; analysis of denervated and reinnervated muscle spindles. *Mailing Add:* Georgetown Univ Med Sch 3900 Reservoir Dr NW Washington DC 20007

NORMANDIN, RAYMOND O, plastics; deceased, see previous edition for last biography

NORMANDIN, ROBERT F, b Laconia, NH, July 22, 27; m 54; c 3. RADIATION BIOLOGY. *Educ:* St Anselm's Col, AB, 50; Univ NH, MS, 53; Ohio State Univ, PhD(zool), 59. *Prof Exp:* Res biologist, NH Fish & Game Dept, 53-54; cur path mus, Ohio State Univ, 56-59; res biologist, US Fish & Wildlife Serv, 59-60; assoc prof biol, 60-70, PROF BIOL, ST ANSELM'S COL, 70- *Concurrent Pos:* Consult pvt fishery, Ohio, 57-59; consult, NH Water Pollution Comn, 60-66; chmn, NH Comn Radiation Control, 62-74; consult lectr, NH Civil Defense Orgn, 64-66; chmn, NH Legis Comn Prof Nursing, 66; mem, NH Adv Comprehensive Health Planning Coun, 69. *Mem:* AAAS; Am Inst Biol Scientists. *Res:* Water pollutional control, especially algal blooms; radiation biology, especially physiology and pathology of radio-sensitivity radio-ecology. *Mailing Add:* St Anselm's Col Manchester NH 03102

NORMANN, RICHARD A, b Oakland, Calif, Feb 8, 43. BIOMEDICAL ENGINEERING. *Educ:* Univ Calif, Berkeley, BS, 65, MS, 67, PhD(elec eng), 73. *Prof Exp:* From asst prof to assoc prof, 79-88, PROF BIOENG, PHYSIOL & OPHTHAL, UNIV UTAH, 88- *Mem:* Biomed Eng Soc. *Mailing Add:* Dept Bioeng 2480 MEB Univ Utah Salt Lake City UT 84112

NORMANN, SIGURD JOHNS, b Cincinnati, Ohio, Oct 24, 35; m 65; c 2. PATHOLOGY. *Educ:* Univ Wash, MD, 60, PhD(path), 66. *Prof Exp:* Intern surg, Univ Calif, San Francisco, 60-61; resident path, Univ Washington, 61-66; US Army Medical Corp, 66-68; from asst prof to assoc prof, 68-76, PROF PATH, UNIV FLA, 76- *Concurrent Pos:* Consult path, Vet Admin Hosp, Gainesville, 68-; Nat Insts Allergy & Infectious Dis res career develop award, Univ Fla, 70-75; vis cardiovasc path, Northwick Park Hosp, England, 75; consult, Swiss Inst Med Res, Davos, Switz, 75-87; sci review comt, Gainesville Vet Admin Hosp, 77-; sci review comt, Fla Div, Am Cancer Soc, 78-88. *Mem:* AAAS; Am Asn Path; Int Acad Path; Reticuloendothelial Soc (secy, 70-73, pres-elect, 78, pres, 79); Am Asn Cancer Res; Int Soc Heart Transplantation; Soc Cardiovasc Path. *Res:* Monocyte and macrophage heterogeneity; macrophage function; host inflammatory reactions; factors mediating cancer induced immune suppression; cardiac pathology; cardiovascular pathology. *Mailing Add:* Dept Path Univ Fla Box J-275 JHMHC Gainesville FL 32601

NORMANSELL, DAVID E, b Smethwick, UK, Jan 13, 38; m 62; c 2. MEDICAL LABORATORY IMMUNOLOGY. *Educ:* Univ Birmingham, UK, BSc, 59, PhD(chem), 63. *Prof Exp:* Res asst chem, Univ Birmingham, UK, 62-63, res fel exp path, 63-67; sr res fel biochem, 67-68, from asst prof to assoc prof, 68-87, PROF PATH & MICROBIOL, UNIV VA, 87- *Concurrent Pos:* Vis investr, Scripps Clin & Res Found, La Jolla, Calif, 83-84. *Mem:* Am Asn Immunologists; Am Soc Microbiol; Asn Clin Lab Physicians & Scientists; AAAS; Brit Soc Immunol. *Res:* Immune response in autoimmune diseases, particularly rheumatoid arthritis; human monoclonal antibodies; candida antigens. *Mailing Add:* Div Clin Immunol Dept Path Med Sch Univ Va Charlottesville VA 22908

NORMARK, WILLIAM RAYMOND, b Seattle, Wash, Jan 21, 43; m 67. OCEANOGRAPHY, MARINE GEOLOGY. *Educ:* Stanford Univ, BS, 65; Univ Calif, San Diego, PhD(oceanog), 69. *Prof Exp:* Res oceanogr, Scripps Inst Oceanog, 69-70; asst prof geol & oceanog, Univ Minn, Minneapolis, 70-74; GEOLOGIST, PAC-ARCTIC BR MARINE GEOL, US GEOL SURV, 74- *Mem:* AAAS; Geol Soc Am; Am Geophys Union; Soc Explor Geophys. *Res:* Continental margin sedimentation, particularly deep-sea turbidites, growth patterns of deep-sea fans, structure and history of continental margins and evolution of lithospheric plate boundaries; erosion of deep-sea sediment. *Mailing Add:* US Geol Surv Mail Stop 999 345 Middlefield Rd Menlo Park CA 94025

NORMENT, BEVERLY RAY, b Whiteville, Tenn, July 23, 41; m 64; c 1. ENTOMOLOGY. *Educ:* Memphis State Univ, BS, 64, MS, 66; Miss State Univ, PhD(entom), 69. *Prof Exp:* Asst zool, Memphis State Univ, 64-66, instr, 66; from res asst to asst prof entom, 66-74, assoc prof, 74-81, PROF ENTOM, MISS STATE UNIV, 81- *Mem:* Am Soc Trop Med & Hygiene; Int Soc Toxinology; Entom Soc Am; Am Mosquito Control Asn. *Res:* Medical entomology; bioassay techniques; enzyme assays; toxinological studies; biological control. *Mailing Add:* PO Box 6343 Mississippi State MS 39762

NORMENT, HILLYER GAVIN, b Washington, DC, Jan 13, 28; m 73; c 3. ATMOSPHERIC SCIENCE, PHYSICAL CHEMISTRY. *Educ:* Univ Md, BS, 51, PhD(phys chem), 56. *Prof Exp:* Group leader x-ray diffraction, Callery Chem Co, 56-59; phys chemist, US Naval Res Lab, 59-62; opers res analyst nuclear fallout, Res Triangle Inst, 62-63; sr scientist nuclear fallout modeling, Tech Opers Inc, 63-67; sr scientist, Arcon Corp, 67-71; prin scientist nuclear fallout cloud physics, Mt Auburn Res Assoc, 71-75; PROPRIETOR NUCLEAR FALLOUT CLOUD PHYSICS AIR POLLUTION, ATMOSPHERIC SCI ASN, 75- *Mem:* Sigma Xi; AAAS; Air Pollution Control Asn; Am Meteorol Soc; Am Geophys Union; Royal Meteorol Soc. *Res:* Cloud physics; modeling of atmospheric transport processes; air pollution modeling; atmospheric turbulence; aerosol physics. *Mailing Add:* 186 Peter Spring Rd Concord MA 01742

NORMILE, HUBERT CLARENCE, b Los Angeles, Calif, Apr 16, 06; m 32; c 3. ANALYTICAL CHEMISTRY. *Educ:* Northeastern Univ, BS, 29. *Prof Exp:* Chem engr, Bethlehem Steel Co, 29-31, US Army, Philadelphia, 31-46; chemist, Vet Admin, 46-51; CHIEF CHEMIST, US AIR FORCE SPACE PROG, CAPE CANAVERAL, FLA, 51- *Mem:* Am Chem Soc; Am Inst Aeronaut & Astronaut; Am Inst Chem Engrs. *Res:* Testing of materials and contamination control for the United States Space Program. *Mailing Add:* 2129 W New Haven Ave Apt 131 Melbourne FL 32904-3849

NORMINTON, EDWARD JOSEPH, b Hensall, Ont, Sept 8, 38; m 63; c 2. APPLIED MATHEMATICS. *Educ:* Univ Western Ont, BA, 61, MA, 62; Univ Toronto, PhD(fluid dynamics), 65. *Prof Exp:* Asst prof math, 65-69, ASSOC PROF MATH, CARLETON UNIV, 69- *Mem:* Can Math Cong. *Res:* Mathematical software. *Mailing Add:* Dept Math Carleton Univ Colonel Dr Ottawa ON K1S 5B6 Can

NORNES, HOWARD ONSGAARD, b Winger, Minn, Apr 27, 31; m 58; c 3. NEUROBIOLOGY, DEVELOPMENTAL BIOLOGY. *Educ:* Concordia Col, BA, 53; Purdue Univ, Lafayette, MS, 63, PhD(biol), 71. *Prof Exp:* Teacher biol, Richfield Pub Sch, 58-66; from instr to lectr, Purdue Univ, Lafayette, 66-72; ASSOC PROF, COLO STATE UNIV, 72- *Mem:* Am Soc Zoologists; Soc Neurosci. *Res:* Developmental neurobiology. *Mailing Add:* Dept of Anat Colo State Univ Ft Collins CO 80523

NORNES, SHERMAN BERDEEN, b Winger, Minn, Jan 10, 29; m 53; c 2. SURFACE PHYSICS. *Educ:* Concordia Col, Moorhead, Minn, BA, 51; Univ NDak, MS, 56; Wash State Univ, PhD(physics), 65. *Prof Exp:* Res engr, Rocketdyne Inc Div, NAm Aviation, Inc, 56-59; assoc prof, 59-61, chmn dept, 67-81, PROF PHYSICS, PAC LUTHERAN UNIV, 65- *Concurrent Pos:* Consult physicist, Lawrence Livermore Lab, 74-81, Rockwell Int, Rockflats Lab, Boulder, Colo, 81-82; vis scientist, Max Planck Inst für Stromungsforschung Göttingen, W Germany, 88-89. *Mem:* Am Vacuum Soc; Am Asn Physics Teachers. *Res:* X-ray photo electron spectroscopy and avger electron spectroscopy studies of surface reactions of Actinides; physics of the interaction of spectroscopically pure gases with ultra clean metal surfaces. *Mailing Add:* Dept of Physics Pac Lutheran Univ S 121st & Park Ave Tacoma WA 98447

NORONHA, FERNANDO M OLIVEIRA, b Portugal, Feb 10, 24; m 60; c 2. VIROLOGY. *Educ:* Lisbon Tech Univ, DVM, 49. *Prof Exp:* WHO fel, Col Med, Univ Montpellier, 50-52; French Acad bursary, Pasteur Inst, Paris, 53-55; researcher, Virus Res Inst, Eng, 56; researcher, Nat Inst Sch Higher Vet Med, Lisbon Tech Univ, 59-63; assoc prof virol, 64-66, PROF VIROL, NY STATE VET COL, CORNELL UNIV, 66- *Concurrent Pos:* Portuguese Govt scholar, Animal Virus Inst, Univ Tubingen, 59-63; resident, Portuguese Inst Vet Res, 59-63. *Honors & Awards:* Chevalier de Merite Agricole, Fr Govt, 50. *Mem:* Am Soc Microbiol; Portuguese Soc Vet Med. *Res:* Virus oncology. *Mailing Add:* Dept Microbiol Cornell Univ Ithaca NY 14853

NORRBOM, ALLEN LEE, b Darby, Pa, Nov 13, 57. SYSTEMATIC ENTOMOLOGY, DIPTEROLOGY. *Educ:* Drexel Univ, BS, 80; Pa State Univ, MS, 83, PhD(entom), 85. *Prof Exp:* Teaching asst, dept entom, Pa State Univ, 80-85; RES ENTOMOLOGIST, SYST ENTOM LAB, USDA, 85- *Honors & Awards:* Asa Fitch Award, Entom Soc Am, 84. *Mem:* Entom Soc Am; Am Entom Soc. *Res:* Systematics of acalyptrate Diptera, especially families Tephritidae (true fruit flies) and Sphaeroceridae (lesser dung flies); biology and evolution of phytophagous and saprophagous Diptera. *Mailing Add:* 106 Patterson Bldg University Park PA 16802

NORRDIN, ROBERT W, b Brooklyn, NY, Oct 2, 37; m 63; c 4. VETERINARY PATHOLOGY. *Educ:* Brooklyn Col, BS, 58; Cornell Univ, DVM, 62, PhD(vet path), 69. *Prof Exp:* Gen pract vet med, Flemington, NJ & Locke, NY, 62-65; asst prof vet med, NY State Vet Col, Cornell Univ, 65-66, NIH trainee nutrit path, 66-69; ASSOC PROF VET PATH, DEPT PATH & COLLABR, RADIOL HEALTH LAB, COL VET MED & BIOMED SCI, COLO STATE UNIV, 69- *Concurrent Pos:* NIH & Med Res foreign fel, Res Unit 18, Hospital Lariboisiere, Paris, 75-76. *Mem:* Am Col Vet Pathologists; Am Vet Med Asn. *Res:* Multifaceted studies of nutritional and metabolic bone diseases in animals; metabolic studies; bone density and composition; bone cell cultures; histomorphometric evaluation of bone and pertinent endocrine organs. *Mailing Add:* Dept Vet Path Colo State Univ Ft Collins CO 80523

NORRED, WILLIAM PRESTON, b Tallassee, Ala, July 11, 45; m 69; c 3. PHARMACOLOGY, TOXICOLOGY. *Educ:* Emory Univ, BA, 66; Univ Ga, BS, 69, PhD(pharmacol), 71. *Prof Exp:* RES PHARMACOLOGIST, RICHARD B RUSSELL AGR RES CTR, USDA-AGR RES SERV, 71- *Concurrent Pos:* Adj assoc prof, Sch Pharm, Univ Ga, 76- & Grad Sch, 78- *Mem:* Am Soc Pharmacol & Exp Therapeut; Soc Toxicol; AAAS; Sigma Xi. *Res:* Drug metabolism; mycotoxins. *Mailing Add:* Toxicol Lab USDA-Agr Res Serv Russell Agr Res Ctr PO Box 5677 Athens GA 30604

NORRIE, D(OUGLAS) H, b Wellington, NZ, Dec 4, 29; m 54; c 4. INDUSTRIAL & MANUFACTURING ENGINEERING, INTELLIGENT SYSTEMS. *Educ:* Univ Canterbury, BE, 51, Hons, 53; Univ Otago, NZ, BSc, 52; Univ Adelaide, PhD(mech eng), 65. *Prof Exp:* Design draftsman, Hamilton & Co, NZ, 53-54; grad apprentice, Rolls-Royce, Derby, UK, 54-55; tech off, Fairey Aviation, SAustralia, 55-57; lectr mech eng, Adelaide, 57-58, sr lectr, 58-66; prof & head dept, 66-71, dir info servs, 71-74, PROF MECH ENG, UNIV CALGARY, 74-, ASSOC HEAD MECH ENG, 90-, HEAD, DIV MFG ENG, 90- *Concurrent Pos:* Mem adv comt eng res, Defence Res Bd Can, 67-74, chmn, 71 & Adv Bd Sci & Tech Info, Nat Res Coun Can, 74-77; chmn, Adv Comt Marine Physics & Eng, Defense Res Bd Can, 72-74 & Human Context Sci & Technol Grant Selection Comt, Social Sci & Humanities Res Coun, 82-83; mem, Nat Sci Eng Res Coun, Mech Eng Grant Sel Comt, 80-83; chmn, Int Conf Comput Assisted Learning, 86-87; co-chmn, Int Steer Comt, Int Conf Series Comput Assisted Learning, 87-; co-chmn prog comt, Int Conf Comput Assisted Learning, 88-90. *Honors & Awards:* Killam Res fel, 79; Inst Comp Assisted Learning fel, 87. *Mem:* Soc Naval Archit & Marine Engrs; Royal Aeronaut Soc; Brit Inst Mech Engrs; Am Asn Artificial Intel; Royal Inst Naval Architects; Soc Mfg Eng. *Res:* Fluid mechanics; finite element methods; computer aided design and manufacturing; computer assisted learning; artificial intelligence. *Mailing Add:* Dept Mech Eng Univ Calgary Calgary AB T2N 1N4 Can

NORRIS, A R, b Meadow Lake, Sask, May 18, 37; m 60. INORGANIC CHEMISTRY, SYNTHETIC INORGANIC CHEMISTRY. *Educ:* Univ Sask, BE, 58, MSc, 59; Univ Chicago, PhD(chem), 62. *Prof Exp:* Res assoc, Univ Chicago, 60-61; Nat Res Coun fel, Univ Col London, 62-64; from asst prof to assoc prof, 63-81, PROF CHEM, QUEEN'S UNIV, ONT, 82- *Concurrent Pos:* Vis scholar chem, Stanford Univ, 71-72; vis res fel chem, Clarkson Col Sci & Technol, 81-82. *Mem:* Assoc Prof Eng Ont; fel Chem Inst Can. *Res:* Metal ion-biomolecule interactions; 199 Hg nuclear magnetic resonance of Hg(II) and CH3 Hg(II)-biomolecule complexes; 13C and 199 Hg nuclear magnetic resonance studies of CH3 Hg(II) exchange reactions; oxidation, reduction and addition reactions of coordinated ligands in transition metal complexes; organometallic chemistry. *Mailing Add:* 55 Jane Ave Kingston ON K7M 3G7 Can

NORRIS, ALBERT STANLEY, b Sudbury, Ont, July 14, 26; m 50; c 3. PSYCHIATRY. *Educ:* Univ Western Ont, MD, 51. *Prof Exp:* Fel, Harvard Med Sch, 55-56; instr psychiat, Med Sch, Queen's Univ, Ont, 56-57; from asst prof to assoc prof, Col Med, Univ Iowa, 57-64; assoc prof, Med Sch, Univ Ore, 64-65; from assoc prof to prof, Col Med, Univ Iowa, 65-72; prof psychiat & chmn dept, Sch Med, 72-84, EMER PROF, SOUTHERN ILL UNIV, 85- *Mem:* AMA; fel Am Psychiat Asn. *Res:* Anatomical and physiological traits which predispose the development of mental illness; prenatal influences affecting intellectual and emotional development; capillary morphology in mental illness; psychosomatic obstetrics and gynecology; investigations of the efficacy of LSD 25 in the treatment of sexual deviation. *Mailing Add:* Cedar River Tower Suite 133 Cedar Rapids IA 52401

NORRIS, ANDREW EDWARD, b Santa Rosa, Calif, Jan 13, 37; m 79; c 1. NUCLEAR CHEMISTRY. *Educ:* Univ Chicago, SB, 58; Wash Univ, PhD(chem), 63. *Prof Exp:* Res assoc chem, Wash Univ, 63-64 & Brookhaven Nat Lab, 64-66; STAFF MEM, LOS ALAMOS NAT LAB, 66- *Concurrent Pos:* Vis guest chemist, Lawrence Berkeley Lab, 73-74; asst for res, Off of Dir, Los Alamos Sci Lab, 75-77; prin investr geochem, Nev Nuclear Waste Storage Invests, Yucca Mountain Proj, 82-90; proj leader, Los Alamos Environ Restoration Prog, 90- *Mem:* Sigma Xi; Am Chem Soc; Am Phys Soc; AAAS. *Res:* Fission yields; heavy ion reaction; nuclear waste management. *Mailing Add:* PO Box 1663 Los Alamos NM 87545

NORRIS, BILL EUGENE, biology, for more information see previous edition

NORRIS, CARROLL BOYD, JR, b New Orleans, La, Apr 28, 41. PHYSICS, ELECTRICAL ENGINEERING. *Educ:* Stanford Univ, BSEE, 63, MSEE, 64, PhD(elec eng), 67. *Prof Exp:* Scientist phys electronics, Lockheed Res Lab, 63; res asst device physics, Stanford Electronics Labs, 64-66, res assoc, 66-70; MEM TECH STAFF SEMICONDUCTOR PHYSICS, SANDIA LABS, 70- *Concurrent Pos:* Consult, Hewlett-Packard Assocs, 67-68, Gen Elec Res Lab, 68 & Watkins-Johnson Co, 68-70. *Mem:* Am Phys Soc. *Res:* Luminescence and the nature of optical transitions in compound semiconductors; physics of charge transport in semiconductors; irradiation and ion implantation effects in semiconductors; electron beam-semiconductor active devices. *Mailing Add:* 1117 Turner Dr NE Albuquerque NM 87123

NORRIS, DALE MELVIN, JR, b Essex, Iowa, Aug 19, 30; m, 90; c 2. NEUROBIOLOGY, PLANT PHYSIOLOGY. *Educ:* Iowa State Univ, BS, 52, MS, 53, PhD(zool, plant path), 56. *Prof Exp:* Asst entomologist, Agr Exp Sta, Univ Fla, 56-57; from asst prof to assoc prof entom, 58-66, PROF ENTOM, UNIV WIS-MADISON, 66- *Concurrent Pos:* Vis lectr various univs and cols, 63-; sci consult, 63- *Honors & Awards:* Founders Mem Award, Entomol Soc Am, 75. *Mem:* AAAS; Entom Soc Am; Asn Chemoreception Sci; Bioelectrochem Soc; Biophys Soc. *Res:* Insect interactions with plants; symbiosis; chemoreception; neurobiology; stress biochemistry; antioxidents; three US patents and over 160 scientific papers. *Mailing Add:* Univ Wis Dept Entom 642 Russell Lab Madison WI 53706

NORRIS, DANIEL HOWARD, b Toledo, Ohio, Dec 29, 33; m 58. BOTANY. *Educ:* Mich State Univ, BS, 54; Univ Tenn, PhD(bot), 64. *Prof Exp:* Instr bot, Univ Tenn, 59-60; asst prof biol, Cent Methodist Col, 61-63; assoc prof, Catonsville Community Col, 64-67; asst prof biol, 67-74, PROF BOT, HUMBOLDT STATE UNIV, 74- *Mem:* AAAS; Am Bryol & Lichenol Soc; Nordic Bryol Soc; Brit Bryol Soc; Bot Soc Am. *Res:* Bryogeography and taxonomy of Dominican Republic, Newfoundland, California and the tropical Pacific Islands; bryoecology of Great Smoky Mountains National Park; phytogeography. *Mailing Add:* Dept of Bot Humboldt State Univ Arcata CA 95521

NORRIS, DAVID OTTO, b Ashtabula, Ohio, Oct 1, 39; m 66; c 2. COMPARATIVE ENDOCRINOLOGY. *Educ:* Baldwin-Wallace Col, BS, 61; Univ Wash, PhD(fish thyroid), 66. *Prof Exp:* From asst prof to assoc prof, 66-77, PROF BIOL, UNIV COLO, BOULDER, 77- *Mem:* Herpetologists League; Am Soc Zoologists; Sigma Xi; Endocrine Soc; Soc Study Amphibians & Reptiles. *Res:* Comparative endocrinology of lower vertebrates; influences of environmental factors on endocrine activity; physiology and behavior in relation to life history events, especially reproduction and metamorphosis. *Mailing Add:* Dept Environ Pop & Organismic Biol Campus Box 334 Univ Colo Boulder CO 80309-0334

NORRIS, DEAN RAYBURN, b Indianola, Iowa, Jan 21, 37; m 61; c 2. PHYCOLOGY, PHYTOPLANKTON. *Educ:* Iowa State Univ, BS, 59; Tex A&M Univ, MS, 67, PhD(oceanog), 69. *Prof Exp:* Discipline scientist, Nat Aeronaut & Space Admin, Johnson Space Ctr, 69-75; assoc prof, 75-85, PROF OCEANOG, FLA INST TECHNOL, 85 - *Concurrent Pos:* Co-investr, Overflight Tektite II, Johnson Space Ctr, NASA, 70; mem, Oceanog Working Group, Shuttle Sortie Workshop, 72; vis scientist, Scripps Inst Oceanog, NASA, 74, aboard Res Vessel, Va Key, 73. *Mem:* Am Soc Limnol & Oceanog; Int Phycological Soc; Sigma Xi; Phycol Soc Am. *Res:* Ecology and taxonomy of marine phytoplankton, in particular the dinoflagellates; effects of various toxic substances on phytoplankton; effects of increased ultraviolet-light on phytoplankton; life-cycle phenomena and ecology of selected dinoflagellates; biology and ecology of benthic dinoflagellates. *Mailing Add:* 5680 Live Oak Ave Melbourne FL 34773

NORRIS, DONALD EARL, JR, b Hammond, Ind, Oct 16, 40; m 64; c 1. PARASITOLOGY. *Educ:* Ind State Univ, Terre Haute, BS, 63; Tulane Univ, MS, 66, PhD(parasitol), 69. *Prof Exp:* From asst prof to assoc prof, 70-81, PROF BIOL, UNIV SOUTHERN MISS, 81- *Mem:* AAAS; Am Soc Parasitol; Am Soc Trop Med & Hyg; Royal Soc Trop Med & Hyg; Wildlife Dis Asn; Sigma Xi. *Res:* Life histories of parasites; helminthology. *Mailing Add:* Dept of Biol Univ of Southern Miss Hattiesburg MS 39401

NORRIS, DONALD KRING, geology, for more information see previous edition

NORRIS, EUGENE MICHAEL, b New York, NY, July 4, 38; m 69; c 3. COMPUTER SCIENCE, SOFTWARE SYSTEMS. *Educ:* Univ SFla, BA, 64, PhD(math), 69. *Prof Exp:* Lectr math, Univ Fla, 65-66 & 68-69; asst prof, WVa Univ, 69-72; asst prof math & comput sci, Univ SC, 72-78; analyst, Ketron, Inc, 78-80; dept chmn, 84-86, ASSOC PROF COMPUT SCI, GEORGE MASON UNIV, 80- *Concurrent Pos:* Consult. *Mem:* AAAS; Sigma Xi; Asn Comput Mach. *Res:* Applications of neural models to cognition problems; user interface design. *Mailing Add:* 5249 Ridge Ct Fairfax VA 22032-2631

NORRIS, FLETCHER R, b Brownsville, Tenn, Sept 2, 34; m 60; c 2. MATHEMATICS, COMPUTER SCIENCE. *Educ:* Vanderbilt Univ, BA, 56; George Peabody Col, MA, 62, PhD(math), 68. *Prof Exp:* Teacher high sch, Tenn, 59-62; instr eng math, Vanderbilt Univ, 62-64, 66-68, asst prof, 68-70; lectr math, Univ NC, Wilmington, 70-71; Vanderbilt fel, Fla State Univ, 71-72; assoc prof, 72-77, PROF MATH, UNIV NC, WILMINGTON, 77- *Mem:* Asn Comput Mach; Am Asn Univ Prof; Nat Coun Teachers Math; Math Asn Am. *Res:* Application of computers and computing to mathematics and statistics. *Mailing Add:* Dept Math Sci Univ NC601 S Col Rd Wilmington NC 28403

NORRIS, FORBES HOLTEN, JR, b Richmond, Va, May 1, 28; m 55; c 3. NEUROLOGY, NEUROPHYSIOLOGY. *Educ:* Harvard Univ, BS, 49, MD, 55. *Prof Exp:* Guest worker electromyography, NIH, 54-55; intern surg, Johns Hopkins Hosp, 55-56; med officer, NIH, 56-61; from sr instr to asst prof neurol, Univ Rochester, 61-63, assoc prof & actg chmn div, 63-66; trustee, Inst Med Sci, Pac Med Ctr, 69-72 & 79-81, assoc dir, Inst Neurol Sci, 66-81; VPRES & CLIN DIR, ALS RES FEDN, 81- *Concurrent Pos:* USPHS spec fel, 61-63; ad hoc consult, NIH, 62; sr res fel, Inst Neurol, Univ London, 66; adj prof neurol, Univ of the Pac, 70-80; ann lectr, Japan Neurol Asn, 89. *Mem:* AAAS; AMA; Am Acad Neurol; Am Asn Electrodiag Med; Am Neurol Asn. *Res:* Clinical and experimental studies of the function of normal and diseased nervous system. *Mailing Add:* 2100 Webster Suite 110 San Francisco CA 94115

NORRIS, FRANK ARTHUR, b Pittsburgh, Pa, July 2, 13; m 39; c 1. BIOCHEMISTRY. *Educ:* Univ Pittsburgh, BS, 35, PhD(biochem), 39. *Prof Exp:* Asst chem, Univ Pittsburgh, 36-39; Rockefeller fel, Univ Minn, 39-41; res chemist, Gen Mills, Inc, 41-44; head oil mill res, Swift & Co, 44-64, head edible oil res, 64-66; sr scientist, Res & Develop Div, Kraftco Corp, 66-69, dir res admin, 69-72, assoc mgr edible oil prod, 72-78; RETIRED. *Concurrent Pos:* Consult, 78- *Honors & Awards:* Bailey Award, Am Oil Chemist Soc, 80. *Mem:* Am Oil Chemists Soc (vpres, 72-73, pres, 73-74); AAAS; Inst Food Technologists; Am Chem Soc. *Res:* Synthetic glycerides; fatty acid chemistry; oilseed processing; vegetable proteins; edible fats and oil processing. *Mailing Add:* 6700 S Brainard No 303 Countryside IL 60525

NORRIS, GEOFFREY, b Romford, Eng, Aug 6, 37; m 58; c 4. GEOLOGY, PALEONTOLOGY. *Educ:* Cambridge Univ, BA, 59, MA, 62, PhD(geol), 64. *Prof Exp:* Sci officer, NZ Geol Surv, 61-64; fel geol, McMaster Univ, 64-65; sr res scientist, Res Ctr, Pan Am Petrol Corp, Okla, 65-67; from asst prof to assoc prof, 67-74, chmn, Dept Geol, 80-90, PROF GEOL, UNIV TORONTO, 74- *Concurrent Pos:* Res assoc, Royal Ont Mus, Toronto, 68-; mem, Int Comn Palynology (secy, treas, 77-); Humboldt Res Fel, Univ Cologne, Ger, 78-; mem, Earth Sci Grant Selection Comt, Nat Sci Eng Res Coun Can, 80-83; secy, Div Earth Space Sci, Royal Soc Can, 90 & Can Nat Comt, Int Union Geol Sci, 90; pres, Can Asn Palynologists, 80. *Mem:* Am Asn Stratig Palynologists (pres, 71-72); fel Geol Asn Can; Paleont Soc; Brit Palaeont Asn; fel Geol Soc Am; fel Royal Soc Can. *Res:* Palynology; stratigraphic and paleoecologic applications; taxonomy of dinoflagellate cysts. *Mailing Add:* Dept Geol Earth Sci Centre Univ Toronto Toronto ON M5S 3B1 Can

NORRIS, H THOMAS, b Johnson City, Tenn, Nov 24, 34; m; c 3. GASTRO-INTESTINAL PATHOLOGY. *Educ:* Univ Southern Calif, MD, 59. *Prof Exp:* PROF & CHMN, DEPT PATH & LAB MED, SCH MED, E CAROLINA UNIV, 83- *Concurrent Pos:* Dept Path, Sch Med, Univ Wash, 67-83. *Mem:* Am Col Physicians; Col Am Pathologists; Int Acad Path; Am Soc Clin Path. *Res:* GI pathology. *Mailing Add:* Dept Path & Lab Med Sch Med E Carolina Univ Greenville NC 27834

NORRIS, JAMES NEWCOME, IV, b Santa Barbara, Calif, Sept 8, 42; m 77; c 1. MARINE PHYCOLOGY. *Educ:* San Francisco State Col, BA, 68, MA, 71; Univ Calif, Santa Barbara, PhD(marine bot), 75. *Prof Exp:* Asst cur, Gilbert M Smith Herbarium Hopkins Marine Sta, Stanford Univ, 69-70; assoc, dept biol sci, Univ Calif, Santa Barbara, 71; sta dir & resident marine biologist, Marine Biol Lab, Puerto Penasco, Mex, Univ Ariz, 72-74; assoc cur, 75-80, CUR, DEPT BOT, NAT MUS NATURAL HIST, SMITHSONIAN INST, 80- *Concurrent Pos:* Res assoc, dept ecol & evolutionary biol, Univ Ariz, 74-80 & Scripps Inst Oceanog, 81-; mem panel, Adv Com Syst Biol Prog, NSF, 79-80; adj prof, George Mason Univ, 83-; vis prof, Univ Hawaii, 85-86. *Mem:* Phycol Soc Am; Int Phycol Soc; Bot Soc Mex; Int Asn Plant Taxon; Asn Trop Biol. *Res:* Biosystematics and ecology of marine benthic algae; chemotaxonomy of marine algae; marine flora of the Gulf of California, Pacific Mexico, Galapagos Islands, Belize and Caribbean Panama. *Mailing Add:* Dept of Bot 166 NHB Smithsonian Inst Washington DC 20560

NORRIS, JAMES SCOTT, b Selma, Ala, Aug 6, 43; m 66; c 2. ENDOCRINOLOGY. *Educ:* Keene State Col, BS, 66; Univ Colo, PhD(biol), 71. *Prof Exp:* Ford Found fel endocrinol, Univ Ill, Urbana, 70-71, Nat Cancer Inst fel, 71, Am Cancer Soc fel, 71-74; instr cell biol, Baylor Col Med, 74-77; asst prof, 77-83, ASSOC PROF, MED SCH, UNIV ARK, 83- *Mem:* Am Soc Microbiol; Endocrine Soc; Am Soc Cell Biol; Tissue Cult Asn. *Res:* Hormonal regulation of oncogene expression and cancer cell growth. *Mailing Add:* Dept Med & Physiol Univ Ark Med Sci 4301 W Markham St Little Rock AR 72205

NORRIS, KARL H(OWARD), b Glen Richey, Pa, May 23, 21; m 48; c 2. SPECTROSCOPY. *Educ:* Pa State Univ, BS, 42. *Prof Exp:* Radio engr, Airplane & Marine Instrument Co, 45-46; electronic engr, Univ Chicago, 46-49; lab dir, Instrumentation Res Lab, Agr Res Serv, USDA, 50-77, chief instrument res lab, sci & educ admin, 77-88; RETIRED. *Honors & Awards:* McCormick Gold Medal, Am Soc Agr Engrs, 74; Alexander von Humboldt Award, 78. *Mem:* Nat Acad Eng; Am Soc Agr Engrs; Soc Appl Spectros. *Res:* Instrumentation for the measurement of quality factors of agricultural products. *Mailing Add:* 11204 Montgomery Rd USDA Beltsville MD 20705

NORRIS, KENNETH STAFFORD, b Los Angeles, Calif, Aug 11, 24; m 53; c 4. ZOOLOGY. *Educ:* Univ Calif, Los Angeles, MA, 51, PhD(zool), 59. *Prof Exp:* Asst ichthyol, Scripps Inst, Calif, 51-53; cur, Marineland of Pac, 54-60; lectr zool, Univ Calif, Los Angeles, 60-65, from assoc prof to prof, 65-72; dir, Long Marine Lab, Univ Calif, Santa Cruz, 72-75, chmn environ studies, 76-78, prof natural hist, 72-90; RETIRED. *Concurrent Pos:* Mem sci adv comt, US Marine Mammal Comn, 72-; mem adv bd, Bur Land Mgt, Dept Interior, 75-; bd gov, Calif Nature Conserv & bd dirs, Monterey Bay Aquarium Res Inst; founding pres, Soc Marine Mammal, 87-88; founder, Univ Calif Natural Reserve Syst. *Honors & Awards:* Mercer Award, Ecol Soc Am. *Mem:* AAAS; Am Soc Ichthyologists & Herpetologists; Soc Study Evolution; Ecol Soc Am; Am Soc Mammalogists; Sigma Xi. *Res:* Echolocation and natural history of cetaceans; natural history of desert reptiles; behavior of spinner dolphins. *Mailing Add:* 1987 Smith Grade Santa Cruz CA 95060

NORRIS, LOGAN ALLEN, b Oakland, Calif, May 23, 36; m 58; c 3. PESTICIDE CHEMISTRY, WATERSHED RESEARCH. *Educ:* Ore State Univ, BS, 61, MS, 64, PhD(plant physiol, biochem), 69. *Prof Exp:* Asst biochem, 61-68, PROF & HEAD DEPT FOREST SCI, ORE STATE UNIV, 83-; chief res chemist & proj leader, Forestry Sci Lab, Pac Northwest Forest & Range Exp Sta, US Forest Serv, 68-83. *Concurrent Pos:* Consult, fate of chemicals on rights of way & in forest environment; chief res chemist & proj leader, Forestry Sci Lab, Pac Northwest Forest & Range Exp Sta, US Forest Serv, 68-83. *Mem:* Soc Am Foresters; Weed Sci Soc Am; fel Soc Am Foresters, 86; hon mem, Western Soc Weed Sci, 88. *Res:* Woody plants; behavior and impact of chemicals in the forest environment; watershed management. *Mailing Add:* 4045 NW Dale Pl Corvallis OR 97330

NORRIS, PATRICIA ANN, b 1932; m 77; c 2. APPLIED PSYCHOPHYSIOLOGY, CLINICAL PSYCHONEUROIMMUNOLOGY. *Educ:* Univ Calif, Santa Barbara, BA, 54; Union Grad Sch, PhD(psychol), 76. *Prof Exp:* Personnel officer, US Naval Ord Test Sta, China Lake, Calif, 56-67; rehab counr, NY Guild Jewish Blind, 58-59; clin psychol intern, Brooklyn Psychiat Ctrs, 68-69; clin psychologist, Kans Reception & Diag Ctr, Topeka, Kans, 70-74; PSYCHOPHYSIOL PSYCHOTHERAPIST, MENNINGER FOUND, TOPEKA, KANS, 74-, CLIN DIR BIOFEEDBACK & PSYCHOPHYSIOL CTR, 79- *Concurrent Pos:* Dir biofeedback seminars & workshops, Vol Controls Prog, Menninger Found, 74-77, assoc dir biofeedback res, 77-79, fac mem biofeedback seminars & workshops, 74-; adj prof, Union Grad Sch, Yellow Springs, Ohio, 82-; assoc ed, Biofeedback & Self Regulation, 82-85; bd dirs, Inst Behav Med, Santa Barbara, 85-, Inst Adjunctive Cancer Ther, 83-, Drake Inst Behav Med, Santa Monica, 82- *Mem:* Biofeedback Soc Am (pres, 85-86). *Res:* Research and writing in psychophysiologic self regulation, clinical psychoneuroimmunology, medical and psychiatric applications of biobehavioral self-regulation; mind/body, conscious/unconscious integration. *Mailing Add:* Biofeedback & Psychophysiol Ctr Menninger Clin Box 829 Topeka KS 66601-0829

NORRIS, PAUL EDMUND, b Detroit, Mich, Nov 9, 18; m 44; c 2. PHARMACEUTICAL CHEMISTRY. *Educ:* Univ Mich, BS, 41, MS, 42, PhD(pharmaceut chem), 52. *Prof Exp:* Asst to F F Blicke, Univ Mich, 46-48, instr pharm, 48-51, asst prof, 51-54; group leader, Procter & Gamble Co, 54-64, toxicologist, 64-76, monitoring systs coordr, 76-82; RETIRED. *Concurrent Pos:* Bomb disposal officer, US Navy, 44-45; mem, Ohio Sch Bd, Finneytown, 60-63; bd dirs, Fulton Cryogenics Co, 64-67. *Mem:* AAAS; Am Chem Soc; Soc Toxicol; Am Pharmaceut Asn. *Res:* Synthetic drugs; synthesis of organic medicinals; concentration in area of potential ergot substitutes; synthesis of esters and amides of beta-amino acids; development of anticaries agents; design of clinical tests; toxicological testing and safety evaluation. *Mailing Add:* 476 Beech Tree Dr Cincinnati OH 45224

NORRIS, RICHARD C, b Schenectady, NY, Nov 2, 35; m 58; c 3. ELECTRICAL ENGINEERING. *Educ:* Harvard Univ, AB, 57, SM, 58; Mass Inst Technol, ScD(elec eng), 62. *Prof Exp:* RES ELEC ENG, ARTHUR D LITTLE, INC, 62-, PRACT LEADER, LOSISTICS, 78- *Mem:* Opers Res Soc Am. *Res:* Operations research; effectiveness analysis; computer applications; transportation. *Mailing Add:* Arthur D Little Inc 35 Acorn Park Cambridge MA 02140

NORRIS, ROBERT FRANCIS, b Buckinghamshire, Eng, July 4, 38; m 63; c 2. PLANT PHYSIOLOGY, WEED SCIENCE. *Educ:* Univ Reading, BSc, 60; Univ Alta, PhD(crop ecol), 64. *Prof Exp:* NIH & USPHS grants, Mich State Univ, 64-67; asst prof bot, 67-74, ASSOC PROF BOT, UNIV CALIF, DAVIS, 74- *Concurrent Pos:* Assoc ed, Weed Sci, 82-86. *Mem:* AAAS; Bot Soc Am; Weed Sci Soc Am; Ecol Soc Am; Sigma Xi; Am Soc Agron; Am Inst Biol Sci. *Res:* Plant growth regulators; cuticle structure and penetration; weed control; ecology of crop-weed association; weed-insect interactions, integrated control; herbicide action and physiology. *Mailing Add:* Dept Bot Univ Calif Davis CA 95616

NORRIS, ROBERT MATHESON, b Los Angeles, Calif, Apr 24, 21; m 52; c 3. GEOLOGY. *Educ:* Univ Calif, Los Angeles, AB, 43, MA, 49; Univ Calif, San Diego, PhD(oceanog), 51. *Prof Exp:* Asst geol, Univ Calif, Los Angeles, 46-49; asst submarine geol, Scripps Inst, 49-51, assoc marine geol, 51-52; from lectr to prof, 52-86, chmn dept, 60-63, EMER PROF GEOL, UNIV CALIF, SANTA BARBARA, 86- *Concurrent Pos:* Mem staff, US Geol Surv, 55-60; geologist, NZ Oceanog Inst, 61-62, 68-69 & 75-76; sr Fulbright scholar, NZ,

61-62. *Honors & Awards:* Neil A Miner Award, Nat Asn Geol Teachers, 81. *Mem:* Nat Asn Geol Teachers; Geol Soc Am; Soc Econ Paleontologists & Mineralogists; Am Geog Soc; Am Asn Petrol Geol; Geol Soc New Zealand. *Res:* Quaternary and marine geology; geomorphology. *Mailing Add:* Dept Geol Sci Univ Calif Santa Barbara CA 93106

NORRIS, ROY HOWARD, b Scammon, Kans, Apr 13, 30; m 60; c 2. REHABILITATION ENGINEERING. *Educ:* Univ Wichita, BSEE, 59, MSEE, 62; Okla State Univ, PhD(elec eng), 72. *Prof Exp:* From instr to assoc prof, 61-78, PROF ELEC ENG, WICHITA STATE UNIV, 78-, CHMN DEPT, 80- *Concurrent Pos:* Res engr, Autonetics, Downey, Calif, 59; res assoc elec eng, Univ Fla, 63-64; NSF sci fac fel, 66-67; res engr, Boeing Airplane Co, 68-72; tech engr, Boeing Co, 73-74; dir technol & eng staff, Wichita State Univ Rehab Eng Ctr, 76-78, co-dir, 78-79 & dir, 79- *Mem:* Am Soc Eng Educr; Inst Elec & Electronics Engrs; Rehab Eng Soc NAm. *Res:* Rehabilitation engineering, including the vocational prospects of the severely disabled. *Mailing Add:* Rehab Eng Ctr Wichita State Univ Wichita KS 67208

NORRIS, STEVEN JAMES, b Apr 8, 51; m; c 2. MICROBIOLOGY, IMMUNOLOGY. *Educ:* Univ Calif, Los Angeles, PhD(microbiol & immunol), 82. *Prof Exp:* Asst prof, 82-89, ASSOC PROF PATH, MICROBIOL & IMMUNOL, UNIV TEX MED SCH, HOUSTON, 89- *Mem:* Am Soc Microbiol; Am Asn Immunologists. *Res:* Syphilis; bacterial infection. *Mailing Add:* Dept Path & Lab Med Univ Tex Med Sch PO Box 20708 Houston TX 77225

NORRIS, TERRY ORBAN, b NC, Apr, 29, 22; m 51; c 2. ORGANIC CHEMISTRY. *Educ:* Univ NC, PhD(chem), 54. *Prof Exp:* Anal res chemist instrumental methods, E I du Pont de Nemours & Co, 49-51, polymer systs, 54-56; asst dir res, Keuffel & Esser Co, 56-57, dir, 58-62, dir corp res, 62-63; res mgr, IBM Corp, 63-66; dir res, Nekoosa Edwards Paper Co, Inc, 66-69, vpres res & develop, 69-87, bd dirs, 72-87; CONSULT, 87. *Mem:* AAAS; Am Chem Soc; Am Inst Chem; Soc Photog Sci & Eng; NY Acad Sci. *Res:* Polymer systems, coating; light sensitive systems and electrophotography; recording materials technology; pulp and paper chemistry; paper coatings. *Mailing Add:* Hound Ears PO Box 188 Blowing Rock NC 28605-0188

NORRIS, THOMAS HUGHES, b Princeton, NJ, Feb 8, 16; m 42; c 1. PHYSICAL INORGANIC CHEMISTRY, CHEMICAL KINETICS. *Educ:* Princeton Univ, AB, 38; Univ Calif, Berkeley, PhD(chem), 42. *Prof Exp:* Chemist, Linde Air Prod Co, NY, 38-39; asst chemist, Gen Elec Co, Mass, 42; res assoc, Nat Defense Res Comt, Univ Calif, Berkeley, 43, instr chem, 44-46; asst prof, Univ Minn, 46-47; from asst prof to prof, 47-81, EMER PROF CHEM, ORE STATE UNIV, 81- *Mem:* Am Chem Soc; AAAS. *Res:* Radioactive tracer studies in physical and inorganic chemistry; exchange reactions and reaction mechanisms; non-aqueous ionizing solvents; complex formation in solution; nuclear magnetic resonance in solution. *Mailing Add:* 3111 NW Norwood Pl Corvallis OR 97330

NORRIS, WILFRED GLEN, b Malmo, Sweden, Apr 21, 32; US citizen; m 55; c 3. CHEMICAL PHYSICS. *Educ:* Juniata Col, BS, 54; Harvard Univ, PhD(chem), 63. *Prof Exp:* Instr physics, 58-59, from asst prof to prof, 59-66, dean, 70-77, WILLIAM I & ZELLA B BOOK PROF PHYSICS, JUNIATA COL, 66- *Concurrent Pos:* NSF sci fac fel, Univ Md, 67-68; vis scholar, Columbia Univ, 84. *Mem:* AAAS; Am Phys Soc; Am Asn Physics Teachers; Optical Soc Am. *Res:* Diffusion of hydrogen in steel; surface reactions of hydrogen on steel; photochemistry; infrared and microwave spectra of small molecules at high temperatures. *Mailing Add:* Dept Physics Juniata Col Huntington PA 16652

NORRIS, WILLIAM C, b 1911; m; c 8. ENGINEERING ADMINISTRATION. *Educ:* Univ Nebr, EE, 32. *Prof Exp:* Chmn & chief exec officer, 57-86, EMER CHMN, CONTROL DATA CORP, 86- *Concurrent Pos:* Chmn bd dirs, William C Norris Inst; dir & mem, Qual & Technol Comt, Control Data Corp; dir & chmn, Greater Minn Corp. *Honors & Awards:* Nat Medal of Technol, 86; Founders Medal, Inst Elec & Electronics Engrs. *Mem:* Nat Acad Eng. *Res:* Computer technology; job creation; new directions for agriculture; raising the level and efficiency of technological innovation. *Mailing Add:* Control Data Corp 8100 34th Ave S PO Box O Minneapolis MN 55440-4700

NORRIS, WILLIAM ELMORE, JR, b Nixon, Tex, Feb 23, 21; m 44; c 3. PLANT PHYSIOLOGY. *Educ:* Southwest Tex State Teachers Col, BS, 40; Univ Tex, PhD(physiol), 48. *Prof Exp:* Instr biol & physiol, Univ Tex, 45-47; instr physiol, biochem & bact, Bryn Mawr Col, 47-48, asst prof biol, 48-49; assoc prof, Southwest Tex State Univ, 49-52, chmn dept, 52-67, dean, Sch Sci, 65-70, Col Arts & Sci, 70-74, vpres acad affairs, 74-80, prof biol, 52-82, dean, Univ, 80-82; RETIRED. *Mem:* Am Soc Plant Physiol. *Res:* Bioelectrics; plant respiration; plant growth substances. *Mailing Add:* 808 W Bluebonnet San Marcos TX 78666

NORRIS, WILLIAM PENROD, b Loogootee, Ind, Sept 2, 20; m 43, 77; c 3. RADIOBIOLOGY. *Educ:* DePauw Univ, AB, 41; Univ Ill, PhD(biochem), 44. *Prof Exp:* Biochemist, Dow Chem Co, Mich, 44 & Manhattan Area Engrs, 44-46; group leader, Biol Div, Argonne Nat Lab, 46-52, assoc biochemist, 52-70, group leader, Div Biol & Med Res, 70-79, scientist, 73-79; RETIRED. *Concurrent Pos:* Chmn, Whitewater-Rice Lakes Mgt Dist, Walworth County, Wis, 86- *Mem:* Radiation Res Soc; Am Soc Biol Chemists; Reticuloendothelial Soc; Health Physics Soc; Am Nuclear Soc. *Res:* Responses of dogs to continuous or protracted gamma-irradiation; sphingolipids; metabolism of phosphorus and alkaline earths; isolation and synthesis of dihydrosphingosine; radioautography; effects of ionizing radiations in animals; measurement and chemistry of radioactive elements; radiation chemistry; paper electrophoresis. *Mailing Add:* 5186 State Park Dr RR 3 Whitewater WI 53190

NORRIS, WILLIAM WARREN, b Choudrant, La, Mar 27; m 49; c 3. ZOOLOGY. *Educ:* La Polytech Inst, BS, 50; La State Univ, MS, 51, PhD, 55. *Prof Exp:* Histologist, Res Labs, Swift & Co, 55-60; assoc prof zool, Western Ky State Col, 60-65; assoc prof, 65-69, PROF BIOL, NORTHEAST LA UNIV, 69- *Mem:* Am Soc Zool. *Res:* Histology, especially as related to proteolytic action of muscle fibers and connective tissue; reproductive physiology as related to pineal gland. *Mailing Add:* Dept Biol Northeast La Univ 700 University Ave Monroe LA 71209

NORSTADT, FRED A, b Sidney, Iowa, Mar 15, 26; m 50; c 2. AGRICULTURAL MICROBIOLOGY, AGRICULTURAL CHEMISTRY. *Educ:* Nebr State Teachers Col, BS, 50; Univ Nebr, MS, 58, PhD(agron), 66. *Prof Exp:* Teacher high schs, Nebr, 49-56, prin, Holmesville High Sch, 51-53; instr exten div, Univ Nebr, 56-64; chemist, Soil & Water Conserv Res, Agr Res Serv, USDA, 64-66, soil scientist, 66-88; RETIRED. *Concurrent Pos:* Collabr, Dept Agron, Colo State Univ, 66-, Agr Res Serv, USDA, 88- & Colo Asn Soil Conserv Dist, 88- *Honors & Awards:* Commendation Award Conserv, Soil & Water Conserv Soc, 76. *Mem:* Emer mem Soil Sci Soc Am; fel Soil & Water Conserv Soc Am; emer mem Am Soc Agron; emer mem Crop Sci Soc Am; emer mem Sigma Xi. *Res:* Microbial crop residue decomposition; phytotoxic substances; mineralization of carbon, nitrogen, phosphorus and sulfur from soil organic matter and animal wastes; soil physical, chemical and biological effects on plant growth and vice versa. *Mailing Add:* 1933 Yorktown Ct Ft Collins CO 80526

NORSTOG, KNUT JONSON, b Grand Forks, NDak, June 11, 21; m 44; c 3. BIOLOGY. *Educ:* Luther Col, Iowa, BA, 43; Univ Mich, MS, 47, PhD(bot), 55. *Prof Exp:* Biologist, Dept Game, Fish & Parks, SDak, 47-49; instr biol, Luther Col, 49-51; assoc prof, Wittenberg Univ, 54-63; assoc prof bot & bact, Univ SFla, 63-66; prof biol sci, Northern Ill Univ, 66-77, RES ASSOC, FAIRCHILD TROP GARDEN, 78- *Concurrent Pos:* NSF fac fel, 59; res fel bot, Yale Univ, 59-60; ed-in-chief, Am J Bot, 80-85. *Mem:* Torrey Bot Club; Bot Soc Am; Sigma Xi. *Res:* Plant morphogenesis; embryogenesis in the cycads; plant tissue culture. *Mailing Add:* 10901 Old Cutler Rd Miami FL 33156

NORTH, CHARLES A, b Kingston, RI, Aug 24, 32; m 62; c 2. ZOOLOGY, ORNITHOLOGY. *Educ:* Univ Mo, BA, 54; Okla State Univ, MS, 62, PhD(zool), 67. *Prof Exp:* Asst prof, 66-76, ASSOC PROF BIOL, UNIV WIS-WHITEWATER, 75- *Concurrent Pos:* Wis State Univ res grant, 69-70; Instnl grants, 68-81; NSF travel grant, Poland, 71. *Res:* Avian ecology; wildlife management. *Mailing Add:* Dept Biol Univ Wis 800 W Main St Whitewater WI 53190

NORTH, EDWARD D(AVID), b Akron, Ohio, Oct 30, 18; m 42; c 2. CHEMICAL ENGINEERING. *Educ:* Univ Mich, MS, 41, PhD(chem eng), 50. *Prof Exp:* Area supt mfg, Mallinckrodt Chem Works, 50-57, plant mgr, 57-62; plant mgr, United Nuclear Corp, 62-64; asst plant mgr, Nuclear Fuel Serv, Inc, 64-65, plant mgr, 65-68, mgr environ protection & licensing, Rockville, 68-72; group leader, Oak Ridge Nat Lab, 76-79, mgr, component develop, 79-84; RETIRED. *Concurrent Pos:* Vchmn, Inst Nuclear Mat Mgr, 60, chmn, 61. *Mem:* Am Chem Soc; fel Am Inst Chem Engrs. *Res:* Economic processing of uranium and plutonium nuclear fuels; chemistry and technology of production of high purity inorganic compounds; nuclear fuel reprocessing and waste management. *Mailing Add:* Rt 3 Brooksview Rd Lenoir City TN 37771

NORTH, HARPER QUA, physics; deceased, see previous edition for last biography

NORTH, HENRY E(RICK) T(UISKU), b Lethbridge, Alta, Nov 18, 31; m 60; c 5. FLUID MECHANICS, OPERATIONS RESEARCH. *Educ:* Queen's Univ, Ont, BSc, 55; Col Aeronaut, Dipl, 57. *Prof Exp:* Spec projs engr, United Aircraft Corp Can Ltd, 57-59; asst prof fluid mech, Univ Man, 59-66; assoc prof fluid mech & chmn sch eng, 66-77, assoc prof, 77-86, PROF MECH ENGR, LAKEHEAD UNIV, 86- *Concurrent Pos:* Proj consult, Bristol Aero Indust Ltd, 62-64 & Hawker Siddeley Canada Ltd, 66-78; mem, comt internal aerodyn ducts, Nat Res Coun Can, 63- *Res:* Internal aerodynamics of straight and curved diffusers; jet interactions on lifting wings; propellor static thrust; parametric linear programming applied to project planning; non-Newtonian fluids processing; mechanical resonant systems in high-power applications; cavitation processing; resonance grinding, comminution; microprocessor control fluid power systems. *Mailing Add:* Sch Eng Lakehead Univ Thunder Bay ON P7B 5E1 Can

NORTH, JAMES A, b Charleston, Utah, Mar 18, 34; m 56; c 4. VIROLOGY, IMMUNOLOGY. *Educ:* Brigham Young Univ, BS, 58, MS, 60; Univ Utah, PhD(microbiol), 64. *Prof Exp:* Sr res assoc virol, Univ Cincinnati, 64-65; assoc prof, 65-72, PROF MICROBIOL, BRIGHAM YOUNG UNIV, 72- *Mem:* Am Soc Microbiol; Sigma Xi. *Res:* Viral purification and physical analysis; immunochemistry; viral etiology of cancer. *Mailing Add:* 775 WIDB Brigham Young Univ Provo UT 84601

NORTH, PAUL, b Coventry, Eng, Nov 5, 40; m 64; c 2. AERONAUTICAL ENGINEERING. *Educ:* Lanchester Col Technol, Eng, BS, 62; Col Aeronaut Eng, MS, 64; Univ Nottingham, PhD(fluid dynamics), 67. *Prof Exp:* Sr engr, Lockheed Ga Co, 67-68; asst prof aerothermopropulsion, Univ W Fla, 68-77; ASSOC SCI, IDAHO NAT ENG LAB, 77- *Mem:* Am Inst Aeronaut & Astronaut; Royal Aeronaut Soc; Brit Inst Mech Engrs. *Res:* Aeronautical systems; performance of conical diffusers with compressible flow; use of metal additives in tri-propellant rocket combustion. *Mailing Add:* Exp Spec & Anal/Tsb Idaho Nat Eng Lab Idaho Falls ID 83401

NORTH, R ALAN, b Halifax, UK, May 20, 44; c 3. CELLULAR-NEUROPHYSIOLOGY. *Educ:* Univ Aberdeen, Scotland, MD, 69, PhD(pharmacol), 73. *Prof Exp:* Assoc prof, Pharmacol, Loyola Univ, Chicago, 75-81; prof, Neuropharmacol, Mass Inst Technol, 81-86; SR SCIENTIST, VOLLUM INST, 87-, PROF NEUROL, OREGON HEALTH

SCI UNIV, 87- *Concurrent Pos:* Vis prof, Flinders Univ, Adelaide, Australia, 82; vis prof, Bogomoletz Inst Physiol, Kiev, USSR, 84; vis prof, Frankfurt Univ, Ger, 88. *Honors & Awards:* Schweppe Prize, 77; Boehringer Prize, 86. *Mem:* Am Physiol Soc; Brit Pharmacol Soc; Soc Neurosci; Physiol Soc; Am Soc Pharmacol. *Res:* Studies of ion channels and receptors on single mamaralian nerve cells, and synaptic connections between cells; studies of drug action on mammalia nervous system particularly opioids and other drugs of abuse antiphychotics etc autonomic and gastrointestinal physiology. *Mailing Add:* The Oregon Health Scis Univ 3181 SW Sam Jackson Park Rd Portland OR 97201

NORTH, RICHARD RALPH, b Hamilton, Ont, Aug 8, 34; m 55; c 4. NEUROLOGY. *Educ:* Queen's Univ, Ont, MD, CM, 59. *Prof Exp:* Fel neurophysiol, Col Med, Baylor Univ, 63-64, fel neurol, 64-66, asst prof, 66-67; from asst prof to assoc prof neurol, 67-75, clin assoc prof neurol, 75-82, CLIN PROF NEUROL, UNIV TEX HEALTH SCI CTR, DALLAS, 82- *Mem:* Am Acad Neurol; Am EEG Soc; Am Epilepsy Soc. *Res:* Clinical electroencephalography; cerebrovascular disease; epilepsy; efficacy of levodopa therapy in parkinsonism. *Mailing Add:* 7777 Forest Lane B410 Dallas TX 75230

NORTH, ROBERT J, b Bathurst, New S Wales, Australia, Aug 22, 35; US citizen; m 61; c 3. IMMUNITY TO TUMORS & BACTERIAL INFECTION. *Educ:* Univ Sydney, New S Wales, 59; Australian Nat Univ, PhD(biol), 67. *Prof Exp:* Tech officer, Electron Micros Unit, Univ Sydney, 59-62; res asst, John Curtin Sch Med Res, Austrlian Nat Univ, 62-67; vis investr, 67-70, assoc mem, 70-74, mem, 74-65, DIR, TRUDEAU INST, NY, 76- *Concurrent Pos:* Mem, bd trustees, Trudeau Inst, 80- *Honors & Awards:* Friedrich-Sasse-Stiftung Prize, 84; Soc Leukocyte Biol Res Award, 90. *Mem:* Am Asn Immunologists; Am Soc Microbiol; AAAS; Soc Leukocyte Biol; Transplantation Soc; Am Asn Cancer Res. *Res:* Cellular and physiological basis of immunological defenses against infectious diseases and cancers; development of models of cellular immunity against intracellular baterial pathogens and syngeneic tumors with the aim of understanding why natural immunity fails to protect under certain circumstances. *Mailing Add:* Trudeau Inst Inc PO Box 59 Saranac Lake NY 12983

NORTH, W(ALTER) PAUL TUISKU, b Vulcan, Alta, Dec 15, 34; m 57; c 4. MECHANICAL ENGINEERING. *Educ:* Queen's Univ, BSc, 58; Univ Sask, MSc, 59; Univ Ill, PhD(appl mech), 65. *Prof Exp:* Lectr mech eng, Univ Sask, 59-61, asst prof, 64-65; asst prof, 65-70, PROF MECH ENG, UNIV WINDSOR, 70- *Concurrent Pos:* Mem nat comt, Inst Union Theoret & Appl Mech, 66-67. *Mem:* Soc Exp Stress Anal; Am Soc Mech Engrs. *Res:* Stress analysis, machine design, metrology. *Mailing Add:* Dept of Mech Eng Univ of Windsor Windsor ON N9B 3P4 Can

NORTH, WHEELER JAMES, b San Francisco, Calif, Jan 2, 22 005 WH; m 53; c 2. OCEANOGRAPHY. *Educ:* Calif Inst Technol, BS, 44 & 50; Univ Calif, PhD(oceanog), 53. *Prof Exp:* Electron engr, US Navy Electron Lab, 47-48; NSF fel, Cambridge Univ, 53-54; Rockefeller fel marine biol, Scripps Inst Oceanog, 55-56; asst res biologist & proj officer, Inst Marine Resources Kelp Prog, Univ Calif, 56-63; sr res scientist, Lockheed Calif Co, 63; assoc prof environ health eng, 63-68, PROF ENVIRON SCI, CALIF INST TECHNOL, 68- *Mem:* AAAS; Soc Gen Physiol; Am Soc Zool; Am Malacol Union; Am Geophys Union; Soc Limnol & Oceanog. *Res:* Marine ecology and agal physiology; resource management. *Mailing Add:* W M Keck Lab Calif Inst Technol Pasadena CA 91125

NORTH, WILLIAM CHARLES, b Chungking, China, Aug 17, 25; m 71; c 9. ANESTHESIOLOGY, PAIN. *Educ:* DePauw Univ, BA, 45; Northwestern Univ, MS, 48, MD, 50, PhD, 52. *Prof Exp:* Intern, Chicago Mem Hosp, 49-50; from instr to asst prof pharmacol, Northwestern Univ, 50-59; asst prof anesthesiol, Sch Med, Duke Univ, 59-62, assoc prof anesthesiol & pharmacol, 63-65; chmn, Dept Anesthesiol, 65-82, med dir, Pain Ctr, 82-**89**, PROF ANESTHESIOL & PHARMACOL, COL MED, UNIV TENN, MEMPHIS, 65- *Concurrent Pos:* Res anesthesia, Chicago Wesley Mem Hosp, 56-59. *Mem:* AAAS; Am Soc Anesthesiol; Am Soc Pharmacol & Exp Therapeut; AMA. *Res:* Neuropharmacology; analgesia; local anesthesia; shock; inhalation anesthesia. *Mailing Add:* Dept Anesthesiol Univ Tenn Col Med Memphis TN 38163

NORTH, WILLIAM GORDON, b Woodstock, Ill, Aug 29, 42; m 64; c 2. STRATIGRAPHY. *Educ:* Carleton Col, AB, 63; Univ Ill, MS, 65, PhD(geol), 69. *Prof Exp:* Petrol geologist, Texaco, Inc, 68-75; mem staff, 77-80, ONSHORE DIST MGR, SANTA FE ENERGY, 80- *Res:* Stratigraphy, particularly subsurface stratigraphy; sedimentary petrology; statistics. *Mailing Add:* Santa Fe Energy 1616 S Voss Suite 555 Houston TX 77057

NORTHAM, EDWARD STAFFORD, b Lansing, Mich, Oct 18, 27; m 61; c 2. MATHEMATICS. *Educ:* Univ Mich, BS, 47, MS, 48; Mich State Univ, PhD(math), 53. *Prof Exp:* Mathematician, Bendix Aviation Corp, 53-54; from instr to asst prof math, Wayne State Univ, 54-64; assoc prof, 64-71, PROF MATH, UNIV MAINE, ORONO, 71-, COOP PROF ENG & SCI, 77- *Mem:* Am Math Soc; Math Asn Am. *Res:* Abstract algebra; lattice theory. *Mailing Add:* Dept Comput Sci Neville Hall Univ Maine Orono ME 04469

NORTHCLIFFE, LEE CONRAD, b Manitowoc, Wis, Mar 20, 26; div; c 2. NUCLEAR PHYSICS. *Educ:* Univ Wis, BS, 48, MS, 51, PhD(physics), 57. *Prof Exp:* Asst physics, Univ Wis, 51-57; from instr to asst prof, Yale Univ, 57-65; assoc prof, 65-70, PROF PHYSICS, TEX A&M UNIV, 70- *Concurrent Pos:* Prin investr, Dept Energy. *Mem:* Am Phys Soc; Am Asn Physics Teachers; Sigma Xi; Am Asn Univ Profs. *Res:* Nucleon-nucleon scattering; penetration of heavy ions through matter; accelerator development and instrumentation for nuclear research; charge distributions of heavy ions; nuclear reactions and scattering. *Mailing Add:* Cyclotron Inst Tex A&M Univ College Station TX 77843-3366

NORTHCOTT, JEAN, b Australia, June 26, 26; nat US. PHYSICAL CHEMISTRY, ORGANIC CHEMISTRY. *Educ:* Univ Sydney, PhD(chem), 53; State Univ NY Buffalo, MLS, 90. *Prof Exp:* Chemist & bacteriologist, Campbell Soup Co, Can, 54-55; patent asst, Nat Aniline Div, Allied Chem Corp, 56-68, head info serv, Spec Chem Div, 68-79, supvr tech info serv, chem sect, 80-88; RETIRED. *Mem:* Am Chem Soc; Spec Libr Asn. *Res:* Chemical literature; patents. *Mailing Add:* 315 Evans St Apt G Williamsville WY 14221

NORTHCUTT, RICHARD GLENN, b Mt Vernon, Ill, Aug 7, 41; m 65. NEUROANATOMY. *Educ:* Millikin Univ, BA, 63; Univ Ill, Urbana, MA, 66, PhD(zool), 68. *Prof Exp:* Asst prof anat, Case Western Reserve Univ, 68-72; assoc prof zool, 72-77, PROF BIOL SCI, UNIV MICH, ANN ARBOR, 77- *Concurrent Pos:* Res assoc, Cleveland Aquarium, 68-72; vis prof neurosci, Univ Calif, San Diego, 79; John Simon Guggenheim mem fel, 79. *Mem:* AAAS; Am Asn Anat; Am Soc Zool; Am Soc Ichthyol & Herpet; Soc Neurosci. *Res:* Evolution of the vertebrate nervous system; vertebrate paleontology, phylogeny and morphology; vertebrate behavior. *Mailing Add:* Dept Neurosci, Scripps Neurobiol A-001, Univ Calif La Jolla CA 92093

NORTHCUTT, ROBERT ALLAN, b Luling, Tex, Sept 30, 37. MATHEMATICS. *Educ:* Univ Tex, Austin, BA, 60, MA, 62, PhD(math), 68. *Prof Exp:* Instr math, San Antonio Col, 62-64; Southwest Tex State Univ, 64-67; NSF fac fel, Univ Tex, 67-68; from asst prof to assoc prof, 68-73, chmn dept, 71-80, PROF MATH, SOUTHWEST TEX STATE UNIV, 73- *Mem:* Am Math Soc; Math Asn Am. *Res:* Differential and integral equations. *Mailing Add:* Dept of Math Southwest Tex State Univ San Marcos TX 78666

NORTHERN, JERRY LEE, b Albuquerque, NMex, Sept 13, 40; c 4. AUDIOLOGY. *Educ:* Colo Col, BA, 62; Gallaudet Col, MS, 63; Univ Denver, MA, 64; Univ Colo, PhD(audiol), 66. *Prof Exp:* Chief audiol clin, Dept Otolaryngol, Brooke Army Med Ctr, San Antonio, Tex, 66-67; asst dir, US Army Audiol & Speech Path Ctr, Walter Reed Army Med Ctr, Washington, DC, 67-70; HEAD AUDIOL SERV, DEPT OTOLARYNGOL, MED CTR, UNIV COLO, DENVER, 70- *Concurrent Pos:* Mem, Nat Registry Interpreters for the Deaf, 68- *Mem:* Am Speech & Hearing Asn; Acoust Soc Am; Am Audiol Soc; Nat Asn Deaf. *Res:* Clinical audiology; acoustic impedance of the ear. *Mailing Add:* Dept Otolaryngol Univ Colo Med Ctr 4200 E Ninth Ave Denver CO 80262

NORTHEY, WILLIAM T, b Duluth, Minn, Aug 10, 28; m 50; c 7. IMMUNOLOGY. *Educ:* Univ Minn, BA, 50; Univ Kans, MA, 57, PhD(immunol), 59. *Prof Exp:* Res asst, Abbott Labs, 50-51; res asst immunol & virol, Naval Med Res Unit 4, 51-55; teaching & res, Univ Kans, 55-59; from asst prof to prof immunol, Ariz State Univ, 59-86; RETIRED. *Concurrent Pos:* Mem, Allergy Found Am, 60-; consult, Latric Corp, 70- *Mem:* Am Soc Microbiol; Am Asn Immunol; Am Asn Univ Profs. *Res:* Immune response in cold exposure; hypothermia; immunological aspects of the delayed response primarily in Coccidioidomycosis; allergenic extracts. *Mailing Add:* 5337 N 46th St Phoenix AZ 85018

NORTHINGTON, DEWEY JACKSON, JR, b New Orleans, La, Jan 31, 46; c 2. ORGANIC CHEMISTRY. *Educ:* La State Univ, BS, 67; Univ Fla, PhD(org chem), 70. *Prof Exp:* Fel chem, La State Univ, 70-71 & Syva Res Inst, 71-72; res scientist, Carnation Co, 72-74; ASST TECH DIR, WEST COAST TECH SERV, INC, 74- *Mem:* Am Chem Soc. *Res:* Analysis of polymers and surfactants; organic analyses; spectral identification of organic compounds. *Mailing Add:* WCAS 9840 Alburtis Ave Santa Fe Springs CA 90670

NORTHOUSE, RICHARD A, b Lanesboro, Minn, Apr 2, 38; m 61; c 2. COMPUTER GRAPHICS, ROBOTICS. *Educ:* Univ Wis, BS, 66, MS, 68; Purdue Univ, MS, 70, PhD(elec eng), 71. *Prof Exp:* Prof elec eng, Univ Wis, Milwaukee, 70-84. *Concurrent Pos:* Fel syst design & anal, NSF, 69-81, elec eng, 70-81; PRES, ICEO COMPCO, 75- *Mem:* Inst Elec & Electronics Engrs; Asn Comput Mach; Pattern Recognition Soc. *Res:* Computers and system theory to solve agricultural problems, particularly dairy cow problems. *Mailing Add:* Compco Computers Inc 13660 Capital Dr Brookfield WI 53005

NORTHOVER, JOHN, b Keynsham, Eng, Nov 7, 37; m 70; c 2. PLANT PATHOLOGY, MYCOLOGY. *Educ:* Bristol Univ, BSc, 60; Univ London, PhD(plant path), 65. *Prof Exp:* Res asst plant path, Agr Res Coun, UK, 60-63; fel, Nat Res Coun Can, 65-66; RES SCIENTIST PLANT PATH, AGR CAN, 66- *Mem:* Can Phytopath Soc; Am Phytopath Soc. *Res:* Mycological diseases of fruit crops; epiphytology; biological and chemical control. *Mailing Add:* Res Sta Agr Can Vineland Sta ON L0R 2E0 Can

NORTHRIP, JOHN WILLARD, b Tulsa, Okla, July 7, 34; m 54; c 4. BIOMECHANICS, TECHNICAL EDUCATION. *Educ:* Southwest Mo State Col, BS, 54; Okla State Univ, MS, 58, PhD(physics), 64. *Prof Exp:* Staff mem phys res, Sandia Corp, 58-61; asst prof physics, Cent Mo State Col, 63-65; Fulbright lectr solid state physics, Sch Eng Sao Carlos, Univ Sao Paulo, 65-67; from asst prof to assoc prof, 67-75, PROF PHYSICS & ASTRON, SOUTHWEST MO STATE UNIV, 75- *Concurrent Pos:* Consult, Sandia Corp, 61-64. *Mem:* AAAS; Am Phys Soc; Am Asn Physics Teachers. *Res:* Study of musculoskeletal motion, especially as applied to athletics; physics of sensory perceptions; technical education in the information age. *Mailing Add:* Dept of Physics Southwest Mo State Univ 901 S National Springfield MO 65804-0094

NORTH-ROOT, HELEN MAY, b Greely, Colo, Apr 7, 47; m 72. IN VITRO TOXICOLOGY METHODS DEVELOPMENT, REGULATORY TOXICOLOGY. *Educ:* Univ Calif, Davis, BS, 69; Univ Calif, San Francisco, PhD(comparative pharmacol & toxicol), 74; Am Bd Toxicol, dipl. *Prof Exp:* Biologist, Stanford Res Inst, 69; asst res pharmacologist, Univ Calif, Davis, 73-76; asst prog coordr, W Contra Costa Community Health Care, 76-78; sr toxicologist, 78-80, group leader toxicol, 80-84, DIR PROD SAFETY & REGULATORY AFFAIRS, DIAL CORP, 84- *Concurrent Pos:* Instr, Chapman Col, Beale Air Force Base, 76 & Peralta Community Col, 78;

consult, NIH, 76-77 & 79-80; adj asst prof, Univ Ariz Col Pharmacy, 84- *Mem:* Soc Toxicol. *Res:* In vitro non-animal alternatives to the traditional irritancy; Draize rabbit eye test; optimization of rabbit corneal epithelial cell line to assess potential irritancy of surfactants and alcohols. *Mailing Add:* 8127 E Via del Sol Scottsdale AZ 85258

NORTHROP, DAVID A, b New Haven, Conn, Feb 4, 38; m 60; c 2. GEOCHEMISTRY, PETROLEUM ENGINEERING. *Educ:* Univ Chicago, BS, 60, MS, 61, PhD(chem), 64. *Prof Exp:* Tech staff mem res, 64-70, DIV SUPVR RES, SANDIA LABS, 71- *Mem:* AAAS; Soc Petrol Engrs. *Res:* Enhanced oil and natural gas recovery; in situ conversion of fossil fuels; carbon composite materials research for aerospace applications; high temperature phase equilibria with emphasis upon vaporization phenomena; oxygen and stable isotope geochemistry. *Mailing Add:* 7207 Harwood Ave NE Albuquerque NM 87110

NORTHROP, JOHN, b New York, NY, Feb 1, 23; m 52; c 3. GEOPHYSICS. *Educ:* Princeton Univ, BA, 47; Columbia Univ, MA, 48; Univ Hawaii, PhD(solid earth geophys), 68. *Prof Exp:* Geol asst marine geol, Oceanog Inst, Woods Hole, 48-49; res assoc, Lamont Geol Observ, Columbia Univ, 49-51; head dept geol, Bates Col, 51-52; geologist, Hudson Labs, 52-61; asst specialist, Scripps Inst, Calif, 61-64; assoc geophysicist, Hawaii Inst Geophys, 64-65; assoc specialist marine geophys, Marine Phys Lab, Scripps Inst Oceanog, Univ Calif, 65-67; geophysicist, US Naval Ocean Systs Ctr, 67-86; consult, Comput Sci Corp, 86-91; RETIRED. *Concurrent Pos:* Consult, Artemis Proj, 58-59. *Mem:* Am Geophys Union; Seismol Soc Am; Acoust Soc Am; NY Acad Sci. *Res:* Submarine geology; marine geophysics; underwater sound; earthquake waves; geoacoustics; hydroacoustics. *Mailing Add:* 7015 Vista Del Mar La Jolla CA 92037

NORTHROP, ROBERT BURR, b White Plains, NY, Jan 11, 35; m 61; c 3. ELECTRICAL & BIOMEDICAL ENGINEERING. *Educ:* Mass Inst Technol, BS, 56; Univ Conn, MS, 58, PhD(physiol), 64. *Prof Exp:* Asst elec engr, Univ Conn, 56-59, asst zool, 59-60, asst prof elec eng, 63-69, assoc prof, 69-78, PROF ELEC ENG, UNIV CONN, 78-, HEAD HON PROG, BIOMED ENG, 91- *Concurrent Pos:* Consult, Conn State Bd Fisheries & Game, 61- & Northeast Utilities, 78. *Mem:* Am Soc Zool; Inst Elec & Electronic Engrs; Am Physiol Soc; Sigma Xi. *Res:* Neural data processing in insect compound eye vision; biomedical instrumentation; biomedical engineering education; environmental engineering; immune system modelling; glucose sensor; cable discharge detection; author of book and holder of patents. *Mailing Add:* 28 Chaplin St Chaplin CT 06235

NORTHROP, ROBERT L, virology, biochemistry, for more information see previous edition

NORTHROP, THEODORE GEORGE, b Poughkeepsie, NY, Dec 15, 24; div; c 5. SPACE PHYSICS. *Educ:* Yale Univ, BS, 44; Cornell Univ, MS, 49; Iowa State Col, PhD(physics), 53. *Prof Exp:* Instr physics, Vassar Col, 46-47; asst, TV tube dept, Labs, Radio Corp Am, NJ, 47; instr physics, Yale Univ, 53-54; mem staff, Theoret Div, Lawrence Radiation Lab, Univ Calif, 54-65; mem staff, Goddard Space Flight Ctr, NASA, 65-67, chief lab space physics, 67-77, mem lab, high energy astrophys, 77-91; RETIRED. *Mem:* Fel Am Phys Soc; Am Geophys Union. *Res:* Biophysics; electronic instrumentation; plasma physics; theory of planetary magnetospheres and rings. *Mailing Add:* Goddard Space Flight Ctr NASA Code 665 Greenbelt MD 20771

NORTHRUP, CLYDE JOHN MARSHALL, JR, b Oklahoma City, Okla, Apr 25, 38; m 60; c 3. SYSTEM ARCHITECTURE. *Educ:* Okla State Univ, BS(math) & BS(physics), 61, PhD(physics), 66. *Prof Exp:* Staff asst, 58 & 59-60, mem tech staff, 67-77, div supvr, 77-84, MEM, FUTURE OPTIONS GROUP, SANDIA NAT LABS, 84- *Concurrent Pos:* Ed, Sci Basis Nuclear Waste Mgt, Mat Res Soc, 80-83. *Mem:* Am Asn Physics Teachers; Mat Res Soc (pres, 82); Am Nuclear Soc; Am Chem Soc. *Res:* Radiation effects in solids; nuclear waste management; materials science; photonics; computerized numerical data bases for the properties of materials. *Mailing Add:* 9200 Crestwood Ave NE Albuquerque NM 87112-3918

NORTHUP, LARRY L(EE), b Audubon, Iowa, Aug 6, 40; m 76; c 2. AEROSPACE ENGINEERING, MECHANICAL ENGINEERING. *Educ:* Iowa State Univ, BS, 62, MS, 63, PhD(aerospace & mech eng), 67. *Prof Exp:* From instr to asst prof aerospace eng, 66-73, from assoc prof to prof, Freshman Eng, 74-91, PROF ENG, IOWA STATE UNIV, 91- *Mem:* Am Soc Eng Educ; Soc Mfg Engrs; Nat Acad Adv Asn. *Res:* Vibrational relaxation in high temperature gas dynamics; compressible gas dynamics; low speed aerodynamics, particularly studies of autorotation characteristics; computational fluid mechanics; fundamental engineering education. *Mailing Add:* Div Eng Fundamentals & Multidisciplinary Design Iowa State Univ 212 Ames IA 50011

NORTHUP, MELVIN LEE, b Floris, Iowa, Oct 11, 41; m 65; c 3. ENVIRONMENTAL SCIENCES. *Educ:* Parsons Col, BS, 63; Purdue Univ, Lafayette, MS, 65; Univ Mo-Columbia, PhD(soil chem), 70. *Prof Exp:* Methods develop chemist, Norwich Pharmacal Co, 65-66; NSF fel soils, Univ Wis-Madison, 70-72; from asst prof to assoc prof environ sci, 72-84, PROF NATURAL RESOURCES MGT, GRAND VALLEY STATE UNIV, 84- *Mem:* Am Soc Agron. *Res:* Interfacing environmental sensors with computerized crop management models; nutrient cycling in nature; computer simulation of ecosystems; computer simulation of agro-ecosystems. *Mailing Add:* Grand Valley State Univ Allendale MI 49401

NORTHUP, SHARON JOAN, 1942; m 71; c 2. BIOCHEMISTRY. *Educ:* Univ Mo, PhD(biochem), 71. *Prof Exp:* RES DIR, BAXTER HEALTHCARE, INC, 76- *Mem:* Soc Toxicol; Am Col Toxicol; Am Soc Pharmacol Exp Therapeut; Am Soc Cell Biol; Soc Biomat; Am Chem Soc. *Res:* Cell biology; biochemistry. *Mailing Add:* Baxter Healthcare Inc PO Box 490 Round Lake IL 60073

NORTHWOOD, DEREK OWEN, b Hitchin, Eng, July 28, 43; Can citizen; m 70; c 2. METALLURGY, ELECTRON MICROSCOPY. *Educ:* Imp Col, Univ London, BSc, 64; Univ Surrey, MSc, 66, PhD(chem physics), 68. *Prof Exp:* Investr melting & casting, Brit Non-Ferrous Metals Res Asn, London, 64-65; fel mat, Univ Windsor, 69-71; metall engr, Atomic Energy Can Ltd, Chalk River, Ont, 71-76; assoc prof mat, Univ Windsor, 76-79, asst dean res, 80-83, head dept, 85-89, PROF ENG MAT, UNIV WINDSOR, 79- *Concurrent Pos:* Vis prof, Univ Queensland, 83-84, 86. *Mem:* Fel Am Soc Metals Int; Micros Soc Can; fel Inst Metals, UK, 87; Metals Soc. *Res:* Structure/property relationships in materials for energy systems; zirconium alloys; mechanisms of creep; hydrogen storage materials; electron microscopy; corrosion. *Mailing Add:* Dept Mech Eng Univ Windsor Windsor ON N9B 3P4 Can

NORTON, CHARLES LAWRENCE, b Neponset, Ill, Dec 20, 17; m 45; c 5. DAIRY SCIENCE. *Educ:* Univ Ill, BS, 40; Cornell Univ, PhD(animal husb), 44. *Prof Exp:* Asst prof animal husb, Cornell Univ, 45-47; prof dairy husb & head dept, Univ RI, 47-50; head dept dairying, Okla State Univ, 50-58; orif daurt & poultry sci & head dept, Kans State Univ, 58-77, prof animal sci, 770; RETIRED. *Concurrent Pos:* Res dairy & poultry scientist, Agr Exp Sta, 70- *Mem:* Am Soc Animal Sci; Am Dairy Sci Asn. *Res:* Vitamin needs of dairy calves; dry calf starters; nutritive value of forages for dairy cattle. *Mailing Add:* 402 Bluemont Circle Manhattan KS 66502

NORTON, CHARLES WARREN, b Scranton, Pa, Aug 2, 44; div. MICROPALEONTOLOGY. *Educ:* Antioch Col, BA, 68; Univ Va, MA, 74; Univ Pittsburgh, PhD(geol), 75. *Prof Exp:* Res geologist, Gulf Res & Develop Corp, 74; coal geologist, WVa Geol Surv, 75-87; ASSOC PROF GEOL, KENT STATE UNIV, 87- *Mem:* AAAS; Paleont Soc; Am Asn Petrol Geologists; Sigma Xi; Soc Econ Paleontologists & Mineralogists. *Res:* Coal resource and reserves study in West Virginia involving mapping, correlation, thickness and quality measures on minable seams; long-term study of interbedded marine horizons for future correlation aid; study of provenance of pennsylvanians sandstones in Appalachian basin; stratigraphic studies of oil producing beds in Ohio for paleogeongraphy as drilling aid. *Mailing Add:* Geology Dept Kent State Univ Stark Campus Canton OH 44720

NORTON, COLIN RUSSELL, seed physiology, tissue culture of plants, for more information see previous edition

NORTON, CYNTHIA FRIEND, b Shelburne Falls, Mass, Aug 18, 40. MICROBIOLOGY. *Educ:* Smith Col, AB, 61; Boston Univ, PhD(marine microbiol), 67. *Prof Exp:* Sci aide, Polaroid Corp, 61-63; res assoc microbial physiol, Univ NH, 67-68; NIH fel microbiol, Sch Med, Yale Univ, 68-71; from asst prof to assoc prof, 71-80, PROF BIOL, UNIV MAINE, AUGUSTA, 80-, CHMN, DIV MATH, SCI & SOCIAL SCI, 80- *Mem:* Am Soc Microbiol. *Res:* Microbial pigments and exoenzymes; interactions of marine organisms; ecological role of soluble exoproducts in sea. *Mailing Add:* Dept of Biol Univ of Maine Univ Heights Augusta ME 04330

NORTON, DANIEL REMSEN, b Brooklyn, NY, Jan 27, 22; m 44; c 3. ANALYTICAL CHEMISTRY. *Educ:* Antioch Col, BS, 44; Princeton Univ, MA & PhD(analytical chem), 48. *Prof Exp:* Asst bacteriologist, Antioch Col, 40-41, asst org chem, 43-44; analyst, Eastern State Corp, NY, 41-42; analytical chemist, Merck & Co, Inc, NJ, 42-43 & Manhattan Proj, Princeton Univ, 44-46; monitor, Radiol Surv Sect, Oper Crossroads, 46; asst prof chem, George Washington Univ, 48-52; res chemist, Bur Anal Labs, US Geol Surv, 52-55 & Sprague Elec Co, Mass, 55-70. *Concurrent Pos:* Res fel, Woods Hole Oceanog Inst, 50-52, assoc, 52-58; instr, Williams Col, 60-68. *Mem:* Am Chem Soc; Sigma Xi. *Res:* Solid state and surface chemistry; ceramics; thermoanalytical techniques; polarography; absorption and emission spectrometry; particle and pore size analysis; silicate rock analysis; coal analysis. *Mailing Add:* 29611 Fairway Dr Evergreen CO 80439

NORTON, DAVID JERRY, b Manhattan, Kans, Oct 23, 40; m 64; c 1. AEROSPACE ENGINEERING. *Educ:* Tex A&M Univ, BS & MS, 63; Purdue Univ, PhD(mech eng), 68. *Prof Exp:* Sr res engr, Jet Propulsion Lab, 68-70; asst prof, 70-75, assoc prof, 75-79, PROF AEROSPACE ENG, TEX A&M UNIV, 79-, ASST DIR, TEX ENG EXP STA, 81- *Concurrent Pos:* Consult, Caudill, Rowlett & Scott, 76-, Exxon Prod Res, 78- & Transworld Drilling Inc, 81- *Mem:* Assoc fel Am Inst Aeronaut & Astronaut; Am Soc Eng Educ; Am Soc Mech Engrs; Sigma Xi. *Res:* Wind engineering; flow in the atmospheric boundary layer; propulsion and gas dynamics. *Mailing Add:* 41 Dovewood The Woodlands TX 77381

NORTON, DAVID L, biological rhythms, for more information see previous edition

NORTON, DAVID WILLIAM, b Mar 4, 44; US citizen; m 67; c 3. PHYSIOLOGICAL ECOLOGY. *Educ:* Harvard Col, AB, 67; Univ Alaska, MS, 70, PhD(zoophysiol), 73. *Prof Exp:* Staff ecologist environ consult, Dames & Moore, Fairbanks, 73-74; state supvr environ surveillance, Joint State-Fed Fish & Wildlife Adv Team for Surveillance Construct Trans-Alaska Oil Pipeline, 74-75; dep proj mgr, Biol, Coastal & Marine Ecol Res, Nat Oceanic & Atmospheric Admin-Outer Continental Shelf Environ Assessment Prog, 75-82; EXEC ED, BIOL PAPERS, UNIV ALASKA, 82-; EXEC SECY, ARCTIC INST NAM, 84- *Concurrent Pos:* Res assoc, Inst Arctic Biol, 73-, adj assoc prof appl sci, Geophys Inst, Univ Alaska, 77-82; mem, Coun Biol Ed & bd gov, Arctic Inst NAm, 83-; assoc ed, Arctic, 87-; acad coodr, Rural Alaska Hon Inst, 88- *Mem:* Sigma Xi; Ecol Soc Am; Arctic Inst NAm. *Res:* Ecological energetics of migrating and breeding in tundra shorebirds of Alaska. *Mailing Add:* 1749 Red Fox Dr Fairbanks AK 99701

NORTON, DENIS LOCKLIN, b Elba, NY, Jan 2, 39; m 60; c 3. GEOCHEMISTRY, GEOLOGY. *Educ:* Univ Buffalo, BA, 60; Univ Calif, Riverside, PhD(geol), 64. *Prof Exp:* Res asst geochem, Univ Calif, 60-64; from res geologist to sr res geologist, Kennecott Copper Corp, 64-69, chief geochem div, 69-73; asst prof, 73-76, ASSOC PROF GEOSCI, UNIV ARIZ,

76- *Mem:* Mineral Soc Am; Glaciol Soc; Am Inst Mining, Metall & Petrol Engrs; Geochem Soc; Mineral Asn Can. *Res:* Solving geological problems by developing a more thorough understanding of processes through geochemical investigations. *Mailing Add:* Dept of Geosci Univ of Ariz Tucson AZ 85721

NORTON, DON CARLOS, b Toledo, Ohio, May 22, 22; m 52; c 4. PLANT NEMATOLOGY. *Educ:* Univ Toledo, BS, 47; Ohio State Univ, MS, 49, PhD(bot), 50. *Prof Exp:* From asst prof to assoc prof plant path, Agr Exp Sta, Agr & Mech Col Tex, 51-59; from assoc prof to prof plant path, 59-89, EMER PROF PLANT PATH, IOWA STATE UNIV, 89- *Mem:* Am Phytopath Soc; fel Soc Nematol (treas, 77-80); Sigma Xi. *Res:* Nematology; root diseases; ecology of plant parasitic nematodes. *Mailing Add:* Dept Plant Path Iowa State Univ Ames IA 50011-1020

NORTON, DONALD ALAN, b Mt Kisco, NY, Mar 15, 20; m 48; c 2. MATHEMATICS. *Educ:* Harvard Univ, BS, 41; Univ Wis, PhD(math), 49. *Prof Exp:* ASSOC PROF MATH, UNIV CALIF, DAVIS, 49- *Mem:* Am Math Soc; Math Asn Am; Asn Comput Math; NY Acad Sci. *Res:* Generalized groups; algorithms. *Mailing Add:* Dept of Math Univ of Calif Davis CA 95616

NORTON, EDWARD W D, b Sommerville, Mass, Jan 3, 22. OPHTHALMOLOGY. *Educ:* Harvard Col, BA, 43; Cornell Univ, MD, 46; Am Bd Ophthal, dipl. *Hon Degrees:* DSc, Thomas Jefferson Univ, Philadelphia, 85. *Prof Exp:* Intern, Cincinnati Gen Hosp, Ohio, 46-47; asst resident neurol, Kingsbridge Vet Hosp, Bronx, NY, 49-50; from asst resident to resident ophthal, New York Hosp-Cornell Med Ctr, 50-53; from instr to asst prof surg, Cornell Univ, 53-58; assoc prof ophthal & chief div, Sch Med, 58-59, PROF OPHTHAL & CHMN DEPT, SCH MED & JACKSON MEM HOSP, UNIV MIAMI, 59- *Concurrent Pos:* Mem retina serv, Mass Eye & Ear Infirmary, Wilmer Inst Ophthal, Johns Hopkins Hosp & Mayo Clin, 53-54; res fels ophthal, Howe Lab, Boston, 53-54; asst attend surgeon, New York Hosp-Cornell Med Ctr, 54-58; consult, Hosp Spec Surg, New York, Mem Hosp Cancer & Allied Dis, Bellevue Hosp & Kingsbridge Vet Hosp, 54-58. *Mem:* Am Ophthal Soc; Retina Soc; Am Acad Ophthal; Asn Res Vision & Ophthal; Asn Univ Profs Ophthal. *Res:* Ophthalmology; retinal diseases; neuroophthalmology. *Mailing Add:* Bascom Palmer Eye Inst PO Box 016880 Miami FL 33101

NORTON, ELINOR FRANCES, b Brooklyn, NY, July 22, 29. ANALYTICAL CHEMISTRY. *Educ:* Wellesley Col, BA, 51. *Prof Exp:* From jr chemist to assoc chemist, 51-74, CHEMIST, BROOKHAVEN NAT LAB, 74- *Mem:* Am Chem Soc. *Res:* Analytical chemistry, particularly trace elements; spectrophotometry; x-ray fluorescence; neutron activation; radiochemical separations and atomic absorption. *Mailing Add:* Chem Dept Brookhaven Nat Lab Upton NY 11973

NORTON, JAMES AUGUSTUS, JR, b Philadelphia, Pa, Jan 3, 21; m 58; c 1. STATISTICS. *Educ:* Antioch Col, AB, 47; Purdue Univ, MS, 49, PhD, 59. *Prof Exp:* Res assoc statist, Div Educ Reference, Purdue Univ, 47-50, Statist Lab, 51-61, from instr to asst prof, 51-61; from asst prof to assoc prof statist, Dept Psychiat, Sch Med, 61-68, RES CONSULT, INST PSYCHIAT RES, IND UNIV, INDIANAPOLIS, 61-, PROF STATIST, DEPT PSYCHIAT, SCH MED, 68- *Res:* Tests of hypotheses in the case of unequal variances. *Mailing Add:* Dept Biostat & Gen Med Med Sch Purdue Univ Ind Univ 1120 South Dr Indianapolis IN 46223

NORTON, JAMES JENNINGS, b Elmira, NY, May 1, 18; m 46. GEOLOGY. *Educ:* Princeton Univ, AB, 40; Northwestern Univ, MS, 42; Columbia Univ, PhD, 57. *Prof Exp:* Geologist, US Geol Surv, 42-84; RETIRED. *Mem:* AAAS; Mineral Soc Am; Geol Soc Am; Soc Econ Geol. *Res:* Geology of Black Hills, South Dakota; pegmatites of Black Hills and elsewhere; industrial minerals; lithium. *Mailing Add:* 3612 Wonderland Dr Rapid City SD 57702

NORTON, JAMES MICHAEL, b Bangor, Maine, July 22, 46; m 78; c 2. ERYTHROCYTE FUNCTION, OXYGEN TRANSPORT. *Educ:* Col Holy Cross, AB, 67; Dartmouth Med Sch, BMS, 69, PhD(physiol), 79. *Prof Exp:* Res asst, Dept Res, Maine Med Ctr, 71-74, res assoc, 78-80; from asst prof to assoc prof, 80-90, PROF PHYSIOL, COL OSTEOP MED, UNIV NEW ENG, 90- *Concurrent Pos:* Prin investr, AHA-Maine Affil, Inc, 81-82 & 84-85, Kroc Found, 81-83; consult, Oper Cork, Dartmouth Med Sch, 79-82 & Nat Bd Examr Osteop Physicians & Surgeons, 83-86; bd mem & res comt chair, AHA-Maine Affil, 86- *Mem:* Am Physiol Soc; Am Col Sports Med; Sigma Xi; AAAS; Int Soc Oxygen Transp To Tissue; NAm Soc Biorheology. *Res:* Function of vascular smooth muscle; control of coronary artery flow; erythrocyte deformability; erythrocyte membrane transport mechanisms during cell development; oxygen transport during adaptation to simulated high altitude. *Mailing Add:* Col Osteop Med Univ New Eng 11 Hills Beach Rd Biddeford ME 04005

NORTON, JOHN LESLIE, b Chanute, Kans, June 1, 45; m 66. MATHEMATICAL PHYSICS. *Educ:* Kans State Col Pittsburg, BA, 66; Univ Kans, MS, 68, PhD(physics), 70. *Prof Exp:* Fel physics, Los Alamos Nat Lab, 70-71, staff mem weapons output, 71-74, staff mem hydrodyn, 74-77, mem staff, asst group leader & alt group leader comput serv, 77-80, mem staff, assoc group leader & dep group leader, Computational Physics, 80-84, assoc div leader, Weapons Computing, 84, staff mem & team leader, Computational Physics, 84-87, group leader, Computer User Serv, 87-88, STAFF MEM, SECT LEADER, COMPUTER USER SERV, LOS ALAMOS NAT LAB, 88- *Mem:* Am Phys Soc. *Res:* Numerical analysis; hydrodynamics; radiation transport; nuclear physics; computer science. *Mailing Add:* 2502 X St Los Alamos NM 87544

NORTON, JOSEPH DANIEL, b Flat Rock, Ala, Oct 14, 27; m 50; c 1. GENETICS, HORTICULTURE. *Educ:* Auburn Univ, BS, 52, MS, 55; La State Univ, PhD(agron, hort), 61. *Prof Exp:* Asst hort, Auburn Univ, 54; asst veg crops specialist, Univ Fla, 54-60; from asst prof to assoc prof hort & from asst horticulturist to assoc horticulturist, 60-73, PROF HORT & HORTICULTURIST, 73- *Concurrent Pos:* Biol scientist, US Air Force, 63-80; consult, 83- *Honors & Awards:* Kelley Mosley Environ Award, 82. *Mem:* Am Soc Hort Sci; Am Genetic Asn. *Res:* Plant breeding, especially plum, muskmelon, watermelon, strawberry and apple; greenhouse tomato studies. *Mailing Add:* Dept Hort Auburn Univ Auburn AL 36849

NORTON, JOSEPH R(ANDOLPH), b Lisbon, Ark, Dec 23, 15; m 39; c 2. ENGINEERING MECHANICS. *Educ:* Okla State Univ, BS, 39, MS, 51; Univ Tex, PhD, 63. *Prof Exp:* Draftsman, W C Norris Mfr, Okla, 39-40, plant engr, 40-46; prof eng res, 46-51, asst dir off eng res, 51-65, head sch gen eng, 63-78, EMER HEAD SCH GEN ENG, OKLA STATE UNIV, 78- *Concurrent Pos:* Consult, Fenix & Scission, 65-80, Grand River Dam Auth, 76-77, BC Power, Hehei Water Conservancy, Nationalist China, 78-83, Triangle Power, Inc, Mesa, Ariz, 84- *Mem:* Nat Soc Prof Engrs; Am Soc Eng Educ; Sigma Xi. *Res:* Mechanics of distributed systems; fluid and solid systems; heat transfer in distributed systems; use of hydraulic air compressors in the production of electrical power; Norton thermodynamic cycle utalizing hydraulic air compressors and gas turbine for production of electric power. *Mailing Add:* 724 Skyline Stillwater OK 74075-8210

NORTON, KARL KENNETH, b London, Eng, Nov 13, 38; US citizen; div; c 2. MATHEMATICS. *Educ:* Yale Univ, BS, 59; Univ Chicago, MS, 61; Univ Ill, Urbana, PhD(math), 66. *Prof Exp:* Asst prof math, Univ Colo, Boulder, 66-73; Independent Res, 73-82; assoc qual engr, NBI, Inc, 82-85; pvt tutor, 85-87; ASSOC PROF MATH, COLBY COL, MAINE, 87- *Concurrent Pos:* Off Naval Res res assoc, Univ Mich, Ann Arbor, 69-70; vis mem, Inst Advan Study, 70-71; vis res math, Univ Geneva, 74; NSF res grant, 75; sr vis fel math, Univ York, Eng, 75; mem-at-large, Coun Am Math Soc, 75-77; vis prof math, Univ Colo, Boulder, 86, 87. *Mem:* Am Math Soc. *Res:* Number theory, particularly the orders of magnitude of arithmetical functions. *Mailing Add:* 94 Thorton Rd Bangor ME 04401

NORTON, LARRY, b Bronx, NY, Apr 9, 47; m 70. CELL KINETICS, BIOELECTROCHEMISTRY. *Educ:* Univ Rochester, AB, 68; Columbia Univ, MD, 72. *Prof Exp:* Intern & resident, Albert Einstein Col Med, 72-74; clin assoc oncol, Nat Cancer Inst, 74-77; ASST PROF ONCOL, MOUNT SINAI SCH MED, 77- *Mem:* NY Acad Sci. *Res:* Treatment of human cancer including theory of tumor growth kinetics with implications in the design of chemotherapy programs; electrochemistry of tumors; enhancing chemotherapy with induced currents. *Mailing Add:* Dept Neoplastic Dis Sch Med Mount Sinai 100 N ST & 5 N Ave New York NY 10029

NORTON, LILBURN LAFAYETTE, b Lenoir City, Tenn, Jan 2, 27; m 46, 86; c 5. ORGANIC POLYMER CHEMISTRY. *Educ:* Carson-Newman Col, BS, 49; Northwestern Univ, MS, 51; Univ Tenn, PhD(chem), 54. *Prof Exp:* Asst chem, Northwestern Univ, 51; sr chemist, 54-59, group supvr, 59-65, from res assoc to sr res assoc, 65-87, RES FEL, RES & DEVELOP, E I DU PONT DE NEMOURS & CO, INC, 87- *Mem:* Am Chem Soc. *Res:* Fiber forming synthetic polymers; antiarthritic and anticarcinogenic chemicals. *Mailing Add:* 402 N Willey St Seaford DE 19973

NORTON, LOUIS ARTHUR, b Gloucester, Mass, Jan 12, 37; m 63; c 2. ORTHODONTICS, CELL BIOLOGY. *Educ:* Bowdoin Col, AB, 58; Harvard Univ, DmD, 62, cert orthod, 64. *Prof Exp:* From asst prof to prof orthod, Univ Ky, 66-74; PROF ORTHOD, UNIV CONN, 74- *Concurrent Pos:* Fulbright res fel, Int Exchange Persons, 71; vis prof orthod, Hadassah Sch Dent Med, Hebrew Univ, Jerusalem, Israel, 72-73 & Royal Dent Col, Univ Aarhus, Denmark, 88-89; Fogarty fel, NIH, 80; vis scientist biophys, Strangeway Res Lab, Cambridge, UK, 80-81; NATO vis fel, NATO/NATION Denmark, 88; chmn, Coun Res Affairs, Am Asn Orthod, 90-91. *Mem:* Am Asn Orthodontists; Am Col Dentists; Int Asn Dent Res; Orthopedic Res Soc; Bioelec Repair & Growth Soc (pres, 87-88); fel AAAS. *Res:* In vitro and in vivo bioelectric studies on connective tissues; cell biology of tooth movement; clinical studies on temporomandibular dysfunction in orthodontics. *Mailing Add:* Dept Orthod Univ Conn Sch Dent Med Farmington CT 06030

NORTON, MAHLON H, b Denver, Colo, Nov 20, 22; m 70; c 5. ELECTRONICS ENGINEERING, COMMUNICATIONS. *Educ:* Univ Colo, BS, 44. *Prof Exp:* Elec engr, Metron Instrument Co, 44-45; electronic engr, Pilot Radio Corp, 46-47; proj engr, Fada Radio & Electronic Co, 47-52; pres & chief engr, Norton Electronics, 52-53; proj engr, Aeronaut Radio Inc, 53-57; res dir, Booz Allen Appl Res Inc, 57-63; proj leaderommun command & control, Ctr Naval Anal, 63-64; sr scientist, Kaman Aircraft Corp, 64-71; staff engr, Telcom, Inc, 71-74, pres, Norton Electronics, Inc, 74-77; eng mgr, Tracor, Inc, Rockville, Md, 77-80; pres, Norton Eng, PC, 80-85; tech dept dir, Tracor Appl Sci Inc, Alexandria, Va, 85-87; PRES, NORTON ENG PC, 87- *Concurrent Pos:* Consult, Petcar Res, 50-52, Walter Kidde & Co, Thomas A Edison Co & Am Mach & Foundry Co, 52-53, Aeronaut Radio, 82-84, Vega Precision, 83, Calif Microwave Inc, 83-84 & Tracor, 87-88. *Mem:* Inst Elec & Electronics Engrs. *Res:* Communications command and control; direction and planning; adaptive electronic circuitry; prototype design and fabrication; electronic warfare; special surveillance systems; high frequency radio propagation; antenna design; power systems design; electro-chemistry interactions; battery applications. *Mailing Add:* 4436 45th St NW Washington DC 20016

NORTON, RICHARD E, b New York, NY, Mar 2, 28; m 66; c 2. THEORETICAL PHYSICS. *Educ:* Lehigh Univ, BS, 52; Univ Pa, PhD(physics), 58. *Prof Exp:* From asst prof to assoc prof, 60-69, PROF PHYSICS, UNIV CALIF, LOS ANGELES, 69- *Concurrent Pos:* John Simon Guggenheim Found fel, 76-77. *Mem:* Am Phys Soc. *Res:* Theoretical research in field theory and elementary particle physics. *Mailing Add:* Dept of Physics 3-174 Knudsen Hall Univ of Calif 405 Hilgard Ave Los Angeles CA 90024

NORTON, RICHARD VAIL, b Hackensack, NJ, Feb 22, 40; m 62, 81; c 3. ORGANIC CHEMISTRY. *Educ:* Rutgers Univ, BS, 61; Univ Maine, MS, 65, PhD(org chem), 67. *Prof Exp:* Chemist, Armstrong Cork Co, 61-63; sr res chemist, Mobay Chem Co, 67-68; from assoc res chemist to sr res chemist, Res & Develop Lab, Sun Oil Co, 68-77; mgr process res & develop, Ashland Chem Co, 77-84; res dir, Bouden Chem Co, 84-87; RES ASSOC, HERCULES CHEM CO, 87- *Concurrent Pos:* Tech asst to vpres, Cryog Vessel Div, Sun Shipbuilding & Drydock, Chester, Pa, 74-75. *Mem:* AAAS; Am Chem Soc. *Res:* Catalytic reactor design; reactor scale up; reaction engineering; polymer blends, adhesives, inks, new polymers. *Mailing Add:* 3207 Swarthmore Rd Wilmington DE 19807

NORTON, ROBERT ALAN, b Hazelton, Pa, Jan 3, 26; m 50; c 6. HORTICULTURE. *Educ:* Rutgers Univ, BS, 50, MS, 51; Mich State Univ, PhD, 54. *Prof Exp:* Asst, Mich State Univ, 51-54; from asst prof to assoc prof hort, Utah State Univ, 54-61; assoc agriculturist, Dept Pomol, Univ Calif, Davis, 61-62; supt & assoc horticulturist, 62-71, supt, 71-88, HORTICULTURIST, RES & EXTEN CTR, WASH STATE UNIV, 71- *Concurrent Pos:* Consult, Kenai Natives Asn, Kenai, Alaska, 75-78. *Mem:* Am Soc Hort Sci; Am Pomol Soc. *Res:* Culture and physiology of horticultural plants. *Mailing Add:* Res & Exten Ctr Wash State Univ Mt Vernon WA 98273

NORTON, ROBERT JAMES, b Fitchburg, Mass, May 22, 14; m 42; c 2. ECONOMIC ENTOMOLOGY, PHYTOPATHOLOGY. *Educ:* Univ NH, MSc, 39; Univ Mass, PhD(entom, plant path), 51. *Prof Exp:* Res entomologist agr chems, US Rubber Co, 40-42, tech rep, 46-47, mgr tech serv, 48; assoc dir, Crop Protection Inst, 51-54, vpres, 65-69, DIR CROP PROTECTION INST, 54-, PRES, 69- *Mem:* Entom Soc Am; Am Phytopath Soc; Weed Sci Soc Am; Royal Entom Soc London. *Res:* Detection and development of biocides as insecticides, fungicides, herbicides and nematicides. *Mailing Add:* 84 Madbury Rd Durham NH 03824

NORTON, SCOTT J, b Marlow, Okla, Oct 21, 36; m 57; c 2. BIOCHEMISTRY, PESTICIDE CHEMISTRY. *Educ:* Abilene Christian Univ, BS, 59; Univ Tex, Austin, PhD(chem), 63. *Prof Exp:* From asst prof to assoc prof, 63-71, PROF CHEM, NTEX STATE UNIV, 71- *Concurrent Pos:* USPHS grant, 63-68; Robert A Welch Found res grant, 63-87. *Mem:* Am Chem Soc; Am Soc Biol Chem. *Res:* Studies in amino acid metabolism; study of the glyoxalase system; membrane studies; synthesis and study of metabolite analogs; synthesis of insecticides. *Mailing Add:* Div Biochem Univ NTex Denton TX 76203

NORTON, STATA ELAINE, b Mt Kisco, NY, Nov 28, 22; m 49. PHARMACOLOGY, TOXICOLOGY. *Educ:* Univ Conn, BA, 43; Columbia Univ, MA, 45; Univ Wis, PhD(zool), 49. *Prof Exp:* Res scientist neuropharmacol, Burroughs Wellcome & Co, 49-62; from asst prof to assoc prof, 62-68, PROF PHARMACOL, UNIV KANS MED CTR, KANSAS CITY, 68- *Mem:* Am Soc Pharmacol & Exp Therapeut; Biomet Soc; Ecol Soc Am; Am Soc Zool; Soc Toxicol. *Res:* Neuropharmacology; animal behavior; brain development; effects of exposure to toxic substances on development and behavior. *Mailing Add:* Dept Pharmacol Univ Kans Med Ctr 39th St & Rainbow Blvd Kansas City KS 66103

NORTON, STEPHEN ALLEN, b Newton, Mass, May 21, 40; m 70; c 3. GEOLOGY. *Educ:* Princeton Univ, AB, 62; Harvard Univ, MA, 63, PhD(geol), 67. *Prof Exp:* From asst prof to assoc prof, 68-77, PROF GEOL, UNIV MAINE, ORONO, 77- *Concurrent Pos:* Dean, Col Arts & Sci, Univ Maine, 84-86. *Mem:* Geol Soc Am; Mineral Soc Am. *Res:* Low temperature and pressure geochemistry. *Mailing Add:* Dept Geol Sci Univ Maine Orono ME 04469

NORTON, TED RAYMOND, b Stockton, Calif, Nov 16, 19; m 44; c 3. ORGANIC CHEMISTRY, PHARMACOLOGY. *Educ:* Univ Pac, BA, 40; Northwestern Univ, PhD(org chem), 43. *Prof Exp:* Dir, agr chem lab, Dow Chem Co, Mich, 53-57, from asst dir to dir, Britton Lab, 62-66, asst dir, Independent Labs, 66-68; prof, 68-85, EMER PROF PHARMACOL, UNIV HAWAII, MANOA, 86- *Concurrent Pos:* Consult pharmacol, toxicol & org chem. *Mem:* Am Chem Soc. *Res:* Isolation and characterization of marine natural products for antitumor and heart stimulant properties. *Mailing Add:* 312 Iliaina St Kailua HI 96734

NORTON, VIRGINIA MARINO, b Memphis, Tenn, Nov 14, 34; m 52; c 3. GERONTOLOGY, PHSYIOLOGY. *Educ:* Memphis State Univ, BS, 69, MS, 71, PhD(biol), 75; Univ Tenn, BSN, 81. *Prof Exp:* Fel & instr physiol, Dept Physiol & Biophys, Ctr Health Sci, Univ Tenn, Memphis, 75-77; asst prof anat & physiol, Dept Biol, Memphis State Univ, 77-81; nursing staff, Nursing Home Care Unit, Vet Admin Hosp, Memphis, 81-82; sr instr Sch, 82-86, DIR RES & DEVELOP NURSING METHODIST HOSP, MEMPHIS, 86- *Concurrent Pos:* Consult, Sci Curriculum Work Shop; instr, Memphis City Schs, 79, Nursing Inserv, 81. *Mem:* Gerontol Soc Am; Sci Res Soc; Am Nurses Asn; Sigma Xi. *Mailing Add:* 5583 Ashley Sq N Memphis TN 38119

NORTON, WILLIAM THOMPSON, b Damariscotta, Maine, Jan 27, 29; m 57; c 2. NEUROCHEMISTRY. *Educ:* Bowdoin Col, AB, 50; Princeton Univ, MA, 52, PhD(org chem), 54. *Prof Exp:* Res chemist, E I du Pont de Nemours & Co, 53-57; instr biochem, 57-58, Albert Einstein Col Med, 57-58, sr fel interdisciplinary prog, 57-60, assoc med & biochem, 58-69, asst prof biochem, 59-64, assoc prof neurol, 64-71, PROF NEUROL, ALBERT EINSTEIN COL MED, 71-, PROF NEUROSCI, 74- *Concurrent Pos:* Vis prof, Charing Cross Hosp Med Sch, London, 67-68; adv comt, Nat Mult Sclerosis Soc, 69-80; mem, neurol A study sect, NIH, 71-76, chmn, BNS-1 study sect, 79-84; chief ed, J Neurochem, 81-85; Jacob Javits neuroscience investr award, 84. *Mem:* AAAS; Am Soc Biol Chem; Am Soc Neurochem (pres, 87-89); Int Soc Neurochem; Soc Neurosci. *Res:* Lipid and myelin chemistry; demyelination; chemistry of isolated brain components; glial cells; cytoskeleton. *Mailing Add:* Dept of Neurol Albert Einstein Col of Med New York NY 10461

NORUSIS, MARIJA JURATE, b Ansbach, Ger, Jan 3, 48; US citizen. BIOSTATISTICS. *Educ:* Univ Ill, BA, 68; Univ Mich, MPH, 71, PhD(biostatist), 73. *Prof Exp:* Res assoc statist & med, Univ Chicago, 73-76; STATISTICIAN, SPSS, INC, 76- *Mem:* Am Statist Asn; Biomet Soc. *Res:* Applications of statistics and computers to biomedical research. *Mailing Add:* Dept Prev Med Rush Univ 600 S Paulina St Chicago IL 60612

NORVAL, RICHARD ANDREW, b Harare, Zimbabwe, Oct 14, 50; m 71; c 2. TICK ECOLOGY & CONTROL, EPIDEMIOLOGY & CONTROL TICK-BORNE DISEASES. *Educ:* Rhodes Univ, SAfrica, PhD(entom), 75, DSc, 85. *Prof Exp:* Chief res officer, Vet Res Lab, Harare, Zimbabwe, 75-87; scientist, Ilrad, Nairobi, Kenya, 87-89; RES SCIENTIST EPIDEMIOL, UNIV FLA, 90- *Concurrent Pos:* Consult, Food & Agr, UN, 81-87. *Res:* Tick biology, ecology and behavior; tick pheromones and their use in control; population modelling; epidemiology of tick-borne disease; geographic information systems. *Mailing Add:* Dept Infectious Dis Univ Fla Mowry Rd Bldg 471 Gainesville FL 32611-0633

NORVELL, JOHN CHARLES, b Jacksonville, Tex, Jan 11, 40; m 66; c 3. MOLECULAR BIOPHYSICS, CRYSTALLOGRAPHY. *Educ:* Rice Univ, BA, 63; Yale Univ, MS, 65, PhD(physics), 68. *Prof Exp:* Fel molecular physics, Res Establishment Riso, Roskilde, Denmark, 68-69; fel biophysics, Biophysics Lab, Univ Wis-Madison, 70-72; res assoc biophysics, Biol Dept, Brookhaven Nat Lab, Upton, NY, 72-75; biophysicist, Lab Molecular Biol, Nat Inst Arthritis, Metab & Digestive Dis, NIH, 75-76; staff officer, Nat Res Coun, Nat Acad Sci, Washington, DC, 76-78; MEM STAFF, NAT INST GEN MED SCI, NIH, 78- *Mem:* AAAS; Biophys Soc. *Res:* Diffraction studies of the structure of proteins and other biological systems. *Mailing Add:* Biophys & Physiol Sci Prog NIH Westwood Bldg Rm 907 Bethesda MD 20892

NORVELL, JOHN EDMONDSON, III, b Charleston, WVa, Nov 18, 29; m 62; c 2. NEUROANATOMY. *Educ:* Univ Charleston, BS, 53; WVa Univ, MS, 56; Ohio State Univ, PhD(anat), 66. *Prof Exp:* Instr biol sci, Johnstown Col, Pittsburgh, 56-60; asst prof, Otterbein Col, 60-62; asst instr anat, Ohio State Univ, 62-65; from asst prof to assoc prof anat, Med Col Va, Va Commonwealth Univ, 66-76; prof & chmn dept anat, Oral Roberts Univ, 76-87, prof dept anat, 87-89; RETIRED. *Concurrent Pos:* Consult, US Naval Hosp, Portsmouth, 66-71; chmn, Okla State Anat Bd, 78-89; chmn, sect anat sci, Am Asn Dent Schs, 79-80; mem, Okla Governor's Mini-Cabinet Health & Human Resources, 80-82; vis prof, Dept Human Anat, Univ Nairobi, Kenya, 82, Sch Med Sci, Univ Benin, Nigeria, 89. *Mem:* AAAS; Am Asn Anat; Soc Neurosci; Transplantation Soc; Int Brain Res Orgn. *Res:* Degeneration and regeneration of nerves in transplanted organs; localization of biogenic amines and transmitters in the nervous system. *Mailing Add:* Dept Anat-Univ Central Del Caribe Call Box 60-327 Bayamon PR 00621-6032

NORVITCH, MARY ELLEN, b Minneapolis, Minn, Feb 4, 55. CLINICAL RESEARCH-ONCOLOGY, MONOCLONAL ANTIBODIES. *Educ:* Univ Minn, PhD(pathobiol), 82. *Prof Exp:* Fel cell & molecular biol, Mayo Clin, 82-86; scientist cell & molecular biol, 86-87, clin res assoc, Div Med Res, 87-89, RES ASSOC BIOTECHNOL, AM CYANAMID CO, 90- *Concurrent Pos:* Am Soc Cell Biol Travel Grant, 80; consult field reader, Off Orphan Prod Develop, Food & Drug Admin, 89- *Honors & Awards:* Bacaner Res Award, 83. *Mem:* Am Chem Soc; Am Soc Cell Biol; Soc Nuclear Med. *Res:* Clinical oncology research in use of monoclonal antibody immunoconjugates for therapy. *Mailing Add:* Am Cynamid Co/MRD Bldg 140 Rm 455 Pearl River NY 10965

NORWICH, KENNETH HOWARD, b Toronto, Ont, May 8, 39; m 63; c 3. BIOPHYSICS, PHYSIOLOGY. *Educ:* Univ Toronto, MD, 63, BSc, 67, MSc, 68, PhD(physics), 70. *Prof Exp:* assoc prof, 70-80, PROF PHYSIOL & APPL SCI, INST BIOMED ENG, UNIV TORONTO, 80-, CROSS-APPOINTED PROF PHYSICS, 85-, ASSOC DIR, INST BIOMED ENG, 89- *Mem:* Biomed Eng Soc; Psychonomics Soc; Can Med Asn; Can Med & Biol Eng Soc; Soc Math Biol. *Res:* Mathematical studies in physiology and medicine, studies of the transport of tracers; mathematical studies of metabolism; theoretical studies of sensory perception; applications of computers in medical decision-making. *Mailing Add:* Dept Physiol Univ Toronto Toronto ON M5S 1A8 Can

NORWOOD, CHARLES ARTHUR, b Crystal City, Tex, Jan 8, 38; m. AGRONOMY. *Educ:* Tex A&I Univ, BS, 61; Okla State Univ, MS, 69, PhD(soil sci), 72. *Prof Exp:* Res technician soils, Agr Res Serv, USDA, Tex, 63-67; res asst soil fertil, Okla State Univ, 67-72; ASSOC PROF DRYLAND SOILS RES, GARDEN CITY EXP STA, KANS STATE UNIV, 72- *Mem:* Am Soc Agron; Soil Sci Soc Am; Soil & Water Conserv Soc. *Res:* Soil fertility; management of dryland soils, dryland cropping systems, reduced tillage research pertaining to efficient water use under dryland conditions; irrigated to dryland transition. *Mailing Add:* Garden City Exp Sta Kans State Univ Garden City KS 67846

NORWOOD, FREDERICK REYES, b Mexico City, Mex, May 13, 39; US citizen; div; c 2. SOLID MECHANICS, APPLIED MATHEMATICS. *Educ:* Univ Calif, Los Angeles, BS, 62; Calif Inst Technol, MS, 63, PhD(appl mech), 67. *Prof Exp:* MEM TECH STAFF, SANDIA LABS, 66- *Mem:* Am Soc Mech Engrs; Soc Indust & Appl Math; Am Geophys Union; Am Acad Mech; Sigma Xi. *Res:* Theoretical study of transient phenomena in liquids, gases and solids. *Mailing Add:* 1819 Ross Pl SE Albuquerque NM 87108

NORWOOD, GERALD, b Tecumseh, Okla, Apr 23, 38. MATHEMATICS. *Educ:* Purdue Univ, PhD(math), 67. *Prof Exp:* prof math, NCarolina Univ, 67-73; PROF MATH, MERCER UNIV, 73- *Mailing Add:* Math Dept Mercer Univ Macon GA 31207

NORWOOD, JAMES S, b Burleson, Tex, Oct 2, 32; m 62. REPRODUCTIVE PHYSIOLOGY, GENETICS. *Educ:* Tex Tech Col, BS, 54; Kans State Univ, MS, 55, PhD(reprod physiol), 63. *Prof Exp:* Instr dairy sci, Southwest Tex State Col, 55-56 & 58-60; asst prof biol, Arlington State Col, 62-63; assoc prof reprod physiol, 63-68, assoc prof biol, 68-71, PROF BIOL, E TEX STATE UNIV, 71- *Mem:* Am Soc Animal Sci; Sigma Xi. *Res:* Post-partum regression of the bovine uterus; conception rates; inhibition of estrus and effect of steroid hormones on the endometrium of the uterus. *Mailing Add:* Agr Dept E Tex State Univ Commerce TX 75428

NORWOOD, RICHARD E(LLIS), b Park Ridge, Ill, May 17, 34; m 58; c 3. MECHANICAL ENGINEERING. *Educ:* Mass Inst Technol, SB, 56, SM, 59, ScD(mech eng), 61. *Prof Exp:* SR TECH STAFF MEM, INFO PROD DIV, IBM CORP, 61- *Mem:* Am Soc Mech Engrs; Sigma Xi. *Res:* Fluid power and automatic control; fluid mechanics and logic devices; machine design. *Mailing Add:* 325 Fox Ct Boulder CO 80303

NOSAL, EUGENE ADAM, b Chicago, Ill, Jan 15, 42. GEOPHYSICS. *Educ:* Ill Benedictine Col, BS, 63; Univ Wyo, MS, 65, PhD(physics), 69. *Prof Exp:* SCIENTIST, DENVER RES CTR, MARATHON OIL CO, 69- *Mem:* AAAS; Am Asn Physics Teachers; Am Geophys Union; Soc Explor Geophysicists; Sigma Xi. *Res:* Analysis of digitized data for geophysical interpretation; application of probability and statistics to geological data. *Mailing Add:* Marathon Oil Co PO Box 3128 Houston TX 77253-3128

NOSANOW, LEWIS H, b Philadelphia, Pa, July 9, 31; m 55; c 2. THEORETICAL PHYSICS, LOW TEMPERATURE PHYSICS. *Educ:* Univ Pa, BA, 54; Univ Chicago, PhD(chem physics), 58. *Prof Exp:* NSF fel, 58-59; res assoc physics, Inst Theoret Physics, Univ Utrecht, 59-60; asst res physicist, Univ Calif, San Diego, 60-62; from assoc prof to prof physics, Univ Minn, Minneapolis, 62-73; prof physics & astron & chmn dept, Univ Fla, 73-74; HEAD CONDENSED MATTER SCI SECT, DIV MAT RES, NSF, 74- *Concurrent Pos:* Guggenheim fel, 66-67; vis prof physics, Univ Sussex, Eng, 71-72, Univ Wash, Seattle & Drexel Univ, Philadelphia, 78-79; assoc provost & prof physics, Univ Chicago, 81-82. *Mem:* Am Phys Soc; AAAS. *Res:* Statistical mechanics and phase transition; macroscopic quantum systems. *Mailing Add:* VChancellor Res Univ California Irvine Irvine CA 92717

NOSEK, THOMAS MICHAEL, b Brooklyn, NY, Apr 25, 47; m 68; c 2. CELLULAR ELECTROPHYSIOLOGY & MECHANICS. *Educ:* Univ Notre Dame, BS, 69; Ohio State Univ, PhD(biophys), 73. *Prof Exp:* Instr physiol, Bowman Gray Sch Med, Wake Forest Univ, 73-76; asst prof, 76-82, ASSOC PROF PHYSIOL, MED COL GA, 82- *Concurrent Pos:* Post-doctoral trainee, NIH, 74-76; prin investr, NIH Res Award, 79-82 & 87-91, Ga affil, Am Heart Asn, 81-85 & 87-89; German academic exchange study fel, 86; co-investr, NIH, 87-92, Navy Res Lab, 91-94. *Mem:* Am Physiol Soc; Soc Neurosci; Biophys Soc; Sigma Xi; Am Heart Asn; NY Acad Sci. *Res:* Excitation-contraction coupling process in the heart and skeletal muscle; influence of altered cellular metabolism during hypoxia and fatigue, inositoltrisphosphate on excitation-contraction coupling. *Mailing Add:* Dept Physiol Med Col Ga 1120 15th St Augusta GA 30912-3000

NOSHAY, ALLEN, b Philadelphia, Pa, Oct 14, 33; m 56; c 2. ORGANIC CHEMISTRY, POLYMER CHEMISTRY. *Educ:* Temple Univ, BA, 55; Univ Pa, MS, 57, PhD(polymer org chem), 59. *Prof Exp:* Chemist, Esso Res & Eng Co, 59-61, proj leader polymer chem, 61-65; res chemist, 65-69, proj scientist, 69-75, res scientist, Plastics Div, 75-79, sr group leader, 79-84, RES ASSOC, UNION CARBIDE CORP, 84- *Mem:* Am Chem Soc. *Res:* Polymer synthesis and modification; block copolymers; epoxy resins; polyolefin catalysis; gas phase polymerization. *Mailing Add:* 66 Wellington Rd East Brunswick NJ 08816

NOSHPITZ, JOSEPH DOVE, b New York, NY, Aug 31, 22; m 56; c 1. CHILD PSYCHIATRY. *Educ:* Univ Louisville, MD, 45; Baltimore-DC Psychoanal Inst, grad psychoanal, 69, grad child psychoanal, 71. *Prof Exp:* Chief children's serv, Topeka State Hosp, Kans, 51-56; chief children's unit, NIMH, 56-60; dir clin inst, Hillcrest Children's Ctr, Children's Hosp, DC, 69-74; dir training, Dept Child Psychiat, Children's Hosp & Nat Med Ctr, Washington, DC, 76-79, sr attend staff psychiatrist, 79-88; RETIRED. *Concurrent Pos:* prof child health & human develop, Med Sch, George Washington Univ; med dir, Florence Crittenton Home, Washington, DC, 73-; vis prof child psychiat, Sch Med, Tel Aviv Univ, 75-76; consult, Community Ment Health Ctr, Washington, DC. *Mem:* Am Psychiat Asn; fel Am Orthopsychiat Asn; fel Am Acad Child Psychiat; Am Asn Children's Residential Ctrs. *Res:* Adolescence; delinquency; treatment of emotionally disturbed children; residential treatment; gender identity in women; the tomboy phenomenon. *Mailing Add:* 3141 34th St NW Washington DC 20008

NOSKOWIAK, ARTHUR FREDRICK, b Galt, Calif, Nov 14, 20. WOOD TECHNOLOGY. *Educ:* Univ Calif, Berkeley, BS, 42, MF, 49; State Univ NY Col Forestry, Syracuse, PhD(wood technol), 59. *Prof Exp:* From instr to asst prof forestry, Colo State Univ, 51-54; instr wood technol, State Univ NY Col Forestry, Syracuse Univ, 58-59; from asst prof to prof forestry & wood technol, 60-83, ASSOC EMER PROF FORESTRY & RANGE MGT, WASH STATE UNIV, 83-; ASSOC EMER PROF FORESTRY & RANGE MGT, WASH STATE UNIV, 83- *Concurrent Pos:* Consult, 83- *Mem:* Soc Am Foresters. *Res:* Spiral grain in trees; strength properties of wood; anatomy of wood. *Mailing Add:* N 5307 Argonne Rd No 3 Spokane WA 99212-4430

NOSS, RICHARD ROBERT, b Chicago, Ill, May 31, 50; m 73. ENVIRONMENTAL ENGINEERING. *Educ:* Harvard Univ, AB, 72; Univ Mich, MSE, 73; Mass Inst Technol, MS, 78, PhD(civil eng), 80. *Prof Exp:* Asst prof environ eng, Dept Civil Eng, Univ Mass, 80-88; CONSULT, BENNETT & WILLIAM INC, 88- *Concurrent Pos:* Pub mem, Mass Water Resources Comn, 76-80; Kellogg Nat fel, 82-85. *Mem:* Am Soc Chem Engrs; Am Geophys Union; Am Water Resources Asn; Water Pollution Control Fedn; Asn Environ Eng Professionals. *Res:* Water quality management; environmental systems analysis; environmental policy analysis. *Mailing Add:* Bennett & Williams Inc 2700 E Dublin Grandville Rd Suite 550 Columbus OH 43291

NOSSAL, NANCY, b Fall River, Mass, Feb 9, 37; m 59; c 3. BIOCHEMISTRY, MOLECULAR BIOLOGY. *Educ:* Cornell Univ, AB, 58; Univ Mich, PhD(biochem), 63. *Prof Exp:* NIH fels, Brussels, 63-64 & Bethesda, Md, 64-65, RES CHEMIST, LAB BIOCHEM PHARMACOL, NAT INST DIABETES & DIGESTIVE & KIDNEY DIS, NIH, 66-, CHIEF, SECT NUCLEIC ACID BIOCHEM, 83- *Mem:* Am Soc Biol Chemists; Am Soc Microbiol. *Res:* Enzymology of nucleic acids; DNA replication. *Mailing Add:* Nat Inst Health Bldg Eight Rm 2A-19 Bethesda MD 20892

NOSSAL, RALPH J, b Brooklyn, NY, Dec 26, 37; m 59; c 3. BIOPHYSICS, PHYSICS. *Educ:* Cornell Univ, BEng Phys, 59; Univ Mich, MS, 61, PhD(nuclear eng), 63. *Prof Exp:* NSF fel statist mech, Brussels, 63-64; Nat Acad Sci-Nat Sci Found res assoc, Nat Bur Standards, 64-66; PHYSICIST, NIH, 66- *Concurrent Pos:* Adj prof chem physics, Univ Md, 83-; vis prof biophys, Univ Col Berkley, 88- *Mem:* Am Phys Soc; Biophys Soc; Am Soc Cell Biol. *Res:* Statistical physics; laser scattering; membrane biophysics; cellular physiology. *Mailing Add:* Phys Sci Lab Div Comp Res Tech Nat Inst of Health Bethesda MD 20892

NOSSAMAN, NORMAN L, b Cherokee, Okla, Jan 21, 32; m 55; c 3. SOIL FERTILITY. *Educ:* Okla State Univ, BS, 53, MS, 57; Kans State Univ, PhD(soils fertil), 63. *Prof Exp:* In chg of dryland soil mgt, Garden City Br Exp Sta, Kans State Univ, 60-65; agronomist, Western Ammonia Corp, 65-69; chief agronomist, Nipak, Inc, 60-78; CONSULT AGR, 78-; AGRONOMIST, TERRA INT, INC, 88- *Mem:* Am Soc Agron; Soil Sci Soc Am. *Res:* Dryland soil management; soil fertility; soil moisture; tillage methods. *Mailing Add:* Terra Int PO Box 20606 Waco TX 76702-0606

NOTARO, ANTHONY, b Queens, NY, Sept 13, 56. REAL TIME COMPUTER SYSTEMS, SYSTEMS ANALYSIS & DESIGN. *Educ:* Hofstra Univ, BS, 78. *Prof Exp:* Programmer, Sperry Corp, 78-80; sr systs analyst, Grumman Corp, 84-85; sr software engr, Satellite Transmission Systs, 85; consult computers, 85-90; nuclear systs designer, 80-84, PLANT ENGR, LONG ISLAND LIGHTING CO, SHOREHAM NUCLEAR POWER STA, 90- *Concurrent Pos:* Consult, 78- *Mem:* Inst Elec & Electronic Engrs; Inst Elec & Electronic Engrs Computer Soc; Asn Comput Mach; Armed Forces Commun & Electronics Asn. *Res:* Computer systems design. *Mailing Add:* 319 Garfield Ave West Hempstead NY 11552

NOTATION, ALBERT DAVID, b Moosomin, Sask, Oct 28, 35; m 63; c 2. ENDOCRINOLOGY, BIOCHEMISTRY. *Educ:* Univ Sask, BE, 58, MSc, 59; McMaster Univ, PhD(org chem), 64. *Prof Exp:* Lab asst fats & oils, Prairie Regional Lab, Nat Res Coun Can, 53-54; NIH Steroid Training Prog fel, Sch Med, Univ Minn, Minneapolis, 64-66, res assoc biochem, 66-67, asst prof, 67-77, asst prof obstet & gynec & lab med, 71-77; chmn dept chem & phys sci, 78-81, assoc prof, 77-81, PROF CHEM, SCH ALLIED HEALTH, QUINNIPIAC COL, 81- *Concurrent Pos:* Ayerst Squibb travel fel, 68; Minn Med Found grant, 71; res affil, Dept Obstet & Gynec, Med Sch, Yale Univ, 80- *Mem:* AAAS; Am Chem Soc; Chem Inst Can; Endocrine Soc; Soc Study Reproduction; Sigma Xi. *Res:* Steroid metabolism and biochemistry; radioimmunoassay of hormones; competitive protein binding assays. *Mailing Add:* Dept Chem Phys Quinnipaic Col Box 370 Hamden CT 06518

NOTEBOOM, WILLIAM DUANE, b East Fairview, NDak, Mar 31, 33; m 67; c 1. BIOCHEMISTRY. *Educ:* Ore State Univ, BS, 55, MS, 61; Univ Ill, PhD(physiol), 65. *Prof Exp:* Fel, McArdle Inst Cancer Res, Univ Wis, 64-67; ASSOC PROF BIOCHEM, UNIV MO-COLUMBIA, 67- *Mem:* Am Chem Soc; Sigma Xi. *Res:* Regulation of cell growth and metabolism. *Mailing Add:* Dept Biochem Univ Mo Columbia MO 65212

NOTHDURFT, ROBERT RAY, b Cape Girardeau, Mo, Nov 13, 39; m 61; c 2. PHYSICS. *Educ:* Wash Univ, St Louis, AB, 61; Univ Mo-Rolla, MS, 64, PhD(physics), 67. *Prof Exp:* Res physicist, US Bur Mines, Mo, 62-67; PROF PHYSICS, NORTHEAST MO STATE COL, 67- *Res:* Kilocycle range dislocation damping, especially in magnesium single crystals. *Mailing Add:* Dept Sci Northeast Mo State Univ Kirksville MO 63501

NOTHNAGEL, EUGENE ALFRED, b Litchfield, Minn, June 5, 52; m 80; c 2. PLANT CELL BIOLOGY, PLANT SENESCENCE. *Educ:* Univ Minn, Morris, BA, 73; Southern Ill Univ, MA, 75; Cornell Univ, MS, 77, PhD(appl physics), 81. *Prof Exp:* Res assoc, dept chem, Univ Colo, Boulder, 81-83; asst prof, 83-89, ASSOC PROF PLANT PHYSIOL, DEPT BOT & PLANT SCI, UNIV CALIF, RIVERSIDE, 89- *Mem:* Am Soc Plant Physiologists; Am Soc Cell Biol; Biophys Soc; Am Soc Hort Sci. *Res:* Dynamics and biochemistry of interactions between the cytoskeleton, plasma membrane and cell wall of plant cells; mechanisms of leaf senescence and fruit ripening; membrane role in chilling sensitivity. *Mailing Add:* Dept Bot & Plant Sci Univ Calif Riverside CA 92521-0124

NOTIDES, ANGELO C, b New York, NY, Dec 11, 36; m 61; c 2. ENDOCRINOLOGY, BIOCHEMISTRY. *Educ:* Hunter Col, BA, 59, MA, 62; Univ Ill, PhD(physiol), 66. *Prof Exp:* Asst prof pharmacol, 68-74, assoc prof toxicol, 74-79, PROF BIOPHYS & TOXICOL, SCH MED & DENT, UNIV ROCHESTER, 79- *Concurrent Pos:* Vis prof, Paris, 86, Norway, 76. *Mem:* Am Soc Biol Chemists; Endocrine Soc; Am Chem Soc; AAAS. *Res:* Mechanism of hormone action; reproductive biochemistry; receptor biochemistry. *Mailing Add:* Dept Biophys Univ Rochester Sch Med & Dent Box #EHSC Rochester NY 14642

NOTKINS, ABNER LOUIS, b New Haven, Conn, May 8, 32; m 69. VIROLOGY, IMMUNOLOGY. *Educ:* Yale Univ, BA, 53; NY Univ, MD, 58. *Prof Exp:* Intern internal med, Johns Hopkins Hosp, 58-59, asst resident, 59-60; res assoc, Nat Cancer Inst, 60-61; investr, Lab Microbiol & Immunol, 61-67, chief viral sect, 67-73, CHIEF LAB ORAL MED, NAT INST DENT RES, NIH, 73-, DIR INTRAMURAL RES, 85- *Concurrent Pos:* Bd dirs, Paul Ehrlich Found; adv comt, res progs, Nat Multiple Sclerosis Soc, Nat Dis Res Interchange. *Honors & Awards:* Meritorious Serv Medal, Dept Health & Human Serv, 73, Distinguished Serv Medal, 81; David Rumbaugh Sci

Award, Juvenile Diabetes Found, 80; Paul E Lacy Res Award, Nat Diabetes Res Interchange, 82; Paul Ehrlich Prize, 86; Rolf Luft Medal, 88; Philip Hench Award in Rheumatology, 89. *Mem:* AAAS; Asn Am Physicians; Am Soc Exp Path; Am Asn Immunol. *Res:* Virology; immunology; endocrinology. *Mailing Add:* Lab Oral Med Bldg 30 Rm 121 NIH Bethesda MD 20892

NOTLEY, NORMAN THOMAS, b Bristol, Eng, Apr 10, 28; Brit & US citizen; c 2. POLYMER CHEMISTRY, PHOTOGRAPHIC SCIENCE. *Educ:* Bristol Univ, BSc, 49, PhD(phys chem), 52. *Prof Exp:* Res assoc polymer chem, Cornell Univ, 52-54; res chemist, E I du Pont de Nemours & Co, 54-59; res supvr, Metal Box Co, Eng, 59-62; chief chemist, Kalvar Corp, La, 62-63, dir res, 63-65, dir res & eng, 65-66; dir chem res, Bus Equip Group, Bell & Howell, 66-69; pres, Photomedia Co, 69-78; consult, 3M Co, 78-85; CONSULT, NOVAMEDIA CO, 85- *Mem:* Am Chem Soc; Soc Photog Sci & Eng; Sigma Xi. *Res:* Solution properties of polymers; polymerization kinetics; photopolymerization; oxidation kinetics of polyolefines; permeability of plastics; unconventional photographic and reprographic systems. *Mailing Add:* 1895 E Orange Grove Blvd Pasadena CA 91104

NOTO, THOMAS ANTHONY, b Tampa, Fla, Dec 27, 31; m 63; c 3. PATHOLOGY. *Educ:* Spring Hill Col, BS, 53; St Louis Univ, MD, 57. *Prof Exp:* Instr path, Sch Med, Univ Miami, 62-64; asst prof, Med Col Ala, 64-65, clin path, 65-68; assoc prof, Med Ctr, Univ Ala, Birmingham, 68-69; assoc dir clin path, 69-77, dir clin path II, 77-78, PROF, SCH MED, UNIV MIAMI, 77-; DIR TRANSFUSION SERV, JACKSON MEM HOSP, 78- *Concurrent Pos:* Asst clin pathologist, Jackson Mem Hosp, 62-64; asst clin pathologist & dir blood bank, Med Ctr, Univ Ala, Birmingham, 64-69. *Mem:* Fel Am Soc Clin Path; Col Am Path; Am Asn Blood Banks; Int Soc Blood Transfusion. *Res:* Immunohematology; transfusion medicine. *Mailing Add:* Dept Path D-33 Univ Miami Sch Med PO Box 016960 Miami FL 33101

NOTTEBOHM, FERNANDO, ETHOLOGY. *Prof Exp:* DIR, FIELD RES CTR ETHOLOGY & ECOL, ROCKEFELLER UNIV. *Mem:* Nat Acad Sci. *Mailing Add:* Rockefeller Univ Field Res Ctr Ethology & Ecol RR 2 Box 38B Millbrook NY 12545

NOTTER, MARY FRANCES, b Johnstown, Pa, Aug 25, 47; m 73. TISSUE CULTURE, NEUROBIOLOGY. *Educ:* Pa State Univ, BS, 69, MS, 71, PhD(physiol), 73. *Prof Exp:* Scholar microbiol, Pa State Univ, 74-76; tech assoc infectious dis, 76-78, res assoc microbiol, 78-80, instr microbiol, 80-81, ASST PROF ANAT & MICROBIOL, UNIV ROCHESTER MED CTR, 81- *Concurrent Pos:* Teaching fel anat, Univ Rochester Med Sch, 73-74. *Mem:* NY Acad Sci; Soc Neurosci; Tissue Cult Asn; Asn Anatomists. *Res:* Tissue culture of the fetal nervous system; viral cell surface receptors; surface markers-neural cell identification, changes with development; viral markers in tumor tissue and their visualization. *Mailing Add:* Dept Anat & Microbiol Univ Rochester Med Ctr 601 Elmwood Rochester NY 14642

NOTZ, WILLIAM IRWIN, b Washington, DC, Oct 16, 51; m 80. OPTIMAL EXPERIMENTAL DESIGN. *Educ:* Johns Hopkins Univ, BS, 73; Cornell Univ, MS, 76, PhD(statist), 78. *Prof Exp:* asst prof statist, Purdue Univ, 78-84; ASSOC PROF STATIST, OHIO STATE UNIV, 84- *Concurrent Pos:* Assoc ed, J Am Statist Asn, 84- *Mem:* Inst Math Statist; Am Statist Asn. *Res:* Robust experimental designs for regression problems; optimal experimental designs; construction of experimental designs. *Mailing Add:* Dept Statist Ohio State Univ Columbus OH 43210

NOUFI, ROMMEL, b Nazareth, Israel, Oct 16, 47. MATERIAL SCIENCE, SEMICONDUCTOR RESEARCH. *Educ:* Southwest Tex State, BSc, 70 & MSc, 73; Univ Tex, PhD(phys & anal chem), 78. *Prof Exp:* Mem tech staff photoelectrochem of semiconductors, Rockwell Int Sci Ctr, 78-80; sr scientist, 80-82, GROUP LEADER & PRIN SCIENTIST MAT SCI PHOTO VOLTAIC RES, SOLAR ENERGY RES INST, 82- *Mem:* Electrochem Soc; Am Chem Soc; Am Phys Soc; Mat Res Soc; Am Asn Advan Sci. *Res:* Fabrication and characterization of new compound semi-conductor materials for photovoltaic applications and thin film superconductors. *Mailing Add:* Solar Energy Res Inst 1617 Cole Blvd Golden CO 80401

NOUJAIM, ANTOINE AKL, b Cairo, Egypt, Feb 26, 37; m 64. NUCLEAR PHARMACY. *Educ:* Cairo Univ, BS, 58; Purdue Univ, MSc, 63, PhD(bionucleonics), 65. *Prof Exp:* Head res & statist dept, Gen Orgn Drugs, Cairo Univ, 59-61; res assoc bionucleonics, Purdue Univ, 65-66; from asst prof to assoc prof bionucleonics, 66-73, chmn dept, 68-74, PROF NUCLEAR PHARM, UNIV ALTA, 73-, CHMN RADIATION CONTROL, 72-, PROF PHARMACEUT STUDIES, UNIV & MEM STAFF, CANCER RES UNIT, 77- *Concurrent Pos:* Nat Res Coun fel, Cairo Univ, 61-65; consult water resources div, Alta Dept Agr, 66-; sr vis scientist, CSIRO, Australia, 74-75; sr res fel nuclear med, Dr W W Cross Cancer Inst, 75- *Mem:* AAAS; NY Acad Sci; Am Pharmaceut Asn; Soc Nuclear Med. *Res:* Radiation effects on biological systems; drug metabolism; isotope dilution methods; activation analysis; research and development of radiopharmaceuticals; clinical applications of radioactive drugs. *Mailing Add:* Dept Pharm Univ Alta Edmonton AB T6G 2E2 Can

NOVACEK, MICHAEL JOHN, b Evanston, Ill, June 3, 48; m 71; c 1. EVOLUTIONARY BIOLOGY, PALEONTOLOGY. *Educ:* Univ Calif, Los Angeles, AB, 71; San Diego State Univ, MA, 73; Univ Calif, Berkeley, PhD(paleont), 77. *Prof Exp:* Res asst vert paleont, San Diego State Univ, 71-72, teaching asst, 72-73; res asst, Univ Calif, Berkeley, 73-75; lectr zool, San Diego State Univ, 76-77, asst prof, 77-79, assoc prof, 79-81; asst cur fossil mammals, 81-85, assoc cur & chmn, Dept Vert PALEONT, 85-89, CUR VERT PALEONT, VPRES & DEAN SCI, AM MUS NATURAL HIST, NEW YORK, 89- *Concurrent Pos:* Res assoc, Univ Calif Mus Paleont, Berkeley, 77-; NSF fel, Am Mus Natural Hist, 79-80; adv, NSF, 86- *Mem:* Soc Study Evolution; Soc Syst Zool (pres, 90-91); Am Soc Mammalogists; Paleont Soc; Soc Vert Paleont. *Res:* Mammalian evolution and phylogeny; vertebrate paleobiogeography; multivariate analysis of community structure; ontogeny; functional morphology of auditory features in bats and other mammals; morphological and molecular systematics. *Mailing Add:* Am Mus Natural Hist Central Park W & 79th New York NY 10024

NOVACK, GARY DEAN, b Oakland, Calif, Nov 21, 53; m 77; c 2. GLAUCOMA, BIOSTATISTICS. *Educ:* Univ Calif, Santa Cruz, AB, 73; Univ Calif, Davis, PhD(pharmacol & environ toxicol), 77. *Prof Exp:* Res fel anesthesiol, Sch Med, Univ Calif, Davis, 77; NIH fel neurophysiol, Ment Retardation Res Ctr & Brain Res Inst, Univ Calif, Los Angeles, 77-79; pharmacologist, Merrell Res Ctr, Richardson-Merrell Inc, Cincinnati, Ohio, 79-81, sr res pharmacologist, Merrell Dow Res Ctr, Merrell Dow, 81-82; clin coordr, 82-84, mgr, Ophthal Clin Res, Allergan, Smith Kline Beckman, Irvine, Calif, 84-88, prin clin scientist, Nelson Res, Irvine, Calif, 88-89; ASST RES NEUROBIOL & PSYCHOBIOL, CTR NEUROBIOL, LEARNING & MEMORY, UNIV CALIF, IRVINE, 84-, PRES & CHIEF SCIENTIST PHARMALOGIC DEVELOP, INC, IRVINE, CALIF, 89- *Mem:* Am Soc Clin Pharmacol & Therapeut; Soc Neurosci; Digital Equip Comput Users Soc; Statist Anal Systs Users Group Int; Am Soc Pharmacol & Exp Therapeut. *Res:* Clinical research of novel anti-glaucoma agents; psychobiology of pharmacologic agents on learning and memory; research design methodology; computer applications. *Mailing Add:* 46 Ashwood Irvine CA 92714-3339

NOVACK, JOSEPH, b Brooklyn, NY, Mar 31, 28; m 53; c 2. ECONOMIC ANALYSES, INDUSTRIAL HYGIENE. *Educ:* City Col New York, BChE, 49; Am Int Col, MBA, 60. *Prof Exp:* Chem engr, Nat Dairy Res Labs, 49-51; chem engr, Monsanto Chem Co, 54-64, eng specialist, 64-70, group leader, sr group supvr, 73-82, PRIN INVESTR, MONSANTO CO, 83- *Mem:* Am Inst Chem Engrs. *Res:* Pilot plant research and development and process scale-up of unit operations and economic evaluations applied to polymeric and organic processes; regulatory compliance/chemical hygiene officer. *Mailing Add:* Technol Dept Monsanto Chem Co 730 Worcester St Indian Orchard MA 01151

NOVACO, ANTHONY DOMINIC, b Orange, NJ, Mar 24, 43; m 66; c 2. CONDENSED MATTER THEORY, SURFACE PHYSICS. *Educ:* Stevens Inst Technol, BS, 64, MS, 66, PhD(physics), 69. *Prof Exp:* Assoc res scientist physics, Hudson Labs, Columbia Univ, 67-69; fel physics, Battelle Mem Inst, 69-71; asst physicist, Brookhaven Nat Lab, 71-73; from asst prof to assoc prof, 73-89, PROF PHYSICS, LAFAYETTE COL, 89- *Concurrent Pos:* Dir, Acad Comput Ctr, Lafayette Col, 78-81. *Mem:* Am Phys Soc; Sigma Xi. *Res:* Theoretical research in condensed matter physics; statistical physics; physisorption and surface physics; quantum liquids and solids. *Mailing Add:* Dept of Physics Lafayette Col Easton PA 18042-1782

NOVAK, ALFRED, b Chicago, Ill, Jan 28, 15; m 44; c 3. HEALTH SCIENCES. *Educ:* Univ Chicago, BS, 36, MS, 42; Chicago Teachers Col, ME, 40; Mich State Univ, PhD, 50. *Prof Exp:* From instr to prof biol, Mich State Univ, 44-60; chief, Div Sci & Math, Stephens Col, 60-74, dir, Div Allied Health, 74-80; prof, Univ Mo-Columbia, 80-81; adminr, Stanley H Kaplan Educ Testing Serv, 81-84; freelance ed work, 86-88; RETIRED. *Concurrent Pos:* NIH spec res fel, Calif Inst Technol, 50-51; Guggenheim fel, Cambridge Univ, 57-58; adv biol ed, Encycl Americana, 59-69; consult biol sci curriculum studies, Am Inst Biol Sci, 60-62; collabr, Nat Sci Adv Bd, Encycl Britannica Films, Inc, 65-; res fel, Inst Path, Med Sch, Univ Bologna, 72. *Res:* Protein synthesis and hormonal influence. *Mailing Add:* 212 E 51st St Kansas City MO 64112

NOVAK, BRUCE MICHAEL, b Flint, Mich; m; c 2. POLYMER CHEMISTRY. *Educ:* Calif State Univ, Northridge, BS, 83, MS, 85; Calif Inst Technol, PhD(chem), 89. *Prof Exp:* ASST PROF CHEM, DEPT CHEM, UNIV CALIF, BERKELEY, 89-, FAC ASSOC MAT SCI, LAWRENCE BERKELEY LAB, 90- *Concurrent Pos:* NSF presidential young investr, 91-; Alfred P Sloan Found fel, 91-93. *Mem:* Am Chem Soc; Sigma Xi; AAAS; Am Soc Plastics Engrs. *Res:* Aqueous ring-opening metathesis polymerizations; transition metal catalyzed living polymerizations; synthesis of organic-inorganic composite materials; photoresists; photodoped conducting polymers; conducting polymers; synthesis of polymers displaying stable helical conformations. *Mailing Add:* Dept Chem Univ Calif Berkeley CA 94720

NOVAK, ERNEST RICHARD, b Szoce, Hungary, Apr 17, 40; US citizen; m 62; c 2. ORGANIC CHEMISTRY. *Educ:* Oberlin Col, BA, 63; Univ Rochester, PhD(org chem), 67. *Prof Exp:* Res chemist, Plastics Dept, Exp Sta, Wilmington, Del, 66-73, sr res chemist, 73-78, RES ASSOC, PLASTICS PROD & RES, E I DU PONT DE NEMOURS & CO, INC, PARKERSBURG, 78- *Mem:* Am Chem Soc. *Res:* Oxonium ions; optically active compounds; monomer and polymer research; polymerization chemistry; polymer stabilization. *Mailing Add:* 23 Valley View Dr Vienna WV 26105

NOVAK, GORDON SHAW, JR, b Ft Collins, Colo, July 21, 47; m 77; c 2. PHYSICS PROBLEM SOLVING, AUTOMATIC PROGRAMMING. *Educ:* Univ Tex, Austin, BS, 69, MA, 71, PhD(comput sci), 76. *Prof Exp:* Systs programmer, Tracor Inc, 66-68, mgr, Systs Prog, 68-76; instr comput sci, Univ Tex, Austin, 76-77; res scientist, SRI Int, 77-78; vis asst prof comput sci, Stanford Univ, 81-83; asst prof, 78-81, 83-84, ASSOC PROF COMPUT SCI, UNIV TEX, AUSTIN, 84-, DIR, ARTIFICIAL INTEL LAB, 84- *Mem:* Asn Comput Mach; Inst Elec & Electronics Engrs; AAAS; Asn Computational Ling; Am Asn Artificial Intel. *Res:* Automatic understanding and solving of physics problems by computer; automatic programming, reusable programs and abstract data types. *Mailing Add:* Dept Comput Sci Univ Tex Austin TX 78712-1188

NOVAK, IRWIN DANIEL, b New York, NY, June 23, 42; div; c 3. GEOMORPHOLOGY. *Educ:* Hunter Col, AB, 66; Univ Fla, MS, 68; Cornell Univ, PhD(geol), 71. *Prof Exp:* Asst prof, 71-75, chmn, dept, 82-88, ASSOC PROF GEOL, UNIV SOUTHERN MAINES, 75- *Concurrent Pos:* Mem adv bd, Maine Land Use Regulation Comn, 73-80,; consult geologist, 75- *Mem:* AAAS; Geol Soc Am; Soc Econ Paleontologists & Mineralogists. *Res:* Landslide mapping; geomorphology and sedimentary processes in coastal and fluvial environments of Maine. *Mailing Add:* Univ Southern Maine College Ave Gorham ME 04038

NOVAK, JAMES LAWRENCE, b Berwyn, Ill, Apr 19, 61. CAPACITIVE SENSORS, SLIP DETECTION. *Educ:* Univ Ill Urbana-Champaign, BS, 83, MS, 85, PhD(elec eng), 88. *Prof Exp:* SR MEM TECH STAFF, SANDIA NAT LABS, 88- *Mem:* Inst Elec & Electronic Engrs; Am Welding Soc. *Res:* Transducers for use in manufacturing process control; sensors for monitoring automated machining systems and welding robots. *Mailing Add:* Sandia Nat Labs MS 1410 PO Box 5800 Albuquerque NM 87185

NOVAK, JOSEF FRANTISEK, b Ceske Budejovice, Czech, Oct 24, 42. CANCER BIOLOGY. *Educ:* Prague Univ, MS, 64; Yale Univ, PhD(biol), 74. *Prof Exp:* Res asst plant physiol, Inst Exp Bot, Czech Acad Sci, 65-68; instr biochem, Dept Biol, Princeton Univ, 73-75; res assoc cancer biol, Dept Clin & Biochem Pharmacol, St Jude Children's Res Hosp, 75-77; res assoc cancer biol, Cancer Res Unit, 77-79, DIR, ORTHOP RES LAB, ALLEGHENY GEN HOSP, 79- *Mem:* Soc Exp Biol Gt Brit; Sigma Xi; Int Soc Differentiation; Am Asn Cancer Res; Orthop Res Soc. *Res:* Molecular and biochemical aspects of tumor metabolism with special emphasis on endogenous and exogenous substances determining the rate of cell division in normal and tumor cells. *Mailing Add:* Orthop Res Lab Allegheny Gen Hosp 320 E N Ave Pittsburgh PA 15212

NOVAK, JOSEPH DONALD, b Minneapolis, Minn, Dec 2, 30; m 53; c 3. BIOLOGY. *Educ:* Univ Minn, BS, 52, MS, 54, PhD(bot, educ), 58. *Prof Exp:* Asst bot, Univ Minn, 52-56, instr, 56-57; asst prof biol, Kans State Teachers Col, 57-59; from asst prof biol to assoc prof biol & educ, Purdue Univ, 59-67; coordr, Shell Merit Progs, 68-72, PROF SCI EDUC & CHMN SCI & ENVIRON EDUC DIV, CORNELL UNIV, 67- *Concurrent Pos:* res assoc, Harvard Univ, 64-65; Fulbright-Hays sr scholar, 80; distinguished vis prof, Univ NC, 81, vis prof, Univ W Fla, 87-88. *Mem:* AAAS; Bot Soc Am; Nat Asn Biol Teachers (vpres, 64); Nat Asn Res Sci Teaching (exec secy, 62-67, pres, 68); Nat Sci Teachers Asn; Sigma Xi; Am Ed Res Asn. *Res:* Biological education; analysis of concept learning; cognitive learning theory; metacognitive learning. *Mailing Add:* Kennedy Hall Cornell Univ Ithaca NY 14853

NOVAK, LADISLAV PETER, b Chlum, Czech, Sept 18, 22; m 50; c 4. PHYSIOLOGICAL ANTHROPOLOGY. *Educ:* Charles Univ, Prague, BSc, 48; Univ Minn, Minneapolis, MA, 61, PhD, 62. *Prof Exp:* Fel physiol, Univ Minn, 62-64, vis asst prof physiol anthrop, 62-63, asst prof physiol anthrop, 63-66, consult, Mayo Grad Sch Med, 66-72; PROF ANTHROP, SOUTHERN METHODIST UNIV, 72- *Concurrent Pos:* NIH grants, 64-72; AMA lectr, 71-72; pres, Int Comt Standardization Phys Fitness Tests, 76-80; NIH proj officer, Foreign Res Prog; res grants, Xerox Co, 76-78 & Sun Gas Co, 78-; exchange scientist, Czech Acad Sci-US Nat Acad Sci, 78-79. *Mem:* Am Asn Phys Anthrop; NY Acad Sci; Brit Soc Study Human Biol; Am Aging Asn; Am Asn Clin Nutrit; Am Asn Cardiopulmonary Rehab. *Res:* Nutrition-cardiac rehabilitation; physiological growth and development; body composition; physiology of exercise. *Mailing Add:* Dept Anthrop Southern Methodist Univ Dallas TX 75275

NOVAK, MARIE MARTA, b Prague, Czechoslovakia, May 10, 40; Can citizen. ZOOLOGY. *Educ:* Univ Man, PhD(zool), 74. *Prof Exp:* From asst prof to assoc prof, 74-85, PROF BIOL, UNIV WINNIPEG, 85- *Mem:* Am Soc Parasitologists. *Res:* Host-parasite relationships governing the growth differentiation and morphology of populations of helminths; NMR and MRI of tapeworms; parasite response to various anthelminthics. *Mailing Add:* Biol Dept Univ Winnipeg 515 Portage Ave Winnipeg MB R3B 2E9 Can

NOVAK, MILOS, b Sumperk, Czech, Mar 29, 25; Can citizen; m; c 1. ENGINEERING SCIENCE. *Educ:* Prague Tech Univ, Ing, 49; Inst Theoret & Appl Mech, Prague, PhD(eng mech), 57. *Prof Exp:* Designer struct, Govt Consult Agency Spec Indust Struct, Bratislava, 49-52; head sect, Govt Consult Agency Heavy Indust, Prague, 52-53; sci worker dynamics, Inst Theoret & Appl Mech, Prague, 56-67; PROF ENG SCI, UNIV WESTERN ONT, 67- *Concurrent Pos:* Consult, Can Westinghouse, 69-74, Golder Assoc, 73-75, Ont Hydro, 74-75, Law Eng, Converse Ward Dan's, Dixon, ATEC Assocs, Harding-Lawson Assocs, Can Industs Ltd & Kraftwerk Union; researcher, Nat Res Coun res grants-in-aid, 70-; chmn task group soils & found, Can Nat Comt Earthquake Eng, 76-82; consult, UN Develop Prog, India, 80, Yugoslavia, 82 & Chinese Develop Proj, 84. *Mem:* Can Nat Comt Earthquake Eng; Am Soc Civil Engrs; Can Geotech Soc; Can Civil Eng Soc. *Res:* Dynamics of structures, foundations and soils; wind engineering; earthquake engineering. *Mailing Add:* Fac Eng Sci Univ Western Ont London ON N6A 5B9 Can

NOVAK, RAYMOND FRANCIS, b St Louis, Mo; m; c 4. NUCLEAR MAGNETIC, RESONANCE SPECTROSCOPY. *Educ:* Univ Mo-St Louis BS, 68; Case Western Reserve Univ, PhD(phys chem), 73. *Prof Exp:* Fel, chem, Case Western Reserve Univ, 73-74; fel pharmacol, Northwestern Univ Med Sch, 74-75, NIH fel, 75-76, assoc, 76-77, from asst prof to assoc prof, 77-86, prof pharmacol, 86-88; DIR, INST CHEM TOXICOL & PROF PHARMACOL, WAYNE STATE UNIV, 88- *Concurrent Pos:* Mem toxicol study sect, NIH, 84-88. *Mem:* Am Soc Pharmacol & Exp Therapeut; Biophys Soc; Sigma Xi; AAAS; Am Soc Biochem & Molecular Biol; Am Asn Cancer Res; Soc Toxicol; Int Soc Study Xenobiotics; Am Chem Soc; Am Soc Hemat. *Res:* Purification of the individual enzymes and related components of the hepatic monooxygenase system and studies on the molecular basis of induction of the cytochromes P450 and the glutathione s-transferases by solvents and other xenobiotics in hepatic and extra hepatic (nasal, renal) tissues and in cultured cells; additional research focuses on the role of the proteosome and the calcium-activated neutral protease, cal pain, in the recognition and degradation of cellular proteins damaged by oxidant stress or by xenobiotics. *Mailing Add:* Inst Chemical Toxicol Wayne State Univ 2727 Second Ave Rm 4000 Detroit MI 48201

NOVAK, ROBERT EUGENE, b Spring Valley, Ill, Apr 30, 49; m 72; c 3. AUDIOLOGY. *Educ:* Univ Iowa, BS, 72, MA, 74, PhD(audiol), 77. *Prof Exp:* Audiologist, St Luke's Hosp, Cedar Rapids, Iowa, 73-74; audiol asstship, Vet Hosp, Iowa City, 76-77; ASST PROF AUDIOL, SAN DIEGO STATE UNIV, 77- *Concurrent Pos:* Mem San Diego City Noise Bd, community noise anal & forensic audiol. *Mem:* Am Speech & Hearing Asn; Sigma Xi; Acad Rehab Audiol. *Res:* Aural rehabilitation, especially assessment and remediation of problems associated with hearing loss in the aged populations and in various ethnic subgroups. *Mailing Add:* 7497 Gorge View Terr San Diego CA 92120

NOVAK, ROBERT LOUIS, b Chicago, Ill, Oct 1, 37; m 65; c 1. BIOCHEMISTRY. *Educ:* Xavier Univ, Ohio, AB, 59; Univ Del, PhD(cell physiol, biochem), 64. *Prof Exp:* NIH fel, Dept Biol Sci, Purdue Univ, 63-65; asst prof biochem, Univ NH, 65-67; res fel, Harvard Univ, 67-69; asst prof, 69-74, ASSOC PROF CHEM & BIOL, DEPAUL UNIV, 74- *Concurrent Pos:* Res grants, Res Corp, 66-67 & 71-72, Am Cancer Soc, 71-72 & NIH, 71- *Mem:* Am Chem Soc. *Res:* Model polypeptides for studying biopolymer interactions; insertion sites for viral information in cancer cells; phosphorous chemistry in polynucleotides; biochemical engineering of human genes. *Mailing Add:* Dept Chem & Biol DePaul Univ 25 E Jackson Blvd Chicago IL 60604

NOVAK, ROBERT OTTO, b Oak Park, Ill, Sept 12, 30; m 59; c 1. MYCOLOGY, PLANT PATHOLOGY. *Educ:* Mich State Univ, BS, 52; Univ Ill, MS, 56; Univ Wis, PhD(bot), 63. *Prof Exp:* Res microbiologist, Lederle Labs, 63-68; asst prof, 68-73, assoc prof, 73-81, PROF BIOL, ORE COL EDUC, 81- *Mem:* AAAS; Mycol Soc Am. *Res:* Ecology and taxonomy of soil microfungi. *Mailing Add:* Dept Biol Western Ore State Col 345 N Monmouth Ave Monmouth OR 97361

NOVAK, ROBERT WILLIAM, b Hoboken, NJ, Aug 2, 39; m 60; c 3. ORGANIC CHEMISTRY. *Educ:* Wagner Col, BS, 60; Purdue Univ, MS, 64, PhD(org chem), 66. *Prof Exp:* Instr chem, Purdue Univ, 64-66; res chemist, 66-72, mgr paper chem tech serv, 72-74, SR RES CHEMIST, AM CYANAMID CO, 74- *Mem:* AAAS; Am Chem Soc; Tech Asn Pulp & Paper Indust. *Res:* Organic synthesis of natural products; reaction of bromine with terminal disubstituted olefins; organic synthesis and product development of paper chemicals; mechanisms of retention of chemical additives for paper application. *Mailing Add:* 1951 Matson Lane Lisle IL 60532-1148

NOVAK, RONALD WILLIAM, b Elmira, NY, Dec 3, 42; m 63; c 2. EMULSION POLYMERIZATION. *Educ:* Univ S Fla, BA, 64; Fla State Univ, PhD(org chem), 68. *Prof Exp:* Chemist, 69-81, sr res assoc, 81-86, RES SECT MGR, ROHM & HAAS CO, 86- *Mem:* Am Chem Soc. *Res:* Solution and emulsion polymers; mechanism of emulsion polymerizations; oxidative polymerizations of reactive diluents. *Mailing Add:* Rohm & Haas Co Bldg 3A Spring House PA 19477

NOVAK, STEPHEN ROBERT, b Homestead, Pa, June 23, 39; m 61; c 3. MECHANICAL ENGINEERING. *Educ:* Univ Pittsburgh, BS, 61, MS, 66, PhD(metall eng, mat sci), 77. *Prof Exp:* Assoc technologist process develop, 61-65, assoc technologist res fracture mech, 65-68, res engr, 68-73, SR RES ENGR FRACTURE MECH, RES LAB, US STEEL CORP, 73- *Concurrent Pos:* Mem adv panel subcritical cracking high strength steels, US Navy, 76-; mem rev panel corrosion fatigue behav steels, Fed Hwy Admin, US Dept Transp, 78-; mem comt, Environ Assisted Cracking Test Methods for High Strength Steel Weldments, Nat Res Coun, Nat Mat Adv Bd, 80- *Mem:* Am Soc Testing & Mat; Am Soc Metals; Nat Asn Corrosion Eng; Am Soc Mech Eng. *Res:* Mechanical metallurgy; fracture mechanics; subcritical crack growth; fatigue; stress corrosion cracking; corrosion fatigue; hydrogen embrittlement; brittle fracture; ductile fracture; crack initiation; stable crack growth; r-curve technology; corrosion; physical metallurgy; engineering mechanics; materials science engineering. *Mailing Add:* 4901 McNulty Rd Pittsburgh PA 15236

NOVAK, THADDEUS JOHN, b Lansford, Pa, Apr 26, 40; m 71. CHEMICAL DETECTION. *Educ:* Drexel Univ, BS, 63; Univ Del, MS, 74. *Prof Exp:* Res chemist, Allied Chem Corp, 63-68, Phys Res Lab, Edgewood Arsenal, 68-71, Chem Lab, 71-74, Chem Systs Lab, US Army, Arradcom, 74-79; RES CHEMIST, CHEM RES, DEVELOP & ENG CTR, AMCCOM, ABERDEEN PROVING GROUND, 80- *Honors & Awards:* Res & Develop Award, Dept Army, 80 & 84. *Mem:* Am Chem Soc; Sigma Xi. *Res:* Detection and identification of chemical warfare agents; development of analytical methods, especially ultra micro methods, for detection and identification of toxic chemicals; synthesis of new analytical reagents; thin-layer chromatography. *Mailing Add:* 706 E Mac Phail Road Bel Air MD 21014

NOVALES, RONALD RICHARDS, b San Francisco, Calif, Apr 24, 28; m 53; c 2. COMPARATIVE VERTEBRATE ENDOCRINOLOGY. *Educ:* Univ Calif, Berkeley, BA, 50, MA, 53, PhD(zool), 58. *Prof Exp:* Assoc zool, Univ Calif, Berkeley, 56; from asst prof to prof biol sci, 58-70, PROF NEUROBIOL & PHYSIOL, NORTHWESTERN UNIV, ILL, 80- *Res:* Comparative vertebrate endocrinology. *Mailing Add:* Dept Neurobiol & Physiol Northwestern Univ Evanston IL 60208

NOVELLO, ANTONIA COELLO, b Fajardo, Puerto Rico, Aug 23, 44; m 70. PEDIATRIC NEPHROLOGY, RESEARCH ADMINISTRATOR. *Educ:* Univ PR, BS, 65; Univ PR Sch Med, MD, 70; John Hopkins Univ, MPH, 82; John F Kennedy Sch Govt, Harvard Univ, cert, 87. *Prof Exp:* Intern, pediat, Univ Mich Med Ctr, 70-71, resident pediat, 71-73, fel adult & pediat nephrol, 73-74,; pediat nephrol, Georgetown Univ Hosp, 74-75; private practice, pediat & nephrol, Springfield, Va, 76-78; proj off, artificial kidney chronic uremia prog, Nat Inst Arthritis, Metab & Digestive Dis, Nat Inst Health, 78-79; staff physician, kidney & urology, chronic renal disease program, Nat Inst Arthritis, Diabetes & Digestive & Kidney Diseases, Nat Inst Health, 79-80; exec secy, renal disease, bone mineral & metabolism, div res grants, 81-86; DEP DIR, GEN PEDIAT & OBSTET & GYNECOL, NAT INST

CHILD HEALTH & HUMAN DEVELOP, NAT INST HEALTH, 86- *Concurrent Pos:* Clin instr pediat, Georgetown Univ Hosp, lieutenant, US Pub Health Serv, active res, 78-79, lieutenant comdr, US Pub Health Serv, 79-80; clin asst prof pediat, Georgetown Univ Hosp, 79-82; comdr, US Pub Health Serv, 80-84; mem comt res, Soc Pediat Nephrol, 81-; co-chmn, Third Int Congress Nutrit & Metab Renal Dis, 82; mem, clin res appl study group, div res grants, Nat Inst Health, 82. *Mem:* Am Acad Pediat; Am Soc Nephrol; Int Soc Nephrol; Am Soc Pediat Nephrol. *Mailing Add:* 1315 31st St NW Washington DC 20007

NOVELLO, FREDERICK CHARLES, b Somerville, Mass, July 27, 16; m 48; c 3. ORGANIC CHEMISTRY. *Educ:* Harvard Univ, SB, 38, MA, 39, PhD(org chem), 41. *Prof Exp:* Res assoc Nat Defense Res Comt, Harvard Univ, 41-43; res chemist, Sharp & Dohme Div, Merck & Co, Inc, 43-69; sr res fel, Merck Sharp & Dohme Res Labs, 69-73, sr investr, 73-81; RETIRED. *Honors & Awards:* Modern Pioneers Award, Nat Asn Mfrs, 65; Albert Lasker Award, Albert & Mary Lasker Found, 75. *Mem:* Am Chem Soc. *Res:* Medicinal chemistry. *Mailing Add:* 786 Bair Rd Berwyn PA 19312

NOVICK, AARON, b Toledo, Ohio, June 24, 19; m 48; c 2. MICROBIOLOGY. *Educ:* Univ Chicago, BS, 40, PhD(chem), 43. *Prof Exp:* Scientist, Manhattan Dist Proj, Univ Chicago, 43-44, AEC Proj, 46-47, asst prof biophys, 48-55, assoc prof biophys & microbiol, 55-58; scientist, Los Alamos Nat Lab, 45-46; dir inst molecular biol, Univ Ore, 59-69, prof biol, 59-70, dean grad sch, 71-80; RETIRED. *Concurrent Pos:* Guggenheim fel, Pasteur Inst, France, 53-54. *Mem:* AAAS; Genetics Soc Am; Am Soc Microbiol; Biophys Soc; Sigma Xi. *Res:* Biosynthetic regulatory mechanisms. *Mailing Add:* 3960 Blanton Road Sch Eugene OR 97405

NOVICK, ALVIN, b Flushing, NY, June 27, 25. AIDS PUBLIC POLICY, BIOETHICS. *Educ:* Harvard Col, AB, 47, MD. 51. *Hon Degrees:* MA, Yale Univ, 82. *Prof Exp:* Teaching fel med, Harvard Univ, 52-53, res fel biol, 53-57; from instr to asst prof zool, 57-65, assoc prof, 65-83, PROF BIOL, YALE UNIV, 83- *Concurrent Pos:* Chmn, Mayor's Task Force on AIDS, 86-; Nat AIDS Network, 87-90; mem bd, Am Asn Physicians for Human Rights, 83-; mem, Antivirals Adv Comt, Food & Drug Asn, 89-; mem adv comt, US Pharmacopeial Conv, 90-; ed, AIDS & Pub Policy J, 91- *Mem:* Am Physiol Soc; Am Soc Zool; Am Soc Mammal; Ecol Soc Am; Animal Behav Soc; Am Asn Physicians for Human Rights (pres, 85-86). *Res:* Biomedical ethics in reference to acquired immune deficiency syndrome, health education; public health measures in reference to acquired immune deficiency syndrome; echolocation in bats; sensory physiology; vertebrate biology. *Mailing Add:* Dept of Biol Yale Univ PO Box 6666 New Haven CT 06511-7444

NOVICK, DAVID THEODORE, materials science, for more information see previous edition

NOVICK, RICHARD P, b New York, NY, Aug 10, 32; m 58; c 2. MICROBIAL GENETICS. *Educ:* Yale Univ, BA, 54; NY Univ, MD, 59. *Prof Exp:* Intern, Yale-New Haven Med Ctr, 59-60; Nat Found fel, Nat Inst Med Res, Eng, 60-62; asst resident, Hosp, Vanderbilt Univ, 62-63; USPHS fel, Rockefeller Univ, 63-65; assoc, 65-69, assoc mem, 69-75, MEM & CHIEF DEPT PLASMID BIOL, PUB HEALTH RES INST OF CITY OF NEW YORK, INC, 75-, DIR, 80- *Concurrent Pos:* Res asst prof, NY Univ, 66-69, res assoc prof, 69-75, res prof, 76-; lectr, Columbia Univ, 66-68; ed-in-chief, Plasmid, 77-87, ed, 87- *Mem:* Am Soc Microbiol; Genetics Soc Am; Harvey Soc; fel AAAS. *Res:* Microbial physiology; control mechanisms in biosynthetic pathways; extrachromosomal resistance factors in Staphylococcus aureus; genetic control of replication; misuse of science; microbiol pathogenesis. *Mailing Add:* Dept Microbiol Pub Health Res Inst 455 First Ave New York NY 10016

NOVICK, ROBERT, b New York, NY, May 3, 23; m 47; c 3. PHYSICS. *Educ:* Stevens Inst Technol, ME, 44, MS, 49; Columbia Univ, PhD(physics), 55. *Prof Exp:* Engr microwaves, Wheeler Labs, Inc, 46-47; instr physics, Columbia Univ, 52-54, res assoc, 54-57, adj asst prof, 57; from asst prof to assoc prof, Univ Ill, 57-60; assoc prof, 60-62, dir radiation lab, 60-68, chmn dept physics, 83-88, PROF PHYSICS, COLUMBIA UNIV, 62-, DIR ASTROPHYS LAB, 68- *Concurrent Pos:* Sloan fel, 55-59 & 65-72; mem adv panel physics, NSF, 62-65; chmn subpanel atomic & molecular physics, Nat Acad Sci, 64-65; mem Nat Acad Sci Panel, adv to Nat Bur Standards Atomic Physics Div, 66-69; consult, Gen Precision Lab, Gen Time, Inc, & Perkin-Elmer Corp, Mitre Corp. *Honors & Awards:* Award for excellence in Space Sci, 83. *Mem:* Fel Am Phys Soc; fel Inst Elec & Electronics Engrs; fel AAAS. *Res:* Atomic physics, collisions and frequency standards; quantum electronics; nuclear moments; x-ray astronomy. *Mailing Add:* 366 W 245th St Bronx NY 10471

NOVICK, RUDOLPH G, b Warsaw, Poland, Dec 16, 10; US citizen; m 37; c 2. PSYCHIATRY. *Educ:* Northwestern Univ, BS, 31, MD, 36; Chicago Inst Psychoanal, 46-50. *Prof Exp:* Jr physician psychiat, Jacksonville State Hosp, Ill Dept Pub Welfare, 37-40; jr physician, Manteno State Hosp, Ill, 40-41; sr physician, Elgin State Hosp, Ill, 41-43; med dir, Ill Soc Ment Health, 43-55; med dir psychiat, 56-77, psychiatrist in chief, Forest Hosp, 77-; AT DEPT PSYCHOL, UNIV HEALTH SCI CHICAGO MED SCH. *Concurrent Pos:* Pvt pract, 45-; psychiat consult, Comt Community Serv, NIMH, 54-57, Chicago State Hosp, 55-56, Munic Tuberc Hosp, Chicago, 55-58 & Ill State Dept Pub Health, 69-; assoc prof psychiat & actg chmn dept, Univ Health Sci/ Chicago Med Sch, 75-77. *Mem:* AMA; fel Am Psychiat Asn; Group Advan Psychiat (chmn, Comt Prev Psychiat). *Mailing Add:* 555 Wilson Lane Des Plaines IL 60016

NOVICK, STEWART EUGENE, b New York, NY, Sept 10, 45; m. PHYSICAL CHEMISTRY. *Educ:* State Univ Stony Brook, BS, 67; Harvard Univ, AM, 68, PhD(chem physics), 73. *Prof Exp:* Fel chem, Harvard Univ, 74-76; res assoc, Joint Inst Lab Astrophysics, Univ Colo & Nat Bur Standards, 76-78; asst prof, 78-85, ASSOC PROF CHEM, WESLEYAN UNIV, 85- *Mem:* Am Phys Soc; Am Chem Soc. *Res:* Spectroscopy, structure and bonding of weakly bound complexes; molecular beams; ions; radicals. *Mailing Add:* Dept Chem Wesleyan Univ Middletown CT 06457

NOVICK, WILLIAM JOSEPH, JR, b Revloc, Pa, Dec 14, 31; m 55; c 4. PHARMACOLOGY, BIOCHEMISTRY. *Educ:* St Francis Col, Pa, BS, 53; Duke Univ, PhD(pharmacol), 61. *Prof Exp:* Technician toxicol, Dept Pub Health, Univ Pittsburgh, 54-55; jr biochemist, Smith Kline & French Labs, 55-58, sr scientist, 61-64, group leader pharmacol, 64-65, from asst sect head to sect head, 65-67; mgr pharmacol dept, William H Rorer, Inc, Pa, 67-70; dir pharmacol dept, 70-77, DIR BIOL SCI, HOECHST ROUSSELL PHARMACEUT, INC, 77-, SR DIR INT PROD DEVELOP, 83- *Mem:* Am Soc Pharmacol & Exp Therapeut; NY Acad Sci. *Res:* Effects of age and thyroid hormone on monamine oxidase; pharmacological activities of steroids; drug metabolism. *Mailing Add:* Hoechst Roussel Pharmaceut Inc Rte 202-206 N Somerville NJ 08876

NOVIKOFF, PHYLLIS MARIE, CELL BIOLOGY, EXPERIMENTAL PATHOLOGY. *Educ:* NY Univ, PhD(cell biol), 72. *Prof Exp:* ASSOC PROF CELL BIOL & CYTOCHEM, ALBERT EINSTEIN COL MED, 70- *Res:* Carcinogenesis. *Mailing Add:* Albert Einstein Col Med Dept Path 1300 Morris Park Ave Bronx NY 10461

NOVITSKI, EDWARD, b Wilkes Barre, Pa, July 24, 18; m 43; c 4. GENETICS. *Educ:* Purdue Univ, BS, 38; Calif Inst Technol, PhD(genetics), 42. *Prof Exp:* Guggenheim fel, Univ Rochester, 45-46, res assoc, Atomic Energy Proj, 46-47; res assoc, Univ Mo, 47-48; sr res fel, Calif Inst Technol, 48-51; assoc prof zool, Univ Mo, 51-56; head biologist, Oak Ridge Nat Lab, 56-58; head dept, 64-67, PROF BIOL, UNIV ORE, 58- *Concurrent Pos:* NSF sr fel, Univs Zurich, 61-62 & Canberra, 67-68; Guggenheim fel, Univ Leiden, 74-75 & Fulbright fel, 75-76. *Mem:* Genetics Soc Am (treas, 62-66); Soc Exp Biol & Med; Am Soc Nat; Soc Human Genetics. *Res:* Chromosome behavior; speciation; statistical analysis of genetic data; use of computing methods in biology. *Mailing Add:* 1690 E 26th Eugene OR 97403

NOVITSKY, JAMES ALAN, b Hazleton, Pa, Nov 14, 51; m 73. MARINE MICROBIOLOGY. *Educ:* Pa State Univ, BS, 73; Ore State Univ, PhD(microbiol), 77. *Prof Exp:* Asst prof, 77-82, ASSOC PROF BIOL, DALHOUSIE UNIV, 82- *Concurrent Pos:* Res assoc oceanog, Dalhousie Univ, 78- *Mem:* Am Soc Microbiol; AAAS; Sigma Xi. *Res:* Sediment microbial ecology. *Mailing Add:* The Jan Group PO Box 148 Wrightsville Beach NC 28480

NOVLAN, DAVID JOHN, b Colorado Springs, Colo, Dec 8, 47; m 74. METEOROLOGY, MATHEMATICS. *Educ:* Univ Colo, BS, 69; Colo State Univ, MS, 73. *Prof Exp:* Meteorologist, 73-77, sr meteorologist, 77-78, CHIEF FORECASTER, ATMOSPHERIC SCI LAB, 78- *Concurrent Pos:* Vis prof, NMex State Univ, 87- *Mem:* Am Meteorol Soc. *Res:* Improvement of operational synoptic and ballistic meteorology; improvement and computerization of regional and local weather forecasting and methods in climatology. *Mailing Add:* Atmospheric Sci Lab White Sands Missile Range NM 88002

NOVOA, WILLIAM BREWSTER, b Havana, Cuba, July 16, 30; US citizen. BIOCHEMISTRY. *Educ:* Univ Fla, BS, 55; Duke Univ, PhD, 59. *Prof Exp:* USPHS res fel, Univ Wash, 59-61; res assoc, McIlvain Lab, Med Ctr, Univ Kans, 61, from instr to asst prof biochem, 62-70; from assoc prof to prof, 70-89, EMER PROF CHEM, CENT CONN STATE UNIV, 89- *Mem:* Am Chem Soc. *Res:* Enzymes. *Mailing Add:* Dept of Chem Cent Conn State Univ New Britain CT 06050

NOVODVORSKY, MARK BENJAMIN, b Moscow, Russia, July 21, 46. PURE MATHEMATICS. *Educ:* Univ Moscow, PhD(physics, math), 71. *Prof Exp:* Asst math, Inst Advan Study, 74-75; asst prof, 75-80, ASSOC PROF MATH, PURDUE UNIV, WEST LAFAYETTE, 80- *Mem:* Am Math Soc. *Res:* Zeta functions associated to automorphic representations of reductive groups over global fields. *Mailing Add:* Dept Math-Sci Dept 1935 Purdue Univ West Lafayette IN 47906

NOVOSAD, ROBERT S, b Chicago, Ill, May 1, 20; m 46; c 3. MATHEMATICS. *Educ:* Ill Inst Technol, BS, 42; Univ Chicago, MS, 48, PhD(math), 52. *Prof Exp:* Instr math, Tulane Univ, 50-53; asst prof, Pa State Univ, 53-60; assoc res scientist, Denver Div, Martin-Marietta Corp, 60-67, chief space systs opers anal, 67-71; PROF PHYSICS & AERONAUT SYSTS, UNIV W FLA, 71-, PROF MATH & STATIST, 77- *Mem:* Am Math Soc; Opers Res Soc Am; Am Astronaut Soc; Am Inst Aeronaut & Astronaut; Sigma Xi. *Res:* Operations research; systems analysis. *Mailing Add:* 2170 Pauline St Cantonment FL 32533

NOVOTNY, ANTHONY JAMES, b Chicago, Ill, Aug 14, 32; m 53; c 2. FISHERIES, AQUACULTURE. *Educ:* Morton Col, BEd, 52; Univ Wash, BSc, 58. *Prof Exp:* Fisheries res biologist, US Bur Com Fisheries, 58-67; fisheries res biologist marine aquacult, Nat Marine Fisheries Serv, Nat Oceanic & Atmospheric Admin, 68-88; OWNER & DIR, MARINKA INT, 88- *Concurrent Pos:* Consult, Nat Marine Fisheries Serv, Ctr Ocean Brest, France, 73-88e, 73-88. *Mem:* Am Fisheries Soc; Nat Shellfisheries Asn; Am Inst Fisheries Res Biologists; Pacific Fishery Biologists (pres, 88). *Res:* Marine aquaculture; fish diseases; salmonid propagation and enhancement; influences of salmonid survival. *Mailing Add:* 1919 E Calhoun Seattle WA 98112

NOVOTNY, CHARLES, b New York, NY, July 27, 36; m 58; c 2. MOLECULAR BIOLOGY. *Educ:* Wis State Col Stevens Point, BS, 59; Univ Pittsburgh, PhD(bact), 65. *Prof Exp:* Fel microbiol, Sch Med, Univ Pittsburgh, 65, res assoc, 65-68; from asst prof to assoc prof, microbiol 68-85, PROF MICROBIOL & MOLECULAR GENETICS, UNIV VT, 84- *Mem:* Am Soc Microbiol. *Res:* Genetics of development. *Mailing Add:* Dept Microbiol & Molecular Genetics Univ Vt Agr Col 85 S Prospect St Burlington VT 05405

NOVOTNY, DONALD BOB, b Cedar Rapids, Iowa, Nov 15, 37; m 67; c 2. PHYSICAL CHEMISTRY. *Educ:* Univ Iowa, BS, 59; Iowa State Univ, PhD(phys chem), 64. *Prof Exp:* Sr res chemist, Mound Lab, Monsanto Res Corp, 64-66; res assoc chem, Mass Inst Technol, 66-67; res chemist, Vacuum Measurement Sect, Nat Bur Standards, Washington, DC, 67-71, gen phys scientist, Fire Technol Div, 71-75, PHYS CHEMIST, SEMICONDUCTER ELECTRON DEVICES DIV, NAT BUR STANDARDS, GAITHERSBURG, MD, 75- *Mem:* Am Phys Soc; Inst Elec & Electronics Engrs. *Res:* Metal physics; metals; alloys; alloy phases; phase stability from thermodynamic considerations; lattice vibrations and energetics; photolithographic processes including photochemistry of resists and measurements utilizing optics; microelectronic devices and integrated circuit fabrication processes. *Mailing Add:* 5101 Moorland Lane Bethesda MD 20814

NOVOTNY, DONALD WAYNE, b Chicago, Ill, Dec 15, 34; m 55; c 2. ELECTRICAL ENGINEERING. *Educ:* Ill Inst Technol, BS, 56, MS, 57; Univ Wis-Madison, PhD(elec eng), 61. *Prof Exp:* Instr elec eng, Ill Inst Technol, 57-58; from instr to assoc prof, Univ Wis-Madison, 58-68, assoc dept, 68-76, chmn, Dept Elec & Comput Eng, 76-80, PROF ELEC ENG, UNIV WIS-MADISON, 68-, ASSOC DIR, UNIV-INDUST RES PROG, 80- *Concurrent Pos:* Consult, A O Smith Corp, 63 & ASD, Inc, 85; res grant, Marathon Elec Mfg Corp, 64, consult, 64-; consult, Hevi-Duty Equip Co, 65-67, Wis Dept Natural Resources, 70- & Borg Warner Res Lab, 73-89; res grant, Allen Bradley Co, 70-71; vis prof, Eindhoven Tech Univ, 73-74 & Leuven, Belgium, 86; consult, Rexnord Corp, 80-82 & Allen Bradley Co & Eaton Corp, 81-; res grants, Gen Elec Co, 78-79 & Allen Bradley, 80; dir, Wis Elec Mach & Power Electronics Consortium, 81; Fulbright lectr, Tech Univ Gent, Belgium, 81; res grant, NASA, 84, Whirlpool, 85, 86, Wis Develop Fund, 85, Emerson, 87, 90, NASA, 88-; chmn EE Prog, Nat Technol Univ, 89- *Mem:* Am Soc Eng Educ; fel Inst Elec & Electronic Engrs. *Res:* Electromechanical devices and control systems; electric machine analysis and design; power electronics. *Mailing Add:* Elec & Comput Eng Dept Univ Wis 1415 Johnson Dr Madison WI 53706

NOVOTNY, EVA, acoustics, hydrodynamics, for more information see previous edition

NOVOTNY, JAMES FRANK, b Washington, DC, May 17, 37; m 41; c 5. MICROBIOLOGY, IMMUNOLOGY. *Educ:* Univ Md, BS, 62, MS, 70, PhD(microbiol), 73. *Prof Exp:* Microbiologist, US Army Res Ctr, 62-65, USDA, 65-69 & US Army Foreign Sci & Technol Ctr, 69-70; res assoc, Washington Hosp Ctr, 70-72; dir microbiol, Woodard Res Corp, 72-75; asst prof, Col Osteop Med & Surg, 75-78; CHMN, MICROBIOL DEPT, NEW ENG COL OSTEOP MED, 78-, PROP MICROBIOL, 88- *Concurrent Pos:* Adj prof, Drake Univ, 75-78. *Mem:* Am Soc Microbiol; Nat Registry Micribiol; Med Lab Microbiol & Pub Health. *Res:* Immunology of viral diseases and cell culture metabolism; environmental virology and microbiology. *Mailing Add:* Chmn, Dept Microbiol, Col Osteop Med Univ New England 605 Pool Rd Biddeford ME 04005

NOVOTNY, JAROSLAV, b Brtnice, Czech, Mar 11, 24; m 56; c 3. MEDICINAL CHEMISTRY. *Educ:* Univ Adelaide, BSc, 59, PhD(org chem), 63. *Prof Exp:* Demonstr, Dept Org Chem, Univ Adelaide, 62-63; res fel org & med chem, State Univ NY Buffalo, 64-65; chemist, 65, SUPVR ORG CHEM, STARKS ASSOCS, INC, 65- *Mem:* Am Chem Soc. *Res:* Synthesis and isolation of carcinogenic hydrocarbons from tars and the use of carbon-14 for them; synthesis of inhibitors of folic reductase and thymidylic synthetase; synthesis of medicinals and other organic compounds; research and development of new antimalarials. *Mailing Add:* 215 Fruitwood Terr Buffalo NY 14221

NOVOTNY, ROBERT THOMAS, mammalian ecology, paleoecology; deceased, see previous edition for last biography

NOVOTNY, VLAD JOSEPH, b Vsetin, Czech, Mar 12, 44; Can citizen; m 66; c 2. SOLID STATE PHYSICS, SURFACE AND INTERFACE CHARACTERIZATION. *Educ:* Czech Tech Univ, BSc, 65, MSc, 66; Univ Toronto, PhD(physics), 72. *Prof Exp:* Res scientist, Inst Physics, Czech Acad Sci, 67-68; Nat Res Coun Can fel, Univ Toronto, 69-73; Med Res Coun Can fel biophysics, Ont Cancer Inst & Univ Toronto, 73-74; sr res scientist physics, Xerox Res Ctr Can, 74-79; mgr device physics, Exxon Enterprises, Calif, 79-83; MGR FILM PROCESSOR & CHMN & RES STAFF MEM, IBM ALMADEN RES CTR, CALIF, 83- *Mem:* Am Chem Soc; Am Phys Soc. *Res:* Physics of surfaces and interfaces; thin film processing, mechanical, tribological, magnetic, acoustic, corrosion, light scattering and electrical characterization; thermodynamics, neutron scattering, polymer conformation; magnetic and optical recording; imaging technologies; display devices. *Mailing Add:* K64/803 IBM Aladen Res Ctr 650 Harry Rd San Jose CA 95120

NOVY, MILES JOSEPH, b Berlin, Ger, Nov 23, 37; c 2. PERINATAL PHYSIOLOGY, REPRODUCTIVE ENDOCRINOLOGY. *Educ:* Yale Univ, BA, 59; Harvard Med Sch, MD, 63. *Prof Exp:* Dir, Infertility Clin, Med Sch, Univ Ore, 73-77, chmn, Steering Comn, Human Seman Bank, 75; mem, Maternal Child Health Res Comn, Nat Inst Child Health Develop, NIH, 76-79; HEAD PERINATAL PHYSIOL, ORE REGIONAL PRIMATE RES CTR, 70-, SCIENTIST, 72-; PROF OBSTET & GYNEC, ORE HEALTH SCI UNIV, 78- *Mem:* Am Physiol Soc; Soc Gynec Invest; Am Fertil Soc; Soc Study Reproduction; Am Col Obsteticians & Gyecologists. *Res:* Endocrine regulation of parturiton in women and in nonhuman primates; perinatal and reproductive physiology and endocrinology; regulation of uteroplacental blood flow and placental hormone production. *Mailing Add:* 3181 SW Sam Jackson Park Rd Portland OR 97201

NOWACK, WILLIAM J, b New York, NY, Jan 15, 45; m. NEUROLOGY, MATHEMATICS & STATISTICS. *Educ:* Brown Univ, AB, 66; Yale Univ, MPhil, 69; Stanford Univ, MD, 73; Am Bd Psychiat & Neurol, cert neurol, 79. *Prof Exp:* Clin instr neurol, Sch Med, Univ Va, 77-79; asst prof, Sch Med, Ind Univ, 79-82; clin asst prof, Col Health Sci & Hosp, Univ Kans, 81-85; assoc prof, Univ Ark Med Sci, 85-90 & Univ Okla, 90-91; ASSOC PROF NEUROL, UNIV S ALA, 91- *Concurrent Pos:* NIH postdoctoral fel develop neurol & attend neurologist, Sch Med, Univ Va, 77-79; asst prof psychiat & attend neurologist, Sch Med, Ind Univ, 79-82; vis prof, Hurley Med Ctr, Flint, Mich, 81 & Chicago Med Sch, 82; consult neurol, Marion Vet Admin Hosp, 81-82; fac mem, Karl Menninger Sch Psychiat, 82-85; actg chief, Neurol Serv, Colmery-O-Neil Vet Admin Med Ctr, 82, chief, 82-85; attend neurologist, Univ Ark Med Sci, 85-90, Okla Mem Hosp, Okla Children's Mem Hosp & O'Donoghue Rehab Inst, 90-; staff neurologist, Oklahoma City Vet Admin Med Ctr, 90- *Mem:* Am Acad Neurol; Am EEG Soc; Southern EEG Soc (secy-treas, 90-); Am Med EEG Soc; Soc Math Biol. *Res:* Electroencephalography and mathematical analysis; author of numerous technical publications. *Mailing Add:* 12701 N Pennsylvania Apt 25 Oklahoma City OK 73120

NOWACZYNSKI, WOJCIECH, endocrinology; deceased, see previous edition for last biography

NOWAK, ANTHONY VICTOR, b Chicago, Ill, Aug 6, 38. ANALYTICAL CHEMISTRY. *Educ:* Loyola Univ, BS, 60; Northern Ill Univ, MS, 63; Univ Ill, Urbana, MS, 65, PhD(chem), 68. *Prof Exp:* Anal chemist, US Army QM Food & Container Inst, 60-62; SR MGR ANAL RES & COMPUT, ATLANTIC RICHFIELD CO, 67- *Concurrent Pos:* Lectr, Chicago sect, Soc Appl Spectros, 71. *Mem:* Am Chem Soc; Soc Appl Spectros. *Res:* Microprocessors in analytical instrumentation; applications of computers to laboratory automation; selective gas chromatographic detectors. *Mailing Add:* 2033 Camino Del Sol Fullerton CA 92633-1317

NOWAK, ARTHUR JOHN, b Erie, Pa, June 25, 37; m 61; c 4. PEDIATRIC DENTISTRY. *Educ:* Univ Pittsburgh, DMD, 61; Columbia Univ, cert (pediat dent), 66, MA, 67; Am Bd Pediat Dent, dipl, 71. *Prof Exp:* Fel pediat dent, Columbia Univ, 64-67; asst prof, Sch Dent Med, Univ Pittsburgh, 67-70; dir pediat dent, Allegheny Gen Hosp, Pittsburgh, 70-73; assoc prof, 73-77, PROF PEDIAT DENT, COL DENT, UNIV IOWA, 77-, PROF PEDIAT, COL MED, 84- *Concurrent Pos:* Consult, President's Comt Ment Retardation, 74-75; pres, Acad Dent Handicapped, 74-75; pres, Nat Found Dent Handicapped, 75-76; exec coun mem, Am Soc Dent Children, 75-78; prof adv coun, Nat Easter Seal Soc Crippled Adults & Children, 75-; assoc ed, Pediat Dent, J Am Acad Pediat Dent; consult comn on accreditation, Am Dent Asn; dir & chmn, Am Bd Pediat Dent; distinguished vis prof, Ohio State Univ. *Honors & Awards:* Preventive Dent Award, Am Dent Asn. *Mem:* Am Dent Asn; Am Acad Pediat Dent; Am Soc Dent Children; Am Asn Dent Sch; Int Asn Dent Res. *Res:* Dental management of effect of drugs on behavior of children in dental setting; effect of long term intubation on oral facial development in low birth weight infants; effect of feeding devices on oral facial development. *Mailing Add:* Dept of Pediat Dent Col Dent Univ of Iowa Iowa City IA 52242

NOWAK, EDWIN JAMES, b Chicago, Ill, Aug 12, 36; m 62; c 3. CHEMICAL ENGINEERING. *Educ:* Northwestern Univ, BS, 58; Princeton Univ, PhD(chem eng), 63. *Prof Exp:* Res engr, Calif Res Corp, 62-63; engr, Process Res Div, Esso Res & Eng Co, NJ, 63-66; asst prof chem eng, Univ NMex, 66-69; MEM TECH STAFF, SANDIA LABS, 69- *Mem:* Am Chem Soc. *Res:* Catalysis; chemical kinetics; surface phenomena; mass transport; electrochemistry. *Mailing Add:* Sandia Labs Orgn 6332 PO Box 5800 Albuquerque NM 87185

NOWAK, ROBERT MICHAEL, b South Milwaukee, Wis, Oct 28, 30; m 57; c 2. ORGANIC CHEMISTRY. *Educ:* Univ Wis, BS, 53; Univ Ill, PhD(org chem), 57. *Prof Exp:* Res chemist, 56-64, group leader, 64-68, asst lab dir, Phys Res Lab, 68-72, res & develop mgr, Plastics Dept, 73-83, dir, Plastics Dept, 83-88, DIR, CENT RES LABS, DOW CHEM USA, 88- *Mem:* Am Chem Soc. *Res:* Grafting vinyl monomers onto polymer backbones and synthesis and study of new monomers and their polymers; new resin systems for reinforced plastics; polymer chemistry. *Mailing Add:* 1212 Bayberry Lane Midland MI 48640

NOWAK, THADDEUS STANLEY, JR, b Bloomington, Ind, Oct 24, 49. NEUROCHEMISTRY, NEUROPATHOLOGY. *Educ:* Mas Inst Technol, BS, 71, PhD(nutrit biochem), 79. *Prof Exp:* staff fel, Lab Neurochem, 79-86, SR STAFF FEL & SPEC EXPERT LAB NEUROPATH & NEUROANAT SCI NAT INST NEUROL & COMMUNICATIVE DISORDERS & STROKE, 86- *Mem:* Am Soc Neurochem; Soc Neurosci. *Res:* Changes in gene expression after ischemia and other insults to brain, specifically utilization of heat shock/stress response as index of ongoing neuronal pathophysiology. *Mailing Add:* Bldg 36 Rm 4D04 NIH Bethesda MD 20892

NOWAK, THOMAS, b Niagara Falls, NY, Nov 25, 42; m 67; c 3. BIOCHEMISTRY, BIOPHYSICS. *Educ:* Case Inst Technol, BS, 64; Univ Kans, PhD(biochem), 69. *Prof Exp:* NIH fel, 69-71, res assoc, Inst Cancer Res, 71-72; PROF CHEM, UNIV NOTRE DAME, 84- *Concurrent Pos:* Res career develop award, NIH, 79-83; vis scientist, Univ Groningen, Netherlands, 81-82; vis prof, Catholic Univ Chile, Santiago, 86. *Mem:* AAAS; Am Chem Soc (counr); Am Soc Biol Chemists; Int Soc Magnetic Resonance; Biophys Soc. *Res:* Mechanism of action of enzymes; function of metal ions in enzymatic catalysis; protein-protein information transfer; applications of magnetic resonance to biological systems. *Mailing Add:* Dept Chem & Biochem Univ Notre Dame Notre Dame IN 46556

NOWAK, WELVILLE B(ERENSON), b Hartford, Conn, Oct 6, 21; m 50; c 2. PHYSICS, MATERIALS SCIENCE. *Educ:* Mass Inst Technol, SB, 42, PhD(physics), 49. *Prof Exp:* Staff mem, Radiation Lab, Mass Inst Technol, 42-45, res assoc surface impedance metals, 47-49, staff mem, 49-52; staff mem, Microwave Assocs, Mass, 52-54; group leader appl physics, Nuclear Metals Div, Textron Corp, 54-57, mgr tech coord, 57-60, dir electronic res, 60-62; assoc prof mat sci, 62-63, PROF MAT SCI, NORTHEASTERN UNIV, 63- *Mem:* Am Phys Soc; fel, Am Soc Metals Int; Am Vacuum Soc; Am Soc Mech Engrs. *Res:* Materials engineering; applied physics; metallurgy; microwaves; thin films; capacitors; energy conversion; electronic materials; plasma processing. *Mailing Add:* 17 Furbush Ave West Newton MA 02165

NOWATZKI, EDWARD ALEXANDER, b Bronx, NY, Feb 23, 36; m 62; c 4. CIVIL ENGINEERING, SOIL SCIENCE. *Educ:* St Joseph's Col, NY, BA, 57; Manhattan Col, BCE, 62; Univ Ariz, MSCE, 65, PhD(civil eng), 66. *Prof Exp:* Res assoc nuclear weapons effects, Univ Ariz, 65-66; res scientist geoastrophys res, Grumman Aerospace Corp, 66-68; assoc prof soil mech & civil eng, Calif State Polytech Col, 68-69; res scientist, Grumman Aerospace Corp, 69-73; tech consult, Joseph S Ward & Assocs, 73-75; assoc prof civil eng, Univ Ariz, 75-89; PROF CIVIL ENG, CALIF POLY STATE UNIV, SAN LUIS OBISPO, 89- *Concurrent Pos:* Pvt consult, var geotech eng consult firms & mining indust. *Mem:* fel Am Soc Civil Engrs; NY Acad Sci; Sigma Xi. *Res:* Plasticity analysis of soils; environmental effect on properties of granular materials; physicochemical aspects of soil mechanics; soil-structure interaction; soil dynamics; off-road vehicle mobility analysis; low level nuclear waste disposal--geotechnical aspects; metastable soil behavior. *Mailing Add:* Dept Civil & Environ Eng Calif Poly State Univ San Luis Obispo CA 93407

NOWELL, JOHN WILLIAM, b Wake Co, NC, Aug 26, 19. CHEMISTRY. *Educ:* Wake Forest Col, BS, 40; Univ NC, PhD(phys chem), 45. *Prof Exp:* Sr physicist, Am Cyanamid Co, Conn, 44-45; from asst prof to prof, 45-88, chmn dept, 63-72, EMER PROF CHEM, WAKE FOREST UNIV, 88- *Mem:* Am Chem Soc; NY Acad Sci; Sigma Xi. *Res:* Membrane permeability to gases; polarographic identification of ions; tracer methods using radioisotopes. *Mailing Add:* 1001 Paschal Dr Winston-Salem NC 27106

NOWELL, PETER CAREY, b Philadelphia, Pa, Feb 8, 28; m 50; c 5. PATHOLOGY. *Educ:* Wesleyan Univ, AB, 48; Univ Pa, MD, 52. *Prof Exp:* Intern, Philadelphia Gen Hosp, 52-53, clin asst lab serv, 56-70; resident path, Presby Hosp, Philadelphia, 53-54; from instr to assoc prof path, Sch Med, Univ Pa, 56-64, chmn dept, 67-73, dir, Cancer Ctr, 73-75, acad coordr, Dept Path & Lab Med, 80-87, PROF PATH, SCH MED, UNIV PA, 64-, DEP DIR, CANCER CTR, 75- *Concurrent Pos:* USPHS sr res fel & career develop award, Sch Med, Univ Pa, 56-61, USPHS res career award, 61-67; mem, Coun Am Soc Exp Path, 67-71, Adv Comt Atomic Bomb Casualty, Nat Res Coun, Nat Acad Sci, 69-74, Path Training Comt, NIH, 70-73, Study Sect Carcinogenesis, Am Cancer Soc, 71-74, Nat Coun Radiation Protection, 73-78, Ad Hoc Study Sect Carcinogenesis, NIH, 77-78, Comt Res Needs Path Effects Fossil Fuel Combustion, Nat Res Coun, 77-80 & Adv Comt to Dir, NIH, 90-; consult lab serv, Philadelphia Gen Hosp & Philadelphia Vet Admin Hosp, 70-77 & consult path, Children's Hosp, Philadelphia, 74-; dir, Am Asn Cancer Res, 70-73 & 90-; NIH outstanding investr grant, 86. *Honors & Awards:* Parke-Davis Award Exp Path, 65; Shubitz Prize, 80; La Madonnina Award, Milan, Italy, 82; Rous-Whipple Award, Am Asn Pathologists, 86; Robert de Villiers Award, Leukemia Soc Am, 87; Cotlove Award, Acad Clin Lab Physicians & Scientists, 87; Philip Levine Award, Am Soc Clin Pathologists, 89. *Mem:* Nat Acad Sci; Inst Med-Nat Acad Sci; Am Soc Exp Path (pres, 70-71); Am Asn Path & Bact; Am Asn Immunologists; Am Asn Cancer Res; Am Asn Pathologists; Acad Clin Lab Physicians & Scientists. *Res:* Growth regulatory mechanisms and cytogenetics of normal and leukemic leukocytes; radiation carcinogenesis; cellular immunology. *Mailing Add:* Dept Path & Lab Med Sch Med Univ Pa Philadelphia PA 19104-6082

NOWELL, WESLEY RAYMOND, b Oakland, Calif, Feb 9, 24; m 46; c 4. MEDICAL ENTOMOLOGY, DIPTERA-DIXIDAE. *Educ:* Stanford Univ, AB, 47, AM, 48, PhD(biol sci), 51. *Prof Exp:* Biomed Sci Corps, US Air Force, 51-78, med entomologist, 5th Air Force, Korea, 51-52, Air Res & Develop Command, 52-55, US Air Force Europe Command, 55-58, Strategic Air Command, 58-62, 4th US Air Force Epidemiol Flight, Turkey, 62-65, US Air Force Epidemiol Lab, Aerospace Med Div, Tex, 65-75 & Pac Air Forces Command, 75-78; assoc chief med entom, Biomed Sci Corps, US Air Force Sch Aerospace Med, 68-74, dep chief, Epidemiol Div, 72-75; Instr Biol, Univ Calif Santa Cruz & Monterey Peninsula Col, 78-81; RETIRED. *Concurrent Pos:* Mem armed forces pest control bd, Dept Defense, 67-74. *Mem:* Sigma Xi; Entom Soc Am; Am Soc Trop Med & Hyg; Am Mosquito Control Asn. *Res:* Global medical entomology; arthropod-associated diseases; vector control program analysis and organization; scientific research and training programs administration; Diptera: Dixidae; mosquito fauna, Mariana Islands. *Mailing Add:* 357 Reindollar Ave Marina CA 93933-3639

NOWER, LEON, b Sosnowiec, Poland, Aug 16, 27; US citizen; m 60; c 2. MATHEMATICS. *Educ:* City Col New York, BS, 53; Stanford Univ, MS, 62, PhD(math), 65. *Prof Exp:* ASSOC PROF MATH, SAN DIEGO STATE UNIV, 63- *Mem:* Am Math Soc; Math Asn Am. *Res:* Harmonic analysis; theory of distributions. *Mailing Add:* Dept Math San Diego State Univ San Diego CA 92182

NOWICK, A(RTHUR) S(TANLEY), b US, Aug 29, 23; m 49; c 4. SOLID STATE PHYSICS, MATERIALS SCIENCE. *Educ:* Brooklyn Col, AB, 43; Columbia Univ, AM, 48, PhD(physics), 50. *Prof Exp:* Jr instr physics, Johns Hopkins Univ, 43-44; physicist, Nat Adv Comt Aeronaut, 44-46; asst physics, Columbia Univ, 46-47; instr, Inst Study Metals, Univ Chicago, 49-51; from asst prof to assoc prof metall, Yale Univ, 51-57; mgr metall res dept, Res Ctr, Int Bus Mach Corp, NY, 57-66; adj prof, 57-66, PROF METALL, COLUMBIA UNIV, 66- *Concurrent Pos:* Consult, Oak Ridge Nat Lab, 65-70; mem, panel physics of condensed matter, Nat Res Coun, 69-; ed mat sci ser, Acad Press, Inc, 72- *Honors & Awards:* A Frank Golick Mem Lectureship, Univ Mo, 70. *Mem:* Fel Am Phys Soc; Am Inst Mining, Metall & Petrol Engrs; fel Metall Soc; Am Ceramic Soc; Sigma Xi. *Res:* Lattice imperfections in solids; anelasticity and internal friction; alloy thin films; ionic crystals; solid electrolytes. *Mailing Add:* Henry Krumb Sch Mines Columbia Univ New York NY 10027

NOWICKE, JOAN WEILAND, b St Louis, Mo; m 63; c 1. PALYNOLOGY, SYSTEMATICS. *Educ:* Washington Univ, AB, 59, PhD(biosyst), 68; Univ Mo-Columbia, AM, 62. *Prof Exp:* Fel, Mo Bot Garden, 68-71; asst prof biol, Univ Mo-St Louis, 71-72; assoc cur, 72-78, CUR, BOT DEPT, SMITHSONIAN INST, 78- *Mem:* AAAS; Bot Soc Am; Am Soc Plant Taxonomists; Int Soc Plant Taxonomists; Torrey Bot Club. *Res:* Pollen morphology, structure, function and use in systematics; classification of Phytolaccaceae; Apocynaceae and Boraginaceae of Central America; pollen morphology of Onagraceae and Ranunculales. *Mailing Add:* NHB 166 Smithsonian Inst Washington DC 20560

NOWINSKI, JERZY L, b Czestochowa, Poland, Mar 2, 05; nat US; m 29; c 1. MECHANICS. *Educ:* Warsaw Tech Inst, MS, 29, DSc(tech sci), 51. *Hon Degrees:* DSc, Warsaw Tech Inst, 51. *Prof Exp:* Asst mech, Warsaw Tech Inst, 30-37; res scientist, Polish Aeronaut Inst, 37-39 & 46-50; prof appl mech, Warsaw, 54-57; vis lectr elasticity, Johns Hopkins Univ, 57-58; prof res, Math Res Ctr, Univ Wis, 58-60; prof, Univ Tex, 60-61; H Fletcher Brown prof, 61-73, EMER PROF MECH ENG, UNIV DEL, 73- *Concurrent Pos:* Dir, Bur Reconstruct, Poland, 45-49; res scientist, Inst Math, Polish Acad Sci, 50-57; ed, Arch Appl Mech, 51-57, Int J Nonlinear Mech, 65-, J Thermal Stresses, 77- *Honors & Awards:* Huber Prize, Polish Acad Sci, 52; Polish State Sci Achievement Prize, 54; Appl Mech Rev Award, Am Soc Mech Engrs, Soc Eng Sci, 72; Res Award, Sigma Xi, 74. *Mem:* Soc Eng Sci; Soc Natural Philos. *Res:* Theory of elasticity, plasticity, viscoelasticity, nonlocal elasticity, thin-walled structures, finite elasticity, thermoelasticity, elastodynamics, variational methods, applied functional analysis. *Mailing Add:* Dept Mech Eng Univ Del Newark DE 19716

NOWLIN, CHARLES HENRY, b Wilmington, Del, Feb 1, 32; m 56; c 2. APPLIED MATHEMATICS, APPLIED PHYSICS. *Educ:* Washington & Lee Univ, BS, 55; Harvard Univ, SM, 56, PhD(appl physics), 63. *Prof Exp:* Appl physicist, 63-77, group leader basic measurement sci group, 77-81, tech consult, 81-85, spec assignment, res instrument sect, instrumentation, 85-87, GROUP LEADER, ELECTRONIC SYSTS DEVELOP, OAK RIDGE NAT LAB, 87- *Mem:* Inst Elec & Electronics Engrs. *Res:* Instrumentation; information theory and signal processing; network analysis and synthesis; acoustic ranging systems; motor current analysis. *Mailing Add:* Instrumentation & Controls Div Oak Ridge Nat PO Box 2008 Oak Ridge TN 37831-6006

NOWLIN, DUANE DALE, b Huron, SDak, Mar 14, 37; m 67; c 3. WATER CHEMISTRY. *Educ:* Macalester Col, BA, 58; Iowa State Univ, PhD(chem), 64. *Prof Exp:* Chief chemist, Lindsay Co, Union Tank Car Co, 64-65; sr chemist, Garrett Res & Develop Co, 65-66; mgr water chem, Econs Lab, Inc, 66-69; dir res & develop, Lindsay Co, Marmon Group, 69-89; PRES, SPECTRUM LABS, INC, 89- *Concurrent Pos:* USEPA Nat Drinking Water adv coun, 83-85; pres, Water Qual Asn, 88-89. *Mem:* Am Water Works Asn; Am Chem Soc; Nat Water Well Asn. *Res:* Ion-exchange technology; isotope separation studies; water treatment and metal ion chelation. *Mailing Add:* Spectrum Labs 301 W County Rd E2 New Brighton MN 55112

NOWLIN, WORTH D, JR, b Smithville, Tex, Oct 1, 35; m; c 2. PHYSICAL OCEANOGRAPHY. *Educ:* Tex A&M Univ, BA, 58, MS, 60 & PhD(physoceanogr), 66. *Prof Exp:* Oceanog technician, Res Found, Tex A&M Univ, 60, res scientist, 61-63; asst prof phys oceanog, Univ, 63-67; oceanogr & prog dir, Ocean Sci & Technol Div, Off Naval Res, DC, 67-69; from assoc prof to prof oceanog, 69-87, head dept, 76-79, ASSOC DEAN GEOSCI & DIR, DIV ATMOSPHERIC & OCEAN SCI, TEX A&M UNIV, 85-, DISTINGUISHED PROF OCEANOG, 87-, VPRES, TEX INST OCEANOG, 90- *Concurrent Pos:* Actg dep head, Off for Int Decade Ocean Explor, NSF, DC, 70-71, consult, 70-74, dep head, 71; mem oceanog panel, Div Environ Sci, NSF, 70-72, Univ Nat Oceanog Lab Syst, 70-72, Fleet Replacement Comt, 84-86; mem oceanog panel, Comt Polar Res, Nat Acad Sci-Nat Res Coun, 72-75, Ocean Sci Bd, 78-81, steering comt, Acad Res Fleet Study, Ocean Sci Bd, 80-81, Comt Evaluate Marine Ecosyst Res in Antarctic, Polar Res Bd-Ocean Sci Bd, 80-81, Ocean Climate Res Strategies Comt, Bd Ocean Sci & Policy, 83-85, Ad Hoc Comt Phys & Chem Oceanog Southern Ocean, Polar Res Bd, 83-86, Comt Nat Strategies Ocean Sci & Policy, Bd Ocean Sci & Policy, 84-86, ComtEarth Sci, Space Sci Bd, 85-87; US nat rep, Southern Oceans Intergovt Oceanog Comn-UNESCO, 74-88; assoc ed, J Phys Oceanog; dir, Tex A&M Sea Grant Col Prog, 77-78; courtesy prof, Ore State Univ, 80-81; chmn, Work Group Sci Comt Oceanic Res, 74, Gen Circulation Southern Ocean, 82-86; co-chmn, Panel World Ocean Circulation Exp, Bd Ocean Sci & Policy, Nat Acad Sci-Nat Res Coun, 83-85; convener & co-chmn prog comt, Ocean Sci 82, Joint Am Geophys Union, Am Soc Limnol & Oceanog, Oceanog Meeting, San Antonio; chmn, Ed Search Comt, J Geophys Res-Oceans, 83; chmn & co-chmn, US Sci Steering Comt, World Ocean Circulation Exp, 85-, co-convener, Global Open Circulation, Vancouver, Can, 87, chmn, US Hydrographic Prog Implementation Panel, 87-, mem, Int Sci Steering Group, 89-; counr, Tex A&M Res Found, 85-; chmn, Univ Nat Ocean Lab Syst Fleet Improv Comt, 86-89, vchmn, 89-90, mem coun, 89- & subcomt Improved Sci Mission Req, 90-; mem, Adv Comt for Polar Prog, Nat Sci Found, 86-87; mem, Space & Earth Sci Comt, NASA Adv Coun, 86-88; mem, Tex Space Grant Adv Panel Tex A&M Univ, 89-; chmn, Ocean Observing Syst Develop Panel, Intergovt Oceanog Comn, UNESCO, World Meteorol Orgn & Int Coun Sci Unions, 89- *Mem:* fel Am Geophys Union; Sigma Xi; Am Meteorol Soc. *Res:* Meso-scale and large-scale oceanic distributions of properties; dynamics of ocean circulation; research planning and management. *Mailing Add:* Dept Oceanog Tex A&M Univ College Station TX 77843

NOWLIS, DAVID PETER, b New Haven, Conn, Feb 12, 37; div; c 4. BEHAVIORAL MEDICINE, FAMILY MEDICINE. *Educ:* Haverford Col, AB, 58; Harvard Univ, AMT, 59, PhD(clin & social psychol), 65. *Prof Exp:* Instr psychol, Claremont Men's Col, 63-64; asst prof educ & child develop, Bryn Mawr Col, 64-67; lectr psychol, Stanford Univ, 67-69; consult life sci, Garrett Corp NASA, 69-74; assoc res psychol, Jewish Gen Hosp, Mc Gill Univ, 74-85; ASSOC CLIN PROF BEHAV SCI, UNIV CALIF MED SCH, FRESNO, 85- *Mem:* Soc Teachers Family Med; Am Psychol Asn. *Res:* Behavioral aspects of family medicine and family health. *Mailing Add:* Dept Family Pract Valley Med Ctr 445 S Cedar Ave Fresno CA 93702-2998

NOWOGRODZKI, M(ARKUS), b Warsaw, Poland, Sept 13, 20; nat US; m 42; c 2. ELECTRICAL ENGINEERING. *Educ:* Polytech Inst Brooklyn, BEE, 48, MEE, 51. *Prof Exp:* Engr microwave equip, Hazeltine Electronics Corp, NY, 48-51; sr engr microwave tubes, Amperex Electronic Corp, 51-54, supv engr magnetron dept, 54-55; eng leader electronic components & devices, Radio Corp Am, 55-57, mgr microwave design & develop, 57-60, mgr prod & equip eng, 60-62, mgr microwave eng prog, 62-64, mgr traveling-wave tubes & solid-state devices oper, 64-69; mgr eng, Hicksville Div, Amperex Electronic Corp, 69-76; mgr, Liasion Div, David Sarnoff Res Ctr, RCA Labs, 76-79, head, Subsysts & Spec Proj, 79-87; RETIRED. *Concurrent Pos:* Consult, Microwave Sensors, 87- *Mem:* Inst Elec & Electronic Engrs; NY Acad Sci. *Res:* Microwave tubes and solid-state devices. *Mailing Add:* Rd #2 Box 722 Sussex NJ 07461

NOWOTNY, ALOIS HENRY ANDRE, b Gyongyos, Hungary, July 30, 22; US citizen; m 60; c 2. IMMUNOLOGY, IMMUNOCHEMISTRY. *Educ:* Pazmany Peter Univ, Budapest, dipl chem, 45, PhD(chem), 47; Temple Univ, MA, 79. *Hon Degrees:* MA, Univ Pa. *Prof Exp:* Asst prof biochem, Med Sch, Univ Budapest, 47-51; res assoc immunochem, Hungarian Blood Serv Ctr, 51-54, vchmn res dept, 54-56; res assoc, A Wander Res Inst, Freiburg, WGer, 57-60; sr res assoc immunochem, City of Hope Med Ctr, Duarte, Calif, 60-62; prof immunochem & immunol, Med Sch, Temple Univ, 62-76; PROF, IMMUNOL, PATH & MICROBIOL, SCH MED, UNIV PA, 76- *Concurrent Pos:* Mem adv bd, Cancer Res Inst, NY, 67-80; guest prof, Med Sch, Univ Heidelberg, 69-70 & 72, Am Univ Beirut, Lebanon, 80; Univ S FL, Tampa, 85; consult, NIH, 73-; Lady Davis guest prof immunol, Hadassah Univ, Jerusalem, Isreal, 79-80; vis prof, Univ Oxford, Oxford, Eng, 85. *Mem:* Am Asn Immunol; Am Microbiol Soc; Ger Immunol Soc; Hungarian Chem Soc. *Res:* Cellular antigens; pathomechanisms of periodontitis; mode of action of bacterial endotoxins; immunology of erythrocyte membranes; immunology of tumor cells. *Mailing Add:* Levy Ctr Oral Health Res 4010 Locust St Res Bldg A2 Philadelphia PA 19104

NOWOTNY, HANS, b Linz, Austria, Sept 27, 11; m 44; c 1. PHYSICAL CHEMISTRY, MATERIALS SCIENCE. *Educ:* Vienna Tech Univ, Dipl Engr, 33, Dr Tech, 34. *Hon Degrees:* Dr Mont, Mining & Metall Col, Austria, 65. *Prof Exp:* Res asst phys chem, Vienna Tech Univ & Karlsruhe Tech Univ, 34-41; sci researcher metall, Max Planck Inst, 41-45; sci dir, Inst Metall Res, Tettnang, Ger, 45-47; assoc prof phys chem, Univ Vienna, 47-52; prof, Vienna Tech Univ, 52-58; prof & head dept, 58-77, EMER PROF PHYS CHEM, UNIV VIENNA, 77-; UNIV PROF METALL, INST MAT SCI, UNIV CONN, 77- *Concurrent Pos:* Vis prof, Univ Calif, 55, Univ Strasbourg, 57, Univ Amsterdam, 60, Univ Ill, 61, Univ Paris, 62 & Univ Conn, 68, 69 & 71; Battelle prof, Ohio State Univ, 63. *Honors & Awards:* Lavoisier Medal, Soc Chim France, 60; Medal Sci & Art, Austrian Fed Ministry Sci & Res, 78; Heyn Medal, Ger Soc Metall, 87. *Mem:* Austria Acad Sci; Hungary Acad Sci; Gottingen Acad Sci; Leopoldina Acad Sci. *Res:* Structural and alloy chemistry. *Mailing Add:* Steinbach 7 Ernstbrunn NO 2115 Austria

NOWOTNY, KURT A, b Vienna, Austria, Apr 8, 31; US citizen; m 60; c 2. INDUSTRIAL CHEMISTRY, INORGANIC CHEMISTRY. *Educ:* Univ Vienna, PhD(natural prod), 59. *Prof Exp:* Fel polyfunctional catalysis, Wash Univ, 59-60; from sr res chemist to proj leader functional fluids res specialist, Monsanto Co, 60-65; supvr packing res & develop, Crown Zellerbach Corp, 65-69, supvr polymer res, 69-70, mgr chem res dept, Cent Res Dept, 70-73; group vpres & dir technol & eng, Evans Prod Co, 73-77; pres, Neudiedler AG Papierfabrikation, Austria, 77-; gen mgr, Disc Media Div, Ampex Corp, San Jose, 84-85; pres, Lightwave Technologies, Inc, 86-89; PRES, TECH SOLUTIONS INT, 89- *Mem:* Am Chem Soc; Ver Oesterreichischer Chemiker; Ges Deutscher Chemiker. *Res:* Natural products; pharmaceutical and physical chemistry; oil additives; functional, aviation and fire resistant fluids; cellulose; adhesives; coatings; physics. *Mailing Add:* PO Box 2759 Vancouver WA 98668-2759

NOYCE, DONALD STERLING, b Burlington, Iowa, May 26, 23; m 46; c 3. ORGANIC CHEMISTRY. *Educ:* Grinnell Col, AB, 44; Columbia Univ, MA, 45, PhD(org chem), 47. *Prof Exp:* NIH fel, Columbia Univ, 47-48; from instr to prof chem, Univ Calif, Berkeley, 48-86, asst dean, Col Chem, 52-60, 66-68 & 75-80, assoc dean, 80-86; RETIRED. *Concurrent Pos:* Guggenheim fel, 57; NSF sr fel, 64. *Mem:* Am Chem Soc; Royal Soc Chem. *Res:* Stereochemistry; mechanisms of organic reactions; organic kinetics. *Mailing Add:* 1504 Olympus Ave Berkeley CA 94708

NOYCE, ROBERT NORTON, physics; deceased, see previous edition for last biography

NOYES, CLAUDIA MARGARET, b Haverhill, NH, Apr 30, 40. ANALYTICAL BIOCHEMISTRY. *Educ:* Univ Vt, BS, 61; Univ Colo, PhD(chem), 66. *Prof Exp:* Res chemist, Armour Grocery Prod Co, 65-67; res assoc med, Sch Med, Univ Chicago, 68-75; res assoc med, Sch Med, Univ NC, Chapel Hill, 75-85, res asst prof med & path, 85-87. *Mem:* AAAS; Am Chem Soc; Sigma Xi; Am Soc Biochem & Molecular Biol. *Res:* Protein primary structure determination; chromatography; structure and function of blood coagulation proteins. *Mailing Add:* 1522 NE 175th St Apt 304 Seattle WA 98155

NOYES, DAVID HOLBROOK, b Hampton, Va, Feb 5, 35; m 60; c 2. PHYSIOLOGY. *Educ:* Rensselaer Polytech Inst, BEE, 59; Univ Ala, Birmingham, PhD(physiol & biophys), 69. *Prof Exp:* Test engr radar systs, Gen Elec Light Mil Dept, NY, 59; supvr electronic model shop & biomed engr, Univ Ala, Birmingham, 62-66, biomed engr & eng design consult, Res Model Shop, 66-69; asst prof, 69-75, ASSOC PROF PHYSIOL, COL MED, OHIO STATE UNIV, 75-, ASST PROF, DEPT ELEC ENG, 69- *Concurrent Pos:* NIH grants, Col Med, Ohio State Univ, 72-75, NSF grant, 73-75. *Mem:* Am Physiol Soc; Int Asn Dent Res. *Res:* Gastrointestinal physiology; stomatology, periodontology and muscle physiology. *Mailing Add:* Dept Physiol Ohio State Univ Col Med 370 W Ninth Ave Columbus OH 43210

NOYES, H PIERRE, b Paris, France, Dec 10, 23; US citizen; m 47; c 3. THEORETICAL PHYSICS. *Educ:* Harvard Univ, BA, 43; Univ Calif, Berkeley, PhD(physics), 50. *Prof Exp:* Mem staff, Radiation Lab, Mass Inst Technol, 43-44; physicist, Radiation Lab, Univ Calif, 50; Fulbright grantee math physics, Univ Birmingham, 50-51; asst prof physics, Univ Rochester, 51-55; physicist & group leader, Lawrence Livermore Lab, Univ Calif, 55-62; assoc prof theoret physics, 62-67, admin head sect, 62-69, PROF THEORET PHYSICS, STANFORD LINEAR ACCELERATOR CTR, 67- *Concurrent Pos:* Leverhulme lectr, Univ Liverpool, 57-58; Avco vis prof, Cornell Univ, 61; vis scholar, Ctr Advan Studies Behav Sci, Stanford Univ, 68-69; consult, Gen Atomic Div, Gen Dynamics Corp, 59-62, Lockheed Aircraft Corp, 62-63, Lawrence Radiation Lab, Livermore, 62-67 & Physics Int, 65-67; chmn comt for a dir attack on the legality of the Vietnam War, 69-72; mem policy comt, US People's Comt Iran, 77-82; Alexander von Humboldt US sr scientist award, 79. *Mem:* AAAS; fel Am Phys Soc; Sigma Xi; Alternative Natural Philos Asn (pres, 79-87). *Res:* Nucleon-nucleon and meson-nucleon interaction; quantum mechanical 3-body problem; applied hydrodynamics and neutronics; computational techniques; foundations of quantum mechanics. *Mailing Add:* Stanford Linear Accelerator Ctr Stanford CA 94305

NOYES, HOWARD ELLIS, b Memphis, Tenn, Apr 5, 22; m 47; c 3. BACTERIOLOGY. *Educ:* Univ Tenn, BS, 47; Ohio State Univ, MS, 49; George Washington Univ, PhD(bact), 55; Am Bd Microbiol, dipl. *Prof Exp:* US Dept Army, bacteriologist, Ft Detrick, Md, 49-51, chief bact sect, Div Surg, Inst Res, Walter Reed Army Med Ctr, 61-63, chief, Dept Bact & Mycol, US Army Med Component, SEATO Med Res Lab, 63-66, dep chief, Dept Lab Serv, Div Surg, Walter Reed Army Inst Res, 66, chief, Dept Surg Microbiol, 66-67, chief, Dept Bact & Mycol, Med Res Lab, US Army Med Component, SEATO, 67-70, assoc dir, Walter Reed Army Inst Res, 70-88; CONSULT, 88- *Mem:* AAAS; Am Soc Microbiol; Soc Exp Biol & Med; NY Acad Sci. *Res:* Medical microbiology as it relates to surgery with emphasis on creation and therapy of experimental wounds, the mechanism of action of bacterial toxins and the evaluation of new antimicrobial agents. *Mailing Add:* 1004 Stedwick Rd Apt 103 Gaithersburg MD 20879

NOYES, PAUL R, b Shreveport, La, Oct 3, 28; m 61; c 2. ORGANIC CHEMISTRY, POLYMER CHEMISTRY. *Educ:* Centenary Col, BS, 49; Univ Tex, MS, 54, PhD(chem), 55. *Prof Exp:* Res chemist, 54-63, staff chemist, 63-84, RES ASSOC, E I DU PONT DE NEMOURS & CO, INC, 84- *Mem:* Am Chem Soc; Tech Asn Pulp & Paper Indust. *Res:* Adhesives; adhesion; paper coatings; teflon coatings. *Mailing Add:* E I du Pont de Nemours & Co Inc PO Box 3886 Philadelphia PA 19146

NOYES, RICHARD MACY, b Champaign, Ill, Apr 6, 19; m 46, 73. PHYSICAL CHEMISTRY, CHEMICAL KINETICS. *Educ:* Harvard Univ, AB, 39; Calif Inst Technol, PhD(phys chem), 42. *Prof Exp:* Instr, Calif Inst Technol, 42-44, res fel, Nat Defense Res Comt Proj, 42-46; from instr to assoc prof chem, Columbia Univ, 46-58; head dept, 66-68 & 75-78, PROF CHEM, UNIV ORE, 58- *Concurrent Pos:* Vis prof, Univ Leeds, 55-56 & Oxford Univ, 71-72; mem subcomt kinetics chem reactions, Nat Res Coun, 60-, mem-at-large, Div Chem & Chem Tech, 63-66; Fulbright fel, Univ Victoria, NZ, 64; NSF sr fel, Max Planck Inst Phys Chem, Gottingen, 65; mem chem adv panel, NSF, 69-71; Alexander von Humboldt fel, Max Planck Inst Biophys Chem, Gottingen, 78-79; assoc ed, J Phys Chem, 80-82; vis prof, Univ Calif, Davis, 85. *Mem:* Nat Acad Sci; Am Chem Soc; Am Phys Soc; Royal Soc Chem (London). *Res:* Thermodynamic properties of ions in solution; mechanisms of chemical reactions including reactions of diatomic molecules, diffusion controlled reactions, isotopic exchange reactions of organic iodides, and reactions oscillating in time and space; nucleation of gases in supersaturated solutions. *Mailing Add:* Dept Chem Univ Oregon Eugene OR 97403

NOYES, ROBERT WILSON, b Winchester, Mass, Dec 27, 34; m 60; c 2. SOLAR AND STELLAR PHYSICS. *Educ:* Haverford Col, BA, 57; Calif Inst Technol, PhD(physics), 63. *Prof Exp:* Lectr astron, 62-73, assoc dir, Ctr Astrophys, 73-80, PROF ASTRON, HARVARD UNIV, 73-; PHYSICIST, SMITHSONIAN ASTROPHYS OBSERV, 62- *Concurrent Pos:* Bd dir, Asn Univ Res Astron, 78, chmn, 86-89. *Mem:* Am Astron Soc; Int Astron Union. *Res:* Solar and stellar spectroscopy, from ground-based and space observatories; solar and stellar seismology; evolution of solar and stellar rotation and magnetic activity. *Mailing Add:* Smithsonian Astrophys Observ 60 Garden St Cambridge MA 02138

NOYES, RUSSELL, JR, b Indianapolis, Ind, Dec 25, 34; m 60; c 3. PSYCHIATRY. *Educ:* DePauw Univ, BA, 56; Ind Univ, MD, 59; Am Bd Psychiat & Neurol, dipl, 66. *Prof Exp:* Rotating intern, Philadelphia Gen Hosp, 59-60; resident psychiat, Inst Living, Conn, 60-61 & Univ Iowa, 61-63; mem staff, US Naval Hosp, Great Lakes, Ill, 63-65; from asst prof to assoc prof, 65-78, PROF PSYCHIAT, UNIV IOWA, 78- *Mem:* Fel Am Psychiat Asn; fel Acad Psychosom Med. *Res:* Anxiety disorders, psychosomatic medicine. *Mailing Add:* Dept Psychiat Univ Iowa Iowa City IA 52242

NOYES, WARD DAVID, b Schenectady, NY, Aug 25, 27; m 50; c 7. INTERNAL MEDICINE, HEMATOLOGY. *Educ:* Univ Rochester, BA, 49, MD, 53. *Prof Exp:* From intern to asst resident med, King County Hosp, Seattle, Wash, 53-56; instr, Univ Wash, 59-61; from asst prof to assoc prof, 61-70, prof med, 70-77, PROF & CHIEF HEMAT, COL MED, UNIV FLA, 77- *Concurrent Pos:* Res fel hemat, Univ Wash, 56-59; USPHS res fel, Oxford & Malmo Gen Hosp, Sweden, 58. *Mem:* Am Fedn Clin Res; Int Soc Hemat; Am Soc Hemat; Am Soc Clin Oncol. *Res:* Problems in erythrokinetics and iron metabolism. *Mailing Add:* Dept Hemat Univ of Fla Col of Med One J Hillis Miller Health Ctr Gainesville FL 32610

NOZ, MARILYN E, b New York, NY, June 17, 39; m. MEDICAL PHYSICS. *Educ:* Marymount Col, NY, BA, 61; Fordham Univ, MS, 63, PhD(physics), 69. *Prof Exp:* From instr to asst prof physics, Marymount Col, 64-69; assoc prof physics, Ind Univ Pa, 69-74; ASSOC PROF RADIOL, NY UNIV, 74- *Concurrent Pos:* Fogarty sr int fel, Karolinska Inst, Sweden, NIH, 83-84; NIH Fogarty Sr Int fel, NDEA fel. *Mem:* Am Phys Soc; Am Asn Physicists Med;

Soc Nuclear Med. *Res:* High energy theoretical physics; covariant harmonic oscillators; quark and parton theory; symmetry schemes for classifying elementary particles; projection operator techniques for calculating generalized vector coupling coefficients. *Mailing Add:* Dept of Radiol NY Univ 550 First Ave New York NY 10016

NOZAKI, KENZIE, b Los Angeles, Calif, June 1, 16; m 44; c 2. CHEMISTRY, PHYSICAL CHEMISTRY. *Educ:* Univ Calif, Los Angeles, BA, 37, MA, 38; Stanford Univ, PhD(chem), 40. *Prof Exp:* Asst chem, Univ Calif, Los Angeles, 37-38; Franklin fel, Stanford Univ, 38-40; instr Univ Calif, 41-42; dir res, War Relocation Authority, Calif, 43; Pittsburgh Plate Glass fel, Harvard Univ, 43-45, res assoc, 43-45; res chemist, Shell Develop Co, 46-72, sr staff res chemist, Shell Oil Co, 72, consult res chemist, 72-83; RETIRED. *Mem:* Am Chem Soc. *Res:* High polymers; free radicals; reaction mechanisms; molecular rearrangements; organic peroxides; kinetics; homogeneous and heterogeneous catalysis; organic nitrogen compounds; carbonylation; coordination chemistry; automotive fuels. *Mailing Add:* 1524 Madera Circle El Cerrito CA 94530

NOZAKI, YASUHIKO, b Yamagata, Japan, June 14, 13; c 2. BIOPHYSICAL CHEMISTRY. *Educ:* Univ Tokyo, BS, 37, PhD(pharm), 45. *Prof Exp:* Asst pharm, Univ Tokyo, 37-39, instr, 39-45; prof chem, Nihon Women's Col, 46-48; prof, Kyoritsu Col Pharm, 48-51; tech off microanal, Nat Inst Hyg Sci, Tokyo, 51-60, sect chief vitamin chem, 60-62; EMER ASSOC BIOCHEM, MED CTR, DUKE UNIV, 62- *Concurrent Pos:* Res assoc, Harvard Univ, 54-55 & Univ Iowa, 57-59. *Mem:* Am Chem Soc; Am Soc Biochem Molecular Biol. *Res:* Naphthoresorcinol reaction of glucuronic acid; interaction of copper and zinc ions with imidazoles; titration of native and denatured proteins; solubility of amino acids in relation to configuration of proteins; structure and function of biological membranes; determination of biopolymer concentration. *Mailing Add:* Dept Biochem Duke Univ Med Ctr Durham NC 27710

NOZIK, ARTHUR JACK, b Springfield, Mass, Jan 10, 36; m 58; c 2. PHYSICAL CHEMISTRY, ENERGY CONVERSION. *Educ:* Cornell Univ, BChE, 59; Yale Univ, MS, 62, PhD(phys chem), 67. *Prof Exp:* Res engr chem eng, Douglas Aircraft Co, Santa Monica, Ca, 59-60; res engr, Cent Res Div, Am Cyanamid Co, 61-64, res chemist, 67-74, sr staff chemist & group leader, Mat Res Ctr, Allied Chem Co, 74-78; sr scientist, Photoconversion Res Br, 78-79, br chief, 79-84, SR RES FEL, SOLAR FUELS RES DIV, SOLAR ENERGY RES INST, 84- *Concurrent Pos:* Instr, Southern Conn State Col, 62-64, lectr, 68-74; US rep & operating agent, Prog on Photocatalytic H2 Prod, Int Energy Agency, 80-; prin investr, US Dept Energy Chem Sci Prog, 80-, US-Yugoslavia & US-Israel Coop Res Prog, 82- & US-Japan Conf Photoconversion Progs, NSF, 83-; Int Organizing Comt, Int Conf on Photochem Conversion & Storage Solar Energy, 82-88; vis fel, Univ Colo, Boulder, 85-88; mem, sci rev comt, US Dept Energy Nat Lab Res Progs, 86-89; NSF sci rev panels, 87-89; chmn, Energy Technol Div, Electrochem Soc, 84; distinguished lectureship, US Dept Energy, Am Western Univ, 89-90. *Honors & Awards:* Outstanding Achievement Award, Solar Energy Res Inst, 84, Van Morris Award, 85; Van Morris Award, Solar Energy Res Inc, 85. *Mem:* Am Phys Soc; Am Chem Soc; AAAS; Mat Res Soc; Soc Photo Optical Instrument Engrs; Electrochem Soc. *Res:* Photoelectrochemistry; solar energy photo conversion; optical, magnetic and transport properties of solids and thin films; hydrogen energy systems; photocatalysis; heterogeneous catalysis; chemical and physical applications of Mossbauer spectroscopy. *Mailing Add:* Solar Energy Res Inst 1617 Cole Blvd Golden CO 80401

NOZZOLILLO, CONSTANCE, b Spencerville, Ont, July 18, 26; m 52. PLANT BIOCHEMISTRY. *Educ:* Queen's Univ (Ont), BA, 49, MA, 50; Univ Ottawa, PhD(plant biochem), 63. *Prof Exp:* Res off plant physiol, Can Dept Agr, 50-53, microbiol, 58-60; from asst prof to assoc prof, 63-85, PROF BIOL, UNIV OTTAWA, 85- *Mem:* Am Soc Plant Physiol; Bot Soc Am; Can Soc Plant Physiol; Phytochem Soc NAm (pres, 81-82); Can Bot Asn. *Res:* Physiology of seedling development; chemotaxonomic aspects of anthocyanins. *Mailing Add:* Dept Biol Univ Ottawa Ottawa ON K1N 6N5 Can

NRIAGU, JEROME OKONKWO, b Ora-eri Town, Nigeria, Oct 24, 42; Can citizen; c 3. GEOENVIRONMENTAL SCIENCE, WATER RESOURCES. *Educ:* Univ Ibadan, Nigeria, BS, 65; Univ Wis-Madison, MS, 67; Univ Toronto, PhD(geol & geochem), 70. *Hon Degrees:* DSc, Univ Ibadan, 87. *Prof Exp:* RES SCIENTIST, ENVIRON CAN, CAN CTR INLAND WATERS, 70-; ADJ PROF, DEPT EARTH SCI, UNIV WATERLOO, 85- *Concurrent Pos:* Mem, numerous int panels & comts on environmental studies; ed, Advan Environ Sci & Technol & Sci of Total Environ. *Honors & Awards:* Frank Rigler Medal, Can Soc Limnologists. *Mem:* Geochem Soc; AAAS; Am Soc Limnol & Oceanog; Royal Can Inst. *Res:* Biogeochemistry of the elements in the environment; stable isotopes as pollutant source and behavior indicators; environmental cycling of trace metals; author of book on lead poisoning in antiquity; edited 23 environmental books. *Mailing Add:* Can Ctr Inland Waters Box 5050 Burlington ON L7R 4A6 Can

NUCCITELLI, RICHARD LEE, b San Francisco, Calif, Feb 18, 48; m 70; c 3. DEVELOPMENTAL BIOLOGY. *Educ:* Univ Santa Clara, BS, 70; Purdue Univ, MS, 72, PhD(biol), 75. *Prof Exp:* NIH fel physiol, Los Angeles, 76-78, from asst prof to assoc prof, 78-88, PROF ZOOL, UNIV CALIF, DAVIS, 88- *Concurrent Pos:* NIH res career develop award, 82-87. *Mem:* Am Soc Cell Biol; Soc Develop Biol; Biophys Soc; AAAS. *Res:* Galvano taxis of neural crest cells; role of the plasma membrane in pattern formation; bioelectric aspects of development; role of pH and calcium in the activation of development. *Mailing Add:* Dept Zool Univ Calif Davis CA 95616

NUCKLES, DOUGLAS BOYD, b Hampton, Va, Mar 7, 31; m 57; c 2. DENTISTRY. *Educ:* Med Col Va, BS, 59, DDS, 60; The Citadel, MAT, 73. *Prof Exp:* Instr oper, Crown & Bridge & Dent Mat, Sch Dent, Med Col Va, 60-64, asst prof oper & dent mat, 64-68, assoc prof restorative dent, 68-71; assoc prof crown & bridge, Col Dent Med, Med Univ SC, 71-72, assoc prof, 72-73, asst dean clin affairs, 81-86, PROF OPER DENT, COL DENT MED, MED UNIV SC, 73-, SPEC ASST TO DEAN, 89-, FACIL COORDR, 90- *Concurrent Pos:* Johnson & Johnson res grant, Med Univ SC, 73-75; consult dept mat sci, Sch Eng, Univ Va, 64-71; mem subcomts dent instruments & hand pieces, Am Nat Standards Comt, 70; dir, Dent Auxiliary Utilization Prog, Med Univ SC, 76-81; coun mem, Dent Mats & Devices, WG4 rotary cutting instruments & dent handpieces, Am Dent Asn, 70- *Mem:* Am Dent Asn; Am Asn Dent Schs. *Res:* Properties of dental materials; clinical evaluation of dental restorations; evaluation of clinical instructors. *Mailing Add:* Col Dent Med 171 Ashley Ave Charleston SC 29425

NUCKOLLS, JOE ALLEN, b Waynesville, NC, Dec 21, 29; m 52; c 3. ENERGY PROCESSING SYSTEM, ELECTRICAL ENGINEERING. *Educ:* NC State Univ, BSEE, 58. *Prof Exp:* Comnd missile specialist, Missiles Radar Airborne Syst, US Air Force, 50-54; reflectivity radar & testing Radiation Inc, 58-60; div develop engr, Outdoor Lightning Dept, Gen Elec, 60-65; mfr Ballast & Control Eng, Elec & Electronic, 66-75, mgr B&C physic tech mkt & sr tech consult, Lighting Syst Dept Gen Elec, 76-83; DIR RES & DEVELOP & ADV ENG, LIGHTING INDUST TECHNOL & NEW PROD DEVELOP, HUBBELL LIGHTING INC, 84- *Res:* Electrical systems; electrical loads; energy processing to arcs; plasma electrical dynamics; materials application. *Mailing Add:* 2000 Electric Way Christianburg VA 24073

NUCKOLLS, JOHN HOPKINS, b Chicago, Ill, Nov 17, 30; m 83; c 2. APPLIED PHYSICS. *Educ:* Wheaton Col, BS, 53; Columbia Univ, MA, 55. *Hon Degrees:* Dr Sci, Fla Inst Tech, 77. *Prof Exp:* Staff physicist, 55-80, assoc div leader, 65-80, assoc prog leader in Laser Fusion Prog, 75-83, X-div leader, Inertial Fusion Target Design, 80-83, assoc dir physics, 83-88, DIR, LAWRENCE LIVERMORE NAT LAB, UNIV CALIF, 88- *Concurrent Pos:* Mem vulnerability task force, Defense Sci Bd, 69-72. *Honors & Awards:* E O Lawrence Award, US AEC, 69; James Clerk Maxwell Prize, Am Phys Soc, 81; Fusion Leadership Award, Fusion Power Assocs, 83. *Mem:* Fel Am Phys Soc; AAAS; Fusion Power Assocs; World Future Soc. *Res:* Nuclear explosives; inertial confinement; underground nuclear explosions; astrophysics; deterrance and arms control; global energy and environmental problems; economic competitiveness; leadership/management of research and development organizations. *Mailing Add:* Lawrence Livermore Nat Lab Univ Calif, PO Box 808 L-1 Livermore CA 94550

NUDELMAN, SOL, b Brooklyn, NY, Aug 14, 22; m 50; c 2. PHYSICS. *Educ:* Union Col, NY, BS, 45; Ind Univ, MS, 48; Univ Md, PhD(physics), 55. *Prof Exp:* Asst physics, Ind Univ, 46-48; instr, Union Col, NY, 48-49; instr, Knox Col, 49-51; physicist, US Naval Ord Lab, 51-56; res physicist, Univ Mich, 56-61; mgr solid state res, IIT Res Inst, 61-64; prof elec eng, Univ RI, 64-73; prof radiol & optical sci, Univ Ariz, 73-83; PROF RADIOL & DIR, DIV RES MED IMAGING, MED SCH, UNIV CONN, 83- *Mem:* AAAS; Am Phys Soc; Inst Elec & Electronics Engrs; Soc Photo-Optical Instrumentation Engrs. *Res:* Influence of electric fields on phosphors; electroluminescent phosphors; luminescent displays; electrical and optical properties of semiconductors; infrared sensitive photodetectors; photoelectronic imaging devices; imaging for diagnostic medicine. *Mailing Add:* Health Ctr Dept Radiol Univ Conn Farmington CT 06032

NUENKE, RICHARD HAROLD, b Bay City, Mich, Sept 3, 32; div; c 2. BIOCHEMISTRY. *Educ:* Univ Mich, BS, 53; Vanderbilt Univ, PhD(biochem), 61. *Prof Exp:* Res assoc biochem, Univ Ill, 60-62; asst prof physiol chem, 62-66, ASSOC PROF MED BIOCHEM, OHIO STATE UNIV, 67-, POWELSON PROF MED, 78- *Concurrent Pos:* Lab coordr, Ohio State Univ, 67-71, admin dir, Independent Study Prog, Col Med, 85-89. *Mem:* AAAS; Am Chem Soc. *Mailing Add:* 366 Hamilton Hall Ohio State Univ 1645 Neil Ave Columbus OH 43210

NUESE, CHARLES J, b Endicott, NY, July 39; m 61; c 3. ELECTRICAL ENGINEERING. *Educ:* Univ Conn, BS, 61; Univ Ill, MS, 62, PhD(elec eng), 66. *Prof Exp:* Mem tech staff, 66-77, head semiconductor device res, 77-84, dir lab, 84-85, vpres mfg, 85-87, VPRES SOLID STATE TECHNOL, SARNOFF RES CTR, RCA LABS, 88- *Mem:* Am Phys Soc; Inst Elec & Electronics Engrs. *Res:* Semiconductor injection lasers and light-emitting diodes in III-V compounds; GaAs bipolar transistors; defects and device degradation; Zn-diffusion; direct-indirect effects; semiconductor contacts and electrolytic etching; heterojunction lasers; vapor-phase epitaxy. *Mailing Add:* Harris Corp Semiconductor Div PO Box 883 Melbourne FL 32901

NUESSLE, ALBERT CHRISTIAN, b Philadelphia, Pa, Feb 24, 15; m 40; c 4. TEXTILE CHEMISTRY. *Educ:* Univ Pa, BS, 36. *Prof Exp:* Textile res chemist, Joseph Bancroft & Sons Co, 37, asst supt finishing, 38-42, textile res chemist, 43-46; head, Textile Appl Lab, Rohm and Haas Co, 47-64, sr fibers res chemist, 65-76; RETIRED. *Mem:* Am Asn Textile Chemists & Colorists. *Res:* Textile finishing agents and auxiliaries, including surfactants; textile application processes and methods, including bleaching, dyeing and finishing; fiber, yarn, fabric properties. *Mailing Add:* 305 W Cnty Li Rd Hatboro PA 19040

NUESSLE, NOEL OLIVER, b St Louis, Mo, June 20, 28; m 56; c 5. PHARMACEUTICS. *Educ:* St Louis Col Pharm, BS, 49; Univ Fla, MS, 55, PhD(pharmaceut chem), 58. *Prof Exp:* Asst, Univ Fla, 54-57; from asst prof to assoc prof, 58-75, prof pharm & coordr externships, 75-90, EMER PROF, UNIV MO-KANSAS CITY, 90- *Mem:* Am Pharmaceut Asn; Acad Pharmaceut Sci; Sigma Xi; Am Asn Pharmaceut Scientists. *Res:* Pharmaceutical formulation; pharmaceutical processing; sterile products research; stability. *Mailing Add:* Dept Pharm Univ Mo 5100 Rockhill Rd Kansas City MO 64110

NUETZEL, JOHN ARLINGTON, b East St Louis, Ill, Feb 16, 25; m 46; c 6. INTERNAL MEDICINE. *Educ:* Wash Univ, MD, 47; Am Bd Internal Med, dipl, 54. *Prof Exp:* Intern, Univ Mich, 47-48; resident, US Vet Admin, Mo, 49-51; CLIN ASSOC PROF INTERNAL MED, ST LOUIS UNIV, 85-

Concurrent Pos: Fel hypertension, Wash Univ & Barnes Hosp, 48-49; med dir, St Mary's Health Ctr; mem courtesy staff, St John's Mercy Med Ctr & St Louis Univ Hosp. *Mem:* AMA; Am Heart Asn; fel Am Col Cardiol; fel Am Col Physicians. *Res:* Hypertension; cardiology. *Mailing Add:* 13309 Conway Rd St Louis MO 63141

NUFFIELD, EDWARD WILFRID, b Gretna, Man, Apr 13, 14; m 39; c 3. MINERALOGY. *Educ:* Univ BC, BA, 40; Univ Toronto, PhD(mineral), 44. *Prof Exp:* Asst, Geol Surv Can, 40-42; from lectr to prof, 43-79, assoc dean, Fac Arts & Sci, 62-64, chmn dept, 64-72, EMER PROF GEOL, UNIV TORONTO, 79- *Concurrent Pos:* Geologist, Ont Dept Mines, 49-51. *Honors & Awards:* Sr Award, Royal Soc Can, 63; Queen's Silver Jubilee Medal, 77; L G Berry Medal, 88. *Mem:* Fel Mineral Soc Am; Am Crystallog Asn; fel Royal Soc Can; Mineral Asn Can (pres, 56-58). *Res:* X-ray crystallography; crystal chemistry; determination of the crystal structures of sulphosalt minerals. *Mailing Add:* 1835 Morton Ave Apt 1603 Vancouver BC V6G 1V3 Can

NUGENT, CHARLES ARTER, JR, b Denver, Colo, Nov 18, 24; m 50; c 3. INTERNAL MEDICINE, ENDOCRINOLOGY. *Educ:* Yale Univ, MD, 51. *Prof Exp:* From intern to asst resident, New Haven Hosp, 51-53; resident, Col Med, Univ Utah, 54-56, from instr to assoc prof, 56-67; prof, Col Med, Univ Hawaii, 67-70; PROF MED, COL MED, UNIV ARIZ, 70- *Concurrent Pos:* Res fel, Col Med, Univ Utah, 54. *Mem:* Am Soc Clin Invest; Endocrine Soc. *Res:* Hypertension; radioimmunoassay of steroid; control of adrenal steroid secretion. *Mailing Add:* Dept Internal Med Univ Ariz Sch Med 1501 N Campbell Ave Tucson AZ 85724

NUGENT, GEORGE ROBERT, b Yonkers, NY, Feb 6, 21; m 47; c 5. NEUROSURGERY. *Educ:* Kenyon Col, AB, 50; Univ Cincinnati, MD, 53. *Prof Exp:* Instr neurosurg, Med Ctr, Duke Univ, 57-58; asst dir div neurosurg, Col Med, Univ Cincinnati, 58-61; from asst prof to assoc prof surg, 61-69, PROF SURG, DIV NEUROSURG, MED CTR, WVA UNIV, 69- *Concurrent Pos:* Chief neurosurg, Cincinnati Vet Admin Hosp, 58-61; chmn div, Wva Univ, 70-85. *Mem:* Am Asn Neurol Surg; Soc Neurol Surg; Cong Neurol Surgeons; Int Soc Res Stereoencephalotomy. *Res:* Treatment of trigeminal neuralgia; sterotaxic brain surgery; teaching methods; microneurosurgery; treatment of pain. *Mailing Add:* Dept Neurosurg Health Sci Ctr N Morgantown WV 26506

NUGENT, LEONARD JAMES, b Chicago, Ill, Oct 1, 30; m 52; c 4. CHEMICAL PHYSICS. *Educ:* Ill Inst Technol, BS, 54, MS, 57; Univ Wis, PhD(phys chem), 59. *Prof Exp:* Res assoc, Nat Res Coun, Nat Bur Standards, 58-59, physicist, 59-61; sr physicist, Gen Dynamics Corp, 61-62; sr physicist, Electro-Optical Systs, Inc, 62-66; sr staff scientist, Chem Div, Oak Ridge Nat Lab, 66-76; mem tech staff, Corp Technol, Inc, 76-79; PRES, ALLWEST TECHNOL, 79- *Mem:* Fel Am Phys Soc; Am Chem Soc; Am Asn Phys Teachers. *Res:* Molecular and atomic spectra; atmospheric, chemical and plasma physics; microwave components and subsystems and systems. *Mailing Add:* 1556 Halford Ave No 181 Santa Clara CA 95051

NUGENT, MAURICE JOSEPH, JR, b Salt Lake City, Utah, Dec 22, 37. ORGANIC CHEMISTRY. *Educ:* Univ Colo, BA, 61; Calif Inst Technol, PhD(chem), 65. *Prof Exp:* NIH res fel chem, Harvard Univ, 65-66; asst prof, 66-73, ASSOC PROF CHEM, TULANE UNIV, 73- *Concurrent Pos:* Vis assoc chem, Calif Inst Technol, 75-76. *Mem:* Am Chem Soc. *Res:* Mechanisms of enzymatic reactions; antibiotics and metabolic control reagents. *Mailing Add:* 7868 Apple Blossom Ln Sebastopol CA 95472

NUGENT, ROBERT CHARLES, b Jersey City, NJ, Sept 22, 36; m 58; c 3. GEOLOGY. *Educ:* Hofstra Univ, BA, 58; Univ Rochester, MS, 60; Northwestern Univ, PhD(geol), 67. *Prof Exp:* Develop geologist, Chevron Oil Co, Standard Oil Co Calif, La, 65-68; asst prof, State Univ NY, 68-70, chmn dept earth sci, 72-76, assoc prof geol, 70-80; SR RES SPECIALIST, SEISMIC STRATIGRAPHY, EXXON PROD RES CO, 80- *Mem:* Geol Soc Am; Soc Econ Paleont & Mineral; Am Asn Petrol Geologists. *Res:* Erosion and deposition along shores of Lake Ontario and the St Lawrence River; bottom sediments of Chaumont Bay, New York; spit and gravel bar formation at mouth of Salmon River, New York. *Mailing Add:* Exxon Prod Res Dept Rm 5T-654 PO Box 2189 Houston TX 77098

NUGENT, SHERWIN THOMAS, b St John's, Nfld, Sept 25, 38; m 62; c 2. ENGINEERING, PHYSICS. *Educ:* Mem Univ Nfld, BSc, 59; Tech Univ NS, BEng, 61; Univ Toronto, MASc, 63; Univ NB, PhD(elec eng), 67. *Prof Exp:* Engr, Maritime Tel & Tel Co Ltd, 61-62; lectr physics, Univ NB, 66-67; Nat Res Coun Can fel, Univ BC, 67-68; from asst prof to assoc prof eng physics, 68-89, PROF ENG, DALHOUSIE UNIV, 89- *Mem:* Inst Elec & Electronic Engrs. *Res:* Digital signal processing; control systems. *Mailing Add:* Dept Eng Dalhousie Univ Halifax NS B3H 3J5 Can

NUITE-BELLEVILLE, JO ANN, b Albany, NY, Oct 30, 45; m 80. PHARMACOLOGY, NEUROBIOLOGY. *Educ:* Col St Rose, BA, 67; Univ NC, Chapel Hill, PhD(pharmacol), 71. *Prof Exp:* Pharmacologist, Drug Abuse Ctr, NIMH/Nat Inst Drug Abuse, 71-74; asst prof pharmacol, Schs Med & Dent, Georgetown Univ, 74-; AT NB ASSOCS, ROCKVILLE, MD. *Concurrent Pos:* Res assoc cent nerv syst pharmacol, Univ NC, Chapel Hill, 71-72; mem, Drug Abuse Adv Comt. *Mem:* AAAS; NY Acad Sci; Int Narcotic Addiction Res Club; Sigma Xi. *Res:* Central nervous system pharmacology and neurobiology; pharmacological and toxicological effects of drugs on adult and developing organism, particularly psychoactive drugs methods of characterizing novel drugs and relating changes in biogenic amine, amine metabolism to possible mechanism of action. *Mailing Add:* NB Assocs 5104 Clavel Terr Rockville MD 20853

NUKI, KLAUS, b Vienna, Austria, May 5, 31; m 63; c 3. HISTOCHEMISTRY, PERIODONTOLOGY. *Educ:* Univ London, BDS, 55, PhD(histochem), 67; Univ Ill, Chicago, MS, 60. *Prof Exp:* House surgeon, Queen Victoria Hosp, East Grinstead, Eng, 55-56; res asst oral path, Univ Ill, Chicago, 57-60; lectr path & periodont, London Hosp Med Col, 60-67; res assoc periodont, Royal Dent Col, Denmark, 63-64; assoc prof dent, Univ Iowa, 67-71, prof dent & head dept oral biol & div histol & histochem, Col Dent, 71-75; prof periodont & assoc dean clin dent educ, 75-89, PROF ORAL DIAG, SCH DENT MED, UNIV CONN, 89- *Concurrent Pos:* Assoc chief of staff, John Dempsey Hosp, Univ Conn, 84-89; assoc ed, J Periodont Res, 72- *Mem:* Am Asn Dent Schs; Int Asn Dent Res; Royal Soc Med. *Res:* Bone resorption mechanisms; pathology; micro-chemistry of inflammation; microcirculation of gingiva; investigations of periodontium; dental implants. *Mailing Add:* Univ Conn Health Ctr Farmington CT 06032

NULL, HAROLD R, b Memphis, Tenn, May 16, 29; m 50; c 3. CHEMICAL ENGINEERING. *Educ:* Univ Tenn, BS, 50, MS, 51, PhD(chem eng), 55. *Prof Exp:* Res engr chem eng, E I du Pont de Nemours & Co, 56-57; asst prof, Univ Dayton, 57-59; res chem engr, Monsanto Co, 59-64, group leader, Eng Res, 64, eng supt, 64-69, eng fel, 69-77, sr eng fel, 77-87; consult chem engr, 87-89; SR TECHNOLOGIST, WSI TECHNOL INC, 90- *Concurrent Pos:* Consult, 57-59; affil prof, Wash Univ, 68- *Mem:* Am Inst Chem Engrs; Sigma Xi. *Res:* Thermodynamics; phase equilibrium; separations technology; liquid-liquid extraction; energy conservation. *Mailing Add:* 14214 Cypress Hill Dr Chesterfield MO 63017

NUMMY, WILLIAM RALPH, b Brooklyn, NY, Oct 2, 21; m 49; c 4. ORGANIC CHEMISTRY. *Educ:* Univ of the South, BS, 47; Univ Rochester, PhD(chem), 50. *Prof Exp:* Anal chemist, Magnus, Mabee & Reynard, Inc, 39-42; res chemist, Arnold, Hoffman & Co, 50-53; res chemist, Dow Chem Co, 53, group leader, 53-56, div leader, 56, asst dir phys res lab, 56-60, asst dir polymer res lab, 60-61, dir plastics dept, res labs, 61-64, mgr plastics develop & serv, 64-67, mgr plant sci bus, 67-68, mgr agr prod dept, 68-69, mgr life sci res, 69-70, dir cent res labs, 70-74, dir res & develop, Dow Lepetit Co, 74-76; dir pharmaceut res & develop, Dow Chem Co, 76-82; RETIRED. *Concurrent Pos:* Consult, res mgt & venture capital. *Mem:* Am Chem Soc. *Res:* Claisen rearrangement mechanism; high polymer synthesis; polyamides; polysulfides and plastic foams. *Mailing Add:* 711 W Meadowbrook Dr Midland MI 48640-3486

NUNAMAKER, RICHARD ALLAN, b Youngstown, Ohio, Aug 4, 51; m 78. ELECTRON MICROSCOPY, IMMUNOCYTO CHEMISTRY. *Educ:* Miami Univ, BA, 74; Univ Northern Colo, MA, 76; Univ Wyo, PhD(entom), 80. *Prof Exp:* Entomologist, Honey Bee Pesticides Dis Res Unit, 77-85, RES ENTOMOLOGIST, ARTHROPOD BORNE ANIMAL DIS RES LAB, AGR RES SERV, USDA, 85- *Concurrent Pos:* Asst adj prof plant sci, Univ Wyo, 82- *Mem:* Entom Soc Am; Int Bee Res Asn; Apimondia; Am Honey Producers Asn; NAm Benthological Soc. *Res:* Biochemical and physiological modifications in honey bees that have been exposed to pesticides, with particular emphasis to enzyme systems and cell ultrastructure; immunogold labelling of arboviruses, especially blue tongue virus and epizootic hemorrhagic disease virus in the vector, culicoides variipennis. *Mailing Add:* PO Box 3965 Univ Sta Laramie WY 82071-3965

NUNAN, CRAIG S(PENCER), b Medford, Ore, Dec 22, 18; m 48; c 4. ELECTRICAL ENGINEERING. *Educ:* Univ Calif, BS, 40, MS, 49. *Prof Exp:* Engr, Pac Gas & Elec Co, 40-41; engr, Elec Sect, Bur Ships, US Dept Navy, DC, 41-46; proj engr, Lawrence Radiation Lab, Univ Calif, 46-53; dir res, Chromatic TV Corp, 53-55; gen mgr, Radiation Div, 55-68, SYSTS SPECIALIST, VARIAN ASSOCS, 68- *Res:* Development and application of medical-electronic equipment for cancer therapy and medical diagnostics. *Mailing Add:* 26665 St Francis Dr Los Altos Hills CA 94022

NUNES, ANTHONY CHARLES, b New Bedford, Mass, Nov 1, 42; m 72; c 3. MAGNETISM, NEUTRON SCATTERING. *Educ:* Mass Inst Technol, BSc, 64, PhD(physics), 69. *Prof Exp:* Res assoc physics, Brookhaven Nat Lab, 69-71, res assoc biol, 71-74; physicist, Inst Laue-Langevin, Grenoble, France, 74-76; assoc prof, 76-82, PROF PHYSICS, UNIV RI, 82- *Concurrent Pos:* Consult small angle neutron scattering, Mass Inst Technol, 76-77; Res Corp grant, 77-; NSF grant, 78-86. *Mem:* Sigma Xi; Am Phys Soc; AAAS. *Res:* Thermal neutron and x-ray scattering applied to problems in solid state physics, magnetic colloids, amorphous magnets and protein structure; development of instrumentation for neutron scattering. *Mailing Add:* Dept Physics Univ RI Kingston RI 02881

NUNES, PAUL DONALD, b New Bedford, Mass, Aug 29, 44; div. GEOCHRONOLOGY. *Educ:* Tufts Univ, BS, 66; Univ Calif, Santa Barbara, PhD(geol), 70. *Prof Exp:* Asst I geochronology res, Swiss Fed Inst Technol, 70-72; nat res coun assoc, US Geol Surv, 72-73, geologist, 73-75; geologist, US Geol Surv, 73-75; res assoc geol, Royal Ont Mus, 75-81; ministry geochronologist, Ont Geol Surv, 75-81; CONSULT GEOSCI, 84- *Mem:* Am Geophys Union. *Res:* Used of uranium-thorium-lead and rubidium-strontium natural decay systems to probe and further clarify our understanding of the evolution of the earth and moon; collaborative investigations, 69-82. *Mailing Add:* 28 Jenny Lind St New Bedford MA 02740

NUNEZ, LOYS JOSEPH, b New Orleans, La, Mar 18, 26. CHEMISTRY, TOXICOLOGY. *Educ:* Tulane Univ, BS, 47; La State Univ, Baton Rouge, MS, 55, PhD(chem), 60. *Prof Exp:* Chemist, Cities Serv Refining Corp & Res & Develop Corp, 47-53, res chemist, 59-60; supvr, Phys Res Sect, Austin Labs, Jefferson Chem Co, Inc, Tex, 60-71; asst prof pharm, 72-77, asst prof dent, 72-83, head, Biomat Sect, Mat Sci Toxicol Labs, 71-85, ASSOC PROF PHARM, UNIV TENN CTR FOR HEALTH SCI, 77-, PROF DENT, 83- *Concurrent Pos:* Fulbright fel, Univ Tuebingen, WGer, 57-58; vis prof, Univ Regensburg, WGer, 84 & 87. *Mem:* Am Chem Soc; NY Acad Sci; Am Asn Adv Sci; Int Fedn Dent. *Res:* Dental biomaterials; toxicity of thermodegradation products of fabrics and plastics; chemical and polymer carcinogenesis; drug-plastic absorption diffusion; mutagenicity. *Mailing Add:* Col Dent Univ Tenn Ctr Health Sci Memphis TN 38163

NUNEZ, WILLIAM J, III, b New Orleans, La, Jan 17, 44; m 65; c 2. IMMUNOLOGY, MICROBIOLOGY. *Educ:* La State Univ, Baton Rouge, BS, 65, MS, 67; NTex State Univ, PhD(immunol), 70. *Prof Exp:* From asst prof to prof biol, Univ Detroit, 70-83, chmn dept, 73-83; DEAN ARTS & SCIS, MO WESTERN STATE COL, 83- *Concurrent Pos:* Tuberc & Health Soc Wayne County fel, 71-73, Univ Detroit, 71-, Nat Multiple Sclerosis Soc, Environ Protection Agency & Scholl Found res grants, 75; res fel, NASA-Ames, Stanford Univ, 77 & 78, vis scholar, NIH sr fac fel award, 79, 80 & 81. *Mem:* AAAS; Am Soc Microbiol; Am Thoracic Soc; Am Asn Univ Admin; Nat Acad Deans Conf; Coun Arts & Sci Cols. *Res:* Allergy, particularly cellular mechanisms of delayed hypersensitivity involved in tuberculin hypersensitivity and experimental autoallergic encephalomyelitis; passive transfer mechanisms of tuberculin, chemical contact dermatitis and experimental allergic encephalomyelitis; fluorescent antibody techniques for rapid and specific detection of human pollution indicators in water. *Mailing Add:* Dean Arts & Scis Mo Western State Col 4525 Downs Dr St Joseph MO 64507-2294

NUNKE, RONALD JOHN, b Kenosha, Wis, Mar 9, 26; m 53; c 3. MATHEMATICS. *Educ:* Univ Chicago, SB, 50, SM, 51, PhD(math), 55. *Prof Exp:* Instr math, Northwestern Univ, 53-54 & Yale Univ, 55-58; from asst prof to assoc prof, 58-69, PROF MATH, UNIV WASH, 69- *Mem:* Am Math Soc; Math Asn Am. *Res:* Abelian groups; homological algebra. *Mailing Add:* Dept of Math Univ of Wash Seattle WA 98195

NUNLEY, ROBERT GRAY, b Quinwood, WVa, Feb 5, 30; m 52; c 4. PHYCOLOGY. *Educ:* Marshall Col, AB, 52, MA, 53; WVa Univ, PhD(bot), 66. *Prof Exp:* Teacher pub schs, WVa, 53-54, teacher & prin, 57-61; teacher, Ohio, 56-57; instr biol, WVa Univ, 63-65; assoc prof, 65-68, PROF BIOL & HEAD DEPT, MORRIS HARVEY COL, 68- *Mem:* AAAS; Am Inst Biol Sci. *Res:* Ecology of freshwater benthic algae; taxonomy of the genus Trachelomonas in West Virginia. *Mailing Add:* Dept Biol Univ Charleston 2300 Maccorkle Ave SE Charleston WV 25304

NUNN, ARTHUR SHERMAN, JR, b Independence, Mo, Nov 9, 22; m 50; c 5. PHYSIOLOGY. *Educ:* Kans State Univ, BS, 55; Univ Iowa, MS, 59, PhD(physiol), 60. *Prof Exp:* Instr physiol, Sch Med, St Louis Univ, 60-62; from asst prof to assoc prof, Sch Med, Univ Miami, 62-67; from assoc prof to prof physiol, Sch Med, Ind Univ, Indianapolis, 67-87; RETIRED. *Concurrent Pos:* US Army res & develop fel, Univ Miami, 62-67. *Mem:* AAAS; Am Physiol Soc. *Res:* Membrane transport of sugars; the role of cations in the transmembrane movement of organic compounds; gastro-intestinal physiology. *Mailing Add:* 7445 Charrington Ct Indianapolis IN 46254

NUNN, DOROTHY MAE, b Cincinnati, Ohio. MICROBIOLOGY. *Educ:* Univ Cincinnati, BS, 55, PhD(microbiol), 62. *Prof Exp:* Asst prof microbiol, Univ Dayton, 62-65; assoc prof, E Tenn State Univ, 65-67; assoc prof biol, Univ Akron, 67-83; RETIRED. *Mem:* AAAS; Am Soc Microbiol; Brit Soc Gen Microbiol. *Res:* Bacterial physiology and bioenergetics; mechanism of action of bacterial toxins. *Mailing Add:* 90 Devon Lane Univ Akron Akron OH 44313

NUNN, ROBERT HARRY, b Tacoma, Wash, Nov 9, 33; m 55; c 3. DYNAMIC SYSTEMS, MODELLING AND SIMULATION. *Educ:* Univ Calif, Los Angeles, BS, 55, MS, 64; Univ Calif, Davis, PhD(mech eng), 67. *Prof Exp:* Naval aviator, US Naval Weapons Ctr, 55-60, res aerospace engr, 60-68; asst prof, 68-71, assoc prof, 71-80, chmn dept, 71-75, PROF MECH ENG, NAVAL POSTGRAD SCH, 80- *Concurrent Pos:* Dep sci dir, Off Naval Res, London, 75-77. *Mem:* Assoc fel Am Inst Aeronaut & Astronaut; Am Soc Mech Engrs; Soc Naval Architects & Marine Engrs. *Res:* Microprocessor control of servomechanisms; power systems; design optimization. *Mailing Add:* 1115 Melton Pl Pacific Grove CA 93950

NUNN, ROLAND CICERO, b Miami, Fla, July 24, 30; m 51; c 4. FUEL TECHNOLOGY, PHYSICAL CHEMISTRY. *Educ:* Duke Univ, BS, 52; Pa State Univ, PhD(fuel technol), 55. *Prof Exp:* Chemist, Esso Res & Eng Co, NJ, 55-59; mgr, 59-77, MGR SPEC PROD, CHEVRON RES CO, 77- *Mem:* Am Chem Soc; Sigma Xi; Tech Asn Pulp & Paper Indust; Am Soc Testing & Mat; Soc Coatings Technol. *Res:* Railroad and marine diesel fuels and lubricants; distillate burner fuels; liquified petroleum gas uses; special products. *Mailing Add:* 23 Las Piedras Orinda CA 94563

NUNN, WALTER M(ELROSE), JR, b New Orleans, La, Sept 16, 25; m 49. PHYSICS, MATHEMATICS. *Educ:* Tulane Univ, BSEE, 50; Okla State Univ, MSEE, 52; Univ Mich, PhD(elec eng), 61; Univ Ill, Urbana, MS, 69. *Prof Exp:* Instr elec eng, Okla State Univ, 50-52 & Rensselaer Polytech Inst, 52-54; res engr, Hughes Aircraft Co, Calif, 54-56; res assoc elec eng, Univ Mich, 56-60; asst prof, Univ Minn, Minneapolis, 60-63; prof, Tulane Univ, 63-65, res prof physics, Sch Med, 65-67, consult appln lasers to cancer res, 65-67; PROF ELEC ENG, FLA INST TECHNOL, 69-, DIR MICROWAVE LAB & ELECTROMAGNETIC CURRIC, 81- *Concurrent Pos:* Consult, Minneapolis-Honeywell Co, Minn, 60-63, Nat Res Coun-Nat Acad Sci, 63-64, US Dept Navy, 65-66, Harris Corp, Melbourne, Fla, 71, Redstone Arsenal, US Army Missile Comnd & DBA Systs, Inc, 78, 89-90; prin investr, NASA contract, 81; res scientist, DBA Systs Inc, Melbourne, Fla, 83-85; sr staff engr, Hughes Aircraft Co, Fullerton, Calif, 85-86; consult, Strategic Defense Initiative Orgn, (DOD), 88- *Mem:* Sr mem, Inst Elec & Electronic Engrs; Am Phys Soc. *Res:* Electromagnetic theory; microwave electronics; accelerator physics; electromagnetic theory, antennas, radar systems. *Mailing Add:* Dept of Elec Eng Fla Inst Technol 150 W University Blvd Melbourne FL 32901-6998

NUNNALLY, DAVID AMBROSE, b Memphis, Tenn, Sept 13, 34; m 56; c 5. ZOOLOGY, BIOLOGY. *Educ:* Univ of the South, BS, 56; Wash Univ, PhD(zool), 61. *Prof Exp:* From instr to asst prof, 60-65, ASSOC PROF BIOL, VANDERBILT UNIV, 65- *Mem:* AAAS; Am Soc Zool. *Res:* Physiological embryology; physiology of parasitic flatworms. *Mailing Add:* Dept Biol Vanderbilt Univ Nashville TN 37240

NUNNALLY, HUEY NEAL, b Atlanta, Ga, Dec 28, 44; m 68; c 2. INDUSTRIAL CONTROLS, ROBOTICS. *Educ:* Ga Inst Technol, BEE, 66, MSEE, 68, PhD(elec eng), 71. *Prof Exp:* Assoc aircraft engr, Lockheed-Ga Co, 65-66; res consult biomed eng, US Army Edgewood Arsenal, 71. *Concurrent Pos:* Asst res prof, Sch Med, Emory Univ, 71-74. *Mem:* Inst Elec & Electronics Engrs. *Res:* Auditory physiology; electrical stimulation of nervous system; electrical stimulation of muscular system; power systems; robotics. *Mailing Add:* Philip Morris Res Ctr PO Box 26583 Richmond VA 23261

NUNNALLY, NELSON RUDOLPH, b Monroe, Ga, Dec 24, 35; m 58; c 4. PHYSICAL GEOGRAPHY. *Educ:* Univ Ga, BS, 58, MA, 61; Univ Ill, PhD(geog), 65. *Prof Exp:* From instr to assoc prof geog, E Tenn State Univ, 60-67; assoc prof, Fla Atlantic Univ, 67-68; asst prof, Univ Ill, 68-71; assoc prof, Univ Okla, 71-74; assoc prof geog, 74-81, ASSOC PROF GEOG & EARTH SCI, UNIV NC, CHARLOTTE, 81- *Mem:* Asn Am Geog; Am Soc Photogram. *Res:* Physical geography; remote sensing. *Mailing Add:* Dept Geog & Earth Sci Univ of NC Univ Sta Charlotte NC 28223

NUNNALLY, STEPHENS WATSON, b Gadsden, Ala, Nov 30, 27; m 57; c 3. CONSTRUCTION ENGINEERING, CIVIL ENGINEERING. *Educ:* US Mil Acad, BS, 49; Northwestern Univ, MS, 58, PhD(civil eng), 66. *Prof Exp:* Officer, US Army Corps Engrs, 49-70; asst prof construct eng, Univ Fla, 71-75; from assoc prof to prof construct eng, NC State Univ, 75-89; RETIRED. *Concurrent Pos:* Consult, Res Triangle Inst, 77-; vis lectr, Univ Stellenbosch, SAfrica, 80; dir training prog, M Binladin Orgn, Saudi Arabia, 81. *Mem:* Am Soc Civil Engrs; Am Soc Eng Educ; Soc Am Mil Engrs; Am Rd & Transp Builders Asn. *Res:* Construction equipment management; project planning and control; construction productivity improvement. *Mailing Add:* 474 St Lucia Ct Satellite Beach FL 32937

NUNNELEY, SARAH A, b New York, NY, Aug 26, 41. AEROSPACE MEDICINE. *Educ:* Mt Holyoke Col, BA, 63; Univ Minn, MD, 67. *Prof Exp:* RES MED OFFICER, US AIR FORCE SCH AEROSPACE MED, BROOKS AFB, 74- *Concurrent Pos:* Dept of Physiol, State Univ NY, Buffalo, 70-74. *Mem:* Aerospace Med Asn; Am Physiol Soc; Am Col Sports Med. *Mailing Add:* US Air Force Sch Aerospace Med Brooks AFB San Antonio TX 78235

NUNNEY, LEONARD PETER, b Greenford, Middlesex, Eng, May 31, 49. EVOLUTIONARY THEORY, POPULATION DYNAMICS. *Educ:* Univ Sussex, BSc, 70; Univ Nottingham, PhD(genetics), 77. *Prof Exp:* Demonstr zool, Univ Edinburgh, 73-78; res fel, Princeton Univ, 78-80; vis lectr, 80-81, asst prof, 81-87, ASSOC PROF BIOL, UNIV CALIF, RIVERSIDE, 87- *Mem:* Am Soc Naturalists; Soc Study Evolution; Genetics Soc Am; Ecol Soc Am. *Res:* Population models; competition in Drosophila; levels of selection; sex ratio in parasitoids; molecular evolution, particularly in relation to codon usage. *Mailing Add:* Dept Biol Univ Calif Riverside CA 92521

NUR, AMOS M, b Haifa, Israel, Feb 9, 38; m 68. EXPLORATION GEOPHYSICS, TECTONOPHYSICS. *Educ:* Hebrew Univ Jerusalem, BS, 62; Mass Inst Technol, PhD(geophys), 69. *Prof Exp:* Res assoc geophys, Mass Inst Technol, 69-70; from asst prof to prof, 70-86, CHMN & PROF, GEOPHYSICS DEPT, STANFORD UNIV, 86- *Concurrent Pos:* Sloan fel, Stanford Univ, 72-74; mem earth sci adv bd, NSF, 74-77; vis prof, Weizmann Inst Sci, 75. *Honors & Awards:* Maclwane Award, Am Geophys Union, 74; Newcomb Cleveland Prize, AAAS, 75. *Mem:* Fel Am Geophys Union; Seismol Soc Am; Soc Petrol Engrs; fel Geol Soc Am. *Res:* Tectonophysics; rock mechanics; physical hydrology; earthquake mechanics; exploration geophysics. *Mailing Add:* Dept Geophysics Stanford Univ Stanford CA 94305-2215

NUR, HUSSAIN SAYID, b Mahmoudia, Iraq, July 1, 39; m 69. MATHEMATICAL ANALYSIS. *Educ:* Univ Baghdad, BSc, 61; Univ Calif, Berkeley, MA, 64, PhD(math), 67. *Prof Exp:* Teaching asst math, Univ Baghdad, 61-62; from asst prof to assoc prof, 67-73, PROF MATH, FRESNO STATE UNIV, 73- *Mem:* Am Math Soc. *Res:* Singular perturbation of differential equations. *Mailing Add:* Dept Math Calif State Univ 6241 N Maple Ave Fresno CA 93740

NUR, UZI, b Ein Harod, Israel, June 28, 28; m 52; c 2. CYTOGENETICS. *Educ:* Hebrew Univ, Israel, MSc, 58; Univ Calif, Berkeley, PhD(genetics), 62. *Prof Exp:* USPHS trainee, 62-63, from asst prof to prof, 63-90, EMER PROF BIOL, UNIV ROCHESTER, 90- *Mem:* Genetics Soc Am; Am Soc Naturalists; Soc Study Evolution. *Res:* Evolution of sexual and parthenogenetic scale insects; maintenance of B chromosomes exhibiting meiotic drive. *Mailing Add:* Dept Biol/Hutchison Univ Rochester Rochester NY 14627

NURMIA, MATTI JUHANI, b Rauma, Finland, Aug 26, 30. NUCLEAR PHYSICS, SEISMOLOGY. *Educ:* Univ Helsinki, MA, 52, PhD(physics), 58. *Prof Exp:* Asst prof physics, Univ Ark, 57-58; assoc prof, Univ Helsinki, 58-62 & 63-66; vis prof, Okla State Univ, 62-63; scientist nuclear chem, Lawrence Radiation Lab, 66-78; SR SCIENTIST NUCLEAR SCI, LAWRENCE BERKELEY LAB, 78- *Concurrent Pos:* Seismologist, Univ Helsinki, 58-62; consult, Finnish Atomic Energy Co, 58-62; vis scientist, Lawrence Radiation Lab, 65-66. *Res:* Synthesis and study of new chemical elements; natural radioactivities. *Mailing Add:* Dept Nuclear Sci Lawrence Berkeley Lab Berkeley CA 94720

NURNBERGER, JOHN IGNATIUS, SR, b Chicago, Ill, Apr 9, 16; m 43; c 5. CYTOLOGY, NEUROSCIENCE. *Educ:* Loyola Univ, Ill, BS, 38; Northwestern Univ, MS, 42, MD, 43. *Hon Degrees:* DSc, Marion Col, Indianapolis. *Prof Exp:* Res neurologist, Neurol Inst, NY, 46-48; res fel, Med Nobel Inst Cell Res & Genetics, Stockholm, Sweden, 49-50; asst prof med & psychiat, Sch Med, Yale Univ, 53-56; actg dean, Sch Med, 63-64, prof psychiat & chmn dept & dir, Inst Psychiat Res, 56-74, DISTINGUISHED PROF PSYCHIAT & CHMN EXEC BD, INST PSYCHIAT RES, MED

CTR, IND UNIV, INDIANAPOLIS, 74- *Concurrent Pos:* Psychiatrist, Inst Living, 48-49, res assoc, 50-56, dir residency training, 50-56; mem, Ment Health Study Sect, NIMH, 59-63, mem exp psychol study sect, USPHS, 60-63; mem bd dir, Found Fund Res Psychiat, 72- *Mem:* Soc Biol Psychiat (pres, 70); Histochem Soc; fel Am Psychiat Asn; Asn Res Nerv & Ment Dis (pres, 71); Am Neurol Asn; Am Asn Chairmen Depts Psychiat (pres, 67). *Res:* Cytochemical and biochemical studies in nitrogen metabolism; pastoral and clinical psychiatry; neural development of neurotransmitters. *Mailing Add:* Inst of Psychiat Res 791 Union Dr Indianapolis IN 46223

NURSALL, JOHN RALPH, b Regina, Sask, Dec 25, 25; m 53; c 3. ZOOLOGY. *Educ:* Univ Sask, BA, 47, MA, 49; Univ Wis, PhD(zool), 53. *Prof Exp:* Instr biol, Univ Sask, 48-49; asst zool, Univ Wis, 49-53; lectr, 53-55, from asst prof to assoc prof, 55-64, chmn dept, 64-69 & 74-78, PROF ZOOL, UNIV ALTA, 64- *Concurrent Pos:* Nuffield travel grant, Gt Brit, 54, Nuffield fel, 62-63. *Mem:* AAAS; Am Soc Ichthyol & Herpet; Soc Study Evolution; Can Soc Zool (pres, 82-83); Am Soc Zoologists; Biol Coun Can (pres 84-85); Zool Soc London. *Res:* Morphology of fish; paleontology; fish behavior. *Mailing Add:* PO Box 37 Whaletown BC V0P 1Z0 Can

NUSBICKEL, EDWARD M, JR, b Philadelphia, Pa, Apr 30, 30; m 54; c 2. ULTRASONIC INSPECTION, REAL TIME DISTRIBUTED CONTROL-STEEL MILLS. *Educ:* Drexel Univ, BSEE, 62. *Prof Exp:* Engr, Burroughs Corp GVL, 60-64; engr, Homer Res Labs, 64-82, SR ENGR, BETHLEHEM STEEL CORP, SPARROWS PT, MD, 82- *Mem:* Am Iron & Steel Inst. *Res:* Nondestructive testing; ultrasonic testing; steel rolling mills; process control; hydraulic servo-control; automatic gauge control; vibration testing and evaluation; awarded five patents. *Mailing Add:* 283 Bowline Rd Severna Park MD 21146

NUSIM, STANLEY HERBERT, b New York, NY, Oct 2, 35; m 60; c 2. CHEMICAL ENGINEERING. *Educ:* City Col New York, BChE, 57; NY Univ, MChE, 60, PhD(chem eng), 67. *Prof Exp:* Asst res engr, Chem Eng, Battelle Mem Inst, Ohio, 56; res chem engr, Chem Eng Res & Develop Dept, Merck Sharp & Dohme Res Labs, 57-60, sr chem engr, 60-64, eng assoc, 66-68, Sect mgr, 68-70, mgr tech serv dept, Merck Chem Div, 70-73, mgr mfg, Merck Chem Mfg Div, 73-81, dir subsid projs, 81-82, exec dir tech opers, Merck Sharp & Dohme Intern Div, 82-83, exec dir opers, Latin Am, Far east, Near East, Merck Sharp & Dohme Intern Div, 83-88, EXEC DIR LICENSEE, LATIN AM, ASIA OPERS, MERCK PHARM MFG DIV, MERCK & CO INC, 89- *Mem:* Am Chem Soc; Am Inst Chem Engrs; NY Acad Sci. *Res:* Synthetic organic process research and development with emphasis in the areas of kinetics, catalysis and diffusional operations. *Mailing Add:* 454 Prospect Ave Apt 165 West Orange NJ 07052-4192

NUSSBAUM, ADOLF EDWARD, b Rheydt, Ger, Jan 10, 25; nat US; m 57; c 2. MATHEMATICS. *Educ:* Columbia Univ, MA, 50, PhD(math), 57. *Prof Exp:* Lectr, Columbia Col, 51-52; staff mem, Electronic Comput Proj, Inst Advan Study, 52-53; instr math, Univ Conn, 53-55; from instr to asst prof, Rensselaer Polytech Inst, 55-58; from asst prof to assoc prof, 58-66, PROF MATH, WASHINGTON UNIV, 66- *Concurrent Pos:* Mem, Inst Advan Study & NSF fel, 62-63; vis scholar, Stanford Univ, 67-68. *Mem:* AAAS; Am Math Soc. *Res:* Analysis and functional analysis. *Mailing Add:* Dept of Math Washington Univ St Louis MO 63130

NUSSBAUM, ALEXANDER LEOPOLD, b Leipzig, Ger, Dec 30, 25; nat US; m 57; c 3. ORGANIC CHEMISTRY, BIOCHEMISTRY. *Educ:* City Col New York, BS, 48; Purdue Univ, MS, 50; Wayne State Univ, PhD(chem), 54. *Prof Exp:* Vis investr, Inst Chem, Nat Univ Mex, 50; chemist, Syntex, SAm, 50-51 & Worcester Found Exp Biol, 54-55; sr chemist, Schering Corp, 55-65; group chief, Hoffmann-La Roche, Inc, 65-76; DIR DEPT BIOORG CHEM, BOSTON BIOMED RES INST, 76-; ASSOC PROF, DEPT BIOL CHEM & MOLECULAR PHARMACOL, HARVARD MED SCH, 77- *Concurrent Pos:* Res assoc, Med Sch, Stanford Univ, 61-63; vis lectr, Stevens Inst Technol, 64-65; assoc res prof, Sch Med, Univ Md; vis lectr, Fairleigh Dickenson Univ, 74-75; vis prof, Dept Biol Chem, Harvard Med Sch, 75. *Mem:* AAAS; Am Chem Soc; NY Acad Sci; Am Soc Biol Chemists; fel Am Inst Chem. *Res:* Chemistry of natural products, especially steroids and nucleotides & peptides; photochemistry; nucleic acid biochemistry; virology; molecular biology of cancer viruses. *Mailing Add:* Dept Biol Chem & Molecular Pharmacol Harvard Univ Med Sch Boston MA 02115

NUSSBAUM, ALLEN, b Phila, Pa, Aug 22, 19; m 45; c 4. SOLID STATE PHYSICS. *Educ:* Univ Pa, BA, 39, MA, 40, PhD(physics), 54. *Prof Exp:* Physicist, Honeywell Inc, 54-62; PROF ELEC ENG & DIR GRAD STUDY, UNIV MINN, MINNEAPOLIS, 62- *Mem:* Am Phys Soc; Inst Elec & Electronics Engrs; Brit Inst Physics. *Res:* Geometrical optics; semiconducting materials device development; theory of junction diodes; electromagnetic and quantum properties of materials. *Mailing Add:* Dept of Elec Eng 139 Elec Eng Univ of Minn 123 Church St SE Minneapolis MN 55455

NUSSBAUM, ELMER, b Monroe, Ind, Sept 2, 20; m 49; c 4. BIOPHYSICS. *Educ:* Taylor Univ, BA, 49; Ball State Univ, MA, 52; Univ Rochester, PhD(radiation biol), 57. *Prof Exp:* instr physics, Taylor Univ, 49-52, asst prof, 53; res assoc biophys, Univ Rochester, 53-57; from assoc prof to prof physics, dir res, 61-85, chmn sci div, 68-85, EMER PROF PHYSICS & PROF IN RESIDENCE, TAYLOR SCI UNIV, 85- *Concurrent Pos:* Consult, Oak Ridge Assoc Univs, 60-85, sr scientist, 62-63; sabbatical leave, Oak Ridge Nat Lab, 79. *Mem:* AAAS; Am Asn Physics Teachers; Sigma Xi; Health Physics Soc. *Res:* Solubility of radon in fatty acids and body tissues; diffusion of radon through semipermeable materials; radionuclides in the biosphere; physics; environmental radioactivity; radiation detectors; solar energy; health physics. *Mailing Add:* Dept of Physics Taylor Univ Upland IN 46989

NUSSBAUM, MIRKO, b Belgrade, Yugoslavia, July 24, 30; US citizen; div. EXPERIMENTAL HIGH ENERGY PHYSICS. *Educ:* Rutgers Univ, AB, 54; Univ Chicago, SM, 56; Johns Hopkins Univ, PhD(physics), 62. *Prof Exp:* Sr physicist, Martin Co, Md, 56-60; instr physics, Johns Hopkins Univ, 61-62; res assoc, Columbia Univ, 62-64; asst prof, Univ Pa, 64-69; assoc prof, 69-71, PROF PHYSICS, UNIV CINCINNATI, 71- *Mem:* Fel Am Phys Soc. *Res:* Fundamental experimental particle research. *Mailing Add:* Dept Physics Univ Cincinnati 11 Cincinnati OH 45221-0011

NUSSBAUM, NOEL SIDNEY, b Brooklyn, NY, Jan 26, 35; m 57; c 3. PHYSIOLOGY, ENDOCRINOLOGY. *Educ:* Brooklyn Col, BA, 56; Williams Col, MA, 58; Yale Univ, PhD(biol), 64. *Prof Exp:* From instr to asst prof biol, Bowdoin Col, 63-65; from asst prof to assoc prof biol, 65-75, ASSOC PROF PHYSIOL, SCH MED, WRIGHT STATE UNIV, 75- *Concurrent Pos:* Consult, Aerospace Med Res Lab, US Air Force, Wright-Patterson AFB, 74- *Mem:* Am Soc Zool; Am Soc Cell Biol; Orthopaedic Res Soc. *Res:* Vertebrate physiology and morphogenesis; mineral metabolism; ultrastructure of calcified tissue; biochemical and biomechanical response of dense connective tissue to stress. *Mailing Add:* Dept Physiol & Biophysics Wright State Univ Sch Med PO Box 927 Dayton OH 45401

NUSSBAUM, ROGER DAVID, b Philadelphia, Pa, Jan 29, 44; m 66; c 2. MATHEMATICAL ANALYSIS. *Educ:* Harvard Univ, AB, 65; Univ Chicago, PhD(math), 69. *Prof Exp:* Asst prof, 69-73, assoc prof, 73-77, PROF MATH, RUTGERS UNIV, NEW BRUNSWICK, 77- *Mem:* Am Math Soc. *Res:* Nonlinear functional analysis, particularly fixed point theorems and their applications to nonlinear problems. *Mailing Add:* Dept Math Rutgers Univ New Brunswick NJ 08903

NUSSBAUM, RONALD ARCHIE, b Rupert, Idaho, Feb 9, 42. HERPETOLOGY. *Educ:* Univ Idaho, BS, 67; Cent Wash State Col, MS, 68; Ore State Univ, PhD(zool), 72. *Prof Exp:* Res assoc ecosyst anal, Sch Forestry, Ore State Univ, 72-73; asst prof & asst cur, 74-80, ASSOC PROF & ASSOC CUR, DIV BIOL SCI & MUS ZOOL, UNIV MICH, ANN ARBOR, 80- *Concurrent Pos:* Assoc ed, Systematic Zoology & J Morphology. *Mem:* Soc Syst Zoologists; Am Soc Naturalists; Am Soc Ichthyologists & Herpetologists; Ecol Soc Am; Soc Study Evolution. *Res:* Evolution, systematics, life history and ecology of amphibians and reptiles; evolution of life history strategies; island biogeography; vertebrate morphology; cytotaxonomy. *Mailing Add:* Div Biol Sci & Mus Zool Univ Mich Main Campus Ann Arbor MI 48109

NUSSBAUM, RUDI HANS, b Furth, Ger, Mar 21, 22; nat US; m 47; c 3. RADIATION HEALTH EFFECTS. *Educ:* Univ Amsterdam, PhD(exp physics), 54. *Prof Exp:* Res assoc nuclear physics, Inst Nuclear Res, Univ Amsterdam, 52-54; UNESCO fel, Nuclear Physics Lab, Liverpool, Eng, 54-55; res assoc, Ind Univ, 55-56; European Orgn Nuclear Res sr fel, Geneva, 56-57; asst prof physics, Univ Calif, 57-59; from assoc prof to prof, 59-87, EMER PROF PHYSICS, PORTLAND STATE UNIV, 88- *Concurrent Pos:* Vis prof, Univ Wash, 65-66; consult, Tektronix Inc; exchange prof, Univ Canterbury, NZ, 71-72; vis prof, Univ Wash, 74, Univ Groningen, Leiden, Delft, Neth, Duisburg, Konstanz, Ger; radiation safety officer, 74-84. *Mem:* Am Phys Soc; Am Asn Physics Teachers; Netherlands Phys Soc; Am Fedn Scientists; Union Concerned Scientists. *Res:* Mossbauer effect applied to impurity lattice dynamics; study of adsorbed monolayers of molecules on graphite surfaces; health effects of low level ionizing radiation. *Mailing Add:* Dept Physics Portland State Univ PO Box 751 Portland OR 97207-0751

NUSSENBAUM, SIEGFRIED FRED, b Vienna, Austria, Nov 21, 19; nat US; m 51; c 2. BIOCHEMISTRY, ORGANIC CHEMISTRY. *Educ:* Univ Calif, BS, 41, MA, 48, PhD(biochem), 51. *Prof Exp:* Anal chemist, Manganese Ore Co & Pan Am Eng Co, 42-43, asst chief chemist, 43-45; chemist, Nat Lead Co, 45; asst, Univ Calif, 48-51, fel, 51-52; from instr to prof chem, Calif State Univ, Sacramento, 52-90, chair, Dept Chem, 58-65; RETIRED. *Concurrent Pos:* Consult biochemist, Sacramento County Hosp, 58-69; dir Master Clin Lab Sci, Univ Cal, San Francisco, 69-87; lecter clin chem, Univ Calif, Davis Med Ctr, 70- *Mem:* AAAS; Am Chem Soc; Nat Acad Clin Biochem; Am Asn Clin Chem. *Res:* Polysaccharide metabolism and chemistry; enzyme, analytical and clinical chemistry. *Mailing Add:* Dept Chem Sacramento CA 95819

NUSSENBLATT, ROBERT BURTON, b New York, NY, June 10, 48; m; c 3. IMMUNOLOGY. *Educ:* City Col NY, BS, 68; State Col NY, MD, 72; Am Bd Ophthal, dipl. *Hon Degrees:* Dr, Bar-Ilan Univ, 88. *Prof Exp:* Med intern, Bellevue Hosp, NY, 72-73, residency internal med, 73-74; teaching asst med, NY Univ Med Ctr, 73-74, residency ophthal, 74-77; clin assoc, Nat Eye Inst, NIH, 77-79, sr staff ophthalmologist, 79-81, chief, Clin Ophthalmic Immunol Sect, 81-87, dept clin dir, 84-87, HEAD, LAB IMMUNOL, NAT EYE INST, NIH, 86-, CLIN DIR, 87- *Concurrent Pos:* Mem, Med Bd Comt, Clin Ctr, NIH, 86-; chmn, Surg Admin Comt, Clin Ctr, NIH, 87-88. *Honors & Awards:* Charles H May Mem Lectr, NY Acad Sci, 87; Mark J Schoenberg Mem Lectr, 90. *Mem:* Sigma Xi; fel Am Acad Ophthal; Am Asn Immunologists; Am Asn Ophthal; Asn Res Vision & Ophthal. *Res:* Ophthalmic immunology. *Mailing Add:* NIH Nat Eye Inst Lab Immunol Bldg 10 Rm 10N202 Bethesda MD 20892

NUSSENZVEIG, HERCH MOYSES, b Sao Paulo, Brazil, Jan 16, 33; m 62; c 3. SCATTERING THEORY, QUANTUM OPTICS. *Educ:* Univ Sao Paulo, BSc, 54, PhD(physics), 57. *Prof Exp:* Asst prof theoret physics, Univ Sao Paulo, 56-57; from asst prof to prof, Brazilian Ctr Physics Res, 57-68; vis prof, Univ Rochester, 65-68, sr res assoc & prof physics, 68-75; prof physics, Univ Sao Paulo, 75-82, dir, Inst Physics, 78- 82; PROF PHYSICS, PONTIFICAL CATH UNIV, RIO DE JANEIRO, 82- *Concurrent Pos:* Nat Res Coun Brazil res fels, Theoret Physics Inst, State Univ Utrecht, 60-61, Dept Math Physics, Univ Birmingham, 61 & Inst Theoret Physics, Swiss Fed Inst Technol, 61; vis mem, Courant Inst Math Sci, NY Univ, 63-64 & Inst Adv Study, 64-65; vis prof, Univ Paris, Orsay, 73; vis scientist, Nat Ctr Atmospheric Res, 79-80; vis lectr, Col France, Paris, 86; vis scientist, NASA, 89-90. *Honors & Awards:* Max Born Award, 86; Elliott Montroll Lectr, Univ Rochester, 88. *Mem:* Am Phys Soc; Asn Math Physics; NY Acad Sci; Brazilian Phys Soc (pres, 81-); Brazilian Acad Sci; fel Optical Soc Am. *Res:* Quantum optics; scattering theory; dispersion relations; complex angular momentum theory. *Mailing Add:* Dept Physics PUC RJ Rua Marques Sao Vincente 225 Rio de Janeiro 22453 Brazil

NUSSENZWEIG, VICTOR, b Sao Paulo, Brazil, Nov 2, 28; m 54; c 3. IMMUNOPARASITOLOGY. *Educ:* Univ Sao Paulo, MD, 53, PhD(parasitol), 57. *Prof Exp:* Asst prof parasitol, Univ Sao Paulo, 53-63; res fel immunochem, Pasteur Inst, Paris, 58-60; res assoc microbiol & immunol, Escola Paulista de Med, Sao Paulo, 61-62; res assoc, 63-65, asst prof, 65-67, assoc prof, 67-71, PROF PATH & IMMUNOL, SCH MED, NEW YORK UNIV, 71- *Concurrent Pos:* Chmn, Chagas Dis Steering Comt, WHO, 75-; prin investr grants, NIH, World Health Orgn & Rockefeller Found, 65-; appointee, Comt Immunodiagnosis Cancer, NIH, 70-73, Organizing Comt, Int Complement Workshop, 79-; ed, J Immunol Methods & Contemp Top Immunol, 78. *Mem:* Am Asn Immunologists; Harvey Soc; Brazilian Soc Immunologists. *Res:* Immunology especially regulatory mechanisms of activation of the complement system and membrane receptors for complement; parasitology especially immune response and protective immunity to parasites, in particular malaria antigens. *Mailing Add:* Dept Path 550 First Ave New York NY 10016

NUSSER, WILFORD LEE, b Sylvia, Kans, Oct 6, 24; m 43; c 5. PARASITOLOGY, PHYSIOLOGY. *Educ:* Bethel Col, Kans, BA, 49; Kans State Univ, MS, 50; Iowa State Univ, PhD(parasitol), 58. *Prof Exp:* Instr microbiol, Col Osteop Med & Surg, 54-57, asst prof physiol, 57-60, actg head dept, 58-60; res fel neuroanat, Emory Univ, 60-63; prof physiol & head dept, Col Osteop Med & Surg, 63-66; grants assoc, 66-67, dir arthritis & orthop prog, Nat Inst Arthritis, Metab & Digestive Dis, 67-74; chief scientist, Prog Br, Nat Eye Inst, NIH, 74-77; assoc dir, Entramaural progs, Nat Inst Environ Health Sci, 77-; RETIRED. *Res:* Electro Electron microscopy; Wallerian degeneration and regeneration of peripheral nerves; endocrinology; site of production of the erythropoietic hormone; neuroanatomy. *Mailing Add:* 4933 Hermitage Dr Raleigh NC 27612

NUSSMANN, DAVID GEORGE, b Burlington, Iowa, May 8, 37. GEOLOGY, STRATIGRAPHIC GEOPHYSICS. *Educ:* Harvard Univ, AB, 59; Univ Mich, MA, 61, PhD(geol, geochem), 65. *Prof Exp:* Geologist, Explor & Prod Res Div, Shell Develop Co, Tex, 64-70, staff geologist, Shell Oil Co, La, 70-74, geol engr, 74-76, sr staff engr, Explor & Prod Res Div, 76-80, mgr geol res, 80-86, mgr interpretive data processing, 86-89, MGR, GEOL-TECH SERV, SHELL DEVELOP CO, 89- *Mem:* Am Asn Petrol Geologists; Geol Soc Am; Soc Petrol Eng; Geochem Soc; Sigma Xi. *Res:* Sediment geochemistry; geostatistics; petroleum geology. *Mailing Add:* Shell Oil Co PO Box 2463 Houston TX 77252-2463

NUSYNOWITZ, MARTIN LAWRENCE, b New York, NY, July 21, 33; m 55; c 3. NUCLEAR MEDICINE, ENDOCRINOLOGY. *Educ:* NY Univ, BA, 54; State Univ NY, MD, 58. *Prof Exp:* Intern med, Letterman Gen Hosp, San Francisco, Med Corps, US Army, 58-59, resident internal med, Tripler Gen Hosp, Honolulu, Hawaii, 59-62, chief radioisotope-endocrine serv, 62-63, chief nuclear-med endocrine serv & dept med res & develop, William Beaumont Gen Hosp, 65-77; prof, Univ Tex Health Sci Ctr, San Antonio, 77-82, PROF NUCLEAR MED & HEAD DIV, UNIV TEX MED BR, GALVESTON, 82- *Concurrent Pos:* Fel mil med & allied sci, Walter Reed Army Inst Res, Washington, DC, 63-64; fel endocrinol & metab, Sch Med, Univ Calif, San Francisco, 64-65; Dorothy Hutton scholar endocrinol, 68; consult, Surgeon Gen, US Army, 71-77. *Mem:* Soc Nuclear Med; Endocrine Soc; fel Am Col Physicians; fel Am Col Nuclear Physicians; Am Fedn Clin Res. *Res:* In vitro thyroid function tests; clinical aspects of thyroid disease; cardiovascular nuclear medicine; diagnostic nuclear medical techniques in clinical problems. *Mailing Add:* Div of Nuclear Med Univ Tex Med Br Galveston TX 77550-2780

NUTAKKI, DHARMA RAO, b Vizayawada, Andhra Pradesh, India, July 18, 37; m 58; c 3. ELECTRICAL ENGINEERING. *Educ:* Andhra Univ, India, BE, 59; Indian Inst Sci, ME, 61, PhD(elec eng), 65. *Prof Exp:* Lectr elec eng, Indian Inst Sci, 62-67; from asst prof to assoc prof, 67-76, PROF ELEC ENG, UNIV CALGARY, 76- *Mem:* Inst Elec & Electronics Engrs. *Res:* Power system stability and analysis; sensitivity analysis; optimal control; network diakoptics; switching surges; state estimation; spreadsheet and database management in power systems; artificial intelligence and expert system applications in power systems. *Mailing Add:* Dept Elec Eng Univ Calgary Calgary AB T2N 1N4 Can

NUTE, C THOMAS, b Troy, Pa, Dec 12, 45; m 68; c 3. COMPUTER ENGINEERING. *Educ:* Univ Calif, San Diego, BA, 68; Tex A&M Univ, MS, 70, PhD(comput sci), 77. *Prof Exp:* Systs analyst space tracking officer, US Air Force, 68-75; instr comput sci, Tex A&M Univ, 75-77; asst prof comput eng, Case Western Reserve Univ, 77-; ASST PROF COMPUT SCI, TEXAS CHRISTIAN UNIV- *Concurrent Pos:* Consult, Bailey Control Co, 78-, Nordson Corp, 78- & Data Basics, Inc, 78- *Mem:* Asn Comput Mach; Inst Elec & Electronics Engrs. *Res:* Computer program behavior; performance measurement and evaluation. *Mailing Add:* Comput Sci Dept Texas Christian Univ PO Box 32886 Ft Worth TX 76129

NUTE, PETER ERIC, b Manchester, NH, Nov 20, 38; m 61; c 1. PHYSICAL ANTHROPOLOGY, BIOCHEMICAL GENETICS. *Educ:* Yale Univ, BS, 60; Duke Univ, PhD(anat), 69. *Prof Exp:* NIH fel med genetics, 69-71, spec fel med & hemat, 71-72, lectr anthrop, 72-74, actg asst prof, 75, from asst prof to assoc prof, 75-84, PROF ANTHROP & ADJ PROF MED, UNIV WASH, 84- *Concurrent Pos:* Res affil, Regional Primate Res Ctr, Univ Wash, 72-, sr fel med, 72-73, res asst prof, 73-77, adj assoc prof, 77-; co-prin investr, USPHS contract, 74-76, prin investr, USPHS grant, 77-80. *Mem:* AAAS; Am Asn Phys Anthropologists; NY Acad Sci; Am Soc Human Genetics; Pop Ref Bur; Human Biol Coun. *Res:* Molecular genetics and evolution, especially as related to medicine and primate evolution. *Mailing Add:* Dept of Anthrop Univ of Wash Seattle WA 98195

NUTLEY, HUGH, b Tacoma, Wash, Jan 30, 32; m 55; c 6. NUCLEAR PHYSICS. *Educ:* Mass Inst Technol, SB, 54; Univ Wash, PhD(physics), 60, BA, 73, MA, 74, MS, 76. *Prof Exp:* Res specialist, Radiation Effects Lab, Boeing Co, 62-66; from asst prof to prof physics, 66-87, PROF ELEC ENG, SEATTLE PAC UNIV, 87- *Concurrent Pos:* Consult, Sundstrand Corp, 80, Jet Propulsion Lab, Calif Inst Tech, 86-87; vis prof, Det Norske Veritas, Oslo, Norway, 81; NASA fac fel, Jet Propulsion Lab, Calif Inst Tech, 85-86; fac fel, Hanscom AFB, USAF, 89. *Mem:* Inst Elec & Electronic Engrs; Am Geophys Union. *Res:* Crack detection in offshore oil platforms; vapor explosions; extreme value analysis of solar-flares; tritium production by solar flare protons. *Mailing Add:* Dept Elec Eng Seattle Pac Univ 3rd Ave W & W Nikkerson Seattle WA 98119

NUTT, RUTH FOELSCHE, b Flensburg, Ger, July 12, 40; US citizen; m 58; c 1. MEDICINAL CHEMISTRY, NATURAL PRODUCT CHEMISTRY. *Educ:* Univ NMex, BS, 62; Univ Pa, PhD, 81. *Prof Exp:* Asst chemist, Merck, Sharp & Dohme Res Labs, 62-65, from res chemist to sr res chemist, 65-76, from res fel to sr res fel, 76- 85, sr investr, 85-89, SR SCIENTIST, MERCK SHARP & DOHME RES LABS, 89- *Mem:* Am Chem Soc. *Res:* Organic synthesis; heterocycles nucleosides; central nervous system-active peptides; design and synthesis of releasing hormone agonists and antagonists; mechanisms of amino acid racemization; amino acid; enzyme and protein synthesis. *Mailing Add:* Merck Sharp & Dohme Res Labs W26A-4044 West Point PA 19486

NUTTALL, ALFRED L, b Lowell, Mass, Mar 27, 43; c 2. SENSORY PHYSIOLOGY, MICROCIRCULATION. *Educ:* Lowell Technol Inst, BS, 65; Univ Mich, MS, 68, PhD(bioeng), 72. *Prof Exp:* Instr, Univ Mich Med Sch, 74-76, from asst prof to assoc prof, Dept Otolaryngol, 76-87, PROF, DEPT OTOLARYNGOL, UNIV MICH MED SCH, 87- *Mem:* Asn Res Otolaryngol; Europ Soc Microcirculation; Soc Neurosci; Acoust Soc Am; AAAS; Inst Elec & Electronic Engrs. *Res:* Physiology of the peripheral auditory system and the inner ear; vascular system of the inner ear; sensory processes leading to transduction and encoding of auditory information. *Mailing Add:* Kresge Hearing Res Inst Univ Mich 1301 E Ann St Ann Arbor MI 48109-0506

NUTTALL, FRANK Q, b May 8, 29; US citizen; c 4. INTERNAL MEDICINE. *Educ:* Univ Utah, MD, 55; Univ Minn, PhD(biochem), 70. *Prof Exp:* Chief admin sect, Vet Admin Hosp, 61-63, chief clin chem, 63-69, assoc prof, Univ, 71-75, PROF INTERNAL MED, SCH MED, UNIV MINN, MINNEAPOLIS, 75-, CHIEF METAB-ENDOCRINE SECT, VET ADMIN HOSP, 70- *Mem:* Am Diabetes Asn; Am Soc Biol Chemists; fel Am Col Physicians; Endocrine Soc. *Res:* Diabetes mellitus; control of glycogen metabolism; glycogen synthetase system. *Mailing Add:* Met Endo Sect 111G Minneapolis Veterans Admin Med Ctr Minneapolis MN 55417

NUTTALL, HERBERT ERICKSEN, JR, b Salt Lake City, Utah, Apr 10, 44; m 67; c 3. CHEMICAL ENGINEERING, GEOCHEMICAL ENGINEERING. *Educ:* Univ Utah, BS, 66; Univ Ariz, MS, 68, PhD(chem eng), 71. *Prof Exp:* Chem engr, Garrett Res & Develop Co, Inc, 71-72; vis asst prof, Univ Tex, Austin, 72-73; from asst prof to assoc prof, 74-83, PROF, DEPT CHEM & NUCLEAR ENG, UNIV NMEX, 83- *Mem:* environmental science. *Res:* Process simulation by digital computer; fossil energy, particularly coal, oil shale and tar sands; disposal of radioactive wastes; personal computers and graphics; colloid science. *Mailing Add:* Dept Chem & Nuclear Eng Univ NMex Albuquerque NM 87131

NUTTALL, JOHN, b Haslingden, Eng, Oct 8, 36; m 62; c 2. THEORETICAL PHYSICS. *Educ:* Cambridge Univ, BA, 57, PhD(theoret physics), 61. *Prof Exp:* Res fel, St John's Col, Cambridge, 61-62; scientist, RCA Victor Co Ltd, Can, 62-64; NSF sr scientist, Tex A&M Univ, 64-65, from assoc prof to prof physics, 65-72; PROF PHYSICS, UNIV WESTERN ONT, 72- *Concurrent Pos:* Vis prof, Univ Western Ont, 72. *Res:* Quantum scattering theory in atomic physics; approximation theory. *Mailing Add:* Dept Physics Univ Western Ont London ON N6A 5B9 Can

NUTTALL, WESLEY FORD, b Regina, Sask, Oct 24, 30. SOIL FERTILITY. *Educ:* Univ Sask, BSA, 58, MSc, 60; McGill Univ, PhD(soil chem), 65. *Prof Exp:* Res scientist, Res Br, Can Dept Agr, 65-91; RETIRED. *Mem:* Can Soc Soil Sci; Am Soc Agron. *Res:* Plant nutrition; statistics; agrometeorology. *Mailing Add:* Res Sta Can Dept Agr Box 1901 Melfort SK S0E 1A0 Can

NUTTER, GENE DOUGLAS, b Columbus, Tex, June 9, 29; m 56; c 2. PHYSICS. *Educ:* Univ Nebr, BS, 51, MS, 56. *Prof Exp:* Physicist, Nat Bur Standards, 52-54; eng supvr, Atomics Int, NAm Rockwell, Inc, 56-67; asst dir measurement sci lab, 67-87, RESEARCHER, INSTRUMENTATION SYSTS CTR, UNIV WIS-MADISON, 87- *Concurrent Pos:* UN Ed, Sci & Cult Orgn consult & lectr, Repub Korea, 68. *Mem:* past sr mem Insrument Soc Am; past sr mem Instr Elec & Electronics Engrs; Am Inst Physics; Am Soc Test & Mat; Optical Soc Am. *Res:* Methodology and instrumentation for high accuracy measurements, especially measurement of temperature and thermal properties by radiometric methods; radiation thermometry from 100 degrees celsius to 4000 degrees celsius. *Mailing Add:* Instrumentation Systs Ctr Univ of Wis-Madison 1500 Johnson Dr Madison WI 53706

NUTTER, JAMES I(RVING), b Lawrence, Mass, Sept 4, 35; m 59; c 4. CHEMICAL ENGINEERING, PHYSICAL CHEMISTRY. *Educ:* Northeastern Univ, BS, 59; Iowa State Univ, MS, 61, PhD(chem eng), 63. *Prof Exp:* Teaching asst chem eng, Iowa State Univ, 59-63; res engr polymers, Wash Res Ctr, W R Grace & Co, 63, 64-66 & process res & develop, 66-68; supvr kinetics group, NASA Marshall Space Flight Ctr, 63-64; supvr process develop, Lord Mfg Co, Lord Corp, 68-70; tech dir design & develop, Robin Industs, Inc, Cleveland, 70-75; mgr res & develop, Scandura, Inc, 75-81; mgr process eng, Reeves Bros, Inc, 81-88; PROJ MGR, DAY ENGR, DUPONT, 88- *Mem:* Am Chem Soc; Am Soc Testing & Mat; Soc Plastic Engrs; Sigma Xi; NY Acad Sci. *Res:* Adsorption, permeation and diffusion processes; polymers; plastics and rubber process design. *Mailing Add:* 8033 Rising Meadow Charlotte NC 28277

NUTTER, JOHN, b Pennsboro, WVa, Jan 15, 30. INFECTIOUS DISEASES. *Educ:* Univ Va, PhD(microbiol), 65. *Prof Exp:* Chief, Planning Off, 81-86, CHIEF, AIDS PROG, PREV BR, NAT INST ALLERGY & INFECTIOUS DIS, NIH, 86- *Mem:* Am Asn Immunologists; Am Soc Microbiol; Sigma Xi. *Mailing Add:* 1416 W 12th St Frederick MD 21701

NUTTER, ROBERT LELAND, b Boston, Mass, Jan 20, 22; m 46; c 2. VIROLOGY, TUMOR IMMUNOLOGY. *Educ:* Andrews Univ, BA, 44; Univ Colo, MS, 49; Iowa State Univ, PhD(biophys), 57. *Prof Exp:* Physicist, Nat Bur Standards, 44-45; asst physics, Univ Colo, 45-46; from instr to asst prof, Pac Union Col, 46-52; asst, Iowa State Univ, 52-53, instr, 53-56; asst prof, Pac Union Col, 56-57; from instr to assoc prof, 57-68, PROF MICROBIOL, SCH MED, LOMA LINDA UNIV, 68- *Concurrent Pos:* Instr, Univ Colo, 48-49; Am Cancer Soc Scholar, Col Med, Hershey Med Ctr & vis prof microbiol, Pa State Univ, 71-72. *Mem:* AAAS; Am Soc Microbiol. *Res:* Use of mice and rats in modification of spontaneous or induced tumor responses by different sources and levels of dietary proteins and fats; mechanisms involved and cell-mediated immunity status. *Mailing Add:* 2317 E 377th Ave Washougal WA 98671

NUTTER, ROY STERLING, JR, b Kingwood, WVa, Apr 28, 44; m 66; c 2. ELECTRICAL ENGINEERING, COMPUTER ENGINEERING. *Educ:* WVa Univ, BSEE, 66, MSEE, 68, PhD(elec eng), 71. *Prof Exp:* Lectr elec eng, WVa Univ, 72; prof engr comput design, NCR Corp, 72-74; from asst prof to assoc prof, 74-82, PROF ELEC ENG, 83-, INTERIM CHMN, WEST VA UNIV, 90- *Concurrent Pos:* Consult coal mine monitoring & control systems, microprocessor and data acquisition. *Mem:* Sr mem Inst Elec & Electronics Engrs; Nat Soc Prof Engrs; Sigma Xi. *Res:* Microprocessor applications to coal mine monitoring and control; computer architecture; multivalued logic; expert systems & AI; neural networks. *Mailing Add:* Dept of Elec & Comput Eng POB 6101 ECE Morgantown WV 26506-6101

NUTTER, WILLIAM ERMAL, b Boomer, WVa, Sept 26, 27; m 52; c 3. BIOCHEMISTRY. *Educ:* WVa Inst Technol, BS, 53; Univ WVa, MS, 57, PhD(biochem), 59. *Prof Exp:* Res assoc biochem, Univ Iowa, 59-61; from asst ed to assoc ed, 61-65, asst head dept biochem ed, 65-71, head dept biochem ed anal, 71-73, ASST MGR, BIOCHEM DEPT, CHEM ABSTR SERV, OHIO STATE UNIV, 73- *Mem:* Am Chem Soc; AAAS; Sigma Xi. *Res:* Bacterial utilization of amino acids and peptides; mammalian metabolism of tryptophan; synthesis of kynurenine; kinetics of d-amino acid oxidase with d-kynurenine; chemical information and documentation. *Mailing Add:* 354 E Schrock Rd Westerville OH 43081-3455

NUTTING, ALBERT DEANE, b Otisfield, Maine, Sept 6, 05; m 40. FORESTRY. *Educ:* Univ Maine, BS, 27. *Hon Degrees:* DSc, Univ Maine, 87. *Prof Exp:* Forester, Finch, Pruyn & Co, 27-31; exten forester, Univ Maine, 31-48; forest comnr, State of Maine, 48-58; dir, Sch Forestry, 58-71, dir sch forest res, 70-71, EMER DIR SCH FOREST RESOURCES, UNIV MAINE, ORONO, 71- *Concurrent Pos:* Chmn, Nat Coop Forestry Res Bd. *Mem:* Fel Soc Am Foresters; Am Forestry Asn. *Res:* Forest management and products. *Mailing Add:* RFD 1 Box 4140 Oxford ME 04270

NUTTING, EHARD FORREST, b Milwaukee, Wis, Oct 4, 29; m 51; c 6. ENDOCRINOLOGY, PHARMACOLOGY. *Educ:* Utah State Univ, BS, 51; Univ Wis, MS, 56, PhD(endocrinol), 62. *Prof Exp:* Res fel genetics, Univ Wis, 54-56 & res fel zool, 58-60; res investr endocrinol testing, 60-64, reprod endocrinol, 65-69, sr res investr reprod physiol, 69-70, res group leader fertility control, 70-71, res dir endocrinol dept, 71-73, dir, Dept Biol Res, Searle Labs Div, 73-75, sr res scientist, Dept Sci, 75-78, dir contraception res, 78-79, DIR CELLULAR & ENDOCRINE DIS, SEARLE LABS DIV, G D SEARLE & CO, 80- *Concurrent Pos:* Consult, Nat Inst Child Health & Human Develop, 71-78. *Mem:* Endocrine Soc; Soc Exp Biol & Med; Soc Study Reprod; Brit Soc Study Fertil; Am Soc Pharmacol & Exp Therapeut. *Res:* Physiology of the female reproductive tract; pharmacological control of fertility; gamete transport; physiology and pharmacology of factors involved in inflammation and cellular growth processes. *Mailing Add:* 8233 Linder Ave Morton Grove IL 60053

NUTTING, WILLIAM BROWN, b Worcester, Mass, Apr 15, 18; m 68; c 4. ACAROLOGY, INVERTEBRATE ZOOLOGY. *Educ:* Univ Mass, BS, 40, MS, 48; Cornell Univ, PhD(zool), 50; FCPP, 78. *Prof Exp:* From instr to assoc prof, 50-64, PROF ZOOL, UNIV MASS, AMHERST, 64- *Concurrent Pos:* Fel, Univ Queensland, 58; vis prof, Wallaceville Res Ctr, NZ, 72; vis prof, Stanford Univ Med Sch, 75 & Univ Queensland, 79. *Mem:* Sigma Xi. *Res:* Biology; ecology; pathogenesis and systematics of mites, especially Demodiciedae; natural history. *Mailing Add:* Dept Zool Univ Mass Amherst MA 01003

NUTTING, WILLIAM LEROY, b Pepperell, Mass, July 26, 22; m 44; c 3. ENTOMOLOGY. *Educ:* Harvard Univ, AB, 43, PhD(biol), 50. *Prof Exp:* Res fel biol, Harvard Univ, 50-55; assoc entomologist, 55-62, from assoc prof to prof, 58-86, EMER PROF ENTOM & ENTOMOLOGIST, UNIV ARIZ, 86- *Concurrent Pos:* Prin res scientist & prog leader, Grassland Termites Res Prog, Int Ctr Insect Physiol & Ecol, Nairobi, Kenya, 79-81. *Mem:* AAAS; Am Soc Zool; Entom Soc Am; Am Inst Biol Sci; Int Union Study Social Insects; Sigma Xi. *Res:* Insect morphology, behavior and ecology, especially Orthoptera, Isoptera and other social insects. *Mailing Add:* 6810 N Nanini Dr Tucson AZ 85704-6130

NUTTLI, OTTO WILLIAM, geophysics; deceased, see previous edition for last biography

NUZZI, ROBERT, b New York, NY, July 7, 42; m 68. ESTUARINE ECOLOGY, PHYTOPLANKTON. *Educ:* Fordham Univ, BS, 63, MS, 65, PhD(microbiol), 69. *Prof Exp:* Teaching asst biol, Fordham Univ, 64-69, asst prof, 69; instr, St Francis Col, NY, 68-70; assoc res scientist microbiol, NY Ocean Sci Lab, Affil Cols & Univs, Inc, 70-74, res scientist, 74-75; marine biologist, Dept Environ Control, 75-78, MARINE BIOLOGIST, DEPT HEALTH SERV, SUFFOLK COUNTY, 78- *Concurrent Pos:* Res assoc, NY Ocean Sci Lab, Affil Cols & Univs, Inc, 75-; adj assoc prof, Suffolk County Community Col, 84-; instr, Nat Swimming Pool Found, 84- *Mem:* AAAS; Am Inst Biol Sci; Am Soc Limnol & Oceanog; NY Acad Sci; Estuarine Res Fedn; Water Pollution Control Fedn. *Res:* Phytoplankton systematics; estuarine ecology; wetland ecology. *Mailing Add:* 66 Bucks Path East Hampton NY 11937

NYBAKKEN, JAMES W, b Warren, Minn, Sept 16, 36; m 60; c 2. MARINE ECOLOGY. *Educ:* St Olaf Col, BA, 58; Univ Wis, MS, 61, PhD(zool), 65. *Prof Exp:* Teaching asst marine biol, Univ Miami, 58-59; cur, Zool Mus, Univ Wis, 61, 62 & 64-65; from asst prof to assoc prof, 65-72, PROF BIOL, CALIF STATE UNIV, HAYWARD, 73-, STAFF MEM, MOSS LANDING MARINE LAB, 66- *Concurrent Pos:* Off Naval Res vis investr, Marine Lab, Univ Ariz, 66; res assoc, Univ Wash, 68-69, 72 & 73. *Mem:* AAAS; Ecol Soc Am; Am Soc Zool; Marine Biol Asn UK; Am Malacological Union (pres, 85-86); Western Soc Naturalists (pres, 86). *Res:* Ecology and systematics of gastropod mollusks of the genus Conus; ecology of nudibranch mollusks; benthic invertebrate ecology. *Mailing Add:* Moss Landing Marine Labs Box 450 Moss Landing CA 95039

NYBERG, DAVID DOLPH, b Vancouver, Wash, June 10, 28; m 51; c 2. POLYMER CHEMISTRY. *Educ:* Ore State Univ, BS, 52, PhD(org chem), 56. *Prof Exp:* Group leader, Shell Chem Co, 56-66; staff mem, 66-72, mgr explor develop, thermofit tech dept, 72-76, mgr advan mat develop, Energy Div, 76-82, tech fel corp res & develop, 82-84, SR STAFF MEM, RAYCHEM CORP, 84- *Mem:* Am Chem Soc. *Res:* Polymer synthesis; plastics and rubber technology; radiation crosslinking; adhesion; specialty formulating. *Mailing Add:* 1426 New Foundland Dr Sunnyvale CA 94087

NYBERG, DENNIS WAYNE, b Oklahoma City, Okla, Feb 11, 44; m 67; c 5. GENETICS, EVOLUTION. *Educ:* Mass Inst Technol, SB, 65; Univ Ill, Urbana-Champaign, MS, 69, PhD(zool), 71. *Prof Exp:* Fel biol, Univ Sussex, 71-72; res assoc, Ind Univ, 72-73; asst prof, 73-77, ASSOC PROF BIOL, UNIV ILL, CHICAGO CIRCLE, 77- *Concurrent Pos:* NSF grant, Univ Ill, Chicago Circle, 76-78; NIH grant, Univ Ill, Chicago, 79-82; fac res leave, Argonne Nat Lab, 80-81. *Mem:* Soc Study Evolution; Soc Protozoologists. *Res:* Evolutionary role of breeding systems; genetics of metal tolerance in protozoa; restoration ecology. *Mailing Add:* Dept of Biol Sci MC 066 Univ of Ill at Chicago Chicago IL 60680

NYBOER, JAN, b Holland, Mich, Apr 21, 06; m 39; c 4. INTERNAL MEDICINE, BIOENGINEERING & BIOMEDICAL ENGINEERING. *Educ:* Univ Mich, AB, 28, MS, 29, MD, 35. *Hon Degrees:* ScD, Univ Mich, 32. *Prof Exp:* Asst biol, Hope Col, 25-27; asst mammal, Univ Mich, 28, asst & demonstr physiol, 28-34, asst electrocardiol, Univ Hosp, 34-35; intern, St Louis Maternity Hosp & Barnes Hosp, Wash Univ, 35-36 & Harvard Med Serv, Boston City Hosp, 36-37; fel med, NY Post-Grad Med Sch, Columbia Univ, 37-41; asst med dir, Conn Mutual Life Ins Co, 41-47; asst prof pharmacol, Dartmouth Med Sch, 47-55; chief cardiovasc physiol & assoc physician, Harper Hosp, 55-65; prof physiol & pharmacol, 59-76, assoc physiol res, 73-76, DIR RES, REHAB INST, WAYNE STATE UNIV, 65-, EMER PROF PHYSIOL, SCH MED, 76- *Concurrent Pos:* Asst attend physician, NY Postgrad Hosp & Clin, 37-47; clin instr, Sch Med, Yale Univ, 42-47; consult, US Vet Admin Hosp, Vt, 48-55; consult, Hitchcock Clin, 48-55, res assoc, Hitchcock Found, 54-55; exchange scientist, Czech Acad Sci, 71 & All India Inst Med Sci 78. *Mem:* Am Physiol Soc; Am Heart Asn; fel Am Col Cardiol; Am Fedn Clin Res. *Res:* Physiology of the heart and respiration; esophagial; electrical impedance plethysmography; development of direct writing electrocardiography; displacement and counterforce ballistocardiography; bioelectrical impedance for total body water during renal dialysis and congestive heart failure; leanness and fatness; hypothermic swim of Lynne Cox between Diomede Islands. *Mailing Add:* 570 Cadieux Grosse Pointe MI 48201

NYBORG, WESLEY LEMARS, b Ruthven, Iowa, May 15, 17; m 45; c 1. BIOPHYSICS, ACOUSTICS. *Educ:* Luther Col, AB, 41; Pa State Univ, MS, 44, PhD(physics), 47. *Prof Exp:* Asst physics, Pa State Univ, 41-43, instr, 43-44 & 47-49, asst, 44-47, asst prof, 49-50; from asst prof to assoc prof, Brown Univ, 50-60; prof, 60-86, EMER PROF PHYSICS, UNIV VT, 86- *Concurrent Pos:* USPHS fel, Sch Adv Study, Mass Inst Technol, 56-57; vis scientist, Oxford Univ, 60-61; mem adv comt radiation bio-effects, Food & Drug Admin, 72-76; mem, Nat Coun Radiation Protection & Measurements, 82-88, hon mem, 88- *Honors & Awards:* Pioneer Award, Am Inst Ultrasound Med, 85, W R Fry Lectr, 89; Silver Medal, Acoust Soc Am, 90. *Mem:* Fel AAAS; fel Acoust Soc Am; Am Phys Soc; Biophys Soc; Am Asn Physics Teachers; fel Am Inst Ultrasound Med. *Res:* Acoustics; ultrasonics; biophysical ultrasound; environmental biophysics; physical mechanisms for biological effects of ultrasound. *Mailing Add:* Dept Physics Cook Phys Sci Bldg Univ Vt Burlington VT 05405

NYBURG, STANLEY CECIL, b London, Eng, Dec 15, 24; m 49; c 2. CRYSTALLOGRAPHY. *Educ:* King's Col, Univ London, BSc, 45, DSc(crystallog, thermodynamics), 73; Univ Leeds, PhD(crystallog), 49. *Prof Exp:* Crystallographer, Brit Rubber Producers' Res Asn, 49-52; from asst lectr to sr lectr chem, Univ Keele, 52-64; prof 64-87, emer prof chem, Univ Toronto, 88; RETIRED. *Concurrent Pos:* Vis prof crystallog, Univ Pittsburgh, 62-63; vis prof, Univ Sydney, Australia. *Mem:* Fel The Chem Soc; assoc Brit Inst Physics & Phys Soc. *Res:* Crystal structure analysis; thermodynamics. *Mailing Add:* Dept Chem King's Col Strand London WC2R 2LS England

NYCZEPIR, ANDREW PETER, b Englewood, NJ, Feb 25, 52; m 83. NEMATOLOGY. *Educ:* Univ Ga, BSA, 74; Clemson Univ, MS, 76, PhD(plant pathol-nematol), 80. *Prof Exp:* Grad res asst, Clemson Univ, 74-76, 77-80; nematologist, res, Wash State Univ, Agr Res Serv, USDA, 80-82; NEMATOLOGIST, SOUTHEAST FRUIT & TREE NUT RES LAB, AGR RES SERV, USDA, BYRON, GA, 82- *Mem:* Soc Nematologist; Orgn Trop Am Nematologists; Sigma Xi; Am Phytopath Soc. *Res:* Host-parasite relationships and control of plant parasitic nematodes. *Mailing Add:* Agr Res Serv USDA PO Box 87 Byron GA 31008

NYDEGGER, CORINNE NEMETZ, b Milwaukee, Wis. MEDICAL ANTHROPOLOGY, GERONTOLOGY. *Educ:* Univ Wis, Madison, BA, 51; Cornell Univ, MA, 70; Pa State Univ, PhD(human develop), 73. *Prof Exp:* Res field-team mem, Six Cultures Proj, Harvard-Yale-Cornell Proj, 54-56;

asst dir res geriatrics, Pa State Univ, 72-73; NIMH fel, Dept Sociol & Inst Human Develop, Univ Calif, Berkeley, 73-74; lectr med anthrop prog, 75-77, adj assoc prof, 77-82, ADJ PROF MED ANTHROP PROG, UNIV CALIF, SAN FRANCISCO, 82- *Concurrent Pos:* USPHS res fel, Human Develop & Med Anthrop Prog, Univ Calif, San Francisco, 74-76; lectr adult develop, Sch Soc Welfare, Univ Calif, Berkeley, 76; mem, Numerous Res Proj in Gerontol, 77-, life course rev comt, NIMH; assoc ed, Res Aging; numerous jours. *Mem:* Fel Geront Soc; fel Soc Appl Anthrop; Soc Cross Cult Res; Soc Study Social Probs; Med Anthrop Asn. *Res:* Timing; effects of chronological age on events and roles; responses to timing deviance; deviance and medical models; group responses to negative labeling; fatherhood, especially deviant patterns. *Mailing Add:* Med Anthrop Prog Univ Calif Box 0850 San Francisco CA 94143

NYE, EDWIN (PACKARD), b Atkinson, NH, Jan 29, 20; m 44; c 3. MECHANICAL ENGINEERING. *Educ:* Univ NH, BS, 41; Harvard Univ, MS, 47. *Prof Exp:* Serv engr controls, Bailey Meter Co, 41-42; instr mech eng, Univ NH, 42-44; proj engr fluid flow, Nat Adv Comt Aeronaut, 44-46; from instr to assoc prof mech eng, Pa State Univ, 47-59; prof eng, Trinity Col, Conn, 58-60, chmn dept, 60-70, dean Fac, 70-81, Hallden prof eng, 60-83, emer Hallden prof eng; RETIRED. *Concurrent Pos:* Chmn bd dir, Univ Res Inst Conn, Inc, 70- *Mem:* Soc Hist Technol; Am Soc Mech Engrs. *Res:* Conversion of stored energies to mechanical and electrical form. *Mailing Add:* PO Box 615 Ashfield MA 01330

NYE, MARY JO, b Nashville, Tenn, Dec 5, 44; m 68; c 1. HISTORY OF PHYSICAL SCIENCES, HISTORY OF FRENCH SCIENCE. *Educ:* Univ Wis, BA, 65, PhD(hist of sci), 70. *Prof Exp:* Andrew Mellon fel Univ Pittsburgh, 74-75; NSF fel, 69-70; PROF HIST OF SCI, UNIV OKLA, 70- *Concurrent Pos:* Mem Sch of Hist Studies, Inst Adv Study, Princeton, 81-82; mem, US Nat Comt, Int Union Hist & Philos of Sci, 86-; vis prof hist & sci, Harvard Univ, 88. *Mem:* Hist Sci Soc (pres, 88-); AAAS; Am Hist Asn; Brit Soc Hist Sci; Europ Asn Study Sci & Technol. *Res:* History of modern physical sciences, particularly atomism; French science and its institutions; science and politics; philosophy of science, especially models and conventions in scientific methodology. *Mailing Add:* 601 Elm St Rm 621 Univ Okla Norman OK 73019

NYE, ROBERT EUGENE, JR, b Cincinnati, Ohio, Feb 6, 22; m 48; c 3. PHYSIOLOGY. *Educ:* Ohio Univ, AB, 43; Univ Rochester, MD, 47. *Prof Exp:* Intern med, Strong Mem Hosp, Rochester, NY, 47-48, asst resident, 48-49; house physician, Hammersmith Hosp, London, Eng, 51; instr, Univ Rochester, 54-56; from instr to assoc prof, 56-73, prof, 73-87, EMER PROF PHYSIOL, DARTMOUTH MED SCH, 87- *Concurrent Pos:* Buswell fel, Univ Rochester, 49-50, univ fel, 51-54; assoc staff, Mary Hitchcock Mem Hosp, 56-65, consult staff, 65- *Mem:* Am Physiol Soc; Am Thoracic Soc; Am Asn Hist Med. *Res:* Cardiovascular and pulmonary physiology, especially the relations between perfusion and ventilation in the lung. *Mailing Add:* PO Box 260 Norwich VT 05055

NYE, SYLVANUS WILLIAM, b Buffalo, NY, Mar 28, 30; m 56; c 2. PATHOLOGY. *Educ:* Hamilton Col, AB, 52; Univ Rochester, MD, 57; Am Bd Path, cert. *Prof Exp:* From intern to chief resident path, Sch Med, Univ NC, Chapel Hill, 57-60, instr path & trainee clin microbiol & path, 60-62; from asst prof to prof path, Sch Med, ECarolina Univ, 69-74, chmn dept, 71-74; MEM STAFF, LENOIR MEM HOSP, 74- *Concurrent Pos:* Pathologist, SEATO Med Res Lab, US Army Component, Bangkok, Thailand, 63-65. *Mem:* AAAS; Col Am Path; Int Acad Path; Am Soc Trop Med & Hyg; Am Soc Clin Path. *Res:* Geographic pathology. *Mailing Add:* Lenoir Mem Hosp Kinston NC 28501

NYGAARD, KAARE JOHANN, atomic physics, quantum optics; deceased, see previous edition for last biography

NYGAARD, ODDVAR FRITHJOF, b Oslo, Norway, Oct 30, 22; nat US; m 46; c 2. RADIOBIOLOGY, CHEMISTRY. *Educ:* Norweg Tech Univ, Sivilingenior, 47; Univ Minn, PhD(physiol chem), 51. *Prof Exp:* Asst tech org chem, Tech Univ Norway, 47; asst physiol chem, Univ Minn, 48-51; fel oncol, McArdle Mem Lab, Univ Wis, 51-52, res assoc, 54-57; res biochemist, Norsk Hydro's Inst Cancer Res, Norweg Radium Hosp, 52-54; Norweg Cancer Soc fel, 54; res chemist, AEC biol effects of irradiation lab, Univ Mich, 57-59; sr inst radiol, 59-62, res assoc, 59-63, asst prof biochem, 59-68, asst prof radiol, 62-65, assoc prof radiol, 65-68, assoc prof biochem & radiol, 68-75, assoc dir div radiation biol, 63-76, PROF RADIOL, CASE WESTERN RESERVE UNIV, 75-, DIR DIV RADIOTION BIOL, 76- *Concurrent Pos:* Ed-in-chief, Radiation Research, Radiation Res Soc, 72-79; special asst to the dir, Low-Level Radiation Effects, Nat Cancer Inst, 79- *Mem:* Environ Mutagen Soc; Am Chem Soc; Soc Develop Biol; Radiation Res Soc; Coun Biol Ed; Sigma Xi. *Res:* Nucleic acid metabolism and control; biological effects of radiation; biochemistry of hypoxic cell radiosensitizers and carcinogens; modification of radiation response in hypoxic and anoxic cells. *Mailing Add:* Dept Radiol Case Western Reserve Univ Cleveland OH 44106

NYGREEN, PAUL W, b Bellingham, Wash, May 15, 25; m 50; c 2. GEOLOGY, OCEANOGRAPHY. *Educ:* Univ Wash, BS, 53; Univ Nebr, MSc, 55. *Prof Exp:* Geologist, Standard Oil Co Tex, 55-62, biostratigrapher, 62-64; biostratigrapher, Calif Oil Co, Okla, 64-65; sr biostratigrapher, Chevron Oil Co, 65-69, staff biostratigrapher, 69-71, staff biostratigrapher, Chevron Overseas Petrol Inc, 71-84; CONSULT GEOLOGIST, 84- *Concurrent Pos:* Supvry paleontologist, West Australian Petrol Proprietary Ltd, 71-74; supvr biostratigraphy, Arabian Am Oil Co, Saudi Arabia, 78-84, asst chief geologist, 80-84, consult geologist, 84- *Mem:* Fel AAAS; fel Geol Soc Am; Paleont Soc; Am Asn Petrol Geol; Am Asn Stratig Palynologists (pres, 67-68). *Mailing Add:* 173 Sylvan Rd Walnut Creek CA 94596

NYGREN, STEPHEN FREDRICK, b Evanston, Ill, Mar 3, 42; m 64. ELECTRICAL ENGINEERING. *Educ:* Carnegie Inst Technol, BS, 64; Stanford Univ, MS, 65, PhD(elec eng), 69. *Prof Exp:* MEM TECH STAFF, BELL TEL LABS, INC, 68- *Mem:* Am Phys Soc. *Res:* III-V compound semiconductor materials, including crystal growth, epitaxial layer growth and diffusion. *Mailing Add:* 3129 Merritt Pkwy Sinking Spring PA 19608

NYHAN, WILLIAM LEO, b Boston, Mass, Mar 13, 26; m 48; c 3. PEDIATRICS, BIOCHEMISTRY. *Educ:* Columbia Univ, MD, 49; Univ Ill, MS, 56, PhD(pharmacol), 58. *Hon Degrees:* DSc, Tokushima Univ, Japan, 81. *Prof Exp:* From asst prof to assoc prof pediat, Sch Med, Johns Hopkins Univ, 58-63; prof pediat & biochem & chmn dept pediat, Sch Med, Univ Miami, 63-69; chmn dept, 68-69, PROF PEDIAT, SCH MED, UNIV CALIF, SAN DIEGO, 69- *Concurrent Pos:* Res fel, Nat Found Infantile Paralysis, 55-58; Am Cancer Soc fac res assoc, 61-63; mem, Prog Proj Comt, Nat Inst Child Health & Human Develop, NIH, 64-67, Advt Comt Teratogenic Effects Certain Drugs, Food & Drug Admin, 64-70, Pediat Panel Coun Drugs, AMA, 64-70; mem, Comt Maternal Nutrit, Nat Res Coun, 66-70, Nat Adv Child Health & Human Develop Coun, 67-71, Med & Sci Adv Comt, Leukemia Soc Am, Inc, 68-72, Basic Res Adv Comt, Nat Found, March of Dimes, 73-81, Clin Cancer Prog Proj Rev Comt, Nat Cancer Inst, NIH, 76-80, Prof Adv Bd, Int Retts Syndrome Asn, 85-; Wellcome guest prof, Royal Alexandria Hosp Children, Sydney, Australia, 80; extraordinary prof, Univ Salvador, Arg, 82. *Honors & Awards:* Borden Award, Am Acad Pediat, 80; Bernard Sachs Lectr, Child Neurol Soc, Ariz, 84; Weinstein-Goldenson Award for Res, United Cerebral Palsy Asn, 87. *Mem:* Fel AAAS; Am Chem Soc; Soc Pediat Res (pres, 70-71); Am Asn Cancer Res; Am Pediat Soc; Sigma Xi; Am Soc Pharmacol & Exp Therapeut; NY Acad Sci; Am Acad Pediat; Am Inst Biol Sci; Soc Exp Biol & Med; Am Soc Clin Invest; Biochem Soc; corresp mem Fr Pediat Soc. *Res:* Amino acid metabolism; biochemical genetics; metabolism of tumors; developmental pharmacology. *Mailing Add:* Dept Pediat Univ Calif San Diego Sch Med La Jolla CA 92093

NYHUS, LLOYD MILTON, b Mt Vernon, Wash, June 24, 23; m 49; c 2. SURGERY, PHYSIOLOGY. *Educ:* Pac Lutheran Col, BA, 45; Univ Ala, MD, 47; Am Bd Surg, dipl, 57. *Hon Degrees:* Dr, Univ Thessoloniki, 69, Univ Uppsala, 74, Chihuahua Univ, 75, Jagielonian Univ, 80, Univ Gama Filho, 83, Univ Louis Pasteur, 84 & Univ Athens, 88. *Prof Exp:* From intern to asst resident surg, King County Hosp, Seattle, 47-50; res assoc, Univ Wash, 52-53; resident, King County Hosp, 54; from instr to prof surg, Sch Med, Univ Wash, 54-67; PROF SURG & EMER HEAD DEPT, COL MED, UNIV ILL, CHICAGO & EMER SURGEON-IN-CHIEF, UNIV HOSP, 89- *Concurrent Pos:* USPHS res fel, 52-54; Guggenheim fel, Sweden & Scotland, 55-56; sr consult, West Side Vet Admin Hosp, Chicago, 67-; sr attend, Cook County Hosp, Chicago, 68- *Mem:* Soc Univ Surg; Am Gastroenterol Asn; fel Am Col Surgeons; Am Surg Asn; Int Soc Surg; hon mem Nat Acad Med Brazil; hon mem Nat Acad Med France; hon mem Brazilian Acad Mil Med; hon mem Fr Asn Surg. *Res:* Gastric physiology and surgery; peptic ulcer; hernia; esophageal physiology. *Mailing Add:* Dept Surg Univ Ill Hosp M/C 958 Chicago IL 60612

NYI, KAYSON, b Chungking, China, Apr 11, 45; m 70. ORGANIC CHEMISTRY. *Educ:* Mass Inst Technol, SB, 65; Univ Chicago, PhD(org chem), 71. *Prof Exp:* SR CHEMIST, RES DIV, ROHM AND HAAS CO, 71- *Mem:* Am Chem Soc. *Res:* Synthetic organic chemistry; photochemistry. *Mailing Add:* 121 Tower Rd Sellersville PA 18960

NYIKOS, PETER JOSEPH, b Salzburg, Austria, Mar 8, 46; US citizen. TOPOLOGY. *Educ:* Washington & Jefferson Col, BA, 67; Carnegie-Mellon Univ, MS, 68, PhD(topology), 71. *Prof Exp:* Mathematician, Biomed Lab, Edgewood Arsenal, US Army, 72-73; NSF fel, Univ Chicago, 73-74; vis lectr math, Univ Ill, Urbana, 74-76; vis asst prof, Auburn Univ, 76-79; mem staff, Inst Med & Math, Ohio Univ, 79; ASSOC PROF, UNIV SC, 79- *Concurrent Pos:* Vis prof, Math Inst, Oxford Univ, 86. *Mem:* Am Math Soc. *Res:* Set-theoretic topology, especially theory of nonmetrizable manifolds, countably compact spaces, zero-dimensional spaces, and Stone-Cech compactification of discrete spaces; theory of partially ordered sets and Boolean algebras. *Mailing Add:* 1729 Fairhaven Dr Columbia SC 29210

NYLAND, GEORGE, b Eastburg, Alta, Apr 3, 19; US citizen; m 41; c 3. PLANT PATHOLOGY. *Educ:* State Col Wash, BS, 40, PhD(plant path), 48; La State Univ, MS, 42. *Prof Exp:* Asst, State Col Washington, 39-40, from instr to asst prof plant path, 46-48; asst, La State Univ, 40-42; instr & jr plant pathologist, 48-50, asst prof & plant pathologist, 50-56, assoc prof & assoc plant pathologist, 56-62, PROF PLANT PATH & PLANT PATHOLOGIST, COL AGR, UNIV CALIF, DAVIS, 62-85. *Concurrent Pos:* Jr plant pathologist, Exp Sta, State Col Washington, 46-48, asst plant pathologist, 47-48. *Honors & Awards:* Calif Asn Nursery Men spec Res Award, 67. *Mem:* AAAS; Am Phytopath Soc; Sigma Xi. *Res:* Stone, pome fruit and ornamental plants virus and mycoplasma diseases; chemotherapy; thermotherapy. *Mailing Add:* Div Plant Path Univ Calif Davis CA 95616

NYLEN, MARIE USSING, b Copenhagen, Denmark, Apr 13, 24; US citizen; m 56; c 3. DENTAL RESEARCH, ELECTRON MICROSCOPY. *Educ:* Royal Dent Col, Denmark, DDS, 47. *Hon Degrees:* DrOdont, Royal Dent Col, Denmark, 73. *Prof Exp:* Pvt pract, 47-48; instr oper dent, Royal Dent Col, Denmark, 48-49; guest worker dent histol, Nat Inst Dent Res, 49-50, asst prof oral diag & res assoc electron micros, Royal Dent Col, Denmark, 51-55; vis assoc biophys, 55-60, biologist, 60-65, actg chief lab histol & path, 65-69, mem dent study sect, 70-74, chief lab biol struct, 69-76, DIR INTRAMURAL RES, NAT INST DENT RES, 76- *Concurrent Pos:* USPHS fel, Nat Inst Dent Res, 50-51; vis investr, Marine Biol Lab, Woods Hole, 69-72; prof lectr, Schs Med & Dent, Georgetown Univ, 70- *Honors & Awards:* Dept Health, Educ & Welfare Super Serv Honor Award, 69; Int Asn Dent Res Award, 70; Federal Woman's Award, 75; Isaac Scheur Mem Award, 77. *Mem:* Fel AAAS; Electron Micros Soc; Am Dent Asn; Am Soc Cell Biol; fel Am Col Dent; Sigma Xi. *Res:* Biophysical studies of developing and mature mineralized tissues and associated cells in normal and pathologic states. *Mailing Add:* Nat Inst Dent Res Westwood Bldg Rm 503 Bethesda MD 20892

NYLUND, ROBERT E, b Lowell, Mass, May 12, 37; m 62; c 1. PHYSICAL CHEMISTRY. *Educ:* Northeastern Univ, BS, 60; Univ Iowa, PhD(phys chem), 64. *Prof Exp:* Asst prof, 64-69, ASSOC PROF CHEM, SUSQUEHANNA UNIV, 69-, CHMN DEPT, 75- *Mem:* Am Chem Soc. *Res:* Physical chemistry of biologically important polymers and proteins. *Mailing Add:* Dept of Chem Susquehanna Univ Selinsgrove PA 17870

NYLUND, ROBERT EINAR, b Ely, Minn, Jan 22, 16; m 40; c 2. HORTICULTURE. *Educ:* Univ Minn, BS, 38, MS, 42, PhD(hort), 45. *Prof Exp:* Asst, Univ Minn, St Paul, 38, horticulturist, Northwest Exp Sta, 39-41, from instr to assoc prof, 39-59, actg head dept hort sci & landscape archit, 75-76, prof hort, 59-78; RETIRED. *Concurrent Pos:* Fulbright lectr, Univ Helsinki, 59-60 & 72; Univ Minn Off Int Prog grant, 65; consult agr res, Govt Philippines, 81. *Mem:* Fel AAAS; Am Soc Hort Sci; Weed Sci Soc Am; European Potato Asn; hon mem Potato Asn Am. *Res:* Physiology of vegetable crops and potatoes; weed control. *Mailing Add:* 2210 Midland Gr Rd No 101 St Paul MN 55113

NYMAN, CARL JOHN, b New Orleans, La, Oct 21, 24; m 50; c 3. INORGANIC CHEMISTRY. *Educ:* Tulane Univ, BS, 44, MS, 45; Univ Ill, PhD(inorg chem), 48. *Prof Exp:* Jr technologist, Shell Oil Co, Inc, Calif, 44; asst chem, Univ Ill, 45-47, instr, 48; from instr to assoc prof, 48-61, assoc provost res, 81-85, dean grad sch, 68-86, vprovost res, 85-86, PROF CHEM, WASH STATE UNIV, 61- *Concurrent Pos:* Vis fel, Cornell Univ, 59-60; vis fel, Imp Col Sci & Technol, London, 66-67, Eidgenossische Technische Hochschule, Zurich, 73, Tulane Univ, 86-87. *Mem:* Sigma Xi; Am Chem Soc. *Res:* Catalytic reduction of sodium sulfate; polarography in liquid ammonia; stability of complex ions; solutions of complex and polynuclear inorganic ions; organometallic complexes; peroxo-complexes of metals; catalytic oxygenation reactions. *Mailing Add:* NW 1320 Orion Dr Pullman WA 99163

NYMAN, DALE JAMES, b Bancroft, Iowa, June 4, 31; m 60; c 2. HYDROGEOLOGIST. *Educ:* Iowa State Univ, BS, 53, MS, 58. *Prof Exp:* Geologist, US Geol Surv, 58-87, hydrologist, 58-87; hydrologist, Ground Water Protection Div, La Dept Environ Qual, 87-88; consult, Woodward-Clyde, 88-89, Geraghty & Miller Inc, 89-90, CONSULT, NYMAN & ASSOCS, 90- *Mem:* Am Geophys Union; Geol Soc Am; Soc Prof Well Log Analysts; Nat Water Well Asn; Nat Water Resources Asn; Am Soc Photogrammetry & Remote Sensing. *Res:* Application of geology to hydrologic problems; ground water-surface water relationships; ground water management and modeling; application of remote sensing to ground water problems. *Mailing Add:* 3168 Sherry Dr Baton Rouge LA 70816

NYMAN, MELVIN ANDREW, b Big Rapids, Mich, Aug 19, 44. COMBINATORICS & FINITE MATHEMATICS. *Educ:* Ferris State Col, BS, 65; Mich State Univ, MS, 67, PhD(math), 72. *Prof Exp:* Instr math, Ferris State Col, 67-69; from asst prof to assoc prof math, Manchester Col, 72-81; ASSOC PROF TO PROF & CHMN MATH & COMPUT SCI, ALMA COL, 81- *Concurrent Pos:* Analyst & mathematician, Solar Energy Res Inst, Golden, Colo, 79-80; vis lectr, Univ Otago, New Zealand, 88. *Mem:* Am Math Soc; Math Asn Am; New Zealand Math Soc; Soc Indust & Appl Math. *Res:* Mathematical models for growth of Macrocystis. *Mailing Add:* Dept Math & Comput Sci Alma Col Alma MI 48801-1599

NYMAN, THOMAS HARRY, b Seattle, Wash, Feb 11, 42. ELECTRONICS, SYSTEMS ENGINEERING. *Educ:* Univ Wash, BS, 64; Mass Inst Technol, SM, 66. *Prof Exp:* Staff mem data transmission, Bell Telephone Labs, 64-66; officer, USS Enterprise Engr Dept, US Navy, 66-67, prog mgr sensor syst develop, Defense Comn Planning Group, 67-69; group leader sensor syst, Mitre Corp, 69-75; staff specialist electronics, Off Dir, Defense Res & Eng, Dept Defense, 75-78; dept dir, Gen Res Corp, 78-80; ASSOC & DEPT MGR SYSTS INTEGRATION, MITRE CORP, 81-, DEPT HEAD, COMMUN ENG- *Concurrent Pos:* Mem navig & commun/positioning/identification working group, Dept Defense, 75-77, mem steering comt embedded comput software, 76-78; mem steering comt avionics, Am Defense Prepardness Asn, 77-78; mem adv comt guid, control & info processing, NASA, 77-78. *Mem:* Inst Elec & Electronics Engrs; Am Inst Navig. *Res:* Systems issues of controlling and stabilizing networks of communications and command and control systems; structuring issues of interoperability by hierarchial layers; dynamic reallocation of computer and communications resources for land battle applications. *Mailing Add:* Mitre Corp 7525 Colshire Prive McLean VA 22102

NYMANN, DEWAYNE STANLEY, b Cedar Falls, Iowa, June 27, 35; m 56; c 1. MATHEMATICS. *Educ:* Univ Northern Iowa, BA, 57; Univ Kans, MA, 59, PhD(math), 64. *Prof Exp:* Asst prof math, Tex Christian Univ, 64-65 & Univ Tex, 65-70; assoc prof, 70-82, PROF MATH, UNIV TENN, CHATTANOOGA, 82- *Mem:* Am Math Soc; Sigma Xi; Math Asn Am. *Res:* Algebra group theory; generalized nilpotent and solvable groups. *Mailing Add:* Dept Math Univ Tenn 615 McCallie Ave Chattanooga TN 37403

NYMANN, JAMES EUGENE, b Cedar Falls, Iowa, Nov 24, 38; m 58; c 3. MATHEMATICS. *Educ:* Univ Northern Iowa, BA, 61; Univ Ariz, MS, 63, PhD(math), 65. *Prof Exp:* Asst prof, Univ Hawaii, 65-67; assoc prof, 67-74, PROF MATH, UNIV TEX, EL PASO, 74- *Concurrent Pos:* Fulbright-Hays lectr, Univ Liberia, 72-73 & Univ Malawi, 77-79. *Mem:* Am Math Soc; Math Asn Am. *Res:* Number theory and analysis. *Mailing Add:* Dept of Math Univ of Tex El Paso TX 79968

NYPAN, LESTER JENS, b Minneapolis, Minn, Oct 30, 29; m 54; c 1. MECHANICAL ENGINEERING. *Educ:* Univ Minn, Minneapolis, BS, 51, MSME, 52, PhD(mech eng), 60. *Prof Exp:* Instr mech eng, Univ Minn, Minneapolis, 54-58; sr engr, Mech Div, Gen Mills, Inc, 56-60 & Lockheed-Calif Co, 60-62; PROF ENG, CALIF STATE UNIV, NORTHRIDGE, 62- *Mem:* Am Soc Mech Engrs; Soc Lubrication Engrs; Am Soc Eng Educ; Nat Soc Prof Engrs. *Res:* Hydrodynamic lubrication; journal bearings; rolling element bearings; measurement; instruments; vibration dynamics; material fatigue. *Mailing Add:* Sch Eng Calif State Univ Northridge CA 91330

NYQUIST, DENNIS PAUL, b Detroit, Mich, Aug 18, 39. ELECTRICAL ENGINEERING. *Educ:* Lawrence Inst Technol, BS, 61; Wayne State Univ, MS, 64; Mich State Univ, PhD(elec eng), 66. *Prof Exp:* Prod engr, Ford Motor Co, 61-63, res engr, Res Lab, 65; from instr to asst prof elec eng, 66-70, sr investr res grant, US Air Force, 67-71, assoc prof elec eng & systs sci, 70-81, PROF ELEC ENG, MICH STATE UNIV, 81- *Mem:* AAAS; Inst Elec & Electronics Engrs; Sigma Xi. *Res:* Applied electromagnetics; radiation and scattering of electromagnetic waves; interaction of electromagnetic waves with plasma; application of electromagnetic theory to electric theory of nerves. *Mailing Add:* Dept of Elec Eng Mich State Univ East Lansing MI 48823

NYQUIST, GERALD WARREN, b Detroit, Mich, Dec 28, 40. HUMAN IMPACT TOLERANCE, VEHICLE CRASH SAFETY. *Educ:* Lawrence Inst Technol, BS, 63; Wayne State Univ, MS, 67; Mich State Univ, PhD(appl mech), 70. *Prof Exp:* Prod test engr, Ford Motor Co, 63-65; res asst, Wayne State Univ, 65-67 & Mich State Univ, 67-68; res assoc, Wayne State Univ, 70-72; sr res engr, Gen Motors Corp, 72-76, staff proj engr, 76-79, sect supvr, 79-80, staff anal engr, 80-82; PRES, GERALD W NYQUIST, INC, 82- *Concurrent Pos:* Res assoc, Wayne State Univ, 83-86. *Honors & Awards:* Arch T Colwell Merit Award, Soc Automotive Engrs, 76. *Mem:* Asn Advan Automotive Med; Soc Automotive Engrs; Sigma Xi. *Res:* Biomechanics; human injury tolerance; mechanical response to impact; development of human surrogates for use in vehicle safety crash testing; vehicle occupant crash protection. *Mailing Add:* 19059 Holbrook St East Detroit MI 48021

NYQUIST, HARLAN LEROY, b Scobey, Mont, Aug 12, 29; m 55; c 3. ORGANIC CHEMISTRY. *Educ:* Mont State Col, BS, 51; Univ Calif, Los Angeles, PhD(chem), 56. *Prof Exp:* From instr to asst prof chem, Univ Calif, Santa Barbara, 56-62; from asst prof to assoc prof, 62-68, PROF CHEM, CALIF STATE UNIV, NORTHRIDGE, 68- *Concurrent Pos:* Distinguished vis prof, Dept Chem, Air Force Acad, Colo, 82-83. *Mem:* Am Chem Soc; Sigma Xi. *Res:* Mechanisms of organic reactions; synthesis and reactions of s-triazines; stereochemistry. *Mailing Add:* Dept Chem Calif State Univ Northridge CA 91326

NYQUIST, LAURENCE ELWOOD, b Tracy, Minn, July 28, 39; m 63; c 2. GEOCHEMISTRY, MASS SPECTROMETRY. *Educ:* Macalester Col, BA, 61; Univ Minn, MS, 63, PhD(physics), 69. *Prof Exp:* Sci asst mass spectrometry, Swiss Fed Inst Technol, 65-68; res assoc physics, Univ Minn, 69-71; aerospace technician geochem, 71-81, SPACE SCIENTIST, NASA JOHNSON SPACE CTR, 81- *Concurrent Pos:* Mem, Lunar Sample Anal Planning Team, 75-76 & Lunar Sci Rev Panel, 76-77; Lunar & Planetary Geosciences Rev Panel, 83-84. *Honors & Awards:* Exceptional Sci Serv, NASA, 76, 83. *Mem:* Am Geophys Union; Meteoritical Soc. *Res:* Chronological and geochemical evolution of the moon and other solid bodies in the solar system; radiometric ages by Rb-Sr and Sm-Nd techniques; use of strontium and neodymium isotopes as petrogenetic tracers. *Mailing Add:* 802 Richvale Houston TX 77062

NYQUIST, RICHARD ALLEN, b Rockford, Ill, May 3, 28; m 56; c 4. MOLECULAR SPECTROSCOPY. *Educ:* Agustana Col, Ill, BA, 51; Okla State Univ, MS, 53. *Prof Exp:* Proj leader infrared spectros, Chem Physics Res Lab, 53-71, from assoc scientist to sr assoc scientist infrared & raman spectros, Analytical Labs, 71-87, ANALYTICAL SCI, DOW CHEM CO, 88- *Honors & Awards:* Williams-Wright Award, 85. *Mem:* Am Chem Soc; hon mem Coblentz Soc; Am Soc Test Mat; Soc Appl Spectrosc. *Res:* Vibrational spectroscopy; elucidation of chemical structure. *Mailing Add:* 3707 Westbrier Terr Midland MI 48640

NYQUIST, SALLY ELIZABETH, b Hutchinson, Minn, Nov 8, 41. CELL BIOLOGY. *Educ:* Wheaton Col, BS, 63; Purdue Univ, Lafayette, PhD(biol), 70. *Prof Exp:* Volunteer, Peace Corps, 63-65; Kettering Found res fel, Charles F Kettering Res Lab, 70-72; ASST PROF BIOL, BUCKNELL UNIV, 72- *Mem:* Am Soc Cell Biol. *Res:* Golgi apparatus membranes, enzymatic activities and composition. *Mailing Add:* Dept of Biol Bucknell Univ Lewisburg PA 17837

NYQUIST, WYMAN ELLSWORTH, b Scobey, Mont, June 13, 28; m 52; c 2. STATISTICAL ANALYSIS, QUANTITATIVE GENETICS. *Educ:* Mont State Univ, BS, 50; Univ Calif, PhD(genetics), 53. *Prof Exp:* Grad asst agron, Univ Calif, 50-52, instr, 53-57, lectr, 57-58, asst prof, 58-63; assoc prof, 63-68, PROF AGRON, PURDUE UNIV, 68- *Concurrent Pos:* Jr agronomist, Exp Sta, Univ Calif, 53-57, asst agronomist, 57-63; NIH spec res fel, 69-70. *Mem:* Fel Am Soc Agron; Genetics Soc Am; Biomet Soc; Coun Agr Sci & Technol; fel Crop ci Soc Am; Sigma Xi. *Res:* Development of statistical models relating to quantitative genetic variation, particularly in plant populations and their utilization in evaluating alternative breeding systems. *Mailing Add:* Purdue Univ 1150 Lilly Hall Life Sci West Lafayette IN 47907-1150

NYQUIST-BATTIE, CYNTHIA, b Burbank, Calif, Dec 19, 47; m 81; c 2. NEUROBIOLOGY. *Educ:* Univ Calif, Irvine, BS, 72; Univ Calif, Los Angeles, PhD(anat), 77. *Prof Exp:* Fel biochem, Mt Sinai Med Sch, 77-79; res assoc neurobiol, Vet Admin, Kansas City, 79-80; asst prof biol sci, Northern Ill Univ, 80-84; ASST PROF BASIC LIFE SCIS, UNIV MO, 84- *Concurrent Pos:* Instr, Univ Calif, Irvine, 76. *Mem:* Soc Neurosci; Soc Cell Biol. *Res:* Cholinergic innervation of the heart; developmental biology of the heart and the central nervous system. *Mailing Add:* Dept Basic Life Sci Univ Mo 5100 Rockhill Rd Kansas City MO 64110

NYREN, PAUL ERIC, b Malvern, Iowa, May 21, 43; m 69; c 1. RANGE SCIENCE & MANAGEMENT. *Educ:* Wash State Univ, BS, 73, MS, 75. *Prof Exp:* Res assoc, bot dept, Land Reclamation Res Ctr, 75, botanist, Dickinson Exp Sta, 75-81, SUPT & RANGE SCIENTIST, NDAK STATE UNIV, CENT GRASSLANDS RES CTR, 81- *Mem:* Soc Range Mgt; Sigma Xi. *Res:* Range improvement (interseeding, fertilization); vegetation, animal interaction (various grazing systems); remote sensing (use of computer enhanced aerial photographs to determine above ground biomass). *Mailing Add:* Box 21 Streeter ND 58483

NYROP, JAN PETER, b Baltimore, Md, Apr 6, 55; m 79; c 2. POPULATION ECOLOGY, DECISION SUPPORT SYSTEMS. *Educ:* Univ Maine, BS, 78; Mich State Univ, MS, 80 & 82 & PhD(entomol), 82. *Prof Exp:* Res assoc, 83-85, ASST PROF INSECT POPULATION, NY STATE AGR EXP STA, CORNELL UNIV, 85- *Mem:* Entomol Soc Am; Am Asn Advan Sci. *Res:* Population ecology of arthropod pest of horticultural crops, including sampling and decision making, biological control, dispersal and host finding; dynamics of insecticide resistance. *Mailing Add:* Dept Entomol Barton Lab NY State Agr Exp Sta Geneva NY 14456

NYSSEN, GERARD ALLAN, b Hattiesburg, Miss, Nov 9, 42; m 65; c 3. ENVIRONMENTAL CHEMISTRY. *Educ:* Olivet Nazarene Col, AB, 65; Purdue Univ, PhD(inorg chem), 70. *Prof Exp:* Teaching asst chem, Purdue Univ, 65-66, res asst, 66-70; from asst prof to assoc prof, 70-76, PROF CHEM, TREVECCA NAZARENE COL, 76- *Concurrent Pos:* Vis prof, Belmont Col, Nashville, Tenn, 70-83, Vanderbilt Univ, 87-88; res assoc, dept chem, Okla State Univ, Stillwater, 71 & Vanderbilt Univ, 76-77 & 84-85; res scientist, anal chem div, Oak Ridge Nat Lab, 74; consult, MoTec Inc, 86-88. *Mem:* Am Chem Soc. *Res:* Investigation of the solubilities of hydrophobic compounds (pesticides, etc) in aqueous-organic solvent mixtures and how such mixtures migrate through compacted soil; improved method of removing dissolved antimony; improved method of removing dissolved pentachlorophenol. *Mailing Add:* Dept Sci & Math Trevecca Nazarene Col 333 Murfreesboro Rd Nashville TN 37217

NYSTED, LEONARD NORMAN, b Marshfield, Wis, May 17, 27; m 47; c 2. PHARMACEUTICAL CHEMISTRY. *Educ:* St Olaf Col, BS, 51. *Prof Exp:* Preparations & res chemist, G D Searle & Co, 51-56; dir chem res, Duraclean Co, 56-58; res chemist, G D Searle & Co, 58-64, sr res investr, 64-86; RETIRED. *Concurrent Pos:* Consult, Dupont Critical Care, Duraclean Co, 86- *Mem:* Am Chem Soc. *Res:* Medicinal chemistry particularly steroids and prostaglandins, insecticides and detergents. *Mailing Add:* 617 Rice St Highland Park IL 60035

NYSTROM, ROBERT FORREST, organic chemistry, radiochemistry; deceased, see previous edition for last biography

NYVALL, ROBERT FREDERICK, b Thief River Falls, Minn, Aug 23, 39; m 62; c 2. PLANT PATHOLOGY. *Educ:* Univ Minn, BS, 65, MS, 66, PhD(plant path), 69. *Prof Exp:* From asst prof to assoc prof, 70-81, exten plant pathologist, 70-85, PROF PLANT PATH, IOWA STATE UNIV, 81-; PROF PLANT PATH & SUPT, N CENTRAL EXP STA, UNIV MINN, 85- *Mem:* Am Inst Biol Sci; Am Phytopath Soc. *Res:* Ecology of soil; root rot and wilt fungi. *Mailing Add:* Univ Minn N Cent Exp Sta 1861 Hwy 169 E Grand Rapids MN 55744

NYYSSONEN, DIANA, b Cambridge, Mass; div; c 2. OPTICS. *Educ:* Boston Univ, BA, 65; Univ Rochester, PhD(optics), 75. *Prof Exp:* Staff scientist optics, Tech Oper, Inc, 64-68; physicist optics, 69-85, group leader, 80-85, GUEST RESEARCHER, NAT BUR STANDARDS, 85-; CONSULT, CD METROL INC, 85- *Mem:* Optical Soc Am; Soc Photo-Optical Instrumentation Engrs. *Res:* Optical metrology, scanning optical microscopy; scanning electron microscopy; integrated-circuit metrology; diffraction and scattering, partially coherent imaging, dimensional measurement standards. *Mailing Add:* IBM E Fishkill Facil Rte 52 Hopewell Junction NY 12533

O

OACE, SUSAN M, b St Paul, Minn, Nov 10, 41. NUTRITION. *Educ:* Univ Minn, St Paul, BS, 63; Univ Calif, Berkeley, PhD(nutrit), 67. *Prof Exp:* Res nutritionist, Univ Calif, Berkeley, 67-68; asst prof nutrit, Univ Calif, Davis, 68-73; asst prof, 73-75, dir, Coord Dietetics Prog, 76-80, ASSOC PROF NUTRIT, UNIV CALIF, BERKELEY, 75- *Concurrent Pos:* Ed, J Nutrit Educ, 79-82; mem joint coun, Foods & Agr Sci, 79-82. *Mem:* Am Inst Nutrit; Am Dietetic Asn; AAAS; Soc Exp Biol & Med. *Res:* Metabolic interrelationships among folic acid, vitamin B-12 and methionine; interaction of intestinal microflora with nutritional and health status of host; dietary fiber. *Mailing Add:* Dept of Nutrit Sci Univ of Calif Berkeley CA 94720

OAKES, BILLY DEAN, b Tulsa, Okla, Sept 5, 28; m 50; c 2. ORGANIC CHEMISTRY. *Educ:* Okla State Univ, BS, 50; Univ Wichita, MS, 52. *Prof Exp:* AEC asst, Univ Wichita, 52 & Ga Inst Technol, 52-53; res chemist, Dowell Div, 53-62 & Chem Dept Res Lab, Mich, 62-64, proj leader, 64-69, res specialist, Resources Res Dept, 69-76, res leader, Process Res & Eng Dep, Tex Div, Dow Chem Co, 76-84; RETIRED. *Mem:* Am Chem Soc; Nat Asn Corrosion Eng; Sigma Xi. *Res:* Corrosion studies; corrosion inhibitors for aqueous solutions, primarily acids, amine gas processing solvents and automotive engine coolants; environmental modifications designed to allow the use of low cost construction materials in sea water desalination plants. *Mailing Add:* 444 Southern Oaks Dr Lake Jackson TX 77566-4514

OAKES, DAVID, b Stockport, Eng, May 8, 47; m 77. SURVIVAL ANALYSIS. *Educ:* Cambridge, BA, 68, MA, 72; London, PhD(statist), 72. *Prof Exp:* Asst prof statist, Harvard Univ, 72-77; sr lectr occup health statist, London Sch Hyg & Trop Med, 77-83; assoc prof, 83-87, PROF STATIST & BIOSTATIST, UNIV ROCHESTER, 87-, CHMN DEPT STATIST, 89- *Concurrent Pos:* Assoc ed, Biometrika, 79-; prin investr, Statist Methods for Multiple Event Time Data, NCI Grant; chief biostatistician, Datatop-Deprenyl & Tocopherol Antioxidative Ther of Parkinsonism Study, 87- *Mem:* Royal Statist Soc; Am Statist Asn; Biometric Soc; Inst Math Statist. *Res:* Statistical theory and methodology; clinical trials; epidemiologic studies; analysis of survival data; random effects (frailty) models for multiple event times. *Mailing Add:* Dept Statist Univ Rochester Rochester NY 14627

OAKES, JOHN MORGAN, b Bridgeport, Conn, Feb 8, 55; m 84; c 2. SPECTROSCOPY. *Educ:* Univ Conn, BS, 77; Univ Colo, PhD(chem physics), 84. *Prof Exp:* Res asst chem, Univ Conn, Storrs, 75-77; chemist, Clairol, Inc, 77-78; res asst chem, Univ Colo, Boulder, 78-84; asst prof chem, Gonzaga Univ, Wash, 84-86; instr, San Diego Col, 86-87; INSTR, UNIV CALIF, SAN DIEGO, 88- *Mem:* Sigma Xi. *Res:* Negative ion photoelectron spectroscopy used to study small organic ions and the hydroperoxyl anion; spectroscopic, geometric and thermodynamic information on the important species and their corresponding neutrals. *Mailing Add:* Bishop Sch La Jolla CA 92137

OAKES, LESTER C, b Knoxville, Tenn, Oct 11, 23. REACTOR CONTROLS & INSTRUMENTATION. *Educ:* Univ Tenn, BS, 49, MS, 62. *Prof Exp:* Staff engr, Reactor Systs Sect, 51-68, head reactor systs sect, 68-70, ASSOC DIR, INSTRUMENTATION & CONTROLS DIV, OAK RIDGE NAT LAB, 70- *Concurrent Pos:* Corp fel, Martin Marietta Corp, 86- *Mem:* Fel Inst Elec & Electronic Engrs; Am Nuclear Soc. *Res:* Development of advanced control systems for nuclear reactors. *Mailing Add:* 5016 Mountaincrest Dr Knoxville TN 37918

OAKES, MELVIN ERVIN LOUIS, b Vicksburg, Miss, May 11, 36; m 63; c 3. PLASMA PHYSICS. *Educ:* Fla State Univ, PhD(plasma physics), 64. *Prof Exp:* Asst physics, Fla State Univ, 58-60 & 60-64; physicist, Army Res Guided Missile Agency, Redstone Arsenal, 60; asst prof physics, Univ Ga, 64; res assoc, 64-65, from asst prof to assoc prof, 65-70, PROF PHYSICS, UNIV TEX, AUSTIN, 75- *Mem:* Am Phys Soc; Am Asn Physics Teachers. *Res:* Electromagnetic interaction with plasmas; plasma waves and radio frequency heating. *Mailing Add:* Dept of Physics Univ of Tex Austin TX 78712

OAKES, ROBERT JAMES, b Minneapolis, Minn, Jan 21, 36. THEORETICAL HIGH ENERGY PHYSICS. *Educ:* Univ Minn, BS, 57, MS, 59, PhD(physics), 62. *Prof Exp:* NSF fel physics, Stanford Univ, 62-64, lectr, 64, asst prof, 64-68; assoc prof, 68-70, PROF PHYSICS, NORTHWESTERN UNIV, 70-, PROF PHYSICS & ASTRON, 76- *Concurrent Pos:* A P Sloan Found res fel, 65-67; vis scientist, Europ Orgn Nuclear Res, 66-67; mem, Inst Advan Study, 67-68; vis scientist, Deutsches Elektron-Synchrotron, 71-72, Los Alamos Sci Lab, 71- & Fermi Nat Accelerator Lab, 75-; mem, Deep Underwater Muon & Neutrino Detection steering comt, 78-85; fac assoc, Argonne Nat Lab, 82- *Mem:* Fel Am Phys Soc; NY Acad Sci; fel AAAS. *Res:* Strong interactions; theoretical, high energy and nuclear physics; field theory; weak interactions; relativity and cosmology. *Mailing Add:* Dept of Physics Northwestern Univ Evanston IL 60208

OAKES, THOMAS WYATT, b Danville, Va, June 14, 50; m 74. ENVIRONMENTAL ENGINEERING, HEALTH PHYSICS. *Educ:* Va Polytech Inst & State Univ, BS, 73, MS, 75; Univ Tenn, MS, 80. *Prof Exp:* Health physicist, Va Polytech Inst & State Univ, 73-74; engr radiation, Nuclear Div, Babcock & Wilcox, 74-75; health physics supvr monitoring group, Oak Ridge Nat Lab, 75-76, group leader environ monitoring, 76-77, sect head, Environ Surveillance Sect, 77-79, environ coordr, 78-79, dept head, 79; corp environ coordr, Martin Marietta Energy Systs, Oak Ridge, Tenn, 79-87; div mgr & asst vpres, Sci Appln Systs, 87-90; MGR ENVIRON & GEOTECH SERV, WESTINGHOUSE, 90- *Mem:* AAAS; Health Physics Soc; Am Nuclear Soc; NY Acad Sci; Am Indust Hyg Asn; Am Qual Assurance Soc. *Res:* Environmental surveillance; quality assurance; determinations of pollutant assessment of environments impact of energy system; biological monitoring neutron activation; health physics; environment health; Cost-benefit analysis. *Mailing Add:* 1208 Buxton Dr Knoxville TN 37922

OAKESHOTT, GORDON B, b Oakland, Calif, Dec 24, 04; m 29, 85; c 3. GEOLOGY. *Educ:* Univ Calif, BS, 28, MS, 29; Univ Southern Calif, PhD(geol), 36. *Prof Exp:* Asst field geologist, Shell Oil Co, 29-30; instr earth sci, Compton Dist Jr Col, 30-48; supv mining geologist, Calif State Div Mines & Geol, 48-56, dep chief, 56-57 & 59-72, actg chief, 58; RETIRED. *Concurrent Pos:* Instr geol, Calif State Univ, Sacramento, 73 & Calif State Univ, San Francisco, 75; consult geologist, 73-85. *Mem:* Fel AAAS; fel & sr mem Geol Soc Am; Asn Prof Geol Scientists; Seismol Soc Am; Webb hon mem Nat Asn Geol Teachers (pres, 71); hon mem Am Asn Petrol Geologists; hon mem Asn Eng Geologists. *Res:* Geology of the San Gabriel Mountains; stratigraphy of California Coast and Transverse Ranges; surface faulting and associated earthquakes; auth geology of California, (2 ed). *Mailing Add:* 3040 Totterdell St Oakland CA 94611

OAKFORD, ROBERT VERNON, b Winfield, Kans, Nov 3, 17; m 48; c 3. INDUSTRIAL ENGINEERING. *Educ:* Stanford Univ, BA, 40, MS, 56. *Prof Exp:* Engr, Northrop Aircraft, Inc, 41-45; secy-treas, Oakford Gas & Appliance Corp, 45-49; engr, US Govt, 51-53; lectr, 55-57, assoc prof, 57-62, prof, 62-77, EMER PROF INDUST ENG, STANFORD UNIV, 77- *Concurrent Pos:* Consult, McKinsey & Co, 57-62, Res Div, State Dept Correction, Calif, 59-62 & Calif Pub Utilities Staff, 85-90. *Honors & Awards:* E L Grant Award, 83; Wellington Award, 85. *Res:* Theory and practice of school scheduling; engineering economy and capital budgeting. *Mailing Add:* 10 Maywood Lane Menlo Park CA 94025

OAKLEY, BERL RAY, b Roxboro, NC, Sept 25, 49; m 77. CELL BIOLOGY, CELL MOTILITY. *Educ:* Duke Univ, BS, 71; Univ London, PhD(cell biol), 74. *Prof Exp:* Instr pharmacol, Med Sch, Rutgers Univ, 77-81; asst prof microbiol, 82-88, ASSOC PROF MOLECULAR GENETICS, OHIO STATE UNIV, 88- *Mem:* Am Soc Cell Biol; Genetics Soc Am; AAAS; Soc Evolutionary Protistology. *Res:* Mechanisms of mitosis and organellar movement; genetic and molecular genetic analyses of the structure and function of microtubule proteins. *Mailing Add:* Dept Molecular Genetics 557 Biosci Ohio State Univ 484 W 12th Ave Columbus OH 43210

OAKLEY, BRUCE, b Philadelphia, Pa, Oct 22, 36; m 58; c 2. NEUROBIOLOGY. *Educ:* Swarthmore Col, BA, 58; Brown Univ, MSc, 60, PhD(psychol), 62. *Prof Exp:* Asst prof psychol, Brown Univ, 62-63; Nat Acad Sci-Nat Res Coun sr fel physiol, Royal Vet Col Sweden, 63-64; USPHS fel zool, Univ Calif, Los Angeles, 64-65, asst res zoologist, 65-66; asst prof, 66-71, assoc prof zool, 71-76, PROF BIOL SCI, UNIV MICH, ANN ARBOR, 76- *Mem:* AAAS; Am Soc Zool; NY Acad Sci; Soc Neurosci; Am Chem Soc. *Res:* Physiology and behavior of taste; neural mechanism of preference and aversion; sensory receptors and coding; trophic functions of neurons; developmental neurobiology. *Mailing Add:* 4042a Nat Sci Biol Univ Mich Ann Arbor MI 48109-1048

OAKLEY, DAVID CHARLES, b Marysville, Calif, July 4, 29; m 52; c 3. UNDERGROUND NUCLEAR TESTING. *Educ:* Calif Inst Technol, BS, 50, MS, 52, PhD(physics & math), 55. *Prof Exp:* Physicist nuclear sci, Lawrence Livermore Lab, 54-73; asst dept dir for testing, Nuclear Sci Admin, Defense Nuclear Agency, 73-76; PHYSICIST NUCLEAR SCI, LAWRENCE LIVERMORE LAB, 76- *Mem:* Am Phys Soc; Sigma Xi. *Res:* Design and testing of nuclear devices; administration of underground nuclear testing of nuclear weapons effects on materials and systems of military interest. *Mailing Add:* Livermore Lab L-41 PO Box 808 Livermore CA 94550-0622

OAKLEY, GODFREY PORTER, JR, b Greenville, NC, June 1, 40; m 61; c 3. BIRTH DEFECTS & DEVELOPMENTAL DISABILITIES, PEDIATRICS. *Educ:* Bowman Gray Sch Med, MD, 65; Univ Wash, MSPH, 72. *Prof Exp:* DIR, DIV BIRTH DEFECTS & DEVELOPMENTAL DISABILITIES, CTR DIS CONTROL, USPHS, 68- *Concurrent Pos:* Clin asst, Div Med Genetics, Emory Univ, 68-70, prof pediat, 72- & prof obstet & gynecol, 81-; mem vis staff, Grady Mem Hosp, Atlanta, 74-; chmn, Int Clearinghouse Birth Defects Monitoring Systs, 81-84. *Mem:* Teratology Soc (pres, 84-85); Am Acad Pediat; Soc Epidemiol Res; Am Epidemiol Soc; Am Pub Health F. *Res:* Environmental health; the prevention of birth defects and developmental disabilities through surveillance and epidemiologic and genetic studies. *Mailing Add:* Ctr Environ Health F-37 1600 Clifton Rd NE Atlanta GA 30333

OAKS, B ANN, b Winnipeg, Man, June 4, 29. PLANT PHYSIOLOGY-MELABOLIC & MOLECULAR ASPECTS. *Educ:* Univ Toronto, BA, 51; Univ Sask, MA, 54, PhD(plant physiol), 58. *Hon Degrees:* FRSCan. *Prof Exp:* Res asst plant physiol, Univ Man, 53-54; Von Humboldt grant, Bact Inst, Feising, Ger, 58-60; res assoc biol sci, Purdue Univ, 60-64; from asst prof to prof biol, McMaster Univ 64-89; res prof, Botany Dept, Univ Guelph, 89. *Honors & Awards:* Gold Medal, Can Soc Plant Physiologists, 89. *Mem:* Am Soc Plant Physiol; Can Soc Plant Physiol. *Res:* Intermediary metabolism in plants; processes regulating development in seedlings; physiological and molecular studies; emphasis on the regulation of nitrate reductase and the hydrolysis of storage proteins in maize and barley. *Mailing Add:* Dept Botany Univ Guelph Guelph ON N1G 2W1 Can

OAKS, EMILY CAYWOOD JORDAN, b Pittsburgh, Pa, Feb 15, 39; div; c 2. VERTEBRATE ZOOLOGY, ANATOMY. *Educ:* Rice Univ, BA, 61; Yale Univ, MS, 64, PhD(biol), 67. *Prof Exp:* Asst prof biol, Utah State Univ, 67-83; asst prof zool, 84-85, ASST PROF BIOL, STATE UNIV NY, OSWEGO, 85- *Mem:* Am Soc Mammal; Sigma Xi; AAAS. *Res:* Anatomy and adaptive function of the middle ear in mammals. *Mailing Add:* Dept Biol State Univ NY Oswego NY 13126

OAKS, J HOWARD, b Camden, NJ, Mar 3, 30; m 57; c 5. DENTISTRY. *Educ:* Wesleyan Univ, BA, 52; Harvard Univ, DMD, 56. *Prof Exp:* Instr oper dent, Sch Dent Med, Harvard Univ, 56-64, lectr oper dent & assoc dean, 64-68, actg dean, 67-68; prof dent med & dean, Sch Dent Med, 68-74, VPRES HEALTH SCI, STATE UNIV NY STONY BROOK, 74- *Concurrent Pos:* Mem dent educ rev comt, Bur Health Manpower Educ, NIH, 67-70, consult, div physician & health professions educ, 70-; vpres for Deans, Am Asn Dent Schs, 72-73. *Mailing Add:* Health Sci State Univ NY Health Sci Ctr Stony Brook NY 11794

OAKS, JOHN ADAMS, b Alma, Mich, Apr 8, 42; m 65; c 2. CELL BIOLOGY, PARASITOLOGY. *Educ:* Colby Col, BA, 64; Tulane Univ, MS, 68, PhD(cell biol), 70. *Prof Exp:* Asst prof parasitol, Sch Pub Health, Tulane Univ, 70-73; asst prof, Dept Anat, Col Med, Univ Iowa, 73-79, assoc prof, 79-81; ASSOC PROF, DEPT COMP BIOSCI, SCH VET MED, UNIV WIS-MADISON, 83-, COORDR, CTR RES & TRAINING PARASITIC DIS. *Mem:* AAAS; Am Soc Cell Biol; Am Soc Parasitol; Am Soc Trop Med Hyg. *Res:* Mechanisms of plasma membrane synthesis; structural and functional aspects of parasitic helminth epithelia; cytoskeleton of cestode tegument; mechanism of host cellular reactions to Toxocara canis; comparative aspects of free-living and parasitic helminth epithelia; parasite-host intestinal interactions. *Mailing Add:* Sch Vet Med Dept Comp Biosci Univ Wis-Madison Madison WI 53706

OAKS, ROBERT QUINCY, JR, b Houston, Tex, Aug 29, 38; div; c 2. GEOLOGY. *Educ:* Rice Univ, BA, 60; Yale Univ, PhD(geol), 65. *Prof Exp:* Res geologist, Jersey Prod Res Co, 64 & Esso Prod Res Co, 65-66; from asst prof to assoc prof, 66-79, PROF GEOL, UTAH STATE UNIV, 79- *Concurrent Pos:* Consult, Magellan Petrol Australia Ltd, 79-86. *Mem:* AAAS; Soc Econ Paleont & Mineral; Geol Soc Am; Am Asn Petrol Geol; Int Asn Sedimentol; Sigma Xi. *Res:* Ordovician quartzites in northern Utah and southen Idaho; the role of fine-grained sediments in promoting erosion in mountain lands; regolith classification for land-management needs; Cambrian carbonates and sandstones in northern Utah and southern Idaho; mud flows and debris flows on alluvial fans in Death Valley and Eureka Valley, Calif; structure and stratigraphy of northeastern Amadeus basin, Central Australia. *Mailing Add:* Dept Geol Utah State Univ Logan UT 84322-4505

OAKS, WILBUR W, b Philadelphia, Pa, Oct 12, 28; m 54; c 3. INTERNAL MEDICINE. *Educ:* Lafayette Col, BS, 51; Hahnemann Med Col, MD, 55. *Prof Exp:* From instr to assoc prof med, 61-69, teaching coordr & dir postgrad educ, 62-70, PROF MED & DIR DIV GEN INTERNAL MED, HAHNEMANN MED COL, 69-, CHMN DEPT MED, 73- *Concurrent Pos:* Staff physician, Hahnemann Hosp, 61- *Honors & Awards:* Christian R & Mary F Lindback Found Award. *Mem:* Am Fedn Clin Res; AMA; Am Col Chest Physicians; Asn Am Med Cols. *Res:* Hypertension. *Mailing Add:* 1320 Race St Philadelphia PA 19107

OALMANN, MARGARET CLAIRE, b Covington, La, Aug 16, 29; m 62; c 3. EPIDEMIOLOGY. *Educ:* La State Univ, BSNEd, 56; Tulane Univ, MPH, 58, DPH(chronic dis, epidemiol), 60. *Prof Exp:* From actg head nurse to head nurse, Charity Hosp La, New Orleans, 49-53, clin instr, 53-57; res assoc, Tulane Univ, 60-62; from instr to asst prof path, 62-70, asst prof, 70-72, ASSOC PROF PATH, PUB HEALTH & PREV MED, MED CTR, LA STATE UNIV, NEW ORLEANS, 72- *Concurrent Pos:* Vis scientist, Charity Hosp, La, 60-; epidemiol consult mortality in nuns, Am Cancer Soc Grant, 63- *Mem:* AAAS; fel Am Heart Asn; Am Pub Health Asn; Asn Teachers Prev Med; Royal Soc Health. *Res:* Epidemiology of cardiovascular disease and cancer. *Mailing Add:* Dept Path La State Univ Med Ctr 1901 Perdido St New Orleans LA 70112

OATES, GORDON CEDRIC, aeronautical engineering; deceased, see previous edition for last biography

OATES, JIMMIE C, b Memphis, Tenn, Apr 14, 33; m 54; c 2. PHYSICS. *Educ:* Memphis State Univ, BS, 58; Vanderbilt Univ, MS, 60, PhD(physics), 63. *Prof Exp:* Assoc prof, 62-70, PROF PHYSICS, QUEENS COL, NC, 70- *Res:* Bioacoustics. *Mailing Add:* Dept of Physics Queens Col 1900 Selwyn Ave Charlotte NC 28274

OATES, JOHN ALEXANDER, b Fayetteville, NC, Apr 23, 32; m 56; c 3. CLINICAL PHARMACOLOGY, INTERNAL MEDICINE. *Educ:* Wake Forest Col, BS, 53; Bowman Gray Sch Med, MD, 56. *Prof Exp:* Intern internal med, New York Hosp, 56-57, asst resident, 57-58; clin assoc exp therapeut, Nat Heart Inst, 58-61; asst resident med, New York Hosp, 61-62; sr investr, 62-63; from asst prof to assoc prof pharmacol & med, 63-69, PROF MED & PHARMACOL, SCH MED, VANDERBILT UNIV, 69-, CHMN DEPT MED, 83- *Concurrent Pos:* Burroughs Wellcome scholar clin pharmacol, 65-70. *Honors & Awards:* Am Soc PHarmacol & Exp Therapeut Award Exp Therapeut, 69. *Mem:* Am Soc Clin Invest; Asn Am Physicians (secy, 70-75, pres, 82); Am Soc Pharmacol & Exp Therapeut. *Res:* Vasoactive amines and peptides; prostaglandins; antihypertensive agents; autonomic pharmacology. *Mailing Add:* Vanderbilt Univ Sch Med Nashville TN 37232

OATES, PETER JOSEPH, b New York, NY, July 26, 47; m 71; c 2. DIABETIC COMPLICATIONS, GASTROINTESTINAL PHYSIOLOGY. *Educ:* Boston Col, BS, 69; Vanderbilt Univ, PhD(molecular biol), 75. *Prof Exp:* Res fel membrane biochem, Dept Molecular Biol, Vanderbilt Univ, 75-79; res scientist gastrointestinal dis, 79-81, sr res scientist, 81, proj leader gastrointestinal dis & diuretic res, 82-86, SR RES INVESTR DIABETIC COMPLICATIONS, CENT RES DIV, PFIZER INC, 86- *Mem:* AAAS; Am Soc Cell Biol; Am Diabetes Asn; Europ Asn Study Diabetes; Am Soc Hypertension. *Res:* Mechanisms of diabetic complications, especially nephropathy, retinopathy and neuropathy; mechanisms of gastrointestinal ulceration; development of experimental model systems for the discovery of new therapeutic agents. *Mailing Add:* Cent Pfizer Inc 16 Ferry View Dr Gales Ferry CT 06335

OATES, RICHARD PATRICK, b Gary, Ind, Mar 17, 37; m 68. BIOSTATISTICS. *Educ:* Purdue Univ, BS, 58; Iowa State Univ, MS, 60, PhD(bact), 64. *Prof Exp:* From asst prof to assoc prof, interim chmn dept prev med, 85-90 , PROF BIOSTATIST, STATE UNIV NY HEALTH SCI CTR, 87-, CHMN DEPT PREV MED, 90- *Concurrent Pos:* NIH fel, Iowa State Univ, 64-65. *Mem:* Sigma Xi; NY Acad Sci; Am Pub Health Asn. *Res:* Medical research; epidemiology. *Mailing Add:* Dept Prev Med State Univ NY Health Sci Ctr Syracuse NY 13210

OATFIELD, HAROLD, b Concord, Ore, Jan 25, 10; m 38; c 1. ORGANIC CHEMISTRY. *Educ:* Rice Inst, BA, 31; Iowa State Univ, MS, 33. *Prof Exp:* Tester, control lab, Crown Zellerbach Corp, Ore, 33-34; chemist, intel div exp sta, E I du Pont de Nemours & Co, Del, 34-45; libr res fel biol, Mass Inst Technol, 45-47; prof assoc, div med sci, Nat Res Coun, 47-51; res chemist, chem res & develop div & head lit res group, Chas Pfizer & Co, Inc, NY, 51-59, head tech info serv, Pfizer Med Res Labs, Conn, 60-69, head comp based info dept, 70-71; RETIRED. *Concurrent Pos:* Guest lectr, Columbia Univ, 57-60. *Mem:* Fel AAAS; Drug Info Asn; fel Am Inst Chem; Am Chem Soc; Med Libr Asn; Am Soc Info Sci (treas, 54). *Res:* Dibenzofuran; embalming; literature of chemical periphery; science documentation. *Mailing Add:* 12705 S E River Rd Milwaukie OR 97222

OATMAN, EARL R, b Sylvester, Tex, Oct 21, 20; m 53; c 3. ENTOMOLOGY. *Prof Exp:* Asst instr entom, Univ Mo, 51-52; res asst entom & parasitol, Univ Calif, Berkeley, 53-56; from asst prof to assoc prof entom, Univ Wis, 56-62; from asst entomologist to assoc entomologist, 62-72, ENTOMOLOGIST, DIV BIOL CONTROL, UNIV CALIF, RIVERSIDE, 72-, PROF ENTOM, 75- *Mem:* Entom Soc Am. *Res:* Population ecology and biological control of insects and mites associated with agronomic and horticultural crops. *Mailing Add:* 565 W Campus View Dr Riverside CA 92507

O'BANNON, JOHN HORATIO, b West Palm Beach, Fla, Sept 23, 26; m 52; c 3. NEMATOLOGY. *Educ:* Univ Ariz, BS & MS, 57; Ariz State Univ, PhD(bot), 65. *Prof Exp:* Nematologist, Cotton Res Ctr, Ariz, 57-65 & nematol invests, Agr Res Serv, USDA, Orlando, 65-78; nematologist, Irrig Agr Res & Ext Ctr, Wash State Univ, 78-84; CHIEF, BUR NEMATOL, 84- *Mem:* Am Phytopath Soc; fel Soc Nematologists (pres, 74-75). *Res:* Nematology concerned with the biology, ecology and control of plant parasitic nematodes; regulatory nematology. *Mailing Add:* Div Plant Indust Fla Dept Agr PO Box 1269 Gainesville FL 32602-1269

O'BARR, RICHARD DALE, b Thorsby, Ala, Apr 26, 32; m 59; c 3. PLANT NUTRITION, PLANT ANALYSIS. *Educ:* Ala Polytech Inst, BS, 52; Univ Ga, MS, 71, PhD(plant sci), 74. *Prof Exp:* Asst prof hort, 74-79, PROF & RESIDENT DIR HORT & ADMIN, PECAN RES & EXTENSION STA, LA AGR CTR, LA STATE UNIV, 79- *Mem:* Am Soc Hort Sci; Res Ctr Adminr Soc; Reserve Officers Asn; Nat Pecan Res & Exten Scientists. *Res:* Physiology of bearing pecan trees; germination and seed sources for rootstocks, toxicity studies and colchicine; new varieties are studied for geographical adaptability and possible release; tissue analysis of leaves, shucks, shells and kernels to determine nutritional status of trees and treatment effects. *Mailing Add:* Pecan Res-Exten Sta La Agr Ctr PO Box 5519 Shreveport LA 71135-5519

O'BARR, WILLIAM MCALSTON, b Sylvania, Ga, Dec 1, 42; m 65; c 2. ANTHROPOLOGY. *Educ:* Emory Univ, BA, 64; Northwestern Univ, MA, 66, PhD(anthrop), 69. *Prof Exp:* Res assoc med sociol, Communicable Dis Ctr, USPHS, 64-67; res assoc social anthrop, Univ Dar es Salaam, 67-68; from asst to prof anthrop, Duke Univ, 69-78, chmn dept, 82-85; JOINT APPOINTMENT, DEPT SOCIOL & ADJ PROF LAW, UNIV NC, CHAPEL HILL, 89- *Concurrent Pos:* Vis assoc prof, Dalhousie Univ, Can, 76, Northwestern Univ, 78. *Mem:* Am Anthrop Asn; African Studies Asn; Am Ethnol Soc; Royal Anthrop Inst Gt Brit; Ling Soc Am; Law & Soc Asn. *Res:* Language of politics and law; advertising language; bilingualism; African ethnology; investment behavior. *Mailing Add:* Dept Anthrop Duke Univ Durham NC 27706

OBBINK, RUSSELL C, b Omaha, Nebr, Sept 29, 24; m 46; c 5. ANALYTICAL CHEMISTRY, ORGANIC CHEMISTRY. *Educ:* Univ Portland, BSc, 53. *Prof Exp:* Analyst, Alcoa, Aluminum Co Am, Wash, 49-53, analytical chemist, 53-57, sr chemist, 57-65, res chemist, Alcoa Res Labs, Aluminum Co Am, 65-67, group leader, 67-70, sect head, 73-78, sci assoc, Analytical Chem Div, Alcoa Tech Ctr, 78-81; RES ASSOC, AM SOC TESTING & MAT, 80- *Mem:* Sigma Xi; Fine Particle Soc; Am Soc Testing & Mat. *Res:* Development of analytical methods for use in aluminum industry. *Mailing Add:* 636 Vance Dr Lower Burrell PA 15068

OBEAR, FREDERICK W, b Malden, Mass, June 9, 35; m 59; c 3. INORGANIC CHEMISTRY. *Educ:* Lowell Tech Inst, BS, 56; Univ NH, PhD(chem), 61. *Hon Degrees:* LHD, Univ Lowell, 85. *Prof Exp:* Asst chem, Univ NH, 56-58, fel, 58-60; asst prof, Oakland Univ, 60-66, dean freshmen, 64-66, asst provost, 65-68, vprovost, 68-70, assoc prof chem, 66-78, vpres acad affairs & provost, 70-81; CHANCELLOR, UNIV TENN, CHATTANOOGA, 81- *Mem:* Am Chem Soc; AAAS; Am Asn Higher Educ. *Res:* Transition metal inorganic chemistry structure and mechanisms; academic administration. *Mailing Add:* Off Chancellor Univ Tenn Chattanooga TN 37403

O'BEIRNE, ANDREW JON, Philadelphia, Pa, Oct 26, 44; m 68; c 2. IMMUNOCHEMISTRY, VIROLOGY. *Educ:* Philadelphia Col Pharm & Sci, BSc, 66; Univ Mich, MPH, 72, DrPH(virol), 73. *Prof Exp:* Co-head virus, 73-74, dir res & develop, 74-78, vpres res & develop, 78-80, vpres opers, res & develop, 80-83, sr vpres, 83-84, GEN MGR DIAG DIV, WHITTAKER M A BIOPRODUCTS, 84- *Concurrent Pos:* Lectr, Hood Col, 83-, mem, sci adv coun, 83-; ed, J Clin Microbiol, Am Soc Microbiol, 83- *Mem:* Am Soc Microbiol; Sigma Xi; Am Pub Health Asn; Am Asn Pathologists; NY Acad Sci. *Res:* Rapid viral diagnostic methods; immunodiagnostic techniques. *Mailing Add:* Whittaker Bioprod Inc 8830 Biggs Ford Rd Walkersville MD 21793

OBEJI, JOHN T, b Aleppo, Syria, Jan 10, 39; US citizen; m 61; c 5. RESEARCH ADMINISTRATION. *Educ:* Fairleigh Dickinson Univ, MS, 75. *Prof Exp:* VPRES RES, SCHER CHEM, INC, 85- *Res:* Surfactants; new class of cationic emulsifiers. *Mailing Add:* 203 Charles St Clifton NJ 07013-3853

O'BENAR, JOHN DEMARION, b Chicago, Ill, Apr 10, 43; m 71. SHOCK RESEARCH, ENVIRONMENTAL NEUROPHYSIOLOGY. *Educ:* Cornell Univ, BA, 64; Univ Ill, MS, 69, PhD(physiol), 71. *Prof Exp:* Res fel neurophysiol, Univ Calif, Berkeley, 73; res physiologist, Naval Weapons Ctr, 73-76 & Naval Aerospace Med Res Lab, 76-79; anal chemist, Army Aviation Ctr, 79-80; RES PHYSIOLOGIST, LETTERMAN ARMY INST RES, 80- *Concurrent Pos:* Consult, Airforce Off Sci Res, 69-71, Transp Systs Ctr, 73-75, Naval Res Lab, 74-75, Brooks AFB, 75-76, Naval Med Res Inst, 76-79 & Army Med Res & Develop Command, 80-86; prin investr, Naval Aerospace Med Res Lab, 76-79 & Letterman Army Inst Res, 80-86. *Mem:* Am Physiol Soc; AAAS; Bioelectromagnetics Soc; NY Acad Sci; Am Acad Optom; Sigma Xi; Shock Soc. *Res:* Neurophysiological effects of electromagnetic radiation; cardiovascular effects of endorphins; biophysics of microcirculation in coronary vasculature; developmental strategies for medical management of hemorrhagic shock; uses of new pharmacologic agents and blood substitutes. *Mailing Add:* 610 Richardson Ave San Francisco CA 94123

OBENCHAIN, CARL F(RANKLIN), b Hailey, Idaho, Apr 25, 35; m 61; c 2. NUCLEAR ENGINEERING, CHEMICAL ENGINEERING. *Educ:* Ore State Univ, BS, 58; Univ Mich, MSE, 59 & 61, PhD(chem eng), 64; Univ Idaho, MBA, 77. *Prof Exp:* Reactor analyst-physicist, Atomic Energy Div, Phillips Petrol Co, 64-67, group leader, 67-69; power reactor safety anal group, Idaho Nuclear Corp, 69-71; group supvr, Nuclear Safety Prog Div, Aerojet Nuclear Corp, 71-74; proj mgr, Reactor Behav Div, Aerojet Nuclear & EG&G, Idaho, 74-77; mgr, Regulatory Support Br, 77-85, MGR, NUCLEAR REGULATORY COMN TECH ASST, EG&G, IDAHO, 85- *Concurrent Pos:* Adj prof, Univ Idaho, 65- *Mem:* AAAS; Am Inst Chem Engrs; Am Nuclear Soc. *Res:* Solute interactions in dilute liquid metal systems; thermal-hydraulic analysis of nuclear reactors; water reactor safety. *Mailing Add:* EG&G Idaho Inc PO Box 1625 Idaho Falls ID 83415

OBENCHAIN, ROBERT LINCOLN, b Indianapolis, Ind, Apr 2, 41; m 69, 76; c 1. STATISTICS, OPERATIONS RESEARCH. *Educ:* Northwestern Univ, BS, 64; Univ NC, Chapel Hill, PhD(statist), 69. *Prof Exp:* MEM TECH STAFF STATIST, BELL TELEPHONE LABS, 69- *Concurrent Pos:* Vis assoc prof statist, Univ Wis-Madison, 79-80; assoc ed, J Am Statist Asn, 80- *Mem:* Am Statist Asn; Inst Math Statist. *Res:* Biased linear regression techniques; multivariate analysis. *Mailing Add:* 5261 Woodfield Dr Carmel IN 46032

OBENDORF, RALPH LOUIS, b Milan, Ind, July 11, 38; m 67; c 2. PLANT PHYSIOLOGY, AGRONOMY. *Educ:* Purdue Univ, BS, 60; Univ California, Davis, MS, 62, PhD(plant physiol), 66. *Prof Exp:* Asst prof field crop sci, 66-71, assoc prof, 71-77, PROF CROP SCI, CORNELL UNIV, 77- *Concurrent Pos:* Vis scientist, Inst Cancer Res, Philadelphia, Pa, 72-73; vis plant physiologist, Plant Growth Lab, Univ Calif at Davis, 83; Int Bus Mach Corp advan educ proj award software develop, 88. *Honors & Awards:* Sower's Medal, Poznan Agr Univ Poland, 87; Res Agri Medal, Polish Ministry Agri, 87. *Mem:* AAAS; Am Soc Plant Physiol; Am Inst Biol Sci; Crop Sci Soc Am; Am Soc Agron; Tissue Culture Asn; Am Asn Cereal Chemists; NY Acad Sci; Am Chem Soc. *Res:* Physiology and biochemistry of seed and regenerated embryo, formation, maturation and germination; seed biology. *Mailing Add:* Dept Soil Crop & Atmospheric Sci Cornell Univ 619 Bradfield Hall Ithaca NY 14853-0144

OBENLAND, CLAYTON O, b Kansas City, Mo, Dec 22, 12; m 41; c 1. INDUSTRIAL CHEMISTRY. *Educ:* Kans State Univ, BS, 35, MS, 50. *Prof Exp:* Asst chemist, Monsanto Chem Co, Ill, 39-41, tech asst plant develop, 41-42 & 46-47, asst supvr chem prod, 47-49; scheduler prod control, Gen Aniline Works, NY, 50-51; chemist, Olin-Mathieson Chem Corp, NY, 52-56, group leader pre-pilot lab, 56-59, sr chemist, Conn, 59-69; patent agent, Carborundum Co, Niagara Falls, 69-75; LAB SUPVR, NIAGARA UNIV, 76-82; RETIRED. *Mem:* Am Chem Soc. *Res:* Synthetic organic and inorganic chemistry, especially boron chemistry. *Mailing Add:* 1300 Birch Dr Rogers AR 72756

OBENSHAIN, FELIX EDWARD, b Pikeville, Ky, Mar 31, 28; m 50; c 4. NUCLEAR PHYSICS. *Educ:* Va Polytech Inst, BS, 52; Univ Pittsburgh, PhD(physics), 59. *Prof Exp:* Physicist, Atomic Power Div, Westinghouse Elec Corp, 52-56; PHYSICIST, OAK RIDGE NAT LAB, 59-; PROF PHYSICS, UNIV TENN, KNOXVILLE, 68- *Concurrent Pos:* Partic, AEC Int Sci Exchange Prog, Cent Inst Nuclear Res, Karlsruhe, Ger, 65-66. *Mem:* Fel Am Phys Soc. *Res:* Heavy-ion nuclear reactions, relativistic heavy-ion reactions, fusion and fission studies; applications of Mossbauer effect in nuclear and solid state physics; positron polarization as related to C and P violation in weak interactions. *Mailing Add:* Oak Ridge Nat Lab PO Box 2008 Oak Ridge TN 37831-6372

OBER, CHRISTOPHER KEMPER, b Magnolia, Ark, Nov 1, 54; Can citizen; m 80; c 2. POLYMER SYNTHESIS. *Educ:* Univ Waterloo, BSc, 78; Univ Mass, MS, 80, PhD(polymer sci & engr), 82. *Prof Exp:* sr mem res staff, Xerox Res Can, 82-86; ASST PROF, CORNELL UNIV, 86- *Mem:* Mat Res Soc; Am Phys Soc; Am Chem Soc. *Res:* Synthesis and characterization of thermoplastic and thermoset liquid crystalline polymers; development of polymers for linear and nonlinear wave guides; new materials for microelectronics packaging. *Mailing Add:* Mat Sci & Engr Dept Cornell Univ Bard Hall Ithaca NY 14853

OBER, DAVID RAY, b Garrett, Ind, Dec 6, 39; m 63; c 3. NUCLEAR PHYSICS. *Educ:* Manchester Col, BA, 62; Purdue Univ, MS, 64, PhD(physics), 68. *Prof Exp:* Teaching asst physics, Purdue Univ, 62-66, res asst, 66-68; asst prof, 68-72, assoc prof, 72-76, PROF PHYSICS, BALL STATE UNIV, 76- *Mem:* Am Asn Physics Teachers; Am Phys Soc; Nat Sci Teachers Asn. *Res:* Low energy nuclear physics. *Mailing Add:* Dept Physics & Astron Ball State Univ Muncie IN 47306

OBER, ROBERT ELWOOD, b Springfield, Ohio, Nov 13, 31; m 55; c 3. DRUG METABOLISM, NEW DRUG DEVELOPMENT. *Educ:* Ohio State Univ, BS, 53, MS, 55; Univ Ill, PhD(biochem), 58. *Prof Exp:* From assoc res biochemist to res biochemist, Res Div, Parke, Davis & Co, 58-66; assoc prof, Col Pharm, Ohio State Univ, 66-69; head biochem pharmacol group, 69-70, supvr drug metab, 70-73, mgr drug metab, Riker Labs, 73-83, lab mgr, Develop Labs, 83-84, RES ADMINR, 3M RIKER, 84-; ASSOC PROF PHARM, UNIV MINN, MINNEAPOLIS, 70- *Mem:* Am Chem Soc; Am Asn Pharmaceut Scientists; AAAS; Sigma Xi; Int Soc Study Xenobiotics. *Res:* Laboratory and clinical drug metabolism and pharmacokinetics; biochemistry of fatty acids; steroids; radiotracer methodology; synthesis of radioisotopically labelled compounds. *Mailing Add:* 3M Pharmaceut 3M Ctr Bldg 270 3A 01 St Paul MN 55144-1000

OBER, WILLIAM B, b Boston, Mass, May 15, 20; m 52; c 2. PATHOLOGY. *Educ:* Harvard Col, AB, 41; Boston Univ, MD, 46. *Prof Exp:* Pathologist, Boston Lying-in Hosp, 53-55; assoc prof & clin prof path, NY Med Col, 60-72; prof path, Mt Sinai Sch Med, 72-78; pathologist, Beth Israel Hosp, 70-78; DIR LABS, HACKENSACK HOSP, HACKENSACK, NJ, 78- *Concurrent Pos:* Instr path, Harvard Med Sch, 53-55; dir labs path, Knickerbocker Hosp, 56-70; consult pathologist, First US Army Med Lab, 58-68, Margaret Sanger Res Bur, 60-73, Lutheran Hosp, Brooklyn, 65-, St Barnabas Hosp, Bronx, 66- & Roger Williams Hosp, Providence, RI, 68-; vis prof path, NJ Col Med, Univ Med & Dent NJ, 78- *Mem:* Royal Micros Soc; Am Asn Pathologists; NY Acad Med; Int Acad Path; Am Asn Path & Bact; Am Chem Soc. *Res:* Experimental production of toxaemia of pregnancy; medical analysis of literary problems. *Mailing Add:* Hackensack Med Ctr 30 Prospect Ave Hackensack NJ 07601

OBERDING, DENNIS GEORGE, b Plum City, Wis, Dec 4, 35; m 56; c 5. ANALYTICAL CHEMISTRY, SPECTROSCOPY. *Educ:* Univ Wis-River Falls, BS, 60. *Prof Exp:* Chemist, 60-73, sr anal chemist, 73-77, supvr instrument anal, 77-88, MGR, RES & DEVELOP, ETHYL CORP RES

LABS, 88- *Mem:* Am Chem Soc. *Res:* Spectroscopy and chromatography specifically, infrared; nuclear magnetic resonance and high pressure liquid chromatography. *Mailing Add:* Ethyl Corp Res Labs 8000 GSRI Ave Baton Rouge LA 70820

OBERDORFER, MICHAEL DOUGLAS, b Athens, Ohio, Aug 29, 42; m 68; c 1. DEVELOPMENTAL NEUROBIOLOGY. *Educ:* Rockford Col, BA, 66; Univ Wis-Madison, PhD(zool), 75. *Prof Exp:* NIH fel neurosci, Dept Anat, Univ Wis, 75-77; asst prof, Dept Neurobiol, Univ Tex Med Sch Houston, 77-82; PANEL DIR DEVELOP NEUROSCI, NSF, 82-; AT NAT EYE INST, NIH. *Mem:* AAAS; Am Soc Cell Biol; Soc Neurosci. *Res:* Anatomical development of mammalian visual systems. *Mailing Add:* Nat Eye Inst NIH 9000 Rockville Pike Bldg 31-Rm 6A48 Bethesda MD 20892

OBERDRSTER, GÜNTER, b Cologne, Ger, Feb 27, 39; m 68; c 3. INHALATION TOXICOLOGY. *Educ:* Univ Giessen, DVM, 64, Dr med vet, 66; Ger Bd Pharmacol & Toxicol, cert, 74. *Prof Exp:* Sci staff mem, Lab Pharmacol, Tropon Co, 66-67; asst prof physiol, Inst Normal & Pathol Physiol, Univ Cologne, 68-71; sci staff mem toxicol, Inst Toxicol & Aerosol Res, 71-79; assoc prof inhalation toxicol, 79-89, PROF TOXICOL, UNIV ROCHESTER, 89- *Concurrent Pos:* Vis asst prof, Univ Rochester, 75-76; appointee, Contact Group Heavy Metals, Europ Commission, 77-79; consult, WHO, 83-84 & 86; subcomt Heavy Metals, Environ Protection Agency, Sci Adv Bd, 84- *Honors & Awards:* Joseph von Fraunhofer Prize, Munich, 82. *Mem:* Ger Physiol Soc; Asn Aerosol Res; Am Thoracic Soc; Int Soc Aersols in Med; Soc Toxicol; Am Conf Govt Indust Hygienists. *Res:* Effects of air pollutants (fibers, heavy metals, oxidants, particles of combustion processes); lung clearance mechanisms (lymphatics, macrophages); fibrogenic and carcinogenic mechanisms of lung injury, inflammatory mediators, effects of complexing agents after inhalation of heavy metal aerosols; development of methods in inhalation toxicology; extrapolation modeling and risk assessment. *Mailing Add:* Dept Biophys, Environ Health Sci Ctr Univ Rochester Rochester NY 14642

OBERENDER, FREDERICK G, b Cambridge, Mass, Feb 6, 33; m 57; c 4. ORGANIC CHEMISTRY, PETROLEUM CHEMISTRY. *Educ:* Trinity Col, Conn, BS, 54, MS, 56; Pa State Univ, PhD(org chem), 60. *Prof Exp:* Chemist, Texaco Inc, 59-60, sr chemist, 60-64, res chemist, 64-69, group leader, 69-73, asst supvr, 73-77, supvr, 77-82, sr coordr, 82-85, sr technologist, 85-89, CONSULT, TEXACO INC, 89- *Mem:* Sigma Xi; Am Chem Soc. *Res:* Synthetic lubricants; lubricant additives; polynuclear aromatic hydrocarbons; chelate polymers; nitrogen containing heterocyclics; fuel and lubricant technology. *Mailing Add:* RR 3 Box 58A Van Voorhis Dr Fishkill NY 12524

OBERHELMAN, HARRY ALVIN, JR, b Chicago, Ill, Nov 15, 23; m 46; c 5. SURGERY. *Educ:* Univ Chicago, BS, 46, MD, 47. *Prof Exp:* From instr to assoc prof, Sch Med, Univ Chicago, 56-60; assoc prof, 60-64, PROF SURG, SCH MED, STANFORD UNIV, 64- *Concurrent Pos:* USPHS grant, Stanford Univ, 60-68; dir, Am Bd Surg, 72-78. *Mem:* Soc Univ Surg; Am Surg Asn; Am Gastroenterol Asn. *Res:* Gastrointestinal physiology, with emphasis on gastric motility and inflammatory diseases of the pancreas and colon. *Mailing Add:* Div Gen Surg Rm S067 Stanford CA 94305

OBERHOFER, EDWARD SAMUEL, b Elizabeth, NJ, May 11, 39; m 67; c 2. NUCLEAR PHYSICS. *Educ:* NC State Univ, BS, 61, MS, 64, PhD(physics), 67. *Prof Exp:* Asst prof, 67-74, chmn physics dept, 77-89, ASSOC PROF PHYSICS, UNIV NC, CHARLOTTE, 74- *Mem:* Am Phys Soc; Am Asn Physics Teachers. *Res:* Low energy nuclear physics; nuclear spectroscopy. *Mailing Add:* Dept Physics Univ NC Charlotte NC 28223

OBERHOLTZER, JAMES EDWARD, b Elizabethtown, Pa, June 18, 42; m 67. ANALYTICAL CHEMISTRY. *Educ:* Elizabethtown Col, BS, 64; Purdue Univ, PhD(analytical chem), 69. *Prof Exp:* Res chemist, Arthur D Little, Inc, 68-85, group leader appln develop, 85-91; SR ANALYTICAL CHEMIST, CHEM QUAL SERV, EASTMAN KODAK CO, 91- *Mem:* AAAS; Am Chem Soc; Sigma Xi. *Res:* Instrumentation and methodology for chemical analyses, especially gas chromatography and mass spectrometry; application of digital computers to scientific research; analytical techniques for monitoring environmental pollution. *Mailing Add:* Eastman Kodak Co 343 State St Rochester NY 14650

OBERLANDER, HERBERT, b Manchester, NH, Oct 2, 39; m 62; c 2. TISSUE CULTURE. *Educ:* Univ Conn, BA, 61; Western Reserve Univ, PhD(biol), 65. *Prof Exp:* NIH fel, Inst Zool, Zurich, 65-66; asst prof biol, Brandeis Univ, 66-71; res physiologist, 71-75, res leader, 75-84, LAB DIR, INSECT ATTRACTANTS, BEHAV & BASIC BIOL RES LAB, AGR RES SERV, USDA, 85- *Concurrent Pos:* Mem grad fac, Dept Entom, Univ Fla, 71- *Mem:* Entom Soc Am; Soc Develop Biol; Tissue Cult Asn; Int Soc Develop Biol. *Res:* Insect physiology; endocrine control of post-embryonic development in insects; cell and organ culture of insect imaginal discs. *Mailing Add:* Behav & Basic Biol Res Lab Agr Res Serv USDA PO Box 14565 Gainesville FL 32604

OBERLANDER, THEODORE M, b Corning, NY, May 5, 33. GEOMORPHOLOGY, ARID PROCESSOR. *Educ:* Syracuse Univ, BA, 54, MA, 56, PhD(geog), 63. *Prof Exp:* PROF GEOG, UNIV CALIF, BERKELEY, 63- *Honors & Awards:* Kirk Brown Award, Geol Surv Am, 86. *Mem:* Am Asn Geog; Geol Soc Am; AAAS. *Mailing Add:* Dept Geog Univ Calif Berkeley CA 94720

OBERLE, THOMAS M, b Mankato, Minn, Mar 10, 30; m 53; c 7. INORGANIC CHEMISTRY. *Educ:* Col St Thomas, BSc, 52. *Prof Exp:* Jr chemist, Ames Lab, Atomic Energy Comn, Iowa State Univ, 52-54; chemist, US Econs Lab, 56-60, group leader prod develop, 60-64, mgr, 64-69, dir, instnl & consumer res & develop, 69-72, asst vpres res & develop, 72-75, vpres res & develop, 75-86. *Mem:* Am Inst Chem. *Res:* Product development of detergent compounds for institutional and consumer needs. *Mailing Add:* 1819 Faro Lane Mendota Heights MN 55118

OBERLEAS, DONALD, b Sheridan, Ind, Feb 14, 33; m; c 3. NUTRITIONAL BIOCHEMISTRY. *Educ:* Purdue Univ, BS, 55; Univ Ky, MS, 59; Univ Mo, PhD(agr chem), 64. *Prof Exp:* Res chemist nutrit biochem, Vet Admin Hosp, Allen Park, Mich, 64-76; from instr to assoc prof biochem med, Wayne State Univ, 64-76; prof Nutrit & Food Sci & chmn dept, Univ Ky, 76-84; chmn, 85-87, PROF NUTRIT, TEX TECH UNIV, 85- *Concurrent Pos:* Actg assoc chief staff res, Vet Admin Hosp, Allen Park, Mich, 73-76; sabbatical leave, Human Nutrit Res Ctr, USDA Beltsville, Md, 84. *Mem:* Am Inst Nutrit; Soc Exp Biol & Med; Sigma Xi; Int Asn Bioinorg Scientists; Inst Food Technologists. *Res:* Trace element bioavailability, metabolism and homeostasis; affect of phytate on mineral absorption, reabsorption and their mechanisms; consequence of mineral deficiencies particularly subclinical deficiencies and how to identify such deficiencies. *Mailing Add:* 3404 88th St Lubbock TX 79423-2706

OBERLEY, TERRY DE WAYNE, b Effingham, Ill, Jan 23, 46; m 68; c 2. EXTRACELLULAR MATRIX. *Educ:* Northwestern Univ, Evanston, BS, 68, PhD(microbiol), 73; Northwestern Univ, Chicago, MD, 74. *Prof Exp:* Intern path, 74-75, fel, 75-77, resident & Am Cancer Soc fel, 77-78, asst prof, 77-83, ASSOC PROF PATH, SCH MED, UNIV WIS, 83-; CHIEF, ELECTRON MICROS, WILLIAM S MIDDLETON VET ADMIN HOSP, MADISON, WIS, 83- *Concurrent Pos:* Co-prin investr, NIH grant- "In Vitro Cult Studies of Renal Dis", 75-78; prin investr, March of Dimes grant, Immunohistochem & Biochem Studies of Hereditary Nephritis, 78-81; prin investr, Vet Admin Rev grant, Studies on Growth of Glomerular Cells in Culture, 84- *Mem:* Am Soc Cell Biol; Am Soc Nephrology; Am Soc Microbiol; AMA; Am Asn Pathologists. *Res:* Regulation of kidney cell growth and differentation by the extracellular matrix and reactive oxygen metabolites. *Mailing Add:* 2500 Overlook Terr Madison WI 53705

OBERLIN, DANIEL MALCOLM, b Tulsa, Okla, Sept 16, 48; m 67; c 1. MATHEMATICAL ANALYSIS. *Educ:* Univ Tulsa, BS, 70; Univ Md, MA, 72, PhD(math), 74. *Prof Exp:* Asst prof, 74-81, ASSOC PROF MATH, FLA STATE UNIV, 81- *Mem:* Am Math Soc. *Res:* Harmonic analysis on locally compact groups. *Mailing Add:* Dept Math Fla State Univ Tallahassee FL 32306

OBERLY, GENE HERMAN, b Palisade, Colo, Apr 27, 25; m 47; c 1. POMOLOGY. *Educ:* Utah State Univ, BS, 49, MS, 50; Mich State Univ, PhD(hort, plant nutrit), 59. *Prof Exp:* Salesman agr chem, C D Smith Drug Co, Colo, 50-51; exten horticulturist, Utah State Univ, 51-54; farm dir & assoc prof hort, Am Univ, Beirut, 54-57; asst hort, Mich State Univ, 57-59; assoc prof pomol, Univ Conn, 59-62; assoc prof, 62-82, PROF POMOL & CHMN DEPT, CORNELL UNIV, 82- *Concurrent Pos:* Vis prof pomol, Ore State Univ, Corvallis, 68-69; vis prof hort, Nat Taiwan Univ & Joint Comn Rural Reconstruction, Taipai, Taiwan, 76. *Mem:* Am Soc Hort Sci; Am Chem Soc. *Res:* Plant nutrition on horticultural crops. *Mailing Add:* Dept Pomol 134 Plant Sci Bldg Cornell Univ Ithaca NY 14850

OBERLY, RALPH EDWIN, b Columbus, Ohio, Feb 13, 41; m 64; c 2. PHYSICS. *Educ:* Ohio State Univ, BS, 63, PhD(physics), 70. *Prof Exp:* Res engr, NAm Aviation, Ohio, 63-64; PROF PHYSICS, MARSHALL UNIV, 70- *Mem:* Am Asn Physics Teachers; Optical Soc Am. *Res:* Infrared molecular spectroscopy of small molecules; optical instruments. *Mailing Add:* Dept Physics & Phys Sci Marshall Univ Huntington WV 25701

OBERMAN, ALBERT, b St Louis, Mo, Feb 9, 34; m 54; c 3. PREVENTIVE MEDICINE, EPIDEMIOLOGY. *Educ:* Wash Univ, AB, 55, MD, 59; Univ Mich, MPH, 66. *Prof Exp:* Investr thousand aviation study, US Naval Base, Pensacola, Fla, 62-65; Nat Heart Inst spec res assoc epidemiol, Sch Pub Health, Univ Mich, 66-67; prof pub health & epidemiol & assoc prof med, 66-81, PROF & CHMN DEPT PREV MED, SCH MED, UNIV ALA, BIRMINGHAM, 81-, DIR, DIV PREV MED, MED CTR, 66- *Concurrent Pos:* Mem med adv bd, Naval Aerospace Med Inst Fla, 65-; mem policy & data monitoring comn, Nat Heart, Lung & Blood Inst, 78-; clin trials review comt, NIH, HEW; Mosby scholar award, Univ Mo; vchmn gen prev med, Am Bd Prev Med, 79-; mem adv panel, Am Inst Biol Sci Oper Med, NASA, 80- *Mem:* Am Pub Health Asn; fel Am Col Physicians; fel Am Col Prev Med; fel Am Heart Asn; Int Soc Cardiol. *Res:* Epidemiology of chronic diseases, especially cardiovascular; exercise and cardiovascular rehabilitation. *Mailing Add:* 1717 11th Ave S Rm 719 Birmingham AL 35294

OBERMAN, HAROLD A, b Chicago, Ill, Oct 21, 32; m 59; c 3. MEDICINE, PATHOLOGY. *Educ:* Univ Omaha, AB, 53; Univ Nebr, MD, 56. *Prof Exp:* Asst chief dept path, Walter Reed Gen Hosp, 61-63; from asst prof to assoc prof path, 63-69, PROF PATH, SCH MED, UNIV MICH, ANN ARBOR, 69-, HEAD SECT CLIN PATH, 80-, DIR, CLIN LAB, UNIV HOSP, 80- *Concurrent Pos:* Consult, Vet Admin Hosp, Ann Arbor, 63- *Honors & Awards:* Elliot Mem Award, Am Asn Blood Banks. *Mem:* Am Soc Clin Path; Col Am Path; Int Acad Path; Am Asn Path; Am Asn Blood Banks. *Res:* Blood banking and blood transfusion; surgical pathology; pathology of breast disease, lymph node disease and neoplasms of head and neck. *Mailing Add:* 2980 Provincial Dr Ann Arbor MI 48104

OBERMAYER, ARTHUR S, b Philadelphia, Pa, July 17, 31; m 63; c 3. ORGANIC CHEMISTRY, PHYSICS. *Educ:* Swarthmore Col, BA, 52; Mass Inst Technol, PhD(chem), 56. *Prof Exp:* Group leader, Tracerlab, Inc, Mass, 56-59; mgr div phys sci, Allied Res Assocs, Inc, 59-61; PRES & CHMN BD, MOLECULON RES CORP, MASS, 61- *Concurrent Pos:* Dir, Strem Chem, Inc, Mass; adv coun exp res & develop incentives prog, NSF, 73-75; dir govt mgt task force, 75-78; cong assessment panel appln sci & technol, 76-; adv coun, NSF, 80-; mem, US Senate Small Business Comt, Nat Adv Coun, 81- *Mem:* Sigma Xi; Fed Am Scientists (treas, 71-73); Am Chem Soc; Asn Tech Professionals. *Res:* Polymer membrane research for controlled release pharmaceuticals, toxic vapor monitoring, and chemical separation processes; science policy options relative to small business innovation. *Mailing Add:* Moleculon Res Corp 239 Chestnut St West Newton MA 02165

O'BERRY, PHILLIP AARON, b Tampa, Fla, Feb 1, 33; m 60; c 6. VETERINARY MICROBIOLOGY. *Educ:* Univ Fla, BS, 55; Auburn Univ, DVM, 60; Iowa State Univ, PhD(vet microbiol), 67. *Prof Exp:* Res vet, Nat Animal Dis Lab, 61-67; res adminr cattle dis, Vet Sci Res Div, USDA, 67-72, asst dir, 72-73, dir, Nat Animal Dis Ctr, Agr Res Serv, 73-88, prin scientist biotech, Off Agr Biotech, 88-90, NAT TECH TRAN COORDR AGR RES SERV, USDA, 88-; ADJ PROF, IOWA STATE UNIV, 73- *Concurrent Pos:* Mem, Food & Agr Orgn Expert Panel on Livestock Infertility, 63-70; mem tech rev res projs, Turkey & Israel, 68; mem, Comt Fed Labs, Fed Coun Sci & Technol, 74-, Steering Comt, World Food Conf of 76, 74-76, Comt Animal Health, World Food & Nutrit Study, Nat Acad Sci, 75-77, Sci Adv Comt, Pan Am Ctr Zoonotic Dis, Buenos Aires, Arg, Pan Am Health Orgn, 76-86 & sci adv, Italian govt, 79; mem, Conf Res Workers Animal Dis; chmn rev panel Animal Dis, US-Israel, BARD, 84-86; tech rev Res Projs, Egypt, 87; consult, Agr Biotech Adv Comt, USDA, 88-90, mem, Patent Rev Comt, 88-, Nat Needs Grad Fel Grants Prog Rev Panel, 89-; mem, Bd Sci Rev, Am J Vet Res, 90-93. *Honors & Awards:* Sci Coop Medal, Rector Warsaw Agr Univ, 85. *Mem:* AAAS; Am Vet Med Asn; Am Soc Microbiol; Am Asn Bovine Practitioners; Am Public Health Asn; Nat Asn Fed Veterinarians; Am Asn Lab Animal Med; US Animal Health Asn; Am Soc Animal Sci; Livestock Conserv Inst. *Res:* Microbiological aspects of Vibrio fetus and Mycoplasma species and their relationship to diseases of livestock. *Mailing Add:* Nat Soil Tilth Lab 2150 Pammel Dr Ames IA 50011

OBERST, FRED WILLIAM, b Falls City, Nebr, July 7, 04; m 28; c 2. DRUG ADDICTION. *Educ:* W Tex State Teacher's Col, BA, 27; State Univ Iowa, MS, 28, PhD(org chem), 30. *Prof Exp:* Researcher biochem, Med Col, State Univ Iowa, 30-35; researcher biochem & drug addiction, USPHS, Lexington, Ky, 35-44; head, dept biochem, William S Merrell Co, Cincinnati, 44-48; researcher biochem, Sch Aviation Med, San Antonio, Tex, 48-50; researcher & head dept pharmacol & toxicol, Med Labs, Edgewood Arsenal, Md, 50-70; RETIRED. *Mem:* Am Chem Soc; Am Soc Biol Chemists; Am Soc Pharmacol & Exp Therapeut; Soc Toxicol. *Res:* Diseases in pregnant women; drug addiction; air-sickness and other aviation health problems; military chemistry on super-toxins and health hazards including protective musks and clothing; health hazards from toxic chemicals. *Mailing Add:* 107 Viola St Avon Park FL 33825-2249

OBERSTAR, HELEN ELIZABETH, b Ottawa, Ill, Aug 29, 23; m 45. COSMETIC CHEMISTRY. *Educ:* Monmouth Col, BS, 43. *Hon Degrees:* LLD, Monmouth Col, 87. *Prof Exp:* Asst food technol, Standard Brands, Inc, NY, 43-45; chemist, Miner Labs, Midwest Div, Arthur D Little, Inc, 46-50; res chemist & supvr, Toni Co Div, Gillette Co, 51-65; group leader, Shulton Inc, 65-71; sect mgr consumer prod div, Am Cyanamid Co, NJ, 72-75; mgr res & develop, Bristol-Myers Co, Int Div, 75-82, DIR RES & DEVELOP, CONSUMER PROD, BRISTOL-MYERS INT GROUP, NEW YORK, 82- *Concurrent Pos:* Exec Comn of CTFA Int Comt. *Mem:* Soc Cosmetic Chem. *Res:* Product development and exploratory research in hair coloring, lightening, hair care products, cosmetics and toiletries; mechanical device developments for cosmetics. *Mailing Add:* 512 Belden Hill Rd Wilton CT 06897-4221

OBERSTER, ARTHUR EUGENE, b Canton, Ohio, July 6, 29; m 55; c 6. ORGANIC CHEMISTRY, POLYMER CHEMISTRY. *Educ:* Mt Union Col, BS, 51; Univ Notre Dame, PhD(org chem), 57. *Prof Exp:* Sr chemist, Merck & Co, Inc, NJ, 55-59; res specialist, 59-67, sr group leader polymerization, 67-75, RES ASSOC, FIRESTONE TIRE & RUBBER CO, 75- *Honors & Awards:* Am Chem Soc Award, 50. *Mem:* Am Chem Soc; Sigma Xi. *Res:* Organic synthesis, particularly steroids, alkaloids, heterocyclics, rubber chemicals, antioxidants, antiozonants and monomers; polymer synthesis, particularly anionic polymerization; elastomer compounding; foam processing; polyurethanes; process and product development and manufacturing of rubber and related products. *Mailing Add:* 6021 Hollydale Ave NE North Canton OH 44721

OBERT, EDWARD FREDRIC, b Detroit, Mich, Jan 18, 10; m 35, 82. MECHANICAL ENGINEERING. *Educ:* Northwestern Univ, BS, 33, ME, 34; Univ Mich, MS, 40. *Prof Exp:* Mfg engr, Western Elec Co, Ill, 29-30; inspector eng mat, Naval Inspection, 35-37; from instr to prof mech eng, Northwestern Univ, 37-59; chmn dept, 63-67, PROF MECH ENG, UNIV WIS-MADISON, 59- *Concurrent Pos:* Consult, US Air Force Acad & Air Force Arctic Medic Lab, 58; mem comt, Nat Res Coun-Nat Acad Sci, 63-6. *Honors & Awards:* Westinghouse Award, Am Soc Eng Educ, 53 & G Edwin Burks Award, 71; Benjamin Smith Reynolds Award, Univ Wis, 73. *Mem:* Hon mem & fel Am Soc Mech Engrs; fel Soc Automotive Engrs. *Res:* Thermodynamics; properties of gases; internal combustion engines; author of three texts. *Mailing Add:* Dept of Mech Eng Univ of Wis Madison WI 53706

OBERT, JESSIE C, b Port Byron, Ill, Mar 26, 11; m 35. NUTRITION. *Educ:* Park Col, AB, 31; Univ Chicago, MS, 42; Ohio State Univ, PhD(nutrit), 51. *Prof Exp:* Nutritionist, Chicago Welfare Dept, 37-42; dir nutrit serv, Maricopa County Chapter, Am Red Cross, 43-47; asst prof home econ, Ohio State Univ, 47-51; instr, Univ Calif, Los Angeles, 52-53; chief nutrit div, Los Angeles Co Health Dept, 53-76; CONSULT, 76- *Mem:* Nutrit Today Soc; Am Dietetic Asn; Soc Nutrit Educ; Am Pub Health Asn; Am Home Econ Asn. *Res:* Activity and weight control; nutritional surveillance; community nutrition; nutrition and exercise. *Mailing Add:* 5122 Bomer Dr Los Angeles CA 90042-4514

OBERTEUFFER, JOHN AMIARD, b Boston, Mass, May 31, 40; m 62; c 2. AUTOMATIC SPEECH RECOGNITION, COMPUTER SCIENCES. *Educ:* Williams Col, BA, 62, MA, 64; Northwestern Univ, PhD(physics), 69. *Prof Exp:* Asst, Northwestern Univ, Ill, 64-69; res assoc neutron physics, Mass Inst Technol, 69-71, mem staff, Francis Bitter Nat Magnet Lab, 71-74; asst dir develop, Sala Magnetics, 74-75, tech dir, 75-76, vpres mkt, 76-80; exec vpres, Sontek Industs, Inc, 80-84; pres, Iris Graphics, 84-86; vpres, Med & Sci Enterprises, 87-89; PRES, VOICE INFO ASSOCS, 90- *Mem:* Inst Elec & Electronic Engrs; Am Phys Soc; Sigma Xi. *Res:* High gradient magnetic separation; basic and applied magnetism in liquids, solids, gases and mixed systems; solid state physics; x-ray and neutron diffraction; acoustics; automatic speech recognition. *Mailing Add:* 14 Glen Rd S Lexington MA 02173

OBEY, JAMES H(OWARD), b Detroit, Mich, Aug 29, 16; m 39; c 3. CHEMICAL ENGINEERING. *Educ:* Lawrence Inst Technol, BChE, 43. *Prof Exp:* Proj leader, Ford Motor Co, 39-43; sr fel, Mellon Inst, 43-56; coordr mkt develop, Consol Coal Co, 56-62; pres, Danville Prod, Inc, 62-64; mgr northeast dist, Blaw-Knox Chem Plants, Inc, 64-73; regional sales mgr, Jacobs Eng, 73-74; pres & gen mgr, Blaw-Knox Food & Chem Equip, 74-86; pres, Wirz Y Machuca, Mexico City, 74-86; RETIRED. *Concurrent Pos:* Consult, Boeckeler, Assocs, 43; bd mem, Blow-Knox, Tokyo, Japan, 74-84. *Mem:* Am Inst Chem Engrs. *Res:* Industrial utilization of proteins; solvent extraction of vegetable oils; pilot plant design and operation; commercialization of new chemical products. *Mailing Add:* 50 Rollingwood Williamsville NY 14270

OBIJESKI, JOHN FRANCIS, b Bridgeport, Conn, Apr 11, 41; m 62; c 2. MICROBIOLOGY, VIROLOGY. *Educ:* Univ Conn, BA, 65; Rutgers Univ, MS & PhD(virol), 71. *Prof Exp:* Res virologist, Virol Sect, Ctr Dis Control, 71-81; SR DIR PROD DEVELOP, GENENTECH, 81- *Mem:* Am Soc Microbiol; Brit Soc Gen Microbiol. *Res:* Molecular and biochemical properties of animal viruses. *Mailing Add:* Genentech 460 Point San Bruno Blvd South San Francisco CA 94080

OBLAD, ALEX GOLDEN, b Salt Lake City, Utah, Nov 26, 09; m 33; c 6. PETROLEUM CHEMISTRY, PETROLEUM REFINING. *Educ:* Univ Utah, BA, 33, MA, 34; Purdue Univ, PhD(phys chem), 37. *Hon Degrees:* DSc, Purdue Univ, 59, Univ Utah, 80. *Prof Exp:* Asst, Purdue Univ, 34-37; res chemist, Standard Oil Co, Ind, 37-42; group leader, Magnolia Petrol Co, Tex, 42-44, sect leader, 44-46, chief chem res, 46; head indust res, Tex Res Found, 46-47; dir chem res, 47-52, assoc mgr res & develop, 52-55, vpres res & develop & dir, Houdry Process Corp, Pa, 55-57; vpres res & develop, M W Kellog Co, NY, 57-60, vpres res & eng develop, process eng, patent & res, 66-69; vpres, Ireco Chem Co, Utah, 69-70; assoc dean col mines & mineral industs, 70-72, dean, 72-75, prof metall & fuels, 70-75, DISTINGUISHED PROF FUELS ENG, PROF CHEM, UNIV UTAH, 75- *Concurrent Pos:* Consult, Atomic Energy Projs, 50-57; managing ed publ, Div Petrol Chem, Am Chem Soc, 54-69; dir, Int Cong Catalysis, 55-64; vpres & dir, Nat Inst Catalysis, 65-69. *Honors & Awards:* E V Murphree Award, Am Chem Soc, 69; Chem Pioneer Award, Am Inst Chemists, 72. *Mem:* Nat Acad Eng; Sigma Xi; Am Chem Soc; AAAS; Am Inst Chem Engrs; Catalysis Soc Am. *Res:* Catalysis; reaction mechanisms; kinetics and thermodynamics of hydrocarbon reactions; heat capacity of glasses; optical methods of analysis; uranium chemistry and processing; petroleum chemistry; petrochemicals; administration and management of research and development; process engineering and patent licensing; petroleum refining; chemicals from petroleum; catalysis. *Mailing Add:* Mineral Sci Bldg Col Mines & Earth Sci Salt Lake City UT 84112

OBLINGER, DIANA GELENE, b Des Moines, Iowa, Feb 11, 54; c 4. COMPUTER TRAINING, APPLICATION OF TECHNOLOGY TO HIGHER EDUCATION. *Educ:* Iowa State Univ, BS, 74, MS, 76, PhD(plant breeding & cytogenetics), 80. *Prof Exp:* Plant breeder, Dekalb AgRes, Inc, 78-80; asst prof hort, Mich State Univ, 80-81; asst prof agron, Univ Mo-Columbia, 81-86, asst dean agr, 85-86, assoc dean, 86-89; ACAD DISCIPLINE SPECIALIST, IBM, 89- *Concurrent Pos:* Adj prof, Agron Dept, Clemson Univ, 89- & Dept Crop Sci, NC State Univ, 90- *Mem:* Am Inst Biol Sci; Crop Sci Soc Am; Coun Agr Sci & Technol; Nat Asn Cols & Teachers Agr. *Res:* Identifying computer applications of value to higher education in the life sciences and agriculture. *Mailing Add:* 2525 Meridian Pkwy Suite 50 Durham NC 27713

OBLINGER, JAMES LESLIE, b Ashland, Ohio, Nov 3, 45; m 68; c 2. FOOD MICROBIOLOGY, FOOD SCIENCE. *Educ:* DePauw Univ, BA, 67; Iowa State Univ, MS, 70, PhD(food technol), 72. *Prof Exp:* From asst prof to assoc prof food microbiol, Univ Fla, 72-84; prof, Univ Mo, Columbia, 84-86; ASSOC DEAN, DIR ACADEMIC PROG, NC STATE UNIV, 86- *Mem:* Inst Food Technologists; Int Asn Milk Food & Environ Sanitarians; Am Soc Microbiol; Sigma Xi; Am Meat Sci Asn. *Res:* Sources and analysis of microbiological aspects of the food supply; food-borne diseases; red meat and poultry microbiology. *Mailing Add:* 115 Patterson Hall Box 7642 NC State Univ Raleigh NC 27607

OBRADOVICH, JOHN DINKO, b Fresno, Calif, May 2, 30. GEOPHYSICS. *Educ:* Univ Calif, Berkeley, BA, 57, MA, 59, PhD(geophys), 64. *Prof Exp:* GEOPHYSICIST, US GEOL SURV, 61- *Mem:* AAAS; Am Geophys Union; Geol Soc Am. *Res:* Isotope geology; K-Ar, K-Ca, Ar-Ar and Rb-Sr geochronology, particularly K-Ar dating related to biostratigraphic and Time Scale studies. *Mailing Add:* US Geol Surv Bldg 21 Fed Ctr Denver CO 80225

OBREMSKI, HENRY J(OHN), b Elmont, NY, Oct 24, 31; m 58; c 2. FLUID DYNAMICS. *Educ:* Polytech Inst Brooklyn, BS, 58; Ill Inst Technol, MS, 62, PhD(mech eng), 66. *Prof Exp:* Engr, United Aircraft Res Lab, 58-60; assoc engr, IIT Res Inst, 60-65; SR RES SCIENTIST, MARTIN MARIETTA LABS, 65- *Mem:* Am Inst Aeronaut & Astronaut; Am Phys Soc; Am Chem Soc. *Res:* Transition in non-steady boundary layers; cavity flows and instabilities; thermal pollution; heat and mass transfer in aluminum cells; ventilation; brine field modelling. *Mailing Add:* 5113 Avoca Ave Ellicott MD 21043

OBREMSKI, ROBERT JOHN, b Brooklyn, NY, Aug 19, 41; m 64; c 2. MOLECULAR SPECTROSCOPY. *Educ:* St John's Univ, NY, BS, 62, MS, 64; Univ Md, College Park, PhD(phys chem), 68. *Prof Exp:* Chemist, Uniroyal, Inc, 68-69 & Spectra-Physics, Inc, 69-71; sr appln chemist, 71-77, prin chemist, 77-85, MGR, ADVAN SYSTS DEVELOP, BECKMAN

INSTRUMENTS, INC, 85- *Mem:* Am Chem Soc; Soc Appl Spectros; Coblentz Soc. *Res:* Application of computers to molecular spectroscopy; design of computerized instrumentation; nuclear counting techniques. *Mailing Add:* Beckman Instr Inc 200 S Kraemer Blvd MS W 345 Brea CA 92621-6209

O'BRIEN, ANNE T, b New York, NY, Apr 11, 36. ORGANIC CHEMISTRY. *Educ:* Marymount Col, NY, BS, 57; Fordham Univ, PhD(org chem), 64. *Prof Exp:* Fac mem parochial sch, 57-59; from instr to assoc prof chem, Marymount Col, NY, 62-73; assoc prof, 73-76, ADJ ASSOC PROF, DEPT MAN-ENVIRON STUDIES, UNIV WATERLOO, 76-, SR RES LIT CHEMIST, MED RES DIV, AM CYANAMID, 76- *Mem:* Am Chem Soc; Am Inst Chem; Sigma Xi; Am Asn Univ Prof; AAAS. *Res:* Halogen-catalyzed autoxidation; porphyrin synthesis; drug syntheses, resolutions and absolute configurations; futuristics; environmental chemistry; alkaloids; literature science. *Mailing Add:* 15 Crest Dr Tarrytown NY 10591

O'BRIEN, BARBARA COONEY, CHOLESTEROL, DIET FAT. *Educ:* Duke Univ, PhD(chem), 59. *Prof Exp:* RES SCIENTIST CHEM, TEX A&M UNIV, 74- *Res:* Dietary effects on lipid metabolism. *Mailing Add:* Dept Biochem & Biophys Tex A&M Univ College Station TX 77843-2128

O'BRIEN, BENEDICT BUTLER, JR, b New Britain, Conn, July 11, 34; c 2. PHYSICS. *Educ:* Mass Inst Technol, BS, 55; Univ Munich, PhD(physics), 65. *Prof Exp:* Mem staff elec eng, Ramo Wooldridge Corp & Space Tech Lab, 56-60; scientist plasma physics, Max Planck Inst for Plasma Physics, 60-67; asst prof elec eng, Univ Southern Calif, 67-71; mgr high power laser, 71-77, mgr millimeter wave technol, 77-84, MGR INTEGRATED MICROSENSORS, NORTHRUP RES & TECHNOL CTR, 84- *Concurrent Pos:* Instr physics, Univ Md, 62-67; consult, Northrop Res & Tech Ctr, 70-71. *Mem:* Am Phys Soc; Inst Elec & Electronics Engrs. *Res:* Millimeter wave; high power laser; plasma physics; silicon microsensors. *Mailing Add:* Northrop Res & Technol Ctr One Research Park Palos Verdes Peninsula CA 90274

O'BRIEN, BRIAN, b Denver, Colo, Jan 2, 1898; m 22, 56; c 1. ELECTRO-OPTICS. *Educ:* Yale Univ, PhB, 18, PhD(physics), 22. *Prof Exp:* Res engr, Westinghouse Elec & Mfg Co, 22-23; res physicist, Buffalo Tuberc Asn, 23-30; prof physiol optics, Univ Rochester, 30-46, dir Inst Optics, 38-53, res prof physics & optics, 46-53; vpres res & trustee, Am Optical Co, 53-58; CONSULT PHYSICIST, BRIAN O'BRIEN ASSOC, 58- *Concurrent Pos:* Mem comt pilot selection & training & vision comt, Nat Res Coun, 39-46, chmn div phys sci, 53-61; mem, Nat Defense Res Comt, 40-46; chmn space prog adv coun, NASA, 70-74; mem sci adv bd, US Air Force, 59-70; chmn, Nat Acad Sci Comt, Adv to Air Force Systs Command, 62-74. *Honors & Awards:* President's Medal for Merit, 48; Ives Medal, 51; Distinguished Pub Serv Medal, NASA, 71; Nat Acad Sci Founder & Develop Award, Air Force Studies, NASA, 79. *Mem:* Nat Acad Sci; Nat Acad Eng; fel Optical Soc Am (pres, 51-53); fel Am Phys Soc; fel Am Inst Elec & Electronics Engrs; Am Geophys Union; Am Acad Arts & Sci; Am Philos Soc. *Res:* Optical properties of metals and thin films; solar ultraviolet and atmospheric ozone; photographic processes; motion picture systems; very high speed photography; photobiochemical effects; flicker phenomena in vision; retinal structure and visual processes; fiber optics. *Mailing Add:* Box 166 Woodstock CT 06281

O'BRIEN, DANIEL H, b Berkeley, Calif, Oct 26, 32; m 59; c 6. ORGANIC CHEMISTRY. *Educ:* Univ Va, BS, 54, PhD(chem), 61. *Prof Exp:* From instr to asst prof chem, Univ Dayton, 60-66; fel, Case Western Reserve Univ, 66-67; ASSOC PROF CHEM, TEX A&M UNIV, 67- *Concurrent Pos:* Res assoc, Chem Br, Aeronaut Res Lab, Wright-Patterson AFB, 62-66. *Mem:* Am Chem Soc; Sigma Xi. *Res:* Organometallic chemistry; organic synthesis and characterization of organosilicon compounds. *Mailing Add:* Dept of Chem Tex A&M Univ College Station TX 77843

O'BRIEN, DEBORAH A, CELL BIOLOGY, REPRODUCTIVE BIOLOGY. *Educ:* Harvard Univ, PhD(physiol), 78. *Prof Exp:* SR STAFF FEL, NAT INST ENVIRON HEALTH SCI, 83- *Mem:* Am Soc Cell Biol; Soc Develop Biol; Soc Study Reproduction; Sigma Xi; AAAS. *Res:* Expression of tissue-specific gene products during mamalian spermatogenesis; interactions between germ cells and Sertoli cells; mannose 6-phosphate receptors. *Mailing Add:* Univ NC MacNider Bldg Campus Box 7500 Chapel Hill NC 27599

O'BRIEN, DENNIS CRAIG, b Great Bend, Kans, July 20, 38; m 60; c 2. GEOLOGY. *Educ:* Cornell Col, AB, 60; Miami Univ, MS, 64; Univ Mass, Amherst, PhD(geol), 71. *Prof Exp:* Instr, 68-71, ASSOC PROF GEOL, DRAKE UNIV, 76- *Mem:* Geol Soc Am; Sigma Xi; Nat Asn Geol Teachers; Soc Econ Paleontologists & Mineralogists. *Res:* Sedimentary petrography and petrology; Precambrian sedimentary rocks; environmental geology. *Mailing Add:* Dept Geog & Geol Drake Univ 25th St Univ Des Moines IA 50311

O'BRIEN, DONOUGH, b Edinburgh, Scotland, May 9, 23; m 50; c 2. PEDIATRICS. *Educ:* Cambridge Univ, BA, 44, MB, BCh, 46, MA, 47, MD, 52; FRCP, 72. *Prof Exp:* House physician, St Thomas' Hosp, Univ London, 47-48, registr, Inst Child Health, 50-52 & Hosp Sick Children & Guy's Hosp, 53-57; from asst prof to assoc prof pediat, Sch Med, 57-64, PROF PEDIAT, UNIV COLO MED CTR, DENVER, 64- *Concurrent Pos:* Markle Found scholar, 60. *Mem:* Am Diabetes Asn; Am Fedn Clin Res. *Res:* Biochemical applications to pediatrics. *Mailing Add:* Univ Colo Med Ctr 4200 E Ninth Ave Denver CO 80220

O'BRIEN, EDWARD E, b Toowoomba, Australia, May 16, 33; m 59; c 6. FLUID MECHANICS. *Educ:* Univ Queensland, BS, 55; Purdue Univ, MSME, 57; Johns Hopkins Univ, PhD(mech), 60. *Prof Exp:* Asst thermodyn, Univ Queensland, 54-55; thermodynamicist, Canadair Aircraft Co, Can, 55; asst heat power, Purdue Univ, 55-57; instr mech, Johns Hopkins Univ, 57-60, fel, 60-61; from asst prof to assoc prof eng, 61-67, PROF ENG, STATE UNIV NY, STONY BROOK, 67-, CHMN DEPT MECH ENG, 83- *Concurrent Pos:* NSF grants, 62-72; mem meteorol group, Brookhaven Nat Lab, 64; US Pub Health Serv grant, 64-66. *Mem:* Fel Am Phys Soc. *Res:* Turbulent diffusion and mixing; chemically reacting turbulent flows; classical incompressible fluid mechanics; geophysical fluid mechanics. *Mailing Add:* Dept Mech Engrs State Univ of NY Stony Brook NY 11794-2300

O'BRIEN, FRANCIS XAVIER, b Quincy, Mass, Sept 6, 35; m 65; c 3. MARINE BIOLOGY. *Educ:* Suffolk Univ, BS, 63; Univ NH, MS, 65, PhD(zool), 72. *Prof Exp:* PROF BIOL & CHMN DEPT, SOUTHEASTERN MASS UNIV, 68- *Concurrent Pos:* Biol consult, US Environ Protection Agency, 75-77. *Mem:* Sigma Xi; Am Malacol Union; Nat Shellfish Asn; Crustacean Soc. *Res:* Ecology of marine benthic invertebrates. *Mailing Add:* Dept Biol Southeastern Mass Univ North Dartmouth MA 02747

O'BRIEN, HAROLD ALOYSIOUS, JR, b Dallas, Tex, May 17, 36; m 58; c 3. NUCLEAR CHEMISTRY, NUCLEAR MEDICINE. *Educ:* Univ Tex, Austin, BA, 59; NMex State Univ, MS, 61; Univ Tenn, Knoxville, PhD(phys chem), 68. *Prof Exp:* Res assoc nuclear chem, Isotopes Develop Ctr, Oak Ridge Nat Lab, 62-68; mem staff, 68-74, assoc group leader, 74-80, group leader, 80-85, MEM STAFF, LOS ALAMOS NAT LAB, 86- *Concurrent Pos:* Adj asst prof, 70-78, adj assoc prof, Sch Med, Univ NMex, 78-; Am Cancer Soc grant, 72-74; mem subcomt radiochem, Nat Acad Sci-Nat Res Coun, 74-78; chmn, State NMex Radiation Tech Adv Coun, 75-85, 90-; vis scientist, Lawrence Berkeley Lab, Lawrence Livermore Lab & Crocker Nuclear Lab, Univ Calif, Davis, 85-86. *Mem:* Am Bd Sci in Nuclear Med; AAAS; Am Chem Soc; Soc Nuclear Med; Sigma Xi. *Res:* Nuclear reactions; cross section studies; radioisotope production and applications; nuclear medicine; high temperature thermodynamics. *Mailing Add:* P-3 MS D449 PO Box 1663 Los Alamos NM 87545

O'BRIEN, JAMES FRANCIS, b Philadelphia, Pa, July 4, 41; m 70; c 2. PHYSICAL CHEMISTRY, INORGANIC CHEMISTRY. *Educ:* Villanova Univ, BS, 64; Univ Minn, Minneapolis, PhD(chem), 68. *Prof Exp:* Fel, Los Alamos Sci Lab, 68-69; from asst prof to assoc prof, 69-79, PROF CHEM, SOUTHWEST MO STATE COL, 79- *Concurrent Pos:* Vis assoc prof, Univ of Del, 75-76; adj prof, Univ Mo-Rolla, 87-88. *Mem:* Am Chem Soc; AAAS. *Res:* Molecular orbital calculations; electrolyte solutions; nuclear magnetic resonance. *Mailing Add:* 3740 E Meadowmere Pl Springfield MO 65809

O'BRIEN, JAMES FRANCIS, b Rochester, NY, Aug 23, 34. CYTOLOGY, MICROBIOLOGY. *Educ:* Spring Hill Col, BS, 60; Fordham Univ, MS, 62, PhD(biol), 65. *Prof Exp:* Res assoc biol, Cancer Res Inst, Boston Col, 69; asst prof, 69-73, ASSOC PROF BIOL, LE MOYNE COL, NY, 73- *Concurrent Pos:* Postdoc fel, St Thomas Inst, Cincinnati, 78-79. *Mem:* AAAS; Nat Asn Biol Teachers; Am Asn Jesuit Sci; Sigma Xi. *Res:* Cytological investigation of the development of intestinal tract of mosquitoes; electrophoretic study of tissues of rats infected with shaychloroma tumors; electron microscopic study of hindgut of mosquito Aedea Aegypti. *Mailing Add:* Dept of Biol Le Moyne Col Syracuse NY 13214

O'BRIEN, JAMES J, b New York, NY, Aug 10, 35; m 58; c 3. METEOROLOGY, OCEANOGRAPHY. *Educ:* Rutgers Univ, BS, 57; Tex A&M Univ, MS, 64, PhD(meteorol), 66. *Prof Exp:* Chemist, Elchem Dept, E I du Pont de Nemours & Co, 57-58, tech rep, 60-62; fel, Adv Study Group, Nat Ctr Atmospheric Res, 66-67, staff scientist, 67-69; assoc prof, 69-74, PROF METEOROL & OCEANOGR, FLA STATE UNIV, 69- *Concurrent Pos:* Co-dir coastal upwelling exp, Int Decade Ocean Explor-NSF; co-ed, Progess Oceanogr, 79-84, assoc ed, J Physical Oceanogr, 71-, ed, J Geophys Res Oceans, 83-; res chmn, Fla State Univ, 85. *Honors & Awards:* Sverdrup Gold Medal, Am Meteorol Soc, 87. *Mem:* Fel Am Meteorol Soc; Am Geophys Union; fel Royal Meteorol Soc; Japan Oceanogr Soc. *Res:* Micrometeorology; air-sea interactions; numerical analysis; applied statistics; numerical modeling of ocean circulation; ecological modeling. *Mailing Add:* Dept of Meteorol Fla State Univ Tallahassee FL 32306

O'BRIEN, JOAN A, veterinary medicine, laryngology; deceased, see previous edition for last biography

O'BRIEN, JOHN C, b Norfolk, Va, Apr 4, 45. CELL BIOLOGY. *Educ:* Univ Southern Calif, PhD(biochem), 76. *Prof Exp:* Dep, 83-84, DIR, GRANTS MGT, UNIFORMED SERV UNIV HEALTH SCI, 84- *Mem:* Am Soc Biol Chemists; Nat Coun Univ Res Adminr. *Mailing Add:* 9805 Greenbrier Lane Walkersville MD 21793

O'BRIEN, JOHN S, b Rochester, NY, July 14, 34; m 57; c 6. PATHOLOGY, MEDICINE. *Educ:* Creighton Univ, MS, 58, MD, 60. *Prof Exp:* From instr to assoc prof path & med, Sch Med, Univ Southern Calif, 62-68, lectr biochem, 62-67, coordr set clin lects & chief div chem path, 64-68; assoc prof neurosci, 68-70, PROF NEUROSCI, SCH MED, UNIV CALIF, SAN DIEGO, 70-, CHIEF DIV NEUROMETAB DISORDERS, 69-85. *Concurrent Pos:* USPHS grants, 63-79; Nat Inst Child Health & Human Develop grant, 66-68; Nat Multiple Sclerosis Soc grant, 66-70; Nat Genet Found grant, 70-72; Nat Cyctic Fibrosis Res Found grant, 70-72; Nat Found March of Dimes grant, 70-73; Nat Inst Gen Med Sci grant, 70-75; vis scientist, Univ Calif, Los Angeles, 62 & Scripps Inst, Univ Calif, 63; consult, Children's Hosp Los Angeles, Pac State Hosp, Pomona, Children's Hosp San Diego, Fairview State Hosp, Costa Mesa, Pasadena Found Med Res & Dept Neurol, Univ Southern Calif, 64-68. *Mem:* AAAS; Am Fedn Clin Res; Am Soc Exp Path; Am Soc Human Genet; NY Acad Sci. *Res:* Relationships between the molecular structure and disease states; role of lipid molecules in membrane structure and stability; brain lipids; myelination and demyelination. *Mailing Add:* Dept Neuroscis M-008 Univ Calif San Diego Sch Med La Jolla CA 92093

O'BRIEN, KERAN, b Brooklyn, NY, Nov 5, 31; m 61; c 2. RADIATION PHYSICS. *Educ:* Fordham Univ, BS, 53. *Prof Exp:* Physicist, Environ Measurements Lab, 53-81, dir, Radiation Physics Div, US Dept Energy, 81-87; PVT CONSULT, 87-; ADJ PROF PHYSICS, DEPT PHYSICS & ASTRON, NORTHERN ARIZ UNIV, 88- *Concurrent Pos:* Mem adv panel accelerator safety, AEC, 65-70 & 77-; mem reference nuclear data panel, Nat Nuclear Data Ctr, 81. *Honors & Awards:* Shielding & Dosimetry Div Award Outstanding Serv, Am Nuclear Soc, 76. *Mem:* Fel Am Nuclear Soc; Radiation Res Soc; Am Phys Soc; Archeol Inst Am. *Res:* Radiation dosimetry associated with particle accelerator shielding with naturally occurring radiation sources; high energy radiation theory and the propagation of atmospheric cosmic rays. *Mailing Add:* 1645 Fabulous Texan Way PO Box 967 Sedona AZ 86336-0967

O'BRIEN, LARRY JOE, b Big Spring, Tex, Sept 14, 29; m 53; c 3. PHYSIOLOGY. *Educ:* Hardin-Simmons Univ, BA, 49; NTex State Col, MA, 54; Univ Tex, PhD(physiol), 57; Med Col Ga, MD, 71. *Prof Exp:* Asst physiol, Univ Tex Med Br, 55-56, instr, 56-57; from instr to asst prof, Albany Med Col, 57-60; assoc prof res physiol, Sch Med, Univ Okla, 60-62; from asst prof to assoc prof physiol, Med Col Ga, 62-72, actg chmn dept, 71-72; prof physiol & chmn dept, 72-75, CLIN PROF PHYSIOL & MED, SCH MED, TEX TECH UNIV, 75- , PROF INTERNAL MED. *Concurrent Pos:* Chief, Circulation Sect, Civil Aeromed Res Inst, 60-62; attend physician internal med, Methodist Hosp, Lubbock, Tex. *Mem:* AAAS; Am Physiol Soc; Am Heart Asn; NY Acad Sci; Soc Exp Biol & Med; Royal Soc Med; Am Fed Clin Res. *Res:* Cardiac and peripheral vascular function. *Mailing Add:* Dept Internal Med Health Sci Ctr Univ Tex 3801 19th St Med-Prof Bldg Suite 401 Lubbock TX 79410

O'BRIEN, MICHAEL, b Melrose, Iowa, Oct 1, 18; m 44; c 3. AGRICULTURAL ENGINEERING, BIOLOGICAL ENGINEERING. *Educ:* Iowa State Univ, BS, 48, MS, 49, PhD(agr eng, biol sci), 51. *Prof Exp:* From instr to assoc prof, 50-70, PROF MAT HANDLING, DEPT AGR ENG, UNIV CALIF, DAVIS, 70- *Concurrent Pos:* Grants, Nat Canners Asn, 58, Cling Peach Adv Bd, 61-64, Canners League, 66-68, Tomato Indust, 59-84 & NSF, 70; consult educ & training, Int Harvester Co, Ill, 50, mat handling, US Steel Co, Pa, 65-66 & mat handling & mech harvesting, Dole Co, Hawaii, 68; consult, Neuman, Williams & Anderson Assocs, 79-83. *Mem:* Fel Am Soc Agr Engrs; Am Soc Eng Educ. *Res:* Automation of materials handling and physical properties and quality control of biological materials. *Mailing Add:* Dept Agr Eng Univ Calif Davis CA 95616

O'BRIEN, MICHAEL HARVEY, b Soperton, Ga, Mar 22, 42; m 63; c 1. ORGANIC CHEMISTRY. *Educ:* Berry Col, BA, 63; NC State Univ, PhD(chem), 69. *Prof Exp:* NASA training grant, 63-67; RES CHEMIST, E I DU PONT DE NEMOURS & CO, INC, 68- *Mem:* Am Chem Soc; Sigma Xi. *Res:* Synthesis, structure and reactions of pentavalent organoarsenic compounds. *Mailing Add:* 112 Lanchester Ct Waynesboro VA 22980-1563

O'BRIEN, MORROUGH P, engineering; deceased, see previous edition for last biography

O'BRIEN, NEAL RAY, b Newark, Ohio, May 25, 37; m 62; c 3. GEOLOGY. *Educ:* DePauw Univ, BA, 59; Univ Ill, MS, 61, PhD(geol), 63. *Prof Exp:* PROF GEOL, STATE UNIV NY COL, POTSDAM, 63- *Concurrent Pos:* Postdoctoral researcher, Kyoto Univ, 69-70 & 77-78. *Mem:* Soc Econ Paleont & Mineral; Clay Mineral Soc. *Res:* Sedimentology; electron microscope study of clay sediment and rocks. *Mailing Add:* Dept Geol Potsdam Col State Univ NY Potsdam NY 13676

O'BRIEN, PAUL J, b Haddonfield, NJ, Feb 11, 33; m 61; c 3. BIOCHEMISTRY. *Educ:* Mt St Mary's Col, Md, BS, 54; St John's Univ, NY, MS, 56; Univ Pa, PhD(biochem), 60. *Prof Exp:* Res chemist, Sect Intermediary Metab, Lab Biochem & Metab, Nat Inst Arthritis & Metab Dis, 60-64; res chemist, Sect Cell Biol, Ophthal Br, Nat Inst Neurol Dis & Blindness, 64-65, from actg chief to chief, 65-71; res chemist, Lab Retinal Cell & Molecular Biol, Nat Eye Inst, 71-81, chief sect cell biol, 81-90, sci dir, 90; CONSULT, HEALTH RES ASSOS, 90- *Concurrent Pos:* Ed, Exp Eye Res, 75 - *Mem:* AAAS; Asn Res Vision & Ophthal (pres, 81); Am Soc Biol Chemists; Am Soc Cell Biol; Int Soc Eye Res (treas, 88-92). *Res:* Biosynthesis of glycoproteins; control mechanisms; visual pigment biosynthesis and photoreceptor renewal; biochemical defects in inherited retinal degenerations. *Mailing Add:* Health Res Assocs 12 Duke St S Rockville MD 20850

OBRIEN, PETER CHARLES, b New York, NY, July 11, 43; m 83; c 1. EPIDEMIOLOGY. *Educ:* Univ Penn, BS, 65; Iowa State Univ, MS, 68, PhD, 70. *Prof Exp:* Consult Biostat, Mayo Clin, 72; from asst prof to assoc prof, 70-88, PROF BIOSTAT, MAYO MED SCH, 88- *Concurrent Pos:* Assoc consult biostat, Mayo Clin, 70-72; consult, Food & Drug Admin, 75-77; assoc ed, Am J Epidemiol, 78-87; primary statist reviewer, J Neurosurg, 89-, J Oral & Maxillofacial Implants, 90- *Mem:* Am Stat Assoc; Biometric Soc; Soc Epidermal Res. *Res:* Development and application of quantitative methods in neurology; development of new statistical procedures. *Mailing Add:* Dept Health Sci Res Mayo Clin Rochester MN 55905

O'BRIEN, PETER J, b London, Eng, July 6, 37. BIOCHEMISTRY. *Educ:* Univ London, BSc, 59; Univ Birmingham, PhD, 63. *Prof Exp:* Res fel med biochem, Univ Birmingham, 63-64, sr res assoc, 64-67; assoc prof, 67-74, PROF BIOCHEM, MEM UNIV NFLD, 74-, MEM CAN RES UNIT, 76-86; Assoc Dean Res, Fac of Pharm, Univ of Toronto 86- *Concurrent Pos:* Assoc ed, Can J Biochem, 71-75. *Mem:* Brit Biochem Soc; Can Soc Cell Biol; Can Biochem Soc; Am Soc Biol Chemists. *Res:* Intracellular formation, molecular effects and function of lipid peroxides, steroid hydroperoxides and hydrogen peroxide; drug and steroid metabolism; functional organization of electron transport in intracellular membranes. *Mailing Add:* Fac Pharm Univ Toronto Toronto ON M5S 1A1 Can

O'BRIEN, REDMOND R, b Quincy, Mass, Oct 27, 31; m 59; c 2. MATHEMATICS. *Educ:* Mass Inst Technol, SB, 53, SM, 54, PhD(math), 57. *Prof Exp:* Adv res engr, Sylvania Elec Prod, 57-60; SR MATHEMATICIAN, IBM CORP, 60- *Mem:* Inst Elec & Electronic Engrs; Soc Indust & Appl Math. *Res:* Mathematical analysis of semiconductor devices. *Mailing Add:* 84 Colburn Dr Poughkeepsie NY 12603

O'BRIEN, RICHARD DESMOND, b Sydenham, Eng, May 29, 29; m 61, 81; c 2. NEUROCHEMISTRY. *Educ:* Univ Reading, BSc, 50; Univ Western Ontario, PhD(chem), 54, BA, 56. *Prof Exp:* Chemist, Pesticide Res Inst, Can, 54-60; from assoc prof to prof entom, Cornell Univ, 60-65, chmn sect biochem, 64-65 & sect neurobiol & behav, 65-70, prof neurobiol, 65-78, dir div biol sci, 70-78; provost, Univ Rochester, 78-84; PROVOST, UNIV MASS, AMHERST, 84- *Concurrent Pos:* Nat Res Coun Can fel, Inst Animal Physiol, Cambridge, 56-57; vis assoc prof, Univ Wis, 58-59; Guggenheim fel, Int Lab Genetics & Biophys, Naples, 67-68; ed & founder, Pesticide Biochem & Physiol, 70-78. *Honors & Awards:* Int Award, Am Chem Soc, 71. *Mem:* Fel AAAS. *Res:* Selective toxicity; modes of toxic action; comparative biochemistry; neuropharmacology. *Mailing Add:* Off Provost Univ Mass Amherst MA 01003

O'BRIEN, RICHARD LEE, b Shenandoah, Iowa, Aug 30, 34; m 57; c 4. ONCOLOGY, CELL BIOLOGY. *Educ:* Creighton Univ, MS, 58, MD, 60. *Prof Exp:* From intern to asst resident med, First Med Div, Bellevue Hosp, Columbia Univ, 60-62; from asst prof to prof path, Sch Med, Univ Southern Calif, 66-82, dep dir, Cancer Ctr, 75-81, dir, 81-82; PROF MED & MED MICROBIOL & DEAN SCH MED, CREIGHTON UNIV SCH MED, 82-, VPRES, HEALTH SCIS, 85- *Concurrent Pos:* Fel biochem, Inst Enzyme Res, Univ Wis-Madison, 62-64; spec fel, Nat Cancer Inst, 67-69; grants, Am Cancer Soc, 67-68 & 75-82, Nat Cancer Inst, 67-70 & 74-88, John A Hartford Found, 69-74 & Wright Found, 76-81; vis prof molecular biol, Univ Geneva, 73-74. *Mem:* AAAS; Am Asn Path; Am Asn Cancer Res; Asn Am Med Col; Am Asn Cancer Educ. *Res:* Mechanisms of oncogenesis; environmental carcinogenesis; control of cell proliferation; molecular biology; cellular immunology; health policy. *Mailing Add:* Creighton Univ Sch Med Calif St & 24 St Omaha NE 68178-0006

O'BRIEN, ROBERT, b Chicago, Ill, April 21, 45. MATHEMATICS. *Educ:* George Washington Univ, PhD(math), 76. *Prof Exp:* Engr, Fairchild Space Corp, 81-85; SYSTS ENGR, OMITRON CORP, 85- *Mailing Add:* Omitron Corp 6411 Ivy Ln Suite 706 Greenbelt MD 20770

O'BRIEN, ROBERT L, b New Bedford, Mass, July 29, 36; m 82; c 5. SPACE PROPULSION, THERMODYNAMICS. *Educ:* Carnegie Inst Technol, BS & MS, 57. *Prof Exp:* Res engr, 55-65, supvr air-breathing propulsion, 65-67, chief, 67-77, mgr eng res, United Aircraft Res Labs, 77-81, MGR DIV COORDR UNITED TECHNOL RES CTR, 81- *Concurrent Pos:* Lectr, Rensselaer Polytech Inst, 64-65; mem, Air Augmented Propulsion Working Group, Joint Army-Navy-Nasa-Air Force Interagency Propulsion Comt, 68-, Naval Aeroballistics Adv Comt, 70 & 71 & US Gas Turbine Sci Deleg to People's Repub China, 79, Shuttle Challenger Failure Rev Team, 86 & Titan Failure Rev & Recovery Team, 86-88. *Mem:* Am Soc Mech Engrs; Am Inst Aeronaut & Astronaut. *Res:* Supersonic inlet technology; gas dynamics; turbulent flow; supersonic boundary layers and heat transfer; turbulent mixing; supersonic compressors; advanced air-breathing propulsion concepts. *Mailing Add:* United Technol Res Ctr East Hartford CT 06108

O'BRIEN, ROBERT NEVILLE, b Nanaimo, BC, June 14, 21; m 52; c 5. PHYSICAL CHEMISTRY. *Educ:* Univ BC, BASc, 51, MASc, 52; Univ Manchester, PhD(metall), 55. *Prof Exp:* Asst physics, Univ BC, 52; fel pure chem, Can Nat Res Coun, 55-57; from asst prof to assoc prof chem, Univ Alta, 57-66; from assoc prof to prof, 66-88, EMER PROF CHEM, UNIV VICTORIA, 88- *Concurrent Pos:* Vis scholar, Univ Calif, Berkeley, 64-65; consult, UniRoyal Res Labs & Bapco Paint, Dupont & Monsanto; mem bd mgt, BC Res Coun; chief consult, Troove Tech Ltd. *Mem:* Am Chem Soc; Electrochem Soc; fel Chem Inst Can; fel Royal Soc Arts. *Res:* Electrode processes; optical studies of working electrodes in metal electrodeposition cells; surface chemistry; electrets; gas-liquid interface. *Mailing Add:* Dept of Chem Univ of Victoria Victoria BC V8W 2Y2 Can

O'BRIEN, ROBERT THOMAS, b Bismarck, NDak, Dec 20, 25; m 48; c 5. BACTERIAL PHYSIOLOGY, BIOCHEMISTRY. *Educ:* Univ NDak, BS, 50, MS, 52; Wash State Univ, PhD(bact), 56. *Prof Exp:* Asst bact, Univ NDak, 50-52, instr, 52; asst, Wash State Univ, 52-53, actg instr, 53-54, asst, 54-56; biol scientist, Gen Elec Co, 56-64, consult microbiologist, 64-66; prof biol, 66-88, chmn dept, 78-88, EMER PROF, NMEX STATE UNIV, 88- *Concurrent Pos:* Consult, Azar Bros, Inc, 67- *Mem:* AAAS; Am Soc Microbiol; Brit Soc Gen Microbiol. *Res:* Survival of animal viruses in aquatic environments; mechanisms of halogen inactivation of animal viruses. *Mailing Add:* 2901 Karen Dr Las Cruces NM 88003

O'BRIEN, RONALD J, b Ottawa, Can, June 3, 32. MATHEMATICS. *Educ:* St Lawrence Univ, MA, 65; NMex State Univ, MA, 67. *Prof Exp:* CHMN MATH DEPT, STATE UNIV NY, CANTON, 61- *Mailing Add:* Math Dept State Univ NY Col Tech Canton Canton NY 13617

O'BRIEN, STEPHEN JAMES, b Rochester, NY, Sept 30, 44; m 68; c 2. GENETICS. *Educ:* St Francis Col, BS, 66; Cornell Univ, PhD(genetics), 71. *Prof Exp:* NIH fel biochem, Frederick Cancer Res Fac, 71-73, GENETICIST, NAT CANCER INST, 73-, CHIEF SECT GENETICS, LAB VIRAL CARCINOGENESIS, FREDERICK CANCER RES & DEVELOP CTR, 80- & CHIEF LAB, 83- *Concurrent Pos:* NIH fel, Geront Res Ctr, 71-72 & Nat Cancer Inst, 72-73. *Mem:* Am Soc Naturalists; Genetics Soc Am; Tissue Cult Asn; Am Genetics Asn. *Res:* Human biochemical genetics; somatic cell genetics; virol oncology. *Mailing Add:* Frederick Cancer Res & Develop Ctr Nat Cancer Inst Frederick MD 21702-1201

O'BRIEN, THOMAS DORAN, b Washington, DC, Mar 31, 10; m 35; c 2. INORGANIC CHEMISTRY. *Educ:* George Washington Univ, BS, 35, MS, 38; Univ Ill, PhD(inorg chem), 40. *Prof Exp:* Res chemist, Naval Res Lab, 40-42 & Barrett Div, Allied Chem & Dye Corp, NJ, 42-43; asst prof chem, Tulane Univ, 43-45; from asst prof to assoc prof, Univ Minn, 45-55; prof & head dept, Kans State Univ, 55-60; prof & dean grad sch, 60-77, res coordr, 70-77, EMER PROF CHEM, UNIV NEV, RENO, 77- *Concurrent Pos:* Prog dir & actg dep div dir, NSF, 67-68. *Mem:* Am Chem Soc. *Res:* Inorganic coordination compounds; heterogeneous catalysis. *Mailing Add:* 174 Sunset Pl Sequim WA 98382-8514

O'BRIEN, THOMAS V, b Cincinnati, Ohio, Apr 30, 37; m 62; c 4. TOPOLOGY. *Educ:* Xavier Univ, BS, 59, MS, 60; Syracuse Univ, PhD(math), 65. *Prof Exp:* Asst prof math, Marquette Univ, 64-69; asst prof, 69-71, ASSOC PROF MATH, BOWLING GREEN STATE UNIV, 71- *Mem:* Am Math Soc; Math Asn Am. *Res:* Dynamical systems; study of continuous flows on manifolds. *Mailing Add:* Dept of Math Statistics Bowling Green State Univ Bowling Green OH 43403

O'BRIEN, THOMAS W, b Rochester, Minn, Sept 17, 38; m 64; c 2. MOLECULAR BIOLOGY, BIOCHEMISTRY. *Educ:* St Thomas Col, BS, 62; Marquette Univ, MS, 63, PhD(physiol), 65. *Prof Exp:* Fel biochem, NJ Col Med, 64-65, instr, 65-66; asst prof, 66-74, assoc prof, 74-81, PROF BIOCHEM & MOLECULAR BIOL, COL MED, UNIV FLA, 81- *Concurrent Pos:* Co-dir, Univ Fla Biotech Ctr, 87-88. *Mem:* Am Chem Soc; Am Soc Biochem & Molecular Biol; Am Soc Cell Biol. *Res:* Structure and function of ribosomes; mitochondrial biogenesis; mitochondrial protein synthesis. *Mailing Add:* Dept Biochem & Molecular Biol Univ Fla Med Col Health Sci Ctr Gainesville FL 32610

O'BRIEN, VIVIAN, b Baltimore, Md, Feb 1, 24. FLUID DYNAMICS. *Educ:* Goucher Col, AB, 45; Johns Hopkins Univ, MA, 50, MS, 54, PhD, 60. *Hon Degrees:* SDc, Goucher Col, 86. *Prof Exp:* Jr aerodynamicist, Martin Co, 45-47; asst aeronaut, 47-55, assoc physicist fluid dynamics, Appl Physics Lab, 55-58, physicist fluid dynamics, Appl Physics Lab, 58-86, lectr, Dept Chem Eng, 79-86, PROF, DEPT MECH ENG, JOHNS HOPKINS UNIV, 86- *Mem:* Fel Am Phys Soc; Am Inst Aeronaut & Astronaut; Soc Women Engrs. *Res:* Theoretical transonic and supersonic aerodynamics; turbulent flow; viscous vortex flow; viscous biological flows; drops and bubbles; porous media flows; rheology of liquids. *Mailing Add:* Dept Mech Engrs Johns Hopkins Univ 3400 W Charles St Baltimore MD 21218

O'BRIEN, WILLIAM DANIEL, JR, b Chicago, Ill, July 19, 42. BIOACOUSTICS, BIOENGINEERING. *Educ:* Univ Ill, Urbana, BS, 66, MS, 68, PhD(elec eng), 70. *Prof Exp:* Res assoc ultrasonic biophys, Univ Ill, Urbana, 70-71; res scientist, Bur Radiol Health, Food & Drug Admin, 71-75; asst prof, 75-80, assoc prof, 80-87, PROF ELEC ENG & BIOENG, UNIV ILL, URBANA, 87- *Concurrent Pos:* ed-in-chief, Trans Ultrasonics, Ferroelecs & Frequency Control, Inst Elec & Electronics Engrs. *Honors & Awards:* Inst Elec & Electronics Engrs Centennial Medal, 84; Am Inst Ultrasound in Med, Pres Recognition Award, 85. *Mem:* Inst Elec & Electronics Engrs; fel Acoust Soc Am; fel Am Inst Ultrasound Med (pres 88-); AAAS; Sigma Xi. *Res:* Examination of the mechanisms by which ultrasonic energy interacts with biological materials including ultrasonic biophysics, bioengineering, dosimetry and bioeffects along with selected ultrasonic instrument development. *Mailing Add:* Dept Elec Eng Bioacoust Res Lab Univ Ill 1406 W Green St Urbana IL 61801

O'BRIEN, WILLIAM E, INBORN ERRORS OF METABOLISM, GENE REGULATION. *Educ:* Univ Ga, PhD(biochem), 71. *Prof Exp:* ASSOC PROF BIOCHEM, BAYLOR COL MED, 83- *Res:* Bio-chemical genetics. *Mailing Add:* Dept Pediatric & Baylor Col Med One Baylor Plaza Houston TX 77030

O'BRIEN, WILLIAM JOHN, b Summit, NJ, Nov 30, 42; m 64; c 3. AQUATIC ECOLOGY, LIMNOLOGY. *Educ:* Gettysburg Col, BA, 65; Cornell Univ, 65-69; Mich State Univ, PhD(aquatic ecol), 70. *Prof Exp:* Teaching fel, Kellogg Biol Sta, Mich State Univ, 70-71; from asst prof to assoc prof, 71-82, PROF SYSTS & ECOL, UNIV KANS, 82-; RES SCIENTIST, ECOSYSTEM CTR, MARINE BIOL LAB, 86- *Concurrent Pos:* Res grants, polar progs, NSF, 75-80, 83-86, 87-88 & 89-90, ecol prog, 78-81, 81-83, 84-86 & 88-90, US Dept Energy, 78-81; sr res assoc, Ctr Northern Studies, 77, Arctic LTER, 88-; distinguished lectr, Kans Acad Sci, 90. *Mem:* Am Soc Limnol & Oceanog; Ecol Soc Am; Int Asn Theoret & Appl Limnol; Am Fisheries Soc; Animal Behav Soc. *Res:* Zooplankton ecology; ecology of fish feeding; artic limnology; ecological effects of nuclear war. *Mailing Add:* Dept Systs Ecol Univ Kans 6008 Haworth Lawrence KS 66045

O'BRIEN, WILLIAM JOSEPH, b New York, NY, July 25, 35; div; c 2. DENTAL MATERIALS, SURFACE CHEMISTRY. *Educ:* City Col New York, BS, 58; NY Univ, MS, 62; Univ Mich, PhD(metall eng), 67. *Prof Exp:* Assoc dir res, J F Jelenko Co, 58-61; from instr to assoc prof mat sci, Marquette Univ, 61-70; assoc prof, 70-73, PROF & DIR, SURV SCI LAB, DENT RES INST, UNIV MICH, ANN ARBOR, 73- *Concurrent Pos:* Consult, WHO, 67-68 & Am Dent Asn, 67-; chmn dept dent mat, Marquette Univ, 67-70; prin investr, USPHS grant; res assoc, Vet Admin Hosp, Wood, Wis, 67-; secy, Dent Mat Group, 69-73, pres, 75-76. *Honors & Awards:* UN Cert, UN, 67. *Mem:* Gen Systs Res; Int Asn Dent Res. *Res:* Biomaterials; noble metals; surface phenomena; capillary phenomena; ceramics-mechanical and optical properties; innovation management. *Mailing Add:* Dent 2203 & Dent 1074 Univ Mich Ann Arbor MI 48109

O'BRIEN, WILLIAM M, b Bethel, Maine, Feb 26, 31; m 57; c 4. RHEUMATIC DISEASES, STATISTICS. *Educ:* Tufts Univ, BS, 52; Yale Univ, MD, 56. *Prof Exp:* Clin assoc, Nat Inst Arthritis & Metab Dis, 58-60; sr registr, Manchester Royal Infirmary, Eng, 60-61; sr clin investr, Nat Inst Arthritis & Metab Dis, 61-64; asst prof, Sch Med, Yale Univ, 64-67; assoc prof, 67-72, PROF INTERNAL MED, SCH MED, UNIV VA, 72- *Concurrent Pos:* Consult, US Food & Drug Admin, US Fed Trade Comn. *Mem:* AAAS; Am Epidemiol Soc; Am Rheumatism Asn; Asn Comput Mach; Heberden Soc. *Res:* Rheumatic diseases; drug testing. *Mailing Add:* Dept Med Univ VA Sch Med Box 395 Charlottesville VA 22908

O'BRYANT, DAVID CLAUDE, b Canton, Ill, Mar 18, 35; m 60; c 1. MECHANICAL ENGINEERING. *Educ:* Univ Ill, Urbana, BS, 58, MS, 61, EdD(voc & tech educ), 70. *Prof Exp:* Teaching asst gen eng, Univ Ill, Urbana, 59-60; designer, Carroll Henneman & Assoc, 60-61; from instr, to asst prof, 60-83, ASSOC PROF GEN ENG, UNIV ILL, URBANA, 83- *Concurrent Pos:* Tech skills coordr, Peace Corps, Kenya, 71-72. *Mem:* Am Tech Educ Asn; Nat Soc Prof Engrs; Tech Design Assocs. *Res:* Vocational and technical education; engineering education and evaluation; teaching improvement. *Mailing Add:* Dept Gen Eng 209 Transp Univ Ill Urbana IL 61801

OBST, ANDREW WESLEY, b Lwow, Poland, Aug 3, 42; US citizen; m 68. PHYSICS. *Educ:* Univ Alta, BSc, 63, MSc, 66; Univ Ky, PhD(exp nuclear physics), 70. *Prof Exp:* Fel nuclear physics, Fla State Univ, 71-73, Univ Tex, Austin, 73-75 & Northwestern Univ, 75-78; STAFF MEM NUCLEAR PHYSICS, LOS ALAMOS SCI LAB, 78- *Res:* Experimental low and medium energy nuclear physics. *Mailing Add:* MS D406 Los Alamos Nat Los Alamos NM 87545

OCAIN, TIMOTHY DONALD, b Waupun, Wis. ENZYME INHIBITORS, ANTIHYPERTENSIVE AGENTS. *Educ:* Univ Wis, BS, 81; Univ Wis-Madison, PhD(pharmaceut chem), 86. *Prof Exp:* Postdoctoral fel, Univ Minn, 86; sr chemist, 86-87, RES SCIENTIST, WYETH-AYERST RES, 87- *Mem:* Am Chem Soc; AAAS. *Res:* Design of novel therapeutic agents for the treatment of hypertension and cardiac disease; synthesis and mechanism of action of enzyme inhibitors. *Mailing Add:* Wyeth-Ayerst Res CN 8000 Princeton NJ 08543-8000

O'CALLAGHAN, DENNIS JOHN, b New Orleans, La, July 26, 40; m 67; c 1. ANIMAL VIROLOGY, MEDICAL MICROBIOLOGY. *Educ:* Loyola Univ, New Orleans, BS, 62; Univ Miss Med Ctr, PhD(microbiol), 68. *Prof Exp:* Fel virol, Dept Biochem, Med Ctr, Univ Alta, 68-70, asst prof biochem, 70-71; from asst prof to assoc prof, 71-77, prof microbiol, Med Ctr, Univ Miss, 77-84; PROF & CHMN, DEPT MICROBIOL & IMMUNOL, LA STATE UNIV MED CTR, 84-, CO-ORDINATOR, MOLECULAR BIOL RES CTR, 87- *Concurrent Pos:* Res grant, Brown-Hazen Fund Res Corp, 72; NIH & NSF res grants, 73-; assoc ed, J Miss Acad Sci, 75-; ad hoc grant reviewer, Am Cancer Soc & NIH; vis prof, Univ Nev Sch Med, 85; vis prof, Johns Hopkins Univ Sch Med, 86; vis prof, La State Univ, Baton Rouge, 88; NIH grant, Molecular Biol Prog USDA; pres-elect, DNA Viruses Div, Am Soc Microbiol, 90-91, pres, 91-92. *Mem:* Am Soc Biol Chemists; Am Asn Cancer Res; Brit Soc Gen Microbiol; Soc Exp Med & Biol; Am Soc Exp Path; Sigma Xi; Am Soc Virol; Am Soc Microbiol; AAAS. *Res:* Biochemistry of herpes virus replication, concerning control of viral DNA replication of defective interfering particles; tumor virology; role of herpes viruses in human cancer. *Mailing Add:* Dept Microbiol Immunol La State Univ Med Ctr 1501 Kings Hwy Shreveport LA 71130

O'CALLAGHAN, JAMES PATRICK, b West Palm Beach, Fla, Mar 6, 49. NEUROCHEMISTRY, NEUROTOXICOLOGY. *Educ:* Purdue Univ, BS, 71; Emory Univ, PhD(pharmacol), 75. *Prof Exp:* Res fel, Res & Testing Lab, NY State Div Substance Abuse Serv, 75-78; res assoc pharmacol, Nat Inst Gen Med Sci & Nat Heart, Lung & Blood Inst, NIH, 78-80, staff fel, 80; GUEST INVESTR, LAB MOLECULAR & CELLULAR NEUROSCI, ROCKEFELLER UNIV, 83-; ADJ ASSOC PROF, NY UNIV MED CTR, 90- *Honors & Awards:* Sci Achievement Award, US Environ Protection Agency, 84, 86, 87, 88 & 89. *Mem:* Am Soc Pharmacol & Exp Therapeut; Soc Neurosci; NY Acad Sci; Soc Toxicol; AAAS. *Res:* The use of nervous system specific proteins as biochemical indicators of neurotoxicity. *Mailing Add:* Neurotoxicol Div MD-74B US Environ Protection Agency Research Triangle Park NC 27711

O'CALLAGHAN, MICHAEL JAMES, b Des Moines, Iowa, Sept 2, 54; m 82; c 1. SPATIAL LIGHTJMODULATORS, FERROELECTRIC LIQUID CRYSTAL ELECTRO-OPTIC MODULATORS. *Educ:* Iowa State Univ, BS, 76; Mich State Univ, MS, 78; Univ Colo, PhD(physics), 87. *Prof Exp:* Mem tech staff physics, Missile Systs Div, Hughes Aircraft Co, 78-81; res asst planetary astrophysics, Lab Atmospheric & Space Physics, 82-83; res asst atomic physics, Joint Inst Lab Astrophysics, 83-87, res assoc non-linear optics, 87-89; SR RES PHYSICIST ELECTRO-OPTIC DEVICES, DISPLAYTECH INC, 89- *Mem:* AAAS; Am Physic Soc; Optical Soc Am. *Res:* Research and development of electro-optic devices based on the use of ferroelectric liquid crystall technology. *Mailing Add:* 964 W Maple Court Louisville CO 80027

O'CALLAGHAN, RICHARD J, b New Orleans, La, Jan 17, 44; m 65; c 2. MICROBIOLOGY, MOLECULAR BIOLOGY. *Educ:* La State Univ, BS, 65; Univ Miss, MS, 66, PhD(med microbiol), 70. *Prof Exp:* Fel biochem, Univ Alta, 70-71, instr, 71; asst prof, 72-77, ASSOC PROF MICROBIOL, MED CTR, LA STATE UNIV, 77- *Concurrent Pos:* E G Schleider Educ Found res grants, 64-81; La Heart Asn res grants, 76-85. *Mem:* Am Soc Microbiol; Am Soc Virol. *Res:* Mechanisms and epidemiology of antibiotic resistance; characterization of influeza C virus. *Mailing Add:* Dept Microbiol La State Univ Sch Med 1542 Tulane Ave New Orleans LA 70112

OCAMPO-FRIEDMANN, ROSELI C, b Manila, Philippines, Nov 23, 37; c 2. ECOLOGY, PHYCOLOGY. *Educ:* Univ Philippines, BSc, 58; Hebrew Univ Jerusalem, MSc, 66; Fla State Univ, PhD(biol), 73. *Prof Exp:* Res assoc limnol & phycol, Nat Inst Sci & Technol Philippines, 58-67; teaching asst bot & lab technologist phycol, 67-73, res assoc phycol, Fla State Univ, 73-75; from asst prof to assoc prof, 74-87, PROF BIOL, FLA A&M UNIV, 87- *Concurrent Pos:* Co-prin investr res grants, NSF & NASA. *Mem:* Int Phycol Soc; Phycol Soc Am; AAAS; Indian Phycol Soc; Sigma Xi; Brit Phycol Soc; Am Soc Microbiol; Planetary Soc. *Res:* Culture and taxonomy of unicellular blue-green algae; microbiology; biology of microorganisms in extreme habitats. *Mailing Add:* Dept of Biol Fla A&M Univ Tallahassee FL 32307

OCCELLI, MARIO LORENZO, b Luino, Italy, Dec 16, 42; US citizen; m 70; c 4. CHEMISTRY. *Educ:* Iowa State Univ, BS, 67, PhD(phys chem), 73. *Prof Exp:* Chemist, Air Prod & Chem, Inc, 73-76; sr res chemist, Davison Div, W R Grace Co, 76-79; sr res chemist, Gulf Oil Corp, 79-85; RES ASSOC, UNION OIL CALIF, 85- *Mem:* Am Chem Soc; Int Zeolite Asn. *Res:* Zeolites synthesis and characterization; preparation and testing of zeolite based catalysts for the upgrading of heavy oils, fluid cracking catalysts, shale oils and raffinates from coal; synthesis of chemicals from zeolite catalyzed reactions. *Mailing Add:* Union Oil Co PO Box 76 Brea CA 92621-6399

OCCOLOWITZ, JOHN LEWIS, b Melbourne, Australia, July 30, 31; m 62; c 2. MASS SPECTROMETRY. *Educ:* Univ Melbourne, BSc, 52, DipED, 53, MSc, 65. *Prof Exp:* Instr sci, Victoria, Australia Educ Dept, 53-55; res scientist chem, Dept Supply Defense Stands Labs, Australian Govt, 55-67; RES ASSOC CHEM, LILLY RES LABS, ELI LILLY & CO, 67- *Mem:* Am Soc Mass Spectrometry. *Res:* Determination of organic structures using mass spectrometry; elucidation of the structure of ions formed in the mass spectrometer, by labelling and measurement of energetics of formation. *Mailing Add:* 6840 Dover Rd Indianapolis IN 46220

OCHIAI, EI-ICHIRO, b Tokyo, Japan, Sept 15, 36; m; c 2. BIOINORGANIC CHEMISTRY, CHEMICAL EVOLUTION. *Educ:* Univ Tokyo, BSc, 59, MSc, 61, PhD(chem), 64. *Prof Exp:* Instr indust chem, Univ Tokyo, 64-69; fel chem, Ohio State Univ, 66-68; fel, Univ BC, 69-71, instr, 71-80; sr res assoc, Univ Md, 80-81; assoc prof, 81-87, DEPT CHMN, 85-, PROF DEPT CHEM, JUNIATA COL, 87- *Mem:* Am Chem Soc; AAAS; Japanese Chem Soc; NY Acad Sci; Int Soc Study Origin Life. *Res:* Bioinorganic chemistry, especially B-12 coenzyme dependent enzyme mechanism, oxygen activation, ribonucleotide reductase mechanism; bioinorganic aspects of chemical evolution and biological evolution. *Mailing Add:* Dept Chem Juniata Col Huntington PA 16652

OCHIAI, SHINYA, b Kofu City, Japan, June 6, 35; m 66; c 1. PROCESS CONTROL, MATHEMATICAL SIMULATION. *Educ:* Waseda Univ, Japan, BS, 60; Rice Univ, MS, 62; Purdue Univ, PhD(control), 66. *Prof Exp:* Instrumentation develop engr, Celanese Fibers Co, 65-67, res & develop engr, 67-68; sr process syst engr, Celanese Chem Co, 68-74; eng assoc, 74-90, SR ENG ASSOC, HOECHST CELANESE, 90- *Concurrent Pos:* Vis assoc prof, Tex A&I Univ, 74-75; prog coordr, nat conf, Instrument Soc Am, 89 & 90. *Honors & Awards:* Schuck Award, Am Control Coun, 74. *Mem:* Fel Instrument Soc Am; Am Inst Chem Engrs; Sigma Xi. *Res:* Inferential methods of measuring process variables; application of advanced control theory; mathematical simulation to improve industrial chemical process operation; author of 16 publications. *Mailing Add:* Hoechst Celanese Box 9077 Corpus Christi TX 78469

OCHOA, SEVERO, b Luarca, Spain, Sept 24, 05; nat; m 31. MOLECULAR BIOLOGY. *Educ:* Malaga Col, Spain, BA, 21; Univ Madrid, MD, 29. *Hon Degrees:* Many from var Am & foreign cols & univs. *Prof Exp:* Lectr physiol & biochem, Sch Med, Univ Madrid, 31-35, head physiol div, Inst Med Res, 35-36; guest res asst physiol, Kaiser-Wilhelm Inst Med Res, 36-37; Lankester investr, Marine Biol Lab, Plymouth, Eng, 37; demonstr biochem & Nuffield res asst, Oxford Univ, 38-40; instr pharmacol & res assoc, Sch Med, Wash Univ, 41-42; res assoc med, Sch Med, NY Univ, 42-45, asst prof biochem, 45-46, prof pharmacol & chmn dept, 46-54, prof biochem & chmn dept, 54-74; distinguished mem, Roche Inst Molecular Biol, 74-85; PROF BIOL, UNIV AUTONOMA, MADRID, 85- *Concurrent Pos:* Fel from Univ Madrid, Kaiser-Wilhelm Inst, Berlin & Heidelberg, 29-31 & Nat Inst Med Res, London, 32-33; mem physiol study sect, USPHS, 47-50, biochem study sect, 52-55; mem biochem panel, US Off Naval Res, 53-55, chmn, 55-57; US rep, Int Union Biochem, 55-61; hon prof, San Marcos Univ, Lima, 57; hon mem fac, Univ Chile, 57; mem sci adv comt, Mass Gen Hosp, 57-60; mem dept biol, Brookhaven Nat Lab, 59-62; mem staff, Jane Coffin Childs Fund Med Res, 61-63, Merck Inst Ther Res, 65-74 & Am Cancer Soc, 69-70. *Honors & Awards:* Nobel Prize in Med, 59; Neuberg Medal Biochem, 51; Price Award, Fr Soc Biol Chem, 55; Borden Award, Asn Am Med Cols, 58; NY Univ Medal, 60; Order of the Rising Sun 2nd Class & Gold Medal, Japan, 67; Carlos Jimenez Diaz Lectr Award, Univ Madrid, 69; Quevedo Gold Medal Award, Spain, 69; Albert Gallatin Medal, NY Univ, 70; Nat Medal Sci, 79. *Mem:* Nat Acad Sci; Am Soc Biol Chemists (pres, 58); Harvey Soc (vpres, 52-53, pres, 53-54); fel Am Acad Arts & Sci; fel NY Acad Sci. *Res:* Biochemistry of muscle and fermentation; respiratory enzymes; enzymatic mechanisms of carbon dioxide assimilation, citric and fatty acid cycles; synthesis of nucleic acid and proteins, genetic code and translation of the genetic message. *Mailing Add:* Ctr Biol Molecular Univ Autonoma Madrid Canto Blanco 28049 Spain

OCHRYMOWYCZ, LEO ARTHUR, b Shaok, Ukraine, May 20, 43; US citizen; m 70. ORGANIC CHEMISTRY. *Educ:* St Mary's Col, Minn, BA, 65; Iowa State Univ, PhD(org chem), 69. *Prof Exp:* Asst prof, 69-74, ASSOC PROF CHEM, UNIV WIS-EAU CLAIRE, 74- *Concurrent Pos:* Res fel, Iowa State Univ, 70-71. *Mem:* Am Chem Soc; Sigma Xi. *Res:* Chemistry of macrocyclic polythioethers; coordination chemistry of post-transitional metals; general organo-sulfur chemistry, especially beta-keto sulfoxides and lipids; development of carbonyl synthesis reagents. *Mailing Add:* Dept Chem Univ Wis Eau Claire WI 54702-4004

OCHS, HANS D, b Sept 29, 36; m; c 2. PEDIATRICS. *Educ:* Montpellier, France, MD, 61. *Prof Exp:* FAC MEM, DEPT PEDIAT, DIV IMMUNOL & RHEUMATOL, SCH MED, UNIV WASH. *Mem:* Soc Pediat Res; Am Asn Immunologists. *Mailing Add:* Dept Pediat Div Immunol & Rheumatol RD 20 Seattle WA 98195

OCHS, RAYMOND S, b Chicago, Ill, July 6, 52; m 83. METABOLIC REGULATION. *Educ:* Purdue Univ, BS, 74; Ind Univ, PhD(biochem), 79. *Prof Exp:* Teaching trainee biochem, Enzyme Inst, Univ Wis, 79-82; staff fel metab, Nat Inst Alcohol Abuse & Alcoholism, 82-83; res assoc biochem, Case Western Reserve Univ, 83-85; ASST PROF BIOCHEM, KANS STATE UNIV, 85- *Mem:* Am Soc Biochem & Molecular Biol. *Res:* Hormonal control of intermediary metabolism of liver; regulation of gene expression for metabolically important enzymes. *Mailing Add:* Dept Biochem Kans State Univ Manhattan KS 66506

OCHS, SIDNEY, b Fall River, Mass, June 30, 24; m 49; c 3. NEUROPHYSIOLOGY, MEDICAL BIOPHYSICS. *Educ:* Univ Chicago, PhD, 52. *Prof Exp:* Res assoc, Ill Neuropsychiat Inst, 53-54; asst prof, Med Ctr, Univ Tex, 56-58; from assoc prof to prof physiol, 58-70, dir med biophys prog, 68-86, PROF PHYSIOL, SCH MED, IND UNIV, INDIANAPOLIS, 70- *Concurrent Pos:* Res fel, Calif Inst Technol, 54-56; NSF sr fel, dept biophys, Univ Col, Univ London, 63-64; ed, J Neurobiol, 68-77, assoc ed, 77-; vis prof, dept neurosci, Ctr Invest & Advan Studies, Nat Polytech Inst, Mexico, 68. *Mem:* Soc Neuroscience; Am Physiol Soc; Int Brain Res Orgn; Am Neurochemical Soc; Biophys Soc. *Res:* Functions of cerebral cortex; axoplasmic transport in nerve; muscle membrane properties. *Mailing Add:* Physiol Dept MS 383 Ind Univ Sch Med 635 Barnhill Dr Indianapolis IN 46202

OCHS, STEFAN A(LBERT), b Frankfurt, Ger, Sept 16, 22; nat US; m 48; c 3. ENGINEERING PHYSICS. *Educ:* Columbia Univ, BS, 43, MA, 49, PhD(physics), 53. *Prof Exp:* Asst, Columbia Univ, 47-50; tutor, City Col New York, 50-51; res engr, RCA Corp, 52-65, sr engr, 65-75; sr mech engr, Int Signal & Control Corp, 75-87; RETIRED. *Mem:* AAAS; Am Chem Soc. *Res:* Mechanical design of electronic equipment. *Mailing Add:* 637 Wyncroft Lane No 2 Lancaster PA 17603

OCHSNER, JOHN LOCKWOOD, b Madison, Wis, Feb 10, 27; m 54; c 4. SURGERY. *Educ:* Tulane Univ La, MD, 52; Am Bd Surg, dipl, 60; Am Bd Thoracic Surg, dipl, 60. *Prof Exp:* From intern to asst resident, Univ Mich Hosp, Ann Arbor, 52-54; resident, Baylor Univ Affil Hosp, Houston, Tex, 56-60; instr, Sch Med, Baylor Univ, 60-61; instr surg, 61-65, clin assoc prof, 65-69, chmn dept surg, 66-87, CLIN PROF SURG, SCH MED, TULANE UNIV, 69-, EMER CHMN, OCHSNER CLIN, 87- *Concurrent Pos:* Chief surg res, Jefferson Davis Hosp, Houston, 58-59 & Tex Children's Hosp, Houston, 59-60; mem staff, Ochsner Clin, 61-66; chief surg, Ochsner Found Hosp, New Orleans; vis surgeon, Charity Hosp La & E A Conway Mem Hosp, Monroe; mem courtesy staff, Sara Mayo Hosp & Flint Goodridge Hosp, New Orleans; consult cardiovasc surgeon, Lafayette Mem Hosp; consult thoracic surgeon, USPHS Hosp, New Orleans; consult heart surgeon, La Dept Health. *Mem:* Soc Vascular Surg; Soc Thoracic Surg; Am Col Surgeons; Am Col Chest Physicians; Int Cardiovasc Soc (secy-gen, 77-87, pres, 89-). *Mailing Add:* 1514 Jefferson Hwy New Orleans KY 70121

OCHSNER, SEYMOUR FISKE, b Chicago, Ill, Nov 29, 15; m 45; c 3. RADIOLOGY. *Educ:* Dartmouth Col, AB, 37; Univ Pa, MD, 47. *Prof Exp:* Staff physician, Stony Wold Sanatorium, 47-49; intern, Johnston-Willis Hosp, 49-50; fel radiol, Ochsner Found Hosp, 50-53; consult, Ochsner Clin, 53-90; assoc prof, 62-67, PROF CLIN RADIOL, SCH MED, TULANE UNIV, 67- *Honors & Awards:* Distinguished Serv Award, Southern Med Asn, 71; Gold Medal, Am Col Radiol, 81, Am Roentgenol Ray Soc, 86. *Mem:* Radiol Soc NAm (vpres, 65); Roentgen Ray Soc (vpres, 66, pres, 76); Am Col Radiol (pres, 72); AMA. *Res:* Clinical radiology and radiation therapy. *Mailing Add:* 107 Holly Dr Metairie LA 70005

OCKERMAN, HERBERT W, b Chaplin, Ky, Jan 16, 32. FOOD CHEMISTRY, STATISTICS. *Educ:* Univ Ky, BS, 54, MS, 58; NC State Univ, PhD(animal husb, statist), 62. *Prof Exp:* PROF ANIMAL SCI, OHIO STATE UNIV, 61- *Mem:* Am Meat Sci Asn; Am Soc Animal Sci; Inst Food Technol; Int Cong Meat Sci & Technol; Am Soc Testing & Mat; Can Soc Meat Sci. *Res:* Lipids and antioxidants; food flavor and analysis; sterile tissue; tissue biochemistry and microbiology; tumbling of tissue; electrical stimulation of tissue. *Mailing Add:* Dept Animal Sci Ohio State Univ Columbus OH 43210-1094

OCKERSE, RALPH, b Brussels, Belg, May 17, 33; US citizen; m 56, 78; c 4. PLANT PHYSIOLOGY, BIOCHEMISTRY. *Educ:* State Teachers Col Neth, BA, 56; Baldwin Wallace Col, BS, 62; Yale Univ, PhD(plant physiol, biochem), 66. *Prof Exp:* Lab asst photosynthesis, Philips Res Labs, Neth, 53-55; res asst tissue culture, Chas Pfizer Co, NY, 57-58; biochemist, Union Carbide Co, Ohio, 62; from asst prof to prof biol, Hope Col, 66-76; PROF BIOL & CHMN DEPT, SCH SCI, PURDUE UNIV, INDIANAPOLIS, 76- *Concurrent Pos:* NSF res grant, Hope Col, 69-70, dir NSF undergrad res partic grant, 71, 72, 73 & 75; res grant, Res Corp, 74-76. *Mem:* AAAS; Am Chem Soc; Am Soc Plant Physiol; Royal Neth Bot Soc; Japanese Soc Plant Physiologists; Sigma Xi. *Res:* Physiology and biochemistry of plant growth regulation; mechanism of hormone action and interactions with macromolecules; plant tissue culture and development. *Mailing Add:* 4716 Laurel Circle N Dr Indianapolis IN 46226

OCKMAN, NATHAN, b New York, NY, Dec 29, 26. PICOSECOND DYNAMICS IN SEMICONDUCTORS. *Educ:* Purdue Univ, BS, 49; Univ Calif, Berkeley, MA, 50; Univ Mich, PhD(physics), 57. *Prof Exp:* Fel chem, Harvard Univ, 57-58; mem tech staff, RCA Labs, Inc, 59-65; mem tech staff, Gen Tel & Electronics Labs, 65-69; fel physiol, Albert Einstein Col Med, 70-81; vis assoc prof, 81-83, RES ASSOC PHYSICS, CITY COL, 83- *Mem:* Am Phys Soc. *Res:* Ultraviolet and visible absorption and reflection spectroscopy applied to the study of phospholipid dispersions, monolayers and bilayers; infrared spectroscopy applied to study of monolayers and films of phospholipids and proteins; picosecond laser spectroscopy applied to study of vibrational relaxation of biological molecules; liquids and electron hole dynamics in semiconductors. *Mailing Add:* 137 Riverside Dr New York NY 10024

OCONE, LUKE RALPH, b Bridgeport, Conn, Mar 10, 25; m 53; c 4. INORGANIC CHEMISTRY. *Educ:* Brooklyn Polytech Inst, BS, 51; Pa State Univ, PhD(chem), 56. *Prof Exp:* Res chemist photoprod dept, E I du Pont de Nemours & Co, 55-58; group leader explor chem, 58-70, GROUP LEADER

COM DEVELOP, PENWALT CORP, 71- *Mem:* Am Chem Soc; Sigma Xi. *Res:* Metal complexes and metallorganics; photographic science and technology. *Mailing Add:* 8315 Widener Rd Windmoor Philadelphia PA 19118

O'CONNELL, EDMOND J, JR, b Providence, RI, Apr 26, 39; m 65; c 5. ORGANIC CHEMISTRY, PHOTOCHEMISTRY. *Educ:* Providence Col, BS, 60; Yale Univ, PhD(chem), 64. *Prof Exp:* Res chemist radiation physics lab, E I du Pont de Nemours & Co, 64-67; from asst prof to assoc prof chem, 67-75, PROF CHEM, FAIRFIELD UNIV, 75-, CHMN DEPT CHEM, 81- *Concurrent Pos:* Petrol Res Fund grant, 68-70 & 72-74; Res Corp grant, 72-73. *Mem:* Am Chem Soc. *Res:* Organic photochemistry, especially reaction mechanisms and the relation of photo-reactivity to molecular structure. *Mailing Add:* Dept Chem Fairfield Univ Fairfield CT 06430

O'CONNELL, FRANK DENNIS, b Lynn, Mass, July 21, 27; m 48; c 3. PHARMACOGNOSY. *Educ:* Mass Col Pharm, BS, 51, MS, 53; Purdue Univ, PhD(pharmacog), 57. *Prof Exp:* Asst prof pharm, 57-58, from asst prof to assoc prof pharmacog, 58-69, actg dean, Sch Pharm, 72-73, asst dean, Sch Pharm, 74-81, prof pharmacog, 69-, ACTG DEAN, WVA UNIV, 81-, ASSOC DEAN, 86- *Mem:* Am Soc Pharmacog (treas, 73-76); Am Asn Col Pharm. *Res:* Isolation and biosynthesis of natural medicinal products; biochemistry; plant tissue cultures; biochemical transformations. *Mailing Add:* Dept Pharm WVa Univ Morgantown WV 26506

O'CONNELL, HARRY E(DWARD), b Glens Falls, NY, Mar 17, 16; m 39. CHEMICAL ENGINEERING. *Educ:* Univ Mich, BS, 38, MS 39, PhD, 42. *Prof Exp:* Chem engr, Ethyl Corp, 41-47, head process design, 47-51, group head process eng, 51, assoc dir process develop, 51-54, staff asst to vpres, 54, proj mgr, 54-60; mgr proj develop, Tenneco Chem Co, Tenn Gas Transmission Co, 60-63, vpres, Tenneco Mfg Co, 63-71, vpres & gen mgr Newport Div, Tenneco Chem, Inc, 71-76; exec vpres, Petro-Tex Chem Corp, 76-78, pres, 78-82; PRES, O'CONNELL ENG CO, 82- *Mem:* Am Chem Soc; fel Am Inst Chem Engrs. *Res:* Distillation; heat transfer; thermodynamics. *Mailing Add:* 163 Litchfield Lane Houston TX 77024

O'CONNELL, JAMES S, b Chicago, Ill, Jan 15, 32. ELECTROMAGNETIC REACTIONS, NEUTRINO REACTIONS. *Educ:* Beloit Col, BS, 53; Univ Ill, PhD(physics), 61. *Prof Exp:* Res asst, Yale Univ, 60-63; nuclear physicist, Nat Bur Standards, 63-89; CONSULT, 89- *Concurrent Pos:* Vis scholar, Stanford Univ, 70-71; Mass Inst Technol, 79-80. *Mem:* Fel Am Physical Soc. *Res:* Theoretical and experimental research on nuclear reactions using photons and electrons; over 60 journal publications. *Mailing Add:* Nat Bur Standards RADP B109 Gaithersburg MD 20899

O'CONNELL, JOHN P, b Morristown, NJ, Sept 19, 38; m 59; c 3. CHEMICAL ENGINEERING. *Educ:* Pomona Col, BA, 61; Mass Inst Technol, SB, 61, SM, 62; Univ Calif, Berkeley, PhD(chem eng), 67. *Prof Exp:* Actg instr chem eng, Univ Calif, Berkeley, 65-66; from asst prof to assoc prof chem eng, Univ Fla, 66-74, actg chmn, 81-82, chmn, 82-84, prof chem eng, 74-88; PROF & CHMN CHEM ENG, UNIV VA, 88- *Concurrent Pos:* Vis prof, Univ Calif, Los Angeles, 73, Inst Chemical Technol, Danish Tech Sch, 85; vis scholar, Stanford Univ, 73-74 & Calif Tech, 73. *Mem:* Fel AAAS; Am Chem Soc; Am Soc Eng Educ; fel Am Inst Chem Engrs. *Res:* Statistical mechanics; molecular thermodynamics; phase equilibria; surfactant systems; micellization and solubilization. *Mailing Add:* Dept Chem Eng Univ Va Charlottesville VA 22903

O'CONNELL, PAUL WILLIAM, b Newark, NY, Aug 5, 22; m 49; c 7. BIOCHEMISTRY. *Educ:* Univ Notre Dame, BS, 43; Univ Rochester, PhD(biochem), 49. *Prof Exp:* Fel chem, Univ Pittsburgh, 49-51; res assoc, 51-64, res sect head, 64-79, res mgr, 79-84, res contract consult, Upjohn Co, 84-87; RETIRED. *Mem:* AAAS; Am Chem Soc. *Res:* Biochemistry of lipids; enzymes; information handling. *Mailing Add:* 2509 Russet Dr Kalamazoo MI 49008

O'CONNELL, RICHARD JOHN, b Helena, Mont, Aug 27, 41; c 1. TECTONOPHYSICS, SOLID EARTH GEOPHYSICS. *Educ:* Calif Inst Technol, BS, 63, MS, 66, PhD(geophysics), 69. *Prof Exp:* Res fel geophysics, Calif Inst Technol, 69-70; res geophysicist, Univ Calif, Los Angeles, 70-71; asst prof geol, 71-74, assoc prof, 74-77, PROF GEOPHYSICS, HARVARD UNIV, 77- *Concurrent Pos:* Consult, Los Alamos Nat Lab, 78-, Rockwell Int Sci Ctr, 84-87, Amoco Tulsa Res Ctr, 84-; mem, Lunar & Planetary Coun, Univs Space Res Asn, 82-; bd dirs, Univs Space Res Asn, 84-; mem comt Earth Sci, Space Sci Bd, 85-86, Geophysics study comt, Nat Res Coun, 85-; mem earth sci rev panel, Nat Sci Found, 85-88; mem, Inst Geophysics & Planetary Physics Review Panel, Los Alamos Nat Lab, 83-, Geophysics Panel, NASA Earth Syst Sci Comt, 85-86, vis sci comt, NASA Goddard Space Flight Ctr, 86-87; chmn steering comt, Unavco, 90- *Mem:* AAAS; fel Am Geophys Union; Sigma Xi; Royal Astron Soc. *Res:* Dynamics and evolution of planetary interiors; plate tectonics; rheology of solids; mechanical properties of rocks and minerals. *Mailing Add:* Dept Earth & Planetary Sci 20 Oxford St Cambridge MA 02138

O'CONNELL, ROBERT F, b Athlone, Ireland, Apr 22, 33; m 63; c 3. SOLID STATE PHYSICS, THEORETICAL PHYSICS. *Educ:* Nat Univ Ireland, BSc, 53, DSc Univ Ireland, 75; Univ Notre Dame, PhD(physics), 62- *Prof Exp:* Asst lectr physics, Univ Col, Galway, 53-54; with telecommun br, Dept Posts & Tel, Ireland, 54-58; res assoc theoretic physics, Inst Advan Studies, Dublin, 62-64; from asst prof to assoc prof, 64-69, PROF PHYSICS, 69-, BOYD PROF, LA STATE UNIV, BATON ROUGE, 86- *Concurrent Pos:* Syst analyst, Int Bus Mach Corp, Ireland, 63-64; Nat Acad Sci-Nat Res Coun sr res assoc, NASA Inst Space Studies, NY, 66-68; consult, Theoret Phys Div, Lawrence Livermore Lab, Univ Calif, Livermore, 73-75; sr res fel, Sci Res Coun, Eng, 76. *Honors & Awards:* Sir J J Larmor Prize, Univ Col, Galway, 54; Distinguished Res Master, La State Univ, 75. *Mem:* Fel Am Phys Soc; Am Astron Soc; Int Astron Union; Int Soc Gen Relativity & Gravitation. *Res:* High-energy astrophysics; gravitation; atomic physics; solid state physics. *Mailing Add:* Dept Phys & Astron La State Univ Baton Rouge LA 70803-4001

O'CONNELL, ROBERT WEST, b San Francisco, Calif, Mar 22, 43; m 76; c 1. EXTRAGALACTIC ASTRONOMY. *Educ:* Univ Calif, Berkeley, AB, 64; Calif Inst Technol, PhD(astron & physics), 70. *Prof Exp:* Res astronomer, Lick Observ, Univ Calif, 69-71; chmn dept & dir, Leander McCormick Observ, 79-85; from asst prof to assoc prof, 71-85, PROF ASTRON, UNIV VA, 86- *Concurrent Pos:* Chmn, Int Joint Sci Working Group Starlab, 80-82; co-investr, Astro Missions, 79- *Mem:* Am Astron Soc; Royal Astron Soc; Int Astron Union; Sigma Xi. *Res:* Extragalactic astronomy (stellar content of normal galaxies, active galaxy nuclei, radio galaxies); ultraviolet and space astronomy; stellar photometry; properties of low mass stars. *Mailing Add:* Dept Astron Univ Va Math-Astron Bldg Rm 314 Charlottesville VA 22903

O'CONNOR, BRIAN LEE, b Lennox, Calif, Sept 17, 44; m 64; c 2. HUMAN BIOLOGY, PHYSICAL ANTHROPOLOGY. *Educ:* Univ Calif, Berkeley, AB, 69, PhD(anthrop), 74. *Prof Exp:* Asst prof, 74-81, ASSOC PROF ANAT, SCH MED, IND UNIV, INDIANAPOLIS, 81- *Mem:* Am Asn Anatomists. *Res:* Articular neurology and its functional, evolutionary and pathological significance. *Mailing Add:* Dept Anat Ind Univ Sch Med 1120 S Dr Indianapolis IN 46223

O'CONNOR, CAROL ALF, b Hamilton, Ohio, Nov 7, 48. STATISTICS, BIOSTATISTICS. *Educ:* Bowling Green State Univ, BS, 70, MA, 72, PhD(math), 75. *Prof Exp:* Asst prof math statist, Wright State Univ, 75-76; ASSOC PROF ENG MATH & COMPUT SCI, UNIV LOUISVILLE, 76- *Mem:* Am Math Soc; Inst Math Statist; Am Statist Asn; Asn Women in Math. *Res:* Biostatistics, simulation, artificial intelligence. *Mailing Add:* 3017 Juniper Hill Rd Louisville KY 40206

O'CONNOR, CECILIAN LEONARD, b Philadelphia, Pa, Oct 23, 22. PHYSICS. *Educ:* Catholic Univ, BS, 45, MS, 51, PhD(physics), 54. *Prof Exp:* Instr physics, De La Salle Col, Catholic Univ, 50-54; assoc prof, 54-67, chmn dept, 60-74, PROF PHYSICS, MANHATTAN COL, 67-, CHMN DEPT RADIOL & HEALTH SCI, 77- *Mem:* AAAS; Acoust Soc Am; Am Asn Physics Teachers; NY Acad Sci. *Res:* Thermodynamic properties of gases and liquids at ultrasonic and hypersonic frequencies. *Mailing Add:* Dept Physics Manhattan Col Bronx NY 10471

O'CONNOR, CHARLES TIMOTHY, b Atlantic City, NJ, Aug 1, 30. MEDICAL ENTOMOLOGY. *Educ:* Rutgers Univ, BS, 53, MS, 55; Ohio State Univ, PhD, 58; Tulane Univ, MPH, 61. *Prof Exp:* Asst dept entom, NJ Agr Exp Sta, 53-55; asst, Ohio Agr Exp Sta, 55-58; med entomologist, USPHS, Tex, 59-60; malaria specialist, US AID, Vietnam, 61-63, Ethiopia, 64-66; specialist malaria eradication br, Ctr Dis Control, USPHS, Haiti, 66-70, Brazil, 70-71; malaria specialist, WHO, Indonesia, 71-79 & Thailand, 80-84; RETIRED. *Mem:* Entom Soc Am; Am Mosquito Control Asn; Sigma Xi. *Res:* Control of livestock pests and of vectors of human diseases; study of the entomological aspects of a large-scale malathion trail in Central Java; cytogenetic studies of several anopheline vectors of malaria. *Mailing Add:* 470-B West Shore Dr Brigantine NJ 08203

O'CONNOR, DANIEL THOMAS, b Chicago, Ill. HYPERTENSION, NEPHROLOGY. *Educ:* Loyola Univ, Los Angeles, BS, 70; Univ Calif, Davis, MD, 74. *Prof Exp:* Intern, 74-75, resident, 75-77, fel nephrology & hypertension, 76-79, asst prof, 79-85, ASSOC PROF MED, UNIV CALIF, SAN DIEGO, 86-; CHIEF HYPERTENSION, VET ADMIN MED CTR, SAN DIEGO, 79- *Concurrent Pos:* Estab investr, Am Heart Asn, 83. *Honors & Awards:* Harry Goldblatt Award, Am Heart Asn, 86. *Mem:* Am Heart Asn; Am Soc Nephrology; Am Fedn Clin Res; Am Physiol Soc. *Res:* Human catecholamine storage vesicle proteins; human hypertension. *Mailing Add:* Vet Admin Hosp Dept Med 3350 La Jolla Village Dr San Diego CA 92161

O'CONNOR, DAVID EVANS, b Ft Ogden, Fla, Apr 16, 32; m 53; c 2. FATS & FATTY ACID CHEMISTRY. *Educ:* Univ Fla, BS, 54, PhD(chem), 61. *Prof Exp:* Res asst fluorine chem, Univ Fla, 57-58; res chemist, Procter & Gamble Co, 61-66; res chemist, Geigy Chem Corp, 66-67, group leader polymer chem, 67-68; res chemist, 68-73, SECT HEAD, PROCTER & GAMBLE CO, 73- *Mem:* Am Chem Soc; Am Oil Chemists Soc. *Res:* Oxidation of unsaturated fats and fatty acids. *Mailing Add:* Procter & Gamble Co Miami Valley Labs PO Box 398707 Cincinnati OH 45239-8707

O'CONNOR, DONALD J, b New York, NY, Nov 7, 22; m 48; c 3. CIVIL ENGINEERING. *Educ:* Manhattan Col, BCE, 44; Polytech Inst Brooklyn, MCE, 47; NY Univ, EngScD, 56. *Prof Exp:* Instr sanit eng, Manhattan Col, 46-47; mem staff, Polytech Inst, Brooklyn, 48-50; from asst prof to assoc prof, 52-64, PROF CIVIL ENG, MANHATTAN COL, 64- *Concurrent Pos:* NSF res grants, 60-64, mem undergrad res participation prog, 60-65, mem adv panel civil & environ eng, 84-; USPHS res grants, 62-66; New York Health Res Coun res grants, 62-65; mem eval comt, Environ Protection Agency, 67-72, res grants, 68-72, mem environ eng, comt, 83-; mem sci adv comt, Nat Oceanic & Atmospheric Admin, 72-77; consult, HydroQual, 76-; co-chmn, Adv Comt Phosphate Mgt, US-Can Joint Comn, Off Technol Assessment, US Cong, mem, Sci & Eng Comt, 78-80; co-chmn, Phosphorus Mgt Strategies for the Int Joint Comn on Great Lakes, 80-82. *Honors & Awards:* Rudolf Hering Award, Am Soc Civil Engrs, 58, 66, 83 & 89, Simon W Freese Environ Eng Award, 90. *Mem:* Nat Acad Eng; Am Soc Eng Educ; Am Soc Limnol & Oceanog; Am Soc Civil Engrs; Int Asn Water Pollution Res; Sigma Xi; Am Asn Univ Prof; Am Asn Environ Eng Prof. *Res:* Mathematical analysis of pollution in all natural bodies of water; water quality modeling toxic and conventional pollutants; transport across the air-water interface; sediment transport; author of several publications. *Mailing Add:* Dept Environ Eng & Sci Manhattan Col Parkway Bronx NY 10471

O'CONNOR, GEORGE ALBERT, b Seymour, Ind, Mar 30, 44; m 68; c 1. SOIL CHEMISTRY. *Educ:* Univ Mass, Amherst, BS, 66; Colo State Univ, MS, 68, PhD(agron), 70. *Prof Exp:* Asst agron, Colo State Univ, 66-70; from asst prof to assoc prof, 70-81, PROF AGRON, NMEX STATE UNIV, 81- *Mem:* Am Soc Agron; Soil Sci Soc Am. *Res:* Salinity; pesticides; heavy metals; hazardous wastes. *Mailing Add:* Dept agron & Hort NMex State Univ Las Cruces NM 88003-0003

O'CONNOR, GEORGE RICHARD, b Cincinnati, Ohio, Oct 8, 28. OPHTHALMOLOGY. *Educ:* Harvard Univ, AB, 50; Columbia Univ, MD, 54. *Prof Exp:* Asst clin prof ophthal, Univ Calif, San Francisco, 62-66, consult ophthal clin, 62-70, from asst dir to dir, Francis I Proctor Found Res Ophthal Med Ctr, 62-84, from assoc prof to prof 70-84, emer prof ophthal, 84; RETIRED. *Concurrent Pos:* NIH spec trainee biochem, Inst Biochem, Univ Uppsala, 60-61 & immunol, State Serum Inst, Copenhagen, Denmark, 61-62. *Honors & Awards:* Doyne Medal, Oxford Ophthal Congress, 84. *Mem:* Asn Res Vision & Ophthal; AMA. *Res:* Microbic immunology; immunologic response of human subjects to toxoplasma infections; antibody formation in the toxoplasma infected eye. *Mailing Add:* Six Mercato Ct San Francisco CA 94131

O'CONNOR, JOEL STURGES, b Auburn, NY, Mar 6, 37; m 78; c 2. ECOLOGY, FISHERIES. *Educ:* Cornell Univ, BS, 58; Univ RI, PhD(oceanog), 65. *Prof Exp:* Chief comput opers br, Div Tech Info Exten, US AEC, 65-68; sr res assoc marine ecol, Biol Dept, Brookhaven Nat Lab, 68-71; res assoc marine sci res ctr, State Univ NY Stony Brook, 71-73; ecologist, NE off, Off Marine Pollution Assessment, 73-85, sr ecologist, Ocean Assessments Div, 85-88, OCEAN POLICY COORDINATOR, WATER MGT DIV, NAT OCEANIC & ATMOSPHERIC ADMIN, 88- *Concurrent Pos:* Adj assoc prof, Marine Sci Res Ctr, State Univ NY, Stony Brook, 73- *Mem:* AAAS; Am Fisheries Soc; Estuarine Res Fedn; Am Soc Limnol & Oceanog; Ecol Soc Am; Sigma Xi; Int Asn Ecol. *Res:* Sampling statistics and computer science; ecology of estuarine and coastal environments, with emphasis on effects of contaminants; marine resource management. *Mailing Add:* USPA Region 2 Water Mgt Div 26 Federal Plaza NY NY 10278

O'CONNOR, JOHN DENNIS, b Chicago, Ill, Mar 20, 42; m 64; c 3. ZOOLOGY, BIOCHEMISTRY. *Educ:* Loyola Univ, Chicago, BS, 63; DePaul Univ, MA, 66; Northwestern Univ, Ill, PhD(biol), 68. *Prof Exp:* NIH fel, Mich State Univ, 68-70; from asst prof to assoc prof zool, Univ Calif, Los Angeles, 68-79, prof develop biol, 79-87, chair, Dept Biol, 79-81, dean, Dept Life Sci, 81-87; vchancellor res & grad studies & dean, Grad Sch, Univ NC, Chapel Hill, 87-88, actg provost, 88, vchancellor acad affairs & provost, 88-91, PROF BIOL, UNIV NC, CHAPEL HILL, 87-; PRES, UNIV PITTSBURGH, 91- *Concurrent Pos:* Vis prof, Roman Cath Univ, Nijmegen, Neth, 75-76, Tissue Cult Asn, 82, Acad Sci, Shanghai, China, 83 & Inst Zool, Peking, 83; mem bd dirs, Coun Res & Technol, 87-, Res Triangle Inst, 88- & NC Sch Sci & Math, 88- *Mem:* Fel AAAS; Am Soc Zool; Soc Develop Biol; Sigma Xi. *Res:* Regulation of metabolism during crustacean molt cycle. *Mailing Add:* Univ Pittsburgh 107 Cathedral of Learning Pittsburgh PA 15260

O'CONNOR, JOHN FRANCIS, b Waterloo, NY, May 24, 35. BIOCHEMISTRY, ENDOCRINOLOGY. *Educ:* St John Fisher Col, BS, 60; Univ Rochester, PhD(org chem), 71. *Prof Exp:* Res fel, Inst Steroid Res, Montefiore Hosp, 69-74, investr, 74-80; ASST PROF, ALBERT EINSTEIN COL MED, 80- *Concurrent Pos:* Instr, Albert Einstein Col Med, 71- *Mem:* AAAS; Am Chem Soc; Royal Soc Chem; Endocrine Soc; AAAS; Sigma Xi. *Res:* Natural products; thyroid physiology; radioimmunoassay techniques; immunology. *Mailing Add:* 781 Pelham Rd 1A New Rochelle NY 10805

O'CONNOR, JOHN JOSEPH, b Bowling Green, Ky, May 7, 16; m 46; c 4. PHYSICS. *Educ:* Western Ky State Univ, BS, 39; Vanderbilt Univ, MS, 40. *Prof Exp:* Asst physics, Western Ky State Univ, 38-39, Vanderbilt Univ, 39-40, Ohio State Univ, 40-42; res physicist, Remington Arms Co, Inc, Conn, 42-44, res physicist, E I du Pont de Nemours & Co, Tenn, 44, Wash, 44-46, res physicist, Remington Arms Co, Inc, Div, 46-51, sr physicist, 51-61, res assoc, 61; proj engr, 61-84, CONSULT, RCA SERV CO, 84- *Mem:* Am Phys Soc; Am Astronaut Soc; Am Astron Soc. *Res:* Artificial radioactivity; Faraday effect; exterior ballistics and accuracy of projectiles; missile trajectory analysis; orbital mechanics. *Mailing Add:* Three Garrison Rd Wellesley MA 02181

O'CONNOR, JOHN THOMAS, b New York, NY, Feb 11, 33; m 66; c 2. CIVIL & SANITARY ENGINEERING. *Educ:* Cooper Union, BCE, 55; NJ Inst Technol, MSCE, 58; Johns Hopkins Univ, EngD, 61. *Prof Exp:* Sanit engr, Elson T Killam Sanit & Hydraul Consult Engrs, 55-56; civil engr, George A Fuller Construct Co, NY, 56-57; sanit engr, Parsons, Brinckerhoff, Quade & Douglas, 57; from asst prof to assoc prof sanit eng, Univ Ill, Urbana, 61-69, prof civil eng, 69-75; chmn dept, 75-89, PROF CIVIL ENG, UNIV MO-COLUMBIA, 75- *Mem:* Am Chem Soc; Am Soc Civil Engrs; Am Water Works Asn; Water Pollution Control Fedn; Am Soc Limnol & Oceanog; Sigma Xi. *Res:* Fate of radionuclides in natural waters; removal of iron and trace metals from ground waters; water and wastewater treatment; ozonation. *Mailing Add:* Dept Civil Eng Univ Mo Columbia MO 65211

O'CONNOR, JOSEPH MICHAEL, b Newark, NJ, Oct 31, 25; m 49; c 2. CAREER CONSULTANT. *Educ:* Seton Hall Univ, BS, 50; Stevens Inst Technol, MS, 55. *Prof Exp:* From jr chemist to sr res scientist, Pharmaceut Div, Ciba-Geigy Corp, 50-86; CONSULT, OC ASSOCS, 87- *Honors & Awards:* Fel, NY Acad Sci. *Mem:* Am Chem Soc. *Res:* Analytical chemistry; chromatography and gas chromatography; synthetic organic chemistry. *Mailing Add:* 1096 Overlook Terrace Union NJ 07083

O'CONNOR, LILA HUNT, b Cleveland, Ohio, Mar 17, 36; c 3. NEW PRODUCTS DEVELOPMENT. *Educ:* St Louis Univ, BS, 58; Wayne State Univ, MS, 70; Cleveland State Univ, 83. *Prof Exp:* Group leader, Union Carbide, 73-85; MAT DEVELOP MGR, R J REYNOLDS TOBACCO CO, 85- *Mem:* Am Chem Soc; Am Indust Hyg Asn. *Res:* Analytical and inorganic chemistry; chromatography and spectroscopy. *Mailing Add:* 3950 Philpark Dr Winston-Salem NC 27106

O'CONNOR, MATTHEW JAMES, b New York, NY, July 31, 40; m 67; c 2. ANALYTICAL CHEMISTRY, PHYSICAL CHEMISTRY. *Educ:* Univ Calif, Long Beach, BS, 70, MS, 72. *Prof Exp:* Mgr anal & qual assurance labs, Apollo Technol, 72-74, dir, 74-76, asst vpres pollution, 76-79, vpres mkt, 79-83; PRES, O'CONNOR ASSOCS, 83- *Mem:* Am Chem Soc; Am Soc Testing & Mat; Air Pollution Control Asn. *Res:* Air pollution. *Mailing Add:* 157 Ironia Rd Mendham NJ 07945

O'CONNOR, MICHAEL KIERAN, b Dublin, Ireland, Sept 19, 52; m 80; c 2. NUCLEAR MEDICINE. *Educ:* Univ Col, Dublin, BS, 74; Trinity Col, Dublin, PhD(med physics), 79. *Prof Exp:* Lectr med physics, Dublin Inst Technol, 75-76, res scientist, 79-80; physicist, Heath Hosp, Dublin, 76-78; sr physicist, St James's Hosp, Dublin, Ireland, 79-86; sr assoc consult, 86-89, CONSULT RADIOLOGIC PHYSICS, MAYO CLIN, 89- *Concurrent Pos:* Asst prof, Mayo Med Sch, 86-88, assoc prof, 88- *Mem:* Soc Nuclear Med; Am Asn Physicists Med. *Res:* Single photon emission computed tomography; whole-body quantitation of radiotracer biodistribution and kinetics; non-invasive measurement techniques for determination of organ blood flow. *Mailing Add:* Sect Nuclear Med Mayo Clin Charlton Bldg 200 First St SW Rochester MN 55905

O'CONNOR, MICHAEL L, b South Bend, Ind, Dec 4, 38; m 65. PATHOLOGY. *Educ:* Rockhurst Col, BS, 58; Univ Wis-Madison, MS, 60; Univ Kans, MD, 64. *Prof Exp:* Asst prof path, Case Western Reserve Univ, 69-71; asst prof path, Univ Iowa, 72-76, dir clin labs, Univ Hosps, 72-76; ASSOC PROF PATH, BOWMAN GRAY SCH MED, 76- *Mem:* Col Am Path; Am Soc Clin Path; Am Asn Clin Chem. *Res:* Computer applications in pathology. *Mailing Add:* Dept Path Bowman Gray Sch Med Winston-Salem NC 27103

O'CONNOR, ROD, b Cape Girardeau, Mo, July 4, 34; m 55; c 4. ORGANIC BIOCHEMISTRY. *Educ:* Southeast Mo State Col, BS, 55; Univ Calif, PhD, 58. *Prof Exp:* Asst prof chem, Univ Omaha, 58-60; assoc prof, Mont State Univ, 60-66; assoc prof chem & dir gen chem, Kent State Univ, 66-67; staff assoc, Adv Coun Col Chem, Stanford Univ, 67-68; prof chem, Univ Ariz, 68-72; vis prof, Wash State Univ, 72-73; prof chem, Tex A&M Univ, 73-86; PRES, TEXAS ROMEC, INC, 83- *Concurrent Pos:* NIH res grants, 61-66; consult, Hollister-Stier Labs, 64-66; Am Chem Soc vis scientist & tour speaker; mem, Col Chem Consult Serv; educ consult, Tucara-4 Media Resources, Inc; mem, nat adv comt, Individualized Sci Instrnl Syst, 72-77; vpres, Romec Environ Res & Develop, Inc, 80-83. *Honors & Awards:* Am Chem Soc Award, 71; Nat Teaching Award, Mfg Chemists Asn, 78. *Mem:* Fel AAAS; Am Chem Soc; Sigma Xi. *Res:* Multi-media instruction; environmental chemistry; chemistry of insect venoms; invention development and patents. *Mailing Add:* 1300 Angelina College Station TX 77840

O'CONNOR, TIMOTHY EDMOND, b Cork, Ireland, Dec 5, 25; nat US; m 52; c 6. ORGANIC CHEMISTRY, BIOCHEMISTRY. *Educ:* Nat Univ Ireland, BSc, 47, MSc, 48, PhD(chem), 51. *Prof Exp:* Chemist, E I du Pont de Nemours & Co, Del, 52-60, res scientist, 60-61; chemist, Nat Cancer Inst, 63-67, head moleuclar virol sect, 66-72, assoc chief viral leukemia & lymphoma br, 67-72, dir molecular control prog, 72-75; dir div biol & med res, Argonne Nat Lab, 75-77; interim dir, Div Educ Progs, Roswell Park Mem Inst, 80-81, assoc dir sci affairs, 80-85; assoc dir res, Biotherapeut, Inc, 86-90; RETIRED. *Concurrent Pos:* Fel, Mayo Found, Univ Minn, 50-51 & Univ Wis, 51-52; USPHS spec res fel, NIH, 61-63. *Mem:* AAAS; Am Chem Soc; Am Asn Cancer Res; Sigma Xi. *Res:* Organic nitrogen compounds; steroids; polymers; refractories; virology; oncology; biophysical characterization of viruses; biochemistry and immunology of chromatins; cytokines and immunology. *Mailing Add:* 201 Cambridge Place Franklin TN 37064

O'CONNOR, WILLIAM BRIAN, b Brattleboro, Vt, Feb 24, 40; m 64; c 3. REPRODUCTIVE PHYSIOLOGY. *Educ:* St Michael's Col, Vt, BS, 62; Purdue Univ, MS, 66, PhD(zool), 68. *Prof Exp:* Asst prof zool, 67-73, ASSOC PROF ZOOL, UNIV MASS, AMHERST, 73- *Concurrent Pos:* chmn, Dept Pre-Med, Univ Mass, Amherst, 74- *Mem:* AAAS; Am Soc Zool. *Res:* Physiology and immunology of relaxin and other protein hormones. *Mailing Add:* Moorrill Sci Ctr Univ Mass Amherst MA 01003

O'CONOR, VINCENT JOHN, JR, b Chicago, Ill, Jan 10, 27; m 52; c 4. UROLOGY. *Educ:* Yale Univ, AB, 49; Northwestern Univ, MD, 53; Am Bd Urol, dipl, 62. *Prof Exp:* From intern to resident, Peter Bent Brigham Hosp, Boston, Mass, 53-58; chief resident, Chicago Wesley Mem Hosp, 58-59; from instr to assoc prof, 59-70, PROF UROL, SCH MED, NORTHWESTERN UNIV, 70-; CHMN DEPT UROL, CHICAGO WESLEY MEM HOSP, 63- *Concurrent Pos:* William Quinby fel urol, Harvard Med Sch, 57-68; attend urologist, Vet Admin Res Hosp & Rehab Inst, Chicago, 59- & Cook County Hosp, 66-; lectr, US Naval Hosp, 65- *Mem:* Fel Am Col Surg; Am Urol Asn; Soc Pelvic Surg; Am Asn Genito-Urinary Surg; Int Soc Urol Endocrinosurg. *Res:* Surgery of the kidney; renal hypertension and transplantation. *Mailing Add:* 251 E Chicago Ave Chicago IL 60611

ODAR, FUAT, b Harbin, China, May 8, 34; US citizen; m 79; c 2. FLUID DYNAMICS, REACTOR SAFETY. *Educ:* Tech Univ Istanbul, Dipl, 56; Northwestern Univ, MS, 58, PhD, 62. *Prof Exp:* Civil engr, Corps Eng, US Army, 58-60, res civil engr, Cold Regions Res Eng Lab, 60-67; sr engr, Bettis Atomic Power Lab, 67-74; MEM STAFF, NUCLEAR REGULATORY COMN, 74- *Mem:* Am Nuclear Soc. *Res:* Flow instabilities in nuclear reactors; emergency cooling and safeguards systems in nuclear reactors; forces exerted on bodies moving arbitrarily in fluid; similitude of drifting snow; reactor accident management; reactor safety analysis code program; advanced thermal and hydrodynamics reactor safety code assessment programs; evaluation of reactor safety analysis. *Mailing Add:* Nuclear Regulatory Comn Mail Stop NLN353 Washington DC 20555

ODASZ, F(RANCIS) B(ERNARD), JR, b Richmond Hill, NY, Oct 8, 22; m 56; c 5. ENGINEERING. *Educ:* Polytech Inst Brooklyn, BChE, 44, MChE, 47. *Prof Exp:* Res engr, Parmelee Motor Fuel Co, 45-47; asphalt res engr, Husky Refining Co, 47-52, asst dir tech serv, Husky Oil Co, 52-56, mgr develop & control, Husky Hi-Power, Inc, 56-60; chief process engr, Southwestern Eng Co, 60-62; asst chief planning engr, Bechtel Corp, San Francisco, 62-74; vpres, Energy Trans Systs Inc, 74-83; CONSULT, CHEM ENGR, CASPER, WY, 83- *Concurrent Pos:* Environ planning comnr, City of Mountain View. *Mem:* Am Chem Soc; Am Inst Chem Engrs; Nat Soc Prof Engrs. *Res:* Asphalt; statistics; rubberized asphalt; refinery and chemical plant design and project management; environmental control; coal slurry pipelines; clean cool technology; hydrology and water resources. *Mailing Add:* 4525 Squaw Creek Rd Casper WY 82604

O'DAY, DANTON HARRY, b Vancouver, BC, Jan 31, 46; m 66; c 1. DEVELOPMENTAL BIOLOGY. *Educ:* Univ BC, BS, 67, MS, 69; Univ Del, PhD(develop biol), 72. *Prof Exp:* From asst prof to assoc prof, 71-84, PROF ZOOL, UNIV TORONTO, 84- *Mem:* Can Soc Zoologists; Soc Protozoologists; Can Soc Cell Biol; Can Micros Soc; Soc Develop Biologist. *Res:* Ca2 and development; characterization and mode of action of sexual pheromones in cellular slime molds; regulation of cell and nuclear fusion; signal transduction during biomembrane fusion. *Mailing Add:* Erindale Col Univ Toronto Mississauga ON L5L 1C6 Can

ODDIS, JOSEPH ANTHONY, b Greensburg, Pa, Nov 5, 28; m 54; c 2. PHARMACY. *Educ:* Duquesne Univ, BS, 50. *Hon Degrees:* DSc, Mass Col Pharm, 75, Philadelphia Col Pharm & Sci, 75, Union Univ, 76, Duquesne Univ, 89 & Long Island Univ, 91. *Prof Exp:* Staff pharmacist, Mercy Hosp, Pittsburgh, Pa, 50-51, asst chief pharmacist, 53-54; chief pharmacist, Western Pa Hosp, 54-56; staff rep hosp pharm, Am Hosp Asn, Chicago, 56-60; secy, Res & Educ Found, 69-87, EXEC VPRES, AM SOC HOSP PHARMACISTS, 60-, PRES, RES & EDUC FOUND, 69- *Concurrent Pos:* Dir, Div Hosp Pharm, Am Pharmaceut Asn, 60-62; deleg, US Pharmacopeial Conv, Inc, 60-80; secy, Nat Coun Patient Info & Educ, 82-85; mem, adv comt, Vet Admin Cent Off Pharm Serv & comn pharmaceut serv ambulant patients, 82-85, bd trustees, Pharmacists Against Drug Abuse, 84-; chmn, USPHS adv group Pub Health Facil & Construct Serv Pharm Manual, 84-; vis prof, Dept Pharm, Howard Univ Col Pharm, Washington, DC; lectr, Duquesne Univ Sch Nursing & Sch Pharm, Mt Mercy Col Sch Nursing & Mercy Hosp Sch Nursing. *Honors & Awards:* Harvey A K Whitney Lectr, Am Soc Hosp Pharmacists, 70, Donald E Francke Medal, 86; Julius Sturmer Mem Lectr, Philadelphia Col Pharm & Sci, 71; Howard C Newton Lectr, Mass Col Pharm, 77; Samuel Melendy Lectr, Univ Minn Col Pharm, 78; Hugo H Schaefer Award, Am Pharmaceut Asn, 83, Remington Honor Medal, 90; Reed & Alice Henniger Lectr, Univ Ill, Chicago Col Pharm, 84. *Mem:* AAAS; Am Pharmaceut Asn; Am Soc Hosp Pharmacists; Int Pharmaceut Fedn (pres, 86-90); Israel Pharmaceut Asn; Pharmaceut Soc Gt Brit; Pharmaceut Soc Nigeria; Soc Hosp Pharmacists Australia. *Res:* Hospital pharmacy. *Mailing Add:* 6509 Rockhurst Rd Bethesda MD 20817

ODDIS, LEROY, b Export, Pa, Aug 30, 31; m 54; c 3. ENDOCRINOLOGY. *Educ:* Utica Col, BA, 59; Rutgers Univ, MS, 63, PhD(zool), 64. *Prof Exp:* USPHS fel, 64; asst prof biol, 64-70, ASSOC PROF BIOL, RIDER COL, 70- *Mem:* AAAS; Brit Soc Endocrinol. *Res:* Skin-pigment cell interaction; hormonal control of pigmentation. *Mailing Add:* Dept of Biol Rider Col 2083 Lawrenceville Rd Lawrenceville NJ 08648

ODDONE, PIERMARIA JORGE, b Arequipa, Peru, Mar 26, 44; c 2. HIGH ENERGY PHYSICS. *Educ:* Mass Inst Technol, BS, 65; Princeton Univ, PhD(physics), 70. *Prof Exp:* Res fel physics, Calif Inst Technol, 69-72; PHYSICIST, LAWRENCE BERKELEY LAB, 72- *Concurrent Pos:* Exp facil coordr, Positron-Electron Proj, Lawrence Berkeley Lab, 75-; spokesman, Positron-Electron Proj Collab, 84-87; head physics div & assoc dir, Lawrence Berkeley Lab, 87- *Mem:* Am Phys Soc; Sigma Xi. *Res:* Elementary particle research in electron-positron annihilation. *Mailing Add:* 50 Lawrence Berkeley Lab Ely Group Univ Calif Berkeley CA 94720

ODDSON, JOHN KEITH, b Selkirk, Man, Nov 30, 35; m 60; c 3. DIFFERENTIAL EQUATIONS, MODELING AND SIMULATION. *Educ:* Univ Toronto, BASc, 57; Mass Inst Technol, SM, 60; Univ Md, PhD(appl math), 65. *Prof Exp:* Lectr math, Univ Waterloo, 57-58 & 60-62; Can NATO res fel partial differential equations, Univ Genoa, 65-66; res asst prof math, Inst Fluid Dynamics & Appl Math, Univ Md, 66-67; asst prof, 67-69, ASSOC PROF MATH, UNIV CALIF, RIVERSIDE, 69- *Concurrent Pos:* Vis prof, Inst Math, Univ Firenze, Florence, Italy, 73-74, 79-80. *Res:* Partial and ordinary differential equations, mathematical modelling of physical, biological, economic processes and the computer simulation of such systems. *Mailing Add:* Dept Math & Comput Sci Univ Calif Riverside CA 92521

ODE, PHILIP E, b Decorah, Iowa, Mar 10, 35; m 61; c 3. POPULATION BIOLOGY, EVOLUTION. *Educ:* Luther Col, BA, 57; Cornell Univ, MS, 63, PhD(entom), 65. *Prof Exp:* From asst prof to assoc prof, 65-75, chmn dept, 68-74, coordr environ prog, 73-79 & hons prog, 81-85, PROF BIOL, THIEL COL, 75-, CHMN DEPT, 79- *Mem:* AAAS; Entom Soc Am; Soc Study Evolution; Am Inst Biol Sci; Sigma Xi; Soc Conserv Biol. *Res:* Biology of Diptera; dispersal of insects and other organisms; animal behavior; population biology; evolution studies. *Mailing Add:* Dept of Biol Thiel Col Greenville PA 16125

ODE, RICHARD HERMAN, b Glendale, Calif, Oct 14, 41; m 69; c 1. ENVIRONMENTAL SCIENCES. *Educ:* Univ Notre Dame, Bs, 63, MS, 65; WVa Univ, PhD(org chem), 68. *Prof Exp:* Vis lectr chem, Duquesne Univ, 68-69; fel natural prod, Ariz State Univ, 69-71, sr res chemist marine natural prod, Cancer Res Inst, 71-73, asst to dir, 73-77; sr res chemist, 77-80, group leader, Environ/Anal Res, 80-85, SECT MGR, ENVIRON TESTING SERV, MOBAY CORP, 85- *Mem:* Am Chem Soc. *Res:* Environmental research associated with water, air and solid wastes; carbon analysis and application; biological treatment of waste waters. *Mailing Add:* Mobay Corp New Martinsville WV 26155-0500

O'DEA, ROBERT FRANCIS, CHROMICAL PHARMACOLOGY, PEDIATRICS. *Educ:* Univ Minn, PhD(pharmacol), 72, MD, 73. *Prof Exp:* ASSOC PROF PHARMACOL, MED SCH, UNIV MINN, 79- *Mailing Add:* Box 404 Univ Minn Hosp Minneapolis MN 55455

ODEGAARD, CHARLES E, b Chicago Heights, Ill, Jan 10, 11; wid; c 1. HISTORY OF MEDICINE, HIGHER EDUCATION. *Educ:* Dartmouth Col, AB, 32; Harvard Univ, AM, 33, PhD, 37. *Hon Degrees:* Various from US cols & univs, 51-63. *Prof Exp:* From instr to prof hist, Univ Ill, 37-48; exec dir, Am Coun Learned Soc, 48-52; dean, Univ Mich, 52-58; pres 58-73, prof, 73-81, EMER PRES, UNIV WASH, 73-, EMER PROF HIGHER EDUC & BIOMED HIST, 81- *Concurrent Pos:* prof higher educ, Univ Wash, 73-81, prof biomed hist, 75-81; US National Comn, UNESCO, 49-55, Int Coun Philos & Humanistic Studies, vpres, 52-59, pres 59-65, dir, Minorities in Med, Josiah Macy Found, 75-76; vpres, Int Coun Philos & Humanistic Studies, 52-59, pres, 59-65; lectr & vis prof, confs in Eng, France, Neth, Denmark, Ger, Switz, Romania, Mex, Japan, Korea, Greece, Spain, Austria, Scotland, Italy, 56-66; dir, minorities in med, Josiah Macy Found, 75-76. *Mem:* Inst Med-Nat Acad Sci. *Res:* Various publications on the history and sociology of medicine and education. *Mailing Add:* Biomed Hist Univ Wash Med Sch Seattle WA 98195

ODEH, A(ZIZ) S(ALIM), b Nazareth, Palestine, Dec 10, 25; US citizen; m 56; c 2. ENGINEERING. *Educ:* Univ Calif, Berkeley, BS, 51; Univ Calif, Los Angeles, MS, 53, PhD(eng), 59. *Prof Exp:* Res technologist petrol prod, Field Res Lab, Socony Mobil Oil Co, 53-56; assoc eng, Univ Calif, Los Angeles, 58-59; sr res technologist petrol prod, Field Res Lab, Mobil Res & Develop Corp, 59-63, res assoc, 63-76, sr res assoc, 76-80, sr scientist, 80-89; RETIRED. *Concurrent Pos:* Mem bd dirs, Abu Dhabi Nat Reservoir Res Found, 80-89. *Honors & Awards:* John Franklin Carll Award, Soc Petrol Engrs, 84 & Outstanding Achievement Award, 89. *Mem:* Nat Acad Eng; Soc Petrol Engrs. *Res:* Petroleum production and reservoir engineering; methods and means to produce oil and natural gas efficiently. *Mailing Add:* 5840 Gallant Fox Lane Plano TX 75093

ODEH, FAROUK M, b Nablus, Palestine; US citizen. APPLIED MATHEMATICS. *Educ:* Cairo Univ, BS, 55; Univ Calif, Berkeley, PhD(appl math), 61. *Prof Exp:* RES STAFF MATH, WATSON RES CTR, IBM CORP, 61- *Concurrent Pos:* Temp mem math, Courant Inst Math Sci, NY Univ, 62-63; assoc prof, Am Univ Beirut, 67-68. *Mem:* Am Math Soc; Soc Indust & Appl Math; Math Asn Am. *Res:* Stability theory of difference schemes; bifurcation theory in mechanics. *Mailing Add:* IBM T J Watson Res Lab Box 218 Yorktown Heights NY 10598

ODEH, ROBERT EUGENE, b Akron, Ohio, Dec 21, 30; m 58; c 3. MATHEMATICAL STATISTICS, COMPUTER SCIENCE. *Educ:* Carnegie Inst Technol, BS, 52, MS, 54, PhD(math), 62. *Prof Exp:* Programmer analogue comput, Goodyear Aircraft Corp, Ohio, 52-53; instr math, Carnegie Inst Technol, 58-59; from instr to asst prof, Univ Ore, 59-64; assoc prof, 64-71, PROF MATH, UNIV VICTORIA, 71- *Concurrent Pos:* Sr investr, Air Force Off Sci Res grant, 61-62; consult, Ore Res Inst, 62-64 & Attorney Gen Dept, 65-68. *Mem:* AAAS; Am Statist Asn; Inst Math Statist; fel Royal Statist Soc; Math Asn Am. *Res:* Computing; non-parametric c-sample rank-sum tests; transformations used in analysis of variance. *Mailing Add:* Dept of Math Univ of Victoria Box 1700 Victoria BC V8W 2Y2 Can

ODELL, ANDREW PAUL, b Galesburg, Ill, May 6, 49. ASTRONOMY. *Educ:* Univ Iowa, BA, 70; Univ Wis-Madison, PhD(astron), 73. *Prof Exp:* Fel meteorol, Univ Wis, 73-74; asst prof astron, Univ Northern Iowa, 74-79; res assoc, Steward Observ, Univ Ariz, 79-81; asst prof astron, 81-85 ASSOC PROF ASTRON, NARIZ UNIV, 86-; assoc prof astron, Univ Vienna, 85-86; ASSOC PROF ASTRON, N Ariz Univ, 86 - *Concurrent Pos:* Assoc prof astron, Univ Vienna, 85-86. *Mem:* Am Astron Soc; Int Astron Union. *Res:* Stellar evolution and pulsation; radiation transfer in planetary atmospheres. *Mailing Add:* Rte 4 Box 962A Flagstaff AZ 86001

O'DELL, AUSTIN ALMOND, JR, b Houston, Tex, Nov 28, 33; m 53; c 2. PHYSICS, NUCLEAR RADIATION. *Educ:* Univ Tex, Austin, BS, 54, MA, 55; Mass Inst Technol, PhD(physics), 61. *Prof Exp:* Physicist, Northrop Corp, 61-65; prof physics, Calif Lutheran Col, 65-68; physicist, EG&G, Inc, 68-72; physicist, Mission Res Corp, 72-74; PHYSICIST, LAWRENCE LIVERMORE NAT LAB, 74- *Mem:* Am Phys Soc; Am Nuclear Soc; Sigma Xi; Am Chem Soc. *Res:* Nuclear radiation interactions, transport and detection; statistical data analysis; computer-based instrumentation and applications; nuclear criticality analysis and monitoring systems. *Mailing Add:* 5258 Blackbird Dr Pleasanton CA 94566

O'DELL, BOYD LEE, b Hale, Mo, Oct 14, 16; m 44; c 2. BIOCHEMISTRY, NUTRITION. *Educ:* Univ Mo, AB, 40, PhD(biochem), 43. *Prof Exp:* Sr chemist, Parke, Davis & Co, 43-46; from asst prof to prof, 45-87, EMER PROF NUTRIT BIOCHEM, UNIV MO-COLUMBIA, 87- *Concurrent Pos:* NIH spec fel, Cambridge Univ, 64-65 & Harvard Med Sch, 72; Fulbright scholar, Commonwealth Sci & Indust Res Orgn, Australia, 73. *Honors & Awards:* Borden Award, Am Inst Nutrit; Spencer, Am Chem Soc. *Mem:* Am Soc Biol Chemists; Am Chem Soc; Soc Exp Biol & Med; fel Am Inst Nutrit. *Res:* Biochemical and physiological functions of micronutrients; vitamins and trace elements; role of trace elements in reproduction and connective tissue metabolism; bioavailability of zinc; biochemical function of copper and zinc in brain; role of zinc in membrane structure and function. *Mailing Add:* 322 Chem Bldg Univ of Mo Columbia MO 65211

O'DELL, CHARLES ROBERT, b Hamilton Co, Ill, Mar 16, 37; m 91; c 2. ASTRONOMY. *Educ:* Ill State Univ, BSEd, 59; Univ Wis, PhD(astron), 62. *Prof Exp:* Carnegie fel, Mt Wilson & Palomar Observs, Calif, 62-63; asst prof astron, Univ Calif, Berkeley, 63-64; from asst prof to assoc prof astron, Univ Chicago, 64-67, prof & chmn dept, 67-72, dir, Yerkes Observ, 66-72; assoc dir astron, Sci & Eng Directorate, Marshall Space Flight Ctr, 72-76 & 80-82, assoc dir sci, 76-80; proj scientist Hubble Space Telescope, Nat Aeronaut &

Space Admin, 72-83; PROF SPACE PHYSICS & ASTRON, RICE UNIV, 82- *Concurrent Pos:* Guest lectr, Univ Col London, 70 & Univ Moscow, 71; US scientist, Copernicus Astron Ctr, 73-78, Int Halley Watch, 81-, Astro Halley Sci Team, 84-86, NASA Hubble Space Telescope Sci Working Group, 83-; mem ed comt, Physics Today, 86- *Mem:* Am Astron Soc (treas, 88-); Int Astron Union; AAAS. *Res:* Physical processes in and evolution of planetary nebulae; dynamical evolution and turbulence in diffuse nebulue; physical processes and origin of comets; characteristics of interstellar grains. *Mailing Add:* 3911 Riley Houston TX 77005

ODELL, DANIEL KEITH, b Auburn, NY, Nov 16, 45; m 69; c 3. VERTEBRATE ZOOLOGY. *Educ:* Cornell Univ, BS, 67; Univ Calif, Los Angeles, MA, 70, PhD(biol), 72. *Prof Exp:* Asst prof, 73-79, assoc prof marine biol, Rosenstiel Sch Marine & Atmospheric Sci, Univ Miami, 79-88; RES BIOLOGIST, SEA WORLD, FLA, 87- *Concurrent Pos:* Sci adv, US Marine Mammal Comn, 78-81. *Mem:* Oceanog Soc; Am Soc Mammalogists; Am Soc Zoologists; Wildlife Soc; Soc Marine Mammal. *Res:* Biology of marine mammals; cetacean stranding phenomena; biology of the Bottle Nose Dolphin, Pygmy Sperm Whale and West Indian Manatee. *Mailing Add:* Sea World 7007 Sea World Dr Orlando FL 32821-8097

ODELL, GEORGE VAN, JR, BIOCHEMISTRY, TARANTULAS VENOM. *Educ:* Tex A&M Univ, PhD(biochem), 66. *Prof Exp:* PROF ORGANIC & BASIC BIOCHEM, OKLA STATE UNIV, 56- *Res:* Isolation and characterization of venom toxins from snakes and spiders. *Mailing Add:* Dept Biochem Okla State Univ Stillwater OK 74075

O'DELL, JEAN MARLAND, b Independence, Mo, June 10, 31; m 56; c 5. NUCLEAR SCIENCE, OPERATIONS RESEARCH. *Educ:* Univ Kans, BSEP, 54, MS, 61, PhD(physics), 65. *Prof Exp:* Asst physics, Univ Kans, 58-62, asst nuclear physics, 62-65; exp physicist, 65-73, SR SCIENTIST, LAWRENCE LIVERMORE NAT LAB, UNIV CALIF, 73- *Concurrent Pos:* staff mem, Sci & Technol Div, IDA, 79-81; actg asst dep proj mgr, Survivability, US Navy, 81-83. *Res:* Nuclear explosives for defense and peaceful uses; compound nucleus formation versus direct interaction investigations bombarding sulphur 32 with deuterons; evaluation of requirements for nuclear weapons for strategic and naval tactical weapon systems. *Mailing Add:* 461 Pismo Ct Livermore CA 94550

ODELL, LOIS DOROTHEA, b Watertown, NY, Sept 25, 15. HUMANISTIC BOTANY, FIELD NATURAL SCIENCE. *Educ:* State Univ NY Albany, AB, 40; Cornell Univ, MA, 45, PhD(sci educ), 51. *Prof Exp:* Teacher pub sch, NY, 41-43; instr biol, 47-62, prof, 62-81, EMER PROF BIOL, TOWSON STATE UNIV, 81- *Concurrent Pos:* Sr lectr, Towson State Univ, 81-, consult, dendrol, 81- *Mem:* AAAS; Am Nature Study Soc. *Res:* Field natural science; botany; science education in environmental concepts. *Mailing Add:* 76 Cedar Ave Baltimore MD 21204

ODELL, NORMAN RAYMOND, b Rochester, NY, Aug 4, 27; m 50; c 3. ORGANIC CHEMISTRY. *Educ:* Whittier Col, BA, 50; Ore State Col, MS, 52, PhD(chem), 55. *Prof Exp:* Chemist, Taylor Instrument Co, 52; chemist grease res, Texaco, 54-59, group leader, 59-60, asst supvr, Lubricants Res Sect, 60-65, res technologist-managerial, 65-67, supvr lubricants res, 67-73, supvr, Lubricants Field Serv, Port Arthur Res Lab, 73-81, asst mgr fuels & lubricants, Beacon Res Labs, 81-82, assoc dir, Petrol Prod Res Div, 82-85; RETIRED. *Concurrent Pos:* Chem res consult, 87-; instr, Dutchess Community Col, 59-60 & Beaufort County Community Col, 86-87. *Mem:* AAAS; Am Chem Soc; fel Am Inst Chemists; Sigma Xi. *Res:* Synthesis of tracer compounds; tracer studies; exploratory research in grease thickening agents; synthetic lubricants; petroleum additives; mechanisms of petroleum additive behavior; additive processing; automotive and industrial lubricants; lubricant application; lubricating oil processing; fuel research, testing and application; lubricant and fuel technical service support. *Mailing Add:* 123 Forecastle Ct Pamlico Plantation Washington NC 27889

ODELL, PATRICK L, b Watonga, Okla, Nov 29, 30; m 58; c 4. MATHEMATICAL STATISTICS. *Educ:* Univ Tex, BS, 52; Okla State Univ, MS, 58, PhD(math statist), 62. *Prof Exp:* Mathematician, Flight Determination Lab, White Sands Missile Range, NMex, 52-53; res scientist, Kaman Nuclear, Inc, 58-59; mathematician, US Navy Nuclear Ord Eval Unit, 59-60, consult mathematician, US Weapons Eval Facility, 60-62; asst prof math, Univ Tex, 62-66; prof, Tex Tech Univ, 66-72; exec dean grad studies & res, 72-75, PROF MATH SCI, UNIV TEX, DALLAS, 72- *Concurrent Pos:* Consult mathematician, Ling-Tempco-Vought, Inc, Tex, 62-64; NSF grants, 63-65 & 77-79; assoc dir, Tex Ctr Res Appl Math & Mech, 64-72; Tex Hwy Res Ctr grant; NASA grant, 70-71, res grant, 72-76; training grant, Environ Protection Agency, 75-76. *Mem:* Soc Indust & Appl Math; fel Am Statist Asn. *Res:* Statistical problems associated with remote sensing from space; statistical problems; statistical analysis of sensitivity data; general theory of matrix inversion; mathematical modelling of the environment. *Mailing Add:* Baylor Univ Waco TX 76798

O'DELL, RALPH DOUGLAS, b Leavenworth, Ind, June 11, 38; m 59; c 4. NUCLEAR CRITICALITY SAFETY. *Educ:* Univ Tex, BS, 61, PhD(mech eng), 65. *Prof Exp:* From asst prof to assoc prof nuclear eng, Univ NMex, 64-73, dir, Los Alamos Grad Div, 69-73; staff mem, Los Alamos Nat Lab, Group TD-5, 73-74, sect leader, Group T-1, 74-76, assoc group leader, Group T-1, 76-79, alternate group leader, Group T-1, 79-80, dep group leader, Group T-1, 80-82, assoc group leader, Group X-6, 82-90, STAFF MEM, GROUP HSE-6, LOS ALAMOS NAT LAB, 90- *Concurrent Pos:* Long term vis staff mem, Los Alamos Sci Lab, 68-69; consult, US Geol Surv, TRIGA Reactor Facility, 69-90 & US Air Force Nuclear Safety Div, Kirtland AFB, 69-72; mem reactor safeguards adv comt, Univ NMex, 73- *Mem:* Am Nuclear Soc; NY Acad Sci; Sigma Xi. *Res:* Nuclear transport and reactor theory; nuclear criticality safety computations. *Mailing Add:* 525 Bryce Ave Los Alamos NM 87544-3607

O'DELL, THOMAS BENIAH, b Cassopolis, Mich, June 19, 20; m 43; c 3. PHARMACOLOGY. *Educ:* Wabash Col, AB, 42; Univ Minn, PhD(pharmacol), 50. *Prof Exp:* Chemist coated fabrics, US Rubber Co, 42-44; anal chemist nutrit, Upjohn Co, 44; asst prof pharmacol, Sch Med, Univ Minn, 49-51; sr res pharmacologist, Irwin, Neisler Labs, Ill, 51-60, dir biol res, 60-64; assoc dir med & sci coordr, William S Merrell Co, 64-68, dir drug regulatory affairs, Merrell-Nat Labs, 68-81; sr dir regulatory affairs, Merrell Dow Pharmaceut, Dow Chem Co, 81-89; RETIRED. *Mem:* AAAS; Am Soc Pharmacol & Exp Therapeut; Soc Toxicol; NY Acad Sci. *Res:* Virus chemotherapy; hypertension; atherosclerosis; antispasmodics; analgesics; central nervous system. *Mailing Add:* Nine Jewel Lane Cincinnati OH 45218

O'DELL SMITH, ROBERTA MAXINE, b May 5, 30; c 1. RENAL PHYSIOLOGY. *Educ:* Duke Univ, PhD(zool), 59. *Prof Exp:* Assoc prof, 76-83, ASSOC PROF PHYSIOL, UNIV NEW ORLEANS, 83- *Mem:* Am Physiol Soc; Am Nephrology Soc. *Res:* Renal and intestinal transport. *Mailing Add:* Univ New Orleans New Orleans LA 70148

ODEN, JOHN TINSLEY, b Alexandria, La, Dec 25, 36; m 65; c 2. ENGINEERING MECHANICS. *Educ:* La State, BS, 59; Okla State Univ, MS & PhD(struct mech), 62. *Hon Degrees:* Dr, Tech Univ Lisbon, Portugal, 86. *Prof Exp:* Asst prof struct eng, Okla State Univ, 62-63; sr struct engr, Gen Dynamics, Ft Worth, 63-64; assoc prof eng mech, Univ Ala, Huntsville, 64-67, prof, 67-77, chmn dept, 70-77; Carol & Henry Groppe prof eng, 79-87, PROF AEROSPACE ENG & ENG MECH, UNIV TEX, 77-, ERNEST & VIRGINIA COCKRELL CHAIR ENG, 88- *Concurrent Pos:* Dir, Tex Inst Comput Mech, 74-; vis prof, Fed Univ Rio, Rio de Janiero, 74; vis sci res fel, Brunel Univ, Uxbridge, Eng, 81; distinguished vis mathematician, Univ Md, 81; chmn exec comt, Eng Mech Div, Am Soc Civil Engrs; mem-at-large, US Nat Comt Theoret & Appl Mech, chmn, Subcomt Res Directions in Mech; US ed, Computer Methods Appl Mech & Eng; assoc ed, Int J Eng Sci & Nonlinear Analysis, Theory, Methods & Appln. *Honors & Awards:* Walter Huber Res Prize, Am Soc Mech Engrs, 73, Worcester Reed Warner Medal, 90; Billy & Claude R Hocott Award for Distinguished Eng Res, Nat Acad Sci, 87; Chevalier dans l'Ordre des Palmes Acad, French Govt, 90; Eringen Medal, Soc Eng Sci, 90. *Mem:* Nat Acad Eng; Soc Indust & Appl Math; fel Soc Eng Sci (pres, 79); Am Soc Eng Sci; fel Am Acad Mech (pres, 85-); fel Am Soc Civil Engrs; US Asn Comput Mech (pres); fel Am Soc Mech Engrs. *Res:* Nonlinear continuum mechanics; approximation theory; numerical analysis of nonlinear problems in continuum mechanics; author or co-author of seventeen books or monographs. *Mailing Add:* WRW 305 ASE/EM Dept Univ Tex Austin TX 78712-1085

ODEN, PETER HOWLAND, b Stuttgart, Ger, July 3, 33; US citizen; m 61; c 1. ELECTRICAL ENGINEERING, COMPUTER SCIENCE. *Educ:* Columbia Col, AB, 55; Columbia Univ, BS, 56, MS, 58, PhD(elec eng), 66. *Prof Exp:* Instr elec eng, Columbia Univ, 59-63; MEM RES STAFF, T J WATSON RES CTR, IBM CORP, 63- *Mem:* Asn Comput Mach; Inst Elec & Electronics Engrs; Am Chem Soc. *Res:* Circuit and system theory; automated computer design; stochastic models of computing systems; programming languages; optimizing compilers. *Mailing Add:* 120 Pine Bridge Rd Ossining NY 10562

ODENCRANTZ, FREDERICK KIRK, b New York, NY, Oct 6, 21; m 59; c 3. PHYSICS. *Educ:* Muhlenberg Col, BS, 43; Rutgers Univ, MS, 49; Univ Utah, PhD(physics), 58. *Prof Exp:* Physicist, New Brunswick Lab, US Atomic Energy Comn, 49-50; physicist, US Naval Weapons Ctr, 50-79; RETIRED. *Mem:* Am Phys Soc; Am Geophys Union. *Res:* Physical optics; atmospheric physics. *Mailing Add:* 3780 St Andrews Dr Reno NV 89502

ODENHEIMER, KURT JOHN SIGMUND, b Regensburg, Ger, May 9, 11; US citizen; m 39; c 4. CLINICAL PATHOLOGY, EXPERIMENTAL PATHOLOGY. *Educ:* Univ Munich, MedDent, 35; Univ Pittsburgh, DDS, 40, MEd, 54; Univ Heidelberg, Ger, DMD, 58; Western Reserve Univ, PhD(path), 64. *Prof Exp:* Asst prof gen path, Univ Pittsburgh, 49-53, head dept, 53-55; assoc prof oral diag & clin path, State Univ NY Buffalo, 61-66; prof path & chief exp path, Loyola Univ, 66-69, chmn diag & roentgenol, 67-69; prof path, 69-80, prof dent, 71-80, prof oral path, 77-80, EMER PROF PATH, GEN DENT & EAR, NOSE & THROAT, LA STATE UNIV MED CTR, NEW ORLEANS, 80- *Concurrent Pos:* Res fel path & fed teachers training grant, Western Reserve Univ, 59-61; res grant viral studies primates, Loyola Univ, 67-; consult, Presby Hosp Tumor Bd, Pittsburgh, 51-55, temporomandibular joint disturbances, NIH, 61, oral diag & path, Meyer Mem Hosp, Buffalo, 63-66, new ed Oral Path by K Thomas, 66 & Gulf South Res Inst, La, 67-; vis res assoc, Tulane Delta Primate Res Ctr, 66-; consult dent med, Touro Infirmary, New Orleans, 68-, stomatologist, 81; dent coordr oncol, Charity Hosp; lectr, Loyala Univ, 80- *Mem:* AAAS; Am Dent Asn; AMA; fel Am Acad Oral Path; fel Am Col Dent. *Res:* Peridontal and temporomandibular joint therapy with myotatic splint; effects of enteric viruses upon embryo, child and adult, in acute or chronic, clinical or subclinical forms, using primates as experimental models; radioactive gold and oral cancers; viral infections in genetics. *Mailing Add:* 4123 Vixen St New Orleans LA 70114

ODENSE, PAUL HOLGER, b Winnipeg, Man, Dec 12, 26; m 54; c 3. BIOCHEMISTRY. *Educ:* Univ Toronto, BA, 50, MA, 54; Univ Okla, PhD(biochem), 59. *Prof Exp:* Asst scientist & biochemist, 54-56, assoc scientist, 59-62, SR SCIENTIST, FISHERIES RES BD CAN, 62-, BIOCHEMIST, 59-; sr res officer, 82-89, GUEST WORKER, NAT RES COUN CAN, 89-; ADJ PROF, DEPT FOOD SCI & TECHNOL, UNIV NOVA SCOTIA, 89- *Mem:* Can Biochem Soc; Chem Inst Can; NY Acad Sci; Electron Microsc Asn Am; Int Electrophoresis Soc; AAAS. *Res:* Marine comparative biochemistry and histology; isoenzymes; electrophoretic separations; light and electron microscope histochemistry. *Mailing Add:* Nat Res Coun Can Inst Marine Biosci 1411 Oxford St Halifax NS B3H 3Z1 Can

ODER, FREDERIC CARL EMIL, b Los Angeles, Calif, Oct 23, 19; m 41; c 3. AEROSPACE SCIENCES. *Educ:* Calif Inst Technol, BS, 40, MS, 41; Univ Calif, Los Angeles, PhD(atmospheric physics), 52. *Prof Exp:* Dir geophys res, Air Force Cambridge Res Labs, 49-52, dir weapons syst ballistic missiles div, Air Res & Develop Command, 56-59, dep comdr space systs, 59-60; asst to dir res & eng, Apparatus & Optical Div, Eastman Kodak Co, NY, 60-61, prog mgr, 61-66; vpres & asst gen mgr, Lockheed Missiles & Space Co, 66-73, VPRES & GEN MGR, SPACE SYSTS DIV, LOCKHEED MISSILES & SPACE CO, INC, LOCKHEED AIRCRAFT CORP, 73- *Concurrent Pos:* Consult, Air Force Studies Bd, Nat Acad Sci, 75-; mem, Defense Intel Agency, Sci Adv Comt, 72- *Mem:* Nat Acad Eng; Soc Photog Sci & Eng; Sigma Xi; fel Am Inst Aeronaut & Astronaut. *Res:* Specialized optical and photographic systems; atmospheric geophysics. *Mailing Add:* 400 San Domingo Way Los Altos CA 94022

ODER, ROBIN ROY, b Jefferson City, Tenn, Sept 26, 34; m 71; c 2. SOLID STATE PHYSICS, ENGINEERING PHYSICS. *Educ:* Mass Inst Technol, BS, 59, PhD(physics), 65. *Prof Exp:* Res physicist magnetism, Francis Bitter Nat Magnet Lab, Mass Inst Technol, 65-70; group leader minerals beneficiation, Clay Div, J M Huber Corp, 70-74; eng specialist process technol, Bechtel Corp, 74-77; res assoc coal & minerals beneficiation, Chemical & Minerals Div, Gulf Sci & Technol Co, 77-82, dir, coal technol, 80-82; PRES, EXPORTECH CO, INC, 82- *Concurrent Pos:* Solid state physics consult, Elec Power Res Inst, 75-77; mem tech adv comts, Prog Coal Gasification, Gulf-TRW Inc, 77-80, Coal Cleaning Test Facil, Elec Power Res Inst, 81-84 & Eriez Mfg Co, Erie, Pa, 81-84. *Mem:* AAAS; Am Phys Soc; Am Chem Soc; Am Inst Chem Engrs; Am Inst Mining, Metall & Petrol Engrs. *Res:* Magnetism; calormetry; thermoelectricity; heat transport; superconductivity; De Haas Van Alphen effect; magnetic separation; clay mineralogy; coal structure; coal desulfurization; economics of coal cleaning; minerals beneficiation; coal beneficiation; particle separation; size reduction. *Mailing Add:* 4919 Simmons Dr Export PA 15632

ODETTE, G(EORGE) ROBERT, b Detroit, Mich, Aug 22, 43. NUCLEAR ENGINEERING, MATERIALS SCIENCE. *Educ:* Rensselaer Polytech Inst, BS, 65; Mass Inst Technol, MS, 68, PhD(nuclear eng), 70. *Prof Exp:* Res asst nuclear eng, Mass Inst Technol, 68-70; assoc prof, 70-81, PROF CHEM & NUCLEAR ENG, UNIV CALIF, SANTA BARBARA, 81- *Concurrent Pos:* Consult, Am Sci & Eng Co, 69-70, Hanford Eng Develop Lab, 70- & Argonne Nat Lab, 74- *Mem:* Am Nuclear Soc; Am Soc Testing & Mat. *Res:* Nuclear metallurgy; radiation effects modeling and data correlation; fundamental microstructure and mechanical behavior; test development; fission, fusion; energy related materials technology. *Mailing Add:* Dept of Chem & Nuclear Eng Univ of Calif Santa Barbara CA 93106

ODIAN, GEORGE G, b New York, NY, July 19, 33; div; c 1. POLYMER CHEMISTRY. *Educ:* City Col New York, BS, 55; Columbia Univ, MA, 56, PhD, 59. *Prof Exp:* Asst, Columbia Univ, 55-57; sr chemist, Thiokol Chem Corp, 58-59; res dir, Radiation Applns, Inc, 59-68; assoc prof, 68-72, PROF CHEM, COL STATEN ISLAND, NY, 72- *Concurrent Pos:* Asst prof, Columbia Univ, 63-68. *Mem:* AAAS; Am Chem Soc. *Res:* Polymer science; radiation chemistry. *Mailing Add:* One Bolivar St Staten Island NY 10314-5603

ODIORNE, TRUMAN J, b Johnson City, Tex, Aug 9, 44; m 70, 83. PHYSICAL CHEMISTRY. *Educ:* Southwestern Univ, Tex, BS, 66; Rice Univ, MA & PhD(chem), 71. *Prof Exp:* NIH trainee, Inst Lipid Res, Baylor Col Med, 71-73; staff chemist, Res Triangle Inst, 73-74; chemist, Monrovia, Calif, 74-77, DEVELOP ASSOC, MED PROD, GLASGOW SITE, E I DU PONT DE NEMOURS & CO, INC, 77- *Mem:* AAAS; Am Chem Soc; Am Phys Soc; Sigma Xi. *Res:* Chemical reaction kinetics; organic mass spectrometry; electron spectroscopy chemical analysis; analytical instrumentation; separation science. *Mailing Add:* E I du Pont de Nemours & Co Inc 20511 Glasgow Wilmington DE 19898

ODIOSO, RAYMOND C, b Pittsburgh, Pa, Apr 17, 23; m 53; c 6. ORGANIC CHEMISTRY. *Educ:* Duquesne Univ, BS, 47; Carnegie Inst Technol, MS, 50, DSc(chem), 51. *Prof Exp:* Asst, Carnegie Inst Technol, 47-49; fel, Mellon Inst, 51-54; res supvr, Gulf Res & Develop Co, 54-61; res mgr, Colgate-Palmolive Co, NJ, 61-67, assoc dir res lab opers, 67-68; VPRES RES & DEVELOP, DRACKETT CO, BRISTOL-MYERS CO, 68- *Concurrent Pos:* Instr, Carnegie Inst Technol, 52-53. *Mem:* AAAS; Am Chem Soc. *Res:* Kinetics of the benzidine rearrangement; monomer synthesis; alkylation; isomerization; dealkylation; dehydrocyclization; olefin reactions; product development, laundry and dishwashing detergents, bleaches, toilet soap, aerosols, paper products; process research, petrochemicals, and detergents. *Mailing Add:* 1588 Beech Grove Dr Cincinnati OH 45233

ODLAND, GEORGE FISHER, b Minneapolis, Minn, Aug 27, 22; m 45; c 3. ANATOMY, DERMATOLOGY. *Educ:* Harvard Med Sch, MD, 46. *Prof Exp:* From intern to resident med, Mass Gen Hosp, Boston, 46-55; asst dermat, Harvard Med Sch, 53-55; clin instr anat, 55-60, clin instr med, 56-60, clin asst prof anat & med, 60-62, from asst prof to assoc prof biol struct & med, 62-69, PROF BIOL STRUCT & MED, SCH MED, UNIV WASH, 69-, HEAD DIV DERMAT, 62- *Concurrent Pos:* Res fel anat, Harvard Med Sch, 49-51, res fel dermat, 51-53; attend staff, Univ Hosp, Harborview Med Ctr, Vet Admin Hosp, & USPHS Hosp, Seattle; mem adv coun, Nat Inst Athritis, Metabolic & Digestive Dis, NIH, 73-76, chmn comt, Nat Inst Arthritis, Digestive Dis & Kidney. *Honors & Awards:* Rothman Award, Soc Invest Dermat. *Mem:* Am Acad Dermat; Soc Invest Dermat; Asn Professors Dermat; Am Dermat Asn. *Res:* Electron microscopy; biosynthesis of epidermal keratins; structure/physiologic correlation in microcirculation; wound repair. *Mailing Add:* Dept Derm Univ Wash Med Ctr MS Rm 14 Seattle WA 98195

ODLE, JOHN WILLIAM, b Tipton, Ind, July 23, 14; m 37; c 3. MILITARY WEAPON SYSTEMS ANALYSIS. *Educ:* Univ Mich, BS, 37, MS, 38, PhD(math), 40. *Prof Exp:* Instr math, Univ Wis, 40-42; asst prof, Pa State Univ, 42-44; opers analyst, US Army Air Force, 44-45; head math div, US Naval Ord Test Sta, 46-55; opers analyst, US Air Force, 55-57; res mathematician eng res inst, Univ Mich, 57-58; dir adv develop dept, Crosley Div, Avco Corp, 58-60; mem staff opers res sect, Arthur D Little, Inc, 60-71; opers res analyst, Naval Ord Lab, 72-79; CONSULT, 79- *Concurrent Pos:* Consult math & opers res, 79- *Mem:* Oper Res Soc Am. *Res:* Applied mathematics; operations research. *Mailing Add:* 9208 Friars Rd Bethesda MD 20817-2321

O'DOHERTY, DESMOND SYLVESTER, b Dublin, Ireland, July 27, 20; m 51; c 2. NEUROLOGY. *Educ:* LaSalle Col, AB, 42; Jefferson Med Col, MD, 45. *Hon Degrees:* DSc, Georgetown Univ, 85. *Prof Exp:* Adj instr, DC Gen Hosp, 50-51; res fel, Med Ctr, 51-52, from instr to assoc prof, 52-61, PROF NEUROL, GEORGETOWN UNIV, 61-, CHMN DEPT, 59- *Concurrent Pos:* Dir, Muscular Dystrophy Clin, 54-; med dir, Georgetown Hosp, 66-67; consult, US Army, US Navy & Vet Admin; dir, Am Bd Psychol & Neurol. *Mem:* Asn Res Nerv & Ment Dis; AMA; fel Am Acad Neurol; Am Neurol Asn; Am Epilepsy Soc. *Res:* Parkinsonism; temporal lobe suppression; syncope; muscular dystrophy; cerebrovascular disease; multiple sclerosis; Huntington's chorea. *Mailing Add:* Dept Neurol 1 W Hosp Georgetown Georgetown Univ Hosp 37th & 0 Sts NW Washington DC 20007

O'DOHERTY, GEORGE OLIVER-PLUNKETT, b Derry, Ireland, Oct 26, 36; US citizen; m 61; c 5. ORGANIC CHEMISTRY. *Educ:* Univ Col, Dublin, Ireland, BSc, 57, BSc, 58, PhD(org chem), 62. *Prof Exp:* Res fel org chem, Sch Pharm, Univ London, 62-64, lectr, 64-66; sr scientist, 66-71, res scientist, 71-76, res assoc, 76-86, SR RES SCIENTIST, LILLY RES LABS, ELI LILLY & CO, 86- *Mem:* Am Chem Soc; NY Acad Sci. *Res:* Animal health vizecto-parasitology, helminthology and coccidiosis; organic chemistry. *Mailing Add:* 3957 S 800 E Greenfield IN 46140

ODOM, GUY LEARY, b Harvey, La, May 20, 11; m 33; c 3. NEUROSURGERY. *Educ:* Tulane Univ, MD, 33; Am Bd Neurol Surg, dipl. *Prof Exp:* Prof, Sch Med, Duke Univ, 50-74, James B Duke prof neurosurg, 74-81; RETIRED. *Concurrent Pos:* Consult to Surgeon Gen, USPHS Neurol Prog Proj Comt, 60-65; mem, Int Cong Neurol Surg; secy-treas, Am Bd Neurol Surg, 64-70, chmn, 70-; mem, Nat Adv Neurol Dis & Stroke Coun, 69-71. *Mem:* AMA; Am Asn Neurol Surg (pres, 71-72); Am Acad Neurol Surg (pres, 67); hon mem Cong Neurol Surg; Soc Neurol Surg (secy-treas, 60-65, pres, 70-71). *Res:* Cerebral circulation and intracranial neoplasms. *Mailing Add:* 2812 Chelsea Circle Durham NC 27707

ODOM, HOMER CLYDE, JR, b Hattiesburg, Miss, Dec 9, 42; m 63; c 1. ORGANIC CHEMISTRY. *Educ:* Univ Southern Miss, BA, 63, MS, 66; Clemson Univ, PhD(chem), 70. *Prof Exp:* Chemist, Pan Am Tung Res, 64-66; PROF CHEM & HEAD DEPT, BAPTIST COL CHARLESTON, 70- *Mem:* Am Chem Soc; Royal Soc Chem; Sigma Xi. *Res:* Natural products; organic residues in nature. *Mailing Add:* Dept Chem Baptist Col PO Box 10087 Charleston SC 29411-0087

ODOM, IRA EDGAR, b Dover, Tenn, June 12, 32; m 57; c 3. MINERALOGY, PETROLOGY. *Educ:* Southern Ill Univ, BA, 56; Univ Ill, MS, 58, PhD(geol), 63. *Prof Exp:* Assoc geologist, Ill State Geol Surv, 57-64; asst prof, 64-67, PROF GEOL, NORTHERN ILL UNIV, 75- *Mem:* Geol Soc Am; Mineral Soc Am; Clay Minerals Soc. *Res:* Mineralogy, petrology of sedimentary rocks. *Mailing Add:* Amar Collod Co 1500 W Shure Dr Arlington Heights IL 60004-1434

ODOM, JEROME DAVID, b Greensboro, NC, Apr 27, 42; m 65; c 2. INORGANIC CHEMISTRY. *Educ:* Univ NC, Chapel Hill, BS, 64; Ind Univ, PhD(chem), 68. *Prof Exp:* NSF fel, Bristol Univ, 68-69; from asst prof to assoc prof, 69-77, PROF CHEM, UNIV SC, 77-, CHMN DEPT, 85- *Concurrent Pos:* Alexander von Humboldt fel, Univ Stuttgart, 75-76. *Mem:* Am Chem Soc; Sigma Xi. *Res:* Chemistry of nonmetals; nuclear magnetic resonance spectroscopy; structure and bonding investigations. *Mailing Add:* Dept Chem Univ SC Columbia SC 29208

O'DONNELL, ASHTON JAY, b Los Angeles, Calif, Apr 7, 21; m 43; c 4. PHYSICS. *Educ:* Whitman Col, AB, 43. *Prof Exp:* Res physicist radiation lab, Univ Calif & Tenn Eastman Corp, 43-44; res physicist, Tenn Eastman Corp, 44-47; chief spec projs br, Hanford Opers Off, US AEC, Wash, 47-51, dir tech opers div, San Francisco Opers Off, 51-54; mgr nuclear econ res, Stanford Res Inst, 55-56, mgr nuclear develops, 56-57, mgr prog develop, 57-61; Dept State sr sci adv, US Mission to Int Atomic Energy Agency, Vienna, Austria, 61-64; mgr develop, Bechtel, Inc, 64-67, mgr, Sci Develop Dept, 67-69, mgr bus develop, 69-72, mgr, Uranium Enrichment Prog, 72-74, gen mgr uranium enrichment assocs, 74-76, vpres & mgr, San Francisco Off, Refinery & Chem Div, 76-78, vpres & mgr, Nuclear Fuel Oper, 78-84, vpres & gen mgr, Advan Technol Div, Bechtel Civil & Minerals, Inc, 85-86, vpres & gen mgr, Res & Develop Div, 85-86, dir, Bechtel Group Inc & Bechtel Investments, Inc, 85-87, CONSULT, BECHTEL, 87- *Concurrent Pos:* Adv to US rep sci adv comts, UN & Int Atomic Energy Agency, 61-64; mem adv comt nuclear mat safeguards, US AEC, 70-72; chmn, US Nat Comt Natural Resources, Pac Basin Econ Coun, 70-73; mem, Atomic Indust Forum; mem bd trustees, Whitman Col, 82-91, vpres, 85-86, chmn & pres, 86-90, emer mem, 91-; mem adv bd, INET, 87- *Mem:* Am Nuclear Soc; Inst Nuclear Mat Mgt. *Res:* Research planning, administration; international nuclear development; industrial applications of technology. *Mailing Add:* 131 Croydon Way Woodside CA 94062-2312

O'DONNELL, C(EDRIC) F(INTON), b Durham Bridge, NB, June 16, 20; nat US; m 44; c 3. ENGINEERING PHYSICS. *Educ:* McGill Univ, BEng, 49; Mass Inst Technol, MS, 51. *Prof Exp:* Supvr digital comput circuitry, Res Div, Burroughs Corp, 51-52; res engr, Autonetics Div, NAm Aviation, Inc, 52-57, group leader guid anal, 57-58, chief systs anal, 58-59, sect chief comput, 59-60, chief engr, Comput & Data Systs Div, 60-62, VPRES RES & TECHNOL, ROCKWELL INT, 63- *Mem:* Sigma Xi; Am Chem Soc. *Res:* Solid state physics as it applies to digital computers; control and stability of sampled data systems; mass data processing; microelectronics. *Mailing Add:* 1301 N Riedel Fullerton CA 92631

O'DONNELL, EDWARD, b New York, NY, Oct 13, 38; m 68; c 3. GEOLOGY. *Educ:* Queens Col, NY, BS, 61; Univ Cincinnati, MS, 63, PhD(geol), 67. *Prof Exp:* Lectr geol, Queens Col, NY, 60-61; hydrologist, US Geol Surv, 61-62; res asst zooplankton ecol, Lamont Geol Observ, 63; lectr geol, Queens Col, NY, 63-65; geologist, Pan Am Petrol Corp, 67-68; asst prof geol, Univ SFla, 68-76; GEOLOGIST, US NUCLEAR REGULATORY COMN, 76- *Mem:* AAAS; Geol Soc Am; Soc Econ Paleont & Mineral; Am Asn Petrol Geologists; Soc Econ Mineralogists & Paleontologists; Am Geophys Union. *Res:* Sedimentation; stratigraphy; structural geology; petroleum geology; tectonics of the Gulf Coast; tectonics of the Caribbean; nuclear reactor siting; high-level radioactive waste disposal; low-level radioactive waste disposal. *Mailing Add:* US Nuclear Regulatory Comn Washington DC 20555

O'DONNELL, JOSEPH ALLEN, III, b Chicago, Ill, May 18, 47; m 76; c 3. RESEARCH ADMINISTRATION. *Educ:* Ill Benedictine Col, BS, 69; Boston Univ, MA, 71; Univ Calif, Davis, PhD(nutrit), 78. *Prof Exp:* Teaching fel, Univ Calif, Davis, 78-79, lectr, 79; assoc staff nutritionist, Quaker Oats Co, 79-81; group leader, Kraft, Inc, 81-83; mgr, 83-85, VPRES, DAIRY RES FOUND, 85- *Concurrent Pos:* Mem, adj fac, Ill Benedictine Col, 80- *Mem:* Inst Food Technologists; Am Dairy Sci Asn; Am Oil Chemists Asn; Am Inst Nutrit; Am Cultured Dairy Prod Asn. *Res:* Metabolism of carbohydrates, lipids and protein; lipoproteins and atherosclerosis; nutritional content of dairy foods; processing of dairy foods. *Mailing Add:* Dairy Found Res Ctr Dairy Coun Calif 1101 National Dr Suite B Sacramento CA 95834-1274

O'DONNELL, MARTIN JAMES, b Iowa Fall, Iowa, July 8, 46; m 72; c 3. ORGANIC CHEMISTRY. *Educ:* Univ Iowa, BS, 68; Yale Univ, PhD(org chem), 73. *Prof Exp:* Fel, Cath Univ Louvain, Belg, 73-75; from asst prof to assoc prof, 75-86, PROF CHEM, IND UNIV-PURDUE UNIV, 84- *Concurrent Pos:* Res grants, Petrol Res Fund, Am Chem Soc, 77-79, Res Corp, 78-80, NATO, 78-86 & NIH, 80-83 & 85- *Honors & Awards:* Res Award, Sch Sci, Purdue Univ, 83. *Mem:* Am Chem Soc; Royal Soc Chem; Belg Chem Soc. *Res:* Development of new synthetic methods for the preparation of amino acid derivatives, especially methods involving the use of phase-transfer reactions; asymmetric synthesis. *Mailing Add:* Dept Chem Ind Univ-Purdue Univ 1125 E 38th St Indianapolis IN 46205

O'DONNELL, RAYMOND THOMAS, b Baltimore, Md, July 14, 31; m 57; c 5. ANALYTICAL CHEMISTRY, PHYSICAL CHEMISTRY. *Educ:* Loyola Col, Md, BS, 54; Mich State Univ, PhD(anal & phys chem), 67. *Prof Exp:* Instr chem, Col St Thomas, 57-59; asst prof, 64-65, ASSOC PROF ANALYTICAL CHEM, STATE UNIV NY COL OSWEGO, 65-; sr res fel, Aberdeen Proving Ground, MD, 85-86; SR STAFF SCIENTIST, SAUDI ARABIAN RES INST, 87- *Concurrent Pos:* Coun, Am Chem Soc, Syracuse, 85-; vis fac, Univ Pittsburg, 87. *Mem:* AAAS; Am Chem Soc. *Res:* Theoretical studies, development and extension of instrumental methods of analysis, especially polarography, cyclic and pulse voltametry, chronopotentiometry, spectrofluorometry and radiochemistry. *Mailing Add:* Dept Chem State Univ NY Oswego NY 13126

O'DONNELL, TERENCE J, b Chicago, Ill, Nov 13, 51. COMPUTER GRAPHICS IN DRUG DESIGN. *Educ:* Univ Ill, Chicago, BS, 73, MS 76, PhD(phys chem), 80. *Prof Exp:* Res fel chem, Nat Resource Comput Chem, Lawrence Berkeley Labs, 80; theoret chemist, Abbott Labs, 82-88; CONSULT, ODONNELL ASSOC, 88- *Mem:* Am Chem Soc; AAAS; Drug Info Asn. *Res:* Application of computational methods which describe molecular and electronic structure to the understanding of the basis of molecular activity, quantum mechanics, molecular mechanics and computer graphics; computational chemistry in drug design. *Mailing Add:* O'Donnell Assoc 1307 Byron St Chicago IL 60613

O'DONNELL, VINCENT JOSEPH, biochemistry, for more information see previous edition

O'DONNELL PHD, JAMES FRANCIS, b Cleveland, Ohio, July 22, 28; m 55; c 3. BIOCHEMISTRY. *Educ:* St Louis Univ, BS, 49; Univ Chicago, PhD(biochem), 57. *Prof Exp:* Asst prof biol chem & res biochemist, Univ Cincinnati, 57-65, asst prof biol chem & assoc prof exp med, 65-68; grants assoc, Div Res Grants, NIH, 68-69, prog dir, Pop & Reprod Grants Br, Ctr Pop Res, Nat Inst Child Health & Human Develop, 69-71, asst dir, 71-76, dep dir, Div Res Resources, 76-81 & 82-89, actg dir, 81-82, DIR, OFF EXTRAMURAL PROG, OFF EXTRAMURAL RES, OFF DIR, NIH, 90- *Mem:* NY Acad Sci; Soc Res Admin; Nat Coun Univ Res Admin. *Res:* Nucleic acid metabolism, particularly liver disease in humans and laboratory animals. *Mailing Add:* Off Extramural Progs NIH Bldg 31 Rm 5B32 Bethesda MD 20892

O'DONOGHUE, AILEEN ANN, b Denver, Colo, Sept 7, 58. RADIO ASTRONOMY, TEACHING SCIENCE TO NON-SCIENTISTS. *Educ:* Ft Lewis Col, BS, 81; NMex Inst Mining & Technol, MS, 87, PhD(physics), 89. *Prof Exp:* ASST PROF PHYSICS, DEPT PHYSICS, ST LAWRENCE UNIV, 88- *Mem:* Am Astron Soc; Astron Soc Pac; Am Phys Soc; Nat Sci Teachers Asn; Am Asn Physics Teachers. *Res:* Plasma flow dynamics of tailed radio galaxies, including radio and optical observations of these galaxies, theoretical work in modelling the fluid and plasma aspects of the flow, and the nature, formation and evolution of the clusters of galaxies in which these radio sources are found. *Mailing Add:* Dept Physics St Lawrence Univ Canton NY 13617

O'DONOGHUE, JOHN LIPOMI, b Lowell, Mass, Apr 12, 47; m 67; c 2. NEUROPATHOLOGY, NEUROTOXICOLOGY, LABORATORY ANIMAL MEDICINE. *Educ:* Univ Pa, VMD, 70, PhD, 79. *Prof Exp:* USPHS trainee & fel path, Univ Pa, 70-74; pathologist, 74-86, DIR TOXICOL SCI, EASTMAN KODAK CO, 86- *Concurrent Pos:* Instr lab animal med, Univ Rochester, 74-85; assoc prof Lab Animal Med & assoc prof toxicol, Univ Rochester, 85- *Mem:* Am Vet Med Asn; Soc Toxicol; Electron Micros Soc Am; Am Bd Toxicol; Am Asn Neuro Pathologists. *Res:* Toxicological pathology, expecially in neurotoxic diseases and in spontaneous neurological disorders. *Mailing Add:* B-320 Kodak Park Eastman Kodak Co Rochester NY 14652-3615

O'DONOHUE, CYNTHIA H, b Washington, DC, Oct 3, 36; div; c 1. ORGANIC CHEMISTRY. *Educ:* Randolph-Macon Woman's Col, AB, 57; Univ Richmond, MS, 67, MBA, 79. *Prof Exp:* Res asst, Am Brands, Inc, 57-61; chemist endocrinol lab dept med, Med Col Va, 61-63; asst scientist, Philip Morris, 65-66, assoc scientist & group leader lit processings, Tech Info Facil, 66-70, res scientist, 70-73, head tech info facil, 73-77, sr scientist & leader planning & econ group, 78-79, mgr biomat sci div, 79-81, sr staff process analyst, tobacco prod standards, 81-83, asst plant mgr, 84-85; dir facil develop, Bio-Solar, Inc, 85-86; chem specialist, Kendall McGraw Co, 86-88; SR STAFF SCIENTIST, MCGAW INC, 88- *Concurrent Pos:* Abstr, Chem Abstracts Serv, 63-68 & 75-77. *Mem:* Am Chem Soc; Am Mkt Asn; AAAS. *Res:* Information storage, retrieval and processing; steroid and synthetic organic chemistry; technical forecasting; acquisition analysis; economic planning; implementation of new technology into manufacturing environment. *Mailing Add:* 14112 Picasso Ct Irvine CA 92714-1824

O'DONOVAN, GERARD ANTHONY, b Cork, Ireland, Nov 28, 37. MICROBIOLOGY, BIOCHEMISTRY. *Educ:* Nat Univ, Ireland, BS, 60, MS, 61; Univ Calif, Davis, Phd(microbiol), 65. *Prof Exp:* Res demonstr microbiol, Univ Col, Cork, 60-61; asst, Univ Calif, Davis, 62-65; lectr bact, Univ Calif, Berkeley, 65-66, Am Cancer Soc fel molecular biol, 66-68; asst prof biochem, biophys & genetics, Tex A&M Univ, 68-71, assoc prof, 72-83; PROF BIOL SCI, UNIV N TEXAS, 83- *Concurrent Pos:* Founder & chmn, Gulf Coast Molecular Biol Conf, 69-; prin investr, Int Collab NATO Res Grant Basic Sci, 70-75; Robert A Welch chem res grant, Tex A&M Univ, 70-78. *Mem:* AAAS; Am Soc Microbiol; Brit Biochem Soc; Genetics Soc Am. *Mailing Add:* Univ N Texas PO Box 5218 Denton TX 76203

ODOR, DAVID LEE, b Washington, DC, Nov 5, 43; m 68; c 2. ENVIRONMENTAL SCIENCE. *Educ:* WVa Wesleyan Col, BS, 65; Univ Okla, MS, 67; Purdue Univ, PhD(bionucleonics), 72. *Prof Exp:* Radiol engr, Potomic Elec Power Co, 72-74; supv engr, 74-81, RES COORDR, PSI ENERGY, 81- *Mem:* Nat Hon Res Soc. *Res:* Manage research activities to derive corporate benefits; optimizing interdepartmental communication to coordinate technological impacts for existing or future company programs. *Mailing Add:* PSI Energy 1000 E Main St Plainfield IN 46168

ODOR, DOROTHY LOUISE, b Washington, DC, May 25, 22. HISTOLOGY, FEMALE REPRODUCTION. *Educ:* Am Univ, BA, 45; Univ Rochester, MS, 48, PhD(anat), 50. *Prof Exp:* Instr, Univ Wash, 50-56; from asst prof to assoc prof, Univ Fla, 56-61; from asst prof to assoc prof, Bowman Gray Sch Med, 61-69; assoc prof anat, Med Col Va, Va Commonwealth Univ, 69-73, prof anat, 73-77; actg chmn, 84-86, PROF ANAT, SCH MED, UNIV SC, 77- *Mem:* AAAS; Am Asn Anat; Am Soc Cell Biol; Soc Study Reprod; Sigma Xi; Am Inst Biol Sci. *Res:* Light microscopy and ultrastructure (TEM and SEM) of oviductal and cervical epithelium under normal and experimental conditions; ciliogenesis; histogenesis of ovary; ultrastructure of ovarian follicles; fertilization and meiotic divisions; changes in oviductal and uterine cytology after hormonal administration; mesothelium. *Mailing Add:* Dept Anat Cell Biol & Neurosci Sch Med Univ SC Columbia SC 29208

ODREY, NICHOLAS GERALD, b Ford City, Pa, Mar 22, 42; m 71. MANUFACTURING CONTROL SYSTEMS, FLEXIBLE AUTOMATION. *Educ:* Pa State Univ, BS, 64, MS, 68, PhD(indust eng), 78. *Prof Exp:* Develop engr flight dynamics, Goodyear Aerospace Corp, 67-70; res asst, Mat Res Lab, Pa State Univ, 70-76; asst prof indust eng, Univ RI, 76-81; assoc prof, WVa Univ, 81-83; assoc prof, 83-91, PROF INDUST ENG, DEPT INDUST ENG, LEHIGH UNIV, 91- *Concurrent Pos:* Dir mat processing & metrol, Univ RI, 76-81; dir, Robotics Int, Soc Mfg Engrs, 87-89, Inst Robotics, Lehigh Univ, 87-91. *Mem:* Am Soc Mech Engrs; Inst Indust Engrs. *Res:* Manufacturing systems engineering, particularly flexible automation, design of automated systems, intelligent control systems for manufacturing including hierarchical architectures; production engineering and precision engineering. *Mailing Add:* Dept Indust Eng Mohler Lab 200 Lehigh Univ Bethlehem PA 18015

O'DRISCOLL, KENNETH F(RANCIS), b Staten Island, NY, July 22, 31; m 54; c 5. POLYMER CHEMISTRY, CHEMICAL ENGINEERING. *Educ:* Pratt Inst, BChE, 52; Princeton Univ, MA, 57, PhD(chem), 58. *Prof Exp:* Develop engr, E I du Pont de Nemours & Co, 52-53; from asst prof to prof chem, Villanova Univ, 58-66; from assoc prof to prof chem eng, State Univ NY, Buffalo, 66-70; PROF CHEM ENG, UNIV WATERLOO, 70- *Concurrent Pos:* Vis prof, Kyoto Univ, 64-65, Univ Mainz, 76-77, Univ Lund, 77 & Univ Lancaster, 84. *Honors & Awards:* Petrol Res Fund Int Award, 64. *Mem:* Am Chem Soc; Can Inst Chem; Can Soc Chem Engrs. *Res:* Kinetics and thermodynamics of polymerization; immobilization of catalysts. *Mailing Add:* Dept Chem Eng Univ Waterloo Waterloo ON N2L 3G1 Can

ODUM, EUGENE PLEASANTS, b Lake Sunapee, NH, Sept 17, 13; m 39; c 1. ECOSYSTEM ECOLOGY, ORNITHOLOGY. *Educ:* Univ NC, AB, 34; Univ Ill, PhD(ecol, ornith), 39. *Hon Degrees:* DLitt, Hofstra Univ, 80; DH, Ferum Col, 86; LHD, Univ NC, 90. *Prof Exp:* Asst zool, Univ NC, 34-36; instr biol & ornith, Western Reserve Univ, 36-37; asst, Univ Ill, 37-39; res biologist, Edmund Niles Huyck Preserve, NY, 39-40; from instr to prof, 40-57, alumni found distinguished prof zool, 57-84, dir inst ecol, 61-84, Fuller

E Callaway prof ecol, 76-84, EMER PROF & DIR, UNIV GA, 84- *Concurrent Pos:* NSF sr fel, 57-58; instr, Marine Biol Lab, Woods Hole, 57-61. *Honors & Awards:* Mercer Award, Ecol Soc Am, 56; Inst de la Vie, 75; Tyler Ecol Award, 77; Crafoord Prize, Royal Swed Acad Sci, 87. *Mem:* Nat Acad Sci; AAAS; Ecol Soc Am (pres, 64-65); Am Soc Limnol & Oceanog; Am Soc Nat; Am Acad Arts & Sci; fel Am Ornithologists Union. *Res:* General principles of ecology; vertebrate populations; productivity and ecosystem energetics; ecology of birds; estuarine and wetland ecology; landscape ecology. *Mailing Add:* Inst Ecol Univ Ga Athens GA 30601

ODUM, HOWARD THOMAS, b Durham, NC, Sept 1, 24; m 47; c 2. ECOLOGY, SYSTEMS ECOLOGY. *Educ:* Univ NC, AB, 47; Yale Univ, PhD(zool), 51. *Prof Exp:* Asst, Univ NC, 46 & Yale Univ, 47-48; asst prof biol, Univ Fla, 50-54; asst prof zool, Duke Univ, 54-56; dir & res scientist, Inst Marine Sci, Univ Tex, 56-63; chief scientist, Rain Forest Proj, P R Nuclear Ctr, 63-66; prof ecol, Univ NC, Chapel Hill, 66-70; GRAD RES PROF ENVIRON ENG SCI, UNIV FLA, 70-, DIR CTR WETLANDS, 72- *Concurrent Pos:* Instr, Trop Weather Sch, CZ, 45; prin investr, Off Naval Res grant, Univ Fla, 52-54; instr, Marine Biol Lab, Woods Hole, 53, 58; grants, NSF, 55-60, Rockefeller Found, 57-58 & 73-77, Off Naval Res, 58-60, Atomic Energy Comn, 58, 63-76 & USPHS, 59-63; mem comn herbicide in Viet Nam, Nat Acad Sci. *Honors & Awards:* George Mercer Award, Ecol Soc, 57; Merit Award, Asn Tech Writers, 71; Cert of Achievement, Soc Tech Commun; Distinguished Serv Award, Indust Develop Res Coun; Prize, Inst la Vie, Paris, 76. *Mem:* AAAS; Am Soc Limnol & Oceanog; Am Meteorol Soc; Ecol Soc Am; Geochem Soc. *Res:* Energy analysis; biological oceanography; biogeochemistry; ecological engineering; tropical meteorology. *Mailing Add:* Dept Environ Eng Univ Fla Gainesville FL 32601

ODUM, WILLIAM EUGENE, b Athens, Ga, Oct 1, 42. ECOLOGY. *Educ:* Univ Ga, BS, 64; Univ Miami, MS, 66, PhD(marine sci), 70. *Prof Exp:* Can Govt res fel, Univ BC, 70-71; asst prof, 71-75, ASSOC PROF ECOL, UNIV VA, 75-, PROF ENVIRON SCI, 80-, CHMN DEPT, 84- *Mem:* Am Soc Limnol & Oceanog; Ecol Soc Am; Am Fisheries Soc; Inst Fisheries Biologists. *Res:* Wetland ecology; plant detritus production; ecology of resource management; ecology of fishes; cycling of heavy metals in ecosystems; estuarine food webs. *Mailing Add:* Dept of Environ Sci Clark Hall 212 Univ of Va Charlottesville VA 22903

O'DŪNLAING, COLM PÁDRAIG, algorithmic motion-planning, for more information see previous edition

O'DWYER, JOHN J, b Grafton, NSW, Australia, Nov 9, 25; m 55; c 5. THEORETICAL SOLID STATE PHYSICS. *Educ:* Univ Sydney, BSc, 45, BE, 47; Univ Liverpool, PhD(physics), 51. *Prof Exp:* Lectr math, Sydney Tech Col, 47-48; sr res officer physics, Commonwealth Sci & Indust Res Orgn, Australia, 51-57; assoc prof, Univ New South Wales, 57-64; sr scientist, Westinghouse Res Labs, 65; prof physics, Univ Southern Ill, 66-70; PROF PHYSICS, STATE UNIV NY COL OSWEGO, 70- *Concurrent Pos:* Vchmn digest comt, Dielectrics Conf, 67, secy, 70-71, chmn, 74-75. *Mem:* Am Phys Soc; Am Asn Physics Teachers; Brit Inst Physics. *Res:* Theory of dielectrics, especially phenomena occurring at high field strengths. *Mailing Add:* Dept of Physics State Univ of NY Col Oswego NY 13126

ODYA, CHARLES E, b Franfurt, W Germany, Oct 3, 47. RECEPTORS, VASAL ACTIVE PEPTIDES. *Educ:* Univ Wis-Madison, PhD(pharmacol), 75. *Prof Exp:* Asst prof, 79-84, ASSOC PROF PHARMACOL, IND UNIV, 84- *Mem:* Sigma Xi; AAAS. *Mailing Add:* 270 Beechwood Dr Ellettsville IN 47429-1053

OECHEL, WALTER C, b San Diego, Calif, Jan 15, 45; m 67. ECOSYSTEMS SCIENCE, PHYSIOLOGICAL ECOLOGY. *Educ:* San Diego State Univ, AB, 66; Univ Calif, Riverside, PhD(physiol ecol), 70. *Prof Exp:* Teaching asst, 66-67, res asst, 67-68, fel, Dept Life Scis, Univ Calif, Riverside, 67-70; from asst prof to assoc prof, Biol Dept, McGill Univ, 70-78; res prof, 78-82, dir, Systs Ecol Res Group, 82-87, PROF, BIOL DEPT, SAN DIEGO STATE UNIV, 83- *Concurrent Pos:* Nat Sci Found Fel, ecol, Orgn Tropical Studies, 67-70; chmn, Physiol Ecol Sect, Ecol Soc Am, 82-83; consult, USDA, Woodward, Okla, 82-85 & County of Inyo, 90; mem, Watershed Mgt Task Force, San Diego, Dept Agr, 82-86; prin investr, Nat Res Coun Can, 71-78, Dept Indian Affairs & Northern Develop, 71-78, USDA Forest Serv, 79-84, NSF, 77-, US Dept Energy, 77-, Southern Calif Edison, 90-, Water Resources Ctr, 90- & US - Spain Joint Comn, 90- *Mem:* Ecol Soc Am; AAAS; Int Asn Ecol; Int Asn Bryologists; Int Assoc Ecol; NY Acad Scis; Sigma Xi; Am Inst Biol Sci. *Res:* The effects of global increases in atmospheric CO2 and climate change on unmanaged ecosystems including the arctic tussock tundra and mediterranean type ecosystems, patterns in and controls on ecosystem function in Mediterranean-type ecosystems; scientific publications in above areas. *Mailing Add:* Biol Dept Ecosyts Res Lab San Diego State Univ San Diego CA 92120

OEGEMA, THEODORE RICHARD, JR, b Jan, 3, 45; m; c 2. CONNECTIVE TISSUE PROTEOGLYCANS, ORTHOPEDIC SURGERY. *Educ:* Univ Mich, PhD(biochem), 72. *Prof Exp:* PROF BIOCHEM, DEPT ORTHOPEDICS & BIOCHEM & DIR ORTHOPEDIC RES LAB, UNIV MINN, 75- *Mailing Add:* 6361 Washington NE Minneapolis MN 55432

OEHLBERG, RICHARD N, b Evanston, Ill, Nov 7, 42; m 68. NUCLEAR REACTOR SAFETY, TRAINING. *Educ:* Loyola Univ, Chicago, BS, 65; Southern Ill Univ, MS, 67; Univ Notre Dame, PhD(physics), 72, Univ Southern Calif, MPA, 75. *Prof Exp:* Physicist, Naval Weapons Ctr, 72-74; prog mgr, US Atomic Energy Corp, 74-76; proj mgr, 76-85, tech asst to vpres nuclear, 85-88, SR PROJ MGR, ELEC POWER RES, INC, 88- *Concurrent Pos:* Treasurer, 80-81, pres, Stanford Fed Credit Union, 81-82; fel, Southern Ill Univ, 66-67, Nat Defense Educ, 67-68; mem prog comt, Nuclear Reactor Safety Div, Am Nuclear Soc, 90-; mem Nuclear Power Marine Purposes Comt, Joint Owners Group Accident Mgt adv comt, 90. *Mem:* Am Phys Soc; Sigma Xi; Sci Res Soc; Am Nuclear Soc. *Res:* Author of textbooks and scientific articles on safety of nuclear power plant. *Mailing Add:* 530 Castano Corte Los Altos CA 94022

OEHLSCHLAGER, ALLAN CAMERON, b Hartford, Conn, Sept 8, 40; m 60; c 3. BIO-ORGANIC CHEMISTRY. *Educ:* Okla State Univ, PhD(org chem), 65. *Prof Exp:* NATO fel org chem, Univ Strasbourg, 65-66; from asst prof to assoc prof, 66-78, PROF CHEM, SIMON FRASER UNIV, 78- *Honors & Awards:* Labatt Award, Chem Inst Can. *Mem:* Am Chem Soc; Royal Soc Chem; Chem Soc France; fel Chem Inst Can. *Res:* Organometallic chemistry; bio-organic chemistry of terpenes; insect attractants. *Mailing Add:* Dept Chem Simon Fraser Univ Burnaby BC V5A 1S6 Can

OEHME, FREDERICK WOLFGANG, b Leipzig, Ger, Oct 14, 33; US citizen; m 60, 81; c 5. CLINICAL TOXICOLOGY, COMPARATIVE MEDICINE. *Educ:* Cornell Univ, BS, 57, DVM, 58; Kans State Univ, MS, 62; Univ Geissen, Dr med vet, 64; Am Bd Vet Toxicol, dipl, 67; Univ Mo-Columbia, PhD(toxicol), 69; Am Bd Toxicol, dipl, 80; Acad Toxicol Sci, dipl, 81. *Prof Exp:* Veterinarian, Md, 58-59; instr surg & med col vet med, Kans State Univ, 59-64, from asst prof to assoc prof, 64-66; instr physiol pharm, Univ Mo-Columbia, 66-67, NIH spec res fel toxicol, 67-69; assoc prof toxicol & med, 69-73, PROF TOXICOL, MED & PHYSIOL, COL VET MED, KANS STATE UNIV, 73-, DIR COMP TOXICOL LAB, 69- *Concurrent Pos:* Vis prof col vet med, Univ Giessen, 63-64; consult, private indust gen public, 69-, Food & Drug Admin, Washington, DC, 70-, Univ Kans, 71-76, Nat Acad Sci, 71, various state & nat govts, 75-, Nat Inst Environ Health Sci, 74- & World Health Org, 79-; ed & publ, Vet & Human Toxicol, 70- *Honors & Awards:* Vet Med Fac Award, Kans State Univ, 77. *Mem:* Fel Am Acad Vet Comp Toxicol (secy-treas, 70-78); Soc Toxicol (pres, 84-85); Am Acad Clin Toxicol (pres, 78-80); World Fed Asn Clin Toxicol Ctrs & Poison Control Ctrs (pres, 85-89); Soc Toxicol Path; NY Acad Sci. *Res:* Biotransformation and biochemical action of toxicants; clinical and diagnostic toxicology; public health aspects of toxicants; comparative toxicology as a research and diagnostic tool; teaching and communication techniques for standards of excellence in science. *Mailing Add:* Comp Toxicol Lab Kans State Univ Manhattan KS 66506

OEHME, REINHARD, b Wiesbaden, Ger, Jan 26, 28; m 52. THEORETICAL PHYSICS. *Educ:* Univ Goettingen, PhD, 51. *Prof Exp:* Res asst, Max Planck Inst Physics, Goettingen, Ger, 50-53; res assoc, Enrico Fermi Inst Nuclear Studies, Univ Chicago, 54-56; mem, Inst Adv Study, 56-58; from asst prof to assoc prof, 58-60, PROF, DEPT PHYSICS & ENRICO FERMI INST, UNIV CHICAGO, 64- *Concurrent Pos:* Vis prof, Inst Theoret Physics, Sao Paulo, Brazil, 52-53 & Univ Md, 57; vis scientist, Brookhaven Nat Lab, 57, 60 & 62, vis sr scientist, 65 & 67; visitor, Europ Orgn Nuclear Res, Switz, 61, 64, 71, 73 & 85; vis prof, Univ Vienna, 61, Imp Col, Lond, 63-64, Guggenheim fel, 63-64; vis prof, Int Ctr Theoret Physics, 66, 68, 69, 70 & 72; vis prof, Univ Karlsruhe, Ger, 74, 75 & 77, Univ Tokyo, 76 & 88, Max Planck Inst, Munich, 78 & 84-87 & Yukawa Inst, Kyoto Univ, 88; fel, Japan Soc Prom Sci, 76 & 88. *Honors & Awards:* Sr US Scientist Award, Alexander von Humboldt Found, 74. *Mem:* Fel Am Phys Soc. *Res:* Elementary particle physics; quantum field theory. *Mailing Add:* Enrico Fermi Inst Univ Chicago Chicago IL 60637

OEHMKE, RICHARD WALLACE, b St Clair, Mich, Dec 19, 35; m 61; c 6. APPLIED CHEMISTRY, RESEARCH ADMINISTRATION. *Educ:* St Joseph's Col, Ind, BS, 58; Univ Ill, Urbana, PhD(inorg chem), 64. *Prof Exp:* Sr chemist, 64-73, supvr com tape div, 73-74, mgr prod develop, 74-78, MGR APPL TECHNOL, COM TAPE DIV, 3M CO, 78- *Mem:* Sigma Xi. *Res:* High temperature polymers; polymer composites; organometallic catalysis; bio-and biologically reactive polymers; chemical reactions of natural polymers; polymer coatings; chromatography; polymer adhesives; polymer films; technical management. *Mailing Add:* 951 Trout Brook Rd Hudson WI 54016-7419

OEHMKE, ROBERT H, b Detroit, Mich, Aug 6, 27; m 50; c 1. MATHEMATICS. *Educ:* Univ Mich, BS, 48; Univ Detroit, MA, 50; Univ Chicago, PhD(math), 54. *Prof Exp:* Instr math, Ill Inst Technol, 53-54; asst prof, Butler Univ, 54-56; instr, Mich State Univ, 52-62; instr, Inst Defense Anal, 62-64; PROF MATH, UNIV IOWA, 64-, CHMN DEPT, 79- *Mem:* Am Math Soc; Math Asn Am; Asn Comput Mach. *Res:* Nonassociative algebra, semigroups and automata. *Mailing Add:* Dept Math Univ Iowa Iowa City IA 52240

OEHSER, PAUL HENRY, b Cherry Creek, NY, Mar 27, 04; m 27; c 2. SCIENCE WRITING, HISTORY OF SCIENCE. *Educ:* Greenville Col, AB, 25. *Prof Exp:* Asst ed bur biol surv, USDA, 25-31; ed, US Nat Mus, 31-50, asst chief ed div, Smithsonian Inst, 46-50, chief ed & pub div, 50-66, ed Sci Pub, Nat Geog Soc, 66-78; RES ASSOC, SMITHSONIAN INST, 66- *Concurrent Pos:* Ed proc, 8th Am Sci Cong, 40-43; gen ed, US Encycl Hist, 67-68. *Mem:* Wilderness Soc; Thoreau Soc Am (pres, 60); Am Ornithologists' Union. *Res:* Science editing and publication; biological editing and bibliography; American naturalists; conservation; Smithsonian history. *Mailing Add:* 9601 Southbrook Dr Jacksonville FL 32256-0489

OEI, DJONG-GIE, b Solo, Indonesia, Apr 18, 31; US citizen; m 63; c 4. PHYSICAL CHEMISTRY, INORGANIC CHEMISTRY. *Educ:* Univ Indonesia, Drs, 58; Univ Ky, PhD(phys chem), 61. *Prof Exp:* Lectr phys & inorg chem, Bandung Inst Tech, Indonesia, 61-63; fel dept chem, Univ Ky, 63-64; staff chemist off prod div, Int Bus Mach Corp, Ky, 64-67; sr res engr, Sci Res Staff, 67-69, ASSOC PRIN RES SCIENTIST, RES STAFF, FORD MOTOR CO, 69- *Mem:* Am Chem Soc; Electrochem Soc; Sigma Xi. *Res:* Chemistry of non-aqueous solvents; coordination compounds; electrochemistry; inorganic sulfur chemistry; chemical vapor deposition. *Mailing Add:* 665 Brentwood Rd Dearborn MI 48124

OEKBER, NORMAN FRED, b Avon, Ohio, Mar 3, 27; m 63; c 9. HORTICULTURE. *Educ:* Ohio State Univ, BS, 49; Cornell Univ, MS, 51, PhD(veg crops), 53. *Prof Exp:* Exten specialist veg crops, Univ Ill, 53-60; EXTEN SPECIALIST VEG CROPS, UNIV ARIZ, 60-, PROF HORT, 71-, RES SCIENTIST HORT, AGR EXP STA, 76- *Concurrent Pos:* Pres, Nat Agr Plastics Cong, 66; res horticulturist, Arid Lands Res Ctr, Abu Dhabi, United Arab Emirates, 75. *Mem:* Am Soc Hort Sci (vpres, 88); Sigma Xi. *Res:* Crop ecology and environmental control; desert vegetable production; post-harvest handling. *Mailing Add:* Dept Plant Sci Univ Ariz Tucson AZ 85721

OELBERG, THOMAS JONATHON, b Waukon, Iowa, July 11, 56; m 79. DAIRY SCIENCE NUTRITION RESEARCH. *Educ:* SDak State Univ, BS, 79, MS, 81; Ohio State Univ, PhD(animal nutrit), 85. *Prof Exp:* Res asst dairy sci, SDak State Univ, 79-81; res assoc dairy sci, Ohio State Univ, 81-85; res scientist, 85-87, MGR ANIMAL RES, INT MULTIFOODS, 87- *Mem:* Am Dairy Sci Assoc; Am Soc Animal Sci; Poultry Sci Assoc. *Res:* Design all experiments and summarize results from all experiments on animal nutrition research for International Multifoods; major research activities include defining nutrient requirements and nutrient availabilities in feed stuffs for swine, poultry and cattle. *Mailing Add:* Supersweet Research Farm PO Box 117 Courtland MN 56021

OELFKE, WILLIAM C, b Kansas City, Mo, May 28, 41; m 64. GRAVITATIONAL RESEARCH, PHYSICS OF MEASUREMENTS. *Educ:* Stanford Univ, BS, 63; Duke Univ, PhD(physics), 69. *Prof Exp:* Res assoc microwave spectros, Duke Univ, summer 69; asst prof physics, Fla Technol Univ, 69-72, assoc prof, 72-80; PROF PHYSICS, UNIV CENT FLA, 80- *Concurrent Pos:* Sr res assoc, La State Univ, 75-76. *Honors & Awards:* Fac Develop Award, Fla Technol Univ, 75. *Mem:* Sigma Xi; Am Asn Physics Teachers; Am Phys Soc; AAAS; Am Chem Soc. *Res:* Detection of gravitational radiation, superconducting, accelerometry, quantum-limited measurements, measurement of universal gravitational constant. *Mailing Add:* 2319 Huron Trail Maitland FL 32751

OELKE, ERVIN ALBERT, b Green Lake, Wis, Dec 14, 33; m 58; c 2. AGRONOMY, PLANT PHYSIOLOGY. *Educ:* Univ Wis, BS, 60, MS, 62, PhD(agron), 64. *Prof Exp:* Agronomist, Rice Exp Sta, Univ Calif, 64-68; from asst to assoc prof, 68-79, PROF AGRON & EXTEN AGRONOMIST, UNIV MINN, ST PAUL, 79- *Concurrent Pos:* Chair, Div C-3, Crop Sci Soc Am; chair, Wheat Indust Resource Comt. *Mem:* Am Soc Agron; Crop Sci Soc Am. *Res:* Ecological factors influencing the seedling development and growth of rice; influence of environment and breeding on chlorophyll and other plant constituents in corn; culture and physiology of wild rice. *Mailing Add:* Dept of Agron & Plant Genetics Univ Minn 1991 Buford Circle St Paul MN 55108

OELS, HELEN C, b Philadelphia, Pa, Apr 13, 31. PATHOLOGY, CYTOLOGY. *Educ:* Chestnut Hill Col, BS, 53; Woman's Med Col Pa, MD, 57; Univ Minn, PhD(path), 71; Am Bd Path, dipl & cert anat path, 65, cert clin path, 67. *Prof Exp:* NIH fel, Sch Med, Temple Univ, 58-60; physician, Whitesburg Mem Hosp, Ky, 60-61; mem staff, Mayo Clin, 61-69; asst prof microbiol, 69-71, ASSOC PROF MICROBIOL & IMMUNOL, SCH MED, TEMPLE UNIV, 71-; DIR ANAT PATH, SMITH KLINE BIOSCIENCE LABS, 75- *Concurrent Pos:* Dir anat path, Lab Procedures, Inc, 75- *Mem:* Am Soc Clin Path; Am Soc Cytology; AMA; Col Am Pathologists. *Res:* Immunopathology; autoimmune diseases; lymphomas; antigens of histoplasma capsulatum; delayed hypersensitivity; virology; viral oncolysis. *Mailing Add:* 131 Elm St Rockledge PA 19111

OEN, ORDEAN SILAS, b Grafton, NDak, June 29, 27; m 53; c 3. PHYSICS. *Educ:* Concordia Col, BA, 49; Univ NDak, MS, 53; Univ Mo, PhD(physics), 58. *Prof Exp:* Instr physics, Univ Mo, 53-55; SUMMER PHYSICIST SOLID STATE THEORY, 57-, RES PHYSICIST, SOLID STATE DIV, OAK RIDGE NAT LAB, 58- *Concurrent Pos:* Vis prof plasma physics, Max Planck Inst, Munich, Ger, 77-78. *Mem:* Am Phys Soc; Sigma Xi; Boehmische Phys Soc. *Res:* Codiscoverer of channeling phenomenon; particle-solid interaction channeling. *Mailing Add:* 119 Lehigh Lane Oak Ridge TN 37830

OERTEL, DONATA (MRS BILL M SUGDEN), b Bonn, Ger, Aug 13, 47; US citizen; m 76; c 1. NEUROBIOLOGY. *Educ:* Univ Calif, Los Angeles, BA, 69, Santa Barbara, PhD(biol), 75. *Prof Exp:* Res asst neurobiol, Marine Biol Lab, Woods Hole, Mass & Univ Calif, Santa Barbara, 72-75; res assoc, Univ Wis-Madison, 75-77; NIH fel, Harvard Med Sch, 77-78; proj assoc, 78-80, asst scientist, 80-81, asst prof, 81-86, ASSOC PROF NEUROPHYSIOL, UNIV WIS-MADISON, 86- *Concurrent Pos:* Mem, Commun Disorders Rev Comt, NIDLD. *Mem:* Soc Neurosci; Sigma Xi; Asn Res Otolaryngol. *Res:* Neurophysiology, in particular, physiology of the auditory and visual systems; membrane biophysics. *Mailing Add:* Dept Neurophys 281 Med Bldg Univ Wis Madison WI 53706

OERTEL, GERHARD FRIEDRICH, b Leipzig, Ger, Apr 22, 20; m 46; c 3. STRUCTURAL GEOLOGY. *Educ:* Univ Bonn, Dr rer nat(geol), 45. *Prof Exp:* Asst geol, Univ Bonn, 46-50, Privatdozent, 50; geologist, Co Petrol Portugal, 51-53; geologist, Portuguese State Overseas Ministry, 53-56; assoc prof geol, Pomona Col, 56-60; from assoc prof to prof geol, 60-90, EMER PROF GEOL, UNIV CALIF, LOS ANGELES, 90- *Concurrent Pos:* John Simon Guggenheim Mem Found fel, Univ Edinburgh, 66-67. *Mem:* Am Geophys Union; Geol Soc Am. *Res:* Structural geology of plutonic, volcanic, metamorphic and sedimentary rocks; deformation and fracture of rocks; slaty cleavage; preferred orientation of phyllosilicate mineral grains in rocks. *Mailing Add:* Dept Earth & Space Sci Univ Calif Los Angeles CA 90024-1567

OERTEL, GOETZ K, b Stuhm, Ger, Aug 24, 34; US citizen; m 60; c 2. ATOMIC & MOLECULAR PHYSICS. *Educ:* Univ Kiel, Ger, Vordipl, 56; Univ Md, College Park, PhD(physics), 64. *Prof Exp:* Researcher, NASA Langley, 63-68, chief solar physics, NASA Wash, 68-74; analyst, Exec Off Pres, 74-75; mem staff, US Dept Energy, 75-87; PRES, ASN UNIVS RES ASTRON, 87- *Concurrent Pos:* Vis prof, Univ Md, 67-68; pres & sci adv, NSF, 74; chair, Tech Rev Group, West Valley, 88-; mem bd, Am Asn Advan Technol, 89- *Mem:* Am Phys Soc; Am Astron Soc; NY Acad Sci; AAAS; Sigma Xi; Am Asn Advan Technol. *Res:* Plasma spectroscopy; broadening of spectral lines in a plasma; solar physics; atomic physics. *Mailing Add:* Asn Univs Res Astron Inc 1625 Massachusetts Ave NW Suite 701 Washington DC 20036

OERTEL, GOETZ KUNO HEINRICH, b Stuhm, Ger, Aug 24, 34; m 60; c 2. ASTRONOMY. *Educ:* Univ Kiel, Vordiplom, 56; Univ Md, PhD(physics), 64. *Prof Exp:* Res asst physics, Univ Md, 57-62; aerospace engr, Langley Res Ctr, NASA, 63-68, staff scientist, Solar Prog Mgt, 68-69, dep chief solar physics prog, 69-71, chief, 71-75; head astron sect, NSF, 75; dir commun & mgt support, Off Asst Adminr Nuclear Energy, ERDA, 75-77, asst dir, Div Waste Prod, Nuclear Waste Mgt, 77-81, dir, Off Defense Nuclear Waste & Prod, 81-83, actg mgr, Savannah River Opers Off, Aiken, SC, 83-84, dep mgr, Albuquerque Opers Off, 84-85, dept asst secy, Safety, Health & Qual Assurance, 85-86; PRES, ASSOC UNIV RES ASTRON, WASH, DC, 86- *Concurrent Pos:* Fulbright grant, 54-55; at dept physics & astron, Univ Md, 67-68; partic, Fed Exec Develop Prog, Off Mgt & Budget & Civil Serv Comn, 74-75; spec analyst, Sci & Technol Policy Off, NSF, 74 & Exec Off of the President. *Honors & Awards:* Silver Medal, US Dept Energy 65, Bronze Medal, 83; Except Serv Medal, NASA. *Mem:* Am Astron Soc; Am Phys Soc; Sigma Xi. *Res:* Plasma spectroscopy; solar physics and astrophysics; program management; science policy; federal budget analysis and formulation; federal management and administration; communications; radioactive waste management. *Mailing Add:* 9609 Windcroft Way Potomac MD 20854

OERTEL, RICHARD PAUL, b New York, NY, Jan 12, 44; m 67. SPECTROCHEMISTRY. *Educ:* Oberlin Col, BA, 64; Cornell Univ, PhD(chem), 68. *Prof Exp:* Asst prof chem, Cornell Univ, 68-69; RES CHEMIST, PROCTER & GAMBLE CO, 69- *Mem:* Am Chem Soc; Sigma Xi. *Res:* Application of spectroscopy, chiefly infrared and Raman, to the elucidation of molecular structure in chemical and biochemical systems. *Mailing Add:* 3191 Minton Rd Hamilton OH 45013-4347

OERTLI, JOHANN JAKOB, b Ossingen, Switz, July 16, 27; US citizen; m 61; c 4. SOIL SCIENCE, PLANT NUTRITION. *Educ:* Swiss Fed Inst Technol, FEng, 51; Univ Calif, MS, 53, PhD(soils), 56. *Prof Exp:* From instr to asst prof soil sci, Univ Calif, Los Angeles, 57-63; from assoc prof to prof, Univ Calif, Riverside, 63-74; head, Inst Bot, Univ Basel, Switz, 74-79; PROF CROP SCI, SWISS FED INST TECHNOL, ZURICH, 79- *Concurrent Pos:* Fulbright fel, Ger, 70-71. *Mem:* German Soc Plant Nutrit; German Soc Soil Sci; Swiss Soc Plant Physiol; Soil Sci Soc Am; Am Soc Agron; Swiss Soc Soil Sci; Crop Sci Soc Am. *Res:* Mineral nutrition and water relations in plants. *Mailing Add:* Inst Plant Sci Swiss Fed Inst Technol Zurich CH 8092 Switzerland

OESPER, PETER, b Cincinnati, Ohio, Sept 25, 17; m 43; c 1. BIOLOGICAL CHEMISTRY. *Educ:* Swarthmore Col, AB, 38; Princeton Univ, MS, 40, PhD(chem), 41. *Prof Exp:* Instr phys chem, Univ Md, 41-42, asst prof, 42-45; res assoc sch med, Univ Pa, 45-51; asst prof, Hahnemann Med Col, 51-56, from assoc prof to prof, 56-68; chmn, 68-77, prof, 68-88, EMER PROF CHEM DEPT, ST LAWRENCE UNIV, 88- *Mem:* Am Chem Soc; Am Soc Biol Chem. *Res:* Dipole moments; kinetics and thermodynamics of enzyme reactions. *Mailing Add:* Dept of Chem St Lawrence Univ Canton NY 13617

OESTERLING, MYRNA JANE, b Butler, Pa, Oct 24, 17. BIOCHEMISTRY. *Educ:* Univ Ill, AB, 39, MS, 41; George Washington Univ, PhD(biochem), 44. *Prof Exp:* Asst chem, Sch Med, Yale Univ, 46-47& Univ Ill, 47-50; asst prof physiol chem, Med Col Pa, 51-67, res assoc prof, 67-71, res assoc prof obstet & gynec(biochem), 71-86; RETIRED. *Concurrent Pos:* Talbot fel, Yale Univ, 44, Coxe fel, 45; Berquist fel, Karolinska Inst, Sweden, 50-51. *Mem:* AAAS; Am Asn Clin Chem; Am Soc Biol Chemists. *Res:* Amino acid nutrition; ascorbic acid method; stability and determination of epinephrine and norepinephrine; excretion of catecholamines in relation to stress; fetoplacental function; perinatology. *Mailing Add:* 3313 Midvale Ave Philadelphia PA 19129

OESTERLING, THOMAS O, b Butler, Pa, Mar 6, 38; m 60; c 3. PHARMACY, CHEMISTRY. *Educ:* Ohio State Univ, BS, 62, MS, 64, PhD(pharmaceut chem), 66. *Prof Exp:* Sr res assoc, Upjohn Co, 66-70, res head, 70-76; dir dermat res & develop, Johnson & Johnson, 76-78, dir pharmaceut res & develop, 78-80; mem staff, Mallinckrodt Inc Res & Develop, 80-; AT COLLABORATIVE RES, INC. *Mem:* Acad Pharmaceut Sci; Am Chem Soc; AAAS; Am Acad Dermat; Sigma Xi. *Res:* Rates and mechanisms of chemical degradation of drugs; physically programmed release of drugs from drug delivery systems. *Mailing Add:* 23420 Commerce Park Rd Beachwood OH 49122-5813

OESTERREICH, ROGER EDWARD, b Chicago, Ill, Feb 7, 30; m 75. PHYSIOLOGICAL PSYCHOLOGY. *Educ:* Univ Mo, AB, 52; Univ Chicago, PhD(biopsychol), 60. *Prof Exp:* Asst psychol, Univ Chicago, 55-59; res assoc, Ill State Psychiat Inst, 60-64; vis asst prof, Ill Inst Technol, 64; asst prof, 64-66, ASSOC PROF PSYCHOL, STATE UNIV NY ALBANY, 66- *Res:* Neural bases of perception; audition; sound localization. *Mailing Add:* RD 1 Esperance NY 12066

OESTERREICHER, HANS, b Innsbruck, Austria, May 16, 39; US citizen; m 69; c 1. SOLID STATE CHEMISTRY. *Educ:* Univ Vienna, PhD(chem), 65. *Prof Exp:* Res assoc solid state chem, Univ Pittsburgh, 65-66 & Brookhaven Nat Lab, 66-67; instr, Cornell Univ, 67-69; asst prof, Ore Grad Ctr, 68-73, assoc prof, 73; assoc prof, 75-81, PROF SOLID STATE CHEM, UNIV CALIF, SAN DIEGO, 81- *Concurrent Pos:* Guest prof, Univ Konstanz, WGer, 78, 79, 80 & 81, Univ Cologne, 78, Max Planck Inst, Stuttgart, 82; vis scientist, Los Alamos, 89 & 90. *Mem:* Am Phys Soc. *Res:* Structural, magnetic, electric and thermal properties metallic and ceramic solids, superconductivity; intermetallic compounds and hydrides thereof. *Mailing Add:* Dept of Chem B-017 Univ of Calif at San Diego Box 109 La Jolla CA 92093

OESTERWINTER, CLAUS, b Hamburg, Ger, Jan 18, 28; US citizen; m 53; c 2. ASTRONOMY, CELESTIAL MECHANICS. *Educ:* Yale Univ, MS, 64, PhD(astron), 65. *Prof Exp:* Chief comput, Western Geophys Co Am, 54-59; astronr, Naval Surface Warfare Ctr, 59-; RETIRED. *Mem:* Am Astron Soc; Int Astron Union; Cospar. *Res:* Determination of artificial satellite orbits in support of satellite geodesy; global solution for orbits of moon and major planets; developing new semianalytical theory for artificial satellite motion. *Mailing Add:* PO Box 431 Dahlgren VA 22448

OESTREICH, ALAN EMIL, b New York, NY, Dec 4, 39; m 73; c 1. RADIOLOGY, PEDIATRIC RADIOLOGY. *Educ:* Princeton Univ, AB, 61; Johns Hopkins Univ, MD, 65. *Prof Exp:* Asst prof radiol, Univ Mo, Columbia, 71-74, asst prof child health, 72-74, assoc prof radiol & child health, 74-79; assoc prof, 80-86, PROF RADIOL & PEDIAT, COL MED, UNIV CINCINNATI, 86-; PEDIAT RADIOL, CINCINNATI CHILDREN'S HOSP, 80- *Concurrent Pos:* Vis prof, Meharry Med Col, 72-77; chmn, sect radiol, Nat Med Asn, 85-87. *Mem:* Nat Med Asn; Soc Pediat Radiol; Am Col Radiol; Radiol Soc NAm; Am Roentgen Ray Soc; Int Skeletal Soc. *Res:* Pediatric radiology, especially orthopedics, and including ultrasound and mathematics applications; applications of computers in bone dysplasias, vertebral column, chest, gastrointestinal. *Mailing Add:* XRay Dept Children's Hosp Med Ctr Elland & Bethesda Aves Cincinnati OH 45229-2899

OESTREICHER, HANS LAURENZ, b Vienna, Austria, Apr 22, 12; m 43; c 1. MATHEMATICS. *Educ:* Univ Vienna, PhD(math), 34. *Prof Exp:* Res mathematician, Helmholtz Inst, Ger, 43-47; RES MATHEMATICIAN & CHIEF MATH & ANALYSIS BR, AEROSPACE MED RES LAB, WRIGHT-PATTERSON AFB, 47- *Mem:* Fel Acoust Soc Am; Am Math Soc; NY Acad Sci; Inst Elec & Electronics Eng. *Res:* Applied mathematics; wave propagation; partial differential equations; theory of sound; information processing. *Mailing Add:* 2923 Green Vista Dr Brook Hollow Fairborn OH 45324

OETINGER, DAVID FREDERICK, b Buffalo, NY, Apr 25, 45; m 70; c 2. HELMINTHOLOGY, PARASITOLOGY. *Educ:* Houghton Col, BA, 67; Univ Nebr-Lincoln, MS, 69, PhD(zool), 77. *Prof Exp:* From asst prof to assoc prof biol, Houghton Col, 81-84; assoc prof, 84-86, PROF BIOL, KY WESLEYAN COL, 87- *Mem:* Am Soc Parasit; Am Micros Soc; Am Soc Microbiol; Am Inst Biol Sci. *Res:* Helminth biology, in particular, taxonomy; morphology and physiology of the Acanthocephala. *Mailing Add:* Dept Biol Ky Wesleyan Col Box 60 Owensboro KY 42302-1039

OETKING, PHILIP, b Madison, Wis, Mar 27, 22; m 45; c 2. MARINE GEOLOGY. *Educ:* Univ Wis, PhB, 46, MS, 48, PhD, 52. *Prof Exp:* Asst geol, Univ Wis, 46-50; res geologist, Sun Oil Co, 51-57; geological consult, 57-61; explor geologist, Scott Hammonds, oil producer, 61-62; sr scientist lunar geol, Chance Vought Corp, 62-63; res scientist, Grad Res Ctr, Southwest, 63-67; dir, Ocean Sci & Eng Lab, Southwest Res Inst, 67-76; DIR PLANNING & ENVIRON AFFAIRS, PORT OF CORPUS CHRISTI, 76- *Mem:* AAAS; Am Asn Petrol Geol; Am Asn Port Authorities. *Res:* Lunar and planetary geological research; light reflectivity measurements; preparation of regional geological highway maps of the United States; Gulf Coast stratigraphic, structural and sedimentation problems; coastal zone, estuarine and Gulf of Mexico environmental geology. *Mailing Add:* 409 Yacht Club Dr Rockwall TX 75087

OETTGEN, HERBERT FRIEDRICH, b Cologne, Ger, Nov 22, 23; m 57; c 3. MEDICINE, IMMUNOLOGY. *Educ:* Univ Bonn, MD, 51; Univ Koeln, MD, 51. *Prof Exp:* Intern, Red Cross Hosp, Neuwied, Ger, 51; resident, dept path, City Hosp, Koln, Ger, 52-54, dept med, 55-58, chief dept med, 62-63; asst prof med, Univ Koeln, Ger, 62-69, appl prof clin & exp cancer res, 69-74; prof immunobiol, Univ Frieburg, Ger, 73-75; from instr to assoc prof med, Cornell Univ Med Col, 66-72, from asst prof to assoc prof biol, 66-81, PROF IMMUNOL, CORNELL UNIV GRAD SCH MED SCIS, 82-, PROF MED, CORNELL UNIV MED COL, 72- *Concurrent Pos:* Vis res fel, Sloan-Kettering Inst, 58-60, res assoc, 60-62, consult, 62-67, assoc mem, 67-69, mem, 70-; fel Dept Med, Mem Hosp, 58-62, clin asst, Dept Med, 65-67, asst attend physician, 67-69, assoc attend physician, 69-71, attend physician & chief, Clin Immunol Serv, 71-89, attend physician, Clin Immunol Serv, 89-, attend physician, Thoraic Oncol Serv, Mem Hosp, 90-; res gynec & obstet, Sch Med, Univ Marburg, 54-55; pvt docent, Univ Cologne, 62-69, vis prof, 69-; clin asst, Mem Hosp Cancer & Allied Dis, 65-67, asst attend physician, 67-69, assoc attend physician, 69-70, attend physician, 70-; asst prof med, Col Med, Cornell Univ, 67-70; mem cancer res ctr rev comt, NIH. *Honors & Awards:* Wilhelm Warner Award for Cancer Res, 70; Lisec-Artz Award, Cancer Res, Univ Bonn, 82. *Mem:* Am Asn Cancer Res; Am Fedn Clin Res; Am Soc Hemat; NY Acad Sci; Am Soc Clin Oncol; Int Soc Hematol; Am Asn Immunologists; fel Am Col Physicians; Harvey Soc; Europ Asn Cancer Res. *Res:* Cancer immunology; clinical oncology. *Mailing Add:* Mem Sloan-Kettering Cancer Ctr 1275 York Ave New York NY 10021

OETTING, FRANKLIN LEE, b Pueblo, Colo, June 21, 30; m 56; c 4. PHYSICAL CHEMISTRY, THERMODYNAMICS. *Educ:* Univ Colo, BS, 52, MS, 54; Univ Wash, Seattle, PhD(phys chem), 60. *Prof Exp:* Instr chem, Regis Col, 54-56; phys chemist, Dow Chem Co, Mich, 60-63, from res chemist to sr res chemist, Rocky Flats Div, Golden, 63-68, assoc scientist, 68-74; mem dept chem, Int Atomic Energy Agency, Vienna, Austria, 74-76; MEM STAFF, ATOMICS INT DIV, ROCKY FLATS PLANT, ROCKWELL INT, 76- *Concurrent Pos:* Mem, Calorimetry Conf, 61- *Mem:* AAAS; Am Chem Soc; Sigma Xi. *Res:* High temperature and isothermal calorimetry; compilation of thermodynamic data. *Mailing Add:* 4125 Eutaw Dr Boulder CO 80303

OETTING, ROBERT B(ENFIELD), b Lee's Summit, Mo, Aug 5, 33; m 60; c 2. AEROSPACE & MECHANICAL ENGINEERING, EDUCATIONAL ADMINISTRATION. *Educ:* Mo Sch Mines, BS, 55; Purdue Univ, MS, 57; Univ Md, PhD(mech eng), 64. *Prof Exp:* Grad res asst aeronaut eng, Purdue Univ, 55-57; instr mech eng, Mo Sch Mines, 58-59 & Univ Md, 59-64; assoc prof, 64-69, PROF MECH & AEROSPACE ENG, UNIV MO-ROLLA, 69- *Concurrent Pos:* Dir, Continuing Educ, Univ Mo-Rolla, 90- *Mem:* Assoc fel Am Inst Aeronaut & Astronaut; Am Soc Eng Educ; Soc Automotive Engrs. *Res:* Gas dynamics; airplane performance, including stability and control; flight simulation. *Mailing Add:* Mech & Aerospace Eng Univ of Mo Rolla MO 65401

OETTING, WILLIAM STARR, b Oak Park, Ill, Apr 16, 55. FLOW-CYTOMETRY, TWO-DIMENSIONAL ELECTROPHORESIS. *Educ:* Univ Kans, BS, 77; Univ Nebr, MS, 79, PhD(genetics), 83. *Prof Exp:* Fel, Univ Calif, Riverside, 83-84; res assoc genetics, Univ Nebr, 84-86; RES ASSOC, CIT INST OF HUMAN GENETICS, 86- *Mem:* Genetic Soc Am; Int Pigment Cell Soc. *Res:* Using melanocytes to study the control and regulation of genes responsible for pigment formation; recombinant DNA methodology, including insertional mutagenesis and 2-dimensional electrophoresis. *Mailing Add:* Dept Med Div Genetics Box 485 UMHE Univ Minn 420 Delaware St SE Minneapolis MN 55455

OETTINGER, ANTHONY GERVIN, b Nuremberg, Ger, Mar 29, 29; nat US; m 54; c 2. APPLIED MATHEMATICS, LINGUISTICS. *Educ:* Harvard Univ, AB, 51, PhD(appl math), 54. *Hon Degrees:* DLitt, Univ Pittsburgh, 84. *Prof Exp:* NSF fel & res fel appl math, 54-55, instr, 55-56, from asst prof to assoc prof appl math & ling, 57-63, prof ling, 63-75, GORDON MCKAY PROF APPL MATH, HARVARD UNIV, 63-, PROF INFO RESOURCES POLICY, 75-, CHMN, PROG INFO RESOURCES POLICY, 73- *Concurrent Pos:* Consult, Arthur D Little, Inc, Cambridge, Mass, 56-80; consult off sci & technol, Exec Off of President of US, 61-73; consult, Bellcomm, Inc, Washington, DC, 63-67; mem, Coun Foreign Rel; mem adv comt automatic lang processing, Nat Acad Sci-Nat Res Coun, Washington, DC, 64-66, chmn comput sci & eng bd, 67-73; mem res adv comt, Syst Develop Corp, Santa Monica, Calif, 65-68; res assoc prog technol & soc, Harvard Univ, 66-72; chmn bd trustees, Lang Res Found, Cambridge, Mass, 70-75; consult comt automation opportunities serv areas, Fed Coun Sci & Technol, 71-72; mem CATV comn, Commonwealth Mass, 72-75, chmn, 75-79, chmn comt regulation, 72- assoc univ sem comput & relation to man & soc, Columbia Univ, 73-78; adv subcomt econ & soc impact of new broadcast media, Comt Econ Develop, New York, 73-75; consult, Nat Security Coun, 75-81; consult, Pres, Foreign Intel Adv Bd, 81. *Mem:* Fel AAAS; Asn Comput Mach; fel Am Acad Arts & Sci; fel Inst Elec & Electronics Engrs; Sigma Xi. *Res:* Information resources policy automatic information processing systems, design and applications, educational technology and policy, programming theory and information sciences. *Mailing Add:* Aiken Comput Lab Harvard Univ Cambridge MA 02138

OETTINGER, FRANK FREDERIC, b New York, NY, Aug 6, 40; m 63; c 3. SEMICONDUCTOR ELECTRONICS. *Educ:* Pratt Inst, BEE, 63; NY Univ, MSEE, 66. *Prof Exp:* Electronics engr semiconductor measurement develop, Appl Sci Lab, US Navy, 63-67; proj leader, Electron Devices Sect, 67-73, chief, 73-80, dep chief, Semiconductor Devices & Circuits Div, 80-85, dep chief, Semiconductor Electronics Div, 85-86, CHIEF, SEMICONDUCTOR ELECTRONICS DIV, NAT INST STANDARDS & TECHNOL, 86- *Honors & Awards:* Silver Medal, Dept Com, 76; Edward Bennett Rosa Award, Nat Inst Standards & Technol, 90. *Mem:* Fel Inst Elec & Electronics Engrs; Int Electronics Packaging Soc. *Res:* Development of improved measurement methods and associated technology to enhance the performance, interchangeability, and reliability of semiconductor devices and integrated circuits; physical electronics; development of techniques to thermally characterize VLSI packages. *Mailing Add:* Semiconductor Electronics Div Bldg 225 Rm B344 Nat Inst Standards & Technol Gaithersburg MD 20899

OETTINGER, PETER ERNEST, b Antwerp, Belg, Apr 14, 37; US citizen; m 67; c 4. PHOTOEMISSION, LASERS. *Educ:* Cornell Univ, BAgrE, 59; Calif Inst Technol, MS, 60; Stanford Univ, PhD(aeronaut & astronaut), 66. *Prof Exp:* Res engr, Lockheed Missile & Space Co, 60-63; res assoc & lectr, Dept Physics & Aero Eng Sci, Joint Inst Lab Astrophys, Univ Colo, 66-68; mem tech staff, Aerospace Corp, 68-71; sr laser scientist, Ctr Res Plasma Physics, Switz, 71-72; dir, Laser Lab, Thermoelectron Corp, 72-88; VPRES & CHIEF OPERS OFFICER, PHOTOELECTRON CORP, 89- *Mem:* Assoc fel Am Inst Aeronaut & Astronaut; Am Phys Soc; Optical Soc Am; Am Soc Mech Engrs; AAAS; Am Electronics Asn. *Res:* Development of high brightness laser-activitated thermionic- and photoemissive-electron sources; E-beam lithography; high definition television; electron microscopes; medical x-ray devices. *Mailing Add:* Four Phlox Lane Acton MA 01720

O'FALLON, JOHN ROBERT, b Princeton, Minn, Aug 20, 37; m 62; c 3. PHYSICS. *Educ:* St John's Univ, BS, 59; Univ Ill, Urbana, MS, 61, PhD(physics), 65. *Prof Exp:* Res assoc high energy physics, Univ Mich, 65-67; from asst prof to prof physics, St Louis Univ, 67-78; asst to pres, Argonne Univ Asn, 76-77, admin dir, 77-79, vpres admin, 79-; ASST VPRES ADVAN STUDIES, UNIV NOTRE DAME. *Concurrent Pos:* Guest scientist, Argonne Nat Lab, 65-81 & Lawrence Berkeley Lab, 70-71; vis scientist, Brookhaven Nat Lab, 66-68 & 80- *Mem:* Am Phys Soc; Sigma Xi; Am Chem Soc. *Res:* Experimental high energy physics; proton-proton and proton-neutron scattering with polarized beams and polarized targets. *Mailing Add:* Div High Energy Physics ER-223-GTN US Dept Energy Washington DC 20545

O'FALLON, NANCY MCCUMBER, b Jackson, Miss, Oct 25, 38; m 62; c 3. RESEARCH ADMINISTRATION, INSTRUMENTATION. *Educ:* St Louis Univ, BS, 60; Univ Ill, Urbana, MS, 61, PhD(physics), 66. *Prof Exp:* Vis asst prof physics, Univ Mo, St Louis, 72-74; asst physicist, Appl Physics Div, 74-76, physicist, 76-83, prog mgr instrumentation & control fossil energy, 78-83, ASST VPRES RES, ARGONNE NAT LAB, UNIV CHICAGO, 83- *Mem:* Asn Women Sci (treas, 84-85); Am Phys Soc; AAAS. *Res:* Development of instruments based on nuclear, acoustic, optical and other advanced techniques for process control in large-scale coal gasification, liquefaction, and fluidized-bed combustion. *Mailing Add:* US Arms Cent & Disarmament Agency SP/SA 4494 320 21st St NW Washington DC 20451

O'FALLON, WILLIAM M, b Princeton, Minn, Mar 7, 34; m 61; c 2. BIOSTATISTICS, MATHEMATICAL STATISTICS. *Educ:* St John's Univ, Minn, BA, 56; Vanderbilt Univ, MAT, 57; Univ NC, PhD(statist), 67. *Prof Exp:* Instr math, St John's Univ, Minn, 57-60; res assoc biostatist, Univ NC, Chapel Hill, 63-67; from asst prof to assoc prof community health sci, Med

Ctr, Duke Univ, 66-74, asst prof math, 66-74; Nat Cancer Inst fel, 74-75; assoc prof, 78-83, CONSULT BIOSTATIST & HEAD SECT BIOSTATISTICS, MAYO CLIN, 75-, PROF BIOSTATIST, MAYO MED SCH, 84- *Mem:* AAAS; Math Asn Am; Am Statist Asn; Biomet Soc; fel Am Statist Asn; Soc Epidemiol Res; Am Pub Health Asn. *Res:* Mathematical and stochastic models of biological phenomena; multivariate analysis of biomedical data; statistical analysis of epidemiological data. *Mailing Add:* Sect Biostatist Mayo Clin Rochester MN 55905

O'FARRELL, CHARLES PATRICK, b Elizabeth, NJ, Oct 30, 37; m 61; c 1. RUBBER CHEMISTRY. *Educ:* St Peter's Col, NJ, BS, 60; Rutgers Univ, PhD(phys & org chem), 65. *Prof Exp:* Res chemist, Esso Res & Eng Co, 64-66; head group mkt tech serv & tire applications, Exxon Chem Co, 77-78, res assoc rubber chem, 68-87; VPRES RES & DEVELOP, SID RICHARDSON CARBON GASOLINE CO, 87- *Concurrent Pos:* Planning assoc, Elastomers Dept, Exxon Chem Americas, 79-85. *Mem:* Am Chem Soc. *Res:* Artificial lattices produced from elastomeric polymers. *Mailing Add:* Sid Richardson Carbon Gasoline Co First City Bank Tower 201 Main St Ft Worth TX 76102

O'FARRELL, MICHAEL JOHN, b Los Angeles, Calif, July 4, 44; m 65; c 3. MAMMALIAN ECOLOGY, BEHAVIORAL BIOLOGY. *Educ:* Univ Nev, Las Vegas, BS, 68; NMex Highlands Univ, MS, 71; Univ Nev, Reno, PhD(zool), 73. *Prof Exp:* Fel & res assoc mammal, Savannah River Ecol Lab, Univ Ga, 73-74; res assoc desert ecol, Univ Nev, Las Vegas, 74-75, asst prof biol, 75-76; cur res, Zool Soc Nev, Las Vegas, 76-; mem staff, Westek Serv, Inc, 89; PRIN ECOLOGIST, O'FAFFELL BIOL CONSULT, 89- *Mem:* Am Soc Mammalogists; Ecol Soc Am; Animal Behav Soc; Sigma Xi. *Res:* Small mammal community ecology; coexistence and social structure in natural populations, temperature regulation and body composition; behavioral ecology. *Mailing Add:* 2912 N Jones Blvd Las Vegas NV 89108

O'FARRELL, THOMAS PAUL, b Chicago, Ill, Mar 18, 36; m 58. MAMMALIAN ECOLOGY. *Educ:* St Mary's Col, Minn, BS, 58; Univ Alaska, MS, 60; Univ Tenn, PhD(zool), 65. *Prof Exp:* Res asst biol, St Mary's Col, Minn, 55-58; res asst wildlife, Dept Wildlife Mgt, Univ Alaska, 58-60; wildlife biologist, Alaska Dept Fish & Game, 60-61; teaching asst zool, Univ Tenn, 61-63; AEC fel ecol, Univ Wash, 65-67; res scientist, Pac Northwest Labs, Battelle Mem Inst, Wash, 67-68, mgr terrestrial ecol sect, 68-71, prog leader, 71-73, sr res scientist, 73-74; assoc res prof, Desert Res Inst, Univ Nev Syst, 74-75, dir, Appl Ecol & Physiol Ctr, 75-78; mgr, Bioenviron Studies Sect, EG&G Energy Measurements, Inc, Santa Barbara, Calif, 78-87; MGR, ENVIRON STUDIES PROJ, 87- *Mem:* Fel AAAS; Am Soc Mammalogists; Ecol Soc Am. *Res:* Mammalian ecology, particularly as affected by ionizing rqdiation; effects of petroleum production activities on flora and fauna; ecological life histories of mammals; physiological ecology; arid lands ecology; environmental impact assessment; endangered species. *Mailing Add:* 611 Ave H Boulder City NV 89005

OFELT, GEORGE STERLING, b Washington, DC, Jan 22, 37; m 62; c 2. SPECTROSCOPY. *Educ:* Col William & Mary, BS, 57; Johns Hopkins Univ, PhD(physics), 62. *Prof Exp:* Instr physics, Johns Hopkins Univ, 62-63; asst prof, Col William & Mary, 63-67; ASSOC PROF OCEANOG, OLD DOM UNIV, 67- *Concurrent Pos:* Consult & lectr, NASA, 63- *Mem:* Optical Soc Am; Am Chem Soc. *Res:* Ultraviolet and visible spectroscopy; optical oceanography. *Mailing Add:* Old Dominion Col Oceanog Bldg Rm 101 Norfolk VA 23508

OFENGAND, EDWARD JAMES, b Taunton, Mass, Aug 15, 32; m 63; c 3. RIBOSOMES, TRANSFER RNA. *Educ:* Mass Inst Technol, BS & MS, 55; Wash Univ, St Louis, PhD(microbiol), 59. *Prof Exp:* NSF fel, Med Res Coun, Cambridge, Eng, 59-61; Rockefeller Inst fel, 61-62, NIH fel, 62; lectr biochem sch med, Univ Calif, San Francisco, 62-67, asst prof, 67-69; assoc mem, 69-77, MEM, ROCHE INST MOLECULAR BIOL, 78- *Concurrent Pos:* Adj prof microbiol, NJ Med Sch, 83-; mem, Physiol Chem Study Sect, NIH, 83-87. *Mem:* AAAS; Am Soc Biol Chem; Am Chem Soc; NY Acad Sci. *Res:* Structure and function of the ribosome as studied by site-specific mutagenesis of ribosomal RNA and by photoaffinity labelling with probe-modified transfer RNA; modified nucleotides in ribosomes. *Mailing Add:* Dept of Biochem Roche Inst of Molecular Biol Nutley NJ 07110

OFFEN, GEORGE RICHARD, US citizen. MECHANICAL ENGINEERING, ENVIRONMENTAL ENGINEERING. *Educ:* Stanford Univ, BS, 61, PhD(mech eng), 73; Mass Inst Technol, MS, 62. *Prof Exp:* Test engr conv bombs, Air Proving Ground Command, Eglin AFB, 62-65; res engr oil field reservoir eng, Chevron Res Co, Standard Oil Calif, 65-66 & France petrol Inst, 66-67; teaching & res asst mech eng, Stanford Univ, 68-73; actg asst prof, Santa Clara Univ, 73-74; staff engr, sect leader & dep mgr eng, Energy & Environ Div, Acurex Corp, 74-85; PROG MGR, AIR QUAL CONTROL, ELEC POWER RES INST, 85- *Mem:* Am Soc Mech Engrs. *Res:* Air pollution control technologies; combustion and postcombustion nitrogen oxide control, particulate control, dry sulphur dioxide control; management systems for environmental programs; management. *Mailing Add:* Elec Power Res Inst 3412 Hillview Ave Palo Alto CA 94303

OFFEN, HENRY WILLIAM, b Uelzen, Ger, Apr 28, 37; US citizen; m 61; c 2. HIGH PRESSURE SPECTROSCOPY. *Educ:* St Olaf Col, BA, 58; Univ Calif, Los Angeles, PhD(chem), 63. *Prof Exp:* Instr chem, Occidental Col, 62-63; from asst prof to assoc prof, 63-73, dean, Off Res Develop, 72-78, PROF PHYS CHEM, UNIV CALIF, SANTA BARBARA, 73- *Mem:* AAAS; Am Phys Soc; Am Chem Soc; Sigma Xi. *Res:* High pressure spectroscopy; aqueous solutions; organic solids; inorganic phosphors. *Mailing Add:* Dept of Chem Univ of Calif Santa Barbara CA 93106

OFFENBACHER, ELMER LAZARD, b Frankfurt am Main, Ger, Sept 29, 23; US citizen; m 52; c 3. PHYSICS. *Educ:* Brooklyn Col, AB, 43; Univ Pa, MS, 49, PhD(physics), 51. *Prof Exp:* Asst instr physics, Amherst Col, 43-44; physicist, Nat Adv Comt Aeronaut, Va, 44; asst instr physics, Univ Pa, 47-51; from asst prof to assoc prof, 51-65, PROF PHYSICS, TEMPLE UNIV, 65- *Concurrent Pos:* Physicist, Frankford Arsenal, US Dept Army, 53-54, consult, 54- *Mem:* Am Phys Soc; Am Asn Physics Teachers. *Res:* Solid state physics; ice physics; biomechanics; electron spin resonance. *Mailing Add:* Dept of Physics Temple Univ Broad & Montgomers Philadelphia PA 19122

OFFENBERGER, ALLAN ANTHONY, b Lintlaw, Sask, Aug 11, 38; m 63; c 2. PLASMA PHYSICS, LASERS. *Educ:* Univ BC, BASc, 62, MASc, 63; Mass Inst Technol, PhD(nuclear eng), 68. *Prof Exp:* From asst prof to assoc prof elec eng, 68-75, PROF ELEC ENG, UNIV ALTA, 75-, DIR, ALTECH, 84- *Concurrent Pos:* Killam res fel, 80-82. *Mem:* Can Asn Physicists (vpres, 88-89, pres, 89-90); Am Phys Soc; Sigma Xi. *Res:* High power KrF laser systems for fusion research; nonlinear optics for laser pulse compression; laser/plasma interaction; laser induced parametric instabilities; nonequilibrium plasma physics; transport. *Mailing Add:* Dept of Elec Eng Univ of Alta Edmonton AB T6G 2G7 Can

OFFENHARTZ, EDWARD, acoustical & mechanical engineering, for more information see previous edition

OFFENHAUER, ROBERT DWIGHT, b Sandusky, Ohio, June 24, 18; m 42; c 3. ORGANIC CHEMISTRY. *Educ:* DePauw Univ, AB, 40; Univ Wis, PhD(org chem), 44. *Prof Exp:* Res chemist, Allied Chem & Dye Corp, 44-48; chemist, 48-59, RES ASSOC LABS, MOBIL OIL CORP, 59- *Mem:* Am Chem Soc. *Res:* Petrochemicals; microbiology; petroleum chemistry; high polymers; textiles and organic synthesis; cyclohexanoic acid and derivatives; homogeneous catalysis; micropropagation of heliconia. *Mailing Add:* 6430 Sun Eagle Lane No 305 Bradenton FL 34210-4100

OFFENKRANTZ, WILLIAM CHARLES, b Newark, NJ, Sept 2, 24; m 85; c 2. PSYCHIATRY, PSYCHOANALYSIS. *Educ:* Rutgers Univ, BS, 45; Columbia Univ, MD, 47; William Alanson White Inst, NY, dipl psychoanal, 57; Chicago Inst Psychoanal, dipl, 66. *Prof Exp:* Prof psychiat, Univ Chicago, 57-79; PROF PSYCHIAT & PSYCHOANAL, MED COL WISCONSIN, 79- *Mem:* Am Psychoanal Asn; Am Psychiat Asn; Group Advan Psychiat. *Res:* Clinical research in psychoanalysis; investigating relationships among dreams of the night. *Mailing Add:* 6623 N Scottsdale Rd Scottsdale AZ 85250

OFFER, DANIEL, b Berlin, Ger, Dec 24, 29; US citizen; m 71; c 2. PSYCHIATRY. *Educ:* Univ Rochester, BA, 53; Univ Chicago, MD, 57. *Prof Exp:* Dir residency prog & assoc dir hosp, 66-73, co-chmn, Dept Psychiat, 74-77, DIR, LAB STUDY ADOLESCENTS, MICHAEL REESE HOSP, 68-, CHMN, DEPT PSYCHIAT, 77-; PROF PSYCHIAT, PRITZKER SCH MED, UNIV CHICAGO, 73- *Concurrent Pos:* Consult, Dept Ment Health, State of Ill, 64-; mem exec comt, Nicholas Pritzker Ctr & Hosp, 68-73; mem prof adv comt, Family Inst Chicago, 69-73; ed in chief, J Youth & Adolescence, 70-; mem prof adv coun, Ment Health Div, Chicago Bd Health, 70-73; fel, Ctr Advan Study Behav Sci, Stanford, Calif, 73-74. *Mem:* AAAS; fel Am Psychiat Asn; AMA; Int Cong Child Psychiat & Allied Profs; Am Soc Adolescent Psychiat (pres, 72-73). *Res:* Study of normal adolescent boys from teenage to young manhood; analysis of data collected over the past few years on juvenile delinquents. *Mailing Add:* Dept Psychiat Univ Chicago Pritzker Sch Med 5841 Maryland Ave Chicago IL 60637

OFFIELD, TERRY WATSON, b Amarillo, Tex, May 27, 33; m 57; c 2. GEOLOGY. *Educ:* Va Polytech Inst, BS, 53; Univ Ill, MS, 55; Yale Univ, PhD(geol), 62. *Prof Exp:* Geologist, Aluminum Co Am, 55; geologist, NY State Geol Surv, 57-59; geologist, US Geol Surv, Ariz, 61-69, coordr uranium geophys prog, 74-77, asst chief, Br Petrophys & Remote Sensing, 75-77, chief br uranium & thorium resources, 77-81, chief off energy & marine geol, 82-87, GEOLOGIST, US GEOL SURV, COLO, 69- *Concurrent Pos:* US govt sr exec serv, 82-87. *Honors & Awards:* Meritorious Serv Award, Dept Interior. *Mem:* AAAS; Geol Soc Am; Am Geophys Union; Sigma Xi. *Res:* Structural geology; Himalayan structure; lunar geology; terrestrial impact structures; remote sensing techniques for geologic mapping and uranium exploration; thermal-infrared studies; uranium geology; energy and marine geology; SE piedmont structural geology. *Mailing Add:* US Geol Surv 954 Nat Ctr Reston VA 22092

OFFNER, FRANKLIN FALLER, b Chicago, Ill, Apr 8, 11; m 56; c 4. BIOPHYSICS. *Educ:* Cornell Univ, BChem, 33; Calif Inst Technol, MS, 34; Univ Chicago, PhD(physics), 38. *Prof Exp:* Asst, Univ Chicago, 35-38; pres, Offner Electronics, Inc, Ill, 39-63; PROF BIOPHYS, NORTHWESTERN UNIV, ILL, 63- *Honors & Awards:* Centennial Medal, Inst Elec & Electronics Engrs. *Mem:* Nat Acad Eng; fel Inst Elec & Electronics Engrs; Am Phys Soc; Am Electroencephalog Soc; Biophys Soc. *Res:* Theory of the excitable membrane. *Mailing Add:* Biomed Eng Dept Northwestern Univ Evanston IL 60207

OFFNER, GWYNNETH DAVIES, b Cleveland, Ohio, Sept 6, 55. BIOCHEMISTRY, PLANT PHYSIOLOGY. *Educ:* Wellesley Col, BA, 76; Sch Med, Boston Univ, PhD(biochem), 84. *Prof Exp:* RES ASSOC, SCH MED, BOSTON UNIV, 84- *Mem:* AAAS. *Res:* Structure and function of proteins; protein sequence. *Mailing Add:* Dept Biochem Boston Univ Sch Med 80 E Concord St Boston MA 02118-2394

OFFORD, DAVID ROBERT, b Ottawa, Ont, Nov 13, 33; m 62; c 3. CHILD PSYCHIATRY. *Educ:* Queens Univ, Ont, MDCM, 57; CRCP(C), 65. *Prof Exp:* Jr rotating intern, Montreal Gen Hosp, 57-58, resident staff psychiat, 58-59 & 59-60; fel child psychiat, Children's Serv Ctr Wyo Valley, Wilkes-Barre, Pa, 60-62; from instr to asst prof psychiat, Div Child Psychiat, Univ Fla, 62-67; assoc prof behav sci, Col Med, Pa State Univ, 67-72; prof psychiat, Univ Ottawa, 72-78; staff psychiatrist, Royal Ottawa Hosp, 72-78, DIR RES & TRAINING CHILDREN'S SERV, 75-; assoc prof behav sci, Milton S Hershey Med Ctr, Pa State Univ, 67-72; PROF PSYCHIAT, MCMASTER UNIV HOSP, 78-; HEAD DIV CHILD PSYCHIAT, MCMASTER UNIV, 78-; CHILD PSYCHIAT, SHEDOKE HOSP, 78- *Concurrent Pos:* Examr psychiat, Royal Col Physicians & Surgeons Can, 75- *Mem:* AMA; Am

Psychiat Asn; Am Acad Child Psychiat; Am Orthopsychiat Asn; Am Pub Health asn. *Res:* The study of the natural histories, including childhood antecedents of the severe psychosocial diseases of adulthood, especially schizophrenia, sociopathy, alcoholism, affective disorder and retardation. *Mailing Add:* Chedoke-McMaster Hosps Box 2000 Sta A Hamilton ON L8N 3Z5 Can

OFFUTT, WILLIAM FRANKLIN, b Slippery Rock, Pa, Feb 24, 19; m 53; c 3. OPERATIONS RESEARCH. *Educ:* Grove City Col, BSc, 40; Univ Pittsburgh, PhD(phys chem), 48. *Prof Exp:* Physicist div war res, Columbia Univ, 42-43; mem staff opers eval group, Off Chief Naval Opers, US Dept Navy, 48-54; sr scientist, E H Smith & Co, 54-56; mem staff weapons systs eval div, Inst Defense Anal, 56-60, sr res assoc, 65-66; sr scientist, Fed Systs Div, IBM Corp, 60-81; RETIRED. *Mem:* Fel AAAS; Opers Res Soc Am (treas, 56-59). *Res:* Thermodynamics of solutions; operations analysis. *Mailing Add:* 11009 Rosemont Dr Rockville MD 20852

O'FLAHERTY, LARRANCE MICHAEL ARTHUR, b Wynyard, Sask, June 14, 41; US citizen; m 64. ALGOLOGY. *Educ:* Western Wash State Col, BA, 63; Ore State Univ, MA, 66, PhD(bot, phycol), 68. *Prof Exp:* From asst prof to assoc prof, 69-79, PROF BIOL, WESTERN ILL UNIV, 79- *Mem:* Am Bryol & Lichenological Soc; Phycol Soc Am; Int Phycol Soc; Am Soc Limnol & Oceanog. *Res:* Blue-green algae in culture; phytoplankton ecology. *Mailing Add:* Dept Biol Sci Western Ill Univ Macomb IL 61455

OFNER, PETER, b Berlin, Ger, June 21, 23; US citizen. ORGANIC CHEMISTRY, ANDROLOGY. *Educ:* Univ London, BSc, 45, PhD(org chem), 50. *Prof Exp:* Res chemist, Wellcome Found, Burroughs Wellcome & Co, Eng, 46-50; asst res chemist, Courtauld Inst Biochem, Middlesex Hosp, London, 52-53; res biochemist, Nat Inst Med Res, London, 53-54 & Hormone Res Lab, Sch Med, Boston Univ, 55-57; CHIEF STEROID BIOCHEM LAB, LEMUEL SHATTUCK HOSP, 57- *Concurrent Pos:* Econ Coop Admin fel enzymol, Univ Chicago, Sloan-Kettering Inst Cancer Res & Univ Utah, Steroid Chem & Biochem, 50-52; res fel med, Harvard Univ, 58-60; res assoc med, Harvard Univ, 60-62 & Sch Dent Med, 63-, asst prof, 69-73, lectr pharm, 73-79, lectr toxicol, Sch Pub Health, 79-; pres, Cancer Res Asn, Boston, 85-86; lectr, Sch Med, Tufts Univ, 69-73, assoc prof urol, 73-; assoc prof, Anat Cellularbiol, 84- *Mem:* Am Chem Soc; Royal Soc Chem; Endocrine Soc; Soc Andrology; Sigma Xi; Asn Toxicol Soc. *Res:* Effects and disposition of androgens and estrogens in male accessory sex organs; prostate tissue culture, hormonal carcinogenesis. *Mailing Add:* 602 Belmont St Watertown MA 02172

O'FOGHLUDHA, FEARGHUS TADHG, b Dublin, Ireland, Jan 8, 27; m 56; c 2. MEDICAL PHYSICS. *Educ:* Nat Univ Ireland, BSc, 48, MSc, 49, PhD(physics), 61; Am Bd Radiol, cert, 65. *Prof Exp:* Asst physics, Univ Col, Dublin, 50-54; sr physicist, St Luke's Hosp, Dublin, 54-63; assoc prof radiation physics & asst prof biophysics, Med Col Va, Richmond, 63-65, prof & chmn div, 65-70; prof & dir div, 70-88, EMER PROF RADIATION PHYSICS, DUKE UNIV MED CTR, DURHAM, NC, 88-; VPRES, QUANTUM RES SERV, INC, DURHAM, NC, 88- *Concurrent Pos:* Consult, McGuire Vet Admin Hosp, Richmond, Va, 63-70; hon vis physicist, St Anne's Hosp, Dublin, 67-75; vis lectr, Univ Col Dublin, 58-63; adj prof physics, Duke Univ, 75-; res assoc, Univ Uppsala, 51, Univ Lund, 52, Univ Mich, Ann Arbor, 57; sr vis scientist, Univ Lund, 61; ed, Med Physics, 85-87; NAm ed, Physics Med & Biol, 76-79; vis scientist, Oak Ridge Inst Nuclear Studies, 56; vpres, Dosimetrix, Inc, Durham, NC, 85-; adj prof radiation ther, E Carol Univ, Greenville, NC, 90- *Mem:* Am Phys Soc; fel Am Asn Physicists Med (pres, 71); Am Col Radiol; Brit Inst physics (fel Phys Soc). *Res:* Ionizing radiation; gamma ray spectrometry. *Mailing Add:* 1513 Pinecrest Rd Durham NC 27705

OFSTEAD, EILERT A, b Minneapolis, Minn, Dec 15, 34; m 62; c 3. POLYMER CHEMISTRY. *Educ:* St Thomas Col, BS, 56; Univ Md, PhD(org chem), 63. *Prof Exp:* Am Dent Asn res asst dent, Nat Bur Standards, 57-58; SR RES & DEVELOP ASSOC, GOODYEAR TIRE & RUBBER CO, 62- *Mem:* Am Chem Soc. *Res:* Reaction mechanisms; ring-opening polymerizations; synthetic elastomers. *Mailing Add:* 610 Brookpark Dr Cuyahoga Falls OH 44223-1497

OFTEDAHL, MARVIN LOREN, b Chicago, Ill, June 21, 31; m 58; c 1. ORGANIC CHEMISTRY. *Educ:* Northwestern Univ, BA, 56; Wash Univ, St Louis, PhD(carbohydrate chem), 60. *Prof Exp:* Sr res chemist org chem div, Indust Chem Co, 60-67, from res specialist to sr res specialist, 67-80; res group leader, Corp Res & Develop Staff, Monsanto Co, 80-83, sr res group leader, Health Care Div, 83-86; LAB DIR, DEPT CHEM, WASH UNIV, ST LOUIS, MO, 86- *Concurrent Pos:* Consult, G D Searle, 86-88, Glycomed, Inc, 89- & Orbital Chemie, 90- *Mem:* Am Chem Soc; Sigma Xi. *Res:* Chemistry of carbohydrates; organosulfur chemistry; oxygen and nitrogen heterocyclics; food and flavor chemistry; pharmaceuticals; pharmaceutical process development. *Mailing Add:* 1511 Andrew Dr St Louis MO 63122

O'GALLAGHER, JOSEPH JAMES, b Chicago, Ill, Oct 23, 39; m 63; c 2. PHYSICS. *Educ:* Mass Inst Technol, SB, 61; Univ Chicago, SM, 62, PhD(physics), 67. *Prof Exp:* Res assoc physics, Enrico Fermi Inst, Univ Chicago, 67-70; asst prof physics, Univ Md, College Park, 71-76; SR RES ASSOC, ENRICO FERMI INST, UNIV CHICAGO, 76- *Concurrent Pos:* Vis fel, Max Planck Inst Physics, 75-76; fel, Alexander Von Humboldt Found, 75; vis scientist, Argonne Nat Lab, 84-; sr lectr & exec officer, dept physics, Univ Chicago, 88- *Mem:* Am Phys Soc; Int Solar Energy Soc; Am Soc Mech Engrs. *Res:* Solar energy; photo-thermal and photovoltaic conversion utilizing non-imaging and non-tracking optical concentrators; technical aspects of alternate and renewable energy sources; space physics; cosmic ray modulation; interplanetary propagation of solar cosmic radiation; detection techniques for energetic charged particles. *Mailing Add:* Enrico Fermi Inst 6640 S Ellis Chicago IL 60637

OGAR, GEORGE W(ILLIAM), b Dorchester, Mass, Nov 24, 18; m 54; c 6. SYSTEMS ANALYSIS, COMPUTER SIMULATION. *Educ:* Col Holy Cross, AB, 40; Boston Col, MA, 41; Harvard Univ, MEngSc, 49. *Prof Exp:* Electronics scientist, Air Force Cambridge Res Ctr, 50-52; electronics engr, Boston Naval Shipyard, 52-54; electronics scientist, Air Force Cambridge Res Ctr, 55; from asst prof to assoc prof elec eng, Air Force Inst Technol, 55-62; sr engr, AC Spark Plug Electronics Div, Gen Motors Corp, 62-64; systs analyst, Dynamics Res Corp, 64-66; PRIN ENGR, RAYTHEON MISSILE SYSTS DIV, 66- *Mem:* Inst Elec & Electronics Engrs. *Res:* Communications; electronic systems; applied mathematics; avionics; radar systems. *Mailing Add:* 50 Apple Hill Dr Tewksbury MA 01876

O'GARA, BARTHOLOMEW WILLIS, b Laurel, Nebr, Mar 21, 23; m 88. ZOOLOGY, FISH & GAME MANAGEMENT. *Educ:* Mont State Univ, BS, 64; Univ Mont, PhD(zool), 68. *Prof Exp:* Leader, 68-78, RES BIOLOGIST, US FISH & WILDLIFE SERV, MONT COOP WILDLIFE RES UNIT, UNIV MONT, 68-, LEADER, 78- *Concurrent Pos:* Affil prof zool, forestry & wildlife biol, Univ Mont, 70-; US Fish & Wildlife Refuge Div grant, Red Rocks Nat Wildlife Refuge, Mont, 71; US Forest Serv grant elk migrations, 73-; res grants coyote-domestic sheep interactions, 74-, coyote-pronghorn interactions, 77, Sheldon Game Range, Nev, 78 & C M Russell Game Range, Mont, 78- *Mem:* Soc Study Reproduction; Am Soc Mammalogists; Wildlife Soc. *Res:* Mammalian reproduction, particularly of game species. *Mailing Add:* Mont Coop Wildlife Res Unit Univ Mont Missoula MT 59812

OGARD, ALLEN E, b Ada, Minn, Dec 9, 31; m 56; c 2. GEOCHEMISTRY & NUCLEAR WASTE. *Educ:* St Olaf Col, BA, 53; Univ Chicago, PhD(chem), 57. *Prof Exp:* Res chemist, Westinghouse Res Labs, 57; staff mem chem, Los Alamos Sci Lab, 57-70; guest scientist, Swiss Inst Reactor Res, 70-72; CHEMIST, LOS ALAMOS NAT LAB, 72- *Concurrent Pos:* Res fel, Inst Transurane, Karlsruhe, Ger, 64-65. *Mem:* Am Chem Soc; Sigma Xi; Am Geophys Union. *Res:* Preparation and properties of high temperature plutonium reactor fuel materials; on-site and laboratory research pertaining to the geochemistry and hydrology of nuclear waste repository areas. *Mailing Add:* 570 Garcia St Santa Fe NM 87501

OGASAWARA, FRANK X, b San Diego, Calif, Nov 10, 13; m 45; c 3. COMPARATIVE PHYSIOLOGY. *Educ:* Univ Calif, BS, 50, PhD, 57. *Prof Exp:* Res physiologist reprod, 55-58, asst prof poultry reprod, 58-66, assoc prof poultry reprod & animal physiol, 66-73, assoc physiologist, Exp Sta, 69-74, prof animal physiol, 73-76, PROF AVIAN SCI, UNIV CALIF, DAVIS, 73-, PHYSIOLOGIST EXP STA, 74- *Mem:* Am Poultry Sci Asn; Am Fertil Soc; World Poultry Sci Asn. *Res:* Poultry reproductive physiology. *Mailing Add:* 630 E St Davis CA 95616

OGATA, AKIO, b Puunene, Hawaii, July 20, 27; m 54; c 2. HYDRAULIC ENGINEERING. *Educ:* Utah Univ, BS, 51; Northwestern Univ, MS, 56, PhD(civil eng), 58. *Prof Exp:* Jr indust engr, Hawaiian Com & Sugar Co, 51-53; hydraul engr, East Maui Irrig Co, 53-55; res assoc fluid mech, Northwestern Univ, 57-58; hydraul engr, US Geol Surv, 58-63, res hydraul engr, 63-72, res hydrologist, Hawaii, 72-77, mem staff, 77-86; CONSULT, 86- *Mem:* Am Geophys Union. *Res:* Transport mechanism in flow of fluids through porous media. *Mailing Add:* 1376 Akiahala St Kailua HI 96734

OGATA, HISASHI, b Tokyo, Japan, June 10, 26; m 58; c 2. NUCLEAR PHYSICS. *Educ:* Tokyo Col Sci, BSc, 55; Tokyo Univ Educ, MSc, 57; Case Western Reserve Univ, PhD(physics), 63. *Prof Exp:* Res assoc nuclear physics nuclear data proj, Nat Acad Sci-Nat Res Coun, 62-63; res assoc nuclear data proj physics div, Oak Ridge Nat Lab, 64-65; asst prof, 65-68, ASSOC PROF NUCLEAR PHYSICS, UNIV WINDSOR, 68- *Concurrent Pos:* Nat Res Coun Can grant, 65-; Ont Dept Univ Affairs res grant, 67-70; vis scientist, Univ Tokyo, 74-75; consult, Cent Res Lab, Hitachi Ltd, Japan, 85-86. *Mem:* Am Phys Soc; Can Asn Physicists. *Res:* Theoretical nuclear physics, particularly nuclear structure and low energy nuclear properties; superconductivity. *Mailing Add:* Dept Physics Univ Windsor Windsor ON N9B 3P4 Can

OGATA, KATSUHIKO, b Tokyo, Japan, Jan 6, 25; m 61; c 1. MECHANICAL ENGINEERING, INSTRUMENTATION. *Educ:* Univ Tokyo, BS, 47; Univ Ill, MS, 53; Univ Calif, PhD(eng sci), 56. *Prof Exp:* Asst combustion, Sci Res Inst, Tokyo, 48-51; fuel engr, Nippon Steel Tube Co, 51-52; from asst prof to assoc prof mech eng, Univ Minn, Minneapolis, 56-60; prof elec eng, Yokohama Nat Univ, 60-61; PROF MECH ENG, UNIV MINN, MINNEAPOLIS, 61- *Concurrent Pos:* Consult, Univac Div, Remington Rand Corp, Minn, 56. *Mem:* Am Soc Mech Engrs; Sigma Xi. *Res:* Automatic controls; nonlinear system analysis; operations research. *Mailing Add:* Mech Engr Univ Minn 307 Mech Engr Minneapolis MN 55455

OGAWA, HAJIMU, b Pasadena, Calif, June 18, 31; m 58; c 2. MATHEMATICS. *Educ:* Calif Inst Technol, BS, 53; Univ Calif, Berkeley, PhD(appl math), 61. *Prof Exp:* Lectr math, Univ Calif, Riverside, 60-61, from asst prof to assoc prof, 61-68; PROF MATH, STATE UNIV NY ALBANY, 68- *Concurrent Pos:* NSF grants, 62-68 & 69-71; vis asst prof, Univ Calif, Berkeley, 63-64; vis mem, Courant Inst Math Sci, NY Univ, 66-67. *Mem:* Am Math Soc; Math Asn Am; Soc Indust & Appl Math. *Res:* Partial differential equations. *Mailing Add:* Dept of Math Statistics State Univ NY 1400 Wash Ave Albany NY 12222

OGAWA, JOSEPH MINORU, b Sanger, Calif, Apr 24, 25; m 54; c 3. DISEASE CONTROL, EPIDEMIOLOGY. *Educ:* Univ Calif, BS, 50, PhD(plant path), 54. *Prof Exp:* Asst specialist plant path, 53-55, lectr, 55-62, from asst prof to assoc prof, 62-68, jr plant pathologist, 55-57, from asst plant pathologist to assoc plant pathologist, 57-68, PROF PLANT PATH & PLANT PATHOLOGIST, UNIV CALIF, DAVIS, 68- *Concurrent Pos:* Mem staff, UN Food & Agr Orgn, 67-68; assoc ed, Phytopath, Plant Dis & Plant Dis Reporter; consult, Nat Inst Agr Technol & Int Confedn Agr Credit, Arg, 72, Univ Tabriz, Iran, 74, Univ Baghdad, Iraq, 79, Inter-Am Inst Coop

Agr EMBRAPA, Brazil, 80, 82 & 84 & Univ Chile, 85; prof, Nat Taiwan Univ, 68; JSPS fel, Hirusaki Univ, Aomori, Japan, 90. *Mem:* AAAS; Am Phytopath Soc; Sigma Xi; fel Am Phytopathol Soc, 86. *Res:* Diseases of deciduous fruit and nut crops; hop diseases; postharvest diseases-fruit crops and tomatoes; fungicides. *Mailing Add:* Dept Plant Path Univ Calif Davis CA 95616

OGBORN, LAWRENCE L, b Richmond, Ind, May 2, 32; m 54; c 1. POWER ELECTRONICS, INSTRUMENTATION. *Educ:* Rose Polytech Inst, BSEE, 54; Purdue Univ, MSEE, 57, PhD(elec eng), 61. *Prof Exp:* Mem tech staff elec eng, Bell Tel Labs, Inc, 54-56; asst, 56-57, from instr to asst prof elec eng, 57-63, assoc prof, 63-78, dir, Elec & Hybrid Vehicle Systs Develop Lab, 78-84, ASSOC PROF ELEC ENG, PURDUE UNIV, W LAFAYETTE, 78-, DIR, LAB PROGS, 87- *Concurrent Pos:* Coordr, Coop Elec Eng Prog, Purdue Univ, W Lafayette, 85- *Mem:* AAAS; Inst Elec & Electronics Engrs; Nat Soc Prof Engrs; Am Soc Eng Educ. *Res:* Electrical properties of materials; electronic circuit analysis and design; instrumentation; power electronics. *Mailing Add:* Dept Elec Eng Purdue Univ West Lafayette IN 47907

OGBURN, CLIFTON ALFRED, b Philadelphia, Pa, Apr 27, 30. IMMUNOLOGY. *Educ:* Univ Pa, PhD(bact), 57. *Prof Exp:* From instr immunol in pediat to asst prof pediat, Sch Med, Univ Pa, 57-69; assoc prof, 69-75, PROF MICROBIOL, MED COL PA, 75-, MEM MED STAFF, 69-, ASSOC DEAN GRAD STUDIES, 76-, ACTG CHMN, 80- *Mem:* Transplantation Soc; Am Asn Immunologists; Am Soc Microbiol; NY Acad Sci. *Res:* Genesis of antibody; serology; transplantation immunity; purification; characterization of interferon tumor immunology. *Mailing Add:* Dept Microbiol & Immunol Med Col Pa 3300 Henry Ave Philadelphia PA 19129

OGBURN, PHILLIP NASH, b Klamath Falls, Ore, Aug 18, 40; m 61; c 2. CARDIOVASCULAR PHYSIOLOGY. *Educ:* Wash State Univ, DVM, 65; Ohio State Univ, PhD(cardiovasc physiol), 71. *Prof Exp:* Intern med & surg, Animal Med Ctr, 65-66; vet pvt pract, 66-67; asst prof, 71-75, ASSOC PROF CARDIOL, COL VET MED, UNIV MINN, ST PAUL, 75- *Concurrent Pos:* Consult, 3M & Daig & Raltech Corp. *Mem:* Am Vet Med Asn; Acad Vet Cardiol; Am Animal Hosp Asn; Am Chem Soc. *Res:* Effects of myocardial hypertrophy on excitation of ventricular myocardium and conduction system; comparative electrophysiology; myocardial disease. *Mailing Add:* C336 Vet Hosp Univ Minn St Paul MN 55108

OGDEN, DAVID ANDERSON, b Westfield, NJ, June 25, 31; m 54; c 4. NEPHROLOGY, INTERNAL MEDICINE. *Educ:* Cornell Univ, BA, 53, MD, 57. *Prof Exp:* From asst prof to assoc prof, Univ Colo, Denver, 63-69; assoc prof, 69-74, PROF MED, MED CTR, UNIV ARIZ, 74-, CHIEF RENAL SECT, 69- *Concurrent Pos:* Am Col Physicians fel, 2nd Med Div, Bellevue Hosp, New York, 59-60, NY Heart Asn fel, 60-61; clin investr, Vet Admin Hosp & Med Ctr, Univ Colo, Denver, 63-66; chief renal sect, Vet Admin Hosps, Denver, Colo, 65-69 & Tucson, Ariz, 69-; mem, Nat Renal Transplant Adv Group, US Vet Admin, 71-77; vpres, Nat Kidney Found, 80-82, pres, 82-84. *Mem:* Int Soc Nephrology; Am Soc Nephrology; Am Soc Artificial Internal Organs; Am Fedn Clin Res. *Res:* Function of the transplanted human kidney; divalent ion metabolism in uremia; problems and innovations in chronic hemodialysis. *Mailing Add:* Health Scis Ctr Univ Ariz Tucson AZ 85724

OGDEN, INGRAM WESLEY, b Henryetta, Okla, Apr 29, 20; m 47; c 3. DENTISTRY. *Educ:* Univ Okla, BS, 41; Univ Mo, DDS, 44; Columbia Univ, cert oral biol, 55. *Prof Exp:* PROF ORAL DIAG, COL DENT, HOWARD UNIV, 66-; MEM STAFF DENT AUXILIARIES, MONTGOMERY COL, 70- *Concurrent Pos:* Ed consult, J Am Dent Asn, 66-; USPHS grant cancer training, Col Dent, Howard Univ, 70-73, USPHS grant dent therapist training, 71-72. *Mem:* Am Acad Oral Path. *Res:* Clinical recognition of oral cancer; improved health care delivery. *Mailing Add:* 9904 Holmhurst Rd Bethesda MD 20817

OGDEN, JAMES GORDON, III, b Martha's Vineyard, Mass, July 6, 28; m 56; c 4. ECOLOGY. *Educ:* Fla Southern Col, BS, 51, BA, 52; Univ Tenn, MS, 54; Yale Univ, PhD(biol), 58. *Prof Exp:* Asst biol, Fla Southern Col, 48-52; lab asst bot, Univ Tenn, 52-54; asst biol, Yale Univ, 54-57; climatologist, Conn Agr Exp Sta, 56; from asst prof to assoc prof bot, Ohio Wesleyan Univ, 58-63, prof & dir radiocarbon dating lab, 63-69; PROF BIOL & DIR RADIOCARBON DATING LAB, DALHOUSIE UNIV, 69- *Concurrent Pos:* Mem comt, NS Environ Control Coun, 73-77; fel, J S Guggenheim Found, 63; pres, Environ Res Asn Ltd, 78- *Mem:* Am Soc Limnol & Oceanog; Am Quaternary Asn. *Res:* Precipitation chemistry; biogeochemistry; biogeography; pollen stratigraphy; paleoecology; climatology; microclimatic ecology; instrumentation for environmental studies; post-glacial history of vegetation and climate. *Mailing Add:* 6259 Coburg Rd Apt 13 Halifax NS B3H 2A2 Can

OGDEN, JOAN MARY, plasma physics, applied mathematics, for more information see previous edition

OGDEN, JOHN CONRAD, b Morristown, NJ, Nov 27, 40; m 69; c 2. MARINE ECOLOGY. *Educ:* Princeton Univ, AB, 62; Stanford Univ, PhD(biol), 68. *Prof Exp:* NIH trainee genetics, Univ Calif, Berkeley, 68-69; Smithsonian Inst fel, Smithsonian Trop Res Inst, CZ, 69-71, Am Philos Soc fel, 71; asst prof marine biol, 71-77, WI Lab, St Croix, VI, 74-77, assoc prof, Rutherford, 77-80, prof marine biol & dir, W Indies Lab, VI, Fairleight Dickinson Univ, Teaneck, NJ, 80-88; DIR, FLA INST OF OCEANOG & PROF BIOL, UNIV S FLA, 88- *Concurrent Pos:* Mem Coastline Zone Comn, VI, 81-88. *Mem:* Am Soc Naturalists; Ecol Soc Am; AAAS; Sigma Xi. *Res:* Ecology and behavior of animals and plants in shallow water marine communities; coastal zone development and management. *Mailing Add:* Fla Inst of Oceanog 830 First St S St Petersburg FL 33701

OGDEN, LAWRENCE, b Maryville, Mo, Nov 9, 19; m 48; c 2. GEOLOGY. *Educ:* Univ Tulsa, BS, 48; Univ Wis, MS, 50; Northwest Mo State Col, BS, 51; Colo Sch Mines, DSc(geol), 58. *Prof Exp:* Instr geol, Colo Sch Mines, 52-58, asst prof, 58-63; assoc prof, 63-66, PROF GEOL, EASTERN MICH UNIV, 66- *Mem:* Geol Soc Am; Nat Asn Geol Teachers. *Res:* Engineering geology; ground water; earth science education. *Mailing Add:* Dept Geog & Geol Eastern Mich Univ Ypsilanti MI 48197

OGDEN, PHILIP MYRON, b Nampa, Idaho, Feb 3, 38; m 62; c 2. PHYSICS. *Educ:* Seattle Pac Col, BS, 59; Univ Calif, PhD(physics), 64. *Prof Exp:* From asst prof to assoc prof physics, Seattle Pac Col, 64-69; assoc prof, 69-76, PROF PHYSICS, ROBERTS WESLEYAN COL, 76-, CHMN DIV NATURAL SCI & MATH, 74- *Mem:* Am Phys Soc; Am Sci Affiliation; Am Asn Physics Teachers. *Res:* High energy or elementary particle physics; cosmic ray physics. *Mailing Add:* Div of Natural Sci & Math Roberts Wesleyan Col 2301 Westside Dr Rochester NY 14624

OGDEN, ROBERT VERL, b Richfield, Utah, Jan 16, 38; m 64; c 3. SENSORY ANALYSIS, FOOD PRODUCT DEVELOPMENT. *Educ:* Utah State Univ, BS, 63, MS, 67; Iowa State Univ, PhD(dairy microbiol), 74. *Prof Exp:* Instr & assoc dairy foods, Dept Food Technol, Iowa State Univ, 66-71; prod develop scientist, Pillsbury Co, 72-77; sr res scientist, W L Clayton Res Ctr, Anderson, Clayton & Co, 77-87; PROD DEVELOP MGR, FRIGO CHEESE CORP, UNIGATE PLC, UK, 87- *Mem:* Am Dairy Sci Asn; Inst Food Technologists. *Res:* Commercial product development of cereal based, dairy, and confectionary products; intermediate moisture food; sensory analysis methods development. *Mailing Add:* Frigo Cheese Corp PO Box 19024 Green Bay WI 54307

OGDEN, THOMAS E, b Lincoln, Nebr, Mar 23, 29; m 75; c 3. PHYSIOLOGY. *Educ:* Univ Calif, BA, 50, MD, 54, PhD(physiol), 62. *Prof Exp:* From res instr to res assoc prof physiol & neurol, Col Med, Univ Utah, 69-72, prof neurol, 72-75; prof physiol, Doheny Eye Found, 75-76, PROF PHYSIOL, UNIV SOUTHERN CALIF, 75- *Mem:* Asn Res Vision Ophthal. *Res:* Retinal neurophysiology and anatomy. *Mailing Add:* Dept of Physiol/ Biophys Univ Southern Calif 2025 Zonal Ave Los Angeles CA 90033

OGDEN, WILLIAM FREDERICK, b Randolph Field, Tex, Sept 11, 42. MATHEMATICS, COMPUTER SCIENCE. *Educ:* Univ Ark, Fayetteville, BS, 64; Stanford Univ, MS, 66, PhD(math), 69. *Prof Exp:* asst prof, 68-76, assoc prof math & comput sci, Case Western Reserve Univ, 76-; AT DEPT COMPUT SCI, OHIO STATE UNIV, COLUMBUS. *Concurrent Pos:* NSF res grant, 71-72. *Mem:* Am Math Soc; Asn Comput Math. *Res:* Automata theory and programming languages. *Mailing Add:* Dept Comput Sci Ohio State Univ 228 Civil Aero Bldg Columbus OH 43210

OGG, ALEX GRANT, JR, b Worland, Wyo, May 3, 41; m 62; c 2. PLANT PHYSIOLOGY, AGRONOMY. *Educ:* Univ Wyo, BS, 63; Ore State Univ, MS, 66, PhD(bot), 70. *Prof Exp:* Aide weed res, Univ Wyo, 59-63; res technician, Wash, 63-66, plant physiologist, Ore, 66-69, PLANT PHYSIOLOGIST, USDA, 69- *Concurrent Pos:* Mem, Coun Agr Sci & Technol. *Mem:* Weed Sci Soc Am. *Res:* Weed control research in dryland field crops in the Pacific Northwest. *Mailing Add:* Wash State Univ, 215 Johnson Hall Pullman WA 99164

OGG, FRANK CHAPPELL, JR, b Champaign, Ill, Jan 22, 30. APPLIED MATHEMATICS. *Educ:* Bowling Green State Univ, BA, 51; Johns Hopkins Univ, MA & PhD(math), 55. *Prof Exp:* Jr instr math, Johns Hopkins Univ, 51-55; res engr, Bendix Corp, 55-57, prin res engr, 57-60; res scientist, Carlyle Barton Lab, Johns Hopkins Univ, 60-68, lectr elec eng, 61-67; ASSOC PROF MATH, UNIV TOLEDO, 68- *Concurrent Pos:* Indust consult. *Mem:* Am Math Soc; Am Math Asn; Sigma Xi. *Res:* Mathematical physics and engineering. *Mailing Add:* 4220 N Holland Sylvania Rd Toledo OH 43606

OGG, JAMES ELVIS, b Centralia, Ill, Dec 24, 24; m 48; c 2. MICROBIAL GENETICS. *Educ:* Univ Ill, BS, 49; Cornell Univ, PhD(bact), 56. *Prof Exp:* With US Army Biol Labs, Md, 50-53 & 56-58; head dept, 67-77, prof microbiol, 58-85, EMER PROF, COLO STATE UNIV, 85- *Concurrent Pos:* Consult-evaluator, NCent Asn Cols & Schs, 74-; Fulbright-Hayes sr lectr, Nepal, 76 & 81; consult genetic eng & biotechnol, UN Develop Prog STAR, Peoples Rep China, 87; consult & acad admin adv, Inst Agr & Animal Sci, Tribhuvan Univ, Nepal, 88-90; consult, Consortium Int Develop, Tucson, Ariz, 91- *Mem:* AAAS; Am Soc Microbiol. *Res:* Microbial and molecular genetics; genetic and biochemical aspects of pathogenicity in bacteria. *Mailing Add:* Dept Microbiology Colo State Univ Ft Collins CO 80523

OGIER, WALTER THOMAS, b Pasadena, Calif, June 18, 25; m 54; c 4. EXPERIMENTAL PHYSICS. *Educ:* Calif Inst Technol, BS, 47, PhD(physics), 53. *Prof Exp:* From instr to asst prof physics, Univ Calif, 54-60; from asst prof to prof physics, 60-89, chmn, Dept Physics & Astron, 72-89, EMER PROF PHYSICS, POMONA COL, 89- *Mem:* Am Phys Soc; Am Asn Physics Teachers. *Res:* X-ray scattering from deformed metals; proton produced x-rays; superfluid helium films. *Mailing Add:* Dept of Physics Pomona Col 333 College Way Claremont CA 91711

OGILBY, PETER REMSEN, b Manila, Philippines, Apr 9, 55; US citizen. SPECTROSCOPY, CHEMICAL DYNAMICS. *Educ:* Univ Wis-Madison, BA, 77; Univ Calif, Los Angeles, PhD(chem), 81. *Prof Exp:* Res fel chem, Univ Calif, Berkeley, 81-83; asst prof, 83-89, ASSOC PROF CHEM, UNIV NMEX, 89- *Mem:* Am Chem Soc; Am Phys Soc; Sigma Xi. *Res:* Time-resolved spectroscopy, with emphasis on singlet molecular oxygen dynamics; near-infrared and infrared spectroscopic techniques; oxygen dynamics and photophysics in solid organic polymers. *Mailing Add:* Dept Chem Univ NMex Albuquerque NM 87131

OGILVIE, ALFRED LIVINGSTON, periodontics, for more information see previous edition

OGILVIE, JAMES LOUIS, b Houston, Tex, Sept 20, 29; m 51; c 2. ANALYTICAL CHEMISTRY. *Educ:* Univ Tex, Austin, BS, 50, MA, 52, PhD(chem), 55. *Prof Exp:* Res chemist, Shell Oil Co, Tex, 54-61; sr res chemist, 61-69, res specialist, 69-74, sr res specialist, 74-79, SR GROUP LEADER, MONSANTO CO, 79- *Mem:* Am Chem Soc; Soc Appl Spectros; Sigma Xi. *Res:* X-ray photoelectron spectroscopy applied to catalyst structure; x-ray emission spectroscopy; x-ray diffraction. *Mailing Add:* 438 Wild Wood Pkwy Ballwin MO 63011

OGILVIE, JAMES WILLIAM, JR, b Orlando, Fla, Oct 20, 25; m 50; c 3. BIOCHEMISTRY, ORGANIC CHEMISTRY. *Educ:* Rollins Col, BS, 50; Johns Hopkins Univ, MA, 52, PhD(chem), 55. *Prof Exp:* USPHS fel biochem, Sch Med, Johns Hopkins Univ, 55-57, from instr to asst prof, 57-67; ASSOC PROF BIOCHEM, SCH MED, UNIV VA, 67- *Mem:* AAAS; Am Soc Biol Chemists; Am Chem Soc; Biophys Soc. *Res:* Mechanisms of enzyme action; biochemical control mechanisms; organic and bioorganic reaction mechanisms. *Mailing Add:* Rt 3 Box 365 Normandy Dr Charlottsville VA 22901

OGILVIE, KEITH W, b Solihull, Eng, Feb 20, 26; m 76; c 2. PHYSICS, SPACE SCIENCE. *Educ:* Univ Edinburgh, BSc, 49, PhD(physics), 53. *Prof Exp:* Physicist, Brit Elec & Appl Indust Res Asn, 54; fel physics, Nat Res Coun Can, 55-57; lectr, Univ Sydney, 57-60; Nat Acad Sci fel, 60-63; physicist, NASA, 63-67, sect head, Goddard Space Flight Ctr, 67-71, BR HEAD, GODDARD SPACE FLIGHT CTR, NASA, GREENBELT, MD, 71- *Honors & Awards:* Except Sci Achievement Medal & Except Serv Medal, NASA. *Mem:* Am Geophys Union; fel Brit Inst Physics. *Res:* Composition and properties of the interplanetary plasma; cosmic rays and solar produced high energy particles. *Mailing Add:* Code 692 Goddard Space Flight Ctr Greenbelt MD 20771

OGILVIE, KELVIN KENNETH, b Windsor, NS, Nov 6, 42; m 64; c 2. BIO-ORGANIC CHEMISTRY. *Educ:* Acadia Univ, BSc, 63, Hons, 64; Northwestern Univ, PhD(chem), 68. *Hon Degrees:* DSc, Acadia Univ, 83, Univ New Brunswick, 91. *Prof Exp:* From asst prof to assoc prof chem, Univ Man, 68-74; from assoc prof to prof chem, McGill Univ, 74-87, Can Pac prof biotech, 84-87, dir, Off Biotech, 84-87; PROF DEPT CHEM & VPRES ACAD, ACADIA UNIV, 87- *Concurrent Pos:* Upjohn Chem Co fel, 74-76 & Steacie Mem fel, Nat Sci & Educ Res Coun Can, 82-84; consult, Upjohn Co, 74-78, Enscor Inc, 79-89, Can Pac lTD, 86-87; bd dirs, Enscor, 80-84, Nova Biotech, 86-87; adv bd, Enscor, 79-84, Allelix Biopharmaceut, 91-; sub-comt Intellectual Prop & Regulatory Affairs, Nat Adv Comt Biotechnol, 89-, sub-comt Human Biopharmaceut & Diag, 90-; bd gov, Plant Biotechnol Inst, 87-90; adv bd & chair, Nat Res Coun Marine Sci, 90- *Honors & Awards:* Huggins Lectr, Acadia Univ, 82; Buck Whitney Award Lect, 83; Buck Whitney Medal, Am Chem Soc, 83. *Mem:* Am Chem Soc; fel Chem Inst Can; AAAS. *Res:* Synthesis of nucleotides; photochemistry of biological systems; phosphate chemistry; antiviral compounds; inventor of "BIOLF-62" = "GANCICLOVIR"; developed chemistry of the Bio Logicals DNA/RNA Synthesizer; developed first total chemical synthesis of large RNA molecules and first total chemical synthesis of a functional TRNA; numerous publications and patents. *Mailing Add:* VP Academic Acadia Univ Wolfville NS B0P 1X0 Can

OGILVIE, MARILYN BAILEY, b Duncan, Okla, Mar 22, 36; div; c 3. HISTORY OF SCIENCE, BIOLOGY. *Educ:* Baker Univ, BA, 57; Univ Kans, MA, 59; Univ Okla, PhD(hist sci), 73, MA, 82. *Prof Exp:* Teacher biol, Phoenix Union High Sch, 59-61; teacher biol & chem, St Andrew's Col, Tanzania, E Africa, 61-62; adj asst prof hist sci, Portland State Univ, Ore, 71-75; prof hist, Oscar Rose Jr Col, Midwest City, Okla, 75-76; vis asst prof hist sci, Univ Okla, 77, adj asst prof hist sci, 77-79; specialist natural sci, 79-80, asst prof, 80-85, ASSOC PROF & CHMN, DIV NATURAL SCI, OKLA BAPTIST UNIV, 85- *Concurrent Pos:* NSF resident pub serv sci, Omniplex Mus, Univ Okla, 77-78; adj assoc prof hist of sci, Div Natural Sci, Okla Baptist Univ, 87. *Mem:* Sigma Xi; Hist Sci Soc. *Res:* Women in science; Robert Chambers and pre-Darwinian evolutionary biology. *Mailing Add:* Div Natural Sci Okla Baptist Univ Shawnee OK 74801

OGILVIE, MARVIN LEE, b Pontiac, Mich, May 3, 35; m 65; c 2. PHYSIOLOGICAL CHEMISTRY. *Educ:* Mich State Univ, BS, 57; Univ Wis-Madison, MS, 59, PhD(biochem), 62. *Prof Exp:* Babcock fel biochem, Univ Wis, 62-63; asst prof biochem & soil sci, Univ Nev, Reno, 63-64; SR RES SCIENTIST BIOCHEM & PHYSIOL, UPJOHN CO, 64- *Mem:* AAAS; Soc Study Reproduction; Am Soc Animal Sci. *Res:* Physiological chemistry relating to the mechanism of drug action; development of growth promotants for ruminants and non-ruminants; rumen function and physiology. *Mailing Add:* 1915 Holiday Kalamazoo MI 49008

OGILVIE, RICHARD IAN, b Sudbury, Ont, Oct 9, 36; m 65; c 2. CLINICAL PHARMACOLOGY. *Educ:* Univ Toronto, MD, 60; FRCP(C), 66. *Prof Exp:* Lectr pharmacol, McGill Univ, 67-68, from asst prof to prof pharmacol & med, 68-83, chmn dept pharmacol & therapeut, 78-83; sr physician & dir, clin pharmacol div, Montreal Gen Hosp, 76-83; dir, Cardiol & Clin Pharmacol Div, Toronto Western Hosp, 83-89; PROF MED & PHARMACOL, UNIV TORONTO, 83- *Concurrent Pos:* Fel, Can Found Adv Therapeut, 67-69; res grants, Que Med Res Coun, Med Res Coun Can, Que Heart Found & Can Found Adv Therapeut, 68-; res asst, Royal Victoria Hosp, Montreal, 68; dir, Can Found Adv Clin Pharmacol, 78- *Mem:* Can Soc Clin Invest; Pharmacol Soc Can; Am Soc Clin Pharmacol; Am Soc Pharmacol & Exp Therapeut; Can Soc Clin Pharmacol (pres, 79-82); Int Union Pharmacol (chmn, Clin Pharmacol Sect, 84-87); fel Am Col Physicians. *Res:* Clinical pharmacology of drugs affecting the cardiovascular system and metabolism; correlation of drug disposition with effect; clinical trials. *Mailing Add:* Dept Med Toronto Western Hosp 399 Bathurst Toronto ON M5T 2S8 Can

OGILVIE, ROBERT EDWARD, b Wallace, Idaho, Sept 25, 23; div; c 3. ELECTRON OPTICS, MATERIALS SCIENCE. *Educ:* Univ Wash, BS, 50; Mass Inst Technol, SM, 52, MetE, 54, ScD, 55. *Prof Exp:* From asst prof to assoc prof, 55-66, PROF METALL, MASS INST TECHNOL, 66- *Concurrent Pos:* Vpres & dir res, Adv Metals Res, Burlington, 63-; vis res lab, Boston Mus Fine Arts, 68- *Mem:* Am Soc Metals; Mat Anal Soc (pres, 70). *Res:* Crystallography and x-ray diffraction; electron optics; phase diagrams, diffusion and phase transformations; metallic meteorites. *Mailing Add:* Rm 13-5065 Mass Inst of Technol Cambridge MA 02139

OGILVIE, T(HOMAS) FRANCIS, b Atlantic City, NJ, Sept 26, 29; m 50; c 3. SHIP HYDRODYNAMICS, PERTURBATION METHODS. *Educ:* Cornell Univ, AB, 50; Univ Md, MS, 57; Univ Calif, Berkeley, PhD(eng sci), 60. *Prof Exp:* Physicist, Underwater Explosions Br, David Taylor Model Basin, Dept Navy, 51-55, ship-wave anal sect, 55-60, head, 60-62, head free surface phenomena br, 62, liaison scientist, Off Naval Res, London Br Off, 62-64, head free surface phenomena br, David Taylor Model Basin, Washington, DC, 64-67; prof fluid mech, Univ Mich, Ann Arbor, 67-81, chmn, Dept Naval Archit & Marine Eng, 73-81; HEAD, DEPT OCEAN ENG, MASS INST TECHNOL, 82- *Concurrent Pos:* Prof lectr, Am Univ, 61-62 & 65-66. *Honors & Awards:* Wm H Webb Medal, Soc Naval Architects & Marine Engrs, 89. *Mem:* Soc Naval Architects & Marine Engrs; Soc Naval Architects of Japan. *Res:* Free surface phenomena, especially interactions between waves, ships or other structures; boundary value problems and perturbation theory in fluid mechanics. *Mailing Add:* 110 Gray St Arlington MA 02174

OGILVY, WINSTON STOWELL, b Le Mars, Iowa, Dec 3, 18; m 46; c 3. FOOD SCIENCE. *Educ:* Iowa State Univ, BS, 41, PhD(food technol), 50. *Prof Exp:* Bacteriologist, Armour & Co, 50-52, head bact sect, 52-56, assoc tech dir, 56-57; dir nutrit prod develop, Mead Johnson & Co, 57-66, VPRES FOOD PROD RES & DEVELOP, MEAD JOHNSON & CO DIV, BRISTOL-MYERS CO, 66- *Mem:* Am Oil Chem Soc; Am Chem Soc; Inst Food Technol. *Res:* Development of infant formulas, therapeutic foods; nutrition; food microbiology; research and development administration. *Mailing Add:* 2501 E Gum St Evansville IN 47713

OGIMACHI, NAOMI NEIL, b Los Angeles, Calif, Oct 10, 25; m 53; c 4. FLUORINE CHEMISTRY, HIGH ENERGY COMPOUNDS. *Educ:* Univ Calif, Los Angeles, BS, 50; Univ Calif, Berkeley, PhD(org chem), 55. *Prof Exp:* Chemist, Naval Weapons Test Sta, 50-52; asst chem, Univ Calif, Davis, 52-55; res chemist, E I du Pont de Nemours & Co, 55-57; res chemist, Naval Weapons Test Sta, 57-59; res specialist, Rocketdyne Div, Rockwell Int, 59-69; res chemist, Halocarbon Prod Corp, NJ, 70-75; staff scientist, Fluorochem Inc, 75; SR RES SCIENTIST, TELEDYNE MCCORMICK SELPH, 75- *Mem:* AAAS; Am Chem Soc; The Chem Soc; Sigma Xi; Int Pyrotech Soc. *Res:* Chemistry of inorganic fluorine compounds; hydrazine and ammonia chemistry; organic synthesis; aromatic molecular complexes; characterization of liquid rocket propellants; analysis of solid propellants and explosives; fluorocarbon chemistry; synthesis of high energy compounds. *Mailing Add:* 1218 Santa Clara St Eureka CA 95501

OGINSKY, EVELYN LENORE, b New York, NY, Apr 6, 19. MICROBIAL PHYSIOLOGY. *Educ:* Cornell Univ, BA, 38; Univ Chicago, MS, 39; Univ Md, PhD(bact), 46. *Prof Exp:* Instr bact, Univ Md, 42-46; res assoc, Merck Inst Therapeut Res, 48-56; assoc prof, Med Sch, Univ Ore, 57-63, prof bact, 63-73; prof microbiol & assoc dean, Grad Sch, Biomed Sci, Univ Tex Health Sci Ctr, San Antonio, 74-83; RETIRED. *Concurrent Pos:* Donner Found fel, Harvard Med Sch, 46-47. *Mem:* AAAS; Am Soc Microbiol; Am Soc Biol Chemists; Am Acad Microbiol; Brit Soc Gen Microbiol. *Res:* Physiology and metabolism of microorganisms. *Mailing Add:* 5528 SW Seymour St Portland OR 97221

OGLE, JAMES D, b Cleveland, Ohio, Feb 16, 20; m 42. BIOCHEMISTRY. *Educ:* Miami Univ, BS, 42; Univ Cincinnati, MS, 49, PhD(biochem), 52. *Prof Exp:* Asst prof, 53-65, ASSOC PROF BIOL CHEM, COL MED, UNIV CINCINNATI, 65- *Mem:* Am Chem Soc. *Res:* Proteolytic enzymes; complement proteins. *Mailing Add:* Dept Biochem Univ Cincinnati Col Med 231 Bethesda Ave Cincinnati OH 45267

OGLE, PEARL REXFORD, JR, b Columbus, Ohio, June 27, 28; m 53; c 6. INORGANIC CHEMISTRY, PHYSICAL CHEMISTRY. *Educ:* Capital Univ, BS, 50; Ohio State Univ, MS, 52; Mich State Univ, PhD(chem), 55. *Prof Exp:* Prin scientist, Goodyear Atomic Corp, 55-64; asst prof, 64-70, ASSOC PROF CHEM, OTTERBEIN COL, 70-, CHMN DEPT, 74- *Concurrent Pos:* Res Corp grant, 65-; Air Force Off Sci Res grant, 66-70. *Mem:* AAAS; Chem Soc; Sigma Xi. *Res:* Mixed solvents; nitrogen dioxide-hydrogen fluoride; nitrosyl fluoride-hydrogen fluoride; nitryl fluoride-hydrogen fluoride; chemical isotope separation; uranium-molybdenum; nitrogen oxide-transition metal fluoride chemistry; selective ion electrodes. *Mailing Add:* Dept Chem Otterbein Col Westerville OH 43081-2006

OGLE, THOMAS FRANK, b St Paul, Minn, Oct 10, 42; c 2. REPRODUCTIVE ENDOCRINOLOGY. *Educ:* Purdue Univ, BS, 66; Wash State Univ, MS, 69, PhD(wildlife biol & zool), 73. *Prof Exp:* Fel endocrinol, Sch Med, Univ Va, 72-74; asst prof, 74-80, ASSOC PROF PHYSIOL, MED COL GA, 80- *Concurrent Pos:* NIH fel, 74. *Honors & Awards:* Nat Wildlife Fedn Fel Award, 70. *Mem:* Sigma Xi; Endocrine Soc; Soc Study Reproduction; Am Physiol Soc; Soc Exp Biol & Med. *Res:* Mechanism of action of progesterone in maintenance of pregnancy. *Mailing Add:* Dept Physiol & Endocrinol Med Col Ga Augusta GA 30912-3000

OGLE, WAYNE LEROY, b Knoxville, Tenn, Dec 23, 22; m 48; c 3. HORTICULTURE. *Educ:* Univ Tenn, BS, 48; Univ Del, MS, 50; Univ Md, PhD(hort), 52. *Prof Exp:* Asst hort, Univ Del, 49-50; from asst to asst prof, Univ Md, 50-54; asst prof, Univ RI, 54-57; assoc prof, 57-66, PROF HORT, CLEMSON UNIV, 66- *Mem:* Am Soc Hort Sci; Inst Food Technologists. *Res:* Mineral nutrition of plants; herbicides and breeding of vegetable crops. *Mailing Add:* Dept of Hort Clemson Univ 201 Sikes Hall Clemson SC 29631

OGLESBY, CLARKSON HILL, b Clarksville, Mo, Nov 9, 08; m 38; c 3. CIVIL ENGINEERING. *Educ:* Stanford Univ, AB, 32, Engr, 36. *Prof Exp:* Draftsman & computer, State Hwy Dept, Ariz, 28-30, off engr & inspector, 32-34, bridge designer, 36-38, from field engr to resident engr, 38-41; from construct engr to chief engr, Vinson & Pringle Construct Co, 41-43; from actg asst prof to prof civil eng, Stanford Univ, 43-74; RETIRED. *Concurrent Pos:* Consult engr, Stanford Res Inst, 58-77 & Systan, Info in Costa Rica, 78; engr, State Div, Bay Toll Crossings, Calif; Fulbright lectr, Imp Col, Univ London, 65-66; lectr at univs in Colombia, Chile, Australia & SAfrica, 74-78; sci writer, 82. *Honors & Awards:* Golden Beaver Award, 64; Hwy Res Bd Award, 69 & 71. *Mem:* Nat Acad Eng; Am Rd Builders Asn; hon mem Am Soc Civil Engrs. *Res:* Highway and public works economics; construction management technology. *Mailing Add:* 850 Webster St Apt 923 Palo Alto CA 94301

OGLESBY, DAVID BERGER, b Charlottesville, Va, Mar 3, 41; m; c 2. ENGINEERING SCIENCE. *Educ:* Va Mil Inst, BS, 63; Univ Va, MAM, 65, DSc, 69. *Prof Exp:* Res engr, Univ Va, 67-68; ASSOC PROF ENG MECH, UNIV MO, 68- *Concurrent Pos:* Res assoc, D K Eng Assoc Inc, 72-74. *Mem:* Am Soc Civil Engrs; Am Soc Eng Educ. *Res:* Stress analysis of engineering structures with primary emphasis on plates and shells. *Mailing Add:* Dept Basic Eng Univ Mo Rolla MO 65401

OGLESBY, GAYLE ARDEN, b McGehee, Ark, Mar 11, 25; m 46; c 2. PETROLEUM GEOLOGY, MINERALOGY. *Educ:* Univ Ark, BS, 51, MS, 52. *Prof Exp:* Staff geologist, Ark State Geol Surv, 52-55 & Ohio Oil Co, 55-56; dist geologist, Oil & Gas Div, Reynolds Mining Corp, 56-57; wellsite geologist, Petrobras Exploracao, Brazil, 57-58; regional geologist, Br Mineral Classification, Gulf Coast Region, Superior Oil, US Geol Surv, 58-73, chief, Br Marine Eval, 73-77, supvry geologist, New Energy Off, US Gen Acct Off, 77-81, joint interest specialist, Offshore Div, 81, eng adv, Mobil Oil, US Geol Surv, 81-89; RETIRED. *Mem:* Am Asn Petrol Geologists; Am Inst Prof Geologists. *Res:* Petroleum geology exploration and development; geochemical and mineralogical analysis to determine stratigraphic correlations; oil, gas, sulfur and salt reservoir and deposit studies and evaluations in the outer continental shelf. *Mailing Add:* 43 Fineza Way Hot Springs Village AR 71909

OGLESBY, LARRY CALMER, b Corvallis, Ore, Mar 26, 36; m 64; c 2. INVERTEBRATE ZOOLOGY. *Educ:* Ore State Col, BA & BS, 58; Fla State Univ, MS, 60; Univ Calif, Berkeley, PhD(zool), 64. *Prof Exp:* Asst prof biol, Reed Col, 64-67; NATO fel, Univ Newcastle, 67-68; from asst prof to assoc prof, 68-71, instr, Marine Biol Lab, 75-78, PROF ZOOL, BIOL DEPT, POMONA COL, 79-, CHMN DEPT ZOOL, 88- *Concurrent Pos:* NSF res grant, 65-67; vis prof, Marine Biol Inst, Univ Ore, 69, 70, 73, 79, 83 & 87, Col William & Mary, 74-75; res grant, Energy Res & Develop Admin, 76-77. *Mem:* Brit Soc Exp Biol; Am Soc Zoologists; Marine Biol Asn UK; fel AAAS; Estuarine Res Fedn. *Res:* Physiology of osmotic and ionic regulation in polychaete annelids, sipunculans and molluscs; annelid parasites and other symbionts; ecological physiology of brackish water invertebrates; life cycles of trematodes; ecology of Salton Sea. *Mailing Add:* Dept Biol Pomona Col Claremont CA 91711

OGLESBY, RAY THURMOND, b Lynchburg, Va, Apr 16, 32; m 56; c 3. AQUATIC SCIENCE. *Educ:* Univ Richmond, BS, 53; Col William & Mary, MA, 55; Univ NC, PhD(environ biol), 62. *Prof Exp:* Res biologist, Bur Com Fisheries, US Fish & Wildlife Serv, 58-59; res asst prof sanit biol, Univ Wash, 62-66, from res asst prof to assoc prof appl biol, 66-68; task group leader aquatic sci, Col Agr & Life Sci, 71-74; assoc prof dept conserv, 68-77, chmn dept, 82-87, PROF DEPT NATURAL RESOURCES, CORNELL UNIV, 77- *Concurrent Pos:* Consult, Mobil Oil Co, Wash, 65, Rockefeller Found, 70-71 & Village of Lake Placid, 71-77; co-prin investr, Pac Northwest Pulp & Paper Asn grant, 65-67; prin investr, USPHS grant, 65-68, res contract, 66-68; prin investr, NSF res grant, 67-68; Int Aluminum Corp res contract, 67-68; Off Water Resources Res res grant, 68-71; Cayuga County res grant, 71-73; co-prin investr, Rockefeller Found res grant, 71-76; sci adv, NY State Assembly Comn on Environ Conserv; mem, Am Inst Biol Sci Life Sci Team Assessment Biol Impacts, 74-; prin investr res grant, Monsanto Chem, 77-78; vis scientist, Water Res Ctr, Stevenace, Eng, 75-76; mem, NY Sea Grant Gov Bd, 83-, chmn, 85-86; actg dir, Cornell Lab Environ Appl Remote Sensing, 90-91 & Global Environ Prog, Cornell, 91-; dir, Cornell Biol Field Sta, Bridgeport, 91- *Mem:* Am Soc Limnol & Oceanog; Am Fish Soc; Sigma Xi. *Res:* Lake and estuarine eutrophication; effects of pollutants on the biota of receiving waters; trophodynamics of freshwater systems. *Mailing Add:* Fernow Hall Cornell Univ Ithaca NY 14853

OGLESBY, SABERT, JR, b Birmingham, Ala, May 14, 21; m 44; c 1. ELECTRICAL ENGINEERING. *Educ:* Auburn Univ, BS, 43; Purdue Univ, MS, 50. *Prof Exp:* Res engr, Southern Res Inst, 46-48; instr elec eng, Purdue Univ, 48-50; head, Spec Eng Proj Sect, 50-57, head, Eng Div, 57-64, dir eng res, 64-74, vpres, 74-80, pres, 80-87, EMER PRES, SOUTHERN RES INST, 87- *Res:* Air conditioning; servomechanisms; general engineering; environmental engineering. *Mailing Add:* Southern Res Inst 2000 Ninth Ave S PO Box 55305 Birmingham AL 35255-5305

OGLIARUSO, MICHAEL ANTHONY, b Brooklyn, NY, Aug 10, 38; m 61; c 1. PHYSICAL ORGANIC CHEMISTRY. *Educ:* Polytech Inst Brooklyn, BS, 60, PhD(chem), 65. *Prof Exp:* Teaching fel chem, Polytech Inst Brooklyn, 61-62, instr phys chem, 64; actg asst prof org chem, Univ Calif, Los Angeles, 65-66, res asst phys org chem, 66-67; from asst prof to assoc prof, 67-78, PROF ORG CHEM, 78-, VA POLYTECH INST & STATE UNIV, ASSOC DEAN, COL ARTS & SCI, 84- *Mem:* Sigma Xi. *Res:* Electron-paramagnetic resonance spectroscopy; observations and reactions of organic radical ions; non-benzenoid aromatic molecules, stable organic anions; dianions and carbonium ions. *Mailing Add:* Çol Arts & Sci 126 Williams Hall Va Polytech Inst & State Univ Blacksburg VA 24061

O'GORMAN, JAMES, b Ireland, Sept 25, 59; m 91. LASER PHYSICS, OPTICAL INSTABILITIES-CHAOS. *Educ:* Nat Univ Ireland, BSc, 79, dipl computer sci, 80; Dublin Univ, PhD(physics), 89. *Prof Exp:* Lectr physics, Dublin Inst Technol, 79-89; res asst optoelectronics, Optronics Ireland, 89-90; POSTDOCTORAL MEM STAFF, AT&T BELL LABS, 90- *Mem:* Inst Elec & Electronics Engrs, Lasers & Electro-Optics Soc; Optical Soc Am. *Res:* Optoelectronics, digital optical communications; laser physics. *Mailing Add:* AT&T Bell Labs Rm 1E-460 600 Mountain Ave Murray Hill NJ 07976

OGRA, PEARAY L, INFECTIOUS DISEASES. *Educ:* Christian Med Col, Ludhiana, MD, 61. *Prof Exp:* PROF PEDIAT & MICROBIOL & CHIEF INFECTIOUS DIS, STATE UNIV NY, BUFFALO & CHILDREN'S HOSP OF BUFFALO, 66- *Res:* Vaccines; mucosal immunology. *Mailing Add:* 219 Bryant St Buffalo NY 14222

O'GRADY, RICHARD TERENCE, b Victoria, BC, Jan 13, 56; c 1. SCIENCE PUBLISHING. *Educ:* Univ BC, Vancouver, BSc, 78; McGill Univ, Montreal, MSc, 81; Univ BC, PhD(zool), 87. *Prof Exp:* Res fel, Nat Sci & Eng Res Coun, Smithsonian Inst, 87-89, res fel, Nat Mus Natural Hist, 87-89, Killam Found, 87-89, RES ASSOC, SMITHSONIAN INST, NAT MUS NATURAL HIST, 89- *Concurrent Pos:* Ed, Smithsonian Inst, 88-90, sci ed, Johns Hopkins Univ Press, 90-; awards comt, Am Soc Parasitologists, 89-90; counr, Soc Syst Biologists, 89-92. *Mem:* Soc Syst Biologists; Am Soc Zool; Am Soc Parasitologists; Sigma Xi; AAAS. *Res:* Biosystematics; phylogenetic reconstruction; cladistics-evolutionary theory and methods of historical reference in biology; organisms have been parasitic helminths; developed and published books in organismal-level biology; earth and space sciences; mathematical sciences. *Mailing Add:* Johns Hopkins Univ Press 701 W 40th St Baltimore MD 21211-2190

O'GRADY, WILLIAM EDWARD, b Longmont, Colo, Oct 8, 39; m 66; c 3. PHYSICAL CHEMISTRY, SURFACE SCIENCE. *Educ:* Colo Sch Mines, BS, 64; Univ Pa, PhD(chem), 73. *Prof Exp:* Chemist anal chem, Shell Chem Co, 64-66; sr res assoc electrochem, Case Western Reserve Univ, 72-77; CHEMIST, DEPT APPL SCI, BROOKHAVEN NAT LAB, 77- *Concurrent Pos:* Ed, J Electrochem Soc, 82- *Mem:* Am Electrochem Soc; Am Vacuum Soc; Am Chem Soc. *Res:* Electrocatalysis and electrode kinetics; applications of surface science to electrochemistry; structure and properties of small particles; single crystal studies; quantum mechanical calculations on small particles. *Mailing Add:* Naval Res Lab Code 6170 Washington DC 20375

OGREN, DAVID ERNEST, geology; deceased, see previous edition for last biography

OGREN, HAROLD OLOF, b Grayling, Mich, Apr 24, 43; m 68; c 2. EXPERIMENTAL HIGH ENERGY PHYSICS. *Educ:* Univ Mich, BS, 65; Cornell Univ, MS, 67, PhD(physics), 70. *Prof Exp:* Vis scientist high energy physics, Nat Lab Frascati, 70-73; fel, Europ Asn Nuclear Res, 73-75; asst prof, 75-78, ASSOC PROF PHYSICS, IND UNIV, BLOOMINGTON, 78- *Mem:* Am Phys Soc. *Res:* Strong interaction physics of fundamental particles. *Mailing Add:* Dept of Physics Ind Univ Bloomington IN 47405

OGREN, HERMAN AUGUST, b Kenosha, Wis, Mar 31, 25; m 51; c 6. ECOLOGY. *Educ:* Univ Wis, BS, 50; Univ Mont, MS, 54; Univ Southern Calif, PhD(zool), 60. *Prof Exp:* State biologist, Mont Fish & Game Comn, 52-54; lab assoc zool, Univ Southern Calif, 54-56; instr, Exten Div, Univ NC, 56-57; state biologist, Dept Game & Fish, NMex, 57-60; asst prof zool, Elmhurst Col, 60-63; assoc prof, 63-70, bd trustees fac res grant, 64-65, PROF ZOOL, CARTHAGE COL, 64- *Concurrent Pos:* NSF grant, 62. *Honors & Awards:* Ann Honorarium, Am Soc Mammal, 55. *Mem:* Am Soc Mammal; Wildlife Soc. *Res:* Fish and wildlife technology; wild sheep of United States, Canada and Mexico, and of Essox Maskinonge in northern Wisconsin; presence of radioactivity in animals; taxonomy African game and smaller mammals. *Mailing Add:* Dept Biol Carthage Col 2001 Alford Dr Kenosha WI 53141

OGREN, PAUL JOSEPH, b Madrid, Iowa, July 3, 41; m 63; c 4. PHYSICAL CHEMISTRY. *Educ:* Earlham Col, BA, 63; Univ Wis, PhD(chem), 68. *Prof Exp:* From asst prof to assoc prof chem, Maryville Col, 67-72; assoc prof chem, Cent Col, Iowa, 72-79; assoc prof, 79-84, PROF CHEM, EARLHAM COL, RICHMOND, IND, 84- *Concurrent Pos:* Advan Study Prog fel, Nat Ctr Atmospheric Res, 71-72. *Mem:* Am Chem Soc. *Res:* Solid state radiation chemistry; pulse radiolysis; photochemistry. *Mailing Add:* Dept Chem Earlham Col Richmond IN 47374

OGREN, ROBERT EDWARD, b Jamestown, NY, Feb 9, 22; m 48; c 2. ZOOLOGY, PARASITOLOGY. *Educ:* Wheaton Col, Ill, BA, 47; Northwestern Univ, MS, 48; Univ Ill, PhD(zool, physiol), 53. *Prof Exp:* Asst zool, Univ Ill, 48-53; asst prof physiol & zool, Ursinus Col, 53-57; asst prof histol & anat, Dickinson Col, 57-63, actg chmn dept biol, 59-60; assoc prof embryol, histol & cell biol, 63-81, prof biol, anat & physiol, 81-86, EMER PROF, WILKES UNIV, 86- *Mem:* Fel AAAS; Am Soc Zoologists; Am Soc Parasitologists; Am Micros Soc; Soc Protozoologists; Ecol Soc Am; Soc Syst Zool. *Res:* Comparative morphology; cytology; development and physiology of tapeworm hexacanth embryos; biology of land planarians; invertebrates; parasitology; histology; embryology. *Mailing Add:* Dept Biol Wilkes Univ Wilkes-Barre PA 18766

OGREN, WILLIAM LEWIS, b Ashland, Wis, Oct 8, 38; m 67; c 3. PLANT PHYSIOLOGY. *Educ:* Univ Wis-Madison, BS, 61; Wayne State Univ, PhD(biochem), 65. *Prof Exp:* Res chemist, Parker Rust Proof Co, 61-62; plant physiologist, 65-79, RES LEADER, USDA, 79- *Concurrent Pos:* From asst prof to assoc prof, 66-77, prof physiol, Univ Ill, Urbana, 77- *Honors & Awards:* Crop Sci Award, 79; C F Kettering Award, 86; Alexander von Humboldt Award, 90. *Mem:* Nat Acad Sci; Am Soc Plant Physiologists; fel Am Soc Agron; fel Crop Sci Soc Am; Am Soc Biochem & Molecular Biol; Japan Soc Plant Physiologists; Am Acad Arts & Sci. *Res:* Biochemistry, physiology and genetics of photosynthesis in soybeans and other crops; photorespiration. *Mailing Add:* USDA/ARS S218 Turner Hall 1102 S Goodwin Ave Urbana IL 61801

OGRYZLO, ELMER ALEXANDER, b Dauphin, Man, Aug 18, 33; m 59; c 4. PHYSICAL CHEMISTRY. *Educ:* Univ Man, BSc, 55, MSc, 56; McGill Univ, PhD(phys chem), 59. *Prof Exp:* Exhibition of 1851 overseas fel, Univ Sheffield, 58-59; from instr to assoc prof chem, 59-71, PROF CHEM, UNIV BC, 71- *Concurrent Pos:* Nat Res Coun Can sr res fel, Univ Amsterdam, 66-67. *Mem:* Am Chem Soc; Chem Inst Can; Royal Soc Chem. *Res:* Kinetics of halogen atom reactions; reactions of electronically excited oxygen molecules; spectroscopy of small molecules; chemiluminescence. *Mailing Add:* Dept Chem Univ BC 2075 Westbrook Mall Vancouver BC V6T 1W5 Can

OGUNNAIKE, BABATUNDE AYODEJI, b Ijebu-Igbo, Nigeria, Mar 26, 56; m 83; c 2. PROCESS CONTROL, MATHEMATICAL MODELING. *Educ:* Univ Lagos, Nigeria, BS, 76; Univ Wis-Madison, MS & PhD(chem eng), 81. *Prof Exp:* Res engr, Shell Develop Co, 81-82; from asst prof to assoc prof chem eng & statistics, Univ Lagos, Nigeria, 82-88; vis assoc prof chem eng, Univ Wis-Madison, 88-89; sr res engr, 89-91, RES ASSOC, E I DU PONT DE NEMOURS & CO, INC, 91- *Concurrent Pos:* Adj prof chem eng, Univ Del, 89- *Mem:* Am Inst Chem Engrs; Soc Indust & Appl Math; NY Acad Sci. *Res:* Theory and application of multivariable process control with emphasis on model predictive control; mathematical modeling with emphasis on industrial chemical and polymerization reactors; applied statistics. *Mailing Add:* E I Du Pont de Nemours & Co Inc PO Box 80101 Bldg 1 Rm 104 Wilmington DE 19880-0101

OGURA, JOSEPH H, otolaryngology; deceased, see previous edition for last biography

OH, BYUNGDU, b Seoul, Korea, May 5, 58; m 85; c 2. SUPERCONDUCTIVITY, SQUIDS & FLUX TRANSFORMERS. *Educ:* Seoul Nat Univ, BS, 81; Stanford Univ, MS, 87, PhD(appl physics), 89. *Prof Exp:* Res asst superconductivity, Ginzton Lab, Stanford Univ, 83-88; POSTDOCTORAL FEL SQUIDS & FLUX TRANSFORMERS, IBM, T J WATSON RES CTR, 88- *Mem:* Am Phys Soc. *Res:* Solid state physics; low-temperature physics; pure and applied superconductivity; superconducting thin films and devices; flux transformers and superconducting quantum interference devices; study of vortex motions in high-Tc superconductors. *Mailing Add:* IBM T J Watson Res Ctr Yorktown Heights NY 10598

OH, CHAN SOO, b Kwangju, Korea, July 4, 38; m 68; c 2. LIQUID CRYSTAL DISPLAYS, CLINICAL DIAGNOSTICS CHEMISTRY. *Educ:* Seoul Nat Univ, BS, 61; St John's Univ, MS, 67; PhD(org chem), 70. *Prof Exp:* Sr chemist, Columbia Pharmaceut Corp, 65-70; mem tech staff, David Sarnoff Res Ctr, 70-72; sr chemist, Helipot Div, Beckman Instruments, 72-74; mem tech staff, David Sarnoff Res Ctr, RCA Labs, 74; sr chemist, Helipot Div, 75-82, MGR, DIAGNOSTICS SYSTS DIV, BECKMAN INSTRUMENTS, 82- *Concurrent Pos:* Consult, Flat Panel Displays. *Mem:* Am Chem Soc; Sigma Xi. *Res:* Organic synthesis; liquid crystal materials; liquid crystal display devices; immunochemistry; clinical chemistry. *Mailing Add:* Beckman Instruments Co 200 S Kraemer Brea CA 92621

OH, HILARIO LIM, b Cagayan de Oro, Philippines, Jan 14, 36; m 68; c 4. MECHANICAL ENGINEERING. *Educ:* Univ Philippines, BS, 60; Purdue Univ, MS, 63, Univ Calif, Berkeley, PhD(mech eng), 67. *Prof Exp:* Mech engr, Calif Packing Corp, 60-62; res asst, Univ Calif, Berkeley, 64-67, NSF res grant, 67-68; sr res scientist, Gen Motors Res Labs, 68-80, staff res scientist, 80-85; CPC GROUP, GEN MOTORS CORP, 85- *Mem:* Am Soc Mech Engrs. *Res:* Fracture of brittle and polymeric solids; mechanical behavior of materials; digital signal processing; robust product design. *Mailing Add:* CPC Group Gen Motors Corp 30001 Van Dyke Warren MI 48090

OH, JANG OK, b Seoul, Korea, Jan 15, 27; m 55; c 1. VIROLOGY, PATHOLOGY. *Educ:* Severance Union Med Col, MD, 48; Univ Wash, PhD(microbiol), 60. *Prof Exp:* Instr microbiol, Severance Union Med Col, 49-51; resident path, Hamot Hosp, Erie, Pa, 53-55; teaching & res asst, Sch Med, Univ Wash, 57-59, res assoc, 59-60, res instr, 60-61; res assoc path, Fac Med, Univ BC, 61-63, asst prof, 63-66; from asst to assoc res microbiologist, 66-71, RES MICROBIOLOGIST, FRANCIS I PROCTOR FOUND, MED CTR, UNIV CALIF, SAN FRANCISCO, 71-, ASSOC DIR, 85- *Concurrent Pos:* Res fel microbiol, Commun Dis Lab, Med Ctr, Ind Univ, 55-56; Lederle med fac award, 63-66; consult, NIH, 79-83, 90- *Mem:* Asn Res Vision & Ophthamol; Am Soc Microbiol; Am Soc Exp Path; Soc Exp Biol & Med. *Res:* Ocular virology, pathogenesis and treatment; nonspecific resistance to viral infection, its mechanism and application to experimental infection; chemotherapy of ocular viral infection. *Mailing Add:* Proctor Found Univ of Calif Med Ctr San Francisco CA 94143-0412

OH, SE JEUNG, b Kimhae, Korea, June 17, 35; US citizen; m 64. ELECTRICAL ENGINEERING, COMPUTER SCIENCES. *Educ:* Univ Colo, BS, 59; Columbia Univ, MS, 63, PhD(elec eng), 66. *Prof Exp:* Instr elec eng, Southern Colo State Col, 59-60; engr, Develop Lab, Data Systs Div, Int Bus Mach Corp, 60-61; lectr elec eng, City Col New York, 61-64; res asst, Columbia Univ, 63-66; mem tech staff, Bell Tel Labs, 66-68; from asst prof to assoc prof, 68-76, chmn dept, 74-78, 81-82, PROF ELEC ENG, CITY COL NEW YORK, 77-, assoc dean grad studies, 84-87. *Concurrent Pos:* Electronics consult, UNESCO, 69; consult, Bell Labs, 72-80. *Mem:* Inst Elec & Electronics Engrs; Assoc Comput Mach. *Res:* Computer architecture and software engineering. *Mailing Add:* Dept Elec Eng City Col New York NY 10031

OH, SHIN JOONG, b Seoul, Korea, Nov 16, 36; US citizen; m 66; c 2. NEUROLOGY. *Educ:* Seoul Nat Univ, Korea, MD, 60, Master Med, 62. *Prof Exp:* Asst prof neurol, Meharry Med Col, 68-70; from asst prof to assoc prof, 70-80, PROF, DEPT NEUROL, SCH MED, UNIV ALA, BIRMINGHAM, 80- *Concurrent Pos:* Res fel, Inst Endemic Dis, Seoul Nat Univ Hosp, 64-66 & Epidemiol & Genetic Unit, Univ Minn Med Ctr, 68; consult, Brookwood Hosp, Birmingham, 75; dir, Electomyogram Lab, Univ Hosp, Birmingham, 70- *Mem:* Am Acad Neurol; Am Asn Electromyography & Electrodiagnosis. *Res:* Neuromuscular disease; electromyography and electrodiagnosis. *Mailing Add:* Dept Neurol Univ Ala Univ Sta Birmingham AL 35294

OH, WILLIAM, b Philippines, May 22, 31; m 60; c 2. PEDIATRICS. *Educ:* Xavier Univ, Philippines, BA, 53; Univ Santo Tomas, Manila, MD, 58. *Hon Degrees:* DSc, RI Col. *Prof Exp:* Res assoc, Karolinska Inst, Sweden, 64-66; asst prof pediat, Chicago Med Sch, 66-68; assoc prof, Univ Calif, Los Angeles, 69-72, prof, 72-74; PROF PEDIAT & OBSTET, BROWN UNIV, 74-, CHMN PEDIAT, 89- *Concurrent Pos:* Fel neonatology, Michael Reese Hosp, Chicago, 62-64; pediatrician-in-chief, Women & Infants Hosp, RI, 74; mem, Study Sect Embryol & Develop, NIH, 80-84. *Mem:* Soc Pediat Res; Am Pediat Soc; Am Soc Clin Res; Perinatal Res Soc (pres, 81). *Res:* Perinatal biology with specific interest on carbohydrate metabolism, fluid and electrolyte balance and respiratory distress syndrome in the newborn. *Mailing Add:* Dept of Pediat Women & Infants Hosp 101 Dudley St Providence RI 02905

O'HALLORAN, THOMAS A, b New York, NY, Apr 13, 31; m 54; c 4. PHYSICS. *Educ:* Ore State Col, BS, 53, MS, 54; Univ Calif, Berkeley, PhD(elem particle physics), 63. *Prof Exp:* Res assoc elem particle physics, Univ Calif, Berkeley, 63-64; res fel, Harvard Univ, 64-66; from asst prof to assoc prof, 66-70, PROF PHYSICS, UNIV ILL, URBANA, 70- *Concurrent Pos:* J S Guggenheim fel, 79. *Mem:* Am Phys Soc. *Res:* Elementary particles. *Mailing Add:* Dept of Physics Univ Ill 1110 W Green St Urbana IL 61801

OHANIAN, HANS C, b Leipzig, Ger, Apr 29, 41; m 66. PHYSICS. *Educ:* Univ Calif, Berkeley, BA, 64; Princeton Univ, PhD(physics), 68. *Prof Exp:* asst prof physics, Rensselaer Polytech Inst, 68-76; assoc prof physics, Union Col, 76-83; ADJ PROF PHYSICS, RENSSELAER POLYTECH INST, 79- *Concurrent Pos:* Vis fel physics, Princeton Univ, 75-76; vis prof physics, Univ Rome, 85-86. *Mem:* Am Phys Soc. *Res:* Gravitation and field theory; writer of physics textbooks. *Mailing Add:* Dept Physics Rensselaer Polytech Inst Troy NY 12180-3590

O'HARA, NORBERT WILHELM, b Youngstown, Ohio, Oct 7, 30; m 67; c 7. BASEMENT GEOLOGY, GEOLOGICAL OCEANOGRAPHY. *Educ:* Mich State Col, BS, 52;; Mich State Univ, MS, 54, PhD(geophysics), 67. *Prof Exp:* Asst prof geol, Eastern Mich Univ, 60-63; asst prof, Mich State Univ, 63-65; asst prof, Grand Valley State Col, 65-67; res assoc, Mich State Univ, 67-68; assoc prof oceanog, World Campus Afloat, Chapman Col, 68-69; res assoc, Univ Mich, 69-71; res geophysicist, Naval Res Ctr, 71-75; PROF OCEANOG & DEPT HEAD, FLA INST TECHNOL, 75- *Concurrent Pos:* Geosci consult, N W O'Hara & Assocs, 75-82; chmn, Fla Intercollegiate Comt Oceanog, 78-80; mem exec comt, Sea Grant Asn, 78-81; geophysicist, NSF Antarctic Field Expedition, 65-66; prin investr, Western Great Lakes Aeromagnetic Surv, 63-67. *Mem:* Soc Explor Geophysicists; Geol Soc Am; Am Geophys Union. *Res:* Basement geology through the use of gravity and magnetic methods; regional tectonics; geopotential mapping; geothermal areas. *Mailing Add:* 400 Seventh Ave Indialantic FL 32903-4338

OHARA, ROBERT J, systematics, evolutionary biology, for more information see previous edition

O'HARE, JOHN MICHAEL, b Des Moines, Iowa, Oct 2, 38; m 64; c 4. SOLID STATE PHYSICS, OPTICS. *Educ:* Loras Col, BS, 60; Purdue Univ, MS, 62; State Univ NY Buffalo, PhD(physics), 66. *Prof Exp:* Instr physics, State Univ NY Buffalo, 65-66; from asst prof to assoc prof, 66-77, PROF PHYSICS, UNIV DAYTON, 77-, CHMN DEPT, 83-, PROF ELECTRO-OPTICS, 83- *Mem:* Am Phys Soc; Optical Soc Am. *Res:* Theoretical atomic and molecular physics; crystal field theory; optical properties of solids, experimental and theoretical; nonlinear optics. *Mailing Add:* Dept Physics Univ Dayton Dayton OH 45409

O'HARE, PATRICK, b Dundalk, Ireland, Aug 6, 36; m 64; c 3. PHYSICAL CHEMISTRY. *Educ:* Nat Univ Ireland, BS, 57, MS, 58; Queen's Univ Belfast, PhD(phys chem), 61, DSc(phys chem), 71. *Prof Exp:* Demonstr chem, Queen's Univ Belfast, 58-61, Imp Chem Indust fel, 61-63; resident res assoc, Argonne Nat Lab, 64-66, assoc chemist, 66-82, sr chemist, 82-89; RES CHEMIST, NAT INST STANDARDS & TECHNOL, 89- *Concurrent Pos:* Vis prof, Univ Toronto, & Int Atomic Energy Agency, Vienna, 76-78; ed, J Chemical Thermodynamics, 82-; chmn, Comn Thermodynamics, Int Union Pure & Appl Chem, 90- *Honors & Awards:* Huffman Award, 86. *Mem:* Am Chem Soc; assoc mem Int Union Pure & Appl Chem. *Res:* Thermochemistry of nuclear materials; bond energies in inorganic molecules; thermodynamies of new materials. *Mailing Add:* Nat Inst Standards & Technol Gaithersburg MD 20899

OHASHI, YOSHIKAZU, b Tokyo, Japan, June 30, 41; m 70; c 3. MINERALOGY, CRYSTALLOGRAPHY. *Educ:* Tokyo Univ, BS, 66, MS, 68; Harvard Univ, PhD(geol), 73. *Prof Exp:* Fel, Geophys Lab, Carnegie Inst Washington, 72-76; asst prof geol, Univ Pa, 76-83; SR RES GEOLOGIST, ARCO E P R RES CTR, 83- *Mem:* Mineral Soc Am; Mineral Asn Can; Am Crystallog Asn; Mineral Soc Great Brit; Inst Elec & Electronics Engrs Comput Soc. *Res:* Crystal structure analysis of rock-forming silicate minerals; crystallographic computing. *Mailing Add:* Arco Explor & Technol 2300 W Plano Pkwy Plano TX 75075

O'HAVER, THOMAS CALVIN, b Atlanta, Ga, Oct 13, 41; m 68; c 2. ANALYTICAL CHEMISTRY. *Educ:* Spring Hill Col, BS, 63; Univ Fla, PhD(chem), 68. *Prof Exp:* From asst prof to assoc prof, 68-78, PROF CHEM, UNIV MD, 78- *Concurrent Pos:* Advan Res Proj Agency, Dept Defense fel, Ctr Mat Res, 68-71, NSF Instnl Sci Equip Prog fel, 69-72 & Petrol Res Found-Am Chem Soc fel, 69-72; NSF res grants, 74-80; res grant, US Dept Agr, Agr Res Serv, 76- *Mem:* Am Chem Soc; Soc Appl Spectros; Sigma Xi. *Res:* Multi-element atomic spectrometry; analytical instrumentation; automation; computer applications. *Mailing Add:* Dept Chem Univ Md College Park MD 20742

O'HEA, EUGENE KEVIN, b Cork, Ireland, Mar 27, 41; m; c 4. PHYSIOLOGY. *Educ:* Univ Col, Dublin, BAgr Sc, 64, MAgrSc, 66; Univ Ill, PhD(nutrit biochem), 69, Univ Western Ont, MD, 76. *Prof Exp:* Asst prof, 70-80, ASSOC PROF PHYSIOL, UNIV WESTERN ONT, 80- *Concurrent Pos:* Pvt pract. *Mem:* Can Physiol Soc; Nutrit Soc Can; Am Physiol Soc. *Res:* Endocrinology; fat synthesis and its regulations; energy balance. *Mailing Add:* 600 Colborne London ON N6A 5C1 Can

O'HEARN, GEORGE THOMAS, b Manitowoc, Wis, Sept 26, 34; m 59; c 3. SCIENCE EDUCATION. *Educ:* Univ Wis-Madison, BS, 57, MS, 59, PhD(sci educ), 64. *Prof Exp:* Asst prof sci educ, Univ & prin investr, Res & Develop Ctr Cognitive Learning, Univ Wis-Madison, 64-68, US Off Educ grant sci literacy, 64-66; assoc prof, 68-70, PROF SCI EDUC, UNIV WIS-GREEN BAY, 70-, DIR, INST RES, 75-, CHAIR EDUC, 88- *Mem:* Fel AAAS; Nat Asn Res Sci Teaching; Nat Sci Teachers Asn. *Res:* Research design, social implications of science, science learning. *Mailing Add:* 202 Warren Ct Green Bay WI 54301

O'HERN, ELIZABETH MOOT, b Richmondville, NY, Sept 1, 13; m 52. MEDICAL MICROBIOLOGY, SCIENCE ADMINISTRATION. *Educ:* Univ Calif, BA, 45, MA, 47; Univ Wash, PhD(microbiol, mycol), 56. *Prof Exp:* Asst parasitol & mycol, Univ Calif, 45-48; instr microbiol, Univ Wash, 56-57 & State Univ NY Downstate Med Ctr, 57-62; asst prof, Sch Med, George Washington Univ, 62-65; sr scientist, Bionetics Res Labs, 65-66, prin investr, 66-68; prog adminr microbiol training prog, Nat Inst Gen Med Sci, 68-75, prog adminr genetics res grants, 74-75, spec asst to dir, 75-76, prog admin trauma, burn, anesthesiol & training grants, 77-86; RETIRED. *Concurrent Pos:* mem Staff Training Extramural Programs comt, NIH, 74-77, Extramural Assocs Rev Comt, 78; mem, Chair Comt Status Women Microbiologists, Am Soc Microbiol, 75-78; mem, Joint Bd Sci & Engr Educ, Wash Acad Sci, 75-78; mem bd dir, Found Microbiol, 76-77; panel mem, US Civil Service Comn, 77-86; consult, Sci Admin, Univ Calif-Berkeley, 90. *Honors & Awards:* Alice Evans Award, Am Soc Microbiol, 86. *Mem:* Am Pub Health Asn; AAAS; Am Soc Cell Biol; Am Soc Trop Med & Hyg; Am Soc Microbiol; Mycol Soc Am. *Res:* Medical mycology; host-parasite relationships; malaria chemotherapy; books on women scientists and the history and philosophy of science. *Mailing Add:* 633 G 25 SW Washington DC 20024

O'HERN, EUGENE A, b Flint, Mich, Jan 20, 27; m 51; c 4. MECHANICAL ENGINEERING. *Educ:* Purdue Univ, BS & BNS, 48, MS, 49, PhD(mech eng), 51. *Prof Exp:* Res engr, Missile & Control Equip Div, N Am Rockwell Corp, 51-54, proj engr, 54-55, anal supvr, 55-57, proj engr autonetics, 57-60, sect chief, 60-62, asst mgr, 62-64, chief scientist, 65-70, mgr syst eng space div, 70-72, mem tech staff, Space Div, Rockwell Int, 72-82; SYSTS ANAL CONSULT 82- *Concurrent Pos:* Lectr, Univ Southern Calif, 59-65. *Mem:* Am Inst Aeronaut & Astronaut; Soc Automotive Engrs; Inst Elec & Electronics Engrs. *Res:* Control system analysis; aircraft dynamics; flight control system synthesis; avionics system analysis; risk analysis. *Mailing Add:* 1233 Longview Dr Fullerton CA 92631

OHI, SEIGO, b Toyama, Japan, Apr 8, 43; m 70; c 3. GENE THERAPY, ADENO-ASSOCIATED VIRUSES. *Educ:* Toyama Univ, BS, 66; Princeton Univ, MS, 70, PhD(molecular biol), 73. *Prof Exp:* Fel molecular biol, Carnegie Inst Washington, Baltimore, 73-75; res assoc biochem, Univ Pittsburgh, 75-78 & Johns Hopkins Univ, Baltimore, 78-83; vis scientist virol, NIH, Bethesda, Md, 83-86; ASST PROF MOLECULAR BIOL, MEHARRY MED COL, NASHVILLE, TENN, 86- *Concurrent Pos:* Vis asst prof, Univ Conn, Farmington, 80-82. *Mem:* NY Acad Sci; AAAS; Am Soc Biochem & Molecular Biol; Am Soc Microbiol; Am Soc Cell Biol; Am Chem Soc; Biophys Soc. *Res:* Gene therapy of hemoglobinopathics and AIDS using human paravoviruses. *Mailing Add:* Dept Biochem-Sch Med Meharry Med Col 1005 18th Ave N Nashville TN 37208

OHKI, KENNETH, b Livingston, Calif, June 13, 22; m 45; c 2. PLANT PHYSIOLOGY. *Educ:* Univ Calif, Berkeley, BS, 49, MS, 51, PhD(plant physiol), 63. *Prof Exp:* Supvr & specialist plant physiol res, Int Minerals & Chem Corp, Ill, 64-70; from asst prof to prof, 71-88, EMER PROF AGRON, GA EXP STA, UNIV GA, 88- *Concurrent Pos:* Res asst, Calif Inst Technol, 50-53. *Mem:* Am Soc Plant Physiologists; Am Soc Agron; Crop Sci Soc Am; Soil Sci Soc Am. *Res:* Plant nutrition and analysis; growth and development of sugar beets; ion absorption in relation to antecedent nutrition; plant growth regulator; micronutrient nutrition of plants. *Mailing Add:* Dept Agron Ga Exp Sta Univ Ga Griffin GA 30223-1797

OHKI, SHINPEI, b Japan, Jan 1, 33; m 71. BIOPHYSICS. *Educ:* Kyoto Univ, BS, 56, MS, 58, PhD(physics), 65. *Prof Exp:* Instr physics, Tokyo Metrop Univ, 61-65; res assoc theoret biol, 65-66, from asst prof to assoc prof biophys, 66-85, PROF BIOPHYS, STATE UNIV NY BUFFALO, 86- *Concurrent Pos:* Grants, Damon Runyon Mem Fund, 68-71, Nat Inst Neurol Dis & Stroke, 69-75 & Nat Inst Gen Med, 78-; fel, Japan Soc Prom Sci; vis prof, Japan Soc Prom Sci, 78-79. *Mem:* Biophys Soc; NY Acad Sci; Am Chem Soc; AAAS. *Res:* Investigation of a mechanism of excitatory and inhibitory phenomena of membranes; adhesion and fusion of model and biological membranes. *Mailing Add:* Dept Biophys Rm 224 SUNY Cary Hall Health Sci Ctr 3435 Main St Buffalo NY 14214

OHL, DONALD GORDON, b Milton, Pa, Apr 13, 15; wid; c 1. MATHEMATICS. *Educ:* Ursinus Col, BS, 36; Bucknell Univ, MS, 47. *Prof Exp:* High sch teacher, Pa, 37-41; instr math, Univ & chmn dept, Olney Undergrad Ctr, Temple Univ, 46; from instr to asst prof, 46-60, assoc prof, 60-81, EMER PROF MATH, BUCKNELL UNIV, 81-; CONSULT, 81- *Mem:* Am Math Soc; Math Asn Am; Nat Coun Teachers Math. *Res:* Mathematics education; general mathematics. *Mailing Add:* 605 Buffalo Rd Lewisburg PA 17837

OHLBERG, STANLEY MILES, b Brooklyn, NY, June 20, 21; m 54; c 3. PHYSICAL CHEMISTRY. *Educ:* Univ Mich, BS, 43; Rutgers Univ, MS, 50, PhD(phys chem), 51. *Prof Exp:* Analytical & indust chemist, Manhattan Proj, Linde Air Prod Co, 43-46; asst inorg chem, Rutgers Univ, 46-51; fel x-ray diffraction anal, Dept Res Chem Physics, Mellon Inst, 51-56; res scientist, Verona Res Ctr, Koppers Co, 56-58; scientist, 58-71, staff scientist, 71-73, SR SCIENTIST, GLASS RES CTR, PPG INDUSTS, INC, 73- *Honors & Awards:* Frank Forest Award, Am Ceramic Soc, 66. *Mem:* Am Chem Soc; fel Am Ceramic Soc. *Res:* Structures of materials; physics and chemistry of glass. *Mailing Add:* PO Box 564 New Paris PA 15554

OHLE, ERNEST LINWOOD, b St Louis, Mo, Dec 17, 17; m 43; c 4. ECONOMIC GEOLOGY. *Educ:* Washington Univ, AB, 38, MS, 40; Harvard Univ, MA, 41, PhD, 50. *Prof Exp:* Geologist, Am Zinc Co, Tenn, 41-42, Ark, 42-43 & Tenn, 44-46, asst mine supt, 46-47; geologist, St Joseph Lead Co, Mo, 48-57; chief geologist, White Pine Copper Co, 57-61; vpres explor, Copper Range Co, 60-61; staff geologist, 61-65, asst chief geologist, 65-68, eval mgr, 68-71, consult geologist, Hanna Mining Co, 71-78; RETIRED. *Concurrent Pos:* Mem bd mineral resources, NSF, 74-77; adj prof, Univ Utah, 73; pres, Soc Econ Geol Found, 78-83 & 86-; adj prof, Univ Ariz, 83- *Honors & Awards:* Marsden Medal, Soc Econ Geol, 87. *Mem:* Soc Econ Geologists (pres, 84); Am Inst Prof Geologists; Am Asn Petrol Geologists; Am Inst Mining, Metall & Petrol Engrs. *Res:* Ore deposition; structural control of ore deposits; limestone permeability as related to ore deposition. *Mailing Add:* 8989 E Escalante Tucson AZ 85750

OHLENBUSCH, ROBERT EUGENE, b Edinburg, Tex, Oct 18, 30; m 63; c 2. DENTISTRY, MICROBIOLOGY. *Educ:* Tex Lutheran Col, BS, 52; Univ Tex, DDS, 56. *Prof Exp:* Dent Corps, US Army, 56-, dentist, 56-58, res asst dent, Walter Reed Army Inst Res, 58-61, res dent officer oral microbiol, US Army Inst Dent Res, 63-67, comdr, 137th Med Detachment, Repub Vietnam, 67-68, exec officer, 102nd Med Detachment, Munich, Ger, 70-71, exec officer prev dent & dir dent educ, Dent Co, Ft Jackson, SC, 71-74, spec projs officer, Hq, Health Serv Command, Ft Sam Houston, Tex, 74-75, CHIEF GARRISON DENT CLIN, DEPT DENT, FT SAM HOUSTON, 75- *Concurrent Pos:* Lectr, Georgetown Univ, 63-67. *Mem:* Am Dent Asn; Am Soc Microbiol; Int Asn Dent Res. *Res:* Preventive dentistry. *Mailing Add:* 11511 Whisper Dew San Antonio TX 78230

OHLENDORF, HARRY MAX, b Lockhart, Tex, Oct 12, 40; m 76; c 3. WILDLIFE RESEARCH. *Educ:* Tex A&M Univ, BS, 62, MS, 69, PhD(wildlife ecol), 71. *Prof Exp:* Wildlife res biologist, 71-73, asst dir ecol res admin, Laurel, Md, 73-80, wildlife res biologist, Patuxent Wildlife Res Ctr, fish & wildlife serv, US Dept Interior, Univ Calif, Davis, 80-90; ENVIRON SCIENTIST, CH2M HILL, SACRAMENTO, 90- *Mem:* Wilson Ornith Soc; Pac Seabird Group; Cooper Ornith Soc; Wildlife Soc. *Res:* Effects of environmental pollutants on wildlife and their habitat; ecological risk assessments of hazardous waste sites. *Mailing Add:* CH2M Hill 3840 Rosin Ct Suite 110 Sacramento CA 95834

OHLINE, ROBERT WAYNE, b St Louis, Mo, Mar 9, 34; m 66; c 2. ANALYTICAL CHEMISTRY. *Educ:* Grinnel Col, AB, 56; Northwestern Univ, MS, 58, PhD, 60. *Prof Exp:* Analytical chemist, Mallinckrodt Chem Works, 60-61; from asst prof to assoc prof, 61-72, chmn dept, 77-81, PROF CHEM, NMEX INST MINING & TECHNOL, 72- *Mem:* Am Chem Soc. *Res:* Instrumental methods; wet analytical methods; gas chromatography; ion production in flames; acid rain equilibrium and dynamics. *Mailing Add:* Dept Chem NMex Inst Mining & Technol Socorro NM 87801

OHLROGGE, ALVIN JOHN, b Chilton, Wis, Sept 19, 15; m 44; c 2. SOILS. *Educ:* Univ Wis, BS, 37; Purdue Univ, PhD(soils), 43. *Prof Exp:* Asst, 42-45, from asst prof to prof, 45-86, EMER PROF AGRON, PURDUE UNIV, WEST LAFAYETTE, 86- *Concurrent Pos:* Soil fertil adv, Pakistan Govt, Food & Agr Orgn, UN, 52; NSF fel, Univ Calif, 58-59; vchmn, Nat Joint Comt Fertilizer Appln, 60, chmn, 61; consult, FDA & indust. *Honors & Awards:* Soil Sci Award, Am Soc Agron, 61, Agron Res Award, 76. *Mem:* Fel AAAS; Soil Sci Soc Am; fel Am Soc Agron; hon mem Am Soybean Asn. *Res:* Soil fertility; plant physiology; nutrient uptake from fertilizer bands; growth regulators. *Mailing Add:* 2619 Peace Dr West Lafayette IN 47906

OHLSEN, WILLIAM DAVID, b Evanston, Ill, June 8, 32; m 56; c 3. SOLID STATE PHYSICS. *Educ:* Iowa State Univ, BS, 54; Cornell Univ, PhD(physics), 62. *Prof Exp:* From asst prof to assoc prof, 61-74, PROF PHYSICS, UNIV UTAH, 74 - *Concurrent Pos:* Vis prof, Munich Tech Univ, 69, Marburg, 89. *Mem:* Am Phys Soc; Am Asn Phys Teachers; Am Asn Univ Professors. *Res:* Magnetic resonance and optical studies of defects in solids. *Mailing Add:* Dept Physics Univ Utah Salt Lake City UT 84112

OHLSON, JOHN E, b Seattle, Wash, May 29, 40; m 62; c 3. SATELLITE COMMUNICATIONS. *Educ:* Mass Inst Technol, BS, 62; Stanford Univ, MS, 63, PhD(elec eng), 67. *Prof Exp:* Asst prof elec eng, Univ Southern Calif, 67-71; assoc prof elec eng, Naval Postgrad Sch, 71-78, prof, 78-81; VPRES, STANFORD TELECOMMUNICATIONS, INC, 81- *Concurrent Pos:* Consult, Jet Propulsion Lab, 68-81; Sigma Xi Mennneken res award, Naval Postgrad Sch, 79. *Mem:* Inst Elec & Electronics Engrs; Sigma Xi. *Res:* Communication and radar theory and applications; satellite communications; radioscience. *Mailing Add:* 2706 Ramos Ct Mountain View CA 94040

OHLSSON, ROBERT LOUIS, b St Paul, Minn, Jan 10, 15. ELECTRICAL ENGINEERING. *Educ:* Univ Mich, BSE, 39, MS, 40. *Prof Exp:* Spec res assoc, Radio Res Lab, Harvard Univ, 43-45; engr, Eng Res Inst, 46-58, asst dir, 58-60, ASSOC DIR INST SCI & TECHNOL, WILLOW RUN LABS, UNIV MICH, ANN ARBOR, 60- *Mem:* Inst Elec & Electronics Engrs. *Res:* Guided missile systems; air defense problems; countermeasures; combat surveillance. *Mailing Add:* 170 Hill St Ann Arbor MI 48104

OHLSSON-WILHELM, BETSY MAE, b Boston, Mass, July 17, 42; m 69. GENETICS, IMMUNOLOGY. *Educ:* Radcliffe Col, AB, 63; Harvard Univ, PhD(bact), 69. *Prof Exp:* Am Cancer Soc fel biophys, Univ Chicago, 68-70; res assoc, Inst Cancer Res, 70-73; asst & assoc prof microbiol, Sch Med & Dent, Univ Rochester, 73-85; ASSOC PROF MED, MICROBIOL & IMMUNOL, DIV ENDOCTRINOL, M S HERSHEY MED, PENN STATE UNIV, 85- *Concurrent Pos:* NIH grant. *Mem:* Soc Analytical Cytol. *Res:* Somatic cell genetics; regulation and control mechanisms in somatic cells; cell biology. *Mailing Add:* MS Hershey Med Penn State Univ PO Box 850 Div Endocrinol Dept Med Hershey PA 17033

OHM, E(DWARD) A(LLEN), b Milwaukee, Wis, July 4, 26. ELECTRICAL ENGINEERING. *Educ:* Univ Wis, BS, 50, MS, 51, PhD(elec eng), 53. *Prof Exp:* Mem tech staff, Bell Tel Labs, Inc, 53-86; RETIRED. *Mem:* Inst Elec & Electronics Engrs. *Res:* Waveguide components; low noise receiving systems; space communications; optical modulators. *Mailing Add:* Four Stonehenge Dr Holmdel NJ 07733

OHM, HERBERT WILLIS, b Albert Lea, Minn, Jan 28, 45; m; c 2. AGRONOMY, GENETICS. *Educ:* Univ Minn, BS, 67; NDak State Univ, MSc, 69; Purdue Univ, PhD(genetics, plant breeding), 72. *Prof Exp:* from asst prof to assoc prof, 72-82, PROF AGRON, PURDUE UNIV, 83- *Concurrent Pos:* Int agron, USAID-Purdue Proj, Burkina Faso, 83-85. *Honors & Awards:* Meritorious Serv Award, Crop Sci Soc Am, 89, Crops & Soils Serv Award, 89. *Mem:* Am Soc Agron; fel Crop Sci Soc Am; Sigma Xi; Coun Agr Sci & Technol. *Res:* Plant disease and insect (pest) resistance; wheat; oats; plant physiological traits; cereal protein; Yellow Dwarf virus; plant breeding systems. *Mailing Add:* Dept Agron Purdue Univ West Lafayette IN 47907-7899

OHM, JACK ELTON, b Milwaukee, Wis, Sept 23, 32. MATHEMATICS. *Educ:* Univ Chicago, BS, 54; Univ Calif, Berkeley, PhD(math), 59. *Prof Exp:* Teaching asst math, Univ Calif, Berkeley, 54-57, teaching assoc, 57-59; NSF fel, Johns Hopkins Univ, 59-60; asst prof, Univ Wis, 60-65; from assoc prof to prof, 65-83, EMER PROF MATH, LA STATE UNIV, BATON ROUGE, 83- *Concurrent Pos:* Wis Alumni Res Found fel & res assoc math, Univ Calif, Berkeley, 64-65; vis prof, Purdue Univ, 71-72 & Univ Wis, Milwakee, 78-79. *Mem:* Am Math Soc; Math Asn Am. *Res:* Algebraic geometry; commutative algebra. *Mailing Add:* Dept of Math La State Univ Baton Rouge LA 70803

OHMAN, GUNNAR P(ETER), b Sweden, Dec 15, 18; nat US; m 48; c 3. ELECTRICAL ENGINEERING. *Educ:* Ill Inst Technol, BS, 43; Univ Md, MS, 48, PhD(elec eng), 59. *Prof Exp:* Elec engr, 43-45, electronics engr & group leader, 45-47, electronic scientist & sect head, 47-67, consult, 67-69, res & develop prog coord, 69-76, ASSOC SUPT TACT ELECTRONIC WARFARE DIV, NAVAL RES LAB, 76- *Concurrent Pos:* Lectr, grad sch, Univ Md, 50- *Mem:* Sci Res Soc Am; sr mem Inst Elec & Electronics Engrs. *Res:* Electronics; radar; navigational aids; microwave radiometry; pulse and microwave circuitry; circuit theory; applied mathematics. *Mailing Add:* 5717 Blackhawk Dr Forest Heights MD 20745

OHMART, ROBERT DALE, b Tatum, NMex, Jan 2, 38; m 58; c 4. ZOOLOGY. *Educ:* NMex State Univ, BS, 61, MS, 63; Univ Ariz, PhD(vert zool), 68. *Prof Exp:* NIH fel, Univ Calif, Davis, 68-70; from asst prof to assoc prof, 70-81, PROF ZOOL, ARIZ STATE UNIV, 81-, ASSOC DIR, CTR ENVIRON STUDIES, 80- *Concurrent Pos:* Mem, Desert Tortoise Coun & Found NAm Wild Sheep. *Honors & Awards:* Frank M Chapman Award, Am Mus Natural Hist, 65, 66; Sigma Xi Res Award, 67. *Mem:* Am Ornith Union; Wildlife Soc; Cooper Ornith Soc; Am Inst Biol Sci; Sigma Xi; AAAS; Am Soc Mammalogists; Ecol Soc Am; Wilson Ornithol Soc. *Res:* Wildlife ecology; avian environmental physiology; Riparian Community ecology. *Mailing Add:* Ctr Environ Studies Ariz State Univ Tempe AZ 85287-1201

OHME, PAUL ADOLPH, b Montgomery, Ala, Nov 4, 40; m 64; c 2. MATHEMATICS. *Educ:* Huntingdon Col, BA, 63; Univ Ala, MA, 64; Fla State Univ, PhD(math), 71. *Prof Exp:* Asst prof math, Franklin & Marshall Col, 71-73; assoc prof math, Miss Col, 73-80; MEM FAC, DEPT MATH, NORTHEAST LA UNIV, 80- *Mem:* Sigma Xi; Math Asn Am; Nat Coun Teachers Math; Am Math Soc. *Res:* Application of differential equations to biological problems. *Mailing Add:* Dept Comput Sci Northeast La Univ 700 Univ Ave Monroe LA 71209-0575

OHMER, MERLIN MAURICE, b Napoleonville, La, Mar 15, 23; m 47; c 3. MATHEMATICS. *Educ:* Tulane Univ, BS, 44, MS, 48; Univ Pittsburgh, PhD(math). *Prof Exp:* Asst physics, Tulane Univ, 42-43, asst math, 47-48; from asst prof to prof, Univ Southwestern La, 48-66; prof & head dept, 66-69, dean, 69-88, EMER DEAN, COL SCI, NICHOLLS STATE UNIV, 69- *Concurrent Pos:* Vis instr, Tulane Univ, 49-50; vis assoc prof, Univ Pittsburgh, 53-54; vis lectr, Math Asn Am, 64-73; consult, La State Dept Educ; mem, Nat Metric Speakers Bur; chief health prof adv, Nichols State Univ. *Mem:* Math Asn Am; Am Math Soc; Nat Coun Teachers Math. *Res:* Teacher training and innovations; game theory; symbolic logic; geometry; promotion of the metric system in the United States; author of 2 articles and 8 books. *Mailing Add:* Dean Col Sci Nicholls State Univ Univ Sta Thibodaux LA 70310

OHMOTO, HIROSHI, b Heijo, Japan, Nov 7, 41; m 65; c 2. GEOCHEMISTRY. *Educ:* Hokkaido Univ, BS, 64; Princeton Univ, AM, 67, PhD(geol), 69. *Prof Exp:* Fel geochem, Univ Alta, 68-69, fel, 68-70, lectr, 69-70; from asst prof to assoc prof, 70-78, PROF GEOCHEM, PA STATE UNIV, UNIVERSITY PARK, 78- *Concurrent Pos:* NSF RES GRANT, 72-; Humboldt fel, 81. *Honors & Awards:* Lindgren Award, Soc Econ Geologists, 70; Clark Medal, Geochem Soc, 73. *Mem:* Soc Econ Geologists; Geochem Soc; Am Geophys Union; Soc Mining Geologists Japan. *Res:* Causes of variation of stable isotopes in geologic processes; geological, geochemical and hydrological processes of the formation of metallic ore deposits. *Mailing Add:* Dept of Geosci Pa State Univ University Park PA 16802

OHMS, JACK IVAN, b Walnut, Iowa, Jan 1, 30; m 54; c 3. ANALYTICAL BIOCHEMISTRY. *Educ:* Iowa State Univ, BS, 53; Mich State Univ, PhD(dairy husb), 61. *Prof Exp:* Res asst, Am Found Biol Res, Wis, 53-57; NIH fel, 61-62; appln specialist, 62-66, SR SCIENTIST, SPINCO DIV, BECKMAN INSTRUMENTS INC, 66- *Mem:* AAAS; NY Acad Sci. *Res:* Physiology of reproduction of domestic animals; endocrine immunochemistry; biomedical instrumentation; biochemical calorimetry; liquid chromatography of biochemicals; automated sequence determination of proteins and peptides. *Mailing Add:* 877 Aspen Way Palo Alto CA 94303

OHMS, RICHARD EARL, b Payette, Idaho, June 13, 27; m 56; c 2. PLANT PATHOLOGY. *Educ:* Univ Idaho, BS, 50, MS, 52; Univ Ill, PhD(plant path), 55. *Prof Exp:* Asst plant path, Univ Idaho, 52-55; asst plant pathologist, SDak State Col, 55-57; assoc horticulturist & exten specialist hort, 57-74, exten & res prof, 73-74, exten prof & exten crop specialist, Esten Serv, Univ Idaho, 74-83; RETIRED. *Concurrent Pos:* Seed potato consult, Jordan & Int Potato Ctr, Lima, Peru. *Mem:* Am Soc Hort Sci; Am Phytopath Soc; Potato Asn Am. *Res:* Cereal. *Mailing Add:* 3920 Country Club Dr Lewiston ID 83501

OHNESORGE, WILLIAM EDWARD, b Acushnet, Mass, Sept 11, 31; m 60. ANALYTICAL CHEMISTRY. *Educ:* Brown Univ, ScB, 53; Mass Inst Technol, PhD(chem), 56. *Prof Exp:* From instr to assoc prof chem, Univ RI, 56-65; assoc prof, 65-71, asst chmn dept, 69-77, PROF CHEM, LEHIGH UNIV, 71- *Concurrent Pos:* Vis asst prof, Mass Inst Technol, 64-65; vis prof clin biochem, Univ Toronto, 74; prog dir chem anal, Nat Sci Found, 80-81. *Mem:* Am Chem Soc. *Res:* Fluorescence spectroscopy; electroanalytical chemistry; clinical chemistry; complex ions; luminescence analysis. *Mailing Add:* Dept Chem Lehigh Univ Seeley Mudd Bldg 6 Bethlehem PA 18015

OHNISHI, TOMOKO, b Kobe, Japan, June 8, 31; m 58; c 2. BIOCHEMISTRY, BIOPHYSICS. *Educ:* Nat Kyoto Univ, BS, 56, MS, 58; Nat Nagoya Univ, PhD(biochem), 62. *Prof Exp:* Johnson Res Found vis asst prof, 67-71, asst prof, 71-77, res assoc prof, 77-84, RES PROF BIOCHEM & BIOPHYS, UNIV PA, 84- *Concurrent Pos:* Invited foreign scholar, Centre National de la Recherche Scientifique, Paris, 83. *Mem:* Am Soc Biochemists; Biophys Soc; AAAS. *Res:* Electron transfer and energy conservation in mitochondria; membrane structure and function; bacterial and yeast electron transfer; metabolism. *Mailing Add:* Dept Biochem & Biophys 606 Univ Pa Sch Med Philadelphia PA 19104

OHNISHI, TSUYOSHI, b Otsu, Japan, Dec 17, 31; m 58; c 2. BIOPHYSICS, BIOCHEMISTRY. *Educ:* Kyoto Univ, BS, 54, MS, 56; Nagoya Univ, PhD(biophys), 60. *Prof Exp:* Japanese Soc Prom Sci fel, Nagoya Univ, 60-62 & res fel physiol, Med Sch, 62-63, res fel physics, Univ, 64-65; assoc prof biophys, Waseda Univ, Japan, 65-67; vis assoc prof, Johnson Found, Univ Pa, 67-68; asst prof biophys, Med Col Pa, 69-72; assoc prof biochem & anesthesia, Hahneman Med Col, 73-84; res prof hemat, Sch Med, Hahneman Univ, 84-85; dir, Membrane Res Inst, 85-89; DIR, PHILADELPHIA BIOMED RES INST, 89- *Concurrent Pos:* Vis res fel, Univ Tokyo, 63-65. *Mem:* Biophys Soc; Am Soc Hemat; Am Soc Biochem & Molecular Biol; Am Soc Neurosci. *Res:* Calcium transport in sarcoplasmic reticulum and erythrocyte; molecular mechanism of maglignant hyperthermia; the role of membranes in sickle cell anemia; development of drugs which protect cell membranes from ischemia, traumatic and chemical injuries; development of chemotherapeutic drugs for malaria and AIDS. *Mailing Add:* Philadelphia Biomed Res Inst 502 King of Prussia Rd Radnor PA 19087

OHNO, SUSUMU, b Tokyo, Japan, Feb 1, 28; m 52; c 3. CYTOGENETICS. *Educ:* Tokyo Univ Agr & Technol, DVM, 49; Hokkaido Univ, PhD(path), 56, DSc, 61. *Hon Degrees:* DHH, Kwansei Gakain Univ, 83, ScD, Univ Pa, 84. *Prof Exp:* Res assoc, Dept Res Staff Path, Tokyo Univ, 50-53; res assoc, Dept Exp Path, City of Hope Med Ctr, 52-62, sr res scientist, dept biol, 62-66, chmn, 66-81, DISTINGUISHED SCIENTIST, CITY OF HOPE MED CTR, 81- *Honors & Awards:* Amory Prize, Am Arts & Sci, 81; Kihara Prize Genetics, 83. *Mem:* Nat Acad Sci; Int Soc Hemat; Genetics Soc Am; Am Asn Cancer Res; AAAS. *Res:* Clinical genetics. *Mailing Add:* Dept Theoret Biol Beckman Res Inst City Hope 1450 E Duarte Rd Duarte CA 91010-0269

OHNUKI, YASUSHI, b Kawasaki City, Japan, July 30, 26; c 2. CYTOGENETICS. *Educ:* Hokkaido Univ, BSc, 54, MSc, 56, DSc(cytogenetics), 61. *Prof Exp:* Tobacco Indust Res Comt fel, Dept Anat, Univ Tex Med Br, Galveston, 59-60 & Dept Cellular Biol, Pasadena Found Med Res, 60-61; res assoc, Makino Lab, Fac Sci, Hokkaido Univ, 61-64; chief cytogenetics sect, 64-66, dir dept cytogenetics, 66-72, PROJ DIR CYTOGENETICS STUDIES STRUCTURE & FUNCTION CHROMOSOMES, HUNTINGTON MED RES, 72-, HEAD DEPT CYTOGENETICS, 77- *Mem:* Am Soc Cell Biol; Tissue Cult Asn; Genetics Soc Japan. *Res:* Studies on structure and function of chromosomes; cytogenetic studies of cellular aging and development of prostatic cancer. *Mailing Add:* Huntington Med Res Inst 99 N El Molino Ave Pasadena CA 91101

OHNUMA, SHOROKU, b Akita, Japan, Apr 19, 28; m 56; c 2. ACCELERATOR PHYSICS. *Educ:* Univ Tokyo, BSc, 50; Univ Rochester, PhD(physics), 57. *Prof Exp:* Res assoc physics, Yale Univ, 56-59; asst prof, Waseda Univ, Japan, 59-62; res assoc, Yale Univ, 62-66, sr res assoc, 67-70; physicist II, Fermi Nat Accelerator Lab, 70-; AT PHYS DEPT, UNIV HOUSTON. *Concurrent Pos:* Consult, Brookhaven Nat Lab, 64-69. *Mem:* Am Phys Soc. *Res:* Accelerator physics theory. *Mailing Add:* Phys Dept Univ Houston University Park Houston TX 77204

OHNUMA, TAKAO, b Sendai, Japan, May 16, 32; m 66; c 3. INTERNAL MEDICINE, CANCER. *Educ:* Tohoku Univ, Japan, MD, 57; Univ London, PhD(biochem), 65. *Prof Exp:* Intern, Naval Hosp, Yokosuka, Japan, 57-58 & Lincoln Hosp, Bronx, NY, 58-59; asst resident, Bird S Coler Mem Hosp, Welfare Island, 59-60; resident, Roswell Park Mem Inst, 60-61; vis scientist chem, Chester Beatty Res Inst, London, 63-65; asst med, Tohoku Univ Hosp, Sendai, Japan, 68; cancer res clinician, Dept Med, Roswell Park Mem Inst,

68-73; assoc prof, 73-79, PROF, DEPT NEOPLASTIC DIS, MT SINAI SCH MED, 80- *Concurrent Pos:* Res fel med, Roswell Park Mem Inst, 61-63; res asst prof med, State Univ NY Buffalo, 70-73; assoc attend physician, Dept Neoplastic Dis, Mt Sinai Hosp, NY, 73-79, attend physician, 80- *Mem:* AAAS; Am Asn Cancer Res; Am Soc Clin Oncol; NY Acad Sci. *Res:* Cancer chemotherapy; preclinical and clinical pharmacology. *Mailing Add:* Dept of Neoplastic Dis Mt Sinai Sch of Med 100 St and 5 Ave New York NY 10029

OHR, ELEONORE A, b New York, NY, Jan 28, 32. CELL PHYSIOLOGY. *Educ:* Univ Rochester, AB, 54, MS, 58, PhD(physiol), 62. *Prof Exp:* Asst prof, 63-68, ASSOC PROF PHYSIOL, SCH MED, STATE UNIV NY BUFFALO, 68- *Concurrent Pos:* NIH Fel biophys, Univ Buffalo, 60-61; Am Heart Asn career investr fel muscle proteins, Cardiovasc Res Inst, Med Ctr, Univ Calif, San Francisco, 62-63; gen res support grant, Sch Med, State Univ NY Buffalo, 65-66; Heart Asn Western NY, Inc res support grant, 67-68; United Health Found Western NY, Inc res support grant, 69-74. *Mem:* AAAS; assoc Am Physiol Soc. *Res:* Chemistry of tricainemethanesulfonate and its effects on epithelial transport systems; inulin space in skeletal muscle in vitro; transport of organic bases by renal cortical slices; active transport; electrophysiology. *Mailing Add:* 85 D Oakbrook Dr Williamsville NY 14221

OHRING, GEORGE, b New York, NY, June 20, 31; m 53; c 3. METEOROLOGY, CLIMATE STUDIES. *Educ:* City Col New York, BS, 52; NY Univ, MS, 54, PhD(meteorol), 57. *Prof Exp:* Res asst, NY Univ, 53-57; atmospheric physicist, Air Force Cambridge Res Lab, 57-60; dir meteorol lab, GCA Corp, 60-71; assoc prof meteorol, Tel Aviv Univ, 71-82; chief, land sci br, 83-84, CHIEF, STATELLITE RES LAB, NAT ENVIRON SATELLITE DATA INFO SYST, NAT OCEAN & ATMOSPHERIC ADMIN, 84- *Concurrent Pos:* Vis prof, Tel-Aviv Univ, 69-70; consult, Master Plan, Israel Meteorol Serv, 72-74; vis scientist, Environ Res & Technol, 74 & Nat Oceanic & Atmospheric Admin, 77-81; vis prof, Univ Md, 76; mem, Int Radiation Comn, 75-; Nat Acad Sci resident res assoc, NASA Goddard Space Flight Ctr, 82-83. *Mem:* AAAS; Am Meteorol Soc; Am Geophys Union; Sigma Xi; Israel Meteorol Soc; Int Asn Meterol & Atmospheric Physics. *Res:* Atmospheric radiation; climate models; atmospheric soundings from satellites; meteorology of planetary atmospheres; ocean and land surfaces; satellite remote sensing of the earth. *Mailing Add:* NOAA-NESDIS Washington DC 20233

OHRING, MILTON, b Stryj, Poland, Apr 6, 36; US citizen; m 60; c 3. METALLURGY, MATERIALS SCIENCE. *Educ:* Queen's Col, NY, BS, 58; Columbia Univ, BS, 58, MS, 61, DSc(metall), 65. *Prof Exp:* From instr to asst prof metall, Cooper Union, 59-64; from asst prof to assoc prof metall, 64-77, PROF MAT & METALL, STEVENS INST TECHNOL, 77- *Concurrent Pos:* Coordr, Stevens OPAP Prog, AT&T Bell Labs. *Mem:* Am Vacuum Soc. *Res:* Thin film technology; electromigration and diffusion. *Mailing Add:* Dept Metall Eng Stevens Inst Technol Castle Point Sta Hoboken NJ 07030

OHRN, NILS YNGVE, b Avesta, Sweden, June 11, 34; m 57; c 2. CHEMICAL PHYSICS, QUANTUM CHEMISTRY. *Educ:* Univ Uppsala, MS, 58, PhD(quantum chem), 63, FD, 66. *Prof Exp:* Res assoc quantum chem, Quantum Chem Group, Univ Uppsala, 63-66; assoc prof chem & physics, 66-70, Air Force Off Sci Res & NSF res grants, 69-87, assoc dir quantum theory proj, 67-77, chmn chem dept, 77-83, PROF CHEM & PHYSICS, UNIV FLA, 71-, EXEC DIR QUANTUM THEORY PROJ, 83- *Concurrent Pos:* Ed, Int J Quantum Chem; Fulbright-Hayes fel, 61-63. *Mem:* Fel Am Phys Soc; Sigma Xi; Am Chem Soc; Finnish Acad Sci; Royal Acad Sci, Sweden; Royal Danish Acad Sci & Lett. *Res:* Electronic structure, properties and spectra of atoms; molecules and solids; quantum mechanical studies of properties of matter. *Mailing Add:* WH 363 Univ Fla Gainesville FL 32611

OHTA, MASAO, b Kobe, Japan, May 4, 19; m 49. PHYSICAL CHEMISTRY. *Educ:* Kyoto Univ, BS, 43; Univ Calif, MS, 56; Univ Akron, PhD(chem), 59. *Prof Exp:* Res chemist polymer chem, Mitsui Chem Co, Japan, 43-52 & Monsanto Co, 59-71; sr assoc ed, Chem Abstr Serv, 71-80, sr ed, 80-84; RETIRED. *Mem:* Am Chem Soc; Sigma Xi. *Res:* Physical chemistry of polymers. *Mailing Add:* 5945 NE 201st St Seattle WA 98155

OHTAKE, TAKESHI, b Chiba, Japan, Jan 22, 26; m 53; c 3. CLOUD PHYSICS, METEOROLOGY. *Educ:* Tohoku Univ, Japan, BSc, 52, DSc(meteorol), 61. *Prof Exp:* Technician meteorol, Cent Meteorol Observ, Tokyo, Japan, 43-44; res assoc, Meteorol Res Inst, 47-49; sr res assoc, Geophys Inst, Tohoku Univ, Japan, 52-64; from assoc prof to prof geophys, Geophys Inst, 64-88, EMER PROF PHYSICS, UNIV ALASKA, FAIRBANKS, 88- *Concurrent Pos:* USPHS res fel, 65-69; NSF res fels, 65-82; vis assoc prof, Dept Atmospheric Sci, Colo State Univ, 69-71; vis prof, Nat Inst Polar Res Tokyo, 79-80, Air Force Geophysics Lab, 85-87; distinguished vis prof, Nagoya Univ, 90-91. *Mem:* Am Geophys Union; fel Am Meteorol Soc; fel Royal Meteorol Soc; Meteorol Soc Japan; Sigma Xi. *Res:* Electron microscopic studies for fog, cloud, ice crystal and ice fog nuclei and cloud condensation nuclei; physical and chemical explanation of fog, cloud, snowfall, polar stratospheric clouds and rainfall formation mechanisms; ice crystal nucleation; weather modification. *Mailing Add:* Dept Geophys Univ Alaska 116 Dunnell Fairbanks AK 99701

OIEN, HELEN GROSSBECK, b Paterson, NJ, July 11, 40; m 68. BIOCHEMISTRY, NUTRITION. *Educ:* Rutgers Univ, New Brunswick, AB, 62; Cornell Univ, MS, 64, PhD(biochem), 68. *Prof Exp:* Sr scientist biochem, Shulton, Inc, NJ, 68-69; sr res biochemist, 69-75, res fel, 75-77, sr proj coordr, 77-82, DIR, DEPT PROJ PLANNING & MGT, MERCK & CO, INC, 82- *Mem:* AAAS; Am Chem Soc; Sigma Xi. *Res:* Prostaglandins, especially mechanisms of action; cyclic adenosinemonophosphate; collagen biosynthesis; mechanisms of skin penetration by drugs; intermediary metabolism of nitrogen bases. *Mailing Add:* Merck & Co Inc PO Box 2000 R70-14 Rahway NJ 07065

OISHI, NOBORU, b Kapaa, Hawaii, Nov 11, 28; m 57; c 1. ONCOLOGY & IMMUNOHEMATOLOGY, MOLECULAR BIOLOGY. *Educ:* Wash Univ, AB, 49, MD, 53. *Prof Exp:* From assoc prof to prof med, Sch Med, Univ Hawaii, Manoa, 65-90; PROF MED, UNIV TEX HEALTH SCI CTR, 91- *Concurrent Pos:* USPHS fel hemat, Univ Rochester, 58-59; consult, Queen's Med Ctr, Kuakini Hosp & Home, Inc & St Francis Hosp, 58-; dir clin lab, Annex Lab, Inc, 67-; vpres, Blood Bank Hawaii; dir clin sci, Cancer Ctr Hawaii, 74-85. *Mem:* Fel Am Col Physicians; Am Soc Hemat; Am Soc Clin Oncol; NY Acad Sci; Asn Advan Med Instrumentation. *Res:* Study of the immune capacity of patients with malignant disorders; tumor markers, invitro growth of tumor cells; molecular biology. *Mailing Add:* Med Cancer Ther & Res Ctr Univ Tex Health Sci Ctr 4450 Medical Dr San Antonio TX 78229

OJA, TONIS, b Tallin, Estonia, Aug 22, 37; US citizen; m 61; c 2. PHYSICS, CHEMISTRY. *Educ:* McGill Univ, BSc, 59; Rensselaer Polytech Inst, PhD(physics), 66. *Prof Exp:* Res asst, RCA Victor Res Labs, Can, 59-61; res assoc physics, Brown Univ, 66-68, asst prof, 68-69; asst prof, Univ Denver, 69-73; assoc prof, Ohio Univ, 73-74; assoc prof physics, Hunter Col, NY, 74-80; Matec Inc, 80-84; consult, 84-86; Coulter Inc, 86-89; MATEC APPL SCI, HOPKINTON, MASS, 89- *Concurrent Pos:* Consult, Matec Inc, RI, 72-76. *Mem:* Am Phys Soc. *Res:* Nuclear magnetic and nuclear quadrupole resonance; glasses and material of biological interest; phase transitions; critical phenomena; acoustics; colloid and interface science. *Mailing Add:* 2 Oakwood Pl Scarsdale NY 10583

OJAKAAR, LEO, b Valga, Estonia, Apr 26, 26; US citizen; m 59; c 2. ORGANIC CHEMISTRY, POLYMER CHEMISTRY. *Educ:* Millikin Univ, BS, 53; Va Polytech Inst, MS, 61, PhD(org chem), 64. *Prof Exp:* Res asst chem, Rutgers Univ, 56-59; RES CHEMIST, E I DU PONT DE NEMOURS & CO, INC, 64- *Mem:* Am Chem Soc; Sigma Xi. *Res:* New high molecular weight aromatic hydrocarbons; new isocyanates and urethanes with industrial importance; synthetic rubbers; new neoprene compounds; elastomer chemistry; fluorocarbon chemistry and perfluoroelastomer chemistry. *Mailing Add:* E I du Pont Polymer Prod Dept PO Box 6098 Newark DE 19714-6098

OJALVO, IRVING U, b New York, NY, Jan 16, 36; m 64; c 2. ACCIDENT RECONSTRUCTION, STRUCTURAL DYNAMICS. *Educ:* City Col New York, BS, 56; Mass Inst Technol, MS, 57; New York Univ, ScD(mech eng), 61. *Prof Exp:* Instr eng mech, New York Univ, 57-60; Fulbright scholar, Delft Tech Inst, Neth, 60-61; eng specialist, Repub Aviation, 61-66; consult eng, Harry Belock Assocs, 66-68; group leader & proj engr struct, Grumman Aerospace, 68-80; supvr appl mech, Perkin-Elmer Corp, 80-83; Bullard prof mech eng & chmn dept, Univ Bridgeport, 83-90; HEAD & CHMN, CONN TECHNOL ASSOCS, 83- *Concurrent Pos:* Investr, NSF grant, 60; Fulbright scholar; sr res scientist, Columbia Univ. *Honors & Awards:* NASA Citation. *Mem:* Fel Am Soc Mech Engrs; Am Acad Mech; Am Asn Univ Profs; Am Inst Aeronaut & Astron; Soc Automobile Engrs; Human Factors Soc. *Res:* Eigenvalue computation scheme; structural dynamics; accident reconstruction; improved correlation methods between analysis and testing; solution of ill-conditioned equations. *Mailing Add:* 350 Mayapple Rd Stamford CT 06903

OJALVO, MORRIS, b New York, NY, Mar 4, 24; m 48; c 4. CIVIL ENGINEERING. *Educ:* Rensselaer Polytech Inst, BCE, 44, MCE, 52; Lehigh Univ, PhD(civil eng), 60; Ohio State Univ, JD, 78. *Prof Exp:* Tutor civil eng, City Col New York, 47-49; instr, Rensselaer Polytech Inst, 49-51; asst prof, Princeton Univ, 51-58; res instr, Lehigh Univ, 58-60; assoc prof, 60-64, prof civil eng, 64-82, EMER PROF, OHIO STATE UNIV, 82- *Concurrent Pos:* vis prof, Univ Texas, Austin, 82-83. *Mem:* Am Soc Civil Engrs; Struct Stability Res Coun. *Res:* Structural mechanics and stability; plasticity; elasticity; limit analysis; thin-walled structural members. *Mailing Add:* 2258 Wickliffe Rd Columbus OH 43221

OJALVO, MORRIS S(OLOMON), b New York, NY, July 6, 23; m 49; c 3. CHEMICAL ENGINEERING. *Educ:* Cooper Union, BME, 44; Univ Del, MME, 49; Purdue Univ, PhD(mech eng), 62. *Prof Exp:* Asst, Eng Exp Sta, Pa State Univ, 46-47; instr physics, Univ Ill, 47-48; asst & instr mech eng, Univ Del, 48-50; from asst prof to assoc prof, Univ Md, 50-55; assoc prof, Univ Del, 55-56; asst & instr, Purdue Univ, 56-60; prog dir, eng div, NSF, 65-90; assoc prof, 60-62, prof eng & appl sci, 62-66, PROF & LECTR, GEORGE WASHINGTON UNIV, 66-; CONSULT, 90- *Concurrent Pos:* Prof lectr, George Washington Univ, 66-; UNESCO mech eng, Nat Polytech Inst, Mex, 67-69; NSF Sci Fac fel, 61. *Mem:* Fel Am Soc Mech Engrs; Soc Automotive Engrs; Am Soc Eng Educ; Fine Particle Soc (vpres, 77-78, pres, 78-79); Am Asn Aerosol Res. *Res:* Combined forced and free convection heat transfer; thermodynamics; fluid mechanics; energy conversion; particulate technology. *Mailing Add:* 6006 Broad Branch Rd N W Washington DC 20015-2504

OJEDA, SERGIO RAUL, b Valdivia, Chile, Apr 19, 46; m 68; c 3. NEUROENDOCRINOLOGY, ENDOCRINOLOGY. *Educ:* Univ Chile, DVM, 68. *Prof Exp:* Instr physiol, Univ Austral, Chile, 68, asst prof, 68-71; Ford Found res fel, 72-74, asst prof, 74-78, assoc prof physiol, 78- PROF PHYSIOL, HEALTH SCI CTR, UNIV TEX. *Concurrent Pos:* Prin investr, Univ Tex grant, 74-75 & NIH grant, 77-81 & 81-85. *Mem:* Endocrine Soc; Am Physiol Soc; Int Soc Neuroendocrinol; Soc Exp Biol Med; Soc Study Reproduction. *Res:* Neuroendocrinology of sexual development; neural and hormonal control of anterior pituitary function. *Mailing Add:* Dept Physiol Univ Tex Health Sci Ctr 5323 Harry Hines Blvd Dallas TX 75235

OJIMA, IWAO, b Yokohama, Japan, June 5, 45; m 71. SYNTHETIC ORGANIC & BIOORGANIC CHEMISTRY. *Educ:* Univ Tokyo, BS, 68, MS, 70, PhD(org chem), 73. *Prof Exp:* Res chemist, Sagami Chem Res Ctr, 70-76, sr res fel & group leader, 76-83; assoc prof, 83-84, PROF, STATE UNIV NY STONY BROOK, 84- *Concurrent Pos:* Lectr, Tokyo Inst Technol, 78-79 & 83 & Tokyo Univ Agr & Technol, 83; prin investr, NSF, 83- & NY Sci, Technol Found, 85-, NIH, 86-, ACS-Prof, 87- *Honors & Awards:*

Progress Award, Chem Soc Japan, 76. *Mem:* Am Chem Soc; AAAS; NY Acad Sci. *Res:* Homogeneous catalysis of transition metal complexes; asymmetric synthesis; organic synthesis by means of organometallic reagents; peptide synthesis; beta-lactam chemistry; organo flourine chemistry; bio-organic chemistry, especially in regard to enzyme inhibitors. *Mailing Add:* Dept Chem State Univ NY Stony Brook NY 11790

OKA, MASAMICHI, PATHOLOGY, ENDOCARDITIS. *Educ:* Nippon Med Sch, Tokyo, Japan, MD, 45. *Prof Exp:* ASSOC PROF PATH, CORNELL UNIV, 71-; ATTEND PATHOLOGIST, NORTH SHORE UNIV HOSP, 71- *Mailing Add:* Five Crossway Scarsdale NY 10583

OKA, SEISHI WILLIAM, b Salinas, Calif, Feb 11, 36. HUMAN GENETICS, DENTISTRY. *Educ:* Univ Calif, Berkeley, BS, 59; Univ Pac, DDS, 63; Univ Pittsburgh, PhD(human genetics), 78. *Prof Exp:* Res fel anthrop, Cowell Mem Hosp, Univ Calif, Berkeley, 64-66; res assoc dent anthrop, Cleft Palate Res Ctr, Univ Pittsburgh, 66-69, trainee human genetics dept biostatist, 69-74; chief gen serv genetics, Lancaster Cleft Palate Clin, 74-, exec dir, 79-; RETIRED. *Concurrent Pos:* Pvt pract dent, 64-65; lectr dent anthrop, Univ Pittsburgh, 67-70, clin instr dept oper dent, Sch Dent Med, 69-74, clin staff, Cleft Palate Res Ctr, 72-74; clin instr div human genetics, Dept Pediat, Milton S Hershey Med Ctr, Pa State Univ, 75-; consult prog develop, Lancaster Cleft Palate Clin, 78- *Mem:* Am Soc Human Genetics; Am Genetic Asn; Int Soc Elec & Electronics Engrs; Am Cleft Palate Asn; Soc Craniofacial Genetics. *Mailing Add:* 100 Hidden Acres Lane Conastoga PA 17516

OKA, TAKAMI, b Tokyo, Japan, Jan 1, 40; m 69; c 1. DEVELOPMENTAL BIOLOGY, PHARMACOLOGY. *Educ:* Univ Tokyo, BS, 63; Stanford Univ, PhD(pharmacol), 69. *Prof Exp:* Vis scientist, Nat Inst Radiol Sci, Japan, 63-64; res fel muscle protein & actin, Med Sch, Stanford Univ, 64-65; vis fel, 69-71, staff fel, 71-73, RES CHEMIST & SECT CHIEF, MAMMARY GLAND, NAT INST DIABETES, DIGESTIVE & KIDNEY DIS, 74- *Res:* Hormonal regulation of cellular function and development; biochemistry. *Mailing Add:* NIADDK NIH Bldg Eight Rm 311 Bethesda MD 20892

OKA, TAKESHI, b Tokyo, Japan, June 10, 32; Can citizen; m 60; c 4. ATOMIC & MOLECULAR PHYSICS. *Educ:* Univ Tokyo, BSc, 55, PhD, 60. *Prof Exp:* Res physicist, Nat Res Council Can, 63-81; PROF, UNIV CHICAGO, 81- *Honors & Awards:* Steacie Prize; Plyler Prize. *Mem:* Am Phys Soc; Am Astro Soc; Optical Soc Am; fel Royal Soc Can. *Res:* Molecular Spectroscopy; Molecular ions; Laser Spectroscopy; Intramolecular interactions. *Mailing Add:* Dept Chem Univ Chicago 5735 S Ellis Ave Chicago IL 60637

OKABAYASHI, MICHIO, b Tokyo, Japan, Dec 10, 39; m 68; c 2. PLASMA PHYSICS. *Educ:* Univ Tokyo, BS, 63, MS, 65, PhD(physics), 68. *Prof Exp:* Fel, 68-71, res staff, 71-75, RES PHYSICIST, PRINCETON UNIV, 75- *Mem:* Am Phys Soc. *Res:* Thermonuclear fusion research. *Mailing Add:* Plasma Physics Lab James Forrestal Campus Princeton Univ Princeton NJ 08540

OKABE, HIDEO, b Naganoken, Japan, Dec 13, 23; m 59; c 3. PHYSICAL CHEMISTRY, PHOTOCHEMISTRY. *Educ:* Univ Tokyo, BS, 47; Univ Rochester, PhD(chem), 57. *Prof Exp:* Nat Res Coun Can fel, 56-58; phys chemist, Nat Bur Standards, 59-83, sr scientist, Div Phys Chem, 74-78 & Div Molecular Spectros, 79-83; RES PROF, DEPT CHEM, HOWARD UNIV, 83- *Concurrent Pos:* Guest prof, Inst Phys Chem, Univ Bonn, 63-65; vis prof, Dept Chem, Tokyo Inst Technol, Japan, 78 & Inst Molecular Sci, Okazaki, Japan, 85; sr res assoc, Goddard Space Flight Ctr, NASA, Greenbelt, Md, 82-83; vis prof, Res Inst Appl Elec, Hokkaido Univ, Sapporo, Japan, 90. *Honors & Awards:* Gold Medal, Dept of Com, 73. *Mem:* Am Chem Soc; Sigma Xi; Chem Soc Japan. *Res:* Vacuum ultraviolet photochemistry in the gas phase; fluorescence; air pollution; planetary atmospheres; laser induced chemical vapor deposition. *Mailing Add:* Dept Chem Howard Univ Washington DC 20059

OKADA, R(OBERT) H(ARRY), b New York, NY, Nov 13, 25; m 56; c 4. ELECTRICAL ENGINEERING. *Educ:* Drexel Inst, BS, 48; Univ Pa, MS, 49, PhD(elec eng), 57. *Prof Exp:* Assoc res engr, Burroughs Corp, 49-52; chief engr, Polyphase Instrument Co, 52-54; from asst prof to assoc prof elec eng, Moore Sch, Univ Pa, 54-61, asst prof elec eng in med, Med Sch, 57-61; sr staff mem, Arthur D Little, Inc, 61-62; tech dir, Epsco Pac, 62-63; PRES & FOUNDER, R O ASSOCS, INC, 63- *Concurrent Pos:* Consult, Providential Mutual Life Ins Co, 56-59, Burroughs Corp, 57- & Air Design, Inc. *Mem:* Sigma Xi; Biophys Soc; Inst Elec & Electronics Engrs. *Res:* Networks; circuits; fields; electrocardiography; instrumentation. *Mailing Add:* PO Box 61419 Sunnyvale CA 94088

OKADA, TADASHI A, b Numazu, Japan, Mar 31, 28; m 56. CYTOGENETICS, MOLECULAR BIOLOGY. *Educ:* Hokkaido Univ, BS, 53, MS, 55, DS(cytogenetics), 59. *Prof Exp:* Asst res scientist, Dept Biochem, City of Hope Med Ctr, Duarte, Calif, 56-62; asst res scientist, DNA denaturation, Univ Edinburgh, 62-63; res scientist, Biol Div, Ciba Ltd, Switz, 63-65; asst res scientist, Dept Biochem, 65-69, assoc res scientist, Dept Med Genetics, 69-85, DEPT MOLECULAR GENETICS, CITY OF HOPE MED CTR, 85-; ASSOC RES SCIENTIST, UNIV CALIF, SAN DIEGO, 79- *Mem:* Am Soc Cell Biol; Electron Micros Soc Am. *Res:* Ultrastructure of meiosis and mitosis; initiation of DNA synthesis; structure of chromatin and DNA; protein complexes. *Mailing Add:* 609 Royal View St Duarte CA 91010

OKAL, EMILE ANDRE, b Paris, France, Aug 7, 50; m 78; c 2. SEISMOLOGY, GEOPHYSICS. *Educ:* Univ Paris, Agrege, 71; Ecole Normale Superiere, Univ Paris, MS, 72; Calif Inst Technol, PhD(geophys), 78. *Prof Exp:* Res fel geophys, Calif Inst Technol, 78; asst prof, Yale Univ, 78-81, assoc prof Geophys, 81-83; assoc prof, 84-89, PROF, NORTHWESTERN UNIV, 89- *Mem:* Am Geophys Union; Seismol Soc Am. *Res:* Seimology; plate tectonics; planetology. *Mailing Add:* Dept Geol Sci, 212 Locy Northwestern Univ Evanston IL 60208

OKAMOTO, K KEITH, b Upland, Calif, Sept 15, 20; m 55; c 3. CHEMICAL ENGINEERING. *Educ:* Univ Tulsa, BS, 57; Univ Tex, MS, 49, PhD(chem eng), 52. *Prof Exp:* Res engr, Calif Res Corp, 52-59; sr res engr, plastics div, 59-62, group leader, hydrocarbon div, 62-65, group supvr, 65-66, SR GROUP SUPVR, HYDROCARBON & POLYMER DIV, MONSANTO CO, 66- *Mem:* Am Chem Soc; Am Inst Chem Engrs; Sigma Xi. *Res:* Hydrocarbon process research; hydrocarbon phase equilibria; ethylene; reforming; methanol process technology. *Mailing Add:* 1317 Plantation Dr Dickinson TX 77539

OKAMOTO, MICHIKO, b Tokyo, Japan, Mar 3, 32; m 59; c 1. PHARMACOLOGY. *Educ:* Tokyo Col Pharm, BS, 54; Purdue Univ, MS, 57; Cornell Univ, PhD(pharmacol), 64. *Prof Exp:* Assoc prof, 71-77, PROF PHARMACOL & ANESTHESIOL, MED COL, CORNELL UNIV, 77- *Concurrent Pos:* Res fel pharmacol, Med Col, Cornell Univ, 64-66, USPHS training fel, 64-67. *Mem:* AAAS; Am Soc Pharmacol & Exp Therapeut; Harvey Soc; NY Acad Sci; Soc Neurosci. *Res:* Neuropharmacology of synaptic transmission and nerves; pharmacology of drugs of abuse; sedative-hypnotics and alcohol. *Mailing Add:* Dept Pharmacol Cornell Univ Med Coll 1300 York Ave New York NY 10021

OKAMURA, JUDY PAULETTE, b Glendale, Calif, Oct 12, 42; m 70; c 2. ANALYTICAL CHEMISTRY. *Educ:* Calif State Univ, Fresno, BS, 64; Calif State Univ, San Diego, MS, 69; Univ Calif, Riverside, PhD(chem), 72. *Prof Exp:* Sr chemist, Analytical Res Labs Inc, 72-73; PROF CHEM, SAN BERNARDINO VALLEY COL, 73- *Mem:* Am Chem Soc. *Res:* Physical basis for separations in all areas of chromatography. *Mailing Add:* Dept of Chem San Bernardino Valley Col San Bernardino CA 92410

OKAMURA, KIYOHISA, b Chuseinan-do, Korea, Feb 8, 35. MECHANICAL ENGINEERING. *Educ:* Kyushu Univ, BS, 57; Univ Tokyo, MSME, 59; Purdue Univ, PhD(automatic control), 63. *Prof Exp:* Mem res staff, Japan Atomic Energy Res Inst, 59-60; sr proj engr, Allison Div, Gen Motors Corp, 63-66; asst prof systs eng, Rensselaer Polytech Inst, 66-68; assoc prof mech eng, N Dak State Univ, 68-85; PROF MECH ENG, BRADLEY UNIV, 85- *Concurrent Pos:* Consult, Japanese Indust, Med & Press agents. *Mem:* Am Soc Mech Engrs; Am Soc Eng Educ. *Res:* Automatic control; bioengineering; microprocessor interfacing; microprocessor applications in biomedical engineering. *Mailing Add:* Dept of Mech Eng Bradley Univ Pvt Univ Sta Peoria IL 61625

OKAMURA, SOGO, b Mie Pref, Japan, Mar 18, 18; m 43; c 3. UNIVERSITY ADMINISTRATION, MICROWAVE ENGINEERING. *Educ:* Univ Tokyo, BEng, 40, DrEng, 51. *Prof Exp:* Prof elec eng, Univ Tokyo, 51-78, dean, Fac Eng, 73-75, sr adv to pres, 75-77, emer prof, 78; dean Grad Sch, 79-83, PROF ELEC ENG, TOKYO DENKI UNIV, 78-, PRES, 90-; SR ADV TO RECTOR, UN UNIV, 88- *Concurrent Pos:* Consult, Elec Commun Labs, Nippen Tel & Tel Pub Corp, 69-73; mem, Broadcast Eng Coun, Japan Broadcast Corp, 71-, Japanese Nat Comn, UNESCO, 83-; sci adv, Ministry Educ, Sci & Cult, 75-78; emer prof, Univ Tokyo, 78; dir, Japanese Nat Rwy, 81-87. *Mem:* Inst Elec & Electronics Engrs; Int Asn Traffic & Safety Sci. *Res:* Measurement of the characteristics of electron devices in microwave and millimeterwave frequencies; science and technology in Japan. *Mailing Add:* Tokyo Denki Univ 2-2 Kanda Nishiki-Cho, Chiyoda-Ku Tokyo 101 Japan

OKAMURA, WILLIAM H, b Los Angeles, Calif, Feb 19, 41; m 70. SYNTHETIC ORGANIC CHEMISTRY, MEDICINAL CHEMISTRY. *Educ:* Univ Calif, Los Angeles, BS, 62; Columbia Univ, PhD(org chem), 66. *Prof Exp:* Nat Acad Sci-Nat Res Coun-Air Force Off Sci Res fel, Cambridge Univ, 66-67; asst prof org chem, 67-74, assoc prof, 74-76, PROF CHEM, UNIV CALIF, RIVERSIDE, 76- *Mem:* Am Chem Soc; Am Soc Photobiol; Sigma Xi. *Res:* Natural products; vitamin D analogs and metabolites; retinoids; chemistry of vision; aromatic molecules; heterocycles; organometallics; allene chemistry. *Mailing Add:* Dept of Chem Univ of Calif 1900 University Ave Riverside CA 92521

O'KANE, DANIEL JOSEPH, b Jackson Heights, NY, June 20, 19; m 46; c 3. BACTERIOLOGY. *Educ:* Cornell Univ, BS, 40, PhD(bact), 47; Univ Wis, MS, 41. *Hon Degrees:* MA, Univ Pa, 71. *Prof Exp:* From vdean to actg dean grad sch arts & sci, Univ Pa, 66-74, dep assoc provost, 74-78, chmn dept biol, 78-80, from asst prof to prof microbiol, 47-89, assoc chmn, 80-89, EMER PROF, UNIV PA, 89- *Concurrent Pos:* Fulbright & Guggenheim fel, Oxford Univ, 55-56; Wis Alumni Res Found scholar & Abbott Labs fel; mem microbiol training comt, Nat Inst Gen Med Sci, chmn microbiol training panel, 68-69. *Mem:* AAAS; Am Soc Microbiol; fel Am Acad Microbiol; NY Acad Sci; Brit Soc Gen Microbiol. *Res:* Microbial metabolism and physiology; control of carbohydrate metabolism. *Mailing Add:* 365 Nansemond St Leesburg VA 22075

O'KANE, KEVIN CHARLES, b Boston, Mass, July 13, 46. BIOMEDICAL COMPUTING, INFORMATION STORAGE & RETRIEVAL. *Educ:* Boston Col, BS, 68; Penn State, PhD(computer Sci), 72. *Prof Exp:* Asst prof computer sci, Penn State Univ, Univ Park, 72-74, Wash Univ, St Louis, 74-75, Ohio State Univ, Columbus, 75-78; sr engr computer sci, Raytheon Co, Norward, Mass, 78-79; from assoc prof to prof computer sci, Univ Tenn, Knoxville, 79-86; PROF COMPUTER SCI, UNIV ALA, TUSCALOOSA, 86- *Concurrent Pos:* Asst prof, Dept OBGYN, Penn State Univ, Hershey, 72-74; asst prof allied med, Ohio State Univ, Columbus, 75-78; head, Computer Sci Dept, Univ Ala, Tuscaloosa, 86-90. *Mem:* Asn Comput Mach; Inst Elec & Electronics Engrs Computer Soc; Mumps User's Group; Sigma Xi. *Res:* Application of computers to biomedicine; computer based information storage and retrieval; mumps language development. *Mailing Add:* Computer Sci Dept Univ Ala Tuscaloosa AL 35487-0290

OKASHIMO, KATSUMI, b Vancouver, BC, Mar 18, 29; m 54; c 4. SOFTWARE, ACADEMIC ADMINISTRATION. *Educ:* McMaster Univ, BA, 52; Univ Toronto, MA, 53, PhD(math), 55. *Prof Exp:* Rep sci comput, Defence Res Bd, 55-59; mgr math & eng comput, Ontario Hydro, 60-67; from assoc prof to prof math & statist, dir, Inst Computer Sci, Univ Guelph, 67-77, prof computer & info sci, 71-87; RETIRED. *Mem:* Asn Comput Mach. *Res:* Management and administration of data processing; educational techniques in computer science; system analysis and design. *Mailing Add:* 80 Oxford St Guelph ON N1H 2N6 Can

OKAYA, YOSHI HARU, b Osaka, Japan, Feb 11, 27; m 50; c 4. CRYSTALLOGRAPHY. *Educ:* Osaka Univ, BS, 47, PhD(crystallog), 56. *Prof Exp:* Res assoc physics, Pa State Univ, 53-56, from asst prof to assoc prof, 56-61; res physicist, Res Ctr, IBM Corp, 61-67; PROF CHEM, STATE UNIV NY STONY BROOK, 67- *Concurrent Pos:* Mem staff, Brookhaven Nat Lab, 58 & Nat Lab High Energy Physics, Tsukuba, Japan, 81. *Mem:* Am Crystallog Asn; Crystallog Soc Japan; Chem Soc Japan. *Res:* Crystal chemistry; crystal structure determination in organic and inorganic compounds; disorder and diffuse scattering; structural basis for physical behaviors; computer controlled experiments, design and concept; solid state chemistry; use of synchrotron radiation; use of x-ray absorbtion spectra in radiology. *Mailing Add:* Dept Chem State Univ NY Stony Brook NY 11794-3400

OKAZAKI, HARUO, b Kochi, Japan, Apr 11, 26; US citizen; m 57; c 3. PATHOLOGY, NEUROPATHOLOGY. *Educ:* Kochi Col, BS, 47; Osaka Univ, MD, 51. *Prof Exp:* Asst instr psychiat, Keio Univ Med Sch, 52-53; resident neurol, Kings County Hosp, Brooklyn, 54-56, resident neuropath, 56-57; assoc, Col Physicians & Surgeons, Columbia Univ, 58-59; from instr to asst prof, State Univ NY Downstate Med Ctr, 59-63; spec appointee, 65, from asst prof to assoc prof path, 67-82, MEM STAFF, MAYO CLIN, UNIV MINN, 66-, PROF PATH, MAYO MED SCH, 82- *Concurrent Pos:* Res fel, Rockland State Hosp, NY, 53-54; vis fel path, Columbia-Presby Med Ctr, NY, 64; Nat Inst Neurol Dis & Blindness spec fel, 65; mem stroke coun, Am Heart Asn. *Mem:* Am Asn Neuropath; Asn Res Nerv & Ment Dis; Am Acad Neurol. *Res:* Human pathology with particular reference to neurologic and psychiatric diseases; anatomic pathology. *Mailing Add:* Dept Path Mayo Med Sch Rochester MN 55901

OKE, JOHN BEVERLEY, b Sault Ste Marie, Ont, Mar 23, 28; m 55; c 3. ASTRONOMY. *Educ:* Univ Toronto, BA, 49, MA, 50; Princeton Univ, PhD(astron), 53. *Prof Exp:* Lectr astron, Univ Toronto, 53-56, asst prof, 56-58; from asst prof to assoc prof astron, 58-64, assoc dir, Hale Observ, 70-78, PROF ASTRON, CALIF INST TECHNOL, 64- *Mem:* Am Astron Soc; Int Astron Union; Astron Soc Pac. *Res:* Galaxies and quasars; astronomical instrumentation. *Mailing Add:* 105-24 Calif Inst Technol 1201 E California Blvd Pasadena CA 91125

OKE, TIMOTHY RICHARD, b Devon, Eng, Nov 22, 41; m 67; c 2. URBAN CLIMATOLOGY, MICROCLIMATOLOGY. *Educ:* Bristol Univ, BSc, 63; McMaster Univ, MA, 64, PhD(geog), 67. *Prof Exp:* Res asst microclimat, McMaster Univ, 64-66, lectr, 66-67; asst prof geog, McGill Univ, 67-70; from asst prof to assoc prof, 70-78, PROF GEOG, UNIV BC, 78- *Concurrent Pos:* Res grants, Nat Res Coun Can, 67-68, 68- & Atmospheric Environ Serv, 68-; rapporteur & consult urban climate, World Meteorol Orgn, 71-; vis scientist, Uppsala Univ, 75; vis prof, Bristol Univ, 76; ed, Atmosphere-Ocean, 77-80; accredited consult meteorologist, Can Meteorol & Ocean Soc; vis fel, Keble Col, Oxford, 90-91. *Honors & Awards:* President's Prize, Can Meteorol & Oceanog Soc; Killam Res Prize. *Mem:* Am Meteorol Soc; fel Royal Meteorol Soc; Can Meteorol & Ocean Soc; Can Asn Geog; Am Geophys Union. *Res:* Urban climatology; micrometeorology of the lowest layers of the atmosphere; energy balance; boundary layer climatology. *Mailing Add:* Dept Geog Univ BC Atmospheric Sci Prog Vancouver BC V6T 1W5 Can

O'KEAN, HERMAN C, b New York, NY, Sept 28, 33. MICROWAVE. *Educ:* Columbia Col, BA, 55; Col Sch Eng, PhD(eng), 56. *Prof Exp:* Mgr & engr, Bell Telephone Lab, 56-66; mgr & engr, AIL, 66-77; VPRES, GOV SYST, LNR COMMUN LAB, 71- *Mem:* Fel Inst Elec Electronics Engrs. *Mailing Add:* Govt Syst Div LNR Comm Inc 180 Marcus Blvd Hauppauge NY 11788

O'KEEFE, DENNIS D, CARDIAC PHYSIOLOGY. *Educ:* Cornell Univ, MD, 62. *Prof Exp:* ASSOC PROF SURG, MED SCH, HARVARD UNIV, 72-; ASSOC PHYSIOLOGIST, MASS GEN HOSP, 72- *Concurrent Pos:* Mem sr staff, emergency dept, Lawrence Gen Hosp, 80- *Res:* Interoperative myocardial preservation. *Mailing Add:* 259 Granville Lane North Andover MA 01845

O'KEEFE, DENNIS ROBERT, b Ottawa, Ont, Nov 15, 39; m 65; c 2. PHYSICS, MICROELECTRONICS ENGINEERING. *Educ:* Univ Ottawa, BSc, 62; Univ Toronto, MASc, 64, PhD(appl sci & eng), 68. *Prof Exp:* staff scientist exp physics & chem, Gen Atomic Co, 68-85; STAFF ENGR, TECHNOL CTR, HUGHES AIRCRAFT CO, 85- *Concurrent Pos:* Nat Res Coun scholar, Can. *Mem:* Am Phys Soc. *Res:* Surface physics and chemistry; rarefied gas dynamics; gas-surface interactions; molecular and atomic beam techniques; ultrahigh vacuum technology; advanced energy concepts; nuclear reactor engineering; thermochemical water splitting for hydrogen production; microelectronics process engineering. *Mailing Add:* 5045 Park West Ave San Diego CA 92117

O'KEEFE, EDWARD JOHN, b Washington, DC, Jan 29, 41; c 2. BIOCHEMISTRY. *Educ:* Yale Univ, BA, 62, MD, 66. *Prof Exp:* Asst prof med, John Hopkins Univ Sch Med, 74-75; assoc prof, 78-83, PROF DERMAT, UNIV NC, 83- *Concurrent Pos:* Mem gen med study sect, NIH; assoc ed, J Invest Med, 87-; vis scientist, Howard Hughes Med Inst, Dept Biochem, Duke Univ, 87-88. *Mem:* Am Soc Cell Biol; Soc Invest Dermat; Am Acad Dermat. *Res:* epithelial junctions and purification and properties of desmosomal proteins; structural proteins of epithelium. *Mailing Add:* Dept Dermat Univ N C Sch Med Chapel Hill NC 27514

O'KEEFE, EUGENE H, PROSTAGLANDINS, THROMBOXANE ANTAGONISTS. *Educ:* Univ Bridgeport, BA, 52. *Prof Exp:* asst res investr, 61-87, LAB SUPVR, SQUIBB INST FOR MED RES, 87- *Mailing Add:* Bristol-Myers/Squibb Pharm Res Inst PO Box 4000 Princeton NJ 08540

O'KEEFE, J GEORGE, b Averill Park, NY, Feb 6, 31; m 64; c 2. PHYSICS. *Educ:* St Bernardine of Siena Col, BS, 52; Rensselaer Polytech Inst, MS, 56; Brown Univ, PhD(physics), 61. *Prof Exp:* Res assoc physics, Brown Univ, 61-62; from asst prof to assoc prof, 62-71, PROF PHYSICS, RI COL, 71- *Mem:* Fel AAAS; Am Asn Physics Teachers; Am Phys Soc; Nat Sci Teachers Asn; Sigma Xi. *Res:* Nuclear magnetic resonance. *Mailing Add:* 20 Maple Crest Dr Greenville RI 02828

O'KEEFE, JOHN ALOYSIUS, b Lynn, Mass, Oct 13, 16; m 41; c 9. TEKTITES, PLANETARY RINGS. *Educ:* Harvard Univ, AB, 37; Univ Chicago, PhD(astron), 41. *Hon Degrees:* Dsc, Alfred Univ, 85. *Prof Exp:* Prof math & physics, Brenau Col, 41-42; mathematician, Army Map Serv, 45-58; GEOPHYSICIST, GODDARD SPACE FLIGHT CTR, NASA, 58- *Concurrent Pos:* Vis Prof Ceramics, Alfred Univ, 83. *Mem:* Am Astron Soc; fel Am Geophys Union; Int Astron Union; Am Ceramic Soc; fel Meteoritical Soc. *Res:* Tektites; planetary rings; mass extinctions; origin of the moon. *Mailing Add:* Code 681 NASA Goddard Space Flight Ctr Greenbelt MD 20771

O'KEEFE, JOHN DUGAN, b Anaconda, Mont, Nov 7, 37; m 68; c 2. PLANETARY IMPACT, LASER PHYSICS. *Educ:* Calif State Univ, Long Beach, BS, 62; Univ Southern Calif, MS, 65; Univ Calif, Los Angeles, PhD(planetary physics), 76. *Prof Exp:* Res physicist, Space Div, Rockwell Int, 62-69; sr scientist & prog mgr, 69-83, MGR OPTICS & DIR ENERGY LAB, TRW INC, 83- *Concurrent Pos:* Vis res assoc, Calif Inst Technol, 76- *Mem:* Optical Soc Am; Am Geophys Union; Asn Planetary Sci. *Res:* Role of hypervelocity impact on the evolution of the solar system, including the effect on extinction of biota on earth; high energy lasers and phenomenology. *Mailing Add:* 418 Prospect Ave Manhattan Beach CA 90266

O'KEEFE, MICHAEL ADRIAN, b Melbourne, Australia, Sept 8, 42. SOLID STATE PHYSICS, ELECTRON MICROSCOPY. *Educ:* Univ Melbourne, BSc, 70, PhD(physics), 75. *Prof Exp:* Sci officer catalysis, Commonwealth Sci & Indust, 61-68; exp officer, Res Orgn, 68-69, exp officer electron micros, 69-75; FEL ELECTRON MICROS, ARIZ STATE UNIV, 76- *Mem:* AAAS; Electron Micros Soc Am. *Res:* High resolution; computation of electron microscope lattice images of minerals and defect structures for comparison with experimental images as an aid to structure determination. *Mailing Add:* Ctr Solid State Sci Ariz State Univ Tempe AZ 85287

O'KEEFE, THOMAS JOSEPH, b St Louis, Mo, Oct 2, 35; m 57; c 5. METALLURGICAL ENGINEERING. *Educ:* Mo Sch Mines, BS, 58; Univ Mo-Rolla, PhD(metall eng), 65. *Prof Exp:* Process control metallurgist metall eng, Dow Metal Prod, 59-61; res dir, Metal System Div, Air Prod & Chem, 77-78; PROF METALL ENG, MAT RES CTR, UNIV MO-ROLLA, 78- *Concurrent Pos:* Prof metall eng, Univ Mo-Rolla, 65-77; comt mem, Electrolytic Technol Adv Comt, Dept Energy, 77-80; Jefferson-Smurfit fel, 84-85. *Mem:* Am Inst Mining & Metall Engrs; Sigma Xi. *Res:* Electrodeposition processes of nonferrous metals and dental materials. *Mailing Add:* 5 Crestview Dr Rolla MO 65401

O'KEEFFE, DAVID JOHN, b New York, NY, July 5, 30; m 56; c 6. PHYSICS. *Educ:* St Peter's Col, NJ, BS, 53; Univ Md, MS, 62; Cath Univ Am, PhD(physics), 69. *Prof Exp:* Physicist, Naval Surface Warfare Ctr, White Oak Lab, 55-86; prof physics, US Naval Acad, Md, 86-87; ASST PROF PHYSICS & CHMN DEPT, ST PETERS COL, 87- *Mem:* Am Phys Soc. *Res:* Theoretical investigations of the behavior of solids under high pressure utilizing the methodologies of statistical mechanics, lattice dynamics and the nature of the cohesive energy of the solid in question. *Mailing Add:* 745 Dianne Ct Rahway NJ 07065

O'KEEFFE, LAWRENCE EUGENE, b Walhalla, NDak, May 12, 34; m 54; c 6. AGRICULTURAL ENTOMOLOGY. *Educ:* NDak State Univ, BS, 56, MS, 58; Iowa State Univ, PhD(econ entom), 65. *Prof Exp:* Port entomologist, Wis Dept Agr, 60-62; asst state entomologist, Iowa Dept Agr, 62-65; exten entomologist, 65-69, from asst prof & asst entomologist to assoc prof & assoc entomologist, 69-81, asst head dept, 82-87, PROF ENTOM & ENTOMOLOGIST, UNIV IDAHO, 81-, HEAD DEPT, 87- *Mem:* Entom Soc Am; Sigma Xi. *Res:* Host-plant resistance to insects and host-plant selection; applied and regulatory entomology; control of insects of grain legumes and soil arthropods; intergrated pest management. *Mailing Add:* Dept of Plant Soil & Entom Sci Univ of Idaho Moscow ID 83843

O'KEEFFE, MICHAEL, b Bury St Edmunds, Eng, Apr 3, 34. SOLID STATE CHEMISTRY. *Educ:* Bristol Univ, BSc, 54, PhD(chem), 58, DSc, 77. *Prof Exp:* Chemist, Mullard Res Labs, 58-59; res assoc chem, Ind Univ, 60-62; from asst prof to assoc prof, 63-69, PROF CHEM, ARIZ STATE UNIV, 69- *Concurrent Pos:* Fel, Ind Univ, 60-62. *Res:* Chemistry of solids. *Mailing Add:* Dept of Chem Ariz State Univ Tempe AZ 82587

O'KELLEY, GROVER DAVIS, b Birmingham, Ala, Nov 23, 28; m 50; c 2. NUCLEAR CHEMISTRY. *Educ:* Howard Col, AB, 48; Univ Calif, Berkeley, PhD(chem), 51. *Prof Exp:* Chemist nuclear chem res, Radiation Lab, 49-51; lead chemist, Calif Res & Develop Co, 51-54; sr chemist, Oak Ridge Nat Lab, 54-59, group leader, 59-74, sr res staff mem, Chem Div, 74-89, GROUP LEADER, CHEM TECHNOL DIV, OAK RIDGE NAT LAB, 89- *Concurrent Pos:* Prof chem, Univ Tenn, Knoxville, 64-82; chmn, Gordon Res Conf on Nuclear Chem, 67; mem, Lunar Sample Preliminary Exam Team, 69-70; mem comt on nuclear sci & chmn subcomt on radiochem, Nat Res Coun, 74-78. *Honors & Awards:* Apollo Achievement Award, NASA, 70 & Group Achievement Award, 73. *Mem:* AAAS; fel Am Phys Soc; Am Geophys Union; Am Chem Soc; Sigma Xi; fel Am Inst Chemists. *Res:* Chemical behavior and separations chemistry of transuranium elements; radiochemistry and nuclear chemistry; packaging and transportation of radioactive materials. *Mailing Add:* Oak Ridge Nat Lab PO Box 2008 MS-6385 Bldg 7930 Oak Ridge TN 37831-6385

O'KELLEY, JOSEPH CHARLES, b Unadilla, Ga, May 9, 22; m 51; c 4. PLANT PHYSIOLOGY. *Educ:* Univ NC, AB, 43, MA, 48; Iowa State Col, PhD(plant physiol), 50. *Prof Exp:* From asst prof to prof biol, Univ Ala, 51-90, chmn dept, 70-73,; RETIRED. *Concurrent Pos:* NSF fel, Univ Wis, 54-55; NIH fel, Johns Hopkins Univ, 65-66. *Mem:* AAAS; Am Soc Photobiol; Phycol Soc Am; Bot Soc Am; Am Soc Plant Physiologists. *Res:* Algal and cell physiology; photobiology. *Mailing Add:* 553 17th St Tuscaloosa AL 35401

OKEN, DONALD, b New York, NY, Jan 21, 28; m 66; c 2. PSYCHIATRY, PSYCHOSOMATIC MEDICINE. *Educ:* Syracuse Univ, 44-45; Harvard Univ, MD, 49; Am Bd Psychiat & Neurol, dipl psychiat, 58. *Prof Exp:* Res assoc, Psychosomatic & Psychiat Inst, Michael Reese Hosp, Chicago, 56-58, from asst dir to assoc dir, 58-65; chief clin res br, NIMH, 66-68, actg dir div extramural res progs, 66-67; prof psychiat, State Univ NY Upstate Med Ctr, 68-83, chmn dept, 68-82; PROF PSYCHIAT, UNIV PA, 83- *Concurrent Pos:* Buswell fel med & psychiat, Sch Med, Univ Rochester, 51-52; NIMH training fel, Neuropsychiat & Psychiat Inst, Univ Ill, 55-56; Found Fund Res in Psychiat fel, Psychosom & Psychiat Inst, Michael Reese Hosp, 57-59; examr, Am Bd Psychiat & Neurol, 61-; mem, Ment Health Bd, Onondaga County, NY, 69-76; vis prof psychiat, Col Med, Univ NC, 77-78; Ed-in-chief, Psychosomatic Med, 82-; dir, Consult-Liaison Serv, Dept Psychiat, Pa Hosp, 83- *Mem:* AAAS; Am Psychiat Asn; Am Psychosom Soc. *Res:* Psychosomatic medicine. *Mailing Add:* Dept Psychiat Pa Hosp 800 Spruce St Philadelphia PA 19107-6192

OKERHOLM, RICHARD ARTHUR, b Woburn, Mass, Nov 10, 41; m 65; c 2. BIOCHEMISTRY. *Educ:* Lowell Technol Inst, BS, 64; Boston Univ, PhD(biochem), 70. *Prof Exp:* NIH fel, Boston Univ, 69-70; from res biochemist to sr res biochemist, Parke-Davis & Co, 70-77; sect head, 77-79, DEPT HEAD DRUG METAB, MARION MERRELL DOW RES INST- *Honors & Awards:* Sigma Xi Res Award, 64. *Mem:* Am Chem Soc; Am Soc Mass Spectrometry; NY Acad Sci; Am Soc Pharmacol & Exp Therapeut; Acad Pharmaceut Sci. *Res:* Analysis, pharmacokinetics and metabolism of drugs; application of mass spectrometry to the analysis of drugs in biological fluids. *Mailing Add:* Drug Metab Dept Marion Merrell Dow Res Inst 2110 E Galbraith Cincinnati OH 45215

OKIISHI, THEODORE HISAO, b Honolulu, Hawaii, Jan 15, 39; m 63; c 4. TURBOMACHINE FLUID DYNAMICS. *Educ:* Iowa State Univ, BS, 60, MS, 63, PhD(mech eng & eng mech), 65. *Prof Exp:* From asst prof to assoc prof, 67-77, PROF MECH ENG, IOWA STATE UNIV, 77-, CHAIR, DEPT MECH ENG, 90- *Honors & Awards:* Cert of Recognition, NASA, 75; Ralph R Teetor Award, Soc Automotive Engrs, 76; Melville Medal, Am Soc Mech Engrs, 89. *Mem:* Am Soc Mech Engrs; Am Inst Aeronaut & Astronaut. *Res:* Turbomachinery fluid mechanics. *Mailing Add:* Dept Mech Eng Iowa State Univ Ames IA 50011

OKINAKA, YUTAKA, b Osaka, Japan, Jan 22, 26; m 50; c 2. ELECTROCHEMISTRY, ANALYTICAL CHEMISTRY. *Educ:* Tohoku Univ, Japan, BS, 48, DSc, 59; Univ Minn, MS, 57. *Prof Exp:* Assoc anal chem, Tohoku Univ, Japan, 48-54, asst prof electrochem, 60-63; fel, Univ Minn, 56-60; mem tech staff, Bell Tel Labs, Inc, 63-83, distinguished mem prof staff, Bell Commun Res, 84-86, distinguished mem tech staff, 86- 90, CONSULT, AT&T BELL LABS, 90- *Honors & Awards:* Res Award, Electrochem Soc; Sci Achievement Award, Am Electroplaters & Surface Finishers Soc. *Mem:* Am Chem Soc; Electrochem Soc; Am Electroplaters & Surface Finishers Soc; Electrochem Soc Japan; Metal Finishing Soc Japan. *Res:* Electrochemical and chemical metal deposition process research and development. *Mailing Add:* 45 Spring Garden Dr Madison NJ 07940

OKITA, GEORGE TORAO, b Seattle, Wash, Jan 18, 22; m 58; c 3. PHARMACOLOGY. *Educ:* Ohio State Univ, BA, 48; Univ Chicago, PhD(pharmacol), 51. *Prof Exp:* From instr to asst prof pharmacol, Univ Chicago, 53-63; from assoc prof to prof, 63-91, actg chmn, 68-70 & 76-77, EMER PROF PHARMACOL, MED SCH, NORTHWESTERN UNIV, 91- *Concurrent Pos:* USPHS fel, Univ Chicago, 52. *Mem:* Cardiac Muscle Soc; AAAS; Am Soc Pharmacol & Exp Therapeut; Int Soc Biochem Pharmacol; Am Heart Asn. *Res:* Metabolism and mechanism of action of digitalis; metabolic effect and mode of action of environmental toxicants. *Mailing Add:* 4300 Waialae Ave No B-803 Honolulu HI 96816

O'KONSKI, CHESTER THOMAS, b Kewaunee, Wis, May 12, 21; m 48; c 4. CHEMISTRY, BIOPHYSICS. *Educ:* Univ Wis, BS, 42; Northwestern Univ Ill, MS, 46, PhD(phys chem), 49. *Prof Exp:* From instr to assoc prof, 48-60, PROF CHEM, UNIV CALIF, BERKELEY, 60- *Concurrent Pos:* Guggenheim fel, 55; Knapp Mem lectr, Univ Wis, 58; Miller Sci Found res prof, Univ Calif, Berkeley, 60; NIH fel, Princeton Univ & Harvard Univ, 62-63; Nobel guest prof, Univ Uppsala, 70; Wis Alumni Res Found fel; Nat Res Coun fel. *Mem:* AAAS; Am Chem Soc; Am Phys Soc; Biophys Soc. *Res:* Physical chemistry of macromolecules; electro-optics; electronic structure of molecules; proteins and mechanisms of muscle and membranes; nucleic acids. *Mailing Add:* Dept Chem Univ Calif 2120 Oxford St Berkeley CA 94720

OKOS, MARTIN ROBERT, b Toledo, Ohio, Sept 30, 45; m 70; c 4. CHEMICAL ENGINEERING, FOOD SCIENCE. *Educ:* Ohio State Univ, BS, 67, MS, 72, PhD(chem eng), 75. *Prof Exp:* Res assoc food sci, Ohio State Univ, 67-68; lab instr dept mech, US Military Acad, 69-70; teaching & res assoc chem eng, Ohio State Univ, 70-75; prof biochem eng, 75-77, AT DEPT CHEM ENG, PURDUE UNIV,77- *Mem:* Am Soc Agr Eng; Am Inst Chem Eng; Am Soc Cereal Chem; Am Soc Dairy Sci; Am Chem Soc; Sigma Xi. *Res:* Food process design; heat and mass transfer in food; energy conservation in food processing; immobilized microbiology cells and enzymes for food processing; protein diffusion and adsorption; drying of pasta and other foods. *Mailing Add:* Dept Chem Eng Purdue Univ West Lafayette IN 47907

OKRASINSKI, STANLEY JOHN, b Wilkes-Barre, Pa, Oct 29, 52; m 79; c 1. KINETICS, CATALYSIS. *Educ:* Kings Col (Pa), BS, 74; Princeton Univ, MA, 76, PhD(chem), 78. *Prof Exp:* Fel, Univ Chicago, 79-81; asst prof organometallic chem, Univ Nebr-Lincoln, 81-86, EASTMAN CHEM CO, 86- *Mem:* Am Chem Soc. *Res:* Organometallic reaction mechanisms; synthesis of organometallics; organometallics applied to organic synthesis; metal cluster chemistry; homogeneous catalysis; carbonylation processes. *Mailing Add:* Eastman Chem Co PO Box 1972 Kingsport TN 37662-1972

OKREND, HAROLD, b New York, NY, Mar 25, 34; m 63. MICROBIOLOGY. *Educ:* City Col New York, BS, 55; Syracuse Univ, MS, 57; Rutgers Univ, PhD(environ sci), 63. *Prof Exp:* Fel environ sci, Rutgers Univ, 63-65; asst prof microbiol, 65-71, ASSOC PROF MICROBIOL, HOWARD UNIV, 71- *Mem:* AAAS; Am Soc Microbiol. *Res:* Pollution microbiology; adenosine triphosphate firefly assay; falrine mononucleotide luciferase assay; nutrition of bacteria. *Mailing Add:* Dept Bot Howard Univ 2400 Sixth St NW Washington DC 20059

OKRENT, DAVID, b Passaic, NJ, Apr 19, 22; m 48; c 3. NUCLEAR PHYSICS. *Educ:* Stevens Inst Technol, ME, 43; Harvard Univ, MA, 48, PhD(physics), 51. *Prof Exp:* Mech engr, Nat Adv Comt Aeronaut, 43-46; from assoc physicist to sr physicist, Argonne Nat Lab, 51-71; PROF ENG & APPL SCI, UNIV CALIF, LOS ANGELES, 71- *Concurrent Pos:* US deleg, Int Conf Peaceful Uses Atomic Energy, Geneva, 55, 58, 64 & 71; consult adv comt reactor safeguards & reactor hazards, Eval Br, Atomic Energy Comn, 59-87, mem adv comt reactor safeguards, 63-87, vchmn, 65, chmn, 66; Guggenheim fel, 61-62 & 67-68; vis prof, Univ Wash, 63 & Univ Ariz, 70-71. *Honors & Awards:* Tommy Thompson Award, Am Nuclear Soc, Glenn Seaborg Medal, 87. *Mem:* Nat Acad Eng; fel Am Phys Soc; fel Am Nuclear Soc. *Res:* Nuclear reactor physics, safety and fuels; neutron cross sections; high temperature materials; fast reactor technology; fusion reactor technology; risk-benefit; society risks and environmental risks; expert systems; diagnostic techniques for complex processes. *Mailing Add:* 48-121 Eng IV Univ Calif Los Angeles CA 90024

OKRESS, ERNEST CARL, physics, electronics engineering; deceased, see previous edition for last biography

OKTAY, EROL, b Safranbolu, Turkey, Aug 3, 38; m 65; c 2. NUCLEAR ENGINEERING, PLASMA PHYSICS. *Educ:* Univ Mich, BS, 63, MS, 64, PhD(nuclear eng), 69. *Prof Exp:* Asst res eng, Gas Dynamics Lab, Univ Mich, 63-68, teaching fel nuclear radiation, 69, asst res engr, 69-70; asst prof nuclear eng, Mass Inst Technol, 70-71; asst prof physics & astron, Univ Md, College Park, 71-74; PHYSICIST, US DEPT ENERGY, 74- *Mem:* AAAS; Am Phys Soc; Am Nuclear Soc; NY Acad Sci. *Res:* Exploding wires; plasma physics and spectroscopy; gas dynamics. *Mailing Add:* 5233 Windmill Lane Columbia MD 21044

OKUBO, AKIRA, b Tokyo, Japan, Feb 5, 25. OCEANOGRAPHY, MATHEMATICAL BIOLOGY. *Educ:* Tokyo Inst Technol, BE, 47, MA, 49; Johns Hopkins Univ, PhD(oceanog), 63. *Prof Exp:* Lab asst, Japan Meteorol Agency, Tokyo, 50-54, chief chem sub-sect, 54-58; res asst oceanog, Johns Hopkins Univ, 58-63, res assoc, Chesapeake Bay Inst, 63-68, res scientist, 68-74; PROF PHYS ECOL, MARINE SCI RES CTR, STATE UNIV NY STONY BROOK, 74- *Concurrent Pos:* Vis prof, Kyoto Univ, 78-79; adj prof, Cornell Univ, 83- *Honors & Awards:* Medal Oceanog, Soc Japan, 83. *Mem:* AAAS; Am Soc Naturalists; Ecol Soc Am; Ecol Soc Japan; Am Soc Limnol & Oceanog; Soc Indust Appl Math. *Res:* Mathematical ecology; turbulent diffusion in the sea; mathematical modeling for animal dispersal. *Mailing Add:* Marine Sci Res Ctr State Univ NY Stony Brook NY 11794-5000

OKUBO, SUSUMU, b Tokyo, Japan, Mar 2, 30; US citizen; m 65. PARTICLE PHYSICS. *Educ:* Univ Tokyo, MS, 52; Univ Rochester, PhD(physics), 68. *Prof Exp:* Res assoc physics, Univ Rochester, 67-69; res assoc, Univ Napoli, Italy, 69-70; vis physicist, CERN, Geneva, Switz, 70-71; sr res assoc, 72-74, PROF PHYSICS, UNIV ROCHESTER, 74- *Honors & Awards:* Nishina Mem Award, Nishina Found, Japan, 76. *Mem:* Am Physics Soc. *Res:* Particle physics, mostly on symmetry principle. *Mailing Add:* Dept Physics Univ Rochester Rochester NY 14627

OKULITCH, ANDREW VLADIMIR, b Toronto, Ont, Dec 1, 41; m 65; c 2. GEOLOGY, STRUCTURAL GEOLOGY. *Educ:* Univ BC, BSc, 64, PhD(geol), 69. *Prof Exp:* RES SCIENTIST GEOL, GEOL SURV CAN, 74- *Mem:* Geol Asn Can; Geol Soc Am. *Res:* Structure and stratigraphy of southern Cordillera, British Columbia; tectonics; metamorphism and evolution; proterozoic fold belts of the Canadian shield; structure and stratigraphy of Canadian arctic archipelago; compilation of Geological Atlas of Canada. *Mailing Add:* Geol Surv of Can 3303 - 33rd St Northwest Calgary AB T2L 2A7 Can

OKULITCH, VLADIMIR JOSEPH, b St Petersburg, Russia, June 18, 06; nat Can; m 34; c 2. GEOLOGY, INVERTEBRATE PALEONTOLOGY. *Educ:* Univ BC, BASc, 31, MASc, 32; McGill Univ, PhD(geol, paleont), 34. *Hon Degrees:* DSc, Univ BC, 72. *Prof Exp:* Surveyor, Atlin Ruffner Mines, 29, geologist, 30-31; asst geol, Univ BC, 31-32 & Univ McGill, 33-34; Royal Soc Can fel, Mus Comp Zool, Harvard Univ, 34-36; instr, Univ Toronto, 36-69, lectr gen & hist geol, 39-42; from asst to assoc prof geol, Univ BC, 42-49, prof paleont & stratig, 49-53, chmn div geol, 53-59, R W Brock prof geol & head dept, 59-63, dean fac sci, 63-71, EMER DEAN SCI, UNIV BC, 71- *Concurrent Pos:* Vis prof, Univ Southern Calif, Univ Calif, Los Angeles & Univ Hawaii; sr asst, Que Geol Surv, 33-34; geologist, Shawinigan Chem, Ltd, 37; consult geologist, Calif Standard Co, Shell Oil Co & Sproule & Assocs; mem, Nat Adv Comt Astron. *Mem:* Fel Geol Soc Am; fel Paleont Soc; fel Royal Soc Can; Royal Astron Soc Can. *Res:* Lower Cambrian fauna; Archaeocyatha; corals and sponges of the Paleozoic era; lower Cambrian fossils; Paleozoic stratigraphy; Cordilleran geology. *Mailing Add:* 4504 49th St NW Calgary AB T3A 1X4 Can

OKULSKI, THOMAS ALEXANDER, b New Brunswick, NJ, 1943. RADIOLOGY. *Educ:* Jefferson Med Col, MD, 69; Am Bd Radiol, dipl, cert diag radiol, 75. *Prof Exp:* Intern, Temple Hosp, 69-70, resident diag radiol, 70-71 & 73-75, fel angiography & ultrasound, 75-76; asst prof radiol, Univ SFla, 76-81, clin asst prof, 81; RADIOLOGIST, SHEER AHEARN, 81- *Mem:* Am Inst Ultrasound Med. *Mailing Add:* 8600 Hidden River Pkwy Suite 900 Tampa FL 33637-1013

OKUMURA, MITCHIO, b Sept 1, 57; US citizen. LASER SPECTROSCOPY. *Educ:* Yale Univ, BS & MS, 79; Cambridge Univ, CPGS, 80; Univ Calif, Berkeley, PhD(chem), 86. *Prof Exp:* Postdoctoral res assoc, Univ Chicago, 87-88; ASST PROF CHEM PHYSICS, CALIF INST TECHNOL, 88- *Concurrent Pos:* Camille & Henry Dreyfus Found distinguished new fac award, 88; NSF presidential young investr award, 89. *Mem:* Am Phys Soc; Am Chem Soc; Optical Soc Am; Am Soc Mass Spectrometry. *Res:* Laser spectroscopy of molecular ions and ionic clusters; microscopic aspects of solvation; molecular and ion beam studies of atmospheric chemistry. *Mailing Add:* Chem Dept 127-72 Calif Inst Technol Pasadena CA 91125

OKUN, D(ANIEL) A(LEXANDER), b New York, NY, June 19, 17; m 46; c 2. ENVIRONMENTAL ENGINEERING. *Educ:* Cooper Union, BS, 37; Calif Inst Technol, MS, 38; Harvard Univ, ScD(sanit eng), 48. *Prof Exp:* Asst sanit engr, USPHS, DC, Ohio, NJ & NY, 40-42; from asst to assoc, Malcolm Pirnie Inc, NY, Conn, Fla, Va, Israel & Venezuela, 48-52; assoc prof, 52-55, prof, 55-73, chmn dept, 55-73, dir int prog, 62-83, chmn fac, 70-73, Kenan prof environ eng, 73-82, EMER PROF, UNIV NC, CHAPEL HILL, 82- *Concurrent Pos:* Consult, WHO, World Bank, UNDP, AID & Environ Protection Agency, 52-; NSF sr fel, 60; vis prof, Int Course Sanit Eng, Delft Technol Univ, 60-61, Tlanjin Univ, People's Repub China, 81, Univ Col, Univ London, 66-67 & 73-74; Fulbright Award 73-74; dir, US Dept Com Environ Control Sem, Bangkok, 70, Rotterdam, Warsaw, Prague & Bucharest, 71; chmn, Comt Metrop Wash Water Supply Studies, Nat Res Coun, 76-80; vis lectr, Duke Univ, Asian Inst Technol, Thailand & Tongs Univ, People's Repub China, 87; mem, Bd Sci & Technol for Int Develop, Nat Res Coun, 78-81, res grants comt, 83-86; vchmn, Environ Studies Bd, 80-83; Water Sci & Technol Bd, water resource res comt, 85-; consult, Camp, Dresser & McKee, Inc, 66-, US, Turkey, Egypt, Thailand, Colombia, Indonesia Morocco; Comn Human Rights, Nat Acad Sci, 88-; mem environ adv coun, Rohm & Haas Co, 85-; consult, C Lott Assoc, People's Repub China, 88-90 & SAFEGE. *Honors & Awards:* Eddy Medal, Water Pollution Control Fedn, 50; Gordon Maskew Fair Award, Am Acad Environ Engrs, 72; Thomas Jefferson Award, Univ NC, Chapel Hill, 73; Gordon Y Billard Award, NY Acad Scis, 75; Simon W Freese Award, Am Soc Civil Engrs, 77; Gordon M Fair Medal, Water Pollution Control Fedn, 78; Abel Walman Award of Excellence, Am Water Works Asn, 91. *Mem:* Nat Acad Eng; Inst Med-Nat Acad Sci; fel Am Soc Civil Engrs; hon mem Water Pollution Control Fedn; hon mem Am Acad Environ Engrs (pres, 69-70); fel AAAS; hon mem Am Water Works Asn. *Res:* Water quality management, including international, institutional, financial, and technical issues. *Mailing Add:* 900 Linden Rd Chapel Hill NC 27514

OKUN, LAWRENCE M, b Toledo, Ohio, Apr 12, 40; m 70; c 2. FLUORESCENCE MICROSCOPY. *Educ:* Wesleyan Univ, BA, 62; Stanford Univ, PhD(genetics), 68. *Prof Exp:* Nat Res Coun fel biol chem, Harvard Med Sch, 69-70, USPHS fel, 70-71; from asst prof to assoc prof, 71-88, PROF BIOL, UNI UTAH, 88- *Concurrent Pos:* Estab investr, Am Heart Asn, 75-80. *Res:* Development and physiology, of neurons cultured in vitro; design of fluorescent labels for specific cell types and activities. *Mailing Add:* Dept Biol Univ Utah 201 S Biol Bldg Salt Lake City UT 84112

OKUN, RONALD, b Los Angeles, Calif, Aug 7, 32; m 58; c 2. CLINICAL PHARMACOLOGY. *Educ:* Univ Calif, Los Angeles, BA, 54; Univ Calif, San Francisco, MD & MS, 58. *Prof Exp:* Teaching asst, Univ Calif, San Francisco, 58; intern & resident, Vet Admin, 58-61; from asst prof to assoc prof med, med pharmacol & therapeut, Univ Calif, Irvine Calif Col Med, 63-90; RETIRED. *Concurrent Pos:* Fel clin pharmacol, Johns Hopkins Hosp, 61-63; res pharmacologist, Vet Admin, 64- *Res:* Toxicology; internal medicine. *Mailing Add:* 12381 Ridge Circle Los Angeles CA 90049

OKUNEWICK, JAMES PHILIP, b Chicago, Ill, Apr 30, 34; m 57; c 4. HEMATOLOGY, MEDICAL SCIENCE. *Educ:* Loyola Univ, Calif, BS, 51; Univ Calif, Los Angeles, MS, 62, PhD(biophys, nuclear med), 65. *Prof Exp:* Mem res staff radiobiol div, Atomic Energy Proj, Univ Calif, Los Angeles, 55-57, US Naval Radiol Defense Lab, San Francisco, 57-59, radiobiologist, Labs Nuclear Med & Radiation Biol, 59-65; phys scientist, Rand Corp, Santa Monica, Calif, 65-66; sr proj leader, Armed Forces Radiobiol Res Inst, Bethesda, Md, 66-68; assoc biologist, Cellular & Radiation Biol Labs, 68-72, sr biologist & head viral oncogenesis, Cancer Res Unit, Allegheny Gen Hosp, 72-77, HEAD, EXP HEMAT SECT & SR SCIENTIST, CANCER RES LABS, ALLEGHENY-SINGER RES INST, ALLEGHENY GEN HOSP, PITTSBURGH, PA, 77-, DIR, FLOW CYTOMETRY FACIL, 84-; PROF MED, ALLEGHENY BR, MED COL PA, 89- *Concurrent Pos:* Consult, Oak Ridge Inst Nuclear Studies, 66-68; lectr, Dept Biol, Am Univ, 67; adj assoc prof radiation health, Univ Pittsburgh, 70-80; assoc ed, exp hemat, 73, radiation res, 77-80, transplantation, 87-, bone marrow transplantation, 88-; pres, Transplantation Soc, Int Soc Exp Hemat, 73-74. *Honors & Awards:* Outstanding Scientist Award, Health Res & Serv Found, United Way Pittsburgh, 75. *Mem:* AAAS; Radiation Res Soc; Soc Exp Biol Med; Am Soc Hemat; Int Soc Exp Hemat. *Res:* Radiobiology; hematopoietic system; oncogenic virus; bone marrow transplantation and transplantation immunology. *Mailing Add:* Allegheny-Singer Res Inst 320 E North Ave Pittsburgh PA 15212

OKWIT, SEYMOUR, b Brooklyn, NY, Aug 31, 29. POWER GENERATION. *Educ:* Brooklyn Col, BS, 52; Adelphi Univ, MS, 61. *Prof Exp:* Dep div dir, AIL Div, Cutlers Hammer Corp, 50-71; FOUNDER & PRES, LNR COMMUN, 71- *Concurrent Pos:* Nat lectr, Inst Elec & Electronics Engrs, 74. *Mem:* Fel Inst Elec & Electronics Engrs. *Mailing Add:* 180 Marens Blvd Hauppage NY 11788

OLAFSON, BARRY D, b Valley City, NDak, Oct 12, 49. THEORETICAL CHEMISTRY. *Educ:* Univ NDak, BS, 71; Calif Inst Technol, PhD(chem), 78. *Prof Exp:* Res fel chem, Harvard Univ, 78-81; res fel chem, Calif Inst Technol, 81-85; PRES, BIODESIGN, INC, PASADENA, CA, 85. *Mem:* Am Chem Soc. *Res:* Electronic structure of molecules; macromolecular dynamics; protein-substrate interactions; theoretical studies of biological reaction mechanisms. *Mailing Add:* Biodesign Inc 199 S Los Robles 270 Pasadena CA 91101

OLAFSSON, PATRICK GORDON, b Winnipeg, Man, Aug 21, 20; m 58. ORGANIC CHEMISTRY. *Educ:* McGill Univ, BS, 46; Univ Man, MS, 50; Ore State Univ, PhD(org chem), 59. *Prof Exp:* Metall chemist, Vulcan Iron & Steel Works, Can, 39-44; res chemist, E I du Pont de Nemours & Co, 59-60; assoc prof, 60-64, prof chem, State Univ NY Albany, 64-80; RETIRED. *Mem:* Am Chem Soc. *Res:* Thermal hydrogen shifts across conjugated pi-electron systems; differential thermal analysis of nucleic acids; reactions of singlet oxygen; atmospheric chemistry. *Mailing Add:* 46 Hanes Westmere NY 12203

OLAH, GEORGE ANDREW, b Budapest, Hungary, May 22, 27; m 49; c 2. ORGANIC CHEMISTRY. *Educ:* Budapest Tech Univ, PhD(org chem), 49. *Prof Exp:* Asst & assoc prof chem, Budapest Tech Univ, 49-54; assoc dir & head org chem dept, Cent Res Inst, Hungarian Acad Sci, 54-56; res scientist, Dow Chem Co Can, 57-64 & Dow Chem Co, 64-65; prof chem, 65-69, C F Mabery prof res chem, Case Western Reserve Univ, 69-77; LOKER DISTINGUISHED PROF CHEM, UNIV SOUTHERN CALIF, 77-, DIR, HYDROCARBON RES INST, 80- *Honors & Awards:* Petrol Chem Award, Am Chem Soc, 64, Baekeland Award, 67, Morley Medal, 70 & Synthetic Chem Award, 79; Roger Adams Medal, 89. *Mem:* Nat Acad Sci; Am Chem Soc; fel Chem Inst Can; The Chem Soc; Swiss Chem Soc. *Res:* Organic reaction mechanism; hydrocarbon chemistry; carbocations; Friedel-Crafts reactions; intermediate complexes; biological alkylating agents; organic fluorine and phosphorus compounds. *Mailing Add:* Dept Chem Univ Southern Calif Los Angeles CA 90007

OLAH, JUDITH AGNES, b Budapest, Hungary, Jan 21, 29; m 49; c 2. ORGANIC CHEMISTRY. *Educ:* Tech Univ, Budapest, MS, 55. *Prof Exp:* Res chemist, Cent Res Inst Chem, Hungarian Acad Sci, 55-56; sr res assoc chem, Case Western Reserve Univ, 66-77; adj assoc prof chem, Univ Southern Calif, 77-89; RETIRED. *Res:* Synthetic and mechanistic organic chemistry, electrophilic reactions. *Mailing Add:* 2252 Gloaming Way Beverly Hills CA 90210

OLANDER, D(ONALD) R(AYMOND), b Duluth, Minn, Nov 6, 31; m 56; c 3. NUCLEAR ENGINEERING. *Educ:* Columbia Univ, AB, 53, BS, 54; Mass Inst Technol, ScD(chem eng), 58. *Prof Exp:* Asst prof chem eng, 58-61, from asst prof to assoc prof nuclear eng, 61-70, PROF NUCLEAR ENG, UNIV CALIF, BERKELEY, 70- *Concurrent Pos:* Europ Atomic Energy Comn sr fel, 65-66. *Mem:* Am Nuclear Soc. *Res:* Nuclear materials and chemistry; high temperature reactions and properties. *Mailing Add:* Dept Nuclear Eng Univ Calif Berkeley CA 94720

OLANDER, DONALD PAUL, b Boulder, Colo, June 24, 40; m 66. ANALYTICAL CHEMISTRY. *Educ:* Washburn Univ, Topeka, BSc, 64; Univ Nebr-Lincoln, MSc, 67, PhD(chem), 70. *Prof Exp:* Asst prof, 69-74, ASSOC PROF CHEM, APPALACHIAN STATE UNIV, 74- *Mem:* Sigma Xi. Am Chem Soc. *Res:* Ion-solvent interactions; nonaqueous solvents; analyses of natural waters and soils. *Mailing Add:* Dept of Chem Appalachian State Univ Boone NC 28608

OLANDER, HARVEY JOHAN, b San Francisco, Calif, Nov 19, 32; m 69; c 2. VETERINARY PATHOLOGY. *Educ:* Univ Calif, Davis, BS, 56, DVM, 58, PhD(comp path), 63. *Prof Exp:* Asst prof vet path, Purdue Univ, 62-65; asst prof, Univ Calif, Davis, 65; assoc prof, State Univ NY Vet Col, Cornell Univ, 65-68; prof vet path, Animal Dis Diag Lab, Purdue Univ, West Lafayette, 68-81; PROF VET PATH, DEPT VET PATH, UNIV CALIF, DAVIS, 81- *Concurrent Pos:* Vis prof path, Ontario Vet Col, Univ Guelph, 78; vis res scientist, Royal Danish Vet Sch & St Serum Lab, Copenhagen, 79-80, USDA Animal & Plant Health Inspection Serv, FADDL, Plum Island, NY, 89, Scripps Inst, San Diego, 89; prin investr, USAID Collab Res Support Prog, Animal Health Proj, Sobral, Ceara, Brazil, 81- 86. *Mem:* Am Vet Med Asn; Am Col Vet Path (pres, 76); Conf Res Workers Animal Dis; Comp Gastroenterol Soc (pres 73-74); Am Asn Vet Lab Diagnosticians; US Animal Health Asn. *Res:* Pathology of infectious diseases of domestic animals; comparative gastroenterology; swine dysentery; salmonellosis; caseous lymphadenitis; foot and mouth disease-in situ hybridization. *Mailing Add:* 766 Mulberry Lane Davis CA 95616

OLANDER, JAMES ALTON, b Boulder, Colo, Oct 9, 44. INORGANIC CHEMISTRY, HYDROMETALLURGY. *Educ:* Washburn Univ, Topeka, BS, 65; La State Univ, Baton Rouge, PhD, 70. *Prof Exp:* Asst chem, La State Univ, Baton Rouge, 68-70; res chemist, Deepsea Ventures, Inc, 70-74, res dir, 74-76; res chemist, US Bur Mines, Salt Lake City, 76-78; sr metallurgist, Amoco Minerals, 78-82; chief metallurgist, Cyprus Thompson Creek, 82-85. *Mem:* Am Chem Soc; Am Inst Mining, Metall & Petrol Engrs; Sigma Xi. *Res:* Ion-ion and ion-solvent interactions; electrolytic solutions; nonaqueous solvents; liquid ion exchange. *Mailing Add:* PO Box 192 Winnemucca NV 89445-0192

OLBRICH, STEVEN EMIL, b Chicago, Ill, Nov 24, 38; m 68; c 4. DAIRY SCIENCE. *Educ:* Univ Wis-Madison, BS, 65; Univ Hawaii, MS, 68; Univ Mo-Columbia, PhD(nutrit), 71. *Prof Exp:* From res asst to res assoc nutrit, Univ Mo-Columbia, 68-72; res assoc, dept animal sci, Univ Hawaii, 72-74, dairy specialist, 75-80, asst prof animal sci, 76-80; CONSULT AGR, 80- *Mem:* Am Dairy Sci Asn; Sigma Xi. *Res:* Animal nutrition, dairy science, environmental physiology and aquaculture. *Mailing Add:* PO Box 627 Waianae HI 96792

OLCOTT, EUGENE L, b St Louis, Mo, Apr 18, 18; m 49; c 3. REFRACTORY MATERIALS, GRAPHITE COMPOSITES. *Educ:* Univ Mo, BS, 40. *Prof Exp:* Metall observer, Bethlehem Steel Corp, NY, 40-41; jr metallurgist, J H Williams Co, 41; head high temperature sect, Gen Elec Co, 41-46; metall engr, Mat Sect, Bur Aeronaut, US Dept Navy, Washington, DC, 46-51, head high temperature mat sect, Bur Ships, 51-56; metall engr missile eng group, Atlantic Res Corp, 56-57, dir mat div, 57-75; CONSULT, 75- *Mem:* Am Soc Metals; Am Ceramic Soc. *Res:* High temperature, graphite composite and propulsion materials. *Mailing Add:* Rte 1 Box 44 Shepherdstown WV 25443

OLCZAK, PAUL VINCENT, b Buffalo, NY, May 24, 43; m 73; c 3. PSYCHOLOGY & LAW. *Educ:* Defiance Col, BA, 66; Northern Ill Univ, MA, 69, PhD(psychol), 72. *Prof Exp:* Asst prof psychol, Mercyhurst Col, 71-73; instr, Pa State Univ, Behrend, 72; asst prof psychol, Hartwick Col, 73-75; clin fel, Family Court Psychiat Clin, Buffalo, 75-76, clin psychologist, 76-77; assoc prof psychol & dir, Counselor Training Ctr, 77-90, PROF PSYCHOL, STATE UNIV NY, GENESCO, 90- *Concurrent Pos:* Psychologist, Family Court, Erie, Pa, 72-73, NY State Supreme & Family Courts, 77-83, Psychiat Assoc Western NY, 76- & Dept Youth Serv, Mental Health Clinic, Buffalo, 82-; consult, Dr Gertrude Barber Ctr Erie, 72-73; consult psychologist, Hopevale Inc, 77-79, supvr psychol serv, 79-89; ed, Guilford Press, 90- *Mem:* Am Psychol Asn; Sigma Xi; Psychonomic Soc; Soc Exp Social Psychologists; NY Acad Sci. *Res:* Psychology and law; cognition and interpersonal psychiatry; personality effects in social behavior. *Mailing Add:* Dept Psychol State Univ New York Geneseo NY 14454

OLD, BRUCE S(COTT), b Norfolk, Va, Oct 21, 13; m 39; c 5. METALLURGY. *Educ:* Univ NC, BS, 35; Mass Inst Technol, ScD(metall), 38. *Prof Exp:* Consult, Dewey & Almy Chem Co, Mass, 37-38; res engr, Bethlehem Steel Co, Pa, 38-41; metallurgist, Arthur D Little, Inc, 46-51, dir, 49-51, vpres, 51-60, sr vpres, 60-78; PRES, BRUCE S OLD ASSOCS INC, 79- *Concurrent Pos:* Chief metall & mat br, AEC, 47-49, consult, 49-55; pres, Cambridge Corp, 51-53, chmn bd, 53-55; consult, Sci Adv Comt, Exec Off of the President, 53-56; pres, Nuclear Metals, Inc, 54-57; chmn, Nat Conf Admin Res, 66. *Mem:* Nat Acad Eng; NY Acad Sci; fel AAAS; fel Am Soc Metals; Am Inst Mining, Metall & Petrol Engrs. *Res:* Direct reduction; age hardening; physical chemistry of steel making; blast furnace; metallurgy as related to atomic energy; cryogenic engineering; radioactive waste disposal. *Mailing Add:* Arthur D Little Inc 25 Acorn Park Cambridge MA 02140

OLD, LLOYD JOHN, b San Francisco, Calif, Sept 23, 33. CANCER, IMMUNOLOGY. *Educ:* Univ Calif, BA, 55, MD, 58. *Prof Exp:* Res fel, Sloan-Kettering Inst Cancer Res, 58-59, res assoc, 59-60, assoc, 60-64, assoc mem, 64-67, vpres & assoc dir, 73-76, assoc dir res, Mem Sloan-Kettering Cancer Ctr & Mem Hosp, 73-76, vpres & assoc dir sci develop, 76-83; res assoc, 60-62, from asst prof to assoc prof, 62-69, PROF BIOL, GRAD SCH MED SCI, CORNELL UNIV, 69-; DIR, LUDWIG INST CANCER RES, 88- *Concurrent Pos:* Consult, Nat Cancer Inst, 67-70, mem develop res working group, 69, mem spec virus cancer prog, Immunol Group, 70, mem Virus Cancer Prog Adv Comt, 75-78, mem Bd Sci Counr, Div Cancer Cause & Prev, 78-81; assoc med dir, New York Cancer Res Inst, Inc, 70, med dir, 71-; mem med & sci adv bd & bd trustee, Leukemia Soc Am, 70-73; mem sci adv bd, Jane Coffin Childs Mem Fund Med Res, 70-75; adv ed, J Exp Med, 71-76, Progress in Surface & Membrane Sci, 72-74; assoc ed, Virology, 72-74; Harvey Soc lectr, 72; mem Sci Adv Comt, Ludwig Inst Cancer Res, Inc, 74-; mem Res Coun, Pub Health Res Inst City New York, Inc, 77-80, bd dirs, 79-; mem bd dirs, Am Asn Cancer Res, 80-83; William E Snee chair cancer immunol, Sloan-Kettering Inst Cancer Res, 83-, mem, 67-; attending immunologist, Clin Immunol Serv, Dept Med, Mem Hosp Cancer & Allied Dis, 84-; mem sci adv comt, Ludwig Inst Cancer Res, 74-86, chmn, 88-, mem bd dirs, 89- & dir, NY Unit, 90- *Honors & Awards:* Roche Award, Roche Inst, 57; Award, Alfred P Sloan Found, 62; Lucy Wortham James Award, James Ewing Soc, 70; Louis Gross Award, 72; Award Cancer Immunol, Cancer Res Inst, 75; Rabbi Shai Shacknai Mem Award, 76; Res Recognition Award, Noble Found, 78; G H A Clowes Mem Lectr, 80; Robert Roesler de Villiers Award, 81; Robert Koch Prize, 90. *Mem:* Nat Acad Sci; Inst Med-Nat Acad Sci; Am Asn Cancer Res; Soc Exp Biol & Med; Am Asn Immunologists; Sigma Xi; NY Acad Sci; Am Acad Arts & Sci; Reticuloendothelial Soc; AAAS. *Res:* Nature, specificity and importance of the immune response to cancer in experimental animals and humans; immunogenetics; viral oncology. *Mailing Add:* Ludwig Inst Cancer Res 1345 Ave of the Americas New York NY 10105

OLD, THOMAS EUGENE, b Spokane, Wash, Aug 2, 43; m 63; c 1. ATMOSPHERIC PHYSICS. *Educ:* Gonzaga Univ, BS, 66; Univ Idaho, MS, 69; State Univ NY Albany, PhD(physics), 71. *Prof Exp:* Res assoc physics, State Univ NY Albany, 71-72; TECH STAFF ATMOSPHERIC PHYSICS, MISSION RES CORP, 72- *Mem:* Am Geophys Union. *Res:* Atmospheric nuclear effects; magnetohydrodynamics. *Mailing Add:* 555 Pasa Robles Dr Santa Barbara CA 93108

OLDALE, ROBERT NICHOLAS, b North Attleboro, Mass; m 56; c 2. GEOLOGY. *Educ:* St Lawrence Univ, BS, 53. *Prof Exp:* GEOLOGIST, US GEOL SURV, 55- *Mem:* Fel Geol Soc Am; Am Quaternary Asn. *Res:* Glacial and marine geology, with emphasis on the advance and retreat of the late Wisconsin ice and the postglacial marine transgression in the northeastern United States. *Mailing Add:* US Geol Surv Woods Hole MA 02543

OLDEMEYER, JOHN LEE, b Ft Collins, Colo, Mar 12, 41; m 63; c 3. UNGULATE HABITAT ECOLOGY. *Educ:* Colo State Univ, BS, 63, MS, 66; Pa State Univ, PhD(forest resources), 81. *Prof Exp:* Biometrician, 66-71, mem staff, Denver Wildlife, Res Ctr, US Fish & Wildlife Serv, 71-85; WILDLIFE RES BIOLOGIST, NAT ECOL RES CTR, 86- *Concurrent Pos:* Affil fac, Colo State Univ, 79- *Mem:* Biomet Soc; Wildlife Soc; Soc Range Mgt; Sigma Xi. *Res:* Effects of forest and range management on wildlife and their habitat; dealing with ungulate habitat on National Wildlife Refuges. *Mailing Add:* Nat Ecol Res Ctr 4512 McMurray Ave Ft Collins CO 80525-3400

OLDEMEYER, ROBERT KING, b Brush, Colo, Sept 23, 22; m 44; c 2. HORTICULTURE, AGRONOMY. *Educ:* Colo State Univ, BS, 47; Univ Wis, MS, 48, PhD(genetics), 50. *Prof Exp:* plant breeder, Great Western Sugar Co, 50-60, dir seed develop, 60-68, agr res, 68-71, mgr seed processing & prod, 71-73, mgr variety develop, 73-85; dir opers, Hilleshög Mono-Hy Inc, 85-89; RETIRED. *Concurrent Pos:* Staff affil, Colo State Univ, 73-; consult, 89-; sugarbeet agr ed, Sugar J, 90- *Mem:* Am Soc Agron; hon mem Am Soc Sugar Beet Technologists; Sigma Xi; Int Inst Beet Sugar Res. *Res:* Breeding of sugar beets, dry beans and vegetables; sugar beet diseases; agricultural writing. *Mailing Add:* 530 Gay St Longmont CO 80501

OLDENBURG, C(HARLES) C(LIFFORD), b Blue Island, Ill, Feb 1, 29; m 53; c 3. CHEMICAL ENGINEERING. *Educ:* Ill Inst Technol, BS, 50; Purdue Univ, MS, 51; Univ Tex, PhD(chem eng), 57. *Prof Exp:* Res chem engr, Jefferson Chem Co, 51-57; Calif Res Co, 57-64; sect head, Stauffer Chem Co, 64-65; sr res assoc, Chevron Res Co, 65-70, MEM TECH STAFF, STANDARD OIL CO CALIF, 70- *Mem:* Am Chem Soc; Am Inst Chem Engrs; Sigma Xi. *Res:* Catalysis; kinetics; dialysis; crystallization. *Mailing Add:* 717 Graham Ct Danville CA 94526

OLDENBURG, DOUGLAS WILLIAM, b Edmonton, Alta, Mar 9, 46; m 67; c 2. GEOPHYSICS. *Educ:* Univ Alta, BSc, 67, MSc, 69; Univ Calif, San Diego, PhD(earth Sci), 74. *Prof Exp:* Fel geophysics, Dept Physics, Univ Alta, 74-75, Isaac W Killam fel, 75-76, res assoc, 76-77; from asst prof to assoc prof, 77-87, PROF, DEPT GEOPHYS & ASTRON, UNIV BC, 87- *Concurrent Pos:* Assoc ed, Geophys J, Royal Astron Soc, 79-81. *Mem:* Royal Astron Soc; Soc Explor Geophysicists; Europ Asn Explor Geophys. *Res:* Application of inversion theory to geophysics. *Mailing Add:* Dept Geophys & Astron Univ BC Vancouver BC V6T 1W5 Can

OLDENBURG, THEODORE RICHARD, b Muskegon, Mich, Apr 8, 32; m 59; c 2. PEDODONTICS. *Educ:* Univ NC, DDS, 57, MS, 62. *Prof Exp:* From asst prof to assoc prof dent, 62-69, PROF PEDODONT & CHMN DEPT, SCH DENT, UNIV NC, CHAPEL HILL, 69- *Concurrent Pos:* United Cerebral Palsy clin fel, 60-62; mem USAF sr dent prog, 56-57; consult accreditation comt, Am Dent Asn, 67-77, coun dent educ & US Army Dent Corps; examr, Am Bd Pedodont, 72-79. *Mem:* Am Dent Asn; Am Acad Pedodont (pres, 79-80); Am Soc Dent for Children; Int Asn Dent Res. *Res:* Educational research in clinical restorative pedodontics. *Mailing Add:* 403 Laurel Hill Rd Chapel Hill NC 27515

OLDENDORF, WILLIAM HENRY, b Schenectady, NY, Mar 27, 25; m 45; c 3. NEUROLOGY, PSYCHIATRY. *Educ:* Albany Med Col, MD, 47; Am Bd Psychiat & Neurol, dipl, cert psychiat, 53, cert neurol, 55. *Prof Exp:* Intern med, Ellis Hosp, Schenectady, NY, 47-48; resident psychiat, Binghamton State Hosp, 48-50 & Letchworth Village State Sch, Thiells, 50-52; resident neurol, Univ Minn Hosps, Minneapolis, 54-55; assoc chief neurol, 55-69, med investr, Wadsworth Hosp, 69-75, SR MED INVESTR, BRENTWOOD HOSP, VET ADMIN CTR, 75-; PROF NEUROL, SCH MED, UNIV CALIF, LOS ANGELES, 70-, PROF PSYCHIAT, 76- *Concurrent Pos:* Clin fel, Univ Minn, Minneapolis, 55; from clin instr to asst clin prof, Sch Med, Univ Calif, Los Angeles, 56-65, assoc prof, 65-70, USPHS res grants, 59-82; NIH spec fel physiol, Univ Col, Univ London, 65-66; mem coun stroke, Am Heart Asn, 69. *Honors & Awards:* Silver Medal Award, Soc Nuclear Med, 69; Ziedses des Plantes Gold Medal, Med Physics Soc, W rzburg, 74. *Mem:* Nat Acad Sci; fel Am Acad Neurol; fel Am Acad Arts & Sci; AMA; fel Am Neurol Asn; Soc Nuclear Med. *Res:* Nuclear medicine applied in clinical neurological research, including studies related to human brain isotope uptake studies, cerebral blood flow in man, cerebrospinal fluid, central nervous system instrumentation and development of photographic techniques relating to brain isotope scanning and cerebral angiography in man. *Mailing Add:* B-151C West Los Angeles Vet Admin Med Ctr Los Angeles CA 90073

OLDENKAMP, RICHARD D(OUGLAS), b Conrad, Mont, Aug 23, 31; m 53; c 5. CHEMICAL ENGINEERING. *Educ:* Mont State Col, BS, 56; Univ Pittsburgh, MS, 59, PhD(chem eng), 62. *Prof Exp:* Engr, Bettis Atomic Power Lab, Westinghouse Elec Corp, 56-63; res specialist, Atomics Int Div, 63-67, proj engr, 67-71, proj mgr, 71-75, prog mgr, 75-78, DIR APPLN & PRELIMINARY ENG, ENVIRON & ENERGY SYSTS DIV, ENERGY SYSTS GROUP, ROCKWELL INT CORP, 78- *Mem:* Am Chem Soc; Sigma Xi. *Res:* Air pollution control process development. *Mailing Add:* 1608 Stoddard Ave Thousand Oaks CA 91360

OLDFIELD, DANIEL G, b New York, NY, July 24, 25; m 50; c 2. CELL BIOLOGY, BIOPHYSICS. *Educ:* Columbia Univ, BS, 50; Univ Chicago, MS, 58, PhD(math biol), 65. *Prof Exp:* Jr scientist, Argonne Cancer Res Hosp, Univ Chicago, 53-55, assoc scientist, 55-62, sr biophysicist, Univ Chicago, Toxicity Lab, 62-64, asst prof, Lab Cytol, 65-68; from asst prof to assoc prof biol sci, DePaul Univ, 68-88; RETIRED. *Mem:* Histochem Soc; Am Phys Soc; Radiation Res Soc; Soc Math Biol; Int Soc Stereology. *Res:* Analysis of cell cycle and organelle interactions by cytofluorometry; cytopathology. *Mailing Add:* PO Box 840 East Hempstead NH 03826

OLDFIELD, EDWARD HUDSON, b Mt Sterling, Ky, Nov 22, 47; m 74. NEUROSURGERY, NEUROENDOCRINOLOGY. *Educ:* Univ Kentucky, BS, 69, MD, 73. *Prof Exp:* Intern resident surg, Vanderbilt Univ Hosp, 73-75; registr neurol, Nat Hosp Nervous Dis, Queen Sq, London, England, 75-76; resident neurosurg, Vanderbilt Univ Hosp, Nashville, Tenn, 76-80; sr staff fel neurosurg, 81-83, CHIEF CLIN NEUROSURG SECT, SURG NEUROL BR, 83-, NIH CHIEF, SURG NEUROL BR, NAT INST NEUROL & COMMUN DIS & STROKES, 87- *Mem:* Am Asn Neurol Surg; Cong Neurol Surg; Endocrine Soc; Am Col Surgeons; Am Fedn Clin Res. *Res:* Clinical research of pituitory tumors, arteriovenous malformations of the spinal cord and brain tumors; clinical neuroendocrinology. *Mailing Add:* Bldg 10A Rm 3368 NIH Bethesda MD 20205

OLDFIELD, ERIC, b London, Eng, May 23, 48. MOLECULAR BIOLOGY, BIOPHYSICAL CHEMISTRY. *Educ:* Bristol Univ, BS, 69; Sheffield Univ, PhD(chem), 72. *Hon Degrees:* DSc, 82. *Prof Exp:* Europ Molecular Biol Org fel chem, Ind Univ, 72-74; vis scientist, Mass Inst Technol, 74-75; from asst prof to assoc prof, 75-82, PROF CHEM, UNIV ILL, URBANA, 82- *Honors & Awards:* Meldola Medal; Colworth Medal. *Mem:* Fel Royal Soc Chem; Am Chem Soc. *Res:* Nuclear magnetic resonance spectroscopy; biological membrane structure; lipid-protein interactions; catalyst structure. *Mailing Add:* 47 Noyes Lab Sch Chem Sci Univ Ill 505 Mathews Ave Urbana IL 61801

OLDFIELD, GEORGE NEWTON, b San Jose, Calif, Sept 6, 36; m 59; c 3. ENTOMOLOGY, ACAROLOGY. *Educ:* Fresno State Col, BA, 62; Univ Calif, Riverside, MS, 66, PhD(entom), 71. *Prof Exp:* Agr res technician, USDA, 62-65, entomologist, 65-68, res entomologist, 68-86, res leader, Sci & Educ Admin-Agr Res, USDA, 87-88; POST DOCTORAL RES PLANT PATHOLOGIST, DEPT PLANT PATH, UNIV CALIF, RIVERSIDE, 88- *Mem:* Am Phytopath Soc; Entom Soc Am; Acarological Soc Am; Sigma Xi. *Res:* Arthropod vectors of fruit tree pathogens; reproductive biology and virus vector capabilities of Eriophyoidea. *Mailing Add:* Boyden Entom Lab USDA Univ of Calif Riverside CA 92521

OLDFIELD, JAMES EDMUND, b Victoria, BC, Aug 30, 21; nat US; m 42; c 5. ANIMAL NUTRITION. *Educ:* Univ BC, BS, 41, MS, 49; Ore State Col, PhD(animal nutrit, biochem), 51. *Prof Exp:* Instr animal husb, Univ BC, 48-49; from asst prof to prof animal nutrit, Ore State Univ, 51-87, head dept animal sci, 67-83, dir, Nutrit Res Inst, 87-90; RETIRED. *Concurrent Pos:* Basic res award, Ore Agr Exp Sta, 61, Sigma Xi res award, Ore State Univ, 64; Fulbright scholar, Massey Univ, NZ, 74, Rosenfeld distinguished prof agr sci, 81, 82 & 83; mem Coun Agr Sci & Technol. *Honors & Awards:* Res Travel Award, Nat Feed Ingredients Asn, 69; Morrison Award, Am Soc Animal Sci, 72, Distinguished Serv Award, 74. *Mem:* Am Chem Soc; fel Am Inst Nutrit; hon fel Am Soc Animal Sci (secy-treas, 62-65, vpres, 65, pres, 66); NY Acad Sci; Agr Inst Can. *Res:* Vitamins, minerals, and antibiotics as feed supplements in animal nutrition; nutritional diseases; metabolic diseases; alternate feed sources; relative efficiencies of domestic animal species as human food producers. *Mailing Add:* Dept Animal Sci Ore State Univ Withycombe Hall Corvallis OR 97331

OLDFIELD, THOMAS EDWARD, b Fond du Lac, Wis, Oct 11, 47; m 70; c 2. PHYSIOLOGICAL ECOLOGY. *Educ:* Mich Technol Univ, BS, 70; Utah State Univ, MS, 73, PhD(physiol), 75. *Prof Exp:* Asst prof physiol, Ferris State Col, 75-76; asst prof physiol, Millikin Univ, 76-78; ASST PROF PHYSIOL, FERRIS STATE COL, 78- *Concurrent Pos:* Sigma Xi res grant, 74. *Mem:* AAAS; Am Soc Zoologists; Sigma Xi. *Res:* Development of reliable methods to be used in the determination of energy utilization of free-living animals. *Mailing Add:* Dept Biol Sci Ferris State Univ Big Rapids MI 49307

OLDFIELD, WILLIAM, materials science, metallurgy, for more information see previous edition

OLDHAM, BILL WAYNE, b Paris, Tex, Oct 30, 34; m; c 3. MATHEMATICS. *Educ:* Abilene Christian Univ, BA, 56; Okla State Univ, MS, 63; Univ Northern Colo, EdD(math educ), 72. *Prof Exp:* Teacher math, Ft Sumner Sch, NMex, 56-59; teacher sci, Yale City Sch, Okla, 60-61; PROF MATH, HARDING UNIV, 61- *Mem:* Nat Coun Teachers Math; Am Math Soc. *Res:* Mathematics learning theory; mathematics applications in business and industry. *Mailing Add:* Dept Math Harding Univ Sta A Box 921 Searcy AR 72143

OLDHAM, JAMES WARREN, b Helena, Ark, Jan 1, 50; m 75; c 3. GENETIC TOXICOLOGY, REPRODUCTIVE TOXICOLOGY. *Educ:* Univ Ark, BA, 72, BS, 74, PhD(toxicol), 79; Am Bd Toxicol, dipl. *Prof Exp:* Res asst, Univ Ark Med Sci, 76-77; res biologist, Nat Ctr Toxicol Res, 77-78; res assoc, Ohio State Univ, 78-80; res scientist, McNeil Pharmaceut, 80-81, Sr scientist, 81-82, prin scientist toxicol, 82-87, res fel, 87-88; SECT HEAD, R W JOHNSON PHARMACEUT RES INST, 88- *Concurrent Pos:* Mem bd gov, Genetic Toxicol Asn, 84-87; steering comt, Indust In Vitro Toxical Group, 89- *Mem:* Soc Toxicol; Am Asn Cancer Res; Environ Mutagen Soc; Genetic Toxicol Asn. *Res:* Development and implementation of testing procedures to define the type and degree of toxicity induced by drugs and environmental chemicals on mammalian systems. *Mailing Add:* 409 Grenoble Dr Sellersville PA 18960

OLDHAM, KEITH BENTLEY, b Ashton-under-Lyne, Eng, Feb 4, 29; m 53; c 5. ELECTROCHEMISTRY, PHYSICAL CHEMISTRY. *Educ:* Univ Manchester, BSc, 49, PhD(phys chem), 52, DSc, 70. *Prof Exp:* Res assoc anal chem, Univ Ill, Urbana, 52-55; vis scientist, Rensselaer Polytech Inst, 55; asst lectr phys chem, Imp Col, Univ London, 55-57; lectr chem, Univ Newcastle, 57-64; vis assoc anal chem, Calif Inst Technol, 64-65; lectr chem, Univ Newcastle, 65-66; mem tech staff, Sci Ctr, NAm Rockwell Corp, Calif, 66-70; chmn dept, 72-79, PROF CHEM, TRENT UNIV, 70- *Concurrent Pos:* Consult, Consumers' Asn, 56-66; mem schs exam bd, Univ Durham, 58-64; mem, Int Comt Electrochem Thermodyn & Kinetics, 61-; consult, Royal Aircraft Estab, 63-66; vis scientist, Sci Ctr, NAm Aviation, Inc, 64-65; fac mem, Calif Lutheran Col, 67-69; div ed, J Electrochem Soc, 78-81; mem, Chem Grant Selection Comt, Nat Sci & Eng Res Coun Can, 78-81; adj prof, Queen's Univ, 80-; vis prof, Deakin Univ, Australia, 81-82 & 86-89, Southampton Univ, 90. *Honors & Awards:* Heyrovsky Medal, Czech Acad Sci. *Mem:* Electrochem Soc; Royal Inst Chem; Chem Inst Can. *Res:* Transport processes; various aspects of physical and analytical chemistry; practical and theoretical aspects of electrochemistry; applied mathematics; properties of functions, fractional calculus. *Mailing Add:* Dept Chem Trent Univ Peterborough ON K9J 7B8 Can

OLDHAM, ROBERT KENNETH, b Pocatello, Idaho, Sept 16, 41; c 5. INTERNAL MEDICINE, MEDICAL ONCOLOGY. *Educ:* Univ Mo, MD, 68. *Prof Exp:* Clin assoc radiation, Nat Cancer Inst, 70-71, clin assoc immunol, 71-72; res assoc, ICIG-Hosp Paul Brousse, Villejuif, France, 72-73; sr investr immunodiag, Nat Cancer Inst, 73-75; DIR DIV ONCOL & ASSOC DIR CANCER CTR, VANDERBILT UNIV HOSP, 75- *Concurrent Pos:* Dir biol response modifer prog, Nat Cancer Inst, 80-84, Biol Ther Inst, 84-90. *Mem:* Am Asn Cancer Res; Am Fedn Clin Res; Am Soc Clin Oncol; fel Am Col Physicians. *Res:* In vitro assays in tumor immunology; stimulation of immunoreactivity of host defense systems with adjuvants; culture of human tumors in vitro. *Mailing Add:* Biol Ther Inst Hosp Dr PO Box 1700 Franklin TN 37064

OLDHAM, SUSAN BANKS, b San Francisco, Calif, Aug 6, 42; m 88. ENDOCRINOLOGY, CHEMICAL DEPENDENCY. *Educ:* Univ Calif, Los Angeles, BS, 64; Univ Southern Calif, PhD(biochem), 68. *Prof Exp:* Fel endocrine res, Mayo Clin, 68-72; res assoc endocrinol, 72-73, asst prof med & biochem, 73-79, ASSOC PROF RES MED, SCH MED, UNIV SOUTHERN CALIF, 79- *Concurrent Pos:* Mem, spec grants rev comt, Nat Inst Dent Res, 78-82. *Mem:* Am Soc Bone & Mineral Res; Endocrine Soc; Am Fedn Clin Res; Sigma Xi; Am Soc Biochem & Molecular Biol; Asn Med Educ Res Substance Abuse. *Res:* Regulation of bone adenylate cyclase; effects of ethanol on neurosecretion; regulation of intracellular magnesium; impact of educational interventions on physician skills and attitudes related to substance abuse. *Mailing Add:* Sch of Med Univ of Southern Calif 2025 Zonal Ave RBl-130 Los Angeles CA 90033

OLDHAM, WILLIAM G, b Detroit, Mich, May 5, 38; m 60. ELECTRICAL ENGINEERING. *Educ:* Carnegie Inst Technol, BS, 60, MS, 61, PhD(elec eng), 63. *Prof Exp:* Mem staff physics, Siemens-Schuckertwerke, Erlangen, Ger, 63-64; asst prof elec eng, 64-76, PROF ELEC ENG, UNIV CALIF, BERKELEY, 76- *Mem:* Nat Acad Eng; Electrochem Soc; fel Inst Elec & Electronics Engrs; Am Phys Soc. *Res:* Semiconductor electronics; electrical properties of III-V compounds. *Mailing Add:* Dept of Elec Eng Univ of Calif Berkeley CA 94720

OLDS, DANIEL WAYNE, b Richland Co, Ill, Mar 27, 35; m 60; c 2. PHYSICS. *Educ:* Wabash Col, AB, 56; Duke Univ, PhD, 64. *Prof Exp:* Asst prof, 63-66, actg chmn dept, 63-71, assoc prof, 66-76, mgr comput terminal, 68-75, dir, Acad Comput Ctr, 75-80, CHMN DEPT, WOFFORD COL, 71-, PROF PHYSICS, 76-, DIR COMPUT SERV. *Mem:* Am Asn Physics Teachers. *Mailing Add:* Dept Comput Sci Wofford Col N Church St Spartanburg SC 29301

OLDS, DURWARD, b Conneaut, Ohio, Apr 12, 21; m 47; c 2. DAIRY SCIENCE, VETERINARY MEDICINE. *Educ:* Ohio State Univ, DVM, 43; Univ Ill, MS, 54, PhD(dairy sci), 56. *Prof Exp:* Artificial insemination technician, Clark County Breeder's Coop, Wis, 44-46; asst dairying, Univ Ky, 45-51, from assoc prof to prof, 51-83; RETIRED. *Mem:* Am Soc Animal Sci; Am Dairy Sci Asn; Am Vet Med Asn; Soc Stud Reprod. *Res:* Physiology of reproduction; artificial insemination; causes of infertility in dairy cattle. *Mailing Add:* 1605 Elizabeth St Lexington KY 40503

OLDS-CLARKE, PATRICIA JEAN, b Waseca, Minn, Aug 15, 43; m 75. REPRODUCTIVE PHYSIOLOGY, ANDROGENETICS. *Educ:* Macalester Col, BA, 65; Wash Univ, PhD(biol), 70. *Prof Exp:* Fel, Johns Hopkins Univ, 70-71; res fel, Harvard Med Sch, 71-72, res assoc anat, 72-73, instr, 73-74; asst prof biol, Bryn Mawr Col, 74-80; From asst prof to assoc prof, 80-89, PROF ANAT, SCH MED, TEMPLE UNIV, 89- *Mem:* Am Soc Androl; Am Soc Cell Biol; Soc Develop Biol; Soc Study Reprod; Sigma Xi. *Res:* Analysis of fertile and sterile sperm with abnormal motility, from mice with one or two T haplotypes. *Mailing Add:* Dept Anat Sch Med Temple Univ 3400 N Broad St Philadelphia PA 19140

OLDSHUE, J(AMES) Y(OUNG), b Chicago, Ill, Apr 18, 25; m 47; c 3. FLUID MIXING. *Educ:* Ill Inst Technol, BS, 47, MS, 49, PhD, 51. *Prof Exp:* Chem engr, Los Alamos Sci Lab, 45-46; head develop eng, 50-54, dir res, 54-64, tech dir, 64-71, V PRES MIXING TECHNOL, MIXING EQUIP CO, INC, 71- *Concurrent Pos:* Lectr, Univ Rochester, 54 & 57; continuing educ lectr, Am Inst Chem Engrs & Ctr Prof Advan. *Honors & Awards:* Founders Award, Am Inst Chem Engrs; Ken Rowe Award, Am Asn Eng Socs. *Mem:* Nat Acad Eng; Am Inst Chem Engrs (pres, 79-, treas, 83-88); Am Chem Soc; Am Asn Eng Socs (treas, 83, pres, 85); World Fedn Eng Orgn (vpres, 86). *Res:* Fluid mixing; author or coauthor of over 100 publications. *Mailing Add:* 141 Tyringham Rd Rochester NY 14617

OLDSTONE, MICHAEL BEAUREGUARD ALAN, b New York, NY, Feb 9, 34; m 60; c 3. EXPERIMENTAL BIOLOGY, NEUROBIOLOGY. *Educ:* Univ Ala, BS, 54; Univ Md, MD, 61. *Prof Exp:* USPHS fel rickettsiol & virol, Sch Med, Univ Md, 58, intern med, Univ Hosp, 61, resident, 62-63, resident neurol, 63-64, chief resident, 64-66; fel, Dept Exp Path, 66-69, assoc, 69-71, ASSOC MEM, DEPT IMMUNOPATH & DIV NEUROL, SCRIPPS CLIN & RES FOUND, 71-, HEAD NEUROL RES, FOUND, 69- *Concurrent Pos:* NIH-AID career develop award, 69; adj prof path, Univ Calif, San Diego, 71- & adj prof neurosci, 72-; mem ad hoc sci adv comt, Multiple Sclerosis. *Mem:* Am Acad Neurol; Am Asn Immunol; Am Asn Neuropath; Am Soc Exp Path; Am Soc Microbiol. *Res:* Viral immunopathology; viruses and immunity. *Mailing Add:* Dept Immunol Scripps Clin & Res Found 10666 N Torrey Pines Rd La Jolla CA 92037

O'LEARY, ANNE K, pharmacology, nerve gas; deceased, see previous edition for last biography

O'LEARY, BRIAN TODD, b Boston, Mass, Jan 27, 40; div; c 2. ASTRONOMY. *Educ:* Williams Col, BA, 61; Georgetown Univ, MA, 64; Univ Calif, Berkeley, PhD(astron), 67. *Prof Exp:* Physicist, Goddard Space Flight Ctr, NASA, 61-62; high sch teacher, Washington, DC, 64; scientist-astronaut, Manned Spacecraft Ctr, NASA, 67-68; asst prof astron & space sci,

Cornell Univ, 68-71; vis assoc, Calif Inst Technol, 71; assoc prof interdisciplinary sci, San Francisco State Univ, 71-72; asst prof astron & sci policy assessment, Hampshire Col, 72-75; spec consult on energy, Subcomt on Energy & Environ, US House Interior Comt, 75; res fac mem & lectr, Physics Dept, Princeton Univ, 75-80; sr scientist, Sci Applns Int Corp, 82-87; chmn bd dir, Inst Security & Coop Outer Space, 87-88; COFOUNDER, INT ASN NEW SCI. *Concurrent Pos:* Consult, NASA Ames Res Ctr, 71-72; vis assoc prof, Sch Law, Univ Calif, Berkeley, 72; lectr astron, Univ Pa, 78-80; chmn, Nat Space Coun, Am Geophys Union; writer & lectr, 80-82. *Mem:* Soc Sci Explor; US Psychotronics Asn; Inst Noetic Sci. *Res:* Advanced concepts in space development. *Mailing Add:* 10615 N 11th St Phoenix AZ 85020

O'LEARY, DENNIS PATRICK, b Dec 24, 39; US citizen; m 64; c 2. BIOPHYSICS, NEUROPHYSIOLOGY. *Educ:* Univ Chicago, SB, 62; Univ Iowa, PhD(physiol, biophys), 69. *Prof Exp:* Asst prof surg & anat, Univ Calif, Los Angeles, 71-74; res assoc prof otolaryngol & pharmacol, Univ Pittsburgh, 74-78, assoc prof otolaryngol & physiol, 78-84; PROF, DEPTS OTOLARYNGOL, PHYSIOL, BIOPHYSICS & BIOMED ENG, UNIV SOUTHERN CALIF, 84- *Concurrent Pos:* USPHS res fel, Univ Calif, Los Angeles, 69-70. *Mem:* Am Physiol Soc; AAAS; Biophys Soc; Soc Neurosci; Int Brain Res Orgn. *Res:* Biophysical mechanisms of sensory transduction; stability and control in the vestibular system; applications of time series analysis, stochastic processes and filtering theory to investigations of neuronal information processing. *Mailing Add:* Univ Southern Calif 1420 San Pablo St Parkview Med Bldg C103 Los Angeles CA 90033

O'LEARY, GERARD PAUL, JR, b Bridgeport, Conn, Oct 16, 40. MICROBIOLOGY, BIOCHEMISTRY. *Educ:* Mt St Mary's Col, Md, BS, 62; NMex State Univ, MS, 64; Univ NH, PhD(microbiol), 67. *Prof Exp:* Nat Res Coun Can fel, Macdonald Col, McGill Univ, 67-69; ASSOC PROF BIOL, PROVIDENCE COL, 69- *Mem:* AAAS; Am Soc Microbiol; Am Inst Biol Sci; Can Soc Microbiol; NAm Apiotherapy Soc. *Res:* Isolation and chemical characterization of marine bacterial lipopolysaccharides; effects of bacterial endotoxins on fresh water fish; characterization of an actomyosin-like protein complex from a procaryotic system; the effects of bee venom in adjuvant induced arthritis; high pressure liquid chromatography of steroids and venoms; elemental analysis of body fluids and tissues by DC plasma emission spectroscopy; isoelectric focusing of animal and bacterial proteins. *Mailing Add:* Dept of Biol Providence Col River Ave & Eaton St Providence RI 02918

O'LEARY, JAMES WILLIAM, b Painesville, Ohio, Aug 10, 38; m 63; c 3. PLANT PHYSIOLOGY. *Educ:* Ohio State Univ, BS, 60, MS, 61; Duke Univ, PhD(bot), 64. *Prof Exp:* Asst prof bot, Univ Ariz, 63-66; asst prof biol, Bowling Green State Univ, 66-67; assoc prof biol sci, Univ Ariz, 67-71, plant physiologist, Environ Res Lab, 68-74, prof biol sci, 71-90, res prof biol sci, 74-90, PROF PLANT SCI & PROF ARID LANDS, UNIV ARIZ, 90- *Mem:* AAAS; Am Soc Plant Physiol; Scand Soc Plant Physiol. *Res:* Plant water relations; responses of plants to salinity stress; physiology of halophytes; development of seawater irrigation crops. *Mailing Add:* Dept Plant Scis Univ Ariz Tucson AZ 85721

O'LEARY, JOHN FRANCIS, b Adrian, Pa, Dec 12, 17; m 72; c 7. PHARMACOLOGY. *Educ:* Univ Rochester, MS, 49, PhD(pharmacol), 51; Penn State, BS, 42. *Prof Exp:* Pharmacologist, McNeill Labs, 51-54; chief, Pharmacol Br, Edgewood Arsenal, Aberdeen Proving Ground, Md, 61-73, med sci adminr, 74-79; RETIRED. *Mem:* Am Chem Soc; fel AAAS; Am Soc Pharmacol & Exp Therapeut; Am Soc Exp Biol & Med. *Mailing Add:* Rte 1 Box 213 Queenstown MD 21658

O'LEARY, KEVIN JOSEPH, b Winthrop, Mass, Aug 8, 32; m 54; c 5. POLYMER CHEMISTRY. *Educ:* Boston Col, BS, 55; Case Western Reserve Univ, MS, 62, PhD(eng), 67. *Prof Exp:* Anal chemist, US Army Chem Ctr, Md, 55-57; sr res chemist & anal area supvr, Diamond Alkali Res Ctr, Ohio, 57-64, res assoc, Res Ctr, Diamond Shamrock Corp, 67-69, mgr anal & electrochem res, 69-71, assoc dir res, T R Evans Res Ctr, 71-, assoc dir res electrochem/anal, 71-75, asst dir res phys sci, 75-76, dir technol, Electrolytic Systs Div, 76-78, dir corp res, T R Evans Res Ctr, 78-80; mem staff, Olin Corp, 80-; AT ELTECH SYST CORP. *Concurrent Pos:* Vis prof, Lake Erie Col, 67-68. *Mem:* Am Phys Soc; Am Chem Soc; Soc Appl Spectros; Am Crystallog Asn; Electrochem Soc; Sigma Xi. *Res:* Polymer morphology, deformation of crystalline polymers; small and wide angle x-ray diffraction; electron microscopy; thermodynamics and analytical chemistry; electrochemistry. *Mailing Add:* 21785 High Pine Trail Boca Raton FL 33428

O'LEARY, MARION HUGH, b Quincy, Ill, Mar 24, 41; m 81; c 3. ORGANIC CHEMISTRY, BIOCHEMISTRY. *Educ:* Univ Ill, BS, 63; Mass Inst Technol, PhD(org chem), 66. *Prof Exp:* NIH res fel biochem, Harvard Univ, 66-67; from asst prof to prof, Univ Wis-Madison, 67-80, prof chem & biochem, 80-89; PROF & HEAD, DEPT BIOCHEM, UNIV NEBR, LINCOLN, 89- *Concurrent Pos:* NIH & Res Corp res grants, 67-; Sloan Found fel, 72-74; NSF res grant, 70-; USDA res grant, 79-; Guggenheim fel, 82-83. *Mem:* Fel AAAS; Am Chem Soc; Am Soc Biol Chemists; Sigma Xi; Am Soc Plant Physiologists. *Res:* Mechanisms of action of enzymes; bio-organic chemistry; plant biochemistry; isotope effects. *Mailing Add:* Dept Biochem Univ Nebr Lincoln NE 68583

O'LEARY, TIMOTHY JOSEPH, b Birmingham, Ala, July 14, 52; m 75; c 3. DIAGNOSTIC MOLECULAR PATHOLOGY, BIOPHYSICAL CHEMISTRY. *Educ:* Purdue Univ, BS, 72; Stanford Univ, PhD(chem), 76; Univ Mich, MD, 79. *Prof Exp:* Res assoc physiol, Univ Mich, 76-78; resident path, 79-81, fel anat path, 81-82, ATTEND PATHOLOGIST, NIH, 82-; CHMN CELLULAR PATH, ARMED FORCES INST PATH, 87- *Concurrent Pos:* Consult, Lawrence Livermore Lab, 75-82; sr staff fel, Food & Drug Admin, 86-87. *Mem:* Fel Col Am Pathologists; fel Am Soc Clin Pathologists; Biophys Soc; Am Phys Soc; Am Chem Soc. *Res:* Diagnosis of disease using molecular biologic methods; structure and function of biological membranes using theoretical, infrared spectroscopic, raman spectroscopic and calorimetric methods. *Mailing Add:* Dept Cellular Path Armed Forces Inst Path Washington DC 20306-6000

OLECHOWSKI, JEROME ROBERT, b Buffalo, NY, Jan 10, 31; m 56; c 5. PHYSICAL CHEMISTRY, ORGANIC CHEMISTRY. *Educ:* Canisius Col, BS, 52; Pa State Univ, MS, 55; La State Univ, PhD(chem), 58. *Prof Exp:* Asst chem, Pa State Univ, 52-54 & La State Univ, 54-57; res chemist, Copolymer Rubber & Chem Corp, 57-61 & Cities Serv Res & Develop Co, 61-63; prin res chemist, Columbian Carbon Co, 63-70; res assoc, Petrochem Group, Cities Serv Co, NJ, 70-77, int tech liaison, 71-77; group leader, 77-80, res scientist, Chem Prod Res, Union Camp Corp, 80-87; CONSULT, IPCC RECYCLING GROUP, 87-; ENFORCEMENT OFFICER, US ENVIRON PROTECTION AGENCY, REGION II, 90- *Concurrent Pos:* Prof, McNeese State Col, 63. *Mem:* Am Chem Soc. *Res:* Reactions of ozone with organic substances; Ziegler-Natta type catalysts for polymerization of unsaturated substances; metal-ion olefin complexes and oxidation; rearrangement of trihaloalkenes; organometallic chemistry of transition elements; chemistry of medium ring compounds; chemistry of terpenes, rosin and other silvichemicals. *Mailing Add:* 17 Empress Lane Lawrenceville NJ 08648

OLECKNO, WILLIAM ANTON, b St Charles, Ill, Dec 16, 48; m 75. HEALTH BEHAVIORS & RISK FACTORS IDENTIFICATION. *Educ:* Ind Univ, BS, 71, HSD(health & safety educ), 80; Univ Pittsburgh, MPH, 73. *Prof Exp:* Sanitarian, DuPage County Health Dept, 70-71; sanitarian II, Ill Dept Pub Health, 72; res assoc, Consad Res Corp, 73; coordr & asst prof environ health, Ind Univ Med Sch, 73-80; coordr & assoc prof community health, 80-89, PROF, NORTHERN ILL UNIV, 89- *Concurrent Pos:* Pub health consult, Nat Automatic Merchandising Asn, 75-80, 81-; vis lectr, Ind Univ, 77-78; ed, Hoosier Sanitarian, 78-80; tech trainer hygiene & sanitation, Peace Corps, Niger Nutrit Educ Training Prog, ACTION, 81, sec tech trainer epidemiol & pub health, Peace Corps, Thialand Lab Serv Training Prog, ACTION, 82; tech consult, Nat Educ Comt, Nat Environ Health Asn, 81-82; prin investr, Job Satisfaction among Environ Health Practrs Ill, Ill Environ Health Asn Res Grants, 86; rep, Nat Sanitation, Foundation's Joint Comt Drinking Water Additives, 89- *Mem:* Nat Environ Health Asn; Am Pub Health Asn; fel Royal Soc Health, London. *Res:* Public health epidemiology, especially risk-factor identification and risk-taking behaviors in adolescents and young adults; environmental health, especially risk perception; environmental health education; general environmental health; environmental health, smoking and health behavior. *Mailing Add:* Sch Allied Health Northern Illinois Univ DeKalb IL 60115

OLEESKY, SAMUEL S(IMON), b New York, NY, June 16, 13; m 39; c 5. ELECTRONICS ENGINEERING, PLASTICS. *Educ:* Univ Iowa, BSEE, 35. *Hon Degrees:* ScD, London, 72. *Prof Exp:* Sect head antennas & radomes, US Naval Air Develop Ctr, Pa, 45-51; vpres & chief scientist, Zenith Plastics Co, Calif, 51-56, chief scientist, Zenith Plastics Div, Minn Mining & Mfg Co, 56-60; electronics consult, 60-64; pres & gen mgr, Nasol Corp, 64-65; design eng specialist, Northrop Ventura, 65-66; sr engr & scientist, Missile & Space Systs Div, Douglas Aircraft Co, Calif, 66-70; CONSULT, 65- *Concurrent Pos:* US deleg, UNESCO, 53; chmn, Reinforced Plastics Div, Soc Plastics Indust, 54-56; consult, Stanford Res Inst, 57, Arnold Eng Ctr, US Air Force, 62, Telecomput Corp, Calif, 62-64 & Flexible Tubing Corp, Conn, 63-64; lectr, Univ Calif, Los Angeles, 63-64; consult ed, Microwave J, 65- *Honors & Awards:* Soc Plastics Indust Exec Award, 56. *Mem:* Sr mem Soc Plastics Engrs. *Res:* Radome and antenna design; reinforced plastics; ceramics; inventor multilayer broadband microwave radome. *Mailing Add:* 5438 Saloma Ave Van Nuys CA 91411

OLEINICK, NANCY LANDY, b Pittsburgh, Pa, Feb 26, 41; m 62; c 2. BIOCHEMICAL ONCOLOGY, RADIATION BIOLOGY. *Educ:* Chatham Col, BS, 62; Univ Pittsburgh, PhD(biochem), 66. *Prof Exp:* From instr to assoc prof, 68-85, PROF RADIATION BIOL & BIOCHEM, ENVIRON HEALTH SCI & ONCOL, SCH MED & SCH DENT, 85-, DIR, DIV BIOCHEM ONCOL, DEPT RADIOL, CASE WESTERN RESERVE UNIV, 86- *Concurrent Pos:* Fel biochem, Sch Med, Case Western Reserve Univ, 66-68; Nat Inst Allergy & Infectious Dis fel, 67-69; Nat Cancer Inst res grants, 73-; vis prof, Hebrew Univ Jerusalem, Israel, 78; assoc ed, Radiation Res, 79-82, 85-88; mem Radiation Study Sect, NIH, 84-88. *Mem:* Am Soc Biochem & Molecular Biol; Am Soc Cell Biol; Radiation Res Soc; Am Soc Photobiol. *Res:* Radiosensitivity of transcriptionally active versus inactive chromatin; influence of chromatin structure on DNA repair; role of nuclear matrix in repair of DNA damage; photodynamic effects of phthalocyanines; ataxia telangiectasia. *Mailing Add:* 3727 Meadowbrook Blvd Cleveland OH 44118

OLEM, HARVEY, b Boston, Mass, Aug 23, 51; m 83; c 1. ENVIRONMENTAL ENGINEERING, CIVIL ENGINEERING. *Educ:* Tufts Univ, BS, 73; Pa State Univ, MS, 75, PhD(civil eng), 78. *Prof Exp:* Res asst environ eng, Pa State Univ, 76-78; environ engr, Tenn Valley Auth, 78-88; PRES, OLEM ASSOCS, 88- *Concurrent Pos:* Instr, Munic Training Div Dept Community Affairs, Commonwealth Pa, 77-78. *Mem:* Water Pollution Control Fedn; Am Water Works Asn; Nat Soc Prof Engrs; Sigma Xi; NAm Lake Mgt Soc (treas, 86-89); Alliance Environ Educ (secy, 90). *Res:* Water pollution control; water quality management; wastewater microbiology; water and wastewater disinfection; industrial wastewater treatment; acid mine drainage pollution control; aquatic liming; non point sources of pollution. *Mailing Add:* 1000 Connecticut Ave NW Suite 802 Washington DC 20036

OLENCHOCK, STEPHEN ANTHONY, b Mt Pleasant, Pa, Dec 12, 46; m 69; c 3. OCCUPATIONAL HEALTH. *Educ:* St Vincent Col, BA, 68; WVa Univ, PhD(med microbiol), 75. *Prof Exp:* Immunologist, 75-80, chief, Immunol Sect, 80-91, ASST TO DIV DIR, NAT INST OCCUP SAFETY & HEALTH, 91-; ADJ PROF MICROBIOL & IMMUNOL, WVA UNIV HEALTH SCI CTR, 85- *Concurrent Pos:* Proj dir, several projs, Nat Inst Occupational Safety & Health, 75-, tech adv grants, 76-; instr microbiol, WVa Univ Med Ctr, 75-77, from adj asst prof to adj assoc prof, 77-85; mem critique panel, Agr Dust Res Prog, USDA & mem rev panel, Byssinosis Res Prog, 81; consult, People's Repub China, 81-; adv associateship prog, Nat Res Coun,

85. *Mem:* Am Asn Immunologists; Am Soc Microbiol; Am Thoracic Soc; Am Acad Microbiol; Comn Officers Asn-US Pub Health Serv; Am Acad Allergy & Immunol. *Res:* Immunologic mechanisms involved in the development of pathophysiology in the lungs and respiratory tract due to the interactions of agricultural dusts with the complement system, antibodies and allergic mechanisms, and macrophages; endotoxins. *Mailing Add:* Nat Inst Occup Safety & Health 944 Chestnut Ridge Rd Morgantown WV 26505

OLENICK, JOHN GEORGE, b Throop, Pa, Oct 31, 35; m 59; c 5. BIOCHEMISTRY, MICROBIOLOGY. *Educ:* Univ Scranton, BS, 58; Albany Med Col, MS, 61; Univ Md, College Park, PhD(microbiol), 71. *Prof Exp:* Biochemist, Sect Kidney & Electrolyte Metab, Nat Heart Inst, 61-62; chief chem sect, Dept Cell & Media Prod, Microbiol Assocs, Inc, 62-63; biochemist, Walter Reed Army Inst Res, 63-71, asst chief, Dept Molecular Biol, 71-76, res biochemist, Dept Biol Chem, 76-79, asst chief, 79-81, chief, Dept Appl Biochem, 81-87, RES CHEMIST, DEPT MOLECULAR PATH, WALTER REED ARMY INST RES, 87- *Concurrent Pos:* Consult, Bio-Medium Corp, 73-75; vis res assoc prof, Univ Md Dent Sch Baltimore, 88-89. *Mem:* Am Soc Biochem & Molecular Biol. *Res:* Mode of action of staphylococcal enterotoxins; immunoelectron microscopy. *Mailing Add:* Dept Molecular Path Walter Reed Army Inst of Res Washington DC 20307-5100

OLENICK, RICHARD PETER, b Chicago, Ill, Dec 22, 51. COMPUTATIONAL PHYSICS, MAGNETIC COOPERATIVE PHENOMENA. *Educ:* Ill Inst Technol, BS, 73; Purdue Univ, MS, 75, PhD(physics), 79. *Prof Exp:* from asst prof to assoc prof, 79-90, PROF PHYSICS, UNIV DALLAS, 90- *Concurrent Pos:* Vis assoc, Calif Inst Technol, 83-86. *Mem:* Am Phys Soc; Sigma Xi; NY Acad Sci; Am Asn Physics Teachers. *Res:* Computer simulations of dielectric phenomena; synergistic hysteresis. *Mailing Add:* Dept Physics Univ Dallas Irving TX 75062-4799

OLER, NORMAN, b Sheffield, Eng, July 12, 29; US citizen; m 57; c 2. MATHEMATICS. *Educ:* McGill Univ, BS, 51, MS, 53, PhD(math), 57. *Prof Exp:* Asst prof math, McGill Univ, 57-60; res assoc, Columbia Univ, 60-61, asst prof, 61-63; assoc prof, 63-67, PROF MATH, UNIV PA, 67- *Concurrent Pos:* Nat Res Coun Can overseas fel, 58-59; NSF sci fac fel, 65-66. *Mem:* Am Math Soc; Can Math Soc. *Res:* Geometry of numbers with particular interest in packing and covering problems. *Mailing Add:* Dept Math Univ Pa Philadelphia PA 19104

OLES, KEITH FLOYD, b Seattle, Wash, June 9, 21; m 46; c 2. GEOLOGY. *Educ:* Univ Wash, BS, 43, MS, 52, PhD, 56. *Prof Exp:* Instr geol, Univ Wash, 47-52; asst prof, Washington & Lee Univ, 52-53; mem spec explor group, Union Oil Co, Calif, 53-58, dist geologist, Wyo Dist, 58-61; from assoc prof to prof, 61-89, EMER PROF GEOL, ORE STATE UNIV, 89- *Concurrent Pos:* Mem prof develop panel, Coun Educ in Geol Sci, Am Geol Inst, 68-73, visitor, Vis Geol Sci Prog, 69-70; consult & chief party, North Slope, Alaska, Union Oil Co Calif-Gulf Oil Corp, 69; geol consult, Cairo Univ, Egypt, 75; vis prof fel, Univ Wales, 76; vis scholar, Univ Cambridge, Eng, 85; consult construct indust. *Mem:* Am Asn Petrol Geologists; Int Asn Sedimentologists. *Res:* Paleoenvironments of Cretaceous rocks of Pacific Northwest and British Columbia; stratigraphy and structure of fold belts; history of geology. *Mailing Add:* 18304 68th Ave W Lynnwood WA 98037-4141

OLESEN, DOUGLAS EUGENE, b Tonasket, Wash, Jan 12, 39; m 64; c 2. WATER POLLUTION. *Educ:* Univ Wash, BS, 62, MS, 63, PhD(civil eng), 72. *Prof Exp:* Res engr, Space Res Div, Boeing Aircraft Corp, 63-64; res engr, Water & Waste Mgt Sect, Water & Land Resources Dept, Pac Northwest Labs, Pac Northwest Div, Battelle Mem Inst, 67-68, mgr, 68-69, mgr, Water Resources Systs Sect, 69-71, mgr, 71-75, dep dir res, 75, dir res, 75-79, dir, Pac Northwest Div, 79-84, vpres, 79-84, EXEC VPRES & CHIEF OPERATING OFFICER, BATTELLE MEM INST, 84- *Concurrent Pos:* Affiliate assoc prof, Univ Wash, 73- *Mem:* Fed Water Pollution Control Fedn. *Res:* Development of combined biological and physical/chemical processes for municipal and industrial wastewater treatment. *Mailing Add:* 8597 Finlarig Dr Dublin OH 43017

OLESKE, JAMES MATTHEW, b Hoboken, NJ, Mar 16, 45; m 69; c 3. MEDICINE, IMMUNOLOGY. *Educ:* Univ Detroit, BS, 67; NJ Col Med, MD, 71; Columbia Univ, MPH, 74. *Prof Exp:* Intern, Martland Hosp, 71-74; fel immunol, Emory Univ, 74-76; dir, Div Allergy, Immunol & Infectious Dis, Dept Pediat, 81-87, ASST PROF PEDIAT & PREV MED, COL MED & DENT, 76-; FRANCOIS-XAVIER BOUGNAD PROF PEDIAT, NJ MED SCH, UNIV MED & DENT NJ, 87- *Concurrent Pos:* Dir immunol & attend, St Michael's Med Ctr, 76- *Mem:* Am Acad Pediat; Am Acad Allergy & Immunol; Am Pub Health Asn; Am Soc Microbiol. *Res:* Clinical human immunology; infectious diseases and allergy. *Mailing Add:* Dept Pediat NJ Med Sch Univ Med & Dent NJ 185 S Orange Ave Newark NJ 07103

OLESON, NORMAN LEE, b Detroit, Mich, Aug 19, 12; m 39; c 3. PLASMA PHYSICS. *Educ:* Univ Mich, BS, 35, MS, 37, PhD(physics), 40. *Prof Exp:* Asst, Univ Mich, 36-40; instr physics, US Coast Guard Acad, 40-46; res physicist, Gen Elec Co, 46-48; prof physics, Naval Postgrad Sch, 48-69; chmn dept, 69-78, prof physics, 78-90, DEPT ELEC ENG, UNIV S FLA, 90- *Concurrent Pos:* Vis prof, Queen's Univ, Belfast, 55-56 & Mass Inst Technol, 67-68; consult, Lawrence Radiation Lab, Univ Calif, 58-69. *Mem:* AAAS; Sigma Xi; fel Am Phys Soc; Inst Elect & Electronics Engr. *Res:* Gamma ray studies of light radioactive elements; multiple scattering of fast electrons; electrical and optical studies of positive column of gas mixtures; plasmas in magnetic fields; non-linear plasma waves; plasma heating. *Mailing Add:* Dept Elec Eng Univ S Fla Box 118 Tampa FL 33620

OLEXA, STEPHANIE A, b Bethlehem, Pa, May 13, 50. BIOTECHNOLOGY. *Educ:* Temple Univ, PhD(biochem), 76. *Prof Exp:* Mgr biotechnol, Rohm & Haas, Philadelphia, Pa, 82-84; mgr venture develop, Air Prod & Chem Inc, Allentown, Pa, 84-86, mgr strategic prod, 86-89; PRES, BENCH MARK ANALYSIS, 89- *Mem:* AAAS; Am Chem Soc; Am Heart Asn. *Mailing Add:* Bench Mark Anal 1776 Main St Hellertown PA 18055

OLEXIA, PAUL DALE, b McKeesport, Pa, July 31, 39; m 68; c 1. MYCOLOGY, TAXONOMY. *Educ:* Wabash Col, BA, 61; State Univ NY Buffalo, MA, 65; Univ Tenn, Knoxville, PhD(bot), 68. *Prof Exp:* Vis asst prof biol, Colgate Univ, 67-68; from asst prof to assoc prof 68-74, ASSOC PROF, 74-84, PROF BIOL, KALAMAZOO COL, 84- *Concurrent Pos:* chair, Div Natural Sci, Kalamazoo Col, 86-89, assoc provost, 89-91. *Mem:* Mycol Soc Am; AAAS; British Mycological Soc; Sigma Xi. *Res:* Ecology of vesicular-arbuscular mycorrhizal fungi. *Mailing Add:* Biol Dept Kalamazoo Col MI 49007

OLF, HEINZ GUNTHER, b Wetzlar, Ger, Nov 1, 34. POLYMER PHYSICS. *Educ:* Munich Tech Univ, Vordiplom, 57, Diplom, 60, Dr rer nat(physics), 69. *Prof Exp:* Physicist, Res Triangle Inst, 61-78; ASSOC PROF, DEPT WOOD & PAPER SCI, SCH FOREST RESOURCES, NC STATE UNIV, 78- *Mem:* AAAS; Am Phys Soc; Am Chem Soc; Ger Phys Soc. *Res:* Solid state physics of high polymers; application of nuclear magnetic resonance techniques, x-ray scattering, infrared and Raman spectroscopy to polymers; mechanical properties of polymers. *Mailing Add:* 109 Flora Macdonald Lane Cary NC 27511

OLFE, D(ANIEL), b St Louis, Mo, Feb 4, 35; m 64; c 2. ENGINEERING SCIENCE. *Educ:* Princeton Univ, BSE, 57; Calif Inst Technol, PhD(eng sci), 60. *Prof Exp:* Assoc prof aeronaut, NY Univ, 60-64; assoc prof, 64-69, PROF ENG PHYSICS, UNIV CALIF, SAN DIEGO, 69- *Mem:* Am Inst Aeronaut & Astronaut; Am Meteorol Soc. *Res:* Gas dynamics; radiative transfer; geophysical fluid mechanics. *Mailing Add:* Dept Ames Mail R 011 Univ Calif San Diego La Jolla CA 92093

OLHOEFT, GARY ROY, b Akron, Ohio, Feb 15, 49; m 72. GEOTECHNICAL GEOPHYSICS, ELECTROCHEMISTRY. *Educ:* Mass Inst Technol, BSEE & MSEE, 72; Univ Toronto, PhD(physics), 75. *Prof Exp:* Scientist, Lockheed Electronics Co, Houston, 72-73; res engr, Univ Toronto, 73-75; RES GEOPHYSICIST, US GEOL SURV, 75- *Concurrent Pos:* Adj prof, Univ Southern Calif, 77-; vis prof, Univ Colo, 80-; mem, US Nat Comn Rock Mech, 83-86; assoc ed, Soc Explor Geophysicists, 85-87; mem, Sci Adv Comt, Deep Observ & Sampling Earth Continental Crust, NSF, 83-88, Chief Geologists Sci Adv Comt, US Geol Surv, 84-87 & Downhole Measurement Panel, Ocean Drill Prog, Joint Oceanog Insts, 81-88, Mars Sci Working Group, 89-; chmn, Nat Acad Sci, Nat Res Coun, US Nat Comt Rock Mech Ann Rev Panel, 85; adj prof, Colo Sch Mines, 85-; partic scientist, USSR Phobos & Mars '94 Space Mission. *Honors & Awards:* Lunar Sci Group Achievement Award, NASA, 75; Meritorious Serv Award, US Dept Interior, 89. *Mem:* Am Geophys Union; AAAS; Soc Explor Geophysicists; Am Soc Testing & Mat; Soc Prof Well Log Analysts; Inst Elec & Electronics Engrs; Am Astron Soc. *Res:* Physical properties and chemical processes in rocks and minerals for applications to geothermal energy, oil and gas exploration, minerals, permafrost, hazardous waste disposal, volcanic hazards and geotechnical applications of geophysics; nonlinear complex resistivity as a technique to monitor geochemical processes in the laboratory at high temperature and pressure and in situ via borehole and surface measurements; geotechnical applications and computer processing of ground penetrating radar. *Mailing Add:* US Geol Surv PO Box 25046 DFC MS 964 Denver CO 80225-0046

OLIEN, NEIL ARNOLD, b Lemmon, SDak, Mar 3, 35; m 58; c 2. THERMODYNAMICS, CRYOGENICS. *Educ:* SDak Sch Mines & Technol, BS, 60; Stanford Univ, MA, 61. *Prof Exp:* Physicist, Cryog Eng Lab, Nat Bur Standards, 61-65, proj leader, 65-72, sect chief, Cryog Div, 72-80, div chief, Thermophys Properties Div, 80-82 & Chem Eng Sci Div, 82-84, prog analyst, Dir Off, 84-85, div chief, Thermophys Div, 85-89, dir, Ctr Chem Technol, 89-90; RETIRED. *Honors & Awards:* Silver Medal, Dept Com, 81. *Mem:* Am Inst Chem Engrs; Am Soc Testing & Mat; AAAS. *Res:* Critical evaluation of thermophysical data for fluids and mixtures; development of predictive models for thermodynamic data; thermophysical properties; cryogenic engineering. *Mailing Add:* Thermophys Div Nat Bur Standards 325 Broadway Boulder CO 80303

OLIET, SEYMOUR, b Perth Amboy, NJ, July 12, 27; m 49; c 2. DENTISTRY. *Educ:* Univ Pa, DDS, 53; Am Bd Endodontics, dipl, 65. *Prof Exp:* Instr oral med, 53-56, assoc, 56-61, from asst prof to assoc prof clin oral med, 61-71, chmn dept, 71-80, PROF ENDODONT, SCH DENT, UNIV PA, 71- *Concurrent Pos:* Clin asst, Southern Div, Albert Einstein Med Ctr, 53-55, chief endodontics, 55-84, adj, 55-60, sr attend, 60-84; consult endodontics, US Army, Ft Dix, NJ, 61 & Vet Admin Hosp, Philadelphia, Pa, 67-84; dir & pres, Am Bd Endodont, 80. *Honors & Awards:* Am Acad Dent Med Award, 53. *Mem:* Fel AAAS; Am Dent Asn; Int Asn Dent Res; Am Acad Dent Med; hon mem Brazilian Endodontic Asn. *Res:* Use of salt and glass beads for rapid resterilization of root canal instruments and armamentarium; development of a torsional tester for root canal instruments and detailed studies on physical properties of root canal instruments; Gutta Percha; clinical research on one-visit endodontics; treating vertical tooth fractures. *Mailing Add:* 532 Putnam Rd Marion Station PA 19066

OLIGER, JOSEPH EMMERT, b Greensburg, Ind, Sept 3, 41; m 66; c 2. NUMERICAL ANALYSIS, APPLIED MATHEMATICS. *Educ:* Univ Colo, BA, 66, MA, 71; Univ Uppsala, PhD(comput sci), 73. *Prof Exp:* Prog, Nat Ctr Atmospheric Res, 65-73, mgr appl prog, 73-74; asst prof comput sci, 74-80, ASSOC PROF COMPUT SCI, STANFORD UNIV, 80- *Concurrent Pos:* Vis prof, Univ Stockholm, 71-72; vis staff mem, Los Alamos Sci Lab, 75-; consult, Lawrence Livermore Lab, 76-; ed J Numerical Analysis, Soc Indust & Appl Math, 76; vis asst prof, Math Res Ctr, Univ Wis-Madison, 78-79; vis prof, Univ Paris-Sud, 79. *Mem:* Am Math Soc; Soc Indust & Appl Math; Math Asn Am. *Res:* Numerical methods for partial differential equations; computational fluid dynamics and numerical weather prediction, ordinary and partial differential equations. *Mailing Add:* Dept Comput Sci Stanford Univ Bldg 460 Rm 308 Stanford CA 94305

OLIKER, VLADIMIR, b USSR, Oct 7, 45. MATHEMATICS. *Educ:* Leningrad Univ, BA, 67, PhD(math), 71. *Prof Exp:* Prof math, Univ Iowa, 77-84; PROF, EMORY UNIV, 84- *Mailing Add:* Dept Math & Comput Sci Emory Univ Atlanta GA 30322

OLIN, ARTHUR DAVID, b New York, NY, July 5, 28; m 54; c 3. ORGANIC CHEMISTRY. *Educ:* St Peter's Col, NJ, BS, 49; Rutgers Univ, PhD(org chem), 56. *Prof Exp:* Asst chem, Rutgers Univ, 50-54; res chemist polymer chem, Nat Starch & Chem Corp, 55; res chemist org chem, Toms River Chem Corp, 56-81, sr chemist, Ciba-Geigy Corp, 81-84. *Concurrent Pos:* Asst res specialist, Rutgers Univ, 55-56; co adj assoc prof, Ocean County Col, 70-71, adj instr, 90-; chem teacher, Cent Regional High Sch, Ocean County, NJ, 85-88; consult, org indust intermediates & dyes, 85-; lectr, Georgion Ct Col, 88-89. *Mem:* Am Chem Soc; Sigma Xi. *Res:* Process development and synthesis of vat and anthraquinone dyes and organic intermediates. *Mailing Add:* 1222 Tuxedo Terr Lakewood NJ 08701

OLIN, JACQUELINE S, b Lansford, Pa, Nov 27, 32; m 55; c 2. BIOCHEMISTRY, ANALYTICAL CHEMISTRY. *Educ:* Dickinson Col, BS, 54; Harvard Univ, MA, 55. *Prof Exp:* Teaching fel chem, Univ Pa, 55-56; res chemist, NIH, 56-57; instr chem, Dickinson Col, 59-60; res chemist, Cornell Univ, 60-61; res chemist, 62-83, adminr, Dept Archaeometry, 83-88, ASST DIR ARCHAEOMETRIC RES, CONSERV ANALYSIS LAB, SMITHSONIAN INST, 88- *Concurrent Pos:* Res collabr, Brookhaven Nat Lab, 66-83; guest worker, Nat Bur Standards, 68-; Nat Endowment Humanities res grant, 70; collab res grant, US-Spain Joint Comt Cult & Educ Coop, Madrid Spain, 87. *Mem:* Fel Int Inst Conserv Hist & Artistic Works; fel Am Inst Conver Hist & Artistic Works (secy, 81-82); Am Chem Soc; Soc Hist Archaeol; Soc Archaeol Sci. *Res:* Analysis of ceramic artifacts using neutron activation analysis for determination of provenance on basis of trace element constituents, extended to glasses with special attention to medieval stained glass; neutron induced autoradiography of paintings. *Mailing Add:* 9506 Watts Rd Great Falls VA 22066

OLIN, PHILIP, b Winnipeg, Man, Nov 21, 41; m 66. MATHEMATICAL LOGIC. *Educ:* Univ Man, BS, 63; Cornell Univ, PhD(math logic), 67. *Prof Exp:* Res assoc & lectr math, Cornell Univ, 67-69; asst prof, McGill Univ, 69-70; asst prof math, 70-74, assoc prof, 74-, AT DEPT COMPUT SCI, YORK UNIV. *Mem:* Am Math Soc. *Res:* Model theory; recursion theory. *Mailing Add:* Dept Math York Univ 4700 Keele St Toronto ON M3J 1P6 Can

OLIN, ROBERT, b Evanston, Ill, Oct 8, 48; m; c 2. MATHEMATICS. *Educ:* Univ Ind, PhD(math), 75. *Prof Exp:* PROF MATH, VA TECH STATE UNIV, 75- *Concurrent Pos:* NSF res grants, 75- *Mem:* Math Asn Am; Sigma Xi; NY Acad Sci. *Res:* Functional analysis; function and operator theory. *Mailing Add:* Math Dept Va Tech Inst & State Univ Blacksburg VA 24601

OLIN, WILLIAM (HAROLD), SR, b Menominee, Mich, Mar 7, 24; m 50; c 3. ORTHODONTICS. *Educ:* Marquette Univ, DDS, 47; Univ Iowa, MS, 48. *Prof Exp:* From asst prof to assoc prof, 48-64, PROF OTOLARYNGOL & MAXILLOFACIAL SURG, COL MED, UNIV IOWA, 64- *Mem:* Am Dent Asn; Am Cleft Palate Asn; Int Dent Fedn; Am Asn Orthod; Angle Orthod Soc. *Res:* Clefts of lip and palate; growth of the facial bones in cleft lip and palate patients, orofacial deformities. *Mailing Add:* Dept Otolaryngol Univ Iowa Iowa City IA 52246

OLINE, LARRY WARD, b Stafford, Kans, July 8, 37; m 59; c 3. ENGINEERING MECHANICS. *Educ:* Sterling Col, BA, 59; Univ Kans, BS, 61; Univ NMex, MS, 63; Ga Inst Technol, PhD(eng mech), 68. *Prof Exp:* Mech engr, Sandia Corp, 61-63; res asst, Ga Inst Technol, 63-67; assoc prof solid mech, 67-77, PROF ENG, UNIV S FLA, 77- *Concurrent Pos:* NSF grant strain-rate behavior of particulate filled epoxy, 70; NASA grant elastic plate spallation, 71 & 72. *Mem:* Soc Exp Stress Anal. *Res:* Stress wave propagation in solids; particulate-filled composites. *Mailing Add:* Dept Civil Eng-Mech Univ S Fla 4202 Fowler Ave Tampa FL 33620

OLINER, ARTHUR A(ARON), b Shanghai, China, Mar 5, 21; US citizen; m 46; c 2. MICROWAVES, ANTENNAS. *Educ:* Brooklyn Col, BA, 41; Cornell Univ, PhD(physics), 46. *Prof Exp:* Asst physics, Cornell Univ, 41-45; res assoc elec eng, 46-53, assoc prof, 53-57, head dept electrophys, 66-71, dir, Microwave Res Inst, 67-82, head dept elec eng & electrophys, 71-74, PROF ELECTROPHYS, POLYTECH INST NEW YORK, 57- *Concurrent Pos:* Chmn adv panel, Nat Bur Standards, 60-64; dir, Merrimac Industs, NJ, 61-; tech consult to indust; Guggenheim fel, 65-66; Walker-Ames vis prof, Univ Wash, Seattle, 64, vis prof, Cath Univ Rio de Janeiro, 73, Cent China Inst Technol, Wuhan, 80 & Univ Rome, Italy, 82; vis res scholar, Tokyo Inst Technol, Japan, 78. *Honors & Awards:* Inst Premium, Brit Inst Elec Engrs, 63; Microwave Prize, Inst Elec & Electronics Engrs, 67, Microwave Career Award, 82, Centennial Medal, 84; Citation for Distinguished Res, Sigma Xi, 74; van der Pol Gold Medal, Int Union Radio Sci, 90. *Mem:* Nat Acad Eng; fel Brit Inst Elec Engrs; fel Inst Elec & Electronics Engrs; Int Union Radio Sci; Optical Soc Am; hon mem Microwave Theory & Tech Soc; fel AAAS. *Res:* Microwave theory and techniques; electromagnetic guided waves, surface and leaky wave antennas; phased array antennas; periodic structures; plasmas; microwave networks and structures; electromagnetic radiation and diffraction; microwave acoustics; integrated optics; millimeter wave antennas. *Mailing Add:* 545 Westmnstr Rd Brooklyn NY 11201

OLINICK, MICHAEL, b Detroit, Mich, May 29, 41; m 63; c 4. MATHEMATICS. *Educ:* Univ Mich, Ann Arbor, BA, 63; Univ Wis-Madison, MA, 64, PhD(math), 70. *Prof Exp:* Asst lectr math, Univ Col Nairobi, Kenya, 65-66; instr, Univ Wis-Madison, 69-70; from asst prof to assoc prof, 70-81, PROF MATH, MIDDLEBURY COL, 81- *Concurrent Pos:* Res assoc, Univ Calif, Berkeley, 75-76; vis assoc prof, San Diego State Univ, 79, vis prof, Wesleyan Univ, 81 & Univ Lancaster, 88. *Mem:* Am Math Soc; Soc Indust & Appl Math; Math Asn Am; Opers Res Soc Am. *Res:* Topology of manifolds; monotone and compact mappings on euclidean spaces; mathematical modeling in social sciences. *Mailing Add:* Dept Math & Comput Sci Middlebury Col Warner Hall Middlebury VT 05753

OLINS, ADA LEVY, b Tel Aviv, Israel, Mar 5, 38; US citizen; m 61; c 2. CELL BIOLOGY, ELECTRON MICROSCOPY. *Educ:* City Col New York, BS, 60; Harvard Univ, MA, 62; NY Univ, PhD(biochem), 65. *Prof Exp:* Fel biochem, Dartmouth Col, 65-67; consult, Biol Div, Oak Ridge Nat Lab, Univ Tenn, 67-77, res asst prof biol, 77-80, res assoc prof, 80-82, RES PROF, OAK RIDGE GRAD SCH BIOMED SCI, 82- *Concurrent Pos:* USPHS spec fel cell biol, King's Col, Univ London, 70-71; res assoc, Oak Ridge Grad Sch Biomed Sci, Univ Tenn, 72-75, res assoc & lectr, 75-77; vis prof, German Cancer Res Ctr, Heidelberg, 79-80; vis scientist, Inst Pasteur, Paris, France & Max Planck Inst Biophys Chem, Goettingen, WGer; mem rev panel, NSF, 90-93. *Honors & Awards:* Theodor Boveri Award, Univ Wurzberg, 87. *Mem:* Am Soc Cell Biol; Electron Microscopy Soc Am. *Res:* DNA-protein interaction; chromosomal organization; nuclear ultrastructure; 3-dimensional image reconstruction; telomere structure. *Mailing Add:* Oak Ridge Grad Sch Biomed Sci Univ Tenn PO Box 2009 Oak Ridge TN 37831-8077

OLINS, DONALD EDWARD, b New York, NY, Jan 11, 37; m 61; c 2. BIOCHEMISTRY. *Educ:* Univ Rochester, AB, 58; Rockefeller Univ, PhD(biochem), 64. *Prof Exp:* Instr microbiol, Dartmouth Med Sch, 66-67, res fel moleculear biol, 64-65, Whitney fel, 65-67; from asst prof to assoc prof, 67-76, PROF, OAK RIDGE GRAD SCH BIOMED SCI, UNIV TENN, 76- *Concurrent Pos:* Vis scientist, King's Col, Univ London, 70-71 & German Cancer Res Ctr, Heidelberg, 79-80; Humboldt US sr scientist award, 79-80; hon prof, Northeast Normal Univ, Changchun, Peoples Repub China. *Mem:* Am Soc Cell Biol; Fed, Am Socs Exp Biol; fel AAAS. *Res:* Chemical structure of chromosomes and DNA-nucleoproteins. *Mailing Add:* Univ Tenn Grad Sch Biomed Sci Oak Ridge Nat Lab Biol Div Box 2009 Oak Ridge TN 37831-8077

OLIPHANT, CHARLES WINFIELD, b Oklahoma City, Okla, Mar 13, 20; m 42; c 2. GEOLOGY. *Educ:* Harvard Univ, BS, 41, MA, 47, PhD(geophys), 48. *Prof Exp:* Res assoc & asst to dir, Radio Res Lab, Harvard Univ, 42-46; PRES, OLIPHANT LABS, INC, 48-; CHMN, CEJA CORP, 66- *Mem:* AAAS; Soc Explor Geophysicists; Seismol Soc Am; Opers Res Soc Am; Am Asn Petrol Geologists. *Res:* Petroleum exploration; elastic properties of rocks; stratigraphy; seismic signal processing. *Mailing Add:* 4400 One Williams Ctr Tulsa OK 74172

OLIPHANT, EDWARD EUGENE, b San Francisco, Calif, May 31, 42; m 66. REPRODUCTIVE BIOLOGY, BIOLOGICAL CHEMISTRY. *Educ:* Univ Redlands, BS, 64; Calif State Col, Long Beach, MS, 66; Univ Calif, Davis, PhD(biochem), 70. *Prof Exp:* Fel biochem, Univ Calif, Davis, 70-71; fel reprod biol, Univ Pa, 71-73; from asst prof to assoc prof, Dept Obstet & Gynec, Sch Med, Univ Va, 73-88; CONSULT, 88- *Mem:* Soc Study Reprod; Am Soc Cell Biol; Am Asn Anat. *Res:* The biochemistry of sperm penetration into ova and early embryo development. *Mailing Add:* 3734 Park Dr El Dorado Hills CA 95630

OLIPHANT, LYNN WESLEY, b Highland Park, Mich, Dec 21, 42; m 64; c 2. CELL ULTRASTRUCTURE, RAPTOR ECOLOGY. *Educ:* Wayne State Univ, Detroit, BS, 64, MSc, 67; Univ Wash, Seattle, PhD(zool), 72. *Prof Exp:* From asst prof to assoc prof, 71-85, PROF MICROS ANAT, ELECTRON MICROS TECH & CELL ULTRASTRUCTURE, DEPT VET ANAT, WESTERN COL VET MED, UNIV SASK, 85-, DIR COL, 71-, ASSOC MEM, DEPT BIOL, UNIV SASK, 81- *Concurrent Pos:* Vis prof, Cornell Univ, 79-80 & Univ Ariz, 85-86. *Res:* Cell ultra structure, primarily pigment cells and contractile cells; avian histology and fine structure; pigmentation and musculature of the iris; ecology and management of falcons. *Mailing Add:* Dept Vet Anat Univ Sask Saskatoon SK S7N 0W0 Can

OLIPHANT, MALCOLM WILLIAM, b Chicago, Ill, Apr 15, 20; m 43; c 2. MATHEMATICS. *Educ:* Georgetown Univ, BS, 47; Johns Hopkins Univ, MA, 48; Cath Univ Am, PhD(math), 57. *Prof Exp:* Instr math, Johns Hopkins Univ, 47-48; mathematician, Nat Bur Standards, 48-50; prof math & chmn dept, Georgetown Univ, 49-66; dean acad affairs, Hawaii Loa Col, 66-80, Prof math, 66-85; vpres, Response Hawaii, 79-86; RETIRED. *Concurrent Pos:* Mem, Int Cong Mathematicians, Edinburgh, 58. *Mem:* AAAS; Am Math Soc; Math Asn Am. *Res:* Mathematical analysis; modern measure theory. *Mailing Add:* 1606 Ulupii St Kailua HI 96734

OLIVE, AULSEY THOMAS, b Mount Gilead, NC, May 23, 31; m 53; c 2. ENTOMOLOGY. *Educ:* Wake Forest Col, BS, 53; NC State Univ, MS, 55, PhD(entom), 61. *Prof Exp:* From asst prof to assoc prof biol, Wake Forest Univ, 61-88; RETIRED. *Concurrent Pos:* NSF grant, 65-67. *Mem:* Entom Soc Am. *Res:* Taxonomic and biosystematic study of aphids of the eastern United States; cytogenetic approach to the species problem in the aphid genus Dactynotus Rafinesque. *Mailing Add:* 1440 Shangrila Dr Louisville NC 27023

OLIVE, DAVID WOLPH, b Weeping Water, Nebr, Nov 6, 27; m 56; c 1. ELECTRICAL ENGINEERING. *Educ:* Univ Nebr, BS, 50; Ill Inst Technol, MS, 54; Univ Wis, PhD(elec eng), 60. *Prof Exp:* Electronics engr, naval ord div, Eastman Kodak Co, NY, 51-52; power engr, switchgear div, Gen Elec Co, Pa, 54-55; instr elec eng, Univ Nebr, 55-58, asst prof, 60-63; sr engr, elec utility eng, Westinghouse Elec Corp, Pa, 63-65; assoc prof elec eng, Univ Nebr, 65-69; L F Hunt Prof elec power eng, Univ Southern Calif, 69-76; PROF ELEC ENG, UNIV NEBR, 76- *Res:* Electromechanical energy conversion; control systems; power system analysis and operations. *Mailing Add:* 194-W Ne Hall Univ Nebr Lincoln NE 68503

OLIVE, GLORIA, b New York, NY, June 8, 23. MATHEMATICS. *Educ:* Brooklyn Col, BA, 44; Univ Wis, MA, 46; Ore State Univ, PhD(math), 63. *Prof Exp:* Asst math, Univ Wis, 44-46; instr, Univ Ariz, 46-48 & Idaho State Col, 48-50; asst, Ore State Col, 50-51; prof, Univ Wis-Superior, 68-71; sr lectr pure math, Univ Otago, NZ, 72-89; from asst prof to prof & head dept, 52-68, VIS PROF, ANDERSON UNIV, IND, 89- *Concurrent Pos:* Mathematician, US Dept Defense, 51; Wis Bd Regents res grant, Univ Wis-Superior, 70-71; convenor Nat Comt Math, NZ, 85-89. *Mem:* Am Math Soc; Math Asn Am; hon mem NZ Math Soc. *Res:* Generalized powers; b-transform; binomial functions; Catalan numbers. *Mailing Add:* Dept Math Univ Otago Dunedin New Zealand

OLIVE, JOHN H, b Glenford, Ohio, Apr 16, 29; m 58; c 4. CELL PHYSIOLOGY, AQUATIC BIOLOGY. *Educ:* Ohio State Univ, BS, 53; Kent State Univ, MA, 61, PhD(biol), 64. *Prof Exp:* From assoc prof to prof biol, Ashland Col, 64-70; assoc prof, 70-78, PROF BIOL, UNIV AKRON, 78- *Mem:* AAAS; Am Inst Biol Sci; NAm Benthological Soc; Am Soc Limnol & Oceanog; Sigma Xi. *Res:* Phytoplankton photosynthesis; nutrition of blue-green algae; biological indicators of water quality; physiological ecology of plectonema wollei; rate of growth measurement under controlled experimental and natural conditions; organism's metabolism of phosphorus. *Mailing Add:* 1671 Meadow Lane SE North Canton OH 44709

OLIVE, JOSEPH P, b Israel, Mar 14, 41; US citizen; m 79; c 3. COMMUNICATIONS SCIENCE. *Educ:* Univ Chicago, BS & MS, 64, MA & PhD(physics), 69. *Prof Exp:* SUPVR, AT&T BELL LABS, NJ, 69- *Concurrent Pos:* Nat Endowment Arts grant, 74. *Mem:* Acoust Soc Am. *Res:* Production of speech by computers. *Mailing Add:* AT&T Bell Labs Inc 2D513 600 Mountain View Murray Hill NJ 07940

OLIVE, LINDSAY SHEPHERD, botany; deceased, see previous edition for last biography

OLIVE, PEGGY LOUISE, b Montreal, Que, May 30, 48; m 74. RADIATION BIOLOGY, BIOCHEMISTRY. *Educ:* Bishop's Univ, BSc, 69; Univ Western Ont, MSc, 72; McMaster Univ, PhD(biochem), 76. *Prof Exp:* Res asst, Ont Cancer Found, 69-72 & Dept Biochem, McMaster Univ, 72-76; res assoc dept human oncol, Univ Wis, 76-77; from instr to asst prof oncol, Environ Health Sci, Oncol Ctr, Johns Hopkins Univ, 77-83; assoc prof, 83-91, PROF PATH, UNIV BC, 91-, STAFF BIOPHYSICIST, BC CANCER RES CENTER, 83- *Concurrent Pos:* Mem, clin sci study sect, NIH, 82-86, ratiation res study sect, 88- *Mem:* Biophys Soc; Radiation Res Soc; Am Asn Cancer Res; Environ Mutagen Soc. *Res:* Cellular resistance to radiation damage; role of DNA conformation in repair; methods for quantifying hypoxic tumor cells. *Mailing Add:* British Columbia Cancer Res Ctr 601 W 10th Ave Vancouver BC V5Z 1L3 Can

OLIVER, ABE D, JR, b Castleberry, Ala, Dec 3, 25; m 50; c 2. ENTOMOLOGY, FORESTRY. *Educ:* Auburn Univ, BS, 53, MS, 54; La State Univ, PhD(entom), 63. *Prof Exp:* Asst entomologist, Miss State Univ, 54-55; asst entomologist, 55-58, from asst prof to assoc prof, 58-68, PROF ENTOM, LA STATE UNIV, BATON ROUGE, 68- *Mem:* Entom Soc Am. *Res:* Forest insect research, especially ecology and economics; floriculture insect research, especially economics and systematics; biological control of aquatic weeds with insects. *Mailing Add:* 833 Rodney Dr Baton Rouge LA 70808

OLIVER, BARRY GORDON, b Winnipeg, Man, Feb 21, 42; m 64; c 2. ENVIRONMENTAL CHEMISTRY. *Educ:* Univ Man, BSc, 63, MSc, 65 & PhD(phys chem), 69. *Prof Exp:* Res assoc, Rensselaer Polytech Inst, 68-69, asst prof chem, 69-70; Nat Res Coun Can fel phys chem, Inland Waters Br, Can Dept Environ, 70-72, res scientist, Can Centre Inland Waters, 73-87, vpres, Eli Eco Lab Inc, 87-89; MGR ANAL DEVELOP, ZENON ENVIRON LABS, 89- *Mem:* Chem Inst Can; fel Chem Inst Can. *Res:* Vibrational spectroscopy; analysis of trace metals in sediments and natural waters; water chemistry; water and wastewater treatment methods; chlorination byproducts, sources and pathways of organic pollutants in the aqueous environment; analytical chemistry of trace organics. *Mailing Add:* Zenon Environ Labs 8577 Commerce Ct Burnaby BC V5A 4N5 Can

OLIVER, BENNIE F(RANK), b Monessen, Pa, Oct 19, 27; m 52; c 3. METALLURGY. *Educ:* Pa State Univ, BS, 52, MS, 54, PhD(metall, physics), 59. *Prof Exp:* Instr metall, Pa State Univ, 54-59, asst prof, 59-60; res scientist, E C Bain Lab Fundamental Res, US Steel Corp, 60-67; assoc prog dir eng mat, NSF, 67-68; prof in chg metall, Univ Tenn, Knoxville, 68-85; RETIRED. *Mem:* Am Soc Metals; Am Inst Mining, Metall & Petrol Engrs; Mat Res Soc. *Res:* Process metallurgy; refining; high purity materials; trace elements; solidification and crystal growth; intermetallic compounds; composites. *Mailing Add:* 4211 Holston Hills Rd Knoxville TN 37914

OLIVER, BERNARD M(ORE), b Santa Cruz, Calif, May 27, 16; c 3. DESIGN ENGINEERING, RADIO ASTRONOMY. *Educ:* Stanford Univ, BA, 35; Calif Inst Technol, MS, 36, PhD(elec eng & phys), 39. *Prof Exp:* Mem tech staff, Bell Tel Labs, Inc, 39-52; dir res, Hewlett-Packard Co, 52-57, vpres, 57-81; chief, 83-89, DEPUTY CHIEF, SEARCH EXTRA-TERRESTRIAL INTEL PROG OFF, AMES RES CTR, MOFFETT FIELD, CALIF, NASA, 89-; TECH CONSULT TO PRES, HEWLETT-PACKARD CO, 81- *Honors & Awards:* Lammé Medal, Inst Elec & Electronics Engrs, 76, Nat Medal of Sci, 86; Except Engr Achievement Medal, NASA, 90; Pioneer Award, Int Found Telemetering, 90. *Mem:* Nat Acad Sci; Nat Acad Eng; Am Astron Soc; Astron Soc Pac; Inst Elec & Electronics Engrs (pres, 65-66). *Res:* Electronic circuits and system design; electronic instrumentation; information theory; search for extra-terrestrial intelligence. *Mailing Add:* Hewlett-Packard Co PO Box 10490 Palo Alto CA 94303-0969

OLIVER, CALVIN C(LEEK), b Castleberry, Ala, Apr 5, 32; m 54; c 3. MECHANICAL ENGINEERING. *Educ:* Ga Inst Technol, BME, 57, MSME, 58; Purdue Univ, PhD(mech eng), 63. *Prof Exp:* Instr thermodyn, Ga Inst Technol, 57-59, asst prof heat transfer, 59-61; asst prof fluid mech, Purdue Univ, 63-65, assoc prof heat transfer, 65-67; prof, 67-76, PROF MECH ENG & ENG SCI, UNIV FLA, 76- *Concurrent Pos:* Assoc, Ga Tech Exp Sta, 58-61; consult, Midwest Appl Sci Corp, 64- & Controlled Acoust, Inc, 71- *Mem:* Am Soc Mech Engrs; Am Soc Eng Educ. *Res:* Thermal phenomena; nuclear technology; acoustics; thermophysical properties; heat transfer; thermal radiation; hypersonic flow with chemical reactions; electric arcs. *Mailing Add:* Dept Mech Eng Univ Fla Gainesville FL 32611

OLIVER, CARL EDWARD, b Anniston, Ala, Feb 26, 43; m 65; c 2. MATHEMATICS. *Educ:* Univ Ala, Tuscaloosa, BS, 65, MA, 67, PhD(math), 69. *Prof Exp:* Asst prof math, USAF, Inst Technol, 69-74, mathematician, Weapons Lab, 74-78, prog mgr comput math, USAF Off Sci Res, 78-83, prog mgr appl math sci, HQ Dept Energy, 83-85, dep chief scientist, Weapons Lab, 85-89; DIR, OFF LAB COMPUT, OAK RIDGE NAT LAB, 89- *Concurrent Pos:* Lectr, Ohio State Univ, 70-71, Wright State Univ, 71-74, NMex Highlands Univ, 74-78 & George Mason Univ, 79-85. *Mem:* Am Phys Soc; Math Asn Am; Soc Indust & Appl Math; Sigma Xi; Inst Elec & Electronics Engrs. *Res:* Interrelations and applications of the concepts within integration theory; numerical analysis; computational mathematics; supercomputer research. *Mailing Add:* 134 Balboa Cir Oak Ridge TN 37830

OLIVER, CONSTANCE, b Whittier, Calif, Nov 23, 42. CELL BIOLOGY, CYTOCHEMISTRY. *Educ:* Northwestern Univ, BA, 64; Univ Utah, MS, 67; Univ Tex, Austin, PhD(zool), 71. *Prof Exp:* Fel, Sloan Kettering Inst, 71-72, res assoc, 72-74; staff fel, 74-76, sr staff fel, 76-79, RES BIOLOGIST, NAT INST DENT RES, 79- *Concurrent Pos:* Asst attend cytochemist path, Mem Sloan Kettering Cancer Ctr, 73-74. *Mem:* Am Soc Cell Biol; Histochem Soc; Sigma Xi. *Res:* Signal transduction; mechanisms of cellular secretion. *Mailing Add:* Nat Inst Dent Res Bldg 10 Rm 1N101 NIH Bethesda MD 20892

OLIVER, DAPHNA R, b Tel Aviv, Israel, Aug 10, 45; US & Israeli citizen; c 2. BIOLOGY. *Educ:* Bar Ilan Univ, BSc, 66; Univ Calif, Los Angeles, PhD(med microbiol), 73. *Prof Exp:* Fel biol chem, Univ Mich, Ann Arbor, 72-74, res assoc microbiol, 74-76; vis asst prof, 77-78, ASST PROF BIOL SCI, OAKLAND UNIV, 78- *Mem:* Am Soc Microbiol; AAAS; Am Asn Univ Prof; Am Soc Med Technol; Sigma Xi. *Res:* Relationship between plasmids and pathogenicity in streptococci. *Mailing Add:* Dept Biol Sci Oakland Univ Rochester MI 48309

OLIVER, DAVID JOHN, b Marcellus, NY, Sept 13, 49; m 71; c 2. PLANT PHYSIOLOGY, BIOCHEMISTRY. *Educ:* State Univ NY, BS & BForestry, 71, MS & MF, 73; Cornell Univ, PhD(bot), 75. *Prof Exp:* Res & teaching asst plant physiol & genetics, Col Forestry, State Univ NY, 71-73; res & teaching asst plant physiol, Cornell Univ, 73-75; NSF fel, Conn Agr Res Sta, 75-76, agr scientist biochem, 76-80; asst prof, 80-84, ASSOC PROF, UNIV IDAHO, 84- *Concurrent Pos:* Vis researcher, CNRS, Grenoble, France, 88-89. *Mem:* Am Soc Plant Physiol; AAAS. *Res:* Mechanism of photosynthetic carbon dioxide fixation by green plants including electron transport, photophorylation, photorespiration and oxidative phosphorylation; using biochemical and molecular biology techniques. *Mailing Add:* Dept Bact & Biochem Univ Idaho Moscow ID 83843

OLIVER, DAVID W, b Fairfax Co, Va, Dec 21, 32; m 58; c 4. SOLID STATE PHYSICS, CONTROL THEORY. *Educ:* Va Polytech Inst, BS, 55, MS, 56; Mass Inst Technol, PhD(physics), 61. *Prof Exp:* Physicist, Gen Elec Co, 61-73, actg mgr, Liaison Br, 75-78, mgr, Automation & Control Lab, 78-79, actg mgr, Solid State Commun Br, 79-90, LIAISON SCIENTIST, AIRCRAFT ENGINE BUS GROUP, GEN ELEC CO, 73-, SYST ENGR, INFO SYST LAB, 90- *Concurrent Pos:* Consult electronics technol & mat. *Mem:* Am Phys Soc; Inst Elec & Electronics Engrs. *Res:* Microwave spectroscopy with application to chemical reaction rates for diatomic gases; acoustics at microwave frequencies and above; growth of inorganic single crystals and their characterization; ultrasonic nondestructive evaluation. *Mailing Add:* Corp Res & Develop Gen Elec Co PO Box 8 Bldg K-1 Rm 5c35 Schenectady NY 12301

OLIVER, DENIS RICHARD, b Santa Barbara, Calif, Nov 12, 41. BIOCHEMISTRY. *Educ:* Calif State Univ, Long Beach, BS, 65; Univ Iowa, PhD(zool), 71. *Prof Exp:* NSF Ctr Excellence res fel biochem, 71-72, res assoc, 72-73, instr biochem & assoc dir, 73-77, assoc, 77-80, ASSOC PROF BIOCHEM, UNIV IOWA, 80-, DIR, PHYSICIAN'S ASST PROG, 77- *Mem:* Sigma Xi; Am Soc Allied Health Prof. *Res:* Structure and organization of nucleohistone in terms of the spatial relationship between histone chromosomal proteins and DNA in chromatin and the mechanism by which newly synthesized histone is deposited. *Mailing Add:* Dept Physician's Asst Univ Iowa Iowa City IA 52242

OLIVER, DONALD RAYMOND, b Saskatoon, Sask, Aug 20, 30. ENTOMOLOGY, FRESHWATER BIOLOGY. *Educ:* Univ Sask, BA, 53, MA, 55; McGill Univ, PhD(limnol & entom), 61. *Prof Exp:* RES SCIENTIST ENTOM, BIOSYSTS RES INST, AGR CAN, 62- *Concurrent Pos:* Adj prof biol dept, Carleton Univ, 70- *Mem:* Entom Soc Can; fel Arctic Inst NAm. *Res:* Biosystematics of Chironomidae such as Diptera. *Mailing Add:* BRC K W Neatby Bldg CEF Room 1002 Ottawa ON K1A 0C6

OLIVER, EARL DAVIS, b Douglas, Ariz, Aug 10, 23; m 60; c 3. CHEMICAL ENGINEERING, ECONOMICS. *Educ:* Univ Wash, BS, 44, MS, 47; Univ Wis, PhD(chem eng), 52. *Prof Exp:* Engr, Arabian Am Oil Co, 47-50 & Shell Develop Co, 52-60; assoc prof chem eng, Univ NMex, 60-65; engr, Colony Develop Co, 65-67; sr chem engr, SRI Int, 67-75, sr engr & economist, 75-80; mgr process evaluation, Synthetic Fuels Assocs, 80-84; PRIN, OLIVER ASSOCS, INC, 84- *Concurrent Pos:* Exten teacher, Univ Calif, 55-60; sr res scientist, McAllister & Assocs, 61-62; vis prof, Univ Tex, 64. *Mem:* Am Inst Chem Engrs. *Res:* Perform engineering, economic and market studies for industrial and governmental clients; write reports; plan experimental programs; develop, design and evaluate economics of processes for energy, environment and petrochemicals. *Mailing Add:* 2049 Kent Dr Los Altos CA 93440

OLIVER, EUGENE JOSEPH, b Pawtucket, RI, Jan 28, 41; m 70. MICROBIOLOGY, BIOCHEMISTRY. *Educ:* RI Col, BEd, 62; Univ Mass, PhD(microbiol), 69. *Prof Exp:* Teacher, Scituate Jr-Sr High Sch, 62-64; res assoc biochem, Univ Mass, 69-71; health scientist adminr, Nat Inst Gen Med Sci, 74-80, HEALTH SCIENTIST ADMINR, NAT INST NEUROL COMMUN DIS & STROKE, NIH, 80- *Concurrent Pos:* NIH fel biochem, Nat Heart & Lung Inst, 71- *Mem:* AAAS; Am Soc Microbiol. *Res:*

Enzymology of bacteria and higher organisms; carbohydrate and amino acid metabolism; molecular mechanisms of cryobiological adaptation. *Mailing Add:* Nat Inst Neurol Commun Dis & Stroke NIH Fed Bldg Rm 806 Bethesda MD 20892

OLIVER, G CHARLES, b Gainesville, Fla, Sept 30, 31; m 79; c 1. CARDIOVASCULAR DISEASE. *Educ:* Harvard Univ, AB, 53; Harvard Med Sch, MD, 57; Am Bd Internal Med, dipl, 65; Am Bd Cardiovasc Dis, dipl, 70. *Prof Exp:* Chief, Cardiovasc Div, Jewish Hosp, St Louis, 71-81; from instr to prof med, 66-81, co-dir, Cardiovasc Div, 75-81, CLIN PROF MED, SCH MED, WASHINGTON UNIV, 81-; CHIEF CARDIOL, FAITH HOSP, ST LOUIS, 81- *Concurrent Pos:* Fel cardiol, Med Ctr, Stanford Univ, 60-61; USPHS fel, Guy's Hosp, London, Eng, 63-65; mem policy bd, Clin Trials Thrombolytic Agents, NIH, 73-; mem electrocardiography comt, Am Heart Asn, 74- *Mem:* AAAS; fel Am Col Cardiol; NY Acad Sci; Am Heart Asn. *Res:* Computer applications to cardiology; pharmacology of digitalis. *Mailing Add:* St Charles Clin 2850 Welay St Charles MO 63301

OLIVER, GENE LEECH, b Rockford, Ill, June 7, 29; m 55, 80; c 3. PHOTOGRAPHIC CHEMISTRY, ORGANIC CHEMISTRY. *Educ:* Beloit Col, BS, 50; Northwestern Univ, PhD, 55. *Prof Exp:* Res chemist, Eastman Kodak Co, 54-65, res assoc, 65-74, tech staff assoc, 74-89; RETIRED. *Mem:* Am Chem Soc (counr, 74-88); Soc Photog Scientists & Engrs; Sigma Xi. *Res:* Photographic sensitizing and image dyes; silver halide emulsion technology; organic heterocyclic chemistry; photographic patents. *Mailing Add:* 24 Pine Cone Dr Pittsford NY 14534-3534

OLIVER, JACK ERTLE, b Massillon, Ohio, Sept 26, 23; m 64; c 2. GEOPHYSICS. *Educ:* Columbia Univ, BA, 47, MA, 50, PhD(geophys), 53. *Hon Degrees:* Hamilton Col, 88. *Prof Exp:* Scientist's aide, US Naval Res Lab, 47; physicist, Air Force Cambridge Res Lab, 51; res assoc, Lamont-Doherty Geol Observ, Columbia Univ, 53-55, from asst prof to prof geol, 55-57; chmn, Dept Geol Sci, 71-81, IRVING PORTER CHURCH PROF ENG, CORNELL UNIV, 71- *Concurrent Pos:* Consult, President's Sci Adv Comt Panel Seismic Improv, 58-59, Air Force Tech Appln Ctr, 58-65, Advan Res Projs Agency, 59-63, US Arms Control & Disarmament Agency, 62-73, comt seismol & earthquake eng, UNESCO, 65 & Atomic energy Comn, 69-74; mem, Sci Adv Bd, USAF, 60-63 & 64-69; mem, comt polar res, Nat Acad Sci & comt seismol, 60-72, chmn, 66-69; mem, Comn Geosci Energy Res, 89-91; mem, adv comt, US Coast & Geol Surv, 62-66 & Panel Solid Earth Probs, 62-66; mem, Nat Comt Upper Mantle Prog, 63-71, Site Selection Comt, 65, Carnegie Inst Adv Comt Awards, Gilbert & Wood Fund, 64-68 & Geophys Res Bd, Assembly Math & Appl Sci, 81-84; chmn, dept geol, Columbia Univ, 69-71, adj prof, 71-73; chmn, Geodesy Comt, 75-76; chmn, Geol Sci Bd, Assembly Math & Appl Sci, Nat Res Coun, 81-84, Geophys Res Forum, 84-86, mem, Comn Phys Sci, Math & Resources, 87-; mem, Geodynamics Comt, 79-87 & chmn 84-87; dir, Inst for Study of Continents, 83-88; mem adv bd, Sch Earth Sci, Stanford Univ, 84-87; mem, Comn Geosci Energy Res, 89-91. *Honors & Awards:* Walter Bucher Medal, Am Geophys Union, 81; Shell Distinguished Lectr, 83; Virgil Kauffman Gold Medal Award, Soc Explor Geophysicists, 83; Eighth Medal, Seismol Soc Am, 84; Heaberg Award, Inst Study of Earth & Man, 90; Woollard Award, Geol Soc Am, 90. *Mem:* Nat Acad Sci; fel AAAS; fel Seismol Soc Am (vpres, 62-64, pres, 64-65); fel Am Geophys Union; fel Geol Soc Am (vpres, 86, pres, 87); hon foreign fel Europ Union Geosci, 86; hon mem Geol Soc London. *Res:* Seismology; geotectonics. *Mailing Add:* Dept Geol Sci Cornell Univ Snee Hall Ithaca NY 14853-1504

OLIVER, JACK WALLACE, b Ellettsville, Ind, Jan 6, 38; m 60; c 2. VETERINARY PHARMACOLOGY, ENDOCRINOLOGY. *Educ:* Purdue Univ, BS, 60, MS, 63, DVM, 66, PhD(vet physiol), 69. *Prof Exp:* Nat Inst Arthritis & Metab Dis fel, 67-69; asst prof vet pharmacol, Sch Vet Med, Purdue Univ, 69-70; asst prof vet endocrinol & pharmacol, Col Vet Med, Tex A&M Univ, 70-71; pvt pract, Ind, 71-72; asst prof vet pharmacol, Col Vet Med, Ohio State Univ, 72-75; assoc prof, 75-81, PROF VET PHARMACOL, COL VET MED, UNIV TENN, KNOXVILLE, 81- *Mem:* Am Soc Vet Physiol & Pharmacol; Am Vet Med Asn; Am Col Vet Pharm & Therapeut; Conf Res Workers Animal Dis. *Res:* Thyroid function as affected by dietary factors, infectious organisms and altered physiological states; thyroid-structural tissue interrelationships; diagnostic procedures for hormones; therapeutic drug monitoring; mechanisms of fescue toxicity, including biogenic amine-receptor effects and hormonal interaction. *Mailing Add:* Dept Environ Practice Col Vet Med Univ Tenn Knoxville TN 37901-1071

OLIVER, JAMES HENRY, JR, b Augusta, Ga, Mar 10, 31; m 57; c 2. ACAROLOGY, CYTOGENETICS. *Educ:* Ga Southern Col, BS, 52; Fla State Univ, MS, 54; Univ Kans, PhD(entom), 62. *Prof Exp:* NATO fel, Univ Melbourne, 62-63; from asst prof entom & parasitol to assoc prof entom, Univ Calif, Berkeley, 63-68; assoc prof, Univ Ga, 68-69; CALLAWAY PROF BIOL, GA SOUTHERN UNIV, 69-, DIR, IAP, GA SOUTHERN UNIV, 83- *Concurrent Pos:* Consult, US Naval Med Res Unit 3, 63-, US Agency Int Develop & Trop Med & Parasitol Study Sect, NIH, 79-83. *Mem:* Fel AAAS; fel Entom Soc Am (pres); Am Inst Biol Sci; Acarological Soc Am; Int Soc Invert Reproduction; Am Soc Trop Med & Hyg. *Res:* Cytology, genetics, reproduction and bionomics of arthropods; arthropods as vectors of disease. *Mailing Add:* Dept Biol Ga Southern Univ Statesboro GA 30460-8056

OLIVER, JAMES RUSSELL, b Egan, La, Sept 12, 24; m 45; c 3. PHYSICAL CHEMISTRY, NUCLEAR CHEMISTRY. *Educ:* Univ Southwestern La, BS, 50; Tulane Univ, MS, 51, PhD(chem), 55. *Prof Exp:* From assoc prof to prof chem, 54-70, dir comput ctr, 60-70, dean grad sch, 61-73, PROF COMPUT SCI, UNIV SOUTHWESTERN LA, 70-, DEAN ACAD & FINANCIAL PLANNING & V PRES ADMIN AFFAIRS, 73- *Concurrent Pos:* Consult, Electro-Acid Corp & Silverloy Int Corp, Nev; asst state supt educ for mgt, res & finance, 72-73; exec dir, State Bd Educ, 73; mem data processing corrd & adv coun, State La, 78-; pres, Phoenix Comput Systs, Inc, 77-; mem bd dirs, Southwest Educ Res Lab, 78- *Mem:* AAAS; Am Chem Soc; Am Nuclear Soc; Asn Comput Mach; Soc Indust & Appl Math; Sigma Xi. *Res:* Coordination compounds; radioactive tracers; reactor fuel processing; radioactive waste disposal; reaction kinetics; radiation chemistry; digital computers; science eduation; science talent search and writing; design of computer languages. *Mailing Add:* PO Box 4 0133 Univ Southwestern La Sta Lafayette LA 70504

OLIVER, JANET MARY, b Adelaide, South Australia, Nov 14, 45. CELL PHYSIOLOGY. *Educ:* Univ Adelaide, BSc, 66; Flinders Univ South Australia, BScHons, 67; Univ Alta, MSc, 69; Univ London, PhD(biochem), 72. *Prof Exp:* Asst prof physiol, 73-78, assoc prof physiol & path, Health Ctr, Univ Conn, 78-; PROF PATHOL, UNIV NMEX. *Concurrent Pos:* Leukemia Soc Am fel, Harvard Med Sch, 72-73. *Mem:* Am Asn Path; Am Soc Cell Biol; Brit Biochem Soc. *Res:* Role of microtubules and microfilaments in regulation of cell surface functions; control of microtubule assembly in leukocytes and cultured cells; inherited and acquired defects in cytoskeleton and cell surface properties. *Mailing Add:* Dept Pathol Univ NMex Albuquerque NM 87131

OLIVER, JOEL DAY, b Amarillo, Tex, Dec 27, 45; m 66; c 3. STRUCTURAL CHEMISTRY. *Educ:* WTex State Univ, BS, 68; Univ Tex, Austin, PhD(phys chem), 71. *Prof Exp:* asst prof phys chem, WTex State Univ, 71-79; X-RAY LAB MGR, PROCTER & GAMBLE CO, 79- *Mem:* Am Chem Soc; Am Crystallog Asn; Chem Soc; Sigma Xi. *Res:* X-ray diffraction studies of organic and organometallic compounds; laboratory automation; minicomputer applications to chemistry; variable-temperature x-ray powder diffraction studies; phase transitions. *Mailing Add:* Proctor & Gamble Co PO Box 398707 Cincinnati OH 45239-8707

OLIVER, JOHN EDWARD, b Dover, England, Oct 21, 33; m 57; c 2. PHYSICAL GEOGRAPHY. *Educ:* Univ London, BSc, 56; Univ Exeter, cert educ, 57; Columbia Univ, MA, 66, PhD(geog), 69. *Prof Exp:* Lectr geol & geol, Willesden Tech Col, London, 57-58; teacher geog, Warwick Acad, Bermuda, 58-66; lectr, Columbia Univ, 66-69, from asst prof to assoc prof, 69-73; assoc prof, 73-77, PROF GEOG, IND STATE UNIV, 78- *Mem:* AAAS; Asn Am Geogrs; Am Meteorol Soc. *Res:* Environmental science with special reference to applications of climatology. *Mailing Add:* Dept Geog & Geol Ind State Univ 217 N 6 St Terre Haute IN 47809

OLIVER, JOHN EOFF, JR, b Stephenville, Tex, June 22, 33; m 57; c 3. COMPARATIVE NEUROLOGY. *Educ:* Tex A&M Univ, DVM, 57; Auburn Univ, MS, 66; Univ Minn, St Paul, PhD(neuroanat), 69; Am Col Vet Internal Med, dipl. *Prof Exp:* Instr vet surg, Col Vet Med, Colo State Univ, 57-58; veterinarian, Houston, Tex, 60-63; res asst vet neurol, Sch Vet Med, Auburn Univ, 63-66; USPHS fel neuroanat, Col Vet Med, Univ Minn, St Paul, 66-67, spec fel, 67-68; assoc prof comp neurol, Col Vet Med, Univ Ga, 68-72; prof small animal clin & head dept, Sch Vet Sci & Med, Purdue Univ, Lafayette, 72-75; head dept, 75-90, PROF SMALL ANIMAL MED, COL VET MED, UNIV GA, 90- *Concurrent Pos:* Chief Ed, Vet Med Report, 88-91. *Mem:* AAAS; Am Vet Med Asn; Am Vet Radiol Soc; Am Asn Vet Neurol; Am Col Vet Internal Med (vpres, 74-75, pres, 76-77). *Res:* Neural control of micturition; pathophysiology of neurogenic bladder; models of neural diseases; hydrocephalus. *Mailing Add:* Dept Small Animal Med Univ Ga Col Vet Med H385 Athens GA 30602

OLIVER, JOHN PARKER, b New Rochelle, NY, Nov 24, 39; m 63; c 3. ASTRONOMY. *Educ:* Rensselaer Polytech Inst, BS, 62; Univ Calif, Los Angeles, MA, 68, PhD(astron), 74. *Prof Exp:* Mem tech staff astron, Aerospace Corp, 65-67; asst prof, 70-76, ASSOC PROF ASTRON, UNIV FLA, 76-, DIR, ROSEMARY HILL OBSERV, 81- *Mem:* Am Astron Soc; Int Astron Union. *Res:* Astronomical photoelectric photometry; lunar occultations of stars; eclipsing binary stars; astronomical instrumentation; spectroscopy. *Mailing Add:* Dept Astron Univ Fla Gainesville FL 32611

OLIVER, JOHN PRESTON, b Klamath Falls, Ore, Aug 7, 34; m 56; c 3. INORGANIC CHEMISTRY. *Educ:* Univ Ore, BA, 56; Univ Washington, PhD (chem), 59. *Prof Exp:* From asst prof to assoc prof, 59-67, PROF CHEM, WAYNE STATE UNIV, 67-, CHMN DEPT, 71-, ASSOC DEAN RES & DEVELOP, COL LIBERAL ARTS, 87- *Mem:* Am Chem Soc; Sigma Xi. *Res:* Synthesis of organometallic compounds of Groups I, II, III and IV; alkyl-metal exchange reactions; nuclear magnetic resonance spectra of organometallic compounds; CP/MAS applied to organometallic compounds. *Mailing Add:* 135 Chem Bldg Wayne State Univ Detroit MI 48202

OLIVER, KELLY HOYET, JR, b Roseboro, Ark, June 22, 23; m 47; c 3. AQUATIC BIOLOGY. *Educ:* Southern Methodist Univ, BS, 52, MS, 53; Okla State Univ, PhD(zool), 63. *Prof Exp:* Aquatic biologist, Anderson Fish Farms, 53; teacher pub schs, Ark, 53-61; from asst prof to assoc prof biol, Ark State Col, 62-64; chief biologist, Chem Biol Lab, Vitro Serv, Eglin AFB, 64-68; assoc prof, 68-73, PROF BIOL, HENDERSON STATE COL, 73-; PRES, KOCOMORO, INC, 73- *Concurrent Pos:* Res biologist, Crossett Co, 56-61; mem, Int Cong Limnol, 62; NSF vis scientist, Ark high schs, 63-64; contractor, aquatic biol & ecol, Edgewood Arsenal, Md, 72-79; res biologist, Nat Ctr Toxicol, 79-80; aquatic toxicologist, Battelle Columbus Labs, Ohio, 81- *Mem:* AAAS; Ecol Soc Am. *Res:* Fishery management; pollution biology; effect of industrial waste, especially paper and petroleum, on aquatic organisms and bacterial ecology; ecology of Lake De Gray, Arkansas, especially change in fauna and flora and physiochemical conditions during river-lake transition; toxicity of contaminants on bluegills and annelids. *Mailing Add:* Dept of Biol Henderson State Col Arkadelphia AR 71923

OLIVER, LESLIE HOWARD, b Halifax, NS, Dec 8, 40; m 64; c 2. DIGITAL IMAGE ANALYSIS, MEDICAL IMAGING. *Educ:* Acadia Univ, BSc, 62, MSc, 66; McGill Univ, PhD(computer sci), 79. *Prof Exp:* Programmer, Can Pac Rwy, 62-64, training officer, 68-71; res asst biophys, Dalhousie Univ, 67-68; col master, Confederation Col, 71-75; assoc prof computer sci, Univ Fla, 79-85; lectr math, 66-67, PROF COMPUTER SCI & DIR, JODRER SCH COMPUTER SCI, ACADIA UNIV, 85- *Concurrent Pos:* Lectr, Sir George Williams Univ, Montreal, 69-71 & Lakehead Univ, Thunder Bay, 74-75; assoc prof, Col Med, Univ Fla, 80-85; vis scientist, Santa Theresa Lab,

Int Bus Mach, 88. *Mem:* Inst Elec & Electronics Engrs Computer Soc; Asn Comput Mach; Can Info Processing Soc; Pattern Recognition Soc. *Res:* Image based systems for measurement, assessment, and analysis; general problems associated with scientific programming; university level education of computer scientists. *Mailing Add:* Jodrey Sch Computer Sci Acadia Univ Wolfville NS B0P 1X0 Can

OLIVER, MONTAGUE, genetics, physiology; deceased, see previous edition for last biography

OLIVER, MORRIS ALBERT, b Milford-Haven, Wales, Feb 12, 18; nat US; m 48; c 3. ENGINEERING STATISTICS. *Educ:* Oxford Univ, BA, 48, MA, 52. *Prof Exp:* Head sci dept, Eaglebrook Sch, Mass, 48-50; prof math, Bennington Col, 50-55; mathematician, Amp, Inc, 55-65, sr mathematician & mgr eng comput serv, 66-, mgr, Precision Artwork Lab, 73-77; RETIRED. *Concurrent Pos:* Consult, AMP, Inc, 53-55; mem, Grad Fac, Pa State Univ, 58. *Mem:* Am Math Soc; NY Acad Sci; fel Royal Statist Soc; Am Statist Asn. *Res:* Statistical engineering. *Mailing Add:* 6235 Elm Ave Harrisburg PA 17112

OLIVER, PAUL ALFRED, b Tripoli, Libya, Feb 4, 40; US citizen; m 63. COMPUTER SCIENCE, MATHEMATICS. *Educ:* Univ Md, BS, 62; Ohio State Univ, MS, 64; Univ NC, Chapel Hill, PhD(comput sci), 69. *Prof Exp:* Mathematician, Res & Tech Div, Air Force Syst Command, 65-66; dir comput ctr, Univ NC, Chapel Hill, 68-70; sr staff scientist, Sperry-UNIVAC Div, 70-73; dir, FCCTS, US Govt, 73-79; PRES, EDS WORLD CORP, 79- *Mem:* AAAS; Inst Elec & Electronics Engrs; Asn Systs Mgt; Am Mgt Asn; Asn Comput Mach. *Res:* Software engineering; systems performance and management. *Mailing Add:* 10121 Captain Hickory Pl Great Falls VA 22066

OLIVER, RICHARD CHARLES, b Minneapolis, Minn, Mar 16, 30; m 53; c 4. PERIODONTOLOGY. *Educ:* Univ Minn, BS, 52, DDS, 53; Loma Linda Univ, MS, 62. *Prof Exp:* From assoc prof to prof periodont, Sch Dent, Loma Linda Univ, 62-75; dean, Sch Dent, Univ Southern Calif, 75-77; dean, 77-86, PROF PREV DENT, SCH DENT, UNIV MINN, 86- *Concurrent Pos:* Fulbright-Hays res scholar, Denmark, 67-68. *Mem:* Am Dent Asn; Am Acad Periodont; Int Asn Dent Res; Am Asn Dent Sch. *Res:* Vascularization following periodontal surgery; epidemiology of periodontal disease and therapy. *Mailing Add:* Sch Dent Univ Minn 515 Delaware St SE Minneapolis MN 55455

OLIVER, ROBERT C(ARL), b Bisbee, Ariz, Aug 20, 25; m 53; c 3. CHEMICAL ENGINEERING. *Educ:* Univ Wash, BS, 48, MS, 49; Mass Inst Technol, ScD(chem eng), 53. *Prof Exp:* Asst, Univ Wash, 47-49; asst, Mass Inst Technol, 49-50, res assoc, 51-53; res engr, Union Oil Co, Calif, 53-55; sr res engr, Res Div, Fluor Corp, Ltd, 55-59, prin res engr, 59-60; res scientist, Aeronutronic Div, Ford Motor Co, 60-62, supvr, Thermochem, 62-63; asst dir, Sci & Technol Div, 76-78, 80-84, MEM SR TECH STAFF, INST DEFENSE ANAL, 63- *Mem:* Am Inst Aeronaut & Astronaut; Am Inst Chem Engrs; Am Geophys Union. *Res:* Missile propulsion; fuels; materials; aircraft and rocket vehicle atmospheric effects such as contrails, ozone and climate. *Mailing Add:* Inst Defense Anal 1801 N Beauregard Alexandria VA 22311

OLIVER, ROBERT MARQUAM, b Seattle, Wash, May 5, 31; m 60; c 3. OPERATIONS RESEARCH, APPLIED MATHEMATICS. *Educ:* Mass Inst Technol, BSc, 52, PhD(opers res), 57. *Prof Exp:* Sr engr, Goodyear Aircraft Corp, 53; lectr, Mass Inst Technol, 57; div dir & mem bd dir, Broadview Res Corp, 57-60; assoc prof indust eng, 61-65, chmn dept, 64-68, PROF INDUST ENG & OPERS RES, UNIV CALIF, BERKELEY, 65-, ASST TO VPRES, 68- *Concurrent Pos:* Mem, Nat Conf Solid Waste Mgt; mem comt transp systs, Nat Acad Sci. *Mem:* Opers Res Soc Am; Inst Mgt Sci; Oper Res Soc UK. *Res:* Traffic flow; transportation. *Mailing Add:* Dept Indust Eng & Oper Res Univ Calif 2120 Oxford St Berkeley CA 94720

OLIVER, RON, b Mountain Lake, Minn, Aug 11, 47. COMPUTER COMMUNICATIONS, REAL TIME SYSTEMS ENGINEERING. *Educ:* Morningside Col, BA, 70; Univ Kans, MS, 75; Colo State Univ, Ft Collins, PhD(computer sci), 88. *Prof Exp:* DIR, COMPUTER SYSTS LAB, CALIF POLYTECH STATE UNIV, 89- *Mem:* Inst Elec & Electronics Engrs; Asn Comput Mach (vpres, 90-); Philos Sci Asn. *Res:* Concurrent programming. *Mailing Add:* Dept Computer Sci Calif Polytech State Univ San Luis Obispo CA 93407

OLIVER, THOMAS ALBERT, b Winnipeg, Man, Dec 16, 24. GEOLOGY. *Educ:* Univ Man, MSc, 49; Univ Calif, Los Angeles, PhD(geol), 52. *Prof Exp:* Geologist, Man Mines Br, Dept Mines & Nat Resources, Can, 50-52 & Calif Standard Co, 52-59; from asst prof to assoc prof, 59-67, PROF GEOL, UNIV CALGARY, 67-, DEAN SCI, 78- *Mem:* Geol Soc Am; Sigma Xi. *Res:* Stratigraphy; sedimentation. *Mailing Add:* Dept Geol Univ Calgary 2500 University Dr NW Calgary AB T2N 1N4 Can

OLIVER, THOMAS K, JR, b Hobart Mills, Calif, Dec 21, 25; m 49; c 2. PEDIATRICS. *Educ:* Harvard Med Sch, MD, 49. *Prof Exp:* Instr pediat, Med Col, Cornell Univ, 53-55; from asst prof to assoc prof, Ohio State Univ, 55-63; from assoc prof to prof pediat, Univ Wash, 63-70; prof pediat & chmn dept, Sch Med, Univ Pittsburgh, 70-87; SR VPRES, AM BD PEDIAT, 87- *Concurrent Pos:* Spec fel neonatal physiol, Karolinska Inst, Sweden, 60-61; med dir, Children's Hosp Pittsburgh, 71-78; consult ed, Monographs Neonatology, 75-; co-ed, Seminars Perinatology, 75-85. *Mem:* Soc Pediat Res; Am Pediat Soc; Asn Am Med Cols; Am Acad Pediat; Am Bd Pediat; Inst Med. *Res:* Neonatal biology; pulmonary physiology in childhood; acid-base physiology. *Mailing Add:* 111 Silver Cedar Ct Chapel Hill NC 27514

OLIVER, THOMAS KILBURY, b St Louis, Mo, Sept 24, 22; m 63; c 4. INSTRUMENTATION, CONTROL SYSTEMS. *Educ:* US Mil Acad, BS, 43; Mass Inst Technol, MS, 48, DSc(instrumentation), 53. *Prof Exp:* Proj engr, Off Air Res, US Air Force, 48-51, assoc prof elec eng, Air Force Inst Technol, 53-56, US Air Force Acad, 56-60, chief guidance systs testing, Guidance & Control Div, Holloman AFB, 60-63; aerospace scientist, NAm Aviation, Inc, 63-64; prin staff engr, Honeywell, Inc, 64-67; assoc prof elec eng, 67-74, SR RES ENGR, SDAK SCH MINES & TECHNOL, 74- *Mem:* Am Inst Aeronaut & Astronaut; Am Soc Eng Educ; Inst Elec & Electronics Engrs; Sigma Xi. *Res:* Dynamics of complex feedback systems; energy from humid air. *Mailing Add:* Dept Elec Eng SDak Sch Mines & Technol Rapid City SD 57701

OLIVER, WILLIAM ALBERT, JR, b Columbus, Ohio, June 26, 26; m 48; c 2. PALEONTOLOGY. *Educ:* Univ Ill, BS, 48; Cornell Univ, MA, 50, PhD, 52. *Prof Exp:* From instr to asst prof geol, Brown Univ, 52-57; RES GEOLOGIST-PALEONTOLOGIST, US GEOL SURV, 57- *Concurrent Pos:* Trustee, Paleont Res Inst, 77-89, pres, 84-86; ed, Paleont Soc, 64-69; res prof, George Washington Univ, 69-70; mem, US Nat Comt Geol, 75-79, chmn, 78-79; mem, Int Subcomn Devonian Stratig, 73-92, chmn, 84-89; secy-gen, Int Paloeont Asn, 84-89. *Mem:* Geol Soc Am; Paleont Soc (pres, 75); Palaeont Asn; Am Geol Inst (pres, 77); Int Asn Fossil Cnidaria (pres, 83-88). *Res:* Silurian and devonian corals and stratigraphy, paleoecology, and biogeography. *Mailing Add:* US Geol Surv E305 Nat Hist Bldg Washington DC 20560

OLIVER, WILLIAM J, b Blackshear, Ga, Mar 30, 25; m 49; c 3. PEDIATRICS. *Educ:* Univ Mich, MDO, 48; Am Bd Pediat, cert, 54; Sub-Bd Pediat Nephrol, cert, 74. *Prof Exp:* From instr to assoc prof pediat, 53-65, chmn, dept pediat & commun dis, 67-79, PROF PEDIAT, SCH MED, UNIV MICH, ANN ARBOR, 65- *Concurrent Pos:* Chief pediat serv, Wayne County Gen Hosp, 58-61; dir, Pediat Lab, Med Ctr, Univ Mich, Ann Arbor, 59-67. *Honors & Awards:* Clifford G Grulee Award, Am Acad Pediat, 79. *Mem:* Am Pediat Soc; Am Soc Nephrology; Am Acad Pediat; Soc Pediat Res. *Res:* Renal physiology and pathology; fluid and electrolyte metabolism; adaptation of primitive peoples with their environment, disease patterns, mineral metabolism and growth patterns of Yanomamö Indians of South America and Pygmies of African Congo. *Mailing Add:* F 7814 Mott Children's Hosp Univ Mich Box 0244 Ann Arbor MI 48109-0244

OLIVER, WILLIAM PARKER, b Philadelphia, Pa, Dec 23, 40; m 75; c 3. HIGH ENERGY PHYSICS. *Educ:* Univ Calif, Berkeley, BS, 62, PhD(physics), 69. *Prof Exp:* Res assoc, Lawrence Berkeley Lab, Univ Calif, 69-71; res assoc, Univ Wash, 71-76; asst prof, 77-82, ASSOC PROF PHYSICS, TUFTS UNIV, 82- *Concurrent Pos:* Sr scientist, Y T Li Eng, Inc, 90. *Mem:* Am Phys Soc. *Res:* Experimental high energy physics. *Mailing Add:* Dept Physics Tufts Univ Medford MA 02155

OLIVER-GONZALES, JOSE, b Lares, PR, Aug 21, 12; US citizen; m 35; c 3. PARASITOLOGY. *Educ:* Univ PR, BA, 38; Univ Chicago, MS, 39, PhD(parasitol, bact), 41. *Prof Exp:* Instr parasitol, Sch Trop Med, Univ PR, 40; res assoc, Univ Chicago, 42; from asst prof to prof parasitol, Sch Med, Univ PR, San Juan, 43-73, head dept, 60-73; PROF PARASITOL, SAN JUAN BAUTISTA SCH MED, PUERTO RICO, 73- *Concurrent Pos:* Res assoc, Western Reserve Univ, 47-48; consult, Parasitol & Trop Med Study Sect, NIH, 53-56, mem, US-Japan Coop Med Sci Prog, 65-70, Nat Adv Allergy & Infectious Dis Coun, 66-70; consult, Vet Admin Hosp, San Juan, Presby Hosp, Auxilio Mutuo Hosp & Doctor's Hosp, San Juan, La State Univ Med Ctr & Inst Int Med, New Orleans; Guggenheim fel, 62-63. *Honors & Awards:* Ashford, Am Soc Trop Med & Hyg, 47; Purdue Frederick Prize, PR Med Asn, 57; Martinez Award, PR Comt Bilharzia Control, 59; Perez Award, Personnel Off, Govt Pr, 60. *Mem:* AAAS; fel Am Soc Trop Med & Hyg; Am Soc Parasitol; Soc Exp Biol & Med; hon mem Am Soc Microbiologists; hon mem Soc Biol & Exp Med. *Res:* Immunity to infections with helminth parasites; prevention and control of infections with Schistosoma mansoni; metabolism of Schistosoma mansoni and Trichinella spiralis; chemotherapy, immunology and immunochemistry related to parasitic infections. *Mailing Add:* Bucare 13 Santurce PR 00913

OLIVERIO, VINCENT THOMAS, b Cleveland, Ohio, Dec 7, 28; m 53; c 11. PHARMACOLOGY, ORGANIC CHEMISTRY. *Educ:* Xavier Univ, Cincinnati, Ohio, BS, 51, MS, 53; Univ Fla, Gainesville, PhD(oncol), 55. *Prof Exp:* USPHS fel cancer res, Univ Fla, 54-55; proj assoc oncol, McArdle Mem Lab, 55-59; sr investr lab chem, pharmacol, Nat Cancer Inst, NIH, 59-67, head biochem pharmacol sect, 67-73, chief lab chem pharmacol, 69-73, assoc dir exp ther, 73-77, assoc dir, Develop Therapeut Prog, Dir Cancer Treatment, 76-83, ASSOC DIR, DIV EXTRAMURAL ACTIV, NAT CANCER INST, NIH, 83- *Concurrent Pos:* Mem med technologist, Bd US Civil Serv, 63-64; mem study group III, Bur Drugs, Food & Drug Admin, 74; cancer res emphasis grant, Div Cancer Treat, Nat Cancer Inst, 75-76. *Honors & Awards:* Superior Service Honor Award, NIH, Nat Cancer Inst, 74. *Mem:* Am Asn Cancer Res. *Res:* Physiological disposition of antitumor agents and other drugs in animals and man; influence of combination drug therapy in man and animals on protein binding, renal clearance, metabolism, and therapeutic activity of individual agents. *Mailing Add:* NIH Bldg 31 Rm 10A05 Bethesda MD 20892

OLIVERO, JOHN JOSEPH, JR, b Yonkers, NY, Jan 18, 41; m 61; c 4. AERONOMY, METEOROLOGY. *Educ:* Fla State Univ, BS, 62; Col William & Mary, MS, 66; Univ Mich, Ann Arbor, PhD(aeronomy), 70. *Prof Exp:* Aerospace technologist appl res & develop, Langley Res Ctr, NASA, Va, 62-70; res assoc fel, Physics & Astron, Univ Fla, 70-72; from asst prof to assoc prof, 72-84, PROF METEOROL, PA STATE UNIV, 85- & IONOSPHERE RES LAB/COMMUN & SPACE SCI LAB, 72- *Concurrent Pos:* Vis res assoc, Lab Atmospheric & Space Physics, Univ Colo, 83-85. *Mem:* Am Meteorol Soc; Sigma Xi; Am Geophys Union. *Res:* Minor constituent photochemistry and transport; microwave radiometry; middle atmospheric composition, structure, energetics, circulation and thermal processes; atmospheric measurements; aerosol physics; climatic change, environmental impact. *Mailing Add:* 509 Walker Pa State Univ University Park PA 16802

OLIVER-PADILLA, FERNANDO LUIS, dairy science, animal breeding, for more information see previous edition

OLIVETO, EUGENE PAUL, b New York, NY, Mar 15, 24; m 48; c 2. ORGANIC CHEMISTRY. *Educ:* City Col New York, BS, 43; Purdue Univ, PhD(org chem), 48. *Prof Exp:* Res chemist, Schenley Res Inst, 43-44; asst chem, Purdue Univ, 44-48; res chemist, Schering Corp, 48-50, group leader, 50-58, mgr, Natural Prod Res Dept, 58-64; sr group chief, 64, sect chief, 64-66, dir animal health & fine chem res, 66-68, dir fine chem res, Hoffmann-La Roche, Inc, 68-85; tech dir, SST Corp, 85-90; CONSULT, 85- *Mem:* AAAS; fel Am Inst Chemists; Am Chem Soc; fel NY Acad Sci; Royal Soc Chem; Sigma Xi. *Res:* Nitroparaffins; synthesis, photolysis and pyrolysis of organic nitrites; organic medicinals; partial and total synthesis of steroidal sex and adrenal hormones; analogs; food additives; vitamins. *Mailing Add:* 284 Forest Ave Glen Ridge NJ 07028

OLIVIER, ANDRE, b Hull, Que, Oct 31, 38; c 3. NEUROSURGERY, NEUROANATOMY. *Educ:* Univ Montreal, BA, 59, MD, 64; Univ Laval, PhD(neuroanat), 70; McGill Univ, DNS, 70. *Prof Exp:* From asst prof to assoc prof, 71-84, PROF NEUROSURG, MCGILL UNIV, 84-, NEUROSURGEON, MONTREAL NEUROL INST, 71- *Concurrent Pos:* Lectr neuroanat, Univ Laval, 65-67; examr neurosurg, Royal Col Physician & Surgeons, 76-85, Prof Corp Physicians & Surgeons Can, 79-85; chmn bd examr, Prof Corp Physicians & Surgeons Can, 81-85; mem, Coun Physicians & Dentists. *Mem:* Am Soc Stereotactic & Functional Neurosurg (vpres, 84-87, pres, 87-). *Res:* Neurosurgical treatment of epilepsy; stereotactic technology; radiosurgery of brain lesion with photon beams; cerebral cholinergic mechanisms; anatomy of thalamus. *Mailing Add:* Dept Neurol & Neurosurg McGill Inst Montreal Neurol Inst 3801 Univ St Montreal PQ H3A 2B4 Can

OLIVIER, KENNETH LEO, b Los Angeles, Calif, May 19, 32; m 54; c 6. ORGANIC CHEMISTRY. *Educ:* Loyola Univ, Calif, BS, 54; Univ Calif, Los Angeles, PhD(chem), 58. *Prof Exp:* Res chemist, Plastics Dept, E I du Pont de Nemours & Co, 57-60; res chemist, 60-65, sr res chemist, 65-69, SUPVR INDUST CHEM, UNION OIL CO CALIF, 69- *Mem:* Am Chem Soc. *Res:* Petrochemicals; emulsion polymerization; hot melt adhesives. *Mailing Add:* 1151 Monte Vista D Riverside CA 92507

OLIVO, RICHARD FRANCIS, b Brooklyn, NY, Sept 26, 42; m 71. NEUROBIOLOGY & BEHAVIOR. *Educ:* Columbia Univ, AB, 63; Harvard Univ, AM, 65, PhD(biol), 69. *Prof Exp:* Tutor biol, Harvard Univ, 66-68; vis asst prof, State Univ NY Stony Brook, 70-71; asst prof, Williams Col, 71-73; from asst prof to assoc prof, 73-85, PROF BIOL, SMITH COL, 85- *Concurrent Pos:* Consult, Harper & Row, Inc, 69-71. *Mem:* AAAS; Soc Neurosci. *Res:* Neurobiology: control of eye movements in crustaceans; invertebrate nervous systems; laboratory microcomputing for data acquisition and imaging. *Mailing Add:* Dept Biol Sci Smith Col Northampton MA 01063

OLKEN, MELVIN I, b New York, NY, May 5, 35. POWER GENERATION. *Educ:* City Col, BS, 56. *Prof Exp:* Mgr electronics, Am Electronic Tower, 56-80; consult, Gibbs & Hill, 80-84; DIR ENG SOC SERVS, INST ELEC & ELECTRONICS ENGRS, 84- *Mem:* Fel Inst Elec & Electronics Engrs. *Mailing Add:* Inst Elec & Electronics Engr 445 Hoes Lane PO Box 1331 Piscataway NJ 08855

OLKIN, INGRAM, b Waterbury, Conn, July 23, 24; m 45; c 3. MATHEMATICAL STATISTICS, MULTIVARIATE ANALYSIS. *Educ:* City Col New York, BS, 47; Columbia Univ, MA, 49; Univ NC, PhD(math statist), 51. *Prof Exp:* Asst prof statist, Mich State Univ, 51-55; vis assoc prof, Univ Chicago, 55-56; assoc prof, Mich State Univ, 56-60; prof, Univ Minn, 60-61; chmn dept statist, 74-77, PROF STATIST & EDUC, STANFORD UNIV, 61- *Concurrent Pos:* Assoc ed, J Am Statist Asn, 60-70, J Educ Statist, 77- & J Psychomet, 80-; overseas fel, Churchill Col, Cambridge Univ, 67-68; fel psychomet, Educ Testing Serv, 71-72; ed, Ann Math Statist, 71-74; vis prof, Univ Brit Col, 77; Fulbright fel, Univ Copenhagen, 79; guest prof, Swiss Fed Inst Technol, 81; Lady Davis fel, Hebrew Univ, 84; fel, Ctr for Educ Statist, 88. *Mem:* Fel Inst Math Statist (pres, 83-84); fel Am Statist Asn; Am Math Soc; fel Int Statist Inst; Psychomet Soc; fel Royal Statist Soc. *Res:* Multivariate analysis; mathematical models in the behavioral sciences; biometrics-biostatistics. *Mailing Add:* Dept Statist Stanford Univ Stanford CA 94305

OLLA, BORI LIBORIO, b Jersey City, NJ, Jan 22, 37; m 58; c 2. ANIMAL BEHAVIOR, FISH BIOLOGY. *Educ:* Fairleigh Dickinson Univ, BS, 59; Univ Hawaii, MS, 62. *Prof Exp:* Res asst shark behav, Hawaii Marine Lab, Honolulu, 61-62; instr biol, Chaminade Col Hawaii, 62; asst zool, Univ Md, 62-63; asst fish neurol, NIH, 63; asst neurol, Col Med, Seton Hall Univ, 63; asst prof marine sci, C W Post Col, Long Island Univ, 75-79, chief behav invest, Sandy Hook Lab, 63-82; PROG LEADER, FISHERIES BEHAV ECOL, NAT MARINE FISHERIES SERV/ORE STATE UNIV, 85- *Concurrent Pos:* Vis lectr, Boston Univ-Marine Biol Lab, Woods Hole, 76-78; vis adj prof marine biol, Cook Col, Rutgers Univ, 77-; adj mem, Col Oceanog, Ore State Univ, 78-82 & Zool Dept, Rutgers Univ, 79-82; prof, Col Oceanog, Ore State Univ, 82- *Honors & Awards:* Bronze Medal, US Dept Com, 75. *Res:* Field and laboratory studies on marine fishes and invertebrates; behavioral ecology in relation to fishery biology and social behavior, including schooling, territoriality, aggression, reproduction; biorhythms, feeding, home ranges; chemosensory responses; rule of physical and biological factors on the behavior and distribution of marine fish larvae and juveniles. *Mailing Add:* Coop Inst Marine Resource Studies NOAA/NMFS-Ore State Univ Hatfield Marine Sci Ctr Newport OR 97365

OLLEMAN, ROGER D(EAN), b Cornelia, Ga, Nov 25, 23; m 47; c 3. METALLURGY. *Educ:* Univ Wash, BS, 48; Carnegie Inst Technol, MS, 50; Univ Pittsburgh, PhD, 55. *Prof Exp:* Group leader, Westinghouse Res Lab, 50-55; asst br head, Dept Metall Res, Kaiser Aluminum & Chem Corp, 55-59; from assoc prof to prof mech eng, Ore State Univ, 59-76, head dept metall eng, 69-72; PRES, ACCIDENT & FAILURE INVESTS, INC, 74- *Concurrent Pos:* Consult, Kaiser Aluminum & Chem Corp, 59-60, Albany Metall Res Ctr, US Bur Mines, 62-72 & Lawrence Radiation Lab, 65-66; courtesy prof, Ore State Univ, 76- *Honors & Awards:* Templin Award, Am Soc Testing & Mat, 53. *Mem:* Am Soc Metals; Am Soc Mech Engrs; Am Inst Mining, Metall & Petrol Engrs; Am Soc Eng Educ. *Res:* Physical and mechanical metallurgy; mechanisms of flow and fracture; crystallography and phase transformations; cryogenic properties. *Mailing Add:* Accident & Failure Invests Inc 2107 NW Fillmore Ave Corvallis OR 97330

OLLERENSHAW, NEIL CAMPBELL, b Matlock, Eng, Sept 12, 33; Can citizen; m 60; c 3. GEOLOGY. *Educ:* Univ Wales, BSc, 57; Univ Toronto, MA, 59, PhD(geol), 63. *Prof Exp:* res scientist, 62-82, SR MGR & SCI ED, GEOL SURV CAN, 82 - *Mem:* Can Soc Petrol Geologists; Soc Econ Paleont & Mineral; Asn Earth Sci Ed (dir, 87-89). *Res:* Structural geology; stratigraphy and facies-tectonics relationships of the Rocky Mountain Foothills and Eastern Cordillera. *Mailing Add:* Geol Surv of Can 3303 33rd St NW Calgary AB T2L 2A7 Can

OLLERHEAD, ROBIN WEMP, b Simcoe, Ont, Mar 12, 37; m 59; c 3. ION-SOLID INTERACTIONS. *Educ:* Univ Western Ont, BSc, 59; Yale Univ, MS, 60, PhD(physics), 64. *Prof Exp:* Asst res officer nuclear physics, Atomic Energy Can Ltd, 64-66, assoc res officer, 66-68; assoc prof, 68-71, PROF PHYSICS, UNIV GUELPH, 71-, CHMN DEPT, 82- *Concurrent Pos:* NATO sci fel & Rutherford Mem fel, Oxford Univ, 63-64; vis res assoc, Calif Inst Technol, 74-75. *Mem:* Am Phys Soc; Can Asn Physicists. *Res:* Ion-solid interactions; materials analysis using ion beams and nuclear physics techniques; inner-shell atomic physics. *Mailing Add:* Dept of Physics Univ of Guelph Guelph ON N1G 2W1 Can

OLLERICH, DWAYNE A, b Sioux Falls, SDak, June 30, 34. ANATOMY, NEUROANATOMY. *Educ:* Augustana Col, SDak, BA, 60; Univ NDak, MS, 62, PhD(anat), 64. *Prof Exp:* Asst prof neuroanat & histol, Univ Alta, 65-66; from asst prof to assoc prof, 66-77, chmn dept anat, 72-79, PROF NEUROANAT, SCH MED, UNIV NORTH DAK, 77-, ASSOC DEAN ACAD AFFAIRS, 79- *Concurrent Pos:* Fel anat, Univ Alta, 64-65, Med Res Coun Can fel, 65-66. *Mem:* Am Asn Anat; Am Soc Cell Biol. *Res:* Electron microscopy; cytology; drug toxicity. *Mailing Add:* Dept Anat Univ NDak Sch Med 501 Columbia Rd N Grand Forks ND 58201

OLLIS, DAVID F(REDERICK), b San Francisco, Calif, Sept 28, 41; m 64; c 2. CHEMICAL ENGINEERING. *Educ:* Calif Inst Technol, BS, 63; Northwestern Univ, MS, 64; Stanford Univ, PhD(chem eng), 69. *Prof Exp:* Res engr, Montebello Res Lab, Texaco Inc, Calif, 64-65; from asst prof to prof chem eng, Princeton Univ, 69-80; prof chem eng, Univ Calif, Davis, 80-84; DISTINGUISHED PROF CHEM ENGR, NC STATE UNIV, 84- *Concurrent Pos:* Vis prof, Univ Libre Bruxelles, 82, 86, Univ Technol, Helsinki, 86, Ecole Polytechnique Fed Lousanne, 88. *Mem:* Am Inst Chem Engrs; Am Chem Soc. *Res:* Biochemical engineering & photocatalysis. *Mailing Add:* Dept Chem Eng Box 7905 NC State Univ Raleigh NC 27695-7905

OLLOM, JOHN FREDERICK, b Ward, WVa, Dec 28, 22; m 54; c 3. PHYSICS. *Educ:* Harvard Univ, PhD, 52. *Prof Exp:* Asst prof physics, Univ WVa, 51-56; PROF PHYSICS, DREW UNIV, 56- *Concurrent Pos:* Mem tech staff, Bell Labs, Inc. *Mem:* Am Phys Soc; Am Asn Physics Teachers. *Res:* Magnetism; microwave spectroscopy. *Mailing Add:* 17 Albright Circle Madison NJ 07940

OLMER, JANE CHASNOFF, b St Louis, Mo; wid. MATHEMATICS, COMPUTER SCIENCE. *Educ:* Wellesley Col, BA, 34; Univ Paris, cert, 36; Washington Univ, MS, 37; Univ Chicago, cert, 61; Inst Comput, cert, 65, Systs Prof, 85. *Prof Exp:* Tech translr, Antoine St Exupéry, 37-38; teacher, Am Sch Paris, 37-39; ed & broadcaster, Paris Letter, 39-41; statistician, Drop Forging Asn, 42-43 & Fed Pub Housing Authority, 43-44; consult, May Co, 45; anal statistician, US Navy Electronics Supply Off, 54-55, supvr anal statistician, 55-56, digital comput syst specialist trainee, 56, mathematician, 56-59, tech head adv logistics res & develop, 59-61; sr mathematician, Appl Physics Lab, Johns Hopkins Univ, 61-77; group mgr, Price, Williams & Assocs Inc, 78-82; PRES, JANE OLMER & ASSOCS, 83- *Concurrent Pos:* Permanent deleg, Univac Users Asn, 57-61; mgr, Text & Info Processing & Retrieval Proj, SHARE; Durant scholar, Wellesley Col; pub rel chair, Asn Comput Mach, DC Chapter. *Honors & Awards:* US Navy Superior Accomplishment Award, 61. *Mem:* Asn Women Comput; Sigma Xi; Data Processing Mgt Asn; Asn Comput Mach. *Res:* Operations research in inventory control; mathematical programming on business computers and text processing intelligent systems; information handling and retrieval and high level computer languages in research and development environment using business and scientific computers; design and implementation of integrated data processing and word processing systems; data base management systems. *Mailing Add:* 2510 Virginia Ave NW Washington DC 20037

OLMEZ, ILHAN, b Kastamonu, Turkey, Aug 11, 42; m 70; c 2. NUCLEAR CHEMISTRY, ANALYTICAL CHEMISTRY. *Educ:* Mid E Tech Univ, BSc, 69, MSc, 71, PhD(chem), 76. *Prof Exp:* Asst anal chem, Mid E Tech Univ, 69-71; res sci, Ankara Nuclear Res Ctr, 71-72; res fel, Int Atomic Energy Agency, 72-73; div head, Ankara Nuclear Res & Training Ctr, 75-77, dir, 79-83; res assoc, Univ Md, 77-79 & 83-85; SR SCIENTIST, NUCLEAR REACTOR LAB, MASS INST TECHNOL, 85- *Mem:* Air Pollution Control Asn; AAAS; Am Chem Soc. *Res:* Development and application of nuclear activation analysis techniques; trace element studies in environmental, biochemical and materials sciences. *Mailing Add:* Mass Inst Technol Nuclear Reactor Lab 138 Albany St Cambridge MA 02139

OLMSTEAD, JOHN AARON, b Buffalo, NY, Feb 21, 30; m 60. ELECTRICAL ENGINEERING. *Educ:* Univ Buffalo, BS, 52; Newark Col Eng, MS, 57. *Prof Exp:* Engr electron tube div, Radio Corp Am, 52-55, group leader, 55-57; asst prof elec eng, Univ Buffalo, 57-60; engr, Electron Components & Devices Div, Radio Corp Am, 59-61, LEADER TECH STAFF, ELECTRON COMPONENTS & DEVICES DIV, RCA CORP, 61-

Mem: Sr mem Inst Elec & Electronics Engrs. *Res:* Gaseous electronic conduction phenomena; solid state physics; device characterization and associated physical phenomena; device fabrication techniques. *Mailing Add:* RCA Corp Hwy 202 Southbound Somerville NJ 08876

OLMSTEAD, MARJORIE A, b Glen Ridge, NJ, Aug 18, 58. SURFACE & INTERFACE PHYSICS. *Educ:* Swarthmore Col, BA, 79; Univ Calif, Berkeley, MA, 82, PhD(physics), 85. *Prof Exp:* Mem res staff, Xerox Palo Alto Res Ctr, 85-86; asst prof, physics, univ calif, berkeley, 86-90; ASST PROF PHYSICS, UNIV WASH, SEATTLE, 91- *Concurrent Pos:* Prin Investr, Lawrence Berkeley Lab, 88-; consult, Xerox Corp, 86- *Honors & Awards:* Presidential Young Investr Award, Nat Sci Found, 87- *Mem:* Am Phys Soc; Mat Res Soc; Am Vacuum Soc; Asn Women Sci. *Res:* Initial stages of interface formation between dissimilar materials, including the role of the substrate surface composition, the nature of interfacial bonding and the development of three-dimensional properties during film growth. *Mailing Add:* Dept Physics FM-15 Univ Wash Seattle WA 98195

OLMSTEAD, PAUL SMITH, b Wilton, Conn, Aug 10, 97; wid; c 2. DETECTION OF NONRANDOMNESS. *Educ:* Princeton Univ, BSc, 19, PhD(physics), 23. *Prof Exp:* Eng statistician, Bell Tel Labs, 25, head, statist studies dept, 61-62; sr consult engr, Lockheed Electronics Co, 62-63; CONSULT STATIST, 63- *Honors & Awards:* Shewhart Medal, Am Soc Qual Control, 59. *Mem:* Fel Am Phys Soc; fel Am Soc Qual Control; fel Am Statist Asn; Am Soc Testing & Mat; Opers Res Soc; fel AAAS; Am Math Soc. *Res:* Ionization and radiation potentials; mathematical theory of runs and run related phenomena. *Mailing Add:* 1111 S Lakemont Ave Apt 621 Winter Park FL 32792

OLMSTEAD, WILLIAM EDWARD, b San Antonio, Tex, June 2, 36; m 57; c 2. APPLIED MATHEMATICS, REACTION-DIFFUSION PHENOMENA. *Educ:* Rice Univ, BS, 59; Northwestern Univ, MS, 62, PhD(appl math), 63. *Prof Exp:* Res engr, Southwest Res Inst, 59-60; fel, Johns Hopkins Univ, 63-64; from asst prof to assoc prof, 64-71, dir appl math, 78-80, PROF APPL MATH, NORTHWESTERN UNIV, 71- *Concurrent Pos:* Vis mem, Courant Inst Math Sci, NY Univ, 67-68; chmn, Comt Appl Math, Northwestern Univ, 72-76, mem, Exec Comt, Coun Theoret & Appl Mech, 74-82; regional lectr, Soc Indust & Appl Math, 72-73 & 78-79; vis prof, Univ Col, London, 73, Calif Inst Technol, 87 & 90. *Mem:* Soc Indust & Appl Math; Am Math Soc; Am Phys Soc; Am Acad Mech. *Res:* combustion; visco-elastic and micropolar fluids; bifurcation theory; integral equations; singular perturbations. *Mailing Add:* Technol Inst Northwestern Univ Evanston IL 60201

OLMSTEAD, WILLIAM N(EERGAARD), b Bryn Mawr, Pa, Mar 4, 50; m 81; c 2. PHYSICAL ORGANIC CHEMISTRY. *Educ:* Yale Univ, BS, 72; Stanford Univ, PhD(chem), 77. *Prof Exp:* Fel org chem, Northwestern Univ, 76-78; SENIOR STAFF CHEMIST ORG CHEM, EXXON RES & ENG CO, 78- *Mem:* Am Chem Soc. *Res:* Gas phase and solution chemistry of ions; solvent effects on organic reactions; petroleum, oil, shale and coal chemistry; pyrolysis of organic compounds. *Mailing Add:* Exxon Res & Eng Co Rte 22 E Clinton Township Annandale NJ 08801

OLMSTED, CLINTON ALBERT, b Chicago, Ill, Oct 27, 25; m 52; c 4. PHYSIOLOGY. *Educ:* Univ Calif, AB, 48, MA, 54, PhD(comp physiol), 56. *Prof Exp:* Asst, Univ Calif, 49-51 & 54-56; radiobiologist, US Naval Radiol Defense Lab, 51-54; res neurologist, Med Ctr, Univ Calif, Los Angeles, 57-59; asst prof zool, Univ Wis, 59-63; sr res physiologist, Battelle Mem Inst, 63-64; head, Cell Biol Div, Inst Lipid Res, Berkeley, Calif, 65; ASSOC PROF BIOL SCI, UNIV NEW ORLEANS, 65- *Concurrent Pos:* Mem & head, Lipid Biochem Sect, Div Nutrit Biochem, Commonwealth Sci & Indust Res Orgn, Adelaide, SAustralia, 70-72; sr lectr physiol, Godfrey Huggins Sch Med, Univ Zimbabwe, 77- *Mem:* AAAS; Am Soc Cell Biol; Am Physiol Soc; NY Acad Sci; Australian Biochem Soc. *Res:* Comparative physiology; cellular physiology with emphasis on general mechanisms in neurophysiology; lipid metabolism, lipid transport, and the role of phospholipids in drug transport and sodium transport affected by lipid soluble drugs. *Mailing Add:* 18 W Park Pl New Orleans LA 70124

OLMSTED, FRANKLIN HOWARD, b Los Angeles, Calif, Nov 23, 21; m 55; c 2. GEOLOGY. *Educ:* Pomona Col, BA, 42; Claremont Cols, MA, 48; Bryn Mawr Col, PhD, 61. *Prof Exp:* Lectr & instr geol, Claremont Cols, 48-49; geologist, US Geol Surv, 49-69, hydrologist, Water Resources Div, 69-88; RETIRED. *Mem:* AAAS; Geol Soc Am; Am Geophys Union. *Res:* Ground-water geology; petrology; development of methods of hydrogeologic exploration of geothermal areas. *Mailing Add:* Water Resources Div US Geol Surv Menlo Park CA 94025

OLMSTED, JOANNA BELLE, b Chicago, Ill, Mar 8, 47; m 77; c 1. CELL BIOLOGY. *Educ:* Earlham Col, BA, 67; Yale Univ, PhD(biol), 71. *Prof Exp:* Fel biochem, Lab Molecular Biol, Univ Wis-Madison, 71-74; from asst prof to assoc prof, 75-87, PROF BIOL, UNIV ROCHESTER, 87- *Concurrent Pos:* NIH Molecular Cytol Study Sect, 87-90. *Mem:* Am Soc Cell Biol. *Res:* Control of cell division and differentiation; regulation of synthesis, assembly and organization of cytoskeletal proteins, particularly microtubule-associated proteins. *Mailing Add:* Dept Biol Univ Rochester Rochester NY 14627

OLMSTED, RICHARD DALE, b Bismarck, NDak, Nov 1, 47; m 69; c 1. THEORETICAL CHEMISTRY. *Educ:* Augsburg Col, BA, 69; Univ Wis-Madison, PhD(chem), 74. *Prof Exp:* Fel chem, Univ BC, 74-76; instr, 76-80, ASST PROF CHEM, AUGSBURG COL, 80- *Concurrent Pos:* Vis asst prof, Univ Wis, 77 & 78 & Univ BC, 81 & 82; consult polymer physics, 3M Co, St Paul, 81- *Mem:* Am Chem Soc; Sigma Xi. *Res:* Nonequilibrium statistical mechanics; kinetic theory and transport properties of gaseous systems. *Mailing Add:* Augsburg Col 707 21st Ave S Minneapolis MN 55454

OLNESS, ALAN, b Kenyon, Minn, Sept 22, 41; m 63; c 4. SOIL BIOCHEM, SOIL SCIENCE. *Educ:* Univ Minn, BS, 63, MS, 67, PhD(soil sci), 73. *Prof Exp:* Res asst soil sci, Univ Minn, 63-67; soil scientist, Soil Struct Group, 67-70, Nat Agr Water Qual Mgt Lab, 70-77 & Food Crops Res Team, Regional Develop Off EAfrica, 77-78, SOIL SCIENTIST, NCENT SOIL CONSERV RES CTR, AGR RES SERV, USDA, 78- *Mem:* Am Soc Agron; Soil Sci Soc Am; Int Soc Soil Sci; Am Chem Soc; Soil Water Conserv Soc. *Res:* Influence of agricultural management practices on crop residue nutrient use efficiency with particular emphasis on nitrogen fertilizer applications, transport, and transformations within the agricultural environment. *Mailing Add:* NCent Soil Conserv Res Ctr N Iowa Ave Morris MN 56267

OLNESS, DOLORES URQUIZA, b Kingsport, Tenn, Mar 20, 35; m 57; c 3. ATOMIC & MOLECULAR PHYSICS, SOLID STATE PHYSICS. *Educ:* Duke Univ, AB, 57, PhD(physics), 61. *Prof Exp:* Res assoc molecular physics, Duke Univ, 61 & Univ NC, 61-63; PHYSICIST, LAWRENCE LIVERMORE LAB, UNIV CALIF, 63- *Mem:* Am Phys Soc. *Res:* Organic low temperature molecular spectroscopy; organic photoconductors; lasers, damage to transparent solids; microstructures design, research and development; thermodynamic data base; systems analysis. *Mailing Add:* 4345 Guilford Ave Livermore CA 94550

OLNESS, FREDRICK IVER, b Dayton, Ohio, April 19, 59; m 88. QUANTUM CHROMODYNAMICS, ELECTROWEAK THEORIES. *Educ:* Duke Univ, BS, 80; Univ Wisc-Madison, MS, 82, PhD(physics), 85. *Prof Exp:* Post doctorate physics, Ill Inst Technol, 85-88; POST DOCTORATE PHYSICS, INST THEORETICAL SCI, UNIV ORE, 88- *Mem:* Sigma Xi; Am Phys Soc; Am Inst Physics. *Res:* The QCD-improved parton model, heavy particle production, small-x physics, and mini-jets; electroweak phenomena, factorization of helicity amplitudes; angular correlations; extended electroweak models. *Mailing Add:* Inst Theo Science Univ Ore Eugene OR 97403

OLNESS, JOHN WILLIAM, b Broderick, Sask, Sept 4, 29; US citizen; m 58; c 5. NUCLEAR PHYSICS. *Educ:* St Olaf Col, BA, 51; Duke Univ, PhD(physics), 57. *Prof Exp:* Res assoc nuclear physics, Duke Univ, 57-58; res physicist, Aeronaut Res Labs, Wright-Patterson AFB, Ohio, 58-63; assoc physicist, 63-68, PHYSICIST, BROOKHAVEN NAT LAB, 68- *Res:* Nuclear spectroscopy; relativistic heavy-ion physics. *Mailing Add:* Physics Dept Bldg 901-A Brookhaven Nat Lab Upton NY 11973

OLNESS, ROBERT JAMES, b Milaca, Minn, Jan 22, 33; m 57; c 3. STATISTICAL PHYSICS, THERMODYNAMICS. *Educ:* Mass Inst Technol, BS, 56; Duke Univ, PhD(nuclear physics), 62. *Prof Exp:* Res assoc physics, Univ NC, 61-63; sr physicist, Lawrence Radiation Lab, Univ Calif, 63-68; assoc prof physics, Northern Mich Univ, 68-69; SR PHYSICIST, LAWRENCE LIVERMORE LAB, UNIV CALIF, 69- *Mem:* Am Phys Soc. *Res:* Thermodynamics and statistical mechanics; atomic and molecular physics; plasma physics. *Mailing Add:* 4345 Guilford Ave Livermore CA 94550

OLNEY, CHARLES EDWARD, b Assam, India, Nov 7, 24; US citizen; m 45; c 4. AGRICULTURAL CHEMISTRY. *Educ:* Tufts Col, BS, 45; Univ RI, MS, 53; Univ Conn, PhD(biochem), 67. *Prof Exp:* Teacher, R W Traip Acad, 46-48; agr chemist, Agr Exp Sta, 48-70, PROF FOOD & RESOURCE CHEM, UNIV RI, 70- *Mem:* Sigma Xi. *Res:* Pesticide residues; lipids. *Mailing Add:* Dept Food Sci & Res Chem Univ RI Kingston RI 02881

OLNEY, JOHN WILLIAM, b Marathon, Iowa, Oct 23, 31; m 57; c 3. PSYCHIATRY, NEUROPATHOLOGY. *Educ:* Univ Iowa, BA, 57, MD, 63; Am Bd Psychiat & Neurol, dipl, 70. *Prof Exp:* Intern, Kaiser Permanente Found, San Francisco, Calif, 63-64; resident, 64-68, from instr to assoc prof psychiat, 68-77, PROF PSYCHIAT & NEUROPATH, SCH MED, WASH UNIV, 77- *Concurrent Pos:* NIMH biol sci trainee, Wash Univ, 66-68 & career invest award, 68-; asst psychiatrist, Barnes Hosp, 68-; consult psychiatrist, Malcolm Bliss Ment Health Ctr, 68- *Mem:* Psychiat Res Soc; Am Psychiat Asn; Am Asn Neuropath; Soc Neurosci; Asn Res Nervous & Ment Dis. *Res:* Role of excitatory neurotoxins in disorders of the nervous system. *Mailing Add:* Dept Psychiat Sch Med Wash Univ St Louis MO 63110

OLNEY, ROSS DAVID, b Biloxi, Miss, Aug 8, 51; m 77; c 2. AUTOMOTIVE TECHNOLOGY, SOFTWARE DEVELOPMENT. *Educ:* Calif State Univ, Northridge, BS, 77, MS, 84. *Prof Exp:* PHYSICIST, RES LABS, HUGHES AIRCRAFT CO, 73- *Concurrent Pos:* Physics teacher, Calif State Univ, Northridge, 81. *Mem:* Soc Automotive Engrs. *Res:* Developing microcircuit lithography tools; software development projects in carbon and fortran; ion thruster deep space engines; automotive sensor technology projects. *Mailing Add:* Hughes Res Labs 3011 Malibu Canyon Rd Malibu CA 90265

O'LOANE, JAMES KENNETH, b Walla Walla, Wash, Dec 12, 13; m 43; c 4. CHEMICAL PHYSICS. *Educ:* St Benedict's Col, BSc, 35; Univ Wash, MSc, 44; Harvard Univ, MA, 47, PhD(chem physics), 50. *Prof Exp:* Instr, St Martin's Acad & Jr Col, 37-38; jr chemist, Shell Develop Co, 43-45; from asst prof to assoc prof chem, Univ NH, 48-54; sr anal chemist, Indust Lab, 54-58, proj physicist, Apparatus & Optical Div, Lincoln Plant, 58-59, sr res chemist, Res Labs, 59-62, res assoc, Eastman Kodak Co Res Labs, 63-84; RETIRED. *Mem:* Am Chem Soc; Optical Soc Am; Am Phys Soc. *Res:* Polymer physical chemistry; fourier transform spectroscopy of polymer films; polymer thermodynamics; optical activity. *Mailing Add:* 390 Alameda San Anselmo CA 94960-1231

OLOFSON, ROY ARNE, b Chicago, Ill, Feb 26, 36. ORGANIC CHEMISTRY. *Educ:* Univ Chicago, BS & MS, 57; Harvard Univ, PhD(org chem), 61. *Prof Exp:* Instr org chem, Harvard Univ, 61-64, asst prof, 64-65; assoc prof, 65-71, PROF ORG CHEM, PA STATE UNIV, 71- *Concurrent Pos:* Consult, FMC Corp, 62-73, Armour Pharmaceut Co, 65-70 & McNeil Labs, 74-88; mem, Adv Bd, J Org Chem, 73-77. *Mem:* Am Chem Soc; Royal Soc Chem; Sigma Xi. *Res:* Synthetic organic, heterocyclic and peptide chemistry; reaction mechanisms. *Mailing Add:* Dept Chem Pa State Univ University Park PA 16802

O'LOONEY, PATRICIA ANNE, b Bridgeport, Conn, Dec 2, 54. LIPOPROTEIN METABOLISM, DIABETES. *Educ:* Regis Col, Weston, Mass, BA, 76; George Washington Univ, Washington DC, MS, 78, PhD(biochem), 82. *Prof Exp:* Res asst, George Washington Univ, Washington, DC, 76-82, teaching asst biochem, 78-81, res assoc, 82-84, res scientist, 84-85, sr res scientist, 85-86, asst prof biochem & med, George Washington Med Ctr, 86-88; asst dir res & med progs, 88-90, DIR RES & MED PROGS, NAT MULTIPLE SCLEROSIS SOC, NY, 90- *Concurrent Pos:* Prin investr, New Investr award, NIH, 85-88. *Mem:* Am Soc Biol Chemists; NY Acad Sci; Am Heart Asn; Asn Women Sci; AAAS; Sigma Xi. *Res:* Defective clearance of lipoproteins in experimental diabetes; abnormalities associated with the composition (both lipid and protein analysis) of the lipoprotein particle itself and the activity of the clearing-factor enzyme, lipoprotein lipase. *Mailing Add:* 25 Cartright St Apt 1G Bridgeport CT 06604

O'LOUGHLIN, WALTER K(LEIN), chemical engineering; deceased, see previous edition for last biography

OLSEN, ARTHUR MARTIN, b Chicago, Ill, Aug 27, 09; m 36; c 4. INTERNAL MEDICINE. *Educ:* Dartmouth Col, AB, 30; Rush Med Col, MD, 35; Univ Minn, MS, 38. *Prof Exp:* First asst med, Mayo Grad Sch Med, Univ Minn, 38-39, from asst prof to prof med, 40-76; DIR INT ACTIV, INT ACAD CHEST PHYSICIANS & SURGEONS, AM COL CHEST PHYSICIANS, 76- *Concurrent Pos:* Consult, Mayo Clin, 38-, head sect med, 49-68, chmn div thoracic dis, 68-71; trustee, Mayo Found, 61-68; mem nat heart & lung adv coun, NIH, 70-71. *Mem:* Am Soc Gastrointestinal Endoscopy (secy-treas, Am Gastroscopic Soc, 58-61, pres, 62-63); Am Col Chest Physicians (pres, 70); Am Broncho-Esophagol Asn (pres, 69-70); Am Thoracic Soc; Am Asn Thoracic Surg. *Res:* Esophageal motility; broncho-esophagology. *Mailing Add:* 211 NW Second St Apt 20022 200 First St NW Rochester MN 55901

OLSEN, CARL JOHN, b Oakland, Calif, May 18, 28; m 64; c 2. ORGANIC CHEMISTRY. *Educ:* Univ San Francisco, BS, 50, MS, 52; Univ Southern Calif, PhD(chem), 62. *Prof Exp:* Asst chem, Walter Reed Army Med Ctr, 53-54; instr, Los Angeles Valley Col, 60-61; from asst prof to assoc prof, 61-71, PROF CHEM, CALIF STATE UNIV, NORTHRIDGE, 71- *Concurrent Pos:* Res Corp Cottrell grant, 63-64; NSF inst grants, 65-66. *Mem:* Am Chem Soc; Sigma Xi. *Res:* Organic reaction mechanisms; thermal and photochemical studies on reactions and reactivities of organic free radicals; free radicals in the field of environmental health; synthesis and properties of metalocenes. *Mailing Add:* Dept of Chem Calif State Univ Northridge CA 91330

OLSEN, CHARLES EDWARD, b Dover, NJ, May 17, 43; m 73. IMMMUNOLOGY. *Educ:* Rensselaer Polytech Inst, BS, 65; Univ Calif, Berkeley, PhD(biochem), 76. *Prof Exp:* Fel, Nat Inst Dental Res, NIH, 76-79; SR RES SCIENTIST, JOHNSON & JOHNSON RES, 80- *Mem:* AAAS; Sigma Xi. *Res:* Investigation of the molecular and cellular processes involved in the healing of normal and infected wounds. *Mailing Add:* 150 Calle La Mesa Moraga CA 94556-1602

OLSEN, CLARENCE WILMOTT, b Indianapolis, Ind, Dec 1, 03; m 29; c 2. NEUROLOGY. *Educ:* Ohio State Univ, BA, 23; Univ Mich, MA, 25, MD, 27. *Prof Exp:* Instr med, 30-34, from asst prof to assoc prof nerv dis, 34-52, clin prof, 52-64, PROF NEUROL, SCH MED, LOMA LINDA UNIV, 64- *Concurrent Pos:* Consult, US Vet Admin Hosps, Loma Linda, 78- *Mem:* Fel AMA; Am Psychiat Asn; fel Am Col Physicians. *Res:* Vascular diseases of the brain; respiratory rhythm in nervous diseases. *Mailing Add:* 11210 St Lucas Dr Loma Linda CA 92354

OLSEN, DON B, b Bingham, Utah, Apr 2, 30; m 50; c 5. BLOOD PUMP DEVELOPMENT & RECIPIENT INTERACTIONS. *Educ:* Utah State Univ, BS, 52; Colo State Univ, DVM, 56. *Prof Exp:* Gen pract vet med, 56-63; exten vet appl res, Dept Pathophysiol, Univ Nev, Reno, 63-68; fel path, Univ Colo Sch Med, 68-72; res assoc, Inst Biomed Eng, 72-85, RES PROF PHARMACEUT, UNIV UTAH, 80-, RES PROF & DIR INST BIOMED ENG, 85- *Concurrent Pos:* From res asst prof to res prof surg, Univ Utah, 73-86, prof, 86-, dir Artificial Heart Lab, 76-; vis prof, Von Humboldt Found Award, Munich W Ger, 77-78. *Mem:* Am Soc Artificial Internal Organs; Biomed Eng Soc; Int Soc Artificial Organs (vpres, 77-); Europ Soc Artificial Organs. *Res:* Development and characterization of blood pumps as ventricular assist or replacement; surgical techniques for implantation and post operative management and assessment; research and development on power sources to drive the blood pumps; blood volume and blood flow in normal and experimental animals. *Mailing Add:* 803 N 300 W Salt Lake City UT 84103

OLSEN, DOUGLAS ALFRED, b Minneapolis, Minn, Oct 10, 30; m 58; c 3. PHYSICAL CHEMISTRY. *Educ:* Gustavus Adolphus Col, BA, 53; Univ Iowa, MS, 55, PhD(phys chem), 60. *Prof Exp:* Teaching asst, Univ Iowa, 53-55; develop chemist, Bemis, Inc, 55-57; sr res scientist, Honeywell, Inc, 59-63; proj leader phys chem, Archer Daniels Midland Co, 63-67; sect mgr chem, Appl Sci Div, Litton Systs, Inc, 67-70; vpres, Bio-Medicus Inc, 70-75; pres, 75-85, CONSULT, PMD INC, 85- *Concurrent Pos:* Res assoc, Univ Minn, 73-76; vis prof, Tech Univ Denmark, 74. *Honors & Awards:* IR-100 Award, Indust Res Inst, 72. *Mem:* AAAS; Am Chem Soc; Am Soc Artificial Internal Organs. *Res:* Surface chemistry; electrochemistry; chemical kinetics; radiochemistry; artificial organs; biomedical materials; operations analysis. *Mailing Add:* 5057 Morgan Ave S Minneapolis MN 55419

OLSEN, EDWARD JOHN, b Chicago, Ill, Nov 23, 27; m 54; c 2. GEOLOGY. *Educ:* Univ Chicago, BA, 51, MS, 55, PhD, 59. *Hon Degrees:* LHD, Augustana Col, 79. *Prof Exp:* Geologist, Geol Surv Can, 53, US Geol Surv, 54 & Can Johns-Manville Co, Ltd, 56-59; asst prof mineral & petrol, Western Reserve Univ & Case Inst, 59-60; chmn dept geol, 74-78, CUR MINERAL, DEPT GEOL, FIELD MUS NATURAL HIST, 60- *Concurrent Pos:* Guest researcher, Argonne Nat Lab, 67-80; adj prof, Univ Ill, Chicago Circle, 70-; res assoc prof, Univ Chicago, 77-; assoc ed, Geochim et Cosmochim Acta, 85-91. *Mem:* Fel Mineral Soc Am; Geochem Soc; Mineral Asn Can; fel Meteoritical Soc. *Res:* Thermodynamics of mineral systems; phase equilibria in meteorites; electron microprobe; scanning electron microscopy. *Mailing Add:* Dept of Geol Field Mus of Natural Hist Chicago IL 60605

OLSEN, EDWARD TAIT, b Brooklyn, NY, June 12, 42; m 68. RADIO ASTRONOMY, PLANETARY ATMOSPHERES. *Educ:* Mass Inst Technol, BS, 64; Calif Inst Technol, MS, 67; Univ Mich, PhD(astron), 72. *Prof Exp:* Resident res assoc radio astron, Nat Res Coun, 72-74, sr scientist radio astron & planetary atmospheres, 74-78, MEM TECH STAFF, JET PROPULSION LAB, 78- *Mem:* Am Astron Soc; Sigma Xi; Planetary Soc; AAAS. *Res:* Microwave studies of the Jovian synchrotron emission and of the atmospheres of the major planets; the search for extraterrestrial intelligence. *Mailing Add:* 11530 Pala Mesa Dr Northridge CA 91326

OLSEN, EDWIN CARL, III, b Salt Lake City, Utah, Dec 20, 32; m 57; c 6. IRRIGATION ENGINEERING, ON-FARM WATER MANAGEMENT. *Educ:* Utah State Univ, BS, 59, PhD(irrig eng), 65. *Prof Exp:* Engr, Tipton & Kalmbach, Inc, Pakistan, 64-66; water resources engr, USAID, Laos, 66-68; dir, Int Irrig Ctr, 80-85, ASSOC PROF AGR & IRRIG ENG, UTAH STATE UNIV, 68- *Mem:* Am Soc Civil Engrs; Am Soc Agr Engrs; Am Geophys Union; Int Asn Sci Hydrol. *Res:* On farm water management in sub-humid and arid areas of developing countries, especially drainage, soil salinity control, irrigation system management, evapotranspiration and hydrologic balance. *Mailing Add:* Dept Agr & Irrig Eng Utah State Univ Logan UT 84322-4105

OLSEN, EUGENE DONALD, b La Crosse, Wis, Dec 10, 33; c 3. ANALYTICAL CHEMISTRY, CLINICAL CHEMISTRY. *Educ:* Univ Wis, BS, 55, PhD(anal chem), 60. *Prof Exp:* Instr chem, Univ Wis, 60; asst prof, Franklin & Marshall Col, 60-64; from asst prof to assoc prof, 64-73, PROF CHEM, UNIV S FLA, 73- *Concurrent Pos:* Supvr undergrad res participation, NSF, 60-70, res grants, 64-66; NIH spec fel, Med Sch, 71-72. *Mem:* Am Chem Soc; Am Soc Clin Chem. *Res:* Teaching films; ion exchange separation of radioelements; instrumental methods of analysis; chelometric titrations; electrochemical analysis; automation in analysis; reactions in nonaqueous and mixed solvents; clinical and medicinal chemistry. *Mailing Add:* Dept Chem Univ SFla 4202 Fowler Ave Tampa FL 33620

OLSEN, FARREL JOHN, b Salt Lake City, Utah, Mar 2, 29; m 55; c 3. AGRONOMY. *Educ:* Utah State Univ, BS, 54, MS, 58; Rutgers Univ, New Brunswick, PhD(crop prod), 61. *Prof Exp:* Exten agronomist, WVa Univ, 61-66; plant scientist, Sci Info Exchange, Smithsonian Inst, 66-67, pasture agronomist, WVa Univ, 67-71; assoc prof, 74-78, PROF PLANT & SOIL SCI, SCH AGR, SOUTHERN ILL UNIV, CARBONDALE, 78-, PASTURE AGRONOMIST, 71- *Mem:* Am Soc Agron. *Res:* Planted pastures, their establishment, management and utilization; nutritive value. *Mailing Add:* Dept Plant Sci Southern Ill Univ Carbondale IL 62901

OLSEN, GEORGE DUANE, b DeKalb, Ill, Jan 5, 40; m 65; c 2. CLINICAL PHARMACOLOGY. *Educ:* Dartmouth Col, AB, 62; Dartmouth Med Sch, BMS, 64; Harvard Med Sch, MD, 66. *Prof Exp:* Intern med, Univ Hosps Cleveland, 66-67; med dir, Indian Health Ctr, USPHS, 67-69; instr med, 72-87, from asst prof to assoc prof, 70-89, PROF PHARMACOL, MED SCH, ORE HEALTH SCI UNIV, 89-, ASSOC PROF MED, 87- *Concurrent Pos:* NIH training grant pharmacol, Med Sch, Univ Ore, 69-70; Fogarty sr int fel & NATO sr scientist fel, 81-82; peer chmn review comt, Am Heart Asn, Ore Affil, 88-; mem exec comt, Develop Pharmacol Sect, Am Soc Pharmacol & Exp Therapeut, 85- *Mem:* AAAS; Am Fedn Clin Res; Am Soc Pharmacol & Exp Therapeut; Am Soc Clin Pharmacol & Therapeut. *Res:* Perinatal pharmacology; respiratory pharmacology. *Mailing Add:* Dept Pharmacol Sch Med L221 Ore Health Sci Univ Portland OR 97201-3098

OLSEN, GLENN W, b North Lima, Ohio, Sept 7, 31; m 57; c 2. MATHEMATICS EDUCATION, COMPUTER SCIENCES. *Educ:* Edinboro State Col, BSEd, 53; Pa State Univ, MEd, 58; Cornell Univ, MAT, 66, PhD(math, math educ), 68. *Prof Exp:* Teacher, Randolph East Mead High Sch, 53-59; assoc prof math, Indiana Univ Pa, 60-65; head dept, 68-74, PROF MATH, EDINBORO UNIV PA, 68- *Mem:* Math Asn Am; Nat Educ Asn. *Res:* Problems in the teaching of mathematics. *Mailing Add:* Dept Math & Comput Sci Edinboro Univ PA Edinboro PA 16444

OLSEN, GREGORY HAMMOND, b Brooklyn, NY, Apr 20, 45; div; c 2. SOLID STATE PHYSICS. *Educ:* Fairleigh Dickinson Univ, BS, 66, BS & MS, 68; Univ Va, PhD(mat sci), 71. *Prof Exp:* Teaching asst physics, Fairleigh Dickinson Univ, 66-68; vis scientist electron micros, Univ Port Elizabeth, Repub S Africa, 71-72; mem tech staff crystal growth, RCA Labs, 72-79, res leader, 80-83,; pres, Epitaxx Inc, 84-90; CONSULT, 90- *Mem:* AAAS; fel Inst Elec & Electronic Engrs; Electrochem Soc; Am Phys Soc; Sigma Xi. *Res:* Study of crystal growth and structural defects in semiconductors for electro-optical devices; synthesized the first vapor phase epitaxy gallium arsenic continuous wave laser and the first vapor phase epitaxy indium gallium arsenic phosphorus 1.25, 1.55 and 1.65 micron continuous wave laser; vapor phase epitaxy 1.0-1.7 unit of measure Indium Gallium Arsenic photodetector. *Mailing Add:* Epitaxx Inc 3490 Rte 1 Princeton NJ 08540

OLSEN, JAMES CALVIN, b San Diego, Calif, May 17, 39; m 64; c 2. FISHERIES, BIOMETRICS. *Educ:* Ore State Univ, BS, 61; Univ Wash, MS, 64, PhD(fisheries), 69. *Prof Exp:* Res asst salmon res, Fisheries Res Inst, Univ Wash, 61-63, res asst fisheries res, Col Fisheries, 63-68; math statistician fisheries pop dynamics, 68-77, dep lab dir, 77-85, PROG MGR, AUKE BAY LAB, NAT MARINE FISHERIES SERV, 85- *Concurrent Pos:* Mem, Pac Salmon Comn Res & Statist Comt & Tech Comts. *Mem:* Am Fisheries Soc; Pac Marine Fisheries Biologists. *Res:* Development of analytic methods for estimating abundance and production of marine populations; distribution, migrations, and stock identification of pacific salmon. *Mailing Add:* Auke Bay Fisheries Lab PO Box 210155 Auke Bay AK 99821-0155

OLSEN, JAMES LEROY, b Minneapolis, Minn, June 8, 30; m 55; c 4. INDUSTRIAL PHARMACY. *Educ:* Univ Minn, Minneapolis, BS, 54, PhD(pharm), 64. *Prof Exp:* Pharmacist supvr, Sch Pharm, Univ Minn, 58-63; asst vpres, Clark-Cleveland Div, Richardson-Merrell Corp, 63-69; asst prof pharmaceut, 69-74, ASSOC PROF PHARM & CHMN DIV PHARMACEUT, SCH PHARM, UNIV NC, CHAPEL HILL, 74- *Concurrent Pos:* Food & Drug Admin fel, Sch Pharm, Univ NC, Chapel Hill, 71- *Mem:* Am Pharmaceut Asn; Acad Pharmaceut Sci. *Res:* Pharmaceutical dosage forms; tablets; long-acting dosage forms; dissolution. *Mailing Add:* 304 Beard Hall Univ NC Chapel Hill NC 27514

OLSEN, JOHN STUART, b Chicago, Ill, Mar 10, 50; m 72; c 2. BIOCHEMICAL SYSTEMATICS, CLADISTICS. *Educ:* Univ Ill, Chicago, BS, 71, MS, 73; Univ Tex, Austin, PhD(bot), 77. *Prof Exp:* Teaching asst bot, Univ Tex, Austin, 73-77, instr, 77; asst prof biol, 77-83, ASSOC PROF BIOL, RHODES COL, 83-, CHMN DEPT, 86- *Concurrent Pos:* Instr, Memphis Col Art, 77-84; vis asst prof, Univ Tex Austin, 78-83, Mountain Lake Biol Sta, Univ Va, 81; vis assoc prof, Univ Tex, Austin, 84-; adj prof, Memphis State Univ, 85- *Honors & Awards:* Ralph Alston Award, Bot Soc Am, 77. *Mem:* Int Asn Plant Taxon; AAAS; Sigma Xi; Am Soc Plant Taxonomists; Bot Soc Am; Am Inst Biol Sci. *Res:* Plant systematics and evolution; genus Verbesina within the large plant family Asteraceae. *Mailing Add:* Dept Biol Rhodes Col 2000 N Pkwy Memphis TN 38112-1699

OLSEN, KATHIE LYNN, b Portland, Ore, Aug 3, 52. NEUROENDOCRINOLOGY, PSYCHOBIOLOGY. *Educ:* Chatham Col, BS, 74; Univ Calif, Irvine, PhD(psychobiol), 79. *Prof Exp:* Fel neurobiol, Dept Neuropath, Harvard Med Sch, 79-80; res scientist, Long Island Res Inst, 80-83, ASST PROF, DEPT PSYCHIAT & BEHAV SCI, STATE UNIV NY, STONY BROOK, 83- *Concurrent Pos:* NIH prin investr, 81-; assoc prog dir psychobiol & integrative neural sci, behav & neural sci, NSF, 84-86. *Mem:* Asn Women Sci; Soc Neurosci. *Res:* Developmental and regulatory mechanisms underlying the expression of sexually dimorphic behaviors; developmental psychoneuroendocrinology; hormone-genome interaction. *Mailing Add:* Dept Psychol State Univ NY Stony Brook Main Stony Brook NY 11794

OLSEN, KENNETH, b Bridgeport, Conn, Mar 20, 26. ELECTRICAL ENGINEERING, DIGITAL COMPUTERS. *Educ:* Mass Inst Technol, BS, 50, MS, 52. *Prof Exp:* Mem staff, Digital Computer Lab, Mass Inst Technol, 50-57; PRES & DIR, DIGITAL EQUIP CORP, 57- *Concurrent Pos:* Mem, Pres Comn on Sci & Technol, 71-73; mem bd dirs, Polaroid Corp, Ford Motor Co & Corp of Mass Inst Technol; adv vpres & mem, Joslin Diabetes Found, Inc, Boston, Mass; mem, Computer Sci & Eng Bd, Nat Acad Sci. *Honors & Awards:* Vermilye Medal, Franklin Inst, 80; Founders Medal, Nat Acad Eng, 82; John Ericsson Award, Am Soc Swed Engrs, 88. *Mem:* Nat Acad Eng; fel Inst Elec & Electronics Engrs. *Mailing Add:* Digital Equip Corp 146 Main St Maynard MA 01754

OLSEN, KENNETH HAROLD, b Ogden, Utah, Feb 20, 30; m 55; c 4. GEOPHYSICS, ASTROPHYSICS. *Educ:* Idaho State Col, BS, 52; Calif Inst Technol, MS, 54, PhD(physics), 57. *Prof Exp:* Staff mem, Los Alamos Nat Lab, 57-89, alt group leader, 71-74, group leader, 74-81, LAB ASSOC, LOS ALAMOS NAT LAB, 89- *Concurrent Pos:* Grad asst, Mt Wilson Observ, Calif Inst Technol, 52-57; vis researcher, Appl Seismol Group, Swedish Nat Defence Res Inst, Stockholm, 83; sr vis scientist fel, Norwegian Seismic Array & Inst Geol, Univ Oslo, 83; coordr & ed, Crest Int Res Group on Continental Rifting, 87-; mem, Working Group 3, Inter-Union Comn on the Lithosphere, 87-90; consult, Geophys Consult Serv Int, 89- *Mem:* Royal Astron Soc; Seismol Soc Am; Europ Geophys Soc; Am Geophys Union; fel Geol Soc Am; Am Astron Soc. *Res:* Experimental transition probabilities for atomic spectra; computer calculations of stellar position; underground explosion phenomenology; water waves, tsunamis; seismology; infrared Fourier transform spectroscopy; solar eclipses; lithospheric structures, seismic refraction and reflection profiling; continental magmatism; geophysics and tectonics of continental rift systems. *Mailing Add:* 1029 187th Pl SW Lynwood WA 98037-4920

OLSEN, KENNETH LAURENCE, plant physiology, for more information see previous edition

OLSEN, KENNETH WAYNE, b Chicago, Ill, Dec 19, 44; m 66; c 3. PROTEIN CHEMISTRY, CRYSTALLOGRAPHY. *Educ:* Iowa State Univ, BS, 67; Duke Univ, PhD(biochem), 72. *Prof Exp:* Res assoc protein crystallog, Purdue Univ, 72-75; asst prof biochem, Med Ctr, Univ Miss, 75-83; ASSOC PROF CHEM, LOYOLA UNIV, 83- *Concurrent Pos:* Postdoctoral fel, Am Cancer Soc. *Mem:* Am Crystallog Asn; Am Chem Soc; Am Soc Biochem & Molecular Biol; Biophys Soc; Protein Soc. *Res:* Protein chemistry and crystallog; enzymology; affinity chromatography; hemoglobin; protein stability; plant lectins; prediction of protein 3D structure; protein crosslinking; blood substitutes. *Mailing Add:* Dept Chem Loyola Univ 6525 N Sheridan Rd Chicago IL 60626

OLSEN, LARRY CARROL, b St Joseph, Mo, July 25, 37; m 60; c 3. SOLID STATE SCIENCE. *Educ:* Univ Kans, BS, 60, PhD(physics), 65. *Prof Exp:* Asst prof physics, Univ Kans, 65; sr scientist, McDonnell Douglas Corp, 65-74; assoc prof eng, 74-80, PROF MAT SCI & ENG, JOINT CTR GRAD STUDY, 81- *Concurrent Pos:* Lectr, Joint Ctr Grad Study, 67-74. *Mem:* Am Phys Soc; Sigma Xi. *Res:* Photovoltaic research and development; solar cell studies based on silicon, cuprous-oxide and other materials; transport properties of solids; radiation effects in materials; electron-voltaic, thermoelectric and other forms of energy conversion. *Mailing Add:* 2500 Geo Wash Way 130 Richland WA 99352-1654

OLSEN, ORVIL ALVA, b Biggar, Sask, July 22, 17; m 45; c 3. BOTANY. *Educ:* Univ Sask, BSA, 41; Univ Man, MSc, 48; McGill Univ, PhD(plant path), 61. *Prof Exp:* Res asst potato breeding, Man Dept Agr, 46-49; asst prof plant path, McGill Univ, 51-57; res officer, Can Dept Agr, 57-65, res scientist, 65-68; assoc prof bot, Mem Univ Nfld, 68-; At Dept Biol, Sir Wilfred Grenfell Col, 69-81; RETIRED. *Mem:* Agr Inst Can; Can Phytopath Soc; Can Bot Asn. *Res:* Ecology of vegetation of spray zones created by waterfalls; floristics of Newfoundland. *Mailing Add:* 352 Fourth Ave Box 298 Minitonas MB R0L 1G0 Can

OLSEN, PETER FREDRIC, b Red Bank, NJ, Apr 7, 35; m 70. ECOLOGY. *Educ:* Univ Mich, BS, 56, MS, 57; Auburn Univ, PhD(zool), 65. *Prof Exp:* Res fel muskrat pop dynamics, Wildlife Mgt Inst, Delta Waterfowl Res Sta, 55-56; game biologist, Mich Dept Conserv, 57-58; res assoc dis ecol, Auburn Univ, 60-62; sr res ecologist, head ecol sect & asst res prof wildlife dis ecol, Univ Utah, 63-70; sr res ecologist & vpres, EcoDynamics, Inc, 70-73; sr ecologist, Dames & Moore Consult Engrs, 74-91; RES HAZARDOUS WASTE, GEOWEST GOLDEN INC, 91- *Mem:* AAAS; Wildlife Soc; Am Soc Mammal; Ecol Soc Am; Am Inst Biol Sci. *Res:* Analysis and evaluation of ecological and other impacts upon the environment which may result from proposed development activities. *Mailing Add:* GeoWest Golden Inc 175 W 200 S Salt Lake City UT 84101

OLSEN, RALPH A, b Moroni, Utah, Jan 30, 25; m 49; c 7. SOIL CHEMISTRY. *Educ:* Brigham Young Univ, BS, 49; Cornell Univ, MS, 51, PhD(agron), 53. *Prof Exp:* Soil scientist, USDA, 53-56; assoc prof, 56-64, PROF CHEM, MONT STATE UNIV, 64- *Concurrent Pos:* Fel, Mineral Nutrit Pioneering Res Lab, Beltsville, Md, 61-62 & Plant Stress Lab, 79-80; Inst Biol Chem fel, Univ Copenhagen, 65-66. *Mem:* Soil Sci Soc Am; Am Chem Soc; Sigma Xi. *Res:* Potentiometric measurements in colloidal suspensions; mechanisms involved in the movement of ions from soils to plant roots; inorganic nutrition of green plants; ion movement through living membranes. *Mailing Add:* Dept Soils Sci Mont State Univ Bozeman MT 59717

OLSEN, RICHARD GEORGE, b Independence, Mo, June 25, 37; c 4. VIROLOGY. *Educ:* Univ Mo-Kansas City, BA, 59; Atlanta Univ, MS, 64; State Univ NY Buffalo, PhD(virol), 69. *Prof Exp:* Instr microbiol, Metrop Jr Col, 63-67; from asst prof to assoc prof virol, 69-77, PROF VIROL & IMMUNOL, COL VET MED, OHIO STATE UNIV, 77- *Mem:* Am Soc Microbiol; Am Asn Cancer Res; Tissue Cult Asn; Int Asn Comp Res Leukemia; Am Soc Virol; Am Soc Immunopharm. *Res:* Immunology of the cat and man to oncogenic viruses, including tumor poxviruses and RNA oncornaviruses; immunology of retrovirus diseases; efficacious subunit vaccine that prevents feline leukemia. *Mailing Add:* 2255 Hwy 56 London OH 43140

OLSEN, RICHARD KENNETH, b Provo, Utah, Sept 3, 35; m 54; c 5. ORGANIC CHEMISTRY. *Educ:* Brigham Young Univ, AB, 60; Univ Ill, PhD(org chem), 64. *Prof Exp:* Fel org chem, Stanford Res Inst, 64-65; res assoc, Univ Utah, 65-67; from asst prof to assoc prof chem, 67-77, PROF CHEM & BIOCHEM, UTAH STATE UNIV, 77- *Mem:* Am Chem Soc. *Res:* Synthetic organic chemistry; natural products; peptide antibiotics. *Mailing Add:* Dept Chem Utah State Univ Logan UT 84322-0300

OLSEN, RICHARD STANDAL, b Lansing, Mich, Nov 13, 36; m 65; c 2. CHEMICAL ENGINEERING. *Educ:* Pac Lutheran Univ, BA, 59; Ore State Univ, PhD(chem eng), 67. *Prof Exp:* chem res engr, 66-80, PROJ SUPVR, ALBANY METALL RES CTR, FED BUR MINES, 80- *Mem:* Sigma Xi. *Res:* Chemical processes involved in extractive metallurgy; chlorination processes; chlorination of base metal sulfides. *Mailing Add:* 732 Broadalbin Albany OR 97321

OLSEN, RICHARD WILLIAM, biochemistry; deceased, see previous edition for last biography

OLSEN, ROBERT GERNER, b Brooklyn, NY, Apr 9, 46; m 74; c 3. ELECTROMAGNETISM, ELECTRICAL ENGINEERING. *Educ:* Rutgers Univ, BS, 68; Univ Colo, MS, 70, PhD(elec eng), 74. *Prof Exp:* Sr scientist elec eng, Westinghouse Geores Lab, 71-73; from asst prof to assoc prof, 73-83, PROF ELEC ENG, WASH STATE UNIV, 83- *Concurrent Pos:* Mem, Comn B, US Nat Comt, Int Union Radio Sci, 76-; consult, Jersey Cent Power & Light, 88- & Wash Water Power, 90. *Mem:* Inst Elec & Electronics Engrs; Sigma Xi. *Res:* Antenna theory; underground electromagnetic wave propagation; numerical solution of electrostatics problems; electromagnetic environment of power lines. *Mailing Add:* Dept Elec Eng Wash State Univ Pullman WA 99164-2752

OLSEN, ROBERT THORVALD, b Brookfield, Ill, Mar 10, 15; m 41; c 2. SYNTHETIC ORGANIC CHEMISTRY. *Educ:* Newark Col Eng, BSChE, 36; Columbia Univ, MSChE, 37; Mass Inst Technol, PhD(org chem), 42. *Prof Exp:* Chem engr, Eastman Kodak Co, NY, 37-39; sr res chemist, Gen Aniline & Film Corp, 42-48, Plymouth Cordage Co, 48- 54 & Celotex Corp, 54-56; res mgr, Standard Register Co, 56-61; tech dir fine chem, Ashland Chem Co, 61-76; RETIRED. *Concurrent Pos:* consult, 76-82; health educr, 82-; owner, Deacon's Vineyard (grapevines, hort), 85-; pub health & epidemiol, Mass Health Res Inst, 91- *Honors & Awards:* Am Wine Soc, 86. *Mem:* AAAS; Am Chem Soc; Am Wine Soc chair; Am Soc Enol & Viticult; Soc Wine Educrs. *Res:* Chemical engineering; research management. *Mailing Add:* 300 Country Lane RR-1 Box 48 Eastham MA 02642-3329

OLSEN, RODNEY L, b Duluth, Minn, July 10, 36; m 58; c 2. ANALYTICAL CHEMISTRY. *Educ:* Univ Minn, Duluth, BA, 58; Iowa State Univ, MS, 60, PhD(anal chem), 62. *Prof Exp:* From asst prof to assoc prof, 62-73, PROF CHEM, HAMLINE UNIV, 73- *Mem:* Am Chem Soc; Sigma Xi. *Res:* Fluorometric methods; radiochemistry; analytical separations. *Mailing Add:* Dept Chem Hamline Univ Snelling & Hewitt Ave St Paul MN 55104

OLSEN, RONALD H, b New Ulm, Minn, June 26, 32; m 58; c 6. MICROBIOLOGY. *Educ:* Univ Minn, BA, 57, MS, 59, PhD(microbiol), 62. *Prof Exp:* Asst prof microbiol, Colo State Univ, 62-63; vpres res & develop, Dairy Technics Inc, 63-65; assoc prof, 65-74, PROF MICROBIOL, UNIV

MICH, ANN ARBOR, 74- *Mem:* AAAS; Am Soc Microbiol. *Res:* Bacterial physiology; physiological basis for the minimum temperature of growth. *Mailing Add:* Dept Microbiol M6643 Med Sci 2 Univ Mich 1301 Catherine Rd Ann Arbor MI 48109-0620

OLSEN, STANLEY JOHN, b Akron, Ohio, June 24, 19; m 42; c 1. ZOOARCHAEOLOGY, ARCHAEOLOGY. *Prof Exp:* Lab technician vert paleont, Harvard Univ, 45-56; vert paleontologist, Fla Geol Surv, 56-58; from assoc prof to prof anthropology, Fla State Univ, 68-73; PROF ANTHROP, UNIV ARIZ, 73-; ZOOARCHAEOLOGIST, ARIZ STATE MUS, 73- *Concurrent Pos:* Res assoc, Mus Northern Ariz, 66-; NSF grant, Fla Geol Surv, 64-66, Harvard Univ Guide Found grant, 66; NSF grant, Fla State Univ, 67-69 & 69-70, Am Philos Soc grant, 70, Nat Geog Soc grant, 82-85 & Univ Ariz grant, 85. *Mem:* Soc Vert Paleont (pres, 65-66); Am Soc Mammal; Soc Am Archaeol; Am Soc Ichthyol & Herpet; Soc Syst Zool. *Res:* Analysis of vertebrates from archaeological sites in the Western Hemisphere; beginnings of animal domestication in China and Central Asia; origins of domestic dog, horse, camel & yak. *Mailing Add:* Dept of Anthrop Univ of Ariz Tucson AZ 85721

OLSEN, STEPHEN LARS, b Mar 22, 42; US citizen. PARTICLE PHYSICS. *Educ:* City Col NY, BS, 63; Univ Wis, MS, 65, PhD(physics), 70. *Prof Exp:* Res assoc physics, Univ Wis, 70 & Rockefeller Univ, 70-72; asst prof, 72-75, ASSOC PROF PHYSICS, UNIV ROCHESTER, 75- *Concurrent Pos:* Fel, Alfred P Sloan Found, 73-77; mem prog adv comt, Ferni Nat Accelerator Lab, 75-78. *Res:* Experimental studies of the interactions between elementary particles at high energies. *Mailing Add:* Dept Physics Univ Rochester Wilson Blvd Rochester NY 14627

OLSEN, WARD ALAN, b Holmen, Wis, Sept 13, 34; m 61; c 3. INTERNAL MEDICINE, GASTROENTEROLOGY. *Educ:* Univ Wis-Madison, BS, 56, MD, 59. *Prof Exp:* From intern to resident, Harvard Med Serv, Boston City Hosp, 59-62 & 64-65; instr med, Harvard Med Sch, 67-68; from asst prof to assoc prof, 68-78, PROF MED, UNIV WIS-MADISON, 78- *Concurrent Pos:* Fel gastroenterol, Boston Univ, 65-67; chief gastroenterol, Vet Admin Hosp, Madison, Wis, 68- *Mem:* Am Fedn Clin Res; Cent Soc Clin Res; Am Soc Clin Invest; Am Gastroenterol Asn. *Res:* Physiology, biochemistry and morphology of small intestine mucosa, especially brush border membrane proteins. *Mailing Add:* Vet Admin Hosp 2500 Overlook Terr Madison WI 53705

OLSEN, WILLIAM CHARLES, b Edmonton, Alta, Mar 25, 33; m 57; c 2. NUCLEAR PHYSICS. *Educ:* Univ BC, BASc, 56, MASc, 59; Univ Alta, PhD(nuclear physics), 64. *Prof Exp:* Res officer, Atomic Energy Can Ltd, Ont, 58-60; res assoc, 64, from asst prof to assoc prof, 64-74, PROF NUCLEAR PHYSICS, UNIV ALTA, 74- *Res:* Low and intermediate energy nuclear physics leading to information on the nuclear structure of low and medium mass nuclei. *Mailing Add:* Dept Physics Univ Alberta Edmonton AB T6G 2M7 Can

OLSHEN, ABRAHAM C, b Portland, Ore, Apr 20, 13; m 34; c 2. RESOURCE MANAGEMENT. *Educ:* Reed Col, AB, 33; Univ Iowa, MSc, 35, PhD, 37. *Prof Exp:* Chief statistician, City Planning Comn, Portland, Ore, 33-34; res asst, Math Dept, Univ Iowa, 34-37, biometrics asst, Med Ctr, 36-37; actuary & chief examiner, Ore Ins Dept, 37-42 & 45-46; controller, W Coast Life Ins Co, San Francisco, Calif, actuary, 46-53, vpres, 47-63, chief actuary, 53-63, first vpres, 63-67, sr vpres, 67-68; dir, Home Fed Savings & Loan Asn, San Francisco, 72-86, vchmn bd, 79-85, chmn bd, 85-86; CONSULT, ACTUARIAL & INS MGT, 68-; PRES, OLSHEN & ASSOCS, INC, SAN FRANCISCO, 79- *Concurrent Pos:* Mem, Bd Dirs, W Coast Life Ins Co, San Francisco, Calif, 55-68; mem, Educ Comt, San Francisco Chamber Com, Blanks Comt, Health Ins Asn Am & Actuarial & Statist Comt; mem, Med Care Admin Comt, Univ Calif & Health Ins Coun, Calif Comt; lectr, var univs. *Honors & Awards:* Cert Distinguished Serv, US Off Sci Res & Develop, 45; Presidential Cert Merit, 47. *Mem:* Am Math Soc; Inst Mgt Sci; Inst Math Statist; Opers Res Soc; Soc Indust & Appl Math; fel Sigma Xi; fel AAAS. *Res:* Operations research; corporate organization and management; actuarial science; author of various publications. *Mailing Add:* 760 Market St Suite 739 San Francisco CA 94102

OLSHEN, RICHARD ALLEN, b Portland, Ore, May 17, 42; m 79; c 4. STATISTICS, MATHEMATICS. *Educ:* Univ Calif, Berkeley, AB, 63; Yale Univ, MS, 65, PhD(statist), 66. *Prof Exp:* Lectr statist & res staff statistician, Yale Univ, 66-67; asst prof statist, Stanford Univ, 67-72; assoc prof, Univ Mich, Ann Arbor, 72-75; assoc prof, 75-77, PROF MATH, UNIV CALIF, SAN DIEGO, 77-, DIR BIOSTATIST UNIT, CANCER CTR, 78-, DIR MATH & STATIST LAB, 82- *Concurrent Pos:* Vis asst prof, Columbia Univ, 70-71; vis assoc prof, Stanford Univ, 73-75; vis prof, Harvard Univ, 79-80, Stanford Univ, 87-88; vis scholar, Mass Inst Technol, 79-80, Stamford Univ, 87-; Guggenheim Award, 87-88; Res Scholar CAncer, Am Cancer Soc. *Honors & Awards:* Fel, Inst Math Statist. *Mem:* Fel Inst Math Statist; Am Statist Asn; AAAS. *Mailing Add:* Stanford Univ Sch Med HRP Bldg Rm 110A Stanford CA 94305-5092

OLSON, ALBERT LLOYD, b Mountain View, Calif, Dec 14, 24; m 48; c 3. PATHOLOGY. *Educ:* Col Med Evangelists, MD, 49; Am Bd Path, dipl, 58, cert clin path, 60. *Prof Exp:* From instr to asst prof, 58-64, ASSOC PROF PATH, LOMA LINDA UNIV, 64- *Res:* Medicine. *Mailing Add:* White Mem Med Ctr 1720 Brooklyn Ave Los Angeles CA 90033

OLSON, ALFRED C, b Chicago, Ill, July 18, 26; m 66; c 1. BIOCHEMISTRY. *Educ:* Northwestern Univ, BS, 49; Univ Wis, PhD(phys & org chem), 54. *Prof Exp:* Chemist, Calif Res Corp Div, Standard Oil Co Calif, 54-60; RES CHEMIST, WESTERN REGIONAL RES CTR, AGR RES SERV USDA, 60- *Concurrent Pos:* Res fel, Div Biol, Calif Inst Technol, 60-63. *Mem:* Am Chem Soc; Am Soc Plant Physiol; Tissue Cult Asn; Inst Food Sci & Technol; AAAS; Sigma Xi. *Res:* Bioavailability of nutrients; nutritional problems of dry beans; plant tissue culture; immobilized enzymes. *Mailing Add:* 1811 Arlington Blvd El Sorito CA 94530

OLSON, ALLAN THEODORE, b Cochrane, Ont, July 10, 30; m 64. THERMODYNAMICS, MECHANICS. *Educ:* Queen's Univ, Ont, BSc, 53; Univ Birmingham, MSc, 55. *Prof Exp:* Res officer mech eng, Nat Res Coun Can, 55-57; ASSOC PROF THERMODYN & MECH, DEPT MECH ENG, UNIV WESTERN ONT, 57- *Mailing Add:* Dept Mech Eng Univ Western Ont Engr & Math Sci Bldg London ON N6A 5B9 Can

OLSON, ANDREW CLARENCE, JR, b San Diego, Calif, Nov 10, 17; m 45; c 2. PARASITOLOGY. *Educ:* San Diego State Univ, BA, 39; Univ Idaho, MS, 42; Ore State Univ, PhD, 55. *Prof Exp:* From instr to prof, 46-80, chmn dept, 57-60, EMER PROF ZOOL, SAN DIEGO STATE UNIV, 80- *Mem:* Am Soc Parasitol; Am Soc Mammal; Int Soc Correlative Biol Res. *Res:* Ecology of fish parasites. *Mailing Add:* 3885 Capitol St La Mesa CA 91941-7603

OLSON, ARTHUR OLAF, b Lethbridge, Alta, May 11, 42; m 63; c 2. BIOCHEMISTRY, HORTICULTURE. *Educ:* Univ Alta, BSc, 64, PhD(plant biochem), 67. *Prof Exp:* Exten horticulturist, Alta Dept Agr, 64; res fel biochem, McMaster Univ, 67-68; res biochemist, Atomic Energy Can, Ltd, 68-69; dir, Alta Hort Res Ctr, Alta Dept Agr, 70-74, dir, Plant Indust Div, 74-79, asst dep minister res & oper, 79-86; ASST DEP MINISTER, RES BR, AGR CAN, 86- *Concurrent Pos:* Ed, Can Soc Hort Sci, 71- *Mem:* Agr Inst Can; Can Soc Hort Sci. *Res:* Aging effects of ethylene in plants and animals; mutation damage and its repair in algae and yeast; post harvest physiology and storage biochemistry. *Mailing Add:* Agr Can Sir John Carling Bldg 7th Floor 930 Carling Ave Ottawa ON L1A 0C5 Can

OLSON, ARTHUR RUSSELL, b Lawrence, Mass, Jan 22, 19; m 42; c 4. TEXTILE CHEMISTRY. *Educ:* Mass Inst Technol, BS, 39. *Prof Exp:* Develop chemist, Hunt-Rankin Leather Co, Mass, 39-42; plastics chemist, United Shoe Mach Corp, 46-50; dir & chemist, McMillan Lab, Inc, 50-52; res chemist, Finishing Div, Kendall Co, 52-60, Fiber Prod Div, 60-64, res mgr, Fiber Prod Div, 64-77, res assoc, 77-82; CONSULT, 82- *Mem:* AAAS; Am Chem Soc. *Res:* Plastics technology; synthetic resins and plastics; textiles and textiles materials. *Mailing Add:* Six Eastover Rd Walpole MA 02081

OLSON, AUSTIN C(ARLEN), b Missoula, Mont, Feb 4, 18; m 45; c 2. CHEMICAL ENGINEERING. *Educ:* Mont State Univ, BS, 39; Univ Minn, MS, 41, PhD(chem eng), 48. *Prof Exp:* Asst chem eng, Univ Minn, 39-41, instr, 41-42; assoc res engr, Chevron Res Co Div, 48-51, res engr, 51-59, group supvr, 59-62, supv engr, 62-66, sect supvr, 66-73, chmn proj mgr, Chevron Res Co, Standard Oil Co Calif, 73-83; RETIRED, 83- *Mem:* Am Chem Soc; Am Inst Chem Engrs. *Res:* Petroleum and petrochemical plant and process design and economic evaluation; environmental engineering. *Mailing Add:* 20 Stevenson Ave Berkeley CA 94708

OLSON, BETTY H, b Salt Lake City, Utah, Sept 28, 47; c 1. WATER QUALITY, PUBLIC HEALTH. *Educ:* Univ Calif, Irvine, BS, 69; Univ Calif, Berkeley, MS, 71, PhD(environ health sci), 74. *Prof Exp:* Asst res, Univ Calif, Davis, 73-74, asst prof, Irvine, 74-80; vis scientist, Imperial Col, London, Eng, 81; assoc prof, 80-84, PROF, UNIV CALIF, IRVINE, 84- *Concurrent Pos:* Fel, NIH, 74; prof & chair, Environ Anal & Design, Social Ecol, Univ Calif, Irvine, 89- *Mem:* Fel Am Soc Microbiol. *Res:* Molecular biology to improve the health of the environment; nucleic acid probe research in biodegradation pathways and identification of pathogenic micro-organisms; environmental management and policy. *Mailing Add:* Prog Social Ecol Univ Calif Irvine CA 92717

OLSON, CARL, b Sac City, Iowa, Sept 15, 10; m 34; c 4. COMPARATIVE PATHOLOGY. *Educ:* Iowa State Univ, DVM, 31; Univ Minn, MS, 34, PhD(comp path), 35. *Prof Exp:* Asst prof path, State Univ NY Vet Col, Cornell Univ, 35-37; res prof, Univ Mass, 37-45; prof animal path & hyg & chmn dept, Univ Nebr, 45-56; chmn dept, 62-63, VET SCI, UNIV WIS-MADISON, 56- *Concurrent Pos:* Vis prof, Turkey, Ger, Japan & Brazil. *Mem:* Am Col Vet Path; Am Vet Med Asn; Am Asn Pathologists; Am Asn Avian Path; Conf Res Workers Animal Dis (pres, 58). *Res:* Pathogenesis of animal diseases, especially neoplastic particularly pepillomatosis and bovine leukosis. *Mailing Add:* 921 University Bay Dr Madison WI 53706

OLSON, CARTER LEROY, b Iola, Wis, Jan 13, 35; m 60; c 1. ANALYTICAL CHEMISTRY. *Educ:* Wis State Col, Stevens Point, BS, 56; Univ Kans, PhD(chem), 62. *Prof Exp:* Res assoc & fel anal chem, Univ Wis, 61-63; from asst prof to assoc prof pharmaceut anal, 63-73, ASSOC PROF MED CHEM, COL PHARM, OHIO STATE UNIV, 73- *Mem:* Am Chem Soc; Sigma Xi. *Res:* Continuous methods of analysis; electroanalytical chemistry; analysis based on the rates of chemical reactions including the rates of enzyme catalyzed reactions. *Mailing Add:* Dept Pharm Ohio State Univ 217 Lloyd Parks Hall Columbus OH 43210

OLSON, CLIFFORD GERALD, b Osakis, Minn, July 6, 42; m 65; c 2. SOLID STATE PHYSICS. *Educ:* Hamline Univ, BS, 64; Iowa State Univ, PhD(physics), 70. *Prof Exp:* Fel, 70-71, PHYSICIST, AMES LAB, US DEPT ENERGY, IOWA STATE UNIV, 71- *Mem:* Am Phys Soc; Optical Soc Am. *Res:* Photo emission and optical properties of solids using synchrotron radiation. *Mailing Add:* Synchrotron Radiation Ctr Ames Lab US Dept Energy 3725 Schneider Dr Rte 4 Stoughton WI 53589

OLSON, DALE WILSON, b Mountain Lake, Minn, July 27, 41. MAGNETISM. *Educ:* Carleton Col, BA, 62; Univ Rochester, PhD(physics), 70. *Prof Exp:* Asst prof, 68-73, ASSOC PROF PHYSICS, UNIV NORTHERN IOWA, 73- *Mem:* Am Phys Soc; Am Asn Physics Teachers; Sigma Xi; Optical Soc Am. *Res:* Holography; phase transitions. *Mailing Add:* 3801 Viking Rd Waterloo IA 50701

OLSON, DANFORD HAROLD, b Minneapolis, Minn, Jan 17, 35; m 55; c 4. ORGANIC CHEMISTRY. *Educ:* Univ Minn, BS, 56; Kans State Univ, PhD(chem), 62. *Prof Exp:* Res scientist petrochem, Denver Res Ctr, Marathon Oil Co, 62-66; res scientist, Res Ctr, Wood River, 66-70, supvr, eng lubricants, 70-77, supvr eng lubricants, Develop Res Ctr, Houston, 77-78, Int Petrol Co, Chester, Eng, staff eng, 79-80, staff chemist, 81-85, SR STAFF ENGR, SHELL OIL CO, HOUSTON, 87- *Mem:* Am Chem Soc; Soc Automotive Engrs; Sigma Xi; Am Soc Testing & Mat. *Res:* Synthesis of small ring heterocyclics; use of nitrogen oxides in organic synthesis; liquid and gas phase oxidation reactions; dehydrogenation; computer applications; product development. *Mailing Add:* 12622 Campsite Trail Cypress TX 77429

OLSON, DAVID GARDELS, b Melrose Park, Ill, Jan 25, 40; m 73. OPERATIONS RESEARCH, SYSTEMS ANALYSIS. *Educ:* Purdue Univ, BS, 61; Mass Inst Technol, SM, 63; Northwestern Univ, Evanston, PhD(indust eng), 71. *Prof Exp:* Mem prof staff, Ctr Naval Anal, Inst Naval Studies, 62-64; physicist, US Naval Undersea Res & Develop Ctr, 64-68; sr opers res analyst, Opers Res Task Force, Chicago Police Dept, 68-69; asst prof indust eng, Ctr Large Scale Systs, Purdue Univ, 71-74; opers res analyst, Naval Underwater Systs Ctr, Newport, RI, 74-78; assoc prof, 78-80, ADJ ASSOC PROF INDUST ENG, UNIV RI, 80-; OPERS RES ANALYST, NAVAL UNDERWATER RES SYSTS CTR, 80- *Concurrent Pos:* Lectr, Traffic Inst, Northwestern Univ, Evanston, 70-78; consult, US Naval Underwater Systs Ctr, RI, 72-74; mem panel, Comt Undersea Warfare, Nat Res Coun-Nat Acad Sci, 72-74; sci & Tech adv, asst chief Naval opers, undersea warfare, Attack Submarine Div, 87-88. *Res:* Applications of search theory and Markov processes to problems of surveillance in the areas of pollution and police operations; applications of mathematical programming; military operations research; computer simulation and modeling applied to warfare analysis. *Mailing Add:* Naval Underwater Syst Ctr Attn: David Olson Code 61 Newport RI 02841-5047

OLSON, DAVID HAROLD, b Stoughton, Wis, Apr 27, 37; m 59; c 2. PHYSICAL CHEMISTRY. *Educ:* Univ Wis, BS, 59; Iowa State Univ, PhD(phys chem), 63. *Prof Exp:* Sr res chemist, 63-74, RES SCIENTIST, MOBIL RES & DEVELOP CORP, 74- *Mem:* Am Chem Soc; Am Crystallog Asn; Int Zeolite Assoc. *Res:* Chemical crystallography; zeolite crystal chemistry and catalysis. *Mailing Add:* Cent Res Div Mobil Res & Develop Corp Princeton NJ 08540

OLSON, DAVID LEROY, b Oakland, Calif, Mar 17, 42; m 63; c 4. METALLURGY & WELDING METALLURGY, CORROSION. *Educ:* Wash State Univ, BS, 65; Cornell Univ, PhD(mat sci), 70. *Prof Exp:* Mem tech staff, Semiconductor Res & Develop Lab, Tex Instruments, Inc, 69-70; res assoc metall eng, Ohio State Univ, 70-72; dir, Ctr Welding Res, 81-86, dean & vpres res & develop, 86-89, from asst prof to assoc prof metall eng, 72-78, PROF PHYS METALL, COLO SCH MINES, 78- *Concurrent Pos:* Consult, govt & indust, 72-; vis sr scientist, Norweg Inst Technol, Troudheim, Norway, 79. *Honors & Awards:* Bradley Stoughton Award, Am Soc Metals, 76; Adams Mem Award, Am Welding Soc, 78, Adams lectr, 84, McKay-Helms Award, 85, 86, 89, 90; Savage Award, Am Nuclear Soc, 87; Savage Award, 87. *Mem:* Fel Am Soc Metals; Am Inst Mining, Metall & Petrol Engrs; Am Phys Soc; Am Welding Soc; Am Ceramic Soc; Am Soc Metals; Nat Asn Corrosion Engrs. *Res:* Welding metallurgy; reactive metals; pyrometallurgical reactions in welding; weld metal microstructure-property relationships; phase transformations during welding and the behavior of welding consumables and processes; corrosion; high temperature corrosion. *Mailing Add:* Dept Metals & Metall Eng Colo Sch Mines Golden CO 80401

OLSON, DONALD B, b Greybull, Wyo, May 28, 52; m 74; c 3. PHYSICAL OCEANOGRAPHY. *Educ:* Univ Wyo, BS, 74; Tex A&M Univ, MS, 76, PhD(phys ocean), 79. *Prof Exp:* Res asst, Tex A&M Univ, 75-79; from asst prof to assoc prof, 79-90, PROF, ROSENSTIEL SCH MARINE & ATMOSPHERIC SCI, UNIV MIAMI, 90- *Concurrent Pos:* Asst res scientist, Tex A&M Univ, 79. *Mem:* Am Meteorol Soc; Am Geophys Union. *Res:* General ocean circulation, particularly mesoscale eddies, rings and fronts and their relationship to the mean flow. *Mailing Add:* Div Meteorol & Phys Oceanog Univ Miami 4600 Rickenbacker Causeway Miami FL 33149

OLSON, DONALD RICHARD, b Sargent, Nebr, Dec 26, 17; m 44; c 3. MECHANICAL ENGINEERING. *Educ:* Ore State Univ, BS, 42; Yale Univ, MEng, 44, DEng, 51. *Prof Exp:* Asst prof mech eng, Yale Univ, 51-57, assoc prof, 57-62; head power plants, Ord Res Lab, Pa State Univ, 62-72, prof, 62-83, head dept, 72-83, EMER PROF MECH ENG, PA STATE UNIV, 83- *Mem:* Am Soc Mech Engrs; Soc Automotive Engrs; Sigma Xi. *Res:* Thermodynamics and combustion as related to power systems. *Mailing Add:* 621 Glenn Rd State College PA 16803

OLSON, DOUGLAS BERNARD, b Beeville, Tex, May 14, 45. PHYSICAL CHEMISTRY. *Educ:* Southwest Tex State Col, BS, 67; Univ Tex, MA, 69, PhD(chem), 77. *Prof Exp:* Res chemist physical chem, Aero Chem Res Labs, 77-88; SR RES SCIENTIST, CETR, NMEX TECH, 88- *Mem:* Combustion Inst; Am Chem Soc; Sigma Xi. *Res:* Gas kinetics of combustion reactions; soot formation. *Mailing Add:* CETR Campus Sta NMex Tech Socorro NM 87801

OLSON, EARL BURDETTE, JR, b Oct 2, 39; m; c 4. PULMONARY METABOLISM & DEVELOPMENT. *Educ:* Univ Wis, PhD(biochem), 69. *Prof Exp:* ASSOC PROF PREV MED, UNIV WIS-MADISON, 71- *Concurrent Pos:* Res career develop award, 77-82. *Mem:* Soc Toxicol; Am Physiol Soc; Sigma Xi. *Res:* Monoamine neurotransmitter. *Mailing Add:* Univ Wis 504 N Walnut St Madison WI 53705

OLSON, EDWARD COOPER, b US, June 7, 30; m 59; c 2. ASTROPHYSICS. *Educ:* Worcester Polytech Inst, BS, 52; Ind Univ, PhD(astron), 61. *Prof Exp:* Instr physics, Worcester Polytech Inst, 52; asst prof astron, Smith Col, 60-63 & Rensselaer Polytech Inst, 64-65; from asst prof to assoc prof, 66-81, PROF ASTRON, UNIV ILL, URBANA, 81- *Concurrent Pos:* Adj prof, San Diego State Univ. *Mem:* Am Astron Soc; Int Astron Union; Astron Soc Pac. *Res:* Photometry and spectrophotometry of interacting stars. *Mailing Add:* Dept Astron Univ Ill 1002 W Green St Urbana IL 61801

OLSON, EDWIN ANDREW, b Gary, Ind, May 21, 25; m 54; c 3. GEOLOGY, GEOCHEMISTRY. *Educ:* Univ Pittsburgh, BS, 47, MS, 49; Columbia Univ, PhD(geochem), 63. *Prof Exp:* Lectr math, Univ Pittsburgh, 47-49; develop engr, E I du Pont de Nemours & Co, 49-53; asst prof phys sci, Northwestern Col, Minn, 53-56; asst geochem, Lamont Geol Observ, Columbia Univ, 56-60; from asst prof to assoc prof geol, 60-69, PROF GEOL, WHITWORTH COL, WASH, 69- *Concurrent Pos:* NSF res grant, 61-64 & 72-73. *Mem:* Geochem Soc; Am Sci Affil (pres, 87). *Res:* Theoretical geochemical aspects relating to the accuracy of radiocarbon dating. *Mailing Add:* Dept Physics & Earth Sci Whitworth Col Spokane WA 99251

OLSON, EDWIN S, b Red Wing, Minn, Oct 23, 37; m 63; c 3. ORGANIC CHEMISTRY. *Educ:* St Olaf Col, BA, 59; Calif Inst Technol, PhD(org chem), 64. *Prof Exp:* Asst prof org chem, Idaho State Univ, 64-68; from asst prof to assoc prof org chem, SDak State Univ, 68-77, prof chem, 77-80; res chemist, Grand Forks Energy Technol Ctr, 80-83; RES SUPVR, UNIV NDAK ENERGY RES CTR, 83- *Mem:* Am Chem Soc. *Res:* Organic analytical; high performance liquid chromatography; gas chromatography; mass spectrometry; nuclear magnetic resonance; coal chemistry; transition metal catalysis. *Mailing Add:* Mineral Res Ctr 223 Cirole Hills Dr Grand Forks ND 58201

OLSON, EMANUEL A(OLE), b Holdredge, Nebr, Feb 19, 16; m 39; c 2. AGRICULTURAL ENGINEERING. *Educ:* Univ Nebr, BS, 39. *Prof Exp:* Asst exten engr, Univ Nebr-Lincoln, 39-45, proj leader exten eng, 46-78, from assoc prof to prof agr eng, 52-78; RETIRED. *Concurrent Pos:* Consult, Behlen Mfg Co, Nebr, 64-65, Hart Carter, Ill, 68, Centron Corp, Kans, 69, Farmland Industs, Inc, 70 & Hubbard Milling Co, Minn, 71; Nebr Environ Control Coun, 79-84; pvt consult, 78-86. *Honors & Awards:* Awards, Am Soc Agr Engrs, 48, 53, 62, 66, 70, 71, 74, 75 & Gunlogson Countryside Eng Award, 79. *Mem:* Fel Am Soc Agr Engrs; Nat Soc Prof Engrs. *Res:* Livestock waste management; rural water systems; livestock production facilities; crop conditioning, storage and processing; farmstead mechanization and concrete tilt-up construction. *Mailing Add:* 925 S 52nd St Lincoln NE 68510

OLSON, ERIK JOSEPH, pharmacology; deceased, see previous edition for last biography

OLSON, EVERETT CLAIRE, b Waupaca, Wis, Nov 6, 10; m 39; c 3. VERTEBRATE PALEONTOLOGY. *Educ:* Univ Chicago, BS, 32, MS, 33, PhD(geol, vert paleont), 35. *Prof Exp:* From instr to prof vert paleont, Univ Chicago, 35-69; prof, 69-78, chmn dept, 70-72, EMER PROF ZOOL, UNIV CALIF, LOS ANGELES, 78- *Concurrent Pos:* Secy dept Geology, Univ Chicago, 45-57, chmn, 57-61, assoc dean, Div Phys Sci, 48-60, chmn, Interdiv Comt Paleozool, 48-69; ed, Evolution, 53-58 & J Geol, 62-68. *Mem:* Nat Acad Sci; fel Geol Soc Am; Soc Vert Paleont (secy-treas, 48, pres, 50); Soc Study Evolution (pres, 65); Soc Syst Zool (pres, 79). *Res:* Permian reptiles and amphibians; biometry of fossils. *Mailing Add:* Dept Biol Univ Calif Los Angeles CA 90024-1606

OLSON, FERRON ALLRED, b Tooele, Utah, July 2, 21; m 44; c 5. METALLURGICAL ENGINEERING, PHYSICAL CHEMISTRY. *Educ:* Univ Utah, BS, 53, PhD(fuel technol), 56. *Prof Exp:* Chemist, Shell Develop Co Div, Shell Oil Co, 56-61; assoc res prof, 61-63, from assoc prof to prof metall, 63-68, chmn dept mining, metall & fuel eng, 66-74, PROF METALL, UNIV UTAH, 74- *Concurrent Pos:* Fulbright prof, Univ Belgrade, Bor Campus, 74-75; bd dir, Accreditation Bd, eng & technol, 75-82; consult hydrometall copper, Chilean Res Ctr, 78-; guest lectr, Univ Belgrade, 80; Fulbright distinguished prof, 80. *Mem:* Am Inst Mining, Metall & Petrol Engrs; Fulbright Alumni Asn; Sigma Xi. *Res:* Physics and chemistry of surfaces; hydrometallurgy; interfacial phenomena, hydrometallurgy, kinetics and mechanism of solid state decomposition. *Mailing Add:* 412 W C Browning Bldg Univ of Utah Salt Lake City UT 84112

OLSON, FRANK R, b Uddevalla, Sweden, May 25, 22; nat US; m 50; c 1. MATHEMATICS. *Educ:* Alfred Univ, BA, 47; Kent State Univ, MA, 50; Duke Univ, PhD(math), 54. *Prof Exp:* Instr math, Hamilton Col, 47-48, Kent State Univ, 49-50 & Duke Univ, 53-54; from asst prof to assoc prof, State Univ NY Buffalo, 54-67; PROF MATH & CHMN DEPT, STATE UNIV NY COL FREDONIA, 67- *Mem:* Am Math Soc; Math Asn Am. *Res:* Algebra, especially arithmetic properties of Bernoulli numbers and determinants. *Mailing Add:* 9445 Rte 60 Fredonia NY 14063

OLSON, GARY LEE, b Ancon, Canal Zone, Mar 15, 45; US citizen; m 69; c 1. DRUG DESIGN. *Educ:* Columbia Col, AB, 67; Stanford Univ, PhD(chem), 71. *Prof Exp:* Sr scientist, Hoffmann-La Roche Inc, 71-80, asst res group chief, 80-83, res fel, 83-85, res investr, 85-90, RES LEADER, HOFFMANN-LA ROCHE INC, NUTLEY, NJ, 90- *Concurrent Pos:* Vis exchange scientist, F Hoffmann-La Roche & Co, Ltd, Basle, Switz, 78-79; vis lectr, Jilin Univ, Changchun, China, 81. *Mem:* Am Chem Soc; Am Acad Arts & Sci; NY Acad Sci; Sigma Xi. *Res:* Design and synthesis of conformationally defined drugs as probes of receptor function and structure; drugs affecting the central nervous system; peptide mimetics. *Mailing Add:* 431 Everson Pl Westfield NJ 07090

OLSON, GEORGE GILBERT, b Omaha, Nebr, Nov 6, 24; m 47; c 4. PHYSICAL CHEMISTRY, METALLURGY. *Educ:* Univ Colo, BA, 47, PhD(phys chem), 51. *Prof Exp:* Du Pont fel chem, 50; res dir, Builders Supply Corp, Ariz, 51-53; Ariz Res Consults, Inc, 53-61; assoc dir, Colo State Univ Res Found, 61-63, dir, 63-68; vpres res, Farad Corp, 68-71; prof chem eng & vpres res, Colo State Univ, 71-83; PRES, UNIV FINANCIAL SERV CORP, 83- *Concurrent Pos:* Consult, Webb & Knapp, Inc, NY, 58-64 & Henry J Kaiser, Hawaii, 60-63; consult & dir, Malleable Iron Fitting Co, Conn, 62-69 & Bigelow Mfg Co, 63-73; dir, Chemsearch Corp, Colo, 63-68; dir, Assoc Western Univs, 71-83; pres, Colo State Univ Res Found, 71-83; dir res resources, NIH Adv Coun, 83-; adminr, Consortium Sci Comput, Princeton, NJ, 84-85. *Mem:* Fel AAAS; Inst Mining, Metall & Petrol Engrs; Sigma Xi; Am Chem Soc; Licensing Execs Soc. *Res:* Hydrometallurgy; solvent extraction; biomass processing, coatings and sealants, synthetic fuels; computer applications; research administration. *Mailing Add:* 1909 Mohawk St Ft Collins CO 80525

OLSON, GERALD ALLEN, b Iowa City, Iowa, Dec 14, 44; m 70; c 3. LABORATORY ANIMAL MEDICINE. *Educ:* Auburn Univ, DVM, 68; Univ Fla, MS, 71. *Prof Exp:* NIH fel, Univ Fla, 68-71, from asst prof to assoc prof vet med & microbiol, 71-77,; assoc prof vet med & microbiol, Univ Tenn, 77-88, assoc prof comp med, 88-89; ASSOC VET, WASH UNIV, 89- *Concurrent Pos:* Consult, Am Asn Accreditation Lab Animal Care, 74-78, Memphis VA Hosp, 77-89, St Judes Children's Res Hosp, 78-89, Schering Plough Corp, 80-89, Baptist Hosp, 85-89 & Rhodes Col, 86-89. *Honors & Awards:* Pres Award, Am Soc Microbiol, 71; Res Award, Am Asn Lab Animal Sci, 90. *Mem:* Am Vet Med Asn; Am Asn Lab Animal Sci; Am Col Lab Animal Med; Am Soc Lab Animal Practitioners. *Res:* Oral microbiology and immunology; diseases and pathology of animals; development of animal models for research. *Mailing Add:* Wash Univ 660 S Euclid Ave Box 8061 St Louis MO 63110

OLSON, GORDON LEE, b Clinton, Minn, Oct 1, 51. STELLAR WINDS, STELLAR EVOLUTION. *Educ:* Univ Minn, Minneapolis, BS, 73; Univ Wis-Madison, MS, 75, PhD(astrophys), 77. *Prof Exp:* Scientific collabr, Astrophys Inst, Free Univ, Brussels, 77-79; res assoc, Joint Inst for Lab Astrophys, Univ Colo, Boulder, 79-81; STAFF MEM, LOS ALAMOS NAT LAB, 81- *Mem:* Am Astron Soc. *Mailing Add:* Los Alamos Nat Lab PO Box 1663 MS B257 Los Alamos NM 87545

OLSON, H(ILDING) M, b Crystal Lake, Ill, Dec 14, 23; m 84; c 2. ELECTRICAL ENGINEERING. *Educ:* Northwestern Univ, BSEE, 48; Stevens Inst Technol, MS, 54; Polytech Inst Brooklyn, DEE, 59. *Prof Exp:* Engr, Radio Corp Am, 48-53; mem tech staff, Integrated Circuit Reliability, Bell Labs, 53-60, supvr magnetron develop, 60-63, supvr semiconductor device develop, 63-68, supvr traveling wave-tube & avalanche-diode develop, 68-73, mem tech staff microwave integrated circuits, 73-81, mem tech staff, 81-87; VIS RES SCIENTIST, LEHIGH UNIV, 87- *Concurrent Pos:* Teaching grad elec eng, 87- *Mem:* Inst Elec & Electronics Engrs; Sigma Xi. *Res:* Microwave electron tubes; microwave semiconductor devices and circuits employing them; computer modeling of semiconductor devices; failure analysis of integrated circuits; inter-connections for high-speed integrated circuits. *Mailing Add:* 1306 Alsace Rd Reading PA 19604

OLSON, HAROLD CECIL, dairy bacteriology; deceased, see previous edition for last biography

OLSON, HOWARD H, b Chicago, Ill May 23, 27; m 51; c 4. DAIRY SCIENCE, PHYSIOLOGY. *Educ:* Univ Wis, BS, 48; Univ Minn, MS, 50, PhD(dairy physiol), 52. *Prof Exp:* Mem res staff, Curtiss Breeding Serv, 52-54; from asst prof to prof dairy sci, 54-89, dir int agr, 79-87, EMER PROF DAIRY SCI, UNIV SOUTHERN ILL, CARBONDALE, 89- *Concurrent Pos:* Physiol teaching workshop grant, 60; Fulbright-Hays lectr, Ain Shams Univ, Cairo, 66-67; res grant, Population Dynamics, 74 & 75; Fulbright lectr, Univ Peradeniya, Sri Lanka, 81-82; consult, US Feed Grains Coun, Japan & People's Rep China, 81. *Mem:* Am Dairy Sci Asn; Am Soc Animal Sci. *Res:* Artificial insemination of dairy cattle; physiology of reproduction; dairy cattle nutrition studies on ad libitum grain feeding; development of complete feeds. *Mailing Add:* 30 Hillcrest Dr Carbondale IL 62901-2445

OLSON, J(ERRY) S, b Chicago, Ill, Mar 22, 28; m 50; c 2. ECOLOGY. *Educ:* Univ Chicago, PhB, 47, BS, 48, MS, 49, PhD(bot), 51. *Prof Exp:* Asst geol Univ Chicago, 47-49; fel statist, 51-52; asst forest ecologist, Conn Agr Exp Sta, 52-57, assoc forest ecologist, 57-58; PROF BOT, UNIV TENN, KNOXVILLE, 64-; SYSTEMS ECOLOGIST, OAK RIDGE NAT LAB, 68-, SR ECOLOGIST, ENVIRON SCI DIV, 75- *Concurrent Pos:* Geobotanist, Health Physics Div, Oak Ridge Nat Lab, 58-70, plant ecol, group leader, 64-67; lectr, Univ Tenn, Knoxville, 60-64; Oak Ridge Assoc Univs traveling lectr, 60-; Guggenheim fel, 62-64; univ lectr, Univ London, 63; sr res adviser ecol sci div, 70-73; environ analyst biomed environ res, Atomic Energy Comn/Energy Res & Develop Admin Div, 73-75. *Honors & Awards:* Ecol Soc Am Mercer Award, 58. *Mem:* Am Meteorol Soc; Bot Soc Am; AAAS; Ecol Soc Am; Am Geophys Union. *Res:* Geomorphology; Pleistocene; sedimentology; statistic pedology; sand dunes; eastern hemlock forests; biogeochemistry; radiation; productivity; regulation of ecosystems; computers and systems biology; global ecology; geochemical cycling of carbon and nitrogen; climatic change from CO2; environmental and energy policy. *Mailing Add:* Eblen Cave Rd Rte 2 Lenoir City TN 37771

OLSON, JAMES ALLEN, b Minneapolis, Minn, Oct 10, 24; m 53; c 3. NUTRITIONAL BIOCHEMISTRY. *Educ:* Gustavus Adolphus Col, BS, 46; Harvard Univ, PhD(biochem), 52. *Hon Degrees:* Dr, Univ Ghent, 88. *Prof Exp:* Fel biochem, Int Ctr Chem & Microbiol, Rome, 52-54, Ital Govt spec fel, 54; res assoc, Harvard Univ, 54-56; from asst prof to prof biochem, Col Med, Univ Fla, 56-66; prof & chmn dept, biochem, Mahidol Univ, Thailand, 66-72, grad res prof biochem & nutrit, 72-74; prof biomed sci, Fed Univ Bahia, Brazil, 74-75; prof biochem & chmn dept biochem & biophys, 75-85, DISTINGUISHED PROF, IOWA STATE UNIV, 84- *Concurrent Pos:* Consult, NSF, 62-65; mem, Comt Spec Caroten, Nat Res Coun, 64-68; guest prof, Kyoto Univ, 65; staff mem, Rockefeller Found, 66-75; ed for Asia, J Lipid Res, 66-72; consult, WHO, 69 & 75; mem, Comt Grad Educ Nutrit, Int Union Nutrit Sci, 70-72; guest prof, Univ London, 71; mem comt int nutrit prog, Nat Acad Sci, 76-80; mem US nat comt, Int Union Nutrit Sci, 77-83, chmn, 83-86; consult, NIH, 78-82; counr, Am Inst Nutrit, 83-86, pres, 86-87, mem US-Japan malnutrit panel, NIH, 83-, chmn, 88-; prog mgr, comp res grants, USDA, 84-85; Food & Agr Orgn/World Health Orgn expert on vitamin A, iron, folate & vitamin B12, 85, chmn, com functional consequences of vitamin deficiencies, Int Union Nutrit Sci, 86-90; sr Int res fel, Fogarty, NIH, 85-86; vis prof Univ London & Nat Inst Nutrit, Rome, Italy, 85-86; counr, Soc Exp Biol & Med, 89-92; assoc ed, Mod Nutrit Health Dis, 88- *Honors & Awards:* Wilton Park Award, Iowa State Univ, 77; Borden Award, Am Inst Nutrit, 89; Kullavanijaya Int Lectr, UK, 90; Wellcome vis lectr, Univ Va, 91. *Mem:* Fel AAAS; Am Chem Soc; Am Soc Biochem & Molecular Biol; Am Inst Nutrit; Soc Exp Biol & Med; Sigma Xi. *Res:* Metabolism, nutrition, molecular biology and function of vitamin A and carotenids; absorption and storage of fat soluble vitamins; evaluation of human nutritional status. *Mailing Add:* Dept Biochem & Biophys Iowa State Univ Ames IA 50011

OLSON, JAMES GORDON, b Palo Alto, Calif, Dec 16, 40; m 64; c 2. EPIDEMIOLOGY, VIROLOGY. *Educ:* Univ Calif, Santa Barbara, BA, 64; San Jose State Col, MA, 66; Univ Calif, Berkeley, MPH, 72, PhD(epidemiol), 77. *Prof Exp:* Virologist, US Navy Med Res Unit 2, 75-78, virologist, Jakarta Detachment, 78-80; virologist, Yale Arborvirus Res Unit, Sch Med, Yale Univ, 80-84; asst dir, Infectious Disease Dept & head biotechnol br, US Naval Med Res Inst, Bethesda, Md, 84-86; adj assoc prof epidemiol, Uniformed Ser Univ Health Sci, 84-86; PROG MGR, MOSQUITO VECTOR RES & RAPID EPIDEMIOL ASSESSMENT, NAT ACAD SCI, WASHINGTON, DC, 86- *Mem:* Am Soc Trop Med & Hyg; Soc Epidemiol Res; Am Pub Health Asn; Roy Soc Trop Med & Hyg; Am Soc Rickettsiology; fel Am Col Epidemiol; fel Am Acad Microbiol. *Res:* Epidemiology of arboviral diseases of man, the association of vector bionomics with incidence rates of arthropod-borne diseases and prediction of and control of epidemics. *Mailing Add:* Nat Acad Sci 2101 Constitution Ave Washington DC 20418

OLSON, JAMES PAUL, b Quincy, Mass, Feb 10, 41; m 61; c 2. CIVIL ENGINEERING. *Educ:* Tufts Univ, BSCE, 62, MS, 66; NC State Univ, PhD, 69. *Prof Exp:* Engr, Sikorsky Aircraft, United Aircraft Corp, 62-63 & Sylvania Electronic Systs, Gen Tel & Electronics Corp, 63-64; asst prof civil eng, 69-77, ASSOC PROF CIVIL ENG, UNIV VT, 77- *Mem:* Assoc mem Am Soc Civil Engrs; Am Soc Photogram. *Res:* Soil behavior; foundation engineering; engineering applications of remote sensing and photo interpretation. *Mailing Add:* Dept Civil Eng Agr Col Univ Vt 85 S Prospect St Burlington VT 05405

OLSON, JAMES ROBERT, b Whittier, Calif, July 26, 43; m 64; c 4. WOOD SCIENCE & TECHNOLOGY, FOREST PRODUCTS. *Educ:* Univ Calif, Berkeley, BS, 71, MS, 74, PhD(wood sci technol), 79. *Prof Exp:* Asst specialist timber physics, Forest Prod Lab, Univ Calif, Berkeley, 71-79; ASST PROF WOOD SCI & TECHNOL, DEPT FORESTRY, UNIV KY, 79- *Concurrent Pos:* Lectr, Dept Forestry & Resource Mgt, Univ Calif, 75. *Mem:* Forest Prod Res Soc; Soc Wood Sci & Technol; Soc Am Foresters. *Res:* Methods for better utilization of low grade hardwoods; improved procedures for hardwood drying; effect of tree improvement practices on wood quality. *Mailing Add:* Dept Forestry 00731 Univ Ky Lexington KY 40546

OLSON, JEAN L, b Winchester, Mass, Nov 20, 48. RENAL PATHOLOGY. *Educ:* Univ Rochester, MD, 74. *Prof Exp:* ASSOC PROF PATH, SCH MED, JOHNS HOPKINS UNIV, 80- *Mem:* Am Asn Path; Int Acad Path; Am Soc Nephrology; Int Soc Nephrology. *Mailing Add:* Dept Path Johns Hopkins Hosp Baltimore MD 21205

OLSON, JERRY CHIPMAN, b Los Angeles, Calif, Dec 25, 17; m 43, 68; c 5. GEOLOGY, MINERAL DEPOSITS & RESOURCES. *Educ:* Univ Calif, Los Angeles, AB, 39, MA, 47, PhD(geol), 53. *Prof Exp:* Geologist, US Geol Surv, 39-60, chief, Radioactive Mat Br, 60-69, geologist, 69-79 & 79-82; RETIRED. *Concurrent Pos:* Consult, Southern Interstate Nuclear Bd, Atlanta, Ga, 68-69. *Mem:* Mineral Soc Am; sr fel Geol Soc Am; sr mem Soc Econ Geol. *Res:* Economic geology; metamorphic rocks; geology of pegmatites, alkalic rocks, thorium, uranium and rare-earth mineral deposits and resources; geology of southwestern Colorado; study of mineral deposits and resources. *Mailing Add:* 13556 W Park Dr Magalia CA 95954

OLSON, JIMMY KARL, b Twin Falls, Idaho, Feb 18, 42; m; c 5. MEDICAL & URBAN ENTOMOLOGY. *Educ:* Univ Idaho, BS, 65; Univ Ill, Urbana, PhD(entom), 71. *Prof Exp:* from asst prof to assoc prof, 71-80, PROF ENTOM, TEX A&M UNIV, 80- *Concurrent Pos:* Tech adv, WHO/FAO/UNEP Joint Panel Experts on Environ Mgt; spec adv, Vector Biol & Control Div, WHO. *Honors & Awards:* Distinguished Entomologist Award, Am Registry Prof Entomologists, 88. *Mem:* Am Mosquito Control Asn (pres, 83); Entom Soc Am; Am Registry Prof Entomologists; Soc Vector Ecol. *Res:* Mosquito bionomics and control; role of hematophagous arthropods in the transmission of disease agents affecting man and his domestic animals; biology, ecology and control of insect pests in urban environments. *Mailing Add:* Dept Entom Tex A&M Univ Col Sta TX 77843-2475

OLSON, JOHN BENNET, b Minneapolis, Minn, Feb 13, 17; m 41; c 4. ZOOLOGY, SCIENCE EDUCATION. *Educ:* Beloit Col, BS, 38; Univ Calif, Los Angeles, MA, 41, PhD(zool), 50. *Prof Exp:* Instr biol, Brooklyn Col, 48-50; asst prof, San Jose State Col, 50-53; sr res assoc cardiol, Childrens Hosp, Los Angeles, 53-57; res fel chem, Calif Inst Technol, 57-58; chmn, Div Natural Sci, Shimer Col, 58-64; assoc prof, 64-81, EMER PROF BIOL & EDUC, PURDUE UNIV, WEST LAFAYETTE, 81- *Concurrent Pos:* Vis lectr, Univ Calif, Santa Barbara, 50, Berkeley, 51, 52 & 53; sci consult pub schs, Calif, 52-53; dir summer inst, NSF, Purdue, 64-73. *Mem:* AAAS; Nat Asn Res Sci Teaching; Nat Asn Biol Teachers; NY Acad Sci; Am Inst Biol Sci; Sigma Xi. *Res:* Marine copepods, ecology and taxonomy; congenital heart defects in mammals; protozoan physiology; amphibian development. *Mailing Add:* Dept Biol Sci Purdue Univ West Lafayette IN 47907

OLSON, JOHN BERNARD, b Chicago, Ill, Aug 20, 31; m 61; c 3. ECOLOGY, GENETICS. *Educ:* Univ Ill, Urbana, BS, 58, PhD(zool), 65. *Prof Exp:* From asst prof to prof biol, Winthrop Col, 63-72, dean, Col Arts & Sci, 72-75, prof, 75-90, EMER PROF BIOL, WINTHROP COL, 90- *Mem:* Cooper Ornith Soc. *Res:* Bioenergetic requirements during the annual cycle of migratory and nonmigratory birds. *Mailing Add:* Dept Biol Winthrop Col Rock Hill SC 29733

OLSON, JOHN MELVIN, b Niagara Falls, NY, Sept 18, 29; m 53; c 3. PHOTOBIOLOGY. *Educ:* Wesleyan Univ, BA, 51; Univ Pa, PhD(biophys), 57. *Prof Exp:* USPHS res fel, Biophys Lab, Univ Utrecht & Univ Leiden, 57-58; instr physics & biochem, Brandeis Univ, 58-59, asst prof, 59-61; from asst biophysicist to assoc biophysicist, 61-65, BIOPHYSICIST, BROOKHAVEN NAT LAB, 65- *Concurrent Pos:* USPHS res fel, Lab Chem Biodynamics, Lawrence Berkeley Lab, 70-71; lectr, Univ Calif, Berkeley, 71; adj prof, State Univ NY Stony Brook, 72; Lady Davis vis scholar, Israel Inst Technol, Haifa, 81. *Mem:* AAAS; Biophys Soc; Am Soc Biol Chem; Am Soc

Photobiol; Int Solar Energy Soc; Sigma Xi. *Res:* Energy conversion in photosynthesis; structure and function of photosynthetic membranes in green bacteria; photoproduction of hydrogen; evolution of photosynthesis. *Mailing Add:* Dept 5SU Bldg 002 IBM Corp Boulder CO 80301-9191

OLSON, JOHN RICHARD, b Ferryville, Wis, Sept 14, 32. PHYSICS. *Educ:* Luther Col, BA, 54; Iowa State Univ, PhD(physics), 63. *Prof Exp:* Instr physics, Iowa State Univ, 63; mem tech staff, Bell Tel Labs, 63-65; lectr physics, Bryn Mawr Col, 65-66, asst prof, 66-72; MEM FAC, DEPT PHYSICS, UNIV DENVER, 72- *Concurrent Pos:* Vis asst prof, Univ Denver, 70-71. *Mem:* Am Asn Physics Teachers; Sigma Xi. *Res:* Atomic and molecular physics, especially molecular energy transfer; quadrupole mass spectrometry; ultra-high vacuum systems; balloon-borne experimentation techniques; sound propagation in deep water. *Mailing Add:* Dept of Physics Univ of Denver Denver CO 80208

OLSON, JOHN S, b Evanston, Ill, May 21, 46; m 68; c 3. HEME PROTEINS, OXYGEN TRANSPORT. *Educ:* Univ Ill, Champaign-Urbana, BS, 68; Cornell Univ, PhD(biochem), 72. *Prof Exp:* Fel biochem, Univ Mich, 72-73; from asst prof to assoc prof, 73-83, PROF BIOCHEM & CHMN DEPT, RICE UNIV, 84- *Mem:* Am Soc Biol Chemists; Am Chem Soc; Biophys Soc. *Res:* Biophysical chemistry, rapid reaction techniques, fundamental properties of hemoglobins, myoglobins and model heme compounds; gas transport by red blood cells and capillaries; biosynthesis of hemoglobin; anion transport. *Mailing Add:* Dept Biochem Rice Univ PO Box 1892 Houston TX 77251-1892

OLSON, JOHN VICTOR, b Kibbie, Mich, June 24, 13; m 40; c 1. DENTISTRY. *Educ:* Univ Mich, DDS, 36, MS, 38. *Prof Exp:* Asst prof prosthetics, Sch Dent, St Louis Univ, 47-49, assoc prof, 49-50, dir postgrad dent educ, 48-50; dean, Dent Sch, Univ Tex Health Sci Ctr, San Antonio, 69-72; dean, 52-82. *Concurrent Pos:* Consult, Vet Admin Hosp, Houston, Tex, 52, Univ Tex Med Br & M D Anderson Hosp & Tumor Inst, 52- *Mem:* AAAS. *Res:* Prosthetic dentistry; dental education. *Mailing Add:* 2725 Pemberton Houston TX 77005

OLSON, JON H, b Akron, Ohio, Jan 31, 34; m 57, 83; c 3. CHEMICAL ENGINEERING. *Educ:* Princeton Univ, BS, 55; Yale Univ, DEng(chem eng), 60. *Prof Exp:* Engr, Eng Res Lab, E I du Pont de Nemours & Co, Del, 59-62 & Radiation Physics Lab, 62-63; from asst prof to assoc prof chem eng, 63-68, assoc dean, 77-85, PROF CHEM ENG, UNIV DEL, 68- *Mem:* AAAS; Am Inst Chem Engrs; Am Chem Soc. *Res:* Chemical reactor engineering; process control. *Mailing Add:* Dept of Chem Eng Univ of Del Newark DE 19711

OLSON, KARL WILLIAM, b Canton, Ohio, Mar 27, 36; m 59; c 3. ELECTRICAL ENGINEERING. *Educ:* Ohio State Univ, BSc & MSc, 59, PhD(elec eng), 65. *Prof Exp:* Res assoc, 60-64, res assoc, Commun & Control Systs Lab, 64-65, asst supvr, 65-67, ASST PROF ELEC ENG, UNIV & ASSOC SUPVR COMMUN & CONTROL SYSTS LAB, OHIO STATE UNIV, 67- *Mem:* Inst Elec & Electronics Engrs. *Res:* Automatic highways; designing and testing automatic lateral and longitudinal control systems for automobiles. *Mailing Add:* Dept Elec Eng Ohio State Univ 15th Ave & N High St Columbus OH 43210

OLSON, KENNETH B, b Seattle, Wash, Jan 21, 08; m 37; c 2. MEDICINE, ONCOLOGY. *Educ:* Univ Wash, BS, 29; Harvard Med Sch, MD, 33. *Prof Exp:* Resident path, Boston City Hosp, Mass, 33-34; intern & asst resident & resident surgeon, Presby Hosp, New York, 34-40; instr, Col Physicians & Surgeons, Columbia Univ, 40-41; dir, Firland Sanatorium, Seattle, Wash, 43-45; from instr to prof med, Albany Med Col, 50-71, dir div oncol, 51-72, prof med & actg head med oncol, Div Hematol, J Hillis Miller Health Ctr, Univ Fla, 76, EMER PROF MED, ALBANY MED COL, 72-; EMER PROF MED ONCOL, DIV HEMATOL, J HILLIS MILLER HEALTH CTR, UNIV FLA, 76- *Concurrent Pos:* Mem cancer clin invest rev comt, Nat Cancer Inst, 68-72, mem breast cancer task force, 71-, chief, Diag Br, Div Cancer Biol & Diag, 72-73; consult, 73-81; consult, Fla Cancer Coun, Halifax Med Ctr Hosp; chmn, Am Asn Cancer Educ, 58. *Mem:* Am Asn Cancer Res; AMA; Soc Surg Oncol; Am Soc Clin Oncol (pres, 71-72); fel Am Col Physicians; Am Asn Cancer Educ. *Res:* Lung and breast cancer; liver function; chemotherapy of cancer; biological effects of irradiation; estrogen receptors; blood clotting. *Mailing Add:* 810 Oakview Dr New Smyrna Beach FL 32069

OLSON, LEE CHARLES, b Austin, Minn, June 2, 36; m 61; c 4. PLANT BIOCHEMISTRY, PLANT PHYSIOLOGY. *Educ:* SDak State Univ, BS, 58; Univ Wis, MS, 62, PhD(biochem), 64. *Prof Exp:* Asst prof plant path & physiol, Univ Minn, St Paul, 65-67, agron & plant genetics, 67-70; MEM FAC, DEPT BIOL, CHRISTOPHER NEWPORT COL, 70- *Mem:* Am Chem Soc; Sigma Xi; Am Soc Plant Physiologists; AAAS; Am Asn Univ Profs. *Res:* Amino acid metabolism; plant cell culture. *Mailing Add:* Dept Biol Christopher Newport Col Newport News VA 23606

OLSON, LEONARD CARL, b Marietta, Ohio, July 17, 45; m 69. LABORATORY ANIMAL MEDICINE, PRIMATE MEDICINE. *Educ:* St Olaf Col, BA, 67; Purdue Univ, DVM, 75. *Prof Exp:* Fel, Lab Animal Med, Med Sch, Univ Mich, 77-79; HEAD ANIMAL SCI DEPT PRIMATE MED, ORE REGIONAL PRIMATE RES CTR, 79- *Mem:* Am Col Lab Animal Med; Am Asn Lab Animal Sci; Asn Primate Vet Clinicians; Am Vet Med Asn; Am Soc Vet Clin Path. *Res:* Documenting or investigating diseases of laboratory animals with an emphasis on diseases of non-human primates; establishing normative data for a wide variety of non-human primate species. *Mailing Add:* Dept Comp Med Brody Bldg GE01 E Carolina Univ Sch Med Greenville NC 27834

OLSON, LEROY DAVID, b East Chain, Minn, July 22, 29; m 56; c 1. VETERINARY PATHOLOGY, VETERINARY VIROLOGY. *Educ:* Univ Minn, BS, 54, DVM, 58; Purdue Univ, MS, 62, PhD(vet path), 65. *Prof Exp:* Instr vet path, Purdue Univ, 60-65; assoc prof, 65-81, PROF VET PATH, UNIV MO-COLUMBIA, 81- *Mem:* AAAS; Am Vet Med Asn; Am Soc Microbiol; Am Soc Vet Parasitol. *Res:* Immunopathology, especially experimental allergic encephalomyelitis; avian cancer viruses; viral enteric diseases of swine; Pasteurella multocida infections in turkeys. *Mailing Add:* Sch Vet Med Univ Mo Columbia MO 65211

OLSON, LEROY JUSTIN, b Fargo, NDak, May 28, 26; m 53; c 2. MICROBIOLOGY. *Educ:* Concordia Col, BA, 50; Kans State Univ, MS, 53; Univ Tex, Galveston, PhD(parasitol), 57. *Prof Exp:* Instr biol, Concordia Col, 50-51; asst vet parasitol & zool, Kans State Univ, 51-53; asst prev med & pub health, 53, instr bact & parasitol, 54-56, from asst prof to assoc prof microbiol, 57-66, PROF MICROBIOL, UNIV TEX MED BR, GALVESTON, 66- *Mem:* Am Soc Parasitol; Am Soc Microbiol. *Res:* Host-parasite relationships; immunology; gut allergy and pathology. *Mailing Add:* Dept Microbiol Univ Tex Med Br Galveston TX 77551

OLSON, LLOYD CLARENCE, b Spokane, Wash, Jan 30, 35; m 58; c 3. MICROBIOLOGY, PEDIATRICS. *Educ:* Reed Col, AB, 57; Harvard Med Sch, MD, 61. *Prof Exp:* From intern to resident pediat, Univ Rochester, 61-63, resident, 63-64; virologist, Walter Reed Army Inst Res, 64-67; vis assoc prof microbiol, Mahidol Univ, Thailand, 67-70, vis lectr, 68-70, vis prof microbiol, 70-73; assoc prof microbiol, Sch Med, Ind Univ, 73-76; PROF PEDIAT, SCH MED, UNIV MO, 76-, CHMN DEPT, 82-; MEM STAFF INFECTIOUS DIS, CHILDREN'S MERCY HOSP, 76- *Concurrent Pos:* Virologist, SEATO Med Res Lab, Thailand, 67-70; mem staff, Rockefeller Found, Thailand, 70-73. *Mem:* AAAS; Soc Pediat Res; Soc Exp Biol Med; Am Soc Microbiol; Infectious Dis Soc Am. *Res:* Clinical and epidemiologic aspects of virus diseases; host defense mechanisms to virus infections. *Mailing Add:* The Children's Mercy Hosp Kansas City MO 64108-9898

OLSON, LYNNE E, LUNG STRUCTURE VENTILATION & PROFUSION. *Educ:* Mich State Univ, PhD(physiol), 81. *Prof Exp:* ASST PROF PHYSIOL, OHIO STATE UNIV, 83- *Res:* Pulmonary physiology. *Mailing Add:* Vet Phys-Pharm 353 Sisson Ohio State Univ Columbus OH 43210-1092

OLSON, MAGNUS, b South Fron, Norway, June 29, 09; US citizen; m 38; c 2. ZOOLOGY. *Educ:* St Olaf Col, AB, 32; Univ Minn, AM, 34, PhD(histol), 36. *Prof Exp:* Asst zool, 32-36, asst, Comt Educ Res, 36-37, instr educ & zool, 37-38, from instr to prof zool, 38-77, chmn dept, 66-77, EMER PROF ZOOL, UNIV MINN, MINNEAPOLIS, 77- *Mem:* AAAS; assoc Am Soc Zool; Sigma Xi. *Res:* Histology of invertebrate muscle; ovogenesis in Mammalia; myeloid metaplasia in mammalian thymus; wound healing in rabbits. *Mailing Add:* 1666 Coffman Apt 122 St Paul MN 55108

OLSON, MARK OBED JEROME, b Clarkfield, Minn, Aug 20, 40; m 66; c 3. BIOCHEMISTRY, MOLECULAR BIOLOGY. *Educ:* St Olaf Col, BA, 62; Univ Minn, Minneapolis, PhD(biochem), 67. *Prof Exp:* From asst prof to assoc prof pharmacol, Baylor Col Med, 69-79; assoc prof, 79-83, PROF BIOCHEM, UNIV MISS MED CTR, 83-, CHMN, DEPT BIOCHEM, 90- *Concurrent Pos:* Med Res Coun Can fel, Univ Alta, 67-69; Am Cancer Soc grant pharmacol, Baylor Col Med, 69-70; NSF grant, 73-74; Nat Cancer Inst proj grant, 77-79; vis scientist, Nat Inst Arthritis, Metab & Digestive Dis, NIH, 78; NIH grant, 80-; mem biochem study sect, NIH, 82, 84 & 85, Biomed Sci Study Sect, NIH, 88- *Mem:* AAAS; Am Asn Cancer Res; Am Chem Soc; Am Soc Cell Biol; Am Soc Biol Chemists; Protein Soc. *Res:* Functional and evolutionary aspects of protein structure; proteolytic enzymes; structure and role of nuclear proteins in genetic regulation; structure and organization of the nucleus; role of nonribosomal nucleolar proteins in ribosome biogenesis; protein chemistry. *Mailing Add:* Dept Biochem Univ Miss Med Ctr 2500 N State St Jackson MS 39216

OLSON, MAYNARD VICTOR, b Washington, DC, Oct 2, 43; m 68; c 2. MOLECULAR GENETICS. *Educ:* Calif Inst Technol, BS, 65; Stanford Univ, PhD(chem), 70. *Prof Exp:* Asst prof chem, Dartmouth Col, 70-76; res assoc, Univ Wash, 76-79; ASST PROF GENETICS, WASHINGTON UNIV, 79- *Mem:* AAAS. *Res:* Structure and function of eukaryotic genes. *Mailing Add:* Dept Genetics Washington Univ Sch Med 660 S Euclid Ave St Louis MO 63110

OLSON, MELVIN MARTIN, b Bangor, Wis, Aug 30, 15; m 41; c 3. ORGANIC CHEMISTRY. *Educ:* Wis State Teachers Col, La Crosse, BS, 38; Univ Wis, PhM, 39, PhD(chem), 50. *Prof Exp:* Instr high sch, Wis, 39-41; instr high sch & Maquoketa Jr Col, Iowa, 41-43; asst chem, Univ Wis, 46-50, prof, Exten Div, 50-51; res chemist, Paint Div, Pittsburgh Plate Glass Co, 51-54; res chemist, Tape Div, Minn Mining & Mfg Co, 54-59, res specialist, 56-69, sr res specialist, Indust Tape Div, 3MCO, 69-81, div scientist, 81-82; RETIRED. *Mem:* Am Chem Soc. *Res:* Modification of paint vehicles with silicon compounds; synthesis of silicon compounds; formulation of organic structural adhesives; water dispersible pressure-sensitive adhesives, water-dispersible hot-melt adhesives; ethylene oxide detection; Water dispersible release coatings. *Mailing Add:* 2131 Grand View Blvd Onalaska WI 54650

OLSON, MERLE STRATTE, b Northfield, Minn, Aug 22, 40; m 64. BIOCHEMISTRY. *Educ:* St Olaf Col, BA, 62; Univ Minn, Minneapolis, PhD(biochem), 66. *Prof Exp:* Asst biochem, Univ Minn, Minneapolis, 62-66; assoc prof biochem, Col Med, Univ Ariz, 68-76; PROF BIOCHEM, UNIV TEX, HEALTH SCI CTR SAN ANTONIO, 76-, CHMN DEPT. *Concurrent Pos:* Johnson Res Found res fel biochem, Univ Pa, 66-68. *Mem:* Am Soc Biol Chem; Am Soc Cell Biol. *Res:* Regulation of nultienzyme complexes in complex metabolic systems; complement receptors on cellular membranes. *Mailing Add:* Dept of Biochem Univ Tex Health Sci Ctr 7703 Floyd Curl Dr San Antonio TX 78284

OLSON, NORMAN FREDRICK, b Edmund, Wis, Feb 8, 31; m 57; c 2. FOOD TECHNOLOGY. *Educ:* Univ Wis, BS, 53, MS, 57, PhD(dairy & food industs), 59. *Prof Exp:* From asst prof to assoc prof, 58-71, PROF FOOD SCI, UNIV WIS-MADISON, 71- *Honors & Awards:* Pfizer Award in Cheese Res; Dairy Res Found Award; Macy Award Technol Transfer; Borden Award. *Mem:* Am Dairy Sci Asn (pres, 84-85); Inst Food Technol; Int Dairy Fedn. *Res:* Chemistry, microbiology and enzymology of cheese manufacture and maturation; mechanization of cheese manufacture; rheology of cheese; bacteriological problems of natural cheese; food fermentations, particularly cheese technology; technology of encapsulated and immobilized enzymes. *Mailing Add:* Dept Food Sci Univ Wis Madison WI 53706

OLSON, NORMAN O, b Regan, NDak, June 1, 14; m 40; c 3. VETERINARY MEDICINE. *Educ:* Wash State Univ, DVM, 38. *Prof Exp:* Vet disease control res, USDA, 38-44, res vet, Food & Drug Admin, 44-48; prof animal path, WVa Univ, 48-79; RETIRED. *Honors & Awards:* Am Feed Mfrs Award, Am Vet Med Asn, 72; Upjohn Achievement Award, Am Asn Avian Path, 79. *Mem:* Am Vet Med Asn; Poultry Sci Asn; Am Asn Avian Path; Am Health Asn. *Res:* Virus and bacterial diseases of farm animals; mycoplasma and virology. *Mailing Add:* 1301 Heritage Pl Morgantown WV 26507

OLSON, OSCAR EDWARD, b Sioux Falls, SDak, Jan 19, 14; m 43; c 2. AGRICULTURAL BIOCHEMISTRY. *Educ:* SDak State Col, BS, 36, MS, 37; Univ Wis, PhD(biochem), 48. *Prof Exp:* Sta analyst, Agr Exp Sta, SDak State Col, 37-42; chemist, Fruit & Veg Prod Lab, USDA, 48-49; biochemist, Div Labs, Mich State Dept Health, 49-51; head, Exp Sta Biochem, 51-73, dean, grad div, 58-65, prof, 75-79, EMER PROF CHEM, SDAK STATE UNIV, 79- *Concurrent Pos:* Vis prof, Inst Enzyme Res, Univ Wis-Madison, 62-63; vis scientist, US Plant, Soil & Nutrit Lab, NY, 73-74. *Mem:* Poultry Sci Asn; Am Chem Soc; Am Inst Nutrit; Am Soc Biol Chem; Sigma Xi. *Res:* Selenium and nitrate poisoning. *Mailing Add:* Box 2170 Univ Sta Brookings SD 57007

OLSON, PETER LEE, b Lincoln, Nebr, Aug 8, 50; m 77. GEOPHYSICS, GEOPHYSICAL FLUID DYNAMICS. *Educ:* Univ Colo, BA, 72; Univ Calif, MA, 74, PhD(geophys), 77. *Prof Exp:* ASST PROF GEOPHYSICS, JOHNS HOPKINS UNIV, 77- *Mem:* Am Geophys Union. *Res:* Dynamics of the earth's interior; geomagnetism. *Mailing Add:* Dept Earth & Planetary Sci Johns Hopkins Univ 3400 N Charles St Baltimore MD 21218

OLSON, RANDALL J, b Glendale, Calif, Apr 12, 47; m 70; c 5. BIOMATERIALS, CORNEAL SURGERY. *Educ:* Univ Utah, BA, 70, MD, 73. *Prof Exp:* Intern med, Mary Imogene Bassett Hosp, 73-74; resident ophthal, Jules Stein Eye Inst, Univ CAlif, Los Angeles, 74-77; fel cornea res, Univ Fla, Gainesville, 77; asst prof ophthal, La State Univ Eye Ctr, New Orleans, 77-79; assoc prof, 79-81, PROF OPHTHAL, UNIV UTAH, 81-, CHMN DEPT, 79- *Concurrent Pos:* Mem bd, Inmedica Develop Co, 82-; med dir ophthal, King Khaled Eye Specialist Hosp, Saudi Arabia, 85-86. *Mem:* AMA; Am Acad Ophthal; Asn Res Vision & Ophthal; Asn Univ Prof Ophthal (secy, 82-84). *Res:* Biomaterials and the eye, especially the interaction of intraocular lenses and other newer materials; control of astigmatism in corneal transplantation with newer methods of performing the procedure; new approaches to enhance corneal refractive surgery. *Mailing Add:* Dept Ophthal Univ Utah Salt Lake City UT 84132

OLSON, REUBEN MAGNUS, b Ferryville, Wis, Feb 12, 19; m 43; c 4. FLUID MECHANICS, HEAT TRANSFER. *Educ:* Univ Minn, Minneapolis, BME, 39, MSME, 41, PhD(mech eng), 64. *Prof Exp:* Res asst mech eng, Eng Exp Sta, Univ Minn, Minneapolis, 39-40, instr, 40-42; ord engr, US Naval Ord Lab, Md, 42-50; res assoc hydraul, St Anthony Falls Hydraul Lab, Univ Minn, Minneapolis, 50-55, lectr hydromech, Univ, 55-62 & 63-64; from assoc prof to prof fluid mech, 64-84, chmn dept civil eng, 75-84, EMER PROF FLUID MECH, OHIO UNIV, 85- *Concurrent Pos:* Vis prof, Tech Univ Norway, 70-71; NSF Sci Fac Fel, 62-63. *Mem:* Am Soc Mech Engrs; Int Asn Hydraul Res. *Res:* Revetment hydraulics; cavitation testing; flow development in ducts; mass transfer as related to heat transfer; shear layers; vortex flows. *Mailing Add:* 2800 E Minnehaha Pkwy Minneapolis MN 55406

OLSON, RICHARD DAVID, b Reading, Pa, Oct 10, 44. NEUROSCIENCES. *Educ:* Univ Redlands, BA, 66; St Louis Univ, MS, 68, PhD(psychol), 70. *Prof Exp:* Asst prof psychol, La State Univ, 70-74, assoc prof anat, Sch Med, 76-77; assoc prof, 74-80, chmn psychol, 74-81, PROF PSYCHOL, UNIV NEW ORLEANS, 80-, ASSOC DEAN GRAD SCH, 82- *Concurrent Pos:* Prin investr, New Orleans Parent-Child Develop Ctr, 75-79. *Mem:* Am Psychol Asn; Am Statist Asn; Animal Behav Soc; Soc Neurosci; Sigma Xi. *Res:* Studying the effects of neuropeptides, especially the endogenous opiates, on behavior using animal models to evaluate the clinical potential of effective peptides. *Mailing Add:* Dept Psychol Univ New Orleans New Orleans LA 70148

OLSON, RICHARD HUBBELL, b Meriden, Conn, Nov 25, 28; m 54; c 4. ECONOMIC GEOLOGY. *Educ:* Tufts Univ, BS, 50; Univ Utah, PhD(geol), 60. *Prof Exp:* Field geologist, US AEC, 51-53; geologist & draftsman, Gulf Oil Corp, 55-56; econ geologist, Union Carbide Corp, 57-59; asst prof econ geol, Univ Nev, 59-64; econ geologist, Chas Pfizer & Co, Inc, 65-69; dir, Mineral Resources Dept, Monsanto Co, 69-71; CONSULT GEOLOGIST, 72- *Honors & Awards:* Hal Williams Hardinge Award, Am Inst Mining Metall & Petrol Engrs. *Mem:* Geol Soc Am; Am Inst Mining, Metall & Petrol Engrs; Soc Econ Geol; Explorers Club. *Res:* Industrial mineral deposits; economic geology of mineral deposits. *Mailing Add:* 1536 Cole Blvd No 320 Golden CO 80401-3405

OLSON, RICHARD LOUIS, b El Paso, Tex, Aug 14, 32; m 55; c 2. BOTANY, GENETICS. *Educ:* Univ Utah, BS, 54, MS, 55; Univ Calif, Berkeley, PhD(genetics), 64. *Prof Exp:* Preliminary design engr, AiResearch Mfg Co Div, Garrett Corp, Calif, 62-64; mgr biosci, Boeing Co, 64-74, gen mgr, Best Prog, Resources Conserv Co Div, 74-77, technol mgr, large space systs, 78-91, MAN-SYSTS MGR, SPACE STA FREEDOM, BOEING AEROSPACE & ELECTRONICS CO, 91- *Concurrent Pos:* Chmn scientist comt, Northwest Pollution Control Asn; chmn, life sci & systs tech comt, Am Inst Aeronaut & Astronaut, past gen chmn, Int Conf Environ Systs. *Mem:* Assoc fe Am Inst Aeronaut & Astronaut. *Res:* Technology advancement for large space systems, life support systems and sludge drying equipment development; waste and water management investigations related to manned space flight; toxicology; space craft sterilization and exobiology; urban development; urban and agricultural pollution abatement. *Mailing Add:* Boeing Aerospace & Electronics Co PO Box 240002 MS JR01 Huntsville AL 35824-6402

OLSON, ROBERT ELDON, b Thief River Falls, Minn, May 4, 40; m 65; c 2. FISH DISEASE RESEARCH. *Educ:* Concordia Col, Moorhead, Minn, BA, 62; Mont State Univ, MS, 64, PhD(zool), 68. *Prof Exp:* Res assoc, Ore State Univ, 68-75, asst prof zool, 75-79, asst prof, 79-81, ASSOC PROF FISHERIES & WILDLIFE, ORE STATE UNIV, 81- , ASSOC DIR EDUC, HATFIELD MARINE SCI CTR, 88- *Mem:* Am Soc Parasitol; Am Fisheries Soc; Am Micros Soc; Wildlife Dis Asn; Soc Protozoologists. *Res:* Ecology of protozoans and helminths parasitizing marine fishes and invertebrates. *Mailing Add:* Hatfield Marine Sci Ctr Ore State Univ Newport OR 97365-5296

OLSON, ROBERT EUGENE, b Minneapolis, Minn, Jan 23, 19; m 44; c 5. BIOCHEMISTRY, MEDICINE. *Educ:* Gustavus Adolphus Col, AB, 38; St Louis Univ, PhD(biochem), 44; MD, 51; Am Bd Nutrit, dipl. *Prof Exp:* Asst biochem, Sch Med, St Louis Univ, 38-43; instr biochem & nutrit, Sch Pub Health, Harvard Univ, 46-47, estab investr, Am Heart Asn, 51-52; prof biochem & nutrit, Grad Sch Pub Health & lectr med, Sch Med, Univ Pittsburgh, 52-65; assoc prof med, 65-72, ALICE A DOISY PROF BIOCHEM & CHMN DEPT, SCH MED, ST LOUIS UNIV, 65-, PROF MED, 72- *Concurrent Pos:* Harvard Med Sch, Nutrit Found fel, 47-49, Am Heart Asn fel, 49-51; Guggenheim & Fulbright fel, Oxford Univ, 61-62 & Univ Freiburg, 70-71; house physician, Peter Bent Brigham Hosp, Boston, Mass, 51-52; dir nutrit clin, Falk Clin, Univ Pittsburg Med Ctr, 53-65; consult metab & nutrit study sect, Res Grants Div, USPHS, 53-57, 58-59, biochem training comt, 59-63, training comt, Nat Heart Inst, 64-68; clin assoc, St Margaret Mem Hosp, 54-60, consult, 55-65; mem panel biochem & nutrit, Comt Growth, Nat Res Coun, 54-56; sci adv comt, Nat Vitamin Found, 55-58; from vpres to pres, Am Bd Nutrit, 60-63; consult & dir metab unit, Presby Hosp, 60-65; dir, Anemia Malnutrition Res Ctr, Thailand, 65-77; Atwater Lectureship, USDA, 78. *Honors & Awards:* McCollum Award, Am Soc Clin Nutrit, 65; Goldberger Award, AMA, 74; Noyes Lectureship, Univ Ill, Urbana, 80. *Mem:* Asn Chmn Biochem Med Schs (pres, 79-80); Am Chem Soc; Am Soc Biol Chem; Am Inst Nutrit (pres, 80-81); fel Am Pub Health Asn. *Res:* Cardiac metabolism; role of the fat-soluble vitamins; experimental and clinical nutrition. *Mailing Add:* Dr Olson Lab SUNY Z-8278 Stony Brook NY 11794-8278

OLSON, ROBERT LEROY, b Portland, Ore, Apr 24, 32; m 57; c 2. PHYSIOLOGY, MEDICINE. *Educ:* Ore State Univ, BS, 54; Univ Ore, MD, 58. *Prof Exp:* Resident, 62-64, from instr to assoc prof, 64-74, PROF DERMAT, SCH MED, UNIV OKLA, 74-; CLIN PROF, PHYSICIANS PROF BLDG, 77- *Mem:* Soc Invest Dermat; Am Acad Dermat; Am Dermat Asn. *Res:* Cutaneous physiology; ultraviolet light physiology; electron microscopy. *Mailing Add:* Physicians Prof Bldg 3400 NW Expressway Suite 710 Oklahoma OK 73112

OLSON, ROY E, b Aberdeen, Wash, May 24, 29; m 62; c 3. SOLID STATE PHYSICS. *Educ:* Univ Calif, Berkeley, PhD(physics), 58. *Prof Exp:* PROF PHYSICS, CALIF STATE UNIV, NORTHRIDGE, 62- *Mailing Add:* Dept Physics & Astron Calif State Univ 18111 Nordhoff St Northridge CA 91330

OLSON, ROY EDWIN, b Richmond, Ind, Sept 13, 31; m 85; c 3. GEOTECHNICAL ENGINEERING, CIVIL ENGINEERING. *Educ:* Univ Minn, BS, 53, MS, 55; Univ Ill, Urbana, PhD(civil eng), 60. *Prof Exp:* From instr to prof, Univ Ill, Urbana, 58-70; PROF CIVIL ENG, UNIV TEX, AUSTIN, 70- *Concurrent Pos:* NSF res grants; Air Force Weapons Lab grant; Off Naval Res res grant; mem hwy res bd, Nat Acad Sci-Nat Res Coun. *Honors & Awards:* Huber Prize, Am Soc Civil Engrs, 72; Hogentogler Award, Am Soc Testing & Mat, 73 & 87; Norman Medal, Am Soc Civil Engrs, 75; Croes Medal, Am Soc Civil Engrs, 84. *Mem:* Am Soc Civil Engrs; Am Soc Testing & Mat; US Nat Soc Soil Mech & Found Eng (chmn, 73-74). *Res:* Foundation engineering; physico-chemical properties of clays; numerical analysis of consolidation of clays; apparatus development; marine foundation engineering; research in consolidation of soft clays, water flow around radioactive waste disposal sites, physicochemical phenomena and soil mechanics. *Mailing Add:* Dept Civil Eng Univ Tex Austin TX 78712-1076

OLSON, STANLEY WILLIAM, b Chicago, Ill, Feb 10, 14; m 36; c 3. MEDICINE. *Educ:* Wheaton Col, Ill, BS, 34; Univ Ill, MD, 38; Univ Minn, MS, 43; Am Bd Internal Med, dipl. *Hon Degrees:* LLD, Wheaton Col, Ill, 53; DSc, Univ Akron, 79, NE Ohio Univ, 85, Morehouse Sch Med, 88. *Prof Exp:* Asst to dir, Mayo Found, Univ Minn, 46-47, asst dir, 47-50; prof med & dean col med, Univ Ill, 50-53 & Baylor Univ, 53-66; dir regional med progs serv, Health Serv & Ment Health Admin, 68-70; pres, Southwest Found Res & Educ, 70-73; provost, 73-79, EMER PROVOST & PROF MED, COL MED, NORTHEASTERN OHIO UNIV, 79- *Concurrent Pos:* Consult, State Univ NY, 49 & Comn, 54; mem med adv panel, US Off Voc Rehab Admin, 60-65; mem comt med sch-vet admin rels, 62-66; mem nat adv coun health res facil, NIH, 63-67; mem rev panel construct schs med, USPHS, 64-66; mem, Nat Adv Comn Health Manpower, 66; prof med, Vanderbilt Univ, 67-68; clin prof med, Meharry Med Col, 67-68; dir, Tenn Mid-South Regional Med Prof, 67-68; consult med educ, 79-; vpres Acad Affairs, Morehouse Sch Med, 83-85, dean, 85-87. *Mem:* Am Asn Med Cols (vpres, 60-61); fel Am Col Physicians. *Res:* Medical administration; medical education for national defense. *Mailing Add:* 5901 Church View Dr No 25 Rockford IL 61107

OLSON, STEVEN T, b Aug 14, 48. BLOOD COAGULATION, ENZYME KINETICS. *Educ:* Univ Mich, Ann Arbor, PhD(biochem), 79. *Prof Exp:* STAFF INVESTR, HENRY FORD HOSP, 83. *Mem:* Am Soc Biochem & Molecular Biol. *Res:* Mechanisms of regulation of blood coagulation pathways. *Mailing Add:* Div Biochem Res Henry Ford Hosp 2799 W Grand Blvd Detroit MI 48202

OLSON, STORRS LOVEJOY, b Chicago, Ill, Apr 3, 44; m 81; c 2. ORNITHOLOGY, PALEONTOLOGY. *Educ:* Fla State Univ, BS, 66, MS, 68; Johns Hopkins Univ, ScD(biol), 72. *Prof Exp:* CUR ORNITH, SMITHSONIAN INST, 72- *Honors & Awards:* A B Howell Award, Cooper Ornith Soc, 72; Ernest P Edwards Award, Wilson Ornith Soc, 73. *Mem:* Am Ornithologists Union; Soc Hist of Nat Hist; Biol Soc Wash; Ornith Soc NZ; Sigma Xi. *Res:* Higher systematics, evolution and paleontology of birds. *Mailing Add:* Smithsonian Inst NHB Stop 116 Washington DC 20560

OLSON, WALTER HAROLD, b New Haven, Conn, May 12, 45; m 70; c 4. BIOMEDICAL ENGINEERING. *Educ:* Pa State Univ, BS, 67; Univ Mich, MS, 68, PhD(bioeng), 73. *Prof Exp:* Teaching fel elec eng, Univ Mich, 69-73; asst prof elec eng, Univ Ill, Urbana, 73-77; assoc prof health sci technol, Mass Inst Technol, 77-80; SR RES SCIENTIST, MEDTRONIC, INC, 80- *Mem:* Inst Elec & Electronics Engrs; Asn Advan Med Instrumentation; Bakken Soc. *Res:* Biomedical instrumentation including transducers, electronics and digital signal processing; computer analysis of ambulatory ECG; cardiac pacemakers, sensing, implantable defibrillator detection. *Mailing Add:* Medtronic Inc 3055 Old Hwy 8 PO Box 1453 Minneapolis MN 55440

OLSON, WALTER T, b Royal Oak, Mich, July 4, 17; m 43; c 1. CHEMISTRY. *Educ:* DePauw Univ, AB, 39; Case Inst Technol MS, 40, PhD(chem), 42. *Prof Exp:* Instr chem, Case Inst Technol, 40-42; res chemist, Nat Adv Comt Aeronaut, Lewis Res Ctr, NASA, 42-45, chief, Combustion Br, 45-50, Fuels & Combustion Res Div, 50-58, Propulsion Chem Div, NASA-Lewis, 58-60, Chem & Energy Conversion Div, 60-62, asst dir pub affairs, 62-72, dir tech utilization & pub affairs, 72-81; DIR COMMUNITY CAPITAL INVEST STRATEGY, GREATER CLEVELAND GROWTH ASN, 81- *Concurrent Pos:* Consult, Dept Defense; chmn adv com, Fenn Col Eng, Cleveland State Univ. *Mem:* Fel AAAS; fel Am Inst Aeronaut & Astronaut; Am Chem Soc; Sigma Xi; Combustion Inst. *Res:* High speed combustion for aircraft engines; high energy fuels; organic synthesis; local anesthetics; rocket engines and propulsion and power for space flight; energy conversion systems. *Mailing Add:* 18960 Coffinberry Blvd Cleveland OH 44126-1602

OLSON, WILLARD ORVIN, b Polk Co, Wis, May 9, 33; m 61; c 2. NUCLEAR ENGINEERING, PHYSICS. *Educ:* Univ Wis-River Falls, BS, 56; Vanderbilt Univ, MS, 58; Ore State Univ, PhD(nuclear eng), 74. *Prof Exp:* Instr physics, Mankato State Col, 58-62, asst prof, 64-67; instr nuclear eng, Ore State Univ, 72-73; SR PHYSICIST REACTOR PHYSICS, IDAHO NAT ENG LAB-ENERGY RES DEVELOP ADMIN. *Mem:* Am Nuclear Soc. *Res:* Reactor physics; numerical and computer methods. *Mailing Add:* 3319 April Dr Idaho Falls ID 83401

OLSON, WILLARD PAUL, b Detroit, Mich, Aug 6, 39; m 64; c 2. SPACE PHYSICS, STRATEGIC PLANNING. *Educ:* Univ Calif, Los Angeles, BS, 62, MS, 64, PhD(physics), 68. *Prof Exp:* Res scientist, Astrophysics Res Corp, 64-65; consult, Rand Corp, 65-66; sr scientist, McDonnell Douglas Astronaut Co, 68-74, chief, Environ Effects Br, 74-77, br chief thermal physics & radiation environ & head, Thermodyn & Environ Dept, 77-81, chief, Thermodyn, Environ & Biotechnol Dept, 81-85, sr mgr, Advan Technol Develop, 85-87, Advan Technol Ctr, 87-89; CONSULT, 89- *Concurrent Pos:* Mem, Air Force Sci Adv Bd Comt on Aeronomy, 76-78; chmn, Univ Calif Irvine, Corp Affil, 86-88. *Mem:* Int Asn Geomagnetism & Aeronomy; AAAS; Am Geophys Union; Am Phys Soc; Comt Space Res. *Res:* Coupling of the earth's magnetosphere with solar wind and interplanetary magnetic field; quantitative modelling and prediction of environmental effects and associated computer software development. *Mailing Add:* 4805 Cove Creek Dr Brownsboro AL 35741

OLSON, WILLIAM ARTHUR, b Minneapolis, Minn, Oct 19, 32; m 56; c 3. TOXICOLOGY, BIOCHEMISTRY. *Educ:* Univ Minn, BS, 54, MS, 60, PhD(nutrit), 62. *Prof Exp:* Res supvr nutrit, Chas Pfizer & Co, 62-65; dir animal health res, Smith Kline & French Labs, 65-69; staff scientist, Hazelton Labs, 69-73; OWNER & MGR, CFR SERV, 73- *Mem:* AAAS; Am Soc Animal Sci; Am Dairy Sci Asn; Animal Nutrit Res Coun. *Res:* Determination of the safe use of chemicals as drugs, veterinary drugs, food additives, pesticides or household articles; toxicology; product development. *Mailing Add:* 2347 Paddock Lane Herndon VA 22091

OLSON, WILLIAM BRUCE, b Omaha, Nebr, Dec 28, 30; m 59; c 3. PHYSICAL CHEMISTRY. *Educ:* Univ Wash, BS, 53, PhD(phys chem), 60. *Prof Exp:* Res assoc phys chem, Princeton Univ, 60-61; PHYSICIST, NAT BUR STANDARDS, 61- *Honors & Awards:* Silver Medal, US Dept Com, 73. *Mem:* Optical Soc Am. *Res:* Molecular structure and vibrational-rotational level structure of polyatomic molecules in gas phase via high resolution infrared spectroscopy; high precision infrared spectrophotometric instrumentation; precision infrared wavelength standards. *Mailing Add:* Nat Bur Standards 221 B-268 Gaithersburg MD 20899

OLSON, WILLIAM MARVIN, b Rock Island, Ill, June 15, 29; m 55; c 3. THERMODYNAMICS & MATERIAL PROPERTIES. *Educ:* Augustana Col, Ill, BA, 51; Univ Iowa, PhD(inorg chem), 59. *Prof Exp:* MEM STAFF, LOS ALAMOS NAT LAB, 56- *Concurrent Pos:* Fel, Europ Inst Transuranium Elements, Euratom, Karlsruhe, Ger, 68-69. *Mem:* Am Vacuum Soc. *Res:* Thermodynamics, especially of actinides and their compounds; thin films as protective coating for actinides. *Mailing Add:* 655 Totavi St Los Alamos NM 87544

OLSON, WILMA KING, b Philadelphia, Pa, Dec 1, 45; m 69; c 1. BIOPHYSICAL CHEMISTRY, POLYMER CHEMISTRY. *Educ:* Univ Del, BS, 67; Stanford Univ, PhD(chem), 71. *Prof Exp:* Fel chem, Stanford Univ, 70-71; Damon Runyon Fund fel, Columbia Univ, 71-72; from asst prof to assoc prof, 72-79, prof chem, 79-90, MARY I BUNTING PROF, RUTGERS UNIV, NEW BRUNSWICK, 90- *Concurrent Pos:* Sloan Found fel, 75; USPHS career develop award, 75; J S Guggenheim Found fel, 78-79; guest prof, Univ Basel, 78-79 & Jilin Univ, China, 81. *Honors & Awards:* Merit Award, NIH, 88- *Mem:* Am Chem Soc; Biophys Soc. *Res:* Relationship of structure, conformation and function in biological macromolecules, particularly nucleic acids. *Mailing Add:* Dept Chem Wright/Rieman Labs Rutgers Univ New Brunswick NJ 08903

OLSSON, CARL ALFRED, b Boston, Mass, Nov 29, 38; c 2. UROLOGY, SURGERY. *Educ:* Bowdoin Col, 59; Sch Med, Boston Univ, MD, 63; Am Bd Urol, cert, 72. *Prof Exp:* Intern surg, Mass Mem Hosp, 63-64; resident, Univ Hosp, 64-66; resident urol, Boston Vet Admin & Boston City Hosps, 66-69; from asst prof to prof, Sch Med Boston Univ, 71-80, chmn, Dept Urol, 74-80; chief, Dept Urol, Univ Hosp, 71-80; PROF UROL & CHMN DEPT, COL PHYS & SURG, COLUMBIA UNIV, 80-; DIR UROL, PRESBY HOSP, NY, 80- *Concurrent Pos:* Fel urol, Cleveland Clin Found, 67-68; fel, Nat Kidney Found, 69 & NIH Spec Res, 69; consult urol, US Naval Hosp, 72-74; lectr surg, Sch Med, Tufts Univ, 72-80; consult urol, US Pub Health Hosp, 72-80; chief Urol Dept, Boston City Hosp, 74-77; delegate res comt, Am Urol Asn, 74-77; mem exam comt, Am Bd Urol, 74-79, trustee, 89-; chief Urol Sect, Boston Vet Admin Hosp, 75-80, chief consult Surg Serv, 76-80; trustee, Boston Int Hosp Orgn Bank, 76-80; ed bd, Invest Urol, 77-88, World J Urol, 85-, the Prostate, 85-, Urol, 85-; asst ed, J Urol, 78-88. *Honors & Awards:* Grayson Carroll Award, Am Urol Asn 63 & 65 Gold Gstocope Award, 78. *Mem:* AMA; Am Fertil Soc; Am Soc Artificial Internal Organs; Am Urol Asn; Soc Univ Urol; Am Asn Urol; Transplantation Soc, Clin Soc Urol Surg; Soc Urol Oncol. *Res:* Urologic cancer. *Mailing Add:* 630 W 168th Street New York NY 10032

OLSSON, NILS OVE, b Arvika, Sweden, July 15, 32; Can citizen; m 65; c 2. AGRICULTURAL AND CONSTRUCTION EQUIPMENT. *Educ:* Gothenburg Tech Inst, BS, 54. *Prof Exp:* Mgr prod planning, 77-78, dir eng, 79-80, vpres eng, 81-82, MGR, INT HARVEST WORLDWIDE, 82- *Concurrent Pos:* Mem, Rops Comt, Can, 72-77, Soc Automative Eng, 72-88, Bus Adv Comt Oreo, Paris, 80-82. *Res:* Agricultural and construction equipment design. *Mailing Add:* 7600 Country Line Rd Hinsdale IL 60521

OLSSON, RAY ANDREW, b Livermore, Calif, Nov 30, 31; m 54; c 3. PHARMACOLOGY. *Educ:* George Washington Univ, MD, 56. *Prof Exp:* With US Army, 55-76, resident internal med, George Washington Univ, 58-60, res internist, Dept Cardiorespiratory Dis, Walter Reed Army Inst Res, 61-63, dir div clin res, SEATO Med Res Lab, Bangkok, 63-66, asst chief dept cardiorespiratory dis, Walter Reed Army Inst Res, 67-70, dep dir div med, 70-71, dir div med, 71-76; PROF MED, UNIV SOUTH FLA COL MED, 76- *Concurrent Pos:* Assoc ed, Am J Physiol. *Honors & Awards:* Order of the White Elephant, Royal Thai Govt, 66. *Mem:* AAAS; Am Physiol Soc; Am Col Physicians; Am Heart Assn; Am Coll Cardiol; Int Soc Heart Res; Sigma Xi. *Res:* Cardiovascular physiology; coronary circulation; cardiac metabolism. *Mailing Add:* Dept Internal Med Univ South Fla Col Med Tampa FL 33612

OLSSON, RICHARD KEITH, b Newark, NJ, Mar 23, 31; m 57; c 3. GEOLOGY. *Educ:* Rutgers Univ, BS, 53, MS, 54; Princeton Univ, MA, 56, PhD(geol), 58. *Prof Exp:* Asst field geologist, Mobil Oil Co Div, Socony Mobil Oil Co, Inc, 53; explor geologist, Pinon Uranium Co, 55; asst geol, Rutgers Univ, 53-54 & Princeton Univ, 54-56; from instr to prof geol, 57-77, CHMN GEOL SCI DEPT, RUTGERS UNIV, 77- *Concurrent Pos:* Consult petrol explor & revel geol, NAm Micropaleont Soc. *Mem:* Geol Soc Am; Soc Econ Paleontologists & Mineralogists; Am Asn Petrol Geol; Paleont Soc. *Res:* Stratigraphy; micropaleontology, sedimentalogy; paleoecology; Late Cretaceous and Tertiary section of the geologic rock column; evolution and biostratigraphy of planktonic foraminifera. *Mailing Add:* Dept Geol Sci Rutgers Univ New Brunswick NJ 08903

OLSTAD, ROGER GALE, b Minneapolis, Minn, Jan 16, 34; m 55; c 2. SCIENCE EDUCATION. *Educ:* Univ Minn, BS, 55, MA, 59, PhD(educ), 63. *Prof Exp:* Instr high sch, Minn, 55-56; instr biol educ, Univ Minn, 56-63; asst prof sci educ, Univ Ill, 63-64; from asst prof to assoc prof, 64-71, assoc dean grad studies, 68-81, PROF SCI EDUC, UNIV WASH, 71- *Concurrent Pos:* Bd trustees, Pac Sci Ctr Found, 77-82; chmn, Bd Trustees, Biolog Sci Currie Study, 89- *Mem:* Fel AAAS; Nat Asn Res Sci Teaching (pres, 76-79); Nat Sci Teachers Asn; Nat Asn Biol Teachers; Am Educ Res Asn; Asn Educ Teachers Sci (pres, 91-). *Res:* Science teacher selection and education, organization and evaluation; science tests and testing; teacher evaluation. *Mailing Add:* Col Educ DQ-12 Univ Wash Seattle WA 98195

OLSTER, ELLIOT FREDERICK, b New York, NY, Nov 23, 42; m 64; c 3. MATERIALS ENGINEERING, STRUCTURAL ENGINEERING. *Educ:* Drexel Inst Technol, BS, 65; Mass Inst Technol, MS, 67, ScD, 71. *Prof Exp:* Staff engr mat res, Systs Div, Avco Corp, 71-73; res engr, Plastics Develop Ctr, Ford Motor Corp, 73-75; ROTOR SYSTS ENGR AEROSPACE DEVELOP, SIKORSKY AIRCRAFT DIV, UNITED TECHNOL CORP, 75 - *Mem:* Sigma Xi. *Res:* Crack propagation; fatigue and failure mechanisms in composite materials. *Mailing Add:* 460 Skyline Dr Orange CT 06477

OLSTOWSKI, FRANCISZEK, b New York, NY, Apr 23, 27; m 52; c 5. POLYMER CHEMISTRY. *Educ:* Tex A&I Univ, BS, 54. *Prof Exp:* Res & develop engr, magnesium technol, Dow Chem Co, 54-56, proj engr fluorine res, 56-62, proj engr graphite technol, 62-65, sr res engr, 65-72, res specialist, 72-79, res leader, 79-87; CONSULT, 87- *Mem:* AAAS; Am Chem Soc; Electrochem Soc; NY Acad Sci; fel Am Inst Chemists. *Res:* Electrolytic production of magnesium metal and fluorocarbons; synthesis of fluoro-olefins; inorganic fluoride synthesis; natural graphite technology; catalytic oxidation of olefins; polyurethane technology; reaction molding systems; plastic composite processes. *Mailing Add:* 912 North Ave A Freeport TX 77541

OLSZANSKI, DENNIS JOHN, b Rossford, Ohio, Feb 27, 48; m 68; c 3. CATALYSIS. *Educ:* Univ Dayton, BS, 70; Mich State Univ, PhD(inorg chem), 75. *Prof Exp:* Res assoc inorg chem, Ohio State Univ, 75-77; res chemist, Engelhard Minerals & Chem Corp, 77-79; sr res chemist, Calsicat Div, Mallinckrodt, Inc, 79-86; DIR COM DEVELOP, MOONEY CHEM, INC, 86- *Mem:* Am Chem Soc. *Res:* Coordination chemistry of scandium; bioinorganic chemistry, especially activation of molecular oxygen and nitrogen by transition metal complexes; catalysis by transition metals and zeolites. *Mailing Add:* Mooney Chem Inc 2301 Scranton Rd Cleveland OH 44113

OLTE, ANDREJS, b Skujene, Latvia, Nov 17, 27; nat US; m 63; c 2. ENGINEERING. *Educ:* Univ Calif, BS, 54, MS, 57, PhD(elec eng), 59. *Prof Exp:* Asst elec eng, Univ Calif, 54-59; from asst prof to assoc prof, 59-66, PROF ELEC ENG, UNIV MICH, ANN ARBOR, 66- *Mem:* Inst Elec & Electronics Engrs. *Res:* Electromagnetic diffraction theory; propagation of radio waves; ionized gases. *Mailing Add:* Dept Elec-Comn Engrs Univ Mich Ann Arbor MI 48109

OLTENACU, ELIZABETH ALLISON BRANFORD, b Pontefract, Eng, Sept 5, 47; m 73. ANIMAL SCIENCE. *Educ:* Univ Edinburgh, BSc, 70; Univ Minn, St Paul, MS, 72, PhD(animal sci), 74. *Prof Exp:* Res assoc, 74-79, lectr, 78-79, asst prof, 79-85, ASSOC PROF ANIMAL SCI, CORNELL UNIV, 85- *Mem:* Brit Soc Animal Prod; Inst Biol; Am Genetic Asn; Am Dairy Sci Asn. *Res:* Computer studies of genetic and environmental influences affecting production and behavioral traits in domestic livestock; use of computers in teaching genetics, selection and animal management in extension education programs for youth audiences. *Mailing Add:* 192 Roberto Hall Cornell Univ Ithaca NY 14853-5901

OLTHOF, THEODORUS HENDRIKUS ANTONIUS, b Deventer, Netherlands, July 15, 34; Can citizen. NEMATOLOGY, PLANT PATHOLOGY. *Educ:* State Col Trop Agr, Netherlands, dipl, 55; McGill Univ, BSc, 58, PhD(plant path), 63. *Prof Exp:* Res officer potatoes, 62-64, res officer nematol, 64-73, RES SCIENTIST, RES STA, CAN DEPT AGR, 73- *Mem:* Am Phytopath Soc; Can Phytopath Soc; Soc Nematol; Soc Europ Nematol. *Res:* Nematode-fungus interactions; field ecology of root-lesion nematodes in tobacco; crop loss assessment; population dynamics. *Mailing Add:* Res Sta Agr Can Vineland Station ON L0R 2E0 Can

OLTJEN, ROBERT RAYMOND, b Robinson, Kans, Jan 13, 32; m 56; c 3. RUMINANT NUTRITION. *Educ:* Kans State Univ, BS, 50, MS, 58; Okla State Univ, PhD(animal nutrit), 61. *Prof Exp:* Res animal husbandman, Beef Cattle Res Br, Animal Sci Res Div, USDA-ARS, 62-69, leader nutrit invest, 69-72, chief, Ruminant Nutrit Lab, Agr Res Serv, 72-77, dir, Roman L Hruska US Meat Animal Res Ctr, Clay Ctr, Nebr, 77-88, ASSOC DEP ADMIN, ANIMAL & POST HARVEST SCI, NAT PROG STAFF, USDA-ARS, 88- *Concurrent Pos:* Mem comt animal nutrit, Nat Acad Sci-Nat Res Coun, 72-76, chmn comt animal nutrit, 77-79. *Honors & Awards:* Nutrit Res Award, Am Feed Mfrs Asn, 71; Presidential Citation, 72; Cert Animal Scientist, Am Soc Animal Sci, 75, Animal Indust Serv Award, 87; Moorman Mfg Travel Award, 77. *Mem:* Am Inst Nutrit; Am Soc Animal Sci (pres, 82-83); Am Dairy Sci Asn; AAAS. *Res:* Digestion and metabolism of various nitrogen and energy sources by ruminants; finishing diets for beef cattle; research on cellulose wastes and non-protein nitrogen for ruminants; animal agriculture. *Mailing Add:* USDA-ARS-NPS Rm 134 Bldg 005 BARC-W West Beltsville MD 20705

OLUM, PAUL, b Binghamton, NY, Aug 16, 18; m 42; c 3. MATHEMATICS. *Educ:* Harvard Univ, AB, 40; Princeton Univ, AM, 42; Harvard Univ, PhD(math), 47. *Prof Exp:* Theoret physicist, Manhattan Proj, Princeton Univ, 41-42 & Los Alamos Sci Lab, 43-45; Jewett fel, Harvard Univ, 47-48 & Inst Advan Study, 48-49; from asst prof to prof math, Cornell Univ, 49-74, chmn dept, 63-66; dean, Col Natural Sci, Univ Tex, Austin, 74-76; vpres acad affairs & provost, 76-80, actg pres, 80-81, PRES, UNIV ORE, 81- *Concurrent Pos:* Mem, Inst Advan Study, 55-56 & Nat Res Coun adv comt, Off Ordn Res, 58-61; vis prof, Univ Paris & Hebrew Univ, Israel, 62-63; NSF sr res fel, Stanford Univ, 66-67; vis prof, Univ Wash, 70-71; mem bd trustees, Cornell Univ, 71-75. *Mem:* Am Math Soc; Math Asn Am; Sigma Xi. *Res:* Algebraic topology. *Mailing Add:* 116 B No 350 1011 Valley River Way Eugene OR 97401

OLVER, ELWOOD FORREST, b Connelsville, Pa, Apr 10, 22; m 44; c 3. AGRICULTURAL ENGINEERING. *Educ:* Pa State Univ, BS, 43, MS, 49; Iowa State Univ, PhD(agr eng), 57. *Prof Exp:* Rural engr rural elec, Pa Power & Light Co, 46-48; asst prof agr eng, Pa State Univ, 48-50, assoc prof, 52-57, dir dept security, 57-60; educ dir, Iowa Rural Elec Coop Asn, 50-52; prof dept agr eng, Univ Ill, 60-82, head processing div, 70-82; asst dean, Col Eng, Univ Ill, 85-87; ENG CONSULT, 87- *Concurrent Pos:* Group leader sci team, J Nehru Agr Univ, India, 67-69; consult, Coop Res Sci & Educ Admin, USDA, 77-78 & 80-84; exec secy, Ill Electrification Coun, 70-82; eng consult, 84-85. *Mem:* Fel Am Soc Agr Engrs; Sigma Xi; Am Soc Eng Educ. *Res:* Automation in dairy farming; grain drying and conditioning. *Mailing Add:* 402 Burkwood Ct Urbana IL 61801

OLVER, FRANK WILLIAM JOHN, b Croydon, Eng, Dec 15, 24; wid; c 3. ASYMPTOTICS, COMPUTER ARITHMETIC. *Educ:* Univ London, BSc, 45, MSc, 48, DSc(math anal), 61. *Prof Exp:* Exp officer numerical anal, Nautical Almanac Off, Eng, 44-45; sr prin sci officer, Nat Phys Lab, 45-61, head numerical methods sect, 53-61; mathematician, Nat Bur Stand, DC, 61-69; RES PROF, INST PHYS SCI & TECHNOL, UNIV MD, 69- *Concurrent Pos:* Mathematician, Nat Bur Stand, Washington, DC, 69-86; ed, J Math Anal, Soc Indust & Anal Math, 69-, Appl Math, 64-68; assoc ed, J Res, Nat Bur Standards, 66-78 & Math Comput, 84- *Mem:* Am Math Soc; Soc Indust & Appl Math; Math Asn Am; fel Inst Math & Applns UK. *Res:* Asymptotics; numerical analysis; special functions. *Mailing Add:* Inst Phys Sci & Technol Univ Md College Park MD 20742

OLVER, JOHN WALTER, b Honesdale, Pa, Sept 3, 36; m 59. ANALYTICAL CHEMISTRY. *Educ:* Rensselaer Polytech Inst, BS, 55; Tufts Univ, MS, 56; Mass Inst Technol, PhD(voltametric studies), 61. *Prof Exp:* Actg head dept chem, Franklin Inst, 56-58; instr chem, Mass Inst Technol, 61-62; ASST PROF CHEM, UNIV MASS, AMHERST, 62- *Mem:* Am Chem Soc. *Res:* Electrochemical methods of analysis; chemical separations. *Mailing Add:* 1333 West St Amherst MA 01002

OLVER, PETER JOHN, b Twickenham, Eng, Jan 11, 52; m 76; c 3. LIE GROUPS, DIFFERENTIAL EQUATIONS. *Educ:* Brown Univ, ScB, 73; Harvard Univ, PhD(math), 76. *Prof Exp:* L E Dickson instr math, Univ Chicago, 76-78; res assoc math, Univ Oxford, Eng, 78-80; from asst prof to assoc prof, 80-85, PROF MATH, UNIV MINN, 85- *Mem:* Am Math Soc; Soc Indust & Appl Math. *Res:* Applications of Lie groups to differential equations in fluid dynamics and elasticity, especially in conservation laws and Hamiltonian structures. *Mailing Add:* Dept Math 1127 Vincent Hall Univ Minn Minneapolis MN 55455

OLVERA-DELA CRUZ, MONICA, b Mexico City, Mex, July 22, 58. POLYMER SCIENCE. *Educ:* Nat Independent Univ Mex, BA, 81; Univ Cambridge, Eng, PhD(physics), 84. *Prof Exp:* Res fel, Univ Mass, 84-85; GUEST WORKER, NAT BUR STANDARDS, 84- *Mem:* Am Phys Soc. *Res:* Statistical mechanics problems; phase transitions and critical phenomena in long molecules systems; dynamical properties of long polymers in topologically restricted environments. *Mailing Add:* Dept Mat Sci & Eng Northwestern Univ Tech Inst Evanston IL 60208

OLYNYK, PAUL, b Ymir, BC, Aug 5, 18; m 71; c 4. CHEMISTRY. *Educ:* McGill Univ, BSc, 39; Univ Toronto, PhD(chem), 44. *Prof Exp:* Qual control supvr radar parts, Res Enterprises Ltd, 44-45; res chemist biol mat, Fine Chem of Can Ltd, 45-46; instr chem, Univ Rochester, 46-48; res assoc antibiotic plants, Babies & Children's Hosp, 48-50, res assoc fats & fat metab, 50-52; res chemist paint formulation, Sherwin-Williams Co, 52-57 & Glidden Co, 57-59; assoc dir inst urban studies & dir div environ sci, 69-71, ASSOC PROF CHEM, CLEVELAND STATE UNIV, 59- *Mem:* Am Chem Soc; Fedn Am Scientists; Int Asn Great Lakes Res. *Res:* Antibiotics from plant; radiocarbon C-14 in proteins and fats; fat metabolism; water and air pollution; sediment and environmental chemistry; toxic organics. *Mailing Add:* 3011 Ludlow Rd Cleveland OH 44120

OLYPHANT, MURRAY, JR, b New York, NY, May 2, 23; m 47; c 3. ELECTRICAL ENGINEERING. *Educ:* Princeton Univ, BS, 44, MSE, 47. *Prof Exp:* Instr plastics & dielec, Princeton Univ, 47-51, res assoc, 51-52; res engr elec prod, Minn Mining & Mfg Co, 52-65, group leader elec prod lab, 65-67, supvr, dielec mat & syst lab, 67-70, sr res specialist, 70-76; sr res specialist electronic prod, 76-80, DIV SCIENTIST, 3M CO, 80- *Concurrent Pos:* Bd mem, Conf Elec Insulation & Dielec Phenomena, Nat Res Coun. *Mem:* AAAS; fel Inst Elec & Electronics Engrs. *Res:* Electrical properties of plastics; breakdown and corona resistance of electrical insulation; microwave dielectric measurements; computer connecting systems. *Mailing Add:* 8609 Hidden Bay Trail N Lake Elmo MN 55042

OMACHI, AKIRA, b Sacramento, Calif, Sept 10, 22; m 51; c 2. PHYSIOLOGY. *Educ:* Univ Buffalo, BA, 44; Univ Minn, MS, 48, PhD(physiol), 50. *Prof Exp:* Instr physiol, Univ Minn, 48-49; from instr to asst prof, Med SCh, Loyola Univ, Ill, 49-57; from asst prof to assoc prof, 57-69, PROF PHYSIOL, COL MED, UNIV ILL, CHICAGO, 69- *Mem:* AAAS; Am Physiol Soc; Am Asn Univ Profs; Biophys Soc; Sigma Xi. *Res:* Electrolyte metabolism; cell physiology and metabolism; membrane transport. *Mailing Add:* Dept of Physiol & Biophys Univ of Ill Med Ctr Box 6998 Chicago IL 60680

O'MALLEY, ALICE T, b Clinton, Mass, Apr 1, 29. PHYSIOLOGY. *Educ:* Clark Univ, PhD(biol), 65. *Prof Exp:* ASSOC PROF BIOL, STATE COL, FITCHBURG, MASS, 66- *Mem:* Am Physiol Soc; Soc Neurosci. *Mailing Add:* Dept Biol State Col 160 Pearl St Fitchburg MA 01420

O'MALLEY, BERT W, b Pittsburgh, Pa, Dec 19, 36; m 60; c 4. ENDOCRINOLOGY, MOLECULAR BIOLOGY. *Educ:* Univ Pittsburgh, BS, 59, MD, 63. *Prof Exp:* From intern to resident, Med Ctr, Duke Univ, 63-65; clin assoc, Endocrinol Br, Nat Cancer Inst, 65-67, sr investr, 67-68, head molecular biol sect, 68-69; prof reproductive physiol, Vanderbilt Chair, Sch Med, Vanderbilt Univ, 69-72; PROF CELL BIOL & CHMN DEPT, BAYLOR COL MED, 72- *Mem:* Endocrine Soc; Soc Study Reproduction; Sigma Xi. *Res:* Reproductive physiology; mechanism of hormone action; hormone-mediated cell differentiation; molecular biology of the animal cell. *Mailing Add:* Dept Cell Biol Baylor Col Med One Baylor Plaza Houston TX 77030

O'MALLEY, EDWARD PAUL, b Hudson, NY, May 30, 26. PSYCHIATRY, PHARMACOLOGY. *Educ:* St John's Univ, NY, BS, 49; Loyola Univ, Ill, MS, 53, PhD(pharmacol), 54; State Univ NY Downstate Med Ctr, MD, 58; Am Bd Psychiat & Neurol, dipl, 65. *Prof Exp:* Intern, St Vincent's Hosp, New York, 58-59; resident psychiat, Bronx Vet Admin Hosp & NY State Psychiat Inst, New York, 59-62; sch psychiatrist, Bur Child Guid, NY Bd Educ, 63-67; med dir, West Nassau Ment Health Ctr, 67-68; dir, Community Ment Health Serv Suffolk County, 68-72; comnr, Orange County Ment Health Serv, NY, 72-74; asst clin prof psychiat, NJ Med Sch, 74-80; ASSOC CLIN PROF PSYCHIAT, UNIV CALIF, SAN DIEGO, 80- *Concurrent Pos:* Asst clin prof, State Univ NY Stony Brook; consult psychiatrist, Riverside Hosp, Bronx, NY, 62-63; visitor, New York Dept Corrections, 62-66; consult, Cath Charities Guid Inst, 62-; assoc attend, St Vincent's Hosp, 62-; clin attend, St Francis Hosp, Bronx, 63-; courtesy attend psychiatrist, Arden Hill Hosp, Goshen, NY; assoc attend psychiatrist, Bronx Lebanon Hosp & St Luke's Hosp; clin instr, NY Med Col, 76- *Mem:* AMA; fel Am Psychiat Asn. *Res:* Psychopharmacological and psychiatric research; pharmacological, biochemical and physiological studies at the basic science level. *Mailing Add:* 3711 Alcott St San Diego CA 92106

O'MALLEY, JAMES JOSEPH, b Philadelphia, Pa, Sept 17, 40; m 64; c 3. POLYMER CHEMISTRY, POLYMER PHYSICS. *Educ:* Villanova Univ, BS, 62; State Univ NY Col Forestry, Syracuse Univ, MS, 64, PhD(phys chem), 67. *Prof Exp:* From assoc scientist to sr scientist, Xerox Corp, 67-74, mgr polymer res, 74-79; MGR POLYMER SCI & TECHNOL, EXXON CHEM CO, 79- *Mem:* AAAS; Am Chem Soc; Am Phys Soc. *Res:* Synthesis and characterization of block and graft copolymers; interactions of synthetic and bio-polymers; surface and electrical properties of polymers; chemical modification of polymer, properties of polyolefins. *Mailing Add:* 12 Turnburry Rd Washington NJ 07882

O'MALLEY, JOHN RICHARD, b Huntington, Ind, May 17, 28; m 53; c 6. ELECTRICAL ENGINEERING. *Educ:* Purdue Univ, BS, 51, MS, 52; Georgetown Univ, LLB, 56; Univ Fla, PhD(elec eng), 64. *Prof Exp:* Patent examr, US Patent Off, 52-56; patent atty, Gen Elec Co, 56-59; assoc prof, 59-81, PROF ELEC ENG, UNIV FLA, 81- *Mem:* Inst Elec & Electronics Engrs; Am Soc Eng Educ. *Res:* Patent law; network synthesis; switching circuits. *Mailing Add:* Dept Elec Eng Larsen Hall Univ of Fla Gainesville FL 32611

O'MALLEY, JOSEPH PAUL, medicine, for more information see previous edition

O'MALLEY, MARY THERESE, b Chicago, Ill. MATHEMATICS. *Educ:* Univ Nebr, BA, 55; Cath Univ Am, MA, 63; Columbia Univ, PhD(math), 71. *Prof Exp:* Asst prof, 68-80, ASSOC PROF MATH, ROSARY COL, RIVER FOREST, ILL, 80- *Mem:* Math Asn Am; Am Math Soc; Sigma Xi. *Res:* Algebra. *Mailing Add:* 7900 W Division St River Forest IL 60305

O'MALLEY, MATTHEW JOSEPH, b Miami, Fla, Oct 21, 40; m 64; c 2. ALGEBRA. *Educ:* Spring Hill Col, BS, 62; Fla State Univ, MS, 64, PhD(math), 67. *Prof Exp:* Mathematician, Johnson Spacecraft Ctr, NASA, 67-71; asst prof, 71-74, assoc prof, 74-81, PROF MATH, UNIV HOUSTON, 81- *Mem:* Am Math Soc. *Res:* Commutative ring theory; pattern recognition. *Mailing Add:* Dept Math Univ Houston Univ Park Houston TX 77204

O'MALLEY, RICHARD JOHN, b Jersey City, NJ, Dec 20, 45; m 70. MATHEMATICAL ANALYSIS. *Educ:* Seton Hall Univ, BS, 67; Purdue Univ, MS, 69, PhD(math), 72. *Prof Exp:* ASST PROF MATH, UNIV WIS-MILWAUKEE, 72- *Concurrent Pos:* Univ Wis-Milwaukee res grant, 75. *Mem:* Am Math Soc. *Res:* Analysis of functions of a real variable, particularly approximate continuity and density properties. *Mailing Add:* Dept Math Univ Wis PO Box 413 Milwaukee WI 53201

O'MALLEY, ROBERT EDMUND, JR, b Rochester, NH, May 23, 39; m 68; c 3. APPLIED MATHEMATICS. *Educ:* Univ NH, BS, 60, MS, 61; Stanford Univ, PhD(math), 66. *Prof Exp:* Asst prof math, Univ NC, Chapel Hill, 66-68; assoc prof, NY Univ, 68-73; prof math, Univ Ariz, 73-76, chmn appl math, 76-81; chmn math sci & Ford Found prof, Rensselaer Polytech Inst, 81-90, chmn dept, 81-84; PROF & CHAIR, APPL MATH, UNIV WASH, 91- *Concurrent Pos:* Vis mem, Courant Inst Math Sci, NY Univ, 66-67; vis asst prof, Math Res Ctr, Univ Wis, 67-68; sr vis fel, Univ Edinburgh, 71-72; sabbatical leaves, Stanford, Vienna. *Mem:* Am Math Soc; Soc Indust & Appl Math (pres, 91-92); Math Asn Am; Asn Women Math. *Res:* Singular perturbation problems; asymptotic expansions; differential equations; control theory. *Mailing Add:* Dept Appl Math Univ Wash Seattle WA 98195

O'MALLEY, ROBERT FRANCIS, b Framingham, Mass, Apr 2, 18; m 44; c 5. FLUORINE CHEMISTRY, ELECTROORGANIC CHEMISTRY. *Educ:* Boston Col, BS, 40, MS, 48; Mass Inst Technol, PhD(inorg chem), 61. *Hon Degrees:* DSc, Boston Col, 88. *Prof Exp:* From instr to prof, 47-88, admin asst, 52-56, chmn dept chem, 56-66 & 71-74, EMER PROF INORG CHEM, BOSTON COL, 88- *Concurrent Pos:* Vis assoc prof, Harvard Univ, 65. *Mem:* AAAS; Am Chem Soc; Sigma Xi. *Res:* Fluorination and chlorination via electron transfer; organic electron transfer. *Mailing Add:* Dept Chem Boston Col Chestnut Hill MA 02167-3809

O'MALLEY, THOMAS FRANCIS, b New York, NY, Nov 13, 28; m 62; c 1. QUANTUM SCATTERING THEORY, ELECTRON TRANSPORT GASES. *Educ:* Bellarmine Col, AB, 53 & PhL(philos), 54; NY Univ, PhD(physics), 61. *Prof Exp:* Res scientist, Gen Res Corp, 64-70; assoc prof physics, Univ Conn, 70-72; res scientist, Dolphin Proj, 73-75; vis fel, Univ Paris, 74, Queens Univ, Belfast, 76-77, Australian Nat Univ, 77-78, Univ Western Ont, 79-80; SPECIALIST, GEN ELEC CO, 80- *Mem:* Fel Am Phys Soc. *Res:* Quantum theory of collisions and its application to reactions among electrons, atoms and small molecules and to electron transport in dense gases; large scale computations. *Mailing Add:* 1718 Ocala Ave #135 San Jose CA 95122

OMAN, CARL HENRY, b Camden, NJ, Apr 21, 38; m 62, 81; c 3. EXPLORATORY GEOLOGY. *Educ:* Rutgers Univ, BA, 59, MS, 60; Fla State Univ, PhD(geol), 65. *Prof Exp:* Exp & prod geologist oil, gas & minerals & district geol, Exxon Corp, 65-78; chief geologist, Everest Minerals Corp, 78-89; CHIEF GEOLOGIST, POWER RESOURCES INC, 89- *Mem:* Am Asn Petrol Geologists. *Res:* Locating commercial quantities of oil, gas and uranium using techniques of biostratigraphy, physical stratigraphy, structural geology, geochemistry and geophysics. *Mailing Add:* Power Resources Inc 800 Werner Ct Suite 230 Casper WY 82601

OMAN, CHARLES MCMASTER, b Brooklyn, NY, Feb 22, 44; m 74; c 2. AERONAUTICAL ENGINEERING. *Educ:* Princeton Univ, BSE, 66; Mass Inst Technol, SM, 68, PhD(instrumentation, automatic control), 72. *Prof Exp:* from asst prof to assoc prof aeronaut & astronaut, Mass Inst Technol, 72-79, Hermann von Helmholtz assoc prof, Div Health Sci & Technol, 77-79, prin res scientist, 79-81, actg dir, Man Vehicle Lab, 87-88, SR RES ENGR, MASS INST TECHNOL, 81- *Concurrent Pos:* Investr, NASA Space Shuttle Space Lab Missions 1, D-1 & SLS-1. *Mem:* Soc Neurosci; Barany Soc; Aerospace Med Asn; Human Factors Soc. *Res:* Vestibular function; human spatial orientation and disorientation, motion sickness; control of posture and movement; biomedical instrumentation; automatic control and signal processing; aerospace human factors. *Mailing Add:* Man Vehicle Lab Rm 37-219 Mass Inst Technol Cambridge MA 02139

OMAN, HENRY, b Portland, Ore, Aug 29, 18; m 54; c 3. ENERGY, POWER. *Educ:* Ore State Univ, BSEE, 40, MSEE, 51. *Prof Exp:* Appln engr, Allis Chalmers Mfg Co, 40-48; res engr, 48-63, ENGR MGR, BOEING CO, 63- *Concurrent Pos:* secy space power comt, Am Inst Astronaut. *Mem:* Fel Inst Elec & Electronics Engrs (vpres, 84-88); fel Am Inst Aeronaut & Astronaut; AAAS; Astron Soc Pac. *Res:* Power and energy; carbon-dioxide buildup from burning of fossil fuels; transportation energy and fuel saving strategies; alternative energy for post petroleum era; power sources for space craft and underground installation. *Mailing Add:* 19221 Normandy Park Dr SW Seattle WA 98166

OMAN, PAUL WILSON, b Garnett, Kans, Feb 22, 08; m 31; c 6. ENTOMOLOGY. *Educ:* Univ Kans, AB, 30, AM, 35; George Washington Univ, PhD(entom), 41. *Prof Exp:* Jr entomologist bur entom & plant quarantine, USDA, 30-34, from asst entomologist to prin entomologist, 34-59, chief insect identification & parasite introd res br, 59-60, dir res & tech progs div, Far East Regional Res Off, Agr Res Serv, 60-63, asst dir entom res div, 63-67; actg head dept, 73-74, prof entom, 67-75, EMER PROF ENTOM, ORE STATE UNIV, 75- *Mem:* Fel Entom Soc Am (2nd vpres, 48, pres-elect, 58, pres, 59); fel Royal Entom Soc London. *Res:* Insect systematics; biological control of insects and weeds; medical entomology. *Mailing Add:* Dept Entom Ore State Univ Corvallis OR 97331

OMAN, ROBERT MILTON, b Easton, Mass, Aug 12, 34; m 62; c 1. ELECTRON PHYSICS. *Educ:* Northeastern Univ, BS, 57; Brown Univ, ScM, 60, PhD(physics), 63. *Prof Exp:* From res asst to res assoc physics, Brown Univ, 58-63; res scientist, United Aircraft Corp, Conn, 63-64; tech specialist, Litton Systs, Inc, Minn, 64-66; sr res scientist, Norton Res Corp, Mass, 66-70; prof math/physics, N Shore Community Col, 71-91; AUTH-PUBL, OMAN PUBLISHERS, MASS, 76- *Mem:* Am Vacuum Soc; Am Phys Soc; Sigma Xi. *Res:* Theoretical physical electronics; solid state and experimental solid state physics; electron mirror microscopy; magnetron gauges; electron transport in semiconductors; applied mathematics. *Mailing Add:* 6268 Palma Del Mar Blvd No 502 E St Petersburg FL 33733

O'MARA, JAMES HERBERT, b St Clair, Mich, Oct 11, 36. PHYSICAL CHEMISTRY. *Educ:* George Washington Univ, BS, 57, MS, 61; Duke Univ, PhD(phys chem), 68. *Prof Exp:* Chemist, Polymers Div, Nat Bur Standards, 57-63; SR CHEMIST, ROHM AND HAAS CO, 67- *Mem:* AAAS; Am Chem Soc; Am Soc Testing & Mat; Soc Automotive Engrs. *Res:* Physical chemistry of polymers in dilute solution and bulk; chemistry of oil additives. *Mailing Add:* Rohm and Haas Co Spring House PA 19477

O'MARA, MICHAEL MARTIN, b Lackawanna, NY, Jan 24, 42; m; c 1. ORGANIC ANALYTICAL CHEMISTRY. *Educ:* Canisius Col, BS, 64; Univ Cincinnati, PhD(chem), 68. *Prof Exp:* Develop scientist, B F Goodrich Co, 68-72, sr develop scientist, 72-75, res & develop group leader, 75-77, develop mgr estane thermoplastic polyurethane, 75-78, gen mgr, 78-81, vpres res & develop, 81-85, vpres com develop, 85-86; vpres & adv, Polymers & Tech, 86-88; MGR, CHEM RES CTR, GEN ELEC CORP RES & DEVELOP, 88- *Concurrent Pos:* Chmn Technol Comt Vinyl Inst; panel chmn, Nat Bur Standard; mem comt, res & tech planning, Am Soc Testing Mat. *Honors & Awards:* Merck Award, 64; B F Goodrich Tech Lectr Award, 72. *Mem:* Am Clin Soc; AAAS; Am Soc Testing & Mat; NY Acad Sci. *Res:* Thermal degradation of polymers; flammability, smoke and toxic gas generation characteristics of polymers; gas chromatographic-mass spectrometric analyses. *Mailing Add:* Gen Elec Corp Res & Develop PO Box 8 Bldg K1 Rm 5A68 Schenectady NY 12301

O'MARA, ROBERT E, b Flushing, NY, Dec 8, 33; m 64; c 3. RADIOLOGY, NUCLEAR MEDICINE. *Educ:* Univ Rochester, BA, 55; Albany Med Col, MD, 59; Am Bd Radiol, dipl, 67; Am Bd Nuclear Med, dipl, 72. *Prof Exp:* Intern, St Louis Hosp, Wash Univ, 59-60, resident radiol, St Vincent's Hosp, New York, 63-66; dir nuclear med, Sch Med, Univ Ariz, 71-75; chief nuclear med serv, Vet Admin Hosp, Tucson, 71-75; PROF RADIOL & CHIEF DIV NUCLEAR MED, DEPT RADIOL, UNIV ROCHESTER, SCH OF MED & DENT, 75- *Concurrent Pos:* NIH fel nuclear med, State Univ NY Upstate Med Ctr, 66-67 & clin nuclear med, 67-71. *Mem:* Am Col Nuclear Physicians; Soc Nuclear Med; Radiol Soc NAm; Am Col Radiol; Am Bd Nuclear Med. *Res:* Diagnostic applications of radionuclides and development of radiopharmaceuticals. *Mailing Add:* Dept Radiology & Nuclear Med Univ Rochester Med Ctr Box 648 Rochester NY 14642

OMATA, ROBERT ROKURO, b Hanford, Calif, Nov 3, 20; m 48; c 3. BACTERIOLOGY. *Educ:* Univ Calif, AB, 44; Univ Minn, MS, 46, PhD(bact), 49. *Prof Exp:* Asst bact & immunol, Univ Minn, 45-47; from asst bacteriologist to bacteriologist, Nat Inst Dent Res, NIH, 53-60, mem staff div res grants, NIH, 60-63, scientist adminr, Off Int Res, 63-68, scientist adminr, Off Int Res, Pac Off, Tokyo, 64-67, head int fels sect, Scholars & Fels Prog Br, Fogarty Int Ctr, NIH, 68-74, Int Prog Specialist, Off Int Affairs, Nat Cancer Inst, 74-85; RETIRED. *Concurrent Pos:* Am Dent Asn res fel, Nat Inst Dent Res, 49-53 & USPHS, 53-85; hon mem, Japanese Cancer Asn, 88. *Mem:* AAAS; Am Soc Microbiol; Sigma Xi. *Res:* Oral microbiology; bacterial physiology and biochemistry; physiology of oral fusobacteria, spirochetes and clostridia; biochemistry and nutrition; science administration; general medical sciences. *Mailing Add:* 314 Beech Grove Ct Shipley's Choice Millersville MD 21108

OMAYE, STANLEY TERUO, b Detroit, Mich, Jan 25, 45. BIOCHEMICAL NUTRITION. *Educ:* Calif State Univ, Sacramento, BA, 68; Univ Pac, MS, 72; Univ Calif, Davis, PhD(biochem nutrit), 75. *Prof Exp:* Fel toxicol, Calif Primate Res Ctr, 75-76; res chemist biochem, Letterman Army Inst Res, 76-80; RES NUTRITIONIST BIOCHEM, WESTERN REGIONAL RES CTR, USDA, 80- *Concurrent Pos:* NIH fel, 75-76; adj prof dept pharmacol, Univ Pac, 76- *Mem:* Am Inst Nutrit; Am Chem Soc; Am Soc Pharmacol & Exp Therapeut; Am Col Toxicol; Soc Toxicol. *Res:* Mechanisms of action of vitamins and essential trace elements; influence of nutrients on hepatic/extrahepatic drug/toxicant metabolism; natural toxicants found in foods; nutrient-nutrient interactions; dietary fiber. *Mailing Add:* Letterman Army Inst Res SGRD ULY MT OMAYE Presidio San Francisco CA 94129-9991

OMDAHL, JOHN L, b Des Moines, Iowa, July 29, 40; m 63; c 4. ENDOCRINOLOGY. *Educ:* Colo State Univ, BS, 63, MS, 65; Univ Ky, PhD(physiol & biophys), 69. *Prof Exp:* Res chemist physiol, Vet Admin, 66-69; fel biochem, Univ Wis, 69-72; from asst prof to assoc prof, 72-86, PROF BIOCHEM, UNIV NMEX, 86- *Concurrent Pos:* Vis staff mem, Los Alamos Nat Lab, 80-81; vis assoc prof, Univ Tex Health Sci Ctr, Dallas, 81. *Mem:* Am Soc Biol Chemists; Sigma Xi; Am Soc Bone & Mineral Res. *Res:* Molecular mechanisms of enzyme regulation; cytochrome p-450 hydroxglase enzymes and the bioactivation of vitamin D; calcium homostasis and the modes-of-action for vitamin D and parathyroid hormone. *Mailing Add:* Dept Biochem Univ NMex Sch Med Albuquerque NM 87131

O'MEARA, DAVID LILLIS, b Camden, NJ, May 12, 59; m 86; c 2. SEMICONDUCTOR METAL PROCESS INTEGRATION. *Educ:* Stanford Univ, BS (mech eng) & BS (prod design), 81, MS, 84. *Prof Exp:* Res asst, Appl Physics Dept, Stanford Univ, 81-85; mech designer, Harris Microwave Semiconductor, Inc, 84-85; res assoc, J C Schmacher, 85-86, res engr, Air Prods Subsid, 86-90; STAFF ENGR, HUGHES AIRCRAFT TECHNOL CTR, 90- *Concurrent Pos:* Consult, Appl Physics Dept, Stanford Univ, 86. *Mem:* Mat Res Soc. *Res:* Semiconductor metal process integration; liquid source low pressure chemical vapor desposition of dielectrics, design and processing; continuous Czochralski silicon growth design; fiber optical coupler development; single crystal fiber growth research; computer modeling; design of experiments; four US patents. *Mailing Add:* 632 S Freeman St Oceanside CA 92054

O'MEARA, DESMOND, organic chemistry, statistics, for more information see previous edition

O'MEARA, GEORGE FRANCIS, b Lowell, Mass, Sept 9, 41; m 64; c 5. MEDICAL ENTOMOLOGY, GENETICS. *Educ:* Univ Notre Dame, BS, 64, MS, 67, PhD(biol), 69. *Prof Exp:* PROF, FLA MED ENTOM LAB, UNIV FLA, 69- *Concurrent Pos:* Adj assoc prof zool, Fla Atlantic Univ, 77- *Mem:* AAAS; Entom Soc Am; Am Mosquito Control Asn; Am Genetic Asn. *Res:* Reproductive and population biology of mosquitoes. *Mailing Add:* Fla Med Entomol Lab 200 Ninth St SE Vero Beach FL 32962

O'MEARA, O TIMOTHY, b Cape Town, SAfrica, Jan 29, 28; m 53; c 5. MATHEMATICS. *Educ:* Univ Cape Town, BSc, 47, MSc, 48; Princeton Univ, PhD(math), 53. *Hon Degrees:* LLD, Univ Notre Dame, 87. *Prof Exp:* Lectr math, Univ Otago, NZ, 54-56; asst prof, 58-62; mem, Inst Advan Study, 57-58; dept chmn, 65-66 & 68-72, prof math, 62-76, KENNA PROF MATH, UNIV NOTRE DAME, 76-, PROVOST, 78- *Concurrent Pos:* Sloan fel, 60-63; vis prof, Calif Inst Technol, 68; mem adv panel math sci, NSF, 74-77, bd dir Asn Catholic Col & Univ, 86-, bd gov, Notre Dame, Australia, 90-; Gauss prof, Gottingen Acad Sci, 78; Consult, 60- *Mem:* Am Math Soc. *Res:* Algebra; linear groups; number theory; quadratic forms. *Mailing Add:* Off Provost Univ Notre Dame Notre Dame IN 46556

O'MELIA, CHARLES R(ICHARD), b New York, NY, Nov 1, 34; m 56; c 6. ENVIRONMENTAL ENGINEERING, ENVIRONMENTAL CHEMISTRY. *Educ:* Manhattan Col, BCE, 55; Univ Mich, Ann Arbor, MSE, 56, PhD(sanit eng), 63. *Prof Exp:* Asst engr, Hazen & Sawyer, Engrs, 56-57; asst sanit eng, Univ Mich, 57-60; asst prof, Ga Inst Technol, 61-64; fel appl chem, Harvard Univ, 64-66, lectr, 65-66; assoc prof environ sci & eng, Univ NC, Chapel Hill, 66-70, prof, 70-80, dep chmn, 77-80; PROF ENVIRON ENG, JOHNS HOPKINS UNIV, 80-, CHMN, 90- *Concurrent Pos:* Consult, Eng & Urban Health Sci Study Sect, NIH, 70-; Swiss Fed Inst Technol, 71; eng & urban health sci study sect, Environ Protection Agency, 71-; assoc ed, Environ Sci & Technol, J Am Chem Soc, 75- *Honors & Awards:* Environ Sci Award, Asn Environ Eng Prof, 75; Simon Freese Award, Am Soc Civil Engrs, 85; Publ Award, Am Water Works Asn, 65 & 85, A P Black Res Award, 90; Pergammon Press Medal, Int Asn Water Pollution Res & Control, 88. *Mem:* Nat Acad Eng; Am Soc Civil Engrs; Am Water Works Asn; Water Pollution Control Fedn; Am Soc Limnol & Oceanog; Am Chem Soc. *Res:* Aquatic chemistry; predictive modeling of natural systems; theory of water and wastewater treatment. *Mailing Add:* Dept Geog & Environ Eng Johns Hopkins Univ Baltimore MD 21218

OMENN, GILBERT S, b Chester, Pa, Aug 30, 41; m 67; c 3. GENETICS, SCIENCE POLICY. *Educ:* Princeton Univ, AB, 61; Harvard Med Sch, MD, 65; Univ Wash, PhD(genetics), 72. *Prof Exp:* Intern & asst resident, Mass Gen Hosp, 65-67; res assoc protein chem, Nat Inst Arthritis & Metab Dis, 67-69; fel med genetics, 69-71, from asst prof to prof med, 71-79, chmn, dept environ health, 82-83, PROF INTERNAL MED GENETICS, UNIV WASH, 79- , PROF ENVIRON HEALTH, 81-, DEAN, SCH PUB HEALTH & COMMUNITY MED, 82- *Concurrent Pos:* Teaching fel internal med, Mass Gen Hosp, Harvard Univ, 66-67; Nat Genetics Found fel, 71-72, Nat Inst Gen Med Sci res career develop award, 72-76; White House fel, AEC, 73-74; attend physician, Univ Hosp, Harborview Med Ctr, Providence Hosp & Children's Orthop Hosp, Seattle, Wash, 71-; dir, Robert Wood Johnson Found Clin Scholars Prog, 75-77; consult, AEC & Fed Energy Admin, 74-76; investr, Howard Hughes Med Inst, 76-77; asst dir human resources & social & econ serv, Off Sci & Technol Policy, Exec Off of the President, Washington, DC, 77-78, assoc dir, 78-80, assoc dir human resources, veterans & labor, Off Mgt & Budget, 80-81; chmn, Scientific Adv Bd, Elec Power Res Inst; Agency Toxic Substances & Dis Registry; Dept Ecol, Wash State Superfund; Clean Sites Inc & Bd Environ Sci & Toxicol, Nat Res Coun-Nat Acad Sci; mem, bd dir, Rohm & Haas, Amgen, Biotech Labs, Ecova & Immune Response Corp. *Mem:* Inst Med-Nat Acad Sci; Am Soc Human Genetics; AAAS; fel Am Col Physicians; Am Occup Med Asn; fel Ramazzini Soc. *Res:* Cancer prevention; health promotion in older adults; ecogenetics; risk assessment; genetic screening. *Mailing Add:* Sch Pub Health & Community Med Univ Wash SC-30 Seattle WA 98195

OMER, GEORGE ELBERT, JR, b Kansas City, Kans, Dec 23, 22; m 49; c 2. ORTHOPEDIC SURGERY. *Educ:* Ft Hays Kans State Univ, BA, 44; Univ Kans, MD, 50; Baylor Univ, MS, 55; Am Bd Orthop Surg, dipl. *Prof Exp:* Chief surg serv, Irwin Army Hosp, Ft Riley, Kans, 57-59, consult orthop surg, Eighth US Army, Korea, 59-60, chief hand surg, Fitzsimons Army Hosp, Denver, Colo, 62-64, dir orthop path, Armed Forces Inst Path, Washington, DC, 64-65, chief hand surg ctr, Brooke Army Med Ctr, Ft Sam Houston, Tex, 66-70, chief orthop, 67-70; PROF ORTHOP, CHMN DEPT & CHMN DIV HAND SURG, SCH MED, UNIV N MEX, 70- *Concurrent Pos:* Instr, Univ Kans, 46-47; asst clin prof, Sch Med, Univ Colo, 61-63; assoc clin prof, Georgetown Univ, 64-65, Univ Tex Med Sch San Antonio, 66-70 & Univ Tex Med Br Galveston, 68-70; consult orthop & hand surg, numerous civilian and mil hosps, 62- Indian Health, USPHS, 66- & orthop, Surgeon Gen, US Army, 69-; pres, Sunderland Soc, Int Nerve Study Soc, 83. *Mem:* Am Acad Orthop Surg; Am Asn Surg of Trauma; Am Col Surg; Am Orthop Asn; Am Soc Surg of Hand (pres, 68). *Res:* Hand surgery; peripheral nerve repair and sensibility; tendon transfers for reconstruction of nerve loss; evacuation of injuries and emergency room care; burn injuries. *Mailing Add:* 2211 Lomas Dr NE Albuquerque NM 87131-2357

OMER, GUY CLIFTON, JR, b Mankato, Kans, Mar 20, 12; m 42; c 2. PHYSICS. *Educ:* Univ Kans, BS, 36, MS, 37; Calif Inst Technol, PhD(physics), 47. *Prof Exp:* Instr, Univ Hawaii, 41-43; asst prof, Occidental Col, 47-48 & Univ Ore, 48-49; asst prof phys sci, physics & astron, Univ Chicago, 49-55; chmn dept phys sci, 67-72, PROF PHYSICS & ASTRON, UNIV FLA, 55- *Concurrent Pos:* Res fel, Calif Inst Technol, 47-48; vis prof, Univ Calif, Los Angeles, 64-65; sr vis fel, Inst Astron, Cambridge, Eng, 72; prof, Fla at Utrecht, Univ Utrecht, Holland, 73; sr vis fel, Dept Astrophys, Univ Oxford, 73. *Mem:* Am Phys Soc; Int Astron Union; Seismol Soc Am; Am Asn Physics Teachers; fel Royal Astron Soc. *Res:* Relativity; cosmology. *Mailing Add:* 1080 SW 11 Terr Gainesville FL 32601

OMER, SAVAS, aeronautical engineering, astronautical engineering, for more information see previous edition

OMID, AHMAD, b Abadeh, Iran, May 6, 31; m 57; c 2. AGRONOMY, PHYSIOLOGY. *Educ:* Calif State Polytech Col, BS, 55 & 57; Ore State Univ, MS, 62, PhD(farm crops), 64. *Prof Exp:* PLANT PHYSIOL GROUP LEADER, CHEVRON CHEM CO, 62-, SR PLANT PHYSIOLOGIST, 70- *Mem:* Weed Sci Soc Am. *Res:* Use of herbicides and growth regulators for control and modification of plant growth, including studies of mode of action and movement of these chemicals in plants and soils. *Mailing Add:* 533 Sturbridge Ct Walnut Creek CA 94598

OMIDVAR, KAZEM, b Mashad, Iran, Dec 6, 26. ATOMIC PHYSICS, ASTROPHYSICS. *Educ:* Univ Teheran, BS, 51; NY Univ, MS, 54, PhD(physics), 59. *Prof Exp:* Instr physics, Cooper Union, 54-56 & NY Univ, 56-57; lectr, Rutgers Univ, 57-58; instr, City Col New York, 58-59; assoc prof, Univ Teheran, 59-60; astrophysicist, Theoret Studies Group, 60-77, ASTROPHYSICIST, LAB FOR ATMOSPHERES, GODDARD SPACE FLIGHT CTR, NASA, 77- *Concurrent Pos:* Lectr, Univ Md, 61-64. *Mem:* Fel Am Phys Soc; Am Astron Soc. *Res:* Development of reliable formula for calculating ionization, excitation and charge exchange cross sections in electron-atom and ion-atom collisions; calculation of multiphotom absorption of radiation in atoms; calculation of ozone production in the atmosphere due to the solar ultraviolet radiations. *Mailing Add:* Lab for Atmospheres Code 914 NASA-Goddard Space Flight Ctr Greenbelt MD 20771

OMODT, GARY WILSON, b LaCrosse, Wis, July 30, 29; m 57; c 6. PHARMACEUTICAL CHEMISTRY. *Educ:* Univ Minn, BS, 53, PhD(pharmaceut chem), 59. *Prof Exp:* assoc prof 58-68, PROF PHARMACEUT CHEM, COL PHARM, SDAK STATE UNIV, 68- *Mem:* Am Pharmaceut Asn; Am Asn Cols Pharm; Sigma Xi. *Res:* Synthesis of organic medicinal agents. *Mailing Add:* Dept Pharmaceut Chem Col Pharm SDak State Univ Box 2201 Brookings SD 57007

O'MORCHOE, CHARLES C C, b Quetta, India, May 7, 31; m 55; c 2. ANATOMY. *Educ:* Univ Dublin, BA, 53, MB, BCh & BA Obstet, 55, MA, 59, MD, 61, PhD(physiol), 69, DSc, 81. *Prof Exp:* Intern med, surg, gynec & obstet, Halifax Gen Hosp, Eng, 55-57; lectr anat, Sch Med, Univ Dublin, 57-61; vis lectr physiol, Sch Med, Univ Md, 61-62; instr anat, Harvard Med Sch, 62-63; lectr, Sch Med, Univ Dublin, 63-65, assoc prof physiol, 66-68; from assoc prof to prof anat, Sch Med, Univ Md, Baltimore City, 68-74, actg chmn dept, 71-73; prof anat & chmn dept, Stritch Sch Med, Loyola Univ, Chicago, 74-84; DIR COL & PROF ANAT SCI & SURG, COL MED, UNIV ILL, URBANA, 84- *Concurrent Pos:* Chmn, Anat Bd Med, 71-73. *Mem:* Am Soc Nephrology; Int Soc Lymphology; Am Asn Anat; Dir of Placement Serv, Exec Comm; Past Pres. *Res:* Anatomy and physiology of renal vascular circulation and renal lymphatic system; urinary cytology in reproductive endocrinology; medical education; histology; lymphology; nephrology. *Mailing Add:* Col Med Univ Ill 190 Med Sci Bldg 506 S Mathews Urbana IL 61801

O'MORCHOE, PATRICIA JEAN, b Halifax, Eng, Sept 15, 30; m 55; c 2. HISTOLOGY, CYTOLOGY. *Educ:* Univ Dublin, BA, 53, MB, BCh & BAO, 55, MA & MD, 66. *Prof Exp:* House officer obstet & gynec, surg & med, Brit Nat Health Serv, Halifax Gen Hosp, Eng, 55-57; jr lectr physiol & histol, Univ Dublin, 59-61; instr cytopath, Johns Hopkins Univ, 61-62; res assoc path & surg, Harvard Univ, 62-63; lectr physiol & histol, Univ Dublin, 63-68; instr cytopath, Johns Hopkins Univ, 68-70; asst prof anat & histol, Univ Md, Baltimore City, 70-74; asst prof path, Johns Hopkins Univ, 73-74; from assoc prof to prof path & anat, Stritch Sch Med, Loyola Univ Chicago, 74-84; PROF PATH & CELL & STRUCT BIOL, COL MED, UNIV ILL, URBANA-CHAMPAIGN, 84- *Concurrent Pos:* Vis lectr, Johns Hopkins Hosp & Med Sch, Johns Hopkins Univ, 70-73,; Frank C Bressler Reserve Fund grant anat, Univ Md, Baltimore City, 71-72; NIH grants, 74-77, 78-80 & 81-83; biomed res support grant, Col Med, Univ Ill, 85-86; assoc ed, Lymphology, 87- *Mem:* Am Soc Cytol; Am Asn Anat; Int Acad Cytol; NAm Soc Lymphology (secy, 88-90, treas, 90-); Int Soc Lymphology. *Res:* structure and function of lymphatic system; computer assisted morphometric analysis; computers in teaching. *Mailing Add:* Col Med Univ Ill 506 S Mathews Urbana IL 61801

OMRAN, ABDEL RAHIM, b Cairo, Egypt, Mar 29, 25; m 53; c 3. EPIDEMIOLOGY. *Educ:* Minufia Sch, Egypt, BS, 45; Cairo Univ, MD, 52, DPH, 54; Columbia Univ, MPH, 56, DrPH, 59; Trudeau Sch Tuberc, Nat Tuberc & Respiratory Dis Asn, cert, 59. *Prof Exp:* Lectr, Cairo Univ, 59-63; res scientist & clin assoc prof environ med, NY Univ, 63-66; assoc prof epidemiol, Univ NC, Chapel Hill, 66-71, prof epidemiol, 71-81; DIR, POP HEALTH & DEVELOP PROG, CTR INT DEVELOP, UNIV MD. *Concurrent Pos:* WHO study fel health serv eastern & western Europe, 63; clin assoc prof, Univ Ky, 64-66; Ford Found consult, India, 69; coordr epidemiol studies in Asian countries, WHO, 69-, dir, Int Reference Ctr on Epidemiol of Human Reproduction, 71-; assoc dir pop epidemiol, Carolina Pop Ctr, 70- *Honors & Awards:* Sci Achievement Medal, Egyptian Govt. *Mem:* Royal Soc Health; Asn Teachers Prev Med; Soc Epidemiol Res; Am Thoracic Soc; fel Am Pub Health Asn. *Res:* Epidemiological studies of human reproduction in Asia and the Middle East; population problems and prospects in the Middle East; development of manual on community medicine for developing countries; health and disease patterns associated with demographic change; Muslim fertility. *Mailing Add:* Ctr Int Develop Rm 2125 Mill Bldg Univ Md College Park MD 20742-4515

OMTVEDT, IRVIN T, b Rice Lake, Wis, June 12, 35; m 59; c 2. SWINE GENETICS. *Educ:* Univ Wis-Madison, BS, 57; Okla State Univ, MS, 59, PhD(animal breeding), 61. *Prof Exp:* Res asst, dept animal sci, Okla State Univ, 58-61; exten swine specialist & from asst prof to assoc prof, dept animal sci, Univ Minn, 62-64; from assoc prof to prof, dept animal sci, Okla State Univ, 64-73; assoc dir, Ala Agr Exp Sta, Auburn Univ, 73-75; prof animal sci & head dept, 75-82, DEAN AGR RES, AGR RES DIV, UNIV NEBR, 82- *Concurrent Pos:* Mem, Alexander von Humboldt Award Selection Comt, 82-86, tech adv comt, Binat Agr Res & & Develop US & Israel, 84-86 & Animal Health Sci Adv Coun, USDA, 84-86. *Mem:* Am Soc Animal Sci (secy-treas, 80-83, pres, 84-85); Coun Agr Sci & Tech. *Res:* Crossbreeeding, selection, and environmental physiology of swine. *Mailing Add:* Dept Animal Sci Univ Nebr Lincoln NE 68503

OMURA, JIMMY KAZUHIRO, b San Jose, Calif, Sept 8, 40; div; c 2. ELECTRICAL ENGINEERING. *Educ:* Mass Inst Technol, BS & MS, 63; Stanford Univ, PhD(elec eng), 66. *Prof Exp:* Sci adv commun eng, Stanford Res Inst, 66-69; from asst prof to prof elec eng, Univ Calif, Los Angeles, 59-85; CONSULT, 85- *Concurrent Pos:* Consult, indust & govt. *Mem:* Fel Inst Elec & Electronics Engrs. *Res:* Feedback communication systems; statistical analysis of systems; coding techniques; mathematical programming applications; data compression systems; spread spectrum communication systems; satellite communication systems; rate distortion theory; cryptography. *Mailing Add:* c/o Cylink Corp 130 B Kifer Ct Sunnyvale CA 94086

ONAL, ERGUN, b Dec 7, 46. RESPIRATION, RESPIRATORY DISORDERS. *Educ:* Univ Hacettepe, Ankara, Turkey, MD, 70. *Prof Exp:* PROF RESPIRATORY & CRITICAL CARE MED, UNIV ILL, 82- *Mem:* Am Thoracic Soc; Am Physiol Soc; Cent Soc Res; Am Asn Appl Psychol. *Res:* Regulation of respiration; sleep. *Mailing Add:* Dept Med M C 787 Univ Ill Col Med PO Box 6998 Chicago IL 60680

O'NAN, MICHAEL ERNEST, b Ft Knox, Ky, Aug 9, 43. ALGEBRA. *Educ:* Stanford Univ, BS, 65; Princeton Univ, PhD(math), 69. *Prof Exp:* Asst prof, 70-73, assoc prof, 73-77, PROF MATH, RUTGERS UNIV, NEW BRUNSWICK, 77- *Concurrent Pos:* Vis asst prof math, Univ Chicago, 71-72; Sloan fel, 74. *Mem:* Am Math Soc. *Res:* Classification problems in the study of finite simple groups, those associated with doubly-transitive permutation groups. *Mailing Add:* Dept of Math Rutgers State Univ New Brunswick NJ 08903

ONAT, E(MIN) T(URAN), b Istanbul, Turkey, Jan 28, 25; m 59. MECHANICS. *Educ:* Tech Univ Istanbul, Dipl, 48, DSc, 51. *Prof Exp:* Res fel appl math, Brown Univ, 51-52, Jewett fel, 52-53, res assoc, 53-57, assoc prof eng, 57-60, prof, 60-65; PROF ENG, YALE UNIV, 65- *Concurrent Pos:* Consult, Turkish Elec Authority, 56-57; Guggenheim fel, 63-64. *Mem:* Am Math Soc; Opers Res Soc Am; assoc Am Soc Mech Engrs. *Res:* Applied mechanics of solids; plasticity and viscoelasticity. *Mailing Add:* 434 College St New Haven CT 06511-6632

ONCLEY, JOHN LAWRENCE, b Wheaton, Ill, Feb 14, 10; m 33, 72; c 2. MOLECULAR BIOPHYSICS. *Educ:* Southwestern Col, Kans, AB, 29; Univ Wis, PhD, 33. *Hon Degrees:* MA, Harvard Univ, 46; DSc, Southwestern Col, Kans, 54. *Prof Exp:* Asst chem, Univ Wis, 29-31; Nat Res Coun fel, Mass Inst Technol, 32-34; instr, Univ Wis, 34-35 & Mass Inst Technol, 35-43; from asst prof to prof biol chem, Harvard Med Sch, 43-62; prof, 62-80, dir, Biophys Res Div, Inst Sci & Technol, 72-76, EMER PROF CHEM & BIOL CHEM, UNIV MICH, ANN ARBOR, 80- *Concurrent Pos:* Res assoc, Harvard Sch Med Sch, 39-41, assoc, 41-43; Guggenheim & Fulbright fels, King's Col, Univ London, 53; mem comn plasma fractionation, Ctr Blood Res, 53-80; fel, Coun High Blood Pressure Res, Am Heart Asn; mem sci adv coun, Southwestern Col, 80. *Honors & Awards:* Pure Chem Award, Am Chem Soc, 42; Stouffer Prize, 72. *Mem:* Nat Acad Sci; Am Chem Soc; Biophys Soc; Am Acad Arts & Sci. *Res:* Dielectric properties of gases, liquids and proteins; biophysical chemistry of protein systems; fractionation and interactions of proteins and lipoproteins. *Mailing Add:* Biophys Res Div Univ of Mich Ann Arbor MI 48109

ONCLEY, PAUL BENNETT, b Chicago, Ill, June 22, 11; m 33; c 2. ACOUSTICS. *Educ:* Southwestern Col Kans, AB, 31; Univ Rochester, BMusic, 32, MMusic, 33; Columbia Univ, PhD(music acoustics), 52. *Prof Exp:* Asst prof vocal music, Univ NC, Greensboro, 37-42; electronics engr, Div Phys War Res, Duke Univ, 42-45; mem tech staff, Bell Tel Labs, Inc, 45-49; assoc prof vocal music, Westminster Choir Col, 49-53; asst to dir eng, Gulton Indust, 53-58; res specialist, Aerospace Div, Boeing Co, 58-63, Space Div, 63-67 & Com Airplane Div, 67-74; sr res scientist, Man, Acoustics & Noise, Inc, 74-80; SR SCI ADV, OKB ENTERPRISES, 86- *Concurrent Pos:* Res Corp res grant acoustics of singing, 49-51; design consult, Rangertone, Inc, 49-51 & Kay Elec Co, 53; lectr, Columbia Univ, 50-52; guest prof, Highline Col, 62-63; conductor, Tacoma Youth Symphony, 63-66; guest prof, Pa Lutheran Univ, 75-76. *Mem:* Acoust Soc Am; Am Inst Aeronaut & Astronaut. *Res:* Music; acoustics of speech and music, particularly physics and physiology of singing; electronic music; ultrasonics and underwater sound; systems engineering for aerospace and physics of space exploration; airplane noise propagation; acoustic impedance measurements. *Mailing Add:* 6535 Seaview Dr NW 311-B Seattle WA 98117-6051

ONDETTI, MIGUEL ANGEL, b Buenos Aires, Arg, May 14, 30; m 58; c 2. ORGANIC CHEMISTRY. *Educ:* Univ Buenos Aires, Dr Chem, 57. *Prof Exp:* Sr res chemist, Squibb Inst Med Res, Buenos Aires, 56-60, NJ, 60-66, res supvr, Dept Org Chem, 66-73, sect head, 73-76, dir, Biol Chem Dept, 76-79, assoc dir, 79-81, VPRES BASIC RES, SQUIBB INST MED RES, 81- *Concurrent Pos:* Instr, Univ Buenos Aires, 57-60. *Honors & Awards:* Alfred Burger Award, Am Chem Soc; CIBA Hypertension Award, Am Heart Asn. *Mem:* Am Chem Soc; Arg Chem Soc; Swiss Chem Soc; Am Soc Biol Chemists; Sigma Xi. *Res:* Isolation, structure determination and synthesis of natural products; peptide isolation and synthesis; synthesis and study of enzyme inhibitors. *Mailing Add:* Squibb Inst Med Res PO Box 4000 Princeton NJ 08540

ONDIK, HELEN MARGARET, b New York, NY, Dec 25, 30. DATA COMPILATION, DATABASE FORMATION. *Educ:* Hunter Col, AB, 52; Johns Hopkins Univ, MA, 54, PhD(chem), 57. *Prof Exp:* Fulbright grant, Univ Amsterdam, 57-58; PHYS CHEMIST, NAT INST STANDARDS & TECHNOL, 58- *Honors & Awards:* Silver Medal, Dept Commerce, 71. *Mem:* AAAS; Am Chem Soc; Am Ceramic Soc. *Res:* Solution and crystal chemistry of condensed phosphates; crystallographic data compilation; inorganic structures; phase diagram database formation. *Mailing Add:* 2737 Devonshire Place NW Washington DC 20008

ONDO, JEROME G, b Homestead, Pa, July 25, 39; m 67; c 2. NEUROENDOCRINOLOGY, CANCER. *Educ:* Univ Pittsburgh, BS, 62; John Carroll Univ, MS, 64; Univ Va, PhD(physiol), 70. *Prof Exp:* Asst prof, 72-77, ASSOC PROF PHYSIOL, MED UNIV SC, 77- *Concurrent Pos:* NIH grant, Univ Tex Southwestern Med Sch Dallas, 70-72. *Mem:* Am Physiol Soc. *Res:* Reproduction; mammary cancer. *Mailing Add:* Dept Physiol Med Univ SC Col Med 171 Ashley Ave Charleston SC 29425

ONDOV, JOHN MICHAEL, b Somerville, NJ, Jan 22, 48. AEROSOL CHEMISTRY, NUCLEAR ANALYTICAL CHEMISTRY. *Educ:* Muhlenberg Col, BS, 70; Univ Md, PhD(chem), 74. *Prof Exp:* SR SCIENTIST AEROSOL CHEM, LAWRENCE LIVERMORE LAB, UNIV CALIF, 75- *Mem:* Am Chem Soc; AAAS. *Res:* Physical and chemical properties of aerosols; studies of formation distribution and transport and transformation of physical and chemical anthropogenic atmospheric aerosol; nuclear activation techniques for elemental analysis. *Mailing Add:* 9430 Clocktower Lane Columbia MD 21046-1806

ONDREJKA, ARTHUR RICHARD, b New York, NY, Mar 20, 34; m 59; c 3. ELECTRICAL ENGINEERING, PHYSICS. *Educ:* Univ Colo, AB, 62, MS, 73. *Prof Exp:* Technician, 59-63, PHYSICIST & ENGR ELECTRONICS, NAT BUR STANDARDS, 64- *Mem:* Sigma Xi. *Res:* Measurement of time domain characteristics of systems, in particular antennas and microwave systems and lasers and optical components. *Mailing Add:* 67 Anemone Dr Boulder CO 80302

O'NEAL, CHARLES HAROLD, b Miami, Fla, Feb 18, 36; m 60; c 3. BIOCHEMISTRY. *Educ:* Ga Inst Technol, BS, 57; Emory Univ, PhD(biochem), 63. *Prof Exp:* Asst chem, Ga Inst Technol, 54-56 & Eng Exp Sta, 56-57; asst biochem, Emory Univ, 57-59; scientist, Nat Heart Inst, 63-65; res assoc biochem, Rockefeller Univ, 65-68, asst prof, 68; assoc prof biophys, Med Col Va, 68-; AT DEPT MICROBIOL, VA COMMONWEALTH UNIV, RICHMOND. *Concurrent Pos:* vis scientist, Lab Molecular Biol, Med Res Coun, Cambridge, Eng, 71-72; comnr Sci Manpower Comn & Pres Va Commonwealth Univ Fac Senate, 77-78. *Mem:* AAAS; Am Inst Chem; Biochem Soc; Biophys Soc; Am Chem Soc. *Res:* Biochemical genetics; cytology. *Mailing Add:* 9224 Holbrook Dr Richmond VA 23229

O'NEAL, FLOYD BRELAND, b Fairfax, SC, May 4, 28; m 54; c 5. ANALYTICAL CHEMISTRY. *Educ:* The Citadel, BS, 48; Tulane Univ, MS, 50; Ga Inst Technol, PhD(chem), 59. *Prof Exp:* From instr to prof chem, Univ Tex, El Paso, 54-66; head, Dept Chem, Augusta Col, 66-76, prof, 66-85; CONSULT, 85- *Mem:* Fel AAAS; Am Chem Soc; Sigma Xi. *Res:* Chelate formation; solvent extraction. *Mailing Add:* PO Box 3393 Augusta GA 30914-3393

ONEAL, GLEN, JR, b Great Falls, Mont, Feb 2, 17; m 41; c 1. APPLIED PHYSICS, ELECTRONICS. *Educ:* Mont State Univ, BS, 40; Univ Pa, MS, 47. *Prof Exp:* Assoc physicist, US Naval Ord Lab, 41-45; physicist, Sun Oil Co, 47-55; res physicist, Am Viscose Corp, 55-63; res physicist, FMC Corp Chem Group, Princeton, NJ, 63-; sr physicist, Corp Eng & Construct, Princeton, NJ, 81-82; RETIRED. *Concurrent Pos:* Res assoc, Nat Bur Standards, 68-70; consult, 82- *Honors & Awards:* Citation, Nat Bur Standards, 70. *Mem:* Am Phys Soc; Am Soc Testing & Mat; Sigma Xi; Inst Elec & Electronics Engrs. *Res:* Application of electronic instrumentation to problems in material testing, polymer research and pharmaceutical applications; materials research. *Mailing Add:* 128 Yale Ave Swarthmore PA 19081

O'NEAL, HARRY E, b Cincinnati, Ohio, Apr 2, 31; div; c 2. PHYSICAL CHEMISTRY, CHEMICAL DYNAMICS & THERMODYNAMICS. *Educ:* Harvard Univ, BA, 53; Univ Wash, PhD(chem), 57. *Prof Exp:* Res chemist, Shell Oil Co, 57-59; res assoc, Univ Southern Calif, 59-60; from asst prof to assoc prof, 61-70, PROF CHEM, SAN DIEGO STATE UNIV, 70- *Concurrent Pos:* Alexander von Humboldt Found sr scientist res award, 75. *Mem:* Am Phys Soc. *Res:* Thermodynamics; chemical kinetics. *Mailing Add:* Dept Chem San Diego State Univ 5300 Campanile Dr San Diego CA 92182

O'NEAL, HARRY ROGER, b Kansas City, Kans, Jan 2, 35. PHYSICAL CHEMISTRY. *Educ:* Univ Kansas City, BS, 58; Univ Calif, Berkeley, PhD(chem), 63. *Prof Exp:* Res chemist, Du Pont Co, 62-65 & Douglas Aircraft Co, 65-66; fel chem, Pa State Univ, 66-67; asst prof, Lowell Tech Inst, 67-70; asst prof chem, Pa State Univ, Ogontz Campus, 70-82; PROF, DEPT CHEM, AURORA UNIV, 82- *Mem:* Am Chem Soc. *Res:* Thermodynamic and transport properties of solids at low temperatures; thermodynamics and kinetics of phase transitions. *Mailing Add:* Dept Chem Aurora Univ 347 S Gladstone Ave Aurora IL 60506

O'NEAL, HUBERT RONALD, b Rotan, Tex, Apr 27, 37; m 60; c 2. ORGANIC CHEMISTRY. *Educ:* Tex Tech Col, BS, 59; NTex State Univ, MS, 64, PhD(chem), 67. *Prof Exp:* Chemist, Sherwin-Williams Co, Tex, 59-61; res chemist, W R Grace & Co, Md, 67-68; from res chemist to sr res chemist, 68-74, res group leader, Petro-Tex Chem Corp, 74-77; supt process res, Denka Chem Corp, 77-80; tech dir, Kocide Chem Corp, 80-83; PRES & OWNER, HOCHEM ENT INC, 84- *Mem:* Am Chem Soc; NY Acad Sci; Sigma Xi. *Res:* Polymer research on elastomeric materials; ketene chemistry and aldehyde polymerizations; synthetic rubber chemistry; polymerization kinetics; agricultural fungicides. *Mailing Add:* 11202 Pecan Creek Dr Houston TX 77043

O'NEAL, JOHN BENJAMIN, JR, b Macon, Ga, Oct 15, 34; m 60; c 2. ELECTRICAL ENGINEERING. *Educ:* Ga Inst Technol, BEE, 57; Univ SC, MEE, 60; Univ Fla, PhD(elec eng), 63. *Prof Exp:* Engr, Southern Bell Tel & Tel Co, 57 & Martin Co, Fla, 59-60; mem tech staff, Bell Tel Labs, Inc, 64-66, supv eng, 66-67; from asst prof to prof elec eng, 67-84, DISTINGUISHED PROF ELEC & COMP ENG, NC STATE UNIV, 84- *Concurrent Pos:* Consult for 12 companies. *Mem:* Fel Inst Elec & Electronic Engrs. *Res:* Studies of communications theory to determine efficient use of band-width for television transmission; information theory. *Mailing Add:* Dept Elec Comp Eng NC State Univ Raleigh NC 27695-7911

O'NEAL, LYMAN HENRY, behavior-ethology, genetics, for more information see previous edition

O'NEAL, PATRICIA L, b St Louis, Mo, June 14, 23. PSYCHIATRY. *Educ:* Wash Univ, AB, 44, MD, 48. *Prof Exp:* Intern, Univ Iowa Hosps, 48-49; resident psychiat, Barnes & Allied Hosps, St Louis, Mo, 49-51; from instr to asst prof, 52-61, PROF PSYCHIAT, SCH MED, WASH UNIV, 61- *Concurrent Pos:* Fel psychosom med, Sch Med, Wash Univ, 52. *Res:* Clinical psychiatry, especially geriatric psychiatry. *Mailing Add:* Barnes Hosp Plaza Suite 16-428 Pavillion St Louis MO 63110

O'NEAL, ROBERT MUNGER, b Wiggins, Miss, Oct 7, 22; m 47; c 4. PATHOLOGY. *Educ:* Univ Miss, BS, 43; Univ Tenn, MD, 45. *Prof Exp:* Resident chest dis, State Sanatorium, Miss, 49-52; from instr to assoc prof path, Sch Med, Wash Univ, 54-61; prof & chmn dept, Col Med, Baylor Univ, 61-69; prof, Albany Med Col, 69-73; prof path & chmn dept, Cols Med & Dent, Univ Okla, 73-78; PROF & CHMN, DEPT PATH, UNIV MISS, 78- *Concurrent Pos:* Fel path, Mass Gen Hosp, 52-54; dir, Bender Hyg Lab, 69-73. *Mem:* Am Soc Exp Path; Am Soc Clin Path; Am Thoracic Soc; Am Asn Path & Bact; Col Am Path. *Res:* Cardiovascular and pulmonary diseases; atherosclerosis. *Mailing Add:* Dept Path & Oral Path Univ Miss Med Ctr 2500 N State St Jackson MS 39216

O'NEAL, RUSSELL D, b Columbus, Ind, Feb 15, 14; m 48; c 2. TECHNICAL MANAGEMENT. *Educ:* DePauw Univ, AB, 36; Univ Ill, PhD(physics), 41. *Prof Exp:* Asst physics, Univ Ill, 36-41, instr, 41-42; mem staff, Radiation Lab, Mass Inst Technol, 42-43, sect head, 43-45, group leader, 45; proj physicist, Eastman Kodak Co, 45-48; head aerophysics group, Willow Run Res Ctr, Univ Mich, 49, from asst dir to dir, 50-52; asst div mgr, Consol Vultee Aircraft Corp, 52-53, prog dir, 53-55; dir systs planning, Bendix Corp, 55-57, gen mgr, Systs Div, 57-60, vpres eng & res, 60-63, vpres & group exec aerospace systs, 63-66; asst secy, Army Res & Develop, Washington, 66-68; exec vpres aerospace, Bendix Corp, 68-69, pres, Bendix Aerospace-Electronics Group, 69-72, pres, Group Opers, 72-73; exec vpres, KMS Fusion, Inc, 74-75, chmn & chief exec officer, 75-76, consult, 76-; VICE CHMN, ENVIRON RES INST, 83- *Concurrent Pos:* Mem, Study Group Guided Missiles, Dept Defense, 58-59; consult, Sci Adv Panel, US Army, 69-80; mem, Nat Sci Bd, 72-78; mem, Army Sci Bd, 80-86; mem bd, Varo Inc, 78-88, Sanders Assocs, Inc, 78-86 & Montgomery & Assocs, 87- *Mem:* Fel Am Phys Soc; fel Inst Elec & Electronic Engrs; assoc fel Am Inst Aeronaut & Astronaut. *Res:* Nuclear physics; guided missiles; systems analysis; laser fusion. *Mailing Add:* Environ Res Inst PO Box 8618 Ann Arbor MI 48106

O'NEAL, STEVEN GEORGE, b Peru, Ind, Oct 28, 47; m 70; c 1. MEMBRANE BIOCHEMISTRY, ENZYMOLOGY. *Educ:* Wabash Col, BA, 70; Univ SC, PhD(chem & biochem), 77. *Prof Exp:* Instr biol, Wabash Col, 70-72; NIH fel, Cornell Univ, 77-79, assoc fel, 79-80; ASST PROF CHEM, UNIV OKLA, 80- *Mem:* AAAS; Am Chem Soc; NY Acad Sci; Sigma Xi. *Res:* Determining the reaction mechanisms, membrane biogenesis, and physiological significance of membrane-bound ion translocases and other receptors. *Mailing Add:* 1220 Crossroads Ct Norman OK 73069

O'NEAL, THOMAS DENNY, b Vandalia, Ill, June 5, 41; div; c 1. BIOREGULATORS, AGRICULTURAL CHEMICALS. *Educ:* Southern Ill Univ, Carbondale, BS, 63; Duke Univ, DF(plant physiol), 68. *Prof Exp:* Res assoc plant physiol, Univ Ky, 68-69 & Carleton Univ, 71-73; fel, Brandeis Univ, 69-71; asst prof, Rensselaer Polytech Inst, 73-75; group leader plant growth regulator discovery, Am Cyanamid Co, 75-79; TECH PROD MGR, BASF, 80- *Mem:* Am Soc Plant Physiologists; Sigma Xi; AAAS; Am Soc Agron. *Res:* Laboratory, greenhouse and field testing of plant growth regulants to enchance yield, quality or harvestability of key agronomic crops; regulation of N and C metabolism in plants. *Mailing Add:* BASF Corp PO Box 13528 Durham NC 27709

O'NEAL, THOMAS NORMAN, b Ft Smith, Ark, Sept 17, 38; m 63; c 3. EXPERIMENTAL SOLID STATE PHYSICS. *Educ:* Miss Col, BS, 60; Univ Fla, MS, 62; Clemson Univ, PhD(physics), 70. *Prof Exp:* From asst prof to assoc prof, 63-75, PROF PHYSICS, CARSON-NEWMAN COL, 75- *Concurrent Pos:* Pew fel, Univ Ky, 87-88. *Mem:* Am Asn Physics Teachers; Sigma Xi; Am Solar Energy Soc. *Res:* Electron irradiation damage in metals. *Mailing Add:* Dept of Physics Carson-Newman Col Jefferson City TN 37760

ONEDA, SADAO, b Akita, Japan, June 30, 23; m 50; c 2. THEORETICAL PHYSICS. *Educ:* Tohoku Univ, Japan, BSc, 46, MSc, 48; Nagoya Univ, DrSci(theoret physics), 53. *Prof Exp:* Res assoc theoret physics, Tohoku Univ, Japan, 48-50; from asst prof to prof, Kanazawa Univ, 50-63; from asst prof to assoc prof, 63-67, PROF PHYSICS, UNIV MD, COLLEGE PARK, 67- *Concurrent Pos:* Mem staff, Res Inst Fundamental Physics, Kyoto Univ, 53-55 & 60-63; Imp Chem Industs fel, Univ Manchester, 55-57; mem, Inst Advan Study, 57-58; res assoc, Univ Md, 58-60; vis prof physics, Univ Wis-Milwaukee, 69-70; vis scientist, Int Ctr Theoret Physics, Trieste, 70; vis res mem, Res Inst Fundamental Physics, Kyoto Univ, 77. *Mem:* Fel Am Phys Soc; Phys Soc Japan. *Res:* Physics; theoretical elementary particle physics; field theory. *Mailing Add:* Dept Physics & Astron Univ Md College Park MD 20742

O'NEIL, CAROL ELLIOT, b St Louis, Mo, Jan 29, 51. OCCUPATIONAL ASTHMA. *Educ:* Univ Ariz, BS, 72; Univ New Orleans, MS, 78; Tulane Univ, PhD(biol), 82. *Prof Exp:* From instr to asst prof, 82-87, ASSOC PROF MED, TULANE MED CTR, 87- *Concurrent Pos:* NIH fel, Tulane Med Ctr, 82-84; consult, US Dept Agr, 82-85. *Mem:* Am Acad Allergy & Immunol; Am Thoracic Soc; Sigma Xi. *Res:* Etiologic agents; pathogenic disease mechanisms of immediate hypersensitivity; occupational exposures to air-borne dusts and gases. *Mailing Add:* 1700 Perdido St New Orleans LA 70112

O'NEIL, DANIEL JOSEPH, b Boston, Mass, June 5, 42. POLYMER SCIENCE, CHEMICAL ENGINEERING. *Educ:* Northeastern Univ, BA, 64; S Conn State Univ, MS, 67; Dublin Univ, Ireland, PhD, 72. *Prof Exp:* Sr res chemist, Raybestos-Manhattan Adv Res Ctr, 64-67; unit leader, Hitco, Inc, 67-68; res fel, Dublin Univ, 68-70; tech dir, Euroglas Ltd, UK & Ireland, 70-72; lectr polymer chem, Nat Inst Higher Educ, 72-75, head polymer prog & dir external liaison & co-op educ, 73-75; sr res scientist, Ga Inst Technol, 75-79, head indust chem group, 76-77, chief, Div Chem & Mat Sci, 77-79, sr staff, Off Dir, 79-81, assoc dir, Energy & Mat Sci Lab, 83-90, PRIN RES SCIENTIST, GA INST TECHNOL, 79-, GROUP DIR, GA TECH RES INST, 90- *Concurrent Pos:* Chief exec, Europ Res Inst Ireland, 81-83. *Mem:* Fel Am Inst Chemists; Am Chem Soc; Biomass Energy Res Asn; Soc Advan Mat & Process Eng; AAAS. *Res:* Energy and environmental processes; advanced composites; process economics. *Mailing Add:* 2660 Nesbit Trail Alpharetta GA 30201-5263

O'NEIL, ELIZABETH JEAN, b Palo Alto, Calif, Jan 29, 42; m 65; c 2. APPLIED MATHEMATICS. *Educ:* Mass Inst Technol, SB, 63; Harvard Univ, MA, 64, PhD(appl math), 68. *Prof Exp:* Vis mem appl math, Courant Inst Math Sci, NY Univ, 66-67, asst prof, 67-68; lectr, Mass Inst Technol, 68-70; ASSOC PROF MATH & COMPUT SCI, UNIV MASS, BOSTON, 70- *Concurrent Pos:* Consult, Bolt, Baranek & Newman, Inc. *Mem:* Soc Indust & Appl Math. *Res:* Computer operating systems analysis; numerical algorithms; asymptotic techniques. *Mailing Add:* Seven Whittier Rd Lexington MA 02173-1716

O'NEIL, JAMES R, b Chicago, Ill, July 16, 34. GEOCHEMISTRY, PHYSICAL CHEMISTRY. *Educ:* Loyola Univ Ill, BS, 56; Carnegie-Mellon Univ, MS, 59; Univ Chicago, PhD(chem), 63. *Prof Exp:* Res fel geochem, Univ Chicago, 63 & Calif Inst Technol, 63- 65; res chemist, Br Isotope Geol, US Geol Surv, 65-87; PROF GEOCHEM, UNIV MICH, 88- *Concurrent Pos:* Exchange scientist to USSR, 72; consult prof, Stanford Univ, 75-87; vis prof, Australian Nat Univ, 76 & Universite Louis Pasteur, 84; vis scientist, Univ Paris, 80; panel mem, NSF, 81-84; assoc ed, Geochim Cosmochim Acta, 84-89, Sci Geologiques, 88- & Geol, 91- *Honors & Awards:* Medal, Am Inst Chem, 56. *Mem:* AAAS; Am Chem Soc; Am Geophys Union; Geochem Soc; Geol Soc Am. *Res:* Studies of stable isotope variations in natural materials; laboratory determination of equilibrium constants for isotope exchange reactions of geologic significance; solute-water interactions; structure of water; mantle geochemistry. *Mailing Add:* Dept Geol Sci Univ Mich Ann Arbor MI 48109-1063

O'NEIL, JOHN J, b Chicago, Ill, Feb 9, 35. PULMONARY PHYSIOLOGY. *Educ:* Univ Calif, San Francisco, PhD(physiol), 74. *Prof Exp:* CHIEF CLIN RES BR, US ENVIRON PROTECTION AGENCY, 76- *Mem:* Am Thoracic Soc; Am Physiol Soc. *Mailing Add:* Clin Studies Br Med Res Bldg C-224H US Environ Protection Agency Chapel Hill NC 27514

O'NEIL, JOHN J, b Detroit, Mich, May 28, 28. ELECTROMAGNETIC COMPATIBILITY. *Educ:* Lawrence Inst Technol, BSEE, 43. *Prof Exp:* Chief electronics, US Army, 72-76; CONSULT, TRANS TECH, 76- *Mem:* Inst Elec & Electronic Engrs; Electromagnetic Compatible Soc. *Mailing Add:* Vectra Inc 1600 Detweiller Rd Peoria IL 61615

O'NEIL, LOUIS C, b Sherbrooke, Que, Jan 20, 30; m 55; c 5. ENTOMOLOGY, ZOOLOGY. *Educ:* Univ Montreal, BA, 50; Laval Univ, BSCApp, 54; State Univ NY Col Forestry, Syracuse Univ, MSc, 56, PhD(forest entom), 61. *Prof Exp:* Res off forest entom & path br, Can Dept Forestry, 55-62; assoc prof, 62-72, secy Fac Sci, 67-70, chmn dept biol, 70-72, dean, Fac Sci, 72-78, prof invert zool & entom, 72-78, DIR PERSONNEL, 78-, SECY FACULTY SCI, UNIV SHERBROOKE. *Concurrent Pos:* Res assoc, Sch Forestry, Yale Univ, 65, 66, 67. *Mem:* Entom Soc Can; Can Soc Zool. *Res:* Ecology of land slug introduced in eastern Canada; relationships between sawflies and their host trees; forest sawfly outbreaks and tree growth. *Mailing Add:* Secy Fac Sci Univ Sherbrooke Sherbrooke PQ J1K 2R1 Can

O'NEIL, PATRICK EUGENE, b Mineola, NY, July 19, 42; m 65; c 2. DATABASE PRICE-PERFORMANCE, INFORMATION RETRIEVAL EFFICIENCY. *Educ:* Mass Inst Technol, BS, 63; Univ Chicago, MS, 64; Rockefeller Univ, PhD(math), 69. *Prof Exp:* Systs programmer, Cambridge Sci Ctr, Int Bus Mach Corp, 64-66, mem staff opers res, 68-69; Off Naval Res fel & res assoc combinatorics, Mass Inst Technol, 69-70, asst prof comput sci, 70-72; consult comput sci, 72-74; asst prof comput sci, Univ Mass, Boston, 74-77; sr systs programmer, Keydata Corp, Watertown, 77-80; prin software eng, Prime Comput Inc, Framingham, Mass, 80-83; sr software engr, Comput Corp Am, Cambridge, Mass, 83-87; prin software engr, 87-88, ASSOC PROF COMPUTER SCI, UNIV MASS, BOSTON, 88- *Concurrent Pos:* Mem, Artificial Intel Lab, Mass, 70-72. *Mem:* Math Asn Am; Asn Comput Mach; Inst Elec & Electronic Engrs. *Res:* Price/performance of database and information retreival systems; design of real world benchmarks for such systems; designing new data access algorithms for efficiency; transactional theory; software development methodology. *Mailing Add:* Seven Whittier Rd Lexington MA 02173

O'NEIL, ROBERT JAMES, b Boston, Mass, July 14, 55; m 79; c 3. PREDATOR-PREY THEORY, BIOLOGICAL CONTROL. *Educ:* Univ Mass, Amherst, BS, 77; Tex A&M, MS, 80; Univ Fla, Gainesville, PhD(entom), 84. *Prof Exp:* Grad res asst, dept entom, Tex A&M Univ, 78-80 & Univ Fla, 80-84; ASST PROF INSECT ECOL, DEPT ENTOM, PURDUE UNIV, 84- *Mem:* Entom Soc Am; Int Orgn Biol Control; Sigma Xi. *Res:* Arthropod predation in agricultural systems; predator-prey theory and insect population ecology. *Mailing Add:* Dept Entom Purdue Univ West Lafayette IN 47907

ONEIL, STEPHEN VINCENT, b Pawtucket, RI, May 3, 47. MOLECULAR PHYSICS, APPLIED MATHEMATICS. *Educ:* Providence Col, BS, 69; Univ Calif, Berkeley, PhD(chem), 73. *Prof Exp:* Res fel chem, Harvard Univ, 73-74; res assoc chem, Joint Inst Lab Astrophys, 74-76; res assoc comput chem, Inst Comput Appln Sci & Eng, 77; MEM APPL MATH, JOINT INST LAB ASTROPHYS, UNIV COLO, BOULDER, 77- *Res:* Ab initio calculation of electronic continuum processes and bound electronic states of molecules. *Mailing Add:* Joint Inst Lab Astrophys MS 440 Univ Colo Boulder CO 80301

O'NEIL, THOMAS MICHAEL, b Hibbing, Minn, Sept 2, 40; m; c 1. THEORETICAL PLASMA PHYSICS. *Educ:* Calif State Univ, Long Beach, BS, 62; Univ Calif, San Diego, MS, 64, PhD(physics), 65. *Prof Exp:* Res physicist, Gen Atomic, 65-67; PROF PHYSICS, UNIV CALIF, SAN DIEGO, 67- *Concurrent Pos:* Div assoc ed, Phys Rev Lett, 79-83; corresp, Comments Plasma Physics & Controlled Fusion, 80-84; mem adv bd, Inst Fusion Studies, 80-83 & Inst Theoret Physics, 83- *Mem:* Fel Am Phys Soc. *Res:* Theoretical plasma physics with emphasis on nonlinear effects in plasmas and on nonneutral plasmas. *Mailing Add:* Dept Physics B-019 Univ Calif San Diego Box 109 La Jolla CA 92093

O'NEILL, BRIAN, b Bristol, Eng, Sept 20, 40; m 69; c 3. STATISTICS. *Educ:* Bath Univ, BSc, 65. *Prof Exp:* Statist consult, Unilever, London, Eng, 65-66; statistician, Wolf Res & Develop Corp, 67-69; PRES, INS INST HWY SAFETY, 69-; PRES, HWY LOSS DATA INST, 77- *Concurrent Pos:* Vchmn, Comt Alcohol & Drugs, Nat Safety Coun, 73-; mem adv panel, Nat Accident Sampling Syst, Dept Transp, 78 & 81. *Mem:* Am Statist Asn; Inst Math & Appl, UK; Royal Statist Soc. *Res:* Highway crash loss reduction, including human behavior, vehicle performance and design and highway design. *Mailing Add:* Ins Inst for Hwy Safety 1005 N Glebe Rd Arlington VA 22201

O'NEILL, EDWARD JOHN, b Washington, DC, Feb 15, 42. TOPOLOGY, OPERATIONS RESEARCH. *Educ:* Cath Univ Am, AB, 63; Yale Univ, MA, 65, PhD(math), 76. *Prof Exp:* Instr, 67-69, ASST PROF MATH, FAIRFIELD UNIV, 69-, ASST PROF COMPUTER SCI, 85- *Mem:* Am Math Soc; Opers Res Soc Am; Asn Comput Mach; Math Asn Am. *Res:* Algebraic topology; cohomology operations. *Mailing Add:* Dept Math & Computer Sci Fairfield Univ Fairfield CT 06430

O'NEILL, EDWARD LEO, b Boston, Mass, Nov 29, 27; m 51; c 3. PHYSICS. *Educ:* Boston Col, AB, 49; Boston Univ, MA, 51, PhD(physics), 54. *Prof Exp:* Res assoc physics, Res Lab, Boston Univ, 54-56, from asst prof to assoc prof, 56-66; vis prof physics, Worcester Polytech Inst, 66-; AT PHYSICS DEPT, CLARK UNIV. *Concurrent Pos:* Mem staff, Itek Corp, 58-60; vis prof, Mass Inst Technol, 61-62 & Univ Calif, Berkeley, 64-65. *Honors & Awards:* Lomb Medal, Optical Soc Am, 58. *Mem:* AAAS; Optical Soc Am; Am Asn Physics Teachers. *Res:* Communication theory; quantum and statistical optics. *Mailing Add:* Dept Physics Worcester Polytech Inst 100 Inst Rd Worcester MA 01609

O'NEILL, EDWARD TRUE, b Charlevoix, Mich, July 20, 40; m 76; c 2. INFORMATION SCIENCE, OPERATIONS RESEARCH. *Educ:* Albion Col, BA, 62; Purdue Univ, BSIE, 63, MSIE, 66, PhD(opers res), 70. *Prof Exp:* From asst prof to assoc prof info sci, Sch Info & Libr Studies, State Univ NY Buffalo, 68-85, asst dean, 70-75. *Concurrent Pos:* Vis distinguished scholar, OCLC, Inc, 78-79. *Mem:* Am Soc Info Sci; Am Libr Asn; Opers Res Soc Am. *Res:* Bibliometrics; subject cataloging; information retrieval; rank frequency modeling; database quality. *Mailing Add:* Off Res OCLC 6565 Frantz Rd Dublin OH 43017

O'NEILL, EILEEN JANE, b Rostrop, Germany, July 20, 55; Brit citizen; m 80; c 1. WATER CONTAMINATION. *Educ:* Univ Newcastle-Upon-Tyne, Eng, BS, 77; Univ Aberdeen, Scotland, PhD(soil sci), 80. *Prof Exp:* Lectr soil sci, Seale-Hayne Col, Eng, 80-82; res fel, Environ Toxicol Ctr, Univ Wis-Madison, 83-84; staff scientist, Environ Control Div, Dynamac Corp, Rockville, Md, 84-85; ENVIRON SCIENTIST, V J CICCONE & ASSOCS, INC, WOODBRIDGE, VA, 85- *Mem:* Brit Soc Soil Sci. *Res:* Diagnostic testing for fertilizer requirements of crops; degradation and environmental fate of pesticides; pesticide contamination of groundwater; regulation of pesticides and drinking water additives; removal of organic contaminants from drinking water. *Mailing Add:* Water Pollution Control Fed 601 Wythe St Alexandia VA 22314

O'NEILL, EUGENE F, b Brooklyn, NY, July 2, 18; m 42; c 4. COMMUNICATIONS ENGINEERING. *Educ:* Columbia Univ, BS, 40, MS, 41. *Hon Degrees:* DSc, Bates Col, 63 & St John's Univ, NY, 65; Milan Polytech Inst, DrEng, 64. *Prof Exp:* Mem tech staff, Bell Tel Labs, 41-56, proj head speech interpolation, 56-60, dir, Telstar Satellite Proj, 60-65, exec dir, Transmission Div, 66-78, exec dir proj planning, 78-85; RETIRED. *Honors & Awards:* Int Commun Award, Inst Elec & Electronics Engrs, 72. *Mem:* Nat Acad Eng; fel Inst Elec & Electronics Engrs. *Res:* Development of long haul toll transmission systems. *Mailing Add:* 17 Dellwood Ct Middletown NJ 07748

O'NEILL, FRANK JOHN, b New York, NY, Apr 26, 40; c 2. CELL BIOLOGY, VIROLOGY. *Educ:* Long Island Univ, BA, 63; Hunter Col, MA, 67; Univ Utah, PhD(molecular biol, genetic biol), 69. *Prof Exp:* Asst prof, 71-79, ASSOC PROF MICROBIOL & PATH, UNIV UTAH MED CTR, 79- *Concurrent Pos:* NIH fel microbiol, Milton S Hershey Med Ctr, Pa State Univ, 69-71; Eleanor Roosevelt-Am Cancer Soc Int Cancer fel, Freiburg, WGer, 77-78. *Mem:* Am Soc Human Genetics; Environ Mutagen Soc; Am Soc Microbiol; Am Asn Cancer Res. *Res:* Elucidation and establishment of mechanisms of viral oncogenesis and viral latency in vitro and in vivo; immortalization of human cells; defective viral DNA. *Mailing Add:* Dept Cell Virol & Molecular Biol Univ Utah Salt Lake City UT 84112

O'NEILL, GEORGE FRANCIS, b Yonkers, NY, Sept 27, 22; m 43; c 2. NUCLEAR PHYSICS. *Educ:* Mt St Mary's Col Md, BS, 43; Fordham Univ, MS, 47, PhD(physics), 51. *Prof Exp:* Physicist, Argonne Nat Lab, 50-53; sr res supvr, exp reactor physics div, Savannah Lab, E I du Pont de Nemours & Co, Inc, 53-68; with AEC Combined Opers Planning, Oak Ridge, 69-70; sr physicist, Advan Oper Planning, 70-74, RES STAFF PHYSICIST, SAVANNAH RIVER LAB, E I DU PONT DE NEMOURS & CO, INC, 74- *Concurrent Pos:* Guest lectr, Dept Nuclear Eng, Univ SC, 67-68; US rep, Can-Am D20 Reactor Comt, 67-68. *Mem:* AAAS; Am Nuclear Soc; Am Phys Soc. *Res:* Cosmic rays; heavy water nuclear reactors; exponential pile assemblies; critical pile assemblies; nuclear reactor fuel cycle. *Mailing Add:* 1230 Evans Rd Aiken SC 29801

O'NEILL, GERARD KITCHEN, systems design & systems science, for more information see previous edition

O'NEILL, JAMES A, JR, b New York, NY, Dec 7, 33; m 59; c 3. PEDIATRIC SURGERY. *Educ:* Georgetown Univ, BS, 55; Yale Univ, MD, 59. *Hon Degrees:* MA, Univ Pa. *Prof Exp:* Instr surg, Sch Med, Vanderbilt Univ, 64-65; instr pediat surg, Sch Med, Ohio State Univ, 67-69; from asst prof to assoc prof, Sch Med, La State Univ, 69-71; prof pediat surg, Sch Med, Vanderbilt Univ, 71-81; PROF PEDIAT SURG, SCH MED, UNIV PA, 81- *Concurrent Pos:* USPHS fel pediat oncol, Ohio State Univ, 67-68; consult, US Army Hosp, Ft Campbell, Ky, 71- & US Army Inst Surg Res, 71-; surgeon-in-chief, Children's Hosp Philadelphia, 81- *Mem:* Am Col Surg; Am Burn Asn; Am Acad Pediat; Am Surg Asn; Am Pediat Surg Asn. *Res:* General and thoracic pediatric surgery; various aspects of burn injury; stress ulceration; gastrointestinal effects of injury. *Mailing Add:* Children's Hosp Philadelphia 34th & Civic Ctr Blvd Philadelphia PA 19104

O'NEILL, JAMES F, b Corona, NY, July 3, 23; m 48; c 3. MEDICAL SCIENCES, BACTERIOLOGY. *Educ:* Long Island Univ, BSc, 50, MSc, 62. *Prof Exp:* Dir & owner, 50-82, DIR, CLIN LAB, SHIEL MED LAB, INC, 82- *Mem:* Am Asn Bioanalysts; fel Am Inst Chemists; Am Chem Soc; Am Soc Microbiologists. *Res:* Tissue culture-leukemic cells; means of culture and transport. *Mailing Add:* 30 Waterview Dr Danbury CT 06811

O'NEILL, JOHN CORNELIUS, b Philadelphia, Pa, July 3, 29. APPLIED MATHEMATICS. *Educ:* Cath Univ, BA, 52; Villanova Univ, MA, 59; Univ Pittsburgh, PhD, 67. *Prof Exp:* Teacher high schs, Pa, 52-53 & Md, 53-64; asst prof, 67-71, chmn, Dept Math Sci, 74-82, ASSOC PROF MATH, LA SALLE, 71- *Mem:* Inst Elec & Electronics Engrs; Math Asn Am. *Res:* Computer software systems; computer architecture; computer communications. *Mailing Add:* Dept Math Sci La Salle Univ Philadelphia PA 19141

O'NEILL, JOHN DACEY, b Detroit, Mich, July 9, 30. MATHEMATICS. *Educ:* Xavier Univ, Ohio, AB, 54; Loyola Univ, Ill, MA, PhL, STL & MS, 60; Wayne State Univ, PhD(math), 67. *Prof Exp:* Instr math & theol, high sch, 56-59; PROF MATH, DETROIT UNIV, 61- *Mem:* Math Asn Am; Am Math Soc. *Res:* Abelian groups; Ring theory. *Mailing Add:* Dept Math Lansing-Reilly Hall Univ Detroit 4001 W McNichols Rd Detroit MI 48221

O'NEILL, JOHN FRANCIS, b Peoria, Ill, Feb 9, 37; m 61; c 1. ELECTRICAL ENGINEERING. *Educ:* St Louis Univ, BSEE, 59; NY Univ, MEE, 61, PhD(elec eng), 67. *Prof Exp:* Mem tech staff data systs, Bell Tel Labs, Inc, 59-65, switching systs, 67-77; dir eng, STC Commun Corp, 77-80; pres, One Com Corp, 80-88; CHMN, CALL MGT PROD, BROOMFIELD, COLO, 88- *Mem:* Inst Elec & Electronics Engrs. *Res:* Active network theory; switching control; switching networks. *Mailing Add:* Call Mgt Prod 555 Q Burbank St Broomfield CO 80020

O'NEILL, JOHN H(ENRY), JR, b New Orleans, La, Mar 25, 22; m 50; c 3. CHEMICAL ENGINEERING. *Educ:* La State Univ, BS, 43; Mass Inst Technol, ScD(chem eng), 51. *Prof Exp:* Res assoc soil solidification, Mass Inst Technol, 49-50; res engr process develop, E I du Pont de Nemours & Co, 50-52, tech investr, Res Div, Film Dept, 52-56, res supvr, 56-64; dir eng, 65-71, res dir, 71-87, CONSULT, MILLIKEN RES CORP, 87- *Mem:* Am Chem Soc; Am Inst Chem Engrs. *Res:* Product and process research and development of polymers, polymeric films, chemical manufacturing; textile fabric product and process research, manufacturing, dyeing and finishing; design and development of textile machinery. *Mailing Add:* 113 Eastwood Circle Spartanburg SC 29302

O'NEILL, JOHN JOSEPH, b Queens, NY, Aug 26, 19; m 43; c 3. BIOCHEMISTRY. *Educ:* St Francis Col, NY, BS, 42; Univ Md, MS, 53, PhD(biochem), 55. *Prof Exp:* Res assoc physiol, Sch Med, Cornell Univ, 46-47; biochemist, Physiol Div, Med Lab, US Army Chem Res & Develop Lab, 47-50, org chemist, Biochem Div, Res Directorate, 50-54, biochemist, 54-56, br chief, 56-60; assoc prof biochem pharmacol, Sch Med, Univ Md, Baltimore City, 60-70; prof biochem pharmacol, Col Med, Ohio State Univ, 70-75; PROF & CHMN DEPT PHARMACOL, MED SCH, TEMPLE UNIV, 75- *Concurrent Pos:* NIMH fel pharmacol, Sch Med, Wash Univ, 63-64; res collab, Biochem Dept, Brookhaven Nat Lab, 61-; consult, US Army Chem Res & Develop Lab, Md, 61-, Melpar Div, Westinghouse Elec Co, Va, 62-, NSF, 72-, Environ Protection Agency, 76-, Nat Res Coun Comt, 80- & Ad Hoc Comt, Nat Inst Neurol Disease & Stroke. *Mem:* Am Chem Soc; Am Soc Neurochem; Int Soc Neurochem; Am Soc Pharmacol & Exp Therapeut; Fedn Am Soc Exp Biol. *Res:* Neurochemistry; biochemistry of the central nervous system; glycolysis; enzyme kinetics and control mechanisms; isotopes in intermediary metabolism; neurotoxins and organophosphorous anticholinesterases; neurohumoral substances and drug action. *Mailing Add:* Dept Pharmacol Temple Univ 3400 N Broad St Philadelphia PA 19140

ONEILL, JOSEPH LAWRENCE, b Philadelphia, Pa, Jan 2, 31. PHARMACY. *Educ:* Philadelphia Col Pharm & Sci, BS, 55, MS, 56. *Prof Exp:* Res assoc, Res Labs, Merck Sharp & Dohme, 56-61, pharmacist, 61-62, res pharmacist, 63-64, sr res pharmacist, 64-73, res fel, 73-90; RETIRED. *Mem:* Am Pharmaceut Asn; Acad Pharmaceut Sci. *Res:* Topical and oral liquid dosage forms; topical gel systems. *Mailing Add:* 403 Spruce Circle Lansdale PA 19446-6029

O'NEILL, MICHAEL WAYNE, b San Antonio, Tex, Feb 17, 40; m 72; c 1. CIVIL ENGINEERING. *Educ:* Univ Tex, Austin, BS, 63, MS, 64, PhD(civil eng), 70. *Prof Exp:* Proj mgr civil eng, Southwestern Labs Inc, 71-74; from asst prof to assoc prof, 74-84, PROF CIVIL ENG, UNIV HOUSTON, 84-, CHAIR DEPT, 89- *Concurrent Pos:* Fel, Univ Tex, 70-71; spec consult, Southwestern Labs, Inc, 74- *Honors & Awards:* John Hawley Award, Am Soc Civil Engrs, 76 & 81, Walter L Huber Res Prize, 86. *Mem:* Am Soc Civil Engrs; Int Soc Soil Mech & Found Eng; Transp Res Bd. *Res:* Behavior of deep foundations, expansive clays and machine foundations. *Mailing Add:* Dept Civil Eng Univ Houston Houston TX 77204-4791

O'NEILL, PATRICIA LYNN, b Chicago, Ill, Apr 19, 49; m 77; c 1. BIOMECHANICS, CONNECTIVE TISSUE. *Educ:* Univ Urbana, BS, 71; Univ Calif, Los Angeles, PhD(biol), 75. *Prof Exp:* Lectr biol, Univ Calif, Los Angeles, 76-77; fel, Cocos Found, Duke Univ, 78-79; asst prof biol, Portland State Univ, 79-86; res fel, Univ Western Australia, 88-89; POSTDOCTORAL RES FEL, AUSTRALIAN RES COUN, 91- *Concurrent Pos:* Hon fel, Dept Zool, Univ Cape Town, 77-78; sr marine biologist, VTN Oregon, Inc, 79-80. *Mem:* Sigma Xi. *Res:* Mechanical properties and structure of invertebrate collagenous tissues; structure of echinoderm calcite; torsional stress analysis of ratite bird limbs; age structure and density regulation of benthic invertebrate populations. *Mailing Add:* Dept Zool Univ Western Australia Nedlands 6009 Western Australia

O'NEILL, PATRICK JOSEPH, b Columbus, Ohio, July 12, 49; m 73; c 2. PHARMACOLOGY. *Educ:* Ohio State Univ, BS, 72, PhD(pharmacol), 76. *Prof Exp:* Res assoc, Ohio State Univ, 72-76; res scientist, McNeil Labs Inc, 76-77, sr scientist, 77-79, group leader, McNeil Pharmaceut, 79-82, dir, Dept Drug Metab, 82-85, exec dir, Res & Develop Proj Planning, 85-86, exec dir, New Prods, 86-87, exec dir, Develop Res, McNeil Pharmaceut, Johnson & Johnson, 88-90. *Mem:* AAAS; Am Soc Pharmacol & Exp Therapeut; NY Acad Sci. *Res:* Drug metabolism. *Mailing Add:* Ethicon PO Box 151 Somerville NJ 08876-0151

O'NEILL, R(USSELL) R(ICHARD), b Chicago, Ill, June 6, 16; m 39, 67; c 2. ENGINEERING. *Educ:* Univ Calif, Berkeley, BS, 38, MS, 40; Univ Calif, Los Angeles, PhD, 56. *Prof Exp:* Engr, Dowell, Inc, 40-41; design & develop engr, Dow Chem Co, 41-44; design engr, AiResearch Mfg Co, 44-46; lectr eng & asst head eng exten, 46-56, from asst dean to dean, Sch Eng & Appl Sci, 56-83, prof eng, 56-84, EMER DEAN, SCH ENG & APPL SCI, UNIV CALIF, LOS ANGELES, 83-, EMER PROF ENG, 84- *Concurrent Pos:* Staff engr, Nat Res Coun, 54; chmn, Maritime Transp Res Bd, Nat Acad Sci-Nat Res Coun, 77-81; bd adv, Naval Post Grad Sch, 75-85; trustee, West Coast Univ, 80-90; mem Nat Nuclear Accrediting Bd, 82-88. *Mem:* Nat Acad Eng; Am Soc Eng Educ; Sigma Xi. *Res:* Cargo handling. *Mailing Add:* Sch Eng & Appl Sci Univ Calif Los Angeles CA 90024-1600

O'NEILL, RICHARD THOMAS, b Winnetka, Ill, Dec 7, 33; m 59; c 2. PHYSICAL INORGANIC CHEMISTRY. *Educ:* Loyola Univ Ill, BS, 55; Carnegie Inst Technol, MS, 59, PhD(chem), 60. *Prof Exp:* From asst prof to assoc prof, 59-70, coordr acad comput, 73-79, chmn dept chem, 82-84, PROF CHEM, XAVIER UNIV, OHIO, 70- *Mem:* Am Chem Soc; Soc Appl Spectros; Coblentz Soc; Sigma Xi. *Res:* Chemical spectroscopy; molecular association; charge transfer complexes; hydrogen bonding; educational use of computers. *Mailing Add:* Dept Chem Xavier Univ Cincinnati OH 45207

O'NEILL, ROBERT VINCENT, b Pittsburgh, Pa, Apr 13, 40; m 67. ECOLOGY. *Educ:* Cathedral Col, BA, 61; Univ Ill, PhD(zool), 67. *Prof Exp:* Fel systs ecol, NSF, 67-68; assoc health physicist, 68-69, SYSTS ECOLOGIST, OAK RIDGE NAT LAB, 69- *Mem:* Ecol Soc Am. *Res:* Systems ecology. *Mailing Add:* Environ Sci Div Bldg 1505 Oak Ridge Nat Lab Oak Ridge TN 37831

O'NEILL, RONALD C, b Beverly, Mass, July 20, 29; m 53; c 2. TOPOLOGY. *Educ:* Columbia Univ, AB, 51, MA, 52; Purdue Univ, PhD(math), 62. *Prof Exp:* From instr to asst prof math, Univ Mich, 61-66; asst prof, 66-67, ASSOC PROF MATH, MICH STATE UNIV, 67- *Mem:* Am Math Soc; Math Asn Am. *Mailing Add:* Dept Math-A319 Wells Hall Mich State Univ East Lansing MI 48824

O'NEILL, THOMAS BRENDAN, b Boston, Mass, Oct 16, 23; m 47; c 5. MICROBIOLOGY, BOTANY. *Educ:* Univ Notre Dame, BS, 50, MS, 51; Univ Calif, PhD(bot), 56. *Prof Exp:* Machinist's apprentice, Bethlehem Steel Co, Mass, 41-43; asst, Univ Notre Dame, 50-51 & Univ Calif, 51-54, assoc bot, 54-55; PROF MICROBIOL, VENTURA COL, 55-, CHMN DEPT, 86- *Concurrent Pos:* Res biologist, US Naval Civil Eng Lab, 57-86; NSF fel, Scripps Inst, Calif, 62-63; lectr, Univ Calif, Santa Barbara. *Mem:* AAAS; Bot Soc Am; Sigma Xi. *Res:* Marine microbiology; teaching methods; biodegradation of hydrocarbons; biological control of marine bovers. *Mailing Add:* Dept Biol Ventura Col Ventura CA 93003

O'NEILL, THOMAS HALL ROBINSON, meteorology; deceased, see previous edition for last biography

O'NEILL, WILLIAM D(ENNIS), b Chicago, Ill, Apr 3, 38; m 61; c 4. SYSTEMS ANALYSIS. *Educ:* Col St Thomas, BA, 60; Univ Notre Dame, BS, 61, MS, 63, PhD(elec eng), 65. *Prof Exp:* Instr elec eng, Univ Notre Dame, 60-61; res assoc, Particle Accelerator Div, Argonne Nat Lab, 61-62; from asst prof to assoc prof, 65-73, PROF SYSTS ENG, UNIV ILL, CHICAGO CIRCLE, 73-, RES ASSOC PHYSIOL, UNIV ILL COL MED, 71- *Concurrent Pos:* Consult, Argonne Nat Lab, 61- & Biosysts, Inc, Mass, 65-; res assoc, Presby St Luke's Hosp, 65-71; res grants, Dept Defense, 67-68, Univ Ill, 68-69 & NIH, 68-77. *Mem:* Inst Elec & Electronics Engrs; Tensor Soc. *Res:* Automatic control; socio-economic system modeling; demographic control systems; statistics, estimation. *Mailing Add:* Dept Biol Eng Univ Ill Box 4348 Chicago IL 60680

ONG, CHUNG-JIAN JERRY, b China, June 19, 44; US citizen; m 71. POLYMER PHYSICS. *Educ:* Nat Taiwan Univ, BS, 68; Univ Mass, MS, 72, PhD(polymer sci, eng), 73. *Prof Exp:* Res scientist, 73-80, sr res scientist polymer physics, 80- 83, FORMULATION CHEMIST, AM CYANAMID CO, 83- *Mem:* Am Chem Soc; Am Soc Testing & Mat; Sigma Xi. *Res:* Physical chemistry of polymers, especially in physical, mechanical and rheological studies of polymer blends and elastomers. *Mailing Add:* 26 Glenview Dr Warren NJ 07059

ONG, DAVID EUGENE, b Elkhart, Ind, Aug 16, 43; m 81. NUTRITION. *Educ:* Wabash Col, BA, 65; Yale Univ, PhD(biochem), 70. *Prof Exp:* NIH fel, 70-74, res asst prof, 75-80, res assoc prof, 81-84, assoc prof, 84-87, PROF BIOCHEM, VANDERBILT UNIV, 87- *Concurrent Pos:* Mem, Nutrit Stud Sect, NIH, 85- *Honors & Awards:* Osborne & Mendel Award, Nutrit Found, 83. *Mem:* Am Inst Nutrit; Am Soc Biol Chemists. *Res:* Biochemical mechanisms of the absorption, transport, storage, and non-visual functions of vitamin A. *Mailing Add:* Dept Biochem Vanderbilt Univ Sch Med 21st Ave S & Garland Nashville TN 37232

ONG, ENG-BEE, b Seremban, Malaysia, Aug 9, 33. PROTEIN CHEMISTRY, FIBRINOLYSIS. *Educ:* Univ Miami, BS, 58; Tulane Univ, PhD(biochem), 64. *Prof Exp:* Instr biochem, Med Ctr, Tulane Univ, 64-65; res assoc, Rockefeller Univ, 65-68, asst prof, 68; clin biochemist, Ochsner Clin, 68-70; ASST PROF CLIN PATH, MED CTR, NY UNIV, 70- *Mem:* Am Chem Soc; Sigma Xi. *Res:* Chemistry of amino acids, peptides and proteins; chromatography of amino acids and peptides; structure and function of proteases; fractionation and characterization of blood coagulation factors; protein sequence. *Mailing Add:* 46-05 76th St Elmhurst NY 11373

ONG, JOHN NATHAN, JR, b Rio Tinto, Spain, May 27, 27; wid; c 4. GENERAL ENGINEERING, METALLURGY. *Educ:* Univ Utah, BSChE, 52, PhD(metall), 55. *Prof Exp:* Res metallurgist, Uranium Corp Am, 55-56 & Howe Sound Co, 56; asst prof mech eng, Washington Univ, 56-61; res scientist, Aeronutronic, 61-63; res scientist, Aeronaut Div, Ford Motor Co, Calif, 63-66; assoc prof mat, 66-68, chmn dept, 67-71, prof, 68-73, PROF INDUST & SYSTS ENG, UNIV WIS-MILWAUKEE, 73- *Concurrent Pos:* Consult, McDonnell Aircraft Corp, 56-60, Mallinckrodt Chem Works, 57-60 & Cutler-Hammer Corp, 67. *Honors & Awards:* Gold Medal Award, Am Inst Mining, Metall & Petrol Engrs, 57. *Mem:* AAAS; Am Inst Mining, Metall & Petrol Engrs; Soc Gen Systs Res; Am Soc Eng Educ; Sigma Xi. *Res:* Design; problem solving; general systems theory; oncology; science of creative intelligence; solid state chemistry and kinetics; thermodynamics. *Mailing Add:* PO Box 271 Fairfield IA 52556

ONG, JOHN TJOAN HO, b Jakarta, Indonesia, Sept 5, 37; m 61; c 2. PHARMACEUTICS. *Educ:* Bandung Inst Technol, CPharm, 60, Drs(pharm), 65; Univ Ky, PhD(pharmaceut), 74. *Prof Exp:* Teaching asst pharmaceut, Sch Pharm, Bandung Inst Technol, 63-65; lectr, Sch Pharm, Univ Indonesia, 65-69; pharmacist, Sanitas Pharm, Jakarta, 69-70; res staff pharmaceut chem, Univ Ky, 70-71, teaching asst pharmaceut, 71-74; sr scientist pharmaceut, Lilly Res Labs, Div Eli Lilly & Co, 74-81; RES SECT LEADER PHARMACEUT, SYNTEX RES, 81- *Mem:* Am Asn Pharm Scientists. *Res:* Stability kinetics of drugs; development of dosage forms; physical and chemical factors influencing the bioavailability of drugs from their dosage forms; tablet compaction. *Mailing Add:* Syntex Res 3401 Hillview Ave Palo Alto CA 94304

ONG, NAI-PHUAN, b Penang, Malaysia, Sept 10, 48; m 82. MICROWAVE IMPEDANCE MEASUREMENT, LOW TEMPERATURE PHYSICS. *Educ:* Columbia Col, BA, 71; Univ Calif, Berkeley, PhD(physics), 76. *Prof Exp:* From asst prof to prof physics, Univ Southern Calif, 76-85; PROF PHYSICS, PRINCETON UNIV, 85- *Concurrent Pos:* Vis fel, Max Planck Inst, Stuttgart, 82; consult, David Sarnoff Res Ctr, Princeton, 87-89 & Airtron, Morristown, 88-89. *Mem:* Fel Am Phys Soc. *Res:* High temperature superconductivity; properties of novel low dimensional metals; properties of solids in intense magnetic fields. *Mailing Add:* Dept Physics Princeton Univ Princeton NJ 08544

ONG, POEN SING, b Semarang, Indonesia, Mar 13, 27; m 61; c 1. ELECTRICAL ENGINEERING. *Educ:* Technol Univ Delft, Engr, 56, ScD, 59. *Prof Exp:* Prin scientist electron & x-ray beam, Philips Electronic Instruments, 61-71; assoc prof biophysics, M D Anderson Hosp, 71-76; ASSOC PROF ELEC ENG, UNIV HOUSTON, 76- *Mem:* Inst Elec & Electronic Engrs. *Res:* Electron and x-ray microbeam technology; x-ray microscopy; electron microscopy; x-ray microanalysis. *Mailing Add:* Dept Elec Engrs Univ Houston-Univ Park 4800 Calhoun Rd Houston TX 77204-4793

ONG, TONG-MAN, b Tainan, Taiwan, June 9, 35; US citizen; m 66; c 2. GENETIC TOXICOLOGY, GENETICS. *Educ:* Taiwan Normal Univ, BSc, 60; Ill State Univ, MS, 67, PhD(genetics), 70. *Prof Exp:* Staff fel genetics, 72-77, geneticist, Nat Inst Environ Health Sci, 77-78; SECT CHIEF MICROBIOL, NAT INST OCCUP SAFETY & HEALTH, 78. *Concurrent Pos:* Adj prof, WVa Univ, 85; distinguished vis prof, Guangzhou Med Col, China. *Mem:* Sigma Xi; AAAS; Environ Mutagen Soc. *Res:* Development and application of short-term assays to study the mutagenic activity of complex mixtures from workplace environments and to study the potential health hazards of chemicals to the exposed population. *Mailing Add:* Nat Inst Occup Safety & Health 944 Chestnut Ridge Rd Morgantown WV 26505-2888

ONGERTH, HENRY J, b San Francisco, Calif, June 19, 13. SANITARY ENGINEERING. *Educ:* Univ Calif, BS, 32; Univ Mich, MPH, 34. *Prof Exp:* Sanit engr, Calif Dept Health, 38-42; sanit engr-in-chg sewerage design, San Francisco Dist Corps Engrs, 42-43; sanit engr, Calif Dept Health, 43-47; consult engr, W J O'Connell & Assocs, 47-48; sanit engr, Calif Dept Health, 48-80, chief, Bur Sanit Eng, 68-78; RETIRED. *Concurrent Pos:* Mem, Adv Comt Revision USPHS Drinking Water Standards, 59-61, Water Supply Panel & Comt Water Qual Criteria, 71-72; chmn, Pub Adv Comt Revision Fed Drinking Water Standards, 76-79; mem, Environ Studies Bd, Nat Acad Sci; mem, Environ Eng Intersoc Bd; diver, Am Water Works Asn, 78. *Mem:* Nat Acad Eng; hon mem Am Water Works Asn; Conf State Sanit Engrs; Am Acad Environ Engrs; fel Am Soc Chem Engrs. *Mailing Add:* 905 Contra Costa Ave Berkeley CA 94707

ONIK, GARY M, b New York, NY, Feb 25, 52; m; c 3. INTERVENTIONAL RADIOLOGY, BIOMEDICAL PRODUCT DEVELOPMENT. *Educ:* Harvard Univ, BA, 74; NY Med Col, MD, 78. *Prof Exp:* Med dr, Allegheny Gen Hosp & Allegheny-Singer Res Inst, 86-89; med dr, Univ Pittsburgh, 89-91; MED DR, ALLEGHENY GEN HOSP, 91- *Concurrent Pos:* Adj asst prof radiol, Univ Pa, 87-89; mem bd, Am Col Cryosurg, 88-, vpres, 90; assoc prof radiol & neurosurg, Univ Pittsburgh, 89- *Honors & Awards:* Exec Coun Award, Animal Work Cryosurg, Am Roentgen Ray Soc, 84; Roscoe E Miller Award, Soc Gastrointestinal Radiologists, 86. *Mem:* Roentgenol Soc NAm; Am Col Cryosurg. *Res:* Development of minimally invasive surgical techniques using imaging guidance; imaging directed tumor treatments aimed at alleviating lower back problems. *Mailing Add:* Dept Neurosurg 14th Floor South Tower Allegheny Gen Hosp 320 E North Ave Pittsburgh PA 15212-9986

ONISCHAK, MICHAEL, b Chicago, Ill, June 23, 43; m 67; c 2. CHEMICAL ENGINEERING. *Educ:* Ill Inst Technol, BS, 65, MS, 68, PhD(gas technol), 72. *Prof Exp:* Proj mgr chem reactors, Energy Res Corp, 72-77; res supvr coal gasification, 77-80, MGR GASIFICATION DEVELOP, INST GAS TECHNOL, 80- *Mem:* Am Inst Chem Engrs; Am Chem Soc; Sigma Xi. *Res:* Gas-solid reaction engineering; heat and mass transfer; specific application to energy conversion devices and coal and biomass gasification plants. *Mailing Add:* 3408 Ithaca Rd Olympia Fields IL 60461

ONKEN, ARTHUR BLAKE, b Alice Tex, Aug 13, 35; m 56; c 2. SOIL CHEMISTRY, SOIL FERTILITY. *Educ:* Tex A&I Univ, BS, 59; Okla State Univ, MS, 63, PhD(soil chem), 64. *Prof Exp:* PROF SOIL CHEM, AGR EXP STA, TEX A&M UNIV, 64- *Mem:* Am Soc Agron; Soil Sci Soc Am; Int Soil Sci Soc; Sigma Xi. *Res:* Plant nutrient reactions in soil; plant response to nutrient level; movement of applied plant nutrients and solutes through soil profiles. *Mailing Add:* Agr Res & Exten Ctr Tex A&M Univ Lubbock TX 79401

ONLEY, DAVID S, b London, Eng, July 4, 34; m 57; c 2. THEORETICAL PHYSICS. *Educ:* Oxford Univ, BA, 56, DPhil(theoret physics), 60. *Prof Exp:* Res assoc physics, Duke Univ, 60-61, asst prof, 61-65; assoc prof, 65-69, chmn dept, 73-78, PROF PHYSICS, OHIO UNIV, 69- *Concurrent Pos:* US Army Res Off res grant, 66-73; sabbatical leave, Univ Melbourne, 71-72; US Dept Energy res grant, 79- *Mem:* AAAS; Am Phys Soc. *Res:* Nuclear structure; photonuclear theory and analysis of electron scattering from atomic nuclei; nuclear fission. *Mailing Add:* Dept Physics Ohio Univ Athens OH 45701-2979

ONN, DAVID GOODWIN, b Newark, Eng, Feb 18, 37; m 67; c 1. SOLID STATE PHYSICS, MATERIALS SCIENCE. *Educ:* Bristol Univ, BSc, 58; Oxford Univ, dipl educ, 60; Duke Univ, PhD(physics), 67. *Prof Exp:* Second physics master, Dean Close Sch, Cheltenham, Eng, 59-61; res assoc physics, Univ NC, Chapel Hill, 66-68, instr, 68-69, vis asst prof, 69-70; asst prof physics, 70-73, from asst prof to assoc prof physics & health sci, 73-76, ASSOC PROF PHYSICS, UNIV DEL, 76- *Mem:* Am Inst Physics; Sigma Xi; Am Asn Physics Teachers; Mat Res Soc; Am Carbon Soc. *Res:* Thermal and magnetic properties of metallic glasses; graphite intercalation compounds; electronic states in cryogenic fluids. *Mailing Add:* 41 Holly Lane Meadow Run Newark DE 19711

ONO, JOYCE KAZUYO, invertebrate zoology, neurophysiology, for more information see previous edition

ONO, KANJI, b Tokyo, Japan, Jan 2, 38; m 62; c 3. MATERIALS SCIENCE. *Educ:* Tokyo Inst Technol, BEng, 60; Northwestern Univ, PhD(mat sci), 64. *Prof Exp:* Res asst mat sci, Northwestern Univ, 60-61 & 64-65; from asst prof to assoc prof, 65-76, PROF ENG, UNIV CALIF, LOS ANGELES, 76- *Concurrent Pos:* ed, J Acoust Emissions. *Honors & Awards:* H M Howe Award, Am Soc Metals, 68. *Mem:* Soc Advan Mat & Process Eng; Am Soc Metals; Am Soc Nondestruct Test; Am Soc Testing & Mat. *Res:* Dislocations and mechanical properties of solids; physical metallurgy; nondestructive testing of materials. *Mailing Add:* Dept Mat Sci & Eng Univ Calif Los Angeles CA 90024

ONOE, MORIO, b Tokyo, Japan, Mar 28, 26; c 3. ELECTRICAL ENGINEERING. *Educ:* Tokyo Univ, BS, 47, PhD(eng), 55. *Prof Exp:* Prof, Inst Indust Sci, Univ Tokyo, 62-86, dir, 83-86; sr exec mgr, 86-88, EXEC VPRES, RICOH CO LTD, 88- *Concurrent Pos:* Mem tech staff, Bell Tel Lab, 61-62 & 66. *Honors & Awards:* C B Sawyer Award, 75; Distinguished Award, Inst Electronic Commun Engrs Japan; Asada Award, Iron & Steel Inst, 85. *Mem:* Fel Inst Elec & Electronic Engrs; Acoust Soc Am. *Res:* Image processing by computer especially for medical and industrial applications; non-destructive testing by ultrasonic waves and eddy current; piezoelectric devices including filters. *Mailing Add:* Ricoh Co Ltd Res & Develop Ctr 16 Shinei-Cho Yokohama 223 Japan

ONOPCHENKO, ANATOLI, b Lvov, Poland, Feb 12, 37; US citizen; m 63; c 2. ORGANIC CHEMISTRY. *Educ:* Pa State Univ, BS, 60, MS, 62; Univ Md, PhD(chem), 66. *Prof Exp:* Res asst, Pa State Univ, 60-62; asst, Univ Md, 63-66; res chemist, 66-72, sr res chemist, 72-80, res assoc, Gulf Res & Develop Co, 80-85; SR RES ASSOC, CHEVRON RES CO, 85- *Mem:* Am Chem Soc; Am Inst Chemists; Sigma Xi. *Res:* General organic and petroleum additive chemistry. *Mailing Add:* 1309 David Lane Concord CA 94518-3845

ONSAGER, JEROME ANDREW, b Northwood, NDak, Apr 8, 36; m 58; c 2. ENTOMOLOGY. *Educ:* NDak State Univ, BS, 58, MS, 60, PhD(entom), 63. *Prof Exp:* Res entomologist, Vegetable Insects Invests, Agr Res Serv, USDA, 63-75; RES LEADER, RANGELAND INSECT LAB, AGR RES SERV, USDA, 75- *Mem:* Entom Soc Am; Pan Am Acridol Soc. *Res:* Biology and ecology of insects that affect rangeland, with emphesis on development and integration of selective methods for biological chemical and cultural control. *Mailing Add:* Rangeland Insect Lab USDA ARS Mont State Univ Bozeman MT 59717

ONSTAD, CHARLES ARNOLD, b Spring Grove, Minn, Oct 30, 41; m 61; c 3. AGRICULTURAL ENGINEERING, HYDROLOGY. *Educ:* Univ Minn, BAE, 64, MSAE, 66; SDak State Univ, PhD(agr eng), 72. *Prof Exp:* AGR ENGR HYDROL, NORTH CENT SOIL CONSERV RES CTR, FOOD RES, AGR RES SERV, USDA, 66- *Concurrent Pos:* Asst prof, Univ Minn, St Paul, 72- *Mem:* Am Soc Agr Engrs; Soil Conserv Soc Am. *Res:* Hydrological and erosional modeling for the purpose of assessing non-point pollution potential on agricultural watersheds. *Mailing Add:* E RR 1 No 187 Morris MN 56267

ONSTOTT, EDWARD IRVIN, b Moreland, Ky, Nov 12, 22; m 45; c 4. THERMOCHEMISTRY. *Educ:* Univ Ill, BS, 44, MS, 48, PhD(chem), 50. *Prof Exp:* Chem engr, Firestone Tire & Rubber Co, 44 & 46; RES CHEMIST, LOS ALAMOS NAT LAB, 50- *Mem:* Fel AAAS; fel Am Inst Chem; NY Acad Sci; Am Chem Soc; Electrochem Soc; Sigma Xi; Int Asn Hydrogen Energy. *Res:* Complex ion and rare earth chemistry; electrochemistry; thermochemical hydrogen production. *Mailing Add:* 225 Rio Bravo Los Alamos NM 87544-3848

ONTELL, MARCIA, b New York, NY, June 28, 37. ANATOMY, ELECTRON MICROSCOPY. *Educ:* Temple Univ, BA, 58; Fairleigh Dickinson Univ, MS, 68; NJ Col Med & Dent, PhD(anat), 72. *Prof Exp:* Assoc anat, Mt Sinai Med Sch, 72-76; asst prof, 76-80, ASSOC PROF ANAT, SCH MED, UNIV PITTSBURGH, 80- *Concurrent Pos:* Prin investr, Muscular Dystrophy Asn, 74-76 & 81, NIH Nat Inst Neurol & Commun Disorders & Stroke, 75-84 & NSF Develop Biol, 78-81. *Mem:* Am Asn Anatomists; Am Soc Cell Biol; Tissue Cult Asn; NY Acad Sci. *Res:* Cell maturation; myogenesis and growth of striated muscle; muscle satellite cells; muscle disorders, including muscular dystrophy, denervation atrophy; response of muscle to various pharmacological agents. *Mailing Add:* Dept Anat & Cell Biol Univ Pittsburgh Sch Med Pittsburgh PA 15261

ONTJES, DAVID A, b Lyons, Kans, July 19, 37; m 60; c 4. ENDOCRINOLOGY. *Educ:* Univ Kans, BA, 59; Oxford Univ, MA, 61; Harvard Med Sch, MD, 64. *Prof Exp:* From intern to resident internal med, Harvard Med Serv, Boston City Hosp, 64-66; from asst prof to assoc prof med & pharmacol, 69-76, chief, Div Endocrinol, 72-81, chmn dept med, 81-89, PROF MED & PHARMACOL, MED SCH, UNIV NC, CHAPEL HILL, 76- *Concurrent Pos:* NIH spec res fel, 71; attend physician, NC Mem Hosp, Chapel Hill, 69- *Mem:* Am Soc Pharmacol & Exp Therapeut; Am Soc Clin Invest; fel Am Col Physicians; Endocrine Soc; Am Fedn Clin Res. *Res:* Chemistry and physiology of peptide hormones; peptide hormone receptors; control of adrenal function. *Mailing Add:* Dept of Med Univ of NC Med Sch Chapel Hill NC 27514

ONTKO, JOSEPH ANDREW, b Syracuse, NY, July 30, 32; m 55; c 4. BIOCHEMISTRY. *Educ:* Syracuse Univ, AB, 53; Univ Wis, MS, 55, PhD(biochem), 57. *Prof Exp:* From instr to asst prof biochem, Med Units, Univ Tenn, 59-64, asst prof physiol & biophys, 65-66; assoc prof, 68-74, PROF RES BIOCHEM, SCH MED, UNIV OKLA, 74-; ASSOC MEM, OKLA MED RES FOUND, 68- *Concurrent Pos:* Am Heart Asn res fel, Med Sch, Bristol Univ, 66-67 & Max Planck Inst Cell Chem, Munich, 67-68; consult, Oak Ridge Inst Nuclear Studies, 61-66; alcohol res rev comt, HEW, 76-79; mem, Metab Study Sect, NIH, 88- *Honors & Awards:* NIH Res Career Develop Award, 70-74. *Mem:* AAAS; Biochem Soc; Am Soc Biol Chemists. *Res:* Lipid metabolism; atherosclerosis; regulation of cell metabolism; isolation of liver cells. *Mailing Add:* Cardiovasc Biol Okla Med Res Found 825 NE 13th St Oklahoma City OK 73104

ONTON, AARE, b Tartu, Estonia, Dec 10, 39; US citizen; m 88; c 3. PHYSICS, ELECTRICAL ENGINEERING. *Educ:* Mass Inst Technol, SB & SM, 62; Purdue Univ, PhD(physics), 67. *Prof Exp:* Res staff mem, Thomas J Watson Res Ctr, IBM Corp, NY, 67-73; RES STAFF MEM, IBM RES, ALMADEN RES CTR, 73- *Concurrent Pos:* Alexander von Humboldt scholar, Max Planck Inst Solid State Res, 72-73; prog mgr, magnetic recording, IBM Corp Develop Staff, Purchase, NY, 82-84. *Mem:* Am Phys Soc. *Res:* Optical properties of solids, particularly electronics bandstructure, electronic impurity states, lattice dynamics, and luminescence; structural, electronic, and magnetic properties of amorphous thin films; storage device and systems architecture. *Mailing Add:* IBM Res Ctr Almaden Res Ctr 650 Harry Rd San Jose CA 95120-6099

ONUSKA, FRANCIS IVAN, b Humenne, Slovak, Mar 28, 35; Can citizen; m 61; c 2. ANALYTICAL CHEMISTRY, ORGANIC CHEMISTRY. *Educ:* Slovak Tech Univ, Bratislava, dipl eng, 58; Tech Univ, Prague, MSc, 66; Purkyne Univ, Brno, PhD(anal chem), 67. *Prof Exp:* Res chemist, Dept Tech Develop, Chemko Strazske, 58-66, scientist, Res Base, 66-68; res chemist, Uniroyal Res Labs, 69-73; SCIENTIST TRACE ANALYSIS, NAT WATER RES INST, 73- *Concurrent Pos:* Fel, Lomonosoff State Univ, Moscow, 67, Ecole Polytechnique, Palaiseau, France, 82; vis scientist, Kyoto Univ, Japan, 86. *Mem:* Fel Chem Inst Can; Am Soc Mass Spectrometry. *Res:* Modern separation methods; capillary gas and liquid chromatography; supercritical fluid chromatography and supercritical fluid extractions; trace organic analysis; environmental analytical chemistry and toxicology. *Mailing Add:* Nat Water Res Inst AMRS PO Box 5050 Burlington ON L7R 4A5 Can

ONWOOD, DAVID P, b London, Eng, Sept 13, 36. PHYSICAL CHEMISTRY. *Educ:* Oxford Univ, BA, 60, DPhil(chem), 66. *Prof Exp:* Res assoc chem, Ill Inst Technol, 63-65, instr 65-66; asst prof, 66-69, ASSOC PROF CHEM, PURDUE UNIV, FT WAYNE, 69- *Concurrent Pos:* NSF partic award, 67-69; vis assoc prof, Univ Toronto, 77-80; consult, Case-Western Reserve Univ, 84; vis scientist, Univ Ky, 85. *Mem:* Royal Soc Chem; Am Chem Soc; Nat Sci Teachers Asn. *Res:* Physical chemistry in aqueous solutions; isotope effects; catalytic effects; computational chemistry. *Mailing Add:* Dept Chem Indiana-Purdue Univ Ft Wayne IN 46805-1082

ONYSHKEVYCH, LUBOMYR S, b Ukraine, Feb 3, 33; US citizen; m 61; c 3. ELECTRONICS ENGINEERING. *Educ:* City Col New York, BEE, 55; Mass Inst Technol, MS, 57, EE, 61. *Prof Exp:* Res asst res lab electronics, Mass Inst Technol, 55-56; staff mem comput memories, RCA Labs, 57-59; mem tech staff comput magnets, Lincoln Labs, Mass Inst Technol, 60-63; mem tech staff comput res, 63-78, GROUP HEAD, ELECTRONIC PACKAGING RES, DAVID SARNOFF RES CTR, RCA LABS, 78- *Honors & Awards:* Indust Res 100 Award. *Mem:* Inst Elec & Electronics Engrs; Int Soc Hybrid Microelectronics. *Res:* Electronic packaging; computer memories; transfluxors; twistors; magnetic logic; thin magnetic films; thin film memories; magnetostriction; magnetoresistance; magnetosonic film memories; memory hierarchies; magnetic bubble domains; thick films; microelectronics; encapsulation; microsonics; hybrids. *Mailing Add:* 9 Dogwood Rd Lawrenceville NJ 08648

ONYSZCHUK, MARIO, b Wolkow, Poland, July 21, 30; nat Can; m 59; c 3. EDUCATIONAL ADMINISTRATION. *Educ:* McGill Univ, BSc, 51; Univ Western Ont, MSc, 52; McGill Univ, PhD(phys chem), 54; Cambridge Univ, PhD(inorg chem), 56. *Prof Exp:* Lectr chem, 56-57, from asst prof to assoc prof, 57-68, chmn dept, 79-85, PROF CHEM, McGILL UNIV, 68-, ASSOC DEAN, FAC GRAD STUDIES & RES, 88- *Concurrent Pos:* Vis prof, Univ Southampton, 68-69, Univ Sussex, 77-78 & Paul Sabatier Univ, 85-86. *Mem:* Am Chem Soc; fel Chem Inst Can; Royal Chem Soc. *Res:* Preparations, properties and structures of organometallic and coordination compounds of germanium, tin, lead halides and pseudohalides. *Mailing Add:* Dept Chem McGill Univ Montreal PQ H3A 2K6 Can

OOI, BOON SENG, b Kuala Lumpur, Malaysia, Apr 21, 40; US citizen; m 66; c 2. MEDICINE, IMMUNOLOGY. *Educ:* Univ Singapore, MB, BS, 64; FRACP, 76. *Prof Exp:* Asst prof, Pritzker Sch Med, Univ Chicago, 73; from asst prof to assoc prof, 73-86, PROF MED, COL MED, UNIV CINCINNATI, 86- *Mem:* Am Soc Nephrology; Am Fedn Clin Res; Am Soc Clin Invest. *Res:* Immunology of kidney diseases. *Mailing Add:* Dept Med Nephrol Univ Cincinnati Col Med 231 Bethesda Ave Cincinnati OH 45267

OOKA, JERI JEAN, b Olaa, Hawaii, Jan 22, 45. PHYTOPATHOLOGY. *Educ:* Univ Hawaii, BA, 66, MS, 69; Univ Minn, PhD(plant path), 75. *Prof Exp:* ASSOC PLANT PATHOLOGIST, UNIV HAWAII, 75- *Mem:* Am Phytopath Soc; AAAS. *Res:* Diseases of maize, especially epidemiology of Northern Corn Leaf Blight in tropical regions of the world; Fusarium moniliforme kernel and ear rot; aerobiology of Fusarium species; diseases of taro, especially Colocasia esculenta, and other edible Araceae with emphasis on fungal root and corm diseases; fungal root diseases of alfalfa in tropical and subtropical regions. *Mailing Add:* Kauai Br Sta Univ of Hawaii 7370-A Kuamoo Rd Kapaa HI 96746

OONO, YOSHITSUGU, b Fukuoka, Japan, Jan 21, 48. STATISTICAL PHYSICS. *Educ:* Kyushu Univ, Japan, BEng, 70, MEng, 73, PhD(chem physics), 76. *Prof Exp:* Asst prof phys chem, Res Inst Indust Sci, Kyushu Univ, Japan, 76-81; asst prof, 81-86, ASSOC PROF THEORET PHYSICS & BECKMAN INST, UNIV ILL, URBANA-CHAMPAIGN, 86- *Concurrent Pos:* Res assoc polymer physics, James Franck Inst, Univ Chicago, 79-81; vis scientist, Schlumberger Doll Res, 84; vis prof, Commun Sci & Comput Eng, Kyusha Univ, 90; vis prof, Appl Physics, Tokoy Inst Technol, 90. *Mem:* Phys Soc Japan; Am Phys Soc; Am Math Soc; Fel Japan Soc Prom Sci. *Res:* Nonequilibrium statistical physics; dynamical system; biophysics; polymer physics. *Mailing Add:* Beckman Inst Univ Ill 405 N Matthews Urbana IL 61801

OORT, ABRAHAM H, b Leiden, Neth, Sept 2, 34; m 59; c 3. DYNAMIC METEOROLOGY. *Educ:* Univ Leiden, Drs, 59; Mass Inst Technol, SM, 63; State Univ Utrecht, PhD(meteorol), 64. *Prof Exp:* Res asst energy-cycle lower stratosphere, Mass Inst Technol, 63-64; res scientist, Royal Neth Meteorol Inst, 64-66; res scientist, 66-85, SR RES SCIENTIST, GEOPHYS FLUID DYNAMICS LAB, NAT OCEANIC & ATMOSPHERIC ADMIN, 85-; 609-452-6518. *Concurrent Pos:* Vis prof, Princeton Univ, 71- *Honors & Awards:* Gold Medal, US Dept Com, 79; Victor P Starr Mem Lectr, Mass Inst Technol, 88. *Mem:* Am Meteorol Soc; Am Geophys Union; Royal Meteorol Soc; NY Acad Sci. *Res:* Dynamic climatology atmosphere and oceans; angular momentum cycle of the earth; energy cycle atmospheric and oceanic motions; El Nino and southern oscillation; hydrological cycle. *Mailing Add:* Nat Ocean & Atmospheric Admin Box 308 Princeton Univ Princeton NJ 08542

OOSTDAM, BERNARD LODEWIJK, b Amsterdam, Holland, Aug 13, 32; m 64; c 2. OCEANOGRAPHY, MARINE GEOLOGY. *Educ:* McGill Univ, BSc, 60; Scripps Inst Oceanog, Univ Calif, San Diego, MS, 64; Univ Del, PhD, 71. *Prof Exp:* Res assoc marine geol, Univ Hawaii, 62-63; staff geologist, Ocean Sci & Eng Inc, 63-65; oceanogr, Tex Instruments, 66; FROM ASST PROF TO PROF OCEANOG, MILLERSVILLE STATE UNIV, 66- *Concurrent Pos:* Managing dir, Environ Sci Res Assocs, 67-; pres, Marine Sci Consortium, 68-76; assoc, Inst Develop Riverine & Estuarine Systs, 68-75; consult, Kuwait Inst Sci Res, 76-78; UN-expert, Marine Geol & Coastal Eng Inc, Trinidad & Tobago, 80-82; consult, US Virgin Island Toxic Substances Proj, 86; vis prof E China Norm Univ, Shanghai, 88, US Info Agency Ampart, 89. *Mem:* Marine Technol Soc; Am Geophys Union; Am Arbit Assoc; Island Resources Found. *Res:* Nearshore and estuarine circulation; sediment transport; marine pollution ocean resources. *Mailing Add:* Dept Earth Sci Millersville Univ Millersville PA 17551

OOSTERHUIS, DERRICK M, b Johannesburg, SAfrica, Aug 22, 47; m 72. PLANT WATER RELATIONS, FOLIAR FERTILIZERS. *Educ:* Natal Univ, SAfrica, BSc, 70; Univ Reading, UK, MSc, 73; Utah State Univ, PhD(plant physiol), 81. *Prof Exp:* Res agronomist, Chiredzi Res Inst, Rhodesia, 70-72; sect leader, Cotton Res Inst, Rhodesia, 73-76; asst prof crop sci, Natal Univ, SAfrica, 77; sr plant physiologist, Water Res Comn, 80-84; from asst prof to assoc prof, 85-89, PROF CROP PHYSIOL, DEPT AGRON, UNIV ARK, 89- *Concurrent Pos:* Chmn, Physiol Conf, Nat Cotton Coun, 89. *Mem:* Plant Physiol Soc Am; Crop Sci Soc Am; Soc Exp Bot. *Res:* Crop water relations; methodology of measuring water potential; photosynthesis; foliar fertilizers; root growth; effects of environmental stress; author of over 200 publications. *Mailing Add:* Dept Agron Univ Ark 276 Altheimer Dr Fayetteville AR 72703

OOSTERHUIS, WILLIAM TENLEY, b Mt Vernon, NY, Mar 28, 40; m 64; c 2. CHEMICAL PHYSICS. *Educ:* Wis State Univ-Platteville, BS, 62; Carnegie Inst Technol, MS, 64; Carnegie Mellon Univ, PhD (physics), 67. *Prof Exp:* At Atomic Energy Res Estab, 67-68; at Carnegie-Mellon Univ, 68-74; solid state chem prog dir, NSF, 74-75, assoc prog dir solid state physics, 78-79, staff assoc, 79-82, prog dir, instrumentation mat res, Mat Res Lab Sect, 82-87, actg sect head, condensed matter sci, 87-90, MGR, SCI & TECHNOL CENTERS, NSF, 90- *Concurrent Pos:* Guest scientist, Nat Bur Standards, 84-85. *Mem:* Am Phys Soc; Mat Res Soc. *Res:* Paramagnetic hyperfine interactions in transition metal complexes, including biological molecules; magnetic critical point phenomena; crystalline electric fields; Mossbauer effect; x-ray scattering. *Mailing Add:* Div Mat Res Nat Sci Found Washington DC 20550

OPAL, CHET BRIAN, aeronomy, astrophysics; deceased, see previous edition for last biography

OPARIL, SUZANNE, b Elmira, NY, Apr 10, 41. HYPERTENSION, CARDIOVASCULAR RESEARCH. *Educ:* Cornell Univ, AB, 61; Columbia Univ, MD, 65. *Prof Exp:* Clin res fel med, Mass Gen Hosp, 68-71; from asst prof to assoc prof med, Med Sch, Univ Chicago, 71-77; assoc prof med, Univ Ala Birmingham, 77-81, asst prof physiol & biophysics, 80-81, ASSOC PROF PHYSIOL & BIOPHYSICS, UNIV ALA BIRMINGHAM, 81-, PROF MED, 81-, DIR HYPERTENSION PROG, 85- *Concurrent Pos:* Ed, Am J Med Sci, 84-; assoc ed, Am J Physiol: Renal, 89-; chmn, Pub Policy Affairs Comt, Am Heart Asn, 90- & Budget Comt, Coun Basic Sci, 90. *Honors & Awards:* Young Investr Award, Int Soc Hypertension, 79. *Mem:* Inst Med-Nat Acad Sci; Am Heart Asn. *Res:* Hypertension research; neural control of the circulation; biochemistry, physiology and molecular biology of the renin-angiotensin system; molecular biology of vascular hypertrophy. *Mailing Add:* Zeigler Res Bldg Rm 1034 Univ Ala Sta Birmingham AL 35294

OPAS, EVAN E, b Englewood, NJ, May 31, 53; m 79; c 3. CELL BIOLOGY, MODELS OF INFLAMMATION. *Educ:* Univ Pa, BA, 76; Case Western Reserve Univ, MS, 78. *Prof Exp:* Teaching asst genetics & gen biol, Case Western Reserve Univ, 76-78; res specialist, Univ Med & Dent of NJ, 78-80; res asst, Univ Conn Health Ctr, 80-82; staff immunologist, 82-85, RES IMMUNOLOGIST, MERCK & CO INC, 85- *Mem:* AAAS. *Res:* Develop new therapeutic compounds against inflamation; eicosanoids and proteases in inflammation, especially involving the immune system; phospholipid metabolism and fatty acid synthesis; molecular genetics of chloroplasts of algae. *Mailing Add:* Dept Bone Biol & Osteoporosis MC WP 26A-1000 Summitown Pike West Point PA 19486

OPATRNY, JAROSLAV, b Plzen, Czechoslovakia, Sept 5, 46; Can citizen; c 2. ALGORITHMS, NETWORK MODELS. *Educ:* Charles Univ, Praque, MSc, 68; Univ Waterloo, PhD(comput sci), 75. *Prof Exp:* Lectr math & comput sci, Tech Univ, Plzen, 68-70; asst prof comput sci, Univ Alta, Can, 75-77; assoc prof, 77-85, ASSOC PROF COMPUT SCI, CONCORDIA UNIV, MONTREAL, 85- *Concurrent Pos:* Fel, Univ Waterloo, 75. *Mem:* Asn Comput Mach; Europ Asn Theoret Comput Sci. *Res:* network algorithms; bandwidth problem for graphs; computer network models. *Mailing Add:* Dept Comput Sci-Concordia Univ Montreal PQ H3G 1M8 Can

OPAVA-STITZER, SUSAN C, b New York City, NY, Mar 14, 47; div; c 2. RENAL PHYSIOLOGY, ENDOCRINOLOGY. *Educ:* Col Mt St Vincent, BS, 68; Univ Mich, Ann Arbor, PhD(physiol), 72. *Prof Exp:* Fel, Dartmouth Med Sch, 72-74; from asst prof to assoc prof, 74-86, PROF PHYSIOL, MED SCH, UNIV PR, 86-, CHMN, DEPT PHYSIOL, 89- *Mem:* Am Physiol Soc; Am Soc Nephrology; Int Soc Nephrology; AAAS; Inter-Am Soc Hypertension. *Res:* Renal physiology, specifically mechanisms of concentration and dilution of urine; electrolyte balance; renin-angiotensin system; control of aldosterone secretion; diabetes insipidus. *Mailing Add:* Dept Physiol Med Sch Univ PR GPO Box 5067 San Juan PR 00936

OPDYKE, DAVID F, MUSCLE PHYSIOLOGY, ENDOCRINOLOGY. *Educ:* Univ Ind, PhD(physiol), 42. *Prof Exp:* EMER PROF PHYSIOL, SCH MED, E CAROLINA UNIV, 82- *Mailing Add:* 1912 Quail Ridge Greenville NC 27834

OPDYKE, NEIL, b Frenchtown, NJ, Feb 7, 33; m 58; c 3. GEOLOGY, GEOPHYSICS. *Educ:* Columbia Col, BA, 55; Univ Durham, PhD(geol), 58. *Prof Exp:* Fel, Rice Univ, 58-59; res fel, Australian Nat Univ, 59-60 & Univ Col Rhodesia & Nyasaland, 61-64; res assoc, Columbia Univ, 65-66, res scientist, Lamont-Doherty Geol Observ, 64-81, sr res assoc, Univ, 66-81; PROF & CHMN, DEPT GEOL, UNIV FLA, 81- *Concurrent Pos:* Fulbright travel grant, Australia, 59-60; adj prof, Columbia Univ, 74-81. *Mem:* AAAS; Geol Soc Am; Am Geophys Union. *Res:* Paleomagnetism; paleoclimatology. *Mailing Add:* Dept Geol Univ Fla Gainesville FL 32611

OPECHOWSKI, WLADYSLAW, b Warsaw, Poland, Mar 10, 11; nat Can citizen; m 33; c 1. THEORETICAL PHYSICS. *Educ:* Univ Warsaw, MPhil, 35. *Hon Degrees:* DSc, Univ Wroclaw, Poland, 73. *Prof Exp:* Lorentz-Fonds fel, Leiden & physicist, Teylers Stichting, 39-45; physicist, Philips Res Lab, Eindhoven, 45-48; from assoc prof to prof, 58-76, HON PROF PHYSICS, UNIV BC, 76- *Concurrent Pos:* Lorentz prof, State Univ Leiden, 64-65; vis prof, Roman Cath Univ Nijmegen, 68-69. *Mem:* Am Phys Soc; fel Royal Soc Can; Can Asn Physicists; Neth Phys Soc. *Res:* Quantum theory of magnetism; applications of group theory. *Mailing Add:* Dept of Physics Univ of BC Vancouver BC V6T 2A6 Can

OPELL, BRENT DOUGLAS, b Washington, Ind, June 12, 49. SYSTEMATICS, FUNCTIONAL MORPHOLOGY. *Educ:* Butler Univ, BA, 71; Southern Ill Univ, MS, 74; Harvard Univ, PhD(biol), 78. *Prof Exp:* from asst prof to assoc prof, 78-89, PROF ZOOLOGY, VA POLYTECH INST & STATE UNIV, 89- *Mem:* Am Arachnological Soc; AAAS; Am Soc Zoologists; Brit Arachnological Soc; Soc Syst Zoologists; Am Micros Soc. *Res:* Morphology; systematics; natural history; behavior of cribellate orb-weaving spiders. *Mailing Add:* Dept Biol Va Polytech Inst & State Univ Blacksburg VA 24061

OPELLA, STANLEY JOSEPH, b Summit, NJ, Sept 29, 47. PHYSICAL CHEMISTRY, BIOLOGICAL CHEMISTRY. *Educ:* Univ Ky, BS, 69; Stanford Univ, PhD(chem), 74. *Prof Exp:* Fel chem, Mass Inst Technol, 75-76; from asst prof to assoc prof, 76-83, PROF CHEM, UNIV PA, 83- *Mem:* Am Chem Soc; Am Phys Soc. *Res:* Nuclear magnetic resonance spectroscopy of biological molecules; high resolution solid state and solution nuclear magnetic resonance spectroscopy. *Mailing Add:* Dept Chem 356 Chem D5 Univ Pa Philadelphia PA 19104

OPENSHAW, MARTIN DAVID, b Mesa, Ariz, Oct 10, 40; m 64; c 5. SOIL FERTILITY, PLANT PHYSIOLOGY. *Educ:* Ariz State Univ, BS, 65; Iowa State Univ, MS, 68, PhD(soils), 70. *Prof Exp:* Teaching asst soil sci, Iowa State Univ, 65-67, instr, 67-70; asst prof soil sci, Univ Ariz, 70-76, assoc prof, 76-77, exten soil specialist, 70-77; dir, Am Registry Certified Prog Agron, Crops & Soils, Ltd, Madison, Wis, 77-; PROF, AGR DEVELOP, NCSU. *Mem:* Am Soc Agron; Soil Sci Soc Am; Soil Conserv Soc Am; Sigma Xi. *Res:* Soil fertility work with field crops and the effect of fertilizers on the environment; technology transfer mechanisms in developing countries. *Mailing Add:* PO Box 50345 Am Embassy Provo UT 84605-0345

OPFELL, JOHN BURTON, b Cushing, Okla, July 24, 24; m 54; c 3. BIOCHEMICAL ENGINEERING, NUCLEAR CHEMICAL ENGINEERING. *Educ:* Univ Wis, BS, 45; Calif Inst Technol, MS, 47, PhD(chem eng & math), 54; Stanford Univ, MBA, 51. *Prof Exp:* Engr, Standard Oil Co, Ind, 47-49; fel, Calif Inst Technol, 53-55; engr, Cutter Labs, 55-61, Dynamic Sci Corp, 61-64 & Philco-Ford Corp, 64-69; asst mgr corp planning, Sunkist Growers, 70-73; asst to exec vpres, Henningson, Durham & Richardson, 80-83; ENGR, AIRES MFG CO, 73-80 & 80- *Concurrent Pos:* Consult, Automated Marine Int, 70, Vitaminerals Inc, 73-80 & Meditech Pharmaceut, Inc, 82-84; lectr, Univ Calif, Santa Barbara, 73 & 82, Calif State Univ, 86. *Mem:* Fel AAAS; Sigma Xi; Opers Res Soc Am; Am Inst Chem Engrs. *Res:* Heat, material and momentum transfer in fluids; thermodynamics and volumetric properties of fluids; sterilization of pharmaceuticals and spacecraft; soil microbial ecology; adsorption; mathematical modeling of physical and economic phenomena. *Mailing Add:* PO Box 7574 Torrance CA 90504

OPIE, JOSEPH WENDELL, b Wilmot, SDak, Sept 27, 13; m 40; c 3. ORGANIC CHEMISTRY, FOOD TECHNOLOGY. *Educ:* Univ Minn, ChB, 35, PhD(org chem), 40. *Prof Exp:* Microanalyst, Univ Minn, 35-39; res chemist, Merck & Co, Inc, NJ, 40-42; res assoc, Nat Defense Res Comt, Univ Minn, 42; res chemist, Tex Co, NY, 42-44 & Wyeth, Inc, 44-46; head develop res sect, 46-48, org develop sect, 48-62, formulation concept dept, 62-67, concept develop dept, 67-69 & res & develop, Sperry Div, 69-71, mgr, 71-74, mgr develop, Dept Corp Res, Gen Mills, Inc, 74-78; RETIRED. *Mem:* Am Chem Soc; AAAS. *Res:* Polyalkylgenzenes; polyalkyl-quinones; substituted coumarins; vitamin E; plastics; antibiotics; medicinals; fatty acids product and processes development; mineral flotation starch; plant gums; corrosion; phospholipids; food products. *Mailing Add:* 2515 W52nd St Minneapolis MN 55410

OPIE, THOMAS RANSON, b Staunton, Va, Dec 19, 48; m 80. ORGANIC CHEMISTRY, ORGANOMETALLIC CHEMISTRY. *Educ:* Davidson Col, BS, 71; Cornell Univ, MS, 74, PhD(org chem), 76. *Prof Exp:* RES CHEMIST, ROHM & HAAS CO, 75- *Mem:* Am Chem Soc; Am Sci Affil; Sigma Xi. *Res:* Synthetic reactions involving completely fluorinated organometallic compounds, chemical process research for agricultural chemicals. *Mailing Add:* 8 Lower State Rd N Wales PA 19454-1216

OPIE, WILLIAM R(OBERT), b Butte, Mont, Apr 3, 20; m 44; c 2. PHYSICAL METALLURGY, EXTRACTIVE METAL. *Educ:* Mont Sch Mines, BS, 42; Mass Inst Technol, ScD, 49. *Hon Degrees:* MetE, Mont Sch Mines, 65; ScD, Mont Tech, 80. *Prof Exp:* Foundry metallurgist, Wright Aeronaut Corp, 42-45; design metallurgist, Anaconda Copper Mining Co, 46; asst, Mass Inst Technol, 47-48; res metallurgist, Am Smelting & Refining Co, 48-50; res supvr, Titanium Div, Nat Lead Co, 50-60; dir res & develop, US Metals Refining Div, Amax Inc, 60-68, vpres & tech dir, Amax Base Metals, 68-71, pres, Amax Base Metals Res & Develop, 71-85; CONSULT, 85- *Concurrent Pos:* Chmn tech comt, Int Copper Res Asn, 64-66 & Int Germanium Res Comt, 64-66. *Mem:* Nat Acad Eng; fel Am Soc Metals; fel Am Inst Mining, Metall & Petrol Engrs; fel Metall Soc. *Res:* Nonferrous extractive metallurgy; copper alloy development; powder metallurgy. *Mailing Add:* 119 Crawfords Corner Rd Holmdel NJ 07733

OPIELA, ALEXANDER DAVID, JR, b Falls City, Tex, Dec 26, 29; m 61; c 5. ELECTRICAL & ENVIRONMENTAL HEALTH ENGINEERING. *Educ:* Univ Tex, Austin, BSc, 52, MSc, 55 & 72. *Prof Exp:* Res engr, Defense Res Lab, Univ Tex, 52-56, systs develop specialist, 56-61; sr engr, Textran Corp, 61-62; sr engr, Tracor, Inc, 62-64; mfg engr, 64-65, asst vpres eng & mfg, 65-69, prod line mgr, 69-70; engr, Air Pollution Control Serv, Tex Dept Health, 72-73, Tex Air Control Bd, 73-81, consult; RETIRED. *Mem:* Inst Elec & Electronics Engrs; Air Pollution Control Asn. *Res:* Cathode ray tube indicators and film data recorders; direction finding and field strength measuring equipment; engineering and manufacturing management; air pollution sources, effects and abatement technology; air pollution control. *Mailing Add:* 6105 Hylawn Dr Austin TX 78723

OPILA, ROBERT L, JR, b Elmhurst, Ill, Mar 31, 51; m 79; c 2. SURFACE SCIENCE. *Educ:* Univ Ill, BS, 75. *Prof Exp:* MEM TECH STAFF, AT&T BELL LABS, 82- *Mem:* Am Chem Soc; Inst Elec & Electronics Engrs; Electrochem Soc; Am Vacuum Soc; Am Soc Testing & Mat. *Res:* Determine how molecular level processes determine macroscopic behaviour at complex interfaces; adhesion; corrosion; interdiffusion. *Mailing Add:* AT&T Bell Labs Rm 1C-260 600 Mountain Ave Murray Hill NJ 07974

OPITZ, HERMAN ERNEST, b Kingston, NY, Sept 28, 29; m 54; c 2. INORGANIC CHEMISTRY. *Educ:* Union Col NY, BS, 51; Ind Univ, PhD(chem), 56. *Prof Exp:* Res chemist, Armstrong Cork Co, Pa, 55-69; MGR GLASS RES PACKAGING PROD DIV, KERR GLASS MFG CORP, 69- *Mem:* Am Ceramic Soc. *Res:* Silicon chemistry; inorganic polymers and foams; glass technology. *Mailing Add:* 1784 Coln Mnr Dr Lancaster PA 17603

OPITZ, JOHN MARIUS, b Hamburg, Ger, Aug 15, 35; US citizen; m; c 5. PEDIATRICS, MEDICAL GENETICS. *Educ:* Univ Iowa, BA, 56, MD, 59; Am Bd Pediat, dipl, 64, Am Bd Med Genet, dipl, 81, DSc, Mont State Univ, Bozeman, 81, Kiel Univ Ger, MD, 86. *Hon Degrees:* DSc, Mont State Univ, Bozeman, 83. *Prof Exp:* From intern to resident pediat, Univ Iowa Hosp, 59-61; resident, Univ Wis-Madison, 61-62, from asst prof to assoc prof pediat & med genetics, 64-72, prof & dir, Wis Clin Genetics Ctr, 72-79; coordr, Shodair-Mont Regional Genetics Prog, 79-82, CHMN, DEPT MED GENET, SHODAIR CHILDREN'S HOSP, 83- *Concurrent Pos:* NIH fels pediat & med genetics, Univ Wis-Madison, 62-64; adj clin prof, Univ Wash, Seattle, Mont State Univ, Bozeman & Univ Wis-Madison, 79-; ed-in-chief, Am J Med Genetics. *Honors & Awards:* Tenth Farber Lectr, Soc Pediat Path, 87. *Mem:* AAAS; Am Soc Human Genetics; Am Soc Zool; Am Inst Biol Sci; Int Soc Cranio-Facial Biol; mem Ger Acad of Scientists Leopoldina; corresp mem, Ger Soc Pediat; Am Pediat Soc; Soc Pediat Res; Teratology Soc; Am Acad Pediat; AAAS. *Res:* Hereditary diseases; single and multiple congenital anomalies; cytogenetics; genetic counseling; errors of sex determination and differentiation; clinical and developmental genetics; mental retardation. *Mailing Add:* Shodair Children's Hospital Box 5539 Helena MT 59604

OPLINGER, EDWARD SCOTT, b Jewell, Kans, Apr 25, 43; m 64; c 3. AGRONOMY. *Educ:* Kans State Univ, BS, 65; Purdue Univ, Lafayette, MS, 69, PhD(agron), 70. *Prof Exp:* Asst prof, 70-75, assoc prof, 75-78, PROF AGRON, UNIV WIS-MADISON, 78- *Mem:* Am Soc Agron; Crop Sci Soc Am; Plant Growth Regulator Working Group. *Res:* Crop production; plant growth regulators; plant physiology. *Mailing Add:* 363 Moore Hall Dept Agron Univ Wis Madison WI 53706

OPP, ALBERT GEELMUYDEN, b Fargo, NDak, July 29, 31; m 60; c 2. ASTROPHYSICS. *Educ:* Univ NDak, BS, 53; St Louis Univ, MS, 55; Univ Gottingen, Dr rer nat(geophys), 61. *Prof Exp:* Terrestrial scientist, Air Tech Intel Ctr, US Air Force, 55-57, space systs specialist, Europ Hq, 59-62, instr physics, US Air Force Acad, 62-63; staff scientist, 63-66, dep chief particles & fields prog, 66-70, chief high energy astrophys, 70-83, dep dir astrophys, NASA HQ, 83-85, asst chief, Lab High Energy Astrophys, NASA Goddard Space Flight Ctr, 85-87; SR PRIN STAFF, SPACE PROGS, BDM CORP, 87- *Concurrent Pos:* Consult, Kaman Nuclear Div, 62-63; fel pub affairs, Princeton Univ, 70-71. *Mem:* Am Phys Soc; Am Astron Soc; AAAS. *Res:* Cosmic rays; gamma ray astronomy; x-ray astronomy; magnetic fields and plasmas in space; trapped radiation; planetary magnetospheres. *Mailing Add:* 1811 Stirrup La Alexandria VA 22308

OPPELT, JOHN ANDREW, b Baltimore, Md, Feb 4, 37; m 60; c 1. MATHEMATICS. *Educ:* Loyola Col, Md, AB, 59; Univ Notre Dame, MS, 61, PhD(math), 65. *Prof Exp:* Instr math, Univ Notre Dame, 65; actg asst prof, Univ Va, 65, asst prof, 65-70; assoc prof, 70-74, chmn dept math, 70-80, actg dean, 80-81, PROF MATH, GEORGE MASON UNIV, 74-; VPRES, BELLARMINE COL, 81- , AT DEPT MATH, BELLARMINE COL. *Concurrent Pos:* Sesquicentennial scholar & vis asst prof, Univ Wash, 69-70. *Mem:* Am Math Soc. *Res:* Modern algebra; Abelian p-mixed groups. *Mailing Add:* Dept Math & Comput Sci Bellarmine Col Newburg Rd Louisville KY 40205

OPPELT, JOHN CHRISTIAN, b Baltimore, Md, Dec 11, 31; m 56; c 5. ORGANIC CHEMISTRY. *Educ:* Loyola Col Md, BS, 53; Univ Md, College Park, MS, 58; Rutgers Univ, PhD(org chem), 67. *Prof Exp:* From chemist to res chemist, 58-74, SR RES CHEMIST, AM CYANAMID CO, 74- *Mem:* Am Chem Soc; Sigma Xi. *Res:* Synthesis of plastics additives, especially antioxidants, stabilizers, ultraviolet and near infrared absorbers; synthesis of alkyl phosphines and derivatives. *Mailing Add:* 606 Talamini Rd Bridgewater NJ 08807-1647

OPPENHEIM, A(NTONI) K(AZIMIERZ), b Warsaw, Poland, Aug 11, 15; nat US; m 45; c 1. MECHANICAL ENGINEERING. *Educ:* Warsaw Inst Technol, dipl, 39; Polish Univ, Eng, dipl, 45; Univ London, DIC & PhD(eng), 45. *Hon Degrees:* DSc, Univ London, 76; Dr, Univ Poitiers, 81, Warsaw Inst Tech, 89. *Prof Exp:* Asst lectr, City & Guilds Col, Eng, 45-48; actg asst prof mech eng, Stanford Univ, 48-50; from asst prof to prof, 50-86, EMER PROF MECH ENG, UNIV CALIF, BERKELEY, 86- *Concurrent Pos:* Lectr, Polish Univ, Eng, 42-48; consult, Shell Develop Co, 52-60; NSF sr fel, 60; assoc prof, Univ Poitiers, 73 & 80. *Honors & Awards:* Water Arbit Prize, Brit Inst Mech Engrs, 49; T Pendray, Am Aeronaut & Astronaut, 66; Dionicy Smolelenski Medal, Polish Acad Sci, 87; Alfred C Egerton Medal, Combustion Inst, 88. *Mem:* Nat Acad Eng; fel & hon mem Am Soc Mech Engrs; fel Am Inst Aeronaut & Astronaut; Int Acad Astronaut; Am Phys Soc; Am Chem Soc; fel Soc Automotive Engrs. *Res:* Gas dynamics; wave propagation and unsteady flow phenomena; detonation; radiation; heat transfer, thermodynamics, combustion; internal combustion engines. *Mailing Add:* Prof Eng Col Eng Univ Calif Berkeley CA 94720

OPPENHEIM, ALAN VICTOR, b New York, NY, Nov 11, 37; m 64; c 2. ELECTRICAL ENGINEERING. *Educ:* Mass Inst Technol, SB & SM, 61. *Hon Degrees:* ScD, Mass Inst Technol, 64. *Prof Exp:* From asst prof to assoc prof, 64-76, assoc div head, Lincoln Lab, 78-80, PROF ELEC ENG, MASS INST TECHNOL, 76- *Concurrent Pos:* Consult, MIT Lincoln Lab, 66-, Comput Signal Processors, Inc, 71-, Sanders Assocs, Inc, 79- & Sperry Corp, 84-; Guggenheim fel, 72; ed, Prentice Hall, Inc, 75- *Honors & Awards:* Centennial Medal, Inst Elec & Electronics Engrs, 84 & Educ Medal, 88. *Mem:* Nat Acad Eng; Fel Inst Elec & Electronics Engrs. *Res:* Digital signal processing, speech, image and seismic data processing; knowledge-based signal processing. *Mailing Add:* Dept Elec Eng & Comput Sci Mass Inst Technol Rm 36-625 Cambridge MA 02139

OPPENHEIM, BERNARD EDWARD, b Chicago, Ill, July 5, 37; m 63; c 4. NUCLEAR MEDICINE, RADIOLOGY. *Educ:* Univ Ariz, BS, 59; Univ Chicago, MD, 63. *Prof Exp:* Intern, Michael Reese Hosp, Chicago, 63-64; resident radiol, Univ Chicago, 64-67, instr nuclear med, 70-71, from asst prof to assoc prof radiol, 71-76; assoc prof, 76-81, PROF RADIOL, IND UNIV, 81- *Concurrent Pos:* James Picker Found scholar radiol res, 72-73; consult, Nat Cancer Inst, Nat Acad Sci. *Mem:* Asn Univ Radiologists; Am Roentgen Ray Soc; AMA; Radiol Soc NAm; Soc Nuclear Med; Am Col Nuclear Physicians. *Res:* Application of computers to imaging in nuclear medicine; long term effects of low dose in utero-irradiation. *Mailing Add:* Dept Radiol Ind Univ Sch Med Indianapolis IN 46223

OPPENHEIM, IRWIN, b Boston, Mass, June 30, 29; m 74; c 1. CHEMICAL PHYSICS. *Educ:* Harvard Univ, AB, 49; Yale Univ, PhD(chem), 56. *Prof Exp:* Asst chem, Calif Inst Technol, 49-51; statist mech, Yale Univ, 51-53; physicist, Nat Bur Stand, 53-60; chief theoret physics, Gen Dynamics/Convair, 60-61; assoc prof, 61-65, PROF CHEM, MASS INST TECHNOL, 65- *Concurrent Pos:* Lectr, Univ Md, 54-60; res assoc, State Univ Leiden, 55-56; vis prof, Weizmann Inst, 58-59 & Univ Calif, San Diego, 66-67; Van Der Waals prof, Univ Amsterdam, 67, Lorentz prof, Univ Leiden, 83; ed, Physica A, 84-; assoc ed, Phys Rev A, 90- *Honors & Awards:* Lennard-Jones Lectr, 87. *Mem:* Fel Am Phys Soc; fel Am Acad Arts & Sci. *Res:* Statistical mechanics of irreversible processes. *Mailing Add:* Dept Chem Rm 6-221 Mass Inst Technol Cambridge MA 02139

OPPENHEIM, JOOST J, b Aug 11, 34. CANCER RESEARCH. *Educ:* Columbia Col, AB, 56, MD, 60. *Prof Exp:* Intern, Kings County Hosp, Seattle, Wash, 60-61; asst med resident, Univ Wash Hosp, 61-62; clin assoc, Med Br, Nat Cancer Inst, NIH, Bethesda, 62-65, sr investr, Cell Biol Sect, Lab Biochem, Nat Inst Dent Res, 66-69, chief, Cellular Immunol Sect, Lab Microbiol & Immunol, 70-82, INSTR POSTGRAD PROG, NIH, BETHESDA, 70-, CHIEF, LAB MOLECULAR IMMUNOREGULATION, BIOL RESPONSE, MODIFIERS PROG, NAT CANCER INST, FREDERICK CANCER RES FACIL, 83- *Concurrent Pos:* Vol internist, Anacostia Ctr Med Serv, Washington, DC, 70-74; sabbat, Lab Biochem Immunol, Weizmann Inst, Rehovot, Israel, 74-75; physician, Allergy Clin, Personnel Health Dept, NIH, 75-82; med dir, USPHS, 75-; OPD physician, Prince Georges County Clin Sr Citizens; internist Mobile Med Care, Montgomery County Free Clin; immunol lectr, Dept Zool, Univ Md, 77-83, adj prof, 79-; mem, Comt Cancer Immunodiag, 76-79, Decision Network Comt, BRMP, DCT, Nat Cancer Inst, 86-87, adv bd, Off Naval Res, 89-92. *Honors & Awards:* Ferdin & Von-Hebra Prize, 82; Marie T Bonazinga Res Award, 87. *Mem:* Fel Am Acad Microbiol. *Res:* Biochemical immunology; molecular immunoregulation. *Mailing Add:* Lab Molecular Immunoregulation Nat Cancer Inst Bldg 560 Rm 21-89A Frederick MD 21701-1201

OPPENHEIM, JOSEPH HAROLD, b Connellsville, Pa, Oct 10, 26. ALGEBRA. *Educ:* Univ Chicago, PhB, 48, BS, 54; Univ Ill, MS, 56, PhD(math, algebra), 62. *Prof Exp:* Asst prof math, Rutgers Univ, 60-65; asst prof, 66-71, ASSOC PROF MATH, SAN FRANCISCO STATE UNIV, 71- *Mem:* Math Asn Am; Am Math Soc. *Res:* Representation theory of groups. *Mailing Add:* Dept Math San Francisco State Univ 1600 Holloway Ave San Francisco CA 94132

OPPENHEIM, RONALD WILLIAM, b Des Moines, Iowa, Nov 2, 38; m 68; c 1. PSYCHOBIOLOGY. *Educ:* Drake Univ, BA, 62; Wash Univ, PhD(biopsychol), 67. *Prof Exp:* Res biologist, Sch Med, Wash Univ, 67-68; RES SCIENTIST, RES DIV, NC DEPT OF MENT HEALTH, 68- *Concurrent Pos:* Adj prof, Neurobiol Prog, Univ NC, Chapel Hill, 71- *Mem:* AAAS; Am Inst Biol Sci; Animal Behav Soc; Neurosci Soc; Sigma Xi. *Res:* Neuroembryology; ontogeny of avian embryonic behavior; early posthatching sensory-motor behavior in Aves; neuroanatomical and neuropharmacological development of the spinal cord. *Mailing Add:* 1048 Watson Ave Winston-Salem NC 27103-4548

OPPENHEIMER, CARL HENRY, JR, b Los Angeles, Calif, Nov 13, 21; m 71; c 3. OCEANOGRAPHY, ECOLOGY. *Educ:* Univ Southern Calif, BA, 47, MA, 49; Univ Calif, PhD(marine microbiol), 51. *Prof Exp:* Asst marine biologist, Scripps Inst Calif, 51-55; sr res scientist, Res Ctr, Pan Am Petrol Corp, Okla, 55-57; res scientist & lectr marine microbiol, Univ Tex, 57-61; assoc prof, Inst Marine Sci, Univ Miami, 61-64; prof biol, Fla State Univ, 64-66, chmn dept oceanog, 66-69, prof oceanog & dir, Edward Ball Marine Lab, 66-71; prof microbiol & dir, Marine Sci Inst, 71-73, PROF MARINE SCI, UNIV TEX, PORT ARANSAS, 73-; DEPT MARINE STUDIES, OPPENHEIMER ENVIRON CO, AUSTIN. *Concurrent Pos:* Fulbright fel, Univ Oslo, 52-53; res fel, Sea Fish Hatchery, Arendal, Norway, 55; chmn, Int Symposium Marine Microbiol, 61; mem subcomt marine biol, President's Sci Adv Comt, 65-66; chmn, Conf Unresolved Probs Marine Microbiol, 66; mem, Japan-US Cultural Exchange Prog, 66 & President's Panel Oil Pollution, 68-71; vpres, Gulf Univ Res Corp, 66-70, mem prog planning coun, 70-; mem, Governor's Comn Marine Sci & Technol, Fla, 67-71 & sci comn, NSF, 69-72; panel chmn, Am Assembly-Uses of the Sea, 69; US rep adv panel eco-sci, NATO, 71-74; Univ Tex Rep, Univ Coun Water Resources, 71-; mem, Tex Senate Interim Coastal Zone Study Comn, 71- & adv coun, Southern Interstate Nuclear Bd, 71-; Fulbright fel, Univ Naples, 78. *Honors & Awards:* Gold Medal, Ital Biol Soc, 76. *Mem:* Am Soc Microbiol; Am Soc Limnol & Oceanog; Geochem Soc; Brit Soc Gen Microbiol; Sigma Xi. *Res:* Distribution and function of marine microorganisms; diseases of marine fishes; biological corrosion; marine productivity cycles; effect of detergents in the oceans; marine microbiology; origin and ecology of oil; biological criteria for coastal zone management. *Mailing Add:* Dept Marine Sci Austin TX 78712-1162

OPPENHEIMER, JACK HANS, b Eglesbach, Ger, Sept 14, 27; US citizen; m 53; c 3. ENDOCRINOLOGY, INTERNAL MEDICINE. *Educ:* Princeton Univ, AB, 49; Columbia Univ, MD, 53; Am Bd Internal Med, dipl, 61. *Prof Exp:* Intern med, Boston City Hosp, 53-54; resident med, Duke Univ Hosp, 57-59; asst clin prof neurol, Sch Med, NY Univ, 62-65; from asst prof to prof med, Albert Einstein Col Med, Yeshiva Univ, 66-73; head endocrine serv, Montefiore Hosp & Med Ctr, 68-76; PROF MED & PHYSIOL, HEAD DIV ENDOCRINOL, UNIV MINN, 86-, CECIL J WATSON PROF MED, 86- *Concurrent Pos:* Fel med, Mem Ctr & Thyroid-Pituitary Study Group, Sloan-Kettering Inst, 54-55; vis fel, Col Physicians & Surgeons, Columbia Univ, 59-60; asst physician, Endocrinol Lab, Presby Hosp, New York, 59-60; from asst attend physician to attend physician, Div Med, Montefiore Hosp & Med Ctr, 60-68; prin investr USPHS grant, 61-; career scientist, Health Res Coun, New York, 62-72; mem bd trustees, Windward Sch, White Plains, NY, 69-71. *Honors & Awards:* Van Meter Award, Am Thyroid Asn, 65; Astwood Award, Endocrine Soc, 78; Parke Davis Award, Am Thyroid Soc, 83. *Mem:* Am Asn Physicians; Am Soc Clin Invest; fel Am Col Physicians; Am Physiol Soc; Am Thyroid Asn (pres, 85); Am Soc Biochem & Molecular Biol. *Res:* Peripheral metabolism and action of the thyroid hormones. *Mailing Add:* Div Endocrinol Univ Minn Box 91 Mayo Bldg Minneapolis MN 55455

OPPENHEIMER, JANE MARION, b Philadelphia, Pa, Sept 19, 11. ZOOLOGY. *Educ:* Bryn Mawr Col, BA, 32; Yale Univ, PhD(zool), 35. *Hon Degrees:* ScD, Brown Univ, 76. *Prof Exp:* Sterling fel biol, Yale Univ, 35-36, res fel, Am Asn Univ Women Berliner, 36-37; res fel embryol, Univ Rochester, 37-38; instr biol, 38-42, Guggenheim fel, 42-43, actg dean grad sch, 46-47, from asst prof to prof biol, 43-74, prof, 74-80, EMER PROF BIOL & HIST OF SCI, BRYN MAWR COL, 80- *Concurrent Pos:* Rockefeller Found fel, 50-51; Guggenheim fel, 52-53, 62-63; sr fel, NSF, 59-60; vis prof biol, Johns Hopkins Univ, 66-67; mem, NIH Hist Life Sci Study Sect, 66-70; exchange prof fac sci, Univ Paris, 69. *Honors & Awards:* Wilbur Lucius Cross Medal, Yale Grad Alumni Asn. *Mem:* Soc Develop Biol; Am Soc Zoologists (treas, 57-59, pres, 74); Hist Sci Soc; Am Asn Anat; Am Asn Hist Med; Am Philos Soc (secy, 87- 92). *Res:* Developmental biology; experimental analysis of teleost development; history of embryology. *Mailing Add:* One Independence Pl No 304 Sixth St & Locust Walk Philadelphia PA 19106-3727

OPPENHEIMER, LARRY ERIC, b New York, NY, Aug 18, 42; m 64; c 2. COLLOID CHEMISTRY, PARTICLE SIZE ANALYSIS. *Educ:* Clarkson Col Technol, BS, 63, MS, 65, PhD(phys chem), 67. *Prof Exp:* Sr res chemist, 66-80, RES ASSOC, RES LABS, EASTMAN KODAK CO, 80- *Mem:* AAAS; Am Chem Soc; Sigma Xi. *Res:* Light scattering by liquids and solutions; colloid and surface chemistry; growth of colloidal particles; unconventional imaging systems; particle size analysis; field-flow fractionation. *Mailing Add:* 93 Carverdale Dr Rochester NY 14618

OPPENHEIMER, MICHAEL, b New York, NY, Feb 28, 46; m; c 1. OZONE DEPLETION, GLOBAL WARMING. *Educ:* Mass Inst Technol, SB, 66; Univ Chicago, PhD(chem physics), 70. *Prof Exp:* Res fel, Smithsonian Ctr Astrophys, Harvard Col Observ, 71-73, physicist, 73-81, lectr, Dept Astron, Harvard Univ, 71-81; SR SCIENTIST, ENVIRON DEFENSE FUND, 81- *Mem:* Am Phys Soc; AAAS; Am Geophys Union; Am Meteorol Soc; Air Pollution Control Asn. *Mailing Add:* Environ Defense Fund 257 Park Ave South New York NY 10010-7304

OPPENHEIMER, NORMAN JOSEPH, b Summit, NJ, Aug 20, 46; m 69; c 1. BIOCHEMISTRY. *Educ:* Brown Univ, BS, 68; Univ Calif, San Diego, PhD(chem), 73. *Prof Exp:* Res assoc biochem, Ind Univ, 73-75; asst prof, 75-80, ASSOC PROF PHARMACEUT CHEM, UNIV CALIF, SAN

FRANCISCO, 80- *Concurrent Pos:* NIH career develop award, 79-84. *Mem:* Am Chem Soc; AAAS; Am Soc Biol Chemists; Sigma Xi. *Res:* Small molecule-macromolecule interactions; redox and nonredox chemistry and metabolic roles of pyridine coenzymes; mechanism of anticancer drugs. *Mailing Add:* Dept of Pharmaceut Chem Univ of Calif San Francisco CA 94143

OPPENHEIMER, PHILLIP R, b San Francisco, Calif, July 9, 48; m 72; c 1. CLINICAL PHARMACY. *Educ:* Univ Calif, PharmD, 72. *Prof Exp:* Residency, Sch Pharm, Univ Calif Med Ctr, 73; from instr to asst prof, 73-82, ASSOC PROF PHARM, UNIV SOUTHERN CALIF, 82- *Concurrent Pos:* Chmn comt pharm educ, Calif Pharm Asn, 76-78; consult, Wadsworth Med Ctr, Vet Admin, 77-85; chair, Ed Comt, Am Pharm Asn. *Mem:* Am Soc Hosp Pharmacists; Am Asn Col Pharm; Am Pharmaceut Asn. *Res:* Clinical therapeutics of drugs, with emphasis on pediatric pharmacology. *Mailing Add:* Dept Clin Pharm Univ Southern Calif Pharm 1985 Zonal Ave Los Angeles CA 90033

OPPENHEIMER, STEVEN BERNARD, b Brooklyn, NY, Mar 23, 44; m 71; c 1. DEVELOPMENTAL BIOLOGY, CANCER. *Educ:* City Univ New York, BS, 65; Johns Hopkins Univ, PhD(biol), 69. *Prof Exp:* Am Cancer Soc fel biol, Univ Calif, San Diego, 69-71; from asst prof to prof biol, 71-84, DIR, CTR CANCER & DEVELOP BIOL, CALIF STATE UNIV, NORTHRIDGE (CSUN),84- *Concurrent Pos:* Trustee, Calif State Univ Found, Northridge, 74-; res & educ grants, Nat Cancer Inst Res, 72-, Am Cancer Soc, 76, NSF, 81-, Thomas Eckstrom Trust, 84-, NASA, 88-, Joseph Drown Found, 88-, Eisenhower Grant Prog, 90-; mem, NIH study sect, 87, NSF panel, 85; chmn prog, dir develop biol, Am Soc Zool, 82-84. *Honors & Awards:* Distinguished Res Award, Sigma Xi, 84. *Mem:* AAAS; Soc Develop Biol; Am Soc Cell Biol; Am Soc Zoologists; Sigma Xi; Am Cancer Soc (pres, 77-). *Res:* Molecular basis of intercellular adhesion; the role of cell surface carbohydrates in morphogenesis and malignancy; cancer prevention; enhancement of science education; author 70 publications including nine books in developmental biology, cancer and science education. *Mailing Add:* Dept of Biol Calif State Univ Northridge CA 91330

OPPENLANDER, JOSEPH CLARENCE, b Bucyrus, Ohio, Apr 6, 31; m 54; c 3. TRANSPORTATION AND TRAFFIC ENGINEERING. *Educ:* Case Inst Technol, Cleveland, BSCE, 53; Purdue Univ, MSCE, 57; Univ Ill, Urbana. PhD(civil eng), 62. *Prof Exp:* Instr civil eng, Purdue Univ, 55-57; traffic planner, Allegheny County Planning Dept, Pa, 57-58; res assoc, Univ Ill, 58-62; assoc prof, Purdue Univ, 62-69; chmn dept, 69-79, PROF CIVIL ENG, UNIV VT, 69- *Concurrent Pos:* VPres, Trans-Op Inc, 70- *Mem:* Am Soc Civil Engrs; Inst Transp Eng; Sigma Xi; Am Road Transport Builders Asn; Transp Res Bd. *Res:* Transportation planning, design and operations; urban planning; optimization of civil engineering systems; traffic accident reconstruction. *Mailing Add:* Dept Civil Eng Univ Vt Burlington VT 05405

ORAHOVATS, PETER DIMITER, b Sofia, Bulgaria, Apr 22, 22; nat US; m 45; c 1. PHYSIOLOGY, PHARMACOLOGY. *Educ:* St Klement Univ, Bulgaria, MD, 45. *Prof Exp:* Instr clin physiol, St Klement Univ, Bulgaria, 45-46; mem staff physiol, Col Physicians & Surgeons, Columbia Univ, 46-51; res assoc physiol & pharmacol, Merck Inst Therapeut Res, 51-59; med dir, Bristol Meyers Co, 59-60, med & sci dir, 61-62, vpres & dir res & develop div, 62-66, vpres & dir Sci Div, Bristol-Meyers Prod Div, 66-85; adj prof pharmacol, Div Biol Sci, Univ Calif, Santa Barbara, 80- *Concurrent Pos:* USPHS fel, Col Physicians & Surgeons, Columbia Univ, 49-51; adj prof pharmacol, Div Biol Sci, Univ Calif, Santa Barbara, 80- *Mem:* Am Soc Pharmacol & Exp Therapeut; Harvey Soc; Royal Soc Med; Pan Am Med Asn. *Res:* Medicine; cardiovascular physiology; blood volume and formation; essential hypertension; pharmacology of analgesics and narcotics. *Mailing Add:* PO Box 169 Valley Lee MD 20692

ORAN, ELAINE SURICK, b Rome, Ga, Apr 16, 46; m 69. COMPUTATIONAL PHYSICS. *Educ:* Bryn Mawr Col, AB, 66; Yale Univ, MPh, 68, PhD(solid state physics), 72. *Prof Exp:* Presidential intern plasma physics, 72-73, res physicist atmospheric physics & combustion, 73-78, res physicist comput physics & fluid dynamics, Lab Comput Physics, 78-82, head, Ctr Reactive Flow & Dynamical Systs, 86-88, SR SCIENTIST, NAVAL RES LAB, 88- *Concurrent Pos:* Prin investr, Navy, Air Force, Defense Advan Res Projs Agency, Dept of Defense & NASA, 74- *Honors & Awards:* Arthur S Fleming Award, US Govt, 79; WISE Sci Award, Women in Sci & Eng, 88; fel Am Inst Aeronaut & Astronaut, 90. *Mem:* Am Phys Soc; Am Inst Aeronaut & Astronaut; Am Geophys Union; Combustion Inst; Sigma Xi; AAAS. *Res:* Numerical methods for large-scale computing, fluid dynamics, turbulence, combustion, energetic materials, solar and atmospheric physics. *Mailing Add:* Code 4404 Naval Res Lab Washington DC 20375

O'RAND, MICHAEL GENE, b Carlisle, Eng, Sept 24, 45; US citizen; m 67; c 1. REPRODUCTIVE & DEVELOPMENTAL BIOLOGY, IN VITRO FERTILIZATION & IMMUNOLOGICAL CONTRACEPTIVES. *Educ:* Univ Calif, Berkeley, AB, 67; Ore State Univ, MS, 69; Temple Univ, PhD(biol), 72. *Prof Exp:* NIH fel, Inst Molecular & Cellular Evolution, Univ Miami, 72-75; asst prof anat, Col Med, Univ Fla, 75-79; frpm asst prof to assoc prof anat, 79-87, PROF CELL BIOL & ANAT, UNIV NC, 87- *Concurrent Pos:* NIH res grant, Univ Fla, 77-79 & Univ NC, 80-82, 83-85 & 85-; mem, Reproductive Biol Study Sect, 84-88; assoc ed, J Exp Zool, 86-89. *Mem:* AAAS; Am Soc Cell Biol; Soc Study Reproduction; Int Soc Immunol Reproduction; Soc Develop Biol. *Res:* Fertilization; immunoreproductive biology; sperm antigens. *Mailing Add:* Dept Cell Biol & Anat Lab Cell Biol Univ NC Chapel Hill NC 27599

O'RANGERS, JOHN JOSEPH, b Philadelphia, Pa, June 11, 36; m 65; c 2. IMMUNOCHEMISTRY. *Educ:* St Joseph's Univ, BS, 58; Hahnemann Med Col, PhD(biochem), 72. *Prof Exp:* Instr biochem, Hahnemann Med Col, 72-74; asst prof biochem, Philadelphia Col Osteop Med, 74-75; sr scientist immunochem, Carter Wallace Co, Princeton, NJ, 75-76; SR RES BIOCHEMIST IMMUNOCHEM, FOOD & DRUG ADMIN, 76- *Concurrent Pos:* Lectr, St Joseph's Univ, 73-75; gen referee immunodiagnostics, Asn Off Analytical Chemists. *Honors & Awards:* Commendable Serv Award, Food & Drug Admin, 78 & 85. *Mem:* Asn Off Anal Chemists; NY Acad Sci; Sigma Xi; Nat Immunoassay Soc; Nat Comt Clin Lab Standards. *Res:* Effect of hormones on the immune response; immunochemical methods of analysis for antibiotics; hormones and metabolites in tissues. *Mailing Add:* 19212 Walters Ave Poolesville MD 20837

ORATZ, MURRAY, b New York, NY, Apr 17, 27; m 52; c 2. BIOCHEMISTRY. *Educ:* Long Island Univ, BS, 48; Clarkson Col Technol, MS, 50; NY Univ, PhD(biochem), 57. *Prof Exp:* ASST CHIEF NUCLEAR MED SERV, VET ADMIN HOSP, 55- *Concurrent Pos:* Adj prof biochem, NY Univ, Col of Dent, 67- *Mem:* Sigma Xi; NY Acad Sci; Soc Nuclear Med; Am Asn Study Liver Dis; Am Chem Soc; Am Inst Chem. *Res:* Regulation and stimuli for the production of plasma proteins; dynamics of cardiac metabolism. *Mailing Add:* Nuclear Med Serv Vet Admin Hosp 423 E 23rd St New York NY 10010

ORAVA, R(AIMO) NORMAN, b Toronto, Ont, Sept 22, 35; m 59; c 2. PHYSICAL METALLURGY. *Educ:* Univ Toronto, BASc, 57; Univ BC, MASc, 59; Univ London, DIC & PhD(phys metall), 63. *Prof Exp:* Sci officer, Eng Physics Sect, Phys Metall Res Labs, Can Dept Mines & Tech Surv, 57; lectr math, Univ Waterloo, 59, lectr appl physics, 59-60; sr scientist, Nonferrous Metall Div, Battelle Mem Inst, Ohio, 63-64; res metallurgist, Mat Sci & Eng Div, Franklin Inst, Pa, 64-65, sr res metallurgist, 65-66; res metallurgist, Metall & Mat Sci Div, Denver Res Inst, Univ Denver, 66-74, lectr, 70-74; actg vpres & dean eng, 75-77, prof metall eng, SDak Sch Mines & Technol, 74-87, dean grad div & dir res, 77-87; DEP DIR, ASSOC WESTERN UNIV INC, 87- *Concurrent Pos:* NSF, Army, Navy & Dept Interior Energy grants; dir, Mining, Mineral Resources & Res Inst, 80-87; dir, Masec Corp, 77-82, chmn bd, 80-82; chmn bd, 82-88, trustee, SDak Sch Mines & Technol Found Res Ctr Inc, 82-89; dir Inst Adv Metall & Mat, SDak Sch Mines & Technol, 87. *Mem:* Soc Res Admins; Am Inst Mining, Metall & Petrol Engrs; Am Soc Eng Educ; Nat Coun Univ Res Adminr. *Res:* Mechanical metallurgy; thermomechanical processing; shock effects; explosive forming. *Mailing Add:* Assoc Western Univs Inc 4190 S Highland Dr Suite 211 Salt Lake City UT 84124-2600

ORAY, BEDII, BIOCHEMISTRY, PHARMACEUTICALS. *Educ:* NTex State Univ, Phd(biochem), 80. *Prof Exp:* PROD MGR, PHARMOTEX INC, 84- *Mailing Add:* 7013 Meadow Parks N N Richland Hills Ft Worth TX 76180-7898

ORBACH, RAYMOND LEE, b Los Angeles, Calif, July 12, 34; m 56; c 3. SOLID STATE PHYSICS. *Educ:* Calif Inst Technol, BS, 56; Univ Calif, Berkeley, PhD(physics), 60. *Prof Exp:* NSF fels, Oxford Univ, 60-61, asst prof appl physics, Harvard Univ, 61-63; assoc prof, 63-66, PROF PHYSICS, UNIV CALIF, LOS ANGELES, 66- *Concurrent Pos:* Alfred P Sloan Found fels, 63-67; asst vchancellor acad change & curric develop, Univ Calif, Los Angeles, 70-72, provost, Col Letters & Sci, 82-; Guggenheim Found fel, 73-74; mem, Nat Comn Res, 78-80; assoc div ed, Phys Review Letters, 80-83, assoc ed, Phys Review B, 84-87; counr, Am Phys Soc, 87-90; vchmn elect, Div Condensed Matter Physics, Am Phys Soc, 88- *Mem:* Am Phys Soc; Sigma Xi. *Res:* Paramagnetic resonance and spin-lattice relaxation; collective properties of ordered magnetic systems; lattice vibrations and their interactions; optical properties of solids; thermal conductivity of paramagnetic salts; magnetic resonance properties of dilute magnetic alloys; non-equilibrium superconductivity; transport in random systems, fractals, dynamics of fractal networks. *Mailing Add:* Dept of Physics Univ of Calif Los Angeles CA 90024

ORBAN, EDWARD, b Youngstown, Ohio, Mar 5, 15; m 41; c 2. APPLIED CHEMISTRY. *Educ:* Univ Buffalo, BS, 40; Univ Md, PhD(phys chem), 44. *Prof Exp:* Analytical chemist, E I du Pont de Nemours & Co, NY, 40-41; asst, Univ Md, 41-44; res chemist, Socony-Mobil Co, Inc, NJ, 44-46; res chemist, Monsanto Co, 46-48, group leader, 48-52, chief, Tech & Chem Sect, 52-54, sr develop investr, 54-57, mgr develop, Cent Res Dept, 57-69, planning mgr, indust chems, 69-82,; MAT & BUS CONSULT, 82- *Mem:* Am Chem Soc; Sigma Xi. *Res:* Electrochemistry; metals; instrumentation; electronic materials; single crystal semiconductor materials preparation. *Mailing Add:* 90 Thorncliff Lane Kirkwood MO 63122

ORBAN, JOHN EDWARD, b Cleveland, Ohio, Jan 22, 52; m 78. STATISTICS. *Educ:* Univ Dayton, BS, 74; Ohio State Univ, MS, 76, PhD(statist), 78. *Prof Exp:* Res assoc statist, Nat Bur Standards, 78-80; mem fac, Dept Math Sci, State Univ NY, Binghamton, 80-85; BATTELLE DRUG CO. *Mem:* Am Statist Asn; Inst Math Statist. *Res:* Non-parametric statistics. *Mailing Add:* Battelle 505 King Ave Columbus OH 43201

ORCHARD, HENRY JOHN, b Oldbury, Eng, May 7, 22; m 47, 71; c 1. ELECTRICAL ENGINEERING. *Educ:* Univ London, BSc, 46, MSc, 51. *Prof Exp:* Sr sci officer, eng res dept, Brit Post Off, 51-56, prin sci officer, 56-61; consult network design, Lenkurt Elec Co, Calif, 61-70; PROF ELEC ENG, UNIV CALIF, LOS ANGELES, 70- *Concurrent Pos:* Lectr, Stanford Univ, 67-70; consult, GTE Lenkurt Inc, 70-84. *Mem:* fel Inst Elec & Electronics Engrs. *Res:* Circuit theory; network design; passive, active and digital filters. *Mailing Add:* 828 19th St #E Santa Monica CA 90403-1923

ORCHIN, MILTON, b Barnesboro, Pa, June 4, 14; m 41; c 2. ORGANIC CHEMISTRY. *Educ:* Ohio State Univ, BA, 36, MA, 37, PhD(org chem), 39. *Hon Degrees:* DSc, Univ Cincinnati, 84. *Prof Exp:* Chemist, Food & Drug Admin, USDA, 39-42, org res chemist, Bur Animal Indust, 42-43; chief org chem sect synthetic liquid fuels, US Bur Mines, 43-53; assoc prof appl sci, 53-56, head dept chem, 56-62, dir basic sci lab, 62-71, prof chem, 56-81, EMER PROF CHEM & DISTINGUISHED SERV PROF, UNIV CINCINNATI, 81-, DIR, HOKE S GREENE LAB CATALYSIS, 71- *Concurrent Pos:* Guggenheim fel, Sieff Inst, 47-48; vis prof, Ohio State Univ, 66, Israel Inst

Technol, 68 & Univ Calif, Berkeley, 69; mem bd gov, Ben Gurion Univ of the Negev, Beersheva, Israel, 75-79. *Honors & Awards:* E V Murphree Award, Am Chem Soc, 80; Morley Medal, 80; E J Houdry Award, Catalysis Soc, 83. *Mem:* AAAS. *Res:* Chemistry of carbon monoxide; ultraviolet absorption spectroscopy; catalysis. *Mailing Add:* Dept of Chem Univ of Cincinnati Cincinnati OH 45221

ORCUTT, BRUCE CALL, b New York, NY, May 18, 37. DATABASE SYSTEMS, SOFTWARE ENGINEERING. *Educ:* Cornell Univ, AB, 61; Purdue Univ, MS, 64; Univ Va, PhD(math), 69. *Prof Exp:* Asst prof math & statist, Georgetown Univ, 69-75; sr res scientist, Nat Biomed Res Found, Georgetown Med Ctr, 75-84; assoc prof comput sci, Va Tech, 84-89. *Concurrent Pos:* Lectr, comput sci, Georgetown Univ. *Mem:* Asn Comput Mach. *Res:* Abstract data types and logic programming. *Mailing Add:* 4014 Lees' Corner Rd Chantilly VA 22021

ORCUTT, DAVID MICHAEL, b Oklahoma City, Okla, June 10, 43; m 67; c 1. PLANT PHYSIOLOGY. *Educ:* Univ Okla, BS, 66; Incarnate Word Col, MS, 70; Univ Md, PhD(plant physiol), 73. *Prof Exp:* Technician plant physiol, US Air Force Sch Aerospace Med, 67-70; asst prof, 73-81, ASSOC PROF PLANT PHYSIOL, VA POLYTECH INST & STATE UNIV, 81- *Mem:* Am Soc Plant Physiologists; Phycol Soc Am. *Res:* The role of plant steroids in growth regulation, host parasite interactions and physiology in higher plants and aquatic plants. *Mailing Add:* Dept Plant Path & Physiol Va Polytech Inst Blacksburg VA 24061-0331

ORCUTT, DONALD ADELBERT, b Traverse City, Mich, June 25, 26; m 50; c 4. BIOCHEMISTRY, INDUSTRIAL PHARMACY. *Educ:* Purdue Univ, Lafayette, BS, 50. *Prof Exp:* Chemist, Armour Pharmaceut, 51-52 & Whitehall Pharmacal Co, 52-58; supvr chem, Miles Labs, Inc, 58-60, asst to dir pharm res & develop, 60-65; vpres & dir labs, Standard Pharmacal Corp, 65-77; mgr, Tablet Tech Serv, Parke-Davis, 78-81; mgr pharmaceut res & develop, 81-83, Am Health, 83-85, vpres protein prod, 85-86, Asn Biol Lab Educ Labs, 86-87, MGR PROD DEVELOP, L PERRIGO CO, 87- *Mem:* Am Pharmaceut Asn; Am Chem Soc; Acad Pharmaceut Sci; Inst Food Technologists; Am Asn Pharmaceut Scientist. *Res:* Pharmaceutical formulation, chemistry and instrumentation; food and drug law. *Mailing Add:* 128 Paddock Lane Greer SC 29650-3841

ORCUTT, FREDERIC SCOTT, JR, b Roanoke, Va, Nov 29, 40; m 65; c 2. ETHOLOGY, ORNITHOLOGY. *Educ:* Cornell Univ, BS, 63, MS, 65, PhD(ethology), 69. *Prof Exp:* Lectr biol, Univ Sask, 69-70; sr res assoc avian reproductive physiol, Univ Wash, 70-71; asst prof, 71-78, ASSOC PROF BIOL, UNIV AKRON, 78-, ASST HEAD DEPT, 90- *Mem:* Animal Behav Soc; Am Ornith Union; Cooper Ornith Soc; AAAS; Sigma Xi; Wilson Ornith Soc. *Res:* Avian vocalizations and reproductive isolating mechanisms; hormonal determinants of behavior; avian reproductive physiology; evolution of behavior. *Mailing Add:* Dept Biol Univ Akron Akron OH 44325-3908

ORCUTT, HAROLD GEORGE, b Washington, DC, May 6, 18; m 44; c 2. FISH BIOLOGY. *Educ:* Wilson Teachers Col, BS, 41; Stanford Univ, PhD(fisheries biol), 49. *Prof Exp:* Asst biol, Wilson Teachers Col, 38-39; teacher high sch, DC, 40-41; marine investr, State Fish Comn, NH, 41-42; marine biologist, Hopkins Marine Sta, Calif Dept Fish & Game, 48-49 & Mus Natural Hist, Stanford Univ, 49-58, dir marine biol res lab, Stanford Univ, 58-61, dir marine resources res lab, 61-67, LAB SUPVR, OPERS RES BR, MARINE RESOURCES REGION, CALIF DEPT FISH & GAME, 67- *Concurrent Pos:* Fisheries officer, US Army Military Govt, Japan, 45-46. *Mem:* Am Inst Fishery Res Biol; Sigma Xi. *Res:* Research conservation and administration of marine fish and shellfish. *Mailing Add:* 2368 Santa Ana Palo Alto CA 94303

ORCUTT, JAMES A, b Mobile, Ala, July 11, 18. TOXICOLOGY, NEUROSCIENCE. *Educ:* Univ Rochester, PhD(pharmacol & toxicol), 46. *Prof Exp:* Actg chmn, dept physiol & chmn, dept pharmacol, Univ Osteop Med & Health Sci, 66-72, prof pre-clin sci, 72-83; RETIRED. *Mem:* Am Soc Pharmacol & Exp Therapeut. *Mailing Add:* Univ Osteop Med & Health Sci 662 Marlou Pkwy Des Moines IA 50320

ORCUTT, JOHN ARTHUR, b Holyoke, Colo, Aug 29, 43; m 67; c 2. SEISMOLOGY, INVERSE THEORY. *Educ:* US Naval Acad, BS, 66; Univ Liverpool, MSc, 68; Univ Calif, San Diego, PhD(earth sci), 76. *Prof Exp:* Fel, 76-77, assoc prof, 82-84, RES GEOPHYSICIST, SCRIPPS INST OCEANOG, 77-, PROF GEOPHYS, 84- *Concurrent Pos:* Vis res assoc, Calif Inst Technol, 78-82; chmn, Inc Res Inst Seismol, 88-90; mem, Nat Res Coun Ocean Studies Bd, 86-, chmn Navy panel; mem, NRC Comt Geod. *Honors & Awards:* Newcomb Cleveland Prize, AAAS, 80. *Mem:* Am Geophys Union; Soc Explor Geophysicists. *Res:* Interaction of acoustic and seismic waves at the sea floor; computation of synthetic seismograms and use in inverse problems; ocean bottom seismology. *Mailing Add:* Scripps Inst Oceanog A-025 Univ San Diego La Jolla CA 92093-0225

ORCUTT, JOHN C, b Monson, Mass, Oct 10, 27; m 53; c 2. CHEMICAL ENGINEERING. *Educ:* Worcester Polytech Inst, BS, 50, MS, 51; Univ Del, PhD(chem eng), 60. *Prof Exp:* Res engr, Atlantic Refining Co, 51-57, supvr petrol process develop, 60-61; res engr rate processes group, Res Triangle Inst, 61-65, dir measurement & controls lab, 65-68, mgr environ eng dept, 71-72; mgr develop process eng, Stauffer Chem Co, 72-82. *Concurrent Pos:* Adj prof, Univ NC, 70 & Duke Univ, 71. *Mem:* Assoc Am Inst Chem Engrs; Sigma Xi. *Res:* Behavior of fluidized bed systems; applied mathematics to chemical engineering problems; process optimization and control; chemical reaction kinetics; advanced waste treatment research and development. *Mailing Add:* 2353 Old Augusta Rd Waldoboro ME 04572

ORCUTT, RICHARD G(ATTON), b Wellington, Colo, Mar 9, 24; m 51; c 3. SANITARY & CIVIL ENGINEERING. *Educ:* Univ NMex, BS, 45; Univ Calif, MS, 51, PhD, 56. *Prof Exp:* Jr construct engr, Creole Petrol Corp, Venezuela, 46-48; asst sanit engr, State Dept Pub Health, Calif, 51-52; sr asst sanit engr, USPHS, 52-53; jr res engr, Sanit Eng Res Ctr, Univ Calif, 55-56; prof civil eng, Univ Nev, Reno, 56-85; PVT CONSULT, 85- *Concurrent Pos:* Fel, Univ Fla, 68. *Mem:* Am Soc Civil Engrs; Am Soc Eng Educ. *Res:* Water resources engineering; economics of water quality management; systems analysis; water resources; waste removal. *Mailing Add:* 985 Munley Dr Reno NV 89503

ORD, JOHN ALLYN, b West Elizabeth, Pa, Mar 9, 12; m 37; c 2. PHYSICS, MATHEMATICS. *Educ:* Carnegie Inst Technol, BS, 34, MS, 35, DSc(physics), 39. *Prof Exp:* Instr physics, Carnegie Inst Technol, 37-38; asst prof, NDak State Col, 38-40; dept dir, Southern Signal Corps Sch, US Army, 42-43, asst commandant, 43-44, commandant, 44, exec officer signal sect, Southwest Pac Area, 45, dir training, Sandia Base, NMex, 46-48, from exec officer to dir, Evans Signal Lab, 49-52, from tech liaison officer to dep theater signal officer, US Army Forces, Far East, 52-55, from chief res div to dir Army Res Off, Off Chief Res & Develop, 55-58, mem staff & fac, Army War Col, 59-62, chief res div, Army Material Command, 62-65; mgr adv concept sect, Systs Tech Ctr, Philco Corp, 65-66; chief scientist, US Army Foreign Sci & Tech Ctr, 66-67, dep for tech opers, 67-71, dep dir, 71-81; RETIRED. *Concurrent Pos:* Teaching fel physics, Carnegie Tech, 34-37; res physicist, Flannery Bolt CO, 46. *Mem:* Sigma Xi. *Res:* Infrared spectroscopy; communications electronics; military research administration. *Mailing Add:* 45 Alters Rd Carlisle PA 17013-8969

ORDAHL, CHARLES PHILIP, b Chatahoochie County, Ga, Aug 8, 45. MOLECULAR BIOLOGY, DEVELOPMENTAL BIOLOGY. *Educ:* Univ Colo, Boulder, BA, 67, MA, 70; Univ Colo, Denver, PhD(molecular biol), 74. *Prof Exp:* Fel molecular & develop biol, Case Western Reserve Univ, 74-77; ASST PROF MOLECULAR & DEVELOP BIOL, DEPT ANAT, SCH MED, TEMPLE UNIV, 77- *Mem:* Soc Develop Biol; Soc Cell Biol. *Res:* Gene transcription changes involved in embryonic differentiation of muscle. *Mailing Add:* Dept Anat-S-1334 Univ Calif San Francisco Third & Parnassus Ave San Francisco CA 94143

ORDAL, GEORGE WINFORD, b Seattle, Wash, Aug 5, 43; div; c 3. BIOCHEMISTRY, MOLECULAR BIOLOGY. *Educ:* Harvard Col, AB, 65; Stanford Univ, PhD(biochem), 71. *Prof Exp:* Res assoc biochem, Univ Wis, 71-73; from asst prof to assoc prof biochem, 73-90, PROF BIOCHEM, UNIV ILL, URBANA, 90- *Concurrent Pos:* Am Cancer Soc fel, Univ Wis, 71-72, USPHS fel, 73; NIH prin investr, Inst Gen Med Sci & AI, 73-; res career develop award, NIH, 76-81. *Mem:* Am Soc Microbiol; Am Soc Biochem & Molecular Biol. *Res:* Biochemistry, molecular biology, and genetics of bacterial chemotaxis; cloning of bacillus subtilis DNA; calcium-binding proteins; enzymatic methylation of membrane proteins; methyl transferases and methylestrases protein phosphorylation; chemoreceptors; adaptation. *Mailing Add:* Dept Biochem 190 Med Sci Bldg 506 S Mathews Ave Univ Ill Urbana IL 61801

ORDEN, ALEX, b Rochester, NY, Aug 9, 16; m 46; c 3. OPERATIONS RESEARCH. *Educ:* Univ Rochester, BS, 37; Univ Mich, MS, 38; Mass Inst Technol, PhD(math), 50. *Prof Exp:* Alternate chief anal sect, Nat Bur Stand, 42-47; instr math, Mass Inst Technol, 48-49; head comput theory sect, Linear Prog, US Dept Air Force, 50-52; mgr, Comput Lab, Burroughs Corp, 52-58; prof appl math, Grad Sch Bus, Univ Chicago, 58-86; ADJ PROF, UNIV ILL, CHICAGO, 87- *Concurrent Pos:* Vis prof, London Sch Econ, 64 & 87, Ga Inst Technol, 66, 67 & 71; prog chmn, Nat Comput Conf, 81-; adj prof, univs & cols, 87- *Mem:* Inst Mgt Sci; Asn Comput Mach; Opers Res Soc Am; Math Prog Soc (chmn, 84-86). *Res:* Linear and nonlinear programming; models of computer-based information systems in business organizations. *Mailing Add:* Grad Sch Bus Univ Chicago Chicago IL 60637

ORDER, STANLEY ELIAS, b Vienna, Austria, Nov 1, 34; US citizen; m 58; c 3. RADIOTHERAPY, IMMUNOLOGY. *Educ:* Albright Col, BS, 56; Tufts Univ, MD, 61; Am Bd Radiol, cert therapeut radiol, 70. *Hon Degrees:* ScD, Albright Col, 78 & Elizabethtown Col, 80. *Prof Exp:* Intern, Ind Univ Hosp, 61-62; resident path, Peter Bent Brigham Hosp, 62-63; chief exp studies, US Army Surg Res Unit, 63-65; from instr to asst prof, Harvard Univ, 69-73, assoc prof radiation ther, 73-78; prof, 78-81, WILLARD & LILLIAN HACKERMAN PROF RADIATION ONCOL, JOHNS HOPKINS UNIV, 81- *Concurrent Pos:* Fel radiation ther, Yale Univ, 65-69. *Honors & Awards:* Elis Beiven Lectr, Stockholm; Pfahler Jacobs lectr, Am Radiol Soc. *Mem:* Am Radium Soc; Am Soc Therapeut Radiol; Am Soc Clin Oncol. *Res:* Hodgkins tumor associated antigens; ovarian carcinoma antigens and immunotherapy; radiolabeled antibody therapy in liver cancer; radiation benign diseases. *Mailing Add:* Dept Radiation Oncol Johns Hopkins Univ 720 Ruthland Ave Baltimore MD 21205

ORDILLE, CAROL MARIA, US citizen. FORECASTING, DATA ADMINISTRATION. *Educ:* Temple Univ, BS, 65; Villanova Univ, MS, 67; Ohio State Univ, PhD(biostatist), 75. *Prof Exp:* Statistician, Merck Sharp & Dohme, 67-68; statist consult, 70-72; res statistician, Schering Corp, 72-75; dir statist & data systs, Delbay Res Corp, 75-80; Mgr Biostatist, E I Du Pont de Nemours & Co, 80-84; data base adminr, Dept Environ Protection, NJ, 85-86; SR STAFF ANALYST, GEN PUBL UTILITIES SERV CORP, 86- *Concurrent Pos:* Statist consult, 85-87. *Mem:* Am Statist Asn; Int Inst Forecasters. *Res:* Development of statistical models to describe customer characteristics, behavior and preferences; information retrieval systems for project life cycle methodology. *Mailing Add:* 18 S New Haven Ave Ventnor NJ 08406

ORDMAN, ALFRED BRAM, b Washington, DC, Oct 25, 48; m 86; c 5. BIOCHEMISTRY. *Educ:* Carleton Col, BA, 70; Univ Wis-Madison, PhD(biochem), 74. *Prof Exp:* Res assoc biochem, Univ Minn, 74-77; ASSOC PROF CHEM, BELOIT COL, 77-, CHAIR, BIOCHEM, 80- *Concurrent Pos:* Vis prof, Tel Aviv Univ, 82-83 & 86-87, McArdle Lab, Wis, 83-84; Kellogg Nat Fel, 84-87; prin invest, US Cong Comn Sci Space Tech, 87. *Mem:* AAAS; Sigma Xi; Union of Concerned Scientists. *Res:* Enzyme mechanisms of four electron dehydrogenases; microbial predation; continuous fermentation; chemical carcinogenesis; cell aging. *Mailing Add:* 632 Church St Beloit WI 53511

ORDMAN, EDWARD THORNE, b Norfolk, Va, Sept 10, 44; m 83; c 6. MICROCOMPUTERS, DISTRIBUTED PROCESSING. *Educ:* Kenyon Col, AB, 64; Princeton Univ, MA, 66, PhD(math), 69. *Prof Exp:* Mathematician, Nat Bur Stand, 62-69; asst prof, Univ Ky, 69-74, res assoc & dir comput serv, Bur Govt Serv, 74-76; from asst prof to assoc prof, New Eng Col, 77-83; vis asst prof, 76-77, ASSOC PROF, MEMPHIS STATE UNIV, 81- *Concurrent Pos:* NSF res grant, 70-71 & 85-87; Fulbright-Hays grant, Univ New South Wales, Sydney, Australia, 73; NASA-Am Soc Eng Educ fel, 78; vis positions, Aalborg Univ Ctr, Denmark, 87, Univ Paris-Sud, France, 88, Faroes Acad, Torshavn, Faroes, 88. *Mem:* Asn Comput Mach; Am Math Soc; Math Asn Am; Australian Math Soc. *Res:* Theory of distributed computing; microcomputers; applications of graph theory to distributed processing; resource allocation algorithms, scheduling, distributed agreement; synchronization primitives; hardware and software implications. *Mailing Add:* Dept Math Sci Memphis State Univ Memphis TN 38152

ORDONEZ, NELSON GONZALO, b Bucaramanga, Colombia, July 20, 44; m 87; c 2. IMMUNOPATHOLOGY, ELECTRON MICROSCOPY. *Educ:* Inst Daza Dangond, Bachelors, 62; Nat Univ Colombia, SAm, 70. *Prof Exp:* Intern, Mil Hosp, Bogota, Colombia, 70-71, resident path, 71-72; resident, Univ NC, Chapel Hill, 72-73; resident, Univ Chicago, Ill, 73-77, asst prof path, 77-78; from asst prof to assoc prof, 78-85, PROF PATH, UNIV TEX, M D ANDERSON CANCER CTR, HOUSTON, 85- *Concurrent Pos:* Dir immunocytochem sect, Univ Tex, M D Anderson Cancer Ctr, Houston, 81- *Mem:* Am Asn Pathologists; Int Acad Path; Am Soc Clin Pathologists; Am Soc Nephrology; Int Soc Nephrology; Am Soc Cytol. *Res:* Morphological and biochemical changes in hypocalemic nephropathy; role of C-type viruses in the pathogenesis of systemic lupus erythematosus; application of immunocytochemical markers in the diagnosis of tumors; electron microscopy in tumor diagnosis. *Mailing Add:* Dept Path M D Anderson Cancer Ctr Houston TX 77030

ORDUNG, PHILIP F(RANKLIN), b Luverne, Minn, Aug 12, 19; m 45; c 2. ELECTRICAL ENGINEERING. *Educ:* SDak State Col, BS, 40; Yale Univ, ME, 42, DEng, 49. *Prof Exp:* Lab asst, Yale Univ, 40-42, from instr to prof elec eng, 42-62; chmn dept, Univ Calif, Santa Barbara, 62-68, prof elec eng, 62-90; RETIRED. *Concurrent Pos:* Consult, Rand Corp, 60; vis prof, Tech Univ Norway, 68 & 69. *Honors & Awards:* Centennial Award, Inst Elec & Electronics Engrs, 84. *Mem:* Fel AAAS; fel Inst Elec & Electronics Engrs; NY Acad Sci. *Res:* Network and solid-state theories. *Mailing Add:* Dept Elec & Computer Engr Univ Calif Santa Barbara CA 93106

ORDWAY, ELLEN, b New York, NY, Nov 8, 27. ENTOMOLOGY, POLLINATION. *Educ:* Wheaton Col Mass, BA, 50; Cornell Univ, MS, 55; Univ Kans, PhD(entom), 65. *Prof Exp:* Asst field biol, Dept Trop Res, NY Zool Soc, 50-52; asst zool, Southwestern Res Sta, Am Mus Natural Hist, New York, 55-57; PROF BIOL, UNIV MINN, MORRIS, 65- *Mem:* Ecol Soc Am; Entom Soc Am; Soc Study Evolution; Soc Syst Zool; Int Union Study Social Insects; Am Inst Biol Sci. *Res:* Biology and behavior of native bees and development of social behavior; biology of native prairies, especially insect-plant relationships and pollination; prairie ecology. *Mailing Add:* Univ of Minn Div Sci & Math Morris MN 56267

ORDWAY, GEORGE A, b Rockville Ctr, NY, Sept 29, 43. CARDIOVASCULAR PHYSIOLOGY. *Educ:* Univ Ky, PhD(physiol), 79. *Prof Exp:* ASST PROF PHYSIOL, UNIV TEX HEALTH SCI CTR, 82- *Mem:* Am Physiol Soc; Am Heart Asn; Am Col Sports Med; Sigma Xi; Am Fedn Clin Res. *Mailing Add:* Dept Physiol Ut Southwestern 5323 Harry Hines Rd Dallas TX 75235-9040

ORDWAY, NELSON KNEELAND, b New Brunswick, NJ, Oct 26, 12; wid; c 4. PEDIATRICS, PUBLIC HEALTH. *Educ:* Yale Univ, MD, 38. *Prof Exp:* Intern pediat, New Haven Hosp, Conn, 38-39 & Hosp Univ Pa, Philadelphia, 39-41; asst path, Sch Med, Yale Univ, 41-45, instr pediat, 45-47; from asst prof to assoc prof, Sch Med, La State Univ, 47-52, prof & head dept, 52-54; prof, Schs Med, Univ NC, 54-57 & Yale Univ, 58-67; prof pediat & pub health, Schs Med & Health, Univ Okla, 67-74; chief pediat, Gallup Indian Med Ctr, 74-78; CHILD HEALTH CONSULT, PHOENIX AREA INDIAN HEALTH SERV, 78- *Concurrent Pos:* Prof, Sch Med, Univ Valle, Colombia, 64-65; consult diarrheal dis, WHO; consult med educ, Pan-Am Health Orgn; clin prof pediat, Sch Med, Univ NMex, 75-78; bd sci coun, Children's Nutrit Res Ctr, US Dept Agr & Baylor Col Med. *Mem:* Am Pub Health Asn; Soc Pediat Res; Am Pediat Soc; Am Acad Pediat. *Res:* Diarrhea; acid-base and electrolyte disturbances. *Mailing Add:* PO Box 5845 Carefree AZ 85377

ORE, FERNANDO, b Trujillo, Peru, Apr 26, 26; US citizen; m 57; c 2. CHEMICAL ENGINEERING. *Educ:* San Marcos Univ, Lima, BS, 49; Univ Wash, MS, 54, PhD(chem eng), 59. *Prof Exp:* Sr res engr, Am Potash & Chem Corp, 59-65, res proj engr, Whittier Res Lab, 65-68, head Crystallization & Math Modeling Sect, 68-70; mgr, Process Develop, Garrett Res & Develop Co, Div Occidental Petrol Corp, 70-76; res dir phosphates, 76-83, VPRES LICENSING-TECH, OCCIDENTAL CHEM, DIV OF OCCIDENTAL PETROL CORP, 83- *Honors & Awards:* Chem Engr Personal Achievement Award, McGraw Hill, 78. *Mem:* Assoc Am Inst Chem Engrs; Am Chem Soc; Sigma Xi. *Res:* Mathematical modeling and process optimization, computer applications; solvent extraction; crystallization; phosphate technology; development of the oxg hemihydrate process. *Mailing Add:* 14732 Glenn Dr Whittier CA 90604

ORE, HENRY THOMAS, b Chicago, Ill, Oct 30, 34; m 54; c 2. SEDIMENTOLOGY. *Educ:* Cornell Col, BA, 57; Wash State Univ, MS, 59; Univ Wyo, PhD(geol), 63. *Prof Exp:* From instr to assoc prof, 63-73, chmn dept, 67-70, PROF GEOL, IDAHO STATE UNIV, 73- *Mem:* Geol Soc Am; Nat Asn Geol Teachers; Int Asn Sedimentol; Am Asn Univ Profs; Sigma Xi. *Res:* Modern depositional environments; braided streams; sedimentary petrography; mechanics of sediment transport. *Mailing Add:* Dept of Geol Idaho State Univ Pocatello ID 83201

O'REAR, DENNIS JOHN, b Harrisburg, Pa, Nov 21, 50; m 77. CATALYSIS. *Educ:* Pa State Univ, BS, 72; Stanford Univ, MS, 75, PhD(chem eng), 80. *Prof Exp:* Res engr, Mobil Res & Develop Co, 72-74; res engr, 77-79, SR RES ENGR, CHEVRON RES CO, 79- *Mem:* Am Inst Chem Engrs; Am Chem Soc. *Res:* Development of refining routes for synthetic fuels from coal and shale oil; developing applications for novel zeolites for use in refineries. *Mailing Add:* Chevron Res Co PO Box 1627 Richmond CA 94802

OREAR, JAY, b Chicago, Ill, Nov 6, 25; c 3. PARTICLE PHYSICS. *Educ:* Univ Chicago, PhB, 43, SM, 50, PhD(physics), 53. *Prof Exp:* Res assoc physics, Inst Nuclear Studies, Univ Chicago, 53-54; from instr to asst prof, Columbia Univ, 54-58; assoc prof, 58-64, PROF PHYSICS, CORNELL UNIV, 64- *Concurrent Pos:* NSF sr fel, Europ Orgn Nuclear Res, 64-65; ed, Forum on Physics & Soc, Am Phys Soc, 71-74. *Mem:* Am Phys Soc; Fedn Am Scientists (chmn, 67-68). *Res:* Large-angle, high energy proton-proton and pion-proton scattering; total cross section at very high energies. *Mailing Add:* Dept Physics Cornell Univ Ithaca NY 14853

O'REAR, STEWARD WILLIAM, b Winchester, Va, Jan 13, 19; m 41; c 4. CHEMICAL ENGINEERING, CHEMISTRY. *Educ:* Univ Va, BChE, 40. *Prof Exp:* Chem engr, Joseph E Seagram & Son's, Inc, 40-44; chemist, E I du Pont de Nemours & Co, Inc, 46-51, from asst supvr to supvr, 51-76, chief supvr, Tech Info Serv, Savannah River Lab, 76-84; RETIRED. *Concurrent Pos:* Consult, 84- *Mem:* Am Chem Soc. *Res:* Technical communication and documentation; atomic energy literature; science communications. *Mailing Add:* 208 Bay Tree Ct Aiken SC 29801

OREBAUGH, ERROL GLEN, b Bradenton, Fla, Sept 28, 37; m 61; c 3. PHYSICAL INORGANIC CHEMISTRY. *Educ:* Univ Fla, BSCh, 59; Fla State Univ, PhD(inorg chem), 72. *Prof Exp:* Chemist, E I Du Pont de Nemours & Co Inc, 60-67; RES CHEMIST, SAVANNAH RIVER LAB, WESTINGHOUSE SAVANNAH RIVER CO, 72- *Mem:* Fel Am Inst Chemists; Am Chem Soc. *Res:* Research and development problems of the nuclear fuel cycle, including actinide chemistry, nuclear waste management and environmental chemistry; instrumentation and complexation thermodynamics. *Mailing Add:* 1710 Alpine Dr Aiken SC 29801

OREHOTSKY, JOHN LEWIS, b Hartford, Conn, July 24, 34; m 59; c 3. METALLURGY, SOLID STATE PHYSICS. *Educ:* Mass Inst Technol, BS, 56; Polytech Inst Brooklyn, MS, 61; Syracuse Univ, PhD(solid state sci), 71. *Prof Exp:* Res scientist, Sylvania Elec Prod, 56-61; staff scientist, Int Bus Mach Corp, 61-69; asst prof solid state, 71-77, ASSOC PROF ENG, WILKES COL, 77- *Mem:* Am Inst Mining, Metall & Petrol Engrs. *Res:* Transformations; critical point phenomena; transport properties; theory of metals and alloys; magnetism, electrochemistry. *Mailing Add:* Dept Physics Wilkes Col 170 S Franklin St Wilkes-Barre PA 18766

O'REILLY, JAMES EMIL, b Cleveland, Ohio, Jan 14, 45; m 68; c 2. SPECTROSCOPIC METHODS, ELECTROCHEMISTRY. *Educ:* Univ Notre Dame, BS, 67; Univ Mich, PhD(chem), 71. *Prof Exp:* Assoc chem, Univ Ill, Urbana, 71-73; asst prof, 73-79, ASSOC PROF CHEM, UNIV KY, 79-, DIR GRAD STUDIES, 87- *Concurrent Pos:* Vis scientist, Food & Drug Admin, 80-81; assoc chmn, Univ Ky, 89- *Mem:* Am Chem Soc; Soc Appl Spectros. *Res:* Electrochemistry; gas chromatography; analytical methods development; ion selective electrodes; atomic absorption; food and fossil-fuel analysis. *Mailing Add:* Dept Chem Univ Ky Lexington KY 40506-0055

O'REILLY, JAMES MICHAEL, b Dayton, Ohio, Nov 25, 34; m 54; c 5. POLYMER PHYSICS, POLYMER CHEMISTRY. *Educ:* Univ Dayton, BS, 56; Univ Notre Dame, PhD(phys chem), 60. *Prof Exp:* Develop chemist, Nat Cash Register Co, 54-56; res assoc polymer physics, Gen Elec Res & Develop Ctr, 60-67; mgr polymer physics & phys chem, Res Labs, Xerox Corp, 67-73, technol mgr, 73-75, prin scientist, 75-80; RES ASSOC, RESEARCH LABS, EASTMAN KODAK CO, 80- *Concurrent Pos:* Assoc prof, Univ Rochester, 71-78. *Mem:* AAAS; Soc Rheol; Am Chem Soc; fel Am Phys Soc. *Res:* Microwave spectroscopy; structure and properties of polymers, electrical, mechanical, thermodynamic and rheological properties; biopolymers; electrophotographic and imaging materials. *Mailing Add:* Res Lab Eastman Kodak Co Rochester NY 14650

O'REILLY, RICHARD, b Brooklyn, NY, Apr 29, 43; m 80; c 2. BONE MARROW TRANSPLANTATION, GENE TRANSFER. *Educ:* Col Holy Cross, Worcester, Mass, BS, 64; Sch Med, Univ Rochester, NY, MD, 68. *Prof Exp:* Intern pediat, Univ Minn Hosp, 68-69; resident pediat, Children's Hosp Med Ctr, Boston, Mass, 71-72, fel, 72-73; instr pediat, Med Col, Cornell Univ, NY, 73, asst prof biol, 74, assoc prof pediat, 78-83; asst attend pediatrician, 73, dir, Bone Marrow Transplantation Prog, 75 & 76, asst atttend physician, Clin Immunol, 78, asst attend phys, Hem-Lymph Serv, 78, assoc attend pediatrician, 78, assoc attend pediatrician med, 81, attend pediatrician, Hem-Lymph, 84, CHIEF, MARROW TRANSPLANTATION PROG, MEM HOSP, NY, 81-, ATTEND PEDIATRICIAN, 82- *Concurrent Pos:* Res assoc, Nat Ctr Dis Control, Atlanta, Ga, 69-71; young investr award, Louise & Allston Boyer, 67; assoc, Sloan-Kettering Inst Cancer Res, NY, 73-83, head lab microbiol & immunol, 73-, assoc mem, 83-84, mem, 84-; mem adv bd, Laura Rosenburg Mem Fund, Inc & Aplastic Anemia Found Am; assoc attend pediatrician, NY Hosp, 78-83, attend pediatrician, 83-; mem bd dirs, Damon Runyon-Walter Winchell Cancer Fund, 86; chmn, Dept Pediat, Mem Sloan-Kettering Cancer Ctr, 86; Vincent Astor prof clin res, Am Pediat Soc. *Mem:* Am Asn Immunologists; Am Asn Pathologists; Am Acad Pediat; AAAS; Transplantation Soc; Am Soc Hemat; Am Pediat Soc; Am Asn Cancer Res. *Res:* Development and evaluation of new approaches for transplantation of human bone marrow as curative treatment of lethal genetic and acquired diseases of the hematopoietic and lymphoid systems. *Mailing Add:* Mem Sloan-Kettering Cancer Ctr 1275 York Ave New York NY 10021

OREJAS-MIRANDA, BRAULIO, herpetology; deceased, see previous edition for last biography

O'RELL, DENNIS DEE, b Sacramento, Calif, Nov 28, 41; m 65; c 2. LEAD-ACID BATTERY, GRAPHIC ARTS. *Educ:* Sacramento State Col, BA, 64; Univ Ore, PhD(org chem), 70. *Prof Exp:* Sr res chemist, Ciba-Geigy Corp, 70-72, supvr, Polymer Additives Dept, 72-75; res mgr, Polyfibron Div, W R Grace & Co, 75-77, tech mgr, 77-79, oper mgr cellulose & new prod, 79-81, mgr mkt & tech activ, 81-84, TECH DIR GRAPHIC ARTS, W R GRACE & CO, 84- *Mem:* Am Chem Soc; Flexographic Tech Asn; Tech Asn Graphic Arts. *Res:* Synthetic organic chemistry; heterocyclic chemistry; monomer synthesis; polymer stabilization; polymer properties and processing; porous membranes; rechargeable batteries; nonwovens; elastomers; offset lithography; flexography. *Mailing Add:* W R Grace & Co 55 Hayden Ave Lexington MA 02173

OREM, JOHN, SLEEP, RESPIRATION CONTROL. *Educ:* Univ NMex, PhD(psychol), 70. *Prof Exp:* PROF PHYSIOL, SCH MED, TEX TECH UNIV, 85- *Mailing Add:* Dept Physiol Tex Tech Univ Sch Med Fourth St & Indiana Ave Lubbock TX 79430

OREM, MICHAEL WILLIAM, b Sturgis, SDak, Nov 30, 42; c 2. INTERFACIAL & SURFACE PHYSICAL CHEMISTRY. *Educ:* Colo Col, BS, 64; Univ Southern Calif, PhD(surface chem), 69. *Prof Exp:* Sr res chemist, Res Labs, 69-76, TECH ASSOC, APPL TECHNOL ORGN, EASTMAN KODAK CO, 76- *Mem:* Am Chem Soc; Soc Photog Scientists & Engrs. *Res:* Applied surface science: Experimental studies of surfactant properties and interfacial phenomena in fluid coating technology and liquid-liquid emulsification technology, particularly in the manufacture of photographic film and paper products. *Mailing Add:* 406 Brookview Dr Rochester NY 14617-4313

OREN, SHMUEL SHIMON, b Bucharest, Romania, Feb 20, 44; US & Israeli citizen; div; c 2. OPERATIONS RESEARCH. *Educ:* Israel Inst Technol, BSc, 65, MS, 69; Stanford Univ, MS & PhD(eng econ systs), 72. *Prof Exp:* Tech officer mech eng, Israel Defence Force, 65-69; asst optimization, Stanford Univ, 70-71; res scientist opers res, Palo Alto Res Ctr, Xerox Corp, 72-80; assoc prof eng econ systs, Terman Eng Ctr, Stanford Univ, 80-82; PROF & CHMN INDUST ENG & OPERS RES, UNIV CALIF, BERKELEY, 83- *Concurrent Pos:* Consult assoc prof, Stanford Univ, 72-80, Xerox Corp, 80-81 & Elec Power Res Inst, 81- *Mem:* Opers Res Soc Am; Inst Mgt Sci; Math Programming Soc. *Res:* Mathematical modelling optimization techniques; market analysis; pricing strategies; economics of telecommunication; electric utilities planning and use of quantitative techniques for problem solving and policy analysis. *Mailing Add:* Dept Indust Eng & Operations Res Etchevery Hall Univ Calif Berkeley CA 94720

ORENSTEIN, ALBERT, b New York, NY; m 62. PHYSICS. *Educ:* NY Univ, PhD(physics), 63. *Prof Exp:* Asst prof physics, Queens Col, NY, 58-85; RETIRED. *Concurrent Pos:* Res assoc, NY Univ, 62- *Res:* Solid state physics. *Mailing Add:* Dept Physics Queens Col 65-30 Kissena Blvd Flushing NY 11367-1575

ORENSTEIN, DAVID M, b Plainsfield, NJ, Nov 3, 45; m 69. PEDIATRIC PULMONOLOGY, EXERCISE PHYSIOLOGY. *Educ:* Amherst Col, BA, 67; Case Western Reserve, MD, 73. *Prof Exp:* Asst prof pediat, Case Western Reserve Univ, 78-81 & Ctr Health Sci, Univ Tenn, 81-84; ASSOC PROF PEDIAT & EXERCISE PHYSIOL, UNIV PITTSBURGH, 84- *Concurrent Pos:* Dir, Cystic Fibrosis Ctr, Memphis, 81-84, Cystic Fibrosis Ctr, Pittsburgh, 84-, Pediat Pulmonology, 84-; mem, Comt Sports Med, Am Acad Pediat, 84-; prog comt, Cystic Fibrosis Found, 89 & 90, Soc Pediat Res, 89 & Pediat Assembly, Am Thoracic Soc, 90-; bd dirs, NAm Soc Pediat Exercise Med, 90- *Mem:* Am Thoracic Soc; Soc Pediat Res; Am Acad Pediat; Cystic Fibrosis Found; NAm Soc Pediat Exercise Med. *Res:* Exercise physiology in children, especially children with cystic fibrosis, asthma and other respiratory diseases. *Mailing Add:* Cystic Fibrosis Res Ctr Univ Pittsburgh 3705 Fifth Ave Pittsburgh PA 15213

ORENSTEIN, JAN MARC, b Boston, Mass, Jan 24, 42; m 65; c 2. ULTRASTRUCTURAL MORPHOLOGY. *Educ:* Johns Hopkins Univ, BA, 63; Downstate Med Ctr, State Univ NY, PhD(path), 69, MD, 71. *Prof Exp:* Intern resident path, Presby Hosp New York, 71-73; res assoc cell biol, Lab Biochem, Nat Cancer Inst, NIH, 73-76, resident path, Lab Path, 76-77; from asst prof to assoc prof, 77-89, PROF PATH, MED CTR, GEORGE WASHINGTON UNIV, 89- *Mem:* Am Soc Cell Biologists; Am Asn Pathologists; Electron Miros Soc Am; Int Acad Pathologists; NY Acad Sci. *Res:* Ultrastructure of HIV and AIDS; ultrastructural features of lung cancer and their significance. *Mailing Add:* Dept Path Ross 502 George Washington Univ 2300 Eye St NW Washington DC 20037

ORENSTEIN, WALTER ALBERT, b New York, NY, Mar 5, 48; m 77; c 2. INFECTIOUS DISEASES, PEDIATRICS. *Educ:* City Col NY, BS, 68; Albert Einstein Col Med, MD, 72. *Prof Exp:* Intern & resident pediat, Univ Calif, San Francisco, 72-74; epidemic intel serv, Div Immunization, Ctr Dis Control, 74-76, med epidemiologist, 76-77; resident pediat, Childrens Hosp Los Angeles, 77-78; fel pediat infectious dis, Univ S Calif Med Ctr, 78-80; med epidemiologist, 80-82, chief surveillance & invest sect, 82-85, chief surveillance, invest & res br, 85-88, DIR, DIV IMMUNIZATION, CTR DIS CONTROL, 88- *Concurrent Pos:* Consult smallpox eradication, World Health Orgn, 74-75 & invest polymyelitis, 88; med adv bd, Ctr Dis Control, 81-84; clin assoc prof, dept community health, Emory Univ, 85-; ed, Pediat Infectious Dis J, 87-; adv comt, March Dimes, 84-; consult, Expanded Prog Immunization, Pan Am Health Asn, 85 & 88. *Honors & Awards:* Commendation Medal, USPHS, 84 & Meritorious Serv Medal, 88. *Mem:* Fel Am Acad Pediat; fel Infectious Dis Soc Am; Am Epidemiol Soc; Soc Epidemiol Res; Am Pub Health Asn. *Res:* Controlling vaccine preventable diseases including, measles, rubella, mumps, polio, diptheria, tetanus and haemophilus influenza type b; evaluated different epidemiologic methods for measuring vaccine efficacy; the relationship of antibody levels in the community to vaccination records; the correlation of low levels of measles and rubella antibodies to protection from disease; methods to determine adverse effects caused by vaccines; author of numerous publications. *Mailing Add:* Div Immunization Ctr Dis Control Atlanta GA 30333

OREY, STEVEN, b Berlin, Ger, July 17, 28; nat US; m 50; c 1. MATHEMATICS. *Educ:* Cornell Univ, BA, 49, MA, 51, PhD, 53. *Prof Exp:* Instr math, Cornell Univ, 53-54; instr, Univ Minn, 54-56; vis asst prof, Univ Calif, 57-58; vis asst prof, 58-59, assoc prof, 59-65, PROF MATH, UNIV MINN, MINNEAPOLIS, 65- *Mem:* Am Math Soc; Asn Symbolic Logic. *Res:* Mathematical logic; probability theory. *Mailing Add:* Dept Math Univ Minn Minneapolis MN 55455

ORF, JOHN W, b Jan 21, 50; c 2. BIOCHEMISTRY, GENETICS. *Educ:* Univ Wis, BS, 72; Univ Minn, PhD(genetics, cell biol & biochem), 77. *Prof Exp:* NIH trainee, 77-78, res assoc, Univ Minn, Minneapolis, 78-79; res scientist, 79-81, res & develop proj mgr, 81-84, dir prod develop, 83-86, dir res & develop, 86-88, VPRES RES & DEVELOP, INCSTAR CORP, STILLWATER, MINN, 88- *Mem:* AAAS; NY Acad Sci; Am Chem Soc. *Res:* Enzymeimmunoassay development (liquid or solid phase); radioimmunoassay development (liquid or solid phase); antisera development (polyclonal& monoclonal); solid phase peptide synthesis; peptide protein purification; author of 9 journal articles. *Mailing Add:* 1990 Industrial Blvd Stillwater MN 55082

ORGAN, JAMES ALBERT, b Newark, NJ, Mar 29, 31; m 56; c 2. ZOOLOGY. *Educ:* Rutgers Univ, AB, 56; Univ Mich, MS, 58, PhD(zool), 60. *Prof Exp:* Instr zool, Univ Mich, 60-61; from instr to assoc prof biol, 61-71, PROF BIOL, CITY COL NEW YORK, 71- *Mem:* Ecol Soc Am; Am Soc Ichthyol & Herpet; Soc Study Evolution; Am Soc Zool; NY Acad Sci. *Res:* Population ecology; animal behavior; herpetology. *Mailing Add:* Dept Biol City Col New York Convent Ave & 138th St New York NY 10031

ORGEBIN-CRIST, MARIE-CLAIRE, b Vannes, France, Mar 20, 36. BIOLOGY, ENDOCRINOLOGY. *Educ:* Sorbonne, Lic Natural Sci & Lic Biol, 57; Univ Lyons, DSc, 61. *Prof Exp:* Res assoc, Med Div, Pop Coun, 62-63; res assoc, Vanderbilt Univ, 63-64, res instr, 64-66, from asst prof to assoc prof, 66-73, LUCIUS E BURCH PROF REPROD BIOL & PROF CELL BIOL, SCH MED, VANDERBILT UNIV, 73-, DIR CTR & DIV REPROD BIOL RES, 73- *Concurrent Pos:* NIH career develop award, 68-73, Fogarty Int sr fel, 77; mem contract rev comt, Ctr Pop Res, NICHD, 68-72, Pop Res & Training Comt, 70-74 & 85-89, chmn, 72-74 & 88-89, Reprod Biol Study Sect, 78-82, key consult five yr res plan, 80; mem, World Health Orgn Task Force Methods for Regulation Male Fertility, 72-75, steering comt, 73-75, Comt Resources Res, 85-90; ed-in-chief, J Andrology, 83-89. *Mem:* AAAS; Am Asn Anatomists; Am Soc Andrology; Endocrine Soc; Int Soc Andrology; NY Acad Sci; soc Study Fertility (Eng); Soc Study Reprod. *Res:* Male reproductive physiology. *Mailing Add:* Ctr Reprod Biol Res Vanderbilt Univ Sch Med Nashville TN 37232

ORGEL, LESLIE E, b London, Eng, Jan 12, 27; m 50; c 3. CHEMISTRY. *Educ:* Oxford Univ, BA, 48, PhD(chem), 51. *Prof Exp:* Reader chem, Cambridge Univ, 51-64; SR FEL, SALK INST BIOL STUDIES, 65- *Concurrent Pos:* Adj prof, Univ Calif, San Diego, 65- *Mem:* Nat Acad Sci; fel Royal Soc Chem; Sigma Xi. *Res:* Prebiotic chemistry; the chemistry of the origins of life. *Mailing Add:* Salk Inst Biol Studies PO Box 85800 San Diego CA 92138

ORGELL, WALLACE HERMAN, b Eldora, Iowa, Dec 28, 28; m 60. PLANT PHYSIOLOGY. *Educ:* Iowa State Univ, BS, 50; Pa State Univ, MS, 52; Univ Calif, PhD(plant physiol), 54. *Prof Exp:* Asst plant physiol, USDA, Pa State Univ, 50-52; asst bot, Univ Calif, 52-54; NSF fel org chem, Univ Ill, 54-55; plant biochemist, US Chem Corps, Ft Detrick, 55-57 & dept zool & entom, Iowa State Univ, 57-63; assoc prof biol sci, Fla Atlantic Univ, 63-66; prof biol, Hiram Scott Col, 66-67; assoc prof, 67-73, PROF BIOL, MIAMI-DADE COMMUNITY COL, 73- *Mem:* Nat Sci Teachers Asn; Nat Asn Biol Teachers. *Res:* Biochemical aspects of ecology. *Mailing Add:* Miami-Dade Community Col Dept Biol South Campus 11011 W 104th St Miami FL 33176

ORGILL, MONTIE M, b Mammoth, Utah, Feb 25, 29; m 56; c 5. ATMOSPHERIC PHYSICS, AIR POLLUTION. *Educ:* Brigham Young Univ, BS, 51; Univ Utah, MS, 57; Colo State Univ, PhD, 71. *Prof Exp:* Consult & forecaster, Intermt Weather, Inc, Utah, 56-58; asst prof meteorol, Univ Hawaii, 58-62; meteorologist, Colo State Univ, 62-72; res scientist, 72-80, proj mgr & sr res scientist, Pac Northwest Labs, Battelle Mem Inst, 80-86; NRC SR RES ASSOC, US ARMY ATMOSPHERIC LAB, WHITE SANDS MISSILE RANGE, NMEX, 86- *Concurrent Pos:* Air weather serv, US Army Forces, 51-55; consult, US Army, 62-63, Forest Serv, 66, Los Alamos Nat Labs, 86-87; sr res assoc, Nat Res Comn. *Honors & Awards:* Twenty-Five Year Award, Am Geophys Union. *Mem:* Am Meteorol Soc; Am Geophys Union; Am Pollution Control Asn; Am Radio Relay League; Sigma Xi. *Res:* Synoptic meteorology; fluid mechanics; micrometeorology and mesometeorology; weather modification; atmospheric resuspension and diffusion research; aircraft sampling research; environmental assessment; applied meteorology; complex terrain meteorology; knowledge-based systems (artificial intelligence). *Mailing Add:* 904 Lenox Ave Las Cruces NM 88005

ORIANI, RICHARD ANTHONY, b El Salvador, Cent Am, July 19, 20; nat US; m 49; c 4. PHYSICAL CHEMISTRY, CORROSION. *Educ:* City Col New York, BChE, 43; Stevens Inst Technol, MS, 46; Princeton Univ, MA & PhD(phys chem), 48. *Prof Exp:* Lab asst chem eng, City Col New York, 43; res chemist, Bakelite Corp Div Union Carbide & Carbon Corp, 43-46; asst, Princeton Univ, 46-48; res assoc, Gen Elec Co, 48-59; mgr phys chem, Res Lab, US Steel Corp, 59-75, sr res consult, 75-80; prof & dir Corrosion Res Ctr, 80-87, EMER PROF & DIR, UNIV MINN. *Concurrent Pos:* Vis instr, Rensselaer Polytech Inst, 53-54; mem Nat Acad Sci adv panel, Nat Bur Stand, 67-73; consult, 80- *Honors & Awards:* First Inland Steel-Northwestern Univ Lectr, 84; Alexander von Humboldt Award, 84; W R Whitney Award, 87. *Mem:* Am Phys Soc; fel Am Inst Chem; fel NY Acad Sci; fel Am Soc Metals; Electrochem Soc; Nat Asn Corrosion Engrs. *Res:* Thermodynamics of alloys; corrosion; structure and binding in solids; kinetics; surface reactions; hydrogen embrittlement. *Mailing Add:* Dept Chem Eng & Mat Sci Univ Minnesota Minneapolis MN 55455

ORIANS, GORDON HOWELL, b Eau Claire, Wis, July 10, 32; m 55; c 3. ECOLOGY, EVOLUTION. *Educ:* Univ Wis, BS, 54; Univ Calif, Berkeley, PhD(zool), 60. *Prof Exp:* From asst prof to assoc prof, 60-68, PROF ZOOL, UNIV WASH, 68- *Concurrent Pos:* Dir, Inst Environ Studies, Univ Wash, 76-86. *Honors & Awards:* Brewster Award, Am Ornithologist Union. *Mem:* Nat Acad Sci; Ecol Soc Am; Soc Study Evolution; Animal Behav Soc; fel Am Ornithologists Union; fel Royal Dutch Acad Sci; Am Soc Naturalists (vpres, 87-88); Cooper Ornith Soc; Orgn Trop Studies (pres, 88-). *Res:* Evolution of vertebrate social systems; factors determining the number of species an environment will support on a sustained basis; plant-herbivore interactions; ecology of rare species. *Mailing Add:* Dept Zool Univ Wash Seattle WA 98195

ORIEL, STEVEN S, b New York, NY, Feb 9, 23; m 50; c 3. STRUCTURAL GEOLOGY, STRATIGRAPHY. *Educ:* Columbia Univ, BA, 45; Yale Univ, MS, 47, PhD(geol), 49. *Prof Exp:* Geologist, US Geol Surv, 47-50; res engr geol, Stanolind Oil & Gas Co, 50-53; GEOLOGIST, US GEOL SURV, 53- *Concurrent Pos:* Mem, Am Comn Stratig Nomenclature, chmn, 73-74. *Honors & Awards:* Meritorious Service Award, US Dept Interior, 79 & Distinguished Service Award, 85. *Mem:* AAAS; Geol Soc Am; Am Asn Petrol Geol; Am Geophys Union; Am Inst Prof Geologists. *Res:* Geology of sedimentary rocks; synthesis of stratigraphic information; areal mapping, tectonic analyses and resource appraisals of deformed rocks and Cenozoic sediments in western Wyoming and southeastern Idaho thrust belt and northeastern basin and range, including geothermal resource and earthquake hazard assessments. *Mailing Add:* 14195 Crabapple Rd Golden CO 80401

ORIENT, JANE MICHEL, b Tucson, Ariz, Nov 2, 46. MEDICINE. *Educ:* Univ Ariz, BS & BA, 67; Columbia Univ, MD, 74. *Prof Exp:* Instr med, Univ Ariz Col Med, 77-81; PVT PRACT, 81- *Concurrent Pos:* Bd sci advs, Am Coun Sci & Health. *Mem:* Fel Am Co Physicians. *Res:* Clinical examination; medical ethics; medical economics. *Mailing Add:* Med Sq Suite 9 1601 N Tucson Blvd Tucson AZ 85716

ORIHEL, THOMAS CHARLES, b Akron, Ohio, Feb 10, 29; m 52; c 4. MEDICAL PARASITOLOGY. *Educ:* Univ Akron, BS, 50; Univ Wash, MS, 52; Tulane Univ, PhD(parasitol), 59. *Prof Exp:* NIH fel, Tulane Univ & Univ Miami, 59-61; vector-borne dis specialist, Dept of State, AID Mission to Brit Guiana, 61-63; assoc dir, 75-76, DIR, INT CTR MED RES, 76-, RES SCIENTIST & HEAD PARASITOL, DELTA PRIMATE CTR, TULANE UNIV, 63-, PROF PARASITOL, SCH PUB HEALTH & TROP MED, 72-, WILLIAM G VINCENT PROF TROPICAL DIS, 82- *Concurrent Pos:* Mem study sect trop med & parasitol, Nat Inst Allergy & Infectious Dis, NIH, 73-77, chmn study sect, 75-77; mem expert panel parasitic dis, WHO, 73-78; mem panel parasitic dis, US-Japan Coop Med Sci Prog, 74-; dir, Int Col Inf Dis Res, 80. *Mem:* Am Soc Parasitol; Am Soc Trop Med & Hyg; Royal Soc Trop Med & Hyg; Am Soc Microbiol; Sigma Xi. *Res:* Human and zoonotic filariasis; host-parasite interactions; primate parasites; nematology; parasite morphology and taxonomy. *Mailing Add:* 115 Bertel Dr Covington LA 70433

ORING, LEWIS WARREN, b New York, NY, Apr 18, 38; m 60; c 2. ECOLOGY, ETHOLOGY. *Educ:* Univ Idaho, BS, 60; Univ Okla, MS, 62, PhD(zool), 66. *Prof Exp:* NSF fel ethology, Univ Copenhagen, 66-67; NIH trainee, Univ Minn, Minneapolis, 67-68; from asst prof to prof biol, Univ NDak, 77-90; PROF CONSERV BIOL, UNIV NEV, 90- *Concurrent Pos:* Vis prof, Univ Minn, 68, Cornell Univ, 75-76. *Mem:* Animal Behav Soc; Ecol Soc Am; Am Ornith Union; Wilson Ornith Soc; Soc Cons Biol. *Res:* Ecological and evolutionary aspects of animal behavior, especially the ways in which environment effects mating systems, spacing patterns and communication; conservation biology, population biology and field endocrinology. *Mailing Add:* Ecol Evolution & Conserv Biol Univ Nev RWF 1000 Valley Blvd Reno NV 89512

ORKAND, RICHARD K, b New York, NY, Apr 23, 36; m 60; c 1. PHYSIOLOGY, NEUROBIOLOGY. *Educ:* Columbia Univ, BS, 56; Univ Utah, PhD(pharmacol), 61. *Hon Degrees:* MS, Univ Pa, 74. *Prof Exp:* USPHS fel biophys, Univ Col, Univ London, 61-64; USPHS spec fel neurophysiol, Harvard Med Sch, 64-66; asst prof physiol, Med Ctr, Univ Utah, 66-68; assoc prof zool, Univ Calif, Los Angeles, 68-72, prof biol, 72-74; prof physiol & chmn Dept Physiol & Pharmacol, Sch Dent Med, Univ Pa, 74-86; PROF PHYSIOL, UNIV PR, 86-, DIR, INST NEUROBIOL, 86- *Concurrent Pos:* Exchange fel Czech Acad Scis, 77-83; vis prof, Univ Geneva, 81-82, Heidelberg Univ, 82-83; res assoc marine sci, Univ Calif, Santa Cruz, 81- *Honors & Awards:* Humboldt Prize, Humboldt Found, Ger, 82. *Mem:* Soc Neurosci; Fel AAAS; assoc Am Physiol Soc; Soc Gen Physiologists; Am Soc Pharmacol & Exp Therapeut; Int Brain Res Orgn; Fed Am Scientists. *Res:* The physiological functions of neuroglia and the interactions between neuron and glia in the nervous system. *Mailing Add:* Inst Neurobiol Univ PR Med Sci Campus 201 Blvd del Valle San Juan PR 00901

ORKIN, LAZARUS ALLERTON, medicine, surgery; deceased, see previous edition for last biography

ORKIN, LOUIS R, b New York, NY, Dec 23, 15; m 38; c 1. ANESTHESIOLOGY. *Educ:* Univ Wis, BA, 37; NY Univ, MD, 41. *Prof Exp:* Dir anesthesia, W W Backus Hosp, Conn, 48-49; asst prof anesthesiol, Postgrad Med Sch, NY Univ, 50-55; prof anesthesiol & chmn dept, 55-82, distinguished univ prof, 82-86, EMER DISTINGUISHED UNIV PROF, ALBERT EINSTEIN COL MED, 87- *Concurrent Pos:* Consult, Norwich State Hosp, Conn, 48-49 & Uncas on Thames, 48-49; anesthesiologist, Bellevue Hosp, 50-55; asst attend anesthetist, NY Univ Hosp, 50-54, assoc attend anesthetist, 54-55; consult, US Naval Hosp, St Albans, 54, USPHS Hosp, Staten Island, 54, Triboro Hosp, 55-65 & Montefiore Hosp, New York, 65-; titular prof, Univ Venezuela, 67; vis prof, Univ Calif, San Diego, 71; mem nat res coun, Nat Acad Med, 65-70; sr consult, Bronx Vet Admin Hosp, 55-84. *Mem:* Am Soc Anesthesiol; AMA; fel NY Acad Sci; fel Am Col Chest Physicians; fel NY Acad Med; hon mem Japanese Soc Anesthesiol. *Res:* Physiology and pharmacology related to anesthesiology. *Mailing Add:* 11 Stuyvesant Oval New York NY 10009

ORKIN, STUART H, b New York, NY, Apr 23, 46. PEDIATRICS. *Educ:* Mass Inst Technol, BS, 67; Harvard Univ, MD, 72; Am Bd Pediat, cert, 78. *Prof Exp:* Intern pediat, Children's Hosp Med Ctr, 72-73, resident, 75-76; clin fel, Harvard Med Sch, 72-73 & 75-76; res assoc, Lab Molecular Genetics, Nat Inst Child Health & Human Develop, NIH, 73-75; fel med, Children's Hosp Med Ctr & Dana Farber Cancer Inst, 76-78; from asst prof to assoc prof pediat, 78-87, LELAND FIKES PROF PEDIAT MED, HARVARD MED SCH, 87-; ASSOC MED, CHILDREN'S HOSP MED CTR, 78- *Concurrent Pos:* Basil O'Connor starter res award, Nat Found-March of Dimes, 76-78; NIH young investr award, 76-79; res fel pediat, Harvard Med Sch, 76-78, lectr hemat, 79-; asst physician, Dana Farber Cancer Inst, 78-; attend physician, Div Hemat Oncol, Children's Hosp & Sidney Farber Cancer Inst, 78-, Dept Med, Children's Hosp, 80-; res career develop award, Nat Heart Lung & Blood Inst, NIH, 79-84; lectr postgrad internal med & pediat, Harvard Med Sch, 79-; investr, Howard Hughes Med Inst, 86-; mem, Comt Mapping & Sequencing the Human Genome, Nat Res Coun, 87-88; mem, Mammalian Genetics Study Sect, NIH, 87-; Ben Abelson vis prof, Dept Pediat, Washington Univ, 88; counr, Am Soc Hemat, 88; Robert & Courtney Steel vis prof pediat, Mem Sloan Kettering Cancer Ctr, 89. *Honors & Awards:* Dameshek Award, Am Soc Hemat, 86; Mead Johnson Award, Am Acad Pediat, 87; Karl Meyer Lectr, Univ Calif Med Ctr, 88; Harvey Lectr, 88. *Mem:* Nat Acad Sci; Am Soc Hemat; Soc Pediat Res; Am Soc Clin Invest (pres, 89-90); Asn Am Physicians. *Res:* Molecular genetics and biology of human disease; thalassemia syndromes and hemoglobin synthesis; prenatal diagnosis of genetic disease; molecular biology of coagulation; molecular genetics and biochemistry of the phagocytic cell; gene transfer and expression in hematopoietic cells. *Mailing Add:* Dept Pediat Hemat & Oncol Childrens Hosp 300 Longwood Ave Boston MA 02115

ORKNEY, G DALE, b Raymond, Wash, Dec, 19, 36; m 59; c 3. PHYSIOLOGY & GENETIC REGULATION OF FLOWERING. *Educ:* Northwest Nazarene Col, BA, 59; Univ Idaho, MS, 61, PhD(bot), 71. *Prof Exp:* Asst prof biol, LaVerne Col, 62-63; asst prof, 63-64 & 65-68, assoc prof, 69-75, PROF BIOL, GEORGE FOX COL, 75- *Concurrent Pos:* Vis instr bot, Univ Idaho, Moscow, 64-65 & 65, fac fel, 68-69; NSF fel, Ore State Univ, 74; vis res prof, Univ Calif, Davis, 81. *Mem:* AAAS; Cactus & Succulent Soc Am. *Res:* Physiology and genetic regulation of flowering. *Mailing Add:* Biol Dept George Fox Col Newberg OR 97132

ORLAND, FRANK J, b Little Falls, NY, Jan 23, 17; m 43; c 4. HISTORY OF DENTISTRY, ORAL MICROBIOLOGY. *Educ:* Univ Ill, BS, 39, DDS, 41; Univ Chicago, SM, 45, PhD(bact), 49; Am Bd Med Microbiol, dipl. *Prof Exp:* Intern, Zoller Mem Dent Clin, 41-42, asst dent surg, 42-49, from instr to asst prof bact, 49-54, dir, Zoller Clin, 54-66, assoc prof microbiol, 54-58, res assoc prof, 58-64, from instr to assoc prof dent surg, 49-58, prof, 58-88, EMER PROF DENT SCI, UNIV CHICAGO, 88-, PROF FISHBEIN CTR, STUDY HIST SCI & MED, 80- *Concurrent Pos:* Fel, Inst Med, Univ Chicago; consult & mem dent study sect & prog planning comt, Nat Inst Dent Res, 55-59; Darnell mem lectr, US Naval Dent Sch, 56; chmn comt info & pub, Inst Med; mem panel, Nat Formulary; chmn dent adv bd, Med Heritage Soc; ed, J Dent Res, 58-69. *Honors & Awards:* Hayden Harris Award, 80. *Mem:* Fel AAAS; Am Soc Microbiol; Electron Micros Soc Am; Am Dent Asn; fel Am Acad Microbiol; Int Asn Dent Res (pres, 71-72); Am Acad Hist Dent (pres, 76-77). *Res:* Experimental dental caries; caries and periodontal disease in germfree and germfree inoculated animals; antigenic analysis of lactobacilli; mechanism of fluoride inhibition in oral bacteria and in tooth decay; morphology and physiology of oral protozoa; historical research in dentistry. *Mailing Add:* 519 Jackson Blvd Forest Park Chicago IL 60130

ORLAND, GEORGE H, b Los Angeles, Calif, Oct 8, 24; m 44; c 3. MATHEMATICS. *Educ:* City Col New York, BEE, 45; Univ Chicago, MS, 56; Univ Calif, Berkeley, PhD(math), 62. *Prof Exp:* Instr elec eng, Polytech Inst Brooklyn, 46-47; tutor, City Col New York, 48-49; electronics engr, Dept Otolaryngol, Univ Chicago, 50-57, mathematician, Inst Systs Res, 57-59; instr math, Univ Ill, 62-65; asst prof, Wesleyan Univ, 65-67; ASSOC PROF MATH, HARVEY MUDD COL, 67- *Mem:* Am Math Soc. *Res:* Classical and functional analysis; measure theory; convexity; the bending of polyhedra. *Mailing Add:* Dept of Math Harvey Mudd Col Claremont CA 91711

ORLANDO, ANTHONY MICHAEL, b Brooklyn, NY, Jan 27, 42; m; c 2. BIOSTATISTICS. *Educ:* City Univ New York, BBA, 63; Columbia Univ, MS, 67; Yale Univ, MPhil, 69, PhD(biostatist), 70. *Prof Exp:* Sr sales analyst, Am-Standard, Inc, 64-65; statistician, Monsanto Chem Co, 65; prin biostatistician, Mead Johnson & Co, Evansville, 70-75, dept head statist sci, 75-79, dir clin info & statist, 79-82; mgr data analysis & database admin, NCTR, 82-84; DIR STATIST & CLIN INFO, BOOTS PHARMACEUT, SHREVEPORT, 84- *Mem:* AAAS; Am Statist Asn; Biomet Soc; Sigma Xi; Am Inst Decision Sci; Inst Int Forecasters; Drug Info Asn; World Future Soc. *Res:* Analyzing clinical parameter distributions under a mixture of normal densities; drug comparison hypothesis testing. *Mailing Add:* Boots Pharmaceut Inc 8800 Ellerbe Rd Shreveport LA 71136-6750

ORLANDO, CARL, b Palermo, Italy; US citizen; m 43; c 5. AUTOMATIC MEDICAL ALERTING DEVICES. *Educ:* Columbia Univ, BS, 41. *Prof Exp:* Chief photo optics, US Army Electronic Comn, 45-75; consult, 75-79; vpres res & develop, Anal Res & Develop Inc Consult, 79-88; CONSULT RES & DEVELOP, ENG DEVELOP CO, 86-; PRES & DIR RES & DEVELOP, SENS-O-TECH INDUSTS, INC MED EQUIP, 88- *Concurrent Pos:* Dir, Soc Photog Scientists & Engrs, 78-82 & Soc Imaging Sci & Technol, 86-88. *Honors & Awards:* Serv Award, Soc Photog Scientists & Engrs, 66. *Mem:* Sr mem Soc Photog Scientists & Engrs; Soc Photog Instrumentation & Eng; Armed Forces Electronic Asn. *Res:* Photo optics of aerial cameras; night photo systems using image intensifier; seismic intrusion sensors; image interpretation; active and passive sensors; target recognition and automatic target screening; non-intrusive medical equipments; apnea and cardiac functions monitors; author of various publications. *Mailing Add:* Eng Develop Co 47 Willow Rd Tinton Falls NJ 07724

ORLANDO, CHARLES M, b Tappan, NY, Sept 30, 35; m 64; c 3. ORGANIC CHEMISTRY. *Educ:* Fordham Univ, BS, 57; NY Univ, PhD(org chem), 61. *Prof Exp:* Res chemist, Esso Res & Eng Co, 61-64 & Kay-Fries Chem Co, 64-66; res chemist, 66-72, MGR PROCESS BR, GEN ELEC CORP RES & DEVELOP CTR, 72- *Honors & Awards:* IR 100 Award, 74. *Mem:* Am Chem Soc. *Res:* Organic photochemistry; organic synthesis; heterocyclic chemistry; chemistry of quinones; polymer synthesis; phenol chemistry; bromination chemistry and flame retardants; polymer blends. *Mailing Add:* Res & Develop Ctr Gen Elec Corp PO Box 8 Schenectady NY 12301-0008

ORLANDO, JOSEPH ALEXANDER, b Methuen, Mass, July 16, 29; m 58; c 6. BIOLOGICAL CHEMISTRY. *Educ:* Merrimack Col, BS, 53; NC State Col, MS, 56; Univ Calif, PhD(biochem), 58. *Prof Exp:* Nat Cancer Inst fel biochem, Brandeis Univ, 58-61; ASSOC PROF BIOCHEM, BOSTON COL, 61- *Res:* Cellular biochemistry-studies on the signal transduction mechanisms associated with the bending of growth and maturation factors to cell surface receptors of normal and transformed mammalian cell lines. *Mailing Add:* Dept Biol Boston Col Chestnut Hill MA 02167

ORLANDO, TERRY PHILIP, b Baton Rouge, La, June 3, 52; m 76; c 1. ELECTRICAL ENGINEERING. *Educ:* La State Univ, BS, 74; Stanford Univ, MS, 75, PhD(physics), 81. *Prof Exp:* ASSOC PROF, DEPT ELEC ENG & COMPUT SCI, MASS INST TECHNOL, 81- *Mem:* Am Phys Soc; Inst Elec & Electronic Engrs; Mat Res Soc. *Res:* Superconductors in high magnetic fields; fabrication of thin films of superconductors, superconducting tunneling; electronic transport in semiconducting devices. *Mailing Add:* Mass Inst Technol 13-3026 77 Mass Ave Cambridge MA 02139

ORLANS(MORTON), F BARBARA, b Jan 14, 28; US citizen; m 59, 73; c 2. PHYSIOLOGY, BIOETHICS. *Educ:* Birmingham Univ, BSc, 49; London Univ, MSc, 54, PhD(physiol), 56. *Prof Exp:* Res physiologist hormonal control autonomic nervous syst, Nat Heart & Lung Inst, NIH, 56-60; writer free-lance, 67-73; sr staff scientist pop biol, Med Ctr, George Washington Univ, 73-74; grants assoc, Admin Training, Nat Heart, Lung & Blood Inst, NIH, 74-75, health sci adminr cardiovasc dis, 75-77, exec secy, 77-79, Adv Coun, health scientist adminr, Cardiac Dis Bd, 79-84; dir, Sci Ctr Animal Welfare, 84-88; RES ASSOC, KENNEDY INST ETHICS, GEORGETOWN UNIV, 89- *Concurrent Pos:* Pres, Scientists Ctr Animal Welfare, 78-84; scholars residence, Rockefeller Found Study Ctr, Bellagio, Italy, 89. *Honors & Awards:* Doerenkamp-Zbinden Found Award for Animal Welfare, 86. *Mem:* Am Soc Pharmacol & Exp Therapeut; Nat Sci Teachers Asn; Nat Asn Biol Teachers. *Res:* bioethics. *Mailing Add:* Kennedy Inst Ethics Georgetown Univ Washington DC 20057

ORLANSKI, ISIDORO, b Buenos Aires, Arg, June 6, 39; m 62; c 3. GEOPHYSICS. *Educ:* Univ Buenos Aires, Lic Physics, 65; Mass Inst Technol, PhD(geophys), 67. *Prof Exp:* SR RES SCIENTIST, GEOPHYS FLUID DYNAMICS LAB, NAT OCEANIC & ATMOSPHERIC ADMIN, 67- *Concurrent Pos:* Res assoc, Mass Inst Technol, 67-68; lectr, Princeton Univ, 71- , prof. *Mem:* Am Meteorol Soc; Am Geophys Union. *Res:* Turbulence in stratified flows; dynamics of rotating stratified fluids; atmospheric mesoscale dynamics. *Mailing Add:* Geophys Fluid Dynamics Lab US Dept Commerce, NOAA Princeton Univ PO Box 308 Princeton NJ 08540

ORLIC, DONALD, b Universal, Pa, Dec 11, 32; m 61; c 3. CELL BIOLOGY. *Educ:* Fordham Univ, BS, 59; NY Univ, MS, 63, PhD(biol), 66. *Prof Exp:* Instr anat, Harvard Med Sch, 68-69; from asst prof to assoc prof, 69-79, actg chmn dept, 74-76, PROF ANAT, NY MED COL, 79- *Concurrent Pos:* NIH fel, Inst Cellular Path, France, 66-67; NIH fel anat, Harvard Med Sch, 67-68. *Honors & Awards:* A Cressy Morrison Award, NY Acad Sci, 66. *Mem:* Am Soc Cell Biol; Soc Study Blood; Am Soc Hemat; Am Asn Anat. *Res:* Effects of erythropoietin on stem cell differentiation and red cell maturation; development of human fetal gastrointestinal tract. *Mailing Add:* Dept Anat NY Med Col Elmwood Hall Valhalla NY 10595

ORLICK, CHARLES ALEX, b Milltown, NJ, Sept 30, 27; m 50; c 6. PHYSICAL CHEMISTRY. *Educ:* Rutgers Univ, BSc, 47, MSc, 50, PhD(chem), 52. *Prof Exp:* Res assoc chem, Hercules Inc, 51-59, sr res chemist, 59-60, supvr propellant res group, 60-65, supvr appl res group, 65-68, mkt specialist, 68-76, mgr propulsion tech mkt, Allegany Ballistics Lab, 76-86, mgr, propulsion technol progs, 86-89, CONSULT, HERCULES INC, 90-; CONSULT, ANALYTICAL ENG SERV, 90- *Mem:* Am Chem Soc. *Res:* Nitrate ester decomposition mechanisms; double-base propellant combustion kinetics; propellant thermochemistry. *Mailing Add:* RFD 1, Box 44A LaVale MD 21502

ORLIDGE, ALICIA, CELL BIOLOGY, MICROVASCULAR. *Educ:* Pa State Univ, PhD(physiol), 83. *Prof Exp:* RES ASSOC, CHILDREN'S HOSP MED CTR, BOSTON, 83- *Res:* Endothelial cells. *Mailing Add:* Children's Hosp Med Ctr 300 Longwood Ave Boston MA 02115

ORLIK, PETER PAUL, b Budapest, Hungary, Nov 12, 38; US citizen; m 64; c 2. MATHEMATICS. *Educ:* Norweg Inst Technol, BS, 61; Univ Mich, Ann Arbor, PhD(math), 66. *Prof Exp:* From asst prof to assoc prof, 66-73, PROF MATH, UNIV WIS-MADISON, 73- *Concurrent Pos:* Mem & NSF fel, Inst Advan Study, 67-69; vis prof, Univ Oslo, 71-72. *Mem:* Am Math Soc. *Res:* Topology; transformation groups. *Mailing Add:* Dept Math 615 Van Vleck hall Univ Wis Madison WI 53706

ORLIK-RUCKEMANN, KAZIMIERZ JERZY, b Warsaw, Poland, May 20, 25; Can citizen; m 50; c 2. APPLIED AERODYNAMICS, FLIGHT MECHANICS. *Educ:* Royal Inst Technol, Stockholm, Sweden, MSc, 47, Tekn Lic, 68, Tekn D, 70. *Prof Exp:* Aeronaut scientist, Royal Inst Technol, Stockholm, 47-51; sr aeronaut scientist, Aeronaut Res Inst, Sweden, 51-55; res officer, Nat Res Coun, Can, 55-58, head, High Speed Aerodyn Lab, 58-62, head Unsteady Aerodyn Lab, 63-90, HEAD, APPL AERODYN LAB, NAT RES COUN, CAN, 90- *Concurrent Pos:* Assoc ed, CASI Trans, 69-75; docent, Royal Inst Technol, Stockholm, 71-; mem, Fluid Dynamics Panel, NATO Adv Group Aerospace Res & Develop, 74-, chmn, 80-81; consult, orgn Can, USA, Sweden; assoc ed, Progress Aerospace Sci, 84- *Honors & Awards:* Ground Testing Award, Am Inst Aeronaut & Astronaut, 86; NATO Agard von Karman Medal, 89. *Mem:* Fel Can Aeronaut & Space Inst; fel Am Inst Aeronaut & Astronaut; Supersonic Tunnel Asn (pres, 63-64). *Res:* Stability of aircraft; dynamic wind tunnel experiments; flight mechanics; applied aerodynamics. *Mailing Add:* Appl Aerodyn Lab Nat Res Coun Ottawa ON K1A 0R6 CAN

ORLIN, HYMAN, b New York, NY, July 16, 20; m 43; c 2. GEOPHYSICS, GEODESY. *Educ:* City Col New York, BBA, 42; George Wash Univ, AB, 48, AM, 52; Ohio State Univ, PhD, 62. *Prof Exp:* Mathematician, Coast & Geod Surv, 47-58, supvry geodesist, 58-59, asst chief gravity & astron br, 59-66, tech asst geod div, 66-69, spec asst to dir, Nat Ocean Surv Earth Sci Activities, 69-71; chief scientist, Nat Oceanic & Atmospheric Admin, Nat Ocean Surv, US Dept Commerce, 71-75; exec secy, Comt Geod, Nat Acad Sci, 75-82; CONSULT, NAT ACAD SCI, 82- *Concurrent Pos:* Lectr, George Wash Univ, 54-61, asst prof lectr, 62-63, assoc prof lectr, 64-65, prof lectr and fac adv, 66-82. *Honors & Awards:* Silver Medal, US Dept Com, 52; Heiskanen Award, Ohio State Univ, 70. *Mem:* Fel Am Geophys Union. *Res:* Physical and geometric geodesy; geophysics. *Mailing Add:* 3005 Portfino Isle F-2 Coconut Creek FL 33066

ORLOFF, DAVID IRA, b Brooklyn, NY, Mar 14, 44; m 72; c 3. COMBUSTION ENGINEERING, HEAT & MASS TRANSFER. *Educ:* Drexel Univ, BSME, 66, MSME, 68, PhD(thermal sci), 74. *Prof Exp:* Proj engr, Eng Div, Ford Motor Co, 73-74; asst prof, 74-79, ASSOC PROF MECH ENG, UNIV SC, 79- *Concurrent Pos:* Fac researcher, Savannah River Lab, Reactor Eng Group, 75-76 & 77. *Honors & Awards:* Ralph Teetor Award, Soc Automotive Engrs, 77. *Mem:* Am Soc Mech Engrs; Combustion Inst. *Res:* Mathematical models of industrial combustion processes; hazardous and radioactive waste incineration. *Mailing Add:* Inst Paper Sci & Technol 575 14th St NW Atlanta GA 30318

ORLOFF, HAROLD DAVID, b Winnipeg, Man, Nov 24, 15; nat US; m 41; c 3. ORGANIC CHEMISTRY. *Educ:* Univ Man, BSc, 37, MSc, 40. *Prof Exp:* Chemist, Dom Grain Res Labs, Can, 39-40; instr chem, McGill Univ, 40-41; res chemist wood chem, Howard Smith Papers Mills, Ltd, 41-48; res supvr, Ethyl Corp, 48-61, supvr res planning, 61-83; RETIRED. *Concurrent Pos:* Instr, Univ Detroit, 58. *Honors & Awards:* Weldon Mem Gold Medal, Can Pulp & Paper Asn, 47. *Mem:* AAAS; Am Chem Soc. *Res:* Recovery and utilization of lignin from wood pulping processes; Arborite plastics; alkaline pulping processes; chlorination of benzene; chlorocyclohexane compounds; stereoisomerism of cyclohexane derivatives; phosphorous chemistry; lubricant and fuel additives; antitoxidants; chemistry of phenol, hydrocarbon oxidation; coal combustion in utility boilers; zeolites. *Mailing Add:* 2903 Victoria Circle Apt #D3 Coconut Creek FL 33066

ORLOFF, JACK, physiology, medicine; deceased, see previous edition for last biography

ORLOFF, JONATHAN H, b New York, NY; m 67; c 2. ELECTRON & ION OPTICS. *Educ:* Mass Inst Technol, BS, 64; Ore Grad Ctr, PhD(physics), 77. *Prof Exp:* Vpres, Elektros, Inc, 71-74; consult, Electron Optics, 74-75; instr, 75-77, from asst prof to assoc prof, 77-84, PROF PHYSICS, ORE GRAD CTR, 84- *Concurrent Pos:* Consult, Perkin-Elmer Corp, 79-, Hughes Aircraft Corp, 76-78 & Varian Assocs, 78-79. *Mem:* Am Phys Soc; Electron Microscopy Soc Am. *Res:* High brightness electron and ion sources; development and applications of electron and ion optics. *Mailing Add:* Dept Applied Physics Ore Grad Ctr 19600 NW Von Newmann Dr Beaverton OR 97006

ORLOFF, LAWRENCE, b Brooklyn, NY, Feb 12, 41; m 70; c 3. COMBUSTION. *Educ:* Rensselaer Polytech Inst, BS, 63; Georgetown Univ, MA, 71. *Prof Exp:* Patent examr optics, US Patent Off, 64-65; res scientist, 66-78, SR RES SCIENTIST FIRE RES, FACTORY MUTUAL RES CORP, 78- *Honors & Awards:* Silver Medal, Combustion Inst. *Mem:* Combustion Inst; Inst Elec & Electronics Engrs. *Res:* Basic experimental research on effects of scale, configuration, and fuel on fire behavior. *Mailing Add:* Factory Mutual Res Corp 1151 Boston-Providence Turnpike Norwood MA 02062

ORLOFF, MALCOLM KENNETH, b Philadelphia, Pa, Feb 26, 39; m 60; c 3. QUANTUM CHEMISTRY. *Educ:* Univ Pa, BA, 60, PhD(phys chem), 64. *Prof Exp:* Fel theoret chem, Yale Univ, 64-65; sr res chemist theoret chem, Am Cyanamid Co, 65-74, group leader, Dyes Tech Serv, 74-75, prod mgr dyes, 75-77, mkt mgr, 77-78; V PRES, RES & DEVELOP, BUFFALO COLOR CORP, 78-; VPRES BUS DEVELOP, MOORE & MUNGER INC. *Mem:* Am Asn Textile Colorists & Chemists; Am Chem Soc; Am Phys Soc; Sigma Xi. *Res:* Quantum mechanics of atomic and molecular systems. *Mailing Add:* 18 Claire Pl Trumbull CT 06611

ORLOFF, MARSHALL JEROME, b Chicago, Ill, Oct 12, 27; m 53; c 6. SURGERY. *Educ:* Univ Ill, BS, 49, MD & MS, 51; Univ Colo, Am Bd Surg, dipl, 59; Am Bd Thoracic Surg, dipl, 60. *Prof Exp:* Asst instr pharmacol, Univ Ill, 49-51; intern, Univ Calif Hosp, San Francisco, 51-52; resident surg, Univ Pa, 52-58, from asst instr to instr, 52-58; from instr to asst prof, Univ Colo, 58-61; prof, Univ Calif, Los Angeles, 61-67; PROF SURG & CHMN DEPT, SCH MED, UNIV CALIF, SAN DIEGO, 67- *Concurrent Pos:* Fel, Harrison Dept Surg, Univ Pa, 52-58, Nat Cancer Inst Inst trainee, 56-58; res grants, Univ Pa, Univ Colo & Univ Calif, 55-74; Markle scholar acad med, 59-64; consult, Vet Admin Hosps, Denver & Grand Junction, Colo & Albuquerque, NMex, 58-61; chief surg, Harbor Gen Hosp, Calif, 61-67; lectr, Univs Chicago, Tex & Boston, 65, Univ Colo, 66 & Albert Einstein Col Med, 68; consult, US Naval Hosp, San Diego, 66- & Nat Bd Med Examr, 67-70; mem clin res training comt, NIH, 66-, surg study sect, 68-; Francis M Smith lectr, Scripps Clin & Res Found, 67; Samuel Lillienthal vis chief surg, Mt Zion Hosp, San Francisco, 68; Michael & Janie Miller vis prof, Univ

Witwatersrand, 70; Edward Pierson Richardson vis prof, Harvard Univ & Mass Gen Hosp, 71; vis prof numerous univs, cols & founds, 67-78; ed-in-chief, World J Surg. *Mem:* AAAS; Am Surg Asn; Am Col Surg; Soc Univ Surg (pres, 71-72); Am Gastroenterol Asn. *Res:* Liver physiology and disease; transplantation; gastrointestinal physiology and surgery; adrenal physiology; metabolism; shock; neuropharmacology; cancer and vascular surgery; diabetes; microsurgery. *Mailing Add:* Univ Calif Med Ctr 225 Dickinson San Diego CA 92103

ORLOFF, NEIL, b Chicago, Ill, May 9, 43. SCIENCE POLICY, ENVIRONMENTAL SCIENCES. *Educ:* Mass Inst Technol, BS, 64; Harvard Univ, MBA, 66; Columbia Univ, JD, 69. *Prof Exp:* Opers officer, World Bank, 69-71; dir regional liason, US Environ Protection Agency, 71-73; legal counsel environ law, President's Coun Environ Qual, 73-75; from assoc prof to prof environ law & policy, Cornell Univ, 75-88, dir, Ctr Environ Res, 84-87; coun, 86-89, PARTNER, ENVIRON PRAC GROUP, 89- *Concurrent Pos:* Fac lectr, Environ Law Inst, 76-77, Cornell Law Sch, 77-79 & US Environ Protection Agency, 83; adv, Int Joint Comn, 79-81, NY Dept Environ Conserv, 84-; gov bd, Cornell Ctr Environ Res, 79-82, NY Sea Grant Inst, 84-87. *Res:* The design of government programs for the protection of the environment; analysis of efficiency and effectiveness of alternate approaches for reducing human exposure to toxic substances. *Mailing Add:* Irell & Mandella 1800 Avenue of the Stars Suite 900 Los Angeles CA 90067

ORMAN, CHARLES, b Wichita, Kans, July 16, 44. POST HARBORS CITRUS RESEARCH. *Educ:* Calif State Univ, BS, MS. *Prof Exp:* MGR, FRUIT SCI DEPT, SUNKIST GROWERS, INC, 77- *Concurrent Pos:* Mem bd dirs, Citrus Qual Coun. *Mem:* Am Hort Soc. *Mailing Add:* Sunkist Growers Inc Box 3720 760 E Sunkist St Ontario CA 91761

ORME-JOHNSON, NANETTE ROBERTS, b San Antonio, Tex, Apr 11, 37. PHYSICAL BIOCHEMISTRY, ENDOCRINOLOGY. *Educ:* Univ Tex, Austin, BA, 59; Univ Wis-Madison, PhD(biochem), 73. *Prof Exp:* Prin res scientist, dept chem, Mass Inst Technol, 80-82; asst prof, 82-87, ASSOC PROF BIOCHEM, SCH MED, TUFTS UNIV, 87- *Mem:* Am Chem Soc; Sigma Xi; Endocrine Soc; Am Soc Biol Chemists. *Res:* Delineation of the subcellular signalling mechanisms that allow peptide hormones to control steroid hormone biosynthesis, by controlling the rate of a reaction that occurs in mitochondria of steroidogenic tissues. *Mailing Add:* Dept Biochem Health Sci Campus Tufts Univ 136 Harrison Ave Boston MA 02111

ORME-JOHNSON, WILLIAM H, b Phoenix, Ariz, Apr 23, 38. ENZYMOLOGY, INORGANIC BIOCHEMISTRY. *Educ:* Univ Tex, Austin, BSc, 59, PhD(chem), 64. *Prof Exp:* Fel, Biochem Inst, Univ Tex, Austin, 64-65; fel, Inst Enzyme Res, Univ Wis, 65-67, asst prof enzyme chem, 67-70, from asst prof to prof biochem, 70-79; PROF CHEM, MASS INST TECHNOL, 79- *Mem:* Am Soc Biol Chemists; Am Chem Soc; Am Soc Microbiologists; fel AAAS. *Res:* Molecular mechanisms of enzyme catalysis, especially electron-transfer reactions, including, nitrogen fixation, steroid hormone synthesis and methanogenesis. *Mailing Add:* Dept Chem (18-407) Mass Inst Technol 77 Mass Ave Cambridge MA 02139

ORMES, JONATHAN FAIRFIELD, b Colorado Springs, Colo; m 64; c 3. PARTICLE ASTROPHYSICS. *Educ:* Stanford Univ, BS, 61; Univ Minn, MS, 66, PhD(physics), 67. *Prof Exp:* Nat Acad Sci-Nat Res Coun resident res assoc, NASA, 66-68, astrophysicist & head cosmic ray group, 68-82, actg chief, Lab High Energy Astrophys, 82-83, High Energy Astrophys Br, NASA Hq, 83-84, head, Cosmic Radiations Br, 84-85, head, Nuclear Astrophys Br, 85-87, assoc chief, Lab High Energy Astrophysics, 87-90, CHIEF, LAB HIGH ENERGY ASTRO PHYSICS, GODDARD SPACE FLIGHT CTR, NASA, 90- *Concurrent Pos:* Secy, High Energy Astrophys Mgt Opers Working Group, 74-76; vis assoc prof, Univ Minn, Minneapolis, 77; co-prin investr, Transition Radiation and Ionization Calorimeter for Spacelab, 79-; proj scientist, Space Sta Attached Payloads, 85-; study scientist, Particle Astrophys Mgt Facil, 85- *Honors & Awards:* Except Serv Medal, NASA, 86. *Mem:* Fel Am Phys Soc; Am Geophys Union; Am Astrophys Soc; AAAS. *Res:* Particle astrophysics, studying antiprotons, positrons, electrons and composition and energy spectra of high energy cosmic rays to understand acceleration, propagation and nucleosynthesis of galactic material. *Mailing Add:* Code 660 NASA Greenbelt MD 20771

ORMROD, DOUGLAS PADRAIC, b Langley, BC, May 27, 34; m 57; c 3. PLANT PHYSIOLOGY, AIR POLLUTION. *Educ:* Univ BC, BSA, 56; Univ Calif, PhD(plant physiol), 59. *Prof Exp:* Instr agron, Univ Calif, 59-60; from asst prof to prof plant sci, Univ BC, 60-69; PROF HORT SCI, 69-, DEAN GRAD STUDIES, UNIV GUELPH, 86- *Mem:* Am Soc Hort Sci; Air & Waste Mgt Asn. *Res:* Growth and development of plants, especially as affected by pollution stresses. *Mailing Add:* Dept Hort Sci Univ Guelph Guelph ON N1G 2W1 Can

ORMSBEE, ALLEN I(VES), b Reno, Nev, Aug 20, 26; m 46; c 3. AERONAUTICAL ENGINEERING. *Educ:* Univ Ill, BS, 46, MS, 49; Calif Inst Technol, PhD(aeronaut, math), 55. *Prof Exp:* From instr to assoc prof, Univ Ill, 46-57, head, Aviation Res Lab, 77-79, prof aviation, 72-88, PROF AERONAUT & ASTRONAUT ENG, UNIV ILL, 57-, ASSOC DIR, INST AVIATION, 88- *Concurrent Pos:* Res physicist, Hughes Aircraft Co, 52-54; consult, Gen Dynamics/Convair, 56-57; design specialist, Douglas Aircraft Co, 57-59, consult, Missiles & Space Systs, 57-78; mem bd aeronaut adv, Ill State Dept Aeronaut, 69-73, Eng Accreditation Comn, 84- *Mem:* Assoc fel Am Inst Aeronaut & Astronaut; Am Soc Eng Educ. *Res:* Aerodynamics and dynamics of guided missiles; spacecraft and aircraft. *Mailing Add:* Dept Aeronaut & Astronaut Eng 102 Transp Bldg 104 S Mathews Ave Urbana IL 61801-2997

ORMSBY, ROBERT B, JR, ENGINEERING. *Prof Exp:* RETIRED. *Mem:* Nat Acad Eng. *Mailing Add:* 23961 Wildwood Canyon Rd Newhall CA 91321

ORMSBY, W(ALTER) CLAYTON, b Alfred, NY, Nov 13, 24; m 59; c 2. CERAMICS. *Educ:* Alfred Univ, BS, 48; Pa State Univ, MS, 51, PhD(ceramics), 55. *Prof Exp:* Asst, Pa State Univ, 48-55; ceramic engr & phys chemist, Nat Bur Standards, 55-65; res chemist, 65-76, SUPVRY RES CHEMIST, US DEPT TRANSP, FED HWY ADMIN, 76- *Mem:* Am Ceramic Soc; Mineral Soc Gt Brit & Ireland. *Res:* Stabilization of soils; rheology and mineralogy of clays; surface chemistry of inorganic solids; chemical and mineralogical properties of highway materials. *Mailing Add:* Fed Hwy Admin HNR-30 6300 Georgetown Pike McLean VA 22101

ORNA, MARY VIRGINIA, b Newark, NJ, July 4, 34. PHYSICAL CHEMISTRY, ANALYTICAL CHEMISTRY. *Educ:* Chestnut Hill Col, BS, 55; Fordham Univ, MA, 58, PhD(phys chem), 62; Catholic Univ, MA, 67. *Prof Exp:* Jr chemist, Hoffmann-LaRoche, Inc, NJ, 55-56; teacher, Acad Mt St Ursula, 58-61; instr chem, Bronx Community Col, 61-62; PROF & CHMN DEPT INORG & ANAL CHEM, COL NEW ROCHELLE, 66- *Honors & Awards:* Am Cyanamid Co Award, 60-61; Catalyst Award, Chem Mfrs Asn, 84. *Mem:* Am Chem Soc; Soc Appl Spectros; Sigma Xi. *Res:* Chelation in mixed solvents; inorganic infrared spectroscopy. *Mailing Add:* Dept Chem Col New Rochelle New Rochelle NY 10801

ORNATO, JOSEPH PASQUALE, b New Haven Conn, Dec 30, 47; m 69. MEDICINE. *Educ:* Boston Univ, BA & MD, 71. *Prof Exp:* Internal med resident, Mt Sinai Hosp, NY, 71-74; Cardiol fel, NY Hosp, Cornell Univ Med Ctr, NY, 74-76; maj, US Army Med Corp, Ft Eustis, Va, 77-78; asst prof internal med & dir emergency med serv, Univ Nebr Med Ctr, Omaha, 79-85; assoc prof internal med cardiol & chief Internal Med Sect Emergency Med Serv, 85-87, PROF INTERNAL MED, MED COL VA, 88- *Concurrent Pos:* chmn, advan cardiac life support, Am Heart Asn, 88- *Mem:* Fel Am Col Cardiol; Am Heart Asn; Am Col Emergency Physicians. *Res:* Cardiopulmonary resuscitation; pharmacological and electrical therapy during cardiac arrest; emergency and services delivery systems; prehospital cardiac care. *Mailing Add:* Med Col Va Box 525 Richmond VA 23298

ORNDORFF, ROY LEE, JR, b Washington, DC, Aug 10, 35; m 59; c 5. AUTOMOTIVE ENGINEERING, TRIBOLOGY. *Educ:* Va Polytech Inst, BS, 57; Purdue Univ, MS, 58. *Prof Exp:* Lab instr mech eng, Purdue Univ, 57-58; from engr to sr res engr, B F Goodrich Res Ctr, 60-72, sr prod engr, Eng Systs Div, 72-78, sect leader, Eng Prod Group, 78-82, tech mgr, 82-91, MGR RES & DEVELOP, B F GOODRICH, 91- *Mem:* Am Soc Mech Engrs; Soc Automotive Engrs. *Res:* Rubber and plastic new products; wet and dry non-metallic bearings; dry dynamic seals; rubber friction and wear; composites; applied thermodynamics; inflatable products. *Mailing Add:* 7827 Birchwood Dr Kent OH 44240-6255

ORNDUFF, ROBERT, b Portland, Ore, June 13, 32. BOTANY. *Educ:* Reed Col, BA, 53; Univ Wash, MSc, 56; Univ Calif, Berkeley, PhD(bot), 61. *Prof Exp:* Asst prof biol, Reed Col, 61-62; asst prof bot, Duke Univ, 62-63; from asst prof to assoc prof, 63-71, assoc dir, Univ Bot Garden, 71-73, dir, Jepson Herbarium & Libr, 68-82 & Univ Herbarium, 75-82, PROF BOT, UNIV CALIF, BERKELEY, 69-, DIR, UNIV BOT GARDEN, 73- *Mem:* AAAS; Bot Soc Am; Soc Study Evolution; Am Soc Plant Taxon. *Res:* Biosystematics of angiosperms; reproductive ecology of seed plants. *Mailing Add:* Dept Integrative Biol Univ Calif Berkeley CA 94720

ORNE, MARTIN THEODORE, b Vienna, Austria, Oct 16, 27; US citizen; m 62; c 2. PSYCHIATRY, PSYCHOLOGY. *Educ:* Harvard Col, BA, 48, Harvard Univ, AM, 51, PhD, 58; Tufts Univ, MD, 55. *Prof Exp:* USPHS fel, Boston Psychopath Hosp, 56-57; lectr social rels, Harvard Univ, 58-59, res assoc, 59-60; instr psychiat, Harvard Med Sch, 59-62, assoc, 62-64; assoc prof, 64-67, PROF PSYCHIAT, MED SCH, UNIV PA, 67-; DIR UNIT EXP PSYCHIAT & SR ATTEND PSYCHIATRIST, INST PA HOSP, 64- *Concurrent Pos:* Dir, Studies in Hypnosis & Human Ecol Projs, Mass Ment Health Ctr, 59-64; Fulbright scholar, Univ Sydney, Australia, 60; ed, Int J Clin & Exp Hypnosis, 62; consult, Surg Gen Study Sect, NIMH, 66-77 & 80-84. *Honors & Awards:* Bernard B Raginsky Award, Soc Clin & Exp Hypnosis, 69; Benjamin Franklin Gold Medal, Int Soc Hypnosis, 82; Distinguished Sci Award Applns Psychol, Am Psychol Asn, 86. *Mem:* Am Psychiat Asn; Am Psychol Asn; Int Soc Hypnosis (pres, 76-79); Am Psychosomatic Soc; Soc Clin & Exp Hypnosis (pres, 71-73); hon mem Royal Soc Med. *Res:* Nature of hypnosis and special states of consciousness, objectification of subjective events, effects of the observational context on data in social and experimental psychology, psychotherapy and psychophysiology. *Mailing Add:* Dept Psychiat Inst Pa Hosp Philadelphia PA 19139-2798

ORNITZ, BARRY LOUIS, b Orlando, Fla, July 23, 49; m 81. ELECTRICAL PROPERTIES OF MATERIALS. *Educ:* Clemson Univ, BS, 70, MS, 74, PhD(chem eng), 80. *Prof Exp:* Res chem engr, 80-85, SR RES CHEM ENGR, EASTMAN CHEM DIV, EASTMAN KODAK CO, 85- *Mem:* Am Inst Chem Engrs; sr mem Instrument Soc Am. *Res:* Design and development of instrumentation for process control; inferential electrical measurements of chemical composition in industrial environments; analog electronic circuit design. *Mailing Add:* 4740 Edens View Rd Kingport TN 37664

ORNSTEIN, DONALD SAMUEL, b New York, NY, July 30, 34; m 65; c 3. PURE MATHEMATICS. *Educ:* Univ Chicago, PhD(math), 56. *Prof Exp:* PROF MATH, STANFORD UNIV, 60- *Honors & Awards:* Bocher Prize, Am Math Soc. *Mem:* Nat Acad Sci. *Res:* Ergodic theory. *Mailing Add:* Dept Math Stanford Univ Stanford CA 94305

ORNSTEIN, LEONARD, b New York, NY, Feb 8, 26; m 45; c 4. CELL BIOLOGY, BIOPHYSICS. *Educ:* Columbia Univ, AB, 48, AM, 49, PhD(zool), 57. *Prof Exp:* Asst zool, Columbia Univ, 49-52, asst cytol, 50-52, instr histol, 51-52; res assoc path, 54-67, DIR CELL RES LAB, MT SINAI HOSP, NEW YORK, 54-, PROF PATH, MT SINAI SCH MED, 66- *Concurrent Pos:* Res assoc, Columbia Univ, 52-64; consult, Am Cyanamid Co, 57-58, Canal Indust Corp, 62-69, Space-Gen Corp, 65-66, Airborne

Instrument Corp, 65-67, Farrand Optical Co, 66-67, IBM Watson Labs, 67-68 & Technicon Corp, 69-; vis prof, Harvard Univ, 67-68; mem develop biol panel, NSF, 70-73. *Mem:* Am Soc Cell Biol; Histochem Soc. *Res:* Techniques and instrumentation in microspectrophotometry, phase, interference and electron microscopy; microcinematography; enzyme cytochemistry; microchemistry; fluorescence microscopy; freeze-substitution; microtomy; electrophoresis; information theory and pattern recognition. *Mailing Add:* S Biltom Rd White Plains NY 10607

ORNSTEIN, WILHELM, b Jaroslaw, Poland, Nov 9, 05; US citizen; m 46. APPLIED MATHEMATICS, MATHEMATICAL ANALYSIS. *Educ:* Tech Hochsch, Brunn, MS, 29; Tech Univ, Berlin, DrIng, 33. *Prof Exp:* Tech secy to dir, Gen Motors, Poland, 34-39; prof mech, Turkish Mil Acad, Istanbul, 40-42; tech adv, Mid East Hq, US Army, Cairo, 43-46; lectr mech, Univ Pa, 47-48; from assoc prof to prof mech eng, Newark Col, 48-54; prof, 55-74, EMER PROF APPL MATH, WASH UNIV, 74- *Concurrent Pos:* Vis prof, Stevens Inst Technol, 50-52. *Honors & Awards:* Medal of Freedom, War Dept, 48. *Mem:* Am Math Soc. *Mailing Add:* Dept Syst Sci & Math Wash Univ St Louis MO 63130

ORNSTON, LEO NICHOLAS, b Philadelphia, Pa, Jan 7, 40; m 74; c 2. MICROBIAL PHYSIOLOGY, BIOCHEMISTRY. *Educ:* Harvard Univ, BA, 61; Univ Calif, Berkeley, PhD(comp biochem), 65. *Prof Exp:* Res assoc bact, Univ Calif, Berkeley, 65-66; NIH fel biochem, Univ Leicester, 66-68; fel, Univ Ill, Urbana, 68-69; asst prof, 69-74, assoc prof, 74-81, PROF BIOL, YALE UNIV, 81- *Concurrent Pos:* John Simon Guggenheim Found fel, 73-74. *Mem:* Am Soc Biol Chem; Am Soc Microbiol; Brit Soc Gen Microbiol; Am Chem Soc. *Res:* Regulation of convergent and divergent pathways in bacteria; evolution of microbial enzymes. *Mailing Add:* Dept Biol Yale Univ 836 Kline Biol Tower New Haven CT 06520

ORNT, DANIEL BURROWS, b Rochester, NY, Jan 21, 51; m 76; c 3. NEPHROLOGY, RENAL PHYSIOLOGY. *Educ:* Colgate Univ, BA, 73; Univ Rochester, MD, 76. *Prof Exp:* Res fel nephrol, Univ Mich, 80-81; sr instr med, 81-82, ASST PROF MED, UNIV ROCHESTER, 82-, ASST PROF PEDIAT, 84- *Mem:* Am Soc Nephrol; Int Soc Nephrol; Am Physiol Soc; Am Fedn Clin Res; AAAS; NY Acad Sci. *Res:* Mechanisms which regulate potassium excretion by the kidney; reduction of potassium excretion during dietary restriction of potassium; metabolic changes which occur in the kidney during potassium depletion. *Mailing Add:* Sch Med Box Med Univ Rochester 601 Elmwood Ave Rochester NY 14642

ORO, JUAN, b Lerida, Spain, Oct 26, 23; m 49; c 4. BIOCHEMISTRY. *Educ:* Univ Barcelona, Lic, 47; Baylor Univ, PhD(biochem), 56. *Prof Exp:* From instr to assoc prof chem, 55-63, chmn dept biophys sci, 67-69, prof chem, 63-, AT DEPT BIOPHYS SCI, UNIV HOUSTON. *Concurrent Pos:* NASA grants, 62-, partic, Viking Mars Lander Molecular Anal Team, 69-; hon coun, Highest Coun Sci Res, Spain, 69; prof, Univ Barcelona, 71, pres sci coun, Inst Fundamental Biol, 71. *Mem:* AAAS; Am Chem Soc; Am Soc Biol Chem; Geochem Soc; Span Soc Biochem; Sigma Xi. *Res:* Mechanisms of enzyme action; synthesis of biochemical compounds; neurobiochemistry; application of biological principles for the improvement of the quality of human life; comprehensive study of carbonaceous, organic and organogenic matter in returned lunar samples; biochemical applications of gas chromatography and mass spectrometry; organic cosmochemistry and paleochemistry; molecular and biological evolution; origin of life. *Mailing Add:* Dept Biophys Biochem Sci SR I Univ Houston-University Park 4800 Calhoon Rd Houston TX 77004-8399

OROFINO, THOMAS ALLAN, b Youngstown, Ohio, Nov 12, 30; m 53. PHYSICAL CHEMISTRY, OLFACTORY SCIENCE. *Educ:* Kent State Univ, BS, 52; Cornell Univ, PhD(phys chem), 56. *Prof Exp:* NSF fel polymer chem, Univ Leiden, Holland, 56-57; sr chemist, Rohm & Haas, 57-58; fel, Mellon Inst, 58-64; group leader, Monsanto Co, Durham, NC, 64-70, res mgr, 70-74, res mgr, Dayton, Ohio, 74-79, res mgr, Dalton, Ga, 79-85; CONSULT, SCI & EDUC, 85- *Concurrent Pos:* Chmn, Chattanooga Sect, Am Chem Soc; adj prof, Univ Tenn, Chattanooga, dir, William H Wheeler Ctr Odor Res; technol consult & expert witness. *Mem:* Am Chem Soc. *Res:* Physical chemistry of polymers; thermodynamics of polymer solutions; separations and selective transport in polymer films and hollow fibers; optical fiber technology; recreational surfaces and artificial turf; chemical aspects of olfactory science. *Mailing Add:* Astroturf Indust 809 Kenner St Dalton GA 30720

ORONSKY, ARNOLD LEWIS, b New York, NY, Oct 22, 40; m 63; c 1. BIOCHEMISTRY, PHYSIOLOGY. *Educ:* NY Univ, AB, 61; Columbia Univ, PhD(physiol), 68. *Prof Exp:* Mgr inflammatory & immunol res, CIBA-GEIGY Res Labs, 71-; AT LEDRLE LAB, AM CYANAMID. *Concurrent Pos:* NIH res fel connective tissue biochem, Harvard Med Sch, 68-71, King Trust Found res scholar, 70-71; adj assoc prof orthopedics, Mt Sinai Sch Med, 73- *Res:* Biochemistry of connective tissue; immunology; inflammation. *Mailing Add:* Ledrle Labs N Middletown Rd Pearl River NY 10965

OROS, MARGARET OLAVA (ERICKSON), b Sheyenne, NDak, Oct 5, 12; m 51; c 1. PETROLEUM GEOLOGY. *Educ:* Univ NDak, BA, 39. *Prof Exp:* Lab instr geol, Univ NDak, 38-40; radio assembler, Bendix Aviation Co, Calif, 42-43; lab asst spectrog anal, Aluminum Co Am, 42-48; chem engr, Harvey Aluminum Co, 48-51; geologist, Ill State Geol Surv, 53-62; geologist, State Geol Surv, Kans, 62-78; GEOL CONSULT, 78- *Mem:* Sigma Xi; Am Asn Petrol Geol; fel Geol Soc Am. *Res:* Spectrography; metallurgical control of aluminum alloys; geological cross sections; field studies; petroleum statistical reports; crude oil reserves estimates; pipeline and petroleum industry map. *Mailing Add:* 913 Madeline Lane Lawrence KS 66049

OROSHNIK, JESSE, b New York, NY, May 12, 24; m 55; c 2. PHYSICS. *Educ:* City Col New York, BS, 48; Univ Fla, MS, 50. *Prof Exp:* Aeronaut res scientist, NASA, 54; physicist, Standard Piezo Co, 54-55; sr engr, Gen Tel & Electronics Labs, Inc, 55-61; sr engr res & develop labs, Westinghouse Elec Corp, 61-66; group leader, Adv Fabrication Tech Sect, Electronics Res Lab, Corning Glass Works, 66-68; physicist, Nat Bur Standards, Washington, DC, 68-71; physicist, Naval Weapons Eng Support Activity, 71-75; physicist, Naval Res Lab, US Navy, 75-79; physicist, Navy Primary Standards Lab, 79-87; PHYSICIST, NAT BUREAU OF STANDARDS, 87- *Concurrent Pos:* NY State Regents War Serv scholar, 57. *Mem:* Sigma Xi. *Res:* Adhesion and adhesive fracture of polymers in polymer-metal bonds; thin film adhesion, scanning electron microscopy and failure analysis. *Mailing Add:* 14800 Marlin Terrace Rockville MD 20853

OROSZ, CHARLES GEORGE, b Cleveland, Ohio, Jan 9, 49; m 74; c 3. TRANSPLANTATION, IMMUNOBIOLOGY. *Educ:* Cleveland State Univ, BS, 71, MS, 75, PhD(regulatory biol), 78. *Prof Exp:* Teaching asst, dept biol, Cleveland State Univ, 73-75; res fel, Cleveland Clinic Found & Cleveland State Univ, 75-78; teaching fel, Immunobiol Res Ctr, Univ Wis, 78-80; vis res fel, dept lab med & path, Univ Minn, 80-81, asst prof, 81-83; asst prof, Dept Surg & dir Therapeut Immunol Lab, 83-88, DIR HUMAN HISTOCOMPATIBILITY TESTING LAB, UNIV HOSP, 83-; ASSOC PROF, DEPT SURG & PATHOL, OHIO STATE UNIV, 88-, ASSOC PROF, DEPT PHARMACOL, 89- *Concurrent Pos:* Prin investr, NIH grants, biomed res, 82-; full mem, Comprehensive Cancer Care Ctr, 84-; ad hoc grants rev, div res grants, NIH & March of Dimes, 85-; bd dirs, Leukemia Soc Am Cent Ohio Chap, 84-; ad hoc J rev, J Immunol, Sci, Life Sci & Immunopharm, 85-; exec bd, Midwest Immunol Conf, 87-89. *Mem:* Am Soc Histocompatibility & Immunogenetics; Transplantation Soc; Am Asn Immunologists; InterAm Soc Chemother; AAAS; NY Acad Sci; Asn Med Lab Immunologists. *Res:* In vivo and in vitro immunoregulatory mechanisms which control human and murine T lymphocyte responses to tumors, viruses and grafts; regulation of T cell function by immunosuppressants such as cyclosporine or selected viral components; relationship between antigen-reactive T cell frequency and in vitro immunoreactivity in humans; regulation and behavior of grapt-infitrating lymphoid populations in mice and humans; clinical application of basic research findings to transplantation and cancer immunotherapy. *Mailing Add:* Dept Surg Div Transplantation Ohio State Univ Rm 258 Means Hall 1655 Upham Dr Columbus OH 43210

OROSZIAN, STEPHEN, b Hungary, July 8, 27; nat US; m 54; c 2. IMMUNOCHEMISTRY, BIOCHEMICAL PHARMACOLOGY. *Educ:* Budapest Tech Univ, Chem Eng, 50; Georgetown Univ, PhD(pharmacol), 60. *Prof Exp:* Res assoc, Biol Res Inst, Hungarian Acad Sci, 50-56; asst mem, Albert Einstein Med Ctr, 63-66; asst prof med, George Washington Univ, 66-68; head immunochem, Flow Labs, Inc, 68-78; HEAD, IMMUNOCHEM SECT, FREDERICK CANCER RES CTR, 78-; HEAD DIR, LAB MOLECULAR VIROL & CARCINOGENESIS, 86- *Concurrent Pos:* Res fel, Lab Chem Pharmacol, Nat Cancer Inst, 60-63. *Mem:* AAAS; Am Soc Pharmacol & Exp Therapeut; Int Soc Biochem Pharmacol; Am Soc Microbiol; Am Chem Soc. *Res:* Macromolecular biochemistry and its application to cancer research; viral proteins; immunochemistry of tumor viruses; viral immunol; immunology and chemotherapy. *Mailing Add:* Immunochem Sect Frederick Cancer Res Ctr PO Box B Frederick MD 21701

O'ROURKE, EDMUND NEWTON, JR, b New Orleans, La, Nov 22, 23; m 47; c 1. HORTICULTURE. *Educ:* Southwestern La Inst, BS, 48; Cornell Univ, MS, 53, PhD, 55. *Prof Exp:* Jr horticulturist, USDA, 48-51; asst pomol, Cornell Univ, 51-54; assoc prof, 54-59, PROF HORT, LA STATE UNIV, BATON ROUGE, 59- *Mem:* Am Soc Hort Sci. *Res:* Breeding and improvement of figs, pears, apples and other fruits for the Gulf Coast area; ecology of fruit varieties; commercial floriculture. *Mailing Add:* Dept Hort 231 Agr-Hort Bldg La State Univ Baton Rouge LA 70803

O'ROURKE, JAMES, b Trenton, NJ, Mar 2, 25; m 54; c 4. OPHTHALMOLOGY. *Educ:* Georgetown Univ, MD, 49; Univ Pa, MSc, 54. *Prof Exp:* Resident surgeon, 52-54; clin assoc ophthal, NIH, 54-55, chief clin res, 55-57; assoc prof, Sch Med, Georgetown Univ, 57-65, prof ophthal surg, 65-69; PROF PATH & DIR VISION-IMMUNOL CTR, HEALTH CTR, UNIV CONN, 84- *Concurrent Pos:* Res fel, Wills Eye Hosp, Pa, 51-52; consult, Oak Ridge Nat Lab, 55-57 & NIH, 57-, mem vision res & training comt, Nat Eye Inst, 71-75; adj prof eng, Trinity Col, Conn, 84-; armed forces vision res comt, Nat Res Coun. *Mem:* Asn Res Vision & Ophthal; Am Acad Ophthal & Otolaryngol; Am Ophthal Soc; Soc Nuclear Med; Sigma Xi; Cosmos Club. *Res:* Ocular vascular diseases; vascular biology and medicine. *Mailing Add:* Univ Conn Health Ctr Dept Path 263 Farmington Ave Farmington CT 06030

O'ROURKE, JOHN ALVIN, b Walsenburg, Colo, Feb 4, 24; m 44; c 6. X-RAY DIFFRACTION ANALYSIS, MECHANICAL METALLURGY. *Educ:* Colo Sch Mines, BE & ME, 50. *Prof Exp:* Staff mem, Gen Elec Co, 50-52; STAFF MEM, LOS ALAMOS NAT LAB, 52- *Mem:* AAAS; Am Soc Metals; Microprobe Analysis Soc Am. *Res:* Methods of evaluating crystalline textures of metals, alloys and ceramics; correlating these measurements with their anisotropic mechanical properties; preparation and characterization of new superconducting and magnetic intermetallics. *Mailing Add:* Group MST-5 Los Alamos Nat Lab MS G734 Los Alamos NM 87545

O'ROURKE, JOSEPH, THEORY, GEOMETRY. *Educ:* St Joseph's Univ, BS, 73; Univ Pa, MS, 77, PhD(computer sci), 80. *Prof Exp:* Physicist, Naval Air Develop Ctr, 73-77; grad res fel, Univ Pa, 77-80; from asst prof to assoc prof computer sci, Johns Hopkins Univ, 80-88; PROF COMPUTER SCI, SMITH COL, 88- *Concurrent Pos:* Guggenheim fel, 87; vis assoc prof computer sci, Johns Hopkins Univ, 88- *Honors & Awards:* Presidential Young Investr Award, 84. *Mem:* Asn Comput Mach; Asn Math Soc; Math Asn Am; Soc Indust & Appl Math; Sigma Xi. *Res:* Computational geometry. *Mailing Add:* Dept Computer Sci Smith Col Northampton MA 01063

O'ROURKE, PETER JOHN, b Pueblo, Colo, Apr 17, 46; m 75; c 2. COMPUTER MODELING OF FLUID FLOWS. *Educ:* Cath Univ Am, BA, 69; Stanford Univ, MS, 71; Princeton Univ, MS, 78, PhD(mech eng), 81. *Prof Exp:* STAFF MEM, LOS ALAMOS NAT LAB, 71- *Concurrent Pos:* Vis prof, Univ Marseilles, France, 84-85 & Ecole d'Ete, Paris, France, 89. *Honors & Awards:* Aren T Colwell Mem Prize, Soc Automotive Engrs, 87. *Res:* Fluid dynamics; computer modeling of fluid flows; formulate models and methods and implement these in large scale computations that investigate fluid flow phenomena. *Mailing Add:* Three La Rosa Ct Los Alamos NM 87544

O'ROURKE, RICHARD CLAIR, b Minneapolis, Minn, July 8, 30. PHILOSOPHY OF SCIENCE, BOTANY. *Educ:* Col St Thomas, BS, 53; Univ Minn, Minneapolis, MS, 58, PhD(philos sci, bot), 66. *Prof Exp:* PROF BIOL & PHILOS SCI, WINONA STATE UNIV, 65- *Concurrent Pos:* Staff mem, Univ Minn, 71-72. *Mem:* AAAS; John Dewey Soc; Inst Relig in Age Sci. *Res:* Logic of the empirical sciences; evolution. *Mailing Add:* Dept Biol Winona State Univ 8th & Johnson Winona MN 55987

O'ROURKE, THOMAS DENIS, b Pittsburgh, Pa, July 31, 48; m 78. CIVIL & GEOTECHNICAL ENGINEERING. *Educ:* Cornell Univ, BS, 70; Univ Ill, MS, 73, PhD(civil eng), 75. *Prof Exp:* Engr, Dames & Moore Consult Engrs, 70; res asst eng, Univ Ill, 70-75, asst prof, 75-78; from asst prof to assoc prof, 78-87, PROF ENG, CORNELL UNIV, 87- *Honors & Awards:* C A Hogentogler Award, Am Soc Testing & Mat, 76; Collingwood Prize, Am Soc Civil Engrs, 83 & Huber Res Prize, 88. *Mem:* Am Soc Civil Engrs; Am Soc Mech Engrs; Am Soc Testing & Mat; Earthquake Eng Res Inst. *Res:* Underground construction with emphasis on construction in urban areas; structural interaction with soil and rock; ground failure; buried pipelines; soil and rock instrumentation; earthquake engineering. *Mailing Add:* 265 Hollister Hall Cornell Univ Ithaca NY 14853

OROWAN, EGON, physics; deceased, see previous edition for last biography

ORPHAN, VICTOR JOHN, b Norfolk, Va, Oct 7, 40; m 68; c 2. NUCLEAR PHYSICS & ENGINEERING. *Educ:* Univ Va, BME, 62; Mass Inst Technol, SM, 64, ScD(nuclear eng), 67. *Prof Exp:* Mgr, Nuclear Technol Br, Gulf Radiation Technol, San Diego, 67-80; VPRES, INSTRUMENTATION RES DIV, SCI APPL INC, 80-, CORP VPRES, 90- *Mem:* Am Phys Soc; Am Nuclear Soc. *Res:* Neutron capture gamma-ray spectroscopy; inelastic neutron scattering cross sections. *Mailing Add:* 1138 Eolus St Leucadia CA 92024

ORPHANIDES, GUS GEORGE, b Kew Gardens, NY, Jan 27, 47; m 68; c 3. EMULSION POLYMER CHEMISTRY. *Educ:* Hobart Col, BS, 67; Ohio State Univ, PhD(org chem), 72. *Prof Exp:* First lieutenant, US Army Foreign Sci & Technol Ctr, 72-74; chemist, Elastomer Chem Dept, E I Du Pont de Nemours & Co, Inc, 74-81; sr prin res chemist, 81-83, RESEARCH MGR, POLYMER CHEM DIV, AIR PROD & CHEM, INC, 83- *Concurrent Pos:* Chemist, US Army Tech Working Group Org Mat, 72-74. *Mem:* Am Chem Soc. *Res:* Structure property relationships in organic polymers; elastomer curing chemistry; general organic synthesis and chemistry of divalent carbon; emulsion polymer synthesis and applications. *Mailing Add:* 3460 Highland St Allentown PA 18104

ORPHANOS, DEMETRIUS GEORGE, b Naxos, Greece, Dec 22, 22; US citizen; m 61; c 2. ORGANIC CHEMISTRY. *Educ:* Nat Univ Athens, BSc, 55; McGill Univ, PhD(org chem), 63. *Prof Exp:* Anal chemist, Nemea Gen Chem Lab, Greece, 55-56 & Lorado Uranium Mills, Ltd, Can, 57-58; res assoc org res, Mass Inst Technol, 62-64; group leader, Can Tech Tape Ltd, 64-67; sr org chemist, Tracerlab-Lab for Electronics, Inc, 67; group leader org chem, New Eng Nuclear Corp, 67-74, sr res chemist, 74-80; sr res chemist, E I Du Pont De Nemours & Co, Inc, 80-86; RETIRED. *Mem:* Am Chem Soc. *Res:* Analytical chemistry; synthesis of natural products; heterocyclics; propolymer heterocyclic systems; biologically active compounds; organic peroxides; radiochemicals. *Mailing Add:* 25 Towne Crest Dr Stoneham MA 02180

ORPHANOUDAKIS, STELIOS CONSTANTINE, b Rethymno, Crete, Nov 13, 48; m 74; c 2. MEDICAL IMAGING, DIGITAL IMAGE PROCESSING. *Educ:* Dartmouth Col, BA, 71, PhD(biomed eng), 76; Mass Inst Technol, MS, 73. *Prof Exp:* Med physicist med physics bioeng, Yale-New Haven Hosp, 75-76; asst clin prof diagnostic radiol, Sch Med, Yale Univ, 76-77, asst prof, 77-78, asst prof diagnostic radiol, eng & appl sci, 78-81, assoc prof diagnostic radiol & elec eng, 81-91, RES AFFILIATE, SCH MED, YALE UNIV, 91-; PROF & CHMN COMPUTER SCI, UNIV CRETE, GREECE, 91- *Concurrent Pos:* Mem med staff, Yale-New Haven Hosp, 79- *Mem:* NY Acad Sci; Am Asn Physicists Med; sr mem Inst Elec & Electronic Engrs; AAAS. *Res:* Digital signal processing; image processing; image analysis applications in medical imaging: computed tomography, ultrasound, digital radiography. *Mailing Add:* Dept Diagnostic Radiol Yale Univ 333 Cedar St New Haven CT 06510-3219

ORPURT, PHILIP ARVID, b Peru, Ind, Aug 9, 21; m 45; c 5. MYCOLOGY. *Educ:* Manchester Col, BA, 48; Univ Wis, MS, 50, PhD(bot), 54. *Prof Exp:* Instr bot & zool, Wausau Exten, Univ Wis, 50-51; PROF BIOL, MANCHESTER COL, 54-, CHMN BIOL DEPT, 80- *Concurrent Pos:* NIH fel, Inst Marine Sci, Univ Miami, 62-63. *Mem:* Mycol Soc Am; Am Inst Biol Sci; Int Oceanog Found. *Res:* Soil and marine microfungi; mycotoxins; paleoecology. *Mailing Add:* 303 Damron Dr North Manchester IN 46962

ORR, ALAN R, b Des Moines, Iowa, May 13, 36; div; c 3. BOTANY, CELL BIOLOGY. *Educ:* Simpson Col, BA, 61; Purdue Univ, MS, 64, PhD(bot), 66. *Prof Exp:* From asst prof to assoc prof, 65-78, PROF BIOL, UNIV NORTHERN IOWA, 78- *Mem:* Bot Soc Am; AAAS; Nat Asn Biol Teachers; Sigma Xi; Am Soc Cell Biol. *Res:* Developmental botany, morphogenesis of reproductive organs, particularly biochemical events associated with morphological changes in meristematic tissue. *Mailing Add:* Dept Biol Univ Northern Iowa Cedar Falls IA 50614

ORR, BEATRICE YEWER, b Milwaukee, Wis, Jan 4, 51; m 77. ANATOMY, TERATOLOGY. *Educ:* Univ Denver, BS, 73; Med Col Wis, PhD(anat), 78. *Prof Exp:* Fel teratology, Am Dent Asn Res Inst, 78-80; res asst, Wash Univ, St Louis, 82-84; DIR ANAT, TAHOE FRACTURE CLIN, 90- *Concurrent Pos:* Adj instr histol, Stritch Sch Med, Loyola Univ, 78-90. *Mem:* Teratology Soc; Am Asn Lab Animal Sci. *Res:* Sirenomelia; craniofacial abnormalities; neural crest; scanning electron microscopy. *Mailing Add:* PO Box 731004 South Lake Tahoe CA 96158

ORR, CHARLES HENRY, b Griffin, Ga, May 24, 24; m 52; c 6. COMPUTER SCIENCE. *Educ:* Ohio State Univ, BSc, 48, MSc, 49; Syracuse Univ, PhD(chem), 53. *Prof Exp:* Asst instr, Syracuse Univ, 50-51; res chemist, Procter & Gamble, Co, 53-69, design specialist compt lang, 70-87; RETIRED. *Mem:* Am Chem Soc; Sigma Xi. *Res:* Introduction of new computer languages; computer application to automatic data reduction; use of Boolean logic in computer languages; decision tables. *Mailing Add:* 1831 Greenpine Dr Cincinnati OH 45231

ORR, CLYDE, JR, b Lewisburg, Tenn, Oct 1, 21; m 44; c 4. CHEMICAL ENGINEERING. *Educ:* Univ Tenn, BS, 44, MS, 48; Ga Inst Technol, PhD(chem eng), 53. *Prof Exp:* Chem engr, Wilson Dam, Tenn Valley Authority, 46-47; asst, Eng Exp Sta, 48-51, res engr, 51-58, assoc prof 53-58, res prof, 58-62, prof, 62-66, Regents Prof, 66-80, EMER PROF CHEM ENG, GA INST TECHNOL, 80- *Concurrent Pos:* Chmn, Bd Dirs, Micromeritics Instrument Corp, Ga. *Mem:* Am Chem Soc; Am Inst Chem Engrs. *Res:* Surface chemistry; aerosols; heat transfer; instrumentation. *Mailing Add:* 5091 Hidden Branches Circle Dunwoody GA 30338

ORR, DONALD EUGENE, JR, b Kokomo, Ind, Jan 17, 45; m 67; c 2. ANIMAL SCIENCE. *Educ:* Purdue Univ, West Lafayette, BS, 67; Pa State Univ, MS, 69; Mich State Univ, PhD(animal husb, nutrit), 75. *Prof Exp:* Fac res asst animal sci, Pa State Univ, 67-69; swine res specialist, Feed Res Div, Cent Soya Co, 74-75; asst prof, 75-81, assoc prof animal sci, Tex Tech Univ, 81-; AT LAKE HEAD UNIV. *Mem:* Am Soc Animal Sci; Sigma Xi. *Res:* Nutrition of the newborn and early-weaned pig; nutrition of the gestating and lactating sow; swine housing and management. *Mailing Add:* Box 108 Sheridan IN 46069

ORR, FRANKLIN M, JR, b Baytown, Tex, Dec 27, 46; m 70; c 2. CARBON DIOXIDE FLOODING, PHASE EQUILIBRIA. *Educ:* Stanford Univ, BS, 69; Univ Minn, PhD(chem eng), 76. *Prof Exp:* Res engr, Shell Develop Co, 76-78; sr engr, N Mex Petrol Recovery Res Ctr, N Mex Inst Mining & Tech, 78-85; assoc prof, 85-87, PROF PETROL ENG, DEPT PETROL ENG, STANFORD UNIV, 87- *Concurrent Pos:* Vis res fel, Heriot-Watt Univ, Edinburgh, Scotland, 81 & Univ New S Wales, Sydney, Australia, 85; mem sci adv bd, David & Lucile Packard Fel Sci & Eng, 88-; mem bd dirs, Monterey Bay Aquarium Res Inst & Wolf Trap Found Performing Arts, 88-; distinguished lectr, Soc Petrol Engrs, 90. *Mem:* Soc Petrol Engrs; Am Inst Chem Engrs; Soc Indust & Appl Math; Sigma Xi. *Res:* Measurement and prediction of high pressure phase equilibria of hydrocarbon and carbon dioxide-hydrocarbon mixtures; interactions of phase behavior and multiphase flow in porous media. *Mailing Add:* Petrol Eng Dept Mitchell Bldg Rm 360 Stanford Univ Stanford CA 94305

ORR, FREDERICK WILLIAM, b Vancouver, BC, Nov 1, 42; c 3. CANCER METASTASIS. *Educ:* Univ Alta, MD, 66. *Prof Exp:* Asst prof, Univ Manitoba, 80-84; assoc prof, 84-88, PROF PATH, MCMASTER UNIV, 88- *Mem:* Am Asn Pathologists; Int Acad Path; Can Asn Pathologists; Soc Clin Invest. *Mailing Add:* Dept Path McMaster Univ 1200 Main St W Hamilton ON L8N 3Z5

ORR, HENRY PORTER, b Opelika, Ala, Aug 20, 21. FLORICULTURE. *Educ:* Ala Polytech Inst, BS, 42; Ohio State Univ, MS, 47, PhD, 62. *Prof Exp:* From asst prof to assoc prof, 47-62, PROF HORT, AUBURN UNIV, 62- *Mem:* Am Soc Hort Sci. *Res:* Watering methods; leaf abscission of Azaleas; nutrition of woody ornamental plants; marketing of woody ornamentals. *Mailing Add:* 2811 Frederick Rd Opelika AL 36801

ORR, JACK EDWARD, b Delphi, Ind, Dec 11, 18; m 42; c 1. PHARMACY. *Educ:* Purdue Univ, BS, 40; Univ Wis, PhD(pharmaceut chem), 43. *Prof Exp:* Instr pharm, Ohio State Univ, 43-44, asst prof, 46-47; prof pharmaceut chem, Univ Utah, 47-52; prof pharmaceut chem & dean sch pharm, Mont State Univ, 52-56; Wash State chemist, Univ Wash, 56-78, dean, Sch Pharm, 56-78, from prof to emer prof pharm, 56-90; RETIRED. *Mem:* Am Pharmaceut Asn; Am Asn Cols Pharm (pres, 64-65). *Res:* Pharmaceutical education; history of pharmacy. *Mailing Add:* Sch Pharm Univ Wash SC-69 Seattle WA 98195

ORR, JAMES ANTHONY, b Madison, Wis, Sept 9, 48; m 71; c 1. MAMMALIAN PHYSIOLOGY, CARDIOPULMONARY PHYSIOLOGY. *Educ:* Loras Col, BS, 70; Univ Wis-Madison, PhD(vet sci physiol), 74. *Prof Exp:* Asst scientist cardiopulmonary physiol, Univ Wis-Madison, 74-75; asst prof, 75-80, ASSOC PROF PHYSIOL & CELL BIOL, UNIV KANS, 80- *Concurrent Pos:* Estab investr, Am Heart Asn, 81-86. *Mem:* Am Soc Vet Physiologists & Pharmacologists (pres-elect, 85-86); Am Physiol Soc. *Res:* Chemical control of ventilation, specifically the role of chemoreceptors in health and disease; cerebral circulation and response of this circulatory bed to hyperoxia and hypoxia. *Mailing Add:* Dept of Physiol & Cell Biol Univ of Kans Lawrence KS 66045

ORR, JAMES CAMERON, b Paisley, Scotland, Aug 10, 30; US citizen; m 59; c 2. BIOCHEMISTRY, ORGANIC CHEMISTRY. *Educ:* Univ London, BSc & ARCS, 54; Univ Glasgow, PhD(chem), 60. *Prof Exp:* Asst lectr chem, Univ Glasgow, 55-57; res chemist, Syntex SA, Mex, 59-63; from assoc to assoc prof biol chem, Harvard Med Sch, 63-75, tutor biochem sci, Harvard Univ, 63-75; PROF BIOCHEM & CHEM, BASIC MED SCI, MEM UNIV NFLD, 75- *Concurrent Pos:* Royal Air Force & Air Sea Rescue. *Mem:* Can Biochem Soc; Am Chem Soc; Am Soc Biol Chem; Endocrine Soc. *Res:* Mechanisms of organic chemical reactions, particularly those involving steroids and enzymes. *Mailing Add:* Basic Med Sci Mem Univ Nfld St John's NF A1C 5S7 Can

ORR, JOHN R, b Waukon, Ohio, Jan 16, 33. PHYSICS, HIGH ENERGY PHYSICS. *Educ:* Univ Ohio, BS, 55; Univ Wash, PhD(physics), 65. *Prof Exp:* Res asst, Boeing Airlines, 56-60; res prof high energy physics, Univ Ill, 60-69; HEAD MAGNETIC & TEST, FERMI LAB, 69- *Mem:* Am Phys Soc; AAAS. *Mailing Add:* Fermi Lab PO Box 500 M/S 306 Batavia IL 60510

ORR, LEIGHTON, b Pittsburgh, Pa, Feb 11, 07; m 32; c 2. MECHANICAL ENGINEERING. *Educ:* Univ Pittsburgh, BS, 28. *Prof Exp:* Test engr, Pittsburgh Testing Lab, 30-36; res engr, PPG Industs Inc, 36-50, head phys testing, 50-72; TECH GLASS CONSULT, 72- *Concurrent Pos:* Consult to various industs on glass problems. *Mem:* Fel Am Soc Mech Engrs. *Res:* Analysis of fractures and defects in glass, determine type of stress and stress intensity causing failure; safety and impact performance, rating degree of anneal and temper; selection of glass sizes, type and thickness for wind load on single and double windows in buildings and water loads on multiple laminated windows in aquaria. *Mailing Add:* Box 291 RD 4 Tarentum PA 15084

ORR, LOWELL PRESTON, b Ross Co, Ohio, Dec 11, 30; m 64; c 2. VERTEBRATE ECOLOGY, HERPETOLOGY. *Educ:* Miami Univ, BS, 52; Kent State Univ, MEd, 56; Univ Tenn, PhD(zool), 62. *Prof Exp:* Instr biol, Pub Sch, Ohio, 52-55; instr, 56-57 & 61-63, from asst prof to assoc prof, 63-73, PROF BIOL SCI, KENT STATE UNIV, 73- *Mem:* Soc Study Amphibian Reptiles; Sigma Xi; Am Soc Ichthyol & Herpet. *Res:* Evolutionary ecology of vertebrate populations; competitive interactions and natural history of salamanders. *Mailing Add:* Dept of Biol Sci Kent State Univ Kent OH 44242

ORR, MARSHALL H, b Providence, RI, Dec 7, 42; m 64; c 3. OCEAN ACOUSTICS, OCEANOGRAPHY. *Educ:* Univ RI, BS, 65; Univ Maine, MS, 67; Pa State Univ, PhD(physics), 72. *Prof Exp:* Instr physics, Univ Maine, 67-68; asst prof, Pa State Univ, 72-73, res assoc, Appl Res Lab, 73-74; asst scientist underwater acoust, Dept Ocean Eng, Woods Hole Oceanog Inst, 74-78, assoc scientist underwater acoust, 78-81; sr res geophysicist, Seismol Dept, Gulf Res & Develop Co, 81-83; staff geophysicist, Sohio Petrol Co, 83-87; PROG MGR, OCEAN ACOUST, OFF NAVAL RES, 87- *Concurrent Pos:* Consult, Off Technol Assessment, US Cong, 80-81. *Mem:* Am Asn Petrol Geologists; Am Geophys Union; Acoust Soc Am. *Res:* Ocean acoustics; land and marine hydrocarbon exploration vertical seismic profiling, two and three dimensional seismology and well logging techniques; marine depositional processes; use of high frequency acoustic backscattering systems as remote sensors of physical processes in the ocean environment; mixing processes; internal waves, sediment transport; experimental low energy nuclear structures physics. *Mailing Add:* 305 Roosevelt Ct Vienna VA 22180

ORR, MARY FAITH, b Ashland, Ala, June 29, 20. ANATOMY. *Educ:* Univ Ala, BA, 41; Vanderbilt Univ, MA, 54, PhD(anat), 61. *Prof Exp:* Asst tissue cult, Univ Tex Med Br, 45-50; instr oncol & dir tissue cult, Vanderbilt Univ, 50-60; from instr to asst prof anat, 60-67, ASSOC PROF ANAT, MED SCH, NORTHWESTERN UNIV, CHICAGO, 67-85; RETIRED. *Concurrent Pos:* Sr investr on assoc staff & dir tissue cult lab, Chicago Wesley Mem Hosp, Ill, 60-64. *Mem:* AAAS; Soc Neurosci; Am Asn Anat; Tissue Cult Asn; Am Soc Cell Biol. *Res:* Cancer; development of sensory elements of the inner ear in vivo and in organ culture; differentiation of nerve tissue and organ rudiments in tissue culture; spleen cultures. *Mailing Add:* 3551 N Whipple Chicago IL 60618

ORR, NANCY HOFFNER, b Allentown, Pa, Dec 20, 41; m 84. DEVELOPMENTAL BIOLOGY & GENETICS. *Educ:* Chestnut Hill Col, BS, 63. *Prof Exp:* Res asst embryol, Inst Cancer Res, 63-66; res asst biochem, Sch Med, Univ Pa, 66-67; res asst anat, 67-70, asst, 70-73, res instr, 73-80, RES ASST PROF PHYSIOL, MED COL PA, 80- *Concurrent Pos:* Co-investr, Gen Med Sci Inst, NIH, 83- *Mem:* AAAS; Int Soc Develop Biologists; Int Soc Differentiation; Soc Develop Biol. *Res:* Process of cell specialization; amphibian nuclear transplantation. *Mailing Add:* 2844 Pennsylvania St Allentown PA 18104

ORR, ORTY EDWIN, b Neodesha, Kans, Aug 18, 20; m 47; c 2. VERTEBRATE ZOOLOGY. *Educ:* Kans State Col, Pittsburgh, BS, 51; Okla State Univ, MS, 52, PhD(zool), 58. *Prof Exp:* Biologist, Nebr Game & Parks Comn, 54-62; from asst prof to assoc prof biol sci, San Antonio Col, 62-66; prof biol sci, Mo Southern State Col, 66-86, head dept, 68-86; RETIRED. *Concurrent Pos:* Nat Sci Insts fels, field biol, Colo State Univ, 68, radiation biol, Argonne Nat Lab, 69, math col teachers, Univ Mont, 70 & radiation biol & environ qual, Cornell Univ, 71-72. *Mem:* Am Fisheries Soc; Am Inst Biol Sci. *Res:* Fresh water fishery. *Mailing Add:* 1519 N Florida Ave Joplin MO 64801

ORR, RICHARD CLAYTON, b Oakland, Calif, Mar 28, 41; m 64; c 2. MATHEMATICS. *Educ:* Humboldt State Col, AB, 64; Syracuse Univ, MA, 66, PhD(math), 69. *Prof Exp:* From asst prof to assoc prof math, State Univ NY, Col Oswego, 69-81, chmn dept, 74-81; PRES, CHRONOS CORP, LEXINGTON, MASS, 84- *Res:* Analytic number theory; sieve methods. *Mailing Add:* 34 Meriam St Lexington MA 02173

ORR, ROBERT S, b Philadelphia, Pa, Dec 18, 37. ORGANIC CHEMISTRY. *Educ:* Univ Pa, BA, 59; Univ Del, MS, 62, PhD(chem), 64. *Prof Exp:* From asst prof to assoc prof, 64-72, PROF CHEM, DEL VALLEY COL, 72-, CHMN DEPT, 66-, CHMN, DIV SCI, 80- *Mem:* AAAS; Am Chem Soc; Sigma Xi. *Res:* Teaching of organic chemistry and biochemistry on undergraduate level. *Mailing Add:* 101 Oak Dr Lansdale PA 19446

ORR, ROBERT THOMAS, b San Francisco, Calif, Aug 17, 08; m 72. MAMMALOGY, ORNITHOLOGY. *Educ:* Univ San Francisco, BS, 29; Univ Calif, MA, 31, PhD(zool), 37. *Hon Degrees:* DSc, Univ San Francisco, 76. *Prof Exp:* Asst, Mus Vert Zool, Univ Calif, 32-35; wildlife technician, US Nat Park Serv, 35-36; from asst curator to curator, 36-75, assoc dir acad, 64-75, SR SCIENTIST, DEPT BIRDS & MAMMALS, CALIF ACAD SCI, 75- *Concurrent Pos:* From asst prof to prof, Univ San Francisco, 42-64; vis prof, Univ Calif, Berkeley, 62 & 64, prof-in-res, Univ Exten Prog, San Blas, Mex, 62; trustee, Western Found Vert Zool, 78- *Honors & Awards:* Fel Medal, Calif Acad Sci, 73. *Mem:* Fel AAAS; hon mem Am Soc Mammal (secy, 38-42, 47-59, vpres, 53-55, pres, 58-60); hon mem Cooper Ornith Soc (vpres, 64-67, pres, 67-70); Soc Syst Zool; fel Am Ornith Union; Australian Mammal Soc; Soc Marine Mammal; Friends of Sea Otter. *Res:* Population, behavioral and taxonomic studies on marine mammals; life history and taxonomy of North American bats and rabbits; distributional and taxonomic studies on Mexican birds and mammals; distribution of fleshy fungi in western North America; author of over 200 publications. *Mailing Add:* Calif Acad Sci Golden Gate Park San Francisco CA 94118

ORR, ROBERT WILLIAM, b Evansville, Ind, Nov 24, 40; m 85. GEOLOGY. *Educ:* Univ Ill, BS, 62; Univ Tex, MA, 64; Ind Univ, PhD(geol), 67. *Prof Exp:* Res asst geol, Ill Geol Surv, 63-64; from asst prof to assoc prof, 67-75, chmn dept, 81-87, PROF GEOL, BALL STATE UNIV, 75- *Concurrent Pos:* Fel Ind Geol Surv Natural Resources. *Mem:* Fel Geol Soc Am; Soc Econ Paleont & Mineral. *Res:* Conodont biostratigraphy of the Devonian system, especially Middle Devonian rocks of North America. *Mailing Add:* Dept Geol Ball State Univ Muncie IN 47306

ORR, ROBIN DENISE MOORE, b Sydney, Australia, Feb 7, 34; US citizen; m 58; c 2. NUTRITION. *Educ:* Sydney Tech Col, ASTC, 54; Iowa State Univ, SM, 59; Harvard Sch Public Health, MSc, 72, DSc, 79. *Prof Exp:* Instr nutrit, East Sydney Tech Col, 55-57; therapeut dietitian, Seiler's Dietary Serv, Watham, Mass, 66-70; lectr, 77-79, ASSOC PROF COMMUN NUTRIT, FAC MED, MEM UNIV NEWFOUNDLAND, 79- *Concurrent Pos:* Tutor, Quincy House, Harvard Col, 73-76. *Mem:* Nutrit Soc Can; Am Dietetic Asn; Can Dietetic Asn; Can Public Health Asn; Newfoundland Public Health Asn. *Res:* Dietary patterns and nutritional status of population groups; dietary intake methodology; organization of nutrition services; evaluation of nutrition programmes. *Mailing Add:* Fac Med Health Sci Ctr St John's NF A1B 3V6 Can

ORR, WILLIAM CAMPBELL, b St Louis, Mo, Dec 27, 20; m 46; c 3. NUCLEAR CHEMISTRY. *Educ:* Princeton Univ, AB, 42; Univ Calif, PhD(chem), 48. *Hon Degrees:* ScD, MacMurray Col, 68. *Prof Exp:* Chemist, Radiation Lab, Univ Calif, 48-49; from asst prof to prof chem, 49-78, assoc provost, 65-74, assoc vpres acad affairs, 74-78, EMER PROF CHEM, UNIV CONN, 78- *Mem:* Fel AAAS; Am Chem Soc; Am Phys Soc. *Res:* Tracer studies of diffusion in solids and molten salts. *Mailing Add:* 23 Dunham Pond Rd Storrs CT 06268-2022

ORR, WILLIAM H(AROLD), communications technology, for more information see previous edition

ORR, WILLIAM N, b Sioux Falls, SDak, June 13, 39; m 58; c 1. GEOLOGY. *Educ:* Univ Okla, BS, 61; Univ Calif, MA, 63; Mich State Univ, PhD(geol), 67. *Prof Exp:* Asst prof geol, Eastern Wash State Col, 66-67; asst prof, 67-74, ASSOC PROF GEOL, UNIV ORE, 74- *Concurrent Pos:* Prog assoc, Ocean Sediment Coring, NSF, 78- *Res:* Paleontology; biostratigraphy; paleoecology. *Mailing Add:* Dept of Geol Univ of Ore Eugene OR 97403

ORR, WILSON LEE, b Roswell, NMex, Jan 8, 23; m 44; c 2. ORGANIC GEOCHEMISTRY. *Educ:* Bethany-Peniel Col, AB, 45; Purdue Univ, MS, 47; Univ Southern Calif, PhD, 54. *Prof Exp:* Chemist prod control, Wabash Ord Works, US Dept Army, Ind, 45; asst, Purdue Univ, 45-47; vis asst prof, Sch Aeronaut, Univ Southern Calif, 48-49, staff chemist & res assoc geochem, Allan Hancock Found, 49-55, lectr & res assoc, dept geol, 55-58, res assoc chem, 58-59; res engr & supvr, Atomic Int Div, NAm Aviation, Inc, 60-63; RES ASSOC, FIELD RES LAB, MOBIL RES & DEVELOP CORP, 64- *Mem:* AAAS; Am Chem Soc; Geochem Soc; Sigma Xi. *Res:* Marine and petroleum geochemistry; organic geochemistry of sediments and petroleum. *Mailing Add:* PO Box 3729 443 Allison Dr Dallas TX 75208-2402

ORRALL, FRANK QUIMBY, b Somerville, Mass, Oct 15, 25; m 49; c 3. ASTROPHYSICS. *Educ:* Univ Mass, BS, 50; Harvard Univ, AM, 53, PhD(astron), 55. *Prof Exp:* Astrophysicist, Col Observ, Harvard Univ, 55-56; solar physicist, Sacramento Peak Observ, NMex, 56-64; PROF PHYSICS & ASTRON, INST ASTRON & UNIV HAWAII, MANOA, 64- *Concurrent Pos:* Vis scientist, High Altitude Observ, Colo, 70-71. *Mem:* Am Astron Soc; Inst Astron Union. *Res:* Solar physics; observation and theory of phenomena of the sun's atmosphere. *Mailing Add:* Dept Physics & Astron Ifa C 206 Univ Hawaii Manoa 2500 Campus Rd Honolulu HI 96822

ORREGO, HECTOR, b Santiago, Chile, Dec 4, 23; Chilean & Can citizen; m 48; c 4. GASTROENTEROLOGY, PHARMACOLOGY. *Educ:* Univ Chile, MD, 49; FRCPS(C). *Prof Exp:* Asst prof med, Univ Chile, 50-64, prof med & pathophysiol, 64-73; head gastroenterol prog, Dept Med, Addiction Res Found Clin Inst, 77-91; vis prof med physiol & pharmacol, 74-76, prof pharmacol, Med & Physiol, 76-91, emer prof, Dept Physiol, Univ Toronto, 91-; RETIRED. *Concurrent Pos:* W R Kellog Found & Am Col Physicians int fels, 50-52; head sect gastroenterol, Univ Chile Clin Hosp, Santiago, 53-73; prof function pathophysiol & gastroenterol, Cath Univ Chile, Santiago, 56-73; subrogate dean, Univ Chile, 69; Addiction Res Found grant, 77- *Mem:* Am Col Physicians; Int Asn Study Liver; Am Asn Study Liver Dis; Can Soc Clin Invest. *Res:* Alcohol and the liver; intestinal absorption. *Mailing Add:* 114 Silver Birch Toronto ON M5S 1A8 Can

ORREN, MERLE MORRISON, neuropsychology, physiological psychology, for more information see previous edition

ORRENIUS, STEN, b Motala, Sweden, Feb 14, 37. TOXICOLOGY. *Educ:* Karolinska Inst, BM, 60, MD, 65. *Hon Degrees:* PhD, Univ Stockholm, 82. *Prof Exp:* Res assoc, Wenner-Gren Inst Exp Biol, Stockholm, 65-67, Dept Biochem, Univ Stockholm, 67-71; prof & chmn, Dept Forensic Med, 71-84, dean, Med Sch, 83-90, PROF & CHMN, DEPT TOXICOL, KAROLINSKA INST, 840, DIR, NAT INST ENVIRON MED, 88- *Concurrent Pos:* Res assoc, Dept Pharmacol, Sch Med, Yale Univ, 72; vis scientist, Dept Biochem, Med Sch, Univ Tex, Dallas, 72; Wellcome vis prof basic med sci, 81 & 88; distinguished lectr med sci, Mayo Clin, Rochester, Minn, 88. *Honors & Awards:* Poulsson Medal for Important Contrib in Toxicol, Norweg Soc Pharmacol & Toxicol, 84; John Barnes Prize, Brit Toxicol Soc, 91; Millercomm Lectr, Univ Ill, 91. *Mem:* Foreign assoc mem Inst Med-Nat Acad Sci; hon mem Am Soc Pharmacol & Exp Therapeut; hon mem Am Soc Biochem & Molecular Biol. *Mailing Add:* Karolinska Inst Solnavagen 1 POB 60400 Stockholm 104 01 Sweden

ORROK, GEORGE TIMOTHY, b Boston, Mass, Nov 25, 30. PHYSICS. *Educ:* Harvard Univ, BA, 52, MA, 53, PhD(physics), 59. *Prof Exp:* Mem tech staff, Bell Tel Labs, 59-62; mem tech staff, Bellcomm, Inc, 62-64, suprvr environ sci, 64-65, dept head space sci, 65-72; supvr systs identification, Bell Tel Labs, 72-80, optical mat res, 81, transmission systs eng, 82-88; RETIRED. *Mem:* AAAS; Am Phys Soc; Sigma Xi. *Res:* Solidification of pure metals; semiconductor device application and evaluation; environment and space sciences; quality assurance; optical materials; telecommunications networks. *Mailing Add:* 1030 Ivy Lane Ashland OR 97520

ORS, JOSE ALBERTO, b Havana, Cuba, June 14, 44; m 67; c 1. PHOTOCHEMISTRY. *Educ:* Univ S Fla, BA, 67, MS, 72, PhD(chem), 74. *Prof Exp:* Chemist air & water pollution, Environ Protection Comn, Fla, 68-69; res staff mem org photochem, T J Watson Res Ctr, IBM Corp, 74-80; mem res staff polymeric mat eng, Eng Res Ctr, Western Elec Co, 80-; AT AT&T, INC. *Mem:* Am Chem Soc; NY Acad Sci; Inter-Am Photochem Soc; Sigma Xi. *Res:* Photocycloaddition of aromatic hydrocarbons to olefins; photochemical reactions in terpenes; reactions involving singlet oxygen. *Mailing Add:* AT&T Inc ERC Box 900 Princeton NJ 08540

ORSBORN, JOHN F, b Gary, Ind, Aug 13, 29; m 54; c 4. FLUID MECHANICS, HYDROLOGY. *Educ:* Colo Col, BA, 52; Univ Colo, BSCE, 57; Univ Minn, MSCE, 60; Univ Wis, PhD(civil eng), 64. *Prof Exp:* Eng aid, Soil Conserv Serv, USDA, Colo, 52-55; res fel hydraul models, St Anthony Falls Hydraul Lab, Univ Minn, 57-61, instr civil eng, 58-61; res assoc, Univ Wis, 61-64; asst prof fluid mech & asst hydraul engr, Lab, 64-69, assoc prof civil eng, Univ, 69-77, prof civil eng & chmn dept, 77-81, PROF & HYDRAUL ENGR, WASH STATE UNIV, 81- *Concurrent Pos:* Ford Found fel, Univ Wis, 61-64; assoc hydraul engr & head, R L Albrook Hydraul Lab, 69-74; consult, UN, Environ Protection Agency, Nat Water Comn, US Fish & Wildlife Serv, US Forest Serv & consulting firms. *Mem:* Am Soc Civil Engrs; Am Fish Soc; Am Soc Engr Educ; Nat Soc Prof Engrs. *Res:* Drainage basin influences on hydrologic parameters; dams, appurtenances and hydraulic design; river engineering and mechanics; flood plain management; drainage basin systems and stream flows. *Mailing Add:* Dept Civil-Environ Eng Wash State Univ Pullman WA 99164

ORSENIGO, JOSEPH REUTER, b Barryville, NY, Apr 4, 22; m 52; c 2. WEED SCIENCE. *Educ:* Cornell Univ, BS, 48, PhD(veg crops), 53. *Prof Exp:* Agronomist & mgr, Rice Res Sta, Ibec Res Inst, Venezuela, 53-55; agronomist-horticulturist & mgr cacao ctr, Inter-Am Inst Agr Sci, Costa Rica, 55-57; asst horticulturist, Univ Fla, 57, prof, 57-75, horticulturist, plant physiologist & emer prof herbicide, weed sci & plant growth regulator res, Agr Res & Educ Ctr, Inst Food & Agr Sci, 57-75; dir res, 76-83, VPRES RES, FLA SUGAR CANE LEAGUE, 83- *Concurrent Pos:* Adv, US Agency Int Develop, Costa Rica, 65-67; weed sci consult, Inter-Am Inst Agr Sci, Cent Am, 66; consult, Univ Calif, Multidisciplinary Pest Mgt Surv-Cent Am, 72. *Mem:* Fel AAAS; Am Soc Agron; Am Soc Hort Sci; Plant Growth Regulator Soc Am; Weed Sci Soc Am. *Res:* Weed science and herbicide research in agronomic crops in tropical America; sugarcane, agronomic and vegetable crops on organic soils; cocoa, coffee, corn, rice, sugarcane and general agriculture in Latin America; sucrose enhancement in sugarcane; air and water quality related to sugarcane. *Mailing Add:* Sci & Agr INE PO Box 1089 Belle Glade FL 33430

ORSI, ERNEST VINICIO, b New York, NY, Aug 10, 22; m 49; c 4. CELL BIOLOGY, VIROLOGY. *Educ:* Queen's Col, NY, BS, 44; Fordham Univ, MS, 48; St Louis Univ, PhD(biol), 55. *Prof Exp:* Sr staff mem, Virus Res Sect, Lederle Labs, Am Cyanamid Co, 54-61; prin res scientist virus diag, New York City Health Dept, 61-65; assoc prof, 65-72, PROF BIOL, SETON HALL UNIV, 72- *Concurrent Pos:* Brown-Hazen Fund grant, 65; contract hyperbaric enhancement of virus infection, Off Naval Res, 68-74; Deborah Hosp Found grant, 77-78, Fannie E Ripple Found grant, 79 & Stony-Wold-Herbert Fund grant, 80-81; exchange prof, Virol Dept, Wuhan Univ, 81; NSF grant, 83. *Mem:* AAAS; Am Soc Microbiologists; Harvey Soc; Sigma Xi; Tissue Cult Asn. *Res:* Cell-virus relationships in normal neoplastic and transplantation systems as influenced either by changes in cell properties following sub-cultivation or induced by environmental factors. *Mailing Add:* Dept Biol Seton Hall Univ South Orange NJ 07079

ORSINI, FRANK R, b Buffalo, NY, Apr 4, 38. IMMUNOHEMATOLOGY. *Educ:* State Univ NY, Buffalo, PhD(immunochem), 70. *Prof Exp:* Cancer res scientist exp therapeut & dir clin immunol, 72-79, SCI DIR TRANSFUSION SERV, ROSWELL PARK MEM INST, 79- *Mem:* Am Soc Immunol; Am Asn Blood Banks; Am Soc Histocompatibility & Immunogenetics. *Mailing Add:* Dept Microbiol 203 Sherman Hall SUNY Health Sci Ctr 3435 Main St Buffalo NY 14214

ORSINI, MARGARET WARD (GIORDANO), b LeRoy, NY, Mar 18, 16; wid. EMBRYOLOGY, PHYSIOLOGY OF REPRODUCTIONS. *Educ:* Mt Holyoke Col, BA, 37; Cornell Univ, PhD(zool), 46. *Prof Exp:* Technician & asst path, Sch Med, St Louis Univ, 37-38; med technologist, Lister Clin Lab, 38-40; asst biol, Manhattanville Col, 40-42; asst histol & embryol, Cornell Univ, 42-46; instr zool, Duke Univ, 46-49; NIH fel, 49-52, res assoc, 52-53, prof, 72-87, EMER PROF ANAT, MED SCH, UNIV WIS-MADISON, 87- *Concurrent Pos:* Prin investr NIH grants, 59-76 & 79-82. *Mem:* Fel AAAS; fel NY Acad Sci; Am Asn Anat; Soc Exp Biol & Med; Soc Study Reproduction. *Res:* Physiology of reproduction; estrous cycle of hamster; ovulation; comparative morphogenesis and endocrinology of implantation; giant cells and vascular changes in gestation and post-partum involution; control of decidualization; comparison of pregnancy and pseudopregnancy; immunoreproduction; hormonal and morphology correlates of hamster. *Mailing Add:* 4413 Tokay Blvd Madison WI 53711

ORSZAG, STEVEN ALAN, b New York, NY, Feb 27, 43; m 64; c 3. APPLIED MATHEMATICS. *Educ:* Mass Inst Technol, BS, 62; Princeton Univ, PhD(astrophys), 66. *Prof Exp:* Mem, Inst Adv Study, 66-67; prof appl math, Mass Inst Technol, 67-84; HAMRICK PROF ENG & PROF APPL & COMPUT MATH, PRINCETON UNIV, 84-, DIR APPL & COMPUT MATH, 90- *Concurrent Pos:* Chmn, Cambridge hydro, 75-; J S Guggenheim Fel, 89. *Honors & Awards:* Fluids & Plasma Dynamics Award, Am Inst Aeronaut & Astronaut; Otto Laporte Prize, Am Phys Soc, 91. *Mem:* Am Phys Soc; Soc Appl & Indust Math. *Res:* Fluid dynamics; numerical methods. *Mailing Add:* 218 Fine Hall Princeton Univ Princeton NJ 08544-1000

ORT, CAROL, b Boston, Mass, Sept 14, 47; m 71. ELECTROPHYSIOLOGY, CELL BIOLOGY. *Educ:* Mich State Univ, BS, 69; Univ Calif, Berkeley, PhD(molecular biol, neurobiol), 74. *Prof Exp:* Muscular Dystrophy Asn Am Inc fel neurobiol, Univ Calif, San Francisco, 74-76; lectr zool & neurobiol, San Francisco State Univ, 76-77; asst prof, 77-83, ASSOC PROF BIOL, UNIV NEV, 83- *Concurrent Pos:* Prin investr, Univ Nev Res Adv Bd, 77-79, 86-87, 88-89, NIH, 79-82. *Mem:* Neurosci Soc; AAAS; Am Soc Cell Biol. *Res:* Mammalian central nervous system neurons in tissure culture. *Mailing Add:* Dept Biol Univ Nev Reno NV 89557

ORT, DONALD RICHARD, b Weymouth, Mass, Feb 20, 49; m 71; c 2. BIOCHEMISTRY, PLANT PHYSIOLOGY. *Educ:* Wake Forest Univ, BS, 71; Mich State Univ, PhD(plant biochem), 74. *Prof Exp:* Fel biochem, Purdue Univ, 74-76; sr fel, Univ Wash, 76-78; ASST PROF PLANT BIOCHEM, UNIV ILL, URBANA, 78- *Concurrent Pos:* NIH Nat Serv award, 75-77. *Mem:* Am Soc Plant Physiol; Biophys Soc. *Res:* Photosynthesis; mechanism of energy coupling between photosynthetic electron transport and adenosine triphosphate. *Mailing Add:* Dept Plant Biol Univ Ill Urbana Campus 505 S Goodwin Ave Urbana IL 61801

ORT, MORRIS RICHARD, b Liberty Center, Ohio, Sept 16, 27; m 51; c 3. SYNTHESIS, CATALYSIS. *Educ:* Bowling Green State Univ, BS, 52; Univ Cincinnati, MS, 62. *Prof Exp:* Res chemist, Monsanto Co, 51-66, res specialist, 66-73, sci fel, 73-80, sr sci fel, 80-88; RETIRED. *Mem:* Am Chem Soc. *Res:* Polymer synthesis; catalysis; gas phase polymerization; solid state polymerization; polyolefins; condensation polymers; organometallics; organic synthesis. *Mailing Add:* Six Bittersweet Lane Wilbraham MA 01095

ORTALDO, JOHN R, NATURAL KILLER CELLS, TUMOR IMMUNOLOGY. *Educ:* George Washington Univ, PhD(microbiol), 78. *Prof Exp:* INSTR IMMUNOL, HOOD COL, 82-; CHIEF, LAB EXP IMMUNOL, FREDERICK CANCER RES FACIL, NIH, 85- *Mailing Add:* Chief Lab Exp Immunol NCI-FCRF 560/31-93 Frederick MD 21701-1013

ORTEGA, GUSTAVO RAMON, b El Paso, Tex, Dec 6, 45. PHARMACEUTICAL CHEMISTRY. *Educ:* Univ Tex, Austin, BS, 69, PhD(pharmaceut chem), 76. *Prof Exp:* Asst prof, 75-80, ASSOC PROF PHARMACEUT CHEM, SOUTHWESTERN OKLA STATE UNIV, 80- *Mem:* AAAS; Am Chem Soc; Am Asn Col Pharm; Sigma Xi. *Res:* Application of structure activity relationships to the synthesis of synthetic medicinal agents; analysis of commercially available pharmaceuticals. *Mailing Add:* Sch Pharm Southwestern Okla State Univ 100 Campus Dr Weatherford OK 73096

ORTEGA, JACOBO, b Allende, Mex, Mar 21, 29; m 56; c 3. PLANT PATHOLOGY, PLANT GENETICS. *Educ:* Univ Coah, Mex, BSc, 56; Okla State Univ, MSc, 58; Univ Minn, PhD(plant path & genetics), 60. *Prof Exp:* Asst wheat breeding, Rockefeller Found, Mex, 54-56; head sect, Nat Inst Agr Res SAG, 61-66; chief pathologist & plant breeder, World Seeds, Inc, 66-71; res agronomist, Calif Milling Corp, Los Angeles, 71-73; from asst prof to assoc prof, 73-84, PROF BIOL, PAN AM UNIV, 84- *Mem:* AAAS. *Res:* Wheat diseases and genetics; agro-economical research of the wheat and milling industries of California; production, localization and activities of cellulolytic enzymes of soil fungi; modeling and kinetics of fungal enzymes; cellulolytic activities of plant pathogens. *Mailing Add:* Dept Biol Univ Tex Pan Am 1201 W University Dr Edinburg TX 78539-2999

ORTEGA, JAMES M, b Madison, Wis, June 15, 32; m 57; c 1. NUMERICAL ANALYSIS. *Educ:* Univ NMex, BS, 54; Stanford Univ, PhD(math), 62. *Prof Exp:* Programmer, Sandia Corp, NMex, 56-58, mathematician, 62-63; mathematician, Bellcomm, Inc, Washington, DC, 63-64; asst prof comput sci, Univ Md, College Park, 64-66, assoc prof comput sci & appl math, 66-69; vis prof, Univ Calif, San Diego, 71-72; dir, ICASE, NASA-Langley Res Ctr, 72-77; prof math & head dept, NC State Univ, 77-79; Charles Henderson prof & chmn, appl math & comput sci, Univ Va, 79-84, assoc dean, eng & appl sci, 80-82, chmn, appl math, 84-89, DIR, INST PARALLEL COMPUTATION, UNIV VA, 90- *Concurrent Pos:* From assoc ed to ed, SIAM Rev, Soc Indust & Appl Math, 65-68; prof comput sci, math & appl math, Univ Md, College Park, 69-74; adj prof math & physics, Col William & Mary, 73-77; adj prof appl math & comput sci, Univ Va, 73-77; consult & mem COSERs numerical comput panel, NSF, 75-78. *Mem:* Am Math Soc; Asn Comput Mach; Soc Indust & Appl Math. *Res:* Numerical analysis; applied mathematics; computer science. *Mailing Add:* Appl Math Thornton Hall Univ Va Charlottesville VA 22901

ORTEGO, JAMES DALE, b Washington, La, Sept 23, 41; m 65; c 3. INORGANIC CHEMISTRY. *Educ:* Univ Southwestern La, BS, 63; La State Univ, PhD(chem), 68. *Prof Exp:* Asst prof, 68-75, assoc prof, 75-81, PROF CHEM, LAMAR UNIV, 81- *Concurrent Pos:* Welch Found res grant, Lamar Univ, 69-83. *Mem:* Am Chem Soc; Sigma Xi. *Res:* Coordination chemistry of transition metal and uranium complexes; oxygen complexes of biological interest. *Mailing Add:* 5165 Laurel Beaumont TX 77710

ORTEL, WILLIAM CHARLES GORMLEY, b Spokane, Wash, Sept 28, 26; m 51; c 3. AUTOMATED VOICE TELEPHONE SERVICES, DATA COMMUNICATIONS. *Educ:* Yale Univ, BS, 49, MS, 50, PhD(physics), 53. *Prof Exp:* Asst physics, Yale Univ, 53-54; NSF fel, Inst Theoret Physics, Denmark, 54-55; physicist, Missile Div, Lockheed Aircraft Corp, 55-56; physicist, Bell Labs, 56-82; AT&T Info Systs, 82-87; NYNEX SCI & TECHNOL, 87- *Res:* Assessment of new technologies for computing and communication systems including experiments with components, analysis of performance; proposal, market evaluation and development of speech recognition systems, microprocessor applications, mini and microcomputer software; integrated circuits; data communications; optical transmission. *Mailing Add:* 125 Washington Pl New York NY 10014

ORTEN, JAMES M, b Farmington, Mo, Nov 29, 04; m 32. BIOCHEMISTRY. *Educ:* Univ Denver, BS, 28, MS, 29; Univ Colo, PhD(physiol chem), 32. *Prof Exp:* Asst pharm, Univ Denver, 27-28; instr biochem, Univ Colo, 31-32; from asst prof to assoc prof, 37-57, prof biochem, 57-75, dir grad affairs, 71-75, EMER PROF BIOCHEM & HON DIR ASST DEAN GRAD PROGS, SCH MED, WAYNE STATE UNIV, 75- *Concurrent Pos:* Coxe fel, Yale Univ, 32-33 & 34-35, Nat Res Coun fel, 33-34, Pfizer & Co fel, 35-37; guest, Int Physiol Cong, Switz, 38. *Honors & Awards:* Lawrence M Weiner Award, Wayne State Univ, 88. *Mem:* Fel AAAS; Am Chem Soc; Am Soc Biol Chem; Soc Exp Biol & Med; fel Am Inst Nutrit (secy, 52-55). *Res:* Effect of dietary protein and metallic elements on hemopoiesis; metabolism of organic acids, alcohol and the porphyrins; carbohydrate metabolism in experimental diabetes. *Mailing Add:* Dept Biochem Wayne State Univ Sch Med 540 E Canfield Ave Detroit MI 48201

ORTH, CHARLES DOUGLAS, b Seattle, Wash, June 1, 42; m 64; c 2. FUSION TARGET DESIGN, FUSION REACTOR STUDIES. *Educ:* Univ Wash, BS, 64; Calif Inst Technol, PhD(physics), 70. *Prof Exp:* Res assoc cosmic ray physics, NASA Manned Spacecraft Ctr, 70-72; asst res physicist astrophys & cosmic ray physics, Univ Calif, Berkeley, 72-78; PHYSICIST, LAWRENCE LIVERMORE NAT LAB, 78- *Concurrent Pos:* Mem, HEAMOWG Comt, NASA, 76-78, consult, Space Sci Steering Comt, 79-82. *Mem:* Am Phys Soc; Sigma Xi; Am Nuclear Soc; Am Inst Aeronaut & Astronaut. *Res:* Charge, age, and energy spectra of high energy cosmic rays; anisotropy of microwave blackbody radiation; real time correction of atmospherically degraded telescope images; target and reactor design for inertial confinement fusion; space propulsion using ICF; systems analysis. *Mailing Add:* L-481 Lawrence Livermore Nat Lab PO Box 5508 Livermore CA 94550-0618

ORTH, CHARLES JOSEPH, b Fontana, Calif, Aug 13, 30; m 59; c 1. GEOCHEMISTRY. *Educ:* San Diego State Univ, BA, 54; Univ NMex, PhD(chem), 69. *Prof Exp:* Staff mem radiochem, 56-83, FEL, LOS ALAMOS NAT LAB, 83- *Honors & Awards:* John Dustin Clark Award, Am Chem Soc, 85. *Mem:* Geol Soc Am; Am Phys Soc; Am Chem Soc; Sigma Xi; Am Geophys Union; AAAS. *Res:* Nuclear structure; pion-nucleus and muon-nucleus reactions; muonic x-rays; trace element abundances in sedimentary rocks; mass extinction and meteorite impact phenomena. *Mailing Add:* 281 Chamisa Los Alamos NM 87544

ORTH, GEORGE OTTO, JR, b Seattle, Wash, July 14, 13; m 40; c 3. ORGANIC CHEMISTRY. *Educ:* Univ Wash, BS, 36. *Prof Exp:* Plant chemist, Solvents, Inc, 33-35; plant chemist & engr, Preservative Paint Co, 35-41; res chemist, Seattle Gas Co, 41-42; lab mgr, A J Norton, Consult Chemists, 42-51; vpres, Food, Chem & Res Labs, Inc, 53-58; lab dir, Korry Mfg Co, 63-68; CONSULT, 51- *Concurrent Pos:* Mem comt estab synthetic rubber mfg corp, Tokyo, Japan, 53; private res, 58- *Mem:* Fel AAAS; fel Am Inst Chem; emer mem Am Chem Soc; Asn Consult Chem & Chem Eng; emer mem Inst Food Technol. *Res:* High polymers; wood specialities and chemistry; organic nitrogen derivatives; protective coatings; adhesives and plastics; wood fiber specialties; oil adsorbents as related to ecology; wood cellulose animal nutrition technology; agriculture crop utilization. *Mailing Add:* PO Box 25592 Seattle WA 98125

ORTH, JOANNE M, REPRODUCTIVE BIOLOGY, TESTICULAR DEVELOPMENT. *Educ:* Temple Univ, PhD(anat), 78. *Prof Exp:* Asst prof, 80-85, ASSOC PROF ANAT & CELL BIOL, SCH MED, TEMPLE UNIV, 85- *Mem:* Endocrine Soc; Am Soc Cell Biol; Am Asn Anatomists. *Mailing Add:* Seven Krauser Rd Dowingtown PA 19335

ORTH, JOHN C(ARL), b Sioux City, Iowa, June 20, 31; m 56; c 5. CHEMICAL ENGINEERING. *Educ:* Univ Seattle, BS, 57. *Prof Exp:* Trainee, Allis-Chalmers Mfg Co, Wis, 58-59, appln engr, 59-61, res engr, 61-62, engr-in-charge appl res, 62-64; group leader energy conversion, Monsanto Res Corp, Mass, 64-66; chief advan power sources br, Power Equip Div, US Army Mobility Equip Res & Develop Ctr, 66-69, chief, Advan Develop Div, Electrotechnol Lab, 69-71, chief, Electrotechnol Dept, 71-74; tech dir, Mineral Res Ctr, Butte, Mont, 74-76; dir, Mont Dept Natural Resources, 77-78; dep managing dir, 78-80, VPRES, MONT ENERGY RES & DEVELOP INST, 80- *Mem:* Am Inst Chem Engrs. *Res:* Solids processing and agglomeration; direct reduction of iron ore; electro-chemical energy conversion. *Mailing Add:* 3500 Hannibal Butte MT 59701

ORTH, PAUL GERHARDT, b Chicago, Ill, Apr 18, 29; m 57; c 4. SOIL FERTILITY, IRRIGATION. *Educ:* Univ Ariz, BS, 51; Wash State Univ, MS, 53; Univ Wis, PhD(phosphate availability), 59. *Prof Exp:* Asst soils chemist, 59-77, ASSOC SOILS CHEMIST, AGR RES & EDUC CTR, UNIV FLA, 77- *Mem:* Am Soc Agron; Soil Sci Soc Am; Am Soc Horticultural Sci. *Res:* Slow release fertilizers; irrigation; calcareous soils; climatology; analysis of plant tissue for inorganic plant nutrients; fertilizer efficiency and leaching loss. *Mailing Add:* 25360 SW 182nd Ave Homestead FL 33031

ORTH, ROBERT JOSEPH, b Elizabeth, NJ, Nov 16, 47; m 73. BIOLOGICAL OCEANOGRAPHY. *Educ:* Rutgers Univ, BA, 69; Univ Va, MS, 71; Univ Md, College Park, PhD(zool), 75. *Prof Exp:* From asst marine scientist to assoc marine scientist, Va Inst Marine Sci, 74-83; from instr to asst prof, 77-83, ASSOC PROF MARINE BIOL, SCH MARINE SCI, COL WILLIAM & MARY & SR MARINE SCIENTIST, 83-, ASST DIR, VA INST MARINE SCI, 86- *Concurrent Pos:* Instr marine biol, Dept Marine Sci, Univ Va, 75-77. *Mem:* Int Asn Aquatic Vascular Plant Biologists; Ecol Soc Am; Estuarine Res Fedn (treas, 85-88, pres 89-91). *Res:* Biology of seagrasses; structional and functional aspects of marine benthos. *Mailing Add:* Va Inst Marine Sci Gloucester Point VA 23062

ORTH, WILLIAM ALBERT, b Coatesville, Pa, Sept 28, 31; m 58; c 2. CONTINUUM MECHANICS. *Educ:* US Mil Acad, BS, 54; Purdue Univ, MS, 61; Brown Univ, PhD(appl math), 70. *Prof Exp:* Dep chief staff civil eng, Strategic Air Command Hq, 72-74; asst prof mech, US Air Force Acad, 61-63, asst prof math, 68-70, prof physics & head dept, 74-78, dean fac, 78-83; pres, Trident Tech Col, 83-85; pres, Spartan Sch Aeronaut, 85-87; PRES, ATLANTIC COMMUNITY COL, 87- *Mem:* Am Soc Eng Educ; Am Asn Physics Teachers; Math Asn Am; Am Inst Aeronaut & Astronaut. *Mailing Add:* Atlantic Community Col Mays Landing NJ 08330

ORTHOEFER, JOHN GEORGE, b Columbus, Ohio, Nov 7, 32; m 64; c 3. VETERINARY MEDICINE. *Educ:* Ohio State Univ, BS, 59, DVM, 63; Univ Calif, Davis, MS, 72. *Prof Exp:* Vet, div epidemiol, Fla State Bd Health, 63-64 & Commun Dis Ctr, Kans City, Kans, 64-66; toxicologist path, Food & Drug Admin, Washington, DC, 66-69; vet lab animal, Nat Air Pollution Control Admin, USPHS, 69-71; training path, Univ Calif, Davis, 71-74; pathologist, Health Effects Res Lab, US Environ Protection Agency, 74-81; adj res scientist, Div Comp Path, Univ Fla, 81-84; CONSULT, 84- *Mem:* Am Vet Med Asn; Int Acad Path; Am Col Toxicol. *Res:* Descriptive pathology of the lung with a goal toward the development of more accurate morphometric techniques in lung pathology. *Mailing Add:* 3525 NW 31st St Gainesville FL 32605

ORTHWEIN, W(ILLIAM) C(OE), b Toledo, Ohio, Jan 27, 24; m 48; c 3. ENGINEERING MECHANICS. *Educ:* Mass Inst Technol, BS, 48; Univ Mich, MS, 50, PhD(eng mech), 59. *Prof Exp:* Aerophysicist, Convair Div, Gen Dynamics Corp, Tex, 51-52; res assoc, Univ Mich, 52-58; adv engr, Int Bus Mach Corp, NY, 58-61; assoc prof Univ Okla, 61-63; res scientist, NASA, 63-65; PROF ENG, SOUTHERN ILL UNIV, 65- *Concurrent Pos:* Auth. *Mem:* Am Soc Mech Engrs; Am Soc Civil Engrs; Soc Exp Stress Analysis; Tensor Soc; Nat Soc Prof Engrs; Soc Automotive Engrs. *Res:* Machine design; computer aided design; nonlinear elasticity; vibrations; author of two books. *Mailing Add:* PO Box 3332 Carbondale IL 62902-3332

ORTIZ, ARACELI, b Culebra, PR, Jan 15, 37; m 76; c 1. DENTISTRY, ORAL PATHOLOGY & MEDICINE. *Educ:* Univ PR, BS, 58, DMD, 62; Ind Univ, MSD(oral path), 67; Am Bd Oral Path, dipl; Am Bd Oral Med, dipl; Nat Bd Forensic Odontol, dipl. *Prof Exp:* Assoc prof, McGill Univ, 67-73; PROF ORAL PATH, ORAL MED & FORENSIC ODONTOL, SCH DENT, UNIV PR, 73- *Concurrent Pos:* Mem clin cancer training comt, NIH, 72-73; consult, Vet Admin Hosp, San Juan, PR, 74-84, Inst Forensic Sci, PR, 74- *Honors & Awards:* Cert Appreciation, Fed Bureau Invest, 87; Cert Merit, Am Acad Oral Med, 90; Samuel C Miller Medal Award, 91. *Mem:* Fel Am Acad Oral Path; Am Acad Oral Med; Am Soc Forensic Odontol; Can Soc Forensic Sci; Am Dental Asn; Am Soc Forensic Sci; Int Asn Forensic Sci; Can Asn Forensic Sci; Nat Asn Commun & Educ Technol; Am Asn Dental Schs. *Res:* Herpes virus; cancer education programs; forensic odontology. *Mailing Add:* Dept Dent Univ PR Med Sci GPO Box 5067 San Juan PR 00936

ORTIZ, MELCHOR, JR, b Victoria, Tex, Nov 14, 42; m 69; c 3. APPLIED STATISTICS. *Educ:* Tex A&I Univ, BS, 70, MS, 71; Tex A&M Univ, PhD(statist), 75. *Prof Exp:* Asst prof, 75-80, ASSOC PROF, 80-85, PROF STATIST, NMEX STATE UNIV, 85- *Concurrent Pos:* Statist consult. *Mem:* Am Statist Asn; Biometric Soc; Sigma Xi. *Res:* Nonconvex mathematical programming. *Mailing Add:* 4955 Ocotillo Las Cruces NM 88001

ORTIZ DE MONTELLANO, PAUL RICHARD, b Mexico City, Mex, Sept 6, 42; US citizen; c 2. HEMOPROTEINS, ENZYME INHIBITORS. *Educ:* Mass Inst Technol, BA, 64; Harvard Univ, MA, 66, PhD(chem), 68. *Prof Exp:* NATO fel, Swiss Fed Inst Technol, 68-69; group leader pharmaceut chem, Syntex Res Labs, 69-71; From asst prof to assoc prof, 72-80, PROF PHARMACEUT CHEM, SCH PHARM, UNIV CALIF, SAN FRANCISCO, 80- *Concurrent Pos:* Consult, Panel Vapor Phase Pollutants, Nat Acad Sci, 72-74 & Merck Sharp & Dohme Res Labs, 81-87; assoc prof, Inst Chem, Univ Louis Pasteur, 78-79; A P Sloan Found res fel, 78; hon res fel, Dept Biochem, Univ Col, London. *Mem:* Am Chem Soc; Royal Chem Soc; fel AAAS; Sigma Xi; Soc Toxicol; Am Soc Pharmacol Exp Therap; Am Soc Biochem Molecular Biol. *Res:* Mechanisms of hemoproteins, especially enzymes and peroxidases; the biochemistry and toxicology of carbon radicals; mechanisms of heme-biosynthetic enzymes; development of inhibitors for the HIV protease. *Mailing Add:* Dept Pharm Chem Sch Pharm Univ Calif San Francisco CA 94143

ORTMAN, ELDON EMIL, b Marion, SDak, Aug 11, 34; m 57; c 3. PEST MANAGEMENT. *Educ:* Tabor Col, AB, 56; Kans State Univ, MS, 57, PhD(entom), 63. *Prof Exp:* Asst entom, Kans State Univ, 56-59, instr, 59-61; res entomologist, Northern Grain Insect Res Lab, Entom Res Div, Agr Res Serv, USDA, 61-68, lab dir & invest leader, 68-72; prof entom & dept head, 72-89, ASSOC DIR AGR RES, PURDUE UNIV, W LAFAYETTE, 89- *Concurrent Pos:* Prof, SDak State Univ, 68-72; mem Coun Agr Sci & Tech. *Mem:* Entom Soc Am; Sigma Xi; fel AAAS; Am Inst Biol Sci. *Mailing Add:* Ag Res-AGAD Purdue Univ West LafAyette IN 47906

ORTMAN, HAROLD R, b Buffalo, NY, Dec 19, 17; m 39, 60; c 6. DENTISTRY. *Educ:* Univ Buffalo, DDS, 41; Am Bd Prosthodont, dipl, 56; FACD. *Prof Exp:* From instr to assoc prof prosthodont, Univ Buffalo, 42-62; clin prof, 62-64, PROF PROSTHODONT & CHMN DEPT, SCH DENT, STATE UNIV NY, BUFFALO, 64- *Concurrent Pos:* Consult, Bethesda Navel Hosp Dent Fac, Eastman Dent Ctr, Rochester, Roswell Park Hosp, Buffalo & Vets Admin Hosp, Buffalo & Batavia, NY. *Honors & Awards:* Clin named in honor, Harold R Ortman Postgrad Clin, Sch Dent Med, State Univ Ny, Buffalo. *Mem:* Am Prosthodont Soc (pres, 80); Am Col Prosthodontists; fel Greater NY Acad Prosthodontists; fel Am Col Dentists. *Res:* Prosthodontics; bone physiology; occlusion; denture retention; denture resins; refitting denture bases with todays concepts and material. *Mailing Add:* 3800 Main St Buffalo NY 14226

ORTMAN, ROBERT A, b Detroit, Mich, Mar 4, 26; m 53; c 4. ZOOLOGY. *Educ:* Univ Calif, BA, 51, MA, 53, PhD(zool), 55. *Prof Exp:* Asst zool, Univ Calif, 53-55, jr res zoologist, Mus Vet Zool, 56; res officer, Med Res Coun, Otago, NZ, 55-56; instr zool, Tulane Univ, 56-57; assoc prof biol, Mem Univ, 57-59; asst prof, Boston Col, 59-62; asst prof, 62-70, ASSOC PROF BIOL, CITY COL NEW YORK, 70- *Concurrent Pos:* Res grants, Nat Res Coun Can, 58-59, USPHS, 59-66. *Mem:* AAAS; Am Soc Zool; Am Asn Anatomists. *Res:* Pituitary cytology and endocrinology; comparative endocrinology; vertebrate histology; cytochemistry; electron microscopy. *Mailing Add:* Dept Biol City Col New York Convent & 138th St New York NY 10031

ORTNER, DONALD JOHN, b Stoneham, Mass, Aug 23, 38; m 60; c 3. PHYSICAL ANTHROPOLOGY. *Educ:* Columbia Union Col, BA, 60; Syracuse Univ, MA, 67; Univ Kans, PhD(anthrop), 70. *Prof Exp:* Grant, 68-70, assoc curator, 70-80, CUR ANTHROP, SMITHSONIAN INST, 80-, CHMN, 88- *Concurrent Pos:* NIH grants, 69-72, 72-73 & 84-87; assoc prof, Univ Md, 71-75; hon vis prof, Univ Bradford, Eng, 88. *Mem:* Am Asn Phys Anthrop; Paleopath Asn; Int Skeletal Soc. *Res:* Calcified tissue biology; paleopathology; biological history of ancient near east; human biocultural adaptation. *Mailing Add:* Dept Anthrop Smithsonian Inst Washington DC 20560

ORTNER, MARY JOANNE, b Windsor, Can, 10, 46; US citizen. PHARMACOLOGY, BIOPHYSICS. *Educ:* Calif State Univ, BS, 68, MS, 71; Univ Hawaii, PhD(pharmacol), 76; Univ Conn, MBA, 87. *Prof Exp:* Instr biol, Ramona Convent High Sch, 68-69; res asst zool, Calif State Univ, 69-71; res asst pharmacol, Univ Hawaii, 71-76; NRSA fel molecular pharmacol, Nat Heart, Lung & Blood Inst, 76-77; staff fel, Nat Inst Environ Health Sci, NIH, 77-82; sr pharmacologist, dept pharmacol, Prin Sci, Dept Med Chem, 87-90, MGR, CLIN RES, BOEHRINGER INGELHEIM PHARMACEUT INC, 90- *Mem:* AAAS; Am Soc Photobiol; Skin Pharmacol Soc; Proj Mgt Inst; Am Soc Pharmacol & Exp Ther. *Res:* Effects of drugs on membranes; biophysical techniques, especially ESR, fluorescence spectroscopy and microscopy, circular dichroism; the mechanism of histamine release from mast cells and the molecular mechanism of compound 48/80; influence of microwave radiation on molecular systems; molecular mechanisms of photosensitization; robotics system, high capacity screening immunology; project management; research management. *Mailing Add:* Dept Clin Res Boehringer Ingelheim Pharmaceut Inc 90 E Ridge Rd Box 368 Ridgefield CT 06877

ORTOLEVA, PETER JOSEPH, b Brooklyn, NY, Nov 5, 42; div; c 1. BIOLOGICAL RHYTHMS, BIOPHYSICS. *Educ:* Rensselaer Polytech Inst, BS, 64; Cornell Univ, PhD(appl physics), 70. *Prof Exp:* Fel chem, Mass Inst Technol, 70-75; asst prof, 75-77, ASSOC PROF CHEM, IND UNIV, 77- *Res:* Onset of spatial and temporal patterns that occur in far from equilibrium reacting systems are studied with the theory of irreversible processes, with applications made to physical, chemical, biological and geological systems; rate of chemical reactions in condensed systems studied using statistical mechanics; growth of chemically complex crystals studied theoretically. *Mailing Add:* Dept Chem Ind Univ Bloomington IN 47401

ORTON, COLIN GEORGE, b London, Eng, June 4, 38; m 64; c 3. MEDICAL PHYSICS. *Educ:* Bristol Univ, BSc, 59; Univ London, MSc, 61, PhD(radiation physics), 65. *Hon Degrees:* MA, Brown Univ, 76. *Prof Exp:* Lectr physics, St Bartholomew's Hosp, Univ London, 61-66; from asst prof to assoc prof physics, Med Ctr, NY Univ, 66-75; chief physicist, RI Hosp, 76-78; assoc prof radiation med, Brown Univ, 76-81; PROF RADIATION ONCOL, MED SCH, WAYNE STATE UNIV, 81-; CHIEF PHYSICIST, HARPER HOSP, 81- *Concurrent Pos:* Consult, Morristown Mem Hosp, NJ, 69-75; lectr, St Vincent's Hosp, New York, 70-71 & Metrop Hosp, New York, 71-75; ed bull, Am Asn Physicists in Med, 71-73; adj prof biol, Fairleigh Dickinson Univ, 73-75; ed, Progress Med Radiation Physics, 78- & Med Physics World, 85-; chmn, Am Col Med Physics, 85. *Honors & Awards:* Marie Curie Gold Medal, 87. *Mem:* Am Asn Physicists in Med (pres, 82); Brit Inst Physics; Brit Inst Radiol; Health Physics Soc; Am Inst Physics; Am Soc Therapeut Radiologists; Radiol Soc NAm; Radiat Res Soc; Am Col Radiol; Am Col Med Phys. *Res:* Radiation; radiobiology; radiation oncology; computers in medicine; radiation dosimetry. *Mailing Add:* Radiation Oncol Ctr Harper Grace Hosp 3990 John R Detroit MI 48201

ORTON, GEORGE W, b Hopkinsville, Ky, Oct 31, 19; m 43; c 2. MATERIALS & ENVIRONMENTAL ENGINEERING. *Educ:* Univ Ky, BS, 43; Carnegie Inst Technol, MS, 48; Ohio State Univ, PhD(metall eng), 61. *Prof Exp:* US Air Force, 43-69, sect chief alloy develop, Wright-Patterson AFB, Ohio, 48-52, div chief mfg methods, Ger, 52-55, br chief mech equip, NMex, 55-59, instr metall, US Air Force Acad, 61-62, assoc prof, 62-64, dir res, 64-66, exchange prof, Philippine Mil Acad, 66-68, chief mfg technol div, Air Force Systs Command, Wright-Patterson AFB, Ohio, 68-69; PROF ENG, UNIV PR, MAYAGUEZ, 69- *Concurrent Pos:* Chmn, PR Comt Indust Pollution. *Mem:* Am Soc Eng Educ; Am Soc Metals; Sigma Xi. *Res:* Development of high-strength, low density alloys for aerospace vehicles; adaption of American manufacturing standards to European plants; design of material handling systems. *Mailing Add:* 10618 Emerald Point Sun City AZ 85351

ORTON, GLENN SCOTT, b Fall River, Mass, July 24, 48; m 79; c 2. PLANETARY ATMOSPHERES. *Educ:* Brown Univ, ScB, 70; Calif Inst Technol, PhD(planetary sci), 75. *Prof Exp:* Assoc scientist, Div Geol & Planetary Sci, 75, resident res planetary atmospheres, Space Sci Div, 75-77, sr scientist, Earth & Space Sci Div, 77-79, MEM TECH STAFF, NASA JET PROPULSION LAB, CALIF INST TECHNOL, 79- *Concurrent Pos:* Mem adv missions studies team, Jet Propulsion Lab, 78-; Mem NASA infrared telescope facil, Proposal Review Comt, 81-83 & 88-90; div Planet Sci Comt, Am Astron Soc, 86-88; mem sci teams, Pioneer 10/11, Galileo, Cassini; Planetary Atmospheres Mgt/Opers Working Group, NASA, 85-90. *Mem:* Sigma Xi; Int Astron Union; Am Astron Soc. *Res:* Determination of thermal structure, composition, opacity, energy balance and climatology of planetary atmospheres; remote sensing of thermal radiation and reflected solar radiation from planetary atmospheres. *Mailing Add:* 169-237 Jet Propulsion Lab 4800 Oak Grove Dr Pasadena CA 91109

ORTON, JOHN PAUL, b June 25, 25; Can citizen; m 55; c 2. METALLURGY & PHYSICAL METALLURGICAL ENGINEERING. *Educ:* Imp Col Sci & Technol, BSc, 48. *Prof Exp:* Metallurgist iron & steel making, Stelco Inc, 54-65, mgr metall serv, 64-89; PRES METALL CONSULT, ORTON METALL INC, 90- *Honors & Awards:* E H Gary Medal, Am Iron & Steel Inst, 77. *Mem:* Fel Am Soc Metals; fel Am Soc Testing & Mat; Can Inst Mining & Metall; Am Inst Mining & Metall Engrs. *Mailing Add:* 158 Kent St Hamilton ON L8P 3Z3 Can

ORTON, WILLIAM R, JR, b Texarkana, Tex, May 4, 22; m 50; c 4. MATHEMATICS. *Educ:* Univ Ark, BA, 47; Univ Ill, MA, 48, PhD(math), 51. *Prof Exp:* Fulbright fel, Univ Paris, 51-52; instr math, Oberlin Col, 52-53; from asst prof to assoc prof, 53-61, prof, 61-87, EMER PROF MATH, UNIV ARK, FAYETTEVILLE, 87- *Concurrent Pos:* NSF fac fel, Univ Wash, 65-66; mem NSF staff, India, 67-68; Fulbright lectr, Univ Tehran, Iran, 77-78; dir, Minority Engr Prog, 81-82. *Mem:* Am Math Soc; Math Asn Am. *Res:* Integration; mathematics education. *Mailing Add:* Dept Math SE 301 Univ Ark Fayetteville AR 72701

ORTS, FRANK A, b Gonzales, Tex, May 6, 31; m 53; c 2. MEAT SCIENCE. *Educ:* Tex A&M Univ, BS, 53, MS, 59, PhD(animal sci), 68. *Prof Exp:* Instr meats, 59-63, asst prof animal sci, 63-81, exten meat specialist, Tex A&M Univ, 63-88; RETIRED. *Honors & Awards:* Distinguished Meats Exten-Indust Award, Am Meat Sci Asn, 75. *Mem:* Am Meat Sci Asn; Am Soc Animal Sci; Sigma Xi. *Res:* Carcass composition. *Mailing Add:* Rte 3 Box 1520 Bryan TX 77802

ORTTUNG, WILLIAM HERBERT, b Philadelphia, Pa, June 16, 34; m 63; c 2. PHYSICAL CHEMISTRY. *Educ:* Mass Inst Technol, SB, 56; Univ Calif, Berkeley, PhD(phys chem), 61. *Prof Exp:* Asst prof chem, Stanford Univ, 60-63; from asst prof to assoc prof, 63-79, PROF CHEM, UNIV CALIF, RIVERSIDE, 79- *Mem:* AAAS; Am Chem Soc; Am Phys Soc. *Res:* Theory of molecular polarizability; molecular electrostatics; dielectric theory; electro-optical phenomena; crystal and molecular optics; physical biochemistry. *Mailing Add:* Dept of Chem Univ of Calif Riverside CA 92521

ORTWERTH, BERYL JOHN, b St Louis, Mo, Aug 13, 37. BIOCHEMISTRY, OPHTHALMOLOGY. *Educ:* St Louis Univ, BS, 59; Univ Mo, MS, 62, PhD(biochem), 65. *Prof Exp:* Asst prof, 68-73, assoc prof, 73-81, PROF OPHTHAL, UNIV MO-COLUMBIA, 81- *Concurrent Pos:* Am Cancer Soc grant, Oak Ridge Nat Lab, 66-68; sabbatical leave, Nat Cancer Inst, 76. *Honors & Awards:* Alcon Distinguished Sci Award. *Mem:* Int Soc Eye Res; Am Soc Biol Chem; Asn Res Vision & Ophthal; Am Chem Soc. *Res:* Glycation mechanisms; lens proteinases and inhibitors; cataract biochemistry. *Mailing Add:* Mason Inst Ophthal Univ Mo Columbia MO 65212

ORVIG, CHRIS, b Montreal, Que. BIOINORGANIC CHEMISTRY. *Educ:* McGill Univ, BSc, 76; Mass Inst Technol, PhD(inorg chem), 81. *Prof Exp:* Nat Sci & Eng Res Coun Can postdoctoral fel, Univ Calif, Berkeley, 81-83; postdoctoral, McMaster Univ, 83-84; asst prof chem, 84-90, ASSOC PROF CHEM & PHARMACEUT SCI, UNIV BC, 90- *Mem:* Am Chem Soc; Sigma Xi. *Res:* Bioinorganic chemistry and the study of metallo-drugs. *Mailing Add:* Dept Chem Univ BC Vancouver BC V6T 1Z1 Can

ORVILLE, HAROLD DUVALL, b Baltimore, Md, Jan 23, 32; m 54; c 4. METEOROLOGY. *Educ:* Univ Va, BA, 54; Fla State Univ, MS, 59; Univ Ariz, PhD(meteorol), 65. *Prof Exp:* Res asst meteorol, Univ Ariz, 63-65; from asst prof to prof, 65-73, HEAD NUMERICAL MODELS GROUP, INST ATMOSPHERIC SCI, SDAK SCH MINES & TECHNOL, 65-, HEAD DEPT, 74- *Concurrent Pos:* Mem, Int Comn Cloud Physics, 71-80; consult, Off Environ Modification, Nat Oceanic & Atmospheric Admin, 72-73; ed, J Atmospheric Sci, 78-85; sr sci officer, World Meteorol Orgn, 82; counr, Am Meteorol Soc, 83-86. *Mem:* Am Meteorol Soc; Am Geophys Union; Sigma Xi; Weather Modification Asn. *Res:* Numerical modeling of medium scale atmospheric motions and cloud development; cloud modification; physics and dynamics of clouds; severe storms. *Mailing Add:* Inst Atmospheric Sci SDak Sch Mines & Technol Rapid City SD 57701-3995

ORVILLE, PHILIP MOORE, b Ottawa, Ill, Feb 24, 30; m 57; c 3. GEOLOGY. *Educ:* Calif Inst Technol, BS, 52; Yale Univ, MA, 54, PhD(geol), 58. *Prof Exp:* Fel petrol, Geophys Lab, Carnegie Inst Technol, 57-60; asst prof petrol & mineral, Cornell Univ, 60-62; assoc prof, 62-72, PROF PETROL & MINERAL, YALE UNIV, 72- *Concurrent Pos:* Ed, Am J Sci. *Mem:* Geol Soc Am; Am Geophys Union; Geochem Soc; Mineral Soc Am. *Res:* Experimental petrology; feldspars; equilibrium between volatile and silicate phases. *Mailing Add:* 127 Everit New Haven CT 06511

ORVILLE, RICHARD EDMONDS, b Long Beach, Calif, July 28, 36; m 64; c 2. ATMOSPHERIC PHYSICS, SPECTROSCOPY. *Educ:* Princeton Univ, AB, 58; Univ Ariz, MS, 63, PhD(meteorol), 66. *Prof Exp:* Mgr mfg, Procter & Gamble Co, 59-60; res asst physics, Johns Hopkins Univ, 60-61; res asst meteorol, Univ Ariz, 61-64, res assoc, 64-66; sr res scientist, Westinghouse Elec Corp, 66-68; assoc prof, 68-81, chmn, 83-88, PROF ATMOSPHERIC SCI, STATE UNIV NY ALBANY, 81- *Concurrent Pos:* Assoc prog dir, NSF, 70-71; vis prof, Univ Wis-Madison, 79; mem, Am Meteorol Soc Coun, 75-81, 86-, chmn publ comt, 86-92; pres elect, Atmospheric Sci Sect, Am Geophys Union, 86-88, pres, 88-90; mem bd trustees, Univ Corp Atmospheric Res, 85-90. *Honors & Awards:* Sackler Medal Geophys, Tel Aviv Univ, Israel, 87; First Suomi Lectr, Univ Wisc, 89. *Mem:* fel Am Meteorol Soc; Am Geophys Union. *Res:* Cloud physics; atmospheric electricity; spectroscopy. *Mailing Add:* Dept Atmospheric Sci State Univ NY 1400 Wash Ave Albany NY 12222

ORWIG, EUGENE R, JR, b Los Angeles, Calif, May 19, 19; m 43; c 4. STRUCTURAL GEOLOGY. *Educ:* Univ Calif, Los Angeles, BA, 46, MA, 47, PhD(geol), 58. *Prof Exp:* Res geologist, Mobil Oil, 48-82; RETIRED. *Concurrent Pos:* Consult, 82- *Mem:* Am Asn Petrol Geologists; Fel Geol Soc Am. *Res:* Development of oil & gas project in large areas. *Mailing Add:* PO Box 1682 Rancho CA 92390

ORWIG, LARRY EUGENE, b York, Pa, Jan 8, 44; div; c 1. SPACE SCIENCES, ASTROPHYSICS. *Educ:* Lebanon Valley Col, BS, 65; Univ NH, PhD(physics), 71. *Prof Exp:* Nat Acad Sci-Nat Res Coun fel, 72-74, ASTROPHYSICIST X-RAY ASTRON, GODDARD SPACE FLIGHT CTR, NASA, 74- *Honors & Awards:* Spec Achievement Award, NASA, 81. *Mem:* Am Phys Soc; Am Astron Soc. *Res:* Experimental x-ray and gamma-ray astronomy; solar flare phenomena; pulsars. *Mailing Add:* Code 682 NASA Greenbelt MD 20771

ORWOLL, ROBERT ARVID, b Minneapolis, Minn, Aug 28, 40; m 72; c 2. THERMODYNAMIC PROPERTIES OF LIQUID CRYSTALS. *Educ:* St Olaf Col, BA, 62; Stanford Univ, PhD(chem), 66. *Prof Exp:* Res instr chem, Dartmouth Col, 66-68; inst mat sci fel chem, Univ Conn, 68-69; from asst prof to assoc prof, 69-82, PROF CHEM, COL WILLIAM & MARY, 82-, DIR, APPL SCI PROG, 88- *Concurrent Pos:* Vis scholar, Stanford Univ, 80-81. *Mem:* AAAS; Am Chem Soc. *Res:* Measurement of equation-of-state properties of liquid crystals and study of their phase equilibria; thermodynamic properties of polymer solutions; thermodynamic properties of liquid crystals. *Mailing Add:* Dept Chem Col William & Mary Williamsburg VA 23185

ORY, HORACE ANTHONY, b Amite, La, Dec 16, 32; m 58. PHYSICAL CHEMISTRY. *Educ:* Southeastern La Col, BS, 53; La State Univ, PhD(chem), 57. *Prof Exp:* Phys scientist, Monsanto Chem Co, 57-60, E H Plesset Assocs, 60-63, Astrophys Res Corp, 63-64, Heliodyne Corp, 63-64 & Rand Corp, 64-71; PHYS SCIENTIST, R&D ASSOCS, 71- *Mem:* AAAS; Am Chem Soc; Am Phys Soc; fel Am Inst Chem; Sigma Xi. *Res:* Spectroscopy; reaction kinetics; electrooptics. *Mailing Add:* 1752 Thurber Pl Burbank CA 91501

ORY, MARCIA GAIL, b Dallas, Tex, Feb 8, 50; m 72. SOCIAL GERONTOLOGY, MEDICAL SOCIOLOGY. *Educ:* Univ Tex-Austin, BA, 71; Ind Univ, MA, 73; Purdue Univ, PhD, 76; Johns Hopkins Univ, MPH, 81. *Prof Exp:* Res assoc, Sch Pub Health, Univ NC, Chapel Hill, 75-76, res asst prof, 76-77; res fel, Minn Family Study Ctr, 77-78, vis asst prof, Health Servs Res Ctr, Univ Minn, Minneapolis, 77-78; asst prof, Sch Pub Health, Univ Ala, Birmingham, 78-80; prog dir, Biosoc Aging & Health Sect, Behav Sci Res Prog, 81-86, CHIEF SOCIAL SCI RES AGING, BEHAV & SOC RES PROG, NAT INST AGING, NIH, BETHESDA, MD, 87- *Concurrent Pos:* Nat Inst Mental Health fel, 77, 81; adj asst to assoc prof, Sch Pub Health, Univ NC, Chapel Hill, 78-88; gov coun, Geront Health Sect, Am Pub Health Asn, 86-88, chair-elect, 89-90, chair, 91-92; prog chair, Pub Health Track, Soc Behav Med, 88-89. *Mem:* Am Pub Health Asn; Soc Behav Med; Am Sociol Asn; Geront Soc Am; Asn Health Servs Res. *Res:* Research and research administration in social and behavioral aspects of aging, health, health care and cancer prevention in old age; injury prevention and control; gender differences in health and longevity; burdens of care for Alzheimer's disease. *Mailing Add:* 9810 Parkwood Dr Bethesda MD 20814

ORY, ROBERT LOUIS, b New Orleans, La, Nov 26, 25; m 48, 77; c 2. BIOCHEMISTRY, NUTRITION. *Educ:* Loyola Univ, La, BS, 48; Univ Detroit, MS, 50; Agr & Mech Col, Tex, PhD(biochem, nutrit), 54. *Prof Exp:* Asst, Univ Detroit, 48-50; chemist, Exp Sta, Agr & Mech Col, Tex, 50-54; chemist, Colloidal Properties Unit, Southern Regional Res Ctr, 54, oil seeds sect, 55-57, biochemist, Seed Proteins Pioneering Lab, 58-69, head protein properties res, Oilseed & Food Lab, 69-72, res leader biochem res, 72-85; CONSULT, 85- *Concurrent Pos:* Fulbright res scholar, Polytechnic Inst Denmark, 68-69; chmn, Div Agr & Food Chem, Am Chem Soc, 80, Gulf Coast Sect, Inst Food Technol; adj prof food sci, Va Tech, 75- *Honors & Awards:* Distinguished Serv Award, Div Agr & Food, Am Chem Soc, 86; Award of Merit, Am Oil Chem Soc, 87. *Mem:* Am Chem Soc; Sigma Xi; Am Soc Biol Chemists; Inst Food Technologists; Am Oil Chemists' Soc; Am Asn Cereal Chemists; Am Peanut Res & Educ Soc; Plant Growth Regulator Soc Am. *Res:* Biosynthesis of amino acids in microorganisms; isolation and characterization of proteins and enzymes, including lipases, proteases, phosphatases and amylases, from oilseeds and barley seeds; plant lectins; biosynthesis of aflatoxins; rice dietary fiber; peanut proteins for food uses; antigens in cotton dust and bioassay for byssinosis; plant growth regulators; published 165 papers in referred journals and edited 6 books. *Mailing Add:* 7324 Ligustrum Dr New Orleans LA 70126

ORZECH, CHESTER EUGENE, JR, b Chicago, Ill, Dec 31, 37; m 63; c 4. ANALYTICAL CHEMISTRY, THERMAL ANALYSIS. *Educ:* Ill Inst Technol, BS, 60; Mich State Univ, PhD(org chem), 68. *Prof Exp:* SECT HEAD ANAL CHEM, WYETH-AYERST RES, 65- *Concurrent Pos:* NSF Fel, 62-64. *Mem:* NAm Thermal Anal Soc; Soc Appl Spectros; Sigma Xi. *Res:* Analytical chemistry of pharmaceuticals; infrared spectroscopy; NMR spectroscopy; thermal analysis of pharmaceuticals; x-ray diffraction; x-ray fluorescence. *Mailing Add:* Wyeth-Ayerst Res Rouses Point NY 12979

ORZECH, MORRIS, b Arg, Feb 10, 42; US citizen; m 63. MATHEMATICS. *Educ:* Columbia Col, AB, 62; Cornell Univ, PhD(math), 67. *Prof Exp:* Asst prof math, Univ Ill, Urbana, 66-68; from asst prof to assoc prof, 68-82, PROF MATH, QUEEN'S UNIV, ONT, 82- *Concurrent Pos:* Inst Adv Studies, Princeton Univ, 74-75, Cornell Univ, 81-82. *Mem:* Can Math Soc; Am Math Soc. *Res:* Algebra, primarily Brauer groups of commutative rings. *Mailing Add:* Dept Math-Statis Queen's Univ Kingston ON K7L 3N6 Can

ORZECHOWSKI, RAYMOND FRANK, b Camden, NJ, Feb 28, 38; m 61; c 3. PHARMACOLOGY. *Educ:* Philadelphia Col Pharm, BSc, 59; Temple Univ, MS, 61, PhD(pharmacol), 64. *Prof Exp:* Sr res pharmacologist, Nat Drug Co, Richardson-Merrell, Inc, Philadelphia, 64-70; from asst prof to assoc prof, 70-80, PROF PHARMACOL, PHILADELPHIA COL PHARM & SCI, 80- *Mem:* AAAS; Am Pharmaceut Asn; Sigma Xi; Soc Toxicol; Am Soc Pharmacol. *Res:* cardiovascular and autonomic pharmacology; inflammation and anti-inflammatory agents; peripheral dopamine receptors. *Mailing Add:* 124 Cherry Tree Lane Cherry Hill NJ 08002

OSAWA, YOSHIO, b Tokyo, Japan, May 28, 30; m 55; c 2. ORGANIC CHEMISTRY, ENDOCRINOLOGY. *Educ:* Tokyo Metrop Univ, BS, 53; Univ Tokyo, MS, 55, PhD(org chem), 59. *Prof Exp:* Res scientist, Tsurumi Res Lab Chem, Teikoku Hormone Mfg Co, Japan, 56-58, chief div steroid chem, Div Total Synthesis of steroids, 58-64, prin res scientist, 64-67; assoc scientist, Med Found Buffalo, 67-70, head dept steroid biochem, 71-75, prin res scientist & assoc dir, 70-78, HEAD DEPT ENDOCRINE BIOCHEM, MED FOUND BUFFALO, 75-; asst res prof biochem, 74-81, ASSOC RES PROF, ROSWELL PARK MED DIV, STATE UNIV NY BUFFALO, 81-, ASSOC DIR, 88- *Concurrent Pos:* Fel steroid chem, Dept Med, Roswell Park Mem Inst, 60-63; Am Cancer Soc fac res award, 69-74; consult to clin staff, Roswell Park Mem Inst, 73-; post-doctoral training prog, Med Found Buffalo, 70-78; vis prof, Sch Med, Nihon Univ, Tokyo, 77-85. *Honors & Awards:* NIH Merit Award, Nat Inst Child Health, 86- *Mem:* Am Chem Soc; Chem Soc Japan; Endocrine Soc; NY Acad Sci; Sigma Xi. *Res:* Synthesis and structure-function relationships of estrogens; total synthesis of steroids; steroid reactions; polymer-iron preparation for anemia; estrogen metabolites; stereochemistry and mechanism of steroid biosynthesis; clinical assay of estrogen conjugates; steroid biochemistry and endocrinology; aromatose purification and inhibitors; aromatose in breast cancer. *Mailing Add:* Med Found Buffalo 73 High St Buffalo NY 14203

OSBAHR, ALBERT J, JR, b Nov 4, 31; US citizen; m 57; c 3. PHYSICAL BIOCHEMISTRY, PHYSICAL ORGANIC CHEMISTRY. *Educ:* Fordham Univ, BS, 52, MS, 53; Georgetown Univ, PhD(phys biochem), 59. *Prof Exp:* Chemist, Cent Labs, NY, 52-53 & Quaker Maid Labs, 53-54; biochemist, Naval Med Res Inst, 55-58; res biochemist, Geront Inst, NIH, 58-60, RES INVESTR, NAT INST ARTHRITIS, METAB & DIGESTIVE DIS, 60- *Concurrent Pos:* Prof lectr, Georgetown Univ, 58- *Mem:* AAAS; Sigma Xi. *Res:* Structure of macromolecules and its correlation to function of molecules; relationship of structure to biological activity; protein modification; enzymology and enzyme kinetics; function of metal ions in biological systems; blood clotting mechanism; muscle biochemistry; polymerization of proteins as involved in muscular contraction and blood coagulation. *Mailing Add:* 1000 Smithlevel Rd C1 The Vigs Carrboro NC 27510

OSBERG, G(USTAV) L(AWRENCE), b Alta, Can, July 9, 17; m 44; c 3. CHEMICAL ENGINEERING. *Educ:* Univ Alta, BSc, 39; Univ Toronto, MA, 47, PhD(phys chem), 49. *Prof Exp:* Jr res officer, Nat Res Coun Can, 40-41, from assoc res officer to sr res officer chem eng & head sect, Appl Chem Div, 49-69, chief, Financial Planning & Budgeting, 67-81; RETIRED. *Concurrent Pos:* Ed, Can J Chem Eng, 62-67. *Res:* Fluidized solids applications and theory; catalytic oxidation. *Mailing Add:* 41 Bedford Crescent Ottawa ON K1K 0E6 Can

OSBERG, PHILIP HENRY, b Melrose, Mass, Oct 23, 24; m 48; c 2. GEOLOGY. *Educ:* Dartmouth Col, AB, 47; Harvard Univ, AM, 49, PhD(geol), 52. *Prof Exp:* Asst prof geol, Pa State Univ, 51-52; from asst prof to assoc prof, Colby Col, 52-57; assoc prof, 57-63, head dept geol sci, 66-71, PROF GEOL, UNIV MAINE, ORONO, 63- *Concurrent Pos:* Fulbright sr res fel, Oslo, Norway, 65-66; geologist, US Geol Surv, var times, 68- *Mem:* Geol Soc Am; Mineral Soc Am; Norweg Geol Soc. *Res:* Metamorphic petrology; structural geology; regional geology of the northern Appalachians. *Mailing Add:* PO Box 224 Orono ME 04473

OSBORG, HANS, b Halberstadt, Ger, Oct 18, 01; nat US; m 30; c 2. PHYSICAL CHEMISTRY. *Educ:* Univ Braunschweig, Ger, BS, 24, MS, 25, DrIng, 27. *Prof Exp:* Mem staff, Res Inst, Ger Metal Co, Inc, 27-29; dir res prod & eng, Lithium Metal Dept, Maywood Chem Works, 29-34; exec vpres, Del, 58-66, PRES, CHEMIRAD CORP, DEL, 66-, TECH DIR, 58-, PRES, NY, 50- *Concurrent Pos:* Mgt consult, 34-; tech dir, Metalectro Corp, Md, 50-53, pres, 53-55. *Mem:* Am Chem Soc; Am Inst Mining, Metall & Petrol Eng; NY Acad Sci; Tech Asn Pulp & Paper Indust. *Res:* Ethylene imines; aziridines; hydrazines, especially unsymmetrical dimethyl hydrazine; chemistry and metallurgy of lithium; boron hydrides; especially diborane and its compounds. *Mailing Add:* PO Box 152 80 Longview Rd Port Washington NY 11050-0163

OSBORN, CLAIBORN LEE, b Austin, Tex, Sept 15, 33; m 59; c 3. POLYMER CHEMISTRY, RADIATION CHEMISTRY. *Educ:* Univ Tex, BA, 56, PhD(org chem), 60. *Prof Exp:* Chemist, Celanese Chem Co, 60-62, sr res chemist, 62-63; Robert A Welch Found fel photochem, Univ Tex, 63-65; res chemist, Union Carbide Corp, 65-68, proj scientist radiation chem, 68-72, res scientist, 72-75, sr res scientist, 75-78, res assoc, 79-88; RETIRED. *Mem:* Am Chem Soc; AAAS; Union Concerned Scientists. *Res:* Photochemistry; application of photochemical techniques to polymer processing; radiation processing of electronic materials; polymer characterization; photoresists; photoelectro chemistry; integrated circuits technology. *Mailing Add:* 3905 Staunton Ave SE Charleston WV 25304

OSBORN, DONALD EARL, b Cleveland, Ohio, Mar 25, 52; m 76; c 3. SOLAR ENERGY RESEARCH. *Educ:* Univ Ariz, BS, 76, MS, 85. *Prof Exp:* Physicist, Helio Assocs Inc, 75-77; assoc dir solar energy, Ariz Solar Energy Comn, 77-81; instr, Maricopa Tech Community Col, 80; PRIN CONSULT, SPECTRUM ENERGY ASSOCS, 78-; dir res, 81-88, SR RES SPECIALIST, UNIV ARIZ SOLAR & ENERGY RES FACIL, 88- *Concurrent Pos:* Res engr, Optical Sci Ctr, Univ Ariz, 77; prin investr, Col Eng, Univ Ariz, 81- *Mem:* Int Solar Energy Soc; Am Solar Energy Soc; Am Soc Heating, Refrig & Airconditioning Engrs; Am Soc Testing & Mat; Asn Energy Engrs; Am Asn Artificial Intel. *Res:* Solar energy, energy management, photothermal, photoquantum, solar cooling, computer applications, fiber composites, photovoltaic, expert systems and system simulation, monitoring and evaluation. *Mailing Add:* SERF Eng Bldg #20 Univ Ariz Tucson AZ 85721

OSBORN, ELBURT FRANKLIN, b Winnebago Co, Ill, Aug 13, 11; m 39; c 2. EXPERIMENTAL GEOCHEMISTRY, RESEARCH ADMINISTRATION. *Educ:* DePauw Univ, BA, 32; Northwestern Univ, MS, 34; Calif Inst Technol, PhD(geol), 38. *Hon Degrees:* ScD, Alfred Univ, 65, Northwestern Univ, 72, DePauw Univ, 72 & Ohio State Univ, 72. *Prof Exp:* Instr geol, Northwestern Univ, 37; geologist, Val d'Or, Can, 38; petrologist, Geophys Lab, Carnegie Inst, 38-42; phys chemist, Nat Defense Res Comt, 42-45; res chemist, Eastman Kodak Co, NY, 45-46; prof geochem, Col Mineral Industs, Pa State Univ, 46-71, head dept earth sci, 46-53, dean col, 53-59, vpres res, 59-70; EMER PROF GEOCHEM, COL EARTH & MINERAL SCI, PA STATE UNIV, UNIVERSITY PARK, 71- , EMER VPRES RES, 71- *Concurrent Pos:* Mem adv comt geophys br, Off Naval Res, Nat Acad Sci-Nat Res Coun, 47-50, mem, Mat Adv Bd, 65-69, chmn comt on mineral sci & technol, 66-70, chmn bd mineral resources & mem comn natural resources, Nat Res Coun, 74-77; mem earth sci panel, NSF, 53-55, mem div comt math, phys & eng sci, 55-59, chmn, 57-58, mem adv panels course content improv prog, 60-61, phys sci facil, 60-64, consult sci & technol, Policy Off, 74-76; mem bd dirs, Am Geol Inst, 56-59; NSF sr fel, Cambridge Univ, 58; mem adv panel mineral prod div, Nat Bur Standards, 58-62, chmn, 58-59, metall div, 58-64, mem div chem & chem technol, 60-63; mem bd, Univ Corp Atmospheric Res, 59-67; mem adv comt basic res, US Army Res Off Ceramics, 60-63; vchmn, Sirimar Corp, Italy, 61-63; mem bd, Geisinger Med Ctr, 62-83, emer mem bd, 83-, mem bd dirs, Inst Med Educ & Res, 69-74; mem, Pa Health Res Inst, 66-70; mem bd dirs, Pa Sci & Eng Found, 68-71, 74-76; mem earth sci adv bd, Stanford Univ, 69-72; mem bd dirs, Pa Res Corp, 71-, vpres, 74-80, pres, 80-; dir, Bur Mines, Washington, DC, 70-73; distinguished prof, Geophys Lab, Carnegie Inst, Washington, 73-77; mem adv bd, Col Eng, Univ Calif, Berkeley, 74-78; mem mat adv comt, Off Technol Assessment, Cong of US, 74-79; mem geosci adv panel, Los Alamos Sci Lab, 75-79; chmn, US Nat Comt for Geol, Nat Acad Sci, 75-77; chmn adv comt mining & mineral eng res, Secy of Interior, 78-81; mem, Continental Sci Drilling Comt, Nat Res Coun, 80-83, Panel Geochem Fibrous Mat Related to Health Risks, 79-81 & chmn, Comt Geol Aspects Indust Waste Disposal, 79-81. *Honors & Awards:* Award, Am Iron & Steel Inst, 54; Roebling Medal, Mineral Soc Am, 72; John Jeppson Award, Am Ceramic Soc, 73; Hal Williams Hardinge Award, Soc Mining Engrs, 74; Albert Victor Bleininger Award, Am Ceramic Soc, 76. *Mem:* Nat Acad Eng; Am Inst Metall & Mining Engrs; distinguished mem Am Ceramic Soc (pres, 65); hon mem Can Ceramic Soc; Soc Econ Geologists (pres, 72); fel Geol Soc Am; Geochemical Soc (pres, 67); fel Mineral Soc Am (pres, 60); Distinguished mem Soc Mining Eng; fel AAAS; fel Am Geophys Union. *Res:* Experimental high temperature and high pressure phase equilibrium research on oxide systems; genesis of volcanic magmas and problems of steelplant refractories and blast furnace slags; physical chemistry of refractories; blast furnance and open hearth slags; phase equilibira in mineral systems; industrial minerals; mineral synthesis; mineral and energy resources. *Mailing Add:* 330 E Irvin Ave State College PA 16801

OSBORN, H(ARLAND) JAMES, b Alexandria, Minn, June 24, 32; m 53; c 3. PHOTOGRAPHIC & SYNTHETIC ORGANIC CHEMISTRY. *Educ:* Univ Minn, BS, 54, PhD(org chem), 58. *Prof Exp:* Res chemist, Eastman Kodak Co, 58-88; RETIRED. *Concurrent Pos:* Adj prof org chem, Col Continuing Educ, Rochester Inst Technol, 89- *Mem:* Am Chem Soc. *Res:* Organic synthesis; design of color photographic systems; chemistry and physics of photographic development. *Mailing Add:* 24 Skycrest Dr Rochester NY 14616

OSBORN, HENRY H(OOPER), b Hartford, Conn, Dec 3, 26; m 59; c 2. MECHANICAL ENGINEERING. *Educ:* Tufts Univ, BSME, 50; Purdue Univ, MSME, 56, PhD(mech eng), 58. *Prof Exp:* Anal engr, Pratt & Whitney Aircraft Div, United Aircraft Corp, 50-52; instr mech eng, Bucknell Univ, 52-54; from asst prof to assoc prof, Purdue Univ, 58-64; proj engr, 64-65, mgr appl res, 65-77, CHIEF ENGR, AIR PREHEATER CO, INC, 77- *Mem:* Am Soc Mech Engrs; Am Soc Eng Educ. *Res:* Thermodynamics; heat and mass transfer; air pollution control; water desalination. *Mailing Add:* Rd 4 Box 394 Wellsville NY 14895

OSBORN, HERBERT B, b San Juan, PR, July 9, 29. HYDROLOGY. *Educ:* Stanford Univ, BS, 52, MS, 53, CE, 56; Univ Ariz, PhD, 71. *Prof Exp:* Hydraul engr, Ariz, 56-58, Miss, 58-60, RES HYDRAUL ENGR, SOUTHWEST WATERSHED RES CTR, USDA, 61- *Concurrent Pos:* Staff scientist, Nat Prog Staff, Agr Res Serv, USDA, 74-75; mem steering comt, Bd Agr & Renewable Resources, Nat Res Coun-Nat Acad Sci, 75-76. *Mem:* AAAS; Am Geophys Union; Am Meteorol Soc; Int Asn Statist in Phys Sci. *Res:* Hydrologic research on thunderstorms and thunderstorm runoff on semiarid rangeland watersheds in the southwest. *Mailing Add:* 2341 S Lazy A Pl Tucson AZ 85713

OSBORN, HOWARD ASHLEY, b Evanston, Ill, May 16, 28; m 51; c 4. MATHEMATICS. *Educ:* Princeton Univ, AB, 49; Stanford Univ, PhD, 55. *Prof Exp:* Instr math, Univ Calif, 54-56; from asst prof to assoc prof, 56-65, PROF MATH, UNIV ILL, URBANA, 65- *Concurrent Pos:* Consult, Rand Corp, 54-65; NSF grants, 58-60, 62- *Mem:* Am Math Soc. *Res:* Differentiable manifolds. *Mailing Add:* Dept Math 361 Altgerd Hall Univ Ill Urbana IL 61801

OSBORN, J MARSHALL, JR, b New York, NY, Sept 11, 30; m 59, 75; c 2. MATHEMATICS. *Educ:* Princeton Univ, AB, 52; Univ Chicago, PhD(math), 57. *Prof Exp:* Instr math, Univ Conn, 57; res instr, 57-59, from asst prof to assoc prof, 59-67, PROF MATH, UNIV WIS-MADISON, 67- *Honors & Awards:* Sigma Xi. *Mem:* Am Math Soc; Math Asn Am. *Res:* Nonassociative algebra. *Mailing Add:* Dept of Math Univ of Wis 480 Lincoln Dr Madison WI 53706

OSBORN, J(OHN) R(OBERT), b Kansas City, Mo, Aug 11, 24; m 45; c 3. MECHANICAL ENGINEERING. *Educ:* Purdue Univ, BS, 50, MS, 53, PhD(mech eng), 57. *Prof Exp:* Jr engr, Thiokol Chem Corp, 50-51; asst, Purdue Univ, West Lafayette, 51-57, from asst prof to assoc prof mech eng, 57-61, prof mech eng, 61-70 & 71-79, prof aeronaut & astron, 80-89; RETIRED. *Concurrent Pos:* Br chief, Ballistics Res Lab, Aberdeen Proving Ground, 70-71. *Mem:* Am Inst Aeronaut & Astronaut; Soc Automotive Engrs. *Res:* Combustion instability in rockets; high frequency response instrumentation; combustion in solid rockets and interior ballistics. *Mailing Add:* 2405 Sauk Pl Lafayette IN 47905

OSBORN, JAMES MAXWELL, b Ypsilanti, Mich, Sept 10, 29; m 55; c 4. COMPLEX ANALYSIS. *Educ:* Univ Mich, BS, 51, MS, 52, PhD(math), 55. *Prof Exp:* Instr math, Ohio State Univ, 54-57; asst prof, 57-60, ASSOC PROF MATH, GA INST TECHNOL, 60- *Mem:* Am Math Soc; Math Asn Am. *Res:* Complex variables. *Mailing Add:* Sch Math Ga Inst Technol Atlanta GA 30332

OSBORN, JEFFREY L, RENAL & CARDIOVASCULAR PHYSIOLOGY, HYPERTENSION. *Educ:* Mich State Univ, PhD(physiol), 79. *Prof Exp:* ASST PROF PHYSIOL, MED COL WIS, 81- *Res:* Neural control of circulation. *Mailing Add:* Dept Physiol Med Col Wis 8701 Watertown Plank Rd Milwaukee WI 53226

OSBORN, JEFFREY WHITAKER, b London, Eng, Jan 2, 30; m 73; c 4. DENTAL ANATOMY, PALEONTOLOGY. *Educ:* Univ London, BDS, 55, PhD(anat), 66; FDSRCS, 60. *Prof Exp:* Registr conserv dent, Guy's Hosp, Univ London, 56-57, demonstr dent anat, 57-64, lectr anat, 64-68, sr lectr, 68-71, reader, 71-75, prof, 75-78; PROF ORAL BIOL, FAC DENT, UNIV ALTA, 78- *Concurrent Pos:* Vis prof vert paleont, Mus Comp Zool, Harvard Univ, 71-72. *Res:* Biomechanics of jaws evolution, comparative anatomy, development and structure of teeth and jaws. *Mailing Add:* Dept Oral Biol Rm 5151B Dent Pharm Bldg Univ Alberta Edmonton AB T6G 2E2 Can

OSBORN, JOHN EDWARD, b Onamia, Minn, July 12, 36; m 59; c 3. NUMERICAL ANALYSIS, PARTIAL DIFFERENTIAL EQUATIONS. *Educ:* Univ Minn, Minneapolis, BS, 58, MS, 63, PhD(math), 65. *Prof Exp:* From asst prof to assoc prof, 65-75, chair math dept, 82-85, actg dean, Col Computers, Math & Phys Sci, 89-90, PROF MATH, UNIV MD, 75- *Concurrent Pos:* NSF res grants, 66-68 & 72-89; mem coun, Am Math Soc, 84-86; assoc ed, Math Computations, 81- *Mem:* Am Math Soc; Soc Indust & Appl Math. *Res:* Numerical solution of partial differential equations. *Mailing Add:* Dept Math Univ Md College Park MD 20742

OSBORN, JUNE ELAINE, b Endicott, NY, May 28, 37; m 66; c 3. VIROLOGY, INFECTIOUS DISEASE. *Educ:* Oberlin Col, BA, 57; Case Western Reserve Univ, MD, 61. *Hon Degrees:* DSc, Univ Med & Dent, NJ, 89. *Prof Exp:* Intern & resident pediat, Children's Hosp Med Ctr, Boston, 61-62 & 63-64; resident, Mass Gen Hosp, 62-63; fel infectious dis, Sch Med, Johns Hopkins Univ, 64-65 & Sch Med, Univ Pittsburgh, 65-66; from instr to prof pediat & med microbiol Sch Med, Univ Wis-Madison, 66-84; DEAN, SCH PUB HEALTH, UNIV MICH, ANN ARBOR, 84- *Concurrent Pos:* Mem panel viral & rickettsial vaccines, Food & Drug Admin, 73-79, mem exp virol study sect, NIH Div Res Grants, 75-79; assoc dean, Grad Sch Biol Sci, Univ Wis-Madison, 75; mem, Panel Vaccines & Related Immunol Prod, 80-, chair, Life Sci Associateship Review Panel, Nat Res Coun, 80-, mem, Med Affairs Adv Comt, Yale Univ, 81-86; chmn, AIDS Adv Comt, Nat Heart Lung & Blood Inst, 84-89; mem, Robert Wood Johnson Found Adv Comt on AIDS Health Serv, 86-90, chmn, 88-90; chmn, US Nat Comn AIDS, 89- *Mem:* Am Acad Pediat; Am Asn Immunologists; Soc Pediat Res; Infectious Dis Soc Am; Am Acad Microbiol; Inst Med. *Res:* Pathogenesis of cytomegalovirus infections; nucleic acid homology of human and simian papovaviruses; mechanism of papovavirus oncogenesis; mechanisms of viral persistence and latency; public policy and HIV epidemic. *Mailing Add:* Sch Pub Health Univ Mich 109 Observatory St Ann Arbor MI 48109

OSBORN, KENNETH WAKEMAN, b Lansing, Mich, Feb 4, 37. MARINE BIOLOGY, COASTAL ECOLOGY. *Educ:* Mich State Univ, BS, 60. *Prof Exp:* Marine biologist res, Tex Game & Fish Comn, 61-63; fishery biologist res, US Bur Com Fisheries, 63-67; fishery biologist environ impact assessment, US Bur Sport Fish & Wildlife, 67-69; marine biologist, US Army Corps Engrs, 69-72; fisheries biologist coastal zone mgt, Nat Marine Fisheries Serv, 72-78; FISHERIES ADVISOR FISH DEVELOP, USAID, 78- *Mem:* Am Inst Fishery Res Biologists; Gulf Estuarine Res Soc; Coastal Soc. *Res:* Marine, estuarine and coastal fishery biology, ecology, development and management; environmental conservation of living marine resources and the habitats upon which they depend; coastal zone management. *Mailing Add:* 6009 Bush Hill Dr Alexandria VA 22310

OSBORN, MARY JANE, b Colorado Springs, Colo, Sept 24, 27; m 50. MOLECULAR BIOLOGY, BIOCHEMISTRY. *Educ:* Univ Calif, Berkeley, BA, 48; Univ Wash, PHD(biochem), 58. *Prof Exp:* From instr to asst prof microbiol, Sch Med, NY Univ, 61-63; from asst prof to assoc prof molecular biol, Albert Einstein Col Med, 63-68; PROF MICROBIOL, SCH MED, UNIV CONN HEALTH CTR, 68-, HEAD DEPT, 80- *Concurrent Pos:* USPHS fel, Sch Med, NY Univ, 59-61, career develop award & res grants, Sch Med, NY Univ, ALbert Einstein Col Med & Univ Conn Health Ctr, 62-; NSF res grants, Albert Einstein Col Med, 65-68; mem microbial chem study sect, NIH, 68-72; mem res comt, Am Heart Asn, 72-77; mem comt space biol & med, Space Sci Bd, Nat Acad Sci, 74-77; mem bd sci counrs, Nat Heart &

Lung Inst, 75-79; assoc ed, J Biol Chem, 78-80; mem, Nat Sci Bd, 80-86; mem, Nat Inst Gen Med, Sci Adv Coun, 83-86; bd trustees, Biosis, 86-89. *Honors & Awards:* Harvey Lect, 83. *Mem:* Nat Acad Sci; AAAS; Am Soc Biol Chem (pres, 81-82); Am Chem Soc; fel Am Acad Arts & Sci; Fedn Am Soc Exp Biol (pres, 82-83); Am Soc Microbiol. *Res:* Biosynthesis of bacterial lipopolysaccharides; biogenesis of membranes. *Mailing Add:* Dept Microbiol Univ Conn Health Ctr Farmington CT 06032

OSBORN, ROGER (COOK), b Crystal City, Tex, Mar 14, 20; m 42; c 3. MATHEMATICS. *Educ:* Univ Tex, BA, 40, MA, 42, PhD(math), 54. *Prof Exp:* Teacher high sch, Tex, 40-42; instr appl math, 42-44, 46-53, from instr to assoc prof, 53-67, PROF MATH, UNIV TEX, AUSTIN, 67- *Res:* Number theory; differential equations; mathematics education. *Mailing Add:* Dept Math Univ Tex Austin TX 78712

OSBORN, RONALD GEORGE, b Los Angeles, Calif, May 22, 48; m 77. SYSTEMS ECOLOGY, STATISTICS. *Educ:* Calif State Col, Long Beach, BS, 71; San Diego State Univ, MS, 75. *Prof Exp:* Res asst ecol, US Int Biol Prog, 72, 73 & 74; statistician ornith, Patuxent Wildlife Res Ctr, 75-77, WILDLIFE BIOLOGIST ENVIRON MGT, WESTERN ENERGY & LAND USE TEAM, US FISH & WILDLIFE SERV, 77- *Honors & Awards:* Spec Achievement Award, US Fish & Wildlife Serv, 78. *Res:* Development of simulation modeling, geographical analysis and data base structuring to aid understanding and management of ecological processes; geo-processing. *Mailing Add:* 1619 Sagewood Dr Ft Collins CO 80525

OSBORN, TERRY WAYNE, b Roswell, NMex, May 17, 43. BIOCHEMISTRY, PHYSICAL BIOCHEMISTRY. *Educ:* Univ Calif, Riverside, BS, 69, PhD(biochem), 75, Pepperdine Univ, MBA, 81. *Prof Exp:* Res scientist biochem, McGraw Labs, Am Hosp Supply Corp, 76- , res scientist & proj leader, 76, res scientist & proj team leader, 76-77, group leader & sr res scientist biochem & chem, 78-80, mgr clin res, 78-80, group-prod mgr, 80-82; dir marketing, Ivac Corp, Eli Lilly, 82-83, dir mkt planning, 84-87; VPRES SALES & MARKETING, NICHOLS INST, 87- *Mem:* Sigma Xi; Am Chem Soc; Am Oil Chemists Soc; Inst Food Technologists. *Res:* Sales-marketing; development and use of lipids and amino acids solutions for total parenteral nutrition; studies on essential fatty acids; photochemical degradation of tryptophan; vitamin D and steroid binding proteins. *Mailing Add:* 31841 Via Patito Trabuco Canyon CA 92679

OSBORN, WAYNE HENRY, b Los Angeles, Calif, Oct 8, 42; m 71; c 3. PHOTOMETRY, TEACHING. *Educ:* Univ Calif, Berkeley, AB, 64; Wesleyan Univ, MA, 67; Yale Univ, MS, 67, PhD(astron), 71. *Prof Exp:* Adj prof astron, physics dept, Univ Los Andes, Merida, Venezuela, 71-72; staff astronomer, Nat Sci & Technol Res Coun Astron Proj, Merida, Venezuela, 72-76; PROF ASTRON, PHYSICS DEPT, CENT MICH UNIV, 76- *Concurrent Pos:* Adj prof astron, Univ Los Andes, Venezuela, 72-76; vis astronomer, Lowell Observ, Flagstaff, Ariz, 83. *Mem:* Int Astron Union; Am Astron Soc; Royal Astron Soc. *Res:* Observational astronomy, particularly photoelectric photometry; variable stars, stellar clusters, occultations and standard stars for comet photometry. *Mailing Add:* Physics Dept Cent Mich Univ Mt Pleasant MI 48859

OSBORNE, CARL ANDREW, b Pittsburgh, Pa, Sept 17, 40; m 64; c 3. VETERINARY MEDICINE, PATHOLOGY. *Educ:* Purdue Univ, Lafayette, DVM, 64; Univ Minn, St Paul, PhD(vet med), 70. *Hon Degrees:* DSc, Purdue Univ, 89, Univ Gent, 90. *Prof Exp:* From instr to assoc prof vet med, 64-75, chmn dept small animal clin sci, 76-84, PROF VET CLIN SCI, COL VET MED, UNIV MINN, ST PAUL, 75- *Concurrent Pos:* Mem, Nat Kidney Found. *Honors & Awards:* Vet of Year, Am Animal Hosp Asn, 74; Gaines Award, Am Vet Med Asn, 77; Puna Res Award, 82; Bourgelat Award, 83; Carnation Res Award, 85; Int Prize for Sci Achievement, World Small Animal Vet Asn, 88. *Mem:* Am Soc Vet Clin Path; Am Vet Med Asn; Am Soc Nephrology. *Res:* Pathophysiology of congenital and inherited renal diseases; pathophysiology of glomerulonephritis; pathophysiology of neoplasms of the urinary system; relationship of calcium metabolism to renal disease; urolithiasis. *Mailing Add:* Dept Clin Sci Univ of Minn C306 Vet Tch Hosp St Paul MN 55108

OSBORNE, CHARLES EDWARD, b Mt Croghan, SC, Aug 4, 29; m 55; c 2. NMR. *Educ:* Univ NC, BS, 51; Univ Maine, MS, 53; Northwestern Univ, PhD(chem), 56. *Prof Exp:* Chemist, Eastman Kodak Co, 51-52, chemist, Tenn Eastman Co, 56-59, sr chemist, 59-63, sr chemist, Eastman Chem Prod, Inc, 63-64, sr chemist, 64-71, DEVELOP ASSOC, TENN EASTMAN CO, 71- *Mem:* Am Chem Soc. *Res:* Syntheses of organic compounds. *Mailing Add:* 1540 Belmeade Dr Kingsport TN 37660

OSBORNE, DAVID WENDELL, b Brush, Colo, Jan 11, 35; m 55; c 2. ORGANIC CHEMISTRY, PATENT LAW. *Educ:* Colo Col, BS, 57; Univ Del, PhD(org chem), 60. *Prof Exp:* Chemist, 60-61, res chemist, 61-65, proj leader, 65-66, group leader, 66-69, div leader, 69-70, res mgr, 70-79, sr res mgr, 79-84, sr tech specialist, 84-85, PATENT ASSOC, DOW CHEM CO, 85- *Mem:* Am Chem Soc. *Res:* Synthesis and chemical properties of agricultural chemicals. *Mailing Add:* 139 Mount Everest Ct Clayton CA 94517-1605

OSBORNE, FRANK HAROLD, b San Juan, PR, Dec 20, 44; m 67. MICROBIOLOGY. *Educ:* State Univ NY Albany, BS, 67, MS, 69; Rensselaer Polytech Inst, PhD(biol), 73. *Prof Exp:* ASST PROF BIOL, KEAN COL NJ, 73- *Mem:* AAAS; Am Soc Microbiol; Soc Indust Microbiol. *Res:* Microbiology of arsenic compound oxidation including ecology of arsenite oxidizers and biochemistry and enzymology of arsenite oxidation. *Mailing Add:* 515 Kaplan St Roselle NJ 07203

OSBORNE, FRANKLIN TALMAGE, b Pewee Valley, Ky, Dec 1, 39; m 61; c 3. CHEMICAL ENGINEERING. *Educ:* Tex A&M Univ, BS, 61; NC State Univ, PhD(chem eng), 69. *Prof Exp:* Res engr synthetic fibers, Chemstrand Res Ctr, Inc, NC, 64-66, sr res engr, 69-70, sr res specialist non-woven fabrics, 70-71, sr group leader, Textile Div, 71-77, SUPVR FIBER INTERMEDIATES, MONSANTO CO, 77- *Mem:* Am Inst Chem Engrs. *Res:* Chemical reactor design; textile chemistry; team innovating new chemical processes. *Mailing Add:* Monsanto Co PO Box 12830 Pensacola FL 32575

OSBORNE, GEORGE EDWIN, pharmacy; deceased, see previous edition for last biography

OSBORNE, J SCOTT, III, b Raleigh, NC, July 23, 51; m 76; c 1. CHRONIC DISEASE EPIDEMIOLOGY, SCIENTIFIC REVIEW. *Educ:* Univ Miami, BA, 78; Fla Atlantic Univ, MA, 80; Mich State Univ, PhD(epidemiol & med anthrop), 85; Univ NC, Chapel Hill, MPH, 86. *Prof Exp:* Asst prof med res, Col Human Med, Mich State Univ, 87-90, dir res clerkships, 87-90, dir res, 87-90; fel epidemiol, 86, HEALTH SCI ADMINR, NIH, 90-, SCI REV ADMINR, EPIDEMIOL & DIS CONTROL, 90- *Concurrent Pos:* Prin investr, Univ NC, 86, Mich State Univ, 87-90; consult & panel mem indoor air qual, Environ Protection Agency/Int Life Sci Inst, 90-91; consult epidemiol, NIH, 90, mem, Comt on Epidemiol, 91; chair, Soc Epidemiol Res & Epidemiol Methods, 91. *Mem:* Sigma Xi; Am Pub Health Asn; AAAS; Soc Epidemiol Res; Soc Med Anthrop. *Res:* Health effects of heating with wood; indoor air quality; perception of risk and risk-taking behavior; epidemiologic assessment of excess mortality and morbidity in small populations; chronic disease epidemiology; peer review process. *Mailing Add:* NIH WW203C 5333 Westbard Bethesda MD 20892

OSBORNE, JAMES WILLIAM, b Pana, Ill, Jan 17, 28; m 50; c 2. PHYSIOLOGY, RADIOBIOLOGY. *Educ:* Univ Ill, BS, 49, MS, 51, PhD(physiol), 55. *Prof Exp:* Asst physics, Univ Ill, 50-51, res assoc, Control Systs Lab, 51-55; from asst prof to assoc prof, 55-67, PROF RADIOBIOLOGY, RADIATION RES LAB, UNIV IOWA, 67-, DIR, 78- *Concurrent Pos:* NIH spec fel, Eng, 65-66. *Mem:* Radiation Res Soc; Cell Kinetics Soc; Am Physiol Soc; Soc Exp Biol & Med. *Res:* Effect of x-radiation on mammalian small intestine; means of preventing radiation damage or promoting recovery in irradiated animals; radiation-induced carcinogenesis; identification and characterization of tumor-associated circulating antigens. *Mailing Add:* Radiation Res Lab Univ of Iowa Col of Med Iowa City IA 52242

OSBORNE, JERRY L, b Denver, Colo, June 13, 40; m 62; c 3. CLINICAL NUTRITION. *Educ:* Univ Calif, Davis, PhD(physiol), 71. *Prof Exp:* Asst prof, Irvine Med Sch, Univ Calif, 72-77; mgr, med dept, Am McGaw, 81-84, DIR CLIN & SCI AFFAIRS, KENDALL MCGAW, 84- *Mem:* Am Physiol Soc; Am Soc Parenteral & Enteral Nutrit. *Mailing Add:* 29462 Ana Maria Lane Laguna Niguel CA 92677

OSBORNE, JOHN ALAN, medicine, cardiology; deceased, see previous edition for last biography

OSBORNE, JOHN WILLIAM, dental research, for more information see previous edition

OSBORNE, LANCE SMITH, b Albuquerque, NMex, Nov 8, 49; m 72; c 1. AGRICULTURAL ENTOMOLOGY. *Educ:* Univ Calif, Davis, BS, 74, PhD(entom), 80. *Prof Exp:* Res fel entom, Univ Calif, Davis, 79-80, res entomologist, 80-81; ASST PROF ENTOM, AGR RES CTR APOPKA, UNIV FLA, 81- *Mem:* Entom Soc Am. *Res:* Control of arthropool pests of tropical foliage plants. *Mailing Add:* Cent Fla Res & Educ Ctr 2807 Binion Apopka FL 32703

OSBORNE, LOUIS SHREVE, b Rome, Italy, Sept 8, 23; m 50; c 3. ELEMENTARY PARTICLE PHYSICS. *Educ:* Calif Inst Technol, BS, 44; Mass Inst Technol, PhD(physics), 50. *Prof Exp:* Res assoc physics, 49-50, assoc prof, 59-64, group supvr, Synchrotron Lab, 50-74, PROF PHYSICS, MASS INST TECHNOL, 50-, PROF PHYSICS, UNIV, 64- *Concurrent Pos:* Guggenheim & Fulbright fels, 59-60; Minna James Heineman fel, 66-67. *Mem:* Am Phys Soc; Am Acad Arts & Sci. *Res:* High energy nuclear physics. *Mailing Add:* Dept Physics Mass Inst Technol 77 Mass Ave Cambridge MA 02139

OSBORNE, MELVILLE, b Carteret, NJ, Apr 21, 22; m 59; c 3. PHARMACOLOGY, PHYSIOLOGY. *Educ:* Drew Univ, BA, 52; Princeton Univ, MA & PhD(biol sci), 59. *Prof Exp:* Jr pharmacologist, Ciba Pharmaceut Co, 52-55; group leader cardiovasc pharmacol, Geigy Chem Co, 59-61 & Warner-Lambert Pharmaceut Co, 61-66; cardiovasc pharmacologist, 66-71, res group chief, Sect Cardiovasc Pharmacol, 71-73, SECT HEAD CARDIOVASC-RENAL PHARMACOL, HOFFMANN LA-ROCHE, INC, 73- *Mem:* Am Soc Pharmacol & Exp Therapeut. *Res:* Adrenergic mechanism relative to the control of the cardiovascular system and development of drugs which alter these mechanisms. *Mailing Add:* 14 Sherwood Close Somerville NJ 08876

OSBORNE, MORRIS FLOYD, b Henderson Co, NC, Sept 14, 31; m 55; c 3. CHEMISTRY, MATERIALS SCIENCE ENGINEERING. *Educ:* Univ NC, BA, 53. *Prof Exp:* Res asst mat properties, 54-57, mem res staff, 57-66. MEM RES STAFF REACTOR SAFETY, OAK RIDGE NAT LAB, 66- *Mem:* Am Nuclear Soc. *Res:* Utilization and application of nuclear energy, especially fission product behavior, nuclear fuel performance; high temperature materials; nuclear reactor safety. *Mailing Add:* 3803 24th Ave SE Temple Hills MD 20748

OSBORNE, PAUL JAMES, b Blackwater, Va, Dec 29, 21; m 42; c 2. INVERTEBRATE ZOOLOGY, PHYSIOLOGY. *Educ:* Univ Va, BA, 50, MA, 51; Univ Fla, PhD, 55. *Prof Exp:* Asst biol, Univ Va, 51, technician biochem res, 51-52; asst biol, Univ Fla, 52-54; assoc prof, 55-62, chmn dept, 65-68, PROF BIOL, LYNCHBURG COL, 63-, CHMN DIV LIFE SCI, 71- *Concurrent Pos:* Mem staff, Randolph-Macon Woman's Col, 68-69. *Mem:* AAAS; Am Inst Biol Sci; Sigma Xi. *Res:* Morphology and histochemistry of invertebrates, especially enzymes of platyhelminthes; biochemical analysis of invertebrates; human physiology, genetics and history; cell physiology; histochemistry and electron microscopy; acid phosphates; molluscs; chick embryogenesis. *Mailing Add:* 1432 Northwood Circle Lynchburg VA 24503-1916

OSBORNE, RICHARD HAZELET, b Kennecott, Alaska, June 18, 20; m 44; c 3. HUMAN GENETICS, ANTHROPOLOGY. *Educ:* Univ Wash, BS & BA, 49; Columbia Univ, PhD(genetics), 56. *Prof Exp:* Res assoc, Inst Study Human Variation, Columbia Univ, 53-58; from asst prof to assoc prof, Sloan-Kettering Div, Cornell Univ, 58-64; prof anthrop & prof med, 64-85, EMER PROF ANTHROP, SCH LETT & SCI & EMER PROF MED, GENETICS, SCH MED, UNIV WIS-MADISON, 85- *Concurrent Pos:* Vis asst prof, Albert Einstein Col Med, 57-58; asst div prev med, Sloan-Kettering Inst Cancer Res, 58-60, assoc, 60-62, assoc mem, 62-64; assoc univ sem genetics & evolution of man, Columbia Univ, 58-64; assoc res geneticist, Dept Prev Med, Mem Hosp, 59-63, clin geneticist, Dept Med, 63-65; career scientist, Health Res Coun New York, 62-64; consult cult anthrop fel rev comt, NIMH, 69-73, perinatal res comt, Nat Inst Neurol Dis & Stroke, 70-73, genetics task force & phys growth task force; mem comt epidemiol & vet follow-up studies, Nat Res Coun, 69-75. *Mem:* Fel AAAS; Am Soc Nat; Am Asn Phys Anthrop; Am Soc Human Genetics; Genetics Soc Am. *Res:* Undergraduate and graduate teaching of human biology and evolutionary theory; twin and sibling studies of quantitative variables and of genetic environmental interaction in health and disease. *Mailing Add:* Box 2349 Port Angeles WA 98362-0303

OSBORNE, RICHARD VINCENT, b Manchester, Eng, Mar 12, 36; m 59; c 3. RADIATION PROTECTION. *Educ:* Cambridge Univ, BA, 59; Univ London, PhD(biophys), 62. *Prof Exp:* Fel radiation detection, Dept Indust Med, NY Univ Med Ctr, 62; res officer, Biol & Health Physics Div, 62-82, head environ res br, Chalk River Nuclear Labs, Atomic Energy Can, Ltd, 82-88; exec asst pres, Atomic Energy Can, Head Off, 88-89, DIR, HEALTH SCI DIV, AECL RES, 89- *Concurrent Pos:* Mem, comt 36, Nat Coun Radiation Protection & Measurements, 72-75; mem, expert group on effluent releases, Nuclear Energy Agency, Orgn Econ Coop & Develop, 77-79, expert group uranium mgt, 80-84, expert group long term objectives waste mgt, 82-84 & comt radiation protection & pub health, 83-; dir, Health Physics Soc, 76-79; consult & mem, various adv groups, radiation protection & waste mgt, Int Atomic Energy Agency; mem, comt 4, Int comn Radiol Protection, 81-; exec coun, Int Radiation Protection Asn, 88- *Honors & Awards:* Elda E Anderson Award, Health Physics Soc, 75. *Mem:* Can Radiation Protection Asn (pres, 79); Health Physics Soc. *Res:* High sensitivity alpha particle spectroscopy; transport of radionuclides in the environment; microdosimetry; detection of hazards of and protection against tritium. *Mailing Add:* AECL Research Chalk River Labs Chalk River ON K0J 1J0 Can

OSBORNE, ROBERT HOWARD, b Akron, Ohio, June 29, 39; m 66. STRATIGRAPHY. *Educ:* Kent State Univ, BS, 61; Wash State Univ Univ, MS, 63; Ohio State Univ, PhD(geol), 66. *Prof Exp:* Asst prof, 66-70, assoc prof 70-80, PROF GEOL, UNIV SOUTHERN CALIF, 80- *Concurrent Pos:* Consult, numerous co; expert witness, State Calif. *Mem:* Am Statist Asn; Geol Soc Am; Am Asn Petrol Geol; Soc Econ Paleont & Mineral. *Res:* Quantitative stratigraphy; statistical applications in sedimentary petrology and stratigraphy. *Mailing Add:* Dept Geol Sci Univ Southern Calif Los Angeles CA 90089

OSBORNE, ROBERT KIDDER, b Kansas City, Mo, Feb 9, 21; m 48; c 1. PHYSICS. *Educ:* Mass Inst Technol, SB, 42, PhD(physics), 47. *Prof Exp:* Instr physics, Mass Inst Technol, 45-49; MEM STAFF, LOS ALAMOS SCI LAB, 49- *Mem:* Am Phys Soc; Inst Elec & Electronics Engrs. *Res:* Radioactive disintegration; investigation of nuclear energy levels through study of radioactive disintegration. *Mailing Add:* 299 Crossfield King of Prussia PA 19406

OSBORNE, WILLIAM WESLEY, b Buhl, Idaho, Apr 29, 20; m 50; c 2. BACTERIOLOGY, PUBLIC HEALTH. *Educ:* Wash State Univ, BS, 46. *Prof Exp:* Bacteriologist, Lab Sect, Sta Dept Health, Wash, 47-48; bacteriologist, Ft Lawton Sta Hosp, Seattle, 48-49; microbiol res div, Bur Agr & Indust Chem, USDA, Md, 49-50, microbiol sect, Poultry Prod Unit, Western Utilization Res Br, Agr Res Serv, 50-55; process develop div, Biol Warfare Labs, US Army, 55-56; microbiol unit, Res & Develop Sect, Indust Test Lab, Bur Ships, US Navy, 56-58; chief bacteriologist, Lab Serv, Presby Hosp Hosp, Philadelphia, 58-60; mgr, Microbiol Control Dept, William H Rorer, Inc, 60-85; RETIRED. *Mem:* Am Soc Microbiol. *Res:* Food technology; agriculture bacteriology; clinical, medical and sanitary bacteriology; germicides and disinfectants. *Mailing Add:* 1736 Aidenn Lair Rd Dresher PA 19025

OSBOURN, GORDON CECIL, b Kansas City, Mo, Aug 13, 54; c 2. VISION SCIENCE, MACHINE VISION. *Educ:* Univ Mo, BS, 74, MS, 75; Calif Inst Technol, PhD(physics), 79. *Prof Exp:* Tech staff, 79-83, DIV SUPVR, SANDIA NAT LABS, 83- *Honors & Awards:* E O Lawrence Award, Dept Energy, 85. *Mem:* Fel Am Phys Soc. *Mailing Add:* Div 1155 Sandia Nat Labs PO Box 5800 Albuquerque NM 87185

OSBURN, BENNIE IRVE, b Independence, Iowa, Jan 30, 37; m 60; c 3. VETERINARY PATHOLOGY. *Educ:* Kans State Univ, BS, 59, DVM, 61; Univ Calif, Davis, PhD(comp path), 65; Am Col Vet Path, dipl. *Prof Exp:* From asst prof to assoc prof vet path, Col Vet Med, Okla State Univ, 64-68; Nat Inst Child Health & Human Develop spec res fel, Wilmer Inst, Sch Med, Johns Hopkins Univ, 68-70; assoc prof vet path, 70-74, PROF PATH, SCH VET MED, UNIV CALIF, DAVIS, 74-, ASSOC DEAN RES, 76- *Concurrent Pos:* Researcher immunol, Calif Primate Res Ctr, 74- *Mem:* Am Vet Med Asn; Int Acad Path; Am Soc Exp Path; US Animal Health Asn; Am Col Vet Path (pres, 82). *Res:* Host responses and pathogenesis of congenital infections. *Mailing Add:* Vet-Med-Path Univ Calif Davis CA 95616

OSBURN, RICHARD LEE, b Pensacola, Fla, May 16, 40; m 64; c 1. ENTOMOLOGY, GENETICS. *Educ:* Ga Southern Col, BS, 62, MS, 70; Univ Ga, PhD(entom), 77. *Prof Exp:* Res asst biol, Ga Southern Col, 68-71, instr anat, 71-72, res assoc tick biol, 72-77; res entomologist tick res, Agr Res, Sci & Educ Admin, USDA, 77-82; assoc prof & head, biol dept, GA Southern Col, 82-87; PROF BIOL & DEAN, SCH SCI & MATH, STEPHEN F AUSTIN, STATE UNIV, 87- *Concurrent Pos:* NIH fel, Ga Southern Col, 76-77. *Mem:* Entom Soc Am; Acarological Soc Am; AAAS. *Res:* Physiology and genetics of ticks (Ixodidae) with special emphasis on genetic hybridization of Boophilus sp and reproductive biology of Amblyomma americanum (Lone Star Tick). *Mailing Add:* 1400 Gunnison Dr Wichita Falls TX 76305

OSCAI, LAWRENCE B, b Doylestown, Ohio, July 13, 38; div; c 1. EXERCISE PHYSIOLOGY. *Educ:* La Sierra Col, BS, 61; Univ Colo, MS, 63; Univ Ill, Urbana, PhD(phys educ), 67. *Prof Exp:* Asst prof, 70-72, ASSOC PROF PHYS EDUC, UNIV ILL, CHICAGO CIRCLE, 72-, PROF, UNIV ILL, CHICAGO. *Concurrent Pos:* USPHS res fel, Sch Med, Wash Univ, 67-70. *Mem:* Am Col Sports Med; Am Inst Nutrit; Am Physiol Soc. *Res:* Effect of exercise on lipid metabolism, including appetite and weight gain, weight reduction, body composition, adipose tissue cellularity and the enzyme lipoprotein lipase. *Mailing Add:* Dept Phys Educ-M/C 194 Univ Ill Chicago IL 60680

OSCAR, IRVING S, b New York, NY, July 20, 23. ELECTRONICS ENGINEERING. *Educ:* NY Univ, BA, 47, MSEE, 52. *Prof Exp:* Proj dir tech writing, Coastal Pub Inc, NY, 50-54; sr engr, Davies Labs, Md, 54-57, chief engr, Davies Labs Div, Honeywell Inc, 57-60, chief engr, Data Sect, Systs Div, Pa, 60-61; chief engr, Systs Div, Epsco, Inc, Mass, 61-63; prin engr, Booz Allen Appl Res, Inc, 63-65; consult electronic eng & instrumentation, 65; PRES, I S OSCAR ASSOCS, INC, 65- *Mem:* Inst Elec & Electronics Engrs. *Res:* Computer oriented, data acquisition systems; real time analysis and control; magnetic tape recording systems; digital and analog; developmental engineering in electronic instrumentation. *Mailing Add:* I S Oscar Assocs Inc 8811 Colesville Rd Silver Spring MD 20910

OSDOBY, PHILIP, b Monticello, NY, Dec 8, 49; m 77; c 2. CELL BIOLOGY, BIOCHEMISTRY. *Educ:* Case Western Res, PhD(develop biol), 78. *Prof Exp:* from asst prof to assoc prof growth anat histol, Sch Dept Med, 81-90, ASSOC PROF BIOL, WASH UNIV, 90- *Concurrent Pos:* Oral biol study sect, NIH, 86-90; Postdoctoral Award, Arthritis Found, 79-81; Arthritis Investr Award, 81-83; Res Career Develop Award, NIH, 85-90. *Honors & Awards:* Young Investr Award, Am Soc Bone & Mineral Res, 84. *Mem:* Fel AAAS; Am Soc Bone & Mineral Res; Am Soc Cell Biol. *Res:* Bone cell development and regulation; identifying orteoclast-specific cell surface molecules utilizing monoclonal antibody and molecular techniques; biochemical and functional characterization of the cell surface molecules; complex regulatory mechanisms of bone modeling. *Mailing Add:* Sch Dent Med Wash Univ 45559 Scott Ave St Louis MO 63110

OSEASOHN, ROBERT, b New York, NY, Jan 23, 24; m 48; c 3. EPIDEMIOLOGY. *Educ:* Tufts Univ, BS, 43; State Univ NY, MD, 47. *Prof Exp:* Demonstr prev med & med, Sch Med, Western Reserve Univ, 51-52; asst med, Sch Med, Boston Univ, 55-57; sr instr prev med & med, Sch Med, Western Reserve Univ, 57-60, from asst prof to assoc prof prev med, 60-67, asst prof med, 61-67; prof epidemiol & community med & chmn dept, Univ NMex, 67-72, prof med, Sch Med, 68-72; prof epidemiol & assoc dean, Sch Pub Health & prof med, Sch Med, Univ Tex, Houston, 72-74; prof epidemiol & health, chmn dept & prof med, McGill Univ, 74-83; PROF EPIDEMOL & ASSOC DEAN SCH PUB HEALTH, UNIV TEX, SAN ANTONIO, 83- *Concurrent Pos:* Attend physician, Vet Admin Hosp, Boston, Mass, 55-57; asst physician, Univ Hosps, Cleveland, 57-67; attend physician, Crile Vet Admin Hosp, 59-67; chief epidemiol sect & dep dir, Pak-SEATO Cholera Res Lab, Dacca. EPakistan, 63-65; attend physician, Vet Admin Hosp, Albuquerque, 67-72 & Bernalillo County Med Ctr, Albuquerque, 67-72; attend staff, Royal Victoria Hosp, Montreal, 74-83. *Mem:* Fel Am Col Epidemiol; Am Epidemiol Soc; Int Epidemiol Asn; Can Soc Clin Res; fel Am Col Physicians. *Res:* Disease control. *Mailing Add:* Univ Tex, Sch Pub Health 7703 Floyd Curl Dr San Antonio TX 78284

OSEBOLD, JOHN WILLIAM, b Great Falls, Mont, Jan 9, 21; m 44; c 2. IMMUNOLOGY, MICROBIOLOGY. *Educ:* Wash State Univ, BS, 43, DVM, 44; Ore State Univ, MS, 51; Univ Calif, PhD(comp path), 53. *Prof Exp:* Fed vet in charge brucellosis diag lab, Bur Animal Indust, USDA, 44-46; pvt pract vet med, 46-49; instr vet med & vet microbiol, Ore State Univ, 49-50; lectr vet microbiol, Univ Calif, Davis, 50-53, asst prof vet med, 53-57, assoc prof immunol, 57-62, prof immunol, 62-86; RETIRED. *Mem:* Am Asn Immunol; Am Col Vet Microbiol; Am Soc Microbiol; Am Vet Med Asn; Am Acad Microbiol. *Res:* Dynamics of infection and resistance; cellular immunity, immunoglobulins serology and immunization. *Mailing Add:* 826 Oeste Rd Davis CA 95616

OSEI-GYIMAH, PETER, b Ghana, March 14, 46. PHARMACY. *Educ:* Univ Wis, Platteville, BSc, 72; Ohio State Univ, Columbus, PhD(pharm), 76. *Prof Exp:* Fel res assoc, Rensselaer Polytech Inst, 76-80; RES SCIENTIST, ROHM AND HAAS CO, 81- *Mem:* Am Chem Soc; Sigma Xi. *Res:* Design and synthesis of bronchodilator agents bearing the isoquinoline nucleus; morphine-like compounds as irreversible opiate receptor ligandus; synthetic polymers and their fabrication into ultrafiltration and reverse osmosis membranes. *Mailing Add:* 141 Hunt Dr Horsham PA 19044

OSEPCHUK, JOHN M, b Peabody, Mass, Feb 11, 27; m 56; c 3. PHYSICS, MICROWAVE POWER TECHNOLOGY. *Educ:* Harvard Univ, AB, 49, AM, 50, PhD(appl physics), 57. *Prof Exp:* Res & develop engr microwave tubes, Raytheon Co, 50-53, consult, 53-56, tech liaison, Gen Tel Co, France, 56-57, eng sect head, Spencer Lab, 57-60, prin engr, 60-62; chief microwave engr, Sage Labs, Inc, 62-64; prin res scientist, 64-75, CONSULT SCIENTIST, RES DIV, RAYTHEON CO, 75- *Mem:* Fel Inst Elec &

Electronics Engrs; fel Int Microwave Inst; Bioelectromagnetics Soc. *Res:* Crossed field electron devices; interaction of electromagnetic waves with electron beams or plasmas; secondary emission from dielectrics; biological effects of microwaves; microwave ovens; electron beam and semiconductor hydrid devices; microwave power applications. *Mailing Add:* 248 Deacon Haynes Rd Concord MA 07142

OSER, BERNARD LEVUSSOVE, b Philadelphia, Pa, Feb 2, 1899; m 23; c 2. NUTRITION, TOXICOLOGY. *Educ:* Univ Pa, BS, 20, MS, 25; Fordham Univ, PhD, 27; Am Bd Nutr, dipl; Am Bd Clin Chem, dipl; Am Bd Indust Hyg, dipl, cert Toxicol. *Prof Exp:* Asst physiol chem, Jefferson Med Col, 20-21; biochemist, Philadelphia Gen Hosp, Pa, 22-26; asst dir biol lab, Food & Drug Res Labs, Inc, 26-34, dir, 34-70, vpres, 39-57, pres, 57-90, chmn, 70-74; CONSULT, BERNARD L OSER ASSOCS, INC, 74- *Concurrent Pos:* Instr, Grad Sch Med, Univ Pa, 22-26, biochemist, Philadelphia Hosp, 22-26; pres, Am Coun Independent Labs, 48-49; mem expert comt food additives, Food & Agr Orgn, WHO, 58-63; adj prof, Columbia Univ, 59-70; mem food protection comt, Nat Acad Sci-Nat Res Coun, 64-71; sci ed, Food Drug Cosmetic Law J, 57-83; mem, White House conf Food, Nutrit & Health, 70. *Honors & Awards:* Babcock-Hart Award, Inst Food Technol, 58; Distinguished Food Scientist Award, Inst Food Technologists; Merit Award, Soc Toxicol, 80. *Mem:* Am Chem Soc; Soc Toxicol; NY Acad Med; Am Soc Pharmacol & Exp Therapeut; Am Col Toxicol; fel Am Inst Nutrit; hon mem Soc Flavor Chem; Sigma Xi; Inst Food Technologists (pres, 68-69); Am Indust Hyg Asn. *Res:* Technicolegal; physiological chemistry; author numerous publications. *Mailing Add:* 655 Pomander Walk Apt 554 Teaneck NJ 07666

OSER, HANS JOERG, b Constance, Ger, Dec 7, 29; m 58; c 3. APPLIED MATHEMATICS. *Educ:* Univ Freiburg, Dipl, 54, Dr rer nat(math), 57. *Prof Exp:* Fel, Inst Fluid Dynamics, Univ Md, 57-58; mathematician-analyst, Comput Lab, Nat Bur Standards, 58-62, consult to chief appl math div, 62-66; exec secy commerce tech adv bd, Dept Commerce, 66-67; chief systs dynamics sect, 68-70, chief math analysis sect, Appl Math Div, 70-78, sr sci adv, Nat Bur Standards, 78-86; tech dir, Soc Indust & Appl Math, Philadelphia, 86-88; consult, US Congress, 88-90; STAFF OFFICER, BD MATH SCI, NAT RES COUN, 90- *Concurrent Pos:* Asst prof, Univ Col, Univ Md, 58-62; lectr, Catholic Univ, 62-66; dir comt sci & technol fel prog, 81-; ed, Nat Bur Standards J Res, 85-86. *Mem:* AAAS; Asn Comput Mach; Soc Rheol; Soc Indust & Appl Math. *Res:* Numerical and functional analysis; computer modelling; science administration; science policy. *Mailing Add:* 8810 Quiet Stream Ct Potomac MD 20854

OSGOOD, CHARLES EDGAR, b Washington, DC, Feb 1, 38; m 64; c 2. AGRICULTURAL CHEMICALS, RESEARCH & DEVELOPMENT. *Educ:* Univ Md, BS, 61; Ore State Univ, MS, 64, PhD(entom), 68. *Prof Exp:* Res scientist, Res Sta, Can Dept Agr, 68-76; tech rep agr chem develop, Diamond Shamrock Corp, 76-83; tech rep, SDS Biotech Corp, 83-85; field res rep, 85-90, SR TECH REP, BASE CORP, 90- *Mem:* Entom Soc Am; Am Phytopath Soc; Sigma Xi; Western Soc Weed Sci. *Res:* Behavior and management of the leaf-cutter bee Megachile rotundata; density dependent behavioral changes affecting reproductive rate in insects; pest management of flea beetles on rape crops; evaluation and development of agricultural chemicals. *Mailing Add:* 11134 Chickadee Dr Boise ID 83709

OSGOOD, CHARLES FREEMAN, b New Castle, Pa, Oct 16, 38; m 68; c 1. MATHEMATICS. *Educ:* Haverford Col, BA, 60; Univ Calif, Berkeley, MA, 62, PhD(Diophantine approximation), 64. *Prof Exp:* From instr to assoc prof math, Univ Ill, Urbana, 64-72; RES MATHEMATICIAN, DEPT DEFENSE, 70- *Concurrent Pos:* Res assoc, Nat Bur Standards, DC, 66-68; adj prof math, Univ RI, 80- *Mem:* Am Math Soc; Inst Elec & Electronics Engrs. *Res:* Diophantine approximation of power series by rational functions; complex variables, principally Nevanlinna Theory; numerical integration of singular integrals. *Mailing Add:* 2633 Woodley Pl NW Washington DC 20008

OSGOOD, CHRISTOPHER JAMES, b Winfield, Kans, Jan 3, 49; m 77; c 1. BIOLOGY, GENETICS. *Educ:* Brown Univ, AB, 71, PhD(biol), 77. *Prof Exp:* Res assoc biochem, Dept Genetics, Univ Calif, Davis, 77-80; MEM FAC, T H MORGAN SCH BIOSCI, UNIV KY, 80- *Mem:* Genetics Soc Am; AAAS; Environ Mutagen Soc; Sigma Xi. *Res:* Biochemistry and genetics of DNA-repair in higher organisms. *Mailing Add:* Dept Biol Sci Old Dominion Univ Norfolk VA 23508

OSGOOD, DAVID WILLIAM, b Grant's Pass, Ore, May 19, 40; m 58; c 2. VERTEBRATE ZOOLOGY, ECOLOGY. *Educ:* Portland State Col, BS, 63; Duke Univ, MA, 65, PhD(zool), 68. *Prof Exp:* Instr comp anat, Duke Univ, 67-68; asst prof comp anat, Butler Univ, 68-73, assoc prof, 73-79, prof, 79-82; PRES, MINI-MITTER CO, INC, 82- *Mem:* AAAS; Am Soc Ichthyol & Herpet; Am Soc Mammal. *Res:* Effects of developmental temperature on systematic characters in colubrid snakes; bio-telemetric study of body temperature in reptiles; applications of telemetry to behavioral and ecological problems in small mammals and reptiles. *Mailing Add:* PO Box 3386 Sunriver OR 97707-3386

OSGOOD, ELMER CLAYTON, b Greenfield, Mass, Aug 4, 06; m 34; c 2. ENGINEERING. *Educ:* Rensselaer Polytech Inst, CE, 28, DEng(struct), 31. *Prof Exp:* Jr civil engr, Civilian Conserv Corps, US Forest Serv, 34-35, asst civil engr, 35-41; design sect, Pub Works Dept, Operating Base, Bur Yards & Docks, US Dept Navy, Va, 41-42, assoc struct engr, 42-44, struct engr, 44-46; from asst prof to prof civil eng, 46-73, EMER PROF CIVIL ENG, UNIV MASS, AMHERST, 73- *Mem:* Am Soc Civil Engrs; Sigma Xi. *Res:* Structures; civil engineering. *Mailing Add:* Depot Rd Leverett MA 01054

OSGOOD, PATRICIA FISHER, PAIN RESEARCH. *Educ:* Boston Univ, PhD(pharmacol), 71. *Prof Exp:* ASST PROF ANESTHESIA PHARMACOL, HARVARD MED SCH, MASS GEN HOSP, 81-, STAFF SCIENTIST, SHRINER'S BURN INST, 81- *Res:* Clinical and laboratory investigations in pain. *Mailing Add:* Dept Anesthesia Mass Gen Hosp Fruit St Boston MA 02114

OSGOOD, RICHARD MAGEE, JR, b Kansas City, Mo, Dec 28, 43; m 66; c 3. LASERS. *Educ:* US Mil Acad, BS, 65; Ohio State Univ, MS, 68; Mass Inst Technol, PhD(physics), 73. *Prof Exp:* Res assoc physics, Mass Inst Technol, 69-73, mem res staff, Lincoln Lab, 73-81; from assoc prof to prof elec eng, 81-87, PROF APPL PHYSICS, COLUMBIA UNIV, 82-, DIR MICROELECTRONICS SCI LABS, 86-, HIGGINS PROF ELEC ENG, 87- *Concurrent Pos:* Vis staff, Los Alamos Nat Lab, 78; mem ad hoc Comt Advan Isotope Separation, Energy Res Adv Bd, Dept Energy, 80; mem adv bd, Spectros Lab, Mass Inst Technol, 81-; consult, Los Alamos Nat Lab & Allied Corp, 82-90; co-dir, Columbia Radiation Lab, Columbia Univ, 84-90; dir, Microelectronics Sci Lab, 85-90; mem D Sci Coun, Defense Advan Res Projs Agency, 85-; mem bd, Army Sci & Technol, 87- *Honors & Awards:* Samual Burka Award, US Avionics Lab, 69. *Mem:* Optical Soc Am; fel Inst Elec & Electronic Engrs; Sigma Xi. *Res:* Microelectronic devices and fabrication; laser devices and physics; surface chemistry; atomic and molecular physics. *Mailing Add:* Dept Elec Eng Columbia Univ New York NY 10027

OSGOOD, ROBERT VERNON, b Wilmington, Del, Dec 26, 41; m 64; c 2. WEED SCIENCE, PLANT GROWTH REGULATION. *Educ:* Univ Miami, BS, 64; Univ Hawaii, MS, 67, PhD(hort), 69. *Prof Exp:* From asst agronomist to assoc agronomist, 69-77, agronomist, 77-89, CROP SCI DEPT HEAD, HAWAIIAN SUGAR PLANTERS ASN, 89- *Mem:* Weed Sci Soc Am; Asian Pac Weed Sci Soc; Plant Growth Regulator Soc Am. *Res:* Development of new herbicides and growth regulators for sugarcane; studies relating to reasons for varietal differences in tolerance to herbicides. *Mailing Add:* Hawaiian Sugar Planters' Asn 99-193 Aiea Hts Dr Aiea HI 96701

O'SHAUGHNESSY, CHARLES DENNIS, b Moose Jaw, Sask, Oct 11, 41; m 64; c 2. STATISTICS. *Educ:* Univ Sask, BA, 62, PhD(statist), 68; Univ Chicago, MSc, 65. *Prof Exp:* Asst prof, 68-71, ASSOC PROF MATH, UNIV SASK, 71- *Concurrent Pos:* Nat Res Coun Can grant, Univ Sask, 69-72. *Mem:* Inst Math Statist; Can Math Cong; Statist Sci Asn Can. *Res:* Design of statistical experiments, in particular, block designs. *Mailing Add:* Dept Math & Statist Univ Sask McLan Hall Rm 142 Saskatoon SK S7N 0W0 Can

O'SHEA, DONALD CHARLES, b Akron, Ohio, Nov 14, 38; m 62; c 4. OPTICS. *Educ:* Univ Akron, BS, 60; Ohio State Univ, MS, 63; Johns Hopkins Univ, PhD(physics), 68. *Prof Exp:* Fel physics, Harvard Univ, 68-70; from asst prof to assoc prof, 70-87, PROF PHYSICS, GA INST TECHNOL, 87- *Concurrent Pos:* Prin lectr, Laser Syst Design, Univ Wis Extension, 78-; mem educ comt, Optical Soc Am, 84-91, chair, 89-90; chmn educ comt, Soc Photo-Optical Instrumentation Engrs, 84-85; vis scholar, Optical Sci Ctr, Univ Ariz, 85-86; bd gov, Soc Photo-Optical Instrumentation Engrs, 88-90, tutorial text ed, 90- *Mem:* Sigma Xi; Optical Soc Am; Am Asn Physics Teachers; fel Soc Photo-Optical Instrumentation Engrs. *Res:* Applied optics; optical and laser system design; opto-mechanical design; author of textbooks on lasers and optics; consultant in optical design. *Mailing Add:* Sch Physics Ga Inst Technol Atlanta GA 30332-0430

O'SHEA, FRANCIS XAVIER, b New York, NY, May 14, 32; m 54; c 6. ORGANIC POLYMER CHEMISTRY. *Educ:* St John's Univ, NY, BS, 54; Univ Notre Dame, PhD(org chem), 58. *Prof Exp:* Res chemist, Naugatuck Chem Div, US Rubber Co, 57-65, group leader plastics res, 65-68; develop mgr, 68-72, sr res scientist corp res, 72-78, RES ASSOC, UNIROYAL, INC, 78- *Mem:* Am Chem Soc; Sigma Xi. *Res:* Synthesis of nondiscoloring antioxidants, particularly sterically hindered phenols; stabilization of hydrocarbon polymers; synthesis of new plastics; thermoplastic polyurethane synthesis and morphology; thermoplastic elastomers; polymer blends; ionomers; EPDM polymerization. *Mailing Add:* 211 Wedgewood Dr Naugatuck CT 06770

O'SHEA, TIMOTHY ALLAN, b Lorain, Ohio, Nov 3, 44; m 68; c 2. ENVIRONMENTAL CHEMISTRY, ANALYTICAL CHEMISTRY. *Educ:* Univ Toledo, BS, 66, MS, 69; Univ Mich, MS, 70, PhD(environ health sci), 72. *Prof Exp:* Asst prof chem, Univ Mich-Dearborn, 72-78; assoc prof chem, Tex Wesleyan Col, 78-; AT DEPT APPL TECHNOL, MACOMB COUNTY COMMUNITY COL. *Mem:* AAAS; Am Chem Soc; Sigma Xi. *Res:* Application of electroanalytical techniques to the study of trace metals in natural waters. *Mailing Add:* Dept Appl Technol Macomb County Community Col 14500 12 Mile Rd Warren MI 48093

OSHEIM, YVONNE NELSON, b Dec 27, 44; m; c 2. ELECTRON MICROSCOPY, CHROMATIN SPREADING. *Educ:* Univ Va, PhD(biol), 83. *Prof Exp:* RES ASSOC, MICROBIOL DEPT, UNIV VA, 84- *Res:* Ultrastructure of nascent transcripts. *Mailing Add:* Univ Va Jordan Hall Rm 7-59 Charlottesville VA 22908

OSHER, JOHN EDWARD, b Estherville, Iowa, Oct 15, 29; m 52; c 3. PLASMA PHYSICS. *Educ:* Iowa State Col, BS, 51; Univ Calif, PhD(physics), 56. *Prof Exp:* Res fel exp nuclear physics, Univ Calif, 56-57; staff mem, Los Alamos Sci Lab, 57-63; plasma physics dept head, Aerojet-Gen Nucleonics Div, Gen Tire & Rubber Co, Calif, 63-65; PHYSICIST, LAWRENCE LIVERMORE LAB, UNIV CALIF, 65- *Mem:* Fel Am Phys Soc; Inst Elec & Electronic Engrs. *Res:* High energy nuclear physics. *Mailing Add:* 148 Via Bonita Alamo CA 94507

OSHER, STANLEY JOEL, b Brooklyn, NY, Apr 24, 42; m 87; c 1. MATHEMATICS. *Educ:* Brooklyn Col, BS, 62; NY Univ, MS, 64, PhD(math), 66. *Prof Exp:* Asst mathematician, Brookhaven Nat Lab, 66-67, assoc mathematician, 67-68; asst prof math, Univ Calif, Berkeley, 68-70; assoc prof math, State Univ NY, Stony Brook, 70-76; PROF MATH, UNIV CALIF, LOS ANGELES, 76- *Concurrent Pos:* Sloan fel, SRC fel (Eng), Bi-Nat Sci Found fel(Israel). *Mem:* Am Math Soc; Am Inst Aeronaut & Astronaut; Soc Ind & Appl Math. *Res:* Numerical and analysis scientific computing; partial differential equations. *Mailing Add:* Dept Math 6356 Math Sci Univ Calif 405 Hilgard Ave Los Angeles CA 90024

OSHEROFF, DOUGLAS DEAN, b Aberdeen, Wash, Aug 1, 45; m 70. LOW TEMPERATURE PHYSICS. *Educ:* Calif Inst Technol, BS, 67; Cornell Univ, MS, 69, PhD(physics), 73. *Prof Exp:* Mem tech staff physics res, Bell Labs, AT&T Co, 72-82, head solid state & low temperature, 82-87; PROF PHYSICS, STANFORD UNIV, 87- *Concurrent Pos:* MacArthur Prize fel, MacArthur Found, 81. *Honors & Awards:* Simon Prize, Brit Inst Physics, 76; Oliver E. Buckley Solid-State Physics Prize, Am Phys Soc, 81. *Mem:* Nat Acad Sci; Am Physic Soc; fel Am Acad Arts & Sci. *Res:* Ultralow temperature physics; properties of solid and liquid helium-3; superfluidity; nuclear ordering; glasses, low temperature detectors. *Mailing Add:* Dept Physics Stanford Univ Stanford CA 94305

OSHEROFF, NEIL, b Washington, DC, Apr 23, 52. EUKARYOTIC DNA, TOPOISOMERASES. *Educ:* Northwestern Univ, PhD(biochem & molecular biol), 79. *Prof Exp:* ASST PROF BIOCHEM, SCH MED, VANDERBILT UNIV, 83- *Mem:* Am Soc Biol Chemists; Sigma Xi; AAAS; Am Chem Soc. *Mailing Add:* Dept Biochem Light Hall Vanderbilt Univ Nashville TN 37232-0146

OSHIMA, EUGENE AKIO, b Kaneohe, Hawaii, Jan 2, 34; m 59; c 3. SCIENCE EDUCATION, BIOLOGY. *Educ:* Univ North Col, BA, 55, MA, 56; Okla State Univ, DEd(sci ed), 66. *Prof Exp:* Head sci dept high sch, Kans, 56-64; mem spec staff sci educ, Okla State Univ, 64-66; from asst prof to assoc prof, 66-75, head dept, 67-71, PROF BIOLOGY SCI EDUC, CENT MO STATE UNIV, 75- *Concurrent Pos:* Sci consult, Okla State Univ, 65 & E Cent State Col, 66. *Mem:* Nat Sci Teachers Asn. *Res:* Changes in attitudes and confidence in teaching science of prospective elementary school teachers. *Mailing Add:* Dept Biol Cent Mo State Univ Warrensburg MO 64093

OSHIRO, GEORGE, b Kenora, Ont, Nov 17, 38; m 67; c 2. PHARMACOLOGY. *Educ:* Univ Man, BS, 64, PhD(pharmacol), 72. *Prof Exp:* Can Heart Found fel cardiovasc pharmacol, McGill Univ, 72-75; sr pharmacologist, Wyeth-Ayerst Res, Inc, 75-78, res assoc, 78-80, sect head, hypertension, 80-90, ASSOC DIR, CARDIOVASC & METAB DISORDERS, WYETH-AYERST RES, INC, 91- *Mem:* AAAS; Am Heart Asn; Am Soc Hypertension. *Res:* Discovery and development of new cardiovascular agents for the treatment of hypertension, arterial occlusive disease, arrhythmias and migraine; the use of potassium channel openers in hypertension, arterial occlusive disease, urinary incontinence, asthma and irritable bowel disease. *Mailing Add:* Wyeth-Ayerst Res Inc CN-8000 Princeton NJ 08543-8000

OSHIRO, LYNDON SATORU, b Hilo, Hawaii, Aug 8, 33; m 61; c 3. VIROLOGY, ELECTRON MICROSCOPY. *Educ:* Univ Utah, BS, 56, PhD(microbiol), 63; Univ Hawaii, MS, 60. *Prof Exp:* ELECTRON MICROSCOPIST, VIRAL & RICKETTSIAL DIS LAB, CALIF STATE DEPT HEALTH SERV, 66- *Concurrent Pos:* NIH trainee virol, Viral & Rickettsial Dis Lab, Calif State Dept Health, 63-65; Nat Inst Allergy & Infectious Dis fel electron micros, Col Physicians & Surgeons, Columbia Univ, 65-66. *Mem:* Am Soc Microbiol; Electron Microscopy Soc Am. *Res:* Morphology and development of several selected viral agents. *Mailing Add:* Calif State Dept Health Serv 2151 Berkeley Way Berkeley CA 94704

OSHIRO, YUKI, b Okinawa, Japan, Feb 20, 35; m 71; c 2. BIOCHEMISTRY. *Educ:* Univ Calif, Los Angeles, BS, 62; Univ Southern Calif, PhD(biochem), 67. *Prof Exp:* Sr scientist, Frederick Cancer Res Ctr, Nat Cancer Inst, 73-77; RES SCIENTIST, G D SEARLE & CO, 78- *Concurrent Pos:* Fel biochem, Salk Inst Biol Studies, 67-70, USPHS fel, 68-70; staff fel chem carcinogenesis & etiol, Biol Br, Nat Cancer Inst, 70-72; dipl, Am Bd Toxicol. *Mem:* Am Chem Soc; AAAS; Environ Mutagen Soc; Am Asn Cancer Res. *Res:* Chemical carcinogenesis; toxicology; pharmacology. *Mailing Add:* G D Searle & Co 4901 Searle Pkwy Skokie IL 60077

OSHMAN, M KENNETH, COMPUTERS, ENGINEERING. *Prof Exp:* Pres & chief exec officer, Rolm Corp; PRES, CHIEF EXEC OFFICER & CHMN, ECHELON CORP. *Mem:* Nat Acad Eng. *Mailing Add:* Echelon Corp 4015 Miranda Ave Palo Alto CA 94304

OSIFCHIN, NICHOLAS, b Phillipsburg, NJ, June 3, 24; m 61; c 5. ELECTRICAL ENGINEERING. *Educ:* Rutgers Univ, BS, 51; Stevens Inst Technol, MS, 59. *Prof Exp:* Head eng mech, 61-63, head exploratory develop, 63-68, dir, Electro Mech Lab, 68-71, dir, Loop Transmission Syst, 71-74, dir, Bldg & Energy Syst, 74-76, dir Power Systs, 76-79, dir Info Systs, 79-84, DIR C3I & PROCESSING SYSTS, BELL LABS, AT&T, 84- *Concurrent Pos:* Prof mech eng, Stevens Inst Technol, 65-74. *Honors & Awards:* Inst Elec & Electronic Engrs Regional Award, 84. *Mem:* Inst Elec & Electronics Engrs; Am Soc Mech Engrs; AIAA. *Res:* Command control & communications systems; systems engineering; information processing systems. *Mailing Add:* Dir Energy & Bldg Systs Lab Bell Labs Whippany NJ 07981

OSINCHAK, JOSEPH, b Fountain Springs, Pa, Sept 20, 37; m 60; c 4. ANATOMY. *Educ:* Susquehanna Univ, AB, 59; Duke Univ, PhD(anat), 63. *Prof Exp:* USPHS fel, Duke Univ, 63-64; instr anat, Albert Einstein Col Med, 64-66, assoc, 66-68; assoc prof, 68-80, PROF BIOL, CITY COL NEW YORK, 80- *Concurrent Pos:* NIH fel, 64-66. *Mem:* Am Soc Cell Biol; Histochem Soc; Am Asn Anat; Electron Micros Soc Am. *Res:* Fine structure of neuroendocrine organs; application of cytochemical techniques to electron microscopy. *Mailing Add:* Dept Biol City Col New York Convent & 138th St New York NY 10031

OSINSKI, MAREK ANDRZEJ, b Wroclaw, Poland, May 28, 48; m 87; c 2. OPTOELECTRONICS, SEMICONDUCTOR MATERIALS SCIENCE. *Educ:* Univ Warsaw, Poland, MSc, 71; Polish Acad Sci, PhD (phys sci), 79. *Prof Exp:* Sr asst, Inst Physics, Polish Acad Sci, 71-80; res fel electronics, Univ Southampton, Eng, 80-84; sr res assoc coherent optical commun, Dept Eng, Univ Cambridge, Eng, 84-85; ASSOC PROF, ELEC ENG & PHYSICS, CTR HIGH TECHNOL MAT, UNIV NMEX, 85- *Concurrent Pos:* NTT vis prof telecommun, Res Ctr Advan Sci & Technol, Univ Tokyo, Japan, 87- *Honors & Awards:* Achievement Prize, Polish Acad Sci, 80. *Mem:* Inst Elec & Electronic Engrs; Optical Soc Am; Soc Photo-Optical Instrument Engrs; Mat Res Soc; Sigma Xi. *Res:* Semiconductor lasers; optoelectronics; integrated and fiber optics; coherent optical communications; coupled waveguide systems; materials science of bulk and low-dimensional III-V compounds; optical information processing. *Mailing Add:* Ctr High Technol Mat Univ New Mexico Albuquerque NM 87131-6081

OSIPOW, LLOYD IRVING, b Brooklyn, NY, Feb 23, 19; m 41. PHYSICAL CHEMISTRY. *Educ:* Columbia Univ, BS, 39. *Prof Exp:* Chemist, Chem Warfare Serv, 41-44; res group dir surface chem, Foster D Snell, Inc, 48-69; pres, Omar Res, Inc, 69-81; pres, Aerosol Prod Tech, Inc, 81-85; PRES, COSTECH, INC, 85- *Concurrent Pos:* Lectr, Polytech Inst Brooklyn, 55. *Mem:* Am Chem Soc; Am Inst Chemists; Soc Cosmetic Chem. *Res:* Surface and cosmetic chemistry; fats and oils; detergents. *Mailing Add:* 2 Fifth Ave New York NY 10011-8855

OSIPOW, SAMUEL HERMAN, b Allentown, Pa, Apr 18, 34; m 56; c 4. COUNSELING PSYCHOLOGY, CAREER PSYCHOLOGY. *Educ:* Lafayette Col, BA, 54; Columbia Univ, MA, 55; Syracuse Univ, PhD(psychol), 59. *Prof Exp:* Psychologist, Pa State Univ, 61-67, asst prof psychol, 65-67; chmn, 73-86, PROF PSYCHOL, OHIO STATE UNIV, 67- *Concurrent Pos:* Res assoc, Ctr Res Careers, Harvard Univ, 65; vis prof psychol, Tel Aviv Univ, 72 & Univ Md, College Park, 80-81; lectr, Univ Wis, Madison, 61; ed, Jour Vocational Behav, 70-75 & Jour Coun Psychol, 75-81. *Honors & Awards:* Leona Tyler Award, 89. *Mem:* Fel Am Psychol Asn. *Res:* Processes of career development and adjustment and methods of intervention to facilitate those processes. *Mailing Add:* 330 Eastmoor Blvd Columbus OH 43209

OSKI, FRANK, b Philadelphia, Pa, June 17, 32; m 57; c 3. MEDICINE, PEDIATRICS. *Educ:* Swarthmore Col, BA, 54; Univ Pa, MD, 58; Am Bd Pediat, cert, 63, cert, pediat hemat-oncol, 74. *Hon Degrees:* DSc, State Univ NY, 91. *Prof Exp:* Assoc pediat, Sch Med, Univ Pa, 63-65, from asst prof to assoc prof, 65-72; prof pediat & chmn dept, State Univ NY Upstate Med Ctr, 72-85; PROF PEDIAT & CHMN DEPT, JOHNS HOPKINS UNIV SCH MED, 85-, PHYSICIAN-IN-CHIEF, CHILDREN'S MED & SURG CTR, JOHNS HOPKINS HOSP, 85- *Concurrent Pos:* Res fel pediat, Children's Hosp Med Ctr & Harvard Med Sch, 61-63; attend pediatrician, Sch Med, Univ Pa, 65-72, Upstate Med Ctr, 72-85, Crouse-Irving Mem Hosp, 72-85; mem coun, Human Blood & Transfusion Serv, NY State, 81-82; oral examr, Am Bd Pediat Coun, Am Pediat Soc, 88-; coun, Am Pediat Soc, 88-; med adv comt, Johns Hopkins Hosp, 89- *Honors & Awards:* Mead Johnson Award, 72; St Geme Award for Pediat Leadership, 90. *Mem:* Inst Med-Nat Acad Sci; Soc Pediat Res (pres, 77-78); Am Soc Hemat; Am Soc Clin Invest; Am Fedn Clin Res; AAAS; Am Soc Clin Nutrit; Am Pediat Soc; Asn Am Physicians. *Res:* Problems in pediatric hematology, specifically red cell metabolism and the consequences of nutritional deficiencies. *Mailing Add:* Johns Hopkins Hosp 600 N Wolfe St Baltimore MD 21205

OSLER, MARGARET J, b New York, NY, Nov 27, 42. HISTORY & PHILOSOPHY OF SCIENCE. *Educ:* Swarthmore Col, BA, 63; Ind Univ, MA, 66, PhD(hist & philos sci), 68. *Prof Exp:* Asst prof hist sci, Ore State Univ, 68-70; asst prof hist, Harvey Mudd Col, 70-74; asst prof, Wake Forest Univ, 74-75; asst prof, 75-77, ASSOC PROF HIST, UNIV CALGARY, 77- *Mem:* Hist Sci Soc; Can Soc Hist & Philos Sci (pres, 87-90). *Res:* 17th century science. *Mailing Add:* Dept Hist Univ Calgary 2500 University Dr NW Calgary AB T2N 1N4 Can

OSLER, ROBERT DONALD, b Elsie, Nebr, Feb 6, 24; m 47; c 3. CROP BREEDING, AGRONOMY. *Educ:* Univ Nebr, BS, 47; Colo Agr & Mech Col, MS, 49; Univ Minn, PhD(plant breeding), 51. *Prof Exp:* Asst barley breeding, Colo Agr & Mech Col, 47-49; asst oat breeding, Univ Minn, 49-51; asst soybean breeding, US Regional Soybean Lab, 51-54; assoc geneticist corn breeding, Rockefeller Found, 54-56, geneticist, 56-60, asst dir agr sci, 60-65, assoc dir, 65-82; dep dir resident progs, Int Ctr Corn & Wheat Improv, 67-73, dep dir gen & treas, 74-89; RETIRED. *Mem:* AAAS; Am Soc Agron. *Mailing Add:* 14627 Oak Bend Dr Houston TX 77079

OSMAN, ELIZABETH MARY, b Ottawa, Ill, Apr 29, 15. ORGANIC CHEMISTRY, AGRICULTURAL & FOOD CHEMISTRY. *Educ:* Univ Ill, BS, 37, MS, 38; Bryn Mawr, PhD(org chem), 42. *Prof Exp:* Lit searcher, Hercules Powder Co, 41-42, res chemist, 42-44; res chemist, Corn Prod Ref Co, 44-51; from asst prof to assoc prof foods & nutrit, Mich State Univ, 51-56; assoc prof home econ, Univ Ill, 56-63; prof home econ, Univ Iowa, 63-78; prof, 78-85, EMER PROF FOOD & NUTRIT, IOWA STATE UNIV, 85- *Concurrent Pos:* Adj prof food sci & nutrit, Univ Ill, 85- *Mem:* Am Chem Soc; Inst Food Technol; Am Asn Cereal Chem. *Res:* Starch; carbohydrate chemistry; polymerization; three carbon tautomerism; foods; starch in foods; carbohydrate enzymes. *Mailing Add:* 2004 S Vine Urbana IL 61801-5820

OSMAN, JACK DOUGLAS, b Philadelphia, Pa, Jan 26, 43; m 66, 81; c 5. HEALTH EDUCATION, FAT CONTROL. *Educ:* West Chester State Col, Pa, BS, 65; Univ Md, MA, 67; Ohio State Univ, PhD(health & educ), 71. *Prof Exp:* High sch teacher health, Washington, DC, 67-69; instr, Ohio State Univ, 69-71; PROF HEALTH, TOWSON STATE UNIV, 71- *Mem:* Fel Am Sch Health Asn; Am Alliance Health; Nutrit Today Soc; Soc Nutrit Educ; Am Med Joggers Asn. *Res:* Health behaviors focusing on clarifying values and initiating change in the areas of nutrition, weight control and fitness; spiritual aspects of health and counseling. *Mailing Add:* Dept Health Sci Towson State Univ Towson MD 21204

OSMAN, M(OHAMED) O M, b Cairo, Egypt, Aug 10, 36; Can citizen; m 66; c 2. MECHANICAL DESIGN, MANUFACTURING SYSTEMS. *Educ:* Cairo Univ, BSc, 57; Swiss Fed Inst Technol, DrScTechn, 64. *Prof Exp:* Assoc prof, 66-73, PROF ENG, CONCORDIA UNIV, 73-, CHMN DEPT MECH ENG, 87- *Concurrent Pos:* Consult, var corp, 68- *Mem:* Am Soc Mech Engrs; Soc Mfg Engrs; Can Soc Mech Engrs; Int Fedn Mach & Mechanisms; Prof

Engr Ont & Que. *Res:* Design and production engineering; dynamics and stability of machine tools; kinematics and dynamics of mechanisms and robotics; surface mechanics; deep-hole machining and metal cutting. *Mailing Add:* Dept Mech Eng Concordia Univ 1455 de Maisonneuve Blvd W Montreal PQ H3G 1M8 Can

OSMAN, RICHARD WILLIAM, b Fountain Hill, Pa, Feb 24, 48. MARINE ECOLOGY, EVOLUTIONARY BIOLOGY. *Educ:* Brown Univ, AB, 70; Univ Chicago, PhD(geophys sci), 75. *Prof Exp:* Asst prof geol, Northern Ill Univ, 75-76; asst res biologist, Marine Sci Inst, Univ Calif, Santa Barbara, 76-81; asst cur, 81-87, ASSOC CUR, ACAD NAT SCI, BENEDICT EST RES LAB, 87- *Concurrent Pos:* Vis res scholar, Univ Melbourne, Melbourne, Australia, 85. *Mem:* AAAS; Ecol Soc Am; Soc Study Evolution; Am Inst Biol Sci; Brit Ecol Soc; Soc Am Naturalists. *Res:* Community dynamics in fluctuating environments; larvel recruitment processes in marine communities; life history changes in species interactions; symbiotic relationships of sessile marine invertebrates. *Mailing Add:* Acad Nat Sci Benedict Est Res Lab Benedict MD 20612

OSMENT, LAMAR SUTTON, b Pascagoula, Miss, Apr 9, 24; m 47. DERMATOLOGY. *Educ:* Birmingham-Southern Col, BS, 45; Univ Ala, MD, 51. *Prof Exp:* Intern, Univ Hosp, 51-52, resident dermat, Med Col, 52-53, from asst prof to assoc prof, 57-70, PROF DERMAT, MED COL, UNIV ALA, BIRMINGHAM, 70- *Concurrent Pos:* Fel dermat, Med Col, Univ Ala, Birmingham, 53-55; mem staff, Univ Hosp, 55-, Vet Admin Hosp, 57-, Baptist Hosps, 58-, St Vincent Hosp, 59- & Children's Hosp, 62- *Mem:* AMA; Am Acad Dermat; Am Dermat Asn. *Res:* Systemic and topical agents used in dermatology; acrodermatitis enteropathica; bacteriology and mycology of the skin. *Mailing Add:* 4224 Old Leeds Lane Birmingham AL 35213

OSMER, PATRICK STEWART, b Jamestown, NY, Dec 17, 43; m 73; c 2. ASTRONOMY. *Educ:* Case Inst Technol, BS, 65; Calif Inst Technol, PhD(astron), 70. *Prof Exp:* Res assoc astron, Cerro Tololo Int-Am Observ, 69-70, asst astronr, 70-73, assoc astronr, 73-76, astronr, 77-85, dir, 81-85; ASTRONR, NOAO-KITT PEAK NAT OBSERV, 86-, DEP DIR, 88- *Concurrent Pos:* Mem, Am Astron Soc Coun, 85-88. *Mem:* Am Astron Soc; Am Astron Soc Pac; Int Astron Union. *Res:* Spectroscopy and spectrophotometry of galactic and extra galactic objects; quasars. *Mailing Add:* Box 26732 Tuscon AZ 85726-6732

OSMERS, HERMAN REINHARD, chemical engineering; deceased, see previous edition for last biography

OSMOND, DANIEL HARCOURT, b Jerusalem, Israel, Aug 22, 34; Can citizen; m 57; c 4. HYPERTENSION, PHYSIOLOGY. *Educ:* Univ BC, BSA, 58, MSA, 60; Univ Toronto, PhD(clin biochem), 64. *Prof Exp:* Fel & lectr clin biochem, Univ Toronto, 64-66; res fel hypertension, res div, Cleveland Clin, 66-69; from asst prof to assoc prof physiol, 69-80, PROF PHYSIOL & MED, UNIV TORONTO, 80- *Concurrent Pos:* Mem, Coun High Blood Pressure Res, Am Heart Asn; pres, Can Sci & Christian Affil, 79-81; mem, Ont Grad Scholar panel, 80-83, sci rev comt, Can Heart Found, 82-85 & Coun High Blood Presure Res, Am Heart Asn, gov coun, Univ Toronto, 88-91. *Mem:* Can Physiol Soc; Am Physiol Soc; Am Sci Affil; Am Soc Nephrology; Can Biochem Soc; Am Heart Asn. *Res:* Renin-angiotensin system in relation to fluid, electrolyte and blood pressure regulation; prorenin and its activation and relation to other blood enzyme cascades. *Mailing Add:* Dept Physiol Med Sci Bldg Rm 3334 Univ Toronto 1 King's Col Circle Toronto ON M5S 1A8 Can

OSMOND, DENNIS GORDON, b New York, NY, Jan 31, 30; m 55; c 3. ANATOMY, IMMUNOLOGY. *Educ:* Bristol Univ, BSc, 51, MB, ChB, 54, DSc, 75. *Prof Exp:* House surgeon, Royal Gwent Hosp, Newport, Eng, 54; house physician, Bristol Royal Infirmary, 55; demonstr anat, Bristol Univ, 57-59, lectr, 59-60, 61-64; instr, Univ Wash, 60-61; from assoc prof to prof, 65-74, ROBERT REFORD PROF ANAT, McGILL UNIV, 74-, CHMN DEPT, 85- *Concurrent Pos:* Vis scientist, Walter & Eliza Hall Inst Med Res, 72-73; vis hon res fel, Univ Birmingham, Eng, 79; vis scientist, Basel Inst Immunol, Switz, 80. *Honors & Awards:* Fel Royal Soc, Can, 84. *Mem:* Can Asn Anat; Am Asn Anat; Anat Soc Gt Brit & Ireland; Can Soc Immunol; Reticuloendothelial Soc; Am Soc Immunol. *Res:* Experimental hematology and cellular immunology; life histories and functional properties of lymphocyte populations, especially in bone marrow; B lymphocyte differentiation; lymphocyte cultures; cell separation; kinetics of cellular proliferation; bone marrow microenvironment; control of B lymphocytopoiesis. *Mailing Add:* McGill Univ Dept of Anat 3640 University St Montreal PQ H3A 2B2 Can

OSMOND, JOHN C, JR, b Philadelphia, Pa, Feb 13, 23. PETROLEUM GEOLOGY, REGIONAL GEOLOGY. *Educ:* Univ Tex, BS, 47, Univ Wyo, MA, 50, Columbia Univ, PhD(geol). *Prof Exp:* Pres, Div Pac Gas & Elect, Eureka Energy Co, 73-83; CONSULT GEOLOGIST, 85- *Mem:* Geol Soc Am; Am Petrol Ecologist; Am Inst Prof Geol. *Mailing Add:* 6212 S Galena Ct Englewood CO 80111

OSMOND, JOHN KENNETH, b Janesville, Wis, May 12, 28; m 57; c 3. GEOLOGY. *Educ:* Univ Wis, BA, 50, MA, 52, PhD(geol), 54. *Prof Exp:* Fel geochem, Welch Found, Rice Inst, 57-58 & Petrol Res Fund, Am Chem Soc, 58-59; from asst prof to assoc prof, 59-77, PROF GEOL, FLA STATE UNIV, 77- *Mem:* Geochem Soc; Geol Soc Am. *Res:* Quaternary geology; geochemistry; mineralogy; geology of the radioactive elements; geochronology. *Mailing Add:* Dept of Geol Fla State Univ Tallahassee FL 32306

OSMUN, JOHN VINCENT, b Amherst, Mass, Feb 22, 18; m 42; c 1. ENTOMOLOGY. *Educ:* Univ Mass, BS, 40; Amherst Col, MA, 42; Univ Ill, PhD, 56. *Prof Exp:* Field entomologist, US Dept Army, 42-43; res rep, Merck & Co, Inc, 46-48; asst prof, 48-56, head dept, 56-74, PROF ENTOM, PURDUE UNIV, WEST LAFAYETTE, 56- *Concurrent Pos:* Mem US entom deleg, USSR, 59; mem comt insect pest mgt & control, Nat Acad Sci-Nat Res Coun, 67-68; founder & mem, Govt Interstate Interdist Pesticide Coun, 69-72, chmn, 72; mem, Fed Task Group Training Objectives & Standardization Pesticides, 71-72; mem, Ind Pesticide Rev Bd, 72-76; consult, Coop States Res Serv, 72-73. *Mem:* Entom Soc Am. *Res:* Household and industrial entomology; thermal aerosol insecticide dispersion; pest management; competencies for insecticide use. *Mailing Add:* Dept of Entom Purdue Univ West Lafayette IN 47907

OSOBA, DAVID, b Glendon, Alta, Apr 10, 32; c 2. MEDICAL ONCOLOGY. *Educ:* Univ Alta, BSc, 54, MD, 56; FRCP(C), 61. *Prof Exp:* Clin instr med, Univ BC, 63-65, asst prof, 65-66; from asst prof to prof, Univ Toronto, 66-85; PROF MED, UNIV BC, 86-, BC CANCER AGENCY, 88- *Concurrent Pos:* Fel hemat, Vancouver Gen Hosp, BC, 61-62 & assoc hematologist, 63-65; mem staff, Med Res Coun Can, 65-68, assoc, 68-80; head, Div Med Oncol, Sunnybrook Med Ctr & Toronto-Bayview Regional Cancer Ctr, 78-85; pres, Can Oncol Soc, 85-87; head, Div Communities Oncol, BC Cancer Agency, 88- *Honors & Awards:* Medal, Royal Col Physicians & Surgeons, Can. *Mem:* Can Soc Clin Invest; Royal Col Physicians & Surgeons Can; Am Soc Clin Oncol. *Res:* Immunological aspects of malignancy; functions of thymus; immunological deficiency diseases; cellular and humoral immunity; clinical trials in lung, head and neck cancer; quality of life in patients with cancer; community cancer programs. *Mailing Add:* BC Cancer Agency 600 W Tenth Ave Vancouver BC V5Z 4E6 Can

OSOBA, JOSEPH SCHILLER, b Temple, Tex, Dec 5, 19; m 43; c 1. PHYSICS. *Educ:* Univ Tex, BA, 41; Wash Univ, PhD(physics), 49. *Prof Exp:* Res engr, Humble Oil & Refining Co, 49-52, sr res engr, 52-54, res specialist, 54-58, sr res specialist, Esso Prod Res Co, 58-67; PROF PETROL ENG, TEX A&M UNIV, 67- *Mem:* Am Phys Soc; Am Inst Mining, Metall & Petrol Eng. *Res:* Fluid mechanics; electric and radioactive logging; geochemistry; subsurface flow meters; hydrocarbon analysis; new oil recovery techniques. *Mailing Add:* 2809 Pierre Pl College Station TX 77845

OSOFSKY, BARBARA LANGER, b Beacon, NY, Aug 4, 37; m 58; c 3. ALGEBRA, LOGIC. *Educ:* Cornell Univ, BA, 59, MA, 60; Rutgers Univ, PhD(algebra), 64. *Prof Exp:* Assoc mem tech staff, Bell Labs, Inc, NJ, 60; instr math, Douglass Col, 61-63, from asst prof to assoc prof, 64-71, PROF MATH, RUTGERS UNIV, 71- *Concurrent Pos:* NSF res grant, 65-67, 70-79; mem, Sch Math, Inst Adv Study, 67-68; mem-at-large, Coun Conf Bd Math Sci, 72-74, chmn, Bd Trustees, 80-82; proceedings ed comt, Am Math Soc, 74-77, managing ed, 76-77. *Mem:* Am Math Soc; Math Asn Am; Asn Women Math. *Res:* Associative rings and modules; homological algebra. *Mailing Add:* 1010 S Park Ave Highland Park NJ 08904

OSOFSKY, HOWARD J, b Syracuse, NY, May 24, 35; m; c 3. OBSTETRICS & GYNECOLOGY, PSYCHIATRY. *Educ:* Syracuse Univ, BA, 55, PhD(psychol), 74; NY State Col Med, MD, 58. *Prof Exp:* From asst prof to assoc prof obstet & gynec, State Univ NY Upstate Med Ctr, 64-71; prof obstet & gynec, Health Sci Ctr, Temple Univ, 71-76, res psychiatrist, 79-80, PSYCHIAT DISCIPLINE CHIEF, TOPEKA, 80-; CLIN PROF OBSTET & GYNEC, UNIV KANS MED CTR, 80- *Concurrent Pos:* Ed-in-chief, Clin Obstet & Gynec, 72-, adj prof obstet & gynec, Health Sci Ctr, Temple Univ, 76-; ed, Advan Clin Obstet & Gynec, 80-86; prof & head dept psychol, La State Univ, 86- *Mem:* Fel Am Col Obstet & Gynec; fel Am Orthopsychiat Asn. *Res:* Psychological aspects of obstetrics and gynecology with emphasis upon adjustment patterns of couples; pregnancy, the effects of prenatal nutrition upon pregnancy outcome and infant development and psychological effects of abortion and sterilization. *Mailing Add:* 1915 State St New Orleans LA 70118

OSOL, ARTHUR, analytical chemistry; deceased, see previous edition for last biography

OSRI, STANLEY M(AURICE), b Chicago, Ill, July 30, 16; m 44; c 1. CHEMICAL ENGINEERING, FOOD SCIENCE & TECHNOLOGY. *Educ:* Ill Inst Technol, BS, 38. *Prof Exp:* Supt, Qual Plating Co, 38-40; indust engr, Kraft Foods Co, 40-41, res engr, 45-47, prod mgr, Whey Prod, 47-51, mgr eng res, 51-57; res mgr, Res & Develop Div, Nat Dairy Prod Corp, 57-63, dir prod develop, Kraftco Corp, 63-65, dir eng, Res & Develop Div, 65-71, nat prod sales mgr, Kraft Foods Div, 71-78; RETIRED. *Concurrent Pos:* Pres, Bd Educ, high sch dist 207, Cook County, Ill; vol, Int Exec Serv Corp, 78- *Mem:* Am Inst Chem Engrs; Inst Food Technol; Am Inst Chemists. *Res:* Research administration; new product development; application of chemical engineering unit operations to the food industry. *Mailing Add:* 9620 Oak Lane Des Plaines IL 60016

OSSAKOW, SIDNEY LEONARD, b Brooklyn, NY, Dec 14, 38; m 67; c 3. PLASMA PHYSICS, TECHNICAL MANAGEMENT. *Educ:* Mass Inst Technol, BS, 60; Univ Calif, Los Angeles, MS, 62, PhD(theoret physics), 66. *Prof Exp:* Asst physics, Univ Calif, Los Angeles, 60-63, asst theoret plasma physics, 63-66; res fel plasma physics lab, Princeton Univ, 66-67; assoc res scientist, Lockheed Palo Alto Res Lab, Lockheed Missiles & Space Co, 67-71; res physicist, 71-77, head, Geophys & Plasma Dynamics Br, 78-83, SUPVRY RES PHYSICIST, US NAVAL RES LAB, WASHINGTON, DC, 77-, SUPT, PLASMA PHYSICS DIV, 81- *Concurrent Pos:* Mem tech staff, Hughes Res Lab, 63-64; consult, Rand Corp, 65-67. *Mem:* Am Geophys Union; fel Am Phys Soc; Int Asn Geomagnetism & Aeronomy; Int Union Radio Sci. *Res:* Plasma kinetic and fluid theory; instabilities, non-linear effects, computer simulation and electromagnetic wave propagation; space and ionospheric plasmas; magnetospheric phenomena; nuclear weapons effects. *Mailing Add:* 8407 Cross Lake Dr Fairfax Station VA 22039-2669

OSSERMAN, ELLIOTT FREDERICK, medicine; deceased, see previous edition for last biography

OSSERMAN, ROBERT, b New York, NY, Dec 19, 26; m 76; c 3. GEOMETRY, MATHEMATICAL ANALYSIS. *Educ:* NY Univ, BA, 46; Harvard Univ, MA, 48, PhD, 55. *Prof Exp:* Instr math, Univ Colo, 52-53; actg asst prof, Stanford Univ, 55-57, from asst prof to assoc prof, 57-66, chmn dept, 73-79, Mellon prof interdisciplinary studies, 87-90, PROF MATH, STANFORD UNIV, 66-; DEP DIR, MATH SCI RES INST, BERKELEY, 90- *Concurrent Pos:* Mem Inst Math Sci, NY Univ, 57-58; head math br, Off Naval Res, 60-61; vis lectr & res assoc, Harvard Univ, 61-62; Guggenheim fel, Univ Warwick, 76-77; mem, Math Sci Res Inst, Berkeley, 83-84. *Honors & Awards:* Fulbright lectr, Univ Paris, 65-66. *Mem:* Math Asn Am; Am Math Soc; AAAS. *Res:* Complex variables; minimal and Riemann surfaces; Riemannian geometry. *Mailing Add:* Dept Math Stanford Univ Stanford CA 94305

OSSESIA, MICHEL GERMAIN, b Houston, Pa, May 4, 29; m 55; c 3. MATHEMATICS. *Educ:* Univ Pittsburgh, BS, 50, MLitt, 57, PhD(algebra), 61. *Prof Exp:* Asst prof math, Duquesne Univ, 57-61; prof, Slippery Rock State Col, 61-63; assoc prof, Seton Hall Univ, 63-65; prof, Univ Alaska, 65-66; head dept, 66-74, PROF MATH, CLARION STATE COL, 66- *Mem:* Am Math Soc; Math Asn Am. *Res:* Continuous limits. *Mailing Add:* Dept of Math Clarion State Col Box 150-A RD 1 Sligo PA 16255

OSTAFF, WILLIAM A(LLEN), b Cleveland, Ohio, Nov 29, 14; m 45; c 2. SPACE PHYSICS, ELECTRICAL ENGINEERING. *Educ:* Ohio Univ, BSEE, 40. *Prof Exp:* Elec engr, Parker Appliance Co, 40-41 & Trumbull Elec Mfg Co, 41; electronics engr, Bur Ships, US Dept Navy, 41-52, asst chief engr, Ord Exp Unit, 52-58, lectr, Bur Ord, 56-59, systs engr, Spec Projs Off, 58-63; exp consult, Space Physics Observ, Goddard Space Flight Ctr, NASA, 63-90; RETIRED. *Concurrent Pos:* Instr astron, Montgomery Col, 76; vchair, Inst Elec Electronic Engrs Geosci & Remote Sensing Soc, Washington DC, 88-89, dir, 88- *Mem:* AAAS; sr mem Inst Elec & Electronic Eng; NY Acad Sci; Nat Soc Prof Engrs; Acoust Soc Am. *Res:* Remote sensing; automatic control; astrophysics; space physics instrumentation; systems engineering. *Mailing Add:* 10208 Drumm Ave Kensington MD 20895

OSTAPENKO, A(LEXIS), b Ukraine, Oct 1, 23; nat US; m 58; c 3. CIVIL ENGINEERING, ENGINEERING MECHANICS. *Educ:* Munich Tech Univ, dipl, 51; Mass Inst Technol, ScD, 57. *Prof Exp:* Draftsman, Fay, Spofford & Thorndike, Mass, 52; struct designer, Thomas Worcester, Inc, 52-55, Jackson & Moreland Int, Inc, 55, Badger Mfg Co, 56 & Erdman, Anthony & Hosley, Inc, 57; from asst prof to assoc prof, 57-65, PROF CIVIL ENG, LEHIGH UNIV, 65- *Mem:* Am Soc Civil Engrs; Am Soc Eng Educ. *Res:* Elastic and inelastic behavior of structures and components; plastic design; ultimate strength of plate structures; computer methods in civil engineering. *Mailing Add:* Dept Civil Eng Fritz Eng Lab No 13 Lehigh Univ Bethlehem PA 18015

OSTAZESKI, STANLEY A, b Superior, Wis, Apr 17, 26; m 50; c 5. PLANT PATHOLOGY, MYCOLOGY. *Educ:* Wis State Univ, Superior, BS, 52; Univ Ill, Urbana, MS, 55, PhD(plant path), 57. *Prof Exp:* Plant pathologist, USDA, Fla, 57-60; plant path adv, AID, Ethiopia, 60-62; res plant pathologist, USDA, 62-83; RETIRED. *Mem:* Am Phytopath Soc; Mycol Soc Am; Sigma Xi. *Res:* Diseases of forage legumes. *Mailing Add:* 12179 Preston Dr Lusby MD 20657

O'STEEN, WENDALL KEITH, b Meigs, Ga, July 3, 28; m 83; c 3. NEUROBIOLOGY, RETINAL PATHOLOGY. *Educ:* Emory Univ, BA, 48, MS, 50; Duke Univ, PhD(zool), 58. *Prof Exp:* Asst prof biol, Jr Col, Emory Univ, 48-49, asst, 49-50, instr, 50-51; asst prof, Wofford Col, 51-53; asst zool, Duke Univ, 55-57; from instr to prof anat, Univ Tex Med Br, 58-68; prof anat, Sch Med, Emory Univ, 68-77; PROF & CHMN, NEUROBIOL & ANAT DEPT, BOWMAN GRAY SCH MED, WAKE FOREST UNIV, 77- *Concurrent Pos:* Vis lectr, Sch Med, Univ Miami, 63; pres & chmn, Asn Anat, 90-91. *Mem:* Am Asn Anat (vpres, 90-92); Endocrine Soc; Soc Neurosci; Sigma Xi; Asn Res Vision & Ophthal. *Res:* Neuroendocrinology; retinal pathology; visual system; photoneuroendocrinology; retinal photoreceptor cells gradually are reduced in number as animals age; chronic stress exposure interacts with aging to increase further cell death; pituitary and gonadal hormones are implicated in these changes. *Mailing Add:* Dept Neurobiol & Anat Bowman Gray Sch Med Wake Forest Univ Winston-Salem NC 27103

OSTENDORF, DAVID WILLIAM, b Stamford, Conn, Feb 6, 50; m 78; c 2. CIVIL ENGINEERING. *Educ:* Univ Mich, BSE, 72; Mass Inst Technol, SM, 78, ScD, 80. *Prof Exp:* Engr hydrothermal, Stone & Webster Eng Corp, 73-75; res asst civil eng, Mass Inst Technol, 77-80; ASST PROF FLUID MECH, DEPT CIVIL ENG, UNIV MASS, 80- *Mem:* Am Soc Civil Engrs; Am Geophys Union; Water Pollution Control Fedn. *Res:* Models of momentum, pollutant and sediment transport processes in the groundwater, coastal and estuarine environments. *Mailing Add:* 918 E Pleasant St Amherst MA 01002

OSTENSO, GRACE LAUDON, b Tomah, Wis, Sept 15, 32; m 63. NUTRITION. *Educ:* Stout State Univ, BS, 54; Univ Wis-Madison, MS, 60, PhD(foods, nutrit, indust eng), 63. *Prof Exp:* Dietetic intern, Peter Bent Brigham Hosp, Boston, Mass, 54-55, admin dietitian, 55-57, asst dir dietetics, 57-59; res asst foods & nutrit, Univ Wis-Madison, 59-63, asst prof, 63-67; group head, Nutrit & Tech Serv, US Dept Agr, 70-73, dir nutrit & tech serv staff, Food & Nutrit Serv, 73-78; mem prof staff, 78-87, STAFF DIR, SUBCOMT SCI RES & TECHNOL, COMT SCI, SPACE & TECHNOL, US HOUSE REPS, 87- *Concurrent Pos:* Consult, Hosp Systs Res Group, Dept Indust Eng, Univ Mich, 64-65, div hosp & med facilities, Res Br, US Dept Health, Educ & Welfare, 64-69 & Off of Surgeon Gen, Army Med Specialist Corp, 69-73; assoc ed technol, Encycl Britannica, 67-69. *Honors & Awards:* Medallion Award, Am Dietetic Asn, 78; Cooper Mem Lectr & Award, Am Dietetic Asn, 86. *Mem:* Fel AAAS; Am Pub Health Asn; Am Dietetic Asn; Am Inst Nutrit; Am Soc Clin Nutrit; Inst Food Technologists. *Res:* Nutrition and public policy; food systems analysis and management. *Mailing Add:* 2871 Audubon Terrace NW Washington DC 20008

OSTENSO, NED ALLEN, b Fargo, NDak, June 22, 30; m 63. GEOPHYSICS, MARINE SCIENCES GENERAL. *Educ:* Univ Wis, BS, 52, MS, 53, PhD(geol, geophys), 62. *Prof Exp:* Geophysicist, Arctic Inst NAm, 56-58; proj assoc geol & geophys, Univ Wis, 58-63, asst prof, 63-66; oceanogr, Off Naval Res, Chicago, 66-68, dir, Geol & Geophys Prog, Washington, DC, 68-70; dir, Off Sci & Technol, Exec Off President, 70; dep dir & sr oceanogr, Ocean Sci & Technol Div, Off Naval Res, 70-76; dep asst adminr res & develop, Nat Oceanic & Atmospheric Admin, 78-83, actg asst adminr, 81-83; dir, 76-88, ASST ADM OCEANIC & ATMOSPHERIC RES, NAT SEA GRANT COL PROG, 88- *Concurrent Pos:* Foreign affairs fel, Off Naval Res, 75, Cong fel, 75-76. *Mem:* Am Geophys Union; Geol Soc Am; Soc Explor Geophys; Glaciol Soc; Acad Polit Sci. *Res:* Artic and antarctic research; gravity; magnetism; oceanography; mountain in Antarctica and Seamount in the Arctic Ocean named Mt Ostenso in honor of contributions to Polar Research. *Mailing Add:* 2871 Audubon Terr NW Washington DC 20008

OSTENSON, JEROME E, PHYSICS. *Prof Exp:* FAC MEM, DEPT PHYSICS, IOWA STATE UNIV. *Mailing Add:* Dept Physics Iowa State Univ Ames IA 50011

OSTER, BERND, materials science engineering; deceased, see previous edition for last biography

OSTER, CLARENCE ALFRED, b Madras, Ore, Apr 3, 33; m 58; c 3. COMPUTER SCIENCE, MATHEMATICS. *Educ:* Univ Ore, BS, 57, MS, 58. *Prof Exp:* Assoc res engr comput sci, Jet Propulsion Lab, Calif Inst Technol, 58-59; electronic data processing analyst, Gen Elec Co, Richland, Wash, 59-64, mathematician, 64-65; sr res scientist comput sci & math, PaC Northwest Div, Battelle Mem Inst, 65-87; analyst, 87-88, SR SOFTWARE ENGR, BIOENG COMPUT SERV, RICHLAND INC, 88- *Concurrent Pos:* Lectr math, Joint Ctr Grad Study, Univ Wash, Wash State Univ & Ore State Univ, 65- *Mem:* Sigma Xi; Soc Indust & Appl Math; Asn Comput Mach; Math Asn Am; AAAS. *Res:* Error propagation in digital computation; function evaluation. *Mailing Add:* Boeing Comput Serv Richland Inc PO Box 300 MS K6-77 Richland WA 99352

OSTER, EUGENE ARTHUR, b Cincinnati, Ohio, June 2, 29; m 53; c 2. CHEMICAL ENGINEERING. *Educ:* Mass Inst Technol, BS, 51, MS, 54, DSc(chem eng), 66. *Prof Exp:* Proj engr, Div Sponsor Res, Mass Inst Technol, 52 & 56; sr engr, Small Aircraft Engine Lab, Gen Elec Co, 56-58, supvr, 58-60, mgr, Fuel Cell Lab, 60-65; mgr, Okemos Res Lab, 65-66; dir eng res, Owens/Ill, Toledo-Ohio, 66-69, venture gen mgr plasma flat panel display, 69-72, dir corp eng develop, 72-76, dir electrooptic technol, 76-79, dir corp automation & control, 79-88; VPRES ENG, DANLY MACH, DIV CONNELL LP, 88- *Mem:* Am Chem Soc; Combustion Inst; Sigma Xi. *Res:* Fuel cells; flat-panel displays; machine vision; computer input multiplexing. *Mailing Add:* 431 W Eugenie Unit 1H Chicago IL 60614

OSTER, GEORGE F, b New York, NY, Apr 20, 40. MATHEMATICAL BIOLOGY. *Educ:* US Merchant Marine Acad, BS, 61; Columbia Univ, MS, 63, EngScD, 67. *Prof Exp:* Instr, City Col New York, 64-67; NIH fel biophys, Lawrence Radiation Lab, 68-71, mem fac mech eng, 71-72, mem fac entom, 72-74, assoc prof entom, 74-77, PROF BIOPHYS, UNIV CALIF, BERKELEY, 78- *Concurrent Pos:* Guggenheim fel, J S Guggenheim Found, 75; MacArthur Found fel, 85-90. *Honors & Awards:* Louis E Levy Medal, Franklin Inst, 71 & 74. *Res:* Developmental and cell biology; biomechanics; biomathematics; biophysics; embryology. *Mailing Add:* Dept Molecular & Cellular Biol Univ Calif Berkeley CA 94720

OSTER, GERALD, b Providence, RI, Mar 24, 18; m 73; c 2. BIOPHYSICS, PHYSICAL CHEMISTRY. *Educ:* Brown Univ, ScB, 40; Cornell Univ, PhD(phys chem), 43. *Prof Exp:* Mem staff virus res, Rockefeller Inst Med Res, 45-48; vis scientist x-ray scattering, Univ London, 48-50; vis scientist nucleic acids, Univ Paris, 50-51; prof chem, Polytech Inst Brooklyn, 51-69; PROF BIOPHYS, MT SINAI SCH MED, 69- *Concurrent Pos:* Fel, Mass Inst Technol, 43-44; fel, Princeton Univ, 44-45. *Mem:* AAAS. *Res:* Polymer chemistry and biophysical chemistry as applied to reproductive physiology and to muscle physiology; Moiré optics. *Mailing Add:* 70A Greenwhich Ave Suite 272 New York NY 10011

OSTER, JAMES DONALD, b Hazen, NDak, Nov 24, 37; m 58; c 3. SOIL CHEMISTRY. *Educ:* NDak State Univ, BS, 59; Purdue Univ, MS, 62, PhD(soil chem), 64. *Prof Exp:* Teaching asst soils, Purdue Univ, 59-61, asst soil chem, 61-63; soil scientist, US Salinity Lab, Agr Res Serv, USDA, 65-81; SOIL & WATER SPECIALIST, COOP EXTEN, UNIV CALIF, RIVERSIDE, 81- *Concurrent Pos:* Vis scientist, Agr Res Org, Volcani Ctr, Bet Dagan, Israel, 75-76; ed-in-chief, Appl Agr Res, 84-90; vis scientist, Inst Irrigation & Salinity Res, Tatura, Australia, 88-89. *Mem:* Am Soc Agron; Soil Sci Soc Am. *Res:* Irrigation water management for salinity control; irrigation water quality. *Mailing Add:* Soil & Environ Sci Univ Calif Riverside CA 92521

OSTER, LUDWIG FRIEDRICH, b Konstanz, Ger, Mar 8, 31; US citizen; m 56, 73; c 2. SOLAR PHYSICS. *Educ:* Univ Freiburg, Dipl, 54; Univ Kiel, Dr rer nat, 56. *Prof Exp:* Fel, Univ Kiel, 56-58; asst physics, Yale Univ, 58-60; from asst prof to assoc prof, 65-67; from assoc prof to prof, Joint Inst Lab Astrophys, Univ Colo, Boulder, 67-83; PROG MGR, NSF, WASHINGTON, DC, 83- *Concurrent Pos:* Vis prof, Univ Bonn, 66; vis fel, Joint Inst Lab Astrophys, Univ Colo, 66-67; vis prof, Johns Hopkins Univ, 81; sr res assoc, Goddard Space Flight Ctr, Nat Res Coun, NASA, 81-82. *Honors & Awards:* Sr US Scientist Award, Alexander von Humboldt Found, 74. *Mem:* Am Phys Soc; Am Astron Soc; Ger Astron Soc; Int Astron Union. *Res:* Solar and plasma physics; theoretical radio astronomy; radiation theory. *Mailing Add:* 9302 Parkhill Terr Bethesda MD 20814-3963

OSTER, MARK OTHO, b Chicago, Ill, June 2, 37; m 56; c 3. BIOCHEMISTRY, MICROBIOLOGY. *Educ:* Purdue Univ, BS, 59; Agr & Mech Col Tex, MS, 61; Univ Ill, PhD(chem), 65. *Prof Exp:* Res chemist, Univ Calif, Los Angeles, 65-68; ASST PROF LIFE SCI, IND STATE UNIV, TERRE HAUTE, 68- *Mem:* AAAS; Am Chem Soc; Am Soc Microbiol; Sigma Xi. *Res:* Biosynthesis of diterpenes and gibberellin precursors in plants and microorganisms; regulation of pyruvate breakdown in acetate auxotrophs of Escherichia coli. *Mailing Add:* Dept of Life Sci Ind State Univ Terre Haute IN 47809

OSTER, MARTIN WILLIAM, b New York, NY, Apr 9, 47; m 75; c 3. MEDICAL ONCOLOGY, CANCER CHEMOTHERAPY. *Educ:* Columbia Univ, BA, 67, MD, 71. *Prof Exp:* Med intern, Mass Gen Hosp, 71-72, med resident, 72-73; oncol fel, Nat Cancer Inst, 73-76; asst prof med, 76-80, asst prof clin med, 81-86, ASSOC PROF CLIN MED, COL PHYSICIANS & SURGEONS, COLUMBIA UNIV, 86-, ATTEND PHYSICIAN, COLUMBIA PRESBY MED CTR, 76-, ASSOC PROF MED, CANCER RES CTR, 86- *Mem:* Am Soc Clin Oncol; Am Asn Cancer Researchers; fel Am Col Physicians. *Res:* Clinical cancer chemotherapy for solid and hematologic malignancies especially breast, gastro-intestinal and head-neck cancer. *Mailing Add:* Columbia Presby Med Ctr 161 Fort Washington Ave New York NY 10032

OSTERBERG, ARNOLD CURTIS, b Rochester, Minn, Sept 14, 21; m 46; c 4. PHARMACOLOGY. *Educ:* Univ Iowa, BA, 42; Univ Minn, PhD, 53. *Prof Exp:* Asst pharmacol, Univ Minn, 48-53; res pharmacologist, Lederle Labs, Am Cyanamid Co, 53-84; RETIRED. *Mem:* Am Soc Pharmacol & Exp Therapeut. *Res:* Analgesia; anticholinergics; central nervous system. *Mailing Add:* Rte 2 Box 2133 Spooner WI 54801

OSTERBERG, CHARLES LAMAR, b Miami, Ariz, June 15, 20; m 45; c 3. OCEANOGRAPHY. *Educ:* Ariz State Col, BS, 48, MA, 49; Ore State Univ, MS, 60, PhD(oceanog), 63. *Prof Exp:* Res assoc astron, Lowell Observ, Ariz, 49-53; jr scientist, Atmospheric Res Observ, 53-56; teacher high sch, 56-59; from instr to prof oceanog, Ore State Univ, 62-67; marine biologist, Div Biol & Med, 67-72, from actg chief to chief environ sci br, 69-72, 72, asst dir environ sci, US Atomic Energy Comn, 72-73, mgr environ progs, Div Biomed & Environ Res, Energy Res & Develop Admin, 73-76; dir, Lab Marine Radioactiv, Monaco, 76-79; MARINE SCIENTIST, US DEPT ENERGY, 79- *Concurrent Pos:* Consult, Int Atomic Energy Agency, Vienna, United Nations Environ Prog, Geneva. *Mem:* NY Acad Sci; Sigma Xi. *Res:* Marine radioecology; biological and chemical oceanography; gamma ray spectrometry; fate of artificial radionuclides in sea and their transport through food chains. *Mailing Add:* 9312 Gue Rd Damascus MD 20872

OSTERBERG, DONALD MOURETZ, b Teaneck, NJ, Oct 28, 37; m 62; c 2. ICHTHYOLOGY. *Educ:* Montclair State Col, BA, 59; Mich State Univ, MS, 61; Univ Ottawa, PhD(ichthyol), 78. *Prof Exp:* High sch teacher biol, NJ, 61-68; from asst prof to assoc prof, 68-88, PROF BIOL, STATE UNIV NY, COL POTSDAM, 88- *Concurrent Pos:* Consult & resource adv, NY Dept Environ Conserv, 77-; consult, St Lawrence Seaway Develop Corp, 78-; environ consult, 84-; fisheries scientist, Am Fisheries Soc, 84- *Mem:* Am Fisheries Soc; Sigma Xi; Can Conf Freshwater Fisheries Res. *Res:* Age, growth, food and behavior of the muskellunge and other warmwater fish in Saint Lawrence River & Saint Lawrence County, New York; vertebrates of Saint Lawrence County; originator of St Lawrence Aquarium and Ecological Center. *Mailing Add:* Dept Biol State Univ NY Potsdam NY 13676

OSTERBERG, J(ORJ) O(SCAR), b New York, NY, Jan 18, 15; m 42; c 4. CIVIL ENGINEERING, GEOGRAPHICAL TECHNICAL ENGINEERING. *Educ:* Columbia Univ, BS, 35, CE, 36; Harvard Univ, MS, 37; Cornell Univ, PhD(civil eng), 40. *Prof Exp:* Asst engr, US Waterways Exp Sta, 40-42; assoc, Univ Ill, 42-43; lectr, Northwestern Univ, 43-45, from asst prof to prof, 45-85, dir, Geotech Ctr, 52-85, Walter P Murphy prof, 81-85, EMER PROF CIVIL ENG, NORTHWESTERN UNIV, 85- *Concurrent Pos:* Consult, var firms; adj prof, Univ Colo, Boulder; distinguished lectr, NSF. *Mem:* Nat Acad Eng; hon mem Am Soc Civil Engrs; Am Soc Testing & Mat; Int Soc Soil Mech & Found Engrs; Sigma Xi. *Res:* Soil mechanics and foundations; frost action of soils; holder of six patents. *Mailing Add:* 16416 E Powers Pl Aurora CO 80015

OSTERBERG, ROBERT EDWARD, b Brooklyn, NY, July 4, 42. PHARMACOLOGY, TOXICOLOGY. *Educ:* Brooklyn Col Pharm, BS, 65; Georgetown Univ, MS, 69, PhD(pharmacol), 72. *Prof Exp:* Hosp pharmacist, Metrop Hosp, New York, 66; pharmacologist toxicol, 72-73, consumer prod safety comn,73-78 RY PHARMACOLOGIST, FOOD & DRUG ADMIN, 78- *Mem:* Soc Toxicol; Am Col Toxicol; Environ Mutagen Soc; Asn Govt Toxicologists. *Res:* Petroleum distillates and organic solvents and their toxicological effects; mutagenic, irritant and skin sensitization properties of chemicals in consumer products; cardiovascular and neuropharmacology; ionophores. *Mailing Add:* Div Anti-Infective Drug Prod HFD-520 Food & Drug Admin CDER Rockville MD 20857

OSTERBROCK, CARL H, b Cincinnati, Ohio, Mar 8, 31; m 54; c 2. ELECTRICAL ENGINEERING. *Educ:* Univ Cincinnati, EE, 54; Univ Ill, MS, 55; Mich State Univ, PhD(elec eng), 64. *Prof Exp:* Propagation engr, Crosley Broadcasting Corp, 54; mem tech staff, Bell Tel Labs, 55-56; from instr to assoc prof, 56-71, asst to pres, 72-73, PROF ELEC ENG, UNIV CINCINNATI, 71-, V PROVOST ACAD AFFAIRS, 73- *Mem:* Inst Elec & Electronics Eng; Am Soc Eng Educ. *Res:* Discrete and continuous parameter system theory. *Mailing Add:* Dept Elec Eng & Comp Eng Univ Cincinnati Mail Loc 30 Cincinnati OH 45221

OSTERBROCK, DONALD E(DWARD), b Cincinnati, Ohio, July 13, 24; m 52; c 3. ASTRONOMY. *Educ:* Univ Chicago, PhB, 48, BS, 48, MS, 49, PhD(astron), 52. *Hon Degrees:* DSc, Ohio State Univ, 86. *Prof Exp:* Fel astron, Princeton Univ, 52-53; from instr to asst prof, Calif Inst Technol, 53-58; from asst prof to prof, Univ Wis-Madison, 58-73, chmn dept, 69-72; dir, 72-81, PROF ASTRON, LICK OBSERV, UNIV CALIF, SANTA CRUZ, 72- *Concurrent Pos:* Guggenheim fel, Inst Advan Study, 60-61 & 82-83; vis prof, Univ Chicago, 63-64, Univ Minn, 77-78 & Ohio State Univ, 80 & 86; NSF sr fel, Univ Col, London, 68-69; chmn sect astron, Nat Acad Sci, 71-74, secy, class phys & math sci, 80-83, chmn, 83-85, counr, 85-88; counr, Am Astron Soc, 70-73; Ambrose Monnell fel, Inst Advan Study, 89-90. *Mem:* Nat Acad Sci; Am Acad Arts & Sci; Int Astron Union; Am Astron Soc (vpres, 75-77, pres, 88-90); assoc Royal Astron Soc. *Res:* Gaseous nebulae; quasars and active nuclei of galaxies; galactic structure; interstellar matter. *Mailing Add:* Lick Observ Univ Calif Santa Cruz CA 95064

OSTERBURG, JAMES, b New York, NY, June 1, 44. GRAPHICS. *Educ:* Ind Univ, BA, 66, PhD(math), 71. *Prof Exp:* PROF MATH, UNIV CINCINNATI, 72- *Mailing Add:* Dept Math Univ Cincinnati Cincinnati OH 45221-0025

OSTERCAMP, DARYL LEE, b Garner, Iowa, Feb 5, 32; m 58; c 2. ORGANIC CHEMISTRY. *Educ:* St Olaf Col, BA, 53; Univ Wis, MS, 55; Univ Minn, PhD(chem), 59. *Prof Exp:* Instr chem & math, Luther Col, 54-56; res assoc org chem, Pa State Univ, 59-60; from asst prof to assoc prof, 60-72, PROF CHEM, CONCORDIA COL, 72- *Concurrent Pos:* Fulbright prof, Col Sci, Mosul, Iraq, 64-65; NSF sci fac fel, Univ E Anglia, 69-70; vis prof, Univ Petrol & Minerals, Dhahran, Saudi Arabia, 77-80; adj prof, Univ Fla, 85-86. *Honors & Awards:* Nere Sundet Prof Chem, 73. *Mem:* Am Chem Soc. *Res:* Vinylogues of imides; octahydroquinolines and dacahydroquinolines; conformational analysis of saturated heterocycles; spectroscopic properties of vinylogous systems. *Mailing Add:* Dept Chem Concordia Col Moorhead MN 56562

OSTERHELD, ROBERT KEITH, b Brooklyn, NY, Apr 19, 25; m 52; c 4. PHYSICAL INORGANIC CHEMISTRY. *Educ:* Polytech Inst Brooklyn, BS, 45; Univ Ill, PhD(chem), 50. *Prof Exp:* Instr chem, Cornell Univ, 50-54; from asst prof to assoc prof, 54-65, PROF CHEM, UNIV MONT, 65-, CHMN DEPT, 73- *Mem:* Am Chem Soc; N Am Thermal Analysis Soc; Am Asn Univ Prof. *Res:* Phosphate chemistry; high temperature inorganic chemistry; mechanism of thermal decomposition of solids. *Mailing Add:* 524 Larry Creek Loop Florence MT 59833

OSTERHOLTZ, FREDERICK DAVID, b Charleston, WVa, July 29, 38; m 60; c 2. ORGANIC CHEMISTRY, POLYMER CHEMISTRY. *Educ:* Univ Pa, BSc, 59; Mass Inst Technol, PhD(org chem), 64. *Prof Exp:* CHEMIST, UNION CARBIDE CORP, 63- *Mem:* Am Chem Soc; Sigma Xi. *Res:* Radiation chemistry of organic compounds and nitrogen; polymer radiation chemistry; organosilicon chemistry. *Mailing Add:* 81 Oak Dr Pleasantville NY 10570

OSTERHOUDT, HANS WALTER, b Houston, Tex, Feb 29, 36; m 62; c 2. POLYMER CHEMISTRY. *Educ:* Colo State Univ, BS, 58; Univ Wis-Madison, PhD(phys chem), 64. *Prof Exp:* Res chemist, Armstrong Cork Co, 64-68; sr res chemist, 68-71, res assoc polymer chem, 71-73, lab head, 73-78, RES ASSOC PHYS CHEM & MAT SCI, EASTMAN KODAK CO, 78- *Mem:* Am Chem Soc; Soc Plastics Eng. *Res:* Physical chemistry and materials science; electrochemistry of batteries; physicochemical properties of polymers; polymer membranes; polymer molecular weight determination; electrophoresis; electrophotography. *Mailing Add:* 4090 Canal Rd Spencerport NY 14559

OSTERHOUDT, WALTER JABEZ, b Bramans, Pa, Jan 21, 06; m 35; c 2. GROUND WATERS, TUNNELS. *Educ:* Univ Wis, BA, 30. *Prof Exp:* Technician geophysics, Gulf Oil Corp, 30-49, seismic party chief, Gulf Res & Develop Co, 31-33, seismic parties supvr, 33-35, seismic explor supvr, 35-39, mgr explor geophys, Gulf Prod Div, Houston, 39-49; supvr, Houston Dist, Geophys Serv, Inc, 50-51; mgr explor, Colo, JR Butler and others, Houston, 50-53; head geophysicist, William G Helis, New Orleans, 53-72; CONSULT EXPLOR, 73- *Concurrent Pos:* Explorer, San Juan Basin, Colo & NMex, 49-50. *Mem:* Fel AAAS; Soc Explor Geophysicists; Am Inst Mining Metall & Petrol Engrs; Am Asn Petrol Geologists. *Res:* Exploration for oil, gas, water and some minerals, in the USA, Mexico, Venezuela, Greece & Tunisia; geology; geophysics. *Mailing Add:* 3403 County Rd 250 Durango CO 81301

OSTERHOUT, SUYDAM, b Brooklyn, NY, Nov 13, 25; m; c 3. VIROLOGY. *Educ:* Princeton Univ, BA, 45; Duke Univ, MD, 49; Rockefeller Inst, PhD(microbiol), 59; Am Bd Internal Med, dipl, 59. *Prof Exp:* Intern path, Cleveland City Hosp, 50; intern med, Mass Mem Hosp, 50-51; from jr asst resident to resident, Duke Hosp, 53-56; asst dean admis, 66-87, PROF MICROBIOL & MED, SCH MED, DUKE UNIV, 59- *Concurrent Pos:* Markle scholar, Sch Med, Duke Univ, 59- *Mem:* Fel Am Col Physicians. *Res:* Infectious diseases; internal medicine. *Mailing Add:* Dept Microbiol Med Ctr Duke Univ Box 3007 Durham NC 27712

OSTERKAMP, THOMAS EUGENE, US citizen. GLACIOLOGY. *Educ:* Southern Ill Univ, BA, 62; St Louis Univ, MS, 64, PhD(physics), 68. *Prof Exp:* ASSOC PROF PHYSICS & GEOPHYS, UNIV ALASKA, FAIRBANKS, 68- *Concurrent Pos:* Mem glaciol panel, Comt Polar Res, Nat Acad Sci, 74-77. *Mem:* Glaciol Soc; Am Geophys Union. *Res:* Scientific aspects of engineering problems involving snow, ice and permafrost. *Mailing Add:* 1990 Weston Dr Fairbanks AK 99709

OSTERLE, J(OHN) F(LETCHER), b Pittsburgh, Pa, July 31, 25; m 59; c 2. MECHANICAL ENGINEERING. *Educ:* Carnegie Inst Technol, BS, 46, MS, 49, DSc(mech eng), 52. *Prof Exp:* From instr to assoc prof, 46-58, chmn nuclear sci & eng, 75-85, actg head mech eng dept, 85-87, THEODORE AHRENS PROF MECH ENG, CARNEGIE-MELLON UNIV, 58- *Concurrent Pos:* Vis prof, Delft Univ Technol, Neth, 57-58 & Oxford Univ, 71. *Honors & Awards:* Walter D Hodson Award, Am Soc Lubrication Engrs, 56. *Mem:* Am Soc Mech Engrs; Am Nuclear Soc. *Res:* Fluid mechanics; thermodynamics. *Mailing Add:* Dept Mech Eng Carnegie-Mellon Univ 5000 Forbes Ave Pittsburgh PA 15213

OSTERMANN, RUSSELL DEAN, b Wichita, Kans, Mar 24, 52; m 80; c 1. CHEMICAL ENGINEERING, FERMENTATION ENGINEERING. *Educ:* Univ Kans, BS, 74, PhD(chem eng), 80. *Prof Exp:* Res asst, NSF, Univ Kans, 72; tech serv engr, Sabine River Works, E I Du Pont de Nemours & Co, 74-75; res asst, chem eng, Univ Kans, 75-79; asst prof chem eng, Tex A&M Univ, 79-81; asst prof, 81-86, assoc prof & dept head petrol eng, 86-89, DEAN SCH MINERAL ENG, UNIV ALASKA, 89- *Concurrent Pos:* Planning engr, Northern Natural Gas Co, 75; asst adj instr mat & energy balances, Univ Kans, 77. *Mem:* Am Inst Chem Engrs; Soc Petrol Engrs. *Res:* Kinetic studies of cellulose liquefaction; fermentation technology; petroleum phase behavior; reservoir engineering. *Mailing Add:* Dept Petrol Eng Univ Alaska 435 Duclecring Fairbanks AK 99701

OSTERMAYER, FREDERICK WILLIAM, JR, b New York, NY, Aug 12, 37. PHYSICS. *Educ:* Mass Inst Technol, SB & SM, 59; Lehigh Univ, PhD(physics), 68. *Prof Exp:* Mem tech staff solid state devices, 59-65, MEM TECH STAFF ELECTRONIC DEVICES & MAT, BELL LABS, INC, 68- *Mem:* Am Phys Soc. *Res:* Optical fibers and materials; solid state lasers; luminescence of solids; semiconductor electrochemistry. *Mailing Add:* AT&T Bell Labs Rm 7F-329 600 Mountain Ave Murray Hill NJ 07974

OSTERUD, HAROLD T, b Richmond, Va, May 1, 23; m 49; c 3. PUBLIC HEALTH, PREVENTIVE MEDICINE. *Educ:* Randolph-Macon Col, BS, 44; Med Col Va, MD, 47; Univ NC, MPH, 51. *Prof Exp:* Health officer, Wasco-Sherman Health Dept, 48-50, Coquille County Health Dept, 53-55 & Lane County Health Dept, 56-61; assoc prof pub health, 61-65, PROF PUB HEALTH, MED SCH, ORE HEALTH SCI UNIV, 65-, CHMN DEPT, 67- *Concurrent Pos:* Mem, Prev Med & Dent Rev Panel, 65-70. *Honors & Awards:* Sippy Award, Am Pub Health Asn, 69. *Mem:* Am Pub Health Asn. *Res:* Health manpower; physician study and health care delivery. *Mailing Add:* Dept Pub Health Ore Health Sci Univ 3181 SW San Jackson Park Rd Portland OR 97201

OSTERWALD, FRANK WILLIAM, physical geology; deceased, see previous edition for last biography

OSTERYOUNG, JANET G, b Pittsburgh, Pa, Mar 1, 39; m 69; c 2. ELECTROANALYTICAL CHEMISTRY. *Educ:* Swarthmore Col, BA, 61; Calif Inst Technol, PhD(chem), 67. *Prof Exp:* Res fel chem, Calif Inst Technol, 66-67; asst prof, Mont State Univ, 67-68; res fel, Colo State Univ, 68-69, assoc prof civil eng, 73-79; PROF CHEM, STATE UNIV NY, BUFFALO, 82- *Concurrent Pos:* Vis prof, Calif Inst Technol, 85; prog dir, chem anal, NSF, 77-78; Guggenheim fel, 85-86. *Honors & Awards:* Garran Medal, 87; Anachem Award, 90. *Mem:* Am Chem Soc; Electrochem Soc; AAAS; Sigma Xi. *Res:* Electroanalytical methods development especially for toxic substances; characterization of natural waters; pulse voltammetry; microelectrodes. *Mailing Add:* Dept Chem State Univ NY Buffalo NY 14214

OSTERYOUNG, ROBERT ALLEN, b US, Jan 20, 27; m 69; c 5. ANALYTICAL CHEMISTRY. *Educ:* Ohio Univ, BS, 49; Univ Ill, MS, 51, PhD, 54. *Prof Exp:* Res chemist, Harshaw Chem Co, 51-52; asst, Univ Ill, 52-54; from asst prof to assoc prof chem, Rensselaer Polytech, 54-59; res specialist, Atomics Int Div, NAm Rockwell Corp, 59-62, group leader phys chem, Sci Ctr, 62-67; prof chem, Colo State Univ, 68-79, chmn dept, 68-78; PROF CHEM, STATE UNIV NY AT BUFFALO, 79- *Concurrent Pos:* Vis assoc chem, Calif Inst Technol, 63-68; asst dir phys sci dept, Autonetics Div, NAm Rockwell Corp, 67, dir, Mat & Process Lab & assoc dir, Sci Ctr, 68. *Honors & Awards:* C N Reilley Award Electroanal Chem, 87; Schoelkolof Medal, 90. *Mem:* Fel AAAS; Am Chem Soc; fel Electrochem Soc; Int Soc Electrochem. *Res:* Ionic liquid chemistry and electrochemistry; electroanalytical chemistry; pulse voltommetry; on line use of computers in electrochemistry. *Mailing Add:* Dept Chem State Univ NY Buffalo NY 14214

OSTFELD, ADRIAN MICHAEL, b St Louis, Mo, Sept 2, 26; m 50; c 3. EPIDEMIOLOGY, PUBLIC HEALTH. *Educ:* Wash Univ, MD, 51. *Prof Exp:* Asst resident med, Barnes Hosp, St Louis, Mo, 52-53; instr med, Med Col, Cornell Univ, 55-56; from asst prof to prof prev med, Univ Ill Col Med, 56-66, prof prev med & community health & head dept, 66-68; chem dept epidemiol & pub health, 68-70, LAUDER PROF EPIDEMIOL & PUB HEALTH, YALE UNIV, 68- *Concurrent Pos:* Commonwealth fel, NY Hosp, 53-55; proj dir ecol migrant pop, Int Biol Prog, Nat Acad Sci, 67-; spec consult to Surgeon Gen, USPHS; vpres, Coun Epidemiol, Am Heart Asn, fel, Coun Stroke & chmn, Comt on Hypertension, 70-75; chmn epidemiol & biomet comt, NIH; chmn epidemiol adv comt, Nat Heart & Lung Inst; consult, Inst Med; assoc ed, J Behav Med & Behav Med Ann, 70-75; ed, Am J Epidemiol, 87. *Mem:* Am Psychosom Soc; Am Soc Pharmacol & Exp Therapeut; Am Pub Health Asn; Asn Teachers Prev Med; fel Am Col Angiol. *Res:* Epidemiology of hypertension and cerebrovascular diseases; community surveys; geriatrics; human behavior and psychosomatic medicine; epidemiology of cardiovascular disease. *Mailing Add:* Lab Epidemiol 404 Leph Yale Univ New Haven CT 06520

OSTLE, BERNARD, b Vancouver, BC, June 29, 21; nat US; m 53; c 2. STATISTICS. *Educ:* Univ BC, BA, 45, MA, 46; Iowa State Univ, PhD(statist), 49. *Prof Exp:* Instr statist, Univ Minn, 46-47; from instr to asst prof, Iowa State Univ, 47-52; from assoc prof to prof math, Mont State Col, 52-57, dir statist lab & statistician, Agr Exp Sta, 56-57; mem staff, Reliability Dept, Sandia Corp, NMex, 57-58, supvr statist sect, 58-60; prof eng, Ariz State Univ, 60-63; chmn statist test design, Rocketdyne Div, Rockwell Corp, 63-67; dean, Col Natural Sci, 67-80, actg vpres for res & dean grad studies, 80-81, prof, 81-87, EMER PROF STATIST, UNIV CENT FLA, 87. *Concurrent Pos:* Consult, Humble Oil & Ref Co, 52, Motorola, Inc, 61-63, AvCir, Flight Safety Found, 62 & McElrath & Assocs, Inc, 82- *Mem:* Fel Am Statist Asn; Am Soc Qual Control; Sigma Xi. *Res:* Quality control; reliability; statistical design of experiments; statistical analysis of data. *Mailing Add:* 1174 Winged Foot Circle E Winter Springs FL 32708

OSTLIND, DAN A, b McPherson, Kans, June 19, 36; m 58; c 1. PARASITOLOGY, ENTOMOLOGY. *Educ:* Bethany Col, Kans, BS, 58; Kansas State Univ, MS, 62, PhD(parasitol), 66. *Prof Exp:* Parasitologist, Moorman Mfg Co, 66-67; sr res parasitologist, 67-69, res fel, 69-77, sr res fel, 77-86, SR INVESTR, MERCK INST, 86- *Mem:* Am Soc Parasitol; Am Asn Vet Parasitologists. *Res:* Chemotherapy of parasites; helminthology; radiation. *Mailing Add:* Merck Inst Rahway NJ 07065

OSTLUND, H GOTE, b Stockholm, Sweden, June 26, 23; m 50; c 2. MARINE CHEMISTRY, ATMOSPHERIC CHEMISTRY. *Educ:* Univ Stockholm, Fil Kand, 49, Fil lic, 59. *Hon Degrees:* Dr, Univ Goteborg, Sweden, 84. *Prof Exp:* Asst inorg chem, Royal Inst Technol, Sweden, 48-52; head labs, Svenska Salpeterverken Koping, 52-54; dir, Radioactive Dating Lab, Stockholm, 54-63; vis assoc prof, 60-61, assoc prof geochem, Inst Marine Sci, 63-67, PROF MARINE & ATMOSPHERIC CHEM, ROSENSTIEL SCH MARINE & ATMOSPHERIC SCI, UNIV MIAMI, 67- *Concurrent Pos:* Consult, Adv Panel Oceanog, NSF, 70-74, chmn div chem oceanog, 70-74; coordr proj, Geochem Ocean Sect Study, 76-81, Transient Tracers in the ocean, 79-83. *Mem:* AAAS; Am Geophys Union; Am Meteorol Soc; Royal Soc Arts & Sci Goteborg, Sweden. *Res:* Low level counting techniques; chemical meteorology and air-sea interaction; radioactive isotopes for tracing and dating ocean currents; arctic oceanography. *Mailing Add:* Dept Marine & Atmospheric Chem Univ Miami Marine Div 4600 Rickenbacker Causeway Miami FL 33149

OSTLUND, NEIL SINCLAIR, b Calgary, Alta, Nov 18, 42. THEORETICAL CHEMISTRY. *Educ:* Univ Sask, BA, 63; Carnegie-Mellon Univ, MSc, 68, PhD(theoret chem), 69. *Prof Exp:* Res fel theoret chem, Harvard, 69-71; asst prof, Univ Ark, Fayetteville, 71-75, assoc prof, 75-80; assoc prof, Carnegie-Mellon Univ, 80-82; At Dept Comput Sci, Univ Waterloo, 82-88; CHIEF EXEC OFFICER, HYPERCUBE INC, 88- *Mem:* Am Phys Soc; Am Chem Soc; Inst Elec & Electronics Engrs; Asn Comput Mach. *Res:* computation chemistry; computer architecture; parallel processing; molecular modeling. *Mailing Add:* Hypercube Inc 419 Phillip St No 7 Waterloo ON N2L 3X2 Can

OSTMANN, BERNARD GEORGE, b Washington, DC, Aug 14, 27; m 55; c 5. TEXTILE TECHNOLOGY. *Educ:* Cath Univ Am, BS, 50. *Prof Exp:* From chemist to res chemist, 50-60, from sr res chemist to res assoc, 67-85, RES ASSOC FIBERS MKT CTR TEXTILE RES LAB, E I DU PONT DE NEMOURS & CO, 85- *Mem:* Tech Asn Pulp & Paper Indust. *Res:* Synthetic fiber chemistry and technology. *Mailing Add:* 2212 Pennington Dr Wilmington DE 19810

OSTRACH, SIMON, b Providence, RI, Dec 26, 23; m 75; c 5. FLUID & THERMAL SCIENCES. *Educ:* Univ RI, BS, 44, ME, 49; Brown Univ, ScM, 49, PhD(appl math), 50. *Hon Degrees:* DSc, Technion Israel Inst Technol, 86. *Prof Exp:* Res scientist aeronaut, Nat Adv Comt Aeronaut, 44-47; res assoc appl math, Brown Univ, 47-50; br chief fluid mech, NASA, 50-60; prof eng & head, Div Fluid, Thermal & Aerospace Sci, Case Inst Technol, 60-70; W J AUSTIN DISTINGUISHED PROF ENG, CASE WESTERN RESERVE UNIV, 70- *Concurrent Pos:* Mem res adv comt, NASA, 63-75; distinguished vis prof, City Col of City Univ New York, 66-68, Col Eng Fla A&M Univ/Fla State Univ, 90; Sigma Xi nat lectr, 78-79; Freeman scholar, Am Soc Mech Engrs, 82; Lady Davis fel & vis prof, Technion-Israel Inst Technol, 83-84; Japan Soc Prom Sci fel, 87; consult to numerous corp & industs; mem, Space Processing & Microgravity Appln Comt, Int Astronaut Fedn, NASA Space Sci & Appln Adv Comt, NASA Microgravity Sci & Appln Subcomt; assoc ed, Mfg Rev. *Honors & Awards:* Heat Transfer Mem Award, Am Soc Mech Engrs, 75, Thurston Lectr, 87, 50th Ann Award, Heat Transfer Div, 88; Max Jacob Mem Award, Am Soc Mech Engrs-Am Inst Chem Engrs, 83. *Mem:* Nat Acad Eng; fel Am Soc Mech Engrs; fel Am Inst Aeronaut & Astronaut; fel Am Acad Eng; Sigma Xi; Soc Natural Philos; Am Asn Univ Profs. *Res:* Aerospace science; natural convection; reduced-gravity phenomena; materials processing; biological fluid mechanics; author of 120 scientific articles and papers. *Mailing Add:* Dept Mech & Aerospace Eng Case Western Reserve Univ Cleveland OH 44106

OSTRANDER, CHARLES EVANS, b Jamestown, NY, Oct 30, 16; m 45; c 2. POULTRY SCIENCE. *Educ:* Cornell Univ, BS, 41; Mich State Univ, MS, 60. *Prof Exp:* Teacher pub schs, 41-45; from asst to assoc county agr agent, Onondaga County, NY, 46-51; asst prof poultry sci, Cornell Univ, 51-52; sales mgr, Marshall Bros Hatchery, 52-56; from asst prof to assoc prof, 56-72, prof, 72-81, EMER PROF POULTRY SCI, CORNELL UNIV, 81- *Honors & Awards:* Pfizer Nat Poultry Exten Award, 80. *Mem:* Poultry Sci Asn; World Poultry Sci Asn; Ctr Appl Sci & Technol. *Res:* Poultry management; waste disposal; lighting regimes to conserve energy. *Mailing Add:* 142 W Haven Rd Ithaca NY 14850

OSTRANDER, DARL REED, b Ann Arbor, Mich, Apr 29, 35; m 57; c 2. LIMNOLOGY. *Educ:* Eastern Mich Univ, BS, 61; Univ Mich, MS, 65; Univ Conn, PhD(ecol-fisheries), 74. *Prof Exp:* Asst prof, 65-76, ASSOC PROF BIOL, CENT CONN STATE COL, 76- *Concurrent Pos:* Assoc consult, Environ Mgt Coun, 74- *Mem:* Ecol Soc Am; Am Soc Limnol & Oceanog. *Res:* Analysis of plant and mineral distribution in inland wetlands; recovery of perturbed tidal wetlands. *Mailing Add:* Dept Biol Cent Conn State Univ New Britain CT 06050

OSTRANDER, LEE E, b Glens Falls, NY, Feb 18, 39; c 2. ELECTRICAL ENGINEERING, BIOENGINEERING. *Educ:* Hamilton Col, AB, 61; Univ Rochester, MS, 63, PhD(elec eng), 66. *Prof Exp:* Res assoc eng, Case Western Reserve Univ, 65-66, from asst prof to assoc prof, 66-74; bioengr, S C Johnson & Son Inc, 74-75; ASSOC PROF BIOMED ENG, RENSSELAER POLYTECH INST, 76- *Mem:* Inst Elec & Electronic Engrs. *Res:* Application of electrical engineering and physics to problems in the life sciences, clinical medicine and bioinstrumentation. *Mailing Add:* Biomed Eng Dept Rensselaer Polytech Inst Troy NY 12180-3590

OSTRANDER, PETER ERLING, b Rahway, NJ, Aug 12, 43; m 70; c 1. NUCLEAR PHYSICS. *Educ:* Ill Inst Technol, BS, 65; Pa State Univ, PhD(physics), 70. *Prof Exp:* Asst prof, 69-76, ASSOC PROF PHYSICS, PA STATE UNIV, 76- *Mem:* Sigma Xi. *Res:* Theoretical calculations of photopion cross-sections including final state interactions. *Mailing Add:* Pa State Univ Fayette Campus Rte 119 N PO Box 519 Uniontown PA 15401

OSTRAND-ROSENBERG, SUZANNE T, b New York, NY, July 8, 48; m 71; c 2. IMMUNOLOGY. *Educ:* Barnard Col, Columbia Univ, AB, 70; Calif Inst Technol, PhD(immunol), 75. *Prof Exp:* Fel immunol, Johns Hopkins Univ, 74-77; ASST PROF BIOL SCI, UNIV MD, BALTIMORE COUNTY, 77-, ASSOC PROF BIOL. *Concurrent Pos:* Fac res award, Am Cancer Soc, 83-88. *Mem:* Am Asn Immunologists; AAAS. *Res:* Genetic control of tumor rejection in mammals, with emphasis on the regulation and role of the major histocompatibility complexes in cell mediated tumor immunity. *Mailing Add:* Dept Biol Sci Univ Md Baltimore County 5401 Wilkins Ave Baltimore MD 21228

OSTRIKER, JEREMIAH P, b New York, NY, Apr 13, 37; m 58; c 3. ASTROPHYSICS. *Educ:* Harvard Univ, AB, 59; Univ Chicago, PhD(astrophys), 64. *Prof Exp:* NSF fel, 64-65, lectr & res assoc, 65-66, from asst prof to assoc prof, 66-71, PROF ASTROPHYS, PRINCETON UNIV, 71-, CHMN, DEPT ASTROPHYS SCI & DIR, PRINCETON UNIV OBSERV, 79- *Concurrent Pos:* Alfred P Sloan fel, 70-72. *Honors & Awards:* Helen B Warner Prize, Am Astron Soc, 72; Henry Norris Russell Prize, Am Astron Soc, 80. *Mem:* Nat Acad Sci; Am Acad Arts & Sci; Int Astron Union. *Res:* Structure and stability of nonspherical self gravitating bodies; high energy astrophysics; galaxy origin and evolution. *Mailing Add:* Princeton Univ Observ Princeton NJ 08544

OSTROFF, ANTON G, b Moline, Ill, Nov 6, 25; m 54; c 3. WATER CHEMISTRY. *Educ:* Augustana Col, Ill, BA, 50; Southern Methodist Univ, MS, 52; Univ Iowa, PhD(inorg chem), 57. *Prof Exp:* Res chemist, Dallas, 52-53, sr res technologist, 56-79, MGR ANALYTICAL SERV, RES LAB, MOBIL RES & DEVELOP CORP, 79- *Concurrent Pos:* Consult, Ostroff & Assoc Consults, 85- *Mem:* Am Chem Soc; Am Soc Testing & Mat; Soc Petrol Engrs; Nat Asn Corrosion Engrs. *Res:* Chemistry of subsurface brines; corrosion of oilfield equipment; geochemistry; trace elements in rocks and waters; water pollution; thermal gravimetric analysis and differential thermal analysis of metal sulfates; analysis of phenol carbonation products; analytical chemistry. *Mailing Add:* 619 Misty Glen Lane Dallas TX 75232-1315

OSTROFSKY, BENJAMIN, b Philadelphia, Pa, July 26, 25; m 56; c 2. INDUSTRIAL ENGINEERING, OPERATIONS RESEARCH. *Educ:* Drexel Univ, BS, 47; Univ Calif, Los Angeles, MEng, 62, PhD(eng), 68. *Prof Exp:* Prod test engr elec & mech test, E G Budd Mfg Co, Red Lion, Pa, 48; proj eng consult, George C Lewis Co, Philadelphia, 40-50; prod mgr mfg, Generator Equip Co, Los Angeles, Calif, 53-54; res engr aircraft testing, Douglas Aircraft Co, Santa Monica, 54-57; sr engr advan design, Norair Div, Northrop Corp, 57-61; mem tech staff, Electronics Div, TRW SYSTS, Calif, 61-64, sr staff engr, 64-69; chmn prod logistics mgt, 71-74, PROF PROD & LOGISTICS MGT, COL BUS ADMIN, UNIV HOUSTON, 69-, PROF INDUST & SYSTS ENG, COL ENG, 69- *Concurrent Pos:* Lectr & teaching fel, Ford Found Educ Develop Prog Design, Univ Calif, Los Angeles, 62-65, lectr systs design, 65-69; consult var industs & govt orgn, 70- *Honors & Awards:* Armitage Medal, Soc Logistics Engrs, 78, Eccles Medal, 88. *Mem:* Fel Soc Logistics Engrs; assoc fel Am Inst Aeronaut & Astronaut; Opers Res Soc Am; Nat Soc Prof Engrs; fel AAAS. *Res:* Methods of systems design; information systems for operational performance; quantitative methods for criteria function modeling design and systems optimization; educational methods for teaching systems design; integrated logistics methods and systems theory; reliability and maintainability research. *Mailing Add:* Dept Indust Eng Univ Houston Houston TX 77004

OSTROFSKY, BERNARD, b New York, NY, Jan 14, 22; m 46; c 2. PHYSICS, PHYSICAL CHEMISTRY. *Educ:* City Col New York, BS, 45. *Prof Exp:* Res chemist, Manhattan Proj, Columbia Univ, 42-45; physicist, Res Labs, Nat Lead Indust, 45-53; sr phys chemist, Commonwealth Eng Co Ohio, 53-54; proj supvr mat sci, Eng Res Div, Standard Oil Co, Ind, 54-61; res proj supvr, Res & Develop Dept, Amoco Oil Co, 61-72; sr res assoc & proj mgr, Standard Oil Co, Ind, 72-81; PRES, BERNARD OSTROFSKY ASSOC INC, 81- *Concurrent Pos:* Tech ed, Mat Eval, J Am Soc Nondestructive Testing, 68-88. *Honors & Awards:* John Vaalor Award, 70. *Mem:* Fel Am Soc Nondestructive Testing; Am Crystallog Asn; Am Chem Soc; Electron Probe Soc Am; Am Soc Testing & Mat. *Res:* X-ray and electron diffraction; x-ray microbeam analysis; materials science; ultrasonics, radiography; leak testing; magnetic inspection; liquid penetrants; electronics; crystallography; scanning electron microscopy; acoustic emission. *Mailing Add:* 3041 Fallstaff Rd No 4040 Baltimore MD 21209-2965

OSTROFSKY, MILTON LEWIS, b Bridgeport, Conn, Sept 9, 47. LIMNOLOGY, ECOLOGY. *Educ:* Univ Vt, BA, 69; Univ Waterloo, MSc, 74, PhD(biol), 76. *Prof Exp:* Environ consult limnol, Sheppard T Powell Assocs, 70-75; res assoc, Dept Biol, McGill Univ, 76-78; ASST PROF BIOL, ALLEGHENY COL, 78- *Mem:* Int Asn Theoret & Appl Limnol; Am Soc Limnol & Oceanog; Int Soc Chem Ecol. *Res:* Trophic changes during the maturation of reservoir ecosystems; effect of nutrient concentrations on the qualitative and quantitative aspects of the plankton of freshwater lakes. *Mailing Add:* Dept Biol Allegheny Col Meadville PA 16335

OSTROM, CARL ERIC, b Philadelphia, Pa, May 29, 12; m 41; c 4. FORESTRY. *Educ:* Pa State Univ, BS, 33; Yale Univ, MF, 41, PhD(forestry), 44. *Prof Exp:* Asst technician, Northern Rocky Mt Forest & Range Exp Sta, US Forest Serv, 34-36, silviculturist, Northeastern Forest Exp Sta, 36-44, Southeastern Forest Exp Sta, 44-57, dir, Div Timber Mgt Res, 57-74, assoc dep chief, Biol Res, 74-75; dir sci progs, Soc Am Foresters, 75-77; RETIRED. *Concurrent Pos:* Consult, UN Spec Fund, Thailand, 62. *Mem:* Soc Am Foresters. *Res:* Silviculture; forest management. *Mailing Add:* 2233 Sequoya Dr Prescott AZ 86312

OSTROM, JOHN HAROLD, b New York, NY, Feb 18, 28; m 52; c 2. VERTEBRATE PALEONTOLOGY. *Educ:* Union Col, NY, BS, 51; Columbia Univ, PhD(geol), 60. *Prof Exp:* Res asst vert paleont, Am Mus Natural Hist, 51-56; from instr to asst prof geol, Beloit Col, 56-61; from asst prof to assoc prof geol & asst cur to assoc cur vert paleont, 61-71, PROF GEOL & GEOPHYS, YALE UNIV, 71-, CUR VERT PALEONT, PEABODY MUS NATURAL HIST, 71- *Concurrent Pos:* Guggenheim fel, 66-67; lectr, Brooklyn Col, 55-56; res assoc, Am Mus Natural Hist, 65-; ed, Am J Sci, 67-, Bull, Soc Vert Paleont, 63-73 & Peabody Mus Publ, 70- *Honors & Awards:* Alexander von Hunboldt sr scientist Award, W Ger, 76-77; F V Hayden Mem Geological Medal, 86. *Mem:* AAAS; Soc Study Evolution; Soc Vert Paleont (pres, 69-70); Asn Earth Sci Ed; Sigma Xi (pres, 72-73); Paleontologic Soc; Soc Syst Zool. *Res:* Vertebrate paleontology, especially ancient reptiles; vertebrate evolution; Mesozoic stratigraphy. *Mailing Add:* Dept of Geol & Geophys Yale Univ New Haven CT 06520

OSTROM, MEREDITH EGGERS, b Rock Island, Ill, Nov 16, 30; m 53; c 3. GEOLOGY. *Educ:* Augustana Col, BS, 52; Univ Ill, MS, 54, PhD, 59. *Prof Exp:* Asst geologist, Univ Ill, 52-54; asst geologist coal sect, State Geol Surv, Ill, 54-55, asst geologist, Indust Minerals Sect, 55-59; from asst state geologist to assoc state geologist, Wis Geol & Natural Hist Surv, 59-72, state geologist & dir, 72-90; prof environ sci, Univ Wis-Extension, 76-90; RETIRED. *Concurrent Pos:* Chmn dept environ sci, Univ Wis Exten, 72-76. *Mem:* Geol Soc Am; Soc Econ Paleont & Mineral; Am Asn Petrol Geol; Am Inst Prof Geol; Asn Am State Geol; AAAS. *Res:* Industrial minerals; nonmetallic mineral deposits; stratigraphy; environmental geology. *Mailing Add:* 6802 Forest Glade Ct Middleton WI 53562

OSTROM, THEODORE GLEASON, b Nicollet, Minn, Jan 4, 16; m 49; c 4. COMBINATORICS & FINITE MATHEMATICS. *Educ:* Univ Minn, BA, 37, MA & BS, 39, PhD(math), 47. *Prof Exp:* Instr math, Univ Minn, 46-47; from asst prof to prof math, Univ Mont, 47-60, chmn dept, 54-60; chmn dept, 69-71, prof, 60-81, EMER PROF MATH, WASH STATE UNIV, 81- *Concurrent Pos:* Vis prof, Univ Frankfurt, 66-67, Univ Florence, 82-83, Univ Recife, 86. *Mem:* Am Math Soc; Math Asn Am. *Res:* Finite projective planes. *Mailing Add:* Dept of Math Wash State Univ Pullman WA 99164-2930

OSTROM, THOMAS ROSS, b San Francisco, Calif, May 23, 24; m 49; c 2. SANITARY ENGINEERING, BALLISTICS. *Educ:* Univ Calif, BS, 44; Harvard Univ, MSSE, 48, PhD(sanit eng), 55. *Prof Exp:* US Army, 45-75, mil govt officer, Hq Korea, 45-47, sanit engr, Health Physics Div, Oak Ridge Nat Lab, Tenn, 49-50, engr health physics, Joint Task Force 3, 50-51, sanit engr, Indust Waste Br, Los Alamos Field Off, 51-53, chmn sanit eng div, US Army Environ Health Lab, Army Chem Ctr, Md, 55-58, chmn sanit eng dept, Walter Reed Army Inst Res, Washington, DC, 59-62, sanit engr, Hq 7th Army, 62-64, chief med & biol sci br, Off Chief Res & Develop, Hq, Dept Army, 64-68, commanding officer, Wound Data & Munitions Effectiveness Team, Vietnam, 68-69, mem fac, Indust Col Armed Forces, 69-71, dir res, develop & eng, Munitions Command, 71-72, commanding officer, US Army Ballistics Res Labs, 72-75; RETIRED. *Concurrent Pos:* Consult, US Army Mat Systs Anal Activity, 76- & Gen Elec, TEMPO, 77- *Mem:* Nat Soc Prof Engrs; Am Soc Civil Engrs; Health Physics Soc; Soc Am Mil Engrs; Am Acad Environ Engrs. *Res:* Radio-strontium content of and distribution of radio-stronitum in fallout; disposal of low level radioactive wastes; sewage disposal procedures under extreme cold conditions. *Mailing Add:* 102 Duncannon Rd Bel Air MD 21014

OSTROVSKY, DAVID SAUL, b Cleveland, Ohio, Jan 28, 43. ZOOLOGY. *Educ:* Case Western Reserve Univ, BA, 65; Univ Mich, MS, 67, PhD, 70. *Prof Exp:* Asst prof zool, Ohio Wesleyan Univ, 72-73; ASST PROF BIOL, MILLERSVILLE STATE COL, 73- *Mem:* Am Soc Zool. *Res:* The problem of enzyme development in relation to both structural and functional differentiation. *Mailing Add:* Dept of Biol Millersville Univ Millersville PA 17551

OSTROVSKY, RAFAIL M, US citizen. CRYPTOGRAPHY, DISTRIBUTED COMPUTING. *Educ:* State Univ NY, BA, 84; Boston Univ, MA, 87; Mass Inst Technol, computer sci, 90- *Prof Exp:* Res engr, Digital Equip Corp, 85 & Index Technol Corp, 87-89; res scientist, Math Res Ctr, AT&T Bell Labs & Int Bus Mach T J Watson Res Ctr, 90. *Res:* Computer security; cryptography; distributed computing; theoretical computer science; applied mathematics. *Mailing Add:* Lab Computer Sci NE43-334 Mass Inst Technol Cambridge MA 02136

OSTROW, DAVID HENRY, b Elgin, Ill, June 25, 53; m 78; c 3. CELL BIOLOGY OF LYMPHOCYTES, IMMUNOCHEMISTRY OF ANTIBODIES. *Educ:* Northern Ill Univ, BS, 75, MS, 77; Univ Ill, PhD(immunol), 82. *Prof Exp:* Immunochemist hybridoma res, Coulter Immunol, 82-84; biochemist immunoassay develop, 84-86, sr biochemist, 86-88, SR RES BIOCHEMIST, ABBOTT LABS, 88- *Mem:* AAAS. *Res:* Infection of lymphocytes by viruses, growth and biology of lymphocytes and the purification and characterization of antibodies; hepatitis vaccine; immunoassay development. *Mailing Add:* 985 Manchester Ct Lake Zurich IL 60047

OSTROW, JAY DONALD, b New York, NY, Jan 1, 30; m 56; c 3. GASTROENTEROLOGY. *Educ:* Yale Univ, BS, 50; Harvard Med Sch, MD, 54; Univ London, MSc, 70. *Prof Exp:* Intern med, Johns Hopkins Hosp, 54-55; asst resident, Peter Bent Brigham Hosp, 57-58; instr med, Harvard Med Serv, 61-62; sr instr, Western Reserve Univ & Univ Hosps, 62-64, asst prof, 64-69; from asst prof to assoc prof med, Sch Med, Univ Pa, 71-78; chief, Gastroenterol Sect, 78-87, PROF MED, NORTHWESTERN UNIV, CHICAGO, 78-, CHIEF, GASTROENTEROL RES, 87- *Concurrent Pos:* NIH res fel gastroenterol, Peter Bent Brigham Hosp, 58-59; Nat Inst Arthritis & Metab Dis res fel liver dis & nutrit, Harvard Med Serv & Thorndike Lab, Boston City Hosp, 59-62; Nat Inst Arthritis & Metab Dis res grants, 63-93; med investr, US Vet Admin, 71-78 & 90-95, merit res rev grant, 71-95, chmn merit rev bd gastroenterol, 79-82; asst ed, Liver Dis Am Jour Physiol, 79-86;

sr distinguished scientist award, Alexander von Humboldt Found, Fed Repub Ger, 89-90. *Honors & Awards:* Award Clin Res Gastroenterol, William Beaumont Soc, 79. *Mem:* Am Soc Clin Invest; Am Asn Study Liver Dis, (pres, 86-87); Am Gastroenterol Asn; Am Physiol Soc; Int Asn Study Liver. *Res:* Bilirubin metabolism; gallbladder absorption; gallstones. *Mailing Add:* Vet Admin Lakeside Med Ctr 333 E Huron St Chicago IL 60611

OSTROWSKI, EDWARD JOSEPH, b Phillipston, Pa, Mar 19, 23; m 50; c 2. METALLURGICAL ENGINEERING. *Educ:* Univ Pittsburgh, BS, 64. *Prof Exp:* Mem staff, Follansbee Steel Corp, Ohio, 43-45; mem staff, US Bur Mines, Pa, 45-60, supv prod metallurgist & proj leader, 60; sr res engr, Res Ctr, Nat Steel Corp, 60-63, supvr, 63-67, div chief process metall, Res Ctr, 67-77, mgr qual control, Mineral Resources Div, 77-82, dir qual assurance & technol, 82-88; CONSULT, 88- *Concurrent Pos:* Indust rep, Blast Furnace Res, Inc, 62-; secy, Blast Furnace Comt, Am Iron & Steel Inst, 62-82; trustee, Iron & Steel Soc Found. *Honors & Awards:* J E Johnson, Jr Award, Am Inst Mining, Metall & Petrol Engrs, 63, Iron Making Conf Award, 63; James B Austin Award, Am Inst Mining, Metall & Petrol Engrs, Iron & Steel Soc, 78, Thomas Joseph Award, 82. *Mem:* Distinguished mem Am Inst Mining, Metall & Petrol Engrs, Iron & Steel Soc (pres, 77); Am Soc Testing & Mat; Am Inst Mining, Metall & Petrol Engrs (vpres, 77-79). *Res:* Steel making; blast furnaces; recipient of 5 patents and author of 40 technical papers. *Mailing Add:* 833 Ridgefield Ave Pittsburgh PA 15216-1143

OSTROWSKI, RONALD STEPHEN, b Chicago, Ill, Mar 3, 39; m 68; c 1. GENETICS. *Educ:* Northern Ill Univ, BS, 66, MS, 68; Univ Notre Dame, PhD, 72. *Prof Exp:* Asst prof, 71-77, ASSOC PROF BIOL, UNIV NC, CHARLOTTE, 77-, ASST TO VCHANCELLOR & DIR EVE PROG, 78- *Mem:* AAAS; Genetics Soc Am. *Res:* Developmental genetics. *Mailing Add:* Dept Biol Univ NC Univ Sta Charlotte NC 28223

OSTROY, SANFORD EUGENE, b Scranton, Pa, Dec 28, 39; m 62; c 2. NEUROBIOLOGY. *Educ:* Univ Scranton, BS, 61; Case Inst Technol, PhD(chem), 66. *Prof Exp:* NIH fel biophys chem, Cornell Univ, 66-68; from asst prof to assoc prof, 68-80, PROF BIOL SCI, PURDUE UNIV, 80- *Concurrent Pos:* NIH res career develop award & grant, 70-75; hon res assoc biol, Harvard Univ, 74-75; vis prof biol, Brandeis Univ, 83-84. *Mem:* Fedn Biol Chemists; Am Chem Soc; Biophys Soc; Asn Res Vision & Ophthal. *Res:* Sensory physiology; rhodopsin structure and function; membrane biophysics; vision; chemistry and physiology of visual photoreceptors. *Mailing Add:* Dept Biol Sci Purdue Univ West Lafayette IN 47907-7801

OSTRUM, G(EORGE) KENNETH, b Rockford, Ill, June 7, 38; m 60; c 3. COMPUTER-BASED CHEMICAL INFORMATION RETRIEVAL. *Educ:* Wheaton Col, BS, 60; Northwestern Univ, PhD(org chem), 65. *Prof Exp:* Res chemist, Jet Propulsion Labs, NASA, 65-66 & textile fibers dept, E I Du Pont De Nemours, 66-68; SR TECH SERV REP, CHEM ABSTR SERV, AM CHEM SOC, 68- *Mem:* Am Chem Soc. *Res:* Designing training materials for computerized chemical information retrieval. *Mailing Add:* Chem Abstr Serv 2540 Olentangy River Rd PO Box 3012 Columbus OH 43210

OSTWALD, PETER FREDERIC, b Berlin, Ger, Jan 5, 28; US citizen; m 60; c 2. PSYCHIATRY, COMMUNICATION SCIENCES. *Educ:* Univ Calif, Berkeley, AB, 47; Univ Calif, San Francisco, MD, 50. *Prof Exp:* Asst resident psychiat, Sch Med, Cornell Univ, 51-53 & 55-56; chief resident, Langley Porter Neuropsychiat Inst, 56-57; res psychiatrist, 58-63, from asst prof to assoc prof, 60-70, PROF PSYCHIAT, SCH MED, UNIV CALIF, SAN FRANCISCO, 70- *Concurrent Pos:* USPHS grant, Langley Porter Neuropsychiat Inst, 57-60, Found Fund Res Psychiat grant, 60-63; attend psychiatrist, Langley Porter Neuropsychiat Inst, 60-; consult, Dept Rehab, Psychiat; dir, UCSF Health Prog Performing Artists, 86- *Honors & Awards:* Deems Taylor Award, Am Soc Composers, Authors & Publishers, 85. *Mem:* Am Psychiat Asn; Am Col Psychiatrists; AAAS. *Res:* Acoustical communication; speech disorders; hearing and language problems; psychotherapy; psychiatric care; education; role of sound in human communication; voice changes associated with mental disorder, speech pathology and musical behavior; psychobiography. *Mailing Add:* Sch Med Univ Calif 401 Parnassus Ave San Francisco CA 94143

OSTWALD, PHILLIP F, b Omaha, Nebr, Oct 21, 31; m 56; c 3. INDUSTRIAL ENGINEERING. *Educ:* Univ Nebr, BS, 54; Ohio State Univ, MS, 56; Okla State Univ, PhD(indust eng), 66. *Prof Exp:* Time study engr, Giddings & Lewis Mach Tool Co, Wis, 56-58; proj staff engr & group leader standards sect, Fabrication Div, Collins Radio Co, Iowa, 58-62; asst prof indust eng, Univ Omaha, 62-63; instr, Okla State Univ, 63-66; from asst prof to assoc prof eng design & econ eval, 66-74, PROF MECH AND INDUST ENG, UNIV COLO, BOULDER, 74- *Mem:* Am Soc Eng Educ; Am Inst Indust Engrs; Soc Mfg Eng; Am Asn Cost Eng. *Res:* Manufacturing dynamics in process machining; dynamic chip breaking; memomotion measurement of factory delay allowances; cost engineering analysis of complex engineering systems. *Mailing Add:* Dept Mech Eng Univ Colo Campus Box 427 Boulder CO 80309

OSUCH, CARL, b Chicago, Ill, Feb 2, 25; m 48; c 4. ORGANIC CHEMISTRY. *Educ:* Antioch Col, BS, 50; Univ Pittsburgh, PhD, 55. *Prof Exp:* Res chemist, Monsanto Co, 55-64; from asst prof to assoc prof, 64-70, chmn sci div, 74-83, PROF CHEM, UNIV DUBUQUE, 70- *Concurrent Pos:* Petrol Res Fund grant, 67-70; consult, Dubuque police dept, 72- *Mem:* Sigma Xi; Coblentz Soc; Am Chem Soc. *Res:* Alkyl pyridines; organolithium compounds; bicyclic olefins; organic azides; phosphorus compounds; drug identification. *Mailing Add:* Dept Chem Univ Dubuque Dubuque IA 52001-5099

OSUCH, CHRISTOPHER ERION, b Pittsburgh, Pa, Feb 23, 51. MATERIALS RESEARCH, PHOTO RESISTS. *Educ:* Univ Dubuque, BA, 75; Iowa State Univ, PhD(chem), 80. *Prof Exp:* res chemist, Allied Corp, 80-84; SR RES CHEMIST, ALLIED-SIGNAL, 84- *Res:* Physical and organic chemistry, including work with radical ions and ion pairs; chemistry of sulfur; chemistry of organic materials for photo lithography; polymer structure-property relationships. *Mailing Add:* 151A Randolph Ave Mine Hill NJ 07801

OSUCH, MARY ANN V, b Chicago, Ill, Feb 9, 41; m 78. BIOCHEMISTRY. *Educ:* DePaul Univ, BS, 62; Northwestern Univ, PhD(biochem), 67. *Prof Exp:* Res biochemist, Miles Lab, Inc, 67-68, supvr, Mfg Lab, 68-71, tech serv coordr, 71-72, prod mgr, 72-78, mkt mgr, Res Prod Div, 78-84, sr clin res scientist, 84-87, dir int mfg, 87- 90, DIR MFG PROGS, MILES LAB, INC, 90- *Mem:* Am Chem Soc; Am Asn Clin Chemists. *Res:* Nucleic acids and enzymes; bufadienolides; clinical chemical methodology and instrumentation. *Mailing Add:* 3400 Middlebury St Elkhart IN 46515

O'SULLIVAN, JOHN, b Limerick, Ireland, Oct 30, 42; m 70; c 4. PLANT PHYSIOLOGY, PLANT BREEDING. *Educ:* Univ Col, Dublin, BAgrSc, 68, MAgrSc, 70; Univ Wis, PhD(plant breeding & genetics), 73. *Prof Exp:* RES SCIENTIST AGR, HORT EXP STA, ONT MINISTRY AGR & FOOD, 74- *Mem:* Am Soc Hort Sci; Int Soc Hort Sci; Can Soc Hort Sci; Weed Sci Soc Am. *Res:* Nutrition and irrigation of crops; growth regulators for crop production; weed control. *Mailing Add:* Hort Exp Sta Ont Ministry Agr & Food Simcoe ON N3Y 4N5 Can

OSULLIVAN, JOSEPH, b Ireland, March, 14, 45; m 70; c 2. BIOSYNTHESIS. *Educ:* Nat Univ Ireland, BSc, 66; Univ Dublin, MSc, 67; London Univ, PhD(microbiol), 70. *Prof Exp:* Res asst microbiol, London Univ, 67-72; res fel, Swiss Fed Inst Technol, 72-75, biochem, Univ Oxford, 75-79; GROUP LEADER MICROBIOL & BIOCHEM, SQUIBB INST MED RES, 79- *Mem:* Am Soc Microbiol; Soc Indust Microbiol; Am Chem Soc. *Res:* Biosynthesis of B-lactam antibiotics; biochemistry of natural products. *Mailing Add:* Squibb Inst Med Res PO Box 4000 Princeton NJ 08540

O'SULLIVAN, WILLIAM JOHN, b Springfield, Mass, Apr 21, 31; m 54; c 2. LASERS, FLUID PHYSICS. *Educ:* Univ Pittsburgh, PhD(physics), 58. *Prof Exp:* Res assoc physics, Univ Pittsburgh, 58; asst prof, US Naval Postgrad Sch, 58-59; mem tech staff & consult, Space Tech Labs, Inc, Div Thompson-Ramo-Wooldridge, Inc, 59-65; staff mem, Sandia Corp, 65-68; PROF PHYSICS, UNIV COLO, BOULDER, 68-, DEPT CHAIR, 88- *Mem:* Am Phys Soc. *Res:* Theory of antiferromagnetism; band structure of solids; nuclear magnetic resonance; behavior of Fermi surfaces under pressure; photo autocorrelation studies of critical fluid behavior. *Mailing Add:* Dept Physics DPPA FG21 Univ Colo Boulder CO 80309

OSWALD, EDWARD ODELL, b Newberry, SC, Jan 9, 40; m 61; c 2. BIOCHEMISTRY. *Educ:* Newberry Col, BS, 61; Bowman Gray Med Sch, MS, 63; Univ NC, Chapel Hill, PhD(biochem), 66. *Prof Exp:* NIH res assoc biochem, Univ NC, Chapel Hill, 66-67; dir clin chem, Palms of Pasadena Hosp, 67-68; res biochemist, Nat Inst Environ Health Sci, Environ Protection Agency, 68-79; chmn, 79-88, MEM FAC, DEPT ENVIRON HEALTH SCI, SCH HEALTH, UNIV SC, 79-, DIR, HEALTH & SAFETY PROGS, 85- *Mem:* AAAS; Am Chem Soc; Am Oil Chemists Soc. *Res:* Organic synthesis and metabolism of radioactive labeled lipids and lipid-like natural products. *Mailing Add:* Dept Environ Health Sci Sch Pub Health Univ SC Columbia SC 29208

OSWALD, JOHN WIELAND, b Minneapolis, Minn, Oct 11, 17; m 45; c 3. PLANT PATHOLOGY. *Educ:* DePauw Univ, AB, 38, LLD, 64; Univ Calif, PhD(plant path), 42. *Hon Degrees:* LLD, Cent Col, 66, Univ Louisville, 66, Univ Calif, Davis, 67 & Lehigh Univ, 82; LHD, Juniata Col, 73 & Gettysburg Col, 74; ScD, Temple Univ, 71, Thomas Jefferson Univ, 82 & Univ Ky, 89. *Prof Exp:* From asst prof to prof plant path, Col Agr, Univ Calif, Berkeley, 46-63, vchmn dept, 54-58, from asst vpres to vpres, Univ, 58-63; prof plant path & pres, Univ Ky, 63-68; exec vpres, Univ Calif, Berkeley, 68-70; prof & pres, 70-83, EMER PROF PLANT PATH & EMER PRES, PA STATE UNIV, 83- *Concurrent Pos:* Fulbright res grant, Neth, 53-54. *Mem:* Am Phytopath Soc. *Res:* Potato diseases; potato scab; viruses; cereal root rots; cereal viruses. *Mailing Add:* Ogontz Campus Pa State Univ 1600 Woodland Rd Abington PA 19001

OSWALD, ROBERT B(ERNARD), JR, b Detroit, Mich, May 25, 32; m 62; c 1. PHYSICS. *Educ:* Univ Mich, BSE(mech eng) & BSE(math), 57, MSE, 58, PhD(nuclear eng), 64. *Prof Exp:* Engr, Aircraft Nuclear Propulsion Div, Gen Elec Co, 58-59; res engr, Eng Sci Div, Am Metal Prod Co, 59-60; res asst radiation effect in semiconductors, Univ Mich, 60-64; res physicist, Harry Diamond Labs, Army Mat Command, 64-73, chief, Lab 300, 73-76, assoc tech dir, 76-79; asst dep dir sci technol, Defense Nuclear Agency, 79-; AT US ARMY, ELECTRONICS RES & DEVELOP COMMAND. *Concurrent Pos:* Assoc prof, Dept Nuclear Eng, Univ Mich, 69-70. *Honors & Awards:* Award, Inst Elec & Electronics Engrs Conf, 65; Louis J Hamilton Award, Am Nuclear Soc, 76; USA Spec Serv Award, Harry Diamond Labs, 78. *Mem:* Inst Elec & Electronics Engrs. *Res:* Radiation effects in solids; thermal neutron effects in semiconductor materials. *Mailing Add:* US Army Corps Engrs CERD-ZA 20 Massachusetts Ave Washington DC 20314-1000

OSWALD, THOMAS HAROLD, bioengineering, biochemical genetics, for more information see previous edition

OSWALD, WILLIAM J, b Calif, July 6, 19; m 44; c 4. SANITARY ENGINEERING. *Educ:* Univ Calif, BS, 50, MS, 51, PhD(sanit eng), 57. *Prof Exp:* Eng aid, Sea Water Proj, 49-50, asst sanit eng, 50, jr res engr, 51-52, asst res engr, 53-57, asst prof, Col Eng, 58-59, assoc prof, Col Eng & Sch Pub Health, 59-66, PROF SANIT ENG & PUB HEALTH, UNIV CALIF, BERKELEY, 66- *Concurrent Pos:* Consult to int agencies, US govt & various indust concerns, 53- *Honors & Awards:* Eddy Medal, Water Pollution Control Fedn, 54; Rudolph Hering Medal, 58; James R Croes Medal, Am Soc Civil Engrs, 58, Arthur M Wellington Award, 66. *Mem:* Fel AAAS; Am Soc Civil Engrs; Am Waterworks Asn. *Res:* Bioenvironmental engineering, especially development of complex mass ecosystems of microbes such as algae and bacteria for waste disposal and reclamation; pharmaceutical production, environmental control; microbiology; space technology. *Mailing Add:* 1081 St Francis Concord CA 94518

OSWALT, JESSE HARRELL, b Maben, Miss, May 16, 23; m 51; c 3. INDUSTRIAL ENGINEERING, ACADEMIC ADMINISTRATION. *Educ:* Miss State Univ, BS, 48; Ga Inst Technol, MSIE, 61. *Prof Exp:* Teacher math high sch, Miss, 53-54; process engr, Am Bosch Arma Corp, 54-56; PROF INDUST ENG, MISS STATE UNIV, 56- *Concurrent Pos:* Consult, Voc Rehab Workshop Prog, Miss, 63 & Tech Assistance Prog, Social & Rehab Serv, 66. *Honors & Awards:* President's Citation Meritorious Serv Employ of Handicapped, President's Comt Employ of Handicapped, 72. *Mem:* Am Inst Indust Engrs; Am Soc Eng Educ. *Res:* Hospital management engineering application work methods and training for handicapped workers. *Mailing Add:* 110 James St Greenwood MS 38930

OSWEILER, GARY D, b Sigourney, Iowa, Sept 8, 42; m 66; c 4. VETERINARY TOXICOLOGY. *Educ:* Iowa State Univ, DVM, 66, MS, 68, PhD(vet toxicol), 73. *Prof Exp:* Instr vet path, Iowa Vet Diag Lab, 66-67; instr med, Vet Clin, Iowa State Univ, 68-69, asst prof vet path, 69-70, assoc prof vet toxicol, 71-73; prof vet toxicol, Univ Mo, Columbia, 74-82; PROF VET TOXICOL, IOWA STATE UNIV, AMES, 83- *Concurrent Pos:* Mem exam bd, Am Bd Vet Toxicol, 72-73; consult, 73 & Food & Drug Admin, 87-90; mem task forces, Coun Agr Sci Technol, 87. *Mem:* Am Col Vet Toxicologists; Am Vet Med Asn; Coun Res Workers Animal Dis; Soc Toxicol; Am Soc Animal Sci. *Res:* Reproductive and immunologic effects of myco toxins; bioassay techniques for screening toxins in foods and feeds; toxicology of chlorophenols; behavioral responses to mycotoxins; trace elements in nutrition and toxicology. *Mailing Add:* Vet Diag Lab Iowa State Univ Ames IA 50011

OTA, ASHER KENHACHIRO, b Waianae, Hawaii, Dec 1, 34; m 66; c 1. ENTOMOLOGY. *Educ:* Univ Hawaii, BS, 56, MS, 62; Univ Calif, Berkeley, PhD(entom), 66. *Prof Exp:* Asst entom, Univ Hawaii, 56-61; res entomologist, Entom Res Div, USDA, 66-68; from assoc entomologist to entomologist, 68-74, PRIN ENTOMOLOGIST, EXP STA, HAWAIIAN SUGAR PLANTERS ASN, 74- *Mem:* Sigma Xi; Entom Soc Am. *Res:* Agricultural entomology; insect population ecology; insect resistant varieties; biological control of insects and pests. *Mailing Add:* 46-034 Kumoo Pl Kaneohe HI 96744

OTANI, THEODORE TOSHIRO, b Honolulu, Hawaii, Jan 5, 25; m 47; c 3. BIOCHEMISTRY, SYNTHETIC ORGANIC CHEMISTRY. *Educ:* Univ Hawaii, BS, 47; Univ Denver, MS, 49; Univ Colo, PhD, 53. *Prof Exp:* Fel, Clin Biochem Res Sect, Nat Cancer Inst, 53-55; res assoc chem, Fla State Univ, 55-57; res assoc, Lab Biochem, 57-72, RES BIOCHEMIST, LAB PATHOPHYSIOL, NAT CANCER INST, 72- *Mem:* AAAS. *Res:* Biochemistry of amino acids and peptides; synthesis of amino acids, amino acid analogues and peptides; amino acid antagonists; isolation, characterization and mode of action of antibiotics; isozymes in hepatomas and normal livers; design and chemical synthesis of possible cancer chemotherapeutic compounds. *Mailing Add:* Nat Cancer Inst Div Cancer Treatment Develop Therapeut Prog 7910 Woodmont Ave EPN Bldg Rm 8311 Bethesda MD 20892

OTERO, RAYMOND B, b Rochester, NY, May 8, 38; div; c 3. MICROBIOLOGY, BIOCHEMISTRY. *Educ:* Univ Dayton, BS, 60; Univ Rochester, MS, 63; Univ Md, PhD(microbiol), 68. *Prof Exp:* Control biologist, Lederle Labs Div, Am Cyanamid Co, 63-65; from asst prof to assoc prof microbiol, 68-73, assoc prof biol, 73-76, PROF BIOL, EASTERN KY UNIV, 76- *Concurrent Pos:* Chicago Res Corp grant, 69-; consult path & cytol, Good Samaritan Hosp, 69, St Joseph Hosp, 69- & Cent Baptist Hosp, 80- *Mem:* AAAS; NY Acad Sci; Am Soc Microbiol; Am Thoracic Soc. *Res:* Determination of mode of entry of nucleic acids into whole bacterial cells; bacterial genetics and immunology; clinical microbiology as it applies to patient care. *Mailing Add:* Dept of Biol Sci Eastern Ky Univ Richmond KY 40475

OTEY, FELIX HAROLD, b Melber, Ky, Feb 27, 27; m 50; c 2. ORGANIC CHEMISTRY. *Educ:* Murray State Col, BS, 48; Univ Mo, Columbia, MA, 50. *Prof Exp:* Instr chem, Flat River Jr Col, 50-55 & McMurray Col, 55-56; res chemist, Northern Regional Res Lab, USDA, 56-75, res leader chem, 75-87; RETIRED. *Honors & Awards:* USDA Superior Serv Award, 66. *Mem:* Am Chem Soc. *Res:* The synthesis and characterization of starch derivatives for use as raw materials in surfactants, coatings, rubber, urethane foams, starch-based plastics and films. *Mailing Add:* 1128 E Tripp Ave Peoria IL 61603

OTHERSEN, HENRY BIEMANN, JR, b Charleston, SC, Aug 26, 30; c 3. PEDIATRIC SURGERY. *Educ:* Col Charleston, BSM, 50; Med Univ SC, MD, 53; Am Bd Surg, dipl, 63 & 75; Am Bd Thoracic Surg, dipl, 66. *Prof Exp:* Intern, Philadelphia Gen Hosp, 53-54; resident gen surg, Med Col SC, 57-62; demonstr pediat surg, Children's Hosp & Ohio State Univ, 63-64; from asst prof to assoc prof surg, 65-72, PROF SURG & PEDIAT, MED UNIV SC, 72-, CHIEF PEDIAT SURG, 65- *Concurrent Pos:* Fel oncol, Med Col SC, 59-60, teaching fel surg, 61-62; res fel surg, Mass Gen Hosp, 64-65; Am Cancer Soc advan clin fel, Med Col SC, 66-69; resident pediat surg, Children's Hosp, Columbus, 62-64. *Mem:* Asn Acad Surg; Am Burn Asn; Am Col Surg; Am Acad Pediat; Am Pediat Surg Asn. *Res:* Tracheal and esophageal strictures. *Mailing Add:* Dept Surg/Pediat & Med Univ SC 171 Ashley Ave Charleston SC 29425

OTHMER, DONALD F(REDERICK), b Omaha, Nebr, May 11, 04; m 50. CHEMICAL ENGINEERING, INDUSTRIAL CHEMISTRY. *Educ:* Univ Nebr, BChE, 24; Mich Univ, MChE, 25, PhD(chem eng), 27. *Hon Degrees:* Dr Eng, Univ Nebr, 62, Polytech Univ, Brooklyn, 77, NJ Inst Technol, 78. *Prof Exp:* Chemist, Cudahy Packing Co, 24; chemist & develop engr, Eastman Kodak Co, 27-31; from instr to asst prof, 32-37, prof & head dept, 37-61, DISTINGUISHED PROF, POLYTECH UNIV, NY, 61- *Concurrent Pos:* Consult, US Dept State, Interior, Army, Navy, Cong, many industs & insts in US & foreign countries, 32-; co-founder & ed, Kirk-Othmer Encycl Chem Tech, 47-86; Am Swiss Found Sci exchange lectr tour, 48; hon prof, Conception Univ, Chile, 51; leader, Develop Chem Indust Prog, Burma, 51-53; lectr, US Army War Col, Shri Ram mem, India, 64, Tour of India, 80. *Honors & Awards:* Tyler Award, Am Inst Chem Engrs, 58; Chem Pioneer Award, Am Inst Chemists, 77; Murphree (Exxon) Award, Am Chem Soc, 78; Perkin Medal, Soc Chem Indust, 78; Waddell mem lectr, Royal Mil Col, Can, 81. *Mem:* Hon fel Am Soc Mech Eng; Am Chem Soc; Am Inst Chem Engrs; Am Inst Chemists; NY Acad Sci; Inst Chem Engrs; AAAS; Soc Chem Indust. *Res:* Synthetic fuels; thermodynamics; author or coauthor of over 350 publications; phase equilibria; data correlation; industrial processes for distillation; desalination; refrigeration; acetylene, acetic acid, methanol; petrochemicals; sugar refining; coal, lignite, peat gasification; desulfurization; sewage treatments; industrial chemical manufacturing processes. *Mailing Add:* 140 Columbia Heights Brooklyn NY 11201

OTHMER, EKKEHARD, b Koenigsberg, Germany, Oct 15, 33; m 64. PSYCHIATRY. *Educ:* Univ Hamburg, Dipl, 60, PhD(psychol), 65, MD, 67. *Prof Exp:* Res assoc neurophysiol of sleep, Univ Mo-Columbia, 67-68; asst prof exp psychiat & asst prof psychol, Sch Med, Wash Univ, 68-74, resident psychiat, 71-74; assoc prof psychiat, Med Ctr, Univ Ky, 74-77; PROF PSYCHIAT, MED CTR, UNIV KANS, 77-; MEM STAFF, VET ADMIN HOSP, KANSAS CITY, 77-; MED DIR, N HILLS HOSP, KANSAS CITY, MO, 88- *Concurrent Pos:* Nat Inst Ment Health fel psychiat, Sch Med, Wash Univ, 70-71; NIH fel, Sch Med, Wash Univ & Med Ctr, Univ Ky, 72-74; chief psychobiol unit, Vet Admin Hosp, Lexington, Ky, 74-76. *Mem:* AMA; fel Am Psychiat Asn; Soc Neurosci; Asn Psychophysiol Study Sleep; Am Asn Univ Profs. *Res:* Psychobiology and pathology of sleep; biorhythms; computer analysis of polygraphic sleep recordings; medical model in psychiatry; diagnostic techniques; neurophysiology of sleep. *Mailing Add:* 5744 Woodhaven Lane Parkville MO 64152-4322

OTHMER, HANS GEORGE, b Hastings, Mich, Oct 15, 43; m 69; c 1. BIOMATHEMATICS. *Educ:* Mich State Univ, BS, 65; Univ Minn, PhD(chem eng), 69. *Prof Exp:* Res engr, Exxon Corp, 69-73; from asst prof to assoc prof math Rutgers Univ, 73-79; PROF MATH, UNIV UTAH, 79- *Concurrent Pos:* Assoc ed, Siam J Appl Math, 75-82, Math Biosci, 77-; consult, Math Res Ctr, Univ Wis, 79; vis prof, Univ Minn, 80, Oxford Univ, 87; mem, Spec Study Sect, NIH, 81- *Mem:* Soc Indust & Appl Math; Am Math Soc; Soc Math Biol. *Res:* Theoretical studies of pattern formation in chemically-reacting systems, with application in developmental biology, ecology and chemistry; bifurcation theory and its application in the analysis of nonlinear differential equations. *Mailing Add:* Dept Math Univ Utah Salt Lake City UT 84112

OTHMER, MURRAY E(ADE), chemical engineering; deceased, see previous edition for last biography

OTIS, ARTHUR BROOKS, b Grafton, Maine, Sept 11, 13; m 42; c 1. RESPIRATORY PHYSIOLOGY. *Educ:* Univ Maine, AB, 35; Int YMCA Col, MEd, 37; Brown Univ, MS, 39, PhD(physiol), 41. *Prof Exp:* Res assoc cellular physiol, Univ Iowa, 41-42; from instr to asst prof physiol, Sch Med & Dent, Univ Rochester, 42-51; assoc prof physiol & surg, Sch Med, Johns Hopkins Univ, 52-56; prof physiol, 56-86, head dept, 56-80, EMER PROF, COL MED, UNIV FLA, 86- *Concurrent Pos:* Fulbright res scholar, Cambridge Univ, 50-51 & Univ Nijmegen, 64-65; mem physiol training comt, NIH, 58-60, physiol study sect, 60-64; comt eval post doctoral fel applns, NSF, 58-59; respiration sect ed, Am J Physiol & J Appl Physiol, 62-64; mem rev comt, Fogarty sr post doctoral fel applns, 75-77; mem physiol comt, Nat Bd Med Examrs, 81-84. *Mem:* Am Physiol Soc; Soc Gen Physiol; Sigma Xi. *Res:* Respiratory physiology; physiology of hypoxia; comparative physiology. *Mailing Add:* Dept Physiol Univ Fla Col Med Gainesville FL 32610

OTIS, MARSHALL VOIGT, b New London, Wis, Aug 28, 19; m 42; c 4. CHEMISTRY, PHYSICS. *Educ:* Univ Wis, BS, 42. *Prof Exp:* Anal chemist, Hercules Powder Co, 42-44; prod supvr, Badger Ord Works, 44-45, develop supvr ultrasonics test method, 45; res chemist, Eastman Chem Div, Eastman Kodak Co, 46-59, sr res chemist, 59-83; RETIRED. *Concurrent Pos:* Adv bd, Off Critical Sci Data Tables. *Honors & Awards:* Presidential Medal of Merit Award, 86. *Mem:* Emer mem Am Chem Soc; emer mem Am Inst Chemists. *Res:* Applied spectroscopy; information storage and retrieval by machine methods; high performance liquid chromatography applied densitometry to thin layer chromatograms by computer techniques; invented fluorescent optical brightener for textile fibers. *Mailing Add:* 2009 Greene Lane Kingsport TN 37664-3609

OTKEN, CHARLES CLAY, b Falfurrias, Tex, Nov 6, 27; m 71; c 5. PHYSICAL ORGANIC CHEMISTRY. *Educ:* Tex A&M Univ, BS, 49; Cornell Univ, PhD(biochem), 54. *Prof Exp:* From instr to assoc prof, 54-67, PROF CHEM, PAN AM UNIV, 67- *Mem:* AAAS; Am Chem Soc; NY Acad Sci. *Res:* Nutritional biochemistry. *Mailing Add:* Dept Chem Pan Am Univ 1201 W Univ Dr Edinburg TX 78539

OTOCKA, EDWARD PAUL, b Jersey City, NJ, Mar 11, 40; m 64; c 2. POLYMER CHEMISTRY. *Educ:* Lehigh Univ, BS, 62; Polytech Inst Brooklyn, PhD(polymer chem), 66. *Prof Exp:* Mem tech staff polymer chem, Bell Tel Labs, Inc, 66-73; mem tech staff polymer chem, Pratt & Whitney Aircraft, 73-76; mgr advan mat mfg technol, Xerox Corp, 77-81, UNIT DIR MAT DEVELOP, EASTMAN KODAK, 81- *Concurrent Pos:* Adj prof polymer sci & eng, Univ Mass, 75-87. *Mem:* Am Chem Soc; Am Phys Soc; NY Acad Sci. *Res:* Composite and polymer material science; physical properties of polymers; polymer modification, characterization and stabilization. *Mailing Add:* 104 Topspin Dr Pittsford NY 14534

O'TOOLE, JAMES J, b Peterborough, Ont, Nov 13, 43; m 66; c 3. WEED CONTROL, EDIBLE BEANS. *Educ:* Univ Guelph, Ont, BSc, 67; Univ Western Ont, MSc, 82. *Prof Exp:* Lectr herbicides, Kemptville Col Agr Technol, 67-71; HEAD AGRON DIV CENTRALIA COL AGR TECHNOL, ONT, 71- *Res:* Production management studies in white (navy) and colored beans, rutabagas and canola; weed control research in white and kidney beans; insects and fertility on rutabagas. *Mailing Add:* Dept Agron Centralia College Agr Technol Huron Park ON N0M 1Y0 Can

OTRHALEK, JOSEPH V, b Czech, Feb 16, 22; nat US; m 47; c 4. CHEMICAL ENGINEERING. *Educ:* Univ Mich, BS, 45. *Prof Exp:* Lab technician, US Rubber Co, 42-44; chem engr, Pilot Plant, Armour & Co, 45-46; res chemist & chem engr, BASF Wyandotte Corp, 46-55, res sect head, 55-59, res supvr & chem engr, 59-63, indust & aircraft prod, 63-71, res mgr process develop, Chem Specialties Div, Consumer Prod Group, 71-79; dir res & develop, Indust Chem Specialities Div, Detrex Chem Indust Inc, 79-87; PVT CONSULT, 88- *Mem:* Am Chem Soc; Am Inst Chem Engrs; Am Soc Testing & Mat; Am Ord Asn; Electroplaters Soc. *Res:* Product and process development for detergents, silicates, minerals, metal finishing, soil amendments; food processing; surfactants; corrosion inhibitors; high temperature reactions; agglomeration; drying; metal working, coolants drawing compounds, cleaners, conversion coatings; industrial liquid waste treatment flocculants; paper mill chemical specialities. *Mailing Add:* 5541 Leafwod Ct Milford MI 48042-8736

OTT, BILLY JOE, b Bearden, Okla, Sept 10, 23; m 43; c 1. SOILS. *Educ:* Okla State Univ, BS, 48, MS, 49, PhD(soil chem), 62. *Prof Exp:* Prof soils, Panhandle Agr & Mech Col, 62-67; prof soils & resident dir res, 67-85, PROF SOILS & COORDR FARMING SYSTS RES, TEX A&M UNIV, 85. *Mem:* AAAS; Am Soc Agron; Soil Sci Soc Am. *Res:* Dryland and irrigated soils management; soil fertility; soil-plant-water relationships; soil moisture utilization; plant root energy levels and ion absorption. *Mailing Add:* 3502 76th St Lubbock TX 79423

OTT, COBERN ERWIN, b Osyka, Miss, Jan 15, 41; m 67; c 2. MEDICAL PHYSIOLOGY, NEPHROLOGY. *Educ:* Millsaps Col, BS, 64; Univ Miss, PhD(physiol), 71. *Prof Exp:* Res assoc, Sch Med, Univ Miss, 70-71, NIH fel, 71-72; NIH fel, Mayo Grad Sch Med, 72-75; asst prof, 75-80, ASSOC PROF PHYSIOL, UNIV KY MED CTR, 80- *Mem:* Am Soc Nephrology; Am Physiol Soc; Int Soc Nephrology; Soc Exp Biol & Med. *Res:* Basic renal physiology with emphasis on sodium balance and relationship to fluid volumes and blood pressure. *Mailing Add:* Dept of Physiol Univ of Ky Med Ctr Lexington KY 40506

OTT, DAVID JAMES, b Toledo, Ohio, Apr 19, 46; m 70; c 1. DIAGNOSTIC RADIOLOGY, GASTROINTESTINAL RADIOLOGY. *Educ:* Univ Mich, MD, 71. *Prof Exp:* Intern med, Bowman Gray Sch Med, 71-72, diag radiol resident, 72-75; staff radiologist, Womack Army Hosp, Ft Bragg, 75-77; from instr to assoc prof, 77-86, PROF RADIOL, BOWMAN GRAY SCH MED, 86- *Mem:* Soc Gastrointestinal Radiologists; Am Gastroenterol Asn; Am Radiol Soc N Am; Am Roentgen Ray Soc; Am Col Radiol; AMA. *Res:* Gastrointestinal radiology; endoscopic and manometric correlation in the esophagus; endoscopic correlation in upper gastrointestinal tract and colon; gastroesophageal reflux disease; polypoid disease of colon. *Mailing Add:* Dept Radiol Bowman Gray Sch Med 300 S Hawthorne Rd Winston-Salem NC 27103

OTT, DONALD GEORGE, b Kinsley, Kans, Aug 13, 26; m 48; c 2. CHEMISTRY, SYNTHETIC ORGANIC CHEMISTRY. *Educ:* Colo State Univ, BS, 50; Wash State Univ, PhD(chem), 53. *Prof Exp:* Org res chemist, Dow Chem Co, 53-54; mem staff, 54-63, alternate group leader biomed res, 63-73, group leader org & biochem synthesis, 73-76, MEM STAFF, LOS ALAMOS SCI LAB, 76- *Concurrent Pos:* USPHS-NIH fel, Cambridge Univ, 60-61. *Mem:* AAAS; Am Chem Soc; Brit Chem Soc. *Res:* Organic and biochemical synthesis of isotopic compounds; organic reaction mechanisms; synthesis and chemistry of explosives; synthesis and characterization of oligonucleotides; applications of stable isotopes of carbon, nitrogen and oxygen; nuclear magnetic resonance. *Mailing Add:* 2423 Club Rd Los Alamos NM 87544

OTT, EDGAR ALTON, b Ft Wayne, Ind, Jan 10, 38; m 63; c 2. EQUINE NUTRITION. *Educ:* Purdue Univ, BS, 59, MS, 63, PhD(animal nutrit), 65. *Prof Exp:* Res assoc, Ralston Purina Co, 65-70; assoc prof, 70-79, PROF EQUINE NUTRIT, ANIMAL SCI DEPT, UNIV FLA, 79- *Concurrent Pos:* Mem subcomt horses, Nat Res Coun, Nat Acad Sci, 75-78, chmn, 87-; consult feed co & farms, 70- *Mem:* Equine Nutrit & Physiol Soc; Am Soc Animal Sci. *Res:* Nutrient requirements of horses, with emphasis on amino acid requirements of growing foals; nutrient effects on skeletal development; the relationship between nutrition and reproductive efficiency of the mare; feed ingredient evaluation; pasture utilization. *Mailing Add:* Dept Animal Sci 210 Animal Sci Bld Gainesville FL 32611

OTT, EDWARD, b New York, NY, Dec 22, 41; m 74; c 2. ELECTRICAL ENGINEERING. *Educ:* Cooper Union, BS, 63; Polytech Inst Brooklyn, MS, 65, PhD(electrophys), 67. *Prof Exp:* NSF fel, Dept Appl Math & Theoret Physics, Cambridge Univ, 67-68; prof elec eng, Cornell Univ, 68-79; PROF ELEC ENG & PHYSICS, UNIV MD, COLLEGE PARK, 79- *Mem:* AAAS; Inst Elec & Electronics Engrs; fel Am Physics Soc; Am Geophys Union. *Res:* Plasma physics. *Mailing Add:* 12421 Borges Ave Silver Spring MD 20904

OTT, GRANVILLE E, b San Angelo, Tex, Nov 8, 35; m 63; c 2. ELECTRICAL ENGINEERING. *Educ:* Univ Tex, Austin, BS, 58, MS, 61, PhD(elec eng), 64. *Prof Exp:* Res engr, Defense Res Lab, Tex, 59-62, sr engr, Tex Instruments, Inc, 64-67, mgr adv planning geophys systs, 67-68; assoc prof elec eng, Univ Mo-Columbia, 68-76; MEM TECH STAFF CONSUMER PROD, TEX INSTRUMENTS, INC, 76- *Concurrent Pos:* Lectr, Univ Houston, 64-66; Nat Sci Found res grants, 70- *Mem:* Inst Elec & Electronics Engrs. *Res:* Design and application of digital systems for data handling, processing and control. *Mailing Add:* 1906 Spring Valley Richardson TX 75081

OTT, HENRY C(ARL), b Troy, NY, July 20, 10; m 62; c 2. NUCLEAR ENGINEERING & CHEMICAL ENGINEERING. *Educ:* Rensselaer Polytech Inst, ChE, 32, MChE, 36, PhD(phys chem), 37. *Prof Exp:* From instr to asst prof chem & chem eng, Rensselaer Polytech Inst, 37-48; sect head, Div Res, US Atomic Energy Comn, 48-49, reactor engr, Div Reactor Develop, 49-51, dep chief, Proj Develop Br, 51-52, asst chief, Eval Br, 52-54, asst chief, Civilian Reactors Br, 54-56; CONSULT NUCLEAR ENGR, EBASCO SERVS, INC, 56- *Concurrent Pos:* Chief sanit inspector, Lake George Asn, NY, 41-42; chemist, Clinton Labs, Monsanto Chem Co, 46-47. *Mem:* Am Chem Soc; Am Nuclear Soc; Am Inst Chem Engrs. *Res:* Nuclear power plant design; economics of nuclear power; nuclear reactor systems, safety and licensing; nuclear fuel management; thermodynamics; mass and heat transfer; alternate energy systems. *Mailing Add:* 50 Fort Pl Apt B4B Staten Island NY 10301-2419

OTT, J BEVAN, b Cedar City, Utah, July 21, 34; m 53; c 6. PHYSICAL CHEMISTRY. *Educ:* Brigham Young Univ, BS, 55, MS, 56; Univ Calif, PhD, 59. *Prof Exp:* Asst prof chem, Utah State Univ, 59-60; assoc prof, 60-68, PROF CHEM, BRIGHAM YOUNG UNIV, 68- *Concurrent Pos:* Sabbatical leave as res specialist, Atomics Int Div, NAm Rockwell Corp, 66-67. *Mem:* Am Chem Soc. *Res:* Low temperature and solution thermodynamics; phase equilibria. *Mailing Add:* Dept of Chem Brigham Young Univ Provo UT 84601

OTT, KAREN JACOBS, b Atlanta, Ga, Sept 26, 39; m 59; c 2. PARASITOLOGY, INVERTEBRATE ZOOLOGY. *Educ:* Asbury Col, AB, 59; Univ Ky, MS, 61; Rutgers Univ, PhD(zool, parasitol), 65. *Prof Exp:* Instr microbiol & parasitol, Jefferson Med Col, 64-65, grant, 65; asst res specialist, Bur Biol Res, Rutgers Univ, 65-68; asst prof, 69-77, ASSOC PROF BIOL, UNIV EVANSVILLE, 77- *Mem:* AAAS; Am Soc Parasitol; Am Soc Trop Med & Hyg; Nat Asn Biol Teachers; Am Inst Biol Sci. *Res:* Immunological and physiological aspects of host-parasite relationships. *Mailing Add:* 850 Lodge Ave Evansville IN 47714

OTT, KARL O, b Klein-Auheim, Ger, Dec 24, 25; m 58; c 2. NUCLEAR ENGINEERING. *Educ:* Univ Frankfurt, BS, 48; Univ Gottingen, 52, PhD(physics), 58. *Prof Exp:* Physicist, Univ Karlsruhe, 55-57; physicist, Gesellschaft Kernforschung, 58-61, asst sect head reactor physics, 61-62, sect head, 63-66; sci asst, Univ Karisruhe, 66-67; PROF NUCLEAR ENG, PURDUE UNIV, 67- *Concurrent Pos:* Exchange scientist, Argonne Nat Lab, 65-66, consult, 67-; consult, Gesellschaft Kernforschung, 66-67; mem, Ger Atomic Forum, 67. *Mem:* Fel Am Nuclear Soc. *Res:* Cosmic ray theory; nuclear theory; thermal reactor physics; physics, safety and fuel cycles of fast reactors; risk assessment of large scale technologies. *Mailing Add:* Sch Nuclear Eng Purdue Univ West Lafayette IN 47907

OTT, R(AY) LYMAN, JR, b Kennett Square, Pa, Mar 1, 40; m 64; c 2. TECHNICAL MANAGEMENT OF BIOSTATISTICS, DATA PROCESSING & COMPUTER RESOURCES. *Educ:* Bucknell Univ, BS, 62, MS, 63; Va Polytech Inst, PhD(statist), 67. *Prof Exp:* Sr statistician, Smith Kline & French Labs, 66-68; from asst prof to assoc prof statist, Univ Fla, 68-75; head, dept biostatist, Merrell Dow Res Inst, 75-81, head, Res Data Ctr, 81-83, dir, biomed info systs, 83-90, MANAGING DIR, SYSTS & QUAL IMPROV, MARION MERRELL DOW, INC, 90- *Mem:* Am Statist Asn; Biomet Soc; fel Am Statist Asn; Drug Info Asn. *Res:* Design of experiments and accompanying estimation procedures; process improvement, re-engineering work processes; author or co-author of 5 college textbooks. *Mailing Add:* Marion Merrell Dow Inc Marion Park Dr PO Box 9627 Kansas City MO 64134-0627

OTT, RICHARD L, b Santa Barbara, Calif; m 48; c 2. VETERINARY MEDICINE. *Educ:* Wash State Univ, BS, 44, DVM, 45. *Prof Exp:* Pvt pract, Wash, 45-46 & 48-49; from asst prof to assoc prof med & surg, 49-59, prof & chmn dept, 59-73, PROF VET CLIN MED & SURG, COL VET MED, WASH STATE UNIV, 73-, ASSOC DEAN, 74- *Concurrent Pos:* NIH grant, 60-65, res grant, 67-74. *Honors & Awards:* Gaines Medal, Am Vet Med Asn, 64. *Mem:* AAAS; Am Vet Med Asn; Conf Res Workers Animal Dis; Am Col Vet Internal Med. *Res:* Immunology, pathogenesis and ecology of virus diseases; small animal medicine and surgery. *Mailing Add:* Vet Med McCoy Hall Wash State Univ Pullman WA 99164

OTT, TEUNIS JAN, b Zype, Netherlands, Jan 27, 43; m 68; c 2. OPERATIONS RESEARCH, MATHEMATICAL STATISTICS. *Educ:* Univ Amsterdam, BSc, 65, Drs(math statist), 70; Univ Rochester, PhD(opers res), 75. *Prof Exp:* Asst prof opers res, Case Western Reserve Univ, 74-76; asst prof, Univ Tex, 76-78; MEM TECH STAFF & SUPVR, BELL LABS, 78-, RES MGR, BELL COMMUNICATIONS RES. *Concurrent Pos:* Assoc ed, Mgt Sci. *Mem:* Opers Res Soc Am; Inst Elec & Electronics Engrs; Inst Mgt Sci; Dutch Math Asn; Netherlands Statist Asn; J Asn Comput Mach; Int Fedn Info Processing. *Res:* Operations research in medical systems; probability theory; computer performance analysis. *Mailing Add:* Rm Mre 2P 388 Bell Communications Res Morristown NJ 07960

OTT, WALTER RICHARD, b Brooklyn, NY, Jan 20, 43; m 64; c 3. CERAMIC ENGINEERING. *Educ:* Va Polytech Inst, BS, 65; Univ Ill, Urbana, MS, 67; Rutgers Univ, New Brunswick, PhD(ceramic eng), 69. *Prof Exp:* Process engr, Corhart Refractories, 65-66; staff res engr, Ceramic Div, Champion Spark Plug, 69-70; assoc prof ceramic eng, Rutgers Univ, New Brunswick, 70-80; PROVOST, ALFRED UNIV, ALFRED, NY, 80- *Honors & Awards:* Ralph Teetor Award, Soc Automotive Engrs, 73; Nat Inst Ceramic Engrs Award, 75. *Mem:* Am Ceramic Soc; Brit Soc Glass Technol; Int Confedn Thermal Anal; Am Soc Eng Educ. *Res:* Application of thermal analysis to ceramic systems; kinetics and mechanism of solid state reactions. *Mailing Add:* Provost Alfred Univ Main St Alfred NY 14802

OTT, WAYNE R, b San Mateo, Calif, Feb 2, 40; div. ENVIRONMENTAL STATISTICS,HUMAN EXPOSURE ASSESSMENT. *Educ:* Claremont Mens Col, BA, 63; Stanford Univ, BS, 63, MA, 66, MS, 65, PhD(environ eng), 71. *Prof Exp:* Res engr, Div Air Pollution, US Pub Health Serv, 66-67; sci adv to the dir, Nat Air Pollution Control Admin, 67-68; chief, Extramural Labs Br, Lab Opers Div, US Environ Protection Agency, 71-74, sr systs analyst, Quality Assurance Div, 74-79; vis scholar, Dept Statistics, Stanford Univ, 79-81; sr environ eng, Off Res & Develop, US Environ Protection Agency, 81-84, leader, Air, Toxics & Radiation Monitoring Res Team, 84-90; VIS SCIENTIST, CTR RISK ANAL & DEPT STATIST, STANFORD UNIV, 91- *Concurrent Pos:* Prin investr air pollution res, Dept Statist, Stanford Univ, 79-81; mem, Nat Coop Hwy Res Prog, Nat Acad Sci, 75-76; co chmn, Nat Conf Environ Modeling & Simulation, 75-76; consult to Yugoslavia, World Health Orgn, Switz, 81; consult to Mexico City, Pan Am Health Orgn, 77. *Mem:* Air & Waste Mgt; Am Statist Asn; Am Soc Quality Control. *Res:* Development of mathematical techniques for predicting environmental phenomena, including environmental indices; measurement and modeling of human exposure to air pollution; stochastic and statistical modeling. *Mailing Add:* 1008 Cardiff Lane Redwood City CA 94061

OTT, WILLIAM ROGER, b Philadelphia, Pa, Mar 29, 42; m 67; c 3. EXPERIMENTAL ATOMIC PHYSICS, OPTICAL PHYSICS. *Educ:* St Joseph's Col, BS, 63; Univ Pittsburgh, PhD(physics), 68. *Prof Exp:* NSF assoc plasma spectros, Nat Inst Standards & Technol, 68-70, staff physicist, 70-80, sci asst to dir, 80-81, prog analyst, 82-83, chief, Radiation Physics Div, 83-88, dep dir, Ctr Atomic, Molecular & Optical Physics, 88-90, DEP DIR, PHYSICS LAB, NAT INST STANDARDS & TECHNOL, 91- *Concurrent Pos:* Alexander von Humboldt fel, Univ Dusseldorf, 77-78. *Honors & Awards:* Superior Performance Award, Nat Bur Stand, 75; Dept Com Silver Medal, 76. *Mem:* Fel Optical Soc Am. *Res:* The application of plasma light sources as intensity standards in the ultraviolet; electron atom collisions. *Mailing Add:* Nat Inst Standards & Technol Bldg 221 Rm B160 Gaithersburg MD 20899

OTTE, CAREL, b Amsterdam, Neth, June 29, 22; nat US; m 53; c 4. GEOLOGY. *Educ:* Calif Inst Technol, MS, 50, PhD(geol), 54. *Prof Exp:* Res geologist, Pure Oil Co, 54-55; explor geologist, Shell Oil Co, 55-57; supvr geol res, Earth Energy, Inc, 57-62, vpres & mgr, 62-67; pres, Geothermal Div, Unocal Corp, 72-89; RETIRED. *Mem:* Nat Acad Eng; Geol Soc Am; Am Asn Petrol Geologists. *Res:* Exploration geology. *Mailing Add:* 4261 Commonwealth Ave La Canada CA 91011

OTTE, DANIEL, b Durban, SAfrica, Mar 14, 39; US citizen; m 63; c 2. ZOOLOGY. *Educ:* Univ Mich, BA & PhD(zool), 68. *Prof Exp:* NSF grant & res assoc zool, Univ Mich, 68-69; asst prof, Univ Tex, Austin, 69-75; assoc cur, 75-80, CUR ENTOM, ACAD NATURAL SCI, PHILADELPHIA, 80- *Concurrent Pos:* Ed, Trans Am Entom Soc, Proc Acad Natural Sci Philadelphia. *Mem:* Am Entom Soc; Asn Pac Systematics; Orthopterists' Soc. *Res:* Insect behavior; evolution of behavior; systematics of North American grasshoppers; systematics of Australian, Pacific, and African crickets; communication in grasshoppers. *Mailing Add:* Acad Natural Sci Philadelphia PA 19103

OTTENBACHER, KENNETH JOHN, b Missoula, Mont, Apr 5, 50; m 75; c 2. REHABILITATION, OCCUPATIONAL THERAPY. *Educ:* Univ Mont, Missoula, BS, 72; Univ Cent Ark, BS, 75; Univ Tenn, Knoxville, MS, 76; Univ Mo, Columbia, PhD(spec educ), 82. *Prof Exp:* Prof occup ther, Univ Wis-Madison, 82-90, dir grad studies, Ther Sci, 87-90; PROF REHAB & ASSOC DEAN, STATE UNIV NY, BUFFALO, 90- *Concurrent Pos:* Ed, Occup Ther J Res, 87-91; vis scholar, Cumberland Col Health Sci, Univ Sydney, 88. *Mem:* Fel Am Occup Ther Asn; Am Soc Allied Health Prof; Am Asn Ment Retardation; Coun Except Children. *Res:* Examination of research designs and data analysis procedures appropriate for use in clinical based field of rehabilitation; methods of demonstrating the effectiveness of rehabilitation interventions. *Mailing Add:* State Univ NY 435 Kimball Tower Buffalo NY 14214

OTTENBRITE, RAPHAEL MARTIN, b Claybank, Sask, Sept 20, 36; m 63; c 3. ORGANIC CHEMISTRY. *Educ:* Univ Windsor, BSc, 58, MSc, 62, PhD(chem), 67; Univ Toronto, dipl ed, 64. *Prof Exp:* Instr chem, Western Ont Inst Technol, 60-64 & Univ Windsor, 64-66; US Air Force fel, Univ Fla, 66-67; asst prof, Med Col Va & Richmond Prof Inst, 67-71; assoc prof, 71-76, PROF CHEM, VA COMMONWEALTH UNIV, 76- *Concurrent Pos:* Ed, J Bioactive & Compatible Polymers, 86-; vis prof, Univ Nagasaki, Japan, 85 & 87, Royal Inst Technol, Copenhagen, Sweden, 91; coordr, High Technol Mat Ctr, Va Commonwealth Univ; chmn, polymer chem div, Am Chem Soc, 91. *Mem:* Am Chem Soc (treas, 84-); Chem Inst Can; Sigma Xi; NY Acad Sci; Int Union Pure & Appl Chem; Controlled Release Soc. *Res:* Synthesis of exocyclic diene systems and study of their diels-alder reactivity; the correlation of these properties by MO calculations; preparation and characterization of polyanions as immunomodulators; synthesis of high temperature materials. *Mailing Add:* Dept Chem 1000 Main St Va Commonwealth Univ Richmond VA 23284-2006

OTTENSMEYER, FRANK PETER, b Essen, Germany, July 23, 39; Can citizen; m 69; c 3. ELECTRON OPTICS, CHROMATIN STRUCTURE. *Educ:* Univ Toronto, BASc, 62, MA, 63, PhD(biophys), 67. *Prof Exp:* Res fel electron microscopy, Electronics Optics Lab, Toulouse, France, 67-68; staff physicist molecular biol, Ont Cancer Inst, Toronto, 68-77; from asst prof to assoc prof, 68-77, PROF BIOPHYS, UNIV TORONTO, 77-, CHMN, MED BIOPHYS, 87- *Concurrent Pos:* Vis prof, Electronmicroscopy Inst, Max Planck Soc, Berlin, 77; vis prof path, Univ Capetown, 81; chmn, Fel Panel, Nat Cancer Inst Can, 84-87. *Mem:* Microscopial Soc Can (pres, 81-83); Biophys Soc Can (secy, 85-); Electron Microscopy Soc Am; Biophys Soc. *Res:* High resolution dark field electron microscopy of biological macromolecules; development of microanalysis and elemental mapping by electron energy loss imaging; structure of chromatin. *Mailing Add:* Physics Div Ont Cancer Inst 500 Sherbourne St Toronto ON M4X 1K9 Can

OTTENSTEIN, DANIEL, b Milwaukee, Wis, Feb 16, 30; m 57; c 3. GAS CHROMATOGRAPHY. *Educ:* Col William & Mary, BS, 52, Univ Richmond, MS, 57. *Prof Exp:* Chemist, Allied Chem & Dye Corp, 52-58; mgr appl dept, Fisher Sci Co, 58-61; sr res scientist, Johns Manville Corp, 62-67; mkt group, Corning Glass works, 67-69; vpres res & develop dept, Supelco Inc, 69-86; CONSULT, OTTENSTEIN ASSOCS, 87- *Mem:* Am Chem Soc. *Res:* Development of chromatographic supports and deactiviation of their surface; development of stationary phases for gas chromatography and application of their use. *Mailing Add:* 423 Park Lane State Col PA 16803

OTTER, FRED AUGUST, b West Chester, Pa, Sept 11, 28; m 53; c 2. SOLID STATE PHYSICS. *Educ:* Lehigh Univ, BS, 53; Temple Univ, AM, 55; Univ Ill, PhD(physics), 59. *Prof Exp:* Asst physics labs, Franklin Inst, 53-55; mem staff, Res Labs, United Aircraft Corp, 63-68, mgr, Surface & Film Lab, 68-69; asst prof, Ohio Univ, 59-63; mem staff, Res Labs, United Aircraft Corp, 63-68, mgr surface & film lab, 68-69; prin scientist, 73-81, SR PRIN SCIENTIST SOLID PHYSICS, UNITED TECHNOL RES CTR, 81-; prin scientist physics of solids, 73-81, sr prin scientist, United Technol Res Ctr, 81-87; PROF PHYSICS, UNIV CONN, 87 - *Concurrent Pos:* Vis prof, Univ Conn, 77-87. *Mem:* Am Phys Soc; Sigma Xi; Mat Res Soc. *Res:* Superconductivity; surface science; ion implantation. *Mailing Add:* 21 Woodland Terrace Columbia CT 06237

OTTER, RICHARD ROBERT, b Evanston, Ill, May 17, 20; m 49; c 2. MATHEMATICS. *Educ:* Dartmouth Col, AB, 41; Univ Ind, PhD(org chem), 46. *Prof Exp:* Res chemist, Eastman Kodak Co, NY, 42; mem comt med res, Off Sci Res & Develop, Washington, DC, 44-46; instr math, Princeton Univ, 46-47, fel, Off Naval Res, 47-48, vis asst prof, 53-54; from asst prof to prof, 48-85, EMER PROF MATH, UNIV NOTRE DAME, 85- *Mem:* Am Math Soc; Math Asn Am. *Res:* Theory of probability; combinatorial analysis. *Mailing Add:* 1245 Oak Ridge Dr South Bend IN 46617-1732

OTTER, TIMOTHY, b South Bend, Ind, June 4, 53. CILIARY & FLAGELLA MOVEMENT, CELL MOTILITY. *Educ:* Brown Univ, ScB, 75; Uiv NC, PhD(zool), 81. *Prof Exp:* Res assoc, Albert Einstein Col Med, 81-83; res assoc, Worcester Found Exp Biol, 83-85, sr res assoc, 85-86; ASST PROF PHYSIOL, DEPT ZOOL, UNIV VT, 86- *Concurrent Pos:* Instr, Univ Pa & Marine Biol Lab, 81-83; Am Cancer Soc fel, Albert Einstein Col Med, 82-; NIH fel, Albert Einstein Col Med & Worcester Found Exp Biol, 83-84; prin investr, Worcester Found Exp Biol & Univ Vt, 85-88; prin investr, Nat Sci Found, Univ Vt, 88. *Mem:* Am Soc Cell Biol; Sigma Xi; Soc Gen Physiologists; Biophys Soc. *Res:* The control of ciliary and flagellar movement; the role of calcium and calcium-binding proteins like calmodulin in the control of flagellar bending. *Mailing Add:* Dept Zool Univ Vt Burlington VT 05405-0086

OTTERBY, DONALD EUGENE, b Sioux Falls, SDak, July 2, 32. NUTRITION. *Educ:* SDak State Col, BS, 54, MS, 58; NC State Col, PhD(nutrit), 62. *Prof Exp:* Fel nutrit, NC State Col, 62-63; asst prof nutrit, 63-67, from assoc prof to prof ruminant nutrit, 67-74, PROF ANIMAL SCI, UNIV MINN, ST PAUL, 74- *Mem:* Am Dairy Sci Asn; Am Soc Animal Sci. *Res:* Nutrition of ruminant animals. *Mailing Add:* Animal Sci Univ of Minn 1364 Eckles Ave St Paul MN 55108

OTTERNESS, IVAN GEORGE, b Oct 9, 38; m 65; c 2. INFLAMMATION, ARTHRITIS. *Educ:* Univ Southern Calif, PhD(biophys), 68. *Prof Exp:* RES ADV, PFIZER CENT RES, 71- *Concurrent Pos:* Adj prof, Dept Chem & Biol, Conn Col, 84- & Dept Pharmaceut, Univ RI, 86-; ed, Agents & Actions, 85-; adv ed, Molecular Immunology, 78-82. *Mem:* Inflammation Res Asn (pres, 84-86); AAAS; Am Asn Immunol; NY Acad Sci; Europ Workshop Inflammation; Am Rheumatism Asn. *Res:* Mechanisms of inflammation and cartilage loss in arthritis; design of drugs to treat arthritis. *Mailing Add:* Dept Immunol & Infectious Dis Pfizer Cent Res Groton CT 06340

OTTESEN, ERIC ALBERT, b Baltimore, Md, Nov 13, 43; m 66; c 3. MEDICINE, IMMUNOLOGY. *Educ:* Princeton Univ, BA, 65; Harvard Univ, MD, 70. *Prof Exp:* Intern & resident pediat, Med Ctr, Duke Univ, 70-72; clin assoc infectious dis immunol, 72-75, SR INVESTR IMMUNOL & PARASITOL, NAT INST ALLERGY & INFECTIOUS DIS, 75-, HEAD SEC CLIN PARASITOL, LAB PARASITIC DIS, LAB CLIN INVEST. *Honors & Awards:* Clin Trop Med Res Award, Am Soc Trop Med & Hyg, 78 & 79. *Mem:* Am Asn Immunologists; Am Acad Allergy; Am Soc Trop Med & Hyg; Infectious Dis Soc Am; Sigma Xi. *Res:* Immunology of parasitic diseases, filariasis, schistosomiasis, the cosinophil and immediate hypersensitivity responses. *Mailing Add:* Lab Parasitic Dis NIH Bldg 10 Rm 11C-108 Bethesda MD 20205

OTTESON, OTTO HARRY, b Rexburg, Idaho, June 25, 31; m 57; c 6. NUCLEAR PHYSICS. *Educ:* Utah State Univ, BS, 60, MS, 62, PhD(physics), 66. *Prof Exp:* Teaching asst, 61-62, asst prof, 66-73, ASSOC PROF PHYSICS & ASST DEPT HEAD, UTAH STATE UNIV, 73- *Mem:* Am Phys Soc; Am Asn Physics Teachers. *Res:* Medium energy PI; nuclear interactions. *Mailing Add:* 147 E 200 S Smithfield UT 84335

OTTING, WILLIAM JOSEPH, b St Paul, Minn, Oct 27, 19; m 40; c 3. PHYSICS. *Educ:* George Washington Univ, BS, 46, MS, 49; Cath Univ Am, PhD(physics), 56. *Prof Exp:* Radio engr, Nat Bur Standards, 45-48; physicist, Off Naval Res, 48-52; dir phys sci, Off Sci Res, US Air Force, 52-60; chief scientist, Defense Atomic Support Agency, US Dept Defense, 60-64; asst dean, 64-70, assoc prof physics, Univ Ill, Chicago Circle, 64-83, assoc dean grad col, 70-83; RETIRED. *Mem:* Sigma Xi; fel Am Phys Soc. *Res:* Electrochemistry; electromagnetism; chemical, solid state and nuclear physics. *Mailing Add:* 1215 Cerramar Ct Eagle ID 83616

OTTINGER, CAROL BLANCHE, b Batesville, Ark, Dec 25, 33. MATHEMATICS. *Educ:* Ark Col, BSE, 54; Okla State Univ, MS, 60, EdD(math), 69. *Prof Exp:* Teacher math high schs, Ark, 54-58 & Okla, 60-63; from asst prof to assoc prof, 63-71, chmn dept, 74-83, PROF MATH, MISS

UNIV FOR WOMEN, 71- *Concurrent Pos:* Mem bd govs, Math Asn Am, 83-86. *Mem:* Math Asn Am; Nat Coun Teachers Math; Am Math Soc. *Res:* Topology of decomposition spaces. *Mailing Add:* 510 Dublin Dr Columbus MS 39702

OTTINO, JULIO MARIO, b La Plata, Argentina, May 22, 51; m 76; c 1. FLUID MECHANICS, TRANSPORT PROCESSES. *Educ:* Nat Univ La Plata, dipl, 74; Univ Minn, PhD(chem eng), 79. *Prof Exp:* From asst prof to prof chem eng, Univ Mass, 79-90; WALTER P MURPHY PROF CHEM ENG, NORTHWESTERN UNIV, 91- *Concurrent Pos:* Adj prof polymer sci & eng, Univ Mass, 79-; Presidential Young Investr, NSF-White House, 84; Chevron prof chem eng, Calif Inst Technol, 86-; univ fel, Univ Mass, 88; sr res fel, Ctr Turbulence Res, Stanford Univ-NASA, 89-90. *Honors & Awards:* Colburn Lectr, Univ Del, 87; Merck Sharp & Dohme Lectr, Univ PR; Fifth Stanley Corrsin Lectr, Johns Hopkins Univ, 91. *Mem:* Soc Natural Philos; Am Inst Chem Engrs; Soc Rheology; Soc Indust & Appl Math; Am Chem Soc; Am Physics Soc. *Res:* Fluid mechanics of mixing chaotic dynamics and turbulence, dispersion and reactive mixing; transport in disordered systems, primarily transport in polymer systems. *Mailing Add:* Dept Chem Eng Northwestern Univ Evanston IL 60208-3100

OTTKE, ROBERT CRITTENDEN, b Louisville, Ky, Jan 23, 22; m 43, 58; c 4. ORGANIC CHEMISTRY, BIOCHEMISTRY. *Educ:* Yale Univ, BS, 48, PhD(chem), 50. *Prof Exp:* Nat Cancer Inst fel, Stanford Univ, 50-51; res chemist, Chas Pfizer & Co, Inc, 51-53, prod develop, 53-57; pres, Caribe Chem Corp, 57-58; dir commercial develop, Wallerstein Co Div, Baxter Labs, Inc, 58-61, tech dir, Baxter Int, 61-63; vpres & managing dir labs, Parke Davis, Madrid, Spain, 63-69; partner, Boyden Assocs, 69-76; dir, Europe & MidE, Allergan Pharmaceut, Inc, 76-78; CONSULT, ROBERT OTTKE ASSOCS, MGT CONSULTS, 78- *Mem:* Am Chem Soc; NY Acad Sci. *Res:* Steroid biosynthesis; natural products; enzymology; pharmaceutical chemistry. *Mailing Add:* 2939 Perla Newport Beach CA 92660

OTTMAN, RUTH, b Houston, Tex, May 29, 52. GENETICS. *Educ:* Univ Calif, Berkeley, AB, 75, PhD(genetics), 80. *Prof Exp:* Fel epidemiol, Univ Calif, Berkeley, 80-81; ASST PROF PUB HEALTH, COLUMBIA UNIV, 81- *Concurrent Pos:* Dir, Familial Epilepsy Proj, 80-81; Sergievsky fel, 81-82, Sergievsky scholar, 84-85; prin investr, Epidemiol of Familial Epilepsy, 85-; res comt neuroepidemiol, World Fedn Neurol. *Mem:* Soc Epidemiol Res; Am Pub Health Asn; World Fedn Neurol; Am Soc Human Genetics. *Res:* Role of genetic factors in the etiology of complex diseases, including breast cancer, epilepsy and coronary heart disease; assessment of risk to relatives of affected persons; testing biological hypotheses about genetic and nongenetic mechanisms increasing risk; methodologies for studying familial aggregation. *Mailing Add:* Sergievsky Ctr Columbia Univ 630 W 168th St New York NY 10032

OTTO, ALBERT DEAN, b Marshalltown, Iowa, Nov 5, 39; c 2. MATHEMATICS. *Educ:* Univ Iowa, BA, 61, MS, 62, PhD(math), 65. *Prof Exp:* Asst prof math, Lehigh Univ, 65-69; assoc prof, 69-75, chmn dept, 76-85, PROF MATH, ILL STATE UNIV, 75- *Concurrent Pos:* Consult to Col Bd in Proj Equality. *Mem:* Math Asn Am; Nat Coun Teachers Math. *Res:* Mathematics education and discrete mathematics. *Mailing Add:* Dept Math Ill State Univ Normal IL 61761

OTTO, CHARLOTTE A, b Freeport, Ill, July 12, 46; m 69; c 1. HEART IMAGING AGENT DEVELOPMENT. *Educ:* Univ Wash, Seattle, BS, 68; Univ Ill, Champaign-Urbana, MS, 70, PhD(chem), 76. *Prof Exp:* Fel, dept chem, 76-78, div internal med, 78-79, asst prof, 80-85, ASSOC PROF ORGANIC CHEM, UNIV MICH, DEARBORN, 85- *Concurrent Pos:* Res assoc, Res Ctr, Univ Mich Med Ctr, 80- *Mem:* Am Chem Soc; Soc Nuclear Med; AAAS. *Res:* Development of radiopharmaceutical for the mycocardium and the pituitary; organic synthesis; radioisotopic labelling; pituitary imaging agents. *Mailing Add:* Dept Natural Sci Univ Mich 4901 Evergreen Rd Dearborn MI 48128

OTTO, DAVID A, b Wichita Falls, Tex, Sept 25, 48; m 74; c 4. LIPID METABOLISM. *Educ:* Univ Okla, PhD(biochem), 77. *Prof Exp:* asst prof nutrit, Cook Col, Rutgers Univ, 81-87; ASSOC RES SCIENTIST, DEPT RES, BAPTIST MED CTR, 87- *Mem:* Am Inst Nutrit; Am Soc Biochem & Molecular Biol; Am Diabetes Asn; Am Heart Asn; Int Soc Heart Transplantation. *Res:* Cellular mechanisms for the regulation of hepatic lipid metabolism; the influence of diet, drugs and hormones; regulation of plasma triglyarides and cholesterol through an understanding of lipoprotein metabolism; the role of dietary fat in the immune response as it relates to organ transplantation. *Mailing Add:* Dept Res Baptist Med Ctrs 701 Princeton Ave Birmingham AL 35211

OTTO, DAVID ARTHUR, b Ft Scott, Kans, Feb 19, 34; m; c 3. MARINE BIOLOGY, PALEOBOTANY. *Educ:* Univ Kans, AB, 56; Kans State Teachers Col, MS, 57; Univ Mo, PhD(bot, geol), 67. *Prof Exp:* PROF BIOL, STEPHENS COL, 57- *Concurrent Pos:* Sci Fac Fel, NSF. *Mem:* Am Inst Biol Sci; Bot Soc Am. *Res:* Investigations of Middle Pennsylvanian fossil plants found in coal ball petrifactions; plants fossils; marine mammals and invertebrates. *Mailing Add:* Dept Natural Sci Stephens Col Columbia MO 65215

OTTO, FRED BISHOP, b Bangor, Maine, Aug 17, 34; m 57; c 4. PHYSICS EDUCATION, ENGINEERING. *Educ:* Univ Maine, Orono, BS, 56; Univ Conn, MA, 60, PhD(physics), 65. *Prof Exp:* Elec engr, Sylvania Elec Co, 56-58; grad asst physics, Univ Conn, 58-64; asst prof, Colby Col, 64-68; asst prof elec eng, Univ Maine, 68-74; design engr, Eaton W Tarbell & Assocs, 74-79; qual control engr, PyrAlarm, 79-80. *Concurrent Pos:* Consult engr, 80-84, Inst Maine Maritime Acad, 82- *Mem:* Am Phys Soc; Am Asn Physics Teachers; Illuminating Eng Soc. *Mailing Add:* 430 Col Ave Orono ME 04473

OTTO, FRED DOUGLAS, b Hardisty, Alta, Jan 12, 35; m 60; c 3. CHEMICAL ENGINEERING. *Educ:* Univ Alta, BSc, 57, MSc, 59; Univ Mich, Ann Arbor, PhD(chem eng), 63. *Prof Exp:* From asst prof to assoc prof, 62-70, actg chmn dept, 70-72, PROF CHEM ENG, UNIV ALTA, 70-, CHMN DEPT, 75- *Concurrent Pos:* Dean eng, Univ Alta, 85- *Mem:* Can Soc Chem Eng (pres, 86-87); Am Inst Chem Engrs; Am Soc Eng Educ. *Res:* Natural gas processing; mass transfer; gas-liquid reactions. *Mailing Add:* 5-1 Mech Eng Bldg Univ Alta Edmonton AB T6G 2G8 Can

OTTO, GILBERT FRED, b Chicago, Ill, Dec 16, 01; m 32; c 2. PARASITOLOGY. *Educ:* Kalamazoo Col, AB, 26; Kans State Col, MS, 27; Johns Hopkins Univ, ScD(parasitol), 29. *Prof Exp:* Asst zool, Kans State Col, 26-27; asst helminth, Sch Hyg & Pub Health, Johns Hopkins Univ, 27-29, from instr to assoc prof, 29-42, assoc prof parasitol, 42-53, asst dean sch, 40-47, parasitologist, John Hopkins Hosp, 47-53; head, dept parasitol, Res Div, Abbott Labs, Ill, 53-60, dir, Agr & Vet Res Div, Sci Divs, 60-65, asst dir, Personnel Div, 65-66; univ prof zool, 66-72, adj prof, 72-81, SR RES ASSOC, UNIV MD, 81- *Concurrent Pos:* Consult, USPHS, 41, 46-53, hookworm & malaria div, State Bd Health, Ga, 42-43 & Surgeon Gen, US Army, 58-66; mem & dir field expeds, Nat Res Coun & Am Child Health Asn, 27-32; expert WHO, UN, 52-74, comt parasitic dis, Armed Forces Epidemiol Bd, 53-57, Lobund adv bd, Univ Notre Dame, 58-66 & adv bd parasitol, Univ Md, 65-66; ed, J Am Heartworm Soc, 77-89; consult, Food & Drug Admin, 77-80 & Nat Acad Sci, 83-86. *Mem:* AAAS; Am Soc Parasitol (treas, 37-40, vpres, 55, pres-elect, 56, pres, 57); Am Soc Trop Med & Hyg; Am Heartworm Soc (secy-treas, 74-77, pres, 77-80); Am Epidemiol Soc. *Res:* Epidemiology of parasitic diseases; immunity to animal parasites; chemotherapy of parasitic diseases; pharmacology of arsenic and antimony; ascariasis; trichinosis; hookworm disease; filariasis; amebiasis; trypanosomiasis; malaria; leucocytozoon; coccidiosis; schistosomiasis; canine heartworm disease. *Mailing Add:* 10506 Greenacres Dr Silver Spring MD 20903

OTTO, HARLEY JOHN, b Richfield, Kans, May 5, 28; m 53; c 2. AGRONOMY. *Educ:* Colo State Univ, BSc, 52; Cornell Univ, PhD(plant breeding), 56. *Prof Exp:* Asst prof plant breeding, Cornell Univ, 56-57, asst prof agron, 57-58; asst prof agron, Univ Minn, St Paul, 58-60, assoc prof, 60-63, prof, 63-75; EXEC VPRES, MINN CROP IMPROV ASN, ST PAUL, 75- *Concurrent Pos:* Dir, Asn Off Seed Certifying Agencies, 61- *Mem:* Am Soc Agron; Crop Sci Soc Am; Coun Agr Sci & Technol. *Res:* Extension education and research in production of field crops; quality seed production and distribution. *Mailing Add:* Minn Crop Improv Asn 1900 Hendon Ave St Paul MN 55108-1011

OTTO, JOHN B, JR, b Kingsville, Tex, Nov 4, 18; m 50; c 2. CHEMISTRY. *Educ:* Tex Col Arts & Indust, BS, 41; Univ Tex, MA, 43, PhD, 50. *Prof Exp:* Tutor chem, Univ Tex, 41-42, instr, 42-44; jr chemist, Clinton Lab, Oak Ridge, 44-46; res chemist, Mound Lab, Monsanto Chem Co, Ohio, 50-54; sr res chemist, Dallas Res Lab, Mobil Res & Develop Corp, 54-88; RETIRED. *Mem:* AAAS; Am Chem Soc. *Res:* Chemical reactions in liquid ammonia; radio chemistry; chemical separations and analysis for geochronology research; uranium in situ leach mining and leach solution processing research; heavy oil production research and development. *Mailing Add:* 735 Swallow Dr Coppell TX 75019-3418

OTTO, KLAUS, b Friedrichroda, Ger, Sept 18, 29; US citizen; m 62; c 2. ENVIRONMENTAL CHEMISTRY, PHYSICAL CHEMISTRY. *Educ:* Univ Hamburg, Vordiplom, 53, Dipl, 57, Dr rer nat(phys chem), 59. *Prof Exp:* Asst phys chem, Univ Hamburg, 60; res assoc, Argonne Nat Lab, 60-62; SR SCIENTIST, FORD MOTOR CO, DEARBORN, 62-, PRIN RES SCIENTIST ASSOC, 73-, STAFF SCIENTIST, 81- *Concurrent Pos:* Adj Prof, Mich State Univ, 86- *Honors & Awards:* Parravano Award for Excellence in Catalysis Res & Develop, 86. *Mem:* AAAS; Am Chem Soc; Catalysis Soc; NY Acad Sci; Sigma Xi. *Res:* Hydrides of transition metals; low temperature calorimetry; electrical properties and nuclear magnetic resonance of inorganic glasses; catalysis of air pollutants, especially nitric oxide; coal gasification; automotive fuels and lubricants. *Mailing Add:* 35173 W Six Mile Rd Livonia MI 48152

OTTO, ROBERT EMIL, b Irvington, NJ, Jan 30, 32; m 66; c 2. CHEMICAL ENGINEERING. *Educ:* Rensselaer Polytech Inst, BChE, 53; Univ Del, PhD(chem eng), 57. *Prof Exp:* Staff asst res & develop, US Air Force, 56-59; sr chem engr, Monsanto Co, 59-65; eng supvr, 65-69; develop mgr, Fisher Controls Co, 69-77; SR FEL, MONSANTO CO, 77- *Mem:* Asn Comput Mach; Am Inst Chem Engrs. *Res:* computer control of industrial processes; computational chemistry. *Mailing Add:* 700 Chesterfield Village Pkwy St Louis MO 63198

OTTO, ROLAND JOHN, b Milwaukee, Wis, Aug 29, 46; m 68. NUCLEAR CHEMISTRY, PHYSICAL CHEMISTRY. *Educ:* Valparaiso Univ, BS, 68; Purdue Univ, PhD(chem), 74. *Prof Exp:* Staff mem nuclear chem, 74-78, SCIENTIST PHYS CHEM, LAWRENCE BERKELEY LAB, UNIV CALIF, 78- *Mem:* Am Chem Soc. *Res:* Heavy ion nuclear reaction; thermodynamics of electrolyte solutions. *Mailing Add:* 1037 Silverhill Dr Lafayette CA 94549-1734

OTTO, WOLFGANG KARL FERDINAND, b Ger, June 12, 27;US citizen; m 59; c 3. PHYSICAL CHEMISTRY, TEXTILE CHEMISTRY. *Educ:* Textile Eng Sch, Germany, Textile engr, 49; Aachen Tech Univ, dipl chem, 54, Dr rer nat, 57. *Prof Exp:* Sci asst, Inst Fuel Chem, Aachen Tech Univ, 54-57; trainee & group leader, Esso, Inc, 57, coordr, 59; supvr fiber develop, Hoechst Dye Works, Inc, 60; proj chemist res & develop, United Carbon Co, Inc, 60, sr chemist, 63; sr res chemist, 64-74, res assoc, 74-80, SR SCIENTIST, MILLIKEN RES CORP, 80- *Mem:* Am Chem Soc; Am Asn Textile Chem & Colorists; Fiber Soc; Am Inst Chemists. *Res:* Surface phenomena; colloidal systems; textiles; synthetic fibers; polymers; adhesives; coatings; mechanical and chemical textile processes. *Mailing Add:* Milliken Res Corp M-420 PO Box 1927 Spartanburg SC 29304

OTTOBONI, M(INNA) ALICE, b Perrin, Tex; m 65. TOXICOLOGY, BIOCHEMISTRY. *Educ:* Univ Tex, BA, 54; Univ Calif, Davis, PhD(comp biochem), 59. *Prof Exp:* Res chemist toxicol eval, Dried Fruit Asn, Calif, 59-63; staff toxicologist, Calif State Dept Health, 63-85; CONSULT & AUTH, 85- *Concurrent Pos:* Nat Inst Environ Health Sci res grant, Calif State Dept Pub Health, 67-72; collabr, Dried Fruit Asn Calif & Western Regional Res Lab, Agr Res Serv, USDA, 59-63. *Mem:* Soc Toxicol; NY Acad Sci; Am Conf Govt Indust Hygienists. *Res:* Intermediary metabolism; comparative biochemistry; toxicity and metabolic fate of chemicals; acute and chronic toxicity, mode of action and metabolism of environmental chemicals. *Mailing Add:* 759 Grizzly Peak Berkeley CA 94708

OTTOLENGHI, ABRAMO CESARE, b Torino, Italy, Apr 13, 31; US citizen; m 58; c 4. MEDICAL MICROBIOLOGY. *Educ:* Wilmington Col, Ohio, BS, 50; Rutgers Univ, MS, 52; Univ Pa, PhD(med microbiol), 60. *Prof Exp:* Res asst preparation & potency testing of vaccines, Labs Life, Quito, Ecuador, 50-51 & 53-56 & Univ Pa, 56-59; res microbiologist, Sidney Hillman Med Ctr, Pa, 59-64; from asst prof to assoc prof, 64-72, PROF MED MICROBIOL & IMMUNOL, COL MED, OHIO STATE UNIV, 72- *Concurrent Pos:* Instr, Ogontz Campus, Pa State Univ, 58-59 & 60-64 & Lab Gen Biol, Rutgers Univ, 59-60; consult, South Div, Albert Einstein Med Ctr, 59-63 & Ohio Dept Health Medicare Lab Cert, 66-; mem comt lab pract microbiol, Pub & Sci Affairs Bd, Am Soc Microbiol. *Mem:* Am Soc Microbiol; Sigma Xi. *Res:* Membrane antigens; bacterial exocellular products; phospholipids structure and function relating to host parasite relationships; antiphospholipid antibodies; medical education; membrane structure and function; electron microscopy; Beta lactamase induction. *Mailing Add:* Dept Med Microbiol & Immunol Ohio State Univ 370 W Ninth Ave Columbus OH 43210

OTTOLENGHI, ATHOS, b Pavia, Italy, May 31, 23; US citizen; m 53; c 2. PHARMACOLOGY. *Educ:* Univ Pavia, MD, 46. *Prof Exp:* Assoc prof pharmacol, Univ Bari, 48-53; res assoc, 53-59, asst prof, 59-65, assoc prof, 65-75, PROF PHARMACOL, DUKE UNIV, 75- *Concurrent Pos:* Fel physiol, Univ Pavia, 46-48. *Mem:* Am Oil Chem Soc; Radiation Res Soc; Am Soc Pharmacol & Exp Therapeut. *Res:* Biochemistry of phospholipids and alteration of phospholipase activity in normal and experimental conditions; biochemistry of whole body irradiation in mammals; eosinophilic leukocytes and kinetics of bone marrow. *Mailing Add:* Dept Pharmacol Duke Univ Box 3813 Med Ctr Durham NC 27706

OTTOSON, HAROLD, b Millston, Wis, Mar 21, 30; m 55; c 4. MATHEMATICS. *Educ:* Wis State Univ, BS, 51; Univ Wis, MS, 54. *Prof Exp:* Instr math, Tri-State Eng Col, 54-55; math analyst, Lockheed Aircraft Corp, 55-56; mathematician, Bendix Pac, Bendix Corp, 56-59; sr staff engr, TRW Comput Co, 59-62; mgr syst design dept, Data & Info Syst Div, Int Tel & Tel Corp, 62-64; sr assoc, Planning Res Corp, 64-66; mem res staff, 66-67, subdept head spec projs, 67-68, assoc mgr nat syst analysis dept, 68-69, head data processing systs dept, 69-75, head, Info Systs Dept, 75-78, dept head, Command & Control Eng, 78-84, DEPT HEAD, SYSTS ANALYSIS, MITRE CORP, 84- *Mem:* Asn Comput Mach. *Res:* Operations research; computer simulation and systems design; applied probability. *Mailing Add:* 3712 Camlot Dr Annandale VA 22003

OTVOS, ERVIN GEORGE, b Budapest, Hungary, Mar 14, 35; m 66; c 2. SEDIMENTOLOGY, GEOMORPHOLOGY. *Educ:* Eötvös Loránd Univ, dipl, 58; Yale Univ, MS, 62; Univ Mass, PhD(geol), 64. *Prof Exp:* Co geologist, Mátra Mineral Mines Co, Hungary, 58-60; asst phys, hist & eng geol, Univ Mass, 62-64; geologist, Brit Petrol Ltd, Can, 64; explor geologist, New Orleans Explor Div, Mobil Oil Corp, 65-68; asst prof, La State Univ, New Orleans, 68-71; HEAD GEOL DIV, GULF COAST RES LAB, 71- *Concurrent Pos:* Adj prof, Univ Southern Miss, 71-, adj prof, 74-; mem subcomn Americas, Int Asn Quaternary Res Comn on Shorelines, 73-; spec master, US Supreme Ct, 82; expert witness, var courts contribr, Geol Soc Am DNAG vol ser. *Honors & Awards:* Henry S Kaminski Mem Award, Sigma Xi, 81. *Mem:* Fel Geol Soc Am. *Res:* Sedimentary, geomorphologic and stratigraphic aspects of coastal processes; Northeast Gulf of Mexico Coastal Plain stratigraphy and geomorphology; coastal evolution; sedimentation and pollution problems of recent coastal water bodies; Quaternary stratigraphy of northeast Gulf coast. *Mailing Add:* Geol Div Gulf Coast Res Lab PO Box 7000 703 E Beach Dr Ocean Springs MS 39564-7000

OTVOS, JOHN WILLIAM, b Budapest, Hungary, Nov 26, 17; nat US; m 45; c 3. PHYSICAL CHEMISTRY. *Educ:* Harvard Univ, BS, 39; Calif Inst Technol, PhD(phys chem), 43. *Prof Exp:* Asst chem, Calif Inst Technol, 39-41; chemist, Shell Develop Co, 46-52, asst head spectros dept, 52-57, head chem physics dept, 57-71, mgr anal res, 71-76; STAFF SR SCIENTIST, LAWRENCE BERKELEY LAB, 76- *Mem:* Am Chem Soc. *Res:* Kinetics; reaction mechanisms; radioisotopes; spectroscopy; photochemistry. *Mailing Add:* Lawrence Berkeley Lab Univ of Calif Berkeley CA 94720

OTWAY, HARRY JOHN, b Omaha, Nebr, Apr 23, 35; m 68; c 2. RISK ASSESSMENT, SCIENCE POLICY. *Educ:* NDak State Univ, BS, 58; Univ NMex, MS, 61; Univ Calif, PhD, 69. *Prof Exp:* Instr eng, NDak State Univ, 57-58; staff mem, Los Alamos Sci Lab, Univ Calif, 58-61; sect leader, 61-66; pilot & flight instr, Golden West Airline, Calif, 66-69; staff scientist, Los Alamos Sci Lab, Univ Calif, 69-72; first officer div nuclear safety & environ protection, Int Atomic Energy Agency, sr officer & proj leader, Joint Int Atomic Energy Agency/Int Inst Appl Systs Anal res proj, 74-78; PRIN SCI OFFICER & HEAD TECHNOL ASSESSMENT SECT, JOINT RES CENTRE, COMN EUROPEAN COMMUNITIES, 78- *Concurrent Pos:* Reactor safety consult, A B Atomenergi, Sweden, 71-72; res scholar, Int Inst Appl Systs Anal, 74-; adv, WHO, 78-; risk assessment consult, Asn Reactor Safety, Ger, 78- environmental impact assessment; social and psychological aspects of technological risks; vis prof psychol & eng, Univ Ill, Urbana, 79; vis scholar, Univ Southern Calif, 82 & 85. *Mem:* Soc Risk Anal. *Res:* Regulation and management of hazardous industrial activities. *Mailing Add:* 186 Pierdra Loop White Rock NM 87544

OTWELL, WALTER STEVEN, b Va, Nov 20, 48. FOOD SCIENCE. *Educ:* Va Mil Inst, BS, 71; Univ Va, MS, 73; NC State Univ, PhD(food sci), 78. *Prof Exp:* Res assoc, Dept Food Sci, NC State Univ, 78; ASST PROF, DEPT FOOD SCI & HUMAN NUTRIT, UNIV FLA, 78- *Mem:* Inst Food Technologists. *Res:* Applied food science relative to utilization of seafoods. *Mailing Add:* Dept Food Sci & Nutrit Univ Fla Gainesville FL 32611

OU, JONATHAN TSIEN-HSIONG, b Formosa, July 31, 34; m 62; c 2. MICROBIOLOGY, GENETICS. *Educ:* Nat Taiwan Univ, BS, 57; Univ Pa, PhD(biol), 67. *Prof Exp:* Res assoc microbial genetics, Univ Pa, 67-68; res assoc, 68-74, asst mem, 74-78, assoc mem, Inst Cancer Res, 78-82; NRC sr res assoc, Walter Reed Army Inst Res, 84-86, sr res microbiol, 86-89; PROF, CHANG GUNG MED COL, 90- *Concurrent Pos:* Guest scholar, Inst Virus Res, 81; res fel, Japan Soc Prom Sci. *Mem:* Genetics Soc Am; Am Soc Microbiol. *Res:* The mechanism of bacterial conjugation in Escherichia coli; the interaction between bacteriophage and bacteria; regulation of gene expression; mechanisms of pathogenesis of enteric bacteria; vaccine construction. *Mailing Add:* Chang Gung Med Col 259 Wen-Hwa 1 Rd Kwei-San 33332 Tao-Yuan Taiwan

OU, LO-CHANG, b Shanghai, China, Oct 16, 32; m 61; c 4. PHYSIOLOGY, BIOCHEMISTRY. *Educ:* Peking Univ, BSc, 54; Dartmouth Med Sch, PhD(physiol), 71. *Prof Exp:* Teaching asst biochem, Peking Univ, 54-59, lectr, 59-62; demonstr physiol, Univ Hong Kong, 62-64; res asst, 64-71, Nat Heart & Lung Inst res assoc physiol, 71-77, asst prof, 77-80, res assoc prof, 80-86, RES PROF PHYSIOL, DARTMOUTH MED SCH, 86- *Res:* Pathophysiology of hypoxia. *Mailing Add:* Dept Physiol Dartmouth Med Sch Hanover NH 03755

OUANO, AUGUSTUS CENIZA, b Mandawe, Philippines, Mar 2, 36; US citizen; m 64; c 2. CHEMICAL ENGINEERING, POLYMER PHYSICS. *Educ:* Mapua Inst Technol, BSChE, 57; Purdue Univ, MSChE, 61; Stevens Inst Technol, PhD(chem eng), 69. *Prof Exp:* Develop engr, Continental Can Co, 61-64; res chemist, Rexall Chem Co, 64-66; res & develop engr, Packaging Corp Am, 66-67; mem res staff polymer characterization, 70-81, mgr polymer processing & mat, 81-87, MGR RELIABILITY ENGR, DIGITAL EQUIP, IBM CORP, 87- *Honors & Awards:* Dolittle Award, Am Chem Soc, 77. *Mem:* Am Chem Soc; Am Inst Physics. *Res:* Determination of molecular weight distribution and molecular weight average of polymers and the relationships between molecular characteristics of polymer to its synthesis and end use properties; diffusion and transport phenomenon in liquid and polymeric systems; reliability of advanced high density IC devices. *Mailing Add:* 315 Arroyo Seco Santa Cruz CA 95060

OUDERKIRK, JOHN THOMAS, b Amsterdam, NY, May 21, 31; m 53; c 2. ORGANIC CHEMISTRY, PHYSICAL CHEMISTRY. *Educ:* Hamilton Col, AB, 52; Cornell Univ, PhD(chem), 57. *Prof Exp:* Res chemist, Hooker Chem Corp, 55-58; res chemist, Toms River Chem Corp, 58-61, prod chemist, Chem Specialties Dept, 61-64; asst prod mgr, Ciba Corp, 64-65, sect leader, Textile & Paper Auxiliaries Div, 65-66; tech dir, NY Color & Chem Corp, Belleville, 66-67; res mgr, Chem Div, Sun Chem Corp, 67-71, consult, 71-74; mem staff, Chemetron Corp, 74-80; MEM STAFF, BASF WYANDOTTE CORP, 80- *Mem:* Am Chem Soc; Am Asn Textile Chemists & Colorists; Royal Soc Chem. *Res:* Reaction mechanisms; polyamides; polyesters; dyestuffs; pigments; textile and paper auxiliaries; instrumental analysis; fluorescence. *Mailing Add:* 15 Cross St Westerly RI 02891-2308

OUELLET, HENRI, b Riviere-du-Loup, Que, Jan 29, 38; m 64; c 1. ORNITHOLOGY, BIOSYSTEMATICS. *Educ:* Univ NB, BA, 62; McGill Univ, MSc, 67, PhD(biol), 77. *Prof Exp:* From asst cur to assoc cur vert zool, Redpath Mus, McGill Univ, 66-70; asst cur, 70-76, chief, Div Vert Zool, 77-86, CUR ORNITH, NAT MUS CAN, 77- *Concurrent Pos:* Res minister cult affairs, Govt Que, 65-67; mem, Endangered Wildlife Can Comt, 79-83; secy gen, XIX Cong Int Ornith, 86. *Mem:* Fel Am Ornith Union; Sigma Xi. *Res:* Ornithological research, museum oriented, particularly in the fields of zoogeography and systematics; bird fauna of Holarctic and Neotropical Regions; taxonomy. *Mailing Add:* Can Mus Nature PO Box 3443 Sta D Ottawa ON K1P 6P4 Can

OUELLET, LUDOVIC, b Tingwick, Que, Mar 27, 23; m 50; c 6. PHYSICAL CHEMISTRY. *Educ:* Laval Univ, BSc, 47, DSc(chem), 49. *Prof Exp:* Merck fel, Cath Univ Am, 50-51; USPHS fel, Naval Med Res Inst, 51-52 & trainee, Univ Wis, 52-53; from asst prof to assoc prof chem, Univ Ottawa, 53-58; chmn dept, Laval Univ, 67-72 & 81-85, prof chem, 60-88; RETIRED. *Concurrent Pos:* Pres, Res Comn, 71-78; pres, Univ Res Comn, Univ Coun Quebec, 69-72; mem, Coun Sci Policy Quebec, 71-73; pres, Order Chemists Que, 76-78. *Mem:* Am Chem Soc; Chem Inst Can. *Res:* Chemical kinetics; mechanism of reactions; especially enzymatic reactions; science policy. *Mailing Add:* Five Barc Sameul Holland Apt 1061 Quebec PQ G1S 4S2 Can

OUELLET, MARCEL, b Hebertville, Que, June 3, 40; m 72; c 1. LIMNOLOGY, GEOCHEMISTRY. *Educ:* Ottawa Univ, BSc, 66, MSc, 68, PhD(paleolimnol), 73. *Prof Exp:* Prof biol, Chicoutimi Col, 68-69; PROF WATER SCI, NAT INST SCI RES, 71- *Mem:* Am Quaternary Res Asn; Am Soc Limnol & Oceanog; Can Soc Limnol; Int Soc Limnol; Can Asn Advan Sci. *Res:* Atmospheric pollutants; geochemistry; postglacial evolution of forest, climate and lake productivity by paleolimnological stratigraphical studies of lake sediments; impact of anthropogenic atmospheric pollution on the quality of lakes by studying the geochemistry of water and most recent sediments; impact of industrial pollution on the quality of snow; industrial fluorides on precipitation and vegetation; limnology of meromictic; hypersaline arctic lakes; paleoseismieity, lake stratigraphy earthquake lights; other environmental, earth & marine sciences. *Mailing Add:* Nat Inst Sci Res Eau CP 7500 Sainte-Foy PQ G1V 4C7 CAN

OUELLETTE, ANDRE J, b Brunswick, Maine, July 2, 46. BIOCHEMISTRY. *Educ:* Ind Univ, PhD(microbiol), 72. *Prof Exp:* ASSOC BIOCHEMIST, SHRINERS BURN INST, MASS GEN HOSP, 73- *Mailing Add:* Dept Surg Shriners Burns Inst 51 Blossom St Boston MA 02114

OUELLETTE, ROBERT J, b St Johnsbury, Vt, Dec 13, 38; m 59; c 3. ORGANIC CHEMISTRY. *Educ:* Univ Vt, BS, 59; Univ Calif, Berkeley, PhD(chem), 62. *Prof Exp:* From asst prof to assoc prof, 62-72, PROF CHEM, OHIO STATE UNIV, 72- *Concurrent Pos:* NSF grants, 65-71; Petrol Res Fund grants, 65-67 & 70-72. *Mem:* Am Chem Soc. *Res:* Organometallic chemistry; conformational analysis; nuclear magnetic resonance. *Mailing Add:* Chem Ohio State Univ 88 W 18th Ave Columbus OH 43210-1106

OUGHSTUN, KURT EDMUND, b New Britain, Conn, Jan 12, 49; m 70; c 2. ELECTROMAGNETIC THEORY, QUANTUM OPTICS. *Educ:* Cent Conn State Col, BA, 72; Inst Optics, Univ Rochester, MS, 74, PhD(optics), 78. *Prof Exp:* Res scientist, United Technologies Corp, 76-84; ASST PROF, UNIV WIS-MADISON, 84- *Mem:* Optical Soc Am; Inst Elec & Electronics Engrs; Nat Geog Soc. *Res:* Linear and nonlinear electromagnetic phenomena; dispersive pulse propagation; semiclassical radiation theory; diffraction and scattering theory; guided wave optics. *Mailing Add:* Dept Elec Engr Univ Vt Burlington VT 05405

OUIMET, ALFRED J, JR, b Wilmington, NC, Apr 7, 31; m 54; c 3. INFORMATION SCIENCE, SOLID STATE SCIENCE. *Educ:* Univ Conn, BA, 53, PhD(phys chem), 62. *Prof Exp:* Res chemist, Gen Elec Co, 53-54; asst instr chem, Univ Conn, 56-62; chemist, Esso Res & Eng Co, 62-66; ED CHEM LIT, CHEM ABSTR SERV, 66- *Mem:* Am Chem Soc; Sigma Xi; Sci Res Soc NAm. *Res:* Kinetic study of the thermal decomposition of aluminum trimethyl; basic and products application research on aerospace lubricants. *Mailing Add:* 1655 Doone Rd Columbus OH 43221

OUJESKY, HELEN MATUSEVICH, b Ft Worth, Tex, Aug 14, 30; m 51; c 3. MICROBIOLOGY, RADIATION BIOLOGY. *Educ:* Tex Woman's Univ, BA & BS, 51, PhD(radiation biol), 68; Tex Christian Univ, MA, 65. *Prof Exp:* Teacher high sch, 51-63; asst prof biol, Tex Woman's Univ, 68-73; assoc prof, 73-80, PROF EARTH & LIFE SCI & ACTING DIV DIR, ALLIED HEALTH & LIFE SCI, UNIV TEX, SAN ANTONIO, 80- *Mem:* Soc Indust Microbiol; Am Soc Microbiol; Radiation Res Soc. *Res:* Effects of gases on DNA polymerase and microorganisms; incorporation of calcium 45 in fish scales; survey of plankton; effects of environmental pollutants on microbial metabolism. *Mailing Add:* UTSA Alliance for Educ 310 S St Mary's St No 1516 San Antonio TX 78205

OULMAN, CHARLES S, b Lake Mills, Iowa, Aug 6, 33; m 61; c 2. SANITARY ENGINEERING. *Educ:* Iowa State Univ, BS, 55, PhD(sanit eng), 63. *Prof Exp:* Assoc civil eng, 56-57, from instr to assoc prof, 57-74, PROF CIVIL ENG, IOWA STATE UNIV, 74- *Honors & Awards:* Co-recipient Gold Medal, Filtration Soc, 70. *Mem:* Am Soc Civil Engrs; Am Water Works Asn; Water Pollution Control Fedn; Filtration Soc. *Res:* Water and waste treatment. *Mailing Add:* 1231 Arizona Ames IA 50012

OURSLER, CLELLIE CURTIS, b Cynthiana, Ind, Nov 26, 15; m 37; c 2. MATHEMATICS. *Educ:* Ind Univ, AB, 37; Univ Chicago, SM, 41; Ill Inst Technol, PhD(math), 58. *Prof Exp:* Teacher pub sch, Ill, 37-38, prin, 38-40; teacher, Ind, 40-49; instr math, Ind Univ, Northwest, 49-59; from asst prof to prof, 59-83, EMER PROF MATH, SOUTHERN ILL UNIV, EDWARDSVILLE, 83- *Mem:* Am Math Soc; Math Asn Am. *Res:* Abstract algebra; theory of numbers; finite geometries; quaternions; matrices with integer elements. *Mailing Add:* 912 Blue Ridge Dr Belleville IL 62223-3526

OURTH, DONALD DEAN, b Los Angeles, Calif, 39; m 69; c 3. IMMUNOLOGY, MICROBIOLOGY. *Educ:* Univ Northern Iowa, BA, 61, MA, 66; Univ Iowa, PhD(microbiol), 69. *Prof Exp:* From asst prof to assoc prof, 74-82, PROF BIOL, MEMPHIS STATE UNIV, 82- *Concurrent Pos:* US Army Med Corps, 64-66; res fel, NIH, Dept Microbiol, Harvard Univ, Sch Pub Health, 69-72; res prof, Gade Inst Microbiol, Univ Bergen, Norway, 72-73. *Mem:* Am Asn Immunologists; Am Soc Microbiol; Int Soc Develop & Comparative Immunol; Fedn Am Soc Exp Biol. *Res:* Fish immunology; insect immunity; complement system of fish; molecular mechanisms of bacterial pathogenesis. *Mailing Add:* Dept Biol Memphis State Univ Memphis TN 38152

OUSEPH, PULLOM JOHN, b Kerala, India, Jan 21, 33; m; c 4. NUCLEAR PHYSICS. *Educ:* Univ Kerala, BSc, 53; Univ Saugor, MSc, 55; Fordham Univ, PhD(nuclear physics), 62. *Prof Exp:* Asst physics, Fordham Univ, 57-61; from asst prof to assoc prof, 62-70, chmn dept, 76-83, PROF PHYSICS, UNIV LOUISVILLE, 70- *Mem:* Am Phys Soc; Am Asn Physics Teachers. *Res:* Low energy nuclear physics, especially angular correlation of successive radiations and Mossbauer effect. *Mailing Add:* Dept Physics Univ Louisville Louisville KY 40208

OUSTERHOUT, LAWRENCE ELWYN, nutrition, biochemistry, for more information see previous edition

OUTCALT, DAVID L, b Los Angeles, Calif, Jan 30, 35; m 56; c 4. MATHEMATICS. *Educ:* Pomona Col, BA, 56; Claremont Grad Sch, MA, 58; Ohio State Univ, PhD(math), 63. *Hon Degrees:* Dr, Kyung Hee Univ, Korea, 84. *Prof Exp:* Asst instr math, Ohio State Univ, 60-62; from instr to asst prof, Claremont Men's Col, 62-64; from lectr to assoc prof, 64-72, chmn dept, 69-72, prof math, Univ Calif, Santa Barbara, 72-80; prof math, Univ Alaska, Anchorage, 80-86; PROF NATURAL & APPL SCI, UNIV WIS, GREEN BAY, 86- *Concurrent Pos:* Co-investr, US Air Force Off Sci Res grant, 64-71, prin investr, 71-73; vis assoc prof, Univ Hawaii, 67-68 & vis prof 71-72; acad asst instructional develop, Univ Calif, Santa Barbara, 73-74, dean instructional develop, 75-80; vchancellor acad affairs, Univ Alaska, Anchorage, 80-81, chancellor, 81-86; chancellor, Univ Wis-Green Bay, 86- *Mem:* Math Asn Am; Int Asn Univ Presidents; Sigma Xi. *Res:* Non-associative rings and algebras. *Mailing Add:* 3015 Bayview Dr Green Bay WI 54311

OUTCALT, SAMUEL IRVINE, b Oak Park, Ill, Aug 26, 36; m 59; c 3. PHYSICAL GEOGRAPHY. *Educ:* Univ Cincinnati, BA, 59; Univ Colo, MA, 64; Univ BC, PhD(geog), 70. *Prof Exp:* Actg asst prof environ sci, Univ Va, 68-70; from asst prof to assoc prof, 70-78, PROF GEOG, UNIV MICH, ANN ARBOR, 78- *Concurrent Pos:* Res assoc, Arctic Inst NAm, 70; consult, Geog Applications Prog-US Geol Surv, 70-; res geogr, Infrared & Optics Div, Willow Run Labs, 72; partic, US-Int Biol Prog-Aerobiol Prog, 72- & US-Int Biol Prog-Tundra Biome, 72; mem staff, Activity under US/USSR Environ Protection Agreement, Northern Ecosysts, 76-; mem comt on permafrost, Nat Acad Sci, 76-78; mem staff, US/Yugoslav Smithsonian Limnol Res, 77- *Mem:* Am Geophys Union; Glaciol Soc; Asn Am Geog. *Res:* Interaction between surface climatological conditions and geomorphic evolution in Arctic-Alpine terrain; thermal modeling; computer applications in physical geography. *Mailing Add:* Dept Geog Univ Mich Ann Arbor MI 48109-1063

OUTERBRIDGE, JOHN STUART, b Hunan, China, Sept 20, 36; m 63; c 3. VESTIBULAR SYSTEM, EYE MOVEMENT. *Educ:* Mt Allison Univ, BS, 57, BA, 58; McGill Univ, MD, CM, 62, PhD(physiol), 69. *Prof Exp:* Intern, Royal Victoria Hosp, Montreal, 62-63, jr asst resident med, 63-64; lectr physiol, Biomed Eng Unit, McGill Univ, 67-69, asst prof otolaryngol & biomed eng, 70-73, asst prof physiol, 69-74, dir otolaryngol res labs, 70-87, dir biomed eng unit, 75-85, ASSOC PROF OTOLARYNGOL & BIOMED ENG, McGILL UNIV, 73-, ASSOC PROF PHYSIOL, 74- *Res:* Vestibular and oculomotor systems; nonlinear dynamics. *Mailing Add:* Dept Physiol McGill Univ 3655 Drummond St Montreal PQ H3G 1Y6 Can

OUTERBRIDGE, WILLIAM FULWOOD, b Atlanta, Ga, Jul 20, 30; m 66. COAL GEOLOGY. *Educ:* Brown Univ, BA, 52; Am Univ, MS, 56. *Prof Exp:* GEOLOGIST, US GEOL SURV, 52- *Mem:* Fel Geol Soc Am; Mineral Soc Am. *Res:* Field geology, Pennsylvanian stratigraphy, and geomorphology of the Appalachian Plateaus; coal geology. *Mailing Add:* 11969 Greywing Court Reston VA 22091

OUTHOUSE, JAMES BURTON, b Canandaigua, NY, Sept 20, 16; m 40; c 3. ANIMAL NUTRITION, ANIMAL SCIENCE. *Educ:* Cornell Univ, BS, 38; Univ Md, MS, 42; Purdue Univ, PhD, 56. *Prof Exp:* Instr & asst animal husb, Univ Md, 38-41, from asst prof to assoc prof, 41-52; from instr to prof, 52-82, EMER PROF ANIMAL SCI, PURDUE UNIV, 82- *Concurrent Pos:* Vis prof, Univ Edinburgh, 70-71; US Int Develop prog, Portugal, 82, 84. *Mem:* Hon fel Am Soc Animal Sci; Brit Soc Animal Prod; Sigma Xi. *Res:* Nutrition studies with sheep and beef cattle; breeding studies with Finn sheep; management studies on accelerated lambing; artificial rearing of lambs and confinement programs; forages and pasture for sheep and lambs; animal nutrition; carcass evaluation; performance testing of sheep and cattle. *Mailing Add:* 2643 Duncan Rd Lafayette IN 47904-1044

OUTKA, DARRYLL E, cell biology, protozoology; deceased, see previous edition for last biography

OUTLAW, HENRY EARL, b Pickwick, Tenn, June 17, 37; m; c 1. BIOCHEMISTRY, MOLECULAR BIOLOGY. *Educ:* Delta State Col, BS, 61; Univ Miss, MS, 64, PhD(pharm), 65. *Prof Exp:* Fel pharmacol, Sch Med, Univ Fla, 65-66; asst prof, 66-74, PROF CHEM, DELTA STATE UNIV, 74-, CHMN DEPT PHYS SCI, 69- *Mem:* Am Chem Soc. *Res:* Chemistry of muscle contraction; isolation and biochemical characterization of subcellular constituents. *Mailing Add:* Dept Phys Sci Box 3255 Delta State Univ Cleveland MS 38733

OUTLAW, RONALD ALLEN, b Norfolk, Va, June 6, 37; m 65; c 2. VACUUM PHYSICS, MATERIALS SCIENCE. *Educ:* Va Polytech Inst, BS, 60, MS, 69, PhD(mat sci), 72. *Prof Exp:* AEROSPACE TECHNOLOGIST VACUUM PHYSICS & MAT SCI, LANGLEY RES CTR, NASA, 63- *Concurrent Pos:* Adj prof physics, Old Dominion Univ, 78-; Hugh L Dryden mem fel, Mass Inst Technol, 90. *Mem:* Am Vacuum Soc. *Res:* Physical and chemical adsorption of gases on metals; diffusion and solubility of gases in metals; vacuum instrumentation and low pressure physics; purification of metals; thin film nucleation and growth. *Mailing Add:* MC 493 Langley Res Ctr NASA Hampton VA 23665

OUTWATER, JOHN O(GDEN), b London, Eng, Jan 2, 23; nat US; m 52; c 4. POLYMER ENGINEERING. *Educ:* Cambridge Univ, BA, 43, MA, 48, PhD, 76; Mass Inst Technol, ScD(mech eng), 50. *Prof Exp:* Engr, E I du Pont de Nemours & Co, 50-52; proj engr, Universal Moulded Prod Corp, 52-54; indust liaison officer, Mass Inst Technol, 54-55; chmn dept, 55-63, PROF MECH ENG, UNIV VT, 55- *Concurrent Pos:* Pres, Vt Instrument Co; consult, Nat Acad Sci, Monsanto Res Co, Naval Ord Lab, US Dept Navy, Chemstrand Corp, Whiting Corp, Gen Elec Co, Grumman Aircraft Co & Smithsonian Inst; Leader expeds, Wenner-Gren Found. *Mem:* Fel Am Soc Mech Engrs; Am Soc Testing & Mat; Soc Hist Technol. *Res:* Mechanical properties of non-metallic materials; reinforced plastics; mechanics of rigid bodies; pre-Columbian construction techniques; ski accident research. *Mailing Add:* Dept Mech Eng Univ Vt Burlington VT 05405

OUTZEN, HENRY CLAIR, JR, b Salt Lake City, Utah, Sept 6, 39. ANATOMY, CANCER IMMUNOLOGY. *Educ:* Univ Utah, BS, 61, PhD(anat), 69. *Prof Exp:* Assoc cancer immunol, Inst Cancer Res, 69-78; mem staff, Jackson Lab, Maine, 78-; DIR TRANSPLANTATION & ANIMAL PROD, SYSTEMICS, 89- *Concurrent Pos:* Mem comt care & use of nude mouse, Inst Lab Animal Resources, Nat Acad Sci-Nat Res Coun, 74-76. *Mem:* AAAS; Am Asn Anat; Am Soc Zool; NY Acad Sci. *Res:* Germfree biology. *Mailing Add:* Systemics Inc 3400 W Bayshore Palo Alto CA 94303

OUZTS, JOHNNY DREW, b Indianola, Miss, Nov 29, 34; m 57; c 4. MEDICAL ENTOMOLOGY, AQUATIC WEED CONTROL. *Educ:* Delta State Col, BS, 57; Miss State Univ, MS, 61, PhD(entom), 63. *Prof Exp:* Actg head dept & asst dean, Sch Arts & Sci, 63-74, head dept, 74-79, PROF BIOL SCI, DELTA STATE COL, 63-, ASSOC DEAN, SCH ARTS & SCI, 74-,

DIR, CTR ALLUVIAL PLAINS STUDIES, 80-, AT DEPT BIOL SCI DELTA STATE UNIV. *Mem:* Entom Soc Am; Am Mosquito Asn; Sigma Xi. *Res:* Chlorinated hydrocarbon resistance in fresh and flood-water; breeding mosquitoes; biological control of cotton insects; practical entomology for student agricultural pilots; plant pathology; weed and biological pest control. *Mailing Add:* 551 Hillcrest Circle Cleveland MS 38732

OVADIA, JACQUES, b Vienna, Austria, Nov 16, 23; nat US; m 56; c 2. MEDICAL PHYSICS. *Educ:* Brooklyn Col, BA, 44; Univ Ill, MS, 47, PhD(nuclear physics), 51. *Prof Exp:* Instr radiol, Univ Ill, 51-52; res assoc biophys, Sloan-Kettering Inst, 52-56; chief physicist radiation ther, Michael Reese Hosp, Chicago, 56-72; prof radiol, Med Col Wis, 72; CHMN MED PHYSICS, MICHAEL REESE HOSP, 73-; PROF RADIOL, UNIV CHICAGO, 74- *Concurrent Pos:* Adj prof physics, Ill Inst Technol, 72- *Mem:* Health Physics Soc; Am Soc Ther Radiol; Radiation Res Soc; Radiol Soc NAm; Am Asn Physicists in Med. *Res:* High energy electrons and x-rays for therapy; short-lived isotope tracers; mechanism of action of radiation; pulse radiolysis. *Mailing Add:* Michael Reese Hosp 1722 E 55th St Chicago IL 60615

OVALLE, WILLIAM KEITH, JR, b Ancon, CZ, Mar 18, 44; US citizen. MEDICAL SCIENCE. *Educ:* St Joseph's Col, Pa, BS, 66; Temple Univ, PhD(anat), 71. *Prof Exp:* Muscular Dystrophy Asn Can fel neurophysiol, Univ Alta, 70-72; asst prof anat, 72-77, ASSOC PROF ANAT, FAC MED, UNIV BC, 77- *Mem:* Am Asn Anatomists; Am Soc Cell Biol; AAAS; Can Asn Anatomists; Sigma Xi. *Res:* Neuromuscular ultrastructure and functional morphology; electron microscopy and histochemistry of the peripheral nervous system system and muscle during development in health and disease. *Mailing Add:* Dept Anat Fac Med Univ BC 2075 Wesbrook Mall Vancouver BC V6R 1W5 Can

OVARY, ZOLTAN, b Kolozsvar, Hungary, Apr 13, 07. IMMUNOLOGY. *Educ:* Univ Paris, MD, 35, Lic es Sci, 39. *Prof Exp:* Res assoc microbiol, Sch Med, Johns Hopkins Univ, 55-59; from asst prof to assoc prof path, 59-64, PROF PATH, SCH MED, NY UNIV, 64- *Concurrent Pos:* Res fel, Brazilian Super Coun Res, 52; Fulbright fel, 54-55. *Mem:* Am Soc Immunol; Soc Exp Biol & Med; Am Acad Allergy; Harvey Soc; fel NY Acad Sci. *Res:* Hypersensitivity; anaphylaxis; antibody structure; antigenicity; function of the reticuloendothelial system. *Mailing Add:* Dept Path 550 First Ave NY Univ Sch of Med New York NY 10016

OVE, PETER, b Dellstedt, Ger, Aug 31, 30; US citizen; m 56; c 3. CELL BIOLOGY, BIOCHEMISTRY. *Educ:* Univ Pittsburgh, BS, 63, PhD(cell biol), 67. *Prof Exp:* Asst prof clin pharmacol, Med Ctr, Duke Univ, 68-69; from asst prof to prof cell biol, Sch Med, Univ Pittsburgh, 69-88, dir grad studies 78-88; SCI PROG DIR, AM CANCER SOC, ATLANTA, GA, 88- *Mem:* Am Asn Cancer Res; Am Soc Cell Biol; Am Asn Anat. *Res:* Control of growth of mammalian cells; cancer problems; biology of aging. *Mailing Add:* 1934 N Akin Dr Atlanta GA 30345

OVENALL, DERICK WILLIAM, b Bloxwich, Eng, Sept 26, 30; m 65; c 2. PHYSICAL CHEMISTRY. *Educ:* Univ Manchester, BSc, 52; Univ Birmingham, PhD(chem), 55. *Prof Exp:* Res fel chem, Univ Birmingham, 55-57; res assoc physics, Duke Univ, 57-59; sr sci officer, Nat Phys Lab, Eng, 59-60; physicist, Battelle Mem Inst, Switzerland, 60-61; res chemist, Plastics Dept, 62-71, RES CHEMIST, CENT RES & DEVELOP DEPT, EXP STA, E I DU PONT DE NEMOURS & CO, INC, 71- *Mem:* Am Chem Soc; Sigma Xi. *Res:* Spectroscopy; high resolution nuclear magnetic resonance; electron spin resonance; radiation damage. *Mailing Add:* 1209 Carr Rd Woodside Hills Wilmington DE 19809-1611

OVENFORS, CARL-OLOF NILS STEN, b Stockholm, Sweden, Sept 26, 23; c 2. RADIOLOGY. *Educ:* Karolinska Inst, Sweden, BS, 44, MD, 51, PhD, 64, Docent, 64. *Prof Exp:* Asst dir thoracic radiol, Univ Hosp of Karolinska, Stockholm, 62-69, dir thoracic radiol, 69-70; assoc prof, 70-73, PROF RADIOL, UNIV CALIF, SAN FRANCISCO, 73- *Concurrent Pos:* James Picker Found scholar, 61-62, grant, Karolinska Inst, Sweden, 66-68; radiol res grants, 62-68; consult radiol, Swedish Armed Forces; mem, Coun for Cardiac Radiol; vis prof, Univ Calif, Los Angeles, 79, Karolinska Inst, Stockholm, Sweden, 79 & Univ Lyon, France, 80. *Mem:* Radiol Soc NAm; Asn Univ Radiol; NAm Soc Cardiac Radiol; Am Thoracic Soc; Fleischner Soc. *Res:* Cardiovascular and pulmonary radiology; author of numerous publications. *Mailing Add:* 14210 Arbolitos Dr Poway CA 92064

OVENS, WILLIAM GEORGE, b PATERSON, NJ, July 18, 39; m 63; c 2. MECHANICAL ENGINEERING. *Educ:* Univ Mich, BSE, 64; Univ Conn, MS, 69, PhD(mech metall), 71. *Prof Exp:* Engr, Pratt & Whitney Aircraft, 64-65; res specialist, Univ Conn, 65-71; lectr mech eng, Papua New Guinea Univ Technol, 72-75; asst prof mech eng, Clarkson Univ, 75-81; PROF MECH ENG, ROSE-HULMAN INST TECHNOL, 81- *Concurrent Pos:* Consult, Martin Marietta Orlando Aerospace, 80-, IBM Corp, 83- & Tri-Industs, 83-; appointee, Productivity Comt, Am Soc Mech Engrs, 85-87, chmn Mats Div, 83; sr teaching fel, Nanyang Technol Inst, Singapore, 87-88. *Honors & Awards:* Andrew Kucher Res Award, Univ Mich, 64; Ralph R Teetor Award, Soc Automotive Engrs, 74. *Mem:* Am Soc Mech Engrs; Am Soc Metals; Soc Am Value Engrs. *Res:* Producibility considerations in design; design for lowest cost manufacturing; application of value engineering to production enhancement problems; effect of manufacturing processes on properties of materials and vice-versa; failure analysis and products liability traceable to design and manufacturing defects. *Mailing Add:* 4693 Woodshire Dr Terre Haute IN 47803-2035

OVENSHINE, A THOMAS, b New York, NY, Mar 25, 36; m 59; c 3. GEOLOGY. *Educ:* Yale Univ, BS, 58; Va Polytech Univ, MS, 62; Univ Calif, Los Angeles, PhD(geol), 65. *Prof Exp:* Geologist, 65-70, asst chief, 70-72, chief, Br Alaskan Geol, 76-80, CHIEF, OFF MINERAL RESOURCES, US GEOL SURV, 80- *Mem:* Fel Geol Asn Can; Soc Econ Paleontologists & Mineralogists; Arctic Inst NAm; AAAS; fel Geol Soc Am. *Res:* Pre-Pleistocene glacial deposits; sedimentary structures; Silurian-Devonian conodonts; geology of southeastern Alaska. *Mailing Add:* Chief Off Int Geol US Geol Survey Nat Ctr MSM 917 Reston VA 22902

OVERALL, JAMES CARNEY, JR, b Nashville, Tenn, Sept 27, 37; m 65; c 4. PEDIATRICS, INFECTIOUS DISEASES. *Educ:* Davidson Col, BS, 59; Vanderbilt Univ, MD, 63; Am Bd Pediat, cert, 68. *Prof Exp:* Intern pediat, Vanderbilt Univ Hosp, 63-64; from jr resident to sr resident, Babies Hosp, Columbia Presby Med Ctr, 64-66; res assoc, Nat Inst Neurol Dis & Blindness, NIH, Bethesda, Md, 66-68; instr & fel pediat & microbiol, Sch Med, Univ Rochester, 68-70; from asst prof to assoc prof pediat & microbiol, 70-79, HEAD, DIV INFECTIOUS DIS, DEPT PEDIAT & DIR TRAINING PROG PEDIAT INFECTIOUS DIS, SCH MED, UNIV UTAH, 70-, PROF PEDIAT, 79-, PROF PATH, 81-; MED DIR, DIAG VIROL LAB, ASSOC REGIONAL & UNIV PATHOLOGISTS, 81- *Concurrent Pos:* Head, Clin Epidemiol & Develop Immunol Unit, Infectious Dis Sect, Perinatal Res Br, Nat Inst Neurol Dis & Blindness, 67-68; proj officer, Rubella Vaccine Prog, Nat Inst Allergy & Infectious Dis, 67-68; asst pediatrician, Strong Mem Hosp, Rochester, NY, 68-70; pediatrician, Univ Utah Med Ctr, Salt Lake City, 70-; consult infectious dis, Primary Children's Med Ctr, Salt Lake City, 70-, active staff, 76-; consult infectious dis, Vet Admin Hosp, Salt Lake City, 70- 76; investr, Howard Hughes Med Inst, 74-80; vis res assoc neurol & pediat, Dept Neurol, Univ Calif, San Francisco, 78-79; dir, Virol Portion Med Microbiol Course, Univ Utah, 80-, vchmn acad affairs, Dept Pediat, 83-; mem, Maternal & Child Health Res Comt, Nat Inst Child Health & Human Develop, 82-86, AIDS Res Rev Comt, Nat Inst Allergy & Infectious Dis, 88-92 & Time Ltd Cert Test Comt, Am Bd Pediat, 88- *Mem:* Infect Dis Soc Am; Am Soc Microbiol; Am Fedn Clin Res; Soc Pediat Res; Sigma Xi. *Res:* Virology; herpes simplex virus infections, antiviral chemotherapy; host resistance to viral infections; viral immunology; viral-bacterial or fungal synergism; neonatal infections; hospital acquired infections in pediatric hospitals; rapid viral diagnosis. *Mailing Add:* Pediat Infectious Dis Univ Utah Sch Med 50 N Medical Dr Salt Lake City UT 84132

OVERBECK, HENRY WEST, b Chicago, Ill, Oct 29, 30; m 60; c 2. MEDICAL PHYSIOLOGY, INTERNAL MEDICINE. *Educ:* Princeton Univ, BA, 52; Northwestern Univ, MD, 56; Univ Okla, PhD(physiol), 66; Am Bd Internal Med, cert. *Prof Exp:* Intern, Mary Fletcher Hosp, Univ Vt, 56-57; resident physician, Sch Med, Northwestern Univ, 59-62; instr internal med, Sch Med, Univ Okla, 63-66; from asst prof to prof physiol & internal med, Mich State Univ, 66-76; prof physiol & internal med, Uniformed Servs Univ Health Sci, Bethesda, Md, 76-78; PROF MED & PHYSIOL, UNIV ALA MED CTR, BIRMINGHAM, 78- *Concurrent Pos:* USPHS career develop award, 66-71; Vet Admin clin investr, 63-66; fel cardiol, Emory Univ Sch Med, Atlanta, 73-74; mem hypertension task force, Nat Heart Lung & Blood Inst, 76-78; mem, Exp Cardiovasc Sci Study Sect, NIH, 83-86; fel, Ccoun High Blood Pressure & Coun Circulation, Am Heart Assn. *Mem:* Cent Soc Clin Res; Soc Exp Biol & Med; Am Physiol Soc; fel Am Col Physicians. *Res:* Cardiovascular physiology and pathophysiology, especially hypertension. *Mailing Add:* Dept Med WVa Univ Morgantown WV 26506

OVERBERGER, CHARLES GILBERT, b Barnesboro, Pa, Oct 12, 20; m 45; c 4. POLYMER CHEMISTRY. *Educ:* Pa State Col, BS, 41; Univ Ill, PhD(org chem), 44. *Hon Degrees:* DSc, Holy Cross Col, 66 & L I Univ, 68. *Prof Exp:* Res chemist, Allied Chem Co, Pa, 41; asst org chem, Univ Ill, 41-43; assoc, Rubber Reserve Corp, 44-46; Du Pont fel, Mass Inst Technol, 46-47; from asst prof to assoc prof chem, Polytech Inst Brooklyn, 47-52, prof org chem, 52-66, assoc dir polymer res inst, 51-64, dir, 64-66, head, dept chem, 55-64, vpres res, 64-65, dean sci, 64-66; chmn, dept chem, 67-72, dir, Macromolecular Res Ctr, 68-87, vpres res, 72-82, PROF CHEM, UNIV MICH, 67- *Concurrent Pos:* Consult, Armed Forces; ed, J Polymer Sci, Fortschritte der Hochpolymerenforschung & Macromolecular Synthesis; mem bur, Int Union Pure & Applied Chem, 78- *Honors & Awards:* Witco Award, Am Chem Soc, 68, Polymer Chem Award, 86 & Paul S Flory Polymer Educ Award, 89; Charles Lathrop Parsons Award, 78; Int Award, Soc Plastic Engrs, 79. *Mem:* AAAS; Am Chem Soc (pres, 67); Am Inst Chem; Brit Chem Soc; Brit Soc Chem Indust; Sigma Xi; Brazilian Acad Sci; Int Union Pure & Applied Chem (pres, Macromolecular Div, 75-77). *Res:* The use of polymers as reactants in organic reactions; catalytic properties polymeric nucleophiles; conformation of asymmetric polyamides in solution; synthesis and conformation of substituted polyproline-type materials; grafting of amino acids and purine bases onto hydrophylic polymer chains; effect of side chain interactions on the conformation of polymer molecules. *Mailing Add:* 24 Southwick Ct Ann Arbor MI 48105

OVERBERGER, JAMES EDWIN, b Barnesboro, Pa, July 1, 30; m 58; c 2. DENTISTRY. *Educ:* Univ Pittsburgh, BS, 52, DDS, 54; Univ Mich, MS, 57. *Prof Exp:* Clin instr human resources, Univ Mich, 56-57; asst prof dent mat, Sch Dent, WVa Univ, 57-62, assoc prof prosthodont, 62-67; assoc prof, Sch Dent, Univ NC, Chapel Hill, 67-69; actg dean, 79-80, 81-82, PROF DENT MAT & ASSOC DEAN SCH DENT, WEST VA UNIV, 69- *Mem:* Fel Am Col Dent; Am Dent Asn; Int Asn Dent Res. *Res:* Dental materials; restorative dentistry. *Mailing Add:* WVa Univ Sch Dent Morgantown WV 26506

OVERBY, LACY RASCO, b Model, Tenn, July 27, 20; m 48; c 4. BIOCHEMISTRY, VIROLOGY. *Educ:* Vanderbilt Univ, BA, 41, MS, 48, PhD(chem), 50. *Prof Exp:* Control & prod supvr, E I du Pont de Nemours & Co, 41-43; asst, Vanderbilt Univ, 46-49; sect mgr nucleic acid res, Abbott Labs, 49-62; vis scholar microbiol, Univ Ill, 62-64; mgr, Dept Virol, Abbott Labs, 69-73, dir, Div Exp Biol, 73-81, genetic eng, 81-83; div vpres, Sci & Technol, 81-84, vpres, 84-88, EXEC CONSULT, CHIRON CORP, 88- *Concurrent Pos:* Assoc res fel, Abbott Labs, 64-69; lectr, Northwestern Univ, 68-; assoc ed, J Molecular Virol, 77-88; mem, Int Adv Bd Asian J Clin Sci, 80- *Honors & Awards:* Pasteur Award, Am Soc Microbiol, 86. *Mem:* AAAS; Am Chem Soc; Am Inst Nutrit; Am Soc Biochem & Molecular Biol; Am Soc Microbiol; Sigma Xi. *Res:* Amino acids and proteins; chemistry and biochemistry of vitamins and growth factors; viruses and nucleic acids; hepatitis viruses; molecular biology; genetic engineering. *Mailing Add:* 28 Cherry Hills Ct Alamo CA 94507

OVERCAMP, THOMAS JOSEPH, b Toledo, Ohio, Feb 22, 46; m 69; c 3. AIR POLLUTION & HAZARDOUS WASTE MANAGEMENT. *Educ:* Mich State Univ, BS, 68; Mass Inst Technol, SM, 70, PhD(mech eng), 73. *Prof Exp:* Res assoc, Meteorol Prog, Univ Md, 72-74, vis asst prof, 74- 75; from asst prof to assoc prof, 75-85, PROF ENVIRON SYSTS ENG, CLEMSON UNIV, 85- *Mem:* Air & Waste Mgt Asn; Am Meteorol Soc; Am Soc Mech Engrs; Am Acad Environ Engrs. *Res:* Atmospheric meteorology; air pollution control; environmental engineering. *Mailing Add:* Dept Environ Systs Eng Clemson Univ Clemson SC 29634-0919

OVERCASH, MICHAEL RAY, b Kannapolis, NC, July 17, 44; m 68; c 1. POLLUTION PREVENTION. *Educ:* NC State Univ, BS, 66; Univ NSW, MSc, 67; Univ Minn, PhD(chem eng), 72. *Prof Exp:* PROF CHEM ENG, NC STATE UNIV, 72- *Concurrent Pos:* Dir, Pollution Prev Res Ctr, 89-; mem & chmn, var comts, Am Inst Chem Engrs & Am Soc Agr Engrs, 77-; US Environ Protection Agency distinguished scientist award, 86; consult indust toxic & non-toxic wastes. *Mem:* Sigma Xi; Am Soc Testing Mat; Am Soc Agr Engrs; Am Inst Chem Engrs; Int Water Resources Asn; Water Pollution Control Fedn; Int Humic Substances Soc; Hist Sci Soc. *Res:* Pollution prevention and cost-effective manufacturing improvements; chemical, pulp and paper, textiles, and microelectronics industries. *Mailing Add:* NC State Univ Box 7905 Raleigh NC 27695-7905

OVERDAHL, CURTIS J, b Elk Point, SDak, June 24, 17; m 53; c 3. SOILS. *Educ:* Univ Minn, BS, 47, MS, 49; Purdue Univ, PhD(soils), 57. *Prof Exp:* Exten area agronomist, Iowa State Univ, 49-53; exten agronomist, Purdue Univ, 53-54; EXTEN SPECIALIST SOILS, UNIV MINN, ST PAUL, 56- *Mem:* Am Soc Agron. *Res:* Soil management and fertilizers; agricultural extension service. *Mailing Add:* 1616 Garden Ave St Paul MN 55113

OVEREND, RALPH PHILLIPS, b Eng, Apr 7, 44; Can citizen; m 65. PHYSICAL CHEMISTRY. *Educ:* Univ Salford, MSc, 68; Univ Dundee, PhD(phys chem), 72. *Prof Exp:* Chemist, UKAEA, Winfrith in Dorset, 66-68; Fel, Chem div, Nat Res Coun Can, 72-74, asst res officer, 74-76, biomass energy res & develop prog convenor, 76-80, task coordr, renewable energy, 80-83, bioenergy prog mgr, 83-86, sr res officer, 86-90; BIOMASS POWER PROG MGR, SOLAR ENERGY RES INST, GOLDEN, COLO, 90- *Concurrent Pos:* Assoc prof, dept chem eng, Fac Appl Sci, Univ Sherbrooke, 83- *Mem:* Royal Soc Chem; fel Chem Inst Can; Can Pulp & Paper Asn; Am Chem Soc Tech Sect. *Res:* Gas kinetics of species important to the atmosphere; the chemical conversion of biomass and biomass productivity. *Mailing Add:* Solar Energy Res Inst 1617 Cole Blvd Golden CO 80401-3393

OVERHAGE, CARL F J, b London, Eng, Apr 2, 10; nat US; m 40. ELECTRONICS, PHOTOGRAPHY. *Educ:* Calif Inst Technol, BS, 31, MS, 34, PhD(physics), 37. *Prof Exp:* Physicist, Technicolor Motion Picture Corp, 37-40, actg dir res, 40-42; mem staff, Radiation Lab, Mass Inst Technol, 42-45; res supvr, Color Control Dept, Eastman Kodak Co, 46-48, asst dir, Color Tech Div, 49-54; div head, Lincoln Lab, 55-57, dir, 57-64, prof eng, 61-73, EMER PROF ENG, MASS INST TECHNOL, 73- *Concurrent Pos:* Sci consult, US Army Air Force, 45; mem sci adv bd, US Air Force, 51-57, sci adv comt ballistic missiles, Secy Defense, 59-61; mem, Defense Sci Bd, 62-63; tech adv bd, Fed Aviation Agency, 62-65; mem, Nat Adv Comn on Libr, 66-67; exec dir, Univ Info Technol Corp, 68-73. *Honors & Awards:* Presidential Cert Merit, 48; Exceptional Service Award, US Air Force, 58. *Mem:* Fel Am Acad Arts & Sci (vpres, 64-66); fel Am Phys Soc; fel Optical Soc Am. *Res:* Photography; electronics. *Mailing Add:* 1112 Calle Catalina Santa Fe NM 87501-1016

OVERHAUSER, ALBERT WARNER, b San Diego, Calif, Aug 17, 25; m 51; c 8. THEORETICAL SOLID STATE PHYSICS. *Educ:* Univ of Calif, AB, 48, PhD(physics), 51. *Hon Degrees:* DSc, Univ Chicago, 79. *Prof Exp:* Res assoc, Univ Ill, 51-53; from asst prof to assoc prof physics, Cornell Univ, 53-58; supvr, Solid State Physics, Ford Motor Co, 58-62, mgr, Theoret Sci, 62-69, asst dir, Phys Sci, 69-72, dir, Phys Sci Lab, 72-73; prof, 73-74, STUART DISTINGUISHED PROF PHYSICS, PURDUE UNIV, 74- *Honors & Awards:* Oliver E Buckley Solid State Physics Prize, Am Phys Soc, 75; Alexander von Humboldt US Sr Scientist Award, 79. *Mem:* Nat Acad Sci; Am Acad Arts & Sci; fel Am Phys Soc. *Res:* Many-electron theory; transport theory; neutron interferometry. *Mailing Add:* Dept Physics Purdue Univ West Lafayette IN 47907

OVERHOLSER, KNOWLES ARTHUR, BIOLOGICAL TRANSPORT PHENOMENON. *Educ:* Univ Wis, PhD(chem eng), 70. *Prof Exp:* PROF CHEM & BIOMED ENG, VANDERBILT UNIV, 71- *Res:* Pulmonary fluid dynamics; coronary mass transfer. *Mailing Add:* Dept Chem & Biochem Eng Vanderbilt Univ Nashville TN 37240

OVERHOLT, JOHN LOUGH, b Estherville, Iowa, May 28, 09; m 39; c 3. OPERATIONS RESEARCH. *Educ:* Iowa State Col, BS, 32; Lehigh Univ, MS, 35. *Prof Exp:* Res investr, NJ Zinc Co, 34-55; mem Opers Eval Group, Mass Inst Technol, 55-62; mem, Franklin Inst, 62-67, mem, 67-75,; consult, Ctr Naval Analysis, Univ Rochester, 75-80. *Mem:* Am Chem Soc; Opers Res Soc Am; Am Statist Asn. *Res:* Interior paint formulation; fundamental drying oil research; heavy chemical processes; titanium compounds and processes; statistics and experimental design; Colonial history; military operations research on air, sea and land problems. *Mailing Add:* Rte 1 Box 2240 Kilmarnock VA 22482

OVERLAND, JAMES EDWARD, b Portland, Ore, Oct 30, 47. COASTAL METEOROLOGY AND SEA ICE, PHYSICAL OCEANOGRAPHY. *Educ:* Univ Wash, BS, 70, MS, 71; NY Univ, PhD(oceanog), 73. *Prof Exp:* Phys scientist, Nat Meteorol Ctr, 73-76, SUPVRY OCEANOGR, PAC MARINE ENVIRON LAB, NAT OCEANIC & ATMOSPHERIC ADMIN, 76- *Concurrent Pos:* Affil assoc prof atmospheric sci, Univ Wash, Seattle; ed, JGR Oceans; mem, Comt Coastal Ocean, Nat Res Coun. *Mem:* Am Meteorol Soc; Am Geophys Union. *Res:* Coastal and arctic meteorology, coastal oceanography sea ice modeling. *Mailing Add:* 7330 56th Ave NE Seattle WA 98115

OVERLEASE, WILLIAM R, ecology, botany, for more information see previous edition

OVERLEY, JACK CASTLE, b Kalamazoo, Mich, Aug 23, 32; m 59. ION MICROPROBES. *Educ:* Mass Inst Technol, BS, 54; Calif Inst Technol, PhD(physics), 61. *Prof Exp:* Res assoc physics, Calif Inst Technol, 60-61 & Univ Wis, 61-63; asst prof, Yale Univ, 63-68; asst dean sci serv, Col Lib Arts, 73-75, assoc prof, 68-83, PROF PHYSICS, UNIV ORE, 83- *Concurrent Pos:* Asst dir, A W Wright Nuclear Structure Lab, Yale Univ, 66-68; vis assoc physics, Calif Inst Technol, 75-76; dir, Grad Studies, Univ Ore, 83-, assoc chmn, 84- *Mem:* AAAS; Am Phys Soc; Am Asn Physics Teachers; Sigma Xi. *Res:* Experimental nuclear physics, primarily nuclear spectroscopy and reactions involving charged particles and neutrons; applications of nuclear techniques to MeV-ion microprobe development. *Mailing Add:* Dept Physics Univ Ore Eugene OR 97403

OVERMAN, ALLEN RAY, b Goldsboro, NC, Mar 11, 37; m 64; c 2. AGRICULTURAL ENGINEERING. *Educ:* NC State Univ, BS, 60, MS, 63, PhD(agr eng), 65. *Prof Exp:* Res assoc agron, Univ Ill, Urbana, 65-69; asst prof agr eng & asst agr engr, 69-74, assoc prof, 74-79, PROF AGR & ASSOC AGR ENGR, UNIV FLA, 79- *Mem:* Am Soc Agr Eng; Am Soc Civil Eng; Am Inst Chem Eng; Water Pollution Control Fedn; Am Soc Agron. *Res:* Transport processes in soil and chemical kinetics; wastewater treatment. *Mailing Add:* Dept Agr Eng Univ Fla 9 ROG Bldg Gainesville FL 32611

OVERMAN, AMEGDA JACK, b Tampa, Fla, May 17, 20; m 53. PLANT NEMATOLOGY. *Educ:* Univ Tampa, BS, 42; Univ Fla, MS, 51. *Prof Exp:* Asst in soils chem, 51-56, asst soils microbiologist, 56-68, assoc nematologist, 68-73, NEMATOLOGIST, INST FOOD & AGR SCI, GULF COAST RES & EDUC CTR, UNIV FLA, 73-, EMER PROF, 89- *Honors & Awards:* Ciba-Geigy Award of Excellence, Soc Nematol, 82; Pres Award, Orgn Trop Am Nematol, 85. *Mem:* Soc Nematol; Europ Soc Nematol; Orgn Trop Am Nematol. *Res:* Bionomics, pathogenicity and control of nematodes attacking vegetables, ornamentals and agronomic crops. *Mailing Add:* Gulf Coast Res & Educ Ctr 5007 60th St E Bradenton FL 34203

OVERMAN, DENNIS ORTON, b Union City, Ind, Oct 16, 43; m 66; c 3. TERATOLOGY. *Educ:* Bowling Green State Univ, BA, 65; Univ Mich, MS, 67, PhD(anat), 70. *Prof Exp:* Res assoc develop biol, Univ Colo, 70-71; from instr to asst prof, 71-76, ASSOC PROF ANAT, W VA UNIV, 76- *Mem:* Teratology Soc; Am Asn Anatomists; Soc Develop Biol. *Res:* Mechanisms and inhibition of teratogenesis in orofacial malformations and limb development. *Mailing Add:* Dept Anat WVa Univ Morgantown WV 26506-6203

OVERMAN, JOSEPH DEWITT, b Champaign, Ill, Feb 23, 18; m 45; c 4. PHYSICAL CHEMISTRY. *Educ:* Univ Ill, BA, 41; Univ Rochester, PhD(chem), 45. *Prof Exp:* RES CHEMIST, E I DU PONT DE NEMOURS & CO, INC, PARLIN, NJ, 45- *Mem:* Am Chem Soc; Soc Photo Scientisits & Engrs. *Res:* Reaction kinetics; photochemistry; photographic emulsions; reaction rate of Claisen rearrangement; photochemical decomposition of unsymmetrical dimethylhydrazine. *Mailing Add:* Coffee Run Condos C-2A Hockessin DE 19707

OVERMAN, LARRY EUGENE, b Chicago, Ill, Mar 9, 43; m 66; c 2. ORGANIC CHEMISTRY. *Educ:* Earlham Col, BA, 65; Univ Wis-Madison, PhD(org chem), 69. *Prof Exp:* NIH fel, Columbia Univ, 69-71; from asst prof to assoc prof, 71-79, PROF CHEM, UNIV CALIF, IRVINE, 79-, CHAIR, 80- *Concurrent Pos:* A P Sloan Found fel, 75-77; Camille & Henry Dreyfus teacher-scholar award, 76-81; mem, Med Chem Study Sect, NIH, 83-86; exec comt org div, Am Chem Soc, 86-; Jacob Javits Awardee, Nat Inst Neurological Sci, 85-91; Sr Scientist Award, Alexander von Humboldt Found, 85-87. *Honors & Awards:* Arthur C Cope Scholar Award, 89. *Mem:* AAAS; Am Chem Soc; Japanese Chem Soc; Swiss Chem Soc; Ger Chem Soc; Royal Soc Chem. *Res:* New methods for organic synthesis; natural products synthesis. *Mailing Add:* Dept of Chem Univ of Calif Irvine CA 92717

OVERMAN, RALPH THEODORE, b Clifton, Ariz, Aug 9, 19; m 45, 71; c 2. MEDICAL EDUCATION. *Educ:* Kans State Teachers Col, AB, 39, MS, 40; La State Univ, PhD(phys chem), 43. *Hon Degrees:* DSc, Philadelphia Col Pharm, 59. *Prof Exp:* Plant chemist, Flintkote Co, La, 42; instr chem, La State Univ, 42-43; head dept sci, La Col, 43-44; tech supt, Fercleve Corp, Tenn, 44-45; sr res chemist, Oak Ridge Nat Labs, 45-48, chmn spec training div, Oak Ridge Inst Nuclear Studies, 48-65; consult, Ralph T Overman Consult Serv, 65-68; dir educ, Technicon Corp, 68-69; vpres, Universal Med Servs, 70; planning dir, Bistate Regional Med Prog, 70-74; MEM STAFF REGIONAL MED EDUC CTR, VET ADMIN HOSP, ST LOUIS, 74- *Concurrent Pos:* Assoc prof, Sch Med, St Louis Univ; lectr, Sch Med, Wash Univ; pres, Compuhealth, St Louis. *Mem:* Fel AAAS; fel Am Nuclear Soc; Am Chem Soc; Am Col Nuclear Physicians; Soc Nuclear Med; Sigma Xi. *Res:* Radiochemistry; nuclear medicine; health care planning. *Mailing Add:* 702 N Laffitte Dr Terre du Lac Bonne Terre MO 63628

OVERMAN, TIMOTHY LLOYD, b Cincinnati, Ohio, Dec 9, 43; m 68; c 2. CLINICAL MICROBIOLOGY. *Educ:* Univ Cincinnati, BS, 66, MS, 69, PhD(microbiol), 72; Am Bd Med Microbiol, dipl, 75. *Prof Exp:* Fel pub health & med microbiol, Ctr Dis Control, Atlanta, Ga, 71-73; asst prof path, Col Med, 74-81, asst prof med technol, Col Allied Health Professions, 79-81, ASSOC PROF PATH & MED TECHNOL, UNIV KY, 81-; CLIN MICROBIOL, PATH SERV, VET AFFAIRS MED CTR, LEXINGTON, KY, 74- *Concurrent Pos:* Asst dir, Clinical Microbiol Lab, Univ Ky Hosp, 74-76, consult, 76- *Mem:* Acad Clin Lab Physicians & Scientists; Am Soc Microbiol; Fel Am Acad Microbiol. *Res:* Diagnostic microbiology, especially oxidase-positive fermentative and non-fermentative bacteria; laboratory automation. *Mailing Add:* Path Serv 113CD Vet Affairs Med Ctr Lexington KY 40502

OVERMIER, J BRUCE, b Queens, NY, Aug 2, 38; m 62; c 1. CONDITIONING & LEARNING, BIOPSYCHOLOGY. *Educ:* Kenyon Col, AB, 60; Bowling Green State Univ, MA, 62; Univ Pa, PhD(psychol), 65. *Hon Degrees:* DSc, Kenyon Col, 90. *Prof Exp:* From asst prof to assoc prof,

65-71, PROF PSYCHOL, UNIV MINN, 71- *Concurrent Pos:* NSF sr fel, 71-72, Nat Acad Sci exchange fel, 72, Fulbright Hays fel, 80, Fogarty sr int fel, 84, James McKeen Cattell fel, 85 & Norweg Marshall fel, 87; ed, Learning & Motivation, 73-76; vis prof, Univ Hawaii, 76, 82 & 86; mem adv panel, NSF, 76-79 & NIMH, 88-91, bd gov, Psychonomic Soc, 83-88, Coun Rep, Am Psychol Asn, 87-90 & Steering Comt, 88-90; dir, Ctr Res Learning, Perception & Cognition, 83-89, pres, Physiol & Comp Psychol Div, 90-91 & Exp Psychol Div, 92-93. *Mem:* Fel Am Psychol Asn; fel Am Psychol Soc; Psychonomic Soc (secy, 81-88). *Res:* Psychobiological processes underlying learning; application of learning principles for teaching the mentally retarded; proactive effects of stress on learning and on physiology. *Mailing Add:* 75 E River Rd-Elliott Minneapolis MN 55455

OVERMYER, ROBERT FRANKLIN, b Chicago, Ill, Aug 9, 29; m 59; c 2. ENVIRONMENTAL SCIENCES. *Educ:* DePauw Univ, BA, 50; Ill Inst Technol, MS, 59. *Prof Exp:* Field engr, Schlumberger Well Surv Corp, 54-56; res asst solid state physics, Gen Atomic Div, Gen Dynamics Corp, 58-60, staff assoc, 60-67; mgr radiation effects, Physics Br, Gulf Gen Atomic, Inc, 67-71, mgr spec nuclear effects dept, Gulf Radiation Technol, 71-73; vpres, Intelcom Radiation Technol, 73-76; mgr nuclear prog, 76-80, VPRES, NUCLEAR, ENVIRONMENTAL & GEOTECH PROG, FORD, BACON & DAVIS, UTAH, 80- *Concurrent Pos:* US delegate working group uranium tailings, Orgn Econ Coop & Develop, Nuclear Energy Agency. *Mem:* Am Inst Physics; Am Soc Mining Engrs. *Res:* Semiconductor thermoelectric materials; engineering and environmental assessments of uranium mill tailings; research on radon diffusion; health impact of mill tailings; nuclear waste management; radiation measurement instrumentation; environmental radiation measurements. *Mailing Add:* 16206 San Dominique San Antonio TX 78232

OVERSETH, OLIVER ENOCH, b New York, NY, May 11, 28; div; c 2. EXPERIMENTAL HIGH ENERGY PHYSICS. *Educ:* Univ Chicago, BS, 53; Brown Univ, PhD(physics), 58. *Prof Exp:* Instr physics, Princeton Univ, 57-60; from asst prof to assoc prof, 61-67, PROF PHYSICS, UNIV MICH, ANN ARBOR, 68- *Concurrent Pos:* Sci assoc, CERN, 74-75 & 83- *Mem:* Am Phys Soc. *Res:* Elementary particles. *Mailing Add:* Dept of Physics Univ of Mich Ann Arbor MI 48109

OVERSKEI, DAVID ORVIN, b Brookings, SDak, Dec 26, 48; m 73. PLASMA PHYSICS. *Educ:* Univ Calif, Berkeley, AB, 71; Mass Inst Technol, PhD(physics), 76. *Prof Exp:* Oper mgr, Plasma Physics Ctr Prog, Francis Bitter Nat Magnet Lab Plasma Fusion Ctr, Mass Inst Technol, 76-81; TOKAMAK PHYSICS PROG MGR, FUSION DIV, GEN ATOMIC CO, 81- *Mem:* Am Phys Soc. *Res:* Controlled thermonuclear fusion research. *Mailing Add:* 10235 Rue Chamberry San Diego CA 92131-2239

OVERSTREET, ROBIN MILES, b Eugene, Ore, June 1, 39; m 64; c 2. MARINE PARASITOLOGY, AQUATIC PATHOLOGY. *Educ:* Univ Ore, BA, 63; Univ Miami, MS, 66, PhD(marine biol), 68. *Prof Exp:* Nat Inst Allergy & Infectious Dis fel parasitol, Med Sch, Tulane Univ, 68- 69; parasitologist, Gulf Coast Res Lab, 69-70; HEAD SECT PARASITOL, GULF COAST RES LAB, 70- *Concurrent Pos:* Adj prof biol, Univ Southern Miss, 70-, Univ Miss, 71-, La State Univ, 81-, Auburn Univ, 87- & Univ Ala, 90-; vis prof, Univ Queensland, 84. *Mem:* Am Soc Parasitol (vpres, 91); Am Fisheries Soc; Soc Invert Path; Am Micros Soc; World Aquaculture Soc. *Res:* Systematics and biology of marine and estuarine parasites; pathology of aquatic animals. *Mailing Add:* Gulf Coast Res Lab PO Box 7000 Ocean Springs MS 39564-7000

OVERTON, DONALD A, b Troy, NY, Dec 9, 35; m 58. STIMULUS PROPERTIES OF DRUGS, STATE DEPENDENT LEARNING. *Educ:* Rensselaer Polytech Inst, BS, 58; Harvard Univ, MA, 61; McGill Univ, PhD(physiol psychol), 62. *Prof Exp:* Instr, dept elec eng, Lafayette Col, 58-59; NSF postdoc fel, Ctr Brain Res, Univ Rochester, 62-64; Nat Inst Mental Health postdoc fel neurophysiol & psychiat, dept physiol & psychiat, Albert Einstein Col Med, 64-66; from asst prof to assoc prof, 66-77, PROF, DEPT PSYCHIAT, TEMPLE MED SCH, 77- *Honors & Awards:* Distinguished Sci Contrib Award, Soc Stimulus Properties Drugs, 86. *Mem:* Fel Am Psychol Asn; fel Am Asn Advan Sci; Soc Stimulus Properties Drugs (pres, 78-79); Sigma Xi; Behav Pharmacol Soc; Neurosci Soc; Soc Biol Psychiat. *Res:* Stimulus properties of drugs, especially as regards state dependent learning and drug discriminations; clinical significance underlying mechanisms and applications. *Mailing Add:* Dept Psychiat Temple Med Sch 3401 N Broad St Philadelphia PA 19140

OVERTON, EDWARD BEARDSLEE, b Mobile, Ala, Nov 21, 42. ANALYTICAL CHEMISTRY, INORGANIC CHEMISTRY. *Educ:* Univ Ala, Tuscaloosa, BS, 65, PhD(chem), 70. *Prof Exp:* Asst prof chem, Northeast La Univ, 70-76; PROJ MGR, CTR BIOORG STUDIES, UNIV NEW ORLEANS, 76- *Mem:* Am Chem Soc; AAAS. *Res:* Trace hydrocarbon analysis; data processing high resolution gas chromatographic and gas chromatography/mass spectrometry data. *Mailing Add:* 5536 Riverbend Blvd Baton Rouge LA 70820

OVERTON, JAMES RAY, b Greenville, NC, July 12, 36. POLYMER CHEMISTRY, PHYSICAL CHEMISTRY. *Educ:* E Carolina Univ, BA, 58; Univ SC, PhD(phys chem), 62. *Prof Exp:* Res chemist, 62-64, sr res chemist, 64-72, res assoc, 72-85, SR RES ASSOC, RES LABS, EASTMAN CHEM DIV, EASTMAN KODAK CO, 85- *Mem:* Am Chem Soc. *Res:* Polymer characterization; polymer synthesis. *Mailing Add:* 343 State St No NJ160 Rochester NY 14650-9999

OVERTON, JANE HARPER, b Chicago, Ill, Jan 17, 19; m 41; c 3. BIOLOGY. *Educ:* Bryn Mawr Col, AB, 41; Univ Chicago, PhD(zool), 50. *Prof Exp:* Asst prof natural sci, Col Univ Chicago, 53-60, assoc prof biol sci, col & dept zool, 60-64, assoc prof develop biol, col & dept biol, 64-72, prof develop & cell biol, 72-85, prof, 85-89, EMER PROF DEVELOP & CELL BIOL, DEPT MOLECULAR GENETICS & CELL BIOL, UNIV CHICAGO, 89- *Concurrent Pos:* Investr, Marine Biol Lab, Woods Hole, Mass, 61-62; vis scientist, Cancer Inst, Univ Heidelberg, WGermany, 81; ed assoc, J Morphol, Am Soc Zool. *Mem:* Am Soc Cell Biol (treas, 80-83); Soc Develop Biol; Int Soc Develop Biologists; Am Asn Anatomists; fel AAAS. *Res:* Factors underlying tissue organization; cell-cell and cell substrate contacts and cell junctions. *Mailing Add:* Dept Molecular Genetics & Cell Biol Univ Chicago 1103 E 57 St Chicago IL 60637

OVERTON, WILLIAM CALVIN, JR, b Dallas, Tex, Oct 26, 18; m 46; c 3. PHYSICS. *Educ:* NTex State Univ, BS, 41; Rice Univ, MA, 49, PhD(physics), 50. *Prof Exp:* Proj engr, Missile Guid & Control Systs Res, S W Marshall Co, 46-47; sr res physicist acoustic oil well logging theory, Field Res Labs, Magnolia Petrol Co, 50-51; physicist, US Naval Res Lab, 51-62; physicist, Los Alamos Nat Lab, Univ Calif, 62-87, consult superconductivity, 87-91; RETIRED. *Concurrent Pos:* Lectr, Univ Md Grad Exten Sch at Naval Res Lab, 52, 54-56 & 60. *Mem:* Fel Am Phys Soc; Sigma Xi. *Res:* Solid state and low temperature physics; lattice dynamics; thermodynamics. *Mailing Add:* 1297 47th St Los Alamos NM 87544

OVERTURF, GARY D, b Santa Ana, Calif, Nov 27, 43; m 65; c 1. INFECTIOUS DISEASES, PEDIATRICS. *Educ:* Western NMex Univ, BA, 65; Univ NMex, MD, 69. *Prof Exp:* Fel infectious dis, Univ Southern Calif, 72-74, asst prof pediat, 74-78; DIR, COMMUN DIS SERV, LOS ANGELES COUNTY-UNIV SOUTHERN CALIF MED CTR, 75-; ASSOC PROF PEDIAT, UNIV SOUTHERN CALIF, 78- *Mem:* Soc Pediat Res; Am Soc Microbiol; Am Acad Pediat; Infectious Dis Soc Am. *Res:* Medical microbiology, including antimicrobial susceptibility and mechanisms of action; vaccines and immune responses to infectious agents of humans. *Mailing Add:* Dept Pediat No 3A108 Olive View Med Ctr 14445 Olive View Dr Sylmar CA 91342

OVERTURF, MERRILL L, b Ottumwa, Iowa, Aug 31, 38; m 73. BIOCHEMISTRY. *Educ:* Univ Iowa, BS, 61, PhD(prev med), 70. *Prof Exp:* Instr med parasitol, Univ Iowa, 70-72; from asst prof to assoc prof, 72-82, PROF INTERNAL MED, UNIV TEX MED SCH, HOUSTON, 83- *Concurrent Pos:* Nat Heart & Lung Inst grant & Iowa Heart Asn grant, Univ Iowa, 71-72; Nat Heart & Lung Inst grant, Univ Tex, 72-91. *Mem:* Am Soc Pharmacol & Exp Therapeut; Am Heart Asn; Am Oil Chemists Soc; Am Fedn Clin Res. *Res:* Biochemistry of renovascular hypertension; renin-angiotensin-aldosterone system; atherosclerosis. *Mailing Add:* Dept Internal Med Univ Tex Health Sci Ctr PO Box 20708 Houston TX 77030

OVERWEG, NORBERT I A, b Enschede, Neth; m 59; c 3. HYPERTENSION, NUTRITION. *Educ:* Univ Amsterdam, Med Cand, 54, MD, 57. *Prof Exp:* Rotating intern, Hosp Univ Amsterdam, 58-60; resident, Rochester Gen Hosp, 61-62; fel, dept pharmacol, Col Physicians & Surgeons, Columbia Univ, 62-65, instr, dept pub health, 65-66, res assoc, dept surg, 67-71; asst prof, dept physiol & pharmacol, NY Univ, 71-78. *Concurrent Pos:* Pvt pract, Neth, 60-61 & US, 77-; spec fel, Nat Inst Child Health & Human Develop, NIH, Bethesda, 64-65; asst vis physician, dept med, Harlem Hosp Ctr, NY, 65-66; res collabr & asst attend physician, Brookhaven Nat Lab, 66-67; consult, Lung Res Ctr, Sch Med, Yale Univ, 72-73; prin & co-investr clin pharmacol, 84- *Mem:* Am Physiol Soc; Am Soc Pharmacol & Exp Therapeut; NY Acad Sci; Harvey Soc; Sigma Xi; Am Soc Hypertension. *Res:* Action of neuropeptides, prostaglandins and cyclic AMP on membrane transfer and transport; action of neuropeptides, prostaglandins, catecholamines, beta blockers, calcium antagonists and cyclic nucleotides on uterine, intestinal, vascular and bronchial smooth muscle; regulation of intestinal motility; clinical investigation of new drugs: antihypertension, antidepressant and gastrointestinal drugs. *Mailing Add:* 3777 Independence Ave Riverdale NY 10463

OVERZET, LAWRENCE JOHN, b Oak Park, Ill, Mar 27, 61; m 83; c 3. GASEOUS ELECTRONICS, SEMICONDUCTOR PROCESSING USING PLASMA. *Educ:* Univ Ill, BSEE, 83, MSEE, 85, PhD(elec eng), 88. *Prof Exp:* ASST PROF ELEC ENG, UNIV TEX, DALLAS, 88- *Mem:* Inst Elec & Electronic Engrs; Am Vacuum Soc. *Res:* New discharge diagnostics; glow discharge physics; investigations of particulate generation; negative ion kinetics; plasma processing. *Mailing Add:* Elec Eng Dept BE28 Univ Tex PO Box 830688 Richardson TX 75083-0688

OVIATT, CANDACE ANN, b Bennington, Vt, June 17, 39; div. ECOLOGY. *Educ:* Bates Col, BS, 61; Univ RI, PhD(biol oceanog), 67. *Prof Exp:* Res assoc, Sch Pub Health, Harvard Univ, 65-69; RES PROF & ASSOC DIR, MARINE ECOSYSTS RES LAB, GRAD SCH OCEANOG, UNIV RI, 69- *Mem:* Am Soc Limnol & Oceanog; AAAS. *Res:* Total systems studies of marine environments; the role of different populations in marine systems; the effects of stress and its role in stability; budgets for marine systems; practical application of the above for understanding and/or management. *Mailing Add:* Sch Oceanog Narragansett Marine Lab Univ RI Kingston RI 02881

OVIATT, CHARLES DIXON, b Washington, Pa, Dec 24, 24; m 48; c 5. APPLIED STATISTICS. *Educ:* Tarkio Col, AB, 47; Ohio State Univ, MS, 49, PhD, 54. *Prof Exp:* Prof chem, Tarkio Col, 49-57; staff ed phys sci, McGraw Hill Encyclop of Sci & Technol, 57-60; res chemist, E I du Pont de Nemours & Co, 60-61; sr sci ed, Webster Div, McGraw Hill Book Co, 61-69, exec ed, 69-71, ed-in-chief, 71-72; DIR STATEWIDE ASSESSMENT PROJ, MO STATE DEPT ELEM & SEC EDUC, 74- *Mem:* Am Chem Soc. *Res:* Technical editing; educational publishing, author; educational research. *Mailing Add:* Rte 1 Box 62 Vienna MO 65582-9407

OVSHINSKY, IRIS M(IROY), b New York, NY, July 13, 27; m 59; c 5. SEMICONDUCTORS, BIOCHEMISTRY. *Educ:* Swarthmore Col, BA, 48; Univ Mich, MS, 50; Boston Univ, PhD(biol), 60. *Prof Exp:* Res assoc biochem, Worcester Found Exp Biol, 56-59; VPRES, ENERGY CONVERSION DEVICES, INC, 60- *Mem:* Sigma Xi. *Res:* Amorphous and disordered semiconductors; structural characteristics of information bearing material. *Mailing Add:* Energy Conversion Devices Inc 1675 W Maple Rd Troy MI 48084

OVSHINSKY, STANFORD R(OBERT), b Akron, Ohio, Nov 24, 22; m 59; c 5. SOLID STATE SEMICONDUCTORS, STRUCTURAL CHEMISTRY. *Hon Degrees:* DSci, Lawrence Inst Technol, 80, DEng, Bowling Green State Univ, 81. *Prof Exp:* Pres, Stanford-Roberts Mfg, 46-50; mgr, Ctr Drive Dept, New Brit Mach Co, 50-52; dir res, Hupp Corp, 52-55; pres, Gen Automation, Inc, 55-58 & Ovitron Corp, 58-59; chmn bd, 60-78, PRES, ENERGY CONVERSION DEVICES, INC, 60- *Concurrent Pos:* Adj prof eng sci, Col Eng, Wayne State Univ, 78-; mem bd govs, Cranbrook Inst Sci, 81-; chmn, Inst Amorphous Studies, 82-; hon adv sci & technol, Beijing Inst Aeronaut & Astronaut, China, 84-; mem bd, Metrop Ctr High Technol, 85-; adj prof physics, Univ Cincinnati, 85- *Honors & Awards:* Diesel Gold Medal, 68. *Mem:* Electrochem Soc; Soc Automotive Engrs; fel Am Phys Soc; sr mem Inst Elect & Electronics Engrs; Am Solar Energy Soc; Am Vacuum Soc. *Res:* Amorphous and disordered semiconductors; structural characteristics of information bearing material; high temperature superconductivity. *Mailing Add:* Energy Conversion Devices Inc 1675 W Maple Rd Troy MI 48084

OVUNC, BULENT AHMET, b Samsun, Turkey, Oct 14, 27; m 57; c 1. CIVIL ENGINEERING, COMPUTER SCIENCE. *Educ:* Istanbul Tech Univ, MSc, 54, PhD(elasticity), 63. *Prof Exp:* Asst prof civil eng, Istanbul Tech Univ, 54-65; acad res asst, Univ BC, 65-67; assoc prof, 67-70, prof civil eng, 70-81, M EROI GIRARD PROF CIVIL ENG, UNIV SOUTHWESTERN LA, 81- *Concurrent Pos:* Designer struct, 55-65; Orgn Econ Coop & Develop fel, Istanbul Tech Univ, 63-65; consult comput prog to eng firms, 65-68. *Mem:* Am Soc Civil Engrs; Am Soc Mech Engrs; Am Concrete Inst; Int Asn Bridge & Struct Engrs; Sigma Xi. *Res:* Structures; dynamics of structures; finite element methods; plastic design; elasticity and related computer programmings; plates and shells; soil-structure; boundary element method; general purpose computer code STDYNL. *Mailing Add:* Dept Civil Eng Univ Southwestern La PO Box 4-0172 Lafayette LA 70504

OWADES, JOSEPH LAWRENCE, b New York, NY, July 9, 19; m 69. BIOCHEMISTRY. *Educ:* City Col New York, BS, 39; Polytech Inst Brooklyn, MS, 44, PhD(chem), 50. *Prof Exp:* Chemist, Naval Supply Dept, 40-49 & Fleischmann Labs, 49-53; asst chief chemist & brewing & food consult, Schwarz Labs, 53-56, chief chemist, 56-58, dir labs & brewing res, 59-60; dir res & tech serv, Liebman Breweries, Inc, 60-65; vpres & tech dir, Rheingold Breweries, Inc, 65-68; consult, Karolos Fix Co, Athens, Greece, 69-70; coordr tech ctr, Anheuser-Busch, Inc, 70-71; DIR, CTR BREWING STUDIES, 77- *Concurrent Pos:* Consut brewing indust, 75- *Mem:* Am Chem Soc; Am Soc Brewing Chemists; Master Brewers Asn Am; Inst Food Technologists; NY Acad Sci. *Res:* Yeast derivatives; sterols; brewing; foods. *Mailing Add:* 3097 Wood Valley Rd Sonoma CA 95476

OWCZAREK, J(ERZY) A(NTONI), b Piotrkow, Poland, Nov 2, 26; nat US; m 59; c 5. MECHANICAL ENGINEERING. *Educ:* Polish Univ Col Eng, Dipl Ing, 50; Univ London, PhD, 54. *Prof Exp:* Asst lectr thermodyn, Polish Univ, Col, 50-52; lectr thermodyn & mech, Battersea Col Technol, 52-53, thermodyn & fluid mech, Queen Mary Col, Univ London, 53-54; develop engr, Gen Elec Co, 55-60; assoc prof, 60-65, PROF MECH ENG, LEHIGH UNIV, 65- *Concurrent Pos:* Consult, IMO De Laval, Inc, 67-, Bell Labs, 74- *Mem:* Am Soc Mech Engrs; Am Inst Aeronaut & Astronaut; Sigma Xi. *Res:* Gas dynamics; turbine fluid mechanics; thermodynamics. *Mailing Add:* Dept Mech Eng & Mech Lehigh Univ Bethlehem PA 18015

OWCZARSKI, WILLIAM A(NTHONY), b Adams, Mass, June 13, 34; m 58; c 3. PHYSICAL & PROCESS METALLURGY. *Educ:* Univ Mass, BS, 55; Rensselaer Polytech Inst, MS, 58, PhD(phys metall), 62. *Prof Exp:* Welding engr, Sprague Elec Co, 55-57; metallurgist, Knolls Atomic Power Lab, Gen Elec Co, 58-61; sr res assoc metals joining, Adv Mat Res & Develop Lab, Pratt & Whitney Aircraft Div, United Aircraft Corp, 62-69, res supvr alloy processing, 69-71, tech supvr metal joining, 71-72, asst mgr, Mat Eng & Res Lab, Com Prod Div, 72-80, mgr tech planning, Pratt & Whitney Aircraft Group, 80-87, IRI fel & sr policy analyst, White House Off Sci & Technol Policy, 88-89, DIR TECHNOL DEVELOP, WASH OFF, UNITED TECHNOL CORP, 89- *Honors & Awards:* William Spraragen Award, Am Welding Soc, 66, 69 & 72, Adams Mem Lectr, 73; George Mead Gold Medal, United Technol Corp, 74. *Mem:* fel Am Soc Metals. *Res:* Welding metallurgy; high temperature nickel alloys; manufacturing technology; technology management; science and technology policy. *Mailing Add:* 2681 Marcey Rd Arlington VA 22207

OWCZARZAK, ALFRED, b New York, NY, Jan 21, 23; m 50; c 3. CELL BIOLOGY, BIOLOGICAL STRUCTURE. *Educ:* Cornell Univ, BS, 44; Univ Wis, PhD(bot), 53. *Prof Exp:* Proj asst bot, Univ Wis, 48-49, asst, 49-53, proj assoc, 53-54, proj assoc path, Serv Mem Insts, 54-55; res assoc zool, 55-58, from instr to assoc prof, 55-83, EMER ASSOC PROF ZOOL, ORE STATE UNIV, 83- *Mem:* AAAS; Am Soc Cell Biol. *Res:* Cytological effects of chemical treatments; tissue culture; microtechnique; cytochemistry; comparative vertebrate histology; cell ultrastructure. *Mailing Add:* Dept of Zool Ore State Univ Corvallis OR 97331

OWELLEN, RICHARD JOHN, oncology, medicine; deceased, see previous edition for last biography

OWEN, ALICE KONING, b Kalamazoo, Mich, Jan 20, 30; m 55; c 2. EMBRYOLOGY. *Educ:* Kalamazoo Col, BA, 51; Iowa State Univ, MS, 53, PhD(embryol), 55. *Prof Exp:* Asst prof anat & physiol, Northeast Mo State Univ, 63-65; ASST PROF BIOL & PHYSIOL, UNIV NDAK, 66- *Mem:* AAAS; Am Soc Zool; Am Inst Biol Sci. *Res:* Development of integument in chicks; effect of vitamin E on development in rats. *Mailing Add:* 1118 Reeves Dr Grand Forks ND 58201

OWEN, BERNARD LAWTON, b Presidio, Tex, Nov 10, 29; m 52; c 3. INVERTEBRATE ZOOLOGY, ENTOMOLOGY. *Educ:* Agr & Mech Col, Tex, BS, 51, MS, 54; Auburn Univ, PhD(entom), 59. *Prof Exp:* Entomologist, Agr Res Serv, USDA, 58-59; consult, 59; from asst prof to assoc prof, 59-66, chmn, natural sci & math div, 69-79, 82-85, PROF ZOOL, KANS WESLEYAN UNIV, 66- *Mem:* AAAS; Nat Asn Sci Teachers; Nat Asn Biol Teachers; Soc Col Sci Teachers; Human Anat & Physiol Soc. *Res:* Environmental physiology of arthropods. *Mailing Add:* Dept Biol Kans Wesleyan Univ Salina KS 67401

OWEN, BRUCE DOUGLAS, b Edmonton, Alta, Oct 1, 27; m 51; c 2. ANIMAL NUTRITION, ANIMAL PHYSIOLOGY. *Educ:* Univ Alta, BSc, 50, MSc, 52; Univ Sask, PhD(animal nutrit), 61. *Prof Exp:* Animal nutritionist, Lederle Labs Div, Am Cyanamid Co, NY, 52-54; animal husbandman, Can Dept Agr, 54-57; from lectr to prof animal sci, Univ Sask, 57-77; PROF ANIMAL SCI & CHMN DEPT, UNIV BC, 77- *Mem:* Agr Inst Can; Am Soc Animal Sci; Can Soc Animal Sci; Nutrit Soc Can. *Res:* Passive immunity; fat soluble vitamin transport; vitamin requirements of swine; protein quality evaluation; swine nutrition and physiology, factors influencing composition of feed grains and forages. *Mailing Add:* Dept Animal Sci Univ BC 2357 Main Mall Vancouver BC V6T 1W5 Can

OWEN, CHARLES ARCHIBALD, JR, b Assiut, Egypt, Dec 3, 15; US citizen; m 39; c 3. BIOCHEMISTRY. *Educ:* Monmouth Col, AB, 36, Univ Iowa, MD, 41; Univ Minn, PhD, 50. *Hon Degrees:* DSc, Monmouth Col, 58. *Prof Exp:* Asst path, Univ Iowa, 38-39; from instr to assoc prof clin path, Mayo Found, Univ Minn, 50-60, prof med res, 60-81; RETIRED. *Concurrent Pos:* Consult, Mayo Clin, 50-82, head sect exp biochem, 59-82, head sect clin path, 61-70. *Mem:* AAAS; Am Soc Clin Path; Am Soc Exp Path; Soc Exp Biol & Med; Am Physiol Soc. *Res:* Blood coagulation; biologic aspects of radioisotopes. *Mailing Add:* Mayo Clinic Rochester MN 55905

OWEN, CHARLES SCOTT, b Springfield, Mo, July 4, 42. LYMPHOCYTE ACTIVATION, INTRACELLULAR CALCIUM. *Educ:* Univ Rochester, BS, 64; Univ Pa, PhD(physics), 70. *Prof Exp:* Res assoc physics, Wash Univ, 69-71 & Univ Calif, Santa Barbara, 71-72; res assoc biophysics, 72-77, asst prof path & biochem-biophysics, Univ Pa, 77-81; ASSOC PROF BIOCHEM, JEFFERSON MED COL, 81- *Mem:* Biophys Soc; Fedn Am Soc Exp Biol; AAAS; Am Asn Immunologists. *Res:* Magnetic sorting of cells; physical properties of immunoglobulins and protein-membrane interactions; intracellular calcium measurements; fluorescence spectroscopy. *Mailing Add:* 512 Harvard Ave Swarthmore PA 19081

OWEN, DAVID A, b New York, NY, July 13, 37; c 3. THEORETICAL PHYSICS. *Educ:* Univ Pa, MS, 61; Johns Hopkins Univ, PhD(physics), 70. *Prof Exp:* Physicist appl magnetohydrodyn, Naval Air Develop Ctr, Johnsville, Pa, 61-64; res fel physics, Univ Surrey, 70-72; res assoc, Ctr Theoret Physics, CNRS, Marseille, France, 72-73; vis prof, Racah Inst, Hebrew Univ, Jerusalem, 73-74; sr lectr physics & res, 74-86, ASSOC PROF, BEN GURION UNIV, ISRAEL, 86- *Concurrent Pos:* Gilman fel, Johns Hopkins Univ, 64-67, NASA fel, 67-69, fel, 69-70; res affil, Physics Dept, Yale Univ, 87-88. *Mem:* Am Phys Soc; Sigma Xi. *Res:* Quantum field theory and the bound states occurring in such theories; compton scattering from bound electron; hyperfine structure and lamb shift of positronium and muonium, muonic atoms, gauge theories and Kaluza-Klein theory; Qed phase transitions. *Mailing Add:* Dept Physics Ben Gurion Univ Beer Sheva Israel

OWEN, DAVID ALLAN, b Coronado, Calif, Mar 19, 43; m 77; c 2. INORGANIC CHEMISTRY, ORGANIC CHEMISTRY. *Educ:* Univ Ill, Urbana, BS, 65; Univ Calif, Riverside, PhD(inorg chem), 70. *Prof Exp:* Res chemist & proj leader, Union Carbide Res Inst, Tarrytown, NY, 70-73; res assoc & lectr, Southern Ill Univ, Carbondale, 73-75; vis asst prof chem, Univ Okla, Norman, 75-78; ASSOC PROF ORGANOMETALLIC CHEM, MURRAY STATE UNIV, 78- *Concurrent Pos:* Instr, Shawnee Community Col, Ullin, Ill, 75; consult, micronutrient balance, Compost Corp, Canyon, Tex, 76-78 & Huttrell Corp, Murray, Ky, 79; reviewer chem texts, Holt Rinehart Winston Publ, CBS Inc, NY, 77-78; Col Arts & Sci Res grant, Univ Okla, 77-78; Am Soc Eng Educ vis fac fel, L B Johnson Space Ctr, NASA, 80-81; NSF-ISEP grants, 81 & 85. *Mem:* Am Chem Soc; AAAS; Sigma Xi. *Res:* All aspects of the synthesis, characterization, structural elucidation, electrochemistry and catalytic activity of novel compounds of the transition metals, boron and carbon. *Mailing Add:* Blackburn Sci Hall Murray State Univ Murray KY 42071

OWEN, DAVID R, b San Luis Obispo, Calif, Apr 14, 42; m 68. APPLIED MATHEMATICS. *Educ:* Calif Inst Technol, BSc, 63; Brown Univ, PhD(appl math), 68. *Prof Exp:* From asst prof to assoc prof, 67-77, PROF MATH, CARNEGIE-MELLON UNIV, 77- *Honors & Awards:* Von Humboldt Found Sr US Scientist Award, 74; Bateman Lectr, 79. *Mem:* Math Asn Am; Am Math Soc. *Res:* Continuum thermodynamics; rate-independent materials; materials with elastic range. *Mailing Add:* Carnegie Mellon Univ Pittsburgh PA 15213

OWEN, DONALD BRUCE, b Portland, Ore, Jan 24, 22; m 52; c 4. APPLIED STATISTICS. *Educ:* Univ Wash, BS, 45, MS, 46, PhD(math statist), 51. *Prof Exp:* Assoc math, Univ Wash, 46-50, res assoc statist, 50-51, instr math, 51-52; asst prof math & consult statist lab, Purdue Univ, 52-54; staff mem math statist, Sandia Corp, 54-57, supvr, Statist Res Div, 57-64; head math & stochastic systs div, Grad Res Ctr & adj prof statist, Southern Methodist Univ, 64-66, prof statist, 66-83, dir, Ctr Statist Consult & Res, 87-90, UNIV DISTINGUISHED PROF STATIST SCI, SOUTHERN METHODIST UNIV, 83- *Concurrent Pos:* Consult, Sandia Corp, 64-67; prof, Univ Tex, 65-66; dir, Prog Vis Lectr in Statist, 70-73; coord ed, Marcel Dekker Statist Ser, 70-; ed, Commun in Statist Theory & Methods & Simulation & Computation. *Honors & Awards:* Don Owen Award, Am Statist Asn. *Mem:* Fel AAAS; fel Am Statist Asn; fel Am Soc Qual Control; Am Math Soc; Sigma Xi; fel Inst Math Statist. *Res:* Tabulations of statistical functions; bivariate normal probability distribution and associated integrals; applications of statistics to physics and engineering; double sample procedures; distributions of estimators of process capability indices. *Mailing Add:* Dept Statist Sci Southern Methodist Univ Dallas TX 75275-0332

OWEN, DONALD EDWARD, b San Antonio, Tex, Nov 29, 36; m 81; c 3. STRATIGRAPHY, GEOLOGY. *Educ:* Lamar Univ, BS, 57; Univ Kans, MS, 59, PhD(geol), 63. *Prof Exp:* Res scientist, Bur Econ Geol, Univ Tex, 62-64; from instr to prof geol, Bowling Green State Univ, 64-78, dir geol field camp, 64-68 & 78, NSF res grant, 65-66; geol assoc, 78-79, tech asst to vpres, Eastern Int Area, 79-81, RES ASSOC, CITIES SERV CO, 81- *Concurrent Pos:* Adj asst prof, State Univ NY Binghamton, 65 & 66; sr lectr sch earth sci, Macquarie Univ, Australia, 69-71; consult geologist, Planet Mgt & Res Proprietary Ltd, Australia, 69-71; NMex Bur Mines & Mineral Resources res grants, 74-77; assoc ed, Soc Econ Paleontologists & Mineralogists, 74-; mem, NAm Comn Stratigraphic Nomenclature, 78- *Honors & Awards:* A I Levorsen Mem Award, Am Asn Petrol Geologists, 75. *Mem:* Am Asn Petrol Geologists; Soc Econ Paleontologists & Mineralogists; Geol Soc Australia; Int Asn Sedimentologists. *Res:* Sedimentology; Cretaceous stratigraphy of southwestern United States; Dakota sandstone of San Juan Basin, New Mexico; modern sediments of coastal North Carolina; Devonian of central New South Wales, Australia. *Mailing Add:* Dept Geol Lamar Univ Lamar Univ Sta Beaumont TX 77710

OWEN, DONALD EUGENE, b Emporia, Kans, Oct 20, 28; m 55; c 2. GEOLOGY, PETROLEUM GEOLOGY. *Educ:* Univ Kans, BS, 49; Univ Tex, MA, 51; Univ Wis, PhD(geol), 64. *Prof Exp:* Geophysicist, Calif Co, 51-56, area geologist, Mont, 56-58, dist staff geologist, 58-60; asst prof, 63-66, assoc prof, 66-79, PROF GEOL, IND STATE UNIV, 79- *Mem:* AAAS; Geol Soc Am; Am Asn Petrol Geol; Nat Asn Geol Teachers. *Res:* Cretaceous stratigraphy; paleogeologic and onlap maps of North America; Williston Basin stratigraphy and structure. *Mailing Add:* Dept Geog & Geol Ind State Univ 217 N 6th St Terre Haute IN 47809

OWEN, FLETCHER BAILEY, JR, b Richmond, Va, Nov 6, 29. PHARMACOLOGY. *Educ:* Univ Richmond, BS, 51; Med Col Va, PhD(pharmacol), 56, MD, 59. *Prof Exp:* Intern, Richmond Mem Hosp, 59-60; physician, Med Dept, 60-63, dir, Med Serv, 63-83, ASST VPRES, MED SERV, A H ROBINS CO, INC, 83- *Concurrent Pos:* Instr, Med Col Va, 60-64, lectr, 64-68. *Mem:* AMA; NY Acad Sci; AAAS. *Res:* Clincial pharmacology; drugs. *Mailing Add:* PO Box 1444 Richmond VA 23220

OWEN, FOSTER GAMBLE, b Flat Creek, Ala, Jan 24, 26; m 50; c 4. DAIRY NUTRITION, FORAGE QUALITY. *Educ:* Auburn Univ, BS, 48; Iowa State Univ, MS, 55, PhD, 56. *Prof Exp:* Asst dairy husb, Iowa State Univ, 53-55; asst prof animal husb, Univ Ark, 55-57; assoc prof dairy husb, 57-64, PROF ANIMAL SCI, UNIV NEBR, LINCOLN, 64- *Mem:* Am Soc Animal Sci; Am Dairy Sci Asn; Am Forage & Grassland Coun; Dairy Shrine. *Res:* Dairy cattle nutrition and feeding methods, calf nutrition and feeding programs, forage preservation and nutritional value of by-products. *Mailing Add:* Dept Animal Sci Univ Nebr Lincoln NE 68583-0908

OWEN, FRANCES LEE, immunology, cell biology; deceased, see previous edition for last biography

OWEN, FRAZER NELSON, b Atlanta, Ga, Feb 4, 47. ASTRONOMY. *Educ:* Duke Univ, BS, 69; Univ Tex, PhD(astron), 74. *Prof Exp:* Res assoc astron, 73-75, asst scientist, 75-77, assoc scientist, 77-79, SCIENTIST, NAT RADIO ASTRON OBSERV, 79- *Mem:* Am Astron Soc; Royal Astron Soc; Int Astron Union. *Res:* Extragalactic radio astronomy; clusters of galaxies; radio galaxies; quasars; radio flare stars. *Mailing Add:* Nat Radio Astron Observ 1000 Bullock Ave PO Box 0 Socorro NM 87801

OWEN, GENE SCOTT, b Roswell, NMex, Oct 25, 43; m 77. BIOPHYSICAL CHEMISTRY, COMPUTER SCIENCE. *Educ:* Harvey Mudd Col, BS, 65; Univ Wash, PhD(chem), 70. *Prof Exp:* Fel chem, Ga Inst Technol, 70-72; from asst prof to assoc prof chem, Atlanta Univ, 77-84; assoc prof, 84-89, PROF COMPUT SCI, GA STATE UNIV, 89- *Concurrent Pos:* Chmn, Asst Chief Staff Task Force Comput Chem Educ; chair, Am Comput Mach S/GGRAPH Educ Comt. *Mem:* Am Chem Soc; Inst Elec & Electronic Engrs; Asn Comput Mach. *Res:* Computers in chemical research and education; computer graphics; programming languages. *Mailing Add:* Dept Math & Comput Sci Ga State Univ Atlanta GA 30303

OWEN, GEORGE MURDOCK, b Alliance, Ohio, Feb 16, 30; m 76; c 4. PEDIATRICS, NUTRITION. *Educ:* Hiram Col, AB, 51; Univ Cincinnati, MD, 55. *Prof Exp:* Rotating intern, Univ Iowa, 55-56, resident pediat, 56-58; clin assoc, Nat Inst Allergy & Infectious Dis, 58-60; from asst prof to assoc prof, Univ Iowa, 66; from assoc prof to prof pediat, Col Med, Ohio State Univ, 66-73; prof pediat & dir, clin nutrit prog, Sch Med, Univ NMex, 73-77; prof & dir, human nutrit prog & prof pediat & fel, Ctr Human Growth & Develop, Univ Mich, 77-81; med dir int div, Bristol-Myers Co, 81-90, INT MEAD JOHNSON NUTRIT GROUP, BRISTOL-MYERS SQUIBB CO, 90- *Concurrent Pos:* Res fel, Metab Sect, Univ Iowa, 60-62. *Mem:* Am Acad Pediat; Am Soc Clin Nutrit; Soc Pediat Res; Am Inst Nutrit; Am Pediat Soc. *Res:* Nutrition, growth and body composition of normal infants; nutrition status. *Mailing Add:* Bristol-Myers Squibb Co 2400 W Lloyd Expressway Evansville IN 47721-0001

OWEN, GWILYM EMYR, JR, b Springfield, Ohio, May 5, 33; m 56; c 3. FREE RADICAL CHEMISTRY, ORGANIC PHOTOCHEMISTRY. *Educ:* Antioch Col, BS, 56; State Univ NY, Syracuse, PhD(polymer chem), 67. *Prof Exp:* Asst prof chem, Denison Univ, 65-72; res assoc, Owens-Corning Fiberglas, 72-86; DIR RES & DEVELOP, REICHHOLD CHEM, 86- *Concurrent Pos:* Postdoctoral fel, Ohio State Univ, 64-65. *Mem:* Am Chem Soc; AAAS; Soc Plastics Engrs. *Mailing Add:* PO Box 19129 Jacksonville FL 32245

OWEN, HARRY A(SHTON), JR, b Hawthorn, Fla, Sept 22, 19; m 42; c 2. ELECTRICAL ENGINEERING. *Educ:* Univ Fla, BEE, 48, MSE, 52; NC State Univ, PhD, 63. *Prof Exp:* Instr elec eng, Univ Fla, 48-51; from instr to assoc prof, Duke Univ, 51-64, prof, 64-86, chmn dept, 68-74, EMER PROF ELEC ENG, DUKE UNIV, 86- *Concurrent Pos:* NSF fac fel, 58; consult, Ed Dept, Western Elec Co, 58; NASA sr res assoc, Goddard Space Flight Ctr, 67-68; sr res fel, Europ Space Agency, Noordwijk, The Neth, 75-76. *Mem:* Inst Elec & Electronic Engrs. *Res:* Electronic instrumentation; electronic power conditioning circuits; computer aided design and applications. *Mailing Add:* Elec Eng/16L Eng Duke Univ Durham NC 27706

OWEN, JAMES EMMET, b Cleveland, Ohio, May 23, 35; m 61; c 8. INORGANIC CHEMISTRY. *Educ:* John Carroll Univ, BS, 57; Case Inst Technol, PhD(chem), 61, Case Western Reserve Univ, MBA, 84. *Prof Exp:* Res scientist, US Rubber Co, 61-63; group leader, Harshaw Chem Co, 63-66, dir inorg phys res, 66-67, dir cent res, 67-73, tech dir, Dyestuff Div, 73-76, dir color technol, Harshaw/Filtrol Partnership, 76-85; vpres & gen mgr, Obron Corp, 85-88; pres, Owen & Assocs, 88-90; PRES, AARBOR INT CORP, 91- *Mem:* AAAS; Am Chem Soc; Am Inst Chemists; Sigma Xi. *Res:* Fundamental studies of inorganic pigment systems; pigments based on non-stoichiometric metal oxides; pigments based on II-VI type semiconductors; investigation of adsorption phenomena; structural characteristics of adsorbents as related to adsorption performance. *Mailing Add:* 3002 Manchester Rd Shaker Heights OH 44120

OWEN, JAMES ROBERT, b Leake Co, Miss, Mar 26, 14; m 37. PHYSICAL CHEMISTRY, INORGANIC CHEMISTRY. *Educ:* Okla Agr & Mech Col, BS, 35; Univ Wis, PhM, 37. *Prof Exp:* Asst, Exp Sta, Okla Agr & Mech Col, 33-35; sect chief, Phillips Petrol Co, 37-58, tech supt, Ambrosia Mill, NMex, 58-62, develop engr, Patent Div, 62-69; CONSULT HETEROGENEOUS CATALYSIS, 69- *Mem:* Am Chem Soc. *Res:* Heterogeneous catalysis; technical on-line searching (dialog, etc; 13 US patents. *Mailing Add:* 2413 Summit Rd Bartlesville OK 74006-6342

OWEN, JOEL, b Boston, Mass, Feb 22, 35. STATISTICS, OPERATIONS RESEARCH. *Educ:* Yeshiva Univ, BA, 56; Boston Univ, MA, 57; Harvard Univ, PhD(statist), 66. *Prof Exp:* Res scientist appl math & statist, Waltham Lab, Sylvania Electronic Systs Div, 57-61, adv res scientist math statist, Appl Res Lab, 61-64; sr mathematician, Info Res Asn, Inc, 65-66; asst prof statist & opers res, 66-70, assoc prof statist, 70-73, PROF STATIST, GRAD SCH BUS, NY UNIV, 73-, CHMN DEPT QUANT ANALYSIS, 76- *Res:* Pattern recognition; discrimination analysis; game and decision theory; mathematical formulation of financial accounting theory and portfolio theory. *Mailing Add:* Dept Statist NY Univ Washington Sq New York NY 10003

OWEN, JOHN ATKINSON, JR, b South Boston, Va, Sept 24, 24; m 52; c 2. INTERNAL MEDICINE, METABOLISM. *Educ:* Hampden-Sydney Col, BS, 44; Univ Va, MD, 48. *Prof Exp:* Intern, Cincinnati Gen Hosp, Ohio, 48-49; asst resident internal med, Univ Va Hosp, 50-51; asst prof med, Med Col Ga, 56-58; asst clin prof internal med, Sch Med, George Washington Univ, 58-60; from asst prof to assoc prof internal med, 60-70, dir div clin pharmacol, 64-71, vchmn dept internal med, 72-74, PROF INTERNAL MED, SCH MED, UNIV VA, 70- *Concurrent Pos:* Fel internal med, Univ Va Hosp, 51-52; res fel metab, Duke Univ Hosp, 54-56; Am Diabetes Asn res fel, 55-56; pres, US Pharmacopeial Conv, 75-80; ed-in-chief, Hosp Formulary, 74-83; vchmn, Nat Coun Drugs, 79-80. *Honors & Awards:* Horsley Res Prize, 61. *Mem:* Am Fedn Clin Res; Endocrine Soc; Am Diabetes Asn; AMA; fel Am Col Clin Pharmacol. *Res:* Experimental and clinical diabetes; diabetogenic and anti-diabetic drugs; clinical trials and animal research. *Mailing Add:* Dept Internal Med Univ Va Hosp Box 242 Charlottesville VA 22908

OWEN, JOHN HARDING, b San Francisco, Calif, Dec 22, 31; m 51; c 3. INORGANIC CHEMISTRY, PHYSICAL CHEMISTRY. *Educ:* Hampden-Sydney Col, BS, 53; Fla State Univ, MS, 55. *Prof Exp:* Res chemist, sr res supvr, 68-76, chief supvvr, 76-78, site coordr safeguards, 79-84, PROCESS ASSOC, SAVANNAH RIVER LAB, E I DU PONT DE NEMOURS & CO, INC, 84- *Concurrent Pos:* Instr, Augusta Col, 59-63 & Univ SC, 63-75; traveling lectr, Oak Ridge Inst Nuclear Studies, 63-66. *Res:* Theory and application of diffusion processes in solids; science administration; technical management. *Mailing Add:* 1319 Williams Dr Aiken SC 29801

OWEN, OLIVER ELON, b Roswell, NMex, Dec 24, 34; c 3. MEDICINE, METABOLISM. *Educ:* Cent State Col, Okla, BS, 58; Univ Colo, MD, 62. *Prof Exp:* Asst med, Peter Bent Brigham Hosp, Boston, Mass, 65-68; from asst prof to assoc prof, 68-75, PROF MED, TEMPLE UNIV HOSP, 75-, PROG DIR, GEN CLIN RES CTR, 68- *Concurrent Pos:* Independent investr, Fels Res Inst, Temple Univ Hosp, 68- *Honors & Awards:* George Morris Piersol Award, Am Col Physicians, 68. *Mem:* Am Fedn Clin Res; Am Diabetes Asn; Am Soc Clin Invest; Am Col Physicians; Asn Am Physicians. *Res:* Diabetes. *Mailing Add:* Gen Clin Res Ctr 2MW Temple Univ Hosp Philadelphia PA 19140

OWEN, PAUL HOWARD, b South Gate, Calif, June 6, 44; m 70; c 1. INSTRUMENTATION. *Educ:* Calif State Univ, Long Beach, BA, 67, MA, 72. *Prof Exp:* Res asst toxicol, Calif State Univ, Long Beach, 67-71; res engr, Environics Inc, 71-76; lab mgr, USD Corp, 76-87; ENG SCIENTIST/SPECIALIST, DOUGLAS AIRCRAFT CO, 87- *Mem:* Instrument Soc Am. *Res:* Pesticide metabolism in food chains; electrochemical carbon regeneration in waste water treatment; catalytic reduction of oxides of nitrogen in power plant stack gas; gasoline vapor control systems; underwater diving apparatus; safety equipment for hazardous atmospheres; aircraft structural test methods. *Mailing Add:* 1934 W Beverly Dr Orange CA 92668-2021

OWEN, RAY BUCKLIN, JR, b Providence, RI, July 14, 37; m 62; c 2. WILDLIFE ECOLOGY. *Educ:* Bowdoin Col, AB, 59; Univ Ill, MS, 66, PhD(ecol), 68. *Prof Exp:* Instr biol, Univ Ill, 67-68; from asst prof to assoc prof, 68-77, PROF WILDLIFE RESOURCES, UNIV MAINE, ORONO, 77- *Mem:* Sigma Xi; Am Ornith Union; Wildlife Soc; Ecol Soc Am. *Res:* Ecological energetics; wetlands ecology; animal behavior. *Mailing Add:* Dept Wildlife Univ Maine Orono ME 04469

OWEN, RAY DAVID, b Genesee, Wis, Oct 30, 15; m 39; c 1. GENETICS, IMMUNOLOGY. *Educ:* Carroll Col, Wis, BS, 37; Univ Wis, PhM, 38, PhD(genetics), 41. *Hon Degrees:* ScD, Carroll Col, Wis, 62, Univ Pac, 66 & Univ Wis, 79. *Prof Exp:* Asst genetics, Univ Wis, 40-41, res fel, 41-43, asst prof genetics & zool, 43-47; from assoc prof to prof biol, 47-83, chmn dept, 61-68, vpres student affairs & dean students, 74-80, EMER PROF BIOL, CALIF INST TECHNOL, 83- *Concurrent Pos:* Gosney fel, 46-47; res partic, Oak Ridge Nat Lab, 57-58, consult, 58- *Honors & Awards:* Mendel Medal, Czech Acad Sci. *Mem:* Nat Acad Sci; Am Acad Arts & Sci; Am Soc Human Genetics; Genetics Soc Am (treas, 56-59, vpres, 60-61, pres, 62); Soc Study Evolution; Am Philos Soc. *Res:* Immunogenetics; serology; vertebrate and developmental genetics; genetics and immunology of tissue transplantation. *Mailing Add:* Div Biol Calif Inst Technol Pasadena CA 91125

OWEN, ROBERT BARRY, b Chicago, Ill, Oct 16, 43; m 70; c 2. OPTICS. *Educ:* Va Polytech Inst & State Univ, BS, 66, PhD(physics), 72. *Prof Exp:* Teaching asst physics, Va Polytech Inst, 66-69; aerospace technologist, 66-72, RES PHYSICIST OPTICS, GEORGE C MARSHALL SPACE FLIGHT CTR, NASA, 72-; PRES, OPTICAL RES INST, 82- *Concurrent Pos:* Prin investr, optical measurement systs grants, 79-; consult optical measurement systs, 80-; sci visitor, Nat Ctr Atmospheric Res, Nat Bur Standards, 82, 85-86. *Mem:* Sigma Xi; Optical Soc Am. *Res:* Optical measurement techniques; interferometric, holographic and schlieren/shadowgraph systems for low-gravity use; tracer/opticcal grid and modified schlieren and optical correlation systems; materials processing (metal-models, electrodeposition, crystal growth, surfaces); color schlieren, fluid flow visualization. *Mailing Add:* PO Box 17382 Boulder CO 80308

OWEN, ROBERT L, b Oklahoma City, Okla, July 1, 40. IMMUNOLOGY, INTESTINAL PARASITOLOGY. *Educ:* Harvard Univ, MD, 66. *Prof Exp:* ASST CHIEF CELL BIOL, VA MED CTR, SAN FRANCISCO, CALIF, 72-; PROF MED, UNIV CALIF, SAN FRANCISCO, 73-, PROF EPIDEMIOL & BIOSTATIST, 83- *Mem:* Am Gastroenterol Asn; Am Soc Microbiol; Am Fedn Clin Res; Am Col Physicians. *Res:* Intestinal immunology and infectious disease. *Mailing Add:* Med Ctr 151E 4150 Clement St San Francisco CA 94121-1563

OWEN, ROBERT MICHAEL, b Philadelphia, Pa, Feb 17, 46; m 76; c 2. CHEMICAL OCEANOGRAPHY, MARINE GEOCHEMISTRY. *Educ:* Drexel Univ, BS, 69; Univ Wis-Madison, MS, 74, PhD(oceanog, limnol), 75. *Prof Exp:* Develop chemist, Rohm & Haas Co, 69-72; res asst oceanog, Univ Wis-Madison, 72-75; from asst prof to assoc prof, oceanog, 75-86, PROF MARINE GEOCHEM, DEPT GEOL SCI & PROF OCEANOG, DEPT ATMOSPHERIC, OCEANIC & SPACE SCI, UNIV MICH, 86- *Concurrent Pos:* Dir, Marine Geochem Lab, Univ Mich, 75-; assoc ed, J Marine Mining, 77-, ed, J Geophys Res Lett, 90-; dir, Univ Mich Geol Field Camp, Jackson, Wyo, 90- *Honors & Awards:* Scott Turner Earth Sci Award, 84. *Mem:* AAAS; Int Asn Great Lakes Res; Soc Econ Paleontologists & Mineralogists; Sigma Xi; Am Soc Limnol & Oceanog; fel Geol Soc Am; Am Geophys Union; Geochem Soc; Int Marine Minerals Soc (pres, 88-91). *Res:* Marine geochemical investigations of rare earth cycling; paleoceanographic reconstructions of the history of seafloor hydrothermal activity; development of geostatistical techniques for use in marine exploration. *Mailing Add:* Dept Geol Sci Univ Mich Ann Arbor MI 48109-1063

OWEN, ROBERT W, JR, b Denver, Colo, Jan 29, 35; c 6. OCEANOGRAPHY, MARINE SCIENCE GENERAL. *Educ:* Univ Wash, BSci, 57; Univ Calif, San Diego, MSci, 63, PhD, 72. *Prof Exp:* Asst oceanogr, Univ Wash, 53-59; oceanogr, Nat Marine Fisheries Serv, 61-87, CHIEF EXEC OFFICER, MARINE ENVIRON & RESOURCES, 87- *Concurrent Pos:* Consult, Calif Dept Fish & Game, 65-; oceanogr food & ag, UN Int Comns, Norway. *Mem:* Am Geophys Union; Optical Soc Am; Am Inst Physics; Am Soc Limnol & Oceanog; Zool Soc. *Res:* Dynamics of motion in the sea; consequences in plankton and fisheries. *Mailing Add:* Scripps Inst Oceanog Code A003 La Jolla CA 92093

OWEN, STANLEY PAUL, b Earlie, Alta, Mar 30, 24; US citizen; m 46; c 2. BIOCHEMISTRY, RESEARCH ADMINISTRATION. *Educ:* Univ Alta, BSc, 50, MSc, 52; Univ Wis, PhD(biochem), 55. *Prof Exp:* Res assoc, Fermentation Res & Develop Dept, Upjohn Co, 55-62, biochemist, Biochem Dept, 62-67, head microbiol control, 67-69, mgr prod control II, 69-76, group mgr, Control Lab Opers, 76-81, dir, 81-85, exec dir, 85-86; RETIRED. *Concurrent Pos:* Mem comt revision, US Pharmacopeia, 77-80, exec comt, Drug/Standards Div, 80-85 & 85-90, chmn, Med & Surg Prod Subcomt, 80-85 & 85-90. *Honors & Awards:* Upjohn Award, 75. *Mem:* Fedn Int Pharmaceut; Am Pharmaceut Asn; Int Asn Biol Standardization; Am Inst Biol Sci; Am Soc Microbiol; AAAS; Fedn Int Pharmaceut; Sigma Xi. *Res:* Antibiotic, vitamin and biological product assays; cancer research, tissue culture, organ culture, pathogen and sterility testing; biological availability of drugs; quality control. *Mailing Add:* 1030 S 295 Pl Federal Way WA 98003-3716

OWEN, SYDNEY JOHN THOMAS, b Ironbridge, Eng; m 60; c 3. SEMICONDUCTOR PHYSICS, ELECTROLUMINESCENCE. *Educ:* Univ Nottingham, BS, 57, PhD(elec eng), 61. *Prof Exp:* Lectr elec & electronics eng, Univ Nottingham, 61-71, reader, phys electronics, 71-77; Tektronix chair, electrophys, 77-83, HEAD DEPT ELEC & COMPUT ENG, ORE STATE UNIV, 78-, DEAN ENG, 90-; VCHANCELLOR, ORE CTR ADVAN TECHNOL, 85- *Concurrent Pos:* Vis prof, Univ Ala, 68-69; secy, Europ Solid State Devices Res Conf, 74; consult, Tektronix Labs. *Honors & Awards:* Lord Rutherford Award, Inst Radio & Electronics Engrs, 71. *Mem:* Inst Elec & Electronic Engrs; Inst Elec Eng; Sigma Xi. *Res:* Physics of III-V and II-VI materials; active devices in ternary and quaternary structures; semiconductor heterojunctions; a. c. electroluminescence. *Mailing Add:* Dept Elec & Comput Eng Ore State Univ Corvallis OR 97331

OWEN, TERENCE CUNLIFFE, b Cannock, Eng, Nov 29, 30; m 53; c 2. BIO-ORGANIC MECHANISMS. *Educ:* Univ Manchester, BSc, 51, PhD(chem), 54. *Prof Exp:* Tech officer chem, Nobel Div, Imp Chem Indust, Scotland, 54-57; from lectr to sr lectr, Liverpool Col Technol, Eng, 57-64; assoc prof, 64-70, chmn dept chem, 74-78, PROF CHEM, UNIV SFLA, 70- *Concurrent Pos:* Res fel, Vanderbilt Univ, 60-61; consult, Summers Steelworks, Eng, 58-60 & Bristol Labs, 68-; master contractor, Nat Cancer Inst, 83- *Mem:* Am Chem Soc; Royal Soc Chem; Sigma Xi; fel Royal Inst Chem. *Res:* Meterocyclic synthesis; mechanisms of enzyme catalysis. *Mailing Add:* 2825 Samara Dr Tampa FL 33618

OWEN, THOMAS E(DWIN), b Alexandria, La, Jan 6, 31; m 53; c 2. ELECTRICAL ENGINEERING. *Educ:* Southwestern La Inst, BSEE, 52; Univ Tex, MSEE, 57, PhD, 64. *Prof Exp:* Res engr, undersea warfare div, defense res lab, Univ Tex, 55-60, asst prof elec eng, 57-60; sr res engr, 60, mgr earth sci appln, 60-71, asst dir, Dept Electronic Systs Res, 71-76, DIR, DEPT GEOSCI, ELECTRONIC SYSTS DIV, SOUTHWEST RES INST, 76- *Mem:* AAAS; Acoust Soc Am; Inst Elec & Electronics Engrs; Am Geophys Union; Sigma Xi; Soc Explor Geophysicists; Soc Prof Well Log Analysts (secy, 84-85); Soc Mining Engrs. *Res:* Electronic instrumentation, data acquistion and reduction systems; electro-acoustic transducer design and fabrication; underwater acoustics; ultrasonics; atmospheric acoustics; high-resolution electrical and seismic exploration systems and applications. *Mailing Add:* Southwest Res Inst 6220 Culebra Rd San Antonio TX 78284

OWEN, TOBIAS CHANT, b Oshkosh, Wis, Mar 20, 36; m 60, 91; c 2. ASTROPHYSICS, PLANETARY SCIENCES. *Educ:* Univ Chicago, BA, 55, BS, 59, MS, 60; Univ Ariz, PhD(astron), 65. *Prof Exp:* Res physicist, IIT Res Inst, 64-67, sr scientist, 67-69, sci adv, 69-70; from assoc prof to prof astron, State Univ NY Stony Brook, 72-90; ASTRON, UNIV HAWAII, 90- *Concurrent Pos:* Consult, Space Sci Bd, Nat Acad Sci, 67-; partic, Apollo 15 & 16 Lunar Orbit Exp Team, NASA, 69-72, Viking Mars Landing, 69-77, Voyager Outer Planets Explor, 72-, Galileo Jupiter-Orbiter Probe, 77-; vpres, comn 16, Int Astron Union, 73-76, pres, 76-79, chmn, Outer Solar Syst Task Group, 75-, mem, working group Planetary & Satellite Nomenclature, 75-; res assoc, Calif Inst Technol, 76-; pres, comn D, AAAS, 80; vis prof, Astronome Titulaire Paris Observ, 80; vis assoc prof planetary sci, Calif Inst Technol, 70. *Honors & Awards:* co-recipient Newcomb-Cleveland Prize, AAAS, 77; Vernadsky lectr, USSR Acad Sci, 82; Exceptional Sci Achievement Medal, NASA, 77. *Mem:* Fel AAAS; Am Astron Soc; Int Astron Union; Int Soc Study Origin Life; Int Asn Geochemistry & Cosmochemistry; fel Am Geophys Union. *Res:* Planetary atmospheres and surfaces; satellites; comets; exploration of the solar system; studies of the composition, structures, origin and evolution of planetary atmospheres, exploration of the solar system. *Mailing Add:* Inst Astron 2680 Woodlawn Dr Honolulu HI 96822

OWEN, WALTER S, b Liverpool, Eng, Mar 13, 20; m 53; c 1. MATERIALS SCIENCE, PHYSICAL METALLURGY. *Educ:* Univ Liverpool, BEng, 40, MEng, 42, PhD(metall), 50. *Prof Exp:* Metallurgist, Eng Elec Co, Ltd, 40-46; asst lectr metall, Univ Liverpool, 46-48, lectr, 48-54; mem res staff, Mass Inst Technol, 54-57; Thomas R Briggs prof mat sci, Cornell Univ, 66-70; dean technol inst, Northwestern Univ, Evanston, 70-71, vpres sci & res, 71-73; prof mat sci & eng & head dept, 73-82, prof phys metall, 82-85, EMER PROF & SR LECTR MAT SCI & ENG, MASS INST TECHNOL, 85- *Concurrent Pos:* Commonwealth Fund fel, Mass Inst Technol, 51-52; consult, Int Nickel Ltd, Eng, 61-65, Eng Elec Co, 62-66, Richard Thomas & Baldwin, 63-66 & Manlabs, Mass, 66; Henry Bell Wortley prof metall, Univ Liverpool, 57-66. *Mem:* Nat Acad Eng; Am Soc Metals; Am Inst Mining, Metall & Petrol Engrs; Brit Inst Metall; Brit Iron & Steel Inst. *Res:* Industrial and manufacturing engineering; deformation and fracture of solids; martensitic transformations; metallurgy and physical metallurgical engineering. *Mailing Add:* Mass Inst Technol 77 Massachusetts Ave Rm 13-5114 Cambridge MA 02139

OWEN, WALTER WYCLIFFE, b Pine Bluff, Ark, Sept 19, 06; m 32; c 2. CHEMISTRY. *Educ:* Univ Ark, BS, 29; Univ NC, PhD(phys chem), 39. *Prof Exp:* Supvr treating, Int Creosoting & Construct Co, Tex, 29-30; exp & sales engr, Sowers Mfg Co, NY, 30-32; res chemist, Rayon Div, E I du Pont de Nemours & Co, 32-36, 39-41, chief chemist, Va, 41-45, tech serv engr, Del, 45-49, tech res tire yarn sales, 49-54, mgr indust tech serv, Textile Fibers Dept, 54-60, tech specialist, 60-68; CONSULT PROCESSING & APPLN SYNTHETIC FIBERS, 68- *Mem:* Am Chem Soc. *Res:* Viscose rayon process; permeability of regenerated cellulose film to carbon dioxide; industrial applications of man-made fibers. *Mailing Add:* 1301 New Stine Rd Apt 903 Bakersfield CA 93309-3504

OWEN, WARREN H, b Rock Hill, SC, Jan 8, 27; m 49; c 3. MECHANICAL ENGINEERING. *Educ:* Clemson Univ, BME, 47. *Hon Degrees:* LLD, Clemson Univ, 88. *Prof Exp:* Prin mech engr, Design Eng Dept, 66-71, vpres design eng, 71-78, sr vpres eng & construct, 78-82 & exec vpres, 82-84, EXEC VPRES, ENG CONSTRUCT & PROD GROUP, DUKE POWER CO, 84- *Honors & Awards:* Philip T Sprague Award, Instrument Soc Am, 70; Clyde A Lilly Jr Award, Atomic Indus Forum, 81; James N Landis Medal, Am Soc Mech Eng, 87. *Mem:* Nat Acad Eng; fel Am Soc Mech Eng; Am Nuclear Soc. *Mailing Add:* Duke Power Co 4225 Church St Charlotte NC 28242-0001

OWEN, WILLIAM BERT, b Gilbertsville, Ky, Oct 28, 03; m 32. ENTOMOLOGY. *Educ:* Univ Ky, BA, 27; Univ Minn, MA, 29, PhD(entom), 36. *Prof Exp:* Instr zool, Hamline Univ, 29-30 & Ore State Col, 30-31; from instr to asst prof, Univ Wyo, 31-34; asst, Univ Minn, 35-36; prof, 46-71, EMER PROF ZOOL, UNIV WYO, 71- *Concurrent Pos:* Researcher, Montpellier & Leyden, 64; fac affil, Colo State Univ, 72- *Mem:* Am Soc Parasitol; Entom Soc Am; Am Mosquito Control Asn. *Res:* Taxonomy of Culicidae; biology of mosquitoes of western North America; chemical senses of insects; insect morphology. *Mailing Add:* Dept of Entom Colo State Univ Ft Collins CO 80523

OWENS, AARON JAMES, b Pottsville, Pa, Feb 5, 47; m 70; c 2. ARTIFICIAL NEURAL NETWORKS. *Educ:* Williams Col, BA, 69; Calif Inst Technol, MS, 71, PhD(theoret physics & econ), 73. *Prof Exp:* Asst prof physics, Lake Forest Col, 73-76; physicist, Cent Res Inst Physics, Budapest, 76-77; asst prof, Kenyon Col, 77-78; res physicist, Bartol Res Found, Franklin Inst, 78-80; Sr Consult, E I Du Pont De Nemours & Co, Inc, 80- *Concurrent Pos:* Adj assoc prof, Univ Del, 81- *Mem:* Am Phys Soc; Sigma Xi. *Res:* Mathematical modeling of physical systems; artificial neural networks; applications of super computers; analysis of random data. *Mailing Add:* 23 Lenape Lane Newark DE 19713

OWENS, ALBERT HENRY, JR, b Staten Island, NY, Aug 27, 26; m 49; c 4. INTERNAL MEDICINE, PHARMACOLOGY. *Educ:* Johns Hopkins Univ, BA, MD, 49; Am Bd Internal Med, dipl, 56. *Prof Exp:* Mem staff internal med, Univ Hosp, 49-50, 52-53 & 55-56, physician, 56-76, from instr to assoc prof med, Sch Med, 56-68, instr pathobiol, Sch Hyg & Pub Health, 57-76, PROF ONCOL & MED, SCH MED, JOHNS HOPKINS UNIV, 68-, DIR ONCOL CTR, 76- *Concurrent Pos:* Fel pharmacol & exp therapeut, Sch Med, Johns Hopkins Univ, 53-55; vis physician, Baltimore City Hosp, 57- *Mem:* Am Soc Pharmacol & Exp Therapeut; fel Am Col Physicians; NY Acad Sci; Am Soc Clin Oncol; Am Asn Cancer Insts. *Res:* Neoplastic diseases. *Mailing Add:* 2416 Old Joppa Rd Joppa MD 21085

OWENS, BOONE BAILEY, b Chefoo, China, Dec 13, 32; US citizen; c 4. LITHIUM BATTERY TECHNOLOGY. *Educ:* Whittier Col, BA, 54; Iowa State Univ, PhD(phys & inorg chem, math), 57. *Prof Exp:* Asst chem, Ames Lab, AEC, 54-57; instr, Univ Calif, Santa Barbara, 57-59; sr chemist, Atomics Int Div, NAm Aviation, Inc, 59-63, res specialist, 63-68; mem tech staff, Gen Tel & Electronics Labs, 68-69; prin scientist, Gould Ionics Inc, 69-72 & Energy Technol Dept, Gould Labs, 72-74, mgr battery res & develop sect, Gould Labs-Energy Res, Gould Inc, 74-75; mgr power sources res & develop dept, Medtronic, Inc, 75-78, dir battery develop, 78-85; CONSULT, BOONE B OWENS, INC, 85-; ADJ PROF, DEPT CHEM ENG & MAT SCI, UNIV MINN, 85- *Concurrent Pos:* Consult, Glass Prod Co, Calif, 58-59; ed, Battery Div, J Electrochem Soc, 73-80 & Progress Batteries & Solar Cells, 78-; chmn/mem sci comt, Int Conference Solid Electrolytes, 76-89 & mem sci comt, Int Meeting Lithium Batteries, 82, 84 & 86; prin investr, high energy battery res & develop. *Mem:* AAAS; Am Chem Soc; Sigma Xi; Electrochem Soc; Int Soc Electrochem. *Res:* High energy batteries; solid electrolytes; solid state batteries; lithium pacemaker batteries; batteries for implantable medical devices; medical battery reliability. *Mailing Add:* 4707 Lyndale Ave N Minneapolis MN 55430

OWENS, CHARLES ALLEN, b Bristol, Tenn, Mar 18, 45; m 67; c 2. BOTANY. *Educ:* King Col, AB, 66; Va Polytech Inst & State Univ, MS, 68, PhD(genetics), 71. *Prof Exp:* Asst prof, 70-74, assoc prof, 75-80, PROF BIOL, KING COL, 81- *Mem:* Am Genetic Asn; Am Inst Biol Sci; AAAS; Sigma Xi. *Res:* Development of techniques for studying plant karyotypes to be extended to comprehensive study of the abundant Applachian flora. *Mailing Add:* Dept Biol King Col Bristol TN 37620

OWENS, CHARLES WESLEY, b Billings, Okla, Oct 27, 35; m 55; c 4. PHYSICAL CHEMISTRY. *Educ:* Colo Col, BS, 57; Univ Kans, PhD(phys chem), 63. *Prof Exp:* Res assoc radiochem, Univ Kans, 63; from asst prof to assoc prof, 63-78, PROF CHEM, UNIV NH, 78-, ASSOC VPRES, ACAD AFFAIRS, 83- *Concurrent Pos:* Res fel appl chem, Harvard Univ, 70-71; interim vpres finance & admin, Univ NH, 87- *Mem:* Am Asn of Higher Educ. *Res:* Wood chemistry; hot-atom chemistry; solid state chemistry. *Mailing Add:* VP Res Acad Affairs Radford Univ Radford VA 24142

OWENS, CLARENCE BURGESS, b Smithville, Tex, June 17, 26; m 49; c 5. AGRONOMY, PLANT PHYSIOLOGY. *Educ:* Prairie View Col, BSc, 48; Ohio State Univ, MSc, 51, PhD, 54. *Prof Exp:* Instr agr, US Vet Admin, 48-50; from asst prof to assoc prof, 53-57, PROF AGRON, FLA A&M UNIV, 57- *Concurrent Pos:* Dir summer & in-serv, Earth Sci Insts, NSF, 64-66; campus coordr, US AID-Fla A&M Univ Kenya contract, 72; res agron adv, US ICA, 57-59. *Mem:* AAAS; Am Soc Agron; Weed Sci Soc Am. *Res:* Chemical weed control; crop fertilization; soil fertility. *Mailing Add:* Dept Agr Sci Fla A&M Univ Tallahassee FL 32307

OWENS, CLIFFORD, b Raven, Ky, Jan 5, 31; m 54. PHYSICAL INORGANIC CHEMISTRY. *Educ:* Rutgers Univ, BA, 57; Drexel Univ, MS, 61, PhD(phys chem), 69. *Prof Exp:* Anal chemist petrol, Gulf Oil Corp, 57-60; res chemist org coatings, Campbell Soup Co, 60-63, emission spectroscopist, 63-66; asst chem, Drexel Univ, 66-69; from instr to asst prof, 69-75, ASSOC PROF CHEM, RUTGERS UNIV, 75- *Concurrent Pos:* Consult, Drexel Univ, 74- 80. *Mem:* Soc Appl Spectrog. *Res:* Study of the instability of chloramine compounds; sweat poisoning of activated carbon surfaces; vapor phase equilibria of binary systems where complexation may occur; synthesis and characterization of inorganic complexes involving phosphoryl and nitroxide compounds. *Mailing Add:* Dept Chem Rutgers Univ Camden NJ 08102

OWENS, DAVID KINGSTON, b Milwaukee, Wis, Sept 14, 48; m 69; c 2. PHYSICS. *Educ:* Mass Inst Technol, SM & SB, 71, PhD(physics), 76. *Prof Exp:* RES STAFF PLASMA PHYSICS, PLASMA PHYSICS LAB, PRINCETON UNIV, 76- *Mem:* Am Phys Soc; AAAS. *Res:* Atomic and surface processes in plasmas; studies of high temperature tokamak plasmas. *Mailing Add:* 127 Cedar Lane Princeton NJ 08540

OWENS, DAVID WILLIAM, b Greensburg, Ind, Nov 24, 46; m 71; c 2. COMPARATIVE ENDOCRINOLOGY, MARINE BIOLOGY. *Educ:* William Jewell Col, BA, 68; Univ Ariz, PhD(zool), 76. *Prof Exp:* Marine biologist, Fiji Islands, US Peace Corps, 68-71; fel physiol, Dept Zool, Colo State Univ, 76-78; from asst prof to assoc prof, 78-88, PROF, DEPT BIOL, TEX A&M UNIV, 88- *Concurrent Pos:* Res sabbatical, Great Barrier Reef, 85. *Mem:* AAAS; Herpetologists League; Am Soc Zoologists; Am Soc Ichthyologists & Herpetologist; Animal Behav Soc. *Res:* Endocrine control of reproduction in non-mammalian vertebrates; marine turtle and fish behavior; marine turtle conservation. *Mailing Add:* Dept of Biol Tex A&M Univ College Station TX 77843

OWENS, EDWARD HENRY, b Stamford, Eng, Mar 25, 45. GEOMORPHOLOGY, SEDIMENTOLOGY. *Educ:* Univ Col Wales, BSc, 67; McMaster Univ, MSc, 69; Univ SC, PhD(geol), 75. *Prof Exp:* Res scientist, Geol Surv Can, 71-75; asst prof geol, Coastal Studies Inst, La State Univ, 72-79; vpres, Woodward-Clyde Consults, 79-86; MANAGING DIR, GEOSCI SERV LTD, 86- *Concurrent Pos:* Chmn, Sable Island Dune Restoration & Terrain Mgt Subcomt, Can, 74-75. *Mem:* Geol Soc Am; Geol Asn Can; Soc Econ Paleontologists & Mineralogists; Sigma Xi; Arctic Inst NAm. *Res:* Barrier beach geomorphology and sediments, particularly the mechanics of sediment transport and deposition; action of ice on beaches in mid- and high-latitudes; effects of oil-spills on coasts. *Mailing Add:* Woodward-Clyde Consult Houston TX 77055

OWENS, FRANK JAMES, b Dublin, Ireland, Oct 24, 39; US citizen; m 72. SOLID STATE PHYSICS, CHEMICAL PHYSICS. *Educ:* Manhattan Col, BS, 62; Univ Conn, MS, 64, PhD(physics), 68. *Prof Exp:* Instr physics, Univ Conn, 62-65, res asst, 65-68; solid state physicist, Feltman Res Lab, 68-77; PRES & SOLID STATE PHYSICIST, ENERGETIC MAT LAB, ARDC, 77-; ADJ PROF PHYSICS, HUNTER COL, CITY UNIV, NY. *Concurrent Pos:* Space scientist, Goddard Space Flight Ctr, NASA, 64-65; consult solid state physics, US Army Res Off, 68; lectr, dept physics, Grad Sch Sci & Eng, Fairleigh Dickinson Univ, 78, NJ Inst Technol, 77. *Honors & Awards:* Army Res & Develop Award, Sigma Xi. *Mem:* Am Phys Soc. *Res:* Paramagnetic resonance and raman spectroscopy studies of dynamics processes involved in solid state phase transitions; also effects of shock and radiation on solids; high pressure physics and lattice stability; initiation of detonation in condensed materials; high temperature superconductors. *Mailing Add:* 127 Cedar Lane Princeton NJ 08540

OWENS, FREDERICK HAMMANN, b Royersford, Pa, Dec 29, 28; m 61; c 3. ORGANIC CHEMISTRY, CHEMICAL INFORMATION. *Educ:* Ursinus Col, BS, 53; Univ Ill, MS, 54, PhD(chem), 58. *Prof Exp:* Res chemist, Rohn & Haas Co, 57-73, proj leader plastic res, 73-76, MGR INFO SERV, RES DIV LABS, ROHM & HAAS CO, 76-, COORDR SCI EDUC, 88- *Mem:* Am Chem Soc. *Res:* Macrocyclic compounds; synthesis of monomers and polymers; information science. *Mailing Add:* 480 Steamboat Dr Southampton PA 18966

OWENS, FREDRIC NEWELL, b Baldwin, Wis, Sept 1, 41; m 85; c 2. ANIMAL NUTRITION. *Educ:* Univ Minn, St Paul, BS, 64, PhD(nutrit), 68. *Prof Exp:* Asst prof animal sci, Univ Ill, Urbana, 68-74; from assoc prof to prof, 74-86, REGENTS PROF ANIMAL SCI, OKLA STATE UNIV, 86-, SARKEYS DISTINGUISHED PROF, 90- *Concurrent Pos:* Ed-in-chief, J Animal Sci, 88-90; comt on animal nutrit, Nat Res Coun, 85- *Honors & Awards:* Tyler Award, Okla State, 80; Res Award, Am Feed Mfr Asn, 87. *Mem:* Am Soc Animal Sci; Am Dairy Sci Asn; Am Inst Nutrit; Nutrit Soc. *Res:* Amino acid and nitrogen metabolism of the ruminant animal. *Mailing Add:* Animal Sci Dept Okla State Univ Stillwater OK 74078

OWENS, G(LENDA) KAY, b Durant, Okla, Dec 4, 42. PUBLIC SPEAKING. *Educ:* Southeastern Okla State Univ, BS, 63; Okla State Univ, Stillwater, MS, 65, PhD(math), 70. *Prof Exp:* Prof math, Central State Univ, 65-83, dept chmn math & statist, 83-88; COL DEAN MATH & SCI, UNIV CENT OKLA, EDMOND, 88- *Concurrent Pos:* Chmn, Okla-Ark Sect, Math Asn Am, 89-90. *Mem:* Math Asn Am; Am Math Soc; Sigma Xi. *Res:* Academic planning, leadership, mathematics and society. *Mailing Add:* Univ Cent Okla Edmond OK 73034

OWENS, GARY KEITH, HYPERTENSION. *Educ:* Penn State Univ, PhD(physiol), 79. *Prof Exp:* ASSOC PROF RESPIRATORY & CELL PHYSIOL, UNIV VA, 82- *Res:* Vascular smooth muscle cell growth control and cytodifferentiation. *Mailing Add:* Univ Va Sch Med Med Ctr Box 395 Charlottesville VA 22908

OWENS, GUY, b Amarillo, Tex, Jan 25, 26; m 49; c 2. NEUROSURGERY, ANATOMY. *Educ:* Tufts Univ, BS, 46; Harvard Univ, MD, 50; Am Bd Neurol Surg, dipl, 60. *Prof Exp:* Intern surg, Vanderbilt Univ Hosp, 50-51, asst, Sch Med, 51-52, asst neurosurg, 54-56, from instr to asst prof, 56-60; chief dept, Roswell Park Mem Inst, 60-68; PROF SURG, SCH MED, UNIV CONN, 68- *Concurrent Pos:* Markle scholar, Sch Med, Vanderbilt Univ, 51-52 & 54-56; Rockefeller fel med sci, Nat Res Coun, 57-58; resident, Vanderbilt Univ, 56-57, chief outpatient dept & vis surgeon, 58-60; attend neurosurgeon, Vet Admin Hosp, Nashville, Tenn, 58-60 & Buffalo, 63-; partic brain tumor workshop, NIH, 66- *Mem:* Am Soc Clin Invest; Am Asn Anat; AMA; Am Fedn Clin Res; Am Physiol Soc. *Res:* Neurophysiology; neuroanatomy. *Mailing Add:* 40 Hart St New Britain CT 06052

OWENS, IDA S, b Whiteville, NC, Sept 13, 39. GLUCURONIDATING ENZYMES, DRUG METABOLISM. *Educ:* Duke Univ, PhD(biochem), 67. *Prof Exp:* CHIEF, DRUG BIOTRANSFORMATION SECT, NAT INST CHILD HEALTH & HUMAN DEVELOP, NIH, 82- *Mem:* Am Soc Biol Chemists; Int Soc Toxicol; Am Soc Pharmacol & Exp Therapeut. *Mailing Add:* Bldg 10 Rm 1C660 NIH Warren Grant Magnuson Clin Ctr Bethesda MD 20892

OWENS, JAMES CARL, b Saginaw, Mich, May 24, 37; m 61; c 2. OPTICS, LASER & ELECTRONIC IMAGING SYSTEMS. *Educ:* Oberlin Col,AB, 59; Harvard Univ, AM, 60, PhD(physics), 65. *Prof Exp:* Asst appl physics, Harvard Univ, 63-64; physicist & proj leader atmospheric physics, Cent Radio Propagation Lab, Nat Bur Standards, 64-65 & Inst Telecommun Sci & Aeronomy, Environ Sci Serv Admin, 65-69; head, Phys Optics Lab, Eastman Kodak Co, 69-71, Image Processes Lab, 76-79, res assoc, Physics Div, 69-83, LASER SCANNING SYSTS GROUP LEADER, ELECTRONIC IMAGING RES LABS, EASTMAN KODAK CO, 83- *Concurrent Pos:* Mem comn 1, US Nat Comt Int Sci Radio Union, 65-75; lectr laser sci, Wash Univ, 67-68. *Honors & Awards:* CEK Mees Medal, 87. *Mem:* Am Phys Soc; Optical Soc Am; Inst Elec & Electronic Eng; Soc Photog Sci & Eng; Int Soc Optical Eng. *Res:* Coherent optics and laser applications; photographic science and novel imaging systems; lattice vibrations in crystals; dielectric properties of solids; atmospheric optics. *Mailing Add:* Electronic Imaging Mail Code 01822 Eastman Kodak Co Rochester NY 14650

OWENS, JAMES SAMUEL, b McKinney, Ky, Mar 27, 08; m 34; c 2. CERAMIC ENGINEERING, PHYSICS. *Educ:* Univ Chattanooga, BS, 28; Univ Mich, MS, 30, PhD(physics), 32. *Prof Exp:* Asst physics, Univ Chattanooga, 26-28 & Univ Wis, 28-29; assoc eng res, Univ Mich, 29-31, asst physics & internal med, 32-33; res physicist, Dow Chem Co, 33-39; mem staff, Cent Tech Lab, Armstrong Cork Co, Pa, 39-40, asst chief chemist, Glass & Closure Div, 40-43; chief tech aide, Nat Defense Res Comt, Univ Mich, 43-46; prof univ & exec dir res found, Ohio State Univ, 46-51; asst to mgr, 51-54, Ceramic Div, Champion Spark Plug Co, 51-54, gen mgr, 54-70, vpres & gen mgr ceramic mfg opers, 71-73; RETIRED. *Concurrent Pos:* Mem comt spectros as appl to chem, Nat Res Coun, 37-42; mem panel infrared, Comt Electronics, Res & Develop Bd, 47-52, chmn, 49-52, mem panel metals & minerals, Comt Mat, 52-54; mem adv bd, ceramic eng dept, Univ Ill, 68-71. *Honors & Awards:* Awards, Am Ceramic Soc, 61, 78 & 79. *Mem:* Am Phys Soc; fel Am Ceramic Soc (vpres, 63-64, pres, 67-68); Nat Inst Ceramic Eng. *Res:* Reactions between activated and normal atoms and molecules; quantitative spectrographic analysis of alloys and chemicals; spectrophotometry; photochemistry; physical and chemical properties of glasses; infrared radiation sources, detectors and propagation; refractory, dielectric and high temperature ceramics; spark plug insulators. *Mailing Add:* 79 Oakland Hills Pl Rotonda West FL 33947

OWENS, JOHN CHARLES, b Plainview, Tex, May 30, 44; m 65; c 2. ECONOMIC ENTOMOLOGY. *Educ:* WTex State Univ, BA, 66; Tex Tech Univ, MS, 69; Iowa State Univ, PhD(entom), 71. *Prof Exp:* Asst prof entom, Iowa State Univ, 71-75; assoc prof entom & arthropod biol, Tex Tech Univ, 75-76; Mgr entom, Pioneer Hi-Bred Int, 76-77; ASSOC PROF ENTOM, NEW MEX STATE UNIV, 77- *Concurrent Pos:* Consult entomologist, Corn Prod Syst Inc, 75-; mem invert control task group, Am Inst Biol Sci, 75- *Mem:* Entom Soc Am; Am Soc Agron; Am Registry Prof Entomologists; Crop Sci Soc Am. *Res:* Pest management strategies for field crop and rangeland insects. *Mailing Add:* Dept Entom & Plant Path NMex State Univ PO Box 30003 Las Cruces NM 88003-3003

OWENS, JOHN MICHAEL, b Providence, RI, July 15, 51; m 74. URBAN PEST CONTROL, PESTICIDE RESEARCH ADMINISTRATION. *Educ:* Univ RI, BS, 74; Purdue Univ, MS, 77, PhD(entom), 80. *Prof Exp:* Exten urban entomologist, dept entom, Tex A&M Univ, 80-82; entom group leader, Corp Res Div, 82-89, res assoc, Entom Res Dept, 89-91, SR ASSOC, GOVT AFFAIRS DIV, S C JOHNSON & SON, INC, 91- *Concurrent Pos:* Co-investr urban entom res projs, dept entom, Purdue Univ, 82- *Mem:* Entom Soc Am. *Res:* Behavior, ecology and management of urban insect pests; German cockroach as a domiciliary pest; cockroach movement within structures; population ecology in response to environmental variables; insecticide technology applied to control household insects. *Mailing Add:* S C Johnson & Son Inc 1525 Howe St Racine WI 53402-5011

OWENS, JOHN N, b Portland, Ore, Apr 18, 36; m 54; c 4. BOTANY, FORESTRY. *Educ:* Portland State Col, BS, 59; Ore State Univ, MS, 61, PhD(bot), 63. *Prof Exp:* From asst prof to assoc prof, 63-77, PROF BIOL, UNIV VICTORIA, BC. *Concurrent Pos:* Vis prof, Univ Calif, Davis, 71, Univ Canterbury, NZ, 77, Univ New England, Australia & Northern Foreign Res Sta, Scotland, 88; mem, Plant Biol Comn, Nat Sci & Eng Res Coun Can; dir biol, Univ Victoria. *Mem:* Bot Soc Am; Can Bot Asn; Int Union Res Org. *Res:* Plant anatomy and morphogenesis; developmental anatomy and morphogenesis of conifers; vegetative and reproductive bud and shoot development; pollen development; pollination mechanisms; ovule, embryo and seed development of conifers. *Mailing Add:* Dept Biol Univ Victoria Victoria BC V8W 2Y2 Can

OWENS, JOSEPH FRANCIS, III, b Syracuse, NY, Sept 23, 46; m 70; c 2. THEORETICAL ELEMENTARY PARTICLE PHYSICS. *Educ:* Worcester Polytech Inst, BSc, 68; Tufts Univ, PhD(physics), 73. *Prof Exp:* Res assoc physics, Case Western Reserve Univ, 73-76; res assoc physics, 76-79, res asst prof, 79-80, from asst prof to assoc prof, 80-85, PROF PHYSICS, FLA STATE UNIV, 85- *Mem:* Am Phys Soc. *Res:* Theoretical high energy physics dealing with interactions among various quarks and gluons. *Mailing Add:* Dept of Physics Keen Bldg Fla State Univ Tallahassee FL 32306

OWENS, KENNETH, biochemistry, for more information see previous edition

OWENS, KENNETH EUGENE, b Battle Creek, Mich, Apr 16, 26; m 55; c 2. PHYSICAL CHEMISTRY. *Educ:* Univ Calif, BS, 49; Univ Minn, PhD(phys chem), 55. *Prof Exp:* Microchemist, Univ Calif, 49-50; res chemist, Minn Mining & Mfg Co, 55-64, proj leader, 64-67, res supvr, 67-72, sr res specialist, 72-86; RETIRED. *Mem:* Am Chem Soc; Soc Rheol; Am Inst Chemists; Am Phys Soc; Sigma Xi. *Res:* Chemical reaction kinetics; photochemical processes; surface and physical properties of polymeric materials; reactions and properties of ceramic materials; biological and environmental applications. *Mailing Add:* 8619 Hidden Bay Trail N Lake Elmo MN 55042

OWENS, KEVIN GLENN, b Elizabeth, NJ, Aug 14, 60; m 87. MASS SPECTROMETRY. *Educ:* Univ NH, BS, 82; Ind Univ, PhD(anal chem), 89. *Prof Exp:* Instr chem, 87-89, ASST PROF CHEM, DREXEL UNIV, 89- *Mem:* Am Chem Soc; Am Soc Mass Spectrometry; Soc Appl Spectros; Sigma Xi. *Res:* Development of laser ionization methods for time of flight mass spectrometry; structural characterization of synthetic and biological polymers using mass spectrometry. *Mailing Add:* Dept Chem Drexel Univ Philadelphia PA 19104

OWENS, LOWELL DAVIS, b Wayne, Nebr, Apr 19, 31; m 56; c 3. PLANT PHYSIOLOGY. *Educ:* Univ Nebr, BS, 52; Mich State Univ, MS, 55; Univ Ill, PhD(agron), 58. *Prof Exp:* Soil conservationist, Soil Conserv Serv, USDA, 52-53; asst, Mich State Univ, 53-55 & Univ Ill, 55-58; soil scientist, Agr Res Serv, 58-71, USDA liaison, SC State Col, 71-73, soil scientist, 73-80, PLANT PHYSIOLOGIST, AGR RES SERV, USDA, 80- *Mem:* AAAS; Soil Sci Soc Am; Am Inst Biol Sci; Am Soc Plant Physiol; Am Soc Microbiol. *Res:* Plant tissue culture; plant and microbiology; physiology. *Mailing Add:* Plant Physiol USDA Agr Res Ctr-W Beltsville MD 20705

OWENS, ROBERT HUNTER, b Philadelphia, Pa, Apr 9, 21; m 48; c 4. MATHEMATICS. *Educ:* Webb Inst Naval Archit, BS, 44; Columbia Univ, MA, 48; Calif Inst Technol, PhD(math), 52. *Prof Exp:* Instr math, Stevens Inst Technol, 46-48; asst, Calif Inst Technol, 49-52; res assoc appl math, Brown Univ, 52-53; Off Naval Res phys sci coordr, Calif, 53-54 & mathematician, DC, 54-56; from asst prof to assoc prof math, Univ NH, 56-62; actg head math sci sect, NSF, 63-64; chmn dept, Univ Va, 64-74, prof appl math & computer sci, 64-89, actg chmn dept, 78-79, EMER PROF APPL MATH & COMPUTER SCI, UNIV VA, 89- *Concurrent Pos:* Liaison scientist, Off Naval Res, London, 70-71; vis scientist, Dept Health, Educ & Welfare, 74-75. *Mem:* Math Asn Am; Soc Indust & Appl Math. *Res:* Applied mathematics and numerical analysis; numerical methods. *Mailing Add:* 112 Island Lane Huddleston VA 24104

OWENS, THOMAS CHARLES, b Fargo, NDak, May 8, 41; m 62; c 3. CHEMICAL ENGINEERING. *Educ:* Univ NDak, BSChE, 63; Iowa State Univ, MS, 65, PhD(chem eng), 67. *Prof Exp:* Sr res engr, Exxon Prod Res Co, Tex, 67-68; PROF CHEM ENG, UNIV NDAK, 68- *Mem:* AAAS; Am Inst Chem Engrs; Am Chem Soc. *Res:* Separation and purification; solvent extraction; lignite processing technology; rheology of suspensions; membrane separation techniques; water and waste treatment. *Mailing Add:* 1809 N Third St Grand Forks ND 58201

OWENS, VIVIAN ANN, b Bucksport, SC, Sept 2, 48. DEVELOPMENTAL & STRUCTURAL BOTANY. *Educ:* Howard Univ, BS, 71, MS, 74; Cornell Univ, PhD(bot), 84. *Prof Exp:* Res asst bot, Howard Univ, 71-74; teaching asst, Cornell Univ, 76-77, res asst, 78-81; asst prof bot, Cent State Univ, 82-88; ASSOC PROF BOT, HAMPTON UNIV, 88- *Mem:* Bot Soc Am; Electron Microscopy Soc Am; NY Acad Sci; AAAS; Am Inst Biol Sci. *Res:* Effects of ozone on the role of growth regulators in wound repair and tissue differentiation in callus cultures; effect of aging on the flowering response and apical activity of morning glory. *Mailing Add:* Dept Biol Sci Hampton Univ VA 23668

OWENS, WILLIAM LEO, b Redding, Calif, Jan 13, 30; m 59; c 3. MECHANICAL ENGINEERING. *Educ:* Wash State Univ, BS, 53; Univ Calif, Berkeley, MS, 59; Univ Aberdeen, PhD(mech eng), 61. *Prof Exp:* Res engr, Univ Calif, Berkeley, 57-59; sr res fel, Univ Aberdeen, 59-61; aerothermo specialist, United Technol Ctr, 62-63; staff engr res & develop, Lockheed Missiles & Space Co, 63-85; CONSULT, 85- *Concurrent Pos:* Sr res fel, UK Atomic Energy Auth, 59-61; mem tech adv comt, Heat Transfer Res Inc, 77-81. *Mem:* Am Soc Mech Engrs; Am Inst Aeronaut & Astronaut. *Res:* Heat transfer and fluid dynamics, boiling, condensing and two phase flow; biofouling; ocean thermal energy conversion; chemical and electric propulsion. *Mailing Add:* 47301 Bill Owens Rd PO Box 92 Point Arena CA 95468

OWENS, WILLIAM RICHARD, b Philadelphia, Pa, Aug 9, 41; m 83; c 6. TECHNICAL MANAGEMENT, INFRARED SYSTEMS ENGINEERING. *Educ:* Wesleyan Univ, BA, 63; Case Western Reserve Univ, MS, 65, PhD(physics), 68. *Prof Exp:* Asst physics, Case Western Reserve Univ, 63-68; fel, Tulane Univ, 68-70; assoc prof, Pontifical Cath Univ, Rio De Janeiro, Brazil, 71-77; sr scientist, 78-80, tech mgr infrared anal, Analytics, 80-83; BR HEAD, ELECTRO-OPTICS LAB, GA TECH RES INST, GA INST TECHNOL, 83- *Mem:* Sigma Xi; Am Phys Soc; Inst Elec & Electronic Engrs; Soc Photo-Optical Instrumentation Engrs. *Res:* Infrared modeling and analysis and measurement; nuclear structure physics; radiation damage; x-ray fluorescence analysis; nuclear instrumentation; software development; infrared systems engineering. *Mailing Add:* Electro-Optics Lab Ga Tech Res Inst Ga Inst Technol Atlanta GA 30332-0800

OWENSBY, CLENTON EDGAR, b Clovis, NMex, Mar 17, 40; m 59; c 3. RANGE SCIENCE, ECOLOGY. *Educ:* NMex State Univ, BS, 64; Kans State Univ, PhD(range sci), 69. *Prof Exp:* Res asst, 64-67, from instr to asst prof, 67-74, assoc prof, 74-78, PROF RANGE MGR, KANS STATE UNIV, 79- *Mem:* Soc Range Mgt. *Res:* Native range burning; fertilization; management systems; carbohydrate reserves; soil chemical properties and animal performance; photosynthesis; root exudation. *Mailing Add:* Dept Agron Throckmorton Hall Kans State Univ Manhattan KS 66506

OWERS, NOEL OSCAR, b Nagpur, India, Oct 1, 26; US citizen; m 60; c 4. BIOLOGY, ANATOMY. *Educ:* Nagpur Univ, BSc, 48, PhD(reproduction biol), 57; Argonne Nat Lab, dipl, 58. *Prof Exp:* Asst prof biol, Duquesne Univ, 58-59; res assoc anat, Univ Wis, 61-63; asst prof, Univ Iowa, 63-69; ASSOC PROF ANAT, HEALTH SCI DIV, VA COMMONWEALTH UNIV, 69- *Concurrent Pos:* NSF-AEC fel, Univ Mich, 59; NIH fel, Univ Wis, 59-61; NASA-Am Soc Eng Educ fac fel, Ames Res Ctr, Stanford Univ, 70-71; consult, Univ Wash, 67-68. *Mem:* AAAS; Am Asn Anat; Am Soc Cell Biol; Microcirc Soc. *Res:* Proteolytic activity of normal and reproductive cells, malignant and non-malignant tumor cells, bacteria and fungi and other life forms, in vivo and vitro; methods of detection of proteinases. *Mailing Add:* Dept Anat Va Commonwealth Univ Med Col Box 709 Richmond VA 23298-0001

OWINGS, ADDISON DAVIS, b Hattiesburg, Miss, Feb 8, 36; m 64; c 2. AGRICULTURAL STATISTICS, BIOLOGY. *Educ:* Miss State Univ, BS, 57, MS, 62, PhD(agron), 66. *Prof Exp:* Asst agronomist, Miss Agr Exp Sta, 57-59; from instr to assoc prof, 63-70, head dept, 67-83, prof agr, 70-89, PROF BIOL, SOUTHEASTERN LA UNIV, 89- *Concurrent Pos:* Mem, Coun Agr Sci & Technol; NDEA fel. *Mem:* Am Soc Agron; Am Quarter Horse Asn; Am Farm Bur Fedn. *Res:* Crops and plant breeding. *Mailing Add:* Southeastern La Univ Box 708 Hammond LA 70402-0708

OWINGS, JAMES CLAGGETT, JR, b Baltimore, Md, Feb 8, 40. MATHEMATICAL LOGIC. *Educ:* Dartmouth Col, BA, 62; Cornell Univ, PhD(math), 66. *Prof Exp:* Asst prof, 66-70, ASSOC PROF MATH, UNIV MD, 70- *Mem:* Asn Symbolic Logic; Am Math Soc; Math Asn Am; Soc Exact Philos. *Res:* Metarecursion theory; graph theory; combinatorics; mathematical linguistics. *Mailing Add:* Dept Math Univ Md College Park MD 20742

OWNBEY, GERALD BRUCE, b Kirksville, Mo, Oct 10, 16; m 68; c 1. PLANT TAXONOMY. *Educ:* Univ Wyo, BA, 39, MA, 40; Wash Univ, St Louis, PhD(bot), 47. *Prof Exp:* Instr bot, 47-48, asst prof & actg cur herbarium, 48-49, assoc prof, 51-56, chmn dept bot, 60-62, prof, 56-85, cur herbarium, 49-85, EMER PROF BOT, UNIV MINN, ST PAUL, 86- *Concurrent Pos:* Guggenheim fel, 53-54. *Mem:* Am Inst Biol Sci; Bot Soc Am; Am Soc Plant Taxon; Soc Study Evolution; Am Fern Soc; Int Asn Plant Taxon. *Res:* Taxonomy and cytotaxonomy of Corydalis, Argemone and Cirsium; floristics of Minnesota. *Mailing Add:* Dept Plant Biol Univ of Minn St Paul MN 55108

OWNBY, CHARLOTTE LEDBETTER, b Amory, Miss, July 27, 47. VETERINARY ANATOMY. *Educ:* Univ Tenn, Knoxville, BS, 69, MS, 71; Colo State Univ, PhD(anat), 75. *Prof Exp:* From instr to assoc prof, 74-84, PROF HISTOL & ELECTRON MICROS, OKLA STATE UNIV, 84-, DIR, ELECTRON MICROSCOPE LAB, 77- *Mem:* AAAS; Sigma Xi; Am Asn Vet Anatomists; Electron Micros Soc Am; Am Asn Anatomists; Int Soc Toxinology. *Res:* Light and electron microscopic study of myonecrosis induced by pure myotoxins of rattlesnake venom; improvement of antiserum treatment of snakebite-induced necrosis. *Mailing Add:* 2413 Tanglewood Circle Stillwater OK 74074

OWNBY, DENNIS RANDALL, b Athens, Ohio, July 14, 48; m 70; c 2. IMMUNOLOGY, ALLERGY. *Educ:* Ohio Univ, BS, 69; Med Col Ohio, MD, 72. *Prof Exp:* Intern, Sch Med, Duke Univ, 72-73, resident, 73-74, res fel, 74-77, asst prof pediat immunol, 77-79; STAFF PHYSICIAN, HENRY FORD HOSP, 80- *Concurrent Pos:* Young investr, Nat Inst Allergy & Infectious Dis res grants, 78-81, 87-92. *Mem:* Am Acad Pediat; Am Acad Allergy. *Res:* In vitro IGE measurement; allergen exposure. *Mailing Add:* Henry Ford Hosp 2799 W Grand Blvd Detroit MI 48202

OWNBY, JAMES DONALD, b Sevierville, Tenn, Sept 30, 44; m 69; c 2. PLANT PHYSIOLOGY. *Educ:* Univ Tenn, Knoxville, BS, 66, MA, 71; Colo State Univ, PhD(plant physiol), 74. *Prof Exp:* Res assoc algal physiol, Langston Univ, 74-75; from asst prof to assoc prof, 75-86, PROF BOT, OKLA STATE UNIV, 86- *Mem:* Am Soc Plant Physiologists; Sigma Xi. *Res:* Physiology of aluminum tolerance in plants. *Mailing Add:* 2413 Tanslewood Circle Stillwater OK 74074

OWNBY, P(AUL) DARRELL, b Salt Lake City, Utah, Nov 9, 35; m 61; c 7. CERAMIC ENGINEERING, MATERIALS SCIENCE. *Educ:* Univ Utah, BS, 61; Univ Mo-Rolla, MS, 62; Ohio State Univ, PhD(ceramic eng), 67. *Prof Exp:* Ceramic engr, Sprague Elec Co, 61; res ceramist, Battelle Mem Inst, 63-67; from asst prof to assoc prof, 68-74, res assoc, Indust Res Ctr, 68, PROF CERAMIC ENG, UNIV MO-ROLLA, 74-, RES ASSOC, GRAD CTR MAT RES, 68- *Concurrent Pos:* Consult, Battelle Mem Inst, 68-70, Dynasil Corp Am, 69-80, Eagle Picher, 69-, Monsanto, 79-81, Terratek, 88-, Digital Controls, 87-; vis scientist, Max Planck Inst Mat Sci, Stuttgart, Ger, 74-75; chmn & chief oper officer, MRD Corp, 85- *Mem:* Am Ceramic Soc; Nat Inst Ceramic Engrs; Ceramic Educ Coun. *Res:* Glass ceramics; surface phenomena; field emission; molecular beam; ultra-high vacuum; final sintering; boron compounds; metal-ceramic interfaces; atmosphere control; ceramic matrix composites. *Mailing Add:* Dept Ceramic Eng Univ Mo Rolla MO 65401

OWRE, OSCAR THEODORE, ornithology; deceased, see previous edition for last biography

OWSLEY, DENNIS CLARK, b Los Angeles, Calif, Mar 26, 43; m 63; c 2. SYNTHETIC ORGANIC CHEMISTRY, PROCESS CHEMISTRY. *Educ:* Univ Calif, Riverside, AB, 65, PhD(org chem), 69. *Prof Exp:* Sr chemist, Monsanto Co, 69-75, res specialist, Cent Res Dept, 75-77 & Monsanto Indust Chem Co, 77-80, sr res specialist, Nutrit Chem Div, 80-83, SR RES SPECIALIST, ANIMAL SCI DIV, MONSANTO AGR PROD CO, 83- *Concurrent Pos:* Adj assoc prof, dept of chem, Univ Mo, St Louis, 77-81. *Mem:* Am Chem Soc; Sigma Xi. *Res:* Organic synthesis; photochemistry; organo-sulfur chemistry; kinetics, gas-liquid chromatography; enzyme inhibitors; neurochemistry; protein isolation. *Mailing Add:* 39 Deerfield Lane St Louis MO 63146

OWSTON, PEYTON WOOD, b Pittsburgh, Pa, Feb 13, 38; m 63; c 2. FOREST PHYSIOLOGY. *Educ:* Univ Mich, BSc, 60, MF, 62, PhD(forestry), 66. *Prof Exp:* Res forester, Pac Southwest Forest & Range Exp Sta, 66-69, PLANT PHYSIOLOGIST, PAC NORTHWEST FOREST & RANGE EXP STA, US FOREST SERV, 69- *Mem:* Soc Am Foresters. *Res:* Tree-soil-water relations; forest regeneration. *Mailing Add:* 7405 NW Valley View Dr Corvallis OR 97330

OWYANG, GILBERT HSIAOPIN, b Tientsin, China; US citizen; m 65; c 2. APPLIED PHYSICS, ELECTRICAL ENGINEERING. *Educ:* Ta Tung Univ, China, BSc, 44; Harvard Univ, SM, 50, PhD(appl physics), 59. *Prof Exp:* Engr elec, Shanghai Power Co, China, 44-49; sr designer power syst, Devenco Inc, 51-53; engr, Frank L Capps Co Inc, 54-55; res asst appl physics, Gordon McKay Lab, Harvard Univ, 55-59; assoc res physicist electromagnetics, Radiation Lab, Univ Mich, 59-61; PROF ELEC ENG, WORCESTER POLYTECH INST, 61- *Mem:* Sigma Xi; Inst Elec & Electronics Engrs. *Res:* Wave propagation in ionized media; energy transmission and radiation in homogeneous media and inhomogeneous media; electromagnetics; fiber optics; optical waveguides. *Mailing Add:* Dept of Elec Eng Inst Rd Worcester MA 01609

OXBORROW, GORDON SQUIRES, b Ely, Nev, Dec 7, 39; m 61; c 5. BIOLOGICAL INDICATORS, PARTICULATES. *Educ:* Utah State Univ, BS, 64; Univ Minn, MPH, 68. *Prof Exp:* Microbiologist, Ctr Dis Control, 65-67, res microbiologist, 68-74; res microbiologist, Jet Propulsion Labs, 74-76; res microbiologist, 76-81, DIR MICROBIOLOGIST, US FOOD & DRUG ADMIN, 81-; MGR, MICROBIOL & STERILIZATION SERV, MED & SURG DIV, 3M CO. *Concurrent Pos:* mem, chem indicators comt, Asn Advan Med Instrumentation, 80-, chmn, biol indicators subcomt, 85 -; lectr, Am Soc Microbiol, 85-86. *Honors & Awards:* Apollo Achievement Award, NASA, 69, Award for Contrib to Planetary Quarantine, 74, Viking Achievement Award, 76; Outstanding Serv Award, Food & Drug Admin; Outstanding Serv Award, US Dept Health & Human Serv. *Mem:* Asn Advan Med Instrumentation; Fel Asn Off Analytical Chemists; Inst Environ Sci. *Res:* Development of methods for sterility testing and sterilization of drugs, medical devices and packaging materials; tests for biological and chemical indicators; evaluation of plastics and rubber materials for leachables into parenteral products. *Mailing Add:* 3M Co 3M Ctr Bldg 270-5N-01 St Paul MN 55144-1000

OXENDER, DALE LAVERN, b Constantine, Mich, Aug 30, 32; m 55; c 2. BIOCHEMISTRY. *Educ:* Manchester Col, BA, 54; Purdue Univ, MS, 56, PhD, 59. *Prof Exp:* Res assoc, 58-59, from instr to assoc prof, 59-75, PROF BIOCHEM, MED SCH, UNIV MICH, ANN ARBOR, 75-; VPRES, DEPT BIOTECHNOL, WARNER LAMBERT CO, 90- *Mem:* Am Chem Soc; Am Soc Biol Chem; Am Micros Soc. *Res:* Cloning and expression of amino acid transport genes in Chinese hamster, human cells and microorganisms. *Mailing Add:* Dept Biotechnol Warner Lambert 2800 Plymouth Rd Ann Arbor MI 48106-1047

OXENKRUG, GREGORY FAIVA, b Leningrad, USSR, Oct 25, 41; US citizen; m 72; c 2. PSYCHOPHARMACOLOGY, BIOLOGICAL PSYCHIATRY. *Educ:* First Pavlov Med Sch, Leningrad, MD, 65; Bekhterev Psychoneurol Res Inst, Leningrad, PhD(psychopharmacol), 70. *Hon Degrees:* MA, Brown Univ, 90. *Prof Exp:* From asst prof to assoc prof, Dept Psychopharmacol, Bekhterev Psychoneurol Res Inst, 77-79; assoc prof psychiat, Div Psychiat, Boston Univ, 80-82; from assoc prof to prof psychiat, Dept Psychiat, Wayne State Univ, 82-88, from assoc prof to prof pharamcol, 83-88; PROF PSYCHIAT, DEPT PSYCHIAT & HUMAN BEHAV, BROWN UNIV, 88- *Concurrent Pos:* Assoc psychiat, Dept Neurol, Tufts-New England Med Ctr, Boston, 80-82; res assoc, Lab Neuroendocrine Reg, Dept Nutrit Sci, Mass Inst Technol, 80-82; chief adult inpatient psychiat & psychoendocrine res lab, Lafayette Clin, Detroit, Mich, 82-88; vis prof, Sch Med, Technion Inst Technol, Israel, 87; chief psychiat serv & pineal res lab, Vet Admin Med Ctr, Providence, RI, 88- *Mem:* Am col Neuropsychophamacol; Am Col Neuropsychopharmacol; NY Acad Sci. *Res:* Pineal gland in normal and elderly; dementia and depression of the elderly. *Mailing Add:* Vet Admin Med Ctr 116A 830 Chalkstone Ave Providence RI 02908

OXFORD, C(HARLES) W(ILLIAM), b Texarkana, Tex, Nov 16, 21; m 46; c 2. CHEMICAL ENGINEERING. *Educ:* Univ Ark, BSChE, 44; Univ Okla, PhD(chem eng), 52. *Hon Degrees:* DSc, Univ Ark, Little Rock, 89. *Prof Exp:* Jr engr, Ark-La Gas Co, 40; asst chemist, Koppers Co, 41; control foreman, Dickey Clay Mfg Co, 42; asst prof chem eng, Univ Ark, 48-50 & 51-52, assoc prof, 52-55; res engr, Boeing Airplane Co, 55; consult, Humble Oil Co, 56; vpres acad affairs, 79-87, PROF CHEM ENG & ASSOC DEAN ENG, UNIV ARK, FAYETTEVILLE, 57-, ADMIN VPRES, 68-, EXEC DIR, UNIV ARK FOUND, INC, 88- *Concurrent Pos:* Res partic, Carbide & Carbon Chem Co, 52. *Mem:* Am Soc Eng Educ; Am Inst Chem Engrs. *Res:* Revaporization of hydrocarbons; gas flow at low pressure. *Mailing Add:* 1024 Eastwood Fayetteville AR 72701

OXFORD, EDWIN, b Stamps, Ark, Jun 12, 39. MATHEMATICS. *Educ:* NMex State Univ, PhD(math), 71. *Prof Exp:* Prof math, Univ Southern Miss, 71-82; PROF MATH, BAYLOR UNIV, 82- *Mailing Add:* Baylor Univ Waco TX 76798

OXLEY, JOSEPH H(UBBARD), b Akron, Ohio, Aug 8, 29; m 51; c 6. CHEMICAL ENGINEERING. *Educ:* Carnegie Inst Technol, BS, 52, PhD(chem eng), 56. *Prof Exp:* Asst div chief, Battelle Mem Inst, 56-66, assoc mgr, Battelle Develop Corp, 66-72, mgr fuels & combustion sect, 72-76, ASST MGR, ENERGY & ENVIRON TECH DEPT, BATTELLE-COLUMBUS LABS, 77- *Mem:* Am Chem Soc; Am Inst Chem Engrs; Electrochem Soc; Am Inst Metall Eng; Licensing Exec Soc; Sigma Xi. *Res:* Halide metallurgy; coated particle nuclear fuels; mass transfer phenomenon; catalysis; coal conversion; flue gas treatment. *Mailing Add:* Battelle Mem Inst 505 King Ave Columbus OH 43201

OXLEY, PHILIP, b Utica, NY, Feb 1, 22; m 46; c 5. PETROLEUM GEOLOGY. *Educ:* Denison Univ, BA, 43; Columbia Univ, MA, 48, PhD, 52. *Prof Exp:* Asst, Geol Dept, Columbia Univ, 46-48; instr geol, Hamilton Col, 48-51, asst prof & chmn dept, 51-53; petrol geologist, Calif Co, La, 53-57; dist & div explor supt, Tenn Gas Transmission Co, 57-61; div explor mgr, Signal Oil & Gas Co, 61-65, vpres & mgr domestic explor, 65-69; exec vpres, Tex Crude Oil, 69-71; geol mgr explor, Tenneco Oil Co, 71-72, mgr foreign explor & prof, 72-76, vpres, 72-74, sr vpres, 74-81, PRES, TENNECO OIL EXPLOR & PROD CO, 81- *Concurrent Pos:* Temporary expert, NY State Mus, 50-51; party chief, Nfld Geol Surv, 52. *Mem:* Fel Geol Soc Am; Am Asn Petrol Geol. *Res:* Ordovician Chazyan stratigraphy of New York and Vermont; Ordovician stratigraphy and areal geology of western Newfoundland; petroleum geology of Texas-Louisiana Gulf Coast and offshore. *Mailing Add:* 1010 Milan Ave Houston TX 77225

OXLEY, THERON D, JR, b Randlett, Okla, Apr 3, 31; m 57. MATHEMATICS. *Educ:* Tex Christian Univ, BA, 51; Purdue Univ, MS, 53, PhD(math), 56. *Prof Exp:* Asst, Purdue Univ, 51-56; asst prof math, Kans State Univ, 56-60; asst prof, 60-64, ASSOC PROF MATH, DRAKE UNIV, 64- *Mem:* Am Math Soc; Math Asn Am. *Res:* Complex variable; infinite series. *Mailing Add:* Dept Math & Comput Sci Drake Univ 25th St & Univ Ave Des Moines IA 50311

OXMAN, MICHAEL ALLAN, b Milwaukee, Wis, Nov 7, 35; m 58; c 3. RESEARCH ADMINISTRATION, PHARMACY. *Educ:* Univ Wis, BS, 58, MS, 60, PhD(medicinal chem), 63. *Prof Exp:* Prod chemist, Aldrich Chem Co, Inc, 63; scientist, NIH, 63-65, assoc res chemist, Sterling-Winthrop Res

Inst, 65-68, health scientist adminr, Spec Res Resources Br, Div Res Facilities & Resources, 68-71, prog adminr, Biotechnol Resources Br, Div Res Resources, 71-74, chief tech files implementation br, Spec Info Serv, Nat Libr Med, 74-75, exec secy, review br, Div Extramural Affairs Nat Heart & Lung Inst, 75-78, health scientist adminr, Biomed Res Support Prog, 78-80, asst dir rev, Div Res Resources, 80-90, CHIEF, SCI REV OFF, NAT INST AGING, NIH, 90- *Honors & Awards:* Commendation Medal, US Pub Health Serv, 78, Meritorious Serv Medal, 90. *Mem:* Am Chem Soc. *Res:* Medicinal chemistry; aromatic biogenesis; oxidative phosphorylation; vitamin E; steroids; development of on-line, interactive toxicology data base. *Mailing Add:* Nat Inst Aging NIH Bethesda MD 20892

OXNARD, CHARLES ERNEST, b Durham, Eng, Sept 9, 33; US citizen; m 59; c 2. HUMAN BIOLOGY, EVOLUTIONARY BIOLOGY. *Educ:* Univ Birmingham, BSc, 55, MB, ChB, 58, PhD(med anat), 62, DSc(sci & engr), 75. *Prof Exp:* House physician, Queen Elizabeth Hosp, Birmingham, 58-59, house surgeon, 59; from lectr to sr lectr anat, Univ Birmingham, 62-66; assoc prof, 66-70, master biol col div & assoc dean & div biol sci, 72-73, prof anat, Univ Chicago, 70-78, dean col, 73-77; dean grad sch, 78-83, prof anat & biol sci, Univ Southern Calif, 78-87, univ prof, 83-87; PROF ANAT & HUMAN BIOL, UNIV WESTERN AUSTRALIA, 87-, DIR CTR HUMAN BIOL, 89-, HEAD, DIV SCI, 90- *Concurrent Pos:* Res grants, US Dept Health, Educ & Welfare, 63-66 & 67-70, Agr Res Coun Gt Brit, 64-66, Louis Block Fund, Univ Chicago, 66-70, USPHS, 67-70 & 74-77, NSF, 71-81, NIH, 83-87, Fac Res Fund, 84-87, IBM, 84-87, RAINE FOUND, 88-91, ARC, 87-92, CSIRO, 87-91; overseas assoc, Univ Birmingham, 72-; assoc ed, Am J Anat, 72-78, Am J Phys Anthrop, 74-78 & Am J Primates, 84-; external examr, 75-77, vis prof Univ Hong Kong, 77-; res assoc, Field Mus, Chicago, 79-, Los Angeles Nat Hist Mus, 84- & Page Nat Hist Mus, 86-; author of several published books; bd mem, Software Teacher Inc, 84- *Honors & Awards:* Huang-Chan Mem Medal, Univ Hong Kong, 80. *Mem:* Fel AAAS; Am Asn Anat; Am Asn Phys Anthrop; Am Soc Zool; Anat Soc Gt Brit & Ireland; fel NY Acad Sci; Australian Soc Human Biol (pres, 87-91); Anatomical Soc Australia & NZ (pres, 90-92). *Res:* Evolution of form in mammals, especially primates, utilizing comparative anatomical, biometric and biomechanical techniques; vitamin B-12 deficiency in primates, especially its effects on nervous system, mucous membranes, blood, growth and reproduction; application of diometric techniques to study of behavior, environment and diet; sexual diamorphisms in primates. *Mailing Add:* Dept Anat & Human Biol Univ Western Australia Nedlands Western Australia 6009 Australia

OXTOBY, DAVID W, b Bryn Mawr, Pa, Oct 17, 51; m 77; c 3. PHYSICAL CHEMISTRY. *Educ:* Harvard Univ, BA, 72; Univ Calif, Berkeley, PhD(chem), 75. *Prof Exp:* Res assoc chem, Univ Chicago, 75-76 & Univ Paris, 76-77; from asst prof to assoc prof, 77-86, PROF CHEM, UNIV CHICAGO, 86- *Concurrent Pos:* Alfred P Sloan Found fel, 77-, John Simon Guggenheim fel, 87-88; teacher-scholar, Camille & Henry Dreyfus Found, 80-85; Benjamin Meaker vis prof, Univ Bristol, 87-88; Mellon Found prof, 87- *Honors & Awards:* Marlow Medal, Royal Soc Chem, 83. *Mem:* Am Phys Soc; Royal Soc Chem; Am Chem Soc. *Res:* Theoretical physical chemistry; statistical mechanics of fluids; light scattering; relaxation processes in liquids; phase transitions. *Mailing Add:* James Franck Inst Univ Chicago 5640 S Ellis Ave Chicago IL 60637-1433

OXTOBY, JOHN CORNING, mathematics; deceased, see previous edition for last biography

OYAMA, JIRO, b Los Angeles, Calif, Aug 2, 25; m 66; c 2. PHYSIOLOGY, BIOCHEMISTRY. *Educ:* Northwestern Univ, BS, 49; George Washington Univ, MS, 56, PhD(biochem), 60. *Prof Exp:* Biochemist, NIH, 50-56; res biochemist, US Food & Drug Admin, 56-61; RES SCIENTIST, AMES RES CTR, NASA, 61- *Concurrent Pos:* Lectr & spec consult, San Jose State Univ, 69-74; lectr, Stanford Univ, 70-80. *Mem:* Am Chem Soc; Am Physiol Soc; Sigma Xi; Am Soc Gravitational & Space Biol. *Res:* Gravitational biology; acceleration stress physiology; space sta centrifuge; carbohydrate metabolism; thermoregulation; respiratory metabolism; aging; developmental physiology; bone and muscle. *Mailing Add:* 21435 Holly Oak Dr Cupertino CA 95014-4927

OYAMA, SHIGEO TED, b Tokyo, Japan, Feb 16, 55; m 82; c 1. CATALYSIS, CHEMICAL ENGINEERING. *Educ:* Yale Univ, BS, 76; Stanford Univ, PhD(chem eng), 81. *Prof Exp:* RES ENGR, CATALYTICA ASSOCS, INC, 81- *Mem:* AAAS; Am Inst Chem Engrs; Am Chem Soc. *Res:* Synthesis and characterization of new materials. *Mailing Add:* Six Drumlin Dr Potsdam NY 13676

OYAMA, VANCE I, b Los Angeles, Calif, June 6, 22; m 45; c 3. BIOCHEMISTRY, ORGANIC CHEMISTRY. *Educ:* George Washington Univ, BS, 53, MS, 60. *Prof Exp:* Med technologist, USPHS, Johns Hopkins Univ, 45-47; med technologist, NIH, 47, chemist, 47-60; sr scientist, Jet Propulsion Lab, Calif Inst Technol, 60-63; prin scientist, Ames Res Ctr, NASA, 63-65, chief, Life Detection Systs Br, 65-76, chief, Planetary Explor Off, 76- 80; RETIRED. *Mem:* Sigma Xi. *Res:* Viable organisms in lunar samples, gas exchange exploration, Viking Mission, sounder gas chromatograph Pioneer Venus Mission; pyrolysis; gas chromatography; carbon dioxide fixation; metabolic monitors; life detection sciences; lunar sciences; planetary sciences and biology; chemical evolution; origin of life. *Mailing Add:* Box 6922 Middletown CA 95461

OYEN, ORDEAN JAMES, b Appleton, Minn, June 27, 44; m 65; c 2. SKELETAL BIOLOGY & HISTOLOGY, PHILOSOPHY OF SCIENCE. *Educ:* Univ Minn, BA, 70, PhD(anthrop & anat), 74. *Prof Exp:* Instr anat & anthrop, Univ Minn, 73-74; asst prof phys anthrop & anat, Tex A&M Univ, 74-79; asst prof, 79-82, ASSOC PROF ORAL BIOL, SCH DENT, CASE WESTERN RESERVE UNIV, 82- *Concurrent Pos:* Prin investr morphol & nonphonetric anal, pvt craniofacial growth, NSF, 78-80, bone growth & masticator, function pvt, 83-86, masticator biomech & craniofacial morphol, 86-88; Assoc prof anthrop, Case Western Univ, 81-; res assoc, Cleveland Mus Natural Hist, 81-; chmn, Comt Res & Funding, Am Asn Phys Anthrop, 81-; res assoc, Inst Biomed Engr Res, Univ Akron, 85-; co-prin investr, tooth wear & tooth use in cereopithecus aethropics, NIH (NIDR), 85-88; asst dir, Off Res Admin, Case Western Reserve Univ, 90-; vis scholar & scientist, Nat Taiwan Univ, 89. *Honors & Awards:* Centennial Lectr, Creighton Univ, 78. *Mem:* Am Asn Phys Anthropologists; Int Asn Dent Res; Am Asn Dent Res; fel Am Chem Engrs. *Res:* Evolution biology, with emphasis on skeletal biology, growth and development histology and functional anatomy; theoretical and experimental aspects of craniofacial morphogenesis; philosophy of science; biomedical engineering; research administration; development. *Mailing Add:* Off Res Admin Case Western Reserve Univ Cleveland OH 44106

OYER, ALDEN TREMAINE, b Easton, Pa, July 13, 47; m 69. COMPUTER SCIENCE, ELECTRICAL ENGINEERING. *Educ:* Lafayette Col, BS, 69; Univ Wyo, PhD(physics), 76. *Prof Exp:* MEM STAFF, LOS ALAMOS NAT LAB, 76- *Concurrent Pos:* Consult, Nuclear Regulatory Comn, 77-81. *Mem:* Asn Comput Mach; Inst Elec & Electronic Engrs; Am Phys Soc. *Res:* Engineering evaluations. *Mailing Add:* Los Alamos Sci Lab PO Box 1663 Los Alamos NM 87545

OYER, HERBERT JOSEPH, b Groveland, Ill, Jan 28, 21; m 47; c 1. AUDIOLOGY, PSYCHOLOGY. *Educ:* Bluffton Col, AB, 43; Bowling Green State Univ, MSEd, 49; Ohio State Univ, PhD, 55. *Prof Exp:* Dir, Speech & Hearing Clin, 60-71, prof & chmn, Dept Speech, 64-67, prof & chmn, Dept Audiol & Speech Sci, 67-71, dean, Col Commun Arts & Sci, 71-75, dean, Grad Sch, 75-81, prof audiol & speech sci, Mich State Univ, 81-; EMER PROF DEPT COMMUN, OHIO STATE UNIV, COLUMBUS. *Concurrent Pos:* Am Speech & Hearing Asn fel, 61; vis prof, Fla State Univ. *Mem:* Acoust Soc Am; Am Psychol Asn; Acad Rehab Audiol (pres); Am Speech & Hearing Asn (vpres); Am Speech-Lang Hearing Asn. *Res:* Visual communication; auditory processing; speech intelligibility. *Mailing Add:* 1430 London Dr Columbus OH 43221

OYEWOLE, SAUNDRA HERNDON, b Washington, DC, Apr 26, 43; m 70; c 3. MICROBIOLOGY. *Educ:* Howard Univ, BS, 65; Univ Chicago, MS, 67; Univ Mass, PhD(microbiol), 73. *Prof Exp:* From asst to assoc prof Microbiol, Sch Natural Sci, Hampshire Col, 73-81; from assoc prof to prof, 81-90, CHAIR, BIOL DEPT, TRINITY COL, 90- *Concurrent Pos:* Res assoc, Dept Biochem, Univ Mass, 73- *Mem:* Am Soc Microbiol; AAAS. *Res:* Development, structure and function of bacterial photosynthetic membranes; photochemical apparatus of the green photosynthetic bacteria. *Mailing Add:* Dept Biol Trinity Col Franklin & Mich Ave NE Washington DC 20017-1094

OYLER, ALAN RICHARD, b Key West, Fla, Oct 13, 47; m 70; c 2. HIGH PERFORMANCE LIQUID CHROMATOGRAPHY. *Educ:* Albright Col, BS, 69; Lehigh Univ, PhD(chem), 73. *Prof Exp:* Res specialist chem, Univ Minn, 73-81; MEM STAFF, MOLECULAR SPECTROS DEPT, ORTHO PHARMACEUT CORP, 81- *Mem:* Am Chem Soc; AAAS. *Res:* Pharmaceutical analysis; analytical techniques and chemistry; trace organic analysis; high performance liquid chromatography. *Mailing Add:* Five Wildwood Ct Flemington NJ 08822

OYLER, DEE EDWARD, b Pueblo, Colo, Jan 23, 42; m 62; c 4. MINERAL SUPPLEMENTS, MATHEMATICS. *Educ:* Brigham Young Univ, BS, 65, PhD(phys chem), 71. *Prof Exp:* Fel chem, Univ Lethbridge, 71-72; dir labs, Albion Labs, 72-85; assoc instr math, Univ Utah, 83-89; ASST PROF, UTAH VALLEY COMMUNITY COL, 87- *Concurrent Pos:* Consult, 85-; gen mgr & tech dir, Monarch Nutrit Labs, Ogden, Utah, 87. *Mem:* Am Chem Soc; fel Am Inst Chemists; Am Math Asn Two Year Cols. *Res:* Thermodynamics involving calorimetric investigations of association equilibria and phase equilibria studies of charge-transfer complexes and intermetallic compounds; formulation of mineral supplements and testing their intestinal absorption. *Mailing Add:* 114 W 1900 S Bountiful UT 84010

OYLER, GLENN W(ALKER), b Franklin Co, Pa, Sept 28, 23; div; c 3. METALLURGICAL ENGINEERING. *Educ:* Pa State Univ, BS, 49; Univ Pittsburgh, MS, 51; Lehigh Univ, PhD(metall eng), 53; Univ RI, cert mgt, 63. *Prof Exp:* Welder, Fairchild Aircraft Corp, Md, 41-42; welding engr, Aluminum Co Am, Pa, 49-51; head welding div, Linde Co, Union Carbide Corp, NJ, 53-60; chief welding engr & proj mgr, Metals & Controls, Inc Div, Tex Instruments, Inc, Mass, 60-63; dir appl res & develop, ACF Industs, Inc, NMex, 63-67; mfg mgr, Lockheed Missiles & Space Co, Calif, 67-69; chief advan mfg technol, Martin Marietta Corp, 69-77; tech dir, Am Welding Soc, 77-80; PRES & EXEC DIR, WELDING RES COUN, NY, 80- *Mem:* Am Welding Soc; fel Am Soc Metals; Am Soc Mech Eng. *Res:* Research, development and production support of metallic and non-metallic materials. *Mailing Add:* Welding Res Coun 345 E 47th St New York NY 10017

OYLER, J MACK, b Sapulpa, Okla, Aug 8, 26; m 52; c 2. PHYSIOLOGY. *Educ:* Okla State Univ, DVM, 53, PhD(physiol), 69. *Prof Exp:* Asst prof vet sci, Univ Ark, 54-55; pvt pract, Grove, Okla, 56-65; asst prof physiol, Okla State Univ, 68-69; from asst prof to assoc prof physiol, Univ Ga, 71-73; assoc prof med & surg, Okla State Univ, 73-75; prof vet med, Va Polytech Inst & State Univ, 75-80; PROF & ASSOC DEAN, COL VET MED, OKLA STATE UNIV, 80- *Mem:* Am Vet Med Asn; Am Soc Vet Physiol & Pharmacol. *Res:* Metabolism with special interest in blood-brain barrier and neonatal studies. *Mailing Add:* Col Vet Med Okla State Univ Stillwater OK 74078

OYSTER, CLYDE WILLIAM, b Marietta, Ohio, Apr 29, 40; m 67. VISUAL PHYSIOLOGY. *Educ:* Ohio State Univ, BS, 63; Univ Calif, Berkeley, PhD(physiol optics), 67. *Prof Exp:* From asst prof to assoc prof, 70-84, PROF PHYSIOL OPTICS, SCH OPTOM, MED CTR, UNIV ALA, BIRMINGHAM, 84- *Concurrent Pos:* Fight for Sight res fel, Australian Nat Univ, 68-70; sr res fel visual physiol, Dept Physiol, Med Fac Rotterdam, 70; Nat Eye Inst grant, Univ Ala, Birmingham, 71- mem vision res prog comt, Nat Eye Inst, 74-77. *Honors & Awards:* Res Career Develop Award, Nat Eye Inst, 75-80. *Mem:* Asn Res Vision & Ophthal; Soc Neurosci. *Res:* Information processing in the visual system; neural interactions in retina and visual pathways. *Mailing Add:* Sch of Optom Univ of Ala Med Ctr Birmingham AL 35294

OZA, DIPAK H, b Poona, India, July 20, 54; US citizen; m 79; c 1. ELECTRON-ATOM SCATTERING, ASTRONAUTICS. *Educ:* MS Univ Baroda, India, BS, 74; La State Univ, Baton Rouge, MS, 82, PhD(theoret physics), 83. *Prof Exp:* Res assoc, Univ Va, 83-84, & La State Univ, 84-85; physicist, Nat Bur Standards, 85-88; PRES, PHYSICS CONSULT GROUP, 85-; SR MEM TECH STAFF, COMPUTER SCI CORP, 88- *Mem:* Am Phys Soc; Am Inst Aeronaut & Astronaut. *Res:* Theoretical studies of electron-atom scattering; atom resonance phenomena; spectral-line broadening; astronautics; orbit determination of satellites. *Mailing Add:* Physics Consult Group 13905 Riding Loop Dr North Potomac MD 20878

OZA, KANDARP GOVINDLAL, b Gujarat, India, July 12, 42. ELECTRICAL & SYSTEMS ENGINEERING. *Educ:* Gujarat Univ, India, BEE, 63; Univ Calif, Berkeley, MS, 64, PhD(elec eng), 67. *Prof Exp:* Asst prof elec eng, Univ Okla, 68-69; mem tech staff, AT&T Bell Labs, 69-70; asst prof elec eng, Indian Inst Technol, Kanpur, 70; mem tech staff, AT&T Bell Labs, 71-84; MKT DIR, DIGITAL SOUND CORP, 84- *Concurrent Pos:* Adj prof, Fairleigh-Dickinson Univ, 73-76; Bell Labs vis prof, Tuskegee Inst, Ala, 77-78. *Mem:* Inst Elec & Electronics Engrs. *Res:* Control, communications and computer systems; new telecommunication services; digital switching and transmissions; microprocessor-based systems; engineering economics; voice processing, interactive voice response. *Mailing Add:* Digital Sound Corp 2030 Alameda Padre Serra Santa Barbara CA 93103

OZAKI, HENRY YOSHIO, b Eau Gallie, Fla, Oct 25, 23; m 59; c 2. VEGETABLE CROPS. *Educ:* Univ Fla, BSA, 49; Cornell Univ, MS, 51, PhD, 54. *Prof Exp:* Asst veg crops, Cornell Univ, 49-53; asst horticulturist, Plantation Field Lab, Agr Res Sta, 54-68, assoc horticulturist, 68-69, ASSOC PROF HORT, EVERGLADES RES & EDUC CTR, UNIV FLA, 69- *Mem:* Am Soc Hort Sci; Sigma Xi. *Res:* Vegetable research on organic and sandy soils of Florida. *Mailing Add:* 498 E Conference Dr Boca Raton FL 33432

OZAKI, SATOSHI, b Osaka, Japan, July 4, 29; m 60; c 2. HIGH ENERGY EXPERIMENTAL PHYSICS. *Educ:* Osaka Univ, BS, 53, MS, 55; Mass Inst Technol, PhD(physics), 59. *Prof Exp:* Res asst physics, Mass Inst Technol, 56-59; res assoc, 59-61, asst physicist, 61-63, assoc physicist, 63-66, physicist, 66-72, group leader physics, Brookhaven Nat Lab, 70-, sr physicist, 72-; AT PHYS DIV KEK, OHO-MACHI, JAPAN. *Concurrent Pos:* Vis prof, Osaka Univ, 75-76; vis prof & dir, Dept Physics, Nat Lab High Energy Physics, Japan, 81- *Mem:* Fel Am Phys Soc; Phys Soc Japan. *Res:* Study of high energy particle interactions; particle spectroscopy; high energy physics instrumentation. *Mailing Add:* Accelerator Div Nat Lab High Energy Physics 1-1 Oho Tsukuba Ibaraki-Ken 305 Japan

OZAN, M TURGUT, b Ankara, Turkey, May 8, 19; m 65. INDUSTRIAL ENGINEERING, OPERATIONS RESEARCH. *Educ:* Swiss Fed Inst Technol, BS & MS, 50; Univ Birmingham, MS, 58; Washington Univ, St Louis, DSc, 63. *Prof Exp:* Design engr, Asper Mach Tool Co, Switz, 50; proj engr, MAKKIM, Turkey, 50-52, chief engr, 52-54, dir prod planning, 54-56; asst prof indust eng & chmn dept, St Mary's Univ, 60-63, assoc prof, 63-65, prof indust eng, 65-90, assoc dean eng, 67-90, EMER PROF, DEPT INDUST ENG, ST MARY'S UNIV, SAN ANTONIO, TEX, 90- *Concurrent Pos:* Consult, 60-; consult, Keuffel & Esser Co, Hoboken, 70-; lectr & consult, Inst Technol Y De Estud Superiores De Monterrey, Mex, 72- *Mem:* Am Inst Indust Engrs; Nat Asn Corrosion Engrs; Opers Res Soc Am; Am Soc Eng Educ; Am Soc Nondestructive Testing. *Res:* Probability, statistics, reliability theory and operations research as applied to the design, operation and control of production systems; search theory; scheduling and flow line balancing problems in production planning; design optimization. *Mailing Add:* Dept Eng St Mary's Univ One Camino Santa Maria San Antonio TX 78284

OZARI, YEHUDA, b Alexandria, Egypt, Feb 19, 44; US citizen; m 65; c 3. INORGANIC CHEMISTRY, PHYSICAL CHEMISTRY. *Educ:* Hebrew Univ, Jerusalem, Israel, BSc, 66; Weizmann Inst Sci, Rehovot, Israel, MSc, 69, PhD(polymers), 75. *Prof Exp:* Res chemist, GAF Corp, 76-79, sr res chemist, 79-81; res assoc, Avery Dennison, 81-83, group leader, 83-85, mgr, 85-90, SR MGR, AVERY DENNISON, 90- *Concurrent Pos:* Postdoctoral, polymer chem, Princeton Univ, 76. *Mem:* Soc Advan Mat & Process Engrs; Tech Asn Pulp & Paper Indust; Am Chem Soc. *Res:* Development of new materials in the area of adhesives, additives, plastic films, papers; environmentally friendly office products, new disposable office products; implementing new concepts in industrial operations; development of new concepts with marketing personnel. *Mailing Add:* Avery Int Corp 2900 Bradley St Pasadena CA 91107

OZAROW, VERNON, physical chemistry, for more information see previous edition

OZATO, KEIKO, b Yamagata, Japan; US citizen; m 76. MOLECULAR IMMUNOLOGY, DEMELOPMENTAL BIOLOGY. *Educ:* Kyoto Univ, BA, 65, PhD(biol), 73. *Prof Exp:* Res fel immunol, Carnegie Inst Washington, 73-76; res assoc, Sch Med, Johns Hopkins Univ, 76-78; vis assoc, Nat Cancer Inst, 78-81; sr staff fel immunol, 81-83, RES MICROBIOLOGIST IMMUNOL, NAT INST CHILD HEALTH & HUMAN DEVELOP, 83-, HEAD, UNIT MOLECULAR GENETICS IMMUNITY LDMI, 83- *Concurrent Pos:* Assoc ed, J Immunol, 83-87, chief, molecular genetics immunity, 85-; chief, sect molecular genetics immunity, Nat Inst Child Health & Human Develop, 85- *Honors & Awards:* Super Award, Pub Health Serv, 90. *Mem:* Am Asn Immunol; Int Soc Interferon Res. *Res:* Major histocompatibility antigens; developmental gene regulation during mammalian embryogenesis; transcriptional regulation of major histocompatibility genes and C-fos oncogene via DNA protein interaction; interferon-induced activation of genes and analysis of the underlying mechanisms; major histocompatibility Class I antigen expression; c-fos oncogene expression. *Mailing Add:* Lab Develop & Molecular Immunity NICHD Bldg 6 Rm 1A03 NIH Bethesda MD 20892

OZAWA, KENNETH SUSUMU, b Tokyo, Japan, Nov 22, 31; US citizen; m 63; c 2. FLUID PHYSICS. *Educ:* John Carroll Univ, BS, 59, MS, 60; Univ Kans, PhD(eng mech), 75. *Prof Exp:* Instr physics, John Carroll Univ, 60-63; asst prof, 63-68, assoc prof, 68-75, PROF PHYSICS, CALIF POLYTECH STATE UNIV, 75- *Mem:* Am Phys Soc; Am Assoc Physics Teachers; AAAS; NY Acad Sci. *Res:* Experimental study of wind-induced waves and currents in lakes of limited fetch and depth. *Mailing Add:* Dept Physics Calif Polytech State Univ San Luis Obispo CA 93407

OZBUN, JIM L, b Carson, NDak, Sept 3, 36; m 59; c 2. PLANT PHYSIOLOGY. *Educ:* NDak State Univ, BS, 59, MS, 61; NC State Univ, PhD(soils, plant physiol), 64. *Prof Exp:* From asst prof to assoc prof plant physiol, 64-74, from actg chmn to chmn dept veg crops, 71-75, asst dir res, NY Exp Sta & NY State Col Agr, 67-70, assoc dir exten, NY State Col Agr & Live Sci & prof plant physiol, Cornell Univ, 74-76; prof hort sci & landscape archit & head dept, Univ Minn, 76-81; assoc dean agr & dir res, Kans State Univ, 81-; DEAN AGR & HOME ECON, WASH STATE UNIV, PULLMAN. *Concurrent Pos:* Sabbatical leave, Dept Biochem & Biophys, Univ Calif, Davis, 70-71. *Honors & Awards:* Campbell Award, Am Inst Biol Sci, 65. *Mem:* Am Soc Hort Sci. *Res:* Photosynthesis; effect of light and potassium nutrition on photosynthesis and respiration; influence of ammonia nitrogen on potassium and water use efficiency; vernalization and flowering; yield physiology. *Mailing Add:* 1301 12th Ave N Fargo ND 58105

OZBURN, GEORGE W, b Guelph, Ont; m 39; c 4. ENTOMOLOGY, BIOLOGY. *Educ:* McGill Univ, BScAgr, 55, PhD(entom), 62; Univ London, DIC, 56. *Prof Exp:* Sci officer, Nigerian Stored Prod Res Unit, Nigeria, 56-59; asst prof biol, Northern Mich Univ, 62-65; asst prof, 65-71, assoc prof, 71-76, PROF BIOL, LAKEHEAD UNIV, 76- *Concurrent Pos:* Grants, Nat Res Coun, 65-69, USPHS, 66-68, Dept Univ Affairs, 70-71, Environ Protection, 72-75 & Ministry of Environ, 74-76, 77-80 & 81-90. *Mem:* Can Soc Zool. *Res:* Insecticide resistance; pest control; aquatic entomology and biology related to pollutants and environment changes; aquatic toxicology. *Mailing Add:* Dept Biol Lakehead Univ Oliver Road Thunder Bay ON P7B 5E1 Can

OZER, HARVEY LEON, b Boston, Mass, July 6, 38; m 60; c 1. VIROLOGY, CARCINOGENESIS. *Educ:* Harvard Univ, AB, 60; Stanford Univ, MD, 65. *Prof Exp:* Intern pediat, Children's Hosp Med Ctr, Boston, 65-66; res assoc, Lab Biol of Viruses, Nat Inst Allergy & Infectious Dis, 66-68; sr staff fel, Lab Biochem, Nat Cancer Inst, 69-72; sr scientist, Worcester Found Exp Biol, 72-77; prof, Dept Biol Sci, Hunter Col, 77-89, Thomas Hunter prof sci & math, 83-89; PROF & CHMN, DEPT MICROBIOL & MOLECULAR GENETICS, NJ MED SCH, 88- *Concurrent Pos:* Vis fel, Inst Tumor Biol, Karolinska Inst, Sweden, 64; USPHS fel, Lab Viral Dis, Nat Inst Allergy & Infectious Dis, 68-69; prof lectr, Med Sch, George Washington Univ, 68-71; vis fac microbiol, Med Sch, Univ Mass, 72-77; virol study sect, NIH, 78-82, ad hoc mem, NIH rev panels; vis prof, dept molecular biol & genetics, Johns Hopkins Med Sch, 83-84; vis comt mem & consult, Dept Biol, NY Univ 78-80, SUNY, Stony Brook, grad prog cell & develop biol, 82, Wistar Inst, Univ PA, 83-, Manhattan Col, 87; conf chmn & consult, Mult Res Grants, NIH & NSF, 73-; prog coord, Gene Struct & Function Study Ctr, 86-88; basic res adv comt, NJ Comn Cancer Res, 89- *Honors & Awards:* Borden Award, Med Sch, Stanford Univ, 65. *Mem:* Am Soc Microbiol; Genetics Soc Am; NY Acad Sci; Am Soc Virol; Sigma Xi; Am Soc Biochem & Molecular Biol. *Res:* Animal virology; genetics; mammalian cell biochemistry and genetics; tumor biology; immortalization and transformation of human fibroblasts; exploited human and rodent cells in culture to identify functions essential to cell and viral replication; the relevant genes are being isolated for detailed genetic and biochemical analysis. *Mailing Add:* NJ Med Sch 185 S Orange Ave Newark NJ 07103-2714

OZER, MARK NORMAN, b Cambridge, Mass, Jan 17, 32; m 79; c 5. NEUROLOGY. *Educ:* Harvard Col, BA, 53; Boston Univ, MD, 57. *Prof Exp:* Instr neurol, Sch Med, George Washington Univ, 65-67, from asst prof to assoc prof child health & develop, 67-83; assoc prof neurol, Med Col Va, 83-89; PROF NEUROL, GEORGETOWN UNIV MED SCH, 89- *Concurrent Pos:* NIMH Spec fel, Washington Sch Psychiat, DC, 64-65; consult, US Air Force, 65-71; consult neurologist, Nat Childrens Ctr, Washington, DC, 65-71; prog dir learning studies, Childrens Hosp, Nat Med Ctr, Washington, DC, 65-74; multiple grants, 65-74; assoc prof psychiat & pediat, Univ Md, 67-74; consult neurologist, Prince Georges County Health Dept, Md, 67-70 & 77-82 & Northern Va Training Ctr, Va, 75-77; lectr, Dept Maternal & Child Health, Univ Mich, 68-70; sci fel, McGuire Va Med Ctr, Richmond Va, 82-83, asst chief sci, 83-89; dir stroke prog, Nat Rehab Hosp. *Mem:* Am Acad Neurol; Am Soc Cybernetics (pres, 76-77); Am Acad Aphasia; Int Neuropsychol Soc; Cong Rehab Med; Am Paraplegia Soc; Int Med Soc Paraplegia. *Res:* Development of assessment techniques for use with handicapped individuals that illustrate the sampling of the brain as an adaptive system; the design of delivery systems for the implementation of such techniques and the training of professionals in their use; development of delivery systems for persons with wide range of chronic neurological problems, evaluation systems for such. *Mailing Add:* 1919 Stuart Ave Richmond VA 23220

OZERE, RUDOLPH L, b Winnipeg, Man, Mar 14, 26; m 59; c 2. PEDIATRICS, MICROBIOLOGY. *Educ:* McGill Univ, BSc, 47; Univ Ottawa, MD, 53; Harvard Univ, SM, 67. *Prof Exp:* Lectr bact, 61-70, asst prof prev med, 63-70, from asst prof to assoc prof pediat, 61-70, prof pediat, Dalhousie Univ, 70-, lectr microbiol, 70-, asst prof prev med, 77-; AT DEPT PEDIAT, MEMORIAL UNIV. *Concurrent Pos:* Res fel microbiol, Dalhousie Univ, 59-61. *Mem:* Can Soc Clin Invest; Am Pediat Soc. *Res:* Epidemiology; clinical pediatrics. *Mailing Add:* Dept Pediat Memorial Univ St Johns NF A1C 5S7 Can

OZERNOY, LEONID M, b Moscow, Russia, May 19, 39; m 69; c 2. THEORETICAL ASTROPHYSICS, COSMOLOGY. *Educ:* Moscow State Univ, BS, 61, MS, 63; Sternberg Astron Inst, Moscow, PhD(astrophys), 66, full Doctor Sci, Lebedev Phys Inst, 71. *Prof Exp:* Res scientist & sr res scientist, astrophys, Dept Theoret Physics, Lebedev Phys Inst, USSR Acad Sci, 66-86; distinguished vis scientist astrophys, Harvard-Smithsonian Ctr Astrophys, 86-89; DISTINGUISHED VIS SCHOLAR ASTROPHYS, LOS ALAMOS NAT LAB, 89- *Concurrent Pos:* Asst prof & prof, Moscow Physics & Tech Inst, 68-79; mem, Sci Coun Plasma Astrophys, Presidium USSR Acad Sci, 71-75; vis prof, Boston Univ, 86-87 & Harvard Univ, 87. *Honors & Awards:* Award, Lebedev Phys Inst, 67; Fel, Am Phys Soc. *Mem:* Int Astron Union; Comt Space Res; AAAS; Am Astron Soc; Astron Soc Pac; NY Acad Sci. *Res:* About 200 scientific papers in theoretical astrophysics, high energy astrophysics, radio astronomy, galactic and extra-galactic astronomy and cosmology, as well as several books; work in physics of nonthermal processes in active galaxies and quasars, dynamics of compact stellar systems; astrophysics of black holes, formation of galaxies and large-scale structure of the universe, observational cosmology. *Mailing Add:* K-305 IGPP Los Alamos Nat Lab Los Alamos NM 87545

OZIER, IRVING, b Montreal, Que, Sept 7, 38; m 63; c 3. MOLECULAR SPECTROSCOPY. *Educ:* Univ Toronto, BA, 60; Harvard Univ, AM, 61, PhD(physics), 65. *Prof Exp:* Alfred P Sloan res fel, 72-74; vis res fel, Katholieke Univ, Nijmegen, Netherlands, 76-77; vis res officer, Herzberg Inst Astrophys, Ottawa, Can, 82-83; vis prof, Eidgenossische Technische Hochscule, Zurich, Switz, 88-89. *Mem:* Am Phys Soc; Can Asn Physicists. *Res:* Forbidden rotational transitions, small electric dipole moments; far-infrared and microwave spectroscopy; hindered internal rotation; magnetic resonance in molecules; nuclear hyperfine interactions and rotational magnetic moments; tetrahedral molecules; molecular beams. *Mailing Add:* Dept Physics Univ BC Vancouver BC V6T 2A6 Can

OZIMEK, EDWARD JOSEPH, b West Natrona, Pa, Nov 11, 45; m 68; c 4. SOLID STATE PHYSICS, HYBRID CIRCUIT DESIGN. *Educ:* Mt Union Col, BS, 68; Univ Akron, MS, 72; Colo State Univ, DPhil(physics), 77. *Prof Exp:* Asst prof, Univ Wis-Stout, 76-77; staff physicist piezoelec, Allied Chem Corp, 77-79; adj asst prof, Univ Ariz, 79-81; Staff Engr, IBM Corp, 81-84; AT RES LABS, EASTMAN KODAK CO, 84- *Mem:* Am Phys Soc; Inst Elec & Electronics Engrs. *Res:* Ultrasonic attenuation and velocity dispersion in materials; electron transport in metals; piezoelectric properties of oxide crystals; heat transfer and predictive thermal dosimetry in materials; magnetic properties of thin films; digital magnetic recording heads; optoelectronic image sensor packaging. *Mailing Add:* Res Labs Bldg 81 Eastman Kodak Co Mail Code 02024 Rochester NY 14650-2024

OZIMKOSKI, RAYMOND EDWARD, b New York, NY, July 3, 27. MATHEMATICS. *Educ:* Fordham Univ, BS, 46, MS, 47. *Prof Exp:* From instr to asst prof math, Fordham Univ, 47-59; from asst prof to prof, 59-88, EMER PROF MATH, MERRIMACK COL, 88- *Mem:* Am Math Soc; Math Asn Am. *Mailing Add:* 32H Forest Acres Dr Bradford MA 01835

OZIOMEK, JAMES, b Akron, Ohio, Feb 17, 41; div; c 4. ORGANIC CHEMISTRY, POLYMER CHEMISTRY. *Educ:* Case Inst Technol, BS, 63; Univ Calif, Berkeley, PhD(chem), 68. *Prof Exp:* Res scientist, 68-75, sr res scientist, 75-78, ASSOC SCIENTIST, CENT RES LAB, FIRESTONE TIRE & RUBBER CO, 78- *Concurrent Pos:* Vis scientist, Inst Polymer Sci, Univ Akron, 73-74. *Mem:* Am Chem Soc; Sigma Xi; NY Acad Sci. *Res:* Anionic and Ziegler polymerization for preparation of elastomers; cationic polymerization; urethane chemistry; adhesion; physical organic chemistry. *Mailing Add:* 30 E Mapledale Akron OH 44301

OZISIK, M NECATI, b Istanbul, Turkey, June 17, 23; m 57; c 1. MECHANICAL ENGINEERING. *Educ:* Univ London, BS, 47, PhD(mech eng), 50. *Prof Exp:* Sr res scientist, Am Soc Heating, Refrig & Air Conditioning Engrs Res Lab, 57-59 & Oak Ridge Nat Lab, 59-63; assoc prof, 63-66, PROF MECH ENG, NC STATE UNIV, 66- *Honors & Awards:* F Bernard Hall Prize, Brit Inst Mech Engrs, 52; Western Elec Fund Award, Am Soc Eng Educ, 72; Turkish Sci Award, 80; Distinguished Res Award, Alcoa Found, 84; Reynolds Award, 86. *Mem:* Fel Am Soc Mech Engrs; Int Asn Hydrogen Energy. *Res:* Heat transfer; fluid mechanics; radiative transfer; author of nine books in engineering. *Mailing Add:* Dept Mech NC State Univ Raleigh NC 27695-7910

OZOG, FRANCIS JOSEPH, b Poland, Sept 9, 22; nat US; m 48; c 3. PHYSICAL CHEMISTRY, ORGANIC CHEMISTRY. *Educ:* Univ Detroit, BS, 47; Northwestern Univ, PhD(chem), 50. *Prof Exp:* From instr to assoc prof, 50-63, head dept, 57-69, dir div natural sci & math, 64-66 & 71-75, PROF CHEM, REGIS COL, 64- *Mem:* Am Chem Soc; fel, Asn Off Racing Chemists; fel Am Inst Chemists. *Res:* Toxicology and forensic chemistry. *Mailing Add:* 28908 Cedar Circle Evergreen CO 80439

OZOL, MICHAEL ARVID, b New York, NY, Nov 27, 34; m 58; c 2. GEOLOGY. *Educ:* City Col New York,BS, 57; Univ Pittsburgh, MS, 58; Rensselaer Polytech Inst, PhD(geol), 63. *Prof Exp:* Asst petrologic res, Rensselaer Polytech Inst, 60-63; hwy mat res analyst, Va Hwy Res Coun, 63-71; sr res scientist, Res Inst Advan Studies, 71-78, prin scientist, 78-80, DIR, TECH DEVELOP, MARTIN MARIETTA AGGREGATES,, MARTIN MARIETTA CORP, 80- *Mem:* AAAS; Geol Soc Am; Am Soc Testing & Mat. *Res:* Properties of rocks which determine their suitability for use as aggregate; behavior of aggregate materials, including physical and chemical reactions; petrography of concrete and aggregates. *Mailing Add:* 2403 Kenoak Rd Baltimore MD 21209

OZOLS, JURIS, LIVER MEMBRANE STRUCTURE. *Educ:* Univ Wash, PhD(biochem), 62. *Prof Exp:* PROF BIOCHEM, HEALTH SCI CTR, UNIV CONN, 69- *Mailing Add:* Dept Biochem Univ Conn Health Ctr 1263 Farmington Ave Farmington CT 06032

OZONOFF, DAVID MICHAEL, b Milwaukee, Wis, July 7, 42; m 72; c 2. HAZARDOUS WASTE, TOXIC CHEMICAL EXPOSURES. *Educ:* Univ Wis, Madison, BS, 62; Cornell Univ Med Col, MD, 67; Johns Hopkins Sch Hyg & Pub Health, MPH, 68. *Prof Exp:* Res assoc elec eng, Mass Inst Technol, 68-77, Mellon fel hist pub health, 76-77; asst prof environ health, Sch Med, 77-80, assoc prof pub health, 82-87, CHAIR ENVIRON HEALTH, SCH MED & PUB HEALTH, BOSTON UNIV, 81-, PROF PUB HEALTH, 87- *Concurrent Pos:* Macy fel hist sci, Harvard Univ, 75-76. *Mem:* Am Pub Health Asn; AAAS; Soc Occup & Environ Health; Nat Environ Health Asn; Mass Pub Health Asn (pres, 83-84). *Res:* Epidemiology and health effects of exposure to toxic wastes; history of public health, especially asbestos related diseases. *Mailing Add:* Sch Pub Health Boston Univ 80 E Concord St Boston MA 02118

OZSVATH, ISTVAN, b Kolesd, Hungary, Sept 10, 28; m 50; c 1. MATHEMATICAL PHYSICS. *Educ:* Eotvos Lorand Univ, Budapest, MS, 51; Univ Hamburg, PhD(astron), 60. *Prof Exp:* Res assoc astron, Konkoly Observ, Hungary, 51-56 & Hamburg Observ, Ger, 56-62; vis lectr math, Univ Tex, 62-63; assoc prof, 63-67, PROF MATH & MATH PHYSICS, UNIV TEX, DALLAS, 67- *Res:* Relativistic cosmology; general relativity. *Mailing Add:* Dept Math Univ Tex Dallas Box 830688 Richardson TX 75083-0688

P

PAABO, MAYA, b Estonia, June 11, 29; US citizen; m 51; c 1. ANALYTICAL CHEMISTRY. *Educ:* George Washington Univ, BS, 56, MS, 62. *Prof Exp:* Res chemist, Nat Inst Standards & Technol, 56-90; CONSULT, 90- *Mem:* Am Chem Soc. *Res:* Relative strength of acids and bases in aprotic media; development of acidity scales in alcohol-water solvents and in heavy water; analytical chemical studies of polymeric materials and their combustion products as they relate to toxic hazard from fires. *Mailing Add:* 5324 Colorado Ave NW Washington DC 20011

PAAL, FRANK F, b Budapest, Hungary, Aug 25, 30; US citizen. COMPUTER SCIENCE. *Educ:* McGill Univ, BSc, 59; Stanford Univ, MA, 62; Univ Calif, Los Angeles, MSc, 65, PhD(comput sci), 69. *Prof Exp:* Sr engr, Hoffman Electronics, 61-63; assoc prof, 67-81, PROF ELEC ENG, CALIF STATE UNIV, LONG BEACH, 81- *Mem:* Inst Elec & Electronics Engrs; Am Fedn Info Processing; Int Fedn Info Processing. *Res:* Arithmetic processors; simulation of dynamic systems. *Mailing Add:* Dept Elec Eng Comp Sci Calif State Univ Longbeach 1250 Bellflower Blvd Long Beach CA 90840

PAAPE, MAX J, PHAGOCYTOSIS, MAMMARY PHYSIOLOGY. *Educ:* Mich State Univ, PhD(physiol), 67. *Prof Exp:* RES PHYSIOLOGIST, USDA, 67- *Mailing Add:* Livestock & Poultry Inst USDA Beltsville MD 20705

PAARSONS, JAMES DELBERT, b Los Angeles, Calif, Mar 18, 47; m 69; c 1. SEMICONDUCTOR DEVICE DESIGN, SEMICONDUCTOR DEVICE PROCESS DEVELOPMENT. *Educ:* Univ Calif, Los Angeles, BS, 74, MS, 77, PhD(eng), 81. *Prof Exp:* Engr III, Jet Propulsion Labs, Hughes Aircraft Co, 73-79, sect head res lab, 80-89, dir metal-organic chem vapor deposition res, Advan Technol Mat, 89; ASSOC PROF SOLID STATE PHYSICS, ORE GRAD INST SCI & TECHNOL, 89- *Concurrent Pos:* Prin investr, Hughes Res Labs, Hughes Aircraft Co, 80-89, Hughes Power FET Contract, Classified, 84-89 & Naval Weapon Ctr Contract, 85-89; consult, Tektronix Inc, 90- *Mem:* Inst Elec & Electronics Engrs; Electrochem Soc; Am Vacuum Soc; Am Asn Crystal Growth. *Res:* Relationships of semiconductor properties; solid state device performance; metal-organic chemical vapor deposition growth; new semiconductors and device processing technologies; new solid state device and monolithic IC design concepts in beta-silicon carbide, III-V and II-VI semiconductors. *Mailing Add:* 8485 SW Secretariat Terr Beaverton OR 97005

PAASWELL, ROBERT E(MIL), b Red Wing, Minn, Jan 15, 37; m 58; c 2. TRANSPORTATION, CIVIL ENGINEERING. *Educ:* Columbia Univ, BA, 56, BS, 57, MS, 61; Rutgers Univ, PhD(civil eng), 65. *Prof Exp:* Civil engr, Spencer, White & Prentis, Inc, 57-59; teaching asst soil mech, Columbia Univ, 59-62; from asst prof to assoc prof civil eng, State Univ NY Buffalo, 64-76, prof, 76-80, dir, Ctr Transp Studies & Res, 78-81, prof & chmn, Dept Environ Design & Planning, 80-81; dir urban transp ctr & prof eng, Univ Ill, Chicago, 82-86; EXEC DIR, CHICAGO TRANSIT AUTHORITY, 86- *Concurrent Pos:* State Univ NY Res Found fac grant, 66-; chmn comt, Nat Acad Sci-Nat Res Coun, 66; Greater London Coun res fel polit & econ planning, Eng, 71-72; fac tech consult; energy consult, US Cong, 73-74; fac fel, US Dept Transp, 76-77; foreign expert, Jilin Univ Technol, Changchun, China, 85. *Mem:* AAAS; Am Soc Civil Engrs; Sigma Xi. *Res:* Urban transport studies; special service transportation analysis; problems of the carless; urban transportation; urban planning and systems; environmental systems; investigations of urban transportation systems, focusing on the users and the cost benefits; analysis of interactions of transportation and land use. *Mailing Add:* 167 W Goethe St Chicago IL 60610

PAAVOLA, LAURIE GAIL, b LaCrosse, Wis; m 71. CELL BIOLOGY, ANATOMY. *Educ:* San Diego State Univ, AB, 65; Stanford Univ, PhD(anat), 70. *Prof Exp:* PROF ANAT, SCH MED, TEMPLE UNIV, 72- *Concurrent Pos:* NIH fels, Dept Anat, Stanford Univ, 70-71 & Dept Anat, Temple Univ, 71-72; prin investr, Nat Inst Child Health & Human Develop res grant, 78-80. *Honors & Awards:* Frances Lou Kallman Award, Stanford Univ, 70. *Mem:* Am Asn Anatomists; Am Soc Cell Biol; AAAS. *Res:* Reproductive biology; ovary; cell biology of granulosa cells and luteal cells; smooth membrane biogenesis during luteinization; mechanisms of ovarian lipid metabolism; role of autophagy and heterophagy in luteolysis; electron microscope cytochemistry; transmission and scanning electron microscopy. *Mailing Add:* Dept Anat Sch Med Temple Univ 3400 N Broad St Philadelphia PA 19140

PABST, ADOLF, mineralogy, crystallography; deceased, see previous edition for last biography

PABST, MICHAEL JOHN, b Washington, DC, Oct 19, 45; m 69; c 2. PHAGOCYTES, IMMUNITY. *Educ:* Boston Col, BS, 67; Purdue Univ, PhD(biochem), 72. *Prof Exp:* Staff fel biochem, NIH, 72-74; asst prof oral biol, Sch Dent, 74-77, ASSOC PROF BIOCHEM, SCH MED, UNIV COLO, 76- *Concurrent Pos:* Assoc prof pediat, Nat Jewish Ctr Immunol & Respiratory Med, 77- *Mem:* Am Soc Microbiol; Am Soc Biol Chemists. *Res:* Mechanisms of macrophage and neutrophil killing of microorganisms; macrophages activating drugs. *Mailing Add:* Dent Res Ctr Univ Tenn Health Sci Ctr 894 Union Ave Memphis TN 38163

PACANSKY, THOMAS JOHN, b Erie, Pa, Dec 17, 46; m 72. POLYMER CHEMISTRY. *Educ:* Gannon Col, BS, 68; Univ Mich, Ann Arbor, MS, 70, PhD(chem), 72. *Prof Exp:* Assoc scientist, Xerox Corp, 72-74, scientist polymer chem, 74-79; develop chemist, Exxon Enterprises, 79-80, mgr, Chem Dept, 80-91, MKT PLANNER, EXXON CHEM CO, 91- *Mem:* Am Chem Soc; Sigma Xi. *Res:* Material properties of polymers relating to their imaging by photochemical, thermal and electrical applications. *Mailing Add:* 19 Oak Pl Bernardsville NJ 07924

PACE, CARLOS NICK, b Salt Lake City, Utah, Feb 22, 40; m 59; c 3. PHYSICAL BIOCHEMISTRY. *Educ:* Univ Utah, BS, 62; Duke Univ, PhD(biochem), 66. *Prof Exp:* USPHS fel, Cornell Univ, 66-68; PROF BIOCHEM, TEX A&M UNIV, 68- *Mem:* Am Soc Biol Chemists; Am Chem Soc. *Res:* Physical chemical studies of proteins and enzymes; protein turnover. *Mailing Add:* Med Biochem Dept Col Med Tex A&M Univ College Station TX 77843-1114

PACE, CAROLINE S, b Richmond, Va, May 19, 42; m 61; c 2. PHYSIOLOGY. *Educ:* Va Commonwealth Univ, BS, 67; Med Col Va, PhD(physiol), 73. *Prof Exp:* Teaching asst anat & embryol biol, Va Commonwealth Univ, 65-67, lab instr, 67-68; fel, Wash Univ, 72-74, res asst prof pharmacol, 74-76; from asst prof to assoc prof, 76-84, PROF PHYSIOL, UNIV ALA, 84- *Mem:* Am Diabetes Asn; Endocrine Soc; Am Physiol Soc; Soc Gen Physiologists; Sigma Xi. *Res:* Stimulus-secretion coupling in pancreatic islets of Langerhans employing electrophysiological, biochemical and functional approaches. *Mailing Add:* Dept Physiol Biophysiol Univ Ala Sch Med Univ Sta Birmingham AL 35294

PACE, DANIEL GANNON, b Pittsburgh, Pa, May 1, 45. PHARMACOLOGY. *Educ:* Univ Pittsburgh, BS, 67; Georgetown Univ, PhD(pharmacol), 75. *Prof Exp:* Fel cardiovasc pharmacol, Mt Sinai Hosp, Cleveland, Ohio, 75; asst prof, Case Western Reserve Med Sch, 75-78; sr scientist cardiovasc pharmacol, 78-88, ASST DIR CLIN PHARMACOL, HOFFMAN LA ROCHE, 88- *Mem:* Am Acad Allergy & Immunol. *Res:* Drug effects in man. *Mailing Add:* Dept Clin Pharmacol Hoffmann La Roche Nutley NJ 07110

PACE, HENRY ALEXANDER, b Cleveland, Ohio, Mar 7, 14; m 45; c 2. ORGANIC CHEMISTRY. *Educ:* Univ Ill, BS, 36; Iowa State Univ, PhD(org chem), 40. *Prof Exp:* Asst org chem, Iowa State Univ, 36-40; org res chemist, Goodyear Tire & Rubber Co, 40-65, res sci, 66-79; RETIRED. *Concurrent Pos:* Consult. *Mem:* AAAS; Am Chem Soc; Am Inst Chem. *Res:* Thermosetting and thermoplastic resins; petrochemical by-products; epoxy resins; fabrication of acrylic resins; rigid and elastomeric polyurethane foams and resins; development of advanced equipment for use in electophoresis studies; gradient makers and towers for use in casting polyacrylamide and agarose gels. *Mailing Add:* 1997 Wyndham Rd Akron OH 44313

PACE, HENRY BUFORD, b Thomasville, Tenn, Nov 21, 29; m 54; c 2. PHYSIOLOGY. *Educ:* Univ Tenn, BS, 52, MS, 57; Tex A&M Univ, PhD(physiol), 64. *Prof Exp:* Res asst reproductive physiol, AEC Agr Res Lab, Univ Tenn, 55-60; assoc radiobiologist, Radiation Biol Lab, Tex A&M Univ, 60-65; from asst prof to assoc prof physiol, St Louis Col Pharm, 65-68; assoc prof physiol, 68-74, PROF PHARMACOL & PHYSIOL, SCH PHARM, UNIV MISS, 68- *Concurrent Pos:* NIMH grant teratogenesis of marijuana, Univ Miss, 68-72. *Mem:* Am Asn Lab Animal Sci; Soc Study Reprod; Sigma Xi. *Res:* Pharmacology, toxicology and teratogenesis of drugs and other chemical compounds; male and female reproductive physiology. *Mailing Add:* Dept Pharmacol Univ Miss Sch Pharm University MS 38677

PACE, JUDITH G, b Scranton, Pa, Dec 24, 49; m 72; c 1. ENZYMOLOGY, INTERMEDIARY METABOLISM. *Educ:* Marywood Col, BS, 71; Hood Col, MS, 77; George Washington Univ, PhD(biochem), 80. *Prof Exp:* Med asst, Pvt Physician, 66-71; res asst chem, Marywood Col, 70-71; chemist, US Naval Ship Res & Develop Ctr, 71-73, RES CHEMIST, US ARMY MED RES INST INFECTIOUS DIS, 77- *Concurrent Pos:* Lectr biochem, Hood Col, 77-78; prin investr, 79-; chief, Dept Cell Biol, 86- *Mem:* Am Chem Soc; Am Fed Clin Res; AAAS; Sigma Xi; IST. *Res:* Effect of infectious diseases on fatty acid metabolism, specifically the hepatic carnitine-acylcarnitine system; effect of mycotoxins on host metabolism; the metabolism and detoxication of mycotoxins and other low molecular weight toxins. *Mailing Add:* 5500 Fernwood Ct Frederick MD 21701

PACE, MARSHALL OSTEEN, b Greenville, SC, Sept 10, 41; m 68; c 2. PHYSICAL ELECTRONICS, HIGH VOLTAGE. *Educ:* Univ SC, BS, 63; Mass Inst Technol, MS, 65; Ga Inst Technol, PhD(elec eng), 70. *Prof Exp:* Engr, Columbia Prod Co, 65; from asst prof to assoc prof, 70-78, PROF ELEC ENG, UNIV TENN, KNOXVILLE, 78- *Concurrent Pos:* Res assoc, Environ Systs Corp, 72; prin investr govt & indust res, Univ Tenn, 72-; fac fel, NASA, Huntsville, Ala, 73; consult, Oak Ridge Nat Lab, 74-; mem, Int Conf Grand Res Elec. *Mem:* Inst Elec & Electronic Engrs. *Res:* Gaseous and solid state electronics; direct energy conversion; high voltage phenomena; atomic collisions; statistical communications theory; electromagnetics; power systems; digital signal processing. *Mailing Add:* Dept Elec Eng Univ Tenn Ferris Hall Knoxville TN 37996-2100

PACE, MARVIN M, b Coalville, Utah, Mar 11, 44; m 63; c 5. REPRODUCTIVE PHYSIOLOGY, ANIMAL SCIENCE. *Educ:* Utah State Univ, BS, 66; Univ Minn, MS, 68, PhD(animal sci), 71. *Prof Exp:* Res asst, Univ Minn, 66-71; ASSOC DIR RES, AM BREEDERS SERV DIV, W R GRACE CO, 71- *Mem:* Am Soc Animal Sci; Soc Study Reproduction. *Res:* Study of biochemical changes that take place in spermatozoa during preservation and how these changes affect fertilization. *Mailing Add:* 3749 Vinburn Rd DeForest WI 53532

PACE, NELLO, b Richmond, Calif, June 20, 16; m 39, 59; c 2. PHYSIOLOGY. *Educ:* Univ Calif, BS, 36, PhD(physiol), 40. *Prof Exp:* Asst physiol, Univ Calif, 36-40; res assoc, Med Col Va, 40-41; head, physiol facil, Naval Med Res Inst, Bethesda, Md, 41-46; res assoc med physics, Univ Calif, 46-53, from asst prof to assoc prof physiol, 48-57, chmn dept physiol anat, 64-67, dir White Mountain Res Sta, 50-77, prof, 57-77, EMER PROF PHYSIOL, UNIV CALIF, BERKELEY, 77- *Concurrent Pos:* Officer-in-chg Unit 1, Off Naval Res, 51-53; asst leader & chief scientist, Himalayan Exped to Makalu, 54; leader, Int Physiol Exped to Antarctica, 57-58; consult to adminr, NASA, 63, experimenter on Biosatellite III, 65-70, experimenter on Cosmos 1129, 78-80; chmn panel on gravitational biol, Comt on Space Res, Int Coun Sci Unions, 71-81; mem comn on gravitational physiol, Int Union Physiol Sci, 73-; pres, Galileo Found, 87- *Honors & Awards:* Founders Award, Am Soc Gravitational & Space Biol, 90. *Mem:* AAAS; Am Physiol Soc; Aerospace Med Asn; Int Acad Astronaut. *Res:* Gravitational physiology; environmental physiology; in vivo body composition. *Mailing Add:* 3420 Yosemite Ave El Cerrito CA 94530-3921

PACE, NORMAN R, b Washington, Ind, Sept 20, 42; m 65. BIOCHEMISTRY. *Educ:* Ind Univ, AB, 64; Univ Ill, PhD(microbiol), 67. *Prof Exp:* Prof biol, Res Ctr, Nat Jewish Hosp & Med Ctr, Univ Colo, 69-84; PROF BIOL, IND UNIV, 84- *Concurrent Pos:* USPHS fel molecular genetics, Univ Ill, 67-68; USPHS grants, 69- *Mem:* Nat Acad Sci. *Res:* In vitro DNA transcription; in vitro viral RNA replication; in vivo RNA metabolism; RNA processing; RNA phylogeny; catalytic RNA. *Mailing Add:* Dept Biol Ind Univ Bloomington IN 47401

PACE, SALVATORE JOSEPH, b Trieste, Italy, Apr 23, 44; US citizen; m 67; c 2. CHEMISTRY. *Educ:* Syracuse Univ, AB, 65, PhD(chem), 70. *Prof Exp:* Res fel, Univ Mich, 70-71 & Pa State Univ, 71-72; mem staff electrochem, Kettering Res Lab, 72-73; res scientist, Technicon Inst Corp, 73-80; MEM STAFF, EXP STA, E I DU PONT DE NEMOURS & CO, INC, 80- *Mem:* Am Chem Soc. *Res:* Electrochemical kinetics of fast reactions; electron transfer properties of metalloporphyrins and related compounds of biological importance; development of electrochemical sensors for clinical medicine applications. *Mailing Add:* E I Du Pont de Nemours & Co Inc Glasgow Sight 709 PO Box 6101 Newark DE 19714

PACE, WESLEY EMORY, b Esmont, Va, Nov 15, 24; m 46; c 3. MATHEMATICS. *Educ:* Univ Va, BA, 45, MA, 51. *Prof Exp:* Instr math, Univ Va, 48-53; from instr to asst prof, 53-58, ASSOC PROF MATH, VA POLYTECH INST & STATE UNIV, 58- *Mem:* Math Asn Am. *Res:* Topology, mainly transformation theory. *Mailing Add:* Dept Math PO Box 329 Blacksburg VA 24060

PACE, WILLIAM GREENVILLE, b Columbus, Ohio, Mar 22, 27; m 49; c 3. SURGERY. *Educ:* Dartmouth Col, AB, 48; Univ Pa, MD, 52; Ohio State Univ, MMSc, 59. *Prof Exp:* From instr to prof surg, Col Med, Ohio State Univ, 60-74, asst dean col med postgrad educ, 63-72, asst dean admin, 72-74, CLIN PROF SURG, COL MED, AT OHIO STATE UNIV. *Concurrent Pos:* Adv clin fel, Am Cancer Soc, 61-64. *Mem:* Am Asn Surg of Trauma; Soc Surg Alimentary Tract; Soc Univ Surg; Am Surg Asn; Am Col Surgeons. *Res:* Parathyroid and pancreas interrelationships; effects of chemotherapy in regional administration; cryotherapy; immunology. *Mailing Add:* 1450 Hawthorne Ave Columbus OH 43203

PACE ASCIAK, CECIL, b Malta, Jan 1, 40; Can citizen; m 69; c 2. BIOCHEMISTRY, EICOSANOIDS. *Educ:* Loyola Col, Que, BSc, 62; McGill Univ, PhD(org chem), 66. *Prof Exp:* Med Res Coun fel neurochem, Montreal Neurol Inst, McGill Univ, 66-68, Med Res Coun scholar, 68-72; res scientist, Hosp for Sick Children, 72-78; from asst prof to assoc prof, 78-84, PROF PHARMACOL & PEDIAT, UNIV TORONTO, 84- *Concurrent Pos:* Josiah Macy Found scholar, 80-81. *Mem:* Am Chem Soc; Am Soc Neurochem; Sigma Xi; Chem Inst Can; NY Acad Sci; Can Biochem Soc. *Res:* Natural products; prostaglandins; lipids; prostaglandin biosynthesis and metabolism during animal development. *Mailing Add:* Neurosci 555 University Ave 9th Floor Toronto ON M5G 1X8 Can

PACELA, ALLAN F, b Chicago, Ill, Oct 5, 38; c 3. MEDICAL PUBLISHING. *Educ:* Mass Inst Technol, BS, 60; Univ Miami, MS, 62; Ind Northern Univ, ScD(med eng mgt), 71. *Prof Exp:* Sr scientist, Lear Siegler Med Labs, 63-64; chief res scientist, Beckman Instruments, 64-72; PUBL & ED, QUEST PUBL CO, 70-; PRES & GEN MGR, INTERSCI TECHNOL CORP, 72- *Concurrent Pos:* Ed, J Clin Eng, 71; consult biomed eng, 70- *Mem:* Asn Advan Med Instrumentation; Inst Elec & Electronic Engrs; Am Col Clin Eng. *Res:* Medical engineering and technology; biomedical engineering; inventor of bilateral impedance plethysmograph; author of numerous publications. *Mailing Add:* Quest Publ Co 1351 Titan Way Brea CA 92621

PACER, JOHN CHARLES, b Toledo, Ohio, Apr 23, 47; m 70; c 2. NUCLEAR CHEMISTRY, PHYSICAL CHEMISTRY. *Educ:* Univ Toledo, BS, 69; Purdue Univ, PhD(phys chem), 74. *Prof Exp:* Res asst nuclear chem, Purdue Univ, 69-74; fel nuclear chem, Ames Lab, USERDA, 74-76; sr res scientist, Bendix Field Eng Corp, 76-84; SR SCIENTIST, PA POWER & LIGHT CO, 85- *Mem:* Am Nuclear Soc; Am Phys Soc; Health Physics Soc. *Res:* Isotopic techniques for uranium exploration; helium jet transport of fission products; reactor chemistry enhancements; geochemical characterization of nuclear waste sites; environmental radiation measurements; transport studies of isotopic species; radiation field buildup for reactors; alternate reactor water chemistry; isotopic techniques for fuel performance; uranium exploration; radon measurement techniques; nuclear spectroscopy. *Mailing Add:* 1139 Edward Ave Allentown PA 18103-5353

PACER, RICHARD A, b Toledo, Ohio, Jan 15, 39; m 66; c 2. ANALYTICAL CHEMISTRY, RADIOCHEMISTRY. *Educ:* Univ Toledo, BS, 60, MS, 62; Univ Mich, PhD(analytical chem), 65. *Prof Exp:* From asst to assoc prof, 65-78, PROF CHEM, PURDUE UNIV, FT WAYNE, 78- *Mem:* Am Chem Soc; Am Asn Univ Prof. *Res:* liquid scintillation counting; isotope dilution analysis; analytical chemistry of technetium and rhenium. *Mailing Add:* Dept of Chem Purdue Univ Ft Wayne IN 46805

PACEY, GILBERT E, b Peoria, Ill, May 2, 52; m 75; c 2. ANALYTICAL CHEMISTRY. *Educ:* Bradley Univ, BS, 74; Loyola Univ Chicago, PhD(chem), 79. *Prof Exp:* Asst prof chem, Loyola Univ Chicago, 78-79; from asst prof to assoc prof, 79-89, PROF CHEM, MIAMI UNIV, OHIO, 89- *Concurrent Pos:* Consult, Abbott Labs, 80-86, Tecator Inc, 82-, Metricor, 87-89, Biotope, 87-89, Novatek, 89- *Mem:* Am Chem Soc; Sigma Xi; Soc Appl Spectros. *Res:* Analytical-organic reagents; flow injection analysis; molecular microspectroscopy. *Mailing Add:* Dept Chem Miami Univ Oxford OH 45056

PACEY, PHILIP DESMOND, b Brandon, Man, Oct 8, 41; m 67. HIGH TEMPERATURE KINETICS. *Educ:* McGill Univ, BSc, 63; Univ Toronto, PhD(chem), 67. *Prof Exp:* Res scientist, Imp Oil Enterprises Ltd, 67-69; teaching fel, Univ Col Swansea, Univ Wales, 69-71; from asst prof to assoc prof chem, 71-86, PROF CHEM, DALHOUSIE UNIV, 86- *Concurrent Pos:* Vis assoc prof, Nat Sch Indust Chem, Nancy, France, 77-78. *Mem:* Chem Inst Can; Am Ceramic Soc; Can Ceramic Soc. *Res:* Chemical kinetics, experimental studies of hydrocarbon pyrolysis, unimolecular reactions, hydrogen atom transfers and permeation of gases through ceramics; theoretical studies of curved Arrhenius plots, the tunnel effect and hindered internal rotations. *Mailing Add:* Dept Chem Dalhousie Univ Halifax NS B3H 4J3 Can

PACHA, ROBERT EDWARD, b Payette, Idaho, Sept 4, 32; m 64. MICROBIOLOGY. *Educ:* Univ Wash, BS, 55, MS, 58, PhD(microbiol), 61. *Prof Exp:* Res instr microbiol, Univ Wash, 61-64; from asst prof to assoc prof, Ore State Univ, 64-69; assoc prof, 69-73, chmn Dept Biol Sci, 88-89, PROF BIOL, CENT WASH UNIV, 73-, DIR, CTR MED TECHNOL, 74- *Mem:* AAAS; Am Soc Microbiol; Brit Soc Gen Microbiol; Sigma Xi. *Res:* Bacteriology of pure and polluted water; fish diseases; public health microbiology. *Mailing Add:* Dept Biol Sci Cent Wash Univ Ellensburg WA 98926

PACHECO, ANTHONY LOUIS, b New Bedford, Mass, Sept 12, 32; m 56, 77; c 4. FISH BIOLOGY, FISHERIES. *Educ:* Univ Mass, BS, 54; Col William & Mary, MA, 57. *Prof Exp:* Aquatic res biologist, Va Inst Marine Sci, 56-59; fishery res biologist, Biol Lab, Bur Com Fisheries, US Dept Interior, 59-66 & Sandy Hook Marine Lab, Bur Sport Fisheries & Wildlife, 66-69, FISHERY RES BIOLOGIST, SANDY HOOK LAB, NORTHEAST FISHERIES CTR, NAT MARINE FISHERIES SERV, NAT OCEANIC & ATMOSPHERIC ADMIN, 69- *Mem:* Am Fisheries Soc; fel Am Inst Fishery Res Biologists; Atlantic Estuarine Res Soc; Am Littoral Soc. *Res:* Ecology of estuarine and coastal marine fishes; distribution, age-growth studies and estimation techniques to assess population abundance; coastal fisheries. *Mailing Add:* Sandy Hook Lab Nat Marine Fisheries Serv Highlands NJ 07732

PACHMAN, DANIEL JAMES, b New York, NY, Dec 20, 11; m 35; c 2. PEDIATRICS. *Educ:* Univ NC, AB, 31; Duke Univ, MD, 34; Am Bd Pediat, dipl, 38. *Prof Exp:* Intern pediat, Univ Chicago, 34-35; intern, Med Ctr, Cornell Univ, 35-36; instr pediat, Duke Univ & resident & attend pediatrician, Duke Hosp, 36-37; instr pediat, Univ Chicago, 37-40 & Northwestern Univ, 40-50; clin asst prof, 50-58, clin assoc prof, 58-67, CLIN PROF PEDIAT, UNIV ILL COL MED, 67-; PROF, RUSH MED COL, 71- *Concurrent Pos:* Attend pediatrician, Clins, Univ Chicago, 37-40; mem courtesy staff, Chicago Lying-In Hosp, 49-; attend pediatrician, Ill Res & Educ Hosp, 50-; consult, South Shore Hosp, 54- & Bd Educ, Chicago, 54-; chmn dept pediat, Ill Cent Hosp, 58-70, consult pediatrician, 70-; chmn, Ill Pediat Coord Coun, 69-; consult pediatrician, Presby-St Luke's Hosp, 70-; attend pediatrician, SChicago Hosp, 71-; Am Acad Pediat liaison rep on Nat Adv Health Coun, Nat Cong Parents & Teachers; pvt pract; assoc attend pediatrician, Children's Mem Hosp, 72- *Honors & Awards:* Archibald Hoyne Award, Chicago Pediat Soc, 77. *Mem:* Fel AMA; Am Heart Asn; fel Am Acad Pediat. *Res:* Carbohydrate metabolism; lipoid diseases; clinical approach. *Mailing Add:* 2315 E 93rd St Chicago IL 60617

PACHMAN, LAUREN M, b Durham, NC, Mar 16, 37; c 2. PEDIATRICS, IMMUNOLOGY. *Educ:* Wellesley Col, BA, 57; Univ Chicago, MD, 61. *Prof Exp:* Asst physician pediat, Rockefeller Univ & Columbia Univ Col Physicians & Surgeons, 64-66; from instr to asst prof, Med Sch, Univ Chicago, 66-71; assoc prof, 71-78, PROF PEDIAT, MED SCH, NORTHWESTERN UNIV, 78- *Concurrent Pos:* Dr Astrid Fagraeus sponsor, State Bacteriol Lab, Solna, Sweden, 60; NIH fel, 64-66. *Mem:* Am Asn Immunol; Soc Pediat Res; Am Rheumatism Asn; AAAS. *Res:* Pediatric immunology and rheumatology; pharmacokinetics; cell-mediated immunity; immune complex disease. *Mailing Add:* Childrens Mem Hosp 2300 Childrens Plaza Chicago IL 60614

PACHOLCZYK, ANDRZEJ GRZEGORZ, b Warsaw, Poland, Sept 23, 35; m 63; c 5. HIGH ENERGY ASTROPHYSICS, RADIO ASTROPHYSICS. *Educ:* Univ Warsaw, MSc, 56 & 57, PhD(astrophys), 61. *Prof Exp:* Asst physics, Univ Warsaw, 55-57; sr asst astron, Polish Acad Sci, 58-61, adj, 61-62; vis scientist, Harvard Col Observ, 62-63, res fel, 63-64; vis fel, Joint Inst Lab Astrophys, Boulder, Colo, 64-65; asst prof, 65-68, ASSOC PROF ASTRON, UNIV ARIZ, 68-, ASSOC ASTRONR, STEWARD OBSERV, 70- *Concurrent Pos:* Polish Acad Sci res fel, Astron Inst, 58-59; lectr, Univ Warsaw, 58-59, 61-62; Ital Foreign Ministry res fel, Univ Turin, 59-60; vis scientist, Inst Earth Physics, USSR Acad Sci, 62 & Astron Inst Czech Acad Sci, 62; dir, Pachart Publ House, 70-, ed, Pachart Astron & Astrophys Serv, 73- & Pachart Hist Astron Serv, 83-, pres, Pachart Found, 84-; sr res fel, Univ Sussex, 73; vis prof, Nat Res Coun, Bologna, 73, Univ Florence, Italy, 82; vis scientist, Nat Res Astron Observ, 76, Vatican Observ, 82-84, Physics Inst, Univ Torino, Italy, 84. *Mem:* Int Astron Union; Am Astron Soc; fel Royal Astron Soc; Polish Astron Soc (actg secy, 59-61, secy, 61-63); Polish Phys Soc; Italian Astron Soc. *Res:* Gravitational instability; magnetohydrodynamics; physics of interstellar medium; theoretical radioastronomy; physics of extragalactic radio sources; physics of active galactic nuclei; high energy astrophysics; history and philosophy of astronomy. *Mailing Add:* Dept Astron Univ Ariz Tucson AZ 85721

PACHOLKA, KENNETH, b Ore, May 10, 26. GEOMETRY. *Educ:* Ore Univ, BA, 65. *Prof Exp:* Prof math, Chico State Univ, 60-63; PROF MATH, CALIF STATE UNIV, 65- *Mem:* Am Math Soc. *Mailing Add:* Sacramento Sch Bus Calif State Univ Sacramento CA 95819

PACHTER, IRWIN JACOB, b New York, NY, July 15, 25; m 53; c 2. ORGANIC CHEMISTRY, PHARMACEUTICAL RESEARCH & DEVELOPMENT. *Educ:* Univ Calif, Los Angeles, BS, 47; Univ NMex, MS, 49; Univ Southern Calif, PhD, 51. *Prof Exp:* Fel, Univ Ill, 51-52 & Harvard Univ, 52-53; res chemist, Ethyl Corp, 53-55; assoc res chemist, Smith Kline & French Labs, 55-62, asst sect head, 62; dir med chem, Endo Labs, Inc, NY, 62-66, dir res, 67-70; vpres res & develop, Bristol Labs, 70-82; PHARMACEUT CONSULT, 82- *Concurrent Pos:* Lectr, Adelphi Univ, 63-69; chmn conf med chem, Gordon Res Conf, 71, mem bd trustees, 71-75; chmn task force on drug abuse prev, Pharmaceut Mfrs Asn, 71-73 & Res & Develop Sect, 75-76; mem med chem ad hoc study group, Walter Reed Army Inst Res, 73-75, chmn, 75-77. *Mem:* Am Chem Soc (secy, Div Med Chem, 72-74, chmn, 74-75). *Res:* Heterocyclic relationships of pharmaceuticals; narcotic antagonists; analgetics; psychotherapeutics; diuretics; antibiotics; natural products. *Mailing Add:* 101 Woodberry Lane Fayetteville NY 13066

PACHTER, JONATHAN ALAN, b Philadelphia, Pa, Nov 16, 57; m 87; c 1. DRUG DISCOVERY, SIGNAL TRANSDUCTION. *Educ:* Univ Rochester, BA, 79; Baylor Col Med, MS, 81, PhD(neurosci), 85. *Prof Exp:* Postdoctoral fel pharmacol, Yale Univ Sch Med, 85-89; sr scientist, 89-90, ASSOC PRIN SCIENTIST, SCHERING-PLOUGH CORP, 91- *Concurrent Pos:* Nat Res Serv award fel, 87-88. *Res:* Regulation of neurotransmitter release and muscle contraction by second messengers, such as inositol phosphates and calcium, with the goal of discovering drugs that inhibit these pathways; author of numerous publications. *Mailing Add:* Schering-Plough Res 60 Orange St Bloomfield NJ 07003

PACHUT, JOSEPH F(RANCIS), b Little Falls, NY, Mar 6, 50; m 72; c 3. BRYOZOOLOGY, EVOLUTION. *Educ:* State Univ NY, Oneonta, BA, 72; Mich State Univ, PhD(geol), 77. *Prof Exp:* Grad asst geol, Mich State Univ, 72-77; asst prof, 77-83, ASSOC PROF GEOL, IND UNIV-PURDUE UNIV, INDIANAPOLIS, 83- *Concurrent Pos:* Prin investr, res grants, NSF, 82-85, 84-87, Petrol Res Fund, Am Chem Soc, 86-88, 88-91. *Mem:* AAAS; Int Bryzool Asn; Paleont Soc; Soc Econ Paleontologists & Mineralogists; Sigma Xi; Soc Syst Zool. *Res:* Impact of environmental conditions on genetic and evolutionary strategies of Paleozoic bryozoans; biometrics, developmental patterns and phylogenetic relationships among these organisms. *Mailing Add:* 8924 Sunburst Circle Indianapolis IN 46227-9757

PACIFICI, JAMES GRADY, b Savannah, Ga, May 20, 39. PHYSICAL ORGANIC CHEMISTRY. *Educ:* Univ Ga, BS, 62, PhD(chem), 66. *Prof Exp:* NSF fel, Mass Inst Technol, 66-67; SR RES CHEMIST, TENN EASTMAN CO DIV, EASTMAN KODAK CO, 67- *Mem:* Am Chem Soc; The Chem Soc. *Res:* Photochemistry of molecular and macromolecular systems; mechanism of free radical reactions. *Mailing Add:* 40 E 50th St Savannah GA 31405

PACIOREK, KAZIMIERA J L, b Poland, Feb 18, 31; US citizen; m 57; c 1. POLYMER CHEMISTRY, FLUORINE CHEMISTRY. *Educ:* Univ Western Australia, BSc, 53, hons, 54, PhD(org chem), 57. *Prof Exp:* Fel org chem, Wayne State Univ, 56-57; res chemist, Wyandotte Chem Corp, Mich, 57-61; sr res chemist, US Naval Ordn Lab, Calif, 61-64, MHD Res, Inc, 64-66 & Marquardt Corp, 66-70; sr res chemist, Dynamic Sci Div, Marshall Industs, 70-72; SR SCIENTIST, ULTRASYSTEMS, INC, 72- *Mem:* Am Chem Soc; Royal Australian Chem Inst. *Res:* Natural products, especially triterpenoids and steroids; fluorocarbons; organometallics, particularly the chemistry of phosphorus and silicon; adoption of discharge techniques to various synthetic aspects; oxidative polymer degradation, flammability of materials. *Mailing Add:* 16845 Von Karman Ave Irvine CA 92714

PACIOTTI, MICHAEL ANTHONY, b St Paul, Minn, Aug 18, 42; m 66; c 2. APPLIED PHYSICS, PHYSICS. *Educ:* Univ Calif, Berkeley, BS, 64, PhD(physics), 69. *Prof Exp:* Res asst physics, Lawrence Berkeley Lab, 66-69; MEM STAFF PHYSICS, LOS ALAMOS SCI LAB, 70- *Res:* Practical application of particle accelerators; application of negative pi-mesons to radiation therapy; charged particle beam optics; proton computerized tomography. *Mailing Add:* 340 Potrillo Dr White Rock NM 87544

PACK, ALBERT BOYD, b Kamas, Utah, May 10, 19; m 45; c 3. AGRICULTURAL METEOROLOGY. *Educ:* Brigham Young Univ, BS, 40; Univ Mass, MS, 43; NC State Univ, PhD(plant physiol), 50. *Prof Exp:* Asst prof plant physiol, Conn Agr Exp Sta, 46-47 & 50-55; state climatologist, Nat Weather Serv, US Dept Commerce, Conn & RI, 56-59 & SDak, 59-61; state climatologist, Nat Oceanic & Atmospheric Admin, NY, 62-73, sr res assoc, 73-77, sr ext assoc, 77-81, sr lectr, State Univ NY Col Agr & Life Sci, Cornell Univ, 81-84; RETIRED. *Concurrent Pos:* Asst prof, Cornell Univ, 67-73. *Mem:* Am Meteorol Soc. *Res:* Effect of macro-climate and various weather elements on growth, phenology and yield of crops and on vegetation in general; climate of New York State. *Mailing Add:* 218 Enfield Falls Rd Ithaca NY 14850

PACK, ALLAN I, b, 43; m; c 4. PULMONARY MEDICINE. *Educ:* Univ Glasgow, Scotland, MD & PhD(med), 67. *Prof Exp:* ASSOC PROF MED, UNIV PA, 76- *Mailing Add:* Ctr Sleep & Respiratory Neurobiol Univ Pa 975 Maloney Bldg 3600 Spruce St Philadelphia PA 19104-4283

PACK, JOHN LEE, b Silver City, NMex, June 7, 27; m 50; c 5. PHYSICS. *Educ:* Univ NMex, BS, 50, MS, 52. *Prof Exp:* Res asst electronics, Univ NMex, 49-50, res assoc, 51; int res engr, Westinghouse Res & Develop Ctr, 52-58, res engr, 58-64, sr engr, 64-85; RETIRED. *Mem:* Sr mem Inst Elec & Electronics Eng; Am Phys Soc. *Res:* Drift velocity of electrons in gases; attachment and detachment of electrons in molecular oxygen and molecular oxygen mixtures; negative ions in air; oxygen and gas mixtures; hydration of oxygen negative ions; carbon dioxide lasers; copper halide lasers; excimer lasers; glow to arc transitions. *Mailing Add:* 3853 Newton Dr Murrysville PA 15668

PACK, MERRILL RAYMOND, b Idaho Falls, Idaho, Apr 15, 23; m 49; c 6. PLANT PHYSIOLOGY. *Educ:* Brigham Young Univ, BS, 49; Rutgers Univ, PhD(soils), 52. *Prof Exp:* Asst prof soils, NMex State Univ, 52-56; plant physiologist, US Steel Corp, Utah, 56-63; PROF & PLANT PHYSIOLOGIST, WASH STATE UNIV, 63- *Mem:* AAAS; Am Soc Plant Physiol; Air Pollution Control Asn. *Res:* Soil fertility; plant nutrition; effect of air pollutants on plants; biogenic sources of air pollution; atmospheric analysis. *Mailing Add:* SE 330 Camino Pullman WA 99163

PACK, RUSSELL T, b Grace, Idaho, Nov 20, 37; m 62; c 7. MOLECULAR SCATTERING THEORY, INTERMOLECULAR POTENTIALS. *Educ:* Brigham Young Univ, BS, 62; Univ Wis-Madison, PhD(phys chem), 67. *Prof Exp:* Res fel chem eng, Univ Minn, 66-67; from asst prof to assoc prof chem, Brigham Young Univ, 67-75; res scientist, 75-83, LOS ALAMOS FEL, LOS ALAMOS NAT LAB, 83- *Concurrent Pos:* Vis scientist, Los Alamos Sci Lab, 73-74, consult, 74-75. *Honors & Awards:* Alexander von Humboldt Sr Scientist Award, 81. *Mem:* Am Chem Soc; fel Am Phys Soc; Sigma Xi. *Res:* Theoretical physical chemistry; molecular quantum mechanics; molecular collisions; statistical mechanics; infinite order sudden approximation; centrifugal sudden approximation; coordinate transformations; photodissociation; reactive molecular scattering theory. *Mailing Add:* T-12 MS B268 Los Alamos Nat Lab Los Alamos NM 87545

PACKARD, BARBARA B K, b Uniontown, Pa, Mar 10, 38. MEDICINE, PHYSIOLOGY. *Educ:* Univ Ala, Birmingham, MD 74; WVa Univ, PhD(physiol), 64. *Prof Exp:* ASSOC DIR SCI PROG OPER, NAT HEART LUNG & BLOOD INST, NIH, 80- *Concurrent Pos:* Dir, Div Heart & Vascular Dis, 80-86. *Honors & Awards:* Commendation Medal, USPHS, 78, Outstanding Serv Medal, 87, Meritorious Serv Medal, 88. *Mem:* Am Heart Asn; Am Col Cardiol; Am Physiol Soc. *Mailing Add:* Off Prog Planning & Evaluation Nat Heart Lung & Blood Inst NIH Bethesda MD 20892

PACKARD, BEVERLY SUE, IMMUNOLOGY. *Educ:* Univ Calif, Berkeley, PhD(biophys), 84. *Prof Exp:* Fel, Johns Hopkins Univ, 84-87; staff fel, 87-88, SR STAFF FEL SURG BR, NAT CANCER INST, NIH, 88- *Mem:* Biophys Soc; Am Soc Cell Biol. *Mailing Add:* 29A-3B22-NIH 8800 Rockville Pike Bethesda MD 20892

PACKARD, DAVID, b Pueblo, Colo, Sept 7, 12; wid; c 4. ELECTRICAL ENGINEERING. *Educ:* Stanford Univ, BA, 34, EE, 39. *Hon Degrees:* DSc, Colo Col, 64; LLD, Univ Calif, 66, Cath Univ Am, 70 & Pepperdine Univ, 72; DLitt, Southern Colo State Col, 73; DEng, Univ Notre Dame, 74. *Prof Exp:* Engr, Vacuum Tube Eng Dept, Gen Elec Co, 36-38; co-founder & partner, Hewlett-Packard Co, Calif, 39-47, pres, 47-64, chmn bd & chief exec officer, 64-69; US Dep Secy of Defense, Washington, DC, 69-71; CHMN BD, HEWLETT-PACKARD CO, 72- *Concurrent Pos:* Trustee, Stanford Univ, 54-69, bd pres, 58-60, dir, Stanford Res Inst, 58-69; pres & chmn, David & Lucile Packard Found, 64-, Monterey Bay Aquarium Res Inst, 87-; dir, Caterpillar Tractor Co, 72-83, Standard Oil Calif, 72-85, Alliance to Save Energy, 77-87, Boeing Co, 78-86, Genentech, Inc, 81-, Wolf Trap found, Vienna, Va, 83-89 & Nat Fish & Wildlife Found, 85-87; chmn, Bus Coun, 72-74, mem, 74-; mem, The Trilateral Comn, 73-81, The Wilson Coun, The Bus Roundtable, Indust Adv Comt-The Adv Coun Inc, US-USSR Trade & Econ Coun Comt Sci & Tech, 75-82, bd overseers, Hoover Inst, 72-, Comt on the Present Danger, 75-81, White House Sci Coun, 82- 88, Comn on Exec, Legis & Judicial Salaries, 85-; pres, Bd Regents, Uniformed Serv Univ Health Sci, 75-82; chmn, Monterey Bay Aquarium Found, 78-; chair, US-Japan Adv Comn, 83-85, Priv Sect Coun, Nat Adv Bd, 84-86, President's Comn Defense Mgt, 85-86; vchmn, Calif Nature Conservancy, 83, The Calif Roundtable; Hitachi Found Adv Coun, 86-; trustee, Herbert Hoover Found, Hoover Inst, Am Enterprise Inst & Ronald Reagan Presidential Found, 86-; mem, Dirs Coun Exploratorium, 87- & Pres Coun Adv Sci & Technol, 90-; co-founder & past chmn, Am Electronics Asn. *Honors & Awards:* Gantt Medal, Am Mgt Asn, 70; Crozier Gold Medal, Am Ordinance Asn, 70; James Forrestal Mem Award, Nat Security Indust Asn, 72; Silver Quill Award, Am Bus Press, Inc, 72; Benjamin Fairless Mem Award, Am Iron & Steel Inst, 72; Founders Medal, Inst Elec & Electronics Engrs, 73; Medal of Honor, Electronic Industs Asn, 74; Founders Award, Nat Acad Eng, 79; Vannevar Bush Award, Nat Sci Bd, 87. *Mem:* Nat Acad Eng; Instrument Soc Am; fel Inst Elec & Electronics Engrs; hon mem Nat Acad Pub Admin; fel Am Acad Arts & Sci. *Res:* Design and development of measuring equipment. *Mailing Add:* Hewlett-Packard Co 1501 Page Mill Rd Palo Alto CA 94304

PACKARD, GARY CLAIRE, b Los Angeles, Calif, Oct 25, 38; m 68. PHYSIOLOGICAL ECOLOGY, ZOOLOGY. *Educ:* Univ Ill, Urbana, BS, 60; Univ Kans, MA, 63, PhD(zool), 66. *Prof Exp:* USPHS assoc zoophysiol, Wash State Univ, 66-67; asst prof zool, Clemson Univ, 67-68; asst prof zool, 68-72, assoc prof, 72-77, PROF ZOOL & ENTOM, COLO STATE UNIV, 77- *Mem:* Am Ornith Union; Am Soc Zoologists; Cooper Ornith Soc; Ecol Soc Am; Soc Syst Zool. *Res:* Vertebrate systematics and evolution; comparative physiology of vertebrates. *Mailing Add:* Dept Biol Colo State Univ Ft Collins CO 80523

PACKARD, KARLE SANBORN, JR, b Boston, Mass, July 15, 21; m 44; c 1. MICROWAVE ELECTRONICS. *Educ:* Columbia Univ, AB, 43; NY Univ, MS, 51; Polytech Inst NY, MS, 78. *Prof Exp:* Engr, Div War Res, Columbia Univ, 43-44; physicist, Nat Carbon Co, 44-46; engr, Cutler-Hammer, Inc, 46-55, consult, 55-58, mgr reliability, 58- 61, eng consult, 61-62, dept head, 62-64, dir long range planning, AIL Div, 64-77; dir planning, Govt Systs Opers, Eaton Corp, 78-86; INDEPENDENT CONSULT, 86- *Concurrent Pos:* Adj Prof, Polytech Inst NY, 82-85. *Honors & Awards:* Inst Elec & Electronic Engrs Sect Award, Long Island Sect, 85; Inst Elec & Electronic Engrs Fel Award, 86. *Mem:* Hist Sci Soc; Soc for Hist Technol; Am Phys Soc; fel Inst Elec & Electronics Eng. *Res:* Microwave electronics; technical and business planning; management science. *Mailing Add:* 18 Homestead Path Huntington NY 11743-2428

PACKARD, MARTIN EVERETT, b Eugene, Ore, Mar 10, 21; m 43; c 2. PHYSICS. *Educ:* Ore State Col, BA, 42; Stanford Univ, PhD(physics), 49. *Prof Exp:* Res engr labs, Westinghouse Elec Co, Pa, 42-45; asst physics, Stanford Univ, 45-46, res assoc, 46-48, instr, 49-51; res physicist, 51-53, dir res, Instrument Div, 53-63, vpres anal, Instrument Div, 63-69, asst to bd chmn, 74-86, CORP VPRES, VARIAN ASSOCS, 69- *Concurrent Pos:* Consult. *Honors & Awards:* Morris E Leeds Award, Inst Elec & Electronics Eng, 71. *Mem:* Sigma Xi; Am Phys Soc; Am Geophys Union; Am Inst Aeronaut & Astronaut; Am Chem Soc; BEMS. *Res:* Microwave tubes; nuclear magnetic resonance; molecular spectroscopy; magnetometers atomic frequency standards; clinical immunoassay. *Mailing Add:* 12640 La Cresta Dr Los Altos CA 94022

PACKARD, PATRICIA LOIS, b May 15, 27; US citizen. SYSTEMATIC BOTANY. *Educ:* Col Idaho, BA, 49; Ore State Col, MS, 52; Wash State Univ, PhD(bot), 65. *Prof Exp:* Assoc prof, 55-59, PROF BIOL, COL IDAHO, 59- *Mem:* AAAS; Am Soc Plant Taxon; Torrey Bot Club; Int Asn Plant Taxon; Sigma Xi. *Res:* Constant fertile hybrid between two heterobasic species of Calochortus; evolution of the flora of the Northern Great Basin. *Mailing Add:* 2112 Cleveland Caldwell ID 83605

PACKARD, RICHARD DARRELL, b Livermore Falls, Maine, Apr 21, 28; m 51; c 2. ELECTROMAGNETISM, SURVIVABILITY & VULNERABILITY. *Educ:* Univ Maine, BS, 50; Mass Inst Technol, SM, 51; Yale Univ, MEng, 55; NY Univ, DEngSci, 60. *Prof Exp:* Paper tester, Int Paper Co, 44-45 & 47-48; instr chem eng, NY Univ, 55-57; asst prof, Northeastern Univ, 57-59; device develop engr, Clevite Transistor Prod Div, Clevite Corp, 59-60; proj engr, Electronics Corp Am, 61-63; prin develop engr, Honeywell Radiation Ctr, Mass, 63-66; chief electro-optical device res, Melpar, Inc, 66-67; staff scientist & sect head infrared detectors, NASA Electronics Res Ctr, Cambridge, 67-70; mgr develop, Infrared Indust, Inc, Waltham, Mass & Santa Barbara, Calif, 70-73; PRES, RDP CONSULTS, 73- *Concurrent Pos:* Consult, Martin Co, Arthur D Little Inc. *Mem:* Sr mem Am Chem Soc; sr mem Am Soc Mil Engrs; Sigma Xi. *Res:* Solid state physics; semiconductor devices; infrared detectors, arrays and systems; nuclear and electromagnetic pulse. *Mailing Add:* 102 River St Cambridge MA 02139

PACKARD, ROBERT GAY, b Regina, NMex, Aug 13, 24; m 54. ACOUSTICS. *Educ:* Univ Tex, BS, 49, MA, 50, PhD(physics), 52. *Prof Exp:* Res mathematician, Univ Tex, 47-49, res scientist, 49, defense res lab, 50-52; assoc prof physics, Baylor Univ, 52-54; prof & chmn dept, Col Pac, 54-55; prof, Miss Col, 55-56; assoc prof, 56-60, PROF PHYSICS, BAYLOR UNIV, 60-, CHMN DEPT, 81- *Concurrent Pos:* Vis prof, Calif field staff, Airlangga Univ, Indonesia, 61-62; lectr, US Army-Baylor Sch Hosp Admin, Brooke Army Hosp, Ft Sam Houston, Tex; leader, Southeast Asia Student Conf, 70; vis prof, Univ Sains Malaysia, Penang, 76. *Mem:* Acoustical Soc Am; Nat Asn Physics Teachers; Brit Inst Physics. *Res:* Underwater sound; electrochemistry; medical physics. *Mailing Add:* Dept Physics Baylor Univ Waco TX 76703

PACKARD, THEODORE TRAIN, b Glen Cove, NY, Jan 24, 42; c 6. CHEMICAL OCEANOGRAPHY. *Educ:* Mass Inst Technol, BS, 63; Univ Wash, MS, 67, PhD(chem oceanog), 69. *Prof Exp:* Res assoc chem oceanog, Univ Wash, 69-70, sr res assoc, 70-72, res asst prof, 72-76; RES SCIENTIST, BIGELOW LAB OCEAN SCI, 76- *Concurrent Pos:* Head chem oceanog, Inst Maurice-Lamontagne. *Mem:* Limnol Oceanog; Marine Biol Asn Eng; Catalan Biol Soc; AAAS; Am Chem Soc. *Res:* Deep sea biology and chemistry; plankton respiration. *Mailing Add:* Inst M Lamontagine CP 1000 Mont Joli PQ G5H 3Z4 Can

PACKARD, VERNAL SIDNEY, JR, b Auburn, Maine, June 10, 30; m 57; c 2. FOOD SCIENCE. *Educ:* Univ Maine, BS, 54; Univ Minn, MS, 56, PhD(dairy indust), 60. *Prof Exp:* Assoc prof, 60-75, PROF FOOD SCI & NUTRIT, 75-, EXTEN SPECIALIST DAIRY PRODS, UNIV MINN, ST PAUL, 60- *Mem:* Int Asn Milk, Food & Environ Sanitarians; Am Dairy Sci Asn. *Res:* Hydrolytic rancidity; laboratory and process control; dairy products; food product nutritional evaluation. *Mailing Add:* Dept Food Sci & Nutrit 136 H Meat Sci Lab Col Agr & Human Ecol Univ Minn ABLMS 1354 Eckles Ave St Paul MN 55108

PACKCHANIAN, ARDZROONY (ARTHUR), bacteriology, protozoology; deceased, see previous edition for last biography

PACKEL, EDWARD WESLER, b Philadelphia, Pa, July 23, 41; m 68; c 4. MATHEMATICS ANALYSIS. *Educ:* Amherst Col, BA, 63; Mass Inst Technol, PhD(math), 67. *Prof Exp:* Asst prof math, Reed Col, 67-70; from asst prof to assoc prof, 75-81, PROF MATH, LAKE FOREST COL, 81- *Concurrent Pos:* Sr lectr comput sci, Columbia Univ, 83-85; NSF grants. *Mem:* Math Asn Am; Sigma Xi; Am Math Soc. *Res:* Operator theory; hilbert space and quantum mechanics; game theory; axiomatic social choice theory; theoretical computer science. *Mailing Add:* Dept Math Lake Forest Col Lake Forest IL 60045

PACKER, CHARLES M, b Colorado Springs, Colo, Apr 21, 30; m 53, 71; c 5. MATERIALS SCIENCE. *Educ:* Univ Utah, BSChE, 54; Stanford Univ, MS, 63, PhD(mat sci), 68. *Prof Exp:* Assoc engr, 57-59, scientist, 59-63, sr scientist, 63-71, res scientist, 71-76, STAFF SCIENTIST, LOCKHEED MISSILES & SPACE CO, PALO ALTO, 76- *Mem:* Am Soc Metals; Am

Ceramic Soc. *Res:* Mechanical behavior of metals, particularly relating to superplasticity effects; x-ray diffraction; measurements and development of techniques; evaluation and development of protective coatings for refractory metals. *Mailing Add:* 3961 Brookline Way Redwood City CA 94062

PACKER, KENNETH FREDERICK, b Grand Rapids, Mich, Aug 12, 24; m 48; c 4. METALLURGICAL & INDUSTRIAL ENGINEERING. *Educ:* Univ Mich, BSE, 48-49, MSE, 52; Purdue Univ, PhD(indust eng), 62. *Prof Exp:* Asst metallurgist, Exp Foundry, Am Brake Shoe Co, 50-52; instr prod eng, Univ Mich, 52-57; chief metallurgist, Danly Mach Specialties, 57-62; CHMN & CEO, PACKER ENG INC, 62- *Concurrent Pos:* Faculty rep, Foundry Ed Found; mem Fatigue Welding Joints Comt, Welding Res Coun, 58-62, ord adv comt Welding Armor Res Sub-comt, 62-64. *Mem:* Am Soc Mech Engrs; Am Soc Testing & Mat; Am Soc Agr Engrs; Nat Soc Prof Engrs; fel Am Soc Metals. *Res:* Materials, developing applications and metals manufacturing processes. *Mailing Add:* Packer Eng Inc N Washington at I88 PO Box 353 Naperville IL 60566

PACKER, LEO S, b Roumania, Aug 5, 20; US citizen; m 48; c 3. TECHNICAL MANAGEMENT, SCIENCE POLICY. *Educ:* City Col New York, BME, 41; Harvard Univ, MS, 48; Cornell Univ, PhD, 56. *Prof Exp:* Mech engr res & develop, BG Corp, 41-43; develop engr, Gyroscopics Div, Arma Corp, 47-50; res mech engr, Physics Dept, Cornell Aeronaut Lab, 50-52, head, Instrumentation Sect, Appl Physics Dept, 52-59; dir eng & mgr, Mil Prod Div, Bausch & Lomb, Inc, 59-62; eng mgr & assoc dir, Info Syst Div, Xerox Corp, 62-66; asst postmaster gen in chg, Bur Res & Eng, Post Off Dept, Washington, DC, 66-69; consult, NASA, 69-70; vpres corp res, Recognition Equip, Inc, 69-73; dir technol policy & space affairs, US Dept State, Washington, DC, 73-76; counr sci & technol affairs, US Mission to OECD, Paris, 76-78; consult technol planning & mgt, Washington, DC, 78-81; resident dir, appl sci & technol prog, Nat Acad Sci, Cairo, Egypt, 81-84; TECH ADV, UNION INT TECH ASN, PARIS, 84- *Concurrent Pos:* Prof lectr, Univ Buffalo; consult, 84-; capt, US Army, 43-47; pres, Asn Am Resident Overseas, 88-; consult, Europ Off, NSF, Paris. *Mem:* Am Soc Mech Engrs; Inst Elec & Electronic Engrs; Sigma Xi. *Res:* Mechanics, dynamics and vibrations; aeronautical instrumentation; kinematics; stress analysis; research and development management; technology policy; international science and technology issues. *Mailing Add:* Four Ave Rodin Paris 75116 France

PACKER, LESTER, b New York, NY, Aug 28, 29; m 56; c 3. BIOCHEMISTRY. *Educ:* Brooklyn Col, BS, 51, MS, 52; Yale Univ, PhD(microbiol), 56. *Prof Exp:* Assoc biophys, Johnson Found Med Physics, Pa, 57-59; assoc biochem, Dartmouth Med Sch, 59; asst prof microbiol, Southwestern Med Sch, Tex, 60-61; from asst prof to assoc prof, 61-66, assoc prof, Miller Inst Basic Res Sci, 66-68, PROF PHYSIOL, UNIV CALIF, BERKELEY, 68-, DIR MEMBRANE BIOENERGETICS GROUP, LAWRENCE BERKELEY LAB, 75- *Mem:* Am Chem Soc; Biophys Soc; Am Soc Biol Chemists; Royal Soc Chem; NY Acad Sci; Sigma Xi. *Res:* Membrane bioenergetics; cellular aging. *Mailing Add:* Dept Physiol Univ Calif Berkeley CA 94720

PACKER, LEWIS C, b Wellsville, Ohio, Feb 21, 93. MOTOR ENGINEERING. *Educ:* Ohio State Univ, BEE, 17. *Prof Exp:* Mgr, Motor eng, Westinghouse Corp, 20-58; RETIRED. *Concurrent Pos:* Consulting engr, RCA, 81. *Mem:* Fel Inst Elec & Electronics Engrs. *Res:* Engineering of electrical motors. *Mailing Add:* 2125 SE Lakeview Dr Apt 16 Sebring FL 33870

PACKER, R(AYMOND) ALLEN, b Clemons, Iowa, July 27, 14; m 40; c 5. VETERINARY MICROBIOLOGY. *Educ:* Iowa State Univ, BS & DVM, 40, MS, 42, PhD(vet bact), 47. *Prof Exp:* Asst vet hyg, 40-41, Iowa State Univ, from instr to assoc prof vet hyg, 41-52, head dept, 52-80, prof, 52-85, EMER PROF VET MICROBIOL & PREV MED, IOWA STATE UNIV, 85- *Concurrent Pos:* Secy-treas & ed, Am Vet History Soc. *Mem:* Am Soc Microbiol; Am Vet Med Asn; Am Col Vet Microbiol. *Res:* Veterinary immunology and virology; pathogenic bacteriology. *Mailing Add:* Dept of Vet Microbiol Iowa State Univ Ames IA 50011

PACKER, RANDALL KENT, b Lock Haven, Pa, Nov 5, 45; m 65. COMPARATIVE PHYSIOLOGY. *Educ:* Lock Haven State Col, BS, 67; Pa State Univ, PhD(zool), 71. *Prof Exp:* Asst, Pa State Univ, 67-70; from asst prof to assoc prof, 71-80, PROF BIOL SCI, GEORGE WASHINGTON UNIV, 80- *Concurrent Pos:* Res asst, Univ Bristol, Bristol, Eng, 79; guest worker, Lab Kidney & Electrolyte Metab, Nat Heart Lung & Blood Inst, NIH, 85- *Mem:* AAAS; Am Soc Zoologists. *Res:* Animal physiology, specifically acid-base balance, electrolyte and water balance in mammals and fish as well as respiration in fish. *Mailing Add:* Dept of Biol Sci George Washington Univ Washington DC 20052

PACKETT, LEONARD VASCO, b Concord, Tenn, Feb 22, 32; m 54; c 2. BIOCHEMISTRY, NUTRITION. *Educ:* Berea Col, BS, 54; Tex A&M Univ, MS, 56, PhD(biochem, nutrit), 59. *Prof Exp:* Asst animal husb, Tex A&M Univ, 54-56 & biochem & nutrit, 56-58; asst prof biochem, Purdue Univ, 58-66; from assoc prof to prof nutrit & food sci, Col Home Econ, Univ Ky, 66-87, chmn dept, 66-74; RETIRED. *Concurrent Pos:* NIH fel clin nutrit, Philadelphia Gen Hosp, 65-66; consult foods, dietetics & food sci. *Mem:* Am Inst Nutrit; Am Inst Chem; Am Chem Soc; Soc Nutrit Educ; Am Dietetic Asn. *Res:* Nutrient requirements and metabolism; metabolic disease; nutritional status; urolithiasis; mucopolysaccharides; radioactive tracers. *Mailing Add:* Dept Nutrit & Food Sci Univ Ky Col Home Econ Lexington KY 40506-0501

PACKHAM, MARIAN AITCHISON, b Toronto, Ont, Dec 13, 27; m 49; c 2. BIOCHEMISTRY. *Educ:* Univ Toronto, BA, 49, PhD(biochem), 54. *Prof Exp:* Lectr biochem, 66-67, from asst prof to prof biochem, 67-89, UNIV PROF, UNIV TORONTO, 89- *Honors & Awards:* J Allyn Taylor Int Prize in Med, 88. *Mem:* Can Biochem Soc; Can Soc Clin Invest; Am Soc Hemat; Am Asn Path; Int Soc Thrombosis & Hemostasis. *Res:* Biochemical aspects of the functions of blood platelets in hemostasis and thrombosis. *Mailing Add:* Dept Biochem Univ Toronto Toronto ON M5S 1A8 Can

PACKMAN, ALBERT M, b Philadelphia, Pa, Jan 18, 30; m 61. MEDICINAL CHEMISTRY. *Educ:* Temple Univ, BS, 51; Philadelphia Col Pharm & Sci, MS, 52, DSc, 56. *Prof Exp:* Org chemist, Nat Drug Co, 56-59; sr org chemist, Int Latex Corp, 59-60; res adminstr, Denver Chem Co, 60-61; sr org chemist, 61-67, actg res dir, 67-68, dir tech serv, 68-77, VPRES & TECH DIR, DERMIK LABS, INC, SUB WILLIAM H RORER, INC, 77- *Concurrent Pos:* Lectr, Philadelphia Col Pharm & Sci, 67-69. *Mem:* Am Chem Soc; Am Pharmaceut Asn; Soc Cosmetic Chemists; Am Acad Dermat. *Res:* Synthesis of medicinal compounds particularly cardiovascular agents, analgesics and antibacterial agents; development and clinical testing of dermatological products. *Mailing Add:* 3223 Lenape Dr Dresher PA 19025

PACKMAN, CHARLES HENRY, b Lake Charles, La, Oct 3, 42; m 66; c 2. HEMATOLOGY, CELL BIOLOGY. *Educ:* La State Univ, MD, 67. *Prof Exp:* Intern med, 67-68, asst resident, NC Mem Hosp, 68-69; assoc resident med, Strong Mem Hosp, 72-73; fel hemat, 73-76, asst prof, 76-82, ASSOC PROF MED, SCH MED, UNIV ROCHESTER, 82- *Mem:* fel Am Col Physicians; Am Soc Hemat; Am Fedn Clin Res; Am Asn Immunologists; Am Soc Cell Biol. *Res:* Neutrophil maturation and function; bone marrow structure and function; lymphocyte activation. *Mailing Add:* Dept Med & Hemat Univ Rochester Med Ctr Rochester NY 14642

PACKMAN, ELIAS WOLFE, b Philadelphia, Pa, Mar 13, 30; m 51; c 3. PHARMACOLOGY. *Educ:* Philadelphia Col Pharm, BSc, 51, MSc, 52, DSc(pharmacol), 54. *Prof Exp:* Asst, 51-52, from instr to assoc prof, 52-66, PROF PHARMACOL, PHILADELPHIA COL PHARM & SCI, 66- *Concurrent Pos:* Lectr, Philadelphia Col Osteop, 60; mem, Drug Device Cosmetic Bd, Pa. *Mem:* Am Pharmaceut Asn; Am Asn Cols Pharm; NY Acad Sci; Sigma Xi. *Res:* Reticulo-endothelial-system; histamine and antihistamine pharmacology; fate of white blood cells; techniques involving radioisotopes. *Mailing Add:* Philadelphia Col Pharm & Sci 43rd St & Woodland Ave Philadelphia PA 19104

PACKMAN, PAUL FREDERICK, b Brooklyn, NY, July 30, 36; m 62; c 2. QUALITY CONTROL. *Educ:* Cooper Union Sch Eng, BSME, 60; Syracuse Univ, MS, 62, PhD(solid state), 64. *Prof Exp:* Team leader mech behav res, Lockheed, Ga, 64-68; assoc prof, Vanderbilt Univ, Tenn, 68-71; resident fel, Nat Acad Sci, 71-72; dir mat sci, Vanderbilt Univ, 72-74, prof eng, 74-77; CHMN & PROF, CIVIL & MECH ENG DEPT, SOUTHERN METHODIST UNIV, 78- *Concurrent Pos:* Adj prof, Dept Eng Mech, Ga Inst Technol, 64-68; consult, Failure Anal, 74-82, Nat Mat Adv Bd, 78, 79 & 81 & Hgy Transp Res Bd, 81-82. *Mem:* Am Soc Testing & Mat; Am Soc Metals; Am Soc Non-Destructive Testing; Am Mining, Metall & Petrol Engrs. *Res:* Fracture and fatigue; analysis of failures in structural components; non-destructive inspection and reliability of inspection procedures. *Mailing Add:* Civil Eng Dept Southern Methodist Univ Dallas TX 75275

PACKMAN, PAUL MICHAEL, b St Louis, Mo, Aug 5, 38; m 65; c 2. NEUROENDOCRINOLOGY, NEUROCHEMISTRY. *Educ:* Wash Univ, BA, 59, MD, 63. *Prof Exp:* Intern med, Vanderbilt Univ Hosp, 63-64; staff assoc biochem, Nat Inst Arthritic & Metab Dis, 64-67; fel neuroendocrinol, Dept Anat, Oxford Univ, 67-68, Inst Neurobiol, Univ Goteborg, 68-69 & Dept Pharmacol, Med Sch, Wash Univ, 69-70; asst res, Dept Psychiat, 70-73, asst prof, 73-78, ASSOC PROF PSYCHIAT, MED SCH, WASH UNIV, 78- *Honors & Awards:* Res Award Med, Borden Found, 63; Res Scientist Develop Award, NIMH, 74. *Mem:* Am Psychiat Asn; Psychiat Res Soc; Int Soc Neuroendocrinol; Int Soc Psychoneuroendocrinol; Endocrine Soc; Sigma Xi. *Mailing Add:* Paul M Packman Inc 1034 S Brentwood Blvd Suite 1180 St Louis MO 63117

PACKMAN, SEYMOUR, b Mt Vernon, NY, Feb 3, 43. HUMAN GENETICS, PEDIATRICS. *Educ:* Columbia Col, AB, 63; Wash Univ, MD, 68. *Prof Exp:* ASST PROF PEDIAT, UNIV CALIF, SAN FRANCISCO, 77- *Mem:* Am Soc Human Genetics; AAAS; Am Fedn Clin Res. *Res:* Human biochemical genetics. *Mailing Add:* Dept Pediat Univ Calif San Francisco Med Sch 513 Parnassus Ave San Francisco CA 94143

PACZYNSKI, BOHDAN, b Wilno, Poland, Feb 8, 40. ASTROPHYSICS. *Educ:* Warsaw Univ, MA, 62, PhD(astron), 64, Docent, 67. *Prof Exp:* From asst to prof, N Copernicus Astron Ctr, Polish Acad Sci, Warsaw, 62-82; prof, 82-89, LYMAN SPITZER JR PROF, DEPT ASTROPHYS SCI, PRINCETON UNIV, 89- *Concurrent Pos:* Astron asst, Lick Observ, Univ Calif, 62-63; vis observer, Beograd Astron Observ, Yugoslavia, 64, Haute Provance Observ, France, 65, Meudon Observ, 66; vis prof, Calif Inst Technol, 73, Univ Calif, Berkeley, 79, Harvard Univ, 89; Sherman Fairchild Distinguished Scholar, Calif Inst Technol, 81. *Honors & Awards:* K Schwarzschild lectr, 81; Alfred Jurzykowski Found Award, 82; Eddington Medal, Royal Astron Soc, 87. *Mem:* Polish Astron Soc; Int Astron Union; Am Astron Soc; Royal Astron Soc; Am Phys Soc. *Mailing Add:* Dept Astrophys Sci Princeton Univ Princeton NJ 08544-1001

PADALINO, JOSEPH JOHN, b Newark, NJ, June 26, 22; m 48; c 3. ELECTRICAL ENGINEERING. *Educ:* Newark Col Eng, BSEE, 44; Univ Pa, MSEE, 47; Polytech Inst Brooklyn, PhD(elec eng), 63. *Prof Exp:* Instr elec eng, Newark Col Eng, 47-51; mem staff, Bell Tel Labs, Inc, 51-54; assoc prof, 54-63, PROF ELEC ENG, NJ INST TECHNOL, 63- *Concurrent Pos:* Consult, Bendix, 63, Picatinny Arsenal, 64-65 & Radio Frequency Labs, 66- *Mem:* Am Soc Eng Educ; Inst Elec & Electronics Engrs. *Res:* Servomechanisms; network theory. *Mailing Add:* 500 S Riverside Dr Shark River Hills NJ 07753

PADALINO, STEPHEN JOHN, b NJ, Jan 29, 54; m 79. NUCLEAR PHYSICS. *Educ:* Stockton State Col, BS, 77; Fla State Univ, MS, 83, PhD(nuclear physics), 85. *Prof Exp:* Instr physics, Stockton State Col, NJ, 77-78, Ohio State Univ, 78-79; instr physics, Fla State Univ, 79-80, res assoc, 80-85; ASST PROF PHYSICS, STATE UNIV NY, GENESEO, 85- *Res:* Projectile fragmentation studies in nuclear reactions in the 70 to 100 million electron volt energy range, leading to three-body final states. *Mailing Add:* Dept Physics State Univ NY Geneseo NY 14454

PADARATHSINGH, MARTIN LANCELOT, CELLULAR IMMUNOLOGY, MICROBIOLOGY. *Educ:* George Washington Univ, PhD(immunol), 74. *Prof Exp:* EXEC SECY, PATH B-STUDY SECT, DIV RES GRANTS, NIH, 80- *Mailing Add:* 42 Columbia Ave Takoma Park MD 20912

PADAWER, JACQUES, b Liege, Belg, Sept 3, 25; nat US; m 51; c 3. ANATOMY, CELL BIOLOGY. *Educ:* NY Univ, BA, 50, MS, 52, PhD(physiol), 54. *Prof Exp:* From instr to asst prof biochem, 55-60, from asst prof to assoc prof anat, 60-72, PROF ANAT & STRUCT BIOL, ALBERT EINSTEIN COL MED, 72- *Concurrent Pos:* Am Heart Asn fel, NY Univ, 54-57; Am Heart Asn estab investr, 57-62. *Mem:* Am Asn Anat; Am Soc Cell Biol; Reticuloendothelial Soc; Soc Exp Biol Med; fel NY Acad Sci. *Res:* Behavior and function of mast cell and basophil leukocyte; cellular effects ultrasound; vitamin A and tumors. *Mailing Add:* Dept Anat & Struct Biol Albert Einstein Col Med New York NY 10461

PADBERG, HARRIET A, b St Louis, Mo, Nov 13, 22. MATHEMATICS. *Educ:* Maryville Col, Mo, BA, 43; Univ Cincinnati, MMus, 49; Univ St Louis, MA, 56, PhD(math), 64. *Prof Exp:* Teacher math, Acad Sacred Heart, Cincinnati, 46-47; instr math & music, Acad & Col Sacred Heart, Grand Coteau, 47-48 & 50-55; teacher, Acad Sacred Heart, St Charles, 48-50; from asst prof to assoc prof, 56-68, PROF MATH & MUSIC, MARYVILLE COL, ST LOUIS, 68- *Concurrent Pos:* Organist, 46-; music coordr weekly televised Mass, 65-68; mem, Nat Educ Comt of Nat Asn for Music Therapy, 85; actg chair, Math/Natural Sci Div, Maryville Col, 90-91. *Mem:* Sigma Xi; Am Math Soc; Math Asn Am. *Res:* Aesthetics; computer music; teacher training. *Mailing Add:* Maryville Col 13550 Conway Rd St Louis MO 63141

PADBURY, JOHN JAMES, organic polymer chemistry, for more information see previous edition

PADDACK, STEPHEN J(OSEPH), b Cincinnati, Ohio, Dec 26, 34; m 62; c 4. ASTRODYNAMICS, RADIATION PRESSURE EFFECTS. *Educ:* Cath Univ Am, BAE, 59, MASE, 64, PhD(eng), 73. *Prof Exp:* Assoc engr, Boeing Co, 59-61; aerospace engr, 61-70, proj opers dir, IMP Proj, 70-74, spacecraft mgr, IMP Proj, 72-74, dep tech mgr, ISEE proj, 74-78 & COBE proj, 78-81, CHIEF, ADVAN MISSIONS ANAL OFF, NASA-GODDARD SPACE FLIGHT CTR, 81- *Concurrent Pos:* mem Space Systs Comt, Am Inst Aeronaut & Astronaut, 84-87; mem fac astrodyn, US Naval Acad. *Mem:* Am Geophys Union; AM Inst Aeronaut & Astronaut. *Res:* Technical management and future planning for all aspects of spaceflight projects; astrodynamics, including applied celestial mechanics, orbit and trajectory analysis and mission analysis of space flights, dynamics and mechanics; planetology related to small celestial bodies; space science. *Mailing Add:* NASA/Goddard Space Flight Ctr Code 402 Greenbelt MD 20771

PADDEN, FRANK JOSEPH, JR, b Scranton, Pa, Sept 21, 28; m 55; c 5. POLYMER PHYSICS. *Educ:* Scranton Univ, BS, 50. *Prof Exp:* Res asst, Mellon Inst Indust Res, 51-52, res assoc, 52-54, jr fel, 54-55; physicist, Am Viscose Corp, 57-60, res physicist, 60; MEM TECH STAFF, AT&T BELL LABS, 60- *Honors & Awards:* Polymer Physics Prize, Am Phys Soc, 73. *Mem:* Fel Am Phys Soc. *Res:* Physics of high polymers. *Mailing Add:* AT&T Bell Labs 600 Mountain Ave Rm 1d-201 Murray Hill NJ 07974

PADDISON, FLETCHER C, b Superior, Ariz, Nov 29, 21; m 46; c 3. SYSTEMS ENGINEERING. *Educ:* Cath Univ, AB, 56. *Prof Exp:* Supvr servomech, Auxiliary Power Systs & Gyroscope Develop Proj, Johns Hopkins Univ, 49-56, mem prin prof staff, 56, proj engr intercontinental ballistic missle defense, 56-58, asst proj engr long range Typhon missle, 60-62, proj engr, 62-64, tech adv assoc dir, 62-66, asst advan res proj agency prog mgr, 66-71, asst dir advan res prog off & prog mgr, Regional Opers Res Contrast, US Dept Energy, Appl Physics Lab, Johns Hopkins Univ, 71-88; CONSULT, 88- *Mem:* Am Soc Mech Engrs. *Res:* Servomechanisms; large system interface management techniques; thin material response to nuclear radiation; radar cross section measurement techniques; special instrumentation and mission analysis US Navy Surface Effect Ships; direct applications of geothermal energy in the eastern US and estimated life cycle costs. *Mailing Add:* 10300 La Costa Dr Austin TX 78747-1103

PADDISON, RICHARD MILTON, b Rochester, NY, Aug 20, 19; m 43, 66; c 8. NEUROLOGY. *Educ:* Duke Univ, AB, 43, MD, 45. *Prof Exp:* Instr neurol & neuroanat, Sch Med, Univ Ga, 48-49; from asst prof to assoc prof neurol, 54-59, head dept, 65-77, PROF NEUROL, LA STATE UNIV SCH MED, NEW ORLEANS, 59- *Concurrent Pos:* Dir Continuing Med Educ & Coordr Alumni Affairs, La State Univ Sch Med, 80-; consult, Hotel Dieu Hosp, Southern Baptist Hosp & Charity Hosp, New Orleans, 80- *Mem:* Asn Res Nerv & Ment Dis; fel Am Psychiat Asn; fel Am Col Physicians; fel Am Acad Neurol; Am Acad Cerebral Palsy. *Res:* Vascular disease; epilepsy. *Mailing Add:* 5435 Bellaire Dr New Orleans LA 70124

PADDOCK, ELTON FARNHAM, b Worcester, Mass, Dec 11, 13; m 37; c 2. PLANT CYTOGENETICS. *Educ:* Whittier Col, AB, 36; Univ Calif, PhD(genetics), 42. *Prof Exp:* Asst genetics, Univ Calif, 40; Muellhaupt scholar, Ohio State Univ, 41-43; plant pathologist, Exp Sta, Tex A&M Univ, 43-45; from asst prof to assoc prof, 45-68, prof, 68-85, EMER PROF GENETICS & CYTOL, OHIO STATE UNIV, 85- *Concurrent Pos:* Am consult, Summer Sci Inst, US Agency Int Develop, India, 65 & 67; consult, DNA Plant Tech Inc, 81-86. *Mem:* AAAS; Am Inst Biol Sci; Genetics Soc Am; Bot Soc Am; Am Genetic Asn. *Res:* Cytology of Hevea brasiliensis; cytogenetics of polyploids and interspecific hybrids of Solanum; disease resistance breeding of tomatoes; chiasma failure and adjacent distribution in Rhoeo spathacea; somatic crossing-over in soybean. *Mailing Add:* 391 Glenmont Ave Columbus OH 43214-3209

PADDOCK, GARY VINCENT, b Port Townsend, Wash, July 13, 42; m 66; c 1. MOLECULAR BIOLOGY. *Educ:* Univ Wash, BS, 64; Univ Calif, San Diego, PhD(chem), 73. *Prof Exp:* Officer, Eng Dept, US Navy, 64-69; teaching asst chem, Univ Calif, San Diego, 69-73; fel, Univ Calif, Los Angeles, 73-77; ASSOC PROF MICROBIOL & IMMUNOL, MED UNIV SC, 77- *Concurrent Pos:* Celeste Durand Rogers fel cancer res, 73-74; Helen Hay Whitney fel, 74-77. *Res:* Eucaryotic gene regulation and the technology involved in recombinant DNA construction and nucleotide sequencing; structure and mechanism of action of transfer factor; recombinant DNA vaccines. *Mailing Add:* Dept Microbiol & Immunol 171 Ashley Ave Charleston SC 29425

PADDOCK, JEAN K, RENAL BIOCHEMISTRY, PHYSIOLOGY. *Educ:* Boston Univ, PhD(biochem), 82. *Prof Exp:* DIR, RENAL BIOCHEM LAB & ASST PROF BIOCHEM & MED, JEFFERSON MED COL, THOMAS JEFFERSON UNIV, 82- *Concurrent Pos:* Exec secy, Nat Cancer Inst. *Mailing Add:* NIH-NCI-GRB-DEA Westwood Bldg Westbard Ave Bethesda MD 20892

PADDOCK, ROBERT ALTON, b Port Washington, NY, Dec 10, 42; m 67; c 3. ANALYTICAL PREDICTIONS, DATA ANALYSIS. *Educ:* Washington & Lee Univ, BS, 64; Mich State Univ, MS, 66, PhD(physics), 69. *Prof Exp:* Asst prof physics, Ripon Col, 69-74; MEM STAFF, ENVIRON ASSESSMENT & INFOR SCI DIV, ARGONNE NAT LAB, 75- *Mem:* Am Phys Soc. *Res:* Development of mission planning expert systems to support the US Special Operations Forces. *Mailing Add:* Environ Assessment & Info Sci Div Argonne Nat Lab Bldg 362 Argonne IL 60439-4815

PADEGS, ANDRIS, b Riga, Latvia, Mar 27, 29; nat US; m 54; c 3. COMPUTER SCIENCE, ELECTRICAL ENGINEERING. *Educ:* Dartmouth Col, AB, 53, Thayer Sch Eng, MS, 54; Carnegie Inst Technol, PhD(elec eng), 58. *Prof Exp:* Asst servomechanisms res, Carnegie Inst Technol, 54-56, proj engr, 56-57; assoc engr, data processing mach, IBM Corp, 58-60, staff engr, 60-63, adv engr, 63-65, sr engr, 65-68, mgr, processor archit, 68-69, mgr, archit design & control, 69-71, prog mgr syst archit, 71-87, PROG MGR, ENTERPRISE SYSTS CENT ARCHIT, IBM CORP, 87- *Mem:* Inst Elec & Electronic Engrs; Sigma Xi. *Res:* Computer architecture, planning and design of data processing systems; alternating current servomechanisms. *Mailing Add:* Two Merry Hill Rd Poughkeepsie NY 12603

PADEN, JOHN WILBURN, b Bakersfield, Calif, Dec 24, 33; m 73; c 2. MYCOLOGY. *Educ:* Univ Calif, Berkeley, BS, 55; Univ Idaho, MS, 61, PhD(bot), 68. *Prof Exp:* Asst plant pathologist, Col Agr, Univ Idaho, 60-61; agr res technician soil microbiol, USDA, Wash, 62-63; instr biol, 66-67, asst prof, 67-77, ASSOC PROF BIOL, UNIV VICTORIA, 77- *Mem:* Mycol Soc Am; Can Bot Asn; Brit Mycol Soc; NY Acad Sci. *Res:* Bid systematics of ascomycetes, especially discomycetes. *Mailing Add:* Dept Biol Sci Univ Victoria Box 1700 Victoria BC V8W 2Y2 Can

PADER, MORTON, b New York, NY, June 5, 21; m 45; c 2. INDUSTRIAL CHEMISTRY. *Educ:* City Col New York, BS, 41; NY Univ, MS, 47, PhD(org chem), 49. *Prof Exp:* Chemist, Food Res Labs, Inc, NY, 41-46, chief chemist, 46-47; assoc chemist, Thomas J Lipton, Inc, 50-52; assoc res chemist, Res & Develop Div, 52-60, chief food prod sect, 60-63, VPRES PERSONAL PRODS DEVELOP, RES CTR, LEVER BROS CO, 63- *Mem:* AAAS; Am Chem Soc; Sigma Xi; Soc Cosmetic Chem; Int Asn Dent Res. *Res:* Reaction rates and mechanisms; protein chemistry and technology; nutrition; food technology; industrial chemicals and operations; food chemistry and product development; development of products for oral and personal hygiene; mechanism and inhibition of caries and calculus. *Mailing Add:* 1358 Sussex Rd West Englewood NJ 07666

PADGETT, ALGIE ROSS, b Morocco, Ind, Apr 20, 11; m 33; c 3. ORGANIC CHEMISTRY. *Educ:* Purdue Univ, BS, 32, MS, 36, PhD(org chem), 37; Am Inst Chemists, cert. *Prof Exp:* Lab asst chem, Purdue Univ, 32-35; res chemist, Humble Oil & Ref Co, 37-40, res sect head, 40-41, chemist, Baytown Ordn Works, 41-45, res sect head, 45-47, res dept head, Res & Develop Div, 47-64, asst dir anal res div, Exxon Res & Eng Co, 64-66, sr staff adv, 66-71, ANNUITANT, EXXON RES & ENG CO, 71- *Concurrent Pos:* Coordr equal employ opportunity prog, 66-71. *Mem:* Fel AAAS; fel Am Inst Chemists; Am Chem Soc; Am Inst Chem Eng. *Res:* Pure aromatics from petroleum; synthetic fuels and lubricants; physical, chemical and instrumental analytical technique. *Mailing Add:* 14 Elkins Lake Huntsville TX 77340

PADGETT, BILLIE LOU, b South Gate, Calif, May 23, 30. MEDICAL MICROBIOLOGY. *Educ:* Univ Calif, BA, 52; Univ Wis, MS, 55, PhD(med microbiol), 57. *Prof Exp:* proj assoc med microbiology, Med Sch, Univ Wis-Madison, 57-8 3; RETIRED. *Concurrent Pos:* Alumni Res Found fel, Med Sch, Univ Wis-Madison, 58-59; USPHS fel, 65-67. *Mem:* AAAS; Am Soc Microbiol. *Res:* Viral enhancement by fibroma virus; papova virus associated with progressive multifocal leucoencephalopathy. *Mailing Add:* 3878 Sunset Rd Santa Barbara CA 93110

PADGETT, DAVID EMERSON, b Fayetteville, NC, Jan 27, 45; m 70; c 2. MYCOLOGY, BOTANY. *Educ:* Duke Univ, AB, 67; Ohio State Univ, MS, 73, PhD(bot), 75. *Prof Exp:* Res assoc, Paint Res Inst, 72-75; from asst prof to assoc prof, 75-87, PROF BIOL, UNIV NC, WILMINGTON, 87- *Concurrent Pos:* NSF grant, 78-80; fel, US Atomic Energy Comn. *Mem:* Mycol Soc Am; Brit Mycol Soc. *Res:* Estuarine distribution and salinity tolerance of saprolegniaceaus fungi; decomposition of plant litter beneath salt marsh soils. *Mailing Add:* Dept Biol Univ NC Wilmington NC 28403

PADGETT, DORAN WILLIAM, b Alexandria, Va, Sept 9, 25; m 46; c 3. NUCLEAR PHYSICS. *Educ:* Am Univ, BS, 52; Cath Univ, PhD(meson theory), 64. *Prof Exp:* Nuclear physicist, Nat Bur Stand, 52-55; proj engr, Eng Res & Develop Lab, US Dept Army, 55-56; res admin, Air Force Off Sci Res, 56-63 & Off Naval Res, 63-65; asst prof, Col William & Mary, 65-66; head, Nuclear Phys Br, 66-74, dir radiation & plasma phys progs, 74-78, OFF

NAVAL RES, PHYS CONSULT, 78- *Concurrent Pos:* Vis scientist, Naval Res Lab, 71-72. *Mem:* Am Phys Soc. *Res:* Methods and formalism of quantum field theory; meson, nuclear reaction and atomic and molecular collision theories; experimental nuclear and high energy physics; nuclear structure; transport theory of energetic heavy ion penetration as associated with simulation of fast neutron damage in reactor materials; high energy neutrino beam applications. *Mailing Add:* 624 N Tazwell St Arlington VA 22203

PADGETT, GEORGE ARNOLD, b East Detroit, Mich, Feb 17, 31; m 51. PATHOLOGY, GENETICS. *Educ:* Mich State Univ, BS, 59, MS, 61, DVM, 61. *Prof Exp:* NIH fel, 61-65; from asst to assoc prof vet path, Wash State Univ, 65-77; PROF PATH, MICH STATE UNIV, 77- *Concurrent Pos:* NIH grant, 65-; res assoc, Rockefeller Univ, 66- *Mem:* AAAS; Am Soc Exp Path. *Res:* Experimental pathology of genetic diseases of animals which also occur in man. *Mailing Add:* Dept Path Mich State Univ East Lansing MI 48824

PADGETT, WILLIAM JOWAYNE, b Walhalla, SC, May 15, 43; m 65; c 2. STATISTICS, MATHEMATICS. *Educ:* Clemson Univ, BS, 66, MS, 68; Va Polytech Inst & State Univ, PhD(statist), 71. *Prof Exp:* From asst to assoc prof, 71-77, PROF STATIST, UNIV SC, 77- *Concurrent Pos:* Chmn, dept statist, 85- *Mem:* Am Soc Qual Control; fel Am Statist Asn; fel Inst Math Statist. *Res:* Nonparametric estimation; censored data; reliability theory. *Mailing Add:* Dept of Statist Univ of SC Columbia SC 29208

PADGITT, DENNIS DARRELL, b Malvern, Iowa, Jan 1, 39; m 61; c 2. DAIRY HUSBANDRY. *Educ:* Iowa State Univ, BS, 62; Univ Mo-Columbia, MS, 64, PhD(dairy sci), 67. *Prof Exp:* PROF ANIMAL & DAIRY SCI, NORTHWEST MO STATE UNIV, 67-, PROF AGR. *Mem:* Am Dairy Sci Asn; Nat Asn Col Teachers Agr. *Res:* Effect of rumen buffers on milk production and microbial activity. *Mailing Add:* Dept Agr Northwest Mo State Univ Maryville MO 64468

PADHI, SALLY BULPITT, b Darien, Conn, July 12, 44; m 70; c 3. SOIL MICROBIOLOGY, INSECT PATHOLOGY. *Educ:* Univ Conn, BS, 67; Univ Mass, MS, 69; Rutgers Univ, PhD(microbiol), 74. *Prof Exp:* Fel insect virol, Waksman Inst Microbiol, Rutgers Univ, 74-85, lectr, 77-78; CONSULT, 80- *Mem:* Am Soc Microbiol; Tissue Culture Asn; Sigma Xi. *Res:* The characterization of insect viruses especially those that have potential use as biological control agents; soil engineering through composting of municipal waste. *Mailing Add:* 11 South Dr East Brunswick NJ 08816-1131

PADIAN, KEVIN, b Morristown, NJ, Mar 12, 51; m 73; c 1. PALEOBIOLOGY, MACROEVOLUTION. *Educ:* Colgate Univ, BA, 72, MAT, 74; Yale Univ, MPhil, 78, PhD(biol), 80. *Prof Exp:* ASSOC PROF PALEONT, UNIV CALIF, BERKELEY, 80- *Concurrent Pos:* Consult, NASA, Smithsonian Inst, Calif Acad Sci & NM Mus Nat Hist; grants, NSF, Nat Geog Soc, NATO, Am Chem Soc. *Mem:* Soc Study Evolution; Soc Vert Paleont; Soc Syst Zool; Paleont Soc; Sigma Xi. *Res:* Evolutionary patterns and processes; vertebrate paleontology and zoology; functional morphology; Mesozoic vertebrate faunas. *Mailing Add:* Mus Paleont Univ Calif Berkeley CA 94720-2399

PADILLA, ANDREW, JR, b Honolulu, Hawaii, Apr 21, 37. CHEMICAL ENGINEERING. *Educ:* Col St Thomas, BA, 60; Univ Notre Dame, BS, 60; Univ Mich, Ann Arbor, MS, 62, PhD(chem eng), 66. *Prof Exp:* Sr res engr, Battelle Northwest, Battelle Mem Inst, 66-70, Westinghouse Hanford Co, Westinghouse Elec, 70-71; chem engr, Argonne Nat Lab, 71-73; fel engr, 74-79, adv engr, 79, MGR REACTOR DYNAMICS, WESTINGHOUSE HANFORD CO, WESTINGHOUSE ELEC CORP, 79- *Mem:* Am Nuclear Soc. *Res:* Design (thermal-hydraulic analysis of fuel bundles) and safety (local faults and hypothetical maximum accidents) analyses of fast nuclear reactor cores. *Mailing Add:* 2025 Davison Ave Richland WA 99352

PADILLA, GEORGE M, b Guatemala, May 27, 29; US citizen; m 57; c 3. CELL PHYSIOLOGY, BIOPHYSICS. *Educ:* George Washington Univ, BS, 52, MS, 53; Univ Calif, Los Angeles, PhD(zool), 60. *Prof Exp:* Res fel cell physiol, Nat Cancer Inst, 60-62, biologist, Geront Br, Nat Heart Inst, 62-63; biologist, Biol Div, Oak Ridge Nat Lab, 63-65; from asst prof to assoc prof 65-86, PROF PHYSIOL, MED CTR, DUKE UNIV, 87- *Concurrent Pos:* Dir, Babies Hosp Res Ctr, Wilmington, NC, 65; actg dir, Wrightsville Marine Biomed Lab, 65-66, consult, Biol Div, Oak Ridge Nat Lab, 65-70. *Mem:* Soc Gen Physiol; Am Soc Cell Biol; Am Physiol Soc; Sigma Xi. *Res:* Cell division synchrony; physiology and identification of marine algal toxins; cell growth; cellular control systems in cell cycle; oncology; cell biology; paustate cancer. *Mailing Add:* PO Box 3709 Duke Univ Med Ctr Durham NC 27710

PADLAN, EDUARDO AGUSTIN, b Manila, Philippines, Aug 31, 40; m 61; c 5. MOLECULAR IMMUNOLOGY. *Educ:* Univ Philippines, BS, 60; Johns Hopkins Univ, PhD(biophysics), 68, MS(comput sci), 84. *Prof Exp:* Asst prof physics, Univ Philippines, 68-69; vis scientist molecular biol, NIH, 71-78; res assoc, Johns Hopkins Univ, 69-71, res scientist biophys, 78-83; VIS SCIENTIST, MOLECULAR BIOL, NIH, 83- *Mem:* Am Crystallog Asn; Am Soc Biochem & Molecular Biol; Am Asn Immunol; NY Acad Sci. *Res:* Crystallographic investigation of proteins. *Mailing Add:* Lab Molecular Biol NIDDK NIH Bldg Two Rm 206 Bethesda MD 20892

PADMANABHAN, G R, b Madras, India, July 12, 35; m 64; c 2. ANALYTICAL CHEMISTRY, PHYSICAL CHEMISTRY. *Educ:* Univ Madras, BSc, 56; Univ Pittsburgh, PhD(analytical chem), 63. *Prof Exp:* Jr chemist, Atomic Energy Estab, Bombay, India, 57-60; sr chemist, Hooker Chem Corp, NY, 64-66 & Hoffmann-La Roche, Inc, NJ, 66-70; MGR, PHARM DIV, CIBA-GEIGY CORP, 70- *Mem:* Am Chem Soc; Am Pharmaceut Asn. *Res:* Non-aqueous solvents; electrochemistry; gas, liquid and thin-layer chromatography; thermal methods of analysis microscopy; purification and separation techniques; pharmaceutical analysis with particular emphasis on non-aqueous solvents and chromatographic techniques. *Mailing Add:* Pharma Div Ciba-Geigy Corp Suffern NY 10901

PADMORE, JOEL M, b Macy, Nebr, Dec 8, 38; m 60; c 3. ANALYTICAL CHEMISTRY, AGRICULTURAL CHEMISTRY. *Educ:* Munic Univ Omaha, BA, 60; Mont State Univ, PhD(chem), 64. *Prof Exp:* From asst prof to prof chem, Univ SDak, 64-79; state chemist, SDak State Chem Lab, 79-; AT CONSTABLE LAB. *Mem:* Asn Off Anal Chemists; Am Chem Soc; Asn Am Feed Control Off; Asn Am Fertil Control Off. *Res:* Analytical methods for pesticides; gas chromatography; instrumental analysis; flame spectroscopy. *Mailing Add:* 113 Pen Rite Ct Garner NC 27529

PADNOS, NORMAN, b Brooklyn, NY, Oct 23, 37; c 5. PHYSICAL CHEMISTRY. *Educ:* Brooklyn Col, BS, 57; Univ Rochester, PhD(phys chem), 63. *Prof Exp:* Asst prof chem, NC Col Durham, 63-68; CHEMIST, NEW YORK CITY DEPT AIR RESOURCES, 68- *Concurrent Pos:* Adj prof chem, Cooper Union. *Mem:* Am Chem Soc; NY Acad Sci; Int Solar Energy Soc; Sigma Xi. *Res:* Photochemistry; theoretical chemistry; analytical chemistry of air pollution. *Mailing Add:* 2931 Brighton 1 St Brooklyn NY 11235

PADOVANI, ELAINE REEVES, b Kansas City, Mo, Dec 10, 38; m 63; c 2. METAMORPHIC PETROLOGY, GEOPHYSICS. *Educ:* Vassar Col, AB, 61; Stanford Univ, MS, 63; Univ Tex, Dallas, PhD(petrol geochem), 77. *Prof Exp:* Res assoc, Dept Earth & Planetary Sci, Mass Inst Technol, 77-81; prog dir Petrogenesis & Mineral Res, Earth Sci Div, NSF, 81-85; DEP EXTERNAL RES, OFF EARTHQUAKES, VOLCANOES & ENG, US GEOL SURV, 85- *Mem:* Am Geophys Union; Paleontol Soc; Geol Soc Am. *Res:* Chemistry mineralogy and physical properties of the deep crust beneath the continents; evolution of the continents through time and constraints on mechanisms of melt production, fluid transport and chemical equilibration at depth. *Mailing Add:* Off Earthquakes Volcanoes & Eng US Geol Surv 905 Nat Ctr Reston VA 22092

PADOVANI, FRANCOIS ANTOINE, b Versailles, France, Aug 22, 37; m 63; c 2. PHYSICS. *Educ:* Advan Sch Elec, France, EE, 59; Stanford Univ, PhD(elec eng), 62. *Prof Exp:* Engr, Nat Ctr Study Telecommun, Paris, France, 63-64; mem tech staff, 64-70, eng mgr, 70-76, MGR, RES & DEVELOP, TEX INSTRUMENTS, INC, 76- *Mem:* Am Phys Soc; Sigma Xi; Electrochem Soc. *Res:* Semiconductor physics; laser research and applications to solid state physics; conduction mechanism in Schottky barriers; crystal growth; polysilicon manufacturing; electromechanical controls; metallurgy. *Mailing Add:* 60 Reservoir Rd Westwood MA 02090

PADRINI, VITTORIO ARTURO, biochemistry of connective tissue; deceased, see previous edition for last biography

PADRON, JORGE LOUIS, b Havana, Cuba, July 3, 31; m 57; c 2. BIOCHEMISTRY. *Educ:* Okla Baptist Univ, BS, 52; Univ Okla, MS, 54, PhD(biochem, bact), 56. *Prof Exp:* Asst prof chem, Okla Baptist Univ, 54-56; from assoc prof to prof, 57-64, head dept, 60-74, VPRES ACAD AFFAIRS & DEAN COL, DRURY COL, 74- *Concurrent Pos:* Fulbright lectr, Seville, Spain, 62-63 & Quito, Ecuador, 66-67; NATO fel, Oxford Univ. *Mem:* Am Chem Soc; Am Soc Microbiol; NY Acad Sci. *Res:* Bacterial physiology; altered metabolic pathways accompanying chloramphenicol resistance in Staphylococcus aeurus; glucose acetoacetate condensate as an antidiabetogenic factor. *Mailing Add:* Chem Dept Drury Col Springfield MO 65802

PADRTA, FRANK GEORGE, b Chicago, Ill, Apr 29, 30; m 57; c 2. ORGANIC CHEMISTRY, PHYSICAL CHEMISTRY. *Educ:* Elmhurst Col, BS, 56. *Prof Exp:* Lab technician, Universal Oil Co, 52-56, chemist, 56-64, assoc res coordr, 64-68, res coordr, 68-74, group leader, 74-81, assoc res scientist, 81-86, res scientist, 86-90, RES MGR, UNIVERSAL OIL PROD CO, 90- *Mem:* Am Chem Soc; Am Soc Mass Spectrometry; Am Soc Testing & Mat. *Res:* Mass spectrometry; gas chromatography; liquid solid chromatography; infrared spectroscopy; catalysis research; new petroleum products; air pollution. *Mailing Add:* 425 Florian Dr Des Plaines IL 60016

PADUA, DAVID A, b Caracas, Venezuela, Feb 28, 49. PARALLEL COMPUTING, COMPILER TECHNIQUES. *Educ:* Univ Cent Venezuela, Licenciado Computer Sci, 73; Univ Ill, PhD(computer sci), 80. *Prof Exp:* Vis asst prof computer sci, Univ Ill, Urbana-Champaign, 80-81; asst prof computer sci, Univ Simon Bolivar, 81-85; asst prof, 85-90, ASSOC PROF COMPUTER SCI, 90-, ASSOC DIR, CTR SUPERCOMPUT RES DEVELOP, UNIV ILL, URBANA-CHAMPAIGN, 89- *Mem:* Inst Elec & Electronic Engrs Computer Soc; Asn Comput Mach. *Res:* The organization of parallel computers; languages, compilers and debuggers for such machines. *Mailing Add:* Univ Ill 305 Talbot Lab 104 S Wright Champaign IL 61821

PADUANA, JOSEPH A, b Utica, NY, Sept 9, 17; m 44. SOIL MECHANICS, FOUNDATION ENGINEERING. *Educ:* Cooper Union, BCE, 50; Polytech Inst Brooklyn, MCE, 60; Univ Calif, Berkeley, PhD(civil eng), 66. *Prof Exp:* Struct engr, Corbett, Tinghir & Co, NY, 50-54; proj engr, Moran, Proctor, Meuser & Rutledge, 54-59; sr proj engr, Raymond Int, Inc, 60-62; prof & chmn dept, 65-, EMER PROF ENG, CALIF STATE UNIV, SACRAMENTO. *Concurrent Pos:* NSF grant for instructional sci equip, 67; mem hwy res bd, Nat Acad Sci-Nat Res Coun. *Mem:* Am Soc Civil Engrs; Sigma Xi. *Res:* Strength of soils; vertical drains in clay soils; soil stabilization; cracking and erosion of earth dams; pile capacities; computer programming. *Mailing Add:* Dept Civil Eng Calif State Univ Sacramento CA 95819

PADULO, LOUIS, b Athens, Ala, Dec 14, 36; m 63; c 2. ELECTRICAL ENGINEERING, MATHEMATICS. *Educ:* Fairleigh Dickinson Univ, BS, 59; Stanford Univ, MS, 62; Ga Inst Technol, PhD(elec eng), 66. *Prof Exp:* Engr, Radio Corp Am, 59-60; acting instr elec eng, Stanford Univ, 60-62; asst prof San Jose State Col, 62-63; asst prof math, Ga State Col, 66-67; asst prof math, Morehouse Col, 67-68, assoc prof, 68-69, acting chmn dept, 67-68; vis assoc prof elec eng, Stanford Univ, 69-71, assoc prof, 71-76; dean col eng & prof elec eng, 76-88, assoc vpres, Boston Univ, 86-87; PRES, UNIV ALA, HUNTSVILLE, 88- *Concurrent Pos:* Consult, math, Eng Exp Sta, Ga Inst

Technol, 66-; mathematician, Thomas J Watson Res Ctr, Int Bus Mach Corp, 68; dir dual degree prog, Atlanta Univ Ctr & Ga Inst Technol, 69-; vis assoc prof, Columbia Univ, 69, Harvard Univ, 70 & Stanford Univ, 71; spec asst to pres, Morehouse Col, 70-; vis prof, Tokyo Univ, 86-87, Mass Inst Technol, 87-88. *Honors & Awards:* Walter J Gores Award, 71. *Mem:* Math Asn Am; Inst Elec & Electronics Engrs; fel Am Soc Elec Eng. *Res:* Communication theory and data processing; linear system theory; applied mathematics; automata theory; computation theory; linear algebra; analysis; optimization; discrete mathematics. *Mailing Add:* 4200 Penny St SW Huntsville AL 35805

PADWA, ALBERT, b New York, NY, Oct 3, 37; m 60; c 3. ORGANIC CHEMISTRY. *Educ:* Columbia Univ, BA, 59, MA, 60, PhD(chem), 62. *Prof Exp:* NSF fel, Univ Wis, 62-63; asst prof chem, Ohio State Univ, 63-66; assoc prof chem, State Univ NY Buffalo, 66-69, prof, 69-79; PROF CHEM, EMORY UNIV, 79- *Concurrent Pos:* Guggenheim fel; Alexander Von Humboldt sr scientist. *Mem:* Am Chem Soc (chmn, Org Div, 85-86). *Res:* Mechanistic organic photochemistry; chemistry of small strained rings; free radical reactions. *Mailing Add:* Dept Chem Emory Univ Atlanta GA 30322

PADWA, ALLEN ROBERT, b Amityville, NY, Jan 12, 52. POLYMER CHEMISTRY. *Educ:* Clarkson Univ, BS, 74; Univ Mass, Amherst, MS, 78. *Prof Exp:* Res chemist, 74-87, ASSOC FEL, PLASTICS DIV, MONSANTO CO, 87- *Mem:* Am Chem Soc; Soc Plastics Engrs. *Res:* Multiphase polymer systems including emulsion and anionic polymerization with emphasis on styrenic polymers, polymer grafting, and polymer blends including reactive processing and compatibilization. *Mailing Add:* Monsanto Chem Co Bldg 43 730 Worcester St Springfield MA 01151-1089

PADYKULA, HELEN ANN, b Chicopee, Mass, Dec 27, 24. CELL BIOLOGY. *Educ:* Univ Mass, BS, 46; Mt Holyoke Col, MA, 48; Radcliffe Col, PhD(med sci), 54. *Hon Degrees:* DSc, Mt Holyoke Col, 84. *Prof Exp:* Asst zool, Mt Holyoke Col, 46-48; instr, Wellesley Col, 48-50; from instr to asst prof anat, Harvard Med Sch, 53-64; vis lectr zool, 61-62, prof biol, lab electron micros, Wellesley Col, 64-79; AT MED SCH, UNIV MASS, 79- *Concurrent Pos:* Teaching asst, Marine Biol Lab, Woods Hole, 48-49; vis investr, Med Col, Cornell Univ, 58-59; USPHS res career develop award, 59-64, spec res fel, 71-72; adv to freshmen, Radcliffe Col, 63-64; dir NSF undergrad res participation prog biol sci, Wellesley Col, 66-70; mem reproductive biol study sect, NIH, 68-72; vis investr, Ian Clunies Ross Res Lab, Commonwealth Sci & Indust Res Orgn, Prospect, Australia, 71; Am ed, Histochemie, 71; vis prof, Dept Anat, McGill Univ, 72; assoc ed, Am J Anat, 71-74; mem sci adv comt, Muscular Dystrophy Asn, 72-75; mem bd sci coun, Nat Eye Inst, 72-76; vis prof, Univ NC, Chapel Hill, 75; assoc ed, Anat Record, 75-, pres, 80-81; dir NICHD Prog Proj Dir, Primate Uterus. *Mem:* Am Asn Anat; Histochem Soc; Am Soc Cell Biol; Soc Develop Biol; Am Asn Anatomists (vpres, 78-80); Sigma Xi; fel AAAS; Am Soc Zoologists. *Res:* Histochemistry and electron microscopy; mammalian differentiation, especially uterus, placenta and skeletal muscle. *Mailing Add:* Dept Cell Biol Med Sch Univ Mass 55 Lake Ave N Worcester MA 01605

PAE, K(OOK) D(ONG), b Seoul, Korea. MATERIALS SCIENCE, POLYMER MECHANICS. *Educ:* Missouri Valley Col, BS, 56; Univ Mo, MS, 58; Pa State Univ, PhD(mech), 62. *Prof Exp:* From asst prof to prof, 62-78, DISTINGUISHED PROF MECH & MAT SCI, RUTGERS UNIV, NEW BRUNSWICK, 78- *Mem:* Am Phys Soc; Am Asn Univ Professors; Sigma Xi; fel Am Phys Soc. *Res:* High pressure effects on mechanical and physical properties of polymers and composites. *Mailing Add:* Dept Mech & Mat Sci Rutgers Univ Piscataway NJ 08855-0909

PAECH, CHRISTIAN, b Leipzig, Germany, July 10, 42; m 70; c 4. INDUSTRIAL ENZYMES, PHOTOSYNTHESIS. *Educ:* Univ Freiburg, Germany, MS, 70, PhD(chem), 75. *Prof Exp:* Res assoc biochem, Mich State Univ, 75-78; asst prof res biochem, dept biochem & biophysics, Med Sch, Univ Calif, San Francisco, 78-81; from asst prof to assoc prof biochem, Dept Chem, SDak Univ, 81-86; SR RES SCI, HENKEL RES CORP, 86- *Concurrent Pos:* Prin investr, Am Soybean Asn Grant, 81-83; prin investr, competitive grant, USDA, 82-86, mem, photosynthesis prog, 83; ed, Planta, 85-90. *Mem:* Am Soc Biol Chemists; Sigma Xi; German Chem Soc; NY Acad Sci. *Res:* Enzymological and metabolic basis of photosynthetic carbon dioxide fixation by higher plants; biochemical basis for shedding of soybean flowers; development of industrial enzymes. *Mailing Add:* Henkel Res Corp 2330 Circadian Way Santa Rosa CA 95407-5415

PAEGLE, JULIA NOGUES, b Buenos Aires, Argentina, May 13, 43; m 68; c 3. STATISTICAL METEOROLOGY. *Educ:* Univ Buenos Aires, BS, 63; Univ Calif, Los Angeles, MA, 67, PhD(meteorol), 69. *Prof Exp:* From asst prof to assoc prof, 71-82, PROF METEOROL, UNIV UTAH, 82- *Concurrent Pos:* Vis prof, Univ Buenos Aires, 71; guest investr, Argentine Weather Serv, 72; vis scientist, Nat Ctr Atmospheric Sci, 75 & 77; vis assoc prof, Univ Calif Los Angeles, 78; vis prof, Univ Veracruz, 79; vis scientist, Nat Meteorol Ctr, 80-81; chair, FGGE Adv Panel, Nat Res Coun, 83-87; Mem, Bd Atmospheric Sci & Climate, 85-88; chair, Southern Hemisphere Meteorol Comt, 84-85, coun, 85-88. *Mem:* Am Meteorol Soc; Argentine Meteorological Ctr; Meteorol Soc Japan; Am Meteorol Soc. *Res:* Statistical and dynamical analyses of planetary scale atmospheric circulations; deflection and modification of air currents by mountains; interaction of tropical motions with mid-latitude flows. *Mailing Add:* Dept Meteorol Univ Utah Salt Lake City UT 84112

PAESLER, MICHAEL ARTHUR, b Elgin, Ill, Apr 19, 46; m 80; c 2. AMORPHOUS SEMICONDUCTORS, PRECISION ENGINEERING. *Educ:* Beloit Col, BA, 68; Univ Chicago, MS, 71, PhD(physics), 75. *Prof Exp:* Guest scientist physics, Max Planck Inst, 75-76; res fel appl sci, Harvard Univ, 77-80; from asst prof to assoc prof, 80-90, PROF PHYSICS, NC STATE UNIV, 90- *Mem:* Am Phys Soc; Mat Res Soc. *Res:* Super-resolution optical microscopies and spectroscopies; photoluminescence and Raman spectroscopy of semiconductors; electrical and optical properties of amorphous semiconductors; precision engineering; microscopic spectroscopies. *Mailing Add:* Dept Physics NC State Univ Raleigh NC 27695-8202

PAETKAU, VERNER HENRY, b Rosthern, Sask, July 26, 41; m 65; c 3. IMMUNOLOGY, MOLECULAR BIOLOGY. *Educ:* Univ Alta, BSc, 63; Univ Wis, MSc, 65, PhD(biochem), 67. *Prof Exp:* Fel biochem, Enzyme Inst, Univ Wis, 67-69; from asst prof to assoc prof, 69-78, PROF BIOCHEM, UNIV ALTA, 78- *Concurrent Pos:* Med Res Coun scholar, Univ Alta, 69-73. *Mem:* Am Soc Biol Chemists; Can Biochem Soc; Am Asn Immunologists. *Res:* Regulation of the immune response by cytolcines; regulation of cytolcine gene expression. *Mailing Add:* Dept Biochem Univ Alta Fac Med Edmonton AB T6G 2G3 Can

PAFFENBARGER, RALPH SEAL, JR, b Columbus, Ohio, Oct 21, 22; m 43; c 6. EPIDEMIOLOGY. *Educ:* Ohio State Univ, BA, 44; Northwestern Univ, BM, 46, MD, 47; Johns Hopkins Univ, MPH, 52, DPH, 54. *Prof Exp:* Epidemiologist, USPHS Commun Dis Ctr, Ga, 47-53, Lab Infectious Dis, Nat Microbiol Inst, 53-55, Robert A Taft Sanit Eng Ctr, Ohio, 55-59 & Nat Heart Inst, 59-68; chief bur adult health & chronic dis, Bur Adult Health & Chronic Dis, Calif State Dept Pub Health, 68-70, chief epidemiol sect, 71-78; prof epidemiol, Sch Pub Health, Univ Calif, Berkeley, 69-80; PROF, SCH MED, STANFORD UNIV, 77-; ADJ PROF EPIDEMIOL, SCH PUB HEALTH, HARVARD UNIV, 88- *Concurrent Pos:* Asst, La State Univ, 49-50; res assoc, Sch Hyg & Pub Health, Johns Hopkins Univ, 52-54; asst clin prof, Col Med, Univ Cincinnati, 55-59; clin assoc, Harvard Univ, 60-65, lectr, 65-; vis prof epidemiol, 83-85. *Mem:* AAAS; Sigma Xi; Am Pub Health Asn; AMA. *Res:* Epidemiologic studies of acute and chronic diseases, poliomyelitis, diarrheal disease, mental illness, coronary heart disease, stroke and cancer. *Mailing Add:* 892 The Arlington Berkeley CA 94707

PAFFORD, WILLIAM N, b Camden, Tenn, May 21, 29; m 50; c 2. SCIENCE EDUCATION. *Educ:* George Peabody Col, BS, 57, MA, 58, EdS, 63; Univ Ky, EdD, 67. *Prof Exp:* Instr, Montgomery Bell Acad, Tenn, 58-63; asst prof biol, Radford Col, 63-65; coordr student teaching sci area, Univ Ky, 66-67; chmn dept gen sci & sci educ, 73-78, asst dean, Col Educ, 78-81, assoc dean, 81-82, PROF SCI EDUC, EAST TENN STATE UNIV, 67-, CHMN, DEPT CURR & INST, 82- *Mem:* Nat Sci Teachers Asn; Asn Educ Teachers Sci. *Res:* Adoption of educational innovations in science; effect of personal characteristics on student performance in science courses. *Mailing Add:* Dept Curric & Inst Box 23020A ETenn State Univ Johnson City TN 37614

PAGANELLI, CHARLES VICTOR, b New York, NY, Feb 13, 29; m 54; c 5. BIOPHYSICS, PHYSIOLOGY. *Educ:* Hamilton Col, AB, 50; Harvard Univ, MA, 53, PhD, 57. *Prof Exp:* From instr to assoc prof, 58-71, assoc chmn, 76-80, actg chmn, 80-81, PROF PHYSIOL, SCH MED, STATE UNIV NY BUFFALO, 71-, ASSOC CHMN, 81- *Concurrent Pos:* Polio Found fel, Copenhagen, 56-58; vis prof physiol, Univ Hawaii, 74-75. *Honors & Awards:* Elliott Coues Award, Am Ornithol Union, 81. *Mem:* Undersea Med Soc; AAAS; Am Physiol Soc. *Res:* Permeability of biological and artificial membranes; diffusion; active transport. *Mailing Add:* Dept Physiol State Univ NY Buffalo NY 14214

PAGAN-FONT, FRANCISCO ALFREDO, b Mayaguez, PR, June 18, 40; US citizen; m 65; c 5. FISHERIES MANAGEMENT. *Educ:* Univ PR, BS, 63, MS, 67; Auburn Univ, PhD(fisheries mgt), 70. *Prof Exp:* From asst prof to assoc prof marine & fisheries sci & aquacult, Univ PR, Mayaguez, 70-76, asst dir dept, 71-73, dir, Dept Marine Sci, 73-76; BIOLOGIST, SE FISHERIES CTR, NAT FISHERIES SERV, 76- *Concurrent Pos:* Mem PR & the Sea Comt, Govt PR, 72; US Dept Interior res grants, 71-73; PR Dept Agr & US Dept Com res grant, 71-76; Upjohn Co res grant, 75; fishery resources off & aquaculturist, Food & Agr Orgn, UN, 76-79, res biologist, Brazil, 81-88; exec dir develop & admin marine, lacustrine & flurial resources, PR Corp, 79-81; pres bd dirs, Fajardo Drydock, Inc, 79-81; instr & fisheries consult, Dept Com, Nat Oceanic & Atmospheric Admin, Nat Marine Fisheries Serv; res fisheries biologist, Charleston Lab/Miami Lab. *Mem:* Am Fisheries Soc; World Maricult Soc; Am Inst Fishery Res Biologists; Asn Island Marine Labs Caribbean (1st vpres, 73-74). *Res:* The aquaculture of Tilapia and channel catfish cultures, mono- and polycultures of fishes, hybridization in Tilapias, cage culture of fishes, and mariculture of fishes, shrimp and oysters; artificial reefs; planning aquaculture development and research; training of senior aquaculturist; research administration. *Mailing Add:* 7820 SW 141st St Miami FL 33158

PAGANO, ALFRED HORTON, b New York, NY, July 14, 30; m 53; c 3. ENVIRONMENTAL REGULATION, DYE CHEMISTRY. *Educ:* Cornell Univ, BA, 52; Columbia Univ, MA, 53; Ohio State Univ, PhD(org chem), 60. *Prof Exp:* Res chemist, 60-73, prod supvr, 73-81, sr supvr, E I du Pont de Nemours & Co Inc, 81-88; CONSULT MGR, ENVIRON. *Mem:* Am Chem Soc. *Res:* Hydrogenation of nitrocompounds; amination; action of oxidizing agents on salts of primary nitroalkanes; oxidative dimerization; responsible for production of azo and anthraquinone dyes for textile or carpet use. *Mailing Add:* 18 Mattei Lane Fireside Park Newark DE 19713

PAGANO, JOSEPH FRANK, b Cleveland, Ohio, July 15, 19; m 42; c 2. MICROBIOLOGY. *Educ:* Univ Ill, BS, 42, MS, 47, PhD(microbiol), 49; Am Bd Med Microbiol, dipl. *Prof Exp:* Bacteriologist, Ill State Water Surv, 46-48; res assoc microbiol, Univ Ill, 49-50; head dept chemother res, Squibb Inst Med Res, 50-61; asst dir biol div, Sterling Winthrop Res Inst, 61-62; sect head microbiol, Smith Kline & French Labs, Smith Kline Corp, 62-66, asst dept mgr infectious dis, 66-67, dir microbiol, 67-71, tech dir, Smith Kline Diag, 72-77, DIR ENVIRON HEALTH & QUAL CONTROL PROGS, CORP TECH SERV, SMITH KLINE DIAG DIV, SMITH KLINE CORP, 78-; CONSULT, 79- *Concurrent Pos:* Res assoc prof, Grad Sch, Hahnemann Univ; Wright fel, Univ Ill, 47-48. *Mem:* Fel Am Acad Microbiol; Am Soc Microbiol (pres, 79-81); Soc Indust Microbiol; Am Indust Hyg Asn; fel NY Acad Sci. *Res:* Antibiotics; chemotherapeutics; fermentations; microbial genetics and biochemistry; diagnostics; environmental health quality assurance. *Mailing Add:* 149 Sheldrake Dr Paoli PA 19301

PAGANO, JOSEPH STEPHEN, b Rochester, NY, Dec 29, 31; m 57; c 2. CANCER RESEARCH, TUMOR VIROLOGY. *Educ:* Univ Rochester, AB, 53; Yale Univ, MD, 57. *Prof Exp:* Assoc in med, Univ Pa & mem Wistar Inst, Philadelphia, 62-65; from asst prof to assoc prof med, bacteriol & immunol, 65-73, dir, Virol Lab, 69-74 & Div Infectious Dis, 72-75, PROF MED, MICROBIOL & IMMUNOL, LINEBERGER PROF CANCER RES, UNIV NC SCH MED, 73-, DIR, LINEBERGER COMPREHENSIVE CANCER CTR, 74-; ATTEND PHYSICIAN MED & INFECTIOUS DIS, NC MEM HOSP, 65- *Concurrent Pos:* Resident, Peter Bent Brigham Hosp, Harvard Univ, 60-61; fel, Karolinska Inst, Stockholm, Sweden, 61 & Nat Found, 61-62; Sinsheimer award, 66-71; USPHS res career award, NIH, 68-73; vis prof, Swiss Inst Cancer Res, Lausanne, 70-71; mem, Virol Study Sect, NIH, 73-79; bd assoc ed, Cancer Res, 76-80; consult, Burroughs Wellcome Co, Res Triangle Park, NC, 78-, Am Inst Biol Sci, 85; assoc ed, J Gen Virol, 79-84, Antimicrobial Agents & Chemother, 84-; mem, Newcomb-Anderson Prize Comt, 84-88; pres, Int Assoc Res Epstein-Barr Virus & Assoc Dis, 88; mem, Recombinant DNA adv comt, USPHS, 86-; bd assoc ed, Cancer Res, 76-80. *Mem:* Infectious Dis Soc Am; Am Soc Microbiol; Am Asn Immunol; Am Asn Cancer Res; Am Soc Clin Invest; Am Asn Physicians; Am Soc Virol; Int Soc Antiviral Res; Royal Soc Med. *Res:* Cancer and virus infections; tumor virology and molecular biology of Herpes viruses, especially Epstein-Barr virus; antiviral drugs. *Mailing Add:* 102 Lineberger CRC CB 7295 Sch Med Univ NC Chapel Hill NC 27599

PAGANO, RICHARD EMIL, b New York, NY, June 13, 44; m 66; c 3. CELL BIOLOGY, LIPID BIOCHEMISTRY. *Educ:* Johns Hopkins Univ, BA, 65; Univ Va, PhD(biophys), 68. *Prof Exp:* Staff assoc membrane biophysics, Lab Phys Biol, NIH, 68-70; sr fel membrane biophys, Polymer Dept, Weizmann Inst Sci, Rehovoth, Israel, 70-72; assoc prof biol, 76-82, PROF BIOL, JOHNS HOPKINS UNIV, 82-, PROF BIOPHYSISCS,84-; STAFF MEM, DEPT EMBRYOL, CARNEGIE INST WASHINGTON, 72- *Concurrent Pos:* Sr asst scientist, USPHS, 68-70; mem cell biol study sect, NSF, 77-79, chmn, Membrane Biophys Subgroup, 78-79. *Mem:* AAAS; Am Soc Cell Biologists; Biophys Soc; Am Soc Biol Chemists. *Res:* Organization, dynamics and metabolism of lipids in eukaryotic cells; membrane traffic; membrane fusion; liposomes; fluorescence microscopy. *Mailing Add:* Dept Embryol Carnegie Inst 115 W University Pkwy Baltimore MD 21210

PAGE, ALAN CAMERON, b Lawrence, Mass, May 16, 42; m 66; c 2. FOREST MANAGEMENT. *Educ:* Cornell Univ, BS, 64; Yale Univ, MF, 66; Univ Mass, Amherst, PhD(forestry), 74. *Prof Exp:* Pres forest mgt, 71-84, OWNER, GREEN DIAMOND SYSTS, GREEN DIAMOND FORESTRY SERV, INC, 82- *Mem:* Soc Am Foresters. *Res:* Computer analysis of forest management options both biological and economic. *Mailing Add:* 125 Blue Meadow Rd Belchertown MA 01007

PAGE, ALBERT LEE, b New Lenox, Ill, Mar 19, 27; m 52; c 2. SOIL CHEMISTRY. *Educ:* Univ Calif, Riverside, BA, 56; Univ Calif, Davis, PhD(soil sci), 60. *Prof Exp:* From asst prof to assoc prof, 60-71, head, Div Soil Sci, 71-75, dir, Kerney Found Soil Sci, 75-80, PROF SOIL SCI, UNIV CALIF, RIVERSIDE, 71- *Concurrent Pos:* Guggenheim Mem Found fel & Fulbright-Hays award, 66-67; comt mem, Nat Acad Sci-Subcomt Geochem & Environ Health, 75-81. *Mem:* AAAS; Soil Sci Soc Am; Am Soc Agron; Sigma Xi. *Res:* Chemical and mineralogical properties of soils; chemistry of hydrolyzable metals in colloid systems; ion exchange equilibrium; environmental trace metal contamination. *Mailing Add:* Dept of Soil & Environ Sci Univ of Calif Riverside CA 92521

PAGE, ARTHUR R, b Winona, Minn, Mar 22, 30; m 51; c 3. IMMUNOLOGY, PEDIATRICS. *Educ:* Univ Minn, MD, 56. *Prof Exp:* Asst prof, 61-66, assoc prof, 66-76, PROF PEDIAT, MED SCH, UNIV MINN, MINNEAPOLIS, 76- *Concurrent Pos:* USPHS fel, 59-61, career develop award, 61-71. *Mem:* Am Soc Exp Path; Soc Pediat Res. *Res:* Inflammation. *Mailing Add:* Pediat 1430 Mayo Univ MN Minneapolis MN 55455

PAGE, BENJAMIN MARKHAM, b Pasadena, Calif, May 6, 11; m 35; c 2. PLATE TECTONICS, NEOTECTONICS. *Educ:* Stanford Univ, AB, 33, MA, 34, PhD(geol), 40. *Prof Exp:* Field geologist, 35-37; from instr to asst prof geol, Univ Southern Calif, 37-41; assoc prof, 43-51, exec head dept, 57-69, prof, 51-76, EMER PROF GEOL, STANFORD UNIV, 76- *Concurrent Pos:* Asst geologist, US Geol Surv, 42-45; sci consult, Nat Resources Sect, Allied Occup, Japan, 50-51; geol consult, UN Tech Asst Admin, Yugoslavia, 55; Guggenheim fel, Italy, 59-60; consult, Adv Comt Reactor Safeguards, Nuclear Regulatory Comn, 67-; geol res, Mineral Res & Serv Org, Taiwan, 74 & 77; ed-in-chief, Tectonics, 85-88; ed, Am Geophys Union Bk, 89-90. *Mem:* Fel Geol Soc Am; fel AAAS; fel Am Geophys Union; Sigma Xi. *Res:* tectonics; geology of continental margins; plate interactions. *Mailing Add:* Dept Geol Stanford Univ Stanford CA 94305-2115

PAGE, CARL VICTOR, b Flint, Mich, Apr 26, 38; div; c 2. COMPUTER SCIENCE. *Educ:* Univ Mich, BSE(sci eng) & BSE(eng math), 60, MS, 61, PhD(commun sci), 65. *Prof Exp:* Lectr elec eng, Univ Mich, 66; asst prof info sci, Univ NC, 66-67; from asst prof to assoc prof comput sci, 67-76, PROF COMPUT SCI, MICH STATE UNIV, 76- *Concurrent Pos:* Grad coordr, 86-90. *Mem:* Inst Elec & Electronic Engrs; Asn Comput Mach; Pattern Recognition Soc; Soc Mfg Engrs. *Res:* Artificial intelligence; adaptive systems; automata theory; compiler theory; formal languages; pattern recognition. *Mailing Add:* Dept of Comput Sci Mich State Univ East Lansing MI 48824

PAGE, CHARLES HENRY, b East Orange, NJ, Aug 28, 41; m 63; c 2. NEUROPHYSIOLOGY, NEUROSCIENCE. *Educ:* Allegheny Col, BS, 63; Univ Ill, Urbana, MS, 66, PhD(physiol), 69. *Prof Exp:* NIH fel biol, Stanford Univ, 68-70; asst prof zool, Ohio Univ, 70-74; from asst prof to assoc prof, 74-85, PROF PHYSIOL, RUTGERS UNIV, 85- *Concurrent Pos:* NIH grant, Ohio Univ, 72-74; NIH res grant, 74-; mem adv bd, J Comp Physiol, 85- *Mem:* AAAS; Soc Exp Biol; Am Soc Zool; Soc Neurosci. *Res:* Arthropod neurobiology; proprioceptive reflexes; central control of movement. *Mailing Add:* Dept of Bio Sci Rutgers Univ Piscataway NJ 08855

PAGE, CHESTER H, b Providence, RI, Nov 13, 12. ELECTRICAL ENGINEERING. *Educ:* Brown Univ, AB, ScM, 34; Yale Univ, PhD(physics), 37. *Prof Exp:* Coordr Int Activities, Nat Bur Standards, 41-77; RETIRED. *Honors & Awards:* Gold Medal Award, US Dept Commerce; Harry Diamond Award, 74 & Charles Steinmetz Award, Inst Elec & Electronics Engrs, 74. *Mem:* Inst Elec & Electronics Engrs; fel Am Phys Soc. *Mailing Add:* 1707 Merrifields Dr Silver Springs MD 20906-1251

PAGE, CLAYTON R, III, cell biology, parasitology, for more information see previous edition

PAGE, D(ERRICK) J(OHN), b London, Eng, Jan 20, 37; US citizen; m 66; c 3. ELECTRICAL ENGINEERING. *Educ:* Univ Birmingham, BSc, 59, PhD(elec eng), 63. *Prof Exp:* Sr eng, Res & Develop Ctr, Westinghouse Elec Corp, 63-68, fel eng, 68-70, mgr, semiconductor res, 70-79, mgr, res & develop, Indust Controls Business Unit, 79-82, mgr, VHSIC insertion, Advan Technol Div, Baltimore, 82-83, mgr, solid state sci, 83-86. *Mem:* Inst Elec & Electronics Engrs. *Res:* Research and development of microwave power silicon transistors; light activated semiconductor switches; charge coupled devices; photosensors; biomicroelectronic devices; very large scale integrated circuits; solid state pulse power devices. *Mailing Add:* Energy Comp Res Corp 910 Camino Del Mar Del Mar CA 92014

PAGE, DAVID L, b New York, NY, Jan 20, 41. SURGICAL PATHOLOGY. *Educ:* Johns Hopkins Univ, MD, 66. *Prof Exp:* DIR, ANAT PATH, 76-, PROF PATH, VANDERBILT UNIV, 78- *Mem:* Am Asn Pathologists; Int Asn Pathologists; Nat Acad Cytol; Am Soc Clin Path. *Mailing Add:* Dept Path Sch Med Vanderbilt Univ Nashville TN 37232

PAGE, DAVID SANBORN, b Suffern, NY, June 1, 43; m 64, 77; c 3. BIO-ORGANIC & MARINE ENVIRONMENTAL CHEMISTRY, ANALYTICAL CHEMISTRY. *Educ:* Brown Univ, BS, 65; Purdue Univ, PhD(phys chem), 70. *Prof Exp:* Vis asst prof chem, Purdue Univ, 70-71; asst prof chem, Bates Col, 71-74; assoc prof, 74-82, PROF CHEM, BOWDOIN COL, 82- *Concurrent Pos:* Vis scientist, Plymouth Marine Lab, UK, 87-88. *Mem:* Am Chem Soc. *Res:* Enzyme-substrate and substrate analog interactions; physiologic indicators of pollution stress; trace analysis of hydrocarbons and other pollutants in the marine environment; toxic effects of hydrocarbons and other pollutants on marine organisms. *Mailing Add:* Dept of Chem Bowdoin Col Brunswick ME 04011

PAGE, DEREK HOWARD, b Sheffield, Eng, Nov 22, 29; m 55; c 4. PULP & PAPER TECHNOLOGY. *Educ:* Cambridge Univ, BA, 53, MA, 57, PhD(physics), 68. *Prof Exp:* Scientist, Brit Insulated Callendars Cables, 53-54; head fibre & paper physics sect, Brit Paper & Bd Indust Res Asn, 55-64; DIR RES, PHYS SCIS, PULP & PAPER RES INST CAN, 64- *Concurrent Pos:* Fel, Tech Asn Pulp & Paper Indust, 75. *Honors & Awards:* Res & Develop Div Award, Tech Asn Pulp & Paper Indust USA, 72. *Mem:* Tech Asn Pulp & Paper Indust; fel Royal Microscopic Soc; Fel Inst Physics; Optical Soc Am. *Res:* Structure and physical properties of wood, wood pulp fibres and paper. *Mailing Add:* Pulp & Paper Res Inst Can 570 St John's Blvd Point Claire PQ H9R 3J9 Can

PAGE, EDGAR J(OSEPH), b Putnam, Conn, Nov 7, 19; m 45; c 3. RESINOGRAPHY, TEXTILE MICROSCOPY. *Educ:* Marianapolis Col, BS, 43; Boston Col, MS, 49; PhD(polymer sci), 56. *Prof Exp:* Control chemist, Nat Chromium Co, 42-44; instr chem & phys, Marianapolis Acad, 44-45; from res chemist to chief chemist, 43-79, asst vpres & sr scientist, 79-85, SR TECH CONSULT, BELDING HEMINGWAY CO, 85- *Mem:* Fel Am Soc Testing & Mat; Am Chem Soc; Am Asn Textile Chemists & Colorists; NY Acad Sci; AAAS; fel Am Inst Chem. *Res:* Textile science; dyeing and finishing of textiles; polymerization of nylons; physical and chemical characterization of specialty polyamides. *Mailing Add:* Belding Hemingway Co Putnam CT 06260

PAGE, EDWIN HOWARD, b Glasgow, Ky, Feb 3, 20; m 41; c 2. VETERINARY MEDICINE. *Educ:* Western Ky Univ, BS, 40; Ohio State Univ, DVM, 53. *Prof Exp:* Teacher pub sch, Ky, 40-41; instr mach oper, E I du Pont de Nemours & Co, Inc, 41-42; owner-operator dairy farm, Ky, 46-49; pvt pract vet med, Ky, 53-64; from assoc prof to prof vet clin, Purdue Univ, West Lafayette, 64-73, prof large animal clins, Sch Vet Sci & Med, 73-86, head dept large animals, 76-86; PROF EDUC, COL EDUC & HUMAN SERV, UNIV DETROIT, MERCY, 86- *Mem:* Am Asn Vet Clinicians; Am Vet Med Asn. *Res:* Use of an autogenous vaccine in the treatment of equine sarcoid. *Mailing Add:* Col Educ & Human Serv Univ Detroit Mercy 4001 W McNichols Rd Detroit MI 48221

PAGE, ERNEST, b Cologne, Ger, May 30, 27; nat US. BIOPHYSICS. *Educ:* Univ Calif, AB, 49, MD, 52. *Prof Exp:* Am Heart Asn estab investr biophys, Biophys Lab, Harvard Med Sch, 59-65; assoc prof, 65-69, PROF MED & PHYSIOL, SCH MED, UNIV CHICAGO, 69- *Mem:* Biophys Soc; Am Physiol Soc; Am Soc Cell Biol; Soc Gen Physiologists; Am Asn Anatomists; Asn Am Physicians. *Res:* Membrane transport phenomena and electron microscopy in mammalian heart muscle; cell biology of atrial natrivietic peptide secretion; cardiac gap junctions. *Mailing Add:* Dept Med & Physiol Univ Chicago BH Box 137 5841 S Maryland Ave Chicago IL 60637

PAGE, IRVINE HEINLY, cardiovascular diseases; deceased, see previous edition for last biography

PAGE, JOANNA R ZIEGLER, b Gainesville, Fla, Aug 28, 38; m 66; c 2. PHYCOLOGY. *Educ:* Bucknell Univ, BS, 60; Cornell Univ, MS, 63, PhD(bot), 66. *Prof Exp:* NSF fel bot, Yale Univ, 65-66; asst prof, Univ Conn, 66-68, biol, 68-71; lectr bot, Univ Mass, 73-74, 74-75, 75-76, 77-78 & 78-79. *Concurrent Pos:* Vis asst prof bot, Univ Mass, 79-88; NSF Coop Grad fel. *Mem:* Phycol Soc Am; Int Phycol Soc; Sigma Xi. *Res:* Biological clocks; algal life cycles and reproduction. *Mailing Add:* Blue Meadow Rd Belchertown MA 01007

PAGE, JOHN BOYD, JR, b Columbus, Ohio, Sept 4, 38; m 66; c 2. THEORETICAL SOLID STATE PHYSICS. *Educ:* Univ Utah, BS, 60, PhD(physics), 66. *Prof Exp:* Res assoc physics, Univ Frankfurt, 66-67 & Cornell Univ, 68-69; asst prof, 69-75, assoc prof, 75-80, PROF PHYSICS, ARIZ STATE UNIV, 80- *Concurrent Pos:* Vis scientist, Max Planck Inst Solid State Physics, Stuttgart, Germany, 75-76 & 85 & Inst Theoret Physics, Regensburg, Germany, 85. *Honors & Awards:* Humboldt Sr Scientist Res Award, 90. *Mem:* Fel Am Phys Soc; Am Asn Physics Teachers; Sigma Xi. *Res:* Theoretical solid state physics; resonance light scattering from biochemical molecules and solid state defect systems; impurity-induced raman scattering and phonon physics in imperfect and perfect crystals; light scattering studies of solids, dynamics of strongly anharmonic systems, optical properties of disordered and ordered media. *Mailing Add:* Dept Physics Ariz State Univ Tempe AZ 85287-1504

PAGE, JOHN GARDNER, b Milwaukee, Wis, Sept 14, 40; m 62; c 2. PHARMACOLOGY, TOXICOLOGY. *Educ:* Univ Wis-Madison, BS, 64, MS, 66, PhD(pharmacol), 67, Am Bd Toxicol, dipl, 80. *Prof Exp:* Sr toxicologist, Eli Lilly & Co, 69-77; dir toxicol & path, Rhone Poulenc, Inc, 77-80; dir, Toxicol Dept, Toxigenics, Inc, 80-82; sr res scientist, Battelle Mem Inst, 83-87; HEAD, PRECLIN TOXICOL DIV, SOUTHERN RES INST, 87- *Concurrent Pos:* NIH staff fel, Lab Chem Pharmacol, Nat Heart Inst, 67-69. *Mem:* AAAS; Am Soc Pharmacol & Exp Therapeut; Int Soc Biochem Pharmacol; Sigma Xi; Fed Am Soc Exp Biol; Soc Toxicol. *Res:* Biochemical basis and/or mechanism for pharmacological and toxicological activity of potential therapeutic compounds. *Mailing Add:* 3601 Crosshill Rd Mountain Brook AL 35223

PAGE, LARRY J, b Brookfield, Mo, Dec 14, 41; m 62. SPACE PHYSICS. *Educ:* Colo State Univ, BS, 64; Univ Utah, PhD(physics), 69. *Prof Exp:* Reactor physicist, Pac Northwest Labs, Battelle Mem Inst, 64-66; SR MEM TECH STAFF PHYSICS, ELECTROMAGNETIC SYSTS LABS, 69- *Mem:* Am Phys Soc; Optical Soc Am; Asn Comput Mach. *Res:* Energy band calculations and defect calculations in alkali halides; systems engineering and system design for high speed digital signal processing systems. *Mailing Add:* 1482 Baron Dr Sunnyvale CA 94087

PAGE, LAWRENCE MERLE, b Fairbury, Ill, Apr 17, 44. ICHTHYOLOGY, INVERTEBRATE ZOOLOGY. *Educ:* Ill State Univ, BS, 66; Univ Ill, Urbana, MS, 68, PhD(zool), 72. *Prof Exp:* Res biologist, Ill Natural Hist Surv, 65-72; fish biologist, Sargent & Lundy Engrs, 72; from asst taxonomist to assoc taxonomist, 72-80, ICHTHYOLOGIST, ILL NATURAL HIST SURV, 80- *Concurrent Pos:* Consult, Mo Bot Garden, 73-76; assoc ichthyol, Mus Natural Hist, Univ Kans, 79-; prof, Dept Ecol, Ethol & Evolution, Univ Ill, 83- *Mem:* Am Soc Ichthyologists & Herpetologists; Int Asn Astacology; Soc Syst Zool; Sigma Xi. *Res:* Systematics and life histories of darters (Percidae) and minnows (Cyprinidae), amphipods, isopods and decapods. *Mailing Add:* Natural Hist Surv 607 E Peabody Champaign IL 61820

PAGE, LINCOLN RIDLER, b Lipson, NH, Feb 11, 10; wid; c 2. GEOLOGY, METALLURGY. *Educ:* Darthmouth Col, AB, 31; Univ Minn, MS, 32, PhD(geol & metall), 37. *Prof Exp:* Instr, Dartmouth Col, 32-35; instr, Univ Minn, 35-37; field asst, Minn Geol Surv. 36; jr geologist, Standard Oil Co, Tex, 37-38; asst prof, Univ Colo, 38-39; consult, Slide Mines, Boulder, Colo, 38-39; chief field party, US Geol Surv, 39-48, chief beryllium prog, 48-50, chief uranium reconnaissance group, 49-52, res geologist, 52-54 & 59-60, asst chief geologist, Tepco, 54-59, chief, New Eng Br, 60-73; geologist, State NH, 84-86; CONSULT, BCI GEONETICS, LACONIA, NH, 86- *Concurrent Pos:* Adv, US deleg, UN atoms Peace Conf, Geneva, 55 & 58; New Eng field comn, Dept Interior, 60-73; African sci bd, Nat Acad Sci, 63-70 & 72; mem Egyptian assessment team, Dept Energy, 78; distinguished lectr, Nigerian Geol Soc, 78; mem team outline Egyptian mineral prog, State Dept, 80; mem uranium group, Int Atomic Energy Agency, Vienna, 80-84. *Honors & Awards:* Distinguished Serv Medal, Dept Interior, 67; Centennial Medal, Geol Survey of Finland, 87. *Mem:* Fel Geol Soc Am; Sigma Xi; Fel Mineral Soc Am; Am Asn Petrol Geol; AAAS; Geochem Soc; Norweg Geol Soc; Nigerian Geol Soc; Am Geol Inst; Asn Geol Int Develop. *Res:* Geology. *Mailing Add:* PO Box 171 Melvin Village NH 03850

PAGE, LORNE ALBERT, b Buffalo, NY, July 28, 21; m 46; c 5. EXPERIMENTAL PHYSICS. *Educ:* Queen's Univ, Ont, BSc, 44; Cornell Univ, PhD(physics), 50. *Prof Exp:* Asst physics, Cornell Univ, 45-46, Lab Nuclear Studies, 46-49; from asst prof to assoc prof, 50-58, PROF PHYSICS, UNIV PITTSBURGH, 58- *Concurrent Pos:* Guggenheim fel, Inst Physics, Univ Uppsala, 57-58; Sloan res fel, Univ Pittsburgh, 61-65. *Mem:* Am Phys Soc; Sigma Xi; Optical Soc Am. *Res:* Electron-photon processes; magnetism; electrodynamics. *Mailing Add:* 157 Lloyd Ave Pittsburgh PA 15260

PAGE, LOT BATES, internal medicine; deceased, see previous edition for last biography

PAGE, LOUISE, b Payson, Utah. NUTRITION. *Educ:* Utah State Univ, BS, 45; Univ Wis, MS, 50, PhD(nutrit), 63. *Prof Exp:* Dietitian, foods & nutrit, Women's Med Specialist Corp, US Army, 46-48; home economist foods, Mont Agr Exp Sta, 50-54; nutrit anal foods & nutrit, Consumer & Food Econ Inst, Sci & Educ Admin-Agr Res, USDA, 54-, leader food & diet appraisal res group, 72-; RETIRED. *Concurrent Pos:* Mem, Nat Nutrit Adv Comt Head Start, 66-69, USDA Plentiful Foods Comt, 66-72, Nat Comt Diet Ther, Am Dietetic Asn, 70, Secretariat Tech Comt Nutrit, White House Conf on Aging, 71, Yearbook of Agr Comt, 74 & Rural Health Subcomt, Nat Rural Develop Comt, 75- *Mem:* Am Inst Nutrit; Am Home Econ Asn; Am Dietetic Asn. *Res:* Development of research-based dietary guidance materials; nutritional evaluation of food supplies and diets. *Mailing Add:* 193 W 100 S Payson UT 84651

PAGE, MALCOLM I, b Lavonia Twp, Mich, Apr 28, 30; c 3. INTERNAL MEDICINE, INFECTIOUS DISEASES. *Educ:* Wayne Univ, AB, 52; Univ Chicago, MD, 56; Am Bd Internal Med, dipl, 64. *Prof Exp:* Officer EIS, Nat Commun Dis Ctr, 57-59; epidemiologist, Mary Imogene Bassett Hosp, 59-62; asst chief epidemiol, Nat Commun Dis Ctr, 62-63; asst prof med, Univ Chicago, 63-65; physician-in-chief, Mary Imogene Bassett Hosp, 65-72; PROF MED, MED COL GA, 72- *Concurrent Pos:* Vol instr, Sch Med, Emory Univ, 62-63; assoc clin prof, Col Physicians & Surgeons, Columbia Univ, 65-72. *Mem:* Am Col Physicians; Am Soc Microbiol. *Mailing Add:* 840 S Wood St Chicago IL 60612

PAGE, MICHEL, b Que, Feb 18, 40; m 66; c 4. CANCER. *Educ:* Laval Univ, BA, 60; Ottawa Univ, BSc, 65, PhD(biochem), 69. *Prof Exp:* Asst prof, 74-78, assoc prof, 78-82, PROF BIOCHEM, LAVAL UNIV, 82- *Concurrent Pos:* Clin biochemist, Hotel-Dieu de Que, 71-; res scholar, Nat Cancer Inst, 75-81. *Mem:* AAAS; Am Soc Cell Biol; Int Soc Immunopharmacol; Int Soc Prev Oncol; NY Acad Sci. *Res:* Cancer research, cancer markers, drug targeting, carcinogenesis by mineral fibers and environmental carcinogens. *Mailing Add:* Dept Biochem Laval Univ Quebec PQ G1K 7P4 CAN

PAGE, NELSON FRANKLIN, b Salisbury, NC, Nov 17, 38; m 64; c 2. MATHEMATICS. *Educ:* Univ NC, Chapel Hill, BA, 61, MA, 65, PhD(math), 67. *Prof Exp:* From instr to asst prof math, Univ NC, Greensboro, 66-74; mem fac, 74-77, assoc prof, 77-81, PROF MATH, HIGH POINT COL, 81- *Mem:* Am Math Soc; Math Asn Am. *Res:* Functional Hilbert spaces; analytic function theory. *Mailing Add:* Dept Math High Point Col High Point NC 27262

PAGE, NORBERT PAUL, b Farmersville, Ohio, Dec 29, 32; m 55; c 4. VETERINARY MEDICINE, RADIATION BIOLOGY. *Educ:* Ohio State Univ, DVM, 56; Univ Rochester, MS, 63; Am Bd Vet Pub Health, dipl, Am Bd Toxicol, dipl. *Prof Exp:* Dir rabies control, Dept Health, Columbus, Ohio, 56; officer vet serv, Selfridge AFB, Mich, 56-58; pvt practr vet med, Ohio, 58-59; chief vet serv, Itazuke AFB, US Air Force, Japan, 59-62; fel radiation biol, Univ Rochester, 62-63; dir, Large Animal Radiobiol Prog, Naval Radiol Defense Lab, 63-67, sci adminr mammalian radiobiol, Hq, US AEC, Washington, DC, 67-70; chief prog & data analysis unit, Nat Cancer Inst, 71-73, chief carcinogenic bioassay & prog resources br, 73-76; chief priorities & res analysis br, Nat Inst Occup Safety & Health, US Environ Protection Agency, 76-78, dir health rev div, Off Toxic Substance, 78-79, dir scientific affairs, 79-81; sr toxicologist, Nat Libr Med, 81-84; PRES, PAGE ASSOC, 88- *Concurrent Pos:* Consult, Off Civil Defense & Atomic Energy Comn, 67, & Nat Acad Sci Adv Comt, Inst Lab Animal Resources, 67-70; Atomic Energy Comn liaison, Hemat Study Sect, NIH, 69-71. *Mem:* Am Vet Med Asn; Am Col Toxicol; Radiation Res Soc; Am Pub Health Asn; Am Col Vet Toxicol. *Res:* Biological effects of radiation exposure; radiation and chemical carcinogenesis; toxicology; safety assessment; national and international affairs. *Mailing Add:* 17601 Stoneridge Ct Gaithersburg MD 20878

PAGE, NORMAN J, b Boulder, Colo, Jan 2, 39; m 61; c 2. PETROLOGY. *Educ:* Dartmouth Col, BA, 61; Univ Calif, Berkeley, PhD(geol), 66. *Prof Exp:* Phys sci aide geol, 57-66, GEOLOGIST, US GEOL SURV, 66- *Concurrent Pos:* Teaching asst, Univ Calif, 62-66. *Mem:* Mineral Soc Am; Geol Soc Am; Soc Econ Geol. *Res:* Ultramafic rocks, specificially origin of platinum group and sulfide minerals; serpentization and its relation to emplacement of rocks; details of mineral relations in igneous, metamorphic or ore grade rocks as a means to develop experimental studies to solve geologic problems. *Mailing Add:* US Geol Surv 210 E Seventh St Tucson AZ 85705

PAGE, RECTOR LEE, b Wichita, Kans, Feb 4, 44; div; c 3. MATHEMATICS, COMPUTER SCIENCE. *Educ:* Stanford Univ, BS, 66; Univ Calif, San Diego, PhD(math), 70. *Prof Exp:* Programmer, IBM Sci Ctr, Calif, 65-66; from asst prof to prof comput sci, Colo State Univ, 70-82; RES SUPVR, AMOCO PROD CO, 82- *Mem:* Asn Comput Mach; Inst Elec & Electronics Engrs. *Res:* Programming languages; functional programming; parallel computing; Fourier analysis. *Mailing Add:* Amoco Prod Res PO Box 3385 Tulsa OK 74102

PAGE, ROBERT ALAN, JR, b New York, NY, Nov 28, 38; m 64. SEISMOLOGY, TECTONICS. *Educ:* Harvard Col, AB, 60; Columbia Univ, PhD(geophys), 67. *Prof Exp:* Res assoc seismol, Lamont Geol Observ, Columbia Univ, 67-70; geophysicist, 70-75, chief, Br Earthquake Hazards, 75-78, coordr, Earthquake Hazards Prog, 78-80, GEOPHYSICIST, US GEOL SURV, 80- *Mem:* Seismol Soc Am; Am Geophys Union; Earthquake Eng Res Inst; Geol Soc Am; fel AAAS. *Res:* Seismicity and tectonics, particulary of Alaska; strong-motion seismology; earthquake hazards; volcano seismology. *Mailing Add:* US Geol Surv 345 Middlefield Rd Menlo Park CA 94025

PAGE, ROBERT EUGENE, JR, b Bakersfield, Calif, Nov 12, 49; m 70; c 2. APICULTURE, POPULATION GENETICS. *Educ:* San Jose State Univ, BS, 76; Univ Calif, Davis, PhD(entom), 80. *Prof Exp:* Asst prof entom, Ohio State Univ, 86-89; assoc prof, 89-91, PROF ENTOM, UNIV CALIF, DAVIS, 91- *Mem:* Int Union Study Social Insects; Entom Soc Am; Genetics Soc Am; Soc Study Evolution. *Res:* Evolutionary genetics of social behavior of insects including division of labor, the recognition of kin, and sex determination; author of more than 70 publications and two books. *Mailing Add:* Dept Entom Univ Calif Davis CA 95616

PAGE, ROBERT GRIFFITH, b Bryn Mawr, Pa, Mar 25, 21; m 47; c 3. MEDICINE. *Educ:* Princeton Univ, AB, 43; Univ Pa, MD, 45. *Prof Exp:* Resident med, Hosp Univ Pa, 48-49, instr pharmacol, 49-51, asst, 51-53; from asst prof to assoc prof med, Sch Med, Univ Chicago, 53-68, asst dean med educ, 57-63, assoc dean div biol sci, 63-68; dean, Med Col Ohio, 68-74, provost, 72-74, prof med & pharmacol, 68-78; co-dir, Mountain Valley Health Ctr, Londonderry, 78-80; adj prof, Community Med, Dartmouth Med Sch, 78-89 & clin med, 86-89; adj prof med & pharmacol, 78-86, EMER DEAN & EMER PROF MED PHARMACOL, MED COL OHIO, 86- *Concurrent Pos:* Prof, Univ Rangoon, 51-53; mem bd gov, Inst Med Chicago, 66-68; pvt pract, 80-87. *Mem:* AAAS; Am Soc Pharmacol & Exp Therapeut; Asn Am Med Cols; Am Fedn Clin Res; Am Heart Asn; Am Col Physicians; Am Col Cardiol. *Res:* Cardiac physiology, pharmacology and clinical investigation. *Mailing Add:* Box 524 Londonderry VT 05148-0524

PAGE, ROBERT HENRY, b Philadelphia, Pa, Nov 5, 27; m 48; c 5. MECHANICAL ENGINEERING. *Educ:* Univ Ohio, BSME, 49; Univ Ill, MS, 51, PhD(eng), 55. *Prof Exp:* Instr & res assoc mech eng, Univ Ill, 49-55; vis lectr, Stevens Inst Technol, 56-57, prof, 57-61; prof mech & aerospace eng, Rutgers Univ, 61-79, chmn dept, 61-76; dean eng, 79-83, FORSYTH PROF, DEPT ENG, TEX A&M UNIV, 83- *Concurrent Pos:* Res engr, Esso Res & Eng Co, 55-57, separated flow consult, 60-81; mem, Nat Space Eng Comt, 67-71, chmn, 69-70; hon prof, Ruhr Univ, W Ger, 84. *Honors & Awards:* Centennial Medal, Am Soc Mech Engrs, 80; James H Potter Gold Medal, 83. *Mem:* Fel Am Soc Mech Engrs; fel Am Soc Eng Educ; Am Phys Soc; fel Am Astronaut Soc; fel Am Inst Aeronaut & Astronaut; Sigma Xi; fel AAAS. *Res:* Aerothermodynamics; gas dynamics; separated flows. *Mailing Add:* 1905 Comal Circle College Station TX 77840

PAGE, ROBERT LEROY, b Belgrade, Maine, Mar 20, 31; m 64; c 1. MATHEMATICS, PHYSICAL SCIENCE. *Educ:* Tufts Col, BS, 53; Univ Maine, Orono, MA, 59; Fla State Univ, PhD(math educ), 70. *Prof Exp:* From asst prof to assoc prof math, Nasson Col, 59-65; asst prof, Bates Col, 65-67; from asst prof to assoc prof, 69-74, PROF MATH & PHYS SCI, UNIV MAINE, AUGUSTA, 74-, CHMN, DIV SCI & MATH, 71-, COORDR, DIV SCI & MATH. *Concurrent Pos:* Vis lectr, New Eng Acad Sci, 64-67; coordr dept math, Nasson Col, 64-65. *Mem:* Math Asn Am. *Res:* Pure mathematics, especially number theory; mathematical education, particularly curriculum development and use of audio-visual media. *Mailing Add:* Dept Math & Sci Univ Maine Augusta Univ Heights Augusta ME 04330

PAGE, ROY CHRISTOPHER, b Campobello, SC, Feb 7, 32. EXPERIMENTAL PATHOLOGY, PERIODONTOLOGY. *Educ:* Berea Col, AB, 53; Univ Md, Baltimore, DDS, 57; Univ Wash, Seattle, cert periodont, 63, PhD(exp path), 67. *Hon Degrees:* DSc, Loyola Univ, Chicago, 83. *Prof Exp:* Asst prof, 67-70, assoc prof, 70-74, affil scientist, Ctr Res Oral Biol, 70-76, PROF PATH & PERIODONT, SCHS MED & DENT, UNIV WASH, 74-, DIR, CTR RES ORAL BIOL, 76- *Concurrent Pos:* Nat Inst Gen Med Sci grant, 67-; NIH res career develop award, Univ Wash, 67-72; Nat Heart Inst grant, 67-; Nat Inst Dent Res grant, 70-; mem bd counrs, Nat Inst Dent Res, 73- *Mem:* AAAS; Am Soc Exp Path; Am Soc Microbiol; Am Acad Periodont; Int Asn Dent Res (pres, 87-88). *Res:* Connective tissue maturation and diseases; immunopathology of periodontal disease; long term chronic inflammatory disease. *Mailing Add:* Dept of Path SM-30 Univ of Wash Sch of Med Seattle WA 98195

PAGE, SAMUEL WILLIAM, b Williston, SC, Jan 25, 45; m 81; c 2. ORGANIC CHEMISTRY. *Educ:* The Citadel, BS, 67; Univ Ga, PhD(org chem), 71. *Prof Exp:* Walter Reed Army Reed Inst Res, US Army, 71-74; res chemist, Div Chem Technol, Food & Drug Admin, 74-79, supvry res chemist, Div Chem & Physics, 79-87, CHIEF, NATURAL PRODS & INSTRUMENTATION BR, FOOD & DRUG ADMIN, 88- *Mem:* Am Chem Soc; AAAS; Sigma Xi; Am Soc Pharmacognosy; Int Union Pure & Appl Chem; Asn Off Analytical Chemists. *Res:* Chemistry and toxicology of natural toxins. *Mailing Add:* Div Contaminants Chem HFF-423 Food & Drug Admin 200 C St SW Washington DC 20204

PAGE, STANLEY STEPHEN, b Silverton, Ore, March 14, 37; m 64; c 2. ASSOCIATIVE RINGS. *Educ:* Univ Ore, BSc, 59; Univ Wash, MSc & PhD(math), 67. *Prof Exp:* From asst prof to assoc prof, 67-86, PROF MATH, UNIV BC, 86- *Mem:* Am Math Soc; Math Asn Am. *Res:* Associative rings and algebras; generators and faithful representation. *Mailing Add:* Dept Math Univ BC Vancouver BC V6T 1W5 Can

PAGE, THOMAS LEE, b Lima, Ohio, Mar 5, 41; m 63; c 2. LIMNOLOGY. *Educ:* Ohio Northern Univ, AB, 63; Kent State Univ, MA, 66, PhD(ecol), 71. *Prof Exp:* NIH fel ecol, Ctr Biol Natural Systs, Washington Univ, 70-72; assoc sect mgr, Northwest Labs, Battelle Mem Inst, 72-80, mgr appl ecol sect, Earth Sci Dept, 80-85, mgr environ pathways sect, 85-87, DEP GEN MGR, NORTHWEST LABS, BATTELLE MEM INST, 87- *Mem:* AAAS; Am Soc Limnol & Oceanog; Ecol Soc Am; Am Inst Biol Sci; Int Limnol Soc; Am Inst Fisheries Res Biologists. *Res:* Limnology and ecology of freshwater communities; environmental ecology; pollution biology. *Mailing Add:* Geosciences Dept K5-16 Battelle NW Labs PO Box 999 Richland WA 99352

PAGE, THORNTON LEIGH, b New Haven, Conn, Aug 13, 13; m 48; c 3. ASTROPHYSICS, SPACE PHYSICS. *Educ:* Yale Univ, BS, 34; Oxford Univ, PhD(astrophys), 38. *Hon Degrees:* ScD, Nat Univ Cordoba, 69. *Prof Exp:* Chief asst, Univ Observ, Oxford Univ, 37-38; instr astrophys, Univ Chicago, 38-41 & 46-47, asst prof, 47-50; dep dir opers res off, Johns Hopkins Univ, 50-58; prof astron, Wesleyan Univ, 58-68; Nat Acad Sci res assoc, Manned Spacecraft Ctr, NASA, 68-71; astrophysicist, US Naval Res Lab, 71-76; CONTRACTOR ASTROPHYS RES, NASA, 77- *Concurrent Pos:* Consult, United Aircraft Corp, Conn, 58-71, Grumman Aircraft Eng Corp, 64-69 & Army Officers Sch Rev Bd, 65; vis prof, Univ Calif, Los Angeles, 64; Nat Acad Sci res assoc, Smithsonian Astrophys Observ, 65-66; adj prof, Univ Houston, 69-74 & Wesleyan Univ, 71-74; astrophysicist, Naval Res Lab, 71-76; mem, Adv Bd, Earth Sci Curriculum Proj; consult, Lockheed Electronics Co, 78-81, 88-; prof astron, Univ Houston, 81- *Honors & Awards:* Chapman Res Prize, 37; Exceptional Sci Achievement medal, NASA. *Mem:* AAAS (vpres, 67-70); fel Am Astron Soc; Int Astron Union; fel Royal Astron Soc; Int Statist Inst. *Res:* Physics of gaseous nebulae; masses of galaxies; military operations research; nuclear weapons; communications; space science; far-ultraviolet imagery of stars, and galaxies and comets. *Mailing Add:* 18639 Point Lookout Dr Houston TX 77058-4038

PAGENKOPF, ANDREA L, b Hamilton, Mont, July 28, 42; m 64; c 1. NUTRITION. *Educ:* Univ Mont, BA, 64; Purdue Univ, PhD, 68. *Prof Exp:* Asst prof foods, Purdue Univ, 68 & Univ Ill, 68-69; from asst prof to assoc prof, 69-86, PROF, FOODS & NUTRIT, MONT STATE UNIV, 86- *Mem:* Inst Food Technologists; Am Dietetic Asn; Soc Nutrit Educ; Am Home Econ Asn; Sigma Xi. *Res:* Metabolism of salmonellae in food systems; food quality, both keeping quality and safety aspects of present food supply with emphasis on newer processing methods and convenience foods; nutrition and disease; child nutrition. *Mailing Add:* Coop Exten Serv Mont State Univ Bozeman MT 59717

PAGENKOPF, GORDON K, physical inorganic chemistry; deceased, see previous edition for last biography

PAGER, DAVID, b East London, SAfrica, Mar 26, 35; m 65; c 3. COMPUTER SCIENCE, MATHEMATICS. *Educ:* Univ Cape Town, BSci, 56; Univ London, PhD(math), 61. *Prof Exp:* Mathematician, Data Electronic Comput Group, EMI Indust, Eng, 57-58; consult math, Govt Can, 61-62; mgr sci systs, IBM Corp, 63-64; mgr software develop, Univac, London, 65-68; PROF INFO & COMPUT SCI, UNIV HAWAII, 68-, NSF RES GRANT, 69- *Mem:* Am Math Soc; Asn Symbolic Logic; Asn Comput Mach; Inst Elec & Electronics Engrs; Soc Indust & Appl Math. *Res:* Complexity theory, recursive function theory; software science; communication of complex information; artificial intelligence. *Mailing Add:* Dept Info Univ Hawaii Manoa Honolulu HI 96822

PAGES, ROBERT ALEX, b New York, NY, Oct 10, 41; m 87. BIO-ORGANIC CHEMISTRY. *Educ:* Polytech Inst Brooklyn, BS, 62; Univ Va, PhD(chem), 66. *Prof Exp:* Staff fel, Clin Endocrinol Br, Nat Inst Arthritis, Metab & Digestive Dis, 68-73; res scientist, Phillip Morris USA, 73-77, sr scientist, 77-88, mgr sci & technol, 88-90, DIR SCI & TECHNOL, PHILIP MORRIS MGT CORP, 90- *Mem:* AAAS; Am Chem Soc; Environ Mutagen Soc; Tissue Cult Asn; Soc Risk Anal; Air & Waste Mgt Asn. *Res:* Biological effects of cigarette smoke; environmental mutagenesis; in vitro bioassays; chemistry and thyroid hormones; hormone binding proteins; medicinal chemistry. *Mailing Add:* Phillip Morris Opers Ctr PO Box 26603 Richmond VA 23261

PAGNAMENTA, ANTONIO, b Zurich, Switz, Jan 13, 34; m 54; c 2. PHYSICS. *Educ:* Swiss Fed Inst Technol, BS, 59, MS, 61; Univ Md, College Park, PhD(physics), 65. *Prof Exp:* Res asst, Univ Md, 61-65; asst prof physics, Rutgers Univ, New Brunswick, 65-70; assoc prof, 70-72, PROF PHYSICS, UNIV ILL, CHICAGO CIRCLE, 72- *Concurrent Pos:* Res fel particle physics, Europ Ctr Nuclear Res, Geneva, 66-67. *Mem:* Am Phys Soc. *Res:* Field theory and particle physics; theoretical high energy physics; radiation physics. *Mailing Add:* Dept Physics MC 273 Univ Ill Chicago Circle Chicago IL 60680

PAGNI, PATRICK JOHN, b Chicago, Ill, Nov 28, 42; m 70; c 3. MECHANICAL ENGINEERING. *Educ:* Univ Detroit, BAeE, 65; Mass Inst Technol, SM, 67, MechE, 69, PhD(mech eng), 70. *Prof Exp:* From asst prof to assoc prof, 70-81, vchmn grad study mech eng, 86-89, PROF MECH ENG, UNIV CALIF, BERKELEY, 81-, ACTG ASSOC DEAN, COL ENG, 90- *Concurrent Pos:* NSF, NASA, Nat Bur Standards & Nat Inst Standards & Technol grants, fire physics, Univ Calif, Berkeley, 71-; res fel appl mech, Div Appl Sci, Harvard Univ, 74 & 77. *Mem:* Int Asn Fire Safety Sci; Am Soc Mech Engrs; Combustion Inst. *Res:* Mathematical modeling of fire physics; fire and explosion reconstruction; analysis of glass breaking, excess pyrolyzate generation, and back draft penomena; combustion generated in fires; experimental studies of flame soot and particulates. *Mailing Add:* Dept of Mech Eng Univ of Calif Berkeley CA 94720

PAGNI, RICHARD, b Chicago, Ill, Dec 14, 41; c 2. ORGANIC CHEMISTRY. *Educ:* Northwestern Univ, BA, 63; Univ Wis, PhD(org chem), 68. *Prof Exp:* NIH fel, Columbia Univ, 68-69; asst prof, 69-76, assoc prof, 76-81, PROF CHEM, UNIV TENN, KNOXVILLE, 81- *Concurrent Pos:* Oakridge Nut Lab, 84- *Mem:* AAAS; Am Chem Soc; Royal Soc Chem; Sigma Xi. *Res:* Photochemistry of hydrocarbons; physical organic chemistry; surface organic chemistry. *Mailing Add:* Dept of Chem Univ of Tenn Knoxville TN 37996-1600

PAHL, HERBERT BOWEN, b Camden, NJ, Aug 14, 27; m 51; c 2. PUBLIC HEALTH ADMINISTRATION. *Educ:* Swarthmore Col, BA, 50; Univ Mich, MS, 52, PhD(biochem), 55. *Prof Exp:* Fel, Nat Cancer Inst, Sloan-Kettering Inst, 55-57; asst prof biochem, Vanderbilt Univ, 57-60; exec secy, Grad Res Training Grant Prog, NIH, 60-62, asst chief & chief, Spec Res Resources Br, 62-64, chief, Gen Res Support Br, 64-66; exec secy, Comt Res Life Sci, Nat Acad Sci-Nat Res Coun, 66-69; dep assoc dir sci progs, Nat Inst Gen Med Sci, NIH, 69-71; dep dir Regional Med Prog, Health Serv & Ment Health Admin, USPHS, 71-73, dir, Regional Med Progs, Health Resources Admin, 73-75, asst to adminr, 75; staff dir, Comt Study Nat Needs Biomed & Behav Sci Res Personnel, Nat Acad Sci-Nat Res Coun, 75-82; PROG DIR, CANCER CTR BR, NAT CANCER INST, NIH, 84- *Concurrent Pos:* Instr, City Col New York, 56-57. *Mem:* AAAS. *Res:* Intermediary metabolism of nucleic acids; fractionation of nucleic acids; protein biosynthesis. *Mailing Add:* 626 Warfield Dr Rockville MD 20850

PAHL, WALTER HENRY, engineering management, for more information see previous edition

PAI, ANANTHA M, b Mangalore, India, May 10, 31; m 56; c 4. ELECTRIC POWER SYSTEMS. *Educ:* Univ Madras, India, BS, 53; Univ Calif, Berkeley, MS, 58, PhD (elec eng), 61. *Prof Exp:* Engr, Bombay Elec Supply Co, 53-57; prof elec eng, Indian Inst Technol, Kanpur, India, 63-80; PROF ELEC ENG, UNIV ILL, URBANA, 80- *Mem:* Fel Inst Elec & Electronic Engrs. *Res:* Electrical power systems with focus on dynamics, computation, stability and control; author of four books in electrical engineering. *Mailing Add:* Dept Elec Eng Univ Ill Urbana IL 61801

PAI, ANNA CHAO, b Peking, China, Jan 27, 35; US citizen; m 59; c 2. GENETICS, EMBRYOLOGY. *Educ:* Sweet Briar Col, BA, 57; Bryn Mawr Col, MA, 59; Albert Einstein Col Med, PhD(genetics), 64. *Prof Exp:* Teacher, Moravian Sem for Girls, 59-60; instr genetics, Albert Einstein Col Med, 64-65; from asst prof to assoc prof, 69-84, chmn Dept Biol, 79-82, PROF BIOL, MONTCLAIR STATE COL, 84- *Concurrent Pos:* Bd Overseers, Sweet Briar Col, 84- *Honors & Awards:* Sigma Xi. *Mem:* Soc Develop Biol; Sigma Xi; Am Inst Biol Teachers. *Res:* Developmental genetics, especially the effects of gene action on mammalian development. *Mailing Add:* 223 W Hobart Gap Rd Livingston NJ 07039

PAI, DAMODAR MANGALORE, b Mangalore, India; US citizen; m 65; c 2. MATERIALS SCIENCE, PHYSICS. *Educ:* Indian Inst Sci, DIISc, 57; Univ Minn, MS, 61, PhD(elec eng), 65. *Prof Exp:* Physicist, assoc scientist & scientist physics of solids, 65-73, sr scientist, 73-80, prin scientist org & inorg photoconductors, 80-85, RES FEL, XEROX CORP, 85- *Honors & Awards:* Chester F Carlson Award, Soc Imaging Sci & Technol, 88. *Mem:* Fel Am Phys Soc; Mem Soc Imaging Sci & Technol. *Res:* Design and study of new photoconductive systems for xerography, both organic and inorganic; research of fundamental photoconductor physics; photogeneration, transport, phenomena near glass transition, temperature, etc. *Mailing Add:* 72 Shagbark Way Fairport NY 14450

PAI, DAVID H(SIEN)-C(HUNG), b Kweilin, China, Jan 7, 36; US citizen; m 59; c 2. DESIGN & ANALYSIS OF PRESSURE VESSEL COMPONENT. *Educ:* Va Mil Inst, BS, 58; Lehigh Univ, MS, 60; NY Univ, ScD(eng sci), 65. *Prof Exp:* Develop engr, Res Div, 60-63, sr engr, 64-65, res assoc creep anal & plasticity theory, 66-68, asst head Solid Mech Dept, 69-73, chief Engr Nuclear Dept, 73-77, mgr eng Technol Dept, 77-80, vpres, Foster Wheeler Appln Inc, 80-84, srvpres, 84-88, PRES, FOSTER WHEELER DEVELOP CORP, 88- *Concurrent Pos:* Mem steering comt high temperature struct design technol, US Atomic Energy Comn, 70-84; vpres, Am Soc Mech Eng, 81-83; consult, China Nat Standards Comt Pressure Vessels; sr adv, Chinese Mech Eng Soc, 88- *Honors & Awards:* Pressure Vessel & Piping Medal, Am Soc Mech Engrs, 87. *Mem:* Am Soc Civil Engrs; Am Soc Mech Engrs. *Res:* Creep and plastic behavior of structural and machine components; metal fatigue; pressure vessel design and analysis. *Mailing Add:* 223 W Hobart Gap Rd Livingston NJ 07039

PAI, MANGALORE ANANTHA, b Mangalore, India, Oct 5, 31; m 56; c 4. SCIENCE & TECHNOLOGY POLICY FOR THIRD WORLD COMPANIES. *Educ:* Univ Madras, India BE, 53; Univ Calif, Berkeley, MS, 58, PhD(elec eng), 61. *Prof Exp:* Eng, Bombay Elec Supply & Transport, India, 53-57; asst prof, elec eng, Univ Calif, Berkeley, 61-63; from assoc prof to prof, Indian Inst Technol, Kanpur, India,63-81; PROF, ELEC ENG, UNIV ILL, URBANA, 81- *Concurrent Pos:* vis prof, Mem Univ New Foundland, Can, 69-71, Iowa State Univ, 79-81. *Honors & Awards:* Bhatnagar Award, Coun Sci & Indust Res, Govt India, 74. *Mem:* Fel Inst Elec Electronic Engrs; fel Indian Nat Sci Acad. *Res:* Stability and control of large scale power systems; parallel processing techniques in power system analysis; Ploicy for third world Companies. *Mailing Add:* Dept Elec Eng Univ Ill Urbana Campus 1406 W Green St Urbana IL 61801

PAI, S(HIH) I, b China, Sept 30, 13; nat US; m 60; c 4. FLUID DYNAMICS, APPLIED MATHEMATICS. *Educ:* Nat Cent Univ, China, BS, 35; Mass Inst Technol, MS, 38; Calif Inst Technol, PhD(aeronaut, math), 40. *Hon Degrees:* Dr Techn, Vienna Tech Univ, 68. *Prof Exp:* Prof aerodyn, Nat Cent Univ, China, 40-47; vis prof aeronaut eng, Cornell Univ, 47-48; tech consult, Cornell Aeronaut Lab, Inc, 48-49, res assoc, 49-51; assoc res prof fluid dynamics, 52-56, res prof, 56-83, EMER PROF, INST PHYSICS SCI & TECHNOL & AEROSPACE, UNIV MD, COLLEGE PARK, 83- *Concurrent Pos:* Guggenheim fel, 57-58; vis prof, Inst Theoret Gasdynamics, Aachen, Germany, 57-58, Univ Tokyo, 66, Tech Univ Denmark, 74, Univ Karlsruhe 80-81 & Univ Paris VI, 81 & Vienna Tech Univ, 67-; NSF sr consult, Martin Co, 56-57, Gen Elec Co, 59-, NAm Rockwell Corp, 65- & Sandia Corp, 65-; NSF sr fel, 66-67; vis scientist, Norwegian Defense Res Estab, 73-74; hon prof, Northwestern Polytech Univ, China, 80- & Zhejiang Univ, China, 85- *Honors & Awards:* Sr US Scientist Award, Humboldt Found, Ger, 78. *Mem:* Am Phys Soc; assoc fel Am Inst Aeronaut & Astronaut; Sigma Xi. *Res:* Turbulence; jet and boundary layer flow; stability of flow; supersonic and hypersonic flow; two phase flows; aeroelasticity; magneto-gas dynamics; plasma dynamics; radiation gas dynamics. *Mailing Add:* Inst Phys Sci & Technol Univ Md College Park MD 20742

PAI, SADANAND V, b Cochin, Kerala, India, Aug 13, 37; US citizen; m 69; c 1. ENVIRONMENTAL CHEMISTRY, RESEARCH ADMINISTRATION. *Educ:* Madras Univ, India, BS, 57; Banaras Hindu Univ, India, BS, 61; MS, 62; Wash State Univ, PhD(med chem), 68. *Prof Exp:* Postdoctoral fel pharmacol, Col Med, Univ Ill, 67-69; lectr pharmacol, McGill Univ, Montreal, Can, 69-76; lab dir, Wastex Industs, Pottstown, Pa, 76-79; sr chemist, Gulf S Res Inst, 79-80; mgr environ chem, Allied-Signal Corp, Morristown, NJ, 80-86; SECT HEAD RES & DEVELOP, LYPHOMED, DIV FUJISAWA, MELROSE PARK, ILL, 86- *Mem:* Am Chem Soc; AAAS; Am Asn Pharmaceut Scientists; Parenteral Drug Asn. *Res:* Synthesized new analgesics; adrenergic b-blockers; blockers of neuronal uptake mechanism and antiprotocoal drugs; catecholamine storage and release mechanism from adrenergic nerve terminals; factors regulating the contraction and relaxation of vascular smooth muscle; adrenergic mechanism; new analytical method development; developed stable pharmaceutical formulations. *Mailing Add:* Lyphomed 2045 N Cornell Ave Melrose Park IL 60160

PAI, VENKATRAO K, b Coondapoor, India, Jan 7, 39; m 77; c 1. CHEMICAL ENGINEERING. *Educ:* Univ Bombay, BChE, 61; Northwestern Univ, MS, 63, PhD(chem eng), 65. *Prof Exp:* Res chem engr, Cent Res Labs, 65-80, mgr process develop, Chem Res Div, 80-89, DIR PROCESS DEVELOP, CHEM RES DIV, AM CYANAMID CO, 89- *Mem:* Am Inst Chem Engrs; Am Chem Soc. *Res:* Process development; reactor engineering; heat transfer; rheology of spinning; plant design and economics. *Mailing Add:* Cyanamid Res Labs 1937 W Main St Stamford CT 06904

PAICE, MICHAEL, b Eng, Sept 13, 49; m 70; c 2. ENZYMOLOGY, PAPER SCIENCE. *Educ:* Univ Col, London, Eng, BSc, 71; Univ Alta, Edmonton, MSc, 75. *Prof Exp:* SR SCIENTIST BIOCHEM, PULP & PAPER RES INST CAN, 75- *Concurrent Pos:* Lectr biotechnol, McGill Univ. *Mem:* Am Chem Soc; Can Inst Chem; Tech Asn Pulp & Paper Indust; Can Pulp & Paper Asn. *Res:* Structural characterization of protein and DNA; microbial or enzymic modification of the wood components lignin and hemicellulose. *Mailing Add:* Pulp & Paper Res Inst Can 570 St John's Blvd Pointe Claire PQ H9R 3J9 Can

PAIDOUSSIS, MICHAEL PANDELI, b Nicosia, Cyprus, Aug 20, 35; m 58. APPLIED MECHANICS, FLUID DYNAMICS. *Educ:* McGill Univ, BEng, 58; Cambridge Univ, PhD(eng), 63. *Prof Exp:* Overseas fel nuclear eng, Gen Elec Co, Eng, 58-60; asst res officer hydroelasticity, Chalk River Nuclear Labs, Atomic Energy Can Ltd, 63-65, assoc res officer, 65-67; from asst prof to assoc prof mech eng, 67-76, PROF MECH ENG, MCGILL UNIV, 76-, chmn dept, 77-87, THOMAS WORKMAN PROF MECH ENG, 86- *Honors & Awards:* Brit Asn Medal, 58; George Stephenson Prize, Inst Mech Engrs, London, 75. *Mem:* Fel Am Soc Mech Engrs; fel Brit Inst Mech Engrs; fel Can Soc Mech Engrs; Am Acad Mech; Int Asn Hydraulics Res; fel Royal Soc Can. *Res:* Pressure waves in flexible pipes; vibration of cylinders in flow; dynamics of flexible pipes containing flow and cylinders in axial and cross flow; vibration and stability of shells. *Mailing Add:* Dept Mech Eng 817 Sherbrooke St W Montreal PQ H3A 2K6 Can

PAIGE, DAVID A, b Los Angeles, Calif, May 11, 57. PLANETARY SCIENCE, SOLAR SYSTEM STUDIES. *Educ:* Univ Calif Los Angeles, BS, 79; Calif Inst Technol, PhD(planetary sci), 85. *Prof Exp:* Resident res assoc, Jet Propulsion Lab, 85-86; ASST PROF PLANETARY SCI, UNIV CALIF, LOS ANGELES, 86- *Mem:* Am Astron Soc; Am Geophys Union. *Res:* The surface, atmosphere, polar caps and climate of Mars; spacecraft instrumentation; planetary exploration. *Mailing Add:* Dept Earth & Space Sci 3806 Geol Bldg Univ Calif 405 Hilgard Ave Los Angeles CA 90024-1567

PAIGE, DAVID M(ARSH), b Rochester, NY, Aug 4, 18; m 49. CHEMICAL ENGINEERING, MECHANICAL ENGINEERING. *Educ:* BSChE, Univ Rochester, 41. *Prof Exp:* Engr, E I du Pont de Nemours & Co, 41-50; proj engr atomic energy, Am Cyanamid Co, 50-53; engr & group leader, Phillips Petrol Co, 53-55, sect head chem eng develop, 55-60; assoc staff engr, Idaho Div, Nat Reactor Testing Sta, Argonne Nat Lab, 60-72; sr engr, Allied Chem Corp Idaho Chem Progs, Idaho Nat Eng Lab, 72-78; sr engr, Exxon Nuclear Corp, Idaho, 78-82; RETIRED. *Mem:* Sigma Xi; Am Inst Chem Engrs; Am Nuclear Soc. *Res:* Application of polymers to fibers and films; chemical separations as applied to fuel recovery in atomic energy field; remote equipment design for hot cells; numerical tape controlled equipment for remote hot cell equipment. *Mailing Add:* PO Box 1629 Sun Valley ID 83353-1629

PAIGE, DAVID M, b Brooklyn, NY, Aug 20, 39; m 59; c 2. PEDIATRIC NUTRITION, MATERNAL NUTRITION. *Educ:* Long Island Univ, BS, 60; NY Med Col, MD, 64; Johns Hopkins Univ, MPH, 69. *Prof Exp:* Intern pediat, Univ NY, Downstate, 64-65, resident, 65; pub health officer, USPHS, 65-67; resident, Johns Hopkins Hosp & Univ, 67-69; PROF MATERNAL & CHILD HEALTH, SCH HYGIENE & PUB HEALTH & JOINT APPOINTMENT PEDIAT, SCH MED, JOHNS HOPKINS UNIV, 81- *Concurrent Pos:* Consult pediat, John F Kennedy Inst, 69-; attend pediatrician, Johns Hopkins Hosp, 69-; consult nutrit, USDA, 73-, Fed Trade Comn, 73-79, Dept Health & Mental Hyg, 74- & US Cong Off Technol Assessment, 76-78; mem panel, Nat Cancer Inst, NIH, 76-78; prin investr, USDA, 80-; ed, Manual Clin Nutrit, 81- & Clin Nutrit, 83- *Mem:* Fed Am Soc Exp Biol; Am Soc Clin Nutrit; Am Inst Nutrit; Am Pub Health Asn; Asn Teachers Maternal & Child Care. *Res:* Maternal, infant and child nutrition, including lactose digestion and milk consumption; evaluation of nutritional supplementation of high risk pregnant women and school children; infant feeding patterns and practices. *Mailing Add:* Dept Maternal & Child Health Sch Hyg Johns Hopkins Univ Baltimore MD 21205

PAIGE, EUGENE, b Dallas, Tx, Sept 9, 29. ALGEBRA, ALGEBRAIC CONCEPTS. *Educ:* Univ Chicago, PhD(math), 54. *Prof Exp:* Prof math, Univ Ill, 56-72; PROF MATH, UNIV VA, 72- *Mailing Add:* 1610 Ricky Dr Charlottesville VA 22901

PAIGE, FRANK EATON, JR, b Philadelphia, Pa, Oct 21, 44. ELEMENTARY PARTICLE PHYSICS. *Educ:* Mass Inst Technol, BS, 66, PhD(physics), 70. *Prof Exp:* Res assoc, 70-72, assoc physicist, 72-76, PHYSICIST, BROOKHAVEN NAT LAB, 76- *Concurrent Pos:* Asst res physicist, Univ Calif, San Diego, 75-76. *Res:* Elementary particle theory and phenomenology; theory of hadronic reactions. *Mailing Add:* 11 Harvard Rd Shoreham NY 11786

PAIGE, HILLIARD W, b Oct 2, 19, US citizen. ENGINEERING. *Educ:* Worcester Polytech Inst, BA & ME, 41. *Hon Degrees:* Dr, Worcester Polytech Inst, 71. *Prof Exp:* Sr consult dir, Int Energy Assocs, Ltd, 76-91; RETIRED. *Concurrent Pos:* Mem defense sci bd, Dept Defense, 67. *Mem:* Nat Acad Eng; fel Am Inst Aeronaut & Astronaut; Am Nuclear Soc. *Res:* Orbit control systems. *Mailing Add:* 5163 Tilden St NW Washington DC 20016

PAIGE, RUSSELL ALSTON, b Grand Junction, Colo, May 23, 29; m 53; c 5. ENGINEERING GEOLOGY, GEOMORPHOLOGY. *Educ:* Univ Alaska, BS, 55; Univ Wash, MSc, 59. *Prof Exp:* Eng geologist, US Geol Surv, Alaska, 55-57 & Peter Kiewit Sons Co, Wash, 59-62; mining geologist, 62; eng geologist, Haner, Ross & Sporseen Consult Engrs, Ore, 62-63 & US Naval Civil Eng Lab, 63-72; dept head, Geol Div 170, 72-79, regional geologist, Western US, Harza Engr Co, 83-88. *Concurrent Pos:* Proj geologist, Strontia Springs Dam construct, Harza Engr Co, 79-83; geologist, seismotectonic studies, Front Range, Colo, Cache La Poudre Basin studies; resident geologist, Two Folks Dam; consult, Ertan Arch Dam, Yalong Repub China. *Mem:* Fel Geol Soc Am; Asn Eng Geologists; Am Soc Civil Engrs. *Res:* Engineering and environmental geology of polar regions; engineering geology of large dams. *Mailing Add:* Tech & Mgt Support Sic Proj Valley Bank Ctr Suite 407 101 Convention Center Dr Las Vegas NV 89109

PAIGEN, BEVERLY JOYCE, b Chicago, Ill, Aug 14, 38; m 70; c 2. ENVIRONMENTAL HEALTH. *Educ:* Wheaton Col, Ill, BS, 60; State Univ NY, Buffalo, PhD(biol), 67. *Prof Exp:* Res technician, Roswell Park Mem Inst, Buffalo, 60-62; grad student, State Univ NY, Buffalo, 63-67; fel, Roswell Park Mem Inst, 67-70; lectr, Rachel Carson Col, State Univ NY, Buffalo, 71-73, dir, 73-75; sr res scientist, Roswell Park Mem Inst, 75-78, res scientist V, 78-82; res biochemist, Childrens Hosp Med Ctr, Oakland, Calif, 82- 89; SR STAFF SCIENTIST, JACKSON LAB, BAR HARBOR, ME, 89- *Concurrent Pos:* Consult, Sci Adv Bd, Subcomt Coke Oven Emissions, Environ Protection Agency, 78-80 & Subcomt Polycyclic Org Matter, 78-80 & Carcinogen Assessment Group, 78-81; mem, Admin Toxic Substances Adv Comt, Environ Protection Agency, 77-79. *Mem:* AAAS; Am Cancer Res; Am Asn Human Genetics; Geochem Environ Health; Am Heart Asn. *Res:* Genetic susceptibility to atherosclerosis; environmental health, toxicology, health effects of hazardous waste. *Mailing Add:* Jackson Lab 600 Main St Bar Harbor ME 04609

PAIGEN, KENNETH, b New York, NY, Nov 14, 27; m 47, 70; c 5. BIOCHEMICAL GENETICS. *Educ:* Johns Hopkins Univ, AB, 46; Calif Inst Technol, PhD(biochem), 50. *Prof Exp:* Carnegie fel genetics, Carnegie Inst, 50-52; assoc med, Peter Bent Brigham Hosp, Boston, Mass, 52-53; USPHS fel, Virus Lab, Univ Calif, 53-55; sr cancer res scientist, 55-58, assoc cancer res scientist, 58-62, prin cancer res scientist, 62-67, assoc chief scientist, 67-72, PROF BIOL & CHMN GRAD PROG & DIR, DEPT MOLECULAR BIOL, ROSWELL PARK MEM INST, 72- *Concurrent Pos:* Vis prof genetics, Univ Calif, Berkeley, 81- *Mem:* AAAS; Genetics Soc; Am Chem Soc; Am Soc Biol Chem; Am Asn Cancer Res; Sigma Xi. *Res:* Regulation of gene expression in mammals; intracellular location of enzymes; developmental genetics. *Mailing Add:* Jackson Lab Bar Harbor ME 04609

PAIK, HO JUNG, b Seoul, Korea, Mar 25, 44; m 69; c 2. GRAVITATION PHYSICS, CRYOGENIC ENGINEERING. *Educ:* Seoul Nat Univ, BS, 66; Stanford Univ, MS, 70, PhD(physics), 74. *Prof Exp:* Res assoc physics, Stanford Univ, 74-78; from asst prof to assoc prof, 78-89, PROF PHYSICS, UNIV MD, COLLEGE PARK, 89- *Concurrent Pos:* Alfred Sloan fel, 81-83. *Mem:* Korean Sci & Eng Asn; Am Phys Soc; Am Geophys Union. *Res:* Search for gravitational waves using cryogenic technology; new test of Newton's law of gravity, using superconducting gravity gradiometers; theoretical work on interaction of gravitational waves with antennas of various geometries. *Mailing Add:* Dept of Physics & Astron Univ of Md College Park MD 20742

PAIK, S(UNGIK) F(RANCIS), b Seoul, Korea, Nov 12, 35; US citizen; m 57; c 3. ELECTRICAL ENGINEERING. *Educ:* Northwestern Univ, BS, 58; Stanford Univ, MS, 59, PhD(elec eng), 61. *Prof Exp:* Sr scientist, Raytheon Res Div, 61-64; assoc prof elec eng, Northwestern Univ, 64-68; dir tube res, Microwave Assocs, Inc, Burlington, 68-70; ENG MGR SOLID STATE PRODS, RAYTHEON CO, SMDO, WALTHAM, 70- *Mem:* Inst Elec & Electronics Engrs. *Res:* Solid state microwave amplifiers, circuits. *Mailing Add:* HiHite Microwave Corp 21 Cabot Rd Woburn MA 01801

PAIK, WOON KI, b Naju, Korea, Mar 2, 25; m 59; c 3. BIOCHEMISTRY. *Educ:* Severance Med Col, Korea, MD, 47; Dalhousie Univ, MS, 56. *Prof Exp:* Asst biochem, Severance Med Col, Korea, 47-48; instr, Ewha Women's Univ, 50-53; asst cellular physiol, Dalhousie Univ, 53-56; vis scientist, Nat Cancer Inst, 56-58; res assoc physiol chem, Univ Wis, 58-61; from asst prof to assoc prof biochem, Fac Med, Univ Ottawa, Ont, 61-66; assoc prof, 66-73, SR INVESTR, FELS RES INST, TEMPLE UNIV, 66-, PROF BIOCHEM, 73- *Mem:* Am Soc Biol Chemists; Am Asn Cancer Res; AAAS; NY Acad Sci. *Res:* Enzyme purification; biochemical studies on amphibian metamorphosis; amino acid metabolism; protein modification; gene regulation. *Mailing Add:* Fels Res Inst Temple Univ Philadelphia PA 19140

PAIK, YOUNG-KI, enzymology, eukaryotic gene expression, for more information see previous edition

PAIKOFF, MYRON, b New York, NY, Jan 31, 32; m 55; c 3. PHARMACEUTICAL CHEMISTRY. *Educ:* Columbia Univ, BS, 52; Purdue Univ, MS, 54, PhD, 56. *Prof Exp:* Asst dir pharmaceut res & develop, 71-75, DIR PHARM SCI, STERLING-WINTHROP RES INST, 75- *Mem:* Acad Pharmaceut Sci; Sigma Xi. *Res:* Pharmaceuticals; drug delivery systems; kinetics; general pharmaceutical research. *Mailing Add:* Ten Cloverfield Dr Albany NY 12211

PAILTHORP, JOHN RAYMOND, b Spokane, Wash, Nov 14, 20; m 42; c 3. ORGANIC CHEMISTRY. *Educ:* Mich State Univ, BS, 42; Univ Wis, MS, 47. *Prof Exp:* Res chemist, Res & Develop Div, E I du Pont De Nemours & Co Inc, 47-53, res supvr, 53-56, res div head, 56-59, mgr, Develop Sect, Res & Develop Div, 60-68, asst dir res & develop, Elastomer Chem Dept, 68-79, tech dir, Polymer Prod Div, 79-81; RETIRED. *Mem:* Am Chem Soc; Am Ord Asn; Am Inst Chemists. *Res:* Elastomers; polymers. *Mailing Add:* 852 Cranbrook Dr Wilmington DE 19803

PAIM, UNO, b June 12, 22; Can citizen; m 50. ENVIRONMENTAL PHYSIOLOGY, FISH BIOLOGY. *Educ:* Univ Toronto, BA, 57, PhD(zool), 62. *Prof Exp:* From asst prof to assoc prof, 62-69, prof biol, 69-88, EMER PROF BIOL, UNIV NB, 88- *Mem:* AAAS; Can Soc Zoologists; Entom Soc Can; Am Fisheries Soc. *Res:* Responses of animals to directive factors; biology of salmonids; aquaculture. *Mailing Add:* 17 Spruce Terr Fredericton NB E3B 2S6 Can

PAINE, CLAIR MAYNARD, b Westmoreland, NH, Jan 4, 30; m 55; c 2. BIOCHEMISTRY. *Educ:* Univ NH, BS, 51; Rutgers Univ, MS, 57, PhD(agr, biochem), 59. *Prof Exp:* Asst agr biochem, Rutgers Univ, 54-58; Nat Cancer Inst fel, Sch Med, George Washington Univ, 58-59; sr res fel biochem, May Inst Med Res, Jewish Hosp, Cincinnati, 59-61, biochemist, 61-63; Nat Inst Neurol Dis & Blindness fel, Retina Found, Boston, 63-66; from asst prof to assoc prof, 66-70, PROF BIOL, SALEM STATE COL, 71- *Mem:* AAAS; Am Inst Biol Sci. *Res:* Amino acid transport; electrolyte metabolism; muscle contraction; educational methods. *Mailing Add:* Dept Biol Salem State Col Salem MA 01970

PAINE, DAVID PHILIP, b Dillon, Mont, Dec 27, 29; m 55; c 2. FOREST MENSURATION, AERIAL PHOTOS. *Educ:* Ore State Univ, BS, 53, MS, 58; Univ Wash, PhD(forest mensuration), 65. *Prof Exp:* Forester, Intermountain Forest & Range Exp Sta, US Forest Serv, 55-57 & 58-59; asst prof mensuration aerial photog, Univ WVa, 62; asst prof, 62-67, assoc prof, 67-, PROF FOREST MGT, ORE STATE UNIV. *Honors & Awards:* Aufderheide Award, 73. *Mem:* Soc Am Foresters; Am Soc Photogram. *Res:* Aerial photo mensuration, remote sensing, photo volume tables for trees and stands; coniferous regeneration using aerial photos and double sampling. *Mailing Add:* Sch Forestry Ore State Univ Corvallis OR 97331

PAINE, DWIGHT MILTON, b Albion, NY, Oct 11, 31; m 56; c 4. MATHEMATICS. *Educ:* Univ Rochester, BA, 52; McGill Univ, BA, 56; Fuller Theol Sem, MDiv, 59; Univ Wis, MS, 61, PhD(math), 63. *Prof Exp:* From asst prof to assoc prof math, Wells Col, 63-72; assoc prof, 72-79, PROF MATH, MESSIAH COL, 79- *Concurrent Pos:* Vis res assoc, Cornell Univ, 69-70. *Mem:* Math Asn Am. *Res:* Systems of partitions of sets. *Mailing Add:* Dept Math Messiah Col Grantham PA 17027

PAINE, KENNETH WILLIAM, neurosurgery, for more information see previous edition

PAINE, LEE ALFRED, b Kansas City, Kans, Nov 24, 20. FOREST PATHOLOGY. *Educ:* Univ Idaho, MS, 47; Swiss Fed Inst Technol, PhD(path), 51. *Prof Exp:* Res officer forest path, Forest Biol Div, Can Sci Serv, 53-57; FOREST PATHOLOGIST, PAC SOUTHWEST FOREST & RANGE EXP STA, US FOREST SERV, 57- *Mem:* AAAS; Am Phytopath Soc; Soc Am Foresters; Mycol Soc Am; NY Acad Sci. *Res:* Physiology of fungi; trunk rots; abiotic diseases; hazardous trees in recreational areas. *Mailing Add:* Forest Dis Res Pac SW Forest & Range Exp Sta PO Box 245 Berkeley CA 94701

PAINE, PHILIP LOWELL, b Orlando, Fla, June 11, 45; m 74, 90; c 2. CELL PHYSIOLOGY, INTRACELLULAR TRANSPORT. *Educ:* Calif Inst Technol, BS, 67; Univ Fla Col Med, PhD(anat), 71. *Prof Exp:* Fel physiol, Dept Physiol & Biophysics, Univ Miami Med Sch, 71-73; res assoc, Mich Cancer Found, 73-75, res scientist & head, Intercellular Transp Sect Cell Physiol, Dept Biol, 75-82, assoc mem & chief, Lab Macromolecular Physiol, Dept Physiol & Biophys, 83-89; PROF DEPT BIOL SCI, ST JOHN'S UNIV, 89- *Concurrent Pos:* Fel, Damon Runyon Mem Fund Cancer Res, 71. *Mem:* Biophys Soc; Am Soc Cell Biol; Am Electrophoresis Soc; AAAS. *Res:* Intra movement of molecules, including transport kinetics and equilibrium distributions. *Mailing Add:* St Johns Univ Grand Central & Utopia Pkwys Jamaica NY 11439

PAINE, RICHARD BRADFORD, b Hyannis, Mass, July 16, 28; m 51; c 2. COMPUTER SCIENCES. *Educ:* Walla Walla Col, BS, 52; Univ Wash, MS, 57, PhD, 58. *Prof Exp:* Asst prof math, Cent Mich Col, 57-58; assoc prof, Stephen F Austin State Col, 58-60; asst prof, Walla Walla Col, 60-61; asst prof, 61-63, ASSOC PROF MATH, COLO COL, 63- *Mem:* Asn Comput Mach. *Res:* Software engineering; systems programming. *Mailing Add:* Colo Col Colorado Springs CO 80903-3294

PAINE, ROBERT MADISON, b Yonkers, NY, June 24, 25; m 46; c 3. METALLURGICAL CHEMISTRY. *Educ:* Mich State Univ, BS, 48, MS, 49. *Prof Exp:* Asst phys chem, Mich State Univ, 48-49; res chemist inorg & x-ray chem, Mallinckrodt Chem Works, 49-55; sect supvr phys chem res & develop, Brush Wellman, Inc, 55-85; RETIRED. *Concurrent Pos:* Consult, 86- *Mem:* Metall Soc; Am Soc Lubrication Eng; Am Electroplaters Soc. *Res:* X-ray diffraction and solid gas reactions; intermetallic compounds as high temperature structural materials; corrosion protection and control; wear research; electroplating surface chemistry beryllium copper. *Mailing Add:* 18891 Brewster Dr Aurora OH 44202

PAINE, ROBERT T, b Cambridge, Mass, Apr 13, 33; div; c 3. ZOOLOGY, ECOLOGY. *Educ:* Harvard Univ, AB, 54; Univ Mich, MS, 59, PhD(zool), 61. *Prof Exp:* Sverdrup fel, Scripps Inst Oceanog, Univ Calif, San Diego, 61-62; from asst prof to assoc prof, 62-71, PROF ZOOL, UNIV WASH, 71- *Honors & Awards:* Tansley Lectr, Brit Ecol Soc; Ecol Inst Prize. *Mem:* Nat Acad Sci; Soc Study Evolution; Am Soc Naturalists; Am Soc Limnol & Oceanog; Ecol Soc Am (vpres, 77-78, pres, 79-80). *Res:* Algal ecology and prey-predator relationships; community organization. *Mailing Add:* Dept Zool Univ Wash Seattle WA 98195

PAINE, ROBERT TREAT, JR, b Colorado Springs, Colo, Dec 15, 44; m 67; c 3. INORGANIC CHEMISTRY. *Educ:* Univ Calif, Berkeley, BS, 66; Univ Mich, Ann Arbor, PhD(chem), 70. *Prof Exp:* Res assoc chem, Northwestern Univ, 70-72 & Los Alamos Sci Lab, Univ Calif, 72-74; from asst prof to assoc prof, 72-78, PROF CHEM, UNIV NMEX, 82- *Mem:* Am Chem Soc; AAAS. *Res:* Inorganic chemistry of phosphorus compounds; organometallic chemistry; hydride and fluoride chemistry; structural studies of inorganic compounds. *Mailing Add:* Dept of Chem Univ of NMex Albuquerque NM 87131

PAINE, T(HOMAS) O(TTEN), b Berkeley, Calif, Nov 9, 21; m 46; c 4. PHYSICAL METALLURGY. *Educ:* Brown Univ, AB, 42; Stanford Univ, MSc, 47, PhD(metall), 49. *Hon Degrees:* Numerous Degrees from US, Can & Chinese Univs. *Prof Exp:* Res assoc liquid metals, Stanford Univ, 48-49; res assoc ferromagnetism, Res Lab, Gen Elec Co, 49-51, mgr, Measurements Lab, 51-58, mgr tech anal, Res & Develop Ctr, 59-62, mgr, TEMPO-Ctr Adv Studies, 63-68; from dep adminr to adminr, NASA, 68-70; vpres & group exec power generation, Gen Elec Co, 70-76; pres & chief operating officer, Northrop Corp, 76-82; chmn, US Nat Comn Space, 85-87; DIR, THOMAS PAINE ASSOCS, 87- *Concurrent Pos:* Mem, bd dirs, Eastern Airlines, 81-86, Quotron Systs, 82-, Arthur D Little, 83-86, Radio Corp Am, 83-87, Nat Broadcasting Corp, 83-87, Nike Inc, 83- & Orbital Sci, 87-; consult, 82- *Honors & Awards:* Indust Sci Award, AAAS, 56; John Fritz Medal, United Eng Soc, 71; Faraday Medal, Inst Elec Eng, London, 76; John F Kennedy

Astronaut Award, Am Astronaut Soc, 86; Annual Award, Soc Space Explor, 87; Tsiolkovsky Medal, Moscow, 87; Washington Award, Western Soc Engrs. *Mem:* Nat Acad Eng; AAAS; Am Phys Soc; Am Inst Aeronaut & Astronaut; Inst Elec & Electronics Engrs. *Res:* Magnetic and structural materials; fine particles; instrumentation; interdisciplinary studies; operations research; solid state; government-industrial science; computers; management systems; aerospace. *Mailing Add:* Thomas Paine Assocs 2401 Colorado Ave Suite 178 Santa Monica CA 90404

PAINE, THOMAS FITE, JR, medicine; deceased, see previous edition for last biography

PAINTAL, AMREEK SINGH, b Kanpur, India, Apr 1, 40. HYDRAULICS, HYDROLOGY. *Educ:* Agra Col, BSc, 58; Univ Roorkee, BE, 61, ME, 62; Univ Minn, PhD(fluid mech), 69. *Prof Exp:* Lectr civil eng, Univ Roorkee, 62-65; from asst prof to assoc prof, WVa Inst Technol, 69-73; sr civil engr, 73-79, PRIN CIVIL ENGR, METROP SANIT DIST OF GREATER CHICAGO, 79- *Concurrent Pos:* Instr hydraul, Ill Inst Technol, 74-81, adj asst prof, 81-83, adj assoc prof, 83- *Mem:* AAAS; Am Soc Civil Engrs; Am Acad Environ Engrs; Am Geophys Union; Int Asn Hydraul Res; Am Inst Hydrol. *Res:* Hydraulic engineering, especially sediment transport, stochastic processes, sewer design, hydraulic structure design and urban drainage; hydrological precipitation and runoff relationships. *Mailing Add:* Metrop Sanit Dist Greater Chicago 100 E Erie St Chicago IL 60611

PAINTER, GAYLE STANFORD, b Columbia, SC, Feb 27, 41; m 62; c 2. QUANTUM THEORY OF ATOMIC CLUSTERS. *Educ:* Univ SC, BS, 63, PhD(physics), 67. *Prof Exp:* Fel physics, Univ Fla, 67-69; RES PHYSICIST, METALS & CERAMICS DIV, OAK RIDGE NAT LAB, 69- *Concurrent Pos:* Guest scientist, Int Solid State Res, Juelich, Ger, 74-75 & vis scientist, H H Wills Physics Lab, Univ Bristol, Eng. *Mem:* Fel Am Phys Soc. *Res:* Theory of electronic structure of crystalline compounds, surfaces and atom clusters; theory of chemical phenomena, chemisorption and bonding at surfaces and inhomogeneous systems. *Mailing Add:* Metals & Ceramics Div Oak Ridge Nat Lab Box 2008 Oak Ridge TN 37831-6114

PAINTER, JACK T(IMBERLAKE), b Kincaid, WVa, July 23, 30. CIVIL ENGINEERING. *Educ:* WVa Univ, BSCE, 50, MSCE, 55. *Prof Exp:* Instr civil eng, WVa Univ, 50-51, 53-55; from asst prof to assoc prof, 55-62, PROF CIVIL ENG, LA TECH UNIV, 62- *Mem:* Am Cong Surv & Mapping; Am Soc Civil Eng; Am Soc Eng Educ. *Res:* Structural analysis; history of engineering. *Mailing Add:* Box 6155 Tech Sta Ruston LA 71272

PAINTER, JAMES HOWARD, b Eolia, Mo, Nov 25, 35; m 55; c 2. THERMODYNAMICS. *Educ:* Univ Mo, Rolla, BS, 60. *Prof Exp:* Res scientist, 60-70, sr group engr, Res Labs, 70-80, sr tech specialist, 80-83, sect chief, 83-86, br chief aeromech, McDonnell Douglas Astronaut Co, 86-89, prin mgr, 89-91, CONSULT, MCDONNELL DOUGLAS CORP, 91- *Honors & Awards:* Tech Contrib Award, Am Inst Aeronaut & Astronaut, St Louis Sect, 82. *Mem:* Assoc fel Am Inst Aeronaut & Astronaut. *Res:* Reentry simulation using arc heaters; development of multimegawatt arc heaters for accurate reentry flowfield simulation; development of advanced techniques for reentry simulation using existing reentry arc heater facilities. *Mailing Add:* 1931 Rustic Oak Chesterfield MO 63017

PAINTER, JEFFREY FARRAR, b Munich, Germany, Mar 3, 51; US citizen; m 84; c 2. NUMERICAL ANALYSIS. *Educ:* Harvey Mudd Col, BS, 73; Univ Wis-Madison, MA, 75, MS, 77, PhD(appl math), 79. *Prof Exp:* MATHEMATICIAN, LAWRENCE LIVERMORE NAT LAB, 79- *Mem:* Soc Indust & Appl Math. *Res:* Development of high-level software tools for numerical mathematics and computational science. *Mailing Add:* Lawrence Livermore Nat Lab L-316 PO Box 808 Livermore CA 94550

PAINTER, KENT, b Providence, RI, Sept 23, 42; m 64; c 2. BIONUCLEONICS. *Educ:* Univ RI, BS, 65; Colo State Univ, MBA, 77; Purdue Univ, MS, 67, PhD(bionucleonics), 70. *Prof Exp:* Tech mgr analysis prod, Amersham Searle Corp, 70-73; mgr opers radioimmunoassay, Micromed Diag, Inc, 73-77; MANAGING DIR, WESTERN CHEM RES CORP, 78- *Concurrent Pos:* Affil prof radiol, Colo State Univ, 74-88. *Mem:* Am Chem Soc; Health Physics Soc; Am Asn Clin Chemists; AAAS; Sigma Xi. *Res:* Radioimmunoassay; labeled compound synthesis; liquid scintillation counting. *Mailing Add:* 1305 Lakewood Dr Ft Collins CO 80521

PAINTER, LINDA ROBINSON, b Lexington, Ky, May 4, 40; m 67; c 2. RADIATION PHYSICS. *Educ:* Univ Louisville, BS, 62; Univ Tenn, MS, 63, PhD(physics), 68. *Prof Exp:* Consult physicist & adj res & develop participant, Health & Safety Res Div, Oak Ridge Nat Lab, 67-88; from asst prof to assoc prof, 68-82, asst dean grad sch, 85-88, PROF PHYSICS, UNIV TENN, KNOXVILLE, 82-, ASSOC DEAN GRAD SCH, 88- *Concurrent Pos:* Assoc ed, J Radiation Res, 83-87. *Mem:* Radiation Res Soc. *Res:* Optical and dielectric properties of liquids in the vacuum ultraviolet. *Mailing Add:* Dept Physics Univ Tenn Knoxville TN 37996

PAINTER, RICHARD J, b Greensboro, NC, July 20, 31; m 57; c 2. MATHEMATICS. *Educ:* Univ NC, BA, 53, MA, 56, PhD(math), 63. *Prof Exp:* Programmer math, Martin Co, Colo, 57-58; math analyst, Babcock & Wilcox Co, Va, 58-60; assoc prof, 63-74, PROF MATH, COLO STATE UNIV, 74- *Mem:* Am Math Soc; Math Asn Am. *Res:* Matrix analysis. *Mailing Add:* Dept Math Colo State Univ Ft Collins CO 80523

PAINTER, ROBERT BLAIR, b Columbus, Ohio, Sept 9, 24; m 48; c 5. RADIOBIOLOGY. *Educ:* Ohio State Univ, BSc, 47, MSc, 49, PhD(bact), 55. *Prof Exp:* Mem staff, Chem Warfare Labs, US Dept Army, 49-52; prin bacteriologist, Battelle Mem Inst, 52-54, proj leader, 59-61; asst microbiologist, Brookhaven Nat Lab, 56-59; br chief, Ames Res Ctr, NASA, 61-65; assoc dir lab radiobiol, 74-87, PROF MICROBIOL, UNIV CALIF, SAN FRANCISCO, 65- *Honors & Awards:* Failla Lect Award, Rad Ros Soc, 85. *Mem:* Am Soc Biochem Mol Biol. *Res:* Radiation effects on nucleic acid metabolism; organization of DNA in mammalian cells. *Mailing Add:* Lab of Radiobiol & Environ Health Univ Calif San Francisco CA 94143

PAINTER, ROBERT HILTON, b Eng, Nov 27, 32; m 55; c 2. BIOCHEMISTRY. *Educ:* Liverpool Univ, BSc, 53, PhD(biochem), 56. *Prof Exp:* Biochemist, Blood Prod Unit, Lister Inst, London, 56-57; res asst, Connaught Med Res Labs, Univ Toronto, 57-60, res assoc, 60-68, assoc prof, 68-74, asst dean, Sch Grad Studies, 75-79, PROF BIOCHEM, UNIV TORONTO, 68- *Concurrent Pos:* Provost & vchancellor, Trinity Col, Toronto, 86- *Mem:* Fel Royal Inst Chemists; Am Asn Immunologists; Can Biochem Soc; Can Soc Immunologists; Am Soc Hemat; Sigma Xi. *Res:* Protein chemistry; plasma proteins and enzymes; immunoglobulins; complement; Fc receptors; alpha 2 macroglobulins. *Mailing Add:* Dept Biochem Univ Toronto Toronto ON M5S 1A8 Can

PAINTER, RONALD DEAN, b San Bernardino, Calif, May 8, 45; m 68; c 2. THEORETICAL PHYSICS, SOLID STATE PHYSICS. *Educ:* Univ Redlands, BS, 67; Univ Mich, MS, 68, PhD(physics), 73. *Prof Exp:* Physicist, Naval Res Lab, NAVAIRSYSCOM, 67-68, physicist & oper analyst, Naval Weapon, 69-78, oper analysis, 78-83; DIR OPER ANALYSIS, UNITED TECHNOL, 83- *Mem:* Am Phys Soc. *Res:* Solid state physics; simulation modeling. *Mailing Add:* 2113 Pleasant Grove Encinitas CA 92024

PAINTER, RUTH COBURN ROBBINS, b Bethel, Conn, July 21, 10; m 40; c 2. NUTRITIONAL BIOCHEMISTRY, TOXICOLOGY. *Educ:* Univ Hawaii, BS, 31, MS, 34. *Prof Exp:* Asst nutrit invests, Univ Hawaii, 31-37; assoc chemist, dept home econ, USDA, 37; nutrit chemist, Wash State Col, 37-40; instr chem, NDak State Col, 42-43; technician environ toxicol, Univ Calif, Davis, 60-62, from asst specialist to specialist, 62-76; RETIRED. *Honors & Awards:* Clara Barton Medal, Am Red Cross. *Mem:* AAAS; Am Chem Soc; Entom Soc Am; Inst Food Technologists; Sigma Xi. *Res:* Human and insect nutritional biochemistry; toxicology of food additives especially pesticides; data bank of pesticides used in California. *Mailing Add:* 815 Miller Dr Davis CA 95616

PAIR, CLAUDE H(ERMAN), b Tekoa, Wash, July 20, 11; m 36; c 2. IRRIGATION ENGINEERING. *Educ:* Wash State Univ, BS, 32, MS, 33, BA, 34. *Prof Exp:* Agr engr, Soil Conserv Serv, USDA, 35-48, irrig engr, Soil Conserv Res Serv, 48-54, irrig engr, Agr Res Serv, 54-61, res engr irrig, Snake River Conserv Res Ctr, 61-75; RETIRED. *Concurrent Pos:* Mem nat comt, Int Comt Irrig & Drainage; irrig consult, 75- *Honors & Awards:* John Deere Gold Medal Award, Am Soc Agr Engrs, 81. *Mem:* Fel Am Soc Agr Engrs; Sigma Xi. *Res:* Sprinkler irrigation; irrigation, especially methods of water application and requirements. *Mailing Add:* 215 Hillview Dr Boise ID 83712

PAIRENT, FREDERICK WILLIAM, medical education, biochemistry, for more information see previous edition

PAIS, ABRAHAM, b Amsterdam, Holland, May 19, 18; nat US; m 56; c 1. THEORETICAL PHYSICS. *Educ:* Univ Amsterdam, BSc, 38; Univ Utrecht, MS, 40, PhD(theoret physics), 41. *Prof Exp:* Prof physics, Inst Advan Study, 50-63; prof, 63-81, DETLEV W BRONK PROF PHYSICS, ROCKEFELLER UNIV, 81- *Honors & Awards:* J R Oppenheimer Mem Prize, 79; US Steel & Am Phys Soc Award, 82. *Mem:* Nat Acad Sci; Am Acad Arts & Sci; fel Am Phys Soc; Royal Netherlands Acad Sci; Am Philos Soc; Royal Danish Acad Sci & L ett. *Res:* Physics of fundamental particles; field theory. *Mailing Add:* Dept Physics Rockefeller Univ New York NY 10021

PAISLEY, DAVID M, b Buckhannon, WVa, Feb 26, 35; m 57; c 3. ORGANIC CHEMISTRY. *Educ:* WVa Wesleyan Col, BS, 57; Univ Ill, MS, 59, PhD(org chem), 61. *Prof Exp:* Patent agent, Law Dept, Union Carbide Corp, 61-63, asst to vpres, Develop Dept, 63-65, mgr pharmaceut mkt res, 65-67; dir mkt res, E R Squibb & Sons, Inc, 67-70; drug indust security analyst, Donaldson, Lufkin & Jenrette, Inc, 70-75; vpres & head health res group, Merrill, Lynch, Pierce, Fenner & Smith, Inc, 75-86; PRES, PAISLEY ASSOCS, 86- *Concurrent Pos:* Adj prof chem, Univ Ill, 86- *Mem:* Am Chem Soc. *Res:* New drug research and development; drug marketing and marketing research. *Mailing Add:* 66 RAL Box 38 1209 W Calif Urbana IL 61801

PAISLEY, NANCY SANDELIN, b Duluth, Minn, Feb 20, 36; m 57; c 3. BIOCHEMISTRY. *Educ:* Univ Minn, BS, 57; Univ Ill, MA, 59, PhD(biochem), 61. *Prof Exp:* Res scientist, Esso Res & Eng Co, 61-63, econ analyst, Standard Oil Co, NJ, 63-65; ASST PROF CHEM, MONTCLAIR STATE COL, 66- *Concurrent Pos:* Vis asst prof biochem, NJ Col Med & Dent. *Mem:* AAAS; Nutrit Today Soc. *Res:* Purification and characterization of long lipoprotein lipase. *Mailing Add:* Dept Chem Montclair State Col Upper Montclair NJ 07043

PAJARI, GEORGE EDWARD, b Montreal, Que, Mar 24, 36; m 55; c 2. PETROLOGY, GEOCHEMISTRY & HYDRO-GEOCHEMISTRY. *Educ:* McGill Univ, BSc, 58; Cambridge Univ, PhD(petrol), 66. *Prof Exp:* From lectr to assoc prof petrol, Univ NB, 64-74, from assoc prof to prof geol, 74-81; CHIEF EXEC OFFICER, PAJARI INSTRUMENTS LTD, 81- *Mem:* Mineral Asn Can; Sigma Xi. *Res:* Physio-chemical processes operative in igneous systems; design of surveying instrumentation in remote procedures; instrument design. *Mailing Add:* Box 820 Orillia ON L3V 6K8 Can

PAK, CHARLES Y, b Seoul, Korea, Nov 27, 35; US citizen; m 63. PHYSICAL CHEMISTRY, ENDOCRINOLOGY. *Educ:* Univ Chicago, BS, 58, MD, 61. *Prof Exp:* Intern med, Univ Chicago Clins, 51-62, asst resident, 62-63; vis scientist, Lab Phys Biol, NIH, 63-65, sr investr lab clin endocrinol, Nat Heart Inst, 65-69, head sect mineral metab, Endocrinol Br, 69-71; assoc prof, 72-75, PROF INTERNAL MED & CHIEF SECT MINERAL METAB, UNIV TEX HEALTH SCI CTR DALLAS, 75-, PROG DIR, GEN CLIN RES CTR, 74- *Mem:* Endocrine Soc; Am Soc Nephrology; Am Soc Clin Invest; Biophys Soc; Am Soc Pharmacol & Exp Therapeut. *Res:* Renal stones; calcium and phosphorus metabolism; mechanism and treatment of nephrolithiasis; parathyroid function; calcium absorption; bone disease. *Mailing Add:* Dept Internal Med Med Sch Univ Tex Southwestern 5323 Harry Hines Dallas TX 75235

PAK, RONALD Y S, b Hong Kong, Aug 23, 57; Brit citizen. ELASTICITY, DYNAMICS. *Educ:* McMaster Univ, Can, BS, 79; Calif Inst Technol, MS, 80, PhD(appl mech), 85. *Prof Exp:* Res fel struct soil & dynamics, Calif Inst Technol, 85; ASST PROF STRUCT SOIL & DYNAMICS, UNIV COLO, BOULDER, 85- *Concurrent Pos:* Prin investr, NSF, 86-94; NSF presidential young investr award, 89; consult, Resonant Mach Inc, Tulsa, 90-91. *Mem:* Am Soc Civil Engrs; Am Soc Mech Engrs; Am Acad Mech. *Res:* Structural and soil dynamics; earthquake engineering; theoretical and applied mechanics; dynamic soil-structure interaction; load transfer problems; wave propagation; centrifuge modeling. *Mailing Add:* Dept Civil Environ & Archit Eng Univ Colo Boulder CO 80309-0428

PAK, WILLIAM LOUIS, b Suwon, Korea, Sept 27, 32; US citizen; m 58; c 2. MOLECULAR GENETIC, NEUROBIOLOGY. *Educ:* Boston Univ, AB, 55; Cornell Univ, PhD(physics), 60. *Prof Exp:* Instr physics, Stevens Inst Technol, 60-61, asst prof, 61-65; from asst prof to assoc prof, 65-72, prof biol, 72-87, PAUL F OREFFICE DISTINGUISHED PROF, PURDUE UNIV, 87- *Concurrent Pos:* USPHS trainee biophys, Univ Chicago, 63-64, career develop award, 66-71; NSF fel, 64-65; vis scientist, Med Sch, Keio Univ, Japan, 67 & Max Planck Inst Biol Cybernet, 71; mem, Visual Sci Study Sect, NIH, 72-75; instr neurobiol drosophila, Cold Spring Harbor Lab, 73-75; Herbert Newby McCoy res award, Purdue Univ, 82; Roche Res Found vis prof, Univ Basel, 84, vis prof, Harvard Med Sch, 84; mem, vis res rev comt, 86-90. *Honors & Awards:* Merit Award, NIH, 90. *Mem:* AAAS; Am Physiol Soc; Biophys Soc; Asn Res Vision & Ophthal; Soc Neurosci; Genetics Soc Am. *Res:* Molecular genetic dissection of mechanisms of nerve cell function particularly those pertaining to signal transduction transmembrane signaling using the Drosophila photoreceptor as a model neuron. *Mailing Add:* 1025 Windwood Lane West Lafayette IN 47906

PAKE, GEORGE EDWARD, b Jeffersonville, Ohio, Apr 1, 24; m 47; c 4. PHYSICS. *Educ:* Carnegie Inst Technol, BS & MS, 45; Harvard Univ, PhD(physics), 48. *Hon Degrees:* DSc, Carnegie Inst Technol, 66 & Univ Mo-Rolla, 66; LLD, Kent State Univ, 67. *Prof Exp:* Mem physics adv panel, NSF, 58-60 & 63-66; chmn physics surv comt, Nat Acad Sci-Nat Res Coun, 64-66; mem, President's Sci Adv Comt, 65-69; physicist, Xerox Corp, 69-86, group vpres, Corp Res, 78-86; dir, Inst Res Learning, 86-90; RETIRED. *Concurrent Pos:* Provost & prof physics, Wash Univ, St Louis, 62-70. *Honors & Awards:* Nat Medal Sci, 87. *Mem:* Nat Inst Med-Nat Acad Sci; AAAS; fel Am Phys Soc (pres, 77); Am Acad Arts & Sci. *Res:* Nuclear magnetic resonance; paramagnetic resonance of free radicals; magnetic properties of solids. *Mailing Add:* Inst Res Learning 2550 Hanover St Palo Alto CA 94304

PAKES, STEVEN P, b East St Louis, Ill, Jan 19, 34; m 54; c 4. PATHOLOGY, LABORATORY ANIMAL MEDICINE. *Educ:* Ohio State Univ, BSc, 56, DVM, 60, MSc, 64, PhD(vet path), 72. *Prof Exp:* Vet pathologist, US Army, Ft Detrick, Md, 60-62; chief animal colonies, Pine Bluff Arsenal, Ark, 64-66; chief comp path, Naval Aerospace Med Inst, 66-69; dir lab animal med, Col Vet Med, Ohio State Univ, 69-72; assoc prof, 72-80, PROF COMP MED & CHMN DEPT, SOUTHWESTERN MED SCH, UNIV TEX SOUTHWESTERN MED CTR, DALLAS, 80- *Concurrent Pos:* Mem exam comt, Am Col Lab Animal Med, 68-69; mem exec comt, Inst Lab Animal Resources, Nat Acad Sci-Nat Res Coun; chmn coun accreditation, Am Asn Accreditation of Animal Care, 74-76, treas, Bd Trustees, 83-; adv bd, Vet Specialities, 81-82, chmn, 85; chmn, Lab Guide Rev Comt, Inst Lab Animal Resources, Nat Acad Sci, 83-85; bd trustees, Am Asn Lab Animal Sci, 85-88; mem, Inst Lab Animal Resources Coun, 85-, chm, 87- *Honors & Awards:* Res Award, AAAS, 82. *Mem:* Am Col Lab Animal Med (pres, 73); Am Vet Med Asn; Am Asn Lab Animal Sci; Sigma Xi; Am Soc Microbiol; AAAS. *Res:* Infectious diseases of laboratory animals; effect of spontaneous diseases of laboratory animals on biomedical research. *Mailing Add:* Southwestern Med Sch 5323 Harry Hines Blvd Dallas TX 75235-9037

PAKISER, LOUIS CHARLES, JR, b Denver, Colo, Feb 8, 19; m 39. GEOPHYICS. *Educ:* Colo Sch Mines, GeolE, 42. *Prof Exp:* Geophysicst, Carter Oil Co, 42-49; nat exec dir, Am Vet Comt, 49-52; geophysicist, Geophys Br, 52-60, chief ground surv sect, 54-57, br rep, Denver, 58-60, chief, Crustal Studies Br, 60-66, chief, Off Earthquake Res & Crustal Studies, Nat Ctr Earthquake Res, Calif, 67-71, RES GEOPHYSICIST, US GEOL SURV, DENVER, 71- *Concurrent Pos:* Mem, Ad Hoc Panel Earthquake Prediction, Off Sci & Technol, 64-66; Adv Bd & Steering Comt, Earth Sci Curriculum Proj & Educ Prog, Am Geol Inst, 65-69; chmn, Adv Comt Minority Partic Prog, Am Geol Inst, 73-75; vis scientist, Univ New Orleans, 82-83. *Mem:* Fel AAAS; fel Am Geophys Union; fel Geol Soc Am; Am Indian Sci & Eng Soc; Soc Advan Chicano's & Native Am Sci. *Res:* Geophysics. *Mailing Add:* 2710 S Sandy Ridge Rd Sedalia CO 80135

PAKMAN, LEONARD MARVIN, b Philadelphia, Pa, Apr 8, 33; m 64; c 2. MICROBIOLOGY. *Educ:* Univ Pa, BA, 56, PhD(microbiol), 63. *Prof Exp:* Asst instr microbiol, Univ Pa, 58-63; asst prof med microbiol, Jefferson Med Col, 63-70; from asst prof to prof microbiol, Sch Dent, 70-86. *Concurrent Pos:* Consult microbiol, Johnson Found Med Physics, Univ Pa, 57-61 & Franklin Inst Res Labs, Philadelphia, 68-72; Tech Adv Serv Attorneys, Ft Washington, Pa, 78- *Mem:* AAAS; Am Soc Microbiol; Int Asn Dent Res; Sigma Xi. *Res:* Effects of hyperbaric oxygen on microorganisms, drug potentiation, infectious processes and tumor chemotherapy; effects of chemotherapeutic agents on and interactions of oral microbes; identification of anaerobic microorganisms associated with root canal infections; evaluation of bacterial microleakage from dental restorations. *Mailing Add:* Dept Microbiol & Immunol Temple Univ Med Sch Philadelphia PA 19140

PAKVASA, SANDIP, b Bombay, India, Dec 24, 35; m 78. PHYSICS. *Educ:* Univ Baroda, BSc, 54, MSc, 57; Purdue Univ, PhD(physics), 65. *Prof Exp:* Lectr elec eng, Univ Baroda, 60-61; asst physics, Purdue Univ, 61-65; res assoc, Syracuse Univ, 65-67; assoc physicist, 67-68, from asst prof to assoc prof, 68-74, PROF PHYSICS, UNIV HAWAII, 74- *Concurrent Pos:* Vis scientist, Tata Inst Fundamental Res, 70 & Europ Coun Nuclear Res, Geneva, 82; vis mem, Inst Adv Study, Princeton, 75; vis prof, Univ Wis-Madison, 78 & 86, Tata Inst Fundamental Res, 83, Kappa Eta Kappa Nat Lab High Energy Phys, Japan, 83 & 89, Univ Melbourne, 86; fel, Japan Soc Promotion Sci, 81 & 85. *Mem:* Fel Am Phys Soc. *Res:* Theoretical physics; elementary particle physics. *Mailing Add:* Dept Physics & Astron Univ Hawaii Honolulu HI 96822

PAL, BIMAL CHANDRA, organic chemistry, biochemistry, for more information see previous edition

PAL, DHIRAJ, b India, 1948; m; c 3. SOIL SCIENCE. *Educ:* Agra Univ, BS, 66; Indian Agr Res Inst, MS, 68; Univ Calif, PhD(soil sci), 73. *Prof Exp:* Scholar & scientist, Univ Calif, Davis, 73-75,; res assoc, NC State Univ, 76-80; independent consult & adv, Global Educ Corp, 80-81; sr scientst, Weston, 81-82; contractor, consult & dir, indust & bus, 82-88; SR ENG TO DIR, TECH SUPPORT, 88- *Concurrent Pos:* Fel, Nat Res Adv Coun, New Zealand, 75; res fel, 75. *Mem:* Am Mgt Asn; Am Inst Chem Engrs; Am Chem Soc; Water Pollution Control Fedn; Int Soc Soil Sci. *Res:* Terrestrial and aquatic microbiology and biochemistry; agricultural waste management; design of land treatment systems for a variety of industrial effluents; hazardous waste management alternatives; fate of organic and inorganic (metals) chemicals in the environment; innovation of fixation/stabilization systems (FSS), particularly Maeckite for lead fixation in solid wastes; inventor of Chromtite technology for fixation of chromium in wastes; invent treatment technologies for remediation of hazardous waste sites, soil detoxification, water treatment and environmental problems related to industrial and defense production; direct technical research relevant to waste management. *Mailing Add:* 270 Leonard Ave Chicago Heights IL 60411

PALACAS, JAMES GEORGE, b New York, NY, Nov 12, 30; m 55; c 4. ORGANIC GEOCHEMISTRY. *Educ:* Harvard Univ, BA, 52; Pa State Univ, MS, 57; Univ Minn, PhD(geol, biochem), 59. *Prof Exp:* Geologist, US Geol Surv, 52-53; geologist, Shell Oil Co, 59-63; GEOLOGIST, OIL & GAS RESOURCES BR, US GEOL SURV, 63- *Mem:* AAAS; Geol Soc Am; Am Asn Petrol Geologists; Geochem Soc; Soc Econ Paleontologists & Mineralogists. *Res:* Organic geochemistry of recent and ancient sediments; geochemistry of bitumens and humic acids; petroleum geology; geochemistry and oil source-rock potential of carbonates. *Mailing Add:* Geol Surv Mail Stop 977 Box 25046 Fed Ctr Denver CO 80255

PALADE, GEORGE E, b Jassy, Romania, Nov 19, 12; US citizen; m 41, 70; c 4. CELLULAR & MOLECULAR BIOLOGY. *Educ:* Univ Bucharest, MD, 40. *Hon Degrees:* DSc, Yale Univ, 67, Univ Chicago, 68, Woman's Med Col Philadelphia, 69, Univ Mich, 73, Bristol Univ, 75, Rockefeller Univ, 76, Univ Siena, Italy, 77, Columbia Univ, 86, NY Univ, 86; MD, Univ Bern, Switz, 68, Med Col Ohio, 73, Univ uppsala, Sweden, 77. *Prof Exp:* Intern, Asn Civil Hosps Bucharest, 33-39; prosector asst, Fac Med, Inst Anat, Univ Bucharest, 35-46, from asst prof to assoc prof, 35-46; vis investr, Dept Path & Bact, Rockefeller Inst, NY, 46-48, asst prof, 48-51, from assoc prof to prof, Lab Cell Biol, 51-73, head lab, 61-73; prof & chmn, Sect Cell Biol, Yale Univ Sch Med, 73-83, sr res scientist, Dept Cell Biol, 83-90, spec adv to dean, 83-90; DEAN SCI AFFAIRS, UNIV CALIF, SAN DIEGO SCH MED, 90-, PROF MED RESIDENCE, 90- *Concurrent Pos:* Instr, Queen Mary Sch Nursing, Bucharest, 36-45; vis investr, Dept Biol, NY Univ, 46. *Honors & Awards:* Nobel Prize Physiol or Med, 74; Warren Triennial Prize, 62; Passano Award, 64; Lasker Award, 66; T Duckett Jones Award, 66; Gairdner Spec Award, 67; Louisa Gross Horwitz Prize, 70; Dickson Prize, 71; Brown-Hazen Award, 83; Schleiden Medal, Leopoldina Acad, 85; Henry Gray Award, 86; Nat Medal of Sci, 86. *Mem:* Nat Acad Sci; fel Am Acad Arts & Sci; foreign mem Royal Belg Acad Med; Acad Leopoldina; Royal Soc (London); Pontif Acad Sci; Romanian Acad. *Res:* Cell biology; structural-functional correlations at the subcellular level; biology of cellular membranes; structural basis of blood capillary permeability. *Mailing Add:* Div Cellular & Molecular Med Univ Calif San Diego 9500 Gilman Dr La Jolla CA 92093-0651

PALADINO, FRANK VINCENT, b Brooklyn, NY, April 13, 52; m 88; c 3. COMPARATIVE VERTEBRATE PHYSIOLOGY, SEA TURTLE BIOLOGY. *Educ:* State Univ Col, Plattsburgh, NY, BA, 74; State Univ Col, Buffalo, NY, 76; Wash State Univ, PhD(zoo physiol), 79. *Prof Exp:* Res assoc avian physiol, Wash State Univ, 76-79; vis asst prof biol, Dept Zool, Miami Univ, 79-82; ASSOC PROF PHYSIOL, DEPT BIOL, IND-PURDUE UNIV, 82- *Concurrent Pos:* Adj prof, Sch Med, Ind Univ, 82- & Drexel Univ, Philadelphia, Pa, 88-; mem, Int Union Conserv Nature, 89-; prin investr, Nat Geog Soc grant, 90-, NSF grant, 90- & World Wildlife Fund grant, 91- *Mem:* Am Physiol Soc; Sigma Xi (pres, 88); Am Ornithologist Union; Herpetologists League; Ecol Soc Am; Cooper Ornith Soc. *Res:* Comparative animal physiology, investigating bioenergetics and thermoregulatory mechanisms of sparrows, fish, elephants and sea turtles; thermoregulation and energy use patterns of large and small animals. *Mailing Add:* Dept Biol Sci Ind-Purdue Univ Ft Wayne IN 46805

PALAIC, DJURO, b Ruma, Yugoslavia, July 24, 37; Can citizen; m 63; c 2. PHARMACOLOGY. *Educ:* Univ Zagreb, MD, 62; PhD(pharmacol), 65. *Prof Exp:* Asst pharmacol, Inst Rudjer Boskovic, Zagreb, 62-65; asst prof, 67-73, ASSOC PROF PHARMACOL, UNIV MONTREAL, 73- *Concurrent Pos:* Res fel, Cleveland Clin Found, 65-67; lectr, Fac Med, Univ Zagreb, 64-65; affil, Dept Clin Pharmacol, Montreal Gen Hosp. *Res:* Pharmacology of angiotensin and serotonin. *Mailing Add:* Dept Pharmacol Univ Montreal Montreal PQ H3L 3J7 Can

PALAIS, JOSEPH C(YRUS), b Portland, Maine, Feb 2, 36; m 61; c 2. ELECTRICAL ENGINEERING. *Educ:* Univ Ariz, BSEE, 59; Univ Mich, MSE, 62, PhD(elec eng), 64. *Prof Exp:* Microwave engr, Motorola, Inc, Ariz, 59-60; asst res engr, Cooley Electronics Lab, Univ Mich, 60-64; from asst prof to assoc prof elec eng, 64-73, PROF ELEC ENG, ARIZ STATE UNIV, 73- *Concurrent Pos:* Consult, Stanford Res Inst, 65-67, Motorola, Inc, Ariz, 71 & 78 & Sperry, Inc, Ariz, 72-75; vis assoc prof, Technion-Israel Inst Technol, 73. *Mem:* AAAS; Inst Elec & Electronics Engrs; Optical Soc Am; Soc Photo-optical Instrumentation Engrs. *Res:* Fiber optical communications; laser; holography. *Mailing Add:* Col of Eng & Appl Sci Ariz State Univ Tempe AZ 85287

PALAIS, RICHARD SHELDON, b Lynn, Mass, May 22, 31; m 54; c 3. MATHEMATICS. *Educ:* Harvard Univ, AB, 52, MA, 54, PhD(math), 56. *Prof Exp:* Instr math, Univ Chicago, 56-58; NSF fel & mem, Inst Advan Study, 58-60; from asst prof to assoc prof, 60-65, PROF MATH, BRANDEIS UNIV, 65- *Concurrent Pos:* NSF sr fel, 62-63; Sloan Found res fel, 65-67; ed, Am Math Soc Trans, 66-70; mem, Inst Advan Study, 63-64, 68-69 & 74-75; guest prof, Univ Bonn, 81-82. *Mem:* Am Math Soc; Math Asn Am; Asn Comput Mach; fel AAAS. *Res:* Differential geometry and topology; transformation groups; global analysis. *Mailing Add:* Dept Math Brandeis Univ Waltham MA 02254

PALAMAND, SURYANARAYANA RAO, b Bangalore, India; US citizen; m 61; c 2. HUMAN PHYSIOLOGY, BIOCHEMISTRY. *Educ:* Univ Mysore, BSc, 51; Univ Bombay, MS, 54; Ohio State Univ MSc, 57, PhD(food technol), 60. *Prof Exp:* Tech dir brewing res, United Breweries Ltd, Bangalore, India, 61-64; res group leader flavor chem, 66-80, mgr corp res & develop, 80-85, DIR CORP RES & DEVELOP, ANHEUSER-BUSCH CO, INC, 85- *Concurrent Pos:* Vis prof, dept chem, Southern Ill Univ, 77-80, adj prof, 80-82. *Mem:* Am Chem Soc; Inst Food Technologists. *Res:* Chemistry & sensory properties of aroma & taste compounds in food & beverages; development of analytical methods for flavor measurement; published 34 papers in the area of beverage flavors in a number of scientific journals. *Mailing Add:* 5002 Amberway St Louis MO 63128

PALANISWAMY, PACHAGOUNDER, b Kandisalai, Madras, July 11, 45; Can citizen; m 72; c 3. PLANT PROTECTION, INSECT PLANT RELATIONSHIPS. *Educ:* Univ Madras, BSc, 68, MSc, 71; Univ New Brunswick, Can (PhD), 79. *Prof Exp:* Res asst, entom, 68-69, agr ext officer, Agr Col & Res Inst, Coimbatore, India, 71-72; res scholar, biol, Univ Leuven, Belg, 73-75; vis scientist, etom, Nat Res Coun Can, 80-81, res officer, Plant Biotech Inst, 81-84; prof, res assoc, biol, Univ Sask, 84-85; econ entomologist, 86-87, RES SCIENTIST, ENTOM, ARG CAN, 87- *Concurrent Pos:* Consult, Food & Agr Orgn UN, 83, 86. *Mem:* Entom Soc Am; Entom Soc India; Entom Soc Can. *Res:* Chemical and biological control of insect pests, insecticides, pheromones, hormones, insect neurophysiology and behavior and insect monitoring; published over 35 papers in various scientific journals. *Mailing Add:* 195 Dafoe Rd Winnipeg MB R3T 2M9 Can

PALAS, FRANK JOSEPH, b Libuse, La, Oct 24, 18; m 49. MATHEMATICS. *Educ:* Univ Okla, MEd, 48, PhD(math), 56. *Prof Exp:* Asst math, Univ Okla, 55-56; from asst prof to assoc prof, 56-71, PROF MATH, SOUTHERN METHODIST UNIV, 71- *Mem:* Am Math Soc; Math Asn Am. *Res:* Analysis; special functions; number theory geometry. *Mailing Add:* Dept of Math Southern Methodist Univ Dallas TX 75275

PALASZEK, MARY DE PAUL, b Grand Rapids, Mich, Aug 23, 35. CHEMISTRY. *Educ:* Mercy Col, Mich, AB, 58; Univ Detroit, MS, 60; Mich State Univ, PhD(chem), 65; Univ Windsor, PDD(clinical chem), 78. *Prof Exp:* Instr natural sci, Our Lady of Mercy High Sch, 58; from asst prof to assoc prof chem, Mercy Col, Mich, 60-74, prof, 74-79; MGR & CLIN CHEMIST, SAMARITAN HEALTH CTR, 79- *Concurrent Pos:* Asst to dean, Mercy Col, Mich, 65-66; instr, Our Lady of Mercy High Sch, 65; consortium grant partic, Dept Health, Educ & Welfare, 67-68; Du Pont grant, 69-70; vis assoc, Nat Inst Environ Health Sci, 72-74; chemist, Rex Hosp, Raleigh, NC, 73-74. *Mem:* Am Asn Clin Chemists; Am Bd Clin Chem; Nat Acad Clin Biochem. *Res:* Thiouronium salts; polythienyls; oxetanes; thiosugars; halogenated pesticides; amniotic fluid; analytes; computer applications. *Mailing Add:* Samaritan Health Ctr, Clin Chem 5555 Conner Detroit MI 48213

PALATNICK, BARTON, b Brooklyn, NY, Oct 29, 40. PHYSICS TEACHING. *Educ:* Yale Univ, BA, 62; Columbia Univ, MA, 64, PhD(physics), 68. *Prof Exp:* Asst prof to assoc prof physics, 74-79, actg chmn, 81-82, PROF PHYSICS, CALIF STATE POLYTECH UNIV, POMONA, 79- *Concurrent Pos:* Sr physicist, Gen Dynamics, Pomona, 74-75; vis prof physics, Univ La Verne, 77; mat develop engr, Beckman Instruments, Inc, 79-80; mem tech staff, Jet Propulsion Lab, 82. *Mem:* Sigma Xi; Am Asn Physics Teachers. *Res:* Near ultraviolet magnetic rotation spectrum of carbon disulfide; infrared detectors; holography. *Mailing Add:* Calif State Polytech Univ Dept Physics 3801 W Temple Ave Pomona CA 91768

PALATY, VLADIMIR, b Prague, Czech, Oct 12, 33; m 62; c 1. BIOPHYSICS, PHYSIOLOGY. *Educ:* Col Chem Technol, Prague, Ing, 57, PhD(chem), 61. *Prof Exp:* Asst prof chem, Col Chem Technol, Prague, 57-61; res assoc physiol, Inst Physiol, Czech Acad Sci, 61-69; assoc prof, 69-76, PROF BIOPHYS, UNIV BC, 76- *Concurrent Pos:* Med Res Coun Can fel, Univ BC, 68-69. *Mem:* Can Physiol Soc; Biophys Soc. *Res:* Physiology of vascular smooth muscle. *Mailing Add:* Dept Anat Univ BC Fac Med 2194 Health Sci Mall Vancouver BC V6T 1W5 Can

PALAY, SANFORD LOUIS, b Cleveland, Ohio, Sept 23, 18; div; c 2. NEUROANATOMY, NEUROCYTOLOGY. *Educ:* Oberlin Col, AB, 40; Western Reserve Univ, MD, 43. *Hon Degrees:* MA, Harvard Univ, 62. *Prof Exp:* Intern, New Haven Hosp, Conn, 44; asst resident, Western Reserve Univ Hosps, 45-46; from instr to assoc prof anat, Sch Med, Yale Univ, 49-56; chief sect neurocytol, Lab Neuroanat Sci, Nat Inst Neurol Dis & Blindness, 56-69, chief lab, 60-61; vis investr, Middlesex Hosp, London, Eng, 61; Bullord prof, 61-89, EMER BULLARD PROF, NEUROANAT, HARVARD MED SCH, 89- *Concurrent Pos:* Res fel, Western Reserve Univ Hosp, 45-46; Nat Res Coun fel, Rockefeller Inst, 48, vis investr, 53; Guggenheim fel, 71-72; mem fel bd, NIH, 58-61, cell biol study sect, 59-61, 62-65, assoc, Neuroscience Res Prog, 62-67, hon assoc, 74-; chmn, Gordon Res Conf Cell Struct & Metab, 60; mem anat sci training comt, Nat Inst Gen Med Sci, 68-72; vis prof, Univ Wash, 69, Univ Osaka Med Sch, 78, 88 & Nat Univ Singapore, 83; distinguished scientist lectr, Sch Med, Tulane Univ, 69, 75 & Univ Ark, 77; mem adv comt, HEW Resources, 73-80, mem behav & neurosci study sect, 81-86, chmn, 84-86; distinguished lectr biol structure, Univ Miami, 74; co-managing ed, Anat & Embryol, 78-88; Fogarty scholar-in-residence, NIH, 80-81; ed-in-chief, J Comp Neurol, 80- *Honors & Awards:* Phillips lectr, Haverford Col, 59; Ramsay Henderson Trust lectr, Univ Edinburgh, 62; Rogowski Mem lectr, Yale Univ, 73; Electron Micros Soc Am Award, 81; Centennial Award, Am Asn Anatomists, 87; Biomed Res Award, Asn Am Med Cols, 89; George Bishop Lectr, Wash Univ, 90; Ralph Girard Award, Soc Neurosci, 90; Henry Gray Award, Amn Asn Anatomists, 91. *Mem:* Nat Acad Sci; AAAS; fel Am Acad Arts & Sci; Am Asn Anat (pres, 80-81); Histochem Soc; Soc Cell Biol; Soc Neurosci; hon mem Royal Micros Soc. *Res:* Neurosecretion; electron microscopy of the nervous system. *Mailing Add:* 78 Temple Rd Concord MA 01742

PALAYEW, MAX JACOB, b Montreal, Que, Can, July 21, 30; m 58; c 3. LUNG BIOPSIES. *Educ:* McGill Univ, BA, 51; Univ Montpellier, France, MD, 56, FRCPS, 66. *Prof Exp:* Intern, Jewish Gen Hosp, 56-57, residency, 57-58, asst resident, 58-59; asst resident, Beth Israel Hosp, Harvard Med Sch, 59-60, chief resident, 60-61; teaching fel radiol, Univ Cincinnati, 61-62; assoc prof, McGill Univ, 71-72, prof radiol & chmn dept, 77-87; CHIEF RADIOLOGIST, JEWISH GEN HOSP, 71- *Concurrent Pos:* Vis prof, Univ BC, 76, Univ Man & Johns Hopkins Univ, 78, Univ Cincinnati, 79, Univ Hawaii & Univ Toronto, 80, Univ Sask, Univ Dartmouth, Univ Hawaii, Okinawa, Univ Kamamoto, Japan & Univ Fukuoko, Japan, 81, Univ Ottawa, 82, Cornell Univ & Montefiore Hosp, NY, 83, Albany, NY, 83, Cornell Med Ctr, 84, North Shore Hosp, Manhasset, NY, 84 & Univ Calif & Univ Sherbrook, 85, Univ Laval, Quebec, Univ Vt, Burlington, Univ BC, Vancouver, 86, Univ Alta, Edmonton, 86, Beijing Inst, China, 86, Xi'an Univ, China, 86, Queen's Univ, Kington, 87, Cornell Univ, Quebec, 87; adv med radiol, Royal Victoria Hosp, 86-89; reviewer, J Radiol, Radiol Soc NAm, 86; consult radiol, Jewish Convalescent Hosp, Maimonides Hosp & Home for Aged, Montreal Children's Hosp & Queen Elizabeth Hosp; secy-treas, Med Exec Comt, Sir Mortimer B Davis-Jewish Gen Hosp, mem Planning & Priority comt. *Mem:* Fel Am Col Radiol; Can Asn Radiologists (pres, 83-84); Fr-Can Soc Radiol; Am Col Radiol; Can Med Asn. *Res:* Cancer of biliary tract; right aortic arch; histoplasmosis; arteriography of the lower extremities gas in the portal vein; non-surgical removal of retained common bile duct stone; induced pulmonary disease atelectasis; lung scanning and small cell lung cancer. *Mailing Add:* Jewish Gen Hosp Rm C 212-2 3755 Cote St Catherine Rd Montreal PQ H3T 1E2 Can

PALAZOTTO, ANTHONY NICHOLAS, b Brooklyn, NY, Dec 15, 35; m 60; c 8. SOLID MECHANICS, CIVIL ENGINEERING. *Educ:* NY Univ, BCE, 55, PhD(solid mech), 68; Polytech Inst Brooklyn, MCE, 61. *Prof Exp:* Civil engr, private practice, 55-63; struct engr, Severud Assoc, 63-75; prof eng mech, Univ Bridgeport, 68-75; PROF ENG & SCI, AIR FORCE INST TECHNOL, 75- *Concurrent Pos:* Asst prof, Univ Conn, 63-66; Nat Sci Found initiation grant, 69, AFOSR, 88. *Mem:* Fel, Am Soc Civil Engrs; assoc fel,Am Inst Aeronaut & Astronaut; Sigma Xi. *Res:* Plasticity; use of finite elements; composite materials, shells. *Mailing Add:* 6358 Siena St Dayton OH 45459

PALAZZOLO, MATTHEW JOSEPH, inhalation toxicology, immunotoxicology, for more information see previous edition

PALCHAK, ROBERT JOSEPH FRANCIS, b Braddock, Pa, May 23, 27; m 53; c 5. ORGANIC CHEMISTRY. *Educ:* Univ Pittsburgh, BSc, 48, MSc, 50; Univ Cincinnati, PhD(chem), 53. *Prof Exp:* Asst prof pharm, Univ Pittsburgh, 53-56; sr res chemist, Olin Mathieson Chem Corp, NY, 56-58 & Allegany Ballistics Lab, Hercules Powder Co, 58-61; sr res chemist, Atlantic Res Corp, Va, 61-63, proj dir propulsion chem & polymers, 63-69; prin scientist, Northrop Corp Labs, Hawthorne, 69-71; INDEPENDENT CONSULT, 71- *Concurrent Pos:* Mem, Comt Foamed Plastics, Adv Bd Mil Personnel Supplies, Nat Acad Sci-Nat Res Coun, 63-; pres, Seminars Health Professionals. *Mem:* Am Pharmaceut Asn; Am Chem Soc. *Res:* Chemistry of acetylenes, allenes, thiophenes, boron hydrides, nitrogen-fluorine compounds, polyurethanes, foamed plastics, cationically initiated polymerizations, general organic and organometallic synthesis. *Mailing Add:* 18012 Darmel Pl Santa Ana CA 92705

PALCZUK, NICHOLAS C, b Williamstown, Pa, Nov 11, 27. ALLERGIES, RAGWEED VACCINES. *Educ:* Univ Pa, PhD(microbiol), 58. *Prof Exp:* Prof immunol, Rutgers Univ, 58-83; res prof immunol, Med Ctr, Duke Univ, 83-; RETIRED. *Mem:* Am Asn Immunologists. *Res:* Isolation purification short ragweed antigen and its use as vaccines. *Mailing Add:* c/o Janice Little Div Immunol Duke Univ Med Ctr PO Box 3010 Durham NC 27710

PALDUS, JOSEF, b Bzi, Czech, Nov 25, 35; m 61; c 1. CORRELATION PROBLEM, QUANTUM CHEMICAL METHODOLOGY. *Educ:* Charles Univ, Prague, MSc, 58; Czech Acad Sci, PhD(chem, physics), 61. *Prof Exp:* Jr sci off quantum chem, Inst Phys Chem, Czech Acad Sci, 61-62 & 64-65, sr sci off, 65-68; assoc dean grad studies, Fac Math, 78-80, assoc prof appl math, 68-75, PROF APPL MATH & CHEM, UNIV WATERLOO, 75- *Concurrent Pos:* Prof, Guelph-Waterloo Ctr Grad Work Chem, 74-; vis prof, Catholic Univ Nijmegen, Holland, 81 & Inst Le Bel, Univ Strasbourg, France, 75-76, 82-83, Technion, Haifa, Israel, 83; vis scientist, Inst Phys Chem, Free Univ, Berlin, 81; adj prof, dept chem, Univ Fla, Gainesville, 84-; Inst Advance Study fel, Berlin, 86-87; mem, Adv Ed Bds, Intern J Quantum Chem, John Wiley & Sons, 77-; fel, Killam Res, 87-89. *Honors & Awards:* Annual Prize, Div Chem, Czech Acad Sci, 62 & 67. *Mem:* Am Phys Soc; Can Appl Math Soc; Int Acad Quantum Molecular Sci; Can Chem Soc; Europ Acad Sci, Arts & Letters; fel Royal Soc Can. *Res:* Electronic structure of atomic and molecular systems; correlation problem and the use of the many-body, group theoretical and diagrammatic techniques in theoretical chemistry and physics; advances in quantum chemistry. *Mailing Add:* Dept Appl Math Univ Waterloo Waterloo ON N2L 3G1 Can

PALDY, LESTER GEORGE, b New York, NY, Mar 19, 34; m 59; c 2. SCIENCE COMMUNICATIONS, ARMS CONTROL. *Educ:* State Univ NY Stony Brook, BS, 62; Hofstra Univ, MS, 66. *Prof Exp:* Instr physics, Cold Spring Harbor High Sch, 63-67; assoc ed, The Physics Teacher, 67-72; assoc prog dir, NSF, 72-73; dean, Ctr Continuing Educ, 76-85, PROF TECHNOL

& SOC, STATE UNIV NY STONY BROOK, 76-, DIR, CTR SCI, MATH & TECHNOL EDUC, 85- *Concurrent Pos:* Ed, J Col Sci Teaching, 78-; adv bd, Physics Today, 73-76; sr assoc, NSF, 82-83. *Mem:* Am Asn Physics Teachers; Nat Sci Teachers Asn; Arms Control Asn; Am Phys Soc; fel AAAS. *Res:* Formulation of US science education policy; arms control. *Mailing Add:* Ctr Sci & Technol State Univ NY Stony Brook Stony Brook NY 11794-3733

PALEK, JIRI FRANT, BIOMEMBRANE, RED BLOOD CELL MEMBRANE SKELETON. *Educ:* Charles Univ, Prague, Czechoslovakia, MD, 58. *Prof Exp:* CHMN DEPT BIOMED RES & CHIEF DIV HEMAT & ONCOL, ST ELIZABETH'S HOSP, 78-; PROF MED, MED SCH, TUFTS UNIV, 78- *Mailing Add:* St Elizabeths Hosp 736 Cambridge St Boston MA 02135

PALEN, JOSEPH W, b Springfield, Mo, June 4, 35; m 76; c 6. CHEMICAL ENGINEERING, HEAT TRANSFER. *Educ:* Univ Mo, BS, 57; Univ Ill, MS, 65; Lehigh Univ, PhD, 88. *Prof Exp:* Chem engr, Phillips Petrol Co, 57-61, res engr, 61-63; from asst tech dir to assoc tech dir, 65-86, PRIN STAFF CONSULT, HEAT TRANSFER RES INC, ALHAMBRA, 88- *Honors & Awards:* Invited Keynote Lectr, Int Heat Transfer Conf, 86. *Mem:* Am Inst Chem Engrs; Am Sci Affil. *Res:* Industrial process heat transfer research, especially design methods for process heat transfer equipment. *Mailing Add:* 1221 Pleasantridge Dr Altadena CA 91001

PALENIK, GUS J, b Chicago, Ill, Mar 29, 33; m 59; c 3. CRYSTALLOGRAPHY. *Educ:* Ill Inst Technol, BSc, 53; Univ Southern Calif, PhD(chem), 60. *Prof Exp:* Res chemist, US Naval Weapons Ctr, 59-66; assoc prof chem, Univ Waterloo, 66-70; assoc prof, 70-74, PROF CHEM, UNIV FLA, 74- *Mem:* AAAS; Am Crystallog Asn; Am Chem Soc; Royal Soc Chem. *Res:* X-ray crystallographic studies of various drugs and metal complexes. *Mailing Add:* Dept of Chem Univ of Fla Gainesville FL 32611

PALERMO, FELICE CHARLES, b New Brunswick, NJ, Nov 8, 31; m 74; c 3. ORGANIC POLYMER CHEMISTRY. *Educ:* Rutgers Univ, BA, 59. *Prof Exp:* Prod develop chemist surg dressings, Johnson & Johnson, New Brunswick, 56-60, group leader nonwoven fabrics, Chicago, 60-63, prod develop engr, 63-65, mgr prod develop, 65-70, dir prod develop, 70-73; dir marketing, Avery Int Inc, 74-75, vpres res & develop adhesive tapes, Permacel Div, 73-83; corp dir bus develop, Seton Corp, 83; exec vpres chem & coatings group, United Merchants & Mfrs, 83-88, gen mgr, Decora Div, 88- *Mem:* Am Chem Soc; Soc Chem Indust; Am Soc Testing & Mat; AAAS; Asn Res Dirs. *Res:* Pressure-sensitive, thermoplastic, thermosetting adhesives and coatings for industrial and electrical insulation applications; nonwoven fabrics; surgical dressings. *Mailing Add:* PO Box 172 Asbury Park NJ 08802

PALESTINE, ALAN G, b Middletown, NY, Apr 30, 52; m 74; c 2. OPHTHALMOLOGY. *Educ:* Cornell Univ, BA, 74; Univ Rochester, MD, 78. *Prof Exp:* Res fel immunol, 82-85, SECT CHIEF IMMUNOL, NAT EYE INST, 85- *Mem:* Am Acad Ophthalmol; Assoc Res Vision & Ophthalmol. *Res:* Mechanisms, causes and therapy of ocular inflammatory disease in humans using clinical methods, animal models, and in vitro techniques. *Mailing Add:* B10 R10 D19 9000 Rockville Pike Bethesda MD 20817

PALEVITZ, BARRY ALLAN, b Brooklyn, NY, July 25, 44; m 67; c 3. CELL BIOLOGY, PLANT PHYSIOLOGY. *Educ:* Brooklyn Col, BS, 66; Univ Wis, PhD(bot), 71. *Prof Exp:* Fel, Stanford Univ, 71-74, res assoc biol, 73-74; asst prof biol, State Univ NY Stony Brook, 74-78; assoc prof, 78-83, PROF BOT, UNIV GA, 83- *Concurrent Pos:* Fel, NSF, 71-72 & NIH, 72-73; guest lectr, Int Bot Cong, 81; mem, Peer Rev Panels, NSF, 81-89; vis prof, Univ Siena, Italy, 87; vis fel, Austral Nat Univ, 89; res Marine Biol Lab Woodshole, 84, 86; Sci Cit Classic, ISI; Creative Res Medal, UGA, 89. *Mem:* Am Soc Cell Biol; fel AAAS; Am Soc Plant Physiol. *Res:* Analysis of the basic mechanisms responsible for motility and morphogenesis in plant cells, using optical, electron microscopic and biochemical techniques; plant cytoskeleton. *Mailing Add:* Bot Dept Univ Ga Athens GA 30602

PALEVSKY, GERALD, b New York, NY, Jan 8, 26; m 48; c 3. CIVIL ENGINEERING, ENVIRONMENTAL HEALTH. *Educ:* Va Polytech Inst, BS, 47; Columbia Univ, MS, 49, Eng ScD, 78; NY Univ, MCE, 51; Am Acad Environ Engrs, dipl, 80. *Prof Exp:* Instr civil eng, NC State Col, 47-48; instr civil & sanit engr, NY Univ, 49-51; design engr, Malcolm Pirnie Engrs, NY, 51-53; utilities engr, Brown & Blauvelt, Consult Engrs, NY, 53-55; from asst prof to prof civil eng, 68-90, evening div supvr, 74-84, EMER PROF CIVIL ENG, CITY COL NEW YORK, 90-; CONSULT ENGR, SITE PLANNING, 55- *Concurrent Pos:* From adj asst prof to adj assoc prof, NY Univ, 54-62, from res scientist to sr res scientist, 56-62; consult, USPHS, 58-65; spec lectr, Columbia Univ, 59-60; dir sewage treatment plant operators short course, Grad II, NY, 63-66; vis assoc prof, Manhattan Col, 64-67; lectr, City Col New York, 65-68; adj prof pub health engr, Dept Pub Health, Med Col, Cornell Univ, 77-; designer, aesthetically pleasing water displays for archit & sculptural constructions; expert witness & spec consult, judicial & legis hearings on-site sanit eng probs. *Mem:* Am Soc Civil Engrs; Am Water Works Asn; Water Pollution Control Fedn; Nat Soc Prof Engrs; Am Pub Health Asn; Am Pub Works Asn. *Res:* Environmental impact assessment and health problems relating to environmental control; water supply, treatment and distribution; wastewater collection, treatment and disposal; solid waste handling and resource recovery; industrial hygiene; on site domestic wastewater disposal in sensitive environmental areas of sole source aquifers and wetlands; special effects and weir and spillway sections for controlled and aesthetic requirements. *Mailing Add:* 61 Kalda Ave New Hyde Park NY 11040

PALEY, HIRAM, b Rochester, NY, Sept 9, 33; m 61; c 3. ALGEBRA. *Educ:* Univ Rochester, AB, 55; Univ Wis, MS, 56, PhD(math), 59. *Prof Exp:* Teaching asst, Univ Wis, 58-59; asst prof, 59-66, ASSOC PROF MATH, UNIV ILL, URBANA, 66- *Concurrent Pos:* Mem, Coop Proj Malaysia, Mucia-ITM-Ind Univ, 89-91. *Mem:* Am Math Soc; Math Asn Am; Sigma Xi. *Mailing Add:* 706 W California Urbana IL 61801

PALFFY-MUHORAY, PETER, b Nyiregyhaza, Hungary, May 20, 44; Can citizen; m 85; c 1. LIQUID CRYSTAL PHYSICS, NONLINEAR OPTICS. *Educ:* Univ BC, BASc, 66, MASc, 69, PhD(physics), 77. *Prof Exp:* Res fel, Dept Chem, Univ Sheffield, 81-82; hon asst prof physics, Univ BC, 83-90; instr, Capilano Col, 77-87; ADJ ASSOC PROF, DEPT PHYSICS & SR RES FEL, LIQUID CRYSTAL INST, KENT STATE UNIV, 87-, ASSOC DIR, LIQUID CRYSTAL INST, 90- *Concurrent Pos:* Vis prof, Dept Physics, Kent State Univ & KFKI Hungarian Acad Sci, 87; consult, AT&T Bell Labs, 89-90; chmn, 13th Int Liquid Crystal Conf, Vancouver, 90; guest ed, Molecular Crystals & Liquid Crystals, 90-91; co-chmn, IV Int Topical Meeting Optics Liquid Crystals, 91. *Mem:* Am Phys Soc; Can Asn Physicists; Optical Soc Am; Soc Photo-Optical Instrumentation Engrs; Sigma Xi. *Res:* Liquid crystals; nonlinear optics; pattern formation; nonequilibrium phenomena. *Mailing Add:* Liquid Crystal Inst Kent State Univ Kent OH 44242

PALFREE, ROGER GRENVILLE ERIC, b Nottingham, Eng, Jan 29, 46; Brit & Can citizen; m 83; c 2. DIFFERENTIATION ANTIGENS, LY-SIX MOLECLUES & GENES. *Educ:* Univ London, BSc, 67; London Univ, MSc, 69; McGill Univ, PhD(biol), 78. *Prof Exp:* Postdoctoral fel biochem, Queens Univ, Ont, 78-80, postdoctoral fel immunol, 80-83; asst lab mem immunol, Sloan-Kettering Inst, NY, 83-88, assoc lab mem immunol, 88-89; ASSOC PROF IMMUNOL, MCGILL UNIV, 89- *Mem:* Am Asn Immunol; Can Biochem Soc. *Res:* Structural and functional characterization of mouse differentiation antigens, with special interest in the Ly-6 molecules, the genes encoding them, and their relatives in other species. *Mailing Add:* Endocrine Lab Royal Victoria Hosp 687 Pine Ave W Montreal PQ H3A 1A1

PALFREY, THOMAS ROSSMAN, JR, b Champaign, Ill, Dec 20, 25; m 49; c 3. PHYSICS. *Educ:* Cornell Univ, BA, 49, PhD(physics), 53. *Prof Exp:* Asst physics, Cornell Univ, 49-52; from asst prof to assoc prof, 52-60, asst dean grad sch, 66-70, PROF PHYSICS, PURDUE UNIV, WEST LAFAYETTE, 60- *Mem:* Am Phys Soc. *Res:* Photonuclear effects of high energies; high energy experimental physics. *Mailing Add:* Dept Physics Purdue Univ West Lafayette IN 47907

PALIK, EDWARD DANIEL, b Elyria, Ohio, Sept 21, 28; m 57; c 3. SEMICONDUCTORS. *Educ:* Ohio State Univ, BSc, 50, MSc, 52, PhD, 55. *Prof Exp:* Asst prof, Ohio State Univ, 55-56; fel, Univ Mich, 56-57 & Ohio State Univ, 57-58; res physicist, US Naval Res Lab, 58-88; AT UNIV MD, 88- *Mem:* Optical Soc Am; Am Phys Soc. *Res:* Magnetooptical properties of solids; infrared properties of semiconductors; brillouin scattering. *Mailing Add:* 904 Pocahontas Dr Ft Washington MD 20744

PALIK, EMIL SAMUEL, b Elyria, Ohio, Apr 25, 23; m 53; c 2. PHYSICAL CHEMISTRY. *Educ:* Ohio State Univ, BSc, 49, MSc, 51. *Prof Exp:* Res assoc chem, Ohio State Univ Res Found, 51-53; sr phys chemist, Lamp Div, Gen Elec Co, 53-85; RETIRED. *Mem:* Am Soc Testing & Mat; Fine Particle Soc. *Res:* Methods for the particle size measurement of fine powders of sub-sieve size and for the specific surface area determination of powders and related topics; scanning electron microscopy and energy-dispersive x-ray microanalysis. *Mailing Add:* 863 Woodview Dr Ashland OH 44805-9254

PALILLA, FRANK C, b New York, NY, Feb 10, 25; m 45; c 2. INORGANIC CHEMISTRY. *Educ:* Brooklyn Col, BS, 48; Polytech Inst Brooklyn, MS, 52. *Hon Degrees:* PhD, Brooklyn Col, 74. *Prof Exp:* Res chemist, Sylvania Elec Prod, Inc, 48-59; res engr, Gen Tel & Electronics Labs, Inc, 59-61, eng specialist, 61-64, sr eng specialist, 64-66, mgr luminescent mat, 66-70, mgr optoelectronics mat, 70-72, sr scientist & prin investr explor res, 72-82; prin mem tech staff, Mat Sci Labs, 82-87; CONSULT MAT SCI, 87- *Honors & Awards:* Annual Award, Electrochem Soc, 71. *Mem:* Electrochem Soc; Am Chem Soc; Sigma Xi (pres, 76); Am Inst Chemists; Am Ceramic Soc; Mat Res Soc. *Res:* Electronic and photoconductive luminescent, magnetic and optical materials; research and technical management in high energy lithium batteries and catalytic materials related to energy and pollution problems; development of high temperature ceramics and composites; development of electrical ceramics. *Mailing Add:* 13 Hickory Hill Lane Framingham MA 01701

PALINCSAR, EDWARD EMIL, b Chicago, Ill, May 23, 20; m 57; c 3. BIOLOGY, PHYSIOLOGY. *Educ:* Roosevelt Univ, BS, 52; Northwestern Univ, MS, 54, PhD(biol), 57. *Prof Exp:* Asst, Northwestern Univ, 55-56; from asst prof to prof gen physiol, 57-74, PROF BIOL, LOYOLA UNIV CHICAGO, 74- *Concurrent Pos:* Lalor fel, 59. *Mem:* AAAS; Soc Protozool; Am Soc Cell Biol. *Res:* Ageing and nucleic acid metabolism; physiology of cell division; protein synthesis; eutely; metabolic effects of cell division blockage; action of carcinogenic and carcinostatic agents; ageing and lysosomes; ageing and physiology of cnidaria; biological clocks. *Mailing Add:* Dept Biol Loyola Univ 6525 N Sheridan Rd Chicago IL 60626

PALISANO, JOHN RAYMOND, b Buffalo, NY, Mar 6, 47; m 70; c 1. VIROLOGY, PATHOGENESIS. *Educ:* Univ Tenn, BS, 69, PhD(biol), 75. *Prof Exp:* Fel, Med & Health Sci Div, Oak Ridge Assoc Univ, Oak Ridge, 75-76; res assoc dept path, St Luke Hosp, Cleveland, Ohio, 76-78, dir Virol Lab, Dept Med Res, 78-85; ASST PROF DEVELOP BIOL, MICROBIOL & MOLECULAR BIOL, EMORY & HENRY COL, VA, 85- *Mem:* Sigma Xi; AAAS; Am Soc Cell Biol; Am Microbiol Soc; Electron Micros Soc Am; Va Acad Sci; Nat Asn Adv Health Professions Inc. *Res:* Studying the origin and role of multilamellar endoplasmic reticulum to investigate the dichotomy existing between tumor and normal cells. *Mailing Add:* Dept Biol Emory & Henry Col Emory VA 24327

PALIWAL, BHUDATT R, b Khewra, India, Aug 12, 38; US citizen; m 71; c 3. RADIOLOGICAL PHYSICS. *Educ:* Sri Aurobindo Int Educ Ctr, BSc, 59, MS, 61; Univ Tex, Houston, PhD(biophysics), 73. *Prof Exp:* Asst prof med physics & radiol, 73-77, ASSOC PROF HUMAN ONCOL, UNIV WIS-MADISON, 77- *Concurrent Pos:* Fel Sri Aurobindo Int Educ Ctr, 59-62, WHO, 64-65 & Int Atomic Energy Agency, 70-72; prof human oncol, Dept Radiation Oncol, Univ NC. *Mem:* Am Asn Phys Med; Am Col Radiol. *Res:* High energy radiation applications in radiation therapy of cancer patient; hyperthermia treatment of cancer. *Mailing Add:* Dept Radiation Oncol Univ Wis Madison WI 53792

PALIWAL, YOGESH CHANDRA, plant virology, plant pathology; deceased, see previous edition for last biography

PALKA, JOHN MILAN, b Paris, France, July 29, 39; US citizen; m 60; c 2. NEUROBIOLOGY. *Educ:* Swarthmore Col, BA, 60; Univ Calif, Los Angeles, PhD(zool), 65. *Prof Exp:* Fulbright vis lectr zool, Univ Poona, India, 63, Sri Venkateswara Univ, India, 65-66; asst prof, Rice Univ, 66-69; from asst prof to assoc prof, 69-78, PROF ZOOL, UNIV WASH, 78- *Concurrent Pos:* Guggenheim fel, Cambridge Univ, Eng, 75-76; instr, Marine Biol Lab, Woods Hole, 80, 82, 84, Cold Spring Harbor Lab, 86, 88, 89. *Mem:* Fel AAAS; Am Soc Zool; Soc Neurosci. *Res:* Anatomy and physiology of insect sensory systems; nervous system during development and metamorphosis. *Mailing Add:* Dept Zool Univ Wash Seattle WA 98195

PALKE, WILLIAM ENGLAND, b Youngstown, Ohio, Feb 5, 41; m 62; c 2. CHEMICAL PHYSICS. *Educ:* Calif Inst Technol, BS, 62; Harvard Univ, PhD(chem physics), 66. *Prof Exp:* A A Noyes res fel, Calif Inst Technol, 66-68; asst prof chem, 68-74, assoc prof, 74-80, PROF THEORET PHYS CHEM, UNIV CALIF, SANTA BARBARA, 80- *Concurrent Pos:* Sabbatical leave, prog officer chem physics, NSF, 81-82. *Mem:* Am Phys Soc. *Res:* Theoretical applications of quantum mechanics to problems of chemical interest. *Mailing Add:* Dept Chem Univ Calif Santa Barbara CA 93106

PALL, DAVID B, RESEARCH ADMINISTRATION. *Prof Exp:* FOUNDER & CHMN, PALL CORP. *Honors & Awards:* Nat Med Technol, 90. *Mailing Add:* Pall Corp 30 Sea Cliff Ave Glen Cove NY 11542

PALL, MARTIN L, b Montreal, Que, Jan 20, 42; m 70. BIOCHEMISTRY, GENETICS. *Educ:* Johns Hopkins Univ, BA, 62; Calif Inst Technol, PhD(biochem, genetics), 68. *Prof Exp:* Asst prof biol, Reed Col, 67-72; from asst prof to assoc prof genetics & biochem, 72-82 PROF GENETICS, CELL BIOL & BIOCHEM, WASH STATE UNIV, 82- *Concurrent Pos:* Vis assoc prof, Yale Univ Sch Med, 79-80; mem res comt, Am Heart Asn, Washington, 79-84; vis prof, Univ Calif, San Francisco, 85. *Mem:* Genetics Soc Am; Am Soc Biol Chem. *Res:* Biochemical genetics; biochemistry of regulation; cyclic nucleotides and GTP in cell regulation. *Mailing Add:* Dept Genetics Wash State Univ Pullman WA 99164-4350

PALLADINO, NUNZIO J(OSEPH), b Allentown, Pa, Nov 10, 16; m 45; c 3. NUCLEAR ENGINEERING. *Educ:* Lehigh Univ, BS, 38, MS, 39. *Hon Degrees:* DEng, Lehigh Univ, 64. *Prof Exp:* Steam turbine design engr, Westinghouse Elec Corp, 39-42, 45-46, sect mgr core design, Atomic Power Div, 50-51, subdiv mgr, 52-59; nuclear reactor designer, Oak Ridge Nat Lab, 46-48 & Argonne Nat Lab, 48-50; prof nuclear eng & head dept, Pa State Univ, 59-66, dean, Col Eng, 66-81; chmn, Nuclear Regulatory Comn, 81-86; RETIRED. *Concurrent Pos:* Past mem US Adv Comt Reactor Safeguards; past chmn, Pa Adv Comt Atomic Energy. *Honors & Awards:* Prime Movers Award, Am Soc Mech Engrs, 56; Arthur Holly Compton Award, AM Nuclear Soc, 82. *Mem:* Nat Acad Eng; Am Soc Mech Engrs; Am Soc Eng Educ; Am Nuclear Soc (past pres). *Res:* Experimental and analytical thermal and hydraulic problems encountered in the design of nuclear reactor cores; mechanical problems associated with the design of nuclear reactor cores. *Mailing Add:* 333 W Park Ave State College PA 16803

PALLADINO, RICHARD WALTER, b Cleveland, Ohio, Apr 20, 33; m 60; c 3. PLASMA PHYSICS DIAGNOSTICS, LASER APPLICATIONS. *Educ:* Case Inst Technol, BS, 55. *Prof Exp:* Prof tech staff, 55-80, proj physicist, 81-84, LEAD PHYSICIST, PLASMA PHYSICS LAB, PRINCETON UNIV, 84- *Mem:* Am Phys Soc. *Res:* Diagnostics; spectroscopy; optics; laser technology; far infrared techniques; fiber optics; detector calibration; radiometry. *Mailing Add:* One Hawthorne Ct Hamilton NJ 08690

PALLADINO, WILLIAM JOSEPH, b Camden, NJ, June 9, 47; m 78; c 2. POLYURETHANES, EPOXIES. *Educ:* Tex A&M Univ, BS, 69. *Prof Exp:* Res chemist polyimides, Ciba Geigy, Ardsley, NY, 70-74, develop chemist urethanes, REN Plastics Div, 74-75, tech mgr tooling, epoxies & polyurethanes, 75-76 & tech mgr commun, 76-78; tech dir urethanes & epoxy, 78-83, VPRES RES & DEVELOP, CONAP, INC, OLEAN, NY, 83- *Mem:* Soc Plastic Engrs; Am Soc Testing & Mat; Soc Advan Mat & Process Eng; Am Chem Soc; Prod Develop & Mgt Asn. *Res:* Materials science engineering. *Mailing Add:* Conap Inc 1405 Buffalo St Olean NY 14760

PALLAK, MICHAEL, b Cleveland, Ohio, Sept 15, 42. PSYCHOLOGY. *Educ:* Ohio State Univ, BA, 64; Yale Univ, MA, 66, PhD(psychol), 68. *Prof Exp:* Dept exec dir, Am Psychol Asn, 76-79, exec dir, 79-85; res prof & sr fel, Vanderbilt Inst Public Policy, 86; VPRES RES, BIODYNE INC, 88- *Concurrent Pos:* Vis prof, Georgetown Univ, 86-88. *Mem:* AAAS; Sigma Xi; fel Am Psychol Asn. *Mailing Add:* 705 Winchester Dr Burlingame CA 94010

PALLANSCH, MICHAEL J, b St Joseph, Minn, Nov 17, 18; m 53; c 7. BIOCHEMISTRY. *Educ:* St John's Univ, Minn, BS, 48; Univ Minn, Minneapolis, PhD(biochem), 53. *Prof Exp:* Res asst, Univ Minn, Minneapolis, 49-51, asst scientist surg, 51-53; res fel, Univ Minn, St Paul, 53-55; from chemist to supvry chemist dairy prod, 55-73, staff scientist processing technol, 73-74, asst adminr, 74-77, CHIEF SCIENTIST SCI & EDUC, AGR RES SERV, USDA, 77- *Honors & Awards:* Borden Award, Am Dairy Sci Asn, 63. *Mem:* Am Chem Soc; Sigma Xi; AAAS. *Res:* Food and fiber processing and distribution; nutrition; health and safety. *Mailing Add:* 509 N Kenmore St Arlington VA 22201

PALLARDY, STEPHEN GERARD, b St Louis, Mo, Mar 19, 51; m 81; c 2. WATER RELATIONS, PHYSIOLOGICAL PLANT ECOLOGY. *Educ:* Univ Ill, BS, 73, MS, 75; Univ Wis, PhD(forestry & bot), 78. *Prof Exp:* Asst prof forestry, Kans State Univ, 79-80; asst prof, 80-84, assoc prof, 84-89, PROF FORESTRY, UNIV MO, 89- *Concurrent Pos:* Assoc ed, Forest Science, 83- *Mem:* Am Soc Plant Physiologists; Ecol Soc Am. *Res:* Physiological and ecological responses of plants to water stress; comparative aspects of inter and intra specific variation in water relations of forest trees. *Mailing Add:* Sch Natural Resources 1-31 Agr Bldg Univ Mo Columbia MO 65211

PALLAVICINI, MARIA GEORGINA, b San Francisco, Calif, Sept 20, 52; m; c 3. HEMATOLOGY, TUMOR BIOLOGY. *Educ:* Univ Calif, Berkeley, BA, 73; Univ Utah, PhD(pharmacol), 77. *Prof Exp:* Fel radiation biol, Ont Cancer Inst & Ont Cancer Treat & Res Found, 77-78, BIOMED SCIENTIST, HEMOPOIESIS CELL KINETICS, LAWRENCE LIVERMORE LAB, UNIV CALIF, 78- *Concurrent Pos:* Adj assoc prof, dept lab med, Univ Calif, San Francisco; NIH grantee, 78- *Mem:* Cell Kinetics Soc (treas, 85-87); Asn Advan Cancer Res; Int Soc Exp Hemat; Soc Anal Cytology; AAAS. *Res:* Bone marrow transplantation; in vivo kinetics of hemopoietic stem cells; detection of low frequency subpoulations; in vitro DNA amplification; fluorescence in situ hybridization; multivariate flow cytometriz analysis and sorting. *Mailing Add:* Biomed & Environ Res Div L-452 Lawrence Livermore Nat Lab Livermore CA 94550

PALLER, MARK STEPHEN, b Cleveland, Ohio, Dec 20, 52. NEPHROLOGY. *Educ:* Northwest Univ, BS, 74, MD, 76. *Prof Exp:* Resident internal med, Case Western Reserve Univ, 76-79; fel nephrol, Univ Colo, 79-82; asst prof med, 82-87, ASSOC PROF MED, UNIV MINN, 87- *Mem:* Am Soc Nephrol; Am Fedn Clin Res; Cent Soc Clin Res; Am Col Physicians; Coun High Blood Pressure Res; Int Soc Nephrol; Am Physiol Soc. *Res:* Renal physiology; mechanisms of ischemic renal injury; cyclosporine nephrotoxicity; hypertension. *Mailing Add:* Box 736 UMHC Univ Minn Minneapolis MN 55455

PALLETT, DAVID STEPHEN, b Watertown, Wis, June 14, 38; m 68; c 3. ACOUSTICS. *Educ:* Ripon Col, AB, 61; Pa State Univ, University Park, MS, 64, PhD(eng acoust), 72. *Prof Exp:* Res asst acoust & vibration, Ord Res Lab, Pa State Univ, 64-73; mem appl acoust staff, 72-76, chief appl acoustics, 76-78, chief, Acoust Eng Div, 78-80, physical scientist, 80-87, MGR AUTOMATED RECOGNITION GROUP, INST COMPUT SCI & TECHNOL, NAT BUR STANDARDS, 87- *Concurrent Pos:* Consult, Reliance Elec Co, 70, Jordan-Kitts Music, 84; Dept Com Sci Fel, 77-78. *Mem:* Acoust Soc Am; Inst Elec & Electronics Engrs. *Res:* Acoustic radiation and structural vibration properties of complex structures; noise reduction; speech recognition and synthesis; signal processing. *Mailing Add:* 19348 Hempstone Ave Poolesville MD 20837

PALLMANN, ALBERT J, b Wiesbaden, Ger, Dec 12, 26; m 58; c 1. ATMOSPHERIC PHYSICS, ENVIRONMENTAL SCIENCES. *Educ:* Univ Cologne, Doktorand, 53, PhD(meteorol, physics, math), 58. *Prof Exp:* Res assoc meteorol, Univ Cologne, 54-58; trop meteorologist, Ministry of Defense, El Salvador, 58-59, head anal ctr, 59-62; from asst prof to assoc prof geophys & meteorol, 63-69, PROF METEOROL, ST LOUIS UNIV, 69- *Concurrent Pos:* Consult, Salvadorean Govt, 58-62; prof, Air Force Hq, Nat Naval Sch, El Salvador, 59-62; consult, Univ Corp Atmospheric Res, Colo, 64-67, St Louis Univ sci rep, 67-80; consult, McDonnell Aircraft Corp, 66-67; prin investr res projs sponsored by govt agencies, 67- *Mem:* Am Meteorol Soc; Am Geophys Union; Sigma Xi. *Res:* Radiative transfer physics of planetary atmospheres; air pollution thermal energetics and climate; solar energy meteorological research and training project; encounter of science and technology with theology. *Mailing Add:* 9 Middlesex Dr St Louis MO 63144

PALLONE, ADRIAN JOSEPH, b Lille, France, Apr 8, 28; US citizen; m 54; c 4. APPLIED MECHANICS, AERONAUTICAL ENGINEERING. *Educ:* Polytech Inst NY, BE, 52, ME, 53, PhD(appl mech), 59. *Prof Exp:* Res scientist, Bell Aerospace Syst, 54-55; res assoc, Polytech Inst NY, 55-59; chief, Exp & Theoret Aerodyn Sect, 59-63, mgr, Aerophysics Dept, Avco Syst Div, 63-66; prof aerospace Eng, NY Univ, 66-67; dir technol, Avco Syst Div, 67-77, chief scientist, 77-87; CONSULT, 87- *Mem:* Am Inst Aeronaut & Astronaut; Sigma Xi; NY Acad Sci; fel,Am Inst Aeronaut 7 Astronaut. *Res:* Theoretical and experimental work in gas dynamics; boundary layer theories; reentry physics; plasma chemistry, utilizing electric arc to achieve reaction environment. *Mailing Add:* One Rennie Dr Andover MA 01810

PALLOS, FERENC M, b Hungary, May 24, 33; m 60; c 2. ORGANIC CHEMISTRY, PESTICIDE CHEMISTRY. *Educ:* Swiss Fed Inst Technol, Dipl Ing Chem, 58, DSc(org chem), 62. *Prof Exp:* Res chemist, Fluka, AG, Switz, 58-60; fel, Swiss Fed Inst Technol, 62-63; res chemist, 63-65, sr res chemist, 65-73, RES ASSOC, WESTERN RES CTR, STAUFFER CHEM CO, 73- *Honors & Awards:* IR-100 Award, Indust Res Inst, 74. *Mem:* Am Chem Soc; Swiss Chem Asn; Soc Ger Chem. *Res:* Herbicide antidotes; insect hormones; herbicides; insecticides; drug research. *Mailing Add:* ICI Americas Inc 1200 S 47th St Richmond CA 94804

PALLOTTA, BARRY S, b Brooklyn, NY, Jan 18, 51. PHYSIOLOGY, BIOPHYSICS. *Educ:* State Univ NY Stony Brook, BS, 73; Univ Vt, PhD(physiol & biophysics), 78. *Prof Exp:* Fel physiol & biophysics, Sch Med, Univ Miami, 78-; ASSOC PROF, PHARM DEPT, SCH MED, UNIV NC, 88- *Mem:* Soc Neurosci. *Res:* Synaptic transmission, especially neuromuscular junction; acetylcholine-receptor channel kinetics; mechanisms of transmitter release. *Mailing Add:* Dept Pharm 1020 Flob 231 H Sch Med Univ NC Chapel Hill NC 27514

PALLOTTA, DOMINICK JOHN, b Syracuse, NY, Oct 6, 42; m 67; c 2. BIOLOGY. *Educ:* State Univ NY Col Cortland, BS, 65; State Univ NY Col Buffalo, PhD(biol), 71. *Prof Exp:* Fel biol, Rice Univ, 70-71; asst prof, 71-75, ASSOC PROF BIOL, LAVAL UNIV, 75- *Concurrent Pos:* Nat Res Coun Can fel, 71-73. *Mem:* Am Soc Cell Biol; Can Soc Cell Biol. *Res:* Chromosomal proteins and the regulation of RNA synthesis; developmental biology. *Mailing Add:* 161 Presque Ile St Nicolas PQ G0S 2Z0 Can

PALLUCONI, FRANK DON, b Iron Mountain, Mich, July 8, 39; m 62; c 2. REMOTE SENSING, PLANETARY SCIENCE. *Educ:* Mich Tech Univ, BS, 61; Pa State Univ, MS, 63. *Prof Exp:* Sr engr planetary sci, 69-73, mem tech staff, 74-78, co-investr, 76-78, team leader Mars IR Mapping, 78-80, asst mgr climate res, NASA HQ, 80-81, MISSION SCIENTIST MARS GEOCHEM ORBITER, JET PROPULSION LAB, 81- *Mem:* Am Astron Soc; AAAS; Am Geophys Union. *Res:* Remote sensing of planetary surfaces and atmospheres at visible, infrared and radio wave lengths; interpretation of these observations and spacecraft instrument development. *Mailing Add:* Earth & Space Sci Dev JPL 4800 Oak Grove Dr 183-501 Pasadena CA 91109

PALM, ELMER THURMAN, b Roseburg, Ore, Mar 24, 27; wid; c 4. PLANT PATHOLOGY. *Educ:* Ore State Col, BS, 51, MS, 52, PhD(plant path), 58. *Prof Exp:* Asst plant path, Exp Sta, Ore State Col, 49-52 & 56-57 & Boyce Thompson Inst Plant Res, 52-56; sr plant pathologist, Crop Protection Inst, NH, 57-63; mgr, Biol Res Sect, Velsicol Chem Corp, 63-69; res plant physiologist, Growth Sci Ctr, Int Minerals & Chem Corp, 69-73; consult, 73-74; res & develop rep, Am Hoechst Corp, 74-85; vpres opers, 85-87, PRES, COUNTRY GARDENER ENTERPRISES, 87-; MGR, N CENT FIELD STA, 85- *Mem:* Am Soc Phytopath; Weed Sci Soc Am; Entom Soc Am. *Res:* Development of agricultural products for both domestic and international use; agronomy. *Mailing Add:* 212 Spring Prairie Rd Burlington WI 53105

PALM, JOHN DANIEL, b Missoula, Mont, Sept 5, 24; m 47; c 3. HUMAN GENETICS, BIOLOGY. *Educ:* Gustavus Adolphus Col, BA, 48; Univ Minn, MA, 50, PhD(zool, bact), 54. *Prof Exp:* Res assoc, Dight Inst Human Genetics, Univ Minn, 51-53; sr scientist biol res, Gen Mills, Inc, Minn, 53-56; assoc prof biol, Wis State Univ, Oshkosh, 52-62; assoc prof biol, St Olaf Col, 62, prof; RETIRED. *Concurrent Pos:* Resident res assoc, Biomed Div, Argonne Nat Lab, 66-67; sabbatical, Soya Prod & Res Assoc, 75- *Res:* Developmental physiology and physiological genetics of human neuropsychological diseases. *Mailing Add:* Dept Biol St Olaf Col Northfield MN 55057

PALM, MARY EGDAHL, b Minneapolis, Minn, Jan 27, 54; m 76; c 2. MYCOLOGY. *Educ:* St Olaf Col, BS, 76; Univ Minn, MSc, 79, PhD(plant path/mycol), 83. *Prof Exp:* Postdoctoral fel, Dept Plant Path, Univ Minn, 83-84; MYCOLOGIST, ANIMAL & PLANT HEALTH INSPECTION SERV, USDA, 84- *Concurrent Pos:* Chair, Mycol Comt, Am Phytopath Soc, 88-90, pub comt, Mycol Soc Am, 89-92; rep AAAS, Mycol Soc Am, 88-91, counr, 87-88. *Mem:* Mycol Soc Am (secy, 91-94); Am Phytopath Soc; AAAS. *Res:* Taxonomy and biology of plant pathogenic fungi with emphasis on quarantine significant fungi, especially Coelomycetes. *Mailing Add:* USDA Animal & Plant Health Inspection Serv Rm 329 B-011A BARC-W Beltsville MD 20705-2350

PALM, SALLY LOUISE, CELL BIOLOGY. *Educ:* Univ Minn, PhD(pathobiol), 84. *Prof Exp:* ASSOC, UNIV MINN, 85- *Res:* Extracellular matrix; cellular neurobiology. *Mailing Add:* 6-159 Jackson Univ Minn 321 Church St SE Minneapolis MN 55455

PALM, WILLIAM JOHN, b Baltimore, Md, Mar 1, 44; m 68; c 3. ROBOT ENGINEERING, DYNAMIC SYSTEMS. *Educ:* Loyola Col, Md, BS, 66; Northwestern Univ, Evanston, PhD(mech eng & astronaut sci), 71. *Prof Exp:* From asst prof to assoc prof, 71-87, PROF MECH ENG & APPL MECH, UNIV RI, 87- *Concurrent Pos:* Consult, Naval Syst Eng, 76-; dir, Robotics Res Ctr, Univ RI, 85- *Mem:* Am Soc Mech Engrs; Inst Elec & Electronics Engrs. *Res:* Control systems; optimization techniques; robotics; writings include two texts in control systems engineering. *Mailing Add:* Dept Mech Eng Univ RI Kingston RI 02881-0805

PALMADESSO, PETER JOSEPH, b Newark, NJ, Sept 5, 40; m 67; c 3. PHYSICS. *Educ:* St Peter's Col, NJ, BS, 61; Rutgers Univ, New Brunswick, MS, 63; Stevens Inst Technol, PhD(physics), 70. *Prof Exp:* Nat Res Coun resident res assoc, Goddard Space Flight Ctr, NASA, Md, 70-72; fel, Ctr Theoret Physics, Univ Md, College Park, 72-74; MEM STAFF, NAVAL RES LAB, 74- *Mem:* Am Phys Soc; Am Geophys Union. *Res:* Nonlinear wave phenomena in plasmas; space plasma physics; stability theory. *Mailing Add:* Naval Res Lab-4780 4555 Overlook Ave SW Washington DC 20375

PALMATIER, ELMER ARTHUR, b Tacoma, Wash, Sept 23, 12; m 37; c 3. BOTANY. *Educ:* Univ Nebr, BSc, 35, MSc, 37; Cornell Univ, PhD(bot), 43. *Prof Exp:* Asst bot, Univ Nebr, 35-37; from asst to instr bot, Cornell Univ, 37-42; from instr to assoc prof, 42-59, instr geog, Army Specialized Training Prog, 44-45, dir conserv workshop, 58, PROF BOT, UNIV RI, 59- *Concurrent Pos:* Mem staff, Nebr Dept Conserv, 36-37 & RI Dept Agr & Conserv, 48-52; US State Dept exchange prof, Univ Baghdad, 61-62; assoc prog dir, NSF, 65-66; vis prof Duke Univ, 70. *Mem:* AAAS; Bot Soc Am; Ecol Soc Am. *Res:* Floral morphology and anatomy of Saxifragaceae; marine and fresh water algae; aquatic seed plants; duck food plants; development of vegetation; salt marsh ecology. *Mailing Add:* PO Box 271 Kingston RI 02881

PALMATIER, EVERETT DYSON, b Winnipeg, Man, Mar 31, 17; nat US; m 41; c 2. PHYSICS. *Educ:* Univ Man, BSc, 38; Cornell Univ, PhD(physics), 51. *Prof Exp:* Asst physics, Cornell Univ, 38-40 & 46-48; from asst prof to prof, 49-59, chmn dept physics, 56-65, vchancellor advan studies & res, 65-69, KENAN PROF PHYSICS, UNIV NC, CHAPEL HILL, 59-, ALUMNI DISTINGUISHED PROF FRESHMAN INSTR, 69- *Concurrent Pos:* NSF sr fel, Univ Bristol, 60-61. *Mem:* Am Phys Soc. *Res:* Cosmic radiation; nuclear physics. *Mailing Add:* 208 Glenhill Lane Chapel Hill NC 27514

PALMBLAD, IVAN G, b Gresham, Ore, May 21, 38. ECOLOGY, EVOLUTION. *Educ:* Portland State Col, BS, 60; Univ Wash, Seattle, PhD(ecol), 66. *Prof Exp:* Instr gen sci, Portland State Col, 60-62; assoc biol, Univ Wash, 65-66; from asst prof to assoc prof bot, 66-73, assoc prof biol, 73-80, PROF BIOL, UTAH STATE UNIV, 80- *Mem:* Ecol Soc Am; Audubon Soc Am; Am Soc Naturalists. *Res:* Plant competition; germination polymorphisms; plant-herbivore interaction. *Mailing Add:* Dept Biol Utah State Univ Logan UT 84322

PALMEDO, PHILIP F, b New York, NY, Mar 11, 34; m 61; c 2. NUCLEAR PHYSICS, SYSTEMS ANALYSIS. *Educ:* Williams Col, BA, 56; Mass Inst Technol, MS, 58, PhD(nuclear eng), 62. *Prof Exp:* Instr nuclear eng, Mass Inst Technol, 60-61; vis researcher reactor physics, Saclay Nuclear Res Ctr, France, 62-63; asst physicist, Nuclear Eng Dept, Brookhaven Nat Lab, 64-66, form assoc physicist to physicist, Dept Appl Sci, 66-79, head, Energy Pol Analysis Div, 76-79; CHMN, INT RESOURCES GROUP, 79- *Concurrent Pos:* Assoc ed, Energy Systs & Policy, 74-; chmn bd, Interfin, 86. *Res:* Reactor physics; energy systems and policy analysis; regional energy planning; environmental assessment; energy in developing countries. *Mailing Add:* Four Piper Lane St James NY 17780

PALMEIRA, RICARDO ANTONIO RIBEIRO, b Recife, Brazil, Sept 6, 30; m 61. SPACE SCIENCE, COSMIC RAY PHYSICS. *Educ:* Recife Univ, BS, 53; Mass Inst Technol, PhD(physics), 60. *Prof Exp:* From asst prof to assoc prof cosmic rays, Brazilian Ctr Phys Res, 53-63; Nat Acad Sci-NASA fel, Goddard Space Flight Ctr, 63-66; asst prof cosmic rays, Univ Tex, Dallas, 66-72; prof space sci, Inst Space Sci, Brazil, 72-77; prog supvr, 77-78, RES ASSOC, SEISCOM DELTA, INC, 78- *Concurrent Pos:* Vis prof, Carnegie Inst Dept Terrestrial Magnetism, 61. *Mem:* Am Phys Soc; Am Geophys Union; Phys Soc Brazil; Int Astron Union; Astron Soc Brazil. *Res:* Radioastronomy; high energy physics; astrophysics; gamma ray astronomy. *Mailing Add:* 4045 Linkwood Apt 431 Houston TX 77025

PALMER, ALAN, b Newcastle, Eng, May 2, 36; US citizen; m 63; c 3. EPIDEMIOLOGY, HEALTH SURVEILLANCE. *Educ:* San Francisco State Univ, BA, 64; Univ Calif, Los Angeles, MPH, 68; Univ Utah, PhD(health sci), 75. *Prof Exp:* Physiologist/epidemiologist, Respiratory Dis Control Prog, Bur State Serv, Med Syst Develop Lab, Heart Dis Control Prog & Nat Ctr for Health Statist, Nat Health Surv, 64-69; epidemiologist, Western Area Occup Health Lab, Salt Lake City, 70-74; actg chief morbidity field surv, Med Invest Br, Div Field Studies & Clin Invest, Nat Inst Occup Safety & Health, 74-78; sr epidemiologist & mgr health surveillance group, Ctr Occup & Environ Safety & Health, SRI Int, 78-80; PRES, PALMER ASSOCS, 81- *Concurrent Pos:* Cardio-pulmonary adv, Nat Health Surv, Nat Ctr Health Statist, 64-; consult, Bilateral Occup Health Prog between US, Poland & Yugoslavia, 73-; consult, Emory Univ Respiratory Screening Prog & Stanford Res Inst, Ctr for Occup & Environ Safety & Health on Morbidity Surv, 74-; actg chief, Med Invest Br, Div Field Studies & Clin Invest, Nat Inst Occup Safety & Health, USPHS, Dept Health Educ & Welfare, 74-; asst prof exp med, Pulmonary Div, Univ Cincinnati, 74-; proj dir, Nat Study Prevalence Byssinosis in Cotton Gin & Cotton Warehouse Workers, 78; consult to indust, respiratory dis surveillance. *Mem:* Am Indust Hyg Asn; NY Acad Sci. *Res:* Beauty and household aerosols, particularly hairspray and respiratory disease in beauticians; silicosis and respiratory disease in brickworkers; morbidity of workers in a leadsmelter due to exposures to lead, cadmium arsenic and sulfur dioxide; byssinosis in cotton gin and warehouse workers; mortality study in workers in coal hydrogenation process; respiratory health surveillance; program design and implementation in the paper and pulp, metal and coal mining, and electronics industries; health surveillance & data quality control of chemical workers in 40 plants in US; conducted survey of mine workers for respiratory diseases and established screening program for gold mine workers in Mali WAfrica. *Mailing Add:* Palmer Assoc Box 118 Star Rte 2 La Honda CA 94020

PALMER, ALAN BLAKESLEE, b Syracuse, NY, May 23, 34; m 59. PHYSICAL CHEMISTRY. *Educ:* Syracuse Univ, BA, 56, PhD(phys chem), 63. *Prof Exp:* Res chemist, Exp Sta, Indust & Biochem Dept, 63-68, develop & serv rep, Indust Chem Dept, New York, 71-72, prod mgr, Indust Specialty Div, Del, 72-74, res supvr, Indust Chem Dept, 74-77, tech supt, Chem, Dyes & Pigments Dept, 77-78, prod supt, Chem & Pigments Dept, 78-79, mgr occup safety & health, 79-81, gen supt, 81-82, bus mgr, 82-84, gen supt, 84-85, mgr Safety Health & Environ, Chem & Pigments Dept, 85-90, MGR PROD SAFETY MGT, CHEMICALS, E I DU PONT DE NEMOURS & CO, INC, 90- *Mem:* AAAS; Am Chem Soc; Sigma Xi. *Res:* Infrared spectroscopy; colloid chemistry; high temperature inorganic chemistry; hydrogen peroxide process chemistry; azo-polymerization initiators. *Mailing Add:* 1207 Greenway Rd Wilmington DE 19803

PALMER, ALLISON RALPH, b Bound Brook, NJ, Jan 9, 27; m 49; c 5. INVERTEBRATE PALEONTOLOGY. *Educ:* Pa State Univ, BS, 46; Univ Minn, PhD(geol), 50. *Prof Exp:* Sci aide, Bur Econ Geol, Tex, 47-48; Cambrian paleontologist & stratigrapher, US Geol Surv, 50-66; prof paleont, State Univ NY Stony Brook, 66-80, chmn, Dept Earth & Space Sci, 74-77; COORDR EDUC PROG, GEOL SOC AM, 80-, CENTENNIAL SCI PROG COORDR, 88- *Concurrent Pos:* Pres, Cambrian Subcomn, Int Stratig Comn, 72-84; US Nat Comt Geol, 83-86; pres, Inst Cambrian Studies, 84- *Honors & Awards:* Walcott Medal, Nat Acad Sci, 67. *Mem:* Fel AAAS; Paleont Soc (pres, 83); fel Geol Soc Am; Am Asn Petrol Geologists. *Res:* Cambrian stratigraphy, trilobites and phosphatic brachiopods. *Mailing Add:* Inst Cambrian Studies 445 N Cedarbrook Rd Boulder CO 80304

PALMER, ARTHUR N, b Pittsfield, Mass, Aug 8, 40; m 66. GEOCHEMISTRY. *Educ:* Williams Col, BA, 62; Ind Univ, MA, 65, PhD(hydrogeol), 69. *Prof Exp:* Assoc prof, 67-80, PROF HYDROLOGY, STATE UNIV NY COL ONEONTA, 80- *Concurrent Pos:* Res Found fel & grant-in-aid, State Univ NY, 74; vis prof, Karst Geol, Western Ky Univ, 80-89; hydrol consult, 70-91. *Mem:* Cave Res Found; hon mem Nat Spelol Soc; Geol Soc Am; Brit Cave Res Asn; Nat Water Well Asn. *Res:* Karst hydrology; geomorphology; geochemistry; Mississippian stratigraphy of midwest United States; weathering of limestones; carbonate petrology. *Mailing Add:* Dept Earth Sci State Univ of NY Col Oneonta NY 13820-4015

PALMER, BEVERLY B, b Bay Village, Ohio, Nov 22, 45; m 67; c 1. BEHAVIORAL MEDICINE, HUMAN SEXUALITY. *Educ:* Univ Mich, BA, 66; Ohio State Univ, MA, 69, PhD(psychol), 72. *Prof Exp:* Res asst, Dept Psychiat, Ohio State Univ, 66-68, admin assoc, Admis Dept, 69-70; res psychologist, Health Serv Res Ctr, Univ Calif, Los Angeles, 71-77; PROF PSYCHOL, DEPT PSYCHOL, CALIF STATE UNIV, 73- *Concurrent Pos:* Bd dir, South Bay Ctr Counseling, 79-; Comnr Pub Health, Los Angeles County, 78-81; clin psychologist, 85- *Mem:* Am Psychol Asn; Asn Women Sci. *Res:* Published research in delivery of health services and in community psychology; interpersonal process, love, and sexuality. *Mailing Add:* Dept Psychol Calif State Univ Dominguez Hills Carson CA 90747

PALMER, BRUCE ROBERT, b St Paul, Minn, June 20, 46. SURFACE CHEMISTRY, KINETICS. *Educ:* Colo Sch Mines, BS, 68; Univ Utah, PhD(metall), 72. *Prof Exp:* Prof metall, SDak Sch Mines & Technol, 72-81; sr engr, Tex Gulf Res Lab, 82-83; chem engr, Wendell E Dunn Inc, 83-85; MGR EXPLOR RES, KERR-MCGEE CORP, 85- *Honors & Awards:* A F

Taggart Award, Am Inst Metall Engrs, 78; Bradley Stoughton Award, Am Soc Metals, 81. *Mem:* Am Inst Metall Engrs; Am Soc Metals; Am Ceramics Soc. *Res:* Electrochemical reactions on sulfide compounds; oxide surface chemistry; high temperature corrosion & errosion; light scattering; particle growth by aerosol processes. *Mailing Add:* 1110 Brookhaven Dr Edmond OK 73034

PALMER, BRYAN D, b Logan, Kans, July 5, 36; m 59; c 3. INORGANIC CHEMISTRY. *Educ:* Ft Hays Kans State Col, BS, 58; Univ Ark, MS, 64, PhD(chem), 66. *Prof Exp:* PROF CHEM, HENDERSON STATE UNIV, 66-, CHMN DEPT, 71- *Mem:* AAAS; Am Chem Soc; Health Physics Soc. *Res:* Atmospheric transport of radioactive nuclear weapons debris; environmental radioactivity. *Mailing Add:* Dept Chem 7629 Henderson State Univ Arkadelphia AR 71923

PALMER, BYRON ALLEN, b Dillon, Colo, Aug 27, 49. FOURIER TRANSFORM SPECTROSCOPY. *Educ:* Colo Sch Mines, BS, 71; Purdue Univ, MS, 73, PhD(physics), 77. *Prof Exp:* PHYSICIST, LOS ALAMOS NAT LAB, 77- *Concurrent Pos:* Assoc ed, J Optical Soc Am, 82- *Mem:* Fel Optical Soc Am; Am Phys Soc; Sigma Xi. *Res:* Study of atomic structure of rare earth and actinide elements using emission spectroscopy; generation of wavelength standards for the visible, ultraviolet and infrared regions. *Mailing Add:* 1174 Big Rock Loop Los Alamos NM 87544

PALMER, CATHERINE GARDELLA, b New York, NY, Apr 28, 24; m 54; c 3. CYTOGENETICS. *Educ:* Hunter Col, AB, 47; Smith Col, MA, 49; Ind Univ, PhD(cytogenetics), 53; Am Bd Med, dipl genetics, 82. *Prof Exp:* From asst prof to assoc prof med, 62-66, assoc prof med genetics, 66-70, PROF MED GENETICS, MED CTR, IND UNIV, INDIANAPOLIS, 70- ASST DEAN GRAD AFFAIRS, SCH MED, 81- *Concurrent Pos:* USPHS career develop award, 63-73. *Mem:* AAAS; Genetics Soc Am; Am Soc Human Genetics; Tissue Cult Asn; Genetics Soc Can; Sigma Xi. *Res:* Human cytogenetics. *Mailing Add:* Dept Med Genetics IB 264 Sch Med Ind Univ 975 W Walnut St Indianapolis IN 46202-5251

PALMER, CHARLES HARVEY, b Milwaukee, Wis, Dec 8, 19; c 2. NONDESTRUCTIVE EVALUATION OF MATERIALS & STRUCTURES, OPTICAL INSTRUMENTATION. *Educ:* Harvard Col, SB, 41 & AM , 45; Johns Hopkins, PhD(physics), 51. *Prof Exp:* Staff mem radar res, Radiation Lab, Mass Inst Technol, 42-45; asst & assoc prof physics, Buchnell Univ, 50-54; res assoc, Lab Astrophys & Phys Meteorol, 54-60; prof elect & comput eng, Johns Hopkins Univ, 60-89; RETIRED. *Concurrent Pos:* Consult, Aberdeen Proving Ground, Bendix Radio Corp & Caterpillar Tractor Co; prof elec & comput eng, Johns Hopkins Univ, 89- *Mem:* Fel Optical Soc Am; Am Phys Soc; Inst Elec & Electronic Engrs; AAAS; Sigma Xi. *Res:* Anomalous behavior of diffraction gratings; student experiments in optics; infrared spectroscopy of water vapor; optical interferometry of ultrasonic waves and laser generation; detection of ultrasound for non-destructive evaluation. *Mailing Add:* Elec & Comput Eng Dept Johns Hopkins Univ Baltimore MD 21218

PALMER, CURTIS ALLYN, b Salem, Ore, June 22, 48; m 72; c 2. INORGANIC COAL GEOCHEMISTRY, TRACE ELEMENT ANALYSIS. *Educ:* Principia Col, BS, 71; Ore State Univ, MS, 75; Wash State Univ, PhD(chem), 83. *Prof Exp:* Res technician, Wash State Univ, 74-80; RES CHEMIST, US GEOL SURV, 80-; ASSOC PROF CHEM, PRINCIPIA COL, 89- *Mem:* Am Chem Soc; Int Union Pure & Appl Chem. *Res:* Instrumental neutron activation analysis. *Mailing Add:* Dept Chem Principia Col Elsah IL 62028

PALMER, DARWIN L, b Long Beach, Calif, Dec 20, 30; m 58; c 2. MEDICINE, INFECTIOUS DISEASES. *Educ:* Oberlin Col, BA, 53; Columbia Univ, MA, 54; NY Univ, MD, 60. *Prof Exp:* From intern to resident med, Med Ctr, Univ Colo, 60-64; from instr to assoc prof, 65-76, PROF MED, SCH MED, UNIV NMEX, 76-, CHIEF, DIV INFECTIOUS DIS, DEPT MED, 79-; CHIEF INFECTIOUS DIS, MED SERV, VET ADMIN HOSP, 66- *Concurrent Pos:* NIH fel infectious dis, Univ Colo, 64-65; mem consult staff, Bernalillo County Med Ctr, Vet Admin Hosp, Presby Hosp, St Joseph's Hosp & Bataan Mem Hosp, 65-; vis assoc prof, Johns Hopkins Univ Sch Med, 74-75. *Mem:* AAAS; NY Acad Sci; Am Fedn Clin Res; Infectious Dis Soc Am; Am Soc Microbiol. *Res:* Investigation of host defense mechanisms in the infection prone individual. *Mailing Add:* Vet Admin Hosp Albuquerque NM 87108

PALMER, EDGAR M, b Hartford, Conn, May 17, 34; m 64; c 2. MATHEMATICS. *Educ:* Wesleyan Univ, BA, 56; Trinity Col, Conn, MS, 60; Univ Mich, PhD(math), 65. *Prof Exp:* Numerical analyst comput, Res Labs, United Aircraft Corp, 56-60; teaching asst math, Univ Mich, 60-65, res assoc, 65-67; asst prof, 67-70, assoc prof, 70-76, PROF MATH, MICH STATE UNIV, 76- *Concurrent Pos:* Res assoc, Univ Col, Univ London, 66-67; vis prof, Oxford Univ, 73-74. *Mem:* Am Math Soc; Math Asn Am. *Res:* Combinatorial and graph theory; numerical analyses. *Mailing Add:* Dept Math Mich State Univ East Lansing MI 48824

PALMER, EUGENE CHARLES, b Elmhurst, Ill, Nov 4, 38; m 60; c 1. NEUROCHEMISTRY, PHARMACOLOGY. *Educ:* Tenn Technol Univ, BS, 60; Vanderbilt Univ, PhD(anat), 68. *Prof Exp:* Assoc prof pharmacol, Sch Med, Univ NMex, 70-77; prof pharmacol, Col Med, Univ SAla, 77-82; DIR RES, FIRST-MASSEY NEUROL INST, 82- *Concurrent Pos:* USPHS fel pharmacol, Vanderbilt Univ, 68-70; NIMH & NMex Heart Asn res grants, Univ NMex, 71-; res grants, March Dimes, 74-76, NSF, 77-79, Epilipsy Found Am, 76-77 & Ala Affil, Am Heart Asn, 78-79. *Mem:* Soc Neurosci; Am Soc Neurochem; Am Soc Pharmacol & Exp Therapeut; Neurocirculation Soc. *Res:* Effect of psychotropic drugs on the adenyl cyclase system and other neurochemical processes in the central nervous system. *Mailing Add:* Pennwalt Corp 755 Jefferson Rd Rochester NY 14623

PALMER, FREDERICK B ST CLAIR, b London, Ont, June 5, 38; m 59; c 4. BIOCHEMISTRY. *Educ:* Univ Western Ont, BSc, 60, PhD(biochem), 65. *Prof Exp:* Asst prof, 67-71, ASSOC PROF BIOCHEM, DALHOUSIE UNIV, 71- *Concurrent Pos:* Med Res Coun Can fel, Agr Res Coun Inst Animal Physiol, Babraham, Eng, 65-67. *Mem:* Can Biochem Soc. *Res:* Phospholipid metabolism; lipoprotein structure and metabolism. *Mailing Add:* Dept Biochem Dalhousie Univ Fac Med Sir Chas Tupper Bldg Halifax NS B3H 4H7 Can

PALMER, GLENN EARL, b Maple, Ont, Apr 13, 35; m 62; c 2. ORGANIC CHEMISTRY. *Educ:* Univ Toronto, BSc, 61, MA, 62; Imp Col, Univ London, dipl chem, 63; Queen's Univ, Ont, PhD(chem), 67. *Prof Exp:* Lectr chem, Royal Mil Col Can, 63-66 & Univ Toronto, 66-67; assoc prof, 67-81, PROF CHEM, UNIV PRINCE EDWARD ISLAND, 81- *Concurrent Pos:* Fel, Univ Toronto, 66-67; chmn dept chem, Univ Prince Edward Island, 69-72, 84- *Mem:* Chem Inst Can; Am Chem Soc; Royal Soc Chem. *Res:* Organic reaction mechanisms; molecular photochemistry; chemical education. *Mailing Add:* Dept Chem Univ Prince Edward Island Charlottetown PE C1A 4P3 Can

PALMER, GRAHAM, b Tonyrefail, Eng, Sept 30, 35; m 60; c 2. BIOCHEMISTRY. *Educ:* Univ Sheffield, BSc, 57, PhD(biochem), 62. *Prof Exp:* Asst lectr biochem, Univ Sheffield, 60-61; proj assoc, Inst Enzyme Res, 61-64; asst prof biochem & assoc res biophysicist, Univ Mich, Ann Arbor, 64-70, assoc prof biol chem & res biophsicist, 70-72; PROF BIOCHEM, RICE UNIV, 72-; ASSOC, RICHARDSON COL, 77- *Concurrent Pos:* NIH career develop award, 66- *Mem:* Am Soc Biol Chemists. *Res:* Mechanism of biological oxidations, particularly in applications of spectroscopic techniques to the study of flavoproteins and metal emzymes. *Mailing Add:* Dept Biochem Rice Univ Box 1892 Houston TX 77251

PALMER, GRANT H, b Cleveland, Ohio, Nov 30, 11; m 40. BIOCHEMISTRY, NUTRITION. *Educ:* Univ Redlands, BA, 38. *Prof Exp:* Lab aide, Regional Salinity Labs, 39; lab technician, San Bernardino County Hosp, 39-40; res biochemist, Calif Fruit Growers Exchange, 40-50; res biochemist, Sunkist Growers, Inc, Calif, 50-56, pharmaceut prod develop, 56-63, pharmaceut coordr, 63-69, mgr, Corona Res & Develop Labs, 69-75, asst to div mgr prod res & develop, 75-76; RETIRED. *Mem:* Am Chem Soc. *Res:* Citrus bioflavonoids; pharmaceutical pectin; plasma expanders; toxicity of pesticides; nutritional studies; citrus pharmaceutical products. *Mailing Add:* 817 W Olive St Corona CA 91720-4166

PALMER, H(AROLD) A(RTHUR), b Nampa, Idaho, Dec 22, 19; m 43; c 3. PHYSICAL CHEMISTRY. *Educ:* Col Idaho, BS, 41; Univ Okla, MA, 46, PhD(chem), 50. *Prof Exp:* Physics aide, Nat Bur Standards, 42-43; instr math, Col Idaho, 43-44; chemist, Basic Magnesium, Inc, 44 & Cutter Labs, 44-45; instr math, Univ Okla, 45-49; assoc prof, Midwestern Univ, 50-51; chemist, United Gas Corp, 51-53; chemist, 53-65, supvr prod res, 65-74, SR RES ASSOC, BELLAIRE LAB, TEXACO, INC, 74- *Mem:* AAAS; Am Chem Soc; Am Inst Chem. *Res:* Uranium mining; petroleum production. *Mailing Add:* 2613 Raquet Nacogdoches TX 75961

PALMER, H CURRIE, b Moncton, NB, Oct 25, 35. GEOPHYSICS, GEOLOGY. *Educ:* St Francis Xavier Univ, BSc, 57; Princeton Univ, PhD(geol), 63. *Prof Exp:* Fel, 63-66, from asst prof to assoc prof, 66-81, PROF GEOPHYS, UNIV WESTERN ONT, 81- *Concurrent Pos:* Nat Res Coun Can fel, 64-66. *Mem:* Geol Soc Am; Am Geophys Union. *Res:* Paleomagnetism. *Mailing Add:* Dept Geophys Univ Western Ont London ON N6A 5B7 Can

PALMER, HARVEY EARL, b Inkom, Idaho, Oct 17, 29; m 54; c 5. RADIOLOGICAL PHYSICS, MEDICAL PHYSICS. *Educ:* Idaho State Univ, BS, 55; Univ Idaho, MS, 61. *Prof Exp:* Chemist, Gen Elec Co, 56-59, sr physicist, 59-65; res assoc, Pac Northwest Labs, Battelle Mem Inst, 65-80, staff scientist, 80-90; ASSOC PROF RADIOL, UNIV WASH, 69-; CONSULT, 90- *Mem:* Health Physics Soc. *Res:* Studies of radioactivity in people; metabolism of fallout isotopes in humans; whole body counters and other methods of studying internally deposited radioactivity; radon decay products in people from radon in homes and uranium mines. *Mailing Add:* 1511 Woodbury Richland WA 99352

PALMER, HARVEY JOHN, b New York, NY, Apr 3, 46; m 66; c 3. TRANSPORT PROCESSES, INTERFACIAL PHENOMENA. *Educ:* Univ Rochester, BS, 67; Univ Wash, PhD(chem eng), 71. *Prof Exp:* Res engr, Mixing Equip Co, 66-67; from asst prof to assoc prof, 71-84, assoc dean grad studies, 84-89, PROF CHEM ENG, UNIV ROCHESTER, 84-, CHAIR CHEM ENG, 90- *Concurrent Pos:* Consult, Eastman Kodak Co, 82-, Helios Res Corp, Mumford, NY, 85- *Mem:* Sigma Xi; Am Inst Chem Engrs. *Res:* Interfacial mechanics; phase transfer catalyzed reactions; heat and mass transfer; convective instability; biochemical engineering. *Mailing Add:* Dept Chem Eng Univ Rochester Rochester NY 14627

PALMER, HOWARD BENEDICT, b Indianapolis, Ind, July 10, 25; m 51; c 3. PHYSICAL CHEMISTRY, FUEL SCIENCE. *Educ:* Carnegie Inst Technol, BS, 48; Univ Wis, PhD(chem), 52. *Prof Exp:* Res assoc chem, Brown Univ, 52, instr, 53-55; from asst prof to prof fuel technol, 55-60, prof fuel sci, 66-77, PROF ENERGY SCI, PA STATE UNIV, 77- *Concurrent Pos:* Head, dept fuel technol, Pa State Univ, 59-65, assoc dean, Grad Sch, 77-83, actg dean, 83-85, sr assoc dean, 85-; chmn, Fuel Sci Sect, 69-76; vis scientist, Imp Col, Univ London, 63; vis prof, Univ Pittsburgh, 71; consult, United Technol Res Ctr, 68-, Brookhaven Nat Lab, 75-87, Sandia-Livermore Nat Lab, 78-, Exxon Corp, 80-83, Univ Space Res Asn, 83-, Lawrence Berkeley Nat Lab, 88-90; ed, Combustion & Flame, 72-84. *Mem:* Am Chem Soc; AAAS; Combustion Inst (vpres, 84-85, pres, 85-88); Inst Energy London; Am Phys Soc. *Res:* Gaseous reaction kinetics; combustion processes; chemiluminescent reactions in gases; electronic spectra of small molecules; carbon formation and vaporization. *Mailing Add:* Pa State Univ 114 Kern Bldg Univ Park PA 16802

PALMER, JAMES D(ANIEL), b Washington, DC, Mar 8, 30; m 52; c 3. ELECTRICAL ENGINEERING. *Educ:* Univ Calif, BSEE, 55, MSEE, 57; Univ Okla, PhD, 63. *Prof Exp:* Res engr, Univ Calif, 55-57; from asst prof to prof elec eng, Univ Okla, 57-66, dir sch elec eng, 63-66, proj engr, High-Speed Computer Proj, 58-60, res assoc & proj dir, Res Inst, 58-66, dir systs res ctr, 64-66; proj elec eng & dean sci & eng, Union Col, NY, 66-71; pres, Metrop State Col, 71-78; adminr res & spec admin, US Dept Transp, 78-80; vpres & gen mgr res, Mech Tech Inc, 80-; exec vpres, J J Henry Co, Inc, 82-85; BDM INT PROF INFO TECHNOL, GEORGE MASON UNIV, 85- *Concurrent Pos:* Consult, McDonnell Aircraft Corp, Mo, 59-69, Streeter-Amet Div, Goodman Indust, Ill, 61-66 & Univ Okla Res Inst, 66-68; mem & chmn bd trustees, Hudson Mohawk Asn Cols & Univs, 69-71; mem bd dirs, Auraria Higher Educ Ctr; mem, Colo Fulbright-Hays Scholar Comt; mem bd, Rocky Mt Region Inst Int Engrs; consult & mem bd, Systs Mgt Co, Mass. *Mem:* Inst Elec & Electronic Engrs. *Res:* Plasma dynamics; adaptive control in social systems. *Mailing Add:* George Mason Univ 103 Sci & Tech Bldg 4400 University Dr Fairfax VA 22030

PALMER, JAMES F, b Indio, Calif, Nov 5, 50; m 81; c 2. LANDSCAPE PERCEPTION, ENVIRONMENTAL MEDIATION. *Educ:* Univ Calif, Santa Cruz, BA, 72; Univ Mass, Amherst, MLA, 76, PhD(natural resource mgt), 79. *Prof Exp:* Res asst, Carlozzi, Sinton & Vilkities, 75-78, assoc, 78; res assoc, Environ Inst, Univ Mass, 79-80; asst prof, 80-81, res assoc, 81-85, SR RES ASSOC LANDSCAPE ARCHIT & ENVIRON STUDIES, COL ENVIRON SCI & FORESTRY, STATE UNIV NY, 85- *Concurrent Pos:* Consult var co, 79-; instr, US Army CEngr, 86-87; assoc, Neil Katz & Assocs, 90- *Mem:* AAAS; Asn Am Geographers; Environ Design Res Asn; Int Asn Impact Assessment; Int Asn Landscape Ecol. *Res:* Community perceptions of local environmental quality; cross-cultural studies of landscape perception; landscape planning methods; impact assessment related to visual and cultural resources; environmental conflict resolution and mediation. *Mailing Add:* Col Environ Sci & Forestry State Univ NY One Forestry Dr Syracuse NY 13210-2787

PALMER, JAMES FREDERICK, b Toronto, Ont, Feb 2, 31; m 56; c 3. SAFETY ENGINEERING, ENVIRONMENTAL SCIENCE. *Educ:* Univ BC, BASc, 53; Queen's Univ, Ont, MSc, 55. *Prof Exp:* Nuclear engr reactor res, Atomic Energy Can Ltd, 54-63; liaison officer attached to Phillips Petrol Co, Idaho, 63-65; nuclear safety engr, Atomic Energy Can Ltd, 65-75, Environ Authority, 75-86; RETIRED. *Mem:* Health Physics Soc; Can Radiation Protection Asn. *Res:* Reactor safety, particularly development of methods for detecting and locating fuel failures in nuclear reactors; studies of environmental effects due to releases of radioactivity from nuclear facilities. *Mailing Add:* 4-12901 17th Ave White Rock BC V4A 8T5 Can

PALMER, JAMES KENNETH, b Camden, NJ, Sept 29, 26; m 49; c 6. FOOD BIOCHEMISTRY. *Educ:* Juniata Col, BS, 48; Pa State Univ, MS, 50, PhD(agr & biochem), 53. *Prof Exp:* Asst biochemist, Conn Agr Exp Sta, 52-56; sr res biochemist, Va Inst Sci Res, 56-59; sr biochemist, United Fruit Co, 59-65; prin res scientist, Div Food Preservation, Commonwealth Sci & Indust Res Orgn, Australia, 65-68; assoc prof food biochem, Mass Inst Technol, 68-73; sr res assoc, US Army Natick Labs, 73-75; assoc prof, 75-85, PROF FOOD BIOCHEM, VA POLYTECH INST & STATE UNIV, 85- *Mem:* Am Chem Soc; Inst Food Technologists. *Res:* Biochemistry of fruit ripening; biogenesis of flavor components; phenolases; biochemical analysis; carbohydrates in foods. *Mailing Add:* Dept Food Sci & Tech Va Polytech Inst & State Univ Blacksburg VA 24061

PALMER, JAMES MCLEAN, b Saginaw, Mich, Sept 3, 37; m 63, 81; c 3. OPTICS. *Educ:* Grinnell Col, AB, 59; Univ Ariz, MS, 73, PhD(optics), 75. *Prof Exp:* Sr engr, Hoffman Semiconductor, 59-67, Globe-Union, Centralab Semiconductor, 67-70; res asst, 73-75, res assoc, 76-77, res specialist, 78-80, RES ASSOC, OPTICAL SCI CTR, UNIV ARIZ, 81- *Concurrent Pos:* Pres, Radiometrics, 70-; lectr optical eng, Optical Sci Ctr, Univ Ariz, 81- *Mem:* Optical Soc Am; Inst Elec & Electronic Engrs; Soc Photo-Optical-Instrumentation Engrs; Coun Optical Radiation Measurements. *Res:* Placed Pioneer-Venus solar flux radiometer on Venus; measuring ground-based solar radiation stability; design and build various radiometric instruments. *Mailing Add:* Optical Sci Ctr Univ Ariz Tucson AZ 85721

PALMER, JAY, b Moab, Utah, Nov 18, 28; div; c 2. PHYSICAL CHEMISTRY, INORGANIC CHEMISTRY. *Educ:* Utah State Univ, BS, 50, MS, 56; Northwestern Univ, MA, 76, PhD(inorg chem), 60. *Prof Exp:* Chemist, US Gypsum Co, 56-57 & Portland Cement Asn, 58; dir inorg chem res, Morton Chem Co, Ill, 59-70; mgr res & develop, CF Industs, Inc, 70-73; res assoc & mgr phosphate res, US Gypsum Co, 73-76; ASSOC PROF CHEM, UNIV S FLA, 77- *Concurrent Pos:* Lectr, Roosevelt Univ, 61-70; lectr, Northwestern Univ, 62; chem consult, USGCorp, 76-; vis scholar, Northwestern Univ, 83-; guest scientist, Los Alamos Nat Lab, 85-86, 88-; vpres, Future Great Harvests, Inc, 86-; pres, Jay W Palmer & Assocs, Inc, 87- *Mem:* Am Chem Soc; NY Acad Sci; fel Am Inst Chem; AAAS; Sigma Xi. *Res:* Inorganic reactions in solution and the solid state; surface and colloid chemistry. *Mailing Add:* 1211 A E Seneca Tampa FL 33612

PALMER, JEFFRESS GARY, b Brooklyn, NY, Oct 7, 21; m 51; c 2. INTERNAL MEDICINE, HEMATOLOGY. *Educ:* Emory Univ, BS, 42, MD, 44. *Prof Exp:* Intern med, NC Baptist Hosp, Winston-Salem, 44-45; asst resident & resident, Lawson Vet Admin Hosp, Ga, 47-49; from asst prof to assoc prof, 52-64, PROF MED, SCH MED, UNIV NC, CHAPEL HILL, 64- *Concurrent Pos:* Fel, Col Med, Univ Utah, 49-52; consult, Vet Admin, Fayetteville, NC. *Mem:* AAAS; Am Fedn Clin Res; Am Soc Hemat. *Res:* Antibiotics; hematology; chemotherapy; spleen; leukocytes; steroids. *Mailing Add:* Dept Med CB No 7035 Univ NC Chapel Hill NC 27599

PALMER, JOHN DAVIS, b Sterling, Ill, Oct 31, 31; m 57; c 2. CLINICAL PHARMACOLOGY. *Educ:* Univ Colo, BS, 54, MS, 55; Univ Minn, PhD(pharmacol), 61, MD, 62. *Prof Exp:* Intern internal med, Univ Minn Hosps, 62-63; asst prof pharmacol, Sch Med, Univ Colo, 63-66; DIR MULTIDISCIPLINE TEACHING LABS, COL MED, UNIV ARIZ, 66-, ASSOC PROF PHARMACOL, 69-, ASST PROF INTERNAL MED, 73- *Concurrent Pos:* Grants, USPHS, 64-66, Am Heart Asn, 66-69 & Ariz Heart Asn, 69-; consult, Denver Vet Admin Hosp, Colo, 64-66 & Tucson Vet Admin Hosp, Ariz, 66-; consult physician educ br, Bur Health Manpower Educ, NIH, 71, mem med educ rev comt, 71-73; consult, Bur Health Resources Develop, 73- *Mem:* AAAS; Am Fedn Clin Res; Am Soc Pharmacol & Exp Therapeut. *Res:* Mechanisms and treatment of septic and cardiogenic shock; medical education; design and development of multidiscipline teaching laboratories. *Mailing Add:* Dept Pharmacol Univ Ariz Col Med Tucson AZ 85724

PALMER, JOHN DERRY, b Chicago, Ill, May 26, 32; m 60; c 1. BIOLOGICAL RHYTHMS. *Educ:* Lake Forest Col, BA, 57; Northwestern Univ, MS, 59, PhD(comp physiol), 62. *Prof Exp:* From instr to asst prof biol, Univ Ill, 61-63; NSF fel, Eng, 63-64; from asst prof to prof, NY Univ, 66-74; PROF ZOOL & CHMN DEPT, UNIV MASS, AMHERST, 74- *Concurrent Pos:* Mem corp, Marine Biol Lab, Woods Hole, 64-; chmn dept biol, NY; Univ of Dunedin, 83 & 84; auth Univ, 68-73. *Honors & Awards:* Res Award, Sigma Xi. *Mem:* Fel AAAS; Am Inst Biol Sci; Int Soc Chronobiol; Sigma Xi; Soc Gen Physiologists. *Res:* Biological rhythms in algae, crustaceans and birds. *Mailing Add:* Dept Zool Univ Mass Amherst MA 01003

PALMER, JOHN FRANK, JR, b Cedar Rapids, Iowa, Oct 23, 15; m 70; c 5. INDUSTRIAL ORGANIC CHEMISTRY. *Educ:* Coe Col, BA, 37; Iowa State Univ, PhD(phys chem), 41. *Prof Exp:* Res chemist, Monsanto Co, Mo, 41-47, group leader, 47-53, res liaison, 53-54, group leader, 54-61, res specialist, 61-71; process chemist, Div Ethly Corp, Edwin Cooper Inc, 71-73, plant process res mgr, 73-79, mgr, Process & Anal Technol, 79-84; RETIRED. *Concurrent Pos:* Expert witness for chem processes. *Mem:* Am Chem Soc; Am Soc Testing & Mat. *Res:* Exploratory research on organic chemical processes; resin materials; surface coating; absorption spectroscopy; oil additives for lubricating oils; catalytic hydrogenation and dehydrogenation. *Mailing Add:* 14157 Crosstrails Dr St Louis MO 63017-3308

PALMER, JOHN M, b Yakima, Wash, May 8, 22; m 47; c 2. SPEECH PATHOLOGY, AUDIOLOGY. *Educ:* Univ Wash, BA, 46, MA, 50; Univ Mich, PhD(speech), 52. *Prof Exp:* PROF SPEECH & HEARING SCI, UNIV WASH, 52- *Concurrent Pos:* Consult, Children's Orthop & Hosps, Seattle, 52-, Univ Wash Hosp, 57- *Mem:* Fel Am Speech & Hearing Asn. *Res:* Organic speech disorders; cleft palate; speech pathology, science and physiology. *Mailing Add:* Dept Speech & Hearing Sci Univ Wash Seattle WA 98195

PALMER, JOHN PARKER, b Los Angeles, Calif, July 16, 39; m 60; c 3. ELECTRONICS ENGINEERING. *Educ:* Univ Mich, BSE, 62; Univ Calif, Riverside, PhD(physics), 69. *Prof Exp:* Physicist, Gen Dynamics, Pomona, 62-65; asst prof, 69-72, assoc prof, 72-78, PROF ELEC ENGR, CALIF POLYTECH STATE UNIV, 78-, CHMN DEPT, 81- *Concurrent Pos:* Design specialist, Gen Dynamics, Pomona, 70-80; staff engr, Hughes Aircraft Co, 81- *Mem:* Inst Elec & Electronics Engrs; Optical Soc Am. *Res:* Analog and digital circuit design, electronic noise analysis; management operating system large scale intergration structures; fiber optic devices and communication systems; sampled data signal processing. *Mailing Add:* Dept Elec Eng Calif Polytech State Univ Pomona 3801 W Temple Ave Pomona CA 91768

PALMER, JOHN WARREN, b Marshalltown, Iowa, July 27, 08; m 31; c 3. BIOCHEMISTRY. *Educ:* Univ Iowa, BA, 28, MS, 29; Columbia Univ, PhD(biochem), 33. *Prof Exp:* Asst chem, Univ Iowa, 28-29; asst biochem, Col Physicians & Surgeons, Columbia Univ, 30-36, instr, 36; biochemist, Biol Lab, E R Squibb & Sons, 36-45, head dept protein fractionation, 45-49, dir biochem develop, 48-49, biol develop, 49-52; dir biol prod, Hyland Div, Travenol Labs, Inc, 52-53, dir labs, 53-78, vpres, 55-78; RETIRED. *Concurrent Pos:* Mem, Rev Comt, US Pharmacopoeia, 60-70. *Mem:* AAAS; Am Chem Soc; Am Asn Immunologists; Am Pub Health Asn; NY Acad Sci. *Res:* Chemistry of the skin and of the eye; mucins and mucoids; bacterial carbohydrates and toxins; toxoids; antitoxins and antibodies; serum proteins; bromide metabolism; development and production of biological pharmaceuticals. *Mailing Add:* 3006 La Ventana San Clemente CA 92672

PALMER, JON (CARL), b Galva, Ill, Jan 20, 40; m 64; c 1. IMMUNOLOGY, MOLECULAR BIOLOGY. *Educ:* Knox Col, AB, 62; Univ Calif, Berkeley, PhD(molecular biol), 68. *Prof Exp:* ASST PROF, WISTAR INST ANAT & BIOL, 70-; SCI INFO SERV, SMITHKLINE BEECHAM, 88- *Concurrent Pos:* Nat Inst Allergy & Infectious Dis fel, Walter & Eliza Hall Inst Med Res, Melbourne, 68-70. *Mem:* Am Chem Soc; AAAS. *Res:* Molecular biology of the immune response; cellular cooperation; control of expression of genetic information; tumor-host relationship. *Mailing Add:* SmithKline Beecham Pharmaceut PO Box 1539 King of Prussia PA 19406

PALMER, KEITH HENRY, b Retford, Eng, July 3, 28; m 58; c 1. BIOCHEMICAL PHARMACOLOGY. *Educ:* Univ Nottingham, BPharm, 52, MPharm, 54. *Hon Degrees:* Dr, Univ Paris, 56. *Prof Exp:* Demonstr, Univ Nottingham, 52-53; from asst prof to assoc prof pharmaceut chem, Univ Alta, 58-64; group leader, Div Chem & Life Sci, Res Triangle Inst, NC, 64-71; assoc prof biochem pharmacol, Brown Univ, 71-74; sect leader drug metab & pharmacokinetics, Res Div, Schering Corp, 74-81, sect leader anal biochem, Vet Div, 81-84; RES SCIENTIST TOXICOL, PH&E LABS HEALTH DEPT, STATE NJ. *Concurrent Pos:* Nat Res Coun Can res fel, 56-58; mem alkaloids comt, Brit Pharmacopoeia Comn, 59-64; head, Div Drug Metab, Roger Williams Gen Hosp, Providence, RI, 71-74. *Mem:* Am Soc Pharmacol & Exp Therapeut; Am Chem Soc; Pharmaceut Soc Gt Brit. *Res:* Drug metabolism; analytical biopharmacology; organic chemistry; pharmaceutical chemistry; pharmacognosy; veterinary chemotherapeutics, synthesis, analysis, development of antiviral agents. *Mailing Add:* 33 Woodland Dr East Windsor NJ 08520

PALMER, KENNETH, b Australia, Nov 28, 45. MATHEMATICS. *Educ:* Univ New Castle, BS, 66; Univ Tasmania, PhD(math), 70. *Prof Exp:* Prof math, Fla State Univ, 77-80; PROF MATH, UNIV MIAMI, 81- *Mailing Add:* Dept Math & Comp Sci Univ Miami Coral Gables FL 33124

PALMER, KENNETH CHARLES, b Everett, Mass, Oct 23, 44; m 68; c 2. EXPERIMENTAL PATHOLOGY, PULMONARY DISEASES. *Educ:* Merrimack Col, BA, 66; Villanova Univ, MS, 72; Boston Univ, PhD(path), 77. *Prof Exp:* Res asst bioenergetics, Boston Biomedical Res Inst, 66-67; res asst lung path, Sch Med, Boston Univ, 73-76; res instr path, Sch Med, Boston Univ, 77-79, asst prof path, 79-80; asst prof path & chief, Div Pulmonary Pathobiol, Sch Med, 80-88, ASSOC PROF, WAYNE STATE UNIV, 88- *Concurrent Pos:* NIH fel, Boston Univ, 76-78; NIH young pulmonary invest award, Mallory Inst Path, 78-81, res assoc, 76-80; dipl, Am Bd Toxicology, 86- *Mem:* AAAS; Am Soc Cell Biol; Am Asn Pathologists; Am Thoracic Soc; NY Acad Sci. *Res:* Pathology; electron microscopy; biochemistry. *Mailing Add:* 540 E Canfield Ave Detroit MI 48201

PALMER, KENT FRIEDLEY, b Chicago, Ill, Sept 26, 41; m 68; c 2. MOLECULAR PHYSICS, SPECTROSCOPY. *Educ:* Ohio State Univ, BS, 64, PhD(physics), 72. *Prof Exp:* Acad counselor phys sci, Col Arts & Sci, Ohio State Univ, 71-72; res assoc, Dept Physics, Kans State Univ, 72-74; vis asst prof, Dept Physics, Univ Ky, 74-76; ASSOC PROF PHYSICS & CHMN DEPT, WESTMINSTER COL, 76- *Concurrent Pos:* Summer fac res assoc, US Air Force & Am Soc Eng Educ, 76; subcontractor, ARO Inc & Calspan Inc, Oper Contractors Arnold Eng & Develop Ctr, Arnold Air Force Sta, Tenn, 78-88; res scientist, EG & G Inc, Las Vegas, Nev, 84; res fac, Univ Nev, Las Vegas, 85. *Mem:* Optical Soc Am; Am Asn Physics Teachers. *Res:* High resolution spectroscopy of small molecules; spectroscopy of condensed molecular species found in planetary atmospheres, rocket and jet plumes; time-resolved spectroscopy of rare gas atoms and molecules; optical properties of thin dielectric films and biological materials. *Mailing Add:* Dept of Physics Westminster Col 501 Westminster Ave Fulton MO 65251-1299

PALMER, LARRY ALAN, b Wichita, Kans, Oct 11, 45; m 67; c 1. NEUROPSYCHOLOGY. *Educ:* Drexel Univ, BS, 67; Univ Pa, PhD(physiol), 72. *Prof Exp:* Fel, 72-75, ASST PROF, DEPT ANAT, UNIV PA, 75- *Mem:* Soc Neurosci. *Res:* Physiological and anatomical determination of neuron connectivity in visual cortex of cat. *Mailing Add:* Dept Anat Univ Pa Sch Med Philadelphia PA 19104

PALMER, LAURENCE CLIVE, b Washington, DC, Dec 25, 32; m 56; c 2. ELECTRICAL ENGINEERING. *Educ:* Washington & Lee Univ, BS, 55; Rensselaer Polytech Inst, BEE, 55; Univ Md, MS, 63, PhD(elec eng), 70. *Prof Exp:* Prin engr, Emerson Res Lab, Div Emerson Radio, 57-60; sect mgr engr, Wash Technol Assoc, Inc, 60-61, Radcom-Emertron Div, Litton Syst, 61-63; sr scientist, Comput Sci Corp, 63-74; prin scientist, Systs Tech Div, Comsat Labs, Commun Satellite Corp, 74-90; ADV ENGR, HUGHES NETWORK SYSTS, 90- *Concurrent Pos:* Electron officer, US Army Signal Corps, 55-57; jr engr, Emerson Res Lab, Emerson Radio, 55. *Mem:* Inst Elec & Electronic Engrs. *Res:* Analysis and simulation of the satellite communication systems with emphasis on modulation, multiple-access techniques. *Mailing Add:* 17 River Falls Ct Potomac MD 20854-3885

PALMER, LEIGH HUNT, b Altadena, Calif, May 9, 35; m 58; c 4. PHYSICS, ASTRONOMY. *Educ:* Univ Calif, Berkeley, BA, 57, PhD(physics), 66. *Prof Exp:* Asst prof, 66-87, ASSOC PROF PHYSICS, SIMON FRASER UNIV, 87- *Mem:* Am Asn Physics Teachers; Astron Soc Pac. *Res:* Astronomy; astrophysics. *Mailing Add:* Dept Physics Simon Fraser Univ Burnaby BC V5A 1S6 Can

PALMER, LEONARD A, b Seattle, Wash, Aug 8, 31. GEOLOGY. *Educ:* Univ Wash, BS, 53, MS, 60; Univ Calif, Los Angeles, PhD(geol), 67. *Prof Exp:* Asst prof geol, Univ Hawaii, 65-67; asst prof geol, 67-74, ASSOC PROF EARTH SCI, PORTLAND STATE UNIV, 74- *Concurrent Pos:* Consult, var industs & govt agencies. *Mem:* AAAS; Geol Soc Am; Asn Eng Geol; Int Asn Quaternary Res. *Res:* Environmental geology; applied geomorphology. *Mailing Add:* Dept Geol Portland State Univ PO Box 751 Portland OR 97207

PALMER, MICHAEL RULE, b Amarillo, Tex, Sept 29, 50; c 2. NEUROPHYSIOLOGY, DEVELOPMENTAL NEUROBIOLOGY. *Educ:* Colo Col, BA, 73; Tex A&M Univ, MS, 76; Univ Colo, PhD(pharmacol), 80. *Prof Exp:* Teaching asst mammalian anat, Tex A&M Univ, 74-75, teaching lab coordr physiol, 75-76, res asst & sr lab tech, 76; res fel, 80-81, asst prof, 82-88, ASSOC PROF PHARMACOL, SCH MED, UNIV COLO, 88- *Concurrent Pos:* Co-instr, Med Sch, Univ Colo, 79-80; vis scientist, Karolinska Inst, Stockholm, Sweden, 81-82. *Mem:* Soc Neurosci; AAAS; Brit Brain Res Asn; Europ Brain & Behavior Soc; Am Soc Pharmacol & Exp Therpeut. *Res:* Opiates and opioid peptides; psychotropic drugs; neuropharmacology; brain transplants; drug-neurotransmitter interaction; CNS actions of ethonol; repair of spinal cord damage. *Mailing Add:* Box C236 Dept Pharmacol Univ Colo Health Sci Ctr 4200 ENinth Ave Denver CO 80262

PALMER, PATRICK EDWARD, b St Johns, Mich, Dec 6, 40; m 63; c 3. RADIO ASTRONOMY, IMAGING SCIENCE. *Educ:* Univ Chicago, SB, 63; Harvard Univ, MA, 65, PhD(physics), 68. *Prof Exp:* Radio astronr, Harvard Col Observ, 68; from asst prof to assoc prof, 68-75, PROF ASTRON, UNIV CHICAGO, 75- *Concurrent Pos:* Alfred P Sloan Found fel, 70; vis assoc prof, Calif Inst Technol, 72; vis lectr radio astron, Cambridge Univ, 73; vis res astronomer, Univ Calif, Berkeley, 77 & 86, vis astronomer, Nat Radio Astron Observ, 80-90. *Honors & Awards:* Bart J Bok Prize for res in galactic structure, Harvard Col Observ, 69; Helen B Warner Prize, Am Astron Soc, 75. *Mem:* Royal Astron Soc; Union Scientifique Radio Int; Am Astron Soc; AAAS; Int Astron Union. *Res:* Observation and analysis of radio frequency molecular emission lines from comets and from the interstellar medium. *Mailing Add:* Astron & Astrophysics Ctr 5640 S Ellis Ave Chicago IL 60637

PALMER, RALPH LEE, b Catskill, NY, June, 09; m 40, 69; c 2. ELECTRICAL ENGINEERING. *Educ:* Union Col, BS, 31. *Prof Exp:* Lab asst, IBM, Endicott, NY, 32-35, supvr, Elec Lab, 35-43, sr develop engr, Poughkeepsie, 46-50, mgr Eng Lab, 50-54, mgr prod develop, Data Processing Div, 56-59, dir eng, 59-63, fel, 63-69, vpres, Armonk, 69-70; RETIRED. *Mem:* Nat Acad Eng; Inst Elec & Electronics Engrs. *Res:* Computers. *Mailing Add:* Delray Dunes Country Club 4799 Oak Circle Boynton Beach FL 33436

PALMER, RALPH SIMON, b Richmond, Maine, June 13, 14; c 3. VERTEBRATE ZOOLOGY. *Educ:* Univ Maine, AB, 37; Cornell Univ, PhD(ornith), 40. *Prof Exp:* Asst ornith, Cornell Univ, 38-39; asst, State Conserv Comn, NY, 42; instr zool, Conserv Div, Vassar Col, 42-47, asst prof, 47-49; state zoologist, NY State Mus, 49-76; res assoc, Smithsonian, 79-88; RETIRED. *Mem:* Am Soc Mammal; Arctic Inst. *Res:* Avian and mammalian ecology and life histories. *Mailing Add:* Box 74 Tenants Harbor ME 04860-0074

PALMER, REID G, b Pemberville, Ohio, June 21, 41; m 67. PLANT BREEDING, GENETICS. *Educ:* Ont Agr Col, Toronto, BSA, 63; Univ Ill, MS, 65; Ind Univ, Bloomington, PhD(genetics), 70. *Prof Exp:* Res asst agron, Univ Ill, 63-65; teaching asst bot, Ind Univ, Bloomington, 66; RES GENETICIST, PLANT SCI RES DIV, AGR RES SERV, USDA, IOWA STATE UNIV, 70- *Mem:* Am Soc Agron; Crop Sci Soc Am; Genetics Soc Am; Am Genetic Asn; Sigma Xi. *Res:* Cytogenetics of soybeans. *Mailing Add:* Dept Genetics Iowa State Univ Ames IA 50011

PALMER, RICHARD ALAN, b Austin, Tex, Nov 13, 35; m 62; c 4. INORGANIC CHEMISTRY. *Educ:* Univ Tex, BS, 57; Univ Ill, MS, 62, PhD(inorg chem), 65. *Prof Exp:* NIH fel inorg chem, Copenhagen Univ, 65-66; asst prof, 66-71, assoc prof, 71-78, PROF INORG CHEM, DUKE UNIV, 78- *Mem:* Soc Appl Spectros; Am Chem Soc; Sigma Xi; Coblentz Soc. *Res:* Vibrational spectroscopy of solids, surfaces and interfaces; photoacoustic/photothermal spectroscopy; time-resolved spectroscopy using step-scan FT-IR; transition metal ion complexes; infrared, visible, ultraviolet spectroscopy; single crystal spectroscopy. *Mailing Add:* Dept Chem Duke Univ Durham NC 27706

PALMER, RICHARD CARL, b Walls, Miss, Nov 8, 31; m 53; c 3. RADIATION CHEMISTRY, RADIOLOGICAL PHYSICS. *Educ:* Memphis State Col, BS, 54, MA, 55; Vanderbilt Univ, PhD(chem), 58; Am Bd Radiol, dipl. *Prof Exp:* Lab asst physics, Memphis State Col, 52-54; Oak Ridge Inst Nuclear Studies fel health physics, Vanderbilt Univ, 54-55, res asst chem, 55-58; asst prof, Ga Inst Technol, 58-61; sr scientist, E G & G, Inc, 62, sci specialist nucleonics, 63-65 & accelerator physics, 66; assoc prof radiol & physics, 66-70, PROF ALLIED HEALTH PROFESSIONS, EMORY UNIV, 71-, ASSOC PROF RADIOL, 74-, PROF RADIOL. *Concurrent Pos:* Consult, E G & G, Inc, 67-69. *Mem:* AAAS; Am Chem Soc; Am Asn Univ Prof; Sigma Xi; Am Asn Physicists Med. *Res:* Radiochemistry; radiation effects; radiological health. *Mailing Add:* 6415 Paradise Point Rd Flowery Branch GA 30542

PALMER, RICHARD EVERETT, b New York, NY, Apr 26, 44; m 71. PHYSICS. *Educ:* Mass Inst Technol, SB, 66; Princeton Univ, PhD(physics), 71. *Prof Exp:* MEM STAFF PHYSICS, SANDIA LABS, 71- *Mem:* Am Phys Soc. *Res:* Research and development of high power lasers for laser fusion. *Mailing Add:* Sandia Nat Labs Div 8354 Livermore CA 94550

PALMER, ROBERT B, b London, Eng, Feb 28, 34. PHYSICS. *Educ:* Imperial Col, BS, 56, PhD(physics), 64. *Prof Exp:* Fel Imperial Col, 60-61; RES ASST, BROOKHAVEN NAT LAB, 85- *Mem:* Am Phys Soc; Am Inst Sci. *Mailing Add:* Brookhaven Nat Lab Upton NY 11973

PALMER, ROBERT GERALD, b Phillips, Wis, May 25, 36; m 57; c 3. PROGRAM & CROP MANAGEMENT. *Educ:* Univ Tenn, Knoxville, BS, 60, MS, 62; Iowa State Univ, PhD(soil mgt), 67. *Prof Exp:* Instr soils, Iowa State Univ, 64-66; asst prof soil mgt & exten specialist, Okla State Univ, 66-68; tech adv soils, Univ Tenn-India Agr Progs, 68-70; assoc prof, Western Ill Univ, 70-78, prof agr, 78-81; vpres & treas, Keyager Serv, Inc, Macomb, 77-79, pres, 79-82; mgr, Agron Serv, 82-86, mgr, Agron Serv Support, 86-91, MGR AGRON SERV, SOUTHERN SALES AREA, PIONEER HI-BRED INT, INC, 91- *Concurrent Pos:* Vis assoc prof, Fed Univ Santa Maria, Brazil with Southern Ill Univ & UN Food & Agr Orgn, 75-76; bd dirs, Am Registry Cert Prof Agron, Crops & Soils, 87-, chair bd dirs, 89-90, chair agron sub-bd, 88-87. *Mem:* Am Soc Agron; Soil Sci Soc Am; Soil Conserv Soc Am; Sigma Xi; Nat Alliance Independent Crop Consult; Hiwassee Presidential Assocs (pres, 86-88). *Mailing Add:* PO Box 364 Tipton IN 46072

PALMER, ROBERT HOWARD, b Chicago, Ill, Nov 3, 31; m 60; c 2. INTERNAL MEDICINE, GASTROENTEROLOGY. *Educ:* Oberlin Col, AB, 53; Harvard Med Sch, MD, 57. *Prof Exp:* Intern, Pa Hosp, Philadelphia, 57-58; resident, 58-61, from instr to assoc prof med, Hosps & Clins, Univ Chicago, 69-73; adj prof & med dir, Ctr Prevention of Premature Arteriosclerosis, Rockefeller Univ, 73-79; prof clin med, Col Phys & Surg, Columbia Univ, 79-85; ATTEND PHYSICIAN, PRESBY HOSP, 78- *Concurrent Pos:* Nat Found fel, Karolinska Inst, Sweden, 62-63; Arthritis Found spec investr, Univ Chicago, 66-67; mem gen med A study sect, NIH, 69-73; mem gastroenterol merit rev bd, Vet Admin, 71-; adj prof, Med Sch, Cornell Univ, 73-78; attend physician, New York Hosp, 73-78; adj prof, Col Physicians & Surgeons, Columbia Univ, 86. *Mem:* Am Asn Study Liver Dis; fel Am Col Physicians; Am Fedn Clin Res; Am Gastroenterol Asn; Am Soc Clin Invest. *Res:* Bile acid biochemistry and pharmacology; arteriosclerosis research; gallstone formation; hepatic metabolism. *Mailing Add:* Med Affairs E43 1500 Spring Garden St Philadelphia PA 19101

PALMER, ROGER, b Albany, NY, Sept 23, 31; div; c 4. PHARMACOLOGY, MEDICINE. *Educ:* St Louis Univ, BS, 53; Univ Fla, MD, 60. *Prof Exp:* From intern to resident med, Johns Hopkins Hosp, 60-62; from instr to assoc prof pharmacol, Col Med, Univ Fla, 62-69; chmn dept pharmacol, 70-81, PROF PHARMACOL & CHIEF DIV CLIN PHARMACOL, UNIV MIAMI, 69-, CLIN PROF MED, 81- *Concurrent Pos:* NIH grant, 64-67 & 73-; Markle scholar acad med, 65- *Honors & Awards:* Mosby Scholar Award, 57-60. *Mem:* Am Soc Pharmacol & Exp Therapeut; Am Col Clin Pharmacol; Am Fed Clin Res; Sigma Xi. *Res:* Cardiovascular pharmacology; cardiovascular effects of endorphins; vascular compliances and the nature of pulstile arterial flow in vascular injury; endorphins in smoking addiction; Bendectin and limb defects in humans (case study). *Mailing Add:* 24 W End Dr Key Biscayne FL 33144

PALMER, RUFUS N(ELSON), b Bath, Maine, Mar 7, 02; m 31. ENGINEERING. *Educ:* Mass Inst Technol, SB, 25, ScD(ceramic eng), 38. *Prof Exp:* Res engr & plant engr, Exolon Co, NY & Can, 25-34; indust fel, Mellon Inst, 37-47; consult engr, 47-66; chief engr, Hall Industs, 66-72; RETIRED. *Mem:* AAAS; Am Chem Soc; Am Ceramic Soc. *Res:* Chemical and mechanical engineering; artificial abrasives and refractories; electromagnetic ore separators; ceramics, pearl button machine design. *Mailing Add:* 1290 Boyce Rd B408 Upper St Clair PA 15241

PALMER, RUPERT DEWITT, b Winston Co, Miss, Jan 28, 29; m 54; c 2. WEED SCIENCE. *Educ:* Miss State Univ, BS, 52, MS, 54; La State Univ, PhD(bot), 59. *Prof Exp:* Assoc prof weed sci, Agr Exp Sta, Miss State Univ, 54-66; exten agronomist weed control & assoc prof agron, Tex A&M Univ, 66-74, from assoc prof to prof weed sci, 74-89, exten weed specialist, 74-89; PRES, DRIFT CONTROL ENTERPRISES, 89- *Mem:* Weed Sci Soc Am. *Mailing Add:* 2502 Willow Bend Bryan TX 77802

PALMER, SUSHMA MAHYERA, b Sirhind, India, Jan 13, 44; US citizen; m 67. NUTRITIONAL BIOCHEMISTRY, CLINICAL NUTRITION. *Educ:* Univ Delhi, BSc, 63, MS, 65; Univ Belgrade, DSc, 77. *Prof Exp:* Asst lectr advan nutrit, Lady Irwin Col, Univ Delhi, 65-66; therapeut dietitian, Washington Hosp Ctr, 66-67, res nutritionist, 68; res nutritionist biochem, George Hyman Res Found & George Washington Univ, 68; dir nutrit & asst prof pediat, Prog for Child Develop, Sch Med, Georgetown Univ, 71-75; fel nutrit & biochem & investr, Inst Biochem & Inst Mother & Child Health, Univ Belgrade, 75-78; staff scientist nutrit & biochem, 78-79, proj dir, 78-83, DIR, FOOD & NUTRIT BD, NAT ACAD SCI, 83- *Concurrent Pos:* Consult, NVa Child Develop Ctr, 71-75; instr nutrit, Univ Md, 72-75; mem, Prog Adv Comt, Nat Inst Dent Res, NIH, 79-83; mem, Mayor's Commission Food Nurit & Health, 83-87; mem, Am Pub Health Asn Action Bd, 86-; vis prof, Nat Inst Oncol, Budapes, Hungary, 87- *Mem:* Soc Nutrit Educ; Am Dietetic Asn; Am Inst Nutrit; Am Pub Health Asn. *Res:* Vitamin A, RBP and immune mechanism; subclinical zinc deficiency and growth; pediatric nutrition and developmental disorders; feeding problems in children. *Mailing Add:* Cent Europ Develop Corp Kirfursten Damm 3rd Fl D-1000 Berlin 15 Germany

PALMER, THEODORE W, b Boston, Mass, Oct 19, 35; m 61; c 3. MATHEMATICS. *Educ:* Johns Hopkins Univ, BS & MA, 58; Harvard Univ, AM, 59, PhD(math), 66. *Prof Exp:* Instr math, Tufts Univ, 65-66; vis asst prof, Math Res Ctr, Univ Wis, 66-67; from asst prof to assoc prof, Univ Kans, 67-70; assoc prof, Univ Ore, 70-75, head dept, 80-83, assoc dean, Col Arts & Sci, 86-90, actg dean, 90, PROF MATH, UNIV ORE, 75- *Concurrent Pos:* Vis prof, Univ Calif, Berkeley, 76 & 77, Ind Univ, Bloomington, 83 & 84. *Mem:* AAAS; Am Math Soc; Math Asn Am; Sigma Xi. *Res:* Banach algebras; linear operator theory. *Mailing Add:* Dept of Math Univ of Ore Eugene OR 97403

PALMER, THOMAS ADOLPH, b Salt Lake City, Utah, Oct 12, 35; m 79; c 2. ANALYTICAL CHEMISTRY. *Educ:* Univ Santa Clara, BS, 57; Iowa State Univ, PhD(analytical chem), 61. *Prof Exp:* Sr res analyst, Olin Mathieson Chem Corp, 61-62; sr chemist, Lockheed Propulsion Co, 62-66; staff res chemist, Kaiser Aluminum & Chem Corp, 66-80, sr staff res chem, 80-87; LAB DIR, IMPERIAL WEST CHEM CO, 87- *Concurrent Pos:* Consult, 87- *Mem:* Am Chem Soc; Sigma Xi; Am Water Works Asn. *Res:* Absorption spectroscopy; electrochemistry; environmental chemistry. *Mailing Add:* 704 Thornhill Rd Danville CA 94526

PALMER, TIMOTHY TROW, b Evanston, Ill, Mar 22, 38; m 62; c 3. LABORATORY SAFETY, HAZARDOUS MATERIALS. *Educ:* Carleton Col, BA, 60; Univ Minn, MS, 64, PhD(zool), 74; Univ Southern Calif, MS, 81. *Prof Exp:* Parasitologist, USN, 64-84; asst to dir, Gorgas Mem Lab, 76-79; occup safety & health mgr, Naval Med Res Inst, 79-84; pres, Tepee, Ltd, 84-90; DIR, ENVIRON ASSESSMENTS & TRAINING, PROF SAFETY CONSULTS CO, 87- *Concurrent Pos:* Head, malaria serol, Naval Med Res Inst, 72-76, head, In Vitro Cult Sect, 79-82; dir, Nat Safety Mgt Soc, 85-87 & 89-91. *Mem:* Am Soc Safety Engrs; Nat Safety Mgt Soc (pres, 87-89); emer mem Am Soc Trop Med & Hyg. *Res:* Development of sporozoite-based malaria vaccine; involved tissue and erythrocytic cultures of Avian, rodent and human malarias; immunological responses to mosquito inoculated Ag; occupational safety and health; infectious/tropical disease; zoonoses. *Mailing Add:* 3232 Chrisland Dr Annapolis MD 21403-4367

PALMER, WARREN K, b Oradell, NJ, June 29, 41; m 77. REGULATION OF LIPID METABOLISM. *Educ:* SUNY Cortland, BS 63; Ball State Univ, MA, 66; Univ Iowa, PhD(exercise physiol), 72. *Prof Exp:* assoc prop biochem, 83-88, PROF EXERCISE PHYSIOL, UNIV ILL, 88- *Mem:* Am Soc Biochem & Molecular Biol; Am Col Sports Med. *Res:* Emphasis of this laboratory is to determine the biochemical mechanisms of regulation of the metabolism of endogenous lipid in muscle cells. *Mailing Add:* Dept Health Phys Educ Univ Ill Chicago IL 60680

PALMER, WILLIAM FRANKLIN, b New York, NY, Mar 5, 37; m 63; c 3. PHYSICS. *Educ:* Harvard Univ, AB, 58; Johns Hopkins Univ, PhD(physics), 67. *Prof Exp:* Res assoc physics, Argonne Nat Lab, 67-69; from asst prof to assoc prof, 69-79, PROF PHYSICS, OHIO STATE UNIV, 79- *Mem:* Am Phys Soc. *Res:* High energy physics theory. *Mailing Add:* Dept Physics Ohio State Univ Columbus OH 43210

PALMER, WINIFRED G, b Brooklyn, NY. BIOCHEMISTRY. *Educ:* Brooklyn Col, BS, 62; Univ Conn, PhD(biochem), 67. *Prof Exp:* Res assoc biochem, Oak Ridge Nat Lab, 67-72; res scientist immunochem, Frederick Cancer Res Ctr, 72-78; sr biologist, Enviro Control, 78-80; prin scientist, Tracor Jitco, 80-83; MEM STAFF, BIOMED TOXICOL ASSOCS, 84- *Mem:* Reticuloendothelial Soc; AAAS; Sigma Xi; Soc Toxicol. *Res:* Chemical carcinogenesis; industrial toxicology. *Mailing Add:* Biomed Toxicol Assocs 6184 Viewsite Dr Frederick MD 21701

PALMERE, RAYMOND M, b Irvington, NJ, Aug 20, 25; m 49; c 12. AGRICULTURAL CHEMISTRY. *Educ:* Rutgers Univ, BA, 55; Seton Hall Univ, MS, 63, PhD(chem), 66. *Prof Exp:* Asst, Squibb Inst Med Res, 55-59, chemist, 59-63, res chemist, 63-65; res chemist, 65-67, SR RES CHEMIST, FMC CORP, PRINCETON, 67- *Mem:* Am Chem Soc; Sigma Xi. *Res:* Natural products; synthesis of yohimbine alkaloids, 9-alpha-bromo-11-keto and 11, 12 oxygenated progesterones; enzymatic hydroxylations of beta diketones; insect physiology; juvenile hormone. *Mailing Add:* 17 Morris Rd West Orange NJ 07052

PALMES, EDWARD DANNELLY, b Mobile, Ala, July 5, 16; m 43; c 3. BIOCHEMISTRY. *Educ:* Springhill Col, BS, 38; Georgetown Univ, MS, 39, PhD(biochem), 47. *Prof Exp:* Jr & asst chemist, NIH, 41-44; biochemist, Med Dept, Field Res Lab, 47-48; from asst prof to assoc prof, 48-60, prof, 60-84, EMER PROF ENVIRON MED, MED CTR, NY UNIV, 85- *Concurrent Pos:* Prof agr chem, Ore State Univ, 85- *Mem:* Am Chem Soc; Am Physiol Soc; Am Indust Hyg Asn; Soc Toxicol. *Res:* Toxicology. *Mailing Add:* 2745 Monterey Dr Corvallis OR 97330-3432

PALMGREN, MURIEL SIGNE, b New Orleans, La, Feb 15, 49. INHALATION TOXICOLOGY, MYCOTOXINS. *Educ:* Newcomb Col, Tulane Univ, BS, 71; Southeastern La Univ, MS, 73; Auburn Univ, PhD(mycotoxins), 78. *Prof Exp:* Training fel environ toxicol, Univ Wis-Madison, 77-79; asst prof microbiol & toxicol, Tulane Sch Pub Health & Trop Med, 80-84, adj asst prof, 84-88, RES SCIENTIST, CLIN IMMUNOL SECT, DEPT MED, TULANE MED SCH, 85- *Concurrent Pos:* Vis asst prof microbiol & biol, Univ New Orleans, 79-80 & 85-86, res assoc, Ctr Bio-Org Studies, 85-87; collabr, Southern Regional Res Ctr, Agr Res Serv, USDA, 80-85. *Mem:* Am Soc Microbiol; Am Chem Soc; Am Indust Hyg Asn; Soc Toxicol; Sigma Xi. *Res:* Determining effects of activated phagocytes on lung tissue following toxic gas exposures; determining the effects of mycotoxins on the immune defenses against viral infections (immunotoxicity); metabolism and toxic responses by the lungs to inhaled toxins (in vivo and in vitro). *Mailing Add:* 1224 Helios Ave Metairie LA 70005-1550

PALMIERI, JOSEPH NICHOLAS, b Providence, RI, Aug 24, 32; m 60; c 2. NUCLEAR PHYSICS. *Educ:* Brown Univ, ScB, 54; Harvard Univ, AM, 55, PhD, 59. *Prof Exp:* Instr physics, Harvard Univ, 58-61; from asst prof to assoc prof, 61-67, PROF PHYSICS, 67-, DIR COMPUT, OBERLIN COL, 86- *Concurrent Pos:* NSF sci fac fel, 71-72. *Mem:* Am Phys Soc; Am Asn Physics Teachers; Am Inst Physics; Sigma Xi. *Res:* Nucleon scattering. *Mailing Add:* Dept of Physics Oberlin Col Oberlin OH 44074

PALMISANO, PAUL ANTHONY, b Cincinnati, Ohio, Dec 30, 29. PEDIATRICS, PHARMACOLOGY. *Educ:* Xavier Univ, Ohio, BS, 52; Univ Cincinnati, MD, 56; Am Bd Pediat, dipl, 62; Univ Calif, Berkeley, MPH, 79. *Prof Exp:* Instr pediat, Univ Cincinnati, 61-62; med officer, Food & Drug Admin, 63-66; asst prof pharmacol, 67-68, from asst prof to assoc prof, 68-74, assoc dean, Sch Med, 80-90, PROF PEDIAT, UNIV ALA, BIRMINGHAM, 73-, ASST DEAN, SCH MED, 74- *Concurrent Pos:* Fel clin pharmacol, Univ Ala, Birmingham, 66-67; mem panel pediat scope subcomt, US Pharmacopoeia, 66- *Mem:* Am Acad Pediat; Am Pub Health Asn; Am Soc Pharmacol & Exp Therapeut. *Res:* Placental transfer of drugs, salicylate toxicology; plumbism; poison control; accident prevention. *Mailing Add:* Dept Pediat Univ Ala Med Ctr Birmingham AL 35294

PALMITER, RICHARD, b Poughkeepsie, NY, Apr 5, 42; c 1. MOLECULAR GENETICS. *Educ:* Duke Univ, BA, 64, Stanford Univ, PhD(biol sci), 68. *Prof Exp:* From asst prof to assoc prof, 74-81, PROF BIOCHEM, UNIV WASH, SEATTLE, 81- *Concurrent Pos:* Investr, genetics, Howard Hughes Inst, Univ Wash, 76- *Honors & Awards:* George Thorm Award, 82; Nat Acad Sci Award, 83. *Mem:* Nat Acad Sci; fel AAAS; Am Acad Arts & Sci. *Res:* Regulation of gene expression, especially in transgenic mice. *Mailing Add:* HHMI SL-15 Univ Wash Seattle WA 98195

PALMORE, JULIAN IVANHOE, III, b Baltimore, Md, Sept 26, 38; m 67; c 2. AERONAUTICAL ENGINEERING. *Educ:* Cornell Univ, BEP, 61; Univ Ala, MA, 64; Princeton Univ, MSE, 65; Yale Univ, MS, 66, PhD(astron), 67; Univ Calif, Berkeley, PhD(math), 73. *Prof Exp:* Res assoc, Ctr Control Sci, Univ Minn, 67-68; vis fel math, Princeton Univ, 68-69; actg instr & lectr, Univ Calif, Berkeley, 70-73; instr, Mass Inst Technol, 73-75; vis asst prof, Univ Mich, Ann Arbor, 75-77; from asst prof to assoc prof math, 77-84, prof math & aeronaut & astronaut eng, 84-87, PROF MATH, AERONAUT, ASTRONAUT ENG & GEN ENG, UNIV ILL, URBANA, 87- *Concurrent Pos:* Lilly fel, Mass Inst Technol, 74-75; NSF grants, 74-84 & 87-, prin investr, 78-84 & 87-; Ctr Advan Study fel, Univ Ill, 79; fac judge, Lilly open fels, 79-83; Construct Eng Res Lab, US Army, 89- *Mem:* fel Explorers Club; assoc fel Am Inst Aeronaut & Astronaut. *Res:* Chaos theory and dynamical systems theory; theoretical computer science and computability theory; limits of computers. *Mailing Add:* Dept Math Univ Ill 1409 W Green St Urbana IL 61801

PALMORE, WILLIAM P, b Beatrice, Ala, Dec 13, 34; m 65; c 2. VETERINARY PHYSIOLOGY. *Educ:* Auburn Univ, DVM, 59, MS, 64; Yale Univ, PhD(physiol), 67. *Prof Exp:* NIH res fel, 61-66; Am Vet Med Asn res fel, 66-67; res assoc path, Sch Med, Yale Univ, 67-68; from asst prof to assoc prof physiol, Sch Vet Med, Univ Mo-Columbia, 68-74; assoc prof physiol, Col Vet Med, Univ Fla, 74-89; RETIRED. *Mem:* AAAS; Am Vet Med Asn; Sigma Xi. *Res:* Adrenocortical research. *Mailing Add:* RR 1 Box 550 Alachua FL 32610-9801

PALMOUR, HAYNE, III, b Gainesville, Ga, Feb 27, 25; m 52; c 3. CERAMIC ENGINEERING. *Educ:* Ga Inst Technol, BCerE, 48, MS, 50; NC State Univ, PhD(ceramic eng), 61. *Prof Exp:* Res engr, US Bur Mines, 50-51; mgt trainee, US Dept Interior, 51-52; admin analyst, Nat Capital Parks, US Park Serv, 52-53; from res engr to sr res engr, Am Lava Corp, Tenn, 53-57; instr ceramic eng, NC State Univ, 57-58, res engr, 58-61, res assoc prof, 61-65, res prof, 65-81, assoc head dept, 85-90, PROF CERAMIC ENG, NC STATE UNIV, 81-, ACTG ASSOC HEAD DEPT, 90- *Concurrent Pos:* Consult, Hygrodynamics, Inc, 61-63, US Air Force, 62-64, Am Instrument Co, Inc, 63-67, W R Grace & Co, 68-75, Babcock & Wilcox Co, 72-75, & 81-, US Army, 76-77, EG&G Idaho Corp, 81-, Battelle Columbus Labs, 85-, Exxon Prod Res Co, 85-87, Small Precision Tools, 87-, Coors Ceramic Co, 87-, Akzo Corp, 90, Baikowski Int, 90-, CDB Corp, 90, Gaisor Tool Co, 90, Res Technol Inc, 90-; mem adv comt, Metals & Ceramics Div, Oak Ridge Nat Lab, 75-77, chmn, 77; mem comt dynamic compaction of metal & ceramic powders, Nat Mat Adv Bd, Nat Acad Sci, 80-83; co-ed, Int Conf Mats, 60-89. *Honors & Awards:* Res Award, Sigma Xi, 61; Alcoa Res Award, 81; Fulbright-Distinguished Prof, 84; John Marquis Mem Award, Am Ceramic Soc, 87; Frankel Prize, Int Inst for Sci of Sintering, 89. *Mem:* Fel Am Ceramic Soc; Nat Inst Ceramic Engrs; Int Inst Sci Sintering,. *Res:* Properties and applications of polycrystalline ceramics; densification dynamics and kinetics, rate controlled sintering; microstructural evolution; precision digital dilatometry; computer aided design of optimal paths for sintering. *Mailing Add:* Dept Mat Sci & Eng NC State Univ 229 Riddick Bldg Raleigh NC 27695-7907

PALMOUR, ROBERTA MARTIN, b Tex, Aug 24, 42; c 1. HUMAN GENETICS. *Educ:* Tex Western Col, BA, 63; Univ Tex, Austin, PhD(zool & human genetics), 70. *Prof Exp:* Lectr genetics, Univ Tex, Austin, 70-72; asst prof genetics, Univ Calif, Berkeley, 72-80; NIH postdoctoral fel, Univ Calif, San Diego, 80-82; ASSOC PROF PSYCHIAT & HUMAN GENETICS, MCGILL UNIV, 82- *Concurrent Pos:* Prin investr, many grants, 72-; dir, genetic counseling, McGill Univ, 85-, Psychiat Genetics Clin, 90- *Mem:* Am Soc Human Genetics; Soc Neurosci; Int Soc Primatology. *Res:* Intersection between human genetics and behavioral neuroscience; search for phenotypic and genotypic markers in familial alcoholism, psychiatric genetics in general, behavioral and metabollic studies of Lesch-Nyhan disease and primate models for inherited neurobehavioral and neuroendocrine diseases. *Mailing Add:* McGill Univ 1033 Pine Ave W Montreal PQ H3Z 1W7 Can

PALMQUIST, DONALD LEONARD, b Silverton, Ore, July 2, 36; m; c 5. ANIMAL NUTRITION. *Educ:* Ore State Col, BS, 58; Univ Calif, Davis, PhD(nutrit), 65. *Prof Exp:* Res assoc dairy sci, Univ Ill, Urbana, 65-67; from asst prof to assoc prof, 67-82, PROF DAIRY SCI, OHIO AGR RES & DEVELOP CTR & OHIO STATE UNIV, 82- *Concurrent Pos:* Assoc res biochemist, Radioisotope Res Div, Vet Admin, Wadsworth Vet Admin Hosp, Los Angeles, 73-74; vis prof, Swed Univ Agr Sci, Uppsala, 82, Nat Inst Animal Sci, Denmark, 90; Fulbright travel fel, 90. *Honors & Awards:* Am Cyanamid Award, Am Dairy Asn, 89; Danish Res Acad Lectr, 90. *Mem:* Am Soc Animal Sci; Am Oil Chemists Soc; Am Dairy Sci Asn; Am Inst Nutrit; Brit Nutrit Soc. *Res:* Lipid metabolism of the dairy cow; intermediary metabolism of rumen microorganisms; lipid kinetics; milk synthesis; dairy cattle feeding management. *Mailing Add:* Dept Dairy Sci Ohio Agr Res & Develop Ctr Wooster OH 44691

PALMQUIST, JOHN CHARLES, b Omaha, Nebr, Sept 6, 34; m 56; c 3. GEOLOGY. *Educ:* Augustana Col, AB, 56; Univ Iowa, MS, 58, PhD(geol), 61. *Prof Exp:* Asst geol, Univ Iowa, 56-61; geologist, Calif Co, 61-62; from asst prof to assoc prof geol, Monmouth Col, Ill, 62-68; assoc geol, 68-89, PROF, GEOL, LAWRENCE UNIV, 89- *Concurrent Pos:* consult petroleum explor, 79-; Coun, Undergrad Res. *Mem:* Fel Geol Soc Am; Nat Asn Geol Teachers; Am Asn Petrol Geologists. *Res:* Basement influence on later tectonics; structural analysis of metamorphic tectonites; metamorphic petrology; structural geology. *Mailing Add:* Dept of Geol Lawrence Univ Appleton WI 54912

PALMQUIST, ROBERT CLARENCE, b Chicago, Ill, Aug 8, 38; m 59; c 2. GEOMORPHOLOGY, ENVIRONMENTAL GEOLOGY. *Educ:* Augustana Col, BA, 60; Univ Wis-Madison, PhD(geol), 65. *Prof Exp:* From asst prof to assoc prof geol, Iowa State Univ, 65-81; assoc prof geol, Northwest Col, Powell, Wyo, 81-90; SR GEOLOGIST, APPL GEOTECHNOL, INC, BELLEVUE, WASH, 90- *Mem:* Geol Soc Am; AAAS; Sigma Xi. *Res:* Environmental and temporal influences in the development of mass movements; application of geological data to computerized land use analysis; relationship between depositional landforms and glacial processes. *Mailing Add:* 4926 Donovan Dr SE No C Olympia WA 98501-4873

PALMS, JOHN MICHAEL, b Rijswijk, Neth, June 6, 35; US citizen; m 58; c 3. NUCLEAR PHYSICS, APPLIED PHYSICS. *Educ:* The Citadel, BS, 58; Emory Univ, MS, 59; Univ NMex, PhD(physics), 66. *Hon Degrees:* DSc, The Citadel, 80. *Prof Exp:* Instr physics, US Air Force Acad, 61-62 & Univ NMex, 62-63; staff mem nuclear physics, Los Alamos Sci Lab, 63-66, mem grad thesis prog, 64-66; chmn dept, Emory Col, 69-74, dean, 74-79, vpres arts & sci, 79-82, vpres acad affairs, 82-88, FAC MEM PHYSICS, EMORY UNIV, 66-, CHARLES HOWARD CANDLER PROF PHYSICS, 88- *Concurrent Pos:* Staff mem nuclear physics, Sandia Corp, NMex, 62-63; mem semiconductor detector panel, Nat Res Coun, 65-; consult var nuclear industs and hosps; bd dir, Inst Defense Anal, Accrediation Bd Inst, Nuclear Power Opers, Source Technologies Inc; bd trustees, Brandon Hall Sch, Pace Acad, Wesleyan Col; sci adv, Oak Ridge Nat Lab; adv bd, The Citadel. *Mem:* AAAS; Am Nuclear Soc; Inst Elec & Electronics Engrs; Soc Nuclear Med; Am Physical Soc; Health Physics Soc. *Res:* Neutron inelastic scattering; coulomb excitation; Ge-gamma-ray detectors; semiconductor detectors; medical and radio-environmental physics; x-ray physics, radio-ecology and radiation dose physics. *Mailing Add:* Pres Ga State Univ University Plaza Atlanta GA 30303

PALOCZ, ISTVAN, b Budapest, Hungary, Sept 24, 20; US citizen; m 53; c 2. ELECTROPHYSICS. *Educ:* Budapest Tech Univ, dipl eng, 45, docent, 54; Polytech Inst Brooklyn, PhD(electrophys), 62. *Prof Exp:* Res staff mem, Tungsram Univ, Budapest, 45-50; asst prof elec eng & appl math, Budapest Tech Univ, 50-54, assoc prof, 54-56; res staff mem, Watson Lab, Columbia Univ, 57-61 & Res Ctr, Int Bus Mach Corp, 61-66; assoc prof elec eng, 65-67, prof electrophys, dept elec eng & electrophys, Polytech Inst NY, 67-; AT NY UNIV. *Concurrent Pos:* Consult, Tungsram Univ, Budapest, 50-56, Gen Precision Inc, 66 & Comput Sci Inc, 67-68. *Mem:* Sr mem Inst Elec & Electronics Engrs. *Res:* Applied mathematics; wave propagation; field theory. *Mailing Add:* Dept Electroph Polytech Univ 333 Jay St Brooklyn NY 11201

PALOPOLI, FRANK PATRICK, b Pittsburgh, Pa, Feb 19, 22; m 44; c 5. ORGANIC CHEMISTRY, MEDICINAL CHEMISTRY. *Educ:* Duquesne Univ, BS, 43, MS, 48. *Prof Exp:* Res chemist, Wm S Merrell Co Div, 50-63, dir chem res, Nat Drug Co Div, 63-70, head chem develop, Res Ctr Div, Richardson-Merrell, Inc, 70-81, HEAD CHEM DEVELOP, MERRELL-DOW RES INST, 81- *Mem:* Am Chem Soc; fel Am Inst Chemists. *Res:* Chemistry of synthetic drugs; medicinal agents; anti-inflammatory, fertility-sterility, cardiovascular and hypocholesteremic drugs. *Mailing Add:* 9893 Barnsley Ct Montgomery OH 45242

PALOTAY, JAMES LAJOS, b Los Angeles, Calif, Oct 3, 22; m 42; c 2. COMPARATIVE PATHOLOGY. *Educ:* Kans State Univ, DVM, 50; Colo State Univ, MS, 58; Am Col Vet Path, dipl, 59. *Prof Exp:* Asst prof path & bact, Colo State Univ, 50-51; res vet, Monfort Feed Lots, Colo, 51-55; asst prof path, Wash State Univ, 55-63; sr res scientist, Biol Lab, Gen Elec Co, 63-64; mgr comp toxicol sect, Pac Northwest Labs, Battelle Mem Inst, 64-68; sci dir, Ore Zool Ctr, 68-69; from assoc scientist to scientist, Ore Regional Primate Res Ctr, 69-85; RETIRED. *Mem:* AAAS; Am Vet Med Asn; Am Col Vet Path; Int Acad Path. *Res:* Systematized nomenclature of veterinary medicine. *Mailing Add:* Vet Path Rte 5 Box 434 Hillsboro OR 97124

PALOTTA, JOSEPH LUKE, b Port Sulphur, La, Sept 26, 40; m 78; c 3. PSYCHIATRY, MEDICAL HYPNOSIS. *Educ:* La State Univ, BS, 62, MD, 65. *Prof Exp:* PSYCHIATRIST PVT PRACT, 71- *Mem:* Am Psychiat Asn; Am Soc Clin Hypnosis. *Res:* Author of books on medical hypnosis and on Christian psychiatry. *Mailing Add:* 3801 Houma Blvd Metairie LA 70006

PALOYAN, EDWARD, b Paris, France, Mar 19, 32; US citizen; c 4. SURGERY. *Educ:* Univ Chicago, MD, 56; Am Bd Surg, dipl, 65. *Prof Exp:* Intern, Univ Hosp, Univ Chicago, 56-57, jr asst resident & res asst, 57-58 & 60-61, sr asst resident, 61-62, resident, Univ Hosps & Clins, 62-63, instr & chief resident, 64-65, from asst prof to assoc prof, 64-73, secy dept surg, 70-73; PROF SURG, STRITCH SCH MED, LOYOLA UNIV CHICAGO, 73-; ASSOC CHIEF OF STAFF FOR RES, VET ADMIN HOSP, HINES, 73- *Concurrent Pos:* Am Diabetes Asn fel, Univ Chicago Hosps & Clins, 62-63; mem bd dirs, Chicago Inst Med, 74. *Mem:* Fel Am Diabetes Asn; Am Fedn Clin Res; fel Am Col Surg; Soc Univ Surg; Am Surg Asn; Am Thyroid Asn; Int Soc Surg; Western Surg Asn. *Mailing Add:* Dept Surg Stritch Sch Med Loyola Univ Chicago Maywood IL 60153

PALS, DONALD THEODORE, b Cicero, Ill, Mar 29, 34; m 58; c 2. PHARMACOLOGY. *Educ:* Calvin Col, AB, 56; Univ Ill, MS, 60, PhD(physiol), 63. *Prof Exp:* Sr res pharmacologist, Norwich Pharmacal Co, 63-68, res assoc, 68-74; res assoc, 74-86, SR SCIENTIST, UPJOHN CO, 86- *Concurrent Pos:* Fel, Coun High Blood Pressure Res, Am Heart Asn. *Mem:* Am Soc Pharmacol & Exp Therapeut; Am Chem Soc; Am Heart Asn. *Res:* Cardiovascular pharmacology; mechanism of action and development of antihypertensive drugs. *Mailing Add:* Cardiovasc Dis Res Div Upjohn Co Kalamazoo MI 49001

PALSER, BARBARA FRANCES, b Worcester, Mass, June 2, 16. BOTANY. *Educ:* Mt Holyoke Col, AB, 38, AM, 40; Univ Chicago, PhD(bot), 42. *Hon Degrees:* DSc, Mount Holyoke Col, 78. *Prof Exp:* From instr to prof bot, Univ Chicago, 42-65, examr biol sci, 45-46 & Div Biol Sequence, 47-49; from assoc prof to prof bot, Rutgers Univ, New Brunswick, 65-82, dir grad prog in bot, 72-79; RETIRED. *Concurrent Pos:* Bot adv, Encycl Britannica, 57-59; ed, Bot Gazette, 59-65; vis prof, Duke Univ, 62; Erskine fel, Univ Canterbury, 69; vis res fel, Univ Melbourne, 84-85. *Honors & Awards:* Merit Award, Am Bot Soc, 85. *Mem:* Bot Soc Am (secy, 70-74, vpres, 75, pres, 76); Torrey Bot Club (pres, 68); Int Soc Plant Morphol; Am Inst Biol Sci. *Res:* Anatomy and morphology of Pteridophytes; histological responses to growth-regulating substances; floral morphology and anatomy of angiosperms, particularly Ericales; experimental anatomy. *Mailing Add:* Dept Biol Sci Rutgers Univ PO Box 1059 Piscataway NJ 08855

PALTER, N(ORMAN) H(OWARD), b Brooklyn, NY, May 12, 21; m 44; c 1. ENGINEERING. *Educ:* City Col New York, BEE, 42. *Prof Exp:* Develop engr, Hazeltine Electronics Corp, 45-48; proj engr, Sperry Gyroscope Co Div, 48-50, sr proj engr, 50-52, eng dept head, 52-62, mgr UK Polaris navig prog, 62-64, avionics systs prog mgr, 64-66, chief engr, Info & Commun Div, 66-67, mgr civil & indust systs, Sperry Syst Mgt Div, 68-72, mgr traffic & transp syst, 72-75, mgr ocean & mil syst, 75-80, dir planning, 80-82; RETIRED. *Mem:* Sr mem Inst Elec & Electronic Engrs. *Res:* Design and development of radar, missiles, countermeasures, navigation and avionics systems, vehicular traffic systems, simulation systems; ship positioning systems. *Mailing Add:* 106 Papermill Creek Ct Novato CA 94949

PALUBINSKAS, ALPHONSE J, b Lowell, Mass, Mar 24, 22; m 48; c 2. RADIOLOGY. *Educ:* Oberlin Col, BA, 48; Harvard Med Sch, MD, 52. *Prof Exp:* Intern med, Henry Ford Hosp, Detroit, 52-53; resident radiol, Peter Bent Brigham Hosp, Boston, 53-56, assoc radiologist, 56-57; from asst prof to assoc prof radiol, 59-69, PROF RADIOL, SCH MED, UNIV CALIF, SAN FRANCISCO, 69-, ASSOC RADIOLOGIST, 59- *Concurrent Pos:* Picker fel radiol res, Univ Col Hosp, London, Nat Hosp Nerv Dis, 57-58 & Karolinska Hosp Stockholm, 58-59; Commonwealth fel, 67. *Res:* Clinical diagnostic radiology. *Mailing Add:* Dept Radiol Univ Calif Med Ctr San Francisco 513 Parnassus Ave San Francisco CA 94143

PALUBINSKAS, FELIX STANLEY, b Lowell, Mass, Jan 16, 20; m 46; c 1. MEDICINE, OPTICS. *Educ:* Mass Inst Technol, BS, 44; Harvard Univ, AM, 47; Iowa State Univ, PhD(physics), 52; Tufts Univ, MD, 60. *Prof Exp:* Asst physics, Mass Inst Technol, 40-42 & Spectros Lab, 42-43, physicist, Confidential Instrument Develop Lab, 44-46; instr high sch, Mass, 46-47; instr physics, Iowa State Univ, 47-50; asst, Univ Ill, 50-52; physicist, Gen Elec Co, 52-53; prof electronic eng & head dept, Lowell Tech Inst, 53-56; intern, USPHS Hosp, Baltimore, Md, 60-61; staff to tech dir, Mitre Corp, Bedford, 61-68; PROF PHYSICS, BRIDGEWATER STATE COL, 68- *Concurrent Pos:* Consult biotechnol, Sch Aerospace Med, US Air Force, 79. *Mem:* AAAS; Am Phys Soc; Optical Soc Am; Aerospace Med Soc; Am Asn Physics Teachers; Sigma Xi. *Res:* Teaching of medicine, and physics. *Mailing Add:* 4 Winslow Rd Winchester MA 01890

PALUCH, EDWARD PETER, b New York, NY, Dec 8, 52; c 1. BASIC IMMUNOLOGY, CLINICAL IMMUNOLOGY. *Educ:* NY Univ, BA, 74; Columbia Univ, NY, MA, 76, PhD (path), 78. *Prof Exp:* Asst dir, Clin Immunol Lab, Mass Gen Hosp, 83-86; dir, New England Allergy & Immunol Lab, Inc, 83-86; med res assoc, Dept Clin Res, Boehringer Ingelheim Pharmaceuts Inc, 86-88. *Concurrent Pos:* Nat Res Serv Award, 78-80. *Mem:* Am Acad of Allergy & Immunol; Am Asn of Immunol; Am Soc of Clin Pathologists; Clin Immunol Soc; Asn of Med Lab Immunologists. *Res:* Clinical immunology; allergy; clinical research; cancer immunology. *Mailing Add:* Glaxo Inc Res Clin Group Five Moore Dr Research Triangle Park NC 27709

PALUMBO, FRANCIS XAVIER BERNARD, b Scranton, Pa, June 19, 45; m 71; c 1. HEALTH CARE ADMINISTRATION, PHARMACY. *Educ:* Med Col SC, BS, 68; Univ Miss, MS, 73, PhD(health care admin), 74; Univ Baltimore, JD, 82. *Prof Exp:* Pharmacist, US Army Health Clin, Pentagon, 69-71 & Dart Drug Corp, 69-71; teaching & res asst, Pharmaceut & Pharm Admin, Sch Pharm, Univ Miss, 71-74; actg chmn, Dept Pharm Admin, 78-79, asst prof pharm admin, 74-79, ASSOC PROF PHARM PRAC & ADMIN SCI, SCH PHARM, UNIV MD, 79-, ASSOC DIR, CTR DRUGS & PUB POLICY, 88- *Concurrent Pos:* Prin investr, Hoffmann-La Roche grant, 75-77, Nat Ctr Health Serv res grant, 79-82, Nat Inst aging grant, 84-87; mem, Human Develop & Aging Study Sect, NIH, 84-89. *Mem:* Am Asn Col Pharm; Am Pharm Asn; Am Asn Pharm Scientists. *Res:* Social and economic aspects of health, particularly appropriate drug use by the elderly in long-term care institutions; cost containment in third party drugs payment programs; food and drug law and regulation. *Mailing Add:* Sch Pharm 20 N Pine St Baltimore MD 21201

PALUMBO, PASQUALE JOHN, b Rochester, NY, Sept 24, 32; m; c 5. DIABETES MELLITUS, EPIDEMIOLOGY. *Educ:* Albany Med Sch, MD, 58. *Prof Exp:* CONSULT, INTERNAL MED & ENDOCRINOL, MAYO CLIN, 64-; PROF MED, MAYO MED SCH, 80- *Concurrent Pos:* Ann & Leo Markin Prof Med Educ, Mayo Found. *Mem:* Am Med Asn; Am Diabetes Asn; Endocrine Soc; AAAS; Am Col Physicians; Cent Soc Clin Res. *Res:* Vascular complications. *Mailing Add:* Dept Internal Med Mayo Med Sch 200 First St SW Rochester MN 55905

PALUMBO, SAMUEL ANTHONY, b Oak Park, Ill, June 4, 39; div; c 3. MICROBIOLOGY, FOOD SCIENCE. *Educ:* Loyola Univ, Ill, BS, 61; Univ Ill, Urbana, MS, 63, PhD(food sci), 67. *Prof Exp:* RES MICROBIOLOGIST, MICROBIAL FOOD SAFETY RES UNIT, EASTERN REGIONAL RES CTR, AGR RES SERV, USDA, 67- *Mem:* Am Soc Microbiol; Inst Food Technologists; Am Meat Sci Asn; Nat Registry Microbiologists; fel Am Acad Microbiol. *Res:* Food microbiology; food safety. *Mailing Add:* Eastern Regional Res Ctr Agr Res Serv USDA 600 E Mermaid Lane Philadelphia PA 19118

PALUSAMY, SAM SUBBU, b Tamil Nadu, India, Nov 14, 39; m 65; c 2. SOLID MECHANICS, NUCLEAR ENGINEERING. *Educ:* Univ Madras, BE, 63; Univ Waterloo, MASc, 66, PhD(solid mech), 70. *Prof Exp:* Jr engr, Tamil Nadu Pub Works Dept, India, 63; lectr civil eng, Annamalai Univ, 63-65; res asst solid mech, Univ Waterloo, 65-69, fel, 70; vis asst prof civil eng, Sir George Williams Univ, 70-71, sr res assoc syst bldg, 71-72; sr engr, 72-77, prin engr, 77-80, fel engr 80-84, mgr testing systs, 84-86 MGR STRUCT MAT ENG, WESTINGHOUSE ENERGY SYST, 86- *Concurrent Pos:* Task group chmn, 74-80, subcomt chmn, Pressure Vessel Res Comt, 84-; chmn, Working Group, Am Soc Mech Engrs Boiler Code, 77-85, Subgroup Design Anal, 85-86; tech adv, Mass Inst Technol-Idaho Nuclear Eng Lab, 85- *Mem:* Fel & assoc Am Soc Mech Engrs. *Res:* Pressure vessels; plastic analysis; model testing; system building, structural optimization; safety analysis; fracture mechanics; pipe whip; reactor vessel; diagnostics and monitoring for materials and structural degradation; life extension; nuclear plant startup testing. *Mailing Add:* PO Box 2728 Westinghouse Energy Syst Pittsburgh PA 15230-2728

PALUSINSKI, OLGIERD ALEKSANDER, b Sambor, Poland, Oct 26, 37; m 62; c 2. MICROELECTRONICS, COMPUTER SIMULATION. *Educ:* Tech Univ Silesia, Poland, MSc, 61, PhD(elec eng), 66; Tech Univ Lille, France, Dr, 68. *Prof Exp:* Res assoc, Inst Org Chem, 61; asst prof elec eng controls, Tech Univ Silesia, 61-67, assoc prof, 68-76; vis assoc prof simulation, 76-82, assoc prof elec eng & comput, 82-90, PROF ELEC & COMPUTER ENG, UNIV ARIZ, TUCSON, 90- *Concurrent Pos:* Vis researcher, Tech Univ Lille, France, 68, Oakland Univ, Mich, 72 , Univ Ariz, 72-73 & Univ Heidelberg, 84; ed-in-chief, Trans of Soc Comput Simulation, 84-90; vis prof, Univ Karlsruhe, 86 & Univ Lille, France, 90. *Mem:* Inst Elec & Electronic Engrs; Soc Comput Simulation. *Res:* Very large scale integration electrical modelling and simulation; very large scale integration interconnects electrical modelling; computer aided circuit analysis; electronic packagaing engineering. *Mailing Add:* Dept Elec & Comp Eng Univ Ariz Tucson AZ 85721

PAMER, TREVA LOUISE, b Doylestown, Ohio, Sept 22, 38; m 78. BIOCHEMISTRY. *Educ:* Kent State Univ, BS, 60; City Col New York, MA, 63; New York Med Col, PhD(biochem), 69; Stevens Inst Technol, MS, 85. *Prof Exp:* Chemist, Klett Mfg Co, 60-63; res technologist anal chem, Chas Pfizer & Co, Inc, 63-64; res assoc gastroenterol, New York Med Col, 64-71; lectr org chem, Mercy Col, 67-70; from asst prof to assoc prof, 68-80, PROF CHEM, JERSEY CITY STATE COL, 80-, CHMN, 90- *Mem:* AAAS; Am Chem Soc; Int Inst Conserv Hist & Artistic Works; Am Soc Testing & Mat. *Res:* Sulfated glycoproteins; composition of gastric secretion as related to gastric disorders; chemistry of art materials; conservation of art objects; computers in chemistry. *Mailing Add:* Dept Chem Jersey City State Col Jersey City NJ 07305

PAMIDI, PRABHAKAR RAMARAO, b Bangalore, India, Feb 18, 42; m 74; c 1. STRUCTURAL ANALYSIS, MECHANISMS & KINEMATICS. *Educ:* Univ Mysore, BE, 63; Indian Inst Sci, Bangalore, ME, 65; Okla State Univ, PhD(mech eng), 70. *Prof Exp:* Scientist, Cent Mech Eng Res Inst, Durgapur, India, 65-66; engr, Am Bur Shipping, 70-74; sr prin engr, Comput Sci Corp, 74-82, VPRES ENG, RPK CORP, 82- *Honors & Awards:* Douglas Michel Nastran Achievement Award, NASA & Cosmic, 87. *Mem:* Am Soc Mech Engrs. *Res:* Structural analysis using finite element techniques; applied mechanics; mechanisms and kinematics; vibrations; dynamics; engineering mechanics. *Mailing Add:* 7306 Carved Stone Columbia MD 21045-5226

PAMNANI, MOTILAL BHAGWANDAS, b Rohri, Pakistan, Oct 5, 33; m 70; c 1. PHYSIOLOGY, CARDIOVASCULAR DISEASES. *Educ:* Univ Bombay, MBBS (MD), 57, MS, 67; Mich State Univ, PhD(physiol), 75. *Prof Exp:* Resident med, G T Hosp & Infectious Dis Hosp, India, 57-58; lectr physiol, M P Shah Med Col, 58-61; lectr, Grant Med Col, India, 61-63, asst prof, 63-71; fel, Mich State Univ, 71-74, asst prof physiol, 75-76; asst prof, 76-80, ASSOC PROF PHYSIOL, UNIFORMED SERV UNIV HEALTH SCI, BETHESDA, 80-, ASST PROF MED, 80- *Concurrent Pos:* Grant in aid res, Mich Heart Asn, 76. *Mem:* Am Physiol Soc; Soc Exp Biol & Med; Am Fedn Clin Res. *Res:* Hypertension; studying role of veins, changes in vessel wall composition, membrane sodium potassium pump activity and vasoactive hormones in experimental and essential hypertension. *Mailing Add:* 19221 Seneca Ridge Ct Gaithersburg MD 20879

PAMPE, WILLIAM R, b Parkersburg, Ill, Dec 5, 23; m 49; c 2. GEOLOGY, PALEONTOLOGY. *Educ:* Univ Ill, AB, 47, MS, 48; Univ Nebr, PhD(geol), 66. *Prof Exp:* Explor geologist, Pure Oil Co, 48-60; from asst prof to assoc prof, 66-78, PROF GEOL, LAMAR UNIV, 78- *Mem:* Am Asn Petrol Geologists; Soc Econ Paleontologists & Mineralogists. *Res:* Detailed studies of invertebrate fossils especially on brachiopods; publications on meteorological subjects. *Mailing Add:* Dept Geol Lamar Univ Beaumont TX 77710

PAMPEYAN, EARL H, b Pasadena, Calif, Jan 3, 25; m 54; c 2. ENGINEERING GEOLOGY. *Educ:* Pomona Col, BA, 51; Claremont Col, MA, 53. *Prof Exp:* res geologist, US Geol Surv, Menlo Park, Calif, 52-90; RETIRED. *Mem:* Fel Geol Soc Am; Asn Eng Geol. *Mailing Add:* 747 Los Altos Ave Los Altos CA 94022

PAN, BINGHAM Y(ING) K(UEI), b China, Feb 11, 23; m 49; c 2. CHEMICAL ENGINEERING. *Educ:* Ord Eng Col, China, BS, 47; Va Polytech Inst, MS, 56, PhD(chem eng), 59. *Prof Exp:* teaching asst & instr instrumentation & unit opers, Va Polytech Inst, 55-59; res engr, Monsanto Co, 59-60, sr res engr, 61-65, sr process engr, 65-66, specialist, 66-74; group head, Occidental Res Corp, 74-80; lectr, Calif Polytech Univ, 79-80; res assoc, Dresser Industs, 80-84; PROJ ENGR, NAVAL CIVIL ENG LAB, 84- *Mem:* Am Chem Soc; Am Inst Chem Engrs. *Res:* Thermodynamics; kinetics; reactor design and process optimization. *Mailing Add:* 2442 San Fernando Court Claremont CA 91711

PAN, CHAI-FU, b Loshon, China, Sept 8, 36; US citizen; m 62; c 2. ELECTROCHEMISTRY. *Educ:* Nat Taiwan Univ, BS, 56; Univ Kans, PhD(phys chem), 66. *Prof Exp:* Assoc prof, 66-71, PROF CHEM, ALA STATE UNIV, 71- *Concurrent Pos:* Prin investr on theories of electrolyte solutions, 66- *Mem:* Am Chem Soc; fel Am Inst Chemists; Int Union Pure & Appl Chem. *Res:* Theoretical studies of electrolyte solutions. *Mailing Add:* 2420 Wentworth Dr Montgomery AL 36106

PAN, CHUEN YONG, b Miao-li, Taiwan, Feb 6, 40; Can citizen; m 66; c 2. CHEMICAL ENGINEERING, PHYSICAL CHEMISTRY. *Educ:* Nat Taiwan Univ, BSc, 62; Univ Ottawa, MSc, 66; Univ Toronto, PhD(chem eng), 70. *Prof Exp:* RES ENGR, ALTA RES COUN, 70- *Mem:* Chem Inst Can; Can Soc Chem Eng; Am Inst Chem Eng. *Res:* Physical separation techniques; adsorption; permeation; mass transfer; membrane. *Mailing Add:* Alta Res Coun 250 Karl Clark Rd Edmonton AB T6N 1E4 Can

PAN, CODA H T, b Shanghai, China, Feb 10, 29; US citizen; m 51; c 2. MECHANICAL ENGINEERING, MATHEMATICS. *Educ:* Ill Inst Technol, BS, 50; Rensselaer Polytech Inst, MS, 58, PhD(aeronaut eng, astronaut), 61. *Prof Exp:* Stress analyst combustion eng, Superheater Co, 50; test engr, Gen Elec Co, 50-51, design engr, Meter Instrument Lab, 51-52, rotating assignments, Adv Eng Prog, 52-54, develop engr, Gas Turbine Dept, 54-58, supvr adv eng prog, 58-61, fluid mech engr, 61; mgr res, Mech Technol Inc, 61-70, assoc dir res & technol, 70-73; tech dir, Shaker Res Corp, 73-81; prof mech eng, Columbia Univ, 81-87; SR CONSULT ENGR, DIGITAL EQUIP CORP, 87- *Concurrent Pos:* Adj prof, Rensselaer Polytech Inst, 61-65, 71-; consult, Army Res Off, 70-74; vis prof, Royal Tech Univ Denmark, 71; NIH spec fel microrheology of erythrocytes, Columbia Univ, 71-72; prin investr, Spacelab I, NASA, 78-; mem coun thrombosis, Am Heart Asn. *Honors & Awards:* Indust Res IR-100 Award, 67. *Mem:* Fel Am Soc Mech Engrs; Am Phys Soc; fel Am Soc Lubrication Engrs; Am Acad Mech; AAAS. *Res:* Gas lubrication; fluid film bearings; rotor dynamics; friction excited vibrations; elastohydrodynamics; process fluid lubrication, physics of fluids; mechanics and rheology of erythrocytes; hemodynamics; rheology. *Mailing Add:* Digital Equip Corp 333 South St Shrewsbury MA 01154-4112

PAN, DAVID, b June 6, 44; c 2. ENZYMOLOGY, IMMUNOLOGY. *Educ:* Univ Man, PhD(enzym), 75. *Prof Exp:* RES ASSOC, UNIV WIS, 78- *Mem:* Am Plant Physiol Soc; Cell Biol Soc. *Res:* Biosynthesis of carbohydrates in maize including synthesis of primer, formation of phospho-oligosaccharides and the functioning and structure of multienzyme complex; investigation on the biochemical and molecular genetic regulation or biosynthesis of carbohydrate in maize; study on enzymes involved including synthesis of primer, formation of phospho-oligosaccharides and functioning and structure of multi-enzyme complex; crop improvement. *Mailing Add:* Dept Genetics Univ Wis 14 B Univ Houses Madison WI 53705

PAN, HUO-HSI, b Foochow, China, Nov 11, 18; m 60; c 2. APPLIED MECHANICS, MECHANICAL ENGINEERING. *Educ:* Nat Southwest Assoc Univ, China, BS, 43; Agr & Mech Col Tex, MS, 49; Kans State Col, MS, 50; Univ Calif, Berkeley, PhD(mech eng), 54. *Prof Exp:* Asst engr, Yunnan Copper Smelting Plant, China, 42-43; mem tech staff, 21st Arsenal, 43-44, head inspection dept, 44-47; asst prof eng mech, Univ Toledo, 54-55; asst prof gen eng, Univ Ill, 55-57; asst prof eng mech, NY Univ, 57-59, from asst prof to prof appl mech, 59-71, prof mech eng, 71-73; prof appl mech, Polytech Inst NY, 73-74, prof mech eng, 74- 76, prof mech & aerospace eng, 76-78, prof mech & aero eng, 78-88, prof mech eng, 88-90, EMER PROF MECH ENG, POLYTECH UNIV, 90- *Concurrent Pos:* Res grants, NSF, 64-67 & NASA, 66-68. *Mem:* Am Inst Aeronaut & Astronaut; Soc Indust & Appl Math; Soc Eng Sci; Am Soc Mech Engrs. *Res:* Elasticity; viscoelasticity; vibrations and dynamics of elastic and viscoelastic systems. *Mailing Add:* Dept Mech & Aerospace Eng Polytech Inst NY 333 Jay St Brooklyn NY 11201

PAN, HUO-PING, b Foochow, China, Feb 13, 21; US citizen; m 55; c 1. FOOD SCIENCE, BIOCHEMICAL ENGINEERING. *Educ:* Nat Southwest Assoc Univ, China, BS, 46; Univ Ill, PhD(food sci), 54. *Prof Exp:* Teaching asst analytical chem, Mining & Metall Eng Dept, Nat Yunnan Univ, 46-49; staff mem food chem, Div Indust Corp, Mass Inst Technol, 54-55 & Div Sponsored Res, 55-57, res assoc, Dept Food Technol, 57-58; asst biochemist food chem, Agr Exp Sta, Univ Fla, 58-63, asst res prof, Dept Chem Eng, 63-64; res biochemist drug metab, 64-75, res chemist, Patuxent Wildlife Res Ctr, 75-76, res chemist, Denver Wildlife Res Ctr, US Dept Agr, 76-86, asst dir, Nat Monitoring & Residue Anal Lab, 86-88, RES CHEM, DENVER WILDLIFE RES CTR, US DEPT AGR, 88- *Mem:* AAAS; Am Chem Soc; Int Soc Study Xenobiotics; Am Inst Biol Sciences; fel Am Inst Chemists. *Res:* drug metabolism in birds; toxication-detoxication microsomal enzyme systems; chemical effects in high energy radiation. *Mailing Add:* Denver Wildlife Res Ctr Bldg 16 Denver Fed Ctr Denver CO 80225-0266

PAN, IN-CHANG, b Tokyo, Japan, Mar 28, 29; US citizen; m 54; c 2. IMMUNOPATHOLOGY, VIROLOGY. *Educ:* Nat Taiwan Univ, DVM, 51; Univ Calif, Davis, MS, 60; Purdue Univ, PhD(vet path), 66. *Prof Exp:* Vet med officer microbiol, Taiwan Prov Inst Animal Health, 51-54; instr path, Nat Taiwan Univ, 54-63 & Purdue Univ, 63-66; fel, Ont Vet Col, 66-67, asst prof, 67-68; vet med officer immunopath, Plum Island Animal Dis Ctr, Agr Res Serv, USDA, 68-89; VIS FEL, NAT SCI COUN, TAIWAN, 89- *Concurrent Pos:* Exchange scholar, Univ Calif, Davis, 58-60; mem, Exotic Animal Dis Comt, US Animal Health Asn, 82-89. *Mem:* Am Asn Immunologists; Am Asn Pathologists; Sigma Xi; Conf Res Workers Animal Dis; US Animal Health Asn; AAAS; Am Asn Vet Immunologists; Am Asn Appl Immunologists. *Res:* Immunologic interference of colostral antibody in hog cholera vaccination; persistent viral infection; diagnosis of animal viral infections by per. *Mailing Add:* Taiwan Prov Res Inst Animal Health 376 Chung-Cheng Rd Tansui Taiwan

PAN, KEE-CHUAN, b China, Aug 13, 41; m 71; c 2. PHYSICAL CHEMISTRY. *Educ:* Nat Taiwan Univ, BS, 63; Nat Tsing Hua Univ, Taiwan, MS, 65; State Univ NY, Buffalo, PhD(phys chem), 70. *Prof Exp:* Prof phys chem & chmn dept, Tamkang Univ, Taiwan, 70-78; res assoc, Brandeis Univ, 75-76 & Pa State Univ, 78; res chemist, 78-80, SR RES CHEMIST, EASTMAN KODAK CO, 80- *Mem:* Am Chem Soc; Am Electrochem Soc. *Res:* Electrochemistry. *Mailing Add:* 19 Locke Dr Pittsford NY 14534-4017

PAN, KO CHANG, b Kwangsi, China, Jan 13, 39; m 68. ENGINEERING MECHANICS. *Educ:* Nat Taiwan Univ, BS, 61; Tex A&M Univ, ME, 64; Univ Iowa, PhD, 69. *Prof Exp:* Res asst elasticity, Univ Iowa, 64-69; MECH ENGR, US ARMY ARMAMENT RES & DEVELOP COMMAND, 69- *Mem:* Am Soc Mech Engrs; Sigma Xi. *Res:* Elasticity; plate and shell; dynamics; optimization. *Mailing Add:* Two Eagles Nest Rd Morristown NJ 07960

PAN, POH-HSI, b Hangzhou, China, July 15, 22; US citizen; m 55; c 2. GEOPHYSICS. *Educ:* Chekiang Univ, BS, 44; Colo Sch Mines, MSc, 63; Rice Univ, PhD(geophys), 69. *Prof Exp:* Mech engr, Chinese Petrol Corp, 47-54, seismic & gravity party chief, 54-59, chief geophysicist, 60; geophysicist, 63-73, geophys specialist, 73-77, geophys supvr, 77-81, geophys adv, 82-86, GEOPHYS MGR, MOBIL OIL CORP, 87-; PRES, PAN CONSULT CO, 87- *Concurrent Pos:* Vis prof, Cheng Kung Univ, Taiwan, 58-60; adj lectr, Rice Univ, 76-77. *Mem:* Soc Explor Geophysicists; Sigma Xi. *Res:* Seismic stratigraphic study and gravimetric analysis in petroleum exploration. *Mailing Add:* 16815 Colegrove Dallas TX 75248

PAN, SAMUEL CHENG, biochemistry; deceased, see previous edition for last biography

PAN, STEVE CHIA-TUNG, b Kaohsiung, Formosa, Mar 20, 22; m 51; c 3. MEDICINE. *Educ:* Tokyo Jikeikai Med Col, Japan, MD, 47; Harvard Univ, MPH, 53. *Prof Exp:* Parasitologist, 406th Med Gen Lab, US Army, 46-52; res assoc, 54-55, instr, 55-59, assoc, 59-63, from asst prof to assoc prof, 63-80, PROF TROP PUB HEALTH, SCH PUB HEALTH, HARVARD UNIV, 80- *Concurrent Pos:* Milton res fel, Sch Pub Health, Harvard Univ, 53-54. *Mem:* Am Soc Trop Med & Hyg; Am Soc Parasitol; fel Royal Soc Trop Med & Hyg; Soc Protozool; AAAS. *Res:* In vitro cultivation; nuclear division and histochemistry of intestinal protozoa of man; helminths infections; schistosomiasis and ascariasis; histology and histopathology of fresh water snails; fine structure of human parasites. *Mailing Add:* Dept Trop Pub Health Harvard Univ Sch Pub Health Boston MA 02115-6092

PAN, VICTOR, b Moscow, Soviet Union, Sept 8, 39; US citizen; m 72. DESIGN & ANALYSIS OF ALGEBRAIC & NUMERICAL ALGORITHMS, COMBINATIONAL COMPUTATIONS. *Educ:* Moscow Univ, MS, 61, PhD(math), 64. *Prof Exp:* vis scientist math, IBM Res Ctr, 77-79; mem sch math, Inst Adv Study, Princeton, 80-81; PROF COMPUT SCI, STATE UNIV NY, 79- *Concurrent Pos:* Vis prof comput sci, Stanford Univ, 81, Univ Pisa, Italy, 84, City Univ NY & Lehman Col, 88-91; NSF grants, 80-82, 82-85, 85-88, 88-91. *Mem:* Am Math Soc; Soc Indust & Appl Math; Asn Comput mach; European Asn Theoret Comput Sci. *Res:* Design and analysis of algebraic, numerical and combinatorial algorithms; parallel computing; numerical linear algebra; mathematical programming and operational research. *Mailing Add:* Dept Comp Sci State Univ NY Albany NY 12222

PAN, YUAN-SIANG, b Fuchou, China; US citizen. ENGINEERING. *Educ:* Cheng Kung Univ, Taiwan, BS, 57; Brown Univ, PhD(eng), 64. *Prof Exp:* Res asst fluid mech, Brown Univ, 61-62; mem res staff mech eng, Mass Inst Technol, 62-64, res assoc, 64-65; from res asst prof to res assoc prof aeronaut & astronaut, NY Univ, 65-69; assoc prof aerospace eng, Univ Tenn Space Inst, Tullahoma, 69-73; sr staff scientist, JIFAS, Langley Res Ctr, NASA, 73-75, sr res assoc, Nat Res Coun, 75-78; MECH ENGR PROJ MGR, PITTSBURGH ENERGY TECHNOL CTR, US DEPT ENERGY, 78- *Concurrent Pos:* Consult, Northrop Serv Inc, Huntsville, 71-; Nat Res Coun sr res assoc, Nat Res Coun-Nat Acad Sci, 75-77; NASA fel acoust, JIFAS, NASA Langley Res Ctr & George Wash Univ, 75-77. *Mem:* AAAS; Am Inst Aeronaut & Astronaut; Am Soc Mech Engrs; Combustion Inst; Sigma Xi; Air Pollution Control Asn. *Res:* Fluid mechanics; supersonic, hypersonic and rarefied flow; air pollution control; combustion of coal and coal-derived fuels and combustion technology of coal slurry mixtures; supersonic combustion; noise generation and propagation. *Mailing Add:* Pittsburgh Energy Technol Ctr PO Box 10940 Pittsburgh PA 15236

PAN, YU-CHING E, b Mar 6, 45; m; c 2. BIOCHEMISTRY. *Educ:* Univ Cincinnati, PhD(chem), 76. *Prof Exp:* RES LEADER, HOFFMAN-LA ROCHE, INC, 85- *Concurrent Pos:* NIEHS, NIH; vis prog, NIH. *Mem:* Am Chem Soc; Am Soc Biochem & Molecular Biol; Protein Soc. *Res:* Structural analysis of protein; protein structure/function. *Mailing Add:* Dept Protein Biochem Hoffman-La Roche Inc Nutley NJ 07110

PAN, YUH KANG, b Canton, China, Feb 14, 37; m 64; c 2. THEORETICAL CHEMISTRY. *Educ:* Nat Univ Taiwan, BSc, 59; Mich State Univ, PhD(chem), 66. *Prof Exp:* Fel chem physics, Univ Southern Calif, 66-67; fel theoret chem, Harvard Univ, 67; from asst prof to assoc prof, 67-74, PROF CHEM, BOSTON COL, 74- *Concurrent Pos:* Vis prof, Stuttgart Univ, 74 & Max Planck Inst Radiation Chem, Mülheim, WGer, 75; ed, J Molecular Sci, 81-; hon prof, Chinese Acad Sci, Lanzhou Univ & Jilin Univ. *Mem:* Am Chem Soc; Am Phys Soc; Royal Soc Chem. *Res:* Theory of spectroscopy; quantum dynamics and biology. *Mailing Add:* Dept Chem Boston Col Chestnut Hill MA 02167

PAN, YU-LI, b Cheng-Tu, China, Aug 20, 39. EXPERIMENTAL HIGH ENERGY PHYSICS, LASERS. *Educ:* Univ Okla, BS, 57; Univ Calif, Berkeley, PhD(physics), 64. *Prof Exp:* Res assoc physics, Lawrence Radiation Lab, Univ Calif, 64-65; instr, Univ Pa, 65-68, asst prof, 68-69; SR RES PHYSICIST, LAWRENCE LIVERMORE LAB, UNIV CALIF, 70- *Mem:* Am Phys Soc. *Res:* High energy physics; gas laser; laser fusion. *Mailing Add:* Lawrence Radiation Lab Univ Calif Livermore CA 94550

PANAGIDES, JOHN, b New York, NY, Aug 15, 44; m 67; c 3. PSYCHOPHARMACOLOGY, IMMUNOPHARMACOLOGY. *Educ:* City Col New York, BS, 66; Univ NC, MS, 68; State Univ NY Buffalo, PhD(biol), 72. *Prof Exp:* Res assoc biochem cytol, Rockefeller Univ, 72-73; res biologist, Lederle Labs, 73-80, sr res biologist, 80-83; sr clin monitor, Ayerst Labs, 83-87; GROUP DIR, CNS, ORGANON, 87- *Concurrent Pos:* Adj asst prof, Pace Univ, 78. *Mem:* NY Acad Sci; AAAS; Am Soc Pharmacol & Exp Therapeut. *Res:* Clinical development of psychopharmacological drugs. *Mailing Add:* Organon 375 Mount Pleasant Ave West Orange NJ 07052

PANAGIOTOPOULOS, ATHANASSIOS Z, b Thessoniki, Greece, Mar 27, 60. THERMODYNAMICS, MOLECULAR SIMULATION. *Educ:* Nat Tech Univ, Athens, dipl chem eng, 82; Mass Inst Technol, PhD(chem eng), 86. *Prof Exp:* Res assoc phys chem, Lab Phys Chem, Oxford Univ, 86-87; ASST PROF CHEM ENG, SCH CHEM ENG, CORNELL UNIV, 87- *Concurrent Pos:* NSF presidential young investr, 89. *Mem:* Am Inst Chem Engrs; Am Chem Soc; Am Phys Soc; AAAS. *Res:* Statistical mechanics; molecular thermodynamics; computer simulation methods for the prediction of thermophysical properties of fluid mixtures. *Mailing Add:* Dept Chem Eng 120 Olin Hall Cornell Univ Ithaca NY 14853

PANALAKS, THAVIL, b Bangkok, Thailand, Nov 9, 17; Can citizen; m 58. ANALYTICAL CHEMISTRY, FOOD CHEMISTRY. *Educ:* Univ Hawaii, BS, 43; Univ Toronto, MSA, 48, PhD(chem), 56. *Prof Exp:* Res asst agron, Nebr Agr Exp Sta, 48-50; instr food chem, Univ Toronto, 51-54, sr res asst biochem, Connaught Med Res Labs, 54-56; chemist, Can Dept Nat Health & Welfare, 57-65, res scientist anal chem, 65-79; consult, Can Exec Serv Overseas, 80-81; Baha'I Int Health Agency, 83-86; RETIRED. *Concurrent Pos:* NATO scholar, 65. *Mem:* AAAS; Am Chem Soc; Nutrit Soc Can. *Res:* Detection and determination of toxic substances in foods; nitrosamines, polycyclic aromatic hydrocarbons, mycotoxins, panthenol and vitamin D. *Mailing Add:* 71 Jacqes Cartier Pointe Gatineau PQ J8T 2W3 Can

PANANGALA, VICTOR S, b Jan 6, 39. MONOCLONAL ANTIBODIES. *Educ:* Cornell Univ, PhD(immunol), 81. *Prof Exp:* ASSOC PROF METHOD IMMUNOL, AUBURN UNIV, 81- *Mem:* Am Asn Immunologists; Am Soc Microbiol; Am Vet Med Asn. *Res:* Reproductive diseases (brucellosis) in cattle; hybridoma and molecular biology. *Mailing Add:* Dept Pathobiol Col Vet Med Auburn Univ Rm 254 Green Hall Auburn AL 36849-5519

PANAR, MANUEL, b Edmonton, Alta, Jan 19, 35; m 64. INFORMATION SCIENCE, POLYMER SCIENCE. *Educ:* Univ Alta, BSc, 57; Calif Inst Technol, PhD(chem), 61. *Prof Exp:* Res fel chem, Harvard Univ, 61-63; res chemist, E I du Pont de Nemours & Co, Inc, 64-68, res supvr, 69-80, res mgr, 81-89, SCI DIR, CENT RES DEPT, E I DU PONT DE NEMOURS & CO, INC, 89- *Mem:* Am Chem Soc. *Res:* Physical organic and polymer physical chemistry; structure of anisotropic liquids; solvent-solute interactions; polymer morphology. *Mailing Add:* Cent Res Dept E I du Pont de Nemours & Co Inc Wilmington DE 19898

PANARELLA, EMILIO, b Ferrandina, Italy, Jan 3, 33; m 63; c 3. PHYSICS. *Educ:* Navig Sch, Camogli, Italy, dipl, 51; Univ Navig, Naples, Italy, Dr, 56. *Prof Exp:* With Malpensa Airport, Milano, 58-60; teacher navig, Navig Sch, Savona, 60-61; NATO fel microwave technol, Polytech Inst Brooklyn, 61-62; res officer, Microwave Res Inst, Florence, Italy, 62-64; res officer, Nat Res Coun Can, 64-87; PRES ADV LASER & FUSION TECHNOL, INC, 87- *Concurrent Pos:* Res prof, Univ Tenn, Knoxville; ed, Int J Physics Essays. *Mem:* Am Phys Soc; Europ Phys Soc; Ital Phys Soc; fel NY Acad Sci; sr mem Inst Elec & Electronic Engrs; Sigma Xi. *Res:* Plasma and laser physics; quantum mechanics. *Mailing Add:* Nat Res Coun Ottawa ON K1A 0R6 Can

PANASENKO, SHARON MULDOON, b Oakland, Calif, April 30, 50. BIOCHEMISTRY. *Educ:* Univ Calif, Berkeley, AB, 72; Stanford Univ, PhD(biochem), 77. *Prof Exp:* Fel biochem, Univ Calif, Berkeley, 77-79; ASSOC PROF BIOCHEM, POMONA COL, 79- *Res:* Modification of proteins and phospholipids during chemotaxis and development in myxococcus xanthus. *Mailing Add:* Abbott Lab Diag Div Dept 9NA AP-20 Abbott Park IL 60064

PANASEVICH, ROBERT EDWARD, PHARMACOLOGY. *Educ:* Univ Scranton, BS, 58. *Prof Exp:* ASSOC DIR PHARMACOL, PHARMAKON LABS, 69- *Mailing Add:* Pharmakon Labs Waverly PA 18471

PANAYAPPAN, RAMANATHAN, b Madras, India, Sept 21, 36; US citizen; m 60; c 2. POLLUTION CHEMISTRY. *Educ:* Annamalai Univ Madras, BSc, 59, MSc, 67; Howard Univ, PhD(inorg chem), 73. *Prof Exp:* Lectr chem, Annamalai Univ Madras, 60-67; adj prof polymer chem res, NASA Res Proj, Shaw Univ, 74-75; RES CHEMIST HIGH TEMPERATURE CHEM & POLLUTION, US NAVAL RES LAB, 75- *Mem:* Am Chem Soc. *Res:* Water and air pollution; high temperature solution chemistry. *Mailing Add:* 7508 Buena Vista Terr Rockville MD 20855

PANCELLA, JOHN RAYMOND, b Dunbar, Pa, June 24, 31. SCIENCE TEACHER SUPERVISION, SCIENCE CURRICULUM DEVELOPMENT. *Educ:* Ind Univ Pa, BS, 53; Univ Md, MS, 62, EdD, 72. *Prof Exp:* Instr biol, Penn Hills High Sch, Pittsburgh, Pa, 55-60 & Richard Montgomery High Sch, Rockville, Md, 65-68; instr zool, Univ Md, 60-61, instr sci educ, 62-65; supvr elem sci, 68-71, asst dir testing, 71-76, coordr sec sci, 76-85, SCI EDUC CONSULT, MONTGOMERY COUNTY PUB SCHS, MD, 85- *Concurrent Pos:* Book reviewer, Nat Sci Teachers Asn, 62-, AAAS, 69-, Appraisal-Sci Books Young People, 72- & Nat Asn Biol Teachers, 74-; prof, Univ Md, Baltimore, 69-70; consult, Nat Geog Soc, 73-85, US AID, Cairo, Egypt, 89, Dynamac Corp-Environ Protection Agency, 89-90; lectr, George Washington Univ, 85-86, Sci Resource, Marymount Univ Jr Sch, 85-88. *Mem:* Fel AAAS; Nat Asn Biol Teachers; Nat Sci Teachers Asn. *Res:* Cognitive levels of biology test items; reproduction of fresh water invertebrates; textbook study on teaching; writing objectives; individualized learning; higher level thinking skills. *Mailing Add:* 1209 Veirs Mill Rd Rockville MD 20851-1747

PANCIERA, ROGER J, b Westerly, RI, Sept 30, 29; m 53; c 2. VETERINARY PATHOLOGY. *Educ:* Okla State Univ, DVM, 53; Cornell Univ, MS, 55, PhD, 60; Am Col Vet Path, dipl, 63. *Prof Exp:* From asst to instr vet path, Cornell Univ, 53-56; asst prof, 56-63, PROF VET PATH, OKLA STATE UNIV, 63- *Mem:* Am Vet Med Asn; NY Acad Sci. *Res:* General veterinary pathology; toxicology. *Mailing Add:* Dept Vet Path Col Vet Med Okla State Univ Stillwater OK 74078

PANCOE, WILLIAM LOUIS, JR, b Chester, Pa, Feb 9, 38; m 61; c 3. ENDOCRINOLOGY. *Educ:* Univ Del, BA, 59, MA, 61; Colo State Univ, PhD(physiol), 65. *Prof Exp:* From asst prof to assoc prof zool & physiol, Univ Wyo, 64-76, dir inst radiation biol, 67-70, prof, 76-, assoc dean student prog, Col Human Med, 76-85; PROF & ASSOC DEAN STUDENT AFFAIRS, SCH MED, CREIGHTON UNIV, 85- *Concurrent Pos:* NSF vis lectr, 67; Am Inst Biol Sci vis radiation biologist, 68-76. *Mem:* Fel AAAS; Asn Am Med Cols. *Res:* Thyroidal interactions as applicable to adrenal interactions, atherosclerosis, reproduction and wound healing; basic and therapeutic aspects of cancer; intestinal function as affected by hormones. *Mailing Add:* Sch Med Creighton Univ 2500 California SE Omaha NE 68178

PANDE, CHANDRA SHEKHAR, b Bast, India, Sept 15, 40; m 69; c 2. ELECTRON MICROSCOPY, SUPERCONDUCTIVITY. *Educ:* Delhi Univ, BS, 59, MS, 61; Oxford Univ, PhD(phys sci), 70. *Prof Exp:* In-chg, Electron Micros Lab, Brookhaven Nat Lab, 74-80; SECT HEAD, NAVAL RES LAB, 80- *Mem:* Am Soc Metals Int; Metall Soc; Electron Microscope Soc Am. *Res:* Microstructural-property relationship in superconducting materials; transmission and analytical electron microscopy of materials; modeling of grain growth. *Mailing Add:* Naval Res Lab Code 6320 Washington DC 20375

PANDE, GYAN SHANKER, b Aug 5, 32; US citizen; m 72; c 2. CHEMISTRY. *Educ:* Agra Univ, India, BSc, 51; Lucknow Univ, India, MSc, 53, PhD(chem), 58. *Prof Exp:* Lectr chem, Lucknow Univ, 55-58 & Roorkee Univ, India, 59-60; res assoc chem, Univ Saskatoon, Can, 60-65; res chemist, Uniroyal Res Labs, Guelph, Can, 65-72; mgr prod mkt develop, Conap Inc, Olean, NY, 73-74, dir res & develop, 75-77; mem staff, Cordis Corp, 77-85; PRES, EDNAP MED INC, 85- *Concurrent Pos:* Adj assoc prof, Biomed Eng, Univ Miami, Fla, 83- *Mem:* Am Chem Soc; sr mem Chem Inst Can; Am Soc Testing & Mat; Soc Plastics Engrs; Biomaterials Soc. *Res:* Synthesis and evaluation of polyurethanes; epoxies as adhesives and encapsulants; materials for medical devices. *Mailing Add:* 3308 Island Dr Miramar FL 33023

PANDE, KRISHNA P, b Basti, India; US citizen; m 73; c 2. GALLIUM ARSENIDE TECHNOLOGY, OPTOELECTRONICS. *Educ:* Lucknow Univ, MS, 69; Indian Inst Technol, PhD(solid state physics), 73. *Prof Exp:* Res assoc elec eng, Rensselaer Polytech Inst, 77-79; asst prof, Rutgers Univ, 79-81; mgr, Bendix Aerospace Technol Ctr, 81-86; dir, Unisys Semiconductor Opers, 86-88; ASSOC EXEC DIR, COMSAT LABS, 88- *Concurrent Pos:* Consult, Wright Air Force Avionics Lab, 79-; TAC mem, NSF, 87-; mem, adv comn, Inst Elec & Electronic Engrs & bd eng, Univ Cincinnati, Ohio. *Mem:* Fel Inst Elec & Electronic Engrs; Am Phys Soc; NY Acad Sci; Electromagnetic Soc. *Res:* Development of microwave monolithic integrated circuits and transmit/receive modules for application in satellite communication, radar and missile seekers; author of over 70 publications; five inventions. *Mailing Add:* 12200 Galesville Dr Gaithersburg MD 20878

PANDE, SHRI VARDHAN, b Kanpur, India, Nov 4, 40; m 65; c 2. BIOCHEMISTRY. *Educ:* Univ Lucknow, BSc, 58, MSc, 60; Univ Delhi, PhD(biochem), 65. *Prof Exp:* Jr res officer biochem, V Patel Chest Inst, Delhi, India, 65; res scholar, Lab Nuclear Med & Radiation Biol, Univ Calif, Los Angeles, 66-69; prof res asst, Univ Man, 69-70; sr researcher, 70-72, DIR LAB INTERMEDIARY METAB, CLIN RES INST MONTREAL, 72- *Concurrent Pos:* Tutor, Univ Man, 69-70; from asst prof to assoc prof, Univ Montreal, 71-82, prof, 82-; assoc mem McGill Univ, 73- *Mem:* Am Chem Soc; NY Acad Sci; Am Soc Biol Chemists; Can Biochem Soc. *Res:* Metabolism and functions of carnitine; mitochondrial functions and transports; fatty acid oxidation and ketogenesis; regulation of metabolism. *Mailing Add:* Clin Res Inst Montreal 110 Pine Ave W Montreal PQ H2W 1R7 Can

PANDELL, ALEXANDER JERRY, b San Francisco, Calif, June 19, 42. ORGANIC CHEMISTRY. *Educ:* San Francisco State Univ, BS, 64. *Prof Exp:* Res fel org chem, Harvard Univ, 68-69; asst prof chem, Boston Univ, 69-70; from asst prof to assoc prof, 70-76, chmn dept, 73- 79, PROF CHEM, CALIF STATE UNIV, STANISLAUS, 77- *Concurrent Pos:* Am Chem Soc-Petrol Res Fund grants, 71-74 & 81-83. *Mem:* Am Chem Soc. *Res:* Transition metal catalyzed oxidations of aromatic systems. *Mailing Add:* Dept Chem Calif State Stanislaus 801 W Monte Vista Turlock CA 95380

PANDEY, JAGDISH NARAYAN, b Varanasi, India, Nov 2, 36; m 54; c 5. MATHEMATICS. *Educ:* Banaras Hindu Univ, BSc, 55, MSc, 57; State Univ NY Stony Brook, PhD(math), 67. *Prof Exp:* Lectr math, Banaras Hindu Univ, 57-58 & 59-64; head dept, SC Col, Ballia, India, 58-59; fel, Univ Alta, 67-68; asst prof, 68-72, assoc prof, 72-81, PROF MATH, CARLETON UNIV, 81- *Concurrent Pos:* Nat Res Coun Can grant, 68- *Mem:* Am Math Soc; Can Math Soc. *Res:* Generalized integral transform and functions; abstract differential equations; harmonic analysis. *Mailing Add:* Dept Math Carleton Univ Ottawa ON K1S 5B6 Can

PANDEY, KAILASH N, RESEARCH, TEACHING. *Educ:* Agra Univ, BSc, 66; Kanpur Univ, MSc, 68; Univ Ky, PhD(cell biol), 79. *Prof Exp:* Grad teaching asst biol, Univ Ky, 73-79; res assoc, Tenn State Univ, 79-81; res fel biochem, Vanderbilt Univ, 81-84, res assoc, 84-86, from res instr to res assoc prof, 86-90; ASSOC PROF BIOCHEM & MOLECULAR BIOL, MED COL GA, 90- *Concurrent Pos:* Prin investr, Am Heart Asn, 86-87, 87-89 & 88-91; estab investr, NIH, 89-94, Am Heart Asn, 90-95. *Mem:* Am Soc Biochem & Molecular Biol; Am Heart Asn. *Res:* Studies involving the cellular and molecular mechanisms of action of atrial natriuretic factor and its receptor. *Mailing Add:* Dept Biochem & Molecular Biol Med Col Ga Sch Med Augusta GA 30912-2100

PANDEY, RAGHVENDRA KUMAR, b Bath, Bihar, India, Jan 7, 37; m 67. APPLIED PHYSICS, SOLID STATE MATERIALS. *Educ:* Bihar Univ, BSc, 57; Patna Univ, MSc, 59; Cologne Univ, WGer, Dr rer nat, 67. *Prof Exp:* Sr res physicist, NCR Co, Dayton, 68-72 & Cincinnati Electronics Corp, 73-74; prof electronics, Nat Inst Astrophys, Optics & Electronics, Puebla, Mex, 74-77; ASSOC PROF ELEC ENG, TEX A&M UNIV, 77- *Mem:* Am Phys Soc; sr mem Inst Elec & Electronics Engrs; AAAS. *Res:* Crystal growth, magnetism, ferroelectricity, x-ray diffraction and optical properties of solids; electro-optical devices. *Mailing Add:* Dept Elec Eng Tex A&M Univ College Station TX 77843

PANDEY, RAMESH CHANDRA, b Naugaon, Almora, India, Nov 5, 38. ORGANIC CHEMISTRY, NATURAL PRODUCTS. *Educ:* Univ Allahabad, BSc, 58; Univ Gorakhpur, MSc, 60; Univ Poona, PhD(terpenoids), 65. *Prof Exp:* Jr res fel terpenoids, Nat Chem Lab, Poona, 60-64, res officer med plants, 65-67; res assoc polyene antibiotics, Chem Dept, Univ Ill, Urbana, 67-70; scientist appl res, Nat Chem Lab, Poona, 70-72; vis scientist antibiotics & mass spectros, Chem Dept, Univ Ill, Urbana, 72-77; sr scientist, Antitumor Compounds, NCI-Frederick Cancer Res Facil, Nat Cancer Inst, 77-82, head chem sect, 82-83; sr scientist, Anti-Infective Res Div, Pharmaceut Discovery, Abbott Labs, North Chicago, 83-84; vis prof, Waksman Inst of Micro, Rutgers Univ, 84-86, PRES, XECHEM, INC, SUBSID OF LYPHOMED, 84- *Concurrent Pos:* Consult, Amphotericin B Group, Sch Med, Washington Univ, St Louis, 76-85, Lyphomed Inc, Melrose Park, Ill, 84-86; vis prof, Waksman Inst, Rutgers Univ, Piscataway, NJ. *Mem:* Am Chem Soc; Am Soc Mass Spectrometry; Am Soc Microbiol; NY Acad Sci; Am Soc Pharmacognosy; fel Am Inst Chemists;

Soc Indust Microbiol; AAAS; Indian Sci Cong Asn; Am Asn Cancer Res. *Res:* Antibiotics, terpenoids, applied research, PMR, CMR and mass spectral studies; process developments; biotechnology; biosynthesis; medicinal chemistry. *Mailing Add:* Xechem Inc 100 Jersey Ave Bldg B Suite 310 New Brunswick NJ 08901-3279

PANDEY, RAS BIHARI, b Ballia, India. COMPUTER SIMULATION MODELING, INOHOMOGENEOUS SYSTEMS. *Educ:* Univ Allahabad, BSc, 72, MSc, 74; Univ Roorkee, PhD(physics), 81. *Prof Exp:* Vis asst prof physics, NC State Univ, Raleigh, 81-82; postdoctoral res asst physics, Cologne Univ, Fed Repub Ger, 83, Cavendish Lab, Cambridge Univ, 83-84 & Univ Ga, Athens, 84-85; asst prof physics, Jackson State Univ, 85-88; ASSOC PROF PHYSICS, UNIV SOUTHERN MISS, 88- *Concurrent Pos:* Res scientist, Int Ctr Theoret Physics, Italy, 81; res collabr, Univ Ga, 85, Sandia Nat Labs, Livermore, 86, Cologne Univ, Fed Repub Ger, 87 & Ger Supercomputer Ctr, Juelich, 89; AVH postdoctoral fel physics, KFA Juelich & Cologne Univ, 88. *Mem:* Am Phys Soc; AAAS; Sigma Xi. *Res:* Percolation; growth models; stochastic processes; fractals; transport phenomena and fluid flow; interfacial growth; phase transition and critical phenomena; polymers, conformational and dynamical properties; cellular automata and theoretical immunology. *Mailing Add:* Physics & Astron Univ Southern Miss SS Box 5046 Hattiesburg MS 39406

PANDEY, SURENDRA NATH, b Mirzapur, India, July 3, 45; US citizen; m 67; c 2. EXPERIMENTAL SOLID STATE PHYSICS. *Educ:* Univ Allahabad, India, BSc, 61, MSc, 63; Howard Univ, Washington, DC, PhD(physics), 73. *Prof Exp:* Spectrogr chem-physics, Qual Control Lab, Hindustan Aluminum Corp, Mirzapur, India, 63-64; res physicist, Phys Res Lab, Fertilizer Corp, India, Sindri, India, 64-67; from asst prof to assoc prof, 73-84, PROF PHYSICS, ALBANY STATE COL, ALBANY, GA, 84- *Concurrent Pos:* Mem acad adv comt, Univ Syst Ga; exec coun mem, Ga Acad Sci. *Mem:* Am Asn Physics Teachers; Am Phys Soc; Nat Sci Teachers Asn. *Res:* Low and high temperature study on thermal expansion by x-ray diffraction of solids; electromagnetic radiation absorption by organic dyes and solutions, especially ultraviolet, visible and infrared. *Mailing Add:* Dept Natural Sci Albany State Col Albany GA 31705

PANDEYA, PRAKASH N, b Unnao, India, Dec 30, 43; US citizen; m 69; c 3. POSITIVE DISPLACEMENT COMPRESSORS, REFRIGERATION & AIR-CONDITIONING COMPRESSORS & SYSTEMS. *Educ:* Univ Lucknow, India, BSc, 62; Univ Roorkee, India, BE, 64; City Univ NY, MS, 74; Purdue Univ, PhD(mech eng), 78; NY Inst Technol, MBA, 84. *Prof Exp:* Asst mech engr, Uttar Pradesh State Irrig Dept, 65; lectr mech eng, Univ Roorkee, India, 65-67; asst mgr, Indian Ord Factories, 68-71; design engr, R-M Friction Mat Co, 72-76; sr engr, Carlyle Compressor Div, Carrier Corp, 78-83, sr proj engr, 83-84, proj leader, 84-85; mgr, Prod Develop, 85-87, DIR ENG, BRISTOL COMPRESSORS DIV, YORK INT, 87- *Concurrent Pos:* Grad teaching & res asst, Purdue Univ, 76-78; adj assoc prof mech eng technol, State Univ NY Col Eng Technol, Utica, 79-80. *Mem:* Am Soc Heating Refrig & Air Conditioning Engrs; Am Soc Mech Engrs. *Res:* Mathematical modeling and performance analysis of positive displacement compressors; reciprocating, rolling piston, rotary vane, and scroll type hermetic and semi-hermetic compressors for air-conditioning and refrigeration applications; awarded four US patents. *Mailing Add:* 129 Roscommon Dr Bristol TN 37620

PANDHARIPANDE, VIJAY RAGHUNATH, b Nagpur, India, Aug 7, 40. THEORETICAL PHYSICS, NUCLEAR STRUCTURE. *Educ:* Nagpur Univ, MSc, 61; Bombay Univ, PhD(physics), 69. *Prof Exp:* Res asst, Tata Inst Fundamental Res, 61-73; from asst prof to assoc prof, 73-77, PROF PHYSICS, UNIV ILL, URBANA-CHAMPAIGN, 77- *Mem:* Indian Physics Asn; fel Am Phys Soc. *Res:* Theoretical understanding of the structure and dynamics of nuclei and neutron stars from nuclear forces; study of the relations between properties of quantum liquids and the interparticle forces. *Mailing Add:* Dept Physics Univ Ill Champaign 1110 W Green St Urbana IL 61801

PANDIAN, NATESA G, b Tamilnadu, India, June 21, 47. ECHOCARIOGRAPHY, CARDIOLOGY. *Educ:* Univ Madras, India, MBBS, 70, MD, 78. *Prof Exp:* Fel cardiol, Tufts Univ Sch Med, 76-79; assoc med, Univ Iowa, 79-82; clin asst med cardiol, Mass Gen Hosp, 82-83; instr med cardiol, Harvard Med Sch, 82-83; asst prof med & radiol, 83-85, ASSOC PROF MED & RADIOL, CARDIOL DEPT, TUFTS UNIV SCH MED, 85-; DIR,NON-INVASIVE CARDIAC LAB, NEW ENG MED CTR HOSPS, BOSTON, 83- *Concurrent Pos:* Am Soc Echocardiog rep, Joint Rev Comn Diag Med Sonographers, 88-, mem bd dirs, 85-88; consult, NIH, 83-88, Vet Admin, 83-88; NIH investr award, 82-86. *Mem:* Am Soc Echocardiography; Am Heart Asn; Am Fedn Clin Res; Am Inst Ultrasound Med. *Res:* Ischemic heart disease; echocardiography; pericardial disease; internal medicine. *Mailing Add:* Tufts New Eng Med Ctr Box 32 750 Washington St Boston MA 02111

PANDIT, HEMCHANDRA M, b Chinchani, India, May 4, 24; m 53; c 3. ANIMAL PHYSIOLOGY. *Educ:* Univ Bombay, BSc, 46, MSc, 50; State Univ NY Buffalo, PhD(biol), 67. *Prof Exp:* Jr lectr biol, Elphinstone Col, Bombay, 46-48 & RR Col, Univ Bombay, 49-52; res asst, Indian Cancer Res Ctr, 50-52 & 55-62 & Haffkine Inst Bombay, 52-55; asst biol, State Univ NY Buffalo, 62-66; assoc prof, Villa Marie Col Buffalo, 66-68; from assoc prof to prof, 68-73, EMER PROF BIOL, D'YOUVILLE COL, 90- *Concurrent Pos:* Expert witness, cancer ecol. *Res:* Snake venoms; leprosy; renal physiology and philosophy; effects of non ionizing radiation on the biological system; author of one book. *Mailing Add:* Dept Biol D'Youville Col 320 Porter Ave Buffalo NY 14201

PANDIT, SUDHAKAR MADHAVRAO, b Gherdi, India, Dec 3, 39; m 66; c 2. MODAL & SPECTRUM ANALYSIS, COMPUTER AIDED MANUFACTURING & DESIGN. *Educ:* Univ Poona, India, BE, 61; Pa State Univ, MS, 70; Univ Wis-Madison, MS, 72, PhD(mech eng), 73. *Prof Exp:* Engr, Kirloskar Oil Engines Ltd, Poona, E Asiatic Co Ltd, Bombay, 61-62 & Heavy Eng Corp, 62-68; teaching asst indust eng, Pa State Univ, 68-70; res asst mech eng, Univ Wis-Madison, 70-73, lectr & res assoc, 73-76; assoc prof, 76-81, PROF MECH & INDUST ENG, MICH TECHNOL UNIV, 81- *Concurrent Pos:* Area dir mfg & indust eng, Mich Technol Univ, 88-; assoc ed, J Eng Indust, 89- *Mem:* Am Soc Mech Engrs; Soc Mfg Engrs; Opers Res Soc Am; Am Inst Indust Engrs; Sigma Xi. *Res:* Evolution of new methodology called data dependent systems for system modeling, analysis, forecasting and control from observed data; applications to computer integrated manufacturing and design, quality control, vibration and modal analysis, and machine vision; author of two books and over 100 publications. *Mailing Add:* ME-EM Dept Mich Technol Univ Houghton MI 49931-1295

PANDOLF, KENT BARRY, b Needham, Mass, Feb 24, 45; m 68. ENVIRONMENTAL PHYSIOLOGY, EXERCISE PHYSIOLOGY. *Educ:* Boston Univ, BS, 67; Univ Pittsburgh, MA, 68, MPH, 70, PhD(work physiol), 72. *Prof Exp:* Vis asst fel environ physiol, John B Pierce Found Lab, New Haven, Conn, 73-78; res physiologist, 73-78, chief, physiol br, 78-82, dir, mil ergonomics div, 82-90, DIR, ENVIRON PHYSIOL & MED DIRECTORATE, US ARMY RES INST ENVIRON MED, 90- *Concurrent Pos:* Fel, Sch Med, Yale Univ, 72-73; adj prof, Boston Univ, 83-, Springfield Col, 85- *Honors & Awards:* Honor Award, New Eng Chap, Am Col Sports Med, 89; Commander's Award for Civilian Serv, US Army Res Inst Environ Med, 90. *Mem:* Am Physiol Soc; fel Ergonomics Res Soc; fel Am Col Sports Med; Aerospace Med Asn; Int Soc Adaptive Med. *Res:* Independent, original research into areas of environmental and work physiology pertaining to human performance with some research involving perceptual and motivational aspects of physical work and heat exposure of men. *Mailing Add:* Environ Physiol & Med Directorate US Army Res Inst Environ Med Natick MA 01760-5007

PANDOLFE, WILLIAM DAVID, b Hartford, Conn, Nov 5, 45; m 76; c 2. PHYSICAL CHEMISTRY. *Educ:* Col Holy Cross, BS, 67; Rutgers Univ, MS, 71, PhD(phys chem), 74. *Prof Exp:* MGR RES & DEVELOP, APV GAULIN INC, 74- *Mem:* Am Chem Soc; Inst Food Technologists. *Res:* Colloid science and emulsion technology. *Mailing Add:* 91 Partridge Rd Billerica MA 01821-5610

PANDOLFO, JOSEPH P, b New York, NY, Sept 26, 30; m 52; c 6. METEOROLOGY, OCEANOGRAPHY. *Educ:* Fordham Univ, BS, 51; NY Univ, MS, 56, PhD(meteorol, oceanog), 61. *Prof Exp:* Weather officer, US Air Force, 51-55; from instr to asst prof meteorol & oceanog, NY Univ, 57-62; res scientist, Travelers Res Ctr, Inc, 62-69; res fel, Ctr Environ & Man, Inc, 69-71, vpres, 71-76, pres, 76-86, mem bd dirs, 77-86; ASSOC RES PROF, ST JOSEPH COL, HARTFORD, 86- *Concurrent Pos:* Consult, Barbados Oceanog & Meteorol Exp, Environ Sci Serv Admin, 67-69; NSF res grants, Ctr Environ & Man, Inc, 69-86; consult, Int Field Year Great Lakes, Nat Oceanog & Atmospheric Agency, 71-74 & Int Decade Ocean Explor, NSF, 72-75; vis prof physics, Trinity Col, Hartford, 85-; dir large-scale dynamics & meteorol prog, NSF, Washington, DC, 90-91. *Mem:* AAAS; Am Meteorol Soc. *Res:* Dynamic oceanography; dynamics and physics of planetary boundary layers; computer simulation, atmospheric and oceanic systems; air-land and air-sea interactions; solar energy. *Mailing Add:* 86 Robindale Dr Kensington CT 06037

PANDORF, ROBERT CLAY, b Cincinnati, Ohio. LOW TEMPERATURE PHYSICS. *Educ:* Miami Univ, Ohio, BA, 57, MA, 59; Ohio State Univ, PhD(physics), 67. *Prof Exp:* Fel physics, Mass Inst Technol, 67-70; PROJ PHYSICIST, CHARLES STARK DRAPER LAB, 70- *Concurrent Pos:* Res asst, Ohio State Univ, 59-66, res assoc, 66-67. *Mem:* Am Phys Soc. *Res:* Study of the viscous and other physical properties of liquid helium at ultra-low temperatures and its application to ultra precision inertial instrumentation. *Mailing Add:* 139 Winchester St Newton Highlands MA 02161

PANDRES, DAVE, JR, b Duncan, Okla, Jan 10, 28; m 53; c 4. THEORETICAL PHYSICS. *Educ:* Univ Tex, BS, 49, MA, 56, PhD(physics), 58. *Prof Exp:* Elec engr, Brown & Root, Inc, 49; systs engr, Chance-Vought Aircraft Co, 51-52; adv res scientist, Ohio Oil Co, 58-60; assoc res scientist, Martin Co, 60-62; sr res & develop scientist, Lockheed-Calif Co, 62-64; res scientist, Douglas Advan Res Lab, 64-66, assoc dir math sci, 66-69, dir math sci, 69-71; from asst prof to assoc prof, 71-78, PROF MATH, NORTH GA COL, 78- *Mem:* Am Phys Soc. *Res:* General relativity; unified field theory; foundations of quantum theory. *Mailing Add:* Dept of Math N Ga Col Dahlonega GA 30597

PANDYA, KRISHNAKANT HARIPRASAD, b Mehmedabad, India, Oct 19, 35; m 54; c 2. PHARMACOLOGY, PHARMACOGNOSY. *Educ:* Gujarat Univ, India, BS, 58, MS, 61, PhD(pharmacol), 68. *Prof Exp:* Demonstr, L M Col Pharm, Gujarat Univ, 58-61, tutor, 61-63, jr lectr pharmacol, B J Med Col, 63-69, asst prof, 69-70; from instr to asst prof, 71-77, ASSOC PROF PHARMACOL, KIRKSVILLE COL OSTEOP MED, 77- *Mem:* Am Soc Pharmacol & Exp Therapeut. *Res:* Autonomic nervous system; modifications of the responses, release, evolution, etc of the neurohumoral transmitters in vivo; cardiovascular and respiratory systems. *Mailing Add:* Dept Pharmacol Kirksville Col Osteop Med Kirksville MO 63501

PANEK, EDWARD JOHN, b Grand Rapids, Minn, Sept 27, 41; m 66; c 2. ORGANOMETALLIC CHEMISTRY. *Educ:* Univ Wis-Madison, BS, 63; Mass Inst Technol, PhD(org chem), 68. *Prof Exp:* NIH fel chem, Dept Chem, Iowa State Univ, 68-69; asst prof chem, Tulane Univ, 69-75; res staff chem, 75-81, RES SUPVR, BASF WYANDOTTE CORP, 81- *Mem:* Am Chem Soc; AAAS; Sigma Xi. *Res:* Activation of anions by complexing agents with particular emphasis on carbanions and anionic polymerization catalysts; environmental chemistry. *Mailing Add:* BASF Corp Chem Agr Res PO Box 13528 Research Triangle Park NC 27709-3528

PANEK, LOUIS A(NTHONY), b Boston, Mass, Dec 3, 19; m 43; c 2. MINING ENGINEERING, ROCK MECHANICS. *Educ:* Mich Col Mining & Technol, BS, 41; Columbia Univ, MS, 46, PhD(eng), 49. *Prof Exp:* Mining res engr, US Bur Mines, 49-65, mgr ground control res, 65-78, sr res scientist, 78-84; J S Westwater prof mining eng, Mich Tech Univ, 84-89; RETIRED. *Concurrent Pos:* Warren lectr, Univ Minn, 56. *Honors & Awards:* Peele Award, Am Inst Mining, Metall & Petrol Engrs, 57. *Mem:* Am Statist Asn; Int Soc Rock Mech; Am Inst Mining, Metall & Petrol Engrs; Sigma Xi. *Res:* Support and control of roof and ground in the vicinity of underground openings based on theoretical and experimental stress analysis; application of these principles to the design of mine openings and methods of mining; in-place testing and measurement of rock mass structural behavior. *Mailing Add:* Eight Hillside Dr Denver CO 80215-6609

PANEM, SANDRA, animal virology, research management, for more information see previous edition

PANESSA-WARREN, BARBARA JEAN, b Yonkers, NY, Feb 21, 47; m 77; c 1. CELL BIOLOGY, MICROSCOPY EDUCATION. *Educ:* NY Univ, BA, 68, MS, 71, PhD(cell biol), 74. *Prof Exp:* Sr res technician, Lab Infectious Dis & Immunol, Med Ctr, Bellevue, 68-69, teaching fel, Grad Sch Arts & Sci & Washington Sq Col Arts & Sci, asst prof physiol & pathophysiology, St Vincent's Hosp & Med Ctr, NY, 74-78; res assoc orthop surg & head, Analytical Electron Micros & Cell Biol Lab, 78-80, res asst prof, Dept Anat Sci, Health Sci Ctr, State Univ NY, Stony Brook, 80-85, res asst prof, dept allied health resources, 84-87, RES ASSOC PROF, 87- *Concurrent Pos:* Sr instr, St Vincent's Hosp Sch Nursing, New York, 69-73; consult, Gen Tel & Electronics, NY, 72, Stauffer Chem Co, NY, 73, Esso Res & Eng, NJ, 73 & Latham Publ Co, 74-75; NATO prof, Inst Zool, Univ Siena, Italy, 74 & 75; NIH fel physiol & ophthal, Med Sch, NY Univ, 75-78; guest assoc scientist, Div Instrumentation, Brookhaven Nat Lab, 79-; postdoctoral, NY Univ Med Sch, 74-77; spec study sect micros, NIH, 82-87. *Honors & Awards:* Wilhelm Bernhard Award, Eur Electon Micros Soc, 80; Burton Medal, Electron Micros Soc Am, 81; Excellence Electron Beam Res Award, Microbeam Anal Soc Am, 71. *Mem:* Am Soc Cell Biol; Biophys Soc Am; Electron Micros Soc Am; Microbeam Anal Soc Am; NY Acad Sci. *Res:* Elemental content and distribution in vertebrate eyes (retina-choroid); high resolution imaging and composition of biological subcellular components; x-ray microanalysis; x-ray fluorescence spectroscopy; proton induced x-ray emission spectroscopy; scanning transmission electron microscopy; routine transmission electron microscopy, scanning electron microscope, x-ray contact; microscopy of biomedical samples, synchrotron radiation. *Mailing Add:* 23 Camp Woodbine Rd Port Jefferson NY 11777

PANETH, NIGEL SEFTON, b London, Eng, Sept 19, 46; m 73; c 2. EPIDEMIOLOGY, PEDIATRICS. *Educ:* Columbia Col, AB, 68, MPH, 78; Dartmouth Col, BMS, 70; Harvard Univ, MD, 72. *Prof Exp:* Asst clin prof pediat, Albert Einstein Col Med, 77-78; asst prof pub health & pediat, Columbia Univ, 78-85, assoc prof pediat, 85-89; DIR, PROG EPIDEMIOL, MICH STATE UNIV, 89- *Honors & Awards:* K L Murray Award, Am Acad Cerebral Palsy Develop Med, 86. *Mem:* Am Acad Pediat; Soc Epidemiol Res; Int Epidemiol Asn; Soc Pediat Res; Am Epidemiol Soc. *Res:* Epidemiology of cerebral palsy and mental retardation, in particular the relationship of these conditions to events in the perinatal period; assessment of medical care effectiveness. *Mailing Add:* Prog Epidemiol A206 E Fee Hall Mich State Univ East Lansing MI 48824

PANETTA, CHARLES ANTHONY, b Albany, NY, Sept 12, 32; m 59; c 2. ORGANIC CHEMISTRY. *Educ:* Manhattan Col, BS, 54; Rensselaer Polytech Inst, PhD(org chem), 61. *Prof Exp:* Sr chemist, Bristol Labs, Inc, 60-64, proj supvr, 64-65; res assoc chem, Mass Inst Technol, 65-67; asst prof chem & pharmaceut chem, 67-70, assoc prof chem, 70-73, PROF CHEM, UNIV MISS, 73- *Concurrent Pos:* Res grants, Res Corp, Bristol Labs, NSF & NIH. *Mem:* Am Chem Soc; Sigma Xi. *Res:* Chemistry of penicillin antibiotics; amino acids and peptides; general organic synthesis; electroactive organic materials. *Mailing Add:* PO Box 6792 University MS 38677

PANG, CHAN YUEH, b China, Oct, 47. HIGH ENERGY PHYSICS, SYSTEMS SCIENCE. *Educ:* Univ Chicago, PhD(physics), 73. *Prof Exp:* Res assoc, Univ Ill, Urbana, 73-75, res asst prof physics, 75-77; mem staff, Lawrence Berkeley Lab, 77-80; PHYSICIST, BECHTEL NAT, 81- *Mem:* Am Phys Soc. *Res:* Weak interactions and neutrino physics; nuclear physics. *Mailing Add:* 1641 Ocean View Ave Kensington CA 94707

PANG, CHO YAT, Can citizen; m 76; c 2. SURGICAL RESEARCH, PERIPHERAL VASCULAR PHYSIOLOGY. *Educ:* Univ Man, Can, BSc, 65, MSc, 71, PhD(physiol), 75. *Prof Exp:* Asst prof, dept surg, Southwestern Med Sch, Univ Tex Health Sci Ctr, Dallas, 80-82; asst prof, 82-85, ASSOC PROF, DEPT SURG & PHYSIOL, UNIV TORONTO, 85-; sr scientist, 82-88, HEAD, DIV SURG RES HOSP SICK CHILDREN, TORONTO, 89- *Concurrent Pos:* Mem, fac grad studies, dept physiol & Inst Med Sci, Univ Toronto; mem, Am Plastic Surg Res Coun. *Honors & Awards:* Investr Award, Educ Found, Am Soc Plastic & Reconstructive Surgeons, 84. *Mem:* Can Physiol Soc; Am Physiol Soc; Soc Exp Biol & Med; Plastic Surg Res Coun. *Res:* Peripheral vascular pathophysiology and pharmacology; emphasis on local regulation of skin and muscle blood flow by catecholamines, prostanoids, serotonin, endothelium-derived vascular relaxant factor and endothelium; ischemia induced tissue damage by oxygen free radicals; pharmacology. *Mailing Add:* Hosp Sick Children 555 University Ave Toronto ON M5G 1X8 Can

PANG, DAVID C, b Hong Kong, Jan 20, 42. PHARMACOLOGY. *Educ:* McGill Univ, PhD(biochem), 70. *Prof Exp:* SR SCIENTIST, BERLEX LABS, INC, 83- *Mem:* NY Acad Sci; Am Physiol Soc; Biophys Soc. *Mailing Add:* Seven Clearfield Rd Succasunna NJ 07876

PANG, HENRIANNA YICKSING, b Hong Kong, Dec 23, 52; Brit citizen; m 80; c 1. CHEMISTRY. *Educ:* San Diego State Univ, BSc, 75; Univ Utah, PhD(med chem), 82. *Prof Exp:* ASSOC, MASS INST TECHNOL, 82- *Mem:* Am Chem Soc; Am Soc Mass Spectrometry. *Mailing Add:* 170 Sommerfield Dr Oakville ON L6L 5N8 Can

PANG, KEVIN DIT KWAN, b China; US citizen; m; c 1. SPACE SCIENCES, REMOTE SENSING. *Educ:* Univ Hawaii, BA, 62; Univ Calif, Los Angeles, MS, 63, PhD(space physics), 70. *Prof Exp:* Teaching asst physics, Univ Hawaii, 61-62; geophysicist earth sci, Inst Geophysics, Univ Calif, Los Angeles, 63-69; res assoc atmospheric physics, Lab Atmospheric & Space Physics, Univ Colo, 70-73; mem tech staff, Computer Sci Corp, 73-74; res assoc space physics, Jet Propulsion Lab, Calif Inst Technol, 74-76; staff scientist space sci, Planetary Sci Inst, Sci Appln, Intern, 76-80; CONSULT SPACE SCI, JET PROPULSION LAB, CALIF INST TECHNOL, 80- *Concurrent Pos:* Consult, Planetary Sci Dept, Rand Corp, 64-65; prin investr, var NASA, NSF & other grants, 76- *Honors & Awards:* Group Achievement Award, NASA, 77; Dean Prize; Sigma Xi Award; Herbert C Pollock Award, 88. *Mem:* Sigma Xi; Int Astron Union; Am Astron Soc; Am Meteorol Soc; Am Geophys Union; AAAS. *Res:* Experimental and theoretical investigations of the atmosphere and surface of the earth, planets and satellites, especially by remote sensing techniques; laboratory simulation studies of physical and chemical processes on solar system bodies; history of astronomy and astrophysics. *Mailing Add:* Jet Propulsion Lab T-1182 4800 Oak Grove Dr Pasadena CA 91109

PANG, KIM-CHING SANDY, b Hong Kong. PHARMACOKINETICS, HEPATIC DRUG CLEARANCE. *Educ:* Univ Toronto, BSc, 71; Univ Calif, San Francisco, PhD(pharm chem), 75. *Prof Exp:* Teaching asst pharm dispensing pharmacokinetics, Univ Calif, San Francisco, 71-73, teaching fel pharmacokinetics, 73-74, res asst, 74-75; vis fel metab pharmacokinetics, Lab Chem Pharmacol, NIH, Bethesda, 76-78; asst prof pharmaceut, Univ Houston, 78-82; assoc prof, 82-87, PROF, PHARMACOL DEPT, UNIV TORONTO, 87- *Concurrent Pos:* Consult, Nat Inst Drug Abuse, 77-78; adj asst prof, Baylor Col Med, Houston, 78-; mem, Spec Study Sect, NIH, 79-80 & ad hoc mem, Pharm Study Sect, NIH, 82-86. *Mem:* Am Soc Pharmacol & Exp Therapeut; AAAS; Am Asn Study Liver Dis. *Res:* Hepatic clearance of drugs; pharmacokinetics of drugs and their metabolites; drug metabolism; metabolite kinetics. *Mailing Add:* Fac Pharm Univ Toronto 19 Russel St Toronto ON M5S 1A1 Can

PANG, PETER KAI TO, b Hong Kong, Oct 14, 41; m 69; c 3. COMPARATIVE PHYSIOLOGY, ENDOCRINOLOGY. *Educ:* Univ Hong Kong, BSc, 63, BSc, 64; Yale Univ, MSc, 67, PhD(biol), 70. *Hon Degrees:* DSc, Univ Hong Kong, 81. *Prof Exp:* Res asst, Bingham Lab, Yale Univ, 64-65, instr biol, 70-71; instr pharmacol, Col Physicians & Surgeons, Columbia Univ, 72-73, assoc, 73-74; assoc prof biol, Brooklyn Col, City Col New York, 74-76; from assoc prof to prof pharmacol, Sch Med Tex Tech Univ, 76-85; PROF & CHMN, DEPT PHYSIOL, UNIV ALTA, CAN, 86- *Mem:* Am Physiol Soc; Am Soc Bone & Mineral Res; Soc Endocrinol; Endocrine Soc; AAAS. *Res:* Herbal medicine; hormonal control of calcium regulation and osmoregulation in lower vertebrates; endocrine evolution. *Mailing Add:* Dept Physiol Univ Alta Edmonton AB T6G 2H7

PANG, YUAN, b Taipei, Taiwan, July 24, 47. ENGINEERING MECHANICS, APPLIED MATHEMATICS. *Educ:* Univ Akron, MS, 77; Stanford Univ, PhD(eng), 81. *Prof Exp:* res engr, Homer Res Lab, 81-86; ENGR, GE ELEC, 86- *Mem:* Soc Indust Appl Math; Am Phys Soc; Am Soc Mech Engrs. *Mailing Add:* 21 Foley Dr North Reading MA 01864

PANGBORN, ROBERT NORTHRUP, b New York, NY, May 26, 51; m 76; c 1. HIGH TEMPERATURE MATERIALS COMPOSITES. *Educ:* Rutgers Univ, BS & BA, 74, MS, 77, PhD(mech & mat sci), 79. *Prof Exp:* Res assoc, Electron Micros Lab, Rutgers Univ, 79; PROF ENG SCI & MECH, PA STATE UNIV, 79- *Concurrent Pos:* Lectr appl x-ray methods, Ctr Prof Advan, 78. *Honors & Awards:* Distinguished Serv Award, Am Soc Mech Engrs. *Mem:* Am Soc Metals; Am Crystallographic Asn; Am Soc Eng Educ; Sigma Xi; Am Soc Mech Engrs. *Res:* Characterization and testing of high temperature materials, coatings and composites; failure analysis; nondestructive evaluation. *Mailing Add:* 322 E Irvin Ave State College PA 16801-5415

PANGBORN, ROSE MARIE VALDES, b Las Cruces, NMex, Aug 19, 32; m 56. FOOD TECHNOLOGY, PSYCHOPHYSICS. *Educ:* NMex State Univ, BS, 53; Iowa State Univ, MS, 55. *Prof Exp:* Res asst foods, Iowa State Univ, 53-55; from jr specialist to assoc prof, 55-71, assoc dean, Col Agr & Environ Sci & actg chmn dept consumer sci, 73-75, PROF FOOD SCI & TECHNOL, UNIV CALIF, DAVIS, 71- *Concurrent Pos:* NIH consult, Univ Chile, 66; vis prof, Swed Inst Food Preservation Res, 70-71; vis scientist, Nestle Res Lab, Switz, 78; invited guest, Polish Acad Sci & invited lectr, Univ Guelph, Can, 79. *Mem:* AAAS; Soc Nutrit Educ; fel Inst Food Technologists; Am Inst Nutrit; Asn Chemoreception Sci; Sigma Xi. *Res:* Food psychophysics; behavioral responses to gustatory and olfactory stimuli; oral perception; saliva. *Mailing Add:* Dept Food Sci Univ Calif Davis CA 95616

PANG-CHING, GLENN K, b Hilo, Hawaii, Feb 10, 31; m 61; c 2. AUDIOLOGY. *Educ:* Calif State Univ, Los Angeles, BS, 57; Purdue Univ, MS, 58; Univ Southern Calif, PhD(audiol), 66. *Prof Exp:* Audiologist, Vet Admin, Washington, DC, 58-59 & Los Angeles, 59-65; PROF AUDIOL & SPEECH PATH, DIV SPEECH PATH & AUDIOL, SCH MED, UNIV HAWAII, 65- *Mem:* Am Speech & Hearing Asn. *Res:* Clinical research in audiology. *Mailing Add:* Div Speech Path & Audiol 1410 Lower Campus Dr Honolulu HI 96822

PANICCI, RONALD J, b New York, NY, Oct 20, 41. ORGANIC CHEMISTRY. *Educ:* Col Holy Cross, BS, 63; Univ NH, PhD(org chem), 67. *Prof Exp:* Asst prof, 67-74, assoc prof, 74-77, PROF CHEM, SOUTHERN CONN STATE COL, 77-, CHMN DEPT, 76- *Concurrent Pos:* Consult, Kemtrek Co, Conn, 67- & Baron Consult Co, 68- *Mem:* AAAS; Am Chem Soc. *Res:* Organic reaction mechanisms relative to ring chain tautomerism; rearrangement reactions and organometallic chemistry; qualitative and quantitative use of nuclear magnetic resonance in reaction mechanisms; biochemistry of algae. *Mailing Add:* Dept Chem Southern Conn State Univ 501 Crescent St New Haven CT 06515

PANISH, MORTON B, b New York, NY, Apr 8, 29; m 51; c 3. PHYSICAL CHEMISTRY, MATERIALS SCIENCE. *Educ:* Denver Univ, BS, 50; Mich State Univ, MS, 52, PhD(phys chem), 54. *Prof Exp:* Chemist molten salt chem, Oak Ridge Nat Lab, 54-57; mem tech staff phys chem, Res & Advan Develop Div, Avco Corp, 57-62, sect head phys chem, 62-64; mem tech staff solid state chem, 64-69, dept head mat sci, 69-86, DISTINGUISHED MEM TECH STAFF, BELL LABS, 86- *Honors & Awards:* Electronics Div Award, Electrochem Soc, 73 & Solid State Sci Award, 79; C & C Prize, Japan, 86; Morris Liebruen Award, Inst Elec & Electronic Engrs, 91. *Mem:* Nat Acad Sci; Nat Acad Eng; Electrochem Soc; Fel Am Phys Soc; fel Inst Elec & Electronic Engrs; Mat Res Soc. *Res:* Chemical thermodynamic studies of semiconductor systems; thermodynamics of impurity incorporation in semiconductors; phase chemistry; epitaxial crystal growth of semiconductors; semiconductor devices, particularly injection lasers. *Mailing Add:* AT&T Bell Labs Rm 1C 315 600 Mountain Ave Murray Hill NJ 07974

PANISSET, JEAN-CLAUDE, b Oka, Que, Apr 27, 35; m 57; c 2. PHARMACOLOGY. *Educ:* Univ Montreal, BA, 54, DVM, 59, MSc, 60, PhD(pharmacol), 62. *Prof Exp:* Res assoc autonomic pharmacol, Cleveland Clin, Ohio, 62-63; pharmacologist, Food & Drug Directorate, Can, 63-64; asst prof autonomic pharmacol, 64-70, assoc prof pharmacol, 70-76, PROF PHARMACOL & CHMN DEPT, FAC VET MED, UNIV MONTREAL, 76-; DIR RES, INST BIO-ENDOCRINOL INC, 70- *Mem:* Fel Am Col Vet Toxicologists; fel Am Col Vet Pharmacologists & Therapeut; Can Vet Med Asn; Pharmacol Soc Can; Am Soc Pharmacol & Therapeut. *Res:* Endocrinology; biopharmaceutics; radioimmunology. *Mailing Add:* Dept Indust Med-Hyg Univ Montreal Montreal PQ H3C 3J7 Can

PANITZ, JANDA KIRK GRIFFITH, b Mechanicsburg, Pa, Feb 2, 45; m 67. SURFACE PHYSICS, SOLID STATE PHYSICS. *Educ:* Pa State Univ, BS, 66, MS, 69, PhD(physics), 75. *Prof Exp:* Teaching asst physics, Pa State Univ, 66-68; physicist IR optics, Ballistic Res Lab, Aberdeen Proving Ground, 67; res asst physics, Appl Res Lab, Pa State Univ, 69-75; res assoc elec eng, Bur Eng Res, Univ NMex, 76-77; STAFF SCIENTIST PHYSICS, SANDIA LABS, ALBUQUERQUE, 78- *Concurrent Pos:* Consult, Sandia Labs, 71-73. *Mem:* AAAS; Am Vacuum Soc; Sigma Xi. *Res:* Thin film physics and surface physics; also theoretical solid state physics and lattice dynamics; geometrical optics and infrared optical systems. *Mailing Add:* 228 N Star Rte Edgewood NM 87015

PANKAVICH, JOHN ANTHONY, b Long Island City, NY, Jan 25, 29; m 56; c 4. ANIMAL SCIENCE & NUTRITION. *Educ:* St Francis Col, NY, BS, 51; Syracuse Univ, MS, 53; NY Univ, PhD(parasitol), 66. *Prof Exp:* Biol technician, Rockefeller Inst Med Res, 47; res biologist, Agr Ctr, Am Cyanamid Co, 56-; RETIRED. *Mem:* Am Soc Parasitologists. *Res:* Evaluation and discovery of anthelmintics for parasitisms found in large and small animals; development of anthelmintic screening models for discovering anthelmintics; in vitro growth of parasites; abnormal host-parasite relationships; visceral larva migrans. *Mailing Add:* 20 Sedgwick Rd Hamilton Square NJ 08690

PANKEY, GEORGE ATKINSON, b Shreveport, La, Aug 11, 33; m 72; c 4. INFECTIOUS DISEASES, INTERNAL MEDICINE. *Educ:* Tulane Univ, New Orleans, BS, 54, MD, 57; Univ Minn, Minneapolis, MS, 61; Am Bd Internal Med, cert, 65, cert infectious dis, 72. *Prof Exp:* Intern med, Univ Minn Hosps, Minneapolis, 57-58, resident internal med, 58-60; resident, Vet Admin Hosp & Gen Hosp, Minneapolis, 60-61; instr med, Div Infectious Dis, Sch Med, Tulane Univ, 61-63; MEM STAFF & CONSULT INFECTIOUS DIS, OCHSNER FOUND HOSP, 63-, HEAD SECT INFECTIOUS DIS, OCHSNER CLIN, 72- *Concurrent Pos:* Asst vis physician, Charity Hosp La, New Orleans, 61-62, vis physician, 62-75, sr vis physician, 75-; instr infectious dis, Sch Med, Tulane Univ, 63-65, from clin asst prof to clin assoc prof, 65-73, clin prof med, 73-; clin assoc prof, Dept Oral Diag, Med & Radiol, Sch Dent, La State Univ, New Orleans, 70-83, clin prof, 83-, clin prof med, Sch Med, 79-; consult, Instnl Rels Comt, Am Soc Internal Med, 72-73; med consult, World Health Info Serv, 74-; consult physician, Dept Med, Vet Admin Med Ctr, Biloxi, Miss, 78- *Mem:* Fel Am Col Physicians; fel Am Col Prev Med; fel Infectious Dis Soc Am; fel Am Col Chest Physicians; Am Soc Trop Med & Hyg. *Res:* Antimicrobial evaluation; endocarditis; systemic fungal diseases; mycobacteria, especially in vitro antimicrobial susceptibility and treatment; AIDS. *Mailing Add:* Ochsner Clin 1514 Jefferson Hwy New Orleans LA 70121

PANKIWSKYJ, KOST ANDRIJ, b Lviv, Ukraine, Dec 20, 36; US citizen. GEOLOGY. *Educ:* Mass Inst Technol, BS, 59; Harvard Univ, PhD(geol), 64. *Prof Exp:* Lectr mineral, Univ Mass, 64; ASST PROF MINERAL & PETROL, UNIV HAWAII, 64-, ASSOC PROF GEOL & GEOPHYS, 72- *Concurrent Pos:* Field geologist, Maine Geol Surv, 60- *Mem:* Mineral Soc Am; Geochem Soc; Mineral Soc Gt Brit & Ireland. *Res:* Structural and stratigraphic mapping in northwestern Maine; chemical and mineralogical composition of Hawaiian basalts. *Mailing Add:* Dept Geol Univ Hawaii Manoa Honolulu HI 96822

PANKOVE, JACQUES I, b Russia, Nov 23, 22; nat US; m 50; c 2. SOLID STATE PHYSICS, MATERIALS SCIENCE ENGINEERING. *Educ:* Univ Calif, BS, 44, MS, 48; Univ Paris, PhD(physics), 60. *Prof Exp:* Lab asst, Univ Calif, 46-48; solid state physicist, RCA Labs, 48-71, fel tech staff, 71-85; prof, Dept Elec & Computer Eng, 85-89, HUDSON MOORE JR UNIV PROF, UNIV COLO, 89-; DISTINGUISHED SCIENTIST, SOLAR ENERGY RES INST, 85- *Concurrent Pos:* Vis MacKay Lectr, UNiv Calif, Berkeley, 68-69; assoc ed, J Quantum Electronics, 68-77; vis prof, Univ Campinas, Brazil, 75; distinguished prof, Univ of Mo-Rolla, 84; regional ed, Crystal Lattice Defects & Amorphous Mats, 84- *Honors & Awards:* Ebers Award, Inst Elec & Electronics Engrs, 75. *Mem:* Nat Acad Eng; fel Am Phys Soc; fel Inst Elec & Electronics Engrs; Electrochem Soc; AAAS; Mat Res Soc. *Res:* Semiconductor devices and semiconductor properties; superconductivity; luminescence; lasers; author of numerous articles and a textbook; editor of seven volumes. *Mailing Add:* Campus Box 425 Univ Colo Boulder CO 80309-0425

PANKOVICH, ARSEN M, b Banja Luka, Yugoslavia, Aug 3, 30; US citizen. ORTHOPEDIC SURGERY. *Educ:* Univ Belgrade, MD, 54. *Prof Exp:* Instr orthop surg, Med Col, Cornell Univ, 66-67; from asst prof to assoc prof, Sch Med, Univ Chicago, 67-72; prof orthop surg, Abraham Lincoln Sch Med, Univ Ill Med Ctr & chmn dept orthop surg, Cook County Hosp, 72-; AT DEPT ORTHOP SURG, DOWNSTATE MED CTR, STATE UNIV NY. *Concurrent Pos:* USPHS fel, Univ Chicago, 61-62 & Hosp Spec Surg, New York, 65-67. *Mem:* Am Acad Orthop Surg; Am Col Surg; Orthop Res Soc; Clin Orthopaedic Soc; Sigma Xi. *Res:* Bone physiology, enzymes and calcification; diagnosis and treatment of fractures and dislocations. *Mailing Add:* 201 E 62nd St New York NY 10021

PANKOW, JAMES FREDERICK, b Mexico City, Mex, Jan 14, 51; US citizen; m 79; c 2. ENVIRONMENTAL CHEMISTRY, NATURAL WATER CHEMISTRY. *Educ:* State Univ NY, Bingham, BA, 73; Calif Inst Technol, MS, 76, PhD(environ sci), 78. *Prof Exp:* Instr, 78-79, from asst prof to prof water chem, 80-86, PROF & CHMN, DEPT ENVIRON SCI & ENG, ORE GRAD CTR, 86- *Mem:* Am Chem Soc. *Res:* Behavior of organic and inorganic chemicals in the environment, and their analysis by modern methods including capillary column/mass spectrometry; natural water chemistry. *Mailing Add:* Dept Environ Sci Ore Grad Inst 19600 NW Von Neumann Dr Beaverton OR 97006-1999

PANKRATZ, RONALD ERNEST, b Moundridge, Kans, Apr 19, 18; m 43; c 8. ANALYTICAL CHEMISTRY. *Educ:* Bethel Col, Kans, AB, 41. *Prof Exp:* Chemist, Weldon Springs Ord Works, Atlas Powder Co, Mo, 42-43; res chemist, Presstite Eng Co, 43-45 & Cole Chem Co, 45-48; control chemist, Flint Eaton & Co, Ill, 48-57; chief chemist, Ft Dodge Labs Inc, Am Home Prod Co, 57-62, dir pharmaceut control & res assoc, Diamond Labs, Iowa, 62-64; chemist, Chemagro Corp, Mo, 64-65; res chemist, Abbott Labs, Ill, 65-69; dir qual control, Med Chem Corp, 70-71; MGR CHEM DIV, POLYSCI CORP, NILES, 71- *Mem:* Am Chem Soc. *Res:* Analytical research; tissue residue studies. *Mailing Add:* 3528 Winhaven Dr Waukegan IL 60085

PANLILIO, FILADELFO, b Philippines, Feb 5, 18; US citizen; m 40; c 2. MECHANICS, MECHANICAL ENGINEERING. *Educ:* Univ Philippines, BSME, 38; Univ Mich, MS, 42, PhD(eng mech), 46. *Prof Exp:* Instr mech, Univ Philippines, 38-41; from instr to asst prof, Univ Mich, 45-48; from asst prof to assoc prof, Univ Philippines, 48-53, prof & chmn, 53-55; from assoc prof to prof, 55-87, EMER PROF MECH ENG, UNION COL, 87- *Concurrent Pos:* Consult, Gen Elec Co, 56, ALCO Prod, Inc, 58, Knolls Atomic Power Lab, Gen Elec Co, 60 & Watervliet Arsenal, US Army Weapons Command, 63- *Mem:* Soc Exp Stress Analysis; Am Soc Mech Engrs; Am Soc Metals. *Res:* Deformable body mechanics; continuum mechanics; finite elements. *Mailing Add:* 49 Aspinwall Rd Loudonville NY 12211

PANNABECKER, RICHARD FLOYD, b Bluffton, Ohio, May 21, 22; m 45; c 5. ZOOLOGY. *Educ:* Roosevelt Univ, BS, 48; Univ Chicago, PhD(zool), 57. *Prof Exp:* Asst zool, Univ Chicago, 50-54; from asst prof to assoc prof biol, Ohio Northern Univ, 54-61; assoc prof, 61-64, PROF BIOL, BLUFFTON COL, 64- *Mem:* AAAS; Am Inst Biol Sci; Sigma Xi. *Res:* Vertebrate embryology and endocrinology, particularly sex differentiation in vertebrates. *Mailing Add:* 430 W Elm St Bluffton OH 45817

PANNELL, KEITH HOWARD, b London, Eng Aug 27, 40; c 3. ORGANOMETALLIC CHEMISTRY. *Educ:* Univ Durham, BSc, 62, MSc, 63; Univ Toronto, PhD(organometallic chem), 66. *Prof Exp:* Res assoc, Univ Ga, 66-68; sr res fel, Univ Sussex, 68-70; from asst prof to assoc prof, 71-80, PROF CHEM, UNIV TEX, EL PASO, 80- *Mem:* The Chem Soc; Am Chem Soc; Soc Quim Mexicana. *Res:* Interactions of metals and metalloids with organic molecules; clay/organometallic interactions; physiology of ionophores. *Mailing Add:* Dept Chem Univ Tex El Paso TX 79968

PANNELL, LOLITA, b Millburn, NJ, May 10, 12. BACTERIOLOGY. *Educ:* Brown Univ, PhB, 34; Univ Kans, MA, 47, PhD(bact), 50. *Prof Exp:* Instr phys educ, Essex County Hosp, NJ, 34-39; bacteriologist, Sch Med & Dent, Univ Rochester, 39-41 & Meadowbrook Hosp, 41-43; lab asst bact, hemat & virol, Univ Kans, 46-50; asst prof bact & immunol, Jefferson Med Col, 50-54; from asst prof to prof bact, 54-75, EMER PROF BACT, MED UNIV SC, 75- *Mem:* AAAS; Am Soc Microbiol; Am Pub Health Asn; NY Acad Sci. *Res:* Glanders; melioidosis; tularemia; rheumatic fever. *Mailing Add:* 288 Molasses Lane Mt Pleasant SC 29464

PANNELL, RICHARD BYRON, b Salt Lake City, Utah, May 1, 52; m 77; c 3. CATALYSIS, PHYSICAL CHEMISTRY. *Educ:* Univ Utah, BS, 74; Brigham Young Univ, PhD(chem eng), 78. *Prof Exp:* Sr res engr, Gulf Res & Develop Co, 78-85; STAFF ENGR, EXXON CHEM CO, 85- *Mem:* Am Chem Soc. *Res:* Hydroprocessing and synthesis gas conversion to fuels and chemicals and surface properties of heterogenous catalysts; catalyst preparation, catalyst charcterization and kinetics. *Mailing Add:* Exxon Chem Co PO Box 5200 Baytown TX 77522

PANNELLA, GIORGIO, b Milan, Italy, Feb 26, 34; m 72. INVERTEBRATE PALEONTOLOGY, PALYNOLOGY. *Educ:* Univ Pavia, MA, 58; Univ Colo, PhD(geol), 66. *Prof Exp:* Asst micropaleont, Univ Pavia, 58, asst prof stratig & sedimentology, 58-59; res geologist, Ital Geol Surv, 61 & Peabody Mus Natural Hist, Yale Univ, 65-72; assoc prof geol, Univ PR, Mayaguez, 72-80; PRES, SINCLAIRVILLE PETROL, 81; PRES, PAN ENERGY CO, INC, 81- *Concurrent Pos:* NASA grant, 65-70; NSF grant, 71; NATO fel, 73-75; vis prof, Univ Colo, Boulder, 78-79; geol consult, 80- *Mem:* AAAS; Ital Geol Soc; Paleont Soc; Sigma Xi; Am Asn Petrol Geologists. *Res:* Geological, biological and astronomical cycles; micropaleontology; sedimentology; pre-Quaternary palynology; bivalves; detection and correlation of astronomical, geological and biological rhythms; calcification in biological systems; Caribbean geology. *Mailing Add:* 7311 Mile Strip Orchard Park NY 14127

PANNER, BERNARD J, b Youngstown, Ohio, Oct 9, 28; m 62; c 3. PATHOLOGY. *Educ:* Western Reserve Univ, AB, 49, MD, 53. *Prof Exp:* PROF PATH, MED CTR, UNIV ROCHESTER, 61- *Mem:* AAAS; Am Soc Exp Path; Electron Micros Soc Am; Am Asn Path & Bact. *Res:* Experimental pathology and clinical renal diseases; ultrastructure of smooth muscle. *Mailing Add:* Univ Rochester Med Ctr 601 Elmwood Ave Rochester NY 14642

PANNILL, FITZHUGH CARTER, JR, b Rosemont, Pa, July 4, 21; m 46; c 3. INTERNAL MEDICINE. *Educ:* Yale Univ, BS, 42, MD, 45; Am Bd Internal Med, dipl, 53. *Prof Exp:* From instr to asst prof med, Col Med, Baylor Univ, 51-54; clin instr, Univ Tex Southwestern Med Sch, 59-60, asst prof internal med & asst dean grants & sponsored res, 61-64; assoc prof med, Hahnemann Med Col, 60-61; prof med & dean, Univ Tex Med Sch San Antonio, 65-73; vpres health sci, 73-83, PROF MED, STATE UNIV, NY, BUFFALO, 73- *Concurrent Pos:* Markle Found award acad med, 62-67; pvt pract, Tex, 54-60. *Mem:* Fel Am Col Physicians; AMA; Asn Am Med Col; Am Fedn Clin Res. *Mailing Add:* Dept Med Erie Co Med Ctr 1462 Grider St Buffalo NY 14215

PANNU, SARDUL S, b Wadala Banger, Panjab, India, Nov 1, 35; US citizen; m 66; c 3. ANALYTICAL CHEMISTRY. *Educ:* Khalsa Col, BSc, 56, MSc, 58; George Washington Univ, PhD(anal chem), 65. *Prof Exp:* Lectr chem, Khalsa Col, 58-60; prof, Southern Union State Jr Col, 65-66; from asst prof to assoc prof, Randolph-Macon Col, 66-69; assoc prof, 69-71, NASA res grant, 70-71, PROF CHEM, UNIV DC, 71- *Concurrent Pos:* NIH biomed grant, 74-77. *Mem:* Am Chem Soc; Indian Chem Soc. *Res:* Electroanalytical determination of transition elements. *Mailing Add:* Dept of Chem Univ of DC Washington DC 20008

PANOFSKY, WOLFGANG K H, b Berlin, Ger, Apr 24, 19; nat US; m 42; c 5. PHYSICS. *Educ:* Princeton Univ, AB, 38; Calif Inst Technol, PhD(physics), 42. *Hon Degrees:* Various from US & foreign univs, 63-87. *Prof Exp:* Dir, Off Sci Res & Develop Proj, Calif Inst Technol, Pasadena, 42-43; consult, Manhattan Dist, Los Alamos, NMex, 43-45; physicist, Radiation Lab, Univ Calif, Berkeley, 45-46, from asst prof to assoc prof physics, 46-51; dir, 61-84, PROF PHYSICS, STANFORD LINEAR ACCERATOR CTR, STANFORD UNIV, 51-, EMER DIR, 84- *Concurrent Pos:* Dir high energy physics lab, Stanford Univ, 53-61; mem high energy comn, Int Union Pure & Appl Physics, 58-60; guest prof, Europ Orgn Nuclear Res, Geneva, 59; Guggenheim fel, 59; chmn tech working group high altitude detection, US del, Geneva, 59, vchmn tech working group II, 59; mem, President's Sci Adv Comt, 60-63; Richtmyer lectr, 63; consult, Atomic Energy Comn, US Air Force; consult, Radiation Lab, Univ Calif & NSF; mem, Adv Res Proj Agency. *Honors & Awards:* Ernest Orlando Lawrence Award, Atomic Energy Comn, 61; Richtmyer Lectr, Harvard Univ, 63; Nat Medal of Sci, 69; Franklin Inst Award, 70; Ann Pub Serv Award, Fedn Am Scientists, 73; Officer of French Legion of Honor, 77; Jessie & John Danz Lectr, Univ Wash, Seattle, 79; E Fermi Award, 79; Cherwell-Simon Mem Lectr, Oxford Univ, Eng, 81; Leo Szilard Award, 82; Hilliard Roderick Prize, AAAS, 91; Heisenberg Lectr, Munich, 91. *Mem:* Nat Acad Sci; fel Am Phys Soc (vpres); Am Philos Soc; Am Acad Arts & Sci. *Res:* High energy physics; particle accelerators; co-author of one book; numerous scientific papers in professional journals and arms control and disarmament papers in professional journals. *Mailing Add:* Stanford Linear Accelerator Ctr PO Box 4349 Stanford CA 94305

PANOPOULOS, NICKOLAS JOHN, b Doliana Kynourias, Greece. PLANT PATHOLOGY, MICROBIAL GENETICS. *Educ:* Agr Col Athens, BS, 66; Univ Calif, Berkeley, PhD(plant path), 71. *Prof Exp:* Asst res fel, 71-75, asst plant pathologist & lectr, 75-80, ASSOC PROF PLANT PATH, UNIV CALIF, BERKELEY, 80- *Mem:* Am Phytopath Soc; Am Soc Microbiol; Sigma Xi. *Res:* Genetics of bacterial pathogenicity on plants. *Mailing Add:* 3120 Minna Ave Oakland CA 94619

PANOS, CHARLES, b Pittsburgh, Pa, July 15, 29; m 57; c 3. BACTERIOLOGY, BIOCHEMISTRY. *Educ:* Univ Pittsburgh, BSc, 52, MS, 53, PhD(bact), 56. *Prof Exp:* Asst prof bact, Univ Fla, 56-57; asst prof biochem & res fel, Sch Med, Univ Ill, 57-61; assoc mem res, Albert Einstein Med Ctr, Northern Div, Philadelphia, 61-68, mem, 68-72; PROF MICROBIOL, MED SCH, THOMAS JEFFERSON UNIV, 72- *Concurrent Pos:* USPHS career develop award, 60 & 63-; res assoc prof, Sch Med, Temple Univ, 61-72. *Mem:* Am Soc Microbiol; Am Soc Biol Chem; Brit Soc Gen Microbiol; Int Orgn Mycoplasmology; fel Am Acad Microbiol. *Res:* Microbial metabolism; intermediary lipid metabolism; cell wall biosynthesis; biochemistry of bacterial L-forms and mycoplasmas. *Mailing Add:* Dept Microbiol & Immunol Jefferson Alumni Hall Thomas Jefferson Univ Med Sch Philadelphia PA 19107

PANOS, PETER S, b Chicago, Ill. RESEARCH ADMINISTRATION. *Educ:* Northwestern Univ, BS, 61; Univ Chicago, MBA, 64. *Prof Exp:* Training specialist, Procter & Gamble Mfg Co, 61-63; supvr eng econ, Inst Gas Technol, 63-69; group vpres, Duff & Phelps Inc, 69-82; VPRES, KIDDER, PEABODY & CO, INC, 83- *Concurrent Pos:* Consult, 82-83. *Mem:* Nat Soc Prof Engrs; Am Inst Chem Engrs; Inst Chartered Financial Analysts. *Res:* Utility engineering; liquefied natural gas; substitute gas supply. *Mailing Add:* 145 Attorney St New York NY 10002

PANSKY, BEN, b Milwaukee, Wis, Feb 18, 28; m 53; c 1. ANATOMY. *Educ:* Univ Wis, BA, 48, MS, 50, PhD(anat), 54; New York Med Col, MD, 68; Nat Bd Med Examiners, dipl. *Prof Exp:* Instr histol, neuroanat & anat, Univ Wis, 51-53; from instr to assoc prof anat, New York Med Col, 53-68; resident path, Columbia Presby Hosp, Col Physicians & Surgeons, 68-69, path trainee & supvr autopsy serv, 69-70; dir baccalaurreate, Biol Nursing,76-84, PROF ANAT, MED COL OHIO, 70- *Concurrent Pos:* Affil, AMA. *Mem:* AAAS; Am Asn Anat; Am Heart Asn; Sigma Xi. *Res:* Hematological studies as related to endocrine variations; immunohistochemistry; immunocytochemistry; pituitary gland and hypothalamus; in situ cDNA-mRNA hybridization; extrapancreatic; insulin as related to pituitary gland, retina, brain; molecular biology; medical illustrator and author of 5 textbooks in anatomy, embryology, neuroscience and physiology. *Mailing Add:* Med Col of Ohio Dept of Anat C S 10008 Toledo OH 43699

PANSON, GILBERT STEPHEN, b Paterson, NJ, Apr 11, 20; m 44; c 2. PHYSICAL CHEMISTRY, ORGANIC CHEMISTRY. *Educ:* Brown Univ, ScB, 41; Columbia Univ, MA, 50, PhD(chem), 53. *Prof Exp:* From instr to asst prof chem, Hobart Col, 41-44; res chemist, Tenn Eastman Co, 44-46; from asst prof to assoc prof chem, 46-60, actg dean, Newark Col Arts & Sci, 71-73; chmn dept, 62-72, dean, Grad Sch, 75-81, prof, 60-85, EMER PROF CHEM, RUTGERS UNIV, NEWARK, 85- *Concurrent Pos:* Vis prof chem, Univ Calif, Berkeley, 73-74 & Univ London, 81-82. *Mem:* Am Chem Soc; Sigma Xi. *Res:* Physical organic chemistry; separation of isotopes; reaction kinetics; separation of organic compounds by thermal diffusion. *Mailing Add:* 8A Yorkshire Dr Cranbury NJ 08512

PANTELIDES, SOKRATES THEODORE, b Limassol, Cyprus, Nov 20, 48; m 72; c 2. THEORETICAL SOLID STATE PHYSICS. *Educ:* Northern Ill Univ, BS, 69; Univ Ill, Urbana, MS, 70, PhD(physics), 73. *Prof Exp:* Res assoc physics, Univ Ill, Urbana, 73 & W W Hansen Labs, Stanford Univ, 73-75; sr mgr phys sci, 84-89, tech asst to vpres sci & technol, 89-91, RES STAFF MEM PHYSICS, IBM RES DIV, T J WATSON RES CTR, IBM CORP, 75-, PROG DIR, TECHNOL & SCI MODELING, 91- *Concurrent Pos:* Vis prof, Dept Theoret Physics, Univ Lund, 75; actg chair prof, Dept Theoret Phys, 78. *Mem:* Fel Am Phys Soc; Mat Res Soc; Electrochem Soc; Inst Elec & Electronic Engrs. *Res:* Theory of electronic properties of solids; theory of semiconductor surfaces and interfaces; the systematics of band structures, optical properties and dielectric constants; theory of excitons and photoemission; theory of impurities, defects and diffusion in semiconductors; amorphous solids. *Mailing Add:* T J Watson Res Ctr IBM Corp Yorktown Heights NY 10598

PANTO, JOSEPH SALVATORE, b Lawrence, Mass, Dec 25, 25; m 52; c 2. TEXTILE CHEMISTRY. *Educ:* Lowell Technol Inst, BS, 51, MS, 63. *Prof Exp:* Res assoc textile chem, Burlington Industs Inc, 51-53; asst dir tech serv, Albany Int Res Co, 53-90; RETIRED. *Honors & Awards:* Harold C Chapin Award for Outstanding Serv, Am Asn Textile Chemists & Colorists, 90. *Mem:* Am Chem Soc; Am Asn Textile Chemists & Colorists (vpres, 72); fel Am Inst Chemists; Brit Soc Dyers & Colourists; Am Arbitration Asn; AAAS. *Res:* Color measurement and the application of textile colorants to fibrous materials; application, by novel techniques, of resin pre-condensates to cellulosic fibers to impart resistance to creasing. *Mailing Add:* 13 Valley Rd Dover MA 02030

PANTON, R(ONALD) L(EE), b Neodesha, Kans, Feb 14, 33; m 60; c 3. FLUID MECHANICS. *Educ:* Wichita State Univ, AB & BS, 56; Univ Wis, MS, 62; Univ Calif, Berkeley, PhD(mech eng), 66. *Prof Exp:* Engr, NAm Aviation, Inc, 56-58; asst prof mech & aerospace eng, Okla State Univ, 66-69; assoc prof, 69-76, PROF MECH ENG, UNIV TEX, AUSTIN, 76- *Mem:* Assoc fel Am Inst Aeronaut & Astronaut; fel Am Soc Mech Engrs; Am Phys Soc. *Res:* Fluid mechanics, turbulence, incompressible flow; acoustics. *Mailing Add:* Dept of Mech Eng Univ of Tex Austin TX 78712

PANTUCK, EUGENE JOEL, b Boston, Mass, Feb 8, 38; m 60; c 2. ANESTHESIOLOGY, BIOCHEMICAL PHARMACOLOGY. *Educ:* Tufts Univ, BS, 59, MD, 63. *Prof Exp:* Intern med & surg, Montefiore Hosp, New York, 63-64; resident anesthesiol, Presby Hosp, New York, 64-66; dep chief, USPHS Hosp, Baltimore, 67-68, chief, 68-69; from asst prof to assoc prof, 69-86, PROF ANESTHESIOL, COLUMBIA UNIV, 86-; ATTEND ANESTHESIOLOGIST, PRESBY HOSP, 86- *Concurrent Pos:* Res trainee anesthesiol, Columbia Univ, 66-67; vis scientist, Hoffman-LaRoche Inc, 70-85; from asst attend anesthesiologist to assoc attend anesthesiologist, Presby Hosp, 69-86; vis assoc physician, Rockefeller Univ Hosp, NY, 75-85; vis scientist, Roche Inst Molecular Biol, 85-87; adj prof chem biol & pharmacog, Rutgers Univ, 86- *Mem:* Am Soc Pharmacol & Exp Therapeut; Am Soc Anesthesiol; Int Anesthesia Res Soc. *Res:* Drug metabolism; drug interaction; enzyme induction; chemical carcinogenesis; drug-nutrient interactions. *Mailing Add:* Presby Hosp 622 W 168th St New York NY 10032

PANUSH, RICHARD SHELDON, b Detroit, Mich, Nov 9, 42; m 66; c 3. CLINICAL IMMUNOLOGY, RHEUMATOLOGY. *Educ:* Univ Mich Honors Col, BA, 65, Med Sch, MD, 67. *Prof Exp:* Intern, Duke Univ Med Ctr, 67-68 & jr asst res, 68-69; res fel, Dept Med, Harvard Med Sch, 69-71; chief rheumatology, Dept Med, Silas B Hayes Army Hosp, 71-72 & Fitzsimons Gen Hosp, 72-73; asst prof clin immunol, Col Med, Univ Fla, 73-77; CHIEF RHEUMATOLOGY, GAINESVILLE VET ADMIN MED CTR, 73-; CHIEF CLIN IMMUNOL, COL MED, UNIV FLA, 76-, ASSOC PROF, 77- *Mem:* Am Rheumatism Asn; Am Fed Clin Res; Am Col Physicians; Am Asn Immunologists; Am Acad Allergy. *Res:* Immunologic aspects of rheumatic diseases. *Mailing Add:* J H Miller Health Ctr Gainesville FL 32610

PANUSKA, JOSEPH ALLAN, b Baltimore, Md, July 3, 27. ENVIRONMENTAL PHYSIOLOGY. *Educ:* Loyola Col, Md, BS, 48; St Louis Univ, PhL, 54, PhD(biol), 58; Woodstock Col, STL, 61. *Hon Degrees:* LLD, Scranton Univ, 74. *Prof Exp:* NIH trainee physiol, Sch Med, Emory Univ, 62-63, instr, 63; from asst prof to assoc prof biol, 63-72, Rector Jesuit Comm; acad vpres & prof biol, Boston Col, 79-88; PRES, UNIV SCRANTON, 82- *Concurrent Pos:* NSF res grant, 60-61, St Joseph's Univ, Loyola Col, Boston Col Campus Sch Handicapped, 79-82; vis fel & vis scientist, Dept Path & St Edmund's House, Cambridge Univ, 69-70; mem, Am Found Biol Res, 75-85, pres bd dirs, 74-79, vpres, 79-; pres bd dirs, Jesuit Community Georgetown Univ, Inc, 76-79; provincial, Md Prov, Soc of Jesus, 73-79; mem Pres, Cambridge Ctr Social Studies, 73-79; mem bd dirs, Community Med Ctr, Scranton, 82-; vchmn, Comn Independent Col & Univ, 88; PA assoc, Col & Univ Exec Comn, 90; mem adv comt, Commonwealth PA State Bd Educ Coun Higher Educ, 90. *Mem:* Am Physiol Soc; Soc

Cryobiol; Soc Exp Biol & Med; Sigma Xi. *Res:* Low temperature effects on mammals, including hypothermia and hibernation; stress on cardiovascular system and animal behavior. *Mailing Add:* Pres Univ Scranton Scranton PA 18510

PANUZIO, FRANK L, b Bridgeport, Conn, July 18, 07; m 49; c 1. CIVIL ENGINEERING. *Educ:* Cornell Univ, CE, 30, MCE, 32. *Prof Exp:* Hydraul engr, US Bur Reclamation, Colo, 33-36; struct & hydraul engr, NY Dist Corps Eng, US Army, 36-41, chief airport design sect, 41-44, asst & dept chief eng div, 44-51, chief eng div, Atlantic Dist Corps Eng, 51-53, asst chief eng div, NY Dist Corps Eng, 53-65, dep chief eng div, 65-66, chief, 66-69, sr eng consult, 69-72, chief, Postal Proj Mgt Br, 72-74, chief, Design Br Eng Div, 74; CONSULT CIVIL ENG, 74- *Concurrent Pos:* Adj prof, NY Univ, 40, 46-47, 48 & Polytech Inst Brooklyn, 47-48. *Mem:* Am Soc Civil Engrs; Am Geophys Union; Am Concrete Inst; Sigma Xi. *Res:* Administration and direction of engineering endeavors in connection with water resource planning, civil works construction and military construction activities. *Mailing Add:* 1385 Capitol Ave Bridgeport CT 06604

PANVINI, ROBERT S, b Brooklyn, NY, Apr 22, 37; m 59; c 3. HIGH ENERGY PHYSICS. *Educ:* Rensselaer Polytech Inst, ScB, 58; Brandeis Univ, PhD(physics), 65. *Prof Exp:* Res assoc exp high energy physics, Brookhaven Nat Lab, 65-67, from asst physicist to assoc physicist, 67-71; assoc prof physics, Vanderbilt Univ,71-80, prof physics & astron, 80-90; SR STAFF PHYSICIST, US DEPT ENERGY, 90- *Mem:* Sigma Xi; fel Am Phys Soc. *Res:* Experimental high energy physics, instrumentation, programming and analysis of experimental data; administration. *Mailing Add:* 4317 Flower Valley Dr Rockville MD 20893-1811

PANZER, HANS PETER, b Ratingen, WGermany, July 26, 22; nat US; m 56; c 2. ORGANIC POLYMER CHEMISTRY. *Educ:* Univ Muenster, Germany, Dipl, 54, Dr rer nat, 57. *Prof Exp:* Fel carbohydrate chem, Purdue Univ, 57-58; sr res chemist, Res & Develop Div, Am Machine & Foundry Co, 59-63; res specialist, Technol Ctr, Gen Foods Corp, 63-65; group leader, Indust Chem Div, 66-71, sr res chemist, Indust Chem & Plastics Div, 71-72 & Chem Res Div, 72-74, proj leader polymer res, 74-79, res group leader, 79-81, prin res chemist, 81-86, ASSOC RES FEL, CHEM RES DIV, AM CYANAMID CO, 86- *Mem:* Am Chem Soc; German Chem Soc. *Res:* Organic synthesis; synthetic and natural polymers; polyelectrolytes; organic flocculants; water treating and mining chemicals. *Mailing Add:* 150 Old N Stamford Rd Stamford CT 06905

PANZER, JEROME, b New York, NY, Apr 5, 31; m 54; c 3. PETROLEUM TECHNOLOGY. *Educ:* NY Univ, BA, 52; Cornell Univ, PhD(org chem), 56. *Prof Exp:* Chemist, Esso Res & Eng Co, 56-65, proj leader lubricants res, 65-70; RES ASSOC, EXXON RES & ENG CO, 70- *Mem:* Am Chem Soc; Soc Automotive Engrs. *Res:* Gasoline product quality; automotive emissions; non-aqueous colloidal systems; surface chemistry; coal liquids; alcohol fuels. *Mailing Add:* 3 Mountain View Rd Millburn NJ 07041

PANZER, RICHARD EARL, b Hastings, Nebr, Feb 9, 23; m 48; c 2. PHYSICAL INORGANIC CHEMISTRY. *Educ:* Col of the Pac, BA, 48; Univ Nev, MS, 50; Univ NMex, PhD(chem), 59. *Prof Exp:* Aide phys sci, Indust Lab, US Navy Yard, 43-44; asst, Univ Nev, 48-50; chemist, Titanium Res Unit, US Bur Mines, 51-53; sr chemist, Metall Res Lab, Reynolds Metals Co, 53-55; res asst, Univ NMex, 55-58; res electrochemist, US Naval Ord Lab, Calif, 58-70 & Mare Island Naval Shipyard, 70-73; res chemist, Naval Ship Res & Develop Ctr, 73-78; CONSULT, 78- *Concurrent Pos:* Naval Ord Lab fel chem, Univ Mich, 66-67. *Mem:* Sigma Xi; Nat Audubon Soc. *Res:* Energy conversion; computer applications to chemistry; physical metallurgy; coatings; corrosion; oceanography; materials science. *Mailing Add:* 1315 Loma Vista Dr Napa CA 94558

PAO, CHIA-VEN, b An-Whei, China, Aug 10, 33; US citizen; m 63; c 3. APPLIED MATHEMATICS, NUMERICAL METHODS. *Educ:* Nat Taiwan Univ, BS, 59; Kans State Univ, MS, 62; Univ Pittsburgh, PhD(math), 68. *Prof Exp:* Engr, Westinghouse Elec Corp, 62-67; from asst prof to assoc prof, 69-79, PROF MATH, NC STATE UNIV, 80- *Concurrent Pos:* Fel, NASA, 68-69; Army Res Off grant, 70-73. *Mem:* Am Math Soc; Soc Indust & Appl Math. *Res:* Nonlinear reaction; diffusion systems, neutron transport and radiative transfer problems; biological systems; reactor dynamics; Boltzmann equations; integro-partial differential equations and stability theory; numerical solutions of partial differential equations. *Mailing Add:* Dept Math NC State Univ Raleigh NC 27607

PAO, ELEANOR M, b Smyrna, NY, Feb 1, 23. RESEARCH METHODOLOGY IN FOOD CONSUMPTION. *Educ:* Ohio Univ, PhD(nutrit), 77. *Prof Exp:* NUTRITIONIST, USDA, 69- *Mem:* Am Inst Nutrit; Am Public Health Asn; Soc Nutrit Educr; Am Dietetics Asn. *Mailing Add:* 2146 Silentree Dr Vienna VA 22182

PAO, HSIEN PING, b Ningpo, China, July 1, 35; US citizen; m 65; c 2. FLUID MECHANICS, ATMOSPHERIC SCIENCES. *Educ:* Nat Taiwan Univ, BS, 56; Johns Hopkins Univ, PhD(fluid mech), 63. *Prof Exp:* Res asst fluid mech, Johns Hopkins Univ, 58-63, res assoc, 63-64; from asst prof to assoc prof, 64-70, PROF FLUID MECH, CATH UNIV AM, 70- *Concurrent Pos:* Consult, Appl Sci Div, Litton Syst, Inc, 67-68, Singer Info Serv Co, 70-, Flow Res, Inc, Kent, Wash, Integrated Syst, Rockville, Md & Merdoc Res Labs, US Army, Ft Belvoir, Va. *Mem:* Am Meteorol Soc; Am Phys Soc; Am Soc Civil Engrs. *Res:* Stability and boundary layer problems in fluid mechanics; stratified and rotating fluid flows; magnetohydrodynamic flows; dynamics of cable system; water pollution; oceanic fronts; atmospheric dynamics. *Mailing Add:* Dept Civil Eng Cath Univ Am Washington DC 20064

PAO, RICHARD H(SIEN) F(ENG), b China, Apr 22, 26; nat US; m 61. CIVIL ENGINEERING. *Educ:* St John's Univ, China, BSCE, 49; Univ Ill, MS, 51, PhD, 53. *Prof Exp:* Struct designer, J G White Eng Corp, 53-54; from asst prof to prof civil eng, Rose-Hulman Inst Technol, 54-67, chmn dept, 62-67, dir acad develop coun, 66-67; PROF CIVIL ENG & ENG MECH, CLEVELAND STATE UNIV, 67- *Concurrent Pos:* Fel fluid mech, Harvard Univ, 64-65. *Mem:* AAAS; Int Water Resources Asn; Am Acad Mech; Am Soc Civil Engrs; Int Asn Hydraul Res. *Res:* Fluid mechanics; hydrology; hydraulic engineering; engineering systems design. *Mailing Add:* Dept of Civil Eng Cleveland State Univ Cleveland OH 44115

PAO, Y(EN)-C(HING), b Ninpo, China, Mar 4, 35; US citizen; m 60. ENGINEERING, MATHEMATICS. *Educ:* Nat Taiwan Univ, BS, 56; Univ Utah, MS, 59, MA, 61; Cornell Univ, PhD(mech), 65. *Prof Exp:* Mech engr, Montek Assocs, Inc, 59; design engr, EIMCO Corp, 59-61; assoc res engr, Boeing Co, 61-63; consult, Therm Advan Res Co, 64-65; preliminary design engr, Garrett Corp, 65-66; assoc prof eng mech, 66-71, PROF MECH ENG & ENG MECH, UNIV NEBR, LINCOLN, 71- *Concurrent Pos:* NSF-NIH grants, Univ Nebr, 67-79; collabr, vis scientist & consult, Mayo Clin, 73-88; fel, Am Soc of Mech Eng, 85; vis fac, IBM Watson Lab, 86; consult, var industs. *Honors & Awards:* Outstanding Res Sci Award, Sigma Xi, 84; AT&T Found Award, Am Soc for Engr Educ, 88. *Mem:* Sigma Xi; AAAS; Am Soc Mech Engrs; Soc Biomed Engrs; Am Acad Mech; Am Soc Eng Educ. *Res:* Mechanics; applied mathematics; simulation; computer applications; composite materials; cardiopulmonary mechanics. *Mailing Add:* Dept Eng Mech Univ Nebr 311 Bancroft Hall Lincoln NE 68588

PAO, YIH-HSING, b Nanking, China, Jan 19, 30; m 57; c 3. APPLIED MECHANICS, STRUCTURAL ENGINEERING. *Educ:* Univ Taiwan, BS, 52; Rensselaer Polytech Inst, MS, 55; Columbia Univ, PhD(eng mech), 59. *Prof Exp:* Asst eng mech, Rensselaer Polytech Inst, 53-55 & Columbia Univ, 56-58; asst prof mech & mat, 58-62, assoc prof eng mech, 62-68, prof theoret & appl mech, 68-84, chmn dept, 74-80, JOSEPH C FORD PROF THEORET & APPL MECH, CORNELL UNIV, 84- *Concurrent Pos:* Consult, industs; vis prof, Nat Taiwan Univ, 64-65, 78-79 & 84-86, dir, Inst Appl Mech, 84-86, 89-; assoc ed, J Appl Math, 71-73; vis prof, Stanford Univ, 72-73; adv solid mech prog, Eng Mech Div, NSF, 77-78. *Mem:* Nat Acad Eng; Soc Eng Sci; fel Am Soc Mech Engrs; fel Am Acad Mech; Acoust Soc Am; Chinese Soc Theoret & Appl Mech. *Res:* Mechanical vibrations; magneto-elasticity; visco-elasticity; stress waves in solids; dynamics of structures; non-destructive testing of materials by ultrasound; earthquake engineering; physical acoustics; mechanics of composite materials. *Mailing Add:* Dept Theoret & Appl Mech Thurston Hall Cornell Univ Ithaca NY 14853

PAO, YOH-HAN, b China, July 17, 22; nat US; m 48; c 3. COMPUTER SCIENCES. *Educ:* Lester Inst, China, BS, 45; Syracuse Univ, MS, 49; Pa State Univ, PhD(appl physics), 52. *Prof Exp:* Asst prof eng res, Pa State Univ, 52-53; res physicist, E I du Pont de Nemours & Co, 53-56, res assoc, 56-59, supvr phys res, 59-61; sr res physicist, lab appl sci & res assoc, inst study metals, Univ Chicago, 61-62; mem tech staff, Bell Tel Labs, NJ, 62-67; prof appl physics, 67-76, head, Dept Elec Eng & Appl Physics, 69-78, PROF ELEC ENG, CASE WESTERN RESERVE UNIV, 76-, PROF COMPUT SCI, 80-, DIR CTR INTEL SYSTS RES, 81- *Concurrent Pos:* Div dir, elec, comput & systs eng, NSF, 78-80; pres, Bonschul Int, Cleveland, OH. *Mem:* Fel Optical Soc Am; fel Inst Elec & Electronics Engrs; Comput Soc; Am Asn Artificial Intel. *Res:* Pattern recognition; signal processing; machine intelligence; symbolic processing. *Mailing Add:* Case Western Reserve Univ 1099 Euclid Ave Cleveland OH 44106

PAOLETTI, ROBERT ANTHONY, b Greenfield, Mass, Oct 2, 42; m 64; c 2. DEVELOPMENTAL BIOLOGY. *Educ:* Univ Mass, BS, 64; Johns Hopkins Univ, PhD(biol), 68. *Prof Exp:* From asst prof to assoc prof, 68-78, PROF BIOL, KING'S COL, PA, 78-, EDUC COORDR, PHYSICIAN'S ASST PROG, 74- *Concurrent Pos:* Mem, Hastings Ctr, Inst Soc, Ethics & Life Sci. *Mem:* AAAS; Sigma Xi. *Res:* Biology of isolated chromatin; characterization of chromosomal proteins and their interaction with DNA; genetic engineering; bioethics. *Mailing Add:* Dept Biol King's Col 133 N River St Wilkes-Barre PA 18711

PAOLI, THOMAS LEE, b Springfield, Ill, May 27, 40; m 65. APPLIED PHYSICS, QUANTUM ELECTRONICS. *Educ:* Brown Univ, BSc & BA, 62; Stanford Univ, PhD(appl physics), 67. *Prof Exp:* Mem tech staff, Solid State Device Lab, Bell Labs, Inc, 67-81; mem res staff, 81-85, MGR, OPTO-ELECTRONIC MAT & DEVICES, PALO ALTO RES CTR, XEROX CORP, 85- *Mem:* Am Phys Soc; Inst Elec & Electronics Engrs; Optical Soc Am; Fel IEEE. *Res:* Semiconductor lasers. *Mailing Add:* Xerox-Palo Alto Res Ctr 3333 Coyote Hill Rd Palo Alto CA 94022

PAOLINI, FRANCIS RUDOLPH, b Newburgh, NY, July 13, 30; m 56; c 4. X-RAY SPECTROMETRY. *Educ:* Rensselaer Polytech Inst, BS, 51; Mass Inst Technol, PhD(physics, math), 60. *Prof Exp:* Sr scientist, Am Sci Eng Inc, 60-65, proj dir space res syst div, 65-69; tech dir, Philips Electronic Instruments, Inc, 69-74, vpres eng, 74-84, mgr technol coord, NAm Philips Corp, 85-88; RETIRED. *Concurrent Pos:* Adj prof, Univ Conn, Stamford & Norwalk State Tech Col, 88. *Mem:* AAAS; Am Phys Soc; Inst Elec & Electronic Engrs; Am Mgt Asn. *Mailing Add:* 111 Fishing Trail Stamford CT 06903

PAOLINI, PAUL JOSEPH, b Newburgh, NY, May 4, 42; m 67; c 2. BIOPHYSICS, CELL BIOLOGY. *Educ:* Rensselaer Polytech Inst, BS, 63, MS, 64; Univ Calif, Davis, PhD(zool), 68. *Prof Exp:* Asst prof biol sci, Univ Ga, 68-70; res prof, dept physiol, Univ Pac, San Francisco, 77-82; from asst prof to prof, 70-76, PROF BIOL, SAN DIEGO STATE UNIV, 76-, CHMN DEPT, 84- *Concurrent Pos:* Am Heart Asn res grants, Univ Ga, 69-70, San Diego State Univ, 70-71, San Diego County Heart Asn res grants, 71-73; NSF res grant, 73-75, 81-82; res grants, NIH, 77-82 & Am Heart Asn, 82-; chmn physiol prog, San Diego State Univ, 75-77 & 79-84. *Mem:* AAAS; Biophys Soc; Sigma Xi; Am Soc Cell Biol; Soc Gen Physiol; Am Heart Asn. *Res:*

Physiology and biophysics of cardiac and skeletal muscle; cell substructure, ultrastructure; digital image analysis; computers in biomedical research; microscopy. *Mailing Add:* Dept Biol San Diego State Univ San Diego CA 92182-0057

PAOLINO, MICHAEL A, b Albany, NY, Mar 8, 39; m 61; c 3. MECHANICAL ENGINEERING, EDUCATION ADMINISTRATION. *Educ:* Siena Col, BS, 60; Univ Ariz, MS, 67 & PhD(mech eng), 72. *Prof Exp:* Res & develop coordr, US Army Missile Command Redstone Arsenal, 68-69; assoc prof & dir thermal sci, Dept Mech, US Mil Acad, W Point, 72-83, prof & coordr mech engr prog, 84-86; DIR ENG & PROF MECH ENGR, LAFAYETTE COL, 86- *Honors & Awards:* Ralph R Teetor Award, Soc Automotive Engr, 79. *Mem:* Am Soc Eng Educ; Am Soc Mech Engr. *Res:* Computational fluid mechanics and heat transfer; pedagogical issues involving undergraduate engineering education. *Mailing Add:* Dir Eng Lafayette Col Easton PA 18042

PAPA, ANTHONY JOSEPH, b New York, NY, Aug 27, 30; m 57; c 5. ORGANIC CHEMISTRY. *Educ:* WVa Univ, BS, 55, MS, 57; Wash State Univ, PhD(org chem), 61. *Prof Exp:* Res assoc, Purdue Univ, 61-62; res chemist org & polymer chem, E I du Pont de Nemours & Co, Del, 62-66; MEM STAFF, UNION CARBIDE CORP, 66- *Mem:* Am Chem Soc. *Res:* Polymer synthesis (relationship between chemical structure and physical properties of polymers) and development of catalytic processes. *Mailing Add:* Plastics Div Union Carbide Corp South Charleston WV 25303

PAPA, KENNETH E, genetics; deceased, see previous edition for last biography

PAPAC, ROSE, b Montesano, Wash, Oct 18, 27. INTERNAL MEDICINE. *Educ:* Univ Seattle, BS, 49; St Louis Univ, MD, 53. *Prof Exp:* Intern internal med, St Louis Univ Hosps, 53-54; asst resident, Stanford Univ Hosps, 54-56; asst cancer res, Chester Beatty Inst Cancer Res, Eng, 56-57; res physician, Cancer Inst, Med Ctr, Univ Calif, San Francisco, 58-59, asst clin prof, 59-63; from asst prof to assoc prof, 63-90, PROF INTERNAL MED, YALE UNIV, 90- *Concurrent Pos:* Res fel, Sloan-Kettering Inst, NY, 57-58. *Mem:* Am Soc Hemat; Am Asn Clin Oncol; Am Asn Cancer Res. *Res:* Clinical cancer chemotherapy; culture studies of human leukemia cells. *Mailing Add:* Dept Internal Med Yale Univ Sch Med 333 Cedar St New Haven CT 06510

PAPACHRISTOU, CHRISTOS A, b Chalkis, Greece. FIRMWARE ENGINEERING, MICROPROGRAMMABLE MICROPROCESSING. *Educ:* Nat Polytech Univ, Greece, BS, 64; Philips Technol Inst, Netherlands, MS, 65; Johns Hopkins Univ, Baltimore, PhD(elec eng), 73. *Prof Exp:* Asst prof elec eng, Manhattan Col, 72-73; asst prof elec eng, Drexel Univ, 73-79; assoc prof elec eng & comput eng, Univ Cincinnati, 79-, J Morrow Res Chair, 80; AT DEPT COMPUT ENG & SCI, CASE WESTERN RESERVE UNIV. *Concurrent Pos:* Consult, Naval Air Develop Ctr, 74-75; prin investr, US Army Res Off, 82. *Mem:* Inst Elec & Electronics Engrs; Asn Comput Mach; Sigma Xi. *Res:* Hardware microprogram control scheme directed by a user-oriented firmware language; associative, table look-up, processing for residue arithmetic and multiple-valued logic; state machine implementation using programmable array logic. *Mailing Add:* Dept Comput Eng & Sci Case Western Reserve Univ Cleveland OH 44106

PAPACONSTANTINOU, JOHN, b Philadelphia, Pa, Dec 2, 30; m 67; c 2. BIOCHEMISTRY. *Educ:* Temple Univ, BA, 52, MA, 54; Johns Hopkins Univ, PhD(biochem), 58. *Prof Exp:* Am Cancer Soc res fel embryol, Carnegie Inst, 58-60; instr microbiol, Sch Med, Johns Hopkins Univ, 60-62; asst prof zool & entom, Univ Conn, 62-66; res prof zool, Univ Ga, 66-77; mem sr res staff, Biol Div, Oak Ridge Nat Lab, 77-80; MEM FAC, DEPT HUMAN BIOL, CHEM & GENETICS, UNIV TEX MED BR, GALVESTON, 80- *Concurrent Pos:* Mem cell biol study sect, NIH, 68-71, mem spec study sect estab genetics res ctrs, Nat Inst Gen Med Sci; prof, Oak Ridge Grad Sch Biomed Sci, Univ Tenn; staff mem, Biol Div, Oak Ridge Nat Lab, 66-77. *Honors & Awards:* Newcomb-Cleveland Prize, AAAS. *Mem:* AAAS; Soc Develop Biol (secy-treas, 71-72); Am Chem Soc; Am Inst Biol Sci; Am Soc Biol Chemists. *Res:* Nucleic acid and protein synthesis of cells in tissue culture; chemical embryology; mechanisms of differential gene action in cellular differentiation. *Mailing Add:* Dept Human Biol, Chem & Genetics Univ Tex Med Br Galveston TX 77550

PAPACONSTANTOPOULOS, DIMITRIOS A, b Athens, Greece, Sept 10, 36; m 69; c 2. SOLID STATE PHYSICS. *Educ:* Nat Univ Athens, BS, 61; NATO fel, Univ London, 62-64, Imp Col, dipl, 63, MS, 64, PhD(theoret Physics), 67. *Prof Exp:* From asst prof to prof physics, George Mason Univ, 67-77, chmn dept, 74-77; RES PHYSICIST & SR CONSULT, NAVAL RES LAB, 77- *Concurrent Pos:* Consult, Naval Res Lab, 70- *Mem:* Fel Am Phys Soc; Am Asn Physics Teachers. *Res:* Theoretical solid state physics; energy band theory; superconductivity; amorphous semiconductors. *Mailing Add:* Naval Res Lab Code 6630 Metal Physics Br Washington DC 20375

PAPACOSTAS, CHARLES ARTHUR, b Peabody, Mass, Mar 12, 21; m 48; c 4. PHARMACOLOGY. *Educ:* Mass Col Pharm, BS, 42; Boston Univ, AM, 52, PhD(physiol), 56. *Prof Exp:* Retail pharmacist, 45-50; from instr to assoc prof pharmacol, 55-66, prof, 66-85, vchmn, Dept Pharmacol, 75-85, EMER PROF PHARMACOL, SCH MED, TEMPLE UNIV, 85- *Mem:* AAAS; Am Soc Pharmacol & Exp Therapeut; Am Heart Assoc; Sigma Xi. *Res:* Histamine and histamine liberators; vascular response during adrenocortical suppression; adrenal steroids in essential hypertension; cardiovascular and metabolic effects of nicotine and other autonomic agents. *Mailing Add:* 260 N Bent Rd Wyncote PA 19095

PAPACOSTAS, CONSTANTINOS SYMEON, b Paphos, Cyprus, Sept 26, 46. TRAFFIC & TRANSPORT ENGINEERING. *Educ:* Youngstown State Univ, BE, 69; Carnegie-Mellon Univ, MS, 71, PhD(civil eng), 74. *Prof Exp:* asst prof, 74-80, ASSOC PROF CIVIL ENG, UNIV HAWAII, MANOA, 80- *Concurrent Pos:* Prin investr, var projs; univ rep, Transp Res Bd, Nat Acad Sci, 77- *Mem:* AAAS; Transp Res Bd; Am Soc Eng Educ; Am Asn Univ Professors; Am Soc Civil Eng. *Res:* Transportation demand analysis; population dynamics; environmental impacts; statewide transportation safety data systems; simulation of transportation operations. *Mailing Add:* Dept Civil Eng Univ Hawaii Manoa Honolulu HI 96822

PAPADAKIS, CONSTANTINE N, b Athens, Greece, Feb 2, 46; m 71; c 1. URBAN & RURAL STORMWATER MANAGEMENT, HYDRAULICS IN CONVENTIONAL & NUCLEAR POWER PLANTS. *Educ:* Nat Tech Univ, Greece, dipl civil ing, 69; Univ Cincinnati, MS, 70; Univ Mich, PhD(civil eng), 73. *Prof Exp:* Eng specialist, Geotech Group, Bechtel Inc, 74-76, asst chief engr, 76-81; vpres & dir, Water Resources Div, STS Consults Ltd, 81-84; vpres water & environ resources, Tetra Tech-Honeywell, 84; dept head civil eng, Colo State Univ, 84-86; DEAN ENG, UNIV CINCINNATI, 86- *Concurrent Pos:* Chmn fluid transients comt, Am Soc Mech Engrs, 78-80; dir, Groundwater Res Ctr, Univ Cincinnati, 86-, Center Hill Res Facil, US Environ Protection Agency, 86-; interim pres, Ohio Aerospace Inst, Cleveland, 88-89, Inst Advan Mfg Sci, Ohio Edison Technol Ctr, 89-90; mem bd gov, Edison Mat Technol Ctr, Dayton, 88-; mem, Ohio Coun Res & Econ Develop, 88-, Comn Educ Eng Profession, Nat Asn State Univ & Land Grant Cols, 91-; mem bd dirs, NES Inc-Penn Central Corp, 91- *Mem:* Fel Am Soc Civil Engrs; fel Am Soc Mech Engrs; Am Soc Eng Educ; Int Asn Hydraulic Res; Asn Groundwater Scientists & Engrs. *Res:* Hydraulics, water resources and ground and surface water pollution and control; cooling of nuclear and conventional power plants; transients in water supply and sewerage; storm water and flood plain management; dams and reservoirs; pollutant transport and remedial action; treatment plant retrofitting; small head hydroelectric power plants. *Mailing Add:* Col Eng ML 18 Univ Cincinnati Cincinnati OH 45221-0018

PAPADAKIS, EMMANUEL PHILIPPOS, b New York, NY, Dec 25, 34; m 60; c 4. NONDESTRUCTIVE TESTING, ULTRASONICS. *Educ:* Mass Inst Technol, SB, 56, PhD(physics), 62; Univ Mich, MM, 79. *Prof Exp:* Mem tech staff, Bell Tel Labs, Am Tel & Tel Co, Pa, 62-69; br chief, Panametrics, Inc, Mass, 69-70, dept head, 70-73; prin staff engr, Ford Motor Co, 73-87; ASSOC DIR, CNDE, IOWA STATE UNIV, 88- *Honors & Awards:* Biennial Award, Acoust Soc Am, 68; Mehl Honor lectr, Am Soc Nondestructive Testing, 79. *Mem:* Am Phys Soc; fel Acoust Soc Am; fel Inst Elec & Electronic Engrs; Am Soc Testing & Mat; fel Am Soc Nondestructive Testing; Am Soc Qual Control. *Res:* Ultrasonics; research and development; grain scattering; grain size distribution, microstructure, preferred orientation; broadband transducers; velocity, attenuation measurement methods; elevated temperatures; diffraction; propagation; acoustic emission; sonics; eddy currents; acoustic filters; nondestructive testing; statistical process control; quality concepts; financial justification of research and development. *Mailing Add:* Iowa State Univ 1915 Scholl Rd No 2 Ames IA 50011

PAPADOPOULOS, ALEX SPERO, b Roytsi-Arcadia, Greece, Jan 1, 46. APPLIED STATISTICS, OPERATIONS RESEARCH. *Educ:* Univ RI, BS, 68, MS, 69; Va Polytech Inst & State Univ, MS, 70, PhD(statist), 72. *Prof Exp:* Lectr math, Keene State Col, 72-76; asst prof, Col Charleston, 76-78; ASSOC PROF MATH, UNIV NC, CHARLOTTE, 78- *Mem:* Am Statist Asn; Am Math Soc; Sigma Xi. *Res:* Reliability theory and applications; Bayesian statistics; time series; simulation; water pollution modeling. *Mailing Add:* Dept of Math Univ of NC Charlotte NC 28223

PAPADOPOULOS, ELEFTHERIOS PAUL, b Thessaloniki, Greece, Sept 26, 26; m 68; c 1. ORGANIC CHEMISTRY. *Educ:* Univ Thessaloniki, BSc, 54; Univ Kans, PhD(org chem), 61. *Prof Exp:* From instr to asst prof chem, Am Univ Beirut, 61-65; res fel, Harvard Univ, 65-66; from asst prof to assoc prof, Am Univ Beirut, 66-68; res fel, Univ Ky, 68-69; assoc prof chem, 69-85, PROF CHEM, UNIV N MEX, 85- *Mem:* Am Chem Soc; Sigma Xi. *Res:* Chemistry of heterocyclic compounds; reactions of isocyanates and isothiocyanates. *Mailing Add:* Dept of Chem Univ of NMex Albuquerque NM 87131

PAPADOPOULOS, KONSTANTINOS DENNIS, b Larissa, Greece, Dec 29, 37; US citizen; m 69; c 2. SPACE PLASMA PHYSICS, IONOSPHERIC PHYSICS. *Educ:* Univ Athens, BSc, 60; Mass Inst Technol, MSc, 65; Univ Md, PhD(physics), 68. *Prof Exp:* Res assoc, Inst Fluids Dynamics, Univ Md, 68-69; res physicist, Plasma Dynamics Br, Naval Res Lab, 69-72, sr consult, 72-75, div consult, Plasma Div, 75-79; sci adv, Appl Physics Prog, Dept Energy, 78-79; PROF, PHYSICS & ASTRON DEPT, UNIV MD, 79- *Concurrent Pos:* Vis asst prof, Inst Fluid Dynamics & Appl Math, Univ Md, 70-72, vis assoc prof, 72-75; adj prof, physics dept, Univ Md, 75-79; consult, Sci Appln Inc & Naval Res Lab, 79, Atlantic Riechfield Co & John Hopkins Appl Physics Lab, 85. *Honors & Awards:* EO Hulbert Award, 77. *Mem:* Fel Am Phys Soc; Am Geophys Union. *Res:* Plasma physics, space plasma physics, plasma astrophysics; emphasis on theory and computer simulations; published 150 journal articles. *Mailing Add:* Dept Physics & Astron Univ Md College Park MD 20742

PAPADOPOULOS, NICHOLAS M, b Greece, July 21, 23; nat US; m 56; c 1. BIOCHEMISTRY. *Educ:* Am Univ, BA, 51; George Washington Univ, MS, 53, PhD(biochem), 56. *Prof Exp:* Vis scientist, NIH, 56-57; asst prof biochem, Sch Med, Georgetown Univ, 57-61; assoc prof, Med Col Va, 61-64, vis prof neurol, 64-74; biochemist & chief Protein & Lipid Sect, Div Biochem, Walter Reed Army Inst Res, Walter Reed Med Ctr, 64-74; BIOCHEMIST, NAT CANCER INST, NIH, 74- *Mem:* Am Soc Biol Chemists; fel Soc Exp Biol & Med; Am Asn Clin Chemists; Am Acad Neurol. *Res:* Chemical investigation of nervous diseases; nicotine metabolism; clinical chemistry. *Mailing Add:* Dept Clin Path Wareen Grant Magnuson Clin Ctr Bldg 10 Rm 2C407 Washington DC 20057

PAPADOPULOS, STAVROS STEFANU, b Istanbul, Turkey, Sept 21, 36; US citizen; m 63; c 2. GROUNDWATER HYDROLOGY. *Educ:* Robert Col, Istanbul, BS, 59; NMex Inst Mining & Technol, MS, 62; Princeton Univ, MA, 63, PhD(civil eng), 64. *Prof Exp:* Res hydrologist, US Geol Surv, 63-66;

assoc prof groundwater hydrol, Univ Minn, 66-67; head hydrol dept, Harza Eng Co, Ill, 67-69; assoc prof groundwater hydrol, Univ Ill, Chicago Circle, 69-70; res hydrologist, US Geol Surv, 70-79; PRES, S S PAPADOPULOS & ASSOCS, INC, 79- *Concurrent Pos:* Assoc prof lectr, George Washington Univ, 65-66; vis assoc prof, Univ Ill, Chicago Circle, 68-69; chief groundwater hydrologist, Harza Eng Co, 69-70. *Mem:* Am Soc Civil Engrs; Am Geophys Union; Int Asn Hydrogeologists; Sigma Xi; Asn Groundwater Scientists & Engrs; Geol Soc Am. *Res:* Analysis of groundwater systems through use of mathematical and digital modeling techniques; development of methods for pumping test analyses and aquifer evaluation; contaminent transport in aquifers. *Mailing Add:* 12250 Rockville Pike Suite 290 Rockville MD 20852

PAPAEFTHYMIOU, GEORGIA CHRISTOU, b Athens, Greece. BIOLOGICAL PHYSICS, CHEMICAL PHYSICS. *Educ:* Barnard Col, BA, 68; Columbia Univ, MA, 70, PhD(physics), 74. *Prof Exp:* RES STAFF MOSSBAUER SPECTROS, NAT MAGNET LAB, MASS INST TECHNOL, 74- *Mem:* Am Phys Soc; Sigma Xi. *Res:* Mossbauer spectroscopy of biomolecules and iron containing chemical complexes forming the active sites of proteins and enzymes; electronic structure and hyperfine interactions in solids. *Mailing Add:* Mass Inst Technol Bldg NW14-2219B 170 Albany St Cambridge MA 02139

PAPAGEORGE, EVANGELINE THOMAS, b Istanbul, Turkey, Dec 1, 06; US citizen. CLINCAL CHEMISTRY. *Educ:* Agnes Scott Col, Ga, AB, 28; Emory Univ, MS, 29; Univ Mich, PhD(biol chem), 37. *Prof Exp:* Asst biochem, 29-35, from instr to assoc prof, 49-75, asst dean, Sch Med, 56-57, assoc dean, 57-68, exec assoc dean, 68-75, EMER PROF BIOCHEM, EMORY UNIV, 75- *Concurrent Pos:* Consult clin chem, Commun Dis Ctr, 52-56. *Mem:* Am Soc Biol Chemists. *Res:* Phenylalanine metabolism; thiamine, ascorbic acid, rutin studies; insulin and body lipids; rat and guinea pig adrenal glycogen in stress and various nutrtional states; nutritional factors and growth hormone. *Mailing Add:* 848 Springdale Rd NE Atlanta GA 30306

PAPAGEORGIOU, JOHN CONSTANTINE, b Kallithea, Greece, Nov 22, 35; US citizen; m 69; c 4. FORECASTING, OPERATIONS MANAGEMENT. *Educ:* Athens Sch Econ & Bus Sci, BSc, 57; Univ Manchester, Eng, dipl, 63, PhD(mgt sci), 65. *Prof Exp:* Assoc prof mgt sci, St Louis Univ, 71-72; assoc prof opers anal, Univ Toledo, Ohio, 74-76; assoc prof, 76-78, PROF MGT SCI, COL MGT, UNIV MASS, BOSTON, 78- *Concurrent Pos:* Vis prof, Athens Sch Econ & Bus, 72-74. *Mem:* Fel AAAS; Opers Res Soc Am; Sigma Xi; Inst Mgt Sci; Am Inst Decision Sci; Hellenic Opers Res Soc. *Res:* Management science and operations research applications in manufacturing and service systems; project management; production management; forecasting; quantitative analysis. *Mailing Add:* Dept Mgt Sci Univ Mass Boston MA 02125

PAPAGIANNIS, MICHAEL D, b Athens, Greece, Sept 3, 32; m 61; c 2. ASTROPHYSICS, SPACE PHYSICS. *Educ:* Nat Tech Univ, Athens, MS, 55; Univ Va, MS, 60; Harvard Univ, PhD(physics), 64. *Prof Exp:* Lectr radio astron, Harvard Univ, 64-65; chmn dept astron, 69-82, from asst prof to assoc prof, 65-70, PROF ASTRON, BOSTON UNIV, 70- *Concurrent Pos:* Assoc, Harvard Observ, Harvard Univ, 64-; Air Force Geophys Lab grants, 65-; vis prof, Univ Athens, 71-72. *Mem:* Am Astron Soc; Am Geophys Union; Int Union Radio Sci; Int Astron Union. *Res:* Radio astronomy; active sun; solar-terrestrial relations; astrophysical plasmas; bioastronomy-the search for extraterrestrial life. *Mailing Add:* Dept of Astron Boston Univ 725 Commonwealth Ave Boston MA 02215

PAPAHADJOPOULOS, DEMETRIOS PANAYOTIS, b Patras, Greece, Aug 24, 34; m 59; c 3. BIOCHEMISTRY, BIOPHYSICS. *Educ:* Nat Univ Athens, BS, 57; Univ Wash, PhD(biochem), 63. *Prof Exp:* NIH res fel physiol, Inst Animal Physiol, Babraham, Eng, 64-67, Am Heart Asn estab investr, 67-72; from asst res prof to res prof biochem, biophysiol & cell biol, State Univ NY, Buffalo, 72-78, mem ctr theoret biol, 68-78; PROF PHARMACOL, CANCER RES INST, UNIV CALIF MED SCH, SAN FRANCISCO, 78- *Concurrent Pos:* Vis sr cancer res scientist, Roswell Park Mem Inst, 67-68, assoc cancer res scientist, 68-74, prin cancer res scientist, 74-78; estab investr, Am Heart Asn, 67-78, fel coun arteriosclerosis, mem res study comt, 72-75. *Honors & Awards:* K S Cole Award, Biophys Soc, 80. *Mem:* Am Chem Soc; Am Soc Biol Chemists; fel NY Acad Sci; Biophys Soc; Am Soc Cell Biol. *Res:* Membrane structure and function; phospholipid-protein interactions; blood coagulation; surface phenomena; pharmacological use of lipid vesicles; author or coauthor of over 150 publications. *Mailing Add:* Cancer Res Inst Univ of Calif Med Sch San Francisco CA 94143

PAPAIOANNOU, CHRISTOS GEORGE, b Athens, Greece, Dec 3, 34; US citizen; m 65; c 3. ORGANIC CHEMISTRY. *Educ:* Univ Athens, BS, 58; Mich State Univ, PhD(org chem), 67. *Prof Exp:* res chemist, Democritos Nuclear Res Ctr, Athens, 62-63; res assoc org chem, Columbia Univ, 67-68; res chemist, Am Cyanamide Co, 68-71; group leader organometallic chem, Nat Patent Develop Corp, 71-73; SR RES INVESTR PHARMACEUT, BRISTOL-MYERS SQUIBB CO, INC, 73- *Mem:* Am Chem Soc; Greek Chem Soc. *Res:* Pharmaceutical research; synthetic methods; organometallic chemistry; process development. *Mailing Add:* 32 Brearly Rd Princeton NJ 08540-6767

PAPAIOANNOU, STAMATIOS E, b Athens, Greece, Mar 27, 34; m 68. BIOCHEMISTRY. *Educ:* Nat Univ Athens, BS, 57; Ore State Univ, MS, 62, PhD(biochem), 66. *Prof Exp:* Asst biochem, Ore State Univ, 62-66; fel, Univ Minn, 66-68; res investr, 68-75, res scientist, Dept Biol Res, 75-81, RES SCIENTIST, DEPT CARDIOVASC DIS RES, GD SEARLE & CO, 81- *Mem:* Am Chem Soc; Endocrine Soc; Am Soc Pharmacol & Exp Pharmaceut. *Res:* Biochemical pharmacology, especially molecular endocrinological and enzymological studies in cardiovascular research. *Mailing Add:* G D Searle & Co Cardiovasc Dis Res 4901 Searle Pkwy Skokie IL 60077

PAPAIOANNOU, VIRGINIA EILEEN, b Upperlake, Calif, Oct 22, 45; m 83; c 2. DEVELOPMENTAL GENETICS, CELL BIOLOGY. *Educ:* Univ Calif, Davis, BSc, 67; Univ Cambridge, PhD(genetics), 72. *Prof Exp:* Postdoctoral, Marshall Lab, Univ Cambridge, 71-74; postdoctoral, Dept Zool, Univ Oxford, 75-79, res assoc, Dept Path, 79-80; ASSOC PROF EMBRYOL, DEPT PATH, TUFTS UNIV SCH MED & VET MED, 81- *Concurrent Pos:* Assoc prof, Dept Anat & Cell Biol, Tufts Univ Sch Med, 90- *Res:* Early implantation and embryonic development using mutations and transgenic technology; lineage of cells within the embryo. *Mailing Add:* Dept Path Tufts Univ 136 Harrison Ave Boston MA 02111

PAPANASTASSIOU, DIMITRI A, b Athens, Greece, Nov 15, 42; m 64; c 2. MASS SPECTROMETRY. *Educ:* Calif Inst Technol, BS, 65, PhD(physics), 70. *Prof Exp:* Fel physics, 70, Millikan res fel, 71, instr freshmen physics, 71-72 & 79-80, sr res fel planet sci & physics, 72-73 & planetary sci, 73-76, res assoc, 76-81, SR RES ASSOC GEOCHEM, CALIF INST TECHNOL, 81- *Concurrent Pos:* Co-investr lunar samples, NASA, 71-, mem, Comt Post-Viking Explor Mars, 73-74 & Lunar Sample Anal Planning Team, 74-79; John Simon Guggenheim fel & J William Fulbright fel, 82-83. *Honors & Awards:* F W Clarke Medal, Geochemical Soc, 72; Except Sci Achievement Medal, NASA, 76. *Mem:* Am Phys Soc; fel Am Geophys Union; AAAS; fel Meteoritical Soc. *Res:* Formation and evolution of meteorites and planets; establishment of a time sequence of condensation in the solar nebula; search for exotic nuclear matter preserved in meteorites; precision mass spectrometry and its application to physical-chemical problems. *Mailing Add:* Dept Geo & Plant Sci Calif Inst Technol Mail Code 170-25 Pasadena CA 91125

PAPANICOLAOU, GEORGE CONSTANTINE, b Athens, Greece, Jan 23, 43; m 67; c 3. PHYSICAL MATHEMATICS. *Educ:* Union Col, NY, BEE, 65; NY Univ, MS, 67, PhD(math), 69. *Prof Exp:* From asst prof to assoc prof, 69-76, PROF MATH, NY UNIV, 76- *Concurrent Pos:* Sloan fel, 74-75, Guggenheim fel, 83-84. *Mem:* Am Math Soc; Soc Indust & Appl Math. *Res:* Applied mathematics; mathematical physics. *Mailing Add:* Courant Inst 251 Mercer St New York NY 10012

PAPANICOLAS, COSTAS NICOLAS, b Paphos, Cyprus, Oct 1, 50; m 75; c 2. NUCLEON & NUCLEAR STRUCTURE, ELECTRON SCATTERING. *Educ:* Mass Inst Technol, BSc, 72, PhD(nuclear physics), 79. *Prof Exp:* Res assoc, Nuclear Physics Lab, Univ Ill, 79-81, vis res asst prof, physics dept, 81-82, asst prof, 82-85, assoc prof, 85-89, PROF, PHYSICS DEPT, UNIV ILL, 89-; PROF, UNIV ATHENS, GREECE, 90- *Concurrent Pos:* Vis scientist, Inst Fundamental Studies, Atomic Energy Comn, 79. *Honors & Awards:* A O Beckman Award. *Mem:* Am Phys Soc; NY Acad Sci; Sigma Xi. *Res:* Experimental development and theoretical investigation of coincident electron scattering; investigation of nuclear and nucleon structure using scattering technique. *Mailing Add:* Loomis Physics Lab 1110 W Green St Univ Ill Urbana IL 61801

PAPANIKOLAOU, NICHOLAS E, b Greece, July 2, 37; US citizen; m 67; c 2. ORGANIC CHEMISTRY. *Educ:* St Anselm's Col, BA, 62; Univ NH, PhD(chem), 67. *Prof Exp:* Assoc prof, 67-76, PROF CHEM, SLIPPERY ROCK STATE COL, 76- *Mem:* Am Chem Soc; Sigma Xi. *Res:* Optically active organosulfur and organophosphorus compounds. *Mailing Add:* Dept of Chem Slippery Rock Univ Slippery Rock PA 16057

PAPANTONOPOULOU, AIGLI HELEN, b Port Said, Egypt; Greek citizen; m 76; c 1. ALGEBRAIC GEOMETRY, COMMUTATIVE ALGEBRA. *Educ:* Columbia Univ, BA, 69; Univ Calif, Berkeley, MA, 71, PhD(math), 75. *Prof Exp:* Instr, Univ Pa, 75-77, Bryn Mawr Col, 77-79; ASST PROF MATH, LEHIGH UNIV, 79- *Concurrent Pos:* Vis prof, Univ Crete, Greece, 80-81; reviewer, Am Math Reviews, 80- *Mem:* Am Math Soc. *Res:* Algebraic intersection theories and applied to Grassmann varieties and flag varieties and problems in enumerative geometry. *Mailing Add:* Dept Math 263 Riverside Dr Princeton NJ 08540

PAPARIELLO, GERALD JOSEPH, b New York, NY, Feb 3, 34; m 57; c 4. ANALYTICAL CHEMISTRY, TECHNICAL MANAGEMENT. *Educ:* Fordham Univ, BS, 56; Univ Wis, MS, 58, PhD(pharmaceut chem), 60. *Prof Exp:* Sr chemist anal res, Ciba Pharmaceut Co, 60-63, assoc dir phys chem res, 63-66; mgr anal res dept, 66-76, dir pharmaceut res & develop, 76-83, asst vpres res & develop, Wyeth Labs Inc, 83-87, VPRES DEVELOP, WYETH-AYERST RES, 87- *Concurrent Pos:* Mem comt revision, US Pharmacopoeia, 70-80; mem comt specifications, Nat Formulary, 70-74. *Mem:* Am Chem Soc; Am Crystallog Asn; fel Acad Pharmaceut Sci (vpres, 78-79); Am Asn Pharmaceut Sci. *Res:* Organic and functional group analysis; analysis of pharmaceutical systems; x-ray fluorescence and diffraction; analytical separations; automated analysis; chemical kinetics. *Mailing Add:* Wyeth-Ayerst Res PO Box 8299 Philadelphia PA 19101

PAPARO, ANTHONY A, ELECTRO-MICROSCOPY. *Educ:* Fordham Univ, PhD(anat), 69. *Prof Exp:* PROF MICROS ANAT, SOUTHERN ILL UNIV, 73- *Mailing Add:* Dept Zool Southern Ill Univ Carbondale IL 62901

PAPAS, ANDREAS MICHAEL, b Kato Moni, Cyprus, Oct 29, 42; US citizen; m 69; c 1. ANIMAL NUTRITION & HEALTH. *Educ:* Univ Salonica, Greece, BSc, 65; Univ Ill, MSc, 71 & 72, PhD(nutrit biochem), 73. *Prof Exp:* Asst res officer, Agr Res Inst, Cyprus, 65-69 & 73-75; res assoc, 76, asst prof animal nutrit, Univ Man, Can, 76-77; sr res chemist, Res Lab, 77-83, res assoc, 83-85, mgr develop & tech serv, animal nutrit suppl bus unit, 86-88, TECH ASSOC, TECH SERV & DEVELOP, EASTMAN CHEM DIV, 88- *Mem:* Am Inst Nutrit; Am Soc Animal Sci; econ develop comt, biotechnology prog, Cornell Univ. *Res:* Nutritional supplements for ruminant animals; protein and amino acid nutrition; effect of isoC4 and C-5 volatile fatty acids in dairy cows; biotechnology and its agricultural applications; vitamin E in health and disease. *Mailing Add:* Tech Serv Develop Dept PO Box 431 Kingsport TN 37662

PAPAS, CHARLES HERACH, b Troy, NY, Mar 29, 18; m 51; c 1. ELECTRODYNAMICS. *Educ:* Mass Inst Technol, BS, 41; Harvard Univ, MS, 46, PhD(electromagnetic theory), 48. *Prof Exp:* Res fel, Harvard Univ, 48-50; mem staff, Los Alamos Sci Lab & consult, Radiation Lab, Univ Calif, 50-52; assoc prof, 53-59, PROF PHYSICS, CALIF INST TECHNOL, 59- *Mem:* Am Math Soc; Am Phys Soc; Inst Elec & Electronics Engrs; foreign mem Acad Sci Armenian Soviet Socialist Repub. *Res:* Electromagnetic theory; gravitational electrodynamics. *Mailing Add:* 543 Vallombrosa Dr Pasadena CA 91107

PAPAS, TAKIS S, b Athens, Greece, June 19, 35; US citizen; c 2. BIOCHEMISTRY, VIROLOGY. *Educ:* Univ NH, BS, 63, MS, 66; Marquette Univ, PhD(biochem), 70. *Prof Exp:* Res analyst biochem, Biomolecular Hyg Sect, Univ NH, 62-64, res asst, Univ, 64-66; res asst, Marquette Sch Med, 66-70; staff fel, Nat Heart Inst, NIH, 70-72; sr staff fel molecular virol, 72-75, res chemist molecular biol of tumor viruses, 75-78, HEAD CARCINOGENESIS REGULATION SECT, NAT CANCER INST, 78-, CHIEF LAB MOLECULAR ONCOL, 83-; DEPT BIOL, JOHNS HOPKINS UNIV. *Mem:* Am Chem Soc; Am Soc Biol Chemists; Leukemia Soc Am. *Res:* Enzyme chemistry; enzyme kinetics; protein synthesis; nucleic acid; protein nucleic-acid interactions; lipid metabolism; modification of viral growth and replication by natural and synthetic substances; molecular biochemistry of oncorna virus; molecular cloning of transforming genes of acute leukemia viruses. *Mailing Add:* Nat Cancer Inst Frederick Cancer Res Fac Bldg 8 Baltimore MD 21218

PAPASTAMATIOU, NICOLAS, b Athens, Greece, Nov 18, 39; m; c 1. PHYSICS. *Educ:* Nat Univ Athens, BSc, 61; Oxford Univ, DPhil(theoret physics), 66. *Prof Exp:* Res assoc physics, Syracuse Univ, 66-68; from asst prof to assoc prof, 68-78, PROF PHYSICS, UNIV WIS-MILWAUKEE, 78- *Concurrent Pos:* Vis prof, Nat Univ Athens, 72-73 & Univ Alta, 80-81, 82, 89-90. *Mem:* Am Phys Soc; AAAS. *Res:* Quantum field theory; high energy physics. *Mailing Add:* Dept Physics Univ Wis PO Box 413 Milwaukee WI 53201

PAPASTEPHANOU, CONSTANTIN, b Cairo, Egypt, Dec 30, 45; nat US; m 71; c 1. BIOCHEMISTRY, ANALYTICAL CHEMISTRY. *Educ:* Ain Shams Univ, Cairo, BS, 67; Univ London, MS, 68, DIC, 68; Univ Miami, PhD(biochem), 72. *Prof Exp:* Proj assoc biochem, Univ Wis-Madison, 72-73; res investr anal chem, Squibb Inst Med Res, 73-77; sect head methods develop, E R Squibb & Sons, 77, sect head, mat control, 77-78; mgr quality control, Squibb Puerto Rico, 78-80; dept head chemical control, E R Squibb & Sons, 80-81, dir prod qual control, 81-84, dir worldwide pharmaceut technol & develop, 84-87; VPRES TECH OPERS, SQUIBB/CONVATEC, 87- *Concurrent Pos:* Fel, Univ Wis-Madison, 72-73. *Mem:* Am Pharm Assoc; Am Chem Soc; Brit Biochem Soc. *Res:* Biochemistry of terpenes and carotenoids; enzymology; metabolic pathways; analytical biochemistry and chemistry. *Mailing Add:* Convatec-Squibb PO Box 4000 Princeton NJ 08540

PAPASTOITSIS, GREGORY, b Athens, Greece, Dec 2, 58; m 90. ENZYMOLOGY, PROTEIN CHEMISTRY. *Educ:* State Univ NY, Binghamton, BS, 84, MS, 87, PhD(biol), 90. *Prof Exp:* POSTDOCTORAL ENZYM, BOSTON UNIV SCH MED, 90- *Res:* Purification, characterization and localization of enzymes; ion exchange; affinity; hydrophobic interactions chromatography; inhibition studies; use of radioactive and chromogenic substrates; purification of antibodies, western blotting and Elisa assays; author of Five technical publications. *Mailing Add:* Boston Univ Sch Med 80 E Concord St K-5 Boston MA 02118-2394

PAPAVIZAS, GEORGE CONSTANTINE, b Kriminion, Greece, July 10, 22; US citizen; m 54; c 2. PLANT PATHOLOGY. *Educ:* Univ Thessaloniki, dipl, 47; Univ Minn, MS, 53, PhD(mycol, plant path), 57. *Prof Exp:* Plant breeder, Inst Plant Breeding, Greece, 50-51; from asst to instr, Univ Minn, 52-57; exp mycologist, Plant Sci Res Div, 57-64, microbiologist & leader mushroom & microbiol invests, 64-72, CHIEF, SOILBORNE DIS LAB, PLANT PROTECTION INST, SCI & EDUC ADMIN, AGR RES, USDA, 72- *Honors & Awards:* Campbell Award, Am Inst Biol Sci, 63. *Mem:* Fel Am Phytopathological Soc; Mycol Soc Am. *Res:* Microecology of soilborne diseases of plants in relation to biological control; integrated pest management; soil fungi and their activities in soil. *Mailing Add:* Biol Control Plant Dis Lab Agr Res Ctr-W USDA Rm 274 Bldg 011A Beltsville MD 20705

PAPAY, LAWRENCE T, b NJ, Oct 3, 36; c 3. MECHANICAL ENGINEERING, NUCLEAR ENGINEERING. *Educ:* Fordham Univ, BS; Mass Inst Technol, MS, ScD(nuclear eng). *Prof Exp:* Dir res, 70-78, gen supt, utility bulk power operations, 78-80, vpres res & develop, 80-83, SR VPRES, SOUTHERN CALIF EDISON CO. 83- *Concurrent Pos:* Mem energy res adv bd & chair new prod reactor technol assessment panel, Nat Acad Eng; Nat Sci Found Indust Panel Sci & Technol; special comt pub educ, Am Nuclear Soc; indust adv comt, Calif State Univ; mem comt power distrib, Asn Edison Illum Co; fed power comn, Tech Adv Comt. *Mem:* Nat Acad Engrs. *Mailing Add:* Southern Calif Edison Co PO Box 800 Rosemead CA 91770

PAPAZIAN, LOUIS ARTHUR, b Cambridge, Mass, Jan 25, 31. PHYSICAL CHEMISTRY. *Educ:* Northeastern Univ, BS, 54, MS, 56; Wayne State Univ, PhD(phys chem), 61. *Prof Exp:* Sr res chemist, Mobil Chem Co, NJ, 62-70; sr res scientist, Columbian Div, Cities Serv Co, 70-78; SR RES CHEMIST, AM CYANAMID CO, 78- *Res:* Polymer characterization; all phases of liquid chromatography; processing of carbon-black-rubber systems. *Mailing Add:* Am Cyanamid Co 1937 W Main St Stamford CT 06904

PAPE, BRIAN EUGENE, b St Louis, Mo, Oct 13, 43; m 71; c 4. TOXICOLOGY, PATHOLOGY. *Educ:* Wash Univ, BA, 66; Mich State Univ, MS, 69, PhD(toxicol), 72; Univ Mo, MBA, 82. *Prof Exp:* Res asst toxicol, Mich State Univ, 66-72; res assoc, Med Ctr, Univ Mo-Columbia, 72-74, asst prof, 74-78, assoc prof path, 78-82, dir toxicol lab, 72-82; ASSOC PROF PATH, MED SCH, UNIV MASS-WORCESTER, 86- *Mem:* Am Asn Univ Professors; Am Chem Soc; Sigma Xi. *Res:* Analytical toxicology; clinical toxicology; drugs, pesticides, chemicals; clinical and forensic toxicology. *Mailing Add:* 20 Tiger Rd Hudson NH 03501

PAPEE, HENRY MICHAEL, b Cracow, Poland, Oct 22, 23; nat Can; m 51; c 2. GENERAL ATMOSPHERIC SCIENCES, INORGANIC CHEMISTRY. *Educ:* Univ Rome, Dr Chem, 51. *Prof Exp:* Res chemist, Metall Res Div, Sherritt-Gordon Mines, Ltd, Can, 51-54; res assoc chem, Univ Ottawa, Ont, 54-58; res officer, Nat Res Coun Can, 58-60; dir, inst Aerosol Nucleation, Nat Res Coun Italy, 60-83; CONSULT, EXP STA FORESTRY & AGR, ITALY, 83- *Mem:* Am Chem Soc; Am Meteorol Soc; Am Geophys Union; Chem Inst Can. *Res:* Atmospheric chemistry; chemistry of atmospheric precipitation phenomena. *Mailing Add:* Via Grazioli 51/3 Trento 38100 Italy

PAPENDICK, ROBERT I, b Bridgewater, SDak, May 22, 31; m 50; c 5. AGRONOMY. *Educ:* SDak State Univ, BS, 57, PhD(agron), 62. *Prof Exp:* Exten serv soil specialist, SDak State Univ, 58-59; res soil physicist, Tenn Valley Authority, 63-65; res soil scientist, soil & water conserv res div, 65-70, COLLABR AGRON, SCI & EDUC ADMIN-AGR RES, USDA, 70-; ASSOC PROF SOILS, WASH STATE UNIV, 71- *Mem:* Am Soc Agron; Sigma Xi. *Res:* Soil-plant-fertilizer relationships leading to development of new fertilizer materials; soil water management problems leading to erosion control. *Mailing Add:* USDA Sea Fr Wash State Univ 215 Johnson Hall Pullman WA 99164

PAPERNIK, LAZAR, applied mechanics, for more information see previous edition

PAPERT, SEYMOUR A, b Pretoria, SAfrica, Mar 1, 28. COGNITIVE PSYCHOLOGY, MEDIA TECHNOLOGY. *Educ:* Univ Witwatersrand, BA, 49, PhD(math), 52; Cambridge Univ, PhD(math), 59. *Prof Exp:* Royal Comn res scholar, Exhib 1851, St John's Col, Cambridge Univ, 54-56; researcher, Henri Poincare Inst, Univ Paris, 56-57; researcher child develop, int ctr genetic epistemology, Univ Geneva, 58-63; sr fel, Nat Phys Lab, London, 59-61; asst lectr cybernet, Univ Geneva, 62-63; res assoc elec eng, Mass Inst Technol, 63-67, prof appl math & co-dir, Artificial Intel Lab, 67-81, Cecil & Ida Green Prof educ, 74-81; LEGO PROF LEARNING RES, 89- *Concurrent Pos:* Guggenheim mem fel, 80-81; Marconi Int fel, 81. *Res:* Cybernetics; child development; psychology and mathematics; science of cognition. *Mailing Add:* Dept Math Mass Inst Technol Media Lab 20 Ames St Cambridge MA 02139

PAPIAN, WILLIAM NATHANIEL, electrical engineering, for more information see previous edition

PAPIKE, JAMES JOSEPH, b Virginia, Minn, Feb 11, 37; m 58; c 4. CRYSTALLOGRAPHY, MINERALOGY. *Educ:* SDak Sch Mines & Technol, BS, 59; Univ Minn, PhD(geol), 64. *Prof Exp:* Geologist, US Geol Surv, Washington, DC, 64-69; assoc prof crystallog, State Univ NY Stony Brook, 69-71, chmn dept earth & space sci, 71-74, prof crystallog, 71-; AT DEPT GEOL, SDAK SCH MINES TECHNOL. *Concurrent Pos:* Vis prof, Bd Earth Scis, Univ Calif, Santa Cruz, 75-76; mem, Lunar Sample Analysis Planning Team, 70-73; mem, Lunar & Planetary Sci Coun, 74-76; NASA grant, 75-; US Nat Comt for Geochem, Nat Res Coun, 78-; NSF grant, 78- *Honors & Awards:* NASA Medal for Except Sci Achievement, 73; Mineral Soc Am Award, 74. *Mem:* Geochem Soc; fel Mineral Soc Am; Am Geophys Union; Meteoritical Soc; fel Geol Soc Am. *Res:* Crystal chemistry of rock-forming silicates; intracrystalline equilibria; properties of mineral solid solutions. *Mailing Add:* Inst Meteoritics Dept Geol Univ New Mexico Northrop Hall Albuquerque NM 87131-1126

PAPINI, GIORGIO AUGUSTO, b Cremona, Italy, May 8, 34; m 59; c 4. UNIFIED FIELD THEORIES, GRAVITATIONAL EFFECTS. *Educ:* Univ Pavia, Italy, PhD(physics), 58. *Prof Exp:* Researcher theoret physics, Ctr Nuclear Res, Ispra, Italy, 58; scholar theoret physics, Inst Advan Studies, Dublin, 58-60; asst lectr physics, Math Dept, Univ Leeds, 60-61; prof incaricato physics, Univ Catania, 61-64; res assoc physics, Inst Field Physics, Univ NC, Chapel Hill, 64-66; dept head, 69-79, PROF, DEPT PHYSICS & ASTRON, UNIV REGINA, 66- *Concurrent Pos:* Consult, Sicilian Ctr Nuclear Physics, Catania, 62-64, Nat Health Inst, Rome, 64, Hughes Res Labs, 65; vis scientist, Int Ctr Theoret Physics, Trieste, 67, Eur Space Res Inst, Frascati, 70, Astrophysics Labs, Nat Res Coun, Italy, 73, Univ BC, Vancouver, 80-81. *Mem:* Am Phys Soc; Int Soc Gen Relativity Gravitation; NY Acad Sci; Int Soc Neural Networks. *Res:* The role of gravitation in astrophysics; elementary particle physics; solid state physics. *Mailing Add:* Dept Physics & Astron Univ Regina Regina SK S4S 0A2 Can

PAPKA, RAYMOND EDWARD, b Thermopolis, Wyo, July 11, 45; m 64; c 3. ANATOMY, NEUROBIOLOGY. *Educ:* Univ Wyo, BS, 67; Tulane Univ, PhD(anat), 71. *Prof Exp:* assoc prof anat, Med Ctr, Univ Ky, 71-88; PROF ANAT, MED CTR, UNIV OK, 88- *Concurrent Pos:* vis scientist, Flinders Univ South Australia, 81, Univ Debrecen, Hungary, 85. *Mem:* AAAS; NY Acad Sci; Am Soc Cell Biol; Am Asn Anatomists; Soc Neurosci. *Res:* Neurobiology; autonomic nervous system; developmental neurobiology; neurotransmitters. *Mailing Add:* Dept Anatomical Sci Univ Oklahoma PO Box 26901 Oklahoma City OK 73190

PAPKE, KEITH GEORGE, b Mankato, Minn, Feb 5, 24; m 54; c 2. GEOLOGY. *Educ:* SDak Sch Mines & Technol, BS, 48; Univ Ariz, MS, 52. *Prof Exp:* Mining engr, Phelps Dodge Corp, Ariz, 48-50; geologist, Am Smelting & Refining Co, 52-56 & Southern Pac Co, 56-61; chief geologist, Standard Slag Co, Nev, 61-66; econ geologist, Nev Bur Mines & Geol, 66-86; GEOL CONSULT, 86- *Concurrent Pos:* Emer prof, Univ Nev. *Mem:* Am Inst Mining, Metall & Petrol Engrs; Clay Minerals Soc; Soc Econ Geologists. *Res:* Geology, physical and chemical properties, uses and specifications, and economics of industrial minerals, especially those occurring in the western states; talc; saline minerals; fluorspar; clay mineralogy; zeolites; barite; gypsum. *Mailing Add:* 2180 Schroeder Way Sparks NV 89431

PAPKOFF, HAROLD, b San Jose, Calif, June 11, 25; m 53; c 2. BIOCHEMISTRY. *Educ:* Univ Calif, AB, 50, PhD(biochem), 57. *Prof Exp:* From jr res biochemist to assoc res biochemist, 57-73, from lectr to assoc prof exp endocrinol, 62-72, prof anat, 72-73, PROF EXP ENDOCRINOL & RES BIOCHEMIST, HORMONE RES LAB, MED SCH, UNIV CALIF, SAN FRANCISCO, 73-, PROF OBSTET & GYNEC, 76- *Concurrent Pos:* USPHS spec fel chem path, St Mary's Hosp, London, Eng, 61-62; career develop award, 64. *Honors & Awards:* Ayerst Award, Endocrine Soc, 78. *Mem:* AAAS; Am Soc Biol Chemists; Endocrine Soc; Am Soc Zoologists; Soc Study Reproduction. *Res:* Purification and characterization of pituitary hormones; relation of structure to biological activity of proteins and peptides; comparative endocrinology. *Mailing Add:* Univ Calif 1088 HSW San Francisco CA 94143

PAPOULIS, ATHANASIOS, b Greece, Jan 18, 21; nat US; m 53; c 5. SYSTEMS ENGINEERING, APPLIED MATHEMATICS. *Educ:* Polytech Inst Greece, MechE, 41, EE, 42; Univ Pa, MS, 47, MA, 48, PhD(math), 50. *Prof Exp:* Instr elec eng, Univ Pa, 48-51; asst prof, Union Col, 51-52; prof elec eng, 52-90, UNIV PROF, POLYTECH INST NY, 90-, EMER PROF, 91- *Concurrent Pos:* Prof, NSF fac fel, Darmstadt Tech Univ, 59-60; vis prof, Univ Calif, Los Angeles, 64-65. *Honors & Awards:* Humboldt Sr Scientist Award, 85. *Mem:* Fel Inst Elec & Electronics Engrs. *Res:* Applied mathematics; systems; communications; author of 8 books in signal processing, system theory & random processing. *Mailing Add:* Polytech Univ 333 Jay St Brooklyn NY 11201

PAPOUTSAKIS, ELEFTHERIOS TERRY, b Alexandroupolis, Greece, May 29, 51. BIOCHEMICAL ENGINEERING, REACTION ENGINEERING. *Educ:* Nat Tech Univ, Athens, dipl, 74; Purdue Univ, MS, 76, PhD(chem eng), 79. *Prof Exp:* ASST PROF CHEM ENG, RICE UNIV, 80- *Mem:* Am Inst Chem Engrs; Am Chem Soc; AAAS; Am Soc Microbiol; NY Acad Sci. *Res:* Biochemical and fermentation engineering; biokinetics; transport and bioenergetics of single and mammalian cells; immobilized whole cells; single cell protein production from one-carbon compounds; applied mathematics in transport and reaction engineering. *Mailing Add:* Dept Chem Eng Northwestern Univ Evanston IL 60208

PAPP, F(RANCIS) J(OSEPH), b Chicago, Ill, Aug 30, 42; m 65; c 4. INTEGRAL EQUATIONS, LINEAR ALGEBRA. *Educ:* Univ Notre Dame, BS, 64; Univ Del, MS, 66, PhD(math), 69. *Prof Exp:* from asst prof to assoc prof, dept math sci, Univ Lethbridge, 69-81, chmn, Dept Math Sci, 78-79; ASSOC PROF, DEPT MATH, UNIV MICH, DEARBORN, 81- *Concurrent Pos:* Vis scholar, Univ Mich, Ann Arbor, 75-76; consult ed, Math Reviews, 78, ed, 79-81. *Mem:* Am Math Soc; Math Asn Am; Res Soc North Am; Sigma Xi. *Res:* Number theory; functional equations; integral equations. *Mailing Add:* 2675 Valley Dr Ann Arbor MI 48103

PAPP, KIM ALEXANDER, b Calgary, Alta, Mar 21, 53; m 80. GALAXY DYNAMICS, STELLAR DYNAMICS. *Educ:* Univ Calgary, BSc, 74; York Univ, MSc, 76, PhD(physics), 80; Univ Calgary, MD, 85. *Prof Exp:* NATO fel, Univ Chicago, 80-81; asst prof physics, Univ Waterloo, 81-82; CONSULT, ALGAS ENG, 74-, KAPTECH, 80- *Mem:* Am Astron Soc; Am Phys Soc; Can Astron Soc; Alta Med Asn. *Res:* Clinical medicine. *Mailing Add:* 721 Snowcrest Pl Waterloo ON N2J 3Z5 Can

PAPPADEMOS, JOHN NICHOLAS, b St Louis, Mo, Mar 12, 25; m 51; c 3. THEORETICAL HIGH ENERGY PHYSICS, PHILOSOPHY OF SCIENCE. *Educ:* Iowa State Univ, BS, 45; Washington Univ, MA, 52; Univ Chicago, PhD(physics), 63. *Prof Exp:* Tech consult, Washington Univ, 54-55; instr physics, Bailey Schs, Inc, 55-57; asst prof, Univ Ill, Chicago Circle, 57-63, assoc prof physics, 63-90; RETIRED. *Concurrent Pos:* NSF res grant, 65-66; transl consult, Plenum Press, 71-78. *Mem:* Am Asn Physics Teachers. *Res:* Theoretical nuclear physics; theory of elementary particle interactions. *Mailing Add:* Dept Physics Box 4348 Univ Ill Chicago Chicago IL 60680

PAPPAGIANIS, DEMOSTHENES, b San Diego, Calif, Mar 31, 28; m 56; c 2. MEDICAL MICROBIOLOGY. *Educ:* Univ Calif, AB, 49, MS, 51; PhD(bact), 56; Stanford Univ, MD, 62. *Prof Exp:* Asst & lab technician, Div Food Technol, Univ Calif, 46-51, from jr res bacteriologist to asst res bacteriologist, Naval Biol Lab, 51-59, consult, 59-61, lectr, Univ, 56-58; intern, Walter Reed Gen Hosp, 62-63; assoc prof med microbiol, Sch Pub Health, Univ Calif, Berkeley, 63-67, PROF MED MICROBIOL, SCH MED, UNIV CALIF, DAVIS, 67-, CHMN DEPT, 68- *Mem:* Am Soc Microbiol; Mycol Soc Am; Sigma Xi. *Res:* Medical mycology; experimental pathology; physiology and immunology of Coccidioides immitis; human brucellosis; immunology of infectious diseases. *Mailing Add:* Dept of Med Microbiol Univ of Calif Sch of Med Davis CA 95616

PAPPALARDO, LEONARD THOMAS, b New York, NY, Dec 1, 29; m 55; c 4. ORGANIC CHEMISTRY, POLYMER CHEMISTRY. *Educ:* Fordham Col, BS, 51; NY Univ, MS, 59. *Prof Exp:* Chemist, Am Sugar Refining Co, 51-53 & Eng Ctr, Columbia Univ, 55-57; sr res chemist, Tex US Chem Co, 57-67; ASSOC MEM TECH STAFF, ORG MAT RES & DEVELOP DEPT, BELL TEL LABS, MURRAY HILL, 67- *Mem:* AAAS; Am Chem Soc; Sigma Xi. *Res:* Organic polymer chemistry of elastomers and synthetic resins; synthesis and characterization as well as relating properties to structure; printed circuit materials research, development and consulting. *Mailing Add:* 312 Mountain Way Morris Plains NJ 07950

PAPPALARDO, ROMANO GIUSEPPE, b Genoa, Italy; US citizen. SOLID STATE SCIENCE. *Educ:* Univ Pavia, PhD(solid state physics), 55. *Prof Exp:* Docent physics, Univ Pavia, 55-56; fel cryogenics, Univ Bristol, 56-57; vis res prof spectros, Univ Pittsburgh, 57-59; vis scholar, Bell Tel Labs, 59-61; res scientist chem, Cyanamide Europ Res Inst, Geneva, Switz, 61-66; resident res assoc, Transuranic Lab, Argonne Nat Lab, 66-68; MEM TECH STAFF, GTE LABS, 68- *Concurrent Pos:* Fulbright scholar, 57. *Mem:* Electrochem Soc; Am Phys Soc. *Res:* Spectroscopy of transition-metal and rare-earth ions and its relation to phosphor technology. *Mailing Add:* GTE Labs 40 Sylvan Rd Waltham MA 02254

PAPPANO, ACHILLES JOHN, b Allentown, Pa, Mar 21, 40; m 63; c 4. PHARMACOLOGY. *Educ:* St Joseph's Col, Pa, BS, 62; Univ Pa, PhD(pharmacol), 66. *Prof Exp:* Instr pharmacol, Sch Med, Tulane Univ, 66-67; from asst prof to assoc prof, 68-80, PROF PHARMACOL, SCHS MED & DENT MED, UNIV CONN, 80- *Concurrent Pos:* Nat Heart Inst fel, Sch Med, Univ Va, 67-68. *Mem:* Am Heart Asn; Am Soc Pharmacol & Exp Therapeut; Int Soc Heart Res. *Res:* Drug action in embryonic heart; cardiac electrophysiology; developmental pharmacology; heart cell culture. *Mailing Add:* Dept Pharmacol Univ Conn Health Ctr Farmington CT 06032

PAPPAS, ANTHONY JOHN, b Memphis, Tenn, Jan 21, 40; m 79; c 1. CHEMISTRY. *Educ:* Univ Miami, BS, 63, MS, 65, PhD(chem), 69. *Prof Exp:* Fel, Univ Tex, Austin, 69-71; from asst prof to sr assoc prof, 71-83, PROF, MIAMI-DADE COMMUNITY COL, SOUTH CAMPUS, 83- *Mem:* Am Chem Soc. *Res:* Synthesis, characterization and application of inorganic coordination compounds; new synthetic routes for preparation of inorganic and organic compounds via enhancement by coordination to metals; writing experiments for general-organic chemistry. *Mailing Add:* Dept Chem Miami-Dade Community Col 11011 SW 104 St Miami FL 33176

PAPPAS, DANIEL SAMUEL, b Northampton, Mass, Dec 18, 42; m 76; c 1. PLASMA PHYSICS, GENERAL PHYSICS. *Educ:* Univ Calif, Los Angeles, BS, 69, MS, 71. *Prof Exp:* Physicist, Univ Calif, Los Angeles, 70-71; physicist, Jet Propulsion Lab, Calif Inst Technol, 71-72; physicist, Lawrence Livermore Lab, Univ Calif, 72-73; physicist, Nat Magnet Lab, Mass Inst Technol, 74-80 & Plasma Fusion Ctr, 80-84; PHYSICIST, LOS ALAMOS NAT LAB, 84-; PRES, ADVEC CORP, 88- *Mem:* Am Phys Soc. *Res:* Controlled fusion research, neutron and x-ray measurements from thermonuclear systems; non-electrical applications of fusion, nuclear excited lasers, ion temperature by neutrons and charge exchange; scaling laws of fusion devices, particle confinement, recycling and impurity studies; astrons, particle accelerators, thermionics and hydrodynamics; nuclear pumped lasers, high power laser applications, cryogenics, superconducting technology; two patents. *Mailing Add:* Physics Div Los Alamos Nat Lab Los Alamos NM 87545

PAPPAS, GEORGE DEMETRIOS, b Portland, Maine, Nov 26, 26; m 52; c 2. NEUROBIOLOGY. *Educ:* Bowdoin Col, AB, 47; Ohio State Univ, MS, 48, PhD(zool), 52. *Hon Degrees:* DSc, Univ Athens, Greece, 88. *Prof Exp:* Vis investr cytol, Rockefeller Inst, 52-54; from asst prof to assoc prof anat, Col Physicians & Surgeons, Columbia Univ, 56-66; prof anat, Albert Einstein Col Med, 67-77, prof neurosci, 74-77; PROF ANAT & CELL BIOL & HEAD DEPT, COL MED, UNIV ILL, 77- *Concurrent Pos:* Fel anat & cytol, NY Univ-Bellevue Med Ctr, 54-56; vis prof neurosci, Albert Einstein Col Med, 77- *Mem:* Fel AAAS; Am Asn Anat; Am Soc Cell Biol (pres, 75); fel NY Acad Sci; Electron Micros Soc Am; Sigma Xi; Soc Neurosci. *Res:* Electron microscopy; cytochemistry; neurocytology; synapses; ocular tissue; vascular system; biological membranes; neural transplantation. *Mailing Add:* Dept Anat & Cell Biol Col Med Univ Ill 808 S Wood St Chicago IL 60612

PAPPAS, GEORGE STEPHEN, b Poughkeepsie, NY, Oct 16, 23; m 46; c 3. PHYSIOLOGY. *Educ:* Fordham Univ, BS, 47, MS, 49; NY Univ, PhD(biol), 60. *Prof Exp:* From instr to assoc prof, 48-63, chmn dept, 60-69, vpres, 69-70, actg pres, 70-71, vpres, 71-76, PROF BIOL, IONA COL, 63-, CHMN DEPT, 77- *Concurrent Pos:* Am Physiol Soc grant, 60; Fulbright fel endocrine physiol, Cairo, Egypt, 66. *Mem:* AAAS; fel NY Acad Sci. *Res:* Antimetabolites and vitamin analogs; endocrine physiology; gonadotrophic effects of adrenalectonized animals; thyroid-adrenal function; developmental biology; history of biology. *Mailing Add:* Dept of Biol Iona Col New Rochelle NY 10801

PAPPAS, JAMES JOHN, b Bridgeport, Conn, July 22, 31; m 60; c 3. ORGANIC CHEMISTRY. *Educ:* Mass Inst Technol, BS, 52, MS, 54; Columbia Univ, PhD(org chem), 59. *Prof Exp:* Jr chemist, Merck & Co, NJ, 54-56; chemist, Esso Res & Eng Co, 59-62; sr chemist, Cent Res Labs, Inmont Corp, 62-67; mgr chem prod, 67-70; sr res chemist, Esso Res & Eng Co, 70-71; sr res chemist, 71-84, MGR APPL RES, J M HUBER CORP, 84- *Mem:* Am Chem Soc. *Res:* Reactions of ozone; hydrogenation; phosphorus and sulfur chemistry; synthetic lubricants; lubricant and fuel additives; wax modifiers; polymer synthesis and modification; electrophotography; liquid and solid toners; dyes; pigments; printing inks; coatings. *Mailing Add:* 19 Trouville Dr Parsippany NJ 07054

PAPPAS, LARRY GEORGE, b Chadron, Nebr, Nov 16, 46; c 2. INSECT PHYSIOLOGY. *Educ:* Hiram Scott Col, BS, 69; Univ Wyo, MS, 71; Univ Ill, Urbana, PhD(entom), 75. *Prof Exp:* Res asst zool, Univ Wyo, 69-71; res assoc, Va Polytech Inst & State Univ, 75-77; instr biol, Col St Teresa, Winona, Minn, 77-79; ASST PROF BIOL, PERU STATE COL, NEBR, 79- *Mem:* Sigma Xi; Entom Soc Am; AAAS; Am Mosquito Control Asn. *Res:* Physiological basis of insect behavior. *Mailing Add:* Sci Dept Peru State Col Peru NE 68421

PAPPAS, LEONARD GUST, b Mansfield, Ohio, Aug 29, 20; m 45; c 4. POLYMER CHARACTERIZATION. *Educ:* Western Reserve Univ, BS, 47, MS, 51. *Prof Exp:* Chemist, NY Cent RR Lab, 47-48; tech man, polymer res, BF Goodrich Res Ctr, 52-60, res chemist, 60-71, supvr anal serv, 71-85; INSTR INORG & ORG CHEM, CUYAHOGA COMMUNITY COL, 80- *Concurrent Pos:* Lectr, China Asn Sci & Technol, People's Repub China, 84 & Kent State Univ, 84. *Mem:* Am Chem Soc; Am Inst Chemists. *Res:* Rheological equipment design for polymer testing; radiation polymerization and its effects on polymers; solution properties of polymers; thermal analysis of polymers; analytical chemical techniques; liquid, ion and gel permeation chromatography. *Mailing Add:* 8681 Brecksville Rd Brecksville OH 44141

PAPPAS, NICHOLAS, b Kearny, NJ, Sept 22, 30; m 53; c 4. ORGANIC CHEMISTRY, POLYMER CHEMISTRY. *Educ:* Yale Univ, BS, 52; Brown Univ, PhD(org chem), 57. *Prof Exp:* Res chemist, E I du Pont de Nemours & Co, Inc, 56-62, staff chemist, 62-63, res supvr polymers, 63-65, develop supvr finishes res, 65-66, asst lab mgr, Mich, 66-69, lab mgr, 69-70, asst nat mgr indust finishes, 70-71, nat mgr indust finishes & mkt dir, 71-73, div dir,

73-76, mgr, corp planning activ, 76-78, gen mgr, 78-81, vpres, fabrics & finishes, 81-82, group vpres, polymer prod, 82-90; RETIRED. *Res:* Organic synthesis; polymer synthesis and evaluation; development of finishes. *Mailing Add:* 606 Swallow Hollow Rd Greenville DE 19898

PAPPAS, PETER WILLIAM, b Pasadena, Calif, Dec 9, 44; m 66; c 2. MEMBRANE BIOLOGY. *Educ:* Humboldt State Col, Calif, BA, 66, MA, 68; Univ Okla, PhD(zool), 71. *Prof Exp:* From asst prof to assoc prof, 73-82, PROF ZOOL, OHIO STATE UNIV, 82-, CHMN ZOOL, 89- *Concurrent Pos:* NIH fel, Rice Univ, Houston, 71-73. *Honors & Awards:* Sigma Xi Res Award, Ohio State Univ, 75; Henry Baldwin Ward Medal, Am Soc Parasitologists, 84. *Mem:* Am Soc Parasitologists; Am Soc Trop Med & Hyg; Sigma Xi; AAAS. *Res:* Biochemical aspects of host-parasite relationships; biochemistry and physiology of parasites; membrane transport in parasitic helminths and protozoa. *Mailing Add:* Dept Zool 1735 Neil Ave Ohio State Univ Columbus OH 43210

PAPPAS, S PETER, b Hartford, Conn, Feb 3, 36; m 61; c 2. ORGANIC CHEMISTRY, PHOTOCHEMISTRY. *Educ:* Dartmouth Col, BA, 58; Univ Wis, PhD(chem), 62. *Prof Exp:* Fels, Univ Wis, 62-63 & Brandeis Univ, 63-64; asst prof chem, Emory Univ, 64-68; from assoc prof to prof chem, NDak State Univ, 68-89; SCI FEL & DIR TECHNOL ACQUISITION, LOCTITE CORP, 89- *Concurrent Pos:* Indust consult, var chem co; res assoc, DeSoto, Inc, 73; indust grants, 73-; NATO grants, 81, 82, 85 & 86; vis scientist, Air Force Mat Lab, 81-82; vis prof, Univ Stuttgart, 77, Royal Inst Technol, Stockholm, 82 & 87, Soc Polymer Sci, Japan, 84; ed bd, J Radiation Curing & J Photopolymer Sci. *Honors & Awards:* First Place Roon Awards, Fedn Socs Coatings Technol, 74, 75, 76 & 80; DAAD Award, 77; Pavac lectr, 83. *Mem:* Am Chem Soc; InterAm Photochemical Soc. *Res:* Photochemistry, synthesis, catalysis and reaction mechanisms; coatings and adhesives. *Mailing Add:* Loctite Corp 705 N Mountain Rd Newington CT 06111

PAPPATHEODOROU, SOFIA, b New York, NY, Feb 15, 40. ORGANOMETALLIC CHEMISTRY. *Educ:* Univ Miami, BS, 62, MS, 65, PhD(chem), 78. *Prof Exp:* Lab instr chem, Univ Miami, 65-68 & 70-71; lab instr chem, Miami-Dade Community Col, 76-78; vis prof chem, Dept Phys Sci, Fla Int Univ, 76, adj instr, 76-78; fel chem, Papanicolaou Cancer Res Inst, 79-81; LECTR, CHEM, CALIF STATE UNIV, FRESNO, 81- *Mem:* Am Chem Soc; Sigma Xi. *Res:* Synthesis and product analysis of olefins from alcohols and platinum salts, and product distributions in the reactions of Grignard reagents with conjugated enones. *Mailing Add:* Chem Dept Calif State Univ DH Carson CA 90747

PAPPELIS, ARISTOTEL JOHN, b Cloquet, Minn, May 25, 28; m 51; c 3. PLANT PATHOLOGY, PLANT PHYSIOLOGY. *Educ:* Wis State Col, Superior, BS, 51; Iowa State Univ, PhD(plant physiol), 57. *Prof Exp:* Plant physiologist, Mkt Serv, USDA, 57-58; asst prof biol sci, Western Ill Univ, 58-60; from asst prof to assoc prof, 60-71, PROF BOT, SOUTHERN ILL UNIV, CARBONDALE, 71- *Concurrent Pos:* Shell Merit fel, Stanford Univ, 70. *Res:* Senescence and host-parasite interactions in plants. *Mailing Add:* Dept Bot Southern Ill Univ Carbondale Carbondale IL 62901

PAPPENHAGEN, JAMES MEREDITH, b Alliance, Ohio, Apr 1, 26; m 48; c 3. ANALYTICAL CHEMISTRY, SCIENCE COMMUNICATIONS. *Educ:* Mt Union Col, BS, 49; Purdue Univ, MS, 51, PhD(chem), 53. *Hon Degrees:* DSc, Kenyon Col, 89. *Prof Exp:* From asst prof to prof chem, 52-89, chmn dept, 61-83, DIR HAZARDOUS MAT MGT, KENYON COL, 89- *Concurrent Pos:* Vis prof environ systs eng, Clemson Univ, 66-67; consult, Off Water Progs, Environ Protection Agency; mem, Standard Methods Comt, Am Water Works Asn & Water Pollution Control Fedn. *Honors & Awards:* Wendell R LaDue Citation, 77. *Mem:* AAAS; Am Chem Soc. *Res:* Spectrophotometry and instrumentation; determination of ions in wastes and waters; technical management. *Mailing Add:* Box 229 Gambier OH 43022

PAPPENHEIMER, ALWIN MAX, JR, b Cedarhurst, NY, Nov 25, 08; m 38; c 3. BIOCHEMISTRY. *Educ:* Harvard Univ, SB, 29, PhD(org chem), 32. *Prof Exp:* Instr & tutor biochem sci, Harvard Univ, 30-33; Nat Res Coun fel med, Nat Inst Med Res London, 33-34; instr appl immunol, Sch Pub Health, Harvard Univ, 36-39; asst prof biochem & bact, Sch Med, Univ Pa, 39-41; from asst prof to prof microbiol, Col Med, NY Univ, 41-58; chmn bd tutors in biochem, Harvard Univ, 58-63, prof, 58-79, emer prof biol, 79; RETIRED. *Concurrent Pos:* Bradford fel, Harvard Med Sch, 36-37; sr chemist, State Antitoxin & Vaccine Lab, Mass, 36-39; Guggenheim fel, 66-67; consult, Surgeon Gen Mem Comn Immunization, Armed Forces Epidemiol Bd. *Honors & Awards:* Eli Lilly Award, 42; Paul Ehrlic Prize, 90. *Mem:* Nat Acad Sci; Am Acad Arts & Sci; Am Chem Soc; Am Soc Biol Chemists; Am Soc Microbiol. *Res:* Chemistry and mode of action of bacterial toxins; immunochemistry; immunization; bacterial metabolism. *Mailing Add:* 17 Edgecliff Rd Watertown MA 02172

PAPPENHEIMER, JOHN RICHARD, b New York, NY, Oct 25, 15; m 49; c 4. PHYSIOLOGY. *Educ:* Harvard Univ, BS, 36; Cambridge Univ, PhD(physiol), 40. *Prof Exp:* Demonstr pharmacol, Univ Col, Univ London, 39-40; instr, Col Physicians & Surgeons, Columbia Univ, 41-42; assoc, 45-49, asst prof, 49-53, vis prof, 53-68, George Higginson prof, 68-82, EMER PROF PHYSIOL, SCH MED, HARVARD UNIV, 82- *Concurrent Pos:* Res fel physiol, Col Physicians & Surgeons, Columbia Univ, 40-41; Johnson Res Found fel, Univ Pa, 41-45; overseas fel, Churchill Col, Cambridge, Eng, 71-72; career investr, Am Heart Asn, 53-; George Eastman Univ prof & fel Balliol Col, Oxford, 75-76. *Mem:* Nat Acad Sci; AAAS; Am Physiol Soc (pres, 64-65); Soc Gen Physiol; Harvey Soc; Am Acad Arts & Sci. *Res:* Kidney; respiration; hemodynamics; permeability; cerebrospinal fluid; control of sleep. *Mailing Add:* Concord Field Sta Harvard Univ Old Causeway Rd Bedford MA 01730

PAPPER, EMANUEL MARTIN, b New York, NY, July 12, 15; m 75; c 2. ANESTHESIOLOGY. *Educ:* Columbia Univ, AB, 35; NY Univ, MD, 38; Am Bd Anesthesiol, dipl, 43; Univ Miami, PhD, 90. *Hon Degrees:* MD, Univ Uppsala, 64, Univ Turin, 66, Univ Vienna, 77, Columbia Univ, 88. *Prof Exp:* From intern to resident anesthesiol, Bellevue Hosp, 39-42; from instr to assoc prof, NY Univ, 42-49; prof & chmn dept, Col Physicians & Surgeons, Columbia Univ, 49-69; dean, Sch Med & vpres med affairs, 69-81, PROF ANESTHESIOL, SCH MED, UNIV MIAMI, 69- *Concurrent Pos:* Fel, NY Univ, 40; dir anesthesiol serv, Presby Hosp, 49-69; dir & vis anesthesiologist, Francis Delafield Hosp, 51-69; consult, Huntington Hosp, Div Med Sci, Nat Res Coun, chmn comt US Navy & First Army; mem surg study sect, NIH, 58-62, nat adv heart coun, 62-64; mem, President's Comn Heart Dis, Cancer & Stroke, 64; prin consult, Nat Inst Gen Med Sci, 65-66, chmn gen med res prog-proj comt, 66-67. *Mem:* AAAS; Am Soc Anesthesiol; Am Soc Pharmacol & Exp Therapeut; Am Soc Clin Invest; Am Thoracic Soc. *Res:* Pharmacology of drugs; physiology of circulation. *Mailing Add:* One Grove Isle Dr Miami FL 33133

PAPPIUS, HANNA M, b Lakocin, Poland, July 26, 25; Can citizen; m 50; c 3. NEUROCHEMISTRY. *Educ:* McGill Univ, BSc, 46, MSc, 48, PhD(biochem), 52. *Prof Exp:* From lectr to assoc prof, 54-79, PROF NEUROL & NEUROSURG, MCGILL UNIV, 79-, ASSOC NEUROCHEMIST, MONTREAL NEUROL INST, 53- *Mem:* Can Physiol Soc; Can Biochem Soc; Can Soc Clin Chem; Am Soc Neurochem; Int Soc Neurochem; Soc Cerebral Blood Flow & Metab. *Res:* Cerebral edema; water and electrolyte distribution in cerebral tissues; functional disturbances in injured brain; biogenic amines in injured brain. *Mailing Add:* Dept of Neurol MC Gill Univ Sch of Med 3801 University St Montreal PQ H3A 2B4 Can

PAPSIDERO, LAWRENCE D, b Buffalo, NY, Sep 2, 49; m 72; c 1. IMMUNO PATHOLOGY, HYBRIDOMA RESEARCH. *Educ:* Ohio State Univ, MS, 74; State Univ NY, Buffalo, BA, 71, PhD(exp path), 78. *Prof Exp:* Fel, Roswell Park Mem Inst, 78-79, cancer res scientist, 79-81, sr cancer res scientist, 81-; RES & DEVELOP CELLULAR PROD, INC, 85- *Honors & Awards:* Investr Res Award, Nat Cancer Inst, 81. *Res:* Immuno diagnosis of cancer. *Mailing Add:* R & D Cellular Prod Inc 872 Main St Buffalo NY 14202

PAQUE, RONALD E, b Green Bay, Wis, Apr 29, 48. MICROBIOLOGY, IMMUNOLOGY. *Educ:* Wis State Univ-Oshkosh, BS, 60; Univ Wis, MS, 63; Univ Ariz, PhD(microbiol, biochem), 66. *Prof Exp:* Proj asst virol, Univ Wis, 61, asst bact, 61-62, asst virol, 62-63; asst microbiol, Univ Ariz, 63-64, training fel, 65-66; assoc dir immunol div, Cancer Chemother Dept, Microbiol Assocs, 66-68; asst prof microbiol, Univ Ill Col Med, 70-74; assoc prof, 74-87, PROF MICROBIOL, HEALTH SCI CTR, UNIV TEX, SAN ANTONIO, 87- *Concurrent Pos:* USPHS fel cellular immunol, Univ Ill Col Med, 68-70; mem study sect, NIH; reviewer, NSF; invited partic, 2nd RNA Conf, Peking, China; Presidential travel award, 88. *Mem:* AAAS; Am Asn Immunol; Am Soc Microbiol; Am Asn Cancer Res. *Res:* Tumor, cellular and transplantation immunology; Idiotypic regulation of autoimmune myocarditis; antildiotypic mimicry of bacterial antigens; delayed-type hypersensitivity; immunopathology. *Mailing Add:* Dept of Microbiol Univ of Tex Health Sci Ctr San Antonio TX 78284

PAQUET, JEAN GUY, b Montmagny, Que, Jan 5, 38. ELECTRICAL ENGINEERING, AUTOMATIC CONTROL SYSTEMS. *Educ:* Laval Univ, BApplSc, 59, PhD(elec eng), 63; Ecol Nat Sup Aeronautique, Paris, MSc, 60. *Hon Degrees:* DSc, McGill Univ, 82; LLD, York Univ, 83. *Prof Exp:* From assoc prof to prof, Laval Univ, 67-71, head dept elec eng, 67-69, vdean res, Fac Sci, 69-72; spec asst to vpres sci, Nat Res Coun Can, 71-72; vrector acad, Laval Univ, 72-77, rector, 77-87; PRES, LA'LAURENTIENNE, 87- *Concurrent Pos:* Fels, French govt, 59, Nat Res Coun of Can, 61, NATO, 62, NSF, 64 & Que govt, 65; grants, Nat Res Coun Can, 64-77 & Defence Res Bd Can, 65-76; mem assoc comt on automatic control, Nat Res Coun Can, 63-70; mem, Coun Que Univs, 73-77; mem, Spec Task Force on Res & Develop, Sci Coun Can, 76- *Mem:* Inst Elec & Electronics Engrs; Am Soc Eng Educ; Can Res Mgt Asn; fel AAAS; Innovation Mgt Inst Can; fel Royal Soc Can. *Res:* Control systems engineering. *Mailing Add:* 50 Grand-Allee E Laval Univ Quebec PQ G1R 5M4 Can

PAQUETTE, DAVID GEORGE, b Annapolis, Md, Apr 21, 45; m 70. MATERIALS SCIENCE ENGINEERING. *Educ:* Univ Calif, Santa Barbara, BA, 67, MA, 69, PhD(physics), 73. *Prof Exp:* Res assoc/assoc instr, Univ Utah, 73-76, res asst prof mat sci & eng, 76-77; SR SCIENTIST, FORD AEROSPACE & COMMUN CORP, 77- *Mem:* Am Phys Soc; Am Ceramic Soc. *Res:* Fabrication and characterization of advanced composite ceramic materials; characterization of beta alumina ceramic materials for sodium-sulfur battery development; design and analysis of data from sodium sulfur electrochemical cells; ceramic composites with increased fracture toughness. *Mailing Add:* 767 Allegheny Ave Costa Mesa CA 92626

PAQUETTE, GERARD ARTHUR, b Winooski, Vt, Aug 5, 26; m 55; c 5. COMPUTER SCIENCE EDUCATION, MATHEMATICS EDUCATION. *Educ:* La Mennais Col, BA, 52; Bridgewater State Col, MEd, 57; Boston Col, MA, 61; Pa State Univ, PhD(math educ), 71. *Prof Exp:* Teacher high sch, Que, 45-49, high sch, Mass, 49-53 & Mt Assumption Inst, 53-54; chmn dept math, Middletown High Sch, RI, 54-67; assoc prof, 67-76, PROF MATH EDUC, BOSTON STATE COL, 76-; AT DEPT MATH, UNIV LOWELL. *Concurrent Pos:* Lectr, Univ RI, 61-64 & Bridgewater State Col, 61-64; math consult, RI State Dept Educ, 64-67. *Res:* Theory of instruction; instructional strategies; modes of representation in instruction; implementation of computer assisted instruction; mathematics curricular materials. *Mailing Add:* Dept Math Wang Labs Inc One Industrial Ave 014-790 Lowell MA 01851

PAQUETTE, GUY, b Montreal, Que, Oct 4, 30; m 59; c 2. THEORETICAL PHYSICS. *Educ:* Univ Montreal, BSc, 51; Univ BC, MA, 53, PhD(physics), 56. *Prof Exp:* From asst prof to assoc prof, 56-69, chmn dept, 73-82, PROF PHYSICS, UNIV MONTREAL, 69- *Concurrent Pos:* Nat Res Coun Can res fel, King's Col, Univ London, 57-58 & State Univ Leyden, 58-59; consult, RCA Victor Co, 62-64; vis, Saclay Nuclear Res Ctr, Saclay, France, 66-67. *Mem:* Am Phys Soc; Can Asn Physicists. *Res:* Physics of magnetoplasmas; waves in plasmas. *Mailing Add:* Dept Physics Univ Montreal CP 6128 Montreal PQ H3C 3J7 Can

PAQUETTE, LEO ARMAND, b Worcester, Mass, July 15, 34; m 57; c 5. ORGANIC CHEMISTRY. *Educ:* Col of the Holy Cross, BS, 56; Mass Inst Technol, PhD(org chem), 59. *Hon Degrees:* DSc, Col of the Holy Cross, 84. *Prof Exp:* Res assoc chem, Upjohn Co, 59-63; from asst prof to prof, 63-81, Kimberley prof, 81-84, DISTINGUISHED UNIV PROF CHEM, OHIO STATE UNIV, 87- *Concurrent Pos:* Alfred P Sloan Found fel, 65-67, Guggenheim fel, 76-77; vis prof, Mich State Univ, 68, Univ Iowa, 70, Univ Colo, 74, Univ Calif, Santa Barbara, 75, Univ Groningen, 75, Tex A&M Univ, 79, Northwestern Univ, 81, Univ Heidelberg, 84, Univ Paris, 85, Univ Freiburg, 89; mem chem div adv comt, NSF; mem bd chem sci & technol, Nat Res Coun, 84-87; mem med chem study sect, NIH, chmn, 86-88. *Honors & Awards:* Morley Medalist, Am Chem Soc, 71, Nat Award Creative Work in Synthetic Org Chem, 84; Sr Humboldt Award, West Germany, 89. *Mem:* Am Chem Soc; Royal Soc Chem. *Res:* Synthetic organic chemistry; organosilicon reagents; unusual molecules; molecular rearrangements; transition metal catalysis; author of 800 research papers in organic chemistry. *Mailing Add:* Dept Chem Ohio State Univ 120 W 18th Ave Columbus OH 43210-1173

PAQUETTE, ROBERT GEORGE, b Chippewa Falls, Wis, Feb 5, 15; m 41; c 5. PHYSICAL OCEANOGRAPHY, OCEAN ENGINEERING. *Educ:* Univ Wash, BS, 36, PhD(phys chem), 41. *Prof Exp:* Instr math, US Naval Acad, 41-42, instr chem & elec eng, 42-45; res chemist, Plywood Res Found, Wash, 46; res assoc meteorol physics, Univ Wash, 46-49, res oceanogr, 49-55, res assoc prof & lectr oceanog, 55-61; head phys oceanog group, Gen Motors Lab, Santa Barbara, 61-71; assoc prof, 71-75, PROF OCEANOG, NAVAL POSTGRAD SCH, 74- *Mem:* Am Geophys Union. *Res:* Arctic oceanography; oceanographic instrumentation; direct measurement of ocean currents. *Mailing Add:* Dept Oceanog 68 Pa Naval Postgrad Sch Monterey CA 93940

PAQUETTE, THOMAS LEROY, biochemistry, for more information see previous edition

PARACER, SURINDAR MOHAN, b Lyallpur, India, Jan 25, 41; m 68; c 2. SYMBIOLOGY, HISTORY OF SCIENCE. *Educ:* Panjab Univ, BS, 59; SDak State Univ, MS, 61; Univ Calif, Davis, PhD(nematol), 66. *Prof Exp:* Res assoc, Univ Mass, 65-67; asst prof biol, Nichols Col, 67-69 & Assumption Col, Mass, 69-70; from asst to assoc prof, 70-77, PROF BIOL, WORCESTER STATE COL, 77- *Concurrent Pos:* Vis prof nematol, Univ Fla, 84; adj fac, Clark Univ, Worcester, Mass. *Mem:* AAAS; Sigma Xi; Nat Sci Teachers Asn; Hist Sci Soc; Soc Nematologists; Europ Soc Nematologists. *Res:* Biology of the host-parasite relationships; symbiology and the history of science. *Mailing Add:* Dept Biol Worcester State Col Worcester MA 01602-2597

PARADISE, LOIS JEAN, b Boston, Mass, Oct 26, 28. MEDICAL MICROBIOLOGY, IMMUNOLOGY. *Educ:* Brown Univ, AB, 50; Univ Mich, MS, 55, PhD(bact), 60. *Prof Exp:* From res assoc to asst prof microbiol, Univ Mich, Ann Arbor, 60-72; ASSOC PROF MED MICROBIOL & IMMUNOL, UNIV S FLA, 73- *Mem:* Am Soc Microbiol; Reticuloendothelial Soc; Int Soc Immunopharmacol; Am Venereal Dis Asn; Int Soc Develop & Comp Immunol. *Res:* Pathogenesis of respiratory infections; experimental syphilis; tumor immunology; respiratory carcinogenesis. *Mailing Add:* Dept of Med Microbiol & Immunol Box 10 Univ of S Fla Col of Med 12901 N Bruce B Downs Blvd Tampa FL 33612-4799

PARADISE, NORMAN FRANCIS, b Minneapolis, Minn, Jan 19, 43; c 5. PHYSIOLOGY. *Educ:* Univ Minn, BA, 66, PhD(physiol), 71. *Prof Exp:* USPHS fel, 71-73, Minn Med Found fel, 73-74, Minn Heart Asn fel, Biophys Sci Unit, Mayo Grad Sch Med, Mayo Found, 75-77; asst prof, Col Med, Northeastern Ohio Univ, 77-81, assoc prof, 81-84; MINNEAPOLIS SCH ANESTHESIA. *Concurrent Pos:* Lectr, Am Asn Nurse Anesthetists; adj prof, St Mary's Col; res assoc, St Paul-Ramsey Med Ctr. *Honors & Awards:* Hon Commencement Grand Marshall, Northeastern Ohio Univ Col of Med, 86, 87, 88. *Mem:* AAAS; Am Physiol Soc; Sigma Xi; NY Acad Sci; Am Soc Hypertension; Am Heart Assoc. *Res:* Cardiovascular physiology; cardiac muscle; hypertension; obesity; exercise. *Mailing Add:* Dept Biol Univ Akron Akron OH 44325

PARAKKAL, PAUL FAB, b Alwaye, India, July 21, 31; US citizen; m 60; c 2. BIOLOGY, ELECTRON MICROSCOPY. *Educ:* Kerala Univ, Trivandrum, India, BSc, 52; McGill Univ, MSc, 59; Brown Univ, PhD(biol), 62. *Prof Exp:* Res assoc, Med Sch, Boston Univ, 64-68, asst res prof dermatol, 64-68; assoc scientist biol, Ore Reg Primate Ctr, 68-73; HEALTH SCI ADMINR BIOL, NAT INST DENT RES, NIH, 73-, DIV RES GRANTS, DEPT SURG & BIOENG. *Mem:* Am Soc Cell Biol. *Res:* Control mechanisms of the hair growth cycle. *Mailing Add:* Westwood Bldg Rm 437 Div Res Grants Dept Surg & Bioeng NIH Bethesda MD 20205

PARANCHYCH, WILLIAM, b Drumheller, Alta, Feb 4, 33; m 57; c 1. BIOCHEMISTRY, MICROBIOLOGY. *Educ:* Univ Alta, BSc, 54, MSc, 58; McGill Univ, PhD(biochem), 61. *Prof Exp:* Clin biochemist, Misericordia Hosp, Edmonton, Alta, 54-56; res assoc virol, Wistar Inst, 61-63; from asst prof to prof biochem, 71-89, PROF & CHMN MICROBIOL, UNIV ALTA, 89- *Mem:* Am Soc Biol Chemists; Am Soc Microbiol; AAAS; Can Soc Microbiol; Can Biochem Soc. *Res:* Mechanisms of biochemical and genetic studies on bacterial pili: a comparison of the structure and function of conjugative and nonconjugative systems. *Mailing Add:* Dept Microbiol Univ Alta Edmonton AB T6G 2E9 Can

PARANJAPE, BHALACHANDRA VISHWANATH, b Gondia, India, Dec 2, 22; m 54; c 3. PHYSICS. *Educ:* Univ Nagpur, BSc, 45; Univ Liverpool, PhD, 57. *Prof Exp:* Asst physics, Univ Liverpool, 56-57; res assoc, Purdue Univ, 58-59; asst prof, La State Univ, 59; from asst prof to assoc prof, 61-70, PROF PHYSICS, UNIV ALTA, 70- *Res:* Theoretical aspects of semiconductor physics; dielectric breakdown; Ettingshausen coefficient. *Mailing Add:* Dept Physics Univ Alta Edmonton AB T6G 2M7 Can

PARASCANDOLA, JOHN LOUIS, b New York, NY, July 14, 41; div; c 2. HISTORY OF SCIENCE. *Educ:* Brooklyn Col, BS, 63; Univ Wis-Madison, MS, 68, PhD(hist sci), 68. *Prof Exp:* Josiah Macy Jr Found fel hist sci, Harvard Univ, 68-69; from asst prof to prof hist pharm & hist sci, Univ Wis-Madison, 69-83; CHIEF, HIST MED DIV, NAT LIBR MED, 83- *Concurrent Pos:* Dir, Am Inst Hist Pharm, 73-81; vis worker, Nat Inst Med Res, London, 75; vis assoc prof, Inst hist Med, Johns Hopkins Univ, 79-80; coun chair, Am Inst Hist Pharm, 89- *Honors & Awards:* Edward Kremers Award, 80; Merit Award, NIH, 88; Surgeon General's Exemplary Serv Award, 89. *Mem:* Am Inst Hist Pharm; Hist Sci Soc; Am Asn Hist Med; Int Acad Hist Pharm (treas, 81-90). *Res:* History of pharmacology and drug therapy; history of pharmaceutical chemistry and biochemistry. *Mailing Add:* Hist Med Div Nat Libr Med Bethesda MD 20894

PARASKEVAS, FRIXOS, b Serrai, Greece, Sept 13, 28; Can citizen; m 60; c 2. IMMUNOLOGY. *Educ:* Aristotelian Univ Thessaloniki, MD, 51. *Prof Exp:* Lectr, 64-65, from asst prof to assoc prof, 65-76, PROF MED, UNIV MAN, 76-, PROF IMMUNOL, 78- *Concurrent Pos:* Consult, R H Inst, Winnipeg; dir, Immunoprotein Lab, Health Sci Ctr, Winnipeg. *Mem:* Can Soc Immunol; Am Asn Immunologists; Am Soc Hemat. *Res:* Activation of immune response; regulation of immune response. *Mailing Add:* Dept Immunol Univ Man Fac Med 753 McDermot Ave Winnipeg MB R3E 0W3 Can

PARASKEVOPOULOS, DEMETRIS E, physics, electrical engineering, for more information see previous edition

PARASKEVOPOULOS, GEORGE, b Athens, Greece, Apr 25, 29; Can citizen; m 60; c 1. PHYSICAL CHEMISTRY. *Educ:* Nat Tech Univ, Athens, BEng, 54; Univ Toronto, MSc, 60; McGill Univ, PhD(phys chem), 65. *Prof Exp:* Res chemist, Can Industs Ltd, 59-61; fel chem, 65-67, asst res officer, 67-71, assoc res officer, 71-80, SR RES OFFICER CHEM, NAT RES COUN CAN, 80- *Mem:* Chem Inst Can. *Res:* Reaction kinetics in the gas phase, particularly elementary reactions of atoms and free radicals; photochemistry; atmospheric chemistry, laser ablation of materials. *Mailing Add:* Inst Environ Chem Nat Res Coun Can Ottawa ON K1A 0R9 Can

PARASURAMAN, RAJA, b New Delhi, India, Aug 2, 50. NEUROPSYCHOLOGY, HUMAN FACTORS. *Educ:* Univ London, Eng, BSc, 72; Univ Aston, Eng, MSc, 73, PhD(psychol), 76. *Prof Exp:* Lectr II psychol, Lanchester Polytech, Eng, 76-77; lectr II psychol, Wolverhampton Polytech, Eng, 77-78; res psychologist psychol, Univ Calif, Los Angeles, 78-82; ASSOC PROF, CATHOLIC UNIV, 82- *Concurrent Pos:* Consult, Logicon Inc, 81-82. *Mem:* AAAS; Human Factors Soc; Soc Psychophysiol Res; Psychonomic Soc. *Res:* Neurophysiology; human performance; attention and vigilance; man-machine systems. *Mailing Add:* Dept Psychol Catholic Univ Am Washington DC 20064

PARASZCZAK, JURIJ ROSTYSLAN, b Rochdale, Eng, May 19, 52; m. ELECTRONICS ENGINEERING. *Educ:* Univ Sheffield, BSc, 73, PhD(phys chem), 76. *Prof Exp:* Res assoc, Univ Wis, Madison, 77-79; RES STAFF MEM, IBM RES, 79- *Mem:* Sigma Xi. *Res:* Design, manufacture, evaluation and modelling of resist materials for microlithography; study of plasma processing on dielectrics and resists. *Mailing Add:* IBM Res PO Box 218 Yorktown Heights NY 10598

PARATE, NATTHU SONBAJI, b Nagbhir, India, Oct 17, 36; m 69; c 3. WASTES MANAGEMENT-DISPOSAL, ENVIRONMENTAL ENGINEERING & MANAGEMENT. *Educ:* Saugar Univ, BE, 61; Sheffield Univ, Brit, MEng, 65; Paris Univ, Doctorate, 68. *Prof Exp:* Res engr geomech, Ecole Polytech Lab, Paris, 68-69; postdoctoral fel rock fracture, Laval Univ, Que, Can, 69-70; resident dir, Can Int Develop Agency & Govt Niger, Africa, 70-73; sr engr, PNP Consult, LaSalle, Que, 73-75; vis prof geomech & environ eng, Waterloo Univ, Can, 75-77; mgt res assoc & engr, Pa Pub Utility Comn & NC Pub Utility Comn, 77-82; assoc prof civil eng, Tenn State Univ, 82-86; VIS PROF, TEX A&M UNIV, PRAIRIEVIEW, 86- *Concurrent Pos:* Expert witness, Pa Pub Utility Comn, 77-79, NC Pub Utility Comn, 80-81 & Tex Pub Utility Comn, 89-90; expert consult, UN Indust Develop Orgn Prog Govt Guinea, 80-; prin investr, Tenn Valley Authority, 84-85; vis scientist, Oak Ridge Nat Lab; environ consult, Marshall Space Flight Ctr, NASA & Univ Space Res Asn; vis prof, Univ Ala, Huntsville. *Mem:* Int Soc Rock Mech; Asn Geoscientists Int Develop. *Res:* Environmental geotechnology; waste engineering and management-disposal; rock geomechanics, rock-soils-geomaterials; energy-depreciation and wastes management; man-society-environment: worldwide observations and testimonies; forensics. *Mailing Add:* 7920 Rockwood Lane No 227 Austin TX 78758

PARBERRY, IAN, b Hillingdon, Eng, July 27, 59; Australian & Brit citizen; m 83. COMPLEXITY THEORY, ANALYSIS OF ALGORITHMS. *Educ:* Univ Queensland, BSc, 80; BSc(hons), 81; Univ Warwick, PhD(comput sci), 84. *Prof Exp:* Asst prof computer sci, Pa State Univ, 84-90; ASSOC PROF COMPUTER SCI, UNIV NTEX, 90-; DIR, CTR RES PARALLEL & DISTRIB COMPUT, 90- *Concurrent Pos:* Prin investr, Gen Elec, 87, Air Force Off Sci Res, 87-91 & NSF, 88-91. *Mem:* Asn Comput Mach; Europ Asn Theoret Comput Sci; Int Neural Network Soc. *Res:* Computational complexity of parallel algorithms, sorting networks, neural networks. *Mailing Add:* Dept Comput Sci Univ NTex PO Box 13886 Denton TX 76203-3886

PARCELL, ROBERT FORD, organic chemistry, for more information see previous edition

PARCELLS, ALAN JEROME, b Los Angeles, Calif, Jan 29, 29; m 60; c 2. BIOCHEMISTRY. *Educ:* Univ Calif, Berkeley, AB, 54, PhD(biochem), 58. *Prof Exp:* Res fel biochem, Univ Utah, 58-59; RES ASSOC, UPJOHN CO, 59- *Mem:* AAAS; Am Chem Soc. *Res:* Purification and characterization of protein and polypeptide hormones, particularly growth hormones; chromatographic and electrophoretic purification methods; terminal and total amino acid analysis; purification and in vitro characterization of antilymphocyte globulins; modification of immunoglobulins; development of in vitro assays for prediction of immunosuppressive drugs. *Mailing Add:* 1310 Sussex St Kalamazoo MI 49002

PARCHEN, FRANK RAYMOND, JR, b Clinton, Iowa, Nov 29, 23; m 45; c 3. CHEMICAL ENGINEERING. *Educ:* Iowa State Univ, BS, 49, MS, 51, PhD(phys chem), 55. *Prof Exp:* Res assoc, Vet Med Res Inst, Iowa State Univ, 50-55; res chemist, Iowa, 55-58, develop supvr, Iowa, 58-59 & Tenn, 59-60, area supvr, Tenn, 59-61, tech supt, Iowa, 61-67, prod mgr cellophane, Del, 67-70, process mgr cellophane, 70-77, mfg consult, E I Du Pont de Nemours & Co, Inc, 77-85; CONSULT, 85- *Honors & Awards:* Pres Award, Asn Indust Metallizers, Coaters & Laminators, 90. *Mem:* AAAS; Am Chem Soc; Am Inst Chem Eng; Tech Asn Pulp & Paper Indust. *Res:* Polymer chemistry; packaging flexible films. *Mailing Add:* 28 Whitetail Dr Chadds Ford PA 19317-9242

PARCHER, JAMES V(ERNON), b Drumright, Okla, July 21, 20; m 43; c 5. SOIL MECHANICS. *Educ:* Okla State Univ, BS, 41, MS, 48; Harvard Univ, AM, 67; Univ Ark, PhD, 68. *Prof Exp:* Jr engr, Peters Cartridge Div, Remington Arms Co, 41-42; from instr to prof, 47-85, actg head, 68-69, head sch, 69-83, EMER PROF CIVIL ENG, OKLA STATE UNIV, 85- *Concurrent Pos:* NSF sci fac fel, 65-66; comt mem, Bldg Res Adv Bd, Nat Acad Sci, 78- *Mem:* Fel Am Soc Civil Engrs; Nat Soc Prof Engrs (vpres, 81-82). *Res:* Foundation engineering; slope stability; stabilization; behavior of expansive soils. *Mailing Add:* 1024 W Knapp Stillwater OK 74075

PARCHMENT, JOHN GERALD, b Cumberland City, Tenn, Aug 13, 23; m 57; c 2. ZOOLOGY, ECOLOGY. *Educ:* Mid Tenn State Univ, BS, 44; George Peabody Col, MA, 47; Vanderbilt Univ, PhD(limnol), 61. *Prof Exp:* From instr to asst prof, 49-63, PROF BIOL, MID TENN STATE UNIV, 63- *Concurrent Pos:* Tenn Acad Natural Sci Found lectr, vis scientists prog, 64-67. *Mem:* Fel AAAS; Sigma Xi. *Res:* Stream limnology; plankton and bottom organisms. *Mailing Add:* 1014 E Clark Blvd Murfreesboro TN 37130

PARCZEWSKI, KRZYSZTOF I(GNACY), b Wilno, Poland, Sept 23, 26; US citizen. CHEMICAL ENGINEERING. *Educ:* Nat Univ Ireland, BSc, 51; Univ London, dipl chem eng, 52; Univ Nottingham, PhD(eng), 56. *Prof Exp:* Develop engr, Powell Duffryn Carbon Prod Ltd, 52-53; res asst, Sch Eng, Univ Nottingham, 53-58; res engr, Cent Res Lab, Celanese Corp Am, 58-60; res scientist, Res Div, Am Radiator & Standard Sanit Corp, NJ, 60-66; engr, Knolls Atomic Power Lab, Gen Elec Co, 66-74; reactor engr, 74-80, SR CHEM ENGR, US NUCLEAR REGULATORY COMN, 80- *Mem:* Am Chem Soc; Am Inst Chem Engrs; AAAS. *Res:* Psychrometry; heat and mass transfer; thermodynamics; heat transfer in boiling; two-phase flow; nuclear chemical engineering. *Mailing Add:* US Nuclear Regulatory Comn Washington DC 20555

PARDEE, ARTHUR BECK, b Chicago, Ill, July 13, 21; c 4. BIOCHEMISTRY. *Educ:* Univ Calif, Berkeley, BS, 42; Calif Inst Technol, MS, 43, PhD, 47. *Prof Exp:* From instr to assoc prof biol, Univ Calif, Berkeley, 49-61; prof biochem, Princeton Univ, 61-75; PROF BIOCHEM & PHARMACOL, HARVARD MED SCH, 75- *Concurrent Pos:* Merck fel, Univ Wis, 47-48; NSF sr fel, Pasteur Inst, 57-58; Am Cancer Soc scholar, Imperial Cancer Res Inst, London, 72-73; mem comt sci & pub policy, Nat Acad Sci, 73-76; trustee, Inst Cancer Res, Philadelphia, 73-85, Waksman Found, NY, 76-85 & Friederich Miescher Inst, Basel, 80-88, Ludwig Inst, 88- *Honors & Awards:* Paul Lewis Award, Am Chem Soc, 60; Krebs Medal, Europ Biochem Soc, 73; Rosensteil Medal, Brandeis Univ, 75; 3M Award, Fedn Am Socs Exp Biol, 80. *Mem:* Nat Acad Sci; Am Soc Biol Chemists (pres, 80); hon mem Japanese Biochem Soc; Am Acad Arts & Sci; Nat Inst Med; Am Asn Cancer Res (pres, 85). *Mailing Add:* Dana Farber Cancer Inst 44 Binney St Boston MA 02115

PARDEE, JOEL DAVID, b Mt Morris, Mich, 47. FLUORESCENCE ENERGY TRANSFER, CYTOSKELETON. *Educ:* Colo State Univ, PhD(biochem & biophysics), 78. *Prof Exp:* ASSOC PROF CELL BIOL & ANAT, MED SCH, CORNELL UNIV, 83- *Concurrent Pos:* Postdoctoral fel, Stanford Univ, 78-82. *Mem:* Biophys Soc; Am Soc Cell Biologists. *Res:* Mechanisms of cyto-skeletal assembly; actin filament regulation. *Mailing Add:* Dept Cell Biol & Anat Cornell Univ Med Col 1300 York Ave New York NY 10021

PARDEE, OTWAY O'MEARA, b Seattle, Wash, June 26, 20; m 46; c 3. COMPUTER SCIENCE, MATHEMATICS. *Educ:* Univ Wash, BS, 41; Stanford Univ, PhD(elec eng), 48. *Prof Exp:* From instr to assoc prof math, 48-69, dir comput ctr, 62-69, prof, 69-86, EMER PROF COMPUT & INFO SCI, SYRACUSE UNIV, 86- *Mem:* Am Math Soc; Am Phys Soc; Math Asn Am; Inst Elec & Electronics Engrs; Asn Comput Mach; Sigma Xi. *Res:* Applied mathematics; computing. *Mailing Add:* Sch Comput & Info Sci Suite 4-116 CST Syracuse Univ Syracuse NY 13244-4100

PARDEE, WILLIAM A(UGUSTUS), b Valdosta, Ga, Sept 12, 14; m 41; c 2. CHEMICAL ENGINEERING. *Educ:* Emory Univ, BS, 36; Yale Univ, DEng, 41. *Prof Exp:* Chem engr, Gulf Res & Develop Co, 41-63, tech adv, Gulf Oil Corp, 64-67, commercial develop & licensing, Eng Dept, Gulf Res & Develop Co, 67-73; PRES, WILLIAM A PARDEE, INC, 73-; ASSOC, MARKETING SERVICES ASSOCS, 76- *Mem:* Am Chem Soc; Am Inst Chem Engrs; Sigma Xi. *Res:* Rubber chemicals; alkylation; vapor phase catalytic reactions; catalyst development; petroleum refinery processes; alkylation of benzene; chlorination of metal oxides; petrochemical process development; process licensing. *Mailing Add:* 402 Duke of Kent Dr Gibsonia PA 15044

PARDEE, WILLIAM DURLEY, b New Haven, Conn, July 7, 29; m 55; c 4. PLANT BREEDING, AGRONOMY. *Educ:* Dartmouth Col, AB, 51; Cornell Univ, PhD(plant breeding), 60. *Prof Exp:* Res assoc plant breeding, Cornell Univ, 60-61; from asst prof to assoc prof crops exten, Univ Ill, 61-66; assoc prof, 66-70, chmn, Dept Plant Breeding & Biomet, 78-88, EXTEN LEADER, DEPT PLANT BREEDING & BIOMET, CORNELL UNIV, 66-, PROF PLANT BREEDING, 70- *Concurrent Pos:* Vis prof, Ore State Univ, 73; past dir, Asn Off Seed Cert Agencies, Am Soc Agron & Crops Sci Soc Am. *Honors & Awards:* Fel, Am Soc Agron; Fel, Crop Sci Soc Am. *Mem:* Am Forage & Grassland Coun; Am Soc Agron; Crop Sci Soc Am; AAAS; Am Inst Biol Sci. *Res:* Seed production; trends in farmer seed usage; factors influencing variety choice by farmers; maximizing forage and grain yields; factors in companion crop competition; seed certification. *Mailing Add:* Dept Plant Breeding Cornell Univ Ithaca NY 14850

PARDEE, WILLIAM JOSEPH, b Davenport, Iowa, Nov 6, 44; div; c 2. THEORETICAL SOLID STATE PHYSICS. *Educ:* Iowa State Univ, BS, 66; Univ Ill, Urbana, PhD(physics), 71. *Prof Exp:* Res assoc physics, Univ Wash, 70-72; vis asst prof, Univ Ore, 72-73; res assoc, Ind Univ, Bloomington, 73-74; mem tech staff theoret physics, 74-80, GROUP MGR, SCI CTR, ROCKWELL INT, 80- *Mem:* Am Phys Soc; Metall Soc; Sigma Xi. *Res:* Ion transport in solids; acoustic emission; temper embrittlement; ultrasonic non-destructive evaluation; fracture and fatigue including statistical mechanics of crack initiation and early growth. *Mailing Add:* 298 Lynnmere Dr Thousand Oaks CA 91360

PARDEN, ROBERT JAMES, b Mason City, Iowa, Apr 17, 22; m 55; c 4. ENGINEERING. *Educ:* Univ Iowa, BS, 47, MS, 51, PhD(eng), 53. *Prof Exp:* Assoc prof eng, Ill Inst Technol, 53-54; PROF ENG, UNIV SANTA CLARA, 54-, DEAN SCH ENG, 55- *Mem:* Am Soc Eng Educ; Am Soc Mech Engrs; Am Inst Indust Engrs. *Res:* Industrial engineering. *Mailing Add:* 19832 Bonnie Ridge Way Saratoga CA 95070

PARDINI, RONALD SHIELDS, b San Francisco, Calif, Nov 10, 38; m 61; c 2. BIOCHEMISTRY, PHARMACOLOGY. *Educ:* Calif State Polytech Col, BS, 61; Univ Ill, PhD(food sci, biochem), 65. *Prof Exp:* Fel biomed, Stanford Res Inst, 67-68; from asst prof to assoc prof, 68-76, PROF BIOCHEM, UNIV NEV, RENO, 76-, CHMN DEPT, 78-; DIR, ALLIE M LEE LAB CANCER RES, 76-, DIR NATURAL PROD LAB, 81- *Concurrent Pos:* Am Diabetes Asn grant, Univ Nev, Reno, 71-; mem adv comt, State Nev Clin Lab, 71-82 & NSF grant, 76-80. *Mem:* Am Chem Soc; Sigma Xi. *Res:* Pesticide toxicology; biochemical pharmacology; phytochemistry; cancer research. *Mailing Add:* Dept Biochem Univ Nev Reno Reno NV 89557

PARDO, RICHARD CLAUDE, b Danville, Ky, Apr 26, 47; m; c 3. NUCLEAR PHYSICS, NUCLEAR ASTROPHYSICS. *Educ:* Univ Louisville, BS, 71; Univ Tex, Austin, PhD(physics), 76. *Prof Exp:* Res assoc nuclear physics, Cyclotron Lab, Mich State Univ, 76-78, asst prof nuclear physics & asst dir, Heavy Ion Lab, 78-79; asst scientist, 79-84, SCIENTIST, ARGONNE NAT LAB, 84- *Mem:* Am Phys Soc. *Res:* Design and construction of electron cyclotron resonance ion source for heavy ions; design and construction of superconducting heavy ion linear accelerator for nuclear research; accelerator control, beam diagnostics, and magnet design. *Mailing Add:* Bldg 203 F-153 Argonne Nat Lab Argonne IL 60439

PARDO, WILLIAM BERMUDEZ, b New York, NY, Jan 14, 28; m 53; c 3. PHYSICS. *Educ:* Hunter Col, AB, 49; Northwestern Univ, PhD(physics), 57. *Prof Exp:* Asst physics, Northwestern Univ, 49-53; tech dir nuclear develop group, Westinghouse Air Brake Co, 53-56; res physicist, Reaction Motors, Inc, 56-57; asst prof, 57-63, ASSOC PROF PHYSICS, UNIV MIAMI, 63- *Concurrent Pos:* Consult, thermonuclear div, Oak Ridge Nat Lab. *Mem:* Am Phys Soc. *Res:* Nuclear physics and reactors; space propulsion; plasma physics. *Mailing Add:* 8530 SW 28th St Miami FL 33155

PARDRIDGE, WILLIAM M, ENDOCRINOLOGY. *Educ:* Pa State Univ, MD, 74. *Prof Exp:* PROF MED, UNIV CALIF, LOS ANGELES. *Res:* Blood brain barrier. *Mailing Add:* 10833 Le Conte Ave West Los Angeles CA 90024

PARDUE, HARRY L, b Big Creek, WVa, May 3, 34; m 57; c 1. ANALYTICAL CHEMISTRY. *Educ:* Marshall Univ, BS, 56, MS, 57; Univ Ill, PhD(chem), 61. *Prof Exp:* From asst prof to assoc prof, 61-70, head dept chem, 83-87, PROF CHEM, PURDUE UNIV, WEST LAFAYETTE, 70- *Honors & Awards:* Award, Am Asn Clin Chemists, 79; Chem Instrumentation Award, Anal Chem Div, Am Chem Soc, 82. *Mem:* Am Chem Soc; Am Asn Clin Chemists. *Res:* Instrumentation for chemical research; chemical kinetics. *Mailing Add:* Dept Chem Purdue Univ 1393 Brown Bldg West Lafayette IN 47907-1393

PARDUE, MARY LOU, b Lexington, Ky, Sept 15, 33. CELL BIOLOGY, DEVELOPMENTAL BIOLOGY. *Educ:* Col William & Mary, BS, 55; Univ Tenn, MS, 59; Yale Univ, PhD(biol), 70. *Hon Degrees:* DSc, Bard Col, 85. *Prof Exp:* assoc prof, 72-80, PROF BIOL, MASS INST TECHNOL, 80- *Concurrent Pos:* Instr molecular cytogenetics, Cold Spring Harbor Lab, 71-78; mem, Genetics Study Sect, NIH, 74-78; mem, Wistar Inst Sci Adv Comt, 76-; instr molecular biol, Drosophila Colo Spring Harbor Lab, 79 & 80; mem, Cellular & Molecular Basis Dis Rev Comt, NIH, 80-84, Nat Adv Gen Med Sci Coun, 84-88; adv comt, Dept Energy, Health & Environ Res, 87; Sci Adv Coun, Abbott Lab, 87-90. *Honors & Awards:* Esther Langer Award, 77. *Mem:* Nat Acad Sci; Am Soc Cell Biol(pres, 85-86); fel AAAS; Genetics Soc Am (vpres, 81-82, pres, 82-83); Am Acad Arts & Sci. *Res:* Structure and function of eukaryotic chromosomes; studies of gene activity during development and the organization of the eukaryotic chromosome; development of insect muscle cell biology of stress responses. *Mailing Add:* Dept Biol Rm 16-717 Mass Inst Technol Cambridge MA 02139

PARDUE, WILLIAM M, b Lexington, Ky, Sept 14, 35; m 57; c 2. METALLURGICAL ENGINEERING. *Educ:* Va Polytech Inst, BS, 57; Ohio State Univ, MSc, 60. *Prof Exp:* Prin metallurgist, 57-62, proj leader nuclear metall, 62-64, asst div chief, 64-66, div chief nuclear metall, Battelle Mem Inst, 66-71, DEPT MGR, OFF NUCLEAR WASTE ISOLATION, BATTELLE COLUMBUS LABS, 78- *Mem:* Am Soc Metals; Am Ceramic Soc; Am Nuclear Soc; Am Soc Testing & Mat. *Res:* Nuclear fuel cycle; thermodynamic studies of numerous materials and combinations of materials; nuclear waste management. *Mailing Add:* 2591 Henthorn Rd Columbus OH 43221

PARE, J R JOCELYN, b Mont Laurier, Que, Nov 26, 59. NATURAL PRODUCTS EXTRACTION, MASS SPECTROMETRY. *Educ:* McGill Univ, BSc, 80; Carleton Univ, PhD(chem), 84. *Prof Exp:* Vis fel biotechnol, Natural Sci & Eng Res Coun, 84-85; sect head aromas & extraction, St Hyacinthe Food Res Ctr, Agr Can, 88-89; HEAD ANAL PROG, RIVER R D ENVIRON TECHNOL CTR, 90- *Concurrent Pos:* Pres & chief exec officer, J R J Pare Estab Chem Ltd, 80-91; managing ed, Spectros: An Inst Jour, 82-90; vpres, Nat Adv Comt, Can Broadcasting Corp, 88-90; res assoc, Fac Grad Studies & Res, Univ Monaton, 89-; chmn, Grant Selection Comt, Agr-Food Res Coun, Govt Que, 89- 90; adj prof, Dept Food Sci & Technol, Univ Laval, 89- & Dept Food Sci, MacDonald Col, McGill Univ, 90-; advan res award, NATO, 90; consult ed, Elsevier Sci Publ, Holland, 91- *Honors & Awards:* Most Promising Young Chemist Award, Int Union Pure & Appl Chem, 90. *Mem:* Am Chem Soc; Royal Soc Chem; Am Soc Mass Spectrometry; Can Soc Chem; NY Acad Sci; affil mem Int Union Pure Appl Chem. *Res:* Development of novel, innovative extraction technology; application of microwave-assisted process (MAP) to the extraction of a whole variety of substances; structure elucidation of novel chemical components obtaned from the above work. *Mailing Add:* Environ Can River Rd Environ Technol Ctr Ottawa ON K1A 0H3 Can

PARE, VICTOR KENNETH, b Oak Park, Ill, Aug 3, 28; m 50; c 2. PHYSICS. *Educ:* Cornell Univ, BS, 51, PhD(eng physics), 58. *Prof Exp:* Assoc develop engr, 51-54, physicist, 58-, RES STAFF MEM, FUSION ENERGY DIV, OAK RIDGE NAT LAB. *Mem:* Am Nuclear Soc; Inst Elec & Electronics Engrs; Am Phys Soc. *Res:* Reactor noise analysis; reactor instrumentation; plasma diagnostics; tokamaks; radiation defects in metals. *Mailing Add:* Instrument & Control Div Oak Ridge Nat Lab PO Box 2008 Oak Ridge TN 37831

PARÉ, WILLIAM PAUL, b Manchester, NH, Aug 25, 33; m 64; c 4. PHYSIOLOGICAL PSYCHOLOGY, PSYCHOSOMATICS. *Educ:* Fordham Univ, BS, 55; Carnegie-Mellon Univ, MS, 57, PhD(exp psychol), 60; Univ Del, PhD(clin psychol), 76. *Prof Exp:* From asst prof to assoc prof psychol, Boston Col, 59-65; asst dir, 65-66, CHIEF, PAVLOVIAN RES LAB, VET ADMIN HOSP, PERRY POINT, MD, 66-, CHIEF, EASTERN RES & DEVELOP OFF, 78- *Concurrent Pos:* Prin investr stress & dis, Vet Admin Res Prog, Perry Point, Md, 65-, res coordr, Med Ctr, 75-78; assoc prof psychol, Lincoln Univ, 66-67; coordr, Health Serv Res & Develop Ctr, Perry Point, Md, 76-78; dir, St John's Counseling Ctr, Havre de Grace, Md, 76-; affil staff, Sch Hyg & Pub Health, Johns Hopkins Univ Sch Med, 76-80; chmn, Fourth Int Conf Ulcer Dis, Tokyo, 80 & Nat Conf, Vet Admin Res Serv, 83 & 85; adj assoc prof, dept psychol, Univ Del, 82-86; exec secy, behav sci rev bd, Vet Admin, Washington, DC, 85-; secy, training adv bd, State Md Dept Ment Hyg, 85-; res assoc prof, Univ Pa, 89-; adv bd, Int Brain-Gut Soc. *Mem:* Fel Am Psychol Asn; Geront Asn; Pavlovian Soc Am (secy-treas, 88-); Psychonomic Soc; Soc Exp Biol & Med. *Res:* Psychological stress and susceptibility to disease; chronic stress effects on coping ability in the aged; animal models of stress-ulcer disease and depression. *Mailing Add:* Pavlovian Res Lab Vet Admin Med Ctr Perry Point MD 21902

PAREES, DAVID MARC, b New York, NY, Jan 1, 50; m 71; c 2. CHROMATOGRAPHY, MASS SPECTROMETRY. *Educ:* State Univ NY, Binghamton, BA, 71; Univ Mass, Amherst, PhD(anal chem), 76. *Prof Exp:* Sr res chemist, Air Prod & Chem Inc, 76-78, prin promotion res chemist, 78-81, sr prin res chemist, 81-85, lead anal chemist, 85-90, TECH LEADER, MASS SPECTROMETRY, AIR PROD & CHEM, INC, 87-, RES ASSOC CORP RES SERV ANAL SERV, 90- *Mem:* Am Soc Mass Spectrometry. *Res:* Chromatographic methods development for trace analysis (nitrosamines) and coal liquids characterization utilizing element-selective gas chromatographic dectors and quadrupole mass spectrometry; general mass spectrometric organic characterization; hydrid magnetic/quadrupole MS, MS/MS and fast atom bombardment. *Mailing Add:* Air Prod & Chemicals Inc 7201 Hamilton Blvd Allentown PA 18195

PAREJKO, RONALD ANTHONY, b Chicago, Ill, Oct 21, 40; m 63; c 4. ENVIRONMENTAL SCIENCES, MICROBIOLOGY. *Educ:* Wis State Univ-Eau Claire, BS, 63; Univ Wis-Madison, MS, 67, PhD(bact), 69. *Prof Exp:* Asst prof, 69-73, assoc prof microbiol, 74-78, PROF BIOL, NORTHERN MICH UNIV, 79- *Concurrent Pos:* Adj prof, Mich State Univ, 80- *Mem:* Fel Am Soc Microbiol; Am Soc Microbiol; Sigma Xi. *Res:* Regulation and kinetics of Klebsiella pneumoniae nitrogenase and chlorinated hydrocarbon pesticide analysis in higher trophic level animals. *Mailing Add:* Dept Biol Northern Mich Univ Marquette MI 49855

PARENCIA, CHARLES R, entomology, for more information see previous edition

PARENT, ANDRE, b Montreal, Quebec, Oct 3, 44; m 70; c 3. NEUROANATOMY, IMMUNOHISTOCHEMISTRY. *Educ:* Univ Montreal, BSc, 67; Univ Laval, PhD(physiol), 70. *Prof Exp:* From asst prof to assoc prof, 71-81, PROF, DEPT ANAT, FAC MED, LAVAL UNIV, 81- *Concurrent Pos:* Fel, Max Planck Inst Brain Res, Frankfurt, Germany, 70-71; vis prof, Dept Phychol, Brain Res Inst, Univ Calif, Los Angeles, 75; sci dir, Neurobiol Res Ctr, Laval Univ & Enfant-Jesus Hosp, 85- *Mem:* Soc Neurosci; AAAS; Am Asn Anatomists; Can Asn Anatomists. *Res:* Anatomy of chemically specified neuronal systems in the brain, particularly the monoanivergic and the cholinergic systems; basal ganglia and the limbic system. *Mailing Add:* Dept Anat Laval Univ Fac Med Ste-Foy PQ G1K 7P4 Can

PARENT, JOSEPH D(OMINIC), b Boston, Mass, Aug 4, 10; m 41; c 8. CHEMISTRY, CHEMICAL ENGINEERING. *Educ:* Cath Univ, BS, 29; Rensselaer Polytech Inst, MS, 31; Ohio State Univ, PhD(chem eng), 33. *Prof Exp:* Chemist, DC Paper Co, 29-30; asst instr chem eng, Ohio State Univ, 32-33; lithographer, US Coast & Geod Surv, 34-35; from instr to asst prof chem, Loyola Univ, Ill, 35-42; assoc prof chem eng, Kans State Col, 42-43; from chem engr to ed dir, Inst Gas Technol, 43-53; res consult, Peoples Gas Light & Coke Co, Chicago, 53-62; energy consult, Inst Gas Technol, 62-85. *Mem:* Am Chem Soc; Am Gas Asn; Am Inst Chem Engrs. *Res:* Natural gas and energy supply problems. *Mailing Add:* 531 Linden Ave Wilmette IL 60091

PARENT, RICHARD ALFRED, b Lynn, Mass, Jan 1, 35; m 58; c 3. ORGANIC CHEMISTRY. *Educ:* Univ Mass, BS, 57; Northeastern Univ, MS, 59; Rutgers Univ, PhD(org chem), 63. *Prof Exp:* Chemist, Am Cyanamid Co, NJ, 59-61, res chemist, 63-69; res chemist, Xerox Corp, 69-71, mgr explor develop, 71-73, mgr color technol, 73, staff specialist toxicol, 73-80; mem staff, Food Drug Res Labs, Inc, 80-85; STAFF MEM, CONSULTOX LTD, 85- *Concurrent Pos:* Mem bd dirs, Delta Labs Inc, 69-, consult toxicol, 75-; consult toxicologist, Independent Union Airline Flight Attend, 77- *Mem:* Am Indust Hyg Asn; Am Chem Soc; NY Acad Sci; AAAS; Am Soc Testing & Mat. *Res:* Acute, sub-acute, chronic toxicology; mutagenesis; carcinogenesis; teratology; inhalation toxicity; intra-tracheal studies; tracer studies; percutaneous absorption; computer readible data bases; ozone; colorants; polymers; smoke toxicity pyrolysis studies; xerographic materials; air sampling and analysis; health hazard evaluation; industrial toxicology. *Mailing Add:* Consultox Ltd PO Box 14082 Baton Rouge LA 70898

PARETSKY, DAVID, b Brooklyn, NY, Nov 15, 18; m 42; c 5. MICROBIOLOGY. *Educ:* City Col New York, BS, 39; Iowa State Col, PhD(physiol bact), 48. *Prof Exp:* Res assoc bact, Agr Exp Sta, Iowa State Col, 42-43 & 46-48; asst prof biol, Rensselaer Polytech Inst, 48-51; from asst prof to assoc prof, 51-59, chmn dept bact, 57-76, PROF BACT, UNIV KANS, 59-, UNIV DISTINGUISHED PROF, 76- *Concurrent Pos:* Instr, Iowa State Col, 48; mem, Kans Adv Lab Comn, 58-; vis prof, Univ Wis, 64-65 & Hadassah Med Sch, Jerusalem, 76; mem microbiol fel rev panel, NIH, 64-68; NSF Coop Col-Sch Sci Prog panelist, 66; Found Microbiol lectr, 71-72; mem cancer res training panel, Nat Cancer Inst, 71-74; av, Walter Reed Army Med Ctr, Inst Biol Res, 82-84. *Mem:* Am Soc Microbiol; Am Soc Biol Chemists; fel AAAS; Am Soc Cell Biol. *Res:* Production of ethanol, 2, 3-butylene glycol and acetoin by corn fermentation; preparation of immune polysaccharides from meningococci; metabolism of amino acids by microorganisms; physiology of rickettsiae; pathobiology; biochemistry of bacteria; regulation of transcription during infection. *Mailing Add:* Dept of Microbiol Univ of Kans Lawrence KS 66044

PARFITT, A(LWYN) M(ICHAEL), b Nottingham, Eng, May 10, 30; m 54; c 3. MEDICINE SCIENCES. *Educ:* Cambridge Univ, preclin, 51, MB BChir, 55; Univ London, clin, 54. *Prof Exp:* Clin assoc prof, 73-79, CLIN PROF MED, SCH MED, UNIV MICH, ANN ARBOR, 79-; DIR, BONE & MINERAL RES LAB & PHYSICIAN, BONE & MINERAL DIV, HENRY FORD HOSP, DETROIT, 71- *Concurrent Pos:* Assoc prof med, Univ Queensland, Australia, 62-71; sci adv bd, Nat Osteoporosis Found. *Honors & Awards:* Murchison scholar, RCP, 56. *Mem:* Am Fedn Clin Res; Am Soc Nephrology; Endocrine Soc; Cent Soc Clin Res; Am Soc Bone & Mineral Res. *Res:* Cellular mechanisms of bone remodelling in health and disease. *Mailing Add:* Henry Ford Hosp 2799 W Grand Blvd Detroit MI 48202

PARGELLIS, ANDREW NASON, b Washington, DC, June 11, 52; m 82; c 2. LIQUID CRYSTALS, PHASE TRANSITIONS. *Educ:* Univ Calif, Irvine, BA, 74; Stevens Inst Technol, MS, 77, PhD(physics), 80. *Prof Exp:* Postdoctoral fel physics, Stevens Inst Technol, 80-83; res dir physics, Glenro, Inc, 83-86; MEM TECH STAFF PHYSICS, AT&T BELL LABS, 86- *Mem:* Am Physics Soc. *Res:* Symmetry breaking phase transitions by observing the evolution and dynamics of defects generated in liquid crystals during phase transitions generated by thermal quenches or rapid compression. *Mailing Add:* AT&T Bell Labs Rm 1C-247 600 Mountain Ave Murray Hill NJ 07974-0636

PARGMAN, DAVID, b New York, NY, July 15, 37; m 63; c 3. PSYCHOLOGY, SPORT & EXERCISE. *Educ:* City Col New York, BSEd, 59; Columbia Univ, MA, 59; NY Univ, PhD(health, phys educ), 66. *Prof Exp:* Lectr health & phys educ, City Col New York, 59-66; from asst prof to assoc prof educ, Boston Univ, 66-77, chmn dept, 77-79; FULL PROF EDUC, FLA STATE UNIV, 77- *Concurrent Pos:* Chmn, Acad Sport Psychol, Am Alliance Health, Phys Educ & Recreation, 83-84. *Mem:* Fel Am Col Sports Med; Am Alliance Health, Phys Educ & Recreation; NAm Soc Psychol Sport & Phys Activity; Int Soc Sport Psychol; NAm Soc Social Sport; fel Asn Advan Appl Sport Psychol; Am Psychol Asn. *Res:* Sport psychology-sociology; various principles, models, theories of human psychology as applied to sport behavior; anxiety and stress as related to motor performance. *Mailing Add:* Dept Phys Educ Fla State Univ Tallahassee FL 32306

PARHAM, ELLEN SPEIDEN, b July 15, 38; m; c 2. NUTRITION. *Educ:* Univ Tenn, PhD(nutrit), 67. *Prof Exp:* PROF & COORDR DIETETIC NUTRIT SCI, DEPT HUMAN & FAMILY RESOURCES, NORTHERN ILL UNIV, 66- *Concurrent Pos:* Coordr grad fac, Dept Human & Family Resources, Northern Ill Univ, 86-88. *Mem:* Am Inst Nutrit; Soc Nutrit Educ; Am Dietetic Asn. *Res:* Nutrition education; obesity. *Mailing Add:* Northern Ill Univ Dept Human & Family Resources DeKalb IL 60115-2864

PARHAM, JAMES CROWDER, II, bio-organic chemistry, medicinal chemistry; deceased, see previous edition for last biography

PARHAM, MARC ELLOUS, b Quincy, Mass, Sept 16, 48; m 74. ORGANIC CHEMISTRY. *Educ:* Duke Univ, BS, 70; Cornell Univ, PhD(chem), 77. *Prof Exp:* Fel, Mass Inst Tech, 77-78; mem staff, Ciba-Geigy Corp, 78-82; MEM STAFF, CLIN ASSAYS DIV, TRAVENOL LABS, INC, 82- *Mem:* Am

Chem Soc; fel Am Inst Chemists. *Res:* Development of affinity membranes to produce new medical devices for blood purification and rapid diagnostic assays. *Mailing Add:* Res Div W R Grace & Co 128 Spring St Lexington MA 02173

PARHAM, MARGARET PAYNE, b Chillicothe, Ohio, Aug 10, 42; m 69; c 2. PHARMACOLOGY. *Educ:* Med Col Va, BS, 64; Cornell Univ, PhD(biochem), 70. *Prof Exp:* Instr biochem, Marymount Col, 69-71; pharmacist, Tanglewood Pharm, 72-74; tech writer toxicol, Cyanamid, 74-75, infor scientist metab dis, 75-76, sr info scientist oncol, 80-84, assoc dir, Prof Pharm Serv, 84-85; mgr, Hosp Oncol Commun & Training, Lederle Labs, 85-88; MGR, LEAD PROD INFO, CYANAMID, 88- *Concurrent Pos:* Consult, 76-80. *Mem:* Sigma Xi; NY Acad Sci. *Res:* Preclinical investigation, clinical evaluation and ongoing usage of the Lederle global product line and major drugs in general development. *Mailing Add:* Med Res Div Cyanamid Pearl River NY 10965

PARHAM, WALTER EDWARD, b Minneapolis, Minn, Jan 21, 30; m 55; c 5. GEOLOGY, MINERALOGY. *Educ:* Univ Ill, BS, 56, MS, 58, PhD(geol), 62. *Prof Exp:* Asst geologist, Ill State Geol Surv, 58-63; asst prof clay mineral, Minn Geol Serv, dept geol & geophys, Univ Minn, Minneapolis, 63-71; assoc prof geol & geophys, Minn Geol Surv, Dept Geol & Geophys, Univ Minn, St Paul, 71-76; phys sci officer, Off Sci & Technol, Agency Int Develop, US Dept State, 76-78; sr analyst, 78-80, PROG MGR FOOD & RENEWABLE RESOURCES, OFF TECHNOL ASSESSMENT, US CONG, 80- *Concurrent Pos:* Univ grant, Univ Minn, Minneapolis, 65-66, grad sch & off int progs grants rock weathering, Hong Kong, 67 & 69; Am Cancer Soc grant health & rock weathering, Hong Kong, 75; Alfred P Sloan grant, PLATO comput pop dynamics appl to environ geol problems; Rockefeller Bros Fund grant, 90; res assoc, Bishop Mus, Honolulu, Hawaii. *Mem:* Fel AAAS; Sigma Xi; Fedn Am Scientists; Int Asn Clay Studies; Clay Minerals Soc; assoc mem Geoscientists Int Develop. *Res:* Tropical rock weathering; clay mineral formation; environmental geology and developing countries; tropical deforestation; restoration of hot, wet tropical degraded lands. *Mailing Add:* Off Technol Assessment US Cong Washington DC 20510

PARIKH, HEMANT BHUPENDRA, b Baroda, India, May 11, 51; US citizen; m 77; c 2. PILOT PLANT PROCESS & PRODUCT DEVELOPMENT, PROCESS EQUIPMENT DESIGN & PROJECT ENGINEERING. *Educ:* MS, Univ Baroda, India, BECHE, 73; Polytech Inst NY, MS, 80; NY Univ, MS, 80; Fairleigh Dickinson Univ, MS, 87; Stevens Inst Technol, chem eng, 88- *Prof Exp:* Process engr, Gujarat State Fertilizer Co, 73-78, design engr, 80-83; res asst, Antonio Ferri Lab, 78-80; process/proj engr, Stepan Co, 83-90; SR PROCESS ENGR, HENKEL CORP, 90- *Mem:* Am Inst Chem Engrs. *Res:* Resolution of recemic alcohols using lipase in organic solvents. *Mailing Add:* 35 Oxford Lane Harriman NY 10926

PARIKH, INDU, US citizen. BIOCHEMISTRY. *Educ:* Univ Zurich, PHD(chem), 65. *Prof Exp:* Asst prof pharmacol & med, Med Sch, Johns Hopkins Univ, 70-75; group leader, 75-77, asst dept head, 77-80, ASSOC DEPT HEAD, WELLCOME RES LABS, BURROUGHS WELLCOME CO, 80- *Concurrent Pos:* Fel, Weimann Inst Sci, 66-68; NIH fel, 68-70; mem bd dir, Alopecia Areata Res Found. *Mem:* Am Soc Biol Chemists; Am Soc Pharmacol & Exp Therapeut; AAAS; NY Acad Sci; Am Chem Soc. *Res:* Steroidal hormone action; targeted drug delivery; affinity chromatogrphy. *Mailing Add:* 2558 Booker Creek Rd Chapel Hill NC 27514

PARIKH, JEKISHAN R, b India, Dec 21, 22; US citizen; m 59; c 3. STEROID CHEMISTRY. *Educ:* Univ Bombay, BSc, 43; Univ Calif, Berkeley, MS, 50, PhD(pharm chem), 53. *Prof Exp:* Prod chemist, Chemo-Pharma Labs, Ltd, India, 43-45, works mgr, 45-48; res assoc, Cobb Chem Labs, Univ Va, 53-55 & Nat Cancer Inst, 55-58; sci officer, Glaxo Labs, Ltd, Eng, 58-60, exec officer, India, 60-63, mgr fine chem factory, 63-64; res scientist chem processing res & develop, 64-75, sr res scientist, 75-79, ASST TO THE VPRES, FINE CHEM DIV, UPJOHN CO, 74-, DIR, INT CHEM MFG, 84- *Honors & Awards:* W E Upjohn Award, Upjohn Co, 74. *Mem:* Am Chem Soc. *Mailing Add:* Upjohn Co Unit 1100 Kalamazoo MI 49001

PARIKH, N(IRANJAN) M, b Godhra, India, Jan 14, 29; US citizen; m 54; c 3. METALLURGY, CERAMICS. *Educ:* Univ Bombay, BSc, 48; Alfred Univ, BS, 49, MS, 50; Mass Inst Technol, ScD(ceramics), 54. *Prof Exp:* Res engr, Sci Lab, Ford Motor Co, Mich, 54-57; sr res off, Govt India, Bombay, Atomic Energy Res Estab, 58-59; res metallurgist, Ill Inst Technol Res Inst, 59-61, sr metallurgist, 61-63, res metallurgist, Metal Sci Adv, 63-64, asst dir metal sci, 64-65, dir metals res, 65-75; planning dir res & develop, 75-76, dir metals technol, 76-80, MANAGING DIR MAT TECHNOL, TECH CTR, AM CAN CO, 80- *Mem:* Fel Am Soc Metals; fel Am Ceramic Soc; fel Brit Inst Metals; fel Am Inst Chem; Am Inst Mining, Metall & Petrol Engrs. *Res:* Metal container fabrication and corrosion; electrochemistry; packaging technology; resource recovery; recycling of aluminum, steel, tin, copper, etc; furnace design; mechanical working and forming of metals; tribology; biomaterials; slag processing; deformation and fracture, extractive metallurgy; research management. *Mailing Add:* 1206 20th St No 9 Huntsville TX 77340

PARIKH, ROHIT JIVANLAL, b Palanpur, Gujarat, India, Nov 20, 36; m 68; c 2. THEORETICAL COMPUTER SCIENCE, LOGIC. *Educ:* Harvard Univ, AB, 57, AM, 59, PhD(math), 62. *Prof Exp:* Instr math, Stanford Univ, 61-63; reader, Panjab Univ, 64-65; lectr, Bristol Univ, 65-67; from assoc prof to prof, Boston Univ, 67-82; DISTINGUISHED PROF COMPUT SCI, BROOKLYN COL, CITY UNIV NEW YORK, 82- *Concurrent Pos:* Res assoc, Calif Inst Technol, 67; vis assoc prof, State Univ NY, Buffalo, 71-72; vis prof, Stanford Univ, 74, Tata Inst Fundamental Res, 71 & 79 & NY Univ, 81; vis scientist, Mass Inst Technol, 77-82. *Mem:* Am Math Soc; Asn Symbolic Logic; Asn Comput Mach; Inst Elec & Electronics Engrs, Comput Sci. *Res:* The use of logic to understand computer science. *Mailing Add:* Dept Comput Sci City Univ NY Grad Sch & Univ Ctr 33 W 42nd St New York NY 10036

PARIKH, SARVABHAUM SOHANLAL, b Tarapur, India, July 31, 35; m 60; c 2. PHYSICAL CHEMISTRY, ANALYTICAL CHEMISTRY. *Educ:* Gujarat Univ, India, BSc, 55; Ohio State Univ, MSc, 60; McGill Univ, PhD(nuclear chem), 66. *Prof Exp:* Apprentice textile chem, Ahmedabad Advan Mills Ltd, Tata Textiles, India, 55-57; res asst, dept chem, Ohio State Univ, 58-59; asst ed chem, 60-62, asst ed nuclear chem, 66-68, assoc ed, 68-69, sr assoc indexer, 69-73, SR ED, CHEM ABSTR SERV, AM CHEM SOC, OHIO STATE UNIV, 73- *Concurrent Pos:* Res assoc, Radiochem Lab, McGill Univ, 66. *Mem:* Sr mem Am Chem Soc. *Res:* Qualitative and quantitative chemical analysis; nuclear properties and reactions; fission and spallation; radiochemical separations; chemistry of nuclear reactor fuels; indexing and editing of nuclear phenomena and technology; information and literature chemistry. *Mailing Add:* Chem Abstr Serv Ohio State Univ Columbus OH 43210-0012

PARIS, CLARK DAVIS, b Delaware Co, Iowa, Aug 10, 11; m 72. HORTICULTURE. *Educ:* Univ Iowa, BS, 36; Iowa State Univ, MS, 39; Mich State Univ, PhD(hort), 56. *Prof Exp:* Horticulturist, Iowa Ment Health Hosp, 39-44; asst gardener, San Diego County Hosp, 44-45; plant breeder, W Atlee Burpee Co, 45-52 & Pan-Am Seeds, Inc, 52-54; res instr hort, Mich State Univ, 56-61, ed & proj leader food sci, 61-78; RETIRED. *Concurrent Pos:* Bibliogr, Mich State Univ Libr, 68-78. *Honors & Awards:* Laurie Award, Am Soc Hort Sci, 59. *Res:* Flower color genetics; general genetics; plant breeding; photography; world food problems. *Mailing Add:* Hacienda de Valencia Space 274 201 S Greenfield Rd Mesa AZ 85206

PARIS, DAVID LEONARD, b Glasgow, Scotland, Sept 4, 44; Can & UK citizen; m. SPORTS MEDICINE, ATHLETIC THERAPY. *Educ:* York Univ, Toronto, BA, 74; Ind State Univ, Terre Haute, MA, 75; Nat Athletic Trainers Asn, Athletic Trainer(cert), 75; Univ Ore, Eugene, PhD(phys educ & human anat), 80. *Prof Exp:* Head athlete trainer & lectr sports med, Oberlin Col, Ohio, 75-77; head athletic therapist- asst prof sports med & human anat, Univ NB, 80-83; SPORTS MED COORDR, ASST PROF SPORTS MED & HUMAN ANAT, CONCORDIA UNIV, MONTREAL, 83- *Concurrent Pos:* Athletic Therapist Team Canada, XXI Olympiad, Montreal, 76, XI Maccabiah Games, Tel Aviv, 81, World Student Games, Edmonton, 83 & XII Maccabiah Games, Tel Aviv, 85; athletic therapist, Montreal Supra Prof Soccer Club, 88-90, XIII Maccabiah Games, Tel Aviv, 89, Can Nat Soccer teams, 85- *Mem:* Nat Athletic Trainers Asn; Can Athletic Therapists Asn; Can Acad Sports Med. *Res:* Sports medicine; effects of ankle prophyloxis on range of motion and performance over periods of extended activity; ankle prophylaxis and performance. *Mailing Add:* Dept Exercise Sci Concordia Univ 7141 Sherbrooke St W Montreal PQ H4B 1R6 Can

PARIS, DEMETRIUS T, b Stavroupolis, Greece, Sept 27, 28; US citizen; m 52; c 2. ELECTRICAL ENGINEERING. *Educ:* Miss State Univ, BSEE, 51; Ga Inst Technol, MSEE, 58, PhD(elec eng), 62. *Prof Exp:* Design engr, Westinghouse Elec Corp, 52-58; sr engr, Lockheed-Ga Co, 58-59; from asst prof to assoc prof 59-66, PROF ELEC ENG, GA INST TECHNOL, 66-, DIR SCH ELEC ENG, 69- *Concurrent Pos:* Consult, Lockheed-Ga Co, 62-, Sci Atlanta, Inc, 65-, Control Data Corp, 85- *Honors & Awards:* Pettit Award, Southcon, 85. *Mem:* Fel Inst Elec & Electronics Engrs; Am Soc Eng Educ. *Res:* Electromagnetics. *Mailing Add:* Off Pres Ga Tech Atlanta GA 30332

PARIS, DORIS FORT, b Roanoke, Va, July 3, 24; m 45, 67; c 2. BIOCHEMISTRY, MICROBIOLOGY. *Educ:* Univ Va, BS, 46; Univ Ga, MS, 66. *Prof Exp:* Med technologist toxicol, Univ Va Hosp, Charlottesville, 43-45; res chemist biochem, Environ Res Lab, US Environ Protection Agency, 66-86; RETIRED. *Mem:* Am Soc Microbiol; Am Chem Soc. *Res:* Microbial transformation of toxic substances in aquatic systems; biochemistry of algae and bacteria. *Mailing Add:* 145 Meadowview Rd Athens GA 30606

PARIS, JEAN PHILIP, b Buffalo, NY, Dec 2, 35; m 56; c 4. ANALYTICAL CHEMISTRY, PHYSICAL CHEMISTRY. *Educ:* Univ Mich, BS, 57; Purdue Univ, PhD(chem), 60. *Prof Exp:* Res chemist, Polychem Dept, E I du Pont de Nemours & Co, Inc, 60-62; lectr chem, Juniata Col, 62; res chemist, Radiation Physics Lab, E I du Pont de Nemours & Co, Inc, 62-65 & Elastomers Dept, 65-69; PRES, THERMOELEC UNLIMITED, INC, 69- *Concurrent Pos:* Res chemist, Am Cyanamid Co, Conn, 58 & E I du Pont de Nemours & Co, Inc, 59. *Mem:* Am Chem Soc; Soc Appl Spectros. *Res:* Thermoelectric instrumentation; electronic spectroscopy; photochromism; reaction kinetics. *Mailing Add:* 1202 Harrison Ave Holly Oak Terr Wilmington DE 19809-1910

PARIS, OSCAR HALL, b Greensboro, NC, Mar 22, 31; c 3. RADIATION ECOLOGY. *Educ:* Univ NC, AB, 53, MA, 56; Univ Calif, PhD(zool), 60. *Prof Exp:* Asst prof zool, Univ NC, 60-62; from asst prof to assoc prof, Univ Calif, Berkeley, 62-71; prof zool & head dept zool & physiol, Univ Wyo, 71-76, dir, Jackson Hole Biol Res Sta, 73-76; ENVIRON SCIENTIST & ADMIN JUDGE ATOMIC SAFETY & LICENSING BD PANEL, US NUCLEAR REGULATORY COMN, 76- *Concurrent Pos:* Turtox Scholar Biol, 58-59; NSF fel, 60. *Mem:* Fel AAAS; Ecol Soc Am; Brit Ecol Soc; Am Inst Biol Sci; Planetary Soc; Sigma Xi. *Res:* Population ecology; invertebrate ecology. *Mailing Add:* Atomic Safety & Licensing Bd Panel US Nuclear Regulatory Comn Washington DC 20555

PARISEK, CHARLES BRUCE, b Hibernia, NJ, Nov 4, 31; m 64; c 3. ORGANIC CHEMISTRY, POLYMER CHEMISTRY. *Educ:* Rutgers Univ, New Brunswick, BSc, 52; Univ Kans, PhD(org chem), 62. *Prof Exp:* Asst ed, Chem Abstr Serv, Am Chem Soc, 58-59; res chemist, Reaction Motors Div, Thiokol Chem Corp, 62-65; sr res chemist, Tex-US Chem Co, 65-69; sr res scientist latex technol & med prod, Corp Res Ctr, Int Paper Co, Tuxedo, NY, 69-77; sr res scientist, Johnson & Johnson, 78; sr consult, Booz, Allen & Hamilton, 78-80; MGR SUPPLIES & CHEM, MAGNETOGRAPHY, AM INT PRINTER SYSTS, 80- *Concurrent Pos:* Lectr, Ohio State Univ, 58-59; instr, Park Col, 59-60 & Fairleigh Dickinson Univ, 69- *Mem:* Am Chem Soc; The Chem Soc; Sigma Xi. *Res:* Mechanism of organophosphorus compounds decomposition; synthesis of organic and inorganic polymers; latex technology; electrooptical properties of organic compounds; radiation processing. *Mailing Add:* 27 Sherman Pl Morristown NJ 07960

PARISER, HARRY, b Newark, NJ, Jan 19, 11; m 45; c 2. DERMATOLOGY, SYPHILOLOGY. *Educ:* Univ Pa, BA, 31, MD, 35, DSc(med), 40; Am Bd Dermat, dipl, 41. *Prof Exp:* Instr dermat & syphil, Sch Med, Univ Pa, 37-42; venereal dis control officer, USPHS, 42-46, consult dermat & syphil, 46-73; prof microbiol & cell biol, 73-87, assoc prof med, 75-85, PROF MED, EASTERN VA MED SCH, 85- *Concurrent Pos:* Consult dermat & syphil, US Naval Hosp, Portsmouth, Va, 46-58, Vet Admin Hosp, Kecoughtan, Va, 46-66, Norfolk Gen Hosp & DePaul Hosp, Norfolk, Va, 46-; pvt pract dermat, 46-; spec consult, USPHS, 60-73; mem, Nat Comn Venereal Dis; assoc ed, J Am Venereal Dis Asn, 73-78. *Honors & Awards:* Thomas Parras Award, Am Venereal Dis Asn, 80. *Mem:* AMA; Am Acad Dermat; Am Venereal Dis Asn (pres, 70-71); Am Pub Health Asn. *Res:* Author of approximately 40 publications and contributing author in three textbooks on various aspects of sexually transmitted diseases. *Mailing Add:* 601 Med Tower Bldg Norfolk VA 23507

PARISER, RUDOLPH, b Harbin, China, Dec 8, 23; nat US; m 72. PHYSICAL CHEMISTRY. *Educ:* Univ Calif, BS, 44; Univ Minn, PhD(phys chem), 50. *Prof Exp:* Res chemist, Org Chem Dept, E I du Pont de Nemours & Co, Inc, 50-53, res supvr, 53-57, res supvr, Elastomer Chem Dept, 57-59, div head, 59-63, asst lab dir, 63-67, lab dir, 67-70, dir explor res & mgr res & develop, 70-72, mgr mkt res & develop, 72-74, dir explor res, 74-76, dir pioneering res, elastomer chem dept, 76-79, res dir, polymer prod dept, 79-81, dir polymer sci, 81-86, dir advan mat sci, Cent Res & Develop Dept, 86-89; PRES, R PARISER & CO, INC, 89- *Concurrent Pos:* Assoc ed, J Chem Physics, 66-79, chem phys letters, 67-70, Du Pont Innovation, 69-75, adv bd, J Polymer Sci, 80-89, NRC New Polymeric Mat, 85-, mem comt chem sci, 79-82, co-chmn ad hoc panel polymer sci & eng, 79-81; mem, Int Union Pure & Appl Chem & Chem Res Appl World Fuels, 82-; NRC co-chmn, panel high performance composites, 84, adv bd, Mat Div, Nat Sci Found, 86-89; NRC, Comt Mat Sci & Eng, 86-89; adv bd, Dept Mech Eng, MIT, 86-, Dept Mat Sci, Univ Fla, 87- , Dept Mat Sci, Lehigh Univ, 88-, Dept Chem Eng, Univ Wisc, 87-, Sch Dent Med, Univ Pa, 90- *Mem:* AAAS; Am Chem Soc; Am Phys Soc. *Res:* Research administration; polymer science; quantum chemistry; over 40 publications; materials science and engineering. *Mailing Add:* 851 Old Public Rd Hockessin DE 19707

PARISH, CURTIS LEE, b Ellensburg, Wash, Apr 10, 37; m 72; c 10. PLANT PATHOLOGY, BIOCHEMISTRY. *Educ:* Wash State Univ, BS, 59; Univ Ariz, PhD(plant path), 65. *Prof Exp:* PLANT PATHOLOGIST TREE FRUIT VIROL, USDA-AGR RES SERV, 65- *Concurrent Pos:* Consult. *Mem:* Am Phytopath Soc. *Res:* Tree fruit virus diseases; virology; serology; physiology of parasitism. *Mailing Add:* Fruit Res Lab 1104 N Western Ave Wenatchee WA 98801

PARISH, DARRELL JOE, b Beebe, Ark, Dec 20, 34; m 60; c 2. ORGANIC POLYMER CHEMISTRY. *Educ:* Univ Louisville, BS, 56, PhD(chem), 60. *Prof Exp:* Asst, AEC, Louisville, 56-60; res chemist, 60-72, staff scientist, Film Dept, 72-77, SR RES CHEMIST & RES ASSOC, ELECTRONICS DEPT, E I DU PONT DE NEMOURS & CO, INC, 77- *Res:* Radiation and polymer chemistry; research and development of polyimide and polyester film forming polymers. *Mailing Add:* E I du Pont de Nemours & Co Inc PO Box 89 Circleville OH 43113

PARISH, EDWARD JAMES, b San Marcos, Tex, Sept 23, 43; c 1. STEROID & LIPID CHEMISTRY, DRUG & ISOTOPIC LABEL SYNTHESES. *Educ:* Southwest Tex State Univ, BS, 67; Sam Houston State Univ, MA, 70; Miss State Univ, PhD(org chem & biochem), 74. *Prof Exp:* NIH res fel steroid chem, biochem dept, Rice Univ, 74-76, res scientist, 76-81; ASSOC PROF ORG CHEM, AUBURN UNIV, 81 - *Concurrent Pos:* Vis scientist, Western Regional USDA Res Lab, Berkeley, 84, 85, & 86, R B Russell USDA Res Lab, Athens, Ga, 87 & 88. *Mem:* Am Chem Soc; Am Oil Chemists' Soc. *Res:* Chemical synthesis of biologically active steroids and terpenoids; developing new and novel syntheses of steroids and related natural products with possible value in metabolic and pharmaceutical investigations; drug synthesis. *Mailing Add:* Dept Chem Auburn Univ Auburn AL 36849-5312

PARISH, HARLIE ALBERT, JR, b Great Bend, Kans, Oct 29, 40; m 64; c 1. ORGANIC CHEMISTRY, MEDICINAL CHEMISTRY. *Educ:* Ft Hays State Univ, BS, 62, MS, 65; Univ Mo, PhD(chem), 69. *Prof Exp:* Res chemist, Brown & Williamson Tobacco Corp, 68-73; res assoc biochem, Sch Med, Univ Louisville, 73-77; ASST PROF RES, RHODES COL, 77- *Concurrent Pos:* Consult, Molecular Design Int, 77-, vpres, 82- *Mem:* AAAS; Am Chem Soc. *Res:* Organic synthesis; medicinal chemistry of drugs for topical treatment of skin diseases; synthesis of experimental drugs. *Mailing Add:* 4293 Beechcliff Lane Memphis TN 38128

PARISH, JEFFREY LEE, b Sturgis, Mich, May 23, 45; m 71; c 2. PLASMA PHYSICS. *Educ:* Purdue Univ, BS; Univ Ill, MS, 71; Univ Iowa, PhD(physics), 80. *Prof Exp:* RES ASSOC PROF, UTAH STATE UNIV, 80- *Mem:* Am Math Soc; Am Geophys Union. *Mailing Add:* Parish Res Box 420 Howe IN 46746

PARISH, RICHARD LEE, b Kansas City, Mo, May 31, 45; m 68; c 2. MACHINE DESIGN, ENGINEERING MANAGEMENT. *Educ:* Univ Mo-Columbia, BS, 67, MS, 68, PhD(agr eng), 70. *Prof Exp:* From asst prof to assoc prof agr eng, Univ Ark, Fayetteville, 69-74; sr proj leader, 74-80, mgr, Mech Res & Develop, O M Scott & Sons Co, 80-83; assoc prof, 83-88, PROF AGR ENG, LA STATE UNIV, 88- *Concurrent Pos:* Consult prod develop & eval, expert witness testimony, 74, 86- *Honors & Awards:* O M Scott Res Dirs Award. *Mem:* Am Soc Agr Engrs; Am Soc Hort Sci. *Res:* Farm machinery development and testing; design and development of lawn and garden equipment; engineering management. *Mailing Add:* Dept Agr Eng La State Univ Baton Rouge LA 70803-4505

PARISH, ROGER COOK, b Utica, NY, Jan 20, 40; m 62; c 2. ORGANIC CHEMISTRY, MEDICINAL CHEMISTRY. *Educ:* Utica Col, BA, 60; Univ Chicago, MS, 64, PhD(org chem), 65. *Prof Exp:* Res chemist, E I du Pont de Nemours & Co, Inc, 65-67; sr med chemist, Smith Kline & French Labs, 67-70, assoc dir chem, 71, mgr chem res, 71-78, mgr res opers, 78-89, DIR RES TECHNOL & PLANNING, SMITH KLINE BEECHAM ANIMAL HEALTH 89-, DIR PLANNING & TECH ASSESSMENT. *Mem:* Am Chem Soc; NY Acad Sci; Am Soc Microbiol; Sigma Xi; AAAS. *Res:* Synthetic organic chemistry; antiparasitic chemotherapy; animal nutrition. *Mailing Add:* Smith Kline Corp 1600 Paoli Pike West Chester PA 19380

PARISH, TRUEMAN DAVIS, b Cincinnati, Ohio, Mar 15, 39; m 64; c 2. CHEMICAL ENGINEERING. *Educ:* Univ Mich, BSE, 62; Mass Inst Technol, SM, 63; Rice Univ, PhD(chem eng), 67. *Prof Exp:* Chem engr, Procter & Gamble Co, 63-64; chem engr, Eastman Kodak Co, 67-69, supvry engr, Advan Eng Technol Group, 69-71, res assoc, Process Design & Eval Res Lab, 71-76, coordr Advan Process Technol Group, 76-78, dir eng res, Eastman Chem Div, Eastman Kodak Co, 78-90, DIR ADMIN & TECH SERV DIV, RES LABS, EASTMAN CHEMICAL CO, 90- *Mem:* Am Inst Chem Engrs; Nat Soc Prof Engrs; Sigma Xi. *Res:* Chemical reactor design, mixing and optimization; venture analysis. *Mailing Add:* 3737 Arrowhead Trail Kingsport TN 37664

PARISH, WILLIAM R, b Grinnell, Iowa, Sept 6, 20; m 43; c 2. ELECTRICAL ENGINEERING. *Educ:* Iowa State Col, BS, 44; Univ Idaho, MS, 52. *Prof Exp:* From instr to prof, 47-83, EMER PROF ELEC ENG, UNIV IDAHO, 83- *Concurrent Pos:* Engr, Gen Elec Co, 59-60; sr specialist engr, Boeing Aerospace Co, 77-78. *Mem:* Am Soc Eng Educ; Inst Elec & Electronics Engrs. *Res:* Radio communication systems; radio wave propagation. *Mailing Add:* Dept Elec Eng Univ Idaho Moscow ID 83843

PARISI, GEORGE I, b Newark, NJ, Feb 6, 31; m 54; c 3. PHYSICAL CHEMISTRY. *Educ:* Newark Col Eng, BS, 51, MS, 53; Rutgers Univ, PhD(phys chem), 66. *Prof Exp:* Develop engr, Celanese Corp Am, 51-55; mem tech staff, 55-87, ADJ PROF AT LARGE CHEM, PHYSICS & MATH, BELL LABS, INC, 87- *Mem:* Am Chem Soc; Electrochem Soc; Am Inst Chem Eng; fel Am Inst Chem. *Res:* Rheology of polymers; design and development of high reliability electronic components; Mossbauer effect studies; applied research in thin film technology, including sputtering, vapor deposition, photolithography and anodization; silicon integrated circuit processing and quality control. *Mailing Add:* 379 Mountain Ave Murray Hill NJ 07974-2702

PARISI, JOSEPH THOMAS, b Chicago, Ill, Apr 28, 34; m 63; c 1. MICROBIOLOGY. *Educ:* Loyola Univ Chicago, BS, 56; Ohio State Univ, MSc, 58, PhD(microbiol), 62. *Prof Exp:* Asst prof biol, Duquesne Univ, 62-65; from asst prof to assoc prof, 65-75, PROF MICROBIOL, SCH MED, UNIV MO-COLUMBIA, 75- *Concurrent Pos:* NIH res grant microbiol, Univ Mo-Columbia, 67-69. *Mem:* Am Soc Microbiol; Sigma Xi. *Res:* Microbial genetics; epidemiology. *Mailing Add:* Microbiol M610 Med Sci Univ Mo Columbia MO 65212

PARISSE, ANTHONY JOHN, b Brooklyn, NY, Oct 25, 36; m 59; c 2. COSMETIC CHEMISTRY, PHARMACEUTICAL CHEMISTRY. *Educ:* St John's Univ, BS, 58; Rutgers Univ, MBA, 73. *Prof Exp:* Res chemist pharmaceut, Whitehall Labs, Am Home Prod, 59-61; sr scientist, Johnson & Johnson, 61-65; DIR PROD EVAL RES, CARTER-WALLACE, INC, CRANBURY, 65- *Mem:* Soc Cosmetic Chemists; Am Chem Soc; Am Pharmaceut Soc; Acad Pharmaceut Sci. *Res:* Inhibition of perspiration; structure, chemical composition and chemical reactivity of human hair; inhalation of airborne particles; dermal solar protection and moisturization. *Mailing Add:* Carter Wallace Inc PO Box 1 Half Acre Rd Cranbury NJ 08512

PARIZA, MICHAEL WILLARD, b Waukesha, Wis, Mar 10, 43; m 67; c 3. DIET & CANCER. *Educ:* Univ Wis-Madison, BS, 67; Kans State Univ, MS, 69, PhD(microbiol), 73. *Prof Exp:* Trainee oncol, McArdle Lab Cancer Res, 73-76; from asst prof to assoc prof, 76-84, assoc chmn microbiol & toxicol, 81-82, PROF FOOD MICROBIOL & TOXICOL, UNIV WIS-MADISON, 84-, CHMN DEPT, 82-; DIR, FOOD RES INST, 86- *Concurrent Pos:* Chmn, Food Microbiol Div, Am Soc Microbiol & Coun AgrSci, 85; co-organizer, Role of Essential Nutrients in Carcinogenesis, Nat Cancer Inst; assoc dir, Food Res Inst, 85-86. *Mem:* Am Asn Cancer Res; Am Soc Microbiol; Inst Food Technologists; Fedn Am Soc Exp Biol; Am Inst Nutrit. *Res:* Role of diet and nutrition in carcinogenesis. *Mailing Add:* Food Res Inst Dept Food Microbiol & Toxicol Univ Wis 1925 Willow Dr Madison WI 53706

PARIZA, RICHARD JAMES, b Waukesha, WI, Mar 5, 46; c 1. RESEARCH MANAGEMENT. *Educ:* Univ Wis, Milwaukee, BS, 76; Purdue Univ, PhD(chem), 83. *Prof Exp:* Founder & pres, Willow Brook Labs, Waukesha, WI, 68-75; MGR, CHEM SERVS, PHARMACEUT PRODS DIV, ABBOTT LABS, 83- *Mem:* Am Chem Soc; Sigma Xi. *Res:* Synthesis of pharmaceuticals and intermediates for medicinal research; developing new methodology and reagents for organic chemical synthesis. *Mailing Add:* Abbott Labs D-47B Abbott Park North Chicago IL 60064

PARIZEK, ELDON JOSEPH, b Iowa City, Iowa, Apr 30, 20; m 44; c 4. GEOLOGY. *Educ:* Univ Iowa, BS, 42, MS, 47, PhD(geol), 49. *Prof Exp:* Asst geologist, Iowa Geol Surv, 46-47; instr geol, Univ Iowa, 47-49; from asst prof to assoc prof, Univ Ga, 49-57; from asst prof to assoc prof, 57-63, dean, Col Arts & Sci, 78-85, PROF GEOL, UNIV MO-KANSAS CITY, 63-, CHMN DEPT GEOSCI, 74-,. *Concurrent Pos:* Asst geologist, State Geol Surv, Mo, 48-49 & 57-, Ga, 51-57. *Mem:* Geol Soc Am; Asn Am Geogrs; Nat Asn Geol Teachers. *Res:* General geology of southeast Missouri; metamorphic and igneous geology of east central Georgia Piedmont; geomorphology of Georgia Piedmont; structural geology; volume changes in shales; stratigraphy of northwest Missouri; geomorphology of northwest Missouri and eastern Kansas. *Mailing Add:* 6913 W100 Overland Park KS 66212

PARIZEK, RICHARD RUDOLPH, b Stafford Springs, Conn, Aug 15, 34; m 61; c 1. HYDROGEOLOGY, ENVIRONMENTAL GEOLOGY. *Educ:* Univ Conn, BA, 56; Univ Ill, MS, 60, PhD(geol), 61. *Prof Exp:* Res asst, Ill State Geol Surv, 56-61; assoc prof, 61-71, prof hydrogeol, 71-77, PROF GEOL, PA STATE UNIV, UNIVERSITY PARK, 78-, STAFF GEOLOGIST, 61- *Concurrent Pos:* Consult, Off Water Resources Res, 64-, Pa Coal Res Bd, 65- & Dept Health, Educ & Welfare, 65-; consult, Pa Dept Environ Resources, 68- & Pa Dept Transp, 74-; Pa rep, Pollution from Land Use Activities Ref Group, Int Joint Comn, 73-; examiner, Pa Civil Serv Comn, 74- *Mem:* Geol Soc Am; AAAS; Am Geophys Union; Sigma Xi. *Res:* Environmental geology with special concern for land use, water supply, waste disposal and coal mining; occurrence and groundwater resource evaluation in carbonate and fractured rock terrains; pleistocene geology. *Mailing Add:* 751 McKee St State Col PA 16803

PARK, B J, b Seoul, Korea, Feb 28, 34; US citizen; m 58. MECHANICAL BEHAVIOUR OF TEXTILE ASSEMBLAGE, SAFETY PROPERTIES OF CONSUMER PRODUCTS. *Educ:* RI Sch Design, BS, 58; Mass Inst Technol, MS, 61; Leeds Univ, PhD(textile eng), 66. *Prof Exp:* vpres, Consumer Testing Labs, Inc, 66-84, sr vpres, 84-86; PRES, MERCHANDISE TESTING LABS, INC, 86- *Concurrent Pos:* Res asst, Mass Inst Technol, 61-63; res fel, Leeds Univ, Eng, 64-66. *Mem:* Textile Inst; Am Asn Textile Chemists & Colorists; Am Soc Testing & Mat; Am Asn Textile Technol; Indust Fabrics Asn Int; Nat Fire Protection Asn. *Res:* Mechanical behavior of textile assemblage; safety and consumer satisfaction properties of consumer products. *Mailing Add:* Merchandise Testing Labs Inc 244 Liberty St Sharon MA 02041

PARK, CHAN H, b Seoul, Korea, Aug 26, 37; US citizen; m 69; c 2. NUCLEAR MEDICINE, RADIOLOGY. *Educ:* Yonsei Univ, MD, 64. *Prof Exp:* From instr to asst prof nuclear med, Thomas Jefferson Univ Hosp, 70-74; asst prof, Penn State Univ Hershey Med Ctr, 74-75; assoc prof, 75-79, PROF NUCLEAR MED, THOMAS JEFFERSON UNIV HOSP, 79- *Mem:* AMA; Soc Nuclear Med; Am Col Radiol; Radiol Soc N Am. *Res:* Treatment of malignant tumors with an internal radiation dose. *Mailing Add:* Div Nuclear Med Thomas Jefferson Univ Hosp Philadelphia PA 19107

PARK, CHAN MO, b Seoul, Korea, Apr 3, 35; m 63; c 3. DIGITAL IMAGE PROCESSING SYSTEM SIMULATION. *Educ:* Seoul Nat Univ, BS, 58; Univ Md, MS, 64, PhD(chem eng), 69. *Prof Exp:* Res assoc comput sci, Univ Md, College Park, 64, res programmer, 64-68, fel, 68-69, asst prof, 69-72; assoc prof comput sci, Korea Advan Inst Sci, 73-76; sr res scientist, Nat Biomed Res Found, 76-79; assoc prof, 79-82, PROF & CHMN DEPT COMPUT SCI, CATHOLIC UNIV AM, 82- *Concurrent Pos:* Consult, Comput Ctr, Korea Inst Sci & Technol, 73-76; resident prof systs mgt, Korea Ctrs, Univ Southern Calif, 75-76; vis lectr comput sci, Univ Md, 76-; fac consult, Nat Bur Standards, 83-85; vis prof, Comput Info Systs, Boston Univ Overseas Prog, WGer, 85-86. *Honors & Awards:* Mil Oper Res Soc Award, Korea, 77. *Mem:* Inst Elec & Electronics Comput Soc; Asn Comput Mach; Pattern Recognition Soc; Korean Scientists & Engrs Asn Am (vpres, 77-78, pres, 84-85); Sigma Xi; Soc Comput Simulation. *Res:* Application of computers to engineering and biomedical problems; picture processing; computer systems and digital simulation; computer vision. *Mailing Add:* Dept Comput Sci Cath Univ Am Washington DC 20064

PARK, CHARLES RAWLINSON, b Baltimore, Md, Mar 2, 16; m 53; c 1. PHYSIOLOGY, BIOCHEMISTRY. *Educ:* Harvard Univ, AB, 37; Johns Hopkins Univ, MD, 41. *Prof Exp:* Intern, Johns Hopkins Hosp, 41-42; from med asst resident to resident, Peter Bent Brigham Hosp, 42-44; dir, Howard Hughes Med Inst, 68-81, prof physiol & head dept, 52-84, EMER PROF PHYSIOL, VANDERBILT UNIV, 84- *Concurrent Pos:* USPHS fel biochem, Sch Med, Wash Univ, 47-49, Welch fel internal med, 49-52. *Honors & Awards:* Banting Medal for Res in Diabetes, 78. *Mem:* Nat Acad Sci; Am Soc Clin Invest (vpres, 60- 61); Am Soc Biol Chem. *Res:* Hormonal effects on metabolism. *Mailing Add:* Dept Med Physiol & Biophys Vanderbilt Sch Med Nashville TN 37232

PARK, CHONG JIN, b Korea, Apr 2, 36; m 64; c 3. MATHEMATICAL STATISTICS. *Educ:* Univ Wash, BS, 61, BA, 62; Kans State Univ, MS, 63; Univ Wis-Madison, MS, 64, PhD(statist), 68. *Prof Exp:* Instr statist, Kans State Univ, 66-67; asst prof, Univ Nebr, Lincoln, 67-71; vis asst prof, Univ Wis-Madison, 71-72; assoc prof, 72-78, PROF MATH SCI, SAN DIEGO STATE UNIV, 78- *Concurrent Pos:* Consult, Kelco, San Diego, Calif, 73-; res fel, US Air Force, Edwards, Calif, 80; researcher & math statistician, Naval Oceanic Syst Command, San Diego, Calif, 85- *Mem:* Inst Math Statist; Am Statist Asn; Bernoulli Soc Math Statist & Probabilty. *Res:* Nonparametric statistics via combinatorial methods; spectral analysis of time series data; random allocations. *Mailing Add:* Dept of Math San Diego State Univ San Diego CA 92182

PARK, CHUL, b Taegu, Korea, June 8, 34; US citizen; m 62; c 3. AERONAUTICAL ENGINEERING. *Educ:* Seoul Nat Univ, BS, 57, MS, 60; Univ London, PhD(aeronaut eng), 64. *Prof Exp:* Instr aerodyn, Korean Air Force Acad, 58-61; res assoc magnetohydrodyn, 64-67, RES SCIENTIST FLUID MECH, AMES RES CTR, NASA, 67- *Concurrent Pos:* Vis engr mech eng, Mass Inst Technol, 71-72. *Mem:* Am Inst Aeronaut & Astronaut. *Res:* Fluid mechanics and high temperature thermophysics related to space travel. *Mailing Add:* Ames Res Ctr NASA MS 230-2 Moffett Field CA 94035

PARK, CHULL, b Seoul, Korea, Feb 29, 32; m 64; c 2. MATHEMATICS, AERONAUTICAL ENGINEERING. *Educ:* Seoul Nat Univ, BS, 59; Southern Ill Univ, MS, 64; Univ Minn, PhD(math), 68. *Prof Exp:* Teacher, Sook Myung High Sch, 59-62; asst prof math, Bemidji State Col, 67-68; asst prof math, 68-74, assoc prof, 75-81, PROF MATH & STATIST, MIAMI UNIV, 81- *Mem:* Am Math Soc; Math Asn Am; Inst Math Statist. *Res:* Gaussian stochastic processes including Wiener process (Brownian motion process); absorption probabilities of Wiener process, stochastic intergrals, stochastic differential equations; laws of iterated logarithms. *Mailing Add:* Dept Math Miami Univ Oxford OH 45056

PARK, CHUNG GUN, b Seoul, Korea, Jan 11, 39; US citizen; m 67. ATMOSPHERIC PHYSICS. *Educ:* Stanford Univ, BS, 61, MS, 64, PhD(elec eng), 70. *Prof Exp:* Engr semiconductor, Hewlett Packard Co, 61-65; res engr, 65-66, res assoc, 70-76, SR RES ASSOC SPACE SCI, STANFORD UNIV, 76- *Concurrent Pos:* Secy, Subcomn Planetary & Space Prob of Atmospheric Elec, 75-; assoc ed, J Geophys Res, 76-; mem, US Nat Comt & Exec Comt, Int Sci Radio Union, 77-; mem, Comt Atmospheric Elec, Am Geophys Union, 77-; mem, US Panel Mid Atmospheric Prog, 78- *Honors & Awards:* Antarctic Serv Medal, NSF, 76. *Mem:* Am Geophys Union; AAAS; Int Asn Geomagnetism & Aeronomy. *Res:* Space science and atmospheric electricity. *Mailing Add:* 655 Island Pl Redwood City CA 94065

PARK, CHUNG HO, b Seoul, Korea, July 4, 36; US citizen; c 2. ORGANIC CHEMISTRY, MEDICINAL CHEMISTRY. *Educ:* Univ Ill, Urbana, BS, 60; Mass Inst Technol, PhD(org chem), 62. *Prof Exp:* RES CHEMIST, MED PROD DEPT, E I DU PONT DE NEMOURS & CO INC, 65- *Mem:* Am Chem Soc. *Res:* Reactions involving medium and large ring compounds and stereochemistry; macrobicyclic diamines, crown ethers and biologically active compounds; pharmaceuticals. *Mailing Add:* Five Woodley Cir Twin Oaks Wilmington DE 19803

PARK, CHUNG SUN, b Seoul, Korea, Jan 14, 42; US citizen; m 72; c 2. ANIMAL SCIENCE & NUTRITION, MOLECULAR BIOLOGY. *Educ:* Seoul Nat Univ, BS, 64; Univ Ga, Athens, MS, 72; Va Polytech Inst & Su, PhD(animal nutrit), 75. *Prof Exp:* Res asst animal sci, Univ Ga, 70-72; res asst dairy sci, 72-75, fel lactation; res assoc, Purdue Univ, 77-78; from asst prof to assoc prof animal sci, 78-88, PROF ANIMAL SCI, NDAK STATE UNIV, 88- *Concurrent Pos:* Vis prof, Stanford Univ Med Sch, 85; Univ Calif, 86. *Mem:* Am Soc Animal Sci; Am Dairy Sci Asn. *Res:* Role of compensatory growth in lactation; regulation of mammary gene expression - rat, bovine species. *Mailing Add:* Dept Animal Sci NDak State Univ 169 Hultz Hall Fargo ND 58105

PARK, CONRAD B, b Kannapolis, NC, Nov 6, 19; m 43; c 5. CHEMISTRY. *Educ:* Newberry Col, AB, 41; Univ NC, MA, 43, PhD, 52. *Prof Exp:* Asst chem, Clemson Col, 41-42 & Univ NC, 42-43; res chemist, Tex Co, 43-46; prof chem, Carthage Col, 46-48; from assoc prof to prof, Lenoir-Rhyna Col, 48-56; acad dean, 56-74, PROF CHEM, NEWBERRY COL, 56- *Mem:* Am Chem Soc. *Res:* Surface active agents; ultraviolet spectroscopy; low temperature fractional distillation. *Mailing Add:* 2016 Forrest St Newberry SC 29108

PARK, DAVID ALLEN, b New York, NY, Oct 13, 19; m 45; c 4. THEORETICAL PHYSICS, HISTORY OF SCIENCE. *Educ:* Harvard Univ, AB, 41; Univ Mich, PhD(physics), 50. *Prof Exp:* Instr physics, Williams Col, 41-44; res assoc, Radio Res Lab, 44-45; mem, Inst Advan Study, 50-51; instr physics, Univ Mich, 50; from asst prof to assoc prof, 51-60, prof, 60-88, EMER PROF PHYSICS, WILLIAMS COL, 88- *Concurrent Pos:* Lectr, Univ Ceylon, 55-56 & 72; fel, Cambridge Univ, 62-63; prof, Univ NC, 64. *Mem:* Fel Am Phys Soc. *Res:* History and philosophy of science. *Mailing Add:* Dept of Physics Williams Col Williamstown MA 01267

PARK, DUK-WON, b Kyong-Buk, Korea, Mar 8, 45; m 74; c 2. MINING ENGINEERING, GEOLOGICAL ENGINEERING. *Educ:* Inha Univ, Korea, BS, 67; Univ Mo, Rolla, MS, 71, PhD(geol eng), 75. *Prof Exp:* Asst proj engr, D'Applonia Consult Engrs Inc, 75-76; res assoc, WVa Univ, 76-77, from asst prof to assoc prof, 77-81; from assoc prof to prof mineral eng, Univ Ala, 81-87. *Concurrent Pos:* Fac res partic, US Dept Energy, 79. *Mem:* Am Inst Mining Engrs; Soc Mining Metall Explor; Int Soc Rock Mech. *Res:* Rock mechanics; mine ground control; mine ventilation; applications of holographic interferometry to geotechnical problems; slope stability; ground water. *Mailing Add:* Dept Mineral Eng Univ Ala PO Box 870207 University AL 35487-0207

PARK, EDWARD C(AHILL), JR, b Wollaston, Mass, Nov 26, 23; m 51. ELECTRO-OPTICS, SENSOR SYSTEMS. *Educ:* Harvard Univ, AB, 47; Univ Birmingham, PhD(exp nuclear physics), 56. *Prof Exp:* Res asst, Univ Birmingham, 53-54; teaching intern, Amherst Col, 54-55; staff mem radar tech, Lincoln Lab, Mass Inst Technol, 55-57; staff mem superconductivity, Advan Res Div, Arthur D Little, Inc, Mass, 57, proj engr, Eng Div, 57-60; group leader electronic systs, Santa Monica Opers, Calif, 60-64; head, Laser Syst Sect, Res & Develop Div, Hughes Aircraft Co, Calif, 64-68; mgr, Electro Optical Sensor Sect, Litton Systs, Inc, Calif, 68-70; sr phys scientist, Eng Sci Dept, Rand Corp, Calif, 70-72; sr researcher, R & D Assocs, Calif, 72-86; sr scientist, Space & Strategic Systs Div, Hughes Aircraft Co, Calif, 86-88; SR TECH SPECIALIST, NAM AIRCRAFT, ROCKWELL INT, CALIF, 88- *Concurrent Pos:* Vpres, Optical Sensor Systs, Inc, Calif, 69-70; consult, 86- *Mem:* Sigma Xi; Inst Elec & Electronic Engrs; Optical Soc Am; Armed Forces Commun & Electronics Asn; Asn Old Crows; NY Acad Sci. *Res:* Electromagnetic and electro-optic sensor systems, including microwave, millimeter and submillimeter wave, laser and imaging systems; superconductivity; nuclear instrumentation. *Mailing Add:* 932 Ocean Front Santa Monica CA 90403

PARK, GEORGE BENNET, b Port Arthur, Tex, Jan 26, 46; m 70; c 2. ANALYTICAL CHEMISTRY, ELECTROANALYTICAL CHEMISTRY. *Educ:* Newberry Col, BS, 67; Univ NC, Chapel Hill, MA, 69; Univ Kans, PhD(chem), 73. *Prof Exp:* Asst prof chem, Clemson Univ, 73-80; MEM STAFF, DRUG METAB DEPT, STERLING WINTHROP RES INST, 80- *Mem:* Am Chem Soc; Med Electronics & Data Soc. *Res:* HPLC determination of low levels of drugs of abuse; electroanalytical chemistry of drugs of abuse and porphyrins; electroanalytical methods applied to enzyme assays. *Mailing Add:* 62 Cottage St Trumbull CT 06611

PARK, GERALD L(ESLIE), b Minneapolis, Minn, Feb 7, 33; m 57; c 3. ELECTRIC POWER DISTRIBUTION & USE. *Educ:* Univ Minn, Minneapolis, BME, 55, PhD(elec eng), 64; Stanford Univ, MS, 58. *Prof Exp:* Instr elec eng, Univ Minn, 58-64; from asst prof to assoc prof, 64-72, PROF

ELEC ENG, MICH STATE UNIV, 72- *Concurrent Pos:* Engr, Control Data & Honeywell; elected mem, E Lansing Sch Bd, 75-80 & 89- *Mem:* Sr mem Inst Elec & Electronic Engrs; Int Conf Large Elec Systs. *Res:* Electric power systems with emphasis on wind-electric systems, instrumentation and harmonics; electrical accidents. *Mailing Add:* Dept Elec Eng Mich State Univ East Lansing MI 48824-1226

PARK, HEEBOK, b Seoul, Korea, Nov 25, 33; m 62; c 1. STATISTICS. *Educ:* Seoul Nat Univ, BS, 57; Univ Chicago, MS, 61, PhD(statist), 64. *Prof Exp:* Asst prof statist, Purdue Univ, 64-67; assoc prof, 67-73, chmn dept, 74-78, PROF STATIST, CALIF STATE UNIV, HAYWARD, 73- *Concurrent Pos:* Am Heart Asn statistician, Coronary Heart Dis Res Proj, Chicago, 61-68. *Mem:* Inst Math Statist (treas, 78-82); Am Statist Asn. *Res:* Inference procedures and data analysis. *Mailing Add:* Dept Statist Calif State Univ 25800 Hillary St Hayward CA 94545

PARK, HERBERT WILLIAM, III, b Gowanda, NY, May 31, 20; m 47; c 5. REHABILITATION MEDICINE. *Educ:* Univ NC, BS, 44; Duke Univ, MD, 45. *Prof Exp:* Chief phys med & rehab serv, Vet Admin Hosp, Ft Thomas, Ky, 49-50; asst prof phys med, Univ Va, 50-52; prof phys med & rehab, Med Col Va, 52-60, med coordr rehab serv, 57-59, clin prof phys, med & rehab, 60-81; med dir, West End Med Rehab Ctr, 69-80; dir dept phys med, McGuire Clin, 80-88; ASSOC MED DIR, SHELTERING ARMS REHAB HOSP, 88- *Concurrent Pos:* Baruch fel biol, Mass Inst Technol, 46-47; Baruch fel physiol, Harvard Med Sch, 47; Baruch fel phys med, Mass Gen Hosp, 47-49; med dir, Woodrow Wilson Rehab Ctr, Fishersville, Va, 50-52; med dir, Baruch Ctr Phys Med & Baruch Sch Phys Ther, 52-59, psychiatrist-in-chief, Hosps, 52-59; mem comt prosthetic educ & info, Nat Acad Sci-Nat Res Coun, 58-65; consult, McGuire Vet Admin Hosp, Richmond City Nursing Home; consult med devices related to rehab med, Dept Health, Educ & Welfare, 74-78; dir prosthetics & orthotics clin, Med Col Va, 74-; chmn subcomt on prosthetics & orthotics, Panel on Physical Med Devices, Food & Drug Admin, HEW, 74-78. *Mem:* Am Col Physicians; Am Soc Internal Med; AMA; Am Acad Phys Med & Rehab. *Res:* Body mechanics as it relates to the field of prosthetics and orthotics. *Mailing Add:* 7814 Carousel Lane Richmond VA 23294

PARK, JAE YOUNG, b Chochiwon, Korea, May 4, 30; m 51; c 3. THEORETICAL PHYSICS, NUCLEAR PHYSICS. *Educ:* Seoul Nat Univ, BSc, 52; Rensselaer Polytech Inst, MSc, 56; Univ NC, PhD(physics), 62. *Prof Exp:* Instr physics, Pusan Nat Univ, Korea, 52-53 & Han Yang Inst Tech, Korea, 53-54; from asst prof to assoc prof, 62-74, PROF PHYSICS, NC STATE UNIV, 75- *Concurrent Pos:* NATO sr fel sci, NSF, 73; vis prof, Univ Giessen, Germany, 80; guest scholar, Kyoto Univ, Japan, 82. *Mem:* Am Phys Soc; Sigma Xi; Am Asn Univ Prof; Am Asn Physics Teachers. *Res:* Nuclear reactions; nuclear structure; heavy-ion physics. *Mailing Add:* Dept of Physics NC State Univ Box 8202 Raleigh NC 27695-8202

PARK, JAMES LEMUEL, b Wichita Falls, Tex, Dec 6, 40; m 63; c 2. PHYSICS. *Educ:* Univ Tex, BA, 63; Yale Univ, MS, 64, PhD(physics), 67. *Prof Exp:* From asst prof to assoc prof, 67-77, chmn dept, 77-80, PROF PHYSICS, WASH STATE UNIV, 77- *Concurrent Pos:* Vis prof, Mass Inst Technol, 80-81. *Mem:* Sigma Xi. *Res:* Mathematical and philosophical foundations of physics, especially quantum theory; quantum theory of measurement and quantum thermodynamics. *Mailing Add:* Dept Physics Wash State Univ Pullman WA 99164-2814

PARK, JAMES THEODORE, b Palo Alto, Calif, Aug 3, 22; m 52. MICROBIOLOGY, BIOCHEMISTRY. *Educ:* Cent Mich Univ, AB, 43; Univ Wis, MS, 44, PhD(biochem), 49. *Hon Degrees:* DSc, Cent Mich Univ, 62. *Prof Exp:* Asst biochem, Univ Wis, 43-44 & 46-49; biochemist, US Army Chem Corps, Md, 49-53; res assoc, Walter Reed Army Inst Res, 53-57; assoc prof microbiol, Sch Med, Vanderbilt Univ, 58-62; chmn dept, 62-70, PROF MOLECULAR BIOL & MICROBIOL, SCH MED, TUFTS UNIV, 62- *Concurrent Pos:* NSF sr fel, Cambridge Univ, 57-58; NIH spec fel, Univ Umea, Sweden, 69-70; mem, Study Sect Bact & Mycol, NIH, 64-68, chmn, Study Sect Microbiol, Physiol & Genetics, 85-88; mem, microbiol training comt, Nat Inst Gen Med Sci, 70-73. *Mem:* Am Acad Arts & Sci; Am Soc Biochem; Am Soc Microbiol; Brit Soc Gen Microbiol. *Res:* Mode of action of B-lactam antibiotics; bacterial cell division; biosynthesis of bacterial cell walls. *Mailing Add:* Dept Molecular Biol & Microbiol Tufts Univ 136 Harrison Ave Boston MA 02111

PARK, JANE HARTING, b St Louis, Mo, Mar 25, 25; m 53; c 1. BIOCHEMISTRY, MAGNETIC RESONANCE SPECTROSCOPY. *Educ:* Wash Univ, BS, 46, PhD(biochem), 52. *Prof Exp:* From instr to assoc prof, 54-69, PROF PHYSIOL, SCH MED, VANDERBILT UNIV, 69- *Concurrent Pos:* Am Cancer Soc scholar, 55-58; NIH sr res fel, 58-; mem physiol chem study sect, NIH; bd counr, NIH, Inst Aging & Heart, Lung, Blood, NIH. *Mem:* Am Chem Soc; Am Cancer Soc; Am Soc Biol Chemists; Soc Magnetic Resonance Imaging; AAAS. *Res:* Mechanism of enzymatic catalysis; oxidative phosphorylation; metabolic pathways in muscle; muscle diseases; magnetic resonance imaging; magnetic resonance spectroscopy; ESR. *Mailing Add:* Dept Physiol Light Hall 702 Vanderbilt Univ Sch Med Nashville TN 37232

PARK, JOHN HOWARD, JR, b Washington, DC, May 16, 32; m 51; c 3. ELECTRICAL ENGINEERING. *Educ:* Univ Md, BS, 54; Johns Hopkins Univ, MS, 57, DEng(elec eng), 60. *Prof Exp:* Res asst elec eng, Radiation Lab, Johns Hopkins Univ, 54-57, res staff asst, 57-59; tech specialist, Electronics Div, Gen Mills, Inc, 59-62, lab mgr systs anal, 62-63; lectr elec eng, Univ Minn, Minneapolis, 61-62, assoc prof, 63-79; sr prin engr, 79-81, oper mgr, 81-85, sr mem exec staff, 85-88, DIR ADVAN PROGS, COMPUT SCI CORP, 88- *Concurrent Pos:* Consult, Appl Sci Div, Litton Industs, Inc, 63-64, Univac Div, Sperry Rand, Inc, 64-79, Systs & Res Div, Honeywell Inc, 66, Electroprod Div, 3M Co, 70-74 & PKM Corp, 71-74. *Mem:* Inst Elec & Electronics Engrs. *Res:* Statistical theory of communication; estimation of signal parameters; effect of noise in modulation systems; speech compression; satellite communications; adaptive antenna arrays. *Mailing Add:* Comput Sci Corp 3160 Fairview Park Dr Falls Church VA 22042

PARK, JOHN THORNTON, b Phillipsburg, NJ, Jan 3, 35; m 56; c 2. ATOMIC PHYSICS. *Educ:* Nebr Wesleyan Univ, BA, 56; Univ Nebr, PhD(physics), 63. *Prof Exp:* NSF fel, Univ Col, London, 63-64; from asst prof to assoc prof, 64-71, chmn dept. 77-83, vchancellor, acad affairs, 83-85, interim chancellor, 85-86, PROF PHYSICS, UNIV MO-ROLLA, 71-, VCHANCELLOR ACAD, 86- *Concurrent Pos:* Vis assoc prof physics, NY Univ, 70-71; hons lectr, Mid-Am State Univs Asn, 73-77; mem, prog comt for Div of Electron & Atomic Physics, 74-75, nominating comt, 78, gen comt, Int Conf on the Physics of Acad & Atomic Physics, 75-79 & gen comt, Small Accelerator Conf, 78-; mem, panel on Nat Sci Found Energy related post-doctoral fels, Nat Res Coun, 75 & 76 & NATO post doctoral fel panel, Nat Sci Found, 77; counr, Oak Ridge Assoc Univs, 84-85; mem doctoral fac, Univ Mo-Rolla. *Mem:* Fel Am Phys Soc; Am Asn Physics Teachers. *Res:* Ionic and atomic collisional excitation and energy loss processes. *Mailing Add:* Off of the Vice-Chancellor Univ of Mo-Rolla Rolla MO 65401

PARK, JOON B, b Pusan, Korea, June 20, 44; US citizen; m 63; c 3. BIOMATERIALS, BIOMECHANICS. *Educ:* Boston Univ, BS, 67; Mass Inst Technol, MS, 69; Univ Utah, PhD(mat), 72. *Prof Exp:* Postdoctoral eng, Univ Wash, 72-73; vis asst prof, Univ Ill, Urbana, 73-75; from asst prof to assoc prof eng, Clemson Univ, 75-81; prof, Tulane Univ, 81-83; PROF ENG, UNIV IOWA, 83- *Mem:* Soc Biomat; Orthop Res Soc. *Res:* Interface problems between biomaterials and tissues, soft and hard tissues characterization. *Mailing Add:* Dept Biomed Eng 1208 EB Univ Iowa Iowa City IA 52242

PARK, KISOON, US citizen. PHYSICAL CHEMISTRY. *Educ:* Seoul Nat Univ, BS, 61; Univ Utah, MS, 64, PhD(metall), 66. *Prof Exp:* Chemist, 66-73, proj scientist, 73-78, RES SCIENTIST, UNION CARBIDE CORP, 76- *Concurrent Pos:* Guest lectr, NC State Univ, 73-77. *Mem:* Sigma Xi; Electrochem Soc; Nat Asn Corrosion Engrs; Am Chem Soc. *Res:* Surface and colloid chemistry; radiation polymerization; friction and lubrication of textile fibers; corrosion science and technology. *Mailing Add:* 1470 Longleaf Ct Matthews NC 28105

PARK, KWANGJAI, b Wonsan, Korea, Oct 12, 35; m 57; c 5. ATMOSPHERIC CHEMISTRY & PHYSICS. *Educ:* Harvard Univ, BA, 58; Univ Calif, Berkeley, PhD(physics), 65. *Prof Exp:* Res scientist plasma physics, Giannini Sci Corp, 58-61; mem tech staff nonlinear optics, Bell Tel Labs, 65-66; from asst prof to assoc prof, 66-83, PROF PHYSICS, UNIV ORE, 83- *Concurrent Pos:* Vis prof, Sogang Univ, Seoul, Korea, 72-73. *Mem:* Am Phys Soc. *Res:* Far infrared spectroscopy; study of chaos. *Mailing Add:* Dept of Physics Univ of Ore Eugene OR 97403

PARK, LEE CRANDALL, b Washington DC, July 15, 26; m 53, 85; c 2. PERSONALITY DISORDERS, PSYCHOPHARMACOLOGY. *Educ:* Yale Univ, BS, 48; Johns Hopkins Univ, MD, 52; Am Bd Psychiat & Neurol, dipl, 60. *Prof Exp:* Intern med, Johns Hopkins Hosp, 52-53, resident psychiat, 55-59; lieutenant, surgeon & div psychiatrist, US Navy, 53-55; Fel, 55-59, mem fac, 59-71, ASSOC PROF PSYCHIAT, JOHNS HOPKINS UNIV, 71 -; MEM STAFF PSYCHIATRY & MED, JOHNS HOPKINS HOSP, 59 - *Concurrent Pos:* Co-prin investr, USPHS grant, 59-60, prin investr, 60-68, co-dir res proj, 69-73; physician in charge psychiat serv, Student Health Serv, Johns Hopkins Univ, 61-73; pvt pract psychiat, 64 -; dir, outpatient serv & community psychiat, Johns Hopkins Hosp, 72-74 & mem coun dept, psychiat & behav sci, 74-76; dir, Clin Study Borderline & Narcissistic Conditions, 74 -; pres, Md Psychiat Soc, 78-79; mem, Am Psychiat Asn Gov Assembly, 83- *Mem:* Am Col Neuropsychopharmacol; fel Am Psychiat Asn; fel AAAS; Am Soc Adolescent Psychiat; AMA; Soc Psycother Res. *Res:* The origins, dynamics, life history and response to psychotherapy and to drug therapy of borderline and narcissistic personality disorders; psychopharmacology; schizophrenia research; short term psychotherapy. *Mailing Add:* 308 Tunbridge Rd Baltimore MD 21212

PARK, MYUNG KUN, b Suhung, Korea, Sept 30, 34; m 67; c 3. PEDIATRIC CARDIOLOGY, PHARMACOLOGY. *Educ:* Seoul Nat Univ, dipl, 56, MD, 60. *Prof Exp:* Fel pediat cardiol, Sch Med, Univ Wash, 65-68, instr pediat, 66-68, res fel pharmacol, 71-73; asst prof pediat, Univ Kans Med Ctr, Kansas City, 73-76; assoc prof, 76-82, PROF PEDIAT, UNIV TEX HEALTH SCI CTR, SAN ANTONIO, 82- *Mem:* NY Acad Sci; Am Acad Pediat; Soc Pediat Res; Am Soc Pharmacol & Exp Therapeut; Soc Exp Biol & Med. *Res:* Developmental pharmacology; systematic hypertension; digitalis pharmacology; autonomic pharmacology. *Mailing Add:* Dept Pediat Univ Tex Health Sci Ctr San Antonio TX 78284

PARK, PAUL HEECHUNG, b Seoul, Korea, Mar 15, 41; US citizen; m 70; c 3. CHEMICAL ENGINEERING, MATHEMATICS. *Educ:* Seoul Nat Univ, BS, 65; Mich State Univ, MS, 68, PhD(chem eng), 72. *Prof Exp:* Sr develop engr wet spinning, Monsanto Textiles Co, 73-74; res engr polymer processing & melt spinning, Am Cyanamid Co, 74-77; sr res engr mat, 77-80, supvr new prod & technol, 80-85, MGR PROD RES & DEVELOP, MOBIL CHEM CO, 85- *Concurrent Pos:* Res assoc, Mich State Univ, 72-73. *Mem:* Am Inst Chem Engrs; Am Chem Soc; Soc Rheology. *Res:* Polymer characterization via solution and melt rheology and its processing; molecular orientation, structural interpretation of polymer processes and structure-property correlation for spun fiber and biaxially oriented film; multi phase polymer blends and multi layer biaxially oriented film technology development. *Mailing Add:* Eight Cedarwood Circle Pittsburgh PA 14534

PARK, PAUL KILHO, b Kobe, Japan, Feb 4, 31; m; c 2. OCEANOGRAPHY. *Educ:* Pusan Fisheries Col, BS, 53; Tex A&M Univ, MS, 57, PhD(oceanog), 61. *Prof Exp:* From asst prof to prof oceanog, Ore State Univ, 61-76; OCEANOGR, US DEPT COM, NAT OCEANIC & ATMOSPHERIC ADMIN, 77- *Concurrent Pos:* Prog dir phys oceanog, NSF, 69-70, head oceanog sect, 70-71. *Mem:* AAAS; Am Soc Limnol & Oceanog; Am Geophys Union. *Res:* Carbon dioxide system; biogeochemistry of hydrosphere; marine pollution. *Mailing Add:* Seven Fallswood Ct Rockville MD 20854-5502

PARK, RICHARD AVERY, IV, b Sedalia, Mo, Dec 20, 38; div; c 2. ENVIRONMENTAL SCIENCES, ECOSYSTEM MODELING. *Educ:* La State Univ, BS, 61; Univ Wis, MS, 65, PhD(paleoecol), 67. *Prof Exp:* From asst prof to prof geol, Rensselaer Polytech Inst, 76-87; PROF & COODR, HOLCOMB RES INST, BUTLER UNIV, 85- *Concurrent Pos:* Chief math ecologist, Rensselaer Fresh Water Inst, 72-76; dir, Ctr Ecol Modeling, Rensselaer Polytech Inst, 77-82; sole proprietor, Eco Modeling, 77-; consult, Empire State Elec Energy Corp, 83-85; res ecologist, US Environ Protection Agency, 84-85; adj prof geol, Ind Univ & Purdue Univ, Indianapolis, 86- *Mem:* Int Soc Ecol Modelling. *Res:* Ecosystem toxic-substance and land use simulation; multivariate analysis; ecology of American and European lakes. *Mailing Add:* Holcomb Res Inst Butler Univ Indianapolis IN 46208

PARK, RICHARD DEE, b Payson, Utah, Nov 19, 42; c 4. VETERINARY MEDICINE. *Educ:* Utah State Univ, BS, 65; Colo State Univ, DVM, 68; Univ Calif, Davis, PhD(comp path), 71. *Prof Exp:* Resident radiol, Univ Calif, Davis, 68-71, asst prof, 74-75; asst prof, Univ Ill, 71-74; assoc prof, 75-79, PROF RADIOL, COLO STATE UNIV, 79- *Concurrent Pos:* Dipl radiol, Am Col Vet Radiol, 71- *Mem:* Am Vet Med Asn; Am Col Vet Radiol (pres, 78-79); Soc Vet Ultrasound; Int Vet Radiol Soc. *Res:* Veterinary diagnostic imaging, including radiology, ultrasound and computed tomography; application in diagnosis and investigation of imaging modalities in veterinary medicine. *Mailing Add:* 2648 Shadow Mountain Dr Ft Collins CO 80525

PARK, ROBERT H, b Mar 15, 02. HIGH VOLTAGE TRANSMISSION SYSTEMS. *Prof Exp:* Pres, Fastload Control, Inc; RETIRED. *Mem:* Nat Acad Eng; fel Inst Elec & Electronics Engrs. *Mailing Add:* 101 Highland Ave No 312 Providence RI 02906

PARK, ROBERT L, b Kansas City, Mo, Jan 16, 31; m 51; c 2. PHYSICS. *Educ:* Univ Tex, BS, 58, MA, 60; Brown Univ, PhD(physics), 65. *Prof Exp:* Res assoc physics, Brown Univ, 64-65; res physicist, Sandia Corp, 65-69, supvr, Surface Kinetics Div, Sandia Lab, 69-74; dir, Ctr Mat Res, 75-78, chmn dept, 78-84, PROF PHYSICS & ASTRON, UNIV MD, COL PARK, 74-; EXEC DIR, OFF PUB AFFAIRS, AM PHYS SOC, 84- *Mem:* AAAS; fel Am Phys Soc; Am Vacuum Soc. *Res:* Physics of solid surfaces; surface structure and gas-surface interactions. *Mailing Add:* 3303 Chatham Rd Adelphi MD 20783

PARK, ROBERT LYNN, b Idaho Falls, Idaho, Sept 1, 32; m 62; c 6. ANIMAL SCIENCE, ANIMAL BREEDING. *Educ:* Brigham Young Univ, BS, 56; Cornell Univ, MS, 58, PhD(animal husb), 62. *Prof Exp:* Spec county agent, Univ Ariz, 59, res animal husb, Agr Res Serv, USDA, 61-65; from asst prof to assoc prof, 65-73, PROF ANIMAL SCI, BRIGHAM YOUNG UNIV, 73- *Concurrent Pos:* Mem pasture & livestock comt, Caribbean Comn, 63-65. *Mem:* Sigma Xi; Am Soc Animal Sci; Am Dairy Sci Asn. *Res:* Population genetics and animal breeding; biomathematics and biostatistics in agriculture; microcomputers; ranking superior sires and dams in mink; using linear array realtime ultrasound in selecting superior swine breeding stock, and utilizing swine waste solids; economic and technical feasibility of embryo transfer in dairy cattle. *Mailing Add:* Dept Animal Sci Brigham Young Univ Provo UT 84602

PARK, ROBERT WILLIAM, b Eaton County, Mich, Oct 2, 29; m 49; c 4. PAPER CHEMISTRY. *Educ:* Albion Col, AB, 52; Western Mich Univ, MBA, 68. *Prof Exp:* Staff chemist, Film Div, Am Viscose Corp, 52-55, prod supvr, 55-60, prod admin, 60-62; sr res specialist, Res & Develop Div, Packaging Corp Am, 62-66, res group leader, 66-68, dir res & develop, 68-70; mgr new prod res, Appleton Papers, Inc, 70-71, com prod mgr, 71-81; DIR PROD DEVELOP, SHAWANO PAPER MILLS, 81- *Mem:* Am Chem Soc; Tech Asn Pulp & Paper Indust. *Res:* Development of coated specialty papers for graphic arts, technical specialty papers, and coated plastics; research administration; development of specialty light weight machine glazed and crepe papers. *Mailing Add:* Rte 4 Box 60 Wolf River Dr Shawano WI 54166

PARK, RODERIC BRUCE, b Cannes, France, Jan 7, 32; US citizen; m 53; c 3. PLANT PHYSIOLOGY. *Educ:* Harvard Univ, BA, 53; Calif Inst Technol, PhD(biol), 58. *Hon Degrees:* DSc, York Univ, 80. *Prof Exp:* Chemist, Lawrence Radiation Lab, Univ Calif, Berkeley, 58-60, from asst prof to assoc prof, 60-66, provost & dean col lett & sci, 72-80, vchancellor, 80-90, PROF BOT, UNIV CALIF, BERKELEY, 66- *Concurrent Pos:* Mem, Harvard Bd Overseers, 81-87; chmn, bd trustees, Athenian Sch, 85-; trustee, Univ Calif Berkeley Found, 85-90; chmn, Coun Acad Affairs, NASULGC, 88-89. *Honors & Awards:* NY Bot Gardens Award, 62. *Mem:* Assoc mem AAAS; Am Soc Plant Physiol. *Res:* Carbon isotope fractionation in photosynthesis; membrane structure; photosynthesis as related to plant ultrastructure; phloem proteins. *Mailing Add:* Prof Plant Biol 111 GPBB Univ Calif Berkeley CA 94720

PARK, SAMUEL, b Seoul, Korea, May 26, 36. MATHEMATICAL ANALYSIS. *Educ:* WVa Wesleyan Col, BS, 56; Univ Pittsburgh, MA, 57, PhD(math), 59. *Prof Exp:* Teaching fel math, Univ Pittsburgh, 56-59; res assoc, Columbia Univ, 59-60; asst prof, Rutgers Univ, 60-65; assoc prof, 65-71, PROF MATH, LONG ISLAND UNIV, 71-, CHMN MATH DEPT, 79- *Mem:* Am Math Soc; Math Asn Am. *Res:* Analysis, particularly ordinary differential equations and theory of summability; Fourier analysis. *Mailing Add:* Dept of Math Long Island Univ Brooklyn NY 11201

PARK, SU-MOON, b Korea, Dec 1, 41; m 67; c 3. ELECTROANALYTICAL CHEMISTRY, PHOTOCHEMISTRY. *Educ:* Seoul Nat Univ, Korea, BS, 64; Tex Tech Univ, MS, 72; Univ Tex, Austin, PhD(chem), 75. *Prof Exp:* Analytical chemist, Choong-Ju Fertilizer Corp, Korea, 64-67; lab supvr, Yong-Nam Chem Co, Ltd, 67-70; asst prof, 75-81, assoc prof, 81-87, PROF CHEM, UNIV NMEX, 87- *Concurrent Pos:* Fel, Oak Ridge Assoc Univ Fac Res Participation, 80; consult, Los Alamos Nat Lab, 81- *Mem:* Am Chem Soc; Electrochem Soc, Inc; Korean Chem Soc. *Res:* Electrogenerated chemiluminescence; chemistry of exciplex; energy transfer of the excited states; organic electrochemistry; donor-acceptor complexes of organic compounds; energy conversion employing semiconductor; liquid junction cells. *Mailing Add:* Dept of Chem Univ of NMex Albuquerque NM 87131

PARK, TAISOO, b Milyang, Korea, July 9, 29; m 76; c 1. INVERTEBRATE ZOOLOGY, MARINE BIOLOGY. *Educ:* Pusan Fisheries Col, Korea, BS, 52, MS, 57; Univ Wash, PhD(zool), 65. *Prof Exp:* Instr, Pusan Fisheries Col, 52-60; res assoc, Univ Wash, 65-66; asst prof, Univ Md, 66-67; asst scientist, Woods Hole Oceanog Inst, 67-69; asst prof oceanog, 69-73, assoc prof biol & marine sci, 73-83, PROF MARINE BIOL, TEX A&M UNIV, 83- *Concurrent Pos:* NSF grants, 67-76; vis prof, Seoul Nat Univ, Korea, 79, 84-85 & Pusan Fisheries Col, 84-85. *Mem:* Am Soc Zoologists; Ecol Soc Am; Am Soc Limnol & Oceanog; Crustacean Soc; Sigma Xi. *Res:* Systematics and distribution of marine calanoid Copepoda on a world-wide base; general biology of the calanoid Copepoda including functional anatomy, reproduction, and development; community ecology of marine planktonic copepods in the Gulf of Mexico. *Mailing Add:* Dept Marine Biol Tex A&M Univ Galveston TX 77550

PARK, THOMAS, b Danville, Ill, Nov 17, 08; m 28, 69; c 2. POPULATION ECOLOGY. *Educ:* Univ Chicago, BS, 30, PhD(zool), 32. *Hon Degrees:* ScD, Univ Ill, 73. *Prof Exp:* Cur zool, Univ Chicago, 30-31, asst, 31-33; Nat Res Coun fel, Johns Hopkins Univ, 33-35, instr biol, Sch Hyg & Pub Health, 35-36, assoc, 36-37; from instr to prof zool, 37-68, assoc dean div biol sci, 43-46, prof biol, 68-74, EMER PROF BIOL, UNIV CHICAGO, 74- *Concurrent Pos:* Ed, Ecology, Ecol Soc Am, 40-50; Rockefeller Found fel, Oxford Univ, 48; sci attache, Am Embassy, London, 49; vis prof, Univ Calif, 52- & Univ Ill, Chicago Circle, 74-76; ed, Physiol Zool, 55-75. *Honors & Awards:* Eminent Ecologist Citation, Ecol Soc Am, 71. *Mem:* AAAS (pres, 61); Am Soc Zool; Ecol Soc Am (pres, 59); hon mem Am Soc Naturalists; Soc Study Evolution; emer mem Sigma Xi. *Res:* Experimental population studies with Tribolium species; experimental biology of populations. *Mailing Add:* 5715 S Blackstone Ave Chicago IL 60637

PARK, VERNON KEE, b Austin, Tex, Sept 27, 28; m 55; c 4. POLYMER CHEMISTRY. *Educ:* Tulane Univ, BS, 48; Univ Tex, MA, 58. *Prof Exp:* From chemist to sr chemist, 57-73, asst dep supt, 73-75, develop assoc, 75-78, res assoc, 79-83, DEVELOP ASSOC, TEX EASTMAN CO, 84- *Res:* Polyolefin catalysts; application research on adhesives and emulsifiable polymers; polymer research; chemical engineering. *Mailing Add:* 204 Ramblewood Ct Longview TX 75601

PARK, WILLIAM H, b Carlisle, Pa, Mar 2, 29; m 54; c 4. MECHANICAL ENGINEERING. *Educ:* Pa State Univ, BS, 52, MS, 59; Cornell Univ, PhD(mach design), 66. *Prof Exp:* Design engr, Sanders & Thomas, Inc, 52-53; from instr to prof mech eng, 53-87, head grad progs, 80-87, EMER PROF MECH ENG, PA STATE UNIV, 87- *Concurrent Pos:* Ford Found prin res engr assoc, Ford Motor Co, 68-69; consult, Eriez Magnetics, 57-, Syntron Co, 66- & var law firms. *Mem:* Soc Automotive Engrs; Am Soc Mech Engrs; Am Soc Eng Educ. *Res:* Repetitive impact vibrations; analog-hybrid computer simulation; road roughness-vehicle response; road roughness measurement; biomedical research and design. *Mailing Add:* 537 Kemmerer Rd State College PA 16801

PARK, WON JOON, b Sun-Chun, Korea, Sept 18, 35; US citizen; m 66; c 2. MATHEMATICS, STATISTICS. *Educ:* Seoul Nat Univ, BA, 57; Univ Calif, Riverside, MA, 66; Univ Minn, Minneapolis, PhD(math), 69. *Prof Exp:* From asst prof to assoc prof, 69-81, PROF MATH & STAT, WRIGHT STATE UNIV, 81- *Mem:* Inst Math Statist; Am Statist Asn; Inst Elec & Electronic Engrs. *Res:* Stochastic processes and reliability. *Mailing Add:* Dept Math & Stat Wright State Univ Dayton OH 45435

PARK, YOON SOO, b Raichun, Korea, July 4, 29; US citizen; m 60; c 3. SOLID STATE PHYSICS. *Educ:* Seoul Nat Univ, BS, 52; Univ Alta, MS, 55; Univ Cincinnati, PhD(physics), 63. *Prof Exp:* Lectr physics, Pusan Nat Univ, 52; res physicist, D H Baldwin Co, 56-62; res physicist, Wright-Patterson AFB, 63-75 group leader, Aerospace Res Labs, 75-83, task mgr, Air Force Avionics Lab, 83-86; AT OFF NAVAL RES, ARLINGTON, VA, 86- *Concurrent Pos:* Lectr, Univ Dayton, 66; guest lectr, Tech Univ Berlin, 68-69; vis prof, Seoul Nat Univ, 78; consult, UN Indust Develop Orgn, 80; adj prof, Marquett Univ, 81-83; dir, Air Force Off Sci Res Japan. *Mem:* Fel Am Phys Soc; Korean Phys Soc; sr mem Inst Elec & Electronic Engrs. *Res:* Optical and electrical properties of semiconductors; ion implantation; radiation damage; crystal growth, cryogenic and optical techniques; photoconductivity and luminescence. *Mailing Add:* Code 1212 Office Naval Res 800 North Quincy St Arlington VA 22217

PARK, YOUNG D, b Korea, Dec 29, 32; m 65; c 2. SYSTEMS ENGINEERING, CHEMICAL ENGINEERING. *Educ:* Univ Va, BChE, 57. *Prof Exp:* Univ fel, Polytech Inst Brooklyn, 60, NSF fel, 60-61; proj engr, Buyers Lab Inc, 61-66; comput appln engr, Realtime Systs Inc, 66-68; process analyst, Digital Appln, Inc, 68-69; mgr systs eng, Mauchly Assocs, Inc, 69-70; sr engr, Davis Comput Systs, 70-71 & Metromation, Inc, Princeton, 71-73; sr prin engr, Comput Sci Corp, 73-80; process comput systs mgr, Catalytic Inc, 80-87; SYSTS ENGR, UNITED ENGRS & CONSTRUCTORS, 87- *Concurrent Pos:* Workshop coordr course on comput prog, Engrs Joint Coun, 68; consult, Oyer Prof Comput Serv, Inc, 68. *Mem:* Am Inst Chem Engrs; Am Chem Soc. *Res:* Computer applications, optimization and simulation of chemical processes; process control. *Mailing Add:* 15 Chateau Dr Cherry Hill NJ 08003

PARKA, STANLEY JOHN, b Holyoke, Mass, May 18, 35; m 56; c 4. WEED SCIENCE. *Educ:* Univ Mass, BS, 56; Purdue Univ, MS, 58, PhD(plant physiol), 60. *Prof Exp:* Plant physiologist, Ferry Morse Seed Co, Calif, 59-60; veg exten specialist, Univ Fla, 60-61; plant physiologist, Eli Lilly & Co, 61-65, head, Eastern Field Res, 65-66, head, NAm Field Res, 66-72, head, Greenfield Plant Sci Res, 72-73, res assoc, 73-77, res adv, Greenfield Labs, 77-91; RETIRED. *Mem:* Weed Sci Soc Am; Plant Growth Regulator Working Group; Aquatic Plant Mgt Soc. *Res:* Development of chemicals which regulate plant growth and weed control in agriculture. *Mailing Add:* 203 Oak Blvd S Dr Greenfield IN 46140

PARKANYI, CYRIL, b Prague, Czech, Sept 11, 33; m 60; c 1. PHYSICAL ORGANIC CHEMISTRY. *Educ:* Charles Univ, Prague, MS, 56, Dr rer nat(chem), 66; Czech Acad Sci, PhD(org chem), 62. *Prof Exp:* Phys & anal chemist, Res & Control Inst Food Indust, Prague Czech, 55-56; researcher, Inst Org Chem & Biochem, Czech Acad Sci, 56-59, res scientist, Inst Phys Chem, 60-65; vis scientist, State Univ Leiden, 65; res fel, Calif Inst Technol, 65-67; sr res scientist, Inst Phys Chem, Czech Acad Sci, 67-68; res assoc, Calif Inst Technol, 68-69; from assoc prof to prof chem, Univ Tex, El Paso, 69-90, chmn dept chem, 82-88; PROF CHEM & CHMN, DEPT CHEM, FLA ATLANTIC UNIV, BOCA RATON, 89- *Concurrent Pos:* Vis res prof, Univ d'Aix-Marseille, France, 74 , 77 & 86; vis prof, Univ Kuwait, 76, Univ Groningen, Neth, 78 & Univ Sci Techn de Lille, Villeneuve d'Ascq, France, 80, 81 & 82; fac res award, Univ Tex, El Paso, 80. *Honors & Awards:* Czech Acad Sci Award, 63. *Mem:* Inter-Am Photochem Soc; Am Chem Soc; Sigma Xi; Fedn Am Scientists; fel NY Acad Sci; fel AAAS; Europ Acad Sci. *Res:* Theoretical and quantum organic chemistry; photochemistry; heterocyclic chemistry. *Mailing Add:* 245 NW 69th St Boca Raton FL 33487-2390

PARKE, RUSSELL FRANK, b Kirkland, Ill, Jan 11, 32; m 56; c 2. PHARMACY. *Educ:* Purdue Univ, BS, 54, MS, 56, PhD, 58. *Prof Exp:* Asst prof pharm, Butler Univ, 58-64; SR PHARMACEUT CHEMIST, ELI LILLY & CO, INDIANAPOLIS, 64- *Mem:* Am Chem Soc. *Res:* Drug stability; data storage and retrieval; international product development. *Mailing Add:* Dept IC733 100/1 Eli Lilly & Co Indianapolis IN 46285

PARKE, WESLEY WILKIN, b Wallingford, Conn, Sept 30, 26; m 53; c 2. VERTEBRATE ANATOMY, EMBRYOLOGY. *Educ:* Univ Conn, BS, 51, MA, 54, PhD(vert anat), 57. *Prof Exp:* From instr to assoc prof anat, Jefferson Med Col, Thomas Jefferson Univ, 57-74; prof gross anat, Sch Med, Southern Ill Univ, Carbondale, 74-78; PROF & CHMN DEPT ANAT, SCH MED, UNIV SDAK, 78- *Concurrent Pos:* Consult & lectr, US Naval Hosp, Philadelphia, Pa; intern anat, Col Surgeons; assoc ed, Clin Anat, Spine. *Honors & Awards:* Volvo Award for Basic Res on Spine, 85. *Mem:* AAAS; Am Asn Anat; World Asn Vet Anat; fel Sigma Xi; Cervical Spine Res Soc; Am Soc Clin Anatomists. *Res:* Evolution and development of respiratory system; visceral teratology; vascular anatomy of abdominal and thoracic viscera; lymphangiology and microcirculation; development and anatomy of spine and spinal cord; vascularity of spinal cord and nerve roots. *Mailing Add:* Dept Anat Univ SDak Sch Med University & Clark Sts Vermillion SD 57069

PARKE, WILLIAM C, b Washington, DC, Oct 29, 41; m 69. ELEMENTARY PARTICLE PHYSICS. *Educ:* George Washington Univ, BS, 63, PhD(theoret physics), 67. *Prof Exp:* Nat Acad Sci-Nat Bur Standards res assoc theoret physics, Nat Bur Standards, 67-69; from lectr to asst prof, 66-74, ASSOC PROF PHYSICS, GEORGE WASHINGTON UNIV, 74- *Mem:* Am Phys Soc; Am Inst Physics. *Res:* Field theory; Regge poles; low mass nuclei. *Mailing Add:* 1820 South St NW George Washington Univ Washington DC 20009

PARKENING, TERRY ARTHUR, b Omaha, Nebr, Jan 24, 43; m 65; c 1. REPRODUCTIVE BIOLOGY, ENDOCRINOLOGY. *Educ:* Midland Col, BS, 65; Univ SDak, MA, 68; Univ Ore, PhD(biol), 74. *Prof Exp:* From asst prof to assoc prof, 76-79, PROF ANAT, UNIV TEX MED BR, GALVESTON, 85- *Concurrent Pos:* Res scientist, Worcester Found Exp Biol, NIH, 74-76; NIH grants, Univ Tex, 77-80 & 81-88. *Mem:* Sigma Xi; Am Asn Anat; Soc Study Reprod; Geront Soc; Am Asn Tissue Banks; Am Fertil Soc. *Res:* Reproductive biology and endocrinology, specifically fertilization, implantation and the effects of aging on reproduction. *Mailing Add:* Dept of Anat Univ of Tex Med Br Galveston TX 77550

PARKER, ALAN DOUGLAS, b Wellsville, NY, Sept 21, 45; m 71. MYCOLOGY, BOTANY. *Educ:* Eastern Ill Univ, BS, 69, MS, 71; Univ Ill, PhD(mycol), 76. *Prof Exp:* ASST PROF BOT & MICROBIOL, UNIV WIS CTR-WAUKESHA, 76- *Mem:* Mycol Soc Am; Brit Mycol Soc; Sigma Xi. *Res:* Taxonomy and ecology of coprophilous ascomycetes. *Mailing Add:* Univ Wis Waukesha 1500 University Dr Waukesha WI 53188

PARKER, ALBERT JOHN, b East St Louis, Ill, July 16, 53. BIOGEOGRAPHY, SYNOPTIC CLIMATOLOGY. *Educ:* Mich State Univ, BS, 75; Univ Wis, Madison, MS, 77, PhD(geog), 80. *Prof Exp:* ASST PROF GEOG, UNIV GA, 80- *Mem:* Asn Am Geographers; Ecol Soc Am; Soil Sci Soc Am; Sigma Xi. *Res:* Structure and dynamics of conifer forests in western North America with field experience in southeastern Arizona, the Sierra Nevada, and northern Rocky Mountains. *Mailing Add:* Dept Geog Univ Ga Athens GA 30602

PARKER, ALICE CLINE, b Birmingham, Ala, Apr 10, 48; m 80; c 1. COMPUTER ENGINEERING. *Educ:* NC State Univ, BS, 70, PhD(elec eng), 75; Stanford Univ, MS, 71. *Prof Exp:* Asst prof elec eng, Carnegie-Mellon Univ, 75-80; asst prof, 80-83, ASSOC PROF ELEC ENG, UNIV SOUTHERN CALIF, 83- *Concurrent Pos:* Prin investr, US Army Res Off grant, 77-86; consult, US Army Electronics Command, 78-79, Info Sci Inst, Digital Equip Corp, Hewlett Packard Corp, Nat Inst Educ, 80, Xerox Corp & Aerospace Corp, 81-; nat lectr, Asn Comput Mach, 79-80, GE, 87. *Mem:* Fel Inst Elec & Electronics Engrs; Sigma Xi; Asn Comput Mach. *Res:* Automatic design of digital systems; computer programs which evaluate cost and performance tradeoffs and automatically make design decisions in order to produce an implementation of a digital system from a higher-level specification. *Mailing Add:* Dept Elec Eng Systs Univ Southern Calif Los Angeles CA 90089-0781

PARKER, ARTHUR L (PETE), b Stillwater, Okla, July 29, 50; m 73. PROCESS SIMULATION & OPTIMIZATION. *Educ:* NMex State Univ, BS, 73; Univ Wis, MS, 75, PhD(chem eng), 78. *Prof Exp:* Res engr, Shell Develop Co, 78-81; process engr, 81-88, MGR OPER TECH SUPPORT, SHELL OIL CO, 88- *Mem:* Am Inst Chem Engrs. *Mailing Add:* Shell Oil Co PO Box 3105 Houston TX 77001

PARKER, B(LAINE) F(RANK), b Gaston Co, NC, June 12, 24; c 3. AGRICULTURAL ENGINEERING. *Educ:* Va Polytech Inst, BSAE, 50, MSAE, 52; Mich State Univ, PhD(agr eng), 54. *Prof Exp:* Instr agr eng, Va Polytech Inst, 50-52; asst, Mich State Univ, 52-54; asst prof, NC State Col, 54-57; assoc prof, 57-58, chmn agr eng dept, 59-74, PROF AGR ENG, UNIV KY, 59- *Concurrent Pos:* consult, Agr Solar Energy, Ecuador, Jamaica & Iraq. *Mem:* Fel Am Soc Agr Engrs; Int Solar Energy Soc. *Res:* Environmental design and research on requirements for domestic animals, poultry and animal care facilities; solar thermal energy research; development of air type solar collectors (two patents); development of heat storages and drying agricultural products. *Mailing Add:* Dept Agr Eng Univ Ky Lexington KY 40546-0276

PARKER, BARRY RICHARD, b Penticton, BC, Apr 13, 35; US citizen; m 61; c 1. GENERAL RELATIVITY. *Educ:* Univ BC, BA, 59, MSc, 61; Utah State Univ, PhD(physics), 68. *Prof Exp:* Lectr physics, Weber State Col, 63-65, asst prof, 65-66; from asst prof to assoc prof, 67-75, PROF PHYSICS, IDAHO STATE UNIV, 75- *Honors & Awards:* McDonald Observ Writing Award. *Mem:* Am Phys Soc; Sigma Xi. *Res:* Unified field theories; cosmology and black holes. *Mailing Add:* Dept of Physics Idaho State Univ Pocatello ID 83201

PARKER, BEULAH MAE, b Oak City, NC, May, 9, 43. MEDICAL & VETERINARY ENTOMOLOGY. *Educ:* Elizabeth City State Col, BS, 65; Ill State Univ, MS, 69; Univ Ill, PhD(entomol), 77. *Prof Exp:* Instr biol, Ill State Univ, 69-74, asst prof biol, 77-79; asst prof, 79-85, ASSOC PROF ENTOMOL, NC STATE UNIV, 85- *Concurrent Pos:* Vis scientist, Dept Microbiol, Colo State Univ, 89-90. *Mem:* Entomol Soc Am; Am Mosquito Control Asn; Nat Tech Asn; Sigma Xi. *Res:* Biology of mosquitoes. *Mailing Add:* Dept Entomol NC State Univ Box 7613 Raleigh NC 27695-7613

PARKER, BRENT M, b St Louis, Mo, July 3, 27; m 53; c 3. INTERNAL MEDICINE, CARDIOLOGY. *Educ:* Wash Univ, MD, 52. *Prof Exp:* Sect chief cardiol, Vet Admin Hosp, Portland, Ore, 57-59; from instr to to assoc prof med, Wash Univ, 59-73; prof med, 73-89, assoc dean & assoc hosp dir, 76-82, chief cardiol, Sch Med, 83-89, EMER PROF MED, UNIV MO, COLUMBIA, 89- *Concurrent Pos:* USPHS res fel, Wash Univ, 55-57, USPHS res grant, 60-66; consult, Ft Leonard Wood, Mo, 63-70; fel coun clin cardiol, Am Heart Asn, 69; chief of staff, Univ Mo Hosp, 76-82; award prevent cardiol, Nat Heart, Lung & Blood Inst, 82-87. *Mem:* Fel Am Col Cardiol; Am Fedn Clin Res; Am Col Physicians; Cent Soc Clin Res. *Mailing Add:* Sch of Med Univ Mo MA432 807 Stadium Rd Columbia MO 65212

PARKER, BRUCE C, b Rockingham, Vt, June 20, 33; m 61; c 2. PHYCOLOGY. *Educ:* Tufts Univ, BS, 55; Yale Univ, MS, 57; Univ Tex, PhD(bot), 60. *Prof Exp:* NSF fel phycol, Univ Col, London, 60-61; asst prof bot, Univ Calif, Los Angeles, 61-65; assoc prof, Washington Univ, 65-69; PROF BIOL, VA POLYTECH INST & STATE UNIV, 69- *Concurrent Pos:* NSF res grants, 62-85; historian, Phycol Soc Am, 71-; NIH res grants, 74-90. *Honors & Awards:* Darbaker Prize, Bot Soc Am, 70. *Mem:* Phycol Soc Am (secy-treas, 65-67, vpres, 68, pres, 69); Am Soc Limnol & Oceanog; Int Phycol Soc; Bot Soc Am. *Res:* Algal ecology and physiology; microbiology; Antarctic freshwater ecosystems; chemical composition of glacial ice and snow; nontuberculous mycobacteria in human disease. *Mailing Add:* Dept Biol Va Polytech Inst Blacksburg VA 24061

PARKER, CALVIN ALFRED, b Chicago, Ill, Sept 20, 31; m 56; c 3. GEOLOGY. *Educ:* Northwestern Univ, BS, 53; Univ Wis, MS, 56, PhD(geol), 58. *Prof Exp:* Prod geologist, Shell Develop Co, 57-60, staff prod geologist, Shell Oil Co, Tex, 60-67, Royal Dutch Shell Group, Netherlands & Brunei Shell Petrol Co, Borneo, 67-69, staff geol engr, oil co, Tex, 69-70, sr staff geol engr, Shell Oil Co, La, 70-79, ENGR AD, SHELL OIL, HOUSTON, 79- *Concurrent Pos:* Int mgr prod geologist, Occidental Petrol Co. *Res:* Production geology; paleoecology; geochemistry; stratigraphy. *Mailing Add:* 204 Via Carisma Bakersfield CA 93309

PARKER, CAROL ELAINE GREENBERG, b Fayetteville, NC, Dec 25, 47; m 72. ANALYTICAL CHEMISTRY, MASS SPECTROMETRY. *Educ:* Cornell Univ, BA, 69; Univ NC, Chapel Hill, MS, 73, PhD(chem), 77. *Prof Exp:* Res technician chem, Div Environ Sci & Eng, Univ NC, Chapel Hill, 73-76; vis assoc, 76-77, RES CHEMIST, NAT INST ENVIRON HEALTH SCI, 77- *Concurrent Pos:* Res assoc, Dept Plant Sci, NC A&T State Univ, 76- *Mem:* Am Soc Mass Spectrometry; Am Chem Soc; Am Soc Agron. *Res:* Application of analytical chemistry, especially gas chromatography; mass spectrometry to problems in the analysis of biological and/or environmental materials. *Mailing Add:* Rte Three Box 232 Apex NC 27502

PARKER, CHARLES D, b Missoula, Mont, Aug 22, 24; m 49; c 3. AUDIOLOGY. *Educ:* Mont State Univ, BA, 49; Univ Iowa, MA, 52, PhD, 53. *Prof Exp:* Dir, Speech & Hearing Clin, Univ Mont, 57-84, assoc prof, 58-66, prof speech path & audiol & chmn dept, 66-89; RETIRED. *Concurrent Pos:* Adv, US Dept Army, 53-54. *Mem:* Acoust Soc Am; Am Speech & Hearing Asn. *Res:* Speech pathology; psychoacoustics. *Mailing Add:* Box 335 Lolo MT 59847

PARKER, CHARLES J, JR, b Tonawanda, NY, Apr 3, 30. BIOCHEMISTRY. *Educ:* Univ Buffalo, BA, 51, MA, 56, PhD, 58. *Prof Exp:* Mem fac, 60-70, ASSOC PROF BIOCHEM, COL MED, WAYNE STATE UNIV, 70- *Concurrent Pos:* Res fel med, Retina Found, Mass Gen Hosp, 58-60. *Mem:* AAAS; Am Chem Soc. *Res:* Muscle biochemistry. *Mailing Add:* Dept Biochem 4374 Scott Hall Wayne State Univ Sch Med 540 E Canfield Detroit MI 48201

PARKER, CHARLES W, b St Louis, Mo, Mar 23, 30; m 53; c 5. INTERNAL MEDICINE, IMMUNOLOGY. *Educ:* Wash Univ, MD, 53. *Prof Exp:* Instr, 60, from asst prof to assoc prof, 63-66, PROF MED, SCH MED, WASH UNIV, 66- *Concurrent Pos:* Allergy Found fel, Sch Med, Wash Univ, 59-61, USPHS res fel, 61-62; USPHS res & training grants, 62-, assoc prof microbiol, 73-76, dir div immunol, 62-76. *Mem:* Am Soc Clin Invest; Am Acad Allergy; Am Asn Immunol. *Res:* Human penicillin allergy; immunochemistry. *Mailing Add:* Dept Med Sch Med Wash Univ St Louis MO 63110

PARKER, CLEOFUS VARREN, JR, b Houston, Tex, Dec 20, 37; m 61; c 2. RADIATION PHYSICS. *Educ:* Sam Houston State Univ, BA, 61, MS, 62; Univ Tex, Austin, PhD(physics), 69. *Prof Exp:* Instr physics, Sam Houston State Univ, 62-63; res scientist, Tex Nuclear Div, Nuclear-Chicago Corp, 63-68, proj leader, 68-69; assoc prof, 69-74, head dept, 74-76, PROF PHYSICS, ANGELO STATE UNIV, 74- *Mem:* Am Asn Physics Teachers; Sigma Xi. *Res:* Dosimetry techniques for simulated solar proton radiation; thermoluminescent dosimetry; neutron activation analysis; fast neutron cross section measurements; electron spin resonance; computer techniques; microcomputer applications. *Mailing Add:* 2802 Briargrove Lane San Angelo TX 76904

PARKER, CURTIS LLOYD, cell biology, for more information see previous edition

PARKER, DAVID CHARLES, b New York, NY, Oct 8, 45; m 72; c 2. CELLULAR IMMUNOLOGY, LYMPHOCYTE ACTIVATION. *Educ:* Haverford Col, AB, 66; Univ Calif, Berkeley, PhD(immunol), 71. *Prof Exp:* Asst prof biol, Hampshire Col, 71-72; fel immunol, Univ Calif, Berkeley, 72-73 & Univ Col, London, 73-76; asst prof, 76-80, assoc prof, 80-87, PROF, DEPT MOLECULAR GENETICS & MICROBIOL, UNIV MASS MED SCH, 87- *Concurrent Pos:* Assoc ed, J Immunol, 83-87; mem, immunobiol study sect, NIH, 83-86; vis scientist, Whitehead Inst, 86-87. *Mem:* Am Asn Immunologists. *Res:* Intercellular signals which regulate lymphocyte activation and antibody response; cell surface events. *Mailing Add:* Dept Molecular Genetics & Microbiol Univ Mass Med Sch Worcester MA 01655

PARKER, DAVID GARLAND, b Washington, DC, Nov 6, 40; m 64; c 2. ENVIRONMENTAL & CIVIL ENGINEERING. *Educ:* Va Polytech Inst & State Univ, BS, 65, MS, 68, PhD(civil eng), 71. *Prof Exp:* Instr eng technol, Roanoke Tech Inst, 65-66; asst prof, Wytheville Community Col, 67-69; asst prof civil eng, Lamar Univ, 71-72; asst prof, 72-81, PROF CIVIL ENG, UNIV ARK, FAYETTEVILLE, 81- *Concurrent Pos:* Wastewater consult, Mobil Oil Co, 71-72. *Mem:* Am Soc Civil Engrs; Water Pollution Control Fedn; Am Soc Eng Educ. *Res:* Wastewater treatment; sludge treatment; solid waste disposal; water treatment. *Mailing Add:* Dept of Civil Eng Univ of Ark Fayetteville AR 72701

PARKER, DAVID H, b July 4, 53. CHEMISTRY. *Educ:* Univ Calif, Irvine, BA, 75, Los Angeles, PhD(phys chem), 79. *Prof Exp:* Fel, Columbia Univ, 79-81; asst prof chem, Univ Calif, Santa Cruz, 81-90. *Mem:* Am Chem Soc; Am Phys Soc. *Mailing Add:* c/o Prof J Reuss Laser & Molecular Physics Dept KUN Nijmegen Netherlands

PARKER, DON EARL, b Gainesville, Tex, Aug 31, 32; m 61; c 2. APPLIED STATISTICS. *Educ:* NTex State Univ, BA, 57, MA, 60; Univ Okla, PhD(biostatist), 70. *Prof Exp:* Asst prof math, Austin Col, 60-64; asst prof, 69-74, assoc prof, 74-77, PROF BIOSTATIST, OKLA HEALTH SCI CTR, 77- *Concurrent Pos:* Statist Consult, Okla Med Res Found, 75- & FAA, 73- *Mem:* Am Statist Asn; Am Public Health Asn. *Res:* Application of statistical techniques to biological and medical research problems & development of computer techniques for these applications. *Mailing Add:* 1313 Shalamar Edmond OK 73013

PARKER, DONAL C, b Pittsburgh, Pa, Aug 13, 31; m 56; c 1. ENDOCRINOLOGY, INTERNAL MEDICINE. *Educ:* Univ Pittsburgh, BS, 53, MD, 57. *Prof Exp:* Intern, Cleveland Clin Educ Found, 57-58, from resident to sr resident, 61-63, chief resident internal med, 63-64; assoc endocrinol, Scripps Clin & Res Found, 67-72, assoc mem, 72-73; clin investr, 73-76, STAFF PHYSICIAN, SAN DIEGO VET ADMIN MED CTR, 76- *Concurrent Pos:* Spec fel endocrinol & metab, Cleveland Clin, 64-65; res trainee endocrinol, Scripps Clin & Res Found, 65-67; asst adj prof med, Sch Med, Univ Calif, San Diego, 69-72, asst prof med-in-residence, 73-75, assoc adj prof med, 75-79, clin prof med, 79- *Mem:* Western Soc Clin Res; Am Fedn Clin Res; Endocrine Soc; Int Soc Chronobiol. *Res:* Neuroendocrine control mechanisms; pituitary peptide hormone release; hormone release across the sleep-wake cycle; biologic rhythms. *Mailing Add:* San Diego Vet Admin Med Ctr Rm 2264 3350 La Jolla Village Dr San Diego CA 92161

PARKER, DONALD LESTER, b Mobile, Ala, Dec 19, 44. ELEMENTARY PARTICLE PHYSICS. *Educ:* Spring Hill Col, BS, 66; Mich State Univ, PhD(physics), 71. *Prof Exp:* Res assoc. Mich State Univ, 72-75, instr, 74-75; asst prof physics, Iowa State Univ, 75-88; UNIV ALA, 88- *Mem:* Am Phys Soc; AAAS; Fedn Am Scientists. *Res:* Experimental elementary particle research; studies of the strong force and hadronic states of matter. *Mailing Add:* 6420 Tokeneak Trail Mobile AL 36609

PARKER, DONN BLANCHARD, b San Jose, Calif, Oct 9, 29; m 52; c 2. COMPUTER SECURITY & CRIMINOLOGY. *Educ:* Univ Calif, Berkeley, BA, 52, MA, 54. *Prof Exp:* Sr res eng comput, Gen Dynamics, San Diego, 54-62; sr res consult comput, Control Data Corp, Palo Alto, 62-69; SR MGR CONSULT COMPUT SECURITY, SRI INT, MENLO PARK, 69- *Mem:* Asn Comput Mach (secy, 65-75); Am Soc Indust Security. *Res:* Computer and information crime and security. *Mailing Add:* SRI Int 333 Ravenswood Ave Menlo Park CA 94025

PARKER, DOROTHY LUNDQUIST, b Webster, SDak, Sept 13, 39; m 71. VIROLOGY. *Educ:* Univ SDak, BA, 61; Univ Calif, Berkeley, MA, 64, PhD(molecular biol), 71. *Prof Exp:* Res technician molecular biol, Virus Lab, Univ Calif, Berkeley, 64-66; asst prof, 71-76, ASSOC PROF BIOL, HALSEY SCI CTR, UNIV WIS-OSHKOSH, 77- *Mem:* Sigma Xi; Am Soc Microbiol; AAAS. *Res:* Biology of blue-green algae, cyanobacteria; isolation and characterization of viruses, cyanophages, infecting blue-green algae. *Mailing Add:* 5024 Island View Dr Oshkosh WI 54901

PARKER, E T, b Royal Oak, Mich, July 26, 26; m 55; c 1. MATHEMATICS. *Educ:* Northwestern Univ, BA, 47; Ohio State Univ, PhD(math), 57. *Prof Exp:* Res assoc math, Ohio State Univ, 57; instr, Univ Mich, 57-58; prin mathematician, Univac Div, Sperry Rand Corp, Minn, 58-64; from assoc prof to prof, 64-87, EMER PROF MATH, UNIV ILL, URBANA, 87- *Concurrent Pos:* Lectr, Univ Minn, 60-62. *Mem:* Am Math Soc. *Res:* Orthogonal latin squares and finite projective planes; finite groups; graphs; application of computers. *Mailing Add:* Dept Math Univ Ill 1409 W Green St Urbana IL 61801

PARKER, EARL ELMER, b Malvern, Ark, Sept 22, 18; m 41; c 2. ORGANIC POLYMER CHEMISTRY. *Educ:* Miss State Col, BS, 40; Ind Univ, PhD(chem), 43. *Prof Exp:* Res chemist, PPG Industs Inc, 43-50, res supvr, 50-54, dept head, 54-58, res assoc, 58-63, sr res assoc, 64-65, scientist, 65-81; RETIRED. *Mem:* AAAS; Am Chem Soc. *Res:* Synthesis of amino acids; polyester resins; oil bodying catalysts; polyurethanes; epoxy resins; surface coatings; unsaturated polyester resins. *Mailing Add:* 1940 Wolf Laurel Dr Sun City Ctr FL 33570-6414

PARKER, EARL RANDALL, b Denver, Colo, Nov 22, 12; m 35; c 3. PHYSICAL METALLURGY. *Educ:* Colo Sch Mines, MetE, 35. *Prof Exp:* Res metallurgist, Res Lab, Gen Elec Co, 35-44; from assoc prof to prof, 44-78, chmn div mat sci & engr, 53-57, dir inst eng res, 57-64, emer prof phys metall, Univ Calif, Berkeley, 78-88; RETIRED. *Concurrent Pos:* Guggenheim fel, 60. *Honors & Awards:* Mathewson Gold Medal, Am Inst Mining, Metall & Petrol Engrs, 56; Sauveur Award, Am Soc Metals, 64, Gold Medal, 72; Bendix Gold Medal, Am Soc Eng Educ, 69; Nat Medal Sci, Presented by Pres USA, Jimmy Carter, 79. *Mem:* Nat Acad Eng; fel Am Soc Metals (pres, 68); fel Am Inst Mining, Metall & Petrol Engrs; fel Am Phys Soc. *Res:* Mechanical behavior of materials. *Mailing Add:* Dept Mat Sci & Mining Eng Univ Calif 2210 Hearst Mining Bldg Berkeley CA 94720

PARKER, EHI, b Benin City, Nigeria, Oct 24, 52; US citizen; m 81; c 2. INSTRUMENTATION. *Educ:* Manila Univ, Phillipines, BSc, 76. *Prof Exp:* Technician, MCC Powers, 76-81; ENGR, LANDIS & GYRS POWERS, 82- *Concurrent Pos:* Technician, ITT Telecommun, Des Plaines, Ill, 74-76. *Mem:* Assoc mem Inst Elec & Electronic Engrs London. *Mailing Add:* Landis & Gyrs Powers 1000 Deerfield Pkwy Buffalo Grove IL 60089-4510

PARKER, EUGENE NEWMAN, b Houghton, Mich, June 10, 27; m 54; c 2. ASTROPHYSICS. *Educ:* Mich State Univ, BS, 48; Calif Inst Technol, PhD(physics), 51. *Hon Degrees:* DSc, Mich State Univ, 75; Dr, Univ Utrecht, 86. *Prof Exp:* Instr math & astron, Univ Utah, 51-53, asst prof physics, 53-55; res assoc, Fermi Inst, Univ Chicago, 55-57, from asst prof to assoc prof, 57-62, chmn dept physics, 70-72, chmn dept astron, 72-78, PROF PHYSICS, ENRICO FERMI INST, UNIV CHICAGO, 62-, PROF ASTRON, 67- *Concurrent Pos:* Chmn, Astron Sect, Nat Acad Sci, 83-86. *Honors & Awards:* Space Sci Award, Am Inst Aeronaut & Astronaut, 62; John Fleming Award, Am Geophys Union, 68; Henry Norris Russell Lectr, Am Astron Soc, 69, George Ellery Hale Award, 78; Henry Arctowski Award, Nat Acad Sci, 69; Sidney Chapman Award, Royal Astron Soc, 79; Nat Medal of Sci, 89; William Bowie Medal, Am Geophys Union, 90; Karl Schwarzschild Medal, Astronomischen Gesellschaff, 90. *Mem:* Nat Acad Sci; Am Phys Soc; Am Astron Soc; Am Geophys Union; Norweg Acad Sci & Letts. *Res:* Theoretical plasma physics; magnetohydrodynamics; solar and terrestrial physics; basic physics of the active star; application and extension of classical physics to the active conditions found in the astronomical universe e.g. the stellar x-ray corona, the solar wind and the origin of stellar and galactic magnetic fields. *Mailing Add:* Lab Astrophys & Space Res 933 E 56th St Chicago IL 60637

PARKER, FRANCES LAWRENCE, b Brookline, Mass, Mar 28, 06. MICROPALEONTOLOGY. *Educ:* Vassar Col, AB, 28; Mass Inst Technol, MS, 30. *Prof Exp:* Res asst, Cushman Lab, 30-40; sr paleontologist, Shell Oil Co, 43-45; mem staff micropaleont, Woods Hole Oceanog Inst, 47-50; from asst res geologist to assoc res geologist, Scripps Inst Oceanog, Univ Calif, San Diego, 50-66, res paleontologist, 66-73, res assoc, 73-; RETIRED. *Concurrent Pos:* Fel, Woods Hole Oceanog Inst, 37-40; dir & hon fel, Cushman Found Foraminiferal Res, 64- *Honors & Awards:* Joseph A Cushman Award. *Res:* Fossil and recent foraminifera; stratigraphic and ecological studies. *Mailing Add:* 334 W Bourne La Jolla CA 92037

PARKER, FRANCIS DUNBAR, b Boston, Mass, July 27, 18; m 42; c 4. APPLIED MATHEMATICS. *Educ:* Middlebury Col, AB, 39; Boston Univ, AM, 42; Case Inst Technol, PhD(math), 51. *Prof Exp:* Instr, Cranbrook Sch, 39-40; instr math, Staunton Mil Acad, 40-41 & Montclair Acad, 41-42; mem staff, Radiation Lab, Mass Inst Technol, 42-45; instr math, Case Inst Technol, 45-51; asst prof, St Lawrence Univ, 51-53; from assoc prof to prof, Clarkson Col Technol, 53-57; prof & head dept, Univ Alaska, 57-63, chmn div sci & math, 58-61; prof math & exec off dept, State Univ NY Buffalo, 63-66; PROF MATH, ST LAWRENCE UNIV, 66- *Concurrent Pos:* Vis prof, Univ Aberdeen, Scotland, 73, Univ York, Eng, 74, Calif State Univ, Los Angeles, 75, 77-78 & 81-82 & Plymouth Polytech, Eng, 78-79. *Mem:* Soc Indust & Appl Math; Am Math Soc; Math Asn Am. *Res:* Microwave propagation; radar antenna design; graph theory. *Mailing Add:* Star Rte Box 2 Canton NY 13617

PARKER, FRANK L, b Somerville, Mass, Mar 23, 26; m 54; c 4. HYDROLOGY & WATER RESOURCES. *Educ:* Mass Inst Technol, SB, 48; Harvard Univ, PhD(civil eng), 55. *Prof Exp:* Engr, US Bur Reclamation, 48; civil engr, Rockland Light & Power Co, NY, 49; res assoc, Harvard Univ, 50-55; consult hydraul engr, Howard M Turner, Mass, 55; eng leader, Oak Ridge Nat Lab, 56-60, sect chief radioactive waste disposal, 62-67; assoc prof environ & water resources eng, 67-68, PROF ENVIRON & WATER RESOURCES ENG, VANDERBILT UNIV, 68- *Concurrent Pos:* Mem, Int Atomic Energy Agency, Austria, 61; assoc prof, Univ Tenn; vis lectr, Vanderbilt Univ; consult, Ital, Pakistan & Israeli AEC; lectr, UK Atomic Energy Authority; mem, comt waste isolation pilot plant, Nat Acad Sci, 69, chmn, bd radioactive waste mgt, 78-; consult, WHO, 68-; mem panel water qual, UNESCO, 69; consult, adv comt reactor safeguards, Nat Res Coun, 75-, Dept Energy, Battelle Mem Inst, 79-86 & Westinghouse Hanford, 79- *Mem:* Nat Acad Eng; AAAS; Am Soc Civil Engrs; Health Physics Soc; Am Geophys Union; Am Nuclear Soc; Water Pollution Control Fedn. *Res:* Geological and environmental radioactive and hazardous chemical waste disposal; water resources; thermal pollution; environmental effects of energy production; sanitary and environmental engineering. *Mailing Add:* Dept Environ & Water Resources Eng Vanderbilt PO Box 1596 Sta B Nashville TN 37235

PARKER, FRANK S, b Boston, Mass, Jan 25, 21; m 46; c 1. BIOCHEMISTRY, SPECTROCHEMISTRY. *Educ:* Tufts Col, BS, 42, MS, 44; Johns Hopkins Univ, PhD(chem), 50. *Prof Exp:* Asst chem, Tufts Col, 42-44; jr instr, Johns Hopkins Univ, 46-50; asst prof biol, Bryn Mawr Col, 50-54; from asst prof to assoc prof, State Univ NY Downstate Med Ctr, 54-63; assoc prof, 63-70, PROF BIOCHEM, NEW YORK MED COL, 70- *Concurrent Pos:* Vis scientist, Mass Inst Technol, 82-84 & 87. *Honors & Awards:* Career scientist Award, Health Res Coun New York, 63-69. *Mem:* Fel AAAS; Am Chem Soc; Am Soc Biochem & Molecular Biol; Soc Appl Spectros; Sigma Xi; Can Spectros Soc; Biophys Soc. *Res:* Infrared studies with enzymes, amino acids, peptides and carbohydrates; aqueous solution infrared spectroscopy; attenuated total reflectance spectroscopy; binding studies of drugs with biological materials; hydrogen bonding studies; biophysics; analytical chemistry; Raman spectroscopy. *Mailing Add:* Dept Biochem & Molecular Biol NY Med Col Basic Sci Bldg Valhalla NY 10595

PARKER, FREDERICK JOHN, b East Orange, NJ, June 17, 52. CERAMICS ENGINEERING, MATERIALS SCIENCE. *Educ:* Upsala Col, BS, 74; Rutgers Univ, MS, 81. *Prof Exp:* Technician, Corp Res Labs, Exxon Res & Eng, 74-81, technologist & engr, 81-84; sr ceramic engr, 84-86, RES CERAMIST, TECH CERAMICS DEPT, WR GRACE & CO, 86- *Mem:* Am Ceramic Soc. *Res:* Ceramics processing and tailoring of properties by altering crystal chemistry; low thermal expansion materials; ceramic catalyst supports. *Mailing Add:* 6714 Bushranger Path Columbia MD 21046

PARKER, GARALD G, SR, b Leona, Ore, July 2, 05; m 26; c 7. HYDROLOGY & WATER RESOURCES. *Educ:* Cent Wash Col, AB, 35; Univ Wash, Seattle, MS, 46. *Hon Degrees:* PhD, Univ S Fla, 88. *Prof Exp:* Jr geologist to prin geologist, US Geol Surv, 40-69; chief hydrologist & sr scientist, Southwest Fla Water Mgt Dist, Brooksville, Fla, 69-75; sr hydrologist hydrogeol res, P E Lamoreaux & Assoc, 75-76; sr scientist geol & hydrol res, Geraghty & Miller, 76-78; founder & pres hydrogeol res, Parker & Assoc, 78-87; RETIRED. *Concurrent Pos:* Mem, US Civil Serv, Bd Expert Geol Examr, 50-60; prof affil, Am Univ, Washington, DC, 55-56; mem, Hwy Res Bd, Nat Acad Sci, 65-69, US Water Res Coun, Nat Water Assessment, 66, Int Geophys Yr Comn, 67-69; vis prof, Agr Eng Dept, Univ Fla, 75-80. *Honors & Awards:* Gold Medal, Fla Acad Sci, 45; Presidential Award, Am Water Resources Asn, 68. *Mem:* AAAS; Am Geophys Union; Am Inst Hydrol; Am Inst Mining, Metall & Petrol Engrs; Am Water Resources Asn (pres, 67-68); Geol Soc Am; Int Asn Sci Hydrol. *Res:* Ground-water geology and hydrology; river-basin hydrology and geology; geomorphology; history of hydrogeology; ecology of Southeastern Florida; geophysics (salt water-fresh water relationships control and management); subterranean erosion (piping); conservation of water resources; exploration, development, management and protection of water supplies. *Mailing Add:* 137 Avenue E Winter Haven FL 33880

PARKER, GEORGE RALPH, b Tulsa, Okla, Sept 16, 42; m 65; c 2. FOREST ECOLOGY, SILVICULTURE. *Educ:* Okla State Univ, BS, 64, MS, 67; Mich State Univ, PhD(forest ecol), 70. *Prof Exp:* ASSOC PROF FOREST ECOL, PURDUE UNIV, LAFAYETTE, 70- *Mem:* Ecol Soc Am; Soc Am Foresters; Nature Conservancy; Sigma Xi. *Res:* Structure and functioning of forested ecosystems. *Mailing Add:* 3570 Division Rd Purdue Univ West Lafayette IN 47906

PARKER, GEORGE W, b Ft Worth, Tex, May 23, 39; m 67; c 3. ATOMICS & MOLECULAR PHYSICS. *Educ:* Univ of the South, BA, 61; Univ SC, PhD(physics), 65. *Prof Exp:* From instr to asst prof, 65-75, ASSOC PROF PHYSICS, NC STATE UNIV, 75- *Mem:* Am Phys Soc; Am Asn Physics Teachers. *Res:* Numerical solution of orbital-free density functional models for molecules; physics education. *Mailing Add:* Dept Physics NC State Univ PO Box 8202 Raleigh NC 27695-8202

PARKER, GORDON ARTHUR, b 1936; m 62. ANALYTICAL CHEMISTRY. *Educ:* Univ Mich, BS, 58; Wayne State Univ, MS, 62, PhD(anal chem), 66. *Prof Exp:* From instr to assoc prof, 65-84, prof chem, 84-88, PROF CHEM & PHARM PRAC, UNIV TOLEDO, 88- *Mem:* Am Chem Soc; Soc Appl Spectros; Soc Col Sci Teachers. *Res:* Spectrophotometric and polarographic analysis. *Mailing Add:* Dept Chem Univ Toledo Toledo OH 43606-3390

PARKER, H DENNISON, b Lake Wales, Fla, June 20, 41; m 61; c 2. ENVIRONMENTAL PUBLIC RELATIONS. *Educ:* Colo State Univ, BS, 68, PhD(range ecol), 72. *Prof Exp:* Prin scientist ecol remote sensing, Lockheed Electronics Co, NASA, Houston, 72-73; staff ecologist synthetic fuels develop, Cameran Engrs, Denver, 73-74; mgr ecosystems analysis, Western Sci Serv, Colo, 74-76; remote sensing specialist, US Fish & Wildlife Serv, 76-78, leader, Data Systs Group, Western Energy Team, 78-80; PRES, PARKER SCI, INC, 80- *Mem:* Am Soc Photogram; Sigma Xi; AAAS. *Res:* Application of remote sensing to wildlife census; habitat inventory; mapping and analysis. *Mailing Add:* 1400 Independence Rd Ft Collins CO 80526

PARKER, HAROLD R, b Los Angeles, Calif, Jan 10, 20; m 49; c 5. RENAL PHYSIOLOGY, ANESTHESIOLOGY. *Educ:* Univ Calif, Davis, BS, 50, DVM, 52, PhD(comp path), 61. *Prof Exp:* Res assoc radiobiol, 56-58, res assoc physiol, 58-61, asst prof, 61-67, assoc prof physiol, 67-78, PROF SURG, UNIV CALIF, DAVIS, 78- *Concurrent Pos:* Sacramento-Yolo-Sierra Heart Asn fel, 68-69. *Mem:* AAAS; Am Physiol Soc; Am Soc Vet Physiol & Pharmacol; Am Heart Asn; Sigma Xi. *Res:* Renal function, fluid and electrolyte balance in domestic animals; intensive care medicine, especially renal failure, hemofiltration, hemo and peritoneal dialysis; experimental surgery. *Mailing Add:* Dept of Surg Univ of Calif Sch of Vet Med Davis CA 95616

PARKER, HARRY W(ILLIAM), b Tulia, Tex, June 4, 32; m 54; c 2. CHEMICAL ENGINEERING. *Educ:* Tex Technol Col, BS, 53; Northwestern Univ, MS, 54, PhD(chem eng), 56. *Prof Exp:* Theoret develop engr, Phillips Petrol Co, 56-62, res group leader, 62-70; from assoc prof to prof chem eng, Tex Tech Univ, 70-79; Engr, Eng Soc Comn Energy, 79-81; PROF CHEM ENG, TEX TECH UNIV, 81- *Mem:* Soc Petrol Engrs; Am Inst Chem Engrs; Am Chem Soc. *Res:* Processes for renovation of contaminated soil and water; enhanced oil recovery processes; economic comparisons of conventional and alternative energy sources; non-food use of agricultural products; recipeint of 80 US patents. *Mailing Add:* Dept Chem Eng Tex Tech Univ Lubbock TX 79409-3121

PARKER, HELEN MEISTER, b Chicago, Ill, July 17, 35; m 55; c 1. PHYSICAL BIOCHEMISTRY, PROTEIN CHEMISTRY. *Educ:* Ohio State Univ, BSc, 56, MSc, 57; Univ Minn, PhD(phys chem), 67. *Prof Exp:* Res assoc biochem, Univ Ill, Urbana, 67-70 & 84-89, animal sci, 70-84, res specialist food sci, 89, RES SPECIALIST VET PATHOBIOL, UNIV ILL, URBANA, 89- *Mem:* AAAS. *Res:* Protein physical chemistry; enzyme structure and function; mineral metabolism; parathyroid function; mineral bioavailability; biochemistry of bacterial chemotaxis; carcinogenesis; tissue culture; pulmonary toxicology; lysosomal enzymes. *Mailing Add:* Dept Vet Pathobiol Univ Ill 2001 S Lincoln Urbana IL 61801

PARKER, HENRY SEABURY, III, b Newport, RI, Apr 19, 44; m 70; c 2. AQUACULTURE, PHYCOLOGY. *Educ:* Harvard Univ, BA, 66; Univ RI, MA, 71, PhD(biol oceanog), 79. *Prof Exp:* Mgr aquaculture, Marine Colloids Div, FMC Corp, 71-73; vis lectr aquacult, 79-80, asst prof, 80-86, ASSOC PROF AQUACULT & BIOL, BIOL DEPT, SOUTHEASTERN MASS UNIV, 86-; VPRES, PAC SEA RESOURCES, INC, 87- *Concurrent Pos:* Consult, Nat Alcohol Fuels Comn, 78-80 & Nonquitt Asn, 80-84; NSF co-prin investr, Coastal & Estuarine Marine Biol Pre-Col Sch Teachers, 80-81; chmn ed comt & bd mem, Lloyd Environ Ctr Coastal & Estuarine Studies; exec dir, Northeastern Regional Aquacult Ctr, 90-; bd mem, Coalition Buzzards Bay, Mass. *Mem:* Am Soc Limnol & Oceanog; Phycol Soc Am; World Aquacult Soc; Int Phycol Soc; AAAS; Sigma Xi. *Res:* Physiological ecology of marine macroalgae, particularly interacting effects of nutrients, light and water motion on growth and nitrogen metabolism; cultivation of macroalgae. *Mailing Add:* Biol Dept Southeastern Mass Univ North Dartmouth MA 02747

PARKER, HENRY WHIPPLE, b Concord, NH, May 31, 24; m 52; c 4. CIVIL ENGINEERING. *Educ:* Dartmouth Col, BS, 45, MS, 47. *Prof Exp:* Estimating engr, Winston Bros & Co, W Coast, 47-50, estimator, Off Engr Supt, Colombia, 52-56, asst proj mgr construct, Calif, 59-60, proj engr, 60-62, proj engr, Atlas-Winston Ltd, Que, 56-69; assoc prof, 62-72, prof civil eng, Stanford Univ, 72-82. *Mem:* Am Soc Civil Engrs; Am Soc Eng Educ. *Res:* Methods improvement techniques; safety management. *Mailing Add:* 430 Kingsley Palo Alto CA 94301

PARKER, HERBERT EDMUND, b Springville, Tenn, Oct 21, 19; m 45; c 3. BIOCHEMISTRY. *Educ:* Univ Tenn, BSA, 42; Purdue Univ, MS, 48, PhD(biochem), 50. *Prof Exp:* From asst prof to prof biochem, Purdue Univ, Lafayette, 50-85, asst head dept, 77-85; RETIRED. *Mem:* Am Inst Nutrit; Am Soc Biol Chem. *Res:* Mineral metabolism; inorganic nutrition of plants and animals. *Mailing Add:* Dept Biochem Purdue Univ West Lafayette IN 47907

PARKER, HERBERT MYERS, health physics, for more information see previous edition

PARKER, HOWARD ASHLEY, JR, b Sydney, Australia, Aug 20, 22; m 50; c 4. CHEMICAL ENGINEERING. *Educ:* Cornell Univ, BChE, 44. *Prof Exp:* Process design engr, 44-46, group leader, Process Design Sect, 46-51, tech dir, Destrehan Refinery, Pan-Am Southern Corp, 51-56; asst mgr opers, Mfg Dept, Amoco Oil Co, 56-57, asst to gen mgr mfg, 57-59, mgr coord & supply planning, 59-62, admin mgr, Atlanta Region, Mkt Dept, 62-63, regional mgr, Kansas City Region, 63-65, gen mgr reseller mkt, Gen Off, 65-68, gen mgr mkt staff serv, 68-70, gen mgr com mkt, 70-74, pres ceo & dir, Amoco Pipeline Co, 74-86; RETIRED. *Concurrent Pos:* Dir, Atlas Supply Co, 65-68; exec vpres, Harvam Corp, 65-68; pres, Tempo Designs, Inc, 68-70, Amoco Enterprises, Inc, 68-70 & Apex Terminals, Inc, 70-74; dir, Asphalt Inst, 70-74, Colonial Pipeline Co, 74-86 & Chicap Pipeline Co, 74-86; pres & dir, W Shore Pipeline Co, 74-78, Wyco Pipeline Co, 74-78, Dixie Pipeline Co, 78-86. *Mem:* Am Inst Chem Engrs; Am Petrol Inst; Asn Oil Pipelines. *Res:* Plant design. *Mailing Add:* 2240 Imperial Golf Course Blvd Naples FL 33942-1029

PARKER, JACK LINDSAY, b Springville, Utah, June 15, 30; m 56; c 10. NUCLEAR PHYSICS, APPLIED PHYSICS. *Educ:* Brigham Young Univ, BS, 55, MS, 56; Univ Utah, PhD(physics), 67. *Prof Exp:* Physicist, Wright Air Develop Ctr, 58-60; staff mem, 67-88, FEL, LOS ALAMOS NAT LAB, UNIV CALIF, 88- *Mem:* Inst Nuclear Mat Mgt. *Res:* Low energy nuclear physics; gamma ray spectroscopy; nondestructive assay of special nuclear materials. *Mailing Add:* 342 Kimberly Lane White Rock NM 87544

PARKER, JACK STEELE, b Palo Alto, Calif, July 6, 18; m 46; c 1. AERONAUTICAL ENGINEERING. *Educ:* Leland Stanford Jr Univ, BS, 39. *Hon Degrees:* MBA, Southeastern Mass Univ; LLB, Clark Univ; LLD, Rensselaer Polytech Inst. *Prof Exp:* Mech engr, Western Pipe & Steel Co Calif, San Francisco, 39-40; marine survr, Am Bur Shipping, Seattle, 40-42; asst gen supt, Todd Shipyards, Inc, Houston, 42-44, gen supt outfitting, San Pedro, Calif, 44-46; asst chief engr, Am Potash & Chem Co, Trona, Calif,

46-50; asst mgr design & construct, Gen Elec Co, Richland, Wash, 50-52, opers mgr aircraft nuclear propulsion proj, Cincinnati, 52-53, gen mgr small aircraft engine dept, Lynn, Mass, 53-54, gen mgr aircraft gas turbine div, Cincinnati, 55-57, vpres rels serv, Exec Off, 57-61, vpres & group exec aerospace & defense, 61-68, exec vpres, 68, vchmn bd & exec officer, 68-80. *Concurrent Pos:* Trustee, St Louis Univ, 67-69 & Rensselaer Polytech Inst; chmn conf bd, Grand Central Art Galleries, 68-83; chmn bd overseers, Hoover Inst, 73-76; chmn adv coun, Grad Sch Bus, Stanford Univ, 77-79. *Mem:* Nat Acad Eng; assoc fel Royal Aeronaut Soc; fel Am Inst Aeronaut & Astronaut; Soc Automotive Engrs; fel Am Soc Mech Engrs; Aerospace Industs Asn. *Mailing Add:* Gen Elec Co 3135 Easton Turnpike Fairfield CT 06431

PARKER, JAMES HENRY, JR, b Bakersfield, Calif, Dec 4, 26; m 48; c 2. CRYOGENICS. *Educ:* Univ Calif, AB, 48, PhD(physics), 54. *Prof Exp:* Asst, Univ Calif, 49-53; physicist, 53-70, res admin staff, res labs, 70-85, MGR SUPERCONDUCTIVITY & CRYOG, WESTINGHOUSE RES & DEVELOP CTR, 85- *Mem:* Fel Am Phys Soc; Inst Elec & Electronic Engrs. *Res:* Optical properties solids; electron motion in gases; ultrasonics; low temperature physics. *Mailing Add:* 4890 Cole Rd Murrysville PA 15668

PARKER, JAMES WILLARD, b Richmond, Ind, Feb 24, 45; m 68; c 2. ORNITHOLOGY, ECOLOGY. *Educ:* Earlham Col, AB, 67; Univ Kans, PhD(biol), 74. *Prof Exp:* Curatorial asst birds, Mus Natural Hist, Univ Kans, 72-73; teaching asst biol, Ohio Wesleyan Univ, 73-74, vis asst prof zool, 74-75; asst prof biol, Wilmington Col, 75-76, State Univ NY, Fredonia, 76-81 & Univ Maine, Farmington, 81-86; PVT ENVIRON EDUCATOR, AERIE E ENVIRON EDUC PROG, 86- *Mem:* Am Ornithologists Union; Ecol Soc Am; Am Inst Biol Sci. *Res:* General and vertebrate ecology and evolution with emphasis on the population biology and reproductive strategies of birds. *Mailing Add:* Rte 3 PO Box 3110 Farmington ME 04938

PARKER, JENNIFER WARE, b Berkeley, Calif, Apr 18, 59; m 83; c 1. BIOMEDICAL INSTRUMENTATION. *Educ:* Princeton Univ, BSE, 80; Univ Calif, Los Angeles, MS, 83, PhD(chem eng), 86. *Prof Exp:* Develop engr, Am Pharmaseal, 80-81; sr engr, 87-90, LEAD ENGR, BOC GROUP, 90- *Mem:* Am Inst Chem Engrs; Biomed Eng Soc; Am Chem Soc; NY Acad Sci. *Res:* Fiber optic chemical sensors and temperature sensors for biomedical instrumentation. *Mailing Add:* 25 Farmhouse Lane Furlong PA 18925

PARKER, JERALD D, b Ardmore, Okla, Feb 24, 30; m 52; c 4. THERMAL ENGINEERING. *Educ:* Okla State Univ, BS, 55, MS, 57; Purdue Univ, PhD(mech eng), 61. *Prof Exp:* From instr to prof mech eng, Okla State Univ, 55-88; PROF ENG, OKLA CHRISTIAN UNIV, 88- *Honors & Awards:* Year-in-Indust Prof Du Pont Co, 69-70. *Mem:* Am Soc Mech Engrs; Am Soc Heating, Refrig & Air-Conditioning Engrs; Am Nuclear Soc; Nat Soc Prof Engrs. *Res:* Heat transfer and fluid mechanics; energy conversion and utilization; heating, ventilating and air-conditioning. *Mailing Add:* 14217 Heritage Dr Edmond OK 73013

PARKER, JERALD VAWER, b Portland, Ore, Mar 3, 39; m 63; c 2. PLASMA PHYSICS. *Educ:* Calif Inst Technol, BS, 60, PhD(physics), 64. *Prof Exp:* Mem staff, Electro-Optical Syst, Pasadena, 64-66, Bell Telephone Labs, 66-69, Hughes Res Lab, 69-75; assoc group leader, 75-78, MEM STAFF, LOS ALAMOS NAT LAB, 78- *Mem:* Am Phys Soc; Sr mem Inst Elec & Electronics Engrs; Sigma Xi. *Res:* Inertial fusion including gas laser development for laser inertial fusion and other novel power sources for heating and compressing plasmas; electromagnetic launchers for high velocity research and applications. *Mailing Add:* 60 Loma Del Escolar Los Alamos NM 87544

PARKER, JOHN C, b Washington, DC, Sept 13, 35; m 75; c 3. VIROLOGY. *Educ:* Univ Md, PhD(parasitol), 65. *Prof Exp:* Pres, Microbiol Assoc, Inc, 79-87, vchmn, 87-89; CONSULT, 89- *Concurrent Pos:* Laboratory dir, 73-85, vpres oper, 75-79. *Honors & Awards:* Griffin Award, Am Asn Lab Animal Sci. *Mem:* Am Asn Lab Animal Sci; AAAS; Am Soc Microbiol; Am Asn Biol Sci; Tissue Culture Asn. *Res:* 50 publications dealing with laboratory animals. *Mailing Add:* 1602 Bull River Rd Noxon MT 59853

PARKER, JOHN CURTIS, b Boston, Mass, Nov 19, 35; m 59; c 2. MEMBRANE TRANSPORT, HEMATOLOGY. *Educ:* Yale Univ, MD, 61. *Prof Exp:* From asst prof to assoc prof, 67-73, PROF MED, UNIV NC, CHAPEL HILL, 73- *Mem:* Am Fedn Clin Res; Am Soc Clin Invest; Asn Am Physicians; Soc Gen Physiologists; Am Physiol Soc; Am Soc Hemat. *Res:* Transport of solutes across the plasma membrane of red blood cells; regulated transport that subserves volume homeostasis and pH control. *Mailing Add:* Dept Med CB 7035 Univ NC CH Chapel Hill NC 27599

PARKER, JOHN HILLIARD, b Orlando, Fla, Sept 30, 41; m 67; c 2. PHYSICAL CHEMISTRY, ENVIRONMENTAL SCIENCES. *Educ:* Emory Univ, BS, 63; Univ Calif, Berkeley, PhD(phys chem), 69. *Prof Exp:* Actg instr natural sci, Univ Calif, Berkeley, 66-67; asst prof chem, Univ Col Cape Coast, Ghana, 69-71; asst prof chem, Kans State Univ, 71-72; asst prof, 72-73, assoc dean col arts & sci, 75-76, assoc prof, 73-88, PROF ENVIRON SCI & CHEM, FLA INT UNIV, 88-, DIR, ENVIRON STUDIES, 84- *Concurrent Pos:* Asst dir, FIU-FAU Joint Ctr Environ & Urban Probs, 76-80. *Mem:* Am Chem Soc. *Res:* Molecular energy distribution in the products of exothermic reactions; analysis of indoor air pollution; chemical lasers; energy conservation; solar energy. *Mailing Add:* Chem Dept Fla Int Univ Miami FL 33199

PARKER, JOHN MARCHBANK, b Manhattan, Kans, Sept 13, 20; m 41, 70, 78; c 4. PETROLEUM GEOLOGY, GEOPHYSICAL COORDINATION. *Educ:* Kans State Univ, BS, 41. *Prof Exp:* Geologist, Hwy Comn, Kans, 41-42, US Pub Rd Admin, Alaska Hwy, 42 & Canol proj, Imp Oil Ltd, 43-44; dist geologist, Amoco, 44-52; vpres explor, Kirby Petrol Co, 52-74 & Northwest Explor Co, 74-75. *Concurrent Pos:* Consult geologist, 75- *Honors & Awards:* Hon Mem Award, Am Asn Petrol Geologists, 87. *Mem:* Am Asn Petrol Geologists (vpres, 65-66, pres, 82-83); Soc Explor Geophysicists; fel Geol Soc Am; fel AAAS; sr mem Soc Petrol Engrs; Europ Asn Explor Geophysicists. *Res:* Petroleum exploration in the United States, Canada, Ecuador and British North Sea; integration of sedimentary petrology, structural geology and petroleum geology with seismic data; discovery of oil and gas fields. *Mailing Add:* 2615 Oak Dr No 32 PO Box 15187 Lakewood CO 80215

PARKER, JOHN MASON, III, b Owego, NY, Sept 6, 06; m 41; c 1. GEOLOGY, STRUCTURAL PETROLOGY. *Educ:* Cornell Univ, AB, 28, AM, 33, PhD(struct geol), 35. *Prof Exp:* From instr to asst prof geol, NC State Univ, 35-42; from asst geologist to assoc geologist, US Geol Surv, 42-46; from assoc prof to prof, 46-72, EMER PROF GEOL, NC STATE UNIV, 72- *Concurrent Pos:* Geologist, US Geol Surv, 46-61. *Mem:* Geol Soc Am; Am Inst Mining, Metall & Petrol Engrs; Am Inst Prof Geol. *Res:* Paleozoic structure in central Appalachians; geology of pegmatites and crystalline rocks; regional systematic jointing in gently dipping sedimentary rocks; structural geology; petrology; economic geology of nonmetallic minerals. *Mailing Add:* 3113 Darien Dr Raleigh NC 27607

PARKER, JOHN ORVAL, b Millington, Mich, Nov 13, 30; Can citizen; m 55; c 4. CARDIOLOGY, CARDIOVASCULAR PHYSIOLOGY. *Educ:* Queen's Univ, Ont, MD & CM, 54, MSc, 58. *Prof Exp:* Lectr, Queen's Univ, 61- 62, from asst prof to assoc prof, 62-71, chmn div cardiol, 72-88, PROF MED, QUEEN'S UNIV, ONT, 71-, DIR CARDIOPULMONARY LAB, 64- *Concurrent Pos:* Ont Heart Found fel, Queen's Univ, Ont, 61-, Med Res Coun Can med res grant, 68-; attend physician, Kingston Gen Hosp, 61-; consult, Hotel Drew Hosp, Kingston. *Honors & Awards:* Medal in Med, Royal Col Physicians & Surgeons of Can, 61. *Mem:* Am Soc Clin Invest; Can Soc Clin Invest; Can Cardiovasc Soc; Am Col Cardiol; Am Col Physicians. *Res:* Hemodynamic and metabolic abnormalities in coronary artery disease. *Mailing Add:* Dept Med Queen's Univ Kingston ON K7L 3N6 Can

PARKER, JOHN WILLIAM, b Clifton, Ariz, Jan 5, 31; m 57; c 4. IMMUNO PHENOTYPING, FLOW CYTOMETRY. *Educ:* Univ Ariz, BS, 53; Harvard Med Sch, MD, 57; Am Bd Path, dipl 62. *Prof Exp:* Intern med, Med Ctr, Univ Calif, San Francisco, 57-58, resident path, 58-60; resident, Vet Admin Hosp, San Francisco, 60-62, clin investr, 62, chief clin labs, 62-64; from asst prof to assoc prof, 64-75, assoc dean sci affairs, 87-89, PROF PATH, SCH MED, UNIV SOUTHERN CALIF, 75-, DIR CLIN LABS, 74- *Concurrent Pos:* Am Cancer Soc sr Dernham fel, Univ Southern Calif, 64-69, Nat Cancer Inst spec fel, 72-73; clin instr, Sch Med, Univ Calif, San Francisco, 62-64; attend physician, Los Angeles County-Univ Southern Calif Med Ctr, 64-89; vis res fel, Walter & Eliza Hall Inst Med Res, Melbourne, Australia, 72-73; co-chmn, dept path, Sch Med, Univ Southern Calif, 85-; ed, Hemat Oncol, 83- *Mem:* NY Acad Sci; Am Asn Pathologists; fel Col Am Pathologists; fel Am Soc Clin Pathologists; Int Acad Path; Am Soc Hemat. *Res:* Cellular immunology; mechanisms of lymphocyte transformation; hematopathology; lymphoma/leukemia phenotyping by flow cytometry. *Mailing Add:* Dept Path Sch Med Univ Southern Calif Los Angeles CA 90033

PARKER, JOHNSON, b Boston, Mass, Dec 7, 17; m 59; c 2. ARCHAEOLOGY. *Educ:* Harvard Univ, AB, 41; Yale Univ, MF, 47; Duke Univ, DF(plant physiol), 50. *Prof Exp:* Asst prof plant physiol, Sch Forestry, Yale Univ, 57-63; plant physiologist, Bartlett Tree Res Labs, 63-65; plant physiologist, S Forest Exp Sta, US Forest Serv, USDA, 56-57, Northeast Forest Exp Sta, 65-80; RETIRED. *Mem:* AAAS; Am Soc Plant Physiologists; Ecol Soc Am. *Mailing Add:* 58 Falls Rd Bethany CT 06525

PARKER, JON IRVING, b Danville, Pa, Jan 1, 44; m 66; c 2. AQUATIC ECOLOGY, BIOLOGY. *Educ:* Bloomsburg State Col, BS, 65; Univ Idaho, MS, 72; Univ NH, PhD(bot), 74. *Prof Exp:* Teacher sci, Union Springs Cent Sch Dist, 65-69; mem staff acquatic res, Radiol & Environ Res Div, Argonne Nat Lab, 74-76, asst ecologist impact assessment, Environ Impact Systs Div, 76-77, asst ecologist aquatic res, Radiol & Environ Res Div, 77-81; asst prof, Lehigh Univ, 81-86; CONSULT, 86- *Mem:* Am Soc Limnol & Oceanog; Sigma Xi; Int Asn Theoret & Appl Limnol; Int Asn Great Lakes Res. *Res:* Biogeochemical behavior of pollutants in aquatic ecosystems. *Mailing Add:* RD 2 Box 516 Barto PA 19504

PARKER, JOSEPH B, JR, b Knox Co, Tenn, July 8, 16; m 46; c 2. PSYCHIATRY. *Educ:* Univ Tenn, BS, 39, MD, 41. *Prof Exp:* Instr psychiat, Duke Univ, 48-49, assoc prof, 53-59; asst prof, Col Med, Univ Tenn, 49-53; prof & chmn dept, Med Ctr, Univ Ky, 59-70; prof, 70-83, EMER PROF PSYCHIAT, MED CTR, DUKE UNIV, 84-; CONSULT, 83- *Concurrent Pos:* Lectr, Sch Social Work, Vanderbilt Univ & Univ Tenn, 50-53; chief psychiat serv, Vet Admin Hosp, 53-59; consult, Vet Admin & USPHS Hosps; mem, Gov Adv Coun Ment Health. *Mem:* Soc Biol Psychiat; AMA; Am Psychopath Asn; Asn Res Nerv & Ment Dis; fel Am Psychiat Asn; Am Col Psychiatrists. *Res:* Affective diorders. *Mailing Add:* 24 Stoneridge Circle Box 3837 Durham NC 27705

PARKER, JOSEPH R(ICHARD), b Grand Island, Nebr, May 8, 16; m 48; c 4. ELECTRO-OPTICS, ELECTRONICS. *Educ:* Univ Nebr, BSc, 43; Univ Pa, MSc, 52. *Prof Exp:* Student engr, Radio Corp Am, 43-44, engr, 44-63; staff mem, Los Alamos Sci Lab, 63-82; CONSULT, 82- *Concurrent Pos:* Consult, 82-; rev ed, Rev Sci Instruments, 77-84; chair, Inst Elec & Electronic Engrs, 78-80, chmn, Laser & Electro-Optical Soc, 80-89. *Mem:* Inst Elec & Electronic Engrs; Optical Soc Am (secy, 84-91); Laser Inst Am (pres, 80-84); Laser & Electro-Optical Soc (treas, 89-). *Res:* Environment near nuclear burst; rubidium vapor frequency standard; radiation dosimetry; extremely fast solid-state television camera; photon limited electro-optics. *Mailing Add:* 333 Potrillo Dr Los Alamos NM 87544

PARKER, KATHLYN ANN, b Chicago, Ill, 45. SYNTHETIC ORGANIC CHEMISTRY. *Educ:* Northwestern Univ, BA, 66; Stanford Univ, PhD(org chem), 71. *Prof Exp:* Assoc chem, Columbia Univ, 71-73; from asst prof to assoc prof, 73-82, PROF CHEM, BROWN UNIV, 82- *Mem:* Am Chem Soc. *Res:* Development of synthetic methods and total synthesis of natural products. *Mailing Add:* Dept Chem Box H Brown Univ Providence RI 02912

PARKER, KEITH KROM, b Billings, Mont, Jan 24, 50; m 85; c 4. NEUROCHEMISTRY, NEUROPHARMACOLOGY. *Educ:* Mont State Univ, BS, 72; Univ Calif, San Francisco, PhD(pharmacol), 77. *Prof Exp:* Fel pharmacol, Health Sci Ctr, Univ Colo, 77-79; res assoc chem, Univ Denver, 79-81; asst prof, 81-85, ASSOC PROF CHEM, WESTERN MONT COL, 86-, DEAN FAC, 89- *Concurrent Pos:* Fac affil, Univ Mont, 86-; pres, Mont Acad Sci, 88-89. *Mem:* AAAS; Sigma Xi; Am Soc Pharmacog; Am Chem Soc. *Res:* Effects of drugs and toxic agents in nerve cell culture and the developing nervous system. *Mailing Add:* Div Math & Sci Western Mont Col Dillon MT 59725

PARKER, KENNETH D, b Feb 27, 35; m 59; c 1. FORENSIC CHEMISTRY, TOXICOLOGY. *Educ:* Univ Calif, BS, 58, MS, 60; Am Bd Forensic Toxicol, dipl, 78. *Prof Exp:* Res asst, Univ Calif, 57-62, res assoc toxicologist, Med Ctr, dept pharmacol toxicol activ, 69-70; toxicologist-criminalist, Hine Lab, Inc, 62-73, consult & dir, Probe Div, Hine Inc, 73-81; OWNER & DIR-TOXICOLOGIST, PROBE SCI, 81- *Concurrent Pos:* Relief toxicologist to coroner, City & County of San Francisco, 57-70; CAP/AACC Forensic Urine Drug Testing Lab Inspector Accreditation Prog. *Mem:* Sigma Xi; Am Chem Soc; Am Acad Forensic Sci; Int Asn Forensic Toxicologists. *Res:* Microanalytical chemistry; gas chromatography; abuse drug usage-detection and human factors; clinical and forensic toxicology; evaluation of patient response to alcohol and drugs for DUI/DUID purposes. *Mailing Add:* Probe Scientific 2109 Pinehurst Ct El Cerritoe CA 94530-1879

PARKER, LEE WARD, b Fall River, Mass, Sept 28, 23. MATHEMATICAL PHYSICS. *Educ:* NY Univ, BEE, 47; Univ Southern Calif, PhD(physics), 53. *Prof Exp:* Physicist, Calif Res & Develop Co, 53-54, Lawrence Radiation Lab, Univ Calif, 54-61, Allied Res Assocs Inc, Mass, 61-64 & Mt Auburn Res Assocs, Inc, 64-75; PRIN SCIENTIST, LEE W PARKER, INC, 75- *Concurrent Pos:* Asst prof, St Mary's Col, Calif, 59-61. *Mem:* Am Phys Soc; Am Geophys Union; Soc Indust & Appl Math; Inst Elec & Electronics Eng. *Res:* Computational physics; interaction of satellites and probes with plasma; hydrodynamics; focusing of sonic booms; weak shocks; kinetic theory with electric fields; charging of satellite in geomagnetic field. *Mailing Add:* Tech Staff Mitre Burlington Rd M/S N101 Bedford MA 01730

PARKER, LEONARD EMANUEL, b New York, NY, May 13, 38; m 61; c 3. THEORETICAL PHYSICS, COSMOLOGY. *Educ:* Univ Rochester, AB, 60; Harvard Univ, AM, 62, PhD(physics), 67. *Prof Exp:* Instr physics, Univ NC, Chapel Hill, 66-68; from asst prof to assoc prof, 68-75, PROF PHYSICS, UNIV WIS-MILWAUKEE, 75- *Concurrent Pos:* NSF res grants, 70- *Honors & Awards:* Gravity Found Award, Gravity Found, 74, 80 & 84. *Mem:* AAAS; fel Am Phys Soc; Am Asn Physics Teachers; NY Acad Sci. *Res:* Quantum field theory; relativistic astrophysics and cosmology; general relativity. *Mailing Add:* Dept of Physics Univ of Wis Milwaukee WI 53201

PARKER, LEROY A, JR, b Newark, NJ, Feb 12, 30; m 54; c 2. DENTISTRY, NATURAL SCIENCE. *Educ:* Seton Hall Univ, BA, 62; Georgetown Univ, DDS, 54. *Prof Exp:* Instr restorative dent, Col Med & Dent, Seton Hall Univ, 57-60; from asst prof to assoc prof path & oral diag, 60-66, prof oral radiol & treatment planning & chmn dept, 66-69, asst dean clin affairs, 69-70, assoc dean clin affairs, 70-78; assoc dean acad affairs, Univ Med & Dent NJ, 79-85, actg dean, 85-86, interim dean, 86-87, PROF ORAL PATH, BIOL DIAG SCI, UNIV MED & DENT NJ, 69-, ASSOC DEAN ADMIN AFFAIRS, 88- *Concurrent Pos:* Assoc dir dent res, Johnson & Johnson, 65-66. *Mem:* Am Dent Asn; fel Am Col Dent; Int Asn Dent Res; fel Int Col Dentists. *Res:* Improved methods of oral hygiene; applied radiology; calculus etiology; composition. *Mailing Add:* Sch Dent Dept Oral Path Biol & Diag Sci Univ Med & Dent NJ Newark NJ 07103

PARKER, LESLIE, b Consett, Eng, Apr 28, 39; m 62; c 4. CLINICAL CHEMISTRY. *Educ:* Reading Univ, BSc, 60; Yale Univ, MS, 64 & 66, PhD(molecular biophys), 68. *Prof Exp:* Jr res fel physics, Mass Gen Hosp, 60-62; instr physics & math, Washington & Jefferson Col, 62-63; fel chem, Yale Univ, 68-69; asst prof biochem, Carnegie-Mellon Univ, 69-73; CHIEF BIOCHEMIST, WASHINGTON HOSP, 73- *Mem:* Fel Am Inst Chemists; Am Asn Clin Chem; Sigma Xi. *Mailing Add:* Clin Chem Lab Washington Hosp Washington PA 15301

PARKER, LLOYD ROBINSON, JR, b Rome, Ga, Sept 24, 50. ANALYTICAL CHEMISTRY. *Educ:* Berry Col, BA, 72; Emory Univ, MS, 74; Univ Houston, PhD(chem), 78. *Prof Exp:* ASST PROF CHEM, VASSAR COL, 78-; NOW AT DEPT OF CHEM, EMORY UNIV. *Mem:* Am Chem Soc; Sigma Xi. *Res:* Optimization and experimental design in analytical chemistry; pattern recognition; computer applications in chemistry; atomic absorption spectroscopy and inductively-coupled plasma spectroscopy. *Mailing Add:* Dept of Chem Emory Univ Oxford Col Oxford GA 30267

PARKER, MARGARET MAIER, b Portland, Maine, Dec 8, 50; m 74; c 4. CRITICAL CARE MEDICINE. *Educ:* Brown Univ, ScB, 73, MD, 77. *Prof Exp:* Fel, 80-82, HEAD CRITICAL CARE SECT, NIH, 82-; ASST CLIN INSTR MED, GEORGE WASHINGTON UNIV, SCH MED, 82- *Concurrent Pos:* Chmn internal med sect, Soc Critical Care Med, 88-89. *Mem:* Soc Critical Care Med; fel Am Col Physicians; Am Fedn Clin Res; Am Med Womens Asn. *Res:* Cardiovascular function in patients with septic shock using non-invasive and invasive measures of cardiac function. *Mailing Add:* 1114 Betts Trail Way Rockville MD 20854

PARKER, MARTIN DALE, pharmacy, pharmaceutics, for more information see previous edition

PARKER, MARY LANGSTON, b Inverness, Fla, Nov 14, 24; m 53; c 5. MEDICINE, ENDOCRINOLOGY. *Educ:* Fla State Univ, BS, 46, MS, 49; Wash Univ, MD, 53. *Prof Exp:* Intern, St Louis Children's Hosp, 53-54; res instr, 62-64, RES ASST PROF MED & PEDIAT, SCH MED, WASH UNIV, 65-, DIR HEALTH SERV, 71- *Mem:* Endocrine Soc. *Res:* Growth and development of children; growth hormone metabolism; physiologic actions; specific immunoassay for measurement of human growth hormone. *Mailing Add:* Dept Med Servs Wash Univ St Louis MO 63130

PARKER, MICHAEL, b Detroit, Mich, July 25, 38; m 60. ZOOLOGY, LIMNOLOGY. *Educ:* Univ Mich, BS, 60; Univ Wis, MS, 63, PhD(zool), 66. *Prof Exp:* Res assoc zool, Univ Wash, 66-68; asst prof, 68-72, ASSOC PROF ZOOL, UNIV WYO, 72- *Concurrent Pos:* Asst dir, Jackson Hole Res Sta, 69-71; consult, NSF, 71. *Mem:* AAAS; Am Soc Limnol & Oceanog; Ecol Soc Am; Am Inst Biol Sci; Fedn Am Scientists; Int Soc Theoret & Appl Liminol; Sigma Xi. *Res:* Phytoplankton and zooplankton ecology; vitamins in aquatic environments; alpine limnology; modeling. *Mailing Add:* Dept of Zool & Physiol Univ of Wyo Laramie WY 82070

PARKER, MURL WAYNE, b El Dorado, Ark, June 28, 39; m 61; c 3. INDUSTRIAL ENGINEERING. *Educ:* Univ Ark, BSIE, 62, MSIE, 63, PhD(indust eng), 69. *Prof Exp:* Indust engr, E I du Pont de Nemours & Co, 63-66; instr indust eng, Univ Ark, 66-68; from asst prof to assoc prof, 68-75, PROF INDUST ENG, MISS STATE UNIV, 76- *Mem:* Am Inst Indust Engrs; Am Soc Eng Educ; Nat Soc Prof Engrs. *Res:* Engineering economy; systems simulation. *Mailing Add:* Rte 7 Box 119 Starkville MS 39759

PARKER, N(ORMAN) F(RANCIS), b Fremont, Nebr, May 14, 23; m 49; c 4. ENGINEERING. *Educ:* Carnegie Inst Technol, BS & MS, 47, DSc(eng), 48. *Prof Exp:* Asst, Carnegie Inst Technol, 46-48; asst sect chief guid, NAm Aviation, Inc, 48-56, asst chief engr, Autonetics Div, 56-59, vpres & gen mgr, Comput & Data Systs, 59-62, exec vpres, 62-66, pres, Autonetics Div & vpres, NAm Aviation, Inc, 66-67; exec vpres & dir, Bendix Corp, 67-68; pres & dir, Varian Assocs, 68-81, chief exec officer, 72-81; dir, Syst Develop Corp, 79-80, US Leasing, 80-87, Int Game Technol, 81-86, Syntech Int, 86; RETIRED. *Concurrent Pos:* Buhl Fel. *Mem:* Nat Acad Eng; fel Inst Elec & Electronics Engrs; fel Am Inst Aeronaut & Astronaut. *Res:* Servomechanicsms. *Mailing Add:* 2375 Lagoon View Dr Cardiff CA 92007

PARKER, NANCY JOHANNE RENTNER, b Kenosha, Wis, June 17, 37; div; c 2. DEVELOPMENTAL BIOLOGY. *Educ:* Lawrence Col, BA, 59; Univ Tex, PhD(zool), 65. *Prof Exp:* NSF fel, Univ Tex, 65; asst prof, 65-73, assoc prof zool, 73-76, ASSOC PROF BIOL SCI, SOUTHERN ILL UNIV, EDWARDSVILLE, 77- *Mem:* Soc Develop Biol. *Res:* Cell death in chick masonephros. *Mailing Add:* Dept Biol Sci Southern Ill Univ Edwardsville Edwardsville IL 62026

PARKER, NICK CHARLES, b Blytheville, Ark, June 30, 43; m 63; c 2. AQUACULTURE, FISHERY RESEARCH. *Educ:* Memphis State Univ, BS, 70, MS, 72; Tex A&M Univ, PhD(fisheries sci), 77. *Prof Exp:* Res asst, anat & physiol labs, Memphis State Univ, 71-72 & res asst fisheries, 72-73; res asst fisheries, Tex A&M Univ, 73-77; sect leader fish cult, US Fish & Wildlife Serv, 77-78, actg dir fishery res, 78-80, res coordr, 80-81, lectr, Fisheries Acad, 81-86, sci dir, Southeastern Fish Cult Lab, 81-89; PROF & LEADER, TEX COOP FISH & WILDLIFE RES UNIT, TEX TECH UNIV, 89- *Concurrent Pos:* Adj prof genetics, Judson Col, 78-81, fish res, Memphis State Univ, 79-81, fish physiol, Auburn Univ, 80-84, fish res, Univ Wash, 86-; lectr, Aquacult, Verona, Italy, 84; assoc ed, Trans Am Fisheries Soc, 84-86, Prog Fish-Culturist; chmn, tech adv comt of Striped Bass Stocking Comt, US Fish & Wildlife Serv & Atlantic State Marine Fisheries Coun, 85-86. *Mem:* Am Fisheries Soc; World Maricult Soc; Am Inst Fishery Res Biologists; Water Pollution Control Fedn. *Res:* Culture techniques for warm water and anadromous fishes in ponds and water reuse systems; stress and reproductive physiology of fishes; tagging and marking techniques; pond and tank aeration; gas supersaturation; heated effluents; handling and transport of warm water fishes, especially striped bass. *Mailing Add:* Tex Coop Fish & Wildlife Res Unit Tex Tech Univ Lubbock TX 79409

PARKER, PATRICK LEGRAND, b El Dorado, Ark, Mar 13, 33; m 56; c 3. GEOCHEMISTRY, CHEMISTRY. *Educ:* Univ Ark, BA, 55, MS, 57, PhD(chem), 60. *Prof Exp:* Res scientist, Univ Tex, 59-61; org geochemist, Geophys Lab, Carnegie Inst, 61-63; res scientist assoc V & lectr chem, actg dir, Inst Marine Sci, Univ Tex, Port Aransas, Tex, 63-65; from asst prof to assoc prof chem, Univ Tex, 65-73; PROF, DEPT MARINE STUDIES, UNIV TEX, AUSTIN, 73-; RES SCIENTIST, MARINE SCI INST, UNIV TEX, ARANSAS, 76- *Concurrent Pos:* Prog mgr, Int Decade Ocean Explor, NSF, 72. *Mem:* Am Chem Soc; Geochem Soc; Am Geophys Union. *Res:* Application of chemistry to geological and near shore marine problems; stable isotope geochemistry of. *Mailing Add:* Marine Sci Inst Univ Tex at Austin Port Aransas TX 78373

PARKER, PAUL MICHAEL, b Vienna, Austria, June 24, 28; m 58; c 1. MOLECULAR PHYSICS, OPTICS. *Educ:* City Col New York, BS, 53; Ohio State Univ, MS, 55, PhD(physics), 58. *Prof Exp:* Asst physics, Ohio State Univ, 53-56, instr, 57-58; from asst prof to assoc prof, 58-67, PROF PHYSICS, MICH STATE UNIV, 67- *Concurrent Pos:* NIH fel, 62-63. *Mem:* Fel Am Phys Soc; Am Asn Physics Teachers. *Res:* Theory of infrared, microwave and radio frequency spectra. *Mailing Add:* Dept of Physics Mich State Univ East Lansing MI 48824-1116

PARKER, PETER DONALD MACDOUGALL, b New York, NY, Dec 14, 36; m 58; c 3. NUCLEAR ASTROPHYSICS. *Educ:* Amherst Col, BA, 58; Calif Inst Technol, PhD(nuclear physics), 63. *Prof Exp:* Res fel nuclear physics, Calif Inst Technol, 63; from asst physicist to assoc physicist, Brookhaven Nat Lab, 63-66; from asst prof to assoc prof, 66-76, PROF PHYSICS, YALE UNIV, 76- *Mem:* Am Phys Soc. *Res:* Nuclear astrophysics; structure light nuclei. *Mailing Add:* Dept Physics-NSL Yale Univ New Haven CT 06520

PARKER, PETER EDDY, b Flint, Mich, Nov 1, 44; m 67. CHEMICAL ENGINEERING, COMPUTER SCIENCE. *Educ:* Univ Rochester, BS, 66; Univ Pittsburgh, MBA, 67; Univ Mich, PhD(chem eng), 74. *Prof Exp:* Res engr environ sci, Raytheon Co, 74-75; RES ASSOC ENVIRON SCI, INST PAPER CHEM, 75- *Mem:* Am Inst Chem Engrs; Tech Asn Pulp & Paper Indust. *Res:* Computer simulation of processes; environmental system dynamics. *Mailing Add:* James River Corp 1915 Marathon Ave Nina WI 54956

PARKER, RICHARD ALAN, b Milwaukee, Wis, May 20, 31; m 53; c 7. LIMNOLOGY, COMPUTER SCIENCE. *Educ:* Utah State Univ, BS, 53; Univ Wis, MS, 54, PhD(zool), 56. *Prof Exp:* Asst zool, Univ Wis, 53-56; from instr to assoc prof zool, Wash State Univ, 56-66, assoc prof info sci, 63-66, actg dir comput ctr & actg chmn info sci, 65-66, chmn dept zool, 65-72, chmn environ sci, 68-71, chmn dept comput sci, 84, PROF ZOOL & COMPUT SCI, WASH STATE UNIV, 66- *Concurrent Pos:* NSF sr fel, Oxford Univ, 68-69; Erskine fel, Univ Canterbury, 76; Fulbright scholar, Univ Liege, 80; vis scientist, Inst Marine Environ Res, Plymouth, 83; vis prof, Univ Canterbury, 90. *Mem:* Am Soc Limnol & Oceanog. *Res:* Plankton population dynamics; computer simulation of aquatic ecosystems. *Mailing Add:* Sch Elec Eng & Computer Sci Wash State Univ Pullman WA 99164-2752

PARKER, RICHARD C, b Coleman, Tex, July 17, 39; m 63; c 2. PHYSICAL CHEMISTRY, KINETICS OF REACTIONS IN SOLUTION. *Educ:* Calif State Col Long Beach, BS, 62; Univ Wash, PhD(kinetics), 66. *Prof Exp:* Asst prof, 66-74, assoc prof, 75-80, PROF CHEM, NJ INST TECHNOL, 81-, ASSOC DEAN ENG, 88- *Concurrent Pos:* Environ comnr, Oceanport, NJ, 82-; chmn, Oceanport Environ Comn, 86- *Mem:* Am Chem Soc; Am Phys Soc; AAAS; Sigma Xi; Am Inst Chem Eng; Am Soc Eng Educ. *Res:* Kinetics of aminolysis reactions; effect of ultrasound on reaction rates; degradation of trace toxic pollutants in water. *Mailing Add:* Dept Chem NJ Inst Technol 323 King Blvd Newark NJ 07102

PARKER, RICHARD GHRIST, b New Kensington, Pa, July 17, 41; m 68; c 2. ORGANIC CHEMISTRY. *Educ:* Carnegie-Mellon Univ, BS, 63; Univ Nebr, Lincoln, PhD(chem), 68. *Prof Exp:* Fel, Harvard Univ, 67-68, Calif Inst Technol, 68-69; res chemist, B F Goodrich Co, 69-70, sr res chemist, 71-76, res assoc, 76-81, sr res & develop assoc, Res Ctr, 81-85, proj leader, 85-87, ASSOC DIR, CORP RES, B F GOODRICH CO, 87- *Concurrent Pos:* Guest scientist, Calif Inst Technol, 75. *Mem:* Am Chem Soc; Sigma Xi. *Res:* Organic synthesis; plastics stabilization; nuclear magnetic resonance; polymer alloys; composites. *Mailing Add:* Res Ctr B F Goodrich Co 9921 Brecksville Rd D 8534 Brecksville OH 44141

PARKER, RICHARD H, b Anaconda, Mont, June 16, 32; c 7. INTERNAL MEDICINE, MICROBIOLOGY. *Educ:* Univ Wash, BS, 54, MD, 58. *Prof Exp:* Intern, Duke Univ, 58-59; resident med, Univ Utah, 59-60, 62-63, instr, 65-66; assoc, Med Sch, Northwestern Univ, Chicago, 66-68, asst prof, 68-72; ASSOC PROF MED, MED SCH, HOWARD UNIV, 72-; CHIEF SECT INFECTIOUS DIS, VET ADMIN HOSP, WASHINGTON, DC, 72- *Concurrent Pos:* Fel infectious dis, Univ Utah, 60-65. *Mem:* Am Soc Microbiol; Am Fedn Clin Res; Am Thoracic Soc; Am Col Physicians; Am Soc Clin Pharmacol. *Res:* Rapid methods for diagnosis of infectious diseases; pharmacologic study of antimicrobial agents. *Mailing Add:* 817 Varnum St NE Washington DC 20017

PARKER, RICHARD LANGLEY, veterinary medicine; deceased, see previous edition for last biography

PARKER, ROBERT ALLAN RIDLEY, b New York, NY, Dec 14, 36; m 58, 81; c 5. ASTRONOMY. *Educ:* Amherst Col, BA, 58; Calif Inst Technol, PhD(astron), 62. *Prof Exp:* NSF fel, Univ Wis-Madison, 62-63, from asst prof to assoc prof astron, Washburn Observ, 63-72; astronaut, Manned Spacecraft Ctr, 67-91, DIR DIV POLICY & PLANS OFFSPACE FLIGHT, NASA HQ, 91- *Concurrent Pos:* Mem support crew, Apollo XV & XVII; prog scientist, Skylab; mission specialist, Spacelab I (STS-9); mission specialist, Astro-1 (STS-35). *Mem:* Am Astron Soc. *Res:* Interstellar matter; instrumentation. *Mailing Add:* Code MB NASA Hq Washington DC 20546

PARKER, ROBERT DAVIS RICKARD, b Honolulu, Hawaii, May 6, 42; US citizen. ENVIRONMENTAL HEALTH, INDUSTRIAL HYGIENE. *Educ:* Univ Hawaii, BA, 64, MSPH, 67; Univ Minn, MS, 70, PhD(environ health), 75. *Prof Exp:* Instr pub health, Univ Hawaii, 67; res asst, Univ Minn, 71; tech scuba diver, Environ Protection Agency, 72; instr environ health, Univ Minn, 72-73; ASST PROF BIOL & PUB HEALTH, UTAH STATE UNIV, 73- *Mem:* Am Pub Health Asn; Nat Environ Health Asn; Am Indust Hyg Asn; Int Asn Great Lakes Res; Sigma Xi. *Res:* Heavy metals in atmospheric fallout in intermountain west region; composition and magnitude of periphyton in artificial streams. *Mailing Add:* Dept Biol Utah State Univ Logan UT 84322

PARKER, ROBERT FREDERIC, b St Louis, Mo, Oct 29, 07; m 34; c 2. MEDICINE. *Educ:* Wash Univ, BS, 27, MD, 29. *Prof Exp:* Asst radiol, Wash Univ, 29-30, asst med, 31-32, instr, 32-33; intern, Barnes Hosp, St Louis, 30-31; asst, Rockefeller Inst, 33-36; from instr to assoc prof med, 36-77, from assoc to prof microbiol, 48-77, assoc dean med educ, 64-73, secy gen fac, 64-76 & dean, 73-76, EMER PROF MICROBIOL, CASE WESTERN RESERVE UNIV, 77-, EMER DEAN MED EDUC, 76- *Concurrent Pos:* In-chg microbiol lab & physician, Hosp, Case Western Reserve Univ, 48- *Mem:* Am Soc Clin Invest; Am Soc Microbiol; Soc Exp Biol & Med; Am Asn Immunologists. *Res:* Virus diseases; immunology and quantitative aspects of infection; tissue culture. *Mailing Add:* 2819 Coleridge Rd Cleveland Heights OH 44118

PARKER, ROBERT G, b Detroit, Mich, Jan 29, 25; m 49; c 2. RADIOLOGY. *Educ:* Univ Wis, BS, 46, MD, 48. *Prof Exp:* From asst prof to assoc prof, 59-66, prof radiol, Univ Wash, 68-78; PROF RADIATION ONCOL, UNIV CALIF LOS ANGELES, 78-, CHMN RADIOLONCOL. *Concurrent Pos:* Consult, USPHS Hosp, Seattle, 57, Vet Admin Hosp, 59-77 & Madigan Army Hosp, Ft Lewis, Wash, 59-77; consult, Wadsworth Vet Admin Hosp & Harbor Gen Hosp, 78- *Mem:* Am Soc Therapeut Radioloncol; Radiation Res Soc; Am Radium Soc; Radiol Soc NAm. *Res:* Radiation tissue tolerance; fast neutron therapy. *Mailing Add:* Dept Radiation Oncol Univ Calif Los Angeles Ctr Health Sci Los Angeles CA 90024

PARKER, ROBERT HALLETT, b Springfield, Mass, Feb 14, 22; m 45; c 3. DEEP SEA BIOLOGY, ESTUARINE RESOURCE MANAGEMENT. *Educ:* Univ NMex, BSc, 48, MSc, 49; Copenhagen Univ, Mag Sci & Doctoranden, 62. *Prof Exp:* Asst biol, Univ NMex, 48-49; asst zool, Duke Univ, 49-50; marine biologist, State Game & Fish Comn, Tex, 50-51; geophys trainee, Phillips Petrol Co, 51; res biologist, Scripps Inst, Univ Calif, 51-58, jr res ecologist, 58-63; resident ecologist, Systs Ecol Prog, Marine Biol Lab, Woods Hole, 63-66; assoc prof biol & geol, Tex Christian Univ, 66-70; PRES & CHMN BD, COASTAL ECOSYSTS MGT, INC, 70- *Concurrent Pos:* Consult, Am Mus Natural Hist, 57, Standard Oil Co, NJ, 56-58 & Pneumodyn Corp, 60-61; res scientist, Tex Christian Res Found. *Honors & Awards:* Am Asn Petrol Geol Award, 56. *Mem:* Soc Syst Zool; Ecol Soc Am; fel Geol Soc Am; Am Soc Limnol & Oceanog; assoc Am Asn Petrol Geol; fel NY Explorer's Club; AAAS. *Res:* Using shelled invertebrate ecology to define present day and ancient environments; paleoclimatological and paleoecological problems; terrestrial and aquatic ecology damage assessment; marine population dynamics; geochemistry of noble metals. *Mailing Add:* 3601 Wren Ave Ft Worth TX 76133

PARKER, ROBERT LOUIS, b Ft Dodge, Iowa, July 4, 29; m 63; c 2. SOLID STATE PHYSICS. *Educ:* Mass Inst Technol, BS, 51; Univ Md, MS, 58, PhD(physics), 60. *Prof Exp:* Physicist, Mech Div, 54-57 & Metal Physics Sect, 57-63, chief crystallization metal sect, 63-74, sr scientist, Metall Div, 74-78, physicist, Metall Div, 78-86, CONSULT, NAT BUR STANDARDS, 86- *Concurrent Pos:* Mem organizing & prog comts, Int Conf Crystal Growth, 65-; dep prin assoc ed, J Crystal Growth, 70- *Mem:* Am Phys Soc; Am Asn Crystal Growth. *Res:* Crystallization; phase transformations; nucleation; interface kinetics and morphology; morphological stability; moving boundary problems; vapor phase growth of crystals; freezing of liquids; whisker growth; convection; origin of imperfections in crystals; ultrasonic measurements of solidification. *Mailing Add:* 3503 1/2 Livingston St NW Washington DC 20015

PARKER, ROBERT ORION, b Big Pool, Md, Dec 24, 15; wid; c 1. CHEMICAL ENGINEERING. *Educ:* Carnegie Inst Technol, BS, 36; Columbia Univ, MS, 43; NY Univ, EngScD(chem eng), 59. *Prof Exp:* Jr engr, Freedom Oil Works, 36-37; from engr to dir res, Griscom Russell Co, 37-56; adj prof heat transfer & fluid flow, Hartford Univ, 56-57; from instr to asst prof chem eng, NY Univ, 57-62, from assoc prof to prof chem & nuclear eng, 62-74; prof & chmn, dept gas eng, Algerian Petrol Inst, 80-82; pres, Parker Int Consults, 82-88; RETIRED. *Concurrent Pos:* Consult heat transfer, 57-; lectr, Rutgers Univ, 62 & 63; adj prof, Brooklyn Polytech Inst, 76 & 77; tech adv, Hyundai Heavy Industs, Ulsan, Korea, 82-85. *Mem:* Fel Am Inst Chem; Am Nuclear Soc; Am Inst Chem Engrs; Am Chem Soc; NY Acad Sci. *Res:* Heat transfer; mass transfer; simultaneous heat and mass transfer; saline water conversion; thermal radiation hazard evaluation; cryogenic fluids. *Mailing Add:* 417 Vale Dr Pittsburgh PA 15239-1715

PARKER, ROBERT TARBERT, b Towson, Md, Oct 9, 19; m 42; c 4. INTERNAL MEDICINE. *Educ:* Johns Hopkins Univ, AB, 41, MD, 44; Am Bd Internal Med, dipl, 54. *Prof Exp:* Asst dir div infectious dis, 51-54, dir, 54-57, assoc dean, Sch Med & coordr med educ nat defense prof, 55-59, from instr to asst prof, 51-59, assoc prof med, Sch Med, Univ Md, Baltimore, 59-74, DIR MED, THE GOOD SAMARITAN HOSP, 74-, VPRES MED AFFAIRS, 77-, DEPT MED, JOHN HOPKINS UNIV. *Concurrent Pos:* Vis physician, Univ Hosp, 51-; consult, USPHS Hosp, 57-; attend physician, St Joseph's Hosp, Baltimore, 59-; dir med, S Baltimore Gen Hosp, 59-74, chief staff, 65-74. *Mem:* Fel Am Col Physicians; Am Fedn Clin Res; NY Acad Sci. *Res:* Infectious diseases. *Mailing Add:* 5601 Loch Raven Blvd John Hopkins Univ 720 Rutland Ave Baltimore MD 21239

PARKER, RODGER D, b St Louis, Mo, July 2, 34; m 57; c 2. MATHEMATICS, MATHEMATICAL BIOLOGY. *Educ:* Drury Col, BA, 56; Ind Univ, MA, 60; Johns Hopkins Univ, PhD(oper res), 65. *Prof Exp:* Res assoc, 62-66, from asst prof to assoc prof, 66-74, PROF OPERS RES, SCH HYG & PUB HEALTH, JOHNS HOPKINS UNIV, 74-, PROF HEALTH POLICY & MGT. *Concurrent Pos:* Consult, Food & Drug Admin, 70- & Pan Am Health Orgn, 71- *Res:* Develop methods for computer-aided diagnosis and planning of regional health services. *Mailing Add:* 210 Longwood Rd Baltimore MD 21210

PARKER, ROGER A, b West Union, Iowa, Jan 31, 43; m 64; c 4. MEDICINAL CHEMISTRY. *Educ:* Univ Iowa, BS, 65; Ohio State Univ, PhD(med chem), 69. *Prof Exp:* org res chemist, Merrell Res Ctr, Richardson-Merrell, Inc, 69-80, org res chemist, Merrell Dow Res Inst, Merrell Dow Pharmaceuticals, Inc, 80-90, MARION MERRELL DOW RES INST, 90- *Mem:* Am Chem Soc. *Res:* Steroid chemistry; heterocyclic medicinal agents; Antifungal agents; antiviral agents; hypocholesterolemic and hypolipidemic agents; antihypertensive agents. *Mailing Add:* Marion Merrell Dow Res Inst 2110 E Galbraith Rd Cincinnati OH 45215-6300

PARKER, RONALD BRUCE, b Los Angeles, Calif, Mar 21, 32; m 71; c 1. TELECOMMUNICATIONS, SCIENCE EDUCATION. *Educ:* Univ Calif, Berkeley, AB, 53, PhD(geol), 59. *Prof Exp:* From asst prof to assoc prof, 59-68, asst cur, geol mus, 74-76, prof geol, 68-76, ADJ PROF, UNIV WYO, 76-, RES AFFIL, UNIV NEBR STATE MUS, 76- *Concurrent Pos:* Sigma Xi-Sci Res Soc Am grant, 61; NSF res grant, 62-66; Univ Wyo fac fel, 65; NATO fel, 66-67; adj prof, Fergus Falls Col, 77-; ed, Contrib to Geol, 62-76, Minn Shepherd, 82-90; chief exec officer, Neomodels Assoc; dir, Fiber Connections; museum assoc, Sci Mus Minn, 88- *Mem:* Fel AAAS; fel Geol Soc Am. *Res:* Structural analysis of highly deformed rocks; geochemistry of deformed rocks; x-ray emission analysis of fossils, rocks and minerals; biogeochemistry of vertebrate fossils; chemical study of evolutionary trends in higher organisms; paleobiochemistry; animal nutrition and science; buffers and energy storage in earth processes; animal husbandry and general earth sciences; structure of spontaneous processes; author of science material for non-specialists. *Mailing Add:* Sammen Sheep Farm Rte One Box 153 Henning MN 56551-9740

PARKER, RONALD R, ELECTRICAL ENGINEERING. *Prof Exp:* PROF ELEC ENG, MASS INST TECHNOL, 77- *Mem:* Nat Acad Eng. *Mailing Add:* Mass Inst Technol Rm NW16-288 Cambridge MA 02139

PARKER, RONALD W, b Chicago, Ill, Aug 22, 50. COMPUTER PROGRAMMER, SYSTEM ANALYSIS. *Educ:* Ohio State Univ, MS, 74. *Prof Exp:* Sr audit systs analyst, 74-90, MGR AUDIT TECH, IBM, WHITE PLAINS, NY, 90- *Mem:* Am Math Soc. *Mailing Add:* Wilner Rd Somers NY 10589

PARKER, ROY DENVER, JR, b San Diego, Calif, Jan 18, 43; m 67; c 3. AGRICULTURAL ENTOMOLOGY, EXTENSION ENTOMOLOGY. *Educ:* Kilgore Col, SC, 63; Tex A&M Univ, BS, 66, MS, 68, PhD(entom), 79. *Prof Exp:* US Army Med Serv Corps, 68-71; serv mgr, pest control, Hunter Industs, 71-72; co exten entom cotton pest mgr, 72-75, EXTEN ENTOMOLOGIST, TEX AGR EXTEN SERV, 78- *Mem:* Entom Soc Am; Am Registry Prof Entomologists. *Res:* Entomology research and educational programs for agricultural producers on corn, sorghum, cotton, pecans, livestock and other insects that are pests of man and his possessions. *Mailing Add:* Rte Two Box 589 Corpus Christi TX 78410

PARKER, SHERWOOD, b Chicago, Ill, Mar 31, 32. ELEMENTARY PARTICLE PHYSICS, ELECTRONICS ENGINEERING. *Educ:* Univ Ill, BS, 53; Univ Calif, Berkeley, MA, 55, PhD(physics), 59. *Prof Exp:* Physicist, Lawrence Radiation Lab, Calif, 59; from instr to asst prof physics, Univ Chicago, 59-64; asst prof, Lawrence Berkeley Lab, Univ Calif, 64-71; MEM STAFF, UNIV HAWAII, 71- *Mem:* Am Phys Soc; Inst Elec & Electronic Engrs. *Res:* Elementary particle physics. *Mailing Add:* c/o Dept of Physics Lawrence Berkeley Lab 50B-6241 Berkeley CA 94720

PARKER, SIDNEY G, b Campbell, Tex, Jan 21, 25. SOLID STATE CHEMISTRY. *Educ:* ETex State Col, BS, 46; Univ Tex, PhD(inorg chem), 51. *Prof Exp:* Res chemist, Mobil Oil Co, 51-53 & E I du Pont de Nemours & Co, Inc, 53-57; MEM TECH STAFF, TEX INSTRUMENTS, INC, 57- *Res:* Study of single crystal growth and epitaxial deposition of infrared detector and laser materials and magnetic bubble materials and silicon. *Mailing Add:* 6550 Highgate Dallas TX 75214

PARKER, SIDNEY THOMAS, mathematics; deceased, see previous edition for last biography

PARKER, SYDNEY R(ICHARD), b New York, NY, Apr 18, 23; m 47; c 2. ELECTRICAL ENGINEERING. *Educ:* City Col New York, BEE, 44; Stevens Inst Technol, MS, 48, ScD, 64. *Prof Exp:* Sr engr, Int Resistance Co, 51-52; proj engr, Radio Corp Am, 52-56; assoc prof elec eng, City Col NY, 56-65; prof, Univ Houston, 65-66; prof, Naval Postgrad Sch, 66-75, chmn dept, 70-75; dean, Col Eng, Rutgers Univ, 75-76; PROF ELEC ENG, NAVAL POSTGRAD SCH, 76- *Concurrent Pos:* Consult, United Aircraft Corp, 58, Automation Dynamics Corp, 59-65, NAm Rockwell Corp, 66-75 & Lawrence Livermore Lab, Univ Calif, 78-; mem, Eng Educ Accreditation Bd, 75-; assoc ed, Transactions on Circuits & Systs, 75-77. *Honors & Awards:* Sigma Xi Res Award, Naval Postgrad Sch, 77. *Mem:* AAAS; fel Inst Elec & Electronics Engrs; Am Soc Eng Educ; NY Acad Sci; Sigma Xi. *Res:* Automatic control systems; circuit theory; computers; sensitivity studies of circuits and systems; computer aided circuit design; digital filter theory and practice. *Mailing Add:* PO Box AQ Carmel CA 93921

PARKER, TRAVIS JAY, b Oklahoma City, Okla, Nov 23, 13. GEOLOGY. *Educ:* Tex Tech Col, BS, 33; Univ Tex, MA, 39, PhD(geol), 52. *Prof Exp:* Asst prof geol, NTex Agr Col, 39-40; instr, Univ Tex, 41-42; from asst prof to assoc prof, 47-52, prof, 52-81, EMER PROF GEOL, TEX A&M UNIV, 81- *Mem:* Geol Soc Am; Am Asn Petrol Geol. *Mailing Add:* Dept Geol Tex A&M Univ College Station TX 77843

PARKER, VINCENT EVELAND, b Kuala Lumpur, Fedn Malay States, Sept 18, 14; US citizen; m 38, 67; c 4. EXPERIMENTAL NUCLEAR PHYSICS. *Educ:* Evansville Col, AB, 36; Ind Univ, PhD(phys chem), 40. *Prof Exp:* Asst, Ind Univ, 36-40; assoc prof phys sci & head dept, Cent Norm Col, 40-41; from instr to asst prof chem, Univ Del, 41-44, from asst prof physics & actg head dept to prof, 44-50, chmn dept, 46-50; prof physics & astron & head dept, La State Univ, 50-62; dep dir, Oak Ridge Assoc Univs, 62-67; dean, sch sci, 67-77, PROF PHYSICS CALIF STATE POLYTECH UNIV, 67- *Concurrent Pos:* Asst microchemist, Biochem Res Found, Franklin Inst, 41-43; mem, Comn Col Physics, 60-64 & gov bd, Am Inst Physics, 63-69; vchmn bd trustees, Col Oak Ridge, 64-66; chmn, Interserv Comt Tech Facilities, Southeastern US, 66; mem ad hoc comt develop, Univ Ala, Huntsville, 66-67 & US Nat Comt, Int Union Pure & Appl Physics, 67-70; chmn judges comt, Inst-US Steel Award Sci Writing, 69-73; for ed, Nuclear Energy Engr. *Mem:* Fel AAAS; fel Am Phys Soc; Am Chem Soc; Am Asn Physics Teachers (pres, 63). *Res:* High energy particle accelerators; neutron physics; radioactivity. *Mailing Add:* 1679 N First Ave Upland CA 91786

PARKER, VIRGIL THOMAS, b Houston, Tex, July 15, 51; m 74. PLANT ECOLOGY, ECOLOGY. *Educ:* Univ Tex, Austin, BA, 73; Univ Calif, Santa Barbara, MA, 75, PhD(biol), 77. *Prof Exp:* Asst prof biol, Rider Col, Lawrenceville, 77-80; ASST PROF BIOL, SAN FRANCISCO STATE UNIV, 80-, DIR, SIERRA NEV FIELD CAMPUS, 80- *Mem:* Ecol Soc Am; Bot Soc Am; Sigma Xi; Am Soc Naturalists. *Res:* Plant-plant interactions (allelopathy competition); wetland ecology; reproductive ecology. *Mailing Add:* Dept Biol San Francisco State Univ 1600 Holloway Ave San Francisco CA 94132

PARKER, WILLARD ALBERT, b Bremerton, Wash, Aug 10, 38; m 62; c 2. MATHEMATICS. *Educ:* Univ Ore, BA, 60, MA, 66, PhD(math), 70; Fuller Theol Sem, BD, 64. *Prof Exp:* Asst prof, 70-80, ASSOC PROF MATH, KANS STATE UNIV, 80- *Mem:* Am Math Soc; Math Asn Am. *Res:* Harmonic analysis on compact groups. *Mailing Add:* Dept Math Cardwell Hall Kans State Univ Manhattan KS 66506

PARKER, WILLIAM ARTHUR, b Tacoma, Wash, May 23, 49; m 70; c 2. THERAPEUTICS, HOME HEALTH CARE. *Educ:* Univ Minn, BSc, 72, PharmD(clin pharm), 73; Dalhousie Univ, MBA, 79. *Prof Exp:* Res assoc pharmacog, Univ Minn, 71-72, teaching asst clin pharm, 72-73, instr, 73-74; asst prof clin pharm, Dalhousie Univ, 74-79, coordr clin pharm, 74-85, assoc prof pharm, 79-91; ASSOC, SHOPPERS DRUG MART, HALIFAX, 85- *Concurrent Pos:* Clin pharmacist consult, Nursing Home Drug Utilization, Minn State Welfare Med Rev Team, 71-73; therapeut consult, Minn State Epilepsy Found, 73-74; clin pharmacist consult & instr, Comprehensive Seizure Ctr & Dept Neurol & Psychiat, St Paul-Ramsey Hosp & Med Ctr, 73-74 & Dept Med & Family Pract, Halifax Infirmary, 75-85. *Mem:* Am Soc Consult Pharm; Can Pharmaceut Asn; Can Soc Hosp Pharmacists; Am Col Clin Pharm; Sigma Xi. *Res:* Social and behavioral aspects of pharmacy practice; pharmacy home health care. *Mailing Add:* Shoppers Drug Mart Fenwick Med Ctr Halifax NS B3H 4M2 Can

PARKER, WILLIAM EVANS, b Newark, NJ, June 7, 40; m 69; c 2. INORGANIC CHEMISTRY, SYMMETRY. *Educ:* Haverford Col, BA, 62; Univ NC, Chapel Hill, MS, 65, PhD(inorg chem), 67. *Prof Exp:* From instr to asst prof, 67-78, ASSOC PROF CHEM, GETTYSBURG COL, 78-, CHMN DEPT, 82- *Mem:* Am Chem Soc. *Res:* Bonding theories of coordination compounds; kinetic studies of inorganic complexes. *Mailing Add:* Dept Chem Gettysburg Col Gettysburg PA 17325-1486

PARKER, WILLIAM HENRY, b Greenville, Pa, Oct 3, 41; m 64; c 3. LOW TEMPERATURE PHYSICS, SOLID STATE PHYSICS. *Educ:* Allegheny Col, BS, 63; Univ Pa, MS, 64, PhD, 67. *Prof Exp:* Asst prof, 67-70, assoc prof, 70-76, PROF PHYSICS, UNIV CALIF, IRVINE, 76- *Concurrent Pos:* Alfred P Sloan res fel, 68-70. *Mem:* Am Phys Soc; Am Asn Physics Teachers. *Res:* Superconductivity, tunneling and normal metals. *Mailing Add:* Dept Physics Univ Calif Irvine CA 92717

PARKER, WILLIAM JAMES, b Sutherlin, Ore, Dec 11, 26; m 49; c 2. THERMAL SCIENCES. *Educ:* Univ Ore, BS, 52, MS, 53; George Washington Univ, PhD(mech eng), 88. *Prof Exp:* Physicist, High Temperature Physics Sect, US Naval Radiol Defense Lab, 53-69; physicist, Ctr Fire Res, Nat Bur Standards, 69-90, FIRE TECHNOL CONSULT, NAT INST STANDARDS & TECHNOL, 90- *Res:* Heat transfer; fire research; experimental and theoretical research on the heat release rate, flame spread rate and ignitability of materials. *Mailing Add:* Fire Technol Consult 13135 Dairymaid Dr Apt T-2 Germantown MD 20874

PARKER, WILLIAM LAWRENCE, b Vermillion, SDak, Feb 19, 39; m 69; c 1. ORGANIC CHEMISTRY, ANTIBIOTICS. *Educ:* Columbia Col, BA, 61; Harvard Univ, PhD(chem), 65. *Prof Exp:* Res chemist, Dow Chem Co, 65-69; sr res investr, 69-79, res group leader, 79-87, SR RES FEL, SQUIBB INST MED RES, 88- *Mem:* Am Chem Soc; AAAS. *Res:* Isolation, structure determination and synthesis of natural products. *Mailing Add:* Bristol-Myers Squibb Pharm Res One Squibb Dr PO Box 191 New Brunswick NJ 08903

PARKER, WILLIAM SKINKER, b St Louis, Mo, Aug 28, 42; m 79; c 1. VERTEBRATE ECOLOGY. *Educ:* Wabash Col, Ind, BA, 64; Ariz State Univ, MS, 67; Univ Utah, PhD(biol), 74. *Prof Exp:* From asst prof to assoc prof, 74-83, PROF BIOL, MISS UNIV FOR WOMEN, 83- *Concurrent Pos:* Adj prof biol, Miss State Univ; managing ed, J Herpet, 90- *Mem:* Soc Study Amphibians & Reptiles; Sigma Xi; Am Soc Ichthyol & Herpet; Am Soc Naturalists; Herpetologists' League. *Res:* demography of pond turtles demography of fence lizards; ecology of reptiles and amphibians; author of over 40 publications. *Mailing Add:* Div Sci & Math Miss Univ Women Columbus MS 39701

PARKER, WINFRED EVANS, physical organic chemistry, for more information see previous edition

PARKER, WINIFRED ELLIS, b New Haven, Conn, Apr 19, 60. NUCLEAR STRUCTURE, ASTROPHYSICS. *Educ:* Dickinson Col, BS, 82; Carnegie Mellon Univ, MS, 85, PhD(chem), 89. *Prof Exp:* STAFF MEM, LOS ALAMOS NAT LAB, 89- *Mem:* Am Chem Soc; Am Phys Soc; AAAS; Sigma Xi. *Res:* Medium and low energy nuclear reactions; lifetimes of excited state composite; systems; mechanisms of fission; fission resonance structure especially of actinides. *Mailing Add:* Los Alamos Nat Lab MS H803 Los Alamos NM 87545

PARKES, KENNETH CARROLL, b Hackensack, NJ, Aug 8, 22; m 53. ORNITHOLOGY. *Educ:* Cornell Univ, BS, 43, MS, 48, PhD(ornith), 52. *Prof Exp:* Cur birds, Cornell Univ, 47-52; from asst cur to assoc cur, Carnegie Mus, 53-61, CUR BIRDS, CARNEGIE MUS NATURAL HIST, 62-, chief cur life sci, 75-85, SR CUR BIRDS, 86- *Concurrent Pos:* Res fel, Dept Epidemiol & Microbiol, Grad Sch Pub Health, Univ Pittsburgh, 56, adj mem grad fac, 63-; vis lectr, Pymatuning Field Lab, 57; mem bd trustees, Delaware Mus Natural Hist, 76-91; taxon ed, Avian Biol, 71-75, co-ed, 76-; mem bd trustees, Audubon Soc Western Pa, 82-; Bird Art Consult, Readers Digest Books, 87-90. *Mem:* Wilson Ornith Soc (pres, 73-75); fel Am Ornith Union (2nd vpres, 75-76); Cooper Ornith Soc; Brit Ornith Union; Ornith Soc NZ; Royal Australasian Ornith Soc. *Res:* Systematics and distribution of new world and Philippine birds; molt and plumage cycles of birds; avian hybrids. *Mailing Add:* Sect Birds Carnegie Mus Nat Hist 4400 Forbes Ave Pittsburgh PA 15213

PARKHIE, MUKUND RAGHUNATHRAO, b Wardha, India, Aug 4, 33; US citizen; m 70; c 4. TOXICITY TESTING IN DRUG DEVELOPMENT. *Educ:* Univ Jabalpur, India, DVM, 57; Univ Agra, MVSc, 60; Univ Sask, MSc, 63; Univ Mo-Columbia, PhD(med physiol), 70; Coun for Vet Grad, cert, 72. *Prof Exp:* Vet med officer, Maharastra & Rafasthan Govt , India, 57-58; jr res fel biochem & genetics, Indian Vet Res Inst, 58-60; asst prof prev med, Osmania Univ, India, 60-61; serv scholar reprod physiol, World Univ Serv Can, 61-63; Nat Acad Sci res fel cardio dynamics, Marshall Sch Space Flight Ctr, NASA, 70-72; staff scientist, biomed res mgt, URS Matrix Corp, 72-73;

RES ASSOC, CLIN HEMAT, DEPT MED, UNIV TORONTO, 63-; SR PHARMACOLOGIST/TOXICOLOGIST, US DEPT HEALTH, FOOD & DRUG ADMIN, ROCKVILLE, 74- *Concurrent Pos:* Regulatory toxicologist, Gen Coun, FDA, 77-81, proj adv/officer, toxicol res, 77-81, exec secy, drug develop, 84-; sr vis scientist, Molecular Tox-Div, Med Res Coun Toxicol Lab, Carshalton, Eng, 80-81; ad hoc reviewer, Teratol & J Am Col Toxicol, 82- *Mem:* Fel Am Acad Vet & Comparative Toxicol; Am Vet Med Asn (secy, 84-85); Soc Toxicol; Int Soc Study Xenobiotics; Sigma Xi. *Res:* Pharmacological administration; teratology; toxicity testing; cancer risk assessments; drug metabolism; human food safety; metal toxicity. *Mailing Add:* 15-032 Joshua Tree Rd Gaithersburg MD 20878-2549

PARKHURST, DAVID FRANK, b Pittsburgh, Pa, Mar 3, 42; div; c 1. ECOLOGY, PLANT PHYSIOLOGY. *Educ:* Univ Colo, Boulder, BS, 65; Univ Wis-Madison, MS, 68, PhD(bot), 70. *Prof Exp:* Lab technician, Inst Arctic & Alpine Res, Colo, 65-66; lectr plant ecol, Univ Wis-Madison, 70, vis asst prof, 70; res scientist, Div Atmospheric Physics, Commonwealth Sci & Indust Res Orgn, 70-73; from asst prof to assoc prof pub & environ affairs, 73-88, from asst prof to assoc prof biol, 78-88, PROF BIOL, IND UNIV, BLOOMINGTON, 88-, PROF PUB & ENVIRON AFFAIRS, 88- *Concurrent Pos:* Consult; Fulbright fel, Univ Melbourne, Victoria, Australia, 66; fel, Environ Sci Eng, Environ Protection Agency, 82. *Mem:* AAAS; Ecol Soc Am; Soc Risk Anal; Sigma Xi; Am Soc Plant Physiologists. *Res:* Theoretical and experimental studies of adaptive anatomy and physiology of plants; environmental risk analysis; biostatistics. *Mailing Add:* Sch Pub & Environ Affairs Ind Univ Bloomington IN 47405

PARKHURST, LAWRENCE JOHN, b Kansas City, Mo, Nov 29, 37; m 62; c 2. BIOPHYSICAL CHEMISTRY, BIOCHEMISTRY. *Educ:* Yale Univ, BA, 59, MS, 60, PhD(phys chem), 65. *Prof Exp:* Res assoc biophys, Johnson Found, Univ Pa, 65-66; NIH fel, biophys chem, Cornell Univ, 66-68; from asst prof to assoc prof chem, 69-76, PROF CHEM & LIFE SCI, UNIV NEBR, 76- *Mem:* AAAS; Am Chem Soc; Am Soc Biol Chem; Biophys Soc; Int Photochem Soc. *Res:* Physical chemistry of biological molecules; heme proteins, nucleic acid and ribosome kinetics. *Mailing Add:* Dept of Chem 525 Hamilton Hall Lincoln NE 68588-0304

PARKIN, BLAINE R(APHAEL), b Oakland, Calif, July 25, 22; m 45; c 2. AERODYNAMICS, HYDRODYNAMICS. *Educ:* Calif Inst Technol, BS, 47, MS, 48, PhD(aeronaut), 52. *Prof Exp:* Res engr, Hydrodyn Lab, Calif Inst Technol, 51-54, lectr appl mech, 54-55, res fel eng, 55-56; engr, Rand Corp, 56-62; sr eng specialist, AiResearch Mfg Co, 62-64; mgr aerospace technol & systs anal, Convair Aerospace Div, Gen Dynamics Corp, Calif, 64-67, prog mgr advan aircraft, 67-71; prof aerospace eng & dir Garfield Thomas Water Tunnel Appl Res Lab, 72-83, head hydromech, Dept Appl Res Lab, Pa State Univ, Univ Park, 83-86; chief scientist, 86-89, EMER PROF, AEROSPACE ENG & CONSULT, APPL RES LAB, 89- *Honors & Awards:* Knapp Award, Am Soc Mech Engrs, 65. *Mem:* Am Inst Aeronaut & Astronaut; fel Am Soc Mech Engrs; Soc Naval Archit & Marine Engrs; Sigma Xi. *Res:* Mechanics of cavitation inception in water flows; cavity flows; wave propagation in solids; applied mechanics; aircraft design. *Mailing Add:* Garfield Thomas Water Tunnel Pa State Univ University Park PA 16802

PARKIN, DON MERRILL, b Salt Lake City, Utah, Nov 7, 43; m 62; c 2. SOLID STATE PHYSICS, MATERIALS SCIENCE. *Educ:* Univ Utah, BA, 66, PhD(physics), 70. *Prof Exp:* Fel physics, Brookhaven Nat Lab, 70-72, asst physicist, 72- 74; staff mem physics, Los Alamos Nat Lab, 74-83, dep dir, ctr mat sci, 83-88, DIR, CTR MAT SCI, LOS ALAMOS NAT LAB, 88- *Concurrent Pos:* Mem, ed adv bd, J Nuclear Mat. *Mem:* Am Phys Soc; Sigma Xi; Mat Res Soc; Minerals, Metals & Mat Soc. *Res:* Defects in solids; radiation effects; neutron sources. *Mailing Add:* 1980 Camino Moro Los Alamos NM 87544

PARKIN, JAMES LAMAR, b Salt Lake City, Utah, June 2, 39; m 63; c 4. OTOLOGY, HEAD & NECK SURGERY. *Educ:* Univ Utah, Salt Lake City, BS, 63, MD, 66; Univ Wash, Seattle, MS, 70. *Prof Exp:* Instr otolaryngol, Univ Wash, 71-72; asst prof surg, 72-76, assoc prof otolaryngol, 76-81, actg chmn dept surg, 82-84, PROF OTOLARYNGOL, UNIV UTAH, 81-, CHMN DEPT, 74- *Concurrent Pos:* Chief otolaryngol, Vet Admin Hosp, Salt Lake City, 72-74, consult, 74-; instr, Am Acad Otolaryngol-Head & Neck Surg, 72-85; consult, Shriner's Crippled Children's Hosp, Salt Lake City, 74-; vis prof, Wash Hosp Med Ctr, Walter Reed Army Med Ctr & Bethesda Naval Hosp, 75; vis lectr, Univ Wash, 81 & Univ Ore & Yale Univ, 84; adj prof commun, Col Humanities, Univ Utah, 81-; pres med bd & chief of staff, Univ Hosp, Salt Lake City, 83-85; assoc ed, Arch Otolaryngol-Head & Neck Surg, 83- *Honors & Awards:* Res Award, Am Soc Clin Pathologists, 66; Res Award, Am Acad Ophthal & Otolaryngol, 71; Honor Award, Am Acad Otolaryngol, 80. *Mem:* Soc Univ Otolaryngologists (pres, 84-85); Asn Acad Depts Otolaryngol (secy-treas, 82-84, pres-elect, 84-86, pres, 86-88); Am Acad Otolaryngol-Head & Neck Surg; Am Laryngol, Rhinological & Otol Soc; Am Col Surgeons; Am Otol Soc. *Res:* Development and surgical placement of cochlear implants for the restoration of hearing in the profoundly deaf; application of laser technology in head and neck surgery. *Mailing Add:* Univ Med Ctr 3C 120 50 N Medical Dr Salt Lake City UT 84112

PARKINS, BOWEN EDWARD, b Omaha, Nebr, Sept 7, 34; m 55; c 4. ACOUSTICS. *Educ:* Univ Kans, BS, 57, MS, 61, PhD(elec eng), 65. *Prof Exp:* Res engr, Electronics Res Labs, Univ Kans, 59-62 & Res Ctr, 62-65; MEM TECH STAFF, ELEC ENG, BELL TEL LABS, 65- *Mem:* Acoust Soc Am; Sigma Xi. *Res:* Wave propagation; detection and estimation theory and practice. *Mailing Add:* 12 W Circuit Dr Succasunna NJ 07876

PARKINS, CHARLES WARREN, b Pittsburgh, Pa, Feb 4, 37; m 61; c 4. OTORHINOLARYNGOLOGY, SPEECH & HEARING. *Educ:* Bethany Col, BS, 59; Univ Rochester, MD, 63; Am Bd Otorhinolaryngol, dipl, 72. *Prof Exp:* Intern, Wilford Hall, US Air Force Hosp, 63-64; flight surgeon, US Air Force, Ankara, Turkey, 64-67 & Davis-Montham AFB, Ariz, 67-68; resident otolaryngol, Upstate Med Ctr, State Univ NY, 68-72, NIH fel, 72-73, asst prof otorhinolaryngol, 73-81; assoc prof otorhinolaryngol, Med Sch, Univ Rochester, 81-89; PROF OTORHINOLARYNGOL, MED SCH, LA STATE UNIV, 89- *Concurrent Pos:* Fac mem, Self Improv Prog, Am Acad Ophthalmol & Otolaryngol, 73-78. *Mem:* Fel Am Acad Ophthalmol & Otolaryngol; Asn Res Otolaryngol; Asn Acoust Soc Am. *Res:* Auditory neurophysiology involving processing of speech at the eighth nerve single neuron level; neural and electrical field modeling; cochlear implant development. *Mailing Add:* Dept Otolaryngol La State Univ Med Ctr 2020 Gravier St Suite A New Orleans LA 70112

PARKINS, FREDERICK MILTON, b Princeton, NJ, Sept 8, 35; m 59, 77; c 3. PEDIATRIC DENTISTRY, PHYSIOLOGY. *Educ:* Univ Pa, DDS, 60; Univ NC, MSD, 65, PhD(physiol), 69. *Prof Exp:* Instr pedodont, Sch Dent, Univ NC, 65-67; asst prof, Sch Dent Med, Univ Pa, 67-69, dir pedodont & dent auxiliary utilization, 68-69; assoc prof pedodont, Col Dent, Univ Iowa, 69-72, head dept, 69-75, prof, 72-79, asst dean acad affairs, 74-75, assoc dean, 75-79; dean, 79-84, PROF PEDIAT DENT, UNIV LOUISVILLE SCH DENT, 79- *Concurrent Pos:* Dent consult, Dept Pediat, Hahnemann Med Col, 65-69; consult med staff, Children's Hosp of Philadelphia, 67-70 & Vet Admin, 68-70; consult div dent health, USPHS, 69-72, Coun Dent Educ, 74-85 & Nat Inst Dent Res, 75-85, Bur Health Facil, 78-85; Robert Wood Johnson Health Policy fel, 77-78; staff, Health & Environ Subcomt, US House Rep, 77-78; mem bd gov, Univ Louisville Hosp, 79-82; trustee, Am Fund Dent Health, 79-85; mem med staff, Childrens Hosp, Louisville, 80-; US Pub Health Serv fel, 63-67. *Honors & Awards:* Res Award, Acad Pediat Dent, 68. *Mem:* AAAS; Sigma Xi; Am Dent Asn; Int Asn Dent Res; Am Acad Pediat Dent. *Res:* Fluoride metabolism and therapy; preventive dentistry; dental lasers. *Mailing Add:* Pediat Dent Sch Dent Univ Louisville Louisville KY 40292

PARKINS, JOHN ALEXANDER, b Warrenton, Va, Apr 18, 16; m 42; c 2. PHYSICAL CHEMISTRY. *Educ:* Washington & Lee Univ, AB, 39; Johns Hopkins Univ, AM, 49, PhD(chem), 51. *Prof Exp:* Res chemist, Geo W Bollman Co, 46-48; res chemist, E I du Pont de Nemours & Co, Inc, 51-60, tech investr, Film Dept, 60-76, patent consult, 76-81; RETIRED. *Concurrent Pos:* Patent agent, 76-, Univ Del, 83-89, consult, res off. *Mem:* Am Chem Soc. *Res:* Polymer chemistry; technical liaison between patents and regulatory affairs and research and development; manufacturing. *Mailing Add:* 14 Vassar Dr Newark DE 19711

PARKINS, WILLIAM EDWARD, b Bozeman, Mont, Mar 1, 16; m 48; c 3. ENERGY CONVERSION. *Educ:* Mont State Univ, BS, 37; Cornell Univ, PhD(physics), 42. *Prof Exp:* Asst physics, Cornell Univ, 39-42; res assoc, Radiation Lab, Univ Calif, 42-46; asst prof physics, Univ Southern Calif, 46-48; group leader exp physics, NAm Aviation, Inc, 49-51 & reactor eng, 52-54, chief engr, Atomics Int Div, 54-59, dir org reactors dept, 59-61, div dir res & tech, 61-69, mgr component eng & technol, NAm Rockwell Corp, 69-72, dir res & technol, Energy Syst Group, Rockwell Int Corp, 72-81; CONSULT, 81- *Concurrent Pos:* Mem liquid metals comt, AEC & Off Naval Res, 51-54; spec adv US del, Int Conf Peaceful Use Atomic Energy, 55. *Mem:* AAAS; fel Am Phys Soc; fel Am Nuclear Soc; Am Inst Aeronaut & Astronaut. *Res:* Ion and electron physics; mass spectroscopy; particle accelerators; effects of radiation on solids; liquid metal technology; engineering of nuclear reactor plants; energy conversion. *Mailing Add:* 20120 Well Dr Woodland Hills CA 91364

PARKINSON, ANDREW, b Bristol, Eng, Nov 30, 54; m 76; c 3. DRUG METABOLISM, CYTOCHROME P-FOUR FIFTY. *Educ:* Univ Surrey, BSc, 77; Univ Guelph, PhD(biol chem), 81. *Prof Exp:* Postdoctoral fel drug metab, Hoffmann-La Roche, Nutley, NJ, 81-83; from asst prof to assoc prof, 83-90, PROF PHARMACOL & TOXICOL, UNIV KANS MED CTR, KANSAS CITY, 90-, ASSOC DIR, CTR ENVIRON & OCCUP HEALTH, 89- *Res:* Analysis of the structure, function and regulation of liver enzymes, cytochrome P-450 and esterases, involved in drug, steroid and carcinogen metabolism. *Mailing Add:* Dept Pharmacol Univ Kans Med Ctr Kansas City KS 66103

PARKINSON, BRADFORD W, RESEARCH ADMINISTRATION. *Educ:* US Naval Acad, BS, 57; Mass Inst Technol, MS, 61; Stanford Univ, PhD(guid & control eng), 66. *Prof Exp:* Br chief, Cent Inertial Guid Test Facil, Holloman AFB, 61-64; chief, Simulation Div, USAF Test Pilot Sch, Edwards AFB, 66-68; assoc prof astronaut & actg head, Dept Astronaut & Computer Sci, USAF Acad, 69-71; chief engr, Advan Ballistic Re-entry Prog, 72-73; prog dir, Naustar Global Positioning Syst, 73-78; prof mech eng, Colo State Univ, 78-79; vpres, Space Systs Group, Rockwell Int, 79-80; gen mgr & exec officer, Prod Group, Intermetrics, Inc, 80-84; PROF AERONAUT & ASTRONAUT, PROF RES, W W HANSEN HIGH ENERGY EXP PHYSICS LAB & PROG MGR, GRAVITY PROBE-B, STANFORD UNIV, 84- *Concurrent Pos:* Res assoc, Calif Inst Technol. *Honors & Awards:* Gold Medal, Royal Inst Navig, 83; Kirschner Award, Inst Elec & Electronic Engrs, 86; Thomas L Thurlow, Inst Navig, 86. *Mem:* Nat Acad Eng; fel Am Inst Aeronaut & Astronaut; Inst Elec & Electronics Engrs; Air Force Asn; Inst Navig; fel Royal Inst Navig; Sigma Xi; Am Astronaut Soc. *Res:* Navigation; control theory; solar engineering; author of various publications. *Mailing Add:* High Energy Physics Lab Gravity Pro B Stanford Univ Stanford CA 94305-4085

PARKINSON, BRUCE ALAN, b Rochester, Minn, Mar 1, 51; m 87; c 2. ELECTROCHEMISTRY & SOLID STATE CHEMISTRY, TUNNELING MICROSCOPY. *Educ:* Iowa State Univ, BS, 72; Calif Inst Technol, PhD(chem), 77. *Prof Exp:* Fel chem, Bell Labs, 77-78; res chemist, Ames Lab, Dept Energy, 79-81; sr scientist, Solar Energy Res Inst, 81-85; RES CHEMIST, DUPONT, 85- *Mem:* Electrochem Soc; Am Chem Soc. *Res:* Semiconductor electrochemistry; solar energy conversion; inorganic electrochemistry; interfacial electrochemistry. *Mailing Add:* Dupont Cent Res Exp Sta E328/216 Wilmington DE 19898

PARKINSON, CLAIRE LUCILLE, b Bay Shore, NY, March 21, 48. POLAR SCIENCE, CLIMATOLOGY. *Educ:* Wellesley Col, BA, 70; Ohio State Univ, MA, 74, PhD(climat), 77. *Prof Exp:* Res asst, Nat Ctr Atmospheric Res, 76-78; RES SCIENTIST, GODDARD SPACE FLIGHT CTR, NASA, 78- *Honors & Awards:* Lewis Atterbury Stimson Award in Math, 70; Charles Clifford Huntington Award in Geog, 76. *Mem:* Am Meteorol Soc; Am Polar Soc; Asn Philos Math; Sigma Xi; Int Glaciological Soc; Oceanog Soc. *Res:* Analysis of sea ice from satellite imagery; climate modeling; climate change; examination of sea ice/climate connections; history of science and math; author of history of science and mathematics, climate modeling and sea ice books. *Mailing Add:* Code 971 Goddard Space Flight Ctr NASA Greenbelt MD 20771-0001

PARKINSON, DENNIS, b Bolton, Eng, Sept 1, 27. BOTANY, MICROBIOLOGY. *Educ:* Univ London, BSc, 51; Univ Nottingham, PhD(bot), 56. *Prof Exp:* Asst lectr bot, Royal Holloway Col, London, 53-56; lectr, Univ Liverpool, 56-64; prof biol, Univ Waterloo, 64-67; head dept biol, 68-77, prof biol, 67-77 PROF MICROBIOL, UNIV CALGARY, 78- *Concurrent Pos:* Orgn Econ Coop & Develop sr fel, Univ Mich, 63. *Mem:* Brit Mycol Soc; Brit Soc Soil Sci; Can Soc Microbiol. *Res:* Soil microbiology; nature and activity of microbial populations in plant litter; arctic microbiology including effects of oil spillage. *Mailing Add:* Dept Biol Sci Univ Calgary 2500 Univ Dr Calgary AB T2N 1N4 Can

PARKINSON, G(EOFFREY) VERNON, b Regina, Sask, Dec 13, 24; m 54; c 2. FLUID MECHANICS, DYNAMICS. *Educ:* Univ BC, BASc, 46; Calif Inst Technol, MS, 48, PhD(aeronaut), 51. *Prof Exp:* From asst prof to assoc prof, 51-65, PROF MECH ENG, UNIV BC, 65-, LECTR AERONAUT ENG, 70- *Concurrent Pos:* Nat Res Coun Can grants, 60-; Defence Res Bd Can grant, 61-; Nuffield fel, Nat Phys Lab, UK, 62-63; mem assoc comt aerodyn, Nat Res Coun Can, 65- *Mem:* Fel Can Aeronaut & Space Inst. *Res:* Dynamics and fluid dynamics of flow-induced oscillations of bluff bodies; airfoil and wing theory; unsteady fluid mechanics and gas dynamics theory; biomedical fluid mechanics. *Mailing Add:* Dept Mech Eng Univ BC 2075 Wesbrook Pl Vancouver BC V6T 1W5 Can

PARKINSON, JOHN STANSFIELD, b Buffalo, NY, Feb 17, 44; m 67; c 3. MOLECULAR GENETICS. *Educ:* Haverford Col, AB, 65; Calif Inst Technol, PhD(genetics, biophys), 70. *Prof Exp:* Vis asst prof microbial genetics, Ore State Univ, 69-70; NSF fel, Univ Wis, 70-71, NIH fel, 71-72; asst prof, 72-76, assoc prof, 76-81, PROF BIOL, UNIV UTAH, 81- *Concurrent Pos:* Prin investr, NIH grant, 73-, Microbiol Genetics Rev Group, NIH, 77-80. *Mem:* Am Soc Microbiol; Am Genetics Asn; Genetics Soc Am. *Res:* Genetic and biochemical bases of behavior; transmission and expression of genetic information; mechanism of bacterial chemotaxis and motility. *Mailing Add:* Dept Biol 201 S Biol Bldg Univ Utah Salt Lake City UT 84112

PARKINSON, R(OBERT) E(DWARD), b Mt Carmel, Ill, Nov 30, 09; m 36; c 3. CHEMISTRY, METALS. *Educ:* Northwestern Univ, BS, 31; Univ Syracuse, MS, 33. *Prof Exp:* Asst chem, Univ Syracuse, 31-33; metallurgist, Inland Steel Co, 33-35; res engr, Battelle Mem Inst, 35-45; consult, antitrust div, Dept of Justice, 45; coordr tech info, Owens-Ill Glass Co, 45-48; sr testing engr, Sears Roebuck & Co, 49-53; asst to dir res, Kawneer Co, 53-58; asst mgr, Porce- Alume Co, 58; sr res engr, Appl Res Lab, US Steel Corp, 59-71, assoc res consult, 71-74; RETIRED. *Concurrent Pos:* Consult, 74- *Mem:* Am Chem Soc. *Res:* Adhesive bonding of metals; special coatings; porcelain enameling of aluminum; metal finishing; materials. *Mailing Add:* Painehollow Rd Wellfleet MA 02667

PARKINSON, THOMAS FRANKLIN, b Tampa, Fla, Feb 22, 25; m 48; c 1. PHYSICS. *Educ:* Auburn Univ, BS, 47; Univ Va, PhD(physics), 53. *Prof Exp:* Physicist, Carbide & Carbon Chem Corp, 48-49; res physicist, Savannah River Lab, E I du Pont de Nemours & Co, 52-60; from assoc prof to prof nuclear eng sci, Univ Fla, 60-67; prof nuclear eng & chmn dept, Univ Mo-Columbia, 67-75, chmn dept, Va Polytech Inst & State Univ, 79-83; PROF NUCLEAR AND MECHANICAL ENGRS, VA POLYTECH INST & STATE UNIV, 75- *Concurrent Pos:* Fulbright-Hays res fel, Madrid, Spain, 66-67; consult, Nuclear Regulatory Comn, Washington, 74-75 & Int Atomic Energy Agency, Vienna, 79- *Mem:* AAAS; Am Phys Soc; Am Nuclear Soc. *Res:* Nuclear reactor physics; nuclear engineering; activation analysis; neutron spectrometry. *Mailing Add:* Nine Birkdale Ct E Aiken SC 29802

PARKINSON, WILLIAM CHARLES, b Jarvis, Ont, Feb 11, 18; US citizen; m 44; c 2. BIOPHYSICS, NUCLEAR PHYSICS. *Educ:* Univ Mich, BSE, 40, MS, 41, PhD(physics), 48. *Prof Exp:* Asst nuclear physics, Univ Mich, 36-42; physicist, Appl Physics Lab, Johns Hopkins Univ, 42-46; off sci res & develop, 43-44, res assoc, 46-47, from instr to assoc prof, 47-58, dir cyclotron lab, 62-77, PROF PHYSICS, UNIV MICH, ANN ARBOR, 58- *Concurrent Pos:* Fulbright scholar, 52-53; consult, Oak Ridge Nat Lab, 55-60, Los Alamos Nat Lab, 58-60 & Argonne Nat Lab, 59-60; mem subcomt nuclear structure, Nat Acad Sci-Nat Res Coun, 60-72; mem, adv panel physics, NSF, 66-69; mem, physics rev comn, Argonne Nat Lab, 59-63. *Mem:* Fel Am Phys Soc; Bioelectromagnetics Soc; Bioelectrical Growth & Repair Soc. *Res:* Biophysics. *Mailing Add:* Dept Physics Univ Mich Ann Arbor MI 48109

PARKINSON, WILLIAM HAMBLETON, b Trenton, NJ, June 26, 32; m 54; c 2. SPECTROSCOPY, ASTROPHYSICS. *Educ:* Western Ont Univ, BSc, 56, MSc, 57, PhD(physics), 59. *Prof Exp:* Nat Res Coun Can fel spectros, Imp Col, London, 59-61; lectr spectros & res physicist, Harvard Univ, 61-73, lectr astron, 61-91, assoc dir, ctr Astrophys, 78-88; physicist, 73-88, SR PHYSICIST, SMITHSONIAN INST, 88- *Concurrent Pos:* Consult, Solar Physics Subcomt, NASA, 61-, mem, Space Sci Solar Subcomt, 65-; sr res assoc, Harvard Col Observ, 65-; assoc dir, Ctr Astrophys, Smithsonian Astrophys Observ, 68- & Harvard Col Observ, 73- *Mem:* Int Astron Union; Am Astron Soc; fel Am Phys Soc. *Res:* Experimental atomic and molecular spectroscopy; astrophysics and upper atmospheric physics from observations with laboratory light sources and from solar pointed rocket and satellite borne experiments. *Mailing Add:* Dept of Astron Harvard Univ Cambridge MA 02138

PARKINSON, WILLIAM WALKER, JR, b White Oak, SC, June 30, 19; m 47; c 1. PHYSICAL CHEMISTRY, POLYMER CHEMISTRY. *Educ:* Erskine Col, AB, 40; Univ NC, PhD, 49. *Prof Exp:* Chemist, Reactor Chem Div, Oak Ridge Nat Lab, 49-73, Health Physics Div, 73-79, Gaseous Diffusion Plant, 79-84; CONSULT,84- *Mem:* Am Chem Soc; fel Am Inst Chem; AAAS. *Res:* Radiation-induced processes in high polymers; the relationship of physical properties of plastics and rubbers to their molecular structure; health physics of internal radioactive isotopes. *Mailing Add:* 834 Nelson Dr Kingston TN 37763

PARKISON, ROGER C, b Oakland, Calif, May 19, 49; m 71; c 2. REAL TIME CONTROL SYSTEMS, COMPUTATIONAL LINGUISTICS. *Educ:* Univ Calif, Berkeley, BA, 71; Stanford Univ, PhD(computer sci & artificial intel), 80. *Prof Exp:* Sr programmer, Univ Calif, Los Angeles, 75-84; sr software engr, Isitec Corp, 84-87; PRIN SOFTWARE ENGR, DIGITAL EQUIP CORP, 87- *Mem:* Asn Comput Mach; Am Asn Artificial Intel; Asn Computational Ling. *Res:* Real-time control; process automation; computational linguistics; text understanding; speech generation. *Mailing Add:* PO Box 1179 Felton CA 95018

PARKMAN, PAUL DOUGLAS, b Auburn, NY, May 29, 32; m 55. VIROLOGY, PEDIATRICS. *Educ:* St Lawrence Univ, BS, 57; State Univ NY, MD, 57; Am Bd Pediat, dipl, 62. *Hon Degrees:* DSc, St Lawrence Univ, 70. *Prof Exp:* Intern, Mary Imogene Bassett Hosp, Cooperstown, NY, 57-58; resident pediat, State Univ NY Upstate Med Ctr Hosp, 58-60; chief sect gen virol, Lab Viral Immunol, Div Biologics Standards, NIH, 63-72; dir div virol, 72-73, DEP DIR, CTR FOR DRUGS & BIOLOGICS, US FOOD & DRUG ADMIN, 73- *Honors & Awards:* E Mead Johnson Award, Am Acad Pediat, 67; Max Weinstein Award, United Cerebral Palsy, 69; Int Award Distinguished Sci Res, Joseph P Kennedy, Jr Found, 71. *Mem:* Am Asn Immunologists; Soc Pediat Res; Am Epidemiol Soc. *Res:* Laboratory and clinical research in infectious diseases; clinical pediatric investigation. *Mailing Add:* Bur Biologics US Food & Drug Admin 8800 Rockville Pike Bethesda MD 20205

PARKMAN, ROBERTSON, MICROBIOLOGY, IMMUNOLOGY. *Educ:* Yale Univ, MD, 65. *Prof Exp:* PROF MICROBIOL, CHILDREN'S HOSP OF LOS ANGELES, 83-, DEPT PEDIAT, UNIV SOUTHERN CALIF. *Res:* Pediatrics. *Mailing Add:* 627 S Euclid Ave Pasadena CA 91106

PARKS, ALBERT FIELDING, b Sulphur Rock, Ark, June 27, 09; m 33; c 1. ANALYTICAL CHEMISTRY, INTERNATIONAL TRADE. *Educ:* Ark Col, AB, 29; NY Univ, MSc, 31. *Prof Exp:* Prof, Ark Col, 31-36; chemist, US Customs Lab, La, 36-43, asst chief chemist, 46-49, asst chief, Div Tech Serv Bur Customs, 49-59; asst chief, US Tariff Comn, 60-69, dir off trade & indust, US Tariff Comn, 75; RETIRED. *Mem:* Am Chem Soc; Am Soc Qual Control; Am Inst Chem; NY Acad Sci. *Res:* Analytical methods covering a wide range of imported commodities, such as ores, metals, sugar products, narcotics, drugs, food products and miscellaneous manufactured materials. *Mailing Add:* 1500 Fairway Dr West Memphis AR 72301

PARKS, DONALD E, b Montgomery, WVa, Jan 7, 31; m 53; c 2. PHYSICS. *Educ:* Carnegie Inst Technol, BS, 53, MS, 54, PhD(physics), 58. *Prof Exp:* Instr physics, Carnegie Inst Technol, 58; sr staff mem, John Jay Hopkins Lab Pure & Appl Sci, Gen Atomic Div, Gen Dynamics Corp, 58-65, spec nuclear effects lab, 65-67; physicist, 67-69, SR RES SCIENTIST, SYSTS, SCI & SOFTWARE, 69- *Concurrent Pos:* Consult, Nat Comt Nuclear Energy, Italy, 61, 63-64. *Mem:* Am Phys Soc. *Res:* Atomic physics; plasma physics; slow neutron scattering; neutron thermalization; reactor physics; interaction of electromagnetic radiation with matter; transport theory. *Mailing Add:* Systems, Sci & Software PO Box 1620 La Jolla CA 92037

PARKS, E(DWIN) K(ETCHUM), b Riverbow, Alta, Oct 23, 17; nat US; m 41; c 3. AERONAUTICAL ENGINEERING. *Educ:* Univ Alta, BSc, 49, Univ Toronto, MASc, 50, PhD(aeronaut), 52. *Prof Exp:* Res officer, Nat Res Coun Can, 52-53; assoc prof aeronaut eng, Univ Kans, 53-60; prof, 60-87, EMER PROF AERONAUT & MECH ENG, UNIV ARIZ, 87- *Mem:* Assoc fel Am Inst Aeronaut & Astronaut. *Res:* Nonstationary and viscous aerodynamics; shock tubes; wind tunnels; airplane dynamics. *Mailing Add:* Dept of Aeronaut & Mech Eng Univ of Ariz Tucson AZ 85721

PARKS, ERIC K, b Meriden, Conn, May 3, 40. PHYSICAL CHEMISTRY. *Educ:* Rensselaer Polytech, BS, 62; Univ Calif, Berkeley, PhD(phys chem), 66. *Prof Exp:* NSF fel chem, Mass Inst Technol, 66-67, res assoc, 67-68; fel, 69-70, PERMANENT STAFF CHEM DIV, ARGONNE NAT LAB, 70- *Mem:* Am Phys Soc. *Res:* Chemical kinetics; molecular beams; metal cluster research. *Mailing Add:* Bldg 200 Argonne Nat Lab Argonne IL 60439

PARKS, GEORGE A(LBERT), b Oakland, Calif, May 3, 31; m 56; c 2. SURFACE CHEMISTRY, MINERAL ENGINEERING. *Educ:* Univ Calif, Berkeley, BS, 53, MS, 54; Mass Inst Technol, PhD(metall), 60. *Prof Exp:* From asst prof to assoc prof, Stanford Univ, 59-70, assoc chmn, Appl Earth Sci, 75-79, assoc dean res, Sch Earth Sci, 79-81, PROF APPL EARTH SCI, STANFORD UNIV, 70-, CIVIL ENG, 75-, DONALD & DONALD M STEEL PROF EARTH SCI, 78- *Concurrent Pos:* Vis scientist, Dept Metall, Mass Inst Technol, 66, Swiss Fed Inst Water Supply, Wastewater Treatment & Pollution Control, EAWAG, Zurich, Switz, 76; staff scientist III-chemist, Lawrence Berkeley Lab, Univ Calif, Berkeley, CA; prof geol, Stanford Univ, 80- *Honors & Awards:* Gaudin Award, Soc Mining Engrs, Am Inst Mining Metall Engrs, 85. *Mem:* Am Inst Mining, Metall & Petrol Engrs; Am Chem Soc; Geochem Soc; Mineral Soc Am; Am Geophys Union; Sigma Xi; distinguished mem Am Inst Mining Metall Engrs. *Res:* Inorganic surface and colloid chemistry of oxides and minerals; applications in geochemistry, environmental pollution abatement and extractive metallurgy; process innovation and interpretation. *Mailing Add:* Dept of Appl Earth Sci Stanford Univ Stanford CA 94305-2225

PARKS, GEORGE KUNG, b Shanghai, China, July 17, 35; US citizen; m 58; c 2. GEOPHYSICS, PHYSICS. *Educ:* Univ Calif, Berkeley, BA, 61, PhD(physics), 66. *Prof Exp:* Res assoc, Univ Minn, Minneapolis, 67-69; assoc prof space physics, Univ Toulouse, 69-71; asst prof, 71-73, assoc prof 73-76, PROF GEOPHYS, UNIV WASH, 76- *Concurrent Pos:* NSF grants, 72- *Honors & Awards:* Hon dipl, Acad Sci, USSR. *Mem:* Am Geophys Union; Am Inst Physics Teachers. *Res:* Space physics; plasma and wave particle interactions; origin of energetic particles in the earth's magnetic field; solar flares. *Mailing Add:* Dept Geophys Univ Wash Seattle WA 98195

PARKS, HAROLD GEORGE, b Shelburne Falls, Mass, May 15, 42; m 64; c 3. SEMICONDUCTOR PROCESSING, YIELDS. *Educ:* Lowell Technol Inst, BS, 64; Syracuse Univ, MS, 69; Rensselaer Polytech Inst, PhD, 80. *Prof Exp:* Elec engr circuit design, Int Bus Mach, 64-66, engr analog & digital design, Xerox Corp, 66-69; ELEC ENGR ELECTRON OPTICS LASER ANNEALING, SEMICONDUCTOR PROCESSING, GEN ELEC CO, 69- *Honors & Awards:* IR 100 Award Beamos Memory, 75. *Res:* Electron and ion optics, numerical analysis, lens and deflection system designs; beam memory targets and charged particle and target interactions; semiconductor processing and Very-Large-Scale-Integration yield engineering. *Mailing Add:* Electroncis Engr CRD KW- B1319 Gen Elec Co PO Box Eight Schenectady NY 12301

PARKS, HAROLD RAYMOND, b Wilmington, Del, May 22, 49; m 85; c 2. GEOMETRIC MEASURE THEORY. *Educ:* Dartmouth Col, AB, 71; Princeton Univ, PhD(math), 74. *Prof Exp:* J D Tamarkin instr math, Brown Univ, 74-77; from asst prof to assoc prof, 77-89, PROF MATH, ORE STATE UNIV, 89- *Concurrent Pos:* Vis assoc prof math, Ind Univ, 82-83. *Mem:* Am Math Soc. *Res:* Methods for computing solutions of the least area problem, particularly those solutions whose existence is guaranteed by results in geometric measure theory. *Mailing Add:* Dept Math Ore State Univ Corvallis OR 97331

PARKS, JAMES C, b Altoona, Pa, Aug 9, 42; m 64; c 1. PLANT SYSTEMATICS. *Educ:* Shippensburg State Col, BS, 64; Vanderbilt Univ, PhD(bot), 68. *Prof Exp:* Asst prof, 68-72, assoc prof, 72-77, PROF BIOL, MILLERSVILLE UNIV, 77- *Mem:* Int Asn Plant Taxon; Am Soc Plant Taxonomists. *Res:* Plant systematics; floristics; cytotaxonomy; pteridology. *Mailing Add:* Dept of Biol Millersville Univ Millersville PA 17551

PARKS, JAMES EDGAR, b Morganton, NC, Jan 12, 39; m 62; c 4. PHYSICS. *Educ:* Berea Col, BA, 61; Univ Tenn, MS, 65; Univ Ky, PhD(physics), 70. *Prof Exp:* Res assoc health physics, Oak Ridge Nat Lab, 61-64; instr physics, Berea Col, 64-66; res assoc, Univ Ky, 70; asst prof, 70-74, ASSOC PROF PHYSICS, WESTERN KY UNIV, 74- *Concurrent Pos:* Consult, Health Physics Div, Oak Ridge Nat Lab, 74- *Mem:* Am Phys Soc; Am Asn Physics Teachers; Sigma Xi. *Res:* Radiation physics; atomic and molecular physics; atomic collisions; laser physics. *Mailing Add:* Laser Tech Ctr 10521 Research Dr Suite 300 Knoxville TN 37932

PARKS, JOHN S, b Washington, DC, Oct 14, 39; m 59; c 3. PEDIATRICS, ENDOCRINOLOGY. *Educ:* Amherst Col, AB, 61; Univ Pa, MD, 66, PhD(biochem), 71. *Prof Exp:* USPHS fel biochem, Univ Pa, 66-67; from intern to resident pediat, Children's Hosp of Philadelphia, 67-69; clin assoc endocrinol, Endocrinol Br, Nat Cancer Inst, 69-70 & Lab Molecular Biol, 70-71; fel, 71-73, assoc dir, div endocrinol, Children's Hosp of Philadelphia, 73-77; ASST PROF PEDIAT, SCH MED, UNIV PA, 73- *Mem:* Endocrine Soc. *Res:* Bacterial gene control; pediatric endocrinology; growth hormone action. *Mailing Add:* 2040 Ridgewood Dr 34th St & Civic Ctr Blvd Atlanta GA 30322

PARKS, KENNETH LEE, b Pineville, Ky, Mar 3, 31; m 58. INDUSTRIAL CHEMISTRY. *Educ:* Davidson Col, BS, 54; Univ NC, PhD(anal chem), 59. *Prof Exp:* Chemist, Atlantic Ref Co, 59-60; res chemist, Rohm and Haas Co, 60-61; process chemist, Am Cyanamid Co, 61-64, sr chemist, 64-65; group leader process develop, Fla, 65-67, dir environ & chem control, Tenn, 67-69, coordr, Fla, 69-72, dir res & develop, 72-76; dir tech develop, Agrico Chem Co, 76-86; RETIRED. *Mem:* Am Chem Soc. *Res:* Analytical methods; plant food manufacturing processes and chemistry; pollution control methodology for air and water. *Mailing Add:* 1404 Laurel Court Plant City FL 33566

PARKS, LEO WILBURN, b Wetaug, Ill, Nov 21, 30; m 57; c 2. MICROBIOLOGY, BIOCHEMISTRY. *Educ:* Univ Ill, BS, 52; Ind Univ, MA, 53; Univ Wash, PhD(microbiol), 56. *Prof Exp:* Asst microbiol, Ind Univ, 52-53 & Univ Wash, 53-55; resident res assoc, Argonne Nat Lab, 56-58; from asst prof to assoc prof microbiol, Ore State Univ, 58-85; PROF & HEAD, MICROBIOL, NC STATE UNIV, 85- *Concurrent Pos:* NSF fel, Univ Copenhagen, 65-66; US Pub Health Serv Fel, Univ Louis Pasteur, Strasbourg, France, 72-73. *Mem:* Am Soc Microbiol; Am Soc Biol Chem; AAAS; Am Acad Microbiol; Sigma Xi; Soc Indust Microbiol. *Res:* Microbial physiology; molecular biology; biochemistry of bacteria. *Mailing Add:* PO Box 7615 NC State Univ Raleigh NC 27695-7615

PARKS, LEWIS ARTHUR, b Hutchinson, Kans, Oct 31, 47; m 71; c 2. RADIATION PROCESSING. *Educ:* Kans State Teachers Col, BA, 69, MS, 71; Univ Tex, Austin, PhD(physics), 76. *Prof Exp:* Res Assoc, Physics Dept, Fla State Univ, 76-78; staff physicist, 78-88, MKT MGR, IRT CORP, 88- *Mem:* Am Phys Soc; Inst Elec & Electronic Engrs; Am Soc Testing & Mat. *Res:* Radiation gauging and processing; experimental heavy ion nuclear research; activation analysis. *Mailing Add:* 3030 Callan Rd San Diego CA 92121

PARKS, LLOYD MCCLAIN, b Scottsburg, Ind, Mar 21, 12; m 40; c 1. PHARMACEUTICAL CHEMISTRY. *Educ:* Purdue Univ, BS, 33, MS, 36; Univ Wis, PhD(pharmaceut chem), 38. *Hon Degrees:* DSc, Purdue Univ, 62, Ohio State Univ, 83; DPS, Union Col, 71. *Prof Exp:* From instr to prof pharmaceut chem, Univ Wis, 38-56; dean col pharm, Ohio State Univ, 56-77. *Concurrent Pos:* Mem revision comt, US Pharmacopoeia, 50-62, Am Coun Pharmaceut Educ, 62-68. *Honors & Awards:* Ebert Prize, Am Pharmaceut Asn, 52, Res Achievement Award, 66, Remington Honor Medal, 75. *Mem:* AAAS; Am Chem Soc; Am Pharmaceut Asn (pres, 71-72); Am Inst Hist Pharm. *Res:* Organic pharmaceutical chemistry; medicinal chemistry. *Mailing Add:* 7868 E Chauncey St Tucson AZ 85715

PARKS, NORRIS JIM, b Snyder, Tex, July 5, 43. PHYSICAL CHEMISTRY, NUCLEAR CHEMISTRY. *Educ:* Eastern NMex Univ, BS, 65; Univ Nebr, PhD(phys & nuclear chem), 69. *Prof Exp:* Fel high energy chem dynamics, 69-71, res chemist, Radiobiol Lab, 72-80, RES CHEMIST, LAB FOR ENERGY-RELATED HEALTH RES, UNIV CALIF, DAVIS, 80- *Mem:* Am Chem Soc; Sigma Xi; AAAS. *Res:* Chemical effects of nuclear transformations; high energy gas kinetics; radiation chemistry; radiochemical applications; nuclear applications in medicine; radiobiology; chemical toxicology. *Mailing Add:* Lab Energy-related Health Res Univ of Calif Davis CA 95616

PARKS, PAUL BLAIR, b Erwin, NC, Nov 20, 34; m 55; c 1. EXPERIMENTAL NUCLEAR PHYSICS, REACTOR PHYSICS. *Educ:* Duke Univ, BS, 57, PhD(physics), 63. *Prof Exp:* Res assoc nuclear physics, Duke Univ, 63-64; FEL ENGR, SAVANNAH RIVER LAB, E I DU PONT DE NEMOURS & CO, INC, 64- *Res:* High resolution nuclear structure studies; reactor experiments using pulsed neutron sources; reactor kinetic experiments emphasizing space-time effects; reactor safety research using calculated transients, reactor charge design; neutron radiography applied to medical or biological problems; new production reactor design. *Mailing Add:* Savannah River Lab E I du Pont de Nemours & Co Inc Aiken SC 29801

PARKS, PAUL FRANKLIN, b Opelika, Ala, Nov 9, 33; m 53; c 4. RESEARCH ADMINISTRATION, BIOCHEMICAL NUTRITION. *Educ:* Auburn Univ, BS, 56, MS, 59; Tex A&M Univ, PhD(biochem), 62. *Prof Exp:* Asst prof biochem, Tex A&M Univ, 62-65; assoc prof, 65-74, asst dean 68-72, dean, grad sch, 72-85, PROF ANIMAL SCI, AUBURN UNIV, 74-, VPRES RES, 81- *Concurrent Pos:* Vchmn, Coun Oak Ridge Assoc Univs, 87-88; mem bd trustee, Southeastern Univ Res Assoc; mem, Army Sci Bd; mem,Coun Res Policy & Grad Educ; mem, Nat Coun Univ Res Adminrs. *Mem:* AAAS; Am Inst Nutrit. *Res:* Relationship between lipid metabolism and certain nutritional diseases. *Mailing Add:* Vpres Res Auburn Univ Auburn AL 36849-5112

PARKS, ROBERT EMMETT, JR, b Glendale, NY, July 29, 21; m 45; c 3. BIOCHEMISTRY, PHARMACOLOGY. *Educ:* Brown Univ, AB, 44; Harvard Univ, MD, 45; Univ Wis, PhD(biochem), 54. *Prof Exp:* Intern, Boston Children's Hosp, 45-46; res assoc biochem, Amherst Col, 48-51; from asst prof to prof pharmacol, Med Sch, Univ Wis, 54-63, actg chmn dept, 62-63; chmn sect biochem pharmacol, 68-76, E E BRINTZENHOFF PROF MED SCI, BROWN UNIV, 63-, CHMN SECT BIOCHEM PHARMACOL, 82-; ASSOC DIR, ROGER WILLIAMS CANCER CTR, 73- *Concurrent Pos:* Am Cancer Soc fel, Enzyme Inst, Univ Wis, 51-54; John & Mary Markle Found scholar, 56-61. *Honors & Awards:* Pfizer lectr, 79, 85, 87 & 90; C Chester Stock Award, 85; Gov Sci & Technol Award, 88. *Mem:* Am Soc Biol Chemists; Am Soc Pharmacol & Exp Therapeut; Am Asn Cancer Res; Am Chem Soc. *Res:* Mechanism of action of antitumor agents; drug metabolism; enzyme mechanisms; regulation of carbohydrate metabolism; drug action. *Mailing Add:* Div Biol & Med Brown Univ Providence RI 02912

PARKS, ROBERT J, b Los Angeles, Calif, Apr 1, 22; m 47; c 3. ELECTRICAL ENGINEERING. *Educ:* Calif Inst Technol, BSEE, 44. *Prof Exp:* Engr, Jet Propulsion Lab, 47-51, chief, Guid & Control Sect, 51-56, Guid & Control Div, 56-58 & Guid & Control Dept, 58-60, dir, Sergeant Prog, 57-60 & Planetary Prog, 60-62, asst lab dir lunar & planetary projs, 62-65, surveyor proj mgr, 65-67, asst lab dir flight projs, 67-83, dep dir, 83-87; RETIRED. *Concurrent Pos:* Consult, Tech Adv Panel Aeronaut, Off Secy Defense, 57-58 & Advan Res Projs Agency Ad Hoc Panel Commun Satellite, Inst Defense Anal, 59; mem ad hoc group solid propellants, Bur Naval Weapons, 57-58. *Honors & Awards:* Hill Space Transp Award, 63. *Mem:* Nat Acad Eng; fel Am Inst Aeronaut & Astronaut; fel Inst Elec & Electronics Engrs. *Res:* Automatic guidance and control systems, particularly as applied to ballistic missiles and space vehicles, including the optimization of such systems in the presence of unwanted noise or external disturbances. *Mailing Add:* 1504 S Bay Front Balboa Island CA 92662

PARKS, RONALD DEE, experimental physics; deceased, see previous edition for last biography

PARKS, ROSS LOMBARD, b Asheville, NC, Sept 13, 20; m 48; c 3. ANALYTICAL CHEMISTRY. *Educ:* Davidson Col, BS, 43; Univ NC, PhD, 52. *Prof Exp:* Res chemist, Am Enka Corp, 51-54, chief chemist nylon, 54-60, head tire yarn develop, 60-63; head polyester develop, Textiles Div, Monsanto Co, 63-66, tech supvr, Polyester Dept, Monsanto Textiles Co, 66-77, supvr advan devices, Dept Org Mat Res & Develop, Monsanto Res Corp, 77-85; RETIRED. *Mem:* Am Chem Soc. *Res:* Polymer chemistry; polyamides; polyesters. *Mailing Add:* 595 Concord Rd Fletcher NC 28732-9706

PARKS, TERRY EVERETT, b Satanta, Kans, Feb 20, 41; m 81; c 3. ORGANIC CHEMISTRY, MATHEMATICS EDUCATION. *Educ:* Kans State Teachers Col, BS, 61, MS, 65; Brown Univ, PhD(org chem), 71. *Prof Exp:* Teacher high sch, Kans, 61-66; NIH trainee, Cornell Univ, 70-71; gen dir, Shawnee Mission Pub Schs, Kans, 73-76, supvr fed proj, 71-81, math supvr, 72-90, comput educ supvr, 75-90; DIR, BATTLE CREEK AREA MATH/SCI CTR, 90- *Concurrent Pos:* Consult, Pub Schs, Kans, 71-90 & Gen Educ Testing Serv, 85-; mem, Kans State Metric Educ Task Force, 74-76, adj prof, Kans State Univ, 74-75; instr, Int Grad Sch Educ, 78-79; mem, Kans Assessment Adv Comt, 78-88; chmn, regional serv comt, Nat Coun Teachers Math, 82-83, nominee bd dirs, 84 & 86. *Mem:* Nat Coun Teachers Math; Am Chem Soc. *Res:* Instructional design; computers and calculators and their effect on education; problem solving techniques. *Mailing Add:* 354 N 27th St Battle Creek MI 49015

PARKS, THOMAS WILLIAM, b Buffalo, NY, Mar 16, 39; m 62; c 3. ELECTRICAL ENGINEERING, DIGITAL SIGNAL PROCESSING. *Educ:* Cornell Univ, BEE, 61, MS, 64, PhD(elec eng), 67. *Prof Exp:* Engr, Gen Elec Advan Electronics Ctr, 61-63; from asst prof to assoc prof, 67-77, PROF ELEC ENG, RICE UNIV, 77- *Concurrent Pos:* NSF grants, 70-86; Sr Fulbright Fel, 73; pres, Parks Consults, Inc, 82- *Honors & Awards:* Sr Scientist Award, Alexander von Humboldt Found, 73. *Mem:* Fel Inst Elec & Electronics Engrs; Soc Explor Geophysicists. *Res:* Communication and signal theory; digital signal processing and filtering; geophysical signal analysis. *Mailing Add:* 312 Phillips Hall Cornell Univ Sch Elec Eng Ithaca NY 14853

PARKS, VINCENT JOSEPH, b Chicago, Ill, May 5, 28; m 55; c 7. ENGINEERING MECHANICS. *Educ:* Ill Inst Technol, BS, 53; Cath Univ Am, MCE 63, PhD(mech), 68. *Prof Exp:* Proj engr, Andrew Corp, Ill, 53-55; res engr, Armour Res Found, Ill Inst Technol, 55-61; from asst prof to assoc prof, 65-73, PROF CIVIL ENG, CATH UNIV AM, 73- *Concurrent Pos:* Nat Acad Sci resident res associateship, Naval Res Lab, 71-72. *Honors & Awards:* M Hetényi Award, Soc Exp Stress Anal, 74 & M M Frocht Award, 81. *Mem:* Fel Soc Exp Mech; fel Am Soc Mech Engrs; Am Acad Mech; Sigma Xi. *Res:* Experimental stress analysis using holography, moire, three dimensional photoelasticity and brittle coatings. *Mailing Add:* Dept Civil Eng Cath Univ Am Washington DC 20064

PARKS, WILLIAM FRANK, b Pittsburgh, Pa, June 1, 38; m 56. THEORETICAL PHYSICS. *Educ:* Lehigh Univ, BS, 60; Univ Iowa, PhD(physics), 64. *Prof Exp:* Asst prof, 64-73, ASSOC PROF PHYSICS, UNIV MO-ROLLA, 73- *Mem:* Am Phys Soc. *Mailing Add:* Dept Physics Univ Mo Rolla Rolla MO 65401

PARLETT, BERESFORD, b London, Eng, July 4, 32; m 60; c 2. MATHEMATICS. *Educ:* Oxford Univ, BA, 55; Stanford Univ, PhD(math), 62. *Prof Exp:* Assoc res scientist, Courant Inst Math Sci, NY Univ, 62-64; asst prof math, Stevens Inst Technol, 64-65; asst prof math, 65-68, assoc prof comput sci, 68-73, chmn dept, 68-71, PROF MATH, ELEC ENG & COMPUT SCI, UNIV CALIF, BERKELEY, 73- *Mem:* Am Math Soc; Asn Comput Mach; Soc Indust & Appl Math. *Res:* Numerical analysis; linear algebra; partial differential equations. *Mailing Add:* Dept Math Univ Calif Berkeley CA 94720

PARLI, C(AROL) JOHN, DRUG METABOLISM & DISTRIBUTION, PHARMACOKINETICS. *Educ:* Okla State Univ, PhD(biochem), 66. *Prof Exp:* RES ASSOC, ELI LILLY & CO, 68- *Mailing Add:* Eli Lilly Corp Ctr Indianapolis IN 46285

PARLIMENT, THOMAS H, b Hackensack, NJ, Jan 26, 39; m 65; c 2. FOOD SCIENCE. *Educ:* Lehigh Univ, BS, 61; Univ Mass, PhD(food sci), 65. *Prof Exp:* RES SCIENTIST, GEN FOODS CO, TARRYTOWN, NY, 65- *Mem:* Am Chem Soc; Sigma Xi. *Res:* Major research activities involve the isolation, identification and synthesis of thermally and biologically generated food aromas. *Mailing Add:* Gen Foods Corp 555 S Broadway Tarrytown NY 10951

PARLOW, ALBERT FRANCIS, b Boston, Mass, Jan 27, 33. BIOLOGY, PHYSIOLOGY. *Educ:* Harvard Univ, BA, 55; Princeton Univ, PhD(biol), 58. *Prof Exp:* Res fel anat, Harvard Med Sch, 58-59, res fel physiol in obstet & gynec, 59-62; from instr to asst prof physiol, Emory Univ, 62-65; assoc res physiologist, 65-70, RES PROF OBSTET & GYNEC, SCH MED, UNIV CALIF, LOS ANGELES, 70- *Concurrent Pos:* USPHS res career develop award, 66. *Mem:* Endocrine Soc; Am Physiol Soc; Soc Exp Biol & Med; Soc Study Reprod; Soc Gynecol Invest; Sigma Xi. *Res:* Endocrine regulation of reproduction; bioassay, immunoassay, isolation and structure of pituitary gland hormones. *Mailing Add:* 56 Silver Saddle Rolling Hills Estates CA 90274

PARMA, DAVID HOPKINS, b Santa Barbara, Calif, May 10, 40; m 83; c 1. GENETICS. *Educ:* Univ Calif, Davis, BS, 64, MS, 65; Univ Wash, PhD(genetics), 68. *Prof Exp:* NSF fel molecular biol, Carnegie Inst Wash Genetics Res Unit, 68-70; asst prof, Univ Utah, 70-76, res assoc prof biol, 76-81; SR RES ASSOC, UNIV COLO, 87- *Mem:* AAAS. *Res:* Genetics, especially origin and behavior of chromosomal aberrations; bacteriophage genetics; mechanisms of replication and recombination. *Mailing Add:* 8943 Baseline Rd Lafayette CO 80026

PARMALEE, PAUL WOODBURN, b Mansfield, Ohio, Oct 17, 26; m 49; c 2. ZOOLOGY. *Educ:* Ohio Univ, BSEd, 48; Univ Ill, MS, 49; Agr & Mech Univ, Tex, PhD(wildlife sci), 52. *Prof Exp:* Asst prof biol, Stephen F Austin State Univ, 52-53; cur zool, Ill State Mus, 53-65, Asst Mus Dir, 65-73; prof zooarchaeol, Dept Anthrop, Univ Tenn, 73-89, dir McClung Mus, 78-89; RETIRED. *Mem:* Am Soc Mammal; Soc Vert Paleont; Soc Am Archaeol; Am Ornith Union; Wilson Ornith Soc. *Res:* Vertebrate zoology; ornithology; mammalogy; osteology; identification of faunal materials from archaeological sites. *Mailing Add:* Dept Anthrop Univ Tenn Knoxville TN 37996

PARMAR, SURENDRA S, BIOCHEMICAL PHARMACOLOGY, MEDICAL CHEMISTRY. *Educ:* Lucknow Univ, PhD(chem), 53; McGill Univ, Can, PhD(biochem), 57. *Prof Exp:* PROF PHYSIOL, SCH MED, UNIV NDAK, 72- & HILL RES PROF, 76- *Mailing Add:* Dept Physiol Univ NDak Grand Forks ND 58202

PARME, ALFRED L, b 1909; c 4. ENGINEERING. *Educ:* Cornell Univ, BSCE, 35. *Prof Exp:* Mem staff, Ebasco, 36-38; mem staff design earth & concrete dams, Corps Engrs, 38-40; struct engr & dir struct bur & advan eng dept, Portland Cement Asn, 40-68; partner, J Fruchtbaum, 69; consult engr, 70-86; RETIRED. *Honors & Awards:* Lindau Award, Am Concrete Inst; Martin Korn Award, Prestressed Concrete Inst. *Mem:* Nat Acad Eng; hon mem Am Concrete Inst. *Res:* Design reinforced concrete; seismic design; thin shells and arch dams. *Mailing Add:* 6787 Avenida Andorra La Jolla CA 92037

PARMEGIANI, RAULO, microbial physiology, for more information see previous edition

PARMELEE, CARLTON EDWIN, b Hopkins, Mich, Mar 24, 18; m 43; c 2. FOOD SCIENCE. *Educ:* Mich State Univ, BS, 40; Iowa State Univ, MS, 42, PhD(dairy bact), 47. *Prof Exp:* Asst, Iowa State Univ, 40-42; qual control, Kraft Foods Co, Wis, 42-43; res assoc, Iowa State Univ, 43-47, res asst prof dairy bact, 47-50; from asst prof to assoc prof dairy mfg, 50-53, ASSOC PROF ANIMAL SCI, PURDUE UNIV, WEST LAFAYETTE, 53- *Mem:* Am Dairy Sci Asn. *Res:* Dairy and food microbiology; cheese manufacture; quality control of foods. *Mailing Add:* 541 Dexter Lane West Lafayette IN 47906

PARMELEE, DAVID FREELAND, b Oshkosh, Wis, June 20, 24; m 43; c 1. ORNITHOLOGY. *Educ:* Lawrence Univ, BA, 50; Univ Mich, MS, 52; Univ Okla, PhD(zool), 57. *Prof Exp:* Asst zool, Univ Okla, 52-56, instr, 56-58; from asst prof to prof biol, Kans State Teachers Col, 58-70; chmn field biol prog & prog dir Cedar Creek Natural Hist Area, 70-84, dir, Lake Itasca Forestry & Biol Sta, 70-86, PROF DEPT ECOL, EVOLUTION & BEHAV, UNIV MINN, MINNEAPOLIS, 70- *Concurrent Pos:* Dir field opers bird virus-parasite study, Med Ctr, Univ Okla, 63-65; cur birds, Bell Mus, 85-; lectr, Travel Dyamics NY, aboard M/V Illiria & Polar Circle Antarctica, 88- *Mem:* Wilson Ornith Soc; Cooper Ornith Soc; Orgn Biol Field Stas (pres, 84-85). *Res:* Ornithology, especially the taxonomy, ecology, behavior and breeding biology of arctic and antarctic birds; Antarctic place name, Parmelee Massif. *Mailing Add:* 349 Bell Mus Natural Hist Univ Minn Minneapolis MN 55455

PARMELY, MICHAEL J, b Waterloo, Iowa, June 29, 47; c 1. IMMUNOLOGY. *Educ:* Univ Iowa, Iowa City, PhD(microbiol), 74. *Prof Exp:* Asst prof, 77-81, ASSOC PROF MICROBIOL, MED CTR, UNIV KANS, 82- *Mem:* Am Asn Immunologists; Transplantation Soc; Sigma Xi. *Mailing Add:* Dept Microbiol Med Ctr Univ Kans 39th & Rainbow Blvd Kansas City KS 66103

PARMENTER, CHARLES STEDMAN, b Philadelphia, Pa, Oct 12, 33; m 56; c 3. PHYSICAL CHEMISTRY. *Educ:* Univ Pa, BA, 55; Univ Rochester, PhD(phys chem), 63. *Prof Exp:* Tech rep photo prod, E I du Pont de Nemours & Co, 57-58; NSF res fel chem, Harvard Univ, 62-63, NIH res fel, 63-64; from asst prof to prof, 64-88, DISTINGUISHED PROF CHEM, IND UNIV, BLOOMINGTON, 88- *Concurrent Pos:* Simon H Guggenheim fel, Cambridge Univ, 71-72; vis fel, Joint Inst Lab Astrophys, Nat Bureau Standards & Univ Colo, 77-78; Fulbright sr scholar, Griffith Univ, Australia, 80. *Honors & Awards:* Humboldt Sr Scientist Award, Tech Univ Munchen, 86. *Mem:* AAAS; Am Chem Soc; Am Phys Soc. *Res:* Photochemistry; spectroscopy; energy transfer. *Mailing Add:* Dept Chem Ind Univ Bloomington IN 47405

PARMENTER, ROBERT HALEY, b Portland, Maine, Sept 19, 25; m 51; c 2. SOLID STATE PHYSICS. *Educ:* Univ Maine, BS, 47; Mass Inst Technol, PhD(physics), 52. *Prof Exp:* Guest mem, Brookhaven Nat Lab, 51-52; mem staff, Lincoln Lab & solid state & molecular group, Mass Inst Technol, 52-54; res physicist, RCA Labs, 54-66; chmn dept, 77-83, PROF PHYSICS, UNIV ARIZ, 66- *Concurrent Pos:* Vis lectr, Princeton Univ, 60-61; mem res adv comt electrophys, NASA, 64-68. *Mem:* Fel AAAS; fel Am Phys Soc. *Res:* Electronic energy bands in ordered and disordered solids; group theory of crystals; acoustoelectric effect; superconductivity; space-charge-limited current in insulators. *Mailing Add:* Dept of Physics Univ of Ariz Tucson AZ 85721

PARMENTIER, EDGAR M(ARC), b Waynesburg, Pa, Oct 29, 45. EARTH SCIENCES, FLUID MECHANICS. *Educ:* WVa Univ, BS, 68; Cornell Univ, MEng, 69, PhD(earth sci), 75. *Prof Exp:* Res scientist eng, AVCO-Everett Res Lab, 69-72; res fel geol sci, Oxford Univ, 75-77; asst prof, 77-80, ASSOC PROF GEOL SCI, BROWN UNIV, 80- *Mem:* Am Geophys Union; Sigma Xi. *Res:* Transport processes in the earth; heat and mass transfer. *Mailing Add:* Dept Geol Sci Brown Univ Providence RI 02912

PARMERTER, R REID, b Rochester, NY, June 14, 35. SOLID MECHANICS. *Educ:* Calif Inst Technol, BS, 58, MS, 59, PhD(aeronaut), 64. *Prof Exp:* Stress & vibration engr, AiResearch Mfg Co, Ariz, 56-57; from asst prof to assoc prof aeronaut & astronaut, 63-77, PROF AERONAUT & ASTRONAUT, UNIV WASH, 77- *Concurrent Pos:* Consult, Southwest Engrs, Calif, 57-62, Math Sci Corp, 62-66, Math Sci Northwest, Wash, 66- *Mem:* Soc Exp Stress Analysis. *Res:* Nonlinear shell theory; shell stability; photoelasticity; stress analysis of solid propellant rockets; holography; ice mechanics. *Mailing Add:* 11919 Lakeside Ave Seattle WA 98125

PARMERTER, STANLEY MARSHALL, b Rochester, NY, Oct 24, 20; m 43; c 3. ORGANIC CHEMISTRY. *Educ:* Greenville Col, 41; Univ Ill, MS, 42, PhD(org chem), 44; John Marshall Law Sch, JD, 80. *Prof Exp:* Asst physics, Greenville Col, 40-41; asst physics & chem, Univ Ill, 42-44, fel, 44-45; res chemist, William S Merrell Co, Ohio, 45-47 & Eastman Kodak Co, 47-52; from instr to prof chem, Wheaton Col, Ill, 52-64, chmn sci div, 59-64; sect leader, Corn Prod Co, 64-69, dir admin serv, 69-75, patent adv, 75-77, patent agent, 78-79, PATENT ATTY, MOFFETT TECH CTR, CPC INT INC, 80- *Concurrent Pos:* Vis res assoc, Argonne Nat Lab, 63-64. *Mem:* Am Chem Soc. *Res:* Organic synthesis; carbohydrate chemistry. *Mailing Add:* 114 Windsor Park Dr B301 Carol Stream IL 60188

PARMETER, JOHN RICHARD, JR, b The Dalles, Ore, Sept 16, 27; m 55; c 3. PLANT PATHOLOGY. *Educ:* Ore State Col, BS, 51; Univ Wis, PhD, 55. *Prof Exp:* Asst plant path, Univ Wis, 51-55; plant pathologist, US Forest Serv, 55-57; from asst prof to assoc prof, 57-69, PROF PLANT PATH, UNIV CALIF, BERKELEY, 69- *Mem:* Am Phytopath Soc; AAAS. *Res:* Diseases of forest trees; soil microbiology; mycology. *Mailing Add:* Dept Plant Path Univ Calif Berkeley CA 94720

PARMLEY, WILLIAM W, b Salt Lake City, Utah, Jan 22, 36; m 61; c 4. CARDIOLOGY. *Educ:* Harvard Univ, AB, 57; Johns Hopkins Univ, MD, 63. *Prof Exp:* Intern, Osler Med Serv, Johns Hopkins Hosp, Baltimore, 63-64, asst resident, 64-65; clin assoc, Cardiol Br, Nat Heart Inst, Bethesda, 65-69;

instr med, Harvard Med Sch, 69; assoc prof, Sch Med, Univ Calif, Los Angeles & assoc dir dept cardiol, Cedars-Sinai Med Ctr, 70-74; PROF MED, SCH MED, UNIV CALIF, SAN FRANCISCO & CHIEF, DIV CARDIOL, MOFFITT HOSP, 74- *Concurrent Pos:* Res fel, Cardiovasc Unit, Peter Bent Brigham Hosp, Harvard Med Sch, 67-69, jr assoc med, 69; estab investr, Am Heart Asn, 71-76; chmn, Comt Regulatory Nat Res, 82-83. *Honors & Awards:* Theodore & Susan Cummings Humanitarian Award, Am Col Cardiol, 71. *Mem:* Fel Am Col Cardiol; Am Physiol Soc; Am Fedn Clin Res; Am Soc Clin Invest; Am Asn Physicians. *Res:* Muscle physiology; ventricular function; cardiac pharmacology. *Mailing Add:* 1186 Moffitt Hosp Univ Calif San Francisco CA 94143-0124

PARNELL, DENNIS RICHARD, b Rochester, NY, Dec 30, 39. PLANT SCIENCE. *Educ:* Whittier Col, BA, 61; Univ Calif, Los Angeles, MA, 63, PhD(plant sci), 65. *Prof Exp:* From asst prof to assoc prof, 65-74, PROF BIOL SCI, CALIF STATE UNIV, HAYWARD, 74-, DEAN, SCH SCI, 80- *Mem:* AAAS; Soc Study Evolution; Bot Soc Am; Am Soc Plant Taxonomists; Am Inst Biol Sci; Sigma Xi. *Res:* Plant systematics, cytogenetics and anatomy. *Mailing Add:* Dept Biol Sci Calif State Col Hayward Hayward CA 94542

PARNELL, DONALD RAY, b Grafton, NDak, Aug 27, 42; m 64; c 4. SILICONE RELEASE COATINGS. *Educ:* Univ NDak, BS, 64, PhD (org chem), 72. *Prof Exp:* Prin investr, Army Mat & Mech Res Ctr, 69-71; postdoctoral assoc, Univ Wis-Madison, 72-73; sr res scientist, Corning Glassworks, 73-77; supvr chem technol, Am Can, 77-81; TECH DIR, AKROSIL, 81- *Concurrent Pos:* Gen secy, Third Int Symp Organosilicon Chem, 72. *Mem:* Tech Asn Pulp & Paper Indust. *Res:* Chemistry of silicone curing and the influence of the underlying substrate of silicone coated release liners used to protect pressure sensitive adhesives. *Mailing Add:* Akrosil 206 Garfield Ave Menasha WI 54952-8001

PARNELL, JAMES FRANKLIN, b Timmonsville, SC, May 15, 34; m 62. ORNITHOLOGY, ECOLOGY. *Educ:* NC State Univ, BS, 59, MS, 61, PhD(zool), 64. *Prof Exp:* From asst prof to assoc prof, 64-69, chmn dept, 69-71, PROF BIOL, UNIV NC, WILMINGTON, 69- *Mem:* Am Ornith Union; Wildlife Soc; Wilson Ornith Soc; Soc Wetland Scientists (pres, 80-83). *Res:* Ecology and distribution of vertebrates, particularly the habitat relations of birds and the management of colonial waterbird populations. *Mailing Add:* Dept Biol Univ NC Wilmington NC 28403

PARNELL, THOMAS ALFRED, b Lumberton, NC, Nov 24, 31; m 55; c 2. PHYSICS, COSMIC RAY PHYSICS. *Educ:* Univ NC, Chapel Hill, BS, 54, MS, 62, PhD(physics), 65. *Prof Exp:* Res adj physics, Univ NC, Chapel Hill, 63-65, opers analyst, 65-66; asst prof physics, Marshall Univ, 66-67; PHYSICIST, MARSHALL SPACE FLIGHT CTR, NASA, 67-, CHIEF ASTROPHYSICS BR, 68- *Honors & Awards:* NASA Exceptional Scientific Achievement Medal. *Mem:* Am Phys Soc; Sigma Xi. *Res:* Experimental cosmic ray physics, gamma ray astronomy and nuclear instrumentation; space radiation and dosimetry. *Mailing Add:* 907 Corinth Huntsville AL 35801

PARNES, MILTON N, b Detroit, Mich, Nov 13, 39; m 63; c 3. STATISTICS. *Educ:* Wayne State Univ, BS, 60, PhD(math), 68. *Prof Exp:* Asst prof math, State Univ NY, Buffalo, 67-77; ASST PROF , 77-, ASSOC PROF MATH, STATIST DEPT, TEMPLE UNIV, 77- *Res:* Complex analysis; mathematical theory of gambling; biometry; mathematical statistics; measure theory; approximation theory. *Mailing Add:* Dept Statist Temple Univ Philadelphia PA 19122

PAROCHETTI, JAMES V, b Spring Valley, Ill, Apr 24, 40; m 80; c 4. AGRICULTURAL SCIENCE, PESTICIDES. *Educ:* Univ Ill, BS, 62; Purdue Univ, MS, 64, PhD(plant physiol), 67. *Prof Exp:* PRIN WEED SCIENTIST, COOP STATE RES SERV USDA, WASHINGTON, DC, 66- *Concurrent Pos:* Exten Specialist Res, Assoc Prof, Weed Sci Univ, Md; Exten Specialist & Pesticide Coordr, Extension Serv, USDA, Washington, DC. *Mem:* Coun Agricultural Sci Technol; Weed Sci Soc Am. *Res:* Behavior of herbicides in plants and soils; control of herbicide resistant weeds. *Mailing Add:* Coop State Res Serv USDA JSM Bldg Washington DC 20251

PARODE, L(OWELL) C(ARR), b Los Angeles, Calif, Oct 27, 24; m 46; c 3. ELECTRONICS ENGINEERING. *Educ:* Calif Inst Technol, BS, 45, MS, 47. *Prof Exp:* Mgr, Guid Lab, Hughes Aircraft Co, 47-71, mgr, Missile Eng Labs, 71-73, prog mgr, NASA Systs Div, 73-89; RETIRED. *Mem:* Sigma Xi; sr mem Inst Elec & Electronics Engrs. *Res:* Missile guidance; tracking radar design; communication systems. *Mailing Add:* 1645 Via Lazo Palos Verdes Estates CA 90274

PARONETTO, FIORENZO, b Treviso, Italy, Jan 18, 29; m 61; c 2. PATHOLOGY. *Educ:* Univ Padua, MD, 52. *Prof Exp:* Fel med, County Hosp, Salt Lake City, Utah, 54-55; fel, Mt Sinai Hosp, New York, 55-56, resident, path, 57-59, intern, 59-60, res fel path, 60-61, res assoc, 62-63, asst attend pathologist, 63-65; from asst prof to assoc prof, 65-71, PROF PATH, MT SINAI SCH MED, 71-; CHIEF LAB SERV, VET AFFAIRS MED CTR, BRONX, 73- *Mem:* Am Asn Path & Bact; Am Fedn Clin Res; Am Soc Exp Path; Am Asn Immunol; Transplantation Soc. *Res:* Immunopathology. *Mailing Add:* Lab Serv Vet Affairs Med Ctr 130 W Kingsbridge Rd Bronx NY 10468

PARPIA, JEEVAK MAHMUD, b Bombay, India, July 22, 52; m 74. LOW TEMPERATURE CONDENSED MATTER. *Educ:* Ill Inst Technol, BSLA, 73; Cornell Univ, MS, 77, PhD(physics), 79. *Prof Exp:* Assoc low temperature physics, Cornell Univ, 78-79; from asst prof to assoc prof physics, Tex A&M Univ, 79-86; ASSOC PROF PHYSICS, CORNELL UNIV, 86- *Concurrent Pos:* Alfred P Sloan res fel, Tex A&M Univ, 81-85. *Mem:* Am Phys Soc; Mat Res Soc. *Res:* Study of hydrodynamic and magnetic properties of Liquid Helium Three at ultra low temperatures; finite size effects at low temperatures. *Mailing Add:* Dept Physics Cornell Univ, Clark Hall Ithaca NY 14853

PARR, ALBERT CLARENCE, b Tooele, Utah, June 22, 42; m 65; c 2. PHYSICS. *Educ:* Ore State Univ, BS(physics) & BS(math), 64; Univ Chicago, MS, 65, PhD(physics), 71. *Prof Exp:* Res asst physics, Univ Chicago, 69-70; from asst prof to assoc prof, Univ Ala, 71-80; res physicist ultraviolet physics, Nat Bur Standards, 80-86, GROUP LEADER, SPECTRAL RADIOMETRY, NAT INST STANDARDS & TECHNOL, 86- *Concurrent Pos:* Sabbatical leave, Nat Bur Standards, 78-79. *Mem:* Am Phys Soc; Coun Optical Radiation Measurement. *Res:* Atomic and molecular physics; photoionization processes in gas phase; reactions on surfaces; synchrotron radiation; photoelectron spectroscopy; radiometry. *Mailing Add:* Bldg 221 Rm A221 Nat Inst Stanards & Technol Gaithersburg MD 20899

PARR, CHRISTOPHER ALAN, b Oakland, Calif, May 6, 41; m 85; c 1. THEORETICAL CHEMISTRY, COMPUTER SCIENCE. *Educ:* Univ Calif, Berkeley, BS, 62; Calif Inst Technol, PhD(phys chem), 69. *Prof Exp:* Army Res Off fel theory bimolecular reactions, Univ Calif, Irvine, 68-69; Nat Res Coun Can fel, Univ Toronto, 69-71; asst prof, 71-77, col master, Natural Sci & Math, 80-87, assoc dean sci, 86-87, ASSOC PROF CHEM, UNIV TEX, DALLAS, 77-, DEAN UNDERGRAD STUDIES, 87- *Mem:* Am Chem Soc; Am Phys Soc; AAAS; Sigma Xi. *Res:* Theoretical chemical reaction dynamics; unimolecular dynamics and quantization; inter- and intra-molecular potential energy functions; computer graphics. *Mailing Add:* Univ Tex at Dallas Box 830688 Richardson TX 75083-0688

PARR, JAMES FLOYD, JR, b Seattle, Wash, Feb 20, 29; m 87; c 2. SOIL MICROBIOLOGY, PLANT PHYSIOLOGY. *Educ:* Wash State Univ, BS, 52, Purdue Univ, MS, 57, PhD(soil microbiol), 61. *Prof Exp:* Irrig exten agent, Agr Exten Serv, Wash State Univ, 53-54; agr exten agent, Mont State Univ, 54-55; instr agron, Purdue Univ, 55-57; chemist, Calif Dept Water Resources, 57-58; instr agron, Purdue Univ, 58-61; res assoc bot, Univ Mich, 61-63; res chemist & soil microbiologist, Soils & Fertilizer Res Br, Tenn Valley Auth, 63-67; soil microbiologist, 67-75, chief, Biol Waste Mgt & Org Resources Lab, Agr Res Serv, 75-84, COORDR, USDA-USAID DRYLAND MGT PROJ, USDA, 84- *Mem:* Fel Am Soc Agron; fel Soil Sci Soc Am; Am Soc Microbiol; Am Soc Plant Physiol. *Res:* Water infiltration into soils; soil organic matter decomposition; effects of surfactants on plant growth; ion uptake by plants; sorption and reaction of anhydrous NH-3 in soils; fertilizer evaluation; fate and persistence of pesticides in soil and water; recycling municipal wastes for soil improvement and plant growth; improving the productivity of subsistence-level agriculture in semi-arid developing countries. *Mailing Add:* Agr Res Serv Beltsville Agr Res Ctr US Dept Agr Beltsville MD 20705

PARR, JAMES GORDON, b Peterborough, UK, May 26, 27; Can citizen; c 3. SCIENCE ADMINISTRATION. *Educ:* Leeds Univ, UK, BSc, 47; Univ Liverpool, PhD(metall), 53. *Hon Degrees:* Univ Windsor, LLD, 84. *Prof Exp:* Lectr metall, Univ Liverpool, 48-53, Univ BC, 53-55; from assoc prof to prof, Univ Alta, 55-64; dean eng & prof eng mat, Univ Windsor, 64-72; chmn, Comt Univ Affairs, Ont, 72; dep minister, cols and univs, Ont Govt, 73-79; chmn, chief exec officer, TV Ont, 79-85; dir gen, Ont Sci Ctr, 85-88; RETIRED. *Concurrent Pos:* Pres, Hanson Parr Eng, 56-64, Ind Res Inst, Univ Windsor, 65-72; ed-in-chief, Can Metall Quart, 66-71; fel, Ryerson Polytech Inst, 82. *Honors & Awards:* Centennial Medal; Jubilee Medal. *Mem:* Fel Royal Soc Can; fel Am Soc Metals. *Res:* Transformations in metals and alloys, and related properties. *Mailing Add:* Ten Governor's Rd Toronto ON M4W 2G1 Can

PARR, JAMES THEODORE, b Lebanon, Ind, Oct 29, 34; m 66; c 1. ALGEBRA. *Educ:* Ind Univ, AB, 56, PhD(math), 64. *Prof Exp:* Lectr math, Ind Univ, 64; from instr to asst prof math, Univ Ill, Urbana, 64-70; ASST PROF MATH, ILL STATE UNIV, 70- *Concurrent Pos:* NSF res fel math, 65-66. *Mem:* Am Math Soc; Math Asn Am; Sigma Xi. *Res:* Homological algebra; abelian groups; cohomology of cyclic groups of prime square order. *Mailing Add:* Dept of Math Ill State Univ Normal IL 61761

PARR, PHYLLIS GRAHAM, b Princeton, Ind, Dec 4, 37; m 66; c 1. COMPUTER PROGRAMMING, ANALYSIS. *Educ:* Oakland City Col, BSE, 59; Ind Univ, Bloomington, MA, 61, PhD(math), 64. *Prof Exp:* Asst prof math, Ind State Univ, Terre Haute, 64-66 & Univ Ill, Urbana, 66-71; assoc prof, Ill Wesleyan Univ, 71-73 & 76-77; asst prof math, Ill State Univ, 74-75; data processing analyst, 79-89, SR DATA PROCESSING ANALYST, STATE FARM INSURANCE, BLOOMINGTON, ILL, 89- *Mem:* Asn Comput Mach; Data Processing Mgt Asn. *Res:* Cohomology of finite groups; finite rings. *Mailing Add:* 2009 Castle Ave Bloomington IL 61701

PARR, ROBERT GHORMLEY, b Chicago, Ill, Sept 22, 21; m 44; c 3. THEORETICAL CHEMISTRY. *Educ:* Brown Univ, AB, 42; Univ Minn, PhD(phys chem), 47. *Hon Degrees:* Dr, Univ Leuven, 86. *Prof Exp:* Asst prof chem, Univ Minn, 47-48; from asst prof to prof chem, Carnegie Inst Technol, 48-62, chmn gen fac, 60-61; prof, Johns Hopkins Univ, 62-74, chmn dept, 69-72; William R Kenan, jr prof theoret chem, 74-90, WASSILY HOEFFDING PROF CHEM PHYSICS, UNIV NC, 90- *Concurrent Pos:* Fel, Univ Chicago, 49, res assoc, 57; Guggenheim fel & Fulbright scholar, Univ Cambridge, 53-54; Sloan fel, 56-60; assoc ed, J Chem Physics, 56-58, Chem Revs, 61-63, J Phys Chem, 63-67 & 77-79, Am Chem Soc Monographs, 66-71 & Theoretica Chimica Acta, 66-69; mem chem adv panel, Air Force Off Sci Res, 60-65; chmn, Nat Acad Sci-Nat Res Coun Comt Postdoctoral Fels in Chem, 61-63; vis prof chem & mem ctr adv study, Univ Ill, 62; chmn panel theoret chem, Westheimer Comt for the Surv of Chem, Nat Acad Sci, 64; vis prof, State Univ NY, Buffalo & Pa State Univ, 67; NSF sr fel, Oxford Univ & CSIRO, Univ Melbourne, 67-68; vis prof, Japan Soc Promotion Sci, 68 & 79; assoc, Comt on Int Exchange of Persons & mem adv comt for E Asia, 71-; mem, Gordon Res Confs Coun, 74-76; mem, panel pub affairs, Am Phys Soc, 75-77; Firth prof, Univ Sheffield, 76; vis prof, Israel Inst Technol, 77; mem comn human resources, Nat Res Coun, 79-82; chmn adv comt associateships, Off Sci & Eng Personnel, Nat Res Coun, 83-84, mem adv comt, 84-; res assoc, Inst Theoret Physics, Univ Calif, Santa Barbara, 83; mem adv comt, Off Sci & Eng Personnel, Nat Res Coun, 84-87; mem coun, Inst Molecular Sci, Okazaki, Japan, 86-88; mem bd trustees, Inst Fundamental

Chem, Kyoto, Japan, 88- *Honors & Awards:* Petrol Res Fund Award, Am Chem Soc, 64. *Mem:* Nat Acad Sci; Sigma Xi; Am Chem Soc; fel Am Phys Soc; Int Acad Quantum Molecular Sci (vpres, 73-79, hon pres, 79-); Am Acad Arts & Sci; fel AAAS; Am Asn Univ Profs. *Res:* Electronic structure of molecules; chemical physics. *Mailing Add:* Dept Chem Univ NC Chapel Hill NC 27514

PARR, WILLIAM CHRIS, b Ranger, Tex, Aug 13, 53; m 74; c 1. STATISTICS, MATHEMATICS. *Educ:* Trinity Univ, BA, 74; Southern Methodist Univ, MS, 76, PhD(statist), 78. *Prof Exp:* Teaching asst statist, Southern Methodist Univ, 74-77; asst mgr, Consult Lab, 77-78; asst prof statist, Inst Statist, Tex A&M Univ, 78-; AT DEPT STATIST, UNIV FLA. *Mem:* Am Statist Asn; Inst Math Statist; Am Math Soc; Royal Statist Soc; Psychomet Soc. *Res:* Robust estimation and testing; jackknifing; the foundations of inference. *Mailing Add:* Dept Statist Univ Fla Gainesville FL 32611

PARRATT, LYMAN GEORGE, b Salt Lake City, Utah, May 17, 08; m 44; c 2. PHYSICS. *Educ:* Univ Utah, AB, 28; Univ Chicago, PhD(physics), 32. *Prof Exp:* Asst physics, Univ Utah, 28-29 & Univ Chicago, 30-33; Nat Res Coun fel, 33-35, from instr to prof, 35-73, chmn dept, 59-69, EMER PROF DEPT PHYSICS & LAB NUCLEAR STUDIES & LAB ATOMIC & STATE PHYSICS, 73- *Concurrent Pos:* Physicist & head, Eng Div, Naval Ord Lab, 41-43; group leader, Los Alamos Sci Lab, 43-46. *Mem:* AAAS; fel Am Phys Soc; Fedn Am Sci; Am Asn Physics Teachers. *Res:* X-ray spectroscopy; solid state physics; corrosion of metal surfaces; electronics; underwater ordnance; ship's magnetism; acoustic instruments; torpedoes; submarine detection; probability theory and experimental errors. *Mailing Add:* 513 Wyckoff Rd Ithaca NY 14850-2309

PARRAVANO, CARLO, b Rome, Italy, Dec 28, 45; US citizen; m 70. PHYSICAL CHEMISTRY. *Educ:* Oberlin Col, BA, 67; Univ Calif Santa Cruz, PhD (chem), 74. *Prof Exp:* Instr chem, Eastern Mich Univ, 68-70; asst prof, 74-80, ASSOC PROF CHEM, STATE UNIV NY COL, PURCHASE, 80- *Concurrent Pos:* Assoc, Danforth Found, 78- *Mem:* Am Asn Physics Teachers; Am Chem Soc; Nat Sci Teachers Asn. *Res:* Gas phase reaction kinetics; application of physical techniques to biochemical and environmental problems. *Mailing Add:* Div Nat Sci State Univ NY Purchase NY 10577

PARREIRA, HELIO CORREA, b Rio de Janeiro, Brazil, July 12, 26; m 53; c 2. SURFACE CHEMISTRY. *Educ:* Univ Brazil, Bsc, 49; Cambridge Univ, PhD(colloid sci), 58. *Prof Exp:* Asst prof analytical chem, Univ Rio de Janeiro, 50-52; phys chemist, Biophys Inst, Univ Brazil, 51-55 & Brazilian Atomic Cmn, 58-60; res assoc surface chem, Sch Mines, Columbia Univ, 60-62, asst prof chem metall, 62-63, asst prof appl chem, 63-65, dir instruction chem, NSF Joint Prog Tech Educ, Sch Eng, 60-65; group leader dept phys & inorg chem, Cent Res Labs, Inmont Corp, NJ, 65-67, prin scientist, 67-69; assoc dir res, 69-70, dir res & develop & mem exec comt, 70-72, sr res assoc, 72-82, SR SCIENTIST, JOHNSON & JOHNSON, 82- *Concurrent Pos:* Researcher, Cambridge Univ, 54; ed, Universal Ref Libr, Grosset-Dunlap, Inc, 64; consult, Yeshiva, 64 & Stanley-Thompson Labs, Columbia Univ, 65-66. *Mem:* Am Chem Soc; Sigma Xi. *Res:* Surface phenomena; electrokinetics; electrostatics of interfaces; physical chemistry of colloidal electrolytes; surface infra-red, ESCA, ISS/SIMS; surface chemistry of polymers, metals, silica, adhesives; glass, natural and biological surfaces; transcutaneous drug delivery. *Mailing Add:* 24 Bunker Hill Run East Brunswick NJ 08816

PARRETT, NED ALBERT, b Logansport, Ind, Nov 8, 39; m 63; c 2. MEAT SCIENCE. *Educ:* Purdue Univ, BS, 63; Tex A&M Univ, MS, 68, PhD(animal sci), 72. *Prof Exp:* Instr animal sci, Tex A&M Univ, 65-69; asst prof, 69-77, ASSOC PROF ANIMAL SCI, OHIO STATE UNIV, 77- *Mem:* Am Soc Animal Sci; Inst Food Technologists; Am Meat Sci Asn. *Res:* Carcass evaluation and composition; quality and palatability attributes of carcasses and wholesale cuts; packaging of meat and meat products. *Mailing Add:* Dept Animal Sci 110 Animal Sci Bldg Ohio State Univ Main Campus Columbus OH 43210

PARRILL, IRWIN HOMER, b Kinmundy, Ill, Mar 21, 09; m 32; c 3. PHYSICAL CHEMISTRY, AGRICULTURAL CHEMISTRY. *Educ:* Ill State Norm Univ, BEd, 31; Univ Iowa, MS, PhD(phys chem), 39. *Prof Exp:* Supvr Lake County Dairy Herd Improv Asn, Ill, 28-29; teacher & asst prin high sch, 31-33; asst sci, Univ High Sch, Iowa, 33-34; dean, Eagle Grove Jr Col, 34-36; instr high sch, C Z, 36-39; chemist, Munic Lab, 39-47; assoc prof chem & chmn phys sci div, Farragut Col & Tech Inst, 47-48; instr high sch, Ill, 53-58; dir, Parrill Lab, Southern Ill Univ, 52-83, lectr, 57-58, from asst prof to prof, 58-74, EMER PROF CHEM, SOUTHERN ILL UNIV, EDWARDSVILLE, 74- *Mem:* Am Chem Soc. *Res:* Soils chemistry; chemistry education. *Mailing Add:* RR Two Box 206 Edwardsville IL 62025

PARRISH, ALVIN EDWARD, b Washington, DC, Sept 6, 22; m 45; c 2. MEDICINE. *Educ:* George Washington Univ, MD, 45. *Prof Exp:* Instr physiol, 47-48, lectr, 48-50, clin instr med, 51-57, assoc prof & assoc dean, 57-64, dir div renal dis, 64-79, PROF MED, SCH MED, GEORGE WASHINGTON UNIV, 64-, DIR CLIN RES, 66-, DIR OFF HUMAN RES, 79- *Concurrent Pos:* Asst chief serv, Vet Admin Hosp, Washington, DC, 51-57, consult, 57- *Mem:* AAAS; AMA; Am Diabetes Asn; fel Am Col Physicians; Am Fedn Clin Res; Sigma Xi. *Res:* Renal disease. *Mailing Add:* 2150 Pennsylvania Ave NW 509 Washington DC 20037

PARRISH, CLYDE FRANKLIN, b Skillman, Ky, Sept 9, 38; m 58; c 2. PHYSICAL CHEMISTRY, RADIATION CHEMISTRY. *Educ:* Eastern Ky State Col, BS, 59; Univ Louisville, PhD(phys chem), 62. *Prof Exp:* Chemist, Dow Chem Co, Mich, 62-63; asst prof, 63-69, dir, radiation lab, 66-71, ASSOC PROF CHEM, IND STATE UNIV, TERRE HAUTE, 69-; PRES, RPS INDUST & SF LEASING CORP, 71- *Concurrent Pos:* Consult, Dow Chem Co, 63-68 & Nuclear Div, Union Carbide Corp, 65-69. *Mem:* AAAS; Am Soc Testing & Mat; Am Chem Soc. *Res:* Radiation induced solid state polymerization of vinyl monomers; mechanism of particulate adhesion in ten to one hundred micron size range. *Mailing Add:* 1300 A1A - 619 Jupiter FL 33477

PARRISH, DALE WAYNE, b Franklin Co, Ala, Aug 17, 24. MEDICAL ENTOMOLOGY, BIOMEDICAL SCIENCES. *Educ:* Auburn Univ, BS, 47; Univ Md, MS, 64; Okla State Univ, PhD(med entom), 71; Am Registry Prof Entomologists, cert med entom, 71. *Prof Exp:* Teacher agr, Ala State Dept Educ, 47-48; supvr agr, Farmers Home Admin, USDA, 48-51; med entomologist, US Dept Air Force, 51-67, from sr biomed scientist to chief biomed scientist, US Air Force Med Serv, 67-73, assoc dir biomed sci, US Air Force Biomed Sci Corps, 74; health sci adminr, 74-80, sr scientist adminr, US Environ Protection Agency, 81-86; BIOMED SCI CONSULT, PARRISH CONSULT SERV, 86- *Concurrent Pos:* Bd mem, Armed Forces Pest Control Bd, Dept Defense, 56-59 & 71-72; sr med serv consult, Off Surgeon Gen, US Air Force Hq, 71-74; mil joint chiefs-of-staff rep fed working group pest mgt, Pres Coun Environ Qual, 72-73. *Mem:* Entom Soc Am; Aerospace Med Asn; Sigma Xi; Am Mosquito Control Asn. *Res:* Epidemiology of arthropod and rodent-borne diseases; wild vertebrate host relationships of ticks; control of arthropod vectors of diseases; toxicology of pesticides, epidemiology of pesticide poisonings, especially effects on human health; aerial dispersal of pesticides-disease vector control. *Mailing Add:* 5105 Acorn Dr Camp Springs MD 20748

PARRISH, DAVID JOE, b Knoxville, Tenn, Dec 23, 43; m 68; c 2. CROP PHYSIOLOGY. *Educ:* E Tenn State Univ, BS, 67; Wake Forest Univ, MA, 70; Cornell Univ, PhD(bot), 76. *Prof Exp:* Lectr biol, Cornell Univ, 75; fel, Univ Nebr, Lincoln, 76-77; from asst prof to assoc prof, 77-89, PROF CROP PHYSIOL, VA POLYTECH INST & STATE UNIV, 89- *Mem:* Am Soc Plant Physiologists; Am Soc Agron; Crop Sci Soc Am; Plant Growth Regulator Soc Am; Sigma Xi. *Res:* High resolution plant growth measurement; auxin physiology; energy cropping; seed physiology. *Mailing Add:* Dept Crop & Soil Environ Sci Va Polytech Inst & State Univ Blacksburg VA 24061-0404

PARRISH, DAVID KEITH, b Clinton, Mo, Sept 29, 44; c 3. GEOLOGY. *Educ:* Univ Mo-Columbia, BS, 66, MA, 69; Rice Univ, C Houston, PhD(geol), 72. *Prof Exp:* Res assoc geol, Dept Earth & Space Sci, State Univ NY Stony Brook, 71-74; asst prof geol, Prog & Inst Geosci, Univ Tex, Dallas, 74-75; asst prof geol, Tex A&M Univ, 75-80, asst prof geophysics, 75-80; MEM STAFF, RE/SPEC, INC, RAPID CITY, 80- *Mem:* Am Geophys Union. *Res:* Theoretical analysis of crustal structures using finite element models; experimental determination of high pressure, temperature mechanical behavior of rocks; analysis of finite strain in rocks; stress analysis of engineered structures in rock. *Mailing Add:* RE/SPEC Inc PO Box 725 Rapid City SD 57709

PARRISH, DONALD BAKER, b Ft Scott, Kans, Sept 24, 13; m 36; c 3. NUTRITION, BIOCHEMISTRY. *Educ:* Kans State Col, BS, 35, MS, 38, PhD(biochem, nutrit), 49. *Prof Exp:* Instr high schs, Kans, 35-43; asst chemist, 43-49, from asst prof to assoc prof chem, 49-62, biochemist & nutritionist, Agr Exp Sta, 62-83, prof 62-83, EMER PROF BIOCHEM, KANS STATE UNIV, 83- *Mem:* Am Chem Soc; Am Soc Animal Sci; Am Dairy Sci Asn; Asn Off Analytical Chem; Am Inst Nutrit. *Res:* Vitamin A requirements of farm animals; metabolism and methods of analysis; nutritional requirements of animals; nutritional evaluations of feeds and foods. *Mailing Add:* Dept of Biochem Kans State Univ Manhattan KS 66506-3701

PARRISH, EDWARD ALTON, JR, b Newport News, Va, Jan 7, 37; m 63; c 2. PATTERN RECOGNITION, COMPUTER ENGINEERING. *Educ:* Univ Va, BEE, 64, MEE, 66, DSc(elec eng), 68. *Prof Exp:* Sr programmer software, Amerad Corp, Va, 61-63; from asst prof to prof elec eng, Univ Va, 68-86, chmn dept, 78-86; CENTENNIAL PROF ELEC ENG & DEAN, VANDERBILT UNIV, 87- *Mem:* AAAS; fel Inst Elec & Electronics Engrs; Pattern Recognition Soc; Sigma Xi. *Res:* Pattern recognition; image processing. *Mailing Add:* Sch Eng Vanderbilt Univ Nashville TN 37235

PARRISH, FRED KENNETH, b Durham, NC, Oct 17, 27; m 56; c 5. BIOLOGY. *Educ:* Duke Univ, BA, 53; Univ NC, MA, 59; Emory Univ, PhD(biol), 66. *Prof Exp:* Instr biol, Agnes Scott Col, 60-65; asst prof, 65-70, ASSOC PROF BIOL & MEM URBAN LIFE FAC, GA STATE UNIV, 70- *Res:* Aquatic and marine biology; animal behavior. *Mailing Add:* Ga State Univ Univ Plaza Atlanta GA 30303

PARRISH, FREDERICK CHARLES, JR, b Olney, Mo, July 18, 33; m 53; c 4. ANIMAL SCIENCE, BIOCHEMISTRY. *Educ:* Univ Mo, BS, 59, MS, 60, PhD(animal husb), 65. *Prof Exp:* Asst animal husb, Univ Mo, 59-60, instr, 60-65; from asst prof to assoc prof, 65-76, PROF ANIMAL SCI & FOOD TECHNOL, 76- *Concurrent Pos:* Food scientist, Coop State Res Serv, USDA, Washington, DC, 72-73. *Honors & Awards:* Am Soc Animal Sci Res Award, 86; Am Meat Sci Asn Distinguished Res Award, 87, Teaching Award, 88. *Mem:* Inst Food Technol; Am Soc Animal Sci; Am Meat Sci Asn. *Res:* Muscle biochemistry and enzymology; relationship of chemistry and structure to meat palatability; postmortem biochemical and biophysical properties of muscle. *Mailing Add:* 150 Food Res Lab Iowa State Univ Ames IA 50011

PARRISH, HERBERT CHARLES, b Jacksboro, Tex, Oct 8, 19; m 41; c 3. MATHEMATICAL ANALYSIS. *Educ:* N Tex State Univ, BS, 39, MS, 41; Ohio State Univ, PhD(math), 55. *Prof Exp:* Instr math, Ohio State Univ, 48-49; from asst prof to assoc prof, 49-58, dir dept, 58-65, PROF MATH, NTEX STATE UNIV, 58- *Mem:* Am Math Soc; Math Asn Am (nat gov, 62-65); Nat Coun Teachers Math. *Res:* Analysis; functions of a real variable; measure and integration. *Mailing Add:* Dept Math NTex Univ 1001 Ector St Denton TX 76201

PARRISH, JAMES DAVIS, b Bennettsville, SC, Dec 18, 35; m 61; c 2. MARINE ECOLOGY, FISH BIOLOGY. *Educ:* Univ SC, BS, 56; Univ RI, PhD(oceanog), 72. *Prof Exp:* Scientist I & II marine ecol, PR Nuclear Ctr of USAEC, 72-75; asst leader & asst prof fisheries, Mass Coop Fish Res Unit, 75-77, LEADER & ASSOC PROF ZOOL, HAWAII COOP FISH RES

UNIT, US FISH & WILDLIFE SERV, 77- *Concurrent Pos:* Mem, Interagency Sci Adv Subcomt Ocean Dredging & Spoiling, 75-77, Sci Statist Comt, West Pac Fish Mgt Coun, 81-; mem bd, Asn Island Marine Labs Caribbean, 81-83 & 84- *Mem:* AAAS; Am Fisheries Soc; Am Soc Limnol & Oceanog; Ecol Soc Am; Sigma Xi; Asn Island Marine Labs Caribbean; Wildlife Soc; Western Soc Naturalists. *Res:* Aquatic population and community ecology; trophic structure; analytical models of ecological systems; species interactions; predation; dynamics of multi-species fish populations; shallow tropical marine systems, coral reef ecology. *Mailing Add:* Hawaii Coop Fish Res Unit 2538 The Mall Univ Hawaii Honolulu HI 96822

PARRISH, JOHN ALBERT, b Louisvile, Ky, Oct 19, 39; m 62; c 3. PHOTOBIOLOGY, DERMATOLOGY. *Educ:* Duke Univ, BA, 61; Yale Univ, MD, 65. *Prof Exp:* From asst prof to assoc prof, 72-83, PROF, DEPT DERMAT, HARVARD MED SCH, 83-, CHMN DEPT, 87- *Concurrent Pos:* Prof, Div Health Sci & Technol, Harvard-Mass Inst Technol, 87; chief dermat serv, Mass Gen Hosp, 87-, dir, Wellman Labs, 78-, dermatologist, 75-; consult dermatologist, Beth Israel Hosp, 73- *Honors & Awards:* William Montagna Award, Soc Invest Dermat, 82. *Mem:* Am Soc Lasers Surg & Med (pres, 88); Am Dermat Asn; Soc Invest Dermat; Am Acad Dermat; AAAS; Am Soc Photobiol. *Res:* Basic and applied research in photomedicine, laser medicine and dermatology. *Mailing Add:* Dept Dermat Mass Gen Hosp 32 Fruit St Boston MA 02114

PARRISH, JOHN W, JR, b Dennison, Ohio, Mar 5, 41; m 66; c 2. ANIMAL PHYSIOLOGY, BIOCHEMISTRY. *Educ:* Dennison Univ, BS, 63; Bowling Green State Univ, MA, 70, PhD(biol), 74. *Prof Exp:* Teacher gen sci & earth sci, Northwood Jr High Sch, Norfolk, Va, 67; vis instr biol, Kenyon Col, 73-74; from asst prof to assoc prof biol, Emporia State Univ, 76-88; PROF & CHAIR BIOL, GA SOUTHERN UNIV, 88- *Concurrent Pos:* Teacher's aide, Bowling Green State Univ, 68-69 & predoctoral fel, 69-73; Nat Inst Child Health Dis postdoctoral fel, Univ Tex, Austin, 74-76; vis res assoc prof, Cornell Univ, 86. *Mem:* Sigma Xi; Am Ornithologists Union; Am Inst Biol Sci; Soc Study Reproduction. *Res:* Environmental physiology of vertebrates, including reproductive physiology, ultraviolet photoreception and bioenergetics of animals. *Mailing Add:* Dept Biol Ga Southern Univ Landrum Box 8042 Statesboro GA 30460-8042

PARRISH, JOHN WESLEY, JR, b Dennison, Ohio, Mar 5, 41. COMPARATIVE ANIMAL PHYSIOLOGY, REPRODUCTIVE BIOLOGY. *Educ:* Denison Univ, BS, 63; Bowling Green State Univ, MA, 70, PhD(biol), 74. *Prof Exp:* Teacher earth & gen sci, Norfolk Northwood Jr High Sch, 67; teaching asst, dept biol, Bowling Green State Univ, 68-69, Nonservice fel, 69-73; Fel, dept zool, Univ Tex, Austin, 74-76; from asst prof to prof biol, Div Biol Sci, Emporia State Univ, 76-88; PROF & HEAD, DEPT BIOL, GEORGIA SOUTHERN UNIV, 88- *Concurrent Pos:* Vis instr biol, dept biol, Kenyon Col, 73-74; vis res prof, dept physiol, Cornell Univ, 86; numerous res grants, 68-88. *Mem:* Am Ornithologists' Union; Soc Study Reprod; Sigma Xi; Cooper Ornith Soc; AAAS. *Res:* Near ultraviolet vision in birds; nutrition and bioenergetics of birds; testicular function; pineal function; placental function; comparative animal physiology; author of numerous publications. *Mailing Add:* Div Biol Sci Emporia State Univ Emporia KS 66801

PARRISH, JUDITH TOTMAN, b Stockton, Calif, Sept 25, 50. PALEOCLIMATOLOGY. *Educ:* Univ Calif, Santa Cruz, BA, 72, MA, 76, MS, 77, PhD(earth sci), 79. *Prof Exp:* Res assoc, Dept Geophys Sci, Univ Chicago, 78-81; geologist, US Geol Surv, 82-87; ASSOC PROF, PALEOCLIMATOLOGY, DEPT GEOSCI, UNIV ARIZ, 88- *Concurrent Pos:* Distinguished lectr, US Geol Surv, 83; assoc ed, Am Asn Petrol Geologists Bull, 85-87, Palaios, 89-92, Geol, 90-92; steering comt, NAS/NRC Critical Assessment of Solid-Earth Sci. *Honors & Awards:* J C Cam Sproule Award, Am Asn Petrol Geologists, 84; Don R & Patricia Boyd Lectr Petrol Explor, Univ Tex, 90. *Mem:* Geol Soc Am; Int Geol Correlation Prog; Soc Econ Paleontologists & Mineralogists; Paleontol Soc; Geol Soc London; Am Asn Petrol Geologists. *Res:* Ancient marine high-productivity systems; paleoclimate of the supercontinent Pangaea; climate of the Cretaceous polar regions. *Mailing Add:* Dept Geosci Univ Ariz Bldg 77 Tucson AZ 85721

PARRISH, RICHARD HENRY, b Portola, Calif, 1939; m 62; c 2. FISHERIES BIOLOGY, FISHERIES OCEANOGRAPHY. *Educ:* Univ Calif, Berkeley, BS, 62; Ore State Univ, Corvallis, MS, 66, PhD(fisheries), 77. *Prof Exp:* Lakes mgt res officer, Victoria Dept Fisheries & Wildlife, 62-64; assoc marine biologist, Calif Dept Fish & Game, 66-72; mgr sci res, Del Monte Corp, 72-74; FISHERIES BIOLOGIST, NAT MARINE FISHERIES SERV, 75- *Concurrent Pos:* Lectr, Moss Landing Marine Lab, 68-72. *Res:* Population dynamics of marine fishes and modelling of marine ecosystems especially in relation to exploitation and climate change. *Mailing Add:* PO Box 831 Monterey CA 93942

PARRISH, ROBERT A, JR, b Augusta, Ga, Sept 10, 30; m 54; c 2. SURGERY. *Educ:* Mercer Univ, AB, 51; Univ Ga, MS, 53; Med Col Ga, MD, 56; Am Bd Surg, dipl, 63. *Prof Exp:* From instr to assoc prof, 62-70, PROF SURG, MED COL GA, 70-, CHIEF PEDIAT SURG, 72- *Concurrent Pos:* Am Cancer Soc fel, 60-61; NIH fel, 61-62; consult to many hosps in Ga. *Mem:* Am Acad Pediat; Am Col Surgeons; Soc Surg Alimentary Tract; Am Asn Surg of Trauma. *Res:* Pediatric surgery; gastrointestinal tract. *Mailing Add:* Dept Pediat Surg Med Col Ga Augusta GA 30912

PARRISH, ROBERT G, physical chemistry, for more information see previous edition

PARRISH, WAYNE, b Millersport, Ohio, Dec 18, 20. ZOOLOGY. *Educ:* Ohio State Univ, BS, 48, MS, 55, PhD(radiation biol), 58. *Prof Exp:* Asst biol, Ohio State Univ, 52-56, radiation biol, 56-58, cancer, Univ Hosp, 58-59; res biologist, Miami Valley Hosp, 59-62; asst prof, 62-66, assoc prof , 66- PROF ZOOL AT OHIO STATE UNIV. *Mem:* AAAS. *Res:* Tissue culture of chloromyeloid leukemia cells; metabolism; radiation; histology; chemotherapy; cytology; electron microscopy. *Mailing Add:* Dept Genetics Ohio State Univ Main Campus 963 Biol Sci Bldg Columbus OH 43210-1230

PARRISH, WILLIAM, b Philadelphia, Pa, Apr 16, 14; m 41; c 2. X-RAY CRYSTALLOGRAPHY, MATERIALS SCIENCE. *Educ:* Pa State Univ, BS, 35; Mass Inst Technol, PhD(crystallog), 40. *Prof Exp:* Res assoc, Mass Inst Technol, 39-40; instr, Pa State Col, 40-42; chief technologist, Off Chief Signal Officer, NAm Philips Co, Inc, 42-43, head quartz oscillator res & pilot plant, 43-44, chief x-ray & crystallog sect, Philips Labs Div, NY, 44-68; chief mat characterization br, Electronics Res Ctr, NASA, Mass, 68-70; mgr, crystallog & x-ray anal, 72-86, RES STAFF MEM, RES DIV, IBM CORP, 70-72 & 86- *Concurrent Pos:* Mem nat comt crystallog, 48-57, 65-69, secy-treas, 54-56 & 67-69; mem comn crystallog apparatus, Int Union Crystallog, 51-63, chmn, 57-63; ed, World Dir Crystallographers, 57; mem mat adv bd panel struct, Nat Acad Sci-Nat Res Coun, 65-66. *Honors & Awards:* Citation, US Dept War, 46; J D Hanawalt Award, Int Ctr Diffraction Data, 87; Barrett Award, X-ray Diffraction, Denver X-Ray Confs, 88. *Mem:* Fel Mineral Soc Am; Am Crystallog Asn (secy-treas, 45-49); Mineral Soc Gt Brit; Microbeam Anal Soc; Mat Res Soc. *Res:* X-ray crystallography; materials science and characterization; x-ray diffraction, reflectometry and spectroscopy; synchrontron radiation, instrumentation and computers; instrumentation and computers. *Mailing Add:* IBM Almaden Res Ctr K32/802 San Jose CA 95120-6099

PARROTT, EUGENE LEE, b Menasha, Wis, Feb 3, 25; m 48; c 1. PHARMACEUTICS, INDUSTRIAL PHARMACY. *Educ:* Univ Wis, BS, 49, PhD(pharm), 54. *Prof Exp:* Asst prof pharm, Univ Ariz, 54-55, assoc prof, 55-56; pharmaceutical chemist, Chas Pfizer & Co, 56-57; assoc prof pharm, Univ Nebr, 57-62; assoc prof, 62-73, head div pharmaceut, 75-81, PROF INDUST PHARM, COL PHARM, UNIV IOWA, 73- *Mem:* Am Pharmaceut Asn; fel Acad Pharmaceut Sci; fel Am Asn Pharmaceut Scientists. *Res:* Applications of physiochemical and physiological principles to formulation, design, production and evaluation of dosage forms. *Mailing Add:* Col Pharm Univ Iowa Iowa City IA 52242

PARROTT, MARSHALL WARD, b Roseville, Calif, Nov 28, 27; m 72; c 2. RADIOLOGICAL HEALTH. *Educ:* Univ Calif, Berkeley, AB, 52, MA, 59; Tulane Univ, La, DSc(hyg), 69. *Prof Exp:* Res asst, Lawrence Radiation Lab, Univ Calif, 51-52, physiologist, 52-62; res scientist, Radiobiol Br, NASA Ames Res Ctr, 62-63, biosatellite proj, 63-65; res assoc, Delta Primate Ctr, Tulane Univ, La, 65-69; dir radiation control, Ore State Bd Health, 69-83; RETIRED. *Concurrent Pos:* Chmn, Nat Conf Radiation Control Prog Dir, 80-81; chmn, Comt Radiation Measurements, Nat Bur Standards/Conf Radiol Control Prog Dir, 77-80 & 85-; consult health physics, 83- *Mem:* AAAS; Radiation Res Soc; Health Physics Soc. *Res:* Effects of internal and external radiation in rats and monkeys; endocrine abnormalities; effects of pesticides; environmental health. *Mailing Add:* Radiation Coun 10390 SW Cynthia St Beaverton OR 97005

PARROTT, ROBERT HAROLD, b Jackson Heights, NY, Dec 29, 23; m 50; c 6. VIROLOGY. *Educ:* Fordham Univ, 44; Georgetown Univ, MD, 49. *Prof Exp:* Resident pediat, Children's Hosp, Washington, DC, 50-52; pediatrician, Clin Ctr, Nat Microbiol Inst, 52-56; physician-in-chief & dir, Res Found, 56-62, dir, Children's Hosp Nat Med Ctr, Washington, DC, 62-; DEPT CHILD HEALTH & DEVELOP, GEORGE WASHINGTON UNIV SCH MED & HEALTH SCI, 88- *Concurrent Pos:* Clin prof, Med Sch, George Washington Univ, 56-65, prof, 69-, chmn, Dept Child Health & Develop, 71-85; clin prof, Georgetown Univ, 56-65, prof, 65-69. *Mem:* Am Acad Pediat; Am Pediat Soc; Soc Pediat Res; Int Doctors Soc; Am Acad Med Dir. *Res:* Respiratory viral disease in relation to newly isolated viral agents; general infectious diseases. *Mailing Add:* 111 Mich Ave NW Washington DC 20010

PARROTT, STEPHEN KINSLEY, b Chicago, Ill, Mar 3, 41. MATHEMATICS. *Educ:* Univ Mich, BS, 61, PhD(math), 65. *Prof Exp:* Instr math, Mass Inst Technol, 65-67; from asst prof to assoc prof, 67-78, PROF MATH, UNIV MASS, BOSTON, 78- *Res:* Functional analysis; mathematical physics. *Mailing Add:* Dept Math Univ Mass Harbor Campus Boston MA 02125

PARROTT, STEPHEN LAURENT, b East St Louis, Ill, May 3, 49; m 75; c 2. SURFACE CHEMISTRY, PHYSICAL CHEMISTRY. *Educ:* GA Inst Technol, BS, 72, MS, 73, PhD(chem), 77. *Prof Exp:* RES CHEMIST, PHILLIPS PETROL CO, 78- *Concurrent Pos:* Fel Univ Tex Austin, 77-78. *Mem:* Am Chem Soc; Sigma Xi. *Res:* Statistical thermodynamics of solutions; gas-solid interactions; surface spectroscopy; heterogeneous catalysis. *Mailing Add:* 1812 SE East Dr Bartlesville OK 74006

PARROTT, WILLIAM LAMAR, b Fayetteville, Ga, Oct 7, 30; m 53; c 3. ENTOMOLOGY. *Educ:* Purdue Univ, BS, 59, MS, 61; Miss State Univ, PhD, 67. *Prof Exp:* Instr entom, Exp Sta, Purdue Univ, 59-61; ENTOMOLOGIST, AGR RES SERV, USDA, 61- *Mem:* Entom Soc Am. *Res:* Resistance of crop plants to insect attack. *Mailing Add:* S Montgomery Rd Starkville MS 39759

PARRY, CHARLES J, b Oneida, NY, Nov 30, 42. ALGEBRA, NUMBER THEORY. *Educ:* State Univ NY, Oswego, BS, 66; Mich State Univ, MS, 67, PhD(math), 70. *Prof Exp:* Asst prof, La State Univ, Baton Rouge, 70-71; asst prof, 71-77, assoc prof, 77-84, PROF MATH, VA POLYTECH INST & STATE UNIV, 84- *Mem:* Am Math Soc; Math Asn Am. *Res:* Algebraic number theory and class field theory; class number problems. *Mailing Add:* Dept Math Va Polytech Inst Blacksburg VA 24060

PARRY, EDWARD PETTERSON, analytical chemistry, environmental chemistry, for more information see previous edition

PARRY, GORDON, MEMBRANE BIOCHEMISTRY, PROTEIN SECRETION. *Educ:* Univ London, PhD(biochem), 75. *Prof Exp:* STAFF SCIENTIST, LAWRENCE BERKELEY, 82- *Mailing Add:* One Cyclotron Rd Lawrence Berkeley Lab Berkeley CA 94720

PARRY, HUBERT DEAN, b Ogden, Utah, Apr 7, 09; m 36; c 2. METEOROLOGY. *Educ:* Brigham Young Univ, AB, 34; Calif Inst Technol, MS, 37. *Prof Exp:* Supv forecaster, Weather Sta, US Weather Bur, Utah, 42-44, supv analyst, Nat Weather Anal Ctr, DC, 44-46, 49-54; from liaison officer to chief, Meteorol Br, US Off Mil Govt, Ger, 46-49; chief airport meteorologist, Forecast Ctr, US Weather Bur, Hawaii, 54-60; res & develop meteorologist for Nat Weather Serv, Nat Oceanic & Atmospheric Admin, 60-73; CONSULT FOR THE ATMOSPHERIC SCI, 73- *Honors & Awards:* Spec Achievement Award, US Dept Com, 69, Bronze Medal, 73. *Mem:* Am Meteorol Soc. *Res:* Physical meteorology; meteorological instrumentation and observational methods; environmental impact studies; hydrology; forensic meteorology; atmospheric acoustics. *Mailing Add:* 2549 Military Rd Arlington VA 22207

PARRY, MYRON GENE, b Manhattan, Kans, June 20, 33; m 56; c 3. THEORETICAL MECHANICS, APPLIED MECHANICS. *Educ:* Kans State Univ, BS, 59, MS, 61; Univ Ill, PhD(theoret & appl mech), 68. *Prof Exp:* From instr to asst prof eng mech, Univ Mo-Rolla, 61-66; instr, Univ Ill, 66-68; from asst prof to assoc prof eng mech, 68-82, REGISTR,UNIV MO-ROLLA, 82- *Mem:* Am Soc Eng Educ; Am Asn Col Registr & Admis Officers. *Res:* Post-buckling behavior of thin plates. *Mailing Add:* Dept Eng Mech Univ Mo Rolla Rolla MO 65401

PARRY, ROBERT WALTER, b Ogden, Utah, Oct 1, 17; m 45; c 2. INORGANIC CHEMISTRY. *Educ:* Utah State Agr Col, BS, 40; Cornell Univ, MS, 42; Univ Ill, PhD(inorg chem), 46. *Hon Degrees:* DSc, Utah State Univ, 85. *Prof Exp:* Teaching asst, Cornell Univ, 40-42; asst munitions develop lab, Univ Ill, 42-45; from instr to prof chem, Univ Mich, Ann Arbor, 46-69; DISTINGUISHED PROF CHEM, UNIV UTAH, 69- *Concurrent Pos:* Consult, Stauffer Chem Co, Chem & Metal Ind, AKZO Chem Co; ed, Inorg Chem, 60-63; chmn bd trustees, Gordon Res Confs, 69-70; assoc ed, J Am Chem Soc, 70-73 & 75-82; mem bd dirs, Am Chem Soc, 73-83; chmn chem sect, AAAS, 82; mem, Nat Sci Bd Comt Precol, ed math, sci & tech, 82-86. *Honors & Awards:* Inorg Chem Award, Am Chem Soc, 65, Chem Educ Award, 77; Sr US Scientist Award, Alexander Von Humboldt Found, WGer, 80. *Mem:* AAAS; Am Chem Soc (pres, 82); Sigma Xi. *Res:* Coordination chemistry, particularly of fluorophosphines and their derivatives; phosphorus fluorine chemistry, boron hydrides; reactions of metalcarbonyl compounds containing cationic ligands. *Mailing Add:* Dept Chem Univ Utah Salt Lake City UT 84112

PARRY, RONALD JOHN, b Los Angeles, Calif, June 9, 42. BIO-ORGANIC CHEMISTRY, NATURAL PRODUCTS CHEMISTRY. *Educ:* Occidental Col, BA, 64; Brandeis Univ, PhD(chem), 68. *Prof Exp:* Fel chem, Univ Liverpool, 68-69, Cambridge Univ, 69-70 & Stanford Univ, 70-71; asst prof, Brandeis Univ, 71-78; assoc prof, 78-86, PROF CHEM, RICE UNIV, 86- *Concurrent Pos:* Res career develop award, NIH, 75-80. *Mem:* Am Chem Soc; Royal Soc Chem; AAAS. *Res:* Investigations of natural product biosynthesis and of stereochemical aspects of enzyme mechanisms. *Mailing Add:* Dept Chem Rice Univ Houston TX 77251

PARRY, WILLIAM LOCKHART, b Palatine Bridge, NY, Apr 19, 24; m 52; c 3. UROLOGY. *Educ:* Univ Rochester, MD, 47. *Prof Exp:* Intern surg, Yale Univ Hosps, 47-48; resident surg & urol, Univ Rochester Hosps, 48-52; from asst prof to assoc prof urol, State Univ NY Upstate Med Ctr, 54-62; PROF UROL & HEAD DEPT, MED SCH CTR, UNIV OKLA, 62- *Concurrent Pos:* Fel, Yale Univ Hosps, 47-48; fel, Univ Rochester Hosps, 48-52; mem exec comt & secy urol res group, Cent Off, Vet Admin, Washington, DC, 59-, consult, Cent Off, 61-62. *Mem:* Am Urol Asn; Am Col Surg; AMA; Sigma Xi. *Res:* Experimental studies of acute renal failure; clinical studies of urologic disease and surgery. *Mailing Add:* Dept Urol Univ Okla Med PO Box 26901 Oklahoma City OK 73190

PARRY, WILLIAM THOMAS, b Manti, Utah, May 11, 35; m 58; c 2. MINERALOGY, GEOCHEMISTRY. *Educ:* Univ Utah, BS, 57, MS, 59, PhD(geol eng), 61. *Prof Exp:* Exploitation engr, Shell Oil Co, 61-63; assoc prof geosci, Tex Tech Col, 63-67; assoc prof mining & geol eng, 67-69, chmn dept, 69-71, PROF GEOL & GEOPHYS, UNIV UTAH, 71- *Concurrent Pos:* Gulf Oil Corp grant, 64; NSF res grant, 66-68. *Mem:* AAAS; Mineral Soc Am; Geochem Soc. *Mailing Add:* Dept Geol Univ Utah 717 Browning Sci Bldg Salt Lake City UT 84112

PARS, HARRY GEORGE, b Pawtucket, RI, Feb 7, 28; m 52; c 2. ORGANIC CHEMISTRY. *Educ:* Harvard Univ, BA, 53; Univ Mich, MS, 55, PhD(org chem), 57. *Prof Exp:* Head med & chem sci drug res & develop, Arthur D Little Co Inc, 57-70; FOUNDER, CHMN & CHIEF EXEC OFFICER, DRUG RES & DEVELOP, H G PARS PHARMACEUT LAB INC, 70-; PRES & TRUSTEE BASIC RES, SHEEHAN INST RES INC, 70-; CHMN CHIEF EXEC OFFICER, H G PARS PHARMACEUT LAB INC, 85- *Concurrent Pos:* Founder & chmn, Nat Coun Indust Innovation, 79-; dir, Small Bus Found Am, 84- *Mem:* Asn Res Vision & Ophthal; Am Chem Soc; Licensing Exec Soc. *Res:* Pioneered in the design, synthesis, pharmacological and clinical development of new drugs structurally related to the opoids and cannabinoids; central nervous system stimulants, depressants, analgesics, anti-glaucoma agents, antihypertensives, anticonvulsants, antidiarrheals, antiinflammatories, and antibiotics. *Mailing Add:* H G Pars Pharmaceut Labs Inc 128 Spring St Lexington MA 02173-7802

PARSA, ZOHREH, m 74; c 2. NUCLEAR PHYSICS, PARTICLE & ACCELERATOR PHYSICS. *Educ:* Del State Col, BS, 69; NY Univ, MS, 71; Polytech Inst NY, PhD(physics), 77. *Prof Exp:* Dir physics & eng lab, Essex County Col, 72-74; instr physics, Hunter Col City Univ NY, 74-76; asst prof physics, NJ Inst Technol, 77-84; PHYSICIST, BROOKHAVEN NAT LAB, 85- *Concurrent Pos:* Vis scholar, Northwestern Univ, 80-81; guest assoc physicist, Dept Physics, Brookhaven Nat Lab, 81-83; chmn, Digital Equip Co Users Soc & BNL Computer Local Users Group, 86- *Mem:* AAAS; Am Phys Soc; Am Asn Physics Teachers. *Res:* Subnuclear physics and field theories; nuclear physics; accelerator physics. *Mailing Add:* Dept Physics Brookhaven Nat Lab Bldg 510D Upton NY 11973

PARSCH, LUCAS DEAN, b Wausau, Wis, May 30, 46; m 71. PRODUCTION ECONOMICS, AGRICULTURAL RISK ANALYSIS. *Educ:* Univ Wis, BA, 68, MS, 79; Mich State Univ, PhD(agr econ), 82. *Prof Exp:* Asst prof, 82-88, ASSOC PROF AGR ECON, UNIV ARK, 88- *Mem:* Am Agr Econ Asn. *Res:* Conduct research in an interdisciplinary setting to assess risk in agricultural production; development and refinement of computer simulation models to support decision analysis in crop (soybeans, wheat) and livestock (beef forage) systems; economic analysis includes assessment of risk return tradeoffs for alternative production management strategies at farm-firm level. *Mailing Add:* Dept Agr Econ Rural Soc Univ Ark 225 Agr Bldg Fayetteville AR 72701

PARSEGIAN, V(OZCAN) LAWRENCE, b Van, Armenia, Turkey, May 13, 08; nat US; m 38; c 2. NUCLEAR PHYSICS, ENGINEERING. *Educ:* Mass Inst Technol, BS, 33; NY Univ, PhD(physics), 48. *Prof Exp:* Dir res, Tagliabue Mfg Co, 36-47, Physics Dept, Kellex Corp, 47-50 & Res Div, NY Opers Off, US AEC, 50-54; chmn eng group, Rensselaer Polytech Inst, 54-58, prof nuclear sci & eng, 54-61, dean, Sch Eng, 58-61, distinguished chair Rensselaer Prof, 61-75, dir, Armenian Architect Archives Proj, 80-91, EMER RENSSELAER PROF NUCLEAR SCI & ENG, RENSSELAER POLYTECH INST, 75- *Concurrent Pos:* Instr, NY Univ, 49; lectr, Columbia Univ; consult, Joint Comt Atomic Energy of US Cong, State Univ NY, Mass Inst Technol & World Coun Churches, 76-77; mem bd dirs, Radiation Applns, Inc, 54-90 & Cast Tech, Inc, 66-70; consult ed, Acad Press, 64-; mem, NY Gov's Comt Atomic Energy, 58; mem comt com utilization atomic energy, US Chamber Com, 59-65 & comt sci & tech, 65-67; mem, NY State Comt Radiation Utilization, 59-65. *Mem:* AAAS; Am Nuclear Soc; Am Phys Soc; NY Acad Sci. *Res:* Nuclear science and engineering; nuclear and instrument research; teaching materials and methods that interrelate the physical and life sciences; science education. *Mailing Add:* Rensselaer Polytech Inst Troy NY 12180-3590

PARSEGIAN, VOZKEN ADRIAN, b Boston, Mass, May 28, 39; m 62; c 3. BIOPHYSICS. *Educ:* Dartmouth Col, AB, 60; Harvard Univ, PhD(biophys), 65. *Prof Exp:* Res physicist,Phys Scis Lab,NIH, 67-; HEAD, SECT BIOL PHYSICS, LAB BIOCHEM & METAB NIDDK, NIH, 84- *Concurrent Pos:* ed,Biophys Jour,77-80 & Biophys Discussions, 78- *Mem:* Biophys Soc (pres,83-84,coun,74-77). *Res:* Theory and measurement of intermolecular forces especially those governing the organization of cellular activity; surface and interfacial properties. *Mailing Add:* Rm 2007 Bldg 12AT Nat Insts Health Bethesda MD 20892

PARSHALL, BRIAN J, b Penn Yan, NY, Oct 28, 45; m 78. GROUP THEORY, REPRESENTATION THEORY. *Educ:* Univ Rochester, AB, 67; Yale Univ, PhM, 70, PhD(math), 71. *Prof Exp:* From asst prof to prof math, Univ Va, 72-87; prof math, Univ Ill, Urbana, 87-88; PROF MATH, UNIV VA, 88- *Concurrent Pos:* Mem, Inst Advan Study, 75, 79; vis assoc prof, Northwestern Univ, 80; vis prof, Math Sci Res Inst, Berkeley, 90. *Mem:* Am Math Soc; AAAS. *Res:* Representation theory of semi simple algebraic groups, associated finite groups of Lie type and associated Lie algebra. *Mailing Add:* Dept Math Math Astro Bldg Univ Va Charlottesville VA 22903-3199

PARSHALL, GEORGE WILLIAM, b Hackensack, Minn, Sept 19, 29; m 54; c 3. CHEMISTRY. *Educ:* Univ Ill, PhD(chem), 54. *Prof Exp:* Res chemist, 54-65, res supvr, 65-79, DIR CHEM SCI, E I DU PONT DE NEMOURS & CO, INC, 80- *Concurrent Pos:* Ed-in-chief, Inorg Syntheses, 72-74; ed, J Molecular Catalysis, 77-80. *Honors & Awards:* Inorg Chem Award, Am Chem Soc, 83; Chem Res Mgt Award, Am Chem Soc, 89. *Mem:* Nat Acad Sci; Am Chem Soc; Am Acad Arts & Sci. *Res:* Organic and inorganic chemistry; transition metal chemistry; catalysis. *Mailing Add:* Cent Res Dept E I du Pont de Nemours & Co Inc Wilmington DE 19880-0328

PARSIGNAULT, DANIEL RAYMOND, b Paris, France, Feb 5, 37; m 60; c 3. EARTH PARTICLES AND FIELDS. *Educ:* Univ Wis, BS, 60; Southern Ill Univ, MS, 61; Univ Paris, Dr es Sc(nuclear physics), 64. *Prof Exp:* Res physicist, Saclay Nuclear Res Ctr, France, 61-64; res fel nuclear physics, Calif Inst Technol, 64-66; asst prof physics, Ohio State Univ, 66-68; mem res staff, Am Sci & Eng, Inc, 68-76; mem staff, Harvard-Smithsonian Ctr Astrophysics, 76-78; SR SCIENTIST, PHYSICS RES DIV, EMMANUEL COL, 78- *Concurrent Pos:* NATO fel, 64-65. *Mem:* Am Phys Soc; Am Geophys Union. *Res:* Alpha, beta, and gamma ray spectroscopy; nuclear structure and reactions; x-ray astronomy; nuclear physics. *Mailing Add:* Eight Bellevue Ave Winchester MA 01890

PARSLEY, RONALD LEE, b Madison, Wis, July 14, 37; m 86; c 2. INVERTEBRATE PALEONTOLOGY, GEOLOGY. *Educ:* Univ Calif, Los Angeles, AB, 60; Univ Cincinnati, MS, 63, PhD(geol), 69. *Prof Exp:* From asst prof to assoc prof, 66-79, PROF GEOL, TULANE UNIV, LA, 79- *Concurrent Pos:* Chmn dept, Tulane Univ, La, 71-74; Nat Acad Sci-Czechoslovak Acad Sci exhange scientist, 81, 83, 86 & 89. *Mem:* AAAS; Paleont Soc; Paleont Res Inst; Soc Econ Paleont & Mineral; Geol Soc Am; Int Paleont Asn. *Res:* Lower Paleozoic primitive Echinodermata, their mode and tempo of evolution, systematics, functional morphology and paleoecology. *Mailing Add:* Dept Geol Tulane Univ New Orleans LA 70118-5698

PARSLY, LEWIS F(ULLER), JR, b Philadelphia, Pa, Jan 27, 18; m 51; c 4. CHEMICAL ENGINEERING. *Educ:* Univ Pa, BS, 40, MS, 47, PhD(chem eng), 48. *Prof Exp:* Jr engr, Am Chem Paint Co, 40 & Elec Storage Battery Co, Pa, 40-41; chem engr, Day & Zimmermann, Inc, 48-51; SR DEVELOP ENGR, NUCLEAR DIV, UNION CARBIDE CORP, 51- *Mem:* Am Chem Soc; Am Nuclear Soc; Am Inst Chem Engrs; Sigma Xi; Nat Soc Prof Engrs. *Res:* Gas absorption; fission product release and transport; thoria slurry handling. *Mailing Add:* 108 Hutchinson Pl Oak Ridge TN 37830

PARSON, JOHN MORRIS, b New Orleans, La, May 29, 46; m 69; c 2. PHYSICAL CHEMISTRY. *Educ:* Harvard Univ, AB, 68; Univ Chicago, PhD(chem), 72. *Prof Exp:* Teacher, chem & physics, Francis Parker High Sch, 68-69; assoc chem, Univ Chicago, 72; from asst prof to assoc prof, 72-84, PROF CHEM, OHIO STATE UNIV, 84- *Concurrent Pos:* Fel, Alfred P Sloan Found, 76-80; NSF grant, 76-82 & 86-; Res Corp grant, 73-75. *Honors & Awards:* E R Norton Prize, Univ Chicago, 72. *Mem:* Am Phys Soc. *Res:* Crossed molecular beams chemistry; gas phase reactions and energy transfer studied by electronic chemiluminescence and laser excited fluorescence. *Mailing Add:* Dept Chem 120 W 18th Ave Columbus OH 43210

PARSON, LOUISE ALAYNE, b Boston, Mass, May 29, 47; m 69; c 2. AUTOMORPHIC FORMS AND DISCRETE GROUPS. *Educ:* Radcliffe Col, BA, 69; Univ Ill, Chicago, MS, 70, PhD(math), 73. *Prof Exp:* Teaching asst math, Univ Ill, Chicago, 69-72; lectr, 73-75, asst prof, 76-84, ASSOC PROF MATH, OHIO STATE UNIV, 84- *Mem:* Am Math Soc; Math Asn Am; Asn Women Math. *Res:* Arithmetic and analytic properties of automorphic forms and automorphic integrals with rational period functions. *Mailing Add:* Dept of Math Ohio State Univ Columbus OH 43210

PARSON, ROBERT PAUL, b New Brunswick, NJ, Dec 27, 57. THEORETICAL CHEMICAL DYNAMICS. *Educ:* Brown Univ, ScB, 79; Univ Mich, MS, 80, PhD(phys chem), 84. *Prof Exp:* Postdoctoral res asst, Univ Wash, 84-86; ASST PROF, DEPT CHEM & BIOCHEM, UNIV COLO, 86- *Concurrent Pos:* Fel, Joint Inst Lab Astrophys, Univ Colo, 88- *Mem:* Affil mem Am Phys Soc. *Res:* Theoretical chemical dynamics; energy transfer processes in gas and condensed phases; collision theory; dynamics of highly excited molecules and clusters; relaxation theory. *Mailing Add:* Univ Colo Campus Box 440 JILA Boulder CO 80309

PARSON, WILLIAM WOOD, b Boston, Mass, Dec 1, 39; m 61; c 2. BIOCHEMISTRY. *Educ:* Harvard Col, AB, 61; Case Western Reserve Univ, PhD(biochem), 65. *Prof Exp:* Fel, Univ Pa, 65-67; from asst prof to assoc prof, 67-77, PROF BIOCHEM, UNIV WASH, 77- *Concurrent Pos:* Assoc ed, Biochem, 69- *Res:* Photosynthesis. *Mailing Add:* Dept Biochem Univ Wash Seattle WA 98195

PARSONNET, VICTOR, b Deal, NJ, Aug 29, 24; m 50; c 3. BIOMEDICAL ENGINEERING. *Educ:* NY Univ, MD, 47; Am Bd Surg, cert, 56. *Hon Degrees:* LLD, Kean Col, 78. *Prof Exp:* Asst instr path, Col Med, NY Univ, 48-49; dir med educ, Newark Beth Israel Med Ctr, 55-58; fel vascular surg, Cent Surg Serv, Columbia Presby Hosp Col Physicians & Surgeons, New York, 59-62; assoc prof surg & chief, Vascular Div, Seton Hall Col Med & Dent, Newark, 62-66; dir vascular surg, 60-74, DIR SURG, NEWARK BETH ISRAEL MED CTR, 60-, DIR, PACEMAKER CTR, 88- *Concurrent Pos:* Clin assoc prof surg, Univ Med & Dent NJ, Newark, 66-71, clin prof surg, 71-; AEC, 66-76, Nuclear Regulatory Comn, 76-; consult, Stevens Inst Technol, Hoboken, 74-75, Dept Health, Educ & Welfare & Nat Heart, Lung & Blood Inst, 79, Univ Med & Dent, Newark, 82-84, Dept Health & Human Serv, Washington, DC & numerous hosp; vis prof, Royal Postgrad Med Sch, Hammersmith Hosp, London, 81; bd dirs, Metro-Essex Med Care Inc, 86. *Honors & Awards:* Harrison Martland Award, 53; Melvyn H Motolinsky Award, Robert Wood Johnson Med Sch, 88; Clara Barton Award, 89; Howard W Dayton Award, AMA, 90. *Mem:* Am Asn Med Systs & Informatics; Am Asn Thoracic Surg; Am Col Angiol; Am Col Cardiol; Am Col Surgeons; AMA; Am Med Writers Asn; Am Soc Laser Med & Surg; Am Surg Asn; Am Geriat Soc. *Res:* Vasculary & cardiovascular surgery; pathology; cardiac pacing & electrophysiology; 5 patents; over 550 technical publications including books and book chapters, 13 medical tapes and 18 teaching films. *Mailing Add:* 159 Millburn Ave Millburn NJ 07041

PARSONS, CARL MICHAEL, b Salisbury, Md, May 7, 54; m 78. POULTRY NUTRITION. *Educ:* Univ Md, Eastern Shore, BS, 76; Va Polytech Inst & State Univ, MS, 78, PhD(animal sci), 81. *Prof Exp:* Res asst poultry sci, Va Polytech Inst & State Univ, 76-81; ASST PROF ANIMAL SCI, UNIV ILL, URBANA-CHAMPAIGN, 81- *Mem:* Poultry Sci Asn; Animal Nutrit Res Coun; Sigma Xi. *Res:* Poultry production and management with emphasis in the field of nutrition; improved nutritional efficiency for production of poultry meat and eggs, particularly with respect to protein utilization. *Mailing Add:* Univ Ill 322 Mumford Hall 1301 W Gregory Dr Urbana IL 61801

PARSONS, DAVID JEROME, b Berkeley, Calif, May 18, 47; m 75; c 1. PLANT ECOLOGY, ENVIRONMENTAL MANAGEMENT. *Educ:* Univ Calif, Davis, BS, 69; Stanford Univ, PhD(pop biol), 73. *Prof Exp:* RES SCIENTIST, NAT PARK SERV, US DEPT INTERIOR, 73- *Concurrent Pos:* Assoc, Univ Calif Agr Exp Sta, 74-; vchmn, Veg Sect, Ecol Soc Am. *Mem:* Ecol Soc Am; Natural Areas Asn. *Res:* Global change effects; fire ecology of mixed conifer forests; carrying capacities of wilderness areas; effects of air pollution. *Mailing Add:* Sequoia-Kings Canyon Nat Parks Three Rivers CA 93271

PARSONS, DONALD FREDERICK, b Shoreham-by-Sea, Eng, Nov 28, 28; m 58; c 4. MEDICAL ONCOLOGY, CANCER CELL STRUCTURE. *Educ:* Battersea Col Technol, BSc, 50; Imp Col, dipl, 53; Univ London, PhD(phys chem), 53, MB, BS, 57. *Prof Exp:* Res assoc biophys, Duke Univ, 57-59; biophysicist, Oak Ridge Nat Lab, 59-61; asst prof med biophys, Univ Toronto & mem Staff Ont Cancer Res Inst, 61-66; res prof biophys, State Univ NY Buffalo, 66-76; prin res scientist, Roswell Park Mem Inst, 66-76; dir biotechnol, High Voltage Electron Micros Resource Lab, 81-89, RES PHYSICIAN, DIV LAB & RES, NY STATE DEPT HEALTH, 76- *Concurrent Pos:* Res assoc prof biol, State Univ NY, Albany & Rensselaer Polytech Inst, 77. *Mem:* Am Soc Clin Oncol; Am Crystallog Asn; Electron Micros Soc Am; Biophys Soc; Am Asn Cancer Res. *Res:* Characterization of invasive cells by electron microscopy, cytochemistry, immunofluorescence and biochemistry; application of three dimensional image reconstruction and stereoscopy to study structure of cancer cells using the high voltage electron microscopy; electron diffraction of wet protein microcrystals; chemotherapy specific for the motile apparatus of invasive cancer cells. *Mailing Add:* Div Labs & Res Empire State Plaza Albany NY 12201

PARSONS, JAMES SIDNEY, b Roanoke, Va, Jan 28, 22; m 52; c 2. ANALYTICAL CHEMISTRY. *Educ:* Washington & Lee, BS, 43; Univ Va, MS, 48, PhD(chem), 50. *Prof Exp:* Chemist, Westvaco Chem Div, Food Mach & Chem Corp, 43-46; asst, Univ Va, 47-50; anal chemist, Am Cyanamid Co, 50-53, res chemist, 54-57, sr res chemist, 58-74, prin res sci, Chem Res Div, 74-83; RETIRED. *Concurrent Pos:* Chmn, N Jersey Chromatography Discussion Topical Group, Am Chem Soc, 70-71. *Mem:* Am Chem Soc. *Res:* Electroanalytical chemistry; separation techniques; thermal methods of analysis; gas chromatography; environmental analysis. *Mailing Add:* Box 241 Rte Three Lexington VA 24450

PARSONS, JERRY MONTGOMERY, b Memphis, Tenn, Feb 4, 46; m 66; c 2. HORTICULTURE. *Educ:* Univ Tenn, Martin, BS, 69; Miss State Univ, MS, 71; Kans State Univ, PhD(hort), 74. *Prof Exp:* Res asst hort, Miss State Univ, 69-71 & Kans State Univ, 71-74; VEG SPECIALIST HORT, TEX AGR EXTEN SERV, 74- *Mem:* Am Soc Hort Sci; Am Pomol Soc. *Res:* Educate and inform commercial vegetable growers and home gardeners located in the Texas Agricultural Extension Serv District 13 of new and progressive vegetable culture techniques which might enable a more profitable existence using existing resources. *Mailing Add:* Vegetable Specialist 1143 Coliseum Rd San Antonio TX 78219

PARSONS, JOHN ANDRESEN, radiotherapy, for more information see previous edition

PARSONS, JOHN ARTHUR, b Pittsburgh, Pa, Dec 8, 32; m 60; c 1. CELL BIOLOGY. *Educ:* Washington & Jefferson Col, BA, 54; Pa State Univ, MS, S6; Fla State Univ, PhD(physiol), 64. *Prof Exp:* Instr biol, Henderson State Teachers Col, 58-59 & Vanderbilt Univ, 59-60; asst prof, Univ Redlands, 63-65; from asst prof to assoc prof, 65-74, PROF BIOL, SAN DIEGO STATE UNIV, 74- *Mem:* AAAS; Am Soc Zoologists; Soc Protozool; Am Soc Cell Biol. *Res:* Mitochondrial origin and DNA. *Mailing Add:* Dept of Biol San Diego State Univ San Diego CA 92182

PARSONS, JOHN DAVID, b Gary, Ind, Nov 22, 25; m 49; c 1. LIMNOLOGY. *Educ:* Southern Illinois Univ, BA, 50, MA, 51; Univ Mo, PhD(zool), 56. *Prof Exp:* From asst prof to assoc prof biol sci, Western Ill Univ, 55-62; assoc prof biol sci & dir Pine Hills Field Sta, Southern Ill Univ, Carbondale, 62-73, prof bot, 73-76; dir environ studies, Parsons & Assocs, 76-79; sr fish & wildlife biologist, 79-85, chief, Br Tech Anal & Rev, 85- 87, RES PROG ADMINR, OFF SURFACE MINING, DEPT INTERIOR, WASHINGTON, DC, 87- *Concurrent Pos:* Atomic Energy Comn equip grant, 62. *Mem:* Am Fisheries Soc; Int Asn Theoret & Appl Limnol; Am Soc Limnol & Oceanog; Wildlife Soc. *Res:* Aquatic ecology; inorganic pollution of streams; organic and inorganic factors relating to primary production and energy transfer in wetlands. *Mailing Add:* Dept of Interior Off Surface Mining Rm 5101 W 1951 Constitution Ave Washington DC 20240

PARSONS, JOHN G, b Man, Can, Dec 3, 39; m 63; c 2. DAIRY CHEMISTRY. *Educ:* Univ Man, BSc, 61, MSc, 63; Pa State Univ, PhD(dairy sci), 68. *Prof Exp:* Asst dairy sci, Univ Man, 62-63; asst, Pa State Univ, 63-65, res asst, 65-68; from asst prof to assoc prof, 68-79, PROF DAIRY SCI, SDAK STATE UNIV, 79-, HEAD DEPT, 78- *Mem:* Am Dairy Sci Asn (dir); Inst Food Technol. *Res:* Flavor and lipid chemistry of milk products. *Mailing Add:* Dept Dairy Sci SDak State Univ Box 2104 Brookings SD 57007

PARSONS, JOHN LAWRENCE, b Kans, Feb 4, 24; m 45; c 5. AGRONOMY. *Educ:* Kans State Col, BS, 48, MS, 49; Univ Mass, PhD(agron), 53. *Prof Exp:* Asst prof agron, Kans State Col, 48-49; instr, Univ Mass, 49-53; from asst prof to prof, Ohio Agr Res & Develop Ctr, 53-68; asst coordr int affairs, 68-73, PROF AGRON, OHIO STATE UNIV, 73- *Mem:* Am Soc Agron. *Mailing Add:* 922 Blind Brook Dr Columbus OH 43235

PARSONS, JOHN THOREN, b Detroit, Mich, Oct 11, 13; m 40; c 6. QUALITY ASSURANCE SYSTEMS. *Hon Degrees:* DEng, Univ Mich, 88. *Prof Exp:* Div mgr, Danville Div, F L Jacobs Co, 53; vpres, Parsons Corp, 42-53 & 54-56, pres, 56-68; RETIRED. *Concurrent Pos:* Consult, US Army Air Corps & NAm Aviation. *Honors & Awards:* Jacqvort Award, Am Inst Mfg Technol, 68; Eng Citation, Soc Mfg Engrs, 75; Nat Medal Technol, US Dept Com, 85. *Mem:* Hon mem Soc Mfg Engrs; hon mem Am Inst Mfg Technol. *Res:* Numerical control; adhesive bonded metal aircraft structure; processes for producing composite structures; system for machining metal without warpage; ship propellers, jet fan blades and helicopter rotor blades. *Mailing Add:* 1235 Milliken Ct Traverse City MI 49684

PARSONS, L CLAIRE, b Flora, La, Aug 1, 33. NEUROPHYSIOLOGY. *Educ:* Northwestern State Col, BS, 54; La State Univ, Baton Rouge, 56-58; Univ Houston, MS, 64; Univ Tex Med Br Galveston, PhD(physiol), 68. *Prof Exp:* Dir res & assoc prof, Sch Nursing, 71-74, ASST PROF PHYSIOL, MED SCH, UNIV VA, 71-, PROF, SCH NURSING, 74- *Concurrent Pos:* NIH spec fel, Surg Neurol Br, Nat Inst Neurol Dis & Stroke, 68-70; lectr, Surg Neurol Br, Nat Inst Neurol Diseases & Stroke, 68-70 & NIH, 68-70; guest lectr, Case Western Reserve Univ, 70-71; prof dir & ed, Human Physiol Ser, 76-80. *Mem:* Sigma Xi; AAAS; Soc Neurosci. *Res:* Sleep and head trauma. *Mailing Add:* 4540 N Trocha Alegre Tucson AZ 85715-6368

PARSONS, LAWRENCE REED, b Sacramento, Calif, May 22, 44; m 71; c 1. ENVIRONMENTAL PHYSIOLOGY, HORTICULTURE. *Educ:* Univ Calif, Davis, AB, 67, MS, 68; Duke Univ PhD(bot), 72. *Prof Exp:* Res assoc plant physiol, Duke Univ, 72-74; asst prof hort & plant physiol, Univ Minn, St Paul, 74-79; from asst prof to assoc prof, 79-89, PROF FRUIT CROPS, UNIV FLA, 89- *Mem:* Sigma Xi; Am Soc Hort Sci; Am Soc Plant Physiologists. *Res:* Plant-water relations; effects of cold temperature on plants; plant hardiness; freeze protection; irrigation; stress physiology. *Mailing Add:* Univ Fla Citrus Res & Educ Ctr 700 Experiment Sta Rd Lake Alfred FL 33850

PARSONS, MICHAEL L, b Oklahoma City, Okla, Apr 20, 40; m 72; c 2. ANALYTICAL CHEMISTRY. *Educ:* Pittsburg State Univ, BA, 62, MS, 63; Univ Fla, PhD(anal chem), 66. *Prof Exp:* Res chemist, Phillips Petrol Co, 66-67; from asst prof to assoc prof anal chem, Ariz State Univ, 67-78, prof, 78-84; Los Alamos Nat Lab, 84-88; DIR RES & DEVELOP, HTL/KIN-TECH DIV, PAC SCIENTIFIC, 88- *Honors & Awards:* W F Meggers Award, Soc Appl Spectros, 67 & 76. *Mem:* Am Chem Soc; Soc Appl Spectros (pres, 79); Optical Soc Am; Sigma Xi. *Res:* Atomic spectroscopy, including trace metal analysis; theoretical measurements and calculations; study of excitation sources and spectra; sensors for fire detection. *Mailing Add:* HTL/Kin-Tech Div Pac Scientific Duarte CA 91010

PARSONS, ROBERT HATHAWAY, b Passaic, NJ, May 13, 41. COMPARATIVE PHYSIOLOGY. *Educ:* Del Valley Col, BS, 63; Ore State Univ, MS, 67, PhD(physiol), 69. *Prof Exp:* NIH fel, Case Western Reserve Univ, 69-71, res assoc, 71-72; asst prof, 72-78, assoc chmn, 82-85, actg chmn, 85, ASSOC PROF BIOL, RENSSELAER POLYTECH INST, 78- *Mem:* Am Soc Zool; Soc Gen Physiologists; Am Physiol Soc; Biophys Soc. *Res:* Membrane transport; water transport; non-electrolyte transport. *Mailing Add:* Dept Biol Rensselaer Polytech Inst Troy NY 12180-3590

PARSONS, ROBERT JEROME, pathology; deceased, see previous edition for last biography

PARSONS, ROBERT W(ESTWOOD), b San Diego, Calif, Sept 13, 32; m 54; c 2. CHEMICAL ENGINEERING. *Educ:* Univ Idaho, BS, 54; Univ Ill, MS, 56, PhD(chem eng), 58. *Prof Exp:* Res engr, Ohio Oil Co, 58- 63, adv res engr, Marathon Oil Co, 63-69, sr res engr, 69-86; RETIRED. *Concurrent Pos:* Sabbatical leave, Chem Eng Dept, Univ Colo, 76-77. *Mem:* Am Inst Mining, Metall & Petrol Engrs; Am Inst Chem Engrs; Sigma Xi. *Res:* Enhanced oil recovery; use of carbon dioxide, surfactants, and thermal methods; physical properties of petroleum fluids. *Mailing Add:* 6777 Southridge Lane Littleton CO 80120

PARSONS, RODNEY LAWRENCE, b Southampton, NY, Dec 7, 39; m 61; c 3. PHYSIOLOGY, BIOPHYSICS. *Educ:* Middlebury Col, AB, 62; Stanford Univ, PhD(physiol), 65. *Prof Exp:* From asst prof to assoc prof, 67-73, prof physiol, 73-79, PROF & CHMN ANAT & NEUROBIOL, COL MED, UNIV VT, 79-, DEPT ANAT & PHYSIOL. *Concurrent Pos:* NIH fel physiol, Columbia Univ, 65-67. *Honors & Awards:* Jacob Javits Neurosci Investr Award. *Mem:* Am Physiol Soc; Soc Gen Physiol. *Res:* Comparative electrophysiology and pharmacology of synaptic transmission; regulation of nicotinic acetylcholine receptor channels. *Mailing Add:* Dept Anat & Neurobiol Univ VT Agri Col B5 S Prospect St Burlington VT 05405

PARSONS, ROGER BRUCE, soil genesis, morphology; deceased, see previous edition for last biography

PARSONS, STANLEY MONROE, b Jan 17, 43; c 2. NEUROCHEMISTRY, PHARMACOLOGY. *Educ:* Calif Inst Technol, PhD(biochem), 71. *Prof Exp:* PROF BIOCHEM, UNIV CALIF, SANTA BARBARA, 72- *Concurrent Pos:* Jacob Javits Neurosci Investr. *Mem:* Soc Neurosci; Int Soc Neurochem; Am Soc Biochem & Molecular Biol. *Res:* Pre-synaptic cholinergic structure and function. *Mailing Add:* Dept Chem Univ Calif Santa Barbara CA 93106

PARSONS, THERAN DUANE, b St Maries, Idaho, Dec 26, 22; m 45; c 2. INORGANIC CHEMISTRY. *Educ:* Univ Wash, BS, 49, PhD(chem), 53. *Prof Exp:* Res fel, Univ Chicago, 53-55; from asst prof to assoc prof, 55-68, asst dean, Col Sci, 67-70, actg dean, 70-72 & 79-81, assoc dean, 73-81, PROF CHEM, ORE STATE UNIV, 68-, ACTG VPRES ADMIN, 81- *Honors & Awards:* Carter Award, 59; Mosser Award, 66. *Mem:* Am Chem Soc. *Res:* Chemistry of boron; electron-deficient bonding; organometallic chemistry. *Mailing Add:* Pres S Off Ore State Univ Corvallis OR 97331

PARSONS, THOMAS STURGES, b New York, NY, Sept 1, 30; m 57. VERTEBRATE ZOOLOGY. *Educ:* Harvard Univ, AB, 52, AM, 53, PhD(biol), 57. *Prof Exp:* Instr biol, Harvard Univ, 57-60; from asst prof to assoc prof 60-69, PROF ZOOL, UNIV TORONTO, 69-, DEPT ZOOL, ST MICHAELS COL. *Mem:* Soc Vert Paleont; Soc Study Evolution; Am Ornith Union; Am Soc Zool; Am Soc Ichthyol & Herpet. *Res:* Anatomy, embryology and paleontology of vertebrates, primarily amphibians and reptiles. *Mailing Add:* Dept Zool Univ Toronto Toronto ON M5S 1A1 Can

PARSONS, TIMOTHY F, b Cambridge, Mass, Sept 27, 38; m 60; c 5. QUALITY MANAGEMENT. *Educ:* Boston Col, BS, 60; Vanderbilt Univ, PhD(org chem), 64. *Prof Exp:* Eastman Kodak sr res chemist, Res Labs, 64-68, tech assoc & suprv develop, Paper Sensitizing Div, 69-74, suprv develop, Paper Support Div, 74-75, asst dir, Paper Serv Div, 75-77, prog dir photograph prog develop, US & Can Photog Div, 78-80, asst supt, Prod Finishing Div, 80-81, dir, Paper Serv Div, 81-83, Film Tech Serv, 83-84, qual asst to gen mgr, 84-85, dir qual serv orgn, 85-89, DIR MFG QUAL ASSURANCE ORGN, EASTMAN KODAK, 90- *Concurrent Pos:* Chmn qual adv comt, Eastman Kodak Co, 82-89. *Honors & Awards:* Charles Ives Award, Soc Photog & Eng, 80. *Mem:* Am Chem Soc; Soc Photog Sci & Eng; Am Soc Qual Control. *Res:* Synthesis, structure, and mechanism of antiradiation drugs, especially those capable of thioalkylation; chemistries of photographic systems; photographic process and product development. *Mailing Add:* 270 Manitou Beach Rd Hilton NY 14466

PARSONS, TIMOTHY RICHARD, b Colombo, Ceylon, Nov 1, 32; nat Can; m 58; c 3. OCEANOGRAPHY. *Educ:* McGill Univ, BSc, 53, MSc, 56, PhD(biochem), 58. *Prof Exp:* Sr scientist, Pac Oceanog Group, Fisheries Res Bd Can, 59-62, Off Oceanog, UNESCO, France, 62-64 & Pac Oceanog Group, Fisheries Res Bd Can, 64-71; PROF OCEANOG, DEPT OCEANOG, UNIV BC, 71- *Honors & Awards:* Prize, Oceanog Soc Japan, 88; J P Tully Medal, 89; Killam Prize, 90. *Mem:* Am Soc Limnol & Oceanog; Int Asn Biol Oceanog; Plankton Soc Japan; Oceanog Soc Japan; fel Royal Soc Can; Hon Fel, Nat Bur Oceanog, Peoples Repub of China. *Res:* Biological and chemical oceanography. *Mailing Add:* Dept Oceanog Univ BC Vancouver BC V6T 1W5 Can

PARSONS, TORRENCE DOUGLAS, mathematics; deceased, see previous edition for last biography

PARSONS, WILLARD H, b New York, NY. VOLCANO GEOLOGY. *Educ:* Hamilton Col, BS, 30; Princeton Univ, MA, 35, PhD(geol), 36. *Prof Exp:* Assoc prof, 46-50, chmn dept & prof, 50-76, EMER PROF GEOL, WAYNE STATE UNIV, DETROIT, MICH, 76- *Concurrent Pos:* Scientist, Int Mineral Resource Develop Lab, Japan, 77-78 & Hawaii Volcano Observ, 56-57 & 72-73. *Mem:* Fel Geol Soc Am; fel Mineral Soc Am; Am Geol Petrol Union. *Res:* Volcano rocks in Montana and Wyoming. *Mailing Add:* 33 Mountain Laurel Dr Morgan Chapel Village Columbus NC 28722

PARSONS, WILLIAM BELLE, JR, b Apollo, Pa, Nov 10, 24; m 45, 73; c 5. MEDICINE. *Educ:* Univ Pittsburgh, BS, 47, MD, 48; Am Bd Internal Med, dipl, 56. *Prof Exp:* Intern med, Western Pa Hosp, Pittsburgh, 48-49; res fel res med, Univ Pittsburgh, 49-50; fel med, Mayo Found, 50-53; first asst, Mayo Clin, 55-56; mem staff, Dept Internal Med, Jackson Clin, 56-73; dir clin res, Armour Pharmaceut Co, 74-78; MEM STAFF, SCOTTSDALE MEM HOSP, 77- *Concurrent Pos:* Dir res med, Jackson Found, 57-73; fel coun arteriosclerosis, Am Heart Asn, 57-; pvt pract internal med, 78-; mem staff, Scottsdale Mem Hosp-N, 83- *Mem:* Fel Am Col Physicians; AMA. *Res:* Lipid research; cholesterol-reducing drugs; hypertension; coronary risk factors; hematology; atherosclerosis. *Mailing Add:* 7331 E Osborn Dr Scottsdale AZ 85251

PARTAIN, CLARENCE LEON, b Memphis, Tenn, July 12, 40; m 64; c 3. MAGNETIC RESONANCE IMAGING. *Educ:* Univ Tenn, BS, 63; Purdue Univ, MS, 65, PhD(nuclear eng), 67; Wash Univ Sch Med, MD, 75. *Prof Exp:* Develop engr, Instrumentation & Controls, Oak Ridge Nat Lab, Union Carbide Corp, 66-68; assoc prof nuclear eng, Univ Mo, 68-75; res asst radiol, Mallinckrodt Inst, Sch Med, Wash Univ, 72-75; assoc prof & dir radiol & comput tomography, Univ NC, Chapel Hill, 75-79; prof & dir radiol Nuclear Med Div, 80-85, PROF & DIR, DIV MED IMAGING, SCH MED, VANDERBILT UNIV, 85- *Concurrent Pos:* Fel, Am Col Nuclear Physicians, 81-82; reviewer & abstractor radiol, Radiol Soc NAm, 77-; consult, Site Vis Nat Cancer Inst & Nat Heart Blood & Lung, NIH, 80-; consult, Vet Admin, Dept Med & Surg, 80- & Sci Adv Comn, Am Cancer Soc, 82- *Mem:* Am Roentgen Ray Soc; Asn Univ Radiologists; Inst Elec & Electronics Engrs; Soc Nuclear Med; Radiol Soc NAm; fel Am Col Nuclear Physicians; fel Am Col Radial; Soc Nuclear Med. *Res:* Development and evaluation of medical imaging modalities including magnetic resonance imaging nuclear medicine; position emission tomography; auth of several journals and books. *Mailing Add:* Hosp Radiol Vanderbilt Univ Nashville TN 37232

PARTAIN, GERALD LAVERN, b Little Rock, Ark, Feb 18, 25; m 68; c 2. FOREST ECONOMICS. *Educ:* Univ SC, BS, 45; Ore State Univ, BS, 49; NY State Col Forestry, Syracuse Univ, PhD(forestry econ), 72. *Prof Exp:* Design engr, Kaiser Engrs, Calif, 51-53 & Bechtel Engrs, 53- 54; assoc prof forestry econ, Humboldt State Univ, 54-69, prof forestry, 69-83, chmn dept, 73-83; dir, Calif Dept Forestry, Sacramento, 83-90; RETIRED. *Concurrent Pos:* Consult land develop; legislative aide & consult, US Senate. *Mem:* Soc Am Foresters; Am Forestry Asn; Int Soc Trop Foresters. *Res:* Forestry in economic development; rural subdivisions and the environment. *Mailing Add:* Calif Dept Forestry 1416 Ninth St Rm 1505 Sacramento CA 95814

PARTANEN, CARL RICHARD, b Portland, Ore, Nov 23, 21; m 61; c 2. PLANT CYTOLOGY, PLANT MORPHOGENETICS. *Educ:* Lewis & Clark Col, BA, 50; Harvard Univ, AM, 51, PhD, 54. *Prof Exp:* Am Cancer Soc res fel, Columbia Univ, 54-55 & Harvard Univ, 55-57; res assoc, Children's Cancer Res Found, 57-61; from assoc prof to prof, chmn dept, 64-70, EMER PROF BIOL, UNIV PITTSBURGH, 87- *Concurrent Pos:* Res fel, Univ Edinburgh, 71-72 & Univ Nottingham, 78-79. *Mem:* Bot Soc Am; Am Genetic Asn; Soc Develop Biol; Tissue Cult Asn. *Res:* Plant development; cell nucleus in relation to growth, differentiation and morphogenesis; radiation-induced tumorization in plants; biology of plants, tissues and cells in culture; plant genetic manipulation. *Mailing Add:* Dept Biol Sci Univ Pittsburgh Pittsburgh PA 15260

PARTANEN, JOUNI PEKKA, b Kaavi, Finland, Sept 27, 57; m 81. NONLINEAR OPTICS, LASER PHYSICS. *Educ:* Helsinki Univ, MSc, 82, DTech, 87. *Prof Exp:* Res scientist nuclear eng, Tech Res Ctr Finland, 81-88; res assoc optics, Rutherford Appleton Lab, Eng, 84-87; res assoc, 88-90, RES PROF OPTICS, UNIV SOUTHERN CALIF, 91- *Mem:* Optical Soc Am; Inst Elec & Electronic Engrs. *Mailing Add:* Dept Elec Eng Univ Southern Calif Los Angeles CA 90089-0484

PARTCH, RICHARD EARL, b Long Beach, Calif, Aug 8, 36; m 57; c 3. ORGANIC CHEMISTRY, MEDICINAL CHEMISTRY. *Educ:* Pomona Col, AB, 58; Univ Rochester, PhD(org chem), 62. *Prof Exp:* Asst prof chem, NMex Highlands Univ, 62-65; asst prof, 65-68, assoc prof & exec off dept, 68-71, ROTARY FOUND FEL, CLARKSON UNIV, 71- *Concurrent Pos:* Res grants, Res Corp, 63-65, NIH, 63- & Am Heart Asn, 63-; Nat Acad Sci res fel, 67-68. *Mem:* Am Chem Soc; Am Ceramic Soc; Sigma Xi; Am Asn Aerosol Res. *Res:* Adamantane derivatives of medicinal interest; recycling and detoxification of waste; lead tetraacetate reactions; synthesis and modification of fine particles composed of inorganic and organic materials. *Mailing Add:* Dept Chem Clarkson Univ Potsdam NY 13699-5810

PARTENHEIMER, WALTER, b Chicago, Ill, Jan 10, 41; m 67; c 1. INORGANIC CHEMISTRY. *Educ:* Univ Wis-Whitewater, BS, 63; Univ Iowa, PhD(chem), 68. *Prof Exp:* NIH fel, Univ Ill, 68-70; Petrol Res Fund fel, Clarkson Col Technol, 70-73, asst prof inorg chem, 70-75; res chemist, 75-78, sr chemist, 78-80, RES ASSOC, AMOCO CHEM CORP, 81- *Mem:* Am Chem Soc; Sigma Xi. *Res:* Coordination chemistry; organometallic chemistry; catalysis; catalytic oxidation of organic molecules. *Mailing Add:* 352 Pearsson Circle Naperville IL 60540

PARTER, SEYMOUR VICTOR, b Chicago, Ill, June 9, 27; m 57; c 2. NUMBER THEORY. *Educ:* Ill Inst Technol, BS, 49, MS, 51; NY Univ, PhD(math), 58. *Prof Exp:* Staff mem, Los Alamos Nat Lab, 51-57; asst prof math, Mass Inst Technol, 57-58, Ind Univ, 58-60, Cornell Univ, 60-62; vis asst prof comput sci, Stanford Univ, 62-63; assoc prof, 63-64, PROF MATH & COMPUT SCI, UNIV WIS, 64- *Concurrent Pos:* Consult, Midwest Res Inst, Kans City, 57-60, Lawrence Livermore Lab, Calif, 62-63; collabr, Los Alamos Nat Lab, NMex, 68; managing ed, Soc Indust & Appl Math J Numerical Anal, 77-80; mem panel appl math, Nat Bur Standards, 81-84; mem sci coun, Inst Comput Applns Sci & Eng, 81-83; chmn, Conf Bd Math Sci, 83-85. *Mem:* Soc Indust & Appl Math (pres, 81-82); Math Asn Am; Asn Comput Mach. *Res:* Numerical methods for elliptic partial differential equations emphasizing the solution of large discrete systems; studies of preconditioning and multigrid; toeplitz forms and singular perturbation theory. *Mailing Add:* Five S Rock Rd Madison WI 53705

PARTHASARATHI, MANAVASI NARASIMHAN, b Madras, India, Jan 13, 24; m 44; c 3. TECHNICAL MANAGEMENT. *Educ:* Banaras Hindu Univ, BS, 44; Univ Ill, MS, 57, PhD(metall eng), 60. *Prof Exp:* Mem res staff, Univ Ill, 55-60; asst head res dept, Bird & Co, 60-62; gen mgr, Zinc Devlop Asn & Lead Develop Asn, 62-77; mgr, develop, Int Lead Zinc Res Orgn, 78-82; DIR, MAT TECHNOL, UNIV NEW HAVEN, 83- *Concurrent Pos:* Hon secy, Indian Inst Metals, 62-77, ed, Transaction, 63-72; prof, Bengaleng Col, 63-65. *Honors & Awards:* John Taylor Gold Medal, Indian Inst Metals, 72. *Mem:* Fel Am Soc Metals; fel Soc Die Casting Engrs; fel Inst Metallurgists London; fel Indian Standards Inst. *Res:* Metallurgical research; fundamental, applied and industrial; pilot plant; lead zinc and cadmium; product development. *Mailing Add:* 142 Linsley Lake Rd North Branford CT 06471

PARTHASARATHY, RENGACHARY, b Nagari, India, May 3, 36; m 62; c 4. BIOPHYSICS, CRYSTALLOGRAPHY. *Educ:* Univ Madras, MA, 57, Msc, 58, PhD(physics), 62; State Univ NY, PhD(biophys), 66. *Prof Exp:* Sr res asst physics, Univ Madras, 61-62, Coun Sci & Indust Res sr fel, 62-63; from cancer res scientist to sr cancer res scientist, 63-67, assoc cancer res scientist, 67-76, PRIN CANCER RES SCIENTIST, BIOPHYS, ROSWELL PARK MEM INST, 76- *Concurrent Pos:* Asst res prof, Roswell Div, State Univ NY, 66-72, assoc res prof, 72-76, res prof biophys, 76-; NIH grants, 69-72, 74-77, 78-80 & 81-83. *Mem:* Biophys Soc; AAAS; Am Crystallog Asn. *Res:* X-ray optics; protein crystallography; conformational analysis; biological structures; stereochemistry of nucleic acids. *Mailing Add:* Biol Phys Dept Roswell Park Mem Inst Buffalo NY 14263

PARTHASARATHY, SAMPATH, b Madras, India, Dec 27, 47; m 74; c 2. BIOCHEMISTRY. *Educ:* Univ Madras, BSc, 67, MSc, 69; Indian Inst Sci, PhD(biochem), 75. *Prof Exp:* Asst res officer, Ind Counc Med Res, 71-72; sr res fel, Univ Grants Comn, 72-74; Jap Govt Scholar, Kyoto Univ, 75-76; res assoc, Duke Univ Med Ctr, 76-77; fel, Hormel Inst, Univ Minn, Austin, 77-78, res assoc, 78-81, asst prof, 81-; ASSOC ADJ PROF, DEPT MED, UNIV CALIF, SAN DIEGO. *Mem:* Am Soc Cell Biol; Am Heart Asn. *Res:* Metabolism of phospholipids; cholesterol metabolism; regulation of phospholipid synthesis, phospholipases; membrane structure and assembly; lipid dependent enzymes; lipid in normal and neoplastic cells; lipid peroxidahion; lipoprotein metabolism; lipoxygenases; anhioxidants; atherosclerosis. *Mailing Add:* Dept Med Univ Calif San Diego 0613-D La Jolla CA 92093

PARTHASARATHY, TRIPLICANE ASURI, b Madras, India; m; c 1. STRUCTURE-PROPERTY CORRELATIONS, HIGH TEMPERATURE DEFORMATION. *Educ:* Indian Inst Technol, BTech, 76; Ohio State Univ, MS, 82, PhD(metall eng), 83. *Prof Exp:* Develop engr, Welding Res Inst, Bharat Heavy Elec Ltd, India, 77-80; grad res assoc metall eng, Ohio State Univ, 80-83, researcher, 84; asst prof metall eng, dept mat sci, Univ Ill, Urbana-Champaign, 84-88; SR RES SCIENTIST, UNIVERSAL ENERGY SYSTS, WRIGHT-PATTERSON AFB, DAYTON, 88- *Concurrent Pos:* Mem, comt alloy phases, Metall Soc, Am Inst Mining, Metall & Petrol Engrs, 85-; consult, Cummings Engine Inc, 86. *Mem:* Am Ceramic Soc; Metall Soc-Am Inst Mining, Metall & Petrol Engrs; Mat Res Soc. *Res:* Deformation mechanisms in advanced materials and composites; computer simulation of defects in materials; processing and mechanical behavior of high temperature ceramic composites. *Mailing Add:* Universal Energy Systs 4401 Dayton-Xenia Rd Dayton OH 45432-1894

PARTHENIADES, EMMANUEL, b Athens, Greece, Nov 3, 26; US citizen; m 67. HYDRAULIC & COASTAL ENGINEERING. *Educ:* Nat Tech Univ Athens, dipl civil eng, 52; Univ Calif, Berkeley, MS, 55, PhD(hydraul eng), 62. *Prof Exp:* Designer, G Vingos, Greece, 52-53 & A Mahairas, 53-54; sr engr analyst, Dames & More, Calif, 55-59; sr engr, Cooper & Clark, 59-60; teaching asst hydraul, Univ Calif, Berkeley, 59-61, res engr, 61-62; asst prof hydraul & fluid mech, San Jose State Col, 62-63; asst prof hydraul, Mass Inst Technol, 63-66; assoc prof, State Univ NY Buffalo, 66-68; vis assoc prof coastal & oceanog eng, 68-69, PROF COASTAL & OCEANOG ENG, UNIV FLA, 69-, PROF ENG SCI, 80- *Concurrent Pos:* Ford Found fel, Mass Inst Technol, 63-65; co-investr with Dr J F Kennedy, NSF res grant, 65-66; Fed Water Pollution Control Admin res grant, 67-71; NSF res grants, 70-74, 76-79, 78-81; Environ Protection Agency traineeship grant, 70-; Coastal Eng Res Ctr res contract, 70-73; prof chair hydrol structure, Univ Thessaloniki, Greece, 73-; US Army res grants, 80-82. *Mem:* Am Soc Eng Educ; Am Soc Civil Engrs; Int Asn Hydraul Res. *Res:* Erosion and deposition of fine cohesive sediments; stratified flow and salinity intrusions in estuaries; shoaling of estuarial channels; turbulence; sediment transport by waves; thermo-pollution. *Mailing Add:* Dept Sci Univ Fla Gainesville FL 32611

PARTIDA, GREGORY JOHN, JR, b Pomona, Calif, Sept 14, 42; m 64; c 2. ECONOMIC ENTOMOLOGY. *Educ:* Calif State Polytech Col, Kellogg-Voorhis, BS, 65; Univ Calif, Riverside, MS, 69, PhD(entom), 71. *Prof Exp:* Lab asst agr biol, Calif State Polytech Col, Kellogg-Voorhis, Pomona, 63-65; lab technician entom, Univ Calif, Riverside, 65-66, res asst, 66-70; asst prof entom, Kans State Univ, 71-77; assoc prof, 77-80, PROF PLANT & SOIL SCI, CALIF STATE POLYTECH UNIV, 80- *Mem:* Entom Soc Am. *Res:* Bionomics of insects attacking stored-products; insect ecology. *Mailing Add:* Dept Plant Sci Calif State Polytech Univ Pomona 3801 W Temple Ave Pomona CA 91768

PARTIN, DALE LEE, b Toledo, Ohio, Mar 31, 49; m 75; c 2. SOLID STATE PHYSICS, SOLID STATE ELECTRONICS. *Educ:* Carnegie-Mellon Univ, BS, 71, MS, 73, PhD(elec engr), 78. *Prof Exp:* assoc sr res scientist, 78-80, sr res scientist physics, 80-82, staff res scientist, 82-85, SR STAFF RES SCIENTIST, GEN MOTORS RES LAB, 85- *Mem:* Inst Elec & Electronic Engrs; Am Vacuum Soc; Am Inst Physics. *Res:* Lead salt laser diodes grown by molecular beam epitaxy for tuneable, high resolution infra-red spectroscopy; magnetic field sensors made from indium antimonide. *Mailing Add:* Dept Physics Gen Motors Res Lab Warren MI 48090-9055

PARTIN, JACQUELINE SURRATT, b Mississippi, 34; m 55; c 5. CELL BIOLOGY. *Educ:* Univ Cincinnati, MSc, 70. *Prof Exp:* RES ASSOC ULTRASTRUCT, DEPT PEDIAT, SCH MED, STATE UNIV NY, STONY BROOK, 80- *Mem:* Am Soc Cell Biol; Soc Ped Pathol; Am Asn Study Liver Dis; Electron Micros Soc Am. *Res:* Pediatric liver diseases; drug intoxication affecting the liver. *Mailing Add:* Dept Pediat State Univ NY Stony Brook NY 11790

PARTIN, JOHN CALVIN, b Ft Lewis, Wash, Mar 21, 33; c 5. MEDICINE, CELL BIOLOGY. *Educ:* Univ Ky, BS, 55; Univ Cincinnati, MD, 59; Am Bd Pediat, dipl, 66. *Prof Exp:* Intern med, Cincinnati Gen Hosp, 59-60; resident pediat, Cincinnati Children's Hosp, 60-62, chief resident, 62-63; from asst prof to assoc prof pediat, Univ Cincinnati, 66-74, prof, 74-, dir electron micros lab, Children's Hosp Res Round, 66-, assoc dir div gastroenterol, 68-, dir microbiol lab & bact lab, 73-, clin res ctr, 76-; CHMN DEPT PEDIAT, STATE UNIV NY, STONY BROOK. *Concurrent Pos:* Fel, Med Sch, Johns Hopkins Univ, 65-66; attend pediatrician, Children's & Cincinnati Gen Hosps, 69-; chief pediat, US Trop Res Med Lab, San Juan, 76- *Mem:* AAAS; Am Soc Microbiol; Soc Pediat Res; Am Fedn Clin Res. *Res:* Ultrastructure in childhood gastrointestinal disease affecting lipid metabolism. *Mailing Add:* 10 Childs Lane Old Field NY 11733

PARTINGTON, MICHAEL W, b Birmingham, Eng, Jan 28, 26; Can citizen; wid; c 5. PEDIATRICS, MEDICAL GENETICS. *Educ:* Univ London, MB & BS, 48, PhD(physiol), 54; FRCP(E), 65; FRCP(C), 74, FCCMG, 76. *Prof Exp:* Intern med, St Bartholomew's Hosp, London, Eng, 49-50; res asst physiol, London Hosp Med Col, 51-53; resident med, St Bartholomew's Hosp, 53-54, West Norwich Hosp, 54-56 & Jenny Lindy Children's Hosp, Norwich, 56-68; lectr child health, Univ Sheffield, 58-59; res fel neurol, Res Inst, Hosp Sick Children, Toronto, 59-61; from asst prof to prof pediat, Queen's Univ, Ont, 61-89, head dept, 71-76, chmn, Div Med Genetics, 81-86; HEAD, REGIONAL MED GENETICS UNIT, SUBURBS HOSP, WARATAH, AUSTRALIA, 89- *Concurrent Pos:* Queen Elizabeth II scientist, 61-67. *Mem:* Can Soc Clin Invest; Am Pediat Soc; Can Pediat Soc; Can Col Med Geneticists. *Res:* Effects of ultraviolet light on skin; dysmorphology and medical genetics; mental retardation; phenylketonuria; neonatal tyrosine metabolism; neonatal behavior; serotonin metabolism; fragile X syndrome. *Mailing Add:* Regional Med Genetics Unit Suburbs Hosp Waratah NSW 2298 Australia

PARTON, WILLIAM JULIAN, JR, b Palmerton, Pa, Nov 2, 44. BIOMETEOROLOGY, MATHEMATICAL BIOLOGY. *Educ:* Pa State Univ, BS, 66; Univ Okla, MS, 68, PhD(meteorol), 72. *Prof Exp:* Res asst meteorol, Univ Okla, 66-68, spec instr, 68-71; fel math biol, Natural Resource Ecol Lab, Colo State Univ, 71-74; fel biometeorol, Nat Ctr Atmospheric Res, 74-75; RES ASSOC MATH BIOL, COLO STATE UNIV, 75- *Concurrent Pos:* Consult, Nat Park Serv, 75- *Mem:* Am Meteorol Soc; Sigma Xi; Soc Comput Simulation; Ecol Soc Am. *Res:* Development of computer models of biological systems; determining impact of weather modification on biological systems and studying interactions between the atmosphere and biosphere. *Mailing Add:* Nat Resource Ecol Lab Colo State Univ Ft Collins CO 80523

PARTOVI, AFSHIN, b Tehran, Iran, Dec 5, 63; US citizen. ELECTRO-OPTICS, OPTICAL NONLINEARITIES. *Educ:* Univ Redlands, BS(eng) & BS(physics), 83; Univ Southern Calif, MS, 86, PhD(elec eng), 90. *Prof Exp:* Mem tech staff, Jet Propulsion Labs, 85-87; consult, Hughes Res Labs, 88-90; MEM TECH STAFF, AT&T BELL LABS, 90- *Mem:* Am Phys Soc; Optical Soc Am. *Res:* Nonlinear optical properties of bulk and layered semiconductors for information processing applications. *Mailing Add:* AT&T Bell Labs Rm 1A-155 600 Mountain Ave Murray Hill NJ 07974-2070

PARTRIDGE, ARTHUR DEAN, b Brooklyn, NY, Feb 17, 27; m 90; c 3. PHYTOPATHOLOGY. *Educ:* Univ Maine, BSF, 53; Univ NH, MS, 56, PhD(bot), 57. *Prof Exp:* Pathologist forest & forest prod, US Forest Serv, 57-60; prof forest path, 60-74, prof forestry, 74-77, PROF FOREST RESOURCES, UNIV IDAHO, 77-; PRES, TREAZ INC CONSULT, 87- *Concurrent Pos:* Consult, expert witnessing, diagnostics. *Mem:* AAAS; Am Phytopath Soc; Bot Soc Am; Am Forestry Asn; Sigma Xi. *Res:* Forest and shade tree diseases. *Mailing Add:* Col Forestry Univ Idaho Moscow ID 83843

PARTRIDGE, GORDON R(ADFORD), electrical engineering, for more information see previous edition

PARTRIDGE, JAMES ENOCH, b Riverside, Calif, Nov 24, 42; m 63; c 2. MYCOLOGY. *Educ:* Calif State Polytech Univ, BS, 66; Univ Calif, Riverside, PhD(bot), 73. *Prof Exp:* Fel, Dept Biochem, Univ Calif, Riverside, 72-75; fel, 75-78, asst prof, 78-82, ASSOC PROF, DEPT PLANT PATH, UNIV NEBR, 82- *Mem:* Am Phytopathol Soc; Sigma Xi. *Res:* Host-parasite interactions between fungal parasites and their hosts; fungal ribosomes and protoplast culture/regeneration of corn and sorghum. *Mailing Add:* Dept Plant Pathol Univ Nebr Lincoln NE 68588

PARTRIDGE, JERRY ALVIN, b American Falls, Idaho, Oct 24, 39; m 63; c 6. PHYSICAL CHEMISTRY. *Educ:* Brigham Young Univ, BS, 61, PhD(inorg chem), 65. *Prof Exp:* Sr res scientist, Pac Northwest Labs, Battelle Mem Inst, 65-76; sr scientist, Westinghouse Hanford Co, 76-82; SR CHEMIST, ADVAN NUCLEAR FUELS CO, 82- *Mem:* Am Chem Soc. *Res:* Metal ion complexes in solution; aqueous phase thermodynamics; solvent extraction of metal ions or compounds; ion exchange; solution calorimetry; nuclear fuel reprocessing; nuclear waste management. *Mailing Add:* 1817 Marshall Ct Richland WA 99352

PARTRIDGE, JOHN JOSEPH, b Altanta, Ga, Dec 13, 42; m 67; c 1. ORGANIC CHEMISTRY. *Educ:* Wheaton Col, Ill, BS, 64; Northwestern Univ, PhD(org chem), 68. *Prof Exp:* NIH fel chem, Columbia Univ, 68-69; res leader, Res Div, Hoffmann-La Roche Inc, 69-88; DEPT HEAD, DEPT SYNTHETIC ORG CHEM, GLAXO INC, 88- *Concurrent Pos:* NIH fel; adj prof chem, Seton Hall Univ, SOrange, NJ, 84-88, Univ NC, Chapel Hill, 90- *Mem:* Am Chem Soc; Royal Chem Soc London; NY Acad Aci; AAAS; Sigma Xi. *Res:* Synthetic organic chemistry, asymmetric synthesis, natural products, terpenes, terpene glucosides, prostaglandins, vitamin D3 metabolites; steroids; 1, 3-dipolar additions; enzymatic chemistry, enzyme inhibitors, antibiotics, pharmaceutical research and development; scale-up of organic synthesis. *Mailing Add:* Glaxo Inc 5 Moore Dr Research Triangle Park NC 27709

PARTRIDGE, L DONALD, b Philadelphia, Pa, May 10, 45; m 65, 84; c 2. NEUROPHYSIOLOGY. *Educ:* Mass Inst Technol, BS, 67; Univ Wash, PhD(physiol), 73. *Prof Exp:* Wellcome res fel, Dept physiol, Univ Bristol, Eng, 73-74; fel, Friday Harbor Labs, Univ Wash, 74-76; asst prof, 76-84, ASSOC PROF, DEPT PHYSIOL, UNIV NMEX, 84- *Concurrent Pos:* Fulbright scientist, Max-Planck Inst, Munich, Ger, 85-86. *Mem:* Soc Neurosci. *Res:* Neuron membrane current processes with particular reference to slow currents underlying the encoding functions of neurons. *Mailing Add:* Dept Physiol Univ NMex Albuquerque NM 87131

PARTRIDGE, LLOYD DONALD, b Cortland, NY, Dec 18, 22; m 44; c 3. MOTOR CONTROL ANALYSIS, NERVOUS SYSTEM MODELING. *Educ:* Univ Mich, BS, 48, MS, 49, PhD(physiol), 53. *Prof Exp:* Instr physiol, Univ Mich, 53-56; res asst neurol, Yale Univ, 56-62, from instr to asst prof physiol, 57-62; assoc prof, 62-70, dep chmn dept, 66-73, PROF PHYSIOL & BIOPHYS, UNIV TENN, MEMPHIS, 70- *Concurrent Pos:* Consult, Middletown State Hosp, 56-62; vis prof, Univ Vt, 65 & 66, Med Col, Ohio, 85; assoc ed, Trans Biomed Eng, 76-80; vis scientist, Univ Western Ont, 81; vis scientist, Acad Sci USSR, 87; dir, Bioeng Soc, 85-89; assoc ed Annuals Bio Med Eng, 85- *Mem:* Am Acad Neurol; Am Physiol Soc; Bioeng Soc; Soc Neurosci; Int Brain Res Orgn; Inst Elec Engrs. *Res:* Distributed biological control systems; signal processing by nervous system; unit function in reflexes; reflex regulation and control of movement; muscle dynamics; organization and conversion of neural activity to mechanical response; energy exchange in muscle. *Mailing Add:* Dept Physiol & Biophys Univ Tenn Med Units Memphis TN 38163

PARTRIDGE, ROBERT BRUCE, b Honolulu, Hawaii, May 16, 40; m; c 2. ASTRONOMY. *Educ:* Princeton Univ, AB, 62; Oxford Univ, DPhil(physics), 65. *Prof Exp:* Instr physics, Princeton Univ, 65-67, asst prof, 67-70; assoc prof, 70-76, PROF ASTRON, HAVERFORD COL, 76- *Concurrent Pos:* Rhodes Scholar fel, Alfred P Sloan Res Found, 71-75; Fulbright Award, Norway, 79; Guggenheim fel, 88-; dean col, Haverford Col, 82-85. *Honors & Awards:* Award for Res in Undergrad Inst, Am Phys Soc, 88. *Mem:* AAAS; Am Astron Soc; Int Astron Union. *Res:* Radio astronomy; astrophysics; cosmology. *Mailing Add:* Dept Astron Haverford Col Haverford PA 19041

PARTYKA, ROBERT EDWARD, b Wakefield, RI, Sept 5, 30; m 57; c 3. PLANT PATHOLOGY. *Educ:* Univ R I, BS, 52; Cornell Univ, PhD(plant path), 58. *Prof Exp:* From asst prof to prof plant path, Ohio State Univ, 57-74, exten plant pathologist, 57-74; PLANT PATHOLOGIST, CHEM-LAWN CORP, 74- *Concurrent Pos:* Adj prof, Ohio State Univ, 74- *Mem:* Am Phytopath Soc; Potato Asn Am; Soc Nematol; Am Inst Biol Sci. *Res:* Turf and urban ornamental disease problems. *Mailing Add:* Chem-Lawn Serv Corp 8275 N High St Columbus OH 43085

PARUNGO, FARN PWU, b Ann-King Anwhei, China, June 18, 32; US citizen; m 62; c 3. ORGANIC CHEMISTRY, PHYSICAL CHEMISTRY. *Educ:* Nat Taiwan Univ, BS, 55; Univ Colo, PhD(org chem), 61. *Prof Exp:* Res assoc chem, Cornell Univ, 61-62; scientist, Nat Ctr Atmospheric Res, 62-67; res scientist, chem, EG&G, 67-69; RES CHEMIST, ENVIRON SCI LAB, NAT OCEANIC & ATMOSPHERIC ADMIN, 69- *Res:* Atmospheric chemistry; natural and anthropogenic chemical reactions in the atmosphere, environmental impact of air pollutions; advertent and inadvertent weather modification. *Mailing Add:* 2630 Cornell Circle Boulder CO 80302

PARVEZ, ZAHEER, b Hyderabad, India, Mar 20, 39; US citizen; m 68; c 2. HEMATOLOGY, CYTOLOGY. *Educ:* Osmania Unv, Hyderabad, India, BS, 59 & 62; Okla State, Stillwater, MS, 68; Univ Calif, Berkeley, PhD(immunol), 74. *Prof Exp:* Lectr zool & immunol, Univ Nairobi, Kenya, 74-77; res assoc, 78-81, instr, 81-83, ASST PROF & DIR, LOYOLA UNIV MED CTR, MAYWOOD, ILL, 83- *Concurrent Pos:* NIH fel, Univ Calif, Berkeley, 69-74; adj prof, Columbia Pac Univ, San Raphael, Calif, 78-; vis scientist, Univ Turku, Finland, 85-86; ed Contrast Media vol I & II, CRC Press, 85-86; prin investr, Sickle Cell Ctr Grant, Univ Chicago, 86-; ed consult, Anal Biochem, 86- *Mem:* NY Acad Sci; fel, Am Col Angiology; Am Asn Immunologists; AAAS; Asn Clin scientists. *Res:* Immunopharmacology of radiocontrast agents, their interactions with antineoplastic drugs, platelets and red cells (in relation to sickle cell traits), and their mutagenic potential. *Mailing Add:* 152 Constitution Dr Bloomingdale IL 60108

PARVIN, PHILIP EUGENE, b Manatee, Fla, July 3, 27. FLORICULTURE, PROTEA & NEW CROPS. *Educ:* Univ Fla, BSAg, 50; Miss State Univ, MS, 52; Mich State Univ, PhD(hort), 65. *Prof Exp:* Instr ornamental hort, Miss State Univ, 52; asst prof floricult, Univ Fla, 53-56; res instr, Floricult Mkt, Mich State Univ, 59-61; exten specialist, Univ Calif, Davis, 63-66; gen mgr, Rod McLellan Co, 66-68; HORTICULTURIST & SUPT, MAUI AGR RES CTR, UNIV HAWAII, 68- *Mem:* Fel Am Soc Hort Sci; Int Soc Hort Sci; Int Plant Propagators Soc; Int Protea Asn. *Res:* Introduction, evaluation, culture, management and marketing of Proteas and other southern hemisphere flora. *Mailing Add:* Maui Agr Res Ctr Univ of Hawaii PO Box 269 Kula HI 96790

PARVULESCU, ANTARES, b Ploesti, Romania, Nov 15, 23; US citizen; c 1. MATHEMATICAL PHYSICS, ACOUSTICS. *Educ:* Univ Bucharest, Lic math sci, 43. *Prof Exp:* French govt fel, Univ Paris, 46-47; lectr appl math, Univ Witwatersrand, 47-50; lectr math, Univ Calif, Berkeley, 50-51; asst prof physics, Bard Col, 51-54 & Villanova Univ, 54-55; assoc prof & head dept, Gallaudet Col, 55-60; res scientist, Hudson Labs, Columbia Univ, 60-64, sr res assoc physics, 64-68; PROF OCEAN ENG & GEOPHYS, HAWAII INST GEOPHYS, UNIV HAWAII, 68- *Concurrent Pos:* Lectr & actg dir, Vassar Col Observ, 53-54; res physicist, David Taylor Model Basin, US Navy, 59-60; res prof physics, Gallaudet Col, 60-68; consult, Hollerith Comput Mach Ltd SAfrica, 48-50; chief scientist on 19 oceanog expeds, 61-; consult, US Navy, 75- *Mem:* Fel Acoust Soc Am; sr mem Inst Elec & Electronics Eng; Audio Eng Soc; NY Acad Sci. *Res:* Mathematical logic; theory of evidence and inference; theory of probability; foundations of quantum theory; ocean acoustics; wave equations; theory of information and signal processing; 05814613Xradiation; marine bioacoustics. *Mailing Add:* 1106 Villa May Blvd Alexandria VA 22307

PARYSEK, LINDA M, b Springville, NY, Jan 30, 54. CYTOSKELETON. *Educ:* State Univ NY, Buffalo, PhD(anat sci), 83. *Prof Exp:* Fel biol, Univ Rochester, 83-84, FEL BIOL, NORTHWESTERN UNIV, 84- *Mem:* Am Asn Anatomists; Am Soc Cell Biol. *Mailing Add:* Dept Biol Univ Rochester Rochester NY 14627

PARZEN, EMANUEL, b New York, NY, Apr 21, 29; m 59; c 2. STATISTICS. *Educ:* Harvard Col, AB, 49; Univ Calif, Berkeley, MA, 51, PhD(math), 53. *Prof Exp:* Asst math, Univ Calif, 49-52; res scientist, Hudson Labs, Columbia Univ, 53-56; from asst prof to prof statist, Stanford Univ, 56-70; prof & chmn statist sci div, State Univ NY, Buffalo, 70-78; DISTINGUISHED PROF STATIST, TEX A&M UNIV, 78- *Concurrent Pos:* Guest prof, Imp Col, Univ London, 61-62; vis prof, Mass Inst Technol, 64 & Harvard Univ, 76, 88; fel IBM Systs Res Inst, 69-70 & Ctr Adv Study Behav Sci, Stanford, 83-84. *Mem:* fel AAAS; fel Inst Math Statist; fel Am Statist Asn; Am Math Soc; Soc Indust & Appl Math. *Res:* Probability limit theorems; applied probability; statistical inference on stochastic processes; non-parametric statistical data modeling; time series analysis; spectral analysis; statistical communication and control theory; statistical computation; econometrics; systems identification. *Mailing Add:* Dept Statist Tex A&M Univ Col Sta TX 77843-3143

PARZEN, GEORGE, b New York, NY, Mar 11, 24. THEORETICAL PHYSICS. *Educ:* City Col New York, BEE, 45; Stanford Univ, PhD(physics), 49. *Prof Exp:* Instr physics, Univ Chicago, 49-51; from asst prof to assoc prof, Notre Dame Univ, 51-56; physicist, Midwestern Univs Res Asn, 56-63; PHYSICIST, ACCELERATOR DEPT, BROOKHAVEN NAT LAB, 63- *Mem:* Am Phys Soc. *Res:* Electron scattering; electrodynamics; energy levels in metals; accelerators; computer physics. *Mailing Add:* Bldg 1005 S-4 Brookhaven Nat Lab Upton NY 11973

PARZEN, PHILIP, b Poland; nat US; m 49; c 2. PHYSICAL ELECTRONICS. *Educ:* City Col New York, BS, 39; NY Univ, MS, 46, PhD(math), 53. *Prof Exp:* Develop engr, Fed Telecommun Labs, Int Tel & Tel Corp, 47-53; res scientist radiation lab, Johns Hopkins Univ, 53-56; res assoc prof elec eng, Polytech Inst Brooklyn, 56-58; dir res & develop, Parzen Assocs, 58-60; chief space physics, Repub Aviation, Inc, 60-62; sr mem tech adv staff, Astro- electronics Div, Radio Corp Am, NJ, 62-69; prof, 69-86, EMER PROF INFO ENG, UNIV ILL, CHICAGO CIRCLE, 86- *Mem:* Am Phys Soc; sr mem Inst Elec & Electronic Engrs. *Res:* Plasmas; gas discharges; microwave tubes; solid state devices. *Mailing Add:* 247 E Chestnut Apt 2102 Chicago IL 60611

PAS, ERIC IVAN, b Cape Town, SAfrica, May 12, 48; m 74; c 2. TRANSPORTATION ENGINEERING. *Educ:* Univ Cape Town, BSc, 70, MSc, 74; Northwestern Univ, PhD(transp eng), 80. *Prof Exp:* Engr design, Van Niekerk, Kleyn & Edwards, 71-72; lectr soil mech & transp eng, Univ Cape Town, 72-77; asst prof, 80-84, ASSOC PROF TRANSP ENG, DUKE UNIV, 84- *Concurrent Pos:* Co-ed, J Advan Transp, 87-90; proj mgr patronage forecasting, Dallas Area Rapid Transit, 86. *Mem:* Am Soc Civil Engrs; Int Asn Travel Behav; Transp Res Bd; Int Asn Time Use Res. *Res:* Modeling and analysis of urban travel behavior; forecasting demand for new and modified transportation systems; understanding travel needs and desires. *Mailing Add:* Dept Civil & Environ Eng Duke Univ Durham NC 27706

PASACHOFF, JAY M(YRON), b New York, NY, July 1, 43; m 74; c 2. ASTRONOMY, ASTROPHYSICS. *Educ:* Harvard Univ, AB, 63, AM, 65, PhD(astron), 69. *Prof Exp:* Res physicist, US Air Force Cambridge Res Labs, 68-69; Menzel Res Fel Astron, Harvard Col Observ, 69-70; res fel astrophys, Calif Inst Technol & Hale Observs, 70-72, asst prof & chmn, Dept Astron, 72-77; from assoc prof to prof, 77-84, DIR, HOPKINS OBSERV, WILLIAMS COL, 72-, FIELD MEM PROF ASTRON, DEPT ASTRON, 84- *Concurrent Pos:* Guest investr, NASA, 73-78; mem, US-Australia Coop Sci Prog, Australian Nat Radio Observ, 74; US nat rep, Comn Teaching Astron, Int Astron Union, 76-; adj assoc prof, Univ Mass, Amherst, 77-83, adj prof, 83-; vis assoc prof, Inst Astron, Univ Hawaii, 80-81, vis colleague, 80-81 & 84-85; var eclipse expeds, 59-91; mem, Astron Comt, Am Asn Physics Teachers, 83-85; chmn, Astron Sect D, AAAS, 87-88; vis scientist, Inst Astrophysics, Paris, 88; mem, Inst Advan Study, Princeton Univ, 89-90,

Astron Educ Adv Bd, Am Astron Soc, 90-, Astrophysics Coun, NASA, 90- & Astron News Comt, Am Astron Soc, 91- *Honors & Awards:* Lockhart lectr, Univ Manitoba, 79. *Mem:* Am Astron Soc; Int Astron Union; fel NY Acad Sci; Int Union Radio Sci; fel Am Phys Soc; fel AAAS; fel Royal Astron Soc. *Res:* Solar physics, including the structure of the chromosphere and corona; eclipses; stellar spectroscopy; spectral-line radio astronomy; author of textbooks in astronomy and physics and a field guide in astronomy. *Mailing Add:* Hopkins Observ Williams Col Williamstown MA 01267

PASAMANICK, BENJAMIN, b New York, NY, Oct 14, 14; m 42, 82. PSYCHIATRY. *Educ:* Cornell Univ, AB, 36; Univ Md, MD, 41. *Prof Exp:* Resident psychiat, NY State Psychiat Inst, 32; asst child develop, Sch Med, Yale Univ, 43-45; instr psychiat, Med Sch, Univ Mich, 46-47; asst clin prof, Long Island Med Col, 48-49; asst clin prof, Col Med, State Univ NY, 49-50; from asst prof to assoc prof pub health admin, Sch Hyg & Pub Health, Johns Hopkins Univ, 50-55; prof psychiat, Col Med, Ohio State Univ, 55-65; assoc dir res, Ill Dept Ment Health, 65-67; pres & dean, 67-72, Sir Aubrey & Lady Hilda Lewis Prof Social Psychiat, NY Sch Psychiat, 72-75; res prof, 75-84, EMER PROF PEDIAT, ALBANY MED COL, 84-; RES PROF PSYCHIAT, NY UNIV, 84- *Concurrent Pos:* Chief children's serv, Neuropsychiat Inst, Univ Mich, 40-47; chief psychiat div, Kings County Hosp, NY, 47-50; psychiatrist, Johns Hopkins Hosp, 51-55; consult, Baltimore Bd Sch Comnr, 51-55, Nat Inst Neurol Dis & Blindness, 55-58, NIMH, 55-58 & Milbank Fund, 56-58; dir res, Columbus Psychiat Inst & Hosp, Health Ctr, Ohio State Univ, 55-65; mem comt ment retardation, Coun State Govt, 57-65; chmn subcomt ment health, US Nat Comt Vital & Health Statist, 58-66 & 70-72; Cutter lectr, Harvard Univ, 60; Bailey lectr, Ill Psychiat Inst, 61; adj prof sociol & anthrop, Ohio State Univ, 63-65; clin prof psychiat, Univ Ill, 65-67 & Chicago Med Sch, 66-67; mem eval unit, Proj Head Start, US Off Econ Opportunity, 65-70; expert comt standardization psychiat diag, WHO, 65-70; Kolb lectr, NIMH, 66-; adj prof epidemiol, Sch Pub Health & admin med, Columbia Univ, 67-76; comnr for res, NY State Dept Ment Hyg, 67-76; adj prof psychol, NY Univ, 68-76; adj prof pediat, Albany Med Col, 72-78, res prof, 78; res prof psychiat & behav sci, State Univ NY, Stony Brook, 77-82. *Honors & Awards:* Hofheimer Res prize, Am Psychiat Asn, 49 & 67; Stratton Award, 61, Am Psychopath Asn, Samuel Hamilton Medal, 68; Lapouse Gold Medal, Am Pub Health Asn, 77; McGavin Award, Am Psychiat Asn, 86. *Mem:* AAAS; Am Col Psychiatrists; Soc Biol Psychiat; Soc Res Child Develop; Soc Psychol Study Social Issues; Am Col Epidemiol. *Res:* Child development; ethnic group psychology; epidemiology of neuropsychiatric disorder; measurement and evaluation in psychiatry; sociology; public health. *Mailing Add:* PO Box 2008 Albany NY 12220-0008

PASBY, BRIAN, b London, Eng, June 8, 37; m 59; c 2. OCEANOGRAPHY, ECOLOGY. *Educ:* Univ London, BSc, 59, MSc, 60; Tex A&M Univ, PhD(oceanog), 65. *Prof Exp:* Asst prof oceanog, Univ Hawaii, 64-67; assoc prof, 67-78, PROF BIOL, PACE UNIV, WESTCHESTER CAMPUS, 78-, CHMN DEPT BIOL. *Mem:* Fel Zool Soc London. *Res:* Aquatic ecology. *Mailing Add:* Dept Math & Sci Pace Univ Pleasantville Briarclif 861 Bedford Rd Pleasantville NY 10570

PASCERI, RALPH EDWARD, b Philadelphia, Pa, Aug 27, 37; m 63; c 1. CHEMICAL ENGINEERING, AIR POLLUTION. *Educ:* Villanova Univ, BChE, 59; Johns Hopkins Univ, DEng(atmospheric aerosols), 64. *Prof Exp:* Sr res engr, Hercules Inc, Md, 64-68; RES SCIENTIST, NJ STATE DEPT ENVIRON PROTECTION, 68- *Res:* Atmospheric aerosols; chemistry; gas dynamics and electromagnetic properties of solid propellant rocket exhausts; motor vehicle air pollution; particulate air pollution. *Mailing Add:* 286 Anderson Rd Morrisville PA 19067

PASCHALL, EUGENE F, b Neosho, Mo, Jan 7, 22; m 46; c 3. CARBOHYDRATE CHEMISTRY. *Educ:* Western State Col Colo, AB, 44; Iowa State Univ, PhD(chem), 51. *Prof Exp:* Sect leader, Moffett Tech Ctr, CPC Int, Inc, 56-61, dir prod develop, 61-75, Starch & Sweetner Dept, 75-86; RETIRED. *Mem:* AAAS; Am Chem Soc. *Res:* Synthesis of polysaccharide derivatives for use in food, textile and paper fields; measurement of physical and chemical properties of starch and starch fractions. *Mailing Add:* 7640 W 123rd Pl Palos Heights IL 60643

PASCHALL, HOMER DONALD, b Montgomery Co, Tenn, Aug 29, 26; m 50; c 4. CELL PHYSIOLOGY, ANIMAL PHYSIOLOGY. *Educ:* Trevecca Nazarene Col, AB, 48; Austin Peay State Col, BS, 50; Peabody Col, AM, 50; Iowa State Univ, PhD, 63. *Prof Exp:* Teacher high sch, Tenn, 48-49 & pub sch, Ky, 50-51; instr sci & math, Bethany-Nazarene Col, 51-55; PROF PHYSIOL & HEALTH SCI, BALL STATE UNIV, 55- *Mem:* Sigma Xi; Am Sci Affil. *Mailing Add:* Dept of Physiol & Health Sci Ball State Univ Muncie IN 47306

PASCHER, FRANCES, dermatology; deceased, see previous edition for last biography

PASCHKE, EDWARD ERNEST, b Evergreen Park, Ill, Feb 22, 43; m 69; c 2. POLYMER CHEMISTRY. *Educ:* Univ Ill, Urbana, BS, 65; Univ Minn, Minneapolis, MS, 67; Univ Iowa, PhD(chem), 71. *Prof Exp:* Res chemist, Standard Oil Co, 70-80, sr res chemist, 80-84, res supv, 84-87; RES ASSOC AMOCO CHEM CO, 87- *Mem:* Am Chem Soc. *Res:* Chemistry; organic chemistry. *Mailing Add:* 25 W 681 Coventry Ct Wheaton IL 60187

PASCHKE, JOHN DONALD, b Upland, Calif, Nov 6, 25; m 67; c 6. ENTOMOLOGY, INTERNATIONAL CROP PROTECTION. *Educ:* Univ Calif, PhD(entom), 58. *Prof Exp:* Sr lab technician, Citrus Exp Sta, Univ Calif, 50-53; asst entomologist, Ill Natural Hist Surv, 58-60; from asst prof to assoc prof, 60-68, asst dir, Div Sponsored Progs, 81-84, assoc coordr, 88-90, COORDR, NIGER APPL AGR RES PROJ, PURDUE UNIV, LAFAYETTE, 90-, PROF ENTOM, 68- *Concurrent Pos:* Vis fel, St Cross Col, Oxford Univ, 68-69; NIH grants, 61-64, 70-73, 73-76, NSF, 75-77 & USAID, 79-81. *Mem:* AAAS; Entom Soc Am; Soc Invert Path. *Res:* Insect pathology; invertebrate pathology; insect virology; microbial control of pests; integrated pest management. *Mailing Add:* Int Prog Agr Rm I AgAd Purdue Univ West Lafayette IN 47907-1168

PASCHKE, RICHARD EUGENE, b Chicago, Ill, Mar 13, 37; m 62; c 2. NEUROPSYCHOLOGY. *Educ:* Univ Ill, Urbana, BS, 61; Northern Ill Univ, MA, 64; Purdue Univ, PhD(exp psychol), 69. *Prof Exp:* Teaching fel, Univ Conn, 69-71; asst prof interdisciplinary studies, William James Col, 71-83; PROF PSYCHOL, GRAND VALLEY STATE COL, 83- *Concurrent Pos:* Fels & grants, Nat Inst Ment Health res fel, Inst Biobehav Sci, Univ Conn, 69-71. *Mem:* AAAS; Am Psychol Asn. *Res:* Relationships between biological determinants of behavior and the spatio-temporal environment; cognitive/perceptual processing and cognitive retraining following trauma to the central nervous system. *Mailing Add:* Dept Psychol Grand Valley State Col Allendale MI 49401

PASCHKE, WILLIAM LINDALL, b New York, NY, Aug 19, 46; m 68. MATHEMATICAL ANALYSIS. *Educ:* Dartmouth Col, BA, 67; Univ Ore, MA, 69, PhD(math), 72. *Prof Exp:* From asst prof to assoc prof, 72-84, PROF MATH, UNIV KANS, 84- *Concurrent Pos:* Vis scientist, Div Math Sci, NSF, 88-89. *Mem:* Am Math Soc. *Res:* Functional analysis, in particular algebras of operators on Hilbert space. *Mailing Add:* Dept of Math Univ of Kans Lawrence KS 66045-2142

PASCHOS, EMMANUEL ANTHONY, b Veroia, Greece, July 13, 40; US citizen; m 67; c 2. ELEMENTARY PARTICLE PHYSICS. *Educ:* City Col New York, BS, 62; Cornell Univ, PhD(physics), 67. *Prof Exp:* Res assoc physics, Stanford Linear Accelerator Ctr, 67-69; res assoc, Rockefeller Univ, 69-71; asst physicist, Fermi Nat Accelerator Lab, 71-75; assoc physicist, Brookhaven Nat Lab, 75-79; PROF, UNIV DORTMUND, WGERMANY, 79- *Concurrent Pos:* Adj prof, Univ Wis, 74-75; Res Corp grant, 75; prof theoret physics, Univ Dortmund, WGer, 78. *Mem:* Fel Am Phys Soc; Ger Phys Soc; Europ Phys Soc; NY Acad Sci. *Res:* Elementary particle theory; gauge field theories; elementary particle physics covering a wide range of topics from the structure of hadrons (parton model) to properties of the weak interactions; neutral and charged currents; proton decay. *Mailing Add:* Inst Physik Univ Dortmund Dortmund 50 4600 Germany

PASCIAK, JOSEPH EDWARD, b Pawtucket, RI, July 3, 50; m 78. APPLIED MATHEMATICS, NUMERICAL ANALYSIS. *Educ:* Northeastern Univ, BA, 73; Cornell Univ, PhD(appl math), 77. *Prof Exp:* Programmer, AVCO Everett Res Lab, 69-73; numerical analyst, Far Field Inc, 74-77; RES MATH, BROOKHAVEN NAT LAB, 77- *Mem:* Soc Indust & Appl Math; Am Math Soc. *Res:* Numerical analysis of finite element, finite difference and other competing methods, analytical analysis as well as implementation. *Mailing Add:* Dept Appl Math Brookhaven Natl Lab Upton NY 11973

PASCO, ROBERT WILLIAM, b Salem, Ohio, Feb 22, 47. SURFACE FINISHING PROCESSES, THIN FILMS. *Educ:* Carnegie-Mellon Univ, BS, 69; Syracuse Univ, PhD(solid state sci), 81. *Prof Exp:* Res engr, Youngstown Steel, 69-71; metallurgist, Sharon Steel Corp, 71-74; operating metallurgist, Globe MetallDiv, Interlake, Inc, 74-77; res asst, Syracuse Univ, 78-81, postdoctoral res assoc, 81-83; develop eng mgr, 83-91, SR ENGR, IBM, 91- *Mem:* Am Soc Metals Int; Metall Soc; Soc Plastics Engrs; Am Soc Testing & Mat; Am Mgt Asn; Am Vacuum Soc. *Res:* Ceramic finishing process development; thin film process development; surface finishing processes for thin film application in electronic packaging. *Mailing Add:* One Alpert Dr Wappingers Falls NY 12590

PASCU, DAN, b Arad, Romania, July 20, 38; US citizen; m 65; c 3. ASTROMETRY, CELESTIAL MECHANICS. *Educ:* Western Reserve Univ, Cleveland, BA, 61; Case Inst Technol, Cleveland, MS, 64; Univ Va, Charlottesville, PhD(astron), 72. *Prof Exp:* ASTRONR, NAUTICAL ALMANAC OFF, NAVAL OBSERV, 63- *Honors & Awards:* Newcomb Award, Naval Observ, 83. *Mem:* Int Astron Union; Am Astron Soc; Sigma Xi. *Res:* Astrometric, photometric and dynamical studies of the planetary satellites. *Mailing Add:* Naval Observ Washington DC 20392-5100

PASCUAL, ROBERTO, b Buenos Aires, Arg, Oct 20, 42; m 66; c 2. THIN FILM FABRICATION & CHARACTERIZATION, MECHANICAL PROPERTIES OF METALS. *Educ:* Balseiro Physics Inst, Arg, Lic physics, 65, Dr(physics), 69. *Prof Exp:* Postdoctoral, Physics Div, Nat Res Coun, Ottawa, 69-71, res staff mem, 77-80; from asst prof to assoc prof, Balseiro Physics Inst, Arg, 71-76; prof, Mil Inst Eng, Brazil, 80-88; RES PROF, METALL ENG DEPT, QUEEN'S UNIV, 88- *Concurrent Pos:* Consult, mat res & ins co. *Mem:* Metall Soc; Mat Res Soc. *Res:* Fabrication and characterization of thin films and protective coatings; mechanical properties of metals and ceramics; damage accumulation during fatigue of metals; surface modification by ion implantation. *Mailing Add:* Dept Metall Eng Queen's Univ Kingston ON K7L 3N6 Can

PASELK, RICHARD ALAN, b Inglewood, Calif, July 20, 45; m 67; c 2. EDUCATIONAL SOFTWARE DEVELOPMENT. *Educ:* Calif State Univ, Los Angeles, BS, 68; Univ Southern Calif, PhD(biochem), 75. *Prof Exp:* Res technologist, Univ Southern Calif Sch Med, 68-69; lectr biochem, Calif State Univ, Long Beach, 74-76; from asst prof to assoc prof biochem & anal chem, 76-86, chair, Chem Dept, 86-89, PROF BIOCHEM & ANALYTICAL CHEM, HUMBOLDT STATE UNIV, 86- *Concurrent Pos:* Tech ed, Zymed Corp, 84-85. *Res:* Software development; scientific visualization in chemistry education; Atomic Orbitals program on CD-ROM; commercial software; author of numerous publications. *Mailing Add:* Chem Dept Humbold State Univ Arcata CA 95521

PASFIELD, WILLIAM HORTON, b Brooklyn, NY, Dec 28, 24; m 47; c 5. PHYSICAL CHEMISTRY. *Educ:* Mass Inst Technol, BS, 48; Univ Conn, PhD(phys chem), 55. *Prof Exp:* Res chemist, Metal Hydrides, Inc, 49-50; res chemist, E I du Pont de Nemours & Co, 55-58; from asst prof to assoc prof, 58-65, PROF CHEM, ST JOHN'S UNIV, NY, 65- *Mem:* Am Chem Soc. *Res:* Molecular interactions in gases and liquids; volume changes in solutions. *Mailing Add:* Dept Chem St John's Univ Jamaica NY 11439

PASHLEY, DAVID HENRY, b Seattle, Wash, Apr 24, 39; m 62; c 2. PHYSIOLOGY, DENTIN PERMEABILITY. *Educ:* Univ Portland, BS, 60; Univ Ore, DMD, 64; Univ Rochester, PhD(physiol), 71. *Prof Exp:* From asst prof to prof oral biol, Sch Dent, asst prof physiol, Sch Med, 71-80, assoc prof, 81-86, PROF BIOL, SCH DENT, MED COL GA, 78-, PROF PHYSIOL, SCH MED, 86- *Concurrent Pos:* Prin investr grant, 72-92. *Mem:* AAAS; Am Physiol Soc; Sigma Xi; Int Asn Dent Res. *Res:* Body fluids; renal salivary physiology; renal salivary gland metabolism; permeability characteristics of teeth and oral epithelia; pulp biology. *Mailing Add:* Dept Physiol Sch Med Med Col Ga Augusta GA 30902

PASHLEY, EMIL FREDERICK, JR, b Lakewood, Ohio, Dec 19, 30; m 52; c 4. GEOLOGY, HYDROLOGY. *Educ:* Ohio State Univ, BSc, 52, MSc, 56; Univ Ariz, PhD(geol), 66. *Prof Exp:* Geologist & hydrologist, US Geol Surv, 59-68; assoc prof, 68-74, PROF GEOL, WEBER STATE COL, 74-, CHMN DEPT, 88- *Mem:* Geol Soc Am. *Res:* Geohydrology and environmental geology of the Wasatch Front. *Mailing Add:* Dept of Geol Weber State Univ 3750 Harrison Blvd Ogden UT 84408

PASIK, PEDRO, b Buenos Aires, Arg, Jan 16, 26; US citizen; m 52; c 3. NEUROANATOMY, NEUROPHYSIOLOGY. *Educ:* Univ Buenos Aires, MD, 51. *Prof Exp:* Asst resident neurol, Bellevue Med Ctr, 53-54; asst resident, Mt Sinai Hosp, 55-56, from res asst to res assoc, 56-66; assoc prof, 66-70, PROF NEUROL, MT SINAI SCH MED, 70-, PROF ANAT, 84- *Concurrent Pos:* Abrahamson fel neurol, 55-56; Nat Inst Neurol Dis & Stroke res career develop award, 63-73; mem grad fac, City Univ New York, 69-; adj prof, Queens Col, NY, 70- *Mem:* Am Neurol Asn; Soc Neurosci; Harvey Soc; Am Physiol Soc; Am Asn Anat; Int Basal Ganglia Soc. *Res:* Synaptology of the visual and basal ganglia systems; relationship between brain structure and function and visually guided behavior, through measurement of the effect of experimental cerebral lesions and the description of the anatomic pathways involved. *Mailing Add:* Mt Sinai Sch Med Dept Neurol Fifth Ave & 100th St New York NY 10029

PASIK, TAUBA, b Yasi, Rumania, Mar 18, 27; US citizen; m 52; c 3. NEUROPHYSIOLOGY, NEUROANATOMY. *Educ:* Univ Buenos Aires, MD, 51. *Prof Exp:* Asst resident neurol, Bellevue Med Ctr, 53-54; asst resident, Mt Sinai Hosp, 55-56, res asst, 56-59, res assoc, 59-66; from assoc prof to prof, 66-86, PROFESSORIAL LECTR NEUROL, MT SINAI SCH MED, 86- *Concurrent Pos:* Sugarmen fel neurol, 55-56; City of New York Health Res Coun career scientist award, 71; mem fac, City Univ New York, 69-; adj prof, Queens Col, NY, 70- *Mem:* AAAS; Am Acad Neurol; Am Neurol Asn; Soc Neurosci; Harvey Soc. *Res:* Synaptology of visual and basal ganglia systems in primates; relationship between brain structures and function and visually guided behavior, through measurement of the effect of experimental cerebral lesions and the description of the anatomic pathways involved. *Mailing Add:* Mt Sinai Sch of Med Dept Neurol Fifth Ave and 100th St New York NY 10029

PASIPOULARIDES, ARES D, b Athens, Greece, May 15, 43; US citizen; m 67. BIOMEDICAL ENGINEERING, CARDIOVASCULAR MECHANICS. *Educ:* Univ Athens, Greece, BA, 65; Univ Minn, MD, 71, PhD(math hemodynamics & chem eng), 72. *Prof Exp:* Res assoc med physiol, Dept Physiol, Univ Minn Med Sch, 71-72; instr physiol, Artificial Organs Lab, Brown Univ, 72-74, investr med sci, Div Biol & Med, 74-76, asst prof fluid mech, Div Eng, 76-79; res assoc med, Dept Med, Harvard Med Sch, 79-81; med officer, Dept Med, Brooke Army Med Ctr, Tex, 81-85, dir res, 85-88; ASSOC PROF BIOMED ENG & MED, DUKE UNIV & MED CTR, 88- *Concurrent Pos:* Med res & develop dir, Med Inc, Minn, 72; investr, Div Biol & Med, Brown Univ, 76-79; assoc med, Peter Bent Brigham Hosp, Boston, 79-80; res assoc, Cardiac Unit, Mass Gen Hosp, Boston, 80-81; clin assoc prof cardiol, Dept Med, Univ Tex, San Antonio, 83-88 & Cardiovasc Div, Dept Med, Duke Med Ctr, 88-; fel, Coun Circulation, Am Heart Asn, 88- *Mem:* Fel Am Col Cardiol; Am Heart Asn; sr mem Biomed Eng Soc; Int Soc Heart Res; Asn Advan Med Instrumentation. *Res:* Cardiovascular mechanics in health and disease; unsteady hemodynamics of intracardiac and arterial flows studied in humans by solid-state micromanometric and velocimetric catheterization; doppler and angiocardiography; computer simulations of ventricular systolic and diastolic function. *Mailing Add:* Dept Biomed Eng & Med Duke Univ & Med Ctr 136 Eng Bldg Durham NC 27706

PASK, JOSEPH ADAM, b Chicago, Ill, Feb 14, 13; m 38; c 2. CERAMIC ENGINEERING, MATERIALS SCIENCE. *Educ:* Univ Ill, BS, 34, PhD(ceramic eng), 41; Univ Washington, Seattle, MS, 35. *Prof Exp:* Ceramic engr, Willamina Clay Prod Co, Ore, 35-36; asst nonmetallic eng, Northwest Exp Sta, US Bur Mines, 35; asst ceramic eng, Univ Ill, 38, instr, 38-41; asst ceramic engr, Electrotech Lab, US Bur Mines, 41, assoc ceramic engr, Northwest Exp Sta, 42-43; res ceramist lamp div, Westinghouse Elec Corp, NJ, 43-46, res engr ceramic sect, 46-48; assoc prof, 48-53, chmn dept mineral technol, 57-61, assoc dean, Col Eng, 69-80, prof, 53-80, EMER PROF CERAMIC ENG, UNIV CALIF, BERKELEY, 80- *Concurrent Pos:* Asst prof ceramic eng & head dept, Col Mines, Univ Wash, Seattle, 41-43; mem clay mineral comt, Nat Res Coun, 54-64, mat adv bd, 64-68, chmn ad hoc comt ceramic processing, Nat Acad Sci, 64-67; sr scientist, Mat & Molecular Res Div, Lawrence Berkeley Lab, Univ Calif, 68-; pres, Pask Res & Eng, 81- *Honors & Awards:* John Jeppson Medal, Am Ceramic Soc, 67, Ross Coffin Purdy Award, 79. *Mem:* Nat Acad Eng; fel AAAS; Am Ceramic Soc (vpres, 53); fel Mineral Soc Am; fel Acad Dent Mat; Int Acad Ceramics. *Res:* Solid state reactions-diffusion studies; ceramic microstructures; mechanical properties of nonmetallic materials; electrochemistry of glass-metal systems; wetting and bonding; stable and metastable phase equilibria. *Mailing Add:* 994 Euclid Ave Berkeley CA 94720

PASKAUSKY, DAVID FRANK, b Waukegan, Ill, Mar 1, 38; m 67; c 3. PHYSICAL OCEANOGRAPHY, PHYSICS. *Educ:* Univ Chicago, SB, 60; DePaul Univ, MS, 64; Tex A&M Univ, PhD(oceanog), 69. *Prof Exp:* Instr physics, Aquinas Col, 62-64; res scientist, Nat Video Corp, 64-65; instr, Milwaukee Sch Eng, 65; asst prof oceanog, Univ Conn, 69-75; sci officer, Off Naval Res, 75-78; head, Oceanog Br, 78-90, CHIEF, SURVEILLANCE SYSTS BR, US COAST GUARD RES & DEVELOP CTR, 90- *Concurrent Pos:* Consult, Sun Oil Co, 70-71, Stone & Webster Eng Corp, 72-73 & Gilbert Assocs, Inc, 72-75; vis assoc prof, Univ Conn, 78-86. *Mem:* Am Geophys Union; Am Meteorol Soc; Am Asn Physics Teachers; fel Explorers Club; Sigma Xi. *Res:* Real-time data to forecast and to develop numerical models to predict ocean surface circulation for search and rescue, oil spill movement and iceberg drift applications. *Mailing Add:* Nine Laurel Dr Groton CT 06340

PASKE, WILLIAM CHARLES, b Tacoma, Wash, Oct 17, 44; m; c 1. MOLECULAR PHYSICS, SCIENCE EDUCATION. *Educ:* Alaska Methodist Univ, Anchorage, BA, 66; Univ Okla, Norman, MS, 70, PhD(physics), 74. *Prof Exp:* Instr physics, Univ Okla, 74-75; asst prof physics, Kans State Univ, 75-77; asst prof phys, Fort Hays State Univ, 77-78; res assoc, Univ Okla, 78-81; res physicist, Seismograph Serv Corp, Tulsa, 81-84; RES PHYSICIST, SPERRY-SUN DRILLING SERV RES & DEVELOP, HOUSTON, 84- *Mem:* Sigma Xi; Am Asn Physics Teachers; Am Phys Soc; Soc Petrol Well Logging Analysts; Soc Petrol Engrs. *Res:* Molecular physics, transition probabilities, quenching processes of polyatomic molecules; physics education, non-traditional approach to lab and lecture format, application of Piaget concepts of learning to undergraduate course work; modeling and redesign of nuclear well logging tools; computer modeling and design of measurement while drilling; oil field research tool design; holder of five patents. *Mailing Add:* 1011 Margate Ct Pearland TX 77584

PASKIEVICI, WLADIMIR, b Bucharest, Rumania, Mar 7, 30; Can citizen; m 57; c 3. NUCLEAR PHYSICS, NUCLEAR ENGINEERING. *Educ:* Univ Strasbourg, LSc, 55, PhD(nuclear physics), 57. *Prof Exp:* Lectr atomic physics, Polytech Sch, Montreal, 58-59, asst prof atomic & nuclear physics, 59-63, assoc prof mod physics, 63-69, assoc prof nuclear eng, 69-77; prof, 77-90, EMER PROF NUCLEAR ENG, UNIV MONTREAL, 91-,; CONSULT, 91- *Concurrent Pos:* Nat Res Coun Can fel, 58-59; head, Nuclear Eng Div, Univ Montreal, 67-70, Nuclear Eng Dept, 81-82, dean res, 82-85, dean res grad studies, 85-90; consult, Atomic Energy Control Bd, 76-83, Hydro-Que, 75-80, Ont-Hydro Energy Mine & Resources, Environ Can. *Mem:* Am Nuclear Soc; Can Nuclear Asn; AAAS; French Soc Nuclear Eng; Can Res Mgt Asn. *Res:* Reactor physics; reactor control; reactor safety licensing; comparative risks; author of 10 articles and journals and 46 technical reports. *Mailing Add:* 4874 Cote des Neiges Montreal PQ H3V 1H4 Can

PASKIN, ARTHUR, b Brooklyn, NY, Feb 15, 24; m 53; c 3. PHYSICS. *Educ:* SDak Sch Mines & Technol, BS, 48; Iowa State Univ, PhD(physics), 53. *Prof Exp:* Asst physics, Inst Atomic Res, Iowa State Univ, 48-53; sr physicist, Sylvania Elec Prod Inc, 53-55; solid state physicist, US Army Mat Res Lab, 55-63; physicist, Brookhaven Nat Lab, 63-68; PROF PHYSICS, QUEENS COL NY, 68- *Concurrent Pos:* Secy Army fel, 60-61. *Honors & Awards:* US Dept Energy Award, Outstanding Res in Metall & Ceramics. *Mem:* Fel Am Phys Soc; Am Soc Metals; Sigma Xi; Soc Automotive Engrs. *Res:* Theoretical and solid state physics; thermodynamics of magnetic systems; x-ray diffraction diffuse scattering, radiation damage; superconductivity; liquid state of metals; computer simulations and shock waves in solids, fracture and accident reconstruction; forensic computer graphics. *Mailing Add:* Dept of Physics Queens Col Flushing NY 11367

PASKINS-HURLBURT, ANDREA JEANNE, b Southampton, Eng, Apr 26, 43; US citizen; m 68. TUMOR ANGIOGENESIS, COLLATERAL VESSEL DEVELOPMENT. *Educ:* McGill Univ, BSc, 65, MSc, 70, PhD(exp surg), 74. *Prof Exp:* Res asst, Montreal Gen Hosp, 65-66; technician gastrointestinal res, McGill Univ, 67-74, res assoc, 74-77; RES ASSOC, DEPT RADIOL, HARVARD UNIV, 77- *Mem:* AAAS. *Res:* Factors which promote the development of a collateral arterial supply to the heart and kidney when the main artery has been occluded; relationship between tumor growth, tumor perfusion and the reactivity of the tumor vasculature; national medical research studies on interrelationships between circulatory and metabolic adaptations to ischemic hindlimb musculature. *Mailing Add:* 12 Kenilworth St Newton MA 02158

PASKO, THOMAS JOSEPH, JR, b Reading, Pa, Mar 1, 37; m 61; c 2. CONCRETE-PORTLAND CEMENT TECHNOLOGY, HIGHWAY PAVEMENTS & STRUCTURES. *Educ:* Pa State Univ, BS, 59, MS, 61. *Prof Exp:* Soils engr, Pa Dept Transp, 60; mat engr, Mat Testing Inc, State College-Pa, 60-61; hwy res engr, Fed Hwy Admin-McLean, Va, 61-76, div chief, Paving & Struct Mat Div, 76-82, Mat Technol & Chem Div, 82-85, Pavements Div, 85-87, DIR OFF RES, FED HWY ADMIN, MCLEAN, VA, 87- *Concurrent Pos:* Liaison, Strategic Hwy Res Prog, 87-92; US Rep, Flexible Pavements Comt, Permanent Int Asn Road Congresses, 89-; adv, Pa Transp Inst, 90-, Advan Cement Based Mat Prog, Northwestern Univ, 90-; bd dirs, Am Concrete Inst, 84-87, mem, numerous comts. *Honors & Awards:* Kennedy Award, Am Concrete Inst, 89; Super Achievement Res Awards, Fed Hwy Admin, 81, 84, 86. *Mem:* Fel Am Concrete Inst; fel Am Soc Civil Engrs; Am Soc Testing & Mat; Nat Soc Prof Engrs; Transp Res Bd; Nat Asn Corrosion Engrs. *Res:* Development of highway technology; low energy binders; alternatives to asphalt; use of wastes; protection for steel; non-corroding deicers; improved pavement markings; non-destructive testing of bridges & pavements; protecting bridge decks; prestressed pavements. *Mailing Add:* 3901 Arnheim St Annandale VA 22003

PASKUSZ, GERHARD F, b Vienna, Austria, Jan 21, 22; US citizen; m 43; c 1. ELECTRICAL ENGINEERING. *Educ:* Univ Calif, Los Angeles, BS, 49, PhD(eng), 61. *Prof Exp:* Teaching asst, Univ Calif, Los Angeles, 50-52, assoc eng, 52-60, res specialist, 60-61; assoc prof elec eng, 61-68, assoc, Dean Cullen Col Eng, 68-75, MEM GRAD FAC, 61-, PROF ELEC ENG, UNIV HOUSTON 68- *Concurrent Pos:* Fac consult & lectr, Nat Sci Found Inst Use of Comput Elec Eng, 62-64, Nat Sci Found lectr, 65; vis assoc prof, Baylor Col Med, 65-67; consult, Houston Speech & Hearing Ctr, 61-62, SIE-Dresser Electronics, 63 & Inst Bus Mach Corp, Calif, 64; prog dir, Minority Eng, 73-

Mem: Am Soc Eng Educ; Asn Comput Mach; Inst Elec & Electronics Engrs; Sigma Xi. *Res:* Bioengineering; circuits and systems; computer applications; linear circuit analysis; teaching methods. *Mailing Add:* Dept Elec Eng Univ Houston Bldg D Houston TX 77004-4793

PASLAY, PAUL R(OBERT), mechanics, for more information see previous edition

PASLEY, JAMES NEVILLE, b Jefferson City, Mo, Sept 19, 39; m 66; c 2. GASTROINTESTINAL & REPRODUCTIVE ENDOCRINOLOGY, BIOLOGICAL RHYTHMS. *Educ:* Westminster Col, Mo, AB, 61; Univ Mo, MA, 65; Ore State Univ, PhD(physiol), 69. *Prof Exp:* From asst prof to assoc prof, 70-83, PROF PHYSIOL, MED CTR, UNIV ARK, LITTLE ROCK, 83- *Concurrent Pos:* NIH fel pop, endocrines & behav, Albert Einstein Med Ctr, Philadelphia, 69-70. *Mem:* Am Gastroenterol Asn; Am Physiol Soc; NY Acad Sci; Soc Exp Biol & Med; Endocrine Soc; Int Soc Chronobiology. *Res:* Effects of environmental factors on endocrines, gastrointestinal peptides and sexually transmitted diseases. *Mailing Add:* Dept of Physiol Univ of Ark Med Ctr Little Rock AR 72205

PASQUA, PIETRO F, b Englewood, Colo, May 30, 22; m 45; c 4. NUCLEAR ENGINEERING. *Educ:* Univ Colo, BS, 44; Northwestern Univ, MS, 47, PhD(mech eng), 52. *Prof Exp:* Instr, Univ Colo, 44-45; from asst to instr mech eng, Northwestern Univ, 45-52; assoc prof, 52-56, prof & actg head dept, 56-57, PROF NUCLEAR ENG & HEAD DEPT, UNIV TENN, KNOXVILLE, 57- *Concurrent Pos:* Consult reactor eng exp div, Union Carbide Nuclear Co, Tenn, 54- *Mem:* Am Nuclear Soc; Am Soc Eng Educ. *Res:* Heat transfer. *Mailing Add:* 319 W Hunt Rd Alcoa TN 37701

PASS, BOBBY CLIFTON, b Cleveland, Ala, Nov 4, 31; m 53; c 1. INSECT PEST MANAGEMENT. *Educ:* Auburn Univ, BS, 52, MS, 60; Clemson Univ, PhD(entom), 62. *Prof Exp:* Res asst entom, Auburn Univ, 58-60 & Clemson Univ, 60-62; from asst prof to assoc prof, 62-71, PROF ENTOM, UNIV KY, 71-, CHMN DEPT, 68- *Concurrent Pos:* Pres, Am Registry Prof Entomologists, 79; pres, Entom Soc Am, 87. *Mem:* Fel Entom Soc Am; Can Entom Soc; AAAS. *Res:* Biology, ecology and control of insect pest of forage crops. *Mailing Add:* Dept of Entom Univ of Ky Lexington KY 40546-0091

PASS, ROBERT FLOYD, b Birmingham, Ala, Sept 2, 47; m 71; c 2. INFECTIOUS DISEASES, VIROLOGY. *Educ:* Univ Ala, BS, 69; Univ Ala-Birmingham, MD, 73. *Prof Exp:* Resident & intern pediat & infectious dis, Sch Med, Stanford Univ, 73-76; fel infectious dis, 76-79, asst prof pediat & infectious dis, 79-82, assoc physicist GCRC, 80-81, ASST PROF MICROBIOL, UNIV ALA, BIRMINGHAM, 81-, ASSOC PROF PEDIAT & INFECTIOUS DIS, 82- *Mem:* Am Soc Microbiol; Transplantation Soc; Soc Pediat Res; Pediat Infectious Dis Soc; Am Soc Virol; Am Fedn Clin Res. *Res:* Epidemiology and host immune response to human cytomegalovirus infection, particularly congenital and perinatal infections. *Mailing Add:* Tower Suite 752 Children's Hosp 1600 Seventh Ave Birmingham AL 35294

PASSANANTI, GAETANO THOMAS, b Wilkes-Barre, Pa, Sept 23, 25; m 50; c 2. ENZYMOLOGY. *Educ:* Pa State Univ, BS, 49, MS, 53, PhD(biochem & org chem), 57, Am Bd Forensic Toxicol, dipl. *Prof Exp:* Instr biochem, Univ Mich, 56-60; biochemist, Dept Path, Hurley Hosp, Flint, Mich, 60-63; asst prof biochem, Ohio State Univ, 63-67; biochemist, Inst Path, Harrisburg Hosp, 67-70; res asst biochem, 51-56, ASST PROF PHARMACOL, SCH MED, PA STATE UNIV, 70- *Mem:* Am Soc Pharmacol & Exp Ther; fel Am Col Clin Pharmacol. *Res:* Biochemical pharmacology and drug metabolism. *Mailing Add:* Dept of Pharmacol Hershey Med Ctr Pa State Univ Hershey PA 17033

PASSANITI, ANTONINO, TUMOR METASTASIS, SELF-SURFACE CARBOHYDRATES. *Educ:* Univ Va, PhD(biochem), 82. *Prof Exp:* Fel, Sch Med, Johns Hopkins Univ, 85-89; SR STAFF, NIH, 89- *Res:* Cell membranes. *Mailing Add:* NIH 4940 Eastern Ave Baltimore MD 21224

PASSANO, LEONARD MAGRUDER, b Staten Island, NY, Dec 16, 24; m 81; c 1. ZOOLOGY. *Educ:* Harvard Univ, AB, 48; Yale Univ, PhD(zool), 52. *Prof Exp:* NSF fel, Cambridge Univ, 52-53; instr zool, Univ Wash, 53-55; from instr to asst prof, Yale Univ, 55-64; assoc prof, 64-69, PROF ZOOL, UNIV WIS-MADISON, 69- *Concurrent Pos:* Instr, Woods Hole Marine Biol Lab, 54-55; Fulbright res fel, Univ West Indies, 60-61; Guggenheim fel, 63-64; vis prof, French Nat Univ Syst, Univ Lyons, 70. *Mem:* Am Soc Zool; Brit Soc Exp Biol; Int Soc Neuroethology. *Res:* Coelenterate neurophysiology; behavioral physiology; crustacean endocrinology; invertebrate physiology. *Mailing Add:* Dept Zool Univ Wis 426 Birge Hall Madison WI 53706

PASSCHIER, ARIE ANTON, b The Hague, Netherlands, Mar 21, 40; US citizen; m 62; c 2. PHYSICAL CHEMISTRY, ANALYTICAL CHEMISTRY. *Educ:* Calif State Col, Long Beach, BS, 61, MA, 63; Univ Wash, PhD(phys chem), 68. *Prof Exp:* Mem tech staff, Rocketdyne Div, 68-69, mem tech staff, 69-86, MGR, AUTONETICS DIV, ROCKWELL INTERNATIONAL CORP, 86- *Mem:* AAAS; Am Chem Soc; Am Vacuum Soc. *Res:* Analysis of gas phase contaminants in inertial instruments; high vacuum systems; mass spectrometry; determination of thermodynamic properties of materials; gas chromatography and gas chromatography/mass spectrometry. *Mailing Add:* 2459 Shady Forest Lane Orange CA 92667

PASSELL, LAURENCE, b Cleveland, Ohio, Mar 23, 25; m 46; c 2. SOLID STATE PHYSICS. *Educ:* US Merchant Marine Acad, BS, 45; Harvard Univ, AB, 50; Univ Calif, MA, 52, PhD(physics), 55. *Prof Exp:* Port engr, US War Shipping Admin, DC, 45-46; asst, Univ Calif, 50-54, sr scientist, Lawrence Radiation Lab, 55-61; scientist, Danish Atomic Energy Comn Res Estab, Riso, 61-63; SCIENTIST, BROOKHAVEN NAT LAB, 63- *Mem:* Fel Am Phys Soc. *Res:* Neutron physics; low temperature physics; neutron scattering study of condensed phases. *Mailing Add:* Bldg 510 Brookhaven Nat Lab Upton NY 11973

PASSELL, THOMAS OLIVER, b Chicago, Ill, Nov 24, 29; m 52; c 6. NUCLEAR CHEMISTRY. *Educ:* Okla State Univ, BS, 51; Univ Calif, PhD(chem), 54. *Prof Exp:* Asst chem, Univ Calif, 51-52, Radiation Lab, 52-54; nuclear res chemist, Atomic Energy Div, Phillips Petrol Co, 54-55; physicist, Stanford Res Inst, 55-69 & Physics Int Co, 69-71; staff scientist, Lockheed Palo Alto Res Labs, 71-75; PROJ MGR, ELEC POWER RES INST, 75- *Mem:* Am Chem Soc; Am Phys Soc; Am Nuclear Soc; Am Geophys Union; fel Am Inst Chemists. *Res:* Beta ray spectroscopy; neutron cross sections; levels in light nuclei; nuclear power reactor technology; corrosion technology; controlled thermonuclear research; plasma physics; nuclear detonation effects; x-ray technology; deuterated metals. *Mailing Add:* 3825 Louis Rd Palo Alto CA 94303

PASSENHEIM, BURR CHARLES, b St Louis, Mo, Dec 15, 41; div; c 2. EXPERIMENTAL PHYSICS. *Educ:* Univ Calif, Berkeley, AB, 63; Univ Calif, Riverside, MA, 65, PhD(physics), 69. *Prof Exp:* Technician physics, Gen Atomic, 58-63; res asst, Univ Calif, Riverside, 63-68; physicist, Gulf Energy & Environ Systs, 68-71; physicist, IRT Corp, 71-78; group leader, Mission Res Corp, 78-83; SR PRIN SCIENTIST, JAYCOR, 83- *Concurrent Pos:* Ed, J Radiation Effects, Res & Eng, 86, 87; session chmn, Inst Elec & Electronic Engrs/NSRE conf, 88, 91. *Mem:* Am Phys Soc; Inst Elec & Electronic Engrs. *Res:* Effects of nuclear, electromagnetic, thermal and directed energy on electronics, optical systems, material and structures. *Mailing Add:* JAYCOR PO Box 85154 San Diego CA 92138-9259

PASSERELLO, CHRIS EDWARD, b New Jersey, Apr 12, 44; m 65; c 2. ENGINEERING, MECHANICS. *Educ:* Univ Cincinnati, BSASE, 67, MSASE, 68, PhD(mech), 72. *Prof Exp:* Assoc prof aerospace eng & mech, Univ Cincinnati, 76-78; assoc prof eng mech, 78-86, PROF ENG MECH, MICH TECHNOL UNIV, 86- *Mem:* Am Soc Mech Engrs; Soc Eng Mech. *Res:* Dynamics and vibrations; structures; finite elements. *Mailing Add:* Dept of Mech Eng & Eng Mech Mich Technol Univ Houghton MI 49931

PASSEY, RICHARD BOYD, b Mesa, Ariz, Aug 5, 37; m 60; c 6. CLINICAL CHEMISTRY, BIOCHEMISTRY. *Educ:* Utah State Univ, BS, 65; Colo State Univ, PhD(biochem), 69. *Prof Exp:* Asst prof pathol, Sch Med, Univ Tex Med Br, 69-76; asst prof, Sch Med, Univ Ky, 76-77; PROF PATHOL, UNIV OKLA HEALTH SCI CTR, 77- *Concurrent Pos:* Dir clin chem & sci dir clin labs, State Okla Teaching Hosps, Oklahoma City, Okla, 77- *Mem:* Am Asn Clin Chem; AAAS; Am Soc Clin Pathologists. *Res:* Application of clinical chemistry methods to problems in clinical medicine; diagnostic usefulness or urinary LDH isoenzymes in predicting pyelonephritis; evaluation of clinical laboratory instrumentation; fetal monitoring. *Mailing Add:* Dept Pathol Univ Okla Health Sci Ctr Box 26307 Oklahoma City OK 73190

PASSINO, DORA R MAY, b Portland, Ore, Mar 22, 40; div; c 1. FISHERY BIOLOGY, AQUATIC TOXICOLOGY. *Educ:* Portland State Univ, BA, 66; Univ Wash, MS, 69, PhD(fisheries), 73. *Prof Exp:* Res & teaching fel, Univ Wash, 67-73; instr, Green River Community Col, Wash, 73; res assoc oceanog, Nat Acad Sci, 73; PROJ LEADER FISHERIES, US FISH & WILDLIFE SERV, 73- *Concurrent Pos:* Mem, pollution abatement comt, Am Fisheries Soc, 71, resolutions comt, 77 & publ awards comt, 81; mem, outer continental shelf task force, US Fish & Wildlife Serv, 73 & Lacy act task force, 74; adj prof, Univ Mich, 80-82; assoc ed, Bulletin Environ Contamination Toxicol, 84-; expert reviewer, Am Fish Soc, 78, Int Joint Comt Can, 79, Mich Dept Natural Resource, 82, US Environ Protection Agency, 82 & 85, Nat Res Coun Can, 84 & Ecol Serv Fish Wildlife Serv, 87; invited lectr, Gulf Sea Frontier, Nat Lab, Munich, WGer, 86-; invited speaker, Ont Ministry Environ, 86, Oakridge Nat Lab & Univ Tenn, 88. *Honors & Awards:* James W Moffett Publication Award, 80. *Mem:* Am Chem Soc; Am Fisheries Soc; Soc Environ Toxicol Chem; Am Inst Fisheries Res Biologists; Int Asn Great Lakes Res. *Res:* Hazard assessment of environmental contaminants to fishery resource; quantitative structure-activity relationships of environmental contaminants. *Mailing Add:* Nat Fisheries Res Ctr 1451 Green Rd Ann Arbor MI 48105

PASSINO, NICHOLAS ALFRED, b Ft Wayne, Ind, May 24, 40; m 64; c 1. MOLECULAR SPECTROSCOPY, INFRARED PHYSICS. *Educ:* St Procopius Col, BS, 62; Ariz State Univ, MS, 65, PhD(physics), 68. *Prof Exp:* Fac assoc physics, Ariz State Univ, 62-68; res physicist, Brown Eng Co, Inc, 68-69, prin res physicist, 69-71, mgr optics technol br, 71-74, DEP MGR, OPTICS DEPT, TELEDYNE BROWN ENG CO, 74- *Mem:* AAAS; Optical Soc Am. *Res:* Infrared optics and semiconductor detectors; nuclear effects on optical materials and semiconductors; infrared instrumentation and calibration; infrared absorption in gases, vibrational and rotational analysis; military optical systems applications. *Mailing Add:* 414 Owens Dr SE Huntsville AL 35801

PASSMAN, DONALD STEVEN, b New York, NY, Mar 28, 40; m 63; c 2. ALGEBRA. *Educ:* Polytech Inst Brooklyn, BS, 60; Harvard Univ, AM, 61, PhD(math), 64. *Prof Exp:* Asst prof math, Univ Calif, Los Angeles, 64-66 & Yale Univ, 66-69; assoc prof, 69-71, PROF MATH, UNIV WIS-MADISON, 72- *Concurrent Pos:* Mathematician, Inst Defense Analysis, Princeton, NJ, 69-70. *Mem:* Am Math Soc. *Res:* Infinite group rings; crossed products and Galois theory. *Mailing Add:* Dept Math Univ Wis 480 Lincoln Dr Madison WI 53706

PASSMAN, FREDERICK JAY, b Philadelphia, Pa, Aug 22, 48; m 68; c 1. ENVIRONMENTAL MICROBIOLOGY. *Educ:* Ind Univ, Bloomington, AB, 70; Univ NH, PhD(microbiol), 77. *Prof Exp:* Microbiologist, Energy Resources Co, Inc, 77-78, dir microbiol lab, 78-80; vpres prod serv, Erco Petrol Serv Inc, 80-81; consult, microbiologist, KVM Eng, Inc, 81-86; BUS MGR, BIOCIDES, ANGUS CHEM CO, 86- *Mem:* Am Soc Microbiol; Am Soc Limnol & Oceanog; Sigma Xi; Oceanic Soc; Soc Indust Microbiol; Soc Lubrication Engrs; Nat Asn Corrosion Engrs. *Res:* Dynamics of marine microbial communities of surface films and sediments; elemental composition of marine bacteria; microbiology of municipal and industrial waste recovery, particularly composting; microbiology of metalworking fluids; biodeterioration of fuels. *Mailing Add:* 1219 W North Shore 2W Chicago IL 60626

PASSMAN, SIDNEY, b Brooklyn, NY, Aug 5, 27; m 51; c 2. PHYSICS, RESEARCH ADMINISTRATION. *Educ:* Columbia Univ, AB, 48, AM, 49, PhD, 52. *Prof Exp:* Asst high energy physics, Nevis Cyclotron Lab, Columbia Univ, 50-52; res physicist guided missile syst analysis, Hughes Aircraft Co, 52-55; group leader infrared physics, Rand Corp, 55-63; phys sci off, Sci & Technol Bur, US Arms Control & Disarmament Agency, 63-70; head sci policy res sect, NSF, Wash, DC, 70-73; dir, Div Sci Res & Higher Educ, UNESCO, Paris, 73-81; special asst planning & policy analysis, Nat Sci Found, Washington, DC, 81-84; PRES, ORGN INT SCI, 84-; CONSULT, INT SCI & TECH. *Concurrent Pos:* Optical Soc Am rep, Nat Res Coun, 66-72; mem nat comt, Infrared Info Systs; ed, J Infrared Physics; exec secy, NSB Comt Planning, Policy & Int Sci, secy, Gen Int Conf Strategies & Policies Informatics. *Mem:* Fel AAAS; fel Optical Soc Am; Sigma Xi; Diplomatic & Consular Officers Retired. *Res:* Scientific research and higher education; science policy; infrared physics and technology; arms control and weapons system research; information retrieval; informatics. *Mailing Add:* One Carderock Ct Bethesda MD 20817

PASSMAN, STEPHEN LEE, b Suffolk, Va, Sept 3, 42; m 65; c 4. MECHANICS. *Educ:* Ga Inst Technol, BSEM, 64, MSEM, 66, PhD(eng mech), 68. *Prof Exp:* Instr, US Naval Acad, 68-70; res assoc & fel, Johns Hopkins Univ, 70-71; from asst prof to assoc prof eng sci & mech, Ga Inst Technol, 71-78; TECH STAFF, SANDIA NAT LABS, 78-; ADJ PROF, UNIV PITTSBURGH, 89- *Concurrent Pos:* Asst prof lectr, George Washington Univ, 69-70; vis scientist, Pittsburgh Energy Technol Ctr, 88-90. *Honors & Awards:* Monie A Ferst Res Award, 68. *Mem:* Am Acad Mech; Sigma Xi; Soc Rheol; Soc Natural Philos; Am Phys Soc. *Res:* Continuum mechanics; multiphase flows, granular media. *Mailing Add:* Div 6212 Sandia Nat Labs Albuquerque NM 87185

PASSMORE, EDMUND M, b Somerville, Mass, July 28, 31; m 54; c 4. METALLURGY, CERAMICS. *Educ:* Mass Inst Tech, SB, 53, SM, 54, ScD(metall), 57. *Prof Exp:* Mem res & develop staff, Avco Space Systs Div, 57-59, ManLabs Inc, 59-63 & Avco Space Systs Div, 63-66; mem res & develop staff, GTE Lighting Prod, 67-90; RETIRED. *Honors & Awards:* Ross-Coffin-Purdy Award, Am Ceramic Soc, 67. *Mem:* Am Inst Mining, Metall & Petrol Engrs; Am Ceramic Soc. *Res:* Publications and patents covering incandescent lamp and evaporation source technology and chemistry; also research and development of materials and components for lighting and other applications; lamps and lighting products; high intensity metal haloid lamps. *Mailing Add:* 41 Craft Rd Glouchester MA 01930

PASSMORE, HOWARD CLINTON, b Drexel Hill, Pa, Sept 12, 42; m 64; c 2. IMMUNOGENETICS. *Educ:* Franklin & Marshall Col, AB, 64; Rutgers Univ, MS, 66; Univ Mich, PhD(immunogenetics), 70. *Prof Exp:* NSF fel, Univ Calif, San Francisco, 70-71; asst prof, 71-77, assoc prof, 77-88, PROF BIOL SCI, RUTGERS UNIV, NEW BRUNSWICK, 88- *Mem:* Am Asn Immunologists; AAAS; Genetics Soc Am; Am Soc Human Genetics; Am Genetic Asn. *Res:* Immunogenetics molecular mechanism of meiotic recombination. *Mailing Add:* Dept Biol Sci Rutgers Univ Piscataway NJ 08855

PASSMORE, JACK, b North Devon, UK, Nov 16, 40; Can citizen; m 62; c 3. INORGANIC CHEMISTRY, FLUORINE CHEMISTRY. *Educ:* Bristol Univ, BS, 63, DSc, 84, Univ BC, PhD(inorg chem), 67. *Prof Exp:* Fel, McMaster Univ, 68-69, Chem Inst Can, 86; from asst prof to assoc prof, 69-78, PROF CHEM, UNIV NB, FREDERICTON, 78- *Mem:* Am Chem Soc; Royal Soc Chem; Chem Inst Can. *Res:* Preparation, characterization and chemistry of simple compounds of interest in terms of current theories of bonding and stereochemistry; sulfer-nitrogen chemistry. *Mailing Add:* Dept Chem Univ New Brunswick Bag Serv 45222 Fredericton NB E3B 6E2 Can

PASSMORE, JOHN CHARLES, SHOCK, HYPERTENSION. *Educ:* Univ NDak, PhD(physiol), 70. *Prof Exp:* ASSOC PROF PHYSIOL, SCH MED, UNIV LOUISVILLE, 70- *Res:* Renal blood flow. *Mailing Add:* Dept Physiol Biophysics Univ Louisville Health Sci Ctr Louisville KY 40292

PASSNER, ALBERT, b Bronx, NY, Aug 30, 38; m 62; c 3. POSITRON TRAP, MAGNET DESIGN. *Educ:* City Col NY, BS, 60; NY Univ, MS, 66. *Prof Exp:* Engr, Astro Div, RCA, 61-63; staff, Princeton-Penn Accelerator, 63-69; MEM TECH STAFF, AT&T BELL LABS, 69- *Concurrent Pos:* Mem, Adv Comt, Queensborough Community Col, 87. *Res:* Producing and trapping the first laboratory antimatter plasma of positrons; making the first semiconductor optical bistable device; high pulsed magnetic fields for research into two dimensional systems. *Mailing Add:* AT&T Bell Labs 600 Mountain Ave Murray Hill NJ 07974

PASSOJA, DANN E, b Chicago, Ill, Jan 28, 41; m 63; c 1. FRACTURE, ANALYTIC OPTICS. *Educ:* Purdue Univ, BS, 63; Rensselaer Polytech Inst, PhD(mat sci), 68. *Prof Exp:* Res fel, Univ Manchester, UK, 68-69; dir technol, EBTEC Corp, Bedford, Mass, 69-70; staff scientist, Kennecott Copper, Cleveland, Ohio, 70-71; proj scientist, Union Carbide, Tarrytown, NY, 71-73, scientist, 73-78, sr scientist, 78-83, sr develop assoc, 83-85; RETIRED. *Concurrent Pos:* Consult, Fusion Labs, Chicopee, Mass, 64-68, EBTEC Corp, Agawan, Mass, ARCO Chem, Grenville, NC, Battelle, Res Triangle, NC, 82-; adj prof mat sci, Pa State Univ, 85- *Mem:* Mat Res Soc; Am Phys Soc. *Res:* Fracture and general physics; optics; physics. *Mailing Add:* 410 E 73rd St New York NY 10021

PASSONNEAU, JANET VIVIAN, biochemistry, for more information see previous edition

PASSOW, ELI (AARON), b Bronx, NY, Sept 4, 39; m 71; c 3. APPROXIMATION THEORY. *Educ:* Mass Inst Technol, BS, 62; Yeshiva Univ, MA, 65, PhD(math), 66. *Prof Exp:* Math analyst, Airborne Instruments Lab, NY, 66-67; instr math, Yeshiva Univ, 67-68; lectr, Bar-Ilan Univ, Israel, 68-70; from asst prof to assoc prof, 70-80, PROF MATH, TEMPLE UNIV, 80- *Concurrent Pos:* Vis assoc prof, Technion, Israel, 79-80. *Mem:* Am Asn Univ Professors; Math Asn Am. *Res:* Approximation theory; approximation and interpolation with constraints. *Mailing Add:* Dept Math 30 N Highland Ave Bala Cynwyd PA 19004

PASSWATER, RICHARD ALBERT, b Wilmington, Del, Oct 13, 37; m 64; c 2. BIOCHEMISTRY, NUTRITION. *Educ:* Univ Del, BS, 59: Bernadean Univ, Las Vegas, PhD, 76. *Prof Exp:* Lab supvr, Allied Chem Corp, 59-64; prod mgr, Baxter-Travenol Labs, 64-77; res dir, Lifesci Labs, 78-79; RES DIR, SOLGAR NUTRIT RES CTR, SOLGAR CO, INC, 79- *Honors & Awards:* Indust Achievement Award, 89. *Mem:* Am Chem Soc; NY Acad Sci; Am Aging Asn; AAAS; Am Geriat Soc; Am Inst Chemists. *Res:* Free radical pathology; role of antioxidant nutrients; trace-elements and amino acids in cancer; isolation of unidentified growth factors. *Mailing Add:* 11017 Manklin Ct Berlin MD 21811-9342

PAST, WALLACE LYLE, b Great Falls, Mont, Feb 15, 24; m 49; c 4. PATHOLOGY. *Educ:* Univ Pa, AB, 44; NY Univ, MD, 48. *Prof Exp:* Resident path, Henry Ford Hosp, Detroit, Mich, 52-55; assoc pathologist, Wilson Mem Hosp, Johnson City, NY, 55-56; PROF PATH, SCH MED, UNIV LOUISVILLE, 56- *Concurrent Pos:* Pathologist, St Edward Hosp, New Albany, Ind, 56-61; consult, Vet Admin Hosp, 56-61. *Mem:* Col Am Path. *Res:* Formation of bone mineral. *Mailing Add:* Floyd Mem Hosp New Albany IN 47150

PASTAN, IRA HARRY, b Winthrop, Mass, June 1, 31; m 53; c 3. MOLECULAR BIOLOGY, CELL BIOLOGY. *Educ:* Tufts Univ, BS, 53, MD, 57. *Prof Exp:* Intern, Grace-New Haven Hosp, Sch Med, Yale Univ, 57-58, asst resident med, 58-59; clin assoc, Nat Inst Arthritis & Metab Dis, 59-61, sr investr endocrine biochem, NIH, 62-69, sr investr, Nat Inst Arthritis & Metab Dis, 62-69, chief, Sect Molecular Biol, Endocrinol Br, Nat Cancer Inst, 69-70, CHIEF, LAB MOLECULAR BIOL, NAT CANCER INST, 70- *Concurrent Pos:* Fel, Lab Cellular Physiol, NIH, 61-62; Burroughs Wellcome vis prof, 78-79; assoc ed, Biochem Int, 80-87; ed, Exp Cell Res, 80-88; spec vis prof, Stanford Med Sci Training Prog, 80; chmn, Gordon Conf Cell Adhesion, Recognition & Movement, 81. *Honors & Awards:* Van Meter Prize, Am Thyroid Asn, 71; Boxer Lectr, Rutgers Med Sch, 72; G Burroughs Mider Lectr, NIH, 73; Windsor C Cutting Lectr, Stanford Univ, 80; Judith Segal Mem Lectr, Hebrew Univ, 80; Stanley Wright Mem Lectr, Western Soc Pediat Res, 82; Rennebohm Lectr, Univ Wis, 88; Pierce Immunotoxin Award, 88. *Mem:* Nat Acad Sci; AAAS; Am Soc Microbiol; Am Soc Biol Chemists; Am Soc Clin Invest; Am Thyroid Asn; Am Asn Phys. *Res:* Regulation of the expression of genetic information in animal cells; development and new agents for cancer treatment. *Mailing Add:* Nat Cancer Inst Bldg 37 Rm 4E 16 Bethesda MD 20892

PASTEELNICK, LOUIS ANDRE, b Newark, NJ, June 22, 29; m 63; c 2. QUALITY ASSURANCE, STATISTICS. *Educ:* NJ Inst Technol, BS, 51; Rutgers Univ, MS, 58. *Prof Exp:* Chem engr drug mfg, Merck & Co, 51-53, chem warfare, US Army, 53-55, plastics mfg, M W Kellogg, Minn Mining & Mfg Co, 55-58; quality control statistician, container mfg, Am Can Co, 58-65; MGR QUALITY ASSURANCE, FOOD DRUG & COSMETIC MFG, WARNER CHILCOTT DIV, WARNER-LAMBERT CO, 65- *Honors & Awards:* E R Ott Award, Am Soc Qual Control, 75. *Mem:* Fel Am Soc Qual Control; Am Inst Chem Eng; Am Chem Soc. *Res:* Pharmaceutical applications of quality control statistics. *Mailing Add:* Warner Lambert Co 182 Tabor Rd Morris Plains NJ 07950

PASTELL, DANIEL L(OUIS), b Seattle, Wash, May 22, 22; wid; c 2. CHEMICAL ENGINEERING. *Educ:* Univ Wash, Seattle, BS, 44, BS & MS, 47. *Prof Exp:* Group leader combustion res, E I Du Pont de Nemours & Co, Inc, 50-53, head res div, 53-54, eng div, 54-60, tech mgr, Western Region, 60-82; RETIRED. *Concurrent Pos:* Chmn motor combustion chamber deposits group, Coord Res Coun USA, 42-43. *Honors & Awards:* Horning Mem Award, Soc Automotive Engrs, 50. *Mem:* Am Chem Soc; Soc Automotive Engrs; Inst Elec & Electronics Engrs; Air Pollution Control Asn. *Res:* Combustion in reciprocating and gas turbine engines. *Mailing Add:* 2904 Via de LaGuerra Palos Verdes Peninsula CA 90274

PASTER, DONALD L(EE), hydrodynamics, acoustics, for more information see previous edition

PASTERCZYK, WILLIAM ROBERT, b Chicago, Ill, Feb 4, 17; m 48; c 3. ANALYTICAL CHEMISTRY, BIOCHEMISTRY. *Educ:* DePaul Univ, BS, 40, MS, 43; Loyola Univ Ill, PhD, 54. *Prof Exp:* From instr to prof chem, DePaul Univ, 43-87; RETIRED. *Mem:* Am Chem Soc; Am Asn Univ Professors; Sigma Xi; Soc Appl Spectros. *Mailing Add:* 8219 N Karlov Ave Skokie IL 60076

PASTERNACK, BERNARD SAMUEL, b Brooklyn, NY, June 18, 32; m 67. EPIDEMIOLOGY, BIOSTATISTICS. *Educ:* Brooklyn Col, BA, 54; NC State Univ, MS, 56, PhD, 59. *Prof Exp:* Asst prof biostatist, Univ NC, 59-61; from asst prof to assoc prof, 61-74, PROF ENVIRON MED, SCH MED, NY UNIV, 74- *Concurrent Pos:* Mem, Epidemiol & Dis Control Study Sect, Nat Inst Health, 80-83. *Mem:* Fel Am Statist Asn; Biomet Soc; fel Am Pub Health Asn; Soc Epidemiol Res; fel Am Col Epidemiol; Int Statist Inst. *Res:* Biomedical statistics; cancer epidemiology. *Mailing Add:* Inst Environ Med NY Univ Med Ctr 550 First Ave New York NY 10016

PASTERNACK, ROBERT FRANCIS, b New York, NY, Sept 20, 36; div; c 2. PHYSICAL INORGANIC CHEMISTRY, BIOINORGANIC CHEMISTRY. *Educ:* Cornell Univ, BA, 57, PhD(phys inorg chem), 62. *Prof Exp:* Res assoc, Univ Ill, 62-63; from asst prof to prof chem, Ithaca Col, 63-76, Charles A Dana Prof, 76-87; EDMUND ALLEN PROF, SWARTHMORE COL, 84- *Concurrent Pos:* Consult, Nat Bur Standards, 63-64 & NY State Dept Educ, 66-67; NSF col teacher res partic grant, 65-68; Petrol Res Fund grant, 67-74; NSF sci fac fel, Univ Rome, 69-70; Sci Course Improv Prog grant, NSF, 69-72; NIH grant, 71-; Res Corp grants, 74-75 & 78-; NSF sci manpower develop grant, Univ London, 77-78; Danforth assoc, 78-84. *Mem:* AAAS; Am Chem Soc; NY Acad Sci; Sigma Xi. *Res:* Rapid reactions by relaxation spectrometry; porphyrin chemistry; nomenclature of inorganic compounds; kinetics of reactions in solution. *Mailing Add:* Dept Chem Swarthmore Col Swarthmore PA 19801

PASTERNAK, GAVRIL WILLIAM, b Brooklyn, NY, June 29, 47; m 77; c 2. NEUROLOGY, NEUROPHARMACOLOGY. *Educ:* Johns Hopkins Univ, BA, 69, MD, 73, PhD(pharmacol), 74. *Prof Exp:* From asst prof to assoc prof, 79-89, PROF NEUROL & PHARMACOL, CORNELL UNIV MED COL, 89-; MEM, MEM SLOAN KETTERING CANCER CTR, 89- *Concurrent Pos:* From asst mem to assoc mem, Mem Sloan Kettering Cancer Ctr, 79-89, from asst attend neurologist to assoc attend neurologist, 79-90, attend neurologist, 90-; attend neurologist, NY Hosp, 89- *Honors & Awards:* S Weir Mitchell Award, Am Acad Neurol; Louise and Allston Boyer Young Investr Award, Bd Sci Coun Addiction Res Ctr Nat Inst Drug Abuse; Granito Mem Lectr; Eino Nelson Mem Lectr. *Mem:* Am Acad Neurol; Am Soc Pharmacol & Exp Therapeut; Soc Neurosci; Am Neurol Asn. *Res:* Central analgesic mechanisms; opiate receptor heterogeneity and correlating the different subpopulations of binding sites with pharmacological and physiological functions. *Mailing Add:* Dept Neurol Sloan Kettering Cancer Ctr 1275 York Ave New York NY 10021

PASTINE, D JOHN, theoretical physics, for more information see previous edition

PASTO, ARVID ERIC, b Ithaca, NY, Dec 17, 44; m 66; c 2. CERAMIC SCIENCE, NUCLEAR MATERIALS. *Educ:* State Univ NY Col Ceramics, Alfred Univ, BS, 67, MS, 69, PhD(ceramics), 72. *Prof Exp:* Res assoc ceramic mat, Oak Ridge Nat Lab, 72-77, subtask leader, Develop Nuclear Fuels, Metals & Ceramics Div, 77-80; mem staff, 80-91, DEPT MGR, GEN TEL & ELECTRONICS LAB, 91- *Honors & Awards:* Indust Res-100 Award, 79. *Mem:* Am Ceramic Soc; Ceramic Educ Coun. *Res:* Synthesis, characterization and fabrication of refractory ceramic materials and their development into final shapes possessing useful properties; oxides, borides, nitrides and cermets. *Mailing Add:* GTE Labs 40 Sylvan Rd Waltham MA 02254

PASTO, DANIEL JEROME, b Elmira, NY, Jan 21, 36; m 71; c 2. ORGANIC CHEMISTRY. *Educ:* Rochester Inst Technol, BS, 58; Iowa State Univ, PhD(org chem), 60. *Prof Exp:* NSF fel chem, Harvard Univ, 60-61; from asst prof to assoc prof, 61-69, PROF CHEM, UNIV NOTRE DAME, 69- *Concurrent Pos:* NSF sr fel, Stanford Univ; consult, Miles Labs, Elkhart, Ind, 75-82; NATO fel, Univ Louis Pasteur, Strasbourg, France, 76. *Mem:* Am Chem Soc. *Res:* Physical organic reaction mechanisms; electrophilic addition and cycloaddition reactions; application of molecular orbital theory to structure and reactivity. *Mailing Add:* Dept Chem & Biochem Univ of Notre Dame Notre Dame IN 46556

PASTORE, PETER NICHOLAS, b Bluefield, WVa, Nov 8, 07; m 39; c 5. OTORHINOLARYNGOLOGY. *Educ:* Univ Richmond, AB, 30; Med Col Va, MD, 34; Univ Minn, MS, 39. *Prof Exp:* Intern, Med Col Va, 34-35, resident, 35-36; instr, Mayo Clin, 36-39, staff, 39-42; prof otol, rhinol & laryngol & chmn dept, Med Col, Va, 42-76; chief otolaryngol, 63-76, DIR, CONTINUING MED EDUC, MCGUIRE VET HOSP, 77- *Concurrent Pos:* Fel, Mayo Clin, Minn, 36-39; consult, US Naval Hosp, Portsmouth, Va, 58 & McDonald Army Hosp, Ft Eustis, 64; mem staff, Crippled Children's Hosp, Richmond, Richmond Eye, Ear, Nose & Throat Hosp, Va Commonwealth Univ Hosp, Richmond Mem Hosp & Med Col Va Hosp; scholar-in-residence, Med Col Va, Va Commonwealth Univ, 80- *Mem:* Am Otol Soc; Am Laryngol, Rhinol & Otol Soc; fel AMA; Pan-Am Asn Oto-Rhino-Laryngol & Broncho-Esophagol; fel Am Col Surgeons. *Res:* Hearing; speech. *Mailing Add:* 5503 Riverside Dr Richmond VA 23225

PASTUSZYN, ANDRZEJ, b Krakow, Poland, July 31, 40; m 68. BIOCHEMISTRY. *Educ:* Jagiellonian Univ, Poland, MS, 62; Univ Vienna, PhD(biochem), 66. *Prof Exp:* Asst anal chem, Univ Agr, Vienna, 66-68; from res assoc to sr res assoc biochem, 68-79, RES ASST PROF, UNIV NMEX, ALBUQUERQUE, 79- *Mem:* Am Chem Soc; Am Soc Biochem & Molecular Biol. *Res:* Protein biosynthesis; cholesterol biosynthesis and regulation. *Mailing Add:* Dept of Biochem Univ of NMex Albuquerque NM 87131

PASZNER, LASZLO, b Koszeg, Hungary, Aug 19, 34; Can citizen; m 60; c 2. WOOD SCIENCE, PULP SCIENCE. *Educ:* Univ BC, BSF, 58, MF, 63, PhD(wood chem), 66. *Prof Exp:* Res technician panel prod, Can Forest Prod Ltd, 58-60; res technician wood technol, Fac Forestry, Univ BC, 60-61; res technician pulping, Harmac Div, MacMillan Bloedell Ltd, 63; fel wood chem, Nat Res Coun Overseas, Ger, Austria, 66-68; res assoc radiation chem, 68-73, asst prof forestry, 73-75, ASSOC PROF FORESTRY, WOOD SCI & INDUST, UNIV BC, 75-, HEAD WOOD SCI & INDUST DIV, 78- *Concurrent Pos:* vchmn wood chem comn, Tech Asn Pulp & Paper Indust, 75- *Mem:* Tech Asn Pulp & Paper Indust; Forest Prod Res Soc; NY Acad Sci; Can Pulp Paper Asn; Am Chem Soc. *Res:* Forest products chemistry; panel products; pulping chemistry; forest industrial pollution and control; polymer chemistry; radiation chemical processing of polymers; chemical conversion of biomass; organosol pulping and saccharification; biochemical technology. *Mailing Add:* Fac Forestry Univ BC Vancouver BC V6T 1W5 Can

PASZTOR, VALERIE MARGARET, b London, Eng, Feb 28, 36; div. INVERTEBRATE NEUROBIOLOGY. *Educ:* Univ Birmingham, BSc, 57; McMaster Univ, PhD(zool), 61. *Prof Exp:* Lectr zool, McMaster Univ, 60-61; from lectr to asst prof zool, 61-70, ASSOC PROF BIOL, McGILL UNIV, 70- *Concurrent Pos:* Dir, Redpath Mus, 90- *Mem:* Soc Neurosci; Soc Exp Biol. *Res:* Neuromodulation of mechanoreception in crustacea. *Mailing Add:* Dept Biol McGill Univ 1205 Ave Dr Penfield Montreal PQ H3A 1B1 Can

PASZYC, ALEKSY JERZY, b St Petersburg, Russia, Aug 9, 12; US citizen; m 43; c 3. SYSTEMS ENGINEERING. *Educ:* Warsaw Polytech, MEE, 35, MMechE, 37; Univ London, DIC & PhD(mech eng), 46. *Prof Exp:* Asst prof mech eng, Warsaw Polytech, 37-39; Polish Univ Col, London, Eng, 43-46; tech dir process plant eng & mfg, Hughes & Lancaster, Ltd, Eng, 46-50; chief engr plate & boiler div, John Ingles Co, Ltd, Can, 50-55; proj engr pulp & paper div, H H Simons Ltd, 55-56 & Ralph M Parsons Co, 56-58; chief engr, J H Pomeroy & Co, Inc, 58-60; mgr & chief engr, Pac Automation Prods, Inc, 60-61; div mgr ground systs group, Hughes Aircraft Co, 61-63; head Mech & Elec Eng Dept, Naval Civil Eng Lab, 63-82. *Concurrent Pos:* Consult, 82- *Mem:* Am Soc Mech Engrs; Asn Prof Engrs Ont; Brit Inst Elec Engrs. *Res:* Scientific management; energy conversion, conservation and recovery; solid state applications to energy control; environmental protection systems; life support systems. *Mailing Add:* 642 Creekmont Ct Ventura CA 93003

PATALAS, KAZIMIERZ, b Rakoniewice, Poland, Aug 24, 25; m 51; c 2. LIMNOLOGY, FISHERIES. *Educ:* Wroclaw Univ, MS, 49, DSc(limnol), 52. *Prof Exp:* Res scientist, Forestry Inst, Poland, 49-50; res scientist, Freshwater Fisheries Inst, Poland, 50-54, head labor, 54-67; RES SCIENTIST, FRESHWATER INST, FISHERIES RES BD CAN, 67- *Concurrent Pos:* Docent habil, Olsztyn Univ, 60; Rockefeller Found fel, Univ BC, Fisheries Res Bd Can & Univ Colo, Boulder, 61; prof, Olsztyn Univ, 67; mem, Int Joint Comn-Int Ref Group on Upper Lakes Pollution, 73-; adj prof zool, Univ Man, 75- *Mem:* Int Asn Theoret & Appl Limnol; Am Soc Limnol & Oceanog; Polish Hydrobiol Asn. *Res:* Crustacean planktonic communities of lakes in relation to eutrophication; eutrophication phenomena in lakes; zoogeographical distribution of planktonic crustaceans. *Mailing Add:* Freshwater Inst 501 University Crescent Winnipeg MB R3T 2N6 Can

PATANELLI, DOLORES J, b Elkhart, Ind, July 20, 32. REPRODUCTIVE PHYSIOLOGY. *Educ:* NY Univ, BA, 55, MS, 58, PhD, 62. *Prof Exp:* Asst to med dir, Pop Coun, Inc, Rockefeller Inst, 56-62; res fel, Merck Inst Therapeut Res, 63-72; REPRODUCTIVE PHYSIOLOGIST, CONTRACEPTIVE DEVELOP BR, CTR POP RES, NAT INST CHILD HEALTH & HUMAN DEVELOP, 72- *Mem:* AAAS; Endocrine Soc; Am Fertil Soc; Am Asn Anat; NY Acad Sci; Soc Andrology; Sigma Xi. *Res:* Endocrinology; control of fertility; developmental biology; development of improved methods of fertility regulation; barrier methods and potential male contraceptives. *Mailing Add:* NICHD Ctr Pop Res 6130 Executive Blvd Rockville MD 20892

PATCHETT, ARTHUR ALLAN, b Middletown, NY, May 28, 29; m 62; c 2. ORGANIC CHEMISTRY. *Educ:* Princeton Univ, AB, 51; Harvard Univ, PhD(org chem), 55. *Prof Exp:* Asst scientist, NIH, 55-57; sr res chemist, Synthetic Chem Dept, Merck & Co, Inc, 57-60, from asst dir to sr dir, 60-71, sr dir lead coord, 71-72, sr dir, New Lead Discovery Dept, 72-76, exec dir, 76-87, VPRES EXP CHEM, MERCK SHARP & DOHME RES LABS, 87- *Concurrent Pos:* Chmn, Div Med Chem, Am Chem Soc, 71. *Mem:* Am Chem Soc; Am Soc Biochem & Molecular Biol; AAAS. *Res:* Medicinal chemistry. *Mailing Add:* Merck Sharp & Dohme Res Labs R5OG-302 Rahway NJ 07065

PATCHETT, JOSEPH EDMUND, ceramics, for more information see previous edition

PATCHICK, PAUL FRANCIS, earth science, for more information see previous edition

PATCHING, THOMAS, b Etzikom, Alta, Oct 2, 15; m 40; c 5. MINING ENGINEERING. *Educ:* Univ Alta, BSc, 36. *Prof Exp:* Mining engr, Int Nickel Co Can Ltd, 36-42 & Hudson Bay Mining & Smelting Co, Ltd, 42-47; from asst prof to assoc prof mining eng, 47-72, prof, 72-80, EMER PROF MINING ENG, UNIV ALTA, 80- *Concurrent Pos:* Tech officer, Fuels & Mining Practice Div, Dept Mines & Tech Surv, Can, 56-60. *Mem:* Can Inst Mining & Metall (pres, 71-72). *Res:* strata stress studies; outbursts of coal and gas. *Mailing Add:* 7322 118th St Edmonton AB T6G 1S4 Can

PATE, BRIAN DAVID, b London, Eng, Feb 11, 28; Can citizen; m 51; c 5. POSITRON OMISSION TOMOGRAPHY. *Educ:* Univ London, BSc, 49, MSc, 51; McGill Univ, PhD(chem), 55. *Prof Exp:* Res assoc chem, McGill Univ, 55-56; from res assoc chem to assoc chemist, Brookhaven Nat Lab, 56-59; asspc prof chem, Wash Univ, 59-64; head dept, Simon Fraser Univ, 64-68, prof chem, 64-77; prof pharm chem, 78-87, PROF MED, UNIV BC, 87- *Concurrent Pos:* Triumf assoc dir appl prog, Univ BC, 77-87, dir Triumf Positron Emmission Tomography, 81-87. *Mem:* Fel Chem Inst Can. *Res:* Radiopharmaceutical chemistry; positron emission tomography. *Mailing Add:* Health Sci Ctr Hosp Univ BC 2211 Wesbrook Mall Vancouver BC V6T 2B5 Can

PATE, FINDLAY MOYE, b Davisboro, Ga, Jan 24, 41; m 61; c 4. ANIMAL NUTRITION. *Educ:* Univ Ga, BS, 65, PhD(animal nutrit), 70; Ore State Univ, MS, 67. *Prof Exp:* Asst prof & asst animal nutritionist, assoc prof & assoc animal nutritionist, Inst Food & Agr Sci, Agr Res & Educ Ctr, Univ Fla, Belle Glade, 70-83, CTR DIR, PROF ANIMAL NUTRIT & ANIMAL NUTRITIONIST, AGR RES & EDUC CTR, UNIV FLA, ONA, 83- *Mem:* Am Soc Animal Sci. *Res:* Beef cattle nutrition; mineral metabolism; sugarcane by-product utilization; beef cattle management. *Mailing Add:* Agr Res & Educ Ctr Univ Fla Ona FL 33865

PATE, JAMES WYNFORD, b Wedowee, Ala, Aug 28, 28; m 48; c 3. MEDICINE. *Educ:* Med Col Ga, MD, 50. *Prof Exp:* Resident surgeon, Med Col, Univ Ala, 53-55; resident thoracic surgeon, Vet Admin Hosp, Memphis, Tenn, 55-57, asst chief thoracic surg, 57-59; asst prof surg, 59-65, chmn dept thoracic surg, 65-74, PROF SURG, COL MED, UNIV TENN, MEMPHIS, 65-, CHMN DEPT, 74- *Concurrent Pos:* Dir training, City Hosps, Memphis, 59-; consult, Vet Admin Hosp, Memphis, 59- & Naval Hosp, 61- *Mem:* Am Asn Thoracic Surgeons; Soc Vascular Surg; Am Col Surgeons; Am Col Chest Physicians; Soc Thoracic Surg; Sigma Xi. *Res:* Thoracic and heart surgery; homografts of tissues and organs. *Mailing Add:* Dept Surg Suite G226 Univ Physicians Found 956 Court Ave Memphis TN 38163

PATEK, ARTHUR JACKSON, JR, medicine, for more information see previous edition

PATEL, ANIL S, b Baroda, India, June 28, 39. OPHTHALMIC ADVANCED PRODUCTS RESEARCH FOR CATARACT & REFRACTIVE SURGERY. *Educ:* Univ Baroda, India, BEE, 60; Purdue Univ, MS, 63; Northwestern Univ, PhD(biomed eng), 66. *Prof Exp:* Jr engr, Koyna Hydro-Elec Proj, 60-61; fel vision res, Biomed Engr Ctr, Northwestern Univ, 66-67; sr res bioengr, Baxter-Travenol Labs Inc, 68-74; prin scientist & biomed engr, Cavitron Corp, 74-79; chief scientist, Cooper Vision Systs, Irvine, Calif, 79-80, chief scientist & mgr, advan prod develop, 80-83; chief scientist & dir, Advan Prod Res, Cilco Div, Cooper Vision, Inc, Bellevue, Wash, 83-89; DIR RES, IOL, ALCON LABS, INC, FT WORTH, 89- *Concurrent Pos:* NIH Postdoctoral fel, 66-67. *Mem:* AAAS; Inst Elec & Electronic Engrs; Asn Advan Med Instrumentation; Asn Res in Vision & Ophthal; Int Soc Refractive Keratoplasty; Soc Biomaterials; Sigma Xi. *Res:* Intraocular lenses; keratorefractive surgery; cardiac assist devices; dialyzers; cataract and vitreous surgery instruments; perimetry for glaucoma diagnostic; lasers in medicine and surgery; cardio-respiratory and neonatal special care medical devices; pulmonary function testing; ultra sound; ultraviolet and infrared technologies in medicine and dentistry. *Mailing Add:* Alcon Labs Inc R5-12 6201 S Freeway Ft Worth TX 76134-2099

PATEL, APPASAHEB RAOJIBHAI, b Baroda, India, Sept 05, 31; m 64; c 5. PHARMACEUTICAL CHEMISTRY. *Educ:* Univ Baroda, BSc, 52, MSc, 55; Univ Calif, PhD(pharmaceut chem), 60. *Prof Exp:* Fel chem, Univ Va, 61-63 & 65-68; res chemist, Leo Pharmaceut Prod, Denmark, 63-64; sr org chemist, Life Sci Div, Melpar, Inc, Va, 68-70; supvr org & anal chem lab, Life Sci Div, Meloy Labs, Inc, 70-77; prog mgr info & resources segment, 77-78, chemist, Toxicol Br, Carcinogenesis Testing Prog, 78-79, PROG DIR, DIET & NUTRIT, DIV CANCER CAUSE & PREV, NAT CANCER INST, BETHESDA, 79- *Res:* Synthesis of quinoline-methanols as potential therapeutics; cigarette smoke condensate production, fractionation and analysis; tobacco and marijuana analysis; synthesis and analysis of carcinogens. *Mailing Add:* 4626 Star Flower Dr Chantilly VA 22021

PATEL, ARVINDKUMAR MOTIBHAI, b Isnav, India, Oct 19, 37; m 63; c 3. COMPUTER SCIENCE, SYSTEMS ENGINEERING. *Educ:* Sardar Vallabhbai Vidyapith, India, BE, 59; Univ Ill, Urbana, MS, 61; Univ Colo, Boulder, PhD(elec eng), 69. *Prof Exp:* Lectr elec eng, Birla Eng Col, India, 59-60; sr assoc engr, IBM Pok, 62-65, staff engr IBM Bldr, 65-69, sr engr, magnetic rec tech, IBM, Pok, 72, sr tech staff, IBM, San Jose, 83; Fel, IBM Corp, 86- *Mem:* Fel Inst Elec & Electronic Engrs. *Res:* Error correcting codes, signal processing, magnetic recording and computer systems reliability. *Mailing Add:* IBM Corp 5600 Cottle Rd Dept H18, Bldg 025 San Jose CA 95193

PATEL, BHAGWANDAS MAVJIBHAI, b Surat, India, Nov 24, 38; m 64; c 3. TRACE & ULTRA-TRACE ANALYSIS, SPECTROSCOPY & CHROMATOGRAPHY. *Educ:* Gujarat Univ, BSc, 61, MSc 63; Univ Bombay, PhD(sci), 80. *Prof Exp:* Res & develop scientist, Dept Atomic Energy, Bhabba Atomic Res Ctr, 63-84; postdoctoral res fel, Dept Chem, Univ Fla, Gainesville, 84-87, res consult, Dept Food Sci, 87-90; TECH COORDR, RES & DEVELOP, TEXACO INC, PORT ARTHUR, 91- *Concurrent Pos:* Vis scientist, Dept Chem, Univ Fla, Gainesville, 71-72, adj postdoctoral res assoc, scientist & consult, 87- *Mem:* Am Chem Soc; Soc Appl Spectros. *Res:* Analytical chemistry; atomic and molecular spectroscopy; chromatography; ultra-trace analysis of environmental and industrial applications; author of numerous publications. *Mailing Add:* Res & Develop Texaco Inc PO Box 1608 Port Arthur TX 77641

PATEL, CHANDRA KUMAR NARANBHAI, b Baramati, India, July 2, 38; US citizen; m 61; c 2. PHYSICS, LASER RESEARCH. *Educ:* Univ Poona, BE, 58; Stanford Univ, MS, 59, PhD(elec eng), 61. *Hon Degrees:* DSc, NJ Inst Technol, 88. *Prof Exp:* Mem tech staff, 61-67, head, Infrared Physics & Electronics Res Dept, 67-70, dir, Electronics Res Labs, 70-76, dir, Phys Res Lab, 76-81, exec dir, Physics & Acad Affairs Div, Bell Labs, 81-87, EXEC DIR MAT SCI, ENG & ACAD AFFAIRS DIV, 87- *Concurrent Pos:* Mem bd trustees, Aerospace Corp, Los Angeles, 79-; mem bd dir, Newport Corp, Calif, 86, Cal Micro Devices, Milpitas, Calif, 90- *Honors & Awards:* Adolph Lomb Medal, Optical Soc Am, 66; Ballantine Medal, Franklin Inst, 68; Honor Award, Assoc Indians in America, 75; Zworykin Award, Nat Acad Eng, 76; Lamme Medal, Inst Elec & Electronic Engrs, 76, Medal of Honor, 89; Texas Instruments Found Founders Prize, 78; Charles Hard Townes Award, Optical Soc Am, 82, Ives Medal, 89; NY Sect Award, Soc Appl Spectrum, 82; Arthur H Schawlow Award, Laser Inst Am, 84; NJ Gov, Thomas Alva Edison Award, 87; George E Pake Prize, Am Phys Soc, 88; Hon Mem, Gynec Laser Surg Soc. *Mem:* Nat Acad Sci; Nat Acad Eng; fel Am Phys Soc; fel Optical Soc Am; fel Am Acad Arts & Sci; fel Inst Elec & Electronics Engrs; AAAS; Am Soc Laser Med & Surg; fel Indian Nat Sci Acad; foreign fel, Inst Elec & Telecommunication; assoc fel, Third World Acad Sci. *Res:* Gas lasers; molecular lasers in the infrared; high power lasers; nonlinear optics; tunable lasers in the infrared; high resolution spectroscopy; pollution detection in the atmosphere and the stratosphere; use of lasers in surgery; pulsed opto-acoustic spectroscopy. *Mailing Add:* Mat Sci Eng & Acad Affairs Div AT&T Bell Labs Murray Hill NJ 07974

PATEL, DALI JEHANGIR, b Agra, India; US citizen; m 57, 81; c 4. MEDICINE, BIOPHYSICS. *Educ:* Grant Med Col, Bombay, MB, BS, 48; Univ Utah, MS, 55; Univ Western Ont, PhD(biophys), 57. *Prof Exp:* Intern, White Mem Hosp, Col Med Evangelists, Los Angeles, 50-51; resident pediat cardiol, Irvington House, NY Univ, 51-52; res asst biophys, Univ Western Ont, 55-57; vis scientist, Nat Heart Inst, 58-62, med officer, Sect Exp Atherosclerosis, 62-79; PROF DEPT PHYSIOL & BIOPHYS, COL MED, HOWARD UNIV, 79- *Concurrent Pos:* Fel med, Med Ctr, Univ Calif, San Francisco, 52-53; fel cardiol, Col Med, Univ Utah, 53-55, Am Heart Asn fel, 57-58. *Mem:* Am Physiol Soc. *Res:* Cardiac physiology; circulatory physics; atherosclerosis; preventive cardiology. *Mailing Add:* Dept Physiol & Biophys Col Med Howard Univ 520 West St NW Washington DC 20059

PATEL, GIRISHCHANDRA BABUBHAI, b Mombasa, Kenya, Nov 10, 48; Can citizen; m 74; c 2. MICROBIAL PHYSIOLOGY. *Educ:* Sardar Patel Univ, India, BSc, 69; Univ Sask, Can, MSc, 72, PhD(agr microbiol), 74. *Prof Exp:* Lectr dairy technol, Dairy Sci Col, Sardar Patel Univ, India, 69-70; res asst, Dairy & Food Sci Dept, Univ Sask, Can, 70-71, qual control analyst, 70-72, lab instr, 73-74; fel, Inst Biol Sci, Nat Res Coun Can, 74-75, res assoc food microbiol, 75-76, from asst res officer to assoc res officer, 76-88, SR RES OFFICER, INST BIOL SCI, NAT RES COUN CAN, OTTAWA, 89- *Honors & Awards:* Bapuji Desai Gold Medal. *Mem:* Can Soc Microbiologists; Am Soc Microbiol. *Res:* Isolation, characterization and metabolic studies of the microflora present in anaerobic ecosystems. *Mailing Add:* Inst Biol Sci Nat Res Coun of Can Ottawa ON K1A 0R6 Can

PATEL, GORDHAN L, b Lourenco Marques, Mozambique, June 12, 36; US citizen; m 64; c 2. MOLECULAR BIOLOGY. *Educ:* Wash Univ, AB, 59, PhD(zool, nuclear proteins), 64. *Prof Exp:* Res assoc biol, State Univ NY Buffalo, 64-67; from asst prof to assoc prof, 67-81, PROF ZOOL & BIOCHEM, UNIV GA, 81-, HEAD ZOOL, 81- *Mem:* AAAS; Am Soc Cell Biol. *Res:* Role of eukaryotic nuclear proteins in chromatin structure and function, and in regulation of nuclear functions. *Mailing Add:* Dept Zool Univ Ga Athens GA 30601

PATEL, GORDHANBHAI NATHALAL, b Manund, Gujarat, India, Feb 2, 42; c 2. PHYSICAL CHEMISTRY, POLYMER SCIENCE. *Educ:* Sardar Patel Univ, India, BS, 64, MS, 66, PhD(polymers), 70. *Prof Exp:* Res asst polymers, Univ Bristol, 70-73; res assoc radiation, Baylor Univ, 73-74; STAFF CHEMIST POLYMERS, ALLIED CHEM CORP, 74- *Mem:* Am Chem Soc. *Res:* Polymer synthesis, characterization and uses; radio chemical changes in polymer; indicators for time, temperature and radiation dosages. *Mailing Add:* J P Labs Inc 26 Howard St Piscataway NJ 08854-1436

PATEL, JAMSHED R(UTTONSHAW), b Calcutta, India, Mar 1, 25; US citizen; m 51; c 3. PHYSICAL METALLURGY. *Educ:* Mass Inst Technol, SB, 49, SM, 51, ScD(metall), 54. *Prof Exp:* Asst mech eng, Mass Inst Technol, 49-51, asst phys metall, 51-53; sr engr, Electronics Div, Sylvania Elec Prod, Inc, 53-55; mem res staff, Res Div, Raytheon Co, 55-61; MEM TECH STAFF, BELL TEL LABS, 61- *Concurrent Pos:* Lectr, NATO Summer Sch, Ghent, Belg, 79 & Durham, UK, 79; vis prof, Univ Pierre et Marie Curie, Paris, France, 84-85; assoc, Harvard Univ, Cambridge, Mass, 89- *Mem:* Am Phys Soc. *Res:* Surface structure; x-ray standing waves; synchrotron radiation; x-ray topography; plastic deformation; dislocations; defects in crystals. *Mailing Add:* 600 Mountain Ave-Rm 1E-328 Murray Hill NJ 07974-2070

PATEL, JITENDRA BALKRISHNA, b Apr 18, 42. CNS PHARMACOLOGY, PSYCHOTROPICS DRUGS. *Educ:* St John's Univ, PhD(pharmacol & toxicol), 77. *Prof Exp:* PRIN PHARMACOLOGIST, ICI AM INC, 78- *Mailing Add:* Biomed Res Dept Attn Jitendra B ICI Am Inc Wilmington DE 19897

PATEL, KALYANJI U, b Kothamdi, India, Dec 23, 25; US citizen; m 57; c 3. ORGANIC CHEMISTRY, CHEMICAL ENGINEERING. *Educ:* Univ Bombay, BSc, 49; Univ Mich, BS, 51; Mich State Univ, MS, 53, PhD(chem eng), 54. *Prof Exp:* Chemist & chem engr, Micarta Div Eng Lab, Westinghouse Elec Corp, 54-55 & Cent Res Lab, Bordon Chem Co, 56-59; sr chem engr & chemist, Chem Div, 3M Co, 59-87; RETIRED. *Res:* Surface catalysis; organic synthesis; polymers and plastics; organic coatings; thermoplastics; thermoset polymers; air flotation; synthesis of functional monomers and polymers; fluorochemicals, elastomers and plastics. *Mailing Add:* 3037 Bartelmy Lane St Paul MN 55109

PATEL, MAYUR, b Kampala, Uganda, EAfrica, Nov 8, 55; Brit citizen; m; c 2. RESEARCH ADMINISTRATION, TECHNICAL MANAGEMENT. *Educ:* Teesside Univ, Eng, UK, BS, 81, PhD(chem eng), 84. *Prof Exp:* Consult eng, Bayfern Contractors, UK, 84-85; technologist eng, Gestetner Mfg, UK, 85-88; sr proj engr, A H Marks & Co, UK, 88-90; SR RES ENGR, NALCO CHEM CO, TEX, 90- *Mem:* Am Inst Chem Engrs. *Res:* Process automation and its adaptability to modern day high tech chemical manufacture to produce high quality chemicals repetitively; developed, designed and constructed new chemical plants; upgraded and modified chemical plants; skid mounted fully automatic process suitable for use on trucks. *Mailing Add:* 13502 Catalina Village Dr Houston TX 77083

PATEL, MULCHAND SHAMBHUBHAI, b Sipor, India, Sept 9, 39. BIOCHEMISTRY. *Educ:* Gujarat Univ, India, BSc, 61; M S Univ Baroda, MSc, 64; Univ Ill, Urbana, PhD(animal sci), 68. *Prof Exp:* Teacher sci, Ahmedabad, India, 61-62; chemist, Alembic Chem Works, Ltd, Baroda, 62; res asst microbiol, M S Univ Baroda, 64-65; res asst nutrit biochem, Univ Ill, Urbana, 65-68; asst prof res pediat, St Christopher's Hosp Children, 70-72; res asst prof biochem, Temple Univ, 70-75, asst prof res med, 72-75, res assoc prof med & biochem, Sch Med, 75-78; assoc prof, 78-86, PROF BIOCHEM, CASE WESTERN RESERVE UNIV, 86- *Concurrent Pos:* NIH fel biochem, Fels Res Inst, Sch Med, Temple Univ, 68-69; NIH res grants, St Christopher's Hosp Children 71-72, Temple Univ, 72-78 & Case Western Reserve Univ, 78-; mem, Biochem Study Sect 2, NIH, 84-88. *Honors & Awards:* Fulbright Res Scholar Award, 87. *Mem:* Brit Biochem Soc; Am Inst Nutrit; Am Soc Biochem & Molecular Biol; Am Soc Neurochem; Int Soc Neurochem. *Res:* Regulation of carbohydrate and lipid metabolism in developing mammalian brain and liver; inborn errors in pyruvate metabolism; molecular biology of pyruvate dehydrogenese complex. *Mailing Add:* Dept Biochem Case Western Reserve Univ Med Sch Cleveland OH 44106

PATEL, NAGIN K, b Navsari, India, June 17, 32, Can citizen; m 62; c 3. PHYSICAL & INDUSTRIAL PHARMACY. *Educ:* Gujarat Univ, India, BPharm, 55; Temple Univ, MS, 57; Univ Md, PhD(pharm), 61. *Prof Exp:* Jr instr pharm, Univ Md, 57-61; instr & fel, Temple Univ, 61-62; asst prof, Duquesne Univ, 62-63 & Univ Man, 63-66; from asst prof to assoc prof, Univ Alta, 66-69; res pharmacist, Frank W Horner, Ltd, 69-70, supvr pharm develop, Res Labs, 70-80; group leader explor res, McNeil CPC, 80-81; DIR,

PAR PHARM, INC, 88- *Concurrent Pos:* Nat Res Coun Can grant, 64-67; pharmaceut consult, 81- *Mem:* Am Pharmaceut Asn; Soc Cosmetic Chem; Indian Pharmaceut Asn; Am Asn Col Pharm; Soc Controlled Release. *Res:* Drug-macromolecular proteins, plastics, surfactants, interaction; controlled release dosage, tableting, membrane and coating technologies. *Mailing Add:* Dept Pharmaceut LIU, Arnold & Marie Schwartz Col Pharm Brooklyn NY 11201

PATEL, NARAYAN GANESH, b Sinaj-na-pura, India, May 5, 28; m 57; c 2. TOXICOLOGY. *Educ:* Univ Bombay, BSc, 50; Univ Poona, MSc, 52; Univ Minn, PhD(toxicol), 59. *Prof Exp:* Entom officer, Bombay Govt, 51-55; res asst entom, Univ Minn, 56-59, res fel toxicol, 59-63, res assoc, 63-64; sr res assoc, Case Western Reserve Univ, 64-69; BIOLOGIST & BIOCHEMIST, SR RES SCI AGR PROD DEPT, EXP STA, E I DU PONT DE NEMOURS & CO, INC, 69- *Concurrent Pos:* Adj prof, Univ Del, 71-; prin investr, Med Plant Expeds, 83, 85-86. *Mem:* AAAS; Entom Soc Am; Soc Develop Biol; Am Soc Cell Biol; Sigma Xi; Earthwatch. *Res:* Physiology, toxicology, endocrinology, developmental biology and biochemistry of insect and invertebrates; natural products. *Mailing Add:* 100 E Sutton Pl Wilmington DE 19810

PATEL, NUTANKUMAR T, b Cambay, Gujarat State, India; m; c 1. BIOCHEMISTRY, MOLECULAR BIOLOGY. *Educ:* Univ Baroda, BS, 56, MS, 58; Univ Bombay, PhD(biochem), 70. *Prof Exp:* Res asst biochem, Univ Baroda, 58-59, demonstr, 59-61; sci officer biochem, Cancer Res Inst, Bombay, India, 62-71; fel biochem, 71-79, res scientist, 79-82, ASST PROF, UNIV TEX MED BR, 82- *Concurrent Pos:* Prin investr, Am Cancer Soc Institnl Grant Comt, 76-77. *Mem:* AAAS; Sigma Xi; The Protein Soc. *Res:* Molecular biology of cancer and chemical carcinogenesis, mRNA processing and transport; enzymology; nucleic acids and proteins; chemistry and metabolism. *Mailing Add:* Dept Human Biochem & Genetics Biochem Div Univ Tex Med Br Galveston TX 77550

PATEL, POPAT-LAL MULJI-BHAI, b Nairobi, Kenya, July 7, Can citizen; m 60; c 3. HIGH ENERGY PHYSICS. *Educ:* Manchester Univ, BSc, 55, MSc, 56; Harvard Univ, PhD(physics), 62. *Prof Exp:* Res assoc physics, Mass Inst Technol, 62-65, asst prof, 62-68; assoc prof, 68-81, PROF PHYSICS, MCGILL UNIV, 81- *Mem:* Am Phys Soc; Can Asn Physicists (secy, 71-76); Can Inst Particle Physics. *Res:* Nucleon-nucleon polarization; nucleon Compton effect; pi-zero photoproduction; boson spectroscopy; electromagnetic interactions of nucleons at high energies; e+e- collider physics. *Mailing Add:* Dept Physics Rutherford Lab McGill Univ 3600 Univ Montreal PQ H3A 2T8 Can

PATEL, PRAFULL RAOJIBHAI, b Nadiad, India, Nov 27, 37; m 69; c 1. DENTAL RESEARCH. *Educ:* Univ Baroda, BSc, 58; Univ Bombay, BSc, 60; Columbia Univ, MS, 62; Univ Mich, PhD(pharmaceut chem), 65; Univ Md, DDS, 78. *Prof Exp:* Asst chem, Col Pharm, Columbia Univ, 60-62; asst phys pharm, Univ Mich, 62-63, res asst, 63-65; res assoc, Res Unit, Am Dent Asn Health Found, 65-74 & 79-81. *Concurrent Pos:* Consult; micro-computer programmer. *Mem:* Am Dent Asn. *Res:* Solubility properties of synthetic and biological calcium phosphate minerals; clinical evaluation of biomaterials. *Mailing Add:* 19140 St Johnsbury Lane Germantown MD 20874

PATEL, RAJNIKANT V, b Nairobi, Kenya, Mar 26, 48; Can citizen; m 77. ROBOTICS, CONTROL SYSTEMS. *Educ:* Univ Liverpool, BEng, 69; Univ Cambridge, PhD(elec eng), 73. *Prof Exp:* Fel, elec eng, Univ Cambridge, 72-75; res assoc, control systs, NASA Ames Res Ctr, Calif, 75-77; sr res scientist, control systs, Delft Univ Technol, Holland, 79-80; lectr, control systs, Inst Sci & Technol, Univ Manchester, 80-81; assoc prof, 82-86, PROF, ELECT ENG, CONCORDIA UNIV, CAN, 86- *Concurrent Pos:* Vis scientist, Lund Inst Technol, Sweden, 73-74; vis asst prof, Univ Waterloo, Can, 77-79; assoc ed, Automatica; assoc ed, IEEE Transactions on Automatic Control; mem, Chem Eng, Inst Elec Eng, UK. *Mem:* Sr mem Inst Elec & Electronic Engrs; Asn Computing Mach; Soc Indust & Appl Math. *Res:* Numerical and computational issues in robotics and control; adaptive control of robot manipulators; neural network applications in robotics and control; redundant manipulators. *Mailing Add:* Dept Elec Eng Concordia Univ Montreal PQ H3G 1M8 Can

PATEL, RUTTON DINSHAW, b Darjeeling, India, June 26, 42; US citizen. FLUID MECHANICS, REACTOR ENGINEERING. *Educ:* Indian Inst Technol, Kharagpur, BTech, 63; Purdue Univ, MS, 66, PhD(chem eng), 67. *Prof Exp:* Resident student assoc chem eng, Argonne Nat Lab, 66-67; vis asst prof, Purdue Univ, 67-68; from asst prof to assoc prof, Polytech Inst New York, 68-78; from hon sr eng to sr staff eng, 78-81, ENG ASSOC & GROUP HEAD, EXXON RES & ENG CO, 81- *Honors & Awards:* Acharya P C Ray Award, Indian Inst Chem Engrs, 64. *Mem:* Am Inst Chem Engrs; Am Chem Soc; AAAS. *Res:* Heat and mass transfer; non Newtonian flow and transfer; thermal fouling and coking; reactor engineering. *Mailing Add:* Exxon Res & Eng Co PO Box 101 Florham Park NJ 07932-0101

PATEL, SHARAD A, b Ahmedabad, India, Aug 29, 25; US citizen; m 58; c 3. ELECTRICAL & MECHANICAL ENGINEERING. *Educ:* Benares Hindu Univ, BSc, 49; Polytech Inst Brooklyn, MAEE, 51, PhD(appl mech), 55. *Prof Exp:* Res assoc aerospace eng & appl mech, Polytech Inst, New York, 54-56, from asst prof to prof appl mech, 56-70, prof mech & aerospace eng, 76-; RETIRED. *Mem:* Assoc fel Am Inst Aeronaut & Astronaut. *Res:* Elasticity; plasticity; creep; structural dynamics; applied and solid mechanics. *Mailing Add:* 290 E Lincoln Ave Mt Vernon NY 10552

PATEL, SIDDHARTH MANILAL, b Santi-Niketan, WBengal, India, Oct 26, 33; m 61; c 3. ANALYTICAL CHEMISTRY, POLLUTION CHEMISTRY. *Educ:* Univ Bombay, BSc, 54, MSc, 57; Ohio State Univ, PhD(analytical chem), 66. *Prof Exp:* Teaching asst analytical chem, Ohio State Univ, 59-66; sr res chemist, 66-73, GROUP LEADER ENVIRON GROUP, QUANT ANALYTICAL SECT, GOODYEAR TIRE & RUBBER CO, 73- *Concurrent Pos:* Lectr, Univ Akron, 67-68, 70-71 & 84-85. *Mem:* Am Chem Soc; Am Soc Mass Spectrometry. *Res:* Environmental analytical methodology; analytical methods of industrial hygiene; process analysis systems; chromatography; spectroscopy and mass spectrometry. *Mailing Add:* 735 Cliffside Dr Akron OH 44313

PATEL, TARUN B, b Nairobi, Kenya, Nov 7, 53. ENDOCRINE PHARMACOLOGY, REGULATION OF METABOLISM. *Educ:* Univ London, Eng, PhD(biochem), 79. *Prof Exp:* Res asst prof pharmacol, Health Sci Ctr, Univ Tex, San Antonio, 79-84; ASST PROF PHARMACOL, CTR HEALTH SCI, UNIV TENN, 84- *Mem:* Am Soc Biol Chemists. *Mailing Add:* Dept Pharmacol Univ Tenn Col Med 800 Madison Ave Memphis TN 38163

PATEL, VIRENDRA C, b Mombasa, Kenya, Nov 9, 38; m 66; c 2. FLUID MECHANICS, HYDRAULICS. *Educ:* Imp Col, Univ London, BSc, 62; Univ Cambridge, PhD(fluid mech), 65. *Prof Exp:* Asst res fluid mech, Univ Cambridge, 65; asst prof aeronaut eng, Indian Inst Technol, Kharagpur, 66-67; sr asst res fluid mech, Univ Cambridge, 67-69; consult aeronaut, Lockheed-Ga Co, 69-70; from asst prof to assoc prof mech & hydraul, 71-75, chmn ,Energy & Mech Eng Div, 76-82, PROF MECH ENG, UNIV IOWA, 75- *Concurrent Pos:* Res engr, Iowa Inst Hydraul Res, 71-; consult in field, Iowa Gov's Sci Adv Coun, 77-83; mem resistance comt, Int Towing Tank Conf, 78-87; vis prof, Univ Fridericiana, Karlsruhe, WGer, 80-81, & Sch Nat Super Mech, Univ Nantes, France, 85; jubilee prof, Chalmers Univ Technol, Gothenburg, Sweden, 88. *Honors & Awards:* US Sr Scientist Award, Alexander von Humboldt Found, WGer, 80-81. *Mem:* Assoc fel Am Inst Aeronaut & Astronaut; fel Am Soc Mech Engrs; Sigma Xi; Am Soc Eng Educ; Soc Naval Architects & Marine Engrs. *Res:* Viscous fluid mechanics; turbulence; industrial aerodynamics; ship hydrodynamics. *Mailing Add:* Dept Mech Eng Univ Iowa Iowa City IA 52242-1585

PATEL, VITHALBHAI AMBALAL, b Dec 26, 36; US citizen; m 60; c 2. APPLIED MATHEMATICS. *Educ:* VP Col, Vallabh Vidyanagar, India, BSc, 57, MSc, 59; Univ Calif, Berkeley, PhD(eng sci), 70. *Prof Exp:* Tutor math, V P Col, Vallabh Vidyanagar, India, 57-59, lectr, 59; lectr, Sardar Patel Univ, 61-63 & MG Sci Inst, Ahmedabad, 63-65; asst, Univ Calif, Berkeley, 65-69; asst prof, Calif State Univ, Humboldt, 69-73; assoc prof, 73-78, PROF MATH, HUMBOLDT STATE UNIV, 78- *Honors & Awards:* Golden Dozen Award, 82. *Mem:* Soc Indust & Appl Math; Soc Eng Sci. *Res:* Numerical solutions of the Navier-Stokes equations for the flows around the obstacles; solutions of the simultaneous equations. *Mailing Add:* Dept of Math Humboldt State Univ Arcata CA 95527

PATEL, VITHALBHAI L, b Samathiara, India, Mar 31, 35; US citizen; m 52; c 2. PLASMA PHYSICS, SPACE PHYSICS. *Educ:* Univ Baroda, BSc, 56; Univ Md, MS, 60; Univ NH, PhD(physics), 64. *Prof Exp:* Res asst physics, Phys Res Lab, Ahmedabad, India, 56-58 & Univ Md, 58-61; res asst, Univ NH, 61-63, instr, 64-65; sr res assoc, Rice Univ, 65-66; from asst prof to prof physics, Univ Denver, 66-90,; ASSOC SUPT, PLASMA PHYSICS DIV, NAVAL RES LAB, WASH, DC, 89- *Concurrent Pos:* NSF & NASA res grants, 67-90, AFOSR, 82-88; vis assoc prof, Dept Physics & Space Sci Ctr, Univ Minn, Minneapolis, 70-71; mem, Int Asn Geomag & Aeronomy; consult, NASA, 71-75; vis scientist, Dept Physics, Mass Inst Technol, 77-78; prog dir, Magnetospheric Physics, NSF, Wash, DC, 88-89; mem coun, Univ Space Res Asn, 75-90, Panel Geomagnetic Res Initiative, Nat Acad Sci, 90 & bd mem, Inst Advan Physics Studies, La Jolla, Calif, 90- *Mem:* Univs Space Res Asn; fel Am Phys Soc; Am Geophys Union. *Res:* Interplanetary magnetic field and plasma; plasma physics of the magnetosphere; space and astrophysical problems; nonlinear plasma physics. *Mailing Add:* Plasma Physics Div Code 4701 Naval Res Lab Washington DC 20375-5000

PATENT, GREGORY JOSEPH, b Hong Kong, May 5, 39; US citizen; m 64; c 2. COMPARATIVE ENDOCRINOLOGY. *Educ:* Univ Calif, Berkeley, 61, MA, 65, PhD(zool), 68. *Prof Exp:* Actg asst prof zool, Univ Calif, Berkeley, 68-69; NIH fel, Zool Sta, Naples, Italy, 69-70; asst prof biol, ECarolina Univ, 70-72; from asst prof to assoc prof, 72-82, PROF ZOOL, UNIV MONT, 82- *Concurrent Pos:* NIH trainee, Div Res, Sinai Hosp, Detroit, Mich, 68-69; res grant, NIH, 75; travel grant, NSF, 75. *Mem:* AAAS; Am Soc Zoologists; NY Acad Sci. *Res:* Endocrinology of carbohydrate metabolism in fishes; factors affecting insulin release in vertebrates; cytology and innervation of pancreatic islets; neural regulation of insulin release. *Mailing Add:* Dept Zool Univ Mont Missoula MT 59801

PATERA, J, b Zdice, Czech, Oct 10, 36. PHYSICS. *Educ:* Prague Univ, PhD(physics), 64. *Prof Exp:* PROF PHYSICS & MATH, UNIV MONTREAL, 69- *Mailing Add:* Ctr Res Math Univ Montreal Montreal PQ H3C 3J7 Can

PATERNITI, JAMES R, JR, b New York, NY, Mar 5, 48. ATHEROSCLEROSIS, LIPID METABOLISM. *Educ:* Queens Col, NY, BA, 69; City Univ New York, PhD(biol), 75. *Prof Exp:* Lectr biol, Queens Col, NY, 69-71; instr biol, Brooklyn Col, 71-75; fel biochem, Mt Sinai Sch Med, 75-78, instr, 78-79, asst prof med & biochem, 79-83; vis fel med, Columbia Univ Col Physicians & Surgeons, 83-84; FEL & SECT HEAD, LIPID & LIPOPROTEIN RES, SANDOZ RES INST, NJ, 84- *Concurrent Pos:* Mem, Coun Arteriosclerosis, Am Heart Asn, 83; prin investr, Nat Heart, Blood & Lung Inst grant, 84-87; adj asst prof med, Mt Sinai Sch Med, 84-87. *Mem:* Sigma Xi; Am Heart Asn; NY Acad Sci; Am Fedn Clin Res. *Res:* Biochemical studies relating to regulation of plasma lipid metabolism and its role in atherosclerosis; effect of pharmacologic agents on heart disease. *Mailing Add:* Dept Lipid & Lipoprotein Metab Sandoz Res Inst Rte 10 East Hanover NJ 07936

PATERSON, ALAN ROBB PHILLIPS, b Nanaimo, BC, Nov 14, 23; m 45; c 2. BIOCHEMISTRY. *Educ:* Univ BC, BA, 50, MA, 52, PhD(biochem), 56. *Prof Exp:* Asst prof biochem, Univ BC, 58-62; hon prof pharm & dir, McEachern Labs, 77-89, PROF BIOCHEM, CANCER RES UNIT, UNIV ALTA, 62- *Mem:* AAAS; Am Asn Cancer Res; Am Soc Biol Chemists; Can Biochem Soc. *Res:* Metabolism of nucleotides and nucleosides. *Mailing Add:* Dept Biochem Univ Alta Edmonton AB T6G 2E2 Can

PATERSON, ARTHUR RENWICK, b Dumfriesshire, Scotland, June 15, 22; nat US; m 48; c 4. ANALYTICAL CHEMISTRY. *Educ:* Davidson Col, BS, 43; Rutgers Univ, MS, 50, PhD(chem), 52. *Prof Exp:* Sect supvr, Phys & Analytical Chem Sect, Allied Chem Corp, 51-71, mgr, Chem Physics Dept, 71-78, dir res, Cent Res Lab, 78-80, dir, Analytical Sci Lab, Corp Res & Develop, 80-83; RETIRED. *Mem:* AAAS; Am Chem Soc; Soc Appl Spectros; Am Soc Testing & Mat; Coblentz Soc. *Res:* Mass spectrometry; chromatography; absorption spectroscopy; polymer characterization. *Mailing Add:* 3105 Wilderness Blvd W Parrish FL 34219-9346

PATERSON, DONALD ROBERT, b Burke, NY, Feb 25, 20; m 43; c 3. HORTICULTURE. *Educ:* Cornell Univ, BS, 47; Univ Calif, MS, 50; Mich State Univ, PhD, 52. *Prof Exp:* Asst county agr agent, NY Agr Exten Serv, 47-49; asst horticulturist, 52-56, assoc prof hort, Univ, 70-76, ASSOC HORTICULTURIST, TEX AGR RES & EXTEN CTR, TEX A&M UNIV, 56-, PROF HORT, UNIV, 76- *Mem:* AAAS; Am Soc Hort Sci; Am Soc Plant Physiol; Sigma Xi. *Res:* Plant growth regulators and mineral nutrition. *Mailing Add:* 205 Garden Club Dr Tex A&M Univ Res & Exten Ctr Overton TX 75684

PATERSON, JAMES LENANDER, b Minneapolis, Minn, Oct 20, 41; m 72; c 2. LOW TEMPERATURE PHYSICS. *Educ:* Yale Univ, BE, 63; Univ Calif, Berkeley, PhD(physics), 73. *Prof Exp:* Asst prof physics, Univ Va, 73-79; RES MGR, NON-VOLATILE MEMORY PROCESS, TEX INSTRUMENTS, 79- *Mem:* Am Phys Soc; sr mem Inst Elec & Electronic Engrs. *Res:* Nonequilibrium superconductivity, studying the characteristic response times. *Mailing Add:* Tex Instruments 13536 N Cent Expressway Dallas TX 75265

PATERSON, JAMES MCEWAN, b Falkirk, Scotland, Dec 1, 37; m 62. PARTICLE PHYSICS, ACCELERATORS. *Educ:* Univ Glasgow, BSc, 59, PhD(physics), 63. *Prof Exp:* Sr res assoc physics, Cambridge Electron Accelerator, Harvard Univ, 63-72; mem staff physics, Stanford Linear Accelerator Ctr, 72-80, PROF APPL RES, STANFORD UNIV, 80- *Mem:* Am Phys Soc. *Res:* Design and development of particle accelerators, and their use in elementary particle physics research. *Mailing Add:* SLAC Bin No 24 PO Box 4349 Stanford Univ Stanford CA 94309

PATERSON, PHILIP Y, b Minneapolis, Minn, Feb 6, 25; m 47; c 3. INFECTIOUS DISEASES, MICROBIOLOGY. *Educ:* Univ Minn, BS, 46, BM, 47, MD, 48. *Prof Exp:* Instr med, Tulane Univ, 50-51; asst resident & co-resident, Sch Med, Univ Va, 53-55, asst prof microbiol & instr med, 55-57; from asst prof microbiol to assoc prof exp med, Sch Med, NY Univ, 57-65; Sackett prof med & chief infectious dis-hypersensitivity sect, McGaw Med Ctr, 66-75, prof microbiol, 75-87, PROF MICROBIOL & IMMUNOL, MED & DENT SCHS, NORTHWESTERN UNIV, CHICAGO, 87-, CHMN DEPT, 75- *Concurrent Pos:* Am Heart Asn res fel, Sch Med, Tulane Univ, 49-50; Am Heart Asn res fel, 53; estab investr, Am Heart Asn, 55-57; med officer, NIH, 57-60; fel coun, AAAS, 85-; prof neurobiol & physiol, Northwestern Univ, Evanston, 83- *Honors & Awards:* Smadel Gold Medal & lectr, Infectious Dis Soc Am, 78. *Mem:* Am Asn Physicians; Am Soc Microbiol; Infectious Dis Soc Am; Am Asn Immunologists; Am Soc Clin Invest; Am Clin Climat Asn. *Res:* Infectious diseases and immunology; experimental tissue damage; autoimmune disease; neuroimmunology. *Mailing Add:* Northwestern Univ Med Sch 303 E Chicago Ave Chicago IL 60611

PATERSON, ROBERT ANDREW, b Reno, Nev, Jan 23, 26; m 56; c 3. BOTANY. *Educ:* Univ Nev, BA, 49; Stanford Univ, MA, 52; Univ Mich, PhD(bot), 57. *Prof Exp:* From instr to assoc prof bot, Univ Md, 56-67; prof bot & head, Dept Biol, 67-79, assoc dean, Col Arts & Sci, 79-84, PROF BOT, VA POLYTECH INST & STATE UNIV, 84-, DIR, CTR for the study of Sci in Soc and Sci and Technol Interdisciplinary Grad Prog, 87- *Concurrent Pos:* NSF grants, 60-67 & antarctic fungi, McMurdo Sta, Antarctica, 71 & 72; investr, Mich Biol Sta, 60-67, 85, vis prof, 67; consult, NSF, McMurdo Area, Antarctica, 73 & 74; investr Alaskan soil fungi, Dept Energy, 79-81; vis prof, Univ Alaska, 85. *Mem:* Sigma Xi; Mycol Soc Am. *Res:* Mycology; limnology; fungal parasites of plankton; lacustrine fungal saprophytes; aquatic Phycomycetes; Antarctic fungi; Antarctic ecosystems; Alaskan soil fungi. *Mailing Add:* Dept Biol Rm 2119 Derring Hall Va Polytech Inst & State Univ Blacksburg VA 24061

PATERSON, ROBERT W, b Jan 19, 39; m 67; c 1. AERODYNAMICS, AEROACOUSTICS. *Educ:* Princeton Univ, BSE, 60; Harvard Univ, MA, 65, PhD(fluid mech), 69. *Prof Exp:* Nuclear power engr, Atomic Energy Comn, Naval Reactors Br, Washington, DC, 60-62, supvr 62-64; res asst, Div Eng & Appl Physics, Harvard Univ, 66-69; res engr, United Technol Res Ctr, 69-73, supvr, Aeroacoust Group, 73-75, mgr, Aeroacoust & Exp Gas Dynamics Group, 75-88, Gas Dynamics & Thermophysics Dept, 88-90, MGR, AERODYN & ENVIRON SCI DEPT, UNITED TECHNOL RES CTR, 90- *Concurrent Pos:* Mem, Tech Comt on Aeroacoust, Am Inst Aeronaut & Astronaut, 74-77. *Mem:* Fel Am Inst Aeronaut & Astronaut; Am Soc Mech Engrs; Sigma Xi. *Res:* Experimental gas dynamics; aeroacoustics. *Mailing Add:* Aerodynamics & Environ Sci Dept United Technologies Res Ctr East Hartford CT 06108

PATERSON, WILLIAM GORDON, b Montreal, Can. CHEMISTRY. *Educ:* Univ Man, BSc, 56, MSc, 57; McGill Univ, PhD(chem), 60. *Prof Exp:* NATO sci fel & Ramsay Mem fel, Oxford Univ, 60-61; asst prof chem, Univ Alta, 61-63; sr res chemist, 3M Co, 63-68, proj coordr, New Bus Ventures Div, 68-71, prog mgr, 71-73, proj mgr, 73-76, proj mgr, Com Tape Div, 76-77, div sales mkt mgr, Chem Res Div, 77-82, strategic planning mgr, 82-86, DIR, BUS DEVELOP CTR, 3M CO, 86- *Concurrent Pos:* Chmn, MDA Mfg, 91- *Res:* Electrochemistry; synthetic inorganic chemistry; nuclear magnetic resonance; electrophotography; aquatic biology; plant nutrition; encapsulation; insect control. *Mailing Add:* 3M Ctr Bldg 223-65-04 St Paul MN 55144

PATES, ANNE LOUISE, b Monongahela, Pa, Mar 23, 13. BACTERIOLOGY. *Educ:* Fla State Col Women, BS, 36; Univ Mich, MS, 38, PhD(bact), 49. *Prof Exp:* Bacteriologist, Presby Hosp, Philadelphia, Pa, 38-44; instr, Sch 9, Univ Mo, 44-45; from asst prof to assoc prof bact, 49-74, assoc prof biol sci, 74-81, EMER PROF, FLA STATE UNIV, 81- *Mem:* AAAS; Am Soc Microbiol; fel Am Pub Health Asn. *Res:* Antigens of hisoplasma capsulation; avian botulism. *Mailing Add:* 2305 Don Patricio Dr Tallahassee FL 32304

PATHAK, MADHUKAR, b Baroda, India, July 29, 27; m 53; c 3. BIOCHEMISTRY, PHOTOBIOLOGY. *Educ:* Univ Baroda, BSc, 48; Univ Bombay, BSc, 50, MB, 53; Univ Ore, MS, 58, PhD, 60. *Prof Exp:* Lectr & biochemist, Med Col, Univ Nagpur, 53-56; res asst biochem & dermat, Med Sch, Univ Ore, 56-60; res fel dermat, Harvard Med Sch, 60-61, from res assoc to assoc, 61-67, asst prof, 67-70, prin assoc dermat, 70-77; res assoc, 61-63, asst biochemist, 63-69, ASSOC BIOCHEMIST, MASS GEN HOSP, 69-; SR ASSOC DERMAT & RES PROF, HARVARD MED SCH, 78- *Mem:* AAAS; Soc Invest Dermat; Am Fedn Clin Res; Biophys Soc; NY Acad Sci. *Res:* Skin Cancer; cutaneous photobiology, photochemistry and photosensitization; melanin pigmentation in mammals; porphyrin metabolism; normal and abnormal reactions of man to light; photo medicine. *Mailing Add:* Dept Dermat Warren Bldg Rm 519 Harvard Med Sch-Mass Gen Hosp 32 Fruit St Boston MA 02114

PATHAK, PRABHAKAR H, b Poona, India, Mar 21, 42; m 72; c 2. ELECTROMAGNETIC THEORY, ANTENNAS & SCATTERING. *Educ:* Univ Bombay, India, BSc, 62; La State Univ, BS, 65; Ohio State Univ, MSc, 70, PhD(elec eng), 73. *Prof Exp:* Instr, Dept Elec Eng, Univ Miss, 65-66; teaching asst, 67-70, grad res assoc, 70-74, res assoc, 73-77, sr res assoc, 77-81, asst prof, 82-86, RES SCIENTIST, ELECTROSCI LAB, OHIO STATE UNIV, 80-, ASSOC PROF, 86- *Concurrent Pos:* Engr, in Electronics Unit, Boeing Airplane Co, 66. *Mem:* Sigma Xi; Fel Inst Elec & Electronics Engrs. *Res:* Development of uniform asymptotic solutions which improve and extend the geometrical theory of diffraction solutions for solving antenna and scattering problems associated with complex structures such as aircraft, spacecraft and ships; diffraction by discontinuities in geometrical and electrical properties of surfaces; analysis of electromagnetic penetration and scattering; numerous publications, books and articles. *Mailing Add:* Electrosci Lab 1320 Kinnear Rd Columbus OH 43212

PATHAK, SEN, b Azamgarh, India, July 13, 40; m 61; c 4. CYTOGENETICS. *Educ:* Banaras Hindu Univ, BS, 61, MS, 63, dipl Ger lang, 65, PhD(cytogenetics), 67. *Prof Exp:* Jr res fel zool, Coun Sci & Indust Res, New Delhi, 63-67; sr res asst cytogenetics, Banaras Hindu Univ, 68-70; fel path, Baylor Col Med, 70-72; prof investr cell biol, 72-73, res assoc, 73-74, asst prof & asst biologist, 78-86, assoc prof, 81-86, GENETICIST & PROF, M D ANDERSON CANCER CTR, 86- *Concurrent Pos:* Vis prof, Univ de Sao Paulo, Brazil, 81-; vis prof exp biol, Baylor Col Med, Houston, 85-; clin prof, Sch Allied Health Sci, 86; mem fac, Grad Sch Biomed Sci, Univ Tex Health Sci Ctr, 74. *Mem:* AAAS; Tissue Cult Asn Am; Am Asn Human Genetics; Am Asn Cancer Res; Nat Acad Sci India. *Res:* Human and mammalian cytogenetics using various chromosomal banding techniques, autoradiography and drug effects on normal and malignant mammalian and human cells; studies on fine structures of mammalian meiotic chromosomes; chromosome analysis in cancer families with special reference to solid tumors; predisposition to human neoplasma and genetic mechanism of tumor metastasis in human and murine systems. *Mailing Add:* Dept Cell Biol M D Anderson Cancer Ctr Houston TX 77030

PATHANIA, RAJESHWAR S, b Kangra, India; Can citizen. STRESS CONTROL CRACKING, CORROSION CONTROL. *Educ:* Delhi Univ, India, BSc, 63; Indian Inst Sci, BEng, 65; Univ BC, PhD(metall eng), 70. *Prof Exp:* Res assoc corrosion, Univ BC, 70-72; res & develop engr corrosion, Atomic Energy Can, Chalk River, 72-79, sect head, 79-85; consult, 85-87, PROJ MGR CORROSION CONTROL, ELEC POWER RES INST, CALIF, 87- *Mem:* Nat Asn Corrosion Engrs. *Res:* Corrosion control in nuclear power plants; increase plant reliability by improvements in materials and water chemistry in current and future designs. *Mailing Add:* 548 Hubbard Ave Santa Clara CA 95051

PATHRIA, RAJ KUMAR, b Ramdas, India, Sept 19, 33; m 58; c 3. THEORETICAL PHYSICS, STATISTICAL MECHANICS. *Educ:* Panjab Univ, BSc, 53, MSc, 54; Univ Delhi, PhD(physics), 57. *Prof Exp:* Lectr physics, Univ Delhi, 58-61, reader, 61-64; vis prof, McMaster Univ, 64-65 & Univ Alta, 65-67; prof theoret physics, Panjab Univ, 67-69; vis prof physics, Univ Waterloo, 69-70 & Univ Windsor, 70-71; assoc prof, 71-74, PROF PHYSICS, UNIV WATERLOO, 74- *Concurrent Pos:* Sr res fel, Nat Inst Sci, India, Delhi Univ, 60-61; vis prof fel, Univ Wales, 70; vis scholar, Univ Calif San Diego & La Jolla Inst, 84-85; vis prof, San Diego State Univ, 88. *Mem:* Am Phys Soc; Can Asn Physicists; Indian Physics Asn; Int Astron Union. *Res:* Statistical mechanics of finite systems; lattice sums; phase transitions; relativity and cosmology. *Mailing Add:* Dept of Physics Univ of Waterloo Waterloo ON N2L 3G1 Can

PATI, JOGESH CHANDRA, b Baripada, India, Apr 3, 37; m 61; c 4. THEORETICAL PHYSICS. *Educ:* Ravenshaw Col, Utkal, India, BSc, 55; Univ Delhi, MSc, 57; Univ Md, PhD(physics), 60. *Prof Exp:* Tolman fel physics, Calif Inst Technol, 60-62; mem staff, Inst Advan Study, 62-63; from asst prof to assoc prof, 63-72, PROF PHYSICS, UNIV MD, COLLEGE PARK, 72- *Concurrent Pos:* Vis scientist, Tata Inst Fundamental Res, India, Int Ctr Theoret Physics, Trieste, 66-67 & Europ Orgn Nuclear Res, 66-67 & 71; vis prof, Univ Delhi, 71-72, Europ Orgn Nuclear Res, 74 & Int Ctr Theoret Physics, Trieste, 74 & 75; John Simon Guggenheim Mem fel, 79. *Mem:* Fel Am Phys Soc; fel Nat Acad Sci India. *Res:* Theoretical elementary particle physics. *Mailing Add:* Dept Physics & Astron Univ Md College Park MD 20740

PATIENCE, JOHN FRANCIS, b Thamesford, Ont, Nov 13, 51; m 74; c 3. DIETARY ELECTROLYTE BALANCE, POST MORTEM MUSCLE METABOLISM. *Educ:* Univ Guelph, Bsc(Agr), 74, MSc, 76; Cornell Univ, PhD(nutrit biochem), 85. *Prof Exp:* Swine exten specialist, Sask Agr, 75-78; head nutritionist, Feed Dept, Federated Coop Ltd, 78-82; vis fel, Animal Res Ctr Agr Can, 85-87; dir, Prairie Swine Ctr, Univ Sask, 87-91; PRES, PRAIRIE SWINE CTR, INC, 91- *Concurrent Pos:* Adj prof, Dept Animal & Poultry Sci, Univ Sask, 87-89, assoc prof, 89- *Mem:* Sigma Xi; Am Soc Animal Sci; Can Soc Animal Sci; Can Soc Nutrit Sci; Am Inst Nutrit. *Res:* Interrelationship among acid-base balance nutrient metabolism and dietary undetermined anion; impact of environmental stresses such as thermal stress; sulphate metabolism, especially epithelial transport mechanisms. *Mailing Add:* 65 Murphy Crescent Saskatoon SK S7J 2T5 Can

PATIL, GANAPATI P, b Sunasgaon, India, Feb 2, 34; m 59; c 4. STATISTICAL ECOLOGY, ENVIRONMENTAL STATISTICS. *Educ:* Univ Poona, BSc, 54, MSc, 55; Univ Mich, MS & PhD(math), 59; Indian Statist Inst, DSc, 75. *Hon Degrees:* DSc, Univ Parma, Italy; DLitt, Univ Poona, India. *Prof Exp:* Assoc statist, Indian Statist Inst, Calcutta, 55-57; lectr math, Univ Mich, 59-61, res assoc, Res Inst, 59-60; from asst prof to assoc prof, McGill Univ, 61-64; prof, 64-90, DISTINGUISHED PROF MATH STATIST, PENN STATE UNIV, UNIVERSITY PARK, 91- *Concurrent Pos:* Fac consult, Cooley Electronics Lab, Univ Mich, 60-61 & Ord Res Lab, Pa State Univ, 64-66; consult, US Forest Serv, Environ Protection Agency, Nat Marine Fisheries Serv & NIH; chmn, Int Statist Ecol Prog, 69-; vis prof, Univ NSW, 67, Univ Wis, 70 & Harvard Univ, 86-88; dir, Ctr Statist Ecol & Environ Statist, Pa State Univ; mem, Statist Rev Comt, Environ Protection Agency, 74, Comt Statist & Environ, Am Statist Asn, 83-88, Rev Comt on Acid Deposition, Environ Protection Agency-Am Statist Asn, 84-85 & Chesapeake Bay Stock Assessment Comt, Nat Oceanic & Atmospheric Admin, 85-90; chmn, Prog & Publ, Am Statist Asn Sect, Statist & Environ, 90- & Comt on Statist Distrib in Sci Work, 72-84; vchmn, Int Conf Biomath, Xian, China, 88; ed, Statist Ecol Ser & Statist Distrib in Sci Work Ser, Int Cooper Publ House, Fairland, Md, 79-; prin investr, Cooper Res Prog, Penn State, Environ Protection Agency & Nat Oceanic & Atmospheric Admin, 76-; bd mem, Int Ctr Theoret & Appl Ecol, Italy, 86- *Honors & Awards:* First Distinguished Statist Ecologist Award, Int Asn Ecol, 86. *Mem:* Fel AAAS; fel Am Statist Asn; fel Inst Math Statist; fel Royal Statist Soc; Int Statist Inst; Ecol Soc Am; Int Asn Ecol. *Res:* Mathematical and applied statistics; statistical ecology; environmental statistics; theory of statistical distributions; spatial statistics; survey design and sampling; encountered data analysis. *Mailing Add:* Dept Statist Penn St Univ University Park PA 16802

PATIL, KASHINATH ZIPARU, b Maharashtra, India; m 55; c 3. INDUSTRIAL CATALYSTS, CATALYTIC PROCESSES. *Educ:* Univ Poona, BS, 51, MS, 53; Indian Inst Technol, MTech, 55. *Prof Exp:* Res asst, Indian Inst Sci, 55-56; sr res asst, Sindri Fertilizers & Chem, 57-58; catalyst chemist, W R Grace & Co, 61-67; sr res chemist, Mallinckrodt Chem, 67-73; sr res chemist, Catalysis, Res & Develop Ctr, M W Kellogg Co, 74-82; CATALYST CONSULT, 83- *Concurrent Pos:* Abstractor, Chem Abstracts, Am Chem Soc, 65-75. *Mem:* Am Chem Soc; Catalysis Soc. *Res:* Catalytic and high pressure reactions; adsorption; catalysis and industrial catalysts; oxidation; hydrogenation; esterification; methanation; steam reforming processes; industrial catalysts. *Mailing Add:* 2914 Kevin Lane Houston TX 77043-1309

PATIL, POPAT N, b Chinchkhede, India, Oct 26, 34; m 64; c 3. PHARMACOLOGY, PHYSIOLOGY. *Educ:* Gujarat Univ, India, BPharm, 56; Ohio State Univ, MS, 60, PhD(pharmacol), 63. *Prof Exp:* From res asst to res assoc, 59-65, asst prof, 65-68, assoc prof, 68-72, PROF PHARMACOL, OHIO STATE UNIV, 72- *Concurrent Pos:* Us sr scientist award, Alexander von Humboldt Found, 80. *Mem:* Am Soc Pharmacol & Exp Therapeut. *Res:* Steric aspects of adrenergic drugs; tachyphylaxis; molecular pharmacology; melanins. *Mailing Add:* Lloyd M Parks Hall Col Pharm 500 West 12th Ave Columbus OH 43210

PATIL, SURESH SIDDHESHWAR, b Sholapur, India, May 2, 35; m 65. PLANT PATHOLOGY. *Educ:* Univ Poona, BSc, 55; Ore State Univ, MS, 59, PhD(plant path), 62. *Prof Exp:* Asst plant path, Ore State Univ, 62-63; asst plant pathologist & Nat Inst Allergy & Infectious Dis res fel, Conn Agr Exp Sta, 63-69; assoc prof, 69-73, PROF PLANT PATH, UNIV HAWAII, 73-, PROG DIR, BIOTECHNOL PROG, 85- *Concurrent Pos:* Vis fel, Dept Biochem, Cambridge Univ, 75-76; vis prof, Kewalo Marine Lab, Univ Hawaii, 83. *Mem:* AAAS; Am Phytopath Soc; Am Soc Plant Physiol; Int Soc Plant Molecular Biol. *Res:* Biochemistry and molecular biology of host-pathogen interactions; bacterial phytotoxins; molecular biology of post-harvest diseases; disease resistance. *Mailing Add:* Dept Plant Path Univ Hawaii 3190 Maile Way Honolulu HI 96822

PATINKIN, SEYMOUR HAROLD, b Chicago, Ill, Mar 25, 26; m 56; c 1. ORGANIC CHEMISTRY. *Educ:* Ill Inst Technol, BS, 49, PhD(chem), 54. *Prof Exp:* Asst US Naval Ord contract, Ill Inst Technol, 50-51 & 52-53; res chemist, Sinclair Res, Inc, 53-66; assoc prof, 66-72, PROF CHEM, ROOSEVELT UNIV, 72-, DEPT CHAIR, 89- *Mem:* AAAS; Sigma Xi; Am Chem Soc. *Res:* Hydrocarbon chemistry; instrumental analysis; free radical chemistry; nuclear magnetic resonance; free radical reactions of organosilicons with aromatic compounds catalyzed with certain metal salts. *Mailing Add:* 4610 W Dempster St Skokie IL 60076

PATMORE, EDWIN LEE, b Connelsville, Pa, Dec 26, 34; m 58; c 2. ORGANIC CHEMISTRY, INORGANIC CHEMISTRY. *Educ:* Eastern Nazarene Col, BS, 58; Univ Conn, MS, 61, PhD(org chem), 63. *Prof Exp:* Chemist, Chem Res Sect, Beacon Res Labs, 62-63, sr chemist, 63-67, res chemist, Lubricants Res Sect, 67-72, group leader, 72-77, asst supvr, 77-80, supvr, 80-82, mgr, Lubricants, 82-85, mgr, New Technol, 85-87, MGR, CATALYSIS RES, TEXACO, INC, 87- *Mem:* Am Chem Soc; Sigma Xi. *Res:* Organic synthesis; free radical chemistry of coordination compounds; oxidation; petrochemicals; carbanion reactions. *Mailing Add:* PO Box 174 Fishkill NY 12524

PATNAIK, AMIYA KRISHNA, b Orissa, India, Mar 19, 30; m 55; c 4. VETERINARY PATHOLOGY, ONCOLOGY. *Educ:* Madras Vet Col, DVM, 51, MS, 62. *Prof Exp:* Res officer salmonellosis, Govt Orissa, India, 62-63, res officer stephanofilariasis, 63-64, vet pathologist, State Vet Labs, 63-64; resident path, Animal Med Ctr, 64-67; pathologist, Food & Drug Res Lab, Inc, 66-68; STAFF PATHOLOGIST, ANIMAL MED CTR, 68- *Honors & Awards:* Ralston Purina Small Animal Res Award For Work In Oncol, 84; Beecham Award Res Excellence, 87. *Mem:* Am Asn Vet Clinicians; Am Vet Med Asn. *Res:* Comparative oncology dealing with histology and biologic behavior and veterinary oncology; immunologic and therapeutic aspects of neoplasms; small animal diseases. *Mailing Add:* Animal Med Ctr 510 E 62nd St New York NY 10021

PATNODE, ROBERT ARTHUR, b Mankato, Minn, May 8, 18; m 46; c 3. IMMUNOLOGY. *Educ:* Univ Minn, BA, 44, MS, 46, PhD(bact), 53. *Prof Exp:* Bacteriologist, Tuberc Unit, Commun Dis Ctr, USPHS, Ga, 47-50, actg chief, 50-53; chief tuberc res lab, Vet Admin Hosp, Washington, DC, 53-60; assoc prof, 60-66, prof, 66-83, EMER PROF MICROBIOL & IMMUNOL, HEALTH SCI CTR, UNIV OKLA, 84- *Concurrent Pos:* Consult grad coun, George Washington Univ, 59-60; vis prof microbiol, Univ Otago, New Zealand, 79. *Mem:* Am Thoracic Soc; Am Soc Microbiol; Am Asn Immunologists; fel Am Acad Microbiol. *Res:* Mechanisms of host resistance in tuberculosis; allergy and immunology in tuberculosis; delayed hypersensitivity. *Mailing Add:* Health Sci Ctr PO Box 26901 Oklahoma City OK 73190

PATON, BRUCE CALDER, b Coonoor, India, Aug 28, 25; US citizen; m 55; c 3. CARDIOVASCULAR SURGERY. *Educ:* Univ Edinburgh, BM & ChB, 51; FRCS, 58; Am Bd Surg, dipl, 64. *Prof Exp:* Chief resident surg, Vet Admin Hosp, Denver, Colo, 59-60; dir, Halstead Lab Exp Surg, 60-74, from instr to prof, 60-70, chief cardiac surg serv, 62-79, actg dean, 78-79, CLIN PROF SURG, COLO MED CTR, DENVER, 79- *Concurrent Pos:* Fulbright scholar, 53-55; NIH fel, 60-61; prin investr, USPHS grants, 60-74 & Am Heart Asn grants, 64-69; vis prof, Univ Med Sci, Bangkok, 64; mem exec comt coun cardiovasc surg, Am Heart Asn, 70-74; mem surg study sect, NIH, 71-75; pres, Denver Acad Sci, 79, Rocky Mountain Traumatic Soc, 80; lectr, Mountain Med Coupes, CA Sierra Club, 85-88; chmn, bd trustees, Colo Outward Bound Sch, 86-88; mem bd & vchmn, Snake River Health Serv Inc, Keystone Co, 81-; sr med officer, Mountain Med Inst, 88- *Honors & Awards:* William Leslie Prize in Med, 56. *Mem:* Am Surg Asn; Am Asn Thoracic Surg; Soc Vascular Surg; Soc Univ Surgeons; Int Cardiovasc Soc. *Res:* Physiology of extracorporeal circulation, hypothermia, prosthetic valves and cardiac valve grafts; accidental hypothermia; frostbite. *Mailing Add:* 5380 E Mansfield Ave Denver CO 80237

PATON, DAVID, b Baltimore, Md, Aug 16, 30; m 85; c 1. MEDICINE, OPHTHALMOLOGY. *Educ:* Princeton Univ, BA, 52; Johns Hopkins Univ, MD, 56; Am Bd Ophthal, dipl, 64. *Hon Degrees:* DSc, Bridgeport Univ, 84 & Princeton Univ, 85. *Prof Exp:* Intern med, New York Hosp-Cornell Med Sch, 56-57; sr asst surg, Nat Inst Neurol Dis & Blindness, 57-59; resident ophthal, Wilmer Inst, Johns Hopkins Hosp, 59-62 & 63-64, from asst prof to assoc prof, 64-70; chmn dept, Baylor Col Med, 71-81, prof ophthal, 71-82; med dir, King Khaled Eye Specialist Hosp, Riyadh, Saudi Arabia, 83-84; founder, chmn & chief med officer, Ocusystems, Inc, 84-86; PROF CLIN OPHTHAL, CORNELL UNIV COL MED, 86- *Concurrent Pos:* Markle scholar acad med, 67-71; past mem, Am Bd Ophthal, past chmn; chmn, dept ophthal, Cath Med Ctr, Brooklyn & Queens, 86- *Honors & Awards:* Medal, Kingdom of Jordan, 2nd Order, 84; Presidential Citizens Medals, 87; Legion of Honor, Chevalier, Fr Govt, 88. *Mem:* Am Acad Ophthal; fel Am Col Surgeons; Pan-Am Asn Ophthal. *Res:* Opthalmology practise management; model system of new techniques for total eye care. *Mailing Add:* St Joseph's Hosp 158-40 79 Ave Flushing NY 11366

PATON, DAVID MURRAY, b East London, SAfrica, Feb 26, 38; m 62; c 2. PHARMACOLOGY. *Educ:* Univ Cape Town, MB & ChB, 61; Univ Witwatersrand, MD, 74, DSc, 81; FRCPC, 78; FRACP, 82; FI BIOL, 85; Moore Theol Col, CTh, 89; Australian Col Theol, ThDIP, 89. *Prof Exp:* Intern internal med & surg, Groote Schuur Hosp, Cape Town, 62-63; lectr pharmacol, Univ Cape Town, 63-64; from asst prof to assoc prof, 66-73, actg chmn dept, 74-75, asst dean med, 77-78, prof pharmacol, Univ Alta, 73-78; prof & head pharmacol, Univ Auckland, 78-88; PROF & CHMN, ORAL BIOL, UNIV ALBERTA, 88- *Concurrent Pos:* James Hudson Brown fel pharmacol, Yale Univ, 65-66; Can Heart Found res fel, Univ Alta, 66-68, sr res fel, 68-72, scholar, 72-74; vis prof pharmacol, Univ Wurzburg, 71-72, Univ Cape Town, 84; vis fel, Corpus Christi Col, Oxford, 84. *Mem:* Int Brain Res Orgn; Germ Pharmacol Soc; Brit Pharmacol Soc; Am Soc Pharmacol Exp Ther. *Res:* Pharmacology of adenosine; mechanisms of transport and release of catecholamines. *Mailing Add:* Dept Oral Biol Univ Alberta Edmonton AB T6G 2N8 Can

PATON, NEIL (ERIC), b Auckland, New Zealand, Dec 3, 38; m 64; c 2. PHYSICAL METALLURGY, MATERIALS SCIENCE. *Educ:* Univ Auckland, BE, 61, ME, 62; Mass Inst Technol, PhD(metall), 69. *Prof Exp:* Res metallurgist, Atlas Steels, Can, 62-65; group leader struct mat dept, 69-77, actg dir struct mat dept, 77-78, dir mat synthesis & processing dept, 78-81, dir mat & struct, 81-83, DIR MAT ENG & TECHNOL, ROCKETDYNAMIC DIV, ROCKWELL INT, 83- *Concurrent Pos:* Mem, Solid State Sci Comt, Nat Acad Sci, 74-78. *Mem:* Sigma Xi; Am Inst Mining, Metall & Petrol Engrs; Am Soc Metals; AAAS. *Res:* Materials synthesis and processing; alloy properties; microstructure analysis. *Mailing Add:* 297 Upper Lake Rd Thousand Oaks CA 91361

PATRASCIOIU, ADRIAN NICOLAE, b Romania; US citizen. THEORETICAL PHYSICS. *Educ:* Mass Inst Technol, PhD(physics), 73. *Prof Exp:* Mem physics, Inst Advan Study, Princeton, 73-75; asst res physicist, Univ Calif, San Diego, 75-77; asst prof physics, 77-80, ASSOC PROF PHYSICS, UNIV ARIZ, 80-, PROF PHYSICS. *Concurrent Pos:* A P Sloan fel, 80. *Res:* Elementary particles. *Mailing Add:* Dept Physics Univ Ariz Tucson AZ 85721

PATRIARCA, PETER, b Utica, NY, Jan 7, 21; m 45; c 3. HIGH-TEMPERATURE NUCLEAR REACTOR COOLING SYSTEMS. *Educ:* Rensselaer Polytech Inst, BME, 48, MME, 50. *Prof Exp:* Mgr, 50-85, CONSULT, OAK RIDGE NAT LAB, 85- *Mem:* Sigma Xi; fel Am Soc Metals; hon mem Am Welding Soc. *Res:* Materials research in support of high-temperature nuclear reactor cooling systems and heat exchangers including mechanical behavior, structural design, joining, nondestructive testing, corrosion and alloy development. *Mailing Add:* 107 Picea Dr Austin TX 78734

PATRIARCHE, MERCER HARDING, b Waltham, Mass, Apr 18, 16; m 46; c 2. FISH BIOLOGY. *Educ:* Mich State Univ, BS, 37, MS, 48. *Prof Exp:* Fisheries biologist, State Conserv Comn, Mo, 48-56; biologist in chg, Rifle River Fisheries Res Sta, State Dept Conserv, Mich, 56-66; fisheris res biologist, Inst Fisheries Res, 66-78, biologist in chg, 78-81; RETIRED. *Concurrent Pos:* Ed, North Am J Fisheries Mgt, 81- *Mem:* Am Fisheries Soc. *Res:* Fish population dynamics; biology and life histories of fish; fish management techniques; basic productivity of lakes and streams. *Mailing Add:* 2677 Esch Ave Ann Arbor MI 48104

PATRIC, JAMES HOLTON, b Rockville, Conn, Dec 24, 22; m 50; c 4. FOREST HYDROLOGY. *Educ:* Univ Conn, BS, 47; Harvard Univ, MF, 57. *Prof Exp:* Work unit conservationist, Soil & Conserv Serv, USDA, 48-55, res forester, San Dimas Exp Forest, US Forest Serv, Calif, 57-60, Coweeta Hydrol Lab, NC, 60-65 & Inst North Forestry, Alaska, 65-67, proj leader watershed mgt res, Parsons Timber & Watershed Lab, Northeastern Forest Exp Sta, US Forest Serv, 67-81; RETIRED. *Concurrent Pos:* Adj prof, WVa Univ, 68-; Bullard fel advan res, Harvard Univ, 72. *Mem:* Soc Am Foresters; Am Geophys Union. *Res:* Forest hydrology; evaporative losses from forest land; managing forest land to produce more and better water. *Mailing Add:* Rt Eight Box 233 L Greenville TN 37743

PATRICK, CHARLES RUSSELL, b Brookfield, Ga, Mar 30, 40; m 69; c 1. ENTOMOLOGY. *Educ:* Clemson Univ, BS, 63; Auburn Univ, MS, 67; Miss State Univ, PhD(entom), 70. *Prof Exp:* Entomologist, Tenn Dept Agr, 70-80; EXTEN ENTOMOLOGIST, AGR EXTEN SERV, UNIV TENN, 80- *Mem:* Assoc Sigma Xi; Entom Soc Am. *Res:* Field crop insects (corn, soybeans, small grain, stored grain). *Mailing Add:* Rte 5 Jackson TN 38301

PATRICK, EDWARD ALFRED, b Wheeling, WVa, Oct 7, 37; m 60; c 4. COMPUTER SCIENCE, MEDICINE. *Educ:* Mass Inst Technol, BS, 60, MS, 62; Purdue Univ, PhD(elec eng), 66; Ind Univ, Indianapolis, MD, 74. *Prof Exp:* From asst prof to assoc prof elec eng, Purdue Univ, 66-74, prof, 74-81; assoc prof community health sci, Sch Med, Ind Univ, Purdue, 70-80; pres, Patrick Consult Inc, West Lafayette, 77-88; RES PROF, ELEC & COMPUT ENG, UNIV CINCINNATI, 85- *Concurrent Pos:* Staff mem, Instrument Lab, Mass Inst Technol, 59-62, instr, Inst, 60-62; grants, US Air Force, 66-69, Naval Ships Syst Command, 66-70, Naval Air Syst Command, 68-70 & NSF, 68-70 & 74-76; consult, Sylvania Appl Res Lab, 60-63, Tex Instrument Corp, 66-70, Dupont Corp, 67-69, Regenstrief Inst Health Care Delivery, 70-, Univ Ill, 74- & Jewish Hosp, Cincinnati, Ohio, 74-; assoc ed, Computers in Biol & Med, 69- *Mem:* AMA; Pattern Recognition Soc; Sigma Xi; Inst Elec & Electronics Engrs. *Res:* Application of engineering to clinical medicine. *Mailing Add:* 5947 Coquina Key Dr Apt C Indianapolis IN 46224

PATRICK, GRAHAM ABNER, b Maysville, NC, Apr 30, 46; m 64; c 2. PHARMACOLOGY, NEUROBIOLOGY. *Educ:* Univ NC, BS, 69, PhD(pharmacol), 73. *Prof Exp:* From instr to asst prof,73-85, ASSOC PROF PHARMACOL, MED COL VA, 85- *Concurrent Pos:* Res grant, Nat Inst Neurol & Commun Dis & Stroke, 76-79, Environ Protection Agency, 79-81, Nat Inst Drug Abuse, 89- *Mem:* Sigma Xi; Am Soc Pharmacol & Exp Therapeut. *Res:* Drug dependence and tolerance development; central neurotransmitters. *Mailing Add:* Dept Pharmacol Med Col Va Box 613 Richmond VA 23298

PATRICK, JAMES BURNS, b Kansas City, Mo, Oct 16, 23; m 44; c 4. ORGANIC CHEMISTRY. *Educ:* Mass Inst Technol, BS, 49; Harvard Univ, MA, 50, PhD(chem), 52. *Prof Exp:* Res chemist, Nat Heart Inst, 52; res chemist, Lederle Labs, Am Cyanamid Co, NY, 53-63, res group leader, 63-67; chmn dept, 67-74, PROF CHEM, MARY BALDWIN COL, 67- *Mem:* Am Chem Soc. *Res:* Chemistry of natural products; structure of alkaloids and antibiotics. *Mailing Add:* Dept of Chem Mary Baldwin Col Staunton VA 24401

PATRICK, JAMES EDWARD, b Westfield, Mass, Oct 25, 44; m 75; c 1. DRUG METABOLISM, PHARMACOKINETICS. *Educ:* Marist Col, BA, 66; Pa State Univ, PhD(org chem), 71. *Prof Exp:* Fel steroid biochem, Worcester Found Exp Biol, 71-73; group leader biotransformation, Drug Metab Sect, Ortho Pharmaceut Corp, 73-80; mgr analytical biochem/drug metab & pharmacokinetics, 80-83, ASSOC DIR DRUG METAB, SCHERING CORP, 84- *Concurrent Pos:* Adj instr, Somerset County Col, 74-76. *Mem:* Am Chem Soc; Am Pharmaceut Asn; Int Soc Study Xenobiotics; Am Asn Pharm Sci. *Res:* Drug disposition and assay development; absorption, distribution, excretion and biotransformation studies of drugs in animals and man; development of analytical methods for drugs in biological fluids. *Mailing Add:* PO Box 494 Bellemead NJ 08502

PATRICK, JAMES R, b Athens, Ohio, Feb 10, 31; m 55; c 3. PEDIATRICS, PATHOLOGY. *Educ:* Ohio Univ, AB, 52; Yale Univ, MD, 56. *Prof Exp:* Intern pediat, Yale-New Haven Med Ctr, 56-57, asst resident, 57-58, chief resident, 58-59; instr pediat, Yale Univ, 58-59; from instr to asst prof path, Univ Southern Calif, 61-65; from asst prof to assoc prof, Georgetown Univ, 65-69; chmn dept path, 69-78, PROF PATH & PEDIAT, MED COL OHIO, 69- *Concurrent Pos:* Fel path, Yale Univ, 59-61; assoc pathologist, Children's Hosp, Los Angeles, 61-65; dir labs & chief pathologist, Children's Hosp, Washington, DC, 65-69. *Mem:* AAAS; Am Asn Path & Bact; Int Acad Path. *Mailing Add:* Med Col Toledo Ohio Toledo OH 43699

PATRICK, MERRELL LEE, b Cynthiana, Ky, July 2, 33; m 59; c 2. APPLIED MATHEMATICS, COMPUTER SCIENCE. *Educ:* Eastern Ky Univ, BS, 55; Univ Ky, MS, 56; Carnegie Inst Technol, PhD(math), 64. *Prof Exp:* Assoc engr, Lockheed Missiles & Space Co, 56-59, sr scientist, 59-62; chmn comput sci, 78-82, PROF COMPUT SCI,DUKE UNIV, 77-; PROG DIR, NEW TECHNOL PROG, DIV ADV SCIENTIFIC COMPUT, NSF, 90- *Concurrent Pos:* Consult, NASA Langley Res Ctr, Inst Comput Applns Sci & Eng, 82-90, assoc mem, 86-90. *Mem:* Soc Indust & Appl Math; Asn Comput Mach; Inst Elec & Electronic Engrs. *Res:* Equations; parallel algorithms and architectures for scientific computations. *Mailing Add:* Dept of Comput Sci Duke Univ Durham NC 27706

PATRICK, MICHAEL HEATH, b Chicago, Ill, Mar 31, 36; m 76; c 4. AGRICULTURAL BIOTECHNOLOGY. *Educ:* Univ Calif, Santa Barbara, AB, 58; Univ Chicago, PhD(biophys), 64. *Prof Exp:* Res assoc biochem, Johns Hopkins Univ, 64-66; from asst prof to assoc prof, 66-80, prof biol, Univ Tex, Dallas, 80-86; CONSULT & LECTR, 86- *Mem:* AAAS; Biophys Soc; Sigma Xi. *Res:* Systems approach to sustainable agriculture, educational programs in biotechnology & bioethics. *Mailing Add:* Rte 1 Box 41A Pine River WI 54965

PATRICK, RICHARD MONTGOMERY, b Rockford, Ill, Sept 24, 28; m 58; c 3. PHYSICS, PHYSICAL CHEMISTRY. *Educ:* Purdue Univ, BS, 50, MS, 52; Cornell Univ, PhD(aero eng), 56. *Prof Exp:* Scientist plasma physics, 56-69, VPRES PLASMA PHYSICS, PRES PLASMA ADVAN PROD & DIR ENERGY TECHNOL, AVCO EVERETT RES LAB, 69- *Mem:* Fel Am Phys Soc. *Res:* Atomic and molecular physics; aerodynamics; fluid flows; plasma physics. *Mailing Add:* 94 Arlington St Winchester MA 01890

PATRICK, ROBERT F(RANKLIN), b Helmetta, NJ, Oct 10, 21; m 49; c 3. CERAMICS. *Educ:* Rutgers Univ, BS, 43, MS 47, PhD(ceramics), 49. *Prof Exp:* Asst, Rutgers Univ, 46-47; sr res ceramist, Pemco Corp, 49-52, asst dir res, 52-60; mgr steel refractory res, Corning Glass Works, 60-62; mgr tech serv, 62-64, tech mgr-bonded, 65-66, DIR TECH SERV, CORHART REFRACTORIES CO, 66- *Mem:* Fel Am Ceramic Soc; Am Soc Testing & Mat; Nat Inst Ceramic Engrs; Am Inst Mining, Metall & Petrol Engrs. *Res:* Physical and chemical properties of inorganic glasses and coatings for metals; porcelain enamel technology; color measurement; properties of refractory oxide systems; crystallization from melts; research and development refractories; refract; application engineering. *Mailing Add:* 2222 Village Dr Louisville KY 40205

PATRICK, ROBERT L, NEUROPHARMACOLOGY, NEURO TRANSMITTERS. *Educ:* Duke Univ, PhD(biochem), 72. *Prof Exp:* ASSOC PROF NEURO MED SCI PHARMACOL, BROWN UNIV, 77- *Mailing Add:* Dept Neurobiol Brown Univ Prog Med 97 Waterman Ave Providence RI 02912

PATRICK, RUTH (MRS CHARLES HODGE IV), b Topeka, Kans, Nov 26, 07; m 31; c 1. BOTANY. *Educ:* Coker Col, BS, 29; Univ Va, MS, 31, PhD(bot), 34. *Hon Degrees:* Numerous from US cols & univs. *Prof Exp:* Asst, Coker Col, 29; asst res, Temple Univ, 34; asst cur micros, 39-47, chmn bd, 73-76, chmn dept limnol, 47-73, CUR LIMNOL, ACAD NATURAL SCI, 47-, FRANCIS BOYER RES CHAIR, 73-, HON CHMN BD, 76- *Concurrent Pos:* Cur, Leidy Micros Soc, 39-47; trustee, Coker Col, 41-; with Am Philos Soc exped, Mex, 47; dir, Henry Found Bot Res, 48-; lectr, Univ Pa, 52-70, adj prof, 70-; US deleg, Int Cong Limnol, 53; leader, Catherwood Exped, Amazon River, 55; trustee, Chestnut Hill Acad, 57-72; mem bd, E I du Pont de Nemours & Co & Pa Power & Light Co, 72-; mem, comt sci & pub policy, Nat Acad Sci, 73-75, mem, subcomt geochem environ, comt geochem, 79-, chmn, sect pop biol, evolution & ecol, 80-83; mem, exec adv comt, Environ Protection Agency, 74-79, chmn, panel ecol, 74-76. *Honors & Awards:* Tyler Ecol Award, 75; James McGirr Kelly Award, Nat Asn Water Co, 90. *Mem:* Nat Acad Sci; Bot Soc Am; Am Soc Limnol & Oceanog; Phycol Soc Am (pres, 54-57); Am Inst Biol Sci; Am Soc Naturalists (pres, 75-77); AAAS; Am Acad Arts & Sci; Am Philos Soc; Sigma Xi; Ecol Soc Am; Am Soc Plant Taxonomists. *Res:* Taxonomy, ecology and physiology of diatoms; limnology; biodynamic cycle of rivers; forces that determine the diversity of ecosystems. *Mailing Add:* Limnol Dept Acad Nat Sci 19th & Parkway Philadelphia PA 19103

PATRICK, TIMOTHY BENSON, b Huntington, WVa, Dec 25, 41; m 64; c 2. ORGANIC CHEMISTRY. *Educ:* Marshall Univ, BS, 63; WVa Univ, PhD(chem), 67. *Prof Exp:* NASA trainee, WVa Univ, 65-67; fel chem, Ohio State Univ, 67-69; asst prof, 69-72, assoc prof, 72-76, PROF CHEM, SOUTHERN ILL UNIV, 76- *Mem:* Am Chem Soc. *Res:* Organic synthesis; physical organic chemistry; heterocyclic chemistry; organofluorine; carbene chemistry; steroid chemistry. *Mailing Add:* Chem Dept Southern Ill Univ Edwardsville IL 62026

PATRICK, WESLEY CLARE, b North Branch, Mich, Dec 29, 51; m 73; c 2. MINING ENGINEERING. *Educ:* Mich Technol Univ, BS, 74; Univ Mo-Rolla, MS, 75, PhD(mining eng), 78. *Prof Exp:* Staff engr, St Joe Minerals Corp, 74; grad asst rock mech, Univ Mo-Rolla, 74-78; 1st lieutenant, US Army Chem Corps, 78; staff engr, Lawrence Livermore Nat Lab, 78-80, proj scientist, 80-83, task leader, 83-85; exp prog mgr, IT Corp, 85-89; TECH DIR, SOUTHWEST RES INST, 87- *Mem:* Soc Mining Engrs; Soc Exp Stress Anal. *Res:* Deep geologic storage of high level wastes from commercial nuclear reactors and transuranic nuclear wastes from national defense activities; radiation damage, heat transfer, and geomechanical response of test repositories. *Mailing Add:* 9611 Alisa Brooke San Antonio TX 78250-6134

PATRICK, WILLIAM H, JR, b Johns, Miss, Nov 9, 25; m 51; c 4. MARINE SCIENCES. *Educ:* La State Univ, Baton Rouge, BS, 50, MS, 51, PhD(soils), 54. *Hon Degrees:* Ghent Univ, Belgium, Dr. *Prof Exp:* From asst prof to prof agron, 53-76, prof, 77-78, BOYD PROF MARINE SCI, LA STATE UNIV, BATON ROUGE, 78- *Mem:* Fel Soil Sci Soc Am; Am Soc Agron; Int Soc Soil Sci; fel AAAS. *Res:* Physicochemical properties of and reactions in soils, particularly wetland soils. *Mailing Add:* Ctr Wetlands Res La State Univ Baton Rouge LA 70803

PATRICK, ZENON ALEXANDER, b Montreal, Que, June 3, 24; m 49; c 3. PLANT PATHOLOGY. *Educ:* McGill Univ, BSc, 48; Univ Toronto, PhD(plant path), 52. *Prof Exp:* Plant pathologist, Can Dept Agr, 52-65; PROF PLANT PATH, UNIV TORONTO, 65- *Concurrent Pos:* Vis prof, Univ Calif, Berkeley, 59-60; mem comt biol control soil borne plant pathogens, NSF, 63-65; consult, Hort Exp Sta, Ont Dept Agr, 65- *Mem:* AAAS; Am Phytopath Soc; Can Phytopath Soc. *Res:* Biology of soil borne pathogens; influence of soil organic matter on root diseases; soil toxins; biological control of plant pathogens. *Mailing Add:* Dept Bot Univ Toronto Toronto ON M5S 1A1 Can

PATRONIS, EUGENE THAYER, JR, b Quincy, Fla, Feb 26, 32; m 51; c 2. PHYSICS. *Educ:* Ga Inst Technol, BS, 53, PhD(physics), 61. *Prof Exp:* Instr physics, Ga Inst Technol, 52-58; res assoc, Brookhaven Nat Lab, 57-59; from asst prof to assoc prof, 59-68, PROF PHYSICS, GA INST TECHNOL, 68- *Concurrent Pos:* Consult, McGraw-Hill Bk Co, 59-61; Indust consult, 61- *Mem:* AAAS; Am Phys Soc; Soc Motion Picture & TV Engrs; Audio Eng Soc. *Res:* Nuclear physics; electronics, electronic instruments and acoustics. *Mailing Add:* Sch Physics Ga Inst Technol Atlanta GA 30332-0430

PATSAKOS, GEORGE, b New York, NY, Mar 24, 42. THEORETICAL PHYSICS. *Educ:* Columbia Col, AB, 62; Stanford Univ, PhD(physics), 69. *Prof Exp:* Res assoc physics, Ind Univ, Bloomington, 68-70; vis asst prof, 70-71, asst prof, 71-76, ASSOC PROF PHYSICS, UNIV IDAHO, 76- *Concurrent Pos:* Vis fac, New Sch Soc Res, NY, 79; assoc prof astron, Wash State Univ, 81; staff engr, Optics Lab, Hughes Aircraft Co, 85-86; fac fel, Jet Propulsion Lab, Am Soc Eng Educ, 84-85. *Mem:* AAAS; Am Asn Physics Teachers; Am Phys Soc; Astron Soc Pac. *Res:* Electromagnetic theory, foundations of quantum mechanics, gravitation. *Mailing Add:* Dept Physics Univ Idaho Moscow ID 83843

PATSCH, WOLFGANG, b Wels, Austria, Aug 10, 46; m 70; c 3. ATHEROSCLEROSIS, METABOLISM. *Educ:* Univ Innsbruck, Austria, MD, 71; Univ Uppsala, Sweden, cert biochem, 78. *Prof Exp:* Resident med, Univ Innsbruck, Austria, 71-74, res intern, 75-77, asst prof, 77-81; vis asst prof prev med, Wash Univ, St Louis, 78-80, asst prof prev med & med, 80-82, dir, Lipid Lab, 79-82; asst prof, 82-85, PROF, DEPT MED, BAYLOR COL, 85- *Concurrent Pos:* mem, Atherosclerosis Coun, Am Heart Asn, 80; specialist internal med, NIH, steering comt study immunol, 85-; grants, Am Heart Asn, 84 & NIH, 85, 86 & 88. *Mem:* Am Fedn Clin Res; Europ Lipoprotein Club; Am Soc Clin Nutrit; Europ Soc Clin Investr; Am Heart Asn. *Res:* Structure, function, and metabolism of lipoproteins; synthesis of apoproteins and lipoproteins by hepatocyte cultures. *Mailing Add:* Dept Med Baylor Col Med MS A601 6565 Fannin Houston TX 77030

PATSIGA, ROBERT A, b New Brighton, Pa, Sept 8, 34; m 56; c 4. ORGANIC CHEMISTRY, POLYMER CHEMISTRY. *Educ:* Geneva Col, BS, 57; State Univ NY Col Forestry, Syracuse Univ, PhD(org chem), 62. *Prof Exp:* Res chemist, Koppers Co, Inc, 61-63; fel polymer chem, Villanova Univ, 63-64; asst prof org chem, Bethany Col, WVa, 64-68; from asst prof to assoc prof, 68-73, PROF CHEM, IND UNIV PA, 73- *Mem:* Am Chem Soc; Sigma Xi. *Res:* Synthesis of new polymers, polymerization mechanisms, free radical reactions and mechanisms. *Mailing Add:* Dept Chem Ind Univ Indiana PA 15705

PATSIS, ANGELOS VLASIOS, b Athens, Greece, May 7, 28; nat US; m 54; c 4. PHYSICAL CHEMISTRY, CHEMICAL ENGINEERING. *Educ:* Athens Univ, BS, 53; Texas A&M Univ, BS, 55; Western Reserve Univ, MS, 58, PhD(chem), 59. *Prof Exp:* Asst anal, Pvt Testing Lab, 47-53; res technologist, Res Lab Agr Tech, Greece, 53-54; exchange vis scientist, Tex Petrol Res Comt, Agr & Mech Univ Tex, 54-55; asst, Off Naval Res, 55 & Western Reserve Univ, 55-59; res chemist, Calif Res Corp, Stand Oil Co Calif, 59-60 & Finishes Dept, E I du Pont de Nemours & Co, Inc, NY, 60-66; assoc prof, 66-70, PROF CHEM, STATE UNIV NY COL NEW PALTZ, 70-, CHMN DEPT. *Mem:* Am Chem Soc; NY Acad Sci; Am Asn Univ Professors; Electrochem Soc; AAAS; Sigma Xi. *Res:* Physical properties of hydrocarbons; oil and gas reservoirs; ultrasonics; colloids; polymers; plastics; coatings; elastomers. *Mailing Add:* Dept of Chem State Univ of NY Col New Paltz NY 12561

PATT, LEONARD MERTON, b Quincy, Mass, Apr 27, 48; m 72; c 4. NATURAL PRODUCT PURIFICATION, HIGH PRESSURE LIQUID CHROMATOGRAPHY. *Educ:* Northeastern Univ, BA, 71; Univ Ariz, PhD(biochem), 76. *Prof Exp:* Fel, Div Biochem Oncol, Fred Hutchinson Cancer Res Ctr, 76-79; sr fel, dept path, Sch Pub Health & Community Med, Univ Wash, 76-79; sr res fel, Virginia Mason Res Ctr, 79-83; vpres, immunogenics corp, 82-86; RES MGR, PROCYTE CORP, 87- *Concurrent Pos:* Lab mgt, bus & tech doc writing & review. *Mem:* AAAS; Am Soc Biol Chemists. *Res:* Control of cell proliferation and function by physiological factors and pharmacological agents; purification and characterization of growth modulating factors from tissues and other natural products; study of biochemical agents for wound healing. *Mailing Add:* 12016 40th Ave NE Seattle WA 98125

PATT, YALE NANCE, b Medford, Mass, June 29, 39. COMPUTER ARCHITECTURE, COMPUTER MICROARCHITECTURE. *Educ:* Northeastern Univ, BS, 62; Stanford Univ, MS, 63, PhD(elec eng), 66. *Prof Exp:* Assoc prof comput sci, NC State Univ, 69-76; prof comput sci, San Francisco State Univ, 76-88; PROF ELEC ENG & COMPUTER SCI, UNIV MICH, 88- *Concurrent Pos:* Consult, Digital Equip Corp, 77-, NCR Corp, 86-; co-dir, Berkeley High Performance Comput Group, 84-88; vis prof comput sci, Univ Calif, Berkeley, 79-88. *Mem:* Inst Elec & Electronic Engrs; Asn Comput Mach. *Res:* Microarchitecture for very high performance computers; developing mechanisms for obtaining performance improvement of several orders of magnitude over what is available today. *Mailing Add:* Dept Elec Eng & Computer Sci 1301 Beal Ave Ann Arbor MI 48109-2122

PATTABHIRAMAN, TAMMANUR R, b Kancheepuram, India, Nov 3, 34. CHEMISTRY, MARINE BIOLOGY. *Educ:* Univ Madras, India, BSc, 56, MA, 57; Univ Hawaii, PhD(chem), 65. *Prof Exp:* Lectr, Univ Madras, India, 58-61; scientist, Coun Sci & Indust Res, India, 65-68; NIH grants & res fel chem, Univ Okla, 68-71; NIH GRANT & RES FEL, UNIV SOUTHERN CALIF, 71- *Mem:* Am Chem Soc; Int Soc Toxinol. *Res:* Natural products, plant and marine origin, isolation, structure elucidation and synthesis; chemotaxonomy. *Mailing Add:* Lab Neurol Res Box 323 1200 N State St Los Angeles CA 90033

PATTEE, HAROLD EDWARD, b Phoenix, Ariz, June 27, 34; m 56; c 7. PLANT PHYSIOLOGY, PLANT BIOCHEMISTRY. *Educ:* Brigham Young Univ, BS, 58; Utah State Univ, MS, 60; Purdue Univ PhD(agron), 62. *Prof Exp:* Fel plant biochem, Univ Calif, Los Angeles, 62-63; biochemist, 63-65, RES CHEMIST, SOUTH ATLANTIC AREA, AGR RES SERV, USDA, 65- *Concurrent Pos:* Symposium chmn, Am Chem Soc, 73, 78 & 82; assoc ed, Peanut Sci, 74-76, ed, 76-; co-ed, Peanut Sci & Technol, 82; ed, Evaluation of Quality of Fruits & Vegetables, 85; adv bd, J Agr Food Chem, 84- *Honors & Awards:* Golden Peanut Res Award, Nat Peanut Coun, 77; Distinguished Serv Award, Div Agr Food Chem, Am Chem Soc, 88. *Mem:* Sigma Xi; fel, Am Peanut Res Educ Soc; fel Am Chem Soc; Coun Biol Eds. *Res:* Influence of seed maturation processes on the physiological processes occurring in the seed; isolation and characterization of flavor components from peanut kernels and of seed enzymes. *Mailing Add:* USDA Agr Res Serv South Atlantic Area NC State Univ Box 7625 Raleigh NC 27695-7625

PATTEE, HOWARD HUNT, JR, b Pasadena, Calif, Oct 5, 26; m 54; c 2. THEORETICAL BIOLOGY, SYSTEMS THEORY. *Educ:* Stanford Univ, BS, 48, PhD(physics), 53. *Prof Exp:* Res assoc physics & biophys, Stanford Univ, 53-60, res biophysicist & lectr biophys, 60-71; vis prof biophys sci, State Univ NY Buffalo, 71-75; prof, Sch Advan technol, 75-83, PROF, SYSTS SCI, T J WATSON SCH OF ENG, STATE UNIV NY BINGHAMTON, 83- *Concurrent Pos:* Consult, Res Lab, IBM Corp, Calif, 56; NSF sr fel med physics, Karolinska Inst Sweden, 59; consult, Ampex Corp, 60. *Honors & Awards:* Fel, AAAS, 84. *Mem:* AAAS. *Res:* Origin of life; complex organization theory; physics of symbol systems; systems theory. *Mailing Add:* T J Watson Sch Eng State Univ of NY Binghamton NY 13901

PATTEE, OLIVER HENRY, b Dinuba, Calif, June 16, 48; m 77; c 1. ENVIRONMENTAL CONTAMINANTS, ENDANGERED SPECIES. *Educ:* Calif State Univ, Fresno, BA, 70, MA, 74; Tex A&M Univ, PhD(wildlife ecol), 77. *Prof Exp:* RES BIOLOGIST, PATUXENT WILDLIFE RES CTR, LAUREL, MD, 78- *Mem:* Wildlife Soc; Soc Conserv Biol; Ecol Soc Am; Nature Conservancy. *Res:* Impacts on contaminants in wildlife, especially heavy metals on raptors; conservation of natural resources with an emphasis on endangered species. *Mailing Add:* Patuxent Wildlife Res Ctr Laurel MD 20723

PATTEE, PETER A, b Brooklyn, NY, Nov 15, 32; m 58; c 4. MICROBIAL GENETICS. *Educ:* Univ Maine, BS, 55; Ohio State Univ, MSc, 57, PhD(bact), 61. *Prof Exp:* From asst prof to assoc prof, 61-69, PROF BACT, IOWA STATE UNIV, 69- *Mem:* AAAS; Am Soc Microbiol; Genetics Soc Am. *Res:* Staphylococcus aureus, especially genetic exchange phenomena, transposons, mutagenesis, mechanisms of antibiotic resistance; genetic and physical mapping of the staphyloccoccus aureus chromosome; interrelationships in the species and genus. *Mailing Add:* Dept Microbiol Iowa State Univ Ames IA 50011

PATTEN, BERNARD CLARENCE, b New York, NY, Jan 28, 31; m 53; c 1. ECOLOGY. *Educ:* Cornell Univ, AB, 52; Rutgers Univ, MS, 54, PhD(bot), 59; Univ Mich, AM, 57. *Prof Exp:* Assoc prof marine sci, Col William & Mary & assoc marine scientist, Va Inst Marine Sci, 59-63; ecologist, Oak Ridge Nat Lab, 63-68; assoc prof bot, Univ Tenn, 64-68; prof, 68-85, REGENTS PROF ZOOL, UNIV GA, 85- *Concurrent Pos:* Pres, Ecol Simulations, Inc, 71- *Mem:* AAAS; Soc Gen Syst Res; Am Soc Limnol & Oceanog; Ecol Soc Am; Am Inst Biol Sci; Int Soc Ecol Modelling N Am (pres, 81-). *Res:* Mathematical and theoretical ecology; limnology; oceanography. *Mailing Add:* Dept of Zool Univ of Ga Athens GA 30601

PATTEN, DUNCAN THEUNISSEN, b Detroit, Mich, Oct 13, 34; m 57; c 4. PLANT ECOLOGY, ENVIRONMENTAL BIOLOGY. *Educ:* Amherst Col, BA, 56; Univ Mass, MS, 59; Duke Univ, PhD(bot), 62. *Prof Exp:* Asst prof bot, Va Polytech Inst, 62-65; from asst prof to assoc prof, 65-73, asst acad vpres, 72-76, chmn, Dept Bot-Microbiol, 77-81, PROF BOT, ARIZ STATE UNIV, 73-, DIR, CTR ENVIRON STUDIES, 80- *Concurrent Pos:* Mem comt environ res assessment, Nat Acad Sci-Nat Res Coun Comn Natural Resources, 75-76, chmn, Mono Lake Ecosyst Study Comt, 85-87; mem, Ecol Sci Adv Panel, NSF, 75-78 Panel, Nat Sci Found, 75-78; mem, Bd Environ Studies & Toxicol, Nat Acad Sci, 87-90, Comt Western Water Mgt, 88-91, Comt Sci in Nat Parks, 89-91, Comn Geosci, Environ & Resources, 90-; Bus mgr Ecol Soc Am, 79; sr scientist, Glen Canyon Studies, Dept Interior (Bur Reclamation), 89- *Mem:* AAAS; Ecol Soc Am; Bot Soc Am; Brit Ecol Soc; Am Inst Biol Sci. *Res:* Ecology of montane and subalpine zones of the northern Rocky Mountains; autecology of desert plants, heat and water flux within desert ecosystems; man's impact on desert ecosystems; reclamation in the Southwest; riparian ecosystem processes. *Mailing Add:* Ctr Environ Studies Ariz State Univ Tempe AZ 85287-1201

PATTEN, GAYLORD PENROD, b Provo, Utah, Nov 30, 38; m 59; c 12. SOIL PHYSICS. *Educ:* Brigham Young Univ, BS, 63; Ohio State Univ, MSc, 65, PhD(soils), 69. *Prof Exp:* From asst prof to assoc prof, 69-77, PROF SOIL SCI, CALIF STATE POLYTECH UNIV, POMONA, 77-, CHMN DEPT PLANT & SOIL SCI, 85- *Mem:* Am Soc Agron; Soil Sci Soc Am. *Res:* Soil-water management; practical applications of soil-water-plant physical relationships. *Mailing Add:* Dept of Plant & Soil Sci Calif State Polytech Univ Pomona CA 91768

PATTEN, RAYMOND ALEX, b Fall River, Mass, Mar 28, 37; m 57; c 2. PHYSICS. *Educ:* Mass Inst Technol, BS, 58; Duke Univ, PhD(physics), 62. *Prof Exp:* Instr & res assoc physics, Duke Univ, 61-62; asst prof, Univ Mass, 62-70; MEM STAFF, NAVAL RES LAB, 70- *Mem:* Am Phys Soc. *Res:* Paramagnetic and cyclotron resonance; electron-nuclear double resonance. *Mailing Add:* Naval Res Lab-Code 6530 Washington DC 20375

PATTENGILL, MERLE DEAN, physical chemistry; deceased, see previous edition for last biography

PATTERSON, ANDREW, JR, b Texarkana, Tex, July 23, 16; m 40; c 4. PHYSICAL CHEMISTRY. *Educ:* Univ Tex, AB, 37, AM, 38, PhD(phys chem), 42. *Hon Degrees:* MA, Yale Univ, 69. *Prof Exp:* Instr chem, Univ Tex, 38-41; asst prof, NTex Agr Col, 41-42; actg assoc prof, Univ NC, 42-43; spec res assoc, Harvard Univ, 43-45; physicist, Underwater Sound Lab, Navy Dept, Conn, 45-46; instr chem & Am Chem Soc fel, 46-48, from asst prof to assoc prof, 48-69, PROF CHEM, YALE UNIV, 69- *Concurrent Pos:* Dir, Edwards St Lab, Yale Univ, 54-56; Master, Morse Col, 61-67; fel, Phys Chem Lab, Oxford, 66-67; mem, consult & chmn mine adv comt, Nat Acad Sci. *Mem:* Am Chem Soc; Sigma Xi. *Res:* Alkali metal-liquid ammonia; ionic association; high field conductance measurements. *Mailing Add:* 175 E Rock Rd New Haven CT 06511

PATTERSON, ARCHIBALD OSCAR, b Monroe, Ga, Dec 26, 08; m 33; c 2. CIVIL ENGINEERING. *Educ:* Ga Sch Technol, BSCivEng, 29. *Prof Exp:* Civil engr, Interstate Com Comn, DC, 29-30; hydraul engr, Water Resources Div, US Geol Surv, Maine & NH, 30-32, WVa & Ky, 32-35, Tenn, Ky & Ala, 35-42, in charge, Knoxville, Tenn, 42-47, dist engr, Surface Water Br, Ocala, Fla, 47-64; dir div water resources, Fla Bd Conserv, 64-67; prof environ eng & dir Fla Water Resources Res Ctr, Univ Fla, 67-70; CONSULT ENGR, 70 - *Concurrent Pos:* Chmn comt surface water, Fla Water Resources Study Comn, 57. *Mem:* Am Soc Civil Engrs (pres, 54). *Res:* Investigation of surface water supply; hydrology; water resources planning; water conservation. *Mailing Add:* 1444 S E Eighth St PO Box 1901 Ocala FL 32678-1901

PATTERSON, C(LEO) MAURICE, b Fairfield, Nebr, Dec 24, 13; m 36; c 2. HEALTH PHYSICS, ENVIRONMENTAL ENGINEERING. *Educ:* Univ Nebr, BS(pharm), 30, BS, 34. *Prof Exp:* Sr supvr radiation protection, Hanford Works, E I du Pont de Nemours & Co, Inc, 44-45, chief supvr, 45-46; supt, Hanford Works, Gen Elec Co, 46-51; supt health physics, Savannah River Plant, E I Du Pont de Nemours & Co, Inc, 51-65, res mgr radiol sci, Savannah River Lab, 65-71, supt health physics dept, Savannah River Plant, 71-78; RETIRED. *Concurrent Pos:* Ed, Health Physics J, 66-76; bd mem, So Carolina Dept Health & Environ Control, 77-80. *Honors & Awards:* Health Physics Founder Award, 84. *Mem:* Am Indust Hyg Asn; fel Health Physics Soc (pres, 62-63); Am Acad Indust Hyg; Am Bd Health Physics. *Res:* Radiological health; planning, development and inauguration of radiation protection programs; fates of radionuclides in the environment. *Mailing Add:* 521 Highland Park Dr Aiken SC 29801

PATTERSON, CHARLES MEADE, b Waynesburg, Pa, June 24, 19; m 67. GEOLOGY. *Educ:* Col Wooster, BA, 40; Columbia Univ, AM, 42, PhD(geol), 47; Calif Inst Technol, MS, 43. *Prof Exp:* Petrographer & geologist, Gulf Res & Develop Co, 47-53 & Nat Rifle Asn, 53-57; commodity specialist, US Bur Mines, 57-63, sci ed, 64-67; geol oceanogr, Ocean Anal Div, US Naval Oceanog Off, Suitland, 67-75, Wash 75-78; RETIRED. *Mem:* Geol Soc Am; Sigma Xi. *Res:* Submarine geology; areal geology; alteration in ore deposits; clays; petrography and petrology; weather forecasting and synoptic meteorology; lime; calcium and calcium compounds. *Mailing Add:* PO Box 784 Hyattsville MD 20783

PATTERSON, CHRISTOPHER WARREN, b Los Angeles, Calif, Aug 1, 46; m 72; c 2. MOLECULAR SPECTROSCOPY, ATOMIC SPECTROSCOPY. *Educ:* Univ Southern Calif, MS, 68, PhD(physics), 74. *Prof Exp:* Asst prof, Univ Campinas, Brazil, 74-77; MEM STAFF, LOS ALAMOS NAT LAB, UNIV CALIF, 77- *Concurrent Pos:* NSF fel, 68-71; NASA fel, 71-72. *Honors & Awards:* Coblentz Award, 82. *Mem:* Am Phys Soc; Optical Soc Am; Am Astron Soc. *Res:* Theoretical atomic and molecular spectroscopy; quantum optics, planet formation. *Mailing Add:* Los Alamos Nat Lab MS J569 PO Box 1663 Los Alamos NM 87545

PATTERSON, CLAIR CAMERON, b Des Moines, Iowa, June 2, 22; m 44; c 4. GEOCHEMISTRY, ENVIRONMENTAL CHEMISTRY. *Educ:* Grinnell Col, BA, 43; Univ Iowa, MS, 44; Univ Chicago, PhD(chem), 50. *Hon Degrees:* DSc, Grinnell Col, 73; Dr Honoris Causa, Univ Paris, 80. *Prof Exp:* Spectroscopist emission & mass, Manhattan Proj, 44-46; res assoc geochem & geochronol & fel, Univ Chicago, 51; fel, Calif Inst Technol 52, sr fel, 53-71, res assoc geochem, 71-72, GEOCHEMIST, CALIF INST TECHNOL, 72- *Concurrent Pos:* Teaching fel, Univ Chicago, 51. *Honors & Awards:* J L Smith Medal, Nat Acad Sci, 73; Goldschmidt Medal, Geochem Soc Am, 80. *Mem:* Nat Acad Sci; Geochem Soc Am; Inst Andean Studies. *Res:* Esotopic evolution of lead and age on earth; biogeochemistry of lead in marine and terrestrial ecosystems; marine, mammalian and atmospheric lead pollution; archaeology of South American metallurgy; history of ancient metal production. *Mailing Add:* Div Geol & Planetary Sci Calif Inst Technol Pasadena CA 91125

PATTERSON, DAVID, b Medford, Mass, Aug 24, 44; m 67, 75; c 2. SOMATIC CELL GENETICS, CANCER. *Educ:* Mass Inst Technol, BS, 66; Brandeis Univ, PhD(biol), 71. *Prof Exp:* Fel, 71-73, res assoc, 73, from asst prof to assoc prof biochem, biophys & genetics, 74-83, assoc prof, dept med, 81-84, PRES, ELEANOR ROOSEVELT INST CANCER RES, UNIV COLO HEALTH SCI CTR, 78-, PROF DEPT BIOCHEM, BIOPHYS & GENETICS, 83-, PROF, DEPT MED, 84- *Concurrent Pos:* Dir, Cytogenetic Core, Univ Colo Cancer Ctr, Denver, 87-89; mem, Geront & Geriatrics Rev Comt, 87-90; chmn workshop, Nat Inst Child Health & Human Develop, 89; chair elect, 1991 Int Workshop Chromosome 21, 90-; chmn workshop, Nat Inst Child Health & Human Develop, 89. *Mem:* Am Soc Cell Biol; AAAS; Am Soc Human Genetics; Sigma Xi. *Res:* Biochemical genetics of somatic mammalian cells grown in culture with emphasis on the relevance of gene regulation to cancer, aging and birth defects. *Mailing Add:* Eleanor Roosevelt Inst Cancer Res 1899 Gaylord St Denver CO 80206-1210

PATTERSON, DAVID THOMAS, b Durham, NC, July 18, 46; m 79. ENVIRONMENTAL PHYSIOLOGY, WEED SCIENCE. *Educ:* NC State Univ, BS, 68; Duke Univ, MA, 70, PhD(bot), 73. *Prof Exp:* Res assoc environ physiol, Dept Bot, Duke Univ, 73-76; PLANT PHYSIOLOGIST, USDA, 76-; ADJ FAC, NC STATE UNIV, 86- *Concurrent Pos:* Adj fac, NC State Univ, 86- *Mem:* Ecol Soc Am; Weed Sci Soc Am. *Res:* Physiological plant ecology, chemical and biological plant interactions, ecology and comparative physiology of weeds and crops, biology of exotic weeds; carbon oxygen enrichment; climate change. *Mailing Add:* Bot Dept USDA-ARS Duke Univ Durham NC 27706

PATTERSON, DENNIS BRUCE, b Chicago, Ill, July 20, 41. POLYMER CHEMISTRY, ORGANIC CHEMISTRY. *Educ:* Univ Chicago, BS, 63; Univ Calif, Berkeley, PhD(org chem), 69. *Prof Exp:* Res assoc org chem, Univ Tex, 69-71; mem tech staff, Bell Labs, 71-73; scholar, polymer chem, Pa State Univ, 74-76; sr res chemist, Goodyear Tire & Rubber Co, 76-81, res & develop assoc, Polymer Chem, 81-88; GROUP LEADER POLYMER CHEM, REVLON SCI INST, 88- *Mem:* Am Chem Soc; AAAS; Sigma Xi. *Res:* Synthesis and chemical modification of polymers; organic synthesis and physical-organic chemistry; homogenous and heterogenous catalysis; nuclear quadrupole resonance spectroscopy. *Mailing Add:* Revlon Sci Inst 4575 Eastgate Mall San Diego CA 92121-1911

PATTERSON, DENNIS RAY, b Cairo, Ill, Sept 6, 46; m 68; c 3. ORGANIC CHEMISTRY. *Educ:* Univ Mo-St Louis, BS, 68; Univ Chicago, PhD(chem), 74. *Prof Exp:* Fel, Univ Sherbrooke, 73-74; RES CHEMIST, ROHM & HAAS CO, 74- *Mem:* Am Chem Soc. *Res:* Synthesis of agriculturally related chemicals. *Mailing Add:* 405 W Prospect Ave North Wales PA 19454

PATTERSON, DONALD DUKE, b Montreal, Que, Oct 13, 27; m 64; c 2. POLYMER CHEMISTRY. *Educ:* McGill Univ, BSc, 49, MSc, 51. *Prof Exp:* Nat Res Coun fel, 53-55; asst prof chem, Univ Montreal, 55-64; assoc prof, Univ Strasbourg, 64-66; assoc prof, 66-69, PROF CHEM, MCGILL UNIV, 69- *Res:* Thermodynamics of polymer solutions; statistical thermodynamics; surface chemistry. *Mailing Add:* Dept Chem McGill Univ Montreal PQ H3A 2K6 Can

PATTERSON, DONALD FLOYD, b Maracaibo, Venezuela, Feb 2, 31; US citizen; m 53; c 2. VETERINARY MEDICINE, MEDICAL GENETICS. *Educ:* Okla State Univ, DVM, 54; Univ Pa, DSc, 68. *Prof Exp:* Intern vet med, Angell Mem Animal Hosp, 54-55; instr, Okla State Univ, 55-56; from instr to asst prof cardiol, 58-66, from assoc prof to prof med, 66-73, CHARLOTTE NEWTON SHEPPARD PROF MED, SCH VET MED, UNIV PA & CO-DIR COMP CARDIOVASC STUDIES UNIT, 66-, NIH PROG PROJ GRANT, 71-, CHIEF SECT MED GENETICS, 71-, PROF HUMAN GENETICS, 74- *Concurrent Pos:* NIH spec fel med genetics, Johns Hopkins Univ, 64-66; grant, 67-72; Ralston-Purina Res Award, Am Vet Med Asn, 81. *Honors & Awards:* Gaines Award, Am Vet Med Asn, 72; Ralston-Purina Res Award, Am Vet Med Asn, 87; Merit Award, NIH, 89. *Mem:* AAAS; fel Am Col Cardiol; Am Vet Med Asn; Am Soc Human genetics. *Res:* Comparative cardiology and medical genetics, particularly the etiology and pathogenesis of congenital heart disease. *Mailing Add:* Sch Vet Med Univ of Pa Philadelphia PA 19104

PATTERSON, DOUGLAS REID, b Port Arthur, Tex, July 30, 45; m 69; c 2. PHARMACEUTICAL RESEARCH & DEVELOPMENT, COMPARATIVE MEDICINE & PATHOLOGY. *Educ:* Tex A&M Univ, BS, 68, DVM, 69; Univ Mo, Columbia, PhD(path), 76. *Prof Exp:* Res assoc path, Univ Mo Vet Sch, 71-75; sr staff pathologist, Hazleton Labs Am, 75-78, head path, Europe, 78-80, dir med, 80-81; supvr, Shell Develop Co, 81-84; dir path/toxicol, 84-87, dir drug safety, 87-90, VPRES, PHARMACEUT PROD DIV, ABBOTT LABS, 90- *Concurrent Pos:* Asst nat prog dir, C L Davis, DVM Found, 81-86; adj asst prof, Dept Comp Med, Univ Tex Med Sch, 83-84; vchmn, Drug Safety Subsect, Pharmaceut Mfrs Asn, 86-91. *Mem:* Pharmaceut Mfrs Asn; Am Col Lab Animal Med; Am Col Vet Pathologists; Soc Toxicol Pathologists; Soc Toxicol. *Res:* Metabolic and pathologic perturbations associated with diabetes mellitus; pathogenesis of spontaneous and induced diseases and lesions, especially changes induced by potential health care products and therapeutics; pharmacodynamics, pharmacokinetics and metabolism of new drug candidates and the chemical characterization; ultrastructural pathology. *Mailing Add:* Dept 46G AP-10 One Abbott Park Rd Abbott Park IL 60064

PATTERSON, EARL BYRON, b Reynolds, Nebr, July 21, 23; m 54; c 1. GENETICS. *Educ:* Univ Nebr, BSc, 47; Calif Inst Technol, PhD(genetics), 52. *Prof Exp:* Fel biol, Calif Inst Technol, 52-53; res assoc bot & agron, 53-55, asst prof agron, 55-73, ASSOC PROF AGRON, COL AGR, UNIV ILL, URBANA-CHAMPAIGN, 73- *Mem:* Genetics Soc Am. *Res:* Maize genetics; maize cooperative collection of genetic traits. *Mailing Add:* Dept Agron S-116 Turner Hall Univ Ill Urbana IL 61801

PATTERSON, EARL E(DGAR), b Oklahoma City, Okla, Apr 14, 23; m 56; c 2. CHEMICAL ENGINEERING. *Educ:* Univ Okla, BS, 44, MChE, 47; Mass Inst Technol, ScD(chem eng), 50. *Prof Exp:* Asst, Mass Inst Technol, 47-50; chem engr titanium res & develop, E I du Pont de Nemours & Co, Inc, 50-54; lab mgr parts div, Reynolds Metals Co, 54-57, asst res dir, Metall Res Labs, 57-58, mem staff corporate planning dept, 58-60, dir res eval, 60-64, exec asst to exec vpres res & develop, 64-71, gen dir metall res div, 71-72, spec asst to exec vpres res & develop, 72-78, gen dir spec projs, 78-80, asst to gen mgr, 80-82; pres, Exp Asn, Inc, 82-86; RETIRED. *Mem:* Am Chem Soc; Am Soc Metals; Am Inst Chem Engrs; Int Transactional Analysis Asn; Int Solar Energy Soc. *Res:* Research administration; economic analysis; titanium metal process metallurgy; high polymers. *Mailing Add:* 8318 Whitewood Rd Richmond VA 23235

PATTERSON, EDWARD MATTHEW, b Savannah, Ga, Dec 11, 43; m 71; c 2. ATMOSPHERIC SCIENCE. *Educ:* Ga Inst Technol, BS, 65, MS, 67, PhD(physics), 74. *Prof Exp:* Fel, Nat Ctr Atmospheric Res, 74-75, scientist, 75-78; RES SCIENTIST, SCH GEOPHYS SCI, GA INST TECHNOL, 78- *Mem:* Am Meterol Soc; Optical Soc Am; Am Geophys Union; Sigma Xi. *Res:* Atmospheric chemistry; aerosol physics and chemistry; radiative effects of aerosols; visibility studies. *Mailing Add:* Sch of Geophys Sci Ga Inst of Technol Atlanta GA 30332

PATTERSON, ELIZABETH CHAMBERS, b Clarksville, Tex; m 40; c 4. HISTORY OF DNA & SUBSEQUENT APPLICATION. *Educ:* Univ Tex, Austin, BA, 37, MA, 40; Oxford Univ, Eng, DPhil, 80. *Hon Degrees:* LHD, Albertus Magnus Col, 87. *Prof Exp:* Asst Prof Chem, NTex Agr Col, 41-42; lectr chem, 47-48, from instr to prof phys sci, Albertus Magnus Col, 57-86; CONSULT ED, AM SCIENTIST, 70- *Concurrent Pos:* corresp mem, Manchester Lit & Philos Soc, Eng, 81. *Mem:* Sigma Xi; AAAS; Am Asn Physics Teachers; Am Physics Soc; Hist Sci Soc; Brit Soc Hist Sci; Hist Sci Soc. *Res:* American and British scientific autodidacts and scientific education; women in science; life and work of Mary Somerville; scientific society, organizations and institutions of the 19th century; C19 American science; scientific publications. *Mailing Add:* 175 E Rock Rd New Haven CT 06511

PATTERSON, ELIZABETH KNIGHT, b Pittsburgh, Pa, Sept 11, 09; m 35. BIOCHEMISTRY, CYTOCHEMISTRY. *Educ:* Wellesley Col, BA, 30; Bryn Mawr Col, MA & PhD(cytol), 40. *Prof Exp:* Tech asst, Rockefeller Inst, 30-34; spec asst, Bur Ord, US Navy Proj, Bryn Mawr Col, 42-43, demonstr biol, 43-44; res assoc, 44-50, assoc mem, 50-77, MEM, INST CANCER RES, 78- *Mem:* Am Asn Cancer Res; Am Chem Soc; NY Acad Sci; Soc Develop Biol; Am Soc Biol Chemists. *Res:* Purification and characterization of dipeptidases. *Mailing Add:* Inst for Cancer Res 7701 Burholme Ave Philadelphia PA 19111

PATTERSON, ERNEST LEONARD, b Crawford, Colo, Nov 18, 18; m 45; c 3. TOXICOLOGY, COMPUTER SCIENCE. *Educ:* Western State Col Colo, AB, 41; Calif Inst Technol, MS, 43; Univ Wis, PhD(biochem), 49. *Prof Exp:* Asst, Western State Col Colo, 39-41 & Calif Inst Technol, 42; explosives chemist, Woods Hole Oceanog Inst, 42-46; asst, Univ Wis, 46-49; res biochemist, Lederle Labs, Am Cyanamid Co, 49-84; RETIRED. *Mem:* Am Chem Soc; Am Soc Biol Chem. *Res:* Isolation, characterization and function of new growth factors; nutrition; antibiotics; fermentation biochemistry. *Mailing Add:* 182 Old Middletown Rd Pearl River NY 10965

PATTERSON, FRED LA VERN, b Reynolds, Nebr, Apr 6, 16; m 43; c 3. PLANT BREEDING, GENETICS. *Educ:* Univ Nebr, BS, 42; Kans State Col, MS, 47; Univ Wis, PhD(plant breeding), 50. *Hon Degrees:* ScD, Univ Nebr, 79. *Prof Exp:* Asst plant breeding, Kans State Col, 46-47 & Univ Wis, 47-50; from asst agronomist to agronomist, 50-76, prof, 70-78, LYNN DISTINGUISHED PROF AGRON, PURDUE UNIV, 78- *Mem:* Am Soc Agron; Genetics Soc Am; Am Phytopath Soc; Genetics Soc Can. *Res:* Plant breeding and genetics of small grains. *Mailing Add:* Dept of Agron Life Sci Bldg Purdue Univ Lafayette IN 47907

PATTERSON, GARY DAVID, b Honolulu, Hawaii, July 31, 46; m 67; c 1. POLYMER PHYSICS, STATISTICAL MECHANICS. *Educ:* Harvey Mudd Col, BS, 68; Stanford Univ, PhD(chem), 72. *Prof Exp:* Mem tech staff, AT&T Bell Labs, 72-84; PROF CHEM, CARNEGIE-MELLON UNIV, 84- *Honors & Awards:* Award for Initiatives in Res, Nat Acad Sci, 81. *Mem:* Am Chem Soc; fel Am Phys Soc; fel Royal Soc Chem; Mat Res Soc. *Res:* Application of chemical physics to polymer science; molecular dynamics of liquids and solutions; light scattering spectroscopy of amorphous media; nature and dynamics of the glass transition; nonlinear optical properties; biopolymers; light scattering; nonlinear optics. *Mailing Add:* Dept Chem Carnegie-Mellon Univ 4400 Fifth Ave Pittsburgh PA 15213

PATTERSON, GARY KENT, b Springfield, Mo, Dec 10, 39; m 60; c 3. CHEMICAL ENGINEERING. *Educ:* Mo Sch Mines, BS, 60; Univ Mich, MS, 61; Univ Mo-Rolla, PhD(chem eng), 66. *Prof Exp:* Engr, Esso Res Labs Div, Humble Oil Co, La, 61-63; from asst prof to assoc prof, Univ Mo-Rolla, 66-78, prof chem eng, 78-; AT DEPT CHEM ENG, UNIV ARIZ. *Concurrent Pos:* Humboldt fel, Max Planck Inst, Gottingen, WGermany, 71-72; sect leader, Chem Eng Div, Lawrence Livermore Labs, Livermore, Calif, 77-78. *Mem:* Am Inst Chem Engrs; Am Chem Soc; Soc Rheol. *Res:* Fluid mechanics; turbulence; fluidization; drag reduction; polymer rheology; viscoelasticity; dilute solutions; melts; turbulent mixing in chemical reactors. *Mailing Add:* 1609 Lincoln Lane Rolla MO 65401

PATTERSON, GEORGE HAROLD, b Uniontown, Pa, Aug 17, 17; m 41; c 2. ORGANIC CHEMISTRY. *Educ:* Juniata Col, BS, 39; Mass Inst Technol, PhD(org chem), 42. *Prof Exp:* Res chemist, E I Du Pont de Nemours & Co, Inc, 42-51, supvr, 51-53, div head, Jackson Lab, 53-80; RETIRED. *Concurrent Pos:* Ed, Chem Bulletin, Am Chem Soc. *Honors & Awards:* Brumbaugh Sci Prize, 39. *Mem:* Am Chem Soc. *Res:* Organic pigments; fluorine and petroleum chemicals; detergents; tanning agents; refrigerants; polymerization of butadiene with organosodium compounds; analytical chemistry. *Mailing Add:* 1501 Woodsdale Rd Wilmington DE 19809-2246

PATTERSON, GLENN WAYNE, b China Grove, NC, Mar 9, 38; m 61; c 3. PLANT PHYSIOLOGY, PLANT BIOCHEMISTRY. *Educ:* NC State Univ, BS, 60; Univ Md, MS, 63, PhD(plant physiol), 64. *Prof Exp:* From asst prof to assoc prof, 64-73, PROF PLANT PHYSIOL, UNIV MD, COLLEGE PARK, 73-, CHMN DEPT, 78- *Mem:* AAAS; Phycol Soc Am; Am Soc Plant Physiol; Am Chem Soc; Am Oil Chem Soc. *Res:* Plant lipid biochemistry, especially lipids of algae. *Mailing Add:* Dept Bot Univ Md College Park MD 20742

PATTERSON, GORDON DERBY, JR, b Columbus, Ohio, May 6, 23. ANALYTICAL CHEMISTRY, POLYMER SCIENCE. *Educ:* Allegheny Col, BS, 44; Purdue Univ, MS, 49, PhD(chem), 52. *Prof Exp:* Asst, Purdue Univ, 46-48, asst instr, 48-50; res chemist, E I Du Pont De Nemours & Co, Inc, 51-58, analytical supvr, Film Dept, 58-73, analytical consult, Film Dept, 73-76, res chemist, Plastic Prod & Resins, 76-81, occupational health coordr, Polymer Prods Dept, 81-83; RETIRED. *Mem:* AAAS; Am Chem Soc; Am Soc Testing & Mat. *Res:* Physical and mechanical properties of plastics and polymeric films; instrumentation; chemical analysis; development of national and international technical standards for plastics and polymeric films; adoption of SI-metric system by industry and standards organizations; coordination of occupational health safeguards in chemical and physical research facilities. *Mailing Add:* 1403 Shallcross Ave Apt 110 Wilmington DE 19806-3038

PATTERSON, GREGORY MATTHEW LEON, b Riverside, Calif, May 7, 54. PHYCOLOGY, CYANOBACTERIOLOGY. *Educ:* Ore State Univ, Corvallis, BS, 76; Univ Ky, Lexington, PhD(biol), 80. *Prof Exp:* Jr researcher, 81-83, ASSOC RESEARCHER, UNIV HAWAII, 83- *Mem:* Phycol Soc Am; Am Soc Microbiol. *Res:* Isolation and cultivation of microscopic algae for evaluation as potential producers of pharmaceuticals or agrochemicals, and characterization of such products after discovery. *Mailing Add:* Dept Chem Univ Hawaii at Manoa 2545 The Mall Honolulu HI 96822

PATTERSON, HARRY ROBERT, b Long Beach, Calif, May 7, 21; m 47; c 4. PHARMACY, BACTERIOLOGY. *Educ:* Univ Southern Calif, BS, 44, MS, 45, PharmD, 51. *Prof Exp:* Retail & hosp pharmacist, 44-48; from instr to assoc prof, Calif State Univ, San Jose, 48-61, prof microbiol, 61-81; DIR PHARM, HILO HOSP, 82- *Mem:* AAAS; Am Soc Microbiol; Am Soc Hosp Pharmacists. *Res:* Synthesis of vanillin; toxicity of synthetic aromatics; hematology; pathogenic microbiology; immunology; electrophoretic characteristics of fish hemoglobins; antigenic characteristics of salmon erythrocytes; biology curriculum development. *Mailing Add:* 100 Honolii Pali Hilo HI 96720

PATTERSON, HOWARD HUGH, b Los Angeles, Calif, Nov 6, 38; m 69. INORGANIC CHEMISTRY, ANALYTICAL CHEMISTRY. *Educ:* Occidental Col, BA, 61; Mass Inst Technol, MS, 64; Brandeis Univ, PhD(inorg chem), 68. *Prof Exp:* From asst prof to assoc prof, 68-79, PROF CHEM, UNIV MAINE, ORONO, 79- *Concurrent Pos:* Vis scientist, Brookhaven Nat Lab; vis prof, Mass Inst Technol. *Mem:* Am Chem Soc; Am Phys Soc; Sigma Xi. *Res:* Optical spectroscopy of unusual transition metal systems such as interacting gold systems, high-temperature-superconductors, and tungsten oxide films used for detection of H2S; use of luminescence to study aluminum-ligand kinetics and speciation; study of polyaniline, polyimide and simple gases on surfaces by Raman, luminescence and neutron scattering. *Mailing Add:* Dept Chem Univ Maine Orono ME 04469

PATTERSON, IAN D(AVID), chemical engineering; deceased, see previous edition for last biography

PATTERSON, JAMES DEANE, b Gillette, Wyo, Sept 1, 34; m 73; c 2. PHYSICS. *Educ:* Univ Mo, AB, 56; Univ Chicago, MS, 57; Univ Kans, PhD(physics), 62. *Prof Exp:* Asst prof physics, Idaho State Univ, 61-63; assoc prof, 63-72, prof physics, SDak Sch Mines & Technol, 72-84; prof, 84-87, PROF & HEAD PHYSICS & SPACE SCI, FLA INST TECH, 87- *Concurrent Pos:* Vis assoc prof, Univ Notre Dame, 69-70; vis prof, Univ Nebr, Lincoln, 75-76; adj prof, SDak Sch Mines & Technol, 89- *Mem:* Am Phys Soc; Am Asn Physics Teachers. *Res:* Solid state theory. *Mailing Add:* 1271 Creel Rd NE Palm Bay FL 32905

PATTERSON, JAMES DOUGLAS, b Chicago, Ill, feb 2, 64. RESEARCH & DEVELOPMENT. *Educ:* Univ NC, BS, 87; NC State Univ, MS, 90. *Prof Exp:* Elec engr, Gen Elec-Nuclear, 88-89; ELECTRONICS ENGR, NASA LANGLEY RES CTR, 91- *Mem:* Inst Elec & Electronic Engrs; Soc Photo-Optical Instrumentation Engrs. *Res:* Photo acoustics; cw lasers; electromagnetics-nondestructive testing; solid state-device physics and processing; optical sensors and networks for aircraft applications. *Mailing Add:* NASA Langley Res Ctr MS 130 Hampton VA 23665

PATTERSON, JAMES FULTON, b Xenia, Ohio, Apr 11, 18; m 42; c 5. INTERNAL MEDICINE. *Educ:* Antioch Col, BS, 41; Harvard Univ, MD, 44; Am Bd Internal Med, dipl, 52. *Prof Exp:* Intern internal med, Boston City Hosp, 44-45; asst resident, Boston Dispensary, 45-46; resident, New Eng Ctr Hosp, 46-48; from instr to assoc prof, 48-66, PROF INTERNAL MED, MED SCH, TUFTS UNIV, 67- *Concurrent Pos:* Asst physician, New Eng Ctr Hosp, 48-54, physician, 54-, chief gastroenterol, 60-72, chief ambulatory med, 72-; clin & res fel, Mass Gen Hosp & Harvard Univ, 57-58. *Mem:* AAAS; AMA; fel Am Col Physicians. *Res:* Clinical research in gastroenterology and ambulatory medicine. *Mailing Add:* 107 Sunrise Ave Medford OR 97504

PATTERSON, JAMES HOWARD, b Kansas City, Mo, Jan 22, 21; m 44; c 5. ACTINIDE CHEMISTRY. *Educ:* Univ Omaha, BA, 43; Iowa State Col, PhD(chem), 50. *Prof Exp:* Anal chemist, Omaha Test Labs, 43-44; res asst & jr chemist, Inst Atomic Res, Iowa State Col, 44-50; assoc chemist, Argonne Nat Lab, 50-72; staff mem, Los Alamos Sci Lab, 72-86; RETIRED. *Concurrent Pos:* Testing engr, Metcalf-Hamilton-Kansas City Bridge Co, Can, 43. *Res:* Actinide element chemistry; interaction of actinides and the environment; chemical analysis of actinides. *Mailing Add:* 2117B 43rd St Los Alamos NM 87544

PATTERSON, JAMES REID, b Charlotte, NC, Jan 15, 18; m 50; c 3. NUCLEAR PHYSICS. *Educ:* Davidson Col, BS, 39; Duke Univ, MA, 41, PhD, 55. *Prof Exp:* Lab instr, Duke Univ, 39-41, asst, 51-52, res assoc, 52-54; instr physics, NC State Col, 46-51; from asst prof to assoc prof, Furman Univ, 54-57; assoc prof, Clemson Col, 57-61; prof & head dept, Rockford Col, 61-67; assoc prof, Univ Wis-Whitewater, 67-69, prof physics, 69-88. *Concurrent Pos:* Researcher, La State Univ, 64, 66 & vis prof, 65. *Res:* Cosmic ray; magnetics; total neutron cross section measurements; beta ray spectrometer. *Mailing Add:* 231 N Esterly Ave Whitewater WI 53190

PATTERSON, JAMES W(ILLIAM), b Montgomery, Ala, Nov 8, 40; m 64; c 2. ENVIRONMENTAL ENGINEERING. *Educ:* Auburn Univ, BS, 64, MS, 67; Univ Fla, PhD(environ eng), 70. *Prof Exp:* Res engr hydraul, Waterways Exp Sta, US Corps Engrs, 64-66; from asst prof to assoc prof, 70-76, chmn dept, 71-89, PROF ENVIRON ENG, ILL INST TECHNOL, 76-; CHIEF EXEC OFFICER, PATTERSON SCHAFER, INC, 90- *Concurrent Pos:* Dir, Pritzker Environ Studies Ctr, 71-75, Indust Waste Elimination Res Ctr, Environ Protection Agency, 80-; pres & chief exec officer, Patterson Assoc, Chicago, 73-88. *Mem:* AAAS; Int Asn Water Pollution Control Res; Am Chem Soc; Am Inst Chem Engrs; Am Water Works Asn. *Res:* Chemical and biological processes in polluted environments; biological waste treatment processes, including monitoring and control techniques, chemical processes for water and wastewater purification. *Mailing Add:* Patterson Schafer Inc 39 S LaSalle St Chicago IL 60603

PATTERSON, JOHN LEGERWOOD, JR, b Roanoke Rapids, NC, Feb 5, 13. INTERNAL MEDICINE. *Educ:* Princeton Univ, AB, 35; Med Col Va, MD, 39; Univ Va, MS, 43; Am Bd Internal Med, dipl. *Prof Exp:* From instr to asst prof physiol & med, Sch Med, Emory Univ, 47-52; from asst prof to assoc prof, 53-60, chmn div cardiopulmonary labs & res, 60-78, RES PROF MED, MED COL VA, 60- *Honors & Awards:* John L Patterson Jr Lectr, Med Col Va, 78. *Mem:* Am Physiol Soc; Am Soc Clin Invest; AMA; Am Heart Asn; Am Fedn Clin Res; Am Col Physicians; Am Col Cardiol; Int Acad Astronaut. *Res:* Mechanisms of control of cerebral circulation and blood flow in the extremities; comparative physiology of the circulation; metabolism of the brain in head injury; determinant mechanisms of breath sounds; metabolism of blood vessels. *Mailing Add:* Box 282 MCV Sta Med Col Va Richmond VA 23298-0282

PATTERSON, JOHN MILES, b Vineland, NJ, Nov 5, 26; m 48; c 1. ORGANIC CHEMISTRY. *Educ:* Wheaton Col Ill, BS, 49; Northwestern Univ, PhD(chem), 53. *Prof Exp:* Res assoc chem, Northwestern Univ, 52-53; from instr to assoc prof, 53-67, PROF CHEM, UNIV KY, 67- *Mem:* Am Chem Soc. *Res:* Aliphatic nitrogen compounds; heterocyclic chemistry and compounds; free radicals in solution; high temperature reactions; photochemistry. *Mailing Add:* Dept Chem Univ Ky Lexington KY 40506

PATTERSON, JOHN W(ILLIAM), b Cleveland, Ohio, Mar 14, 36; m 60; c 2. METALLURGY, CERAMICS. *Educ:* Ohio State Univ, BEM & ME, 62, PhD(metall eng), 66. *Prof Exp:* From asst prof to assoc prof metall, 66-76, PROF MAT SCI & ENG, IOWA STATE UNIV, 76- *Concurrent Pos:* Prin investr solid electrochem, Eng Res Inst, Iowa State Univ, 66-90. *Mem:* Am Asn Physics Teachers; Am Soc Metals; Am Soc Quality Control; Am Soc Eng Educ; Sigma Xi. *Res:* Mass and charge transport in ceramics at elevated temperatures; thermodynamics; statistics; design of experiments. *Mailing Add:* Dept Mat Sci & Eng Iowa State Univ Ames IA 50011

PATTERSON, JOHN WARD, b Baldwin, Kans, Dec 6, 16; m 40; c 4. PHYSIOLOGY, MEDICINE. *Educ:* Ohio Wesleyan Univ, AB, 39; Ohio State Univ, MS, 41, PhD(org chem), 42; Western Reserve Univ, MD, 49. *Hon Degrees:* DSc, Ohio Wesleyan Univ, 65. *Prof Exp:* Asst chem, Ohio State Univ, 39-42; instr, Univ Vt, 42-43; from instr to assoc prof anat, Western Reserve Univ, 47-56, assoc dean, 53-56; prof anat & dean, Univ BC, 56-58; dean sch med, Vanderbilt Univ, 58-62, dir med affairs, 58-59, vchancellor med affairs, 59-62, prof physiol, 62-63; dean, 65-71, exec dir health ctr, 67-75, vpres health affairs, 70-75, PROF PHYSIOL, SCH MED, UNIV CONN, 63- *Mem:* AAAS; Am Chem Soc; Soc Exp Biol & Med; Asn Res Vision & Ophthal; Am Physiol Soc. *Res:* Ultraviolet absorption spectroscopy; stereochemistry; cytochemistry; experimental diabetes; cataracts; medical education. *Mailing Add:* Univ Conn Sch Med Farmington CT 06032

PATTERSON, JOSEPH GILBERT, b Jackson, Miss, Apr 25, 26; m 57; c 3. GEOLOGY. *Educ:* Miss Col, BA, 46; Southern Baptist Sem, BD, 49; US Army Language Sch, Russ, 51; Univ Va, MS, 58. *Prof Exp:* Dist geologist, US Army Engrs, Va, 57-60; dist geologist & asst chief, Mat Testing Lab, Hawaii, 60-62; mat engr, US Bur Pub Rds, Cambodia, 62-64, eng adv soils & geol, USAID, Thailand, 64-70; geologist & real estate appraiser, Miss State Hwy Dept, Jackson, 70-86; CHIEF APPRAISER, VET ADMIN, HARTFORD, CONN. *Mem:* Geol Soc Am. *Res:* Materials testing; engineering geology; ground water. *Mailing Add:* 380 S Lane One Granville MA 01034

PATTERSON, LARRY K, b Marysville, Kans, Feb 14, 37; m 59; c 2. RADIATION CHEMISTRY, PHOTOCHEMISTRY. *Educ:* Kans State Univ, BS, 59, PhD(chem), 67. *Prof Exp:* Asst prof chem, St Benedicts Col, 65-66; fel photochem, Royal Inst Gt Brit, 67-70; fel radiation chem, Carnegie-Mellon Univ, 70-74; res scientist radiobiol, Michael-Reese Hosp, 74-76; PROF SPECIALIST RADIATION CHEM, RADIATION LAB, UNIV NOTRE DAME, 76-, FAC FEL, DEPT CHEM, 80-, ASST DIR, RADIATION LAB, 83- *Concurrent Pos:* Vis scientist, Nat Mus Natural Hist, 81; assoc prof, Mus Natural Hist, Paris, 84. *Mem:* Am Chem Soc. *Res:* Photochemical and radiation chemical kinetics in homogeneous aqueous systems and in organizates (micelles, monolayers, vesicles) associated with aqueous media; laboratory automation. *Mailing Add:* Radiation Lab Univ Notre Dame Notre Dame IN 46556

PATTERSON, LOYD THOMAS, b Dekalb Co, Ala, Feb 23, 30; m 47; c 2. IMMUNOLOGY, ANIMAL VIROLOGY. *Educ:* Auburn Univ, BS, 59, MS, 60, PhD(microbiol), 63. *Prof Exp:* James W McLaughlin fel immunol, Med Br, Univ Tex, 63-65; from asst prof to prof microbiol, 65-86, EXP PROF, UNIV ARK, FAYETTEVILLE, (DISABLED, 86-); PROF EXP PROF, UNIV ARK, (DISABLED, 86-). *Concurrent Pos:* Consult, Pelfreez Biol & Vet Admin Hosp, Fayetteville. *Mem:* AAAS; Reticuloendothelial Soc; Am Soc Microbiol; Poultry Sci Asn; Sigma Xi. *Res:* C-reactive protein in nonspecific resistance and its occurrence in different species; immune response in tumor induction and in resistance to tumors; ontogeny of immunity; regulation of the immune response. *Mailing Add:* 116 Dupree Circle Rainsville AL 35986

PATTERSON, MALCOLM ROBERT, b Waverly, Tenn, Nov 21, 35; m 89; c 2. NUMERICAL ANALYSIS, ANALYTICAL REPRESENTATION OF PHYSICAL SYSTEMS. *Educ:* Ga Inst Technol, BME, 58; Univ Tenn, Knoxville, MS, 64, PhD(physics), 67. *Prof Exp:* Assoc engr, flight test, Lockheed Missiles, Van Nuys, Calif, 58-59; assoc engr, fluid flow, Scott Paper Co, Philadelphia, Pa, 62-63; comput analyst physics, 63-72, GROUP LEADER APPL PHYSICS, OAK RIDGE NAT LAB, TENN, 72- *Concurrent Pos:* Second lieutenant, ordanance, US Army, Aberdeen, Md, 59-60. *Mem:* Am Soc Comput Mach; Math Asn Am. *Res:* Interdiciplinary modeling studies of large systems, such as hydrologic and atmospheric transport models; more finite but useful systems such as Lithium Bromide-Water heat pumps. *Mailing Add:* 8008 Ball Camp Pike Knoxville TN 37931

PATTERSON, MANFORD KENNETH, JR, b Muskogee, Okla, Aug 20, 26; m 53; c 1. CELL CULTURE MODELS. *Educ:* Okla Univ, BS, 53, MS, 54; Vanderbuilt Univ, PhD(biochem), 61. *Prof Exp:* Sr res chemist, 61-66, VPRES & DIR, BIOMED DIV, SR NOBLE FOUND, 73- *Concurrent Pos:* Consult, ICNND, Nutrit Surv, Jordan, 62; head nutrit, SR Noble Found, 66-; co-ed, Tissue Culture: Methods & Appln-Acad Press, 73; adj prof, Univ Okla Dent Sch, 86-; adj prof, Dept Biochem & Molecular Biol, Okla Univ Health Sci Ctr, 87-; chmn, Eminent Scholar Comt, Okla Ctr Advan Sci & Technol, 88- *Mem:* Tissue Culture Asn (treas, 72-76); Soc Biochem & Molecular Biol; fel AAAS; Am Asn Cancer Res; Am Chem Soc; NY Acad Sci. *Res:* Relationship of protein-protein interactions and cellular growth; molecular control of enzymes involved in the reaction as it relates to diseases e.g. cancer, lupus, cataracts. *Mailing Add:* Noble Found PO Box 2180 Ardmore OK 73402

PATTERSON, MARIA JEVITZ, b Berwyn, Ill, Oct 23, 44; m 70; c 2. INFECTIOUS DISEASES, CLINICAL MICROBIOLOGY. *Educ:* Col St Francis, BS, 66; Registry Med Technologists, cert, 66; Northwestern Univ, PhD(microbiol), 70; Mich State Univ, MD, 84. *Prof Exp:* From asst prof to assoc prof, 72-90, PROF MICROBIOL & PUB HEALTH, MICH STATE UNIV, 90- *Concurrent Pos:* NIH trainee clin microbiol, Univ Wash, 70-72; staff microbiologist, Dept Path, Lansing Gen Hosp, Mich, 72-75; fel clin microbiol, Univ Wash, 70-72; staff microbiologist, Clin Labs, Mich State Univ, 78-82; infectious dis fel, Univ Mass Med Ctr, 85-86. *Mem:* Am Soc Microbiol; Am Soc Clin Pathologists; SCent Asn Clin Microbiol; Am Acad Pediat; Pediat Infectious Dis Soc; Infectious Dis Soc Am. *Res:* Host response to infectious disease; viral pathogenesis. *Mailing Add:* Dept Microbiol & Pub Health Mich State Univ East Lansing MI 48824-1101

PATTERSON, MARK ROBERT, b North Tonawanda, NY, Apr 3, 57; US & Can citizen; m 86. BIOFLUID MECHANICS, UNDERWATER INSTRUMENTATION & IMAGE PROCESSING. *Educ:* Harvard Col, AB, 79; Harvard Univ, AM, 82, PhD(biol), 85. *Prof Exp:* ASST PROF PHYS BIOL & MARINE BIOL, DIV ENVIRON STUDIES, UNIV CALIF, DAVIS, 86- *Concurrent Pos:* Prin investr, NSF, 87- & Nat Oceanic & Atmospheric Admin, 87-; chief scientist & aquanaut team leader, Nat Oceanic & Atmospheric Admin, 88; distinguished speaker marine biol, Univ NH, 89. *Mem:* AAAS; Sigma Xi; Am Soc Limnol & Oceanog; Am Soc Zoologists; Ecol Soc Am. *Res:* Apply chemical engineering theory to the physiological ecology and population biology of aquatic organisms; effects of water motion or gas exchange and particle capture; theoretical models developed to complement in situ measurements; computers and electronics; invertebrate zoology. *Mailing Add:* Div Environ Studies Univ Calif Davis CA 95616

PATTERSON, MAX E, b Bourbon, Ind, July 8, 23; m 44; c 2. PLANT PHYSIOLOGY, HORTICULTURE. *Educ:* Purdue Univ, BSA, 49, PhD(plant physiol, biochem, hort), 59; Cornell Univ, MSA, 53. *Prof Exp:* Res assoc plant physiol & hort, NY State Agr Exp Sta, 49-53; instr, Purdue Univ, 53-58; from asst prof to assoc prof, 58-70, PROF PLANT PHYSIOL & HORT, WASH STATE UNIV, 70- *Concurrent Pos:* Vis scientist, East Malling Res Sta, England, 79-80. *Honors & Awards:* Woodbury Award, Am Soc Hort Sci, 63; Dow Chem Co Award, Am Soc Hort Sci, 68; J H Gourley Award, Am Soc Hort Sci, 78. *Mem:* AAAS; Am Soc Hort Sci; Am Soc Plant Physiol; Sigma Xi. *Res:* Post harvest horticulture. *Mailing Add:* Dept Hort Wash State Univ Pullman WA 99164

PATTERSON, MICHAEL MILTON, b Muscatine, Iowa, Mar 17, 42; m 66; c 2. PSYCHOBIOLOGY. *Educ:* Grinnell Col, BA, 64; Univ Iowa, PhD(psychol), 69. *Prof Exp:* Teaching fel psychobiol, Univ Calif, Irvine, 69-71; from asst prof to assoc prof physiol, Kirksville Col Osteop Med, 71-77; dir res, 77-90, PROF BIOMED SCI, COL OSTEOP MED, OHIO UNIV, 77- *Concurrent Pos:* Vis prof, Univ Calif, Irvine, 77; consult, NIH, 77-; ed, J Am Osteop Asn, 88; vis scientist, Am Acad Osteop, 90- *Honors & Awards:* Louisa Burns Lectr, Am Osteop Asn, 80. *Mem:* Fel Am Psychol Asn; Sigma Xi; Psychonomic Soc; Am Psychol Soc. *Res:* Alteration of spinal reflex function in response to altered input and brain substrates of simple learning situations. *Mailing Add:* Col Osteop Med Ohio Univ Athens OH 45701

PATTERSON, OMAR LEROY, physics, for more information see previous edition

PATTERSON, PAUL H, b Chicago, Ill, Oct 22, 43. DEVELOPMENTAL NEUROSCIENCE. *Educ:* Grinnell Col, BA, 65; Johns Hopkins Univ, PhD(biochem), 70. *Prof Exp:* From asst prof to assoc prof neurobiol, Harvard Med Sch, 73-83; PROF BIOL, CALIF INST TECHNOL, 83-, EXEC OFFICER BIOL, 90- *Concurrent Pos:* Mem, Neurol Study Sect, Div Res Contracts, NIH, 77 & 87, Cellular & Molecular Basis of Dis Rev Comt, Nat Inst Gen Med Sci, 82, Res Briefing Panel, Inst Med, Nat Acad Sci, 88 & Sci Adv Bd, Hereditary Dis Found, 91-; panel mem, Long-term strategies on Inflammatory, Demyelinating & Degenerative Dis, NIH, 78; scholar, Rita Allen Found, 79-84; assoc ed, J Neurosci, 80-85 & 88-, Am Rev Physiol, 82-83 & Neuron, 86-; Javits neurosci investr award, Nat Neuro & Commun Disorders & Stroke Coun, 88. *Mem:* AAAS; Soc Neurosci; Am Soc Cell Biol; Soc Develop Biol. *Res:* Influence of target cells, glia, limphokines and hormones on phenotypic decisions of neural crest derivatives; analysis of the function of Thy-1, proteoglycans, proteases and positional information cues in axon outgrowth, regeneration and guidance; production of appropriate cells for neural grafting in neurodegenerative disease models. *Mailing Add:* 216-76 Div Biol Calif Inst Technol Pasadena CA 91125

PATTERSON, RICHARD L, b Brooklyn, NY, Sept 12, 32; m 53; c 2. RESOURCE MANAGEMENT, BIOMETRICS. *Educ:* Univ Mich, BSF, 54, MS, 59, PhD(indust eng), 63. *Prof Exp:* Engr opers anal, Bendix Corp, 60-63; res assoc, Indust Systs Lab, Univ Mich, 63; from asst prof to prof indust & systs eng, Univ Fla, 63-70; PROF NATURAL RESOURCES, UNIV MICH, ANN ARBOR, 70- *Concurrent Pos:* Consult environ anal. *Mem:* Inst Mgt Sci; Sigma Xi. *Res:* Natural resource management; biomathematics; population dynamics; ecology systems; operations research. *Mailing Add:* Sch of Natural Resources Univ of Mich Ann Arbor MI 48109

PATTERSON, RICHARD SHELDON, b Waltham, Mass, Apr 27, 32; m 57; c 2. MUSCOID FLY-BIOLOGICAL CONTROL, COCKROACHES & FLEAS CONTROL. *Educ:* Univ Mass, BS, 54, MS, 55; Cornell Univ, PhD(entom), 62. *Prof Exp:* Entomologist, Midge Res Lab, Fla, 62-64, dir lab, Fla State Bd Health, 64-66; entomologist, 66-78, RES LEADER, INSECTS AFFECTING MAN & ANIMAL BR, GAINESVILLE LAB, USDA, 78- *Concurrent Pos:* Prof, Univ Fla, 68-; WHO scientist genetic control mosquitoes, India, 70-73. *Honors & Awards:* Distinguished Serv Award, USDA, 88; US Fed Lab Award, 89. *Mem:* Am Mosquito Control Asn; Entom Soc Am; Int Orgn Biol Control Noxious Animals & Plants. *Res:* Control of insect pests of truck-crops and ornamentals; biology and control of aquatic insects, primarily chironomid midges; biology and control of mosquito; fly biology and control, particularly biocontrol of noxious flies; biology, ecology, physiology, and control of muscoid flies which attack livestock; control of household pests, cockroaches, ants and fleas; pest ants especially imported fire ant. *Mailing Add:* Agr Res Serv USDA PO Box 14565 Gainesville FL 32604

PATTERSON, ROBERT ALLEN, b Lock Haven, Pa, Nov 23, 27; m 56; c 4. ZOOLOGY. *Educ:* Univ Mich, BSc, 50; Ohio State Univ, MSc, 55, PhD(entom), 57. *Prof Exp:* From asst prof to prof zool, Ariz State Univ, 57-89; RETIRED. *Concurrent Pos:* NIH res grant, 60-66 & 76-77. *Mem:* Fel AAAS; Int Soc Toxinology. *Res:* Physiology and pharmacology of animal venoms. *Mailing Add:* 2031 S Ventura Tempe AZ 85282

PATTERSON, ROBERT PRESTON, b Statesville, NC, July 13, 39; m 61; c 4. CROP PHYSIOLOGY, CROP PRODUCTION. *Educ:* NC State Univ, BS, 61, MS, 63; Cornell Univ, PhD(agron), 68. *Prof Exp:* From asst prof to assoc prof, 68-79, DISTINGUISHED PROF CROP SCI, NC STATE UNIV, 72- *Mem:* Am Soc Plant Physiol; Sigma Xi; fel Am Soc Agron. *Res:* Plant nutrition; environmental and physiological aspects of crop production. *Mailing Add:* NC State Univ Crop Sci 4124 Williams Hall Raleigh NC 27695-7620

PATTERSON, RONALD BRINTON, b Wichita, Kans, Apr 20, 41; m 71; c 2. ANALYTICAL & ORGANIC CHEMISTRY. *Educ:* Hastings Col, AB, 63; Univ Nebr, Lincoln, PhD(org chem), 70. *Prof Exp:* Res chemist, 69-73, supvr prod develop, 73-80, MGR, PROD DEVELOP, LORILLARD RES CTR, LOEW'S, 80- *Mem:* Am Chem Soc. *Res:* Flavors; high performance liquid chromatography; gas chromatography; tobacco chemistry. *Mailing Add:* Lorillard Res Ctr 420 English St Greensboro NC 27420

PATTERSON, RONALD JAMES, b Pittsburgh, Pa, Apr 16, 43; m 70. IMMUNOBIOLOGY. *Educ:* Washington & Jefferson Col, BA, 65; Northwestern Univ, Evanston, PhD(microbiol), 70. *Prof Exp:* Asst prof, 72-77, ASSOC PROF MICROBIOL, MICH STATE UNIV, 77- *Concurrent Pos:* NIH fel, Univ Wash, 70-72; NIH res grant, Mich State Univ, 72- *Mem:* Am Soc Cell Biol. *Res:* Cell biology. *Mailing Add:* Dept Microbiol & Pub Health Mich State Univ East Lansing MI 48824

PATTERSON, ROSALYN MITCHELL, b Madison, Ga, Mar 25, 39; m 61; c 3. CYTOGENETICS. *Educ:* Spelman Col, BA, 58; Atlanta Univ, MS, 60; Emory Univ, PhD(biol), 67. *Prof Exp:* From instr to prof biol, Spelman Col, 60-70; consult, Bur Reclamation, Dept Interior, 70-71, coordr nat environ educ develop prog, Nat Park Serv, 71-72; fel cell biol & Nat Inst Gen Med Sci fel, Div Biol Standards, Lab Path, NIH, 72-73; assoc prof biol, Ga State Univ, 74-76; from assoc prof to prof biol & chairperson dept, Atlanta Univ, 77-86; prof biol, Spelman Col, 86-87; DIR RES CAREERS OFF & ADJ PROF BIOL, MOREHOUSE COL, 88- *Concurrent Pos:* Southern Fel Fund fel, Ga Inst Technol, 69-70, NIH fel, 72-73, Macy fel, Marine Biol Lab, Woods Hole, Mass, 80, Nat Res Coun Ford Found fel, 83-84, Nat Inst Health, NIGMS-MARC fac fel, 84-85. *Mem:* AAAS; Am Soc Cell Biol; Soc Develop Biol; Sigma Xi. *Res:* Determination of potential free radical activity induced by specific environmental agents in vitro; evaluation of induced free radical activity by electron spin resonance and cytogentic analyses. *Mailing Add:* Dept Biol Morehouse Col Atlanta GA 30314

PATTERSON, ROY, b Ironwood, Mich, Apr 26, 26; m 48; c 3. INTERNAL MEDICINE. *Educ:* Univ Mich, BS, 50, MD, 53; Am Bd Internal Med, dipl; Am Bd Allergy, dipl. *Prof Exp:* Asst prof internal med, Univ Mich, 59-60; from asst prof to prof internal med, 60-67, assoc chmn dept med, 71-73, IRVING S CUTTER PROF MED, MED SCH, NORTHWESTERN UNIV, CHICAGO, 64-, CHMN DEPT MED, 74- *Concurrent Pos:* USPHS res career develop award, 63-; consult, US Vet Admin Hosps, Ann Arbor, Mich, 58-60 & Chicago, Ill, 60-; attend physician, Vet Admin Res & Passavant Mem Hosps, 61, Cook County Hosp, 63- & Chicago Wesley Mem Hosp, 64- *Mem:* AAAS; fel Am Col Physicians; fel Am Acad Allergy; Am Asn Immunologists; Am Soc Clin Invest. *Res:* Allergy; immunology. *Mailing Add:* Northwestern Univ Med Sch 320 E Superior St Chicago IL 60611

PATTERSON, RUSSEL HUGO, JR, b New York, NY, Apr 1, 29; m 55; c 3. NEUROSURGERY. *Educ:* Stanford Univ, BA, 48; Cornell Univ, MD, 52; Am Bd Neurol Surgeons, cert, 63. *Prof Exp:* Asst, 55-59, instr, 59-61, clin instr, 61-63, from asst prof to assoc prof, 63-71, PROF SURG, MED COL, CORNELL UNIV, 71- *Concurrent Pos:* Mem, Am Bd Neurol Surgeons, 76- *Mem:* AMA; Am Col Surgeons; Am Acad Neurol Surg (pres, 84); Soc Neurol Surgeons; Am Asn Neurol Surgeons (pres, 85-86). *Res:* Cerebrovascular disease. *Mailing Add:* 525 E 68th St New York NY 10021

PATTERSON, SAM H, b Marion, Iowa, Aug 14, 18; m 42; c 2. GEOLOGY. *Educ:* Coe Col, BA, 40; Univ Iowa, MS, 47; Univ Ill, PhD(geol), 55. *Prof Exp:* Geologist, Sci & Educ Admin-Agr Res, US Geol Surv, 47-86; RETIRED. *Mem:* Geol Soc Am; Mineral Soc Am; Clay Minerals Soc; Am Inst Mining, Metall & Petrol Eng; Soc Econ Geologists. *Res:* Bentonite deposits of South Dakota, Wyoming and Montana; refractory clay deposits of eastern Kentucky; bauxite deposits of Hawaii; Fuller's earth deposits of Georgia and Florida; world bauxite. *Mailing Add:* 2515 Fowler Lane Reston VA 22091

PATTERSON, SAMUEL S, b Indianapolis, Ind, Mar 8, 17; m 58; c 2. DENTISTRY, ENDODONTICS. *Educ:* Ind Univ, DDS, 40, MSD, 60; Am Bd Endodont, dipl. *Prof Exp:* From instr to prof oper dent, 49-74, dir grad endodont, 68-74, prof & chmn dept, 74-85, EMER PROF ENDODONT, SCH DENT, IND STATE UNIV, 85- *Concurrent Pos:* Consult, Vet Admin. *Honors & Awards:* Edgar G Coolidge Award, 89; Maynard K Hine Award, 90. *Mem:* Am Dent Asn; fel Am Asn Endodontists (past vpres, pres, 68); fel Am Col Dent; fel Int Col Dent; Int Asn Dent Res. *Res:* Effect of therapeutic cobalt 60 radiation on dental structures and investing tissues; effect of an apical dentinal plug in monkey teeth. *Mailing Add:* St Vincents Hosp Prof Bldg 8402 Harcourt Rd 405 Indianapolis IN 46260

PATTERSON, STEVEN LEROY, b Waco, Tex, Oct 2, 47. PHYSICAL OCEANOGRAPHY. *Educ:* Tex A&M Univ, BS, 70, MS, 72, PhD(oceanog), 78. *Prof Exp:* res assoc, Dept Oceanog, Tex A&M Univ, 78-79; sr scientist, Sci Applns Inc, McLean, Va, 79-81; asst res scientist, Dept Oceanog, Tex A&M Univ, 81-85; PHYSICAL OCEANOGRAPHER, NAT OCEANOG DATA CTR, 86- *Concurrent Pos:* Sr scientist, Sci Applns Inc, McLean, Va, 79-81. *Mem:* Am Geophys Union. *Res:* Descriptive physical oceanography, especially of the Southern Ocean; water mass formation, circulation, and distribution; structure and motion of Southern Ocean frontal zones and cross frontal property fluxes; kinetic energy distributions from drifting buoys. *Mailing Add:* Nat Oceanog Data Ctr 1825 Connecticut Ave NW Washington DC 20235

PATTERSON, TIM J, b Independence, Mo, Dec 20, 57. SIGNAL PROCESSING, REMOTE SENSING. *Educ:* Brigham Young Univ, BS & MS, 82, PhD(elec eng), 84. *Prof Exp:* Lead scientist, Eyring Res Inst, 84-85; assoc prof elec eng, Brigham Young Univ, 84-85; sr systs engr, TRW, 85-89; SR RES ENGR, ADVAN DECISION SYSTS, 89- *Concurrent Pos:* Reviewer, Micro Prog, Univ Calif, 87-89. *Mem:* Inst Elec & Electronic Engrs; Int Soc Optical Engrs. *Res:* Development of image processing algorithms for enhancement and exploitation of remote sensing imagery; utilization of SAR, IR, visible, and multispectral images. *Mailing Add:* Advan Decision Systs 1500 Plymouth St Mountain View CA 94043

PATTERSON, TROY B, b Columbus, Miss, Dec 5, 23; m 48; c 4. ANIMAL BREEDING, ANIMAL GENETICS. *Educ:* Miss State Univ, BS, 47; Tex A&M Univ, MS, 48, PhD(animal breeding), 56. *Prof Exp:* From instr to assoc prof animal husb, Miss State Univ, 48-57; asst genetics, Tex A&M Univ, 53-56; assoc prof animal sci, 57-65, PROF ANIMAL SCI, AUBURN UNIV, 65- *Mem:* Am Soc Animal Sci; Am Genetic Asn; Sigma Xi. *Res:* Genetic-environment interaction in beef cattle; genetic and environmental parameters of beef cattle; crossbreeding among British breeds. *Mailing Add:* Animal/Dairy Sci Auburn Univ Auburn AL 36849

PATTERSON, TRUETT CLIFTON, b Greenville, SC, Oct 10, 37; m 59; c 2. INORGANIC CHEMISTRY, ANALYTICAL CHEMISTRY. *Educ:* Furman Univ, BS, 59; Univ Tenn, PhD(chem), 66. *Prof Exp:* Asst prof, 64-66, assoc prof, 67-71, PROF CHEM, CARSON-NEWMAN COL, 72- *Mem:* Sigma Xi; Am Chem Soc. *Res:* Coordination chemistry. *Mailing Add:* Carson-Newman Col Carson-Newman Col Box 1906 Jefferson City TN 37760

PATTERSON, VERNON HOWE, metallurgy; deceased, see previous edition for last biography

PATTERSON, WILLIAM ALEXANDER, b Shankhouse, Eng, July 16, 15; m 46; c 2. PHYSICAL CHEMISTRY. *Educ:* Univ NB, BA, 36; Univ Toronto, 38, PhD(phys chem), 40. *Prof Exp:* Asst chemist, NB Int Paper Co, 36-37; asst chem & demonstr electrochem, Univ Toronto, 37-40; res chemist, Res & Develop Lab, Can Industs, Ltd, 40-51; tech consult instrument applns, Baird Assocs, Inc, 51-55; res group leader, Cryovac Div, W R Grace & Co, Duncan, 55-62, res gr, 62-66, sr scientist for advan technol, 66-70, sr scientist & coordr environ res control, Cryovac Div, 70-80; RETIRED. *Concurrent Pos:* Vis prof, Univ W Indies, 68-70. *Mem:* Am Chem Soc; AAAS. *Res:* Diffusion constants of copper; spectroscopy; process control instrumentation; irradiation chemistry; polymer and film technology; environmental problems. *Mailing Add:* 422 Forest Ave Spartanburg SC 29302

PATTERSON, WILLIAM BRADFORD, b New Rochelle, NY, June 25, 21; m 43; c 4. SURGERY, ONCOLOGY. *Educ:* Harvard Univ, AB, 43, MD, 50. *Prof Exp:* Chemist explosives, E I du Pont de Nemours & Co, Inc, NJ, 42-44; intern & resident surg, Peter Bent Brigham Hosp, 50-55; asst dir, Sears Surg Lab, Boston City Hosp, 56-59; chief prof serv, Pondville Hosp, Walpole, Mass, 59-63; asst clin prof surg, Harvard Med Sch, 61-70; prof oncol in surg, Sch Med & Dent, Univ Rochester, 70-78; DIR CANCER CONTROL, DANA FARBER CANCER INST, 78- *Concurrent Pos:* Nat Cancer Inst trainee, Peter Bent Brigham Hosp, Boston, 51-53; Am Cancer Soc clin fel, New Eng Deaconess Hosp, Boston, 53-54; pvt pract, 63-70; surgeon, Strong

Mem Hosp, Rochester, NY, 70-78; consult, Park-Ave Hosp, Rochester, 71-78; chmn surg, Monroe Community Hosp, Rochester, 71-75; mem cancer clin invests rev comt, NIH, 71-75; vis prof surg, Harvard Med Sch, 78-; mem Harvard surg serv, New Eng Deaconess Hosp, Boston, 78- *Mem:* Am Col Surgeons; Am Soc Clin Oncol; Soc Surg Oncol. *Res:* Clinical cancer research, including chemotherapy and surgical technics. *Mailing Add:* Dana Farber Cancer Inst 44 Binney St Boston MA 02115

PATTERSON, WILLIAM CREIGH, JR, b Royalton, Pa, Mar 26, 21; m 47; c 4. VETERINARY MEDICINE. *Educ:* Pa State Col, BS, 43; Univ Pa, VMD, 49. *Prof Exp:* Vet, Southeast Pa Artificial Breeding Coop, 49-51; asst to dir animal dis & parasite res div, 51-61, dir, Southeast Poultry Res Lab, 61-72, ASST AREA DIR, AGR RES SERV, USDA, 72- *Mem:* Am Vet Med Asn; US Animal Health Asn; Am Asn Avian Path; Poultry Sci Asn; World Vet Poultry Asn. *Res:* Veterinary virology; vesicular stomatitis of horses, cattle and swine; vesicular exanthema of swine; poultry diseases. *Mailing Add:* 736 Riverhill Dr Athens GA 30606

PATTERSON, WILLIAM JERRY, b Memphis, Tenn, Aug 8, 39; m 66; c 2. ORGANIC POLYMER CHEMISTRY. *Educ:* Miss State Univ, BS, 61; MS, 63; Univ Ala, PhD(org chem), 74. *Prof Exp:* Chemist, Entom Res Lab, USDA, 61-63; RES CHEMIST, MAT & PROCESSES LAB, MARSHALL SPACE FLIGHT CTR, NASA, 63- *Mem:* Am Chem Soc; Sigma Xi. *Res:* Synthesis, characterization and structure-property correlations of organosilicon and organometallic condensation polymers; reactivity-ratio studies of oxazoline addition copolymers. *Mailing Add:* 300 Quincy Dr SW Huntsville AL 35801

PATTIE, DONALD L, b Volt, Mont, Nov 22, 33; m 56; c 1. BIOLOGY, ECOLOGY. *Educ:* Concordia Col, Minn, BA, 55; Mont State Univ, MA, 60; Univ Mont, PhD, 67. *Prof Exp:* From instr to asst prof biol, Pac Lutheran Univ, 63-69; assoc prof, Camrose Lutheran Col, 69-73; INSTR BIOL SCI, NORTHERN ALTA INST TECHNOL, 73- *Concurrent Pos:* NSF study grants, 67-69; Nat Res Coun & Int Biol Prog grant energy flow in avian high Arctic ecosysts, Devon Island, Northwest Territories, Can, 70-72 & 78-81; mgr, High Arctic Field Sta, Arctic Inst N Am, Devon Island, 78-86. *Mem:* Am Soc Mammal; Cooper Ornith Soc; Arctic Inst NAm. *Res:* Ecology of arctic and alpine mammals and birds; mammalian taxonomy; long term populations of high arctic birds. *Mailing Add:* Dept Biol Sci Northern Alta Inst Technol Edmonton AB T5G 2R1 Can

PATTILLO, ROBERT ALLEN, b Atlanta, Ga, Nov 22, 51; m 72; c 2. REFRACTORIES. *Educ:* Ga Inst Technol, BS, 75, MS, 80. *Prof Exp:* Ceramic engr, Repub Steel Corp, 75-82; mgr tech serv, 82-83, dir res, 83-90, VPRES RES, RIVERSIDE REFRACTORIES, INC, 90- *Mem:* Am Ceramic Soc; Iron & Steel Soc. *Res:* High performance materials for molten metal contact; product development and application in order to create the best possible solution to a problem; finite element modeling of refractory structures. *Mailing Add:* 4939 Meadow Brook Rd Birmingham AL 35242

PATTILLO, WALTER HUGH, JR, b Bayboro, NC, Jan 1, 30; m 58; c 2. PARASITOLOGY. *Educ:* Hampton Inst, BS, 52; Iowa State Univ, MS, 54, PhD(parasitol), 56. *Prof Exp:* Asst zool & entom, Iowa State Univ, 53-56; asst prof biol, Tuskegee Inst & res assoc, Carver Found, 58-61; chmn dept, 64-76, asst undergrad dean, 76-78, PROF BIOL, NC CENT UNIV, 61-, DEAN UNDERGRAD SCH, 78- *Mem:* Sigma Xi; Nat Inst Sci; Soc Protozool. *Res:* Protozoology; Coccidia; Eimeria. *Mailing Add:* 310 E Alton St Durham NC 27707

PATTON, ALTON DEWITT, b Corpus Christi, Tex, Feb 1, 35; m 59; c 2. ELECTRIC POWER-ENERGY SYSTEMS, RELIABILITY OF ELECTRIC POWER SYSTEMS. *Educ:* Univ Tex, BS, 57; Univ Pittsburgh, MS, 61; Tex A&M Univ, PhD(elec eng), 72. *Prof Exp:* Engr, Westinghouse Elec Corp, 57-65; prof elec eng, Tex A&M Univ, 65-79; pres, Assoc Power Anals Inc, 79-83; PROF ELEC ENG, TEX A&M UNIV, 83- *Concurrent Pos:* Dir, Elec Power Inst, Tex A&M Univ, 76-79 & 85- & Ctr Space Power, Tex Eng Exp Sta, 87- *Mem:* Nat Soc Prof Engrs; fel Inst Elec & Electronic Engrs. *Res:* Electric power and energy systems as found in electric utility, industrial plant and aerospace applications; the reliability and operational modeling, assessment and evaluation of electric power systems. *Mailing Add:* Elec Eng Dept Tex A&M Univ College Station TX 77843

PATTON, BRUCE RILEY, b Pittsburgh, Pa, May 30, 44. MANY BODY THEORY, PHASE TRANSITIONS. *Educ:* Swarthmore Col, BA, 66; Cornell Univ, PhD(physics), 71. *Prof Exp:* Res assoc physics, Cornell Univ, 71; lectr, Univ Calif, San Diego, 71-73; asst prof, Mass Inst Technol, 73-78; ASSOC PROF PHYSICS, OHIO STATE UNIV, 78- *Concurrent Pos:* Fel, Woodrow Wilson Found, 66-67, NSF, 71-72 & Alfred P Sloan Found, 74-78; Prin investr, NSF; consult, IBM. *Mem:* Am Phys Soc. *Res:* Many-body phenomena in condensed matter, including superconductivity in metals, superfluidity in liquid 3helium, thermodynamics and transport properties of quasi-one-dimensional conductors, and the properties of inhomogeneous and submicron physical systems. *Mailing Add:* Physics Dept Ohio State Univ Columbus OH 43210

PATTON, CARL E, b San Antonio, Tex, Sept 14, 41. SOLID STATE PHYSICS. *Educ:* Mass Inst Technol, SB, 63; Calif Inst Technol, MS, 64, PhD(elec eng), 67. *Prof Exp:* Asst scientist, Jet Propulsion Lab, Calif Inst Technol, 63; sr res scientist, Res Div, Raytheon Corp, 67-69; vis scientist, Inst Solid State Physics, Univ Tokyo, 69-70; sr res scientist, Res Div, Raytheon Corp, 70-71; assoc prof, 71-76, PROF PHYSICS, COLO STATE UNIV, 76- *Concurrent Pos:* Japan Soc Promotion Sci fel, 69-70; Nat Acad Sci exchange fel, Czech Acad Sci, 73 & 79; vis scientist, Inst Fur Angewandte Festkorperphysin, WGer, 77-78; Humboldt Found res fel, 77-78; ed, Trans Magnetics, Inst Elec & Electronics Eng. *Mem:* Fel Am Phys Soc; Inst Elec & Electronics Eng. *Res:* Magnetic thin films; domain wall motion and ferromagnetic resonance; microwave magnetic materials; spin wave instability; relaxation processes; magnetic metals and alloys; induced anisotrophy; Brillouin light scattering. *Mailing Add:* Dept Physics Colo State Univ Ft Collins CO 80523

PATTON, CHARLES C(LIFFORD), b Cushing, Okla, July 10, 36; m 61; c 2. PETROLEUM ENGINEERING, CORROSION ENGINEERING. *Educ:* Univ Okla, BS, 59; Univ Tex, PhD(petrol eng), 64. *Prof Exp:* Exploitation engr, Shell Oil Co, 59-60; sr res petrol engr, Monsanto Co, 63-64; res scientist corrosion, Continental Oil Co, 64-67, res group leader, 67-69; staff engr, Hudson's Bay Oil & Gas Co, Ltd, 69-70; assoc prof petrol & geol eng, Univ Okla, 70-72, dir sch petrol & geol eng, 72-74; vpres, Petrotech Ltd, 74-80; PRES, C C PATTON & ASSOCS, 80- *Concurrent Pos:* Nat Sci Found res fel sci & pub policy, 71-72. *Mem:* Soc Petrol Engrs; Nat Asn Corrosion Engrs. *Res:* Oilfield corrosion; water treatment for subsurface injection. *Mailing Add:* C C Patton & Assocs Inc 1555 Valwood Pkwy Suite 100 Carrollton TX 75006

PATTON, CURTIS LEVERNE, b Birmingham, Ala, June 13, 35; m 63; c 1. PARASITOLOGY, CELL BIOLOGY. *Educ:* Fisk Univ, BA, 56; Mich State Univ, MS, 61, PhD(microbiol), 66. *Prof Exp:* Asst microbiol, Mich State Univ, 60-63, from asst instr to instr, 63-67; guest investr, Rockefeller Univ, 67-70; asst prof microbiol, 70-74, dir grad studies, 72-74, asst prof epidemiol, pub health & microbiol, 74-76, ASSOC PROF EPIDEMIOL, PUB HEALTH & MICROBIOL, SCH MED, YALE UNIV, 76- *Concurrent Pos:* Biomed Sci Support grant, 66-67; fel parasitol, Rockefeller Univ, 67-70; USPHS training grant, 67-69, USPHS res grants, 72-77, 78-82 & 82-85; dir, Interdisciplinary Parasitol Training Prog, USPHS grants, 77-80 & 86; mem, Minority Access Res Careers & Nat Inst Gen Med Sci, 78-82; mem, Nat Res Coun Comn Human Resources Eval Panel, 79-81; consult, US Army Med Res & Develop Command, 79-82. *Mem:* AAAS; Am Soc Parasitologists; Soc Protozoologists. *Res:* Cell and molecular biology of trypanosomes; membrane transport in parasitic protozoa; physiology of parasitic protozoa; humoral and cellular responses to parasitic protozoa. *Mailing Add:* 61 Whittier Rd New Haven CT 06515

PATTON, DAVID ROGER, b 1934; m 52; c 3. WILDLIFE RESEARCH, GENERAL FORESTRY. *Educ:* Univ WVa, BS, 60; Va Polytech Inst, MS, 63; Univ Ariz, PhD(watershed mgt), 74. *Prof Exp:* Forester, Cleveland Nat Forest, US Forest Serv, 60; biologist, US Fish & Wildlife Serv, US Forest Serv, 60-61, Santa Fe Nat Forest, US Forest Serv, 63-64, res biologist, Rocky Mountain Forest & Range Exp Sta, 64-73, proj leader, Forest Hydrol Lab, Ariz State Univ, Rocky Mountain & Range Exp Sta, 73-; AT DEPT FORESTRY, NORTHERN ARIZ UNIV. *Concurrent Pos:* Wildlife expert to Govt Repub Zambia, Food & Agr Orgn, UN, 66-67; forestry & wildlife res in Mex. *Honors & Awards:* Gulf Conserv Award, 82. *Mem:* AAAS; Wildlife Soc; Soc Am Foresters. *Res:* Influence of forest management practices on distribution and abundance of wildlife populations; ecological relationships of animals and their habitat; develop habitat criteria for game, non-game and endangered species on national forests in the Southwest. *Mailing Add:* 1231 E Bishop Tempe AZ 85282

PATTON, DENNIS DAVID, b Oakland, Calif, Aug 4, 30; m 65; c 2. RADIOLOGY, NUCLEAR MEDICINE. *Educ:* Univ Calif, Berkeley, BA, 53; Univ Calif, Los Angeles, MD, 59. *Prof Exp:* Assoc med physics, Univ Calif, Berkeley, 52-54, sr lab technician, 54-57; mgr biomed group, Planning Res Corp, 58-65; resident radiol, Col Med, Univ Calif, Irvine, 65-68, asst prof, 68-70, chief nuclear med, Orange County Med Ctr, 68-70; assoc prof radiol, Med Ctr, Vanderbilt Univ, 70-75; actg chief nuclear med, Vet Admin Hosp, Tucson, 75-77; DIR, DIV NUCLEAR MED, MED CTR, UNIV ARIZ, 75-, PROF RADIOL, 80- *Concurrent Pos:* Teaching fel, Univ Calif, Irvine, 67-68, grant, 69-70; consult, Long Beach Vet Admin Hosp, Calif, 68-70 & St Joseph's Hosp & Children's Hosp, Orange County, Calif, 69-70. *Mem:* Soc Nuclear Med; Radiol Soc NAm; Am Col Radiol; Am Col Nuclear Med; Asn Univ Radiologists. *Res:* Development of short-life radiopharmaceuticals for diagnostic clinical studies; development of medical imaging systems; operations research and systems analysis in medicine. *Mailing Add:* Univ Ariz Health Sci Ctr Univ Ariz Med Ctr Tucson AZ 85724

PATTON, ELIZABETH VANDYKE, b Omaha, Nebr, June 24, 44; m 67; c 2. PHYSICAL CHEMISTRY. *Educ:* Univ Mich, BS, 66; Univ Wis, PhD(phys chem), 72. *Prof Exp:* Fel biochem, Sch Med & Dent, Univ Rochester, 71-74; LECTR, ROCHESTER INST TECHNOL, 74- *Mem:* Am Chem Soc. *Res:* Physical properties of biological macromolecules and polypeptides. *Mailing Add:* Four Fellview Dr Pittsford NY 14534-4014

PATTON, ERNEST GIBBES, b Greenville, SC, Nov 30, 24; m 50; c 4. ECOLOGY. *Educ:* Yale Univ, BA, 48; Univ NC, MA, 50; Duke Univ, PhD, 55. *Prof Exp:* From asst prof to assoc prof biol, Univ Ala, 53-63, dir arboretum, 58-63; ASSOC PROF BIOL, WOFFORD COL, 63- *Mem:* Soc Study Evolution; Nat Sci Teachers Asn. *Res:* Plant communities; ecology of native shrubs; biological evaluation of land-use. *Mailing Add:* Dept of Biol Wofford Col Spartanburg SC 29301

PATTON, HARRY DICKSON, b Bentonville, Ark, Mar 10, 18; m 43; c 2. PHYSIOLOGY. *Educ:* Univ Ark, BA, 39; Yale Univ, PhD, 43, MD, 46. *Hon Degrees:* LLD, Univ Ark, 84. *Prof Exp:* Res asst physiol, Sch Med, Yale Univ, 43-46; instr psychology, Johns Hopkins Univ, 46-47; from asst prof to assoc prof, 47-56, actg chmn dept, 64-66, PROF PHYSIOL & BIOPHYS, SCH MED, 56-, chmn dept, 66-83, EMER PROF & CHAIR, UNIV WASH, 83- *Concurrent Pos:* Mem, Physiol Test Comt, Nat Bd Med Exam, 64-68, chmn, 65-68; mem, Physiol Study Sect, USPHS, 62-66, chmn, 65-66, ad hoc cerebrovasc training rev comt, 66-67, nat adv neurol dis & blindness coun, Nat Inst Neurol Dis & Blindness, 66-70 & chmn Neurol Dis Review Comt B, 75-79. *Mem:* Am Physiol Soc; Soc Neurosci. *Res:* Thalamocortical representation of taste; neuronal organization of sympathetic chain; neural factors in endocrine secretion; cortical activation of pyramidal tract. *Mailing Add:* 1717 Evergreen Pl Seattle WA 98122

PATTON, HUGH WILSON, b Lebanon, Tenn, Dec 2, 21; m 50; c 4. PHYSICAL CHEMISTRY. *Educ:* Middle Tenn State Col, BS, 45; Vanderbilt Univ, PhD(chem), 52. *Prof Exp:* Prof chem, Ark State Teachers Col, 50-53; from res chemist to sr res chemist, Tenn Eastman Co, 53-62, res assoc, 62-66, res div head, 66-70, asst dir res, 70-73, vpres, Eastman Chem Prod, Inc, 73-78 & dir res, 78-79, dir res, Tenn Eastman Co, 79-82 & vpres, 79-85, dir res, Eastman Chem Div, Eastman Kodak Co, 82- 85; RETIRED. *Mem:* AAAS; Am Chem Soc. *Res:* Heats of vaporization; gas chromatography; x-ray diffraction; nuclear magnetic resonance. *Mailing Add:* 939 Lookout Dr Kingsport TN 37663

PATTON, JAMES LLOYD, b St Louis, Mo, June 21, 41; m 66. MAMMALOGY. *Educ:* Univ Ariz, BA, 63, MS, 65, PhD(zool), 69. *Prof Exp:* Asst prof & asst cur, 69-74, assoc prof & assoc cur, 74-79, PROF ZOOL & CUR MAMMALS, MUS VERT ZOOL, UNIV CALIF, BERKELEY, 79- *Mem:* AAAS; Am Soc Mammal; Soc Study Evolution; Genetics Soc Am; Soc Syst Zool; Sigma Xi. *Res:* Population genetics and historical evolution of heteromyid and geomyid rodents in North America; biosystematics of Neotropical mammals. *Mailing Add:* Univ Calif Zool Life Sci Bldg Berkeley CA 94720

PATTON, JAMES WINTON, b Okemah, Okla, Nov 10, 29; m 61; c 3. ORGANIC CHEMISTRY, GEOCHEMISTRY. *Educ:* Univ NMex, BS, 51, MS, 54; Univ Wis-Madison, PhD(chem), 61. *Prof Exp:* Res assoc chem, Univ Southern Calif, 61-62; res chemist, Marathon Oil Co, 62-67, advan res chemist, 67-79, sr res chemist, 79-86; RETIRED. *Concurrent Pos:* Participated in Leg 77 of deep sea drilling proj, Gulf of Mex, 80-81. *Mem:* AAAS; Am Chem Soc; Geochem Soc; Sigma Xi. *Res:* Petrochemicals; organic chemistry; origin and migration of crude oil; autoxidation processes; stable isotopes in geological interpretation. *Mailing Add:* 737 W Davies Way Littleton CO 80160

PATTON, JOHN BARRATT, geology; deceased, see previous edition for last biography

PATTON, JOHN F, b St Louis, Mo, Feb 8, 39; m 68; c 1. PHYSIOLOGY. *Educ:* Wake Forest Univ, BS, 61; Univ Mo, Columbia, Mo, 65, PhD(physiol), 69. *Prof Exp:* Res physiologist, US Army Res Inst Environ Med, Natick, Mass, 69-76; proj leader physiol & toxicol, Patuxent Wildlife Res Ctr, Dept Interior, Laurel, Md, 76-77; RES PHYSIOLOGIST, US ARMY RES INST ENVIRON MED, 77- *Concurrent Pos:* Exchange scientist, Army Personnel Res Estab, Farnborough, UK, 83-84. *Mem:* Am Physiol Soc; Sigma Xi; fel Am Col Sports Med. *Res:* Environmental physiology; exercise physiology; cold acclimation; hypothermia. *Mailing Add:* Usariem Kansas State Natick MA 01760-5007

PATTON, JOHN STUART, b Columbus, Ohio, Sept 9, 46; m 68; c 3. PROTEIN PHARMACOLOGY, LIPID BIOCHEMISTRY. *Educ:* Pa State Univ, BS, 68; Univ RI, MS, 73; Scripps Inst Oceanog, PhD(marine biol), 76. *Prof Exp:* Swed Med Res Coun fel, Univ Lund, 76-77; Boston Med Found fel, Harvard Med Sch, 77-79; asst prof, Univ Ga, 79-85; scientist, Genentec, Inc, 85-90; FOUNDER & CHIEF EXEC OFFICER, INHALE INC, 90- *Concurrent Pos:* RRS Career Develop Award, NIH, 83-88. *Res:* Lipid digestion; coral-sea anemone physiology; lipid flow and partition in biological systems; delivery and targeting of recombinant proteins; epithelial permeability to proteins; aerosol drug delivery. *Mailing Add:* Inhale Inc 330 Emerald Ave San Carlos CA 94070

PATTON, JOHN THOMAS, JR, b Jonesboro, Ark, Jan 30, 17; m 45; c 2. ORGANIC POLYMER CHEMISTRY. *Educ:* Ark State Univ, BS, 38; Purdue Univ, PhD(org chem), 47. *Prof Exp:* Teacher high sch, Ark, 38-39; res chemist, Ford Motor Co, Mich, 39-42 & Penick & Ford, Inc, Iowa, 42-43; asst org chem, Purdue Univ, 45; res chemist, Wyandotte Chem Corp, 47-53, from actg sect head to sect head, 53-56, supvr org res, 56-63, mgr org res, 63-64, dir res urethane chem res & develop, 64-69; dir advan polymer technol, BASF, 79-82, dep dir cent res corp, 82-83; CONSULT, 83- *Concurrent Pos:* Consult, patent litigations. *Honors & Awards:* Thomas Midgley Award, Am Chem Soc, 75; Chem Pioneer Award, Am Inst Chemists, 78. *Mem:* Am Chem Soc; Am Inst Chemists; Sigma Xi. *Res:* Vegetable proteins; starch; textile assistants; detergents; nitroparaffins; alkylene oxides; organic synthesis and plastics; basis for reported optical activity of salts of aliphatic nitro compounds; polyols; polyglycols; heterocyclic nitrogen compounds; halogenation; polyethers; isocyanates; urethanes; cellular plastics; polymers. *Mailing Add:* 2358 18th St Wyandotte MI 48192-4124

PATTON, JOHN TINSMAN, b Ft Worth, Tex, May 9, 31; m 53; c 6. CHEMICAL ENGINEERING, PETROLEUM ENGINEERING. *Educ:* Okla State Univ, BS, 53, MS, 58, PhD(chem eng), 59. *Prof Exp:* Chem engr, Tex Eastman Co, 53-56; res engr, Jersey Prod Res Co Div, Standard Oil Co, NJ, 59-61, sect head drilling res, 61-63; petrol engr, Int Petrol Co, 63-64; sect head petrol recovery, Esso Prod Res Co, 64-65; res adv coal processing, Humble Oil & Refining Co, 65-68; prof chem eng, Mich Technol Univ, 68-77; PROF & HEAD CHEM ENG DEPT, N MEX STATE UNIV, 77- *Concurrent Pos:* Consult to oil & chem indust, 67-; pres, Comput/Bioeng Inst Inc, 68-; prin investr enhanced oil recovery by carbon dioxide foam flooding, 77- *Mem:* Soc Petrol Engrs; Am Chem Soc; Am Inst Chem Engrs. *Res:* Biochemical engineering, especially kinetics of biosynthesis; improved methods of oil recovery and drilling techniques; pollution control processes with emphasis on pulp and paper industry. *Mailing Add:* Dept Chem Eng NMex State Univ PO Box 3805 Las Cruces NM 88003

PATTON, JON MICHAEL, b North Canton, Ohio, Sept 10, 42. INDUSTRIAL MANUFACTURING & ENGINEERING, CIVIL ENGINEERING. *Educ:* Ohio State Univ, BS, 65; Univ Ala, Huntsville, MA, 68; Purdue Univ, MS, 76, PhD(math), 81. *Prof Exp:* Assoc engr struct anal, Boeing Co, 65-67; instr math, Univ Ala, Huntsville, 68; teaching asst, Purdue Univ, 69-79; systs engr, Pritsker & Assocs, 79-81 & Computer Task Group, 81-82; asst prof systs anal, 83-89, APPLICATIONS CONSULT, ACAD COMPUTER SERV, MIAMI UNIV, 89- *Concurrent Pos:* Adj asst prof, Decision Sci Dept, Miami Univ, 89-, adj asst prof, Systs Anal Dept, 91- *Mem:* Math Asn Am; Soc Indust & Appl Math; Inst Mgt Sci; Opers Res Soc Am. *Res:* Numerical research on the stability of solar system; co-authored publications on operations research covering quality control, decision support systems and flexible manufacturing systems. *Mailing Add:* Miami Univ 80 Charleston Dr Oxford OH 45056

PATTON, LEO WESLEY, b Sublette, Kans, July 23, 19; m 49; c 2. ORGANIC CHEMISTRY, ANALYTICAL CHEMISTRY. *Educ:* Southwestern Col, Kans, BA, 41; Univ Kans, MS, 48; Kans State Col Manhattan, PhD(chem), 50. *Prof Exp:* Asst prof chem, McPherson Col, 49; res chemist, 51-72, FEL PETROCHEMICALS, SABINE RIVER LAB, E I DU PONT DE NEMOURS & CO, INC, 72- *Mem:* Am Chem Soc; Am Acad Arts & Sci. *Res:* Analytical methods and processes associated with the production of nylon intermediates. *Mailing Add:* 2332 Manley Circle Orange TX 77630

PATTON, NANCY JANE, b Springfield, Ohio, Dec 29, 36. NEUROSCIENCES. *Educ:* Ohio State Univ, BS, 58; Univ Mich, MS, 62; Univ Wis, PhD(anat), 69. *Prof Exp:* Instr anat & physiol, Sargent Col, Boston Univ, 62-65; instr functional neurophysiol, Sch Phys Ther, Univ Wis, 68; assoc prof phys ther & asst prof anat, Col Allied Health Professions, Univ Ky, 69-74, assoc dir curric phys ther, 72-74; assoc prof phys therapy & asst prof anat, Col Allied Health Sci, Med Univ SC, 74-78, prof phys therapy & dir phys therapy prog, 78-81; PROF & CHMN HEALTH SCI, COL ARTS & SCI & DIR PHYS THERAPY, CLEVELAND STATE UNIV, 81- *Mem:* Am Soc Allied Health Professions; Am Phys Ther Asn; Am Asn Univ Prof. *Res:* Proprioceptive reflexes; patterns of motor activity in the central nervous system; sensory-motor systems; physical therapy; normal development of posture. *Mailing Add:* Health Sci Cleveland State Univ Euclid Ave at E 24th St Cleveland OH 44115

PATTON, PETER C(LYDE), b Wichita, Kans, June 11, 35; m 57; c 5. COMPUTER HARDWARE SYSTEMS, SOFTWARE SYSTEMS, APPLIED MATHEMATICS. *Educ:* Harvard Univ, AB, 57; Univ Kansas, MA, 59; Univ Stuttgart, PhD(aerospace eng), 66. *Prof Exp:* Jr engr, Wichita Div, Boeing Co, 57; res asst math, Univ Kans, 57-59; assoc engr, Midwest Res Inst, Mo, 59-61; prin programmer, Fed Syst Div, Univac, Minn, 61, sci consult comp appln, Int Div, Lausanne, Switz, 61-63, sci syst mgr, Univac Ltd, London, 63-65; comput group mgr aerospace eng, Univ Stuttgart, 65-67; eng mgr syst design, Data Processing Div, Univac, Minn, 67-68; mgr tech staff, Analysts Inst Corp, 68-71; assoc prof comput Sci & Dir Comput Ctr, Univ Minn, Minneapolis, 71-83; prin scientist, Parallel Processing Prog, Adv Comput Archit, Microelec & Comput Tech Corp, Tex, 83-85; DIR, MINN SUPERCOMPUTER INST, UNIV MINN, 85- *Concurrent Pos:* Lectr, Univ Kansas City, 58-59; instr, Kansas City Jr Col, 60-62; res assoc, Univ Stuttgart, 62-65, lectr, 65-67; assoc dir, Ctr Ancient Studies, Univ Minn, 77-78, dir, 78-81; prof, Ctr Ancient Studies, Univ Minn, 85- *Mem:* Sr mem Inst Elec & Electronics Engrs; Asn Comput Mach; fel Brit Inst Math & Appln. *Res:* New applications of computer technology to science and engineering; new computer architectures to meet the needs of advanced applications; computer applications to the study of ancient languages and literature; automatic generation of computer applications software. *Mailing Add:* Spec Consult Serv 7900 International Dr Minneapolis MN 55425

PATTON, PETER C, b May 29, 49; m 76; c 2. GEOLOGY. *Educ:* Franklin & Marshall Col, BA, 71; Colo State Univ, MS, 73; Univ Tex, PhD(geol), 77. *Prof Exp:* From asst prof to assoc prof, 76-89, chmn dept, 86-89, PROF, DEPT EARTH & ENVIRON SCI, WESLEYAN UNIV, 89- *Concurrent Pos:* Consult, Nature Conserv & Conn River Watershed Coun. *Mem:* Fel Geol Soc Am; AAAS; Sigma Xi; Am Geophys Union; Brit Geomorphol Res Group. *Mailing Add:* Dept Earth & Environ Sci Wesleyan Univ Middletown CT 06457

PATTON, ROBERT FRANKLIN, b Albuquerque, NMex, Oct 28, 19; m 46; c 2. FOREST PATHOLOGY. *Educ:* Univ Mich, BS, 40; Univ Idaho, MS, 42; Univ Wis, PhD(plant path), 52. *Prof Exp:* Res asst med mycol & antibiotics, Parke, Davis & Co, 46-47; res asst plant path, 47-50, from instr to assoc prof forest path, 50-65, PROF FOREST PATH, UNIV WIS-MADISON, 65- *Concurrent Pos:* NATO sr fel sci, Cambridge Univ, 71. *Mem:* Soc Am Foresters; Am Phytopath Soc. *Res:* Resistance to white pine blister rust; forest plantation root diseases; forest tree rusts; fluorescence microscopy; plantation diseases, especially needle infection processes. *Mailing Add:* Dept Plant Path Univ Wis Russell Lab Madison WI 53706

PATTON, ROBERT LYLE, b Stockton, Calif, Nov 19, 43; m 72; c 2. PHYSICAL CHEMISTRY. *Educ:* Univ of the Pac, BS, 64; Univ Calif, Berkeley, PhD(inorg chem), 69. *Prof Exp:* SR RES ASSOC, MOLECULAR SIEVE DEPT, UNION CARBIDE CORP, 69- *Mem:* Am Chem Soc; Int Zeolite Asn. *Res:* Synthesis and properties of new crystalline molecular sieves including zeolites, aluminophosphates and silicoaluminophosphates; ozone chemistry. *Mailing Add:* Union Carbide Linde Bldg Tarrytown Tech Ctr Tarrytown NY 10591

PATTON, SHARON, b Watertown, Tenn, Sept 14, 47; m 80; c 1. PARASITOLOGY. *Educ:* Middle Tenn State Univ, BS, 69; Univ Ky, MS, 71, PhD(parasitol), 75. *Prof Exp:* Scholar Vet Parasitol, dept vet sci, Univ Ky, 75-77; asst prof parasitol, 77-82, ASSOC PROF, DEPT PATHOBIOL, UNIV TENN, 82- *Concurrent Pos:* Exec Comt, Asn Southeastern Biologists, 81-84. *Mem:* Sigma Xi; Am Soc Parasitol; Am Soc Trop Med & Hyg; AAAS; Southeastern Soc Parasitol (pres-elect, 80-81, pres, 81-82, secy treas, 87-); Am Soc Vet Parasitologists. *Res:* Veterinary parasitology, helminths; Toxoplasma Gondii; Zoonotic disease. *Mailing Add:* Dept Pathobiol Col Vet Med Univ Tenn PO Box 1071 Knoxville TN 37901-1071

PATTON, STUART, b Ebenezer, NY, Nov 2, 20; m 45; c 7. FOOD BIOCHEMISTRY. *Educ:* Pa State Univ, BS, 43; Ohio State Univ, MS, 47, PhD(dairy sci), 48. *Prof Exp:* Chemist, Borden Co, 43-44; from assoc prof to prof food sci, Pa State Univ, 49-66, Evan Pugh res prof agr, 66-80; CONSULT, 80- *Concurrent Pos:* Mem nat adv comt for res vessel Alpha Helix, 71-72; adj prof neurosci, Univ Calif San Diego, 80-; Alexander von Humboldt Sr Scientist Award, 81. *Honors & Awards:* Borden Award, 57; Agr & Food Chem Award, Am Chem Soc, 75. *Mem:* AAAS; Am Chem Soc; Am Dairy Sci Asn; Am Oil Chem Soc; Am Soc Biochem & Molecular Biol; Am Soc Cell Biol. *Res:* Chemistry and biology of milk; lipid biochemistry; membrane biology. *Mailing Add:* Dept Neurosci 0634 J Sch Med Univ Calif San Diego La Jolla CA 92093-0634

PATTON, TAD LEMARRE, b Wichita Falls, Tex, June 21, 25; m 52; c 2. ORGANIC CHEMISTRY, POLYMER SYNTHESIS. *Educ:* Baylor Univ, BS, 46; Univ Notre Dame, MS, 50; Univ Tex, PhD(org chem), 52. *Prof Exp:* Res chemist, Monsanto Chem Co, Ohio, 51-53; res fel, Univ Tex M D Anderson Hosp & Tumor Inst, 53-54, from asst biochemist to assoc biochemist, 54-61; res chemist, Spencer Chem Co, 61-63; res specialist, 63-65, res assoc, 65-70, SR RES ASSOC, EXXON RES & ENG CO, 70- *Mem:* Am Chem Soc; Royal Soc Chem; fel Am Inst Chemists; Sigma Xi. *Res:* Steroid synthesis, particularly those related to the androgens and estrogens; relationship of molecular structure to physiological activity; organic reaction mechanisms; synthetic polymers and fuels. *Mailing Add:* 5010 Glenhaven Dr Baytown TX 77521

PATTON, THOMAS HUDSON, b New Orleans, La, May 24, 34; m 58; c 5. VERTEBRATE PALEONTOLOGY. *Educ:* La State Univ, BS, 60; Univ Tex, MA, 62, PhD(geol), 66; Univ Fla, JD, 79. *Prof Exp:* Asst prof zool, Univ Fla, 65-72, asst prof geol, 70-72, assoc prof zool & geol, 72-77; assoc cur, Fla State Mus, 72-77; PRES, PATTON & ASSOCS, INC, 79- *Concurrent Pos:* Assoc cur, Fla State Mus, 64-72; NSF grant, Jamaica, 66-68; Univ Fla & NIH grant & proj leader, Coop Biol Invests, Jamaica, 68-72; consult, local state & fed agencies, private law & eng firms, 79- *Mem:* Soc Vert Paleont; Am Soc Mammal; Am Soc Study Evolution; Geol Soc Am; Am Asn Petrol Geologists. *Res:* Environmental, karst, and historical geology; stratigraphy and geomorphology of Gulf Coast and circum-Caribbean regions; geology of wetlands and associated boundary determinations; environmental law. *Mailing Add:* 2830-D NW 41 St Gainesville FL 32606

PATTON, WENDELL KEELER, b Utica, NY, Sept 29, 32; m 61; c 3. INVERTEBRATE ZOOLOGY. *Educ:* Hamilton Col, AB, 54; Ohio State Univ, MS, 56; Univ Queensland, PhD(zool), 60. *Prof Exp:* Instr zool, Duke Univ, 60-62; from asst prof to assoc prof, 62-71, PROF ZOOL, OHIO WESLEYAN UNIV, 71- *Mem:* AAAS; Am Soc Zoologists; Crustacean Soc; Sigma Xi. *Res:* Animal associations; decapod Crustacea associated with reef corals; ecology of coral reefs. *Mailing Add:* Dept Biol Sci Ohio Wesleyan Univ Delaware OH 43015

PATTON, WILLIAM HENRY, b Albany, NY, Apr 15, 25; m 63; c 2. VETERINARY PARASITOLOGY. *Educ:* Mich State Univ, DVM, 50; Univ Wis, MS, 60, PhD(vet sci), 63. *Prof Exp:* Vet diagnostician, Animal Dis Diag Lab, Wis Dept Agr, 50-55; clin trials vet, Am Cyanamid Co, 55-57; res asst vet sci, Univ Wis, 57-62; ASST PROF VET SCI, PA STATE UNIV, 62- *Mem:* AAAS; Am Asn Avian Path; Wildlife Dis Asn; Tissue Cult Asn; Am Soc Parasitol. *Res:* Intracellular parasites of poultry and livestock; host-parasite relationship at level of parasitized cell, relying on systems of animal cells cultured in vitro. *Mailing Add:* 427 Orlando Ave State College PA 16803

PATTON, WILLIAM WALLACE, JR, b Vancouver, BC, May 25, 23; US citizen; m 51; c 3. GEOLOGY. *Educ:* Cornell Univ, BA, 45, MS, 48; Stanford Univ, PhD(geol), 59. *Prof Exp:* GEOLOGIST, BR ALASKAN GEOL, US GEOL SURV, 48- *Mem:* Fel AAAS; Geol Soc Am; Am Asn Petrol Geol; Am Geophys Union. *Res:* Geology of western Alaska and the Bering Sea region, particularly stratigraphy and tectonics; ophiolites of Alaska. *Mailing Add:* Alaskan Geol Br 345 Middlefield Rd Menlo Park CA 94025

PATTY, CLARENCE WAYNE, b Ringgold, Ga, Oct 7, 32; c 3. TOPOLOGY. *Educ:* Univ Ga, BS, 54, MA, 58, PhD(math), 60. *Prof Exp:* Res instr math, Univ NC, 60-61, from asst prof to assoc prof, 61-67; PROF MATH, VA POLYTECH INST & STATE UNIV, 67-, HEAD DEPT, 70- *Mem:* Am Math Soc; Math Asn Am. *Mailing Add:* Dept Math Va Polytech Inst Blacksburg VA 24061

PATTY, RICHARD ROLAND, b Jonesboro, Ark, Sept 28, 33; m 55; c 3. PHOTOACOUSTIC SPECTROSCOPY. *Educ:* Furman Univ, BS, 55; Vanderbilt Univ, MA, 57; Ohio State Univ, PhD(physics), 60. *Prof Exp:* Res assoc physics, Ohio State Univ, 60-61; sr scientist, Philco Res Lab, 63-64; from asst prof to assoc prof, 64-72, PROF PHYSICS, NC STATE UNIV, 72-, HEAD DEPT, 78- *Mem:* AAAS; fel Optical Soc Am; Am Asn Physics Teachers; Am Phys Soc. *Res:* Absorption and emission of infrared radiation by atmospheric gases as related to temperature, optical thickness and total pressure; airglow originating in upper atmosphere; photoacoustic spectroscopy. *Mailing Add:* Dept Physics NC State Univ Raleigh NC 27695-8202

PATWARDHAN, BHALCHANDRA H, b Amraoti, India, Dec 25, 47; m 75; c 1. BIOTECHNOLOGY, DOWNSTREAM RECOVERY. *Educ:* Nagpur Univ, BS, 67, MS, 69, PhD(phys org chem), 73. *Prof Exp:* Res scientist, Sep Chem Res, Miles, Inc, Elkhart, Ind, 82-83, staff scientist, 83-85, sr staff scientist process develop, 85-89, mgr process recovery process develop, 89; DIR PROCESS DEVELOP, HAARMANN & REIMER CORP, A BAYER USA CO, ELKHART, IND, 89- *Concurrent Pos:* Swiss Nat Fund postdoctoral fel, Inst Org Chem, Univ Bern, Switzerland, 74-75; Minna James Heinnmann postdoctoral fel, Univ Louvain, Belg, 75-76; Sci Res Coun postdoctoral fel, Univ Leicester, Eng, 76-77; postdoctoral fel, Univ Syracuse, NY, 77-79, State Univ NY, Syracuse, 79-82. *Mem:* Sigma Xi; Am Chem Soc; Royal Soc Chem. *Res:* Bioseparations, characterization of r DNA proteins; downstream processing of r DNA proteins; liquid liquid extractions; synthetic heterocyclic chemistry; organometallic chemistry; quantitative structure activity relationships; mechanism of organic reactions. *Mailing Add:* Haarmann & Reimer Corp PO Box 932 Elkhart IN 46515

PATZ, ARNALL, b Elberton, Ga, June 14, 20; c 4. OPHTHALMOLOGY. *Educ:* Emory Univ, BS, 42, MD, 45. *Hon Degrees:* DSc, Univ Pa, 82, Emory Univ, 85 & Thomas Jefferson Univ, 85. *Prof Exp:* SEEING EYE RES PROF OPHTHAL, JOHNS HOPKINS HOSP, 70-, PROF OPHTHAL, WILMER OPHTHAL INST, 69-, CHMN & DIR, DEPT OPHTHAL, 79- *Honors & Awards:* Albert Lasker Award, Am Pub Health Serv, 56; Mead Johnson Award, Am Acad Pediat, 56; Billings Silver Medal, AMA, 72; Friedenwald Award, Asn Res & Vision Ophthal, 80; Jules Stein Award, Res Prevent Blindness, 81. *Mem:* Am Acad Ophthal (pres elect, 86, vpres 84 & 85). *Res:* Retinal blood vessel diseases; laser techniques in ophthalmology. *Mailing Add:* Wilmer Ophthal Inst Baltimore MD 21205

PAU, LOUIS F, b Copenhagen, Denmark, May 29, 48; French citizen; m 83; c 1. COMPUTER VISION, SENSOR & DATA FUSION. *Educ:* Univ Paris, MSc, 69, PhD(computer sci), 72, DSc(control & econ), 75; ENS Adronautique et Espace, MSc, 70; Inst d'études Politiques, MA, 71. *Prof Exp:* From asst prof to assoc prof, Tech Univ, Denmark, 72-74, res prof artificial intel, 86-90; prof computers & electronics, ENS Telecommun, Paris, 79-82; sr scientist, Battelle Mem Inst, 82-86; prof computer sci, Tokyo Univ, 88-91; TECHNOL DIR, DIGITAL EQUIP CORP EUROPE, 91- *Concurrent Pos:* Vis prof, Mass Inst Technol, 77-78; adj prof, Univ Md, College Park, 80-82; vpres, Tech Bd, Int Fedn Automatic Control, 84-87; mem bd, Nordic Technol Corp, 88- & rev bd, Inst Systs Sci, Singapore, 90- *Mem:* Fel Inst Elec & Electronic Engrs; Soc Photo-Optical Instrumentation Engrs; Asn Comput Mach; Soc Mfg Engrs; Soc Econ Dynamics & Control; Am Asn Artificial Intel; Mach Vision Asn; fel Japan Soc Prom Sci. *Res:* Applications of artificial intelligence and image processing in industry, economics and aerospace engineering; failure diagnosis and test systems for aerospace, telecommunication and integrated circuits; economic modeling and games, macroeconomics and finance; international political affairs; science and technology administration. *Mailing Add:* Digital Equip Europe PO Box 129 Sophia Antipolis Valbonne Cedex F06561 France

PAUCKER, KURT, virology; deceased, see previous edition for last biography

PAUDLER, WILLIAM W, b Varnsdorf, Czech, Feb 11, 32; US citizen; m 55; c 3. ORGANIC CHEMISTRY. *Educ:* Univ Ill, BS, 54; Ind Univ, PhD(org chem), 59. *Prof Exp:* Res chemist, Procter & Gamble Co, 58-60; res assoc org chem, Princeton Univ, 60-61; distinguished prof chem, Ohio Univ, 61-65, from assoc prof to prof, 65-69, distinguished prof, 69-72; prof chem & chmn dept, Univ Ala, 72-; dean, Col Sci, 81-82, DEAN, COL LIB ARTS & SCI, PORTLAND STATE UNIV, 82- *Concurrent Pos:* Grants, NSF, 62-69, 71-73, & 75-76 & NIH, 64-69 & 71-74. *Mem:* AAAS; Am Chem Soc; Royal Soc Chem; NY Acad Sci. *Res:* Synthetic and theoretical aspects of heterocyclic chemistry; photochemistry; alkaloid chemistry; mass spectroscopy; nuclear magnetic resonance spectroscopy. *Mailing Add:* Dean, Col Lib Arts & Sci Portland State Univ, PO Box 751 Portland OR 97207

PAUERSTEIN, CARL JOSEPH, b New York, NY, May 31, 32; m 55; c 3. OBSTETRICS & GYNECOLOGY. *Educ:* Lehigh Univ, BA, 54; Hahnemann Med Col, MD, 58; Am Bd Obstet & Gynec, dipl, 66. *Prof Exp:* Assoc obstetrician & gynecologist-in-chief, Sinai Hosp, Baltimore, Md, 65-68; assoc prof, 68-72, prof obstet & gynec, 72-79, CHMN OBSTET & GYNEC, UNIV TEX HEALTH SCI CTR, 79- *Concurrent Pos:* Fel gynec path, Sch Med, Johns Hopkins Univ, 65-66, Am Cancer Soc adv clin fel, 66-68; asst prof gynec & obstet, Johns Hopkins Univ, 66-68. *Mem:* Am Col Obstet & Gynec; Am Fertil Soc; Soc Study Reprod; Soc Gynec Invest. *Res:* Gynecological pathology; reproductive physiology. *Mailing Add:* 7703 Floyd Curl Dr San Antonio TX 78284

PAUKEN, ROBERT JOHN, b Maumee, Ohio, Aug 30, 39; div; c 2. GEOLOGY, STRUCTURAL. *Educ:* Bowling Green State Univ, BSEd, 62, MA, 64, Univ Mo-Columbia, PhD(geol), 69. *Prof Exp:* Lab asst biol, Bowling Green State Univ, 63-64; instr biol, Univ Mo-Columbia, 64-69; sr res geologist, Mobil Res & Develop Corp, 69-79, SR GEOLOGIST, NEW EXPLOR VENTURES CO, 79- *Mem:* Geol Soc Am; Am Asn Petrol Geologists. *Res:* Quantitative geology; paleoecology of Middle Devonian ostracodes; population study of the Pleistocene terrestrial gastropod faunas in the loess deposits of Missouri; fracture analysis using remote sensing imagery; uranium exploration; basin stratigraphy; hydrocarbon exploration; extensional tectonics. *Mailing Add:* New Explor Ventures Co PO Box 650232 Dallas TX 75265-0232

PAUKER, STEPHEN GARY, b New York, NY, Nov 21, 42; m 67; c 2. CLINICAL DECISION MAKING, CARDIOLOGY. *Educ:* Harvard Col, AB, 64; Harvard Med Sch, MD, 68. *Prof Exp:* Intern, Boston City Hosp, 68-69; resident, Mass Gen Hosp, 69-70, fel cardiol, 71-72; from asst prof to assoc prof, 72-83, PROF MED, & CHIEF DIV CLIN DECISION MAKING, TUFTS UNIV, NEW ENG MED CTR, 83- *Concurrent Pos:* Res affil, Mass Inst Technol, 80-; chmn Bd Scientific Counslrs, Nat Libr Med, 81-85; fel cardiol, New Eng Med Ctr, 70-71. *Mem:* Soc Med Decision Making (pres 87-88); Am Col Physicians; fel Am Heart Asn; fel Am Col Med Informatic; Am Soc Clin Invest. *Res:* Application of formal methods of decision making to clinical medicine; development of a set of decision support tools which have enhanced the ability of investigators to analyze clinical choices. *Mailing Add:* 171 Harrison Ave Boston MA 02111

PAUKSTELIS, JOSEPH V, b Linkuva, Lithuania, Nov 25, 39; US citizen. ORGANIC CHEMISTRY. *Educ:* Univ Wis, BS, 60; Univ Ill, Urbana, PhD(org chem), 64. *Prof Exp:* Asst, Univ Ill, Urbana, 60-64; NIH fel, Mass Inst Technol, 64-66; asst prof, 66-74, assoc prof, 74-81, PROF ORG CHEM, KANS STATE UNIV, 81- *Concurrent Pos:* Prog dir, NSF, 79-81. *Mem:* Am Chem Soc; Royal Soc Chem. *Res:* Trans-coplanar rearrangements of glycol monotosylates; synthesis and reactions of N-cyanoammonium salts; synthesis of natural products containing medium sized rings; nuclear magnetic resonance spectroscopy. *Mailing Add:* Dept Chem Kans State Univ Manhattan KS 66504

PAUL, ARA GARO, b New Castle, Pa, Mar 1, 29; m 62; c 2. PHARMACOGNOSY. *Educ:* Idaho State Univ, BS, 50; Univ Conn, MS, 53, PhD(pharmacog), 56. *Prof Exp:* Asst prof pharmacog, Butler Univ, 56-57; from asst prof to assoc prof, 57-69, PROF PHARMACOG, COL PHARM, UNIV MICH, ANN ARBOR, 69-, DEAN, 75- *Concurrent Pos:* Vis prof indust microbiol, Tokyo Univ, 65-66; Am Found Pharm Educ Pfeiffer Mem res fel, 65-66; consult, Argonne Nat Lab, 55; vis fac mem, Univ Calif, Berkeley, 72-73, Eli Lilly Found fel, 51-53, Am Found Pharm Educ fel, 54-56, del, US Pharm Conv, 80 & 90; NIH spec fel, 72-73. *Mem:* Sigma Xi; Am Soc Hosp Pharmacists; Am Soc Pharmacog; Am Pharmaceut Asn; fel AAAS; Am Asn Pharmaceut Scientists; Acad Pharmaceut Sci; Am Asn Col Pharm. *Res:* Biosynthesis of alkaloids; phytochemistry of fungi; phytochemistry of cacti. *Mailing Add:* Univ Mich Col Pharm Ann Arbor MI 48109-1065

PAUL, AUGUSTUS JOHN, III, b Oneida, NY, July 19, 46. BIOENERGETICS, BIOLOGICAL OCEANOGRAPHY. *Educ:* Univ Mass, BS, 69; Univ Alaska, MS, 73; Hokkaido Univ, Japan, PhD(fisheries), 87. *Prof Exp:* Res assoc, 71-88, ASSOC PROF MARINE SCI, INST MARINE SCI, UNIV ALASKA, 89- *Mem:* Crustacean Soc; Nat Shellfisheries Asn; Am Soc Limnol & Oceanog; Pac Sci Asn. *Res:* Describing the factors which regulate the flow of energy through marine communities, especially those modifying survival of fish and crustacean larvae; describing the natural history of Alaskan decapods. *Mailing Add:* Inst Marine Sci Univ Alaska PO Box 730 Seward AK 99664

PAUL, B(URTON), b Jersey City, NJ, June 11, 31; m 58; c 2. MECHANICS. *Educ:* Princeton Univ, BSE, 53; Stanford Univ, MS, 54; Polytech Inst Brooklyn, PhD(appl mech, aeronaut eng), 58. *Prof Exp:* Engr, Bulova Res & Develop Labs, Inc, 54-56; res assoc, Polytech Inst Brooklyn, 56-58; asst prof eng, Brown Univ, 58-60; supvr eng mech, Bell Tel Labs, 61-63; head solid mech res, Ingersoll-Rand Res Ctr, 63-69, PROF MECH ENG, UNIV PA, 69- *Mem:* Fel Am Soc Mech Engrs; Am Soc Eng Educ. *Res:* Elasticity; plasticity; dynamics; vibrations; fracture and flow; structural and machine design; vehicle dynamics; mechanical engineering. *Mailing Add:* Dept Mech Eng & Appl Mech Univ Pa Philadelphia PA 19104

PAUL, BENOY BHUSHAN, b India, Jan 1, 36; m 67; c 1. BIOCHEMISTRY, RADIOBIOLOGY. *Educ:* Comilla Col, Pakistan, BSc, 55; Univ Dacca, MSc, 57; McGill Univ, PhD(biochem), 63. *Prof Exp:* Res asst antibiotics, Univ Dacca, 57-58, lectr phys biochem, 58-60; res biochemist, St Louis Univ, 63-64; res assoc, State Univ NY Downstate Med Ctr, 64-66; biochemist, 66-73, SR BIOCHEMIST & SUPVR MED RES & CLIN LABS, ST MARGARETS HOSP, 74- *Concurrent Pos:* Can Nat Res Coun scholar, 61-63; res fel obstet & gynec, Sch Med, Tufts Univ, 66-70, asst prof, 70-75, assoc prof, 75- *Mem:* Am Asn Clin Chemists; Am Soc Exp Path; Am Soc Microbiol; Reticuloendothelial Soc. *Res:* Biochemical, antimicrobial and immunological aspects of host-parasite interactions during pregnancy, leukemia and after radiation; biochemical mechanisms of antimicrobial activities of leukocytes, catalatic and peroxidatic functions of hemoproteins in host- defense; simplified Helena L/S pregnancy lung maturity test; published peroxidase assay ten times more sensitive than conventional methods; isolated peroxidase from lymphcytes from various species including human. *Mailing Add:* Med Res & Labs St Margarets Hosp Boston MA 02125

PAUL, DAVID I, ferromagnetism, properties of metals, for more information see previous edition

PAUL, DAVID LOUIS, HISTOLOGY. *Educ:* Harvard Univ, PhD(cell & develop biol), 83. *Prof Exp:* Fel, 83-86, INSTR HISTOL, MED SCH, HARVARD UNIV, 86- *Res:* Intercellular communication; molecular biology of gap junction. *Mailing Add:* Dept Anat Harvard Med Sch 25 Shattuck St Boston MA 02115

PAUL, DEREK (ALEXANDER LEVER), b Brussels, Belg, Oct 1, 29; m 51; c 4. PHYSICS. *Educ:* Cambridge Univ, BA, 50; Queen's Univ, Ont, PhD(physics), 58. *Prof Exp:* Lectr physics, Royal Mil Col, Ont, 53-58, from asst prof to assoc prof, 58-63; assoc prof, 64-75, PROF PHYSICS, UNIV TORONTO, 75- *Concurrent Pos:* Dir, Sci for Peace, Can. *Mem:* Can Asn Physicists; Am Phys Soc; Inst Elec & Electronic Engrs. *Res:* Electron and positron interactions with atoms and molecules; positronium; nuclear beta decay. *Mailing Add:* Dept Physics Univ Toronto Toronto ON M5S 2R5 Can

PAUL, DILIP KUMAR, b Rangoon, Burma, July 11, 40; US citizen; m 64; c 1. OPTICAL COMMUNICATION SYSTEMS, INTEGRATED OPTICS. *Educ:* Jadavpur Univ, India, BTeleE, 61; Indian Inst Sci, MS, 62; Univ RI, PhD(elec eng), 74. *Prof Exp:* Lectr elec eng, Univ Roorkee, India, 65-68; postdoctoral fel, Univ Md, 74-76; asst prof, Indian Inst Technol, 76-81; res assoc/scientist coordr, Harvard Univ, 79-81; MGR, OPTICAL COMMUN DEPT, COMSAT LABS, CLARKSBURG, 81- *Concurrent Pos:* Adj prof, Elec Eng Dept, Howard Univ, 85 & adj prof fiber optics, Whiting Sch Eng, Johns Hopkins Univ, 87-; mem, Standards Coord Comt Photonics, Inst Elec & Electronic Engrs, 89. *Mem:* Fel Soc Photo-Optical Instrumentation Engrs; Am Phys Soc; Optical Soc Am; fel Inst Electronics & Telecommun Engrs. *Res:* Prototype development of fiber optic and laser com systems for applications in undersea, terrestrial and space craft; optical signal processing, beam forming of phased array antenna, organic optical materials and optical intersatellite links; amorphous solar cells and magnetic thin films. *Mailing Add:* 8006 Thornley Ct Bethesda MD 20817

PAUL, DONALD ROSS, b Yeatesville, NC, Mar 20, 39; m 64; c 2. CHEMICAL ENGINEERING, POLYMERS. *Educ:* NC State Col, BS, 61; Univ Wis, MS, 63, PhD(chem eng), 65. *Prof Exp:* Instr chem eng, Univ Wis, 64-65; res chem engr, Chemstrand Res Ctr, Inc, Monsanto Chem Co, 65-67; from asst prof to assoc, 67-73, chmn dept, 77-85, PROF CHEM ENG, 73-, MELVIN H GERTZ CHAIR CHEM ENG, UNIV TEX AUSTIN, 85- *Concurrent Pos:* Part-time indust consult, 68-; dir, Ctr Polymer Res, 81-; mem bd dirs, Coun Chem Res, 81-84; mem, Nat Mat Adv Bd, 88-; res award, Soc Plastics Engrs, 82, educ award, 89. *Honors & Awards:* Arthur K Doolittle Award, Am Chem Soc, 73, Phillips Award, 84; Mat Sci Engr Award, Am Inst Chem Engrs, 85. *Mem:* Nat Acad Engr; Am Chem Soc; Soc Plastics Engrs; NY Acad Sci; Fiber Soc; Am Inst Chem Engrs. *Res:* Polymer blends and transport properties of polymers; polymer processing. *Mailing Add:* Dept Chem Eng Univ Tex Austin TX 78712

PAUL, EDWARD GRAY, b Salt Lake City, Utah, Jan 3, 31; m 63. ORGANIC CHEMISTRY. *Educ:* Univ Utah, BSc, 58, PhD(org chem), 62. *Prof Exp:* NIH fel chem, Univ Utah, 61-65; assoc prof, 65-69, asst prof, 69-78, PROF CHEM, BRIGHAM YOUNG UNIV, 78- *Mem:* Am Chem Soc. *Res:* Organic synthesis; natural products; nuclear magnetic resonance spectroscopy. *Mailing Add:* Dept Chem Brigham Young Univ Provo UT 84602

PAUL, EDWARD W, b Newark, NJ, July 16, 44; m 70; c 2. PHASE DIAGRAMS. *Educ:* Brandeis Univ, BA, 66; Univ Ore, PhD(chem physics), 70. *Prof Exp:* Researcher biophys chem, Univ Calif, Santa Cruz, 70-71; researcher biophysics, NIH, 71-72; PROF CHEM TEACHING, STOCKTON STATE COL, 72- *Concurrent Pos:* Vis res assoc, Nat Bur Standards, 84-85. *Mem:* Am Chem Soc; AAAS. *Res:* Alloy phase diagrams; mathematical modeling; thermodynamics; science curricula. *Mailing Add:* Stockton State Col Pomona NJ 08240

PAUL, ELDOR ALVIN, b Edmonton, Alta, Nov 23, 31; m 55; c 2. MICROBIOLOGY. *Educ:* Univ Alta, BSc, 54, MSc, 56; Univ Minn, PhD, 58. *Prof Exp:* Asst soils, Univ Minn, 56-58; from asst prof to assoc prof soil sci, Univ Sask, 59-70, prof, 70-80; prof & chair plant & soil biol, Univ Calif, Berkeley, 80-85; CHAIRPERSON & PROF, CROP & SOIL SCI, MICH STATE UNIV, 85- *Mem:* Fel Can Soc Soil Sci; Am Soc Microbiol; Agr Inst Can; Int Soc Soil Sci; fel Am Soc Agr; Ecol Soc Am; fel Soil Sci Soc Am; AAAS. *Res:* Microbiology and biochemistry of soil organic matter; effect of microorganisms on soil fertility; microbial ecology. *Mailing Add:* Dept Crop & Soil Sci Mich State Univ East Lansing MI 48824-1325

PAUL, FRANK W(ATERS), b Jersey Shore, Pa; c 2. MECHANICAL ENGINEERING. *Educ:* Pa State Univ, BS, 60, MS, 64; Lehigh Univ, PhD(mech eng), 68. *Prof Exp:* Anal engr thermodyn, Hamilton Stand, United Aircraft Corp, 60, controls, 61-64; instr mech eng, Lehigh Univ, 64-68; from asst prof to assoc prof, Carnegie-Mellon Univ, 68-77; from assoc prof to prof, 77-83, MCQUEEN QUATTLEBAUM PROF MECH ENG & DIR, CTR ADVAN MFG, CLEMSON UNIV, 83- *Concurrent Pos:* Mem tech staff, Bell Tel Labs, 65; hon prof eng, Hull Univ, Eng, UK. *Mem:* Am Soc Mech Engrs; Am Soc Eng Educ; Robots Int Soc Mfg Eng. *Res:* Dynamic systems modeling; applied control systems; computer aided manufacturing; automated systems; design; robotics. *Mailing Add:* Dept Mech Eng Clemson Univ 300D Riggs Hall Clemson SC 29634-0921

PAUL, GEORGE, JR, b Houston, Tex, Nov 9, 37; m 69. ELECTRICAL ENGINEERING, MATHEMATICS. *Educ:* Rice Univ, BA, 61, MS, 66, PhD(elec eng), 67. *Prof Exp:* Staff mem, Eng Dept, Houston Sci Ctr, 67-69, mgr eng dept, 69-75, mem staff, Sci Comp Dept, DPD Hq, 75-77, MGR VECTOR ARCHIT, SYST TECHNOL DEPT, T J WATSON RES CTR, IBM CORP, 77- *Mem:* Inst Elec & Electronics Engrs; Soc Indust & Appl Math; Asn Comput Mach; NY Acad Sci. *Res:* Application of computers to large scale scientific problems; computer architecture and design; numerical analysis; Fourier analysis; algebraic coding theory. *Mailing Add:* 96 Pierce Dr Pleasantville NY 10570

PAUL, GEORGE T(OMPKINS), materials science engineering, chemical engineering, for more information see previous edition

PAUL, GILBERT IVAN, b Ont, May 31, 22; m 45; c 4. STATISTICS. *Educ:* Univ Alta, BSc, 51, MSc, 52; Univ NC, PhD(statist), 56. *Prof Exp:* Asst prof genetics & statist, McGill Univ, 55-59; from asst prof to assoc prof, 59-67, PROF STATIST, UNIV MAN, 67- *Mem:* Biomet Soc; Sigma Xi. *Res:* Statistics and population genetics. *Mailing Add:* Dept Statist Univ Man Box 69 Starbuck MB R0G 2P0 Can

PAUL, HARBHAJAN SINGH, b India, June 8, 37; US citizen. NUTRITIONAL BIOCHEMISTRY. *Educ:* Univ Gauhati, India, BVSc, 59; Univ Minn, MS, 66, PhD(nutrit biochem), 69; Am Bd Nutrit, dipl. *Prof Exp:* Fel biochem, State Univ NY Buffalo, 69-70, Univ Hohenheim, Ger, 71-73; res assoc cell biol, 74, res instr nutrit biochem, 74-75, res asst prof, 75-81, res asst prof biochem, 79-81, asst prof biochem, Med Sch, Univ Pittsburgh, 81-85; asst prof med, 81-85, RES ASSOC PROF MED & BIOCHEM, MONTEFIORE UNIV HOSP, 85- *Mem:* Am Inst Nutrit; AAAS; Sigma Xi; Am Soc Biochem & Molecular Biol. *Res:* Regulation of protein and amino acid metabolism in altered nutritional and hormonal states; metabolic effects of the hypolipidemic drug, clofibrate; metabolism of carnitine; metabolism in peroxisomes. *Mailing Add:* Dept Med Montefiore Univ Hosp 3459 Fifth Ave Pittsburgh PA 15260

PAUL, IAIN C, b Glasgow, Scotland, Oct 1, 38; m 76; c 2. PHYSICAL CHEMISTRY, ORGANIC CHEMISTRY. *Educ:* Glasgow Univ, BSc, 59, PhD(phys chem), 62. *Prof Exp:* Res fel, Harvard Univ, 62-64; asst prof phys chem, 64-68, assoc prof chem, 68-74, PROF CHEM, UNIV ILL, URBANA, 74- *Concurrent Pos:* A P Sloan Found fel, 68-70. *Mem:* Am Crystallog Asn; Am Chem Soc; Royal Soc Chem. *Res:* X-ray crystal structure analysis; three dimensional structure of molecules, particularly those of interest in biological and organic chemistry; structure and function of proteins; solid state chemistry. *Mailing Add:* Dept Chem W A Noyes Lab Univ Ill 505 S Mathews Urbana IL 61801

PAUL, IGOR, b Kharkov, USSR, Oct 28, 36; US citizen; m 63; c 3. BIOMECHANICS. *Educ:* Mass Inst Technol, SB, 60, SM, 61, ME, 62, ScD(mech eng), 64. *Prof Exp:* Res asst mech eng, Mass Inst Technol, 60-61, design engr instrumentation, 61, from instr to asst prof, 61-69, ASSOC PROF MECH ENG, MASS INST TECHNOL, 69- *Concurrent Pos:* Ford fel, 65-67; engineering consult, Mass Gen Hosp & Childrens Hosp Med Ctr. *Honors & Awards:* Ralph R Teetor Award, Soc Automotive Engrs, 66. *Mem:* Assoc Am Soc Mech Engrs; Am Soc Eng Educ; Soc Automotive Engrs; Biomed Eng Soc; Orthopedic Res Soc; Biomechanics Soc Am. *Res:* Design and controls; biomedical engineering and biomechanics; transportation. *Mailing Add:* Dept Mech Eng Rm 3-461 Mass Inst Technol Cambridge MA 02139

PAUL, JEDDEO, b Georgetown, Guiana, June 2, 29; m 59; c 3. BIOCHEMISTRY, ANALYTICAL CHEMISTRY. *Educ:* Univ London, BS, 53, MS, 55; Univ Birmingham, PhD(biochem), 58. *Prof Exp:* Imp Chem Industs fel, Univ Birmingham, 58-59; analyst, Govt Lab, Guiana, 59-62; head, Dept Chem & Soils, Ministry Agr, 62-66; from asst prof to assoc prof, 66-72, PROF CHEM, UNIV BRIDGEPORT, 72- *Concurrent Pos:* Consult, Govt Guiana, 62-66; vis fel, Univ Aberdeen, 64-65; sr Fulbright award, Brazil, 84.

Mem: Am Chem Soc; fel Royal Soc Chem. *Res:* Application of selective solvent extraction to simultaneous determination of metals and nonmetals; chromatography of organo-phosphorus insecticides. *Mailing Add:* 50 Champion Terr Stratford CT 06497

PAUL, JEROME L, b St Paul, Minn, Sept 3, 37. GRAPH THEORY. *Educ:* Univ Minn, BS, 59, MS, 62. *Prof Exp:* Prof math, 70-84, PROF COMPUT SCI, UNIV CINCINNATI, 84- *Mem:* Am Math Soc. *Mailing Add:* Comput Sci Dept Univ Cincinnati Mail Loc 8 Cincinnati OH 45221

PAUL, JEROME THOMAS, b Chicago, Ill, Sept 22, 12; m 41; c 2. MEDICINE. *Educ:* Loyola Univ, Ill, MD, 37; Univ Ill, MS, 41. *Prof Exp:* Asst, 39-40, from instr to asst prof, 41-52, ASSOC PROF MED, UNIV ILL COL MED, 53- *Concurrent Pos:* Mem assoc staff, St Francis Hosp, Evanston, Ill, 40-47, mem sr staff, 48-, chmn dept med, 58-65; asst attend physician, Ill Res Hosp, 47-, attend physician, Med Clin, 47-; asst attend physician, Presby-St Luke's Hosp, 57-62; asst attend physician, St Joseph Hosp, 62, chief sect metab dis, 64- *Mem:* Am Diabetes Asn; Am Col Physicians; Am Fedn Clin Res; Int Soc Hemat. *Res:* Metabolic diseases; hematology. *Mailing Add:* 18260 Aceituno St San Diego CA 92128-1560

PAUL, KAMALENDU BIKASH, b Karimganj, India, Apr 1, 37; m 70. HORTICULTURE, PLANT PHYSIOLOGY. *Educ:* Univ Calcutta, BS, 57, BS, 60; Tuskegee Inst, MS, 65; Univ Ottawa, PhD(biol), 71. *Prof Exp:* Res asst biol, Univ Ottawa, 65-70; res scientist plant physiol, Tuskegee Inst, 70-75; asst prof & researcher soybean fertilization, 75-80, ASSOC PROF PLANT & SOIL SCI & PROG LEADER CROP RES, LINCOLN UNIV, 75- *Concurrent Pos:* USDA grant, Tuskegee Inst, 70-, coop state res serv grants, 72-, USDA grant, 75-78 & Sci & Educ Admin-Agr Res grant, 78- *Mem:* Am Soc Hort Sci; Am Soc Agron; Crop Sci Soc Am. *Res:* Seed physiology; various treatments to break seed dormancy; physiological and biochemical changes occurring in seeds during stratification and germination; seed pre-treatment on germination, emergence, growth and yield of vegetable crops; foliar fertilization of soybeans. *Mailing Add:* Dept of Agr Lincoln Univ Jefferson City MO 65101

PAUL, LAWRENCE THOMAS, b Lykens, Pa, June 17, 33. PHYSIOLOGY. *Educ:* Muhlenberg Col, BSc, 55; Ohio State Univ, MSc, 60, PhD(physiol), 63. *Prof Exp:* From instr to asst prof, 63-72, ASSOC PROF PHYSIOL, OHIO STATE UNIV, 72- *Concurrent Pos:* Bremer Found res fel, Ohio State Univ, 63-65. *Mem:* AAAS; Am Physiol Soc. *Res:* Cardiovascular physiology; coronary blood flow; hemodynamics and peripheral circulation. *Mailing Add:* 1114 Rockport Lane Columbus OH 43235

PAUL, LEENDERT CORNELIS, b Zevenhuizen, Neth, Jan 31, 46; m 76; c 2. TRANSPLANT IMMUNOLOGY, NEPHROLOGY. *Educ:* Leiden State Univ, Neth, MD, 69, PhD(immunol), 79. *Prof Exp:* Fel transplantation, Harvard Med Sch, Boston, 79-80; from asst prof to assoc prof, Leiden State Univ, Neth, 81-87; PROF MED, UNIV CALGARY, CAN, 87-; SCIENTIST, ALTA HERITAGE FOUND MED RES, 88- *Concurrent Pos:* Head, Div Nephrology, Univ Calgary & Foothills Hosp, 90-; counr, Can Soc Nephrology, 90-; chmn, Immunol Sci Res Group, Univ Calgary, 91-; dir, Clin Organ Transplantation, Human Organ Procurement Prog, Foothills Hosp, 91- *Mem:* Transplantation Soc; Int Soc Nephrology; Am Soc Nephrology; Can Soc Immunol. *Res:* Organ transplantation and mechanisms of rejection; histocompatibility antigens; immunosuppressive drugs, mechanism of actions, side effects; mechanisms of progressive loss of renal function; author of numerous technical publications. *Mailing Add:* Foothills Hosp 1403-29th St NW Calgary AB T2N 2T9 Can

PAUL, MILES RICHARD, b Portland, Ore, June 25, 40; m 65. DEVELOPMENTAL BIOLOGY. *Educ:* Harvard Univ, BA, 62; Makerere Univ Col, E Africa, Dip Ed, 63; Stanford Univ, PhD(biol), 70. *Prof Exp:* Educ Off Sci, Mzumbe Sec Sch, Tanzania, 63-65; fel biol, Harvard Univ, 70-72; asst prof, 72-78, ASSOC PROF BIOL, UNIV VICTORIA, BC, 78- *Mem:* Am Soc Cell Biol; Am Soc Zoologists; AAAS. *Res:* Responses of marine invertebrate eggs to fertilization; silkmoth eggshell protein synthesis and eggshell structure. *Mailing Add:* Dept Biol Sci Univ Victoria Box 1700 Victoria BC V8W 2Y2 Can

PAUL, MILTON HOLIDAY, b Philadelphia, Pa, Jan 15, 26; m 48; c 3. MEDICINE. *Educ:* Univ Pa, MD, 49; Am Bd Pediat, dipl & cert pediat cardiol, 62. *Prof Exp:* Assoc dir, 58-63, DIR CARDIOL, CHILDREN'S MEM HOSP, 63-; PROF PEDIAT, MED SCH, NORTHWESTERN UNIV, CHICAGO, 67- *Concurrent Pos:* USPHS res fel biochem, Univ Wis, 50-51, cardiol, Harvard Univ, 53-54, pediat cardiol, 54-56; mem & examr, Am Bd Pediat, 62-; assoc prof, Med Sch, Northwestern Univ, 63-67. *Mem:* Am Heart Asn; Am Soc Pediat Res. *Res:* Pediatric cardiology; cardiovascular physiology; biomedical engineering; medical information processing. *Mailing Add:* 2300 Childrens Plaza Chicago IL 60614

PAUL, OGLESBY, b Villanova, Pa, May 3, 16; m 43, 81; c 2. MEDICINE. *Educ:* Harvard Univ, AB, 38; Harvard Med Sch, MD, 42; Am Bd Internal Med, dipl, 54. *Prof Exp:* Asst med, Harvard Med Sch, 46-49; clin assoc prof med, Col Med, Univ Ill, 52-63; prof med, Med Sch, Northwestern Univ, 63-77, vpres health sci, 74-76; dir admission, 77-82, PROF MED, HARVARD MED SCH, 78- *Concurrent Pos:* Consult, US Naval Hosp, Great Lakes, 59-70; chmn, subspecialty bd cardiovasc dis, Am Bd Internal Med, 62-67; chmn adv comt, Ill Regional Med Prog Heart Dis, Cancer & Stroke, 66-70. *Mem:* Fel AMA; Am Heart Asn (pres, 60-61); Am Clin & Climat Asn; fel Am Col Cardiol; Am Epidemiological Soc. *Res:* Epidemiologic study of coronary heart disease. *Mailing Add:* Harvard Med Sch Boston MA 02115

PAUL, PAULINE CONSTANCE, food chemistry, educational administration; deceased, see previous edition for last biography

PAUL, PETER, b Dresden, Ger, Nov 24, 32; m 63; c 3. NUCLEAR PHYSICS. *Educ:* Univ Freiburg, BA, 54, PhD(physics), 59; Aachen Tech Inst, MSc, 56. *Prof Exp:* Res assoc nuclear physics, Univ Freiburg, 57-59, asst physics, 62-63; fel, Stanford Univ, 60-62, res assoc physics, 63-64, lectr, 64-65, asst prof, 65-67; from asst prof to prof physics, 67-80, chmn dept, 86-90, LEADING PROF PHYSICS, STATE UNIV NY STONY BROOK, 80- *Concurrent Pos:* A P Sloan fel, 67-71; Humboldt Award, 83, 84; chmn, Nuclear Safety Analysis Ctr, 89- *Mem:* Fel Am Phys Soc. *Res:* Level structure of nuclei; research on nuclear structure and reactions with accelerators; accelerator design. *Mailing Add:* Dept Physics State Univ NY Stony Brook NY 11790

PAUL, PREM SAGAR, b Jullunder, Panjab, India, Oct 5, 47; US citizen; c 2. VETERINARY MICROBIOLOGY, VIROLOGY & IMMUNOLOGY. *Educ:* Panjab Agr Univ, BVSc, 69; Univ Minn, PhD(vet microbiol), 75; Am Col Vet Microbiologists, dipl, 77. *Prof Exp:* Res asst vet microbiol, Univ Minn, 69-75, res assoc, 75-78; vet med officer, virol immunol, Nat Animal Dis Ctr, USDA, 78-85; assoc prof, 85-89, PROF, VET MED RES INST, IOWA STATE UNIV, AMES, 89- *Mem:* Sigma Xi; Am Vet Med Asn; Am Asn Vet Immunol; Conf Res Workers in Animal Dis; Am Col Vet Microbiol; Am Soc Virol. *Res:* Bovine viral enteric disease; transmissible gastroenteritis virus of swine; viral subunit vaccines; monoclonal antibodies; porcine, bovine and equine rotaviruses; DNA probes; vaccines. *Mailing Add:* Vet Med Res Inst Iowa State Univ Ames IA 50011

PAUL, REGINALD, b Lucknow, India, Oct 17, 36; m 65; c 2. THEORETICAL CHEMISTRY. *Educ:* Univ Lucknow, BSc, 56, MSc, 57; Univ Alta, MSc, 62; Univ Durham, PhD(chem), 66. *Prof Exp:* Sessional instr, 66-67, from asst prof to assoc prof, 67-82, PROF CHEM, UNIV CALGARY, 82- *Res:* Applications of methods of second quantization to chemical kinetics and electrolyte transport. *Mailing Add:* 6008 Dalford Rd NW Calgary AB T3A 1L6 Can

PAUL, RICHARD JEROME, b Chicago, Ill, Apr 4, 44; m 67; c 2. BIOPHYSICS. *Educ:* St Mary's Col, BA, 66; Harvard Univ, PhD(biophysics). *Prof Exp:* Res fel, 72-73, instr, Harvard Univ, 73-74; angestellter, Univ Heidelberg, Ger, 74-75; hon res asst, Univ Col, London, 75-77; from asst prof to assoc prof, 77-87, PROF PHYSIOL, COL MED, UNIV CINCINNATI, 87- *Concurrent Pos:* Vis prof, Univ Heidelberg, Ger, 82; res fel, British Am Heart Asn, 75-77; estab investr, Am Heart Asn, 81-86; mem, Hypertension Task Force, NIH & Coun Basic Sci, American Heart Asn; NIH physiol study sect; Fogarty Sr Int fel, 88. *Mem:* AAAS; Biophys Soc; Am Physiol Soc. *Res:* Muscle energetics and physiology, with particular emphasis on the relations between metabolism and contractility in vascular smooth muscle. *Mailing Add:* Dept Physiol Rm 4259 Univ Cincinnati Col Med 231 Bethesda Ave Cincinnati OH 45267-0576

PAUL, ROBERT E, JR, b Austin, Tex, Oct 14, 27; c 2. RADIOLOGY. *Educ:* Baylor Univ, MD, 50; Temple Univ, MS, 57. *Prof Exp:* Resident radiol, Temple Univ Hosp, 53-56, instr, 56-57; radiologist, New Eng Med Ctr Hosp, 57-65; from asst prof to assoc prof, 57-65, PROF RADIOL & CHMN DEPT, SCH MED, TUFTS UNIV, 65-; RADIOLOGIST-IN-CHIEF, NEW ENG MED CTR HOSP, 65- *Mem:* Fel Am Col Radiol; AMA; Am Roentgen Ray Soc; Asn Univ Radiologists; Radiol Soc NAm. *Res:* Gastrointestinal radiology. *Mailing Add:* New England Med Ctr Hosp 171 Harrison Ave Boston MA 02111

PAUL, ROBERT HUGH, b Muskogee, Okla, Apr 21, 27; m 49; c 2. PHYSICS, ELECTRICAL ENGINEERING. *Educ:* Univ Houston, BS, 51; NMex State Univ, MS, 55, PhD(physics), 65. *Prof Exp:* Electronic scientist, Flight Determination Lab, White Sands Missile Range, 51-55, supvry electronics engr, 55-56, electronic scientist, Systs Eng Br, Integrated Range Mission, 56-57, supvry electronic scientist, Electronic Trajectory Br, Range Instrumentation Develop Directorate, 57-60, supvry electronics engr, US Army Res & Develop Activity, 60-62 & Electronic Trajectory Br, Instrumentation Develop Directorate, 62-64, res physicist, Range Eng, 64-73, TECH DIR, INSTRUMENTATION DEVELOP DIRECTORATE, WHITE SANDS MISSILE RANGE, 73- *Mem:* AAAS; Inst Elec & Electronics Engrs; Am Phys Soc; Sigma Xi. *Res:* Radar signal processing, pattern recognition; propagation of radio waves in the earth's atmosphere; theoretical determination of radar cross section; molecular theories of light scattering by crystals and gases. *Mailing Add:* 2007 Crescent Dr Las Cruces NM 88005

PAUL, ROBERT WILLIAM, JR, b Jersey City, NJ, Nov 27, 46; m 72; c 1. COMMUNITY ECOLOGY, AQUATIC BIOLOGY. *Educ:* Westminster Col, BA, 69; St Louis Univ, MS, 75; VA Polytech Inst & State Univ, PhD(zool), 78. *Prof Exp:* Grad res asst ecol, Va Polytech Inst & State Univ, 75-77; from instr to asst prof, 77-86, ASSOC PROF, BIOL, ST MARY'S COL OF MD, 86-, HEAD, DIV NAT SCI & MATH, 84- *Concurrent Pos:* Teaching asst, Va Polytech Inst & State Univ, 74-75, St Louis Univ, 72-74; res investr, Biol Sta, Univ Mich, 81-82 & Andrew Mellon Found fel, 81-82; vis prof biol, Evergreen State Col, 81-82; adj prof biol, Univ Md, 88-89. *Mem:* Sigma Xi; Int Asn Appl & Theoret Limnol; Ecol Soc Am; AAAS. *Res:* Decomposition in aquatic ecosystems and the effects of perturbation on decomposition processes; thermal pollution of aquatic communities and its effect; insect ecology; aquatic toxicology. *Mailing Add:* Div Nat Sci & Math St Mary's Col Md St Mary's City MD 20686

PAUL, ROLF, b Neuss, Ger, Oct 31, 30; nat US; m 58; c 2. PHARMACEUTICAL CHEMISTRY. *Educ:* Purdue Univ, BS, 52, PhD(org chem), 56. *Prof Exp:* ORG CHEMIST, LEDERLE LABS, AM CYANAMID CO, 56- *Mem:* Am Chem Soc. *Res:* Medicinal chemistry. *Mailing Add:* 687 Montgomery Lane River Vale Westwood NJ 07675

PAUL, RONALD STANLEY, b Olympia, Wash, Jan 19, 23; m 44; c 3. PHYSICS. *Educ:* Univ Ore, BS, 47, MS, 49, PhD(physics), 51. *Prof Exp:* From physicist to sr physicist, Hanford Atomic Prod Oper, Gen Elec Co, 51-56, mgr testing methods, 56-59, instrument res & develop, 59-62, physics & instruments lab, 62-64; mgr phys & instruments dept, Battelle-Northwest, 65, assoc dir, 66-67, dep dir, 67-68, sr mem corp staff, Battelle Mem Inst, 68-70, dir pac northwest div, 71-73, vpres opers, 73-76, sr vpres, 76-78, exec vpres, 79-81, pres, Battelle Mem Inst, 82-87; RETIRED. *Concurrent Pos:* Lectr, Gen Elec Sch Nuclear Eng, 52-58 & Richland Ctr Grad Studies, Univ Wash, 58-63; Int Atomic Energy Agency consult, Japan, 62. *Mem:* Am Phys Soc; Am Nuclear Soc. *Res:* Nuclear reactor physics and instrumentation; nondestructive testing; general physical, life and environmental sciences; research and development management. *Mailing Add:* 7706 173rd St SW Edmonds WA 98020

PAUL, STEVEN M, MENTAL HEALTH. *Prof Exp:* DIR, INTRAMURAL RES PROG, NIMH, NIH, 88- *Mailing Add:* NIH NIMH Ment Health Intramural Res Prog Bldg 10 Rm 4N224 9000 Rockville Pike Bethesda MD 20892

PAUL, WILLIAM, b Toronto, Ont, Mar 5, 18; m 45; c 4. BIOPHYSICS. *Educ:* Univ Toronto, BA, 41, PhD(pharmacol), 48. *Prof Exp:* Fel, Cambridge Univ, 48-49; asst prof pharmacol, 49-53, from asst prof to prof path chem, 53-83, EMER PROF CLIN BIOCHEM, UNIV TORONTO, 83- *Concurrent Pos:* Commonwealth Med fel, Univ London, 70, actg dep chmn, Clin Biochem, 71. *Mem:* Can Asn Physicists; Inst Elec & Electronics Eng; Pharmacol Soc Can. *Res:* Oximetry; radioisotope safety; radioimmunoassays. *Mailing Add:* Dept of Clin Biochem Univ of Toronto Toronto ON M5G 1L5 Can

PAUL, WILLIAM, b Scotland, Mar 31, 26; m 52; c 2. SOLID STATE PHYSICS. *Educ:* Aberdeen Univ, MA, 46, PhD(physics), 51. *Hon Degrees:* AM, Harvard Univ, 60. *Prof Exp:* Asst lectr natural philos, Aberdeen Univ, 46-51, lectr, 51-52; Carnegie fel, Harvard Univ, 52-53, res fel solid state physics, 53-54, from lectr to assoc prof, 54-63, GORDON MCKAY PROF APPL PHYSICS, HARVARD UNIV, 63-, PROF PHYSICS, 85- *Concurrent Pos:* Guggenheim res fel, 59; assoc prof, Univ Paris, 66-67; vis prof, Cavendish Lab, Cambridge Univ, Eng, 74-75; Ripon Prof, Calcutta, 84. *Mem:* Fel Am Phys Soc; fel Brit Inst Physics; fel NY Acad Sci; AAAS; fel Royal Soc Edinburgh. *Res:* Semiconductor physics; optical spectra, especially at infrared wave lengths; high pressure physics; study of properties of amorphous semiconductors and glasses. *Mailing Add:* 229 Pierce Hall Harvard Univ Cambridge MA 02138

PAUL, WILLIAM ERWIN, b Brooklyn, NY, June 12, 36; m 58; c 2. IMMUNOLOGY. *Educ:* Brooklyn Col, AB, 56; State Univ NY Downstate Med Ctr, MD, 60. *Prof Exp:* Intern med, Mass Mem Hosp, Boston Univ, 60-61, asst resident, 61-62; clin assoc, Endocrinol Br, Nat Cancer Inst, 62-64; instr med, Sch Med, NY Univ, 64-68; med officer, 68-70, LAB CHIEF, LAB IMMUNOL, NAT INST ALLERGY & INFECTIOUS DIS, 70- *Concurrent Pos:* USPHS spec res fel path, Sch Med, NY Univ, 64-66, trainee med, 66-67; clin asst vis physician, Bellevue Hosp, New York, 66-68; mem, directing group, Nat Inst Allergy & Infectious Dis Task Force Immunol Dis, 73-74, Int Fel Rev Panel, NIH, 74-75 & transplantation & immunol adv comt, Nat Inst Allergy & Infectious Dis, 74-77; mem fel subcomt, Arthritis Found, 76-81; mem adv group personnel for res, Am Cancer Soc, 78-83; co-chmn, Study Group Immunol, Nat Inst Allergy & Infectious Dis, 80; mem sci rev bd, Howard Hughes Med Inst, 79-85, 87-; US chmn, immunol bd, Vis Okla Med Res Found, 79-85 & US-Japan Coop Med Sci Prog, 81-84; mem bd dirs, Found Adv Educ Sci, 81-90; mem, bd sci adv, Jane Coffin Childs Fund Med Res, 82-90; mem, bd sci consults, Mem Sloan-Kettering Cancer Ctr, 84-; chmn, sci rev comt, Cambridge Br, Ludwig Inst Cancer Res, 83-86; chmn, adv comt, Harold C Simmons Arthritis Res Ctr, Univ Tex Health Sci Ctr, Dallas, 84-91; mem, Sci Adv Coun, Cancer Res Inst, 85-; mem, comt res opportunities biol, Nat Acad Sci-Nat Res Coun, 85-88; mem, Bd on Basic Biol, Nat Res Coun 86-; Alfred P Sloan Jr selection comt, Gen Motors Cancer Res Fedn, 86-87; mem adv comt, PEW Scholars Prog Biomed Sci, 88-92. *Honors & Awards:* 3M Life Sci Award, Fedn Am Soc Exp Biol, 88. *Mem:* Inst Med-Nat Acad Sci; Am Asn Immunologists (vpres, 86-87); Am Fedn Clin Res; Am Soc Clin Invest (pres, 80-81); Am Asn Path; Am Acad Allergy; Asn Am Physicians; hon mem Scand Soc Immunol. *Res:* Recognition and regulatory mechanisms in immune responses. *Mailing Add:* Lab Immunol 10/11N311 Nat Inst Allergy & Infectious Dis 9000 Rockville Pike Bethesda MD 20892

PAUL, WOLFGANG, b Aug 10, 18; m 40, 79. PHYSICS. *Educ:* Univ G06ttingen, dozent, 44. *Hon Degrees:* Dr, Univ Uppsala & Univ Aachen. *Prof Exp:* EMER PROF, UNIV BONN. *Concurrent Pos:* Dir, Physics Div, Europ Orgn Nuclear Res, 64-67, deleg to coun, 75-78, chair, Sci Policy Comt, 75-78; dir, Ger Electron-Synchrotron, Hamburg, 70-75; pres, Alexander von Humboldt Found, 79-89. *Honors & Awards:* Nobel Prize in Physics, 89; Robert W Pohl Prize. *Mailing Add:* Stationsweg 13 Bonn 5300 Germanyp

PAULAY, GUSTAV, b Budapest, Hungary, May 13, 57; US citizen. BIOGEOGRAPHY, INVERTEBRATE ZOOLOGY. *Educ:* Yale Univ, BSc, 79; Univ Wash, PhD(zool), 88. *Prof Exp:* Lectr zool, Friday Harbor Labs, 88-89; postdoctoral fel, Nat Mus Natural Hist, 90-91; ASST PROF MARINE BIOL, MARINE LAB, UNIV GUAM, 91- *Concurrent Pos:* Vis lectr zool, Univ NC, 90; res assoc, Bishop Mus. *Mem:* Am Soc Zoologists; Western Soc Naturalists. *Res:* Biogeography and evolution of invertebrates especially on Pacific Islands; effects of sea level fluctuations on insular faunas and reef morphology. *Mailing Add:* Marine Lab Univ Guam Mangilao GU 96923

PAULE, MERLE GALE, b Sacramento, Calif, June 28, 52; m 82; c 1. BEHAVIORAL PHARMACOLOGY & TOXICOLOGY, PSYCHOPHARMACOLOGY & TOXICOLOGY. *Educ:* Univ Calif, Davis, BS, 76, PhD(pharmacol-toxicol), 83. *Prof Exp:* Teaching asst animal physiol, Univ Calif, Davis, 77-78, technician pharmacol, Sch Med, 78-81, assoc, 81-82; fel pharmacol, Univ Ark Med Sci, 82-83, instr, 83; staff fel, 83-86, PHARMACOLOGIST, NAT CTR TOXICOL RES, DEPT HEALTH & HUMAN SERV-FOOD & DRUG ADMIN, 86- *Concurrent Pos:* Adj asst prof, Univ Ark Med Sci, 84-; travel fel, Am Soc Pharmacol & Exp Therapeut, 81 & 87, Comn Probs Drug Dependence, 86; pres, Ark Chap, Soc Neuroscience, 91; pres elect, S Cent Chap, Soc Toxicol, 90. *Mem:* Am Soc Pharmacol & Exp Therapeut; Soc Toxicil; NY Acad Sci; Soc Neurosci; Sigma Xi. *Res:* Behavioral pharmacology and toxicology; cognitive function deficiencies as indicators of toxicity; pharmacology of learning and memory; sites and mechanism of action of psychotropic agents; epilepsy; developmental pharmacology and toxicology. *Mailing Add:* Pharmacodynamics Br HFT-132 Div Reproductive Develop Toxicol Primate Res Facil Nat Ctr Toxicol Res Jefferson AR 72079-9502

PAULE, ROBERT CHARLES, b St Louis, Mo, Jan 4, 32; m 54; c 2. STATISTICAL CONSULTATION, PHYSICAL CHEMISTRY. *Educ:* Univ Fla, BS, 53, MS, 55; Univ Wis, PhD(phys & high temp chem), 62. *Prof Exp:* Chemist, Esso Res Labs, 62-66; PHYSICAL SCIENTIST, NAT INST STANDARDS & TECHNOL, 66- *Honors & Awards:* Bronze Medal, Dept Com. *Mem:* Am Statist Asn; Am Soc Testing & Mat; Sigma Xi. *Res:* Statistics; development of standard reference materials; high temperature chemistry; Knudsen and Langmuir vapor pressure studies; mass spectrometry; cracking and hydrocracking catalysts. *Mailing Add:* 2 Owens Ct Rockville MD 20850

PAULE, WENDELIN JOSEPH, b Toledo, Ohio, Aug 6, 27; m 59; c 2. ANATOMY. *Educ:* Univ Toledo, BS, 50; Ohio State Univ, MS, 51, PhD, 57. *Prof Exp:* Asst biol, Univ Toledo, 48-50; asst instr anat, Ohio State Univ, 53-56; instr, Univ Buffalo, 56 & 59; asst prof, 59-66, ASSOC PROF ANAT, SCH MED, UNIV SOUTHERN CALIF, 66- *Concurrent Pos:* USPHS fel, Univ Buffalo, 58. *Mem:* Electron Micros Soc Am; Am Asn Anat; AMA; Am Soc Cell Biol. *Res:* Electron microscopy of blood vessels, lipid exocrine glands and endocrines; histochemistry of normal tissues. *Mailing Add:* Dept Anat Univ Southern Calif 2025 Zonal Ave Los Angeles CA 90033

PAULEY, GILBERT BUCKHANNAN, b Klamath Falls, Ore, June 18, 39; m 64; c 1. IMMUNOLOGY, FISHERIES. *Educ:* Univ Wash, BS, 62, MS, 65; Univ Calif, Irvine, PhD(biol), 71; Univ Puget Sound, MBA, 85. *Prof Exp:* Res asst, Col Fisheries, Univ Wash, 62-65; res scientist, Battelle Mem Inst, 65-68; res asst biol, Univ Calif, Irvine, 70-71; fishery biologist, Nat Marine Fisheries Serv, 71-72; immunologist, Eastern Fish Dis Lab, US Fish & Wildlife Serv, 72-74; ASSOC PROF, COL FISHERIES, UNIV WASH, 74- *Concurrent Pos:* US Fish & Wildlife Serv res grant, Battelle Mem Inst, 66-68, consult, US Fish & Wildlife Serv, 68-69; asst unit leader, Wash Coop Fishery Res Unit, 74-84, unit leader, 84- *Honors & Awards:* Thurlow C Nelson Award, Nat Shellfisheries Asn, 64. *Mem:* Am Soc Microbiol; Am Micros Soc; Soc Invert Path; Am Fisheries Soc; Nat Shellfisheries Asn; Aquatic Plant Mgt Soc. *Res:* Immunology, pathology and cell biology of invertebrates and fish; recreational fisheries; freshwater habitat alteration warm water fish population ecology. *Mailing Add:* Wash Coop Fishery Res Unit Univ Wash Seattle WA 98195

PAULEY, JAMES DONALD, b Mason City, Iowa, Nov 15, 41; m 69; c 1. BIOENGINEERING. *Educ:* Iowa State Univ, BS, 63; Univ Colo, MS, 65; Univ Kans, PhD(elec eng), 71. *Prof Exp:* Instr physiol, Univ Colo Med Ctr, Denver, 71-74, bioengr, 70-80, asst prof psychiat, 72-80; MEM STAFF, WYO BIOTELEMETRY, INC, 80- *Mem:* Inst Elec & Electronics Eng; Nat Soc Prof Eng. *Res:* Biotelemetry, especially physiological data from long term implants. *Mailing Add:* Paul Eng Inc 401 E Elkhorn Ave Box 3376 Estes Park CO 80501

PAULEY, JAMES L, b Powersville, Mo, Dec 5, 25; m 48; c 4. PHYSICAL CHEMISTRY. *Educ:* Cent Col, Mo, AB, 48; Univ Nebr, MS, 50; Univ Arkansas, PhD(chem), 54. *Prof Exp:* Instr chem, Ft Hays Kans State Col, 49-50; assoc prof, 54-61, PROF CHEM, PITTSBURG STATE UNIV, 61- *Mem:* Am Chem Soc; Sigma Xi. *Res:* Surface and polymer chemistry; ion exchange; solutions of electrolytes. *Mailing Add:* 1941 Stoneway Grapevine TX 76051

PAULEY, THOMAS KYLE, b Ansted, WVa, Oct 26, 40; m 64; c 2. ECOLOGY. *Educ:* Univ Charleston, BS, 62; Marshall Univ, MS, 66; WVa Univ, PhD(ecol), 77. *Prof Exp:* Elem prin, Putnam County Sch, 62-63; jr high teacher sci, Kanawha County Sch, 63-65, high sch teacher biol, 65-66; prof biol, Dept Natural Sci, Salem Col, 66-82; prof biol, Div Natural Sci, Univ Pittsburgh, Bradford, 82-; PROF BIOL, MARSHALL UNIV. *Concurrent Pos:* Prof, WVa Univ, 72-76; consult, US Forest Serv, 76-79, US Off Surface Mining & WVa Heritage Trust, 80 & Environ Energy, Inc, 81. *Mem:* Soc Study Amphibians & Reptiles. *Res:* Range and distribution of amphibians and reptiles in West Virginia; ecology of the Cheat Mountain salamander; range and distribution of the Mountain earth snake, smallmouth salamander, and jefferson salamander. *Mailing Add:* Dept Biol Marshall Univ Huntington WV 25701

PAULIE, M CATHERINE THERESE, b Greenbush, Kans. MATHEMATICS, STATISTICS. *Educ:* Marquette Univ, BS, 67; Univ Ill, MS, 68; St Louis Univ, PhD(math), 71. *Prof Exp:* Elem teacher, Our Lady Lourdes, Pittsburg, 59-64, St Paul Sch, Lyons, 64-66 & St Theresa Sch, Hutchison, 66-67; from asst prof to assoc prof math, 71-75, acad dean, 75-80, assoc prof math, 81-84, VPRES ACAD AFFAIRS, ST MARY OF THE PLAINS COL, 87- *Concurrent Pos:* Assoc prof math, Wichita State Univ, 84-87. *Mem:* Math Asn Am; Nat Coun Teachers Math. *Mailing Add:* St Mary of the Plains Col 240 San Jose Dr Dodge City KS 67801

PAULIK, FRANK EDWARD, b Chicago, Ill, Sept 11, 35; m 70. ORGANOMETALLIC CHEMISTRY. *Educ:* Univ Ill, BS, 57; Purdue Univ, MS, 60; Univ Cincinnati, PhD(chem), 64. *Prof Exp:* Res assoc, Univ Chicago, 64-65; sr res chemist, Monsanto Polymers & Petrochem Co, 65-75, from res specialist to sr res specialist, 75-84, FEL, MONSANTO INDUST CHEMS CO, 84- *Concurrent Pos:* Ed, Tech Community Monsanto Newsletter, 79, chmn-elect, 80, chmn, 81. *Mem:* Am Chem Soc; Sigma Xi (chmn-elect, 85). *Res:* Physical organometallic chemistry; synthesis; transition metal complexes; homogeneous catalysis; process research. *Mailing Add:* 13150 Amiot Dr St Louis MO 63146

PAULIKAS, GEORGE A, b Pagegiai, Lithuania, May 14, 36; US citizen; m 57; c 1. SPACE PHYSICS. *Educ:* Univ Ill, Urbana, BS, 57, MS, 58; Univ Calif, Berkeley, PhD(physics), 61. *Prof Exp:* Mem tech staff, 61-68, dir, Space Sci Lab, 68-81, vpres, 81-85, GROUP VPRES, AEROSPACE CORP, 86- *Concurrent Pos:* Consult, Lawrence Radiation Lab, Univ Calif, 61-66; assoc ed, J Geophys Res, 72-74; AIAA Committee on Space Sci & Astron, 73-77, chmn, 76-77, mem, Univ Calif Adv Coun Geophys, 74-75, US Air Force Sci Adv Bd, 75-81, comt space physics, Nat Acad Sci, 77-80 & adv coun, Space Sci Lab, Univ Calif, Berkeley, 77-81; consult, NASA, 75-82; mem ad hoc comts, Nat Acad Sci, 70, 73, 79, 80, 84-86 & 88, Los Alamos Physics Div, 83-, NAS/NRC Naval Studies Bd, 89-, NAS/NRC Defense Space Tech Comt, 89-; mem, Comt Space Sci & Astron, Am Inst Aeronaut & Astronaut, 73-77, chmn, 76-77. *Mem:* Fel Am Phys Soc; Am Geophys Union; assoc fel Am Inst Aeronaut & Astronaut; Sigma Xi. *Res:* Space and radiation belt physics; space science; plasma physics; atomic physics; solar cosmic rays; effects of space environment on space systems. *Mailing Add:* Aerospace Corp PO Box 92957 Los Angeles CA 90009

PAULIN, GASTON (LUDGER), b Ste-Anne de Beaupre, Que, Feb 19, 34; m 55; c 4. DYNAMIC METEOROLOGY. *Educ:* Laval Univ, BA, 53; Can Govt, Forecasting Dipl, 57; Univ Montreal, BSc, 60; Univ Toronto, MA, 61; McGill Univ, PhD(meteorol), 68. *Prof Exp:* Res asst ballistics, Can Armament Res Defense, 55-56; forecaster meteorol, Can Meteorol Serv, 57-59, meteorologist, 61-64, res scientist meteorol, 68-72; dir, Sch Meteorol, Univ Que, Montreal, 72-74; dir, Que Meteorol Serv, 74-80; dir res, Que Ministry Environ, 80-84; sci prog coord, Atmospheric Environ Serv Can, 84-91; RETIRED. *Concurrent Pos:* Lectr math, Armed Serv Educ Exten, Univ Md, 61-65 & Sir George Williams Univ, 67-72; lectr, Univ Montreal, 72. *Honors & Awards:* Prize, Que Meteorol Soc, 81. *Mem:* Can Meteorol Soc; Meteorol Soc Japan; Am Meteorol Soc. *Res:* Numerical weather prediction models; parameterization of physical effects in atmospheric simulation models; simulation of solar energy at the ground level; climatology of solar energy in the Province of Quebec; climate simulation modelling; long range transport of atmospheric pollution (acid rain). *Mailing Add:* Serv L'Environ Atmospherique Environ Can 10 Wellington St Hull PQ K1A 0H3 Can

PAULIN, JEROME JOHN, b Milwaukee, Wis, Jan 18, 36; m 61; c 2. PROTOZOOLOGY, ELECTRON MICROSCOPY. *Educ:* Univ Wis, Whitewater, BE, 62; Univ Ill, Urbana, MS, 63, PhD(zool), 67. *Prof Exp:* Res asst phycol, Univ Ill, Urbana, 64-66; res assoc, protozool, Univ Ill, 66-67; from asst prof to assoc prof, 68-77, PROF ZOOL, UNIV GA, 77- *Concurrent Pos:* Fel dept zool, Univ Ga 67-68; NIH res fel, Univ Wis, 73-74; Nat Acad Sci Exchange Scientist, Prague, Czech, & Leningrad, USSR, 78. *Mem:* Soc Protozool; Am Microscop Soc; Electron Microscopy Soc Am. *Res:* Ultrastructure of the trypanosomatid flagellates and three-dimensional reconstructions of their chondrion. *Mailing Add:* Dept of Zool Univ Ga Athens GA 30602

PAULING, EDWARD CRELLIN, b Pasadena, Calif, June 4, 37; m 56, 70; c 4. MOLECULAR BIOLOGY. *Educ:* Reed Col, BA, 59; Univ Wash, PhD(genetics), 64. *Prof Exp:* USPHS fel biophys, Stanford Univ, 64-66; from asst prof to prof biol, Univ Calif, Riverside, 66-82; PROF & CHMN BIOL, SAN FRANCISCO STATE UNIV, 82- *Mem:* AAAS; Genetics Soc Am; Am Soc Microbiol. *Res:* Mechanisms and metabolic significance of repair of radiation induced damage to DNA of bacteria; physiology of phage infection in halobacteria. *Mailing Add:* Dept Biol Univ Calif Riverside CA 92521

PAULING, LINUS CARL, b Portland, Ore, Feb 28, 01; m 23; c 4. CHEMISTRY, PHYSICS. *Educ:* Ore State Col, BS, 22; Calif Inst Technol, PhD(chem), 25. *Hon Degrees:* Thirty from US & foreign univs. *Prof Exp:* Nat Res Coun fel quant anal, 25-26; Guggenheim fel, Univs Munich, Zurich & Copenhagen, 26-27; from asst prof to assoc prof theoret chem, Calif Inst Technol, 27-31, prof chem, 31-63, head div chem & chem eng & dir, Gates & Crellin Chem Labs, 37-58; res prof phys & biol sci, Ctr Study Dem Insts, 63-67; prof chem, Univ Calif, San Diego, 67-69; prof chem, 69-74, EMER PROF CHEM, STANFORD UNIV, 74-; FEL, LINUS PAULING INST SCI & MED, 73- *Concurrent Pos:* George Eastman prof, Oxford Univ, 48; vis prof, Univ Calif, Cornell Univ, Univ Ill, Mass Inst Technol, Harvard Univ, Princeton Univ, plus other univs & cols. *Honors & Awards:* Nobel Prize in Chem, 54; Nobel Peace Prize, 63; Langmuir Prize, Am Chem Soc, 31, Nichols Medal, 41 & Linus Pauling Medal, 66; Davy Medal, Royal Soc, 47; Pasteur Medal, Biochem Soc France, 52; Addis Medal, Nat Nephrosis Fedn, 55; Phillips Mem Award, Am Col Physicians, 56; Avogadro Medal, Ital Acad Sci, 56; Fermat Medal, Sabatier Medal & Int Grotius Medal, 57. *Mem:* Nat Acad Sci; AAAS; Am Phys Soc; Am Philos Soc (vpres, 51-54); Harvey Soc. *Res:* Determination of structure of crystals and molecules; application of quantum mechanics to chemistry; rotation of molecules in crystals; sizes of ions; theory of stability of complex crystals; chemical bond; line spectra; immunochemistry; structure of proteins; molecular abnormality in relation to disease; sickle cell anemia; orthomolecular medicine; vitamin C and cancer; metals and alloys; ferromagnetism. *Mailing Add:* 440 Page Mill Rd Palo Alto CA 94306-2025

PAULISH, DANIEL JOHN, b Suffern, NY, Aug 20, 49; m 79; c 2. ELECTRICAL ENGINEERING. *Educ:* Manhattan Col, BS, 71; Polytech Inst Brooklyn, MS, 72, PhD(elec eng), 76. *Prof Exp:* Staff engr, Burroughs Corp, 75-81; mgr syst software, Gen Elec Co, 81-85; direct engr, Data Card Corp, 85-87; direct software engr, Siemens Med Electronics, 87-90; PROJ LEADER, SIEMENS AG, 90- *Mem:* Inst Elec & Electronic Engrs; Asn Comput Mach. *Res:* Application of metrication techniques in software engineering; espirit pyramid consortium. *Mailing Add:* Siemens AG Otto-Hahn-Ring Six D-8000 Munich 83 Germany

PAULISSEN, LEO JOHN, b Kankakee, Ill, Nov 8, 15; m 56; c 4. MEDICAL MICROBIOLOGY, IMMUNOLOGY. *Educ:* Bradley Polytech Inst, BS, 41; Univ Chicago, SM, 49; Wash Univ, PhD(bact, immunol), 54. *Prof Exp:* Bacteriologist, State Dept Pub Health, Ill, 46-48, serologist, 49-50; asst radiation biol & bact, Sch Med, Wash Univ, 50-54, res assoc, 54-55; res specialist, 55-56, from asst prof to assoc prof, 56-68, prof, 68-86, EMER PROF, BACT, UNIV ARK, FAYETTEVILLE, 86- *Mem:* Fel AAAS; Am Soc Microbiol; Sigma Xi. *Res:* Study of various microbiological and immunological aspects of mouse response to experimental infection with Salmonella enteritidis. *Mailing Add:* Dept of Bot & Bact Univ of Ark Fayetteville AR 72701

PAULK, JOHN IRVINE, b Bufords, Tenn, July 27, 28; m 52; c 2. NUCLEAR ENGINEERING. *Educ:* US Naval Acad, BS, 52; NC State Univ, PhD(nuclear eng), 62. *Prof Exp:* Assoc serv engr, E I du Pont de Nemours & Co, 56-57; reactor physicist, Tenn Valley Authority, 61-63; from assoc prof to prof nuclear eng, 63-81, head dept, 64-81, ASSOC DEAN ENG & DIR, ENG & INDUST RES STA, MISS STATE UNIV, 81- *Concurrent Pos:* Comnr, Southeast Compact Comn Low-Level Radioactive Waste Mgt; chmn, Miss Radiation Adv Coun. *Mem:* Sigma Xi. *Res:* Reactor core analysis; radioactive tracer techniques. *Mailing Add:* Drawer DE Miss State Univ Mississippi State MS 39762

PAULL, ALLAN E, b Regina, Sask, Mar 5, 18; m 53; c 2. STATISTICS. *Educ:* Univ Man, BA, 38; Univ NC, PhD(exp statist), 48. *Prof Exp:* Staff consult statist & math, Grain Res Lab, Bd Grain Comnrs, Can, 39-41, 48-50; chief statistician, Abitibi Power & Paper Co, Ltd, 50-61; dir opers res, Kimberly-Clark Corp, 61-65; prof statist, Dept Math, Fac Arts & Sci & Sch Bus, 65-71, prof statist, Fac Mgt Studies, 71-86, EMER PROF, FAC MGT STUDIES, UNIV TORONTO, 86- *Mem:* Am Statist Asn; Opers Res Soc Am; Can Opers Res Soc; Royal Statist Soc; Statist Soc Can. *Res:* Statistical methodology in marketing and business. *Mailing Add:* 39 Fenn Ave Willowdale Toronto ON M2L 1M7 Can

PAULL, BARRY RICHARD, b Chicago, Ill, June 1, 47; m 73; c 2. ALLERGY, IMMUNOLOGY. *Educ:* Univ Wis, BSc, 69; Univ Miami, MD, 73; Univ Minn, MS, 78. *Prof Exp:* Intern, Mayo Grad Sch Med, 73-74, resident, 74-75, fel pediat allergy, 75-77, allergy & immunol res fel, 77-78; lectr, 78-80, ASST PROF ALLERGY & IMMUNOL, TEX A&M UNIV, 80- *Concurrent Pos:* Assoc consult, Mayo Clin, 77-; instr, Mayo Med Sch, 77-78. *Mem:* Am Acad Pediat; Am Acad Allergy. *Res:* Insect venoms of the class hymenoptera; identification of allergens in honeybee venom and fire ant venom; identification of allergenic components of the major allergen of ragweed pollen, antigen E and fire ant allergy. *Mailing Add:* 2706 Osler Blvd Allergy,Assoc/Bratos Uly Bryan TX 77801

PAULL, KENNETH DYWAIN, b Winslow, Ariz, Aug 22, 42; m 68; c 2. MEDICAL CHEMISTRY, COMPUTER SYSTEMS. *Educ:* Fresno State Univ, BS, 65;Ariz State Univ, PhD(org chem), 69. *Prof Exp:* Fel, NIH, 69-71; sr scientist, Midwest Res Inst, 71-74; vpres, Starks C P Inc, Subsid Starks Assocs, Inc, 74-78; head acquisition sect, Drug Synthesis & Chem Br, Div Cancer Treat, Nat Cancer Inst, 79-88, detailed to co-supervise, Info Tech Br, 88-89, CHIEF INFO TECH BR, DEVELOP THERAPEUT PROG, NAT CANCER INST, 89- *Mem:* AAAS; Am Asn Cancer Res. *Res:* Antitumor evaluation of new substances; detailed analysis and display in vitro antitumor data. *Mailing Add:* Nat Cancer Inst 811 EPN Bethesda MD 20892

PAULL, RACHEL KREBS, b Milwaukee, Wis, Jan 31, 33; m 54; c 3. CONODONT BIOSTRATIGRAPHY, CARBONATE SEDIMENTOLOGY. *Educ:* Univ Wis, BS, 54, MS, 70, PhD(geol), 80. *Prof Exp:* Lectr, geol, Alverno Col, 72-78; lectr, 80-81, from adj asst prof to adj assoc prof, 81-89, ADJ PROF GEOL, UNIV WIS-MILWAUKEE, 89- *Mem:* Fel Geol Soc Am; Paleont Soc; Soc Econ Paleonts & Mineralogists; Am Geol Instit. *Res:* Lower Triassic conodont biostratigraphy, sedimentology and stratigraphy in Rocky Mountains and Great Basin, Western US. *Mailing Add:* Dept Geosci Univ Wis-Milwaukee PO Box 413 Milwaukee WI 53201

PAULL, RICHARD ALLEN, b Madison, Wis, May 20, 30; m 54; c 3. GEOLOGY. *Educ:* Univ Wis, BS, 52, MS, 53, PhD(geol), 57. *Prof Exp:* Geologist, Pan Am Petrol, Standard Oil Ind, 55-57, res geologist, Jersey Prod Res Co, NJ, 57-60, res group leader, 60-62; assoc prof, 62-65, chmn dept, 62-66, PROF GEOL, UNIV WIS-MILWAUKEE, 65- *Concurrent Pos:* Hon curator, Milwaukee Pub Mus, 62-; dir, inst & short Courses, NSF, 65-68; partic, NSF Conf, 65 & 68; consult. *Mem:* Fel Geol Soc Am; Am Asn Petrol Geol; Soc Econ Paleont & Mineral; Nat Asn Geol Teachers (pres, 77-78); Am Geol Inst (secy, 86-88). *Res:* Classification and techniques for describing clastic rocks; depositional environments of sedimentary rocks; Cretaceous stratigraphy of the western United States; Paleozoic stratigraphy of Idaho; Lower Triassic of western Cordillera. *Mailing Add:* Dept Geosci Univ Wis Milwaukee WI 53201

PAULL, WILLIS K, JR, b Butte, Mont, Dec 25, 44; c 3. ANATOMY, NEUROENDOCRINOLOGY. *Educ:* Mont State Univ, BS, 66, MS, 67; Univ Southern Calif, PhD(anat), 73. *Prof Exp:* Asst prof anat, Univ Vt Sch Med, 72-76; assoc prof anat, Med Col Ga, 76-; CHMN, DEPT ANAT, UNIV MO. *Mem:* Am Asn Anat. *Res:* Development of neuroendocrine mechanisms; electron microscopy and immunocytochemistry of neurosecretory systems. *Mailing Add:* Dept Anat M303 Univ Mo Sch Med Columbia MO 65212

PAULLING, J(OHN) R(ANDOLPH), JR, b Doniphan, Mo, Jan 8, 30; m 52; c 3. ENGINEERING. *Educ:* Mass Inst Technol, BS, 52, MS, 53, NavArch, 54; Univ Calif, Berkeley, DEng(naval archit), 58. *Prof Exp:* Assoc prof, 54-68, PROF NAVAL ARCHIT, UNIV CALIF, BERKELEY, 68- *Concurrent Pos:* Nat Sci Found sci faculty fel, 62-63. *Honors & Awards:* David W Taylor Gold Medal, Soc Naval Architects & Marine Engrs, 85. *Mem:* Nat Acad Eng; Fel Soc Naval Archit and Marine Engrs (vpres, 76-77); fel Royal Inst Naval Archit. *Res:* Experimental ship hydrodynamics, steering and control; numerical analysis of ship structures. *Mailing Add:* Dept of Naval Archit Univ of Calif Berkeley CA 94720

PAULNOCK, DONNA MARIE, IMMUNOLOGY. *Educ:* Stanford Univ, PhD(immunol), 81. *Prof Exp:* ASST PROF IMMUNOL, SCH MED, UNIV WIS, MADISON, 83- *Res:* Mechanism regulation of macrophage function. *Mailing Add:* Dept Med Microbiol Med Sch Univ Wis 1300 University Ave Madison WI 53706

PAULOS, JOHN ALLEN, b Denver, Colo, July 4, 45; m 72; c 2. LOGIC, SCIENCE EDUCATION. *Educ:* Univ Wis, PhD(math), 74. *Prof Exp:* ASSOC PROF MATH, TEMPLE UNIV, 74- *Mem:* Am Asn Advan Sci; Asn Symbolic Logic; Am Philos Asn. *Res:* Philosophy of math; mathematical model theory. *Mailing Add:* Dept Math Temple Univ Broad & Montgomery Sts Philadelphia PA 19122

PAULSEN, CHARLES ALVIN, b Portland, Ore, May 3, 24; m 49; c 5. INTERNAL MEDICINE, ENDOCRINOLOGY. *Educ:* Univ Ore, BA, 47, MD, 52. *Prof Exp:* Instr med, Col Med, Wayne State Univ, 57-58; clin instr, 58-61, from asst prof to assoc prof, 61-70, PROF MED, SCH MED, UNIV WASH, 70- *Concurrent Pos:* USPHS fel med, Col Med, Wayne State Univ, 53-55; consult, US Marine Hosp, Windmill Point, Mich, 57-58; dir labs, Pac Northwest Res Found, 58-61; consult, Vet Admin Hosp, Seattle, Wash, 58- & USPHS Hosp, 61-71. *Mem:* AAAS; Am Fedn Clin Res; Am Fertil Soc; NY Acad Sci; fel Am Col Physicians; Am Endocrine Soc; Am Androl Soc. *Mailing Add:* Dept Med Univ Wash Seattle WA 98195

PAULSEN, DOUGLAS F, b Baltimore, Md, Oct 7, 52; m 84. EXPERIMENTAL EMBRYOLOGY. *Educ:* Western Md Col, BA, 74; Bowman Gray Sch Med, Wake Forest Univ, PhD(anat), 79. *Prof Exp:* Instr anat, Nurse Anesthesia Dept, Bowman Gray Sch Med, 79; fel cell adhesion, Nat Cancer Inst, Calif State Univ, Northridge, 79-80; asst prof anat, 80-88, ASSOC PROF ANAT, MOREHOUSE SCH MED, 88- *Concurrent Pos:* Prin investr, Am Heart Asn grant, 81-83; course dir, Med Micros Anat, 82-; vis researcher biol, Univ Iowa, 86-87; NIH & MBRS subproject, 83-90. *Mem:* Soc Develop Biol; Am Asn Anatomists; Am Soc Cell Biol; Tissue Culture Asn. *Res:* In vitro and in vivo studies of the cellular and molecular bases of differentiation and morphogenesis in limb and craniofacial development. *Mailing Add:* Dept Anat Morehouse Sch Med 720 Westview Dr SW Atlanta GA 30310

PAULSEN, DUANE E, b Fargo, NDak, May 4, 37; m 58; c 1. PHYSICAL CHEMISTRY. *Educ:* NDak State Univ, BS, 58, MS, 59; Mass Inst Technol, PhD(phys chem), 65. *Prof Exp:* Res chemist, 64-68, res chemist, Aeronomy Lab, Spectros Studies Br, Air Force Geophys Lab, 68-83. RES CHEMIST, OPTICAL ENVIRON DIV, GEOPHYS LAB, WENTWORTH INST, 83- *Mem:* Sigma Xi. *Res:* Atmospheric chemistry; spectroscopy and spectrophotometry of atmospheric species; gaseous kinetics. *Mailing Add:* 33 Farwell Rd PO Box 202 Tyngsborough MA 01879-0202

PAULSEN, ELSA PROEHL, b Clinton, Iowa, Oct 10, 23; m 55; c 2. PEDIATRICS, ENDOCRINOLOGY. *Educ:* Univ Ill, AB, 43; Ind Univ, MA, 45; Univ Minn, MD, 54. *Prof Exp:* From instr to assoc prof pediat, Albert Einstein Col Med, 58-69; ASSOC PROF PEDIAT, MED SCH, UNIV VA, 69- *Concurrent Pos:* USPHS fel, 57-59; Nathan Hofheimer fel, 59-64; career scientist, Health Res Coun, NY, 66-71; guest investr, Rockefeller Univ, 68-69. *Mem:* Lawson Wilkins Pediat Endocrine Soc; Soc Pediat Res; Am Diabetes Asn. *Res:* Pediatric endocrinology; diabetes mellitus occurring in childhood. *Mailing Add:* Dept Pediat Box 386 2-11 Univ Va Med Sch Charlottesville VA 22903

PAULSEN, GARY MELVIN, b Frederic, Wis, Mar 23, 39; m 66; c 2. PHYSIOLOGY. *Educ:* Univ Wis, BS, 61, MS, 63, PhD(agron, biochem), 65. *Prof Exp:* From asst prof to assoc prof, 65-75, PROF AGRON, KANS STATE UNIV, 75- *Concurrent Pos:* Rockefeller Found grant & vis scientist, Int Rice Res Inst, Philippines, 71-72. *Mem:* AAAS; Am Soc Plant Physiol; Japanese Soc Plant Physiol; Am Soc Agron; Crop Sci Soc Am; Sigma Xi. *Res:* Physiology of grain plants, primarily nitrogen and protein metabolism and crop hardiness. *Mailing Add:* Dept of Agron Kans State Univ Manhattan KS 66506

PAULSEN, MARVIN RUSSELL, b Minden, Nebr, July 17, 46; m 70; c 2. AGRICULTURAL ENGINEERING. *Educ:* Univ Nebr, BS, 69, MS, 72; Okla State Univ, PhD(agr eng), 75. *Prof Exp:* Engr, Chevrolet Div, Gen Motors Corp, 69; res asst, Agr Eng Dept, Univ Nebr, 71-72; res & teaching asst, Okla State Univ, 72-75; res assoc, 75-77, from asst prof to assoc prof, 77-86, PROF, AGR ENG DEPT, UNIV ILL, 86- *Concurrent Pos:* Mem, NC-151 Regional Comt Grain Quality, 78-91, Ill-Iowa Moisture Meter Task Force, 81-82. *Mem:* Am Soc Agr Engrs; Sigma Xi; Am Asn Cereal Chemists; Am Soc Eng Educ. *Res:* Determination of grain quality of corn and soybeans in export shipment, and from combine harvesting; determination of corn varietal effects on corn hardness and milling; testing of grains with near infrared reflectance; computer image analysis to detect grain quality; grain drying and handling. *Mailing Add:* 360B Agr Eng Sci Univ Ill 1304 W Pennsylvania Urbana IL 61801

PAULSEN, PAUL, b Denison, Iowa, June 21, 35; m 60; c 4. ANALYTICAL CHEMISTRY. *Educ:* Univ Calif, Riverside, BA, 57; Cornell Univ, PhD(analytical chem), 62. *Prof Exp:* ANALYTICAL CHEMIST, NAT BUR STAND, 62- *Mem:* Am Chem Soc; Am Soc Mass Spectrometry. *Res:* Trace element determinations in biological matrices, with isotope-dilution spark source mass spectrometry; analysis of high purity materials with spark source mass spectrograph. *Mailing Add:* 17612 Parkridge Dr Gaithersburg MD 20878

PAULSHOCK, MARVIN, b Springfield, Ill, Jan 11, 23; m 51; c 3. ORGANIC CHEMISTRY. *Educ:* Univ Ill, BS, 43; Harvard Univ, AM, 47, PhD(org chem), 48. *Prof Exp:* Res chemist, Grasselli Div, E I du Pont de Nemours Co, Inc, 48-57, res scientist, Indust & Biochem Dept, 57-62, clin res assoc, Pharmaceut Div, Biochem Dept, 63-74, assoc med dir, 74-85, assoc dir prof serv, Biomed Prod Dept, 85-88, mgr med affairs, 88; CONSULT, PAULSHOCK ASSOCS, 89- *Mem:* Am Soc Clin Pharmacol & Therapeut; Drug Info Asn. *Res:* Naphthoquinones; phosphorous chemistry; chemotherapy; clinical evaluation of new drugs. *Mailing Add:* 1306 Marsh Rd Wilmington DE 19803

PAULSON, BOYD COLTON, JR, b Providence, RI, Mar 1, 46; m 70; c 2. CONSTRUCTION ENGINEERING, CONSTRUCTION MANAGEMENT. *Educ:* Stanford Univ, BS, 67, MS, 69, PhD(civil eng), 71. *Prof Exp:* Asst prof civil eng, Univ Ill, Urbana-Champaign, 72-73; from asst prof to assoc prof civil eng, 74-82, assoc chmn dept, 80-85, PROF CIVIL ENG, STANFORD UNIV, 82- *Concurrent Pos:* Secy, Proj Mgt Inst, 74-77; prin investr, NSF projs, US Dept Transp & Bus Roundtable, 74-; vis prof, Univ Tokyo, 78, Tech Univ, Munich, 83, Univ Strathclyde, Glasgow, Scotland, 90-91; Fulbright scholar, 90-91, Brit Coun Scholar, 90-91. *Honors & Awards:* Huber Res Prize, Am Soc Civil Engrs, 80; Construct Mgt Award, Am Soc Civil Engrs, 84; Distinguished Contrib Award, Proj Mgt Inst, 86. *Mem:* Am Soc Civil Engrs; Am Soc Eng Educ; Asn Comput Mach; Inst Elec & Electronic Engrs, Computer Soc. *Res:* Computer applications in construction engineering and management; urban tunneling and deep excavations for transportation and wastewater projects; automation and robotics in construction. *Mailing Add:* Dept Civil Eng 4020 Stanford Univ Stanford CA 94305-4020

PAULSON, CARLTON, b Barrett, Minn, Dec 9, 34; m 56; c 4. PARASITOLOGY. *Educ:* Concordia Col, Moorhead, Minn, BA, 56; Kans State Univ, MS, 58, PhD(parasitol), 61. *Prof Exp:* From asst prof to assoc prof, 61-73, PROF BIOL, CONCORDIA COL, MOORHEAD, MINN, 73- *Mem:* AAAS; Am Soc Parasitol. *Res:* Metabolism of animal parasites; surface antigens of nematodes. *Mailing Add:* Dept Biol Concordia Col Moorhead MN 56560

PAULSON, CHARLES MAXWELL, JR, b Camden, NJ, Sept 15, 36; m 75; c 4. PHYSICAL CHEMISTRY, POLYMER SCIENCE. *Educ:* Drexel Univ, BS, 59; Univ Calif, Berkeley, PhD(chem), 65. *Prof Exp:* SR RES ASSOC PHYS CHEM, E I DU PONT DE NEMOURS & CO INC, 65- *Mem:* Am Chem Soc; Am Phys Soc. *Res:* Molecular electro optics; molecular orientation in polymers; optics of yarns. *Mailing Add:* PO Box 80357 E I du Pont de Nemours & Co Inc Wilmington DE 19880-0357

PAULSON, CLAYTON ARVID, b Fergus Falls, Minn, Oct 10, 38; m 63; c 2. OCEANOGRAPHY, METEOROLOGY. *Educ:* Augsburg Col, BA, 60; Univ Wash, PhD(atmospheric sci), 67. *Prof Exp:* From res assoc to res asst prof atmospheric sci, Univ Wash, 67-70; asst prof, 71-74, assoc prof, 74-79, PROF PHYS OCEANOG, ORE STATE UNIV, 79- *Concurrent Pos:* NATO assoc meteorol, Danish Atomic Energy Comn, Riso, Denmark, 70-71. *Mem:* AAAS; Am Geophys Union; Am Meteorol Soc; Royal Meteorol Soc; Sigma Xi. *Res:* Air-sea interaction; mechanics of turbulence; dynamics of the upper ocean. *Mailing Add:* Dept of Oceanog Ore State Univ Corvallis OR 97331

PAULSON, DAVID F, b Washington, DC, Mar 29, 38; m 61; c 3. UROLOGY, ONCOLOGY. *Educ:* Duke Univ, AB, 60, MD, 64. *Prof Exp:* Assoc prof, 75-80, PROF & CHMN, DIV UROL, MED CTR, DUKE UNIV,80- *Concurrent Pos:* Consult urol, Vet Admin Med Ctr, Oteen, NC & Cabarrus Mem Hosp, Concord, NC, 72-; dir urol res, Med Ctr, Duke Univ, 72-83; nat chmn, Urol-Oncol Res Group, 73-; consult urol, Durham Count Gen Hosp, 77- *Mem:* Am Asn Genito-Urinary Surgeons; Am Urol Asn; Europ Soc Exp Surg; Am Col Surgeons; Soc Univ Surgeons. *Res:* Genitourinary oncology, etiology, diagnosis and treatment; identification of multimodality programs for control of genitourinary malignancy; tumor-specific androgens. *Mailing Add:* 2808 Chelsea Circle Durham NC 27707

PAULSON, DAVID J, RESEARCH ADMINISTRATION. *Prof Exp:* MGR RES & DEVELOP, OSMONICS INC. *Mailing Add:* Osmonics, Inc 5951 Clearwater Dr Minnetonka MN 55343

PAULSON, DENNIS ROY, b Chicago, Ill, Nov 29, 37. ORNITHOLOGY, EVOLUTIONARY BIOLOGY. *Educ:* Univ Miami, BS, 58, PhD(zool), 66. *Prof Exp:* Instr zool, Univ NC, Chapel Hill, 64-65, USPHS fel, 66; res assoc, Univ Wash, 66-69 & 74-76, asst prof zool, 69-74, adj cur, Burke Mus, 76-89; DIR, SLATER MUS, UNIV PUGET SOUND, 90- *Concurrent Pos:* NSF res grants, 71-74. *Mem:* Cooper Ornith Soc; Wilson Ornith Soc; Am Ornith Union; Soc Int Odonata. *Res:* Systematics and biology of neotropical Odonata; adaptive strategies of migratory shorebirds; animal coloration. *Mailing Add:* Slater Mus Univ Puget Sound Tacoma WA 98416

PAULSON, DONALD LOWELL, b St Paul, Minn, Sept 14, 12; m 38; c 3. THORACIC SURGERY. *Educ:* Univ Minn, BS & MB, 35, MD, 36, MS, 37, PhD(surg), 42; Am Bd Surg, dipl; Am Bd Thoracic Surg, dipl. *Prof Exp:* Instr anat, Univ Minn, 36-37; clin instr thoracic surg, 46-48, from clin asst prof to clin assoc prof surg, 48-56, clin assoc prof thoracic surg, 56-62, CLIN PROF THORACIC SURG, UNIV TEX HEALTH SCI CTR, DALLAS, 62- *Concurrent Pos:* Mem, Am Bd Thoracic Surg, 64-73, chmn, 71-73; chief sect thoracic surg, Baylor Univ Med Ctr, 62-77, chmn med bd, 76-77; consult, Parkland Mem. *Mem:* Fel Am Col Surgeons; fel Am Surg Asn; Am Asn Thoracic Surg (pres, 81); Soc Thoracic Surgeons. *Res:* Bronchogenic carcinoma; hiatal hernia. *Mailing Add:* 5359 Drane Dr Dallas TX 75209

PAULSON, DONALD ROBERT, b Oak Park, Ill, Sept 6, 43; m 66; c 2. ORGANIC CHEMISTRY. *Educ:* Monmouth Col, BA, 65; Ind Univ Univ, PhD(chem), 68. *Prof Exp:* NIH fel, Univ Chicago, 68-70; from asst prof to assoc prof, 70-79, PROF CHEM, CALIF STATE UNIV, LOS ANGELES, 79- *Concurrent Pos:* Vis prof, Univ BC, 77-78 & Univ Sussex, 84-85. *Mem:* AAAS; Am Chem Soc; The Chem Soc; Int Photochem Soc; Sigma Xi. *Res:* Organic photochemistry; model systems for biological oxygenation; conducting polymers. *Mailing Add:* Dept of Chem & Biochem Calif State Univ Los Angeles CA 90032

PAULSON, EDWARD, b Grantwood, NJ, July 21, 15; m 58; c 1. MATHEMATICAL STATISTICS. *Educ:* Brooklyn Col, BA, 36; Columbia Univ, MA, 38, PhD(math statist), 48. *Prof Exp:* Statist clerk, M R Scharff, Consult Engr, NY, 39; statist clerk, US Bur Census, Washington, DC, 40; Carnegie asst, Columbia Univ, 41, math statistician, Div War Res, 42-45; instr math statist, Univ NC, 47; asst prof math, Univ Wash, 47-53; from asst prof to prof, 53-81, EMER PROF MATH, QUEENS COL, NY, 81- *Concurrent Pos:* Vis mem, Courant Inst Math Sci, NY Univ, 65-66, Guggenheim fel, 68. *Mem:* Am Statist Asn; AAAS. *Res:* Testing of statistical hypotheses; multiple decision problems; sequential analysis. *Mailing Add:* 72-10 41 Ave Woodside NY 11377

PAULSON, GAYLORD D, b Castlewood, SDak, June 23, 37; m 60; c 2. BIOCHEMISTRY. *Educ:* SDak State Univ, BS, 63; Univ Wis, MS, 65, PhD(biochem), 67. *Prof Exp:* RES CHEMIST, RADIATION & METAB LAB, SCI & EDUC ADMIN-AGR RES, USDA, 67- *Mem:* Am Chem Soc. *Res:* Animal nutrition; metabolism of natural feed stuffs and pesticides. *Mailing Add:* Biosci Res Lab PO Box 5674 State Univ Sta Fargo ND 58105-5674

PAULSON, GLENN, b Sycamore, Ill, Sept 14, 41. ENVIRONMENTAL TOXICOLOGY & POLICY. *Educ:* Northwestern Univ, BA, 63; Rockefeller Univ, PhD(environ sci), 71. *Hon Degrees:* ScD, Long Island Univ, 72. *Prof Exp:* Staff scientist, Natural Resources Defense Coun, Inc, 71-73, adminr, Sci Support Prog, 73-74; asst comnr, Sci & Res, NJ Dept Environ Protection, 74-79; vpres, Sci & Sanctuaries, Nat Audubon Soc, 79-84, sr vpres, 84-85; vpres, Tech Rev & Compliance, Clean Sites, Inc, 85-88; dir, Ctr Hazardous Waste Mgt, 88-90, RES PROF, PRITZKER DEPT ENVIRON ENG, ILL INST TECHNOL, 88- *Concurrent Pos:* Mem, bd dirs, NY Scientists Comt Pub Info, 65-74, Coun Advan Sci Writing, 69-, Scientists' Inst Pub Info, 70-78 & Rene Dubos Ctr for Human Environ, 85-; lectr, New Sch Social Res, 67-69 & 70-71; vis instr, Col of the Atlantic, 71; adj assoc prof, State Univ NY, Purchase, 71-73 & City Col of NY, 73-74; vis lectr, Southampton Col, 72; mem, Outer Continental Shelf Adv Bd, Dept Interior, 76-79; mem, Secy Energy Adv Bd. *Mem:* AAAS; Am Chem Soc; Sigma Xi; Am Soc Environ Toxicol & Chem; fel AAAS; fel Am Inst Chemist. *Res:* Environmental toxicology, toxic chemicals and hazardous wastes; air and water pollution; environmental policy (national and international). *Mailing Add:* Pritzker Dept Environ Eng Ill Inst Technol Chicago IL 60616

PAULSON, JAMES CARSTEN, b Ashland, Wis, Feb 28, 48; m 70; c 3. BIOCHEMISTRY. *Educ:* MacMurray Col, AB, 70; Univ Ill, Urbana, MS, 72, PhD(biochem), 74. *Prof Exp:* NIH fel biochem, Med Ctr, Duke Univ, 74-78; from asst prof to assoc prof, Univ Calif, Los Angeles, 78-85, prof & vchair biol chem, 85-90; VPRES RES, CYTEL CORP, LA JOLLA, CALIF. *Concurrent Pos:* Fac Res Award, Am Chem Soc, 81-86. *Mem:* AAAS; Fedn Am Scientists; Sigma Xi; Am Soc Biochem & Molecular Biol; Am Chem Soc. *Res:* Molecular biology of glycosyl transferases; biosynthesis of carbohydrate prosthetic groups of glycoproteins and glycolipids and biological function of same. *Mailing Add:* 209 Torrey Pines Terr Del Mar CA 92014

PAULSON, JAMES M(ARVIN), b Wausau, Wis, Jan 1, 23; m 46; c 2. CIVIL ENGINEERING. *Educ:* The Citadel, BS, 47; Ill Inst Technol, MSCE, 49; Univ Mich, PhD(civil eng), 58. *Prof Exp:* Draftsman, Wausau Iron Works, 46; asst instr eng, Ill Inst Technol, 47-48; engr, C S Whitney, 48-49; from instr to assoc prof civil eng, Wayne State Univ, 49-60, chmn dept, 67-72, prof civil eng, 60-90, assoc dean, 73-90; RETIRED. *Mem:* Am Soc Civil Engrs; Am Concrete Inst. *Res:* Structural engineering. *Mailing Add:* PO Box 23-2703 US 23 Greenbush MI 48738

PAULSON, JOHN FREDERICK, b Providence, RI, Oct 29, 29; m 55; c 2. PHYSICAL CHEMISTRY. *Educ:* Haverford Col, AB, 51; Univ Rochester, PhD(chem), 58. *Prof Exp:* Proj assoc chem, Univ Wis, 58-59; RES CHEMIST, AIR FORCE GEOPHYS LAB, 59- *Mem:* Am Chem Soc; Am Phys Soc; Sigma Xi; Am Soc Mass Spectrometry; Am Geophys Union. *Res:* Chemical kinetics; photochemistry; ion-neutral reactions; aeronomy. *Mailing Add:* Geophys Directorate Hanscom AFB MA 01731

PAULSON, LARRY JEROME, b Willmar, Minn, Nov 1, 45; m 66. LIMNOLOGY, FISH BIOLOGY. *Educ:* Univ Nev, Las Vegas, BS, 72; Univ Calif, Davis, MS, 75, PhD(ecol), 77. *Prof Exp:* Res asst fisheries, Univ Nev, Las Vegas, 69-71, res assoc limnol, 71-73; res asst, Univ Calif, Davis, 74-77; res asst prof limnol, Univ Nev, Las Vegas, 77-; DIR, LAKE MEAD LIMNOL RES CTR. *Concurrent Pos:* Limnol consult, Ecol Res Assoc, 74-77. *Mem:* Am Fisheries Soc; Am Inst Biol Sci; Am Soc Limnol & Oceanog; Int Asn Theoret & Appl Limnol; Sigma Xi. *Res:* Factors regulating phytoplankton productivity; nutrient cycling and trophic interactions in lakes and reservoirs. *Mailing Add:* Dept Biol Sci Univ Nev Las Vegas NV 89154

PAULSON, MARK CLEMENTS, b Rossville, Ill, Oct 29, 13; m 40; c 2. ORGANIC CHEMISTRY. *Educ:* Univ Ill, BS, 40; Univ Rochester, PhD(org chem), 43. *Prof Exp:* Instr chem, Univ Rochester, 43-44; res chemist, E I du Pont de Nemours & Co, 44-49; from asst prof to prof chem, Bradley Univ, 49-65, head dept, 64 & 65; prof chem, Chatham Col, 65-79, chmn dept, 65-75; RETIRED. *Mem:* Am Chem Soc; AAAS. *Res:* Attempted asymmetric syntheses in the Grignard reaction; synthesis of some substituted thiocarbazones. *Mailing Add:* 616 Tenth St Oakmont PA 15139

PAULSON, OSCAR LAWRENCE, b El Dorado, Ark, Oct 2, 30; m 68. PETROLEUM GEOLOGY, ENVIRONMENTAL GEOLOGY. *Educ:* Miss State Col, BS, 54, MS, 55; La State Univ, PhD(geol), 60. *Prof Exp:* Instr geol, Miss State Univ, 55; computer, United Geophys Corp, 55; instr geol, La State Univ, 59-60; res geologist, La Geol Surv, 60-61; petrol geologist, Gulf Coast Venture, 61-64; independent geologist, 64-66; from asst prof to assoc prof, 66-76, chmn dept, 73-81, PROF GEOL, UNIV SOUTHERN MISS, 76- *Concurrent Pos:* NSF grant, 69-70; consult, NSF In-Serv Inst in Earth Sci, 68-69; fac res award, Gulf Oil Found, 80. *Mem:* Am Inst Prof Geol; Am Asn Petrol Geol. *Res:* Stratigraphy; salt tectonics; research and teaching environmental geology; environmental effects of dead-end canals in coastal zones. *Mailing Add:* 3975 I 55 N-X7 Jackson MS 38216

PAULSON, WAYNE LEE, b LaCrosse, Wis, Sept 5, 34; m 57; c 3. ENVIRONMENTAL ENGINEERING. *Educ:* Univ Wis-Madison, BS, 59, MS, 60; Univ Iowa, PhD(sanit eng), 65. *Prof Exp:* From instr to assoc prof, 60-72, PROF CIVIL ENG, UNIV IOWA, 72- *Concurrent Pos:* Consult, Can-Tex Industs, Tex, 66, Penberthy Div, Houdaille Industs, Inc, 67-70; Environ Protection Agency process consult, 69-, Norton, 71 & Dravo, 71- *Mem:* Water Pollution Control Fedn; Am Water Works Asn; Am Soc Eng Educ; Am Soc Civil Engrs; Am Asn Prof Sanit Engrs. *Res:* Biochemical aspects of wastewater treatment processes, oxygen transfer; stream quality, activated sludge process applications. *Mailing Add:* 4110 Eng Bldg Univ of Iowa Iowa City IA 52242

PAULSON, WILLIAM H, b Del Norte, Colo, Aug 8, 26; m 58; c 1. PASTURE & CORP MANAGEMENT. *Educ:* Colo State Univ, BS, 51; Univ Wis-Madison, MS, 61, PhD(agron), 71. *Prof Exp:* Agronomist, San Juan Basin Br Sta, Colo State Univ, 55-59; res asst agron, 59-63, SUPT, LANCASTER AGR RES STA, UNIV WIS-MADISON, 63- *Mem:* Am Soc Agron; Am Forage & Grassland Coun; Soil & Water Conserv Soc. *Res:* Pasture, crop (small grains, alfalfa, corn) management; tillage for corn and crop rotations. *Mailing Add:* 301 S Harrison St Lancaster WI 53813

PAULUS, ALBERT, b Glendo, Wyo, Feb 28, 27. PLANT PATHOLOGY. *Educ:* Univ Wyo, MS, 51; Univ Wis, PhD(plant path), 54. *Prof Exp:* PLANT PATHOLOGIST, UNIV CALIF, RIVERSIDE, 54-, LECTR PLANT PATH, 70- *Mem:* fel Am Phytopath Soc; Neth Soc Plant Path; Australasian Plant Path Soc. *Res:* Identification and control of diseases of vegetable, field, ornamental and oil palm crops; subtropical diseases; development of new fungicides; integrated pest management control through the testing of resistant or tolerant cultivars; oil palms. *Mailing Add:* 2791 Shenandoah Rd Riverside CA 92506

PAULUS, HAROLD JOHN, environmental health, air pollution; deceased, see previous edition for last biography

PAULY, JOHN EDWARD, b Elgin, Ill, Sept 17, 27; m 49; c 4. ANATOMY. *Educ:* Northwestern Univ, BS, 50; Loyola Univ, Ill, MS, 52, PhD(anat), 55. *Prof Exp:* Res asst anat, Chicago Med Sch, 52-54, res instr, 54-55, instr gross anat, 55-57, assoc, 57-59, asst prof anat, 59-63, asst to pres, 60-62; assoc prof anat, Sch Med, Tulane Univ, 63-67; prof physiol & biophys & head dept, Col Med, 78-80, prof anat & head dept, 67-83, VCHANCELLOR ACAD AFFAIRS & SPONSORED RES, & ASSOC DEAN GRAD SCH, UNIV ARK MED SCI, LITTLE ROCK, 83- *Concurrent Pos:* Tech adv, Encyclop Britannica Films; mem safety & occup health study sect, Nat Inst Occup Safety & Health, 75-79; ed, Anat News, 72-80, ed, Am J Anat, 80- & co-managing ed, Advances Anat, Embryol & Cell Biol, 80- *Honors & Awards:* Cert of Merit, AMA, 53 & 59; Bronze Award, Ill State Med Soc, 59; Lederle Med Fac Award, 66; Centennial Award, Am Asn Anat, 87. *Mem:* Asn Anat Chmn (secy-treas, 69-71); Int Soc Chronobiol; Am Asn Anat (secy-treas, 73-80, pres, 82-83); fel AAAS; Am Physiol Soc. *Res:* Electromyography, histology and comparative anatomy of adrenals; structural change in human adrenal cortex associated with systemic diseases; chronobiology. *Mailing Add:* Off VChancellor Univ Ark Med Sci Little Rock AR 72205

PAUR, SANDRA ORLEY, b Bismarck, NDak, Nov 4, 46; m 67; c 2. MATHEMATICS. *Educ:* Univ NDak, BS, 68; Ind Univ, MA, 70, PhD(math), 73. *Prof Exp:* ASST PROF MATH, NC STATE UNIV, 73- *Mem:* Am Math Soc; Math Asn Am; Sigma Xi; Asn Women Math. *Res:* Geometric analysis; approximation theory. *Mailing Add:* Dept of Math NC State Univ Raleigh NC 27695-8205

PAUSCH, JERRY BLISS, b Leesburg, Ohio, Jan 21, 39; m 63; c 2. PROCESS ANALYTICAL CHEMISTRY. *Educ:* Ohio State Univ, BChE, 61; Purdue Univ, PhD(analytical chem), 69. *Prof Exp:* Chem engr glass fiber res, Hercules, Inc, 61-63; res chemist gas chromatog, B F Goodrich, 69-73, sect leader analytical res, 73-75, mgr analytical res, 75-88, RES & DEVELOP FEL, B F GOODRICH, 88- *Concurrent Pos:* Dir, Indust Res Analysis Group, Am Chem Soc; mem, Presidents Private Sector Surv, Grace Comn. *Mem:* Am Chem Soc. *Res:* Process analytical applications; laboratory information management systems; pyrolysis mass spectroscopy; solid headspace gas chromatography. *Mailing Add:* 9921 Brecksville Rd Brecksville OH 44141

PAUSTIAN, FREDERICK FRANZ, b Grand Island, Nebr, Nov 24, 26; m 53; c 4. INTERNAL MEDICINE, GASTROENTEROLOGY. *Educ:* Univ Nebr, BSc, 52, MD, 53. *Prof Exp:* Intern, Grad Hosp, Univ Pa, 53-54, resident internal med, 54-56, fel gastroenterol, 56-58, asst instr, Div Grad Med, 57-58; instr, 58-60, assoc, Univ Nebr, 60-61, from asst prof to assoc prof, 61-67, PROF INTERNAL MED & PHYSIOL, COL MED, UNIV NEBR, OMAHA, 67-, MEM GRAD FAC, 65-, ASSOC DEAN CONTINUING & GRAD MED EDUC, 80- *Concurrent Pos:* Mem staff, Univ Nebr Hosp & Douglas County Hosp, 58-; attend physician gastroenterol & radioisotopes, Omaha Vet Admin Hosp, 58-; consult, Immanuel Deaconnes Hosp, 58-, Nebr Methodist Hosp, 59- & Children's Mem Hosp, 63-; assoc, Bishop Clarkson Mem Hosp, 59-; fel trop med & parisitol, Cent Am, Panama, Mex & Sch Med, La State Univ, 60; grants, Lic Beverage Indust, Inc, Sci Adv Coun, 62-64, Dorsey Labs, 63-64 & Off Comprehensive Health Planning, State Nebr, 72-73; mem, Residency Rev Comt Internal Med, 71-, vchmn, 75, chmn, 77 & Residency Rev Comt Allergy-Immunol, 74-; internal med specialist site visitor, 79- *Mem:* AMA; fel Am Col Physicians; Am Gastroenterol Asn; NY Acad Sci; Bockus Int Soc Gastroenterol; Am Soc Internal Med; Am Soc Gastrointestinal Endoscopy; Sigma Xi. *Res:* Gastrointestinal physiology, specifically secretory and motility research; clinical investigation regarding gastric ulcer, regional enteritis, ulcerative colitis and obscure gastrointestinal blood loss. *Mailing Add:* Sect Digestive Dis & Nutrit Univ Nebr Med Ctr 600 S 42nd St Omaha NE 68198-2000

PAUSTIAN, JOHN EARLE, b Grand Island, Nebr, Mar 19, 28; m 51; c 3. ORGANIC CHEMISTRY. *Educ:* Univ Nebr, BSc, 50; Stevens Inst Technol, MSc, 57. *Prof Exp:* Chemist, US Naval Ord Test Sta, 51-53; chemist, Reaction Motors, Inc, 53-57, group leader, Reaction Motors Div, Thiokol Chem Corp, 57-67; sr process res specialist, 67-77, sr res chemist, 77-80, PRIN RES CHEMIST, LUMMUS CO, 80- *Mem:* Catalysis Soc; Am Chem Soc; NY Acad Sci. *Res:* Photochemistry; process development and scale-up; organo-metallic and polymer research and development; petrochemicals. *Mailing Add:* Lummus Crest Inc 1515 Broad St Bloomfield NJ 07003

PAUTLER, EUGENE L, b Alden, NY, Aug 10, 31; m 66; c 1. PHYSIOLOGY, BIOPHYSICS. *Educ:* Univ Buffalo, BA, 56, PhD(physiol psychol), 60. *Prof Exp:* Nat Acad Sci fel neuropharmacol, Aerospace Med Lab, Wright-Patterson Air Force Base, Ohio, 60-61; lab dir life sci res, Goodyear Aerospace Corp, 61-65; NIH spec fel electrophysiol, Keio Sch Med, Japan, 65-67; assoc prof physiol & biophys, 67-78, PROF PHYSIOL & BIOPHYS, COL VET MED & BIOMED SCI, COLO STATE UNIV, 78- *Mem:* Am Physiol Soc. *Res:* Electrophysiological and ultrastructural cytochemistry studies of the retina. *Mailing Add:* Dept Physiol & Biophysics Colo State Univ Ft Collins CO 80523

PAUW, PETER GEORGE, CELL BIOLOGY, SOMATIC CELL GENETICS. *Educ:* Univ Mo, Columbia, PhD(biol), 80. *Prof Exp:* RES ASSOC, UNIV UTAH, 84- *Res:* Gene amplification. *Mailing Add:* S 2112 Lincoln Spokane WA 99203

PAVAN, CRODOWALDO, b Campinas, Sao Paulo, Brazil, Nov 29, 19; m 46; c 3. GENETICS. *Educ:* Univ Sao Paulo, BA, 41, PhD(biol), 44. *Prof Exp:* Monitor biol, Univ Sao Paulo, 40-41, asst, 41-52, prof, 52-69; prof zool & dir cytogenetics, pop & theoret genetics proj, Univ Tex Austin, 69-75; prof biol, Univ Sao Paulo, 76-77; PROF GENETICS, UNIV CAMPINAS, BRAZIL, 78- *Concurrent Pos:* Brazilian del sci comt effects atomic radiation, UN, 56-68. *Honors & Awards:* Moinho Santista Prize Biol, 81. *Mem:* Brazilian Acad Sci; Brazilian Soc Advan Sci (vpres, 75-77, pres, 81-86); Brazilian Biol Soc; Brazilian Genetics Soc (pres, 58-59); Vatican Acad Sci; Third World Acad Sci; Chilean Acad Sci; Physiography Soc. *Res:* Population genetics; cytogenetics and chromosomal physiology; biological control of pests. *Mailing Add:* Dept Evolutionary Genetics Univ Campinas Sao Paulo 13081 Brazil

PAVEK, JOSEPH JOHN, b Waubun, Minn, Oct 3, 27; m 57; c 7. PLANT BREEDING, PLANT GENETICS. *Educ:* Univ Minn, BS, 55, MS, 63; Univ Wis, PhD(cytogenetics of oats), 65. *Prof Exp:* RES GENETICIST, AGR RES SERV, USDA, 65- *Mem:* AAAS; Crop Sci Soc Am; Potato Asn Am. *Res:* Variety development and genetics of economically important characteristics of potato; genetics of reaction in oats to rust; cytogenetics of oat aneuploids. *Mailing Add:* Agr Res Serv USDA Br Exp Sta Aberdeen ID 83210

PAVELIC, VJEKOSLAV, b Sisak, Yugoslavia, June 20, 29; US citizen; m 54; c 2. MECHANICAL ENGINEERING. *Educ:* Univ Zagreb, Dipl Eng, 55; Univ Wis, MS, 61, PhD(mech eng), 68. *Prof Exp:* Design engr, Badger Meter Corp, 55-56; proj engr, Neodyne Corp Consult Engrs, 56-57; staff engr, Aqua-Chem Co, 57-60; sr res & develop Engr Cent Res, A O Smith Corp, 60-68; res asst educ, Univ Wis, Madison, 64-68; from asst prof to assoc prof, 68-78, PROF, DEPT MECH ENG, UNIV WIS-MILWAUKEE, 78- *Concurrent Pos:* NSF fel, Univ Wis, Milwaukee, 69-71, Am Welding Soc fel, 71-73; indust consult; assoc ed, J Mech Design; chmn, Fastening & Joining Comt, Am Soc Mech Engrs & mem, Bd Prof Develop; Javier Barros Sierra prof, Nat Autonomous Univ Mex, 84; vpres, Region VI, Am Soc Med Engrs, 87-89. *Honors & Awards:* Western Elec Fund Award, Am Soc Eng Educ, 77-78; James F Lincoln, ARC Welding Found, 77. *Mem:* Am Soc Mech Engrs; Am Welding Soc; Am Soc Eng Educ; Am Soc Med Engrs. *Res:* Machine design; mechanical system analysis; statistical experiment design; welding and plasma arc; mechanical reliability and probabilistic design. *Mailing Add:* Col Eng & Appl Sci Univ Wis Milwaukee WI 53201

PAVEY, ROBERT LOUIS, food science, for more information see previous edition

PAVGI, SUSHAMA, b Varanasi, India, Jan 8, 55. COMPARATIVE ENDOCRINOLOGY. *Educ:* Banaras Univ, BSc, 74, MSc, 76, PhD(zool), 79. *Prof Exp:* Asst prof endocrinol, Dept Zool, Banaras Univ, 80-85; assoc prof endocrinal & physiol, Dept Life Sci, Manipur Univ, 85-86; RES ASSOC ENDOCRINOL, DEPT INTEGRATIVE BIOL, UNIV CALIF, 89- *Concurrent Pos:* Sr vis scientist, Fulbright Found, 86-89. *Honors & Awards:* Young Scientist Award, Indian Nat Sci Acad, 81. *Mem:* Am Soc Zoologists; Int Fedn Endocrinol. *Res:* Evolution of regulatory mechanisms of hormone secretion from various endocrine glands; organizational levels from organismal to cellular and molecular. *Mailing Add:* Dept Integrative Biol Univ Calif Berkeley CA 94720

PAVIA, DONALD LEE, b Portland, Ore, Jan 25, 41. ORGANIC CHEMISTRY, PHOTOCHEMISTRY. *Educ:* Reed Col, AB, 62; Yale Univ, MS, 64, PhD(chem), 68. *Prof Exp:* NIH fel, Univ Wis-Madison, 68-70; asst prof, 70-75, ASSOC PROF CHEM, WESTERN WASH UNIV, 75- *Mem:* Am Chem Soc; Royal Soc Chem; Sigma Xi. *Res:* Autoxidation reactions; organic photochemistry. *Mailing Add:* Dept of Chem Western Wash Univ Bellingham WA 98225

PAVILANIS, VYTAUTAS, b Kaunas, Lithuania, June 7, 20; Can citizen; m 47; c 4. MEDICINE, VIROLOGY. *Educ:* Univ Kaunas, Lithuania, MD; FRCP (C), 82. *Prof Exp:* Asst prof path, Univ Kaunas, Lithuania, 42-44; asst virol, Pasteur Inst, Paris, 45-48; assoc prof virol, Univ Montreal, 48-65, head virus dept, 48-71, sci dir, Inst Armand-Frappier, 71-75, res coordr, 75-78, asst dir, 75-85, prof virol, 65-88, EMER PROF VIROL, UNIV MONTREAL, 88- *Concurrent Pos:* Mem, Can Nat Tech Adv Comt Live Poliovirus Vaccine, 61-64; consult, Inst Armand-Frappier, 85- *Honors & Awards:* Ann Award, Can Soc Microbiol, 84. *Mem:* NY Acad Sci; Royal Soc Can; Can Soc Microbiol; Can Med Asn. *Res:* Quality control of viral, bacterial vaccine and culture media. *Mailing Add:* 4742 The Blovlevard Westmount PQ H3Y 1V3 Can

PAVKOVIC, STEPHEN F, b Highland Park, Mich, Oct 29, 32; m 61; c 4. INORGANIC CHEMISTRY. *Educ:* Wayne State Univ, BS, 55, MS, 61; Ohio State Univ, PhD, 64. *Prof Exp:* PROF & CHMN CHEM, LOYOLA UNIV, CHICAGO. *Mem:* Am Crystallog Asn; Am Chem Soc. *Res:* X-ray crystallographic structural investigations of transition metal coordination complexes. *Mailing Add:* Chemistry Dept Loyola Univ 6525 N Sheridan Rd Chicago IL 60626-5311

PAVLASEK, TOMAS J(AN) F(RANTISEK), b London, Eng, July 15, 23; Can citizen; m 54; c 3. ELECTRICAL ENGINEERING. *Educ:* McGill Univ, BE, 44, MEng, 48, PhD(elec eng), 58. *Prof Exp:* Res assoc, 48-52, from asst prof to assoc prof, 52-62, secy of fac, 66-67, PROF, MCGILL UNIV, 62-, ASSOC DEAN PLANNING & DEVELOP, FAC ENG, 67- *Mem:* Inst Elec & Electronics Engrs; Am Soc Eng Educ; Eng Inst Can. *Res:* Microwave measurements; automatic control; antennas and electromagnetic wave propagation. *Mailing Add:* Dept Elec Eng McGill Univ 3480 University St Montreal PQ H3A 2A7 Can

PAVLATH, ATTILA ENDRE, b Budapest, Hungary, Mar 11, 30; US citizen; m 51; c 2. FLUORINE CHEMISTRY, GLOW DISCHARGE CHEMISTRY. *Educ:* Tech Univ Budapest, MS, 52; Hungarian Acad Sci, PhD(org fluorine chem), 55. *Prof Exp:* Res assoc fluorine chem, Org Chem Inst, Budapest Tech Univ, Hungary, 51-52, asst prof, 52-54; sr group leader, Chem Cent Res Inst, Hungarian Acad Sci, 54-56; res chemist, Dept Chem, McGill Univ, 57-58; sr group leader, Res Dept, Stauffer Chem Co, Calif, 58-67; sr res scientist, 67-72, proj leader, 72-80, RES LEADER, WESTERN REGIONAL RES LAB, USDA, 80- *Concurrent Pos:* Vis prof, Budapest Tech Univ, 54-56; counr, Am Chem Soc, 73-, dir, 91-93. *Honors & Awards:* Award, Hungarian Acad Sci, 52-54; Outstanding Contrib Chem, Am Chem Soc, 76, Henry Hill Award, 89. *Mem:* Am Chem Soc; Royal Soc Chem; Am Inst Chem. *Res:* Organic and inorganic fluorine chemistry; theoretical organic chemistry; aromatic substitution; high energy oxidizers; low temperature plasma chemistry; natural fibers; biomass gasification and liquefaction; carbohydrate chemistry; edible films and coatings; application of computers in chemistry. *Mailing Add:* Western Regional Res Lab USDA 800 Buchanan St Albany CA 94710

PAVLIDES, LOUIS, b Annapolis, Md, May 11, 21; m 50; c 1. GEOLOGY. *Educ:* Brooklyn Col, AB, 43; Columbia Univ, AM, 48. *Prof Exp:* Geologist, NC, 47-48, NJ, 48, Fla, 48-49, Maine, 49-68, Va, 68-70, Eastern Environ Geol Br, 70-82, GEOLOGIST, EASTERN REGIONAL GEOL BR, US GEOL SURV, 80- *Mem:* Geol Soc Am; Soc Econ Geol. *Res:* Resources of manganese deposits; structure, stratigraphy and petrology of rocks of northeast Maine; geology of the outer Piedmont and northeast Virginia; geology of New England and the southeast Piedmont. *Mailing Add:* 7518 Creighton Dr College Park MD 20740

PAVLIDIS, THEO, b Salonica, Greece, Sept 8, 34; US citizen. INTELLIGENT SYSTEMS. *Educ:* Tech Univ Athens, dipl, 57, Univ Calif Berkeley, MS, 62 & PhD(elec eng), 64. *Prof Exp:* Prof elec eng & comput sci, Princeton Univ, 64-80; tech staff comput sci, AT&T Bell Labs, 80-86; LEADING PROF COMPUT SCI, STATE UNIV NY, STONY BROOK, 86- *Concurrent Pos:* Consult, RCA Corp, 78-80, Datacopy Corp, 86-88, AT&T Bell Labs, 86-90, Symbol Technol, 86- & Ricoh Res & Develop, 90- *Mem:* Fel Inst Elec & Electronics Engrs; Asn Comput Mach. *Res:* Image processing and analysis; computer graphics; pattern recognition. *Mailing Add:* 18 Brewster Hill Rd Setauket NY 11733

PAVLIK, EDWARD JOHN, b Cleveland, Ohio, Dec 16, 46; m 75. BIOCHEMISTRY, CELL BIOLOGY. *Educ:* Univ Denver, BS, 69, MS, 73; Univ Tenn Knoxville, PhD(cell biol), 75. *Prof Exp:* Teaching asst biol sci, Univ Denver, 69-71; biol, zool, Univ Tenn, Knoxville, 74-75; trainee reprod, 76-77, vis asst prof physiol, Univ Ill, Urbana, 78-80; dir res gynec & oncol, 80-84, FROM ASST PROF TO ASSOC PROF, UNIV KY, LEXINGTON, 80- *Concurrent Pos:* Nat Cancer Inst fel, 77-78. *Mem:* Am Soc Cell Biol; Endocrine Soc; Sigma Xi; Am Asn Cancer Res; Am Chem Soc. *Res:* Mechanisms of gene activation in eukaryotes; mechanism of steroid hormone action; identification of hormone dependent neoplasias and the endocrinology of these neoplasias; identification of biological control proteins; ovarian cancer treatment; tumor imaging with radiohalogenated ligands; antiestrogen resistance; developer of publication source of scientific software and information. *Mailing Add:* Dept Obstet & Gynec Univ Ky Lexington KY 40536-0840

PAVLIK, JAMES WILLIAM, b Chicago, Ill, Sept 22, 37; m 59; c 3. PHOTOCHEMISTRY, HETEROCYCLIC CHEMISTRY. *Educ:* Carthage Col, AB, 59; Va Polytech Inst & State Univ, MS, 61; George Washington Univ, PhD(chem), 70. *Prof Exp:* Asst prof org chem, Addis Ababa Univ, 67-69; res scientist, George Washington Univ, 69-70; assoc prof, Univ Wis-River Falls, 70-74; PROF ORG CHEM & HEAD DEPT, WORCESTER POLYTECH INST, 74- *Mem:* Sigma Xi; Am Chem Soc. *Res:* Organic photochemistry; phototransposition reactions of heterocyclic compounds. *Mailing Add:* Dept Chem Worcester Polytech Inst Worcester MA 01609

PAVLIN, EDWARD GEORGE, b Dauphin, Man, June 17, 40. ANESTHESIOLOGY. *Educ:* Univ Man, BSc, 61, BSc & MD, 68. *Prof Exp:* Res fel, 72-73, instr, 73-75, asst prof, 75-80, ASSOC PROF ANESTHESIOL, SCH MED, UNIV WASH, 80- *Concurrent Pos:* Med dir, Highline Col Sch Respiratory Ther, Midway, Wash, 74-; asst dir respiratory ther, Harborview Med Ctr, Seattle, 74- *Mem:* Am Soc Anesthesiol. *Res:* Control of respiration and determinants of brain extracellular fluid hydrogen ion concentration; effects of anesthesia on physiologic function, pulmonary respiration, respiratory physiology. *Mailing Add:* Dept of Anesthesiol Univ of Wash Sch of Med Seattle WA 98195

PAVLIN, MARK STANLEY, b Wilmington, Del, Nov 6, 51; m 73; c 3. ORGANIC CHEMISTRY. *Educ:* Lehigh Univ, BS, 72; Univ Ill, Urbana, PhD(org chem), 77. *Prof Exp:* Res scientist, 77-81, GROUP LEADER TERPENES, UNION CAMP CORP, 81- *Mem:* Am Chem Soc. *Res:* Process research in terpene synthesis and turpentine conversion. *Mailing Add:* 142 Lawn Park Ave Lawrenceville NJ 08648

PAVLISKO, JOSEPH ANTHONY, b Bethesda, Md, April 10, 53. BIOMIMETICS, BIOMATERIALS. *Educ:* Rensselaer Polytech Inst, 75; Univ Conn, PhD(polymer chem), 78. *Prof Exp:* Fel polymer chem, Macromolecular Res Inst, Univ Mich, 78-79; SR CHEMIST POLYMER CHEM, RES TRIANGLE INST, 79- *Mem:* Sigma Xi; NY Acad Sci; Am Chem Soc; AAAS. *Res:* Polymer synthetic chemistry; biomimetic systems; catalysis; thermotropic polymer liquid crystals; modifications of polymers. *Mailing Add:* 433 W Bloomfield Rd Pittsford NY 14534

PAVLOPOULOS, THEODORE G, b Thouria-Kalamata, Greece, Aug 20, 25. OPTICAL PHYSICS. *Educ:* Univ Gottingen, dipl, 51, Dr rer nat(phys chem), 53. *Prof Exp:* Res fel, Max-Planck Inst for Phys Chem, 53; fel physics, Univ Toronto, 53-54; fel phys chem, BC Res Coun, 54-55; lectr physics, Univ BC, 55-56; res assoc, Biophys Prog, Tulane Univ, 56-58; sr res engr, Electronics Dept, Convair Div, Gen Dynamics Corp, 58-62; fel chem, Univ Calif, Los Angeles, 62-65; PHYSICIST, NAVAL OCEAN SYSTS CTR, 65- *Mem:* NY Acad Sci; Sigma Xi. *Res:* Spectroscopy; solid state physics; optics; electrochemistry; integrated optics. *Mailing Add:* 4603 Coronado Ave San Diego CA 92107

PAVLOS, JOHN, b Cleveland, Ohio, Dec 29, 27; m 53; c 1. ORGANIC CHEMISTRY, BIOCHEMISTRY. *Educ:* Western Reserve Univ, BS, 53, MS, 54, PhD(org chem), 60. *Prof Exp:* Fel biochem, Albert Einstein Col Med, 59-63, res asst prof, 63; from asst prof to assoc prof chem, Davis & Elkins Col, 63-67; ASSOC PROF CHEM, MANHATTANVILLE COL, 67- *Mem:* AAAS; Am Chem Soc. *Res:* Steroid and peptide synthesis; free radical aromatic substitution. *Mailing Add:* Manhattanville Col Purchase NY 10577

PAVLOVA, MARIA T, medicine, microbiology, for more information see previous edition

PAVLOVIC, ARTHUR STEPHEN, b Bedford, Ohio, Dec 2, 25; m 47; c 4. SOLID STATE PHYSICS. *Educ:* Yale Univ, BE, 46; Case Inst Technol, MS, 49; Pa State Univ, PhD(physics), 56. *Prof Exp:* Prin physicist, Battelle Mem Inst, 51-52; jr engr, Goodyear Aircraft Corp, 52; jr physicist, Union Carbide Metals Co, 56-58, res physicist, 58-59; from asst prof to assoc prof, 59-66, chmn dept, 68-75, PROF PHYSICS, WVA UNIV, 66- *Mem:* Am Phys Soc; Am Asn Physics Teachers. *Res:* Ferromagnetics; paramagnetism; dielectric materials. *Mailing Add:* Dept Physics WVa Univ Morgantown WV 26506

PAVLOVIC, DUSAN M(ILOS), b Dobric, Yugoslavia, July 9, 21; nat US; m 59; c 1. NUCLEAR METALLURGY, NUCLEAR TECHNOLOGY. *Educ:* Sch Mines Freiberg, MetE, 45; Univ Stuttgart, DrEng, 49. *Prof Exp:* Res asst, Univ Pittsburgh, 49-50 & Carnegie-Mellon Univ, 50; res engr, Gibson Elec Co, 50-52; metall engr, Mat Labs, 52-58, resident eng rep in Ger, 59-63, adv scientist, Aerospace Elec Div, Lima, Ohio, 63-69, FEL ENGR, BETTIS ATOMIC POWER LAB, WESTINGHOUSE ELEC CO, WEST MIFFLIN, 69- *Mem:* Am Soc Metals; Am Inst Mining, Metall & Petrol Engrs; Inst Elec & Electronics Engrs. *Res:* Physical and mechanical metallurgy of nuclear, composite, magnetic and high-temperature materials; fracture mechanics; powder metallurgy; arc welding; magnetic phenomena; international liaison in electrical technology. *Mailing Add:* 413 Cheri Dr Bridgeville PA 15017-1705

PAVLOVICH, RAYMOND DORAN, b Rocky Point, Wyo, July 14, 34; m 55; c 3. MATERIALS ENGINEERING, CIVIL ENGINEERING. *Educ:* Univ Wyo, BS, 59, MS, 65; Purdue Univ, PhD(civil eng), 75. *Prof Exp:* Off mgr munic eng, Holder Eng Serv, 59-60; engr design & construct, J T Banner & Assoc, Inc, 60-65; res assoc, Wyo Hwy Dept, Univ Wyo, 65-67; grad instr res, Purdue Univ, 67-75; engr-analyst, Asphalt Inst, 75-77; hwy res eng waste utilization, Fed Hwy Admin, Eng Testing Lab Inc, 77, mgr res & develop, 77-81; assoc prof civil eng, Ariz State Univ, 81-88; EXXON CHEM, 88- *Mem:* Nat Acad Sci; Asn Asphalt Paving Technol; Am Soc Testing & Mat; Am Soc Civil Eng; Nat Soc Prof Eng. *Res:* Construction materials developed from recovered resources such as fly ash, reclaimed tire rubber, incinerator residue and mine wastes. *Mailing Add:* Exxon Chem 2470 S Dairy Ashford No 250 Houston TX 77077

PAVLOVSKIS, OLGERTS RAIMONDS, b Riga, Latvia, Apr 29, 34; m 67; c 2. MICROBIOLOGY, BACTERIOLOGY. *Educ:* Ore State Univ, Corvallis, BS, 58; Univ Wash, BS, 61, MS, 64; Northwestern Univ, Chicago, PhD(microbiol), 70. *Prof Exp:* Sr lab technician genetics, Dept Med, Sch Med, Univ Wash, 58-61; qual control chem & bact, Western Farmers Asn, Seattle, 64-66; res assoc microbiol, Evanston Hosp, Evanston, Ill, 66-68; MICROBIOLOGIST, NAVAL MED RES INST, BETHESDA, MD, 70- *Concurrent Pos:* Res assoc, Nat Res Coun & Bur Med & surg res associateship, Naval Med Res Inst, Bethesda, Md, 70-72. *Mem:* Am Soc Microbiol; Am Acad Microbiol. *Res:* Microbial toxins; Pseudomonas aeruginosa exotoxin; kinetics of exotoxin synthesis in vitro and in patients, its mode of action, role in clinical infections; prophylaxis and treatment of Pseudomonas infections; antibody production; vaccine; mechanisms of pathogenesis; enteric diseases; enteric pathogenesis. *Mailing Add:* Dept of Microbiol Naval Med Res Inst Bethesda MD 20889

PAWEL, JANET ELIZABETH, b Oak Ridge, Tenn, Dec 24, 62. ION IMPLANTATION, SURFACES & INTERFACES. *Educ:* Univ Tenn, BS, 85, MS, 87; Vanderbilt Univ, PhD(eng), 91. *Prof Exp:* GRAD RES ASST, OAK RIDGE NAT LAB, 84- *Mem:* Am Soc Metals Int; Sigma Xi; Am Vacuum Soc. *Res:* Growth of thin films; mechanical adhesion testing of thin films; ion implantation and ion beam mixing, and surface analysis. *Mailing Add:* PO Box 2008 Mail Stop 6118 Oak Ridge TN 37831

PAWEL, RICHARD E, b Glens Falls, NY, Mar 12, 32; m 55; c 2. OXIDATION, DIFFUSION. *Educ:* Univ Tenn, BS, 53, MS, 54, PhD(metall), 56. *Prof Exp:* METALLURGIST, OAK RIDGE NAT LAB, MARTIN MARIETTA ENERGY SYSTS, INC, 59- *Mem:* Am Soc Mat Int; Electrochem Soc; Sigma Xi. *Res:* Reaction mechanisms in metals; thermal properties and measurements; thermoelectric materials; electron microscopy; gas-metal reactions; surface phenomena; diffusion in refractory metals and oxides; properties of thin films; high-temperature gas corrosion; zirconium and zirconium alloy oxidation; sulfidation of Fe-alloys; oxidation of ordered intermetallic alloys; thermal conductivity of fibrous insulators; corrosion of aluminum alloys under high heat flux conditions. *Mailing Add:* Metals & Ceramics Div Oak Ridge Nat Lab PO Box 2008 Oak Ridge TN 37831-6156

PAWELEK, JOHN MASON, b Baltimore, Md, Apr 15, 42; m 64; c 3. BIOCHEMISTRY, GENETICS. *Educ:* Gettysburg Col, AB, 63; Brown Univ, PhD(develop biol), 67. *Prof Exp:* Asst prof, 70-77, SR RES ASSOC PROF SCI DERMATOL, SCH MED, YALE UNIV, 77- *Concurrent Pos:* Am Cancer Soc fel biochem, 67-70, NIH fel genetics, 70-71. *Mem:* Soc Develop Biol; Soc Invest Dermat. *Res:* Biochemical and genetic controls of cell division and the expression of differentiated functions. *Mailing Add:* Dept Dermat Yale Univ Sch Med New Haven CT 06510

PAWLICKI, ANTHONY JOSEPH, b Detroit, Mich, Mar 21, 44. EXPERIMENTAL HIGH ENERGY PHYSICS. *Educ:* Univ Detroit, BS, 65; Cornell Univ, MS, 68, PhD(physics), 70. *Prof Exp:* Res assoc physics, Physics Dept, Ind Univ, 70-72; appointee physics, High Energy Physics Div, Argonne Nat Lab, 72-76; res assoc, Univ Toronto, 76-77; WRITER, 77- *Mem:* Am Phys Soc. *Mailing Add:* Box 878 Woodland Park CO 80866

PAWLISCH, PAUL E, b Madison, Wis, Jan 9, 31; c 2. AGRONOMY. *Educ:* Univ Wis-Madison, BS, 55, MS, 57, PhD(agron & plant path), 59. *Prof Exp:* Res asst cereal breeding, Univ Wis-Madison, 55-59; asst prof, Tex A&M Univ, 59-64; res agronomist, Malting Barley Improv Asn, 64-69, exec dir, 69-80, pres, 80-83; RETIRED. *Mem:* Am Soc Agron; Am Soc Asn Exec. *Mailing Add:* 5456 N Lydell Ave Milwaukee WI 53217

PAWLOWICZ, EDMUND F, b Toledo, Ohio, Apr 2, 41; div; c 3. GEOPHYSICS. *Educ:* Ohio State Univ, BEE, 64, MSc, 65, PhD(geophys), 69. *Prof Exp:* Sr proj scientist, Marine Geophys, Naval Civil Eng Lab, Calif, 68-70; assoc prof geophys & dir, Seismol Observ, Bowling Green State Univ, 70-81; STAFF GEOPHYSICIST, AMOCO PROD CO, DENVER, COLO, 81- *Mem:* Am Geophys Union; Soc Explor Geophys; Can Soc Explor Geophysicist; Europ Asn Explor Geophysicists. *Res:* Exploration geophysics; reflection seismology; vertical seismic profiling; potential fields. *Mailing Add:* 17144 E Prentice Dr Aurora CO 80015

PAWLOWSKI, ANTHONY T, b North Abington, Mass, Nov 11, 22; m 49; c 4. PHYSICAL CHEMISTRY. *Educ:* Gannon Col, BS, 51; Boston Col, MS, 56; Rutgers Univ, PhD(thermal diffusion), 56, PhD(phys chem), 65. *Prof Exp:* Instr chem, Calumet Ctr, Purdue Univ, 58-60; instr, Rutgers Univ, 60-65; asst prof phys chem, Providence Col, 65-69; PROF CHEM & CHMN DEPT, ALLEGANY COMMUNITY COL, 69- *Concurrent Pos:* NIH res grant, 65-; consult, Precious Metals Inc, Mass, 65- & Basic Sci Inc, Conn, 70; pres, Theotron Co, Cumberland, Md, 74-; mem, Chem Week Mgt Adv Panel, 85-; coordr, Western Maryland Union of Concerned Scientists, 85- *Mem:* Am Chem Soc; Am Phys Soc. *Res:* Thermal diffusion; transport phenomena; electrochemistry, especially electrode processes; Faraday effect; combustion processes. *Mailing Add:* Bedford Rd No R3 Cumberland MD 21502

PAWLOWSKI, NORMAN E, b Lynnwood, Calif, Aug 22, 38; c 3. ORGANIC CHEMISTRY. *Educ:* Southern Ore Col, BS, 61; Ore State Univ, PhD(chem), 65. *Prof Exp:* Res assoc chem, Univ Mich, 65-66; asst prof, Ill State Univ, 66-68; asst prof food protection, Ore State Univ, 68-74, assoc prof food sci & technol, 74-85; HEWLETT-PACKARD, 85- *Mem:* Am Chem Soc. *Res:* Kinetics; toxicology; free-radicals; spectroscopic identification of metabolites; cyclopropene chemistry. *Mailing Add:* 1455 NW 13th Ore State Univ Corvallis OR 97330

PAWLOWSKI, PHILIP JOHN, b Dunkirk, NY, July 15, 43; m 68; c 1. CELL BIOLOGY. *Educ:* St Bonaventure Univ, BS, 65; State Univ NY Buffalo, MS & PhD(biol), 69. *Prof Exp:* Fel biochem, Univ Pittsburgh, 70-73; res assoc cell biol, Dept Biol, Wesleyan Univ, 73-80, vis asst prof, 80-; RES FEL, DEPT MOLECULAR GENETICS, PFIZER RES LAB. *Mem:* Soc Develop Biol. *Res:* Regulation of cell growth and development, specifically the regulation of nucleic acid and protein synthesis and degradation. *Mailing Add:* 662 Duanesburg Rd Schenectady NY 12806

PAWLUK, STEVE, b Egremont, Alta, Sept 21, 30; m 60; c 3. SOIL SCIENCE. *Educ:* Univ Alta, BSc, 53, MSc, 55; Univ Minn, PhD, 57. *Prof Exp:* Asst res officer soil surv, Res Coun Alta, 57-59; from asst prof to assoc prof soil sci, Univ Alta, 59-69, assoc chmn dept forest sci, 71-73, chmn dept, 74-79, PROF SOIL SCI, UNIV ALTA, 69-, ASSOC DEAN RES. *Honors & Awards:* Gold Medal, Agr Inst Can. *Mem:* Am Soc Agron; fel Can Soc Soil Sci; Int Soc Soil Sci. *Res:* Soil pedology, especially soil mineralogy and chemistry; forest soils; soil micromorphology. *Mailing Add:* Dept Soil Sci Univ Alta Edmonton AB T6G 2G7 Can

PAWSON, BEVERLY ANN, b East Orange, NJ, June 20, 34; m 77. ORGANIC CHEMISTRY. *Educ:* Smith Col, BA, 56; Mass Inst Technol, PhD(org chem), 66. *Prof Exp:* Res asst, Mellon Inst, 56-57; res chemist, Esso Res & Eng Co, 57-62; sr chemist, Hoffman-La Roche Inc, 66-73, res fel, 74, res group chief, 75-80, asst dir med chem, 81, assoc dir chem res, 81-85; dir, Strategic Res Planning, Cytogen Corp, 87, dir, Prod Develop, 88-89; ASST DIR PRECLIN RES ADMIN, BERLEX LABS, INC, 89- *Mem:* Am Chem Soc; Sigma Xi; AAAS. *Res:* Synthetic organic chemistry; natural products; medicinal chemistry, retinoids and chemotherapeutic agents. *Mailing Add:* 7 Belleclaire Pl Montclair NJ 07042

PAWSON, DAVID LEO, b Napier, NZ, Oct 5, 38; m 62; c 2. MARINE ZOOLOGY. *Educ:* Victoria Univ, NZ, BSc, 60, MSc, 61, PhD(zool), 64. *Prof Exp:* Demonstr zool, Victoria Univ, NZ, 59-62, teaching fel, 62-64, lectr, 64; assoc cur, Div Marine Invert, 64-65, cur-in-charge, Div Echinoderms, 65-71, chmn dept invert zool, 71-75, CUR ECHINODERMS, NAT MUS NATURAL HIST, SMITHSONIAN INST, 75- *Concurrent Pos:* Adj lectr, George Washington Univ, 65-; assoc invert, Harvard Univ, 74- *Mem:* Soc Syst Zool; Biol Soc Wash (pres, 83). *Res:* Systematics; zoogeography and evolution of echinoderms; deep-sea biology. *Mailing Add:* Nat Mus Natural Hist Smithsonian Inst Washington DC 20560

PAWULA, ROBERT FRANCIS, b Chicago, Ill, May 17, 36. COMMUNICATIONS. *Educ:* Ill Inst Technol, BS, 60; Mass Inst Technol, SM, 61; Calif Inst Technol, PhD(elec eng), 65. *Prof Exp:* Teaching asst elec eng, Calif Inst Technol, 61-62; mem tech staff, Hughes Aircraft Co, 62-65; from asst prof to assoc prof aerospace eng, Univ Calif, San Diego, 65-75; PRES, RANDOM APPLICATIONS, INC, 75- *Concurrent Pos:* Commun & radar consult, 65- *Mem:* Inst Elec & Electronics Engrs; Exp Aircraft Asn. *Res:* Applications of probability theory and stochastic processes to problems in communication theory; information theory and statistical control theory. *Mailing Add:* Random Applns Inc 4611 Chateau Dr San Diego CA 92117

PAX, RALPH A, b Celina, Ohio, May 9, 34; m 61; c 3. PHYSIOLOGY. *Educ:* Univ Dayton, BS, 60; Purdue Univ, MS, 62, PhD(physiol), 64. *Prof Exp:* From asst prof to assoc prof, 64-72, PROF ZOOL, MICH STATE UNIV, 72- *Mem:* AAAS; Am Soc Zool; Soc Neurosci. *Res:* Comparative physiology; neurophysiology; electrophysiology. *Mailing Add:* Dept Zool Mich State Univ Nat Sci Bldg East Lansing MI 48824

PAXSON, JOHN RALPH, b Waco, Tex, Sept 22, 42; m 73; c 2. ANALYTICAL CHEMISTRY, INORGANIC CHEMISTRY. *Educ:* St Mary's Univ, San Antonio, BS, 65; Iowa State Univ, Ames, PhD(inorg chem), 70. *Prof Exp:* Instr chem, North Tex State Univ, Denton, 70-72, supvr instrumentation, 72; res assoc chem, Univ Tex, Austin, 72-74; supvr molecular structure, 77-81, supvr inorg analysis, 81-85, MGR ANALYSIS SERVS, PHILLIPS PETROL CO, BARTLESVILLE, OK, 85- *Concurrent Pos:* Res assoc, Ames Lab USAEC, Ames, Iowa, 67-70; teaching fel, NTex State Univ Fac grant, 70-72; res fel, Robert A Welch Found, Tex, 72-74; adv bd, Ctr Process Analytical Chem, 85-; mem, Dir Indust Res Analysis Group, 85- *Mem:* Am Chem Soc; The Chem Soc; Coblent Soc; Soc Appl Spectros; Am Asn Artificial Intel; Am Soc Qual Control. *Res:* Fourier transform infrared spectroscopy; analysis, molecular structure and interactions of petroleum and petroleum products, of polymers and heterogeneous catalytic surfaces; expert systems; quality assurance. *Mailing Add:* 2316 Chapel Hill Ctr Bartlesville OK 74006

PAXTON, H(AROLD) W(ILLIAM), b Eng, Feb 6, 27; nat US; m 53; c 4. PHYSICAL METALLURGY. *Educ:* Univ Manchester, BSc, 47, MSc, 48; Univ Birmingham, PhD(metall), 52. *Prof Exp:* Prof & head dept metall & mat sci & dir metals res lab, Carnegie-Mellon Univ, 53-74; vpres res, US Steel Corp, 74-86; USS PROF, CARNEGIE MELLON, 86- *Concurrent Pos:* NSF sr fel, Imp Col, Univ London, 62-63; adj sr fel, Mellon Inst; dir Div Mat Res, NSF, 71-73; vis prof, Mass Inst Technol, 70. *Mem:* Nat Acad Eng; fel Am Soc Metals; Am Inst Mining, Metall & Petrol Engrs (pres, 82); fel Metall Soc; hon mem Iron & Steel Soc Japan. *Res:* Plastic deformation of solids; phase transformations; materials policy. *Mailing Add:* 115 Eton Dr Pittsburgh PA 15215

PAXTON, HUGH CAMPBELL, b Los Angeles, Calif, Apr 29, 09; m 37; c 2. NUCLEAR PHYSICS. *Educ:* Univ Calif, Los Angeles, AB, 30; Univ Calif, PhD(physics), 37. *Prof Exp:* Mem tech staff, Bell Tel Labs, 30-32; res assoc, Lab Nuclear Chem, Col France, 37-38; instr physics, Columbia Univ, 38-41; physicist, SAM Labs, Manhattan Dist, 42-44; sr admin physicist, Carbide & Carbon Chem Corp, Tenn, 44-45; physicist, Res labs, Sharples Corp, 45-48; mem staff, 48-76, CONSULT, LOS ALAMOS SCI LAB, 76- *Concurrent Pos:* Consult, Nuclear Regulatory Comn Atomic Safety & Licensing Bd Panel, 63-84. *Mem:* Fel Am Phys Soc; fel Am Nuclear Soc. *Res:* Critical assemblies; criticality and reactor safety; reactor physics. *Mailing Add:* 1229 41st St Los Alamos NM 87544

PAXTON, JACK DUNMIRE, b Oakland, Calif, Feb 17, 36; m 60; c 2. PLANT PATHOLOGY, BIOCHEMISTRY. *Educ:* Univ Calif, Berkeley, BS, 58; Univ Calif, Davis, PhD(plant path), 64. *Prof Exp:* Res assoc, 64-65, asst prof, 65-70, ASSOC PROF PLANT PATH, UNIV ILL, URBANA, 70- *Concurrent Pos:* Mem, Advan Study Insts, NATO, 70, 75 & 80; Fulbright award, 72, Salk Inst, 88. *Honors & Awards:* Fulbright Award, 72, Salk Inst, 88. *Mem:* Am Phytopath Soc; Am Soc Plant Physiol. *Res:* Physiology and biochemistry of host-parasite interactions. *Mailing Add:* Dept of Plant Path Univ of Ill 1102 S Goodwin Urbana IL 61801

PAXTON, K BRADLEY, b Norwich, NY, Dec 31, 38; m 62; c 2. MATHEMATICS, ELECTRICAL ENGINEERING. *Educ:* Rensselaer Polytech Inst, BEE, 60; Univ Rochester, MS, 65, PhD(elec eng), 71. *Prof Exp:* Develop engr, Apparatus Div, Eastman Kodak Co, 60-64, sr develop engr, 64-67, proj engr, 67-69, proj physicist, 69-72, res supvr, 72-75, mgr math modeling & anal, Spec Projs Develop, 75-78, supvr subsyst equip develop, 78-86, gen mgr & vpres, Electronic Photog Div, 86-90, DIR ELECTRONIC IMAGING RES LABS, EASTMAN KODAK CO, 90- *Mem:* Soc Indust & Appl Math; Soc Photog Scientists & Engrs; Inst Elec & Electronic Engrs. *Res:* Analysis of novel imaging systems; electrophotography. *Mailing Add:* Eastman Kodak Co 343 State St Rochester NY 14650

PAXTON, R(ALPH) R(OBERT), b Zion, Ill, Mar 4, 20; m 43; c 2. CHEMICAL ENGINEERING. *Educ:* Univ Ill, BS, 43; Mass Inst Technol, ScD, 49. *Prof Exp:* Jr Chem engr, Res Dept, Standard Oil Co, Ind, 43-46; instr chem eng, Univ Colo, 46-47; asst, Mass Inst Technol, 47-49; asst prof, Stanford Univ, 49-55; engr advan process develop, Gen Elec Co, 55-58; chief engr, Pure Carbon Co, 58-81; vpres eng & planning, Pure Indust Inc, 81-88; RETIRED. *Mem:* Am Chem Soc; Sigma Xi; Soc Tribologists & Lubrication Engrs; Am Inst Chem Engrs. *Res:* Reaction kinetics; research management; friction and wear. *Mailing Add:* 627 Sherry Rd St Mary's PA 15857

PAXTON, RALPH, b Cincinnati, Ohio, Aug 10, 50; m 74; c 1. AMINO ACID METABOLISM, ENZYMOLOGY. *Educ:* Miami Univ, Ohio, BA, 72; Univ Cincinnati, PhD(physiol), 80. *Prof Exp:* Teaching asst physiol, dept biol sci, Univ Cincinnati, 74-80; instr anat & physiol, dept biol, Northern Ky State Univ, 76-77; res assoc, dept biochem, Sch Med, Inc Univ, 81-86, asst prof, 84-86; asst prof biochem, Tex Tech Univ, 86-88, assoc prof, 89; ASSOC PROF PHYSIOL & BIOCHEM, AUBURN UNIV, 89- *Concurrent Pos:* Estab investr, Am Heart Asn, 87-92. *Mem:* Am Soc Zoologists; Sigma Xi; Am Physiol Soc; Am Soc Biol Chemists; Biochem Soc. *Res:* Regulation of amino acid and carbohydrate metabolism; enzymology; environmental and comparative biochemistry and physiology; biochemistry and physiology of fish. *Mailing Add:* Dept Physiology & Pharmacology Col Veternary Medicine, Auburn Univ Auburn AL 36849

PAYER, ANDREW FRANCIS, b Pittsburgh, Pa, Aug 20, 43; m 69. HUMAN ANATOMY. *Educ:* Edinboro State Col, BS, 65; Loyola Univ, PhD(anat), 73. *Prof Exp:* ASST PROF ANAT, UNIV TEX MED BR GALVESTON, 73- *Mem:* Am Asn Anat; Sigma Xi. *Res:* Electron microscopy of human and animal gonads. *Mailing Add:* 25 Adler Circle Galveston TX 77550

PAYER, JOE HOWARD, b Cleveland, Ohio, Oct 6, 43; m 65; c 1. CORROSION, ELECTROCHEMISTRY. *Educ:* Ohio State Univ, BS, 66, PhD(metal eng). *Prof Exp:* Sr res engr corrosion, Res Lab, Inland Steel Corp, 71-74; Sr res & assoc mgr corrosion, Columbus Labs, Battelle, 74-83, mgr mat technol, Houston Opers, 83, mgr, Corrosion Sect, Columbus Labs, 84-85; PROF MAT SCI, CASE WESTERN RESERVE UNIV, 85- *Concurrent Pos:* Instr, Corrosion Course, Am Soc Metals, 75-; assoc dir, Case Ctr Electrochem Sci, 86-; deleg, Int Corrosion Cong, 90- *Honors & Awards:* Sam Tour Award, Am Soc Testing & Mat, 79. *Mem:* Nat Asn Corrosion Engrs (pres, 85-86); Am Soc Metals; Electrochem Soc. *Res:* Corrosion; electrochemistry; environmental degradation of materials; reliability; life prediction; materials selection; failure analysis; surface relativity. *Mailing Add:* Dept Mat Sci & Eng Case Western Reserve Univ 10900 Euclid Ave Cleveland OH 44106

PAYET, MARCEL DANIEL, b Meursac, France, Jan 30, 47; m 73; c 2. CARDIAC ELECTROPHYSIOLOGY, BIOPHYSICS. *Educ:* Univ Poitiers, France, MSc, 70, DrPhysiol, 73; Univ Montreal, PhD(biol), 77. *Prof Exp:* Fel, 77-80, asst prof, 80-85, ASSOC PROF, DEPT BIOPHYSICS, FAC MED, UNIV SHERBROOKE, 85- *Concurrent Pos:* Fel, Can Heart Found, 77-80, res scholarship, 80- *Mem:* NY Acad Sci; Biophys Soc. *Res:* Electrical properties of myocardium; patch clamp; fast inward current of rat heart; K current of rat heart; molecular pharmacology; cultered cells; ionic current of glomerulosa cells from rat adrenals. *Mailing Add:* Dept Biophys Univ Sherbrooke Fac Med Sherbrooke PQ J1H 5N4 Can

PAYLORE, PATRICIA PAQUITA, b Roswell, NMex, Sept 27, 09. SCIENTIFIC BIBLIOGRAPHY, SCIENTIFIC DOCUMENTATION. *Educ:* Univ Ariz, BA, 29, MA, 30. *Prof Exp:* Accession asst, Libr, Univ Ariz, 31-37, serials librn, 37-42, acquisition librn, 42-46, asst librn, 46-64, actg librn, Univ, 46, 47-48, 52 & 63-64, res assoc, Off Arid Lands Studies, 65-68, bibliographer, 68-70, asst dir, Off Arid Lands Studies, 71-79; ED, ARID LANDS NEWSLETTER, UNIV ARIZ, 75- *Concurrent Pos:* Mem, Am Libr Asn Coun, 56-65. *Honors & Awards:* Award for Distinguished Contributions Arid Zone Research, AAAS, 74; Creative Sci Award, Univ Ariz Found, 80. *Mem:* AAAS. *Res:* Natural resources of deserts; arid lands. *Mailing Add:* 2306 E Waverly Tucson AZ 85719

PAYNE, ANITA H, b Karlsruhe, Germany, Nov 24, 26; US citizen; wid; c 2. BIOCHEMISTRY, ENDOCRINOLOGY. *Educ:* Univ Calif, Berkeley, AB, 49, PhD(physiol), 52. *Prof Exp:* Lab technician med physics, Donner Lab, Univ Calif, 49-52, res physiologist, 52-53, res assoc biochem, 61-69, assoc res biochemist, 69-71, from asst prof to assoc prof, 71-81, PROF BIOCHEM, STEROID RES UNIT, DEPT OB-GYN & BIOL CHEM, MED CTR, UNIV MICH, ANN ARBOR, 81- *Concurrent Pos:* Nat Cancer Inst fel, 54-55; mem NIH Biochem Endocrinol Study Sect, 79-83; mem coun, Endocrine Soc, 88-91; mem, Pop Res Comt, Nat Inst Child Health Dis, 89- *Mem:* Am Soc Biol Chemists; Soc Study Reprod; Endocrine Soc (secy, 86-89, pres-elect, 89-90, pres, 90-91); Am Soc Andrology; Am Soc Cell Biol. *Res:* Regulation of expression and mouse chromosomal mapping of the genes that encode the enzymes necessary for steroid hormone biosynthesis. *Mailing Add:* Steroid Res Unit Dept Obstet Gyn Univ Mich Med Ctr Ann Arbor MI 48109-0278

PAYNE, ANTHONY, b Delta, Utah, Dec 25, 27. ECONOMIC GEOLOGY. *Educ:* Univ Utah, BS, 49, MS, 50; Stanford Univ, PhD(geol), 59. *Prof Exp:* Resident geologist, Northern Peru Mining & Smelting Co, 50-51; explor geologist, Am Smelting & Ref Co, Ariz, 51-52; geologist, Minas de Matahambre, Cuba, 52-54; consult mining geologist, Shenon & Full, Utah, 54-59; assoc prof mining, Univ Nev, 59-68, prof geol, 68-79; CONSULT GEOLOGIST, 79- *Mem:* Soc Econ Geol; Am Inst Mining, Metall & Petrol Eng; Am Asn Petrol Geologists. *Res:* Ore deposits; metallogeny; regional geology. *Mailing Add:* 99 Brownstone Dr Reno NV 89512

PAYNE, BERTRAM R, Brit citizen. SENSORY PHYSIOLOGY. *Educ:* Univ Durham, BSc, 74, PhD(zool), 78. *Prof Exp:* Fel, Univ Kassel, WGer, 77-78; res assoc, 78-80, INSTR MED COL, PA, 80- Now at Boston Univ Sch of Med Anat Dept. *Concurrent Pos:* Vis prof, Med Col Pa, 81-82. *Mem:* Soc Exp Biol; Soc Neurosci; Asn Res Vision and Ophthalmol; Royal Soc Europ fel. *Res:* Use of electrophysiological and anatomical techniques to study the normal organization, development, plasticity and cell death within the visual system. *Mailing Add:* Dept of Anat Boston Univ Sch of Med 80 E Concord St Boston MA 02118

PAYNE, DAVID GLENN, b Ont, Can, July 26, 50. HIGH-ENERGY ASTROPHYSICS. *Educ:* Univ Waterloo, BSc, 74; Yale Univ, MPhil, 79, PhD(astrophysics), 79. *Prof Exp:* Res fel, Calif Inst Technol, 79-81; res fel astrophysics, Ctr Astrophysics, Harvard Univ, 81-; AT JET PROPULSION LAB, CALIF. *Mem:* Am Astron Soc. *Res:* Dynamical and stochastic problems in high-energy and relativistic astrophysics. *Mailing Add:* 1111 Blanche St Pasadena CA 91106

PAYNE, DEWITT ALLEN, b Pasadena, Calif, Mar 1, 44; m 68; c 2. ELECTROCHEMISTRY, ATOMIC SPECTROSCOPY. *Educ:* Calif Inst Technol, BS, 65; Univ Tex, Austin, PhD(chem), 70. *Prof Exp:* RES CHEMIST, TENN EASTMAN CO, 71- *Res:* Electroanalytical chemistry; electroorganic synthesis; trace element analysis by AAS, ICP-OES and XRF. *Mailing Add:* Tenn Eastman Co PO Box 1972 Kingsport TN 37662

PAYNE, DONNA L, CARBOHYDRATE ABSORPTION, LIPID METABOLISM. *Educ:* Univ Mo, Columbia, PhD(human nutrit), 72. *Prof Exp:* ASSOC PROF HUMAN NUTRIT, COL ARTS & SCI, OKLA STATE UNIV, 72- *Mailing Add:* 1407 E Connell St Stillwater OK 74074

PAYNE, FRED R(AY), b Mayfield, Ky, Jan 26, 31; m 58; c 3. AERONAUTICAL ENGINEERING. *Educ:* Univ Ky, BS, 52; Pa State Univ, MSAE, 64, PhD(aeronaut eng), 66. *Prof Exp:* Vis res assoc aeronaut eng, Pa State Univ, 66, asst prof, 67; design specialist, Gen Dynamics Corp, Tex, 68-69; from asst prof to assoc prof, 69-70, PROF AEROSPACE ENG, UNIV TEX, ARLINGTON, 75-, JOINT PROF MATH, 85- *Concurrent Pos:* Am Soc Eng Educ-NASA fel, 88 & 89. *Mem:* Am Phys Soc; NY Acad Sci; Am Acad Mech; Am Math Soc; Soc Indust Appl Math. *Res:* Atmospheric turbulence; large eddy structure of turbulence; integral methods; computer mathematics; physical mathematics. *Mailing Add:* Dept Aerospace Eng Univ Tex 315G Eng I Arlington TX 76019

PAYNE, GERALD LEW, b Columbus, Ohio, Mar 11, 38; m 63; c 3. THEORETICAL NUCLEAR PHYSICS, THEORETICAL PLASMA PHYSICS. *Educ:* Ohio State Univ, BS & MS, 61; Univ Calif, San Diego, PhD(physics), 67. *Prof Exp:* Res assoc nuclear physics, Univ Md, 67-69; from asst prof to assoc prof, 69-80, PROF PHYSICS, UNIV IOWA, 80- *Mem:* Fel Am Phys Soc; Am Asn Physics Teachers; Am Asn Univ Professors; Sigma Xi; AAAS. *Res:* Theoretical descriptions of nuclear scattering processes and few body systems. *Mailing Add:* Dept Physics & Astron Univ Iowa Iowa City IA 52242

PAYNE, HARRISON H, b Palmer, Mass, Apr 14, 25; m 44. ZOOLOGY, WILDLIFE CONSERVATION. *Educ:* State Univ NY Col Forestry, Syracuse, BS, 50; St Lawrence Univ, MEd, 55; Cornell Univ, EdD(conserv ed), 63. *Prof Exp:* Teacher, Pawling Cent Sch, NY, 52-63; assoc prof zool, 64-69, PROF WILDLIFE CONSERV, COL ENVIRON SCI & FORESTRY, STATE UNIV NY SYRACUSE, 69- *Concurrent Pos:* Ed consult. *Mem:* Soc Am Foresters; Wildlife Soc; Asn Interpretive Naturalists; Am Asn Univ Prof; Nat Asn Student Personnel Admin. *Res:* Conservation education; field biology; natural history. *Mailing Add:* Dept of Environ & Forest Biol State Univ NY Syracuse NY 13210

PAYNE, HOLLAND I, b Johnstown, Ohio, Sept 29, 18; m 42; c 4. RESEARCH ADMINISTRATION, NATURAL SCIENCE. *Educ:* Colo State Col, BA, 49, MA, 52; Okla State Univ, MS, 59, EdD(sci educ), 63. *Prof Exp:* Teacher, Swayne Indian Sch, Nev, 47-50; teacher, Cassia Schs, Idaho, 50-51; teacher pub schs, Nebr, 51-56, prin, 56-58, supvr, 59-62; staff asst math & consult to elem sch teachers, Okla State Univ, 62-63; prog specialist math, Sacramento City Unified Sch Dist, 63-69, dir develop serv, 69-74, dir educ eval & qual control dept, 74-82; RETIRED. *Concurrent Pos:* Mem, Calif Statewide Math Adv Comt, 67-72 & 76-; mem, Math Assessment Comt, Calif State Dept of Educ, 73- *Res:* Comparative analysis in the achievement of students studying a modern program of mathematics when compared to the achievement of students studying a traditional program; factual history of the Civilian Conservation Corps; US government program for youth from 1933 to 1942; language handicap of Indian school children; mathematics education. *Mailing Add:* 4985 Helen Way Sacramento CA 95822

PAYNE, IRENE R, b Ft Morgan, Colo, Oct 30, 21; c 2. NUTRITION. *Educ:* Colo State Univ, BS, 48, MS, 51; Cornell Univ, PhD(animal nutrit), 60. *Prof Exp:* Res asst agr biochem, Univ Wyo, 51-55, nutrit, 55-57, asst prof, 59-60; asst prof biochem, Pa State Univ, 60-65; from assoc prof to prof nutrit, Southern Ill Univ, Carbondale, 65-84, actg dir, Human Develop Dept, Col Human Resources, 80-84; RETIRED. *Mem:* Am Dietetic Asn. *Res:* Metabolic lesions caused by nutritional deficiencies; metabolic and nutritional bases of mental illnesses. *Mailing Add:* 7878 Hwy 34 Granby CO 80446

PAYNE, JAMES, b Athens, Ala, Mar 27, 41; m 63; c 2. ZOOLOGY, ECOLOGY. *Educ:* Univ Tenn, Martin, BS, 62; Memphis State Univ, MS, 65; Miss State Univ, PhD(zool), 68. *Prof Exp:* Instr zool, Memphis State Univ, 63-65; res asst zool, Miss State Univ, 66-67, NSF trainee, 67-68; from asst prof to assoc prof, 68-80, asst & assoc dean arts & sci, 74-81, actg dean, 77-78, PROF BIOL, MEMPHIS STATE UNIV, 80-, CHMN DEPT, 81- *Concurrent Pos:* Consult, Tech Div, Tenn River Pulp & Paper Co, 80- *Mem:* Sigma Xi; Int Asn Astacologists (treas & secy, 81-84, pres elect, 84-87, pres 87-89); Crustacean Soc; NAm Benthological Soc; Am Soc Zoologists. *Res:* Ecology and life histories of freshwater invertebrates; crayfish life histories; influence of stress (environmental disturbance) on marcoinvertebrates in streams. *Mailing Add:* Dept Biol Memphis State Univ Memphis TN 38152

PAYNE, JAMES EDWARD, b Lynchburg, Va, Feb 10, 44; m 69. EXPERIMENTAL SOLID STATE PHYSICS. *Educ:* Hampden-Sydney Col, BS, 66; Clemson Univ, MS, 69, PhD(physics), 71. *Prof Exp:* Asst prof physics, Clemson Univ, 71-72; asst prof, 72-80, ASSOC PROF PHYSICS, SC STATE COL, 80- *Mem:* Am Phys Soc. *Res:* Low temperature solid state physics; superconductivity. *Mailing Add:* Rte 4 Box 1280 Orangeburg SC 29115

PAYNE, JERRY ALLEN, b Winchester, Va, Dec 19, 37; m 81; c 3. ENTOMOLOGY, ECOLOGY. *Educ:* Univ Tenn, BS, 61; Clemson Univ, MS, 63, PhD(entom), 67. *Prof Exp:* Assoc health physicist, Radiation Ecol Sect, Health Physics Div, Oak Ridge Nat Lab, 63-65; res entomologist, Coastal Plain Exp Sta, Mkt Qual Res Div, 67-69, RES ENTOMOLOGIST, SOUTHEASTERN FRUIT & TREE NUT RES LAB, AGR RES, USDA, 69- *Concurrent Pos:* Adj prof, Dept Hort, Univ Ga, Athens; ed & mem bd dirs, Northern Nut Growers Asn, 82- *Mem:* Am Soc Mammal; Ecol Soc Am; Entom Soc Am; NAm Fruit Explorers. *Res:* Entomology, particularly those of biological and ecological nature; ecology of carrion; medical-legal entomology; biology and ecology of stored-products insects; fruit and tree nut insects; edible, native fruits and nuts. *Mailing Add:* Southeastern Fruit & Tree Nut Res Lab Agr Res USDA PO Box 87 Byron GA 31008

PAYNE, KENYON THOMAS, b Amherst, Mass, Jan 3, 18; m 42; c 3. AGRONOMY. *Educ:* Kans State Col, BSc, 39; Univ Nebr, MSc, 41; Univ Minn, PhD(plant breeding), 48. *Prof Exp:* Agent, Bur Plant Indust, USDA, Nebr, 39-41; from asst prof to assoc prof agron, Purdue Univ, 48-52; prof crop sci & chmn dept, 52-68, prof crop & soil sci, 68-88, EMER PROF CROP & SOIL SCI, MICH STATE UNIV, 88- *Concurrent Pos:* Dean fac agr, Univ Nigeria, 64-66. *Mem:* Fel AAAS; fel Am Soc Agron; fel Crop Sci Soc Am. *Res:* Plant breeding; turf grasses; turf grass breeding. *Mailing Add:* Dept of Crop & Soil Sci Mich State Univ East Lansing MI 48824

PAYNE, LAWRENCE EDWARD, b Enfield, Ill, Oct 2, 23; m 48; c 5. APPLIED MATHEMATICS. *Educ:* Iowa State Col, BS, 46, MS, 48, PhD(applied math), 50. *Hon Degrees:* Dr, Nat Univ Ireland, 90. *Prof Exp:* Jr engr res & develop, Linde Air Prod, 46-47; instr math, Iowa State Col, 48-50; asst prof, Univ Ariz, 50-51; res assoc, Inst Fluid Dynamics & Appl Math, Univ Md, 51-52, from asst res prof to res prof, 52-65; dir, Ctr Appl Math, 67-71, PROF MATH, CORNELL UNIV, 65- *Concurrent Pos:* NSF sr fel, King's Col, Univ Newcastle, 58-59; consult, Nat Bur Stand, 58, 59-65; vis prof, Swiss Fed Inst Technol, 68-69 & 79-80, Univ Newcastle, 72, Univ Col Dublin & Univ Tenn, 86-87. *Honors & Awards:* Steele Prize, Am Math Soc, 72. *Mem:* Am Math Soc; Soc Natural Philos; Sigma Xi; Soc Eng Sci; Am Acad Mech; Soc Indust & Appl Math. *Res:* Elasticity; isoperimetric inequalities; partial differential equations; 111 posed problems. *Mailing Add:* Dept Math Cornell Univ Ithaca NY 14853

PAYNE, LINDA LAWSON, b Spartanburg, SC, Oct 12, 46; m 69; c 2. SOLID STATE PHYSICS. *Educ:* Converse Col, SC, AB, 68; Clemson Univ, SC, MS, 70, PhD(physics), 74. *Prof Exp:* Asst prof, 72-80, PROF PHYSICS, SC STATE COL, ORANGEBURG, 80- *Concurrent Pos:* NASA/Am Soc Elec Engrs fac fel, 81, 82; US Army res, 83. *Mem:* Am Asn Physics Teachers; AAAS. *Res:* Study of quantum size effects in thin films of bismuth as the bismuth films are electrically charged. *Mailing Add:* Rte 4 Box 1280 Orangeburg SC 29115

PAYNE, MARVIN GAY, b Barnardsville, NC, Apr 25, 36; m 61; c 2. THEORETICAL PHYSICS. *Educ:* Berea Col, BA, 58; Univ Ky, PhD(theoret physics), 65. *Prof Exp:* From instr to assoc prof physics, Berea Col, 60-69; THEORET PHYSICIST, OAK RIDGE NAT LAB, 71- *Concurrent Pos:* Fel, Yale Univ, 69-70; adj prof, Univ Ky, 73- & Univ Tenn, 90- *Mem:* Am Asn Physics Teachers; Am Phys Soc. *Res:* Radiative transport in planetary atmospheres; atom-atom collision theory; interaction of laser beams with matter-multiphoton processes; theory of spectral line shapes; many body theory; radiation transport theory; biological modeling; theory of superconductivity; nonlinear optics. *Mailing Add:* X-10 Area Bldg 5500 Mail Stop 6378 Oak Ridge Nat Lab Oak Ridge TN 37831

PAYNE, MYRON WILLIAM, b Red Wing, Minn, June 10, 45; m 68. SEDIMENTOLOGY, PETROLEUM GEOLOGY. *Educ:* Clemson Univ, BS, 68; Univ SC, MS, 70; Tex A&M Univ, PhD(geol), 73. *Prof Exp:* Geologist, Tenneco Oil Co, 73-74; asst prof geol, Univ Alaska, 75-81; explor strategist, Sabine Corp, 82-88; VPRES, ALTANA ROAN, 89- *Honors & Awards:* J C Sproule Award. *Mem:* Soc Econ Paleontologists & Mineralogists; Geol Soc Am; Am Asn Petrol Geologists. *Res:* Long range exploration strategy for western United States, Alaska and Canada; stratigraphic hydrocarbon traps; paleoenvironments of deposition; tectono-stratigraphic setting of east-central Alaska. *Mailing Add:* Altana Roan Cadillac Fairview Bldg 1300 311 Sixth Ave SW Calgary BC T2P 3H2 Can

PAYNE, NICHOLAS CHARLES, b Wrotham, Eng, Apr 16, 42; m 81; c 3. CHEMISTRY. *Educ:* Imp Col, Univ London, BSc, 64; Univ Sheffield, PhD(chem), 67. *Prof Exp:* Res asst, Northwestern Univ, Evanston, 67-69; from asst prof to assoc prof, 69-79, PROF INORG CHEM, UNIV WESTERN ONT, 79- *Concurrent Pos:* Vis prof, Osaka Univ, Japan, 81-82. *Mem:* Chem Inst Can; Royal Soc Chem; Am Chem Soc; Am Crystallog Asn. *Res:* Determinations of absolute configurations by the anomalous dispersion of x-rays; design of chiral ligands for homogeneous asymmetric synthesis; crystal structures of catalytically important transition metal complexes; organometallic and phosphine chemistry. *Mailing Add:* Dept Chem Univ Western Ont London ON N6A 5B7 Can

PAYNE, PHILIP WARREN, b New Castle, Ind, Feb 26, 50; m 78. COMPUTATIONAL METHODS, PROTEIN STRUCTURE. *Educ:* Pomona Col, BA, 71; Princeton Univ, MA, 73, PhD(chem), 76. *Prof Exp:* Res assoc chem, Univ NC, Chapel Hill, 76-77; asst prof chem, Univ Hawaii, Manoa, 77-84; sr computational chemist, SRI Int, 85-88; STAFF SCIENTIST, PROTEIN DESIGN LABS, 88- *Concurrent Pos:* NSF fel Univ NC, Chapel Hill, 76. *Mem:* Am Chem Soc; Am Phys Soc; AAAS; NY Acad Sci; Protein Soc. *Res:* Computational study of structure and function of biological macromolecules, using molecular mechanics, molecular dynamics and electronic structure theory; statistical mechanics, conformational analysis; software; biochemistry. *Mailing Add:* Protein Design Labs 2375 Garcia Ave Mountain View CA 94043

PAYNE, RICHARD EARL, b Holyoke, Mass, Apr 2, 36; c 2. SEA SURFACE METEOROLOGICAL MEASUREMENTS, SOFTWARE SYSTEMS. *Educ:* Bowdoin Col, BS, 58; Univ Md, MS; Univ RI, PhD(phys oceanog). *Prof Exp:* NATO fel, Univ Southampton, Eng, 72-73; fel, 71-72, RES ASSOC, WOODS HOLE OCEANOG INST, 73- *Mem:* Am Geophys Union; Am Meteorol Soc; Oceanog Soc. *Res:* Techniques of measuring meteorological parameters in the surface boundary layer over the ocean and methods of deteming air-sea heat fluxes; time series data processing systems and techniques; sensor calibration. *Mailing Add:* Woods Hole Oceanog Inst Woods Hole MA 02543

PAYNE, RICHARD N, horticulture, for more information see previous edition

PAYNE, RICHARD STEVEN, b Worcester, Mass, Mar 8, 43; m 65; c 3. PHYSICS, ELECTRICAL ENGINEERING. *Educ:* Dartmouth Col, AB, 64; Yale Univ, PhD(physics), 70. *Prof Exp:* Mem tech staff, Bell Tel Labs, 70-74, supvr, IC Process Develop, 74-; DIR, WAFER FABRICATION, ANALOG DEVICES. *Mem:* Int Elec & Electronics Engrs. *Res:* Applications of ion implantation to silicon ICs and devices; development of new complimentary mos IC technology; bipolar process development for ICs and discrete devices. *Mailing Add:* Analog Devices 804 Woburn St Wilmington MA 01887

PAYNE, ROBERT B, b Niles, Mich, July 24, 38; m 65; c 1. ZOOLOGY. *Educ:* Univ Mich, BS, 60; Univ Calif, Berkeley, PhD(zool), 65. *Prof Exp:* Res assoc zool, Inst African Ornith, Univ Cape Town, 65-67; asst prof, Univ Okla, 67-70; asst prof, 70-74, ASSOC PROF ZOOL, UNIV MICH, ANN ARBOR, 74-, ASSOC CUR BIRDS, BIRD DIV, MUS ZOOL, 74- *Concurrent Pos:* NSF fel, 65-67 & res grant, 68-73. *Mem:* Soc Study Evolution; Ecol Soc Am; Am Ornith Union; Brit Ornith Union; Animal Behav Soc. *Res:* Reproductive physiology; mechanisms of speciation in vertebrates; behavioral aspects of evolution. *Mailing Add:* Bird Div Mus Zool Univ Mich 1109 Washtenow Ann Arbor MI 48109

PAYNE, ROSE MARISE, b Lake Bay, Wash, Aug 5, 09; m 42. IMMUNOLOGY, HEMATOLOGY. *Educ:* Univ Wash, BS, 32, MS, 33, PhD(bact), 37. *Prof Exp:* Asst prof bact, Okla Agr & Mech Col, 37-38; lectr, Seattle Col, 39-42; res asst & assoc, Med Sch, Stanford Univ, 48-64, sr scientist, 64-72, prof med, 72-90, emer prof med, 85-90; RETIRED. *Concurrent Pos:* McDermott Found Tuberc res fel, Univ Wash, 38-39; mem adv comt leukocyte antigen terminology, WHO, 67-, mem expert adv panel immunol, 69-74; counr, I-VII Int Histocompatability Workshop Conf, 69-; mem comt organ transplantation & tissue typing, Am Asn Blood Banks, 70- *Mem:* Am Fedn Clin Res; Am Soc Hemat; Transplantation Soc; Int Soc Hemat; Int Soc Blood Transfusion. *Res:* Immunohematology; leukocytes; serology; genetics. *Mailing Add:* Dept Med Hemat 800 Welch Rd Rm 303 Stanford Univ Med Sch Stanford CA 94304

PAYNE, STANLEY E, b Chicago, Ill, Sept 26, 39; m 61; c 4. MATHEMATICS. *Educ:* Hastings Col, BS, 61; Fla State Univ, MS, 63, PhD(math), 66. *Prof Exp:* From asst to instr math, Fla State Univ, 61-66; from asst prof to assoc prof, 66-74, PROF MATH, MIAMI UNIV, 74- *Mem:* Am Math Soc; Math Asn Am. *Res:* Existence, uniqueness and combinatorial properties of finite geometries, graphs and designs; number theory; finite groups. *Mailing Add:* 4187 S Sebring Ct Denver CO 80237

PAYNE, THOMAS GIBSON, geology; deceased, see previous edition for last biography

PAYNE, THOMAS LEE, b Bakersfield, Calif, Oct 17, 41; m 63. ENTOMOLOGY. *Educ:* Univ Calif, Santa Barbara, BA, 65; Univ Calif, Riverside, MS, 67, PhD(entom), 69. *Prof Exp:* From asst prof to assoc prof, 69-77, PROF ENTOM, TEX A&M UNIV, 78- *Mem:* Sigma Xi; AAAS; Entom Soc Am. *Res:* Insect behavior; pest management; pheromones; antennal olfactory electrophysiological response. *Mailing Add:* Dept Entom Va Polytech Inst Blacksburg VA 24061

PAYNE, TORRENCE P B, PATHOLOGY. *Educ:* McGill Univ, PhD(path), 50. *Prof Exp:* dir labs, St Luke's Hosp, 50-87; RETIRED. *Mailing Add:* 840 S Collier Blvd No 306 Marco Island FL 33937

PAYNE, WILLARD WILLIAM, b Hastings, Mich, Jan 1, 34; m 53. PLANT TAXONOMY. *Educ:* Alma Col, Mich, AB, 55; Ohio Univ, MS, 57; Univ Mich, PhD(bot), 62. *Prof Exp:* Assoc res botanist, Univ Mich, 62-64; from asst prof to assoc prof bot, Univ Ill, Urbana, 64-73, from assoc cur to cur, Herbarium, 65-73; prof taxon & chairperson dept bot, Univ Fla, 73-77, dir, Div Biol Sci, 75-77; v pres, New York Bot Gardens & dir, Cary Arboretum, 77-84; CONSULT, 85- *Concurrent Pos:* Assoc prog dir syst bot, NSF, 69-70; trustee, Fairchild Trop Garden, 77-80. *Honors & Awards:* George R Cooley Award Plant Taxon, 64. *Mem:* Int Orgn Biosyst; Am Soc Plant Taxonomists (pres, 77-79); Bot Soc Am; Int Asn Plant Taxon. *Res:* Biosystematics, taxonomy, morphology and evolution of genus Ambrosia; the tribe Ambrosieae and the family Asteraceae; use of stomatal data for taxonomy and evolution of embryophytes; indument evolution; pollen wall evolution. *Mailing Add:* 2596 Roosevelt Pl Sanibel FL 33957

PAYNE, WILLIAM JACKSON, b Chattanooga, Tenn, Aug 30, 25; m 49; c 3. MICROBIOLOGY. *Educ:* Col William & Mary, BS, 50; Univ Tenn, MS, 52, PhD(bact), 55. *Prof Exp:* Instr bact, Univ Tenn, 53-54; from asst prof to prof, Univ Ga, 55-67, head dept bact, 58-67, prof microbiol & head dept, 67-77, actg dean, Col Arts & Sci, 77-78, dean, 78-88, ALUMNI FOUND DISTINGUISHED PROF, UNIV GA, 82- *Concurrent Pos:* Consult, Univ Ala, 59, 68, 70 & 85, Iowa State Univ, 88, Howard Univ, 89; chmn, Nat Registry Microbiol, 67-71; consult, Water Qual Prog, Environ Protection Agency, 71-72; lectr, Am Soc for Microbiol Found, 72-73; biol oceanog panel mem, Nat Sci Found, 76-77; vis prof fel biochem, Univ Wales, Cardiff, 75, hon prof fel, 77-87; chmn, Am Soc Microbiol Found Comn, 77-82; chmn, Am Soc Microbiol Comn Undergrad & Grad Educ, 74-80; nitrogen fixation panel mem, USDA Comp Res Group, 82. *Honors & Awards:* M G Michael Res Award, 60; P R Edwards Award, Am Soc Microbiol, 72. *Mem:* Am Soc Microbiol; Am Acad Microbiol; Brit Soc Gen Microbiol; Am Soc Biol Chemists; Sigma Xi. *Res:* Physiology and biochemistry of denitrification; bacterial alkyl sulfatases. *Mailing Add:* Dept Microbiol Univ Ga Athens GA 30602

PAYNE, WILLIAM WALKER, b Calverton, Va, June 1, 13; m 36; c 3. PUBLIC HEALTH. *Educ:* Univ Va, BSE, 35; Univ Mich, MSE, 47; Univ Pittsburgh, MPH, 56, ScD(hyg), 59; Am Asn Environ Eng, dipl, 59. *Prof Exp:* Engr, County Comnrs for Montgomery County, Md, 35-43; sanit engr, USPHS, 43-60, sanit eng dir, 60-63; dep sci dir, Nat Cancer Inst, 63-67, dep dir, Nat Inst Environ Health Sci, 67-73, sci coordr, Frederick Cancer Res Ctr, 73-81; RETIRED. *Mem:* Am Indust Hyg Asn; Am Asn Cancer Res. *Res:* Sanitary and public health engineering; environmental cancer research and investigations. *Mailing Add:* 8623 Pinecliff Dr Frederick MD 21701

PAYNTER, CAMEN RUSSELL, b Plankinton, SDak, Dec 23, 16; m 49; c 7. INTERNAL MEDICINE, GASTROENTEROLOGY. *Educ:* Univ Ill, BS, 44, MD, 46; Univ Chicago, MBA, 69; Am Bd Internal Med, dipl, 54, recert, 77, Bd Gastroenterol, dipl, 56; Bd Life Ins Med, dipl, 75. *Prof Exp:* Asst, SDak Exp Sta, 41-43; intern, Harper Hosp, 46-47; asst med, Univ Chicago, 53-54, instr gastroenterol, 54-55; pvt pract, 55-63; from assoc med dir to med dir, Western Region, Continental Ins Co, 63-84. *Concurrent Pos:* Fel x-ray, Harper Hosp, Detroit, Mich, 49; fel internal med, Mayo Found, Minn, 50-53; asst clin prof, Stritch Sch Med, Loyola Univ Chicago, 63-71, assoc clin prof, 71-83; attend physician gastroenterol, Vet Admin Hosp, Hines, Ill, 65-82. *Mem:* Am Asn Study Liver Dis; fel Am Acad Occup Med; Int Soc Hemat; Am Soc Gastrointestinal Endoscopy; Am Gastroenterol Asn; fel Am Col Physicians. *Res:* Selenium toxicity; liver disease; nutrition. *Mailing Add:* PO Box 2027 Arlington Heights IL 60006

PAYNTER, GERALD C(LYDE), b Savanna, Ill, Oct 24, 38; m 65; c 2. MECHANICAL ENGINEERING, FLUID MECHANICS. *Educ:* Univ Wash, BS, 60, PhD(fluid mech), 65. *Prof Exp:* Actg asst prof mech eng, Univ Wash, 65-66; res engr, 66-67, lead engr, 67-71, sr engr, 71-75, RES ENGR, MIL AIRPLANE DEVELOP, BOEING CO, 75- *Mem:* Am Inst Aeronaut & Astronaut. *Res:* Boundary layer theory; shock-boundary layer interactions. *Mailing Add:* Boeing Com Airplane Group M-S 6M-98 PO Box 3707 Seattle WA 98124-2207

PAYNTER, HENRY M(ARTYN), b Evanston, Ill, Aug 11, 23; m 44; c 6. SYSTEMS ENGINEERING. *Educ:* Mass Inst Technol, SB, 44, SM, 49, ScD, 51. *Prof Exp:* Jr engr, Puget Sound Power & Light Co, Wash, 44-46; asst civil eng, Mass Inst Technol, 46-48, from instr to asst prof civil eng, 48-54, from asst prof to prof mech eng, 54-85, head Systs Dynamics & Control Div, 63-66, EMER PROF & SR LECTR, MASS INST TECHNOL, 86- *Concurrent Pos:* Consult, Stone & Webster Eng Corp, 51-66, Chas A Maguire & Assocs, 52-80, Foxboro Co, 54-62, Jackson & Moreland, 58-60 & A D Little, 64-65; pres, Pi-Square Eng Co, Inc, 54-59; bd chmn, Telos Corp, 71-72 & Dynacycle Corp, 72; dir, UST Capital Corp, 71-80, HyComp, Inc, 72-75 & Lord Corp, 80-88. *Honors & Awards:* Noble Prize, Joint Eng Socs, 53. *Mem:* Am Soc Civil Engrs; Am Soc Mech Engrs; Inst Elec & Electronics Engrs. *Res:* Systems engineering with emphasis on system dynamics, modelling and simulation, machine computation and automatic control. *Mailing Add:* Dept Mech Eng Bldg 3-449 Mass Inst Technol 77 Mass Ave Cambridge MA 02139

PAYNTER, HOWARD L, b West Allis, Wis, Jan 3, 31; m 49; c 3. MECHANICAL ENGINEERING, FLUID MECHANICS. *Educ:* Univ Wis, BS, 55; Univ Calif, San Diego, 56-60; Univ Denver, MS, 65. *Prof Exp:* Jr engr, Gen Dynamics/Convair, 55, sr engr, 55-60; design specialist gas dynamics, Martin Marietta Corp, 60-66, staff engr, 66-69, chief subsysts technol, propulsion res lab, 69-72, chief thermodynamics & fluid mech, propulsion eng, res & develop dept, 71-74; chmn dept, 77-80, assoc prof, 75-, PROF MECH ENG, DEPT TECHNOL, METROP STATE COL. *Mem:* Am Soc Mech Engrs; Nat Soc Prof Engrs; Am Soc Eng Educ. *Res:* Fluid mechanics, including low-gravity behavior associated with aerospace vehicles; technical writing; thermodynamics. *Mailing Add:* Three Meadowbrook Rd Littleton CO 80120

PAYNTER, JOHN, JR, b Brooklyn, NY, Oct 3, 36; m 59; c 6. LUMINESCENT MATERIALS. *Educ:* Queens Col, NY, BS, 58; Columbia Univ, MA, 59, PhD(electrochem), 64. *Prof Exp:* Phys chemist, Gen Elec Res & Develop Ctr, NY, 64-68; tech ldr mat develop, 75-78, mgr phosphors eng, 78-85, phys chemist, 68-75, TECH LDR PHOSPHORS, GEN ELEC CO, OHIO, 85- *Mem:* Electrochem Soc. *Res:* Low pressure mercury discharge lamps; phosphors; phosphor coatings. *Mailing Add:* 10221 Stonehedge Dr Painesville OH 44077

PAYNTER, MALCOLM JAMES BENJAMIN, b Dudley, Eng, Oct 9, 37; m 61; c 2. MICROBIOLOGY, BIOCHEMISTRY. *Educ:* Univ Sheffield, BSc, 59, MSc, 62, PhD(microbiol), 64. *Prof Exp:* Fel bact, Univ Calif, Davis, 64-66; asst prof environ systs eng, 66-68, bot & bact, 68-69, from asst prof to assoc prof, 69-74, sect chmn, 69-71, PROF MICROBIOL, CLEMSON UNIV, 74-, HEAD DEPT, 71- *Mem:* AAAS; Am Soc Microbiol; Sigma Xi; Brit Soc Gen Microbiol; fel Explorers Club. *Res:* Intermediary metabolism of strict anaerobes; rumen microbiology; methane bacteria; mixed culture systems; microbial ecology; bacteriophage ecology. *Mailing Add:* Dept Microbiol Long Hall Clemson Univ Clemson SC 29631

PAYNTER, RAYMOND ANDREW, JR, b New York, NY, Nov 29, 25; m 60; c 2. ZOOLOGY. *Educ:* Bowdoin Col, BS, 46; Yale Univ, MS, 48, PhD, 54. *Prof Exp:* Field dir sci sta, Bowdoin Col, 46-48; leader, Yale Peabody Mus expeds, Yucatan, 48-49 & 50-51, 52; from asst curator to assoc curator birds, Mus Comp Zool, 53-60, lectr biol, Harvard Univ, 63-75, CURATOR BIRDS, MUS COMP ZOOL, HARVARD UNIV, 61-, SR LECTR BIOL, 75- *Concurrent Pos:* Leader, Mus Comp Zool exped, Chiapas, Mex, 54; leader, Harvard-Yale exped, Nepal, Pakistan & India, 57-59 & mus comp zool exped, Ecuador, 65. *Mem:* Soc Syst Zool; Ecol Soc Am; fel Am Ornith Union; Brit Ornith Union. *Res:* Avian biology and systematics. *Mailing Add:* Mus Comp Zool Harvard Univ Cambridge MA 02138

PAYSON, HENRY EDWARDS, b New York, NY, May 12, 25; m 58; c 4. FORENSIC PSYCHIATRY. *Educ:* Harvard Univ, BS, 48; Columbia Univ, MD, 52; Yale Univ Law Sch, MSL, 78; Am Bd Psychiat & Neurol, dipl, 61; Am Bd Forensic Psychiat, dipl, 80. *Hon Degrees:* AM, Dartmouth Col, 81. *Prof Exp:* From instr to asst prof psychiat & med, Sch Med, Yale Univ, 58-63; from asst prof to assoc prof, 63-78, PROF PSYCHIAT, DARTMOUTH MED SCH, 78- *Mem:* Fel Am Psychiat Asn; Am Acad Psychiat & Law. *Res:* Legal abuse of medical systems. *Mailing Add:* Dept Psychiat Dartmouth Med Sch Nine Maynard St Hanover NH 03755

PAYTON, ALBERT LEVERN, b Hattiesburg, Miss, Feb 8, 44; m 65; c 2. ORGANIC CHEMISTRY. *Educ:* Alcorn State Univ, BS, 65; Southern Univ, Baton Rouge, MS, 69; Univ Southern Miss, PhD(chem), 76. *Prof Exp:* Teacher chem, phys & math, Hattiesburg Pub High Sch, 65-67; teaching asst chem, Southern Univ, 67-69; instr chem & math, Dillard Univ, 69-71; teaching asst chem, Univ Southern Miss, 71-74; ASSOC PROF CHEM, MISS VALLEY STATE UNIV, 74- *Concurrent Pos:* Southern fel Fund, Atlanta, 72-73; adv coun mem, Brookhaven Lab, 77-, NSF Grant, 78-81. *Mem:* Am Chem Soc; Nat Inst Sci. *Res:* Lithium-amine reductions of carboxamides, carboxylic acids, and heterocyclic aromatics; synthesis of long-chain aldynoic acids and alkyn-1-ols via coupling of 1-alkynes and dioc acid anhydrides or lactones. *Mailing Add:* Broward Comm Col 3501 SW Davie Rd Ft Lauderdale FL 33314

PAYTON, ARTHUR DAVID, b Chicago, Ill, Sept 19, 35; m 74. ELECTROCHEMISTRY. *Educ:* Ill Inst Technol, BS, 56; Yale Univ, MS, 57, PhD(phys chem), 60. *Prof Exp:* Assoc prof, 62-77, RES PROF CHEM, WILLAMETTE UNIV, 77- *Concurrent Pos:* NIH fel, Bethesda, Md, 60-61; Alexander Von Humboldt Found fel, Bad Godesberg, WGer, 61-62. *Mem:* Am Chem Soc; Royal Soc Chem; Am Asn Physics Teachers; Ger Bunsen Soc Phys Chem; Electrochem Soc. *Res:* Thermoelectric powers and transported entropies in aqueous acids. *Mailing Add:* Dept of Chem Willamette Univ Salem OR 97301

PAYTON, BRIAN WALLACE, b London, Eng. PHYSIOLOGY. *Educ:* Univ London, MB & BS, 57, PhD(pharmacol), 65. *Prof Exp:* Jr lectr, St Bartholomews Hosp Med Sch, Univ London, 58-65; vis asst prof physiol, Col Physicians & Surgeons, Columbia Univ, 65-67; NIH res fel anat, Albert Einstein Med Col, 67-69; PROF PHYSIOL & DIR MED AUDIO-VISUAL SERV, MED SCH, MEM UNIV NFLD, 69- *Concurrent Pos:* Med Res Coun Can res grant, Mem Univ Nfld, 69-72. *Mem:* Biophys Soc; Can Physiol Soc; Asn Biomed Commun Dirs; Inst Med & Biol Illus; Health Sci Commun Asn. *Res:* Synaptic transmission and neural mechanisms; leech neurophysiology. *Mailing Add:* Dept Physiol Mem Univ Nfld Fac Med St John's NF A1C 5S7 Can

PAYTON, CECIL WARREN, b Orangeburg, SC, Mar 9, 42; m 67; c 2. MICROBIOLOGY. *Educ:* Morris Col, BS, 65; Atlanta Univ, MS, 70; Univ Md, Baltimore, PhD(microbiol), 78. *Prof Exp:* Med lab specialist bact, US Army, 65-68; ASSOC PROF MICROBIOL & BIOL, MORGAN STATE UNIV, 72- *Concurrent Pos:* Instr microbiol, Community Col Baltimore, 73- *Mem:* AAAS; Am Soc Microbiol; Environ Mutagen Soc. *Res:* Microbial physiology and biochemistry. *Mailing Add:* Dept Biol Morgan State Univ Baltimore MD 21239

PAYTON, DANIEL N, III, b Lamar, Mo, July 2, 40; m 60; c 1. PHYSICS. *Educ:* Mo Sch Mines & Metall, BS, 62, MS, 64; Univ Mo, PhD(physics), 66. *Prof Exp:* Instr physics, Univ Mo-Rolla, 64; US Atomic Energy Comn fel & staff mem physics, Los Alamos Sci Lab, 65-67; physicist, Air Force Weapons Lab, US Air Force, 67-71, sci adv plasma physics, 71-74, tech dir nuclear tech, 74-84; DIR STRATEGIC TECHNOL, E O S TECHNOLOGIES, INC, 84- *Concurrent Pos:* Consult, Sci Applns Inc, 72-74; adj prof nuclear eng, Univ NMex, 74- *Mem:* Sigma Xi; Am Phys Soc; AAAS. *Res:* Analysis of systems survivability; applications of plasma physics research; statistical mechanics of traffic flow and modeling; nonstandard fusion techniques; applications of emerging technologies. *Mailing Add:* 4516 Andrew NE Albuquerque NM 87109

PAYTON, OTTO D, b Elk City, Okla, Dec 30, 29; m 55; c 1. REHABILITATION, EDUCATION. *Educ:* Univ Kans, BS, 56; Ind Univ, MS, 64; Univ Md, PhD(higher educ), 71. *Prof Exp:* Chief phys therapist, Dixon State Sch, Ill, 56-58; chief phys therapist, Rehab Ctr, Elkhart, Ind, 58-66, asst dir, 62-66; instr rehab, Sch Med, Univ Md, 66-70, asst prof phys ther & actg chmn dept, 70-71; assoc prof, Med Col Va, Va Commonwealth Univ, 71-76, dir grad div, dept phys ther, Sch Allied Health Professions, 71-83, chmn, 83- 88, PROF PHYS THER, MED COL VA, VA COMMONWEALTH UNIV, 76- *Concurrent Pos:* A D Williams res grant, Med Col Va, 72, Med Col Va Found res grant, 73-74; consult, Div Phys Ther, Md State Dept Health, 69-71; chmn grad educ comt, Am Phys Ther Asn, 74 & 77; res fel, Am Phys Ther Asn, 80; Kellog fel, Western Australia Inst Technol, 84. *Honors & Awards:* Golden Pen Award, Am Phys Ther Asn, 81. *Mem:* Am Phys Ther Asn. *Res:* Education. *Mailing Add:* Sch Allied Health Professions Dept Phys Ther Box 224 Richmond VA 23298-0234

PAYTON, PATRICK HERBERT, b New Eagle, Pa, June 28, 41. PHYSICAL CHEMISTRY, NUCLEAR CHEMISTRY. *Educ:* Waynesburg Col, BS, 64; Univ Calif, Los Angeles, PhD(phys chem), 74. *Prof Exp:* Fel nuclear & lunar chem, Univ Calif, Los Angeles, 74-75, fel stable isotope geochem, 75; Robert A Welch fel nuclear chem, Marine Biomed Inst, Univ Tex Med Br, Galveston, 75-76; fel geochem, Univ Calif, Los Angeles, 76-77; res scientist, Teledyne Wah Chang Albany, 77-81; sr res chemist, Occidental Res Corp, 81-83; SR SCIENTIST, TRW CORP, 83- *Mem:* Am Chem Soc; Am Inst Chem Engrs; Assoc Inst Mech Engrs; Am Vacuum Soc. *Res:* Planetary and lunar geochemistry; stable isotope geochemistry; radioactive fallout; analytical methods; neutron and photon activation analysis. *Mailing Add:* 25421 Mina Ct El Toro CA 92630

PAYTON, ROBERT GILBERT, b Louisville, Ky, Jan 1, 29; m 71. APPLIED MATHEMATICS. *Educ:* Univ Louisville, BEE, 52; Yale Univ, ME, 53; Harvard Univ, PhD(appl math), 59. *Prof Exp:* Staff scientist appl mech, Avco Corp, 59-61, res group leader, 62-64; from asst prof to assoc prof, 64-73, PROF MATH, ADELPHI UNIV, 73- *Concurrent Pos:* Temporary mem, Courant Inst Math Sci, 61-62; vis prof, Univ Strathclyde, 71-72; sabbatical leave, Imperial Col, London, 80. *Mem:* Am Math Soc. *Res:* Linear elastic wave propagation. *Mailing Add:* Dept Math & Comput Sci Adelphi Univ Garden City NY 11530

PAZ, MARIO MEIR, b Quezaltenango, Guatemala, Mar 21, 24; m 51; c 4. CIVIL ENGINEERING, STRUCTURAL DYNAMICS. *Educ:* Univ Chile, Civil Eng, 54; Iowa State Univ, MS, 57, PhD(theoret & appl mech), 60. *Prof Exp:* Head dept statist, Chilean Serv Statist, 54-58; instr eng, Iowa State Univ, 58-60; from asst prof to assoc prof civil eng, 60-69, PROF CIVIL ENG, UNIV LOUISVILLE, 69-, CHMN DEPT, 80- *Concurrent Pos:* Instr, Univ Chile, 54-57; consult, Rex Chain Belt Co, Ky, 62-; fallout shelter analyst & blast designer, George Washington Univ, 65. *Mem:* Am Soc Eng Educ; Seismol Soc Am; Am Soc Civil Engrs. *Res:* Mechanical vibrations. *Mailing Add:* 1712 Sutherland Dr Louisville KY 40205

PAZ, MERCEDES AURORA, b Buenos Aires, Arg, Apr 7, 28; US citizen. BIOCHEMISTRY. *Educ:* Univ Buenos Aires, BPharm, 50, MBiochem, 52, PhD(biochem), 55. *Prof Exp:* Chief lab clin anal, Ctr Rheumatic Dis, Univ Buenos Aires, 52-60, instr biol chem, Sch Pharm & Biochem, 57-61, res assoc, Nat Coun Res, 64-65, asst prof, Sch Pharm & Biochem, 65-66; asst prof biochem, Albert Einstein Col Med, 67-72; asst prof, 72-77, ASSOC PROF ORAL BIOL & PATHOPHYSIOL, SCH DENT MED, HARVARD UNIV, 77-; RES ASSOC, CHILDREN'S HOSP MED CTR, 72- *Concurrent Pos:* Arg Nat Coun Res grants, Univ Buenos Aires, 60-61 & Albert Einstein Col Med, 62-63. *Mem:* NY Acad Sci; AAAS; Geront Soc; Protein Soc. *Res:* Cellular aging protein chemistry; salivary proteins; guinones-quinoproteins. *Mailing Add:* Lab Human Biochem Children's Hosp Med Ctr Enders 12 Boston MA 02115

PAZDERNIK, THOMAS LOWELL, b Detroit Lakes, Minn, Jan 3, 43; m 67; c 2. NEUROPHARMACOLOGY, NEUROTOXICOLOGY. *Educ:* Univ Minn, BS, 67; Univ Kans, PhD(med chem), 71. *Prof Exp:* Postdoctoral fel, 71-73, from asst prof to assoc prof, 73-84, PROF PHARMACOL, UNIV KANS MED CTR, 84- *Concurrent Pos:* Vis scientist, Univ Helsinki, Finland, 76; assoc dean, Univ Kans Med Ctr, 86-89, actg chmn, 90- *Mem:* Am Soc Pharmacol & Exp Therapeut; Soc Neurosci; Soc Toxicol; Sigma Xi; AAAS; Int Soc Immunopharmacol. *Res:* Neuropharmacology/toxicology: studies on chemical-induced seizures and brain damage, using local cerebral glucose use, quantitative autoradiographic receptor analysis and brain microdialysis; immunotoxicology: effects of chemicals on hemato/immunopoietic systems. *Mailing Add:* Dept Pharmacol Univ Kans Med Ctr 39th & Rainbow Blvd Kansas City KS 66103

PAZICH, PHILIP MICHAEL, b Sewickly, Pa, Jan 12, 47; m 73. SPACE PHYSICS. *Educ:* Va Mil Inst, BS, 68; Rice Univ, MS, 71, PhD(space physics), 73. *Prof Exp:* Res physicist, US Air Force Geophys Lab, 73-76; proj mgr, US Air Force Space & Missile Systs Orgn, 76-79; PROJ DIR, GEN RES CORP, 79- *Mem:* Am Geophys Union. *Res:* Ionospheric effects of magnetospheric substorms and the properties of the solar wind and its interaction with the earth's magnetosphere. *Mailing Add:* 4559 Atascadero Dr Santa Barbara CA 93110

PAZOLES, CHRISTOPHER JAMES, b Chicago, Ill, Jan 17, 50; m 72; c 2. MOLECULAR BIOLOGY, NEUROBIOLOGY. *Educ:* Oberlin Col, AB, 71; Univ Notre Dame, PhD(microbiol), 75. *Prof Exp:* Staff fel membrane molecular biol, Nat Inst Child Health & Human develop, 75-77, sr staff fel, Clin Hemat Br, Nat Inst Arthritis, Metab & Digestive Dis, NIH, 77-81; ASST DIR, PFIZER CENT RES, PFIZER INC, GROTON, CONN, 81- *Mem:* Soc Neurosci; AAAS; Int Soc Study Pain; Am Pain Soc; Am Soc Pharmacol & Exp Therapeut. *Res:* Discovery and development of new therapeutic agents for the treatment of inflammatory disorders. *Mailing Add:* Dept Immunol & Infectious Dis Pfizer Cent Res Groton CT 06340

PAZUR, JOHN HOWARD, b Czech, Jan 17, 22; US citizen; m 50; c 4. BIOCHEMISTRY. *Educ:* Univ Guelph, BSA, 44; McGill Univ, MSc, 46; Iowa State Univ, PhD(chem), 50. *Prof Exp:* Instr chem, Iowa State Univ, 50-51; asst prof biol chem, Univ Ill, 51-52; from asst prof to prof biochem & nutrit, Univ Nebr, 52-66; PROF BIOCHEM, PA STATE UNIV, UNIVERSITY PARK, 66- *Mem:* Am Chem Soc; Am Soc Biol Chem; Soc Complex Carbohydrates. *Res:* Biochemistry of carbohydrates; mechanism of enzyme action; structure of enzymes; immunochemistry. *Mailing Add:* Penn State Univ University Park PA 16802

PE, MAUNG HLA, b Mandalay, Burma, Nov 14, 20; m 57. PHYSICS. *Educ:* Univ Rangoon, BSc, 46; Lehigh Univ, MS, 51, PhD(elec eng), 57; Polytech Inst Brooklyn, MS, 52; NY Univ, MNuclearEng, 57. *Prof Exp:* Lectr elec eng, City Col New York, 57-61; ASSOC PROF PHYSICS, MANHATTAN COL, 61- *Mem:* Assoc mem Physicists in Med; NY Acad Sci; AAAS; Sigma Xi. *Res:* Magnetohydrodynamics; plasma physics; biophysics. *Mailing Add:* 240 W 261 St Bronx NY 10471

PEABODY, DWIGHT VAN DORN, JR, b Elyria, Ohio, July 19, 24; m 74; c 4. WEED SCIENCE. *Educ:* Ohio State Univ, BS, 49; Wash State Univ, MS, 51. *Prof Exp:* Asst, Wash State Univ, 50-51; asst agronomist, 51-66, assoc agronomist & exten weed scientist, 66-85, EMER WEED SCIENTIST, NORTHWESTERN WASH RES & EXTEN UNIT, WASH STATE UNIV, 85- *Concurrent Pos:* Consult, 71- *Honors & Awards:* Weed Warrior Award, Wash State Weed Asn. *Mem:* Weed Sci Soc Am; Int Weed Sci Soc. *Res:* Weed control in agronomic and horticultural crops. *Mailing Add:* Rte 1 Box 1917 Lopez WA 98261

PEABODY, FRANK ROBERT, b Birmingham, Mich, Oct 13, 20; m 47; c 2. MICROBIOLOGY, ENVIRONMENTAL HEALTH. *Educ:* Mich State Univ, BS, 42, MS, 48, PhD(microbiol), 52. *Prof Exp:* From instr to assoc prof, 48-75, prof microbiol, 75-87, assoc chmn dept, 78-86, EMER PROF MICROBIOL, MICH STATE UNIV, 87- *Concurrent Pos:* Consult, Nat Automatic Merchandising Asn, 57-; adv to Repub SKorea, Int Coop Admin, US Dept State, Washington, DC, 59-60; mem, Coun Pub Health Consults, Nat Sanit Found, 79-85. *Mem:* Am Soc Microbiol; Int Asn Milk, Food & Environ Sanitarians. *Res:* Antiseptics and disinfectants; hospital environment; microbial indicators of water pollution. *Mailing Add:* 3430 Salling Beach Gaylord MI 49735-9464

PEACE, GEORGE EARL, JR, b Norfolk, Va, Feb 4, 45; m 71; c 2. ANALYTICAL CHEMISTRY. *Educ:* Lafayette Col, BS, 66; Univ Ill, Urbana, MS, 68, PhD(analytical chem), 71. *Prof Exp:* Res asst analytical chem, Univ Ill, Urbana, 68-71; from asst prof to assoc prof analytical chem, Lafayette Col, 71-79, chief adv health professions, 75-79; ASSOC PROF ANALYTICAL CHEM, COL OF THE HOLY CROSS, 79- *Concurrent Pos:* Fac fel res grant, 71-72 & 74-75; environ sci & eng fel, AAAS & Environ Protection Agency, 79; vis assoc prof, Colo St Univ, 85-86; fel, Am Coun Educ, 90-91. *Mem:* Soc Appl Spectros; AAAS. *Res:* Teaching analytical chemistry; trace metal analysis. *Mailing Add:* Dept of Chem Col of the Holy Cross Worcester MA 01610-2395

PEACEMAN, DONALD W(ILLIAM), b Miami, Fla, June 1, 26; m 52; c 2. CHEMICAL ENGINEERING, MATHEMATICS. *Educ:* City Col New York, BChE, 47; Mass Inst Technol, ScD, 52. *Prof Exp:* Instr chem eng, Mass Inst Technol, 50; sr res adv, Exxon Prod Res Co, 51-86; CONSULT, 86- *Honors & Awards:* Robert Earll McConnell Award, Am Inst Mining, Metall & Petrol Engrs, 79, Lucas Award, 91; Reservoir Eng Award, Soc Petrol Engrs, 85. *Mem:* Asn Comput Mach; Soc Indust & Appl Math; Soc Petrol Engrs. *Res:* Liquid-side resistance in gas absorption; petroleum production; numerical analysis; petroleum reservoir simulation. *Mailing Add:* 4907 Glenmeadow Houston TX 77096

PEACH, MICHAEL EDWIN, b Nottingham, Eng, May 7, 37; m 64; c 2. INORGANIC CHEMISTRY. *Educ:* Cambridge Univ, BA, 59, PhD(chem), 62, MA, 64. *Prof Exp:* NATO fel, Graz Tech Univ, 62-63; fel, Univ Gottingen, 63-64, res assoc chem, 64-65; asst prof, Dalhousie Univ, 65-66; lectr, Loughborough Univ Technol, 66-67; from asst prof to assoc prof, 67-76, dir grad studies & res, 76-90, PROF CHEM, ACADIA UNIV, 76- *Concurrent Pos:* Guest prof, Univ Wurzburg, 74-75 & 80-81; Alexander Von Humboldt Found fel, 74-75 & 80-81; vis prof, Univ Calgary, 87-88. *Mem:* Chem Inst Can. *Res:* Chemistry of divalent sulfur compounds, particularly derivative of halogenated thiols and the chemistry of non-aqueous solvent systems. *Mailing Add:* Dept Chem Acadia Univ Wolfville NS B0P 1X0 Can

PEACH, MICHAEL JOE, b Morgantown, WVa, Aug 22, 40; m 66; c 1. PHARMACOLOGY, PHYSIOLOGY. *Educ:* Shepherd Col, BS, 63; WVa Univ, MS, 65, PhD(pharmacol), 68. *Prof Exp:* Instr pharmacol, WVa Univ, 67; from asst prof to assoc prof, 68-76, PROF PHARMACOL, SCH MED, UNIV VA, 76- *Concurrent Pos:* Nat Heart & Lung Inst fel, Res Div, Cleveland Clin, 67-68; assoc dean res, Sch Med, Univ Va. *Mem:* AAAS; Am Soc Pharmacol & Exp Therapeut; Soc Neurosci. *Res:* Physiology and pharmacology of the reninangiotensin system and interaction with the sympathoadrenal system; angiotensin and control of aldosterone; arterial hypertension. *Mailing Add:* Dept Pharmacol Box 448 Health Sci Ctr Univ Va Sch Med Charlottesville VA 22908

PEACH, PETER ANGUS, geology; deceased, see previous edition for last biography

PEACH, ROY, b Runcorn, Eng. HISTOLOGY. *Educ:* Manchester Univ, Eng, BS, 56, MS, 57, PhD(biophys), 60. *Prof Exp:* Spec fel rheumatism res, Manchester Univ, 59-61, asst lectr biophys, Dept Anat, 61-64, lectr, 64-67; asst prof oral biol, Sch Dent, 67-69, asst prof anat, Sch Med, 68-74, ASSOC PROF ANAT, SCH MED, UNIV NC, CHAPEL HILL, 74-, ORAL BIOL, SCH DENT, 69- *Mem:* Am Asn Anatomists; Anat Soc Gt Brit & Ireland; Electron Micros Soc Am; Int Asn Dent Res; Royal Micros Soc. *Res:* Development and growth, particularly of facial region, in vitro and in vivo, using mainly microscopic techniques, light and electron. *Mailing Add:* Sch Dent Univ NC Chapel Hill NC 27514

PEACHEY, LEE DEBORDE, b Rochester, NY, Apr 14, 32; m 58; c 3. CELL BIOLOGY, PHYSIOLOGY. *Educ:* Lehigh Univ, BS, 53; Rockefeller Univ, PhD(biophys), 59. *Hon Degrees:* MA, Univ Pa, 71. *Prof Exp:* From asst prof to assoc prof zool, Columbia Univ, 59-65; assoc prof biochem & biophys, 65-71, PROF BIOL, UNIV PA, 70- *Concurrent Pos:* NSF grants, Columbia Univ, 60-65 & Univ Pa, 65-72; Muscular Dystrophy Asn grant, Univ Pa, 73-90 & NIH grant, 73-; Guggenheim & Fulbright-Hayes fels, Cambridge Univ, 67-68; mem molecular biol study sect, NIH, 69-73; adj prof molecular, cellular & develop biol, Univ Colo, 69-; Fogarty sr int fel, Univ Col London, 79-80, guest res fel, Royal Soc London, Cambridge, 86; mem coun, Int Union Pure & Appl Biophys, 78-84 & vpres, 84-87, pres, 87-90 & chmn, Comn Cell & Membrane Biophys, 81-84; mem, Mayor's Sci & Technol Adv Coun, Philadelphia, 72-; mem, Physiol Soc, Eng, 87- *Mem:* Fel AAAS; Am Soc Cell Biol; Biophys Soc (pres, 80-81); Electron Micros Soc Am (pres, 82); Int Union Pure & Appl Biophysics; Soc Gen Physiologists; Int Soc Stereology. *Res:* Structure and function of muscle cells; electron microscopy; muscle physiology. *Mailing Add:* Dept Biol Univ Pa Philadelphia PA 19104-6018

PEACOCK, ERLE EWART, b Durham, NC, Sept 10, 26; m 54; c 2. SURGERY. *Educ:* Univ NC, cert, 47; Harvard Univ, MD, 49; Am Bd Surg, dipl. *Prof Exp:* Lab asst zool, Univ NC, 44-45; assoc plastic surg, Sch Med, Wash Univ, 55-56; from instr to prof surg, Sch Med, Univ NC, 56-69; prof & chmn surg fac, Col Med, Univ Ariz, 69-74; PROF SURG, TULANE UNIV, 77- *Concurrent Pos:* Dir hand rehab ctr, Univ NC, Chapel Hill, 65-69; mem surg study sect, NIH, mem surg training grants study sect; chmn, Plastic Surg Res Coun, 67; consult, Watts Hosp, Durham, NC, Vet Admin Hosp, Fayetteville & Womack Army Hosp, Ft Bragg, NC; chmn, Am Bd Plastic Surg, 75-76; US Army Surgeon Gen's adv panel consult in trauma. *Mem:* Soc Plastic & Reconstruct Surg; Soc Surg of Hand; Am Col Surgeons; Am Surg Asn; Soc Univ Surgeons. *Res:* Tissue transplantation and wound healing; connective tissue research; plastic and reconstructive surgery. *Mailing Add:* 109 Conner Dr Suite 7204 Chapel Hill NC 27514

PEACOCK, HUGH ANTHONY, b Cairo, Ga, May 30, 28; m 49; c 3. PHYTOPATHOLOGY, PLANT BREEDING & GENETICS. *Educ:* Univ Fla, BSA, 52, MSA, 53; Iowa State Univ, PhD(plant breeding), 56. *Prof Exp:* Asst breeding, Iowa State Univ, 55-56; asst agronomist, Univ Fla, 57-58; res agronomist, Exp Sta, Univ Ga & USDA, 59-73; AGRONOMIST & DIR AGR RES & EDUC CTR, AGR EXP STA, UNIV FLA, 73- *Honors & Awards:* Sigma Xi. *Mem:* Am Soc Agron; Am Genetics Asn; Am Soybean Asn; Crop Sci Soc Am. *Res:* Cotton, soybean, corn, grain sorghum and peanut culture and management. *Mailing Add:* Agr Res & Educ Ctr Rt 3 Box 575 Jay FL 32565

PEACOCK, JOHN TALMER, b Madison, Ga, Aug 5, 31; m 57; c 2. PLANT ECOLOGY. *Educ:* Maryville Col, BS, 53; Univ Ala, MS, 55; Univ Tex, PhD(bot), 63. *Prof Exp:* Asst bot, Univ Ala, 54-55; instr biol, Tex Col Arts & Indust, 55-60; res assoc plant ecol, Univ Tex, 61-62; from asst prof to assoc prof, 62-66, chmn dept, 68-76, dean, Col Arts & Sci, 76-89, PROF BIOL, TEX A & I UNIV, 66- *Mem:* AAAS; Am Inst Biol Sci; Bot Soc Am; Ecol Soc Am; Sigma Xi. *Res:* Physiological ecology of southwestern shrubs, particularly mesquite. *Mailing Add:* Dept Biol Tex A & I Univ Kingsville TX 78363

PEACOCK, LELON JAMES, b Brevard, NC, May 25, 28; m 45; c 3. COMPARATIVE PSYCHOLOGY. *Educ:* Berea Col, AB, 50; Univ Ky, MS, 52, PhD(psychol), 56. *Prof Exp:* Psychophysiologist, US Army Med Res Lab, 54-56; res assoc psychol, Yerkes Labs Primate Biol, 56-58, actg dir, 58-59; from asst prof to assoc prof, 59-65, prof, 66-90, EMER PROF PSYCHOL, UNIV GA, 90- *Concurrent Pos:* Pres, Southern Soc Philos & Psychol, 74. *Mem:* AAAS; Am Psychol Asn; Soc Psychophysiol Res; Soc Neurosci; Psychonomic Soc; Am Psychol Soc. *Res:* Psychophysiology of learning; radiation effects on behavior; measurement of general activity; instrumentation; neuropsychology. *Mailing Add:* Dept of Psychol Univ of Ga Athens GA 30602

PEACOCK, MILTON O, b West Monroe, La, Aug 31, 16; m 56; c 4. BIOCHEMISTRY, NUTRITION. *Educ:* La State Univ, BS, 43, MS, 45; Univ Miss, BS, 52; Univ Ala, PhD(biochem), 63; Anglo-Am Inst Drugless Ther, Scotland, ND(naturopathy), 80. *Prof Exp:* Mathematician, Ballistics Res Lab, Md, 45-47; asst prof math & physics, Evansville Col, 47-49; assoc prof biol, Miss Southern Col, 49-51; prof pharmacol & chmn, Sch Pharm, Northeast La State Col, 54-57; prof chem & head dept, Ark A&M Col, 63-67; prof chem & pharmacol & chmn, Div Sci & Math, Whitworth Col, Miss, 67-72 & 75-76; assoc prof pharmacol, La Tech Univ, 76-78; PRES & DIR RES & INSTR MATH & SCI, INST SCI & MATH, LA, 72-, SCI HEALTH COUNSELING, 75-; NUTRIT BIOCHEMIST & CONSULT, NUTRIT SCI CLIN, 80-; PHYSICIAN, 80- *Concurrent Pos:* NSF fel, 53; radiol defense officer, 74-; nutrit scientist & naturopathic med consult, Nutrit Sci Metab Ctr, 87; pres, Evangel Christian Univ Am & dean, Sch Med Sci & Naturopathic Med, 88. *Honors & Awards:* Zinn Award, 43. *Mem:* Fel Am Inst Chemists; Sigma Xi; Am Chem Soc; Am Pharmaceut Asn. *Res:* Mathematical analysis of interferometer of n plates; mathematical description and correlation of physical phenomena; chromogenic reactions of divalent S compounds with ammonia, amines, amides, and others; magnetotropism of germinating plant embryos; mathematical models of serial dilution; gradient change in concentration, extraction and countercurrent exchange phenomena; mathematical models for natural phenomena and analytical chemical processes; nutritional therapy; definition and description of non-drug medicinal substances and locus of healing (biological repair); causal levels of disease and of action of medicinal substances; design of medical regimens for elimination of in-depth cause of disease and thereby for therapeutic intervention and in preventive maintenance of health; medical doctor. *Mailing Add:* Inst Sci & Math 1125 Red Cut Loop Rd West Monroe LA 71292

PEACOCK, PETER N B, b Nairobi, Kenya, Nov 3, 21; m 49; c 3. EPIDEMIOLOGY. *Educ:* Univ Cape Town, MB, ChB, 45, dipl pub health, 47; Royal Col Physicians & Surgeons, dipl indust health, 53; Univ Witwatersrand, DTM&H, 57, MD, 69; Samford Univ, MA, 70; FRCP(C), 71. *Prof Exp:* Med inspector, Dept Health, Union SAfrica, 47-49, asst chief health officer, 49-52; regional med health officer, Govt Sask, 53-55; prof pub health, Univ Witwatersrand, 56-61; asst prof pub health, Univ Sask, 61-65; prof pub health & epidemiol, Med Ctr, Univ Ala, Birmingham, 65-76; CLIN PROF MED, UNIV S ALA, 73- *Concurrent Pos:* Med officer health, City of Germiston, 56-61; regional med health officer, Govt Sask, 61-65; mem, Nat Adv Comt Stoke Epidemiology, Cancer Epidemiol & Bioradiation; clin prof pub health, Med Col, Cornell Univ, 72-79; chief epidemiol, Am Health Found, New York, 72-75; dist med officer, Refugee Prog, Malawi, E Africa, 87; chief prof serv, Ala Medicaid, 90-91. *Mem:* Am Pub Health Asn; Can Med Asn; Can Pub Health Asn; fel Royal Soc Trop Med & Hyg; SAfrican Pub Health Asn (pres, 60-61). *Res:* Medical epidemiology; rheumatic fever; bilharzia; typhoid; occupational health; tick paralysis; poliomyelitis; physical fitness; mercury poisoning; the distribution of health services; statistical epidemiology relating to community health; air pollution; stroke. *Mailing Add:* 2150 Cahaba Valley Rd Indian Springs AL 35124

PEACOCK, RICHARD WESLEY, b Scottsburg, Ind, Sept 24, 39. PHYSICS. *Educ:* Tulane Univ, BS, 61, MS, 66, PhD(physics), 70. *Prof Exp:* Instr & head dept math, Fishburne Mil Sch, 73-76; physicist, US Navy Eastern Standards Lab, 76-77; tech info specialist, Langley Res Ctr, NASA, 77-85; TECH INFO SPECIALIST, NAVAL RES LAB, 85- *Mem:* Am Phys Soc; AAAS. *Res:* Three particle system theory; neutron-deutron scattering as a three particle system; weak interactions and basic nuclear systems. *Mailing Add:* 2240 White Cornus Lane Reston VA 22091

PEACOCK, ROY NORMAN, b Sandpoint, Idaho, Mar 19, 30; m 57; c 2. INSTRUMENTATION. *Educ:* Univ Ore, BA, 52, MA, 53; Univ Ill, PhD(physics), 58. *Prof Exp:* From asst prof to assoc prof physics, Univ Ill, Urbana, 62-71; scientist, Granville-Phillips Co, 71-72, vpres res & eng, 72-76; CO-FOUNDER & VPRES DEVELOP, HPS DIV MKS INSTRUMENTS, 76- *Mem:* Am Vacuum Soc; Am Phys Soc; Am Soc Metals; Am Welding Soc. *Res:* Vacuum techniques; solid surfaces; thin films. *Mailing Add:* 8845 Elgin Dr Lafayette CO 80026

PEACOCK, SAMUEL MOORE, JR, b Philadelphia, Pa, May 8, 22. NEUROPHYSIOLOGY. *Educ:* Princeton Univ, AB, 44; Univ Pa, MD, 48. *Prof Exp:* Intern, Bryn Mawr Hosp, Pa, 48-49; from instr to asst prof neurol, Tulane Univ, 49-57; sr med res scientist neurophysiol, Dept Clin Res, Eastern Pa Psychiat Inst, 57-80; asst prof psychiat, Univ Pa, 59-81; assoc prof psychiat, Jefferson Med Col, 75-83; DIR SLEEP LABS, UNIV SERV, 86- *Concurrent Pos:* Am Col Physicians fel, Tulane Univ, 49-50; USPHS fel, 50-51, USPHS fel, Nat Found Infantile Paralysis, 51-53; Grass Trust fel, Marine Biol Lab, Woods Hole, 51; res assoc, Dept Neurosurg, Pa Hosp, Philadelphia, 61-63. *Mem:* AAAS; Am Physiol Soc; Am Acad Neurol; Soc Biol Psychiat; AMA; Am Acad Clin Neurophysiology. *Res:* Physiology of motor system; neurophysiology of behavior; electroencephalography and evoked response averaging; sleep disorders. *Mailing Add:* Lower Pine Creek Rd Chester Springs PA 19425

PEACOCK, VAL EDWARD, b Sioux City, Iowa, Oct 25, 51; m 78. ORGANIC CHEMISTRY. *Educ:* Iowa State Univ, BS, 73; Univ Wis-Madison, PhD (org chem), 78. *Prof Exp:* Res assoc org chem, Dept Agr chem, Ore State Univ, 78-80; mem staff, Phillip Morris Res Ctr, 80-; AT SEVEN-UP CO; SECT MGR, ANHEISER BUSCH CO INC, 89- *Mem:* Am Chem Soc. *Res:* Physical organic chemistry. *Mailing Add:* Anheiser Busch Co Inc One Busch Pl St Louis MO 63118

PEACOR, DONALD RALPH, b Somerville, Mass, Feb 15, 37; m 60; c 3. MINERALOGY. *Educ:* Tufts Univ, BS, 58; Mass Inst Technol, SM, 60, PhD(crystallog), 62. *Prof Exp:* From instr to assoc prof geol, 62-71, PROF GEOL, UNIV MICH, 71- *Mem:* Mineral Soc Am; Am Crystallog Asn. *Res:* X-ray crystallography of minerals; general mineralogy; high temperature x-ray diffraction. *Mailing Add:* Dept Geol Sci Univ Mich Ann Arbor MI 48109

PEAK, DAVID, b Brooklyn, NY, Nov 28, 41; m 83. THEORETICAL PHYSICS. *Educ:* State Univ NY Col New Paltz, BS, 65; State Univ NY Albany, PhD(physics), 69. *Prof Exp:* Res assoc physics, State Univ NY Albany, 69-71, instr, 71-75; from asst prof to prof, 75-89, BAILEY PROF PHYSICS, UNION COL, 89- *Concurrent Pos:* Res fel, Inst Study Defects in Solids, Albany, NY, 75-; vis fel, Princeton Univ, 78-79; NSF prof develop award, 78-79; vis scientist, Argonne Nat Lab, 83-84; founding mem & secy, Coun Undergrad Res, 85-89; bd mem, Nat Conf Undergrad Res, 88-; mem, Sci, Math & Eng Educ Comt, Sigma Xi, 89- *Mem:* Am Phys Soc; Am Asn Physics Teachers; Sigma Xi. *Res:* Irreversible thermodynamics; ion-solid interactions, condensed phase kinetics-fluorescence quenching in liquids; impurity complexing in semiconductors; corrosion; formation and processing of cosmic dust. *Mailing Add:* Dept Physics Union Col Schenectady NY 12308

PEAK, MEYRICK JAMES, b Southend, Eng, June 29, 37; m 61; c 2. RADIATION BIOLOGY, DNA DAMAGE & REPAIR. *Educ:* Univ Cape Town, SAfrica, BSc, 60; Univ Calif, PhD(biol), 72. *Prof Exp:* Lectr biol, Univ Cape Town, SAfrica, 63-70; sr lectr biochem, Rhodes Univ, SAfrica, 73-79; scientist, 80-86, SR SCIENTIST, BIOL MED RES DIV, ARGONNE NAT LAB, 87- *Concurrent Pos:* Vis prof, Univ Rio de Janeiro, 76; fel, 70-72, vis scientist, Div Biol Med Res, Argonne Nat Lab, 78; vis prof, Univ Cape Town, 86. *Mem:* Am Soc Photobiol; Radiation Res Soc; Europ Soc Photobiol. *Res:* Photobiology; radiobiology; algal biochemistry; mutagenesis; DNA damage and repair. *Mailing Add:* Biol Ecol Res Div Argonne Nat Lab 9700 S Cass Ave Argonne IL 60439

PEAK, WILFERD WARNER, b Los Angeles, Calif, Jan 17, 24; m 51; c 4. GEOLOGY, ENGINEERING GEOLOGY. *Educ:* Univ Calif, Los Angeles, BA, 48. *Prof Exp:* From jr geologist to assoc geologist, Planning Div, Calif State Dept Water Resources, 48-56, assoc to sr geologist, Design & Construct Div, 56-64, chief geologist, Div Safety of Dams, 64-84; COMNR, CALIF SEISMIC SAFETY COMN, 84- *Concurrent Pos:* Mem, Calif State Bd Registrn Geologists, 69-76, pres, 69-72. *Mem:* Fel Geol Soc Am; Asn Eng Geologists (secy, 64); Earthquake Eng Res Inst. *Res:* Methods of utilization of new techniques and tools to investigate foundations of dams and appurtenant hydraulic structures so as to ensure competency of those structures. *Mailing Add:* 6360 Eichler St Sacramento CA 95831

PEAKALL, DAVID B, b Purley, Eng, Mar 17, 31; m 59; c 4. ENVIRONMENTAL SCIENCES. *Educ:* Univ London, BSc, 52, MSc, 54, PhD(chem), 56, DSc, 79. *Prof Exp:* Sci officer, Ministry Supply, Eng, 56-57; Cabot fel surface chem, State Univ NY Agr & Tech Col, Alfred Univ, 57-59; chemist, Distillers Co, Ltd, Eng, 59-60; res assoc protein chem, State Univ NY Upstate Med Ctr, 60-62, asst prof pharmacol, 62-66; Am Heart Asn estab investr, 66-71; sr res assoc, Langmuir Lab, Div Ecol & Systs, Cornell Univ, 68-75; chief wildlife toxicol div, Can Wildlife Serv, 75-91; RETIRED. *Concurrent Pos:* Consult, US-USSR Environ Health Prog, Nat Insts Environ Health Sci, 72-76; mem expert group, Orgn Econ Coop & Develop Ecotoxicol, 78- *Mem:* Am Ornithologists Union; Am Physiol Soc; Ornithologists Union; Soc Environ Toxicol & Chem. *Res:* Effect of pollutants on avian physiology and reproduction; regulation of protein synthesis in spiders. *Mailing Add:* 17 Saint Mary's Rd Wimbledon London SW19 7BZ England

PEAKE, CLINTON J, b Hancock, NY, Sept 11, 32; m 55; c 3. ORGANIC CHEMISTRY, BIOLOGICAL CHEMISTRY. *Educ:* Harpur Col, AB, 58; Univ Notre Dame, PhD(org chem), 62. *Prof Exp:* Res chemist, FMC Corp, 62-63, res chemist, Niagara Chem Div, 63-68, sr res chemist, 68-80, SR RES CHEMIST, AGR CHEM GROUP, 80- *Mem:* Am Chem Soc; Soc Nematol. *Res:* Pesticide research in the nematicides area; blood chemistry, especially hemostasis. *Mailing Add:* 29 Fran Ave Trenton NJ 08628

PEAKE, EDMUND JAMES, JR, b Omaha, Nebr, July 5, 38; m 64; c 2. ALGEBRA. *Educ:* NMex State Univ, BS, 60, MS, 62, PhD(math), 63. *Prof Exp:* ASST PROF MATH, IOWA STATE UNIV, 63- *Mem:* Math Asn Am. *Res:* Abelian groups; homological and universal algebra; applications to computer science. *Mailing Add:* Dept Math Iowa State Univ Ames IA 50011

PEAKE, HAROLD J(ACKSON), b Norton, Va, Dec 7, 20; m 42; c 1. ELECTRONICS. *Educ:* Va Polytech Inst, BS, 42; Univ Md, MS, 53; George Washington Univ, MEA, 69. *Prof Exp:* Radio engr, Radio Div, US Naval Res Lab, Washington, DC, 42-49, electronics scientist, Solid State Div, 49-58; res engr & head, Flight Systs Br, Goddard Space Flight Ctr, NASA, Greenbelt, Md, 58-65, assoc chief, Spacecraft Technol Div, 65-69, chief, Electronics Div, 69-74, chief, Off Nat Needs, 74-80; RETIRED. *Concurrent Pos:* Lectr, George Washington Univ, 43-44, 57-59, prof lectr, 78-; lectr mgt, Am Univ, 75-78, prof lectr, 78- *Mem:* Sci Res Soc Am; fel Inst Elec & Electronics Engrs; Sigma Xi. *Res:* Radio communications systems radio wave propagation; circuit analysis; cathode-ray oscillography; satellite and space probe instrumentation and telemetry; space electronics; management and supervision. *Mailing Add:* Dept of Eng Mgt George Washington Univ Washington DC 20052

PEAKE, ROBERT LEE, b Evansville, Ind, July 6, 35; m 57; c 2. INTERNAL MEDICINE, ENDOCRINOLOGY. *Educ:* Ind Univ, AB, 57, MD, 60. *Prof Exp:* Intern med, Sch Med, Ind Univ, 60-61; resident internal med, 61-62 & 64-65; asst prof, 68-73, ASSOC PROF INTERNAL MED, UNIV TEX MED BR, GALVESTON, 73- *Concurrent Pos:* NIH trainee fel endocrinol & metab, Med Ctr, Ind Univ, 65-68. *Mem:* Am Fedn Clin Res; Endocrine Soc; Am Col Physicians; Am Thyroid Asn. *Res:* Thyroglobulin proteolysis; thyroid hormone release. *Mailing Add:* 1501 Broadway Galveston TX 77550

PEAKE, THADDEUS ANDREW, III, b Louisville, Ky, Jan 6, 48; m 76; c 3. AIR POLLUTION & AIR TOXICS ENGINEERING. *Educ:* Univ Louisville, BS, 78, MEng, 83; Kennesaw State Col, MBA, 89. *Prof Exp:* Engr, US Environ Protection Agency, 79-91; PRES, PEAKE ENG, INC, 88- *Concurrent Pos:* Eng duty officer, USNR, 67- *Mem:* Am Acad Environ Engrs; Am Consult Eng Coun. *Res:* Air pollution and air toxics control engineering for industrial facilities; dispersion modeling; control technology evaluation; emission inventories. *Mailing Add:* 3111 Vandiver Dr NW Marietta GA 30066

PEAKE, WILLIAM TOWER, b Oak Park, Ill, Nov 26, 29; m 52; c 2. ELECTRICAL ENGINEERING. *Educ:* Mass Inst Technol, SB, 51, SM, 53, ScD, 60. *Prof Exp:* Asst elec eng, Mass Inst Technol, 51-53; proj engr, Wright Air Develop Ctr, US Air Force, 53-56; from instr to assoc prof, 56-76, PROF ELEC ENG, MASS INST TECHNOL, 76- *Concurrent Pos:* Res assoc otol, Mass Eye & Ear Infirmary, 63- *Mem:* Inst Elec & Electronics Eng; Acoust Soc Am; Sigma Xi. *Res:* Sensory communication, particularly signal transmission in the ear. *Mailing Add:* Rm 36-825 Mass Inst of Technol Cambridge MA 02139

PEALE, STANTON JERROLD, b Indianapolis, Ind, Jan 23, 37; m 60; c 2. SOLAR SYSTEM PHYSICS, ASTROPHYSICS. *Educ:* Purdue Univ, BS, 59; Cornell Univ, MS, 62, PhD(eng physics), 65. *Prof Exp:* Res assoc space res, Cornell Univ, 64-65; asst prof astron & geophys, Univ Calif, Los Angeles, 65-68; from asst prof to assoc prof, 68-76, PROF PHYSICS, UNIV CALIF, SANTA BARBARA, 76- *Concurrent Pos:* Res Corp Cottrell res grant, 69-; NASA res grant planetary geophys, 70-; consult sci adv group explor outer solar syst, Jet Propulsion Lab, 71-72; vis fel, Joint Inst for Lab Astrophys, Univ Colo, 72-73, 79-80; NASA-Ames Univ Consortium grant, 76-81; mem, NASA Lunar & Planetary Rev Panel, 79-81, 86-89, NASA Planetary Systs Sci Working Group, 88- & Nat Acad Sci Space Sci Bd Comt on Planetary & Lunar Exp, 80-84. *Honors & Awards:* Newcomb Cleveland Prize, AAAS, 80; Except Sci Achievement Medal, NASA, 80; James Craig Watson Award, Nat Acad Sci, 81. *Mem:* Fel AAAS; Am Astron Soc; fel Am Geophys Union; Int Astron Union. *Res:* Small particles in space; dynamics of planetary spins; tidal evolution; nature of the planetary interiors; star formation and origin of the solar system; small bodies in the solar system; asteroid named Peale 3612. *Mailing Add:* Dept Physics Univ Calif Santa Barbara CA 93106

PEANASKY, ROBERT JOSEPH, b Menominee, Mich, Oct 18, 27; m 53; c 5. BIOCHEMISTRY, ENZYMOLOGY. *Educ:* Marquette Univ, BS, 51, MS, 53; Univ Wis, PhD(biochem), 57. *Prof Exp:* NSF fel & trainee, Sch Med, Marquette Univ, 57-60, asst prof biochem, 60, assoc prof, 64-67; assoc prof, 67-70, chmn, 86-89, PROF BIOCHEM, SCH MED, UNIV SDAK, 70- *Concurrent Pos:* USPHS career develop award, 60-67. *Mem:* Am Chem Soc; Am Soc Biochem & Molecular Biol; Soc Exp Biol & Med; Sigma Xi. *Res:* Purification and properties of proteins; mechanism of interaction between protein inhibitors and proteolytic enzymes; biochemistry of protease inhibitors from Ascaris. *Mailing Add:* Dept Biochem & Molecular Biol Univ SDak Sch Med Vermillion SD 57069-2390

PEARCE, CHARLES WALTER, b Philadelphia, Pa, May 22, 47; m 66; c 3. ELECTRICAL ENGINEERING. *Educ:* Univ Nebr, BS, 69; Lehigh Univ, MS, 74. *Prof Exp:* SR STAFF ENGR SILICON MAT, AT&T TECHNOL SYSTS, 69- *Mem:* Electrochem Soc; Am Soc Test & Mat. *Res:* Measurement of electrical, optical and microperfection properties of silicon; mos fabrication technology. *Mailing Add:* 127 E St Joseph Easton PA 18044

PEARCE, DAVID ARCHIBALD, b Montreal, Que, Nov 12, 20; US citizen; m 50; c 2. AGRICULTURAL CHEMISTRY. *Educ:* Sir George Williams Univ, BS, 52. *Prof Exp:* Chief chemist, Green Cross Div, Sherwin-Williams Co Can, 52-56; res chemist, Chemagro Corp, Mo, 57-65; formulation res chemist, Mobil Chem Co, 65-67, sr res chemist, res, develop & eng div, 67-81; RETIRED. *Mem:* Am Chem Soc. *Res:* Apparatus for comparing emulsions; pesticide preparation and formulation; formulation research, insecticides, fungicides, herbicides; product development. *Mailing Add:* 22 Finley Rd Edison NJ 08817

PEARCE, DAVID HARRY, b Newport News, Va, July 20, 43; m; c 1. BIOMEDICAL ENGINEERING, ELECTRICAL ENGINEERING. *Educ:* Va Polytech Inst & State Univ, BS, 66; Univ Va, PhD(biomed eng), 72. *Prof Exp:* NIH cardiovasc training grant, Univ Miss, 72-74, NIH young investr pulmonary award, 74-76, instr physiol, 73-74, asst prof physiol, Med Ctr, 74-, biomed engr, Miss Methodist Rehab Ctr, 76-82; vpres, Bobby J Hall & Assoc,

82-87; DIR, BIOMED ENG, MISS METHODIST REHAB CTR, 87- *Mem:* Biomed Eng Soc; Inst Elec & Electronic Engrs; Sigma Xi; Am Soc Hosp Eng; Asn Advan Med Instrumentation. *Res:* Biomedical engineering as applied to medical instrumentation, respiratory and exercise physiology, mathematical modeling and simulation, biological control theory; real time data processing of physiological signals; rehabilitation engineering. *Mailing Add:* Miss Methodist Rehab Ctr 1350 E Woodrow Wilson Dr Jackson MS 39216-5112

PEARCE, ELI M, b Brooklyn, NY, May 1, 29; m 80; c 2. POLYMER CHEMISTRY. *Educ:* Brooklyn Col, BS, 49; NY Univ, MS, 51; Polytech Inst NY, PhD(polymer chem), 58. *Prof Exp:* Res chemist biochem, NY Univ, Bellevue Med Ctr, 49-53, Army Med Res Lab, 53-55 & E I du Pont de Nemours & Co, Carothers Res Lab, 58-62; sect head, polymer chem, J T Baker Chem Co, 62-68; mgr tech supvr consult, polymer chem, Allied Chem Corp, 68-73; dir polymer sci, Camille Dryfus Lab, 73-74; head chem dept, 76-82, PROF POLYMER CHEM & CHEM ENG, 71- , DEAN ARTS & SCI, 82-, DIR POLYMER RES INST, POLYTECH UNIV NY, 81- *Concurrent Pos:* From assoc ed to ed, J Polymer Sci, 66-88; adv bd, Macromolecular Synthesis, 70, Nat Mat Adv Bd, 75-78 & Petrol Res Fund, 82-84; mem, Comt Polymer Sci & Eng, Nat Res Coun, Nat Acad Sci, 80-81; adv comt, Nat Bur Standards Fire Res, 84- & Mat Adv Comt, Los Alamos Nat Lab, 83-; dir, Polymer Res Inst, Polytech Univ NY, 76-78 & NSF-MRG, 85-; coun, Am Chem Soc, 82- *Honors & Awards:* Int Educ Award, Soc Plastics Eng, 88. *Mem:* Fel AAAS; Am Chem Soc; Fel Am Inst Chem; Fel NY Acad Sci; Fel Soc Plastics Eng; Fel NAm Thermal Anal Soc. *Res:* Polymer synthesis; structure property relationships; polymer degradation; polymer flammability; polymers as chemical reagents; polyamides, polyesters, epoxy resins; polymer compatibility; photo-resists; polymer compatability. *Mailing Add:* Polytech Univ 333 Jay St Brooklyn NY 11201

PEARCE, FRANK G, b Terre Haute, Ind, Sept 17, 18; m 42; c 3. CHEMICAL ENGINEERING. *Educ:* Rose Polytech Inst, BS, 40; Mass Inst Technol, 46. *Prof Exp:* Chem engr, Am Mach & Foundry Co, 46-47; supvr, Res Sect, Pan Am Petrol Corp Div, Standard Oil Co, Inc, 47-58, dir proj eng, Amoco Chem Corp Div, 58-60, div dir res, 60-62, coordr res & develop, 62-64, dir data systs planning, Ill, 64-67, dir info serv & comput sci, 67-70, gen mgr, Info Serv & mgt sci, 70-79; RETIRED. *Mem:* Am Chem Soc; Am Inst Chem Engrs. *Res:* Process design and economics; economics in petrochemicals; industrial and information services management. *Mailing Add:* PO Box 2567 El Macero CA 95618

PEARCE, FREDERICK JAMES, PHYSIOLOGY, SHOCK. *Educ:* Univ Rochester, PhD(physiol), 77. *Prof Exp:* Asst prof surg & physiol, Sch Med, Univ Rochester, 80-90; ASST PROF, UNIFORM SERV HEALTH SCI, 90-; CHIEF, DEPT EXP SURG, DIV SURG, WALTER REED ARMY INST RES, 90- *Res:* Cardiovascular aspect of hemorrhagic septic shock & metabolic aspects. *Mailing Add:* Bldg 40 Rm 3007 Washington DC 20307

PEARCE, GEORGE WILLIAM, b St John's, Nfld, Dec 27, 42; m 69; c 1. GEOPHYSICS, GEOLOGY. *Educ:* Mem Univ Nfld, BSc, 65, MSc, 67; Univ Toronto, PhD(geophys), 71. *Prof Exp:* Sci officer geophys, Geol Surv Can, 67-69; fel, Lunar Sci Inst, Houston, 70-73; lectr, 73-74, asst prof, 74-77, ASSOC PROF GEOL, ERINDALE COL, UNIV TORONTO, 77- *Mem:* Am Geophys Union; Geol Asn Can; Can Geophys Union. *Res:* Paleomagnetism; at present concentration on sedimentary deposits of Paleozoic age; space geophysics; magnetic properties of natural materials, terrestrial and extraterrestrial. *Mailing Add:* Dept Geol Univ Toronto Erindale Col 3359 Mississauga Mississauga ON L5L 1C6 Can

PEARCE, JACK B, b Dearborn, Mich, Sept 20, 30; m 53; c 2. MARINE ECOLOGY, ZOOLOGY. *Educ:* Humboldt State Col, BA, 57; Univ Wash, MS, 60, PhD(zool), 62. *Prof Exp:* NIH fel, Marine Lab, Denmark & Marine Sta, Scotland, 62-63; res assoc, Systs Ecol Prog, Marine Biol Lab, Woods Hole Oceanog Inst, 63-65; asst prof biol, Humboldt State Col, 65-67; marine ecologist, Sandy Hook Lab, 67; dir environ invests, Northeast Fisheries Ctr, Nat Oceanic & Atmospheric Admin, 67-84, actg dir, Estuarine Prog Off, Washington, DC, 84-85; DEP DIR SCI RES, NORTHEAST FISHERIES CTR, WOODS HOLE, MA, 85- *Concurrent Pos:* US mem & chmn environ qual comt, Int Coun Explor Sea; adj assoc prof, Livingston Col, Rutgers Univ & Lehigh Univ; postdoctoral fel, Univ Copenhagen, Denmark; US mem, UN Environ Prog, GESAMP wg 26. *Honors & Awards:* US Dept Commerce Gold Medal. *Mem:* Marine Biol Asn UK; Scottish Marine Biol Asn; Estuarine Res Fedn; Sigma Xi. *Res:* Biology of symbiotic crabs of the family Pinnotheridae; synecological study of epibenthic mytilid communities; the effect of man's activities on the marine environment; the effects of contaminants on habitat quality and mariculture. *Mailing Add:* Buzzards Bay Lab 54 Upland Rd Falmouth MA 02540

PEARCE, KEITH IAN, b London, Eng, Nov 7, 27; Can citizen; m 54; c 4. PSYCHIATRY, MEDICINE. *Educ:* Univ London, MBBS, 54; Univ Sask, MD, 64; FRCPS(C), 61. *Prof Exp:* Resident med officer, Univ Col Hosp, Univ London, 54, casualty surg officer, 55; registr surg, Royal Cancer Hosp, 55-56; registr surg, Whip's Cross Hosp, 56-57; resident psychiat, Univ Sask, 57-61, consult hosp, 61-62, clin dir hosp, 64-66; dir dept psychiat, Foothills Hosp, Calgary, 66-81; head div, Fac Med, 69-81, prof, 69-88, EMER PROF PSYCHIAT, UNIV CALGARY, 88- *Concurrent Pos:* McLaughlin fel, Med Res Coun, Eng, 62-64; chmn Calgary & Region Ment Health Planning Coun, 70-81; Ment Health Rev Bd, 88-; Human Rights Comn, 88- *Mem:* Can Psychiat Asn; Royal Col Psychiat; Int Fedn Med Electronics & Biol Eng; Am Col Forensic Psychiat. *Res:* Psychopharmacology and addiction; forensic psychiatry; CAT scan and psychiatry; diagnosis and treatment of sexual offenders. *Mailing Add:* Box 892 Univ Calgary 2500 University Dr NW Cochrane AB T2N 1N4 Can

PEARCE, MORTON LEE, b Chicago, Ill, Aug 22, 20; m 44; c 3. CARDIOLOGY. *Educ:* Univ Chicago, BS, 41, MD, 44. *Prof Exp:* Intern & resident med, Los Angeles County Hosp, 45-48; res assoc hemat, Atomic Energy Proj, Univ Calif, Los Angeles, 48-51; res fel cardiol, Johns Hopkins Univ, 51-52; res fel & instr, Vanderbilt Univ, 52-53; from asst prof to assoc prof med, 53-63, PROF MED, SCH MED, UNIV CALIF, LOS ANGELES, 63- *Concurrent Pos:* Fel coun atherosclerosis & fel coun clin cardiol, Am Heart Asn; vis investr, Cardiovasc Res Inst, Univ Calif, San Francisco, 62-63; consult cardiol, Vet Admin Hosp, Los Angeles; consult, Rand Corp, Santa Monica. *Mem:* Am Fedn Clin; Am Physiol Soc; fel Am Col Cardiol; Asn Univ Cardiologists. *Res:* Cardiology, including hemodynamics, atherosclerosis and electrocardiography. *Mailing Add:* Dept Med Univ Calif Sch Med Los Angeles CA 90024

PEARCE, RICHARD HUGH, b London, Ont, Apr 30, 24; c 3. EXPERIMENTAL PATHOLOGY. *Educ:* Univ Western Ont, BSc, 46, MSc, 48, PhD(path chem), 51; Am Bd Clin Chem, cert; Can Soc Clin Chem, cert. *Prof Exp:* Res assoc bact, Sch Med, Yale Univ, 49-50; lectr path chem, Univ Western Ont, 50-52, from asst prof to assoc prof, 52-61; from assoc prof to prof, 61-89, EMER PROF PATH, FAC MED, UNIV BC, 89- *Concurrent Pos:* Assoc dir, Dept Clin Path, Meek Mem Labs, Victoria Hosp, Ont, 52-61; consult, St Thomas-Elgin Gen Hosp, St Thomas; St Paul's Hosp, Vancouver, BC; hon prof oral biol, Univ BC, 82. *Honors & Awards:* Ames Award, Can Soc Clin Chem. *Mem:* Fel AAAS; Biochem Soc UK; Orthop Res Soc; Can Biochem Soc; Sigma Xi; Can Acad Clin Biochem; hon mem Can Soc Clin Chem. *Res:* Connective tissue; intervertebral disc aging and degeneration. *Mailing Add:* 2211 Westbrook Mall Vancouver BC V6T 1W5 Can

PEARCE, ROBERT BRENT, b Pendleton, Ore, June 5, 36; m 58; c 4. AGRONOMY, PLANT PHYSIOLOGY. *Educ:* Univ Calif, Davis, BS, 63; Va Polytech Polytech Inst, MS, 65, PhD(agron), 67. *Prof Exp:* Plant physiologist, Forest & Range Br, Plant Indust Sta, USDA, Md, 66-69; CROP PHYSIOLOGIST, DEPT AGRON, IOWA STATE UNIV, 69-, PROF AGRON, 80- *Mem:* Am Soc Agron; Crop Sci Soc Am. *Res:* Increasing crop production through understanding crop growth and its relationship to photosynthesis, plant morphogenesis and the environment of the crop community. *Mailing Add:* Dept Agron Iowa State Univ Ames IA 50011

PEARCE, THOMAS HULME, b Ottawa, Ont, Mar 20, 38; m 87; c 1. PETROLOGY, GEOCHEMISTRY. *Educ:* Carleton Univ, BSc, 61; Univ Western Ont, MSc, 63; Queen's Univ, Ont, PhD(geol), 67. *Prof Exp:* Nat Res Coun Can fel, Univ Manchester, 67-69 & Univ Western Ont, 69-70; asst prof geol, Univ Ga, 70-72; from asst prof to assoc prof, 72-90, PROF GEOL, DEPT GEOL SCI, QUEEN'S UNIV, ONT, 90- *Concurrent Pos:* Inventor, Mult Frequency Laser Interference Microscope. *Mem:* AAAS; Geol Asn Can; Mineral Asn Can; fel Geol Soc Am; Mineral Soc Am. *Res:* Petrogenesis; major element igneous geochemistry; variation diagrams; mechanisms of differentiation; liquid lines of descent; laser interference microscopy (new invention); laser holography; plagioclase zoning. *Mailing Add:* Dept Geol Sci Queen's Univ Kingston ON K7L 3N6 Can

PEARCY, CARL MARK, JR, b Beaumont, Tex, Aug 23, 35. MATHEMATICS. *Educ:* Tex A&M Univ, BA, 54, MS, 56; Rice Univ, PhD(math), 60. *Prof Exp:* Fel math, Rice Univ, 60-61; res engr, Humble Oil & Refining Co, 61-63; Hildebrant res instr math, Univ Mich, 63-64, asst prof, 64-65; vis assoc prof, Univ Miami, 65-66; assoc prof, 66-68, PROF MATH, UNIV MICH, ANN ARBOR, 68- *Concurrent Pos:* Sloan Found res fel math, 66-68; distinguished vis prof, Bucknell Univ, 72-; ed, Am Math Soc Surv Vol 13, 74, J Operator Theory, 78- & managing ed, Michigan Math J, 78-83. *Mem:* Am Math Soc. *Res:* Theory of linear operators on Hilbert space; numerical solution of partial differential equations. *Mailing Add:* Dept of Math Univ of Mich Ann Arbor MI 48104

PEARCY, ROBERT WOODWELL, b Boston, Mass, Aug 28, 41; m 65. PLANT ECOLOGY, PLANT PHYSIOLOGY. *Educ:* Univ Mont, BS, 63, MS, 65; Colo State Univ, PhD(bot), 69. *Prof Exp:* Fel plant physiol, Dept Plant Biol, Carnegie Inst, 69-71; asst prof biol sci, State Univ NY Albany, 71-75; asst prof, 76-78, ASSOC PROF BOT, UNIV CALIF, DAVIS, 78- *Mem:* AAAS; Ecol Soc Am; Am Soc Plant Physiologists; Sigma Xi. *Res:* Physiological and biochemical basis for adaptation to environment in plants; comparative physiology of photosynthesis and respiration in ecospecies and ecotypes. *Mailing Add:* Dept of Bot Univ of Calif Davis CA 95616

PEARCY, WILLIAM GORDON, b Evanston, Ill, Oct 14, 29; m 57; c 3. ANIMAL ECOLOGY, BIOLOGICAL OCEANOGRAPHY. *Educ:* Iowa State Univ, BS, 51, MS, 52; Yale Univ, PhD(ecol, oceanog), 60. *Prof Exp:* From asst prof to assoc prof, 60-70, PROF BIOL OCEANOG, ORE STATE UNIV, 70 - *Concurrent Pos:* Vis prof, Univ Tokyo, 78 & Univ Tromso, 78; chmn, working group 52, Sci Comt Oceanic Res, 77-82; mem, Ocean Sci Bd, 82-83; dir, Coop Inst Marine Resources Studies, 82-85. *Mem:* AAAS; Am Soc Limnol & Oceanog; Am Soc Zoologists; Am Soc Naturalists; Int Assoc Biol Oceanog (vpres). *Res:* Ecology, behavior and distribution of marine nekton, juvenile salmonids, fishes and squids; estuarine ecology; marine biology. *Mailing Add:* Dept Oceanog Ore State Univ Corvallis OR 97331

PEARD, WILLIAM JOHN, b Chicago, Ill, Nov 28, 28; m 51; c 4. INORGANIC CHEMISTRY, ANALYTICAL CHEMISTRY. *Educ:* Northwestern Univ, BS, 51; Univ Iowa, MS, 55, PhD(anal chem), 57. *Prof Exp:* Anal chemist, US Steel Corp, 51; res chemist, Dow Chem Co, 56; res chemist, 57-60, group leader inorg res, 61-64, anal supvr res & develop, 64-67, asst dir res, 67-73, lab mgr, 73-77, MGR, LAB SERV & ENVIRON AFFAIRS, PPG INDUSTS, 77- *Mem:* Am Chem Soc. *Res:* Inorganic solution chemistry; coordination compounds; transition metal chemistry. *Mailing Add:* 5009 W St Charles Ave Lake Charles LA 70601

PEARDON, DAVID LEE, b Eagle, Wis, Mar 26, 32; m 63; c 2. VETERINARY SCIENCE. *Educ:* Univ Wis, BS, 58, MS, 60, PhD(vet sci), 62. *Prof Exp:* Res parasitologist, Ralston Purina Co, 62-66; res assoc, Whitmoyer Labs, Inc, Pa, 66-69; lab head, Rohm & Haas, 69-73, proj leader, 73-78, sr res assoc, 78-79, sr scientist res labs, 79-82; RETIRED. *Mem:* Am Soc Parasitologists; Am Asn Vet Parasitologists. *Res:* Development of veterinary parasiticides and experimental application of such products. *Mailing Add:* Two Deerpath Rd Chalfont PA 18914

PEARINCOTT, JOSEPH V, b Travancore, India, May 26, 29; US citizen; m 58; c 1. PHYSIOLOGY, ZOOLOGY. *Educ:* Univ Travancore, India, BSc, 49; Aligarh Muslim Univ, MSc, 51; Fordham Univ, PhD(physiol), 59. *Prof Exp:* Instr biol, Fordham Univ, 52-56; fel physiol, Col Physicians & Surgeons, Columbia Univ, 59-61; res assoc physiol & pharmacol, NY Med Col, 61-62; asst prof, 62-68, ASSOC PROF PHYSIOL, NORTHEASTERN UNIV, 68- *Mem:* AAAS; Am Soc Zool; Entom Soc Am; NY Acad Sci; Sigma Xi. *Res:* Lipid metabolism; neurophysiology; cardiovascular physiology; experimental production and prevention of arteriosclerosis and atherosclerosis; carbohydrate, protein and fat metabolism during insect metamorphosis. *Mailing Add:* 61 Webb St Lexington MA 02173

PEARL, GARY STEVEN, b New York, NY, Dec 4, 49; m 74; c 2. NEUROSCIENCES, NEOPLASIA. *Educ:* Oberlin Col, BA & MA, 71; Emory Univ, PhD(anat), 76, MD, 77, Am Bd Path, dipl. *Prof Exp:* Fel neurosci, dept anat, Emory Univ, 71-77, resident, 77-81, assoc, 81-82, from instr to asst prof, 83-85; PATHOLOGIST, ORLANDO REGIONAL MED CTR, 85-, DIR MED EDUC PATH, 87- *Concurrent Pos:* Consulting Neuropathologist, Fla Dist 9 Med Exam Off, 89- *Mem:* Col Am Path; Am Asn Neuropath; Am Asn Path. *Res:* Analysis of cerebrospinal fluid in the diagnosis of neurologic disease. *Mailing Add:* Dept Path Orlando Regional Med Ctr Orlando FL 32806-2093

PEARL, IRWIN ALBERT, b Seattle, Wash, Dec 25, 13; m 38; c 2. WOOD CHEMISTRY. *Educ:* Univ Wash, BS, 34, MS, 35, PhD(org chem), 37. *Hon Degrees:* MS, Lawrence Univ, 77. *Prof Exp:* Res assoc org chem, Univ Wash, 37-41; res assoc, 41-55, sr res assoc org chem, Inst Paper Chem, 55-76; CHEM CONSULT, FOREST PROD INDUST, 77- *Concurrent Pos:* Anal chemist, NCent Labs, Wash, 37; supvr lab, State Dept Conserv & Develop, Wash, 38-40; consult, Environ Protection Agency. *Mem:* Am Chem Soc; Tech Asn Pulp & Paper Indust; Electrochem Soc; fel NY Acad Sci; Phytochem Soc NAm. *Res:* Utilization of sulfite waste liquor; chemistry of vanillin and its derivatives; pulp mill pollution; chemicals from wood; chemistry of hardwoods; hardwood extractives and bark; pulp and paper mill effluent analysis. *Mailing Add:* 2115 N Linwood Appleton WI 54914

PEARL, JOHN CHRISTOPHER, b Ann Arbor, Mich, Dec 31, 38; m 63; c 2. PLANETARY SATELLITES. *Educ:* Univ Mich, BSE, 61, MS, 63, PhD(physics), 70. *Prof Exp:* Asst res physicist, Space Physics Res Lab, Univ Mich, Ann Arbor, 59-61 & 62-70; AEROSPACE TECHNOLOGIST SPACE SCI, GODDARD SPACE FLIGHT CTR, NASA, 70- *Mem:* Am Phys Soc; Am Geophys Union; Am Astron Soc. *Res:* Determination of thermal and reflective characteristics; chemical composition; evolution of planets and satellites. *Mailing Add:* Mail Code 693 2 Goddard Space Flight Ctr NASA Greenbelt MD 20771

PEARL, JUDEA, b Tel-Aviv, Israel, Sept 4, 36; US citizen; m 60; c 3. HEURISTIC PROGRAMMING, PROBABILISTIC REASONING. *Educ:* Israel Inst Technol, BSc, 60; Newark Col Eng, MSc, 61; Rutgers Univ, MSc, 65; Polytech Inst Brooklyn, PhD(elec eng), 65. *Prof Exp:* Res engr, Dent Sch, NY Univ, 60-61; mem tech staff, RCA Res Labs, 61-65; dir advan memory devices, Electronic Memories, Inc, Calif, 66-69; PROF ENG SYSTS & COMPUT SCI, UNIV CALIF, LOS ANGELES, 69- *Concurrent Pos:* Instr, Newark Col Eng, 61; consult, Rand Corp, 72, Integrated Sci Corp, 75 & Hughes Aircraft, 89. *Honors & Awards:* Outstanding Achievement Award, RCA Labs, 65. *Mem:* Fel Inst Elec & Electronic Engrs; fel Americal Asn Artificial Intel. *Res:* Superconductivity; computer memories; artificial intelligence; simulation of perceptual, cognitive and decision making processes; pattern recognition; complexity of computations; computer-based decision-aids; reasoning under uncertainty; probabilistic causality. *Mailing Add:* Univ Calif 4731 Boelter Hall Los Angeles CA 90024

PEARL, MARTIN HERBERT, mathematics; deceased, see previous edition for last biography

PEARL, RONALD G, b New York, NY, Oct 5, 49; m 81; c 2. PULMONARY HYPERTENSION. *Educ:* Yale Univ, BA, 71; Univ Chicago, PhD(physiology, pharmacol), 75, MD, 77. *Prof Exp:* Resident int med, 77-80, fel critical care, 80-81, asst prof med, 81-83, resident anesthesiol, 83-85, ASST PROF & ASSOC DIR INTENSIVE CARE, STANFORD UNIV, 85- *Mem:* Am Soc Anesthesiologists; Am Col Physicians; Soc Critical Care Med; Soc Cardiovasc Anesthesiol; AAAS; Am Soc Critical Care Anesthesiologists. *Res:* Pharmacology and physiology of pulmonary hypertension. *Mailing Add:* 853 Canning Ave Palo Alto CA 94301

PEARL, W(ESLEY) L(LOYD), b Seattle, Wash, July 10, 21; m 42; c 4. CHEMICAL ENGINEERING. *Educ:* Univ Wash, BS, 42; Inst Paper Chem, Lawrence Col, MS, 48, PhD(paper chem), 51. *Prof Exp:* Proj engr, Longview Fibre Co, Wash, 51-52; chem engr, Hanford Works, Gen Elec Co, 52-55, mgr chem eng develop subsect, Atomic Power Equip Dept, 55-74; PRES, NWT CORP, 74- *Concurrent Pos:* Consult. *Mem:* Am Soc Mech Engrs; Am Nuclear Soc; Nat Asn Corrosion Engrs; Am Inst Chem Engrs. *Res:* Research and development in corrosion, water chemistry, water treatment and liquid, solid and gaseous radioactive wastes. *Mailing Add:* NWT Corp 7015 Realm Dr San Jose CA 95119

PEARL, WILLIAM, b Hungary, Apr 15, 20; nat US; m 49; c 2. PHYSIOLOGY, PHARMACOLOGY. *Educ:* NY Univ, PhD(biol), 60. *Prof Exp:* Res scientist physiol, Warner-Lambert Res Inst Div, Warner-Hudnut, Inc, 59-61; Nat Heart Inst res fel path, Sch Med, NY Univ, 61-63; res scientist, Lederle Labs, Am Cyanamid Co, Pearl River, 63-76, med ed, dept med commun, 76-85; RETIRED. *Mem:* AAAS; Am Physiol Soc; Histochem Soc; NY Acad Sci. *Res:* Cellular physiology; homogenization techniques; fractionation of subcellular particles by differential centrifugation; cytochemical analysis; distribution and interaction of nucleoprotein particles; biochemical pathology and pharmacology; enzyme parameters of disease; metabolic regulators. *Mailing Add:* 38 Lyncrest Dr Monsey NY 10952

PEARLE, PHILIP MARK, b Bronx, NY, Sept 24, 36; m 59; c 2. THEORETICAL PHYSICS. *Educ:* Mass Inst Technol, BS, 57, MS, 58, PhD(physics), 63. *Prof Exp:* Instr physics, Harvard Univ, 64-66; asst prof physics, Case Western Reserve Univ, 66-69; assoc prof, 72-77, from asst prof to assoc prof, 69-77, Kenan prof, 76-79, PROF PHYSICS, HAMILTON COL, 77- *Concurrent Pos:* Fel, Univ Geneva, 73-74; res fel, Oxford Univ, 81-82, Univ S Carolina, Univ Cambridge, Univ Trieste, 87-88, Univ Trieste, 91. *Mem:* Am Phys Soc. *Res:* Foundations of quantum mechanics. *Mailing Add:* Dept Physics Hamilton Col Clinton NY 13323

PEARLMAN, ALAN L, b Des Moines, Iowa, June 30, 36; m 90; c 3. NEUROBIOLOGY, NEUROLOGY. *Educ:* State Univ Iowa, AB, 58; Wash Univ, MD, 61. *Prof Exp:* USPHS physiologist neurophysiol, NIMH, Bethesda, 62-64; resident neurol, Harvard Med Sch & Mass Gen Hosp, 64-67; asst prof & asst neurologist, Barnes Hosp, St Louis, 69-73, assoc prof & physiol & assoc neurologist, 73-79; PROF NEUROL & CELL BIOL, SCH MED, WASH, 79-; NEUROLOGIST, BARNES HOSP, ST LOUIS, 79- *Concurrent Pos:* USPHS & Nat Inst Neurol Dis & Stroke res fel neurobiol, Harvard Med Sch, 67-69; USPHS & Nat Eye Inst res grant neurol & physiol, Sch Med, Wash Univ, 70- *Mem:* Am Neurol Asn; Am Soc Cell Biol; Am Acad Neurologists; Asn Res Vision & Ophthal; Am Physiol Soc; Soc Neurosci. *Res:* Neural organization and development of mammalian cerebral cortex. *Mailing Add:* Dept Cell Biol & Neurol Sch Med Wash Univ 660 S Euclid Ave Box 8228 St Louis MO 63110

PEARLMAN, BRUCE A, b Feb 18, 49. SYNTHETIC ORGANIC CHEMISTRY. *Educ:* Harvard Univ, AB, 71; Columbia Univ, MPhil(chem), 74, PhD(chem), 76. *Prof Exp:* NIH res fel, Dept Chem, Harvard Univ, 76-79; SR RES SCIENTIST CHEM PROCESS RES & DEVELOP UNIT, UPJOHN CO, 79- *Mem:* Am Chem Soc. *Res:* New methods for synthesis of pharmaceuticals. *Mailing Add:* 1500-91-2 Upjohn Co Kalamazoo MI 49001

PEARLMAN, MICHAEL R, b Boston, Mass, Aug 21, 41; m 64; c 2. INSTRUMENTATION, ELECTRONICS. *Educ:* Mass Inst Technol, SB, 63, SM, 80; Tufts Univ, PhD(physics) 63. *Prof Exp:* SCIENTIST, SMITHSONIAN ASTROPHYS OBSERV, CAMBRIDGE, 68-; RES ASSOC, HARVARD COL OBSERV, 69- *Concurrent Pos:* Vis scientist, Off Geod Satellites, NASA, 71-72. *Mem:* Inst Elec & Electronics Engrs; Am Phys Soc; Am Geophys Union. *Res:* Optics and electronics; applications to satellite tracking for geodesy and geophysics; atmospheric effects on optical and radio wave propagation; solid state physics and lasers. *Mailing Add:* Smithsonian Astro Observ 60 Garden St Cambridge MA 02138

PEARLMAN, NORMAN, b New York, NY, Aug 2, 22; m 46; c 3. SOLID STATE PHYSICS. *Educ:* City Col New York, BS, 42; Univ Chicago, SM, 49; Purdue Univ, PhD(physics), 52. *Prof Exp:* Jr mathematician, Signal Corps Radar Labs, NJ, 42-43; assoc mathematician, Metall Lab, Univ Chicago, 44-46, asst physics, 47-52; from asst to assoc prof, 54-69, PROF PHYSICS, PURDUE UNIV, WEST LAFAYETTE, 69-, ASST HEAD, 82- *Mem:* AAAS; Am Phys Soc; Am Asn Physics Teachers; Sigma Xi. *Res:* Low temperature calorimetry; electrical and thermal transport phenomena. *Mailing Add:* Dept Physics Purdue Univ West Lafayette IN 47907-1396

PEARLMAN, RODNEY, b Melbourne, Australia, May 29, 51. PHARMACEUTICAL CHEMISTRY. *Educ:* Victorian Col Pharm, Melbourne, BPharm, 73; Univ Kans, MS, 76, PhD(pharmaceut chem), 79. *Prof Exp:* Sr pharmaceut chemist, Eli Lilly & Co, 78-81; ASST PROF PHARMACEUT, COL PHARM, UNIV TEX, AUSTIN, 81- *Mem:* Acad Pharmaceut Sci. *Res:* Synthesis and evaluation of drugs; molecular orbital analysis of pharmaceutical agents; physical and organic chemistry of drug degradation; lymphatic absorption of drugs; pharmaceut formulation; peptide stability. *Mailing Add:* PO Box 2108 El Granada CA 94018

PEARLMAN, RONALD C, b Brooklyn, NY, Aug 8, 44; div; c 2. AUDIOLOGY, SPEECH PATHOLOGY. *Educ:* L I Univ, BA, 67, MS, 70; Univ Mo-Columbia, PhD(audiol), 74. *Prof Exp:* Speech pathologist, West Islip Sch Syst, 67-71; teaching asst speech path & audiol, Univ Mo, 71-73; trainee audiol, Vet Admin Hosp, Columbia, Mo, 73-74; asst prof speech path & audiol, 77-78, ASSOC PROF AUDIOL, HOWARD UNIV, 78- *Mem:* Am Speech & Hearing Asn; Am Audiol Soc. *Res:* Acoustic impedance of the tympanic membrane; clinical audiology; hearing aids; intraoperative monitoring. *Mailing Add:* Dept Commun Arts & Sci Howard Univ Washington DC 20059

PEARLMAN, RONALD E, b Calgary, Alta, Dec 22, 41. MOLECULAR BIOLOGY. *Educ:* McGill Univ, BSc, 61; Harvard Univ, MSc & PhD(biochem), 66. *Prof Exp:* Nat Res Coun Can res fel biochem, Biol Inst, Carlsberg Found, Copenhagen, Denmark, 66-68; asst prof, 68-72, assoc prof, 72-82, PROF BIOL, YORK UNIV, 82- *Concurrent Pos:* Vis prof, Biochem Inst B, Univ Copenhagen, 75-76. *Res:* Enzymology of nucleic acid metabolism; replication, transcription and translation of genetic information and its control; structure, function and organization of eukaryotic genes. *Mailing Add:* Dept Biol Sci York Univ 4700 Keele St Downsview ON M3J 1P6 Can

PEARLMAN, SHOLOM, public health and epidemiology, for more information see previous edition

PEARLMAN, WILLIAM HENRY, b New York, NY, Mar 2, 14; m 43. BIOCHEMISTRY. *Educ:* Brooklyn Col, BS, 34; Columbia Univ, PhD(biochem), 40. *Prof Exp:* Res assoc biochem, Clark Univ, 40-44; asst prof, Princeton Univ, 44-45; res assoc, Jefferson Med Col, 45-46, from asst prof to assoc prof, 46-53; res assoc biochem, Harvard Med Sch, 59-61, asst prof biol chem, 61-68; mem staff, Lab Reproductive Biol, 70-81, prof, 69-81, EMER PROF PHARMACOL, SCH MED, UNIV NC, CHAPEL HILL, 81-, ADJ PROF BIOCHEM & NUTRIT, 82- *Concurrent Pos:* Brit Med Res Coun grant chem & biochem, Guy's Hosp Med Sch, London, 54-57; assoc, Peter Bent Brigham Hosp, 59-68; mem adv panel rev contraceptives & other vaginal prod, Food & Drug Admin, 73-78. *Mem:* AAAS; Am Soc Biol Chemists; Endocrine Soc; Brit Biochem Soc; Brit Soc Endocrinol. *Res:* Cancer; steroid-protein interactions; biochemistry of steroid hormones. *Mailing Add:* Dept Biochem CB #7260FLOB Univ NC Sch Med Chapel Hill NC 27599-7260

PEARLMUTTER, ANNE FRANCES, b Chelsea, Mass, Oct 28, 40; wid; c 2. BIOCHEMISTRY. *Educ:* Tulane Univ La, BS, 62; Case Western Reserve Univ, MA, 67, PhD(chem), 69. *Prof Exp:* Teaching assoc biochem, Med Col Ohio, 69-72, instr, 72-74, from asst prof to assoc prof, 74-88; SCI FAC, HORACE MANN BARNARD SCH, 89- *Mem:* Endocrine Soc; Am Soc Biol Chemists; AAAS; Am Chem Soc. *Res:* Endocrinology; fast kinetic processes in biochemistry; role of metal ions in biological systems; hormone-protein and hormone-receptor interactions; brain hormone mechanisms of action. *Mailing Add:* 208 Victory Blvd New Rochelle NY 10804

PEARLSON, WILBUR H, b Los Angeles, Calif, Oct 19, 15; m 44; c 4. CHEMISTRY. *Educ:* Univ Calif, BS, 38; Pa State Col, PhD(org chem), 43. *Prof Exp:* Res assoc, USDA, Univ Calif, 38-40; from instr to asst prof chem, Pa State Col, 43-47; chemist, Minn Mining & Mfg Co, 47-85; CONSULT, 85- *Mem:* AAAS. *Res:* Organofluorine compounds; mechanism of alkylation reaction; epoxy and phenolic resins; aziridine; urethane resins; patent liaison; regulatory affairs. *Mailing Add:* 5261 Bald Eagle Blvd W St Paul MN 55110

PEARLSTEIN, ARNE JACOB, b Los Angeles, Calif, Mar 18, 52. PHOTOCHEMICAL REACTION ENGINEERING. *Educ:* Univ Calif, Los Angeles, BS & MS, 77, Engr, 79, PhD(eng), 83. *Prof Exp:* ASST PROF AEROSPACE & MECH ENG, 83- & ASST PROF CHEM ENG, UNIV ARIZ, 86- *Concurrent Pos:* Prin investr, NSF grant, 85- *Honors & Awards:* Presidential Young Investr Award, NSF, 85. *Mem:* Am Soc Mech Engrs; Am Inst Chem Engrs; Am Chem Soc; Am Phys Soc; Am Inst Aeronaut & Astronaut. *Res:* Stability of chemically reacting and other nonhomogeneous fluids; photochemical engineering; fluid mechanics and mass transfer in materials processing. *Mailing Add:* Aerospace & Mech Eng Dept Univ Ariz Tucson AZ 85721

PEARLSTEIN, ARNE JACOB, b Los Angeles, Calif, Mar 18, 52; m 91. HYDRODYNAMIC STABILITY, INCOMPRESSIBLE FLOW. *Educ:* Univ Calif, Los Angeles, BS & MS, 77, PhD(eng), 83. *Prof Exp:* Asst prof, Dept Aerospace & Mech Eng, Univ Ariz, 83-89; ASSOC PROF, DEPT MECH & INDUST ENG, UNIV ILL, URBANA-CHAMPAIGN, 89- *Concurrent Pos:* Prin investr, NSF grant, 85-91, Air Force Off Sci Res grant & NASA grants, 89-; NSF presidential young investr award, 85; asst prof, Dept Chem Eng, Univ Ariz, 86-89. *Mem:* Am Soc Mech Engrs; Am Inst Chem Engrs; Am Phys Soc; Am Inst Aeronaut & Astronaut; Am Soc Eng Educ; Am Chem Soc. *Res:* Hydrodynamic stability; stability of chemically reacting and other nonhomogeneous fluids; fluid mechanics and mass transfer in materials processing; oscillatory electrodissolution. *Mailing Add:* Dept Mech & Indust Eng Univ Ill 1206 W Green St Urbana IL 61801

PEARLSTEIN, EDGAR AARON, b Pittsburgh, Pa, Mar 19, 27; div; c 1. SOLID STATE PHYSICS. *Educ:* Carnegie Inst Technol, BSc, 47, DSc(physics), 50. *Prof Exp:* Res assoc, Univ Ill, 50-52; res physicist, Carnegie Inst Technol, 52-56; from asst prof to assoc prof, 56-63, PROF PHYSICS, UNIV NEBR-LINCOLN, 63- *Concurrent Pos:* Consult, Westinghouse Elec Corp, 56-57. *Mem:* AAAS; fel Am Phys Soc; Am Asn Physics Teachers; Fedn Am Scientists. *Res:* Radiation effects in solids; electrical and optical properties of insulating crystals. *Mailing Add:* Dept of Physics Univ of Nebr Lincoln NE 68588-0111

PEARLSTEIN, LEON DONALD, b Los Angeles, Calif, Dec 10, 32; m 59; c 2. THEORETICAL PHYSICS. *Educ:* City Col New York, BS, 53; Univ Pa, PhD(physics), 59. *Prof Exp:* Instr physics, Fla State Univ, 58-60; staff mem, Gen Atomic Div, Gen Dynamics Corp, Calif, 60-70; MEM STAFF, LAWRENCE LIVERMORE LAB, 70- *Mem:* Am Phys Soc. *Res:* Theoretical plasma physics applied to controlled thermonuclear research and space physics. *Mailing Add:* Lawrence Livermore Nat Lab PO Box 5511 L630 Livermore CA 94550

PEARLSTEIN, ROBERT DAVID, b Gary, Ind, July 20, 49; m 78. INSTRUMENTATION DESIGN. *Educ:* Univ Mo, BA, 71; Univ NC, MS, 77, PhD(bioeng), 82. *Prof Exp:* Lab instr, Dept Elec Eng, Univ Mo, 71-72; res asst physiol, 75-78, RES ASSOC MED, DEPT SURG, DUKE MED CTR & RES ASSOC BIOPHYSICS, SPETNAGEL LAB, DUKE UNIV, 78- *Concurrent Pos:* Consult, Leigh Biotechnol, Cleveland, OH. *Mem:* AAAS; Sigma Xi; Biophysical Soc. *Res:* The role of polypeptide and arachadonic acid metabolites in the propagation of free radical mediated cellular injury; analytical/clinical instrumentation development and laboratory automation. *Mailing Add:* 2901 Arnold Dr Durham NC 27707

PEARLSTEIN, ROBERT MILTON, b New York, NY, Oct 16, 37; m 60; c 2. BIOPHYSICS, CHEMICAL PHYSICS. *Educ:* Harvard Univ, BA, 60; Univ Md, PhD(physics), 66. *Prof Exp:* Physicist, Marine Biol Lab, Woods Hole, Mass, 62-63; NSF fel photosynthesis, Biol Div, Battelle Columbus Labs, 66-67, res staff mem, 67-75, group leader photosynthesis, 72-75, res staff mem, Chem Div, 75-78, coordr solar energy res, Oak Ridge Nat Lab, 75-76, prin res scientist, 78-79, sr res scientist, Chem Dept, 79-82; chmn, 83-90, PROF PHYSICS, INDIANA UNIV-PURDUE UNIV, INDIANAPOLIS, 83- *Concurrent Pos:* Lectr, Oak Ridge Biomed Grad Sch, Univ Tenn, 69-78; mem hon ed bd, Photochem & Photobiol, 71-73; assoc ed, Biophys J, 81-84. *Mem:* Fel Am Phys Soc; Biophys Soc; fel AAAS; Am Soc Photobiol; Sigma Xi. *Res:* Natural and artificial photosynthesis, photochemical conversion of solar energy, chlorophyll chemistry, electronic excited-state energy transfer and exciton transport in bio-organic systems. *Mailing Add:* 3333 Eden Way Pl Carmel IN 46032

PEARLSTEIN, SOL, b Brooklyn, NY, Feb 21, 30; m 51; c 3. NUCLEAR PHYSICS. *Educ:* Polytech Inst Brooklyn, BS, 51; NY Univ, MS, 53; Rensselaer Polytech Inst, PhD(physics), 64. *Prof Exp:* Physicist, Knolls Atomic Power Lab, Gen Elec Co, 52-64; physicist, 64-68, DIR NAT NUCLEAR DATA CTR, BROOKHAVEN NAT LAB, 68- *Concurrent Pos:* Chmn, Cross Sect Eval Working Group, 66-83; mem adv comt reactor physics, US Dept Energy, 70-, mem nuclear data comt, 74- *Mem:* Am Nuclear Soc; Am Phys Soc; Sigma Xi. *Res:* Nuclear data; reactor physics. *Mailing Add:* Nuclear Energy Dept Brookhaven Nat Lab Upton NY 11973

PEARMAN, G TIMOTHY, b Cape Giradeau, Mo, June 27, 40; m 62; c 1. ACOUSTICS. *Educ:* Cent Methodist Col, BA, 62; Univ Mo, MS, 64, PhD(physics), 67. *Prof Exp:* MEM TECH STAFF, BELL LABS, 67- *Mem:* Inst Elec & Electronics Engrs; Am Phys Soc; Sigma Xi. *Res:* Piezoelectric devices. *Mailing Add:* Bell Telephone Labs 555 Union Blvd Allentown PA 18103

PEARS, COULTAS D, b Clarksburg, WVa, Feb 7, 25; m 49; c 3. MECHANICAL ENGINEERING. *Educ:* Tulane Univ, BEngMech, 46. *Prof Exp:* Asst head eng, Appalachian Exp Sta, US Bur Mines, 47-57, supt underground gasification, Gorgas Exp Sta, 57; head, Dept Mech & Mat Eng, 57-89, VPRES, MECH & MAT ENG RES CTR, SOUTHERN RES INST, 89- *Concurrent Pos:* Spec consult, Ala Power Co, 58-63. *Res:* Analysis and evaluation of materials for special application under extreme environments of temperature, vacuum, pressure and chemistry; thermal and stress analysis of systems. *Mailing Add:* Mech & Mat Eng Res Southern Res Inst 2000 Ninth Ave S Birmingham AL 35255-5305

PEARSALL, GEORGE W(ILBUR), b Brentwood, NY, July 13, 33; m 62. MATERIALS SCIENCE, METALLURGY. *Educ:* Rensselaer Polytech Inst, BMetE, 55; Mass Inst Technol, ScD(metall), 61. *Prof Exp:* Res engr metall, Dow Chem Co, Mich, 55-57; asst prof, Mass Inst Technol, 60-64; assoc prof, 64-66, actg dean, 69-71, dean, Sch Eng, 71-74 & 82-83, PROF MECH ENG, DUKE UNIV, 66-, PROF MAT SCI, 80-, PROF PUB POLICY, 82- *Concurrent Pos:* Bd dirs, Duke Univ; dir, Duke-IBM Product Safety Inst; consult, corporations, attorneys, partnerships in materials development, failure analysis and safe-product design. *Mem:* AAAS; Soc Plastic Engrs; Am Soc Metals; Am Soc Mech Engrs; Am Soc Testing & Mat; Soc Risk Anal. *Res:* Development of materials and processes; deformation behavior of materials; microstructural mechanics; nonlinear properties of alloys and polymers; failure analysis; biomechanics; product design and safety. *Mailing Add:* 2941 Welcome Dr Durham NC 27705

PEARSALL, S(AMUEL) H(AFF), b Guthrie, Ky, July 17, 23; m 46; c 4. ELECTRICAL ENGINEERING. *Educ:* Vanderbilt Univ, BE, 48, MS, 58. *Prof Exp:* Engr, WSM, Inc, 48-56; assoc prof elec eng, Vanderbilt Univ, 56-64; vpres, R W Benson & Assoc, 64-71; vpres, Bonitron, Inc, 66-71; V PRES, CUTTERS EXCHANGE & CUTTERS ELECTRONICS INT, 71- *Concurrent Pos:* Consult, Temco Corp, Avco Corp, E I du Pont de Nemours & Co, Inc, Army Ballistic Missile Agency & NASA; pres, White Owl Systs, Inc. *Mem:* Sr mem Inst Elec & Electronic Engrs; sr mem Soc Mfg Engrs; Sigma Xi. *Res:* Industrial control systems; 4 United States patents. *Mailing Add:* 118 Spring Valley Rd Nashville TN 37214-2822

PEARSALL, THOMAS PERINE, optics, electronic transport, for more information see previous edition

PEARSE, GEORGE ANCELL, JR, b Stoneham, Mass, May 18, 30; m 53; c 5. ANALYTICAL CHEMISTRY, INORGANIC CHEMISTRY. *Educ:* Univ Mass, BS, 52; Purdue Univ, MS, 56; Univ Iowa, PhD(chem), 59. *Prof Exp:* Sr chemist, E I du Pont de Nemours & Co, Inc, 59-60; from asst prof to assoc prof, 60-70, actg chmn dept, 65-66, chmn dept, 66-72, PROF CHEM, LE MOYNE COL, NY, 70- *Concurrent Pos:* Vis res prof, Stockholm Univ, Sweden, 78-79 & Cambridge Univ, Eng, 85-86. *Mem:* Am Chem Soc. *Res:* Coordination chemistry; absorption spectrophotometry; synthesis of analytic-organic reagents and ligands; use of nonaqueous solvents; instrumental methods of analysis. *Mailing Add:* Dept Chem Le Moyne Col Syracuse NY 13214

PEARSE, JOHN STUART, b Boise, Idaho, May 28, 36; m 70; c 1. MARINE BIOLOGY. *Educ:* Univ Chicago, BS, 58; Stanford Univ, PhD(biol), 65. *Prof Exp:* Asst, Antarctic Deep Freeze, 60-62; asst prof biol, Am Univ Cairo, 65-68; res fel, Calif Inst Technol, 68-71; from asst prof to assoc prof, 71-78, PROF BIOL, UNIV CALIF, SANTA CRUZ, 78- *Concurrent Pos:* Jr scientist, Te Vega Exped, 65, sr scientist, 68; ed, Marine Biol, 74-76; vis prof, Harvard Univ, 78; course dir, Bermuda Biol Sta, 80-81, bd rev ed, Sci, 86-88. *Honors & Awards:* Antarctic Serv Award. *Mem:* Fel AAAS; Am Soc Zoologists; Am Inst Biol Sci; Int Soc Invert Reproduction; Western Soc Naturalists. *Res:* Reproduction of marine invertebrates; echinoderm biology; ecology of kelp forests; Antarctic biology; tropical reef biology. *Mailing Add:* Inst Marine Sci Univ Calif Santa Cruz CA 95064

PEARSE, VICKI BUCHSBAUM, b Dec 17, 42; US citizen; m 70; c 1. INVERTEBRATE ZOOLOGY, MARINE SCIENCES. *Educ:* Stanford Univ, AB, 63, PhD(biol), 68. *Prof Exp:* NIH res fel zool, Univ Calif, Los Angeles, 68-69; instr, Univ Calif, Irvine, 69-70; res fel, Calif Inst Technol, 70; ASSOC RES MARINE BIOLOGIST, UNIV CALIF, SANTA CRUZ, 72- *Mem:* Am Soc Zoologists; Western Soc Naturalists; Sigma Xi. *Res:* Development, maintenance and physiology of symbiotic relationships; calcification; reproduction of marine invertebrates; photoperiodism; placozoans. *Mailing Add:* Inst Marine Sci Univ of Calif Santa Cruz CA 95064

PEARSE, WARREN HARLAND, b Detroit, Mich, Sept 28, 27; m 50; c 4. OBSTETRICS & GYNECOLOGY. *Educ:* Mich State Univ, BS, 48; Northwestern Univ, MB, 50, MD, 51; Am Bd Obstet & Gynec, dipl, 60. *Prof Exp:* Resident obstet & gynec, Univ Mich, 50-53, 55-56; instr, Col Med, Univ Nebr, Omaha, 59-61, from asst prof to prof, 61-71, chmn dept, 62-71, asst dean, 63-71; dean, Med Col Va, 71-75; EXEC DIR, AM COL OBSTETRICIANS & GYNECOLOGISTS, 75- *Mem:* Am Col Obstetricians & Gynecologists. *Res:* Obstetric manpower; physiology of labor. *Mailing Add:* 600 Maryland Ave SW Suite 300 Washington DC 20004

PEARSON, ALBERT MARCHANT, b Oakley, Utah, Sept 3, 16; m 46; c 5. FOOD SCIENCE. *Educ:* Utah State Univ, BS, 40; Iowa State Univ, MS, 41; Cornell Univ, PhD(animal husb), 49. *Prof Exp:* Asst, Iowa State Univ, 40-41 & Cornell Univ, 46-49; from asst prof to assoc prof animal husb, Univ Fla, 49-54; from assoc prof to prof, 54-89, EMER PROF FOOD SCI & HUMAN NUTRIT, MICH STATE UNIV, 89- *Concurrent Pos:* Fulbright fel, Meat Indust Res Inst NZ, 71-72. *Honors & Awards:* Morrison Award, Am Soc Animal Sci, 72; R C Pollock Award, Am Meat Sci Asn. *Mem:* Am Soc Animal Sci; Inst Food Technol; Am Inst Nutrit; Am Meat Sci Asn. *Res:* Methods of measuring body composition; isolation and identification of meat flavor components; chemical changes in pre- and post-rigor muscle. *Mailing Add:* Dept Animal Sci Oregon State Univ Withycomb Hall 112 Corvallis OR 97331-6702

PEARSON, ALLAN EINAR, b Minneapolis, Minn, June 18, 36; m 61. ENGINEERING. *Educ:* Univ Minn, BS, 58, MS, 59; Columbia Univ, PhD(control systs), 63. *Hon Degrees:* MA, Brown Univ, 66. *Prof Exp:* From asst prof to assoc prof, 63-66, PROF ENG, BROWN UNIV, 70- *Concurrent Pos:* Fulbright grant, 78. *Mem:* Am Soc Mech Engrs; Inst Elec & Electronics Engrs. *Res:* Automatic control systems theory, particularly the mathematical aspects of adaptive optimal control and system identification. *Mailing Add:* Div of Eng Brown Univ Providence RI 02912

PEARSON, ALLEN MOBLEY, b Mobile, Ala, Jan 14, 09; wid; c 1. BIOLOGY. *Educ:* Ala Polytech Inst, BS, 31; Iowa State Col, MS, 32, PhD(entom), 36. *Prof Exp:* Asst biologist, Soil Conserv Serv, USDA, 35-37; assoc biologist, US Fish & Wildlife Serv, 37-43; exten specialist, Auburn Univ, 43-47; from assoc prof to prof, Auburn Univ, 47-71, emer prof Zool & Entom, 72-; RETIRED. *Mem:* Wildlife Soc; assoc Am Soc Mammal; assoc Am Ornithologists Union. *Res:* Entomology; game management. *Mailing Add:* 1433 Lee Rd 97 Opelika AL 36801

PEARSON, ARTHUR DAVID, b Darlington, Eng, Apr 19, 32; nat US; m 55; c 3. CHEMISTRY. *Educ:* Univ Durham, BSc, 53; Mass Inst Technol, PhD(Inorg Chem), 57. *Prof Exp:* Supvr, AT&T Bell Labs, Inc, 57-88; RETIRED. *Honors & Awards:* Forrest Award, Am Ceramic Soc, 61. *Mem:* Fel Am Ceramic Soc; Optical Soc Am. *Res:* Chemistry and physics of glasses, including chemical, physical, electrical and optical properties; properties of materials for optical waveguide applications. *Mailing Add:* PO Box 11747 Charlotte Amalie VI 00801

PEARSON, BENNIE JAKE, b Austin, Tex, Aug 28, 29; m 59. MATHEMATICS. *Educ:* Univ Tex, BA, 50, PhD(math), 55. *Prof Exp:* Instr math, Ill Inst Technol, 55-57; asst prof, Kans State Univ, 57-59; from asst prof to assoc prof, 59-66, PROF MATH, UNIV MO-KANSAS CITY, 66- *Mem:* Am Math Soc. *Res:* Ordered topological spaces; continua; curves; dendrites. *Mailing Add:* Dept Math Univ Mo Kansas City Col Arts & Sci 109 Haag Hall Kansas City MO 64110

PEARSON, CARL E, b Vitaby, Sweden, Mar 17, 22; nat US; m 53; c 3. ENGINEERING SCIENCE. *Educ:* Univ BC, BASc, 44; Brown Univ, PhD(appl math), 49. *Prof Exp:* Res assoc mech eng, Harvard Univ, 49-50, from instr to asst prof, 51-57; staff scientist, Arthur D Little, Inc, 57-62; vis prof appl math, Tech Univ Denmark, 62-63; head eng sci group, Sperry Rand Res Ctr, 63-65; chief math anal unit, Airplane Div, Boeing Co, Wash, 65-68; PROF AERONAUT & ASTRONAUT & APPL MATH, UNIV WASH, 68- *Mem:* Am Phys Soc; Soc Indust & Appl Math. *Res:* Applied mathematics and mechanics; numerical analysis; thermodynamics; acoustics; electromagnetism; fluid mechanics; elasticity. *Mailing Add:* Dept Appl Math Univ Wash FS-20 Seattle WA 98195

PEARSON, COLIN ARTHUR, b Armidale, Australia, Sept 1, 35; c 2. PHYSICS. *Educ:* Univ Sydney, BS, 56, PhD, 61. *Prof Exp:* Res assoc, Univ Sydney, 61-62; group leader nuclear physics, Australian AEC, 62-64; vis prof physics, Mass Inst Technol, 64-65; vis prof, Niels Bohr Inst, Copenhagen, 65-67; vis lectr, Univ Ariz, 67-69; assoc prof, Univ Ala, Birmingham, 69-71, clin assoc prof radiation oncol, 71-72, prof 72-78, chmn dept, 78-84; SELF EMPLOYED, 84- *Res:* Neutron physics; cosmic rays; high energy photonuclear reactions; relativistic deuteron production; neutron absorption in reactors; nuclear stripping reactions; three-body break-up and scattering; nuclear fission. *Mailing Add:* PO Box 439 Chelksea AL 35043

PEARSON, DALE SHELDON, b Omaha, Nebr, Oct 4, 42; m 64; c 2. POLYMER SCIENCE, MATERIALS SCIENCE. *Educ:* Iowa State Univ, BS, 64; Univ Akron, MS, 70; Northwestern Univ, PhD(mat sci & eng), 78. *Prof Exp:* Res scientist & group leader, polymer sci & radiation res, Firestone Tire & Rubber Co, 65-74; res asst, polymer sci, Northwestern Univ, 74-78; mem tech staff, polymer physics, Bell Labs, 78-; RES ASSOC, POLYMER PHYSICS, EXXON RES & ENG CO, 83- *Mem:* Am Chem Soc; Am Phys Soc; Soc Rheology. *Res:* Polymer physics, particularly molecular structure and mechanical properties of rubber networks and polymer melts. *Mailing Add:* Dept Chem Eng Univ Calif Eng II Rm 3337 Santa Barbara CA 93106

PEARSON, DANIEL BESTER, III, b Perryton, Tex, May 25, 49. PETROLEUM GEOLOGY, ORGANIC GEOCHEMISTRY. *Educ:* Rice Univ, BA, 71, MA, 73 & PhD, 81; Calif Inst Technol, MS, 74. *Prof Exp:* Petrol geochemist, source rock geochem, Superior Oil Co, 74-78; res geologist org geochem, Oil & Gas Resources Br, US Geol Surv, 78-80; mem staff, Mobil Oil Co, 80-81; CONSULT, 81- *Mem:* AAAS; Am Asn Petrol Geol; Soc Econ Paleont & Mineral; Geochem Soc; Geol Soc Am. *Res:* Mechanism of origin of petroleum and applications to predicting character of oil and gas in petroleum exploration; sour cerock geology. *Mailing Add:* 1940 Mayflower Dallas San Antonio TX 75208-3113

PEARSON, DAVID D, b Norwich, Conn, Aug 3, 38; m 74; c 4. IMMUNOBIOLOGY, BIOCHEMISTRY. *Educ:* Univ Conn, BA, 60, MS, 63; Univ Kans, PhD(zool), 66. *Prof Exp:* Asst prof zool, Univ Kans, 65-66; Herbert L Spencer assoc prof, 71-77, asst prof, 66-71, ASSOC PROF BIOL, BUCKNELL UNIV, 71- *Concurrent Pos:* USPHS fel immunol, 65-66; consult res, Geisinger Med Ctr, actg dir res, 73-77. *Mem:* AAAS; NY Acad Sci; Sigma Xi. *Res:* Glycoproteins; immunobiology; protein biochemistry; serology; autoimmune diseases; oligosaccharides. *Mailing Add:* Dept Biol Bucknell Univ Lewisburg PA 17837

PEARSON, DAVID LEANDER, b Fargo, NDak, Nov 11, 43; m 69. ECOLOGY, ORNITHOLOGY. *Educ:* Pac Lutheran Univ, BS, 67; La State Univ, Baton Rouge, MS, 69; Univ Wash, PhD(zool), 73. *Prof Exp:* Asst prof, 74-80, ASSOC PROF BIOL, PA STATE UNIV, 80- *Mem:* Am Ornithol Union; Cooper Ornith Soc; Ecol Soc Am; Soc Study Evolution; Sigma Xi; Entom Soc Am. *Res:* Comparative study of bird community structure in tropical lowland forests of the world and ecology and community structure of tiger beetles of the family Cicindelidae. *Mailing Add:* Dept Biol Bucknell Univ Lewisburg PA 17837

PEARSON, DONALD A, b Los Angeles, Calif, Oct 16, 21; m 45; c 3. CHEMICAL ENGINEERING. *Educ:* Univ Wash, BS, 44, MS, 48. *Prof Exp:* Chem engr, Puget Sound Pulp & Timber Co, 48-51; process engr, Ketchikan Pulp Co, 51-53, tech dir, 53-62, gen supt, 62-66, vpres mfg, Alaska, 66-68, vpres, Wash, 68-70, pres, 70-74; pres & managing prin, Rubens & Pratt, 74-75, pres & managing prin, Rubens, McClure, Pearson Co, Consult Engrs, 75-79, Pres, Pearson, Pape, Allen, Huggins, Inc, 80-; sr vpres, Shucart & Assoc, Inc, 86; RETIRED. *Mem:* Am Chem Soc; Tech Asn Pulp & Paper Indust. *Res:* Sulfite pulping of wood to produce wood pulps for use as regenerated cellulose fibers and films. *Mailing Add:* 8001 Sand Pt Wy NE Seattle WA 98124-3855

PEARSON, DONALD A, b Boone Co, Ind, Oct 7, 31; m 52; c 2. METALLURGY. *Educ:* Purdue Univ, BS, 58. *Prof Exp:* Res metallurgist, FMC Corp, 59-69, chief mat engr, 69-72, mgr qual control, 72-81, mgr qual assurance, Link-Belt Chain Div & Automotive Prod Div, 81-85, DIR QUAL ASSURANCE & RELIABILITY, LINK-BELT CHAIN DIV, REXNORD CORP, 85- *Concurrent Pos:* Legal coordr & prod liability coordr, Link-Belt Div, PT Components, Inc. *Mem:* Am Inst Mining, Metall & Petrol Engrs; Am Foundrymen's Soc; Am Soc Qual Control; affil Am Soc Testing & Mat; affil Soc Automotive Engrs. *Res:* Ferrous and semiconductor metallurgy. *Mailing Add:* Link Belt Roller Chain Oper Rexnord Corp PO Box 346 Indianapolis IN 46206

PEARSON, DONALD EMANUAL, b Madison, Wis, June 21, 14; m 50; c 3. CHEMISTRY. *Educ:* Univ Wis, BS, 36; Univ Ill, PhD(org chem), 40. *Prof Exp:* Res chemist, Pittsburgh Plate Glass Co, Wis, 40-42; tech aide, Nat Defense Res Comt Contract, Univ Chicago, 44-45; res assoc, Mass Inst Technol, 45-46; from asst prof to assoc prof, 46-57, PROF CHEM, VANDERBILT UNIV, 57- *Honors & Awards:* Pearson Award for Best Sr Res. *Mem:* Am Chem Soc. *Res:* Organic chemistry, synthesis and analysis; Beckmann rearrangement; anti-malarial synthesis; drugs; environmental. *Mailing Add:* Dept of Chem Vanderbilt Univ Nashville TN 37235

PEARSON, EARL F, b Scottsville, Ky, Oct 12, 41; m 62; c 3. SPECTROSCOPY, LASERS. *Educ:* Western Ky Univ, BS, 63, MA, 64; Vanderbilt Univ, PhD(chem), 69. *Prof Exp:* Res assoc, Univ Southern Calif, 68-69; chemist, Shell Develop Co, 69-70; PROF CHEM, WESTERN KY UNIV, 70- *Concurrent Pos:* Vis prof, State Univ NY, Binghamton, 77. *Mem:* Am Chem Soc; Sigma Xi. *Res:* Optical properties; electronic spectroscopy; propellant aging; chemical education. *Mailing Add:* 1080 Trammel Boyce Rd Scottsville KY 42164

PEARSON, EDWIN FORREST, b St Louis, Mo, June 3, 38; m 63; c 2. MOLECULAR SPECTROSCOPY, CHEMICAL PHYSICS. *Educ:* Yale Univ, BA, 60; Duke Univ, PhD(physics), 68. *Prof Exp:* Asst prof physics, Southern Ill Univ, 68-73; vis scientist, chem, Univ Ill, 73-74; asst prof physics, Univ Mo-Rolla, 76-78; staff scientist, physics, Laser Anal, Inc, 78-80; sr eng, physics, Electrooptics, Raytheon Co, 80-86; TECH STAFF, TASC, 86- *Concurrent Pos:* Consult Univ Ill, 72-73; Humboldt fel, Alexander von Humboldt Stiftung, Justis Liebig Univ, Ger, 74-76. *Mem:* AAAS; Am Phys Soc; Am Asn Physics Teachers; NY Acad Sci; Optical Soc Am. *Res:* Infrared, millimeter wave and microwave molecular spectroscopy; molecues of interstellar and atmospheric interest; chemical physics; laser radar. *Mailing Add:* 18 Fort Pond Rd Acton MA 01720

PEARSON, ERMAN A, sanitary engineering; deceased, see previous edition for last biography

PEARSON, FREDERICK JOSEPH, b Bethlehem, Pa, July 4, 35. GEOCHEMISTRY, GROUNDWATER GEOLOGY. *Educ:* Harvard Univ, AB, 58; Univ Tex, Austin, MA, 62, PhD(geol), 66. *Prof Exp:* Res sci assoc carbon 14 dating & geochem, Carbon 14 Lab, Univ Tex, Austin, 62-66; hydrologist, US Geol Surv, NY, 66-69, res hydrologist, 69-79; CONSULT, 87- *Mem:* AAAS; Geochem Soc; Geol Soc Am; Am Geophys Union. *Res:* Carbon 14; tritium; light stable isotopes to indicate sources and rates of movement of groundwater; geochemistry of water-rock interactions. *Mailing Add:* Groundwater Geochem 130 Walnut Hill Lane Suite 210 Irving TX 75038

PEARSON, GARY RICHARD, b Livingston, Mon, May 3, 38; m 66; c 3. VIROLOGY, IMMUNOLOGY. *Educ:* Univ Chicago, BS, 60, MS, 63; Stanford Univ, PhD(med microbiol), 67. *Prof Exp:* Fel pub health serv virol, Karolinska Inst, Sweden, 67-69; res assoc, Children's Hosp Philadelphia, 69-70; sr staff fel, 70-72, microbiologist, 72-74, head, microbiol sect, Nat Cancer Inst, NIH, Bethesda, 74-75; consult & assoc prof microbiol, Univ Minn, Mayo Clin & Med Sch, 75-78, consult & prof microbiol & virology, 78-, head sect microbiol, 79-; CHMN & PROF MICROBIOL, MED CTR, GEORGETOWN UNIV. *Concurrent Pos:* Asst prof pediat, Univ Pa, 69-70; vchmn Nat Cancer Prog, 73-75; mem, Sci Rev Comt, Virus Cancer Prog, 76-80. *Mem:* Am Soc Microbiol; AAAS; Am Asn Cancer Res; Am Asn Immunol; Int Asn Comp Res Leukemia & Related Dis. *Res:* Role of herpesviruses in the etiology of human cancers; immunoprevention of herpesvirus-induced diseases; association of herpes- viruses with neurological diseases; definition of immune responses patterns to viruses and associated diseases. *Mailing Add:* Dept Microbiol Georgetown Univ Sch Med 3900 Reservior NW Washington DC 20007

PEARSON, GEORGE DENTON, b Oakland, Calif, May 10, 41; m 66; c 3. ANIMAL VIROLOGY. *Educ:* Stanford Univ, BS, 64, PhD(pharmacol), 69. *Prof Exp:* Calif Div, Am Cancer Soc Dernham jr fel, Stanford Univ, 68-71; asst prof biochem, 71-77, ASSOC PROF BIOCHEM & BIOPHYS, ORE STATE UNIV, 77- *Concurrent Pos:* Am Cancer Soc grants, 72-74 & 75-; Nat Cancer Inst res grant, 75-; NATO sr fel, NSF, 75- *Mem:* Biophys Soc; Am Soc Microbiol; Sigma Xi. *Res:* Structure and function of eukaryotic DNA; viral carcinogenesis; mechanism of anticancer and antiviral agents; developmental biochemistry. *Mailing Add:* 477 NW Survista Corvallis OR 97330

PEARSON, GEORGE JOHN, b Burlington, Iowa, June 16, 28; m 52, 81; c 4. APPLIED PHYSICS, PHOTOGRAPHIC SCIENCE. *Educ:* Grinnell Col, AB, 50; Iowa State Univ, MS, 52; Univ Conn, PhD(physics), 63. *Prof Exp:* Jr res assoc thermal measurement, Inst Atomic Res, Iowa State Univ, 52-55; asst instr physics, Univ Conn, 57-58; sr proj engr superconductivity, Westinghouse Res Lab, 63-64; sr proj engr instrumentation, Res Labs, J & L Steel Co, 64; proj mgr & head appl physics, Fisher Sci Co, 65-66; proj & develop engr, Apparatus & Optical Div, Eastman Kodak Co, 66-69, sr photog engr, Photo Technol Div, 69-74, tech assoc, 75-83, supvr photo sect, 84-89; PRES, THE SOLUTION ENG, 90- *Mem:* AAAS; Soc Photog Scientists & Engrs. *Res:* Photographic instrumentation and systems. *Mailing Add:* 3283 Lake Rd W Williamson NY 14589

PEARSON, GLEN HAMILTON, b Seattle, Wash, June 8, 48; m 71. CHEMICAL ENGINEERING, POLYMER RHEOLOGY. *Educ:* Va Polytech Inst & State Univ, BS, 71; Univ Mass, PhD(chem eng), 76. *Prof Exp:* Res chemist, Polymer Phys Chem, Eastman Kodak Co, 75-81, lab head & chem process engr, Res Lab, 81-83, lab head, Polymer Res, 83-87, lab head, Electronic Interconnect Systs, 87-88, LAB HEAD, PHYSICAL PERFORMANCE LAB, EASTMAN KODAK CO, 88- *Mem:* Soc Rheology; Am Chem Soc; Soc Plastic Engr. *Res:* Fluid mechanics; polymer rheology; polymer processing; physical properties of polymers. *Mailing Add:* 85 Raspberry Patch Dr Rochester NY 14652-3701

PEARSON, HANS LENNART, b Vancouver, BC, Mar 6, 27; US citizen; m 61; c 1. MATHEMATICS. *Educ:* Univ BC, BASc, 49, MA, 51; Ill Inst Technol, PhD(math), 57. *Prof Exp:* From instr to asst prof, 54-63, actg dean grad sch, 75-77, ASSOC PROF MATH, ILL INST TECHNOL, 63-, DEAN GRAD SCH, 88- *Res:* Special functions; partial differential equations. *Mailing Add:* Dept of Math Ill Inst of Technol Chicago IL 60616-3793

PEARSON, HENRY ALEXANDER, b Heidenheimer, Tex, Oct 26, 33; m 58; c 3. RANGE SCIENCE. *Educ:* Blinn Col, AA, 54, Tex A&M Univ, BS, 58, MS, 59; Utah State Univ, PhD(range sci), 68. *Prof Exp:* Asst range mgt, Tex A&M Univ, 58-59; asst range sci, Utah State Univ, 59-62; range scientist, Rocky Mountain Forest & Range Exp Sta, Ariz, 62-69, proj leader range res, Southern Forest Exp Sta, La, 69-75, prin range/wildlife scientist forest environ res, US Forest Serv, Washington, DC, 75-77; PROJ LEADER RANGE RES, SOUTHERN FOREST EXP STA, LA, 77- *Concurrent Pos:* US Forest Serv rep, Range Livestock Nutrit Tech Comt, 64-69. *Honors & Awards:* Chapline Res Award, Soc Range Mgt, 87. *Mem:* Soc Range Mgt; Wildlife Soc; Am Forage & Grassland Coun. *Res:* Nutrition; ruminology; forest grazing systems. *Mailing Add:* 2500 Shreveport Hwy US Forest Serv Pineville LA 71360

PEARSON, HOWARD ALLEN, b Ancon, CZ, Nov 4, 29; US citizen; m; c 5. PEDIATRICS, HEMATOLOGY. *Educ:* Dartmouth Col, AB, 51, dipl, 52; Harvard Med Sch, MD, 54; Am Bd Pediat, dipl, cert pediat hemat-oncol. *Prof Exp:* Res fel hemat, Children's Med Ctr, Harvard Med Sch, 57-58; asst head clin hemat & asst chief pediat, US Naval Hosp, Bethesda, Md, 58-62; from asst prof to prof pediat, Col Med, Univ Fla, 62-68; chmn dept, 74-87, PROF PEDIAT, SCH MED, YALE UNIV, 68- *Concurrent Pos:* Clin instr, Sch Med, Georgetown Univ, 59-62; clin asst prof, Col Med, Howard Univ, 61-62. *Mem:* Am Fedn Clin Res; Am Soc Hemat; Soc Pediat Res; Am Acad Pediat. *Res:* Pediatric hematology. *Mailing Add:* Dir Pediat Yale Univ Sch Med New Haven CT 06510

PEARSON, J(OHN) RAYMOND, b Providence, RI, July 6, 12; m 43; c 2. MECHANICAL ENGINEERING. *Educ:* Univ RI, BSc, 35; Mass Inst Technol, MSc, 46. *Prof Exp:* Trainee eng, Brown & Sharpe Mfg Co, RI, 35-37; instr sci vocations, Am Bd Comnr Foreign Missions, Turkey, 38-41, from asst prof to assoc prof mech eng, Robert Col, Istanbul, 41-45, prof & chmn dept, 46-54, assoc dean eng, 47-56, actg dean fac, 55-56; from vis prof to prof mech engr, 56-81, prof, Dept Phys Med & Rehab, Sch Med, 64-81, assoc chmn, 66-75, chmn, Mech Eng Dept, 75-88, EMER PROF MECH ENG, COL ENG, UNIV MICH, ANN ARBOR, 81- *Concurrent Pos:* Vis prof, Univ Uppsala, 70-72; prof mech eng, Syracuse Univ, 49-50. *Mem:* Fel Am Soc Mech Engrs; Am Soc Eng Educ; Biomed Eng Soc. *Res:* Mechanisms, dynamics of machinery, mechanical vibrations, automatic control; mechanical design; biomechanics; research, design and development of orthotic devices; bioengineering. *Mailing Add:* 334 Sumac Lane Ann Arbor MI 48105

PEARSON, JAMES BOYD, JR, b McGehee, Ark, June 3, 30; m 57; c 6. ELECTRICAL ENGINEERING. *Educ:* Univ Ark, BS, 58, MS, 59; Purdue Univ, PhD(elec eng), 62. *Prof Exp:* Instr elec eng, Univ Ark, 58-59; from instr to asst prof, Purdue Univ, 59-65; assoc prof, 65-70, chmn dept, 74-79, PROF ELEC ENG, RICE UNIV, 70-, J S ABERCROMIE PROF ENG, 79- *Mem:* Fel Inst Elec & Electronics Engrs. *Res:* Analysis and design of control systems; control and system theories. *Mailing Add:* Dept Elec & Comp Eng Rice Univ Houston TX 77251-1892

PEARSON, JAMES ELDON, b Newton, Mass, Oct 8, 26; m 53; c 5. NEPHROLOGY, PHARMACOLOGY. *Educ:* Loyola Univ, La, BS, 56, MS, 58. *Prof Exp:* Prof asst angiol, Touro Res Inst, 64-68; res assoc pharmacol, LSU Med Ctr, 68-71; res assoc ecol, Pac Biomed Res, Univ Hawaii, 71-72; res assoc nephrol, LSU Med Ctr, 72-74 & Univ Miss Med Ctr, 74-76; RES ASSOC NEPHROL, LSU MED CTR, 76- *Honors & Awards:* Honor Achievement Award, Angiol Res Found, 68. *Mem:* Am Soc Pharmacol & Exp Therapeut; Am Soc Nephrol; Int Soc Nephrol; Am Soc Artificial Internal Organs. *Res:* Nephrology, as it concerns the artificial kidney and drug effects on kidney. *Mailing Add:* Dept of Med 1542 Tulane Ave New Orleans LA 70112

PEARSON, JAMES GORDON, medicine, radiotherapy, for more information see previous edition

PEARSON, JAMES JOSEPH, b Kansas City, Mo, Sept 23, 34; m 57, 69; c 2. SOLID STATE PHYSICS. *Educ:* Yale Univ, BS, 56; Stanford Univ, MS, 57; Univ Pittsburgh, PhD(physics), 61. *Prof Exp:* Res assoc physics, Univ Pittsburgh, 61; jr res physicist, Univ Calif, San Diego, 61-62; res scientist, 63-71, staff scientist, 71-74, MGR SIGNAL PROCESSING LAB, LOCKHEED PALO ALTO RES LAB, 74- *Concurrent Pos:* NSF fel, Saclay Nuclear Res Ctr, France, 62-63. *Mem:* Soc Photo Optical Instrumentation Engrs; Am Phys Soc; Inst Elec & Electronics Engrs. *Res:* Theory of ferromagnetism and antiferromagnetism; spin waves; super-exchange; magnetic anisotropy; crystal field theory; electromagnetic scattering; image data processing. *Mailing Add:* ORGN 61-90 Bldg 107 Lockheed Miss & Space Co Inc PO Box 3504 Sunnyvale CA 94088

PEARSON, JAMES MURRAY, b Aberdeen, Scotland, Nov 22, 37; m 62; c 2. PHYSICAL CHEMISTRY, POLYMER CHEMISTRY. *Educ:* Aberdeen Univ, PhD(polymer chem), 62. *Prof Exp:* NSF fel, State Univ NY Col Forestry, Syracuse Univ, 62-64; lectr, Aberdeen Univ, 64-66; asst prof chem, State Univ NY Col Forestry, Syracuse Univ, 66-68; res chemist, Xerox Corp, 68-72, mgr polymer physics & chem, Res Labs, 72-79; LAB HEAD, POLYMER RES, EASTMAN KODAK CO, 79- *Mem:* Am Chem Soc; AAAS. *Res:* Kinetics; mechanism polymerization; ionic polymerization; electrical and photochemical properties of polymers; optical recording in polymers. *Mailing Add:* 1485 Highland Ave Rochester NY 14618-1001

PEARSON, JEROME, b Texarkana, Ark, Apr 19, 38; m 65; c 3. SPACE VEHICLES, SPACEFLIGHT. *Educ:* Wash Univ, BS, 61; Wright State Univ, MS, 77. *Prof Exp:* Aerospace Engr, Langley Res Ctr, NASA, 62-66, Ames Res Ctr, 66-71; CHIEF, STRUCT DYNAMICS BRANCH, WRIGHT LAB, US AIR FORCE, 71- *Concurrent Pos:* Air Force rep, Tech Adv Group Shock & Vibration, Dept Defense, 76-; tech consult, The Brubaker Group, 76- & WED Enterprises, Walt Disney Epcot Ctr, 83-; mem, Mat & Struct Comt, Int Astronaut Fedn, 88- *Honors & Awards:* Apollo Achievement Award, NASA, 69; Sci Achievement Award, US Air Force, 75. *Mem:* Assoc fel Am Inst Aeronaut & Astronaut; Am Astronaut Soc; fel Brit Interplanetary Soc; AAAS. *Res:* Research into structural dynamics, acoustics and vibration of aircraft and spacecraft; new methods for spaceflight. *Mailing Add:* 5491 Corkhill Dr Dayton OH 45424-4707

PEARSON, JOHN, b Leyburn, Eng, Apr 24, 23; nat US; m 44; c 3. EXPLOSIVE ORDNANCE, HIGH-STRAIN-RATE BEHAVIOR. *Educ:* Northwestern Univ, BS, 49, MS, 51. *Prof Exp:* Res engr, 51-55, head warhead res br, 55-58, head solid dynamics br, 58-59, head detonation physics group, 59-67, head detonation physics div, 67-83, SR RES SCIENTIST, US NAVAL WEAPONS CTR, 83- *Concurrent Pos:* Lectr, Univ Calif, Los Angeles, 57-67 & Univ Calif, Santa Barbara, 71-72; consult, Lockheed Aircraft Corp, 58-62 & US Air Force, Ohio, 58-62. *Honors & Awards:* L T E Thompson Medal, 65; William B McLean Medal, 79; Haskell G Wilson Award, 85. *Mem:* Fel Am Soc Mech Engrs; Am Soc Metals; Am Phys Soc; Sigma Xi; Am Inst Mining, Metall & Petrol Engrs. *Res:* Elasticity; plasticity; fracture dynamics; behavior of metal-explosive systems; high velocity impact; explosive ordnance; high energy rate forming of materials; high speed photography. *Mailing Add:* Eng Sci Div US Naval Weapons Ctr Code 38906 China Lake CA 93555

PEARSON, JOHN MICHAEL, b Halifax, Eng, Aug 27, 33; m 59, 81; c 3. THEORETICAL PHYSICS. *Educ:* Univ London, BSc, 54; McMaster Univ, PhD(physics), 59. *Prof Exp:* Instr physics, Western Reserve Univ, 59-60; mem res staff, 60-61, from asst prof to assoc prof, 61-70, PROF PHYSICS, UNIV MONTREAL, 70- *Concurrent Pos:* Nat Res Coun fel, Univ Paris, 65-66. *Mem:* Am Phys Soc; Can Asn Physicists. *Res:* Low energy nuclear physics; nuclear astrophysics. *Mailing Add:* Dept of Physics Univ of Montreal CP 6128 Montreal PQ H3C 3J7 Can

PEARSON, JOHN RICHARD, b Colorado Springs, Colo, Oct 5, 38; m 61; c 2. CLINICAL CHEMISTRY, BIOCHEMISTRY. *Educ:* Colo State Univ, BS, 60, PhD(chem), 66. *Prof Exp:* Sr res assoc chem, Colo State Dept Pub Health, 66-67, prin res assoc, 67-68; instr, 68-69, ASST PROF CHEM, UNIV COLO HEALTH SCI CTR, DENVER, 69-, DIR CLIN CHEM, 70- *Concurrent Pos:* from asst mgr to proj mgr, Lab Infor Syst, 80-83. *Mem:* Am Asn Clin Chemists; Int Soc Clin Enzym; Am Chem Soc; Am Soc Clin Pathol. *Res:* Clinical laboratory management; cost effective laboratory operations; bringing specialized lipid testing into the clinical labratory; use of clinical laboratory to prevent heart disease. *Mailing Add:* Univ Colo Health Sci Ctr Box A022 Denver CO 80262

PEARSON, JOHN W(ILLIAM), b St Paul, Minn, June 2, 18; m 40; c 2. CHEMICAL ENGINEERING. *Educ:* Univ Minn, BChE, 39. *Prof Exp:* Chem engr, Minn Mining & Mfg Co, 39-44 & Manhattan Dist, Los Alamos, NMex, 45-46; develop engr, 46-53, exec engr eng res, 53-55, mgr, New Prod Div, 55-62 & Indust Finishing Dept, 62-68, group eng mgr photographic prod, 68-74, dir div eng, 74-80, vpres develop, 80-82, EXEC DIR DIV ENG, MINN MINING & MFG CO, 80-; CONSULT, PEARSON CONSULT CO, 82- *Mem:* Am Chem Soc; Am Inst Chem Engrs; Vacuum Soc. *Res:*

Construction of magnetic sound recording media; non-woven textiles; pressure sensitive tape coating; producing pressure sensitive tapes; metal plating; interactive video laser disks. *Mailing Add:* 201 Crestway Lane West St Paul MN 55118

PEARSON, JOHN WILLIAM, b Livingston, Mont, June 29, 35; m 64; c 2. MICROBIOLOGY, ONCOLOGY. *Educ:* Mont State Col, BS, 58; George Washington Univ, MS, 63; Rutgers Univ, PhD(microbiol), 68. *Prof Exp:* Microbiologist, Ft Detrick, Md, 63-64; staff fel viral oncol, 68-75, RES MICROBIOLOGIST, NAT CANCER INST, 75- *Mem:* Am Asn Cancer Res; Am Soc Microbiol. *Res:* Viral oncology; design of experiments toward therapy on oncogenic viruses; chemotherapeutic approach through the use of drugs used singly or in combination; immunotherapy and the use of interferon inducers. *Mailing Add:* Frederick Cancer Res Facil Nat Cancer Inst Bldg 537 Frederick MD 21701

PEARSON, JOSEPH T(ATEM), b Portsmouth, Va, Dec 7, 33; m 56; c 3. HEAT TRANSFER, FLUID MECHANICS. *Educ:* NC State Univ, BSME, 56, MS, 61; State Univ NY Stony Brook, PhD(eng), 67. *Prof Exp:* Assoc engr, Douglas Aircraft Co, 56-57; mgr, J T Pearson & Co, 57-59; teaching asst mech eng, NC State Univ, 59-61; instr eng, State Univ NY Stony Brook, 61-67; asst prof, 67-72, ASSOC PROF MECH ENG, PURDUE UNIV, LAFAYETTE, 72- *Concurrent Pos:* Heat transfer res grants, Purdue Univ, 68-; consult heat transfer appln, serveral co, 61-; Am Soc Heating, Refrig & Air-Conditioning Engrs grant, Purdue Univ, 71-72; Fulbright fel, 73-74. *Mem:* Am Soc Mech Engrs; Am Soc Heating, Refrig & Air-Conditioning Engrs; Sigma Xi. *Res:* Experimental and analytical studies on heat transfer equipment. *Mailing Add:* Sch Mech Eng Purdue Univ Lafayette IN 47907

PEARSON, KARL HERBERT, inorganic chemistry; deceased, see previous edition for last biography

PEARSON, KEIR GORDON, b Geeveston, Tasmania, Feb 19, 42; m 76; c 3. NEUROPHYSIOLOGY. *Educ:* Univ Tasmania, BE, 64; Oxford Univ, DPhil(physiol), 69. *Prof Exp:* Jr res fel, Merton Col, Oxford Univ, 67-69; from asst prof to assoc prof, 69-78, PROF PHYSIOL, UNIV ALTA, 78- *Mem:* Soc Exp Biol; Soc Neurosci. *Res:* Control of movements in invertebrates; structure of neurons in insects. *Mailing Add:* Dept Physiol Univ Alta Fac Med Edmonton AB T6G 2G3 Can

PEARSON, LONNIE WILSON, b Jackson, Miss, Mar 15, 46; m 75; c 2. NUMERICAL METHODS, ASYMPTOTIC ANALYSIS. *Educ:* Univ Miss, BSEE, 68, MS, 73; Univ Ill, PhD(elec eng), 76. *Prof Exp:* Electronic engr, Naval Surface Weapons Ctr, 71-74; res asst, Univ Ill, 74-76; asst prof elec eng, Univ Ky, 76-80; assoc prof elec eng, Univ Miss, 80-84; sr scientist, McDonnell Douglas Res Labs, 84-87, prin scientist, 87-90; PROF & HEAD ELEC & COMPUTER ENG, CLEMSON UNIV, 90- *Concurrent Pos:* Mem, Admin Comt, Antennas & Propagation Soc, Inst Elec & Electronics Engrs, 90-92. *Mem:* Fel Inst Elec & Electronics Engrs; Inst Elec & Electronics Engrs Microwave Theory & Technol Soc; US Nat Comt Union Radio Sci. *Res:* Analytically augmented numerical methods in electromagnetic scattering, diffraction and guided waves. *Mailing Add:* Dept Elec & Computer Eng Clemson Univ Clemson SC 29634-0915

PEARSON, LORENTZ CLARENCE, b American Fork, Utah, Jan 28, 24; m 52; c 6. PHYSIOLOGICAL ECOLOGY, LICHENOLOGY. *Educ:* Utah State Univ, Logan, BS, 52; Univ Utah, Salt Lake City, MS, 52; Univ Minn, PhD(plant genetics), 58. *Prof Exp:* Teaching asst biol & genetics, Univ Utah, 51-52; instr bot & agron, Ricks Col, 52-55; teacher sci & math, Swedish Sec Schs, 55-56; res asst plant genetics, Univ Minn, 56-58; from assoc prof to prof agron & bot, 57-64, prof & chmn div bot & life sci, 64-74, PROF BOT, RICKS COL, 74- *Concurrent Pos:* Prin investr, Desert Ecol Res Proj, Ricks Col & NSF, 60-62 & 65-66; researcher, Univ Minn Lake Itasca Res Sta, 63, 66, 67 & 70, Desert Biome Proj, US-Int Biol Prog, 68-72; NSF res fel, Uppsala Univ, 63-64 & 75-76; counr & adminr, Latter-Day Saints Church Lang Training Mission, Sweden, 68-75; consult, Idaho Opers, Idaho Nat Eng Lab, Dept of Energy, 84- , US Park Serv, Craters of the Moon Nat Monument, 88- & Organ Pipe Nat Monument, 89; pres, Idaho Acad Sci, 74-75 & 90-91. *Mem:* Fel AAAS; Bot Soc Am; Am Bryol & Lichenological Soc; Sigma Xi; Int Lichenological Soc. *Res:* Plant genetics, especially the genetics of autopolyploids; physiological ecology and productivity of ecosystems, especially deserts; microclimate research; conservation of natural resources, especially in reference to atmospheric pollution. *Mailing Add:* Dept Biol LSA 284 Ricks Col Rexburg ID 83460-1100

PEARSON, MARK LANDELL, b Toronto, Ont, June 2, 40; m 62; c 2. MOLECULAR GENETICS. *Educ:* Univ Toronto, BA, 62, MA, 64, PhD(med biophys), 66. *Prof Exp:* Asst prof med genetis, Univ Toronto, 69-74, asst prof med biophys, 69-75, assoc prof med genetics, Fac Med, 74-80, assoc prof med biophys, 75-80; asst dir, Cancer Biol Prog, Frederick Cancer Res Ctr, 79-82, dir, Lab Molecular Biol, 82-83; assoc dir, E I Du Pont de Nemours & Co, Inc, 83-84, dir molecular biol, Cent Res & Develop, 85-90; DIR CANCER & INFLAMMATORY DIS, DUPONT MERCK PHAMACEUT CO, DUPONT EXP STA, 91- *Concurrent Pos:* Helen Hay Whitney Found fel biochem, Sch Med, Stanford Univ, 66-69; mem fel panel, Nat Cancer Inst Can, 76-77; Josiah Macy Found fel biol sci, Stanford Univ, 77-78; mem, Mam Genetics study sect, NIH, 85-88; mem adv panel, Off Technol Assess, 85-87; mem bd, Life Sci Res Found, 88-; mem, Prog Adv Comm Human Genome, NIH, 88-; mem bd, Alliance Aging Res, 89-; mem bd, Univ Del Res Found, 90- *Mem:* AAAS; Am Chem Soc; Am Soc Biochem & Molecular Biol; Am Soc Cell Biol; Am Soc Human Genome; Am Soc Microbiol; Am Soc Virol; Biophys Soc; Can Biochem Soc; Can Soc Cell Biol. *Res:* Molecular basis of gene expression; mapping and sequencing the human genome. *Mailing Add:* 914 Fairthorne Ave Wilmington DE 19807

PEARSON, MICHAEL J, b Eng, Sep 25, 38; US citizen; m 62; c 2. ALUMINA CHEMISTRY, ALUMINA CATALYSTS. *Educ:* Univ Leeds, BSc, 59, PhD(chem), 62. *Prof Exp:* Fel chem, Univ Wash, 62-64; chemist, Chevron Chem Co, 64-67; STAFF RES CHEMIST, KAISER ALUMINUM & CHEM CORP, 67- *Mem:* Am Chem Soc. *Res:* Reaction mechanisms for the catalytic decomposition of hydrogen sulfide and carbon sulfur compounds; specialist in Claus Process and related catalytic reactions; alumina chemistry and applications. *Mailing Add:* 599 Sky Farm Dr Castro Valley CA 94552

PEARSON, MYRNA SCHMIDT, b Philadelphia, Pa, July 7, 36; m 61; c 2. ORGANIC CHEMISTRY. *Educ:* Univ Pa, AB, 58; Columbia Univ, AM, 59, PhD(chem), 63. *Prof Exp:* Asst prof chem, Clark Univ, 63-64; from asst prof to assoc prof, 64-82, PROF CHEM, WHEATON COL, MASS, 82- *Mem:* Am Chem Soc; NY Acad Sci. *Res:* Reactions and mechanisms of organophosphorus compounds; free radical chemistry; keto-enol tautomerism; environmental chemistry. *Mailing Add:* Dept Chem Wheaton Col Norton MA 02766-0930

PEARSON, OLOF HJALMER, b Boston, Mass, Feb 7, 13; m 42; c 4. MEDICINE. *Educ:* Harvard Univ, AB, 34, MD, 39. *Prof Exp:* Intern, Mass Gen Hosp, Boston, 39-41, asst med resident, 41-42; flight surgeon, Pan Am Airways, Africa, 42, med dir, 42-45; from asst prof to assoc prof med, Med Col, Cornell Univ, 48-60; assoc prof, 60-68, PROF MED, SCH MED, CASE WESTERN UNIV, 68-, AM CANCER SOC PROF CLIN ONCOL, 73- *Concurrent Pos:* Fel, Mass Gen Hosp, Boston, 46-48; assoc mem, Sloan-Kettering Inst Cancer Res, 48-60; assoc attend physician, Mem Hosp, New York; assoc attend physician, Univ Hosps, Cleveland, 60- *Mem:* AAAS; Am Soc Clin Invest; Am Soc Biol Chem; Asn Am Physicians; Am Fedn Clin Res. *Res:* Clinical endocrinology; endocrine treatment of cancer. *Mailing Add:* 14111 Larcamere Blvd Shaker Heights OH 44120

PEARSON, PAUL (HAMMOND), b Bolenge, Belgian Congo, Feb 18, 21; US citizen; m 68; c 4. PEDIATRICS, ADOLESCENT MEDICINE. *Educ:* Northwestern Univ, Chicago, BS, BMed & MD, 47; Univ Calif, Los Angeles, MPH, 63. *Prof Exp:* Fels & pediatrician, Johns Hopkins Hosp, 51-53; from clin instr to asst clin prof pediat, Univ Southern Calif, 53-62; from actg chief to chief, Ment Retardation Br, Div Chronic Dis, USPHS, 63-65, asst prog dir, Nat Inst Child Health & Human Develop, 65-66, spec asst child health to surgeon gen, 66-67; C Louis Meyer prof child health & pediat & med dir, Meyer Ther Ctr Handicapped Children, Col Med, Univ Nebr Med Ctr, Omaha, 67-, prof prev med & dir Meyer Children's Rehab Inst, 68-81; MCGAW PROF OF PEDIAT, DIR SEC ADOLESCENT MED, UNIV NE COL MED, 82- *Concurrent Pos:* Attend physician, Los Angeles Children's Hosp, 53-62; attend staff, Valley Presby Hosp, 58-62, chmn pediat comt, 58-59, mem exec comt, 60-61; neurol consult pub sch dist, San Fernando Valley, 60-62; from clin assoc prof to clin prof pediat, Sch Med, Georgetown Univ, 64-67; consult, United Cerebral Palsy Asns, Inc, 69; consult, Div Develop Disabilities, Dept Health, Educ & Welfare, 70, mem, Nat Adv Comt, 71-75; mem spec adv comt accessible environ for disabled, Nat Acad Sci; resident child psychiat, dept psychiat, Univ BC, 76-77; vis lectr pediat & adolescent med, Harvard Univ & Childrens Hosp Med Ctr, Boston, 81; med dir, Univ NE Hosp Clins Eating Disorder Prog, 82- *Mem:* Am Acad Pediat; Am Pub Health Asn; Am Acad Ment Deficiency; Am Acad Cerebral Palsy (secy, 74-76, pres, 81-82); Am Asn Univ Affil Progs for Develop Disabled (pres, 76-77); Soc Adolescent Med; Int Soc Adolescent Med. *Res:* Physiologic aspect of eating disorders; delivery of adolescent health services. *Mailing Add:* Pediat Univ Nebr Col Med Omaha NE 68105

PEARSON, PAUL BROWN, b Oakley, Utah, Nov 28, 05; m 33; c 2. NUTRITION. *Educ:* Brigham Young Univ, BS, 28; Mont State Col, MS, 30; Univ Wis, PhD(biochem, nutrit), 37. *Prof Exp:* Asst prof, Mont State Univ, 30-31; res assoc, Univ Calif, 32-35; prof animal nutrit, Tex A&M Univ, 37-41, distinguished prof, 41-47, dean grad sch & head dept biochem & nutrit, 47-49; chief biol br, AEC, 49-58; with Ford Found Prog Sci & Eng, 58-63; pres & sci dir, Nutrit Found, Inc, NY, 63-72; chmn dept nutrit & food, Drexel Univ, 72-74; PROF, DEPT NUTRIT & FOOD SCI & DEPT FAMILY & COMMUNITY MED, UNIV ARIZ, 74- *Concurrent Pos:* Collabr, Bur Animal Indust, 49-57; prof, Johns Hopkins Univ, 51-58; mem exec comt, Div Biol & Agr, Nat Res Coun, 50-52; US/Japan Conf Radiobiol, Tokyo, 54, World Conf on Peaceful Uses of Atomic Energy, Geneva, 55, White House Conf on Food, Nutrit & Health, 69; int symposium, biochem sulphur, Rascoff, France, 56; consult, Energy Res & Develop Admin, 58-69, Off Sci & Technol, AID, sci adv to Pres, 62-70; prog comt, Int Cong Nutrit, 60; deleg, White House Conf on Int Coop, 65; trustee, Food, Law & Drug Law Inst, 69-85; liaison, Food & Nutrit Bd; adv comt, PR Nuclear Ctr, McCollum-Pratt Inst & Secy Agr; mem, Agr Res Inst; vis prof, Thomas Jefferson Univ, 72-74; chief dept nutrit, Sch Med, Univ Autonoma de Guadalajara, 74-82. *Mem:* Fel AAAS; fel Am Soc Animal Sci; Am Chem Soc; fel Am Inst Nutrit; Am Soc Biol Chemists; Brit Nutrit Soc. *Res:* Mineral metabolism and micronutrients; B vitamins; metabolism of sulfur and enzymes; utilization of proteins and amino acids malnutrition and taste in malnourished children; author of more than 170 scientific publications on nutrition and biochemistry. *Mailing Add:* Dept Nutrit & Food Sci Univ Ariz Tucson AZ 85721

PEARSON, PAUL GUY, b Lake Worth, Fla, Dec 5, 26; m 51; c 3. ECOLOGY. *Educ:* Univ Fla, BS, 49, MS, 51, PhD, 54. *Prof Exp:* Lectr, Univ Fla, 52-54; asst prof zool, Univ Tulsa, 54-55; from asst prof to prof, 55-81, chmn dept, 67-72, assoc provost, 72-77, actg pres, 78, exec vpres, Rutgers Univ, 77-81; PROF & PRES, MIAMI UNIV, 81. *Concurrent Pos:* Chmn, NJ Noise Control Coun, 75-76. *Mem:* Ecol Soc Am (secy, 61-64, vpres, 70, treas, 74-77); Am Inst Biol Sci (pres, 78). *Res:* Population dynamics; vertebrate natural history; ecosystem analysis; impacts of pesticides and other pollutants. *Mailing Add:* Pres/Chancellor Miami Univ Main Campus Oxford OH 45056

PEARSON, PHILIP RICHARDSON, JR, b Newburyport, Mass, Apr 15, 27; m 57; c 2. PLANT ECOLOGY. *Educ:* Dartmouth Col, AB, 50; Univ Mass, MS, 56; Rutgers Univ, PhD(ecol), 60. *Prof Exp:* Instr bot, Rutgers Univ, 60-61; asst prof biol, Temple Univ, 61-67; assoc prof, 67-71, PROF BIOL, RI COL, 71- *Mem:* AAAS; Ecol Soc Am; Torrey Bot Club; Am Inst Biol Sci. *Res:* Woodland and forest communities of eastern United States. *Mailing Add:* 185 Pine Hill Rd North Scituate RI 02857

PEARSON, PHILLIP T, b Story Co, Iowa, Nov 21, 32; m 54; c 4. VETERINARY SURGERY, BIOMEDICAL ENGINEERING. *Educ:* Iowa State Univ, DVM, 56, PhD(path, surg), 62. *Prof Exp:* Intern vet med, Angell Mem Animal Hosp, Boston, Mass, 56-57; from instr to assoc prof vet med & surg, Iowa State Univ, 57-64; prof & assoc clin dir, Univ Mo, 64-65; chmn, Small Animal Clin, 67-72, dean, Col Vet Med, 72-89, PROF VET CLIN SCI, IOWA STATE UNIV, 64- *Honors & Awards:* Gaines Award, 66. *Mem:* Am Vet Med Asn; Am Asn Vet Clinicians; Am Col Vet Surg; Asn Am Vet Med Col (pres, 78-79). *Res:* Canine nephritis, orthopedics and hyperparathyroidism; surgical and pathological aspects of artificial heart work. *Mailing Add:* 1136 Vet Med Col Vet Med Iowa State Univ Ames IA 50011

PEARSON, R(AY) L(EON), b New York, NY, Mar 31, 30; m 54; c 5. MATERIALS SCIENCE, PHYSICAL CHEMISTRY. *Educ:* Willamette Univ, BS, 51; Univ Utah, MA, 54; PhD(metall), 56. *Prof Exp:* Sr scientist, Oak Ridge Nat Lab, 56-59; prin metallurgist, Aerojet-Gen Nucleonics Div, Gen Tire & Rubber Co, 59-65; PRIN SCIENTIST, OAK RIDGE NAT LAB, 65- *Res:* Nuclear waste management. *Mailing Add:* Oak Ridge Nat Lab Bldg K1037 MS 7358 PO Box 2003 Oak Ridge TN 37831

PEARSON, RALPH GOTTFRID, b Chicago, Ill, Jan 12, 19; m 41; c 3. CHEMISTRY. *Educ:* Ill Inst Technol, BS, 40; Northwestern Univ, PhD(chem), 43. *Prof Exp:* From instr to prof chem, Northwestern Univ, 43-76; PROF CHEM, UNIV CALIF, SANTA BARBARA, 76- *Concurrent Pos:* Guggenheim fel, 51-52. *Honors & Awards:* Midwest Award, Am Chem Soc, 66 & Inorg Award, 70. *Mem:* Nat Acad Sci; Am Chem Soc. *Res:* Kinetics of organic reactions; theories of organic chemistry; mechanisms of inorganic reactions; theories of inorganic chemistry. *Mailing Add:* Dept of Chem Univ of Calif Santa Barbara CA 93106

PEARSON, ROBERT BERNARD, ELECTROCARDIOGRAPHY, ELECTROMYOGRAPHY. *Educ:* Loma Linda Univ, MD, 49. *Prof Exp:* STAFF PHYSICIAN, RANCHO LOS AMIGOS HOSP, DOWNEY, CALIF, 62- *Mailing Add:* PO Box 4550 Downey CA 90241

PEARSON, ROBERT MELVIN, b Klamath Falls, Ore, Feb 17, 30; m 59; c 3. PHYSICAL CHEMISTRY, ANALYTICAL CHEMISTRY. *Educ:* Univ Nev, Reno, BS, 57, MS, 59; Univ Calif, Davis, PhD(chem), 65. *Prof Exp:* Chemist, Aerojet-Gen Corp, 59-61 & 64-67; chemist, Kaiser Aluminum & Chem Corp, 67-86; RETIRED. *Concurrent Pos:* Consult, solid state & indust magnetic resonance. *Mem:* Am Chem Soc; Am Oil Chemists Soc; Mat Res Soc. *Res:* High resolution nuclear magnetic resonance; wide line nuclear magnetic resonance of solids; process control. *Mailing Add:* 3590 Churchill Ct Pleasanton CA 94588

PEARSON, ROBERT STANLEY, b Lawrence, Kans, Aug 12, 27; m 51; c 2. INORGANIC CHEMISTRY. *Educ:* Kans State Teachers Col Pittsburg, BS, 50; Kans State Univ, MS, 53, PhD(chem), 64. *Prof Exp:* Chemist, Midwest Solvents Co, Kans, 50-51; instr gen chem, Colo Sch Mines, 55-61; asst prof chem, Idaho State Univ, 61-68; assoc prof chem, 68-70, chmn dept, 71-78, PROF CHEM, UNIV ARK, MONTICELLO, 70- *Mem:* Sigma Xi; Am Chem Soc. *Res:* Application of trace element analysis in geochemistry and geochemical prospecting; ceramic glaze design. *Mailing Add:* Box 3095 Univ Ark Monticello AR 71655

PEARSON, RONALD EARL, b Worcester, Mass, Dec 21, 44; m 65; c 2. DAIRY SCIENCE, ANIMAL BREEDING. *Educ:* Univ Mass, Amherst, BS, 66; Iowa State Univ, MS, 70, PhD(animal breeding), 71. *Prof Exp:* Res geneticist, Genetics & Mgt Lab, Beltsville Agr Ctr, Sci & Educ Admin-Agr Res, 71-79, assoc prof, 79-85, PROF, DEPT DAIRY SCI, VA POLYTECH INST & STATE UNIV, BLACKSBURG, VA, 85- *Mem:* Am Dairy Sci Asn; Biomet Soc. *Res:* Dairy cattle breeding and management, including economic evaluation of various breeding and management systems. *Mailing Add:* Dept Dairy Sci Va Tech 2100 Animal Sci Bldg Blacksburg VA 24061-0315

PEARSON, TERRANCE LAVERNE, b Berwick, NS, Dec 19, 37; m 61; c 1. PURE MATHEMATICS. *Educ:* Acadia Univ, BSc, 60; Univ Sask, MSc, 63, PhD(math), 65. *Prof Exp:* Fel, Univ Wis, Madison, 62; lectr, 60-61, from asst prof to assoc prof, 65-78, PROF MATH, ACADIA UNIV, 78-, DEAN PURE & APPLIED SCI, 85- *Concurrent Pos:* NATO fel, Univ Calif, Santa Barbara, 66-67; fel, Birbeck Col, Univ London, 73-74. *Mem:* London Math Soc; Math Asn Am. *Res:* General topology; real analysis. *Mailing Add:* Dept of Math Acadia Univ Wolfville NS B0P 1X0 Can

PEARSON, THOMAS ARTHUR, b Berlin, Wis, Oct 21, 50; m 76; c 2. PREVENTIVE MEDICINE, LIPID METABOLISM. *Educ:* Johns Hopkins Univ, BA, 73, MD, 76, MPH, 76 & PhD(epidemiol), 83. *Prof Exp:* Fel cardiol, Johns Hopkins Sch Med, 81-83, from asst prof to assoc prof med, 83-88, from asst prof to assoc prof epidemiol, 83-88; PROF EPIDEMIOL, COLUMBIA UNIV, 88-; DIR, MARY IMOGENE BASSETT RES INST, 88- *Concurrent Pos:* Chmn, Monitoring Bd, CARDIA Proj, Nat Heart, Lung & Blood Inst, 87-, Res Comt Md Heart Asn, 86-88, Nat Res Comt, Am Heart Asn, 87-88, mem Coun Epidemiol, 87-; comnr, Md Coun Phys Fitness, 85-88; mem, Clin Applns & Prev Adv Comn, NIH, 87-91. *Honors & Awards:* Res Prize, Soc Epidemiol Res, 78. *Mem:* Soc Epidemiol Res; Am Fedn Chem Res; Am Col Epidemiol; Am Col Prev Med; Asn Teachers Prev Med. *Res:* The etiology and pathogenesis of atherosclerosis. *Mailing Add:* One Atwell Rd Cooperstown NY 13326

PEARSON, WALTER HOWARD, b Troy, NY, Mar 25, 46; div; c 2. MARINE BIOLOGY, ANIMAL BEHAVIOR. *Educ:* Bates Col, BS, 67; Univ Alaska, MS, 70; Ore State Univ, PhD(oceanog), 77. *Prof Exp:* Fisheries biol, behav ecol, Nat Marine Fish Serv, Sandy Hook Lab, 75-77; sr res scientist, behav ecol, 77-87, TECH GROUP LEADER, BATTELLE PAC NORTHWEST LAB, MARINE SCI LAB, 87- *Mem:* AAAS; Animal Behav Soc; NY Acad Sci; Crustacean Soc; Asn Chemoreception Sci. *Res:* Ecological and behavioral effect of pollution; chemoreception in crustaceans and fish, estuarine and intertidal ecology; tidal marsh ecology; fisheries biology of crabs; behavioral ecology of marine invertebrates; food habits; predator-prey relationships; ocean disposal; OCS development; design of monitoring programs; oil spill effects; NRDA studies. *Mailing Add:* Battelle Marine Sci Lab 439 W Sequim Bay Rd Sequim WA 98382

PEARSON, WESLEY A, b Red Wing, Minn, July 19, 32; m 71; c 2. ORGANIC CHEMISTRY. *Educ:* St Olaf Col, BA, 54; Univ Minn, PhD(org chem), 58. *Prof Exp:* Assoc prof, 58-75, PROF CHEM, ST OLAF COL, 75- *Concurrent Pos:* NSF fac sci fel, Mass Inst Technol, 64-65. *Mem:* Am Chem Soc. *Res:* Organic reaction mechanisms; stereochemistry of cyclic compounds. *Mailing Add:* Dept of Chem St Olaf Col Northfield MN 55057

PEARSON, WILLIAM DEAN, b Moline, Ill, Dec 6, 41; m 66; c 2. FISHERIES. *Educ:* Iowa State Univ, BS, 63; Utah State Univ, MS, 67, PhD(fisheries, biol), 70. *Prof Exp:* Biologist aid, Iowa State Conserv Comn, 61; fishery aid, US Fish & Wildlife Serv, Yankton, 62-63, fishery biologist, Logan, 64-66; res asst fisheries, Utah State Univ, 67-70; asst prof aquatic ecol, NTex State Univ, 70-75; from asst prof to assoc prof, 75-85, PROF RES, WATER RESOURCES LAB, UNIV LOUISVILLE, 85- *Concurrent Pos:* Actg dir, Water Resources Lab, 86-88. *Mem:* AAAS; Ecol Soc Am; Am Fisheries Soc; NAm Benthological Soc; Sigma Xi; Am Soc Ichthyologists & Herpetologists. *Res:* Drift of stream invertebrates; larvae fish ecology; population dynamics and production of fish and invertebrates; large river fisheries management. *Mailing Add:* Water Resources Lab Univ of Louisville Louisville KY 40292

PEART, ROBERT MCDERMAND, b Kewanee, Ill, Nov 11, 25; m 48; c 2. KNOWLEDGE SYSTEMS, PROCESS SIMULATION. *Educ:* Iowa State Univ, BS, 49; Univ Ill, MS, 57; Purdue Univ, PhD(agr eng), 60. *Prof Exp:* Agr engr, Eastern Iowa Light & Power Corp, 49-54; from instr to asst prof agr eng, Univ Ill, 54-61; from assoc prof to prof agr eng, Purdue Univ, 61-85; GRAD RES PROF AGR ENG, UNIV FLA, 85- *Concurrent Pos:* Vis agr engr, Agr Res Serv, USDA, Cornell Univ, 69-70, energy res coordr, Peoria, Ill, 80-81; dir, Am Soc Agr Engrs, 78-80; vis prof, Clemson Univ, 83. *Mem:* Fel Am Soc Agr Engrs; Am Soc Eng Educ; Am Asn Artificial Intel; Soc Computer Simulation. *Res:* Computer simulations of agricultural processes such as solar grain drying, farming operations, crop and insect growth and the development of expert systems for decision support. *Mailing Add:* Agr Eng Dept Rogers Hall Univ Fla Gainesville FL 32611

PEARTON, STEPHEN JOHN, b Hobart, Tasmania, Australia, Jan 15, 57; Australian. ION SOLID INTERACTIONS, PHYSICS OF DEFECTS SEMICONDUCTORS. *Educ:* Univ Tasmania, BS, 78, BS, 79, PhD(physics), 81. *Prof Exp:* Tech officer, Australian Atomic Energy Comn, 81-82; postdoctoral fel, Univ Calif Berkeley, 82-83; MEM TECH STAFF, AT&T BELL LABS, 84- *Mem:* Mat Res Soc; Am Physical Soc. *Res:* Physics of defects and impurities in semiconductors; heteroepitaxial growth of III V's on Si; beam processing, hydrogen in semiconductors. *Mailing Add:* AT&T Bell Labs Murray Hill NJ 07974

PEASCOE, WARREN JOSEPH, b San Pedro, Calif, Jan 15, 43; m 70; c 2. ORGANIC CHEMISTRY, POLYMER CHEMISTRY. *Educ:* Calif Inst Technol, BS, 65; Univ Ill, Urbana, PhD(chem), 70. *Prof Exp:* Res chemist, Oxford Mgr & Res Ctr, Uniroyal Inc, 70-84; ADVAN POLYMER CHEMIST, G E PLASTICS, 84- *Mem:* Am Chem Soc. *Res:* Polymerization chemistry; emulsion polymerization; engineeing plastics; weatherable plastics; blowing agents. *Mailing Add:* 5607 Greenmont Pl Vienna WV 26105

PEASE, BURTON FRANK, b Reliance, SDak, Jan 25, 28; m 53; c 5. ANALYTICAL CHEMISTRY. *Educ:* Pac Univ, BS, 50; Ore State Univ, PhD(analytical chem), 57. *Prof Exp:* Chemist, Surg Res Team & Med Serv, Grad Sch, Walter Reed Inst Med Res, 51-53; instr chem, Ore State Col, 56-57; res technologist, Shell Oil Co Calif, 57-59; from asst prof to assoc prof, 59-67, head dept, 67-70, PROF CHEM, CALIF STATE UNIV, CHICO, 67- *Mem:* Am Chem Soc; Sigma Xi. *Res:* Chemical microscopy; spectrophotometry. *Mailing Add:* Chico State Univ Chico CA 95927

PEASE, JAMES ROBERT, b Halifax, Vt, June 5, 37; c 1. RESOURCE MANAGEMENT. *Educ:* Univ Mass, BA, 60, MS, 70, PhD(resource planning), 72. *Prof Exp:* Consult & prin partner, Terra Planning Assocs, 71-73; PROF RESOURCE PLANNING, DEPT GEOG & EXTEN SERV, ORE STATE UNIV, 73- *Concurrent Pos:* Sr scientist award, US-Spain Joint Friendship Comt, 85-86. *Mem:* Asn Am Geographers; Am Soc Planning Officials. *Res:* Methodology for environmental impact analysis and for land use planning; ecological criteria for planning policy; land use and land tenure in Latin America. *Mailing Add:* Dept Geog Ore State Univ Corvallis OR 97331

PEASE, MARSHALL CARLETON, III, b New York, NY, July 30, 20; div; c 3. PHYSICS, MATHEMATICS. *Educ:* Yale Univ, BS, 40; Princeton Univ, MA, 43. *Prof Exp:* Res assoc, Radio Res Lab, Harvard Univ, 43-45; from sr engr to eng specialist & mgr tube develop, Sylvania Elec Prod, Inc, 46-60; sr engr, Stanford Res Inst, 60-64, staff scientist, SRI International, 64-84; CONSULT, 84- *Concurrent Pos:* Lectr, Stanford Univ, 60-70. *Mem:* Fel Inst Elec & Electronics Engrs. *Res:* Computer design; theory of information storage and processing; application of modern algebra to system problems. *Mailing Add:* 151 Carmel Way Portola Valley Menlo Park CA 94028

PEASE, PAUL LORIN, b New Britain, Conn, Aug 6, 43; m 74; c 2. PHYSIOLOGICAL OPTICS. *Educ:* Pa Col Optom, BS, 65, OD, 67; Univ Calif, Berkeley, PhD(physiol optics), 75. *Prof Exp:* Dir vision sci, 76-80, assoc prof vision, New England Col Optom, 73-; ASSOC PROF, UNIV HOUSTON COL OPTOM, 82- *Concurrent Pos:* Vis scientist, Univ Calif, Berkeley, 79; consult, Nat Bd Examr Optom, 80- *Mem:* Asn Res Vision & Ophthal; Optical Soc Am; AAAS; Am Acad Optom. *Res:* Neurophysiology and psychophysiology of color vision, color testing and spatial aspects of vision. *Mailing Add:* Univ Houston Col Optom Houston TX 77004

PEASE, ROBERT LOUIS, b Fitchburg, Mass, July 13, 25; m 70; c 1. PHYSICS. *Educ:* Miami Univ, AB, 43; Mass Inst Technol, PhD(physics), 50. *Prof Exp:* Asst prof math, Univ NH, 50-51; asst phys scientist, Rand Corp, 51-53; asst prof physics, Tufts Univ, 53-56; res physicist, Hughes Aircraft Co, 56-57; staff mem, Lincoln Lab, Mass Inst Technol, 57-60; assoc prof physics, Brooklyn Col, 60-64; vis res physicist, Princeton Univ, 64-65; physicist, Brookhaven Nat Lab, 65-67; PROF PHYSICS, STATE UNIV NY, COL NEW PALTZ, 67- *Concurrent Pos:* Consult, Lincoln Lab, Mass Inst Technol, 60-63, Princeton-Pa Accelerator, 64, Airborne Instrument Labs, 79, & GE Astro-space Div, 82- *Mem:* Am Phys Soc; Inst Elec & Electronic Engrs; Opers Res Soc Am. *Res:* Nuclear and elementary particle theory; electromagnetic theory; electrodynamics and particle accelerator theory. *Mailing Add:* Dept Physics State Univ NY New Paltz NY 12561

PEASE, ROBERT WRIGHT, b Evanston, Ill, Mar 30, 17; m 38; c 2. PHYSICAL GEOGRAPHY. *Educ:* Univ Calif, Los Angeles, BA, 38, MA, 46, PhD(geog), 60. *Prof Exp:* Tool engr, Douglas Aircraft Co, 40-45; teacher, Los Angeles City Sch Dist, 44-63; lectr geog, Univ Calif, Los Angeles, 63-64; instr, Los Angeles Jr Col Dist, 66-67; assoc prof, 67-77, PROF GEOG, UNIV CALIF, RIVERSIDE, 77- *Concurrent Pos:* Consult, Calif State Dept Educ, 63 & Geog Prog, US Geol Surv, 73- *Mem:* Sigma Xi; Asn Am Geog; Am Soc Photogram. *Res:* Remote sensing, radiation and crop climatology; utilization of remotely acquired data to measure surface energy-exchange phenomena; systems of filters and application for use with color infrared film for earth resources purposes. *Mailing Add:* Dept Geog Univ of Calif Riverside CA 92521

PEASE, ROGER FABIAN WEDGWOOD, b Cambridge, Eng, Oct 24, 36; m 60; c 3. ELECTRICAL ENGINEERING. *Educ:* Cambridge Univ, BA, 60, MA & PhD(elec eng), 64. *Prof Exp:* Res fel, Trinity Col, Cambridge Univ, 63-67; mem tech staff, Bell Tel Labs, Inc, 67-78; PROF ELEC ENG, STANFORD UNIV, 78- *Concurrent Pos:* Consult, IBM Corp, 64-67; from asst prof to assoc prof, Univ Calif, Berkeley, 64-69. *Mem:* Inst Elec & Electronics Engrs. *Res:* Electron microscopy and electron beam technology; digital encoding of television signals; microstructures and their applications; high density electronic circuitry. *Mailing Add:* Stanford Elec Labs Mcc 204 Stanford Univ Stanford CA 94305

PEASLEE, ALFRED TREDWAY, JR, b Dubuque, Iowa, June 25, 30; m 59; c 3. THEORETICAL PHYSICS. *Educ:* Harvard Univ, AB, 52, AM, 53, PhD(physics), 55. *Prof Exp:* STAFF MEM, LOS ALAMOS SCI LAB, 55- *Mem:* AAAS; Am Phys Soc; Sigma Xi. *Res:* Quantum electrodynamics; classical theoretical physics. *Mailing Add:* 114 El Viento Los Alamos NM 87544

PEASLEE, DAVID CHASE, b White Plains, NY, July 23, 22; m 47, 73; c 3. PARTICLE & THERMAL PHYSICS, NUCLEAR STRUCTURE. *Educ:* Princeton Univ, AB, 43; Mass Inst Technol, PhD(physics), 48. *Prof Exp:* Analyst, Opers Res Group, Washington, DC, 44-46; res assoc physics, Mass Inst Technol, 46-48; assoc physicist, Kellex Corp, 48-49; fel, AEC, Zurich, 49-50; asst prof physics, Wash Univ, 50-51; res assoc, Columbia Univ, 51-54; from assoc prof to prof, Purdue Univ, 54-59; Fulbright fel, Australian Nat Univ, 58, reader theoret physics, 59-61, prof theoret physics, 61-76; vis prof physics, Brown Univ, 76-77, adj prof physics, 77-81; VIS PROF PHYSICS, UNIV MD, 81- *Concurrent Pos:* Mem staff, Dept Energy, Washington, DC. *Mem:* Fel Am Phys Soc. *Res:* Structure and interactions of nucleons and elementary; Hadron spectroscopy and structure; high energy spin physics; low energy antiproton physics; theoretical physics. *Mailing Add:* Dept Physics Univ Maryland College Park MD 20742

PEASLEE, DOYLE E, b Stockton, Kans, Feb 24, 30; m 53; c 2. SOIL FERTILITY. *Educ:* Kans State Univ, BS, 52, MS, 56; Iowa State Univ, PhD(soil fertil), 60. *Prof Exp:* From asst soil scientist to assoc soil scientist, Conn Agr Exp Sta, 60-66; assoc prof, 66-71, PROF AGRON, UNIV KY, 71-, DIR, DIV REGULATORY SCI, RES & TECHNOL, 80- *Mem:* Am Soc Agron; Soil Sci Soc Am; Crop Sci Soc Am; Int Soil Sci Soc; Sigma Xi. *Res:* Fertilizer reactions in soils; availability of fertilizer and soil nutrients to plants; analytical methods for elements in soils; effects of nutrients on photosynthesis. *Mailing Add:* Div Regulatory Serv Univ Ky Lexington KY 40506

PEASLEE, MARGARET H, b Chicago, Ill, June 15, 35; m 57; c 1. ENDOCRINOLOGY. *Educ:* Fla Southern Col, BS, 59; Northwestern Univ, MS, 64, PhD, 66. *Prof Exp:* Asst prof, Fla Southern Col, 66-68; from asst prof to prof biol, Univ SDak, Vermillion, 68-76; prof head, Dept Zool, 76-90, PROF BIOL SCI, COL LIFE SCI, LA TECH UNIV, 76-, ASSOC DEAN & DIR RES & GRAD STUDIES, 90- *Concurrent Pos:* Acad opportunity liaison officer, Univ SDak, Vermillion, 74-76. *Mem:* Fel AAAS; Am Inst Biol Sci; Am Soc Zoologists; Sigma Xi. *Res:* Vertebrate endocrinology; melanocyte-stimulating hormone; pigment cell physiology; hormonal relationships in development. *Mailing Add:* Col Life Sci PO Box 10198 La Tech Univ Ruston LA 71272-0045

PEATMAN, JOHN B(URLING), b Port Chester, NY, Nov 28, 34; m 56; c 3. ELECTRICAL ENGINEERING. *Educ:* Swarthmore Col, BSEE, 56; Case Inst Technol, MSEE, 60, PhD(digital systs), 65. *Prof Exp:* Assoc eval engr, Minneapolis-Honeywell Regulator Co, 56-59; asst prof elec eng, Univ Mo-Rolla, 60-62; from asst prof to assoc prof, 64-77, PROF ELEC ENG, GA INST TECHNOL, 77- *Mem:* Inst Elec & Electronics Engrs. *Res:* Digital systems engineering; algorithmic processes; time-oriented digital systems design techniques. *Mailing Add:* Sch Elec Eng Ga Inst Tech 225 N Ave NW Atlanta GA 30332

PEATMAN, WILLIAM BURLING, b Port Chester, NY, July 15, 39; m 80; c 3. PHYSICAL CHEMISTRY. *Educ:* Harvard Univ, BA, 61; Northwestern Univ, MS, 63, PhD(chem), 69. *Prof Exp:* Instr chem, Elmhurst Col, 63-65; NSF grant, Univ Chicago, 69-70; asst prof chem, Vanderbilt Univ, 70-73, assoc prof, 73-79; DEP DIR RES, BERLINER ELECTRONENSPEICHERRING GES FUER SYNCHROTRONSTRAHLUNG MBH, 79- *Concurrent Pos:* Prof, Free Univ Berlin, W Germany, 80- *Honors & Awards:* A P Sloan, 73. *Mem:* Am Phys Soc; Am Inst Physics. *Res:* Energy transfer processes in plasmas; photoionization studies of molecules. *Mailing Add:* Berliner Elektronenspeicherring GMBH Lentzeallee 100 Berlin 1 33 Germany

PEAVY, HOWARD SIDNEY, b Brookhaven, Miss, Sept 11, 42; m 62; c 2. SANITARY & ENVIRONMENTAL ENGINEERING. *Educ:* La State Univ, BS, 69; Duke Univ, MS, 70; Okla Univ, PhD(environ eng), 74. *Prof Exp:* Chief engr, Okla Dept Pollution Control, 72-74; from asst prof to assoc prof, 74-84, PROF ENVIRON ENG, MONT STATE UNIV, 85-, DIR, WATER RESOURCES RES CTR, 82- *Concurrent Pos:* Vis prof, Rogaland District Shogshole Stavanger, Norway, 80-81. *Mem:* Am Soc Civil Engrs; Am Water Works Asn; Water Pollution Control Fedn; Am Soc Eng Educ. *Res:* Environmental effects of energy production and groundwater quality; municipal and industrial water and wastewater treatment. *Mailing Add:* Dept Civil Eng Mont State Univ Bozeman MT 59715

PEBLY, HARRY E, b Sharpsville, Pa, Feb 1, 23; m 49; c 1. MATERIALS ENGINEERING, INFORMATION TECHNOLOGY. *Educ:* Pa State Univ, BS, 44; Stevens Inst Technol, MS, 57. *Prof Exp:* Chief, plastics fabrication unit, Army Ord, Picatinny Arsenal, 47-60, dir, Plastics Tech Eval Ctr, Army Armament Res & Develop Command, 60-86, chief, Org Mat Br, Eng Command, 86-89; CONSULT, PLASTICS, COMPOSITES & ADHESIVES, 89- *Concurrent Pos:* Trustee, Plastics Inst Am, 71-77 & Eng Index, 70-80, ed, adv bd, Int Plastics Selector, 78-81, mem, Fed Coun Sci Technol, Panel Info Anal Ctr, 67-72. *Honors & Awards:* Outstanding Serv Award, Soc Plastics Indust, 65; Comdr's Award Civilian Serv, 89. *Mem:* Soc Plastics Eng; Soc Advan Mat & Process Eng; Am Defense Preparedness Asn. *Mailing Add:* 198 Center Grove Rd Randolph NJ 07869

PECCEI, ROBERTO DANIELE, b Torino, Italy, Jan 6, 42; m 62; c 2. ELEMENTARY PARTICLE PHYSICS, THEORETICAL PHYSICS. *Educ:* Mass Inst Technol, BS, 62, PhD(physics), 69; NY Univ, MS, 64. *Prof Exp:* Res assoc physics, Univ Wash, 69-71; asst prof physics, Stanford Univ, 71-78; staff mem physics, Max Planck Inst, Munich, 78-84; head theory group physics, Ger Electron-Synchrotron, Hamburg, Ger, 84-89; PROF & CHMN PHYSICS, PHYSICS DEPT, UNIV CALIF, LOS ANGELES, 89- *Concurrent Pos:* Vis, Rutherford Lab, UK, 73-76; Schroedinger prof, Univ Vienna, Austria, 83; mem, LEP Exp Comt, CERN, 82-90, Physics Res Comt, Ger Electron-Synchrotron, 86-89, Physics Adv Comt, Fermilab, 90-, Sci Policy Comt, SLAC, 90-, Bd Overseers, Fermilab, 91- *Mem:* Fel Am Phys Soc. *Res:* Elementary particle theory, particularly on electroneak interactions and in the interface of elementary particles and cosmology. *Mailing Add:* Dept Physics Univ Calif 405 Hilgard Ave Los Angeles CA 90024

PECCI, JOSEPH, b Boston, Mass, Nov 20, 30; m 56; c 5. CLINICAL CHEMISTRY, BIOCHEMISTRY. *Educ:* Mass Col Pharm, BS, 56, MS, 58. *Prof Exp:* Biol chemist, City of Boston Police Dept, 58-59; res chemist, Air Force Cambridge Res Labs, 59-69; res chemist, Vet Admin Hosp, Boston, 69-76, SUPVRY CHEMIST, VET ADMIN MED CTR, BOSTON, 76- *Mem:* AAAS; Am Asn Clin Chemists; Am Chem Soc; AMA. *Res:* Drug detection in biological fluids; clinical chemistry. *Mailing Add:* PO Box 376 Holbrook MA 02343-0376

PECHACEK, TERRY FRANK, b Flatonia, Tex, Dec 14, 47; m 74. CHRONIC DISEASE PREVENTION, SMOKING PREVENTION & CONTROL. *Educ:* Univ Tex, BA, 70, MA, 73, PhD(psychol), 77. *Prof Exp:* Teaching asst math, Univ Tex, Austin, 70-71; clin psychol intern, Palo Vet Admin Hosp, Palo Alto, Calif, 75-76, res assoc psychol, 76-77; postdoctoral fel, Div Epidemiol, Sch Pub Health, Univ Minnesota, Minn, 77-80, from asst prof to assoc prof, 80-86; chief pub health, Smoking Tobacco & Cancer Br, Nat Cancer Inst, Dept Health & Human Serv, NIH, 88-91; ASSOC PROF EPIDEMIOL, DEPT SOCIAL & PREV MED, BUFFALO MED SCH, STATE UNIV NY, 91- *Concurrent Pos:* Actg intervention dir, Div Epidemiol, Mult Risk Factor Intervention Trial, Univ Minn, 77, dir smoking res, Sch Pub Health, 79-86; mem, NIH/ADAMA Study Sect, 80-86 & Prev Comt, Nat Cancer Control, Objectives for year 2000, Nat Cancer Inst, NIH, 84; prin or sr investr, NIH grants, 80-86; expert, Div Cancer Prev & Control, Nat Cancer Inst, Dept Health & Human Serv, NIH, 86-91; chairperson, Eval Adv Comt, Tobacco Control Prog, Calif Dept Health, 89-91; tech consult, Nat Patriotic Health Campaign Comt, Ministry Pub Health, People's Repub China, 90-91; mem, Cardiovasc Epidemiol Coun, Am Heart Asn. *Honors & Awards:* Spec Act Serv Award, Nat Cancer Inst, 88; Award of Merit, NIH, USPHS, 89. *Mem:* Am Pub Health Asn; Soc Epidemiol Res; Am Heart Asn; Am Soc Prev Oncol; Int Fedn Cardiol. *Res:* Design and implementation of chronic disease prevention trials; role of smoking cessation in prevention of cardiovascular, cancer and chronic obstructive lung disease; evaluation of preventive services and health promotion programs. *Mailing Add:* Dept Social & Prev Med Buffalo Med Sch State Univ NY 2211 Main St Buffalo NY 14214

PECHAN, MICHAEL J, b Avoca, Wis, Jan 13, 50; m 78; c 3. CONDENSED MATTER, MAGNETISM. *Educ:* Wis State Univ, BS, 71; Iowa State Univ, PhD(physics), 77. *Prof Exp:* Instr physics, Iowa State Univ, 79-81; asst prof, 81-87, ASSOC PROF PHYSICS, MIAMI UNIV, 87- *Concurrent Pos:* Vis scientist, Argonne Nat Lab, 83, 85 & 86; vis asst prof, Univ Ill, Champaign-

Urbana, 83 & 84. *Mem:* Sigma Xi; Am Phys Soc. *Res:* Superlattice, low dimensional and interface effects in magnetic multilayer materials; use of ferromagnetic resonance, torque magnetometry and squid magnetometry to characterize anisotropies in magnetic properties. *Mailing Add:* Dept Physics Miami Univ Oxford OH 45056

PECHENIK, JAN A, b Jamaica, NY, May 5, 50; m 82; c 1. INVERTEBRATE ZOOLOGY, PHYSIOLOGICAL & LARVAL ECOLOGY. *Educ:* Duke Univ, BA, 71; Mass Inst Technol, MS, 75; Univ RI, PhD(biol oceanog), 78. *Prof Exp:* Marine biologist, Environ Res Lab, 76-78; lectr biol, 78-79, asst prof, 79-84, ASSOC PROF, TUFTS UNIV, 84- *Mem:* Am Soc Zool; Am Inst Biol Sci. *Res:* Reproduction and development of marine invertebrates; larval development; metamorphosis of marine mollusks; author of one book. *Mailing Add:* Dept Biol Tufts Univ Medford MA 02155

PECHET, LIBERTO, b Braila, Rumania, July 6, 26; m 50; c 3. MEDICINE. *Educ:* Hebrew Univ, MD, 52. *Prof Exp:* Intern, Hadassah Hosp, Jerusalem, Israel, 51-52; village physician, Civil Serv, Ministry of Health, 52-53; res fel med, Harvard Med Sch, 57-59, instr med, 59-61, assoc, 61-66; asst prof path, Sch Med, Univ Colo, 66-69; asst chief, Lab Serv, Vet Admin Hosp, Boston, 69-70; assoc prof, 70-74, PROF MED & PATH & CHIEF SECT HEMAT, MED SCH, UNIV MASS, 74- *Concurrent Pos:* Res fel med, Beth Israel Hosp, 57-60, resident, 60-61, assoc, 61-65, asst vis physician, 65-66; attend hemat, Denver Vet Admin Hosp, 66-69. *Mem:* Am Soc Hemat; Am Fedn Clin Res; Soc Exp Biol & Med; Int Soc Hemat; Am Col Physicians. *Res:* Clinical, physiological and biochemical aspects of blood coagulation. *Mailing Add:* Univ Mass Med Sch 55 Lake Ave Worcester MA 01655

PECHUKAS, PHILIP, b Akron, Ohio, Oct 30, 42. THEORETICAL CHEMISTRY. *Educ:* Yale Univ, BS, 63; Univ Chicago, PhD(chem physics), 66. *Prof Exp:* Nat Acad Sci-Nat Res Coun fel theoret chem, Nat Bur Standards, Washington, DC, 66-67; from asst prof to assoc prof, 67-78, PROF CHEM, COLUMBIA UNIV, 78- *Concurrent Pos:* Sloan Found fel, 70-74; Guggenheim fel, 75. *Mem:* Am Chem Soc; fel Am Phys Soc. *Res:* Chemical dynamics; semiclassical approximation. *Mailing Add:* Dept Chem Columbia Univ New York NY 10027

PECHUMAN, LAVERNE LEROY, b Lockport, NY, Oct 18, 13; m 39; c 2. ENTOMOLOGY. *Educ:* Cornell Univ, BS, 35, MS, 37, PhD(entom), 39. *Prof Exp:* Asst entom, Cornell Univ, 35-39; entomologist, Ortho Div, Chevron Chem Co, 39-46, dist mgr, 46-61, sr res scientist, 61-62; from assoc prof to prof, 62-82, cur, 62-82, EMER PROF ENTOM, CORNELL UNIV, 82- *Mem:* Fel AAAS; Entom Soc Am; Entom Soc Can. *Res:* Biogeography and insect distribution patterns; taxonomy and biology of Diptera, especially Tabanidae; archeology of New York State. *Mailing Add:* 16 Lakeview Dr Lansing NY 14882

PECINA, RICHARD W, b Cedar Rapids, Iowa, Mar 2, 35; m 56; c 3. CHEMICAL & INDUSTRIAL ENGINEERING. *Educ:* Univ Iowa, BS, 56, MS, 57, PhD(indust & mgt eng), 62. *Prof Exp:* Res engr, Abbott Labs, 59-61; fel recovery potable water from urine, Univ Iowa, 62; res engr, Abbott Labs, North Chicago, 62-63, mgr mat & packaging res, 63-66, mgr plastics prods res & develop, 66-68, opers mgr plastics prods mfg, 68-69, plant mgr hosp prods opers, 69-71, mgr int hosp prod res & develop, 71-72; vpres sci affairs, Respiratory Care, Inc, Ill, 72-75; PRES, RICHARD W PECINA & ASSOC, INC, 75- *Mem:* Soc Plastics Engrs. *Res:* Packaging and materials research; water vapor permeability of plastic packaging films; research and development of plastic disposable devices for intravenous feeding solutions. *Mailing Add:* 2348 N Lewis Ave Waukegan IL 60087

PECK, AMMON BROUGHTON, IMMUNOGENETICS, TRANSPLANTATION. *Educ:* Univ Wis, PhD(med microbiol), 72. *Prof Exp:* ASSOC PROF PATH, UNIV FLA, 79- *Mailing Add:* Dept Path Col Med Univ Fla Gainesville FL 32610

PECK, CARL CURTIS, b Emporia, Kans, Mar 28, 42; m 72; c 2. EXPERIMENTAL BIOLOGY, MEDICINE. *Educ:* Univ Kans, BA, 63, MD, 68. *Prof Exp:* Intern med, Tripl Army Hosp, & resident phys int med, Letterman Army Hosp, 68-72; res fel clin pharmacol, Univ Calif Med Ctr, 72-74; res clin pharmacol, Div Blood Res, Letterman Army Inst Res, 74-77, chief res & develop, 77-80; prof med & pharmacol & dir, Div Clin Pharmacol, Uniformed Serv Univ Health Sci, 80-87; DIR, CTR DRUG EVAL & RES, FOOD & DRUG ADMIN, 87- *Concurrent Pos:* Asst prof med, Univ Calif Med Ctr, 75- *Honors & Awards:* Res Achievement Award, US Army Med Res & Develop Command, 78. *Mem:* AAAS; Am Soc Clin Pharmacol & Therapeut; Am Col Phys; Am Fedn Clin Res. *Res:* Medicine; biomathematics; biostatistics; experimental design; kinetics; modeling; clinical pharmacology. *Mailing Add:* 5600 Fishers Lane Rockville MD 20857

PECK, CHARLES WILLIAM, b Freer, Tex, Nov 29, 34; m 80; c 4. EXPERIMENTAL HIGH ENERGY PHYSICS. *Educ:* NMex Col Agr & Mech Arts, BS, 56; Calif Inst Technol, PhD(physics), 64. *Prof Exp:* Res fel, 64-65, from asst prof to assoc prof, 65-77, PROF PHYSICS, CALIF INST TECHNOL, 77- *Mem:* Am Inst Physics; Sigma Xi. *Res:* Meson photoproduction; bubble chamber physics; e-plus, e-minus storage ring experiments. *Mailing Add:* Lauritsen Lab Calif Inst Technol Pasadena CA 91125

PECK, D STEWART, b Grand Rapids, Mich, Oct 19, 18; m 44; c 2. ELECTRICAL ENGINEERING. *Educ:* Univ Mich, BS, 39, MS, 40. *Prof Exp:* Design engr, Electron Tubes, Gen Elec, 40-47; dept head, Reliabilities Testing, Bell Labs, 47-80; RETIRED. *Concurrent Pos:* Seminar instr Univs US & Europe, 81. *Mem:* Fel Inst Elec & Electronics Engrs. *Res:* Electrical Engineering. *Mailing Add:* 3646 Highland St Allentown PA 18104

PECK, DALLAS LYNN, b Cheney, Wash, Mar 28, 29; m 51; c 3. GEOLOGY. *Educ:* Calif Inst Technol, BS, 51, MS, 53; Harvard Univ, PhD, 60. *Prof Exp:* Asst field geol, Calif Inst Technol, 51-52; asst struct geol, Harvard Univ, 52-53; geologist, 51-66, asst chief geologist, Off Geochem & Geophys, 67-72, res geologist, 72-77, chief geologist, 77-81, DIR, US GEOL SURV, 81- *Concurrent Pos:* Mem vis comt, Dept Geol Sci, Harvard Univ, 71-75; mem geosci adv comt, Los Alamos Sci Labs, 75-77; mem earth sci adv bd, Stanford Univ, 81- *Mem:* Fel Geol Soc Am; fel, Am Geophys Union; Soc Econ Geologists; Mineral Soc Am; fel AAAS. *Res:* Igneous petrology; Hawaiian lava lakes; Sierra Nevada Batholith. *Mailing Add:* 2524 Heathcliff Lane Reston VA 22091

PECK, DAVID W, b Whitwell, Tenn, Sept 17, 25; m 52; c 3. PHYSICAL ORGANIC CHEMISTRY. *Educ:* Emory & Henry Col, BS, 49; Univ Va, MS, 51, PhD, 52. *Prof Exp:* Res chemist, Union Carbide Corp, 52-62, res proj leader org chem, 62-67, group leader, 67-70, develop scientist, 70-78, sr develop scientist org chem, 78-85; RETIRED. *Mem:* Am Chem Soc. *Res:* Synthetic organic chemistry; pesticides. *Mailing Add:* 102 Foxridge Ct Chapel Hill NC 27514

PECK, EDSON RUTHER, b Evanston, Ill, Oct 29, 15; m 46; c 5. PHYSICS. *Educ:* Northwestern Univ, BA, 36, MS, 37; Univ Chicago, PhD(physics), 45. *Prof Exp:* Res physicist, Nat Defense Res Comt, Northwestern Univ, 42, instr physics, 42-46, asst prof, 46-49, assoc prof, 49-62; prof, 62-78, EMER PROF PHYSICS, UNIV IDAHO, 78- *Concurrent Pos:* Am Asn Physics Teachers-Am Inst Physics regional counr, 63-66. *Mem:* AAAS; Math Soc Am; Am Phys Soc; Optical Soc Am; NY Acad Sci. *Res:* Spectroscopy; interferometry; optical dispersion of gases. *Mailing Add:* 200 James St No 304 Edmonds WA 98020

PECK, EMILY MANN, b Ft Myers, Fla, Sept 5, 46; m 73. MATHEMATICS, FUNCTIONAL ANALYSIS. *Educ:* NC State Univ, Raleigh, BS, 67; Univ Ill, MS, 68, PhD(math), 72. *Prof Exp:* Asst prof math, Vassar Col, 72-73; asst dean math, 73-80, Col Lib Arts & Sci, 80-88, ASST PROF MATH, UNIV ILL, 73-, ASSOC DEAN, COL LIB ARTS & SCI, 88- *Mem:* Am Math Soc; Math Asn Am; Sigma Xi. *Res:* Spaces of continuous functions; Banach lattices. *Mailing Add:* 202 E Pennsylvania Ave Urbana IL 61801

PECK, ERNEST JAMES, JR, b Port Arthur, Tex, July 26, 41; c 2. BIOCHEMISTRY, NEUROCHEMISTRY. *Educ:* Rice Univ, BA, 63, PhD(biochem), 66. *Prof Exp:* Res assoc biol sci, Purdue Univ, 66-68, Am Cancer Soc fel, 67-69, asst prof, 68-73; from asst prof to prof cell biol, Baylor Col Med, 73-82; PROF & CHMN BIOCHEM, UNIV ARK MED SCI, 82- *Mem:* Am Chem Soc; Am Soc Biol Chemists; Am Soc Neurochem; Biophys Soc; Soc Neurosci; Endorine Soc. *Res:* Chemical processes of cell communication, including steroids, peptide hormones and neurotransmitters; detection and mechanism of action. *Mailing Add:* Col Sci/Math Univ 4505 Maryland Pkwy Las Vegas NV 89154

PECK, EUGENE LINCOLN, b Kansas City, Kans, June 2, 22; m 46; c 4. HYDROLOGY, METEOROLOGY. *Educ:* Univ Utah, BS, 47, MS, 51; Utah State Univ, PhD(civil eng), 67. *Prof Exp:* Res hydrologist, Western Region, US Weather Bur, Nat Oceanic & Atmospheric Admin, 48-67, chief res br, Hydrol Res & Develop Lab, 67-72, asst dir, Hydrol Res Lab, 73, dir, 74-80; PRES, HYDEX CORP, FALLS CHURCH, VA, 80- *Concurrent Pos:* Lectr, Univ Utah, 56- *Honors & Awards:* Silver Medal, US Dept Com, 59 & 75. *Mem:* Am Meteorol Soc; Am Geophys Union; Am Water Resources Asn. *Res:* Hydrometeorology; precipitation; snow. *Mailing Add:* 2203 Lydia Pl Vienna VA 22181

PECK, GARNET E, b Windsor, Ont, Feb 4, 30; US citizen; m 57; c 4. PHARMACY. *Educ:* Ohio Northern Univ, BS, 57; Purdue Univ, MS, 59, PhD(indust pharm), 62. *Prof Exp:* Pharmaceut technician, Strong Cobb & Co, Ohio, 47-51 & 53; instr pharmaceut chem, Purdue Univ, 59-62; sr scientist, Mead Johnson Res Ctr, 62-65, group leader, 65-67; assoc prof, 67-73, PROF INDUST PHARM, PURDUE UNIV, WEST LAFAYETTE, 73-, DIR INDUST PHARM LABS, 75-, ASSOC DEPT HEAD, 89- *Honors & Awards:* Cath Nat Acad Sci, US. *Mem:* Fel AAAS; Am Chem Soc; Am Pharmaceut Asn; fel Am Inst Chemists; NY Acad Sci; fel Acad Pharmaceut Res & Sci; Soc Mfg Engrs; Parenteral Drug Asn; Am Asn Pharmaceut Scientists; fel Am Asn Pharmaceut Scientists. *Res:* Pharmaceutical product development; drug stability; application of radioisotopes to pharmaceutical processing and analysis; flow of solids; solid surface studies; solid coating techniques. *Mailing Add:* Purdue Univ Sch Pharm West Lafayette IN 47906

PECK, HARRY DOWD, JR, b Middletown, Conn, May 18, 27; m 75; c 5. MICROBIAL PHYSIOLOGY, ENZYMOLOGY. *Educ:* Wesleyan Univ, BA, 50, MA, 52; Western Reserve Univ, PhD(microbiol), 56. *Prof Exp:* Res fel biochem, Mass Gen Hosp, 56-57; vis investr, Rockefeller Inst, 57-58; assoc biochemist, Oak Ridge Nat Lab, 58-64; NSF sr fel, Nat Ctr Sci Res, Ministry Educ, France, 64-65; head, Dept Biochem, 65-85, dir, Sch Chem Sci, 85-91, PROF BIOCHEM, UNIV GA, 65- *Concurrent Pos:* Mem, adv panel metabolic biol, NSF & adv panel fels in microbiol, NIH, 67-70; found lectr, Am Soc Microbiol, 73-74 & 81-82; dir res, Cent Nuclear Res Acad, France. *Mem:* Am Soc Biol Chemists; Am Soc Microbiol; Soc Gen Microbiol; Am Chem Soc. *Res:* Enzymology of respiratory sulfate reduction, hydrogen metabolism and electron transfer in sulfate reducing bacteria; bioenergetics of anaerobic bacteria; biochemistry and molecular biology of nickel-containing hydrogenases. *Mailing Add:* Dept Biochem Univ Ga Life Sci Bldg Athens GA 30602

PECK, JOHN F, b Phoenix, Ariz, May 16, 36; m 63; c 2. METALLURGY. *Educ:* Univ Ariz, BS, 58; Mass Inst Technol, MS, 61, PhD(metall), 63. *Prof Exp:* Sr engr, Motorola, Inc, Ariz, 63-69; PROJ ENGR, ENVIRON RES LAB, INST ATMOSPHERIC PHYSICS, UNIV ARIZ, 69- *Mem:* Nat Asn Corrosion Engrs; fel AAAS. *Res:* Saline water corrosion; green house design for arid climates; solar space heating; evaporative cooling. *Mailing Add:* 4145 E Fourth St Tucson AZ 85711

PECK, JOHN H, b Concord, MA, Jul 20, 37; m 62; c 3. DRILLING TECHNOLOGY, SITE CHARACTERIZATION. *Educ:* Univ MA, BS, 60, Dartmouth Col, MA, 62. *Prof Exp:* Geologist US Geol Surv, 62-74; sr geologist, Stone & Webster Eng Co, 74-76, consult geologist, 76-85. *Concurrent Pos:* Chmn, Asn Eng Geologist, 82-83. *Mem:* Fel Geol Soc Am; Am Asn Petroleum Geologist; Panhandle Geol Soc. *Res:* Geological disposal of high level radioactive waste; management of field investigation. *Mailing Add:* 5117 Bromley Ave Las Vegas NV 89107

PECK, JOHN HUBERT, b Rochester, NY, Oct 4, 42; m 64; c 4. MEDICAL ENTOMOLOGY, ENVIRONMENTAL SCIENCE. *Educ:* Clark Univ, BA, 64; Univ Calif, Berkeley, PhD(parasitol), 68. *Prof Exp:* Assoc prof, 68-80, PROF BIOL, ST CLOUD STATE UNIV, 80- *Res:* Ecology and natural control of filth flies; impact of camping on wilderness ecology. *Mailing Add:* Dept of Biol St Cloud State Univ St Cloud MN 56301

PECK, LYMAN COLT, b Lebanon, Ohio, Dec 10, 20; m 42; c 2. MATHEMATICS. *Educ:* Yale Univ, BS, 42; Univ Chicago, SM, 47; Ohio State Univ, PhD(math educ), 53. *Prof Exp:* Instr math, Ohio Univ, 47-49; asst prof educ, Fla State Univ, 51-52; asst prof math, Iowa State Teachers Col, 52-56; assoc prof, Ohio Wesleyan Univ, 56-61; chmn dept, Miami Univ, 71-73, prof math, 61-86; RETIRED. *Mem:* Math Asn Am. *Res:* College and high school mathematics curriculum; writing software for the MacIntosh Computer to help children learn mathematical concepts and problem-solving. *Mailing Add:* 223 Field Crest Ave Oxford OH 45056

PECK, MERLIN LARRY, b Boise, Idaho, Mar 1, 40; m 63; c 2. ORGANIC CHEMISTRY. *Educ:* Col of Idaho, BS, 62; Mont State Univ, PhD(chem), 71. *Prof Exp:* Asst prof chem, Lake Superior State Col, 66-69; lectr, Univ Ariz, 69-72; asst dept head educ activity, Am Chem Soc, 72-74; asst prof, 74-78, ASSOC PROF CHEM, TEX A&M UNIV, 78- *Mem:* Am Chem Soc; Nat Sci Teachers Asn; Sigma Xi. *Res:* Development of interactive instructional programs and instructional aids; identification and synthesis of small, naturally occurring amines. *Mailing Add:* Dept Chem Tex A&M Univ College Station TX 77843

PECK, NATHAN HIRAM, b Phelps, NY, Feb 21, 23; m 52; c 6. HORTICULTURE. *Educ:* Cornell Univ, BS, 51, PhD(soils), 56. *Prof Exp:* Soil scientist, Agr Res Serv, USDA, 56-57; res agronomist, Bird's Eye Div, Gen Foods Corp, 57-59; from asst prof to assoc prof, 59-74, PROF HORT SCI, NY STATE COL AGR, CORNELL UNIV, 74- *Concurrent Pos:* Assoc prof, Ore State Univ, 65-66; sabbatical, Univ Calif, Davis, 76. *Mem:* Am Soc Agron; Am Soc Hort Sci. *Res:* Soil fertility, cultural practices and mineral nutrition for vegetable crops. *Mailing Add:* 1463 Lyons Rd Phelps NY 14532

PECK, NEWTON TENNEY, b Honolulu, Hawaii, Feb 3, 37; m 73. FUNCTIONAL ANALYSIS. *Educ:* Haverford Col, BA, 59; Univ Wash, PhD(math), 64. *Prof Exp:* Instr math, Yale Univ, 64-66; lectr, Univ Warwick, 67-68; from asst prof to assoc prof, 68-82, PROF MATH, UNIV ILL, URBANA, 82- *Concurrent Pos:* Alexander von Humboldt Found res fel, Univ Frankfurt, 66-67; Australian Res Coun fel, Monash Univ, 83. *Mem:* Am Math Soc; Math Asn Am; London Math Soc; Sigma Xi. *Res:* Functional analysis, especially non locally convex topological linear spaces. *Mailing Add:* Dept Math Univ Ill 1409 W Green St Urbana IL 61801

PECK, RALPH B(RAZELTON), b Winnipeg, Man, June 23, 12; US citizen; m 37; c 2. CIVIL ENGINEERING. *Educ:* Rensselaer Polytech Inst, CE, 34, DCE, 37. *Hon Degrees:* DEng, Rensselaer Polytech Inst, 74; DSC, Laval Univ, 87. *Prof Exp:* Detailer, Am Bridge Co, Pa, 37-38; lab asst, Arthur Casagrande, Mass, 38-39; lectr, Armour Inst Technol, 39-41; chief testing engr, Holabird & Root, Scioto Ord Plant, Ohio, 41; prof, 41-74, EMER PROF FOUND ENG, UNIV ILL, URBANA, 74- *Concurrent Pos:* Asst subway engr, Chicago, Ill, 39-42; consult, 41- *Honors & Awards:* Norman Medal, Am Soc Civil Engrs, 43, Wellington Prize, 65, Karl Terzaghi Award, 69 & President's Award, 88; Nat Soc Prof Engrs Award, 72, Nat Medal Sci, 74, Washington Award, 76; John Fritz Medal, 88. *Mem:* Nat Acad Eng; hon mem Am Soc Civil Engrs; Nat Soc Prof Engrs; Geol Soc Am; Int Soc Soil Mech & Found Eng (pres, 69-73); Am Acad Arts & Sci. *Res:* Behavior of soil masses under stress; foundations; dams; tunnels. *Mailing Add:* 1101 Warm Sands Dr SE Albuquerque NM 87123

PECK, RAYMOND ELLIOTT, geology; deceased, see previous edition for last biography

PECK, RICHARD MERLE, b Cleveland, Ohio, Mar 1, 21; m 50, 63; c 3. ORGANIC CHEMISTRY. *Educ:* Univ Md, BS, 43, PhD(org chem), 47. *Prof Exp:* Asst chem, Univ Md, 43-45, res chemist, 45-47; res assoc, Res Found, Ohio State Univ, 47-49; res assoc, Inst Cancer Res, 49-56, assoc mem, 56-82; RETIRED. *Concurrent Pos:* Am Cancer Soc fel, Royal Cancer Hosp, London, 55-56. *Mem:* Am Chem Soc; Am Asn Cancer Res. *Res:* Antimalarial synthesis; sulfonamides; quinolines; carcinogen-protein conjugates; alkylating antitumor, mutagenic, and carcinogenic agents with mixed bifunctionality; immunization as protection against chemical carcinogenesis in the rat. *Mailing Add:* 141 Zane Ave Philadelphia PA 19111

PECK, ROBERT E, b Pasadena, Calif, Oct 16, 47; m 70; c 2. COMBUSTION, THERMAL SCIENCES. *Educ:* Univ Calif, Berkeley, BS, 69; Univ Calif, Irvine, MS, 72, PhD(eng), 76. *Prof Exp:* Assoc engr & scientist, McDonnell Douglas Astro Co, 69-71; res asst, Univ Calif, Irvine, 71-76, teaching asst eng, 72-74; ASST PROF MECH ENG, UNIV KY, 76- *Honors & Awards:* Ralph R Teetor Award, Soc Automotive Engrs, 77. *Mem:* Combustion Inst; Am Soc Mech Engrs; Air Pollution Control Asn; Am Soc Eng Educ; Sigma Xi. *Res:* Combustion processes; air pollution control; thermodynamics; kinetic and transport processes affecting pollutant production in pulverized-coal combustion. *Mailing Add:* Dept Mech Eng Ariz State Univ Tempe AZ 85287

PECK, RUSSELL ALLEN, JR, b New Haven, Conn, May 31, 24; m 52. PHYSICS. *Educ:* Yale Univ, BS, 44, MS, 45, PhD(physics), 47. *Prof Exp:* Asst, Yale Univ, 47-48; from instr to assoc prof, 48-59, PROF PHYSICS, BROWN UNIV, 59- *Concurrent Pos:* Guggenheim fel, Univ Birmingham, 56-57; consult, Los Alamos Sci Lab, 63-64; NIH fel radiol, Univ Pa Hosp, 70-71. *Mem:* Fel Am Phys Soc; Am Asn Physics Teachers; Am Asn Physicists in Med. *Res:* Nuclear physics; fast neutron reactions with low energy Cockcroft-Walton accelerator; biomedical applications of neutrons. *Mailing Add:* Dept of Physics Brown Univ Providence RI 02912

PECK, STEWART BLAINE, b Davenport, Iowa, Aug 14, 42; m 70; c 2. ENTOMOLOGY, EVOLUTIONARY BIOLOGY. *Educ:* Univ Ky, BS, 64; Northwestern Univ, Evanston, MS, 66; Harvard Univ, PhD(biol), 71. *Prof Exp:* Lectr biol, 70-71, from asst prof to assoc prof, 71-84, PROF BIOL, CARLETON UNIV, 84- *Concurrent Pos:* Fel, Carleton Univ, 70-71. *Mem:* Fel Nat Speleol Soc; Soc Syst Zool; Soc Study Evolution; Ecol Soc Am; Am Soc Naturalists. *Res:* Evolutionary biology of cave-inhabiting arthropods and staphylinoid beetles. *Mailing Add:* Dept Biol Carleton Univ Ottawa ON K1S 5B6 Can

PECK, THEODORE RICHARD, b Spring Green, Wis, June 16, 31; m 70; c 2. SOIL FERTILITY, SOIL CHEMISTRY. *Educ:* Univ Wis, BS, 57, MS, 58, PhD(soils), 62. *Prof Exp:* From asst prof to assoc prof, 62-74, PROF SOIL CHEM, UNIV ILL, URBANA, 74- *Concurrent Pos:* Sabbatical, Univ Rio Grande do Sul, Brazil, 70-71, Netherlands, 85, China, 86, Pakistan, 89; coun soil Testing & Plant Anal; coun Agr Sci & Technol. *Mem:* AAAS; Soil Sci Soc Am; Am Soc Agron; Sigma Xi; AOAC; Int Soil Sci Soc. *Res:* Soil testing and plant analysis methods, correlation and calibration; soil fertility and pedology; availability of chemical elements in the soil for plants; chemical composition of plants. *Mailing Add:* Dept Agron Univ Ill 1102 S Goodwin Urbana IL 61801

PECK, WILLIAM ARNO, b New Britain, Conn, Sept 28, 33; c 3. ENDOCRINOLOGY. *Educ:* Harvard Univ, AB, 55; Univ Rochester, MD, 60. *Prof Exp:* Intern ward med, Barnes Hosp, St Louis, 60-61, asst resident, 61-62; fel med metab, Sch Med, Washington Univ, 62-63; clin assoc metab dis, NIH, Bethesda, 63-65; chief resident & instr med, Sch Med, Univ Rochester, 65-66, sr instr, 66-67, from asst prof to assoc prof, 67-73, prof med & biochem, 73-76; JOHN E & ADALINE SIMON PROF MED & CO-CHMN DEPT, SCH MED, WASHINGTON UNIV, 76-; PHYSICIAN-IN-CHIEF, JEWISH HOSP ST LOUIS, 76- *Concurrent Pos:* Med officer, USPHS, 63-65; fel, Univ Rochester, 66-67; assoc physician, Strong Mem Hosp, Rochester, 67-69, head endocrine unit & sr assoc physician, 69-73, physician, 73-76; NIH res career prog award, 70-75, mem metab adv comt, Food & Drug Admin, 74-78, chmn, 76-78; mem gen med study sect, NIH, 77-81, chmn, 79-81; chmn, Gordon Conf Chem, NIH, 77, Osteoporosis Consensus Develop Conf, 84; pres, Nat Osteoporosis Found, 85- *Honors & Awards:* Doran J Stephens Award, 60; Lederle Med Fac Award, 67. *Mem:* Endocrine Soc; Am Physiol Soc; Asn Am Physicians; Am Soc Clin Invest; Am Soc Biol Chemists; Am Soc Bone & Mineral Res (pres, 83-84). *Res:* Mechanisms of hormone action; regulation of bone and mineral metabolism. *Mailing Add:* Two Appletree Lane St Louis MO 63124

PECK, WILLIAM B, b Neosho, Mo, Apr 27, 20; m 52. ARACHNOLOGY. *Educ:* Iowa State Univ, BS, 42; Cent Mo State Col, MA, 63; Univ Ark, PhD(zool), 68. *Prof Exp:* Mem staff, US Civil Serv, 53-65; from asst prof to assoc prof, 67-75, prof biol, 75-83, at dept zool, cent mo state univ; RETIRED. *Concurrent Pos:* Collabr Arachnida, Entom Res Div, Agr Res Serv, USDA, 69-87. *Mem:* AAAS; Centre Int de Doc Arachnologique (vpres, 77-80); Am Arachnol Soc (pres, 75-77); Brit Arachnological Soc. *Res:* Biology and systematics of arachnids; human population and behavioral mores. *Mailing Add:* 337 Xanthisma Ave McAllen TX 78504

PECKA, JAMES THOMAS, b Binghamton, NY, May 19, 32; m 55; c 4. TECHNICAL CUSTOMER SERVICE, INDUSTRIAL CHEMISTRY. *Educ:* St Bonaventure Univ, BS, 54; Univ Buffalo, MA, 58, PhD(chem), 61. *Prof Exp:* STAFF SCIENTIST, FILM DEPT, SR CHEMIST, PLASTIC PROD & RESINS DEPT, TECH CONSULT ELECTRONICS DEPT, E I DU PONT DE NEMOURS & CO, INC, 60- *Mem:* Am Chem Soc; Am Nat Standards Inst. *Res:* Reactions of cyanogen with amines, aminophenols and aminothiophenols; weatherable films applications; polymeric nonlubricated bearings; high temperature film application; polyester film development and applications. *Mailing Add:* E I du Pont de Nemours & Co Inc PO Box 100543 Florence SC 29501-0543

PECKARSKY, BARBARA LYNN, b Milwaukee, Wis, Aug 18, 47; m 78. STREAM ECOLOGY, COMMUNITY ECOLOGY. *Educ:* Univ Wis-Madison, BS, 69, MS, 71, PhD(zool), 79. *Prof Exp:* ASST PROF ENTOM, CORNELL UNIV, 79- *Concurrent Pos:* Fel, NSF, 80-81; instr, Rocky Mountain Biol Lab, 79, 81. *Mem:* Ecol Soc Am; NAm Benthological Soc; Entom Soc Am; Am Soc Limnol & Oceanog; Sigma Xi. *Res:* Experimental analysis of the biological factors that influence the distribution and abundance of benthic invertebrates in streams; effects of invertebrate predation and competition among invertebrates on their community structure; behavioral interactions among predators and prey, and among competitors in situ, as well as the results on enclosure experiments. *Mailing Add:* 40 Comstock Hall Cornell Univ Ithaca NY 14853

PECKHAM, ALAN EMBREE, b Boise, Idaho, Aug 11, 31. HYDROLOGY. *Educ:* Earlham Col, BA, 53; Univ Nebr, MS, 55. *Prof Exp:* Geologist, US Geol Surv, 55-63; hydrogeologist, Int Atomic Energy Agency, 63-65; staff geologist, US Geol Surv, 65-66; res assoc, Ctr Water Resources Res, Desert Res Inst, 66-72, asst dir, Inst, 68-71, asst dir, Ctr, 71-72; HYDROLOGIST, US ENVIRON PROTECTION AGENCY, 72- *Concurrent Pos:* Leader task force indust waste injection of working group on protection of underground sources of drinking water, US Environ Protection Agency, 75- *Mem:* AAAS; Am Asn Petrol Geologists; fel Geol Soc Am; Am Geophys Union; Int Asn Sci Hydrol. *Res:* Groundwater geology and hydrology; ground disposal of radioactive wastes; stable and radioactive isotope techniques in hydrogeological studies; protection of underground water sources; water supply and ground disposed of wastes; impacts of waste disposed on groundwater quality; hazardous waste site investigations; underground waste injection control guidance. *Mailing Add:* 8985 W Jefferson Ave Denver CO 80235

PECKHAM, DONALD CHARLES, b Bainbridge, NY, Sept 11, 22; m 49; c 6. PHYSICS. *Educ:* Oberlin Col, AB, 48; Univ Mich, MA, 49; Pa State Univ, PhD(physics), 54. *Prof Exp:* Instr physics, Norwich Univ, 49-51; from asst prof to prof, 54-71, Alvinza Hayward Prof, 71-86, EMER PROF, PHYSICS, ST LAWRENCE UNIV, 86- *Concurrent Pos:* NSF sci fac fel, Colo State Univ, 66-67; vis physics fac mem, Ga Inst Technol, 79-80. *Mem:* AAAS; Am Phys Soc; Am Asn Physics Teachers; Optical Soc Am. *Res:* Optics. *Mailing Add:* Dept Physics St Lawrence Univ Canton NY 13617

PECKHAM, P HUNTER, b Elmira, NY, June 23, 44; m 66; c 2. REHABILITATION ENGINEERING. *Educ:* Clarkson Col Technol, BS, 66; Case Western Reserve Univ, MS, 68, PhD(biomed eng), 72. *Prof Exp:* Res assoc, Case Western Reserve Univ, 72-74, instr, Div Orthapedic Surg, 74-78; RES BIOMED ENGR, MED CTR, CLEVELAND VET ADMIN, 76-; ASST PROF ORTHOPEDICS, CASE WESTERN RESERVE UNIV, 78-, ASSOC PROF BIOMED ENG, 79- *Concurrent Pos:* Res Career Develop Award, NIH, 78; mem staff, Cuyohoga Coun Hosp, 81- *Mem:* Rehab Eng Soc NAm; Int Med Soc Paraplegia; Inst Elec & Electronics Engrs. *Res:* Technology for rehabilitation of the severely disabled; restoration of movement of the arm and hand using functional neuromuscular stimulation. *Mailing Add:* Dept Biomed Engr Case Western Reserve Univ 2040 Adelbert Rd Cleveland OH 44106

PECKHAM, RICHARD STARK, b Concord, NH; Mar 1, 24; m 50; c 9. ZOOLOGY. *Educ:* Univ NH, BS, 48; Univ Notre Dame, MS, 52, PhD(zool), 55. *Prof Exp:* USPHS jr asst sanitarian, Pan Am Sanit Bur, Guatemala, 48-50; instr biol sci, Holy Cross Cent Sch Nursing, Ind, 54-55, Del Mar Col, 55-59 & Dutchess Community Col, 59-63; PROF BIOL SCI, MT ST MARY COL, NY, 63- *Concurrent Pos:* Mem Int Audio-Tutorial Cong. *Mem:* Am Inst Biol Sci; Sigma Xi. *Res:* Plankton studies; ichthyology. *Mailing Add:* Mt St Mary col Newburg NY 12550

PECKHAM, WILLIAM DIEROLF, b Wichita Falls, Tex, Sept 18, 22; m 50; c 4. BIOCHEMISTRY, ENDOCRINOLOGY. *Educ:* Colo Col, AB, 46; Univ Pittsburgh, PhD(biochem), 55. *Prof Exp:* Res biochemist, Schering Corp, 55-59, sr res biochemist, 59-60, head dept biochem, 60-61; RES ASSOC PHYSIOL, SCH MED, UNIV PITTSBURGH, 61- *Mem:* AAAS; Am Chem Soc; Endocrine Soc; NY Acad Sci. *Res:* Peptide synthesis; regulation of pituitary hormone secretion; mechanisms of hormone action; isolation and purification of primate protein hormones; immunoassay of primate protein hormones. *Mailing Add:* Pituitary Hormone Ctr Harbour/UCLA Med Centre 1000 W Carson St Torrance CA 90509

PECKNOLD, PAUL CARSON, b Brandon, Man, Dec 21, 42; US citizen; m 68; c 2. PLANT PATHOLOGY. *Educ:* Calif State Col, Hayward, BS, 67; Univ Calif, Davis, PhD(plant path), 72. *Prof Exp:* Plant pathologist, Calif State Dept Food & Agr, 72-73; assoc prof exten, 73-85, PROF BOT & PLANT PATH, PURDUE UNIV, WEST LAFAYETTE, 85- *Mem:* Am Phytopath Soc; Apple & Pear Dis Workers. *Res:* Fire blight (Erwinia amplovora); fungicide testing for control of fruit and ornamental diseases. *Mailing Add:* Dept Bot & Plant Path Purdue Univ West Lafayette IN 47907

PECORA, ROBERT, b Brooklyn, NY, Aug 6, 38. PHYSICAL CHEMISTRY. *Educ:* Columbia Univ, AB, 59, AM, 60, PhD(chem), 62. *Prof Exp:* Nat Acad Sci-Nat Res Coun fel, Brussels, 63; res physicist, Columbia Univ, 64; from asst prof to assoc prof, 64-77, PROF CHEM, STANFORD UNIV, 77- *Concurrent Pos:* Vis prof, Victoria Univ Manchester, 70-71; vis prof Univ Nice, 78. *Honors & Awards:* Humboldt Sr Scientist Award. *Mem:* AAAS; Am Chem Soc; fel Am Phys Soc. *Res:* Statistical mechanics of equilibrium and non-equilibrium processes especially applications to liquids and polymers; inelastic scattering of neutrons and light from condensed systems; biophysics. *Mailing Add:* Dept Chem Stanford Univ Stanford CA 94305-5080

PECORARO, VINCENT L, b Freeport, NY, Aug 31, 56. BIOINORGANIC CHEMISTRY. *Educ:* Univ Calif, Los Angeles, BS, 77; Univ Calif, Berkeley, PhD(chem), 81. *Prof Exp:* NIH fel biochem, Univ Wis- Madison, 81-84; ASST PROF CHEM, UNIV MICH, ANN ARBOR, 84- *Concurrent Pos:* Mem, Inst Protein Struct & Design, Univ Mich, Ann Arbor, 85-, Ctr Molecular Genetics; Horace H Rackham fel, 85; Eli Lilly teaching fel, 85; G D Searle Biomed Res scholar, 86. *Mem:* Am Chem Soc; NY Acad Sci; Sigma Xi; AAAS; Am Soc Biol Chemists. *Res:* Elucidation of the role of metal ions in biological systems through modelling of enzyme active site structure with small molecule inorganic complexes or through modification of active site structure using site directed mutagenesis. *Mailing Add:* Dept Chem Univ Mich Ann Arbor MI 48109

PECORINI, HECTOR A(NDREW), b New York, NY, Oct 26, 24; m 49; c 3. CHEMICAL ENGINEERING. *Educ:* Cooper Union, BChE, 44; Univ Mich, MS, 50, PhD(chem eng), 54. *Prof Exp:* Chemist, US Indust Chem Co, 46-49; chem engr, 53-70, PROCESS ENGR, E I DU PONT DE NEMOURS & CO, INC, 70- *Mem:* Am Chem Soc; Am Inst Chem Engrs. *Res:* Kinetics of the homogeneous liquid-phase reaction between propylene oxide and methyl alcohol. *Mailing Add:* 2426 Owen Dr Wilmington DE 19808

PECSOK, ROBERT LOUIS, b Cleveland, Ohio, Dec 18, 18; m 40; c 7. ANALYTICAL CHEMISTRY, CHROMATOGRAPHY. *Educ:* Harvard Univ, BS, 40, PhD(chem), 48. *Prof Exp:* Prod foreman, Procter & Gamble Co, 40-43; instr chem, Harvard Univ, 48; from asst prof to prof, Univ Calif, Los Angeles, 48-71; prof chem & chmn dept, Univ Hawaii, Honolulu, 71-80, dean natural sci, 81-90; RETIRED. *Concurrent Pos:* Guggenheim fel, 56-57; Am Chem Soc-Petrol Res Fund int fel, 63-64; sci adv, US Food & Drug Admin, 66-69, dir chem technician curric proj, 69-72. *Honors & Awards:* Tolman Medal, Am Chem Soc, 71. *Mem:* Am Chem Soc; AAAS. *Res:* Polarography; complex ions; chemistry of transition metals; gas chromatography; principles and practice of chromatographic separations. *Mailing Add:* 6009 Haleola St Honolulu HI 96821

PEDDICORD, RICHARD G, computer science, for more information see previous edition

PEDEN, CHARLES H F, b Honolulu, HI, Apr 4, 53; US citizen; m 80; c 4. SURFACE CHEMISTRY, HETEROGENOUS CATALYSIS. *Educ:* Calif State Univ, Chico, BA, 78; Univ Calif, Santa Barbara, MA, 81 & PhD(phys chem), 83. *Prof Exp:* Fel, Surface Sci Div, 83-85, MEM TECH STAFF, INORG MAT CHEM DIV, SANDIA NAT LAB, 85- *Mem:* Am Chem Soc; Am Vacuum Soc; Mats Res Soc. *Res:* fundamental chemistry and physics of ceramic surfaces; surface chemical mechanisms of heterogeneous catalytic reactions. *Mailing Add:* Sandia Nat Lab PO Box 5800 Albuquerque NM 87185

PEDEN, IRENE C(ARSWELL), b Topeka, Kans, Sept 25, 25; m 62; c 2. ELECTRICAL ENGINEERING, RADIO SCIENCE. *Educ:* Univ Colo, BS, 47; Stanford Univ, MS, 58, PhD(elec eng), 62. *Prof Exp:* Jr engr, Del Power & Light Co, 47-49; jr engr, Aircraft Radio Systs Lab, Stanford Res Inst, 49-50, res engr, 50-52, antenna res group, 54-57; res engr, Midwest Res Inst, Mo, 53-54; res asst, Hansen Lab, Stanford Univ, 58-61, actg instr elec eng, 59-61; from asst prof to assoc prof, 61-71, assoc dean eng, 73-77, assoc chair, Elec Eng Dept, 83-86, PROF ELEC ENG, UNIV WASH, 71- *Concurrent Pos:* Mem, Policy Adv Comt Eng & Appl Sci, 76-81, Eng Adv Comt, NSF, 84-; mem adv bd, Alaska Geophys Inst, Univ Alaska, 77-80, US Merchant Marine Acad, 78-84; mem, Army Sci Bd, 78-, vchmn, 83-85, chmn, 86-87, Blue Ribbon Selection Panel Presidential Young Investr Awards, NSF, 83, Comt Educ Utilization Eng, Panel Grad Educ & Res, Nat Res Ctr-Nat Acad Eng, 83-85, bd dir, BDM, Int Corp, 84-, bd visitors, Univ Calif, Davis, 84-, Eng Develop Coun, Univ Colo, 84-, Defense Sci Bd, 88-, vpres, Inst Elec & Electronic Engrs, Antennas & Propagation Soc, 88, mem, US Nat Comt, URSI, 88-91; accreditation bd, Eng & Technol ABET, eng accrediation comt, 75-82, chmn, 81-82, bd dirs, 82-88; dir, BDM Int Corp, 88- *Honors & Awards:* Centennial Medal, Inst Elec & Electronics Engrs, 84, Haraden Pratt Award, 88. *Mem:* Fel AAAS; fel Inst Elec & Electronics Engrs; Am Geophys Union; Int Union Radio Sci; Soc Women Engrs; NY Acad Sci; fel Explorers Club. *Res:* Microwave measurements, networks and periodic circuits; radio science with applications to the polar regions; antennas, wave propagation and scattering; electromagnetic subsurface remote sensing. *Mailing Add:* 409 Elec Eng Bldg FT-10 Univ Wash Seattle WA 98195

PEDERSEN, CHARLES JOHN, macrocyclic crown polyethers; deceased, see previous edition for last biography

PEDERSEN, FRANKLIN D, b St Paul, Nebr, Feb 22, 33; m 61; c 2. MATHEMATICS. *Educ:* Peru State Col, BA, 59; Tulane Univ, MS, 62, PhD(math), 67. *Prof Exp:* Asst prof, 65-76, ASSOC PROF MATH, SOUTHERN ILL UNIV, CARBONDALE, 76- *Concurrent Pos:* Vis lectr, Univ Natal, 70. *Mem:* Am Math Soc; Math Asn Am; Int Coun Teachers Math. *Res:* Algebra; ordered structures. *Mailing Add:* Dept Math Southern Ill Univ Carbondale IL 62901

PEDERSEN, KATHERINE L, b Connersville, Ind, Oct 14, 37; m 61; c 2. TOPOLOGY. *Educ:* St Louis Univ, BS, 59; Tulane Univ, MS, 62, PhD(math), 69. *Prof Exp:* Asst math, Tulane Univ, 60-61; instr math, Newcomb Col, 61-65; instr, 65-69, ASST PROF MATH, SOUTHERN ILL UNIV, CARBONDALE, 69-, ASSOC PROF MATH. *Mem:* Am Math Soc; Math Asn Am; Nat Coun Teachers Math. *Res:* Preparing materials for pre-service education of elementary and secondary mathematics teachers. *Mailing Add:* Dept Math Southern Ill Univ Carbondale IL 62901

PEDERSEN, KNUD B(ORGE), b Odense, Denmark, Nov 26, 32; US citizen; m 53; c 3. NUCLEAR & MECHANICAL ENGINEERING. *Educ:* Iowa State Univ, BS, 58, MS, 64, PhD(nuclear eng), 67. *Prof Exp:* Develop engr, Outboard Marine Corp, 58-60; jr engr, Ames Lab, Atomic Energy Comn, Iowa State Univ, 60-64, instr nuclear eng, univ, 64-67; from asst prof to assoc prof nuclear eng, Univ PR, 67-76, assoc scientist, 67-70, scientist II, 70-76, sr scientist, Ctr Energy & Environ Res, Atomic Energy Comn, 76-80, head, Nuclear Tech Div, 77-80, prof mech & nuclear eng, 76-87; ENGR, LAWRENCE LIVERMORE NAT LAB, 87- *Concurrent Pos:* Consult energy & accident invest. *Mem:* Am Nuclear Soc. *Res:* Reactor safety and kinetics; energy conversion and conservation; methods and policies. *Mailing Add:* Lawrence Livermore Nat Lab PO Box 808 MS 231 Livermore CA 94550

PEDERSEN, LEE G, b Oklahoma City, Okla, June 15, 38; m 64; c 2. PHYSICAL CHEMISTRY, CHEMICAL PHYSICS. *Educ:* Univ Tulsa, BCh, 61; Univ Ark, PhD(phys chem), 65. *Prof Exp:* NSF res assoc theoret chem, Columbia Univ, 65-66; NIH res fel, Harvard Univ, 66-67; from asst prof to assoc prof, 67-76, PROF PHYS CHEM, UNIV NC, CHAPEL HILL, 76- *Concurrent Pos:* Vis scientist, NIEHS, 85- *Mem:* Sigma Xi; Am Asn Univ Professors; Am Phys Soc. *Res:* Theoretical chemistry and biology; rotational barriers; H-bond; free radicals; molecular dynamics. *Mailing Add:* Dept Chemistry Univ North Carolina Chapel Hill NC 27514

PEDERSEN, LEO DAMBORG, b Soroe, Denmark, June 11, 46; m 64; c 3. FOOD SCIENCE & TECHNOLOGY. *Educ:* Tech Univ, MSc, 68; Copehagen Univ, MBA, 72; Univ Calif, PhD(ag engr), 78. *Prof Exp:* Proj leader Biotech Inst, Denmark, 72-75; vis scientist, USDA Western Regional Res Lab, 75-78; Res sci, 78-84, DIV DIR, NAT FOOD PROCESSORS ASSOC, 84-88. *Mem:* Am Soc Heart Engr. *Res:* Process development and equipment manufacturing. *Mailing Add:* 605 Thornhill Rd Danville CA 94526

PEDERSEN, PEDER CHRISTIAN, b Kalundborg, Denmark, Sept 28, 43. PHYSICAL ACOUSTICS. *Educ:* Aalborg Eng Col, Denmark, BS, 71; Univ Utah, ME, 74, PhD(bioeng), 76. *Prof Exp:* ASST PROF ELEC ENG, DEPT ELEC & COMPUT ENG, DREXEL UNIV, 76- *Concurrent Pos:* Proj leader, Sandoz Hosp Supplies Res, Salt Lake City, Utah, 76; Coordr, Clin Eng Prog, Biomed Eng & Sci Inst, Drexel Univ, 80- *Mem:* Inst Elec & Electronics Engrs; Acoust Soc Am; Am Inst Ultrasound Med; AAAS; Sigma Xi. *Res:* Study of diagnostic use of low-intensity microwaves for detecting changes in lung water; development of techniques for ultrasonic characterization of lung tissue. *Mailing Add:* Dept Elec Eng Worcester Polytech Inst Worcester MA 01609

PEDERSEN, PETER L, b Muskogee, Okla, Oct 31, 39; div; c 4. BIOCHEMISTRY, MOLECULAR BIOLOGY. *Educ:* Univ Tulsa, BA, 61; Univ Ark, PhD(chem), 64. *Prof Exp:* From instr to asst prof, 67-72, PROF BIOCHEM, SCH MED, JOHNS HOPKINS UNIV, 75- *Concurrent Pos:* USPHS fel biochem, Sch Med, Johns Hopkins Univ, 64-67; USPHS res grant, 68-; Nat Cancer Inst res awards, 69-; NSF res grant, 76. *Mem:* Am Soc Biol Chemists; Am Cancer Soc; Biophys Soc. *Res:* Bioenergetics of normal and pathological states; mechanism of ATP synthesis and utilization in biological systems; mechanism of glucose catabolism in normal, neoplastic, and parasitic cells; transport processes, particularly phosphate, which is required for both ATP synthesis and glucose catabolism; parasitology; oncology. *Mailing Add:* Dept Biol Chem Johns Hopkins Univ Sch Med Baltimore MD 21205

PEDERSEN, ROGER ARNOLD, b San Bernandino, Calif, Aug 1, 44; m 76; c 2. DEVELOPMENTAL GENETICS, TERATOGENESIS. *Educ:* Stanford Univ, AB, 65; Yale Univ, PhD(biol), 70. *Prof Exp:* USPHS fel, Sch Hyg & Pub Health, Johns Hopkins Univ, 70-71; asst prof radiol, 71-79, asst prof anat, 75-79, assoc prof, 79-85, PROF RADIOL & ANAT, UNIV CALIF, SAN FRANCISCO, 85- *Concurrent Pos:* Ser ed, Current Topics Develop Biol, 90- *Honors & Awards:* Int Cancer Res Technol Transfer Award, 82. *Mem:* AAAS; Am Soc Cell Biol; Soc Develop Biol. *Res:* Mammalian embryology; mechanisms of cell differentiation and commitment; induction and repair of genetic damage in mammalian germ cells and embryos; cell lineage and cell fate in pre- and post-implantation mouse embryo development. *Mailing Add:* Lab of Radiobiol Univ of Calif San Francisco CA 94143-0750

PEDERSON, DARRYLL THORALF, b Valley City, NDak, Aug 12, 39; c 2. HYDROGEOLOGY. *Educ:* Valley City State Col, BSEd, 61; Univ NDak, MST, 66, PhD(geol), 71. *Prof Exp:* Asst prof geol, Minot State Col, 71-73; asst prof geol, Appalachian State Univ, 73-75; ASSOC PROF HYDROGEOL, UNIV NEBR, LINCOLN, 75-, PROF GEOL. *Mem:* Am Inst Prof Geologists; Geol Soc Am; Am Geophys Union; Nat Water Well Asn; AAAS. *Res:* Characterization of the hetereogeneous aquifer systems and the movement of water from the surface into these aquifers; surface geophysical techniques, electrical. *Mailing Add:* Dept Geol Univ Nebr Lincoln NE 68588

PEDERSON, DONALD O(SCAR), b Hallock, Minn, Sept 30, 25; m 50, 78; c 6. ELECTRICAL ENGINEERING, COMPUTER AIDED DESIGN. *Educ:* NDak State Col, BS, 48; Stanford Univ, MS, 49, PhD(elec eng), 51. *Hon Degrees:* DSc, Kathielke Univ, Belgium, 79. *Prof Exp:* Res assoc, Electronic Res Lab, Stanford Univ, 51-53; mem tech staff, Bell Tel Labs, Inc, 53-55; dir, Electronic Res Lab, 60-64, chmn dept, 83-85, PROF ELEC ENG, UNIV CALIF, BERKELEY, 55- *Concurrent Pos:* Lectr, Newark Col Eng, 53-55; Guggenheim fel, 64. *Honors & Awards:* Educ Medal, Inst Elec & Electronics Engrs, 69, Solid State Circuit Coun Outstanding Develop Award, 85. *Mem:* Nat Acad Sci; Nat Acad Eng; fel Inst Elec & Electronics Engrs. *Res:* Electronic circuits. *Mailing Add:* Dept Elec Eng & Computer Sci Univ Calif Berkeley CA 94720

PEDERSON, ROGER NOEL, b Minneapolis, Minn, Dec 28, 30; m 57; c 2. MATHEMATICS. *Educ:* Univ Minn, BS, 52, MS, 53, PhD(math), 57. *Prof Exp:* Temporary mem, Inst Math Sci, NY Univ, 57-58; Moore instr math, Mass Inst Technol, 58-60; from asst prof to assoc prof, 60-70, PROF MATH, CARNEGIE-MELLON UNIV, 70- *Concurrent Pos:* NSF fel, 61-; vis assoc prof, Stanford Univ, 68-69. *Res:* Partial differential equations; real and complex analysis. *Mailing Add:* 332 Sharon Dr Pittsburgh PA 15221

PEDERSON, THORU JUDD, b Syracuse, NY, Oct 10, 41; m 66; c 2. CELL BIOLOGY, MOLECULAR BIOLOGY. *Educ:* Syracuse Univ, BS, 63, PhD(zool), 68. *Prof Exp:* Res fel cell biol, Albert Einstein Col Med, 68-71; staff scientist, Worcester Found, 71-73, co-dir, Cancer Ctr, 75-81, sr scientist, 73-83, DIR CANCER CTR, WORCESTER FOUND EXP BIOL, 81-, PRIN SCIENTIST CELL BIOL, 83-, PRES & SCI DIR, 85- *Concurrent Pos:* Nat scholar, Leukemia Soc Am, 72-77; mem, Cell Biol Study Sect, NIH, 75-79, Molecular Biol Study Sect, 80-83 & Med Res Comt, Am Cancer Soc, Mass Div, 80-; ed, J Cell Biol, 78-83; fac physiol course, Marine Biol Lab, Woods Hole, 81-; nat lectr, Sigma Xi, 84-86. *Honors & Awards:* Hudson Hoagland Award, 83. *Mem:* Am Soc Biol Chemists; Am Soc Cell Biol; fel AAAS. *Res:* Structure and function of the cell nucleus; molecular biology of gene transcription; cell growth; cell motility; cytoskeleton; developmental biology. *Mailing Add:* Pres & Sci Dir Worcester Found Exp Biol Shrewsbury MA 01545

PEDERSON, VERNON CLAYTON, b Nashua, Minn, Dec 26, 29; m 52; c 3. PHYSIOLOGY, ENDOCRINOLOGY. *Educ:* Southwest Mo State Col, BS, 55; Univ Mo, MA, 60, PhD(zool), 63. *Prof Exp:* Instr biol, Southwest Mo State Col, 56-58; asst zool, Univ Mo, 58-61, asst instr, 61-63; asst prof biol, 63-74, assoc prof biol sci, 74-86, PROF BIOL SCI, WESTERN ILL UNIV, 87- *Mem:* AAAS; Am Soc Zool. *Res:* Reproductive physiology and endocrinology of wild animal populations. *Mailing Add:* Dept Biol Sci Western Ill Univ Macomb IL 61455

PEDIGO, LARRY PRESTON, b Great Bend, Kans, Oct 8, 38; m 61; c 2. ENTOMOLOGY. *Educ:* Ft Hays Kans State Col, BS, 63; Purdue Univ, MS, 65, PhD(entom), 67. *Prof Exp:* From asst prof to assoc prof, 67-75, PROF ENTOM, IOWA STATE UNIV, 75- *Honors & Awards:* Am Soybean Asn Res Recognition Award; C V Riley Achievement Award & J E Bussart Mem Award, Entom Soc Am. *Mem:* Entom Soc Am. *Res:* Population ecology and management of insect pests with emphasis on sampling procedure; population characteristics; modeling, bioeconomics and plant stress from insect injury. *Mailing Add:* Dept of Entom Iowa State Univ Ames IA 50011

PEDLOSKY, JOSEPH, b Paterson, NJ, Apr 7, 38; m 75. OCEANOGRAPHY, DYNAMIC METEOROLOGY. *Educ:* Mass Inst Technol, BSc & MSc, 60, PhD(meteorol), 63. *Prof Exp:* From asst prof to assoc prof math, Mass Inst Technol, 64-68; assoc prof, 68-72, prof geophys fluid dynamics, Univ Chicago, 72-79; SR SCIENTIST & DOHERTY PROF PHYS OCEANOG, WOODS HOLE OCEANOG INST, 79- *Concurrent Pos:* Fel, Int Inst Meteorol, Stockholm, 61-62; Sloan Found Fel, 67-68; mem sci adv comt, Nat Ctr Atmospheric Res, 74- *Honors & Awards:* Meisinger Award, Am Meteorol Soc, 71. *Mem:* Nat Acad Sci; Am Meteorol Soc; fel Am Geophys Union. *Res:* Dynamics of ocean and atmospheric flows, particularly the nonlinear mechanics of geophysically important fluid instabilities. *Mailing Add:* Woods Hole Oceanog Inst Woods Hole MA 02543

PEDOE, DANIEL, b London, Eng, Oct 29, 10; m 33, 66; c 2. MATHEMATICS. *Educ:* Univ London, BSc, 30; Magdalene Col, BA, 33, PhD(math), 37. *Prof Exp:* Lectr math, Univ Southampton, 36-42 & Univ Birmingham, 42-47; reader, Univ London, 47-52; prof, Khartoum Univ, 52-59, Singapore Univ, 59-62 & Purdue Univ, 62-64; prof math, Univ Minn, Minneapolis, 64-80, emer prof, 80-84; RETIRED. *Concurrent Pos:* Leverhulme res fel, 46-48; sr math consult, Minn Sch Geom Proj, 64-66. *Mem:* Math Asn Am. *Res:* Algebraic geometry; mathematics education. *Mailing Add:* 704 14th Ave SE Minneapolis MN 55414

PEDRAJA, RAFAEL R, b Sagua La Grande, Cuba, Oct 21, 29; US citizen; m 53; c 2. FOOD SCIENCE, MICROBIOLOGY. *Educ:* Super Sch Arts & Trades, Havana, BS, 47; Univ Havana, MS, 50, Dr(agr eng), 52. *Prof Exp:* Auditor chem, Chas Martin Co, Cuba, 53; chemist & bacteriologist spec prod div, Borden Co, 53-56; res chemist, Griffith Labs, Inc, Ill, 56-62; tech dir, Am Dry Milk Inst, 62-65; tech dir food prod div, Super Tea & Coffee Co, 65-67; vpres res & develop & qual assurance, Booth Fisheries Div, 67-83, DIR QUAL ASSURANCE & REGULATORY AFFAIRS & EXEC DIR, SCI-TEK LABS, DIV SARA LEE BAKERY, SARA LEE CORP, 83- *Honors & Awards:* Ronald Reagan Presidential Award. *Mem:* AAAS; Am Chem Soc; Am Dairy Sci Asn; Am Asn Cereal Chem; Inst Food Technol; NY Acad Sci. *Res:* Industrial hygiene; oxidative rancidity of fats and oils; mechanisms of the curing of meats; microbiology and composition of dry milks; microbiology and biochemistry of fish and seafoods; development of convenience foods. *Mailing Add:* Div Sara Lee Bakery Sara Lee Corp 500 Waukegan Rd Deerfield IL 60015

PEDROTTI, LENO STEPHANO, b Zeigler, Ill, May 21, 27; m 51; c 8. LASERS, OPTICAL PHYSICS. *Educ:* Ill State Univ, BS, 49; Univ Ill, MS, 51; Univ Cincinnati, PhD, 61. *Prof Exp:* From instr to prof physics, US Air Force Inst Technol, 51-82, chmn dept, 64-82; tech ed, Tech Educ Res Ctrs Southwest, US Off Educ, 72-80; chief tech ed, author & consult, 80-82, VPRES-PROG MGR, CTR OCCUP RES & DEVELOP, 82- *Concurrent Pos:* Mem, bd dirs, Laser Inst Am, 75-82, author & lectr, 75-; pres, Lasop, Inc. *Mem:* Am Soc Eng Educ; Am Nuclear Soc; Am Asn Physics Teachers; Optical Soc Am; Am Asn Univ Profs. *Res:* Electrical and optical properties of semiconductor crystals at low temperatures; development of laser scalpels for corneal surgery. *Mailing Add:* 11006 Trailwood Dr Waco TX 76710

PEDROZA, GREGORIO CRUZ, b Pearson, Tex, May 3, 41; m 62; c 3. INDUSTRIAL ORGANIC CHEMISTRY. *Educ:* St Mary's Univ, Tex, BS, 63; WVa Univ, PhD(org chem), 68. *Prof Exp:* SR ENGR-MGR, ADVAN CIRCUIT PACKAGING, IBM CORP, 69- *Mem:* Am Chem Soc. *Mailing Add:* 4 Deborah Dr Apalachin NY 13732

PEEBLES, CHARLES ROBERT, b Oak Park, Ill, May 31, 29; c 2. PARASITOLOGY. *Educ:* Cornell Univ, AB, 51; Univ Ill, MS, 52, PhD(zool), 57. *Prof Exp:* Asst, Univ Ill, 52-57; assoc parasitologist, Agr Exp Sta, Univ PR, 57-63; asst prof biol, Oberlin Col, 63-65; from asst prof to assoc prof, 65-77, dir student sci training prog, 69-85, PROF NATURAL SCI, MICH STATE UNIV, 77- *Mem:* AAAS; Am Micros Soc; Sigma Xi; Wilderness Soc. *Res:* Ultrastructure of nematodes; biology gastrointestinal parasites of ruminants; helminths of Great Lakes fishes. *Mailing Add:* Dept Natural Sci Mich State Univ East Lansing MI 48824-1031

PEEBLES, CRAIG LEWIS, b Cleveland, Ohio, Apr 14, 50; m 77, 90; c 1. TRNA PROCESSING, RNA SPLICING. *Educ:* Col Wooster, BA, 72; Univ Chicago, PhD(biophys), 78. *Prof Exp:* Postdoctoral fel, Dept Chem, Univ Calif, San Diego, 78-82; asst prof, 82-88, ASSOC PROF, DEPT BIOL SCI, UNIV PITTSBURGH, 88- *Concurrent Pos:* Res grants, USPHS-NIH, 83-89, Am Cancer Soc, 90-91. *Mem:* AAAS; Genetics Soc Am. *Res:* Chemistry of RNA splicing; control of stable RNA synthesis, particularly; mechanism of self-splicing by group II introns from yeast; RNA accumulation in yeast cells. *Mailing Add:* Dept Biol Sci 357 Crawford Hall Univ Pittsburgh Pittsburgh PA 15260

PEEBLES, EDWARD MCCRADY, b Greenwood, Miss, June 26, 24; m 49; c 7. ANATOMY. *Educ:* Univ of the South, BS, 49; Tulane Univ, PhD(anat), 54. *Prof Exp:* Jr instr biol, Johns Hopkins Univ, 49-50; asst, 52-54, from instr to assoc prof, 54-68, asst dean Sch Med, 67-76, assoc dean Sch Med, 76-77, PROF ANAT, TULANE UNIV, 68- *Mem:* AAAS; Am Asn Anat; Am Osteopathic Asn. *Res:* Experimental neurology; nerve regeneration; gross anatomy; mast cells; teratology. *Mailing Add:* 2701 Gallinghouse St New Orleans LA 70131

PEEBLES, HUGH OSCAR, JR, b Kountze, Tex, May 4, 33; m 55; c 4. PHYSICS. *Educ:* Univ Tex, Austin, BS, 55; Okla State Univ, MS, 60, PhD(physics), 64. *Prof Exp:* Teacher pub schs, Tex, 55-59; asst prof, 63-65, ASSOC PROF PHYSICS, LAMAR UNIV, 65- *Mem:* Am Astron Soc. *Res:* Stellar atmospheres. *Mailing Add:* Box 715 Kountze TX 77625

PEEBLES, PEYTON Z, JR, b Columbus, Ga, Sept 10, 34; c 2. ELECTRICAL ENGINEERING. *Educ:* Evansville Col, Indiana, BSEE, 57; Drexel Inst Technol, Pa, MSEE, 63; Univ Pa, PhD, 67. *Prof Exp:* Trainee engr, RCA, Moorestown, NJ, 57, radar & systs engr, 58-64, systs engr, 66-69; officer, US AF, Bakalar, 58; David Sarnoff fel, Univ Pa, 64-66; assoc prof eng, Univ Tenn, Knoxville, 69-75, prof, 75-76 & 77-81; vis prof elec eng, Univ Hawaii, Honolulu, 76-77; assoc chmn, acad affairs, elec eng dept, 84-90 PROF ELEC ENG, UNIV FLA, GAINESVILLE, 81-,. *Concurrent Pos:* Chmn, spec cont, Elec Eng Prog, Univ Tenn, 80-81; consult, RCA, Union Carbide Corp, Oak Ridge Nat Lab, BDM Corp, ARO, Inc, Systs Dynamics, Inc. *Mem:* Fel Inst Elec & Electronics Engrs; Sigma Xi. *Res:* Author of over 50 articles and four books. *Mailing Add:* Dept Elec Eng Univ Fla Gainesville FL 32611

PEEBLES, PHILLIP J, b Winnipeg, Man, Apr 25, 35; m; c 3. ASTRONOMY. *Educ:* Univ Man, BS, 58; Princeton Univ, PhD, 62. *Hon Degrees:* DSc, Univ Toronto, 86, Univ Chicago, 86, McMaster Univ, 89, Univ Man, 89. *Prof Exp:* Instr, Princeton Univ, 61-62, res assoc, 62-64, res staff mem, 64-65, from asst prof to prof, 65-84, ALBERT EINSTEIN PROF SCI, PRINCETON UNIV, 84- *Honors & Awards:* A C Morrison Award in Nat Sci, NY Acad Sci, 77; Eddington Medal, Royal Astron Soc, 81; Heineman Prize, Am Astron Soc, 82. *Mem:* Foreign assoc Nat Acad Sci; fel Am Phys Soc; fel Am Acad Arts & Sci; fel Royal Soc; Am Astron Soc; AAAS; Int Astron Union. *Res:* Physics. *Mailing Add:* Dept Physics Jodwin Hall Princeton Univ Princeton NJ 08544-0708

PEEK, H(ARRY) MILTON, b Blackwell, Okla, Feb 8, 28; m 53; c 2. PHYSICAL CHEMISTRY, ATMOSPHERIC PHYSICS. *Educ:* Univ Okla, BA, 47; Univ Rochester, PhD(chem), 50. *Prof Exp:* Asst chem, Univ Rochester, 47-48; mem staff phys chem, Los Alamos Sci Lab, Univ Calif, 50-62, assoc group leader optical physics, 63-71, alt group leader, 71-72, group leader, 72-76, mem staff field test div, 76-77; asst dep dir testing, Defense Nuclear Agency, 77-78; mem staff field test, Los Alamos Sci Lab, Univ Calif, 79-84; mgr, Nat Defense Sci, Holmes & Narver Inc, 84-86; Sr scientist, Geo-Centers, Inc, 86-90; OWNER & SR SCIENTIST, AGEX CONSULTS, 90- *Mem:* Sigma Xi. *Res:* Time resolved absorption spectroscopy of hot gases; radiometric optical studies of nuclear tests; rocket spectroscopy of aurora; ionospheric and magnetospheric plasma injections; weapons systems reliability and operational testing; artificial noctilucent clouds. *Mailing Add:* PO Box 15130 Rio Rancho NM 87174-0130

PEEK, JAMES MACK, b Unionville, Mo, Sept 5, 33; m 62; c 2. ATOMIC & MOLECULAR PHYSICS. *Educ:* Western Ill State Col, BSEd, 55; Ohio State Univ, MS, 58, PhD(chem), 62. *Prof Exp:* mem tech staff theoret physics, Sandia Nat Labs, 62-86; TECH STAFF MEM, LOS ALAMOS NAT LAB, 86- *Concurrent Pos:* Vis fel, Joint Inst Lab Astrophys, Univ Colo, 71-72. *Mem:* Fel Am Phys Soc; Am Math Soc; AAAS; Sigma Xi. *Res:* Theory of collisions involving electrons, atoms and molecules with emphasis on molecule formation and dissociation; theoretical study of the structure of simple atoms and molecules. *Mailing Add:* Group T-Four Los Alamos Nat Lab Los Alamos NM 87545

PEEK, JAMES MERRELL, b Helena, Mont, Sept 18, 36; m 64; c 3. WILDLIFE ECOLOGY. *Educ:* Mont State Univ, BS, 58, MS, 61; Univ Minn, St Paul, PhD(wildlife), 71. *Prof Exp:* Mgt biologist, Mont Fish & Game Dept, 61-63; res biologist, Bur Sport Fisheries & Wildlife, 64; res biologist, Mont Fish & Game Dept, 65-67; res fel wildlife, Univ Minn, St Paul, 67-70; from instr to assoc prof, 70-75, PROF WILDLIFE, UNIV IDAHO, 76- *Mem:* Am Soc Range Mgt; Wildlife Soc; Am Soc Mammal; Ecol Soc Am. *Res:* Ecology and behavior of ungulates. *Mailing Add:* Col Forest Wildlife & Range Sci Univ of Idaho Moscow ID 83843

PEEK, NEAL FRAZIER, b Chico, Calif, Jan 28, 29; m 51; c 4. EXPERIMENTAL NUCLEAR PHYSICS. *Educ:* Univ Calif, Berkeley, BA, 51; Univ Calif, Davis, MA, 59, PhD(exp nuclear physics), 66. *Prof Exp:* Lab technician, Univ Calif, Berkeley, 51-53; sr lab technician, Univ Calif, Davis, 53-55; chemist, US Army, 55-57; lectr physics & res physicist, 59-71, asst prof physics, 71-78, SR LECTR & VCHMN, UNIV CALIF, DAVIS, 78- *Concurrent Pos:* Fel, Assoc Western Univs, 73; vis scientist, Swiss Ints Nuclear Res Würenlingen, Switzerland, 76, Kernforschungsanlage, Jülich, Germany, 80. *Mem:* Sigma Xi; Am Phys Soc. *Res:* Applied nuclear experimental physics, including isotope production and in vivo studies with short lived isotopes; spallation isatope production. *Mailing Add:* Dept of Physics Univ of Calif Davis CA 95616

PEEL, JAMES EDWIN, b Lonaconing, Md, Dec 26, 24; m 55; c 2. CHEMICAL ENGINEERING, MARINE ENGINEERING. *Educ:* Carnegie Inst Technol, BS, 53. *Prof Exp:* Develop engr, Pittsburgh Coke & Chem Co, 53-59; proj mgr, Pittsburgh Plate Glass Co, 59-67; div engr, US Chem Co, 67-69; res engr, US Steel Corp, 69-71; process engr, PPG Indust, Inc, 77-87; MANAGING DIR, PEEL ENG & CONSTRUCT CO, 65- *Res:* Methods of floatation of sunken vessels by in-situ polyurethane foaming and underwater habitation for mammals by silicone membrane osmosis of carbon dioxide and oxygen. *Mailing Add:* Peel Eng & Construct Co 204 Chadwick St Sewickley PA 15143

PEELER, DUDLEY F, JR, b Booneville, Miss, June 17, 31; m 59; c 3. NEUROPSYCHOLOGY, BEHAVIOR GENETICS. *Educ:* Vanderbilt Univ, AB, 53, AM, 62, PhD(psychol), 63. *Prof Exp:* Asst psychol, Vanderbilt Univ, 57-59, res assoc, 60-61; trainee, Vet Admin Hosp, Murfreesboro, Tenn, 61-62; from instr to asst prof, 62-73, chief, Exp Behav Lab, 65-80, CHIEF NEUROSURG LAB, MED CTR, UNIV MISS, 80-, RES PROF NEUROSURG, 87- *Concurrent Pos:* Mem adj fac psychol, Millsaps Col, 64-71, 82; assoc psychol, Univ Miss, 65-76. *Mem:* AAAS; Soc Neurosci. *Res:* Neuropsychology of development; behavioral genetics; neuropsychology of motivation and emotion. *Mailing Add:* Dept of Neurosurg Univ of Miss Med Ctr Jackson MS 39216

PEELLE, ROBERT W, b Toledo, Ohio, Jan 13, 29; m 55; c 2. NUCLEAR PHYSICS. *Educ:* Univ Rochester, BS, 49; Princeton Univ, MA, 51, PhD(physics), 58. *Prof Exp:* PHYSICIST, ENGR MATHS DIV, OAK RIDGE NAT LAB, 54-, SECT HEAD, 76- *Mem:* AAAS; Am Phys Soc; fel Am Nuclear Soc. *Res:* Spectra associated with fission by thermal neutrons; techniques of scintillation and semiconductor spectrometry; space and reactor shielding; continuum reactions in 100 region; fast neutron cross sections of materials important to fusion and fission reactors. *Mailing Add:* Eng Physics & Maths Div Oak Ridge Nat Lab PO Box 2008 Oak Ridge TN 37831-6354

PEEPLES, EARLE EDWARD, b West Palm Beach, Fla, Aug 23, 29; m 65; c 2. HUMAN GENETICS EDUCATION, HANDWRITING RESEARCH. *Educ:* Univ Fla, BS, 51; Southeastern Baptist Sem, BD, 54; Stetson Univ, MS, 62; Univ Tex, Austin, PhD(genetics, zool), 66. *Prof Exp:* Asst prof biol, Mary Hardin-Baylor Col, 60-63; res scientist assoc IV, Univ Tex, Austin, 66-69; assoc prof genetics, Southwestern State Univ, 69-73; assoc prof, 73-80, PROF BIOL SCI, UNIV NORTHERN COLO, 80- *Concurrent Pos:* Consult, Int Found Rat Genetics & Rodent Pest Control, 69-71; ed, J Graphological Sci, 87-88. *Mem:* Am Soc Human Genetics; Sigma Xi; Behav Genetics Asn; Int Graphonomics Soc. *Res:* Genetic control of enzymes in development of reproductive organs of Drosophila, rat and man; genetic regulation of human external ear; inheritance of handwriting factors. *Mailing Add:* Dept Biol Sci Univ Northern Colo Greeley CO 80639

PEEPLES, JOHNSTON WILLIAM, b Estill, SC, July 30, 48; m 70; c 2. ELECTRICAL ENGINEERING. *Educ:* The Citadel, BS, 70; Univ SC, MS, 76, PhD(eng), 78. *Prof Exp:* Elec engr, tech serv, Westvaco Corp, 70- & commun, US Air Force, 70-73; res asst elec mat, Univ SC, 73-76; elec engr, NCR Corp, 76-78, proj engr, 78-80, mgr qual anal, 80-83, mgr engr, 83-86, dir oper, 86- 90, DIR MFG TECHNOL RES CTR, NCR CORP, 90- *Concurrent Pos:* Indust prof eng, Univ SC, 78-, adj prof bus,86-- *Mem:* Int Soc Hybrid Microelectron; Inst Elec & Electronics Engrs. *Res:* Microelectronic failure analysis and reliability; environmental testing of microelectronic packaging; silicon crystal growth for photovoltaic power production; expert system in manufacturing; advanced manufacturing technology; application intergrated circuit design. *Mailing Add:* 3325 Platt Springs Rd West Columbia SC 29169

PEEPLES, WAYNE JACOBSON, b Corder, Mo, Dec 18, 40; m 65. GEOPHYSICS. *Educ:* William Jewell Col, BA, 63; Wichita State Univ, MSc, 65; Univ Alta, PhD(physics), 69. *Prof Exp:* Fel, Univ Alta, 69-70; res geophysicist, Kennecott Explor, Inc, 70-71; asst res prof geophysics, Univ Utah, 71-76; asst prof geol sci, Southern Methodist Univ, 76-81; ASST PROF GEOL SCI, UNIV TEX, EL PASO, 82- *Mem:* Am Geophys Union; Soc Explor Geophys. *Res:* Electromagnetic scattering in geophysics; mathematical inversion of geophysical data; data processing and analysis; solid earth geophysics; lunar geophysics. *Mailing Add:* Dept of Geol Sci Univ Tex El Paso TX 79968

PEEPLES, WILLIAM DEWEY, JR, b Bessemer, Ala, Apr 19, 28; m 56; c 4. MATHEMATICS. *Educ:* Samford Univ, BS, 48; Univ Wis, MS, 49; Univ Ga, PhD(math), 51. *Prof Exp:* Asst math, Samford Univ, 46-47, Univ Wis, 47-49 & Univ Ga, 49-51; from asst prof to assoc prof, Samford Univ, 51-56; asst prof, Auburn Univ, 56-59; PROF MATH, SAMFORD UNIV, 59-, HEAD DEPT, 67- *Concurrent Pos:* Consult, Hayes Inst Corp. *Mem:* Am Math Soc; Math Asn Am. *Res:* Valuation theory, elliptic curves; algebraic geometry. *Mailing Add:* Dept Math Samford Univ Birmingham AL 35229

PEERCY, PAUL S, b Monticello, Ky, Nov 26, 40. SOLID STATE PHYSICS. *Educ:* Berea Col, BA, 61; Univ Wis-Madison, MS, 63, PhD(physics), 66. *Prof Exp:* Mem tech staff, Bell Labs, 66-68; mem tech staff solid state physics res, 68-76, supvr, Ion Implantation Physics Div, 76-82, mgr, Ion Implantation & Radiation Physics, 82-86, MGR, COMPOUND SEMICONDUCTOR & DEVICE RES, 86- *Concurrent Pos:* Vpres, Mat Res Soc, 88, chmn, Prog Comt, 90-91; mem, Solid State Sci Comt, Nat Res Coun, 87-99, Dept Energy Coun Mat Sci, 87-; secy, Electronic Mat Comt, 90-91; dir, Sandia Ctr Compound Semiconductor Technol. *Mem:* Fel Am Phys Soc; fel AAAS; sr mem Inst Elec & Electronics Engrs; The Metall Soc. *Res:* Solid state physics with primary emphasis on structural phase transitions in solids; ion implantation and ion beam analysis of solids rapid melt and solidification and compound semiconductors. *Mailing Add:* Sandia Nat Labs Org Apt 1140 PO Box 5800 Albuquerque NM 87185

PEERSCHKE, ELLINOR IRMGARD BARBARA, b Braunschweig, WGermany, May 7, 54. HEMATOLOGY, HEMOSTASIS & COAGULATION. *Educ:* Rutgers Univ, BA, 75; NY Univ, PhD(path), 80. *Prof Exp:* Asst prof, 80-86, ASSOC PROF PATH, SCH MED, STATE UNIV NY, STONY BROOK, 86-, HEAD CLIN HEMAT LABS, UNIV HOSP, 80- *Concurrent Pos:* Mem, lab & genetics task force, Region II Hemophilia Ctrs, Dept Health & Human Serv, 85-, Am Soc Hemat, Coun Platelets, 87-; prin investr, Nat Heart Lung Blood Inst, 81-, prin investr grant, Am Heart Asn. *Mem:* Am Heart Asn; Am Soc Hemat; Am Fedn Clin Res; NY Acad Sci; Soc Exp Biol & Med; Int Soc Thrombosis & Hemostasis; Sigma Xi. *Res:* Human blood platelet physiology; identification of platelet membrane receptors involved in physiologic and pathologic platelet responses; examination of intraplatelet calcium fluxes contributing to stimulus-response coupling; characterization of biochemical and biophysical requirements for platelet cohesion and adhesion to surfaces. *Mailing Add:* Clin Labs Univ Hosp Level 3 State Univ NY Stony Brook NY 11794-7300

PEERY, CLIFFORD YOUNG, b Coeburn, Va, Sept 24, 34; m 55; c 2. INDUSTRIAL ORGANIC CHEMISTRY. *Educ:* Univ Va, BS, 55; Ohio State Univ, PhD(org chem), 62. *Prof Exp:* Chemist, Merck & Co, Ltd, 55-59; teaching asst, Ohio State Univ, 59-62; chemist, Wyeth Labs, 62-63; res chemist, 63-71, staff chemist, Chem Div, 71-73, prod mgr, 73-85, DIR PROD, FERMENTATION PROD, UPJOHN CO, 85- *Mem:* Am Chem Soc; Am Inst Chem Engrs. *Res:* Separation and purification of antibiotics and chemicals produced by fermentation processes. *Mailing Add:* Upjohn Co 1200-38-1 Kalamazoo MI 49001

PEERY, LARRY JOE, b Moberly, Mo, Dec 24, 41; m 64; c 2. PHYSICS. *Educ:* Univ Mo-Rolla, BS, 64; Okla State Univ, MS, 67, PhD(physics), 70. *Prof Exp:* Coop trainee physics & eng, McDonnell-Douglas Corp, 59-64; res asst physics, Okla State Univ, 64-67; asst prof, 67-69, PROF PHYSICS & CHMN DEPT, CENT METHODIST COL, 69- *Mem:* Am Asn Physics Teachers; Am Phys Soc. *Res:* Plasma physics; lasers; mathematical modeling; innovation in instruction. *Mailing Add:* Dept Physics Cen Methodist Col Fayette MO 65248

PEERY, THOMAS MARTIN, b Lynchburg, Va, Aug 24, 09; m 36; c 3. PATHOLOGY. *Educ:* Newberry Col, AB, 28, DMS, 66; Med Col SC, MD, 32. *Prof Exp:* Intern, Metrop Hosp, NY, 32-33, res physician, 33-34; instr path, Med Col SC, 34-38; from asst prof to prof, 38-74, EMER PROF PATH, GEORGE WASHINGTON UNIV, 74- *Concurrent Pos:* Dir labs, Alexandria Hosp, 41-46; mem bd dirs, Technicon Corp, Tarrytown NY, 69-80; consult, Armed Forces Inst Path, Vet Admin Hosp, Wash, DC, Nat Cancer Inst & Random House Dict English Lang. *Honors & Awards:* Ward Burdick Award, 74. *Mem:* Am Soc Clin Pathologists (vpres, 61 & 66-68, pres, 68-69); fel AMA; Am Asn Pathologists & Bacteriologists; fel Col Am Pathologists; Int Acad Path. *Res:* Brucellosis and heart disease; clinical pathology tests in health evaluation; mortality trends in United States; curriculum planning in medicine. *Mailing Add:* 2115 Belle Haven Rd Alexandria VA 22307-1117

PEES, SAMUEL T(HOMAS), b Meadville, Pa, Nov 16, 26. REMOTE SENSING, SUBSURFACE MAPPING TECHNIQUES. *Educ:* Allegheny Col, BS, 50; Syracuse Univ, MS, 59. *Prof Exp:* Surv recorder topog, US Geol Surv, 48; foreign explor rep, Skelly Oil Co, 62-66; res geologist, Magellan Petrol Pty Ltd, 69-71; sr explor geologist, Geo-Informacoe Ltd, Brazil, 73-76; geol proj leader, Jamestown A Lewis Eng Argentina, 76-78; SR EXPLOR GEOLOGIST & MGR SAMUEL T PEES & ASSOC, 78- *Concurrent Pos:* Lectr, Allegheny Col, 72-73; vis petrol geologist, Oswego Col, 87. *Honors & Awards:* Chauncy Homes Lectr, Syracuse Univ, 87. *Mem:* Am Asn Petrol Geologist; Geol Soc Am; Pateont Res Inst; Am Soc Photogametry & remote sensing. *Res:* Inovative exploration methodology for oil, gas and other minerals; detection of morphotectonic features via remote sensing; sandstone and limestone geometry and porosity of appalachian reservoir; history of geology. *Mailing Add:* 889 Porter St Meadville PA 16335

PEET, MARY MONNIG, b Washington, DC, Aug 4, 47; m 71; c 2. PLANT PHYSIOLOGY. *Educ:* Hiram Col, BA, 69; Univ Wis-Madison, MS, 72; Cornell Univ, PhD(plant physiol), 75. *Prof Exp:* Teaching asst bot, Univ Wis-Madison, 69-71; res asst, Cornell Univ, 72-75; res assoc environ physiol, Duke Univ, 75-80; asst prof, 80-86, ASSOC PROF, NC STATE UNIV, 86- *Concurrent Pos:* Adapter sci papers, Biol Sci Curric Study, 73-74. *Mem:* Am Asn Plant Physiologists; Agron Soc; Am Soc Hort Sci; Sigma Xi. *Res:* Production and physiology of greenhouse-crops, especially as related to carbon dioxide enrichment, nutrition and temperature. *Mailing Add:* Dept Hort Sci NC State Univ Box 7609 Raleigh NC 27695-7609

PEET, NORTON PAUL, b Fargo, NDak, June 14, 44; m 67; c 3. ORGANIC CHEMISTRY, MEDICINAL CHEMISTRY. *Educ:* Univ Minn, Minneapolis, BA, 66; Univ Nebr, Lincoln, PhD(org chem), 70. *Prof Exp:* Instr chem, Concordia Col, 67; res assoc org chem, Mass Inst Technol, 70-71; res assoc & instr, Univ SC, 71-72; res chemist, Dow Chem Co, 72-81, GROUP LEADER, MERRELL DOW, 81. *Concurrent Pos:* Vis scientist, Med Univ SC, 80. *Mem:* Am Chem Soc; Int Soc Heterocyclic Chem. *Res:* Selective adenosine receptor agents; computer modeling in drug design; proteolytic enzyme inhibitors; inhibitors of steroid biosynthetic processing enzymes; new methods for introducing fluorine into molecules of biological interest; synthesis of heterocycles; antiallergic agents; glycohydrolase inhibitors; synthesis and rearrangements of benzotriazepines. *Mailing Add:* Merrell Dow Res Inst 2110 E Galbraith Rd Cincinnati OH 45215-6300

PEET, ROBERT G(UTHRIE), b Bristol, Pa, Apr 16, 33; m 60; c 2. CHEMICAL ENGINEERING. *Educ:* Pa State Univ, BS, 54; Purdue Univ, PhD(chem eng), 63. *Prof Exp:* Process engr, Am Cyanamid Co, 54-57; res engr, Marathon Oil Co, 62-65; staff engr, E I du Pont de Nemours & Co, Inc, 65-69, res supvr, 69-72, develop supvr, 72-77; MGR PROCESS RES & DEVELOP, MOBIL CHEM CO, 77- *Mem:* Am Chem Soc; Am Inst Chem Engrs. *Res:* Mass transfer; polymer processing relationships; petrochemical processing; radiation induced reactions. *Mailing Add:* Films Dept Mobil Chem Co Res & Develop Tech Ctr Macedon NY 14502

PEET, ROBERT KRUG, b Beloit, Wis, Feb 14, 47; m 71; c 2. PLANT ECOLOGY. *Educ:* Univ Wis-Madison, BA, 70, MS, 71; Cornell Univ, PhD(ecol), 75. *Prof Exp:* From asst prof to assoc prof, 75-88, PROF BIOL, UNIV NC, CHAPEL HILL, 89- *Mem:* Ecol Soc Am; Brit Ecol Soc; Int Asn Ecol; Int Asn Veg Sci; fel AAAS. *Res:* Plant succession; diversity of ecological communities; community composition and structure; plant population dynamics; plant geography. *Mailing Add:* Dept of Biol CB # 3280 Univ of NC Chapel Hill NC 27599-3280

PEETE, CHARLES HENRY, JR, b Warrenton, NC, Apr 30, 24; m 54; c 3. MEDICINE. *Educ:* Harvard Med Sch, MD, 47; Am Bd Obstet & Gynec, dipl, 57. *Prof Exp:* Asst prof, 56-68, PROF OBSTET & GYNEC, MED CTR, DUKE UNIV, 68- *Mem:* Am Fertil Soc; AMA; Am Col Obstetricians & Gynecologists. *Res:* Obstetrics and gynecology. *Mailing Add:* Duke Univ Med Ctr Durham NC 27710

PEETE, WILLIAM P J, b Warrenton, NC, Mar 29, 21; m 60; c 1. SURGERY. *Educ:* Univ NC, AB, 42; Harvard Univ, MD, 47. *Prof Exp:* Intern & resident surg, Mass Gen Hosp, 47-54; asst prof surg & asst to dean, 55-66, prof surg, 66-70, PROF GEN & THORACIC SURG & DIR SCH MED, DUKE UNIV, 70- *Concurrent Pos:* Fel surg, Harvard Univ, 53-55, Moseley fel, 54-55; consult, US Army. *Mailing Add:* Dept of Surg Duke Univ Sch of Med Durham NC 27706

PEETERS, RANDALL LOUIS, b San Bernandino, Calif, Dec 9, 45; m 65; c 2. SOLID PROPELLANT, HIGH EXPLOSIVES. *Educ:* Calif State Polytech Univ, BS, 67; Univ Wash, MS, 69, PhD(solid mech), 73. *Prof Exp:* Assoc eng, Boeing Co, 67-68; mat eng, Air Force Astronaut Lab, 72-76; staff mem, Los Alamos Nat Lab, 76-80; mgr, Xerox Corp, 80-81; res assoc, Eastman Kodak Co, 81-82; DIR RES & DEVELOP, AEROJET PROPULSION DIV, 82- *Mem:* Am Inst Aeronaut & Astronaut. *Res:* Development of new concepts for application to solid rocket motors. *Mailing Add:* 202 Baurer Circle Folsom CA 95630

PEETS, EDWIN ARNOLD, b New York, NY, June 26, 29; m 61; c 1. DERMATOPHARMACOLOGY, BIOCHEMISTRY. *Educ:* St John's Univ, NY, BS, 50; Brooklyn Col, MA, 54; NY Univ, PhD(biol), 66. *Prof Exp:* Anal chemist, Ledoux & Co, NJ, 50-55; res assoc geochem, Lamont Observ, Columbia Univ, 55-58; res biochemist, Lederle Labs, Am Cyanamid Co, NY, 58-66; HEAD DEPT BIOCHEM, SCHERING CORP, 66-, ASSOC DIR BIOL RES, 75-, DIR CLIN RES DERMAT, 80-, SR DIR, DERMAT & CLIN PHARMACOL, 85-, SR DIR & PRESIDENTIAL FEL, DERMAT & GEN MED, 89- *Concurrent Pos:* Guest lectr, 71-77, adj assoc prof, Fairleigh Dickinson Univ, 77-88; Founders Day scholar, NY Univ. *Mem:* AAAS; Am Chem Soc; Am Soc Pharmacol & Exp Therapeut; fel Am Inst Chemists; Am Acad Dermat. *Res:* Dermatopharmacology; metabolic diseases; control processes of intracellular metabolism and macromolecules biosynthesis; metabolic and dermatologic diseases and their control by pharmaceutical agents. *Mailing Add:* Schering Corp 2000 Galloping Hill Rd Kenilworth NJ 07033

PEEVY, WALTER JACKSON, soils; deceased, see previous edition for last biography

PEFFER, JOHN ROSCOE, b Natrona Heights, Pa, Aug 13, 28; m 51; c 3. POLYMER CHEMISTRY. *Educ:* Allegheny Col, BS, 50; Mich State Univ, MS, 52; Carnegie Inst Technol, PhD(org chem), 58. *Prof Exp:* Sr res chemist, Coatings & Resins Div, 58-65, proj leader resin develop, 65-72, res assoc 72-79, sr res assoc, PPG Industs, Inc, 72-87; CONSULT, 87- *Mem:* Am Chem Soc; Sigma Xi. *Res:* Development of coatings resins; polymer synthesis, including addition, condensation, ring-opening and urethane polymers; polyether polyols; siloxane polymers; epoxy curing agents; unsaturated polyester resins. *Mailing Add:* 151 Glenfield Dr Pittsburgh PA 15235

PEFLEY, RICHARD K, b Sacramento, Calif, June 17, 21; m 47; c 3. MECHANICAL ENGINEERING. *Educ:* Stanford Univ, BA, 44, MS, 51, ME, 60. *Prof Exp:* Testing engr, Pac Mfg Co, 46-47; res engr, Babcock & Wilcox Co, 47-49; chmn dept, 51-81, prof mech eng, 81-86, EMER PROF MECH ENG, UNIV SANTA CLARA, 86- *Concurrent Pos:* Consult, Enterprise Diesel Eng Co, 53-58, Broadview Res Corp, 57-59, Int Bus Mach Corp & Stanford Res Inst. *Mem:* Am Soc Mech Engrs; Am Soc Eng Educ; Am Soc Heat, Refrig & Air-Conditioning Engrs. *Res:* Heat transfer; fluid mechanics; thermodynamics. *Mailing Add:* 2169 Bohanon Dr Santa Clara CA 95050

PEGG, ANTHONY EDWARD, b Derbyshire, Eng, Apr 13, 42; m 65; c 2. BIOCHEMISTRY, PHYSIOLOGY. *Educ:* Univ Cambridge, BA, 63, MA, 65, PhD(biochem), 66. *Prof Exp:* Fel pharmacol, Med Sch, Johns Hopkins Univ, 66-69; lectr biochem, Middlesex Hosp Med Sch, Univ London, 69-74; from assoc prof to prof, 75-85, EVAN PUGH PROF PHYSIOL, MED SCH, HERSHEY MED CTR, PA STATE UNIV, 85- *Concurrent Pos:* Michael Sobell fel cancer res, Brit Empire Cancer Res Campaign, 69-74; estab investr, Am Heart Asn, 76-81, mem, Pathobiol Chem Study Sect; dir res grants, NIH, Bethesda, 77-81, chmn, 79-81; assoc ed, Cancer Res, 80-, Am J Physiol, 81-84 & Biochem J, 85-; mem cancer preclinical prog proj rev comt, Nat Cancer Inst, Bethesda, 82-86. *Mem:* Biochem Soc; Am Asn Cancer Res; Am Physiol Soc; Am Soc Biol Chemists; Am Heart Asn; Environ Matagenesis Soc. *Res:* Synthesis and function of polyamines in mammalian cells; mechanism of carcinogenesis by nitrosamines; nucleic acid methylation; DNA repair. *Mailing Add:* Dept Cellular & Molecular Physiol Pa State Univ Hershey PA 17033-0850

PEGG, DAVID JOHN, b London, Eng, Sept 2, 40; US citizen; m 65; c 2. ACCELERATOR-BASED SPECTROSCOPY OF ATOMS AND IONS. *Educ:* Univ Manchester, BSc, 63; Univ NH, MS, 68, PhD(physics), 70. *Prof Exp:* Asst prof, 70-75, assoc prof, 75-80, PROF PHYSICS, UNIV TENN, 80- *Concurrent Pos:* Co-ed, Beam-Foil Spectroscopy, 75-76; vis prof, Univ Lund, Sweden, 78, Univ Lyons, France, 78-79; prin investr, NSF res grants, 73-80 & Dept Energy res grants, 83-; consult, Oak Ridge Nat Lab, 70- *Mem:* Am Phys Soc. *Res:* Accelerator-based studies of the structure of atoms and ions using photon and electron spectroscopic techniques--ions of the beam are excited collisionally or selectively by lasers directed perpendicular or collinear to the beam axis; photo detachment of negative ions. *Mailing Add:* Dept Physics Univ Tenn Knoxville TN 37996

PEGG, PHILIP JOHN, b Chester, Eng, Mar 14, 36; m 61; c 3. CLINICAL PATHOLOGY. *Educ:* Univ London, MB, BS, 60; FRCPath, 80. *Prof Exp:* Registr hemat, St George's Hosp, London, 63-64; registr chem path, Westminster Med Sch, 64-65; lectr, Univ West Indies, 65-69; asst prof path, Univ Pa, 69-87; CHMN, DEPT PATH, MED CTR DEL, 87- *Concurrent Pos:* Dir clin chem, Dept Path, Med Ctr Del, 73- *Mem:* Col Am Pathologists; Royal Col Path; Am Asn Clin Chemists; Am Soc Clin Pathologists; Acad Clin Lab Physicians & Scientists; AAAS. *Res:* Radioimmunoassay, interactions of plasma binding proteins and hormones and effect of drugs on binding proteins; concepts biochemical normality. *Mailing Add:* Christiana Hosp PO Box 6001 Newark DE 19718-6001

PEGLAR, GEORGE W, b Independence, Mo, Sept 2, 22; m 47; c 3. MATHEMATICS. *Educ:* Cent Mo State Col, BS, 42; Univ Chicago, SM, 49; Univ Iowa, PhD(math), 53. *Prof Exp:* From instr to assoc prof, 52-70, PROF MATH, IOWA STATE UNIV, 70- *Concurrent Pos:* Opers analyst, US Air Force, 56-66. *Mem:* Fel AAAS; Am Math Soc; Math Asn Am. *Res:* Algebra; number theory. *Mailing Add:* Dept of Math Iowa State Univ Ames IA 50011

PEGLER, ALWYNNE VERNON, mining engineering, rock mechanics, for more information see previous edition

PEGOLOTTI, JAMES ALFRED, b Arcata, Calif, Dec 17, 33. ORGANIC CHEMISTRY. *Educ:* St Mary's Col, Calif, BS, 55; Univ Calif, Los Angeles, PhD(org chem), 59. *Prof Exp:* From instr to assoc prof chem, St Peter's Col, NJ, 59-73, prof chem, 73-81; DEAN, SCH ARTS & SCI, WESTERN CONN STATE UNIV, DANBURY, 81- *Mem:* Am Chem Soc. *Res:* Chemistry of allylic compounds. *Mailing Add:* 158 Brushy Hill Rd Danbury CT 06810

PEGRAM, GEORGE VERNON, JR, b Nashville, Tenn, Feb 7, 37; m 74; c 4. NEUROSCIENCE, PSYCHOPHYSIOLOGY. *Educ:* Univ of the South, BS, 59; Univ NMex, PhD(psychol), 67. *Prof Exp:* Chief, Bioeffects Div, Holloman AFB, NMex, 68-70; assoc dir, 73-77, DIR NEUROSCI, UNIV ALA MED CTR, 77- *Concurrent Pos:* Fel, Nat Res Coun, 67-68; intern, clin psychol, Univ Ala Med Ctr, 76-77; clin polysomnographer, Asn Sleep Dis Ctrs, 78- *Mem:* AAAS; Asn Psychophysiol Study Sleep; Soc Neurosci; Biofeedback Soc Am; Int Soc Chronobiol. *Res:* Sleep research particularly, effects of protein synthesis inhibition, neurochemistry of sleep, the effects of hypnotics on sleep patterns; biofeedback research into headaches, chronic pain and neuromuscular re-education. *Mailing Add:* Neurosci Prog Univ Ala Birmingham 905 Cdld Univ Sta Birmingham AL 35294

PEHL, RICHARD HENRY, b Raymond, Wash, Nov 27, 36; m 80. NUCLEAR & SEMICONDUCTOR PHYSICS. *Educ:* Washington State Univ, BS, 58, MS, 59; Univ Calif, Berkeley, PhD(nuclear chem), 63. *Prof Exp:* Fel nuclear physics, 63-65, PHYSICIST, LAWRENCE BERKELEY LAB, UNIV CALIF, 65- *Concurrent Pos:* Mem, Instrument Develop Sci Team, NASA; consult, Semiconductor Radiation Detectors; prin investr on many proj. *Res:* Development of semiconductor detector spectometers, especially silicon and germanium, for measuring radiation; fabrication and application of detectors made from high-purity germanium. *Mailing Add:* Lawrence Berkeley Lab Univ Calif 1 Cycoltron Rd Berkeley CA 94720

PEHLKE, ROBERT DONALD, b Feb 11, 33; m 56; c 3. METALLURGICAL ENGINEERING, MATERIALS SCIENCE ENGINEERING. *Educ:* Univ Mich, BSE, 55; Mass Inst Technol, SM, 58, ScD(metall), 60. *Prof Exp:* From asst prof to assoc prof, 60-68, chmn dept, 73-84, PROF METALL ENG, UNIV MICH, ANN ARBOR, 68- *Concurrent Pos:* Res engr, Gen Motors Res Labs, 52-54 & Ford Sci Lab, 55-57. *Honors & Awards:* Howe Mem lectr, 80; Gold Medal Sci Extractive Metall, Metall Soc, Am Inst Mining, Metall & Petrol Engrs. *Mem:* Fel Metall Soc; fel Am Soc Metals; Am Foundrymen's Soc; Am Soc Eng Educ; Am Inst Mining, Metall & Petrol Engrs. *Res:* Iron and steelmaking; chemical and extractive metallurgy; melting; computer control of metallurgical processes. *Mailing Add:* Dept Mat Sci & Eng Univ Mich Ann Arbor MI 48109-2136

PEI, DAVID CHUNG-TZE, b Shanghai, China, June 16, 29; Can citizen; m 60; c 2. HEAT TRANSFER, CHEMICAL ENGINEERING. *Educ:* McGill Univ, BEng, 55, PhD(chem eng), 61; Queen's Univ, Ont, MSc, 57. *Prof Exp:* Lectr chem eng, Royal Mil Col, 56-57; from asst prof to assoc prof, 61-70, PROF CHEM ENG, UNIV WATERLOO, 70- *Concurrent Pos:* Nutfield Found travel grant, 65-66. *Mem:* Am Inst Chem Engrs; Chem Inst Can. *Res:* Effects of mass transfer on the rate of heat transfer under combined convection; particle dynamics in a solid-gas conveyer system; high temperature metallurgical reaction initiated by microwaves; investigation of heat transfer rates from a continuous moving surface. *Mailing Add:* Dept Chem Eng Univ Waterloo Waterloo ON N2L 3G1 Can

PEI, RICHARD YU-SIEN, b Suzhou, Jiangsu, China, July 24, 27; US citizen; m 51; c 3. ELECTRO-MECHANICAL ENGINEERING. *Educ:* Univ l'Aurore, BS, 47, MS, 48; Rensselaer Polytech Inst, PhD(mech), 64. *Prof Exp:* Apprentice engr, power, Taiwan Power Co, 48-49 & Shanghai Power Co, 49-50; lab supt, China Light & Power Co, 50-51; chief engr, United Prod Corp, 51-55; design engr, Copes-Vulcan Div, Blaw-Knox Co, 55-56; supvr eng serv, Gen Elec Co, 56-62; chief scientist instrumentation, Neptune Res Lab, 62-64; mem staff, Bellcomm, Inc, 64-67 & defense, Inst Defense Anal, 67-69; tech dir transp, TRW Syst Group, 69-71; DIR, CIVIL TECHNOL PROG, THE RAND CORP, 71- *Concurrent Pos:* Adj prof, Cath Univ Am, 64-69. *Mem:* Fel AAAS; Am Soc Mech Eng; Oper Res Soc AM; Sigma Xi. *Res:* Advanced weapon systems technologies; systems analysis; energy conversion technologies; mathematical analysis; applied mechanics; electromagnetics; applied mathematics. *Mailing Add:* The RAND Corp 2100 M St NW Washington DC 20037-1270

PEI, SHIN-SHEM, b Kweilin, China, Jan 28, 49; m 74; c 2. ELECTRONICS ENGINEERING, COMPOUND SEMICONDUCTOR. *Educ:* Nat Taiwan Univ, BS, 70; State Univ NY, Stony Brook, PhD(solid state physics), 77. *Prof Exp:* Fel, State Univ NY, Stony Brook, 77; mem tech staff, 78-84, supvr, 84-86, DEPT HEAD, BELL LABS, 86- *Mem:* Inst Elec & Electronics Engrs; Am Vacuum Soc. *Res:* Ultra high speed compound semiconductor devices and integrated circuits on molecular beam epitaxy or metal-organic chemical vapor deposition grown heterostructures. *Mailing Add:* 15 Ethan Dr New Providence NJ 07974

PEIERLS, RONALD F, b Manchester, Eng, Sept 8, 35; m 59; c 2. NUMERICAL METHODS, HIGH ENERGY PHYSICS. *Educ:* Cambridge Univ, BA, 56; Cornell Univ, PhD(physics), 59. *Prof Exp:* NATO fel physics, Univ Birmingham, 59-61; mem, Inst Advan Study, 61-62; vis asst prof, Cornell Univ, 62-63; assoc physicist, Brookhaven Nat Lab, 63-66, physicist, 66-72, sr physicist, 72-79, chmn appl math dept, 79-88, HEAD ANAL SCI DIV, DEPT APPL SCI, BROOKHAVEN NAT LAB, 88- *Concurrent Pos:* Adj prof, Stevens Inst Technol, 67-74; vis lectr, Harvard Univ, 69; adj prof Stony Brook, 88- *Mem:* Fel Am Phys Soc; Soc Indust & Appl Math; AAAS; Fedn Am Scientists. *Res:* Computational techniques; Monte Carlo methods; asynchronous parallel algorithms. *Mailing Add:* Brookhaven Nat Lab Upton NY 11973

PEIFER, JAMES J, b Nanticoke, Pa, Apr 29, 24; m 48; c 3. BIOCHEMISTRY. *Educ:* Ursinus Col, BS, 48; Rutgers Univ, PhD(biochem), 54. *Prof Exp:* Asst biochem, Sharp & Dohme Pharm Co, Merck & Co, Inc, 48-49; instr biochem & physiol, Rutgers Univ, 53-54; fel chem & metab of essential fatty acids, Hormel Inst, Minn, 54-56, div leader & res assoc chem, 56-67; actg head dept foods & nutrit, 71-74, ASSOC PROF NUTRIT & BIOCHEM, UNIV GA, 67- *Concurrent Pos:* Am Heart Asn fel, 59-60. *Mem:* AAAS; Am Chem Soc; Am Oil Chem Soc; Soc Exp Biol & Med; Sigma Xi. *Res:* Chemistry and metabolism of lipides; interrelationships in the metabolism of essential fatty acids; protein-lipid interrelationships related to brain development. *Mailing Add:* Dawson Hall Univ Ga Athens GA 30601

PEIFFER, HOWARD R, b Reading, Pa, Feb 8, 31. METALLURGY, PHYSICS. *Educ:* Albright Col, BS, 52; Pa State Univ, MS, 54, PhD(metall), 56. *Prof Exp:* Sr scientist metal physics, Res Inst Advan Study, Martin Co, 56-64; res assoc, 64-78, dir mat eng & res, 78-90, DIR METALS TECHNOL, AMP INC, HARRISBURG, 90- *Concurrent Pos:* Lectr, Loyola Col, Md, 58-64. *Honors & Awards:* David Ford MacFarlane Award, Am Soc Metals. *Mem:* Am Soc Metals; Am Inst Mining, Metall & Petrol Engrs; Am Phys Soc. *Res:* Defect solid state; composite materials; intermetallic compounds; electrical alloys; interconnection systems; electrical resistivity measurements; explosive deformation. *Mailing Add:* Res Div AMP Inc PO Box 3608 Harrisburg PA 17105

PEIFFER, ROBER LOUIS, JR, b Chester, Pa, Dec 5, 47; m 77; c 3. COMPARATIVE OPHTHALMOLOGY. *Educ:* Univ Minn, BS, 69, DVM, 71, PhD(comp ophthal), 80. *Prof Exp:* Intern, small animal med, Iowa State Univ, 71-72; practr, Blue Cross Animal Hosp, 72-73; resident, small animal surg, Purdue Univ, 73-74; comp ophthal, NIH, Univ Minn, 74-76; asst prof comp ophthal, Univ Fla, 76-78; from asst prof to assoc prof, 78-88, PROF OPHTHAL & PATH, UNIV NC, CHAPEL HILL, 89- *Concurrent Pos:* Burroughs-Wellcome, 78-; vis prof ophthal, NC State Univ, 80-85, adj prof path, 85- *Honors & Awards:* Gaines Award, Am Animal Hosp Asn, 88. *Mem:* Am Col Vet Ophthal; Am Acad Ophthal; Asn Res Vision & Ophthal; Am Vet Med Asn; Am Animal Hosp Asn. *Res:* Glaucoma; experimental ophthalmic pathology; ocular implants. *Mailing Add:* 617 Bldg 229 H Univ NC Chapel Hill NC 27599-7040

PEIGHTEL, WILLIAM EDGAR, b Huntingdon, Pa, July 27, 27; m 49; c 3. BIOLOGY. *Educ:* Juniata Col, BS, 49; Univ Va, MA, 54, PhD(biol), 61. *Prof Exp:* Assoc prof, 56-62, PROF BIOL, SHIPPENSBURG STATE COL, 62-, CHMN DEPT, 64- *Mem:* Sigma Xi. *Res:* Shell regeneration in mollusks. *Mailing Add:* 517 Franklin Heights Shippensburg PA 17257

PEIKARI, BEHROUZ, b Hamadan, Iran, Mar 16, 38; m 65; c 3. ELECTRICAL ENGINEERING. *Educ:* Univ Tehran, BEng, 61; Univ Ill, Urbana, MS, 64; Univ Calif, Berkeley, PhD(elec eng), 69. *Prof Exp:* Instr elec eng, Univ Tehran, 61-62; res asst, Univ Ill, Urbana, 64-65; asst, Univ Calif, Berkeley, 66-69; from asst prof to assoc prof, 69-79, PROF ELEC ENG, SOUTHERN METHODIST UNIV, 80- *Concurrent Pos:* Consult, Arj Mfg Co & Mepsal Mfg Co, Tehran, 61-62, Xerox Corp, 75-76, Texas Instruments, 79-80; NASA multidisciplinary grant, Southern Methodist Univ, 71; NSF grant 86-88. *Mem:* Inst Elec & Electronics Engrs; Sigma Xi. *Res:* Analysis and design of nonlinear electronic circuits and communication networks; digital signal processing; development and design of efficient adaptive filters that can be used in echo cancellation; development of algorithms for the control of precision positioning robot arms. *Mailing Add:* Dept Elec Eng Southern Methodist Univ Dallas TX 75275

PEIL, KELLY M, US citizen. ENGINEERING. *Educ:* Okla State Univ, BA, 68, MS, 69, PhD(environ eng), 72. *Prof Exp:* At US Army Med Bioeng Res & Develop Lab, 72-75; proj engr & mgr, 75-, DEPT MGR, WESTON, 85- *Mem:* Am Soc Civil Eng; Water Pollution Control Fedn; Sigma Xi. *Res:* Wastewater treatment conceptual process development and design; treatment plant evaluations; advanced wastewater treatment process evaluation; facilities and regional planning; sewer system evaluation studies; hazardous waste RI/FS. *Mailing Add:* 14201 Vista Ct NE Albuquerque NM 87123

PEINADO, ROLANDO E, b Colombia, SAm, Nov 17, 38; US citizen; div; c 2. ALGEBRA, BIOMATHEMATICS. *Educ:* Union Col, BA, 58; Univ Nebr, MA, 60, PhD(math), 63. *Prof Exp:* Asst math, Univ Nebr, 59-60; instr, San Diego State Col, 60-61 & Univ Nebr, 61-63; asst prof, Univ Iowa, 63-66; vis prof, Univ PR, Mayaguez, 65-66, from assoc prof to prof, 66-73, RES FEL, MED CAMPUS, UNIV PR, 73- *Mem:* Am Math Soc; Math Asn Am; NY Acad Sci. *Res:* Ring theory; group theory; semigroup theory; applications of semigroup to biology, especially to neurobiology; brain research and application; mathematical linguistics to biomedicine research; computer science. *Mailing Add:* Dept Math Univ PR Mayaguez PR 00709

PEIRCE, EDMUND CONVERSE, II, b Montclair, NJ, Oct 9, 17; m 66; c 8. PHYSIOLOGY, SURGERY. *Educ:* Harvard Univ, SB, 40, MD, 43. *Prof Exp:* Res fel surg, Harvard Med Sch, 47-48; asst prof anat, Sch Med, Johns Hopkins Univ, 48; trainee surg, Nat Heart Inst, 50, staff mem, 53-54, instr, Sch Med, Georgetown Univ, 54; chief, Acuff Clin, Knoxville, Tenn, 54-61; fel surg & vis assoc prof physiol, Emory Univ, 62-64, assoc prof physiol, 64-70, prof surg, 66-70; HENRY KAUFMAN PROF HYPERBARIC SURG, MT SINAI SCH MED, 70- *Concurrent Pos:* Attend, Mt Sinai Hosp, 70-; mem, Mt Desert Island Biol Lab, chief dept surg, Bronx Vet Admin Hosp, 71-74. *Honors & Awards:* Gibbon Award, Am Soc Extracorporeal-Technol, 84. *Mem:* Am Soc Artificial Internal Organs (secy-treas, 60-62, pres, 63); Soc Thoracic Surgeons; Am Physiol Soc; Undersea & Hyperbaric Med Soc; Soc

Vascular Surg; Asn Advan Medical Instrumentation. *Res:* Artificial organs; especially lung; hypothermia; acid base balance; tissue transplantation; hyperbaric oxygen; venous disease. *Mailing Add:* Mt Sinai Sch of Med Box 1259 New York NY 10029-6574

PEIRCE, JAMES JEFFREY, b Easton, Md, Mar 27, 49; m 71; c 2. RESOURCE RECOVERY, HAZARDOUS WASTE CONTROL. *Educ:* Johns Hopkins Univ, BES, 71; Univ Wis-Madison, MSCE, 73, PhD(environ eng), 77. *Prof Exp:* Res asst civil eng, Inst Environ Studies, Univ Wis-Madison, 71-73, res asst environ eng, 73-77; sr engr energy & environ, Booz, Allen & Hamilton, Inc, 77-79; asst prof, 79-88, ASSOC PROF CIVIL & ENVIRON ENG & DIR GRAD STUDIES, DUKE UNIV, 88- *Concurrent Pos:* Instr chem, Chesapeake Col, 73; spec consult, Solid & Hazardous Waste, US Environ Protection Agency, 79-, Res Triangle Inst, 82-; prin investr, Solid Waste Planning, Triangle J Coun, NC Govt, 80-81 & Waste-to-Energy Basic Res, US Dept Energy, 82- *Honors & Awards:* Pres Reagan's Investr Award, 84. *Mem:* Sigma Xi; Asn Environ Eng Prof; Am Soc Eng Educ. *Res:* Solid and hazardous waste management; wastewater treatment sludge processing and disposal; waste-to-energy technologies; particle movement in fluids; fluid movement through particles. *Mailing Add:* 906 Kings Mill Rd Chapel Hill NC 27514

PEIRCE, JOHN WENTWORTH, b Boston, Mass, Oct 17, 46; m 77; c 3. MARINE GEOPHYSICS, EXPLORATION GEOPHYSICS. *Educ:* Dartmouth Col, AB, 68; Mass Inst Technol, PhD(oceanog), 77. *Prof Exp:* Asst prof geol, Dalhousie Univ, 76-78; geophys specialist, Explor Geophysics, Petrocan, 78-91; MAN PARTNER, GEDCO, 91- *Concurrent Pos:* Vis res assoc, marine geophysics, Mass Inst Technol, 78; chmn, ocean drilling prog, Site Surv Panel, 85-88; co-chief scientist, Ocean Drilling Prog, Leg 121, 88. *Mem:* Am Geophys Union; Geol Asn Can; Geol Soc Am; Soc Explor Geophys; Sigma Xi; Can Geophys Union; Can Soc Explor Geophys; Can Soc Petrol Geol. *Res:* Non-seismic exploration geophysics, especially in frontier international areas; plate tectonics as applied to continental margins; basin modelling; evolution of Arctic Ocean basin and margins; Himalayan orogeny. *Mailing Add:* Gedco 717 Seventh Ave SW Suite 1500 Calgary AB T2P 0Z3 Can

PEIRCE, LINCOLN CARRET, b Newburyport, Mass, May 10, 30; m 53; c 2. PLANT BREEDING. *Educ:* Cornell Univ, BS, 52; Univ Minn, PhD, 58. *Prof Exp:* From asst prof to assoc prof veg breeding, Iowa State Univ, 58-64; chmn dept, 64-78, actg dean, 78-79, PROF PLANT BIOL, COL LIFE SCI & AGR, UNIV NH, 64- *Concurrent Pos:* Sci ed, Am Soc Hort Sci; consult, USAID, 70, vis prof, Univ Hawaii, 71. *Mem:* Fel Am Soc Hort Sci; Sigma Xi. *Res:* Basic and applied genetic research of vegetable crops. *Mailing Add:* Dept Plant Biol Univ NH Durham NH 03824

PEIRENT, ROBERT JOHN, b Lowell, Mass, Apr 12, 21; m 55; c 6. CHEMISTRY, COLOR SCIENCE. *Educ:* Lowell Tech Inst, BS, 49, MS, 53. *Prof Exp:* From instr to prof textile chem, Lowell Technol Inst, 49-75; prof chem, Univ Lowell, 75-88; RETIRED. *Mem:* Am Asn Textile Chemists & Colorists (vpres, 66-67); Am Chem Soc; Brit Soc Dyers & Colourists. *Res:* Mechanisms of chemical dyeing of textile fibers; dyeing reactions in nonaqueous liquids; rate study and acceleration of the dyeing of hydrophobic synthetic fibers; diffusion of dyes in synthetic polymeric fibers via non aqueous liquids and color science measurements. *Mailing Add:* 1197 Ansover St Univ Lowell Tewksbury MA 01876

PEIRSON, DAVID ROBERT, b Hamilton, Ont, Can, Sept 23, 39; c 2. PLANT PHYSIOLOGY. *Educ:* Univ Waterloo, BSc, 63, MSc, 69; Univ BC, PhD(plant physiol), 72. *Prof Exp:* Teacher sec sch, Baden, Ont, 63-65; teacher biol, Galt Col Inst, 65-67; vis asst prof, Simon Fraser Univ, 72-74; asst prof, 74-80, ASSOC PROF BIOL, WILFRID LAURIER UNIV, 80- *Mem:* Can Soc Plant Physiologists; Am Soc Plant Physiologists. *Res:* Chemical regulation of plant growth; nitrogen metabolism. *Mailing Add:* Dept Biol Sci Wilfrid Laurier Univ Waterloo ON N2L 3C5 Can

PEISACH, JACK, b New York, NY, Aug 23, 32; m 64; c 2. BIOCHEMISTRY. *Educ:* City Col New York, BS, 53; Columbia Univ, AM, 54, PhD(chem), 58. *Hon Degrees:* PhD, Univ Padova, 90. *Prof Exp:* Instr chem, Manhattan Col, 58-59; sr res asst biochem, 59-62, res asst prof pharmacol, 62-65, asst prof, 65-69, PROF MOLECULAR PHARMACOL & MOLECULAR BIOL, ALBERT EINSTEIN COL MED, 69- *Concurrent Pos:* USPHS fel, Albert Einstein Col Med, 59-61; asst prof chem, Yeshiva Col Med, 61-62; vis scientist, Bell Tel Labs, 64-88. *Mem:* Biophys Soc; Am Chem Soc; Am Soc Biol Chemists; NY Acad Sci. *Res:* Role of transition metals in biological systems, especially biological oxidation; biochemistry and biophysics of copper- and iron-containing proteins and porphyrins; electron paramagnetic resonance and electronic spin echo spectroscopy of transition metals. *Mailing Add:* Dept of Molecular Pharmacol Albert Einstein Col of Med Bronx NY 10461

PEISER, HERBERT STEFFEN, b Grunewald, Ger, Aug 19, 17; nat US; m 49; c 3. METROLOGY, INTERNATIONAL RELATIONS. *Educ:* Cambridge Univ, BA, 39, MA, 43. *Hon Degrees:* DSc, Chungnam Nat Univ. *Prof Exp:* Tech officer, Imp Chem Indust, Eng, 40-47; sr lectr physics, Univ London, 47-48; asst res mgr, Hadfields, 48-57; phys chemist, Nat Bur Standards, 57-69, chief, Mass & Scale, 58-62, chief, Crystal Chem Sect, 64-69, chief, Off Int Rels, 69-79; CONSULT, METROL & INT STANDARDIZATION, 79- *Concurrent Pos:* Res fel & lectr appl physics, Harvard Univ, 65-66; mem, Comn Atomic Weights, Int Union Pure & Appl Chem, 69-91; consult, World Bank, Univ Petrol & Minerals, Raytheon Serv Co; vis scientist, Joint Res Ctr, Comn Europe Communities, 89-91. *Res:* X-ray crystallography; nondestructive testing; precision mass determinations; physical measurement standards; crystal growth. *Mailing Add:* 638 Blossom Dr Rockville MD 20850

PEISS, CLARENCE NORMAN, b Ansonia, Conn, Jan 3, 22; m 49; c 2. PHYSIOLOGY. *Educ:* Stanford Univ, AB, 47, AM, 48, PhD(physiol), 49. *Prof Exp:* Asst, Stanford Univ, 47-48, res assoc, 48-49; fel, Johns Hopkins Univ, 49-50; sr instr, Sch Med, St Louis Univ, 50-52, asst prof, 52-54; from asst prof to prof physiol, 54-85, dean, Sch Med, 78-82, EMER PROF PHYSIOL & EMER DEAN, SCH MED, LOYOLA UNIV, 85- *Concurrent Pos:* Markle Found scholar, Loyola Univ Chicago, 53-58, Lederle med fac award, 58-61; vis lectr, Blackburn Col, 53-54; liaison scientist, Off Naval Res, London, 65-66; mem physiol study sect, NIH, 66-70, consult, artificial heart prog, prog proj comt & cardiovasc training comt; assoc dean, Grad Sch Med Ctr, Loyola Univ, 72-78, Acad Affairs, Sch Med, 75-78. *Mem:* Soc Neurosci; Am Heart Asn; Am Physiol Soc. *Res:* Brain metabolism; environmental physiology; temperature regulation; circulation; neurophysiology. *Mailing Add:* 360 E Randolph Dr Apt 1057 William M Scholl Col 1001 N Dearborn St Chicago IL 60601

PEISSNER, LORRAINE C, b Philadelphia, Pa, Jan 15, 19. PHYSIOLOGY. *Educ:* Pa State Univ, BS, 52; Univ Okla, MS, 54, PhD(zool & biochem), 61. *Prof Exp:* Res asst gastroenterol, Okla Med Res Found, 58-62; assoc prof biol, Cent State Univ, Okla, 62-67; ASST PROF PHYSIOL, KIRKSVILLE COL OSTEOP MED, 67-, ASST TO DEAN STUDENTS, 80- *Mem:* AAAS; Shock Soc; Am Physiol Soc; Am Inst Biol Sci. *Res:* Influence of gonadal hormones on anatomy and function of rodent submaxillary gland; mechanisms in regulation of blood flow in peripheral vasculature; search for gastric secretion inhibitor in human saliva and gastric juice; effect of estrogens on parathyroid hormone metabolism. *Mailing Add:* Dean Students Off Kirksville Col Osteop Kirksville MO 63501

PEITHMAN, ROSCOE EDWARD, b Hoyleton, Ill, Feb 26, 13; m 36; c 2. PHYSICS, ELECTRONIC INSTRUMENTATION. *Educ:* Southern Ill Univ, BEd, 35; Univ Ill, MS, 39; Ore State Univ, EdD(educ, physics), 55. *Prof Exp:* Teacher high schs, Ill, 35-42; from asst prof to prof, 46-77, EMER PROF PHYSICS, HUMBOLDT STATE UNIV, 77- *Concurrent Pos:* Actg chmn, Div Natural Sci, Humboldt State Univ, 57-58, chmn, Div Phys Sci, 60-69, Dean, Sch Sci, 69-70. *Mem:* Am Asn Physics Teachers. *Res:* Electrical measurments and electronics; air navigation. *Mailing Add:* 2704 Sunny Grove Ave McKinleyville CA 95521-9226

PEITZ, BETSY, b Pittsburgh, Pa, Dec 11, 48. REPRODUCTIVE PHYSIOLOGY. *Educ:* Carlow Col, BA, 70; Case Western Reserve Univ, PhD(biol), 75. *Prof Exp:* NIH fel reproductive physiol, Med Col, Cornell Univ, 75-77; asst prof, 77-81, ASSOC PROF BIOL, CALIF STATE UNIV, LOS ANGELES, 81- *Mem:* Am Soc Zoologists; AAAS; Soc Study Reproduction; Sigma Xi. *Res:* Physiology of the male reproductive tract and the adaptive and evolutionary significance of different reproductive patterns. *Mailing Add:* Dept Biol Calif State Univ LA 5151 State University Dr Los Angeles CA 90032

PEKALA, PHILLIP H, MOLECULAR BIOLOGY, IMMUNOLOGY. *Educ:* Va Polytech Inst & State Univ, PhD(biochem), 78. *Prof Exp:* ASST PROF BIOCHEM, SCH MED, E CAROLINA UNIV, 81- *Mailing Add:* 1512 Hollybriar Lane Greenville NC 27858

PEKAREK, ROBERT SIDNEY, b Berwyn, Ill, May 1, 40; c 2. MICROBIOLOGY, BIOCHEMISTRY. *Educ:* Knox Col, Ill, BA, 62; Loyola Univ Chicago, MS, 64, PhD(microbiol), 67. *Prof Exp:* Res microbiologist, US Army Med Res Inst Infectious Dis, 69-74; staff mem, Human Nutrit Lab, USDA, 74-77; RES SCIENTIST, LILLY RES LABS, 77- *Mem:* Am Soc Microbiol; Fedn Am Socs Exp Biol; Soc Exp Biol & Med. *Res:* Study of alterations in host metabolism during infection and stress, including trace metals, amino acids, serum proteins and endogenous humoral mediating factors; nutritional biochemistry; development of analytical technique for measuring trace element in biological materials; study of pathophysiologic mechanisms during acute and chronic infections; design models for testing antibiotic efficacy; effect of immunostimulating defense mechanisms; effect of nutritional deficiencies on the immune response; antibiotic evaluation in infectious disease models; antibiotic pharmacokinetics. *Mailing Add:* 725 St Johns Ave Apt 11 Highland Park IL 60035

PEKAS, JEROME CHARLES, b Mott, NDak, Aug 13, 36; m 57; c 5. ANIMAL PHYSIOLOGY, ANIMAL NUTRITION. *Educ:* NDak State Univ, BS, 57; Univ Fla, MS, 58; Iowa State Univ, PhD(animal physiol & nutrit), 61. *Prof Exp:* Asst prof metab, UT-AEC Agr Res Lab, Univ Tenn, 61-64; res scientist, Pac Northwest Labs, Battelle Mem Inst, 64-66; res physiologist, Agr Res Serv, 66-80, GASTROINTESTINAL PHYSIOLOGIST, US MEAT ANIMAL RES CTR, USDA, 80- *Concurrent Pos:* Nat Acad Sci travel grant, Int Cong Physiol, Tokyo, 65; partic, Int Cong Physiol, Munich, 71. *Mem:* Am Physiol Soc; Am Soc Animal Sci; Am Gastroenterol Asn; NY Acad Sci; Sigma Xi. *Res:* Function and regulation of the exocrine pancreas; interaction between the dietary and functions of the gastrointestinal glands, particularly the pancreas and small intestine. *Mailing Add:* USDA US Meat Anim Res Ctr Box 166 Clay Center NE 68933

PEKAU, OSCAR A, b Can, Feb 15, 41. EARTHQUAKE ENGINEERING, STRUCTURAL DYNAMICS. *Educ:* Univ Toronto, BASc, 64; Univ London, MSc, 65; Univ Waterloo, PhD(structural eng), 71. *Prof Exp:* Structural engr, Morrison, Hershfield, Millman & Huggins, Toronto, 65-67; res asst, Univ Waterloo, 67-70; PROF CIVIL ENG, CONCORDIA UNIV, 71- *Concurrent Pos:* Vis scholar, Univ Calif, Berkeley, 78-79. *Mem:* Am Soc Civil Engrs; Asn Prof Engrs. *Res:* Analysis of building structures under static and dynamic loading; non-linear behaviour of coupled shear walls, precast panal structures, steel frames, and lateral-torsional coupling of the response of buildings during earthquakes. *Mailing Add:* Dept Civil Eng Concordia Univ Sir G Williams 1455 de Maisonneuve Montreal PQ H3G 1M8 Can

PEKERIS, CHAIM LEIB, b Alytus, Lithuania, June 15, 08; nat US; m 33. GEOPHYSICS, ASTROPHYSICS. *Educ:* Mass Inst Technol, BS, 29, DSc(meteorol), 33. *Prof Exp:* Fel, Rockefeller Found, 34-35 & Cambridge Univ, 35-36; assoc geophys, Mass Inst Technol, 36-41; mem sci staff, Div War Res, Columbia Univ, 41-45, dir math physics group, 45-50; prof appl math, 50-73, DISTINGUISHED INST PROF, WEIZMANN INST SCI, 73- *Concurrent Pos:* Guggenheim fel, 47. *Honors & Awards:* Rothschild Prize, 66; Vetlesen Prize, C Unger Vetlesen Found, 74; Gold Medal, Royal Astron Soc, 80. *Mem:* Nat Acad Sci; fel Royal Astron Soc; Israel Nat Acad Sci; Ital Nat Acad Sci; Am Philos Soc. *Res:* Seismology; underwater sound; atomic spectroscopy; hydrodynamics; electromagnetic wave propagation; applied mathematics; stellar hydrodynamics. *Mailing Add:* Dept Appl Math PO Box 26 Weizmann Inst Sci Rehovot 76100 Israel

PEKOZ, TEOMAN, b Apr 16, 37; US citizen; m 63; c 2. STRUCTURAL ENGINEERING, CIVIL ENGINEERING. *Educ:* Robert Col, Istanbul, BS, 58; Harvard Univ, MS, 59; Cornell Univ, PhD(civil eng), 67. *Prof Exp:* Engr, struct eng, various eng co, 59-69; PROF STRUCT ENG, CORNELL UNIV, 69- *Concurrent Pos:* Consult, Am Iron & Steel Inst, 69- & Mat Handling Inst, 71-; vis prof, Royal Inst Technol, Sweden, 76-77 & Neth; Gordon McKay fel, Harvard Univ, 58-59. *Mem:* Struct Stability Res Coun; Am Soc Civil Eng; Int Asn Bridge & Struct Eng. *Res:* Developing specifications on thin walled structures; cold-formed steel design. *Mailing Add:* Hollister Hall Cornell Univ Ithaca NY 14850

PELAN, BYRON J, mechanical design, mechanical engineering education, for more information see previous edition

PELCOVITS, ROBERT ALAN, b New York, NY, Jan 25, 54; m 78; c 2. PHASE TRANSITIONS, SUPERCONDUCTIVITY. *Educ:* Univ Pa, BA & MS, 74; Harvard Univ, PhD(physics), 78. *Prof Exp:* Res assoc, Univ Ill, Urbana-Champaign, 78-79, Brookhaven Nat Lab, 79-80; asst prof, 79-86, ASSOC PROF PHYSICS, BROWN UNIV, 86- *Concurrent Pos:* A P Sloan Found fel, 83-87. *Honors & Awards:* Bergmann Mem Award, 83. *Mem:* Am Phys Soc. *Res:* Condensed matter theoretical physics including studies of phase transitions in liquid crystals, amorphous systems and superconductors. *Mailing Add:* Dept Physics Brown Univ Providence RI 02912

PELCZAR, FRANCIS A, b Lawrence, Mass, June 13, 39; m 64; c 2. ORGANIC CHEMISTRY. *Educ:* Merrimack Col, BS, 60; Univ Conn, MS, 63; Univ NH, PhD(org chem), 68. *Prof Exp:* Asst chem, Univ Conn, 60-63; asst, Univ NH, 64-65; from asst prof to assoc prof, 67-77, PROF ORG CHEM, GANNON COL, 77- *Mem:* Am Chem Soc. *Res:* Organometallic and heterocyclic chemistry. *Mailing Add:* Dept Chem Gannon Col On Perry Sq Erie PA 16541

PELED, ABRAHAM, b Suceava, Romania, Sept 21, 45; US citizen; m 67; c 1. COMPUTER SYSTEMS. *Educ:* Technion-Israel Inst Technol, BSc, 67, MSc, 71; Princeton Univ, MA, 72, PhD(elec eng), 74. *Prof Exp:* Res staff mem, IBM, T J Watson Res Ctr, 74-76, res staff mem, IBM Sci Ctr, Israel, 76-78, mgr digital signal processing, T J Watson Res Ctr, 78-80, mgr comput sci, Almaden Res Ctr, 80-83, dir tech planning, T J Watson Res Ctr, 83-85, VPRES SYSTS, IBM, T J WATSON RES CTR, 85- *Concurrent Pos:* Adj prof elec eng, Technion-Israel Inst Technol, 76-78; bd gov, IBM Tokyo Res Lab, 85-; bd trustees, Univ Space Res Asn, 86-; mem, Computer Sci & Telecommun Bd Nat Res Coun, 88- *Mem:* Fel Inst Elec & Electronic Engrs; NY Acad Sci. *Res:* The evolution of computer systems and the information society. *Mailing Add:* IBM Res Ctr PO Box 704 Yorktown Heights NY 10598

PELICCI, PIER GUISEPPE, oncogenes, for more information see previous edition

PELIKAN, EDWARD WARREN, b Chicago, Ill, June 15, 26; m 55; c 2. PHARMACOLOGY. *Educ:* Univ Ill, BS, 48, MS, 50, MD, 51. *Prof Exp:* From instr to asst prof pharmacol, Col Med, Univ Ill, 51-53; asst prof, Grad Sch Med, Univ Pa, 55-57; assoc prof, 57-62, chmn dept, 62-88, PROF PHARMACOL, SCH MED, BOSTON UNIV, 62- *Concurrent Pos:* House officer, Presby Hosp, Chicago, Ill, 51-53. *Mem:* AAAS; Am Soc Pharmacol & Exp Therapeut; NY Acad Sci. *Res:* Methodology; biometrics, autonomic drugs; neuromuscular blocking agents; central nervous system depressants; drugs affecting behavior. *Mailing Add:* Dept Pharmacol Boston Univ Sch Med Boston MA 02118

PELINE, VAL P, b Hooversville, Pa, July 12, 30. ENGINEERING ADMINISTRATION. *Educ:* Univ Pittsburgh, BS, 52, MS, 54; Ohio State Univ, PhD(aeronaut eng), 58. *Prof Exp:* From res specialist to mgr astrodyn, Lockheed Missile & Space Co, 58-66, mgr space systs technol, Space Systs Div & systs anal & preliminary design, chief systs engr & dep prog mgr, 66-78, vpres advan prog & develop, 78-84, vpres & gen mgr, Space Systs Div & vpres corp, 84-86, pres, Space Systs Div, 86-87, GROUP PRES, ELECTRONIC SYSRS, LOCKHEED CORP, 87- *Concurrent Pos:* Mem bd dirs, Stanford Telecommun Inc & Electronic Industs Asn. *Mem:* Nat Acad Eng; fel Am Astronaut Soc; Am Inst Aeronaut & Astronaut; Nat Space Club. *Mailing Add:* Electronic Systs Group Lockheed Corp Daniel Webster Hwy S CS2050 Nashua NH 03061-2050

PELKA, DAVID GERARD, b San Diego, Calif, Apr 25, 43; m 69; c 2. PHYSICS, ENERGY SCIENCE. *Educ:* Calif State Univ, Los Angeles, BS, 65; Univ Calif, Riverside, MA, 68, PhD(physics), 71. *Prof Exp:* Res assoc superconductivity, Univ Calif, Riverside, 68-70; prof physics, Northrop Univ, 71-, dir, Energy Res Ctr, 79-; CONSULT SOLAR ENERGY, L M DEARING ASSOCS, INC. *Concurrent Pos:* Energy consult, various govt & pvt co, 75-; mem, Calif Post Second Comt, HEW grant, 76-78 & Los Angeles Solar City Comt, 77- *Mem:* Int Solar Energy Soc; Am Phys Soc; Am Soc Heating Refrig & Air Conditioning Engrs; Sigma Xi. *Res:* Current research concerned with developing a new optical concentrator which can be used in applications to medium-high temperature thermal energy generation, its characteristics exceed those of fresnel or parabolic concentrators. *Mailing Add:* 8315 Kenyon Ave Los Angeles CA 90045

PELL, ERIK MAURITZ, b Rattvik, Sweden, Sept 22, 23; nat US; m 51; c 3. SOLID STATE PHYSICS. *Educ:* Marquette Univ, BEE, 44; Cornell Univ, PhD(physics), 51. *Prof Exp:* Asst, Cornell Univ, 47-51; res assoc, Gen Elec Res Lab, NY, 51-61; res mgr, Webster Sci Labs, Xerox Corp, 61-71, staff consult, 71-74, dir res planning, 74-81, 81-84, assoc mgr, Webster Res Ctr, 85-86, mgr, tech adv panels, 87-88; RETIRED. *Honors & Awards:* Edward G Acheson Medal, 86. *Mem:* Fel Am Phys Soc; Inst Elec & Electronics Engrs; Electrochem Soc (pres, 80-81); Soc Photog Sci & Eng; Sigma Xi. *Res:* Semiconducting properties of germanium, silicon, and selenium; photoconductivity; electrophotography. *Mailing Add:* 697 Summit Dr Webster NY 14580

PELL, EVA JOY, b New York, NY, Mar 11, 48; m 69; c 2. PLANT PATHOLOGY. *Educ:* City Col New York, BS, 68; Rutgers Univ, PhD(plant biol), 72. *Prof Exp:* Adj asst prof plant biol, Rutgers Univ, 72-73; from asst prof to prof, 73-90, DISTINGUISHED PROF PLANT PATH, PA STATE UNIV, 91- *Mem:* Am Phytopath Soc; Am Soc Plant Physiol; AAAS. *Res:* Air pollution effects on vegetation. *Mailing Add:* 211 Buckhout Lab University Park PA 16802

PELL, KYNRIC M(ARTIN), b Toronto, Ont, Apr 5, 38; US citizen; m 62; c 3. COMPUTER AIDED DESIGN & MANUFACTURE. *Educ:* Univ Fla, BAsE, 62, MS, 63, PhD(aerospace), 67. *Prof Exp:* Design engr, United Aircraft Corp, 62; res assoc aerospace eng, Univ Fla, 63-67, asst prof, 67-68; asst prof, Auburn Univ, 68-71; from asst prof to assoc prof, 71-79, PROF MECH ENG, UNIV WYO, 79-, DEPT HEAD, 89- *Concurrent Pos:* Consult, Northrop Corp, 70-71, US Army Missile Command, 77-82, US Navy, 87-; secy, IDES, 86-; partner, Mountain Technol Assocs, 89- *Honors & Awards:* Transp Res Bd Award, 75. *Mem:* Am Soc Mech Engrs; Sigma Xi. *Res:* Computer workstations and networks for computer aided design and manufacturing; software to guide a designer in piece part design optimized for specific manufacturing process; heat pipe applications; instrumentation. *Mailing Add:* Dept Mech Eng Univ Wyo PO Box 3295 Laramie WY 82071

PELL, MEL, b New York NY, July 11, 42; m 67; c 2. CHEMICAL ENGINEERING. *Educ:* City Univ NY, PhD(chem eng), 71; Duquesne Univ, MBA, 77. *Prof Exp:* Res engr, Sci Des, Halcon Int, 64-66; res engr, Yardney Elec Co, 66-67; process engr, Conoco, Conoco Coal Res, 71-82; SR CONSULT, DU PONT ENG, 82- *Concurrent Pos:* Lectr, Am Inst Chem Engrs Continuing Educ, 91. *Mem:* Am Inst Chem Engrs. *Res:* Fluidization and fluid particle systems. *Mailing Add:* E I du Pont de Nemours & Co Louviers PO Box 6090 Newark DE 19714

PELL, SIDNEY, b New York, NY, Dec 13, 22; m 50. BIOSTATISTICS, EPIDEMIOLOGY. *Educ:* City Col New York, BBA, 47, MBA, 52; Univ Pittsburgh, PhD(biostatist), 56. *Prof Exp:* Statistician, NY Univ-Bellevue Med Ctr, 51-52; mgr epidemiol sect, Med Div, E I Du Pont de Nemours & Co, 55-85; CONSULT EPIDEMIOL, 85- *Mem:* Fel Am Col Epidemiol; fel Am Pub Health Asn; fel Am Heart Asn. *Res:* Epidemiological studies of cardiovascular disease, diabetes, alcoholism and occupational disease. *Mailing Add:* 1416 Emory Rd Green Acres Wilmington DE 19803-5120

PELLA, JEROME JACOB, b Pierz, Minn, Mar 13, 39; m; c 3. FISHERIES. *Educ:* Univ Minn, BS, 61; Univ Washington, MS, 64, PhD(fisheries), 67. *Prof Exp:* Scientist, Inter-Am Trop Tuna Comn, 65-69; MATH STATISTICIAN, NAT MARINE FISHERIES SERV, 69- *Mem:* Am Inst Fishery Res Biologists. *Res:* Applications of stochastic models in fisheries; development of statistical methods for assessing composition of population mixtures; US and Canadian salmon fisheries problems. *Mailing Add:* Biol Lab Nat Mar Fish Serv PO Box 210155 Auke Bay AK 99821

PELLA, MILTON ORVILLE, b Wilmot, Wis, Feb 13, 14; m 44. SCIENCE EDUCATION. *Educ:* Milwaukee State Teachers Col, BE, 36; Univ Wis, MS, 40, PhD(educ, bot), 48. *Prof Exp:* Teacher, Wyler Mil Acad, Wis, 37-38, grade sch, 38-39, Wis High Sch, 39-42; teacher, 39-42 & 46-50, asst prof to assoc prof, 46-57, prof sci edu, 57-80, EMER PROF, SCH EDUC, UNIV WIS-MADISON, 80- *Concurrent Pos:* Consult, Ministries Educ, Turkey, 59, Iran, 62 & Jordan, 63, Am Univ Beirut, 62 & 64, Univ Jordon, 65 & 66, Aleppo, Syria, 66, Egypt, 72-80, Univ Damascus Syria, 80 & Yarmouk Univ Jordon, 80-81. *Mem:* Fel AAAS; Nat Sci Teachers Asn; Nat Asn Res Sci Teaching (pres, 65). *Res:* Teaching of science in elementary schools, its status and development; science concepts and their relationships to factors of educational achievement, culture and experience. *Mailing Add:* 5518 Varsity Hill Madison WI 53705

PELLEG, AMIR, b Palestine, Sept 14, 44; US citizen; m 79; c 2. CARDIAC ELECTROPHYSIOLOGY. *Educ:* Tel-Aviv Univ, Israel, BSc, 69; Polytech Inst Brooklyn, MS, 74; La State Univ, PhD(physiol), 77. *Prof Exp:* Head, Cardiac Metab Unit, Ichilov Hosp, Tel-Aviv Med Ctr, 81-83; assoc investr cardiol, 83-87, SR SCIENTIST, LANKENAU MED RES CTR, PHILADELPHIA, PA, 87- *Concurrent Pos:* Vis asst prof, Med Col Pa, 86- *Mem:* Am Physiol Soc; Cardiac Electrophysiol Soc; Int Soc Heart Res Am Chap. *Res:* Cardiac electrophysiology; mechanisms of arrhythmias; anti-arrhythmic drugs; electrophysiology of purine compounds in the heart. *Mailing Add:* Likoff Cardiol Inst Dept MS 110 Hahneman Univ Med Broad & Vine St Philadelphia PA 19102

PELLEGRINI, CLAUDIO, b Rome, Italy, May 9, 35; m 61; c 3. PARTICLE ACCELERATORS, FREE ELECTRON LASERS. *Educ:* Univ Rome, Laurea physics, 58. *Prof Exp:* Physicist, Lab Nat Frascati, 59-65, sr physicist, 65-76, sr physicist & div leader, 76-78; sr physicist, Brookhaven Nat Lab, 81-89; PROF PHYSICS, UNIV CALIF, LOS ANGELES, 89- *Concurrent Pos:* Assoc chmn physics, Nat Synchrotron Light Source. *Mem:* Fel Am Phys Soc. *Res:* Particle beam physics; high energy accelerators and colliders; free electron lasers. *Mailing Add:* Dept Physics Univ Calif Los Angeles CA 90024

PELLEGRINI, FRANK C, b Brooklyn, NY, July 10, 40; m 67; c 2. ORGANIC CHEMISTRY. *Educ:* St John's Univ, NY, BS, 62, MS, 64, PhD(org chem), 70. *Prof Exp:* Teaching asst chem, St John's Univ, NY, 62-66, lectr pharmaceut chem, Col Pharm, 66-69; from asst prof to assoc prof, 69-79, chmn dept, 74-77, PROF CHEM, STATE UNIV NY AGR & TECH COL FARMINGDALE, 79- *Mem:* Am Chem Soc. *Res:* Heterocyclic chemistry; synthesis of non-benzenoid aromatic ring systems; complex metal hydride reductions. *Mailing Add:* Dept Chem State Univ Tech Col Far Melville Rd Farmingdale NY 11735

PELLEGRINI, JOHN P, JR, b New Haven, Conn, Jan 23, 26; m 53; c 2. ORGANIC CHEMISTRY. *Educ:* Yale Univ, BS, 46; Univ Ill, MS, 47, PhD(chem), 49; Univ Pittsburgh, MBA, 78. *Prof Exp:* Asst gen chem, Univ Ill, 46-49; res chemist, 49-50, group leader explor res, 50-54, sect head prod res, 54-61, res assoc, Gulf Res & Develop Co, 61-82; OWNER, JE-PELLE VENTURES CONSULTS, 82- *Mem:* Am Chem Soc. *Res:* Chemical control and clean up of oil spills; additives for petroleum products; metallocene chemistry; polymer coatings; chemical properties of long-chain olefins; formulation and marketing of surface active chemicals; properties of gly col based sufactants. *Mailing Add:* 617 Orchard Hill Dr Pittsburgh PA 15238

PELLEGRINI, MARIA C, b New Orleans, La, June 3, 47. MOLECULAR BIOLOGY, ORGANIC CHEMISTRY. *Educ:* Conn Col, AB, 69; Columbia Univ, PhD(chem), 73. *Prof Exp:* Asst prof, 77-83, ASSOC PROF MOLECULAR BIOL, UNIV SOUTHERN CALIF, 83- *Concurrent Pos:* Alfred P Sloan Found fel, 79. *Mem:* Sigma Xi; Am Chem Soc; Am Soc Microbiologists; Soc Cell Biologists. *Res:* Molecular biology of ribosome biosynthesis; eukaryotic gene expression; ribosome structure-function relationships, especially antibiotic binding sites. *Mailing Add:* Molecular Biol Univ Southern Calif Los Angeles CA 90089

PELLEGRINI, ROBERT J, b Worcester, Mass, Oct 21, 41. PSYCHOLOGY. *Educ:* Clark Univ, BA, 63; Univ Denver, MA, 66, PhD(psychol), 68. *Prof Exp:* PROF PSYCHOL, SAN JOSE STATE UNIV, 67- *Mem:* Am Psychol Asn. *Res:* Interpersonal-evaluative judgments; attitude and personality measurements. *Mailing Add:* Dept Psychol San Jose State Univ San Jose CA 95192

PELLEGRINO, EDMUND DANIEL, b Newark, NJ, June 22, 20; m 45; c 7. MEDICAL & HEALTH SCIENCES. *Educ:* St John's Univ, NY, BS, 41; NY Univ, MD, 44. *Hon Degrees:* Thirty-five from US cols & univs. *Prof Exp:* Intern med, Bellevue Hosp, New York, 44-45; resident, Goldwater Mem Hosp, 45-46; fel, NY Univ, 48, fel & asst, 49-50; asst prof clin med, 53-59; supv tuberc physician, Homer Folks Tuberc Hosp, Oneonta, NY, 51-53; prof med & chmn dept, Med Ctr, Univ Ky, 59-66; prof med, chmn dept & dir med ctr, State Univ NY Stony Brook, 66-73, dean sch med, 68-72, vpres health sci & dir ctr, 68-73; prof med & med humanities & chancellor, Univ Tenn Ctr Health Sci, 73-75; prof med, chmn bd & dir, Yale-New Haven Med Ctr, Yale Univ, 75-78; pres, Cath Univ Am, 78-82; JOHN CARROLL PROF MED & MED HUMANITY, GEORGETOWN UNIV, 82-, DIR, CTR ADVAN STUDY ETHICS, 89- *Concurrent Pos:* Asst clin vis physician, Bellevue Hosp, NY, 49-50, from asst vis physician to assoc vis physician, 53-59; dir internal med, Hunterdon Med Ctr, Flemington, NJ, 53-59; consult, Vet Admin Hosp, Lexington, Ky, 59-66, USPHS Hosp, 60-66, Dept Med, Div Internal Med & Nephrol, Nassau County Med Ctr, NY, 68-73 & St Charles Hosp, Port Jefferson, NY, 69-73; sr vis scientist & vis attend physician, Brookhaven Nat Lab, 66-73. *Honors & Awards:* Outstanding Contrib Allied Health Educ & Accreditation, Am Med Asn, 80. *Mem:* Inst Med-Nat Acad Sci; Am Bd Internal Med; Asn Am Physicians; Am Col Physicians. *Res:* Cardiovascular-renal diseases; electrolyte metabolism; bone physiology; medical education; medical humanities; author of two books and many articles and numerous reviews. *Mailing Add:* Ctr Advan Study Ethics Georgetown Univ Washington DC 20057

PELLEGRINO, MICHELE A, b Bovalino, Italy, Nov 14, 40. IMMUNOCHEMISTRY, CANCER THERAPEUTICS. *Educ:* Pavia Univ, Italy, PhD(chem), 63. *Prof Exp:* Sci invest, 83-89, ASST DIR, DEPT SCI IMMUNOBIOL, HYBRITECH, INC, 89- *Mem:* Am Asn Immunol. *Mailing Add:* Hybritech Inc 11085 Torreyana Rd San Diego CA 92121

PELLENBARG, ROBERT ERNEST, b Charleston, WVa, Feb 5, 49; m 73; c 2. CHEMICAL OCEANOGRAPHY, CHEMISTRY. *Educ:* George Washington Univ, BS, 71; Univ Miami, MS, 73; Univ Del, PhD(marine sci), 76. *Prof Exp:* Head chem lab, US Naval Oceanog Off, 76-77; fel, 77-79, RES CHEMIST, US NAVAL RES LAB, 79- *Mem:* Sigma Xi; Am Chem Soc; Oceanogr Soc. *Res:* Geochemistry of trace metals; marine chemistry and biochemistry of trace elements and organics; analytical methods development as used for saline waters; trace substances as markers for materials movement in geochemical context. *Mailing Add:* Code 6182 Naval Res Lab Washington DC 20375

PELLER, LEONARD, b New York, NY, July 22, 28. PHYSICAL CHEMISTRY. *Educ:* Univ Calif, BS, 51; Princeton Univ, MA, 53, PhD, 58. *Prof Exp:* Jr res phys chemist, Calif Res Corp, 51; teaching asst, Princeton Univ, 51-53 & 55-57; proj assoc phys chem, Univ Wis, 57-60, instr, 60, lectr, 61; res chemist, Nat Inst Arthritis & Metab Dis, 62-63; vis lectr chem, Univ Calif, Berkeley, 63; ASSOC RES BIOPHYSICIST, MED CTR, UNIV CALIF, SAN FRANCISCO, 63-, ASSOC PROF BIOPHYS, 73- *Mem:* Am Chem Soc. *Res:* Physical chemistry of biological macro-molecules; enzyme kinetics. *Mailing Add:* HSW 841 Univ Calif Med Ctr San Francisco CA 94143

PELLERIN, CHARLES JAMES, JR, b Shreveport, La, Dec 11, 44; c 2. ASTROPHYSICS. *Educ:* Drexel Univ, BS, 67; Cath Univ Am, MS, 70, PhD(physics), 74; Harvard Bus Sch, PMD, 82. *Prof Exp:* Physicist, Sounding Rocket Br, 67-70, astrophysicist, Lab High Energy Astrophys, Goddard Space Flight Ctr, 70-75, DEP DIR, SPACELAB FLIGHT DIV, NASA, 75-, DIR, ASTROPHYS DIV, 82- *Mem:* Am Phys Soc; Am Astron Soc; Int Astron Union. *Res:* Astrophysical problems through measurements of cosmic rays, including solar cosmic rays, galactic positrons and negatrons and gamma rays. *Mailing Add:* 11018 West Ave Kensington MD 20895

PELLETIER, CHARLES A, b New Britain, Conn, Jan 10, 32; m 61; c 2. HEALTH PHYSICS, ENVIRONMENTAL HEALTH. *Educ:* Rensselaer Polytech Inst, BCE, 56; Univ Rochester, MS, 57; Univ Mich, PhD(environ health), 66. *Prof Exp:* Consult health physics, Astra, Inc, 57-58; radiation control engr, Bethlehem Steel Co, 58-60; lectr radiol health, Univ Mich, 60-66, asst prof environ health, 66; chief, Environ Br, Health Serv Lab, US AEC, 67-71, chief environ radiation sect, Div Compliance, 71-73; DEPT MGR, SCIENCE APPLICATIONS INT CORP, 73- *Mem:* Health Physics Soc; Am Nuclear Soc. *Res:* Evaluation of waste management and effluent and environmental monitoring practices of US Nuclear Regulatory Commission licensed facilities; contract research, occupational and environmental health and safety for nuclear power plants; research and development and services in power plant air and water leak detection; radiation instrumentation development. *Mailing Add:* 41 W Main St Niantic CT 06357-2329

PELLETIER, GEORGES H, b June 18, 39; Can citizen; m 68; c 3. ENDOCRINOLOGY. *Educ:* Laval Univ, BA, 60, MD, 65, PhD(endocrinol), 68. *Prof Exp:* Med Res Coun Can fel endocrinol, McGill Univ, 68-69, electron micros, Univ Paris, 69-70 & cytochem, Albert Einstein Col Med, 70-71; asst prof, 71-74, assoc prof, 74-78, prof physiol, Laval Univ, 78-; AT MOLECULAR ENDOCRINOL LAB, LAVAL UNIV HOSP CTR. *Concurrent Pos:* Mem, Group Med Res Coun Can Molecular Endocrinol, 75- *Mem:* Am Soc Cell Biol; Endocrine Soc; Can Physiol Soc; Can Micros Soc; Can Soc Endocrinol & Metab. *Res:* Ultrastructural identification of the structures which produce the hypothalamic regulatory hormones involved in the regulation of anterior pituitary secretion. *Mailing Add:* Dept Physiol Univ Laval Fac Med Ste-Foy PQ G1K 7P4 Can

PELLETIER, GERARD EUGENE, b Ottawa, Ont, June 18, 30; m 61; c 2. CHEMISTRY. *Educ:* Univ Ottawa, BSc, 53, MSc, 55; Laval Univ, PhD(chem), 60. *Prof Exp:* Chemist, OPW Paints, Ltd, 53-54 & Imperial Oil Res Labs, 60-62; vdean fac sci, 64-67, dir dept, 67-76, PROF CHEM, UNIV SHERBROOKE, 62- *Mem:* Chem Inst Can; Am Chem Soc. *Res:* Physical chemistry; enzyme kinetics. *Mailing Add:* Dept Chimie Fac Des Sci Univ Sherbrooke Sherbrooke PQ J1K 2R1 Can

PELLETIER, JOAN WICK, b Northampton, Mass, Nov 13, 42; c 2. PURE MATHEMATICS. *Educ:* Smith Col, BA, 64; McGill Univ, MS, 67, PhD(math), 70. *Prof Exp:* Asst prof math, Concordia Univ, 70-72; assoc prof, 72-80, chmn math dept, 85-89, PROF MATH, YORK UNIV, 80-, ASSOC VPRES RES, 90- *Concurrent Pos:* Vis prof, Univ Mass, 78-79 & Mc Gill Univ, 84-85. *Mem:* Am Math Soc; Can Math Soc; Asn Women Math. *Res:* Applications of the theory of categories to analysis. *Mailing Add:* Dept Math York Univ 4700 Keele St Downsview ON M3J 2R3 Can

PELLETIER, OMER, b May 31, 29; Can citizen; m 58; c 4. ANALYTICAL BIOCHEMISTRY, NUTRITIONAL BIOCHEMISTRY. *Educ:* Univ Montreal, BSc, 53; Univ Ottawa, BSc, 64, PhD(biochem), 70. *Prof Exp:* Technologist, Bur Med Biochem, Can, 55-64, res scientist, Health Protection Br, 64-90, mgr, 70-90; RETIRED. *Mem:* Chem Inst Can; Nutrit Soc Can; NY Acad Sci; Am Soc Clin Nutrit; Am Soc Clin Chem. *Res:* Analytical, biochemical nutrition and metabolism and clinical chemistry: vitamins, sugars and lipids; automated, high performance liquid chromatography and global environmental monitoring system reference methods and materials for vitamins, sugars and lipids; metabolism of ascorbic and isoascorbic acid; vitamin C metabolism of smokers. *Mailing Add:* Five Bernier Hull PQ J8Z 1E7 Can

PELLETIER, R MARC, b Feb 10, 46; div. CELL BIOLOGY OF TESTICULAR CELL MEMBRANE. *Educ:* Univ Sherbrooke, Quebec, PhD(anat & cell biol), 76. *Prof Exp:* ASST PROF GROSS ANAT, DEPT ANAT, SCH MED, UNIV OTTAWA, 86- *Res:* Andrology. *Mailing Add:* Dept Anat Sch Med Univ Ottawa 451 Smyth Rd Ottawa ON K1H 8M5 Can

PELLETIER, S WILLIAM, b Kankakee, Ill, July 3, 24; m 49; c 6. ORGANIC CHEMISTRY. *Educ:* Univ Ill, BS, 47; Cornell Univ, PhD(org chem), 50. *Prof Exp:* Asst org chem, Cornell Univ, 47-50; instr, Univ Ill, 50-51; res asst chem pharmacol, Rockefeller Inst, 51-54, res assoc, 54-57, from asst prof to assoc prof, 57-62; prof & head dept, 62-69, provost, 69-76, ALUMNI FOUND DISTINGUISHED PROF CHEM, UNIV GA, 69-, UNIV PROF & DIR INST NATURAL PROD RES, 76- *Concurrent Pos:* Radio broadcast engr, Sta WKAN, 47; Gordon lectr, 55, 59 & 69; lectr, Ger Acad Agr Sci, Berlin, 59 & Am-Swiss Found Sci Exchange, 60; commemorative dedication lectr, Shionogi Res Lab, Osaka, Japan, 61; session chmn, Natural Prod Symp, Int Union Pure & Appl Chem, Kyoto, Japan, 64, Riga, Latvia, 70 & Varna, Bulgaria, 78. mem, Undergrad Equip Panel, NSF, 65-67, Med Chem Study Sect, NIH, 68-72 & Adv Bd, Ga Mus Art, 68-; mem, Bd Dirs, Story Chem Corp, 69-73 & Ctr Res Libraries, 75-81. *Honors & Awards:* Herty Medalist, Am Chem Soc, 71, Southern Chemists' Award, 72; Victor A Coulter Lectr, Univ Miss, 65; Nason-Preston Lectr, Boston Pub Libr, 82. *Mem:* AAAS; fel Am Chem Soc; The Chem Soc; fel Royal Soc Arts; Am Soc Pharmacog. *Res:* Synthetic organic chemistry; structure and stereochemistry of diterpenoid alkaloids; diterpenes; triterpenes; application of carbon-13 nuclear magnetic resonance to structure elucidation; X-ray crystallographic structures of natural products. *Mailing Add:* Dept of Chem Univ of Ga Athens GA 30602

PELLETT, DAVID EARL, b Topeka, Kans, July 2, 38; m 67. PHYSICS. *Educ:* Univ Kans, AB, 60, MA, 62; Univ Mich, Ann Arbor, PhD(physics), 66. *Prof Exp:* Res assoc particle physics, Univ Mich, Ann Arbor, 66-67; asst prof, 67-74, ASSOC PROF PHYSICS, UNIV CALIF, DAVIS, 74- *Mem:* Am Phys Soc. *Res:* Experimental high energy particle physics. *Mailing Add:* 922 Eureka Davis CA 95616

PELLETT, HAROLD M, b Atlantic, Iowa, Feb 17, 38; m 60; c 6. BREEDING, COLD HARDINESS. *Educ:* Iowa State Univ, BS, 60, MS, 61, PhD(hort), 64. *Prof Exp:* Asst prof hort, Univ Nebr, 64-66; from asst prof to assoc prof, 66-80, supt, Hort Res Ctr, 74-84, PROF HORT, UNIV MINN, ST PAUL, 80- *Concurrent Pos:* Chmn, Woody Landscape Plants, Crop Adv Comt, 86- *Mem:* Am Soc Hort Sci; Int Plant Propagators Soc. *Res:* Breeding of stress tolerant woody landscape plants, cold hardiness and propagation research. *Mailing Add:* Dept Hort Sci Univ Minn Col Agr St Paul MN 55101

PELLETT, NORMAN EUGENE, b Atlantic, Iowa, June 26, 34; m 56; c 3. ORNAMENTAL HORTICULTURE. *Educ:* Iowa State Univ, BS, 58; Univ Minn, MS, 64, PhD(hort), 65. *Prof Exp:* Mgr, Pellett Gardens, 58-61; res asst hort, Univ Minn, 61-64; asst prof, State Univ NY Agr & Tech Col Cobleskill, 64-67; asst prof hort, 67-72, assoc prof, 72-85, PROF PLANT & SOIL SCI, COL AGR, UNIV VT, 85- *Mem:* Am Soc Hort Sci; Int Plant Propagators Soc. *Res:* Winter hardiness of ornamental plants. *Mailing Add:* Dept Plant & Soil Sci Col Agr Univ Vt Burlington VT 05405

PELLEY, RALPH L, organic chemistry, for more information see previous edition

PELLICCIARO, EDWARD JOSEPH, b Beaver Falls, Pa, Mar 20, 21; m 46; c 3. MATHEMATICS. *Educ:* Wagner Col, BS, 49; Univ NC, PhD(math), 53. *Prof Exp:* Instr math, Univ Del, 53-54; res fel, Duke Univ, 54-56; from asst prof to assoc prof, 56-70, PROF MATH, UNIV DEL, 70- *Mem:* Am Math Soc; Math Asn Am; Soc Indust & Appl Math; Sigma Xi. *Res:* Ordinary differential equations. *Mailing Add:* Dept Math Univ Del Newark DE 19711

PELLICER, ANGEL, b Tarragona, Spain, Aug 5, 48; m 80. GENETIC ENGINEERING, MOLECULAR ONCOLOGY. *Educ:* Univ Valencia, Spain, MD, 71; Univ Madrid, Spain, PhD(biochem), 76. *Prof Exp:* Fel res assoc, Columbia Univ, 76-80; ASSOC PROF PATH, NY UNIV MED CTR, 80- *Honors & Awards:* Irma Hirschl Award, Irma Hirschl Trust, 82. *Mem:* Am Soc Microbiol; AAAS; Am Asn Professions; Am Asn Cancer Res. *Res:* Study of mechanisms of activation of oncogenes in animal model systems of induced carcinogenesis; Gene transfer and recombinant DNA as approaches to understanding genetic alterations in cancer. *Mailing Add:* Dept Path NY Univ Med Ctr 550 First Ave New York NY 10016

PELLIER, LAURENCE (DELISLE), b Paris, France, US citizen. METALLURGY. *Educ:* City Col New York, BChE, 39; Stevens Inst Technol, MS, 42. *Prof Exp:* Metallurgist, Sylvania Elec Prod, Inc, 46-51, Am Cyanamid Co, 51-57 & Int Nickel Co, 58-61; metallurgist, Burndy Corp, 62-64, consult, 64-67; OWNER, PELLIER-DELISLE METALL LAB, 67- *Concurrent Pos:* Achievement Award, Soc Women Engrs, 62. *Honors & Awards:* Micrography Prize, Am Soc Metals, 49; Micrography Prize, Am Soc Testing & Mat, 52. *Mem:* Electron Micros Soc Am; Am Soc Metals; Fr Soc Metall; Am Soc Testing & Mat; Soc Women Engrs. *Res:* Powder metallurgy; corrosion; optical and electron metallography; physical metallurgy. *Mailing Add:* Pellier-Delisle Metall Lab 45 Clapboard Hill Rd Westport CT 06880

PELLINEN, DONALD GARY, b Duluth, Minn, Sept 19, 39; m 71; c 4. PLASMA PHYSICS. *Educ:* Univ Calif, Berkeley, BA, 65. *Prof Exp:* Technician, Space Sci Lab, Univ Calif, Berkeley, 61-65; physicist, Physics Int Co, 65-80, GCA Corp, Plasma Prop, 80-82; prog eng, 82-86, PHYSICIST, AMD, 86- *Mem:* Am Physics Soc; Inst Elec & Electron Eng. *Res:* Pulse power technology with emphasis on developing, transporting & diagnosing high power electron particle beams. *Mailing Add:* Pulse Scis Inc 600 McCormick San Leandro CA 94577

PELLINI, WILLIAM S, structural liability; deceased, see previous edition for last biography

PELLIS, NEAL ROBERT, b Greensburg, Pa, Mar 24, 44; m 67; c 2. IMMUNOLOGY. *Educ:* Washington & Jefferson Col, BA, 66; Miami Univ, MS, 68, PhD(microbiol), 72. *Prof Exp:* From instr to asst prof immunol, Med Sch, Northwestern Univ, 73-77; asst prof, 77-80, ASSOC PROF IMMUNOL, MED SCH, UNIV TEX, HOUSTON, 80-, DIR GRAD PROG IMMUNOL, 84- *Concurrent Pos:* Fel, Med Sch, Stanford Univ; mem staff, Exp Tumor Immunol Lab, Vet Admin, Lake Side Hosp, Chicago, Ill, 73-77. *Mem:* Am Asn Immunologists; Am Asn Cancer Res; Soc Exp Biol & Med. *Res:* Immunity to neoplastic disease; chemical characterization of tumor-associated antigens; immunity to allografts; potentiation of the immune response. *Mailing Add:* Dept Gen Surg Univ Tex Med Cancer Ctr 1515 Holcombe Blvd Box 174 Houston TX 77030

PELLIZZARI, EDO DOMENICO, b Orland, Calif, Mar 17, 42; m 71; c 6. ANALYTICAL CHEMISTRY, ANALYTICAL BIOCHEMISTRY. *Educ:* Calif State Univ, Chico, AB, 63; Purdue Univ, West Lafayette, PhD(anal biochem), 69. *Prof Exp:* Res asst biochem, Mich State Univ, 63-65; res asst Purdue Univ,Lafayette, 65-67, teaching asst analy biochem, 67-68; fel anal chem, Tex Res Inst Ment Sci, 69-71; div dir, 71-83, VPRES, RES TRIANGLE INST, 83- *Concurrent Pos:* Fulbright-Hayes fel, Univ Repub, Uruguay, 67; USPHS fel, Tex Res Inst Ment Sci, 69-71. *Mem:* Am Chem Soc; AAAS; Sigma Xi; Soc Appl Spectros; Air Pollution Control Asn. *Res:* Development and application of techniques and instrumental methods of analysis for biomedical and environmental problems. *Mailing Add:* Res Triangle Inst PO Box 12194 Research Triangle Park NC 27709

PELLON, JOSEPH, b Barre, Vt, Mar 9, 28; m 50; c 2. ORGANIC CHEMISTRY. *Educ:* Univ Vt, BS, 50; Columbia Univ, PhD(chem), 57. *Prof Exp:* Mem staff, Am Cyanamid Co, 50-53; asst chem, Columbia Univ, 53-54; res chemist, Cent Res Div, Polymer Flocculants, 57-59, res leader, 59-60, group leader, 60-65 & Fibers Res, 66-69, mgr, Fibers Res, 69-78, prod liaison polyacrylamide emulsions, Water Treating, Mining & Paper Chem, 78-85; PRIN RES CHEMIST, POLYMER FLOCCULANTS, 85- *Mem:* Am Chem Soc. *Res:* Polymer synthesis and reaction mechanism; free radical chemistry; polymer structure versus polymer properties; acrylic polymers and fibers; polyacrylamide emulsions for EOR, water treating, mining and paper chemicals. *Mailing Add:* Am Cyanamid Co 1937 W Main St Stamford CT 06902

PELLOUX, REGIS M N, b Passy, France, Dec 24, 31; US citizen; m 56; c 2. METALLURGY. *Educ:* Ecole Centrale des Arts et Manufactures, Paris, France, dipl, 55; Mass Inst Technol, MS, 56, ScD(metall), 58. *Prof Exp:* Res specialist, Sci Res Labs, Boeing Co, 61-64, head mat unit, Turbine Div, 64-66, res specialist phys metall, Sci Res Labs, Wash, 66-68; assoc prof, 68-76, PROF METALL, MASS INST TECHNOL, 76- *Concurrent Pos:* Vis assoc prof metall, Mass Inst Technol, 66-67. *Mem:* Am Soc Metals. *Res:* Physical metallurgy; electron microscopy; fatigue-fracture. *Mailing Add:* Dept Mat Sci & Eng Mass Inst Technol Rm 8-237 Cambridge MA 02139

PELOQUIN, STANLEY J, b Barron, Wis, July 22, 21. CYTOGENETICS, EVOLUTION. *Educ:* Univ Wis, River Falls, BS, 42; Marquette Univ, MS, 48; Univ Wis-Madison, MS, 49, PhD(genetics), 52. *Prof Exp:* Assoc prof biol, Marquette Univ, 52-56; assoc prof genetics, 56-62, prof hort & genetics, 62-82, CAMPBELL BASCOM PROF HORT & GENETICS, UNIV WIS-MADISON, 82- *Honors & Awards:* Genetic & Breeding Award, Nat Coun Com Plant Breeders, 85. *Mem:* Nat Acad Sci; Am Potato Asn; Europ Potato Ans; Genetic Soc Am; Can Soc Genetics. *Mailing Add:* Dept Hort Univ Wis Madison WI 53706

PELOSI, EVELYN TYMINSKI, b Dec 6, 38; US citizen; m 65; c 2. ORGANIC CHEMISTRY. *Educ:* Univ Mass, BS, 60; Univ NH, MS, 63, PhD(org chem), 65. *Prof Exp:* Asst prof chem, State Univ NY Col Oneonta, 67-68; TEACHER CHEM, NORWICH SR HIGH SCH, 77- *Mem:* Sigma Xi. *Res:* Synthetic and stereo chemistry. *Mailing Add:* 111 Hillview Ct Norwich NY 13815

PELOSI, LORENZO FRED, b San Michele di Serino, Italy, Sept 11, 44; US citizen; m 77; c 1. ORGANIC POLYMER CHEMISTRY, FLUORINE CHEMISTRY. *Educ:* Montclair State Col, BA, 67; Cornell Univ, PhD(org chem), 73. *Prof Exp:* Jr anal chemist, Hoffmann-La Roche, Inc, 67; teaching asst org chem, Cornell Univ, 67-70; fel synthetic org chem, Syntex Res, Inc, 72-74; res chemist, Elastomers & Polymer Prod Depts, 74-81, sr res chemist, Med Prod Dept, 81-89, RES ASSOC FIBERS, BIOMAT TECH CTR, E I DU PONT DE NEMOURS & CO, INC, 89- *Mem:* Sigma Xi; Am Chem Soc; Soc Biomat. *Res:* Elastomer characterization and vulcanization; process development; electrochemical sensors development; solid phase DNA extraction; protein supports for extracorporal blood therapy; biomaterials development. *Mailing Add:* Bldg 302 Du Pont Co Exp Sta 80302 Wilmington DE 19880-0302

PELOSI, STANFORD SALVATORE, JR, b Revere, Mass, Oct 3, 38; m 65; c 2. ORGANIC CHEMISTRY. *Educ:* Boston Col, BS, 60; Univ NH, PhD(org chem), 65. *Prof Exp:* Org chemist, US Army Aviation Mat Labs, Va, 65-67; sr res chemist, 67-79, unit leader, 79-84, SECT HEAD, NORWICH EATON PHARMACEUT, 84- *Concurrent Pos:* Alt counr, Am Chem Soc, 82-84. *Mem:* Am Chem Soc; Sigma Xi. *Res:* Synthetic organic chemistry; heterocyclic compounds; medicinal chemistry; peptide chemistry. *Mailing Add:* Norwich Eaton Pharmaceut Inc PO Box 191 Norwich NY 13815-0191

PELT, ROLAND J, b Marianna, Fla, July 10, 31; m 54. CHEMICAL ENGINEERING, MATHEMATICS. *Educ:* Troy State Univ, BS, 52; Univ Miss, BS & MS, 60; Univ Pittsburgh, PhD(chem eng), 64. *Prof Exp:* Engr, Westinghouse Atomic Power Labs, 60-62; process engr, Mobay Chem Co, 62-64, sr process engr, 64-65, tech asst to dir eng, 65-66; develop group leader, McIntosh, 66-67, proj mgr eng, 67-68, proj eng supvr, 68-69, prod mgr, 69-70, asst plant mgr, 70-74, plant mgr, ciba-geigy chem corp, 74-85; RETIRED. *Concurrent Pos:* Pres, Delvan Develop Corp, 76-85. *Mem:* Am Chem Soc. *Res:* Chemical process design; chemical plant construction; project and production management. *Mailing Add:* 29230 Rosemary Lane Elberta AL 36530

PELTIER, CHARLES FRANCIS, b Boston, Mass, Oct 6, 45; m 69; c 2. TOPOLOGY. *Educ:* Col of the Holy Cross, BS, 67; Univ Notre Dame, MS, 70, PhD(math), 73. *Prof Exp:* Vis asst prof math, Univ Notre Dame, 73-74; ASST PROF MATH, ST MARY'S COL, 74- *Mem:* Am Math Soc; Math Asn Am. *Res:* Fiber bundles and classifying spaces, especially with reference to problems in embedding and immersion of manifolds; finite transformation groups on manifold, diameter and degree of finite graphs. *Mailing Add:* Dept Math 1329 Wall St South Bend IN 46615

PELTIER, EUGENE J, b Concordia, Kans, Mar 28, 10; wid; c 4. ENGINEERING. *Educ:* Kans State Univ, BS, 33, LLD, 61. *Prof Exp:* Instrumentation engr, Kans Hwy Comn, 34-40; asst pub works officer, USN, Great Lakes, Ill, 40-42, sr asst supt civil eng, Boston, Mass, 42-44, officer var assignments, Pensacola, Fla, 45-46, Memphis, Tenn, 46-49, Jacksonville, Fla, 49-51, dis pub works officer, 14th Naval Dist, 51-53, asst chief maintenance & mat, Bur Docks, Washington, 53-56, cmndg officer, Pt Hueneme, 56-57, chief, Bur Yarks & Docks, Navy Dept, Washington, 57-62, chief of civil engrs, 57-62; vpres, Sverdrup & Parcel & Assocs, Inc, St Louis, Mo, 62-64, sr vpres, 64-66, exec vpres, 66-67, pres & dir, 67-75, chief exec officer, 72-75, consult, 75-82; RETIRED. *Concurrent Pos:* Vpres & dir, ARO, Inc, Tullahoma, Tenn, 66-75; dir, Sverdrup & Parcel Int, Inc, 67-75; consult, Environ Protection Agency, 76-80. *Honors & Awards:* Award of Merit, Consult Engrs Coun, 62. *Mem:* Nat Acad Eng; Consult Engrs Coun; hon mem Am Soc Civil Engrs; Am Pub Works Asn; Soc Mil Engrs (pres, 60-61); Am Concrete Inst; Am Road & Transp Builders Asn (pres, 72-73); Nat Soc Prof Engrs; Pub Works Hist Soc (pres, 77-78). *Mailing Add:* 543 Middleton Ct St Louis MO 63122-1554

PELTIER, HUBERT CONRAD, b New York, NY, Apr 6, 25; m 52; c 6. MEDICINE. *Educ:* Ind Univ, Bloomington, MD, 48; Am Bd Pediat, dipl, 54. *Prof Exp:* Intern, Ind Univ Med Ctr, 49; resident pediat, James Whitcomb Riley Hosp Children, 52; pvt pract, 52-56; res physician, Clin Res, Upjohn Co, Kalamazoo, 56-59; chief clin develop, 59-62, mgr, 62-64; vpres & med dir, Bristol Labs, Syracuse, 64-68; sr dir med res & med affairs, 68-70, exec dir domestic med affairs, 70, vpres med affairs, 71-78, vpres res, 78-79, sr vpres develop, 79-81, PRES, MERCK SHARP & DOHME RES LABS, 81- *Mem:* Fel Am Acad Pediat; Am Diabetes Asn; AMA; Pharmaceut Mfrs Asn. *Res:* Clinical pharmacology. *Mailing Add:* 1042 Kennett Way West Chester PA 19380

PELTIER, LEONARD FRANCIS, b Wisconsin Rapids, Wis, Jan 8, 20; m 43; c 2. ORTHOPEDIC SURGERY. *Educ:* Univ Nebr, AB, 41; Univ Minn, MD, 44, PhD(surg), 51. *Prof Exp:* Intern, Sect Orthop Surg, Univ Hosps, Univ Minn, 44-45, resident, 45-46, 48-50, clin instr, 51-53, clin asst prof, 53-56, clin assoc prof surg, 56-57; prof surg & head sect orthop surg, Med Ctr, Univ Kans, 57-71; PROF SURG & HEAD SECT ORTHOP SURG, MED CTR, UNIV ARIZ, 71- *Concurrent Pos:* Markle scholar, 52-56. *Mem:* Soc Univ Surgeons; Am Col Surgeons; Am Acad Orthop Surgeons. *Res:* Orthopedics; traumatology. *Mailing Add:* Univ Hosp Univ Ariz Tucson AZ 85721

PELTIER, WILLIAM RICHARD, b Vancouver, BC, Can, Dec 31, 43. DYNAMIC METEOROLOGY, GEOPHYSICS. *Educ:* Univ BC, BSc, 67; Univ Toronto, MSc, 69, PhD(physics), 71. *Prof Exp:* Vis scientist fluid dynamics & geophys, Coop Inst Res Environ Sci, 72-73; vis prof, 73-77, from asst prof to assoc prof, 74-79, PROF PHYSICS, UNIV TORONTO, 79- *Concurrent Pos:* Consult, Coop Inst Res Environ Sci & Inst Arctic & Alpine Res, 74-; Sloan fel, 77-79; Steacie fel, 78; Killam fel, 80-82; Guggenheim fel, 86-88; fel Clare Hall, Ulc, 88- *Honors & Awards:* Kirk Bryan Award, Geol Soc Am, 80. *Mem:* Can Asn Physicists; Can Meteorol & Oceanographic Soc; fel Am Geophys Union; fel Royal Soc Can; fel Am Meteorol Soc; Am Geophys Union. *Res:* Atmospheric dynamics; thermal convection and atmospheric waves; visco-elastic structure of the planetary interior; mantle convection; glacio-isostatic rebound; numerical modeling; nonlinear dynamical systems. *Mailing Add:* Dept of Physics Univ of Toronto Toronto ON M5S 1A7 Can

PELTON, JOHN FORRESTER, b Los Angeles, Calif, Mar 15, 24; m 48; c 1. PLANT ECOLOGY. *Educ:* Univ Calif, Los Angeles, BS, 45; Univ Minn, MS, 48, PhD(bot), 51. *Prof Exp:* Asst bot, Univ Minn, 46-48, instr, 48-49; instr forestry, Univ Calif, 51-52 & bot, Oberlin Col, 52-53; from asst prof to prof bot, Butler Univ, 53-85, actg head dept, 55-56, head dept, 62-85; RETIRED. *Concurrent Pos:* Treas, Rocky Mountain Biol Lab, 57. *Mem:* Bot Soc Am; Ecol Soc Am; Sigma Xi. *Res:* Ecological life history of seed plants. *Mailing Add:* Box 404 Rte 2 Nashville IN 47448

PELTON, MICHAEL RAMSAY, b Trion, Ga, July 25, 40; m 62; c 2. WILDLIFE BIOLOGY, MAMMALOGY. *Educ:* Univ Tenn, BS, 62; Univ Ga, MS, 65, PhD(wildlife biol), 69. *Prof Exp:* Asst prof, 68-76, ASSOC PROF FORESTRY, UNIV TENN, KNOXVILLE, 76- *Honors & Awards:* Stoddard-Sutton Award, 69. *Mem:* Am Soc Mammal; Wildlife Soc. *Res:* Game mammal ecology; physiological response of small mammals to environmental conditions; reproductive aspects of game mammal biology. *Mailing Add:* Dept Forestry Wildlife Fish Univ Tenn Knoxville TN 37916

PELTZMAN, ALAN, b New York, NY, Dec 25, 37; m 63. CHEMICAL ENGINEERING. *Educ:* City Col New York, BChE, 60; NY Univ, MChE, 64; City Univ New York, PhD(chem eng), 67. *Prof Exp:* Air Force Off Sci Res res asst chem eng, Eng Res Div, NY Univ, 60-62; res asst chem eng, City Col New York, 62-63, Nat Sci Found asst, 63-66, instr, 64; chem engr, Halcon Int, Inc, 66-68, sr chem engr, 68-75; proc mgr res & develop, Halcon Int, Inc & Halcon S D Corp, 75-78, tech dir, Halcon S D Corp, 78-87; TECHNOL MGR, H-R INT INC, 87- *Mem:* Am Inst Chem Engrs; Am Chem Soc. *Res:* Chemical kinetics; process development; mass transfer operations; chemical economics; heat transfer; unit processes of chemical engineering. *Mailing Add:* H R Intl Inc 2045 Lincoln Hwy Edison NJ 08817

PELUS, LOUIS MARTIN, immunobiology, experimental hematology, for more information see previous edition

PELUSO, ADA, b Antwerp, Belg, Feb 8, 41; US citizen. MATHEMATICS. *Educ:* Hunter Col, BA, 60; NY Univ, ScM, 63, PhD(math), 66. *Prof Exp:* Asst prof, 66-70, ASSOC PROF MATH, HUNTER COL, 71- *Mem:* Am Math Soc; Math Asn Am; AAAS; Sigma Xi. *Res:* Group theory. *Mailing Add:* 38 E 85th St Apt 4B New York NY 10028

PELZER, CHARLES FRANCIS, b Detroit, Mich, June 5, 35; m 72; c 1. HUMAN GENETICS, MOLECULAR GENETICS. *Educ:* Univ Detroit, BS, 57; Univ Mich, PhD(human genetics), 65. *Prof Exp:* Kettering Found fel, Wabash Col, 65-66; instr biol, Univ Detroit, 66-68; from asst prof to assoc prof, 69-79, PROF BIOL, SAGINAW VALLEY STATE UNIV, 79- *Concurrent Pos:* Res assoc, Mich State Univ, 76-77; vis scientist, Am Inst Biol Sci/Energy Res & Develop Admin, 75-78; res fel, Henry Ford Hosp, 82-83; grant reviewer, US Dept Educ, 84-86; prin investr, Biochem & Molecular Genetic Studies, Kellogg Found, Monsanto Co & Saginaw Valley State Univ Found, NIH; res fel, Henry Ford Hosp, 89- *Mem:* Am Inst Biol Sci; NY Acad Sci; Soc Human Genetics; Genetics Soc Am; Electrophoresis Soc. *Res:* Genetic screening for blood proteins and enzymes, especially alpha-1 antitrypsin and glutamic-pyruvate transferase as related to such human diseases as emphysema and breast cancer; genetic control of isozymes; mapping of multiple endocrine neoplasia gene; isoelectric focusing; molecular genetics of tumor suppressor genes. *Mailing Add:* Dept Biol Saginaw Valley State Univ University Center MI 48710

PEMBLE, RICHARD HOPPE, b Indianola, Iowa, Aug 16, 41; m 65; c 3. ECOLOGY, PHYTOGEOGRAPHY. *Educ:* Simpson Col, BA, 63; Univ Mont, MA, 65; Univ Calif, Davis, PhD(bot), 70. *Prof Exp:* Teaching asst biol, Univ Mont, 63-65; res asst bot, Univ Calif, Davis, 65-68, teaching asst, 68-69; from asst prof to assoc prof, 69-80, PROF BIOL, MOORHEAD STATE UNIV, 80- *Concurrent Pos:* Chairperson, dept biol, Moorhead St Univ, 85- *Mem:* Ecol Soc Am. *Res:* Phytogeography and ecology of the Red River Drainage Basin in North Dakota and Minnesota. *Mailing Add:* Dept Biol Moorhead State Univ Moorhead MN 56560

PEMRICK, RAYMOND EDWARD, b Troy, NY, Jan 6, 20; m 48. TEXTILE CHEMISTRY. *Educ:* Siena Col, BS, 43; Rensselaer Polytech Inst, MS, 52. *Prof Exp:* Res chemist, Cluett, Peabody Co, Inc, 46-50; res chemist, Behr-Manning Co div, Norton Co, NY, 51-54, asst to res dir, 54-59, dir lab serv, Tech Dept, 59-62, mgr lab serv, Res Dept, 62-63, asst dir res, 63-70, lab mgr, Coated Abrasives Div, 70-71, mgr res & develop serv & group leader cloth finishing res, 71-75, mgr tech serv, Tech Dept, Coated Abrasives Div, 75-76, mgr tech serv & proj mgr, Synthetics Backings, 76-77, mgr tech serv, 78-81, res assoc, 81-85, CONSULT, NORTON CO, NY, 86- *Mem:* AAAS; Am Chem Soc; Am Asn Textile Chemists & Colorists; Sigma Xi. *Res:* Coated abrasives; synthetic resins; textiles, including application and fabric design; patents and patent liaison; agreements and licenses; industrial health and safety. *Mailing Add:* Norton Co Coated Abrasive Div 10 Ave & 25th St Watervliet NY 12189

PEMSLER, J(OSEPH) PAUL, b New York, NY, July 9, 29; m 54; c 3. PHYSICAL CHEMISTRY, MATERIALS SCIENCE. *Educ:* NY Univ, BS, 49, PhD(phys chem), 54. *Prof Exp:* Teaching fel chem, NY Univ, 50-53; phys chemist, Goodyear Atomic Corp, 53-56; phys chemist, Nuclear Metals, Inc, 56-59, group leader chem metall, 59-62; chem metallurgist, Ledgemont Lab, Kennecott Copper Corp, 62-66, staff scientist, 67-74, group leader chem, 74-75, mgr exploratory res, 75-77; dir new technol, EIC Corp, 77-79; PRES, CASTLE TECHNOL CORP, 79- *Concurrent Pos:* Chmn, Gordon Conf Corrosion, 73; vis scientist, NSF-CSIR, India, 83. *Honors & Awards:* Francis Mills Turner Award, Electrochem Soc, 58; Extractive Metall Sci Award, Am Inst Mining, Metall & Petrol Engrs, 78. *Mem:* Am Chem Soc; Electrochem Soc; Am Inst Mining, Metall & Petrol Engrs; Am Inst Chemists. *Res:* Electrochemistry; extractive metallurgy; materials science; energy related research and development; new concepts in batteries, electrodeposition, corrosion and hydrometallurgy, including proof of concepts, publications and patents, and applied research toward commercialization. *Mailing Add:* 6 Castle Rd Lexington MA 02173

PENA, HUGO GABRIEL, b Tarma, Peru, Mar 24, 28; m 57; c 5. RADIOBIOLOGY, NUCLEAR MEDICINE. *Educ:* Agr Univ, Lima, Peru, BS, 51; Purdue Univ, MS, 66, PhD(bionucleonics), 69. *Prof Exp:* Gen mgr point four prog, Exp Ranch, Int Coop Admin, Lima, Peru, 52-55; lab technician, Gen Hosp, Indianapolis, Ind, 57-60; chief technician, Metab Res Lab, Radioisotope Serv, Vet Admin Hosp, Indianapolis, 60-64; res asst bionucleonics dept, Pharm Sch, Purdue Univ, 64-68; ASSOC CHIEF NUCLEAR MED SERV, VET ADMIN HOSP, 68- *Concurrent Pos:* Lectr, Med Sch, Univ NMex, 69-, asst prof radiopharm, Pharm Sch, 77- *Mem:* AAAS; Health Physics Soc; Am Soc Animal Sci; Int Radiation Protection Asn; Nat Asn Agr Eng, Peru. *Res:* Thyroid physiology, especially radioiodinated compounds metabolism; fallout pollutants in the food chain, particularly their metabolism and translocation in farm animals; radioactive lipid dyes in the assay of atherosclerosis; whole-body counting and nuclear medicine research. *Mailing Add:* 1712 Ross Pl SE Albuquerque NM 87108

PENA, JORGE AUGUSTO, b Buenos Aires, Arg, May 6, 19; m 46; c 2. AIR POLLUTION, ATMOSPHERIC CHEMISTRY. *Educ:* Univ Buenos Aires, PhD(chem), 46. *Prof Exp:* Chemist at several factories & labs, Arg, 45-56; instr phys chem, Univ Buenos Aires, 56-59, instr meteorol, 59-67, adj prof, 67; from res asst to res assoc, 67-74, res meteorologist, 74-77, RES ASSOC METEOROL, PA STATE UNIV, 77- *Mem:* Am Meteorol Soc. *Res:* Aerosol particle generation by gas reactions in atmospheric conditions. *Mailing Add:* 827 Wheatfield Dr State College PA 16801

PENCE, HARRY EDMOND, b Martins Ferry, Ohio, Feb 4, 37; m 59; c 3. ENVIRONMENTAL CHEMISTRY. *Educ:* Bethany Col, WVa, BS, 58; Univ WVa, MS, 62; La State Univ, PhD(chem), 68. *Prof Exp:* Instr chem, Washington & Jefferson Col, 61-65, asst prof, 65-66; assoc prof, 67-69, PROF CHEM, STATE UNIV NY COL ONEONTA, 69- *Concurrent Pos:* Pres, Fac Senate, State Univ NY, 75-77. *Mem:* Am Chem Soc; Hist Sci Soc; AAAS; Nat Sci Teachers Asn. *Res:* Writing articles and abstracts in the fields of chemistry, computer applications in chemistry, the history of chemistry and chemical education. *Mailing Add:* Dept Chem State Univ NY Oneonta NY 13820-1381

PENCE, IRA WILSON, JR, b Pontiac, Mich; c 3. TECHNOLOGY TRANSFER, INDUSTRIAL LOGISTICS. *Educ:* Univ Mich, BS, 62, MS, 66, PhD(elec eng), 70. *Prof Exp:* Microwave engr res & develop, Gen Elec Co, 70-72, prog mgr, 72-74, br mgr, 74-76, liaison scientist, 76-78, mgr eng, 78-80, mgr advan electronics, 80-82; vpres eng, Unimation, Inc, 82-87; DIR MAT HANDLING, GA INST TECHNOL, 87- *Concurrent Pos:* Consult, Westinghouse Elec, 82-87, Denon Digital, 90-91, Brown Bourer Asea, 90-91; prin investr, NSF grant, 87- *Mem:* Inst Elec & Electronics Engrs; Soc Mfg Engrs. *Res:* Application of robotics to manufacturing and material handling; storage, movement and control of material flow in manufacturing enterprises; production efficiency resulting from application of industrial logistics. *Mailing Add:* Mat Handling Res Ctr Ga Inst Technol Atlanta GA 30332-0205

PENCE, LELAND HADLEY, b Kearney, Mo, Oct 1, 11; m 38; c 3. BIOORGANIC CHEMISTRY. *Educ:* Univ Fla, BS, 32; Univ Mich, MS, 33, PhD(org chem), 37. *Prof Exp:* Lectr, demonstr & teaching fel, Univ Mich, 33-37; org res chemist, Biochem Res Found, Franklin Inst, 37-39; instr org chem, Reed Col, 39-42, asst prof, 42-45; org res chemist, Difco Labs, Inc, Ore, 40-45, sr scientist, Detroit, 45-83; RETIRED. *Concurrent Pos:* Org res chemist, Mayo Clin, 40; res fel Calif Inst Technol, 43. *Mem:* Fel AAAS; Am Soc Microbiol; emer mem Am Chem Soc; Tissue Cult Asn; Sigma Xi. *Res:* Preparation of fluorescent antibodies; isolation and purification of bile acids, phospholipids, lectins, mitogens, phytohemagglutinin, and concanavalin A; synthesis of diagnostic reagents; cytogenetics of human and animal chromosomes; tissue and chromosome culture reagents. *Mailing Add:* 972 Alberta Ave Ferndale MI 48220-1627

PENCHINA, CLAUDE MICHEL, b Paris, France, Jan 25, 39; US citizen; m 64; c 2. SEMICONDUCTORS, PHOTOCONDUCTIVITY. *Educ:* Cooper Union, BEE, 59; Syracuse Univ, MS, 61, PhD(physics), 64. *Prof Exp:* Res assoc physics & elec eng, Univ Ill, 63-65; from asst prof to assoc prof, 65-76, PROF PHYSICS, UNIV MASS, AMHERST, 76- *Concurrent Pos:* Vis scientist, Kings Col, Univ London, 71, 77, 78 & Max Planck Inst Solid State

Res, Stuttart, Ger, 72; Nat Acad Sci exchange scientist, Inst Solid State Physics, Czech Acad Sci, Prague & Inst Tech Physics, Budapest, Hungary, 79; fel, Lady Davis Found, Technion Israel Inst Technol, Haifa, 79-80 & 88; consult, US Naval Ocean Systs Ctr & Qantix Corp; vis scientist, Bull SA, Les Clayes Sous Bois, France, 87-88; res fel, Inst Indust Sci, Tokyo Univ, 88; grad fel, NSF. *Mem:* Am Phys Soc; Sigma Xi; Optical Soc Am; NY Acad Sci; Inst Elec & Electronic Engrs. *Res:* Theoretical and experimental studies of optical and electronic properties of semiconductors; intrinsic properties and properties of deep impurities; numerical modeling of semiconductor devices; photoconductivity; interface traps. *Mailing Add:* Hasbrouck Lab Dept Phys & Astron Univ Mass Amherst MA 01003

PENDERGAST, DAVID R, CARDIOVASCULAR PHYSIOLOGY, EXERCISE. *Educ:* State Univ NY, Buffalo, EdD, 73. *Prof Exp:* Assoc prof, 74-88, PROF PHYSIOL, STATE UNIV NY, BUFFALO, 88- *Res:* Environment. *Mailing Add:* Dept Physiol 124 Sherman Hall State Univ NY Buffalo NY 14214

PENDERGRASS, LEVESTER, b Columbia, SC, May 30, 46; m 68; c 2. PLANT TAXONOMY. *Educ:* Morris Col, BS, 69; Atlanta Univ, MS, 72, PhD(biol & bot), 76. *Prof Exp:* Range aid, US Forest Serv, 73-75, biol technician, 75-76, botanist, 76-77, regional botanist, 77-87, BOT/RANGE PROG MGR, US FOREST SERV, USDA, 87- *Mem:* Soc Range Mgt. *Res:* Management of lesser vegetation endangered and sensitive plant species. *Mailing Add:* US Forest Serv 1720 Peachtree Rd NW Atlanta GA 30367

PENDERGRASS, ROBERT NIXON, b Ark, Dec 22, 18; m 44; c 3. STATISTICS. *Educ:* Southwest Mo State Col, BS, 46; Univ Mo, MEd, 51, AM, 52; Va Polytech Inst & State Univ, PhD(statist), 58. *Prof Exp:* Instr math, Univ Mo, 51-53; from assoc prof to prof, Radford Col, 53-62; PROF MATH, SOUTHERN ILL UNIV, EDWARDSVILLE, 62-, CHMN DEPT MATH, STATIST & COMPUT SCI, 85- *Mem:* Math Asn Am; Inst Math Statist; Am Statist Assoc; Nat Coun Teachers Math. *Res:* Statistical inferencel rank analysis. *Mailing Add:* 837 Troy Rd Edwardsville IL 62025

PENDERGRASS, THOMAS WAYNE, b Kansas City, Mo, Sept 23, 45; m 73; c 1. PEDIATRICS, HEMATOLOGY. *Educ:* Univ Ariz, BA, 67; Univ Tenn, MD, 71; Univ Wash, MSPH, 78. *Prof Exp:* Pediat resident, Children's Mem Hosp, Northwestern Univ, 71-73; staff assoc epidemiol, Nat Cancer Inst, 73-75; sr fel pediat hemat & oncol, Univ Wash & Children's Orthop Hosp & Med Ctr, 77-78; sr fel epidemiol, 77-78, asst prof pediat & adj asst prof epidemiol, 79-83, ASSOC PROF PEDIAT & ADJ ASSOC PROF EPIDEMIOL, UNIV WASH, 83-; ASST MEM PEDIAT ONCOL & EPIDEMIOL, FRED HUTCHINSON CANCER RES CTR, 78- *Mem:* Soc Epidemiol Res; Am Asn Cancer Educ; Am Soc Clin Oncol; Am Pub Health Asn; Am Acad Med Dirs. *Res:* Etiology and epidemiology of cancer, especially of childhood cancer; educational methods for dealing with stress; oncology; post graduate medical education. *Mailing Add:* Childrens Hosp Med Ctr 4800 Sandpoint Way NE PO Box C5371 Seattle WA 98105

PENDLAND, JACQUELYN C, INSECT PATHOLOGY, ELECTRON MICROSCOPY. *Educ:* Univ Fla, PhD(agron), 76. *Prof Exp:* BIOL SCIENTIST, UNIV FLA, 80- *Res:* Insect immunology. *Mailing Add:* 11801 SW 3rd Ave Gainesville FL 32607

PENDLETON, HUGH NELSON, III, b Gallipolis, Ohio, Aug 14, 35; m 58; c 2. MATHEMATICAL PHYSICS. *Educ:* Carnegie Inst Technol, BS, 56, MS, 58, PhD(physics), 61. *Prof Exp:* From instr to assoc prof, 60-78, PROF PHYSICS, BRANDEIS UNIV, 78- *Mem:* AAAS; Am Phys Soc; Math Asn Am; Am Asn Univ Profs; Int Asn Math & Physics. *Res:* Supergeometry; bound states in quantum electrodynamics. *Mailing Add:* Dept of Physics Brandeis Univ Waltham MA 02254-9110

PENDLETON, JOHN DAVIS, b Elizabeth City, NC, Sept 20, 12; m 46; c 4. SOIL CHEMISTRY. *Educ:* NC State Col, BS, 35; Rutgers Univ, MS, 39; Cornell Univ, PhD(soil chem), 51. *Prof Exp:* Ed mgr, Chilean Nitrate Co, NJ, 39-41; asst prof agron, Va Polytech Inst & State Univ, 46-47, assoc prof, 50-81, emer assoc prof, 81-83; RETIRED. *Concurrent Pos:* Asst prof, Tenn Valley Auth, 46-47. *Mem:* Am Soc Agron. *Res:* Soils; soil physics. *Mailing Add:* 330 Nautilus Ct Ft Myers FL 33908-1610

PENDLETON, ROBERT GRUBB, b Kansas City, Mo, Apr 24, 39; div; c 3. PHARMACOLOGY. *Educ:* Univ Mo, AB, 61; Univ Kans, PhD(pharmacol), 66. *Prof Exp:* Sr scientist, dept pharmacol, Smith Kline & French Labs, 66-69, sr investr, 69-74, asst dir, 74-77, assoc dir pharmacol, 77-80, dir, neuropharmacol, 80-81; dir, gastroenterol, Merck Inst Therapeut Res, 81-86; DIR PHARMACOL, RORER GROUP INC, 86- *Concurrent Pos:* Lieutenant Colonel, MSC, USAR. *Mem:* Am Soc Pharmacol & Exp Therapeut; Am Chem Soc. *Res:* General pharmacology, including biochemical pharmacology and drug receptor interactions. *Mailing Add:* 859 Corinthian Ave Philadelphia PA 19130

PENDLETON, WESLEY WILLIAM, b Providence, RI, Apr 2, 14; m 39; c 4. ELECTRICAL ENGINEERING. *Educ:* Univ RI, BS, 36; Mass Inst Technol, MS, 40. *Prof Exp:* Student engr, Gen Elec Co, 36-38; mem staff, Mass Inst Technol, 38-41; res engr, Westinghouse Elec Corp, 41-49; res engr, Gen Cable Corp, NJ, 49-50; res engr, Anaconda Wire & Cable Co, 50-65, mgr elec sect, Magnet Wire Res Lab, 65-70, mgr res & develop, 70-77, sr engr, 77-80; consult elec insulation, 80-85; RETIRED. *Honors & Awards:* Avant Guarde, Inst Elec & Electronic Engrs, 90. *Mem:* Sigma Xi; fel Inst Elec & Electronics Engrs. *Res:* Insulation development for high voltage; viscometry; semiconductors; corona studies; magnet wire development; thermal stability of insulation; high temperature and inorganic insulation; electrical-thermal testing of magnet wire enamels; electrical insulation; thermal degradation; magnet wire evaluation test equipment. *Mailing Add:* 1542 Clinton St Muskegon MI 49442-5023

PENDSE, PRATAPSINHA C, b Poona, India; US citizen. CYTOGENETICS, BOTANY. *Educ:* Univ Bombay, BS, 47; Univ Poona, MS, 51; Utah State Univ, MS, 59, PhD(plant sci), 65. *Prof Exp:* Instr agron, Univ Bombay, 47-50; lectr, Univ Poona, 50-56; asst prof biol, Colgate Univ, 65-66; asst prof biol, 66-74, assoc prof biol sci, 74-76, MEM FAC BIOL SCI, CALIF POLYTECH STATE UNIV, 76- *Mem:* Am Soc Hort Sci; Asn Trop Biol. *Res:* Developmental biology; androgenesis. *Mailing Add:* Dept Biol Calif State Polytech Col San Luis Obispo CA 93401

PENE, JACQUES JEAN, b Algiers, Algeria, July 19, 37; US citizen; c 1. MICROBIOLOGY, MICROBIAL GENETICS. *Educ:* Univ Calif, Los Angeles, BA, 59, PhD(microbiol), 63. *Prof Exp:* NIH res fel biochem, Albert Einstein Col Med, 63-66; asst prof develop biol, Univ Colo, Boulder, 67-73; ASSOC PROF BIOL SCI, COL ARTS & SCI, UNIV DEL, 73- *Concurrent Pos:* Mem, Microbial Physiol Study Sec, NIH, 80-83. *Mem:* Am Soc Microbiol. *Res:* DNA structure and function; cloning in bacillus; genetic engineering of extremely thermophilic bacteria. *Mailing Add:* Cell & Molecular Biol Sect Sch Life & Health Sci Newark DE 19716

PENFIELD, MARJORIE PORTER, b Mt Pleasant, Pa, May 28, 42; m 72. SENSORY SCIENCE. *Educ:* Pa State Univ, BS, 64, MS, 66; Univ Tenn, PhD(food sci), 73. *Prof Exp:* Exten specialist food & nutrit, Va Polytech Inst & State Univ, 66-71; asst prof food sci, Univ Ky, 74; from asst prof to prof food sci, 74-86, PROF FOOD TECHNOL & SCI, UNIV TENN, KNOXVILLE, 86- *Mem:* Inst Food Technologists; Am Meat Sci Asn; Am Diet Asn; Am Home Econ Asn; Am Asn Cereal Chemists; Am Soc Test Mat. *Res:* Sensory evaluation of foods; bakery products. *Mailing Add:* Dept Food Technol & Sci Univ Tenn PO Box 1071 Knoxville TN 37901-1071

PENFIELD, PAUL, JR, b Detroit, Mich, May 28, 33; m; c 7. ELECTRICAL ENGINEERING, PHYSICS. *Educ:* Amherst Col, BA, 55. *Hon Degrees:* ScD, Mass Inst Technol, 60. *Prof Exp:* From asst prof to assoc prof, 60-69, PROF ELEC ENG, MASS INST TECHNOL, 69- *Concurrent Pos:* Ford Found fel, 60-62; NSF sr fel, 66-67. *Mem:* Fel Inst Elec & Electronics Engrs; Am Phys Soc; Sigma Xi. *Res:* Varactors; solid-state microwave devices and circuits; noise theory; theory of noise and frequency conversion; electrodynamics of moving media; computer-aided circuit theory. *Mailing Add:* 17 Bradford Rd Weston MA 02193-2104

PENFIELD, ROBERT HARRISON, b Oswego, NY, Nov 25, 21; m 45; c 7. PHYSICS. *Educ:* Syracuse Univ, PhD(physics), 50. *Prof Exp:* Asst physics, Syracuse Univ, 48-50; from asst prof to assoc prof, 50-60, PROF PHYSICS, HARPUR COL, STATE UNIV NY BINGHAMTON, 60- *Concurrent Pos:* NSF grants, 64-68. *Mem:* Am Phys Soc. *Res:* Classical field theory; quantum field theory; quantum theory of the gravitational field. *Mailing Add:* Dept Physics State Univ NY Binghamton NY 13901

PENG, ANDREW CHUNG YEN, b Peiping, China, Feb 14, 24; m; c 2. FOOD CHEMISTRY. *Educ:* Wash State Univ, BS, 61; Mich State Univ, MS, 62, PhD(lipid chem), 65. *Prof Exp:* Food scientist, Res & Develop Ctr, Swift & Co, 65-67; asst prof, 68-72, assoc prof, 72-78, PROF FOOD TECHNOL, DEPT HORT, OHIO STATE UNIV, 78- *Honors & Awards:* MacGee Award, Am Oil Chem Soc, 65. *Mem:* Inst Food Technologists; Am Oil Chem Soc; Am Asn Cereal Chemists. *Res:* Lipids and soybean proteins. *Mailing Add:* Dept Hort 152 Howlett Hall Ohio State Univ Main Campus Columbus OH 43210-1096

PENG, FRED MING-SHENG, b Taiwan, China, July 15, 36; US citizen; m 64; c 2. POLYMER SYNTHESIS & STRUCTURE, MORPHOLOGY. *Educ:* Tunghai Univ, BS, 59; Syracuse Univ, MS, 64, PhD(chem eng), 67. *Prof Exp:* Sr res engr, 66-74, res specialist polymer sci, 75-79, SR TECHNOL SPECIALIST, MONSANTO CO, 79- *Mem:* Am Inst Chem Engrs. *Res:* Polymerization kinetics and processes, specifically of mass and emulsion; one phase and two phase polymer systems. *Mailing Add:* 132 Brookwood Dr Longmeadow MA 01106

PENG, JEN-CHIEH, b Canton, China, Jan 14, 49; m; c 1. LEPTON-PAIR PRODUCTION IN HADRON INTERACTION, HYPERNUCLEAR PHYSICS. *Educ:* Tunghai Univ, BS, 70; Univ Pittsburgh, MS, 72, PhD(physics), 75. *Prof Exp:* Res assoc nuclear physics, Nuclear Res Ctr, Saclay, French Atomic Energy Res, 75-77; sr res assoc nuclear physics, Univ Pittsburgh, 77-78; STAFF MEM NUCLEAR & PARTICLE PHYSICS, LOS ALAMOS NAT LAB, 78- *Concurrent Pos:* Vis scientist, Europ Lab Particle Physics, 82-83 & Fermi Nat Accelerator Lab, 89-91. *Mem:* Am Phys Soc; Oversea Chinese Phys Asn. *Res:* Pion-induced eta meson production in nuclei; hypernuclear physics experiment with pion beam; Drell-Yan and Quarkonium production in proton-nucleus collision; production and decay of B-mesons. *Mailing Add:* Mail Stop D456 Los Alamos Nat Lab Los Alamos NM 87545

PENG, SHI-KAUNG, b Taiwan, China, Aug 3, 41; m 68; c 2. PATHOLOGY. *Educ:* Nat Taiwan Univ, MD, 66; Northwestern Univ, PhD(path), 71. *Prof Exp:* Intern, St Francis Hosp, Wichita, Kans, 67-68; resident path, Northwestern Univ, 68-71; AT ALBANY VET ADMIN MED CTR. *Honors & Awards:* Res Prize, Evanston Hosp, 70. *Mem:* Int Acad Path; Am Soc Clin Path; Am Heart Asn; Electron Micros Soc Am. *Res:* Cholesterol metabolism and atherosclerosis. *Mailing Add:* Path Harbor UCLA Med Ctr Torrance CA 90509

PENG, SONG-TSUEN, b Taiwan, China, Feb 19, 37. ELECTRICAL ENGINEERING. *Educ:* Cheng-Kung Univ, BS, 59; Chiao-Tung Univ, MS, 61; Polytech Inst, PhD, 68. *Prof Exp:* PROF, NY INST TECHNOL, 82- *Mem:* Chinese Eng Soc; fel Inst Elec & Electronics Engrs. *Mailing Add:* Ten Kendrick Dix Hills NY 11746

PENG, SYD SYH-DENG, b Miaoli, Taiwan, Jan 27, 39; US citizen; m 68; c 2. MINING ENGINEERING, GROUND CONTROL. *Educ:* SDak Sch Mines & Technol, MS, 67; Stanford Univ, PhD(mining eng), 70. *Prof Exp:* Asst mine supt mining eng, Chinese Coal Develop Corp, Taiwan, 59-65; head, Rock Physics Labs, US Bur Mines, Minneapolis, Minn, 70-74; assoc prof mining eng, 74-78, CHMN & PROF MINING ENG DEPT, WEST VA UNIV, 78- *Concurrent Pos:* Pres, Penfel Co, Morgantown, WVa, 74-; mining engr, US Bur Mines, 74-; prin investr, US Bur Mines & US Dept Energy, 74- *Honors & Awards:* Rock Mech Award, 87, Soc Mining Engrs, Asn Inst Mining Engrs. *Mem:* Am Inst Mining & Metall Engrs; Am Soc Testing & Mat; Int Soc Rock Mech. *Res:* Ground control in coal mining; longwall mining. *Mailing Add:* WVa Univ Col Mineral & Energy Resources Box 6070 Morgantown WV 26506-6070

PENG, TAI-CHAN, b Vietry, Vietnam, Feb 28, 28; m 54; c 2. PHARMACOLOGY. *Educ:* Univ Geneva, BMedSc, 56, MD, 59. *Prof Exp:* Intern, Hartford City Hosp, Conn, 60; res fel pharmacol, Sch Dent Med, Harvard Univ, 61-63; res assoc pathophysiol, Sch Med, Univ Geneva, 63-64, resident internal med, 64-65; instr pharmacol, Sch Dent Med, Harvard Univ, 65; from instr to assoc prof, 65-89, PROF PHARMACOL, SCH MED, UNIV NC, CHAPEL HILL, 90- *Concurrent Pos:* Mem rev panel drug interactions proj, Am Pharmaceut Asn, 71-75; mem, working group, Health Sci Coun Aging, Univ NC, 80. *Mem:* AAAS; Endocrine Soc; Am Soc Pharmacol & Exp Therapeut; Am Asn Univ Profs; Am Soc Bone & Min Res; Sigma Xi (pres, 82-83). *Res:* Hormonal and nonhormonal agents affecting calcium homeostasis; endocrine pharmacology and morphology; effects of alcohol on bone metabolism. *Mailing Add:* Dept of Pharmacol Univ of NC Sch of Med Chapel Hill NC 27514

PENG, YEH-SHAN, b Taipei, Taiwan, Feb 11, 36; m 69; c 2. NUTRITION. *Educ:* Nat Taiwan Univ, BS, 61; Univ Calif, Los Angeles, MS, 65; Univ Wis, PhD(biochem & nutrit), 71. *Prof Exp:* Res assoc nutrit, Univ Ariz, 71-83, res scientist, 83-85, head, Nutrit Sect, 85-89, RES ASSOC, ARIZ CANCER CTR, UNIV ARIZ, 89- *Mem:* Am Inst Nutrit; Inst Food Technol. *Res:* Shrimp nutrition and feed; amino acid imbalance; exploring new food technology for manufacturing practical shrimp feed based on the nutritional need of shrimp; investigating efficient use of chalophytes in animal feed. *Mailing Add:* Environ Res Lab Univ Ariz Tucson AZ 85721

PENGELLEY, DAVID JOHN, b Toronto, Ont, Aug 30, 52; m 78. ALGEBRAIC TOPOLOGY, HOMOTOPY THEORY. *Educ:* Univ Calif, Santa Cruz, BA, 73; Univ Wash, PhD(math), 80. *Prof Exp:* C L E Moore instr math, Mass Inst Technol, 80-82; asst prof, 82-85, ASSOC PROF, NMEX STATE UNIV, 85- *Mem:* Am Math Soc; Math Asn Am; Asn Women Math; Fedn Am Scientists. *Res:* Cobordism Thom spectra, classifying spaces for vector bundles. *Mailing Add:* Dept Math Sci NMex State Univ Las Cruces NM 88003

PENGELLEY, ERIC T, b Toronto, Ont, July 18, 19; US citizen; m 48; c 2. PHYSIOLOGY. *Educ:* Univ Toronto, BA, 54, PhD(physiol), 59. *Prof Exp:* Asst prof zool, Univ Calif, Davis, 59-60, lectr biol, Santa Barbara, 60-61; asst prof, Col William & Mary, 61-62; asst prof, 62-, EMER PROF BIOL, UNIV CALIF, RIVERSIDE. *Res:* Hibernation and estivation of small mammals; biological rhythms; history of biology. *Mailing Add:* Dept Zool Univ Calif Davis CA 95616

PENGRA, JAMES G, b Eugene, Ore, Apr 27, 33; m 56; c 4. NUCLEAR PHYSICS. *Educ:* Univ Ore, BS, 55, MS, 57, PhD(nucleus-electron interaction), 63. *Prof Exp:* From asst prof to assoc prof, 62-76, PROF PHYSICS, WHITMAN COL, 76-, GARRETT FEL, 74- *Mem:* Am Phys Soc; Am Asn Physics Teachers; Sigma Xi. *Res:* Interaction of the atomic electrons with the nucleus and their behavior following beta decay of the nucleus. *Mailing Add:* Dept Physics Whitman Col Walla Walla WA 99362

PENGRA, ROBERT MONROE, b Rapid City, SDak, Jan 20, 26; m 51; c 3. MICROBIAL PHYSIOLOGY, SOIL MICROBIOLOGY. *Educ:* SDak State Univ, BS, 51, MS, 53; Univ Wis, PhD(bact, biochem), 59. *Prof Exp:* Res asst biochem, SDak Exp Sta, 51-52, 53; from asst prof to assoc prof, 57-68, PROF MICROBIOL, DEPT S DAK STATE UNIV, 68- *Mem:* Am Soc Microbiol; Can Soc Microbiol; Soil Sci Soc Am. *Res:* Mechanism of biological nitrogen fixation and ecology and distribution of nitrogen fixing bacteria in soil as related to nutrient availability and toxic materials. *Mailing Add:* Dairy Microbiol Bldg SDak State Univ Box 2104 Brookings SD 57007

PENHALE, POLLY ANN, b St Louis, Mo, Dec 18, 47. MARINE BIOLOGY, AQUATIC BOTANY. *Educ:* Earlham Col, BA, 70; NC State Univ, MS, 72, PhD(zool), 76. *Prof Exp:* Res assoc marine bot, Rosenstiel Sch Marine & Atmospheric Sci, Univ Miami, 75-76; aquatic bot, W K Kellogg Biol Sta, Mich State Univ, Hickory Corners, 77-79; asst prog dir, biol oceanog, NSF, 82-84, assoc prog dir, 84-85; asst prof, 79-82, ASSOC PROF BIOL, VA INST MARINE SCI, GLOUCESTER PT, 85- *Mem:* Am Soc Limnol & Oceanog (secy, 85); Int Asn Theoret & Appl Limnol; Am Geophys Union; Phycological Soc Am. *Res:* Macrophyte-epiphyte productivity in seagrass communities; nutrient cycling; macrophyte-epiphyte interactions; seagrass ecosystems. *Mailing Add:* OCE/OSRS Nat Sci Found 1800 G St NW Washington DC 20550

PENHOET, EDWARD ETIENNE, b Oakland, Calif, Dec 11, 40; m 62; c 2. BIOCHEMISTRY, VIROLOGY. *Educ:* Stanford Univ, BA, 63; Univ Wash, PhD(biochem), 68. *Prof Exp:* Actg asst prof biol, Univ Calif, San Diego, 70-71; asst prof, 71-77, ASSOC PROF BIOCHEM, UNIV CALIF, BERKELEY, 77-; CHIEF EXEC OFFICER, CHIRON CORP, 81- *Concurrent Pos:* Nat Inst Child Health & Human Develop fel, Univ Calif, San Diego, 69-72. *Mem:* AAAS; Am Chem Soc; Am Soc Biol Chemists. *Res:* Control of macromolecular synthesis in normal and virus-infected animal cells. *Mailing Add:* Dept Biochem Univ Calif Berkeley CA 94720

PENHOLLOW, JOHN O, b Tama, Iowa, Aug 26, 34; m 59; c 2. COMPUTER SCIENCE, GEOPHYSICS. *Educ:* State Univ Iowa, BS, 56; Univ Ill, MS, 59, PhD(elec eng), 62. *Prof Exp:* Mem tech staff comput design, Bell Tel Lab, 56-58; teaching asst commun elec eng, Univ Ill, 58-59; res asst comput design, Univ Ill Comput Lab, 59-63; res eng, Exxon Prod Res Co, 63-67, res supvr, 67-68, res mgr, 68-71, asst div geophys mgr, Exxon Co USA, 71-73; geophys mgr, Esso Explor Inc, London & Singapore, 73-77; sr tecnol adv, Dept Sci & Technol, Exxon Corp, 77-79, mgr technol & serv, 79-81, mgr appl develop & coord, 81-83, mgr planning, Dept Commun & Comput Sci, 83-85; mgr commun & comput sci, Esso Explor Inc, Houston, 85-86; independent info systs consult, Houston, 86-87; DIR ELECTRONIC DATA GATHERING, ANAL, & RETRIEVAL PROJ, US SECURITIES & EXCHANGE COMN, WASH, DC, 87- *Mem:* Inst Elec & Electronic Engrs; Soc Explor Geophys. *Res:* Computer design, data processing software systems, geophysical (seismic) data enhancement, information storage and retrieval systems, communication and computer networks, business and technical computer applications, workstation information networks, text management systems, distributed processing and database systems. *Mailing Add:* US Securities & Exchange Comn 450 Fifth St NW Washington DC 20549

PENHOS, JUAN CARLOS, b Buenos Aires, Arg, Feb 12, 18; m 44; c 1. PHYSIOLOGY, ENDOCRINOLOGY. *Educ:* Univ Buenos Aires, BA, 35, MD, 42. *Prof Exp:* Asst biochem res, Inst Physiol, Med Sch, Univ Buenos Aires, 45-47; chief instr med, Modelo Inst, 46-52; chief res exp diabetes & lipid metab, Inst Biol & Exp Med, Univ Buenos Aires, 52-64; assoc prof med, NY Med Col, 64-67, dir exp diabetes unit, 65-67; assoc prof med, George Washington Univ, 68-83,; prof, 83-85, EMER PROF PHYSIOL & BIOPHYS, GEORGETOWN UNIV, 85- *Concurrent Pos:* Arg Asn Advan Sci fel, 53-54; Lederle Labs fel, 56-58; Lederle Int, Chicago, 60-62; Nat Coun Sci & Tech Res, Toronto, 62-63; carrier scientist, Nat Coun Sci & Tech Res, 62-64; NIH grants, 65-75; Vet Admin grants, 68-71; chief endocrine res, Vet Admin Hosp, Washington, DC, 71-83, consult diabetes res, 84-85. *Mem:* AAAS; Am Diabetes Asn; Endocrine Soc; Am Physiol Soc; Am Fedn Clin Res. *Res:* Experimental diabetes; hormones and diabetes; mechanism of insulin and glucagon release; lipid metabolism and hormones in the liver; prostaglandins. *Mailing Add:* Dept Biochem Molecular Biol Miami Univ Med Sch PO Box 016129 Miami FL 33101

PENICK, GEORGE DIAL, b Columbia, SC, Sept 4, 22; m 47; c 5. VASCULAR PATHOLOGY, DERMATOPATHOLOGY. *Educ:* Univ NC, BS, 44; Harvard Med Sch, MD, 46. *Prof Exp:* Intern path, Presby Hosp, Chicago, 46-47, mem, C D Antaller Med Lab, 47-49; from instr to prof path, Sch Med, Univ NC, 49-70; head dept, 70-81, PROF PATH & DERMATOL, UNIV IOWA, 70- *Concurrent Pos:* Consult pathologist, Watts Hosp, 49-70; asst attend pathologist, NC Mem Hosp, 52-56, assoc attend pathologist, 56-63, attend pathologist, 63-70; Markle scholar, 53-58; pathologist, Rex Hosp, 60-62; dir NIH prog-proj on thrombosis & hemorrhage, Univ NC, 61-70; consult pathologist, Vet Admin Hosp, Iowa City, 70-72, chief lab serv, 72-75. *Mem:* AMA; Am Asn Pathologists; Soc Exp Biol & Med; Col Am Pathologists; Am Soc Clin Pathologists. *Res:* Vascular diseases. *Mailing Add:* Dept Path Univ Iowa Iowa City IA 52240

PENICO, ANTHONY JOSEPH, b Philadelphia, Pa, June 11, 23; m 48; c 2. MATHEMATICAL PHYSICS. *Educ:* Univ Pa, AB, 45, PhD(math), 50. *Prof Exp:* Asst instr math, Univ Pa, 44-50, asst, 47-48; from instr to assoc prof, Tufts Univ, 50-56; adv res engr, Microwave Physics Lab, Sylvania Elec Prod, Inc, 56-59, specialist, Labs, Gen Tel & Electronics Corp, 59-62; sr res mathematician, Stanford Res Inst, 62-66; PROF MATH, UNIV MO-ROLLA, 66- *Concurrent Pos:* Consult, Lab Phys Electronics, Tufts Univ, 50-56; mathematician, Air Force Cambridge Res Labs, 52-53; lectr, Eve Grad Eng Sch, Northeastern Univ, 54-56. *Mem:* AAAS; Am Math Soc; Am Phys Soc; Soc Indust & Appl Math; Math Asn Am. *Res:* Equations of mathematical physics; wave theory; abstract algebra. *Mailing Add:* Dept Math Univ Mo Rolla MO 65401

PENK, ANNA MICHAELIDES, b Thessaloniki, Greece, May 4, 28; US citizen; m 50; c 4. MATHEMATICAL LOGIC. *Educ:* Whitman Col, BA, 50; Reed Col, MAT, 64; Univ Ore, PhD(math), 73. *Prof Exp:* Instr, Pub Schs, Ore, 63-66; asst prof math, Ore Col Educ, 73-77, assoc prof, 77-; AT MATH DEPT, WESTERN ORE STATE COL. *Res:* Elementary topos theory, in particular notions of infinity and axioms of choice; relationships of two set-theoretic equivalents of the choice axiom. *Mailing Add:* Dept Math 309 19th St NE Salem OR 97301

PENLAND, JAMES GRANVILLE, b Dallas, Tex, Mar 1, 51; m 77; c 3. COGNITIVE PSYCHOLOGY, COMPUTER APPLICATIONS. *Educ:* Metrop State Col Denver, BA, 77; Univ NDak, MA, 79, PhD(cognitive psychol), 84. *Prof Exp:* From instr to asst prof psychol & statist, Psychol Dept, Univ NDak, 78-87; statistician, 81-84, psychologist, 84-85, RES PSYCHOLOGIST, PSYCHOL RES GROUP, GRAND FORKS HUMAN NUTRIT RES CTR, AGR RES SERV, USDA, 85- *Concurrent Pos:* Adj prof, Psychol Dept, Univ NDak, 87- *Mem:* Am Psychol Asn; Am Statist Asn; Am Inst Nutrit; Sigma Xi. *Res:* Effects of trace element nutrition on neurophysiological function; cognitive performance and behavior in healthy adult humans and animals; psychometric methods; computer technology to behavioral assessment. *Mailing Add:* Grand Forks Human Nutrit Res Ctr Agr Res Serv USDA PO Box 7166 Univ Sta Grand Forks ND 58202-7166

PENLIDIS, ALEXANDER, b Kozani, Greece, Feb 12, 57; Can citizen. MATHEMATICAL MODELLING, REACTOR DESIGN POLYMER SYSTEMS. *Educ:* Univ Thessaloniki, Greece, dipl eng, 80; McMaster Univ, Can, PhD(chem eng), 86. *Prof Exp:* Asst engr prod, MISSR Petrol Co, Egypt, 78; res assoc res, Polymer Prod Technol, McMaster Inst, 85-86; lectr teaching, Dept Chem Eng, McMaster Univ, 85-87; asst prof, 86-90, ASSOC PROF RES & TEACHING, DEPT CHEM ENG, UNIV WATERLOO, 90- *Concurrent Pos:* Lectr, several Indust Intensive Short Courses, 84-; consult, several cos worldwide, 85-; founding co-ed, Polymer Reaction Eng J, 90-

Mem: Chem Inst Can; Am Inst Chem Engrs; Am Chem Soc. *Res:* Kinetics, mathematical modelling and simulation of polymer production processes; sensors for polymerization systems; reactor design, optimization and computer control. *Mailing Add:* Dept Chem Eng Univ Waterloo Waterloo ON N2L 3G1 Can

PENMAN, PAUL D, b Williston, NDak, Sept 25, 37; m 86; c 3. REACTOR MATERIALS, COMPUTERIZED INSPECTIONS. *Educ:* Univ Colo, BS, 59; Univ Louisville, MS, 65. *Prof Exp:* Group mgr, Process Control & Anal, Advan Naval Reactor Cores, 72-77; dept mgr, Core Mfg Develop, 77; proj mgr, Develop Shops, 77-81, proj mgr, Develop Labs Oper, 78-82, PROJ MGR, CORE MFG, BETTIS ATOMIC POWER LAB, WESTINGHOUSE ELEC CORP, 82- *Mem:* US Naval Inst. *Res:* Nuclear reactor manufacturing development; computer scanning systems; irradiation behavior of fuel and poison materials. *Mailing Add:* Bettis Atomic Power Lab PO Box 79 West Mifflin PA 15122

PENMAN, SHELDON, b New York, NY, July 24, 30. CELLULAR & MOLECULAR BIOLOGY. *Educ:* Pa State Univ, BS, 54; Columbia Univ, PhD(physics), 58. *Prof Exp:* Instr physics, Columbia Univ, 58-59; Asst prof physics, Univ Chicago, 59-61; assoc, Columbia Univ & Bell Tel Labs, 61-62; res assoc biol, Mass Inst Technol, 62-64; asst prof biochem, Albert Einstein Col Med, 64-65; from asst prof to assoc prof biophys & microbiol, 65-70, PROF CELL BIOL, MASS INST TECHNOL, 70- *Mem:* Nat Acad Sci; Am Acad Arts Sci; Am Soc Cell Biol. *Res:* Author of over 180 technical journal articles. *Mailing Add:* Mass Inst Technol Rm 56-535 Cambridge MA 02139

PENN, ARTHUR, b Brooklyn, NY, Nov 1, 42. EXPERIMENTAL ATHEROSCLEROSIS. *Educ:* Univ Pa, PhD(molecular biol), 76. *Prof Exp:* RES PROF ENVIRON MED, MED CTR, NY UNIV, 84- *Mailing Add:* Dept Environ Med NY Univ Med Ctr 550 First Ave New York NY 10016

PENN, BENJAMIN GRANT, b Martinsville, Va, July 6, 47. MATERIALS SCIENCE. *Educ:* Winston-Salem State Univ, BS, 69; Rensselaer Polytech Inst, MS, 73; NC State Univ, PhD(fiber & polymer sci), 78. *Prof Exp:* Res assoc, Va Polytech Inst & State Univ, Blacksburg, 78-79; sr res chemist, Res Triangle Inst, NC, 79-80; POLYMER SCIENTIST, MARSHALL SPACE FLIGHT CTR, NASA, 80- *Mem:* Am Chem Soc; Mats Res Soc; Soc Advan Mat & Process Eng (secy-treas, 82-83). *Res:* Polymer and organic synthesis, characterization and crystallization; conductive organic and polymeric materials; environmental effects on composites; nonlinear optical organic and polymeric materials; materials processing in space. *Mailing Add:* Space Sci Lab NASA Marshall Space Flight Ctr ES74 Bldg 4481 Huntsville AL 35812

PENN, HOWARD LEWIS, b Bayonne, NJ, Nov 09, 46. COMPUTERS, MATHEMATICS. *Educ:* Univ Ind, BA, 68; Mich Univ, PhD(Math), 73. *Prof Exp:* Inst math, Northeastern Univ, 72-73; PROF MATH, US NAVAL ACAD, 73- *Mailing Add:* Dept Math US Naval Acad Annapolis MD 21402

PENN, LYNN SHARON, b Iowa City, Iowa, June 18, 43; m 67; c 1. ADHESIVE PERFORMANCE IN COMPOSITES. *Educ:* Univ Pa, BA, 66; Bryn Mawr Col, MA, 70, PhD(phys org chem), 74. *Prof Exp:* Chemist, Lawrence Livermore Nat Lab, 73-78; sr scientist, Textile Res Inst, 78-80 & Ciba Geigy Corp, 80-83; prin scientist, Midwest Res Inst, 83-86; RES PROF, POLYTECH UNIV, 86- *Concurrent Pos:* Consult & lectr, Am Soc Metals, 84-; lectr, Fiber Soc, 85; adj prof, Univ Mo, 86. *Mem:* Adhesion Soc (secy, 84-); Am Soc Testing & Mat; Am Chem Soc; Fiber Soc. *Res:* Filamentary composite materials; fiber surface effects and the interface in composites; materials studies of composites with modified matrices; processing of composites; adhesion. *Mailing Add:* 31 Division Ave South Nyack NY 10960

PENN, THOMAS CLIFTON, b Placid, Tex, Jan 17, 29; m 50; c 1. ELECTRONICS ENGINEERING. *Educ:* Tex Technol Univ, BS, 50; Southern Methodist Univ, MS, 57. *Prof Exp:* Lead systs engr, Chance Vought Aircraft, 54-57; sr mem tech staff, 57-, RES BR MGR, TEX INSTRUMENTS, INC. *Mem:* Inst Elec & Electronics Engrs; Electrochem Soc. *Res:* Computer memories, electronic circuits, semiconductor fabrication techniques; magnetic bubble memories and plasma fabrication of semiconductor devices. *Mailing Add:* 911 Northlake Richardson TX 75080

PENN, WILLIAM B, b Lawrence, Mass, May 2, 17; m 42; c 2. ELECTRICAL ENGINEERING. *Educ:* Mass Inst Technol, BS, 37, MS, 38. *Prof Exp:* Consult engr, DC Motor & Generator Dept, Gen Elec Co, Erie, Pa; CONSULT, 82- *Concurrent Pos:* Chmn, insulation subcomt, rotating mach comt, Power Eng Soc & working group 5 & 6, Int Electrotech Comn. *Mem:* Fel Inst Elec & Electronic Eng; Sr emer mem Am Chem Soc. *Mailing Add:* 4616 Sunnydale Blvd Erie PA 16509

PENNA, MICHAEL ANTHONY, b Buffalo, NY, June 5, 45; m 68; c 2. MATHEMATICS. *Educ:* Union Col, BS, 67; Univ Ill, Urbana, MS, 68, PhD(math), 73. *Prof Exp:* From asst prof to assoc prof, 73-88, PROF MATH & COMPUTER SCI, IND UNIV-PURDUE UNIV, INDIANAPOLIS, 88- *Mem:* Am Math Soc; Math Asn Am; Soc Photo-Optical Instrumentation Engrs; Int Soc Optical Eng; Inst Elec & Electronic Engrs; Asn Comput Mach. *Res:* Applications of geometry to computer graphics, computer vision and image processing. *Mailing Add:* Dept Math Sci Ind Univ-Purdue Univ 1125 E 38th St Indianapolis IN 46223

PENNA, RICHARD PAUL, b Palo Alto, Calif, Sept 7, 35; m 58; c 3. PHARMACY. *Educ:* Univ Calif, BS, 58, PharmD, 59. *Prof Exp:* Pharmacist, Ryan Pharm, 58-66; asst clin prof pharm, Univ Calif, 61-66; prof staff pharm, Am Pharmaceut Asn, 66-84; ASSOC EXEC DIR, AM ASN COLS PHARM, 85- *Mem:* AAAS; Am Pharmaceut Asn. *Mailing Add:* 1426 Prince St Alexandria VA 22314-2841

PENNAK, ROBERT WILLIAM, b Milwaukee, Wis, June 13, 12; m 35; c 2. ZOOLOGY. *Educ:* Univ Wis, BS, 34, MS, 35, PhD(zool), 38. *Prof Exp:* Asst limnol, Univ Wis, 34-36, zool, 36-38; from instr to prof, 38-74, EMER PROF BIOL, UNIV COLO, BOULDER, 74 - *Concurrent Pos:* Stream & lake consult, ecology, mining & land develop firms. *Mem:* Ecol Soc Am; Am Soc Limnol & Oceanog (pres, 63); Am Soc Zoologists; Soc Syst Zool (pres, 64); hon mem Am Micros Soc (vpres, 55, pres, 56). *Res:* Limnology; stream biology; freshwater invertebrates; animal ecology at high altitudes; interstitial microscopic Metazoa of sandy beaches. *Mailing Add:* 14215 E Marine Aurora CO 80014

PENNDORF, RUDOLF, b Chemnitz, Ger, Nov 29, 11; nat US; m 42; c 3. ATMOSPHERIC PHYSICS. *Educ:* Univ Leipzig, PhD(geophys), 36, PhD(habilitation), 44. *Prof Exp:* Asst meteorol, Univ Leipzig, 34-36, asst prof, 36-42; assoc prof, Univ Strassburg, 42-45; chief forecaster, US Weather Cent, Wiesbaden, 46-47; asst chief, Atmospheric Physics Lab, Air Force Cambridge Res Ctr, 47-56; prin scientist, Res & Advan Develop Div, Avco Corp, Wilmington, 56-61, sect chief geophys, Space Systs Div, 61-73; RETIRED. *Concurrent Pos:* Mem, Armed Forces Vision Comt, Nat Res Coun, 52-57; mem, US Nat Comt, Int Sci Radio Union, 59-78; consult, 73-80. *Mem:* Fel Optical Soc Am; Am Geophys Union. *Res:* Physics of the ionosphere; radio wave propagation and communication; atmospheric optics; light scattering; antarctic ionosphere; prediction of high frequency communication conditions; ozone and water vapor. *Mailing Add:* 148 Oakland St Wellesley Hills MA 02181

PENNEBAKER, JAMES W, b Midland, Tex, Mar 2, 50; m 72; c 2. HEALTH PSYCHOLOGY, SOCIAL PSYCHOLOGY. *Educ:* Eckerd Col, BA, 72; Univ Tex, Austin, PhD(psychol), 77. *Prof Exp:* Asst prof psychol, Univ Va, 77-83; assoc prof, 83-87, PROF PSYCHOL, SOUTHERN METHODIST UNIV, 87- *Concurrent Pos:* Vis prof, Stanford Univ, 89; NIH grant, 84-87, NSF grants, 87- *Mem:* Acad Behav Med; Soc Exp Social Psychol; Am Psychophysiol Res; Am Psychol Asn. *Res:* Psychosomatics and health psychology: how individuals cope with traumatic experiences, including the roles of repression and emotional expression; the psychology of physical symptoms and sensations and their links to disease. *Mailing Add:* Dept Psychol Southern Methodist Univ Dallas TX 75275

PENNEBAKER, WILLIAM B, JR, b New Rochelle, NY, Oct 23, 35. PHYSICS. *Educ:* Lehigh Univ, BS, 57; Rutgers Univ, PhD(physics), 62. *Prof Exp:* Mem staff, 62-, MGR EXPLOR TERMINAL TECHNOL, T J WATSON RES CTR, IBM CORP. *Mem:* AAAS; Am Phys Soc; Soc Info Display. *Res:* Solid state physics; thin film physics; display and printing technology; image processing and image compression; this includes theoretical work on fundamental techniques for data compression. *Mailing Add:* T J Watson Res Ctr IBM Corp PO Box 218 Yorktown Heights NY 10598

PENNELL, MAYNARD L, b Skowhegan, Maine, 1910. AERONAUTICAL ENGINEERING. *Educ:* Univ Wash, AeE, 31. *Prof Exp:* Aeronaut struct engr, Douglas Aircraft Co, 33-40; from asst proj engr to proj engr B-29, B-50 & C-97, Boeing Co, 40-70, vpres prod develop, 70-74; RETIRED. *Concurrent Pos:* Mem, various nat comts. *Honors & Awards:* Elmer A Sperry Award, 65. *Mem:* Nat Acad Eng; fel Am Inst Aeronaut & Astronaut. *Res:* Design and development of 707, 720 and 727 transports and supersonic transport. *Mailing Add:* 1545 NE 143rd St Seattle WA 98125

PENNELL, TIMOTHY CLINARD, b Asheville, NC, Oct 31, 33; m 53; c 3. SURGERY, THORACIC SURGERY. *Educ:* Wake Forest Univ, BS, 55, MD, 60; Am Bd Surg, dipl, 67. *Prof Exp:* Intern, 60-61, resident, 61-66, from asst prof to assoc prof, 68-80, PROF SURG & DIR INT HEALTH AFFAIRS, BOWMAN GRAY SCH MED, 80-, CHIEF PROF SERV, WAKE FOREST UNIV. *Concurrent Pos:* Am Thoracic Soc teaching fel, Wake Forest Univ, 67-68. *Mem:* Fel Am Col Surg; Am Thoracic Soc; Am Acad Surg; Soc Univ Surgeons; Soc Surg Alimentary Tract. *Res:* Pulmonary lymphatic system of the lung and tissue transplantation, particularly of the pancreas; clinical surgery. *Mailing Add:* Dept of Surg Bowman Gray Sch Med Winston-Salem NC 27103

PENNEMAN, ROBERT ALLEN, b Springfield, Ill, Feb 5, 19; m 42; c 3. INORGANIC CHEMISTRY. *Educ:* Millikin Univ, AB, 41; Univ Ill, MS, 42, PhD(inorg chem), 47. *Hon Degrees:* ScD, Univ Ill, 61. *Prof Exp:* Res assoc, Metall Lab, Univ Chicago, 42-45 & Clinton Labs, Tenn, 45-46; sect leader, Los Alamos Sci Lab, Univ Calif, 47-50, alt group leader, 50-74, assoc div leader, 74-84; CONSULT, SEPARATIONS CHEM, 84- *Mem:* Am Chem Soc; Sigma Xi. *Res:* Sulfurdioxide and laser chemistry; chemistry of hydrazine; radiochemistry; chemistry of americium and curium; infrared spectroscopy of inorganic solution complexes; structural inorganic chemistry; complex fluorides. *Mailing Add:* 12201 LaVista Grande NE Albuquerque NM 87111-6710

PENNER, ALVIN PAUL, b Arnaud, Man, Dec 1, 47. THEORETICAL CHEMISTRY, OPERATIONS RESEARCH. *Educ:* Univ Man, BSc, 68, MSc, 70, PhD(chem), 74. *Prof Exp:* Fel chem, Univ Laval, 74-75; res assoc chem, Nat Res Coun Can, 76-77; analyst opers res, Fraser Co, Ltd, 78-85; PROCESS SYSTS ANALYST, ONTARIO PAPER, 85- *Concurrent Pos:* Fel, Nat Res Coun Can, 74-75. *Mem:* Can Indust Comput Soc; Can Pulp & Paper Assn. *Res:* Molecular collision theory; stochastic theories of chemical kinetics; operations research; process control theory. *Mailing Add:* Oper Technol Ont Paper Co Thorold ON L2V 3Z5 Can

PENNER, DONALD, b Mt Lake, Minn, Dec 28, 36; m 63. WEED SCIENCE. *Educ:* Univ Minn, BS, 57, MS, 60; Univ Calif, Davis, PhD(plant physiol), 66. *Prof Exp:* Res botanist, Univ Calif, Davis, 66-67; from asst prof to assoc prof plant physiol & herbicide action, 67-76, PROF CROP & SOIL SCI, MICH STATE UNIV, 76- *Honors & Awards:* Outstanding Res Award, Weed Sci Soc Am, 87. *Mem:* AAAS; Weed Sci Soc Am; Am Inst Biol Sci; Am Soc Plant Physiologists; Sigma Xi; Coun Agricult Sci & Technol. *Res:* Herbicide action and metabolism; environmental toxicology. *Mailing Add:* Dept of Crop & Soil Sci Mich State Univ East Lansing MI 48824

PENNER, GLENN H, b Plum Coulee, Man, May 4, 60. QUANTUM CHEMISTRY. *Educ:* Univ Man, BSc, 82, MSc, 84 & PhD(chem), 87. *Prof Exp:* Killam fel chem, Dalhousie Univ, 87-88; ASST PROF CHEM, UNIV GUELPH, 88- *Res:* Investigation of molecular structure and dynamics by solid state nuclear magnetic resonance spectroscopy; dynamic nuclear magnetic resonance studies in solution; quantum chemical calculations of molecular structure and dynamics. *Mailing Add:* Dept Chem Univ Guelph Guelph ON N1G 2W1 Can

PENNER, HELLMUT PHILIP, b Mountain Lake, Minn, Feb 15, 25. CHEMISTRY. *Educ:* Mass Inst Technol, SB, 45, PhD(org chem), 50. *Prof Exp:* Res assoc polymer chem, Mass Inst Technol, 49-56; asst prof chem, Carleton Col, 51-55; vis asst prof, Univ Ky, 56-57; mem res staff, Ocean Spray Cranberries, Inc, 57; asst prof, 62-67, ASSOC PROF CHEM, FISK UNIV, 67- *Mem:* Am Chem Soc. *Res:* Synthesis of heterocyclic compounds of medicinal potential; organometallic chemistry. *Mailing Add:* Fisk Univ PO Box 120456 Nashville TN 37212

PENNER, JOYCE ELAINE, b Fresno, Calif, Oct 3, 48; m 80; c 2. ATMOSPHERIC PHYSICS, APPLIED MATHEMATICS. *Educ:* Univ Calif, Santa Barbara, BA, 70; Harvard Univ, MA, 72, PhD(appl math), 77. *Prof Exp:* PHYSICIST ATMOSPHERIC PHYSICS, LAWRENCE LIVERMORE LAB, UNIV CALIF, 77-, GROUP LEADER, 87- *Mem:* Am Geophys Union; AAAS; Am Meteorol Soc. *Res:* Chemistry of stratosphere and troposphere; physics and microphysics of clouds. *Mailing Add:* Lawrence Livermore Lab PO Box 808 Livermore CA 94550

PENNER, S(TANFORD) S(OLOMON), b Unna, Ger, July 5, 21; nat US; m 42; c 2. AERONAUTICAL & ASTRONAUTICAL ENGINEERING. *Educ:* Union Col, BS, 42; Univ Wis, MS, 43, PhD(phys chem), 46. *Hon Degrees:* Dr, Hochschule Aachen, 81. *Prof Exp:* Res assoc, Alleghany Ballistics Lab, Md, 44-45; res assoc, Esso Res Lab, Standard Oil Develop Co, NJ, 46; sr res engr, Jet Propulsion Lab, Calif Inst Technol, 47-50, from asst prof to prof jet propulsion, 50-64; chmn, dept Aerospace & Mech Eng Sci, Univ Calif, San Diego, 64-68, vchancellor acad affairs, 68-69, dir, Inst Pure & Appl Phys Sci, 68-71, dir, Energy Ctr & Combustion Res, 72-90, PROF ENG PHYSICS, UNIV CALIF, SAN DIEGO, 64- *Concurrent Pos:* Agard-NATO US mem, Propulsion & Energetics Panel, 51-64; vis lectr, US, Italy, Germany, Holland, Belgium, France, Greece & Turkey, 57-66; res adv comt eng sci, Off Sci Res, US Air Force, 61-66; comt high-temperature phenomena, Div Chem & Chem Technol, Nat Acad Sci-Nat Res Coun, 61-71; res adv comt air breathing engines, NASA, 62-64; dir res & eng support div, Inst Defense Anal, 62-64; Guggenheim fel, England, W Germany, Italy, Israel, India, Japan, Australia, 71-72; ed, Energy-The Int J, 75-; Nat Sigma Xi lectr, 77-79; chmn, US Comt, Int Inst Appl Systs Anal, 78-82, US DoE Working Group on Assessment Res Needs for Coal Gasification, 85-87, Innovative Sci & Technol Off, Strategic Defense Initiative Prog, Dept Defense, 85-; dir, NATO Multinational Study Power Sources, Devices Tactical Applications, Brussels, Belgium, 86-87; NASA Microgravity Combustion Sci Disciplinary Working Group; spec guest, Int Coal Sci Conf, 83, 85, 87 & 89; US Dept Energy distinguished assoc award, 90. *Honors & Awards:* Off Sci Res & Develop Award, 45; Spec Citation, Adv Group Aerospace Res & Develop, NATO, 69; G Edward Pendray Award, Am Inst Aeronaut & Astronaut, 75, Thermophysics Award & Energy Systs Award, 83; Numa Manson Medal, Int Colloquia Gasdynamics Explosions & Reactive Systs, 79; Int Columbus Prize, Int Inst Commun Res, Genoa, Italy, 81. *Mem:* Nat Acad Eng; fel AAAS; fel Am Phys Soc; fel Am Inst Aeronaut & Astronaut; fel NY Acad Sci; fel Am Acad Arts & Sci; Int Acad Astronaut; Sigma Xi; fel Optical Soc Am; Am Chem Soc. *Res:* Applied spectroscopy, combustion and propulsion research; energy. *Mailing Add:* Dept Appl Mech & Eng Sci Univ Calif San Diego 9500 Gilman Dr La Jolla CA 92093-0310

PENNER, SAMUEL, b Buffalo, NY, Oct 3, 30; m 52; c 2. EXPERIMENTAL NUCLEAR PHYSICS, ACCELERATOR PHYSICS. *Educ:* Univ Buffalo, BA, 52; Univ Ill, MS & PhD(physics), 56. *Prof Exp:* Asst physics, Univ Ill, 52-55; NSF fel, 56-57; physicist nuclear physics, Nat Bur Standards, 57-66, chief Elecronuclear Physics Sect, 66-72, actg chief, Linac Radiation Div, dep chief, Nuclear Sci Div, 72-76, chief, Nuclear Res Sect, 76-78, chief, Radiation Source & Instrumentation Div, 78-88; CONSULT, 88- *Concurrent Pos:* Guggenheim Found fel, 63-64. *Mem:* Fel Am Phys Soc. *Res:* Electron scattering; photonuclear physics; instrumentation for nuclear physics; accelerator physics. *Mailing Add:* 10500 Pine Haven Terr North Bethesda MD 20852

PENNER, SIEGFRIED EDMUND, b Mt Lake, Minn, Oct 2, 23. OXYCHLORINATION. *Educ:* Mass Inst Technol, SB, 45, PhD(org chem), 48. *Prof Exp:* Res chemist, Pan Am Petrol Corp, 48-60, group leader, Vulcan Chem Div, Vulcan Mat Co, 60-68, sect head org chem, 68-73, asst mgr res & develop, 73-82, asst mgr technol, res & develop, 82-87; RETIRED. *Mem:* Am Chem Soc. *Res:* Synthetic organic chemistry; partial oxidation of hydrocarbons; olefin polymerization; chlorinated hydrocarbons; oxychlorination. *Mailing Add:* 4630 Friar Tuck Lane Sarasota FL 34232

PENNER-HAHN, JAMES EDWARD, b Brunswick, Ga, Aug 27, 57; m 83; c 1. BIOINORGANIC CHEMISTRY, SYNCHROTION RADIATION. *Educ:* Purdue Univ, BS, 79; Stanford Univ, PhD(chem), 84. *Prof Exp:* Res assoc, Stanford Rad Lab, 84-85; assoc prof, 85-90, PROF, PHYS CHEM, UNIV MICH, 90- *Mem:* Am Chem Soc; Am Physical Soc. *Res:* Use of x-ray absorption to characterize structurally the metal sites in metalloproteins. *Mailing Add:* Dept Chem Univ Mich 930 N Univ Ann Arbor MI 48109-1055

PENNEY, CARL MURRAY, b Newport News, Va, Nov 15, 37. OPTICAL PHYSICS. *Educ:* NC State Univ, BS, 59; Univ Mich, MS, 62, PhD(nuclear eng), 65. *Prof Exp:* PHYSICIST, GEN ELEC CORP RES & DEVELOP, 65- *Mem:* Soc Photo-Optical Instrumentation Engrs; Inst Elec & Electronics Engrs; Optical Soc Am; AAAS; Laser Inst Am. *Res:* Laser development and application of lasers to measurements in the atmosphere, combustion gases and on industrial processes. *Mailing Add:* Res & Develop Gen Elec Corp Bldg 37 Rm 625 Schenectady NY 12301

PENNEY, DAVID EMORY, b Decatur, Ga, Jan 26, 38; m 63; c 1. TOPOLOGY. *Educ:* Tulane Univ, BS, 58, PhD(math), 65. *Prof Exp:* Res assoc biophys, Tulane Univ, 55-59; res assoc, Vet Admin Hosp, New Orleans, La, 59-63; instr math, La State Univ, New Orleans, 63-65, asst prof, 65-66; asst prof, 66-71, ASSOC PROF MATH, UNIV GA, 71- *Mem:* Am Soc Ichthyol & Herpet; Am Math Soc; Math Asn Am. *Res:* Number theory; computer utilization; applied mathematics; convexity; geometric topology; knot theory; active transport in biological membranes. *Mailing Add:* Dept Math Univ Ga Athens GA 30602

PENNEY, DAVID GEORGE, b Detroit, Mich, Jan 11, 40; m 63; c 4. CARDIOVASCULAR PHYSIOLOGY. *Educ:* Wayne State Univ, BS, 63; Univ Calif, Los Angeles, MA, 67, PhD(zool), 69. *Prof Exp:* Asst prof physiol, Univ Ill, Chicago Circle, 69-75, assoc prof, 75-77; assoc prof, 77-90, PROF PHYSIOL, WAYNE STATE UNIV, 91- *Mem:* Int Soc Heart Res; Am Physiol Soc; Soc Exp Biol & Med; AAAS; Am Heart Asn. *Res:* Normal and abnormal heart growth; cardiovascular stress; carbon monoxide, physiology exercise; anemia and hypothyroidism; cardiology. *Mailing Add:* Dept Physiol Wayne State Univ 540 E Canfield Ave Detroit MI 48201

PENNEY, DAVID P, b Waltham, Mass, Dec 11, 33; m 56; c 2. HISTOLOGY, ELECTRON MICROSCOPY. *Educ:* Eastern Nazarene Col, AB, 56; Boston Univ, AM, 57, PhD(biol), 62. *Prof Exp:* Instr anat, Sch Med, Yale Univ, 62-64; asst prof, Univ Rochester, 64-69, assoc prof anat, 69-77, from assoc prof to prof oncol in anat, 77-85, PROF ONCOL IN PATH & LAB MED, SCH MED & DENT, UNIV ROCHESTER, 85- *Concurrent Pos:* Assoc ed, The Anat Record, 78-, Stain Technol, 88-91; treas & trustee, Biol Stain Comn, 88- *Mem:* AAAS; Am Asn Anatomists; Am Soc Cell Biol; Histochem Soc; Electron Micros Soc Am; NY Acad Sci; Am Thoracic Soc; fel Royal Microscopial Soc. *Res:* Electron histochemistry; neoplasia; radiation toxicology; drug toxicity; cell biology; pulmonary pathology; fibroblast heterogeneity. *Mailing Add:* Cancer Ctr Box 704 Univ Rochester Sch Med & Dent Rochester NY 14642

PENNEY, GAYLORD W, b Stacyville, Iowa, Nov 20, 98; m 32, 49; c 1. ELECTROSTATIC CLEANING VENTILATING AIR. *Educ:* Iowa State Univ, BS, 23; Univ Pittsburgh, MS, 29. *Hon Degrees:* Carnegie-Mellon Univ, Dr Engr (Hon), 80. *Prof Exp:* Engr trainee, Westinghouse Elec Corp, 23-24, engr power oper, 24-29, res scientist, 29-36, mgr electro-physics, dept res lab, 36-47; prof, 47-69, EMER PROF ELEC ENG, CARNEGIE MELLON UNIV, 69- *Concurrent Pos:* Lectr, Univ Pittsburgh, 28-47; consult electrostatic probs, 47- *Honors & Awards:* Wetherhill Medal, Franklin Inst, 51; Frank A Chambers Award, Air Pollution & Control Asn, 78. *Mem:* Fel Inst Elec & Electronics Engr; fel Am Soc Heating Ventilations & Air Conditioning Engrs; Am Soc Mech Engrs; Air Pollution & Control Asn. *Res:* Electrostatic problems particularly electrostatic precipitation. *Mailing Add:* 216 Paris Rd Pittsburgh PA 15213

PENNEY, RICHARD COLE, b Chicago, Ill, Oct 5, 45; m 71; c 2. COMPLEX ANALYSIS, SOLVABLE LIE GROUPS. *Educ:* Tulane Univ, BS, 68; Mass Inst Technol, PhD(math), 71. *Prof Exp:* PROF MATH, PURDUE UNIV, 71- *Mem:* Am Math Soc. *Res:* Representation theory of lie groups and its interplay with complex analysis; hormonically induced representations and hormonic analysis on nilmanifolds. *Mailing Add:* Dept Math Purdue Univ West Lafayette IN 47907

PENNEY, WILLIAM HARRY, b Rochester, Minn, June 16, 29; m 55; c 4. CHEMICAL ENGINEERING, PHYSICAL CHEMISTRY. *Educ:* Colo Sch Mines, PRE, 51; Univ Minn, PhD(chem eng, phys chem), 57. *Prof Exp:* Chem engr & res supvr, Cent Res Div, 57-63, res supvr, New Prod Div, 63-65, lab mgr, 65-67, recreation prod tech mgr, 67-73, MEM STAFF COM CHEM DIV, 3M CO, 73- *Mem:* Am Inst Chem Engrs; Am Chem Soc. *Res:* Chemical engineering aspects of compound separations; particle research; process design and polymer and synthetic grass design. *Mailing Add:* 1023 London Rd Mendota Heights MN 55150

PENNEYS, RAYMOND, b Philadelphia, Pa, Oct 15, 19; m 46; c 3. MEDICINE. *Educ:* Temple Univ, 40, MD, 43. *Prof Exp:* Intern, Philadelphia Gen Hosp, Pa, 43-44; instr, Vascular Sect, Dept Med, Univ Pa, 50-58, assoc, 58-62, asst prof, 62-70; ASSOC PROF MED, HAHNEMANN MED COL & HOSP, 70- *Concurrent Pos:* Fel pharm & physiol, Grad Sch Med, Univ Pa, 46-47 & med & prev med, Sch Med, Johns Hopkins Univ, 47-50; chief peripheral vascular sect, St Angus Hosp, 51-, Philadelphia Gen Hosp, 58-80; consult vascular dis, Hillman Med Ctr, 51- *Mem:* AAAS; Am Heart Asn; Am Fedn Clin Res; NY Acad Sci. *Res:* Oxygen tension; peripheral circulation and cardiovascular effects of induced hypoxemia in man. *Mailing Add:* 232 Winding Way Merion Station PA 19066

PENNIALL, RALPH, b Southampton, Eng, Dec 25, 22; nat US; m 44; c 2. BIOCHEMISTRY. *Educ:* Knox Col, BA, 47; Univ Iowa, MS, 50, PhD(biochem), 53. *Prof Exp:* Asst prof biochem, Col Med, Baylor Univ, 54-56; res assoc, Inst Enzyme Res, Univ Wis, 56-58; from asst prof to assoc prof, Sch Med, Univ NC, Chapel Hill, 58-68, prof biochem, 68-90; RETIRED. *Concurrent Pos:* Am Heart Asn advan res fel, 58-60; Fogarty Sr Int fel, 80. *Mem:* AAAS; Am Soc Biol Chemists; Brit Biochem Soc; Am Chem Soc. *Res:* Metabolism and function of inorganic polyphosphates. *Mailing Add:* Dept Biochem CB#7260 Univ NC Sch Med Chapel Hill NC 27599-7260

PENNING, JOHN RUSSELL, JR, b Spokane, Wash, Dec 13, 22; m 75; c 1. APPLIED PHYSICS. *Educ:* Univ Wash, BS, 48, PhD(physics), 56. *Prof Exp:* Res assoc physics, Cyclotron-Univ Wash, 55-56; res scientist, Space Tech Labs, Calif, 56-59, Boeing Co, Wash, 59-63 & Northrop Corp, Calif, 63-65; RES SCIENTIST, BOEING CO, SEATTLE, 65- *Mem:* Am Phys Soc. *Res:* Radioactive decay; scintillation detectors; exploding foils; plasma physics; shock propagation; radiation effects on electronics. *Mailing Add:* 32544 36th Ave SW Federal Way WA 98023

PENNINGROTH, STEPHEN MEADER, b Providence, RI, Oct 21, 44; div; c 2. CELL BIOLOGY, CELL MOTILITY. *Educ:* Princeton Univ, PhD(biochem sci), 77. *Prof Exp:* ASSOC PROF PHARMACOL, SCH OSTEOP MED, UNIV MED & DENT NJ, 84- *Concurrent Pos:* Vis assoc prof, Dept Natural Resources, Inst Comp & Environ Toxicol, Cornell Univ, Ithaca, NY, 87-88. *Mem:* Am Soc Cell Biol; Soc Environ Toxicol & Chem. *Res:* Sea urchin sperm motility; dynein enzymiology; pharmacology of microtubule-based motor atpases. *Mailing Add:* Sch Osteop Med Univ Med & Dent NJ 401 S Central Plaza Stratford NJ 08084

PENNINGTON, ANTHONY JAMES, b Newark, NJ, Dec 19, 32; div; c 2. ELECTRICAL ENGINEERING. *Educ:* Princeton Univ, BSE, 57; Univ Mich, MSE, 58, PhD(elec eng), 63. *Prof Exp:* Res engr, Dodco, Inc, NJ, 58-59; appl mathematician, E I Du Pont de Nemours & Co, Del, 59-60; from instr to assoc prof elec eng, Univ Mich, 60-68; assoc prof, Drexel Inst Technol, 68-71; vpres for planning systs, Decision Sci Corp, 71-73; pres, A J Pennington, Inc, Consults, 73-79; prin engr, Franklin Inst, 79-80; sr prin engr, Comput Sci Corp, 80-86; TECH STAFF, LEAD ENG, MITRE CORP, 86- *Honors & Awards:* Henry Russel Award, 65. *Mem:* Inst Elec & Electronics Engrs. *Res:* Space vehicle performance analysis and control; chemical process optimization and control; electrical power system planning, analysis and control; digital computer control systems; project economic analysis and financial planning; energy resource analysis modeling; urban and regional systems analysis; military warfare gaming system; logistics system; strategic defense system; distributed/partitioned assignment algorithm. *Mailing Add:* Mitre Corp Burlington Rd MS T-140 Bedford MA 01730

PENNINGTON, DAVID EUGENE, b Bryan, Tex, Nov 3, 39; m 62; c 2. INORGANIC CHEMISTRY. *Educ:* NTex State Univ, BA, 62, MS, 63; Pa State Univ, PhD, 67. *Prof Exp:* Fel, State Univ NY Stony Brook, 67; from asst prof to assoc prof, 67-81, PROF INORG CHEM, BAYLOR UNIV, 82- *Concurrent Pos:* Chair, premed/predent adv comt, Baylor Univ, 87- *Mem:* Am Chem Soc. *Res:* Mechanisms of inorganic reactions involving coordination compounds, particularly electron transfer and substitution reactions. *Mailing Add:* Dept Chem Baylor Univ PO Box 97348 Waco TX 76798-7348

PENNINGTON, FRANK COOK, b Seattle, Wash, Apr 4, 24; m 47; c 2. ORGANIC CHEMISTRY. *Educ:* Reed Col, BA, 48; Univ Rochester, PhD(chem), 51. *Prof Exp:* Res chemist antibiotic chem, Chas Pfizer & Co, 51-55; prof chem, Coe Col, 55-69, chmn dept, 58-69; dean, Sch Nat Sci, 69-78, PROF CHEM, CALIF STATE UNIV, CHICO, 69- *Concurrent Pos:* Res assoc, Argonne Nat Lab, 62-63. *Mem:* Fel AAAS; Am Chem Soc. *Res:* Chemistry of antibiotics, chlorophyll and cartenoids; synthesis of colchicine analogs; synthesis of heterocycles. *Mailing Add:* Box 63 Chapman Ranch CA 78347

PENNINGTON, JEAN A T, b Los Angeles, Calif, Sept 24, 46; c 1. FOOD COMPOSITION, FOOD CONSUMPTION METHODOLOGIES. *Educ:* Univ Calif, Berkeley, BA, 67, PhD (nutrit), 73. *Prof Exp:* Instr nutrit & physiol, City Col San Francisco, 72-79; ASSOC DIR DIETARY SURVEILLANCE, DIV NUTRIT, FOOD & DRUG ADMN, 79- *Concurrent Pos:* Instr nutrit, Univ Calif, Berkeley Exten, 71-72 & San Francisco State Univ, 77. *Honors & Awards:* Outstanding Serv Award, Am Dietetic Asn, 87. *Mem:* Am Dietetic Asn; Am Inst Nutrit; Am Soc Clin Nutrit; Int Soc Trace Element Res Humans; Soc Nutrit Educ. *Res:* The food and nutrient intake of the US population and various population subgroups; the relationship of food & nutrient intake to health and disease; the nutrient composition of foods; the development and maintenance of food composition databases. *Mailing Add:* FDA 200 C St SW Washington DC 20204

PENNINGTON, KEITH SAMUEL, b West Bromwich, Eng, June 14, 36; m 60; c 3. PRINTING TECHNOLOGIES & PRINTING MATERIALS. *Educ:* Univ Birmingham, BSc, 57; McMaster Univ, PhD(physics), 61. *Prof Exp:* Mem tech staff, Bell Tel Labs, Inc, 61-67; mem prof staff, Info Sci Lab, Gen Elec Res & Develop Ctr, 67; res staff mem, IBM Corp, 67-72, mgr explor displays, 72-73, mgr explor terminal technol, 73-78, mem & dir res staff, I/O & Com, T J Watson Res Ctr, 78-79, SR MGR IMAGE TECHNOLS DEPT, IBM CORP, 79- *Concurrent Pos:* Group leader, Undersea Warfare Comt, Nat Acad Sci, 71. *Honors & Awards:* Charles E Ives Award, Soc Photog Sci & Eng, 87. *Mem:* Soc Info Display; Optical Soc Am; Inst Elec & Electronic Engrs; Soc Photog Sci & Eng. *Res:* Holography; optical information processing; exploratory terminal technologies; printers; image processing. *Mailing Add:* T J Watson Res Ctr IBM Corp PO Box 704 Yorktown Heights NY 10598

PENNINGTON, RALPH HUGH, b Wichita, Kans, Oct 4, 24; m 73; c 2. COMPUTER SCIENCE. *Educ:* US Mil Acad, BS, 46; Stanford Univ, MS, 50, PhD(math), 54. *Prof Exp:* Staff scientist, Proj Matterhorn, US Army, Princeton, 51-52, Radiation Lab, Univ Calif, 53-54, atomic energy specialist, Off Spec Weapons Develop, 55-58, nuclear effects engr, Defense Atomic Support Agency, DC, 58-61, spec asst to dir, Defense Res & Eng, 61-63, chief, Theoret Br, Air Force Weapons Lab, Kirtland AFB, 63-67; chief advan systs studies, Advan Ballistic Missile Defense Agency, 69-71, chief scientist, Defense & Space Div, Syst Develop Corp, 71-75; dir data processing & software, 75-89, SR SCIENTIST, GEN RES CORP, 89- *Concurrent Pos:* Lectr, George Washington Univ, 58-63. *Mem:* Am Math Soc; Asn Comput Mach. *Res:* Nuclear phenomenology; systems analysis; simulation and modelling; numerical analysis; computer system software; computer programming languages; computer architecture. *Mailing Add:* 506 Yankee Farm Rd Santa Barbara CA 93109

PENNINGTON, ROBERT ELIJA, b Brenham, Tex, Nov 22, 26; m 46; c 2. CHEMICAL ENGINEERING. *Educ:* Univ Tex, BS, 48, MS, 50, PhD(chem eng), 56. *Prof Exp:* Res chem engr, Humble Oil & Ref Co, Baytown, 56-58, sr res chem engr, 58-61, res specialist, 62, planning specialist, Houston, 62, sr planning specialist, 63-64, res coordr, 64-65; head, Petrol Dept, Esso Res & Eng Co, 65-66, mgr, Baytown Petrol Res Lab, 66-67, lab mgr, Synthetics Fuels Res Lab, Exxon Res & Eng, 67-82; PRES, PENNINGTON CONSULT & ENG CO, 82- *Mem:* Am Chem Soc; Inst Chem Engrs. *Mailing Add:* 206 W Main Brenham TX 77833

PENNINGTON, SAMMY NOEL, b Dumas, Tex, Mar 19, 41; c 2. ANALYTICAL BIOCHEMISTRY. *Educ:* Kans State Col Pittsburg, BS, 64; Kans State Univ, PhD(chem), 66. *Prof Exp:* Instr biochem, Univ Tex Med Br Galveston, 67-68; asst prof chem, Cent Mo State Col, 68-69; asst scientist biochem, Cancer Res Ctr, Columbia, Mo, 69-70; assoc prof, 70-80, PROF BIOCHEM, ECAROLINA UNIV, 70- *Concurrent Pos:* Consult, Cancer Res Ctr, Columbia, Mo, 68-69; Am Cancer Soc grant, ECarolina Univ, 72-73, NSF res grant, 72-74, alcoholism grant, 76-81 & NIH grant, 79-81. *Mem:* Am Soc Biol Chemists; Sigma Xi. *Res:* Fetal alcohol syndrome; growth and development; molecular mechanism of ethanol induced growth suppression; the role of ethanol-prostaglandin interactions in the pathophysiology of chronic alcoholism. *Mailing Add:* Dept Biochem E Carolina Univ Med Greenville NC 27834

PENNINGTON, WAYNE DAVID, b Rochester, Minn, Dec 19, 50; m 78; c 2. SEISMOLOGY, ROCK PHYSICS. *Educ:* Princeton Univ, AB, 72; Cornell Univ, MS, 76; Univ Wis, Madison, PhD(geophysics), 79. *Prof Exp:* Asst prof geophys, dept geol sci, Univ Tex, Austin, 79-85; ADVAN SR GEOPHYSICIST, MARATHON OIL CO, LITTLETON, COLO, 85- *Mem:* Am Geophys Union; Seismol Soc Am; Soc Explor Geophysicists; Soc Petrol Engrs; Soc Prof Well Log Analysts. *Res:* Elastic properties of rock derived from theoretical, seismic, and engineering observations for use in seismic identification of rocks quantifying hydrocarbon reservoir behavior, and determining strength of rocks. *Mailing Add:* Marathon Oil Co PO Box 269 Littleton CO 80160-0269

PENNISTON, JOHN THOMAS, b St Louis, Mo, Sept 10, 35; m 60; c 2. BIOLOGICAL CHEMISTRY. *Educ:* Harvard Univ, AB, 57, AM, 59, PhD(chem), 62. *Prof Exp:* NIH res fel chem, Harvard Univ, 62; vis asst prof, Pomona Col, 63-64; fel biochem, Inst Enzyme Res, Univ Wis-Madison, 64-66, asst prof, 66-71; assoc prof chem, Univ NC, Chapel Hill, 71-76; assoc prof biochem, 76-79, PROF BIOCHEM & MOLECULAR BIOL, MAYO GRAD SCH MED, 79-, CONSULT BIOCHEM & MOLECULAR BIOL, MAYO CLIN, 76- *Concurrent Pos:* Estab investr, Am Heart Asn, 69-74. *Mem:* AAAS; Am Chem Soc; Am Soc Biol Chemists. *Res:* Membrane structure and biological transducing systems; intracellular calcium metabolism; plasma membrane calcium pumps; erythrocyte metabolism and morphology. *Mailing Add:* Dept Biochem & Molecular Biol Mayo Clin Rochester MN 55905

PENNOCK, BERNARD EUGENE, b Philadelphia, Pa, Jan 30, 38; m 69; c 2. BIOMEDICAL ENGINEERING, PULMONARY PHYSIOLOGY. *Educ:* Drexel Inst, BS, 60, MS, 61; Univ Pa, PhD(biomed eng), 67. *Prof Exp:* Res assoc biomed eng, Presby Hosp, Univ Pa, 67-68; asst prof biophys, Med Col Pa, 68-73; assoc prof med, Univ Okla Health Sci Ctr, 73-80; MEM FAC, UNIV PITTSBURGH, 80- *Concurrent Pos:* Vis lectr, Hosp Univ Pa, 69-73. *Mem:* Biomed Eng Soc; Asn Advan Med Instrumentation; Am Tech Soc; Sigma Xi. *Res:* Electrical properties of biologic materials; pulmonary physiology; excitable membranes. *Mailing Add:* Univ Pittsburgh 440 Scaife Pittsburgh PA 15261

PENNY, JOHN SLOYAN, b Philadelphia, Pa, Aug 3, 14; m 47; c 5. PALEOBOTANY. *Educ:* La Salle Col, AB, 37; Univ Pa, MS, 39, PhD(bot), 42. *Prof Exp:* Herbarium asst & asst instr bot, Univ Pa, 39-42; geologist, US War Dept, 46-47; paleobotanist, Creole Petrol Corp, Venezuela, 47-50; from asst prof to assoc prof, 50-60, prof & chmn dept, 60-79, EMER PROF BIOL, LA SALLE COL, 79- *Concurrent Pos:* Mem teaching fac, Arboretum of the Barnes Found, 65-80, dir, 79-84; bot consult, So Co Servs, 73-89. *Honors & Awards:* Lindback Found Award, 62. *Mem:* AAAS; Geol Soc Am; Bot Soc Am; Int Assoc Plant Taxon. *Res:* Mesozoic paleobotany; stratigraphy; palynology. *Mailing Add:* 504 Portsmouth Court Doyles Town PA 18901

PENNY, KEITH, SR, b Oklahoma City, Okla, Aug 25, 32; m 55; c 4. NUCLEAR PHYSICS. *Educ:* Univ Okla, BS, 54; Univ Tenn, PhD(physics), 66. *Prof Exp:* Physicist, Oak Ridge Nat Lab, 54-62, dir, Radiation Shielding Info Ctr, 63-66; res scientist, Space Sci & Eng Lab, Defense & Space Systs Dept, Union Carbide Corp, NY, 67-68; physicist, Oak Ridge Nat Lab, 68-89; mem staff, Mantech, 89-90; CONSULT, 90- *Res:* Radiation shielding; nuclear structure; computer technology; information handling. *Mailing Add:* 4717 Guinn Rd Knoxville TN 37931

PENNYPACKER, CARLTON REESE, b Covina, Calif, Feb 1, 50; m 78; c 1. ASTROPHYSICS. *Educ:* Univ Calif, Berkeley, BA, 72; Harvard Univ, MA, 74, PhD(physics), 78. *Prof Exp:* Fel, Smithsonian Astrophys Observ, 73-78; RES PHYSICIST, SPACE SCI LAB, LAWRENCE BERKELEY LAB, 78- *Res:* Automated supernova searches, infrared observations of pulsars, infrared astronomy, experimental astrophysics. *Mailing Add:* Bldg 50 Rm 232 Lawrence Berkeley Lab Berkeley CA 94720

PENROSE, WILLIAM ROY, b Hamilton, Ont, Jan 20, 43; m 65; c 3. POLLUTION CHEMISTRY. *Educ:* McMaster Univ, BSc, 65; Univ Mich, Ann Arbor, MS, 67, PhD(biol chem), 69. *Prof Exp:* Fel enzym, McMaster Univ, 69-70; fel molecular biol, Univ Toronto, 70-72; res scientist pollution chem, Fisheries & Marine Serv, Environ Can, 72-80; biochemist & head, Ecol Sect, Radiol & Environ Res Div, Argonne Nat Lab, 80-82, biochemist, environ effects res prog, Biol Environ & Med Res Div, 82-89; SR SCIENTIST, TRANSDUCER RES INC, NAPERVILLE, ILL, 89- *Honors & Awards:* IR100 Invention Award, 84. *Mem:* Sigma Xi; Am Chem Soc. *Res:* Effect of pollutants on survival of aquatic and marine animals; analytical chemistry of pollutants; biochemical transformations of pollutants; gas detection instrumentation. *Mailing Add:* 526 W Franklin Ave Naperville IL 60540

PENSACK, JOSEPH MICHAEL, b Scranton, Pa, Dec 2, 16; m 41; c 6. ANIMAL NUTRITION, ANIMAL PHYSIOLOGY. *Educ:* Pa State Col, BS, 38; Univ NH, MS, 40; Ohio State Univ, PhD(biochem), 48. *Prof Exp:* Asst zool, Univ NH, 38-40, endocrinol, Northwestern Univ, 40-41 & animal sci, Exp Sta, Ohio State Univ, 46-48; res biochemist, Com Solvents Corp, 48-51, dir nutrit res, 51-59; sr biochemist, Am Cyanamid Co, 59-73, chief nutritionist, 73-78, prin scientist, Cyanamid Int, 78-; RETIRED. *Mem:* AAAS; Am Soc Animal Sci; Poultry Sci Asn. *Res:* Antibiotics; vitamins; non-protein nitrogen; enzymes; marine biology; biology; anabolics. *Mailing Add:* Two Sioux Rd Trenton NJ 08635

PENSAK, DAVID ALAN, b Princeton, NJ, Feb 16, 48; m. CHEMISTRY, COMPUTER SCIENCE. *Educ:* Princeton Univ, AB, 69; Harvard Univ, MA, 71, PhD(chem), 73. *Prof Exp:* Fel chem, Harvard Univ, 73-74; res chemist, 74-80, res suprv, Cent Res Dept, 80-87, CORP ADV, COMPUTING TECH, E I DU PONT DE NEMOURS & CO, INC, 87- *Concurrent Pos:* Hon res assoc, Ctr Res Comput Technol, Harvard Univ, 74- *Mem:* Am Chem Soc; Asn Comput Mach; Sigma Xi. *Res:* Application of computer science techniques to chemical problems, especially drug design, molecular modeling and computer-aided synthetic design. *Mailing Add:* Cent Res & Develop Dept Exp Sta Bldg 328 E I du Pont de Nemours & Co Inc Wilmington DE 19898

PENSE, ALAN WIGGINS, b Sharon, Conn, Feb 3, 34; m 58; c 3. PHYSICAL METALLURGY, WELDING. *Educ:* Cornell Univ, BMetE, 57; Lehigh Univ, MS, 59, PhD(metall), 62. *Prof Exp:* From instr to assoc prof, Lehigh Univ, 60-71, chmn dept metall & mat eng, 77-83, assoc dean, Col Eng & Phys Sci, 84-89, dean, 89-90, PROF METALL, LEHIGH UNIV, 71-, PROVOST & VPRES, 90- *Concurrent Pos:* Mem, Pressure Vessel Res Comt; consult, Adv Comt Reactor Safeguards, Nuclear Regulatory Comn, 69-; welding handbk comt, Am Welding Soc, 72- *Honors & Awards:* Sparagan Award, Am Welding Soc, 64, Charles H Jennings Award, 70, Comfort Adams Mem Lectr, 80, William Hobart Medal, 83; Western Elec Award, Am Soc Eng Educ, 73; Stabler Award, 72. *Mem:* Fel Am Soc Metals Int; hon mem Am Welding Soc; Int Inst Welding; Am Soc Testing & Mat. *Res:* Mechanical metallurgy and properties of alloys, particularly low alloy high strength steels; weldability of metals; failure analysis. *Mailing Add:* Alumni Bldg 27 Lehigh Univ Bethlehem PA 18015

PENSKY, JACK, b Canton, Ohio, Aug 25, 24; m 56; c 4. BIOCHEMISTRY. *Educ:* Ohio State Univ, BS, 45; Purdue Univ, MS, 49; Western Reserve Univ, PhD(biochem), 54. *Prof Exp:* Res assoc sulfur metab leucocytes, 54-57, sr instr biochem, Inst Path, 57-60, asst prof exp path, 60-71, ASST PROF BIOCHEM, CASE WESTERN RESERVE UNIV, 71- *Concurrent Pos:* USPHS res career develop award, 64; mem staff, Cleveland Vet Admin Hosp, Ohio, chief, Radioassay Unit, Lab Serv. *Mem:* Fel AAAS; NY Acad Sci; Sigma Xi. *Res:* Protein purification; human complement and hormones; immunochemistry; radioimmunoassay. *Mailing Add:* 7560 Chagrin Rd Chagrin OH 44022

PENSTONE, S(IDNEY) ROBERT, b Winnipeg, Man, Aug 29, 30; m 82; c 2. ELECTRICAL ENGINEERING. *Educ:* Queen's Univ, BSc, 55, MSc, 57. *Prof Exp:* Defence sci serv officer, Defence Res Telecommun Estab, Defence Res Bd, Can, 57-63; from asst prof to assoc prof, 63-73, assoc dean, fac appl sci, 81-84, PROF ELEC ENG, QUEEN'S UNIV, 73- *Mem:* Can Soc Elec Eng; Inst Elec & Electronics Engrs. *Res:* Development of CAD tools for VLSI circuit design; development of interactive computer programs in electrical engineering education. *Mailing Add:* Dept Elec Eng Queen's Univ Kingston Kingston ON K7L 3N6 Can

PENTECOST, JOSEPH L(UTHER), b Winder, Ga, Apr 23, 30; m 49, 69; c 5. CERAMIC ENGINEERING. *Educ:* Ga Inst Technol, BCerE, 51; Univ Ill, Urbana, MS, 54, PhD(ceramic eng), 56. *Prof Exp:* Sr engr, Melpar, Inc, 56-59; chief res engr, Aeronca Mfg Corp, 59-60; assoc prof ceramic eng, Miss State Univ, 60-61; assoc dir res, Res & Develop, Melpar, Inc, 61-68; mgr, Wash Res Ctr, W R Grace & Co, 68-72; PROF CERAMIC ENG, GA INST TECHNOL, 72- *Concurrent Pos:* Proprietor, Matcos Co, 61- *Mem:* Am Ceramic Soc (pres, 86-87); Am Soc Metals; Nat Inst Ceramic Engrs (pres, 76-77). *Res:* Electrical ceramics; fine particle raw materials; crystal growth; glass fracture. *Mailing Add:* Mat Eng Ga Inst Technol Atlanta GA 30332

PENTNEY, ROBERTA PIERSON, b Van Nuys, Calif, Jan 11, 36; m 75; c 1. NEUROANATOMY, MORPHOMETRY. *Educ:* Col Notre Dame, Calif, BA, 60; Univ Notre Dame, PhD(biol), 65. *Prof Exp:* Res assoc, Univ Notre Dame, 65; from instr to assoc prof biol, Col Notre Dame, Calif, 65-74; asst prof, 74-84, ASSOC PROF ANAT, STATE UNIV NY, BUFFALO, 84- *Concurrent Pos:* NSF fel, 61-64; NIH fel, Col Physicians & Surgeons, Columbia Univ, 71-74; prin investr, Nat Inst Alcohol Abuse & Alcoholism, 83- *Mem:* Res Soc Alcoholism; Geront Soc Am; Am Asn Anatomists; Soc Neurosci; Sigma Xi. *Res:* Ethanol and age effects on central nervous system; visual pathways in central nervous system. *Mailing Add:* Dept of Anat Sci State Univ NY Buffalo NY 14214

PENTO, JOSEPH THOMAS, b Masontown, Pa, Sept 1, 43; m 69; c 2. ENDOCRINOLOGY, PHARMACOLOGY. *Educ:* WVa Univ, BA, 65, MS, 67; Univ Mo, PhD(pharmacol), 70. *Prof Exp:* NIH fel, Maimonides Med Ctr, 70-71; asst prof pharmacol, 71-76, assoc prof, 76-80, PROF PHARMACODYNAMICS & TOXICOL, UNIV OKLA HEALTH CTR, 81-, SECT HEAD, 79- *Concurrent Pos:* Exec comt grant, Univ Okla Res Inst, 71-72; NSF grant, 74-78; NIH grant, 86-90. *Mem:* AAAS; Sigma Xi; Am Soc Pharmacol Exp Therapeut; Am Soc Bone Mineral Res; Endocrine Soc. *Res:* Calcium metabolism; calcitonin; parathyroid hormone; radio immunoassay; antiestrogen. *Mailing Add:* Univ Okla HSC 1110 N Stonewall Ave Oklahoma City OK 73190

PENTON, HAROLD ROY, JR, b New Orleans, La, Apr 17, 47; m 69; c 1. POLYMER CHEMISTRY, ORGANIC CHEMISTRY. *Educ:* Fla State Univ, BS, 69; Ga Inst Technol, PhD(org chem), 73. *Prof Exp:* Res chemist polymer chem, Am Enka Co, 73-74, AKZO Res Lab, 74-75; sr res chemist, Am Enka Co, 75-77; sr res chemist polymer chem, 77-81, group leader, Ethyl Corp, 80-81, supvr, 81-87, ASST DIR, 87- *Concurrent Pos:* Adj prof, Univ NC, Asheville, 76-77. *Mem:* Am Chem Soc; Soc Plastic Engrs. *Res:* Anionic and ring opening polymerizations; inorganic polymer synthesis; mechanism of polycondensation catalysis; mechanism of polymer degradation; hydrophilic polymer synthesis; polymer blends; polymer grafting; high temperature resistant polymers, engineering thermoplastics, polyphosphasenes; polyimides; ceramics. *Mailing Add:* 232 Shady Oaks Ct Baton Rouge LA 70810

PENTON, ZELDA EVE, b New York, NY, Jan 15, 39; m 61; c 2. ANALYTICAL CHEMISTRY. *Educ:* City Col New York, BS, 59; Columbia Univ, MA, 60, PhD(chem), 64. *Prof Exp:* Res assoc biochem, Rockefeller Univ, 64-68; chemist, Signetics Corp, Sunnyvale, 72-74; chemist, Calif State Dept Health, 74-78; SR CHEMIST, INSTRUMENT DIV, VARIAN ASSOCS, 78- *Mem:* Am Chem Soc; Am Soc Testing & Mat. *Res:* Analysis of trace metals in biologicals; biological role of trace metals; gas chromatography of biological and environmental samples; capillary gas chromatography. *Mailing Add:* 5649 Greenridge Rd Castro Valley CA 94552

PENWELL, RICHARD CARLTON, b Columbus, Ohio, Apr 2, 42; m 64; c 2. POLYMER SCIENCE, ENGINEERING. *Educ:* Toledo Univ, BSME, 64; Princeton Univ, MSChE, 65; Univ Mass, MSE, 69, PhD(polymer sci), 70. *Prof Exp:* Proj eng polymer processing, Owens Ill Glass Co, 65-67; scientist, Xerox Corp, 72-78, sr scientist polymer sci, 78-79, mgr, spec mat eng, 79-82, mgr, advan mat & mfg processes, 82-86, mgr, mat & mfg technol, 86-89, MGR, FUTURE PROD & MAT ENG, XEROX CORP, 89- *Concurrent Pos:* NSF fel, Strasbourg, France, 70-71; Mat Sci Fel, Northwestern Univ, 71-72. *Mem:* Am Chem Soc. *Res:* Polymer physics, rheology and processing; mechanical and thermal properties of polymers and composites; structure-property relations of polymers. *Mailing Add:* Xerox Corp W147 Webster NY 14580

PENZ, P ANDREW, b Detroit, Mich, July 19, 39; m 61; c 2. ARTIFICIAL NEURAL NETWORKS, LIQUID CRYSTAL DISPLAYS. *Educ:* Brown Univ, ScB, 61; Cornell Univ, PhD(physics), 67. *Prof Exp:* Res scientist sci lab, Ford Motor Co, Mich, 66-72; res scientist, 72-77, mgr Advan Displays Br, 78-84, SR MEM TECH STAFF, TEX INSTRUMENTS INC, 79- *Mem:* Fel Am Phys Soc; Sigma Xi; sr mem Inst Elec & Electronic Eng; fel Soc Info Display; AAAS; Int Neural Network Soc. *Res:* Transport properties and fermi surface studies of pure metals; linear magneto-resistance of potassium demonstration; electrodynamic properties of liquid crystals; pseudo analog LCD; plastic LCD; DSP architecture for neural network accelerators; closeness code for neural networks; neural networks applied to passive radar classification. *Mailing Add:* Tex Instruments Inc PO Box 655936 MS134 Dallas TX 75265

PENZIAS, ARNO A(LLAN), b Munich, Germany, Apr 26, 33; US citizen; m 54; c 3. RADIO ASTRONOMY, PHYSICS. *Educ:* City Col NY BS, 54; Columbia Univ, MA, 58, PhD(physics), 62. *Hon Degrees:* Dr, Observ Paris, 76 & from various US & foreign univs, 79-90. *Prof Exp:* Mem tech staff, AT&T Bell Labs, 61-72, dept head, radio physics res, 72-76, dir radio res lab, 76-79, exec dir res & commun, 79-81, VPRES RES, AT&T BELL LABS, 81- *Concurrent Pos:* Lectr, dept astrophys sci, Princeton Univ, 67-72, vis prof, 72-85; res assoc, Harvard Col Observ, 68-80; adj prof earth & sci, State Univ NY, Stony Brook, 74-84; mem, Comt Concerned Scientists, 75-, vchmn, 76-; trustee, Trenton State Col, 77-79; mem, Astron Advis Panel, NSF, 78-79, Indust Panel Sci & Technol, 81-2; assoc ed, Astrophys J, 78-82; mem, Max-Planck-Inst Radioastron, 78-85; mem bd overseers, Sch Eng & Appl Sci, Univ Pa, 83-86. *Honors & Awards:* Nobel Prize physics, 78; Herschel Medal, Royal Astron Soc, 77; Henry Draper Medal, Nat Acad Sci, 77; Komfner lectr, Standford UNiv, 79; Joseph Handleman Prize in Sci, 83; Grace Adams Tanner lectr, Southern Utah State Col, 87; Gamow Lectr, Univ Colo, 80. *Mem:* Nat Acad Sci; Nat Acad Eng; Am Astron Soc; Am Acad Arts & Sci; Am Phys Soc; World Acad Art & Sci. *Res:* Astrophysics; information systems; contributed over 80 articles to technical journals and granted several patents. *Mailing Add:* AT&T Bell Labs 600 Mountain Ave Rm 6A-411 Murray Hill NJ 07974

PENZIEN, JOSEPH, b Philip, SDak, Nov 27, 24; m 50; c 4. STRUCTURAL ENGINEERING. *Educ:* Univ Wash, Seattle, BS, 45; Mass Inst Technol, ScD, 50. *Prof Exp:* Jr engr, Corps Engrs, 45-46; instr eng, Univ Wash, Seattle, 46-47; res assoc, Mass Inst Technol, 48-50; mem staff, Sandia Corp, NMex, 50-52; sr struct engr, Consol Vultee Aircraft Corp, 52-53; consult, Lawrence Radiation Lab, Univ Calif, Berkeley, 57-67, dir, Earthquake Eng Res Ctr, 67-73 & 77-80, from assoc prof to prof struct eng, 53-88; chmn, Eastern Int Engrs, Inc, Lafayette, 82-90; CHMN, INT CIVIL ENG CONSULTS, INC, 90- *Concurrent Pos:* Postdoctoral fel, NSF, 59, sr sci fel, 73; Inst Advan Study fel, Mass Inst Technol, 59-60; chief tech adv to UNESCO & expert earthquake engr, Int Inst Seismol & Earthquake Eng, Japan, 64-65; mem, Governor's Earthquake Coun, Calif, 72; vchmn, Seismic Adv Bd, Calif Dept Transp; consult, Bay Area Rapid Transit Extensions Prog; mem bd consults, Muni Metro Turnaround Proj, Bechtel Corp; res prize, Am Soc Civil Engrs, 65. *Honors & Awards:* Nathan M Newmark Medal, Am Soc Civil Engrs, 83, Alfred M Freudenthal Medal, 86. *Mem:* Nat Acad Eng; hon mem Am Soc Civil Engrs; Seismol Soc Am; fel Am Acad Mech; Am Concrete Inst. *Res:* Structural mechanics. *Mailing Add:* Int Civil Eng Consults Inc 1995 University Ave Suite 119 Berkeley CA 94704

PEO, ERNEST RAMY, JR, b Watertown, NY, Apr 21, 25; m 43; c 1. ANIMAL NUTRITION. *Educ:* Okla State Univ, BS, 52, MS, 53; Iowa State Univ, PhD(animal nutrition), 56. *Prof Exp:* Res assoc, Iowa State Univ, 53-54, asst, 54-56; from asst prof to assoc prof animal sci, 56-66, PROF ANIMAL SCI, UNIV NEBR, LINCOLN, 66- *Mem:* Am Soc Animal Sci; Am Inst Nutrit; Sigma Xi. *Res:* Monogastric nutrition. *Mailing Add:* Dept Animal Sci Univ Nebr Lincoln NE 68503

PEOPLES, JOHN, JR, b Staten Island, NY, Jan 22, 33; m 56; c 2. HIGH ENERGY PHYSICS. *Educ:* Carnegie-Mellon Univ, BSEE, 55; Columbia Univ, MA, 61, PhD(physics), 66. *Prof Exp:* Engr, Martin-Marietta Corp, 55-59; instr physics, Columbia Univ, 64-66, asst prof, 66-69; from asst prof to assoc prof, Cornell Univ, 69-72; head, Res Div, 75-80, head Antiproton Source Proj, 81-87, DEPUTY HEAD, FERMI NAT ACCELRRATOR LAB, 87- *Concurrent Pos:* Sloan fel, 70-72. *Mem:* Am Phys Soc; AAAS; fel Am Phys Soc, 83. *Res:* High energy particle physics; accelerator technology. *Mailing Add:* Fermilab PO Box 500 Batavia IL 60510

PEOPLES, STUART ANDERSON, b Petaluma, Calif, Nov 3, 07; m 32; c 1. PHARMACOLOGY. *Educ:* Univ Calif, AB, 30, MD, 34. *Prof Exp:* Intern, Univ Hosp, Univ Calif, 33-34, Merck fel, Univ, 34-35; Commonwealth Fund fel, Maudsley Hosp, London, Eng, 35-36; asst prof pharmacol, Univ Louisville, 36-38; assoc prof physiol & pharmacol, Sch Med, Univ Ala, 38-43; prof pharmacol, Col Med, Baylor Univ, 43-47; med consult, Calif Dept Food & Agr, 75-80; prof comp pharmacol, Sch Vet Med, lectr pharmacol & exp therapeut, Sch Med & pharmacologist, Exp Sta, 47-74, EMER PROF PHYSIOL SCI, UNIV CALIF, DAVIS, 74- *Concurrent Pos:* Consult & expert witness toxicol, 80-86. *Mem:* AAAS; Am Soc Pharmacol & Exp Therapeut; Soc Exp Biol & Med; AMA; Am Chem Soc. *Res:* Drug metabolism; toxicology; chemotherapy of tuberculosis; metabolism of arsenic in mammals. *Mailing Add:* 26213 County Rd 96 Davis CA 95616

PEPE, FRANK ALBERT, b Schenectady, NY, May 22, 31. ANATOMY, MACROMOLECULES. *Educ:* Union Col, NY, BS, 53; Yale Univ, PhD(phys chem), 57. *Prof Exp:* Instr, Univ Pa, 57-60, assoc, 60-63, from asst prof to assoc prof, 63-70, chmn dept, 77-90, PROF ANAT, SCH MED, UNIV PA, 70- *Concurrent Pos:* Prin investr res grants, NIH. *Honors & Awards:* Raymond C Truex Distinguished Lectr Award, Hahneman Univ. *Mem:* Fel AAAS; Am Chem Soc; Biophys Soc; Am Asn Anatomists; NY Acad Sci; Electron Microscopy Soc Am; Am Soc Cell Biol; Am Inst Biol Sci. *Res:* Molecular anatomy; studies of the molecular organization of the myofibril of striated muscle using immunochemical genetic and electron microscopy techniques. *Mailing Add:* Dept of Anat Univ of Pa Sch of Med Philadelphia PA 19104

PEPE, JOSEPH PHILIP, b Connellsville, Pa, Nov 10, 47; m 69; c 2. ORGANIC CHEMISTRY, PHOTOGRAPHIC SCIENCE. *Educ:* Univ Pittsburgh, BS, 69; Pa State Univ, PhD(chem), 76. *Prof Exp:* From res chemist to sr res chemist, 76-87, RES ASSOC PHOTOG SCI, RES LAB, EASTMAN KODAK CO, 87- *Mem:* Am Chem Soc. *Res:* Synthesis of organic components for conventional color-sensitized products; investigation of factors affecting dye stability in color print papers; design and precipitation of silver halide photographic emulsions. *Mailing Add:* 64 Wheelock Rd Penfield NY 14526-1430

PEPER, ERIK, b Hague, Netherlands, May 28, 44; US citizen; m; c 2. BIOFEEDBACK, SELF REGULATION. *Educ:* Harvard Univ, BA, 68; Union Grad Sch, PhD(psychol), 75. *Prof Exp:* FAC SELF-REGULATION, SAN FRANSISCO STATE UNIV, 76-; DIR BIOFEEDBACK, BIOFEEDBACK FAMILY INST, 79- *Concurrent Pos:* Sport psychologist, US Rhythmic Gymnastic Team, 81-85. *Honors & Awards:* Recognition Award, Asn Appl Psychophysiol & Biofeedback, 89. *Mem:* Asn Appl Psychophysiol & Biofeedback, (pres 76-77); Soc Behav Med; Soc Psychophysiol Res. *Res:* Optimizations of health; respiratory psychophysiology and self regulation with emphasis on nonpharmacological treatment of asthma and breathlessness. *Mailing Add:* 2236 Derby St Berkeley CA 94705

PEPIN, HERBERT SPENCER, b Birtle, Man, Jan 28, 28; m 62; c 2. PLANT PATHOLOGY. *Educ:* Univ BC, BSA, 54, MA, 56; Univ Ill, PhD(plant path, bact), 59. *Prof Exp:* Asst plant path, Univ Ill, 56-59; res officer, 59-67, RES SCIENTIST, RES BR, CAN DEPT AGR, 67- *Concurrent Pos:* Hon lectr, Univ BC, 83- *Mem:* Am Phytopath Soc; Can Phytopath Soc; Am Mycol Soc. *Res:* Physiology and biochemistry of disease resistance; root rot diseases. *Mailing Add:* Agr Can Res Sta 6660 NW Marine Dr Vancouver BC V6T 1X2 Can

PEPIN, ROBERT OSBORNE, b Wellesley Hills, Mass, Apr 12, 33; m 60. PHYSICS, SPACE SCIENCES. *Educ:* Harvard Univ, AB, 56; Univ Calif, Berkeley, PhD(physics), 64. *Prof Exp:* Res physicist, Univ Calif, Berkeley, 64; res assoc, 65, from asst prof to assoc prof, 66-74, PROF PHYSICS & ASTRON, INST TECHNOL, UNIV MINN, MINNEAPOLIS, 74- *Mem:* AAAS; Am Geophys Union; Sigma Xi. *Res:* History of meteoritic and terrestrial matter by mass spectrometry of rare gases. *Mailing Add:* Univ Minn 148 Physics Minneapolis MN 55455

PEPIN, THEODORE JOHN, b St Paul, Minn, Feb 7, 39; m 60; c 2. PHYSICS. *Educ:* Univ Minn, BS, 62, MS, 68, PhD(physics), 70. *Prof Exp:* Grant, Univ Minn, 70; asst prof, 71-75, ASSOC PROF PHYSICS, UNIV WYO, 75- *Res:* Remote sensing of atmosphere from earth orbiting satellites to determine ozone and aerosol content of stratosphere; atmospheric and astrophysics research. *Mailing Add:* Dept Physics & Astron Univ Wyo Laramie WY 82070

PEPINE, CARL JOHN, b Pittsburgh, Pa, June 8, 41; m 63; c 3. CARDIOVASCULAR DISEASE. *Educ:* Univ Pittsburgh, BS, 62; NJ Col Med, MD, 66; Am Bd Internal Med, dipl, 71, 73. *Prof Exp:* Resident internal med, Jefferson Med Col Hosp, 67-68; resident, Regional Naval Med Ctr, Philadelphia, Pa, 68-69; fel cardiopulmonary physiol & cardiovasc dis, Regional Naval Med Ctr & Jefferson Med Col Hosp, 69-71; from instr to asst prof med, Jefferson Med Col, Thomas Jefferson Univ, Philadelphia, Pa, 71-74; from asst prof to assoc prof, 74-79, PROF MED, DIV CARDIOL, COL MED, UNIV FLA, GAINESVILLE, 79-, ASSOC DIR DEPT, 82- *Concurrent Pos:* Dir, Cardiac Catherization Lab, Naval Regional Med Ctr, Philadelphia, Pa, 71-74; med dir, Cardiac Catherization Labs, Shands Hosp, Univ Fla & Vet Admin Med Ctr, Gainesville, 74-85; chief cardiol, Vet Admin Med Ctr, 78-; prof, Sch Grad Studies, Univ Fla, 79-; prin investr, Nat Heart, Lung, Blood Inst, 85- *Honors & Awards:* Bronze Award, Am Heart Asn, 83; Harold Jeghers Lectr, Univ Med & Dent NJ, 86; William L MacDonald Lectr, Queens Univ, 87; Pioneer Investr Award, Int Soc Holter Monitoring, 90. *Mem:* Asn Univ Cardiologists; Am Soc Clin Invest; fel Am Col Cardiol; Am Heart Asn; fel Am Col Physicians; Am Fedn Clin Res; AAAS; Soc Cardiac Angiography. *Res:* Author of numerous technical publications. *Mailing Add:* Dept Med Div Cardiol Univ Fla Box J-277 Gainesville FL 32610

PEPINSKY, RAYMOND, b St Paul, Minn, Jan 17, 12; m 42; c 2. CRYSTALLOGRAPHY, CHEMICAL PHYSICS. *Educ:* Univ Minn, BA, 33, MA, 34; Univ Chicago, PhD(physics), 40. *Hon Degrees:* Dr, Univ Giessen, 65. *Prof Exp:* Asst physics, Univ Chicago, 36-39, res assoc, 39; res physicist, US Rubber Co, 40-41; from instr to asst prof physics, Ala Polytech Inst, 41-45, from res assoc prof to res prof, 45-49; dir, Crystallog Res Lab, Pa State Univ, 49-63; distinguished prof chem & physics & chmn dept physics, Fla Atlantic Univ, 63-65; Robert O Law prof physics & chem, Nova Univ Advan Technol, 65-68, chmn dept physics, 67-68; vis prof, 68-70, prof physics, metall & mat eng, 70-72, PROF BIOPHYS, PHYSICS DEPT, UNIV FLA, 72- *Concurrent Pos:* Res assoc & mem staff, Radiation Lab, Mass Inst Technol, 42-45; travel grants, Rockefeller Found, Europe, 54 & NSF, Gt Brit, 60; Smith-Mundt fel, Univ Mex, 55; Guggenheim & Smith-Mundt fel, Europe, 58-59; vis prof, Swiss Fed Inst Technol, Zurich, 58-59 & Kyoto Univ, 72; dir Goth Inst, Pa State Univ, 58-63; vis prof & dir instr crystallog, Univ Marburg, Ger, 66; lectr, Japan, 69; consult, Brookhaven Nat Lab, 51-66; mem, Solid State Adv Panel, Off Naval Res, 51-60, Blood Plasma Substitutes Panel, 51-53, Comn Crystallog Apparatus, Int Union Crystallog, 57-66, Crystal Data, 60-63, Gov Bd, Am Inst Physics, 59-62, Adv Comt Data Processing, Nat Bur Standards, 59-61, Comt Use Electronic Comput in Life Sci, 61-63 & Adv Comt Solid State Physics, 60- & Comt Tech Distrib Nuclear Data, Nat Acad Sci, 63-68; secy, Nat Comt Crystallog, Nat Res Coun, 52-54. *Mem:* AAAS; fel Am Phys Soc; Am Crystallog Asn. *Res:* Physical, chemical and surface crystallography; theory of x-ray, neutron and slow electron diffraction; structural mechanisms of crystal transitions and melting; biophysics; intermolecular interactions and assembly of biosystems; biophysics of vision. *Mailing Add:* Dept Physics Univ Fla Gainesville FL 32611

PEPKOWITZ, LEONARD PAUL, b Paterson, NJ, Mar 23, 15; m 41, 69; c 5. CHEMISTRY. *Educ:* Univ Minn, BS, 39; Rutgers Univ, PhD(biochem), 43. *Prof Exp:* Asst plant physiol, Rutgers Univ, 40-43; asst chemist, Exp Sta, RI State Col, 42-43; chemist, Los Alamos Sci Lab, Univ Calif, 43-45, group leader, 45-46; res assoc, Res Lab, Gen Elec Co, 46-47, supvr gen anal unit, Knolls Atomic Power Lab, 47-53, mgr chem & metall sect, 53-56; vpres res, develop & serv, Nuclear Mat & Equip Corp, 57-64, vpres & gen mgr, Boron Isotope Separation Facility, 64-71; from asst prof to prof biol, Erie Community Col, City Campus, 71-85; RETIRED. *Mem:* AAAS; Am Chem Soc; Sigma Xi. *Res:* Isotope and chemical separations; chemistry of nuclear materials; analytical chemistry; radiochemistry; biochemistry; plant metabolism. *Mailing Add:* 929 River Rd Youngston NY 14174

PEPOY, LOUIS JOHN, b Cleveland, Ohio, Feb 8, 38; m 68; c 2. ORGANIC CHEMISTRY. *Educ:* John Carroll Univ, BS, 61, MS, 65; Wayne State Univ, PhD(org chem), 70. *Prof Exp:* Chemist analyst, Repub Steel Res Ctr, 62-65; chemist, E I du Pont de Nemours & Co, Inc, 69-71; electrocoating chemist, Glidden-Durkee Res Ctr, SCM Corp, 71-75; sr res chemist, Chemetron Corp, 75-76, group leader, 76-80, res mgr, 80-; AT BASF WYANDOTTE CORP. *Mem:* Am Chem Soc. *Res:* Organic pigment synthesis and development; relationships between physical properties, chemical structure, and physical forms of pigments; colloid science; electrodeposition of paints; low energy curing of coatings. *Mailing Add:* BASF Chem Div Corp 491 Columbia Ave Holland MI 49423

PEPPAS, NIKOLAOS ATHANASSIOU, b Athens, Greece, Aug 25, 48; m. BIOMATERIALS. *Educ:* Nat Tech Univ Athens, dipl eng, 71; Mass Inst Technol, ScD, 73. *Prof Exp:* From asst prof to assoc prof, 76-81, PROF CHEM ENG, PURDUE UNIV, 81- *Concurrent Pos:* Vis prof sci, Univ Geneva, 82-83, Univ Paris, 86, Univ Parma, 88; consult, 76-; ed, Biomaterials, 83, Polymer News, 84. *Honors & Awards:* Western Elec Award, Am Soc Eng Educ, 80; Zyma Found Award, 82; Mat Award, Am Inst Chem Engrs; Curtis McGraw Award, Am Soc Eng Educ. *Mem:* Controlled Release Soc (pres, 88); Am Phys Soc; NY Acad Sci; Am Inst Chem Engrs; Am Chem Soc; Soc Plastics Engrs. *Res:* Structure of polymers, especially its relation to diffusion of solutes through polymers and membranes; biomedical application of polymers. *Mailing Add:* Sch Chem Eng Purdue Univ W Lafayette IN 47907

PEPPER, EVAN HAROLD, b Windsor, Ont, Aug 14, 27; m; c 6. ENVIRONMENTAL BIOLOGY. *Educ:* Univ Detroit, BS, 54; Mich State Univ, MS, 58, PhD(plant path), 61. *Prof Exp:* Tech coordr, Ladish Malting Co, Wis, 60-62; asst prof plant path & plant pathologist, NDak State Univ, 62-67; assoc prof, 67-71, PROF BOT, BRANDON UNIV, 71- *Concurrent Pos:* Vchmn, Man Res Coun, 71-; mem bd dirs, Biomass Energy Inst, Inc, 72; vis prof, Univ Reading, Eng, 73-74 & Mich State Univ, 83-84. *Mem:* Sigma Xi. *Res:* Landscape plants for northern regions; renewable energy sources; history and philosophy of science. *Mailing Add:* Dept Bot Brandon Univ Brandon MB R7A 6A9 Can

PEPPER, JAMES MORLEY, b Morse, Sask, Mar 30, 20; m 45; c 4. ORGANIC CHEMISTRY. *Educ:* Univ BC, BA, 39, MA, 41; McGill Univ, PhD(chem), 43. *Prof Exp:* Res fel chem, McGill Univ, 43-44, lectr, 44-45; res chemist, Dominion Rubber Co, 45-47; assoc prof chem, 47-55, head dept chem & chem eng, 70-76, prof, 55-85, EMER PROF CHEM, UNIV SASK, 85- *Mem:* Am Chem Soc; Can Pulp & Paper Asn; Chem Inst Can. *Res:* Natural products; lignin chemistry; chemistry of phenolic substances. *Mailing Add:* Dept Chem Univ Sask Saskatoon SK S7N 0W0 Can

PEPPER, PAUL MILTON, b Kendallville, Ind, May 16, 09; m 34; c 1. MATHEMATICS. *Educ:* Ind Univ, AB, 31, AM, 32; Univ Cincinnati, PhD(math), 37. *Prof Exp:* Instr math, Univ Cincinnati, 37-38; from instr to assoc prof, Univ Notre Dame, 38-49; assoc prof indust eng, 49-51, asst to dir res found, 51-53, dir mapping & chart res lab, 53-57, assoc prof math, 53-59, from assoc prof to prof indust eng, 56-74, asst to dir res found, 57-67, assoc dir develop, 67-74, EMER PROF INDUST ENG, OHIO STATE UNIV, 74- *Concurrent Pos:* Fel, Harvard Univ, 47; res scientist, Automet Corp, 59-63; consult, Tech Inc & Chem Abstr, 64-69 & DBA Systs, Inc, 67-70. *Mem:* Fel AAAS; Am Math Soc; Soc Indust & Appl Math; Am Astron Soc; Soc Mfg Engrs. *Res:* Geometry of numbers; abstract metric geometry; potential theory; design of slide rules and nomographs; applied mathematics; photogrammetry; celestial mechanics; geodesy; finite projective planes, by high speed computers. *Mailing Add:* 517 E Schreyer Pl Columbus OH 43214-2254

PEPPER, ROLLIN E, b Glens Falls, NY, June 8, 24; m 53; c 3. MICROBIOLOGY. *Educ:* Earlham Col, BA, 50; Syracuse Univ, MS, 53; Mich State Univ, PhD(microbiol), 63. *Prof Exp:* Assoc scientist, Ethicon, Inc, 51-60; res asst, Mich State Univ, 60-63, res assoc, 63-64; from asst prof to assoc prof, Elizabethtown Col, 64-68, chmn dept, 67-77, prof biol, 68-90; RETIRED. *Concurrent Pos:* Vis prof, Univ Zambia, 72-73; indust consult microbiol. *Mem:* Sigma Xi. *Res:* Radiation and chemical sterilization; bacterial metabolism; food microbiology. *Mailing Add:* 420 N Mt Joy St Elizabethtown PA 17022

PEPPER, THOMAS PETER, b London, Eng, Jan 26, 18; nat Can; m 53; c 3. NUCLEAR PHYSICS. *Educ:* Univ BC, BA, 39, MA, 41; McGill Univ, PhD(nuclear physics), 48. *Prof Exp:* Sr res asst, Nat Res Coun Can, 41-45, asst res officer, 47-52; with Isotope Prod, Ltd & pres, Isotope Prod, Inc, Curtiss-Wright Corp, 52-57; head physics div, 58-67, asst dir, 67-72, dir, 72-80, EXT DIR, SASK RES COUN, 80- *Mem:* Can Asn Physicists; Can Res Mgt Asn. *Res:* Application of radioactivity; operations research. *Mailing Add:* Site 12 C20 RR 2 Nanaimo BC V9R 5K2 Can

PEPPER, WILLIAM DONALD, b Dadeville, Ala, Apr 29, 35; m 68; c 1. FOREST BIOMETRY. *Educ:* Auburn Univ, BS, 59; NC State Univ, MF, 65, PhD(forestry), 75. *Prof Exp:* Res forester, Southeastern Forest Exp Sta,US Forest Serv, 65-66, biometrician, Southern Forest Exp Sta, 77-84; STA BIOMETRICIAN, SOUTHERN FOREST EXP STA, USDA FOREST & SERV, 84- *Mem:* Am Statist Asn; Biomet Soc. *Res:* Design and analysis of studies on biological topics, particularly forest resources; forecasting timber growth and yield. *Mailing Add:* Forestry sci Lab Carlton Green St Athens GA 30602

PEPPERBERG, DAVID ROY, b Chicago, Ill, Nov 10, 44; m 70. NEUROBIOLOGY, BIOPHYSICS. *Educ:* Mass Inst Technol, SB, 66, PhD(biophys), 73. *Prof Exp:* Res fel biol, Harvard Univ, 73-76; asst prof biol sci, Purdue Univ, 77-; AT COL MED OPHTHAL, UNIV ILL. *Concurrent Pos:* Res fel, NIH, 74-76, Fight for Sight, Inc res fel, 76. *Mem:* Asn Res Vision & Ophthal; Am Chem Soc; Soc Gen Physiologists; Biophys Soc. *Res:* Light and dark adaptation of vertebrate photoreceptors; relationship of adaptational processes to the bleaching and regeneration of visual pigment; interconversion of retinoids in the vertebrate eye. *Mailing Add:* 127 Prairie Ave Wilmette IL 60091

PEPPERMAN, ARMAND BENNETT, JR, b New Orleans, La, May 30, 41; div; c 3. ORGANIC CHEMISTRY. *Educ:* La State Univ, New Orleans, BS, 63, PhD(org chem), 73. *Prof Exp:* Chemist, Oilseed Crops Lab, 63-68, res chemist, Spec Finishes Invests, 68-72 & Polymer Finishes Res, 72-74, res chemist, Cotton Textile Chem Lab, Polymer Finishes Res, Southern Regional Res Ctr, Agr Res Serv, USDA, 74-83; res chemist, Crop Protection Res, 83-87, compos & properties res, 87-88, RES LEADER, COMPOS & PROPERTIES RES, USDA, 88- *Concurrent Pos:* Instr, chem & organic chem, 82-89. *Mem:* Am Chem Soc; Sigma Xi; Weed Sci Soc Am. *Res:* Organophosphorus chemistry; flame retardants for cellulosic textiles; nuclear magnetic resonance spectroscopy; synthesis of plant growth regulators; controlled release formulations. *Mailing Add:* Compos & Properties Res 1100 Robert E Lee Blvd New Orleans LA 70124

PEPPERS, RUSSEL A, b Belleville, Ill, Feb 23, 32; m 66; c 3. GEOLOGY, PALYNOLOGY. *Educ:* Univ Ill, BS, 57, MS, 59, PhD(geol, bot), 61. *Prof Exp:* Geologist, Shell Oil Co, 61-63; from asst gelogist to assoc geologist, 63-78, GEOLOGIST, ILL STATE GEOL SURV, 78- *Mem:* AAAS; Geol Soc Am; Sigma Xi; Am Asn Stratigraphic Palynologists; Soc Econ Paleontologists & Mineralogists. *Res:* Palynology of Paleozoic strata; coal geology; stratigraphy. *Mailing Add:* Ill State Geol Surv Champaign IL 61820

PEPPIN, RICHARD J, b Brooklyn, NY, Feb 18, 43; div. ARCHITECTURAL ACOUSTICS, NOISE CONTROL ENGINEERING. *Educ:* City Univ New York, BE, 65; WVa Univ, MS, 66; Rensselaer Polytech Inst, MS, 69. *Prof Exp:* Jr proj engr, Bendix Corp, 65; sr struct analyst, Pratt & Whitney Aircraft, 66-69; consult, Donley Miller & Nowikas, 72-73; sr consult, Kodaras Acoust Labs, 73-75; sr scientist, Sci Applns, Inc, 75-76; county noise control engr, Montgomery County, Md, 76-77; proj dir, Jack Faucett Assoc, 78-79; sr scientist, Dept Labor, Occup Safety & Health Admin, 79-81; actg head safety, Nat Park Serv, 81-82; regional applns engr, Bruel & Kjaer Instruments, 82-83, head acoust & environ groups, 83-85; PRIN, WHITE MOUNTAIN RES, 81-; PRES & CHIEF EXEC OFFICER, SCANTEK, INC, 85- *Concurrent Pos:* Consult, Nat Bur Standards, 82- *Honors & Awards:* Centennial Medal, Am Soc Mech Engrs, 80. *Mem:* Inst Noise Control Engrs (secy, 80-85, vpres, 85-); Am Soc Testing & Mat; Am Soc Mech Engrs; Inst Environ Sci; Acoust Soc Am; Am Soc Heating, Refrigeration & Airconditioning Engrs; Soc Automotive Engrs. *Res:* Environmental, industrial and architectural acoustics; measurement of sound intensity. *Mailing Add:* 5012 Macon Rd Rockville MD 20852

PEPPLER, HENRY JAMES, b Hussenbach, Russia, Nov 29, 11; nat US; m 39; c 2. BIOCHEMISTRY, FOOD FERMENTATIONS. *Educ:* Univ Wis, BS, 36, MS, 38, PhD(bact), 39. *Prof Exp:* Asst, Univ Wis, 36-39; instr bact, Kans State Col, 39-42; lab officer, US War Dept, 42-46; res microbiologist, Res Lab, Carnation Co, 46-51; dir biochem res, Red Star Yeast & Prod Co, 51-54, mgr res, 54-60; mgr res, Universal Foods Corp, 60-66, dir res & develop, 66-72, dir sci affairs, 72-76; RETIRED. *Concurrent Pos:* Consult, 76-91. *Honors & Awards:* Am Chem Soc Award, 66. *Mem:* Am Chem Soc; Am Soc Microbiol; Inst Food Technologists; Acad Am Microbiologists. *Res:* Microbial physiology; zymology; yeast products; food and dairy products. *Mailing Add:* 5157 N Shoreland Ave Whitefish Bay WI 53217-5542

PEPPLER, RICHARD DOUGLAS, b Trenton, NJ, Sept 24, 43; m 68. ANATOMY. *Educ:* Gettysburg Col, BA, 65; Univ Kans, PhD(anat), 69. *Prof Exp:* Asst prof anat & obstet-gynec, La State Univ, New Orleans, 70-74, assoc prof, 74-78; assoc prof, Col Med, East Tenn State Univ, 78-80, prof anat & obstet-gynec, 80-85, assoc chmn anat, 80-85; ASSOC DEAN ACAD & FAC AFFAIRS & PROF ANAT & NEUROBIOL, COL MED, UNIV TENN, MEMPHIS, 85- *Mem:* AAAS; Am Asn Anatomists; Soc Study Reproduction; Endocrine Soc; Soc Study Fertility; Sigma Xi. *Res:* Pituitary-ovarian relationships to include specifically follicular development and control of ovulation; ovarian dynamics in mammals; reproductive physiology. *Mailing Add:* Ctr Health Sci Col Med Univ Tenn 3 N Dunlap Memphis TN 38163

PEQUEGNAT, LINDA LEE HAITHCOCK, b Bedford, Ind, Oct 27, 31; m 57; c 2. BIOLOGICAL OCEANOGRAPHY, TAXONOMY. *Educ:* Pomona Col, BA, 53; Univ Calif, San Diego, MS, 57; Tex A&M Univ, PhD(oceanog), 70. *Prof Exp:* Lab asst zool, Pomona Col, 51-53; lab technician, Univ Tex, 53-54; res asst biol oceanog, Scripps Inst Oceanog, 54-58; res scientist oceanog, Tex A&M Univ, 64-81; res scientist, Tereco Corp, 70-83; INDEPENDANT CONSULT, 83- *Concurrent Pos:* Cur systematic collection marine organisms, 73-81. *Mem:* AAAS; Am Inst Biol Sci; Am Soc Limnol & Oceanog; Soc Syst Zool; Am Soc Zoologists; Crustacean Soc. *Res:* Taxonomy and systematics of deep-sea crustacea (Mysidacea and Decapoda); coral reef Decapoda; bio-fouling studies. *Mailing Add:* 8463 Paseo del Ocaso La Jolla CA 92037

PEQUEGNAT, WILLIS EUGENE, b Riverside, Calif, Sept 18, 14; m 37, 57; c 4. OCEANOGRAPHY. *Educ:* Univ Calif, BA, 36; Univ Calif, Los Angeles, MA, 38, PhD(zool), 42. *Prof Exp:* From instr to prof zool, Pomona Col, 40-59; assoc prog dir, NSF, 60-61, prog dir, 61-63; head dept, Tex A&M Univ, 64-66, prof oceanog, 63-80; mem staff, Tereco Corp, 80-84; CONSULT, OFF TECHNOL ASSESSMENT, INT ASN PORTS & HARBORS, 81-; CONSULT, LGL ECOL RES ASSOCS, INC, 81- *Concurrent Pos:* Ranger naturalist, Nat Park Serv, US Dept Interior, 39; vis asst prof, Univ Chicago, 47; vis assoc prof, Univ Calif, 51, res assoc, Scripps Inst, 54-55; Ford fel, 54; NSF sci fac fel, 58-59; consult, NSF, 63-; pres, Tereco Corp, 69-; vis prof, Univ Aberdeen, 81-83. *Mem:* Fel AAAS; Am Soc Limnol & Oceanog; Ecol Soc Am; Marine Biol Asn UK. *Res:* Biological oceanography; ecology of the deep sea; ecology of epifauna of the continental shelf; taxonomy of Brachyura and Galatheidae. *Mailing Add:* 8463 Paseo del Ocaso La Jolla CA 92037

PERA, JOHN DOMINIC, b Memphis, Tenn, Oct 5, 22; m 44; c 3. MICROBIOLOGY. *Educ:* Rhodes Col, BS, 49; Ind Univ, PhD(org chem), 60. *Prof Exp:* Chemist, Cent Labs, Inc, 42-43; chemist, Buckman Labs Inc, 46-52, prod mgr, 52-53, chief chemist 53-58, vpres res, 58-71, vpres res & develop, 71-88; RETIRED. *Mem:* Am Chem Soc; Am Inst Chemists; Tech Asn Pulp & Paper Indust; Royal Soc Chem. *Res:* Non pressure preservation of wood with pentachlorophenol; formulation of chemical specialty products; synthesis of organic and barium compounds used in industrial microorganism control; organic sulfur chemistry; organic bromine chemistry; polymeric quaternary ammonium compounds. *Mailing Add:* 8838 Eatonwick Cordova TN 38018

PERACCHIA, CAMILLO, b Milan, Italy, Mar 31, 38; m 67; c 3. CELL BIOLOGY, CELL COMMUNICATION. *Educ:* Univ Milan, Italy, MD, 62. *Prof Exp:* PROF PHYSIOL, SCH MED, UNIV ROCHESTER, 83- *Mem:* Biophys Soc; Am Soc Cell Biol; AAAS; Fedn Am Soc Exp Biol; Asn Res Vision & Ophthal. *Res:* Membrane structure and function; regulation of direct cell to cell communication via gap junction channels which are studied electrophysiologically in two cell and reconstituted systems, biochemically and morphologically. *Mailing Add:* Dept Physiol Sch Med Univ Rochester 601 Elmwood Ave Rochester NY 14642

PERACCHIO, ALDO ANTHONY, b Brooklyn, NY, Feb 25, 35; m 59; c 2. ENGINEERING, SPACE SCIENCES. *Educ:* City Col NY, BME, 56; Rensselaer Polytech Inst, MME, 59, PhD(mech eng), 68. *Prof Exp:* Anal engr, Hamilton Standard Div, United Aircraft Corp, 56-61, sr anal engr, 61-69; supvr aeroacoust group, United Technol Res Lab, 69-73; head acoust technol, 73-89, SUPVR COMBUSTION GROUP, PRATT & WHITNEY AIRCRAFT, 89- *Concurrent Pos:* Adj assoc prof, Rensselaer Polytech Inst. *Mem:* Am Inst Aeronaut & Astronaut. *Res:* Fluid mechanics of turbomachinery and nozzles; kinetics theory of gases applied to fluid mechanics problems; prediction of propeller flow field and aerodynamic performance; experimental and theoretical acoustics research on rotating machinery. *Mailing Add:* 102 Fairview Dr South Windsor CT 06074

PERAINO, CARL, b Passaic, NJ, Mar 10, 35; m 58; c 2. BIOCHEMISTRY, ONCOLOGY. *Educ:* Lebanon Valley Col, BS, 57; Univ Wis, MS, 59, PhD(biochem), 61. *Prof Exp:* Nat Cancer Inst fel oncol, 61-64; instr, McArdle Lab, Univ Wis, 64-65; asst biochemist, Argonne Nat Lab, 65-70, biochemist, 70-78, sr biochemist, 78-88; RETIRED. *Mem:* AAAS; Am Soc Biol Chemists; Am Asn Cancer Res. *Res:* Enzyme regulation; hepatocarcinogenesis; enzymology. *Mailing Add:* 4632 Seeley Ave Downers Grove IL 60515

PERALTA, PAULINE HUNTINGTON, virology, for more information see previous edition

PERALTA, RICHARD CARL, b Enid, Okla, Nov 8, 49; m 72; c 4. AGRICULTURE. *Educ:* Univ S Carolina, BS, 71; Utah State Univ, MSAE, 77; Okla State Univ, PhD(agr eng), 79. *Prof Exp:* From asst to assoc prof groundwater modeling & mgt drainage & exp anal, Univ Ark, 80-88; ASSOC PROF GROUNDWATER, MODELING, CONJUNCTIVE WATER MGT, UTAH STATE UNIV, 88- *Concurrent Pos:* Prin Investr, Univ Ark, 80-88, bioenvironmental eng, Air Force Eng & Serv Ctr, Tyndall AFB, 82-, hydrologist, US Geol Surv, Little Rock, Ark, 85-87, Air Force Eng & Serv Ctr, 86, consult, Mid-Am Int Agr Consort Proj, 87-88; Halliburton Found Award for Excellence Res, 84-85 & 87. *Mem:* Am Soc Agr Eng; Am Soc Civil Engrs; Am Soc Water Resources Asn; Am Soc Geophys Union; Sigma Xi. *Res:* Development and application of computer models that optimize groundwater management to assure sustainable agricultural production; maximizing crop production; water utilization; management of groundwater contamination; remedial action. *Mailing Add:* Dept Agr & Irrigation Eng Utah State Univ Logan UT 84322-4105

PERCARPIO, EDWARD P(ETER), b Paterson, NJ, Mar 8, 34; m 56; c 3. MECHANICAL ENGINEERING, RUBBER TECHNOLOGY. *Educ:* Newark Col Eng, BS, 55, MS, 58. *Prof Exp:* Engr, Curtiss-Wright Res Div, 55-58; mech engr, Picatinny Arsenal, 58-60; res engr rubber technol, Uniroyal Res Ctr, Middlebury, Conn, 60-74; res engr, 74-80, SR RES ASSOC & DEPT MGR, BECTON-DICKINSON, 80- *Concurrent Pos:* Adj fac, Community Col Morris, 74- *Mem:* Am Soc Mech Engrs; Am Chem Soc. *Res:* Automobile tire research; skid resistance of tires; rubber friction; rubber technology; sealing; medical product design. *Mailing Add:* 36 Boat St North Haledon NJ 07508

PERCEC, VIRGIL, b Siret, Romania, Dec 8, 46; US citizen; m 68; c 1. NEW POLYMERIZATION REACTIONS & MECHANISMS, SELF ASSEMBLED POLYMER SYSTEMS. *Educ:* Polytech Inst, Lassy, Romania, MSC,69,Inst Macromolecular Chem, Lassy, Romania, PhD(org & polymer chem), 76. *Prof Exp:* From asst prof to assoc prof chem, dept org & macromolecular chem, Polytech Inst, Jassy, Romania, 69-81; from asst prof to assoc prof, 82-86, PROF ORG & POLYMER CHEM, DEPT MACROMOLECULAR SCI, CASE WESTERN RESERVE UNIV, OHIO, 86- *Concurrent Pos:* Res assoc org & polymer chem, Inst Macromolecular Chem, Jassy, Romania, 69-80, sr res assoc, 80-81; vis prof, Inst Macromolecular Chem, Univ Freiburg, W Ger, 81, 84, 85 & 88; consult, 81-; vis scientist, Univ Akron, 81-82. *Mem:* Am Chem Soc; AAAS; NY Acad Sci; Int Union Pure & Appl Chem. *Res:* Polymer synthesis and modification with main emphasis on elucidation of polymerization mechanism of acetylenes by metathesis catalysts; studies of the reactivity of macromonomers; development of new polymerization methods by using phase transfer catalyzed, single electron transfer and ion radical and ionic reaction; special topics in copolymeration, design of novel self-organized polymer systems including liquid crystalline polymers systems; donor acceptor complex polymers. *Mailing Add:* Dept Polymer Sci Case Western Reserve Univ University Circle Cleveland OH 44106-2699

PERCHONOCK, CARL DAVID, b Philadelphia, Pa, Mar 30, 46; m 73; c 2. DEVELOPMENT, CORPORATE RELATIONS. *Educ:* Univ Pa, BA, 67; Univ Wis, PhD(org chem), 72. *Prof Exp:* Assoc sr investr, Smith Kline & French Labs, 72-79, sr investr med chem, 79-87; prin investr, Starks CP, 87-89; DIR, CORP RELS, FOX CHASE CANCER CTR, 90- *Concurrent Pos:* NIH fel, Columbia Univ, 72. *Mem:* Am Chem Soc; AAAS. *Res:* Synthesis of biologically active compounds; total synthesis of beta-lactam antibiotics; design and synthesis of leukotriene receptor antagonists and biosynthsis inhibitors. *Mailing Add:* 421 Wadsworth Ave Philadelphia PA 19119-1130

PERCICH, JAMES ANGELO, b Detroit, Mich, Sept 14, 44; m 82. CROP PEST MANAGEMENT, CHEMICAL CONTROL. *Educ:* Mich State Univ, BS, 67, MS, 71, PhD(plant path), 75. *Prof Exp:* Fel, Univ Wis, 74-76; fel, 76-77, ASST PROF CHEM CONTROL PLANT DIS, DEPT PLANT PATH, UNIV MINN, 77- *Mem:* Am Soc Agron; AAAS; Can Phytopath Soc; Am Phytopath Soc. *Res:* Plant disease control of fungal pathagens of wild rice (zizania aquatica) and sugar beet; various fungicides and adjudcent materials to increase disease control at lower dosages and numbers of applications. *Mailing Add:* Dept Plant Path Univ Minn St Paul MN 55108

PERCIVAL, DOUGLAS FRANKLIN, b Ridgetown, Ont, Jan 8, 26; nat US; m 48; c 5. ORGANIC POLYMER CHEMISTRY. *Educ:* Pomona Col, BA, 50; Mich State Univ, MS, 52, PhD, 55. *Prof Exp:* Res chemist natural prod, Mich State Univ, 52-54; res chemist petrochem, Calif Res Corp, Standard Oil Co, Calif, 55-64; res chemist petrochem, Flexible Packaging Div, Crown Zellerbach Corp, 64-75, dir packaging res & develop, 75-82; CONSULT, 82- *Mem:* Am Chem Soc. *Res:* Polyolefins and copolymers; extrusion coating and laminating. *Mailing Add:* 2475 Harborview Dr San Leandro CA 94577

PERCIVAL, FRANK WILLIAM, b Los Angeles, Calif, June 10, 48; m 69; c 2. PLANT DEVELOPMENT. *Educ:* Occidental Col, BA, 69; Univ Calif, Santa Barbara, PhD(biol sci), 73. *Prof Exp:* Fel plant physiol, Dept Bot & Plant Path, Mich State Univ, 73-75; from asst prof to assoc prof biol, 75-87, PROF BIOL, WESTMONT COL, 87- *Mem:* Am Soc Plant Physiologists; Sigma Xi. *Res:* Regulation of hormone levels in higher plants, focusing on the control of indole-3-acetic acid metabolism; molecular biology of fruit ripening. *Mailing Add:* Dept of Biol Westmont Col 955 La Paz Rd Santa Barbara CA 93108

PERCIVAL, JOHN A, b North Bay, Ont, July 16, 52; m 76. GEOLOGY. *Educ:* Concordia Univ, BSc, 76; Queen's Univ, MSc, 78, PhD(geol), 81. *Prof Exp:* GEOLOGIST, CONTINENTAL GEOSCI DIV, GEOL SURV CAN, 81- *Honors & Awards:* H S Robinson Distinguished Lectr, Geol Asn Can. *Mem:* Fel Geol Asn Can; fel Geol Soc Am; Mineral Asn Can; Mineral Soc Am. *Res:* Field geology; metamorphic petrology; crustal structure and tectonics; deep continental crust. *Mailing Add:* Geol Surv Can 601 Booth St Ottawa ON K1A 0E8 Can

PERCIVAL, WILLIAM COLONY, b Greencastle, Ind, Sept 26, 24; m 50; c 2. ORGANIC CHEMISTRY. *Educ:* Middlebury Col, BA, 46; Pa State Col, MS, 49, PhD(chem), 51. *Prof Exp:* Chemist, NJ, 45-47 & Del, 51-59, qual control supvr, Elastomers Area, Chamber Works, 50-69, div head, 69-80, sr supvr, 80-82, CONSULT REGULATORY AFFAIRS, E I DU PONT DE NEMOURS & CO, INC, 82- *Mem:* Am Chem Soc. *Res:* Addition reactions of Grignard reagents; formaldehyde chemistry; organic analytical chemistry; gas chromatography; spectrophotometric methods. *Mailing Add:* 216 Wellington Rd Fairfax Wilmington DE 19803

PERCUS, JEROME K, b New York, NY, June 21, 26; m 65; c 2. CHEMICAL PHYSICS, MATHEMATICAL BIOLOGY. *Educ:* Columbia Univ, BS, 47, MA, 48, PhD(physics), 54. *Prof Exp:* Instr elec eng, Columbia Univ, 52-54; res assoc physics, Courant Inst Math Sci, NY Univ, 54-55; asst prof, Stevens Inst Technol, 55-58; assoc prof, 58-66, PROF PHYSICS, COURANT INST MATH SCI, NY UNIV, 66- *Honors & Awards:* Pregel Award, 75. *Mem:* Am Math Soc; Am Phys Soc. *Res:* Theory of many-body systems. *Mailing Add:* Courant Inst Math Sci NY Univ 251 Mercer St New York NY 10012

PERCY, JOHN REES, b Windsor, Eng, July 10, 41; Can citizen; m 62; c 1. ASTRONOMY, SCIENCE EDUCATION. *Educ:* Univ Toronto, BSc, 62, MA, 63, PhD(astron), 68. *Prof Exp:* From asst prof to assoc prof, 67-78, PROF ASTRON, ERINDALE COL, UNIV TORONTO, 78-, ASSOC DEAN SCI & VPRIN RES & GRAD STUDIES, 89- *Concurrent Pos:* Leverhulme fel, Cambridge Univ, 72-73. *Honors & Awards:* Royal Jubilee Medal, 77. *Mem:* Int Astron Union; Am Astron Soc; Can Astron Soc; Royal Astron Soc Can (pres, 78-79); Royal Can Inst (pres, 85-86); Am Asn Variable Star Observers (pres, 89-91). *Res:* Theory and observation of variable stars; astronomy education. *Mailing Add:* Dept of Astron Erindale Col Univ of Toronto Mississauga ON L5L 1C6 Can

PERCY, JOHN SMITH, b South Shields, Eng, Aug 8, 38; m 61; c 3. MEDICINE. *Educ:* Univ Durham, MB & BS, 61; Univ Newcastle, Eng, MD, 66; FRCP(C), 68; FRCP(E), 73, FRCP, 79. *Prof Exp:* Fel med, Med Ctr, Univ Colo, 66-68; asst prof, 68-70, assoc prof, 70-77, PROF MED, UNIV ALTA, 77-, DIR, RHEUMATIC DIS UNIT, 68- *Concurrent Pos:* Can Arthritis & Rheumatism Soc fel, Univ Alta, 68-; examr, Royal Col Physicians & Surgeons Can, 71- *Mem:* Am Rheumatism Asn; Can Rheumatism Asn; Can Soc Immunologists. *Res:* Immunopathology of connective tissue diseases. *Mailing Add:* Dept Med Rheumat Univ Alta Fac Med Edmonton AB T6G 2G3 Can

PERCY, JONATHAN ARTHUR, b Penrootyn, Wales, Jan 14, 43; Can citizen; m 68; c 2. MARINE ECOPHYSIOLOGY. *Educ:* Carleton Univ, BSc, 65; Mem Univ Nfld, MSc, 68, PhD(marine biol), 71. *Prof Exp:* Fel, Inst Arctic Biol, Univ Alaska, 71-72; res scientist physiol, 72-86, SECT HEAD BIOL OCEANOG, ARCTIC BIOL STA, FISHERIES & OCEANS, CAN, 86- *Mem:* Arctic Inst NAm; Can Soc Zoologists; Am Soc Limnol & Oceanog; Can Arctic Resources Comt. *Res:* Physiology and ecology of Arctic marine invertebrates (amphipods) and sublethal physiological effects of oil pollution. *Mailing Add:* Arctic Biol Sta Fisheries & Oceans Can 555 Boul St-Pierre Ste Anne de Bellevue PQ H9X 3R4 Can

PERCY, MAIRE EDE, b Toronto, Ont, Dec 19, 39; m 62; c 1. HUMAN GENETICS. *Educ:* Univ Toronto, BSc, 62, MA, 64, PhD(biochem), 72; Royal Conserv Music, ARCT, 67. *Prof Exp:* Teaching fel, Agr Res Coun, Cambridge, UK, 72-73; teaching fel immunol, Hosp Sick Children, Toronto, Ont, 73-75, staff scientist genetics, 75-82; STAFF SCIENTIST OBSTET & GYNEC, MT SINAI HOSP, TORONTO, ONT, 82-; NEUROGENETICIST, SURREY PLACE CTR, TORONTO, 89- *Concurrent Pos:* Nat health res scholar, obstet & gynec, Univ Toronto, 84-89, asst prof, 85- *Mem:* Am Soc Human Genetics; Can Soc Immunol; Can Asn Women Sci. *Res:* Risk factors in human disease; application of techniques of biochemistry, immunology, genetics, study of genetic diseases; muscular dystrophy, hemophilia, Alzheimer's disease, cervical cancer. *Mailing Add:* Surrey Place Ctr Two Surrey Pl Toronto ON M5S 2C2 Can

PERDEW, JOHN PAUL, b Cumberland, Md, Aug 30, 43. DENSITY FUNCTIONAL THEORY. *Educ:* Gettysburg Col, AB, 65; Cornell Univ, PhD(physics), 71. *Prof Exp:* Fel physics, Univ Toronto, 71-74 & Rutgers Univ, 74-77; from asst prof to assoc prof, 77-82, PROF PHYSICS, TULANE UNIV, 82- *Concurrent Pos:* NSF grants, 78-80, 80-85, 85-88 & 88- *Mem:* Am Phys Soc. *Res:* Theoretical condensed matter physics; fundamentals, approximations and applications of the density functional theory of atoms, molecules, solids and surfaces. *Mailing Add:* Dept Physics Tulane Univ New Orleans LA 70118

PERDRISAT, CHARLES F, b Geneva, Switz, July 1, 32; m 62. NUCLEAR PHYSICS, HIGH ENERGY PHYSICS. *Educ:* Univ Geneva, MSc, 56; Swiss Fed Inst Technol, PhD(physics), 61. *Prof Exp:* Res asst, Swiss Fed Inst Technol, 61-62 & Inst Theoret Nuclear Physics, Univ Bonn, 62-63; res asst, Univ Ill, 63-65, res asst prof physics, 65-66; from asst prof, to assoc prof, 66-76, PROF PHYSICS, COL WILLIAM & MARY, 76- *Mem:* Swiss Phys Soc; Am Phys Soc; Europ Phys Soc. *Res:* Nuclear spectroscopy; properties of elementary particles; nuclear reactions induced by intermediate energy particles. *Mailing Add:* Dept of Physics Col of William & Mary Williamsburg VA 23185

PERDUE, EDWARD MICHAEL, b Atlanta, Ga, May 21, 47; m 67; c 3. ENVIRONMENTAL CHEMISTRY, ORGANIC GEOCHEMISTRY. *Educ:* Ga Inst Technol, BS, 69, PhD(chem), 73. *Prof Exp:* Asst prof environ sci & chem, Portland State Univ, 73-77, assoc prof, 77-82; assoc prof, Sch Geophys Sci, 83-88, PROF GEOCHEM, SCH EARTH & ATMOSPHERIC SCI, GA INST TECHNOL, 89- *Mem:* Am Chem Soc; Geochem Soc. *Res:* Thermodynamics of acid-base equilibria; chemical characterization of naturally-occurring polyelectrolytes, such as fulvic and humic acids, in soils and waters. *Mailing Add:* Sch Earth & Atmospheric Sci Ga Inst Technol Atlanta GA 30332

PERDUE, JAMES F, b Chicago, Ill, May 5, 33; m 57; c 3. BIOCHEMISTRY, ENDOCRINOLOGY. *Educ:* Wis State Univ, Oshkosh, BS, 58; Univ Wis-Madison, MS, 61, PhD(zool), 63. *Prof Exp:* Trainee, Inst Enzyme Res, Med Sch, Univ Wis-Madison, 63-65, asst prof oncol, 66-75; staff scientist, Lady Davis Inst, Jewish Gen Hosp, 75-84; mem, Dept Cell Biol, Revlon Biotechnol Res Ctr, 84-86; SR SCIENTIST, AM RED CROSS, 87- *Concurrent Pos:* Assoc mem, Depts Exp Med & Biochem & Molecular Cytol Study Sect, McGill Univ, 77-81. *Mem:* Am Soc Biol Chem; Can Biochem Soc; AAAS; Am Asn Cancer Res; Am Soc Cell Biol; Am Soc Endocrinol. *Res:* Isolation and characterization of insulin-like growth factor II and the receptors which bind this growth factor; regulation of normal and transformed cell replication by growth factors and proteases. *Mailing Add:* Am Red Cross 15601 Crabbs Branch Way Rockville MD 20855

PERDUE, ROBERT EDWARD, JR, b Norfolk, Va, Oct 18, 24. ECONOMIC BOTANY. *Educ:* Univ Md, BS, 49; Harvard Univ, MA, 51, PhD, 57. *Prof Exp:* Botanist, US Geol Surv, 51-53 & Tex Res Found, 53-56; botanist, Beltsville Agr Res Ctr, Agr Res Serv, USDA, 57-89; CONSULT, 90- *Mem:* AAAS; Soc Econ Bot; Am Soc Pharmacognosy; Am Inst Biol Sci; Sigma Xi. *Res:* Vernonia, new crop development. *Mailing Add:* 11000 Waycroft Way Rockville MD 20852

PEREIRA, GERARD P, b Port-au-Prince, Haiti, Mar 22, 37; Can citizen; m 61; c 2. ANATOMY, HISTOLOGY. *Educ:* Columbia Univ, BS, 61; McGill Univ, MS, 65, PhD(anat), 68. *Prof Exp:* Asst prof anat, Fac Med, Laval Univ, 68-70, Col Physicians & Surgeons, Columbia Univ, 70-77; mem staff, Dept Anat, Univ Nebr Med Ctr, Omaha, 77-81. *Concurrent Pos:* Lectr, Fac Med, Sherbrooke Univ, 66-68; Que Med Res Coun grant, Fac Med, Laval Univ, 69-70; NIH res grant, Col Physicians & Surgeons, Columbia Univ, 71- *Mem:* Can Soc Immunologists; Am Asn Anatomists; Am Soc Cell Biol; Electron Micros Soc Am; NY Acad Sci. *Res:* Immunocytology in germ-free animals; cytology in dietary deficiencies; transmission and scanning electron microscopy coupled with x-ray analysis of cellular components. *Mailing Add:* Ctr Med Educ McGill Univ 3655 Drummond St Rm 529 Montreal PQ H3G 1Y6 Can

PEREIRA, JOSEPH, b Brooklyn, NY, Nov 1, 28; m 55; c 3. BIOCHEMISTRY. *Educ:* City Col New York, BS, 51; Univ Conn, MS, 55, PhD(bact, biochem), 58. *Prof Exp:* Res assoc bact, Univ Conn, 53-57; microbiologist & biochemist, 58-68, proj leader metab dis, 68-71, mgr biochem pharmacol, 71-75, MGR GASTROINTESTINAL, IMMUNOL & VIROL RES, PFIZER, INC, 75- *Res:* Microbial metabolism; drugs affecting gastric secretion; biochemistry of drugs influencing lipid and carbohydrate metabolism. *Mailing Add:* 12 Rosemary Lane Quaker Hill CT 06375

PEREIRA, MARTIN RODRIGUES, b Hilversum, Neth, June 7, 20; US citizen; m 60; c 3. ENVIRONMENTAL PHYSIOLOGY. *Educ:* Fla State Univ, BSc, 60; Rutgers Univ, MSc, 61; St Thomas Inst, PhD(biol, exp med), 66. *Prof Exp:* Biologist, Foreign Technol Div, US Air Force Systs Command, Ohio, 66-67; res scientist, Travelers Res Ctr, Inc, 67-69; physiologist, Grumman Aerospace Corp, 69-70; sr microbiologist, New York City Dept Air Resources, 70-74; environ scientist, Gibbs & Hill, Inc, 74-80, sr environ scientist, 80-83; RETIRED. *Concurrent Pos:* Res scientist, Murry & Leonie Guggenheim Inst Dent Res, NY Univ, 71-82. *Res:* Influence of physical environment on man; aeromicrobiology; bioclimatology; microclimatology; synoptic meteorology; magnetobiology; environmental/occupational health. *Mailing Add:* 19 Michael Rd Syosset NY 11791

PEREIRA, MICHAEL ALAN, b New York, NY, May 3, 44; m 67; c 2. BIOCHEMICAL PHARMACOLOGY. *Educ:* Ohio State Univ, BA, 67, PhD(pharmacol), 71. *Prof Exp:* Damon Runyon Mem Fund Cancer Res fel cellular physiol, Nat Heart & Lung Inst, 71-73; res assoc, Gtr New York Blood Ctr, 73-74; res scientist, Dept Environ Med, Med Ctr, NY Univ, 74-78; pharmacologist, Health Effects Res Lab, Us Environ Protection Agency, 78-86; VPRES, TOXICOL ENVIRON HEALTH RES & TESTING, INC, 86- *Mem:* Am Asn Cancer Res, Am Chem Soc; Soc Toxicol; Am Soc Pharmacol & Exp Therapeut; Environ Mutagen Soc; Am Col Toxicol. *Res:* DNA replication in Escherichia coli; chemical carcinogenesis and mutagenesis; tumor promotion and co-carcinogenesis including studies in laboratory animals and cell culture; dose response and interspecties extrapolation of chemical carcinogens; relationship of DNA binding and damage to tumor initiation. *Mailing Add:* 11520 Olde Gate Dr Cincinnati OH 45246

PEREIRA, NINO RODRIGUES, b Amsterdam, Holland, Mar 11, 45; m 77; c 3. NUCLEAR WEAPON SIMULATORS, PLASMA RADIATORS. *Educ:* Amsterdam Univ, MS, 70; Cornell Univ, PhD(plasma physics), 76. *Prof Exp:* Prof, physics, UNESCO, Paraguay, 70-72; res asst, Cornell Univ, 73-76; postdoctoral fel, Lawrence Berkeley Labs, 76-78; scientist, Dynamics Technol, 78-79; scientist, Western Res, 79-80; sr scientist, Maxwell Labs, 80-84; SR SCIENTIST, PHYSICS, BERKELEY RES, 84- *Concurrent Pos:* Consult, Univ Calif, Los Angeles, Pacific Sierra, Physical Dynamics, 78-81. *Mem:* Am Phys Soc. *Res:* Development of nuclear weapon simulators including plasma radiators. *Mailing Add:* PO Box 852 Springfield VA 22150

PEREL, JAMES MAURICE, b Arg, Mar 30, 33; US citizen; m; c 3. CLINICAL PHARMACOLOGY, PSYCHOPHARMACOLOGY. *Educ:* City Col New York, BS, 56; NY Univ, MS, 61, PhD(chem), 64. *Prof Exp:* Assoc res scientist, Dept Med, Med Sch, NY Univ, 64-67; asst prof med & chem, Emory Univ Sch Med, 67-70; assoc res scientist, NY State Psychiat Inst, 70-76; asst prof psychiat, Col Physicians & Surgeons, Columbia Univ, 70-76, assoc prof clin pharmacol, 76-80; chief, Psychiat Res, NY State Psychiat Inst, 76-80; actg chmn, dept pharmacol, Sch Med, 85-88, PROF PSYCHIAT & PHARMACOL, SCH MED & DIR CLIN PHARMACOL, WESTERN PSYCHIAT INST, UNIV PITTSBURGH, 80- *Concurrent Pos:* Fel pharmacol, NY Univ, 62-64; USPHS fel, 63; lectr chem, City Univ New York, 63-67; asst prof chem, Sch Arts & Sci, Emory Univ, 67-70, vis assoc prof med, Sch Med, 70-73; USPHS grants psychopharmacol, Col Physicians & Surgeons, Columbia Univ & NY State Psychiat Inst, 70-80; consult, Food & Drug Admin & NIMH; consult, Res in Mania & Depression, NY City; chief, clin pharmacol, Vet Admin Med Ctr, Pittsburgh; assoc & adv ed, J Psychopharmacol, J Therapeut Drug Monitoring & J Neuro Psychobiol. *Honors & Awards:* Julius Koch Mem Award, 83. *Mem:* Am Soc Pharmacol & Exp Therapeut; NY Acad Sci; Soc Biol Psychiat; Am Fedn Clin Res; fel Am Inst Chemists; Am Soc Clin Pharmacol & Therapeut. *Res:* Clinical pharmacology of psychoactive agents; drug metabolism; enzymatic mechanisms of drug action; molecular pharmacology of anesthetic agents; pharmacokinetics. *Mailing Add:* Western Psychiat Inst Univ Pittsburgh 3811 O'Hara St Pittsburgh PA 15213

PEREL, JULIUS, b New York, NY, Sept 10, 27; m 50; c 4. ATOMIC & MOLECULAR PHYSICS. *Educ:* City Col New York, BS, 51; NY Univ, MS, 52, PhD(physics), 62. *Prof Exp:* Physicist, Anton Electronic Lab, NY, 52 & Electronic Prod Co, NY, 52-54; grad asst & instr physics, NY Univ, 55-62; sr physicist, ElectroOptical Systs, Xerox Corp, 62-65, mgr particle physics dept, 65-74; CO-FOUNDER & SCIENTIST, PHRASOR SCI, INC, 74- *Concurrent Pos:* Lectr, City Col New York, 54-60. *Mem:* Am Phys Soc. *Res:* Atomic collisions-charge transfer; ionic and molecular beams; surface ionization; electric propulsion-colloid propulsion; electrohydrodynamics; spectrometry; increased ion detection sensitivity; fast atom beams. *Mailing Add:* Phrasor Sci Inc 1536 Highland Ave Duarte CA 91010

PEREL, WILLIAM MORRIS, b Chicago, Ill, Oct 17, 27. MATHEMATICS. *Educ:* Ind Univ, AB, 49, AM, 50, PhD, 55. *Prof Exp:* Assoc math, Ind Univ, 49-54; asst prof, Ga Inst Technol, 54-56 & Tex Tech Col, 56-59; assoc prof, La State Univ, New Orleans, 59-62; vis prof, Randolph-Macon Woman's Col, 62-63; prof, Univ NC, Chatlotte, 63-66; chmn dept, 67-76, PROF MATH, WICHITA STATE UNIV, 66- *Mem:* Am Math Soc; Math Asn Am; Sigma Xi. *Res:* Semigroup and ring ideal theory; number theory; urban society and education. *Mailing Add:* Dept Math & Statist Wichita St Univ Wichita KS 67208

PERELSON, ALAN STUART, b Brooklyn, NY, Apr 11, 47; m 68; c 2. BIOMATHEMATICS, IMMUNOLOGY. *Educ:* Mass Inst Technol, BS(elec eng) & BS (life sci), 67; Univ Calif, Berkeley, PhD(biophys), 72. *Prof Exp:* Actg asst prof med physics, Univ Calif, Berkeley, 73; NIH fel & res assoc chem eng, Univ Minn, 73-74; staff mem theoret biol & biophys, Los Alamos Sci Lab, Univ Calif, 74-78; asst prof med sci, Brown Univ, 78-80; MEM STAFF, LOS ALAMOS NAT LAB, 80-; EXTERNAL PROF, SANTA FE INST, 89- *Concurrent Pos:* Adv ed, J Math Biol, 78-, Math Biosci, 81-, Inst Math & Applns J Math Appl Med Biol, 83- & J Theoret Biol, 85-; career res develop award, NIH, 79-84; mem, Sci Bd, Santa Fe Inst, 88-; vis prof physics, Ecole Normale Superieurc, Paris, 90. *Mem:* Inst Elec & Electronic Engrs; Am Asn Immunologists; Soc Indust & Appl Math; Soc Math Biol; Sigma Xi. *Res:* Application of mathematics to chemical and biophysical problems, with particular emphasis on problems in immunology. *Mailing Add:* Theoret Div Los Alamos Nat Lab Los Alamos NM 87545

PERESS, NANCY E, b New York, NY, Feb 2, 43; m 64; c 2. NEUROPATHOLOGY. *Educ:* Hunter Col, MA, 63; State Univ NY, MD, 67. *Prof Exp:* From intern to resident path, Kings County Hosp, New York, 67-69, neuropath, 69-71; asst prof & attend physician path, State Univ NY Downstate Med Ctr & Kings County Hosp, 71-73; asst prof path, 73-80, ASSOC PROF PATH, STATE UNIV NY STONY BROOK & NORTHPORT VET HOSP, 80-, CHIEF NEUROPATH, 73- *Concurrent Pos:* Consult neuropathologist, Nassau County Med Ctr, 74- *Mem:* Am Asn Neuropathologists. *Res:* Pathophysiology of immune disease of the choroid plexus and ciliary body; neuropathology of experimental fetal hypoxia and acidosis. *Mailing Add:* Two Coon Hollow Rd Huntington NY 11743

PERESSINI, ANTHONY L, b Great Falls, Mont, May 30, 34; m 55; c 5. ANALYTICAL MATHEMATICS. *Educ:* Col Great Falls, BS, 56; Wash State Univ, MA, 58, PhD(math), 61. *Prof Exp:* From instr to assoc prof math, 61-73, PROF MATH, UNIV ILL, URBANA, 73- *Mem:* Am Math Soc. *Res:* Functional analysis; ordered topological vector spaces. *Mailing Add:* Dept Math Univ Ill 273 Altgeld Hall Urbana IL 61801

PERETTI, ETTORE A(LEX), b Butte, Mont, Apr 5, 13; m 37; c 2. METALLURGY. *Educ:* Mont Col Mineral Sci & Technol, BS, 34, MS, 35; Germany, DSc(metall eng), 36. *Hon Degrees:* MetE, Univ Mont, 63. *Prof Exp:* From Instr to asst prof metall eng, Mont Col Mineral Sci & Technol, 36-40; asst prof, Sch Mines, Univ Columbia, 40-46; from assoc prof to prof metall eng, Univ Notre Dame, _____46-78, head dept metall eng, 51-69, asst dean col eng, 70-78, asst dir, Nat Consort Grad Degrees, Minorities Eng, 78-90, EMER PROF, UNIV NOTRE DAME, 78-, ASSOC DIR, NAT CONSORT GRAD DEGREES, MINORITIES ENG, 90- *Concurrent Pos:* Consult, 40-46, 59- *Mem:* Fel Am Soc Metals; Am Inst Mining, Metall & Petrol Engrs. *Res:* Metallurgical extractive processes and phase relationships. *Mailing Add:* Box 537 Univ Notre Dame Notre Dame IN 46556

PERETZ, BERTRAM, NEURONAL AGING. *Educ:* Univ Iowa, PhD(zool), 68. *Prof Exp:* PROF MED PHYSIOL, UNIV KY, 69- *Res:* Neural correlates of behavior. *Mailing Add:* Dept Physiol Univ Ky Med Ctr Lexington KY 40536

PEREY, BERNARD JEAN FRANCOIS, b Paris, France, Apr 28, 30; Can citizen; m 57; c 4. SURGERY. *Educ:* Univ Paris, PCB, 49; McGill Univ, MD, CM, 56, MSc, 60, Dipl Surg, 62. *Hon Degrees:* DCL, Bishop's Univ. *Prof Exp:* Asst prof surg, McGill Univ, 64-67; dir clin sci, Univ Sherbrooke, 68-70, prof surg & chmn dept, 67-82, surgeon-in-chief, univ hosp, 67-82, vdean, 73-77; prof surg & chmn dept, Dalhouse Univ, 82-89; SURGEON, VICTORIA GEN HOSP, 82- *Concurrent Pos:* Surgeon, Royal Victoria Hosp, Montreal, 62-67; vpres, Royal Col Physicians & Surgeons Can, 74-76. *Mem:* Am Col Surgeons; Soc Surg Alimentary Tract; Can Asn Gastroenterol (pres); Can Asn Clin Surgeons (pres); Am Surg Asn; Royal Col Physicians & Surgeons Can (pres, 80-82). *Res:* Gastrointestinal physiology; vascular physiology; wound healing. *Mailing Add:* ACC Rm 4115 Victoria Gen Hosp Halifax NS B3H 2Y9 Can

PEREY, FRANCIS GEORGE, b Paris, France, Oct 7, 32; US citizen; c 2. NUCLEAR PHYSICS, PROBABILITY THEORY. *Educ:* McGill Univ, BSc, 56, MSc, 57; Univ Montreal, PhD(physics), 60. *Prof Exp:* PHYSICIST NUCLEAR PHYSICS, OAK RIDGE NAT LAB, 60- *Concurrent Pos:* Mem nuclear data comt, Dept Energy, 70-; US rep nuclear data comt, Nuclear Europ Agency, 78-; corp res fel, Union Carbide Corp, 79. *Mem:* Fel Am Phys Soc. *Res:* Neutron physics. *Mailing Add:* Eng Physics Bldg 6010 MS 6356 Oak Ridge Nat Lab PO Box 2008 Oak Ridge TN 37830

PEREZ, CARLOS A, b Colombia, Nov 10, 34; US citizen; m 88; c 3. RADIATION ONCOLOGY. *Educ:* Univ Antioquia, Colombia, BS, 52, MD, 60. *Prof Exp:* From instr to assoc prof, 64-72, PROF RADIOL, MALLINCKRODT INST RADIOL, SCH MED, WASH UNIV, 72- *Concurrent Pos:* Fel radiother, M D Anderson Hosp, Houston, 63-64; consult, Vet Admin Hosp, St Louis, 67-87, Ellis Fischel State Hosp, Columbia, 67- & Jewish Hosp, St Louis, 70- *Mem:* AAAS; Radiol Soc NAm; Am Soc Ther Radiol Oncol (pres, 81-82); Am Radium Soc; Am Asn Cancer Res; Am Soc Clin Oncol. *Res:* Radiation therapy in combination with surgery or chemotherapy or both in the management of cancer; hyperthermia; gynecological tumors; prostate cancer. *Mailing Add:* 510 S Kings Highway St Louis MO 63110

PEREZ, GUIDO O, b Santa Clara, Cuba, Dec 16, 38; m 66; c 3. NEPHROLOGY, NUTRITION. *Educ:* Univ Miami, MD, 65. *Prof Exp:* Asst prof med, Univ Conn, 70-72; assoc prof, 73-77, PROF, MED, UNIV MIAMI, 77- *Concurrent Pos:* Chief dialysis unit, Miami VA Med Ctr, 72-; vis prof, Univ Geneva, Switz, 81. *Mem:* Am Col Physicians; Am Physiol Soc; Am Soc Nephrol; Int Soc Nephrol; Soc Exp Biol Med. *Res:* Acid-base metabolism; potassium homeostasis in renal failure; uremic toxicity - nutrition in uremia. *Mailing Add:* VA Hosp 1201 NW 16th St Miami FL 33125

PEREZ, HECTOR DANIEL, RHEUMATOLOGY, IMMUNOLOGICAL RESEARCH. *Educ:* Med Sch Buenos Aires, Argentina, MD, 72. *Prof Exp:* ASSOC PROF MED RHEUMATOLOGY, UNIV CALIF, SAN FRANCISCO, 80- *Mailing Add:* Dept Med San Francisco Gen Hosp San Francisco CA 94110

PEREZ, JOHN CARLOS, b Park City, Utah, Apr 29, 41; m 63; c 9. VIROLOGY. *Educ:* Univ Utah, BS, 67; Mankato State Col, MA, 69; Utah State Univ, PhD(immunol), 73. *Prof Exp:* Res assoc virol, Utah State Univ, 71-72; from asst prof to prof, 72-79, PROF BIOL, TEX A&I UNIV, 80- *Concurrent Pos:* Am Soc Microbiol Pres fel, 71. *Mem:* Am Soc Microbiol; Sigma Xi; Int Soc Toxinology. *Res:* Production of a monoclonal antibody against hemorrhagic activity of Crotalus atrox (Western Diamondback Rattlesnake) venom. *Mailing Add:* Dept Biol Tex A&I Univ Santa Gertrudis Kingsville TX 78363

PEREZ, JOSEPH DOMINIQUE, b New Orleans, La, Oct 17, 42; m 72; c 2. THEORETICAL PHYSICS. *Educ:* Loyola Univ, La, BS, 64; Univ Md, College Park, PhD(nuclear physics), 68. *Prof Exp:* Res asst physicist, Inst Pure & Appl Phys Sci, Univ Calif, San Diego, 69-71; asst prof physics, Univ Southern Calif, 71-74; RES SCIENTIST, LOCKHEED PALO ALTO RES LAB, 74- *Mem:* Am Phys Soc; Inst Elec & Electronics Engrs. *Res:* Many-body theory as applied to nuclear structure, shell model and nuclear and atomic scattering; laser plasmas and laser induced fusion. *Mailing Add:* Dept Physics Auburn Univ Auburn AL 36849

PEREZ, RICARDO, b Mex City, Jan 12, 59; US citizen; m 81; c 2. DEVICE PHYSICS, SOFTWARE ENGINEERING. *Educ:* Ill Inst Technol, BS, 81, MS, 83. *Prof Exp:* Analog designer, Ludwig Technol, 78-80; sr analog designer, Zenith Electronics Corp, 81-83; sr develop engr, Teledyne Corp, 84-86; MGR ADVAN MAT TECHNOL, COMPAQ COMPUTER CORP, 86- *Concurrent Pos:* Sanco Consults, 84-87; tech mem, IPC Packaging Comt, 88-91, Orgn Tab & Advan Packaging, 89-91; speaker, Surface Mount Asn, 89-90; comt mem, EIA Standards Electronic Prod Packaging, 89-90; packaging chmn, Int Reliability Physics Soc, 90-91. *Mem:* AAAS; Inst Elec & Electronic Engrs; Int Soc Hybrid Microelectronics; Am Soc Mat. *Res:* Applied research in practical computer system applications leading to the development of interactive hardware and modeling software tools for reducing product design cycle of complex computer systems. *Mailing Add:* 11923 Gatesden Dr Tomball TX 77375

PEREZ-ALBUERNE, EVELIO A, b Ranchuelo, Cuba, July 22, 39; US citizen; m 61; c 3. PHYSICAL CHEMISTRY. *Educ:* Villanova Univ, BChE, 61; Univ Ill, Urbana, MS, 63, PhD(chem eng), 65. *Prof Exp:* Sr res chemist, Eastman Kodak Co, 65-69, res assoc, 69-79, sr res assoc, 79-80, sr lab head, 80-84, asst div dir, 84-86, DIV DIR, RES LAB, EASTMAN KODAK CO, 86- *Mem:* AAAS; Inst Elec & Electronic Engrs; Am Vacuum Soc. *Res:* Optical recording; materials science; thin-film coatings. *Mailing Add:* Res Labs Eastman Kodak Co Rochester NY 14650-2022

PEREZ-CRUET, JORGE, b San Juan, PR; m; c 4. PSYCHOPHARMACOLOGY. *Educ:* Univ PR, MD, 57; McGill Univ, Can, dipl psychiat, 76; FRCP(C). *Prof Exp:* Res assoc, NIH & NIMH, Bethesda, Md, 69-73; PROF PSYCHIAT, MED SCH, UNIV PR, 78-; CHIEF PSYCHIAT, SAN JUAN VET ADMIN HOSP, 78- *Concurrent Pos:* Mem staff, Walter Reed Army Inst Res, 60-62. *Mem:* Am Physiol Soc; Am Col Neuropharmacol; Am Psychiat Asn; Am Soc Pharmacol & Exp Therapeut. *Res:* Neuropsychopharmacology; biological psychiatry; neuroscience. *Mailing Add:* Psychiat Serv Vet Admin Hosp One Veterans Plaza San Juan PR 00927-5800

PEREZ-FARFANTE, ISABEL CRISTINA, b Havana, Cuba, July 24, 16; US citizen; m 41; c 2. ZOOLOGY. *Educ:* Univ Havana, BS, 38; Radcliffe Col, MS, 44, PhD(biol), 48. *Prof Exp:* Prof invert zool, Univ Havana, 48-60; biologist, Ctr Fisheries Res, 52-55, dir, 59-60; EMER RES ASSOC, SMITHSONIAN INST, 63-; SYST ZOOLOGIST, SYSTS LAB, NAT MARINE FISHERIES SERV, 66- *Concurrent Pos:* John Simon Guggenheim Mem Found fel, 42-44; Alexander Agassiz fel, oceanog & zool, 44-45; scholar, Radcliffe Inst Independent Study, 63-65; Am Inst Fishery fel, Res Inst, 77- *Mem:* AAAS; Soc Syst Zool; Crustacean Soc. *Res:* Systematics, morphology and distribution of decapod Crustacea, with special reference to western Atlantic and eastern Pacific regions. *Mailing Add:* Systs Lab Nat Marine Fisheries Serv Nat Mus Natural Hist Washington DC 20560

PEREZ-MENDEZ, VICTOR, b Guatemala, Aug 8, 23; nat US; m 49; c 2. EXPERIMENTAL HEAVY ION PHYSICS. *Educ:* Hebrew Univ, Israel, MS, 47; Columbia Univ, PhD, 51. *Prof Exp:* Res assoc, Columbia Univ, 51-53; staff physicist, 53-61, SR SCIENTIST, LAWRENCE BERKELEY LAB, UNIV CALIF, 60- *Concurrent Pos:* Vis lectr, Hebrew Univ, 59-60; prof physics, Dept Radiol, Univ Calif, San Francisco, 68- *Mem:* Fel Am Phys Soc; fel NY Acad Sci; fel Inst Elec & Electronics Engrs; Soc Photo-Optical Instrument Engrs. *Res:* Physics of heavy ions, instrumentation and medical applications. *Mailing Add:* Lawrence Berkeley Lab Univ of Calif Berkeley CA 94720

PEREZ-TAMAYO, RUHERI, b Tampico, Mex, Dec 2, 26; US citizen; m 63; c 5. RADIOTHERAPY. *Educ:* Nat Univ Mex, BS, 45, MD, 52. *Prof Exp:* Rotating intern, Huron Rd Hosp, Cleveland, Ohio, 52-53, resident radiol, 53-55; resident radiother, Penrose Cancer Hosp, Colorado Springs, Colo, 55-56; instr radiother & radiotherapist, Univ Mich Hosp, 56-57; chief of dept radiother, French Hosp, Mexico City, Mex, 57-61; asst prof radiol & head dept radiother, Med Ctr, Univ Colo, 61-63; assoc radiother, Jefferson Med Col Hosp, 63-64; assoc radiother, Penrose Cancer Hosp, 64-66; chief dept radiother, Ellis Fishchel State Cancer Hosp, Columbia, Mo, 66-70; PROF RADIOL & DIR RADIATION THER, LOYOLA UNIV HOSP, 70- *Concurrent Pos:* Cordell Hull fel, 56-57; fel, Armed Forces Inst Path, 57; grants, AEC Mex, 58, USPHS, 61, 62, 65, 66 & 67-70, Milheim Found Cancer Res, 62-63, 65 & 67, Bent Co Cancer Fund, Fluid Res Fund Comt & Am Cancer Soc, 62 & Alpha Phi Omega & Jefferson Med Col, 63; prof, Nat Univ Mex, 58-61; assoc prof, Jefferson Med Col, 63-64; prof, Sch Med, Univ Mo-Columbia, 66-70; consult, Sch Med, Washington Univ, 68-; dir radiation ther, St Joseph's Hosp, 70-71; Gen Res Support grant, Loyola Univ Hosp, Ill, 70-71, 72; consult, Vet Admin Hosp, Hines, Ill, 72-; comt mem cancer educ prog, Nat Cancer Inst, NIH. *Mem:* Am Soc Therapeut Radiol; Radiol Soc NAm; fel Am Col Radiol; Am Asn Cancer Educ; Am Asn Cancer Res; Sigma Xi. *Res:* Time-dose relationships; mathematical models in radiation therapy; radiation therapy oncology group clinical studies. *Mailing Add:* Yucatan 71 MTZ DeLa Torre Veracruz Mexico

PERFETTI, PATRICIA F, b Charleroi, Pa, July 16, 52; m 75; c 2. TOBACCO & CIGARETTE CHEMISTRY, PAPER CHEMISTRY. *Educ:* Ind Univ, Pa, BSEd, 74. *Prof Exp:* Teacher chem, Monongahela Cath High Sch, Monongahela, Pa, 75; teacher chem, Cave Spring High Sch, Roanoke, Va, 75-77; sr lab technician, Appl Food Res, R J Reynolds Foods Co, Winston-Salem, NC, 78-79, assoc scientist, 79-80; jr res & develop chemist, Fundamental Res, 80-81, assoc res & develop chemist, 81-84, res & develop chemist, 84-87, SR RES & DEVELOP CHEMIST, APPL RES & DEVELOP, R J REYNOLDS TOBACCO CO, WINSTON-SALEM, NC, 87- *Mem:* AAAS; Am Chem Soc; fel Am Inst Chemists; NY Acad Sci; Sigma Xi. *Res:* Design, development and implementation of new technologies on cigarette products. *Mailing Add:* R J Reynolds Tobacco Co Bowman Gray Tech Ctr Winston-Salem NC 27102

PERFETTI, RANDOLPH B, b Greensburg, Pa, Sept 11, 49; div. ORGANIC CHEMISTRY, BIOCHEMISTRY. *Educ:* Ind Univ Pa, BS, 71; Va Polytech Inst & State Univ, PhD(chem), 75. *Prof Exp:* Fel organic chem, Johns Hopkins Univ, 75-76; CHEMIST PESTICIDES, US ENVIRON PROTECTION AGENCY, 76- *Concurrent Pos:* NSF grant, 71-75. *Mem:* Am Chem Soc; Sigma Xi. *Res:* Mechanisms of enzyme catalyzed reactions; elucidation of via chemical modification and kinetics; synthetic organic chemistry; development of new synthetic techniques; redox reactions. *Mailing Add:* US Environ Protection Agency TS-769 401 M St NW Washington DC 20460

PERFETTI, THOMAS ALBERT, b Jeannette, Pa, Mar 22, 52; m 75; c 2. ORGANIC CHEMISTRY, ANALYTICAL CHEMISTRY. *Educ:* Ind Univ Pa, BS, 74; Va Polytech Inst & State Univ, PhD(org chem), 78. *Prof Exp:* Teaching asst org chem, Va Polytech Inst & State Univ, 74-77; res chemist, 77-79, sr res chemist, 79-82, sr staff scientist, 82-85, MASTER SCIENTIST TOBACCO CHEM, R J REYNOLDS TOBACCO CO, 85- *Concurrent Pos:* Res fel, NASA, Langley, 75, 76. *Mem:* Sigma Xi; Am Chem Soc (treas, 80-81); AAAS; NY Acad Sci; fel Am Inst Chemists. *Res:* Organic synthesis of polyaromatic hydrocarbons; electronic effects associated with sigmatropic rearrangements; structure elucidations; pyrolytic glass capillary chromatography, tobacco chemistry; cigarette design technology; nicotine and menthol chemistry; sensory evaluation science; materials development. *Mailing Add:* R J Reynolds Tobacco Co Bowman Gray Tech Ctr Winston-Salem NC 27102

PERHAC, RALPH MATTHEW, b Brooklyn, NY, July 29, 28; m 50; c 2. GEOCHEMISTRY. *Educ:* Columbia Univ, AB, 49; Cornell Univ, AM, 52; Univ Mich, Ann Arbor, PhD(geol), 61. *Prof Exp:* Mining engr, Anaconda Co, 52-53; dist geologist, AEC, 53-55; explor geologist, Caltex Oil, Australia, 55-57; sr res scientist, Humble Oil & Refining Co, 60-67; from assoc prof to prof geochem, Univ Tenn, Knoxville, 67-74; prog mgr, NSF, Washington, DC, 74-76; prog mgr, Elec Power Res Inst, 76-80, dept dir, 80-90, consult, 90-91; RETIRED. *Concurrent Pos:* Consult, Oak Ridge Nat Lab, 72-75. *Mem:* Geol Soc Am; Geochem Soc; Am Asn Petrol Geologists; Int Asn Geochem & Cosmochem; Air Pollution Control Asn; Soc Environ Geochem & Health. *Res:* Hydrogeochemistry; geochemical prospecting; lunar soils; environmental chemistry. *Mailing Add:* Box 7412 Menlo Park CA 94026-7412

PERHACH, JAMES LAWRENCE, b Pittsburgh, Pa, Oct 26, 43; m 67; c 2. PHARMACOLOGY, MEDICAL SCIENCES. *Educ:* Univ Dayton, BS, 66; Univ Pittsburgh, MS, 69, PhD(pharmacol), 71. *Prof Exp:* Sr scientist, Mead Johnson Pharmaceut Div, Mead Johnson & Co, Bristol Myers Corp, 71-74, sr investr, 74-76, sr res assoc, 76-78, prin res assoc, 78-80; dir pharmacol, Wallace Labs, Div Carter-Wallace, Inc, 80, exec dir biol res, 80-84, assoc dir clin res, 84-85, dir clin invest, 85-87, VPRES CLIN PHARMACOL & PHARMACOKINETICS, WALLACE LABS, DIV CARTER-WALLACE, INC, 87- *Concurrent Pos:* Instr pharmacol & exp therapeut, dept pharmacol, Univ Pittsburgh Sch Pharm, 68-69, grad teaching asst, 67-69; instr grad pharmacol, Univ Evansville, Col Arts & Sci, 75, lectr grad physiol, 73-79; assoc fac, Ind Univ Sch Med, Evansville Ctr Med Educ, 73-80; consult, Substance Abuse Comt Tri State Area Planning Coun, Evansville, Ind, 72-75, Addictions Med Educ Prog, Evansville Ctr Med Educ, Ind Univ Sch Med, 72-78, Drug Utilization Rev Coun State NJ, 83-; adj prof toxicol, Philadelphia Col Pharm & Sci, 81- *Mem:* Sigma Xi; Soc Neurosci; Am Soc Clin Pharmacol & Therapeut; Am Soc Pharmacol & Exp Therapeut; Am Col Clin Pharmacol; Soc Exp Biol & Med; NY Acad Sci; Am Col Toxicol; Europ Soc Toxicol; AAAS; Drug Info Asn. *Res:* New drug discovery, elucidation of mechanism of action and safety evaluation of new therapeutic agents in cardiovascular pulmonary and CNS area. *Mailing Add:* Wallace Labs Div Carter-Wallace Inc 301B College Rd E Princeton NJ 08540

PERI, BARBARA ANNE, b Richmond, Calif, May 15, 25; m 46; c 3. IMMUNOLOGY. *Educ:* Univ Calif, Berkeley, BA, 46; Univ Wis-Madison, MS, 48; Univ Notre Dame, PhD(microbiol), 70. *Prof Exp:* Lab asst bact, Hooper Res Found, Univ Calif, Berkeley, 46; from asst prof to assoc prof biol, Valparaiso Univ, 64-74; res assoc, Pritzker Sch Med, Dept Pediat, LaRabida Res Inst, Univ Chicago, 74-79; res assoc & assoc prof, Dept Pediat, Pritzker Sch Med, 79-86; asst prof, Harvard Med Sch, 86-88; assoc immunologist, Mass Gen Hosp, 86-88; RETIRED. *Mem:* Am Soc Microbiol; Am Asn Immunologists; Sigma Xi; Asn Gnotobiology (exec bd, 87-89). *Res:* Secretory antibody; immune response to environmental antigens; diphtheria toxin; maternal regulation of immune response. *Mailing Add:* 2 Tortoise Ln Falmouth MA 02540-1639

PERI, JOHN BAYARD, b Stockton, Calif, May 5, 23; m 46; c 3. PHYSICAL CHEMISTRY. *Educ:* Univ Calif, BS, 43; Univ Wis, PhD(chem), 49. *Prof Exp:* Res asst, Univ Wis, 47-49; res chemist, Calif Res Corp, 49-57; proj chemist, Amoco Corp, 57-58, sr proj chemist, 58-60, sr res scientist, 60-62, res assoc, 62-79, sr res assoc, 79-86; ADJ PROF, CHEM ENG DEPT, NORTHEASTERN UNIV, 87- *Honors & Awards:* R L Burwell Lectr, 87. *Mem:* AAAS; Am Chem Soc; Catalysis Soc. *Res:* Characterization of surface chemistry of solid catalysts and adsorbents by infrared spectroscopic techniques and relation of this chemistry to catalytic and adsorptive properties. *Mailing Add:* Two Tortoise Ln Falmouth MA 02540-1639

PERICAK-SPECTOR, KATHLEEN ANNE, b Buffalo, NY, Jan 24, 54. HYPERBOLIC PARTIAL DIFFERENTIAL EQUATIONS. *Educ:* State Univ NY, Buffalo, BA(math), 75, BA(physics), 76; Univ Pittsburgh, MA, 77; Carnegie-Mellon Univ, MS, 78 & PhD(math), 80. *Prof Exp:* Vis asst prof math, Univ Tenn, Knoxville, 79-81; asst prof, 81-87, ASSOC PROF MATH, SOUTHERN ILL UNIV, CARBONSDALE, 87- *Concurrent Pos:* Sr fel, Inst for Math & It's Appln, Univ Minn, 83 & 85; vis lectr, Heriot-Watt Univ, Edinburgh, 86; prin investr, Nat Sci Found Grant, 88-89. *Mem:* Soc Nat Philos; Asn for Women Math; Soc Eng Sci; Soc Rheology. *Res:* Uniqueness and stability of solutions of equations that govern fluid flow through a porous media; continuum mechanics; cavitation in elasticity. *Mailing Add:* Dept Math Southern Ill Univ Carbondale IL 62901-4408

PERIC-KNOWLTON, WLATKA, b Nuremberg, WGer, Oct 29, 55. ANTICOAGULATION THERAPY, PRIMARY CARE. *Educ:* Ariz State Univ, BS, 77, MS, 86. *Prof Exp:* Staff nurse, Gen Surg Operating Rm & Staff & charge nurse, Cardiovasc Operating Rm, St Joseph's Hosp, Phoenix, Ariz, 78-80; staff & charge nurse, Surg Intensive Care Unit, 80-83, ADULT NURSE PRACTR & DIR, ANTICOAGULATION CLINIC & STABLE DIS CLINIC, AMBULATORY CARE, CARL HAYDEN VET ADMIN MED CTR, PHOENIX, ARIZ, 83- *Concurrent Pos:* Adj clin fac, Ariz State Univ Col Nursing Grad Prog, 87-; pres, Ariz Nurse Practr Coun, 89-91; guest ed, Nurse Practr Forum, W B Saunders pub, Spec Issue on Thromboembolism & Anticoagulant Ther, 90-92; Ariz rep, Am Acad Nurse Practr, 90-92. *Mem:* Am Nurses Asn; Am Acad Nurse Practr; Am Asn Diabetes Educ; Am Heart Asn; Int Asn Study Pain. *Res:* Anticoagulation therapy: comparing fingerstick capillary blood protimes with laboratory venous stick protimes, complications of warfarin therapy; cardiomyopathy: patient management; pain: effectiveness of transcutaneous electric nerve stimulation, pain theories, the pain experience and nonpharmacologic pain relief methods. *Mailing Add:* Carl Hayden Vet Admin Med Ctr OPD (11C-1) Phoenix AZ 85012

PERILLIE, PASQUALE E, b Bridgeport, Conn, Sept 2, 26; m 52; c 5. MEDICINE. *Educ:* Univ Conn, BA, 51; NY Med Col, MD, 55. *Prof Exp:* USPHS res fel, 59-61; instr, 62-64, asst prof, 64-77, CLIN PROF MED, SCH MED, YALE UNIV, 77- *Concurrent Pos:* Vet Admin clin investr, 62-65; USPHS career res award, 65-; chmn dept med, Bridgeport Hosp, 69- *Mem:* Am Fedn Clin Res; Am Soc Hemat; NY Acad Sci; Sigma Xi; Int Soc Hemat. *Res:* Resistance to infection in leukemia; abnormal hemoglobins; histochemistry of blood cells. *Mailing Add:* 267 Grant St Bridgeport CT 06610

PERIMAN, PHILLIP, b Memphis, Tex, Dec 5, 38; m 65; c 3. INTERNAL MEDICINE, HEMATOLOGY. *Educ:* Yale Univ, BA, 61; Wash Univ, MD, 65. *Prof Exp:* From intern to asst resident internal med, Sch Med, NY Univ, 65-67; res assoc, Nat Cancer Inst, 67-69, sr staff fel, Lab Path, 71; asst prof med, Sch Med, George Washington Univ, 71-76; assoc prof & assoc chmn med, Sch Med, Tex Tech Univ, 76-81; MED DIR, DON & SYBIL HARRINGTON CANCER CTR, 81- *Concurrent Pos:* USPHS fel, Sir William Dunn Sch Path, Oxford Univ, 70. *Mem:* Am Soc Hemat; Am Soc Clin Oncol. *Res:* Tumor immunology; cell biology; hematology; oncology; cell differentiation; virus carcinogenesis. *Mailing Add:* Harrington Cancer Ctr 1500 Wallace Blvd Amarillo TX 79106

PERINI, JOSE, b Sao Paulo, Brazil, Mar 1, 28; m 55; c 6. ELECTRICAL ENGINEERING. *Educ:* Polytech Sch, Sao Paulo, BS, 52; Syracuse Univ, PhD(elec eng), 61. *Prof Exp:* Mgr radio maintenance, Real Trasportes Aereos, Brazil, 51-54; asst prof elec eng, Polytech Sch, Sao Paulo, 54-58, assoc prof, 61-62; from asst prof to assoc prof, 62-71, PROF ELEC ENG, SYRACUSE UNIV, 71- *Concurrent Pos:* Consult, Gen Elec Co, Syracuse & Sao Paulo, Brazil, 59-, Coencisa, Brazil, 80, Syracuse Res Corp, 82-, Power Technol Inc, 84- *Mem:* Inst Elec & Electronics Engrs; Brazilian Inst Eng; Am Asn Eng Educ. *Res:* Antennas; transmission lines; radio navigation and communication systems; digital signal processing; computers. *Mailing Add:* Dept Elec & Comput Eng Syracuse Univ Link Hall Syracuse NY 13244

PERINO, JANICE VINYARD, b Oklahoma City, Okla, Sept 3, 46; m 68. PLANT ECOLOGY, APPLIED ECOLOGY. *Educ:* Univ Okla, BS, 68; MS, 71; NC State Univ, PhD(bot), 75. *Prof Exp:* Teacher pub schs, Okla, 68-69; ecol res technician bot, NC State Univ, 71-73; asst prof bot, Miami Univ, Ohio, 75-79; STAFF ECOLOGIST, RADIAN CORP, 79- *Mem:* Ecol Soc Am; AAAS; Torrey Bot Club; Sigma Xi. *Res:* Descriptive and applied plant ecology; strategies controlling plant distribution, including reproductive mechanical anisms, resource allocation, and plant-plant interactions; old-field succession and the tall grass prairies; forests, marshes and wetland studies. *Mailing Add:* 900 W Lake Dr Springfield IL 62707

PERISHO, CLARENCE H(OWARD), b Granite City, Ill, Dec 23, 24; m 46; c 3. APPLIED MATHEMATICS, AERODYNAMICS. *Educ:* Purdue Univ, BS, 47; St Louis Univ, MS, 58. *Prof Exp:* Res engr, McDonnell Aircraft Corp, St Louis, 47-51, proj dynamics engr, 51-59, chief dynamics engr, 59-61, sect mgr struct dynamics, 61-64, chief struct dynamics engr, Eng Technol Div, 64-71, mgr, 71-81, chief technol engr, struct dynamics, 81-83, chief technol engr, Struct Dynamics & Loads, Eng Tech Div, 83-87. *Concurrent Pos:* Tech chmn, Symp Flight Flutter Testing, NASA, 75, chmn, Aerospace Flutter & Dynamic Coun, 76-84, mem, NASA aeronaut adv comt, subcomt structures, 82-84. *Mem:* Assoc fel Am Inst Aeronaut & Astronaut. *Res:* Influence of dynamic response, flutter, vibration and acoustics on the design, reliability and performance of aircraft; vertical takeoff vehicles. *Mailing Add:* 270 Greentails Dr N Chesterfield MO 63017

PERISHO, CLARENCE R, b Newberg, Ore, Apr 29, 17; m 41; c 3. BIOCHEMISTRY. *Educ:* William Penn Col, BS, 38; Haverford Col, MA, 39; NY Univ, PhD(sci educ), 63. *Prof Exp:* Instr math & sci, Friendsville Acad, 39-40; prof phys sci, Nebr Cent Col, 40-44; instr physics, chem & math, McCook Jr Col, 44-47; from asst prof to assoc prof chem & math, Nebr Wesleyan Univ, 47-54; from instr to asst prof sci, Mankato State Univ, 54-63, from assoc prof to prof chem, 63-82; RETIRED. *Concurrent Pos:* Acad Year Exten Res Partic Col Teachers, Mankato State Col, 64-66. *Mem:* AAAS; Math Asn Eng; Nat Educ Asn; Sigma Xi. *Mailing Add:* 804 Belgrade Ave North Mankato MN 56003-3602

PERKEL, DONALD HOWARD, theoretical biology; deceased, see previous edition for last biography

PERKEL, ROBERT JULES, b New York, NY, Feb 23, 26; m 50; c 3. POLYMER CHEMISTRY, ORGANIC CHEMISTRY. *Educ:* City Col New York, BS, 48; Polytech Inst Brooklyn, MS, PhD(polymer chem), 59. *Prof Exp:* Chemist, Lehman Bros Corp, NJ, 48-56; PRES, JEMA-AM, INC, DUNELLEN, 56- *Mem:* NY Acad Sci; fel Am Inst Chemists; Am Chem Soc; Soc Vacuum Coaters (past pres). *Res:* Organic finishes and inorganic solutions related to electroless spray deposition of metals and vacuum deposition of metals on a variety of substrates. *Mailing Add:* 19 Lamington Rd Somerville NJ 08876

PERKINS, A THOMAS, b Youngstown, Ohio, Dec 20, 42; m 62; c 3. AGRONOMY. *Educ:* Pa State Univ, BS, 64, PhD(agron), 69. *Prof Exp:* From instr to asst prof agron, Pa State Univ, 66-70; sr plant physiologist, 70-75, mgr specialty prod res, MGR, TECH CHEM SALES & MTK DEVELOP, LILLY & CO, 84- *Mem:* Am Soc Agron. *Res:* Research and development of pesticides for uses on turfgrass ornamental species and general noncropland usage; general turfgrass management. *Mailing Add:* Lilly Corp Ctr 13-2 Indianapolis IN 46285

PERKINS, ALFRED J, b Elgin, Ill, Mar 8, 12; m 44; c 3. INORGANIC COMPLEX COMPOUNDS, INFRARED & RAMAN SPECTRA. *Educ:* Kenyon Col, BS, 33; Johns Hopkins Univ, PhD(chem), 37. *Prof Exp:* Chemist, EI du Pont de Nemours, 36-38, Underwriter's Labs, 38-41, 46-47; from asst prof to prof chem, Col Pharm, Univ Ill, 47-75, assoc dean, Grad Col Med Ctr, 75-78, actg dean, 77; RETIRED. *Concurrent Pos:* Consult, Argonne Nat Labs, 62-79; vis prof, Col Chem, Univ Bristol, Eng, 68-69. *Mem:* Am Chem Soc; Soc Appl Spectros. *Res:* Inorganic complex compounds; infrared and Raman spectroscopy; inorganic fluorides in anhydrous hydrogen fluoride. *Mailing Add:* Box 277 Burlington IL 60109

PERKINS, BOBBY FRANK, b Greenville, Tex, Dec 9, 29; m 54; c 3. INVERTEBRATE PALEONTOLOGY. *Educ:* Southern Methodist Univ, BS, 49, MS, 50; Univ Mich, PhD(geol), 56. *Prof Exp:* Instr geol, Southern Methodist Univ, 50-51 & 53-55; asst prof, Univ Houston, 55-56; res paleontologist, Shell Develop Co, 56-66; prof geol, La State Univ, Baton Rouge, 66-75, dir mus geosci, 69-75, chmn dept geol & dir sch geosci, 73-75; PROF GEOL, DEAN GRAD SCH & ASSOC VPRES RES, UNIV TEX, ARLINGTON, 75- *Concurrent Pos:* Exec dir, Gulf Coast Sect, Soc Econ Paleontologists & Mineralogists Found, 81- *Mem:* Geol Soc Am; Paleont Soc; Soc Econ Paleontologists & Mineralogists; Am Asn Petrol Geologists; Brit Paleont Soc; Sigma Xi. *Res:* Cretaceous invertebrate paleontology, biostratigraphy and paleoecology; Cretaceous corals and rudistid pelecypods; Cretaceous stratigraphy in Texas and northern Mexico; trace fossils; reef organisms, sediments, and petrography. *Mailing Add:* 1416 Creekford Dr Arlington TX 76012

PERKINS, COURTLAND D(AVIS), b Philadelphia, Pa, Dec 27, 12; m 41; c 2. AERONAUTICAL ENGINEERING. *Educ:* Swarthmore Col, BS, 35; Mass Inst Technol, MS, 41. *Prof Exp:* Prof aeronaut eng, Princeton Univ, 45-75, chmn dept, 51-75, assoc dean sch eng, 65-71; pres, Nat Acad Eng, 75-83; EMER PROF AERONAUT ENG, PRINCETON UNIV, 75- *Concurrent Pos:* Chief scientist, US Air Force, 56-57, asst, Sect Res & Develop, 60-61, mem, Sci Adv Bd, 46-, vchmn, 61-68, chmn 68-; chmn adv group aerospace res & develop, NATO, 63-67, US nat deleg, 63-69; mem space sci bd, Nat Acad Sci, 65-70; mem space prog adv coun, NASA, 71-, chmn space systs command, 71- *Mem:* Nat Acad Eng; fel Am Inst Aeronaut & Astronaut (pres, 64); fel Royal Aeronaut Soc. *Res:* Airplane stability and control; airplane dynamics. *Mailing Add:* 400 Hilltop Terr Alexandria VA 22301

PERKINS, DAVID DEXTER, b Watertown, NY, May 2, 19; m 52; c 1. GENETICS, CYTOLOGY. *Educ:* Univ Rochester, AB, 41; Columbia Univ, PhD(zool), 49. *Prof Exp:* Mem fac biol, 48-61, PROF BIOL, STANFORD UNIV, 61- *Concurrent Pos:* Mem genetics training comt, NIH, 61-65; ed, Genetics, 63-67; USPHS res career award, 64-; exec bd, Int Genetics Fedn, 78-83; Guggenheim fel, 83-85. *Mem:* Nat Acad Sci; Genetics Soc Am (pres, 77). *Res:* Genetics; cytogenetics and biology of Neurospora. *Mailing Add:* Dept Biol Sci Stanford Univ Stanford CA 94305-5020

PERKINS, DONALD YOUNG, b Ponchatoula, La, June 27, 23; m 56; c 4. HORTICULTURE. *Educ:* La State Univ, BS, 50, MS, 51; Cornell Univ, PhD(veg crops), 54. *Prof Exp:* Instr & jr olericulturist, Univ Calif, 54; assoc horticulturist, La State Univ, 54-57; prin horticulturist, Coop State Res Serv, USDA, 57-66; head dept hort, Auburn Univ, 66-85; RETIRED. *Mem:* AAAS; Am Soc Hort Sci; Am Genetic Asn; Am Inst Biol Sci. *Res:* Vegetable breeding; mineral nutrition of vegetable crops; chemical weed control; horticultural research administration. *Mailing Add:* 319 Singleton Ave Auburn AL 36849-4201

PERKINS, EDWARD GEORGE, b Canton, Ill, Nov 1, 34; m 57; c 4. ORGANIC CHEMISTRY. *Educ:* Univ Ill, BS, 56, MS & PhD(food chem), 58. *Prof Exp:* Res chemist, Res Div, Armour & Co, 58-62; PROF FOOD SCI, BURNSIDES RES LAB, UNIV ILL, URBANA, 62- *Mem:* AAAS; Am Chem Soc; Am Oil Chemists Soc; Am Inst Nutrit; Sigma Xi. *Res:* Organic and biochemistry of lipids; lipid methodology; mass spectrometry; organic synthesis. *Mailing Add:* Burnside Res Lab Univ Ill 1208 W Pennsylvania St Urbana IL 61801

PERKINS, EUGENE HAFEN, immunology, for more information see previous edition

PERKINS, FLOYD, differential dynamics, physics; deceased, see previous edition for last biography

PERKINS, FRANK OVERTON, b Fork Union, Va, Feb 14, 38; m 61. MARINE BIOLOGY. *Educ:* Univ Va, BA, 60; Fla State Univ, MS, 62, PhD(exp biol), 66. *Prof Exp:* Assoc marine scientist, 66-69, sr marine scientist & head dept microbiol & path, 69-77, head, Div Biol Oceanog & asst dir, 77-81, actg dir, 81-, DIR, VA INST MARINE SCI, 82-; DEAN MARINE SCI, COL WILLIAM & MARY. *Mem:* Soc Protozoologists. *Res:* Marine invertebrate pathology; ultrastructure taxonomy and ecology of estuarine Protozoa and microalgae. *Mailing Add:* Dean Marine Sci Col William & Mary Gloucester Point VA 23062

PERKINS, HAROLD JACKSON, b London, Ont, July 6, 30; m 54; c 4. BIOCHEMISTRY. *Educ:* Univ BC, BA, 51, MSc, 53; Iowa State Col, PhD(plant biochem), 57. *Prof Exp:* Fel plant physiol, Div Appl Biol, Nat Res Coun Can, 57-58; plant biochemist, Res Sta, Can Dept Agr, 58-63; assoc prof biochem, State Univ NY Col Plattsburgh, 63-64, chmn div, 64-66, prof biochem, 64-77, dean fac sci & math, 66-75, dean grad studies & res, 75-77; PRES, BRANDON UNIV, CAN, 77-, PROF BIOCHEM, 77- *Concurrent Pos:* Consult, Teacher Training Prog Univs, State NY Dept Educ, 66-77, accreditation of col & univ sci & math progs; consult, Design Sci Facilities, State of NJ; mem patent policy bd, State Univ NY, mem grad coun, 65-69 & 75-77; mem, Nat Coun Univ Res Adminr, 75-77; mem, Coun Tissue Cult Asn, 76- *Mem:* AAAS; Sigma Xi. *Res:* Biosynthesis of chlorophylls a and b; analytical chemistry of chlorophylls and their degradation products with particular reference to isotopically labelled compounds. *Mailing Add:* 463 13th St Brandon MB R7A 5Z4 Can

PERKINS, HAROLYN KING, b Six Mile, SC, Jan 24, 37; m 59; c 3. PHYSICAL CHEMISTRY. *Educ:* Wake Forest Col, BS, 57; Cornell Univ, PhD(phys chem), 65. *Prof Exp:* Res assoc phys chem, Cornell Univ, 65-66 & Princeton Univ, 66-67; lit chemist, FMC Corp, NJ, 67-69; Sloan fel, 69-73, res staff & lectr chem eng, Princeton Univ, 73-77; SR RES SCIENTIST, AM CAN CO, RES & DEVELOP CTR, PRINCETON, NJ, 78- *Mem:* Am Chem Soc; Am Phys Soc; Am Inst Chem Eng; Am Vacuum Soc; Adhesion Soc. *Res:* Surface characterization of metals and polymers; adhesion of metals to polymers and polymers to polymers; solid state chemistry; rheology of aqueous foams; mechanical properties of polymers; thermonuclear fusion technology. *Mailing Add:* 65 Woodside Lane Princeton NJ 08540

PERKINS, HENRY CRAWFORD, JR, b Miami, Fla, Nov 23, 35; m 60; c 4. MECHANICAL ENGINEERING. *Educ:* Stanford Univ, BS, 57, MS, 60, PhD(mech eng), 63. *Prof Exp:* Res assoc mech eng, Stanford Univ, 62-63, acting asst prof, 63-64; assoc prof, 64-67, PROF MECH ENG, UNIV ARIZ, 67- *Concurrent Pos:* Vis prof, Tech Univ Denmark, 71-72, US Mil Acad, 78-79 & Stanford Univ, 81-82. *Honors & Awards:* Teetor Award, Soc Automotive Engrs, 76. *Mem:* Am Soc Mech Engrs; fel Am Soc Mech Engrs. *Res:* Thermoscience; air pollution. *Mailing Add:* Dept of Aerospace & Mech Eng Univ of Ariz Tucson AZ 85721

PERKINS, HENRY FRANK, b Quitman Co, Ga, June 19, 21; m 45; c 2. SOIL CLASSIFICATION. *Educ:* Univ Ga, BS, 45, MS, 51; Rutgers Univ, PhD(soils), 54. *Prof Exp:* Asst soil surveyor, Soil Conserv Serv, USDA, 45-47; soils analyst, 47-50, asst agronomist, Exp Sta, 50-52, from asst prof to prof agron, 54-84, D W Brooks distinguished prof, 84-87, EMER PROF AGRON, 87- *Concurrent Pos:* Tech ed, Agron J. *Mem:* Fel Am Soc Agron; Soil Conserv Soc Am; Int Soc Soil Sci; fel Soil Sci Soc Am; fel Nat Asn Cols & Teachers Agr. *Res:* Soil genesis, chemistry and fertility. *Mailing Add:* Dept Agron Univ Ga Athens GA 30602

PERKINS, HERBERT ASA, b Boston, Mass, Oct 5, 18; m 42; c 5. HEMATOLOGY. *Educ:* Harvard Univ, AB, 40; Tufts Univ, MD, 43. *Prof Exp:* Asst prof med, Sch Med, Stanford Univ, 53-58; asst prof, Sch Med, Washington Univ, 58-59; dir hemat, Jewish Hosp, St Louis, Mo, 58-59; dir res, 59-80, sci dir, 80-89, med dir, 80-90, EXEC DIR, IRWIN MEM BLOOD BANK, 87- *Concurrent Pos:* Clin prof med, Univ Calif, San Francisco, 59-; consult, Vet Admin Hosp, San Francisco, Calif; mem exec comt, Nat Marrow Donor Prog, 87- *Honors & Awards:* John Elliott Mem Award, Am Asn Blood Banks. *Mem:* Am Soc Hemat; Int Soc Blood Transfusion; Transplantation Soc; Am Asn Blood Banks; Am Asn Clin Histocompatibility Testing. *Res:* Tissue typing and organ transplantation; immunohematology; AIDS; blood banking. *Mailing Add:* Irwin Mem Blood Bank 270 Masonic Ave San Francisco CA 94118

PERKINS, JAMES, b Midland, Pa, Dec 28, 43; m; c 1. CHEMISTRY. *Educ:* Slippery Rock State Col, BS, 65; Univ Pittsburgh, PhD(phys chem), 71. *Prof Exp:* Teacher, Pub Sch, Pa, 65-67; from teaching asst to res asst, Univ Pittsburgh, 67-71; assoc prof, 71-76, PROF CHEM, JACKSON STATE UNIV, 76-, CHMN DEPT, 74-,DEAN SCH SCI & TECH. *Concurrent Pos:* Instnl res grant, Jackson State Univ, 71-73; res grants, NASA, 72-74 & NIH, 74; consult, Inst Serv Educ, 73-74; NSF equip grant, 74-76. *Mem:* Am Chem Soc; Soc Appl Spectros; Soc Advan Black Chemists & Chem Engrs. *Res:* Infrared and Raman spectroscopy. *Mailing Add:* 1920 Gunbarrel Apt 1512 Chattanooga TN 37421

PERKINS, JAMES FRANCIS, b Hillsdale, Tenn, Jan 3, 24; m 49; c 1. QUANTUM PHYSICS. *Educ:* Vanderbilt Univ, AB, 48, MA, 49, PhD(physics), 53. *Prof Exp:* Sr nuclear engr, Convair Div, Gen Dynamics Corp, 53-54; staff scientist, Lockheed Aircraft Corp, 54-59, mgr nuclear proj support dept, 59-60, scientist, 60-61; physicist, Phys Sci Lab, US Army Missile Command, 61-63, res physicist, 63-77; RETIRED. *Concurrent Pos:* Consult physicist, 77- *Mem:* Am Phys Soc; Sigma Xi. *Res:* Atomic structure; autoionization states; scattering theory; numerical techniques. *Mailing Add:* 102 Mountainwood Dr Huntsville AL 35801

PERKINS, JANET SANFORD, b Chicago, Ill, Oct 27, 13; wid. LASER ACTION ON MATERIALS. *Educ:* Wellesley Col, BA, 36; Smith Col, AM, 38; Mass Inst Technol, PhD(chem), 52. *Prof Exp:* Instr chem, Wilson Col, 38-39 & Kendall Hall, 39-40; res chemist, Arthur D Little, Inc, 40-48; res asst chem, Mass Inst Technol, 48-49; res chemist spectros, Barrett Div, Allied Chem & Dye Corp, 51-54; res assoc chem, Wellesley Col, 55-56; spec instr, Simmons Col, 55, asst prof, 56-61; sr staff scientist, Res & Advan Develop Div, Avco Corp, 62-65; consult chemist, 65-66; res chemist, Cabot Corp, 66-68; res chemist, Army Mat & Mech Res Ctr, 69-85, PRIN INVESTR, ARMY MAT LAB, 85- *Honors & Awards:* Henry A Hill Award, 84. *Mem:* Am Chem Soc; Am Carbon Soc; Sigma Xi; Mat Res Soc; Am Inst Aeronaut & Astronaut. *Res:* Structure-reactivity relationships; nature of aromaticity; polymer degradation, charring and ablation mechanisms; laser/materials interactions; carbon fiber morphology; surface and solid state chemistry; composites; polymer chemistry. *Mailing Add:* US Army Mat Technol Lab SLCMT-MEC Arsenal St Watertown MA 02172

PERKINS, JOHN PHILLIP, b Phoenix, Ariz, June 27, 37; m 61; c 3. PHARMACOLOGY, BIOCHEMISTRY. *Educ:* Ariz State Univ, BS, 60, MS, 62; Yale Univ, PhD(pharmacol), 66. *Prof Exp:* USPHS fel, Univ Wash, 66-68; asst prof, 68-73, assoc prof pharmacol, Sch Med, Univ Colo, Denver, 74-77; PROF & CHMN PHARMACOL, SCH MED, UNIV NC, CHAPEL HILL, 77-; PROF & CHMN PHARMACOL, SCH MED, YALE UNIV. *Concurrent Pos:* Res Career Develop Award, USPHS, 72-77. *Mem:* Am Soc Pharmacol & Exp Therapeut; Am Soc Biochem & Molecular Biol. *Res:* Regulation of the expression and control of function of cell-surface receptors for neurotransmitters, hormones, and growth factors; receptor desensitization and down regulation; regulation of cyclic. *Mailing Add:* 12034 Forestwood Circle Dallas TX 75244

PERKINS, KENNETH L(EE), b St Charles, Mo, Sept 7, 24; m 48; c 4. RELIABILITY MANAGEMENT, HYBRID MICROCIRCUITS. *Educ:* St Louis Univ, BSEE, 50, MS, 52, PhD(physics), 56. *Prof Exp:* Instr elec eng, St Louis Univ, 56; sr res engr, Autonetics Div, NAm Aviation, Inc, 56-60; chief scientist solid state physics, Orbitec Corp, 60; res specialist, Rockwell Int Corp, 60-61, eng supvr, 61-66, electromech eng specialist, 66-68, mem tech staff, 68-85; RES ENGR, NORTHROP CORP, 85- *Honors & Awards:* NASA Award Creative Develop Technol, 76, 79, 80 & 83. *Res:* Adhesive systems and organic coating materials for microcircuit applications; methods of determining the hermeticity and analyzing the moisture contents of hybrid microcircuit packages; low cost methods of packaging hybrid microcircuits. *Mailing Add:* 5162 Wendover Rd Yorba Linda CA 92686

PERKINS, KENNETH ROY, b Woburn, Mass, Dec 10, 42; m 72; c 3. MECHANICAL & NUCLEAR ENGINEERING. *Educ:* Tufts Univ, BS, 66; Univ Ariz, PhD(mech eng), 75. *Prof Exp:* Engr reactor safety, Aerojet Nuclear Co, Gen Tire & Rubber, 73-76; NUCLEAR ENGR REACTOR SAFETY, BROOKHAVEN NAT LAB, 76- *Mem:* Am Nuclear Soc; Am Soc Mech Engrs. *Res:* Liquid metal fast breeder reactor safety and accident analysis; thermal reactor fuels behavior; reactor safety applications of heat transfer and fluid dynamics. *Mailing Add:* Long Pond Rd Rte 1 Wading River Long Island NY 11792

PERKINS, KENNETH WARREN, b Pittsfield, Mass, Mar 3, 27; m 47; c 3. PARASITOLOGY. *Educ:* Berea Col, AB, 48; Purdue Univ, MS, 50, PhD(invert zool), 53. *Prof Exp:* Instr invert zool, Purdue Univ, 51-52; assoc prof biol & chem, High Point Col, 53-55; head dept parasitol & genetics, Carolina Biol Supply Co, 54-75, HEAD DEPT GRAPHIC ARTS, 64- *Mem:* AAAS; Am Soc Parasitologists; Am Asn Lab Animal Sci; Sigma Xi. *Res:* Development of teaching aids in parasitology and genetics; visual aids for zoology and botany. *Mailing Add:* Box 396 Elon College NC 27244

PERKINS, PETER, b Rutland, Vt, Oct 19, 35; m 57; c 5. MATHEMATICS. *Educ:* Univ Vt, BA, 57; Dartmouth Col, MA, 59; Univ Calif, Berkeley, PhD(math), 66. *Prof Exp:* From instr to assoc prof, 62-77, PROF MATH, COL OF THE HOLY CROSS, 77- *Concurrent Pos:* Mathematician, Itek Labs, 62-64, NASA, 68. *Mem:* Am Math Soc; Math Asn Am; Asn Symbolic Logic; Asn Comput Math. *Res:* Automata theory; universal algebra; equational theories. *Mailing Add:* Dept Math Col of the Holy Cross Worcester MA 01610

PERKINS, RICHARD SCOTT, b Hammond, La, June 21, 40; m 64; c 2. ELECTROCHEMISTRY. *Educ:* La State Univ, BS, 62; Univ Utah, PhD(chem) 66. *Prof Exp:* Fel chem, Univ Ottawa, Ont, 66-68 & Univ Utah, 68-69; from asst prof to assoc prof, 69-80, PROF CHEM, UNIV SOUTHWESTERN LA, 80- *Mem:* Am Chem Soc. *Res:* Photoelectrochemistry; electrochemical double layer and electrode kinetics; corrosion. *Mailing Add:* Dept Chem Box 4370 Univ Southwestern La Lafayette LA 70504-4370

PERKINS, RICHARD W, JR, b Poughkeepsie, NY, Apr 28, 32; m 54; c 4. MECHANICAL ENGINEERING, APPLIED MECHANICS. *Educ:* Dartmouth Col, BA, 54; State Univ NY Col Forestry, Syracuse, MS, 59, PhD(forestry), 63. *Prof Exp:* From instr to asst prof wood prod eng, State Univ NY Col Forestry, Syracuse, 58-64; from asst prof to assoc prof, 64-75, chmn dept, 81-90, PROF MECH & AEROSPACE ENG, SYRACUSE UNIV, 75- *Concurrent Pos:* Maitre de Conference, Univ de Poitiers, 73-74. *Mem:* Am Soc Mech Engrs; Am Acad Mech; Soc Eng Sci; Tech Asn Pulp & Paper Indust. *Res:* Mechanical behavior of heterogeneous media, solid wood, paper, composite materials; vibrations and dynamics of mechanical systems. *Mailing Add:* Dept of Mech & Aerospace Eng Syracuse Univ Syracuse NY 13244

PERKINS, ROBERT LOUIS, b Bradford, Pa, Feb 20, 31; m 52; c 2. INFECTIOUS DISEASES, INTERNAL MEDICINE. *Educ:* WVa Univ, AB, 53; MS, 54; Johns Hopkins Univ, MD, 56; Ohio State Univ, MMS, 62. *Prof Exp:* Resident med, Col Med, Ohio State Univ, 56-58, NIH Clin res fel med, Div Infectious Dis, 60-62, resident, 62-63, clin instr, 61-62, from instr to assoc prof, 63-73, assoc prof, 71-74, dir, Div Infectious Dis, 71-87, prof med, 73-90, prof med microbiol, 74-90, EMER PROF MED, COL MED, OHIO STATE UNIV, 90-; DIR MED EDUC, GRANT MED CTR, 90- *Concurrent Pos:* Attend physician, Univ Hosps, Columbus, Ohio, 63-; consult infectious dis, Dayton Vet Admin Ctr, Ohio, 65 & Wright-Patterson AFB Hosp, Fairborn, 65-; attend physician & prog dir internal med, Grant Med Ctr, 90- *Mem:* Fel Am Col Physicians; Infectious Dis Soc Am; Am Fedn Clin Res; Am Soc Microbiol; AMA; Am Col Physician Execs. *Res:* Clinical pharmacology, efficacy and tolerance of antimicrobial agents; in vitro activity and mechanisms of action of antibiotics; scanning electron microscopy. *Mailing Add:* Dept Med Educ Grant Med Ctr 111 S Grant Ave Columbus OH 43215

PERKINS, ROGER A(LLAN), metallurgical engineering, materials science, for more information see previous edition

PERKINS, ROGER BRUCE, b Hammond, Ind, Nov 8, 35; m 57; c 3. NUCLEAR PHYSICS. *Educ:* Univ Wis, BS, 55; Princeton Univ, PhD(physics), 59. *Prof Exp:* Res asst physics, Princeton Univ, 59; staff mem, Los Alamos Sci Lab, Univ Calif, 59-64; physicist, Div Res, AEC, 64-65; staff mem, 65-70, alternate div leader physics div, 71-76, div leader, Laser Res & Technol Div, 76-79, dep assoc dir inertial fusion, 79-80, dep assoc dir eng sci, 81-86, asst dir facil & fabrication, 86-88, DEP ASST DIR SUPPORT, LOS ALAMOS NAT LAB, 88- *Concurrent Pos:* Vis prof, Univ Colo, Boulder, 69-70. *Mem:* AAAS; fel Am Phys Soc. *Res:* Experimental nuclear physics; inertial confinement fusion. *Mailing Add:* Los Alamos Nat Lab Stop A145 Los Alamos NM 87545

PERKINS, RONALD DEE, b Covington, Ky, May 18, 35; m 57; c 2. PETROLOGY, MARINE GEOLOGY. *Educ:* Univ Cincinnati, BS, 57; Univ NMex, MS, 59; Ind Univ, PhD(geol), 62. *Prof Exp:* Geologist, Shell Develop Co, 62-63, res geologist, 63-68; chmn, 78-90, PROF GEOL, DUKE UNIV, 68- *Concurrent Pos:* Consult, several major oil co. *Mem:* Geol Soc Am; Am Asn Petrol Geologists; Soc Econ Paleontologists & Mineralogists; Int Asn Sedimentologists. *Res:* Carbonate petrology; stratigraphy; role of microboring organisms in alteration of marine sediments. *Mailing Add:* Dept Geol 206 Old Chem Bldg Duke Univ Durham NC 27706

PERKINS, STERRETT THEODORE, b Oakland, Calif, July 25, 32; m 70; c 6. NUCLEAR PHYSICS, PLASMA PHYSICS. *Educ:* Univ Calif, Berkeley, BS, 56, MS, 57, PhD(nuclear eng), 65. *Prof Exp:* Mech engr, Lawrence Radiation Lab, Univ Calif, 56-57, nuclear engr, 59-60; nuclear engr, Nucleonics Div, Aerojet-Gen Corp, 57-59, from sr engr to prin nuclear engr, 60-65; PHYSICIST, LAWRENCE LIVERMORE NAT LAB, UNIV CALIF, 65- *Concurrent Pos:* Consult, Nucleonics Div, Aerojet-Gen Corp, 59-60. *Mem:* Am Nuclear Soc; Am Physical Soc. *Res:* Transport theory analysis and cross section evaluation, pertaining to neutrons, charged particles and photons. *Mailing Add:* Lawrence Livermore Nat Lab L-298 PO Box 808 Livermore CA 94550

PERKINS, THOMAS K(EEBLE), b Dallas, Tex, Jan 31, 32; m 63; c 2. PETROLEUM PRODUCTION. *Educ:* Agr & Mech Col, Tex, BS, 52, MS, 53; Univ Tex, PhD(chem eng), 57. *Prof Exp:* Res engr, Dow Chem Co, 52; from instr to asst prof chem eng, Univ Tex, 55-57; res engr, Atlantic Ref Co, Atlantic Richfield Co, 57-71, dir process develop res, 71-76, dir well mech res, 76-81; mgr mat & Arctic res, Arco Oil & Gas Co, 81-83, mgr prod res, Arco Resources Technol, 83-85, DISTINGUISHED RES ADV, ARCO OIL & GAS CO, 85- *Concurrent Pos:* Distinguished lectr, Soc Petrol Engrs, 77-78. *Honors & Awards:* Lester C Uren Award, Soc Petrol Engrs, 78. *Mem:* Nat Acad Engr; Am Inst Mining, Metall & Petrol Engrs. *Res:* Oil production technology. *Mailing Add:* 6816 Stichter Dallas TX 75230

PERKINS, WALTON A, III, b Aurora, Ill, Nov 24, 33; m 54, 87; c 2. COMPUTER SCIENCE. *Educ:* Purdue Univ, BS, 55; Univ Calif, Berkeley, MA, 57, PhD(physics), 59. *Prof Exp:* Physicist, Lawrence Livermore Lab, 59-68, Lawrence Berkeley Lab, 68-73; comput scientist, Stanford Artificial Intel Lab, 73-74 & Gen Motors Res Labs, Warren, Mich, 74-81; COMPUT SCIENTIST, LOCKHEED PALO ALTO RES LAB, 81- *Mem:* Asn Comput Mach; Inst Elec & Electronics Engrs; Am Asn Artificial Intel. *Res:* Instabilities of plasmas contained in magnetic fields; diagnostic equipment used in plasma physics; internal structure of photons and pions; artificial intelligence, knowledge representation and expert systems; image understanding. *Mailing Add:* Lockheed Res Lab 0/96-20 B254F 3251 Hanover St Palo Alto CA 94304-1191

PERKINS, WILLIAM CLOPTON, b Lynchburg, Va, Jan 6, 34; m 55; c 5. NUCLEAR CHEMISTRY. *Educ:* Duke Univ, BS, 55; Johns Hopkins Univ, MA, 57, PhD(chem), 60. *Prof Exp:* RES CHEMIST, SAVANNAH RIVER LAB, E I DU PONT DE NEMOURS & CO, INC, 60- *Mem:* Am Chem Soc; Am Nuclear Soc; Am Soc Testing & Mat. *Res:* Peaceful uses of nuclear explosions; recovery of actinides from debris; nuclear fuel processing; chemical effects of nuclear reactions; process safety; systems safety analysis; solvent extraction; probabilistic risk assessment. *Mailing Add:* 946 Magnolia St SE Aiken SC 29802

PERKINS, WILLIAM ELDREDGE, b Paterson, NJ, Mar 16, 38; m 63; c 4. PHYSIOLOGY, PHARMACOLOGY. *Educ:* Lawrence Univ, BA, 60; Univ Ill, Urbana, MS, 62, PhD(physiol), 67. *Prof Exp:* Sr pharmacologist, Pfizer, Inc, 67-71; sr scientist, Warren-Teed Pharmaceut Inc, Rohm and Haas Co, 71-75, group leader, Preclin Res, Rohm and Haas Co, 75-77; sr res scientist, 77-84, prof leader, Adria Labs Inc, 84-86; SR RES SCIENTIST, G D SEARLE, 86- *Concurrent Pos:* Adj asst prof, Div Pharm, Col Pharm, Ohio State Univ, 81- *Mem:* Sigma Xi; Am Motility Soc. *Res:* Gastrointestinal physiology and pharmacology; general pharmacology. *Mailing Add:* 1003 Harms Ave Libertyville IL 43223

PERKINS, WILLIAM HUGHES, b Kansas City, Mo, Feb 21, 23; m 52; c 4. SPEECH PATHOLOGY. *Educ:* Southwest Mo State Col, BS, 43; Univ Mo, MA, 49, PhD(speech path), 52. *Prof Exp:* From asst prof to assoc prof, 52-60, PROF SPEECH PATH, UNIV SOUTHERN CALIF, 60- *Mem:* AAAS; Am Speech & Hearing Asn; Acoust Soc Am; Am Psychol Asn. *Res:* Behavior and physiology of stuttering; clinical treatment of stuttering; onset of stuttering; vocal behavior; laryngeal physiology. *Mailing Add:* 5425 Weatherford Dr Los Angeles CA 90008

PERKINS, WILLIAM RANDOLPH, b Council Bluffs, Iowa, Sept 1, 34; m 57; c 3. ELECTRICAL ENGINEERING. *Educ:* Harvard Univ, AB, 56; Stanford Univ, MS, 57, PhD(elec eng), 61. *Prof Exp:* Instr elec eng, Stanford Univ, 59-60; from asst prof to assoc prof, 61-69, assoc, Ctr Advan Study, 71-72, PROF ELEC ENG & RES PROF, COORD SCI LAB, UNIV ILL, URBANA, 69- *Honors & Awards:* Centennial Medal, Inst Elec & Electronic Engrs. *Mem:* Distinguished mem Inst Elec & Electronic Engrs Control Systs Soc (pres, 85); fel Inst Elec & Electronic Engrs. *Res:* Control systems, especially parameter sensitivity effects, feedback theory and large scale dynamic systems. *Mailing Add:* Dept Elec & Comput Eng Coord Sci Lab Univ Ill 1101 W Springfield Ave Urbana IL 61801

PERKINS, WILLIS DRUMMOND, b Porterville, Calif, Dec 20, 26; wid. CHEMICAL PHYSICS, SPECTROSCOPY FT-IR. *Educ:* Mass Inst Technol, SB, 48; Harvard Univ, AM, 50, PhD(chem physics), 52. *Prof Exp:* Res technologist analytical chem, Shell Oil Co, 51-60; prod specialist, 60-69, asst prod mgr infrared, 69-75, SR PROD SPECIALIST INFRARED SPECTROS, PERKIN-ELMER CORP, 75- *Concurrent Pos:* Lectr, molecular spectroscopy short course, Ariz State Univ, Tempe, 75- *Mem:* Am Chem Soc; Am Phys Soc; Optical Soc Am; Soc Appl Spectros. *Res:* Spectroscopic instrumentation; infrared and Raman spectroscopy; molecular structure. *Mailing Add:* 4081 Petulla Court San Jose CA 95124

PERKO, LAWRENCE MARION, b Pueblo, Colo, May 5, 36; m 62; c 5. APPLIED MATHEMATICS. *Educ:* Univ Colo, BS, 58, MS, 59; Stanford Univ, PhD(math), 65. *Prof Exp:* Res engr, Martin Marietta Co, 59-60; res scientist, Lockheed Res Labs, 63-68; asst prof math, San Jose State Univ, 65-68; assoc prof, 68-80, PROF MATH, NORTHERN ARIZ UNIV, 80- *Concurrent Pos:* Burlington Scholar, Northern Ariz Univ, 87. *Res:* Periodic orbits in the restricted three body problem, especially existence, approximation, bifurcation and stability; bifurcation of limit cycles for quadratic systems of ordinary differential equations; singular perturbation theory; nonlinear dynamical systems. *Mailing Add:* Dept Math Northern Ariz Univ Flagstaff AZ 86011

PERKOFF, GERALD THOMAS, b St Louis, Mo, Sept 22, 26; c 3. INTERNAL MEDICINE. *Educ:* Washington Univ, MD, 48. *Prof Exp:* Res fel med, Salt Lake County Gen Hosp, 49-50; from res instr to res asst prof med, Univ Utah, 54-62, assoc prof, 62-63; chief, Wash Univ Med Serv, St Louis City Hosp, 63-68, from assoc prof to prof med, Sch Med, Wash Univ, 63-68; prof med, prev med & pub health & dir, Div Health Care Res, 68-79, CURATORS PROF & ASSOC CHMN, DEPT FAMILY & COMMUNITY MED & PROF MED, UNIV MO, COLUMBIA, 79- *Concurrent Pos:* Clin instr, Sch Med, Georgetown Univ, 53-54; Markle scholar med sci, 55-60; career res prof, Found Neuromuscular Dis, 61; chief med serv, Vet Admin Hosp, Salt Lake City, Utah, 61-63; mem, Inst Med, Nat Acad Sci, 78-; Henry J Kaiser sr fel, Ctr Advan Study Behav Sci, Palo Alto, Calif, 76- 77, 85-86. *Mem:* Inst Med-Nat Acad Sci; fel Am Col Physicians; Am Fedn Clin Res; Asn Am Physicians; Am Soc Clin Invest. *Res:* Health care research; biomedical ethics; medical education. *Mailing Add:* M228 Med Sci Bldg Univ Mo Columbia MO 65212

PERKOWITZ, SIDNEY, b Brooklyn, NY, May 1, 39; m 67; c 1. SOLID STATE PHYSICS. *Educ:* Polytech Inst Brooklyn, BS, 60; Univ Pa, MS, 62, PhD(physics), 67. *Prof Exp:* Physicist, Gen Tel & Electronics Labs, Inc, 66-69; from asst to assoc prof, 69-79, chmn dept, 80-83, PROF PHYSICS, 79-, DIR, EMORY RAMAN LAB, 84-, CHARLES HOWARD CANDLER PROF, CONDENSED MATTER PHYSICS, EMORY UNIV, 87- *Concurrent Pos:* Vis prof, Univ Calif, Santa Barbara, 83-; consult, Santa Barbara Res Ctr, 83-, Nat Inst Standards & Technol, 90-91; vis sr scientist, Southwestern Univs Res Asn, 90-91. *Mem:* AAAS; Am Phys Soc; Coblentz Soc; Soc Photo-Optical Instrumentation Engrs; Soc Lit & Sci. *Res:* Optical properties of semiconductors, superconductors and microstructures; far infrared, laser and picosecond spectroscopy; semiconductor transport properties; infared detector physics; science writer and essayist. *Mailing Add:* Dept Physics Rollins Res Ctr Emory Univ Atlanta GA 30322

PERKS, ANTHONY MANNING, b Gloucester, Eng; div. PHYSIOLOGY, PHARMACOLOGY. *Educ:* Cambridge Univ, BA, 54, MA, 58; Univ St Andrews, PhD(physiol), 59; Oxford Univ, MA, 64. *Prof Exp:* Fel, Col Med, Univ Fla, 59-61; instr pharmacol, Col Physicians & Surgeons, Columbia Univ, 61-63; res officer med res, Nuffield Inst Med Res, Oxford Univ, 63-65; assoc prof biol sci, 65-72, assoc mem, Dept Obstet & Gynec, PROF ZOOL, UNIV

BC, 72- *Concurrent Pos:* Mem staff, New Eng Inst Med Res, 67-68; hon fac res scholar, Col Med, Univ Fla, Gainesville, 77- *Mem:* Am Physiol Soc; Am Zool Soc; European Soc Comput Endocrinol; Sigma Xi; Can Physiol Soc; Royal Soc Med, London. *Res:* Endocrinology; studies of the neurohypophysis, particularly elasmobranchs and in mammalian foetuses; water metabolism of the fetus. *Mailing Add:* Dept Zool Univ BC Vancouver BC V6T 1W5 Can

PERKS, NORMAN WILLIAM, b Aug 25, 32; US citizen; m 55; c 2. ELECTRICAL ENGINEERING, OPERATIONS RESEARCH. *Educ:* Drexel Inst, BS, 58; Pa State Univ, MS, 63, PhD(elec eng, math), 67. *Prof Exp:* From jr engr to sr engr, HRB-Singer, Inc, 58-63; from instr to asst prof elec eng, Pa State Univ, 63-67; STAFF ENGR, HRB-SINGER, INC, 67- *Res:* System analysis of information collection systems; queueing theory; mathematical programming; digital signal processing. *Mailing Add:* HRB Inc Systs Science Park Rd State College PA 16803

PERL, EDWARD ROY, b Chicago, Ill, Oct 6, 26; m 53; c 3. PHYSIOLOGY. *Educ:* Univ Ill, BS, 47, MD, 49, MS, 51. *Prof Exp:* Asst physiol, Univ Ill, 47-49; intern med, Harvard Med Serv, Boston City Hosp, 49-50; fel physiol, Sch Med, Johns Hopkins Univ, 50-52; from asst prof to assoc prof med, State Univ NY Upstate Med Ctr, 54-57; from assoc prof to prof, Col Med, Univ Utah, 57-71, actg head dept, 64-65 & 68-69; dir, neurobiol prog, 73-78, PROF PHYSIOL & CHMN DEPT, SCH MED, UNIV NC, CHAPEL HILL, 71-, SARAH GRAHAM KENAN PROF, 83- *Concurrent Pos:* USPHS fel neurophysiol, 51-52; NSF sr fel, 62-63; vis fel, Lab Physiol, Fac Med, Univ Toulouse, 62-63; vis prof, fac sci, Univ Paris, 65, 80 & 85, Univ Milan, 67, Fac Sci, Univ Aix Marseille, 70, Col France, 82 & 85; mem study sect, USPHS, 66-70; vis lectr, Sch Med, Univ Calif, Los Angeles, 68; mem physiol test comt, Nat Bd Med Examrs, 68-71; chmn neurol sci, Found Soc Neurosci, 68-69, res brief panel pain, comt sci, Nat Acad Sci, 85; actg pres, Soc Neurosci, 69-70. *Honors & Awards:* Bishop Lectr, Wash Univ, 70; Bousea Lectr, Int Asn Study Pain, 84. *Mem:* AAAS; Soc Neurosci; Am Physiol Soc; Int Brain Res Orgn; Int Asn Study Pain. *Res:* Neuroanatomy, neurophysiology and neurochemistry of sensory mechanisms for the body associated with thin offerent fibers with particular emphasis on pain; reflex function; autonomic function. *Mailing Add:* Dept Physiol 206H Univ NC Chapel Hill NC 27514

PERL, MARTIN LEWIS, b Brooklyn, NY, June 24, 27; c 4. PHYSICS. *Educ:* Polytech Inst Brooklyn, BChemEng, 48; Columbia Univ, PhD, 55. *Hon Degrees:* DSc, Univ Chicago, 90. *Prof Exp:* Engr, Gen Elec Co, 48-50; asst physics, Columbia Univ, 50-55; from instr to assoc prof, Univ Mich, 55-63; assoc prof, 63-64, PROF PHYSICS, STANFORD UNIV, 64- *Honors & Awards:* Wolf Prize Physics, 82. *Mem:* Nat Acad Sci. *Res:* High energy physics. *Mailing Add:* Stanford Linear Accelerator Ctr Stanford Univ Stanford CA 94309

PER-LEE, JOHN H, b Detroit, Mich, June 30, 29; m 57; c 4. MEDICINE. *Educ:* Dartmouth Col, AB, 51; Cornell Univ, MD, 55. *Prof Exp:* Instr surg, Med Col, Cornell Univ, 58-59; NIH fel otol, 61-62; assoc surg, 62-65, asst prof, 65-69, ASSOC PROF OTOLARYNGOL, SCH MED, EMORY UNIV, 69-, ASSOC PROF SURG. *Honors & Awards:* Hon Award, Am Acad Otolaryngol & Head & Neck Surg. *Mem:* Am Laryngol, Rhinol & Otol Soc; Am Acad Otolaryngol & Head & Neck Surg; AMA; Soc Univ Otolaryngol. *Res:* Otology; otolaryngology. *Mailing Add:* Dept Surg Emory Univ Sch Med Atlanta GA 30322

PERLEY, JAMES E, b Hornell, NY, Jan 21, 39; m 60. PLANT PHYSIOLOGY. *Educ:* Univ Mich, AB, 60; Yale Univ, MS, 61, PhD(biol), 65. *Prof Exp:* Asst prof biol, Wayne State Univ, 65-68; assoc prof, 68-74, PROF BIOL, COL WOOSTER, 75- *Concurrent Pos:* Res assoc, Masey Univ, Palmerston, New Zealand, 73-74, Univ BC, Vancouver, 78-79 & Plant Genetics, Inc, Davis, Calif, 86-87. *Mem:* AAAS; Am Soc Plant Physiologists. *Res:* Microbial biochemistry; plant hormone production in bacteria, fungi and higher plants. *Mailing Add:* Dept Biol Col Wooster Wooster OH 44691

PERLGUT, LOUIS E, b New York, NY, Apr 7, 15; m 40; c 2. BIOCHEMISTRY. *Educ:* Rutgers Univ, BSc, 37, MS, 38, PhD(biochem), 64. *Prof Exp:* Asst res specialist, Rutgers Univ, 64-65; from asst prof to assoc prof, 65-74, chmn dept, 70-71, PROF BIOCHEM, CALIF STATE UNIV, LONG BEACH, 74- *Concurrent Pos:* USPHS grant, 65-67. *Mem:* AAAS; Am Chem Soc. *Res:* Magnesium DNA multistrand complexes; phosphorylated high energy intermediates of oxidative phosphorylation and reverse electron transport. *Mailing Add:* Dept of Chem Calif State Univ Long Beach CA 90840

PERLICH, ROBERT WILLARD, b Minneapolis, Minn, Dec 6, 15; m 39; c 3. ANALYTICAL CHEMISTRY. *Educ:* Univ Minn, BCh, 37, MS, 38. *Prof Exp:* Chemist, Rock Anal Lab, Univ Minn, 39 & Mines Exp Sta, 39-42; chemist, Minn Mining & Mfg Co, 42-46, chief chemist, 46-54, tech supvr atomic energy prog, 54-58, proj supvr, 58-60, supvr photog processing chem, 60-73; RETIRED. *Concurrent Pos:* Pres & chmn bd dirs, Fedn Pharmacy Serv, 74-78. *Mem:* Am Chem Soc; Sigma Xi. *Res:* Micro, trace and industrial analysis. *Mailing Add:* 8607 SE Causey Ave Apt 122 Portland OR 97266

PERLIN, ARTHUR SAUL, b Sydney, NS, July 7, 23; m 50; c 5. ORGANIC CHEMISTRY. *Educ:* McGill Univ, BSc, 44, MSc, 46, PhD(chem), 49. *Prof Exp:* Res chemist, Nat Res Coun Can, 49-67; PROF CHEM, McGILL UNIV, 67- *Concurrent Pos:* Res fel, Univ Edinburgh, 51-52; sessional lectr, Univ Sask; prin scientist, Pulp & Paper Res Inst Can, 68- *Honors & Awards:* Merck Lectr, Chem Inst Can, 61; C S Hudson Award, Am Chem Soc, 79. *Mem:* Am Chem Soc; Can Chem Soc; fel Royal Soc Can. *Res:* Chemistry and biochemistry of carbohydrates and natural products; nuclear magnetic resonance spectroscopy. *Mailing Add:* Dept Chem McGill Univ Pulp & Paper Bldg Montreal PQ H3A 2A7 Can

PERLIN, MICHAEL HOWARD, b Calif. BACTERIAL DRUG RESISTANCE, HOST-PATHOGEN INTERACTIONS. *Educ:* Univ Chicago, AB, 78, SM, 80, PhD(microbiol), 83. *Prof Exp:* Postdoctoral fel, Infectious Dis Sect, Univ Chicago, 84; instr genetics, Loyola Univ, 84; asst prof, 84-90, ASSOC PROF GENETICS, UNIV LOUISVILLE, 90- *Mem:* Sigma Xi; AAAS; Am Soc Microbiol; Soc Indust Microbiol. *Res:* Evolution at the molecular level; development of bacterial resistance to aminoglycosides; characterization of the molecular bases for host specificity for a fungal phytopathogen; design of conditional lethal systems for containment of genetically engineered microbes. *Mailing Add:* Dept Biol Univ Louisville Louisville KY 40292

PERLIN, SEYMOUR, b Passaic, NJ, Sept 27, 25; m 58; c 3. PSYCHIATRY. *Educ:* Princeton Univ, BA, 46; Columbia Univ, MD, 50; Am Bd Psychiat & Neurol, dipl. *Prof Exp:* Resident, NY Psychiat Inst, 50-51; intern, Univ Mich Hosp, 51-52; resident, Manhattan Hosp, 52; resident, NY Psychiat Inst, 53-54; asst physician, Presby Hosp, NY, 54; asst psychiatrist, Col Physicians & Surgeons, Columbia Univ, 54; chief sect psychiat, Lab Clin Sci, NIMH, 55-59; chief div psychiat, Montefiore Hosp, NY, 60-63; lectr, Columbia Univ, 63-64; prof psychiat, Sch Med, Johns Hopkins Univ, 65-73, dir clin care & training, Henry Phipps Psychiat Clin Hosp, 65-73, dep dir dept psychiat & behav sci, 70-73; clin prof, 74-77, PROF PSYCHIAT & BEHAV SCI, SCH MED & DIR GRAD EDUC DEPT PSYCHIAT, MED CTR, GEORGE WASHINGTON UNIV, 77- *Concurrent Pos:* USPHS fel, 52-53; fel, Ctr Advan Study Behav Sci, Calif, 59; neuropsychiatrist, Home Aged & Infirm Hebrews, NY, 54; consult, State Dept Ment Hyg, Md, 64-70; chmn ment health study sect B, Div Res Grants, NIH, 64-66; mem clin prog-projs res rev comt, NIMH, 67-; vis fel, Princeton Univ, 73 & Oxford Univ, 74; Joseph P Kennedy Jr fel med, law & ethics, Kennedy Inst Ctr Bioethics, 74-75, sr res scholar, 74-77. *Honors & Awards:* Louis Dublin Award, Am Asn Suicidology, 78. *Mem:* Fel Am Psychiat Asn; Am Psychosom Soc; Am Asn Suicidology (pres, 69-); Am Psychopath Asn. *Res:* Clinical studies of psychiatric patients; clinical and ethical issues in suicide, death and dying. *Mailing Add:* George Washington Univ Med Ctr Psychiat 2150 Pennsylvania Ave NW Washington DC 20037

PERLIS, ALAN JAY, mathematics; deceased, see previous edition for last biography

PERLIS, IRWIN BERNARD, b Detroit, Mich, Feb 26, 25; m 47; c 4. PLANT CHEMISTRY. *Educ:* Univ Calif, AB, 50; Univ Ill, MS, 52, PhD(bot, chem), 54. *Prof Exp:* Asst plant physiol, Univ Ill, 50-54; res fel plant biochem, Calif Inst Technol, 54-55; biochemist & develop supvr, Gen Cigar Co Inc, 55-69, asst dir develop prod & res, 69-73, asst dir, res labs, 73-86, DIR, RES & DEVELOP, GEN CIGAR CO, INC, DIV CULBRO CORP, 86- *Mem:* Sigma Xi; Am Chem Soc; Am Soc Testing & Mat; Asn Off Anal Chemists; AAAS. *Res:* Auxin effects on enzymes, cell wall substances and organic acids during normal growth; chemistry of tobacco; chemistry and technology of tobacco sheets. *Mailing Add:* 2747 Brookfield Rd Lancaster PA 17601

PERLIS, SAM, b Maywood, Ill, Apr 18, 13; m 39; c 2. ALGEBRA. *Educ:* Univ Chicago, BS, 34, MS, 36, PhD(math), 38. *Prof Exp:* Asst math, Univ Chicago, 38-39; instr, Ill Inst Technol, 38-41, Univ Mich, 41 & Ill Inst Technol, 41-42; sr res asst, Lockheed Aircraft Corp, Calif, 42-46; from instr to prof, 46-79, EMER PROF MATH, PURDUE UNIV, LAYFAYETTE, 79- *Mem:* Am Math Soc; Math Asn Am. *Res:* Algebras; radicals; rings; fields; matrices. *Mailing Add:* 704 Sugar Hill Dr West Lafayette IN 47906

PERLMAN, BARRY STUART, b Brooklyn, NY, Dec 5, 39; m 78; c 1. AUTOMATION. *Educ:* City Col NY, BEE, 61; Polytech Inst NY, MSEE, 64, PhD(electro physics), 73. *Prof Exp:* Sr MTS, RCA Surface Communications, RCA Labs, 61-68, mem staff, 68-81, mgr, 81-86, head, 86-88; CHIEF, US ARMY ELECTRONICS TECH & DEVICE LAB, 88- *Mem:* Fel Inst Elec & Electronic Engrs; Sigma Xi; Asn Comput Mach. *Res:* Microwave solid state research; device and circuit research and development; design and test automation research; technology assessment; software and hardware development; conceptual design; modeling and systems integration. *Mailing Add:* SLCET-MP Ft Monmouth NJ 07703

PERLMAN, ELY, b New York, NY, Nov 11, 13; m 40; c 2. ALLERGY, IMMUNOLOGY. *Educ:* Columbia Univ, AB, 34; NY Univ, MD, 38. *Prof Exp:* Littauer pneumonia res fel, Harlem Hosp, 38-40, intern, 40-42; res fel, Dept Allergy New York Hosp & Med Col, Cornell Univ, 42 & Rockefeller Inst Hosp, 42-46; res assoc, 46-48, ASSOC ATTEND PHYSICIAN, MT SINAI HOSP, NEW YORK, 48-; CHIEF ALLERGY SERV, LONG ISLAND JEWISH HOSP, 54- *Concurrent Pos:* Clin asst prof pediat, State Univ NY Downstate Med Ctr, 59-; assoc prof microbiol, Mt Sinai Sch Med, 66-; assoc prof, Sch Med, Health Sci Ctr, State Univ NY Stony Brook, 71- *Mem:* AAAS; Harvey Soc; Am Asn Immunologists; fel Am Col Allergists; fel Am Acad Allergy. *Res:* Nature of cross reactions; theoretical studies on rates of antigen-antibody reactions; purification and electrophoretic analysis of ragweed fractions with immunological studies; C-protein and C-antibody; cold-agglutinins; chemistry and pharmacology of antihistaminics; pollen counting and air pollution; staphycococcal antigens. *Mailing Add:* 118 Crescent Lane Roslyn Heights NY 11577

PERLMAN, ISADORE, nuclear chemistry; deceased, see previous edition for last biography

PERLMAN, KATO (KATHERINE) LENARD, b Budapest, Hungary, July 18, 28; wid. ORGANIC CHEMISTRY. *Educ:* Eotvos Lorand Univ, Budapest, dipl, 50, Dr rer nat(org chem), 61. *Prof Exp:* Res chemist, Chinoin Pharmaceut, Budapest, 50-54; res assoc, Res Inst Pharmaceut Indust, 54-62; res assoc & fel, Princeton Univ, 63-68; assoc scientist, Sch Pharm, 68-81, SR SCIENTIST, BIOCHEM DEPT, UNIV WIS-MADISON, 81- *Mem:* Am Chem Soc; The Chem Soc. *Res:* Synthesis of organic chemicals of medicinal interest, including barbiturates, alkaloids, steroids, peptides, hallucinogenic drugs, tetrahydro-cannabinol and pteridine; antibiotics; vitamin antagonists; vitamin D metabolites, synthesis. *Mailing Add:* One Chippewa Ct Madison WI 53711

PERLMAN, MAIER, b Moineshti, Romania, May 14, 37; Can citizen; m 60; c 1. ENERGY MANAGEMENT, ELECTRICAL ENERGY UTILIZATION. *Educ:* Polytech Inst Jassy, Romania, MASc, 61. *Prof Exp:* Electromech engr, Vegetable Oil Factory Unirea, 61-63; chief engr equip, Construct & Assembly Enterprise, 63-68; asst prof, Polytech Inst Jassy, Romania, 65-73; mgr, Energy Lab, Standards Inst Israel, 73-80; sr res engr, 80-86, HEAD, ENERGY MGT UNIT, RES DIV, ONT HYDRO, 86- *Concurrent Pos:* Mem, Standard Comt Elec Water Heaters, Can Standards Asn, 81-90; tech adv res projs, Can Elec Asn, 81-; chmn, Tech Comt, Am Soc Heating, Refrigerating & Air- Conditioning Engrs, 87-90; mem, Adv Comt, Am Coun Energy Efficient Econ-Summer Study Energy Efficiency in Bldgs, 88-90. *Honors & Awards:* Energy Award, Am Soc Heating, Refrigerating & Air Conditioning Engrs, 87. *Mem:* Can Elec Asn; Can Standards Asn; sr mem Asn Energy Engrs; Am Soc Heating, Refrigerating & Air Conditioning Engrs; Int Solar Energy Soc; Solar Energy Soc Can; Standards Coun Can. *Res:* Energy management and utilization; heat transfer; fluid dynamics; waste heat recovery; solar energy; heat pump technology; building thermal performance; author of various publications. *Mailing Add:* 344 Yorkhill Blvd Thornhill ON L4J 3B6 Can

PERLMAN, MICHAEL DAVID, b Chicago, Ill, Dec 1, 42. MATHEMATICAL STATISTICS. *Educ:* Calif Inst Technol, BSc, 63; Stanford Univ, MSc, 65, PhD(statist), 67. *Prof Exp:* From asst prof to assoc prof statist, Univ Minn, 68-73; assoc prof & chmn dept, Univ Chicago, 73-77; PROF STATIST & CHMN DEPT, UNIV WASH, 79- *Concurrent Pos:* Assoc ed, Ann Statist, 74-77, ed, 83-86. *Mem:* Fel Inst Math Statist; fel Am Statist Asn; Math Asn Am. *Res:* Exact small sample properties of multivariate tests and estimates; probability inequalities for convex regions; monotonicity properties of multivariate power functions. *Mailing Add:* Dept Statist Univ Wash Seattle WA 98195

PERLMAN, MORRIS LEONARD, solid state chemistry; deceased, see previous edition for last biography

PERLMAN, PHILIP STEWART, b Baltimore, Md, July 27, 45; m 80; c 2. GENETICS, MOLECULAR BIOLOGY. *Educ:* Johns Hopkins Univ, BA, 66; Ind Univ, Bloomington, PhD(biochem), 71. *Prof Exp:* Fel biochem with Prof H R Mahler, Ind Univ, Bloomington, 71; from asst prof to prof genetics, 71-87, prof molecular genetics, Ohio State Univ, 87-90; PROF BIOCHEM, UNIV TEX SOUTHWESTERN MED CTR, 90- *Concurrent Pos:* Vis prof, Universite de Paris Sud, 79-80; dir, Molecular, Cellular & Develop Biol Prog, Ohio State Univ, 83-87; actg chair, genetics, 86-87, chair molecular genetics, Ohio State Univ, 87- *Mem:* AAAS; Sigma Xi; Genetics Soc Am; Am Soc Microbiol. *Res:* Molecular mechanisms of mitochondrial recombination; mitochondrial gene structure and regulation; RNA splicing. *Mailing Add:* Dept Biochem Univ Tex Southwestern Med Ctr 5323 Harry Himes Blvd Dallas TX 75235-9038

PERLMAN, RICHARD, b Madison, Wis, Feb 6, 20; m 68. BIOCHEMISTRY. *Educ:* Univ Wis, BA, 41, MSc, 43, PhD(biochem), 45. *Prof Exp:* Biochemist, Hoffmann-La Roche, Inc, NJ, 45; microbiologist, Merck & Co, 45-47; biochemist, E R Squibb & Sons, 47-52 & Squibb Inst Med Div, Olin Mathieson Chem Corp, 52-67; dean sch pharm, 68-75, PROF PHARMACEUT BIOCHEM, SCH PHARM, UNIV WIS-MADISON, 67- *Concurrent Pos:* Knapp vis prof, Sch Pharm, Univ Wis, 58; mem exec comt, Intersci Conf Antimicrobial Agents & Chemother, 63-66, chmn, 66-67; Guggenheim Found fel, 66; chmn vitamins & metab pathways, Gordon Conf, 67; ed, Advan Appl Microbiol, 67-; chmn, Third Int Symp Genetics Indust Microorganisms, 78. *Honors & Awards:* Marvin J Johnson Award, 78; Fisher Sci Award, 79; Charles Thon Award, Res in Appl Microbiol, 79; Pasteur Award, Appl Microbiol, 79. *Mem:* Am Chem Soc; Am Soc Microbiol; Brit Biochem Soc; fel Am Acad Microbiol; fel Acad Pharmaceut Sci. *Res:* Physiology of microorganisms; intermediary metabolism of molds; production of antibiotics by microorganisms; microbial transformation of organic compounds; tissue culture; industrial fermentations; microbial metabolites; antibiotics. *Mailing Add:* Sch Pharm Univ Wis Madison WI 53706

PERLMAN, ROBERT, b Chicago, Ill, Aug 15, 38; m 64; c 2. BIOCHEMISTRY, MEDICINE. *Educ:* Univ Chicago, AB, 57, SB, 58, MD, 61, PhD(biochem), 63. *Prof Exp:* From intern to resident pediat, Bellevue Hosp, New York, 63-65; staff assoc, Nat Inst Arthritis & Metab Dis, 65-67, med officer, 67-71; assoc prof physiol, Harvard Med Sch, 71-81; PROF & HEAD PHYSIOL & BIOPHYS, UNIV ILL COL MED, 81- *Concurrent Pos:* USPHS fel biochem, Univ Chicago, 61-63. *Res:* Synthesis and secretion of catecholamines; regulation of cell metabolism; physiology. *Mailing Add:* 1134 Hinman Evanston IL 60202

PERLMAN, T(HEODORE), mechanical engineering, technical management; deceased, see previous edition for last biography

PERLMUTT, JOSEPH HERTZ, b Savannah, Ga, Dec 8, 18; m 48; c 3. PHYSIOLOGY. *Educ:* Col Charleston, BS, 39; Univ NC, MA, 42; Princeton Univ, PhD(biol), 50. *Prof Exp:* Asst, Princeton Univ, 47-49; asst prof physiol, Sch Med, Univ Okla, 50; res assoc surg, Sch Med, Univ Pa, 51-53; from asst prof to prof, 53-89, EMER PROF PHYSIOL, SCH MED, UNIV NC, CHAPEL HILL, 89- *Mem:* AAAS; Am Physiol Soc; Soc Exp Biol & Med; Am Soc Nephrology. *Res:* Endocrinology; renal physiology. *Mailing Add:* Dept Physiol Univ NC Sch Med Chapel Hill NC 27514

PERLMUTTER, ALFRED, b New York, NY, Dec 7, 14; m 39; c 3. ICHTHYOLOGY. *Educ:* NY Univ, BS, 34; Univ Mich, MS, 36, ScD(zool), 41. *Prof Exp:* Aquatic biologist, US Fish & Wildlife Serv, 40-49; sr aquatic biologist, State Conserv Dept, NY, 50-60; from assoc prof to prof, 60-84, EMER PROF BIOL, WASH SQ COL, NY UNIV, 84- *Concurrent Pos:* Sea fisheries specialist, State Univ NY-US Opers Mission, Israel, 55-56. *Mem:* AAAS; Am Fisheries Soc; Am Soc Ichthyol & Herpet; Am Soc Limnol & Oceanog; Nat Shellfisheries Asn. *Res:* Aquatic biology; sea fisheries and pond fish culture. *Mailing Add:* Dept Biol Wash Square Col NY Univ New York NY 10003

PERLMUTTER, ARNOLD, b Brooklyn, NY, Nov 4, 28; m 49, 80; c 2. PHYSICS. *Educ:* Univ Calif, Los Angeles, BA, 49; NY Univ, MS, 51, PhD(physics), 55. *Prof Exp:* Asst physics, NY Univ, 51-54; instr, Cooper Union, 54-55 & Brooklyn Col, 55-56; from asst prof to assoc prof, 56-68, PROF PHYSICS, UNIV MIAMI, 68-, SECY, CTR THEORET STUDIES, 65- *Concurrent Pos:* Res assoc, Midway Lab, Univ Chicago, 56 & Univ Calif, Los Angeles, 60; Fulbright travel grant, Nat Inst Nuclear Physics, Univ Trieste, 61-62; Israel Atomic Energy Comn fel, Weizmann Inst, 62-63; vis physicist, Argonne Nat Lab, 65, consult, 66-; vis prof, Imp Col Sci & Technol, Univ London, 73-74, Univ Mich, 78-82, Univ Torino, 83-84. *Mem:* Am Phys Soc; Fedn Am Scientists. *Res:* Photoconductivity of phosphors; optical properties of gallium arsenide; nuclear emulsions; elementary particles; potential scattering computations; electromagnetic radiation; polarized nucleon scattering; global energy problems. *Mailing Add:* Ctr for Theoret Studies Univ of Miami Coral Gables FL 33124

PERLMUTTER, ARTHUR, b Lodz, Poland, Feb 26, 30; US citizen; m 57; c 2. CHEMICAL ENGINEERING. *Educ:* Univ Havana, BS, 55; Columbia Univ, MS, 65. *Prof Exp:* Jr chem engr, Rayon Co, Cuba, 56-57, chem engr, 57-59; chem engr, Goodyear Tire Co Cuba, 59-60; chem engr & pilot plant supvr, Rheingold Breweries, Inc, 61-67; res chem engr, Cent Res Div, Am Cyanamid Co, 68-75, sr process engr, Eng & Construct Div, 75-80; sr tech assoc, corp eng dept, 80-83, plant design mgr, 83-86, proj mgr, 86-90, MGR PLANT DESIGN, GAF CHEM CORP, 90- *Mem:* Am Chem Soc; Am Inst Chem Engrs. *Res:* Development of high tenacity rayon tire cord; development of low-carbohydrate beer; design of shellfish culture plant; development of fuel tank inerting system; development of solid waste disposal system; disposal of thermoplastic materials; development and design of a process waste water conveyance and spill control system. *Mailing Add:* Corp Eng Dept GAF Corp 1361 Alps Rd Wayne NJ 07470

PERLMUTTER, DANIEL D, b Brooklyn, NY, May 24, 31; m 54; c 3. CHEMICAL ENGINEERING. *Educ:* NY Univ, BChE, 52; Yale Univ, DEng, 56. *Prof Exp:* Res & develop engr, Esso Res & Eng Co, 55-58; instr, Newark Col Eng, 56-58; asst prof chem eng, Univ Ill, 58-64; assoc prof, 64-66, chmn grad group chem eng, 69-77, PROF CHEM ENG, UNIV PA, 67- *Concurrent Pos:* Guggenheim fel, 64-65; Fulbright fel, Eng, 68-69 & Yugoslavia, 72. *Honors & Awards:* Lectureship Award, Am Soc Eng Educ, 79. *Mem:* Am Inst Chem Engrs. *Res:* Automatic process control; reactor design. *Mailing Add:* Sch of Chem Eng Univ of Pa Philadelphia PA 19174

PERLMUTTER, FRANK, b NJ, June 2, 12; m 42; c 2. HORTICULTURE. *Educ:* NC State Col, BS, 34; Univ Md, MS, 39. *Prof Exp:* Student aide hort, NC State Col, 34; dist supvr plant dis, USDA, NC, 34, inspector, 36, horticulturist, Plant Indust Sta, Md, 38-41, soil conservationist, NC, 41-46; horticulturist, Caswell Training Sch, 35-36, DC Training Sch, MD, 37-38 & US Vet Admin, NY, 46-48; inspector plant quarantine, USDA, 48-60, supvr, 60-68; teacher biol, De Witt Clinton High Sch, 68-82, teacher horticult, 77-82; RETIRED. *Concurrent Pos:* Vol, New York Bot Garden, 76-; consult & vol horticult greenhouse gardening, Time Life Encycl Gardening, 77; substitute teacher, Bronx High Sch Sci, 83-91. *Mem:* Fel AAAS; Am Soc Hort Sci; NY Acad Sci; Am Inst Biol Sci. *Res:* Biology, botany; plant quarantine, plant breeding and pathology; entomology; genetics. *Mailing Add:* 3965 Sedgwick Ave Bronx NY 10463-3104

PERLMUTTER, HOWARD D, b Brooklyn, NY, May 27, 38; c 1. ORGANIC CHEMISTRY. *Educ:* Lehigh Univ, BS, 59; NY Univ, MS, 62, PhD(org chem), 63. *Prof Exp:* Fel org chem, Univ Wis, 63-65; asst prof, 65-71, ASSOC PROF CHEM ENG & CHEM, NJ INST TECHNOL, 71- *Concurrent Pos:* NIH fel, 63-65. *Mem:* Am Chem Soc. *Res:* Physical organic chemistry; stereochemistry and mechanism of organic reactions; thermal reorganization reactions; cyclooctatetraenes; novel aromatic and heterocyclic compounds; solid state photochemistry; diazocines. *Mailing Add:* Dept of Chem Eng & Chem NJ Inst of Technol 323 High St Newark NJ 07102

PERLMUTTER, ISAAC, b Russia, Nov 23, 12; nat US; m 47; c 2. METALLURGY. *Educ:* Mass Inst Technol, SB, 34. *Prof Exp:* Metallurgist, Carnegie-Ill Steel Corp, 36-40 & US Bur Mines, 40; proj engr, Metals & Ceramics Div, US Air Force Mat Lab, Wright-Patterson AFB, 41-46, chief high temperature mat sect, 46-59, chief phys metall br, 59-73; consult aircraft struct metals, 72-; RETIRED. *Mem:* Fel Am Soc Metals. *Res:* Gas turbine materials; beryllium; heat resistant coatings; refractory metals; high strength steels; bearing materials; aircraft structural alloys of titanium and aluminum. *Mailing Add:* 150 Trailwoods Dr Dayton OH 45415

PERLMUTTER, ROGER M, b Denver, Colo, Sept 16, 52. MOLECULAR IMMUNOLOGY. *Educ:* Reed Col, BA, 73; Washington univ, St Louis, MD & PhD(immunology), 79. *Prof Exp:* Sr res fel biochem, Calif Inst Technol, 81-84; from asst prof to assoc prof med & biochem, 84-90, PROF & CHMN, DEPT IMMUNOL, UNIV WASH, 89-, ASSOC INVESTR, HOWARD HUGHES MED INST, 84- *Mem:* Am Asn Immunologists; Am Asn Clin Invest. *Res:* Molecular biology of lymphoid neoplasta and of signal transduction in immune cells. *Mailing Add:* Howard Hughes Med Inst Univ Wash SL15 Seattle WA 98195

PERLOFF, DAVID STEVEN, b Philadelphia, Pa, Mar 31, 42; m 65; c 3. PHYSICS, MATERIALS SCIENCE. *Educ:* Univ Pa, BA, 63; Brown Univ, PhD(physics), 69. *Prof Exp:* Res physicist, Corning Glass Works, 68-71, sr mem tech staff, Res & Develop Lab, 72-77, mgr device & process characterization, Advan Technol Ctr, 78-81, mgr charaterization technol, Philips Res Lab, Signetics Corp, 81-; PRES, PROMETRIX CORP, 84- *Mem:* Am Phys Soc; Electrochem Soc. *Res:* Applications of ion implantation to silicon device and integrated circuit fabrication; process modeling and characterization. *Mailing Add:* Prometrix Corp 3255 Scott Blvd Bldg 2 Santa Clara CA 95051

PERLOFF, WILLIAM H(ARRY), JR, b Philadelphia, Pa, May 2, 36; m 59; c 3. PEDIATRICS, CRITICAL CARE. *Educ:* Swarthmore Col, BS, 57; Northwestern Univ, MS, 58, PhD(civil eng), 62; Case Western Reserve Univ, MD, 77; Am Bd Pediat, dipl, 82. *Prof Exp:* Asst prof civil eng, Ohio State Univ, 62-65; from assoc prof to prof soil mech, Purdue Univ, 65-73; resident pediat, Cleveland Clinic Found, 77-79; fel clin pharmacol & pediat intensive care, Univ Hosps Cleveland, 79-81; asst prof, Case Western Reserve Univ, 81-82; asst prof pediat & anesthesia, Sch Med, Univ Wis, 82-86, assoc prof, 86-90, DIR PEDIAT CRITICAL CARE MED, UNIV WIS HOSPS, MADISON, 82-, PROF PEDIAT & ANESTHESIA, UNIV WIS, 90- *Mem:* Am Acad Pediat; Soc Critical Care Med; Am Trauma Soc. *Res:* Cardiorespiratory function in pediatric intensive care. *Mailing Add:* Univ Wis Children's Hosp 600 Highland Ave Madison WI 53792

PERLOW, GILBERT JEROME, b New York, NY, Feb 10, 16; m 41. EXPERIMENTAL PHYSICS. *Educ:* Cornell Univ, AB, 36, MA, 37; Univ Chicago, PhD(physics), 40. *Prof Exp:* Instr physics, Univ Minn, 40-41; physicist, Naval Ord Lab, 41-42 & Naval Res Lab, 42-52; res assoc, Univ Minn, 52-53; physicist, 53-58, SR PHYSICIST, ARGONNE NAT LAB, 58- *Concurrent Pos:* Vis assoc prof, Univ Wash, 57; vis physicist, Atomic Energy Res Estab, Eng, 61; vis prof, Univ Ill, Chicago, 65, Munich Tech Univ, 68 & 79 & Free Univ Berlin, 69; ed, J Appl Physics, 70-73 & Appl Physics Lett, 70-90; sr Humboldt award, Munich Tech Univ, 74-75. *Mem:* Fel Am Phys Soc. *Res:* X-rays; radio; cosmic rays; nuclear physics; Mossbauer effect. *Mailing Add:* 4919 Northcott Ave Downers Grove IL 60515

PERLOW, MINA REA JONES, b Harrodsburg, Ky; m 41. INORGANIC CHEMISTRY, RADIOCHEMISTRY. *Educ:* Centre Col, AB, 31; Pa State Col, MS, 33; Univ Chicago, PhD(inorg chem), 41. *Prof Exp:* Chemist, WVa Pulp & Paper Co, 33-38; commodity specialist, Inorg Chem Sect, War Prod Bd, 42-45, Civilian Prod Admin, 45-47 & Off Domestic Commerce Chem & Drugs Div, US Dept Commerce, 48-50; consult, USDA, 51; CONSULT CHEMIST, 51- *Concurrent Pos:* Consult, Argonne Nat Lab, 60-75. *Res:* Lithium-lithium hydride-hydrogen equilibrium; radiochemistry; Mossbauer effect. *Mailing Add:* 4919 Northcott Ave Downers Grove IL 60515

PERLSTEIN, JEROME HOWARD, b New York, NY, Mar 25, 41; m 66; c 2. SOLID STATE CHEMISTRY. *Educ:* Brandeis Univ, AB, 61; Cornell Univ, PhD(chem), 67. *Prof Exp:* Asst prof solid state chem, Johns Hopkins Univ, 67-73; SR RES CHEMIST, EASTMAN KODAK CO, 73- *Mem:* Am Chem Soc. *Res:* Electron transport in transition metal complexes; solid state properties of non-stoichiometric inorganic and organic compounds; low temperature electrical properties and magnetic properties of the solid state; photoconductivity of amorphous organic films. *Mailing Add:* Res Labs Eastman Kodak Co Rochester NY 14650

PERMAN, VICTOR, b Greenwood, Wis, Jan 28, 26; c 5. VETERINARY PATHOLOGY. *Educ:* Univ Minn, BS, 53, DVM, 55, PhD(vet path), 62. *Prof Exp:* Instr, 55-58 & 59-62, from asst prof to assoc prof, 62-68, PROF VET PATH, UNIV MINN, ST PAUL, 68-, CHMN, DEPT VET PATHBIOLOGY, 84- *Concurrent Pos:* Med scientist, Brookhaven Nat Lab, 58-59, res collab, 59-62 & 67. *Mem:* Am Col Vet Path; Am Soc Vet Clin Path; Am Vet Med Asn; Am Animal Hosp Asn. *Res:* Comparative hematology; hypoplastic and proliferative diseases of hemic system; veterinary clinical pathology. *Mailing Add:* Col Vet Univ Minnesota St Paul MN 55108

PERMODA, ARTHUR J, chemical engineering, for more information see previous edition

PERMUTT, SOLBERT, b Birmingham, Ala, Mar 6, 25; m 52; c 3. MEDICINE. *Educ:* Univ Southern Calif, MD, 50. *Prof Exp:* Intern, Univ Chicago Clins, 49-50, asst resident med, 52-53, res assoc anat, Univ, 50-52; chief resident pulmonary div, Montefiore Hosp, New York, 54-55, chief resident med div, 55-56; Nat Found fel med & environ med, Johns Hopkins Univ, 56-58; chief cardiopulmonary physiol div, Nat Jewish Hosp at Denver, Colo, 58-61; asst prof physiol, Sch Med, Univ Colo, 60-61; assoc prof environ med, 61-65, PROF ENVIRON HEALTH SCI, SCH HYG & PUB HEALTH, JOHNS HOPKINS UNIV, 65-, PROF MED, SCH MED, 71-, PROF ANESTHESIOL & CRITICAL CARE MED, 78- *Honors & Awards:* Distinguished Achievement Award, Am Heart Asn & Citation for Distinguished Serv Res; George Wills Comstock MD Award, Am Lung Asn; Gold Medal, Am Col Chest Physicians. *Mem:* AAAS; Am Thoracic Soc; Am Fedn Clin; Am Physiol Soc; Am Heart Asn; Asn Am Physicians; Am Lung Asn. *Res:* Respiratory and circulatory physiology; pulmonary medicine; mechanical interactions of circulatory and respiratory systems. *Mailing Add:* Johns Hopkins Med Insts Johns Hopkins Asthma & Allergy Ctr 301 Bayview Blvd Baltimore MD 21224

PERNICK, BENJAMIN J, b New York, NY, June 8, 31; m 56; c 3. PHYSICAL MATHEMATICS, MATH MODELING. *Educ:* City Col NY, BS, 54; Stevens Inst Technol, MS, 58, PhD(physics), 65. *Prof Exp:* Res physicist, Stevens Inst Technol, 56-65; STAFF SCIENTIST, GRUMMAN AEROSPACE CORP, 65-; ADJ ASSOC PROF PHYSICS, COL STATEN ISLAND, 77- *Concurrent Pos:* Adj asst prof math, Brooklyn Col, 64-77; lectr, Nat Sem Laser Technol, NY Univ, 82-84. *Mem:* Optical Soc Am; Soc Photo-Optical Instrumentation Engrs. *Res:* Laser systems; laser applications; optical computing (analog and digital); holography; spatial light modulators; non-destructive testing; plasma physics; hypervelocity phenomena; megagauss magnetic fields; exploding wire phenomena; thin film technology; recipient of numerous patents and author of many journal papers. *Mailing Add:* 110-11 Queens Blvd No 23E Forest Hills NY 11375

PERO, JANICE GAY, b Lowell, Mass, June 11, 43; m 70; c 2. MOLECULAR BIOLOGY. *Educ:* Oberlin Col, BA, 65; Harvard Univ, PhD(biochem, molecular biol), 71. *Prof Exp:* Res fel biol, Harvard Univ, 71-75, res assoc, 75, asst prof, 75-78, assoc prof, 78-83; VPRES, OMNIGENE, INC, 83- *Concurrent Pos:* Am Cancer Soc fel, 71-73; tutor biochem sci, Harvard Univ, 73-77. *Mem:* Am Soc Microbiol. *Res:* Regulation of gene expression; control of transcription of bacteriophage DNA by regulatory subunits of bacterial RNA polymerase. *Mailing Add:* OmniGene Inc 85 Bolton St Cambridge MA 02140

PERONA, JOSEPH JAMES, b Blanford, Ind, May 28, 30; m 55; c 4. CHEMICAL ENGINEERING. *Educ:* Rose Polytech Inst, BS, 52; Northwestern Univ, PhD(chem eng), 56. *Prof Exp:* Sr chem engr, Oak Ridge Nat Lab, 56-68; head dept, 84-90, PROF CHEM ENG, UNIV TENN, KNOXVILLE, 68- *Concurrent Pos:* Consult, Oak Ridge Nat Lab, 68- & Dept Energy Hq, 75-81. *Mem:* Fel Am Inst Chem Engrs; Am Soc Eng Educ. *Res:* Heat transfer; mass transfer and reaction; hazardous waste treatment. *Mailing Add:* Dept of Chem Eng Univ of Tenn Knoxville TN 37996-2200

PERONE, SAMUEL PATRICK, b Rockford, Ill, Oct 1, 38; m 56, 81; c 4. ANALYTICAL CHEMISTRY. *Educ:* Rockford Col, BA, 59; Univ Wis, PhD(analytical chem), 63. *Prof Exp:* From asst prof to prof analytical chem, Purdue Univ, 62-81; staff analytical chem sect, Lawrence Livermore Lab, 81-83, head, 84-86, mgr, analytical sci, 87-89; PROF, ANALYTICAL CHEM, SAN JOSE STATE UNIV, 89-, ASSOC DEAN SCI RES, 90- *Concurrent Pos:* Chmn, ACS Analysis Div, 88-89. *Mem:* Am Chem Soc; Soc Electroanal Chem. *Res:* Information theory; lab computer applications to pattern recognition and instrumental analysis; chemical sensors. *Mailing Add:* Dept Chem San Jose State Univ One Washington Sq San Jose CA 95192

PERONNET, FRANC04OIS R R, b Valence, France, Jan 27, 46; French & Can citizen; m 85. EXERCISE PHYSIOLOGY. *Educ:* Ecole Normale Superieure, d'Educ Physique, Paris, BSc, 69; Univ Montreal, MSc,75, PhD(physiol), 80. *Prof Exp:* PROF EXERCISE PHYSIOL, UNIV MONTREAL, 76- *Honors & Awards:* New Investr Award, Am Col Sports Med, 81. *Mem:* Fel Am Col Sports Med; Can Asn Sports Sci. *Res:* Exercise and the autonomic system; sympathetic response to exercise and role of the sympathetic system in the cardiovascular and metabolic adjustments to exercise and training. *Mailing Add:* Dept Phys Educ Univ Montreal Cp 6128 Succ A Montreal PQ H3C 3J7 Can

PEROT, PHANOR L, JR, b Monroe, La, July 19, 28; m 54; c 5. NEUROSURGERY. *Educ:* Tulane Univ, BS, 49, MD, 52; McGill Univ, dipl neurosurg, 61, PhD(neurol, neurosurg), 63; Am Bd Neurol Surg, dipl, 63. *Prof Exp:* Intern, Philadelphia Gen Hosp, Pa, 52-53; asst resident gen surg, Hosp Univ Pa, 53-54; resident neurol serv, Montreal Neurol Inst, 56-57, asst res neurosurg, 57-59; sr neurosurg res & demonstr neurol & neurosurg, Fac Med, McGill Univ, 60-61, from lectr to asst prof neurosurg, 62-68; PROF NEUROL SURG & CHMN DEPT, MED UNIV SC, 68- *Concurrent Pos:* Fel neurosurg path, Montreal Neurol Inst, 57, fel neurophysiol, 59-60; Nat Inst Neurol Dis & Blindness Spec fels, 59-62; asst neurosurgeon, Montreal Neurol Inst & Hosp, 61-68; clin asst neurosurgeon, Royal Victoria Hosp, 61-65, asst neurosurgeon, 65-68. *Mem:* Fel Am Col Surgeons; Am Epilepsy Soc; Soc Univ Neurosurgeons; Soc Neurol Surgeons; Am Acad Neurol Surgeons. *Res:* Basic neurophysiological studies in epilepsy in animals and man; auditory and visual memory in the human brain; clinical neurosurgery; clinical and experimental studies in spinal cord injury. *Mailing Add:* Dept Neurol Surg Med Univ SC 171 Ashley Ave Charleston SC 29425-2272

PEROUTKA, STEPHEN JOSEPH, b Baltimore, Md, Feb 21, 54; m 79. NEUROLOGY. *Educ:* Cornell Univ, AB, 75; Johns Hopkins Univ, MD, 79, PhD(pharmacol), 80. *Prof Exp:* Intern med, Stanford Univ, 80-81; resident fel neurol, Johns Hopkins Univ, 81-84; ASST PROF NEUROL & PHARMACOL, STANFORD UNIV, 84- *Mem:* Soc Neurosci; Am Acad Neurol. *Res:* Molecular pharmacology of the central nervous system with emphasis on serotonin receptor analysis. *Mailing Add:* Dept Neurol Pharmacol Stanford Univ Hosp 861 Allardice Way Stanford CA 94305

PEROVICH, DONALD KOLE, b Pittsburgh, Pa, Aug 25, 50; m 79; c 3. SEA ICE GEOPHYSICS, OPTICAL PROPERTIES OF ICE & SNOW. *Educ:* Mich State Univ, BS, 71; Univ Wash, MS, 79, PhD(geophysics), 83. *Prof Exp:* Res assoc, Geophysics Prog, Univ Wash, 84-85; GEOPHYSICIST, US ARMY COLD REGIONS RES & ENG LAB, 86- *Concurrent Pos:* Assoc ed, J Geophys Res: Oceans, 89-; mem, Comt Polar Meteorol & Oceanog, Am Meteorol Soc, 89-; mem, Comt on Snow, Ice & Permafrost, Am Geophys Union, 90- *Mem:* Am Geophys Union; Int Glaciol Soc; Electromagnetics Acad. *Res:* Studying the interaction of shortwave radiation with sea ice; relating the optical properties of sea ice to the state and structure of the ice; sea ice thermodynamics. *Mailing Add:* US Army Cold Regions Res & Eng Lab Hanover NH 03755

PEROZZI, EDMUND FRANK, b Camden, NJ, Feb 16, 46; m 72; c 1. LUBRICANT ADDITIVES. *Educ:* Drexel Univ, BS, 69; Univ Ill, Urbana, PhD(org chem), 74. *Prof Exp:* Fel org chem, Univ Mich, Ann Arbor, 73-74; teacher chem & phys sci, Southeastern Christian Col, 74-77, assoc prof chem, 77-79; MEM STAFF, ETHYL CORP, 79- *Concurrent Pos:* Res assoc, Univ Ill, Urbana, 75-78. *Mem:* Am Chem Soc; Soc Tribologists & Lubricating Engrs. *Res:* New organic synthetic reagents; chemistry of hypervalent compounds; x-ray crystallography; organic photochemistry; lubricant additives synthesis. *Mailing Add:* Ethyl Petrol Addit Div Ethyl Corp 1530 S Second St St Louis MO 63104-3896

PERPER, LLOYD, b New York, NY, Apr 23, 21. ELECTRONICS. *Educ:* Mass Inst Technol, SB, 41; Ohio State Univ, MSc, 48. *Prof Exp:* ENGR, 54- *Mem:* Fel Inst Elec & Electronic Engrs; Inst Navig (pres, 73). *Mailing Add:* L J Perper PE 6700 W El Camino Del Cerro Tucson AZ 85745

PERPER, ROBERT J, b New York, NY, Mar 6, 33; m 58; c 5. IMMUNOPATHOLOGY, VETERINARY MEDICINE. *Educ:* Cornell Univ, DVM, 56; Univ Calif, San Francisco, MS, 65, PhD(comp path), 67. *Prof Exp:* Pvt pract vet med, 56-64; lectr pharmacol, Univ of the Pac, 62-67; sr res scientist, Upjohn Co, Mich, 67-69; sect head anti-inflammatory immunol res, Geigy Pharmaceut, 69-71, mgr anti-inflammatory immunol res, Ciba-Geigy Corp, 71-74; sr dir, Dept Inflammation & Arthritis, Merck Inst Therapeut Res, 74-76; MED DIR, FELINE HEALTH, 78- *Concurrent Pos:* Nat Heart Inst fel, Med Ctr, Univ Calif, San Francisco, 65-68, lectr, Dept Surg, Med Ctr, Univ, 66-67; adj prof, New York Med Col, 71-73, adj prof path, 73-; adj prof path, New York Hosp Med Col, Cornell Univ, 78- *Mem:* Transplantation Soc; Am Asn Immunologists; Am Soc Exp Pathologists; Am Acad Allergy; Am Vet Med Asn. *Res:* Transplantation immunology; anti-inflammatory research. *Mailing Add:* RD-2 Hillsdale NY 12529

PERR, IRWIN NORMAN, b Newark, NJ, Mar 4, 28; m 52; c 4. PSYCHIATRY, FORENSIC MEDICINE. *Educ:* Franklin & Marshall Col, BS, 46; Jefferson Med Col, MD, 50; Cleveland State Univ, JD, 61. *Prof Exp:* Resident psychiat, Bellevue Hosp-NY Univ, 51-52; resident, Philadelphia Psychiat Hosp, 52; resident, Med Ctr, Univ Calif, San Francisco, 54-56; asst supt, clin dir & dir educ, Fairhill Psychiat Hosp, 56-61; pvt pract, Cleveland, 61-72; PROF PSYCHIAT & PROF ENVIRON & COMMUNITY MED, ROBERT WOOD JOHNSON MED SCH, UNIV MED & DENT NJ, 72- *Concurrent Pos:* Lectr, Sch Law, Cleveland State Univ, 62-65; lectr psychiat & law, Fairhill Psychiat Hosp & Cleveland Psychiat Inst, 62-72; consult psychiat, Juv Court Cuyahoga County, Cleveland, 62-70; clin prof legal med, Sch Law, Case Western Reserve Univ, 71-72; mem, Med Rev Panel, State NJ, 73-; counr, Am Acad Psychiat & Law, 74-77, pres, 77-79; adj prof law, Rutgers Univ, Newark, 77- *Mem:* Am Psychiat Asn (vpres, 84-86); Am Acad Forensic Sci (vpres, 75-76, secy, 83-85); Am Acad Psychiat Law; Am Col Legal Med; Am Bd Forensic Psychiat (vpres, 76-79, secy, 81-84). *Res:* Legal medicine; legal psychiatry. *Mailing Add:* UMDNJ Robert Wood Johnson Med Sch Piscataway NJ 08854-5635

PERRAULT, GUY, b Amos, Que, Sept 25, 27; m 57; c 3. EARTH SCIENCES. *Educ:* Polytech Sch, Montreal, BSc, 49; Univ Toronto, MSc, 51, PhD(geol sci), 55. *Prof Exp:* Field engr mineral explor, Iron Ore Co, Can, 49-55 & Moneta Porcupine Mines, Ltd, 55-56; prof mineral eng, Polytech Sch, Montreal, 56-75, chmn dept geol eng, 66-72, chmn dept mineral eng, 74-75; vpres res, Quebec Mining Explor Co, 75-77; PROF MINERAL ENG, POLYTECH SCH, MONTREAL, 77- *Concurrent Pos:* Consult engr, St-Lawrence Columbium & Metals, 56-66; mem bd, Mineral Explor Inst, 73-77 & Dighem Ltd, 75-77. *Honors & Awards:* Scientific Prize, Govt Quebec, 70. *Mem:* Mineral Asn Can (pres, 67-68); fel Mineral Soc Am; Can Inst Mining & Metall; French Soc Mineral & Crystallog; Royal Soc Can. *Res:* Crystal structure of minerals; mineralogy applied to exploration and ore dressing; analytical geochemistry. *Mailing Add:* Dept Mineral Engr Universitie De Montreal Cp 6128 Succursale A Montreal PQ H3C 3J7 Can

PERRAULT, JACQUES, b Montreal, Que, June 25, 44. VIROLOGY, GENE EXPRESSION. *Educ:* McGill Univ, BSc, 64; Univ Calif, San Diego, PhD(virol cell biol), 72. *Prof Exp:* Fel virol, McGill Univ, 72, Univ Calif, Irvine, 72-74, Scripps Clin & Res Found, 74-76; asst res biologist, Univ Calif, San Diego, 76-77; ASST PROF MICROBIOL & IMMUNOL, WASHINGTON UNIV SCH MED, 77- *Concurrent Pos:* Actg instr, Univ Calif, Irvine, 72-74; course master, virol, Washington Univ Sch Med, 78-; Res Career Develop Award, NIH, 80. *Mem:* AAAS; Am Soc Microbiol; Soc Gen Microbiol London. *Res:* Molecular biology of defective interfering particles of RNA viruses and their role in controlling virus growth and maintaining persistent infectious of host cells. *Mailing Add:* Dept Biol San Diego St Univ 5300 Campanile Dr San Diego CA 92182

PERRAULT, MARCEL JOSEPH, b St Anaclet, Can, Apr 24, 14. ANIMAL PHYSIOLOGY, ENDOCRINOLOGY. *Educ:* Univ Ottawa, Ont, BA, 35, PhD(physiol), 64; Laval Univ, MA, 49; Cornell Univ, MS, 51. *Prof Exp:* Lectr biol, Univ Ottawa, 46-54, res fel anat, 53-54, asst prof physiol, 54-55, assoc vdean fac sci & eng, 69-76, assoc prof, 55-79; RETIRED. *Concurrent Pos:* Ont Res Found & Nat Res Coun Can res grants. *Mem:* AAAS; Fr-Can Asn Advan Sci; Can Physiol Soc; Am Physiol Soc; Am Inst Biol Sci. *Res:* Physiological and biochemical evaluation of the androgenic function of the testis; action of stress and endocrine relationships. *Mailing Add:* 515 St Laurent Blvd No 1512 Ottawa ON K1K 3X5 Can

PERREAULT, DAVID ALFRED, b Pawtucket, RI, Oct 25, 42; m 65; c 4. ELECTRICAL ENGINEERING. *Educ:* Purdue Univ, Lafayette, BSEE & MSEE, 68, PhD(elec eng), 70. *Prof Exp:* Asst prof elec eng, Clarkson Col Technol, 70-75; ASSOC PROF ELEC ENG, BOSTON UNIV, 75- *Concurrent Pos:* Consult, Magnavox Co, 70-; Nat Sci Found res grant, Clarkson Col Technol, 72-73. *Mem:* Inst Elec & Electronics Engrs; Am Soc Eng Educ. *Res:* Nonlinear networks and computer aided design; Microprocessors and Microcontroller design; Programmable devices. *Mailing Add:* Dept Elec Eng Boston Univ Boston MA 02215

PERRET, GEORGE (EDWARD), neurophysiology; deceased, see previous edition for last biography

PERRET, WILLIAM RIKER, b Newark, NJ, Feb 29, 08; m 38; c 2. GEOPHYSICS. *Educ:* Mass Inst Technol, SB, 30, SM, 56. *Prof Exp:* Proj engr, Waterways Exp Sta, US Corps Engrs, Miss, 35-36; observer interpreter, Schlumberger Well Surv Corp, Tex, 36-37; chief geophysics sect, Waterways Exp Sta, US Corps Engrs, Vicksburg, Miss, 38-43 & 46-51; asst chief vacuum group, Process Improv, Y-12, Clinton Eng Works, Tenn Eastman Corp, 43-46; staff physicist, Underground Physics Div, Sandia Nat Lab, Albuquerque, NMex, 51-73, consult, 73-83; RETIRED. *Concurrent Pos:* Consult, Boeing Co, 62-69; vis scientist & lectr, NMex Acad Sci, 67-68; Redfield Proctor travelling fel, 30-31; Charles A Coffin fel, 32-33; consult, Defense Nuclear Agency, 75-83; consult, R & D Assocs, 77-83. *Mem:* AAAS; emer fel Am Phys Soc; emer mem Soc Explor Geophys; Seismol Soc Am; Am Geophysics Union. *Res:* Vacuum gauging; soil corrosion; electrical and seismic exploration; static and dynamic earth stress and motion instrumentation and analysis; underground effects of nuclear explosions. *Mailing Add:* 6116 Natalie Ave NE Albuquerque NM 87110

PERRI, JOSEPH MARK, b Philadelphia, Pa, June 3, 17; m 43; c 5. COLLOID CHEMISTRY, SYNTHETIC FIBER SCIENCE. *Educ:* Philadelphia Col Pharm, BS, 38; Univ Pa, MS, 39, PhD(chem), 43. *Prof Exp:* Plant chemist, Evanson Soap Co, 39-41; chief chemist, Nat Foam Syst, Inc, 43-49, chem dir, 49-56; res assoc, Textile Fibers Dept, E I du Pont de Nemours & Co, Inc, 56-82; RETIRED. *Mem:* Am Chem Soc; Am Asn Textile Chemists & Colorists. *Res:* Surface active materials; fire fighting foams; fiber science; apparel (automated manufacturers; reinforced composites. *Mailing Add:* 326 Brockton Rd Sharpley DE 19803

PERRILL, STEPHEN ARTHUR, b Dayton, Ohio, Mar 13, 41; m 67; c 2. HERPETOLOGY. *Educ:* Ohio Wesleyan Univ, BA, 63; Southern Conn State Col, MS, 68; NC State Univ, PhD(zool), 73. *Prof Exp:* Teacher high schs, Conn, 64-66 & 68-70; asst prof, 73-80, assoc prof, 80-87, PROF ZOOL, BUTLER UNIV, 88- *Mem:* Am Soc Ichthyologists & Herpetologists; Soc Study Amphibians & Reptiles; Herpetologists League; Animal Behav Soc; AAAS. *Res:* Behavioral ecology of anurans. *Mailing Add:* Dept of Zool Butler Univ Indianapolis IN 46208

PERRIN, CARROL HOLLINGSWORTH, b Peterborough, Ont, Dec 21, 12; m 44; c 3. ANALYTICAL CHEMISTRY. *Educ:* Univ Toronto, BA, 37. *Prof Exp:* Analyst chem, Can Packers, Ltd, 37-40, res investr gen chem, 40-41, plastics adhesives, 41-45, anal methods, 45-47, group leader anal methods, 47-77; RETIRED. *Mem:* Fel Asn Off Anal Chem. *Res:* Development of analytical methodology. *Mailing Add:* Eight Hardwood Gate Rexdale ON M9W 4G1 Can

PERRIN, CHARLES LEE, b Pittsburgh, Pa, July 22, 38; m 64; c 2. ORGANIC CHEMISTRY. *Educ:* Harvard Univ, AB, 59, PhD(chem), 63. *Prof Exp:* NSF fel, Univ Calif, Berkeley, 63; from asst prof to assoc prof, 64-80, PROF CHEM, UNIV CALIF, SAN DIEGO, 80- *Concurrent Pos:* Nat Inst Neurol Dis & Stroke spec res fel, Gothenburg Univ, 72-73; NATO prof, Univ Padua, 86. *Mem:* Fel AAAS; Am Chem Soc; Sigma Xi. *Res:* Kinetics and mechanisms of organic reactions. *Mailing Add:* Dept Chem Univ Calif San Diego La Jolla CA 92093-0506

PERRIN, EDWARD BURTON, b Greensboro, Vt, Sept 19, 31; m 56; c 2. BIOSTATISTICS, HEALTH SERVICES RESEARCH. *Educ:* Middlebury Col, BA, 53; Columbia Univ, MA, 56; Stanford Univ, PhD(biostatist), 60. *Prof Exp:* Asst prof biostatist, Grad Sch Pub Health, Univ Pittsburgh, 59-62; from asst prof to prof prev med, Sch Med, Univ Wash, 62-70, head div biostatist, 65-69, prof biostatist & chmn dept, Sch Pub Health & Community Med, 70-72; dep dir, Nat Ctr Health Statist, Dept Health, Educ & Welfare, 72-73, dir, 73-75; res scientist, Human Affairs Res Ctr, Battelle Mem Inst, 75-77, dir, Health & Pop Study Ctr, 77-83; PROF & CHMN, DEPT HEALTH SERV, SCH PUBLIC HEALTH, UNIV WASH, 83- *Concurrent Pos:* Fulbright scholar, Edinburgh Univ, 53-54; Milbank Mem Fund fac fel, 64-69; clin prof, Dept Community Med & Int Health, Georgetown Univ, 72-75; chair Health Serv Res Study, Dept Health, Educ & Welfare, 76-79; gov coun, Am Pub Health Asn, 82-85; health serv res sci review & eval bd, Vet Admin, bd health care serv & Inst Med, Nat Acad Sci & steering comt, study health care qual, 87-; vis prof, WChina Univ Med Sci, Chengdu, Sichwan Peoples Rep China. *Honors & Awards:* Billings Gold Medal, AMA, 66; Spiegelman Award, Am Pub Health Asn, 70; Lowell Reed Award, Statist Sect, 89. *Mem:* Inst Med-Nat Acad Sci; fel AAAS; fel Am Pub Health Asn; fel Am Statist Asn; Biomet Soc; Asn Health Serv Res. *Res:* Health data, its collection, analysis and applications to health services research. *Mailing Add:* 4900 NE 39th St Seattle WA 98105

PERRIN, EUGENE VICTOR, b Detroit, Mich, Mar 7, 27; m 56; c 4. PEDIATRICS, PATHOLOGY. *Educ:* Wayne State Univ, AB, 48; Univ Mich, MD, 53. *Hon Degrees:* DD, Univ Life Church, 81. *Prof Exp:* Intern, Sinai Hosp, Detroit, Mich, 53-54; asst resident path, Children's Med Ctr, Boston, Mass, 55-56; resident, Boston Lying-In & Free Hosp Women, 56-57; resident, New Eng Deaconess Hosp, 57-58; instr, Col Physicians & Surgeons, 58-59; sr res assoc path & pediat, Col Med, Univ Cincinnati, 59-61, from instr to assoc prof, 59-66; assoc pathologist, Inst Path, Case Western Reserve Univ, 66-72, pathologist, 72-74, assoc prof path, pediat & reproductive biol, Sch Med, 66-74, assoc develop biol, 72-74; assoc prof path, 75-76, PROF PATH & ASSOC PEDIAT, OBSTET & ANAT, WAYNE STATE UNIV, 76-; DIR ANAT PATH, CHILD HOSP MICH, 75- *Concurrent Pos:* Pvt pract, Mich, 54-55; teaching fel, Harvard Med Sch, 56-57; resident, Babies Hosp, Presby Hosp, New York, 58-59; asst pathologist, Children's Hosp, Cincinnati, Ohio, 59-61, assoc pathologist, 63-66, pathologist-in-chg, Babies' & Children's Hosp, 66-74; dir labs, Health Hill Hosp, Cleveland, Ohio, 68-74; pathologist-in-chg, Rainbow Hosp, Cleveland, 66-74; adj assoc prof path, Case Western Reserve Univ, 74-79; pathologist, Sinai Hosp, Detroit, 74-75; assoc attend pathologist, Sinai Hosp, 75- & Hutzel Hosp, 76-; assoc cytopathol, Receiving Hosp, Detroit, 81-; adj prof anthrop, Wayne State Univ, 81 & assoc community med, 88- *Mem:* AAAS; Am Asn Pathologists; Int Asn Great Lakes Res; Am Soc Cell Biol; Tissue Cult Asn; Teratol Soc (pres, 77-78); Soc Pediat Path (pres, 70-71); Soc Marine Mammal. *Res:* Biology of disease in fetus and newborn; placental pathology; metabolic diseases studied in vitro; pediatric oncology; comparative teratology; placental receptor site biology; occupational fetopathy; Great Lakes water quality; toxicology. *Mailing Add:* Child Hosp Mich Wayne State Univ 3901 Beaubien Blvd Detroit MI 48201

PERRIN, JAMES STUART, b Superior, Wis, June 19, 36. MATERIALS SCIENCE. *Educ:* Mass Inst Technol, BS, 58; Univ Ill, Urbana, MS, 60; Stanford Univ, PhD(mat sci), 69. *Prof Exp:* fel, Columbus Labs, Battelle Mem Inst, 66-74, res leader, 74-81; PRES, FRACTURE CONTROL CORP, 81- *Mem:* Fel Am Soc Testing & Mat; Am Inst Mining, Metall & Petrol Engrs; Am Nuclear Soc; Am Soc Metals; Am Soc Mech Engrs. *Res:* Mechanical and physical properties of unirradiated and irradiated nuclear materials; failure studies and development of new materials for nuclear pressure vessels, piping and fuel rod materials. *Mailing Add:* 2041 Willowick Dr Columbus OH 43227

PERRIN, WILLIAM FERGUS, b Oconto Falls, Wis, Aug 20, 38; c 3. MAMMALOGY. *Educ:* San Diego State Univ, BS, 66; Univ Calif, Los Angeles, PhD(zool), 72. *Prof Exp:* FISHERY BIOLOGIST, SOUTHWEST FISHERIES CTR, NAT MARINE FISHERIES SERV, 68- *Concurrent Pos:* Res assoc, Smithsonian Inst, 72- & Los Angeles County Mus, 73-; mem, Small Cetaceans Subcomt, Sci Comt, Int Whaling Comn, 74-, chmn, 79-; mem, Ad Hoc Consult Group II, Adv Comt Marine Resources Res, Food & Agr Orgn, UN, 74-76, convenor, 75-76; mem, Whales Adv Group, Survival Serv Comn, Int Union Conserv Nature & Natural Resources, 75-80, comt sci adv, US Marine Mammal Comn, 80-, bd dirs, Mex Soc Study Marine Mammal, 80-; affiliate prof, Col Fisheries, Univ Washington, 81-; assoc adj prof, Scripps Inst

Oceanogr, 82- *Honors & Awards:* Scientific Res & Achievement Award, Nat Oceanic & Atmospheric Admin, 79. *Mem:* Am Soc Mammalogists; Am Inst Fishery Res Biologists; Soc Marine Mammalogy (pres, 86-87); Soc Conservation Biol. *Res:* Systematics, growth and reproduction, and community ecology of pelagic delphinid cetaceans, with emphasis on the eastern tropical Pacific Ocean. *Mailing Add:* 5776 Derest View Dr La Jolla CA 92038

PERRINE, JOHN W, b Hightstown, NJ, Feb 26, 27; m 55; c 3. BIOLOGY. *Educ:* Brown Univ, BA, 50; NY Univ, MS, 57, PhD(biol), 64. *Prof Exp:* Res scientist biol, Am Cyanamid Co, 54-66; res scientist, 66-72, group leader, 72-77, res scientist, Pharmaceut Div, 77-84, SAFETY OFFICER RES & DEVELOP, SANDOZ INC, 84- *Mem:* Am Soc Pharmacol & Exp Therapeut; NY Acad Sci. *Res:* Cell culture, atherosclerosis, inflammation; application of cell culture to the development of anti-atherosclerotic drugs. *Mailing Add:* 83 Lexington Ave Sandoz Res Inst Westwood NJ 07675

PERRINE, R(ICHARD) L(EROY), b Mountain View, Calif, May 15, 24; m 45; c 2. ENVIRONMENTAL SCIENCE & ENGINEERING. *Educ:* San Jose State Col, AB, 49; Stanford Univ, MS, 50, PhD(phys chem), 53. *Prof Exp:* Asst chem res, Stanford Univ, 50-53; res chemist, Calif Res Corp, Div, Standard Oil Co Calif, 53-59; assoc prof eng, 59-63, chmn dept environ sci & eng, 73-82, PROF ENG & APPL SCI, UNIV CALIF, LOS ANGELES, 63- *Concurrent Pos:* Lectr, Univ Southern Calif, 57-59; mem, Los Angeles County Energy Comn, 74-81; chmn qual assessment comt, Adv Coun South Coast Air Qual Mgt Dist, 77-82. *Mem:* Am Chem Soc; Am Inst Chem Engrs; Soc Petrol Engrs; Air Pollution Control Asn; Am Water Resources Asn; Nat Asn Environ Profs. *Res:* Energy resources; physics of flow through porous media. *Mailing Add:* Civil Eng Dept 2066 Eng I Univ Calif Los Angeles CA 90024-1593

PERRINO, CHARLES T, b Vandergrift, Pa, Jan 27, 38; m 64; c 2. SOLID STATE CHEMISTRY. *Educ:* Indiana Univ Pa, BS, 62; Ariz State Univ, PhD(chem), 66. *Prof Exp:* From asst prof to assoc prof, 66-75, prof & assoc dean sch sci, 75-80, CHAIR CHEM, CALIF STATE UNIV, HAYWARD, 85- *Concurrent Pos:* Res Found grant, 66-68; NSF acad year exten grant, 67-69. *Honors & Awards:* Fulbright lectr, Univ Ceylon, 69-70. *Mem:* Am Chem Soc. *Res:* Defects and electrical properties in hydrogen bonded solids, namely potassium dihydrogen phosphate and potassium dihydrogen arsenate. *Mailing Add:* Dept Chem 25800 Carlos Bee Dr Calif State Univ Hayward CA 94542

PERRITT, ALEXANDER M, b Kearny, NJ, Nov 28, 28; m 55; c 4. MICROBIOLOGY. *Educ:* Syracuse Univ, AB, 51, MS, 57, PhD(microbiol), 61. *Prof Exp:* Bacteriologist, Nat Yeast Corp, NJ, 52-53, 55; jr chemist, Hoffman LaRoche, Inc, 57-58; bacteriologist, Carter-Wallace, Inc, 61-66; dir microbiol, Worthington Biochem Corp, 66-68; res dir microbiol, Foster D Snell, Inc, 68-70; dir microbiol, Affil Med Res, 70-73; proj mgr chemother fermentation, Frederick Cancer Res Ctr, Litton Bionetics, Inc, Md, 74; PRES & DIR RES, PERRITT LABS, INC, 73- *Concurrent Pos:* Mem tech adv comt poison prev packaging, Consumer Prod Safety Comn; chmn, E35 Comt Pesticides, Am Soc Testing & Mat. *Mem:* AAAS; Am Soc Microbiol; Soc Indust Microbiol; Am Soc Testing & Mat; Am Coun Independent Labs. *Res:* Applied microbiology; fermentation technology; evaluation of poison prevention packaging. *Mailing Add:* Perritt Labs Inc PO Box 147 Hightstown NJ 08520

PERRON, PIERRE OMER, metallurgical engineering, nuclear engineering, for more information see previous edition

PERRON, YVON G, b Montreal, Que, Feb 14, 25; m 51; c 5. ORGANIC CHEMISTRY, MEDICINAL CHEMISTRY. *Educ:* Univ Montreal, BSc, 48, MSc, 49, PhD(org chem), 51. *Prof Exp:* Nat Res Coun Can fel, 51; asst prof, Univ Montreal, 51-54; fel, Univ Rochester, 54-55; res scientist, Bristol Labs, NY, 55-62, asst dir chem res, 62-63; asst dir res, 63-66, dir labs, 66-71, dir res, 71-74, VPRES RES, BRISTOL LABS CAN, 74- *Mem:* Am Chem Soc; fel Chem Inst Can. *Res:* Chemistry of natural products; synthesis of heterocyclic compounds of potential medicinal interest; determination of structure and synthesis of antibiotics. *Mailing Add:* Bristol Labs of Can 100 Industrial Blvd Candiac PQ J5R 1J1 Can

PERRONE, NICHOLAS, b New York, NY, Apr 30, 30; m 57; c 5. APPLIED MECHANICS, BIOMECHANICS. *Educ:* Polytech Inst Brooklyn, BAeroEng, 51, MS, 53, PhD(appl mech), 58. *Prof Exp:* Res asst aeronaut eng & appl mech, Polytech Inst Brooklyn, 51-54, res assoc, 55-58; asst prof eng sci, Pratt Inst, 58-59, assoc prof, 60-62; sr scientist, Off Naval Res, 62-68, dir struct mech prog, 68; NIH spec res fel biomech, Georgetown Univ, 69-70; dir struct mech, Off Naval Res, 70-81; vpres, CGI Inc, 82-85, gen mgr advan technol & res, 86-87; PRES, PERRONE FORENSIC CONSULT INC, 87- *Concurrent Pos:* Adj prof mech, Catholic Univ, 64-; consult, US Govt & Nat Acad Sci. *Honors & Awards:* Commendation on Hwy Safety Design, Dept Transp, Fed Hwy Admin, 75; John Curtis Lect Award, Am Soc Eng Educ, 76. *Mem:* Fel AAAS; Am Inst Aeronaut & Astronaut; NY Acad Sci; fel Am Acad Mech; Am Soc Civil Engrs; fel Am Soc Mech Engrs; Soc Automotive Engrs; Soc Mfg Engrs. *Res:* Dynamic plastic, rate-sensitive response of structures; reactor technology; biodynamics and crashworthiness of vehicle impact; finite difference techniques with arbitrary grids; nonlinear relaxation methods to solve problems with geometric and material nonlinearities; vehicle safety. *Mailing Add:* 12207 Valerie Lane Laurel MD 20708

PERROS, THEODORE PETER, b Cumberland, Md, Aug 16, 22; div. FLUORINE CHEMISTRY, FORENSIC SCIENCES. *Educ:* George Washington Univ, BS, 46, MS, 47, PhD(chem), 52. *Prof Exp:* Analyst, Res Div, US Naval Ord Lab, 43-46; from asst prof to assoc prof chem, George Washington Univ, 46-60, chmn dept forensic sci, 71-73, chmn dept chem, 80-88, PROF CHEM, GEORGE WASHINGTON UNIV, 60- *Concurrent Pos:* Co-dir res grants, Atomic Energy Comn, 52-56, Off Res & Develop, US Army Air Force, 60-64; res corp grant, 53-54; consult, US Naval Ord Lab, 53-56; NSF fel, Inst Org Chem, Munich, Germany, 59; secy, Ahepa Educ Found, 70-; fed comnr, Interstate Comn, Potmac River Basin, 83- *Mem:* Am Chem Soc; Sigma Xi; Soc Appl Spectros; fel Am Inst Chemists; fel Am Acad Forensic Sci; Royal Chem Soc; Chem Soc Germany. *Res:* Coordination chemistry; preparation and characterization of fluorine complexes of platinum; stabilities of inorganic coordination polymers; chemical education; forensic chemistry. *Mailing Add:* Dept of Chem George Washington Univ Washington DC 20052

PERROTTA, ANTHONY JOSEPH, b Erie, Pa, Aug 27, 37; m 61; c 2. MINERALOGY, CRYSTALLOGRAPHY. *Educ:* Pa State Univ, BS, 60; Univ Chicago, MS & PhD(mineral, crystallog), 65. *Prof Exp:* Phys scientist ceramics, mineral & crystallog, Union Carbide Corp Res Inst, NY, 67-69; sr res scientist, Gulf Res & Develop Co, 69-85; ALCOA FEL, ALCOA TECH CTR, 85- *Mem:* Mineral Soc Am; Am Ceramic Soc; Sigma Xi. *Res:* Crystal structure, phase equilibria, thermal and catalytic properties of mineralogical and ceramic materials. *Mailing Add:* 950 Harvard Rd Monroeville PA 15146

PERROTTA, JAMES, b Trenton, NJ, May 19, 19; m 47; c 1. CHEMISTRY. *Educ:* Univ NC, BS, 40. *Prof Exp:* Field rep, Venereal Dis Div, USPHS, 45-46; technician toxoid vaccine mfg, E R Squibb & Sons, Inc, 46-47, tech supvr, 47-50, tech supvr bio-control dept, 50, sect head endocrine mfg, 50-56, asst dept head biochem mfg, 56-61, dept head, 61-64, dept head packaging, 64-65, prod mgr, NY, 65-68, plant mgr, NJ, 68-70, bulk mfg dir, 70-75; asst to pres food div, 75-79, vpres opers, 79-84, sr vpres, 84-87, MFG OPERS CONSULT, INGREDIENT TECHNOL DIV, CROMPTON & KNOWLES CORP, 87- *Mem:* Am Pharmaceut Asn; NY Acad Sci; Am Inst Chemists; Int Soc Pharmaceut Engrs. *Res:* Isolation and crystallization of natural products; insulin; thyroxin; vitamin B12; human blood products; estrogens; development and manufacture of toxoids and vaccines; manufacturing of creams, ointments, liquids, tablets and capsules; fermentation and isolation of antibiotics; manufacture of radiopharmaceuticals; manufacture of tabletting excipients; manufacture of sugar based food ingredients and pharmaceutical excipients. *Mailing Add:* J P Enterprizes Six Brookhill Rd East Brunswick NJ 08816

PERRY, ALBERT SOLOMON, b Salonika, Greece, Apr 15, 15; nat US; m 46; c 3. INSECT TOXICOLOGY, ENVIRONMENTAL SCIENCES. *Educ:* Univ Calif, BS, 41, PhD(entom), 50. *Prof Exp:* Res entomologist, Calif Spray-Chem Co Div, Stand Oil Co Calif, 50-51; res scientist, Tech Develop Lab, USPHS, 51-74; prof, 74-85, EMER PROF INST NATURE CONSERV RES, TEL AVIV UNIV, 85- *Concurrent Pos:* Consult, Malaria Eradication Div, WHO, 58. *Honors & Awards:* Chaim Resnick Award, Advan Pesticide Res, Israel. *Mem:* AAAS; Am Chem Soc; Entom Soc Am; Sigma Xi; Israel Entom Soc; Soc Environ Toxicol & Chem. *Res:* Insect toxicology and biochemistry; mechanisms of resistance to insecticides; toxicity of pesticides to nontarget organisms; bioaccumulation, magnification, and metabolic fate of pesticides in aquatic food-chain organisms; pesticide monitoring. *Mailing Add:* Inst Nature Conserv Res Tel Aviv Univ 202 Sherman Bldg Ramat Aviv 69978 Israel

PERRY, ALFRED EUGENE, b Pendleton, Ore, Jan 12, 31; m 54; c 3. ZOOLOGY, ECOLOGY. *Educ:* Walla Walla Col, BA, 53, MA, 58; Okla State Univ, PhD(zool), 65. *Prof Exp:* Instr biol, Union Col, Nebr, 60-62; res asst zool, Okla State Univ, 62-65; asst prof, Memphis State Univ, 65-69; assoc prof biol, Walla Walla Col, 69-85, assoc develop, 78-80, prof indust tech, 80-85; dir develop, Hinsdale Hosp Found, 85-87, DIR DEVELOP, ATLANTIC UNION COL, 87- *Concurrent Pos:* Consult, Memphis & Shelby County Health Dept, 65-69. *Mailing Add:* Off of Develop Atlantic Union Col South Lancaster MA 01561

PERRY, BILLY WAYNE, b Portland, Tenn, Sept 15, 37. BIOCHEMISTRY, CLINICAL CHEMISTRY. *Educ:* Tenn Technol Univ, BS, 59, MA, 61; Univ Tenn, Memphis, PhD(biochem), 67. *Prof Exp:* From instr to asst prof clin chem, Med Col Ala, 67-70; ASST PROF PATH, MED COL WIS, 70-; ALLIED PROF STAFF, MILWAUKEE COUNTY MED COMPLEX, 70-; ALLIED HEALTH STAFF MEM, FROEDTERT MEM LUTHERAN HOSP, 80- *Concurrent Pos:* Consult scientist, Univ Ala Hosps & Clins, 67-70; consult, Vet Admin Hosp, Birmingham, Ala, 69-70; clin asst prof, Univ-Wis, Milwaukee, 84-86. *Mem:* AAAS; Am Asn Clin Chemists; Am Chem Soc; Sigma Xi; Am Inst Chemists. *Res:* Research and development of methods and instrumentation in clinical chemistry; mechanisms of bilirubin reactions as related to methods for its measurement. *Mailing Add:* Dept Path Med Col Wis 8700 W Wisconsin Ave Milwaukee WI 53226

PERRY, CHARLES LEWIS, b Culver, Ind, Dec 9, 33; div; c 1. PHOTOELECTRIC PHOTOMETRY. *Educ:* Ind Univ, Bloomington, BA, 55; Univ Calif, Berkeley, PhD(astron), 65. *Prof Exp:* Res assoc, Kitt Peak Nat Observ, 63-65; res fel, Mt Stromlo Observ, 65-66; from asst prof to assoc prof, 66-79, PROF PHYSICS & ASTRON, LA STATE UNIV, 79- *Concurrent Pos:* NSF res grants, 67, 69, 77, 87 & 89. *Mem:* Am Astron Soc; Int Astron Union. *Res:* Photoelectric photometry with applications to galactic structure. *Mailing Add:* Dept Physics & Astron La State Univ Baton Rouge LA 70803-4001

PERRY, CHARLES RUFUS, JR, b Throckmorton, Tex, Oct 1, 36. MATHEMATICS, STATISTICS. *Educ:* Tex Tech Univ, BS, 67, MS, 69, PhD(math), 71. *Prof Exp:* assoc prof math, Tex Lutheran Col, 71-82; MATH STATISTICIAN, STATIST REPORTING SERV, US DEPT AGR, 81- *Concurrent Pos:* Nat Res Coun sr res assoc, Earth Observ Div, Johnson Space Ctr, NASA, 78-80. *Mem:* Math Asn Am; Am Statist Asn. *Res:* Linear algebra; probability and statistics; sample theory. *Mailing Add:* 12402 Washington Bridge Rd Fairfax VA 22033

PERRY, CHARLES WILLIAM, b Taunton, Mass, Oct 1, 10; m 43; c 2. CHEMICAL ENGINEERING, PETROLEUM ENGINEERING. *Educ:* Northeastern Univ, BS, 34; Mass Inst Technol, MS, 35; Johns Hopkins Univ, DrEng, 41. *Prof Exp:* Jr technologist, Shell Oil Co, 35-37; develop engr, US

Indust Chem, Inc, 37-40; design engr, Merck & Co, Inc, 41-42; div chief, Rubber Div, War Prod Bd, 43-44; mgr, Chem Process Div, Phillips Petrol Co, 44-51; chief process engr, Phillips Chem Co, 51-52; asst mgr cent eng, Olin Corp, 52-59, assoc dir plant develop, Energy Div, 59-60; vpres mfg, Witco Chem Co, 60-71, sr vpres planning, 71-73; staff petrol engr, Fed Energy Admin, 74; sect head planning, Off Coal Res, 75; sect head enhanced oil recovery, Energy Res & Develop Admin, 76-77; sr staff engr fossil fuels, Extraction Div, Dept Energy, 78-82; heavy oil & tarsands technologist, US Synthetic Fuels Corp, 83-86; SR CONSULT, ENVIRON PROTECTION ENERGY, 86- *Mem:* AAAS; Am Inst Mining, Metall & Petrol Engrs; Am Petrol Inst; Am Inst Chem Engrs. *Res:* Continuous distillation; petroleum refining; operation of synthetic rubber plants; chemicals from petroleum; electrolytic caustic soda and chlorine; chemical fertilizers; ammonia; rocket fuels; fuels and lubricants; enhanced oil recovery. *Mailing Add:* 19724 Greenside Terr Gaithersburg MD 20879

PERRY, CLARK WILLIAM, b Jersey City, NJ, May 14, 36; m 80; c 3. SYNTHETIC ORGANIC CHEMISTRY, RADIOSYNTHESIS. *Educ:* Univ Rochester, BS, 58; Mass Inst Technol, PhD(org chem), 62. *Prof Exp:* Res chemist, Maywood Div, Stepan Chem Co, 62-64, dir res, 64-66; sr chemist, 66-76, asst group chief, Hoffman-La Roche, Inc, 77-85; SR PRIN SCIENTIST, BOEHRINGER INGELHEIM PHARM INC, 85-, RADIATION SAFETY OFFICER, 88- *Mem:* Am Chem Soc; Health Physics Soc; Int Isotope Soc. *Res:* Organic synthesis; labelling of drugs and metabolites with radioisotopes; radiation safety. *Mailing Add:* 32 Pond Crest Rd Danbury CT 06811

PERRY, CLIVE HOWE, b Merton, Eng, May 9, 36; m 60; c 2. SOLID STATE PHYSICS. *Educ:* Univ London, BSc, 57, PhD(solid state physics) 60. *Prof Exp:* Sloan foreign fel infrared spectros & low temperature physics, Sch Advan Study, Mass Inst Technol, 60-61, res assoc, 61-62, asst prof solid state physics, 62-68; assoc prof, 68-72, PROF SOLID STATE PHYSICS, NORTHEASTERN UNIV, 72- *Concurrent Pos:* Consult, Arthur D Little Corp, 63-64, Air Force Cambridge Res Labs, 65-, Borders Electronics Res Corp, 66-, Melpar Space Sci Labs, 66-, PTR Optics, 70-, Norcon Instruments, 72- & Gilford Instruments, 80; vis guest prof, Max Planck Inst Solid State Physics, 72-73; Nat Ctr Sci Res res assoc, Lab Solid State Physics, Univ Paris, 75; Humboldt Found Sr Am scientist award, Max Planck Inst High Magnetic Fields, Grenoble, France, 76; consult, Bellcore, 83-; dir, Optometrics USA, 85-; vis guest prof, Technische Universitat Munchen, 87; resident res assoc, Air Force Systs Command/Nat Res Coun, RADC, Hanscom AFB, 90-91. *Honors & Awards:* Coblentz Soc Award Molecular Spectros, 71. *Mem:* Am Phys Soc; fel Brit Inst Physics. *Res:* Low temperature physics; far infrared and Raman spectroscopy; high-pressure physics; phase transitions; internal molecular and lattice vibrations; antiferromagnetic resonances; ferroelectrics; semiconductors; inelastic/polarized neutron scattering; high magnetic fields; two dimensional electronic systems; superionic conductors; multiple quantum wells and superlattices; heterostructures semiconductor interfaces; fiber optics; multiplexers; optical systems; guided wave devices; high temperature superconductors. *Mailing Add:* Physics Dept Northeastern Univ Boston MA 02115

PERRY, DALE LYNN, b Greenville, Tex, May 12, 47. MATERIALS SCIENCE, SURFACE SCIENCE. *Educ:* Midwestern State Univ, BS, 69; Lamar Univ, MS, 72; Univ Houston, PhD(chem), 74. *Prof Exp:* Welch fel, Rice Univ, 75-76 & Nat Sci Found fel, 76-77; Miller fel chem, Univ Calif, Berkeley, 77-79, staff scientist, 79-87, PRIN INVESTR, LAWRENCE BERKELEY LAB, UNIV CALIF, 81-, STAFF SCIENTIST, 81- *Honors & Awards:* Nat Res Award, Sigma Xi, 74. *Mem:* Am Chem Soc; Soc Appl Spectroscopy; Mat Res Soc. *Res:* Inorganic chemistry, materials chemistry and surface chemistry; basic and applied resin catalysis, corrosion, chemisorption and thin film development; polymers, semi- and superconducts and ceramics. *Mailing Add:* 2216 Lupine Road Hercules CA 94547-1109

PERRY, DAVID ANTHONY, b Kansas City, Kans, Sept 19, 38; wid; c 2. ECOLOGY. *Educ:* Univ Fla, BS, 61; MS, 66; Mont State Univ, MS, 71, PhD(ecol), 74. *Prof Exp:* Res assoc, Dept Biol, Mont State Univ, 74; range ecologist, Mont Dept Natural Resources, 74-75; res forester, Intermountain Forest & Range Exp Sta, US Forest Serv, 75-77; PROF, DEPT FOREST SCI, ORE STATE UNIV, 77- *Mem:* Ecol Soc Am; Sigma Xi. *Res:* Sustainable resource management; rhizosphere dynamics; soil biology and nutrient cycling; host-pest relations; role of mycorrhizal fungi and other mutualistic soil organisms in ecosystem productivity and stability; reclamation of degraded ecosystems; strategies for sustainable agriculture & forestry. *Mailing Add:* Dept Forest Sci Ore State Univ Corvallis OR 97331

PERRY, DAVID CARTER, b Sacramento, Calif, Sept 10, 48; m 80; c 3. NEUROPHARMACOLOGY, MOLECULAR PHARMACOLOGY. *Educ:* Harvard Univ, BA, 70; Univ Calif, San Francisco, PhD(pharmaceut chem), 81. *Prof Exp:* Teaching fel pharmacol, dept neurosci, Johns Hopkins Univ, 81-83; asst prof, 83-88, ASSOC PROF PHARMACOL, GEORGE WASHINGTON UNIV MED CTR, 88- *Concurrent Pos:* Vis assoc prof, Dept Pharmacol, Georgetown Univ, 90-91. *Mem:* Soc Neurosci; Am Asn Pharmaceut Scientists; AAAS. *Res:* Neurotransmitter receptors; receptor autoradiography; opiate receptors; neuropeptides; molecular biology of receptors; stress and depression. *Mailing Add:* Dept Pharmacol Med Ctr George Washington Univ 2300 I St NW Washington DC 20037

PERRY, DENNIS, b West Point, Miss, Dec 25, 32; m 58; c 4. MICROBIOLOGY, BACTERIAL PHYSIOLOGY. *Educ:* Southern Ill Univ, BA, 57, MA, 58; Northwestern Univ, PhD(microbiol), 62. *Prof Exp:* From instr to asst prof, 62-69, ASSOC PROF MICROBIOL, NORTHWESTERN UNIV, CHICAGO, 69- *Mem:* Am Soc Microbiol. *Res:* Genetic transformation of streptococci. *Mailing Add:* Dept Microbiol Northwestern Univ Med Sch Chicago IL 60611

PERRY, DENNIS GORDON, b Bakersfield, Calif, July 8, 42; m 64; c 2. COMPUTER NETWORKS. *Educ:* Westmont Col, AB, 64; Univ Wash, PhD(chem), 70; Univ NMex, MBA, 81. *Prof Exp:* Fel, Brookhaven Nat Lab, 70-72; mem staff, Los Alamos Nat Lab, 72-78, asst group leader, 78-81, dep group leader, 81-82, group leader, 82-85; prog mgr, Defense Advan Res Projs Agency, 85-87; DIR TECHNOL, UNISYS DEFENSE SYSTS, 87- *Mem:* Inst Elec & Electronics Engrs; Am Sci Affil; Asn Comput Mach. *Res:* Fission; charged particle reactions; meson induced reactions; medium and high energy interactions; neutron reactions; cosmic ray interactions; on-line computer applications; network engineering. *Mailing Add:* 3389 Monarch Lane Annandale VA 22003

PERRY, DONALD DUNHAM, b New York, NY, June 9, 22; m 58; c 3. ORGANIC CHEMISTRY, POLYMER CHEMISTRY. *Educ:* Harvard Univ, AB, 43, AM, 48, PhD(org chem), 51. *Prof Exp:* Fel, Ohio State Univ, 50-52; res chemist, Polychem Dept, Exp Sta, E I du Pont de Nemours & Co, Del, 52-55; res chemist, Reaction Motors Inc, NJ, 55-58, unit & sect supvr, Reaction Motors Div, Thiokol Chem Corp, 58-66; mgr cent res, Riegel Paper Corp, 66-70; mgr appl res, Personal Prod Co, 70-72; supvr polymer lab, Polychrome Corp, 72-75; mgr mat sci, Elastimold Div, Amerace Corp, 76-91; RETIRED. *Concurrent Pos:* Meteorologist, USAAF, 43-46. *Mem:* Am Chem Soc; hon mem Sigma Xi; Soc Plastics Engrs; Inst Elec & Electronic Engrs. *Res:* Synthesis of high energy materials; synthesis and application of polymers for coatings, films, adhesives and fiber systems; radiation-curable polymers; application of plastics and elastomers in electrical insulation. *Mailing Add:* 45 Dogwood Dr Milford NJ 08848

PERRY, EDMOND S, b New York, NY, Sept 26, 12; m 39. PHYSICAL CHEMISTRY. *Educ:* Univ Ill, BS, 35; Univ Wis, PhD(phys chem), 38. *Prof Exp:* Res chemist, Nat Aniline Div, Allied Chem & Dye Corp, 38; res chemist, Eastman Kodak Co, NY, 38-42; res chemist, Distillation Prod, Inc, 42-54; res assoc, Eastman Kodak Co, 54-62, sr res assoc, 62-65, asst head, Photomat Div, 65-73, tech asst to dir, Kodak Res Labs, 73-77; RETIRED. *Concurrent Pos:* Chmn Gordon Res Conf Separations & Purifications, 59. *Mem:* Am Chem Soc; Sigma Xi. *Res:* Electrochemistry; colloids; high vacuum engineering and distillation; physical and chemical separations; tall oil chemistry; chemical engineering; science of image formation and signal recording in chemical and physical systems. *Mailing Add:* 310 Oakridge Rochester NY 14617

PERRY, EDWARD BELK, b Oxford, Miss, Sept 29, 39; m 66; c 3. SOIL MECHANICS, FOUNDATION ENGINEERING. *Educ:* Univ Miss, BS, 62; Miss State Univ, MS, 68; Tex A&M Univ, PhD(civil eng), 73. *Prof Exp:* RES CIVIL ENGR, SOIL MECH, WATERWAYS EXP STA, US ARMY CORP ENGRS, 63- *Mem:* Am Soc Civil Engrs; Am Soc Testing & Mats; Am Soc Eng Educ. *Res:* Reinforced earth; dispersive clays; erosion of soils; streambank protection. *Mailing Add:* US Army Corp Engrs Waterways Exp Sta 3909 Halls Ferry Rd Vicksburg MS 39180-6199

PERRY, EDWARD MAHLON, b Providence, RI, Aug 9, 28; m 52; c 5. ORGANIC CHEMISTRY. *Educ:* Brown Univ, ScB, 50; Univ Conn, MS, 55, PhD(org chem), 56. *Prof Exp:* Chemist, ICI Organics, Inc, 55-68; sr develop engr, Owens-Corning Fiberglas Corp, 68-73; lab dir, 73-82, PLANT MGR, SOLUOL CHEM CO, WEST WARWICK, 82- *Mem:* Am Asn Textile Chemists & Colorists; Am Chem Soc. *Res:* Textiles; fiber; paper coatings. *Mailing Add:* 54 Clark Rd Barrington RI 02806

PERRY, ERIK DAVID, b Dallas, Tex, Oct 17, 52; m 74; c 3. MECHANICAL ENGINEERING, TECHNICAL MANAGEMENT. *Educ:* Cornell Univ, BSME, 74; Univ Mich, MSE, 75, Mercer County Comm Col, Cert Mgt, 82. *Prof Exp:* Teaching asst aerospace eng, Univ Mich, 74-75; engr mech-struct design, Antenna Lab, Lockheed Missiles & Space Co, 75-76; engr prof tech staff, Design Fusion Reactors, Plasma Physics Lab, Princeton Univ, 77-79, head gen fabrication & design sect, 79-84, head diags & gen fabrication sect, 84-85, head eng serv sect, 85-88, HEAD TOKAMAK FUSION TEST REACTOR SHUTDOWN MGT BR, PRINCETON UNIV, 88-, HEAD TOKAMAK ENG BR & DT PREP PROJ MGR, 90- *Mem:* Am Inst Aeronaut & Astronaut; Am Soc Mech Engrs; Am Nuclear Soc. *Mailing Add:* Plasma Physics Lab PO Box 451 Princeton NJ 08543

PERRY, ERNEST JOHN, physical chemistry, for more information see previous edition

PERRY, EUGENE ARTHUR, b Baraboo, Wis, Mar 14, 38; m; c 3. MICROBIOLOGY, MYCOLOGY. *Educ:* Northland Col, BA, 60; Ind Univ, PhD(mycol), 67. *Prof Exp:* From asst prof to assoc prof, 67-88, PROF BIOL, KNOX COL, ILL, 88- *Mem:* Am Soc Microbiol; Am Inst Biol Sci; Sigma Xi. *Res:* Morphology and physiology of fungi, particularly differentiation of fungal reproductive structures and the physiology of host-parasite relationships in fungi. *Mailing Add:* Dept of Biol Knox Col Galesburg IL 61401-4999

PERRY, EUGENE CARLETON, JR, b Mar 23, 33; US citizen; m 62; c 2. GEOCHEMISTRY. *Educ:* Ga Inst Technol, BS, 54; Mass Inst Technol, PhD(geol), 63. *Prof Exp:* Res assoc geochem, Minn Geol Surv, Univ Minn, Minneapolis, 64-65, asst prof, 65-72; actg chmn dept, 72-77, ASSOC PROF GEOL, NORTHERN ILL UNIV, 72-, PROF GEOL. *Mem:* AAAS; Geochem Soc; Geol Soc Am; Mineral Soc Am. *Res:* Stable isotope chemistry of metamorphic rocks and ancient sediments. *Mailing Add:* Dept of Geol Northern Ill Univ De Kalb IL 60115

PERRY, FRANK ANTHONY, b Lake Charles, La, Dec 16, 21; m 48; c 2. SURGERY. *Educ:* Meharry Med Col, MD, 45; Am Bd Surg, dipl, 53. *Prof Exp:* Intern, Meharry Med Col, 45-46, resident surg, 47-51; fel & resident cancer surg, Mem Ctr Cancer & Allied Dis, 51-54; surg consult, Sloan-Kettering Inst, 55-56; assoc prof, 56, 58-68, coordr, Regional Med Prog, 67-72, dir, Learning Resources ctr Prog, 73-81, PROF SURG, MEHARRY MED COL, 68-, DIR SURG RES, 58- *Mem:* Am Soc Head & Neck Surg; Am Col Surgeons. *Res:* Fluids and electrolytes. *Mailing Add:* 4223 Drake Hill Dr Nashville TN 37218

PERRY, GEORGE, b Lompoc, Calif, Apr 12, 53; m 83; c 1. CELL BIOLOGY, CYTOSKELETON. *Educ:* Univ Calif, Santa Barbara, BA, 74; Univ Calif, San Diego, PhD(marine biol), 79. *Prof Exp:* Asst cell biol, Baylor Col Med, 79-82; asst prof path, 82-89, ASSOC PROF PATH & NEUROSCI, CASE WESTERN RESERV UNIV, 89- *Mem:* AAAS; Am Soc Cell Biol; Electron Microscope Soc Am; Soc Neurosci; Am Asn Neuropathologists. *Res:* Structural and biochemical changes in Alzheimer, Parkinson and other neurodegenerative diseases; cytoskeletal changes in disease; myloidosis. *Mailing Add:* Dept Path Case Western Reserve Univ 2085 Adelbert Rd Cleveland OH 44106-4901

PERRY, HAROLD, b Hamtramck, Mich, June 26, 24; m 48; c 3. RADIATION ONCOLOGY. *Educ:* Howard Univ, MD, 48. *Prof Exp:* Intern, Freedmens Hosp, Washington, DC, 48-49, resident radiol, 49-52; resident radiation ther, Mem Hosp Cancer & Allied Dis, New York, 52 & 55; from asst prof to assoc prof radiol, Univ Cincinnati, 57-66; clinical assoc prof radiol, 66-79, CLINICAL PROF RADIATION ONCOL, WAYNE STATE UNIV, 82- *Concurrent Pos:* Kress fel, Mem Hosp & Dept Biophys, Sloan-Kettering Inst Cancer Res, 56-57; attend physician, Cincinnati Gen & Daniel Drake Mem Hosps, 57-; Nat Cancer Inst grant, 63-68; consult, Vet Admin Ctr, Dayton, 65-; radiotherapist, Vet Admin Hosp, Allen Park, Mich, 68-72, consult, 72-; mem exec comt, Mich Cancer Found, 70-; attend physician, Sinai Hosp Detroit; vchmn, Mich Cancer Found, 76-79; dir, Abraham & Anna Srere Radiation Ther Ctr, Sinai Hosp, Detroit, 66-81, chmn, Dept Radiation Ther, 81; chairperson, Radiation Ther Adv Panel, Mich Cancer Found, 76-; chmn, Radiation Ther Tech Adv Panel, Comprehensive Health Planning Coun Southeastern Mich, 78-; spec lect, 7th Int Conf on Use of Comput in Radiation Ther, Japan, 80; regional rep, Coun Affil Regional Radiation Oncol Socs, 81- *Honors & Awards:* William E Allen Jr Lectr, Nat Med Asn, 79; Hon, A Cressy Morrison Awards, Nat Sci, NY Acad Sci, 63. *Mem:* NY Acad Sci; Am Soc Clin Oncol; Am Asn Physicists Med; Am Radium Soc; Am Asn Cancer Res; Am Soc Therapeut Radiol & Oncol. *Res:* Radiation therapy for cancer; computer utilization in radiation treatment planning; combination radiation and chemotherapy in Hodgkin's lymphoma; carcinoma of pancreas; pre-operative adjuvant therapy in rectal carcinoma and combination adjuvant therapy in carcinoma of the lung; dosimetry of high-energy electrons and x-rays; retrospective analysis of environmental contacts of patients with respiratory cancer, other cancers and other diseases; dose response and effect of radiation on bioluminescent bacteria; automation of a multi-leaf collimator for a linear accelerator; the development of a three-dimensional radiation therapy treatment unit. *Mailing Add:* Sinai Hosp Detroit 6767 W Outer Dr Detroit MI 48235

PERRY, HAROLD OTTO, b Rochester, Minn, Nov 18, 21; m 44; c 4. DERMATOLOGY. *Educ:* Univ Minn, BSc, 44, MB, 46, MD, 47, MS, 53. *Prof Exp:* Consult staff dermat, St Mary's & Rochester Methodist Hosps, 53; from instr to asst prof, 54-69, PROF DERMAT, MAYO GRAD SCH MED, UNIV MINN, 69- *Concurrent Pos:* Chmn, Dept Dermat, 53-86, head Dept Dermat, Mayo Clinic, 76- 85. *Mem:* Soc Invest Dermat; Am Dermat Asn; AMA; Am Acad Dermat; Sigma Xi. *Res:* Publication of more than 250 scientific articles. *Mailing Add:* Mayo Clin Grad Sch Med Rochester MN 55901

PERRY, HAROLD TYNER, JR, b Bismarck, NDak, Jan 26, 26; m 52; c 2. PHYSIOLOGY. *Educ:* Northwestern Univ, DDS, 52, MS, 54, PhD, 61. *Prof Exp:* Res assoc, 52-54, instr, 54-61, assoc prof, 61-65, PROF ORTHOD, SCH DENT, NORTHWESTERN UNIV, CHICAGO, 65-, CHMN DEPT, 61- *Concurrent Pos:* USPHS fel, Nat Inst Dent Res, 52-54; consult, Vet Admin Hosp, 71. *Mem:* AAAS; Am Dent Asn; Am Asn Orthod; Int Asn Dent Res. *Res:* Electromyography of the head and neck musculature; crania; cranial, facial and dental development. *Mailing Add:* 100 E Chicago St Elgin IL 60120

PERRY, HORACE MITCHELL, JR, b Reading, Pa, June 11, 23; m 45; c 4. MEDICINE. *Educ:* Wash Univ, MD, 46. *Prof Exp:* From intern to asst resident med, Barnes Hosp, 46-48, asst, 47-48; res fel med, Washington Univ, St Louis, 50-51, from instr to assoc prof med, 51-72, prof, 72-, dir, Hypertension Div, 57- & Hypertension Clin, 64-; CHIEF HYPERTENSION SECT, VET ADMIN MED CTR, ST LOUIS. *Concurrent Pos:* Life Ins Med Res fel, 51-53; Am Heart Asn estab investr, 56-61; sr physician, St Louis Vet Admin Hosp, 71-, chief med serv, 63-75; assoc physician, Barnes Hosp, 72-; physician coordr hypertension, Vet Admin, Washington, DC, 77-; consult, Hypertension & Kidney Dis Br, Nat Heart, Lung & Blood Inst, 79- *Mem:* Am Soc Clin Invest; Am Physiol Soc; fel Am Col Physicians; fel Am Col Cardiol. *Res:* Hypertension. *Mailing Add:* 1983 Karlin Dr St Louis MO 63131

PERRY, J WARREN, b Richmond, Ind, Oct 25, 21. MEDICAL EDUCATION. *Educ:* De Pauw Univ, BA, 44; Northwestern Univ, MA, 52, PhD, 55. *Prof Exp:* Asst prof psychol, Univ Ill, 53-56; asst prof orthop surg, Northwestern Univ Med Sch, 57-61; asst chief rehab admin, Dept Educ & Welfare, 61-64, asst comnr res & training, 64-66; prof, 66-85, dean, 77-85, EMER PROF HEALTH SCI ADMIN & EMER DEAN, SCH HEALTH RELATED PROF, STATE UNIV NY, BUFFALO, 85- *Concurrent Pos:* Chmn, NY State Coalition Smoking or Health, 89-91. *Mem:* Inst Med-Nat Acad Sci. *Mailing Add:* 83 Bryant Buffalo NY 14209

PERRY, JACQUELIN, b Denver, Colo, May 31, 18. ORTHOPEDIC SURGERY, KINESIOLOGY. *Educ:* Univ Calif, Los Angeles, BEd, 40; Univ Calif, San Francisco, MS, 50. *Prof Exp:* CHIEF PATHOKINESIOLOGY, RANCHO LOS AMIGOS HOSP, 55-; PROF ORTHOP SURG, UNIV SOUTHERN CALIF, 72- *Concurrent Pos:* Assoc chief surg serv, Rancho Los Amigos Hosp, 55-75; assoc clin prof, Univ Calif, San Francisco, 66-73, clin prof, 73-; consult, Surgeon Gen, US Air Force, 69-80. *Honors & Awards:* Woman of Year in Med, Los Angeles Times, 59; Golden Pen Award, J Am Phys Ther Asn, 65; Award for Orthop Res, Am Acad Orthop Surg, 77; Isabelle R Lenard Goldensen Award, United Cerebral Palsy Asn, 81. *Mem:* Hon mem Am Phys Ther Asn; Am Orthop Asn; Am Acad Orthop Surgeons. *Res:* Rehabilitation; objective measurement of gait; upper extremity function; stroke rehabilitation. *Mailing Add:* Dept Orthop Univ Southern Calif 2025 Zonal Ave Los Angeles CA 90033

PERRY, JAMES ALFRED, US citizen; m; c 3. INTERNATIONAL DEVELOPMENT, RESOURCE MANAGEMENT. *Educ:* Colo State Univ, BS, 68; Western State Col, MA, 73; Idaho State Univ, PhD(environ mgt), 81. *Prof Exp:* Sr water qual specialist, Div Environ, State of Idaho, 74-82; area mgr, Cent Assocs, 82; asst prof, 82-85, ASSOC PROF WATER QUAL, FOREST RESOURCES, UNIV MINN, 85-, DIR GRAD STUDIES WATER RESOURCES, 88-; DIR, CTR NATURAL RESOURCE POLICY & MGT, 83- *Concurrent Pos:* Sr res fel, Am Inst Indian Studies, New Delhi & Madras, 85-86; prin investr, many grants; exec bd, NAm Benthol Soc, 90-91; vis scholar, Oxford Univ, Eng, 90-91; Bush sabbatical fel, 90-91; deleg, Univ Coun water resources. *Mem:* Int Asn Crenobologists (pres, 78-84); Soc Am Foresters; Sigma Xi; Ecol Soc Am; NAm Benthol Soc; Int Soc Theoret & Appl Limnol; Am Water Res Asn; Int Water Resource Asn. *Res:* Water quality management and the role of water quality in watershed management; microbial approaches to assessing water quality; value and management of biodiversity; environmental monitoring in developing countries; policies for water quality management and sustainable development; published more than 75 science and technology papers. *Mailing Add:* Dept Forest Resources Univ Minn 1530 N Cleveland St Paul MN 55108

PERRY, JAMES ERNEST, b Washington, DC, Jan 30, 23; m 56; c 3. ELECTROOPTICS. *Prof Exp:* Elec engr res infrared, Optics Div, Naval Res Lab, 47-64; elec engr res & develop infrared systs, Night Vision & Electrooptics Lab, US Army, 64-80; CONSULT, 80- *Concurrent Pos:* US deleg to study group 7, NATO Panel IV (Infrared), 72-73, US Army rep to NATO AC/243, 72-73. *Honors & Awards:* Res Develop Achievement Award, US Army, 71; Award for Invention of Thermal Imaging Systs, Night Vision & Electro Optics Lab, 72. *Mem:* Sr mem Inst Elec & Electronics Engrs; AAAS; Infrared Info Symp. *Res:* Infrared, including radiometry and thermal imaging; radiometry associated with nuclear research; development of systems for tactical applications for Army. *Mailing Add:* PO Box 527 Mayo MD 21106

PERRY, JAMES WARNER, b Tulsa, Okla, Oct 25, 48; m 77. ULTRASTRUCTURE OF PHLOEM TISSUE, HISTOPATHOLOGY. *Educ:* Univ Wis, Madison, BS, 71, MS, 74, PhD(bot & plant path), 82. *Prof Exp:* ASSOC PROF BIOL, FROSTBURG STATE UNIV, 83- *Mem:* Am Phytopath Soc; Asn Biol Lab Educ; Electron Micros Soc Am. *Res:* Ultrastructure of phloem tissue; histopathology; laboratory exercises for introductory biology and plant morphology. *Mailing Add:* Dept Biol Frostburg State Univ Frostburg MD 21532-1099

PERRY, JEROME JOHN, b Wilkes Barre, Pa, Oct 15, 29; m 56; c 2. MICROBIOLOGY. *Educ:* Pa State Univ, BS, 51; Univ Tex, PhD(microbiol), 56. *Prof Exp:* Asst microbiol, Merck & Co, 51-52; res assoc, Upjohn Co, 56-58; bacteriologist, Am Meat Inst Found, Chicago, 58-59; res scientist, Kitchawan Res Lab, Brooklyn Bot Garden, 59-61 & Univ Tex, 61-64; from asst prof to assoc prof, 64-71, PROF MICROBIOL, NC STATE UNIV, 71- *Concurrent Pos:* Vis prof, Univ Wash, Seatle, 89. *Mem:* AAAS; Am Chem Soc; Am Soc Microbiol; Am Acad Microbiol; Brit Soc Gen Microbiol. *Res:* Degradation of hydrocarbons by microorganisms; lipid synthesis; biodegradation in marine areas; thermophilic bacteria. *Mailing Add:* Dept Microbiol NC State Univ Raleigh NC 27695-7615

PERRY, JOHN ARTHUR, b Ridgefield, Conn, Nov 11, 21; m 42, 64; c 1. ANALYTICAL CHEMISTRY. *Educ:* Univ Rochester, BS, 42; La State Univ, MS, 52, PhD(chem), 54. *Prof Exp:* Jr chemist, Shell Develop Corp, 43; chemist, Vallejo Naval Yard, 44-45, Interchem Res Labs, 45 & Monsanto Chem Corp, 47-50; proj chemist, Stand Oil Co Ind, 54-57; sr res chemist, Sinclair Res Inc, 57-66; lab mgr instrumental analysis, Perkin-Elmer Corp, Downers Grove, 66-69; dir instrument div, McCrone Assocs, 69-70; consult, 71-78; vpres res, Regis Chem, 79-91; CONSULT, 91- *Mem:* Am Chem Soc; AAAS. *Res:* Instrumental analysis; gas chromatography; infrared spectrophotometry; emission spectrography; mass spectrometry; distillation control; dielectrophoresis; instrument design; new forms for chromatography. *Mailing Add:* c/o Casper Bye Four Løkenåsen Fetsund 1900 Norway

PERRY, JOHN E(DWARD), b Roanoke Rapids, NC, Mar 2, 24; m 45; c 3. CHEMICAL ENGINEERING. *Educ:* High Point Col, BS, 44; Purdue Univ, MS, 53. *Prof Exp:* Asst engr, Oak Ridge Plant, Tenn Eastman Co, 44-47; engr, Nat Lab, Monsanto Chem Co, 47-48 & Argonne Nat Lab, 48-51; supv engr, Pilot Plant, Union Carbide Nuclear Co, 53-62, mgr, Electroclad Dept, Parma Tech Ctr, Stellite Div, Union Carbide Corp, Ohio, 62-69; chief chem eng, Metals Universal, Inc, 69-70; proj dir fused salt process, Gen Metals Technol Co, 70-72, pres, 72-84; chem engr, Defense Gen Supply Ctr, 85-86; tech consult, Fike Metal Prods, 85-86, prod mgr, 86-90; TECH CONSULT, PERRY SERV INC, 90- *Mem:* Nat Asn Corrosion Engrs; Am Inst Chem Engrs; Am Soc Metals. *Res:* Design and development of equipment for fused salts processes, particularly electrodeposition of refractory metals niobium, molybdenum, tungsten and tantalum. *Mailing Add:* Perry Serv Inc 4305 Knob Rd Richmond VA 23235

PERRY, JOHN FRANCIS, JR, surgery; deceased, see previous edition for last biography

PERRY, JOHN MURRAY, b Springfield, Vt, July 19, 25; m 49; c 4. MATHEMATICS. *Educ:* Middlebury Col, AB, 45; Harvard Univ, AM, 51; Univ Rochester, PhD, 60. *Prof Exp:* From instr to prof math, Clarkson Tech Univ, 46-65, chmn dept, 56-63, dir comput ctr, 61-64; NSF fac sci fel, Univ Md, 64-65; prof math & chmn dept, 65-69, dean, 69-75, chmn div phys & math sci, Wells Col, 75-77; assoc provost, 77-83, VICE PROVOST, VA POLYTECH INST & STATE UNIV, 83- *Mem:* Math Asn Am. *Res:* Partial differential equations; numerical analysis. *Mailing Add:* 201 Burruss Hall Va Polytech Inst & State Univ 1102 Kam Dr Blacksburg VA 24060

PERRY, JOHN STEPHEN, b Lynbrook, NY, Oct 18, 31; m 53; c 2. METEOROLOGY, COMPUTER SCIENCE. *Educ:* Queens Col, NY, BS, 53; Univ Wash, BS, 54, MS, 60, PhD(meteorol), 66. *Prof Exp:* Forecaster, Air Weather Serv, US Air Force, Korea, 54-55 & La, 55-58, systs analyst, Global Weather Cent, 60-63, Hq, Scott AFB, 66-70 & Air War Col, 70-71, prog mgr, Advan Res Projs Agency, 71-74; exec scientist, US Comt Global Atmospheric Res Prog, 74-76, sci officer, World Meteorol Orgn, 76-78, STAFF DIR, BD ATMOSPHERIC SCI & CLIMATE, NAT ACAD SCI, 78- *Honors & Awards:* Commendation Medal, 63. *Mem:* Fel Am Meteorol Soc; Am Geophys Union; AAAS. *Res:* Science policy; atmospheric sciences, general; computer sciences, general; environmental, earth and marine science. *Mailing Add:* Nat Acad Sci MH 594 2101 Constitution Ave NW Washington DC 20418

PERRY, JOHN VIVIAN, JR, b Danville, Va, July 10, 24; m 46; c 3. MACHINE DESIGN. *Educ:* Va Polytech Inst, BS, 48; Tex A&M Univ, MS, 54, PhD, 63. *Prof Exp:* From asst prof to assoc prof, 48-74, PROF MECH ENG, TEX A&M UNIV, 74- *Concurrent Pos:* Propulsion & struct engr, Gen Dynamics, 51-53; vis prof, US Mil Acad, West Point, NY, 79-80. *Mem:* Am Soc Mech Engrs; Am Gear Mfrs Asn. *Res:* Applied mechanics; vibrations of foundations in soils; soil mechanics. *Mailing Add:* Dept Mech Eng Tex A&M Univ College Station TX 77843

PERRY, JOSEPH EARL, JR, b Belmont, Mass, Oct 29, 17; m 44, 82; c 2. PHYSICS. *Educ:* Mass Inst Technol, BS, 39; Univ Rochester, PhD(physics), 48. *Prof Exp:* Asst physics, Univ Rochester, 40-42, instr, 45-46; physicist, Radiation Lab, Univ Calif, 42-45; physicist, Appl Physics Lab, Johns Hopkins Univ, 45; res fel, Calif Inst Technol, 47-50; group leader, Los Alamos Nat Lab, Univ Calif, physicist staff mem, 50-82; RETIRED. *Mem:* Fel Am Phys Soc. *Res:* Photoelectricity of semiconductors; spectroscopy of light nuclei; absolute differential cross sections for scattering of protons by protons; van de Graaff studies of interactions of hydrogen nuclides; monoenergetic neutron flux measurement; controls for nuclear reactor rocket engines; weapon location by sound ranging; construction of and experimentation with large neodymium/glass laser system; solar heating and cooling of buildings; passive solar heating of buildings. *Mailing Add:* 9632 Galatea Lane Escondido CA 92026

PERRY, KENNETH W, cell physiology, molecular biology; deceased, see previous edition for last biography

PERRY, L JEANNE, protein engineering, recombinant dna technology, for more information see previous edition

PERRY, LLOYD HOLDEN, b Nashua, NH, Mar 8, 16; m 43; c 4. CHEMISTRY. *Educ:* Wesleyan Univ, BA, 38; Univ NH, MS, 40; Mass Inst Technol, PhD(org chem), 46. *Prof Exp:* Instr chem, Univ NH, 39-40; asst, Mass Inst Technol, 42-46; res dir, Union Bay State Chem Co, Fla, 46-48; tech dir, Union Bay State Labs, Inc, 48-59; vpres & tech dir, UBS Chem Co, A E Staley Mfg Co, 59-67; mgr, Chattanooga-Dalton Plants, GAF Corp, 67-81; RETIRED. *Mem:* Am Chem Soc. *Res:* Polymerizations; peroxides; rubber compounding and applications. *Mailing Add:* 203 N Palisades Ave Signal Mountain TN 37377

PERRY, LORIN EDWARD, b Twin Falls, Idaho, July 24, 14; m 38; c 4. SALMON FISHERIES. *Educ:* Utah State Univ, BS, 39; Univ Mich, PhD, 43. *Prof Exp:* Fishery res biologist, US Fish & Wildlife Serv, 38-41 & State Inst Fisheries Res, Mich, 41-43; entomologist & limnologist, USPHS, Ga, 43-46; res biologist, US Fish & Wildlife Serv, Ore, 46-47, biologist & supvr river basins studies, 48-50, chief biologist, Columbia River Fishery Develop Prog, 50-59; prog dir, Columbia Fisheries Prog, US Bur Commercial Fisheries, 59-67; fisheries staff asst, Bur Sport Fisheries & Wildlife, 67-71, dep regional dir Pac region, 71-74; PRES, BIOL SERV, INC, 75- *Concurrent Pos:* Consult, Pac Northwest Regional Comn on Columbia River salmon, 75-78 & Pac Fishery Mgt Coun on Pac Salmon, 78-81. *Mem:* Am Fisheries Soc; Am Inst Fishery Res Biol. *Res:* Impounded waters and water development projects, especially salmon and steelhead; life history of fresh water fish; salmon management. *Mailing Add:* 6350 SW Spruce Beaverton OR 97005-3545

PERRY, MALCOLM BLYTHE, b Birkenhead, Eng, Apr 26, 30; m 56; c 2. BIOCHEMISTRY. *Educ:* Bristol Univ, BSc, 53, PhD(org chem), 57, DSc, 70. *Prof Exp:* Asst prof chem, Queen's Univ, Ont, 56-62, McLaughlin res prof, 62-63; RES OFFICER BIOL SCI, NAT RES COUN CAN, 63- *Honors & Awards:* Can Soc Microbiol Award, 91. *Mem:* Am Soc Microbiol; Royal Soc Can; Can Soc Microbiol. *Res:* Natural products; structural investigation of carbohydrates, bacterial and fungal glycans, glycoproteins; biosynthesis; gas-liquid chromatography; synthetic organic chemistry; immunochemistry. *Mailing Add:* Inst Biol Sci Nat Res Coun Can Ottawa ON K1A 0R6 Can

PERRY, MARGARET NUTT, b Waynesboro, Tenn, Apr 23, 40; m 65; c 2. ACADEMIC ADMINISTRATION, FOOD SCIENCE. *Educ:* Univ Tenn, Martin, BS, 61; Univ Tenn, Knoxville, MS, 63, PhD(nutrit, food sci), 65. *Prof Exp:* Instr food sci, Univ Tenn, Knoxville, 63-64, asst prof, Col Home Econ, 66-68, asst to dean, 67, from asst to assoc dean, 68-73, dean grad studies, 73-79; assoc vpres acad affairs, Tenn Technol Univ, Cookeville, 79-86, CHANCELLOR, UNIV TENN, MARTIN, 86- *Concurrent Pos:* Exec Comn, Coun Grad Sch US & Conf Southern Grad Sch, 74-77; Mem, Comn Women, Am Coun Educ, 79-81, Comn Minorities, 90 & Comn Cols, Southern Asn Cols & Schs, 90- *Mem:* Inst Food Technologists; Am Home Econ Asn. *Res:* Development of interdisciplinary graduate programs and research; application of food science and nutrition research to societal needs; student outcomes assessments. *Mailing Add:* Univ Tenn Box C Martin TN 38238

PERRY, MARY HERTZOG, b Bethlehem, Pa, Oct 24, 22; m 47; c 3. ANALYTICAL CHEMISTRY, ORGANIC CHEMISTRY. *Educ:* Russell Sage Col, BA, 44; Lehigh Univ, MS, 46, PhD(chem), 49. *Prof Exp:* Instr chem, Cedar Crest Col, 47-48 & Muhlenberg Col, 65-68; from asst prof prof, 68-83, chmn dept, 70-72, EMER PROF CHEM, CEDAR CREST COL, 83- *Concurrent Pos:* Tech ed, Chemist-Analyst, J T Baker Chem Co, 58-69. *Mem:* Am Chem Soc. *Res:* Use of organic reagents in analytical chemistry. *Mailing Add:* 76 W Laurel St Bethlehem PA 18018

PERRY, MARY JANE, b New York, NY, Mar 30, 48. BIOLOGICAL OCEANOGRAPHY. *Educ:* Col New Rochelle, BA, 69; Univ Calif, San Diego, PhD(oceanog), 74. *Prof Exp:* Res asst oceanog, Scripps Inst Oceanog, Univ Calif, San Diego, 69-74; asst prof marine ecol, Univ Ga, 75; res instr marine chem, Dept of Pharmacol, Sch Med, Washington Univ, 75-76; RES ASST PROF OCEANOG, DEPT OF OCEANOG, UNIV WASH, 76-, ASSOC PROF. *Concurrent Pos:* Assoc prog dir oceanic biol, Nat Sci Found, 80-82. *Mem:* AAAS; Am Soc Limnol & Oceanog; Phycol Soc Am; Am Geophys Union. *Res:* Marine phytoplankton and bacteria, with emphasis on nutrient dynamics, autotrophy and physiological response to environmental variables and physical water circulation in the ocean. *Mailing Add:* Sch Oceanog Univ of Wash Seattle WA 98195

PERRY, MICHAEL PAUL, b Crystal Falls, Mich, Aug 24, 47. ELECTRICAL ENGINEERING. *Educ:* Mass Inst Technol, ScB, 69; Colo State Univ, MS, 73, PhD(elec eng), 76. *Prof Exp:* Programmer, Inforex Inc, Burlington, 70-71; elec engr, Corp Res & Develop, Gen Elec Co, 76-87; ELEC ENGR, KNOLLS ATOMIC POWER LAB, 87- *Concurrent Pos:* Instr elec eng, Colo State Univ, 75-76; adj prof elec eng, Union Col, NY, 77-83. *Mem:* Inst Elec & Electronics Engrs. *Res:* Measurement and calculation of electric and magnetic fields; electromechanics; computer aided electromagnetic design; total on-line searching and cataloging activities. *Mailing Add:* Knolls Atomic Power Lab Bldg A1-212 PO Box 1072 Schenectady NY 12301

PERRY, NELSON ALLEN, b Louisville, Ky, Mar 26, 37; m 56; c 2. RADIATION HORMESIS. *Educ:* Univ Louisville, BA, 62; Univ Okla, MS, 66. *Prof Exp:* RADIOL SAFETY OFF, UNIV SOUTH ALA, 76-; RADIOL PHYSICIST, PERRY RADIOL CONSULT INC, 70- *Mem:* Am Asn Physicist Med; Health Physics Soc. *Mailing Add:* 1150 Byronell Dr Mobile AL 36693

PERRY, PAUL, flavor chemistry, organic chemistry; deceased, see previous edition for last biography

PERRY, PETER M, b New York, NY, Sept 4, 41; m 67; c 2. IMAGE PROCESSING. *Educ:* City Univ New York, BA, 64; Adelphi Univ, MS, 70; Univ Pa, PhD(astrophys), 74. *Prof Exp:* Instr physics, State Univ NY Maritime Col, 64-70; mem tech staff, 74-77, sect mgr, 77-78, dept mgr, 78-82, PROJ MGR, SYST SCI DIV, COMPUT SCI CORP, 82- *Concurrent Pos:* Prin investr, Int Ultraviolet Explorer Proj, NASA, 78-; proj mgr, Sci Inst space telescope contract, Comput Sci Corp, 83- *Mem:* Am Astron Soc; Am Phys Soc; Astron Soc of the Pac; Sigma Xi. *Res:* Digital image processing as applied to astronomical objects; pseudo color images are created with the echelle spectrograph onboard the Ultraviolet Explorer Spacecraft. *Mailing Add:* Computer Sci Corp 10000 A Aerospace Rd Lanham MD 20706

PERRY, RANDOLPH, JR, b Needham, Mass, Aug 1, 23; m 53; c 2. INDUSTRIAL CHEMISTRY. *Educ:* Harvard Univ, BS, 48; Univ Mich, MS, 49, PhD(chem), 53. *Prof Exp:* Res group leader, Monsanto Co, 52-82; RETIRED. *Mem:* AAAS; Am Chem Soc. *Res:* Industrial chemicals of phosphorus, nitrogen and sulfur; process development. *Mailing Add:* 18 Thorncliff Lane Kirkwood MO 63122

PERRY, RANDY L, b Ashland City, Tenn, Apr 15, 40; m 65; c 2. ENGINEERING ADMINSTRATION. *Educ:* Univ Tenn, Knoxville, BS, 64, MS, 71, PhD(civil eng), 79. *Prof Exp:* Tests engr, Dept Transportation, State of Tenn, 66-74; asst dir, Transportation Ctr, Univ Tenn, 74-79; assoc prof transp eng, 79-81, asst dean, 81-87, DEAN, SCH ENG TECHNOL, UNIV TENN-MARTIN, 87- *Concurrent Pos:* Consult, Flatt & Jared, Attorneys, 81- *Mem:* Am Soc Eng Educ; Am Soc Civil Engrs; Inst Transportation Engrs; Sigma Xi. *Res:* Transportation safety; student retention in engineering. *Mailing Add:* Sch Eng Technol Univ Tenn Martin TN 38238

PERRY, REEVES BALDWIN, b Greenville, Tex, Feb 12, 35; m 64; c 2. PHYSICAL CHEMISTRY. *Educ:* E Tex State Univ, BS, 54; Univ N Tex, MS, 56; Univ Tex-Austin, PhD(phys chem), 66. *Prof Exp:* Chemist, Texaco Inc, 56-59, sr chemist, 65-66; asst prof chem, 66-69, assoc prof, 69-79, PROF CHEM, SOUTHWEST TEX STATE UNIV, 79- *Mem:* Am Chem Soc; Am Inst Chemists. *Res:* Adsorption from solutions. *Mailing Add:* Dept of Chem Southwest Tex State Univ San Marcos TX 78666-4616

PERRY, RICHARD LEE, b Portland, Ore, Jan 22, 30; m 52; c 4. ELECTRON PHYSICS, HOLOGRAPHIC INTERFEROMETRY. *Educ:* Linfield Col, BA, 52; Ore State Univ, MS, 55, PhD(physics), 61. *Prof Exp:* Assoc physicist, Linfield Res Inst, 56-61; from asst prof to assoc prof, 61-71, actg chmn dept, 67-68, 78, PROF PHYSICS, UNIV PAC, 71-, CHMN DEPT, 87- *Concurrent Pos:* Consult, Thompson-Ramo-Wooldridge, Inc, 62-63, Tektronix, Inc, 63-64; consult, US Army Nuclear Defense Lab, 68, res physicist, 69; res assoc, Ames Res Ctr, NASA, 78-85. *Mem:* Am Phys Soc; Am Asn Physics Teachers; Sigma Xi; Soc Photo-Optical Instrument Engrs. *Res:* Field emission from silicon; electron tunneling through thin insulating films; photoelectric effect; holographic interferometry. *Mailing Add:* Dept Physics Univ Pac Stockton CA 95211-0197

PERRY, ROBERT HOOD, JR, b Temple, Tex, Apr 5, 28; m 51. ORGANIC CHEMISTRY, RESEARCH ADMINISTRATION. *Educ:* Baylor Col Med, BS, 48; Univ Tex, PhD(chem), 52. *Prof Exp:* Res chemist, Humble Oil & Ref Co, 52-57, sr res chemist, 57-61, sect head, Exxon Res & Eng Co, NJ, 61-64, asst dir cent basic res lab, 66-67; mgr chem develop, Polaroid Corp, Mass, 67-69; mgr, 69-70, DIR ADVAN DEVELOP, MAGNETIC TAPE LAB, AMPEX CORP, 70- *Mem:* Am Chem Soc; AAAS; NY Acad Sci; Tech Asn Pulp & Paper Indust. *Res:* Chemistry of petroleum hydrocarbons; carbene chemistry; polymerization; autoxidation; synthetic reactions of ozone; photographic chemicals; magnetic materials; synthetic high polymers; magnetic recording; surface coatings; analytical chemistry. *Mailing Add:* 3324 Melendy Dr San Carlos CA 94070

PERRY, ROBERT LEONARD, b New York, NY, Dec 23, 41; m 64; c 2. APPLIED STATISTICS. *Educ:* State Univ NY, Binghamton, BA, 63; Mich State Univ, MS, 65; Rutgers Univ, PhD(statist), 70. *Prof Exp:* Statistician, Ctr Dis Control, USPHS, 67-69; statist consult, Comput Ctr, Rutgers Univ, 69-71; group leader & statistician, Procter & Gamble Co, 71-83; dir diagnostics, McDonnell Douglas Electronics Co, 84-87, prog mgr, quant prob solving, McDonnell Aircraft Co, 87-90; AT PILLSBURY CO, 90- *Concurrent Pos:* Adj fac, Univ Cincinnati, Linderwood Col & Univ Mo, St Louis, Wash Univ, St Louis. *Mem:* Am Statist Asn; fel Am Soc Qual Control; Biomet Soc; Sigma Xi. *Res:* Acceptance sampling, particularly lot acceptance and skip-lots; applied statistical methodology pertaining to real problems in industry; quality and productivity improvements. *Mailing Add:* 3060 Loon Lane Eagan MN 55121

PERRY, ROBERT NATHANIEL, III, b New Orleans, La, Aug 17, 42; m 63, 73, 84; c 3. DESIGN OF CALIBRATION PROCEDURES. *Educ:* Talladega Col, BA, 64; Howard Univ, MS, 67, PhD(physics), 72. *Prof Exp:* Asst prof physics, Univ DC, 68-74; assoc prof physics & math, Dillard Univ, 75-76; mem tech staff, Aerospace Corp, 78-82; sr staff engr, Gen Dynamics, 83-84; dir eng, Gillex Systs, Inc, 84-85; SCIENTIST & ENGR, HUGHES AIRCRAFT CO, 85- *Concurrent Pos:* Lectr electronics, Col Dessert, 82-83. *Mem:* Black Physicist Soc. *Res:* Electro-optical systems; experiment design for optical measurement systems; radiation measurement systems; mathematical analysis. *Mailing Add:* 945 S Osage No 101 Inglewood CA 90301

PERRY, ROBERT PALESE, b Chicago, Ill, Jan 10, 31; m 57; c 3. GENETICS. *Educ:* Northwestern Univ, BS, 52; Univ Chicago, PhD(biophys), 56. *Hon Degrees:* Dr, Univ Paris, 82. *Prof Exp:* Res assoc, Oak Ridge Nat Lab, 56-57; from res assoc to sr mem, Inst Cancer Res, 60-69, assoc dir, 71-74; assoc prof, 66-73, PROF BIOPHYS, UNIV PA, 73-, MEM GRAD GROUP MOLECULAR BIOL, 61-, MEM GRAD GROUP MICROBIOL, 73-, MEM GRAD GROUP GENETICS, 78- *Concurrent Pos:* Fel biophys, Univ Pa, 57-58; Am Cancer Soc fel, Univ Brussels, 59-60; assoc med physics, Johnson Found & instr molecular biol, Univ Pa, 60-65; UN tech assistance expert & vis lectr, Univ Belgrade, 65; ed, J Cellular Physiol, 67-73; mem adv panel molecular biol, Div Res Grants, NSF, 68-71; mem sci adv comt, Damon Runyon Mem Fund Cancer Res, 70-74; mem vis comt, Dept Embryol, Carnegie Inst Wash, 73-76; ed, J Cell Biol, 70-73; ed, Cell, 74-87; Guggenheim Found fel, Univ Paris & Univ Zurich, 74-75. *Mem:* Nat Acad Sci; Am Soc Cell Biol; Int Cell Res Orgn. *Res:* Synthesis and processing of RNA; biosynthesis and function of ribosomes; control mechanisms of growth and division; interrelationships between macromolecular biosyntheses. *Mailing Add:* Inst for Cancer Res 7701 Burholme Ave Fox Chase Philadelphia PA 19111

PERRY, ROBERT RILEY, b Temple, Tex, Dec 7, 34; m 59; c 2. NUCLEAR PHYSICS, COMPUTER SCIENCES. *Educ:* Rice Inst, BA, 57, MA, 58, PhD(physics), 60; Am Bd Radiol, cert radiol physics, 72. *Prof Exp:* Sr res assoc physics, Rice Univ, 60-63; sr physicist, Texaco Inc, Tex, 63-67; asst prof radiol, Univ Tex Med Br, Galveston, 67-77, radiol physicist, 77-; RADIOL PHYSICIST, DICKINSON CO. *Mem:* Am Col Radiol; Am Asn Physicists in Med. *Res:* Application of physics to medicine; use of computers in medicine. *Mailing Add:* Dickinson Co 2710 Mt Vernon Dr Dickinson TX 77539

PERRY, ROBERT W(ILLIAM), b Niagara Falls, NY, Apr 2, 21; wid; c 1. GASDYNAMICS. *Educ:* Cornell Univ, BME, 43, MME, 47, PhD(eng), 51. *Prof Exp:* With Pioneering Res Lab, E I du Pont de Nemours & Co, 50-53; mgr hypervelocity & res brs, Arnold Eng Develop Ctr, 53-59; chief re-entry simulation lab, Repub Aviation Corp, 59-65; prof aerospace eng, Polytech Inst Brooklyn, 65-67; sr staff consult, Liquid Metal Eng Ctr, 68-70; DISTINGUISHED PROF MECH ENG, UNIV LOUISVILLE, 71- *Concurrent Pos:* Mem res adv comt fluid mech, NASA, 59-60; consult, NATO, 66. *Mem:* Am Soc Mech Engrs. *Res:* High temperature gasdynamics; autonomous vehicles. *Mailing Add:* Indust Design Corp 2020 SW Fourth Ave 3rd Fl Portland OR 97201

PERRY, SEYMOUR MONROE, b New York, NY, May 26, 21; m 52; c 3. MEDICAL TECHNOLOGY ASSESSMENT. *Educ:* Univ Calif, Los Angeles, BA, 43; Univ Southern Calif, MD, 47, Am Bd Inter Med dipl, 55. *Prof Exp:* Intern, Los Angeles County Hosp, 46-48, resident med, 48-51; sr asst surgeon, Indian Gen Hosp, Phoenix, Ariz, USPHS, 52; in chg internal med, Outpatient Clin, USPHS, 52-54; fel hemat, Med Ctr, Univ Calif, Los Angeles, 54-55, asst prof med, 56-60, asst res physician, Atomic Energy Proj, 55-57, in chg hemat training, 57-60, asst clin prof med, 60-61; sr investr med br, Nat Cancer Inst, NIH, 61-65, chief med br, 65-68, assoc sci dir clin trials, 66-71, chief human tumor cell biol br, 68-71, assoc sci dir prog, Div Cancer Treatment, 71-73, dep dir, 73-74, actg dir, 74, spec asst to dir, NIH, 74-78, assoc dir med appl res, actg dep asst secy, health technol & actg dir, 78-80; dir, Nat Ctr Health Care Technol, US Dept Health & Human Serv, 80-82; med dir, USPHS, 61-80, asst surgeon gen, 80-82; dir, div biomed res & fac develop, & dep dir dept acad affairs, Asn Am Med Col, 82; sr fel, Dep Dir Inst Health Policy Anal & Prof Med & Community & Family Med, 83-89 interim chmn, 89-90, CHMN, DEPT COMMUNITY & FAMILY MED, GEORGETOWN UNIV MED CTR, PROF MED & COMMUNITY & FAMILY MED, 90- *Concurrent Pos:* Instr, Col Med Evangelists, 51-57; attend specialist, Wadsworth Vet Admin Hosp, 58-61; vis prof, Univ Utah Sch Med, 69, Hebrew Hadassah Univ Med Sch, Jerusalem, 71, Univ PR Sch Med, 75, Univ Wis Sch Med, 82, Swedish Planning & Rationalization Inst Health & Social Serv & Karolinska Inst, 85; chmn, Nutrit Coord Comt, NIH, 77-80 & Inventions & Patents Bd, 78-80; mem Comt Eval Med Technol Clin Use, Inst Med, Nat Acad Sci, 81-84; mem Adv Panel Med Technol & Costs Medicare Prog, Off Technol Assessment, US Cong, 82-84; chmn Criteria Working Group, NASA, 84; consult, Nat Ctr Health Serv Res & Health Care Technol Assessment, 85-90, Agency for Healthcare Policy & Res, Dept Health & Human Serv, 90-; consult, Nat Libr Med, 86-88; mem, Eval Panel, Coun Health Care Technol, Inst Med, Nat Acad Sci, 87-90; gov, Pub Health Serv, Am Col Physicians, 80-82, chmn, Health & Pub Policy Comt, DC Metrop Area, 87-; mem, Health Serv Res Adv Comt, Acad Health Ctrs; chmn, Project Adv Comt Hip Fracture, Univ Md, 90-; Instit Universitaire De Med Soc Et Preventive, Lausanne, 90; Dept Bioeng, Univ Utah, 90. *Honors & Awards:* Commendation Medal, Pub Health Serv, 67; Meritorious Serv Medal, Pub Health Serv, 80; Comendador Ordem Al Merito, Condecoracion, Govt of Peru, 84. *Mem:* Inst Med Nat Acad Sci; Am Pub Health Asn; Am Fedn Clin Res; fel Am Col Physicians; Int Soc Technol Assessment Health Care (pres, 85); Asn Health Serv Res. *Res:* Health care technology assessment. *Mailing Add:* 3750 Reservoir Rd NW Suite 215 Washington DC 20007-2111

PERRY, THOMAS LOCKWOOD, b Asheville, NC, Aug 10, 16; m 41; c 4. BIOCHEMISTRY, NEUROSCIENCES. *Educ:* Harvard Univ, AB, 37, MD, 42; Oxford Univ, BA, 39. *Prof Exp:* Pvt pediat pract, 47-57; res assoc chem, Calif Inst Technol, 57-62; assoc prof, 62-66, PROF PHARMACOL, UNIV BC, 66- *Res:* Biochemical basis of mental and neurological disease; metabolic disorders of childhood; prevention of mental deficiency; development of methods for diagnosis and for treatment of genetically determined diseases; experimental studies of Huntington's disease and Parkinson's disease. *Mailing Add:* Dept of Pharmacol & Therapeut Univ of BC Vancouver BC V6T 1W5 Can

PERRY, THOMAS OLIVER, b Cleveland, Ohio, May 31, 25; m 49; c 5. FOREST GENETICS. *Educ:* Harvard Univ, BS, 49, MA, 50, PhD, 52. *Prof Exp:* From asst prof to assoc prof forestry, Univ Fla, 52-58; NSF res fel, Calif Inst Technol, 59-60; assoc prof, 60-69, prof forestry, NC State Univ, 69-88; CONSULT, 88- *Concurrent Pos:* Charles Bullard fel, Harvard Univ, 68-69; consult nat resource mgt. *Mem:* AAAS; Genetics Soc Am; Soc Am Foresters; Sigma Xi. *Res:* Plant physiology; genetics and physiology of trees; competition. *Mailing Add:* Nat Systs Assoc 5048 Avent Ferry Rd Raleigh NC 27606

PERRY, TILDEN WAYNE, b Timewell, Ill, June 21, 19; m 43; c 2. ANIMAL NUTRITION. *Educ:* Western Ill Univ, BEd, 40; Iowa State Univ, BS, 42; Purdue Univ, MS, 48, PhD(animal nutrit), 50. *Prof Exp:* Teacher high sch, 40-41; res chemist, E I du Pont de Nemours & Co, 42-46; PROF ANIMAL SCI, PURDUE UNIV, WEST LAFAYETTE, 46- *Honors & Awards:* Bohstedt Minerals Award, Am Soc Animal Sci, 78; Am Feed Mfgs Asn Award, 80; Lifetime Serv Award, Ind Beef Cattle Asn, 88. *Mem:* Fel Am Soc Animal Sci; Am Registry Prof Animal Scientists. *Res:* Animal nutrition, especially with beef cattle, swine and sheep. *Mailing Add:* Dept Animal Sci Purdue Univ West Lafayette IN 47907

PERRY, VERNON G, b Boaz, Ala, May 8, 21; m 48; c 2. PLANT NEMATOLOGY. *Educ:* Ala Polytech Inst, BS, 43, MS, 49; Univ Wis, PhD(plant path), 58. *Prof Exp:* Asst nematologist, Agr Res Serv, USDA, Fla, 49-54, nematologist, Wis, 55-58; actg chmn dept entom & nematol, 75-76, PROF NEMATOL & ASST CHMN DEPT ENTOM & NEMATOL, AGR EXP STA, UNIV FLA, 59-, ASST DEAN RES, 76- *Mem:* AAAS; fel Soc Nematol; Orgn Trop Am Nematol. *Res:* Feeding habits and pathogenicity to plants of nematology; taxonomy of nematodes and control of soil nematodes; nematode parasites of insects. *Mailing Add:* Ifas Admin Univ Fla Gainesville FL 32603

PERRY, VERNON P, biology; deceased, see previous edition for last biography

PERRY, WILLIAM DANIEL, b Bradenton, Fla, Oct 8, 44; m 65; c 1. INORGANIC CHEMISTRY. *Educ:* Fla State Univ, BS, 65; Univ Ill, Urbana, PhD(inorg chem), 70. *Prof Exp:* Res assoc chem, Fla State Univ, 70-71; asst prof, 71-78, ASSOC PROF CHEM, AUBURN UNIV, 78- *Mem:* Am Chem Soc. *Res:* Study of the structure, properties and kinetics of transition metal complexes. *Mailing Add:* Dept of Chem Auburn Univ Auburn AL 36830

PERRY, WILLIAM JAMES, b Vandergrift, Pa, Oct 11, 27; m 47; c 5. MATHEMATICS. *Educ:* Stanford Univ, BS, 49, MS, 50; Pa State Univ, PhD(math), 57. *Prof Exp:* Instr math, Univ Idaho, 50-51 & Pa State Univ, 51-54; eng mgr, Sylvania Elec Prod, Inc, 54-64, dir electronic defense lab, 61-64; pres, ESL, Inc & dir, Electromagnetic Systs Lab, Sunnyvale, 64-77, dir defense res & eng, 77; under secy defense, Res & Eng, Off Secy Defense, 77-81; partner, Hambrecht & Quist Investment Bankers, 81-85, PRES, TECH STRATEGIES & ALLIANCES, INC, 85- *Concurrent Pos:* Sci consult, US Dept Defense, 63-; lectr, Santa Clara Univ, 69-77; prof, Sch Eng, Stanford Univ; co-dir, Ctr Int Security & Arms Control; trustee, Mitre Corp. *Honors & Awards:* Medal of Achievement, Am Electronics Asn, 80. *Mem:* Nat Acad Eng; Am Math Soc; Sigma Xi; fel Am Acad Arts & Sci. *Res:* Electromagnetic systems analysis; partial differential equations. *Mailing Add:* 10701 Mora Dr Los Altos CA 94024

PERRY, WILLIAM LEON, b Trenton, Mo, Dec 3, 45; m 67; c 2. NONLINEAR DIFFERENTIAL EQUATIONS. *Educ:* Park Col, BA, 67; Univ Ill, MA, 68, PhD(math), 72. *Prof Exp:* From asst prof to assoc prof, 71-87, PROF MATH, TEX A&M UNIV, 87- *Mem:* Am Math Soc; Math Asn Am. *Res:* Nonlinear differential equations; integral equations and transforms. *Mailing Add:* Dept Math Tex A&M Univ College Station TX 77843

PERRYMAN, CHARLES RICHARD, b Elliot, Iowa, Sept 21, 16; m 40; c 2. RADIOLOGY. *Educ:* Dartmouth Col, BA, 38; Cornell Univ, MD, 42; Univ Pa, DSc, 47. *Prof Exp:* Intern, Bellevue Hosp, NY, 42-43; instr radiol, Sch Med, Univ Pa, 46-47, from assoc to asst prof, 47-49; asst prof, 51-59, assoc prof, 59-78, EMER PROF RADIOL, SCH MED, UNIV PITTSBURGH, 78- *Concurrent Pos:* Mem staff, Dept Radiol, Hosp Univ Pa, 46-51; pres, M D

O'Donnell Diag Clin, 75- *Mem:* Am Radium Soc; Am Roentgen Ray Soc; Radiol Soc NAm; Am Nuclear Soc; fel Royal Soc Health. *Res:* Fundamental biological effects of radiation and their modification; roentgen diagnostic aspects of head and nervous system. *Mailing Add:* 640 Osage Ave Pittsburgh PA 15243

PERRYMAN, ELIZABETH KAY, b Greenwood, Miss, Apr 11, 40; m 83. ENDOCRINOLOGY, TEACHING. *Educ:* Memphis State Univ, BS, 64; Tex Tech Univ, MS, 67; Univ Ariz, PhD(zool), 72. *Prof Exp:* Instr biol, Victoria Col, 67-69; from asst prof to assoc prof, 72-81, PROF BIOL, CALIF POLYTECH STATE UNIV, SAN LUIS OBISPO, 81- *Concurrent Pos:* Sigma Xi res grant-in-aid, 72; NSF grants small col fac, 76-79; Dept Energy fac fel, 79; Burrough Wellcome Award, 81; assoc, Danforth Found. *Mem:* Am Soc Zoologists; fel AAAS; assoc mem Danforth Found. *Res:* Ultrastructure of the pituitary gland in relation to the control of pigmentation; study of the nongranulated (stellate) cells of the pituitary gland. *Mailing Add:* Dept of Biol Sci Calif Polytech State Univ San Luis Obispo CA 93407

PERRYMAN, JAMES HARVEY, b Kansas City, Mo, Aug 18, 18; m 50. NEUROPHYSIOLOGY, BIOLOGY. *Educ:* Stanford Univ, AB, 41; Univ Calif, MA, 43, PhD(physiol), 55. *Prof Exp:* Mem biomech group, Univ Calif, 50-52, mem pain study group, 52-53, asst physiol, 53-54; asst prof biol, Univ San Francisco, 54-56; from asst prof physiol & biophys to assoc prof physiol & pharmacol, Col Dent, NY Univ, 56-72; prof oral biol, Col Med & Dent, NJ, 74-80, prof biodent sci, 80-, chmn occlusion & res dir, 72-; RETIRED. *Concurrent Pos:* Res physiologist, Univ Calif, 55-56; USPHS res grant, NY Univ, 57-67; res assoc ophthal, NY Univ, 57-67, consult med & periodontia, 64-, consult grad fac, 58-, vis prof neuroanat, 70 & 71; consult med & pain, Bellevue Hosp, New York, 66-68; res prof psychol, Queens Col, 66-70; consult neurosurg, Brooklyn Hosp, 68-69; NSF inst award, Col Med & Dent NJ, 73- *Mem:* Fel AAAS; Am Physiol Soc; Am Psychol Asn; Am Asn Anatomists; Soc Neurosci. *Res:* Nonspecific reflex reactions of tactile, pain and stress stimulation; microelectrode recording spinal and brain stem respiratory potentials; extraocular muscle physiology, neural differentiation and coordination. *Mailing Add:* 90 Manchester Pl Newark NJ 07104

PERRYMAN, JOHN KEITH, b Midland, Tex, Oct 16, 35; m 55; c 3. APPLIED MATHEMATICS. *Educ:* Union Col, Nebr, BA, 58; Univ Tex, MA, 60, PhD(math), 63. *Prof Exp:* Spec instr math, Univ Tex, 60-63; asst prof, 63-65, ASSOC PROF MATH, UNIV TEX, ARLINGTON, 65- *Mem:* Am Math Soc; Math Asn Am; Sigma Xi. *Res:* Mathematical analysis; development of integral transforms; mathematical modeling of medically oriented problems. *Mailing Add:* PO Box 1352 Mena AR 72902

PERSAD, EMMANUEL, b Trinidad, Dec 25, 35; Can citizen; m 67; c 2. MOOD DISORDER, PSYCHIATRIC EDUCATION. *Educ:* Univ Durham, Eng, MB BS, 64; Univ Toronto, dipl psychiat, 69; FRCP(C), 69. *Prof Exp:* Psychiatrist, Clarke Inst Psychiat, Toronto, 71-89, dir res, 84-86, head clin res, 86-89; DIR, POSTGRAD EDUC PSYCHIAT, UNIV WESTERN ONT, 89- *Concurrent Pos:* Prin investr, several pharmaceut co, 80-; consult, Mood Dis Prog, London Psychiat Hosp, 89- *Mem:* Can Psychiat Asn. *Res:* Psychiatry; genetics; psychiatric nosology; pharmacology. *Mailing Add:* London Psychiatric Hospital 850 Highbury Ave London ON N6A 4H1 Can

PERSANS, PETER D, b Hempstead, NY, Mar 25, 53. SEMICONDUCTORS, RAMAN SCATTERING. *Educ:* Polytech Inst NY, BS, 75; Univ Chicago, SM, 77, PhD(physics), 82. *Prof Exp:* Res physicist, Exxon Res & Eng, 81-83, sr physicist, 83-86; asst prof, 86-89, ASSOC PROF PHYSICS, DEPT PHYSICS, RENSSELAER POLYTECH INST, 89- *Concurrent Pos:* Consult, Dow-Corning, 87-88, Gen Elec Co, 89-91; vchair, Acad Affairs Comt, Mat Res Soc, 90-91. *Mem:* Mat Res Soc; Am Phys Soc; Metall Soc; Optical Soc Am. *Res:* Size, disorder and interface effects on semiconductor materials; quantum dots and nanocrystals; amorphous semiconductors. *Mailing Add:* Dept Physics 4223-CII Rensselaer Polytech Inst Troy NY 12180-3590

PERSAUD, TRIVEDI VIDHYA NANDAN, b Port Mourant, Guyana, Feb 19, 40; Can citizen; m 66; c 3. ANATOMY, PATHOLOGY. *Educ:* Univ Rostock, Ger, MD, 65, DSc, 74; Univ West Indies, Jamaica, PhD(anat), 70; Royal Col Pathologists, London, MRCPath, 72, FRCPath, 84; Royal Col Physicians Ireland, FFPath, 84. *Prof Exp:* Intern, Kleinmachnow Hosp, 65-66; govt med officer, Guyana, 66-67; lectr anat, 67-70, sr lectr, Univ West Indies, 70-72; from assoc prof to prof, 72-77, PROF ANAT & HEAD DEPT, UNIV MAN, 77-, DIR TERATOLOGY RES LAB, 75-, ASSOC PROF OBSTET, GYNEC & REPROD SCI, 79-, PROF PEDIAT & CHILD HEALTH, 89- *Concurrent Pos:* Ed, West Indian Med Jour, 70-73; consult teratology & clin genetics, Children's Ctr, Winnipeg, 73- & path, Health Sci Ctr, Winnipeg, 73-; vis fel, Wolfson Col, Cambridge Eng, 88. *Honors & Awards:* Carveth Jr Scientific Award, Can Asn Pathologists, 74; Rh Inst Award, Univ Man, 75; Albert Einstein Centennial Medal, Ger, 81; Andreas Vesalius Medal, Italy, 86; Twelfth Raymond C Truex Distinguished Lectureship Award, Hahnemann Univ, 90. *Mem:* Royal Col Pathologists, London; Royal Col Physicians, Ireland; Am Asn Anat; Can Asn Anatomists (pres, 81-83); hon mem Anat Soc Ger. *Res:* Teratology; fetal physiology and reproduction. *Mailing Add:* Dept Anat Fac Med & Dent Basic Med Sci Bldg Univ Man 730 William Ave Winnipeg MB R3E 0W3 Can

PERSELL, RALPH M(OUNTJOY), b New Iberia, La, Sept 17, 08; c 1. CHEMICAL ENGINEERING. *Educ:* Tulane Univ, BEChE, 32. *Prof Exp:* Chemist, State Hyg Dept, Miss, 34-37; field engr, Bristol Co, 37-38; inspector, USFDA, 38-39; chem engr southern regional lab, USDA, 46-59, asst dir prog appraisal, southern utilization res & develop div, Sci & Educ Admin-Agr Res, 59-72, dir Ala & northern Miss res area, 72-75; RETIRED. *Mem:* Am Chem Soc; Am Inst Chem Engrs. *Res:* Improved production of farm crops. *Mailing Add:* The Towers 801 Myrtle Ave Natchez MS 39120

PERSHAN, PETER SILAS, b Brooklyn, NY, Nov 9, 34; m 57; c 2. SOLID STATE PHYSICS. *Educ:* Polytech Inst Brooklyn, BS, 56; Harvard Univ, AM, 57, PhD(physics), 60. *Prof Exp:* Res assoc, 60-61, from asst prof to assoc prof, 61-68, GORDON McKAY PROF SOLID STATE PHYSICS, HARVARD UNIV, 68- *Concurrent Pos:* Sloan Found res grant, 63-67; NSF sr fel, Univ Paris, 71-72; vis prof, Mass Inst Technol, 78-79; vis scientist, Brookhaven Nat Lab, 85-86. *Mem:* Fel Am Phys Soc; NY Acad Sci. *Res:* Cross-relaxation phenomena in magnetic resonance; microwave modulation of light; applications of electron spin resonance; basic theory of nonlinear optics; physical properties of liquid crystals; x-ray scattering using synchrotron radiation. *Mailing Add:* 205 Pierce Hall Harvard Univ Cambridge MA 02138

PERSIANI, PAUL J, b Brooklyn, NY, Oct 24, 21; m 52; c 2. PHYSICS, NUCLEAR ENGINEERING. *Educ:* Clarkson Col Technol, BME, 48; St Louis Univ, PhD(physics), 56. *Prof Exp:* Assoc mech engr, 49-52, PHYSICIST, ARGONNE NAT LAB, 56- *Concurrent Pos:* Asst prof, Sch Technol, Northwestern Univ. *Honors & Awards:* Argonne Pacesetters Award. *Mem:* Sigma Xi; Am Nuclear Soc; AAAS; NY Acad Sci; Am Phys Soc. *Res:* Fission and fusion reactor; neutron, nuclear and plasma physics; physics engineering. *Mailing Add:* 2455 Charles Naperville IL 60540

PERSKY, GEORGE, b Brooklyn, NY, Apr 26, 38; m 64; c 1. SOLID STATE PHYSICS, ELECTRICAL ENGINEERING. *Educ:* Rensselaer Polytech Inst, BSEE, 59; Polytech Inst Brooklyn, MSEE, 61, PhD(physics), 68. *Prof Exp:* Mem tech staff, Bell Labs, Murray Hill, 67-78; sr proj engr, Newport Beach Res Ctr, 78-80, sect head, 80- 81, asst dept mgr, 81-83, DEPT MGR, DESIGN & TEST DEPT, IEG TECHNOL CTR, HUGHES AIRCRAFT CO, CARLSBAD, CALIF, 83-, ASST LAB MGR, 86- *Mem:* Am Phys Soc; sr mem Inst Elec & Electronics Engrs; Sigma Xi. *Res:* High field transport properties of semiconductors; propagation and attenuation of ultrasound and helicon waves in metals; computer aided design; artificial neural networks. *Mailing Add:* 24751 Doria Way Mission Viejo CA 92691

PERSKY, HAROLD, b Chicago, Ill, Aug 11, 17; m 41, 76; c 1. PSYCHOENDOCRINOLOGY, ENDOCRINOLOGY. *Educ:* Univ Chicago, BS, 36, PhD(chem), 41. *Prof Exp:* Asst, Univ Chicago, 41-42; res assoc metab paralyzed muscle, Res Inst, Michael Reese Hosp, 42-44, asst dir res, Div Neuropsychiat, 46-51, dir biochem lab, Inst Psychosom & Psychiat Res & Training, 51-56; from asst prof to assoc prof biochem, Sch Med, Ind Univ, 56-62; adj res assoc prof, 63-71, PROF PSYCHIAT, SCH MED, UNIV PA, 71- *Concurrent Pos:* Lalor Found fel, Marine Biol Lab, Woods Hole, 48; USPHS res scientist awards, 62-81; assoc mem & mem div encocrinol & reproduction res labs, Albert Einstein Med Ctr, 62-71. *Mem:* AAAS; Am Psychosom Soc; Endocrine Soc. *Res:* Psychoendocrinology of human sexual behavior. *Mailing Add:* 340 Spruce Philadelphia PA 19106

PERSKY, LESTER, b Cleveland, Ohio, Mar 4, 19; m 49; c 3. SURGERY, UROLOGY. *Educ:* Univ Mich, BS, 41; Johns Hopkins Univ, MD, 44. *Prof Exp:* From instr to assoc prof, 53-63, prof urol, Sch Med, 63-, clin prof surg, EMER PROF UROL, CASE WESTERN RESERVE UNIV, 90. *Concurrent Pos:* Teaching fel surg, Harvard Univ, 49-50; teach fel, Tufts Univ, 49-50; prof urol, Univ SFla, 88. *Mem:* AAAS; AMA; Am Urol Asn; Am Col Surg; Am Soc Clin Invest. *Res:* Physiology of obstruction to kidney; homotransplantation of organs. *Mailing Add:* 15 Bluebill Ave Apt 803 Naples FL 33963

PERSON, CLAYTON OSCAR, genetics; deceased, see previous edition for last biography

PERSON, DONALD AMES, b Fargo, NDak, July 17, 38; m 62; c 2. PEDIATRIC RHEUMATOLOGY, VIROLOGY. *Educ:* Univ NDak, BS, 61; Univ Minn, MD, 63. *Prof Exp:* Resident neurosurg, Mayo Clin & Mayo Found, 67-68, fel microbiol, 68-71; instr virol & epidemiol, Balor Col, 71-73, asst prof virol, epidemiol & internal med, 73-78, resident pediat, 78-80, asst prof, 80-87; ASSOC CLIN PROF PEDIAT, JOHN A BURNS SCH MED, UNIV HAWAII MANOA, 88- *Concurrent Pos:* Fel, Arthritis Found, 71-73; res fel, Arhritis Found, 72-74, sr investr, 75-77; asst attend pediatrician, Harris County Hosp Dist, 80-87; rheumatologist, Tex Children's Hosp, 80-, attend pediatrician, 82-87; lectr, Tex Women's Univ, 83-87; Consult, St Luke's Episcopal Hosp, Methodist Hosp & Shriner's Hosp Crippled Children, Houston, 83-87; sr investr, Arthritis Found, 75-77; col, Med Corps, US Army; asst chief, dept pediat, Tripler Army Med Ctr, chief out patient pediat & rheumatol serv, 87- *Honors & Awards:* Philip Hench Award, Asn Mil Surgeons US, 90. *Mem:* Fel Am Acad Pediat; fel Am Col Rheumatology; AMA; Asn Mil Surgeons US; Soc Exp Biol & Med; Soc Pediat Res. *Res:* Microbial rheumatoid arthritis; immunopathogenesis of rheumatoid arthritis; juvenile rheumatoid arthritis and related diseases. *Mailing Add:* Dept Pediat Tripler Army Med Ctr Honolulu HI 96859-5000

PERSON, JAMES CARL, b Portland, Ore, Sept 22, 36; m 60; c 3. PLASMA PHYSICS. *Educ:* Willamette Univ, BA, 58; Univ Calif, Berkeley, PhD(phys chem), 64. *Prof Exp:* Asst chemist, Argonne Nat Lab, 63-68, chemist, 68-81; asst prof, St Xavier Col, 81-82; programmer, SPSS Inc, 82-84; PRIN SCIENTIST, PHYS SCI, INC, 84- *Mem:* Am Phys Soc; Radiation Res Soc; Am Chem Soc. *Res:* Modeling chemical kinetics; plasma simulations; laser ablation plume phenomena; ionic environment of space shuttle; photo absorption and photo-ionization measurements; ion-ion recombination kinetics; relating optical data to radiation chemistry yields. *Mailing Add:* 20 New England Business Ctr Andover MA 01810

PERSON, LUCY WU, nuclear chemistry, physics, for more information see previous edition

PERSON, STANLEY R, b US, Apr 14, 28; c 3. BIOPHYSICS, MOLECULAR BIOLOGY. *Educ:* Lafayette Col, BS, 51; Yale Univ, PhD(biophys), 57. *Prof Exp:* Asst res biophysicist, Univ Calif, Los Angeles, 57-65; from assoc prof to prof biophys, Pa State Univ, 66-81, biophys grad prog chmn, 75-78, dir, Indust-Univ Coop Prog Recombinant DNA Technol, 82-83, PROF BIOPHYS & MOLECULAR BIOL, PA STATE UNIV, 81-; DIR, BIOTECHNOL THRUST AREA, ADVAN TECHNOL CTR CENT & NORTHERN PA, 83- *Concurrent Pos:* Biophys Grad Prog Chr, 75-78, dir, Indust Univ Coop Prog, 82-83. *Mem:* AAAS; Biophys Soc. *Res:* Structure, expression and regulation of herpes simplex virus genes; recombinant DNA; expression of eukaryotic genes in bacteria; in vitro mutagenesis of DNA; herpes virus entry; genetic recombination; genetic code; subunit vaccine; gene evolution. *Mailing Add:* Dept MGB Univ Pittsburgh Med Sch E1246 Bio Med Sci Tower Pittsburgh PA 15261

PERSON, STEVEN JOHN, b Albert Lea, Minn, Mar 14, 44; m 65; c 3. MAMMALIAN PHYSIOLOGY. *Educ:* Iowa State Univ, BS, 66, MS, 68; Univ Alaska, PhD(zoophysiol), 75. *Prof Exp:* Instr anat & physiol, Lake Superior State Col, 73-74; pipeline surveillance biologist fish & wildlife mgt, Alaska Dept Fish & Game, 74-75; asst anat & physiol, Lake Superior State Univ, 75-80, from asst prof to assoc prof biol, 80-89, head dept biochem, 86-89, PROF BIOL, LAKE SUPERIOR STATE UNIV, 89- *Mem:* Nat Asn Adv Health Professions. *Res:* Digestibility of forages consumed by Rangifer tarandus with emphasis on the inhibition of rumen microbes caused by dietary lichens. *Mailing Add:* Dept Biol Lake Superior State Univ Sault Ste Marie MI 49783

PERSON, WILLIS BAGLEY, b Salem, Ore, Apr 23, 28; m 85; c 2. MOLECULAR SPECTROSCOPY, BIOPHYSICS. *Educ:* Willamette Univ, BS, 47; Ore State Col, MS, 49; Univ Calif, PhD(chem), 53. *Prof Exp:* Asst, Ore State Col, 47-49 & Univ Calif, 49-52; Du Pont instr phys chem, Univ Minn, 52-53, res fel, 53-54; instr, Harvard Univ, 54-55; from asst prof to assoc prof, Univ Iowa, 55-66; PROF PHYS CHEM, UNIV FLA, 66- *Concurrent Pos:* Guggenheim fel, Univ Chicago, 60-61, vis assoc prof & NSF sr fel, 65-66; collabr, Los Alamos Nat Lab, 75-89; vis sr fel, Royal Holloway Col, Univ London, 78; UNESCO consult, State Univ Campinas, Brazil, 80; assoc mem, Comn I-5 Molecular Struct & Spectros, Int Union Pure & Appl Chem, 81-89, mem Spectros sub-comt of Comn I-5, 89-; vis prof Inst Molecular Sci, Japan, 84 & Univ Pierre & Marie Curie, Paris, 85. *Mem:* AAAS; Am Chem Soc; Optical Soc Am; Royal Soc Chem; Coblentz Soc; Soc Appl Spectros. *Res:* Molecular spectroscopy and structure; vibrational spectroscopy, emphasizing infrared intensities, laser applications and biophysics; theoretical and experimental studies of molecular complexes; studies of tautomerism in nucleic acid bases. *Mailing Add:* Dept Chem Univ Fla Gainesville FL 32611-2046

PERSONEUS, GORDON ROWLAND, b Grand Gorge, NY, Mar 10, 22; m 48; c 3. ORGANIC CHEMISTRY. *Educ:* Union Col, BS, 48. *Prof Exp:* Asst penicillin res, Lederle Labs Div, Am Cyanamid, 40-42, cancer res, 48-52, vet bact, 52-55, head bact & pharmacol test dept qual control, 55-73, dir tech serv regulatory affairs, 73-91; RETIRED. *Concurrent Pos:* Consult, 91- *Mem:* Soc Indust Microbiol; Parenteral Drug Asn. *Res:* Biological and bacteriological sciences; immunology; test development; quality control; drug regulatory. *Mailing Add:* Lederle Labs Am Cyanamid Co Pearl River NY 10965

PERSONICK, STEWART DAVID, b Brooklyn, NY, Feb 22, 47; m 86; c 2. TELECOMMUNICATIONS, OPTICAL COMMUNICATIONS. *Educ:* City Col New York, BEE, 67; Mass Inst Technol, SM, 68, ScD(elec eng), 70. *Prof Exp:* Engr, Bell Commun Res, 67-75, supv engr, 75-78, dept head, Bell Telephone Labs, 83, div mgr, 84-85, ASST VPRES, BELL COMMUN RES, 85- *Concurrent Pos:* Dept mgr, Vidar Div, 78-81, lab mgr, Military Electronics Div, TRW, Inc, 81-83; Consult, 83; comt mem, Nat Res Coun, 84-; vis lectr, Mass Inst Technol, 84. *Honors & Awards:* Centennial Medal, Inst Elec & Electronics Engrs, 89. *Mem:* Fel Inst Elec & Electronic Engrs; fel Optical Soc Am. *Res:* Optical fiber transmission and communications systems; photonic and electronic switching; advanced telecommunications services and technology. *Mailing Add:* Bell Commun Res Rm 2A213 445 South St Morristown NJ 07962

PERSSON, SVEN ERIC, b Lethbridge, Alta, Oct 16, 45; m 68; c 2. ASTRONOMY. *Educ:* McGill Univ, BS, 66; Calif Inst Technol, PhD(astron), 72. *Prof Exp:* Lectr & res fel astron, Harvard Col Observ & Harvard Univ, 72-75; staff mem astron, Hale Observ, Calif Inst Technol, 75-89; STAFF MEM ASTRON, CARNEGIE OBSERV, 89- *Mem:* Am Astron Soc. *Res:* Stellar populations of elliptical galaxies from infrared photometry; star formation. *Mailing Add:* Carnegie Observ 813 Santa Barbara St Pasadena CA 91101

PERSSON, SVERKER, b Karlshamn, Sweden, Aug 3, 21; m 50; c 4. POWER MACHINERY. *Educ:* Chalmers Technol Univ, Sweden, MS, 45; Mich State Univ, PhD(agr eng), 60. *Prof Exp:* Engr, Bolinder-Munktell, Eskilstuna, 45-47; res engr, Swed Inst Agr Eng, 47-49; assoc prof agr eng, Royal Agr Col Sweden, 47-62, acting head dept, 62-63; assoc prof, Mich State Univ, 63-68; assoc prof, 68-73, PROF AGR ENG, PA STATE UNIV, UNIVERSITY PARK, 73- *Honors & Awards:* John Ericson Medal, 45. *Mem:* Am Soc Agr Engrs; Max Eyth Ges; Int Soc Terrain Vehicles Systs. *Res:* Mushroom and greenhouse production systems; tractors; soil-machine systems; methane from agricultural residue; basic principles of harvesting machines. *Mailing Add:* Dept of Agr Eng Pa State Univ University Park PA 16802

PERT, CANDACE B, b Manhattan, NY, June 26, 46; c 3. NEUROSCIENCE. *Educ:* Bryn Mawr Col, AB, 70; Sch Med, Johns Hopkins Univ, PhD(pharmacol), 74. *Prof Exp:* NIH fel, Johns Hopkins Univ, 74-75; staff fel, NIMH, 75-77, sr staff fel, 77-78, res pharmacologist, 78-82, chief sect brain chem, 82-88, GUEST RESEARCHER, NIMH, 88- *Honors & Awards:* Arthur S Fleming Award, 79. *Mem:* Am Soc Pharmacol & Exp Therapeut; Am Soc Biol Chemists; Soc Neurosci; Int Narcotics Res Conf. *Res:* Brain peptides and their receptors: chemical characteristics, brain distribution and function. *Mailing Add:* Nat Inst Ment Health NIH Bldg 10 Rm 3N256 9000 Rockville Pike Bethesda MD 20892

PERTEL, RICHARD, b Tallinn, Estonia, Mar 22, 28; nat US; m 55; c 3. PHYSICAL CHEMISTRY. *Educ:* Univ Wash, BS, 52; Ill Inst Technol, PhD(chem), 60. *Prof Exp:* From asst prof to assoc prof chem, Univ Houston, 59-65; vis prof, Univ Alta, 65-68; phys chemist, 68-69, SR SCIENTIST, INST GAS TECHNOL, IIT CTR, 69- *Mem:* Am Chem Soc; Am Phys Soc. *Res:* Photochemistry; gas-phase kinetics; combustion and flames; mass spectrometry; optical spectroscopy. *Mailing Add:* 1210 S Salem Lane Arlington Heights IL 60005

PERTICA, ALEXANDER JOSÉ, b Buenos Aires, Arg, Dec 7, 61; US citizen; m 90. LASER PHYSICS. *Educ:* Cornell Univ, BA, 84; Univ Southern Calif, MS, 86. *Prof Exp:* Assoc mem tech staff, Aerospace Corp, 85-86; ENGR, LAWRENCE LIVERMORE NAT LABS, 86- *Mem:* Optical Soc Am. *Res:* Optical properties of solid state laser materials; instrumentation for remote spectroscopy; remote sensing. *Mailing Add:* Lawrence Livermore Nat Labs L-407 PO Box 808 Livermore CA 94550

PERUMAREDDI, JAYARAMA REDDI, b Gudur, India, Oct 15, 36; m 63; c 2. INORGANIC CHEMISTRY, PHYSICAL CHEMISTRY. *Educ:* Andhra Univ, India, BS, 56, MS, 57; Univ Southern Calif, PhD(chem), 62. *Prof Exp:* Fel chem, Mellon Inst, 63-67; from asst prof to assoc prof, 67-73, chmn dept, 77-83, PROF CHEM, FLA ATLANTIC UNIV, 73- *Concurrent Pos:* Consult, Inst Explor Res, US Army Electronics Command, Ft Monmouth, NJ, 70-72. *Mem:* AAAS; Am Chem Soc; Royal Soc Chem; Chem Soc Japan; Am Phys Soc. *Res:* Transition-metal chemistry and physics, especially the electronic energy levels of transition-metal systems by electronic spectra; ligand field and molecular orbital theories; photochemistry; structural and magnetic properties. *Mailing Add:* Dept of Chem Fla Atlantic Univ Boca Raton FL 33431

PERUN, THOMAS JOHN, b Auburn, NY, Sept 28, 37; div; c 3. PHARMACEUTICAL CHEMISTRY. *Educ:* Rensselaer Polytech Inst, BS, 59; Univ Rochester, PhD(org chem), 63; Lake Forest Sch Mgt, MBA, 81. *Prof Exp:* Sr chemist, 63-71, assoc res fel, 71-75, MGR CHEM RES, ABBOTT LABS, 75- *Mem:* Am Chem Soc; Am Soc Microbiol; Sigma Xi. *Res:* Chemistry and structure-activity relationships of antibiotics; conformational analysis; enzyme inhibitors. *Mailing Add:* Abbott Labs D-460 Abbott Park IL 60064

PERUZZOTTI, GEORGE PETER, b New York, NY, Jan 11, 35; m 61; c 4. MICROBIAL BIOCHEMISTRY. *Educ:* St John's Univ, NY, BS, 56, MS, 59; Univ Wis-Madison, PhD(microbiol), 73. *Prof Exp:* Asst res biologist biochem, Sterling-Winthrop Res Inst, 59-64; res scientist, Miles Lab Inc, 73-77, sr res scientist chem, 77-81; sr res assoc, 81-89, RES ASSOC, EASTMAN KODAK CO, 89- *Mem:* AAAS; Am Chem Soc; Soc Indust Microbiol; Am Soc Microbiol; NY Acad Sci; Sigma Xi. *Res:* Research and development in microbial transformation of synthetic organic compounds; natural products and microbial enzymes; fermentation and medium optimization of biocontrol agents; agricultural research in biocontrol. *Mailing Add:* 1217 Wildflower Dr Webster NY 14580

PERVIN, WILLIAM JOSEPH, b Pittsburgh, Pa, Oct 31, 30; m 81; c 4. MATHEMATICS, COMPUTER SCIENCE. *Educ:* Univ Mich, BS & MS, 52; Univ Pittsburgh, PhD(math), 57. *Prof Exp:* Sr scientist, Atomic Power Div, Westinghouse Elec Corp, 54-55; asst prof math, Univ Pittsburgh, 55-57; from asst prof to assoc prof, Pa State Univ, 57-64; prof math, Univ Wis-Milwaukee, 64-67, chmn dept, 65-66; prof math, Drexel Univ, 67-73, dir, Comput Ctr, 71-73, dir regional comput ctr, 73-78, chmn dept, 83-85; PROF MATH & COMPUT SCI, UNIV TEX, DALLAS, 73-, MASTER, SCH ENG & COMPUT SCI, 87- *Concurrent Pos:* Vis prof, Univ Heidelberg, 63-64. *Mem:* AAAS; Am Math Soc; Asn Comput Mach; Soc Indust & Appl Math; Math Asn Am; Inst Elec & Electronics Engrs Comput Soc. *Res:* Topology; computer sciences. *Mailing Add:* Univ Tex at Dallas Box 830688 Sta MP31 Richardson TX 75083-0688

PERZ, JOHN MARK, b Paris, France, Mar 23, 40; Can citizen; m 66; c 2. METAL PHYSICS. *Educ:* Univ Toronto, BASc, 60, MASc, 61; Univ Cambridge, PhD(physics), 64. *Prof Exp:* From asst prof to assoc prof, 66-80, PROF PHYSICS, UNIV TORONTO, 80- *Mem:* Am Phys Soc; Can Asn Physicists. *Res:* Ultrasonics in metals; fermi surfaces. *Mailing Add:* Dept of Physics Univ of Toronto Toronto ON M5S 1A7 Can

PERZAK, FRANK JOHN, b Pittsburgh, Pa, July 27, 32. PHYSICAL CHEMISTRY, EXPLOSIVES. *Educ:* Carnegie Inst Technol, BSc, 54; Univ Pittsburgh, PhD, 79. *Prof Exp:* Chemist br health & safety, 57-62, chemist, Explosive Res Ctr, 62-64, res chemist explosives chem sect, 64-66, res chemist, Spec Res Group, Explosive Res Ctr, 66-72, RES CHEMIST, PITTSBURGH RES CTR, US BUR MINES, 72- *Mem:* AAAS; Am Phys Soc; Combustion Inst; Am Inst Chemists. *Mailing Add:* 161 Temona Dr Pittsburgh PA 15236

PESCE, AMADEO J, b Everett, Mass, June 30, 38; m 64; c 3. CLINICAL CHEMISTRY. *Educ:* Mass Inst Technol, BS, 60; Brandeis Univ, PhD(biochem), 64; Am Bd Clin Chem, cert. *Prof Exp:* Dir biochem res, Renal Div, Michael Reese Hosp & Med Ctr, 67-73; assoc prof, 73-79, PROF EXP MED, PATH & LAB MED, UNIV CINCINNATI MED CTR, 79-; DIR, TOXICOL LAB, UNIV HOSP, CINCINNATI, OHIO, 82- *Concurrent Pos:* NIH fel, Univ Ill-Urbana, 64-67; adj asst & assoc prof, Dept Biol, Ill Inst Technol, 67-73; estab investr, Am Heart Asn, 68-73; mem, bd dirs, Nat Acad Biochem, 79-85. *Mem:* Am Asn Clin Chemists; Soc Exp Biol Med; Am Chem Soc; Nat Acad Clin Biochem; Asn Clin Scientists. *Res:* Clinical chemistry toxicology; immunology of antigenic fragments; application of Elisa technology; urinary protein excretion. *Mailing Add:* Univ Cincinnati Med Ctr 226 Lab Med Bldg 231 Bethesda Ave Cincinnati OH 45267-0714

PESCE, MICHAEL A, b New York, NY, July 3, 42; m 69; c 2. CLINICAL CHEMISTRY, LABORATORY MEDICINE. *Educ:* St John's Univ, BS, 63, MS, 67, PhD(synthetic org chem), 71. *Prof Exp:* Asst dir, Micro Chem Lab, 71-77, asst prof, 76-82, ASSOC PROF CLIN PATH, COLUMBIA-PRESBY MED CTR, 82-, DIR, SPEC CHEM LAB, 78- *Concurrent Pos:* Mem, var roundtables, Am Asn Clin Chem, 82-90, Adaptation Immunoassay Tech Centrifugal Analyzers workshop, 83 & 84, chmn Pediat Div, 88-89, mem, Recent Advances Physician Off Labs, 89; centrib ed, Pediat Clin Chem Ref (Normal) Values, 89; ed spec sect, Pediat Lab Med, Clin Chem News, 89; moderator, selected topic session, Respiratory Distress Newborns-Assessment Clin Problems & New Lab Tech, Am Asn Clin Chem, 89. *Mem:* Am Chem Soc; NY Acad Sci; Am Asn Clin Chem; fel Nat Acad Clin Biochem; fel Asn Clin Scientists. *Res:* Use of biochemical markers for rapid detection of myocardial infarction; feasibility of performing laboratory testing in physician's offices, clinics, emergency rooms or at the patient's bedside; use of non-radioiosotopic immunochemical assays for determining reproductive hormones in serum; establishment of clinical chemistry reference values for neonated, newborns and adolescents. *Mailing Add:* Columbia-Presby Med Ctr 622 W 168th St New York NY 10032

PESCH, PETER, b Zurich, Switz, June 29, 34; US citizen; m 55; c 2. ASTRONOMY. *Educ:* Univ Chicago, BS, 55, MS, 56, PhD(astron), 60. *Prof Exp:* Res assoc, Case Western Reserve Univ, 60-61, from asst prof to prof astron, 61-84, chmn dept & dir, Warner & Swasey Observ, 75-84; prog dir, Stars & Stellar Evolution Prog, Div Astron, NSF, 84-86; CHMN DEPT ASTRON & DIR, WARNER & SWASEY OBSERV, CASE WESTERN RESERVE UNIV, 86- *Mem:* Am Astron Soc; felAAAS. *Res:* Stellar photometry and spectroscopy. *Mailing Add:* Dept Astron Case Western Reserve Univ Cleveland OH 44106

PESCHKEN, DIETHER PAUL, b Cologne, Ger, Apr 3, 31; Can citizen; m 60; c 4. BIOLOGICAL CONTROL OF WEEDS. *Educ:* Univ Man, BSA, 59, MSc, 60; Univ Gottingen, DScAgr, 64. *Prof Exp:* Res scientist I biol control, Can Dept Agr Res Inst, Belleville, Ont, 64-72; RES SCIENTIST II BIOL CONTROL WEEDS, RES STA AGR CAN, REGINA, 72- *Mem:* Entom Soc Can; Weed Sci Soc Am; Int Orgn Biol Control Noxious Animals & Plants. *Res:* Investigations of insects for their suitability and safety as biological control agents against weeds; released several and studied their development and effect on the weed in the field; biological control of weeds. *Mailing Add:* Res Sta Agr Can Box 440 Regina SK S4P 3A2 Can

PESEK, JOHN THOMAS, JR, b Hallettsville, Tex, Nov 15, 21; m 52; c 3. SOIL FERTILITY, FERTILIZER USE. *Educ:* Tex Agr & Mech Col, BS, 43, MS, 47; NC State Col, PhD(agron), 50. *Prof Exp:* Asst agron, NC State Col, 47-50; res agronomist, 50-64, PROF AGRON & HEAD DEPT, IOWA STATE UNIV, 64- *Concurrent Pos:* Consult, Int Basic Econ Corp Res Inst, Brazil, 54 & Mid-Am State Univs, Colombia, 66; tech ed soils, Agron J, Am Soc Agron, 65-71; mem, Seventh Soil Sci Cong, Southern Africa. *Mem:* Fel AAAS; fel Am Soc Agron (pres, 79); fel Soil Sci Soc Am; Crop Sci Soc Am; Soil Conserv Soc Am. *Res:* Effect of edaphic and climatological factors on response of crops to fertilizer and the economics of fertilizer use. *Mailing Add:* Dept Agron Iowa State Univ Ames IA 50011

PESEK, JOSEPH JOEL, b Chicago, Ill, June 29, 44; m 67; c 1. SEPARATION METHODS. *Educ:* Univ Ill, BS, 66; Univ Calif Los Angeles, PhD(chem), 70. *Prof Exp:* Postdoctoral fel, Univ Calif Los Angeles, 70-71; asst prof chem, Northern Ill Univ, 71-78; from asst prof to assoc prof, 78-85, PROF CHEM, SAN JOSE STATE UNIV, 85-, CHMN DEPT CHEM, 88- *Concurrent Pos:* Vis scientist, Ecole Polytechnique, 84-85; vis prof chem, Univ d'Aix-Marseille, 88; Camille & Henry Dreyfus scholar, 91-93. *Mem:* Am Chem Soc. *Res:* Chemical modification of oxide surfaces for use in high performance liquid chromatography and high performance capillary electrophoresis. *Mailing Add:* Dept Chem San Jose State Univ San Jose CA 95192

PESELNICK, LOUIS, b New York, NY, Mar 31, 24; m 61. SOLID STATE PHYSICS. *Educ:* Catholic Univ, BEE, 50, MS, 52, PhD(physics), 57. *Prof Exp:* Physicist, Power Condenser & Electronics Corp, 52-55; physicist, Catholic Univ, 55-57; physicist, Nat Ctr Earthquake Res, US Geol Surv, 57-62, res physicist, 62-84; RETIRED. *Concurrent Pos:* Consult, Power Condenser & Electronics Corp, 55-57; lectr, Catholic Univ, 58-65. *Mem:* Am Phys Soc; Am Geophys Union; Soc Explor Geophys; Inst Elec & Electronics Engrs. *Res:* Ultrasonic relaxation in gases and liquids; elastic and anelastic properties of single crystals and condensed matter using infrasonic, sonic and ultrasonic techniques; pressure-temperature phase transitions in solids using ultrasonics; variational methods and average elastic constants. *Mailing Add:* 319 Waverly Menlo Park CA 94025

PESETSKY, IRWIN, b New York, NY, Feb 17, 30; m 56; c 1. ANATOMY, NEUROEMBRYOLOGY. *Educ:* NY Univ, BA, 52; State Univ Iowa, MS, 54, PhD(zool), 59. *Prof Exp:* From instr to asst prof, 59-70, ASSOC PROF ANAT, ALBERT EINSTEIN COL MED, 70- *Concurrent Pos:* NIH res grants, Albert Einstein Col Med, 61-76, NIH fel behav & neurol sci, 62-64; vis asst prof biol, Yeshiva Col, Yeshiva Univ, 66-67. *Mem:* AAAS; Soc Develop Biol; Am Asn Anatomists; Am Soc Cell Biol; Am Soc Zoologists; Sigma Xi. *Res:* Developmental neurobiology; neuroendocrinology; histochemistry. *Mailing Add:* Dept Anat 656 Forch Albert Einstein Col Med 1300 Morris Park Ave Bronx NY 10461

PESHKIN, MURRAY, b Brooklyn, NY, May 17, 25; m 55; c 3. PHYSICS. *Educ:* Cornell Univ, BS, 47, PhD(physics), 51. *Prof Exp:* From instr to asst prof, Northwestern Univ, 51-59; assoc physicist, 59-64, assoc dir, 64-83, SR PHYSICIST, PHYSICS DIV, ARGONNE NAT LAB, 67- *Concurrent Pos:* Weizmann Inst, Israel, 59-60, 68-69 & 84. *Mem:* AAAS; Sci Res Soc Am; Am Phys Soc. *Res:* Quantum theory. *Mailing Add:* Physics Div Argonne Nat Lab 9700 S Cass Ave Argonne IL 60439-4843

PESKIN, ARNOLD MICHAEL, b Paterson, NJ, Mar 15, 44; m 65; c 2. ELECTRICAL ENGINEERING, COMPUTER SCIENCE. *Educ:* NJ Inst Technol, BS, 65; Polytech Inst NY, MS, 71. *Prof Exp:* Elec engr, Int Bus Mach, 65-67; digital systs engr, 67-69, group leader eng, 69-75, div head tech support, 75-78, actg chmn, 78-79, DEPUTY CHMN APPL MATH, BROOKHAVEN NAT LAB, 79- *Concurrent Pos:* Lectr math statist, Columbia Univ, 72; adj prof, State Univ NY, Stony Brook, 79-; comnr, Technol Accreditation Comn, Accreditation Bd Eng & Technol. *Mem:* Sr mem Inst Elec & Electronics Engrs. *Res:* Data communications; design automation of digital systems. *Mailing Add:* Dept Appl Math Brookhaven Nat Lab Assoc Univ Comjunting & Commun Div Bldg 515 Upton NY 11232

PESKIN, MICHAEL EDWARD, b Philadelphia, Pa, Oct 27, 51; m 78; c 2. QUANTUM FIELD THEORY, SYMMETRIES OF QUARK & LEPTON INTERACTIONS. *Educ:* Harvard Univ, AB, 73; Cornell Univ PhD(physics), 78. *Prof Exp:* Jr fel physics, Harvard Univ, 77-80; vis asst prof, Cornell Univ, 80-82; assoc prof, 82-86, PROF PHYSICS, STANFORD LINEAR ACCELERATOR CTR, STANFORD UNIV, 86- *Concurrent Pos:* Vis scientist Ctr Nuclear Studies, Saclay, France, 79-80. *Mem:* Fel Am Phys Soc; AAAS. *Res:* Construction of models to explain the masses of quarks, leptons and vector bosons; mathematical aspects of symmetry in local quantum field theory and in superstring theory; new experiments which probe physics at distances less than 10-16 cm. *Mailing Add:* Stanford Linear Accelerator Ctr Stanford Univ Theory Group Bin 81 Stanford CA 94309

PESKIN, RICHARD LEONARD, b Cambridge, Mass, May 31, 34; div; c 2. ENGINEERING. *Educ:* Mass Inst Technol, BS, 56; Princeton Univ, MS, 58, PhD(eng), 60. *Prof Exp:* Eng consult, Budd Co, 56-57; asst eng, Princeton Univ, 57-59, mem res staff plasma physics, 59-61; assoc prof mech & aerospace eng, 61-68, prof mech & aerospace eng, Rutgers Univ, New Brunswick, 68-; AT MECH & AEROSPACE ENG DEPT, RUTGERS COL. *Concurrent Pos:* Res engr, Budd Co, 55-57; consult, Macrosonics Corp & Inst Defense Anal, 64 & Esso Res & Eng Corp, 66. *Mem:* Am Phys Soc; AAAS; Am Soc Mech Eng. *Res:* Fluid mechanics, turbulence, computational fluid dynamics, gas-solid suspension flows and aerosol mechanics; geophysical fluid dynamics; ignition combustion theory; statistical, parallel computing, thermodynamics; pollution dispersion; parallel computing, software. *Mailing Add:* Dept Mech & Aerospace Eng Rutgers Col PO Box 909 Piscataway NJ 08854

PESSEN, DAVID W, b Berlin, Ger, Nov 23, 25; US citizen; m 59; c 3. MECHANICAL ENGINEERING. *Educ:* Univ Pa, BS, 49, MS, 50; Israel Inst Technol, DSc(mech eng), 62. *Prof Exp:* Res asst thermodyn, Univ Pa, 49-50; res engr, Honeywell, Inc, 50-55; lectr mech eng, Israel Inst Technol, 55-62; eng analyst, Wiedemann Mach Co, Pa, 62-64; assoc prof mech eng, Israel Inst Technol, 64-69; sect supvr, Res & Develop Ctr, Gulf & West, 69-70; assoc prof, 70-90, PROF, MECH ENG, ISRAEL INST TECHNOL, 91- *Concurrent Pos:* Consult, Twinworm Assocs, 60-; vis prof mech eng, Cornell Univ, 84-85. *Honors & Awards:* Blackall Award, Am Soc Mech Engrs, 60. *Mem:* Am Soc Mech Engrs. *Res:* Machine dynamics; automatic control; automation. *Mailing Add:* Dept Mech Eng Israel Inst Technol Haifa Israel

PESSEN, HELMUT, b Berlin, Ger, Sept 6, 21; US citizen; m 66. PHYSICAL CHEMISTRY, BIOCHEMISTRY. *Educ:* Drexel Inst Technol, BSChE, 49; Temple Univ, PhD(chem), 61. *Prof Exp:* Chemist, Fred Whitaker Co, 43 & 46; res asst, Am Viscose Corp, 48; inspector, US Food & Drug Admin, 50; res chemist, US Army Qm Pioneering Res Labs, 50-57; tech translator, 61-63; RES SCIENTIST, EASTERN REGIONAL RES CTR, AGR RES SERV, USDA, 63- *Concurrent Pos:* Patent agent, regist US Patent Off, 63. *Mem:* AAAS; Am Chem Soc; Am Crystallog Asn; Biophys Soc. *Res:* Fine structure of cellulose; microbial degradation of plasticizers; chemical kinetics; physical properties of enzymes; protein structure and interactions; small-angle x-ray scattering; nuclear magnetic resonance; solution properties and hydrodynamics of biomacromolecules. *Mailing Add:* USDA Eastern Regional Res Ctr 600 E Mermaid Lane Philadelphia PA 19118

PESSL, FRED, JR, b Detroit, Mich, Nov 18, 32. GLACIAL GEOLOGY, ENVIRONMENTAL GEOLOGY. *Educ:* Dartmouth Col, AB, 55; Univ Mich, MS, 58. *Prof Exp:* Geologist, US Geol Surv, Conn, 63-71, proj geologist, Conn Valley Urban Pilot Proj, 71-76; CO-DIR, US GEOL SURV, PUGET SOUND EARTH SCI APPLN PROJ, SEATTLE, WASH, 76-; AFFIL PROF, DEPT GEOL SCI, UNIV WASH, 77- *Concurrent Pos:* Arctic Inst NAm res grants glacial geol, E Greenland, 59-61; vis prof, Wesleyan Univ, 71; proj chief, Conn Valley Urban Proj; vis prof geol, Wesleyan Univ, 73-74 & Yale Univ, 75-76. *Mem:* Fel Geol Soc Am; Glaciol Soc; Arctic Inst NAm; Int Glaciol Soc; Sigma Xi. *Res:* Glacial chronology in East Greenland; distribution and stratigraphy of glacial deposits in New England; thermal expansion properties of lake ice; geology and hydrology for use in land-use planning and resource management. *Mailing Add:* 402 Detwiller Lane Bellevue WA 98004

PESSON, LYNN L, b New Iberia, La, Oct 15, 27; m 50; c 3. AGRICULTURAL EDUCATION. *Educ:* La State Univ, BS, 48, PhD(agr educ), 60; Univ Md, MEd, 55. *Prof Exp:* Assoc county agent, La State Univ, Baton Rouge, 48-54, assoc state club agent, 54-60, prof exten educ & training specialist, 60-71, prof exten & int educ & head dept, 71-73, head dept exten educ & coordr int progs, 69-71, asst vchancellor, 73-74, vchancellor admin, 74-81, vchancellor student affairs, 81-; RETIRED. *Concurrent Pos:* Nat 4-H fel, USDA, DC, 53-54, specialist, Fed Exten Serv, USDA, 65; head dept exten, Col Agr, Malaya, 67-68. *Res:* Extension youth programs; effectiveness of extension teaching; diffusion of technology; international agriculture. *Mailing Add:* 5846 S Pollark Pkwy Baton Rouge LA 70808

PESTANA, CARLOS, b Canary Islands, Spain, June 10, 36; m 66. SURGERY. *Educ:* Nat Univ Mex, BS, 52, MD, 59; Univ Minn, PhD(surg), 65. *Prof Exp:* From asst prof to assoc prof, 68-74, assoc dean acad develop, 71-73, PROF SURG, UNIV TEX MED SCH, SAN ANTONIO, 74-, ASSOC DEAN

STUDENT AFFAIRS, 73- *Concurrent Pos:* Edward J Noble Found Award, 65; Piper Found Award, 72. *Mem:* Asn Acad Surgeons; Am Col Surgeons; NY Acad Sci; Soc Surg Alimentary Tract; Sigma Xi. *Mailing Add:* 10123 N Mantox San Antonio TX 78213

PESTANA, HAROLD RICHARD, b Honolulu, Hawaii, Dec 20, 31; m 53; c 2. PALEONTOLOGY. *Educ:* Univ Calif, Berkeley, BA, 57, MA, 59; Univ Iowa, PhD(geol), 65. *Prof Exp:* From instr to asst prof, 59-73, PROF GEOL, COLBY COL, 73- *Mem:* AAAS; Paleont Soc; Geol Soc Am; Nat Asn Geol Teachers; Hist Earth Sci Soc; Soc Econ Paleont & Mineral. *Res:* Morphology and ecology of Paleozoic corals and stromatoporoids; history of geology; origin of carbonate sediments. *Mailing Add:* Dept of Geol Colby Col Waterville ME 04901

PESTKA, SIDNEY, b Drobnin, Poland, May 29, 36; US citizen; m 60; c 3. BIOCHEMISTRY, MEDICAL SCIENCES. *Educ:* Princeton Univ, AB, 57; Univ Pa, MD, 61. *Prof Exp:* Intern pediat & med, Baltimore City Hosps, Md, 61-62; med officer, Nat Heart Inst, 62-66 & Nat Cancer Inst, 66-69; head sect cell regulation, 69-80, head lab molecular genetics, 80-86, MEM, ROCHE INST MOLECULAR BIOL, 69-; PROF & CHMN, DEPT MOLECULAR GENETICS & MICROBIOL, ROBERT WOOD JOHNSON SCH MED, UNIV MED & DENT NJ, 86- *Concurrent Pos:* Adj prof path, Col Physicians & Surgeons, Columbia Univ, 73-; secy, Int Soc Interferon Res. *Honors & Awards:* Selman A Waksman Award Microbiol, 77. *Mem:* NY Acad Sci; Am Soc Microbiol; Am Soc Biol Chemists; Am Asn Cancer Res; Int Soc Interferon Res. *Res:* Interferon synthesis, cloning, purification and action; cell receptors and signals; protein biosynthesis; mode of action of antibiotics; cellular regulatory mechanisms; immunology; cancer. *Mailing Add:* Dept Molec Genetics & Microbiol UMDNJ Robert Wood Johnson Med Sch 675 Hoes Lane Piscataway NJ 08854-5635

PESTRONG, RAYMOND, b New York, NY, Apr 20, 37; m 62; c 2. ENVIRONMENTAL GEOLOGY, GEOMORPHOLOGY. *Educ:* City Col New York, BS, 59; Univ Mass, Amherst, MS, 61; Stanford Univ, PhD(geol), 65. *Prof Exp:* Asst eng geol, Linear Accelerator Ctr, Stanford Univ, 62-63; eng geologist, Soils Eng Firm, 65-66; dir, NSF Earth Sci Inst, 68-72 & 78-81, chmn dept geol, 69-73, PROF GEOL, SAN FRANCISCO STATE UNIV, 66- *Concurrent Pos:* Mem, Coun Educ Geol Sci; Off Naval Res res contractr, 64; dir earthquake educ, NSF Inst; mem, Calif Seismic Safety Educ Comt. *Mem:* Fel Geol Soc Am; Asn Eng Geol; Nat Asn Geol Teachers; Sigma Xi. *Res:* Processes and mechanisms of sedimentation within the tidal flat and tidal marsh environment; multi-media in education; engineering geologic studies of landslides and other slope stability problems; coastal geomorphic phenomena; alternative approaches to introductory geologic education. *Mailing Add:* Dept Geol San Francisco State Univ 1600 Holloway Ave San Francisco CA 94132

PESTRONK, ALAN, b Cambridge, Mass, Jan 19, 46. NEUROMUSCULAR DISEASE. *Educ:* Princeton Univ, AB, 66; Johns Hopkins Sch Med, MD, 70. *Prof Exp:* Resident, Johns Hopkins Sch Med, 71-74, from asst prof to assoc prof neurol, 84-89; PROF NEUROL & DIR NEUROMUSCULAR DIV, WASH UNIV SCH MED, 89- *Honors & Awards:* Lawrence C McHenry Award, 87. *Mem:* Am Neurol Asn; Am Asn Neurol. *Res:* Trophic interactions between nerve and muscle; neuromuscular disease. *Mailing Add:* Dept Neurol Wash Univ Sch Med 660 S Euclid Ave Box 8111 St Louis MO 63110

PESYNA, GAIL MARLANE, health care & marketing, for more information see previous edition

PETAJAN, JACK HOUGEN, b Evanston, Ill, Apr 2, 30; m 52; c 3. NEUROLOGY, PHYSIOLOGY. *Educ:* Johns Hopkins Univ, BA, 53; Univ Wis, PhD(physiol) & MD, 59; Am Bd Psychiat & Neurol, dipl, 66. *Prof Exp:* Asst prof neurol & physiol, Univ Wis, 63-65; vis assoc prof physiol, Univ Alaska & chief physiol sect, Arctic Health Res Lab, Inst Arctic Biol, Univ Alaska, 65-69; assoc prof, 69-73, PROF NEUROL, MED SCH, UNIV UTAH, 73- *Concurrent Pos:* For exchange fel, 62-63; consult, Vet Admin, 63-65 & US Army, 66- *Mem:* Fel Am Acad Neurol; AMA; Int Soc Biometeorol; Am Phys Soc. *Res:* Study of motor unit control; neuromuscular disease; environmental health; toxicology. *Mailing Add:* 2233 Melodie Ann Way Salt Lake City UT 84117

PETCH, HOWARD EARLE, b Agincourt, Can, May 12, 25; m 49, 76; c 3. NUCLEAR MAGNETIC RESONANCE. *Educ:* McMaster Univ, BSc, 49, MSc, 50; Univ BC, PhD(physics), 52. *Hon Degrees:* DSc, McMaster Univ, 74. *Prof Exp:* Fel, McMaster Univ, 52-53; Rutherford fel, Cavendish Lab, Cambridge Univ, 53-54; asst prof physics, McMaster Univ, 54-57, assoc prof physics & metall, 57-60, prof metall & metall eng, 60-67, chmn dept metall & metall eng, 58-62, dir res, 61-67, prin, Hamilton Col, 63-67, chmn interdisciplinary mat res unit, 64-67; prof physics & vpres acad, Univ Waterloo, 67-74, pres pro-tem, 69-70; prof physics, 75, PRES, UNIV VICTORIA, 75- *Concurrent Pos:* Mem Sci Coun Can, 66-72; asst secy, Ministry of State for Sci & Technol, 72; mem Defense Res Bd Can, 73-78 & Royal Comn Air Transp Needs of the Toronto Area, 73-74. *Honors & Awards:* Centennial Medal, 67. *Mem:* Fel Royal Soc Can; Can Asn Physicists (vpres, 66-67, pres, 67-68); Am Phys Soc; Int Union Crystallog. *Res:* Crystallography; solid state physics. *Mailing Add:* Off of the President PO Box 1700 Univ Victoria Victoria BC V8W 2Y2 Can

PETER, JAMES BERNARD, b Omaha, Nebr, June 27, 33; m 54; c 7. LABORATORY MEDICINE, CLINICAL CHEMISTRY. *Educ:* Creighton Univ, BS, 54; St Louis Univ, MD, 58; Univ Minn, PhD(biochem), 63. *Prof Exp:* Lectr chem, 63, from asst prof to assoc prof, 65-76, PROF, UNIV CALIF, LOS ANGELES, 76-; PRES & DIR, SPECIALTY LABS INC, 76- *Concurrent Pos:* Fel med, Univ Minn Hosp, 59-60; attend specialist, Wadsworth Vet Hosp, Los Angeles, 65- *Mem:* Am Chem Soc; Am Soc Biol Chemists; Am Soc Clin Invest; Am Soc Clin Pathologists; Am Asn Immunologists. *Res:* Laboratory diagnosis of human disease. *Mailing Add:* Specialty Labs Inc 2211 Michigan Ave Santa Monica CA 90404

PETER, RICHARD ECTOR, b Medicine Hat, Alta, Mar 7, 43; m 65; c 2. NEUROENDOCRINOLOGY, COMPARATIVE ENDOCRINOLOGY. *Educ:* Univ Alta, BSc, 65; Univ Wash, PhD(zool), 69. *Prof Exp:* Med Res Coun Can fel, Bristol Univ, 69-70; from asst prof to assoc prof, 71-79, PROF ZOOL, UNIV ALTA, 79-, CHMN ZOOL, 83- *Concurrent Pos:* vis scientist, Inst Nat Res Agron, France, 77, 78, 81 & Acad Agr, Poland, 81; Steacie fel, Nat Sci Eng Res Coun, Can, 80-82; proj specialist, Chinese Univ Develop Proj. *Honors & Awards:* Pickford Medal, 85; Geschwind Mem Lectr, 88. *Mem:* Fel AAAS; Am Soc Zoologists; Can Soc Zoologists; Int Soc Neuroendocrinol; Soc Study Reproduction; fel Royal Soc Can; Endocrine Soc. *Res:* Neuroendocrine control of pituitary function in teleost fishes. *Mailing Add:* Dept Zool Univ Alta Edmonton AB T6G 2E9 Can

PETERING, DAVID HAROLD, b Peoria, Ill, Sept 16, 42; m 66; c 2. BIOINORGANIC CHEMISTRY, BIOCHEMICAL PHARMACOLOGY. *Educ:* Wabash Col, BA, 64; Univ Mich, PhD(biochem), 69. *Prof Exp:* Fel, Northwestern Univ, Am Cancer Soc, 69-71; from asst prof chem & biochem to assoc prof, 71-82, PROF CHEM, UNIV WIS, 83- *Concurrent Pos:* Vis sr fel, Nat Inst Environ Health Sci, 81- *Mem:* Am Chem Soc; Sigma Xi; Am Soc Biol Chemists. *Res:* Metabolism of essential transition metals zinc, iron and toxic metals and their complexes; role of zinc in normal and tumor cell proliferation, of cadmium in biological toxicity and various metal complexes in cancer chemotherapy; biochemistry of metallothionein. *Mailing Add:* 7229 N Santa Monica Blvd Milwaukee WI 53217

PETERING, HAROLD GEORGE, b Laporte, Ind; m 39; c 3. BIOCHEMISTRY, BIOCHEMICAL PHARMACOLOGY. *Educ:* Univ Chicago, SB, 34; Univ Wis, PhD(chem, biochem), 38. *Prof Exp:* Asst prof biochem, Mich State Univ, 38-41; sr biochemist, E I du Pont de Nemours & Co, 41-45; sr scientist & group leader biochem pharmacol, Upjohn Co, 45-66; from assoc prof to prof, 66-78, EMER PROF ENVIRON HEALTH & BIOCHEM TOXICOL, COL MED, UNIV CINCINNATI, 78- *Honors & Awards:* Chemist Award, Am Chem Soc, 75. *Mem:* Am Soc Biol Chemists; Am Chem Soc; Am Asn Cancer Res; Am Inst Nutrit; Am Col Nutrit. *Res:* Biochemistry and toxicology of trace metals; chemotherapy of cancer; biological activity and metabolism of chelating agents; nutrition of zinc, copper and iron. *Mailing Add:* 2484 Heather Way Lexington KY 40503

PETERJOHN, GLENN WILLIAM, b Cleveland, Ohio, June 30, 21; m 46; c 2. ZOOLOGY, PHYSIOLOGY. *Educ:* Univ Wis, BS, 47, MS, 49, PhD(zool), 56. *Prof Exp:* From instr to asst prof, Baldwin-Wallace Col, 49-61, head biol, 82-85, prof, 61-86; RETIRED. *Concurrent Pos:* Vis prof, Inst High Sch Sci Teachers, William Jewell Col, 59; NIH fel, 60; NSF res grant, 63. *Mem:* AAAS; Am Soc Parasitol; Am Inst Biol Sci; Nat Audubon Soc. *Res:* Physiology; parasitology; nucleic acids in migrating slime mold; fluke parasites of snails. *Mailing Add:* 20359 Mercedes Rocky River OH 44116

PETERKOFSKY, ALAN, b Mass, Aug 29, 30; m 56; c 2. BIOCHEMISTRY. *Educ:* City Col NY, BS, 53; Union Col NY, MS, 55; NY Univ, PhD(biochem), 60. *Prof Exp:* Asst scientist, State Dept Health, NY, 53-55; asst scientist, US Pub Health Serv Off, 59-61, CHEMIST, NIH, 61- *Mem:* AAAS; Am Soc Microbiol; Am Soc Biol Chem. *Res:* Enzymology; molecular biology; nucleic acids; neuropeptides. *Mailing Add:* Lab Biochem Genetics NIH Bldg 36 Rm C09 Bethesda MD 20892

PETERLE, TONY J, b Cleveland, Ohio, July 7, 25; m 49; c 2. BIOLOGY, ECOLOGY. *Educ:* Utah State Univ, BS, 49; Univ Mich, MS, 50, PhD(wildlife ecol, zool), 54. *Prof Exp:* Res biologist, Dept Conserv, Mich, 51-54; Fulbright scholar natural history, Aberdeen Univ, 54-55; res biologist, Dept Conserv, Mich, 55-59; leader, Ohio Coop Wildlife Res Unit, 59-63; chmn fac pop & environ biol, Col Biol Sci, 68-70, chmn prog environ biol fac zool, 69-71, chmn dept, 71-81, PROF ZOOL, OHIO STATE UNIV, 62- *Concurrent Pos:* Mem panel, NSF, 65 & Adv Conf Pesticides, NATO, 65; ed, J Wildlife Mgt, 69-70; co-organizer, XIII Int Cong Game Biol, 77; mem, Ecol Comt, Sci Adv Bd, US Environ Protection Agency. *Honors & Awards:* Aldo Leopold Mem Award, Wildlife Soc, 90. *Mem:* Int Union Game Biol; Wildlife Soc (vpres, 70, pres-elect, 71, pres, 72); AAAS; Ecol Soc Am; Sigma Xi; Nat Audubon Soc; Soc Environ Toxicol & Chem. *Res:* Ecology of wildlife, translocation and bioaccumulation of pesticides in natural environments; use of radioisotopes to study pesticides; natural regulation of animal populations; population dynamics and censusing; motivation in hunting. *Mailing Add:* Dept Zool Ohio State Univ 1735 Neil Ave Columbus OH 43210

PETERLIN, ANTON, b Ljubljana, Yugoslavia, Sept 25, 08; m 41; c 2. POLYMER PHYSICS. *Educ:* Univ Ljubljana, MS, 30; Univ Berlin, DSc(physics), 38. *Hon Degrees:* DSc, Univ Mainz, 79; Univ Ljubljana, 88. *Prof Exp:* Prof physics, Univ Ljubljana, 39-60, pres, J Stefan Inst, 49-59; prof physics, Munich Tech Univ, 60-61; dir Camille Dreyfus Lab, Res Triangle Inst, 61-73; adj prof polymer sci, Duke Univ, 61-73; asst chief Polymers Div, Nat Bur Standards, Washington DC, 73-84; RETIRED. *Concurrent Pos:* Vis prof, Case Western Reserve Univ, 73-84. *Honors & Awards:* Bingham Medal, Soc Rheol, 70; Ford High Polymer Physics Prize, Am Phys Soc, 72; Silver Medal Award, Dept Com, 81. *Mem:* Am Phys Soc; Am Chem Soc; Soc Rheol; Slovenian Acad Sci Art; Austrian Acad Sci. *Res:* Solution properties; solid state of polymers; plastic deformation of crystalline polymers; acoustic emission; spin resonance; IR; neutron and x-ray scattering. *Mailing Add:* Polymers Div Nat Inst Standards & Technol Gaithersburg MD 20899

PETERLIN, BORIS MATIJA, b Ljubljana, Yugoslavia, July, 47; m 47, 84; c 1. RHEUMATOLOGY, IMMUNOGENETICS. *Educ:* Duke Univ, BS, 68; Harvard Univ, MD, 73. *Prof Exp:* Intern, Stanford Univ, 73-74, res med, 74-75; clin assoc immunogenetics, NIH, 75-77; sr res med, Stanford Univ, 77-78, fel rheumatol, 78-81; ASST PROF MED MICROBIOL IMMUNOL, UNIV CALIF, SAN FRANCISCO, 81-, ASST INVESTR, HOWARD HUGHES MED INST, 85- *Concurrent Pos:* Assoc, Howard Hughes Med Inst, 82-85. *Mem:* AAAS; Am Asn Immunol; Am Fedn Clin Res; Am Soc Clin Invest. *Res:* Major histocompatibility genes-regulation of expression, disease association, bare lymphocyte syndrome; human immunodeficiency virus-regulation, mechanism of activation, trans-activation by virally encoded trans-activators (tat, rev, nef) interaction with other viruses. *Mailing Add:* U-426 Howard Hughes Med Inst Univ Calif San Francisco CA 94143-0724

PETERMAN, DAVID A, b Port Arthur, Tex, Sept 21, 35; m 59; c 6. MECHANICAL ENGINEERING. *Educ:* Lamar State Col, BS(mech eng) & BS(math), 56; Rice Univ, MS, 58, PhD(mech eng), 64. *Prof Exp:* Instr mech eng, Lamar State Col, 58-59; mem tech staff, Tex Instruments Inc, 61-66, br mgr corp res & eng, 67-69, process technol res & develop, 69-76, mgr advan front end, 76-80, electron beam systs, 80-84, strategic develop mgr, 84-86, regional educ laison, 86-88, site mgr, 89, EXTEN TECHNOL MGR, TEX INSTRUMENTS, INC, 90- *Mem:* Am Soc Mech Engrs; Inst Elec & Electronics Engrs; Technol Transfer Soc; Soc Adv Engr Educ. *Res:* Semiconductor device manufacturing science and technology. *Mailing Add:* Tex Instruments Inc PO Box 225621 Mail Sta 385 Dallas TX 75265

PETERMAN, KEITH EUGENE, b Elliottsburg, Pa, Dec 27, 47; m 71; c 4. INORGANIC CHEMISTRY, FLUORINE CHEMISTRY. *Educ:* Shippensburg State Col, BS, 69, MEd, 72; Univ Idaho, PhD(inorg chem), 75. *Prof Exp:* Teacher chem, S Middleton Sch Dist, 69-71; teaching asst chem, Shippensburg State Col, 71-72; res asst chem, Univ Idaho, 72-75; teaching fel chem, Millikin Univ, 75-76; from asst prof to assoc prof chem, 76-89, PROF CHEM & CHMN PHYS SCI DEPT, YORK COL PA, 90- *Concurrent Pos:* Fulbright scholar & guest prof, Ruhr Univ, Bochum, West Germany, 84-85; Vis scholar, Guangxi Inst, Nationalitico, Nanning, PRC, 86; Eastern European Exchange Scholar, Polish Acad Sci, Warsaw, 89. *Mem:* Am Chem Soc; Sigma Xi; Am Asn Univ Prof; Am Chem Soc. *Res:* Synthesis, reaction chemistry, photochemistry, mechanistic studies, and spectroscopic studies of fluorine containing compounds; model for enzyme assay utilizing nuclear magnetic resonance. *Mailing Add:* Dept Phys Sci York Col Pa York PA 17405

PETERMAN, ZELL EDWIN, b Cass County, Iowa, Nov 29, 34; m 60; c 2. GEOLOGY. *Educ:* Colo Sch Mines, GeolE, 57; Univ Minn, MS, 59; Univ Alta, PhD(geol), 62. *Prof Exp:* Geologist, Br Geochem Census, 62-64, geologist, Br Isotope Geol, 64-71, chief, Br Isotope Geol, 71-76, GEOLOGIST, BR ISOTOPE GEOL, US GEOL SURV, 76- *Mem:* Geol Soc Am; Mineral Soc Am; Soc Econ Geologists; Geol Asn Can; Geochem Soc. *Res:* Geochronology; Precambrian geology; isotope tracer studies. *Mailing Add:* US Geol Surv Fed Ctr Box 25046 MS 963 Lakewood CO 80225

PETERS, ALAN, b Nottingham, Eng, Dec 6, 29; m 55; c 3. ANATOMY. *Educ:* Bristol Univ, PhD(zool), 54. *Prof Exp:* Lectr anat, Univ Edinburgh, 58-66; PROF ANAT & CHMN DEPT, SCH MED, BOSTON UNIV, 66- *Concurrent Pos:* Res fel anat, Univ Edinburgh, 57-58; grants, Med Res Coun Gt Brit, 65-66, NIH, 66-; vis lectr, Harvard Med Sch, 63-64; mem, Anat Test Comt, Nat Bd Med Examrs, 72-75; assoc ed, Anat Rec, 72-81, J Neurocytol, 72-90, J Comp Neurol, Anat Embryol, Cerebral Cortex & Studies Brain Function, ed, Cerebral Cortex; mem, Neurol B Study Sect, NIH, 75-79; pres, Am Asn Anat chmn, 76-77; exec comt, Am Asn Anatomists, 86-90. *Honors & Awards:* Symington Prize, Anat Soc Gt Brit & Ireland, 62. *Mem:* Am Soc Cell Biol; Soc Neurosci; Am Asn Anat; Int Brain Res Orgn; Cajal Club. *Res:* Fine structure of the nervous system. *Mailing Add:* Dept Anat & Neurobiol Boston Univ Sch Med Boston MA 02118

PETERS, ALAN WINTHROP, b Cincinnati, Ohio, Oct 9, 37; m 61; c 3. CATALYSIS, ZEOLITES. *Educ:* Univ San Francisco, BS, 63; Univ Cincinnati, PhD(chem), 75. *Prof Exp:* Group leader, Mobil Res & Develop, 70-79; mgr res, 79-86, RES ASSOC, WR GRACE & CO, 86- *Mem:* Am Chem Soc; Sigma Xi. *Res:* Determination of selecting and structure relationships for catalysts. *Mailing Add:* PO Box 247 Ashton MD 20861

PETERS, ALEXANDER ROBERT, b Pender, Nebr, Nov 17, 36; m 59; c 2. MECHANICAL ENGINEERING. *Educ:* Univ Nebr, BS, 59, MS, 63; Okla State Univ, PhD(mech eng), 67. *Prof Exp:* Engr, Space & Info Div, N Am Aviation, Inc, 63-64; from instr to prof mech eng, Univ Nebr, Lincoln, 66-85, chmn dept, 75-85; mgr design & develop eng, Defense Div, Brunswick Corp, 85-86; PROF MECH ENG, UNIV NEBR, LINCOLN, 86- *Concurrent Pos:* NSF res initiation grant, 68-69, instr sci equip grant, 68-70; Am Soc Eng Educ-Ford Found residency fel, Ford Motor Co, 70-71; forensic eng expert witness, 71-; prin investr, NIH Interdisciplinary res grant & NSF instrnl sci equip grant, 75-77; mem bd educ, Soc Automotive Engrs, 84-87; mem bd dirs, Accreditation Bd Eng & Technol, 83-86, prog evaluator, 87- *Mem:* Am Soc Mech Engrs; Soc Automotive Engrs; Am Inst Aeronaut & Astronaut; Am Soc Eng Educ; Soc Advan Mat & Process Eng; Am Soc Heating, Refrig & Air Conditioning Engrs; Am Soc Metals Int; Am Soc Agr Engrs. *Res:* Boundary layer separation; aerodynamic heat transfer; automotive climate control; combustion; aerodynamics; energy; fluidized bed heat transfer; slurry flows and filtration; pneumatic conveying. *Mailing Add:* 3235 S 29th St Lincoln NE 68502

PETERS, B(RUNO) FRANK, b Drumheller, Alta, Jan 24, 33; m 57; c 5. METALLURGY. *Educ:* Univ BC, BASc, 55, MASc, 58; Univ Leeds, PhD(metall), 68. *Prof Exp:* Mem metall & nondestructive test group, Defence Res Estab Pac, 55-56 & 58-65; lectr metall, Univ Leeds, 65-68; head ,metall & nondestructive testing group, 68-82, head, Dockyard Lab, Defence Res Estab Atlantic, 82-84, actg dir, Technol Div, 84-87, HEAD MAT ENG, DEFENSE RES ESTAB PAC, 88- *Mem:* Am Soc Metals. *Res:* Metal failure analysis; electron fractography; recrystallization; engineering applications of nondestructive testing. *Mailing Add:* Defence Res Estab Pac HMC Dockyard Victoria BC V0S 1B0 Can

PETERS, BRUCE HARRY, b Ft Dodge, Iowa, July 18, 37; m 64; c 3. NEUROLOGY. *Educ:* State Univ Iowa, BA, 59, MS & MD, 63. *Prof Exp:* Intern med, Univ Calif, San Francisco, 63-64; resident neurol, Univ Iowa, 66-69; ASSOC PROF NEUROL, UNIV TEX MED BR, GALVESTON & MARINE BIOMED INST, 69- *Concurrent Pos:* Consult, St Mary's Hosp, Galveston, USPHS Hosp & Glaveston County Mem Hosp, 69-; mem, Awards Comt, Am Acad Neurol, 70-72; examr, Am Bd Neurol & Psychiat, 74- *Mem:* Am Acad Neurol; Soc Neurosci; Am Fedn Clin Res. *Res:* Insulinogenesis in periodic paralysis, cellular immunity in myasthenia gravis and polymyositis, physiology of hyman pharyngeo-lingual reflexes, pain and depression in chronic headache treated with antidepressants. *Mailing Add:* 2667 N Chelton Roadway Colorado Springs CO 80909

PETERS, CHARLES WILLIAM, b Pierceton, Ind, Dec 9, 27; c 5. NUCLEAR PHYSICS. *Educ:* Ind Univ, BA, 50. *Prof Exp:* Nuclear physicist, US Naval Res Lab, 50-71; physicist, Environ Protection Agency, 71-76; mgr advan systs, Consol Controls Corp, 76-89; MGR, ADVAN SYSTS DIV, NUCLEAR DIAG SYSTS, INC, 90- *Mem:* Am Phys Soc; Inst Elec & Electronic Eng. *Res:* Diagnostic measurements of nuclear weapons; experimental photodiodes and photomultipliers; nuclear radiation detectors; ultra-sensitive neutron detectors; radiation detection systems for military applications; applications of nuclear technology; neutron diagnostic probe. *Mailing Add:* Nuclear Diag Systs Inc PO Box 726 Springfield VA 22150

PETERS, CLARENCE J, b Midland, Tex, Sept 23, 40. VIROL IMMUNOLOGY. *Educ:* Johns Hopkins Univ, MD, 66. *Prof Exp:* Chief, Med Div, 82-84, DEP COMDR, DIS ASSESSMENT DIV, US ARMY MED RES INST INFECTIOUS DIS, 83-, CHIEF, 84- *Mem:* Am Bd Internal Med; Am Asn Immunologists; Am Soc Trop Med & Hyg; Am Soc Immunologists; Infectious Dis Soc. *Mailing Add:* US Army Med Res Inst Infectious Dis Ft Detrick Bldg 1425 Fredrick MD 21702-5011

PETERS, DALE THOMPSON, b Cincinnati, Ohio, Dec 5, 34; m 59; c 2. METALLOGRAPHY, BATTERIES. *Educ:* Univ Cincinnati, BS, 58; Ohio State Univ, PhD(metall), 62. *Prof Exp:* Res asst, Ohio State Univ, 58-62; res metallurgist, Int Nickel Co, Inc, 62-67, sect supvr magnetic mat, 67-74, mgr phys/anal sect, 74-79, mgr secondary battery res sect, 80-81, res fel, Energy Systs Dept, Inco Res & Develop Ctr, 81-84; tech dir, Int Copper Asn, 84-89; CONSULT, 89- *Mem:* Metall Soc; Am Soc Metals Int; Am Powder Metall Inst; Am Soc Mech Engrs; Mat Res Soc; Am Soc Testing Mat. *Res:* Internal friction, particularly grain boundary relaxation; mechanisms of strengthening of maraging steels; magnetic properties of nickel containing materials; metallography and electron beam analytical techniques; lead-acid and nickel alkaline battery research; copper alloys and products; technical management. *Mailing Add:* 18 Tartan Rd Mahwah NJ 07430

PETERS, DAVID STEWART, b Danville, Pa, July 5, 41; m 59; c 4. FISH BIOLOGY, ECOLOGY. *Educ:* Utah State Univ, BS, 64; NC State Univ, MS, 68, PhD(zool), 71. *Prof Exp:* FISHERY BIOLOGIST, NAT MARINE FISHERIES SERV, 70- *Mem:* Am Fisheries Soc; AAAS; Ecol Soc; Sigma Xi. *Mailing Add:* Nat Marine Fisheries Serv US Dept Commerce Beaufort NC 28516

PETERS, DENNIS GAIL, b Los Angeles, Calif, Apr 17, 37. ANALYTICAL CHEMISTRY, ELECTROCHEMISTRY. *Educ:* Calif Inst Technol, BS, 58; Harvard Univ, PhD(analytical chem), 62. *Prof Exp:* From instr to prof, 62-75, HERMAN T BRISCOE PROF CHEM, IND UNIV, BLOOMINGTON, 75- *Concurrent Pos:* NSF grant, 66-69, 73-75; Petrol Res Fund grants, 69-70, 78-81 & 81-84. *Honors & Awards:* Chem Mfrs Asn Nat Catalyst Award, 88. *Mem:* Am Chem Soc; Electrochem Soc; Int Soc Electrochem; Am Inst Chemists; NY Acad Sci; Sigma Xi. *Res:* Chronopotentiometry; polarography; coulometry; chemistry of noble metals; kinetics and mechanisms of electrode reactions; chemical instrumentation; organic electrochemistry. *Mailing Add:* Dept Chem Ind Univ Bloomington IN 47405

PETERS, DON CLAYTON, b Corn, Okla, Sept 1, 31; m 53; c 3. ENTOMOLOGY. *Educ:* Tabor Col, AB, 53; Kans State Col, MS, 55, PhD, 57. *Prof Exp:* Asst prof biol, Tabor Col, 57; asst prof entom, Univ Mo, 57-59; from asst prof to prof zool & entom, Iowa State Univ, 59-70; head dept, 71-84, PROF ENTOM, OKLA STATE UNIV, 84- *Concurrent Pos:* USAID consult, WAfrica, 79 & Pakistan, 88. *Mem:* Entom Soc Am; Am Registry Prof Entomologists. *Res:* Host plant resistance to insects; insect population management; Schizaphis graminum damage to wheat; causes of aphid biotypes. *Mailing Add:* Dept Entom Okla State Univ Main Campus Stillwater OK 74078

PETERS, DOYLE BUREN, b Chester, Tex, Nov 23, 22; m 50; c 2. SOIL PHYSICS. *Educ:* Agr & Mech Col, Tex, BS, 49, MS, 51; Univ Calif, PhD(soil physics), 54. *Prof Exp:* Asst soils, Agr & Mech Col, Tex, 49-51 & Univ Calif, 51-54; agent, Soil & Water Mgt, USDA, Ill, 54-64; assoc prof agron, 64-70, prof agron & soil scientist, 70-77, PROF SOIL PHYSICS & RES LEADER, SCI & EDUC ADMIN-AGR RES, USDA, UNIV ILL, URBANA, 77-, PROF DEPT AGRON, 80- *Mem:* Am Soc Agron. *Res:* Soil physical conditions and plant growth; energy status of soil moisture during compression. *Mailing Add:* 401 Evergreen Urbana IL 61801

PETERS, E(RNEST), b Can, Jan 27, 26; m 49; c 2. EXTRACTIVE METALLURGY, ELECTROCHEMISTRY. *Educ:* Univ BC, BASc, 49, MASc, 51, PhD(metall), 56. *Prof Exp:* Metallurgist, Geneva Steel Co, Utah, 49-50; jr res engr, Consol Mining & Smelting Co, Can, 51-53; instr metall, Univ BC, 55-56; res engr, Metals Res Lab, Union Carbide Corp, 56-58; from asst prof to assoc prof metall, 58-67, PROF METALL, UNIV BC, 67- *Concurrent Pos:* Consult, Cominco, Ltd, Can, 58- & Kennecott Corp, 64-; Nat Res Coun Can travel award, Univ Gottingen, 65; vis prof, Univ Calif, Berkeley, 71 & 83; Killam fel, Can Coun, 85-87. *Honors & Awards:* Authors Award, Am Inst Mining, Metall & Petrol Engrs, 57 & Chem Inst Can, 62; Extractive Metall Lectr, Am Inst Mining, Metall & Petrol Engrs, 76, James Douglas Gold Medal, 86; Alcan Award, Can Inst Mining & Metall, 83; Distinguished Lectr, Univ Utah, 85. *Mem:* Am Inst Mining, Metall & Petrol Engrs, Metall Soc; Can Soc Chem Engrs; fel Can Inst Mining & Metall; BC Asn Prof Engrs; fel Royal Soc Can. *Res:* Corrosion and metal oxidation; extractive metallurgy; hydrometallurgy; pressure processes; electrochemistry of sulfide minerals; metallurgical thermodynamics and kinetics; reactor and plant design; coal liquefaction. *Mailing Add:* 2708 W 33rd Vancouver BC V6N 2G1 Can

PETERS, EARL, b Leipzig, Ger, Dec 28, 27; US citizen; m 57; c 3. CHEMISTRY, TEXTILES. *Educ:* Oberlin Col, AB, 47; Univ Buffalo, MA, 50, PhD(org chem), 58. *Prof Exp:* Chemist polymers, Sprague Elec Co, 52-54; guest worker, Nat Bur Standards, 54-56; sr res chemist textiles, Milliken Res Corp, 60-67; mgr tech liaison, Burlington Industs, 67-73; assoc prof textiles,

73-77, EXEC DIR DEPT CHEM, CORNELL UNIV, 78- *Mem:* AAAS; Am Chem Soc; Sigma Xi; Fiber Soc. *Res:* Educational administration; laboratory safety; textiles. *Mailing Add:* Dept Chem Baker Lab Cornell Univ Ithaca NY 14853-1301

PETERS, EDWARD JAMES, b Milwaukee, Wis, Aug 5, 44; m 65; c 3. AQUATIC ECOLOGY, ICHTHYOLOGY. *Educ:* Wis State Univ, Stevens Point, BS, 67; Brigham Young Univ, MS, 69, PhD(zool), 74. *Prof Exp:* Instr biol, 72-74, asst prof, Mt Mercy Col, 74-75; asst prof fisheries, 75-80, ASSOC PROF, UNIV NEBR-LINCOLN, 80- *Mem:* Am Fisheries Soc; Am Soc Ichthyol & Herpetol; NAm Benthological Soc; Sigma Xi. *Res:* Effects of silt and sedimentation on fish growth and reproduction; ecology and distribution of freshwater fishes; trophic dynamics of freshwater communities. *Mailing Add:* 6501 NW 105th St Rte 1 Box 49 Malcolm NE 68402

PETERS, EDWARD TEHLE, b Evanston, Ill, July 26, 35; m 62; c 2. MATERIALS CHARACTERIZATION, LITIGATION EXPERT ON MATERIALS. *Educ:* DePauw Univ, BA, 57; Purdue Univ, BS, 57, Univ Wis, MS, 58; Mass Inst Technol, ScD(metall), 63. *Prof Exp:* Instr metall, Univ Wis, 58-59; instr x-ray crystallog, Mass Inst Technol, 59-63; res scientist, Manlabs, Inc, 63-69; SR SCIENTIST, ARTHUR D LITTLE, INC, 69- *Concurrent Pos:* Adj prof, Northeastern Univ, 65-75; pres, XCO Inc, 70-76, consult, 76- *Mem:* Am Soc Metals; Am Indust Hyg Asn; Am Ceramics Soc; Int Soc Optical Eng; AAAS. *Res:* Physical, chemical and structural properties of materials; product and process development; technology assessment; contract research and development; expert witness; metals, ceramics, high temperature materials and thin films/coatings; laser processing; asbestos; environment; failure analysis; optical materials. *Mailing Add:* Arthur D Little Inc 15-161 Acorn Park Cambridge MA 02140

PETERS, ELROY JOHN, b Kaukauna, Wis, June 10, 22; m 53; c 4. AGRONOMY, WEED SCIENCE. *Educ:* Univ Wis, BS, 52, MS, 53, PhD(agron), 56. *Prof Exp:* Asst, Univ Wis, 53-56; agronomist, USDA, 56-85, USAID, 85; from asst prof to assoc prof field crops, 66-71, prof agron, Univ Mo, Columbia, 71-85; CONSULT AGRON, 85- *Concurrent Pos:* Res coordr Liberia, USAID, 85-87. *Mem:* Am Soc Agron; Weed Sci Soc Am. *Res:* Weed science, especially pastures and range land; pasture management. *Mailing Add:* 210 Waters Hall Univ of Mo Columbia MO 65201

PETERS, ESTHER CAROLINE, b Greenville, SC, May 9, 52; m 84; c 1. INVERTEBRATE, ONCOLOGY. *Educ:* Furman Univ, BS, 74; Univ S Fla, MS, 78; Univ RI, PhD(biol oceanog), 84. *Prof Exp:* Assoc biologist histopath, Marine Serv Div, JRB Assoc, 84-85; fel systematics, Smithsonian Inst, 85-86; res fel histopath, registry tumors lower animals, Nat Cancer Inst, 87-90. *Concurrent Pos:* Resident res assoc, Dept Invert Zool, NMNH, Smithsonian Inst, 86- 89; courtesy asst prof, Dept Marine Sci, Univ S Fla, 87- *Mem:* Soc Invertebrate Path; Int Soc Reef Studies; Nat Soc Histotechnol; Am Soc Zoologists; Sigma Xi; AAAS. *Res:* Comparative invertebrate and vertebrate histopathology, especially the study of neoplasms and related disorders in lower organisms; ecology of coral reefs, and how diseases of corals and other reef organisms may affect the reef environment; fates and effects of toxic organic and inorganic compounds on the health of marine organisms. *Mailing Add:* Registry Tumors Lower Animals Nat Mus Natural Hist Washington DC 20560

PETERS, GERALD ALAN, b Monroe, Mich, Mar 3, 43; m 65; c 2. PLANT PHYSIOLOGY. *Educ:* Eastern Mich Univ, BS, 66; Univ Mich, MS, 69, PhD(bot), 70. *Prof Exp:* Res asst, Univ Mich, 66-69; fel photosynthesis, C F Kettering Lab, 70-72, staff scientist, 72-76, investr, 76-80, sr investr, 80-83, res leader, Battelle- Kettering Lab, 83-87; PROF BIOL, VA COMMONWEALTH UNIV, 87- *Concurrent Pos:* Prog mgr, BNF Prog, US Olympic Asn-Competitive Res Grants Off, 82. *Mem:* Am Soc Plant Physiologists; AAAS. *Res:* Symbiotic nitrogen fixation, particularly morphological and physiological investigations on the association which occurs between the water-fern, Azolla and the blue-green algal symbiont, Anabaena azollae. *Mailing Add:* Biol Dept Va Commonwealth Univ 816 Park Ave Box 2012 Richmond VA 23284

PETERS, GERALD JOSEPH, b Baltimore, Md, Sept 29, 41; m 67; c 2. PHYSICS. *Educ:* Loyola Col, BS, 63; Univ Toledo, MS, 65. *Prof Exp:* Electronic engr physicist & res physicist, White Oak Lab, Naval Surface Weapons Ctr, 65-80; physicist & spec asst to assoc dir high energy & nuclear physics, 80-86, PHYSICIST & PROG MGR, ADVAN TECHNOL RES & DEVELOP BR, DIV HIGH ENERGY PHYSICS, OFF ENERGY RES, US DEPT ENERGY, 86- *Concurrent Pos:* NASA trainee physics, Univ Toledo, 66-68. *Mem:* Inst Elec & Electronics Engrs; Am Phys Sci. *Res:* Accelerators; particle beam technology and applications; nuclear radiation instrumentation. *Mailing Add:* Div High Energy Physics US Dept Energy ER 224 GTN Washington DC 20585

PETERS, GERALDINE JOAN, b San Diego, Calif; m 68. ASTRONOMY, ASTROPHYSICS. *Educ:* Calif State Univ, Long Beach, BS, 65; Univ Calif, Los Angeles, MA, 66, PhD(astron), 73. *Prof Exp:* Res assoc astron, Univ Calif, Los Angeles, 74-76; physicist, Div High Energy Astrophys, Ctr Astrophys, Smithsonian Astrophys Observ, Cambridge, Mass, 76-78; ASSOC RES SCIENTIST & ADJ ASST PROF, DEPT ASTRON & PHYSICS, SPACE SCI CTR, UNIV SOUTHERN CALIF, LOS ANGELES, 78- *Concurrent Pos:* Prin investr grants, NASA; mem bd dirs, Astron Soc Pac, 84-87. *Mem:* Am Astron Soc; Int Astron Union; Astron Soc Pac. *Res:* Observation and analysis of the spectra of B-type stars; studies of the spectral variations observed in Be stars; observational studies of binary mass transfer; ultraviolet astronomy. *Mailing Add:* 20621 Septo St Chatsworth CA 91311

PETERS, HENRY A, b Oconomowoc, Wis, Dec 31, 20; m 54; c 4. NEUROLOGY, PSYCHIATRY. *Educ:* Univ Wis, BA, 43, MD, 45; Am Bd Psychiat & Neurol, cert psychiat, 52, cert neurol, 55. *Prof Exp:* Intern, Germantown Hosp & Dispensary, Philadelphia, 45-46; resident neuropsychiat, 48-51, from instr to asst prof, 51-58, assoc prof neurol, 58-70, PROF NEUROL & REHAB MED, MED SCH, UNIV WIS-MADISON, 70- *Concurrent Pos:* Mem bd dirs, Dane County Res Prog & Comn Pub Policy, State Med-Soc; dir muscular dystrophy clin, Univ Hosps, 56-; mem adv comt, March Dimes; consult, Muscular Dystrophy Asn Am Inc, Cerebral Palsy Clin, Univ Hosps & Vet Admin Hosp, Tomah, Wis; examr neurol, Am Bd Neurol & Psychiat; mem tech adv comt drug abuse, Atty Gen; mem med adv bd, Nat Muscular Dystrophy Asn. *Honors & Awards:* Cert Merit, AMA, 68. *Mem:* AMA; fel Am Col physicians; fel Am Psychiat Asn; fel Am Acad Neurol; Soc Clin Neurol; Sigma Xi. *Mailing Add:* 4128 Hiawatha Dr Madison WI 53711

PETERS, HENRY BUCKLAND, b Oakland, Calif, Nov 2, 16; c 5. OPTOMETRY. *Educ:* Univ Calif, AB, 38; Univ Nebr, MA, 39. *Hon Degrees:* Dr Ocular Sci, Southern Col Optom, 71 & New Eng Col Optom, 85; DSc, State Univ NY. *Prof Exp:* Lectr optom, Los Angeles Col Optom, 39-40; lectr, Sch Optom, Univ Calif, Berkeley, 47-48, from asst clin prof to assoc clin prof, 48-62, assoc prof & dir clins, 62-69, asst dean, 65-69; prof optom & pub health, 69-86, dean sch optom, 69-86, DEAN EMER, SCH PUB HEALTH, MED CTR, UNIV ALA, BIRMINGHAM, 87- *Concurrent Pos:* chmn optom adv comt, Vet Admin Dept Med & Surg; pres, Asn Schs & Cols Optom, 67-69; pres, Nat Health Coun, 78-79, Am Acad Optom, 73-74. *Honors & Awards:* Carel C Koch Medal, 74; Thomas P Carpenter Award, Nat Health Coun, 83. *Mem:* Am Optom Asn; fel Am Acad Optom (pres, 72-74). *Res:* Relation of vision to epidemiology of vision problems; vision performance. *Mailing Add:* 508 Sch Optom Univ of Ala Birmingham AL 35294

PETERS, HOWARD AUGUST, b Council Bluffs, Iowa, May 5, 26; m 48; c 3. ENVIRONMENTAL HEALTH. *Educ:* Univ Nebr, Omaha, BA, 51; Univ NC, Chapel Hill, MPH, 58, PhD(environmental health), 65. *Prof Exp:* Instr environ health, Univ NC, Chapel Hill, 61-62, res assoc, 62-64; asst prof pub health & dir environ health & safety, 64-71, ASSOC PROF PUB HEALTH, UNIV MASS, AMHERST, 71-, CHMN ENVIRON HEALTH PROG, 79- *Concurrent Pos:* NIH training grants air pollution & health & safety, 64-71. *Mem:* AAAS; Am Pub Health Asn; Air Pollution Control Asn; Am Indust Hyg Asn; Nat Environ Health Asn. *Res:* Urban environment; physical, chemical and biological contaminants in air, water and the industrial environment. *Mailing Add:* 669 East Pleasant St Amherst MA 01002

PETERS, HOWARD MCDOWELL, b Beech Creek, Pa, Oct 13, 40; m 64; c 2. ORGANIC CHEMISTRY, TRADE SECRET LAW. *Educ:* Geneva Col, BS, 62; Stanford Univ, PhD(org chem), 67; Univ Santa Clara, JD, 78. *Prof Exp:* Res chemist, Dow Chem Co, 66-69; res chemist, SRI Int, 69-75, asst dir chem lab, 75-77, indust economist, 77-78; atty, Hexcel Corp, 78-80; patent atty, Syntex Corp, 80-84 & Burns & Doane Law Firm, 84-85; PARTNER, PHILLIPS, MOORE, LEMPIO & FINLEY LAW FIRM, SAN FRANCISCO, 85- *Honors & Awards:* Mosher Award, Am Chem Soc, 84. *Mem:* Am Chem Soc; Am Int Property Law Asn; Am Bar Asn. *Res:* Synthetic organic chemistry; organic fluorine chemistry; nitration chemistry; chemically related patents; pharmaceutical chemistry; materials chemistry. *Mailing Add:* 3469 Kenneth Dr Palo Alto CA 94303

PETERS, JACK WARREN, b Denver, Colo, Jan 7, 16; m 40; c 3. GEOPHYSICS. *Educ:* Colo Sch Mines, GeolE, 38. *Prof Exp:* Chief gravity interpreter, Magnolia Petrol Co, 38-53; regional geophysicist, Mobil Producing Co, 53-59, sr staff geophysicist, Mobil Oil Co, 60-66, div regional geophysicist, Mobil Oil Corp, 66-69, geophys adv, 69-80; CONSULT, 80- *Concurrent Pos:* Consult, 80- *Mem:* Soc Explor Geophys; Europ Asn Explor Geophys. *Res:* Interpretation of gravity and magnetic data. *Mailing Add:* 13726 Peyton Dr Dallas TX 75240-3715

PETERS, JAMES, b Mt Clemens, Mich, Aug 30, 34; m 56; c 1. COLLOID CHEMISTRY, POLYMER CHEMISTRY. *Educ:* Wayne State Univ, BS, 56, PhD(chem), 64. *Prof Exp:* Chemist, Ditzler Color Div, Pittsburgh Plate Glass Co, 56-57; RES CHEMIST, DOW CHEM CO, 64- *Mem:* Am Chem Soc; Sigma Xi. *Res:* Polymer characterization; water soluble polymers and polyelectrolytes; colloid coagulation; secondary oil recovery; flow through porous media; emulsion polymerization; cement reinforcement. *Mailing Add:* Dow Chem Co Cent Res 1712 Bldg Midland MI 48640

PETERS, JAMES EMPSON, b Seymour, Ind, Sept 11, 54; m 80; c 3. COMBUSTION, TWO-PHASE FLOW. *Educ:* Purdue Univ, BS, 76, MS, 78, PhD(mech eng), 81. *Prof Exp:* Asst prof, 81-86, ASSOC PROF MECH ENG, DEPT MECH & INDUST ENG, UNIV ILL, URBANA, 86- *Concurrent Pos:* Consult, 81- *Mem:* Am Soc Mech Engrs; assoc fel Am Inst Aeronaut & Astronaut; Combustion Inst; Soc Automotive Engrs; Am Soc Eng Educ. *Res:* Combustion process including studies of fuel spray characteristics; air pollution control; coal combustion; gas turbine combustion and reciprocating engine performance. *Mailing Add:* Dept Mech & Indust Eng Univ Ill 1206 W Green St Urbana IL 61801

PETERS, JAMES JOHN, b Ft Wayne, Ind, May 4, 41; m 64; c 5. PARAMAGNETISM, MAGNETIC IMPURITIES. *Educ:* Ind Inst Technol, BS, 63; Univ Detroit, MS, 65; Univ Ill, PhD(metall eng), 71. *Prof Exp:* Asst prof physics, Tri-State Col, 70-71; PROF PHYSICS, HILLSDALE COL, 71- *Mem:* Sigma Xi; Am Asn Physics Teachers; Am Phys Soc. *Res:* Magnetic susceptibilities of transition metals in non-magnetic host liquid alloys. *Mailing Add:* 16 Armstrong St Hillsdale MI 49242

PETERS, JAMES MILTON, biochemistry, for more information see previous edition

PETERS, JEFFREY L, b San Francisco, Calif, Nov 23, 39. ANESTHESIOLOGY, PHYSIOLOGY. *Educ:* San Jose State Col, BA, 62; Univ Calif, Los Angeles, MS, 66; Baylor Col Med, PhD(physiol & biomed physics), 70; Univ Utah, MD, 74. *Prof Exp:* RES PROF BIOENG, UNIV UTAH, 76-, CLIN PROF ANESTHESIOL, MED CTR, 88- *Concurrent Pos:* Asst res prof surg, Univ Utah, 70-, clin assoc prof path, 88-; vpres res, Rocky Mountain Res Inc, 85- *Mem:* Biol Eng Soc; AMA; Am Fedn Clin Res; Am Physiol Soc; Am Soc Artificial Internal Organs; Am Heart Asn. *Mailing Add:* Dept Anesthesiol Med Ctr Univ Utah Salt Lake City UT 84112

PETERS, JOHN HENRY, b Cincinnati, Ohio, May 31, 24; m 47; c 2. BIOCHEMISTRY, BIOCHEMICAL PHARMACOLOGY. *Educ:* Univ Cincinnati, BS, 49; Univ Fla, MS, 50; Univ Minn, PhD(physiol chem), 55. *Prof Exp:* Biochemist, Radioisotope Serv, Vet Admin Hosp, Minneapolis, 52-56; res assoc, Christ Hosp Inst Med Res, Cincinnati, 56-63; assoc res biol chemist, Nat Ctr Primate Biol, Univ Calif, Davis, 63-64; sr biochem pharmacologist, SRI Int, 65-68, dir, biochem pharmacol prog, 68-89; RETIRED. *Concurrent Pos:* Asst prof, Sch Med, Univ Cincinnati, 60-63; mem, US Leprosy Panel, Nat Inst Allergy & Infectious Dis, 69-73. *Mem:* AAAS; Am Chem Soc; Am Asn Cancer Res; Soc Exp Biol & Med; NY Acad Sci; Am Soc Pharmacol Exp Therap. *Res:* Chemical mutagenesis and carcinogenesis; intermediary and drug metabolism; mechanism of vitamin and drug action; comparative biochemistry and pharmacology; pharmacogenetics; metabolism of amino acids in uremia of man; drug and amino acid metabolism in subhuman primates; metabolism of antileprotic and anticancer drugs in animals and man. *Mailing Add:* 11087 Linda Vista Cupertino CA 95014

PETERS, JOSEPH JOHN, b Chicago, Ill, Aug 5, 07. BIOLOGY. *Educ:* St Louis Univ, MA, 34; Univ Detroit, MS, 36; Fordham Univ, PhD(zool), 46. *Prof Exp:* From asst prof to prof biol, Xavier Univ, Ohio, 45-89; RETIRED. *Mem:* Am Physiol Soc; Soc Develop Biol; Am Soc Zool. *Res:* Electroencephalography and electrical activity of the nervous, sensory and muscular systems of the developing chick during various behavioral states such as sleep and attention and while under the influence of such environmental changes as temperature, anoxia and some drugs. *Mailing Add:* Colombiere Ctr PO Box 139 Clarkston MI 48016

PETERS, JUSTIN, b Bryn Mawr, Pa, May 23, 46; m. MATHEMATICS. *Educ:* Minn Univ, PhD(math), 73. *Prof Exp:* from asst prof to assoc prof, 76-86, PROF MATH, IOWA STATE UNIV, 86- *Concurrent Pos:* Wiss asst, Tech Univ Munich, 73-76. *Mem:* Am Math Soc; Math Asn Am. *Mailing Add:* Dept Math Iowa State Univ Ames IA 50011

PETERS, KEVIN SCOTT, b Ponca City, Okla, May 5, 49; m 78. PHYSICAL ORGANIC CHEMISTRY. *Educ:* Univ Okla, BS, 71; Yale Univ, PhD(chem), 75. *Prof Exp:* asst prof chem, Harvard Univ, 78-; AT DEPT CHEM, UNIV COLO, BOULDER. *Mem:* Am Chem Soc. *Res:* Picosecond laser spectroscopy applied to organic photochemistry and protein dynamics. *Mailing Add:* Dept Chem Univ Colo Boulder Box 215 Boulder CO 80309

PETERS, LEO CHARLES, b Smith Co, Kans, Sept 21, 31; m 57; c 9. DESIGN, PRODUCT SAFETY. *Educ:* Kans State Univ, BS, 53; Iowa State Univ, MS, 63, PhD(mech eng & eng mech), 67. *Prof Exp:* Jr engr, John Deere Waterloo Tractor Works, Iowa, 57-59, engr, 59-61; from instr to assoc prof, 61-78, PROF MECH ENG, IOWA STATE UNIV, 78- *Concurrent Pos:* Eng consult, 62- *Mem:* Am Soc Mech Engrs; Am Soc Agr Engrs; Soc Automotive Engrs. *Res:* Design; products liability; product safety. *Mailing Add:* Dept Mech Eng Black Eng Div Iowa State Univ Ames IA 50011

PETERS, LEON, JR, b Columbus, Ohio, May 28, 23; m 53; c 7. ELECTRICAL ENGINEERING. *Educ:* Ohio State Univ, BEE, 50, MSc, 54, PhD(elec eng), 59. *Prof Exp:* Asst, Antenna Lab, 50-51, res assoc, 51-56, asst supvr, 56-59, from asst prof to assoc prof, 59-67, dir tech area, 69-73, PROF ELEC ENG, OHIO STATE UNIV, 67-, ASSOC SUPVR, ANTENNA LAB, 59-, DIR, 83- *Mem:* Sigma Xi; fel Inst Elec & Electronics Engrs. *Res:* Properties of radar targets and the environment in which they may be observed; antennas. *Mailing Add:* 1320 Kinnear Rd Electrosci Lab Ohio State Univ Columbus OH 43212

PETERS, LEROY LYNN, b Deerfield, Mo, June 21, 31; m 54; c 3. ENTOMOLOGY. *Educ:* Kans State Univ, BS, 55, MS, 56; Univ Mo-Columbia, PhD(entom), 71; Am Registry Prof Entom, cert. *Prof Exp:* Admin asst agr chem, Indust Fumigant Co, 58; surv entomologist, Kans State Bd Agr, 59-64; exten entomologist, Univ Mo-Columbia, 64-72; assoc prof, entom, 72-82; CONSULT, BARLE ASSOCS, 91- *Concurrent Pos:* Vis prof, Ore State Univ, Corvallis, 82-83. *Mem:* Entom Soc Am. *Res:* Control of corn and sorghum insects. *Mailing Add:* 411 Walnut Wamego KS 66547

PETERS, LESTER JOHN, b Brisbane, Australia, Aug 23, 42; m 66; c 2. RADIOTHERAPY & RADIOBIOLOGY. *Educ:* Univ Queensland, MB, BS, 66; Univ New S Wales Med Sch, MD; FRACR, 71; FRCR, 75; MACR, 78. *Prof Exp:* Resident radiother, Queensland Radium Inst, 68-70, staff radiotherapist, 71; res fel radiobiol & oncol, Cancer Res Campaign, Eng, 72-74; res fel, Dimbleby Res Found, Eng, 74-75; asst prof, 75-78, assoc prof radiother, assoc radiotherapist, 75-79, PROF & HEAD, DIV RADIOTHER, UNIV TEX M D ANDERSON CANCER CTR, HOUSTON, 81- *Concurrent Pos:* Sr radiol oncologist, Inst Oncol & Radiother, Prince of Wales Hosp, Sydney Australia, 79-81; John G & Marie Stella Kennedy Mem Found Chair, Univ Tex M D Anderson Cancer Ctr. *Mem:* Am Col Radiol; Am Asn Cancer Res; Am Soc Therapeut Radiologists & Oncol; Brit Inst Radiol; Radiation Res Soc; Am Radium Soc. *Res:* Application of radiobiology research into clinical practice; optimization of time-dose relations in radiotherapy, fast neutron radiotherapy; application of tests to predict the response of individual patients tumors to radiation treatment. *Mailing Add:* 1515 Holcombe Box 97 Univ Tex M D Anderson Cancer Ctr Houston TX 77030

PETERS, LEWIS, b Evanston, Ill, Apr 10, 32; m 61, 80; c 6. PARASITOLOGY. *Educ:* DePauw Univ, AB, 54; Purdue Univ, MS, 56, PhD(zool), 60. *Prof Exp:* Instr zool, Univ Wis-La Crosse, 58-61; from asst prof to assoc prof, 61-70, dept head, 65-75, PROF BIOL, NORTHERN MICH UNIV, 70- *Mem:* Am Soc Parasitol. *Res:* Parasitic helminths; echinococcus; allocreadium. *Mailing Add:* Dept Biol Northern Mich Univ Marquette MI 49855

PETERS, LYNN RANDOLPH, b Defiance, Ohio, Jan 25, 25. INDUSTRIAL ORGANIC CHEMISTRY. *Educ:* Oberlin Col, AB, 44; Univ Mich, MS, 49, PhD(chem), 52. *Prof Exp:* Asst, Univ Mich, 52; sr org chemist, Eli Lilly Co, 52-67, res scientist, 67-68, res assoc, 68-83; RETIRED. *Mem:* Am Chem Soc. *Res:* Process research; antibiotics; heterocyclic compounds. *Mailing Add:* 110 Jefferson Ave Defiance OH 43512

PETERS, MARVIN ARTHUR, b Saginaw Co, Mich, June 23, 33; m 57; c 3. BIOCHEMICAL PHARMACOLOGY. *Educ:* Ferris State Col, BS, 57; Loma Linda Univ, MS, 65; Univ Iowa, PhD(pharmacol), 69. *Prof Exp:* Hosp pharmacist, Hinsdale Hosp, Ill, 57-63; from res asst to asst instr, 63-66, asst prof, 69-74, assoc prof, 74-79, PROF PHARMACOL, LOMA LINDA UNIV, 79- *Concurrent Pos:* NIMH grant, Div Narcotic Addiction & Drug Abuse, Loma Linda Univ, 72-78. *Mem:* Sigma Xi; Am Soc Pharmacol & Exp Therapeut. *Res:* Effect of methadone on perinatal development; effect of totigestational exposure to environmental chemicals on postnatal development; metabolism and distribution of drugs, and alterations in these processes which result in effects ranging from therapeutic ineffectiveness to extreme toxicity. *Mailing Add:* R-2 Box 455 Redlands CA 92373

PETERS, MAX S(TONE), b Delaware, Ohio, Aug 23, 20; m 47; c 2. CHEMICAL ENGINEERING. *Educ:* Pa State Univ, BS, 42, MS, 47, PhD(chem eng), 51. *Prof Exp:* Prod supvr, Nitric Acid Plant, Hercules Powder Co, 42-43; tech supvr, prod org chem, George I Treyz Chem Co, 47-49; from asst prof to prof chem eng & head dept, Univ Ill, 51-62; dean, Col Eng, 62-78, prof, 78-87, EMER PROF CHEM ENG, UNIV COLO, 88- *Concurrent Pos:* Consult ed, McGraw-Hill ser chem eng, 60-; mem, President's Comt Nat Medal of Sci, 67-69, chmn, 69. *Honors & Awards:* Westinghouse Award, Am Soc Eng Educ, 59, Lamme Gold Medal, 73; Founders Award, Am Inst Chem Engrs, 74. *Mem:* Nat Acad Eng; Am Soc Eng Educ; fel Am Inst Chem Engrs (pres, 68); Am Asn Cost Engrs; Am Chem Soc. *Res:* Reaction kinetics and catalysis; reduction reactions for nitrogen oxides; alternative energy sources. *Mailing Add:* Dept Chem Eng OT 3-6 Eng Ctr Box 424 Univ Colo Boulder CO 80309

PETERS, MICHAEL WOOD, b Midland, Tex, Mar 28, 38; m 77; c 3. PHYSICS. *Educ:* Calif Inst Technol, BS, 59; Univ Wis, PhD(physics), 64. *Prof Exp:* Proj assoc elem particle physics, Univ Wis, 64-66; from asst prof to assoc prof, 66-82, PROF PHYSICS, UNIV HAWAII, 82- *Mem:* Am Phys Soc. *Res:* High energy elementary particle phenomena; data collection and processing for particle physics; numerical analysis; neutrino physics; holography; proton-antiproton collisions. *Mailing Add:* Dept Physics & Astron Univ Hawaii Honolulu HI 96822

PETERS, PAUL CONRAD, b Kokomo, Ind, Dec 5, 28; m 55; c 4. SURGERY, UROLOGY. *Educ:* Ind Univ, AB, 50, MD, 53. *Prof Exp:* Intern med, Philadelphia Gen Hosp, 53-54; resident urol, Ind Univ, 54-57; from asst prof to assoc prof, 63-72, PROF UROL, UNIV TEX HEALTH SCI CTR DALLAS, 72-, CHMN DIV, 71- *Concurrent Pos:* Sr attend urologist, Parkland Mem Hosp, 63-; consult, Baylor Med Ctr Hosp, Dallas, 63-, Vet Admin Hosp, 64-, Children's Med Ctr, 64- & Presby Hosp, 66. *Mem:* AMA; Am Col Surg; Am Asn Genitourinary Surg; Am Soc Nephrology; Am Geriat Soc; Am Soc Transplant Surg. *Res:* Detection and management of genitourinary defects during intrauterine life and in newborn; renal failure and transplantation. *Mailing Add:* 5303 Harry Hines Blvd Dallas TX 75235

PETERS, PENN A, b Frederic, Wis, June 7, 38; m 59; c 3. TIMBER HARVESTING. *Educ:* Univ Minn, BS(bus admin) & BS(aero eng), 62; Univ Wash, MS, 69. *Prof Exp:* Res engr, stability & control, Boeing Co, 62-69; assoc prof forest eng, Ore State Univ, 75-79; res engr, 69-75, PROJ LEADER LOGGING SYSTS, USDA FOREST SERV, 79- *Concurrent Pos:* Co-prin investr, Smallwood Harvesting Res Coop, Ore State Univ, 75-79; assoc ed, Forest Sci & Northern J Appl Forestry; consult, Dennis Henninger, Atty, 76, Arthur Claflin, Atty, 78; prof logging engr, State Ore, 78-; chmn forest eng, Am Soc Agr Engrs. *Mem:* Am Soc Agr Engrs; Forest Prod Res Soc. *Res:* Timber harvesting systems for steep terrain applications in the Eastern United States. *Mailing Add:* USDA Forest Serv PO Box 4360 Morgantown WV 26505

PETERS, PHILIP BOARDMAN, b Baltimore, Md, May 17, 35; m 66; c 3. ATOMIC COLLISION PHYSICS & NEUTRON ACTIVATION ANALYSIS. *Educ:* Va Mil Inst, BS, 57; Univ NC, Chapel Hill, PhD(physics), 68. *Prof Exp:* Instr physics, Va Mil Inst, 57-58; electronic engr, Bendix Corp, Md, 58-60; instr physics, Va Mil Inst, 60-62; asst, Univ NC, Chapel Hill, 62-67; from asst prof to assoc prof physics, 67-74, chmn dept, 84-89, PROF PHYSICS, VA MIL INST, 74- *Mem:* Am Nuclear Soc; Am Asn Physics Teachers; Am Phys Soc. *Res:* Point defects in metals, introduced by electron bombardment; defects in solids by internal friction measurements, xray production in atomic collision experiments; neutron activation. *Mailing Add:* Dept Physics Va Mil Inst Lexington VA 24450

PETERS, PHILIP CARL, b Berwyn, Ill, Apr 22, 38; div; c 4. GENERAL RELATIVITY, RELATIVISTIC ASTROPHYSICS. *Educ:* Purdue Univ, BS, 60; Calif Inst Technol, PhD(physics), 64. *Prof Exp:* Res asst prof, 64-66, from asst prof to assoc prof, 66-77, PROF PHYSICS, UNIV WASH, 77- *Honors & Awards:* Distinguished Serv Citation, Am Asn Physics Teachers. *Mem:* Am Asn Physics Teachers; Am Phys Soc. *Res:* Gravitation; relativity; astrophysics. *Mailing Add:* Dept Physics FM-15 Univ Wash Seattle WA 98195

PETERS, PHILIP H, b Cleveland, Ohio, Jan 19, 21; m 45; c 1. ELECTRICAL ENGINEERING, PHYSICS. *Educ:* Case Inst Technol, BS, 42; Union Col, NY, MS, 57; Rensselaer Polytech Inst, PhD(elec eng), 62. *Prof Exp:* Engr, Advan Eng Prog, Gen Elec Co, 42-45, res assoc, Res Lab, 45-50, microwave engr, 50-59, res physicist, 59-61, consult engr, Light Mil Electronics Dept, 61-63, prog mgr process equip res & develop, Advan Technol Labs, 63-65, physicist microwave processes prog, Res & Develop Ctr, 66-68, mgr electrophys process equip, 68-69; sr scientist, Environ Technol Inc, 69-72; vpres & dir res, Environ/One Corp, 72-76; consult cool-top induction cooking, 76-77; elec engr, Gen Elec Co Corp Res & Develop, 77-83; CONSULT, 83- *Concurrent Pos:* Lectr, Union Col, NY, 47-48, 50-51. *Mem:* Sr mem Inst Elec & Electronic Engrs; Sigma Xi. *Res:* Thermal mining technology; electron accelerators; magnetrons; microwave heating; high voltage direct current power line digital instrumentation; fiber-optic technology and applications; plasma torch; getter and ion vacuum pumps; sub-micron particle detection; land use planning and pollution control methods; solid state discharge lamp ballast circuits; pulse energization of electrostatic precipitators. *Mailing Add:* RD 1 Box 373 Greenwich NY 12834

PETERS, RALPH I, b Tulsa, Okla, June 30, 47. NEUROCHEMISTRY. *Educ:* Univ Tulsa, BS, 69; Wash State Univ, PhD(zoophysiol), 75. *Prof Exp:* Res assoc, Tex A&M Univ, 75-76; fel, Wash Col Vet Med, 76-77; asst prof, Bates Col, 77-80; asst prof, 80-85, ASSOC PROF, WICHITA STATE UNIV, 85- *Mem:* Am Soc Zoologists; AAAS; Sigma Xi; Soc Neurosci. *Res:* Tryptophan metabolism as it relates to serotonergic neuronal function in the mammalian central nervous system, and the importance of these in locomotor control. *Mailing Add:* Dept Biol Sci Wichita State Univ Wichita KS 67208

PETERS, RANDALL DOUGLAS, b Big Stone Gap, Va, Feb 5, 42; m 65; c 4. MECHANICS. *Educ:* Univ Tenn, BS, 64, PhD(physics), 68. *Prof Exp:* From asst prof to assoc prof physics, Univ Miss, 68-77; mem staff, Bell Helicopter, LTV Aerospace, 77-86; ASSOC PROF PHYSICS, TEX TECH UNIV, 86- *Mem:* Sigma Xi. *Res:* Nonlinear systems studies; deterministic chaos; interface between classical and quantum dynamics. *Mailing Add:* Dept of Physics Texa Tech Univ Lubbock TX 79409

PETERS, RICHARD MORSE, b New Haven, Conn, Feb 21, 22; m 46; c 4. SURGERY. *Educ:* Yale Univ, BS, 43, MD, 45; Am Bd Surg, dipl, Am Bd Thoracic Surg, dipl. *Prof Exp:* Asst, Washington Univ, 48-50; asst prof surg, Sch Med, Univ NC, 52-55, from assoc prof to prof surg in chg thoracic & cardiovasc surg, 55-65, thoracic & cardiovasc surg & biomath & bioeng, 65-69; head div thoracic surg, 69-77, co-head div cardiothoracic surg, 77-83, PROF SURG & BIOENG, UNIV CALIF, SAN DIEGO, 69-, DIR SURG RES & EDUC, 83- *Concurrent Pos:* Fel thoracic surg, Washington Univ, 50-52; head thoracic surg, Vet Admin Hosp, San Diego, mem surg, Merit Rev Bd; pres gov body, Health Systs Agency, San Diego & Imperial Counties, 76-78; Am Bd Thoracic Surg, 81-88; gov, Am Col Surgeons, 85- *Mem:* Am Asn Thoracic Surg; Am Col Surgeons; Am Surg Asn; Am Asn Surg Trauma; Soc Univ Surgeons. *Res:* Cardiopulmonary pathophysiology; bioengineering; biomathematics; computers in patient care. *Mailing Add:* Univ Calif San Diego Med Ctr 225 W Dickinson St San Diego CA 92103-9981

PETERS, ROBERT EDWARD, b Jackson, Ohio, June 17, 40; m 64; c 2. ELECTROMAGNETICS. *Educ:* Miami Univ, BS, 62; Purdue Univ, MS, 65, PhD(high energy physics), 67. *Prof Exp:* Instr physics, Purdue Univ, 67-68; PHYSICIST, FERMI NAT ACCELERATOR LAB, 68- *Mem:* Am Phys Soc. *Res:* Particle accelerator design; superconducting magnet technology. *Mailing Add:* 861 Geneva Rd St Charles IL 60174

PETERS, ROBERT HENRY, b Toronto, Ont, Aug 2, 46; m 74; c 2. LIMNOLOGY, ALLOMETRY. *Educ:* Univ Toronto, BSc, 68, PhD(limnol), 72. *Prof Exp:* Fel hydrobiol, Italiano Inst Hydrobiol, 72-74; asst prof, 74-86, PROF BIOL, MCGILL UNIV, 86-; FEL ZOOL, UNIV MUNICH, 74- *Concurrent Pos:* Vis scientist, Italian Inst Hydrobiol, 80-81 & 87-88; dir, NAm Lake Mgt Soc, 90- *Mem:* Int Asn Theoret & Appl Limnol; Am Soc Limnol & Oceanog; Can Soc Zool; Am Soc Naturalists; NAm Lake Mgt Soc. *Res:* Ecological implications of body size; cycling of phosphorus and polychlorobiephenyls; empirical patterns in ecology and limnology; philosophy of biology; ectrophication; contamination. *Mailing Add:* McGill Univ Dept Biol 1205 Ave Docteur Penfield Montreal PQ H3A 1B1 Can

PETERS, ROGER PAUL, b Washington, DC, Oct 29, 43; c 1. ANIMAL BEHAVIOR. *Educ:* Univ Chicago, BA, 65; Univ Mich, Ann Arbor, PhD(psychol), 74. *Prof Exp:* Instr math, New Col, 66-69; lectr psychol, Univ Mich, Ann Arbor, 74-75; asst prof, 75-80, ASSOC PROF PSYCHOL, FT LEWIS COL, 80-, PROF PSYCH. *Concurrent Pos:* Res intern, US Forest Serv, 72-74. *Mem:* Western Psychol Asn. *Res:* Educational assessment; cognitive mapping; cognitive bias; evolution of cognition; mammalian communication; canine behavior. *Mailing Add:* Dept Psychol Ft Lewis Col Durango CO 81301

PETERS, STEFAN, b Posen, Poland, June 27, 09; US citizen; m 52; c 2. MATHEMATICS. *Educ:* Univ Erlangen, PhD(math), 31. *Prof Exp:* Asst actuary, NY Compensation Ins Rating Bd, 38-43; res mathematician, West Coast Life Ins Co, Calif, 46-48; assoc prof, Univ Calif, Berkeley, 49-51; actuary, Calif Inspection Rating Bur, 50-52; consult actuary, Morss & Seal, NY, 52-53 & Cornell & Price, Mass, 53-58; self employed, 58-60; opers res consult, Arthur D Little, Inc, Mass, 60-73; chief actuary, Div Ins, Commonwealth of Mass, 74-; RETIRED. *Mem:* Fel Soc Actuaries; Opers Res Soc Am; Am Math Soc; Int Math Statist; fel Casualty Actuarial Soc. *Res:* Operations research; numerical analysis. *Mailing Add:* 93 Upland Rd Cambridge MA 02144

PETERS, THEODORE, JR, b Chambersburg, Pa, May 12, 22; m 45; c 4. BIOCHEMISTRY. *Educ:* Lehigh Univ, BS, 43; Harvard Univ, PhD(biol chem), 50; Am Bd Clin Chem, dipl. *Prof Exp:* Asst chem eng, Mass Inst Technol, 43-44; instr physiol chem, Sch Med, Univ Pa, 50-51; assoc biol chem, Harvard Med Sch, 53-55; res biochemist, 55-88, EMER RES SCIENTIST, MARY IMOGENE BASSETT HOSP, 88- *Concurrent Pos:* Biochemist, Boston Vet Admin Hosp, 53-55; from adj asst prof to adj assoc prof biol chem, Col Physicians & Surgeons, Columbia Univ, 55-; Commonwealth Fund vis scientist, Carlsberg Lab, Copenhagen, 58-59; adj prof path, Albany Med Col, 69-84, adj prof biochem, 84-; guest worker, NIH, 71-72. *Honors & Awards:* Gold Medal Award, Electron Microscope Soc Am, 66; Nat Award, Am Asn Clin Chem, 77. *Mem:* Am Chem Soc; Am Soc Biol Chemists; Am Asn Clin Chemists; Am Soc Cell Biol. *Res:* Protein synthesis in animal tissues; plasma protein structure and function. *Mailing Add:* Mary Imogene Bassett Hosp Cooperstown NY 13326

PETERS, THOMAS G, b Cincinnati, Ohio, Oct 3, 45; m 77; c 3. TRANSPLANTATION SURGERY, SURGICAL EDUCATION. *Educ:* Miami Univ, Ohio, AB, 66; Univ Cincinnati, MD, 70. *Prof Exp:* Chief resident gen surg, Med Col Wis, 76-77; fel transplantation, Univ Colo, 77-78; from asst prof to assoc prof, 78-87, PROF SURG, UNIV TENN, MEMPHIS, 87- *Concurrent Pos:* Actg chief transplantation serv, Walter Reed Army Med Ctr, 81-; vis prof surg, Univ Miss, 84, Walter Reed Army Med Ctr, 85, 86, Charlotte Mem Hosp Med Ctr, 88; travelling fel, Am Col Surg, Australian-N* chap; vis lectr surg, Med Col Wis, 87; mem, Abstractor Int Abstr Surg for Surg, Gynec & Obstet. *Mem:* Am Col Surgeons; AMA; Am Soc Transplant Surgeons; Int Surg Soc; The Transplantation Soc; Soc Surg Alimentary Tract. *Res:* Investigation of post-splenectomy sepsis; study of outcomes in hepatic and renal transplantation; utilization of multiple organ donor resources; trends in surgical education. *Mailing Add:* 580 W Eighth St Jacksonville FL 32209

PETERS, THOMAS MICHAEL, b Inglewood, Calif, Oct 7, 37; m 57. ENTOMOLOGY, SYSTEMATICS. *Educ:* Long Beach State Col, BS, 59; Univ Minn, MS, 61, PhD(entom), 64. *Prof Exp:* From instr to assoc prof, 64-75, head dept, 68-75, PROF ENTOM, UNIV MASS, AMHERST, 75- *Mem:* Entom Soc Am; Can Entom Soc; Soc Syst Zool; Sigma Xi. *Res:* Taxonomy of cujlicoidea, especially dixid flies; insect fauna of North American crop ecosystems. *Mailing Add:* Dept of Entom Univ of Mass Amherst MA 01003

PETERS, TILL JUSTUS NATHAN, b Berlin, Ger, Jan 27, 34; US citizen; m 79; c 2. INORGANIC CHEMISTRY. *Educ:* Pa State Univ, BS, 57; Univ Wis, Madison, MS, 61; Western Reserve Univ, MS, 64, PhD, 66. *Prof Exp:* Teacher high sch, NJ, 57-63; res scientist, Sherwin-Williams Co, Ohio, 66-67; advan develop engr, GTE Sylvania, Inc, 67-71; chmn bus & occup div, Berkshire Community Col, 72-74, chmn div occup studies, 74-75, asst dean fac, 75-77; DEAN OCCUP EDUC, GRAND RAPIDS JR COL, 77- *Mem:* Am Chem Soc; Am Voc Asn. *Res:* Coordination chemistry of high oxidation states of molybdenum and tungsten; synthesis of soluble tungstates; extractive metallurgy of heavy metals. *Mailing Add:* 1454 Benjamin Ave NE Grand Rapids MI 49505

PETERS, WILLIAM CALLIER, b Oxford, Ohio, July 12, 20; m 42; c 1. GEOLOGY. *Educ:* Miami Univ, BA, 42; Univ Colo, MS, 48, PhD(geol), 57. *Prof Exp:* Engr, NJ Zinc Co, 42, geologist, Empire Zinc Div, 48-49; asst prof geol, Idaho State Col, 49-54; geologist, Mineral Develop Dept, FMC Corp, 54-60; div geologist, Utah Copper Div, Kennecott Copper Corp, Utah, 60-64; prof, 64-82, EMER PROF MINING & GEOL ENG, UNIV ARIZ, 82-; MINING CONSULT, 82- *Concurrent Pos:* Vis prof, Univ Geneva & Sch Mines, Leoben, Austria, 71, 80 & Univ San Juan, Argentina, 79. *Mem:* Soc Econ Geol; Geol Soc Am; Am Inst Mining, Metall & Petrol Eng; Can Inst Mining & Metall; Brit Inst Mining & Metall. *Res:* Exploration for mineral deposits; mining engineering and geology applied to industrial mineral resources; economics and conservation of natural resources. *Mailing Add:* 5702 E Seventh St Tucson AZ 85711

PETERS, WILLIAM LEE, b Leavenworth, Kans, June 27, 39; m 64; c 1. AQUATIC ENTOMOLOGY. *Educ:* Univ Kans, BA, 60; Univ Utah, MS, 62, PhD(entom), 66. *Prof Exp:* PROF ENTOM, FLA AGR & MECH UNIV, 66- *Concurrent Pos:* Prin investr, NSF grant & Coop State Res Serv, 66-; adj prof, Fla State Univ, 67; res assoc, Fla State Collection Arthropods, 67-; entomologist, Univ Fla, 67-; res dir agr sci, Fla Agr & Mech Univ, 70-85. *Mem:* AAAS; Entom Soc Am; Am Inst Biol Sci; fel Royal Entom Soc London. *Res:* Higher classification of Ephemeroptera, especially Leptophlebiidae. *Mailing Add:* Dept Entom Fla Agr & Mech Univ Tallahassee FL 32307

PETERSDORF, ROBERT GEORGE, b Berlin, Ger, Feb 14, 26; US citizen; m 51; c 2. MEDICINE. *Educ:* Brown Univ, BA, 48; Yale Univ, MD, 52. *Hon Degrees:* DSc, Albany Med Col, 79, State Univ NY, Brooklyn, 87, Med Col Ohio, Toledo, 87, Chicago Med Sch, 87, St Louis Univ, 88; MA, Harvard Univ, 80; DMSc, Med Col Pa, 82, Brown Univ, 83, Wake Forest Univ, 86; LHD, NY Med Col, 86, Med Col Hampton Rds, 88. *Prof Exp:* Instr med, Sch Med, Yale Univ, 57-58; asst prof, Sch Med, Johns Hopkins Univ, 58-60; assoc prof, Sch Med, Univ Wash, 60-62, prof, 62-79, chmn dept, 64-79; prof med, Harvard Med Sch, 79-81; dean & vchancellor health sci, Univ Calif, San Diego, 81-86; CLIN PROF, DEPT MED, GEORGETOWN UNIV SCH MED, 86-; PRES, ASN AM MED COLS, 86- *Concurrent Pos:* Physician, Johns Hopkins Hosp, 58-60; physician-in-chief, Univ Wash Hosp, 64-79, Harborview Med Ctr, 60-64, attend physician, 64-79; attend physician, Seattle Vet Admin & Univ Wash Hosp, 60-64, Brigham & women's Hosp, 79-81; consult, USPHS Hosp, Madigan Army Hosp & US Vet Admin Hosp; mem, Training Grant Comt, Nat Inst Allergy & Infectious Dis, 65-69; mem, Training Grant Comt, Nat Inst Allergy & Infectious Dis, 65-69, coun mem,79-81; mem, Adv Comt to Dir, NIH, 72-77; rep, Intersci Conf Antimicrobial Agents & Chemotherapy, Am Soc Microbiol, 68-70; chmn, Comt Eval Clin Competence, Am Bd Int Med, 71-75, mem exec comt, 72-77, chmn bd gov, 76-77; mem, Comt Nat Med Policy, Am Soc Clin Invest, 72-78; mem adv comt to dir, NIH, 72-77, mem, Centennial Comt, 86-87; mem, Nat Sci Adv Comt, Nat Jewish Hosp & Res Ctr/Nat Asthma Ctr, 72-, chmn, 85-; mem, Steering Comt to Study Impact of Social Security Regulations, Nat Acad Sci-Inst Med, 74-76, chmn, Steering Comt Task Force Clin Invest in Developing Countries, 78-80; mem, Coun Med Affairs, 86-; mem, Comt Med Affairs, Yale Univ Coun, 88-; mem bd dirs, Nat Med Fel, Inc, 89-92, Res Am, 90- *Honors & Awards:* Lilly Medalist, Royal Col Physicians, 78; Alfred E Stengel Mem Award, Am Col Physicians, 80; Robert H Williams Distinguished Chmn of Med Award, Asn Prof Med, 83; Distinguished Internist, Am Soc Internal Med, 87. *Mem:* Inst Med-Nat Acad Sci; master Am Col Physicians (pres, 75-76, emer pres, 85-); Asn Am Physicians (vpres, 75-76, pres, 76-77); Am Soc Clin Invest; Asn Profs Med (pres, 70-71); fel AAAS; Am Asn Immunol; sr mem Am Fedn Clin Res; AMA; Sigma Xi; fel Royal Soc Med; fel Royal Col Physicians. *Res:* Infectious diseases; over 400 publications on bacterial infections, pathogenesis of infections, epidemiology, organization of academic health centers, public policy and physician manpower. *Mailing Add:* Asn Am Med Cols One Dupont Circle Washington DC 20036

PETERSEN, BENT EDVARD, b Copenhagen, Denmark, July 31, 42; Can citizen; m 65; c 3. PARTIAL DIFFERENTIAL EQUATIONS, SEVERAL COMPLEX VARIABLES. *Educ:* Univ BC, BSc, 64; Mass Inst Technol, PhD(math), 68. *Prof Exp:* From asst prof to assoc, 68-80, PROF ORE STATE UNIV, 80- *Concurrent Pos:* Mem, Inst Advan Study, Princeton, NJ, 73-74. *Mem:* Am Math Soc; Math Asn Am; Sigma Xi. *Res:* Partial differential equations; pseudo-differential operators; cohomology with bounds. *Mailing Add:* Dept Math Ore State Univ Corvallis OR 97331

PETERSEN, BRUCE H, b Logan, Vt, Nov 21, 37. MICROBIOLOGY. *Educ:* Indiana Univ, MS, 67, PhD(immunol & microbiol), 69. *Prof Exp:* RES SCIENTIST, LILLY LAB CLIN RES, 69- *Concurrent Pos:* Adj prof biol, Purdue Sch Sci & Sch Med, Indiana Univ, 77- *Mem:* Am Asn Immunol; Am Fedn Clin Res; Int Soc Immuno-Pharmacol. *Mailing Add:* Lilly Lab Clin Res Wishard Mem Hosp 1001 W 10th St Indianapolis IN 46202

PETERSEN, BRUCE WALLACE, b Minneapolis, Minn, Dec 2, 36; m 64; c 2. ZOOLOGY. *Educ:* Univ Nebr, Omaha, BA, 58, BS, 62; State Univ Iowa, MS, 60; Univ Colo, Boulder, PhD(zool), 68. *Prof Exp:* Sci teacher pub schs Nebr, 60-62; asst prof biol, Atlantic Christian Col, 65-66 & Meramec Community Col, St Louis, 67-68; asst prof zool, Southern Ill Univ, Carbondale, 68-84; SCI TEACHER, LOS ANGELES SCHS, 90- *Res:* Human ecology, especially biological implication of the human population increase. *Mailing Add:* 838 N Forbes Dr Brea CA 92621

PETERSEN, CARL FRANK, b Santa Rosa, Calif, Oct 24, 37; m 58; c 2. GEOPHYSICS. *Educ:* Stanford Univ, BS, 59, MS, 61, PhD(geophys), 69. *Prof Exp:* Peace Corps volunteer, 63-66; scientist, Dept Sci & Indust Res, NZ, 69-70; geophysicist shock studies, Stanford Res Inst, 62-63, 67-69 & 70-75; mgr shock physics, 75-83, MGR TECHNOL, S-CUBED, 83- *Mem:* AAAS; Am Geophys Union; Soc Explor Geophysicists. *Res:* Dynamic properties of geologic materials; rock mechanics; shock waves in solids. *Mailing Add:* S-Cubed Div Maxwell Labs PO Box 1620 La Jolla CA 92038-1620

PETERSEN, CHARLIE FREDERICK, b Buhl, Idaho, Aug 5, 15; m 40; c 3. POULTRY NUTRITION. *Educ:* Univ Idaho, BSA, 40, MSA, 46. *Prof Exp:* Field serv agent, Albers Milling Co, Wash, 40-41; asst prof poultry nutrit & asst poultryman, 43-47, assoc prof & assoc poultryman, 47-57, PROF & POULTRYMAN, UNIV IDAHO, 57-, HEAD DEPT POULTRY SCI, 61-, HEAD DEPT ANIMAL SCI, 79- *Mem:* Am Poultry Sci Asn; Am Inst Nutrit; World Poultry Sci Asn. *Res:* Environmental factors related to energy requirements; protein biological value; vitamin requirements and unidentified growth factors. *Mailing Add:* Dept Animal Sci Univ of Idaho Moscow ID 83843

PETERSEN, DENNIS ROGER, TOXICOLOGY, PHARMACOGENETICS. *Educ:* Univ Wyo, PhD(biochem pharmacol), 74. *Prof Exp:* ASSOC PROF PHARMACOL & TOXICOL, UNIV COLO, 78- *Mailing Add:* Pharm Univ Colo Boulder Box 297 Boulder CO 80309

PETERSEN, DONALD E, b Pipestone, Minn, Aug 6, 26. ENGINEERING ADMINISTRATION. *Educ:* Univ Wash, BS, 46; Stanford Univ, MBA, 49. *Hon Degrees:* DSc, Univ Detroit, LLD, Art Ctr Col Design. *Prof Exp:* Exec vpres, 77, Bd dirs, 77-80, pres & chief opers, 80-85, CHMN & CHIEF EXEC OFFICER, FORD MOTOR CO, 85- *Concurrent Pos:* adv bd, Univ Wash Bus Sch & Stanford Univ Bus Sch. *Mem:* Nat Acad Eng; Motor Vehicle Mfrs Asn; Soc Automotive Engrs. *Mailing Add:* Ford Motor Co The American Rd Dearborn MI 48121

PETERSEN, DONALD FRANCIS, b Brookings, SDak, May 20, 26; m 49; c 4. PHARMACOLOGY. *Educ:* DePauw Univ, AB, 47; SDak State Col, MS, 50; Univ Chicago, PhD(pharmacol), 54. *Prof Exp:* Assoc chemist, Exp Sta, SDak State Col, 47-51; from asst to instr pharmacol, Univ Chicago, 51-56; mem staff, 56-77, alternate div leader, 77-91, LAB ASSOC, LOS ALAMOS NAT LAB, 91- *Mem:* AAAS; Am Chem Soc; Am Soc Pharmacol & Exp Therapeut; Am Indust Hyg Asn; NY Acad Sci. *Res:* Influence of ionizing radiations on enzymatic processes in mammalian cells; analytical biochemistry; metabolism of chemotherapeutic agents and isotopically labelled intermediates. *Mailing Add:* Biomed Res Br NSP-DST MSF616 Los Alamos Sci Lab Univ Calif Los Alamos NM 87545

PETERSEN, DONALD H, b Hillsboro, Ore, Mar 17, 34; m 61, 84; c 5. PHYSICAL CHEMISTRY, MATERIALS SCIENCE. *Educ:* Univ Portland, BS, 56; Univ Notre Dame, PhD(phys chem), 61. *Prof Exp:* Res scientist, Pioneering Res Lab, Weyerhaeuser Co, 61-63; res scientist, Ling-Temco-Vought, Inc, 63-65, sr scientist, 65-69, mgr pollution res prog, Advan Technol Ctr, 70-72, supvr, 72-74, mgr structures & mats res, LTV Missiles & Electronics Group, 74-88, CONSULT ENGR, LTV AIRCRAFT PROD GROUP, 88- *Concurrent Pos:* Past chmn, Tech Comt Mat, Am Inst Aeronaut & Astronaut; chmn, Mat Panel Space Technol Workshop III; Tech Info Tech Comt, Am Inst Aeronaut & Astronaut. *Mem:* Soc Advan Mat & Eng; Am Inst Aeronaut & Astronaut. *Res:* Spectroscopy; chromatography; chemistry of high temperature materials including carbon composites and refractory alloys; graphite-epoxy composites; carbon fibers development; non-destructive testing; laminar structures; adhesive bonding; armor and anti-armor. *Mailing Add:* 3654 Park Ridge Dr Grand Prairie TX 75051

PETERSEN, DONALD RALPH, b Wis, Apr 14, 29; m 52, 79; c 5. STRUCTURAL CHEMISTRY. *Educ:* Lawrence Col, BA, 51; Calif Inst Technol, PhD(phys chem), 55. *Prof Exp:* Res chemist, Chem Physics Res Lab, Dow Chem Co, 55-65, group leader, 65-69, div leader, 69-70, res mgr, Comput Res Lab, 70-78, sr res staff, 78-86; SR TECH CONSULT, OMNI TECH INT, 86- *Concurrent Pos:* Ed bd, J Testing & Eval, Am Soc Testing Mat, 85- *Mem:* Am Chem Soc; Am Phys Soc; Am Crystallog Asn; Am Inst Chemists; Am Soc Testing Mat; Classification Soc. *Res:* Molecular structures; properties of catalytic surfaces, x-ray and electron-beam techniques; computer applications to chemical information processing; experimental statistics and data handling. *Mailing Add:* Omni Tech Int PO Box 2237 Midland MI 48641

PETERSEN, EDWARD LELAND, b Myrtle Point, Ore, Aug 12, 32; m 59; c 1. PHYSICS. *Educ:* Ore State Col, BS, 54, MS, 56; Univ Calif, Los Angeles, PhD(physics), 66. *Prof Exp:* Asst prof physics, San Fernando Valley State Col, 63-68 & Oberlin Col, 68-69; RES PHYSICIST, NAVAL RES LAB, 69- *Mem:* Am Phys Soc; AAAS. *Res:* Nuclear reactions; single event upset phenonema; single event upsets in computers in the space environment; satellite vulnerability and survivability. *Mailing Add:* 9502 Babson Ct Fairfax VA 22030

PETERSEN, EDWARD S, b Chicago, Ill, Nov 19, 21; m 44; c 2. INTERNAL MEDICINE. *Educ:* Harvard Med Sch, MD, 45. *Prof Exp:* Intern, St Lukes Hosp, Chicago, 45-46; resident med, Univ Chicago, 48-51; pvt practr, 51-53; asst dir prof serv, Vet Admin Res Hosp, 53-54; dir med clins & assoc prof med, Med Sch, Northwestern Univ, Chicago, 54-72; asst dir, AMA, 72-76, dir, 76-87, asst dir, Dept Undergrad Med Educ, 87-88; RETIRED. *Concurrent Pos:* Med educ nat defense coordr, Northwestern Univ, 58-69, from asst dean to assoc dean, Med Sch, 60-72; pres, Inst Med Chicago, 75; co-secy, Liaison Comt Med Educ, 76-87. *Mem:* Fel Am Col Physicians. *Res:* Diabetes; medical education and administration. *Mailing Add:* 1350 Astor St 3A Chicago IL 60610

PETERSEN, EUGENE E(DWARD), b Tacoma, Wash, Mar 2, 24; m 48; c 2. CHEMICAL ENGINEERING. *Educ:* Univ Wash, BS, 49, MS, 50; Pa State Univ, PhD(fuel sci), 53. *Prof Exp:* Asst fuel technol, Pa State Univ, 50-52; from instr to assoc prof, 53-64, PROF CHEM ENG, UNIV CALIF, BERKELEY, 64- *Concurrent Pos:* Consult, Chevron Res Corp, 61-66, Stauffer Chem Co, 65-67, Gen Motors, 75-81, Lockheed Missiles & Space Co, 77-78 & W R Grace Co, 84-; Solvay prof, Free Univ Brussels, 66; exchange scientist, Acad Sci, USSR, 73. *Honors & Awards:* King lectr, Johns Hopkins Univ, 62; Reilly lectr, Notre Dame Univ, 76; R H Wilhelm Award, Am Inst Chem Engrs, 85. *Mem:* Am Chem Soc; Am Inst Chem Engrs. *Res:* Synthesis of heterogenous catalysts; the dynamics of impregnationg; the characterization of microscopic and microscopic distribution of active elements within catalysts; the characterization of chemical activity and deactivation properties of catalysts. *Mailing Add:* Dept Chem Eng Univ Calif Berkeley CA 94720

PETERSEN, FREDERICK ADOLPH, b Chicago, Ill, Aug 19, 13; m 40; c 3. CERAMIC ENGINEERING. *Educ:* Univ Ill, BS, 37; Ohio State Univ, MSc, 39. *Prof Exp:* Eng trainee, Frigidaire Div, Gen Motors Corp, Ohio, 37-38; millroom operator, Rundle Mfg Co, Wis, 38; develop engr, Ingram-Richardson Mfg Co, Ind, 39-41; spec res prof, Univ Ill, 41-51; trade asn mgt, Thomas Assocs, Inc, 51-80, pres, 61-80; RETIRED. *Concurrent Pos:* Consult, 43-; chmn, Comt Safety Standards, Am Nat Standards Inst, 68-89. *Mem:* Fel Am Ceramic Soc; Nat Inst Ceramic Engrs; Sigma Xi. *Res:* Porcelain enamels development and research. *Mailing Add:* 3224 E Monmouth Rd Cleveland Heights OH 44118

PETERSEN, GARY WALTER, b Frederic, Wis, Aug 17, 39; m 64; c 3. SOIL MORPHOLOGY, REMOTE SENSING. *Educ:* Univ Wis-Madison, BS, 61, MS, 63, PhD(soil genesis, soil morphol), 65. *Prof Exp:* Res asst soils, Univ Wis-Madison, 61-64, teaching asst, 64-65; from asst prof to assoc prof, 65-76, PROF SOIL GENESIS & MORPHOL, PA STATE UNIV, 76-, CO-DIR, OFF REMOTE SENSING EARTH RESOURCES, 70- *Concurrent Pos:* Vis prof land use, Water Resources Ctr, Univ Wis, 74-75. *Mem:* Fel Am Soc Agron; fel Soil Sci Soc Am; Soil & Water Conserv Soc Am; Am Soc Photogram; Int Soc Soil Sci. *Res:* Soil interpretations for proper land use decisions; on site disposal of wastes; development and use of geographic information systems; analysis and interpretation of remotely sensed data. *Mailing Add:* Off Remote Sensing Earth Resources Pa State Univ University Park PA 16802

PETERSEN, GENE, b Salt Lake City, Utah, Aug 6, 44; m 72; c 2. BIOTECHNOLOGY. *Educ:* Univ Utah, BA, 70; Northwestern Iniv, PhD(chem), 76. *Prof Exp:* Grad student & instr, Northwestern Univ, 71-76; res fel, Cal Inst Technol, 76-78; TECH GROUP LEADER, BIOTECHNOL GROUP, JET PROPULSION LAB, 78- *Concurrent Pos:* Res fel, Caltech 76-78; res asst, Kennecott Cooper Corp, 66-71. *Mem:* Am Chem Soc. *Res:* Rheology of polysaccharides; microgravity operational bioreactions; polysaccharide chemistry. *Mailing Add:* Cal Inst Tech, Jet Propulsion Lab 4800 Oak Grove Dr Pasadena CA 91109

PETERSEN, HAROLD, JR, b Natick, Mass, Sept 18, 40; m 66; c 2. INFORMATION MANAGEMENT SYSTEMS. *Educ:* Univ Mass, BS, 62; Univ Ill, PhD(chem), 66. *Prof Exp:* Teaching asst chem, Univ Ill, 62-66; fel, Univ Southern Calif, 66-67; from asst prof to prof chem, Univ RI, 67-84; Battelle Ocean Sci Dept, 84-89; ASST VCHANCELLOR, INFO SYSTS, UNIV TENN, MEMPHIS, 89- *Mem:* Sigma Xi; Am Chem Soc; Asn Computer Mach; AAAS. *Res:* Data management and information system design; chemical oceanography; investigations of drug-receptor interactions by molecular orbital theory; environmental studies. *Mailing Add:* 8464 Glen Ridge Cove Germantown TN 38138

PETERSEN, INGO HANS, b Davenport, Iowa, July 27, 30; m 52; c 4. FLUORANTHENE CHEMISTRY. *Educ:* Iowa State Univ, BS, 52; Univ Iowa, MS & PhD(org chem), 61. *Prof Exp:* Res chemist, Union Carbide Corp, 60-64; from asst prof to assoc prof, 64-66, chmn dept, 66-70 & 80-83, PROF CHEM, STATE UNIV NY COL, BROCKPORT, 66- *Concurrent Pos:* Res assoc, Queen's Univ, Belfast, 70-71; vis prof, Duke Univ, 76-77. *Mem:* Am Chem Soc. *Res:* Synthesis of fluoranthene derivatives; substituent-orienting effects and products; synthesis and polymerization of thiiranes. *Mailing Add:* Seven Keystone Ct Brockport NY 14420

PETERSEN, JEFFREY LEE, b Racine, Wis, Nov 19, 47; m 73. PHYSICAL INORGANIC CHEMISTRY. *Educ:* Carthage Col, BA, 69; Univ Wis-Madison, MS, 71, PhD(chem), 74. *Prof Exp:* Res assoc chem, Argonne Nat Lab, 74; vis scholar, Northwestern Univ, Evanston, 74-75; from asst prof to assoc prof, 75-85, PROF CHEM, WVA UNIV, 85- *Mem:* Am Chem Soc; Sigma Xi; Am Crystallog Asn. *Res:* Systematic investigation by structural and spectroscopic techniques into the nature of the chemical bonding in transition metal complexes of current chemical interest. *Mailing Add:* Dept Chem WVa Univ PO Box 6045 Morgantown WV 26507

PETERSEN, JOHN DAVID, b Glendale, Calif, Nov 21, 47; m 70; c 2. INORGANIC CHEMISTRY. *Educ:* Calif State Univ, Los Angeles, BS, 70; Univ Calif, Santa Barbara, MA & PhD(chem), 75. *Prof Exp:* Asst prof chem, Kans State Univ, 75-80; from asst prof to assoc prof, 80-85, assoc dean sci,

83-88, PROF CHEM, CLEMSON UNIV, 85-, DEPT HEAD, 90- *Concurrent Pos:* Alexander Von Humboldt res fel, Univ Regensburg, Ger, 86-87. *Mem:* Am Chem Soc; InterAm Photochem Soc; Soc Appl Spectros. *Res:* Photochemistry and electron transfer reactions of transition metal complexes to include synthetic techniques, ligand substituent effects and spectroscopic studies; synthesis and conducting behavior of novel metallopolymers. *Mailing Add:* Dept Chem Clemson Univ Clemson SC 29634-1905

PETERSEN, JOHN ROBERT, b LaCrosse, Wis, June 26, 29; m 52; c 5. MEDICAL ADMINISTRATION. *Educ:* Univ Wis-Madison, BMSc, 51, MD, 54. *Prof Exp:* From asst instr to asst prof, 57-68, asst dean, 65-67, ASSOC PROF MED, MED COL WIS, 68-; DIR MED SERV, 67-, ASSOC DEAN MILWAUKEE COUNTY MED COMPLEX, 80- *Concurrent Pos:* Fel internal med, Frank Bunts Educ Inst, 55; consult, Vet Admin Hosp, Wood, Wis, 61-; asst dir dept med, Milwaukee County Gen Hosp, 61-65; mem, State of Wis Health Policy Coun, 80-88. *Mem:* NY Acad Sci; Asn Am Med Cols; Am Fedn Clin Res; Med Admin Conf. *Res:* Endocrinology wth emphasis on Diabetes Mellitus. *Mailing Add:* 960 N Mayfair Rd Wauwatosa WI 53226

PETERSEN, JOSEPH CLAINE, b Fielding, Utah, Feb 14, 25; m 49; c 4. ASPHALT CHEMISTRY, FOSSIL FUEL CHEMISTRY. *Educ:* Univ Utah, BS, 52, PhD(org chem), 56. *Prof Exp:* Res chemist, Am Gilsonite Co, Utah, 56-61; sr res chemist, New Prod Div, Textile Fibers Dept, Exp Sta, E I du Pont de Nemours, Del, 61-64; Sect Supvr, Asphalt Res & Prod Utilization, Laramie Energy Technol Ctr, Dept Of Energy, 64-83; distinguished scientist, Western Res Inst, 84-90; RETIRED. *Concurrent Pos:* Chmn, Transp Res Bd Comt Characteristics Bituminous Mat, 74-80; mem adv panel, Nat Res Coun, 71-; asphalt adv comt, Strategic Highway Res Prog, 87- *Mem:* Am Chem Soc; Asn Asphalt Paving Tech. *Res:* Petroleum, asphalt and polymer chemistry; fossil fuel heavy liquids; asphalt functional group analysis, infrared spectroscopes, chemical composition - physical property relationships. *Mailing Add:* 1072 Colina Dr Laramie WY 82070

PETERSEN, KARL ENDEL, b Tallinn, Estonia, July 11, 43; US citizen; m 69; c 2. ERGODIC THEORY, DYNAMICAL SYSTEMS. *Educ:* Princeton Univ, AB, 65; Yale Univ, MA, 67, PhD(math), 69. *Prof Exp:* from asst to assoc prof, 69-81, PROF MATH, UNIV NC, CHAPEL HILL, 81- *Concurrent Pos:* Vis assoc prof, Dept Math, Yale Univ, 79; prof assoc, Laboratoire de Calcul des Probabilities, Univ de Paris VI, 81; vis researcher, Univ Rennes I, France, 88- *Mem:* Am Math Soc. *Res:* Ergodic theory, probability and analysis, especially questions of mixing, almost everywhere convergence and maximal functions and information and coding. *Mailing Add:* 2719 Jones Ferry Rd Chapel Hill NC 27516

PETERSEN, KENNETH C, b Chicago, Ill, Mar 17, 36. ORGANIC POLYMER CHEMISTRY. *Educ:* Northwestern Univ, BS, 60, MS, 63. *Prof Exp:* Sr chemist condensation polymers, Acme Resin Corp, 58-64; res mgr addn & condensation polymers, 64-76, mgr chem/tech develop & chem mfg, 77-78, mgr chem div, 78-79, mgr, mfg & chem develop, 79-80, vpres mfg, 80, exec vpres, 81, PRES, SCHENECTADY CHEM, INC, 81- *Mem:* Am Chem Soc. *Mailing Add:* 1301 Rosehill Blvd Schenectady NY 12309

PETERSEN, MORRIS SMITH, b West Jordan, Utah, Feb 21, 33; m 54; c 6. PALEONTOLOGY. *Educ:* Brigham Young Univ, BS, 55, MS, 56; Univ Iowa, PhD(geol), 62. *Prof Exp:* Asst prof earth sci, Am Univ, 62-66; asst prof, 66-72, chmn dept, 75-81, PROF GEOL, BRIGHAM YOUNG UNIV, 72- *Mem:* Geol Soc Am; Paleont Soc. *Res:* Devonian stratigraphy and paleontology. *Mailing Add:* Dept Geol ESC 144 Brigham Young Univ Provo UT 84602

PETERSEN, NANCY SUE, b Paris, Tex, Mar 11, 43; m 65; c 2. DEVELOPMENTAL BIOLOGY. *Educ:* Harvey Mudd Col, BS, 65; Brandeis Univ, MA, 68; Univ Calif, Irvine, PhD(biol), 72. *Prof Exp:* Fel bacteriol, Univ Calif, Los Angeles, 72-74; fel biol, City Hope Med Ctr, Duarte, 74-76; SR RES FEL BIOL, CALIF INST TECHNOL, 76-, DEPT BIOCHEM, UNIV WYO. *Mem:* Genetics Soc; AAAS. *Res:* Gene expression in differentiating Drosophila tissues and the mechanisms by which heat shock interrupts normal gene expression and induces developmental defects. *Mailing Add:* Dept Molecular Biol Univ Wyo Laramie WY 82071

PETERSEN, QUENTIN RICHARD, b Bridgeport, Conn, Mar 10, 24; m 46; c 2. ORGANIC CHEMISTRY. *Educ:* Antioch Col, BS, 48; Northwestern Univ, PhD(chem), 52. *Prof Exp:* Instr org chem, Northwestern Univ, 52; from instr to asst prof, Wesleyan Univ, 52-57; assoc prof, Wabash Col, 57-61, prof, 62-66; prof chem & chmn dept, Simmons Col, 66-69 & Monmouth Col, Ill, 69-73; PROF CHEM & CHMN DEPT, CENT MICH UNIV, 73- *Concurrent Pos:* Consult dir res & develop, Kelite Prod, Inc, 50-55; vis lectr, Trinity Col, Conn, 53-54; Fulbright lectr, Univ Valencia & Univ Barcelona, 61-62; vis scientist, Am Chem Soc, 62-74, nat counr, 73-76; cor res fel, Inst Chem, Acad Sinica, Taiwan, 63-; corp dir, H-C Industs Inc, 65-; consult, NSF-USAID, Karnatak Univ, India, 69; vis prof, Tamkang Univ, Rep China, 79-80, Univ Auckland, 86-87. *Mem:* AAAS; Am Chem Soc; Royal Soc Chem; Am Asn Univ Prof. *Res:* Stereochemistry of alicyclic compounds; derived steroids; napthyridines; corrosion inhibition studies; molecular model design; reaction mechanism; stereochemistry of heterogenous catalysis; irreversible photochemical transformations; herbal medicine; insect hormones. *Mailing Add:* Dept Chem Cent Mich Univ Mt Pleasant MI 48858

PETERSEN, RAYMOND CARL, b Ware, Mass, July 24, 29; m 54; c 2. PHYSICAL CHEMISTRY, ANALYTICAL CHEMISTRY. *Educ:* Amherst Col, BA, 51; Brown Univ, PhD(chem), 56. *Prof Exp:* Chemist, Sprague Elec Co, 56-62, sr res scientist, 62-70, dept head phys chem, 66-69; res scientist, Res Inst Advan Studies, Martin Marietta Corp, 71-73, lab dir & toxicologist, City Baltimore, 73-79; MGR, MAT & CHEM, SOLAREX CORP, 79- *Concurrent Pos:* Ed, The Chesapeake Chemist, 79- *Mem:* Sigma Xi; Am Chem Soc; Electrochem Soc. *Res:* Electrochemistry; physical-organic and analytical chemistry; solar photovoltaics. *Mailing Add:* 9329 Joey Dr Ellicott City MD 21043

PETERSEN, RICHARD RANDOLPH, b Astoria, Ore, Mar 9, 40; m 64. LIMNOLOGY. *Educ:* Univ Wash, BS, 65; Duke Univ, PhD(zool), 70. *Prof Exp:* Res biologist, Rayonier Inc, 65-66; from asst prof to assoc prof, 70-82, PROF BIOL, PORTLAND STATE UNIV, 82- *Mem:* Am Soc Limnol & Oceanog; Am Soc Naturalists; AAAS. *Res:* Ecology of fresh water phytoplankton. *Mailing Add:* Dept of Biol PO Box 751 Portland State Univ Portland OR 97207

PETERSEN, ROBERT J, b Hillsboro, Ore, Sept 29, 37; m 65; c 2. ORGANIC CHEMISTRY. *Educ:* Univ Portland, BS, 59; Pa State Univ, PhD(chem), 65. *Prof Exp:* Sr res chemist, Tape Div, Minn Mining & Mfg Co, 64-66; res chemist, North Star Res & Develop Inst, 66-69, sr chemist, 69-74, assoc dir, Chem Div, 74-75, prin chemist, North Star Div, Midwest Res Inst, 75-77; dir res, 78-84, VPRES SCI & TECHNOL, FILMTEC CORP, 84- *Mem:* AAAS; Am Chem Soc; Am Filtration Soc; N Am Membrane Soc. *Res:* Gas phase reactions of organic halides with potassium vapor; epoxy and polyester thermosetting adhesives; plating of plastics; development of new reverse osmosis, hemodialysis and blood oxygenator membranes. *Mailing Add:* 5936 Emerson Ave S Minneapolis MN 55419

PETERSEN, ROBERT VIRGIL, b South Jordan, Utah, Apr 21, 26; m 50; c 4. PHARMACEUTICAL CHEMISTRY. *Educ:* Univ Utah, BS, 50; Univ Minn, PhD(pharmaceut chem), 55. *Prof Exp:* Asst prof pharmaceut chem, Ore State Col, 55-57; from asst prof to assoc prof pharm, 57-66, chmn dept appl pharmaceut sci, 65-68, PROF PHARM, UNIV UTAH, 66-, CHMN DEPT PHARMACEUT, 78- *Mem:* Am Chem Soc; Am Pharmaceut Asn; Am Asn Cols Pharm (vpres, 71-72, pres, 72-73); Inst Hist Pharm; Sigma Xi. *Res:* Drug stability studies, especially of glycosides; biodegradable drug delivery systems; emulsion formation and stability, especially nonaqueous emulsions; toxicity of plastics. *Mailing Add:* Col Pharm Univ Utah Salt Lake City UT 84112

PETERSEN, ROGER GENE, b Essington, Pa, July 22, 24; m 48; c 1. BIOMETRICS. *Educ:* Iowa State Col, BS, 49, MS, 50; NC State Col, PhD(soil fertil, statist), 54. *Prof Exp:* Asst soil fertil, Iowa State Col, 49-50; asst soil fertil, NC State Col, 50-53, asst statistician, 53-55; assoc prof statist, Ore State Col, 55-62, statistician, Exp Sta, 55-62; from assoc prof to prof statist, NC State Col, 62-65; prof, 65-90, EMER PROF STATIST, ORE STATE UNIV, 90- *Concurrent Pos:* Biometrician, Int Ctr Agr Res Dry Areas, Syria, 78-80, 82-86; mem, Int Gov Coun, 84-87; consult, Nat Agr Res Ctr, Islamabad, Pakistan, 88. *Mem:* Biomet Soc; Western NAm Region (pres, 83); Am Soc Agron. *Res:* Experimental designs for agricultural research. *Mailing Add:* Dept Statist Ore State Univ Corvallis OR 97331

PETERSEN, ULRICH, b Negritos, Peru, Dec 1, 27; m 82. ECONOMIC GEOLOGY. *Educ:* Nxat Sch Eng, Lima, EM, 54; Harvard Univ, MA, 55, PhD(geol), 63. *Prof Exp:* Asst geologist, Geol Inst Peru, 49-50; geologist, Nat Inst Mining Res & Develop, 50-51; asst geologist, Cerro de Pasco Corp, 51, geologist, 51-54, asst chief geologist, 56-57, chief geologist, 58-63, consult geologist, 63; lectr geol, 63-66, assoc prof, 66-69, prof mining geol, 69-81, H C DUDLEY PROF ECON GEOL, HARVARD UNIV, 81- *Concurrent Pos:* Consult geologist. *Mem:* AAAS; Geol Soc Am; Mineral Soc Am; Am Inst Mining, Metall & Petrol Eng; Soc Econ Geol. *Res:* Ore deposition; phase equilibria and geochemistry applied to formation of ore deposits; exploration for economic mineral deposits; mineral economics and policies; zoning of hydothermal ore deposits. *Mailing Add:* Dept Earth & Planetary Sci 310 Hoffman Lab Harvard Univ 20 Oxford St Cambridge MA 02138

PETERSEN, WALLACE CHRISTIAN, b Kansas City, Mo, Feb 10, 43; m 67; c 2. CHEMISTRY. *Educ:* DePauw Univ, BA, 65; Northwestern Univ, Evanston, PhD(chem), 70. *Prof Exp:* Res chemist, 69-80, sr res chemist, 80-90, RES ASSOC, E I DU PONT DE NEMOURS & CO, INC, 90- *Mem:* Am Chem Soc. *Res:* Photochemistry; dye chemistry; organophosphorous chemistry; synthesis. *Mailing Add:* 413 Topsfield Rd Hockessin DE 19707

PETERSON, ALAN HERBERT, b Moline, Ill, Aug 27, 32; m 55; c 2. PETROLEUM CHEMISTRY. *Educ:* Augustana Col, AB, 55; Univ Ill, PhD(org chem), 60. *Prof Exp:* Asst inorg chem, Univ Ill, 55-59; RES CHEMIST, MARATHON OIL CO, 59- *Mem:* Soc Automotive Eng; Am Chem Soc; Sigma Xi; Am Soc Testing & Mat. *Res:* Petroleum refining and petrochemicals. *Mailing Add:* Instrumentation & Chem Dept Marathon Oil Co PO Box 269 Littleton CO 80160

PETERSON, ALAN W, physics, astronomy, for more information see previous edition

PETERSON, ALLEN MONTGOMERY, b Santa Clara, Calif, May 22, 22; m 42; c 4. ELECTRICAL ENGINEERING. *Educ:* Stanford Univ, BS, 48, MS, 49, PhD(elec eng), 52. *Prof Exp:* Res assoc elec eng, Stanford Univ, 52-56, from asst prof to assoc prof, 56-61; asst dir, Electronics & Radiation Div, Stanford Res Inst Int, 62-89; PROF ELEC ENG, STANFORD UNIV, 62- *Concurrent Pos:* Head spec tech group, Stanford Res Inst Int, 53-58, mgr, Commun & Propagation Lab, 59-62; mem bd dirs, Granger Assocs, Palo Alto, 54-, Silicon Gen Corp; mem comn III & IV, US Comn Union Radio Sci Int, 56-; mem panel, NSF, 57-60; consult, Press Sci Adv Comt, Washington, DC, 58-, Adv Res Proj Agency, 58-, Defense Atomic Support Agency, 58-, Inst Defense Analysis, 60- & NAm Aviation, Inc, Calif, 62-63; tech adv, US Deleg to Geneva Conf on Discontinuation Nuclear Tests, 59-; mem sci adv comt, Geophys Inst, Univ Alaska, 60-; mem panel, Adv Group Aeronaut Res & Develop, Avionics, 60-64; chief telecommun, Naval Opers Exec Panel, 70-83; mem panel on SDI, White House Sci Coun; mem, Comt Emerging Space Technol, Bd Army Sci & Technol, Nat Res Coun, 85-88, Navy 21 Study Group, Naval Studies Bd, 86-88; chief scientist, Sci Appln Int, Inc; consult, SRI Int; co-dir, Ctr Radar Astron, Stanford Univ. *Mem:* Nat Acad Eng; fel Inst Elec & Electronic Engrs; Am Geophys Union; Sigma Xi; Int Union Radio Sci. *Res:* Digital signal processing; algorithms; system architecture; very large scale integration design; hardware implementation; communication systems; radar systems; remote sensing systems; space systems; cochlear implants. *Mailing Add:* Dept Elec Eng Stanford Univ Stanford CA 94305

PETERSON, ANDREW R, b Calcutta, India, Nov 29, 42; UK citizen; m 61; c 3. GENETIC TOXICOLOGY. *Educ:* Univ Manchester, Eng, MIB, 68, MSc, 69, PhD(biochem), 71. *Prof Exp:* res assoc prof, Inst Toxicol & Cancer Res Ctr, Univ Southern Calif, 76-89; INSTR, LAUSD, 86- *Concurrent Pos:* lectr, Am Chem Soc. *Mem:* Inst Biol; AAAS; Sigma Xi. *Res:* Mammalian cell mutagenesis; carcinogenesis; toxicology. *Mailing Add:* Inst Toxicol Univ Southern Calif 1985 Zonal Ave Los Angeles CA 90033

PETERSON, ARNOLD (PER GUSTAF), b DeKalb, Ill, Aug 7, 14; m 43; c 3. ELECTRONICS ENGINEERING, ACOUSTICS. *Educ:* Univ Toledo, BEng, 34; Mass Inst Technol, SM, 37, ScD(elec eng), 41. *Prof Exp:* Asst elec eng, Mass Inst Technol, 36-40; develop engr, Gen Radio, Inc, 40-68, eng staff consult, 68-74, sr prin engr, 74-77, staff scientist, 77-79; RETIRED. *Mem:* AAAS; fel Acoust Soc Am (vpres, 58-59); Inst Elec & Electronics Engrs; Audio Eng Soc. *Res:* Acoustical instruments and measurements. *Mailing Add:* 4252 Rockaway Beach Rd NE Bainbridge Island WA 98110-3152

PETERSON, ARTHUR CARL, b Everett, Wash, Mar 30, 23; m 57; c 3. BACTERIOLOGY, IMMUNOLOGY. *Educ:* Univ Minn, BChE, 47, MS, 50, PhD(bact, immunol), 56. *Prof Exp:* Instr quant anal chem, Hamline Col, 49; asst dairy industs, Univ Minn, 50-51 & 55-56, Hormel Inst, 51-55; res technologist, Bact Res Dept, Campbell Soup Co, 56-62, sr res microbiologist, 62-66, div head, Microbiol Res Dept, 66-72, mgr qual control frozen food, 72-73, dir inspection serv, 73-76, dir tech admin labs, 77-88, mgr residue systs, 88-90; CONSULT, PETERSON CONSULT, 90- *Mem:* Am Pub Health Asn; Am Chem Soc; Am Soc Microbiol; Inst Food Technologists; Nat Acad Microbiol. *Res:* Bacterial metabolism and physiology; psychrophilic bacteria and their enzymatic activities; lipid and protein metabolism; protein biochemistry and chromatography; Salmonella epidemiology; public health; food and public health microbiology. *Mailing Add:* 149 Avon Terr Moorestown NJ 08057

PETERSON, ARTHUR EDWIN, b Curtiss, Wis, Mar 11, 23; m 44; c 3. SOILS. *Educ:* Univ Wis, BS, 47, MS, 48, PhD, 50. *Prof Exp:* Exten soils specialist, 50-64, PROF SOIL & WATER CONSERV, UNIV WIS-MADISON, 64- *Concurrent Pos:* Chief resident consult, Rockefeller-Ford Found Coop Maize Improv Prog, Egypt, 66-68, 80; consult, Syria, 79, Indonesia, 82. *Mem:* Fel AAAS; Am Soc Agron; Crop Sci Soc Am; Soil Sci Soc Am; fel Soil Conserv Soc Am. *Res:* Soil tillage and water movement; nutrient and sediment runoff; sludge and waste water renovation. *Mailing Add:* Soils Dept Univ Wis 1525 Observation Dr Madison WI 53706

PETERSON, BARRY WAYNE, b Abington, Pa, June 25, 42; m 67. NEUROPHYSIOLOGY. *Educ:* Calif Inst Technol, BS, 64; Rockefeller Univ, PhD(neurophysiol), 69. *Prof Exp:* Res assoc neurophysiol, Rockefeller Univ, 69-71, from asst prof to assoc prof, 71-80; MEM FAC, DEPT PHYSIOL, MED SCH, NORTHWESTERN UNIV, 80- *Mem:* Soc Neurosci; AAAS. *Res:* Study of central nervous structures involved in motor behavior. *Mailing Add:* Dept Physiol Med Sch Northwestern Univ 303 EChicago Ave Chicago IL 60611

PETERSON, BOBBIE VERN (ROBERT), b Price, Utah, Dec 16, 28; m 50; c 3. ENTOMOLOGY. *Educ:* Univ Utah, BS, 51, MS, 53, PhD(entom), 58. *Prof Exp:* Regist sanitarian, Salt Lake City Bd Health, 51-52; neuropsychiat aide, Vet Admin Hosp, 53-54; teaching asst, Univ Utah, 54-57, lectr, 57-58; entomologist, 58-63, RES SCIENTIST, BIOSYSTS RES INST, CENT EXP FARM, RES BR, CAN AGR, 63- *Mem:* Entom Soc Am; Am Mosquito Control Asn; Soc Syst Zool; Entom Soc Can; Can Soc Zool. *Res:* Systematic entomology; systematics of Diptera. *Mailing Add:* Head Diptera Sect Biosysts Res Inst Cent Exp Farm Can Agr Ottawa ON K2C 3N7 Can

PETERSON, BRADLEY MICHAEL, b Minneapolis, Minn, Nov 26, 51; m 78; c 4. EXTRAGALACTIC ASTRONOMY, QUASARS & ACTIVE GALAXIES. *Educ:* Univ Minn, BPhys, 74; Univ Ariz, PhD(astron), 78. *Prof Exp:* Postdoctoral assoc, Dept Physics, Univ Minn, 79; postdoctoral fel, 79-80, from asst prof to assoc prof, 80-91, PROF ASTRON, OHIO STATE UNIV, 91- *Concurrent Pos:* Prin investr, NSF, 81-, NASA, 86-; dir, Asn Univs Res Astron, 87-90; vis astronr, Lick Observ, 89. *Mem:* Am Astron Soc; Int Astron Union. *Res:* Observational studies of active galactic nuclei; reverberation mapping of line-emitting regions in active nuclei; quasar absorption spectra and applications to observational cosmology. *Mailing Add:* Dept Astron Ohio State Univ 174 W 18th Ave Columbus OH 43210

PETERSON, BRUCE BIGELOW, b Boston, Mass, Mar 26, 35; m 56; c 3. MATHEMATICS. *Educ:* Middlebury Col, BA, 56; Syracuse Univ, MA, 58, PhD(math), 62. *Prof Exp:* Asst math, Syracuse Univ, 56-60, instr, 60-62; from instr to prof, 63-80, chmn dept, 68-80, provost & vpres acad affairs, 85-89, CHARLES A DANA PROF MATH, MIDDLEBURY COL, 80- *Concurrent Pos:* Vis scholar, Univ Wash, 70-71. *Mem:* Am Math Soc; Math Asn Am. *Res:* Convex figures; Euclidean topology and geometry. *Mailing Add:* Dept of Math Warner Sci Bldg Middlebury Col Middlebury VT 05753

PETERSON, BRUCE JON, b Chicago, Ill, Apr 9, 45; c 2. LIMNOLOGY, MARINE ECOLOGY. *Educ:* Bates Col, BS, 67; Cornell Univ, PhD(aquatic ecol), 71. *Prof Exp:* Res assoc limnol, Cornell Univ, 71-74; res assoc marine ecol, NC State Univ, 75-76; asst scientist, 76-80, assoc scientist, 80-86, SR SCIENTIST ECOL, MARINE BIOL LAB, 87- *Mem:* Am Soc Limnol & Oceanog; AAAS. *Res:* Aquatic primary productivity, nutrient cycling, phosphorus cycle, carbon cycle, nitrogen cycle and sulfur cycle. *Mailing Add:* Ecosysts Ctr Marine Biol Lab Woods Hole MA 02543

PETERSON, CHARLES FILLMORE, b Indianapolis, Ind, May 11, 20; m 44; c 4. PHARMACY. *Educ:* Univ Southern Calif, BS, 43; Purdue Univ, MS, 49, PhD(pharm), 52. *Prof Exp:* Asst prof pharm, Univ Kans, 50-56; from assoc prof to prof, 56-88, chmn dept, 59-70, EMER PROF PHARM, SCH PHARM, TEMPLE UNIV, 88- *Concurrent Pos:* Vis prof, Univ Panama, 72. *Mem:* Fel Am Soc Consult Pharmacists. *Res:* Effects of formulation on the biopharmaceutics of drugs; Latin American pharmacy practice; drug information in professional pharmacy practice. *Mailing Add:* Temple Univ Sch Pharm 3307 N Broad St Philadelphia PA 19140

PETERSON, CHARLES HENRY, b Lawrenceville, NJ, Feb 18, 46; m 72; c 2. MARINE ECOLOGY, POPULATION BIOLOGY. *Educ:* Princeton Univ, AB, 68; Univ Calif, Santa Barbara, MA, 70, PhD(biol), 72. *Prof Exp:* Instr biol, Univ Calif Exten, 70-72; teaching assoc biol sci, Univ Calif, Santa Barbara, 71-72; asst prof biol sci, Univ Md, 72-76; assoc prof, 76-82, PROF MARINE SCI, UNIV NC, CHAPEL HILL, 83- *Concurrent Pos:* Fel, Ford Found grant, 72; biol oceanog panel, NSF, 80, 85-87; vis res fel, Univ Western Australia, 83; Japan Soc Promotion Sci fel, Univ Nageseki, 90. *Mem:* Am Inst Biol Sci; Ecol Soc Am; Paleontol Soc Am; Paleontol Res Inst; Sigma Xi; Brit Ecol Soc; Am Soc Limol Oceanogs; Int Asn Ecol. *Res:* Population biology and community ecology, especially of marine benthic invertebrates and barrier island plants; invertebrate fisheries biology. *Mailing Add:* Inst Marine Sci Univ NC-Chapel Hill Morehead City NC 28557

PETERSON, CHARLES JOHN, b Seattle, Wash, Oct 13, 45; m 67; c 1. GALAXIES, GLOBULAR CLUSTERS. *Educ:* Univ Wash, BS, 67; Univ Calif, Berkeley, MA, 68, PhD(astron), 75. *Prof Exp:* Carnegie fel astron, Carnegie Inst Wash, 74-76; res assoc, Cerro Tololo Inter-Am Observ, 76-78; asst prof, 78-83, ASSOC PROF ASTRON, DEPT PHYSICS, UNIV MO-COLUMBIA, 83- *Mem:* Am Astron Soc; Astron Soc Pac; Int Astron Union; Sigma Xi. *Res:* Globular clusters; structure and dynamics of galaxies. *Mailing Add:* Dept Physics Univ Mo Columbia MO 65211

PETERSON, CHARLES LESLIE, b Bradner, Ohio, Dec 23, 24; m 48; c 3. PHYSICAL CHEMISTRY, CORROSION. *Educ:* Bowling Green State Univ, BA, 48; Ohio State Univ, MSc, 50. *Prof Exp:* Res chemist, Diamond Alkali Co, 50-53; sr res chemist, Battelle Mem Inst, 53-63; proj mgr mat sci, 63-75, staff mem high temp chem, 75-81, STAFF MEM WEAPON SUB SYSTS, LOS ALAMOS SCI LAB, UNIV CALIF, 81- *Res:* Kinetics of metal-gas reactions; corrosion of metals and materials science; development of thermochemical processes for hydrogen production. *Mailing Add:* Group WX-5 MS-780 Los Alamos Nat Lab PO Box 1663 Los Alamos NM 87545

PETERSON, CHARLES MARQUIS, b New York, NY, Mar 8, 43; m 77. DIABETES, HEMOGLOBIN. *Educ:* Carleton Col, BA, 65; Columbia Univ Sch Physicians & Surgeons, MD, 69. *Prof Exp:* Intern, Harlem Hosp, 69-70, residency, 70-72, chief researcher, 72-73; guest investr, biochem, Rockefeller Univ, 72-73, asst prof, 73-78, assoc prof med biochem, 78-; DIR RES, SANSUM MED RES FOUND. *Mem:* Am Soc Clin Invest; Am Soc Pharmacol; Am Soc Hematol; Soc Exp Biol & Med; Am Diabetic Asn. *Res:* Diabetes mellitus; sickle cell disease; nephrolithiasis; basic anac of metabolism. *Mailing Add:* Sansum Med Res Found 2219 Bath St Santa Barbara CA 93105

PETERSON, CLARE GRAY, b Scobey, Mont, Nov 24, 17; m 41; c 4. SURGERY, PHYSIOLOGY. *Educ:* Univ Ore, BA, 39, MD, 43, MS, 45; Am Bd Surg, dipl, 51. *Prof Exp:* Instr physiol, 44-45, from instr to asst prof, 48-49, from instr to assoc prof, 48-58, PROF SURG, MED SCH, UNIV ORE, 58-, CHIEF SURG SERV, UNIV HOSP, 65- *Concurrent Pos:* Sr surg consult, Vet Admin Hosp, 65- *Mem:* AAAS; fel Am Col Surgeons; NY Acad Sci. *Mailing Add:* Dept Surg Univ Ore Health Sci Portland OR 97201

PETERSON, CLARENCE JAMES, JR, b Park City, Utah, Aug 23, 28; m 56; c 3. AGRONOMY. *Educ:* Univ Idaho, BS, 56, MS, 59; Ore State Univ, PhD(plant breeding), 70. *Prof Exp:* Res agronomist, wheat, genetics, qual, Physiol & Dis Res, Agr Res Serv, USDA, 59-87; AGRONOMIST, WASH STATE UNIV, 88- *Mem:* Am Soc Agron; Crop Sci Soc Am. *Res:* Wheat improvement; development of new winter wheat varieties for the Pacific Northwest. *Mailing Add:* 209 Johnson Hall Wash State Univ Pullman WA 99164

PETERSON, CLINTON E, horticulture; deceased, see previous edition for last biography

PETERSON, CURTIS MORRIS, b Fargo, NDak, Jan 16, 42; m 64; c 3. BOTANY. *Educ:* Moorhead State Univ, BS, 66; Univ Ore, PhD(biol), 70. *Prof Exp:* NDEA fel biol, Univ Ore, 66-69, teaching fel, 69-70; from asst prof to assoc prof, 76-84, PROF BOT, AUBURN UNIV, 84- *Concurrent Pos:* Plant physiologist, USDA Exp Sta, Pendleton, Ore, 79-80. *Mem:* Bot Soc Am; Am Soc Plant Physiologists; Agron Soc; Sigma Xi; Crop Sci Soc. *Res:* Ultrastructure, physiological and environmental factors affecting reproductive development in soybeans, Glycine max, particularly the factors causing abscission of flowers and pods; abiotic, biotic and symbiotic factors influencing the rhizosphere. *Mailing Add:* Dept Bot & Microbiol Auburn Univ Auburn AL 36849-5407

PETERSON, CYNTHIA WYETH, b Philadelphia, Pa, Apr 28, 33; m 57; c 2. CONDENSED MATTER, ELECTRONIC PROPERTIES. *Educ:* Bryn Mawr Col, BA, 54; Cornell Univ, PhD(exp physics), 64. *Prof Exp:* Res asst atmospheric physics, Harvard Univ, 60-62; from instr to asst prof physics, Wesleyan Univ, 63-66; spec fel, Yale Univ, 66-68; from asst prof to assoc prof, 68-83, PROF PHYSICS, UNIV CONN, 83- *Concurrent Pos:* Vis assoc prof, Dept Molecular Biophysics, Yale Univ, 78-79; Am Asn Univ Women fel, Marie Curie Endowment; vis prof, Dept Molecular Biophysics, Yale Univ, 87-88. *Mem:* AAAS; Am Asn Physics Teachers; Am Astron Soc. *Res:* Optical and electrical properties of condensed matter: semiconductors, biomaterials and alloys; ultraviolet photoelectric emission; vacuum ultraviolet spectroscopy; metalloenzymes; spectroscopy of rare gas ion-atom collisions. *Mailing Add:* Shipyard Rd Middle Haddam CT 06456

PETERSON, DALLAS ODELL, b Oakley, Idaho, Aug 17, 25; m 51; c 1. GEOLOGY. *Educ:* Brigham Young Univ, BS, 52, MS, 53; Wash State Univ, PhD, 59. *Prof Exp:* Asst geol, Wash State Univ, 54-57; asst prof, Chico State Col, 57-60; from asst prof to prof, Weber State Col, 60-69, chmn dept, 68, 69; dir acad prog develop & asst to vpres, 69-71, asst vpres acad affairs, 71-72, ASSOC VPRES ACAD AFFAIRS, SYST ADMIN, UNIV WIS-MADISON, 72- *Concurrent Pos:* Field geologist, Shell Oil Co, summers, 53, 56, 58-62; eng geologist, US Forest Serv, summers 66 & 67. *Mem:* Soc Econ Paleont & Mineral; Am Asn Petrol Geol. *Res:* Upper Paleozoic stratigraphy in the Rocky Mountain area. *Mailing Add:* Univ Wis 1544 Van Hise Madison WI 53706

PETERSON, DARRYL RONNIE, b Oak Park, Ill, Aug 25, 43; m 68; c 5. CELL PHYSIOLOGY, CELL BIOLOGY. *Educ:* Wheaton Col, Ill, BS, 66; Southern Ill Univ, MA, 69; Univ Ill, PhD(physiol), 73. *Prof Exp:* Asst prof physiol, Northwestern Univ Med Sch, 74-80; assoc prof physiol, Univ Ill Col Med, 80-83; assoc prof, 83-89, PROF PHYSIOL, CHICAGO MED SCH, 89-, VCHMN, 88- *Mem:* Am Physiol Soc; Am Soc Cell Biologists. *Res:* Regulation of the Na/H antiporter; regulation of cellular growth; mechanism of nephrotoxicity; peptide biology. *Mailing Add:* Chicago Med Sch UHS 3333 Green Bay Rd North Chicago IL 60064

PETERSON, DARWIN WILSON, b Redmond, Utah, Mar 23, 38; m 59; c 5. CARDIOVASCULAR PHYSIOLOGY. *Educ:* Univ Nev, BS, 66, MS, 67; Univ Ala, Birmingham, PhD(physiol), 72. *Prof Exp:* Fel physiol, Sch Med, Univ Ala, Birmingham, 72-73; ASST PROF PHYSIOL, BOWMAN GRAY SCH MED, WAKE FOREST UNIV, 73- *Concurrent Pos:* NIH trainee, Sch Med, Univ Ala, Birmingham, 68-72; NIH fel, Heart, Lung & Blood Inst, 72-73. *Mem:* Sigma Xi; Am Physiol Soc. *Res:* Pathophysiology of dietary lipids, diabetes and hypertension. *Mailing Add:* 432 Cages Bend Rd Gallatin TN 37066

PETERSON, DAVID ALLAN, b Hayward, Wis, Nov 29, 38; m 64; c 2. VIROLOGY, MICROBIOLOGY. *Educ:* Wis State Univ-Stevens Point, BS, 66; Ind Univ, MS, 70, PhD(microbiol), 71. *Prof Exp:* Asst prof microbiol, Rush Med Col, 72-81; asst scientist microbiol & chief diag virol, Rush-Presby-St Luke's Med Ctr, 72-81; clin projs mgr, 81-82, sr virologist, 82-89, RES INVESTR, ABBOTT LABS, INC, NORTH CHICAGO, ILL, 89- *Concurrent Pos:* Fel microbiol, Rush-Presby-St Luke's Med Ctr, 71-72, Nat Inst Allergy & Infectious Dis fel, 71-78; JJ Reingold Trust for congenital Rubella studies, 72-81. *Mem:* AAAS; Am Soc Microbiol; Soc Exp Biol Med; Sigma Xi. *Res:* Virologic aspects of hepatitis; congenital Rubella and slow degenerative diseases of the central nervous system; viral diagnostic systems. *Mailing Add:* Dept 90D Abbott Labs Inc North Chicago IL 60064

PETERSON, DAVID MAURICE, b Woodward, Okla, July 3, 40; m 65; c 2. PLANT PHYSIOLOGY. *Educ:* Univ Calif, Davis, BS, 62; Univ Ill, Urbana, MS, 64; Harvard Univ, PhD(biol), 68. *Prof Exp:* Res biologist, Allied Chem Corp, 70-71; PLANT PHYSIOLOGIST, CEREAL CROPS RES UNIT, AGR RES SERV, USDA, 71-, RES LEADER, 83- *Concurrent Pos:* Asst prof, Univ Wis-Madison, 71-75, assoc prof, 75-80, prof, 80-; assoc ed, Crop Sci, 75-78; assoc ed, Cereal Chem, 88- *Mem:* FEL AAAS; Am Soc Plant Physiol; Am Soc Agron; Crop Sci Soc Am; Am Asn Cereal Chemists. *Res:* Protein synthesis; nitrogen metabolism; plant physiology as related to quality improvement in oats and barley; seed development. *Mailing Add:* Dept of Agron Univ of Wis Madison WI 53706

PETERSON, DAVID OSCAR, b Portland, Ore, June 22, 50. GENE EXPRESSION, MOLECULAR GENETICS. *Educ:* Pomona Col, BA, 72; Harvard Univ, PhD(chem), 78. *Prof Exp:* Fel, Dept Biochem & Biophysics, Univ Calif, San Francisco, 78-81; asst prof, 81-87, ASSOC PROF, DEPT BIOCHEM & BIOPHYSICS, TEX A&M UNIV, 87- *Mem:* Am Soc Biochem & Molecular Biol; Am Soc Microbiol. *Res:* Transcriptional regulation of eukaryotic genes; glucocorticoid-mediated regulation of mouse mammary tumor virus gene expression. *Mailing Add:* Dept Biochem & Biophysics Tex A&M Univ Col Sta TX 77843

PETERSON, DAVID T, b Blue Earth, Minn, Nov 29, 22; m 45; c 2. METALLURGY. *Educ:* Iowa State Univ, BS, 47, PhD(phys chem), 50. *Prof Exp:* Prof metall, Iowa State Univ & sr metallurgist, Ames Lab, 50-87; RETIRED. *Mem:* Am Soc Metals; Am Inst Mining, Metall & Petrol Engrs. *Res:* Chemical metallurgy; kinetics and equilibria of metallic reactions. *Mailing Add:* 405 24th Ames IA 50010

PETERSON, DAVID WEST, b Schenectady, NY, Sept 3, 40; m 61; c 2. STATISTICAL ANALYSIS. *Educ:* Univ Wis, BS, 62; Stanford Univ, MS, 63, PhD(elec eng), 65. *Prof Exp:* Mem tech staff, Ground Systs Group, Hughes Aircraft Co, 62-63; res asst systs group, Stanford Univ, 63-65; from asst prof to assoc prof quant methods, Grad Sch Mgt, Northwestern Univ, Evanston, 67-73; prof bus admin, Fuqua Sch Bus, Duke Univ, 73-84, sr lectr, Duke Law Sch, 82-86, adj prof, Fuqua Sch Bus, 84-89, ADJ PROF, INST STATIST & DECISION SCI, DUKE UNIV, 87-; PRES, PRI ASSOCS, INC, 79- *Concurrent Pos:* Consult, US Army Res Off, NC, 67-68 & USPHS, 70-71; res fel, Int Inst Mgt, Berlin, 71-72; vis lectr systs eng, Univ Ill, Chicago Circle, 73; statist consult, var corps, law firms & govt units, 74- *Mem:* Am Statist Asn; Economet Soc; Inst Elec & Electronic Eng; Inst Mgt Sci. *Res:* Statistical methods in litigation; applications of mathematical models to socio-economic systems; statistical frameworks for determining the fairness of employment practices; methods for detecting computer software copying. *Mailing Add:* 2738 Sevier St Durham NC 27705

PETERSON, DEAN EVERETT, b Aledo, Ill, Apr 14, 41; m 66; c 2. HIGH TEMPERATURE CHEMISTRY. *Educ:* Monmouth Col, BA, 64; Univ Kans, PhD(phys chem), 72. *Prof Exp:* Mem staff chem, Argonne Nat Lab, 63-64 & Savannah River Lab, 66-67; mem staff chem, 72-84, dep group leader, 86-89, CHEM SECT LEADER, LOS ALAMOS SCI LAB, 84- *Concurrent Pos:* Mem staff, Explor Res & Develop Ctr, 88- *Honors & Awards:* Medal, Am Inst Chemists, 64. *Mem:* Am Chem Soc; Sigma Xi. *Res:* Thermodynamics, kinetics, vaporization processes, vapor pressures and mass spectrometry; high temperature superconductors. *Mailing Add:* MST-5 G730 Los Alamos Nat Lab Los Alamos NM 87545

PETERSON, DEAN F(REEMAN), JR, civil & agricultural engineering; deceased, see previous edition for last biography

PETERSON, DONALD BRUCE, b Erie, Pa, Dec 16, 31. PHYSICAL CHEMISTRY. *Educ:* Pa State Univ, BS, 54; Carnegie Inst Technol, MS, 57, PhD(chem), 58. *Prof Exp:* NSF fel radiation chem, Univ Leeds, 58-60; asst prof chem, McMaster Univ, 60-61; res scientist, Univ Notre Dame, 61-64; PROF CHEM, UNIV SAN DIEGO, 64- *Concurrent Pos:* NSF fel, CA State Energy Comn, 77-78; staff, House Subcomt Fossil & Synthetic Fuels, Washington, DC, 84-85. *Mem:* AAAS; Am Chem Soc. *Res:* Photochemistry, radiation chemistry, energy technologies. *Mailing Add:* Dept Chem Univ San Diego Alcala Park San Diego CA 92110-2492

PETERSON, DONALD FREDERICK, physiology, for more information see previous edition

PETERSON, DONALD I, b Moscow, Idaho, July 26, 22; m 42; c 3. NEUROLOGY, PHARMACOLOGY. *Educ:* Walla Walla Col, BA, 44; Loma Linda Univ, MD, 47. *Prof Exp:* Med dir, Jengre Mission Hosp, 54-60; from asst prof to assoc prof neurol, 69-82, ASSOC PROF PHARMACOL, 60-, PROF NEUROL, LOMA LINDA UNIV HOSP, 82- *Mem:* Sigma Xi; Am Acad Neurol; fel Am Col Physicians. *Res:* Toxicity; metabolism; movement disorders. *Mailing Add:* 24418 Lawton Ave Loma Linda CA 92354

PETERSON, DONALD J, b Ladysmith, Wis, Nov 19, 35; m 57; c 3. ORGANIC CHEMISTRY. *Educ:* Wis State Univ, Superior, BS, 57; Iowa State Univ, PhD(org chem), 62. *Prof Exp:* DIR, CORP RES DIV, MIAMI VALLEY LABS, PROCTER & GAMBLE CO, 62- *Honors & Awards:* C D Hurd Lectr, N Western Univ, 83. *Mem:* Am Chem Soc. *Res:* Organometallic chemistry, especially organolithium, organosilicon, organophosphorus and organotin chemistry. *Mailing Add:* Miami Valley Labs Procter & Gamble Co PO Box 39175 Cincinnati OH 45247

PETERSON, DONALD LEE, b Reno, Nev, Apr 6, 30; m 52; c 3. PHYSICAL CHEMISTRY. *Educ:* Univ Nev, BS, 52; Univ Wash, PhD(chem), 56. *Prof Exp:* Chemist, Shell Develop Co, 56-62; NSF fel, Imp Col, Univ London, 62-63; chemist, Shell Develop Co, 63-66; assoc prof, 66-71, chmn dept, 76-82, PROF CHEM, CALIF STATE UNIV, HAYWARD, 71- *Concurrent Pos:* Vis prof chem, Univ Newcastle, NSW, 72-73. *Mem:* Am Chem Soc. *Res:* Adsorption and diffusion in zeolites; theory of gas chromatography; capillary chromatography; GC/MS analysis. *Mailing Add:* Dept Chem Calif State Univ Hayward CA 94542

PETERSON, DONALD NEIL, b Detroit, Mich, Oct 14, 41; m 68. GEOPHYSICS. *Educ:* Wittenberg Univ, BS, 63; Ind Univ, Bloomington, MA, 65; Ohio State Univ, PhD(geol & geophys), 69. *Prof Exp:* Res assoc glaciol & geophys, Inst Polar Studies, Ohio State Univ, 64-68; res assoc geophys and paleomagnetism, Case Western Reserve Univ, 69-70; asst prof, State Univ NY Col Fredonia, 70-75, assoc prof, 75-; MEM STAFF, EXXON PROD RES. *Concurrent Pos:* NSF grant, Case Western Reserve Univ, 69-70; vis assoc, Calif Inst Technol, 78. *Mem:* Am Geophys Union; Geol Soc Am; Soc Explor Geophysicists. *Res:* Paleomagnetism and continental drift; establishment of paleomagnetic stratigraphic sequence for Paleozoic time. *Mailing Add:* Stratig Explor Div Exxon Prod Res Co PO Box 2189 Houston TX 77252-2189

PETERSON, DONALD PALMER, b Bremerton, Wash, June 27, 29; m 52; c 4. MATHEMATICS. *Educ:* Wash State Univ, BA, 52, MA, 54; Univ Ore, PhD(math), 57. *Prof Exp:* Asst, Wash State Univ, 52-54; asst, Univ Ore, 54-55, instr, 56-57; mem staff, Sandia Corp, 57-61; eng specialist, Gen Tel & Electronics Labs, Inc, 62-63; mem staff, 63-68, supvr, 68-73, mem tech staff, 73-83, DISTINGUISHED MEM TECH STAFF, SANDIA NAT LAB, 83- *Concurrent Pos:* Adj prof, Univ NMex, 87- *Mem:* Soc Indust & Appl Math; Oper Res Soc Am; Asn Comput Mach. *Res:* Computer science; computer-aided design; computer representation of solids; computational geometry. *Mailing Add:* Sandia Nat Lab Org 2814 Albuquerque NM 87185

PETERSON, DONALD RICHARD, b Portland, Ore, Jan 12, 21; m 45; c 4. EPIDEMIOLOGY. *Educ:* Ore State Univ, BA, 44; Univ Ore, MS, 46, MD, 47; Univ Calif, Berkeley, MPH, 57. *Prof Exp:* Dir adult health div, Dept Pub Health, Seattle, King County, Wash, 57-61; dir epidemiol div, 61-71, PROF EPIDEMIOL & CHMN DEPT, SCH PUB HEALTH & COMMUNITY MED, UNIV WASH, 71- *Concurrent Pos:* WHO traveling fel, 71. *Mem:* Int Epidemiol Asn; Soc Epidemiol Res; Am Epidemiol Soc. *Res:* Sudden infant death syndrome; disease seasonality. *Mailing Add:* Pub Health Univ Wash Sch Comm Med Seattle WA 98195

PETERSON, DONALD WILLIAM, b San Francisco, Calif, Mar 3, 25; m 48; c 3. GEOLOGY. *Educ:* Calif Inst Technol, BS, 49; Wash State Univ, MS, 51; Stanford Univ, PhD, 61. *Prof Exp:* Asst, Wash State Univ, 49-51; GEOLOGIST, US GEOL SURV, 52- *Concurrent Pos:* Asst, Stanford Univ, 55-56; scientist-in-chg, Hawaiian Volcano Observ, US Geol Surv, 70-75, Cascades Volcano Observ, 80-85. *Honors & Awards:* Meritorious Serv Award, US Dept Interior, 83. *Mem:* AAAS; fel Geol Soc Am; Am Geophys Union. *Res:* Geology and petrology of volcanic rocks; geology and volcanology of Hawaiian and cascade volcanoes. *Mailing Add:* US Geol Surv MS-910 345 Middlefield Rd Menlo Park CA 94025

PETERSON, EARL ANDREW, b Puyallup, Wash, Jan 8, 40. HIGH ENERGY PHYSICS. *Educ:* Univ Wash, BA, 62; Stanford Univ, MA, 67, PhD(physics), 68. *Prof Exp:* Res assoc, 67-73, asst prof, 73-78, ASSOC PROF PHYSICS, UNIV MINN, 78- *Mem:* Am Phys Soc. *Res:* Experimental high energy physics. *Mailing Add:* Sch Physics Univ Minn 116 Church St SE Minneapolis MN 55455

PETERSON, EDWARD CHARLES, b Duluth, Minn, Apr 12, 29; m 58; c 4. PHYSICS. *Educ:* Univ Minn, Duluth, BA, 51; Univ Wis, MS, 56; Mich State Univ, MA, 63. *Prof Exp:* From assoc physicist to physicist, 58-63, from assoc res physicist to res physicist, 63-74, SR RES PHYSICIST, WHIRLPOOL CORP, 74- *Honors & Awards:* Elisha Gray II Award, Sigma Xi, 84. *Mem:* Am Asn Physics Teachers; Am Vacuum Soc; Sigma Xi; Electrostatic Soc Am. *Res:* Electronics; mechanics; high vacuum techniques; instrumentation; particle accelerators; textile test work; development of test methods and instrumentation for evaluating home appliances; electrostatics; ultrasonics. *Mailing Add:* Whirlpool Corp Res Lab Monte Rd Benton Harbor MI 49022

PETERSON, ELBERT AXEL, b Chicago, Ill, June 16, 18; m 43; c 2. BIOCHEMISTRY. *Educ:* Univ Chicago, BS, 41; Univ Calif, PhD(biochem), 51. *Prof Exp:* Chemist, Shell Develop Co, 41-46; fel, 50-52, CHEMIST, NAT CANCER INST, 52- *Mem:* Am Chem Soc; Am Soc Biol Chemists. *Res:* Biosynthesis of proteins; chromatography of proteins; maturation of leukocytes. *Mailing Add:* 4405 Cambria Ave Garrett MD 20896

PETERSON, ELLENGENE HODGES, b Abilene, Tex, Oct 20, 40. NEUROANATOMY, NEUROBIOLOGY. *Educ:* Radcliffe Col, BA, 62; Calif State Univ, Los Angeles, MA, 70; Univ Calif, Riverside, PhD(psychol), 76. *Prof Exp:* Fel neuroanat, Univ Calif, San Diego, 75-76 & Sch Med, Univ PR, 76-77; res assoc neuroanat, Univ Chicago, 77-80; MEM FAC, SCH ANAT, UNIV NEW SOUTH WALES, 80- *Concurrent Pos:* Alfred P Sloan fel, 75-77. *Mem:* Soc Neurosci; Am Soc Zool; AAAS; Sigma Xi. *Res:* Neural control of head movement. *Mailing Add:* Dept Zool Ohio S Univ Athens OH 45701

PETERSON, ELMOR LEE, b McKeesport, Pa, Dec 6, 38; m 66; c 2. MATHEMATICS, OPERATIONS RESEARCH. *Educ:* Carnegie Inst Technol, BS, 60, MS, 61, PhD(math), 64. *Prof Exp:* Sr mathematician, Westinghouse Res & Develop Ctr, 63-66; asst prof math, Univ Mich, 67-69; assoc prof math & indust eng/mgt sci, Northwestern Univ, 69-73, prof, 73-79; PROF MATH & INDUST ENG, NC STATE UNIV, 79- *Concurrent Pos:* Lectr, Carnegie Inst Technol, 63-66; vis assoc prof math res ctr, Univ Wis, 68-69; Air Force Off Sci Res grants, 73-77; vis prof oper res, Stanford Univ, 76-77; NSF res grant, 85-87. *Mem:* Am Math Soc; Soc Indust & Appl Math; Math Asn Am; Opers Res Soc Am. *Res:* Geometric programming with applications to optimal engineering design, location and resource allocation; regression analysis; structural analysis and optimization; network analysis and design and variational analysis in the physical sciences and economics. *Mailing Add:* 3717 Williamsbourgh Ct NC State Univ Raleigh NC 27609

PETERSON, ERNEST A, b New York, NY, June 16, 31; m 52, 87; c 2. COMPUTERS IN MEDICINE, ACOUSTICS. *Educ:* Rutgers Univ, BA, 59; Princeton Univ, MA, 61, PhD(psychol), 62. *Prof Exp:* Asst prof, 64-70, ASSOC PROF OTOLARYNGOL, ANESTHESIOL, PSYCHOL & BIOMED ENG, SCH MED, UNIV MIAMI, 70- *Concurrent Pos:* NIH fel sensory psychol, Princeton Univ, 62-64. *Mem:* Acoust Soc Am. *Res:* Computerized medical records and computer aided instruction; auditory evoked responses; non-auditory effects of noise. *Mailing Add:* Med Computer Systs Lab D-55 PO Box 016960 Miami FL 33101

PETERSON, ERNEST W, b Long Beach, Calif, Dec 10, 38; div; c 2. METEOROLOGY. *Educ:* Univ Calif, Los Angeles, BA, 62; Pa State Univ, PhD(meteorol), 69. *Prof Exp:* from asst prof to assoc prof, 69-82, ASSOC COURTESY PROF ATMOSPHERIC SCI, ORE STATE UNIV, 83- *Concurrent Pos:* Res assoc, Risoe Nat Lab, Roskilde, Denmark, 73, adj scientist, 73-80, vis scientist, 82-83 & 90; res meteorologist, Corvallis Environ Res Lab, Environ Protection Agency, 75-81; vis prof, Inst fir Meereskunde, Univ Kiel, Ger, 85; vis lectr, Linfield Col, Ore, 89. *Mem:* Sigma Xi. *Res:* Boundary layer meteorology; climatology. *Mailing Add:* 1410 SW 53rd Corvallis OR 97333

PETERSON, EUGENE JAMES, b Evergreen Park, Ill, Nov 18, 49; m 79. INORGANIC CHEMISTRY. *Educ:* St Procopius Col, BS, 71; Ariz State Univ, PhD(inorg chem), 76. *Prof Exp:* Staff mem, Los Alamos Sci Lab, 76-78; staff mem, Argonne Nat Lab, 78-79; STAFF MEM, LOS ALAMOS NAT LAB, 79- *Mem:* Am Chem Soc; AAAS. *Res:* Identification of water quality concerns associated with shale oil extraction and research of strategies for mitigation of health and environmental impacts. *Mailing Add:* Los Alamos Nat Lab PO Box 1663 Inc DO MSJ 519 Los Alamos NM 87544

PETERSON, FRANCIS CARL, b Brooklyn, NY, Dec 13, 42; m 66; c 3. PHYSICS. *Educ:* Rensselaer Polytech Inst, BEE, 64; Cornell Univ, PhD(physics), 68. *Prof Exp:* From assoc physicist to physicist, Ames Lab, 68-74, from asst prof to assoc prof, 68-90, PROF PHYSICS, IOWA STATE UNIV, 90- *Mem:* Am Asn Physics Teachers. *Res:* Educational development research. *Mailing Add:* Dept of Physics Iowa State Univ Ames IA 50011

PETERSON, FRANK LYNN, b Klamath Falls, Ore, May 8, 41; m 67. GROUNDWATER GEOLOGY, ENGINEERING GEOLOGY. *Educ:* Cornell Univ, BA, 63; Stanford Univ, MS, 65, PhD(geol), 67. *Prof Exp:* Asst prof geol & asst geologist, 67-71, assoc prof geol & assoc geologist, Water Resources Res Ctr, 71-76, PROF GEOL & HYDROL & HYDROLOGIST, RES CTR, UNIV HAWAII, 76-, GEOL & GEOPHYSICS DEPT CHMN, 83- *Concurrent Pos:* Consult hydrologist, Hawaii & Pacific Basin. *Mem:* Am Water Resources Asn; fel Geol Soc Am; Am Geophys Union; Nat Water Well Asn; Asn Eng Geol; Am Inst Hydrol; Int Asn Hydrologists. *Res:* Hydrologic cycle; occurrence of and exploration for groundwater; groundwater occurrence in volcanic rocks; salt water intrusion; fluid flow and groundwater storage; groundwater contamination; engineering geology problems. *Mailing Add:* Dept Geol & Geophys Univ Hawaii Honolulu HI 96822

PETERSON, FRANKLIN PAUL, b Aurora, Ill, Aug 27, 30; m 59. MATHEMATICS. *Educ:* Northwestern Univ, BS, 52; Princeton Univ, PhD, 55. *Prof Exp:* NSF fel, Univ Chicago, 55-56; Higgins lectr math, Princeton Univ, 56-58; from asst prof to assoc prof, 58-65, PROF MATH, MASS INST TECHNOL, 65- *Concurrent Pos:* Sloan fel, Oxford Univ, 60-61; Fulbright res grant, Kyoto Univ, 67; chmn comt pure math, Mass Inst Technol. *Mem:* Am Math Soc (treas, 73-). *Res:* Algebraic topology, particularly homotopy and cohomology theory. *Mailing Add:* Dept Math Mass Inst Technol Cambridge MA 02139

PETERSON, FRED, b St Johns, Mich, June 29, 33. GEOLOGY. *Educ:* San Diego State Univ, BS, 60; Stanford Univ, PhD(geol), 69. *Prof Exp:* GEOLOGIST, US GEOL SURV, 60- *Mem:* Am Asn Petrol Geol; Geol Soc Am; Soc Econ Paleont & Mineral; Int Asn Sedimentologists. *Res:* Stratigraphy; sedimentation; paleotectonics. *Mailing Add:* US Geol Surv Mail Stop 939 Box 25046 Denver CO 80225

PETERSON, FREDERICK FORNEY, b Madison, Wis, Dec 21, 28; m 73; c 4. SOIL CLASSIFICATION, SOIL GENESIS. *Educ:* Univ Wis, BS, 50; Cornell Univ, MS, 53; Wash State Univ, PhD(soils), 61. *Prof Exp:* Res asst soil surv, Wis Natural Hist & Geol Surv & Univ Wis, 48-49; res asst, Cornell Univ, 50-52; soil genesis, Wash State Univ, 56-59; res soil scientist, Soil Conserv Serv, USDA, 59-62; asst prof soil morphol & classification & asst chemist, Univ Calif, Riverside, 62-67; assoc prof, 67-73, PROF, PLANT, SOIL & WATER SCI DIV, UNIV NEV, RENO, 73- *Mem:* Soil Sci Soc Am; Soc Range Mgt; Sigma Xi. *Res:* Soil classification and survey; geomorphology-soils interrelations. *Mailing Add:* 1430 Peavine Rd Reno NV 89503

PETERSON, GARY A, b Holdrege, Nebr, Apr 30, 40; m 65; c 2. SOIL SCIENCE, STATISTICS. *Educ:* Univ Nebr, BS, 63, MS, 65; Iowa State Univ, PhD(agron), 67. *Prof Exp:* Assoc prof soil sci, Univ Nebr, Lincoln, 67-74, prof, 67-; AT DEPT AGRON, COLO STATE UNIV. *Concurrent Pos:* Vis prof, Colo State Univ, 80-81. *Honors & Awards:* CIBA-Geigy Award, Am Soc Agron, 74; Appl Res Award, Soil Sci, 87. *Mem:* fel Am Soc Agron; fel Soil Sci Soc Am; Soil Conserv Soc Am; Sigma Xi. *Res:* Soil-plant relationships and management of soils under semiarid dry land conditions, particularly effects of cultural practices on soil nitrogen and phosphorus chemistry and fertility. *Mailing Add:* Dept Agron Colo State Univ Ft Collins CO 80523

PETERSON, GARY LEE, b Fargo, NDak, June 24, 36. GEOLOGY. *Educ:* Univ Colo, BA, 59; Univ Wash, MS, 61, PhD(geol), 63. *Prof Exp:* From asst prof to assoc prof, 63-69, chmn dept, 73-76, PROF GEOL, SAN DIEGO STATE UNIV, 69-, ACTG CHMN GEOL DEPT, 84- *Concurrent Pos:* Vis prof geol, Univ Mont, 77. *Mem:* AAAS; Am Asn Petrol Geol; Geol Soc Am; Soc Econ Paleont & Mineral; Sigma Xi. *Res:* Cretaceous stratigraphy of the West Coast; late Mesozoic and Cenozoic stratigraphy of southwestern California and northwestern Baja California; Neogene stratigraphy of Imperial Valley, California; Phanerozoic paleoclimatology of North America. *Mailing Add:* Dept of Geol Sci San Diego State Univ San Diego CA 92182

PETERSON, GEORGE EARL, b Pittsburgh, Pa, June 7, 34. PHYSICS. *Educ:* Univ Pittsburgh, BS, 56, PhD(physics), 61. *Prof Exp:* Res asst physics, Univ Pittsburgh, 56-61; MEM STAFF CRYSTAL PHYSICS, BELL LABS, 61- *Mem:* Optical Soc Am; Am Phys Soc; Am Crystallog Asn; Am Asn Crystal Growth; Am Ceramic Soc. *Res:* Laser materials research; rare earth fluorescence; nuclear magnetic resonance and nuclear quadrupole resonance in ferroelectrics and other nonlinear dielectric materials. *Mailing Add:* Bell Labs Crystal Chem Res Dept Murray Hill NJ 07971

PETERSON, GEORGE HAROLD, b San Francisco, Calif, Apr 11, 31. MICROBIOLOGY. *Educ:* Univ Calif, BS, 53, PhD(soil microbiol), 57. *Prof Exp:* Asst prof bot & plant path, Purdue Univ, 57-63; asst prof, 63-66, assoc prof & assoc dean instr, 66-71, dean acad planning, 70-78, PROF BIOL SCI, CALIF STATE COL, HAYWARD, 71-, ASSOC VPRES ACAD RESOURCES, 78- *Mem:* Fel AAAS; Am Soc Microbiol; NY Acad Sci. *Res:* Physiology of soil microorganism; bacteriology; bacterial and cellular physiology. *Mailing Add:* Dept Acad Planning Calif State Col 25800 Hillary Hayward CA 94542

PETERSON, GEORGE P, New York, NY, Mar 30, 30; m; c 3. AEROSPACE MATERIALS TECHNOLOGY. *Educ:* Columbia Univ, BSME, 51. *Prof Exp:* Mem staff Non-Metallic Mat Div, US Air Force Mat Lab, Wright Aeronaut Labs, Wright-Patterson AFB, Ohio, 51-52, chief West Coast Off, Space Systs Div, El Segundo, Calif, 61-65, chief Advan Composites Div, Ohio, 65-72, chief Mfg Technol Div, 72-74, dir Mat Lab, 74-77, dep dir Wright Aeronaut Labs, 77-80, dir Mat Lab, 80-85; CONSULT, GEORGE PETERSON RESOURCES, INC, 85- *Concurrent Pos:* Mem bd dirs, Aracor, Sunnyvale, Calif, 86-, Howmet Corp, Greenwich, Conn, 87- *Honors & Awards:* Struct Dynamics & Mat Award, Am Inst Aeronaut & Astronaut, 73; Eng Citation, Soc Mfg Engrs, 78. *Mem:* Nat Acad Eng; hon mem Soc Advan Mat & Process Eng; hon mem Am Soc Metals; fel Soc Mfg Engrs. *Mailing Add:* 9877 Washington Church Rd Miamisburg OH 45342-4511

PETERSON, GERALD A, b Minneapolis, Minn, Jan 25, 32; m 57; c 2. THEORETICAL PHYSICS, SOLID STATE PHYSICS. *Educ:* Univ Minn, BA, 52; Cornell Univ, PhD(theoret physics), 59. *Prof Exp:* Res fel appl physics, Harvard Univ, 60-61, lectr, 62-64; consult, 62-64, sr theoret physicist, 64-67, PRIN SCIENTIST THEORET PHYSICS, RES CTR, UNITED TECHNOL CORP, 67- *Concurrent Pos:* Vis lectr, Wesleyan Univ, 64-65, vis assoc prof, 71- *Mem:* Am Phys Soc; Sci Res Soc Am. *Res:* Electronic and optical properties of solids; metals and semiconductors; band theory; impurity levels; current instabilities in semiconductors; nonlinear optics; laser physics. *Mailing Add:* Shipyard Rd Middle Haddam CT 06456

PETERSON, GERALD ALVIN, b Chesterton, Ind, Apr 12, 31; m 53; c 3. PHYSICS. *Educ:* Purdue Univ, BS, 53, MS, 55; Stanford Univ, PhD(physics), 62. *Prof Exp:* Physicist, Dept Defense, 57; res assoc/lectr physics, Yale Univ, 62-64, asst prof, 64-67; vis scientist, Amsterdam, Netherlands, 67; assoc prof, 68-73, PROF NUCLEAR PHYSICS, UNIV MASS, AMHERST, 73- *Concurrent Pos:* NATO fel, Scotland, 69-70; United Kingdom Res Coun sr fel, 70; Japan Soc for Promotion of Sci vis prof, Tohoku Univ, Japan, 72; vis prof, Univ Mainz, West Germany, 75; US-Israel Binat Sci Found fel, 83; vis prof, Shizuoka Univ, Japan, 89. *Mem:* Am Phys Soc. *Res:* Electron scattering; nuclear structure; instrumentation. *Mailing Add:* Dept Physics & Astron Univ Mass Amherst MA 01003

PETERSON, GERALD E, b Ephraim, Utah, Aug 9, 38; m 62; c 7. EQUATIONAL REASONING, FORMAL METHODS. *Educ:* Univ Utah, BS, 61, MA, 63, PhD(math), 65. *Prof Exp:* Res engr, Jet Propulsion Lab, 63-64; asst prof math, Univ Utah, 65-66 & Brigham Young Univ, 66-68; asst prof math, Univ Mo, St Louis, 68-71, assoc prof, 71-83; prof computer sci, Southern Ill Univ, Edwardsville, 83-86; SR SCIENTIST, MCDONNELL DOUGLAS CORP, 86- *Concurrent Pos:* NSF res contract, 70-71; software eng, McDonnell Douglas Corp, 81-86. *Mem:* Math Asn Am; Asn Comput Mach; Am Asn Artificial Intel; Sigma Xi. *Res:* Automated mathematics; especially automatic theorem proving for first-order logic with equality. *Mailing Add:* 2211 Polo Parc Court St Louis MO 63146

PETERSON, GLENN WALTER, b Shell Rock, Iowa, Oct 26, 22; m 57; c 3. PLANT PATHOLOGY. *Educ:* Iowa State Teachers Col, BA, 49; Iowa State Univ, MS, 51, PhD(plant path), 58. *Prof Exp:* Teacher high sch, Iowa, 49-56; PLANT PATHOLOGIST, US FOREST SERV, UNIV NEBR, LINCOLN, 58- *Mem:* Am Phytopath Soc; Sigma Xi. *Res:* Diseases of forest tree seedlings in great plains nurseries and diseases of trees in plains plantations. *Mailing Add:* Forestry Sci Lab East Campus Univ of Nebr Lincoln NE 68503

PETERSON, HAROLD A(LBERT), b Essex, Iowa, Dec 28, 08; m 34; c 2. ELECTRICAL ENGINEERING. *Educ:* Univ Iowa, BS, 32, MS, 33. *Prof Exp:* Elec engr, Works Lab, Gen Elec Co, Mass, 34-37, elec engr, Anal Div, 37-46; prof elec eng, Univ Wis-Madison, 46-75, Wis Utilities Asn prof, 67-74, chmn dept, 47-67, Edward Bennett prof, 74-75, Emer Edward Bennett prof elec eng, 75-88; RETIRED. *Concurrent Pos:* Mem, Int Conf Large Elec Systs. *Honors & Awards:* Educ Medal, Inst Elec & Electronic Engrs, 78, Centennial Medal, 84. *Mem:* Nat Acad Eng; Am Soc Eng Educ; Am Soc Mech Engrs; Nat Soc Prof Engrs; fel Inst Elec & Electronic Engrs. *Res:* Analysis of power system engineering problems; transients in power systems; applied superconductivity and energy storage in power systems; D-C power transmission. *Mailing Add:* 121 Montana Jack Green Valley AZ 85614

PETERSON, HAROLD ARTHUR, b St Paul, Minn, Sept 11, 26; m 50; c 4. SPEECH PATHOLOGY, PSYCHOLINGUISTICS. *Educ:* Minot State Col, BA, 50, BS, 58; State Univ Iowa, MA, 62; Univ Ill, PhD(speech path), 67. *Prof Exp:* Speech therapist, Ward County Pub Sch, 58-61; trainee speech path, Univ Iowa, 61-62; instr & speech clin coordr, Univ Ill, 62-67; assoc prof, dir speech practicum & admin asst to dept head speech path, 67-72, dir, Hearing & Speech Ctr, 73-88, PROF DEPT AUDIOL & SPEECH PATH, UNIV TENN, 72- *Mem:* Am Speech & Hearing Asn; fel Am Speech, Lang & Hearing Asn. *Res:* Appraisal of speech and language disorders and psycholingustic variables in speech-language development and disorders. *Mailing Add:* Hearing & Speech Ctr Yale at Stadium Dr Knoxville TN 37966-2500

PETERSON, HAROLD LEROY, b Mayville, NDak, Mar 24, 46; div; c 2. SOIL MICROBIOLOGY. *Educ:* Mayville State Col, BA, 68; Iowa State Univ, MS, 71, PhD(soil microbiol), 75. *Prof Exp:* Asst prof & asst agronomist, 75-80, assoc prof agron & assoc agronomist, Miss Agr & Forestry Exp Sta, 80-85, PROF & AGRONOMIST, MISS STATE UNIV, 85- *Concurrent Pos:* Vis prof agron & genetics, Iowa State Univ, 81. *Mem:* Sigma Xi; Am Soc Microbiol; Am Soc Agron; Brit Soc Soil Sci; Soil Sci Soc Am; Coun Agr Sci & Technol. *Res:* Ecology of microorganisms in soil; dinitrogen fixation in plants and soil; relationships of soil microorganisms to food and fiber production; increased capture of solar energy; rhizobium DNA research. *Mailing Add:* Dept Agron Box 5248 Miss State Univ Mississippi State MS 39762

PETERSON, HAROLD LEROY, b Stromsburg, Nebr, Apr, 13, 38. MATHEMATICS. *Educ:* Stanford Univ, BS, 60; Univ Ore, MS, 63, PhD(math), 66. *Prof Exp:* Assoc engr, Lockheed Missiles & Space Co, 60-61; asst prof math, Univ Conn, 66-70; asst prof, 70-74, ASSOC PROF MATH, IND UNIV NORTHWEST, 74- *Mem:* Math Asn Am. *Res:* Topological groups; measure theory; general topology; subgroups of finite index, extensions of Haar measure. *Mailing Add:* Dept of Math Ind Univ Northwest Gary IN 46408

PETERSON, HAROLD OSCAR, b Dalbo, Minn, Apr 13, 09; m 34; c 4. MEDICINE, RADIOLOGY. *Educ:* Univ Minn, BS, 30, BM, 33, MD, 34; Am Bd Radiol, dipl, 38. *Prof Exp:* Intern, Kansas City Gen Hosp, Mo, 33-34; resident radiol, Mass Gen Hosp, Boston, 35-36; instr, 37-40, from clin instr to clin prof, 40-57, head dept, 57-70, prof, 57-77, EMER PROF RADIOL, UNIV MINN, ST PAUL, 77- *Concurrent Pos:* Lectr, var univs & socs, 62-; radiologist, Interstate Clin, Red Wing, Minn, 40-57, St Joseph's Hosp, St Paul, 41-43 & Bethesda Hosp, 41-44; radiologist & head dept, Charles T Miller Hosp, 41-57 & Children's Hosp, 48-57; trustee, Am Bd Radiol, 58-71; mem neurol study sect, NIH, 68-72; vis prof radiol, Univ Tex Med Br, Galveston, 79, 80 & 81 & Univ Tex Health Sci Ctr, San Antonio, 80-85. *Honors & Awards:* Henry K Pancoast lectr, Philadelphia, 56; Freedman lectr, Univ Cincinnati, 59; Gold Medal, Am Roentgen Ray Soc, 61; Fred J Hodges Lectr, Ann Arbor, Mich, 62; Award, Am Acad Neurol, 63; Gold Medal, Am Col Radiol, 71. *Mem:* AAAS; Am Soc Neuroradiol (pres, 68); Am Roentgen Ray Soc (pres-elect, 63, pres, 64); fel Am Col Radiol (vpres, 63-64); fel Am Col Chest Physicians. *Res:* Neuroroentgenology. *Mailing Add:* 1995 W County Rd B St Paul MN 55113

PETERSON, HARRY C(LARENCE), b Greeley, Colo, Feb 23, 31; m 50; c 4. VIBRATION DYNAMICS, VEHICLE CRASHWORTHINESS. *Educ:* Colo State Univ, BSME, 53; Cornell Univ, MS, 56, PhD(eng mech), 59. *Prof Exp:* Instr eng mech, Cornell Univ, 53-55 & 57-59, asst prof, 59-60; asst res scientist, Martin Co, 60-62, chief aeroelasticity res sect, 62-66, mgr aeromech & mat dept, 66-67; prof eng mech, Univ Denver, 67-74; prof basic eng, Colo Sch Mines, 74-87; RETIRED. *Concurrent Pos:* Expert witness in motorcycle accident litigation, 72-; sr res engr, Res Inst, Univ Denver, 74-78. *Mem:* Am Inst Aeronaut & Astronaut; Am Soc Eng Educ; Sigma Xi; NY Acad Sci; AAAS. *Res:* Safety systems for mine conveyances; dynamics of motorcycle crashes, rider injuries and protection; nonstationary random vibrations; educational materials and apparatus for undergraduate education in engineering mechanics and mechanical design. *Mailing Add:* 6061 S Aberdeen St Littleton CO 80120

PETERSON, HAZEL AGNES, b Houston, Tex, Apr 7, 16. HYDROGEOLOGY. *Educ:* NY Univ, BA, 39; Univ Tex, MA, 42. *Prof Exp:* Subsurface geologist, Tex Co, 42-44; subsurface geologist, Sun Oil Co, 44-52; supv geologist, Seaboard Oil Co, 52-54; instr, 58-67, asst prof geol, E Tex State Univ, 67-78; CONSULT, 78- *Concurrent Pos:* Consult geologist, 54-; grant, Nat Park Serv, 60-61; fac res grant, ETex State Univ, 65-66, & 68. *Mem:* Am Asn Petrol Geologists; Am Asn Petrol Geol; Sigma Xi. *Res:* Pleistocene vertebrates; stratigraphy of northeast Texas; geohydrology; water resources and environmental geology of Texas; petroleum geology of the Southwest; geoscience audio-visual education. *Mailing Add:* 820 Hillcrest Denton TX 76201

PETERSON, HOWARD BOYD, b Redmond, Utah, Oct 10, 12; m 40; c 3. SOIL CHEMISTRY. *Educ:* Brigham Young Univ, BS, 35, AM, 37; Univ Nebr, PhD(soils), 40. *Prof Exp:* From instr to assoc prof agron, 40-50, chmn dept, 59-64, head dept agr & irrig eng, 71-73, prof agron, 50-80, EMER PROF AGR & IRREGATION ENG, UTAH STATE UNIV, 80- *Mem:* AAAS; Am Soc Agron; Soil Sci Soc Am. *Res:* Irrigation; water quality. *Mailing Add:* 1126 N 16 E Logan UT 84321

PETERSON, IDELLE M(ARIETTA), b Forest City, Iowa, Feb 25, 38; m 60; c 3. CHEMICAL ENGINEERING. *Educ:* Iowa State Univ, BS, 60, MS, 64, PhD(chem eng), 68. *Prof Exp:* Res scientist surface chem, McDonnell-Douglas Res, 68-69; engr aeronaut struct & mat, US Army Aviation Systs Command, 69-77, AEROSPACE ENGR, COMPUT AIDED DESIGN, US ARMY AVIATION RES & DEVELOP COMMAND, 77- *Mem:* Am Helicopter Soc. *Res:* Development of computer aided design software. *Mailing Add:* 1405 Whitney Lane Rolla MO 65401

PETERSON, IRVIN LESLIE, b Earlimart, Calif, Feb 25, 26; m 51; c 3. VETERINARY MEDICINE, POULTRY SCIENCE. *Educ:* Univ Calif, Davis, BS, 51, DVM, 63. *Prof Exp:* Partner, Kerners' Turkey Farms & Hatchery, Calif, 51-54; farm adv, Agr Exten Serv, Univ Calif, 54-59; poultry pathologist, Kimber Poultry Breeding Farm Inc, 63-68; mgr poultry & livestock serv, Western Farmers Asn, 68-71; vet coordr, Nat Poultry Improv Plan, 71-77, Agr Res Serv, 71-77, chief staff vet, animal & plant inspection serv, USDA, 77-84; SR COODR, NAT POULTRY IMPROV PLAN, VET SERV, ANIMAL & PLANT HEALTH INSPECTION SERV, USDA, 84- *Mem:* Am Asn Avian Path; Am Vet Med Asn; Poultry Sci Asn; World Poultry Sci Asn; World Vet Poultry Asn; US Branch-World's Poultry Sci Asn (secy-treas). *Res:* Poultry egg transmitted and hatchery disseminated diseases; coordination of disease control measures through state agencies. *Mailing Add:* Fed Ctr Bldg Rm 771 Nat Poultry Improv Plan USDA Hyattsville MD 20782

PETERSON, JACK EDWIN, b Bremerton, Wash, Feb 7, 28; m 52; c 2. INDUSTRIAL HYGIENE. *Educ:* Wash State Univ, hons BS, 51; Univ Mich, MS, 52, PhD(indust health), 68. *Prof Exp:* Chem engr, Spec Assignments Prog, Dow Chem Co, Mich, 52-53, environ res engr, Biochem Res Lab, 53-65; asst prof, 68-73, ASSOC PROF CIVIL ENG, MARQUETTE UNIV, 73-; PROF OCCUP & ENVIRON MED, SCH PUB HEALTH, UNIV ILL, 77-; CLIN PROF PREVENTIVE MED, MED COL WIS, 80- *Concurrent Pos:* Asst prof environ health eng, Med Col Wis, 68-73; indust hyg consult, Peterson Assocs, 75-; pres, Exigency Group, Inc, 85-86; prof allied health, Univ Wis Parkside, 84-86. *Honors & Awards:* Authorship Award, Am Indust Hyg Asn, 62 & 70; Henry F Smyth Jr Award, Am Acad Indust Hyg, 87. *Mem:* Am Indust Hyg Asn; NY Acad Sci; Sigma Xi; Am Acad Indust Hyg. *Res:* Human biothermal stress-strain; inhalation exposure integration; effects of carbon monoxide on human performance; absorption-excretion of carbon monoxide and organic solvents by man; air pollution; industrial hygiene and toxicology. *Mailing Add:* Peterson Assocs 660 Forest Grove Circle Brookfield WI 53005

PETERSON, JACK KENNETH, b Chicago, Ill, Dec 3, 32; m 56; c 3. POLYMER CHEMISTRY, PHYSICAL CHEMISTRY. *Educ:* Purdue Univ, BS, 54; Ohio State Univ, PhD(phys chem), 61. *Prof Exp:* Sr res chemist, Mobil Chem Co, 61-63, sect leader, 63-68; mgr hot melts develop, Stein Hall & Co, Inc, 68-71; mgr tech corrd, USS Chem, 71-78; mgr, environ control, US Steel Corp, 78-86, ENVIRON SPECIALIST, USX ENGRS & CONSULT, INC, 86- *Concurrent Pos:* Res consult, USS Chem, 75-78. *Honors & Awards:* Merck Index Award, 54. *Mem:* Am Chem Soc. *Res:* Solution, viscoelastic and solid state properties of polymers. *Mailing Add:* USX Engrs & Consults Inc 600 Grant St Pittsburgh PA 15219

PETERSON, JACK MILTON, b Portland, Ore, Apr 25, 20; m 46; c 3. NUCLEAR PHYSICS. *Educ:* Harvard Univ, SB, 42; Univ Calif, PhD(physics), 50. *Prof Exp:* Res assoc & ed, Microwave Techniques, Radiation Lab, Mass Inst Technol, 42-45; from tech investr to asst dir vacuum tube develop comt, Columbia Univ, 43-46; physicist, Lawrence Berkeley Lab, Univ Calif, 50-52, head nuclear physics div, Livermore, 52-61, physicist, B-div, 61-64, physicist, 64-85, sr scientist advan accelerator design group & sr accelerator physicist, super collider cent design group, Univ Res Asn, Univ Calif, Berkeley, 85-89; SR SCIENTIST & PHYSICIST, UNIV RES ASN, DALLAS, TEX, 89- *Concurrent Pos:* Lectr, Univ Calif, Berkeley, 50-53; Fulbright fel, Bohr Inst Theoret Physics, Copenhagen, 60-61; leader beam-transport & injection system Positron-Electron Proj, Stanford Linear Accelerator Ctr, 74-80; mem, Nuclear Cross Sect Adv Group, AEC. *Mem:* Am Phys Soc. *Res:* Design of microwave components; high energy neutron cross sections; excitation curves for high energy deuteron and alpha particle reactions; meson production by x-rays; x-ray fluorescence; design and operation of betatrons, cyclotrons, synchrotrons, storage rings, beam-transport systems; collective acceleration; free electron lasers; design of kicker, septum, iron and superconducting magnets; beam dynamics; beam optics. *Mailing Add:* Lawrence Radiation Lab 90-2148 Univ of Calif Berkeley CA 94720

PETERSON, JAMES ALGERT, b Berrian Co, Mich, Apr 17, 15; m 44; c 3. GEOLOGY. *Educ:* St Louis Univ, BS, 48; Univ Minn, MS, 49, PhD(geol), 52. *Prof Exp:* Geologist paleont, US Geol Surv, 48-50, 75-88; instr geol, State Col Wash, 51; geologist, Shell Oil Co, 52-57, from div stratigrapher to sr geologist, 57-65; assoc prof, 65-67, prof, 67-75, ADJ PROF GEOL, UNIV MONT, 75-; GEOLOGIST, US GEOL SURV, 75- *Mem:* Fel AAAS; fel Geol Soc Am; fel Explorers Club; hon mem Soc Econ Paleont & Mineral; Am Asn Petrol Geol. *Res:* Jurassic, Cretaceous and upper Paleozoic stratigraphy; paleontology and geologic history; petroleum geology; ground water resources; carbonate petrology; world energy resources. *Mailing Add:* Dept Geol Univ Mont Missoula MT 59812

PETERSON, JAMES DOUGLAS, b St Louis, Mo, May 19, 48; m 74; c 2. MANAGEMENT, TELECOMMUNICATIONS. *Educ:* Univ Chicago, BA, 71; PhD(biochem), 74. *Prof Exp:* Res biol chemist, Univ Calif, Los Angeles, 74-75; fel chem, Univ Del, 75-77; res assoc biochem, Univ Kans Med Ctr, Kansas City, 77-79; res chemist, Vet Admin Med Ctr, 78-79; mem tech staff, Bell Labs, Naperville, Ill, 79-85; PROJ & PROG MGR, GTE COMMUN SYSTS, PHOENIX, 85- *Concurrent Pos:* Adj asst prof, Univ Kans Med Ctr, 77-79; fel, Nat Inst Gen Med Sci, 76-77; chmn, TISI & Telecommun Standards Group. *Mem:* Am Chem Soc; Sigma Xi. *Res:* Testing and validation of telephone switching software; planning of new network features; management of telecommunication systems development projects and programs. *Mailing Add:* 4909 E Emile Zola Scottsdale AZ 85254

PETERSON, JAMES LOWELL, b Weston, WVa, Feb 15, 42; m 64; c 2. INORGANIC CHEMISTRY. *Educ:* WVa Wesleyan Col, BA, 64; WVa Univ, MS, 66; Ohio State Univ, PhD(inorg chem), 73. *Prof Exp:* Instr chem, 66-70, from asst prof to assoc prof, 73-77, DEAN ACAD AFFAIRS, GLENVILLE STATE COL, 77- *Mem:* Am Chem Soc. *Mailing Add:* 957 J Mineral Rd Glenville WV 26351

PETERSON, JAMES MACON, b Elizabethtown, NC, Apr 8, 37; m 66; c 3. POLYMER SCIENCE, MATERIALS SCIENCE. *Educ:* Wake Forest Univ, BS, 58; Calif Inst Technol, PhD(chem), 63. *Prof Exp:* Res chemist, Chemstrand Res Ctr, 63-69; RES SCIENTIST, BOEING COMMERCIAL AIRPLANE CO, 69- *Mem:* Am Chem Soc; Am Phys Soc; Sigma Xi. *Res:* Polymer structure; materials flammability; chemical physics of fire. *Mailing Add:* 420 SW 183rd St Seattle WA 98166

PETERSON, JAMES OLIVER, b St Louis, Mo, Feb 26, 37; m 68; c 2. ORGANIC CHEMISTRY, CHEMISTRY OF SEMICONDUCTOR DEVICES. *Educ:* Northwestern Univ, BA, 59; State Univ NY Buffalo, PhD(org chem), 65. *Prof Exp:* Res Chemist, Gen Chem Div, Allied Chem Corp, 64-68, res chemist, Specialty Chem Div, 69-71, sr res chemist, Specialty Chem Div, 71-76, suprv, Prog Res, Thermoset Prods Lab, Specialty Chem Div, 76-79; mgr, Prog Res, Plaskon Prod, Inc, 79-84; mgr res plaskon electronic mat, 84-90, SR RES SPECIALIST, ROHM & HAAS CO, 90- *Mem:* Am Chem Soc; Sigma Xi; Inst Elec & Electronic Engrs. *Res:* Development of encapsulation products for electronic semiconductor devices; development of epoxy, urea, melamine and phenolic molding compounds; synthesis of new and industrial organic compounds; fluorination of organic compounds; reaction of fluorinated compounds; preparation of fluorocarbon polymers; industrial hygiene and environmental regulations. *Mailing Add:* Rohm & Haas Res Labs 727 Norristown Rd Spring House PA 19477

PETERSON, JAMES RAY, b Hollywood, Calif, June 2, 24; m 52; c 4. ATOMIC PHYSICS. *Educ:* Univ Calif, Los Angeles, AB, 48; Univ Calif, PhD, 56. *Prof Exp:* Physicist, Radiation Lab, Univ Calif, 50-56; physicist, 56-75, assoc dir, Molecular Physics Lab, 75-80, prog mgr, 80-89, SR STAFF SCIENTIST, SRI INT, 89- *Concurrent Pos:* Vis assoc prof, Univ Wash, 66-67; vis fel, Joint Inst for Lab Astrophysics, Univ Colo, 74-75; vis scientist, FOM Inst, Amolf, Amsterdam, 86, 90. *Mem:* Fel Am Phys Soc; Am Geophys Union; Am Asn Physics Teachers; AAAS. *Res:* Low energy atomic, molecular, and photon interactions. *Mailing Add:* Molecular Physics Lab PN013 SRI Int 333 Ravenswood Ave Menlo Park CA 94025

PETERSON, JAMES ROBERT, soil fertility, for more information see previous edition

PETERSON, JAMES ROBERT, b St Paul, Minn, Apr 16, 32; m 54; c 1. MAN-MACHINE SYSTEM DEVELOPMENT, HUMAN FACTORS. *Educ:* Univ Minn, BA, 54, MA, 58; Univ Mich, PhD(eng psychol), 65. *Prof Exp:* Develop engr, Honeywell, Inc, 61-65, sr develop engr, 65-67, staff engr, 67-90, SR PROJ STAFF ENGR, HONEYWELL, INC, 90- *Concurrent Pos:* Honeywell corp sponsor rep, 2 Shuttle Student Involvement Progs, 82 & 84. *Mem:* assoc fel Am Inst Aeronaut & Astronaut; Human Factors Soc; Am Defense Preparedness Asn. *Mailing Add:* 3303 San Gabriel St Clearwater FL 34619-3341

PETERSON, JAMES T, b St James, Minn. ATMOSPHERIC & ENVIRONMENTAL SCIENCES. *Educ:* Univ Minn, BS, 63; Univ Wis, MS, 65, PhD(meteorol), 68. *Prof Exp:* Meteorologist, Meteorol Div, Air Resources Lab, 68-77, chief, Analysis & Interpretation Group, 77-81, dir, geophys monitoring climatic change div, 81-90, DEPT DIR, CLIMATE MONITORING & DIAGNOSTICS LAB, US DEPT COM, NAT OCEANIC & ATMOSPHERIC ADMIN, 90- *Concurrent Pos:* Adj prof, dept geosci, NC State Univ, Raleigh, 70-76. *Mem:* Am Meteorol Soc; Am Geophys Union. *Res:* Impact of human activities on weather and climate; atmospheric aerosols; solar radiation; trace gases. *Mailing Add:* US Dept Com Nat Oceanic & Atmospheric Admin 325 Broadway Boulder CO 80303

PETERSON, JANE LOUISE, b Hamilton, Mont, May 20, 47. HUMAN GENOME RESEARCH. *Educ:* Western Col, BA, 69; Univ Colo, PhD, 75. *Prof Exp:* Postdoctoral fel, Dept Human Genetics, Yale Univ Sch Med, 75-78; sr staff fel, Lab Biochem, Nat Cancer Inst, NIH, 78-81; asst prog dir, Develop Biol Prog, NSF, Washington, DC, 81-85; prod adminr, Genetics Prog, Nat Inst Gen Med Sci, NIH, 85-89, CHIEF, RES CTRS BR, NAT CTR HUMAN GENOME RES, NIH, 89- *Concurrent Pos:* NIH postdoctoral trainee, 75-77; alt proj officer, GenBank, 86-89; vchmn, Women in Cell Biol, Am Soc Cell Biol, 88-89, chair, 89-90; mem, Adv Comt Women's Health Issues, NIH, 90- *Mem:* AAAS; Am Soc Cell Biol; Am Soc Microbiol. *Res:* Cellular and developmental biology; numerous technical publications. *Mailing Add:* Nat Ctr Human Genome Res NIH Res Ctr Br Bldg 38A Rm 6N615 Bethesda MD 20892

PETERSON, JANET BROOKS, b Brooklyn, NY, Mar 15, 24. ORGANIC CHEMISTRY. *Educ:* Univ Mich, BS, 45; Univ Ill, MA, 48, PhD(chem), 51. *Prof Exp:* Asst chem, Johns Hopkins Univ, 45-46 & Univ Ill, 46-49; admin & tech asst, Gen Aniline & Film Corp, 50-52; res chemist & lab supvr, Geigy Chem Corp, 52-70, sr res chemist & group leader, 70-79, staff scientist, 79-86, sr staff scientist, Ciba-Geigy Corp, 86-89; RETIRED. *Mem:* Am Chem Soc; Fel Am Inst Chem; NY Acad Sci; Sigma Xi. *Res:* Synthetic organic chemistry, including heterocyclic nitrogen compounds; amino acids and derivatives; hindered phenol antioxidants; scale-up synthesis and process research; corrosion inhibitors and hindered amine light stabilizers. *Mailing Add:* 275 Powder Point Ave Duxbury MA 02332-3926

PETERSON, JANET SYLVIA, b McMinnville, Tenn. APPLIED MATHEMATICS. *Educ:* Calif State Univ, Los Angeles, BS, 73; Univ Tenn, MS, 77, PhD(math), 80. *Prof Exp:* Analyst, Jet Propulsion Lab, Calif Inst Technol, 73-75; ASST PROF MATH, UNIV PITTSBURGH, 81- *Concurrent Pos:* Vis scientist, Inst Comput Appln Sci & Eng, Langley Res Ctr, NASA, 81- *Mem:* Soc Indust & Appl Math; Asn Women Math. *Res:* Numerical solution of partial differential equations from both a theoretical and computational stand point; finite element and finite difference methods. *Mailing Add:* Los Alamos Nat Lab MSB 265 Univ Pittsburgh 4200 Fifth Ave Los Alamos NM 87545

PETERSON, JOHN BOOTH, b Salem, Ore, July 18, 05; m 30; c 2. AGRONOMY. *Educ:* Ore State Col, BS, 28; Iowa State Col, MS, 29, PhD(soil fertil), 36. *Hon Degrees:* DAgr, Purdue Univ, 82. *Prof Exp:* From instr to assoc prof agron, Iowa State Univ, 29-46, prof soils, 46-48; head dept agron, Purdue Univ, Lafayette, 48-71, prof, 48-81, assoc dir, Lab Appln Remote Sensing, Purdue/NASA, 71-81; consult, 81-86; RETIRED. *Concurrent Pos:* Nat Res Coun fel, Univ Calif, 39-40; guest prof, Univ Ariz, 66; consult, Latin Am Fel Prog, Rockefeller Found, 61-62, Orgn Nat Agr Res Prog, Greek Govt, 63, Ford Found Agr Prog for Arg, 65, Food & Agr Orgn, Arg, 70, Int Fertilizer Develop Ctr, 75- & Inter-Am Develop Bank, Cent Am, 76; sr scientist, Jecor, Saudi Arabia, 77; consult, plant & soil sci, 81-86; res consult, Purdue Agr Alumni Crop Improv Asn, 84-86. *Honors & Awards:* Stevenson Award, Am Soc Agron, 48, Agron Serv Award, 78; H H Bennett Award, Soil Conserv Soc Am, 84. *Mem:* Fel AAAS; fel Am Soc Agron (pres, 58-59); fel Soil Sci Soc Am; fel Soil Conserv Soc Am. *Res:* Soil microscopy, fertility, conservation, genesis, microbiology and chemistry; administration of research and teaching; remote sensing of natural resources; land use; development of genetic male sterile corn; contamination of soil and water with toxic materials; contamination of underground aquifers. *Mailing Add:* 2741 N Salisbury West Lafayette IN 47906

PETERSON, JOHN CARL, b US. FLORICULTURE, PLANT PHYSIOLOGY. *Educ:* Univ RI, BS, 74; Rutgers Univ, PhD(hort), 78. *Prof Exp:* Res asst hort, Rutgers Univ, 74-78; ASST PROF, OHIO STATE UNIV, 78- *Mem:* Am Soc Hort Sci. *Res:* Applied physiology and production problems relating to growth and development of floral crops and tropical foliage plants in indoor environments. *Mailing Add:* Dept of Hort Ohio State Univ 2001 Fyffe Ct Columbus OH 43210

PETERSON, JOHN EDWARD, JR, b Myrtle, Ill, June 19, 21; m 45; c 2. MYCOLOGY, MICROBIOLOGY. *Educ:* Northern Ill Univ, BE, 42; Mich State Univ, MS, 52, PhD(bot), 57. *Prof Exp:* Instr biol, Jackson Jr Col, 46-50; from instr to prof bot, Univ Mo, Columbia, 53-71; dean, Sch Lib Arts & Sci, 71-83, PROF BIOL SCI & DIR, UNIV HONS PROG EMPORIA STATE UNIV, 83- *Mem:* AAAS; Mycol Soc Am; Bot Soc Am; Soc Indust Microbiol; Am Soc Microbiol. *Res:* Myxobacteria; genera Fusarium and Micromonospora; microflora of the bark of living trees. *Mailing Add:* Div Biol Sci Emporia State Univ Emporia KS 66801

PETERSON, JOHN ERIC, b Norwalk, Ohio, Oct 26, 14; m 38; c 2. INTERNAL MEDICINE. *Educ:* Col Med Evangelists, MD, 39; Am Bd Internal Med, dipl, 47. *Prof Exp:* Resident med, Henry Ford Hosp, Detroit, Mich, 39-42; from instr to assoc prof, 42-54, PROF MED & CHMN DEPT & ASSOC DEAN, SCH MED, LOMA LINDA UNIV, 54- *Concurrent Pos:* Res assoc, Harvard Med Sch, 60-61. *Mem:* AMA; Am Col Physicians; Am Diabetes Asn. *Res:* Atheromatous disease; fat, lipoprotein metabolism. *Mailing Add:* Loma Linda Univ Med Ctr Loma Linda CA 92354

PETERSON, JOHN IVAN, b Syracuse, NY, Mar 29, 28; m 50; c 5. INSTRUMENT DEVELOPMENT SENSORS. *Educ:* Syracuse Univ, BA, 50, PhD(anal chem), 54. *Prof Exp:* Sr scientist, Chem Div, Sylvania Elec Prod, Inc, 54-55; sr scientist, Res Div, Jones & Laughlin Steel Corp, 55-58; sr engr, Textile Fibers Div, E I du Pont de Nemours & Co, 58-59; group leader anal develop, Melpar, Inc, 59-60; supvr chem br, Chem Warfare Sect, 60, supvr anal br, Res Div, 60-62; mgr chem div, Woodard Res Corp, 62-64; mgr lab, Clin Serv Div, Bionetics Res Labs, Inc, 64-65; STAFF CHEMIST, BIOMED ENG BR, NIH, 65- *Mem:* Instr Soc Am; Am Chem Soc. *Res:* Instrumental methods of analysis and instrument development; fiber optic sensors; clinical chemistry; radiotracer analysis; electrochemistry; biochemical analysis. *Mailing Add:* Biomed Eng & Instrumentation Br NIH Rm 3W13 Bldg 13 Bethesda MD 20892

PETERSON, JOHN M, b Fillmore, Utah, Feb 14, 38; m 60; c 7. MATHEMATICS. *Educ:* Utah State Univ, BS, 62, MS, 64, EdD, 65. *Prof Exp:* Teacher pub schs, Utah, 61-63; from asst prof to assoc prof, 65-74, PROF MATH, BRIGHAM YOUNG UNIV, 74- *Mem:* Math Asn Am; Am Math Soc; Nat Coun Teachers Math. *Res:* Efficiency of different methods of teaching mathematics at various academic levels. *Mailing Add:* Dept Math TMCB 366 Brigham Young Univ Provo UT 84602

PETERSON, JOHNNY WAYNE, b Gilmer, Tex, Aug 30, 46; m 65; c 2. MICROBIOLOGY. *Educ:* Univ Tex, Arlington, BS, 67; NTex State Univ, MS, 69; Univ Tex Southwestern Med Sch, PhD(microbiol), 72. *Prof Exp:* ASSOC PROF MICROBIOL, UNIV TEX MED BR GALVESTON, 77- *Concurrent Pos:* NIH contract, Univ Tex Med Br Galveston; Found for Microbiol lectr, 78-79. *Mem:* Am Soc Microbiol. *Res:* Pathogenic mechanisms of enteric infections and bacterial immunology; salmonellas; cholera. *Mailing Add:* Dept Microbiol Univ Tex Med Br Galveston TX 77550

PETERSON, JOSEPH LOUIS, b Pleasant Grove, Utah, May 16, 29; m 53; c 2. PLANT PATHOLOGY, MYCOLOGY. *Educ:* Utah State Univ, BS, 55, MS, 57; Univ Wis, PhD(plant path), 59. *Prof Exp:* Asst plant path, Utah State Univ, 53-56; agr aide hort crops res br, USDA, 56-57; asst prof plant path & res specialist, 59-63, assoc prof, 63-71, PROF PLANT PATH, RUTGERS UNIV, NEW BRUNSWICK, 71- *Mem:* Am Phytopath Soc; Mycol Soc Am. *Res:* Diseases of ornamental crops; mycology; biological control. *Mailing Add:* Dept of Plant Path Cook Col Rutgers Univ New Brunswick NJ 08903

PETERSON, JOSEPH RICHARD, b Wilmington, Del, Sept 16, 42; wid; c 2. PHYSICAL INORGANIC CHEMISTRY, MICROCHEMISTRY. *Educ:* Swarthmore Col, AB, 64; Univ Calif, Berkeley, PhD(chem), 67. *Prof Exp:* Res asst, Lawrence Radiation Lab, Univ Calif, 64-67; from asst prof to assoc prof, 67-79, PROF CHEM, UNIV TENN, KNOXVILLE, 79- *Concurrent Pos:* Consult, Transuranium Res Lab, Oak Ridge Nat Lab, 67-77, adj Res & Develop Partic II, 77-; NATO fel nuclear chem, Univ Liege, 69-70; consult, Lawrence Livermore Lab, Univ Calif, 73-79; treas, Am Chem Soc, Div Nuclear Chem & Technol, 78-87, vchmn elect, 88, vchmn, 89, chm, 90; guest scientist, Inst Heisse Chemie, Kernforschungszentrum Karlsruhe, 81-82, Europ Inst for Transuranium Elements, Karlsruhe, 81-82, 83, 84, 86, 87, 88 & 89. *Mem:* Am Chem Soc; Am Nuclear Soc; Sigma Xi. *Res:* Determination, correlation, and interpretation of the basic physical and chemical properties of the transuranium elements and their compounds. *Mailing Add:* Dept Chem Univ Tenn Knoxville TN 37996-1600

PETERSON, JULIAN ARNOLD, b Detroit, Mich, Oct 3, 39; m 64. BIOCHEMISTRY. *Educ:* Wittenberg Univ, BS, 61; Univ Mich, Ann Arbor, MS, 65, PhD(biochem), 67. *Prof Exp:* Teaching asst org chem, Wittenberg Univ, 61-62; asst prof, 68-72, assoc prof biochem, 72-79, PROF BIOCHEM & MED COMPUT SCI, UNIV TEX HEALTH SCI CTR, DALLAS, 79- *Concurrent Pos:* NIH fel, Johnson Res Found, Univ Pa, 67-68; Nat Inst Arthritis & Metab Dis res grant, Univ Tex Health Sci Ctr, Dallas, 68-71; Robert A Welch Res Found res grant, 69- & Nat Inst Gen Med Sci res grant, 71-; mem, Phys Biochem Study Sect, NIH, 80- *Mem:* Am Soc Biol Chemists; AAAS; Am Chem Soc; Am Soc Pharmacol Exp Therapeut; Sigma Xi. *Res:* Electron transport and the biological activation of molecular oxygen for hydroxylation reactions; control of microbial reactions involving molecular oxygen; computer based education. *Mailing Add:* 3339 Darbyshire Dallas TX 75229

PETERSON, KENDALL ROBERT, atmospheric physics; deceased, see previous edition for last biography

PETERSON, KENNETH C(ARL), b Superior, Wis, Nov 2, 21; m 65; c 2. CHEMICAL ENGINEERING. *Educ:* Univ Wis, BS & MS, 47. *Prof Exp:* Chem engr, Res Dept, Standard Oil Co, Ind, 47-53, group leader, 53-59, sect leader, 59-65; div dir, Res & Develop Dept, Amoco Chem Corp, 65-69, mgr polymers & plastics res & develop, 69-76, mgr polymer properties & process design, 76-81; RETIRED. *Mem:* AAAS; Am Inst Chem Engrs; Am Chem Soc; Soc Plastics Engrs; Am Mgt Asn. *Res:* Petrochemicals. *Mailing Add:* 1033 Anne Rd Naperville IL 60540

PETERSON, LANCE GEORGE, b Duluth Minn, Jan 1, 40; m 63; c 3. INSECT PHYSIOLOGY, TOXICOLOGY. *Educ:* Univ Minn, BA, 62; Univ Ill, PhD(insect physiol & toxicol), 67. *Prof Exp:* Sr entomologist, Eli Lilly & Co, 67-73, plant sci rep, 73-76, tech adv to agr mkt, 76-77; AREA RES MGR, LILLY INT, 77- *Mem:* Entom Soc Am; Sigma Xi; Am Registry Prof Entomologists. *Res:* Mechanisms of control of insects that are economically important pests of agriculture; developing agricultural chemicals for international agricultural markets, including insecticides, herbicides and fungicides. *Mailing Add:* Eli Lilly & Co PO Box 708 Greenfield IN 46140

PETERSON, LARRY JAMES, b Winfield, Kans, Apr 23, 42; m 69; c 2. ORAL SURGERY. *Educ:* Univ Kans, BS, 64; Univ Mo-Kansas City, DDS, 68; Georgetown Univ, MS, 71. *Prof Exp:* From asst prof to assoc prof oral surg, Med Col Ga, 71-75; from assoc prof to prof oral surg, 75-81, PROF MAXILLOFACIAL SURG, HEALTH CTR, UNIV CONN, 81-; DEPT ORAL SURR, OHIO STATE UNIV. *Concurrent Pos:* Adj prof, Col Eng, Clemson Univ, 74. *Mem:* Am Asn Oral & Maxillofacial Surgeons; Am Dent Asn; Int Asn Dent Res; Asn Acad Surgeons; Am Asn Dent Schs. *Res:* Use of alloplastic implant materials in dentistry and oral surgery; use of prophylactic antibiotics in major and surgical procedures; wound healing and revascularization following bone graft ridge augmentation. *Mailing Add:* Dept Oral Surg Ohio State Univ 305 W 12th Ave Columbus OH 43210

PETERSON, LAUREN MICHAEL, b Minneapolis, Minn, June 11, 43; m 68; c 3. OPTICAL PHYSICS, QUANTUM ELECTRONICS. *Educ:* Univ Minn, BPhys, 66; Pa State Univ, MS, 68, PhD(physics), 72. *Prof Exp:* RES PHYSICIST OPTICAL PHYSICS, ENVIRON RES INST MICH, 72- *Concurrent Pos:* Adj assoc prof elec eng, Univ Mich, 81- *Mem:* Optical Soc Am. *Res:* Nonlinear optics, especially stimulated light scattering, self induced transparency; electro- and acousto- optic modulation of infrared lasers; remote sensing; molecular spectroscopy. *Mailing Add:* 1160 Morehead Ct Ann Arbor MI 48103-6181

PETERSON, LAURENCE E, b Grantsburg, Wis, July 26, 31; m 56; c 4. HIGH ENERGY ASTROPHYSICS. *Educ:* Univ Minn, Minneapolis, BS, 54, PhD(physics), 60. *Prof Exp:* Res assoc physics, Univ Minn, Minneapolis, 60-62; res physicist, 62-63, from asst prof to assoc prof, 63-71, PROF PHYSICS, UNIV CALIF, SAN DIEGO, 71- *Concurrent Pos:* Mem physics subcomt, NASA Space Sci Steering Comt, 64-; NSF fel, 58-59; Guggenheim fel, 73-74; assoc dir sci, Astrophys Div, NASA, 86-88. *Honors & Awards:* Space Sci Award, Am Inst Aeronaut & Astronaut, 78. *Mem:* AAAS; fel Am Phys Soc; Am Astron Soc; Int Astron Union. *Res:* Space astronomy; galactic and solar cosmic rays; solar gamma and x-rays; x-ray and gamma ray astronomy and high energy astrophysics; related balloon and satellite instrumentation. *Mailing Add:* Ctr Astrophys & Space Sci 0111 Univ Calif San Diego La Jolla CA 92093-0111

PETERSON, LAVERNE E, b Warren, Pa, Feb 19, 25; m 47; c 3. MICROBIOLOGY. *Educ:* Pa State Univ, BS, 49, MS, 50. *Prof Exp:* Microbiologist, Eli Lilly & Co, 50-64, sr rep, 64-81, prod regist mgr, 81-85; RETIRED. *Mailing Add:* 4218 Briarwood Dr Indianapolis IN 46250

PETERSON, LENNART RUDOLPH, b Kearny, NJ, Sept 27, 36; m 56, 83; c 4. ATMOSPHERIC PHYSICS, ATOMIC PHYSICS. *Educ:* NC State Univ, BS, 58; Mass Inst Technol, SM, 63, PhD(theoret nuclear physics), 66. *Prof Exp:* Fel atmospheric physics, 66-67, interim asst prof physics & phys sci, 67-68, asst prof, 68-72, assoc prof physics & phys sci, 72-79, PROF PHYSICS, UNIV FLA, 79- *Mem:* Am Phys Soc. *Res:* Nuclear structure and reaction theory; electron impact cross sections and energy degradation as well as aurora and airglow. *Mailing Add:* Dept Phys Univ Fla Gainesville FL 32611

PETERSON, LEROY ERIC, b Anaconda, Mont, July 22, 17; m 41; c 3. PHYSICS. *Educ:* Oberlin Col, AB, 38; Univ Notre Dame, MS, 40, PhD(physics), 42. *Prof Exp:* Instr physics, Univ Notre Dame, 42-45; physics sect leader, Am Viscose Corp, 46-55; res physicist, Polychem Dept, E I du Pont de Nemours & Co, 55-60, sr res physicist, Textile Fibers Dept, 60-61; assoc prof physics, Drexel Inst Technol, 61-67; assoc prof physics, Villanova Univ, 67-83; RETIRED. *Concurrent Pos:* Res physicist, Manhattan Proj, Univ Notre Dame, 43-45; res physicist, SAM Labs, Carbide & Carbon Chem Corp, 45-46. *Res:* Thermodynamics of natural and synthetic rubber; physical behavior of cellulosic and synthetic fibers; equation of state of some synthetic rubbers; optical and physical properties of cellulose and rayon; mechanical properties of thermoplastics; test development and physical testing of plastics; administration and teaching of physics. *Mailing Add:* 217 E Col St No 12 Oberlin OH 44074-1314

PETERSON, LOWELL E, b Keokuk, Iowa, Apr 20, 26; m 48; c 5. PHYSICAL CHEMISTRY. *Educ:* Univ Minn, BS, 50, PhD, 54. *Prof Exp:* Proj leader, Cent Res Labs, Gen Mills, Inc, 54-60, res assoc, 60-64, head chem res dept, 64-68, mgr chem res, Chem Div, 68, dir appl res, Gen Mills Chem Inc, 68-74; DIR BASIC DEVELOP, HENKEL CORP, 75- *Mem:* Am Chem Soc. *Res:* Instrumental methods of organic analysis; resins; water soluble polymers. *Mailing Add:* Gen Mills Inc 2010 E Hennepin Minneapolis MN 55413

PETERSON, LYNN LOUISE MEISTER, b NJ, Jan 3, 41; m 64; c 1. INTELLIGENT COMPUTER-BASED EDUCATION. *Educ:* Wittenberg Univ, BA, 62; Duke Univ, MA, 63, Univ Mich, Ann Arbor, MA, 66; Univ Tex Health Sci Ctr, Dallas, PhD(math sci), 78. *Prof Exp:* Instr math, Kent State Univ, 63-64; programmer & analyst, Human Performance Ctr, Univ Mich, 66-67 & Johnson Res Found, Univ Pa, 67-68; programmer & analyst, Sci Comput Ctr, 68-71, asst dir, Med Comput Res Ctr, 71-79, asst prof med comput sci, Univ Tex Health Sci Ctr, Dallas, 78-82; asst prof, 82-87, ASSOC PROF COMPUT SCI ENG, UNIV TEX ARLINGTON, 87- *Concurrent Pos:* Prog dir, Grad Prog Math Sci, 80-81; vchmn, Spec Interest Group Biomed Comput-Asn Comput Mach, 80-82; treas, Health Educ Network, 81-; managing ed, Med Comput Sci J, 81-82; prin investr subcontract, Nat Libr Med, 81-82. *Mem:* Sigma Xi (pres, 80-81); Soc Med Decision Making; Asn Comput Mach; AAAS. *Res:* Application of artificial intelligence techniques to medicine, manufacturing and computer-based instruction. *Mailing Add:* 3339 Darbyshire Dallas TX 75229

PETERSON, LYSLE HENRY, b Minneapolis, Minn, Jan 21, 21; m 43; c 4. PHYSIOLOGY, CARDIOVASCULAR REHABILITATION. *Educ:* Univ Minn, BA, 43; Univ Pa, MD, 50. *Prof Exp:* Asst physiol chem, Med Sch, Univ Minn, 42-43; from asst to instr physiol, Univ Pa, 46-51, intern, Hosp, 51-52, from asst prof to assoc prof physiol & surg res, 52-60, from vpres to pres, Univ City Sci Ctr, 65-71, prof physiol, 60-75, dir, Bockus Res Inst, 61-75, chmn med bd, Grad Hosp, 69-75; vpres acad affairs, 75-77, PROF PHYSIOL, UNIV TEX HEALTH SCI CTR, HOUSTON, 75-; DIR HOUSTON CARDIOVASC REHAB CTR, 78- *Concurrent Pos:* Estab investr, Am Heart Asn, 52-56; mem sci adv bd, US Air Force, 61-77, chmn aerospace biosci panel, 70-77, consult, US Navy, NIH, Nat Acad Sci & Inst Defense Anal. *Honors & Awards:* Bordon Res Prize; Bell Res Prize; John Clark Res Award; Am Physiol Soc Award. *Mem:* Fel AAAS; Am Physiol Soc; Am Heart Asn (vpres, 65-68); fel Am Col Cardiol; Am Soc Clin Invest; Sigma Xi. *Res:* Control and regulation of circulatory system in health and disease; systems analysis and computer applications to biological systems. *Mailing Add:* 3711 San Felipe 8G Houston TX 77027

PETERSON, MALCOLM LEE, community medicine; deceased, see previous edition for last biography

PETERSON, MAURICE LEWELLEN, b Lyons, Nebr, Dec 30, 13; m 38; c 3. AGRONOMY. *Educ:* Univ Nebr, BS, 38; Kans State Col, MS, 40; Iowa State Col, PhD(agron), 46. *Hon Degrees:* DSc, Univ Nebr, 81. *Prof Exp:* Asst, Div Forage Crops & Dis, USDA, Kans State Col, 38-40; jr agronomist, S Great Plains Field Sta, USDA, Okla, 40-42, asst agronomist, 42-43; res assoc agron, Iowa State Col, 43-46, res asst prof, 46-48; from asst prof to assoc prof agron, 48-67, chmn dept, 52-59, dir agr exp sta, 62-65, dean col agr, 63-67, prof, 67-81, EMER PROF AGRON, UNIV CALIF, DAVIS, 81- *Concurrent Pos:* Fulbright adv res scholar, Welsh Plant Breeding Sta, Aberystwyth, Wales, 56-57; consult, State Exp Stas Div, Agr Res Serv, USDA, 58-; fel, Am Soc Agron; AAAS. *Mem:* Crop Sci Soc Am (pres, 59); Am Soc Agron; Am Soc Plant Physiol. *Res:* Physiology of rice. *Mailing Add:* 1922 Amador Ave Davis CA 95616

PETERSON, MELBERT EUGENE, b Moline, Ill, July 30, 30; m 67; c 1. ORGANIC CHEMISTRY. *Educ:* Augustana Col, Ill, AB, 53; Univ Ill, MS, 55; Okla State Univ, PhD, 67. *Prof Exp:* Res chemist, Morton Chem Co, 56-58; from asst prof to assoc prof, 58-77, PROF CHEM, AUGUSTANA COL, ILL, 77- *Concurrent Pos:* Dir, Sky Ridge Observ, Ill, 58-64; dir, John Deere Planetarium, 87- *Mem:* Am Chem Soc. *Res:* Nucleophilic displacement reactions on organophosphorus compounds. *Mailing Add:* Dept Chem Augustana Col 639 3th St Rock Island IL 61201

PETERSON, MELVIN NORMAN ADOLPH, b Evanston, Ill, May 27, 29; m 58; c 4. MARINE GEOLOGY. *Educ:* Northwestern Univ, BS, 51, MS, 56; Harvard Univ, PhD(geol), 60. *Prof Exp:* Asst res geologist, Scripps Inst Oceanog, Univ Calif, San Diego, 60-62, from asst prof to assoc prof oceanog, 63-87, chief scientist, 68-70, prin investr & proj mgr, Deep Sea Drilling Proj, 70-75; chief scientist, Nat Oceanic & Atmospheric Admin, Washington, DC, 75-89; SCIENTIST, PAC FORUM, 91- *Concurrent Pos:* NSF res grants, 62-66; Am Chem Soc grant, 64-66; mem adv comt to Secy Comn, Marine Petrol & Minerals, 74-78; emer prof, Scripps Inst Oceanog, Univ Calif, San Diego, 87- *Mem:* AAAS; fel Geol Soc Am; NY Acad Sci. *Res:* Origin of minerals and mineral assemblages in sediments; kinetics of crystallization of low temperature phases; volcanism and hydrothermal emanations in ocean basins; deep sea sedimentary and tectonic processes; deep water drilling and scientific exploration technology; management of national and international phases of ocean drilling; rates of dissolution of calcium carbonate in oceans. *Mailing Add:* Pacific Forum 1001 Bishop St Suite 1150 Panahi Towers Honolulu HI 96813

PETERSON, MILLER HARRELL, b Darlington, SC, Jan 3, 25; m 50; c 3. ELECTROCHEMISTRY, CORROSION. *Educ:* Clemson Univ, BS, 44; Univ NC, MA, 51. *Prof Exp:* Instr chem, Clemson Univ, 46-47; phys chemist, US Naval Ord Lab, 51-54, phys chemist, Naval Res Lab, 54-61, res chemist, 61-63, HEAD MARINE CORROSION SECT, US NAVAL RES LAB, 63- *Mem:* Am Chem Soc; Electrochem Soc; Nat Asn Corrosion Eng; AAAS; Sigma Xi. *Res:* Marine corrosion; stress corrosion cracking; corrosion fatigue; cathodic protection; corrosion in the deep ocean; materials for design of marine structures and ships. *Mailing Add:* 12807 Pine Tree Lane Ft Washington MD 20744

PETERSON, NEAL ALFRED, b San Francisco, Calif, Sept 24, 29; m 58; c 2. BIOCHEMISTRY. *Educ:* Univ San Francisco, BA, 52, MA, 54; Stanford Univ, PhD(physiol), 59. *Prof Exp:* Res asst physiol, Stanford Univ, 59-61; USPHS res fel & res asst neurochem, Univ Calif, Berkeley, 61-66; res biochemist, Dept Ment Hyg, 66-73, assoc res biochemist, 73-78, res biochemist, 78-91, EMER PROF, LANGLEY PORTER NEUROPSYCHIAT INST, UNIV CALIF, SAN FRANCISCO, SONOMA STATE HOSP, 91- *Mem:* Soc Neurosci. *Res:* Neurochemistry, including investigation of biochemical processes which take place in nerve endings isolated from the central nervous system. *Mailing Add:* Brain Behav Res Ctr Sonoma State Hosp Eldridge CA 95431

PETERSON, NORMAN CORNELIUS, b Joliet, Ill, May 3, 29; m 55; c 4. PHYSICAL CHEMISTRY. *Educ:* Mass Inst Technol, SB, 51; Iowa State Univ, PhD(chem), 56. *Prof Exp:* Assoc prof phys chem, NDak State Univ, 55-59; res assoc chem, Polytech Inst NY, 59-61, vis asst prof, 61-62, from asst prof to assoc prof, 62-72, PROF CHEM, POLYTECH UNIV, 72- *Concurrent Pos:* Vis scientist, US Nat Bur Standards, 70-71 & 79-82, consult, 71- *Mem:* Am Chem Soc; Am Phys Soc. *Res:* Kinetics of inorganic reactions; chemical instrumentation; molecular beam scattering. *Mailing Add:* 17732 Caddy Dr Derwood MD 20855-1002

PETERSON, NORMAN L(EE), physical metallurgy, ceramics science & technology; deceased, see previous edition for last biography

PETERSON, OTIS G, b Galesburg, Ill, Nov 17, 36; m 75; c 5. OPTICS, LASER PHOTOCHEMISTRY. *Educ:* Univ Ill, BS, 58, MS, 60, PhD(solid state physics), 65. *Prof Exp:* Staff mem thin films, Eastman Kodak Co, 65-68, sr staff mem org dye lasers, 69, group leader org dye lasers, 69-73; mem staff, Lawrence Livermore Lab, 73-75; res assoc, Allied Chem Corp, 75-78, mgr optical prod, 78-79; assoc group leader, Appl Photochem Div, Los Alamos Nat Lab, 79-82, group leader, Chem & Laser Sci Div, 82-89; SCI ATTACHE, DEPT STATE, US EMBASSY, LONDON, 89- *Honors & Awards:* Indust Res-100 Award, 71, 80. *Mem:* Am Phys Soc; Inst Elec & Electronic Engrs; Optical Soc Am; Sigma Xi. *Res:* Thermodynamic properties of noble-gas solids; opto-electronic properties of evaporated thin films of II-IV compounds; tunable lasers, including organic dye lasers; laser isotope separation; laser induced chemistry; Alexandrite lasers; nonlinear optics including stimulated Raman scattering; solid state lasers; free electron lasers; science policy. *Mailing Add:* US Embassy London Box 38 FPO New York NY 09509

PETERSON, PAUL CONSTANT, b Kewanee, Ill, Mar 16, 40; m 61; c 2. ACAROLOGY, SYSTEMATICS. *Educ:* Gustavus Adolphus Col, BS, 68; Univ Nebr, PhD(entom), 68. *Prof Exp:* Instr & cur, Entom Div, Nebr State Mus, Univ Nebr, 67-68; from asst prof to assoc prof, 68-76, PROF BIOL, YOUNGSTOWN STATE UNIV, 76- *Mem:* AAAS; Entom Soc Am; Acarol Soc Am; Am Inst Biol Sci. *Res:* Biology, ecology and systematics of mites parasitic on birds; reproductive strategies among analgoid mites. *Mailing Add:* Dept Biol & Sci Youngstown State Univ Youngstown OH 44555

PETERSON, PAUL E, b Denison, Tex, Aug 27, 29. ORGANIC CHEMISTRY. *Educ:* NTex State Col, BS, 51, MS, 52. *Prof Exp:* NSF fel, Calif Inst Technol, 56-58; asst prof chem, Purdue Univ, 58-59; from asst prof to prof chem, St Louis Univ, 59-71; PROF CHEM, UNIV SC, 71- *Mem:* Am Chem Soc. *Res:* Organosilicon chemistry; chiral reagents; vitamin D. *Mailing Add:* Dept Chem Univ SC Columbia SC 29208

PETERSON, PAUL W(EBER), engineering mechanics; deceased, see previous edition for last biography

PETERSON, PETER ANDREW, b Bristol, Conn, Mar 17, 25; m 48; c 2. GENETICS. *Educ:* Tufts Col, BS, 47; Univ Ill, PhD(bot), 53. *Prof Exp:* Asst, Dept Genetics, Carnegie Inst, 47-48 & Univ Ill, 49-52; asst geneticist, Univ Calif, 53-56; assoc prof genetics, 56-68, PROF AGRON & GENETICS, IOWA STATE UNIV, 68- *Concurrent Pos:* NIH fel, 69; vis scientist, Univ Vienna, 73; Plant Breeding Inst, Cambridge, Eng, 73; co-ed, Maydica, 75-; vis prof, Max Planck Inst, Koln, WGer, 76-; assoc ed, Genetical Res. *Mem:* AAAS; Am Naturalist; Genetics Soc Am. *Res:* Mutation; maize genetics; cytogenetics; linkage; male sterility; controlling regulatory elements; bacteria-insertion sequences. *Mailing Add:* Dept Agron Iowa State Univ Ames IA 50011

PETERSON, RALPH EDWARD, b Paola, Kans, Aug 21, 18; m 44; c 6. ENDOCRINOLOGY. *Educ:* Kans State Col, BS, 40, MS, 41; Columbia Univ, MD, 46. *Prof Exp:* Dir biol lab, Standard Brands, Inc, 42-43; intern, Univ Minn, 46-47, instr med, 47-48; asst chief biochem, Army Med Serv Grad Sch, Walter Reed Army Med Ctr, 50-52; asst med, Peter Bent Brigham Hosp, 52-53; sr clin investr, Nat Inst Arthritis & Metab Dis, 53-58; assoc prof, 58-68, dir Div Endocrinol, 58-63, PROF MED, MED COL, CORNELL UNIV, 68- *Concurrent Pos:* Fulbright Award, 64-65; assoc attend physician, New York Hosp, 58-70, attend physician, 70-83; dir Med Res Serv, Dept Vet Affairs, 83-90. *Mem:* AAAS; Am Soc Clin Invest; Am Fedn Clin Res; Asn Am Physicians; Endocrine Soc. *Res:* Steroid metabolism. *Mailing Add:* 14 Oyster Landing Club Hilton Head Island SC 29928

PETERSON, RAYMOND DALE AUGUST, b Minneapolis, Minn, Oct 5, 30; m 53; c 3. PEDIATRICS, IMMUNOLOGY. *Educ:* Univ Minn, BA, 52, BS, 53, MD, 55. *Prof Exp:* Pediatrician, Univ Minn Hosps, 58-60, Am Heart Asn estab investr, 63-65; Am Heart Asn estab investr, Univ Uppsala, 65-67; from assoc prof to prof pediat, Univ Chicago & La Rabida Inst, 67-73, from actg dir to dir, Inst, 68-73, dir, La Rabida Children's Hosp, 70-73; PROF PEDIAT, MED SCH, UNIV S ALA, 73- *Concurrent Pos:* Am Rheumatism Asn fel immunol, Univ Minn Hosps, 60-61, NIH fel, 61-63; Guggenheim fel, Univ Uppsala, 65-67. *Mem:* AAAS; Am Asn Immunologists; Am Soc Exp Path; Soc Pediat Res. *Res:* Development of the immune system; pathogenesis of malignancies of the lymphoid tissues; cellular differentiation. *Mailing Add:* Univ SAla 394 Cancer Ctr Bldg Mobile AL 36688

PETERSON, RAYMOND GLEN, b Denver, Colo, Feb 12, 36; m 59; c 3. ANIMAL BREEDING, STATISTICS. *Educ:* Univ Wyo, BS, 62; Univ Ill, MSc, 65, PhD(animal breeding), 68. *Prof Exp:* Res asst animal breeding, Univ Ill, 62-68; ASST PROF ANIMAL BREEDING, UNIV BC, 68- *Res:* Estimation of environmental and genetic parameters of biological traits in cattle. *Mailing Add:* 4291 W 16th Vancouver BC V6R 3E5 Can

PETERSON, REIDER SVERRE, b Minot, NDak, May 17, 39; m 63; c 2. STATISTICS. *Educ:* Northern Ariz Univ, BS, 61; Univ Maine, MA, 65; Mont State Univ, PhD(statist), 74. *Prof Exp:* Teacher math, Idaho Falls High Sch, 61-62; instr math & statist, Northern Ariz Univ, 65-68; statistician, Union Carbide Corp, 74-75; ASST PROF MATH & STATIST, SOUTHERN ORE STATE COL, 75- *Mem:* Am Statist Asn. *Res:* Ratio extimation in randomized response designs. *Mailing Add:* Dept Math & Comput Sci Southern Ore State Col Ashland OR 97520

PETERSON, RICHARD BURNETT, b Omaha, Nebr, Apr 12, 49; m 81; c 2. PHOTORESPIRATION, PHOTOSYNTHESIS. *Educ:* Unvi Nebr, BA, 71; Univ Wis, Madison, PhD(biochem), 76. *Prof Exp:* Res assoc, Plant Res Lab, Mich State Univ, 76-78; res assoc, C F Kettering Res Lab, 78-79; asst scientist, 79-82, ASSOC SCIENTIST, CONN AGR EXP STA, 82- *Mem:* Am Soc Plant Physiologists; AAAS; Crop Sci Soc Am. *Res:* Photosynthesis and nitrogen fixation by filamentous cyanobacteria; regulation of photorespiration in higher plants and relationship of photosynthesis to crop yield; relationships between fluorescence and quantum yield of photosynthesis. *Mailing Add:* Dept Biochem & Genetics Conn Agr Exp Sta PO Box 1106 New Haven CT 06504

PETERSON, RICHARD CARL, b Duluth, Minn, Feb 10, 48; m 77. MECHANICAL ENGINEERING, INTERNAL COMBUSTION ENGINES. *Educ:* Univ Minn, Duluth, BA, 70; Purdue Univ, MS, 72, PhD(mech eng), 81. *Prof Exp:* Res asst, Purdue Univ, 74-81; sr res engr, 81-88, STAFF RES ENGR, GEN MOTORS CORP, 88- *Mem:* Am Soc Mech Engrs; Combustion Inst; Soc Automotive Engrs; Sigma Xi. *Res:* Combustion and combustion diagnostics; chemical kinetics; heat transfer; internal combustion engines. *Mailing Add:* Dept 57 Gen Motors Res Labs 30500 Mound Rd Warren MI 48090-9055

PETERSON, RICHARD CHARLES, b Los Angeles, Calif, Feb 25, 31; m 53; c 1. GEOLOGY, ENVIRONMENTAL SCIENCES. *Educ:* Univ Calif, Berkeley, AB, 53; Univ Ariz, MS, 63, PhD(geol), 68. *Prof Exp:* Geophysicist, Pan Am Petrol Corp, 53-60; instr geol, Univ Ariz, 63-68; assoc prof, 68-71, PROF GEOL & DIR PLANETARIUM, ADAMS STATE COL, 71-, HEAD DEPT, 85- *Mem:* Geol Soc Am; Am Asn Petrol Geol; Nat Asn Geol Teachers. *Res:* Structural geology of the Santa Catalina Mountains, Arizona; stratigraphy of the Sangre de Cristo Mountains, Colorado. *Mailing Add:* Dept Geol Adams State Col Alamosa CO 81102

PETERSON, RICHARD ELSWORTH, b Seattle, Wash, Aug 27, 21; m 55; c 5. INTERNAL MEDICINE, NUCLEAR MEDICINE. *Educ:* Univ Wash, BS, 42; Northwestern Univ, MD, 46; Am Bd Internal Med, dipl; Am Bd Nuclear Med, dipl. *Prof Exp:* Clinician, Radioisotope Unit, Vet Admin Hosp, Long Beach, Calif, 50-52; chief radioisotope serv, Vet Admin Hosp, Iowa City, 52-70; from clin assoc to clin asst prof med, 53-60, clin assoc prof internal med, 60-63, assoc prof med & radiol, Univ Iowa, 63-66, prof internal med & radiol, 66-; chief nuclear med serv, Vet Admin Hosp, Iowa City, 70-; AT UNIV WIS-MADISON. *Mem:* Soc Nuclear Med (secy, 65-68, treas, 68-69); Am Bd Nuclear Med; Asn Am Med Cols; NY Acad Sci; Am Col Nuclear Physics. *Res:* Endocrinology; hematology; metabolism; genetics and human radiobiology; diagnostic applications of radioisotope techniques; body composition in health and disease; kinetics of radioactive metabolites as tracers of physiologic processes. *Mailing Add:* Squire Lane RR 2 Box 316 North Liberty IA 52317

PETERSON, RICHARD GEORGE, b Milwaukee, Wis, June 12, 41; m 64; c 2. NEUROBIOLOGY. *Educ:* Bethel Col, Minn, BA, 64; Univ NDak, MS, 67, PhD(anat), 69. *Prof Exp:* Asst res anatomist, Univ Calif, Los Angeles, 69-71; asst prof neurobiol, Med Sch, Univ Tex, Houston, 71-77; assoc prof, 77-85, PROF ANAT, SCH MED, IND UNIV, 85- *Mem:* Am Asn Anat; Am Diabetes Asn; NAm Asn Study Obesity; Sigma Xi. *Res:* Neural ultrastructure; histochemistry of neural myelin; developmental neural structure; diabetic neuropathy; fatty diabetic animal models. *Mailing Add:* Dept Anat Ind Univ Sch Med Indianapolis IN 46202-5120

PETERSON, RICHARD WALTER, b Chewelah, Wash, Mar 6, 33; m 51, 83; c 4. ENGINEERING, MATERIALS SCIENCE. *Educ:* Univ Idaho, BS, 57. *Prof Exp:* Process engr, Alcoa, Wenatchee, Wash, 57-64, Alcoa, Tenn, 64-65, STAFF ENGR RES & DEVELOP, ALCOA LABS, NEW KENSINGTON, PA, 65- *Mem:* Am Chem Soc. *Res:* Improvement in materials and processes for producing carbon electrodes for aluminum smelting by the Hall-Heroult process. *Mailing Add:* Alcoa Labs Box 772 New Kensington PA 15068

PETERSON, ROBERT C, b Suffern, NY, May 18, 36; m 63; c 2. PULP CHEMISTRY, PAPER CHEMISTRY. *Educ:* Mich Technol Univ, BS, 57; State Univ NY Col Forestry, BS, 60; NY Univ, MS, 67; State Univ NY Col Environ Sci & Forestry, PhD(paper sci), 71. *Prof Exp:* Res chemist, Eastern Res Div, ITT Rayonier, 60-62; tech serv engr, St Regis Tech Ctr, 62-65, res & develop engr wood chem, 65-67; asst prof, 71-74, ASSOC PROF PAPER TECHNOL, MIAMI UNIV, 74- *Concurrent Pos:* Fulbright-Hays vis lectr, Coun Int Exchange Scholar, Malaysia, 74-75. *Mem:* Am Inst Chem Engrs; Tech Asn Pulp & Paper Indust. *Res:* Solid waste disposal, wastepaper recycling; water and air pollution abatement within the paper industry; sources of non-wood fibers for paper and the application of novel pulping methods to agricultural residues. *Mailing Add:* 101 E Central Ave Oxford OH 45056

PETERSON, ROBERT H F, b Middletown, Pa, Dec 1, 35. MICROBIOLOGY. *Educ:* Elizabethtown Col, BS, 59; Univ Kans, PhD(microbiol), 70. *Prof Exp:* Asst prof cell biol, Sloan-Kettering Div, Grad Sch Med Sci, Cornell Univ, 75-82; res assoc cancer res, 72-74, 82, asst lab mem, 72-89, ASST LAB MEM, DEPT HEMATOPOIETIC DEVELOP, SLOAN-KETTERING INST CANCER RES, 89. *Concurrent Pos:* Fel, Roche Inst Molecular Biol, 70-72; instr microbiol, Sloan-Kettering Div, Grad Sch Med Sci, Cornell Univ, 73-74. *Mem:* Am Soc Microbiol; Biochem Soc; Am Soc Cell Biol; Am Asn Cancer Res. *Res:* Biochemical changes of plasma membranes accompanying the malignant state. *Mailing Add:* 29 Beck Ave Rye NY 10580

PETERSON, ROBERT HAMPTON, b Jamestown, NDak, Oct 29, 22; m 48; c 8. ORGANIC CHEMISTRY. *Educ:* St John's Univ, Minn, BA, 48; NDak State Univ, MS, 50; Univ Utah, PhD(chem), 62. *Prof Exp:* From instr to prof chem, NDak State Univ, 50-68; vpres acad affairs, 70-76, PROF CHEM & CHMN DIV SCI & MATH, ST LEO COL, 68- *Mem:* Am Chem Soc. *Res:* Phenol-formaldehyde polymers; reaction mechanisms; correlation of properties with molecular structure; chemical modification of high polymers; aromatic azo compounds. *Mailing Add:* St Leo Col Box 2188 St Leo FL 33574

PETERSON, ROBERT LAWRENCE, b Strathmore, Alta, Jan 20, 39; m 63; c 2. BOTANY. *Educ:* Univ Alberta, BEd, 62, MSc, 64; Univ Calif, Davis, PhD(plant morphogenesis), 68. *Prof Exp:* Assoc prof, 68-80, PROF BOT, UNIV GUELPH, 80- *Concurrent Pos:* Lansdowne fel, 84. *Mem:* Bot Soc Am; Can Bot Asn; Sigma Xi; Royal Micros Soc; Can Micros Soc; fel Royal Soc Can. *Res:* Experimental and anatomical studies on plant roots; particularly development and structure of mycorrhizae. *Mailing Add:* Dept Bot Univ Guelph Guelph ON N1G 2W1 Can

PETERSON, ROBERT LEE, b Brady, Nebr, May 21, 30; m 56; c 5. THEORETICAL SOLID STATE PHYSICS, SUPERCONDUCTIVITY. *Educ:* Colo Sch Mines, MetE, 52; Lehigh Univ, MS, 54, PhD(physics), 59. *Prof Exp:* From instr to asst prof physics, Case Inst Technol, 59-63; PHYSICIST, NAT BUR STANDARDS, 63- *Concurrent Pos:* Bd dirs, Boulder Fed Credit Union & World Christian Fel. *Mem:* AAAS; Am Phys Soc; Sigma Xi. *Res:* Semiconductor transport; solid state optics; statistical mechanics and transport theory; spin relaxation; magnetism; superconductivity. *Mailing Add:* Nat Inst Standards & Technol Div 724 03 325 Broadway Rm 2137 Boulder CO 80303

PETERSON, ROBERT W, b Elizabeth, NJ, July 14, 25; m 65; c 1. NUCLEAR PHYSICS. *Educ:* Rutgers Univ, BS, 50; Calif Inst Technol, PhD(physics), 54. *Prof Exp:* Asst physics, Calif Inst Technol, 50-54; staff mem, 54-67, alt group leader, 67-71, group leader weapons test, 71-75, staff mem, 75-78, staff mem theoret design, 78-81, ASST DIV LEADER, INT TECHNOL DIV, LOS ALAMOS SCI LAB, UNIV CALIF, 81- *Mem:* Am Phys Soc; Am Geophys Union; Sigma Xi. *Res:* Low energy nuclear reactions; nuclear spectroscopy; nuclear weapons testing; cosmic ray physics; conjugacy of visual auroras. *Mailing Add:* PO Box 83 Los Alamos NM 87544

PETERSON, ROGER SHIPP, b Ann Arbor, Mich, June 23, 31; m 53; c 3. BOTANY. *Educ:* Harvard Univ, AB, 53; Univ Mich, MA, 57, PhD(bot), 59. *Prof Exp:* Forest pathologist, US Forest Serv, 59-66 & Rocky Mt Forest & Range Exp Sta, Colo, 59-62, leader Native Rust Proj, Intermountain Forest & Range Exp Sta, Utah, 62-66; TUTOR, ST JOHN'S COL, NMEX, 66- *Mem:* AAAS; Mycol Soc Am. *Res:* Ecology and taxonomy of rust fungi; forest tree diseases; ecology of New Mexico. *Mailing Add:* St John's Col Santa Fe NM 87501

PETERSON, ROLF EUGENE, b Minneapolis, Minn, Mar 10, 21; m 44; c 2. NUCLEAR PHYSICS. *Educ:* St Olaf Col, BS, 43; Univ Wis, PhD(physics), 50. *Prof Exp:* Asst, Univ Wis, 43-44; jr scientist, Metall Lab, Univ Chicago, 44; jr scientist, Manhattan Proj, Univ Calif, 44-46; asst, Univ Wis, 46-50; mem staff, 50-57, group leader, 58-64, ALT DIV LEADER, LOS ALAMOS SCI LAB, 64- *Mem:* Am Phys Soc; Am Nuclear Soc. *Res:* Neutron scattering; critical assemblies; development of homogeneous reactors and fast breeder power reactors. *Mailing Add:* 615 Meadow Lane Los Alamos NM 87544

PETERSON, ROLF OLIN, b Minneapolis, Minn, Apr 5, 49; m 70; c 2. WILDLIFE ECOLOGY. *Educ:* Univ Minn, Duluth, BA, 70; Purdue Univ, PhD(wildlife ecol), 74. *Prof Exp:* Fel wildlife ecol, Purdue Univ, 70-74, res assoc, 74-75; asst prof, 75-84, assoc prof biol, 84-87, PROF WILDLIFE ECOL, MICH TECHNOL UNIV, 87- *Mem:* Wildlife Soc; Am Soc Mammal; AAAS. *Res:* Population ecology of northern mammals; long-term study of wolves and moose in Isle Royale National Park, a Lake Superior island; variations in predator-prey relationships in response to environment and density. *Mailing Add:* Sch Forestry & Wood Prod Mich Technol Univ Houghton MI 49931-1295

PETERSON, RONALD A, b Wayland, Mich, May 17, 37; c 2. AVIAN PHYSIOLOGY. *Educ:* Mich State Univ, BS, 60, MS, 62, PhD(avian physiol), 66. *Prof Exp:* Asst prof, 66-71, ASSOC PROF POULTRY PHYSIOL, W VA UNIV, 71- *Mem:* Poultry Sci Asn; World Poultry Sci Asn. *Res:* Physiology involved in feather release in birds; environmental physiology related to poultry. *Mailing Add:* Div of Animal & Vet Sci WVa Univ Morgantown WV 26506-6108

PETERSON, RONALD M, b Parkers Prairie, Minn, Apr 8, 22; m 56; c 3. HORTICULTURE, PLANT BREEDING. *Educ:* Colo State Univ, BS, 47; Univ Calif, Davis, MS, 49; Univ Minn, PhD(plant breeding), 53. *Prof Exp:* From asst prof to assoc prof, 53-66, head dept, 66-82, PROF HORT, SDAK STATE UNIV, 66- *Mem:* Am Soc Hort Sci; Am Pomol Soc. *Res:* Breeding of improved varieties of fruit, especially apples, pears and grapes. *Mailing Add:* Box 2207-C Univ Sta Brookings SD 57007-0996

PETERSON, ROY JEROME, b Everett, Wash, Oct 18, 39; m 62; c 4. NUCLEAR PHYSICS. *Educ:* Univ Wash, BS, 61, PhD(physics), 66. *Prof Exp:* Instr physics, Princeton Univ, 66-68; res assoc, Yale Univ, 68-70; res assoc, 70-71, lectr, 71-74, from asst prof to assoc prof, 74-78, PROF PHYSICS & ASTROPHYS, UNIV COLO, BOULDER, 78- *Concurrent Pos:* Prog dir, Intermediate Energy Physics, NSF, 78-79. *Mem:* Am Phys Soc. *Res:* Direct nuclear reaction studies; medium energy physics. *Mailing Add:* Nuclear Physics Lab Univ Colo Campus Box 446 Boulder CO 80309

PETERSON, ROY PHILLIP, b Alexandria, La, Nov 16, 34; m 66; c 4. ENDOCRINOLOGY, REPRODUCTIVE PHYSIOLOGY. *Educ:* Southern Univ, Baton Rouge, BS, 57; Univ Ore, MA, 61; Univ Iowa, PhD(endocrinol), 67. *Prof Exp:* Asst, Univ Ore, 59-61; instr biol, Southern Univ, New Orleans, 61-63; res asst, Univ Iowa, 63-64, US Pub Healths fel, 64-67; asst prof biol, Southern Univ, Baton Rouge, 67-69; from asst prof to assoc prof biol sci, Southern Ill Univ, Edwardsville, 69-78, asst dean grad sch, 71-75; assoc dir acad & health affairs, Ill Bd Higher Educ, 75-80; interim pres, Tenn State Univ, Nashville, 85-; DEP EXEC DIR ACAD AFFAIRS, KY COUN HIGHER EDUC, 80- *Concurrent Pos:* Consult biol prog, Grambling Col, 61-63, undergrad biol, Fisk Univ, 71 & minorities program Univ Kans Med Sch, 71; fel, Am Coun Educ, 73-74; tech task force life-long learning, Educ Comn of the States, 79-80; vpres, Nat Bd Dir, Compassionate Friends, Inc, 80-81, treas, 79 & pres, 81-85; mem, Nat Comn Higher Educ Issues, 81- *Mem:* AAAS; Soc Study Reproduction; Am Soc Zool; Sigma Xi. *Res:* Uptake, distribution and utilization of serum proteins in the estrogen-stimulated uterus; fate of uterus during drug inhibition of metabolism. *Mailing Add:* 1050 US 127 S Suite 101 Frankfort KY 40601

PETERSON, ROY REED, b Kansas City, Mo, June 26, 24; m 46; c 4. ANATOMY. *Educ:* Univ Kans, AB, 48, PhD, 52. *Prof Exp:* From instr to assoc prof, 52-70, PROF ANAT, WASHINGTON UNIV, 70- *Concurrent Pos:* USPHS spec res fel, 60; vis prof anat, Stanford Univ, 75-76. *Mem:* Am Asn Anat; Sigma Xi. *Res:* Placental permeability; cross-sectional anatomy; thyroid and reproduction in guinea pigs; pituitary cytology. *Mailing Add:* Dept Anat Washington Univ St Louis MO 63110

PETERSON, RUDOLPH NICHOLAS, b New York, NY, June 6, 32. BIOCHEMISTRY, PHARMACOLOGY. *Educ:* St John's Univ, NY, BS, 57; Brooklyn Col, MA, 62; Univ Fla, PhD(biochem), 65. *Prof Exp:* Res assoc chem, Lever Bros Res Labs, 57-61; res assoc biochem, Univ Fla, 65-66; from asst prof to assoc prof, pharmacol, New York Med Col, 66-76; assoc prof, 76-78, PROF PHARMACOL, SOUTHERN ILL UNIV, 78- *Concurrent Pos:* Prin investr, sperm-egg interaction, NIH, 79-89, calcium transport by sperm, 81-89. *Mem:* Soc Exp Biol & Med; Am Soc Cell Biol; Am Soc Pharmacol & Exp Therapeut; Harvey Soc; Soc Develop Biol. *Res:* Energy metabolism in mammalian spermatozoa; molecular mechanisms of fertilization. *Mailing Add:* Sch of Med Southern Ill Univ Carbondale IL 62901

PETERSON, SELMER WILFRED, b Owatonna, Minn, Sept 8, 17; m 41; c 4. PHYSICAL CHEMISTRY. *Educ:* St Olaf Col, BA, 38; Univ SDak, MA, 40; Univ Md, PhD(phys chem), 42. *Prof Exp:* Indust fel, Mellon Inst, 42-43; instr chem, La State Univ, 43-45, asst prof, 45-46; asst prof phys chem, Vanderbilt Univ, 46-49; prin chemist, Oak Ridge Nat Lab, 49-61; prof chem & chemist nuclear reactor, Wash State Univ, 61-67; sr scientist, Chem Div, Argonne Nat Lab, 67-79; RETIRED. *Concurrent Pos:* Fulbright advan res scholar, Netherlands, 54-55; res assoc, Atomic Energy Res Estab, Harwell, Eng, 66; consult, Pac Northwest Lab, Battelle Mem Inst. *Mem:* AAAS; Am Chem Soc; Am Phys Soc; Am Crystallog Asn; Royal Soc Chem; Sigma Xi. *Res:* Neutron diffraction; crystal structure; hydrogen bonding. *Mailing Add:* 6222 Wehner Way San Jose CA 95135-1445

PETERSON, SIGFRED, chemistry, for more information see previous edition

PETERSON, SPENCER ALAN, b Sioux Falls, SDak, Jan 6, 40; m 61; c 2. LIMNOLOGY. *Educ:* Sioux Falls Col, BS, 65; Univ NDak, MS, 67, PhD(limnol), 71. *Prof Exp:* Res asst limnol, Univ NDak, 68-71; aquatic biologist water qual res, Mich Water Resources Comn, 71; RES AQUATIC BIOLOGIST EUTROPHICATION RES, NAT ENVIRON RES CTR, US ENVIRON PROTECTION AGENCY, 71-, SR STAFF ECOLOGIST, 74-, RES ECOLOGIST & COORDR, NAT PROG TO EVAL LAKE RESTORATION TECHNIQUES, 75-, LEADER, HAZARDOUS MAT ASSESSMENT TEAM, 81- *Concurrent Pos:* Chief, Hazardous Waste & Water Br, US Environ Protection Agency, 84, Watershed Br, 88, regional scientist, 89, leader, Regional Effects Team, 90-; Nat Defense grad fel, 65-68, Nat Wildlife Found fel, 69- *Honors & Awards:* Tech Excellence Award, NAm Lake Mgt Soc, 90. *Mem:* NAm Soc Mgt Soc; Sigma Xi. *Res:* Eutrophication control methods; aquatic ecosystem nutrient budgets; algae macrophyte dynamics; lake restoration; regional ecological risk assessment. *Mailing Add:* Nat Environ Res Ctr 200 SW 35th Corvallis OR 97330

PETERSON, STEPHEN CRAIG, b Salt Lake City, Utah, Nov 19, 40; m 64; c 2. PHYSIOLOGY. *Educ:* Univ Calif, Berkeley, BA, 62; NMex Highlands Univ, MS, 64; Wash State Univ, PhD(physiol), 69; Univ Utah, MEEE, 84. *Prof Exp:* Res assoc physiol, Yale Univ, 68-69; from asst prof to assoc prof biophys, Fac Med, Mem Univ Nfld, 69-76; sr scientist, Fluids, Biodyn & Biomat Sect, UBTL/Div, Univ Utah Res Inst, 76-79; mgr bioeng, TRA, 83-89; PRES, SUMMITEK INC, 89- *Mem:* AAAS; Acoust Soc. *Res:* Mechanisms and modeling of cell movement; acoustics; bioengineering. *Mailing Add:* 3791 S 1860 E Salt Lake City UT 84102

PETERSON, STEPHEN FRANK, b Lafayette, Ind, Mar 2, 42; m 68; c 1. ANALYTICAL CHEMISTRY. *Educ:* Ind Univ, Bloomington, BS, 64; Cornell Univ, PhD(anal chem), 69. *Prof Exp:* Res chemist, 68-73, res supvr, 73-80, RES STAFF CHEMIST, SAVANNAH RIVER LAB, E I DU PONT DE NEMOURS & CO, INC, 80- *Mem:* Am Chem Soc. *Res:* Neutron activation analysis; in-line radiochemical analysis; nuclear fuel cycle separations processes; application of small computers to analytical chemistry; automated control of nuclear fuel cycle separations processes; instrumental methods of analysis. *Mailing Add:* 807 Valley View Aiken SC 29801

PETERSON, STEVEN LLOYD, NEUROPHARMACOLOGY. *Educ:* Univ Calif, Davis, PhD(pharmacol & toxicol), 80. *Prof Exp:* ASST PROF MED PHARMACOL, COL MED, TEX A&M UNIV, 82- *Mailing Add:* Dept Pharmacol Col Med Tex A&M Univ College Station TX 77843

PETERSON, TIMOTHY LEE, b Minneapolis, Minn, Aug 20, 46; m 84; c 3. HIGH TEMPERATURE SUPERCONDUCTORS, SEMICONDUCTORS. *Educ:* Sioux Falls Col, BA, 68; Cornell Univ, MS, 75. *Prof Exp:* Mat engr, Air Force Mat Lab, 73-80, MAT RES ENGR, WRIGHT LAB, WRIGHT PATTERSON AFB, 80- *Mem:* Mat Res Soc. *Res:* Electrical and magnetic properties of bulk and thin film high temperature superconductors; electrical transport properties of semiconductors. *Mailing Add:* Wright Lab WL-MLPO Wright Patterson AFB OH 45433

PETERSON, VERN LEROY, b Gothenburg, Nebr, Nov 8, 34; m 61; c 3. AERONOMY, METEOROLOGY. *Educ:* Univ Colo, BS, 56; Ind Univ, MA, 60, PhD(astrophys), 63. *Prof Exp:* Res physicist, Nat Oceanic & Atmospheric Admin, 63-69; assoc prof physics, Utah State Univ, 69-74; adj prof, Univ Sao Paulo, 74-77; sr scientist, Res Anal & Develop, Inc, 77-78; pres, Centennial Sci, Inc, 78-80; sr tech dir, Ocean Data Systs, Inc, 80-86; chief scientist, Tycho Technol, Inc, 86-90; SR RES SPECIALIST, EG&G, 90- *Mem:* Am Meteorological Soc; Am Geophys Union; Int Union Radio Sci. *Res:* Atmospheric radar; atmospheric turbulence; nucleosynthesis; meteorology; digital signal processing; physics of the ionosphere and airglow; numerical modeling of upper atmosphere; satellite image processing; ocean currents. *Mailing Add:* 4616 Ipswich Boulder CO 80301-4219

PETERSON, VICTOR LOWELL, b Saskatoon, Sask, Can, June 11, 34; US citizen; m 55; c 3. AERONAUTICS, ASTRONAUTICS. *Educ:* Ore State Univ, BS, 56; Stanford Univ, MS, 64; Mass Inst Technol, MS, 73. *Prof Exp:* Res scientist, 56-68, asst chief hypersonic aerodyn br, 68-69, dep chief spaceshuttle off, 69-71, chief aerodyn br, 71-74, chief, Thermo & Gas Dynamics Div, 74-84, dir aerophys, 84- 90, DEP DIR AEROPHYS, AMES RES CTR, NASA, 90- *Concurrent Pos:* Sloan fel, Mass Inst Technol, 72-73. *Mem:* Fel Am Inst Aeronaut & Astronaut, 86. *Res:* Assists with direction of research in fields of aeronautics, space sciences, life sciences, earth sciences, aerophysics, and the advancement of large-scale scientific computational systems. *Mailing Add:* NASA-Ames Res Ctr Moffett Field CA 94035

PETERSON, VINCENT ZETTERBERG, b Galesburg, Ill, June 18, 21; m 48; c 4. EXPERIMENTAL HIGH ENERGY PHYSICS. *Educ:* Pomona Col, BA, 43; Univ Calif, PhD(physics), 50. *Prof Exp:* Physicist, Naval Res Lab, 43-46 & Radiation Lab, Univ Calif, 47-50; res fel physics, Calif Inst Technol, 50-53, sr res fel, 53-58, asst prof, 58-62; PROF PHYSICS, UNIV HAWAII, 62- *Concurrent Pos:* Fulbright res scholar & Guggenheim travel grant, Italy, 59-60; prin investr high energy physics prog, Univ Hawaii; consult, AEC. *Mem:* AAAS; fel Am Phys Soc; Sigma Xi. *Res:* Elementary particle physics; multiwire proportional chambers and bubble chamber experiments; high energy neutrino interactions. *Mailing Add:* Dept Physics & Astron Univ Hawaii 2505 Correa Rd Honolulu HI 96822

PETERSON, WARD DAVIS, JR, b Ann Arbor, Mich, May 10, 27; m 50; c 5. MEDICAL MICROBIOLOGY. *Educ:* Univ Mich, Ann Arbor, BA, 47, MS, 51, PhD(epidemiol sci), 60. *Prof Exp:* Virologist, Mich State Dept Health, 51-55; res assoc, 59-62, SR RES ASSOC MICROBIOL, CHILD RES CTR MICH, 62- *Concurrent Pos:* Instr, Med Sch, Wayne State Univ, 62-76, assoc, dept immunol & Microbiol, 73-76, assoc prof, dept pediat, 79- *Mem:* Tissue Cult Asn; AAAS. *Res:* Development of mammalian cell lines in vitro; identification and characterization of cells in vitro; cytogenetics; cell transformation; tumor antigens. *Mailing Add:* 11 Norwich Pleasant Ridge Detroit MI 48201

PETERSON, WARREN STANLEY, b Worcester, Mass, Nov 1, 27. CHEMICAL METALLURGICAL PROCESSES & SAFETY ASPECTS. *Educ:* Clark Univ, AB, 39; Va Polytech Inst, MS, 40;Polytech Inst Brooklyn, PhD(inorg chem). *Prof Exp:* Res engr labs, Aluminum Co Am, 42-45, res suprvr, 45-46, asst chief, 46-47; asst proj supt, Air Reduction Co, 47-48; head phys chem & prcess metall dept, Kaiser Aluminum & Chem Corp, 48-56, reres supv, 56-57; assoc dir metals res labs, Olin Mathieson Chem Corp, 57-71, tech asst to pres Aluminum Group, 71-72; mgr reduction technol aluminum, Martin Marietta Corp, 72-80; mgr product devt, Cabot Corp, 80-82; CONSULT, 82- *Mem:* Am Chem Soc; Am Soc Metals; Am Inst Mining, Metall & Petrol Engrs; fel Am Inst Chem; Am Soc Safety Engrs. *Res:* Process metallurgy of aluminum and copper alloys; reduction and smelting of aluminum; safety in metal processing. *Mailing Add:* 2113 E 37nd St Spokane WA 99203

PETERSON, WILBUR CARROLL, b White Rock, SDak, Aug 20, 13; m 46. ELECTRICAL ENGINEERING. *Educ:* Univ Minn, BEE, 39; Mich State Univ, MS, 51; Northwestern Univ, PhD, 57. *Prof Exp:* Elec engr, Gen Elec Co, 39-48; from instr to asst prof elec eng, Mich State Univ, 48-58; assoc prof elec eng, NC State Univ, 58-79; RETIRED. *Mem:* Inst Elec & Electronic Engrs. *Res:* Circuit theory and automatic control systems; automatic control and heat transfer. *Mailing Add:* 211 Willow Valley Sq Bldg C-215 Lancaster PA 17602

PETERSON, WILLIAM ROGER, b Racine, Wis, May 22, 27; m 46; c 4. ORGANIC CHEMISTRY. *Educ:* Carroll Col, Wis, AB, 49; Univ Ill, PhD(org chem), 52. *Prof Exp:* Asst inorg chem, Univ Ill, 49-50; asst high polymers, Off Rubber Reserve, 50-52; res chemist polymer chem, E I du Pont de Nemours & Co, 52-57; sr chemist & div consult, Cent Res, Continental Can Co, Ill, 57-59; supvr polymer res, United Carbon Co, Inc, 59-61, res div mgr, 61-63; res dir & exec vpres, Southeast Polymers Inc, Tenn, 63-69; tech dir, Chem Div, 69-74, mgr res & develop, 74-79, TECH DIR, TEXTILE RUBBER & CHEM, GAF CORP, 79- *Concurrent Pos:* Spec sci ed, McGraw-Hill Co, 59-62; adj prof & consult, Univ Tenn. *Mem:* Am Chem Soc; AAAS. *Res:* Polymer chemistry, including monomer synthesis; synthetic fibers; vinyl resins, elastomers and emulsion polymerization. *Mailing Add:* 1508 Dalewood Dr Chattanooga TN 37411

PETERSON, WILLIAM WESLEY, b Muskegon, Mich, Apr 22, 24; m 72; c 5. ELECTRICAL ENGINEERING. *Educ:* Univ Mich, AB, 48, BSE, 49, MSE, 50, PhD(elec eng), 54. *Prof Exp:* Assoc engr, Eng Lab, Int Bus Mach Corp, NY, 54-56; assoc prof elec eng, Univ Fla, 56-63; vis prof, Chiao Tung Univ, 63-64; prof elec eng, 64-66, PROF ICS, UNIV HAWAII, 66- *Concurrent Pos:* Vis assoc prof, Mass Inst Technol, 59-60; vis prof, Osaka Univ, 71 & 79. *Honors & Awards:* Centenial Medal, Inst Elec & Electronics Engrs. *Mem:* Inst Elec & Electronics Engrs; fel Asn Comput Mach. *Res:* Information theory; reliability and error control; computers. *Mailing Add:* Dept Info & Comput Sci Univ Hawaii 2565 The Mall Honolulu HI 96822

PETERSON-FALZONE, SALLY JEAN, b Paxton, Ill, Feb 22, 42; m 75. SPEECH, HEARING SCIENCE. *Educ:* Univ Ill, BS, 64, MA, 65; Univ Iowa, PhD(speech path), 71. *Prof Exp:* Speech pathologist, Inst Phys Med & Rehab, Peoria, Ill, 70-71; res assoc audiol, Cleft Palate Clin, 65-67, ASSOC PROF SPEECH PATH, CTR CRANIOFACIAL ANOMALIES, UNIV ILL, 71- *Mem:* Am Speech & Hearing Asn; Am Cleft Palate Asn; AAAS; Sigma Xi. *Res:* Craniofacial malformation syndromes; experimental phonetics; speech production in anomalous supralaryngeal vocal tracts. *Mailing Add:* Ctr Cranofacial Anomalies Univ Calif 747-S Med Sci San Francisco CA 94143-0442

PETERSON-KENNEDY, SYDNEY ELLEN, b Euclid, Ohio, Feb 12, 58; m 83. ELECTRON TRANSFER. *Educ:* Georgetown Univ, BS, 80; Northwestern Univ, MS, 81, PhD(chem), 85. *Prof Exp:* ASST PROF INORG CHEM, WHITWORTH COL, SPOKANE, WASH, 85- *Mem:* Am Chem Soc; Biophys Soc. *Res:* Photoinitiated electron transfer; full temperature dependence within hemoglobin hybrids; effects on electron transfer resulting from a change in the insulating matrix in which the redox centers are imbedded. *Mailing Add:* Chem Dept Whitworth Col Spokane WA 99251-0002

PETERSSON, GEORGE A, b New York, NY, July 6, 42; m 72; c 1. THEORETICAL CHEMISTRY. *Educ:* City Col New York, BS, 64; Calif Inst Technol, PhD(chem), 70. *Hon Degrees:* MA, Wesleyan Univ. *Prof Exp:* Res fel chem, Harvard Univ, 70-71; J W Gibbs instr, Yale Univ, 71-73; from asst prof to assoc prof, 73-85, PROF CHEM, WESLEYAN UNIV, 85- *Concurrent Pos:* Prin investr, Off Adv Sci Comput grants, PRF & NSF. *Mem:* Am Chem Soc; Am Phys Soc. *Res:* Theory of electron correlation in atoms, molecules and especially in transition states for chemical reactions; development of ab-initio and semiempirical methods for the calculation of molecular interaction energies in solids and liquids. *Mailing Add:* Hall-Atwater Labs Wesleyan Univ Middletown CT 06457

PETES, THOMAS DOUGLAS, b Washington, DC, Mar 27, 47; m 73; c 2. GENETICS, BIOCHEMISTRY. *Educ:* Brown Univ, ScB, 69; Univ Wash, PhD(genetics), 73. *Prof Exp:* Jane Coffin Childs fel, Div Microbiol, Nat Inst Med Res, 73-75; fel, Dept Biol, Mass Inst Technol, 75-77; from asst prof to assoc prof, Dept Microbiol, 77-84, prof, Dept Molecular Genetics & Cell Biol, Univ Chicago, 84-88; PROF, DEPT BIOL, UNIV NC, CHAPEL HILL, 88- *Concurrent Pos:* Res grants, Jane Coffin Childs Mem Fund, 77-78, NIH, 77- & March of Dimes, 78-79, Am Chem Soc, 89- *Mem:* Genetics Soc Am. *Res:* Chromosome structure and replication in the yeast Saccharomyces cerevisiae; genetic behavior of repeating DNA genes in eucaryotes. *Mailing Add:* Dept Biol Univ NC Chapel Hill NC 27599-3280

PETHICA, BRIAN ANTHONY, b London, Eng, July 9, 26; m 52, 78; c 5. SURFACE CHEMISTRY, COLLOID SCIENCE. *Educ:* London Univ, BSc, 46, PhD(phys chem), 50, DSc, 62; Cambridge Univ, PhD, 53, ScD, 71. *Hon Degrees:* DSc, Clarkson Univ, 75. *Prof Exp:* Lectr pharmacol, Med Sch, Univ Birmingham, 49-53; sr res asst, Dept Colloid Sci, Univ Cambridge, 53-58; lectr phys chem, Inst Sci & Tech, Manchester Univ, 58-61, sr lectr, 61-63 & hon reader, 63-71; dean arts & sci, Clarkson Col, 76-81, prof biophys & chem, 81-83; CORP VPRES, ELECTROBIOL INC, 84- *Concurrent Pos:* Vis lectr, Columbia Univ, 58; div mgr chem physics, Unilever Res Lab, Port Sunlight, UK, 63-65, head lab, 65-75 & chmn, Technol Policy Study, 75-76; Chmn, Chem Interfaces, Gordon Conf, 79. *Mem:* Fel Royal Soc Chem; Am Chem Soc; Faraday Soc (vpres, 74). *Res:* Surface chemistry; membrane biophysics; solution and surface thermodynamics; colloid science; biocolloids; biological effects of magnetic and electric fields. *Mailing Add:* Electrobiol Inc PO Box 345 6 Upper Pond Rd Parsippany NJ 07054

PETHICK, CHRISTOPHER JOHN, b Horsham, Eng, Feb 22, 42. THEORETICAL PHYSICS. *Educ:* Univ Oxford, BA, 62, DPhil(physics), 65. *Prof Exp:* Res assoc, 66-68, res asst prof, 68-69, assoc prof, 70-73, prof physics, Univ Ill, 73-; PROF PHYSICS, NORDITA, COPENHAGEN, DENMARK, 75- *Concurrent Pos:* Fel physics, Magdalen Col, Oxford Univ, 65-70; vis prof physics, Nordita, Copenhagen, Denmark, 75- *Mem:* Fel Am Phys Soc. *Res:* Theoretical low-temperature and solid state physics; theoretical astrophysics. *Mailing Add:* Nordita Blegdamsvej 17 DK-2100 Copenhagen Denmark

PETICOLAS, WARNER LELAND, b Lubbock, Tex, July 29, 29; m 55; c 5. PHYSICAL BIOCHEMISTRY. *Educ:* Tex Technol Col, BSc, 50; Northwestern Univ, PhD(phys chem), 54. *Prof Exp:* Res chemist, E I du Pont de Nemours & Co, 54-55, 57-59, res assoc, 59-60; sr asst scientist, NIH, 55-57; staff scientist, Res Lab, Int Bus Mach Corp, Calif, 60-65, mgr chem phys group, 66-67; vis assoc prof mat sci, Calif Inst Technol, 65-66; PROF CHEM, UNIV ORE, 67- *Concurrent Pos:* Guggenheim fel, Inst Laue-Langevin, Grenoble, France, 73-74; vis prof, Univ Paris, 80-81. *Honors & Awards:* Alexander Von Humboldt Award, 84/85. *Mem:* Biophys Soc; Am Soc Biol Chemists; Am Chem Soc; Royal Soc Chem; fel Am Phys Soc; Sigma Xi. *Res:* Biomembranes; proteins; nucleic acids; Raman spectroscopy; inelastic laser light scattering. *Mailing Add:* Dept Chem Univ Ore Eugene OR 97403

PETILLO, PHILLIP J, b Jersey City, NJ, Sept 4, 45; m 71; c 5. MEDICAL & HEALTH SCIENCES. *Educ:* Columbia Univ, BS, 67; Graham Inst, MS, 82. *Prof Exp:* Res & develop, Columbia Univ, 65-67; CONSULT, 71- *Concurrent Pos:* Prod developer, Site Microsurgical Systs, 82-89, Codman, 87-89; lectr, Am Chem Soc, 85; prod develop, Am Design Coun, 86, Johnson & Johnson, 88; instrumentation, Columbia Univ, 88. *Res:* Product development for medical devices, microsurgical devices, all industries, and prototypes of all types; engineering all types for new products. *Mailing Add:* 1206 Herbert Ave Ocean NJ 07712

PETITT, GUS A, b Birmingham, Ala, Apr 7, 37. PHYSICS. *Educ:* Mass Inst Technol, BS, 60; Duke Univ, PhD(physics), 65. *Prof Exp:* Mem tech staff, Northrop Space Labs, 64-65; ASSOC PROF PHYSICS, GA STATE UNIV, 65- *Concurrent Pos:* Consult, Northrop Space Labs, 65- *Mem:* Am Phys Soc. *Res:* Neutron scattering; Mossbauer effect; heavy ion nuclear reactions. *Mailing Add:* Dept Physics Univ Plaza Ga State Univ Atlanta GA 30303

PETKE, FREDERICK DAVID, b Cincinnati, Ohio, May 12, 42; m 67; c 2. SURFACE CHEMISTRY. *Educ:* Lehigh Univ, BA, 64; Wash State Univ, PhD(phys chem), 68. *Prof Exp:* Res chemist, Tenn Eastman Co, 68-70, sr res chemist, Res Labs, 70-81, sr res chemist, Eastman Chem Div, Res Labs, 81-85, res assoc, 85-89, TECH ASSOC, TENN EASTMAN CO, 81- *Mem:* Am Chem Soc; Sigma Xi. *Res:* Surface chemistry of adhesion, polymers and fundamentals of hot melt adhesives. *Mailing Add:* 1912 Birchwood Rd Kingsport TN 37660-5022

PETRACEK, FRANCIS JAMES, medicinal chemistry, for more information see previous edition

PETRACK, BARBARA KEPES, b New York, NY, Aug 21, 27; m 49; c 2. BIOCHEMISTRY. *Educ:* Hunter Col, BA, 47; Polytech Inst Brooklyn, MS, 51; NY Univ, PhD(biochem), 57. *Prof Exp:* Guest investr biochem res, Rockefeller Inst, 58-61; biochemist, Geigy Res Labs, Geigy Chem Corp, 61-71, mgr basic biochem, 71-75, SR RES FEL, CIBA-GEIGY CORP, 75- *Concurrent Pos:* Dazian Found fel, 58; Runyon fel, 58-60; adj prof, Dept Biochem, New York Med Col. *Mem:* NY Acad Sci; Am Chem Soc; Am Soc Biol Chem; Sigma Xi. *Res:* Enzymes of catecholamine metabolism; nicotinamide-adenine dinucleotide and nicotinamide metabolism; photosynthetic phosphorylation; arginine metabolism; cyclic nucleotide metabolism; diabetes research; insulin and glucagon secretion; hormone receptors; benzodiazepine receptors; cholesterol metabolism. *Mailing Add:* Molecular Biol/Enzym CIBA-Gelgy Corp 556 Morris Ave Summit NJ 07901

PETRAKIS, LEONIDAS, b Sparta, Greece, July 23, 35; US citizen; m 59; c 2. PHYSICAL CHEMISTRY. *Educ:* Northeastern Univ, 58; Univ Calif, Berkeley, PhD(phys chem), 61. *Prof Exp:* Nat Res Coun Can res fel, 61-62; asst prof phys chem, Univ Md, 62-63; res chemist, E I du Pont de Nemours & Co, 63-65; res chemist, Gulf Res & Develop Co, 65-66, supvr molecular spectros, 66-85, res assoc, 73-78, sr res assoc, Gulf Sci & Technol Co, 78-85; SR SCIENTIST, CHEVRON RES CO, RICHMOND, CALIF, 85- *Concurrent Pos:* Sr lectr chem, Carnegie-Mellon Univ, 72-73; adj prof, Univ Pittsburgh, 81-87; vis prof, Univ Paris, 85. *Mem:* AAAS; Am Phys Soc; Am Chem Soc. *Res:* Catalysis, magnetic resonance; surface chemistry and solid state; fossil energy and the environment; strategic studies in fossil fuels and non-energy and energy minerals; analytical spectroscopy; characterization of materials; relaxation of nuclear spins and intermolecular forces; chemistry and physics of fossil fuels. *Mailing Add:* Brookhaven Nat Lab Bldg 179 A2 Center St Upton NY 11973

PETRAKIS, NICHOLAS LOUIS, b Bancroft, Iowa, Feb 6, 22; m 47; c 3. EPIDEMIOLOGY, ONCOLOGY. *Educ:* Augustana Col, BA, 43; Univ SDak, BS, 44; Washington Univ, MD, 46. *Prof Exp:* Intern, Minneapolis Gen Hosp, 46-47; asst med officer, Naval Radiol Defense Lab, 47-49; sr asst surgeon, USPHS Lab Exp Oncol, Univ Calif, 50-54; from asst prof to assoc prof med & prev med, 54-66, from vchmn to actg chmn dept prev med, 60-66, vchmn dept epidemiol & int health, 66-69, assoc dir, G W Hooper Found, 70-74, actg dir, G W Hooper Found, 74-77, PROF PREV MED, DIV AMBULATORY & COMMUNITY MED, SCH MED, UNIV CALIF, SAN FRANCISCO, 66-, RES ASSOC, CANCER RES INST, 54-, CHMN DEPT EPIDEMIOL & INT HEALTH, 78- *Concurrent Pos:* Eleanor Roosevelt Int Cancer fel, Rome, 62-63; USPHS spec fel, Galton Lab, Dept Human Genetics & Biomet, Univ Col, Univ London, 69-70; consult, Vet Admin Hosps, Martinez, Calif; mem, Epidemiol Comt, Breast Cancer Task Force & consult, Nat Cancer Inst, 73-76, mem, Biometry & Epidemiol Contract Rev Comt, 78-81; consult, Calif State Dept Pub Health, 73-; mem, Bd Scientific Counsellors, Div Cancer Etiology, Nat Cancer Inst, 85- *Mem:* AAAS; Am Asn Cancer Res; Soc Epidemiol Res; Am Epidemiol Soc; Am Soc Prev Oncol; Soc Prospective Med. *Res:* Genetic epidemiology of cancer; sickle cell disease; physical anthropology; hematology; oncology; breast cancer epidemiology. *Mailing Add:* Dept Epidemiol Univ Calif 1699 HSW San Francisco CA 94143

PETRALI, JOHN PATRICK, b Fairview, NJ, July 30, 33; m 63. ANATOMY, IMMUNOCYTOCHEMISTRY. *Educ:* Davis & Elkins Col, BS, 55; Univ Md, PhD(human anat), 69. *Prof Exp:* Biologist cytol, Med Res Lab, Edgewood Arsenal, 62-63, res biologist ultrastruct, Biomed Lab, 64-69, group leader, Electron Micros Sect, Biomed Lab, 69-79; RES ANATOMIST & TEAM LEADER, ELECTRON MICROS, COMP PATH, US AM MED RES INST CHEM DEFENSE, 79- *Concurrent Pos:* Adj asst prof anat, Sch Med, Univ Md, 69-75, adj asst prof, Univ Md, 75-79. *Mem:* Am Asn Anatomists; Electron Micros Soc Am. *Res:* Diagnostic pathology. *Mailing Add:* 204 Goucher Way Churchville MD 21028-1220

PETRARCA, ANTHONY EDWARD, b Providence, RI, July 24, 29; m 55; c 6. DATA BASE INFORMATION SYSTEMS, INFORMATION ANALYSTS. *Educ:* RI Col, BEd, 53; Univ RI, MS, 55; Univ NH, PhD(org chem), 59. *Prof Exp:* Instr chem, Univ NH, 55-59; from instr to asst prof, Seton Hall Univ, 59-64; info scientist, Chem Abstr Serv, 64-68; asst prof, 68-71, actg chairperson, 84-85, ASSOC PROF COMPUT & INFO SCI, OHIO STATE UNIV, 71- *Concurrent Pos:* Vis sr res scientist, Battelle Columbus Labs, 81-82. *Mem:* AAAS; Am Chem Soc; Am Soc Info Sci; Asn Comput Mach. *Res:* Information analysis and representation, classification query enhancement techniques; automatic synonym recognition and vocabulary control; automatic indexing and classification; biomedical and chemical applications. *Mailing Add:* Dept Comput & Info Sci Ohio State Univ Columbus OH 43210

PETRAS, MICHAEL LUKE, b Windsor, Ont, Aug 4, 32; m 56; c 4. ZOOLOGY, GENETICS. *Educ:* Assumption Col, BSc, 54; Univ Notre Dame, MS, 56; Univ Mich, PhD(zool), 65. *Prof Exp:* Lectr biol, Essex Col, Ont, 56-62; from asst prof to assoc prof, 62-70, PROF BIOL, UNIV WINDSOR, 70-, DEPT HEAD BIOL, 88- *Concurrent Pos:* Genetics trainee, Univ Mich, 61-62; grants, Nat Soc Eng Res Coun Can, 60-90; vis investr, Jackson Lab, 79. *Mem:* Genetics Soc Am; Soc Study Evolution; Genetics Soc Can; Environ Mutagen Soc. *Res:* Biochemical variants in mammals; the genetics of mammalian populations, especially mouse populations; sister chromatid exchanges and the monitoring of genotoxicity. *Mailing Add:* Dept Biol Univ Windsor Windsor ON N9B 3P4 Can

PETRASEK, EMIL JOHN, b Passaic, NJ, Jan 22, 40; m 62; c 2. HAZARDOUS WASTE & CHEMICAL MANAGEMENT, SALES & MARKETING MANAGEMENT. *Educ:* Wilkes Col, BS, 61. *Prof Exp:* Tech mgr, Sonneborn Div, De Soto Chem, 61-71; res mgr, CPD WR Grace & Co, 71-82; VPRES, MKT & SALES, COATING SYST INC, 83- *Concurrent Pos:* Consult, mkt & sales, 82-83. *Mem:* Am Chem Soc; Am Soc Testing & Mat; Nat Paint & Coatings Group. *Mailing Add:* 55 Crown St Nashua NH 03060

PETRASSO, RICHARD DAVID, b Portland, Ore, Dec 22, 44. PHYSICS, PLASMA PHYSICS. *Educ:* Oregon State Univ, BS, 67; Brandford Univ, PhD(physics), 72. *Prof Exp:* res solar & lab plasma, Am Sci Int High Tech, 72-83, RES RADIATION & LAB PLASMA, MASS INST TECHNOL PLASMA FUSION CTR, 83- *Mem:* Am Phys Soc. *Res:* Radiation plasma; Solar plasma; laboratory plasma. *Mailing Add:* Mass Inst Technol Plasma Fusion Ctr NW 16-132 167 Albany St Cambridge MA 02139

PETRELLA, RONALD VINCENT, b Youngstown, Ohio, Feb 27, 35; m 57; c 3. PHYSICAL CHEMISTRY. *Educ:* Univ Youngstown, BS, 56; Western Reserve Univ, MS, 58, PhD(phys chem), 63. *Prof Exp:* Chemist, 61-65, res chemist, 65-68, sr res chemist, 68-73, RES SPECIALIST, DOW CHEM CO, 73- *Mem:* Combustion Inst; Am Chem Soc; Sigma Xi. *Res:* Thermodynamics; thermochemistry; molecular spectroscopy; flash pyrolysis; flash photolysis; kinetics; polymer combustion; flammability. *Mailing Add:* 3712 Hillgrove Ct Midland MI 48640

PETRELLA, VANCE JOHN, b Joliet, Ill, July 23, 47; m 72; c 2. TOXICOLOGY, BIOCHEMISTRY. *Educ:* Va Mil Inst, BS, 69; Va Polytech Inst & State Univ, PhD(biochem), 73; Am Bd Toxicol, cert, 80, 86. *Prof Exp:* Res asst, 69-72, res fel biochem toxicol, Va Polytech Inst, 72-73; res biochemist toxicol natural toxins, US Army Med Res Inst Infectious Dis, 74-77, med review officer, 77-81, SR MED REVIEW OFFICER TOXICOL, GILLETTE MED EVAL LABS, 81- *Concurrent Pos:* Adj asst prof, Hood Col, 76. *Mem:* Am Col Toxicol; Soc Toxicol. *Res:* Biochemical toxicology of natural and synthetic toxins; industrial toxicology. *Mailing Add:* 5344 Pommel Dr Mt Airy MD 21771

PETRI, LEO HENRY, b Garland, Nebr, Sept 24, 14; m 42; c 1. PARASITOLOGY. *Educ:* Nebr State Teachers Col, Peru, AB, 37; Univ Nebr, MA, 41; Kans State Col, PhD(parasitol), 51. *Prof Exp:* Prin pub sch, 37-39; asst, Univ Nebr, 39-41; instr & technician, Kans State Col, 41-42, 46-52; assoc prof, 52-57, prof, 57-79, EMER PROF BIOL, WARTBURG COL, 80- *Mem:* AAAS; Am Soc Parasitol; Am Micros Soc. *Res:* Trematodes and nematodes. *Mailing Add:* 105 Iowa Waverly IA 50677

PETRI, WILLIAM HENRY, III, b Hopkinsville, Ky, Nov 8, 38. WOUND RESEARCH, TEACHING. *Educ:* Univ Louisville, BA, 61, DMD, 65; Antioch Univ, MA, 83; Union Exp Col & Univ, PhD(biomed sci), 85; Am Bd Oral & Maxillofacial Surg, dipl, 77. *Prof Exp:* Resident oral & maxillofacial surg, Naval Regional Med Ctr, Naval Hosp, Norfolk, Va, 72-75, instr, 75-76, chief, dent serv, 76-77, staff consult, Naval Hosp, Nat Capital Region, 81-85, staff consult oral & maxillofacial surg, Mid-Atlantic Region, 85-89, mem staff, Naval Dent Clin, 85-89; ASSOC PROF, DEPT ORAL MAXILLOFACIAL SURG, SCH DENT, LA STATE UNIV, 89- *Concurrent Pos:* Prin investr,

Naval Med Res Inst, 81-85. *Mem:* Am Asn Oral & Maxillofacial Surgeons; Am Dent Asn; Am Asn Dent Res; Int Asn Dent Res. *Res:* Wound healing; bone grafts. *Mailing Add:* Dept Oral Maxillofacial Surg Sch Dent La State Univ 1100 Florida Ave Box 220 New Orleans LA 70119

PETRI, WILLIAM HUGH, b San Francisco, Calif, Dec 30, 44; m 76; c 3. DEVELOPMENTAL GENETICS, DROSOPHILA GENETICS & DEVELOPMENT. *Educ:* Univ Calif, Berkeley, AB, 66, PhD(genetics), 72. *Prof Exp:* Res fel, Harvard Univ, 72-76; asst prof, 76-82, ASSOC PROF GENETICS, BOSTON COL, 82- *Concurrent Pos:* Vis scientist, Mass Inst Technol, 85. *Mem:* Genetics Soc Am; Soc Develop Biol. *Res:* Genetics and molecular biology of eggshell development in Drosophila Melanogaster; Drosophila genomic sequences which are difficult to clone. *Mailing Add:* Dept Biol Boston Col Chestnut Hill MA 02167

PETRICCIANI, JOHN C, b Sacramento, Calif, Sept 4, 36. RESEARCH ADMINISTRATION, PATHOLOGY. *Educ:* Rensselaer Polytech Inst, BS, 58; Univ Nev, MS, 60; Stanford Univ, MD, 67. *Prof Exp:* House staff pediatrician, Buffalo Children's Hosp, NY, 67-68; NIH staff assoc, 68-72, chief, Exp Cytol Sect, Lab Path, 72-74; dep dir, Div Path, Bur Biologics, Food & Drug Admin, 74-78, asst dir, 78-81; asst dir, Off Protection from Res Risks, Ctr Drugs & Biologics, NIH, 81-82; VPRES MED & REGULATORY AFFAIRS, PHARM MFRS ASN, 88- *Concurrent Pos:* Adj prof genetics & prof lectr biol sci, George Washington Univ, 78- *Mem:* AAAS; NY Acad Sci; Tissue Cult Asn; AMA. *Res:* Tumor cell biology; Cytogenetics. *Mailing Add:* Pharmaceut Mfrs Asn 1100 15th St SW Washington DC 20005

PETRICH, ROBERT PAUL, b Barberton, Ohio, Jan 11, 41; m 64; c 3. POLYMER SCIENCE. *Educ:* Mass Inst Technol, BS, 63, MS, 64; Univ Akron, PhD(polymer sci), 68. *Prof Exp:* Chemist, Res Div, Rohm & Haas Co, 64-66, group leader, 68-72, mgr, Int Mkt Area, 72-75, head, Europ Plastics Lab, 75-79, Europ Res & Develop Mgr, Int Div, 79-90, PLANT MGR, PHILADELPHIA PLANT, 90- *Mem:* Am Chem Soc; Soc Plastics Engrs; Plastics & Rubber Inst. *Res:* Physical properties of polymers, especially multi-phase polymer systems. *Mailing Add:* c/o Rohm & Haas Co 5000 Richmond St Philadelphia PA 19137

PETRICK, ERNEST N(ICHOLAS), b Taylor, Pa, Apr 9, 22; m 46; c 4. MILITARY ENGINEERING. *Educ:* Carnegie Inst Technol, BS, 42; Univ Mich, cert, 45; Purdue Univ, MS, 48, PhD, 55. *Prof Exp:* Instr mech, Aeronaut, Chem & Metall Eng & co-supvr gas turbine lab, Purdue Univ, 46-53; head heat transfer sect, Wright Aeronaut Div, Curtiss-Wright Corp, 53-56, chief adv propulsion systs, Res Div, 55-60; chief res engr, Kelsey-Hayes Co, 60-65; chief scientist, Hq, US Army Mobility Command, 65-66; chief scientist & tech dir, Hq, US Army Tank-Automotive Command, 66-82; chief scientist, Land Systs Div, Gen Dynamics Land, 82-87; CONSULT, 87- *Concurrent Pos:* Mem bd vis, Oakland Univ; mem adv bd eng, Wayne State Univ; adj prof, Wayne State Univ & Univ Mich; mem adv comt advan automotive power systs, Coun Environ Qual, Exec Off of Pres; eng consult, 82-; mem, NATO Panel combat vehicles; US Army Sci Bd, 83- *Mem:* Sigma Xi; Soc Automotive Engrs; Am Defense Preparedness Asn. *Res:* Transportation systems; surface and flight propulsion; gas turbines; automotive and truck engines; commercial and military vehicles; main battle tanks. *Mailing Add:* 1540 Stonehaven Rd Ann Arbor MI 48104

PETRIDES, GEORGE ATHAN, wildlife management, ecology, for more information see previous edition

PETRIE, JOHN M(ATTHEW), chemical engineering; deceased, see previous edition for last biography

PETRIE, WILLIAM LEO, b Assiut, Egypt, Jan 16, 23; US citizen; m 46; c 2. GEOLOGY. *Educ:* Monmouth Col, AB, 49; Univ Iowa, MS, 51. *Prof Exp:* Asst geol, Monmouth Col, 47-49 & Univ Iowa, 49-51; oceanogr, US Navy Hydrographic Off, 51-52, panel coordr geophys & geol, Off Secy Defense, Res & Develop Bd, 52-54; analyst geophys, US Govt, 54-61; exec secy, Nat Acad Sci-Nat Res Coun, Adv Comt Mohole Proj, 61-64 & Comt Alaska Earthquake, 64-73, exec secy, US Nat Comt Geochem, Bd Earth Sci, Nat Acad Sci-Nat Res Coun, 72-87; exec secy, US Nat Comt Int Geog Union & US Nat Comt, Int Union Quaternary Res, 74-87, staff consult, Nat Acad Sci-Nat Res Coun, 87-89; RETIRED. *Concurrent Pos:* Staff officer, Bd Mineral & Energy Resources, Nat Acad Sci-Nat Res Coun, 77-85. *Mem:* Fel Geol Soc Am; Am Geophys Union; Nat Speleol Soc. *Res:* Earth science administration; marine geology; solid-earth geophysics; deep drilling for scientific purposes; scientific and engineering studies of 1964 Alaska Earthquake; environmental geochemistry related to health and disease. *Mailing Add:* 15001 Wannas Dr Accokeek MD 20607-9406

PETRIE, WILLIAM MARSHALL, b Louisville, Ky, Oct 19, 46; m 68; c 3. PSYCHOPHARMACOLOGY, GERIATRIC PSYCHIATRY. *Educ:* Vanderbilt Univ, BA, 68, MD, 72. *Prof Exp:* Res psychiatrist psychopharmacol, NIMH, 75-77; clin instr psychiat, Georgetown Univ Med Ctr, 76-77; from asst prof to assoc prof, 77-82, ASSOC CLIN PROF PSYCHIAT, DEPT PSYCHIAT, VANDERBILT MED CTR, 82- *Concurrent Pos:* Consult psychopharmacol, NIMH, 77-79 & psychiat Nashville Vet Admin Hosp, 85-86; chief, geriat res, Tenn Neuropsychiat Inst, 80-82; mem, rev comt, Ctr Aging, NIMH, 82-86; med dir, Memory Study Ctr. *Mem:* Fel Am Psychiat Asn; Am Asn Geriat Psychiat; AMA; Am Col Psychiatrists. *Res:* Drug treatments for dementing illnesses. *Mailing Add:* 310 25th Ave N Nashville TN 37203

PETRIELLO, RICHARD P, b Jersey City, NJ, July 6, 42; m 71; c 4. NEMATOLOGY. *Educ:* Iona Col, BS, 64; Seton Hall Univ, MS, 66; Rutgers Univ, PhD(nematol), 70. *Prof Exp:* Res asst nematol, Dept Entom & Econ Zool, Rutgers Univ, 66-68, res assoc, 68-69, asst prof biol, Livingston Col, 69-78; assoc prof biol, St Peter's Col, 77-79, chmn dept, 77-79, 86-87, dean enrollment mgt, 87-90, PROF BIOL, ST PETER'S COL, JERSEY CITY, NJ, 81- *Concurrent Pos:* NIH grant, 71-72, NIH-NSF grant, 78, NSF & LOCI grants, 81 & NJ Dept Higher Educ grants, 85, Retention Initiative grants, 88-89; asst to acad vpres, St Peter's Col, 91. *Mem:* Am Pub Health Asn; Soc Nematol; Soc Parasitol. *Res:* Nematology, especially in relationship to parasites affecting man; parasitology and invertebrate zoology. *Mailing Add:* Biol Dept St Peter's Col Jersey City NJ 07306

PETRILLO, EDWARD WILLIAM, b Neptune, NJ, Nov 4, 47; m 69. SYNTHETIC ORGANIC CHEMISTRY. *Educ:* Princeton Univ, AB, 69; Yale Univ, MPhil, 71, PhD(chem), 73. *Prof Exp:* Res fel chem, Calif Inst Technol, 73-74; res investr, 74-79, RES res group leader chem, 79-83, ASST DIR CHEM & CARDIOVASC, E R SGUIBB & SONS INC, 85- *Mem:* Am Chem Soc. *Res:* Synthetic, organic and medicinal chemistry; cardiovascular drugs; enzyme inhibitors. *Mailing Add:* Squibb Inst Med Res PO Box 4000 Princeton NJ 08544

PETRO, ANTHONY JAMES, b Waterbury, Conn, Nov 28, 30; m 59; c 1. PHYSICAL CHEMISTRY. *Educ:* Trinity Col, Conn, BS, 52, MS, 54; Princeton Univ, PhD(phys chem), 57. *Prof Exp:* Chemist, Esso Res & Eng Co, 57-66; chemist, Mearl Corp, 66-78; chemist, Avon Prods, Inc, 78-88; CHEMIST, MONA INDUSTS, 88- *Mem:* Am Chem Soc; Soc Cosmetic Chemists; Sigma Xi. *Res:* Monodisperse sulfur hydrosols; microwave dielectric constant and loss; dipole moments; rubber latex; corrosion; nacreous pigments; particle size; surface area; porosity; thixotropic nail enamels; rheology; sunscreens; non-aqueous gels; hydrotropes; phosphate esters. *Mailing Add:* 3719 Briarhill St Mohegan Lake NY 10547

PETRO, JOHN WILLIAM, b Vinton, Iowa, Nov 5, 30; m 71; c 2. NUMERICAL AND COMPUTATIONAL MATHEMATICS. *Educ:* Univ Iowa, BA, 52, MS, 59, PhD(math), 61. *Prof Exp:* From asst prof to assoc prof, 61-71, PROF MATH, WESTERN MICH UNIV, 71- *Concurrent Pos:* Vis assoc prof, Ohio State Univ, 66-67; vis prof, Univ Calif, Riverside, 79-80. *Mem:* Am Math Soc; Math Asn Am. *Res:* Theory of pseudovaluations in commutative algebra and number theory; theory of filtrations in commutative algebra; theory of boolean structures in commutative algebra. *Mailing Add:* Dept Math & Statist Western Mich Univ Kalamazoo MI 49008

PETRO, PETER PAUL, JR, b Mobile, Ala, Apr 13, 40; m 65; c 3. ANALYTICAL CHEMISTRY. *Educ:* Spring Hill Col, BS, 62; Univ Ala, MS, 64, PhD(chem), 66. *Prof Exp:* Res chemist, Res & Develop Lab, Lone Star Gas Co, 68-71, mgr chem & plastics sect, Corp Develop & Res Div, 71-75, dir plastics opers, Nipak, Inc, 75-77; dir tech serv, 78-79, vpres Tech Serv & Mat, 79-87, PRES, PLEXCO, 87- *Mem:* AAAS; Am Chem Soc; Soc Plastics Eng; Chem Soc; Coblentz Soc. *Res:* Gas chromatography; infrared spectroscopy; thermal analysis; ultraviolet and visible spectrophotometry and polarography; plastics fabrication; chemical product development. *Mailing Add:* Plexco 3240 N Mannheim Rd Franklin Park IL 60131

PETROF, ROBERT CHARLES, b Beloit, Wis, June 7, 37; m 67; c 1. MECHANICAL ENGINEERING, APPLIED MECHANICS. *Educ:* Northwestern Univ, BSME, 60, MS, 62, PhD(mech eng, astronaut sci), 65. *Prof Exp:* Sr res engr, 65-68, PRIN RES ENGR, SCI RES STAFF, FORD MOTOR CO, 68- *Mem:* Am Inst Aeronaut & Astronaut; Am Soc Mech Engrs; Soc Automotive Engrs; Soc Mfg Engrs. *Res:* Heat transfer; thermal stress analysis; viscoelasticity; finite element structural analysis; mechanics of brakes; composite materials; machine design; vibration analysis; computer analysis methods. *Mailing Add:* Ford Motor Co Sci Res Staff 24500 Glendale Ave Detroit MI 48239

PETROFF, PIERRE MARC, b Paris, France, Apr 26, 40; m 64; c 4. MATERIALS SCIENCE, SOLID STATE ELECTRONICS. *Educ:* Sch Mines, France, BS, 63; Univ Calif, Berkeley, MS, 64, PhD(mat sci), 67. *Prof Exp:* Mil researcher mat sci, Physics Lab, Sci Fac, Orsay, France, 67-69; res fel, Dept Mat Sci, Cornell Univ, 69-71; MEM TECH STAFF, BELL TEL LABS, 71- *Concurrent Pos:* Int Atomic Energy Agency consult, Bhaba Atomic Res Ctr, Bombay, India, 73. *Mem:* Sigma Xi; Am Phys Soc; Am Soc Testing & Mat. *Res:* Crystallographic and electrical properties of defects in semiconductors and optoelectronics materials; study of radiation effects on semiconductors. *Mailing Add:* Eng Mater Dept Univ of Calif Santa Barbara CA 93106

PETRONE, ROCCO A, b Amsterdam, NY, Mar 31, 26; m 55; c 4. AEROSPACE ENGINEERING & TECHNOLOGY. *Educ:* US Mil Acad, BS, 46; Mass Inst Technol, MS, 52. *Hon Degrees:* DSc, Rollins Col, 69. *Prof Exp:* Mem gen staff, Dept Army, 56-60; mgr, Apollo Prog, Kennedy Space Ctr, NASA, 60-66, dir launch opers, 66-69, dir, Wash, 69-73 & Marshall Space Flight Ctr, Hunstville, Ala, 73-74, assoc adminr, Wash, 74-75, pres & chief exec officer, Nat Ctr Res Recovery, 75-81; exec vpres, Space Transp Systs Group, Rockwell Int, 81-82, pres, 82-89; RETIRED. *Mem:* Nat Acad Eng; fel Am Inst Aeronaut & Astronaut; Sigma Xi. *Mailing Add:* 1329 Granvia Altamira Palos Verdes Estates CA 90274

PETRONIO, MARCO, b New York, NY, Aug 15, 12; m 41; c 1. CHEMICAL ENGINEERING. *Educ:* City Col New York, BChE, 37. *Prof Exp:* Sales engr, Calo & Lydon, Inc, NY, 37-38; chem engr, Frankford Arsenal, US Dept Army, 40-47, chemist, 47-49, chem engr, 49-54, chemist, 54-59, chief fluids & lubricants br, 59-72; consult, E M Kipp Assoc, 72-74; vpres & consult, Hangsterfer's Lab, Inc, 72-80; RETIRED. *Concurrent Pos:* Consult. *Honors & Awards:* Deutsch Mem Award, Am Soc Lubrication Engrs, 60. *Mem:* Am Chem Soc; Am Soc Lubrication Engrs; Am Soc Testing & Mat; Sigma Xi; Am Ord Asn. *Res:* Lubricants research and development for the shaping and forming of metals; cleaners for the removal of lubricants; structural adhesives research for the bonding of metals, glass and plastics; hermetic sealing of instruments. *Mailing Add:* Chandler Hall Buck Rd & Barclay St New Town PA 18940

PETROPOULOS, CONSTANTINE CHRIS, b South Norwalk, Conn, May 8, 31; m 56; c 4. ORGANIC POLYMER CHEMISTRY. *Educ:* Brown Univ, BS, 54; Fla State Univ, MS, 57. *Prof Exp:* Sr res chemist, Nat Cash Register Co, 59-65 & Loctite Corp, 65-67; sr res chemist, 67-77, res assoc, 77-79, TECH ASSOC, EASTMAN KODAK CO, 79- *Res:* Synthesis of photographic and photoimaging applications; photochemical rearrangement reactions, photoinitiator and sensitizers; organic reaction mechanisms. *Mailing Add:* Eastman Kodak Co 343 State St Rochester NY 14650

PETROSKI, HENRY J, b New York, NY, Feb 6, 42; m 66; c 2. APPLIED MECHANICS & DESIGN, HISTORY OF ENGINEERING. *Educ:* Manhattan Col, BME, 63; Univ Ill, Urbana-Champaign, MS, 64, PhD(mech), 68. *Hon Degrees:* DSc, Clarkson Univ, 90. *Prof Exp:* Instr, Dept Theoret & Appl Mech, Univ Ill, Urbana, 66-68; asst prof mech, Dept Aerospace Eng & Eng Mech, Univ Tex, Austin, 68-74; mech engr, Reactor Anal & Safety Div, Argonne Nat Lab, 75-80; assoc prof, 80-87, dir grad studies, 81-86, PROF CIVIL ENG, DEPT CIVIL & ENVIRON ENG, DUKE UNIV, 87- *Concurrent Pos:* Prin investr, NSF grants appl mech, 73 & struct dynamics, 82-85, design theory, 89-92; res civil engr, Waterways Exp Sta, 84-86; fel, Nat Humanities Ctr, 87-88; Guggenhein fel, 90-91. *Mem:* Am Soc Civil Engrs; Soc Natural Philos; Am Acad Mech; Hist Sci Soc; Soc Hist Technol. *Res:* Fracture mechanics; structural dynamics of systems containing cracks, including large plastic deformations; books on engineering, history and design philosophy and essays on technology. *Mailing Add:* Dept Civil & Environ Eng Duke Univ Durham NC 27706

PETROU, PANAYIOTIS POLYDOROU, b Nicosia, Cyprus, Nov 20, 55; US citizen. UNDERWATER ACOUSTICS, ANTI-SUBMARINE WARFARE. *Educ:* Univ RI, BSEE, 80. *Prof Exp:* Elec engr, Test & Eval Dept, 81-85 & Advan Systs Concepts Div, 85-90, ELEC ENGR, PLATFORM ANALYSIS DIV, COMBAT SYSTS ANALYSIS DEPT, NAVAL UNDERWATER SYSTS CTR, 90- *Concurrent Pos:* Prin investr, Test & Eval Dept, Naval Underwater Systs Ctr, 82-85 & Combat Systs Analysis Dept, 86- *Res:* Underwater acoustics; undersea ranges and systems; submarine-anti-submarine warfare, combat systems and weapons; acoustic-non-acoustic systems and technologies; low intensity conflict and contingency and limited objective warfare or regional conflict. *Mailing Add:* 2121 W Main Rd, Apt No 607 Portsmouth RI 02871

PETROVIC, LOUIS JOHN, b Cleveland, Ohio, Oct 6, 40; m 67; c 2. CHEMICAL ENGINEERING. *Educ:* Case Inst Technol, BS, 62; Northwestern Univ, MS, 64, PhD(chem eng), 67; Boston Col, MBA, 73. *Prof Exp:* Group leader, Cabot Corp, 67-69; mgr coal prog, Ledgemont Lab, Kennecott Copper Corp, 69-78; pres, 78-88, CHMN, RESOURCE ENG INC, 88-; ASST DEAN ENG, UNIV LOWELL, 88- *Mem:* Am Chem Soc; Am Inst Chem Engrs; Soc Mining Engrs; Int Asn Energy Economists. *Res:* Flame technology; desulfurization of coal; new coal mining technologies; liquefaction and gasification of coal; hydrometallurgy of mineral ores. *Mailing Add:* 26 Dunster Rd Sudbury MA 01776

PETROVICH, FRED, b Binghamton, NY, July 16, 41; m 62, 86; c 3. NUCLEAR PHYSICS. *Educ:* Clarkson Univ, BS, 63; Mich State Univ, MS, 66, PhD(nuclear physics), 71. *Prof Exp:* Instr physics, Mich State Univ, 69-71, res assoc, 71; res assoc physics, Lawrence Berkeley Nat Lab, 71-73; from asst prof to assoc prof, 73-81, PROF PHYSICS, FLA STATE UNIV, 81- *Concurrent Pos:* NSF res grant, 73-; consult, Lawrence Livermore Nat Lab, 76- & Los Alamos Nat Lab, 80-; assoc ed, Int Reviews Nuclear Physics; ed, Spin Excitations in Nuclei; mem, Prog Comt, Div Nuclear Physics, Am Phys Soc, 86-88. *Mem:* Fel Am Phys Soc; Sigma Xi. *Res:* Development of theoretical microscopic models for describing scattering of electrons, nucleons, pions and heavy- ions from nuclei; various aspects of nuclear structure; core polarization effects and spin excitations in nuclei. *Mailing Add:* Dept Physics Fla State Univ Tallahassee FL 32306

PETROWSKI, GARY E, b LaCrosse, Wis, May 18, 41; m 67. FOOD CHEMISTRY, COMPUTER SCIENCE. *Educ:* Loras Col, BS, 63; Univ Calif, Los Angeles, PhD(chem), 69. *Prof Exp:* Res assoc, Univ Colo, Boulder, 69-70; sr res scientist, Carnation Res Lab, 70-76; sect head food develop, Kelco Div, Merck Inc, 76-79; CONSULT, 79- *Concurrent Pos:* Instr, Exten, Univ Calif, Los Angeles, 75-76. *Mem:* Am Chem Soc; Am Oil Chemists Soc; Inst Food Technologists. *Res:* Food emulsion stability; utilization of gums in foods. *Mailing Add:* 115 Taormina Lane Ojai CA 93023-3629

PETRUCCI, RALPH HERBERT, b Saratoga, NY, Jan 21, 30; m 55; c 3. CHEMICAL EDUCATION. *Educ:* Union Col, NY, BS, 50; Univ Wis, PhD(phys chem), 54. *Prof Exp:* Asst phys chem, Univ Wis, 50-54; from instr to assoc prof chem, Western Res Univ, 54-64; PROF CHEM, CALIF STATE UNIV, SAN BERNARDINO, 64- *Mem:* AAAS; Am Chem Soc; Sigma Xi. *Res:* Heterogeneous equilibrium; chemical education. *Mailing Add:* Dept of Chem Calif State Univ San Bernardino CA 92407

PETRUCCI, SERGIO, b Rome, Italy, Feb 7, 32; m 57; c 1. PHYSICAL CHEMISTRY. *Educ:* Univ Rome, PhD(chem), 55, DrSci (phys chem), 66. *Prof Exp:* Instr phys chem, Univ Rome, 55-56; Fulbright fel, Yale Univ, 56-57; asst prof phys chem, Univ Rome, 57-61; NATO fel, Tech Univ Norway, 61-62; res assoc, Princeton Univ, 62-64; vis asst prof, Univ Md, College Park, 64-65; from asst prof to assoc prof, 65-70, PROF PHYS CHEM, POLYTECH UNIV, 70- *Mem:* Am Chem Soc; Sigma Xi. *Res:* Ultrasonic and microwave dielectric relaxation in electrolytic solutions containing macrocytic and polymeric ligands; electrical conductance of ionic solutions; infrared spectra of ionic macrocyclic and polymeric complexes in solution. *Mailing Add:* Dept Chem Polytech Univ Rte 110 Farmingdale NY 11735

PETRUCELLI, LAWRENCE MICHAEL, b Bridgeport, Conn, July 25, 32; m 62; c 1. NEUROPHARMACOLOGY, NEUROANATOMY. *Educ:* Fordham Univ, BS, 54; Ohio State Univ, MS, 58; Georgetown Univ, PhD(pharmacol), 63. *Prof Exp:* Staff scientist, Litton Bionetics, 63-68; asst prof neuropharm, Sch Med, Univ Pittsburgh, 68-70; health scientist, Nat Inst Neurol & Commun Dis & Stroke, 70-71; exec secy, Pharmacol Study Sect, Div Res Grants, NIH, 71-74; ARTHRITIS PROG DIR, NAT INST ARTHRITIS, METAB & DIGESTIVE DIS, 74- *Concurrent Pos:* Res assoc, Sch Med, Univ Pittsburgh, 68-69; exec secy arthritis interagency coord comt, Nat Inst Arthritis, Metab & Digestive Dis, 77- *Mem:* Soc Neurosci. *Res:* Somatosensory physiology neuropharmacology; electrophysiology. *Mailing Add:* Westwood Bldg Rm 405 5333 Westbard Ave Bethesda MD 20205

PETRUK, WILLIAM, b Norquay, Sask, June 30, 30; m 59; c 2. GEOLOGY. *Educ:* Univ Saskatchewan, BEng, 54, MSc, 55; McGill Univ, PhD(geol), 59. *Prof Exp:* Sr geologist, Can Johns-Manville Co, 59-60; RES SCIENTIST, CAN DEPT ENERGY, MINES & RESOURCES, 60- *Concurrent Pos:* Appl mineralogist, res in image anal. *Honors & Awards:* Boldy Award, Can Inst Mining & Metall, 87, Wright Award, 89, Distinguished Lectr Award, 88. *Mem:* Fel Mineral Soc Am; Mineral Asn Can; fel Geol Asn Can; Can Inst Mining & Metall; Am Inst Mining, Metall & Petrol Engrs/The Metall Soc. *Res:* Ore microscopy, image analysis studies related to process mineralogy, application of ore microscopy to mineral beneficiation and exploration; interpretation of textures and correlation of synthetic studies to naturally occurring minerals; semi-quantitative analysis by means of x-ray diffractometer. *Mailing Add:* Can Dept Energy Mines & Resources 555 Booth St Ottawa ON K1A 0G1 Can

PETRUSKA, JOHN ANDREW, b Winnipeg, Man, Feb 6, 33; m 58; c 1. CHEMICAL PHYSICS. *Educ:* Bishop's Univ, Can, BSc, 53; McMaster Univ, MSc, 54; Univ Chicago, PhD(chem physics), 60. *Prof Exp:* Asst molecular physics, Univ Chicago, 54-55, asst phys & anal chem, 56-58; from res fel to sr res fel biol, Calif Inst Technol, 60-68; ASSOC PROF BIOL, UNIV SOUTHERN CALIF, 68- *Res:* Theory and interpretation of molecular spectra; structure and interaction of biological molecules; properties of the chemical bond; molecular biology. *Mailing Add:* Dept Biol Sci Univ Southern Calif Los Angeles CA 90089

PETRY, ROBERT FRANKLIN, b Hebron, Ind, Sept 13, 36; m 59; c 2. NUCLEAR PHYSICS. *Educ:* Univ Ind, BS, 58, MS, 60, PhD(physics), 63. *Prof Exp:* Res assoc nuclear physics, Princeton Univ, 64-67; asst prof, Univ Okla, 67-71, chmn, Dept Physics & Astron, 70-74, assoc prof, 71-80, PROF PHYSICS, UNIV OKLA, 80-, ASSOC DEAN, COL ARTS & SCI, 91- *Mem:* Am Phys Soc. *Res:* Beta-ray and gamma-ray spectroscopy; production and study of nuclides far from stability. *Mailing Add:* Dept Physics & Astron Univ Okla Norman OK 73019

PETRY, ROBERT KENDRICK, b Trenton, NJ, May 11, 12; m 34; c 4. BIOLOGY. *Educ:* Rutgers Univ, BS, 33. *Prof Exp:* Chemist, Sloane-Blabon Corp, 33-36; from chemist to chief chemist, Del Floor Prods, Inc, 36-45, dir res, 45-53; supvr res & develop, Congoleum Industs, Inc, 53-56, mgr res, 56-61, dir res & develop, 61-77; RETIRED. *Mem:* Am Chem Soc. *Res:* Plastics applications in surface coverings. *Mailing Add:* 89 Powder Mill Rd Morris Plains NJ 07950

PETRYKA, ZBYSLAW JAN, b Warsaw, Poland, Feb 12, 30; US citizen; m. TETRAPYRROLE METABOLISM. *Educ:* Univ Warsaw, MSc, 60; Univ London, England, PhD(org chem), 64. *Prof Exp:* Asst prof nuclear chem, Univ Warsaw, Poland, 60-61; fel org chem, Univ Minn, 64-66, res assoc biochem, 66-68, chief res chemist biochem, 69-83; chief res chemist, 83-86, VPRES, PORPHYRIN PRODUCTS, 86- *Concurrent Pos:* Inspector chem safety, Inst Safety Chem Factories, 60-61. *Mem:* Royal Soc Chem; Am Chem Soc; Polish Chem Soc; Am Soc Biol Chemists; AAAS; Sigma Xi. *Res:* Metabolism of tetrapyrroles (porphyrins and bile pigments) particularly in relation to porphyria and jaundice and to chemically induced porphyria; bile pigments formation; hematin treatment of porphyria; chromatographic methods of porphyrins and bile pigments separation; synthesis of photosensitizers for photodynamic therapy. *Mailing Add:* PO Box 31 Porphyrin Products Logan UT 84321-0031

PETRYSHYN, WALTER VOLODYMYR, b Lviv, Ukraine, Jan 22, 29; US citizen; m 56. MATHEMATICS. *Educ:* Columbia Univ, BA, 53, MS, 54, PhD(appl math), 61. *Prof Exp:* Instr math, Notre Dame Col, NY, 54-56; lectr, City Col New York, 59-61; fel, Courant Inst Math Sci, NY Univ, 61-64; from asst prof to assoc prof math, Univ Chicago, 64-67; PROF MATH, RUTGERS UNIV, NEW BRUNSWICK, 67- *Mem:* Am Math Soc; Soc Indust & Appl Math. *Res:* Functional and numerical analysis, especially solution of linear and nonlinear operator equations. *Mailing Add:* Math Dept Rutgers Univ New Brunswick NJ 08903

PETSCHEK, ALBERT GEORGE, b Prague, Czech, Jan 31, 28; nat US; m 49; c 4. PHYSICS. *Educ:* Mass Inst Technol, BS, 47; Univ Mich, MS, 48; Univ Rochester, PhD(physics), 53. *Prof Exp:* Jr res physicist, Carter Oil Co, Okla, 48-49; mem staff, Los Alamos Sci Lab, 53-66; prof physics, NMex Inst Mining & Technol, 66-68; sr scientist, Systs Sci & Software, Calif, 68-71; PROF PHYSICS, NMEX INST MINING & TECHNOL, 71- *Concurrent Pos:* Vis asst prof, Cornell Univ, 60-61; consult, Sandia Corp, 66-70, Los Alamos Sci Lab, 68-80 , 86- & Systs Sci & Software, 71-77; vis prof, Dept Geophys & Planetary Sci, Tel Aviv, 78; fel Los Alamos Nat Lab, 80- *Mem:* AAAS; Am Phys Soc; Am Astron Soc; Sigma Xi. *Res:* Mathematical physics; supernova theory; fracture; cloud physics; mathematical physics applied to a variety of problems including high energy astrophysics, thunderstorms, and inertially confined fusion; numerical analysis. *Mailing Add:* 122 Piedra Loop Los Alamos NM 87544

PETSCHEK, HARRY E, b Prague, Czech, Sept 12, 30; nat US; c 4. PHYSICS. *Educ:* Cornell Univ, BEngPhys, 52, PhD, 55. *Prof Exp:* Instr, Princeton Univ, 55-56; prin res scientist, 56-68, assoc dir, 68-73, chmn & chief exec officer, 78-81, PRES, AVCO-EVERETT RES LAB INC, 73- *Mem:* Am Geophys Union; Am Phys Soc; Am Inst Aeronaut & Astronaut; Am Soc Artificial Internal Organs. *Res:* High temperature gas dynamics; magnetohydrodynamics; plasma dynamics; space plasma; biomedical engineering. *Mailing Add:* 1314 Mass Ave Lexington MA 02173

PETSCHEK, ROLFE GEORGE, b Los Alamos, NMex, Aug 25, 54; m 80. THERMAL PHYSICS. *Educ:* Mass Inst Technol, BS, 75, BS, 75; Harvard Univ, PhD(physics), 81. *Prof Exp:* Fel, Univ Calif, Santa Barbara, 80-82; asst prof, 83-89, ASSOC PROF, CASE WESTERN UNIV, 89- *Concurrent Pos:* Vis lectr, Col Creative Studies, Univ Calif, Santa Barbara, 81-82; fel, Univ Calif, San Diego, 82-83. *Mem:* Am Phys Soc. *Res:* Behavior of materials near phase transitions, particularly dynamic behavior; polymers; liquid crystals. *Mailing Add:* Dept Physics Case Western Reserve Univ Cleveland OH 44106-2623

PETSKO, GREGORY ANTHONY, b Washington, DC, Aug 7, 48; m 71. BIOPHYSICS, BIOCHEMISTRY. *Educ:* Princeton Univ, BA, 70; Oxford Univ, PhD(molecular biophys), 73. *Prof Exp:* Instr, Sch Med, Wayne State Univ, 73-76, asst prof biochem, 76-78; assoc prof, 78-83, PROF CHEM, MASS INST TECHNOL, 83- *Honors & Awards:* Pfizer Award, Am Chem Soc, 87. *Mem:* Am Crystallog Asn; Biophys Soc; Am Chem Soc. *Res:* Structure-function relations in biological macromolecules; protein structure determination by x-ray crystallography, enzyme crystallography at sub-zero temperatures; nervous system structure; site-directed mutagenesis; allergy. *Mailing Add:* Rosenstiel Basic Med Scis Ctr Brandeis Univ Waltham MA 02254

PETTEGREW, RALEIGH K, b July 7, 31; US citizen; m 58; c 2. MAMMALIAN PHYSIOLOGY. *Educ:* Baldwin-Wallace Col, BA, 53; Kent State Univ, PhD(physiol), 68. *Prof Exp:* Teacher pub sch, Ohio, 56-63; instr Eng, Kent State Univ, 58-62; asst prof, 63-74, ASSOC PROF BIOL, DENISON UNIV, 74- *Concurrent Pos:* Nat Defense Educ Act fel, 65-68. *Mem:* AAAS; Soc Neurosci. *Res:* Mammalian temperature regulation. *Mailing Add:* Dept of Biol Denison Univ Granville OH 43023

PETTENGILL, GORDON HEMENWAY, b Providence, RI, Feb 10, 26; m 67; c 2. PHYSICS, ASTRONOMY. *Educ:* Mass Inst Technol, BS, 48; Univ Calif, Berkeley, PhD(physics), 55. *Prof Exp:* Res asst health physics, Los Alamos Sci Lab, 48-50; staff mem, Lincoln Lab, Mass Inst Technol, 54-58, assoc group leader, 58-62, group leader, 62-63; assoc dir, Arecibo Ionospheric Observ, Cornell Univ, 63-65, dir, 68-70; dir, Ctr Space Res, 84-90, PROF PLANETARY PHYSICS, MASS INST TECHNOL, 70- *Concurrent Pos:* Pres, comn 16, Int Astron Union, 70-73; mem, Comt Planetary & Lunar Explor, US Space Sci Bd, 74-77; Guggenheim fel, 80-81; Thomas Gold lectr, Astron, Cornell Univ, 91-92. *Mem:* Nat Acad Sci; Am Phys Soc; Int Union Radio Sci; fel Am Acad Arts & Sci; Am Astron Soc. *Res:* Nuclear and radio physics; planetary and radar astronomy. *Mailing Add:* Ctr Space Res Rm 37-641 Mass Inst of Technol Cambridge MA 02139

PETTENGILL, OLIVE STANDISH, b Newport News, Va, Apr 27, 24. CELL CULTURE. *Educ:* Temple Univ, AB, 45; Brown Univ, MS, 48; Boston Univ, PhD(biol), 60. *Prof Exp:* Instr oncol, Mec Sch, Tufts Univ, 60-61; res assoc anat, Med Sch, Univ Pa, 61-63; res instr, Washington Univ, 63-67, asst prof pathol, 67-68; asst prof pathol, Med Sch, St Louis Univ, 68-71; from asst to res assoc prof pathol, 72-82, RES PROF PATHOL, DARTMOUTH MED SCH, 82- *Mem:* Am Soc Pathologists; AAAS; Am Asn Cancer Res; Tissue Culture Asn. *Res:* Human lung cancer; biologic properties of tumor cells under in vitro conditions and in vivo in nude, athymic mice. *Mailing Add:* Dept Pathol Dartmouth Med Sch Hanover NH 03756

PETTERS, ROBERT MICHAEL, b Wilmington Del, June 8, 50; m; c 2. GENETICS, DEVELOPMENTAL BIOLOGY. *Educ:* Univ Del, BA, 72; NC State Univ, MS, 74, PhD(genetics), 76. *Prof Exp:* Res assoc genetics, NC State Univ, 76; assoc develop genetics, Yale Univ, 76-78; asst prof biol, Pa State Univ, 78-83; ASSOC PROF, DEPT ANIMAL SCI, NC STATE UNIV, 83- *Concurrent Pos:* USPHS trainee, dept biol, Yale Univ, 76-77. *Mem:* AAAS; Am Genetic Asn; Genetics Soc Am; Soc Develop Biol; Am Soc Animal Sci; Soc Study Reproduction. *Res:* Animal biotechnology; embryo culture & gene transfer. *Mailing Add:* Dept Animal Sci NC State Univ Box 7621 Raleigh NC 27695

PETTERSEN, HOWARD EUGENE, b Everett, Wash, Jan 8, 22; m 44; c 3. SOLID STATE PHYSICS. *Educ:* Whitman Col, AB, 43; Univ Minn, MA, 47; Mich State Univ, PhD(physics), 68. *Prof Exp:* Instr physics, Col Training Prog, Army Air Force, State Col Wash, 43-44; instr physics, Whitman Col, 44-45; asst, Univ Minn, 45-47; from asst prof to assoc prof, 47-69, prof, 69-87, EMER PROF PHYSICS, ALBION COL, 88- *Concurrent Pos:* Fac fel, Inst Sci & Technol, Univ Mich, 62-64; Fulbright res prof, Inst Crystallog, Univ Cologne, 72-73. *Mem:* Am Asn Physics Teachers. *Res:* Acousto-optics; strain-optical constants. *Mailing Add:* Dept Physics Albion Col Albion MI 49224

PETTERSEN, JAMES CLARK, b Winona, Minn, Aug 5, 32; m 57; c 3. ANATOMY. *Educ:* St Olaf Col, BA, 54; Univ NDak, MS, 61, PhD(anat), 63. *Prof Exp:* Instr anat, Univ NDak, 62-63; from instr to assoc prof gross anat, Med Sch, Univ Wis-Madison, 63-80, prof, 80-; AT DEPT ANAT, BARDEEN MED LABS. *Mem:* AAAS; Am Asn Anat; Teratol Soc; Sigma Xi. *Res:* Spleen; cellular aspects of the immune response; gross anatomy; anatomical syndromes associated with trisomy. *Mailing Add:* Dept Anat Med Sci Ctr Madison WI 53706

PETTERSON, ROBERT CARLYLE, b Waterville, Maine, Jan 6, 23; m 51. ORGANIC CHEMISTRY. *Educ:* Univ Maine, BS, 47; Univ Southern Calif, PhD(org chem), 57. *Prof Exp:* Res chemist, Purex Corp, Ltd, Calif, 51-55, supvr basic res, 55-58, res coordr, 58-59; USPHS fel, Imp Col, London, 59-60 & Mass Inst Technol, 60-62, res assoc, 62; from asst res prof to assoc res prof, 62-68, RES PROF CHEM, LOYOLA UNIV, LA, 68- *Mem:* Am Chem Soc; Royal Soc Chem; Int Soc Heterocyclic Chem. *Res:* Photorearrangements of nitrogen haloimides, propenes and cyclopropanes; photolysis of aryl diazonium compounds; transannular radical reactions; singlet molecular oxygen; paint and oil removal; surface-active agents; photooxidation. *Mailing Add:* 525 Kenmore Dr Harahan LA 70123

PETTEY, DIX HAYES, b Salt Lake City, Utah, Mar 16, 41. TOPOLOGY, OPERATIONS RESEARCH. *Educ:* Univ Utah, BS, 65, PhD(math), 68. *Prof Exp:* From asst prof to assoc prof, 68-83, PROF MATH, UNIV MO, 83- *Concurrent Pos:* Vis prof math, US Mil Acad, 74; sr scientist, Presearch Inc, 85. *Mem:* Math Asn Am; Soc Indust & Appl Math; Opers Res Soc Am. *Res:* General topology; computer memory systems; military applications; discrete mathematics. *Mailing Add:* Dept of Math Univ of Mo Columbia MO 65211

PETTIBONE, MARIAN HOPE, b Spokane, Wash, July, 08. ZOOLOGY. *Educ:* Linfield Col, BS, 30; Univ Ore, MS, 32, Univ Wash, PhD(zool), 47. *Prof Exp:* Asst zool, Univ Ore, 32-33; teacher high sch & jr col, 35-36; instr jr col, 36-42; actg assoc, Univ Wash, 45-46, assoc, Lab Embryol & Comp Anat, 46-47, instr marine plankton & comp anat, 47-49; res assoc, Johns Hopkins Univ, 49-53; from asst prof to assoc prof zool, Univ NH, 53-63; cur, 63-78, EMER CUR DEPT INVERT ZOOL, SMITHSONIAN INST, 78- *Mem:* AAAS; Soc Syst Zool; Am Soc Zool; Am Inst Biol Sci; Am Asn Zool; Nomenclature Biol Soc Wash. *Res:* Nutrition; growth curves using planarian worms as test animals; taxonomy and ecology of polychaete worms. *Mailing Add:* Dept Invert Zool Smithsonian Inst Washington DC 20560

PETTIGREW, JAMES EUGENE, JR, b Morris, Ill, Sept 17, 45; m 80. ANIMAL NUTRITION, SWINE. *Educ:* Southern Ill Univ, BS, 67; Iowa State Univ, MS, 69; Univ Ill, PhD(animal nutrit), 75. *Prof Exp:* Mgr swine res, Moorman Mfg Co, 74-80; ASSOC PROF ANIMAL NUTRIT, UNIV MINN, 80- *Concurrent Pos:* Vis scientist, Inst Grassland & Animal Prod, Shinfield, Reading, UK, 87-88. *Mem:* Am Soc Animal Sci; British Soc Animal Prod; Coun Agr Sci & Technol. *Res:* Dietary/metabolic effects on sow reproduction milk production; diets for weanling pigs; mathematical modelling of swine metabolism. *Mailing Add:* Dept Animal Sci Univ Minn St Paul MN 55108

PETTIGREW, NORMAN M, FLOW CYTOMETRY, CELL SURFACE MARKERS. *Educ:* Univ Glasgow, Scotland, MBChB, 64. *Prof Exp:* ASSOC PROF PATH, MED COL, UNIV MAN, 81- *Mailing Add:* Dept Path Univ Man Fac Med 753 McDermot Ave Winnipeg MB R3E 0W3 Can

PETTIJOHN, DAVID E, b Cordova, Alaska, Dec 28, 34; c 2. CHROMOSOMIC STRUCTURE, CANCER BIOLOGY. *Educ:* Wash State Univ, BS, 56, MS, 61; Stanford Univ, PhD(biophys), 64. *Prof Exp:* Res scientist human eng, Boeing Co, 59-61; USPHS fel molecular biol, Geneva, 64-66; asst prof, 66-71, assoc prof biophys, 71-77, PROF BIOCHEM, BIOPHYS & GENETICS, UNIV COLO MED CTR, DENVER, 78- *Mem:* AAAS; Am Soc Biol Chemists; Biophys Soc. *Res:* Nucleic acids; structure and synthesis; chromosomic structure with emphasis on DNA packaging; non-histone chromosomal proteins; cancer antigens; mucin structure. *Mailing Add:* Dept Biochem Univ Colo Sch Med 4200 E Ninth Ave Denver CO 80262

PETTIJOHN, FRANCIS JOHN, b Waterford, Wis, June 20, 04; c 3. GEOLOGY. *Educ:* Univ Minn, BA, 24, MA, 25, PhD(geol), 30. *Hon Degrees:* LHD, Johns Hopkins Univ, 78; ScD, Univ Minn, 86. *Prof Exp:* Instr, Macalester Col, 24-25; instr geol, Oberlin Col, 25-29; from instr to prof, Univ Chicago, 29-52; prof, 52-73, chmn dept, 63-68, EMER PROF GEOL, JOHNS HOPKINS UNIV, 73- *Concurrent Pos:* Geologist, US Geol Surv, 43-53; ed, J Geol, 47-52. *Honors & Awards:* Twenhofel Medal, Soc Econ Paleont & Mineralogists, 74; Wollaston Medal, Geol Soc London, 74; Penrose Medal, Geol Soc Am, 75; Sorby Medal, Int Asn Sedimentologists, 82. *Mem:* Nat Acad Sci; fel Geol Soc Am; hon mem Soc Econ Paleont & Mineral (vpres, 54, pres, 55); Am Asn Petrol Geol; fel Am Acad Arts & Sci; Int Asn Sedimentologists. *Res:* Precambrian geology; sedimentary petrology; sedimentation and geochronology; history of geology. *Mailing Add:* Dept Earth & Planetary Sci Johns Hopkins Univ Baltimore MD 21218

PETTIJOHN, RICHARD ROBERT, b San Francisco, Calif, Oct 29, 46; m 70. RADIOCHEMISTRY. *Educ:* Univ Calif, BS, 69; Univ Nebr, PhD(chem), 73. *Prof Exp:* Guest res assoc chem, Brookhaven Nat Lab, 72-73; res assoc, Univ Calif, Davis, 73-75; radiochemist, Stanford Res Inst, 76-77, MGR RADIATION PHYSICS, SRI INT, 77- *Mem:* Am Chem Soc. *Res:* Applications of fluorescence, phosphorescence and chemiluminescence to photography, electrophotography, electron multiplier array and glass capillary array applications, production and control of nonreflecting surfaces of silicon solar cells; night vision techniques; development of photographic image enhancement procedures. *Mailing Add:* 1216 Trancos Rd Menlo Park CA 94025

PETTIJOHN, TERRY FRANK, b Wyandotte, Mich, June 7, 48; m 70; c 3. ANIMAL BEHAVIOR, TEACHING. *Educ:* Mich State Univ, BS, 70; Bowling Green State Univ, MA, 72, PhD(psychol), 74. *Prof Exp:* From asst prof to assoc prof, 74-89, PROF PSYCHOL, OHIO STATE UNIV, MARION, 89- *Concurrent Pos:* Ed, Micro Psychol Network, 87- *Mem:* Animal Behav Soc; Psyconomic Soc; Soc Comput Psychol; Am Psychol Asn; Assembly Sci & Appl Psychol; Am Psychol Soc. *Res:* Comparative development and organization of animal social systems, including study of genetic, social and physical environmental influences on aggression, attachment, reproductive, social and learning behaviors in mammals; human memory; human stress; computer aassisted instruction. *Mailing Add:* Dept Psychol Ohio State Univ Marion OH 43302-5695

PETTINATO, FRANK ANTHONY, b Whitefish, Mont, May 19, 21; m 46; c 2. PHARMACEUTICAL CHEMISTRY. *Educ:* Univ Mont, BS, 49, MS, 54; Univ Wash, PhD(pharmaceut chem), 58. *Prof Exp:* Asst pharmaceut chem, Univ Wash, 54-58; from asst prof to assoc prof, 58-70, PROF PHARMACEUT CHEM, SCH PHARM, UNIV MONT, 70- *Mem:* Fel Am Found Pharmaceut Ed; Sigma Xi. *Res:* Chemistry and medicinal applications of plant constituents. *Mailing Add:* Sch Pharm State Univ Mont Missoula MT 59801

PETTINGA, CORNELIUS WESLEY, b Mille Lacs, Minn, Nov 10, 21; m 43; c 5. BIOCHEMISTRY. *Educ:* Hope Col, AB, 42; Iowa State Col, PhD(chem), 49. *Hon Degrees:* DSc, Hope Col, 84, Ind Univ, 87. *Prof Exp:* Assoc biochemist, Argonne Nat Lab, 49-50; res biochemist, Eli Lilly & Co, 50-53, head biochem res, 53-59, asst to vpres res, develop & control, 60-64, vpres, 64-71, pres, Elizabeth Arden, Inc, 71-72, exec vpres, 72-86; RETIRED. *Concurrent Pos:* Dir, Sci Corp, 87-, Collagen Corp, 87-, Atrix Corp, 87- & Celltrix, 89- *Honors & Awards:* Charles H Best Medal, Am Diabetes Asn, 87. *Mem:* Am Chem Soc; NY Acad Sci; AAAS. *Res:* Peptide synthesis; proteolytic enzymes; antibiotics; insulin. *Mailing Add:* 445 Somerset Dr W Indianapolis IN 46260

PETTINGILL, OLIN SEWALL, JR, b Belgrade, Maine, Oct 30, 07; m 32; c 2. ORNITHOLOGY. *Educ:* Bowdoin Col, AB, 30; Cornell Univ, PhD(ornith), 33. *Hon Degrees:* DSc, Bowdoin Col, 56, Colby Col, 79, Univ Maine, 82. *Prof Exp:* Teaching fel biol, Bowdoin Col, 33-34; instr, Westbrook Jr Col, Portland, Maine, 35-36; instr zool, Carleton Col, 36-41, from asst prof to assoc prof, 41-54; dir, Lab Ornith, Cornell Univ, 60-73. *Concurrent Pos:* Leader & co-leader expeds, Canada, Iceland, Antarctica, Argentina, New Zealand, Mexico and the Falkland Islands, 41-; res assoc, Cranbrook Inst Sci, 40-45; lectr, Audubon Screentours, 43-; trustee, Kents Hill Sch, 75-; vis prof biol, Va Polytech Inst & State Univ, 78. *Honors & Awards:* Eugene Eisemann Medal, Linnaean Soc NY, 75; Ludlow Griscom Award, Am Birding Asn, 82. *Mem:* Wilson Ornith Soc (secy, 37-41, 42-47, pres, 48-50); Cooper Ornith Soc; Wildlife Soc; Nat Audubon Soc (secy, 57-59, 63-66); Am Ornith Union (secy, 46-51). *Res:* Life histories, behavior, distribution and ecology of birds in the United States, Canada, Mexico and the Falkland Islands; photography of wildlife. *Mailing Add:* Gott Rd Wayne ME 04284

PETTIT, DAVID J, b Loma Linda, Calif, Sept 27, 36; m 56; c 3. CARBOHYDRATE CHEMISTRY. *Educ:* Univ Calif, Riverside, BA, 60, PhD(org chem), 64. *Prof Exp:* Res chemist, Textile Fibers Dept, Pioneering Res Div, E I du Pont de Nemours & Co, Inc, 64-66; sr res chemist, 66-67, sect head org res, 68-70, asst tech dir, 71-74, tech dir, 74-80, EXEC DIR, KELCO CO, 81- *Mem:* AAAS; Am Chem Soc; Sci Res Soc Am. *Res:* Chemistry of episulfides; stable episulfonium ions; new trialkyloxonium compounds; preparation, polymerization of small ring monomers for textile fibers; synthesis of polysaccharide derivatives; graft copolymers; fermentation biopolymers; polysaccharide structure and properties. *Mailing Add:* Res Dept Kelco Co 8225 Aero Dr San Diego CA 92123

PETTIT, FLORA HUNTER, NUTRITION, ENZYMOLOGY. *Educ:* Univ Tex, PhD(biochem), 61. *Prof Exp:* RES SCIENTIST, UNIV TEX, 74- *Mailing Add:* 4605 Lantana Hollow Austin TX 78731

PETTIT, FREDERICK S, b Wilkes-Barre, Pa, Mar 10, 30; m 58; c 4. METALLURGY, PHYSICAL CHEMISTRY. *Educ:* Yale Univ, BE, 52, MEng, 60, DEng(metall), 62. *Prof Exp:* Jr scientist, Westinghouse Atomic Power Div, Pa, 52-54 & Lycoming Div, Avco Mfg Corp, Conn, 57-58; res asst metall, Yale Univ, 58-62; NSF fel thermodyn & reaction kinetics, Max Planck Inst Phys Chem, Gottingen, WGer, 62-63; res assoc, Adv Mat Res & Develop Lab, United Aircraft Corp, North Haven, 63-64, sr res assoc, Middletown, 64-68 & East Hartford, 68-72, sr staff scientist, Pratt & Whitney Aircraft Div, 72-79; chmn, 79-88, PROF, MATS SCI & ENG DEPT, UNIV PITTSBURGH, 79- *Mem:* Am Inst Mining, Metall & Petrol Engrs; fel Am Soc Metals Int; Electrochem Soc. *Res:* Oxidation of metals and alloys; thermodynamics and kinetics of solid state reactions; electrochemical processes. *Mailing Add:* 201 Ennerdale Lane Pittsburgh PA 15237

PETTIT, GEORGE ROBERT, b Long Branch, NJ, June 8, 29; m 53; c 5. ORGANIC CHEMISTRY, CANCER. *Educ:* Wash State Univ, BS, 52; Wayne State Univ, MS, 54, PhD(org chem), 56. *Prof Exp:* Asst chem, Wash State Univ, 50-52, lect demonstr, 52; asst, Wayne State Univ, 52-56; sr res chemist, Morton-Norwich Co, 56-57; from asst prof to prof chem, Univ Maine, 57-65; vis prof, Stanford Univ, 65, SAfrican Univs, 78; chmn org div, Ariz State Univ, 66-68, dir, Cancer Res Lab, 74-75, prof, 65-85, DALTON PROF CHEM, ARIZ STATE UNIV, 85-, REGENTS PROF, 90- *Concurrent Pos:* Consult, Smith-Kline-French Labs, 59-63, Nat Cancer Inst, 65-76, mem med chem fel panel, NIH, 67-70, Nat Cancer Inst, Div Cancer Treat Adv Bd, 71-74; mem, Special Study Sect, Nat Cancer Inst & NIH, 76-, Wash State Univ Found Bd, 81-; distinguished res prof, Arizona State Univ, 78-79; dir, Cancer Res Inst, Ariz State Univ, 75- *Honors & Awards:* Chem Pioneer Award, Am Inst of Chemists, 88; Outstanding Investr Award, Nat Cancer Inst, 90. *Mem:* Am Asn Cancer Res; Am Chem Soc; The Chem Soc; Am Soc Pharmacog; Sigma Xi; AAAS. *Res:* Cancer chemotherapy; chemistry of natural products especially steroids, peptides, nucleotides, anticancer constituents, from marine animals, arthropods and plants; general organic synthesis. *Mailing Add:* Dept Chem & Cancer Res Inst Ariz State Univ Tempe AZ 85287-1604

PETTIT, JOHN TANNER, theoretical physics, for more information see previous edition

PETTIT, RAY HOWARD, b Canton, Ga, May 12, 33; m 58; c 3. ELECTRICAL ENGINEERING. *Educ:* Ga Inst Technol, 54, MSEE, 60; Univ Fla, PhD(elec eng), 64. *Prof Exp:* Appln engr, Westinghouse Elec Corp, 54-55, 57-58; sr engr, Orlando Div, Martin Co, 60-61, design specialist, 63; scientist, Lockheed-Ga Res Lab, 63-66; from assoc prof to prof elec eng, Ga Inst Technol, 66-78; PROF ELEC ENG, CALIF STATE UNIV, NORTHRIDGE, 78- *Concurrent Pos:* Lectr, Ga State Col, 64; consult, US Air Force, 75- 81. *Mem:* Inst Elec & Electronics Engrs. *Res:* Communication theory; modulation theory; applied decision theory; information theory; adaptive systems; spread spectrum systems; communications electronic countermeasures and electronic counter-countermeasures. *Mailing Add:* Dept Elec & Comp Eng Calif State Univ Northridge CA 91330

PETTIT, RICHARD BOLTON, b Lockport, NY, Dec 4, 44; m 66; c 2. SOLID STATE PHYSICS. *Educ:* Univ Mich, BS, 66; Cornell Univ, PhD(appl physics), 72. *Prof Exp:* Mem tech staff physics, 71-86, SUPVR, ELEC STANDARDS DIV, SANDIA LABS, 86- *Concurrent Pos:* Ed mat, Solar Energy Eng. *Honors & Awards:* Dept Energy Solar Thermal Award, 81. *Mem:* AAAS; Am Phys Soc; Int Solar Energy Soc; Am Vacuum Soc; Soc Photo-optical Instrumentation Engrs. *Res:* Optical property measurement of solar energy materials including emittance and solar absorptance of solar selective coatings, specular reflectance properties of mirrors and solar transmittance of glazings. *Mailing Add:* Div 7342 Sandia Labs Albuquerque NM 87185-5800

PETTIT, ROBERT EUGENE, b Edina, Mo, Dec 19, 28; m 60; c 4. PLANT PATHOLOGY & PHYSIOLOGY. *Educ:* Univ Mo-Columbia, BS, 55, MA, 60, PhD(plant path), 66. *Prof Exp:* Technician, Wolfe Elec Co, Kans, 51-52, Cessna Aircraft Co, 52, Beech Aircraft Corp, 52-53; teacher pub schs, Mo, 55-56 & 60-61; instr electronics, US Air Force, 56-58; res asst plant physiol, Univ Mo-Columbia, 58-60; instr plant path, Univ Mo-Columbia, 61-66; asst prof, 66-72, assoc prof plant path, Tex A&M Univ, 72-88. *Mem:* Am Phytopath Soc; Am Inst Biol Sci; Am Peanut Res & Educ Asn; Sigma Xi. *Res:* Biochemical factors within plants which impart frost hardiness; soil microbiology as related to plant root ecological balances; diseases of peanuts and their control; mycotoxins and the fungi which produce these metabolic compounds; correlation studies of molecular biology, genetics of disease resistance and mechanisms of plant-pathogen interactions. *Mailing Add:* Dept Plant Sci Tex A&M Univ College Station TX 77843

PETTIT, RUSSELL DEAN, b Burr Oak, Kans, Feb 17, 41; m 70. RANGE ECOLOGY. *Educ:* Ft Hays Kans State Col, BS, 63; Tex Tech Univ, MS, 65; Ore State Univ, PhD(range mgt), 68. *Prof Exp:* Asst prof range mgt, Tex A&I Univ, 68-69; asst prof, 69-74, ASSOC PROF RANGE MGT, TEX TECH UNIV, 74- *Mem:* Am Soc Agron; Soc Range Mgt; Soil Sci Soc Am. *Res:* Phytosociological relationships between plant communities; soil-plant-water relationships; grassland ecosystems. *Mailing Add:* Dept Range & Wildlife Mgt Tex Tech Univ Lubbock TX 79409

PETTIT, THOMAS HENRY, b Salt Lake City, Utah, Jan 20, 29; m 53; c 6. OPHTHALMOLOGY. *Educ:* Univ Calif, Los Angeles, AB, 49; Univ Pa, MD, 55. *Prof Exp:* From asst prof to assoc prof, 63-69, PROF OPHTHAL, SCH MED, UNIV CALIF, LOS ANGELES, 69-, ASSOCDIR, JULES STEIN EYE INST & CHIEF, CORNEA EXTERNAL OCULAR DIS DIV, HEALTH SCI CTR, 80- *Concurrent Pos:* NIH spec fel ophthal, Wash Univ, 61-62; NIH res fel, Proctor Found, San Francisco, 62-63; consult ophthal, Vet Admin Hosps, Long Beach, 63-70, Wadsworth, 63-70 & Sepulveda, 69-90; consult, Nat Eye Inst, NIH, 71-75. *Mem:* Asn Res Vision & Ophthal; Am Acad Ophthal. *Res:* Herpes simplex infections in man; corneal and external ocular diseases; infectious keratitis. *Mailing Add:* Jules Stein Eye Inst Univ Calif Sch Med Los Angeles CA 90024-7003

PETTITT, BERNARD MONTGOMERY, b Houston, Tex, Sept 3, 53; m 82. STATISTICAL MECHANICS OF FLUIDS, POLYMER BIOPHYSICS. *Educ:* Univ Houston, BS(chem) & BS(math), 76, PhD(chem), 80. *Prof Exp:* Teaching fel, Univ Tex, 80-83; NIH fel, Harvard Univ, 83-85; ASST PROF CHEM, UNIV HOUSTON, 85- *Mem:* Am Phys Soc; AAAS. *Res:* Connection between the physical properties of solutions and the properties of their constituent atoms; development and application of theories to the structure and thermodynamics of fluids, aqueous solutions and biomolucules in solution. *Mailing Add:* Dept Chem Univ Houston Univ Park Houston TX 77004

PETTOFREZZO, ANTHONY J, b Bayonne, NJ, Oct 12, 31; m 54; c 4. NUMBER THEORY. *Educ:* Montclair State Col, BA, 53, MA, 57; NY Univ, PhD(math), 59. *Prof Exp:* Instr pub sch, NJ, 53 & 55-56; sr programmer comput math, Wright Aeronaut Corp, NJ, 56-57; asst prof math, Newark Col Eng, 57-60; assoc prof, Montclair State Col, 60-66; assoc prof math educ, Fla State Univ, 66-69; PROF MATH, UNIV CENT FLA, 69- *Mem:* Math Asn Am. *Res:* Linear algebra and its applications. *Mailing Add:* Dept Math Sci Univ Central Fla Alafaya Trail Orlando FL 32816

PETTRY, DAVID EMORY, b Beckley, WVa, Aug 1, 35; m 60; c 3. SOIL MORPHOLOGY. *Educ:* Univ Fla, BS, 62, MS, 65; Va Polytech Inst, PhD(soil sci), 69. *Prof Exp:* Soil scientist, Soil Conserv Serv, USDA, 60-69; from asst prof to assoc prof soil sci, Va Polytech Inst & State Univ, 69-75; PROF SOIL SCI, MISS STATE UNIV, 75- *Concurrent Pos:* NASA & Va State Health Dept grants, Va Polytech Inst & State Univ, 70-; consult, Va State Health Dept, 70-; res grants, Phillips Coal Co, 80-81; consult, UN. *Mem:* Soil Sci Soc Am; Clay Minerals Soc; Am Soc Photogram Eng; Am Soc Agron; Soil Conserv Soc Am. *Res:* Soil genesis, classification; urban soils; clay mineralogy; remote sensing of soils and plant species and conditions. *Mailing Add:* Agron Dept Miss State Univ PO Box 5248 Mississippi State MS 39762

PETTUS, DAVID, b Goliad, Tex, Nov 28, 25; m 47; c 5. VERTEBRATE ZOOLOGY, GENETICS. *Educ:* Ariz State, BA, 51, MA, 52; Univ Tex, PhD, 56. *Prof Exp:* From assoc prof to prof, 56-89, EMER PROF ZOOL, COLO STATE UNIV, 89- *Concurrent Pos:* Vis prof, Tulane Univ, 69. *Honors & Awards:* Krause Award; Harris T Guard Award. *Mem:* Soc Study of Evolution; Am Soc Ichthyol & Herpet; Genetics Soc Am; Sigma Xi; AAAS; Soc Study Amphib & Reptiles. *Res:* Ecology, genetics and evolution of amphibians and reptiles. *Mailing Add:* Dept Biol Colo State Univ Ft Collins CO 80521

PETTUS, WILLIAM GOWER, b Lynchburg, Va, Aug 6, 25; m 47; c 1. FISSION, FUSION. *Educ:* Lynchburg Col, BS, 49; Univ Va, MS, 53, PhD(physics), 56. *Prof Exp:* Instr physics, Lynchburg Col, 52-54; instr, Univ Va, 54-56; ADV PHYSICIST, BABCOCK & WILCOX CO, 56- *Concurrent Pos:* Vis res scientist, Plasma Physics Lab, Princeton Univ, 78; vis prof nuclear eng, Va Polytech Inst & State Univ, 81. *Mem:* Sigma Xi; Am Phys Soc. *Res:* Neutron physics; nuclear reactor physics; fusion; space power and propulsion; advanced energy systems. *Mailing Add:* Space & Nuclear Syst Babcock & Wilcox Co PO Box 11165 Lynchburg VA 24506-1165

PETTY, CHARLES SUTHERLAND, b Lewistown, Mont, Apr 16, 20; m 57. MEDICINE. *Educ:* Univ Wash, BS, 41, MS, 46; Harvard Med Sch, MD, 50; Am Bd Path, dipl, 56. *Prof Exp:* From instr to asst prof path, Sch Med, La State Univ, 55-58; from asst prof to assoc prof forensic path, Sch Med, Univ Md, 58-67; prof, Med Ctr, Ind Univ, 67-69; PROF FORENSIC SCI, UNIV TEX HEALTH SCI CTR DALLAS, 69-, DIR SOUTHWESTERN INST FORENSIC SCI, 69- *Concurrent Pos:* Teaching fel path, Harvard Med Sch, 52-55; vis physician, Charity Hosp La, New Orleans, 55-58; clin pathologist, Md Gen Hosp, 58-61; lectr, Sch Hyg & Pub Health, Johns Hopkins Univ, 59-65, assoc, 65-67; asst state med examr, Md, 58-67; chief med examr, Dallas County, 69- *Mem:* Am Asn Pathologists & Bacteriologists; Am Col Physicians; Am Acad Forensic Sci. *Res:* Pathology; pathologic physiology. *Mailing Add:* PO Box 35728 Dallas TX 75235

PETTY, CLINTON MYERS, b Des Moines, Iowa, June 4, 23; m 57; c 4. GEOMETRY. *Educ:* Univ Southern Calif, AB, 48, MA, 49, PhD(math), 52. *Prof Exp:* Instr math, Princeton Univ, 52-53; res instr, Duke Univ, 53-54; instr, Purdue Univ, 54-55; consult scientist, Lockheed Missile & Space Co, 55-66; assoc prof, 66-74, PROF MATH, UNIV MO-COLUMBIA, 74- *Mem:* Am Math Soc; Math Asn Am. *Res:* Theory of convex sets; geometry of metric spaces. *Mailing Add:* Dept Math Univ Mo Columbia MO 65211

PETTY, HOWARD RAYMOND, b Toledo, Ohio, Aug 1, 54; m 80; c 3. MEMBRANE BIOLOGY. *Educ:* Manchester Col, BS, 76; Harvard Univ, PhD(biophys), 79. *Prof Exp:* Fel, Stanford Univ, 79-81; from asst prof to assoc prof, 81-89, PROF BIOL, WAYNE STATE UNIV, 89- *Concurrent Pos:* Fel, Damon Runyon-Walter Winchel Cancer Res. *Mem:* NY Acad Sci; Biophys Soc; Am Asn Immunologists; Am Soc Cell Biol; Reticuloendothelial Soc; fel AAAS. *Res:* Role of the macrophage and neutrophil cell surface in phagocytosis and cytolysis. *Mailing Add:* Dept Biol Sci Wayne State Univ Col Liberal Arts Detroit MI 48202

PETTY, ROBERT OWEN, ecology; deceased, see previous edition for last biography

PETTY, THOMAS LEE, b Boulder, Colo, Dec 24, 32; m 54; c 3. PULMONARY DISEASES. *Educ:* Univ Colo, BA, 55, MD, 58. *Prof Exp:* Intern, Philadelphia Gen Hosp, 58-59; resident, Med Ctr, Univ Mich, 59-60; resident med, 60-62, fel pulmonary dis, 62-63, chief resident med, 63-64, from instr to assoc prof, 62-74, PROF MED, UNIV COLO MED CTR, 74- HEAD DIV PULMONARY MED, 71- *Mem:* Am Thoracic Soc; fel Am Col Chest Physicians (pres, 81-82); fel Am Col Physicians; Sigma Xi. *Res:* Basic and clinical research in pulmonary sciences; applied physiology. *Mailing Add:* Dept Acad Affairs P/SL Ctr for Health Sci 1719 E 19th Ave Denver CO 80218

PETTY, WILLIAM CLAYTON, b Cedar City, Utah, Aug 8, 38; m 61; c 7. ANESTHESIA MACHINES, RISK MANAGEMENT. *Educ:* Univ Utah, BS, 62, MD, 65. *Prof Exp:* Chief anesthesia, 24th Evacuation Hosp, Vietnam, 69-70; chmn, Dept Anesthesiol, Aultman Hosp, 87-90; prof anesthesiol, Col Med, Northeastern Ohio Univ, 87-90; staff anesthesiologist, Nat Naval Med Ctr, 90-91; PROF ANESTHESIOL, UNIFORM SERV UNIV HEALTH SCI, 88-; PROF ANESTHESIOL, UNIV UTAH, 81-87 & 91-; STAFF ANESTHESIOLOGIST, ST MARKS HOSP, 91- *Concurrent Pos:* Consult anesthesiol, Intermountain Healthcare Inc, 81-87; mem, Prof Liability Comt, Am Soc Anesthesiol, 84-88; ed, Utah Soc Anesthesiologists Newsletter, 85-87, Wellcome Trends Anesthesiol, 87-, Ohio Soc Anesthesiologists Newsletter, 88-90; mem, Comt Technol, Anesthesiologists Patient Safety Found, 88- *Mem:* Am Soc Anesthesiologists; Asn Univ Anesthetists; World Fedn Soc Anesthesiologists; Am Soc Zoologists. *Res:* Anesthesia machines; safety features of anesthesia machines; risk management in anesthesia; history of anesthesia. *Mailing Add:* 2065 E Lincoln Lane Salt Lake City UT 84124

PETTYJOHN, WAYNE A, b Portland, Ind, Aug 4, 33; div; c 2. HYDROLOGY. *Educ:* Univ SDak, BA & MA, 59; Boston Univ, PhD(geol), 64. *Prof Exp:* Instr geol, Bradford Jr Col, 60-63; hydrologist, US Geol Surv, 63-67; prof geol, Ohio State Univ, 67-80; PROF GEOL, OKLA STATE UNIV, 80- *Concurrent Pos:* Groundwater consult; attorney-at-law. *Mem:* Geol Soc Am; Sigma Xi; Am Inst Prof Geologists; Nat Water Well Asn. *Res:* Surface and ground-water relationships and quality of water, including natural quality and pollution; legal aspects of water law; ground-water contamination. *Mailing Add:* Geol 151 PSI Okla State Univ Stillwater OK 74078

PETURA, JOHN C, US citizen. WASTE TREATMENT & MANAGEMENT. *Educ:* Bucknell Univ, BS, 68; Drexel Univ, MS, 73. *Prof Exp:* Chem & environ engr, Gulf Oil Corp, 68-73; sr proj engr & process mgr, Calgon Corp, 73-78; PRIN PROJ MGR & SR PROJ ENGR, ROY F WESTON, INC, 78- *Mem:* Am Inst Chem Engrs; Water Pollution Control Fedn; Am Water Works Asn; Am Acad Environ Engrs. *Res:* Water and wastewater treatment; industrial/solid/hazardous waste management; sludge handling and disposal; sanitary landfill design and surveillance; control and removal of hazardous toxic substances; potable water supply and treatment; contaminated groundwater recovery. *Mailing Add:* Roy F Weston Inc Weston Way West Chester PA 19380

PETZOLD, EDGAR, b Grand Rapids, Mich, Oct 10, 30; m 52; c 4. BIOCHEMISTRY. *Educ:* Mich State Univ, BS, 52; Purdue Univ, MS, 55, PhD(biochem), 59. *Prof Exp:* Res chemist, Grain Processing Corp, 58-61; dept mgr, Salsbury Labs, 61-66; assoc scientist, Biochem Residue Anal Dept, 66-72, ASSOC SCIENTIST, DRUG METAB RES, METAB SECT, UPJOHN CO, 72- *Concurrent Pos:* Comput technol. *Mem:* Am Chem Soc; Am Diabetes Asn. *Res:* Drug absorption; bioavailability; distribution; metabolism excretion; mechanism of action; analytical methods for drugs in biological fluids; biochemistry of carotenoids; isolation and properties of peptide antibiotics; industrial enzymes; carbohydrate chemistry and production; metabolism and residue methods in animals; herbicide methods and metabolism; drug metabolism; metabolism of anti-diabetic agents. *Mailing Add:* 159 S Lake Doster Dr Plainwell MI 49080

PEURA, ROBERT ALLAN, b Worcester, Mass, Jan 26, 43; m 64; c 4. BIOMEDICAL ENGINEERING, ELECTRICAL ENGINEERING. *Educ:* Worcester Polytech Inst, BS, 64; Iowa State Univ, MS, 67, PhD(biomed & elec eng), 69. *Prof Exp:* Asst prof elec & biomed eng, 68-73, ASSOC PROF ELEC ENG & BIOMED ENG, WORCESTER POLYTECH INST, 73-, SITE DIR, WORCESTER POLYTECH INST-ST VINCENT HOSP INTERNSHIP CTR, 72- *Concurrent Pos:* Lectr biomed eng, Med Sch, Univ Mass, 74-; vis assoc prof, Health Sci & Technol Div, Mass Inst Technol, 81-82. *Mem:* Am Heart Asn; Int Elec & Electronics Engrs; Am Soc Eng Educ; Asn Advan Med Instrumentation. *Res:* Noninvasive measurement of circulation; impedance plethysmography; model of electrocardiogram conduction system; computers in biomedicine; models of biological systems; biomedical instrumentation systems; blood pressure measurements; engineering education. *Mailing Add:* Dept Biomed Eng Worcester Polytech Inst 100 Institute Rd Worcester MA 01609

PEURIFOY, PAUL VASTINE, b Wortham, Tex, June 10, 27; m 53; c 4. ANALYTICAL CHEMISTRY. *Educ:* Fla Southern Col, BS, 49; Univ Miami, MS, 51; Kans State Univ, PhD(chem), 56. *Prof Exp:* Instr chem, Kans State Univ, 54-56; res chemist, Houston Res Lab, Shell Oil Co, 56-66, sr res chemist, 67-72, SR RES CHEMIST, SHELL DEVELOP CO, 72- *Mem:* Am Chem Soc; fel Am Inst Chem. *Res:* Polarography; gas-liquid chromatography; spot tests; electrochemistry; high-pressure liquid chromatography; paper and thin layer chromatography; organic functional group analysis; coal analysis. *Mailing Add:* 11715 Glenway Dr Houston TX 77070

PEVEHOUSE, BYRON C, b Lubbock, Tex, Apr 5, 27; m 51; c 3. MEDICINE. *Educ:* Baylor Univ, BS, 49, MD, 52; McGill Univ, MSc, 60; Univ London, cert, 60. *Prof Exp:* From asst prof to assoc prof, 60-76, prof, 76-80, CLIN PROF NEUROL SURG, SCH MED, UNIV CALIF, SAN FRANCISCO, 80-, CHIEF DEPT, PAC MED CTR, 67- *Concurrent Pos:* NSF fel neuropath, 59; USPHS trainee neurophysiol, 59-60. *Mem:* Am Col Surgeons; Am Acad Neurol; Am Acad Neurol Surg; Am Asn Neurol Surg; fel Royal Soc Med. *Res:* Birth defects of the central nervous system; use of hypothermia in the treatment of injuries and diseases of the brain. *Mailing Add:* 2351 Clay St San Francisco CA 94115

PEVERLY, JOHN HOWARD, b Danville, Ill, Oct 15, 44; m 66; c 4. PLANT BIOCHEMISTRY, SOIL CHEMISTRY. *Educ:* Purdue Univ, BS, 66; Univ Ill, MS, 68, PhD(plant physiol), 71. *Prof Exp:* Asst prof, 71-78, ASSOC PROF AGRON, CORNELL UNIV, 78- *Mem:* Am Asn Aquatic Plant Mgt; Int Asn Aquatic Vascular Plant Biol; AAAS; Am Soc Agron; Sigma Xi. *Res:* Influence of agricultural wastes on growth of aquatic plants; nutrient cycling in wetlands; control and biochemistry of aquatic plants. *Mailing Add:* 1016 Bradfield Hall Cornell Univ Ithaca NY 14850

PEVSNER, AIHUD, b Palestine, Dec 18, 25; nat US; m 49; c 2. PHYSICS. *Educ:* Columbia Univ, AB, 47, AM, 48, PhD(physics), 54. *Prof Exp:* Asst, Columbia Univ, 48-53; instr physics, Mass Inst Technol, 53-56; prof, 56-75, JACOB L HAIN PROF PHYSICS, JOHNS HOPKINS UNIV, 75- CHMN DEPT, 74- *Mem:* Am Phys Soc. *Res:* High energy physics, properties of fundamental particles. *Mailing Add:* Dept Physics Johns Hopkins Univ Baltimore MD 21218

PEW, WEYMOUTH D, physiology, for more information see previous edition

PÉWÉ, TROY LEWIS, b Rock Island, Ill, June 28, 18; m 44; c 3. POLAR GEOMORPHOLOGY. *Educ:* Augustana Col, AB, 40; Univ Iowa, MS, 42; Stanford Univ, PhD(geol), 52. *Hon Degrees:* DSc, Univ Alaska, 91. *Prof Exp:* Instr geol & head dept, Augustana Col, 42-46; instr geomorphol, Stanford Univ, 46; assoc prof, Univ Alaska, 54-58, prof & head dept, 58-65; chmn dept, 65-76, PROF GEOL, ARIZ STATE UNIV, 65-, DIR MUS GEOL, 76-; GEOLOGIST, US GEOL SURV, 46- *Concurrent Pos:* Consult, Corps Eng, US Army, 45, geologist, 46; chief glacial geologist, US Nat Comt-Int Geophys Year, Antarctica, 57-58; US mem periglacial comt, Int Geog Comt; mem coun Alaska earthquake comt, Nat Acad Sci-Nat Res Coun, 65, glaciol panel on polar res, 71-73; chmn joint US planning comt, Sec Int Permafrost Conf, Nat Acad Sci-Nat Acad Eng-Nat Res Coun, 72-74; chmn, Permafrost Comn, Nat Acad Sci-Nat Res Coun, 75-81; leader US deleg, Third Int Permafrost Conf, 78, chmn US comt, Fourth Int Permafrost Conf, 80-84; vpres, Int Permafrost Assoc, 83-88, pres, 88- *Honors & Awards:* US Cong Antarctic Medal, 66; USSR Nat Acad Medal, 88. *Mem:* Fel AAAS; fel Geol Soc Am; Am Asn Geol Teachers; Glaciol Soc; fel Arctic Inst NAm. *Res:* Geomorphology; permafrost; glaciation; Quaternary geology of Arctic, Subarctic and Antarctic areas; environmental geology of desert regions. *Mailing Add:* Dept of Geol Ariz State Univ Tempe AZ 85287

PEWITT, EDWARD GALE, b Tenn, July 12, 32; m 56; c 4. EXPERIMENTAL HIGH ENERGY PHYSICS. *Educ:* Vanderbilt Univ, BEEE, 54; Carnegie Inst Technol, MS, 57, PhD, 61; Univ Chicago, MBA, 74. *Prof Exp:* Res physicist & lectr physics, Carnegie Inst Technol, 61-63, asst prof, 63-66; assoc physicist, Argonne Nat Lab, 66-71, actg lab dir & actg dep lab dir, 79, dep lab dir, 80-84, sr physicist & dir High Energy Facil Div, 71-73, assoc lab dir Energy & Environ Technol, 73-79, chief opers officer, 84-89; SCIENTIST III, FERMI NAT ACCELERATE LAB, 89- *Mem:* Fel Am Phys Soc. *Res:* Particle detector development, particularly bubble chambers. *Mailing Add:* 1129 Big Foot Lane Naperville IL 60563

PEWS, RICHARD GARTH, b Leamington, Ont, Mar 25, 38; m 60; c 3. ORGANIC CHEMISTRY. *Educ:* Univ Western Ont, BSc, 60, PhD(org chem), 63. *Prof Exp:* Fel with Prof R W Taft, Pa State Univ, 63-65; res chemist, Govt Res Labs, Esso Res & Eng Co, NJ, 65-66; res chemist, Polymer & Chem Labs, 66-67, proj leader, Hydrocarbons & Monomers Res Lab, 68-71, sr res chemist, Halogens Res Lab, 70-71, group leader, 71-75, sr res specialist, 76-77, assoc scientist, 77-81, sr assoc scientist, Cent Process, 81-85, RES SCIENTIST, DOW CHEM CO, 85- *Mem:* Am Chem Soc. *Res:* Organic reaction mechanisms; molecular rearrangements; reductions with complex metal hydrides; Hammett-Taft correlations; transmission of conjugation in three membered rings and biaryl systems; hexahalocyclopentadienes; propellant chemistry; nucleophilic aromatic substitution; synthesis of heterocycles; chemistry of sulfonyl cyanides; aromatic fluorene chemistry. *Mailing Add:* Dow Chem Co 1776 Bldg 1 Midland MI 48640

PEYGHAMBARIAN, NASSER, b Iran, Mar 26, 54; US citizen. OPTICAL SCIENCES, SOLID STATE OPTICS. *Educ:* Pahlavi Univ, Iran, BS, 76; Ind Univ, Bloomington, MS, 79, PhD(physics), 82. *Prof Exp:* Postdoctoral optical sci, Optical Sci Ctr, Univ Ariz, 82-83, res asst prof, 83-85, from asst prof to assoc prof, 85-91, PROF OPTICAL SCI, OPTICAL SCI CTR, UNIV ARIZ, 91-, DIR, OPTICAL CIRCUITRY COOP, 91- *Concurrent Pos:* 3M young fac award, 87 & 88; TRW young fac award, 89; vis prof, NTT, Japan, 90. *Mem:* Am Phys Soc; Optical Soc Am. *Res:* Femtosecond laser spectroscopy of semiconductors; high-speed optical switching; nonlinear photonics; optical nonlinearities of quantum confined semiconductor microstructures; nonlinear optics of organic polymers; co-author of over 100 publications. *Mailing Add:* Optical Sci Ctr Univ Ariz Tucson AZ 85721

PEYRONNIN, CHESTER A(RTHUR), JR, b New Orleans, La, July 26, 25; m 53; c 3. MECHANICAL ENGINEERING. *Educ:* Tulane Univ La, BE, 47; Ill Inst Technol, MS, 50. *Prof Exp:* From asst prof to assoc prof, 48-64, PROF MECH ENG, TULANE UNIV LA, 64- *Concurrent Pos:* Consult, Corps Engrs, US Army, Chrysler Corp, Gen Dynamics Corp & Port of New Orleans. *Honors & Awards:* Guerrard Mackey Award, Am Soc Mech Engrs. *Mem:* AAAS; Am Soc Mech Engrs; Am Soc Eng Educ. *Res:* Engineering medical research; tidal hydraulics; aerospace design; fire protection; port development. *Mailing Add:* Sch Eng Tulane Univ New Orleans LA 70118

PEYTON, LEONARD JAMES, b Compton, Calif, Feb 8, 24; m 60; c 2. ORNITHOLOGY. *Educ:* Utah State Univ, BS, 51. *Prof Exp:* Biologist, Arctic Health Res Ctr, USPHS, Anchorage, 55-62; asst zoophysiologist, Inst Artic Biol, Univ Alaska, Fairbanks, 62-77, zoophysiologist, 77-90, coordr environ serv, 62-90; RETIRED. *Mem:* AAAS; Am Ornith Union; Cooper Ornith Soc. *Res:* Geographic related song patterns of birds and their relationship within a given species; migration of birds and the use of song patterns in its study; distribution of birds in Alaska. *Mailing Add:* 1790 Red Fox Dr Fairbanks AK 99701

PEZ, GUIDO PETER, b Fiume, Italy, Feb 10, 41; US citizen; m 66; c 3. CATALYSIS. *Educ:* Univ New South Wales, BSc, 62; Monash Univ, Australia, PhD(chem), 67. *Prof Exp:* Fel, McMaster Univ, 67-69; res chemist, Allied Chem Corp, 69-74, sr res chemist, 74-78, res assoc, 78-81; SR RES ASSOC, AIR PROD & CHEMICALS, INC, 81- *Mem:* Am Chem Soc. *Res:* Synthetic inorganic and organometallic chemistry applied to the development of new homogeneous and heterogeneous catalysts, selective gas absorption materials, and gas separation membranes. *Mailing Add:* Air Prod & Chem Inc 7201 Hamilton Blvd Allentown PA 18195-1501

PEZOLET, MICHEL, b Montreal, Que, Jan 30, 46; m 69; c 2. BIOPHYSICAL & POLYMER CHEMISTRY. *Educ:* Laval Univ, BS, 68, PhD(chem), 71. *Prof Exp:* Res assoc physics, Univ BC, 71-72; res assoc chem, Univ Ore, 72-73; asst prof, 73-77, ASSOC PROF CHEM, LAVAL UNIV, 77- *Concurrent Pos:* Fel, Nat Res Coun Can, 71-73; sabbatical leave, vis prof, Ctr Paul Pascal, Talence, France. *Mem:* Chem Inst Can; Biophys Soc. *Res:* Spectroscopic studies of natural and synthetic macromolecules; structure of proteins and membranes determined by Raman and Fourier transform spectroscopy. *Mailing Add:* Dept Chem Laval Univ Quebec PQ G1K 7P4 Can

PFADT, ROBERT E, b Erie, Pa, May 22, 15; m 48; c 4. ENTOMOLOGY. *Educ:* Univ Wyo, BA, 38, MA, 40; Univ Minn, PhD(entom), 48. *Prof Exp:* Field asst, State Fish Comn, Wyo, 38; field supvr, Bur Entom & Plant Quarantine, USDA, 39; field asst, 40-42, from asst prof to prof, 42-85, EMER PROF ENTOM, UNIV WYO, 85- *Concurrent Pos:* Chief, Univ Wyo team, Kabul, Afghanistan, 66-67. *Mem:* Entom Soc Am; Entom Soc Can; fel Royal Entom Soc; Orthopterist's Soc; fel AAAS. *Res:* Ecology and control of insects. *Mailing Add:* Dept Plant Soil & Insect Sci Box 3354 Univ Wyo Laramie WY 82071

PFAENDER, FREDERIC KARL, b Long Beach, Calif, Aug 5, 43; m 72; c 2. MICROBIOLOGY, ENVIRONMENTAL SCIENCE. *Educ:* Calif State Univ, Long Beach, BS, 66, MS, 68; Cornell Univ PhD(microbiol), 71. *Prof Exp:* From asst prof to prof environ microbiol, Dept Environ Sci & Eng, 71-89, DIR & ASSOC CHMN ENVIRON CHEM & BIOL PROG, UNIV NC, CHAPEL HILL, 89- *Concurrent Pos:* Chmn, Gordon Res Conf, Appl Environ Microbiol, 87, Am Soc Microbiol, Bd Educ & Training, 88- *Mem:* AAAS; Am Soc Microbiol; Sigma Xi; Soc Environ Toxicol & Chem; Soc Indust Microbiol; Am Acad Microbiol. *Res:* Environmental distribution and microbial degradation of organic compounds, especially molecules considered as pollutants; microbial activities in salt-marsh estuarine ecosystems, oceans and ground water. *Mailing Add:* Dept of Environ Sci Sch of Pub Health Univ of NC Chapel Hill NC 27599-7400

PFAFF, DONALD CHESLEY, b Los Angeles, Calif, Nov 4, 36; m 64; c 2. MATHEMATICS. *Educ:* Univ Calif, Berkeley, AB, 57, MA, 59, PhD(math), 69. *Prof Exp:* From lectr to asst prof, 61-72, ASSOC PROF MATH, UNIV NEV, RENO, 72- *Concurrent Pos:* Lectr, Math Asn Am, 70-75. *Mem:* Nat Coun Teachers Math; Math Asn Am. *Res:* Measure theory; discontinuous functions. *Mailing Add:* Dept Math Univ Nev Reno NV 89557

PFAFF, DONALD WELLS, b Rochester, NY, Dec 9, 39; m 63; c 3. NEUROPHYSIOLOGY, ANIMAL BEHAVIOR. *Educ:* Harvard Col, AB, 61; Mass Inst Technol, PhD(psychol), 65. *Prof Exp:* Res asst, Harvard Col, 59-61; asst, Mass Inst Technol, 62-65; res assoc, 65-66; NSF fel, 66-68, staff scientist, Pop Coun, 68-69, asst prof, 69-71, assoc prof physiol & psychol, 71-77, PROF NEUROBIOL & BEHAV, ROCKEFELLER UNIV, 77- *Mem:* AAAS; Am Asn Anat; Am Psychol Asn; Soc Neurosci; Am Physiol Soc. *Res:* Brain mechanisms of behavior; effects of hormones on brain electrical activity; brain chemistry and behavior. *Mailing Add:* Rockefeller Univ York Ave & 66th St New York NY 10021

PFAFF, WILLIAM WALLACE, b Rochester, NY, Aug 14, 30; m 60; c 4. SURGERY. *Educ:* Harvard Univ, AB, 52; Univ Buffalo, MD, 56. *Prof Exp:* Intern & jr resident surg, Univ Chicago, 56-58; clin assoc cardiac surg, NIH, 58-60; resident surg, Med Ctr, Stanford Univ, 60-64, instr & chief resident, 64-65; from asst prof to assoc prof, 65-71, PROF SURG, UNIV FLA, 71- *Concurrent Pos:* Consult, Vet Admin Hosps, Gainesville & Lake City & Univ Hosp, Jacksonville, 65-; pres, Southeastern Orgn Procurement Found. *Mem:* Am Col Surg; Asn Acad Surg; Transplantation Soc; Am Surg Asn. *Res:* Organ transplantation, vascular physiology. *Mailing Add:* Col Med Univ Fla Gainesville FL 32610

PFAFFENBERGER, WILLIAM ELMER, b Cleveland, Ohio, Mar 16, 43; m 67; c 1. MATHEMATICS. *Educ:* Univ Ore, BA, 64, MA, 66, PhD(math), 69. *Prof Exp:* Asst prof, 69-80, ASSOC PROF MATH, UNIV VICTORIA, BC, 80- *Concurrent Pos:* Vis colleague, Univ Hawaii, 78; vis scholar, Univ Ore, 82. *Mem:* Am Math Soc; Can Math Soc; Sigma Xi. *Res:* Banach algebras and operator theory. *Mailing Add:* Dept of Math Univ of Victoria Victoria BC V8W 2Y2 Can

PFAFFLIN, JAMES REID, b Connersville, Ind, Dec 3, 30; m 57. WASTEWATER TREATMENT, COASTAL ENGINEERING. *Educ:* Ind State Univ, BS, 52; Johns Hopkins Univ, BES, 56, MS, 57; Univ Windsor, PhD(elec eng), 72. *Prof Exp:* Instr civil eng, Cooper Union, 61-65; asst prof, Polytech Inst Brooklyn, 65-70; assoc prof, Stevens Inst Technol, 73-77 & NJ Inst Technol, 77-83; prof marine eng, US Merchant Marine Acad, 84-89; ED, GORDON & BREACH SCI PUBL, 70- *Mem:* Inst Elec & Electronic Engrs; Can Soc Civil Engrs; Inst Energy; Royal Soc Health. *Res:* Environmental engineering; space heating dynamics; statistical prediction of infrequent natural phenomena. *Mailing Add:* 173 Gates Ave Gillette NJ 07933-1719

PFAFFMAN, MADGE ANNA, b Mobile, Ala, Feb 18, 39. PHARMACOLOGY. *Educ:* Judson Col, BA, 61; Univ Miss, MS, 63, PhD(pharmacol), 65. *Prof Exp:* asst prof pharmacol, Med Sch, Univ Miss, 67-; RETIRED. *Concurrent Pos:* NIH fel physiol, Med Sch, Duke Univ, 65-67; vis prof, Univ Southern Miss, 70-85. *Mem:* AAAS; Am Soc Pharmacol & Exp Therapeut; Biophys Soc. *Res:* Physiology and pharmacology of gastrointestinal and vascular smooth muscle involving the excitation-contraction (E-C) coupling and ionic basis of their function in the presence of drugs. *Mailing Add:* TRT-2 Box 250 Point Washington FL 32454

PFAFFMANN, CARL, b New York, NY, May 27, 13; m 39; c 3. NEUROSCIENCES, PHYSIOLOGICAL PSYCHOLOGY. *Educ:* Oxford Univ, Eng, BA, 37; Brown Univ, PhB, 33, MSc, 35; Cambridge Univ, Eng, PhD(physiol), 39. *Hon Degrees:* DSc, Brown Univ, 65, Bucknell, 66 & Yale, 72. *Prof Exp:* Res assoc biophysics, Johnson Found, Univ Pa, 39-40; instr psychol, Brown Univ, 40-42; from asst prof to prof psychol, Brown Univ, 45-60; vis prof, Yale Univ, 59 & Harvard Univ, 62-63; prof, Florence Pierce Grant Univ, Brown Univ, 60-65; vpres & prof, 65-78, Vincent & Brooke Astor prof physiol psychol, 78-83, EMER PROF, ROCKEFELLER UNIV, 83- *Concurrent Pos:* Nat lectr, Sigma Xi, 63; chmn, Div Behav Sci, Nat Res Coun, 62-64; mem exec comt, Int Union Psychol Sci, 66-72; Kenneth Craik Res Award, St John's Col, Cambridge, Eng, 68-69; mem, Bd Fel, Brown Univ Corp, 69- *Honors & Awards:* Howard Crosby Warren Medal, Soc Exp Psychologists, 60. *Mem:* Nat Acad Sci; Soc Exp Psychol; Am Physiol Soc; Am Philos Soc; AAAS. *Res:* Neurophysiology and psychology of chemical senses, taste and smell; brain mechanisms and behavior. *Mailing Add:* Rockefeller Univ New York NY 10021

PFAHLER, PAUL LEIGHTON, b Essex Co, Ont, Can, Nov 3, 30; nat US; m 67; c 2. GENETICS, AGRONOMY. *Educ:* Univ Mich, AB, 52; Mich State Univ, MS, 54; Purdue Univ, PhD(genetics, plant breeding), 57. *Prof Exp:* Asst, Mich State Univ, 53-54; fel, Carnegie Inst, Purdue Univ & Stanford Univ, 55-56; asst agron, Purdue Univ, 56-57, interim asst prof agron, 57, fel & interim asst prof hort, 57-58; from asst agronomist to assoc agronomist, 58-71, PROF & AGRONOMIST, UNIV FLA, 71- *Concurrent Pos:* Res prof, Univ Nijmegen, Netherlands, 69 & 77; vis scientist, Czechoslovak Acad Sci, 73, 83 & 85 & Hungarian Acad Sci, 88. *Mem:* Am Soc Agron; Crop Sci Soc Am; Am Genetic Asn. *Res:* Population and physiological genetic studies involving small grains, maize, and sesame. *Mailing Add:* Dept Agron Univ Fla 304 Newell Gainesville FL 32611-0311

PFAHNL, ARNOLD, b Austria, June 25, 23; m; c 2. PHYSICS. *Educ:* Graz Univ, PhD(physics), 48; Univ Paris, ScDr, 56. *Prof Exp:* Asst, Graz Univ, 48; res engr, French X-ray Firm, 50-57; res asst, Stanford Univ, 57; mem tech staff, Bell Tel Labs, Inc, NJ, 57-69, supvr, Film Technol Dept, Bell Labs Inc, Pa, 69-80, supvr, Integrated Circuit Packaging Dept, 80-89; RETIRED. *Mem:* Int Soc Hybrid Microelectronics; Inst Elec & Electronics Engrs. *Res:* X-ray tubes; luminescent materials; light emitting diodes; GaAs lasers, photoconductors; image pick-up tubes; thin and thick film hybrid circuit development; integrated circuit packaging development; component qualification and reliability; very-high-speed integrated circuit qualification program. *Mailing Add:* Seven Marshview Dr St Augustine FL 32084

PFALTZGRAFF, JOHN ANDREW, b Huntingdon, Pa, Nov 1, 36; m 60; c 3. MATHEMATICS. *Educ:* Harvard Univ, AB, 58; Univ Ky, MS, 61, PhD(math), 63. *Prof Exp:* Asst prof math, Univ Kans, 63-65; vis asst prof, Ind Univ, Bloomington, 65-67; from asst prof to assoc prof, 67-76, PROF MATH, UNIV NC, CHAPEL HILL, 76- *Mem:* Am Math Soc; Math Asn Am. *Res:* Complex variables; conformal mapping and variational methods; spaces of analytic functions; potential theory; analytic functions in Banach spaces. *Mailing Add:* Dept Math Univ NC Chapel Hill NC 27599

PFALZNER, PAUL MICHAEL, b Vienna, Austria, Aug 18, 23; nat Can; m 50; c 1. MEDICAL PHYSICS. *Educ:* Univ Toronto, BA, 46; McGill Univ, MSc, 51. *Prof Exp:* Res officer, Nat Res Coun Can, 46-50; instr physics, McGill Univ, 50-51; physicist, Ont Cancer Found, 52-64; head dosimetry sect, Dept Res & Isotopes, Int Atomic Energy Agency, 67-68; sr physicist, Ont Cancer Treatment & Res Found, 68-88; RETIRED. *Concurrent Pos:* Ont

Cancer Treatment & Res Found traveling fel, Eng & Scand, 55-56; lectr, Univ Western Ont, 51-66; Int Atomic Energy Agency tech assistance adv med physics, Govt Thailand, 64-65, first officer, Int Atomic Energy Agency, Vienna, 66-68; secy-gen, Fourth Int Conf Med Physics; consult, Pan Am Health Orgn, WHO & Int Atomic Energy Agency. *Mem:* Can Asn Physicists; Brit Hosp Physicists Asn; Brit Inst Radiol; Am Asn Physicists in Med; Ger Soc Med Physics. *Res:* Physical aspects of applications of radiation in biology and medicine; radioisotopes brachytherapy and teletherapy; use of computers in life sciences. *Mailing Add:* 380 Hamilton Ave S Ottawa ON K1Y 1C7 Can

PFANDER, WILLIAM HARVEY, b Lamar, Mo, Aug 9, 23; m 53; c 3. INTERNATIONAL NUTRITION. *Educ:* Univ Mo, BS, 48; Univ Ill, MS, 49, PhD(nutrit), 51. *Prof Exp:* Asst, Univ Wis, 51; assoc prof, 52-54, chmn dept, 75-77, PROF ANIMAL SCI, UNIV MO-COLUMBIA, 54-, ASSOC DEAN & ASSOC DIR, AGR EXP STA, 77- *Concurrent Pos:* Fulbright res fel, Rowett Res Inst, Aberdeen Univ, 51-52; NSF sr fel, Clunies Ross Lab, Univ Sydney, 58-59; Fulbright fel, Univ Alexandria, 66-67; Moorman fel nutrit, 67; mem Nat Acad Sci-Nat Res Coun comt nitrate accumulation; US nat comt, Int Union Nutrit Sci; comt animal nutrit, Nat Res Coun, subcomt horse nutrit, subcomt nutrient & toxic properties water; mem Coun Agr Sci & Technol; nutrit res award, Am Soc Animal Sci. *Mem:* Fel AAAS; Nutrit Soc Gt Brit; Am Chem Soc; Am Inst Nutrit; Am Soc Animal Sci. *Res:* Experimental nutrition; ruminant metabolism and nutrition; carbohydrate utilization; metabolic disorders; appetite control; nutritional environmental interactions; mineral and amino acid requirements; comparative nutrition. *Mailing Add:* 11201 I-70 Dr NE Rte 6 Columbia MO 65202

PFANNKUCH, HANS OLAF, b Berlin, Ger, Nov 23, 32; m 61; c 3. HYDROGEOLOGY. *Educ:* Aachen Tech Univ, MS, 59; Univ Paris, PhD(fluid mech), 62. *Prof Exp:* Res engr, French Petrol Inst, 61-65; res asst prof transport in porous media, Univ Ill, 65-66; asst prof technol, Southern Ill Univ, Carbondale, 66-68; assoc prof, 68-84, PROF HYDROGEOL, UNIV MINN, MINNEAPOLIS, 84- *Mem:* Am Inst Mining, Metall & Petrol Eng; Am Geophys Union; Geol Soc Am; Am Water Resources Asn; Sigma Xi; Nat Water Well Asn. *Res:* Fluid mechanics and transport processes in porous media; ground water geology; analytical geohydrology; watershed modeling and analysis; analysis; surface water (lake)-groundwater interaction; subsurface propagation of oil spills; environmental geology; technical dictionaries. *Mailing Add:* Geol 108 Pillsbury Hall Minneapolis MN 55455

PFAU, CHARLES JULIUS, b Troy, NY, Sept 29, 35; m 58; c 3. VIROLOGY, IMMUNOLOGY. *Educ:* Rensselaer Polytech Inst, BS, 56; Ind Univ, MA, 58, PhD(bact), 60. *Prof Exp:* USPHS fel biophys, Yale Univ, 60-62 & Rickettsia & Virus Dept, State Serum Inst, Copenhagen, Denmark, 62-64; asst prof microbiol, Univ Mass, Amherst, 64-71; assoc prof, 71-76, chmn dept, 82-85, PROF BIOL, RENSSELAER POLYTECH INST, 76- *Concurrent Pos:* Career develop award, USPHS, 65-70; chmn arenavirus study group, Int Comt Taxon Viruses, 74-80; intergovt fel, Nat Inst Neurol & Commun Disorders & Stroke, NIH, 79-80; vis fel, Pasteur Inst, 83, Inst Path, Univ Zurich, 84, Inst Med Microbiol, Univ Copenhagen, 85; Fogarty Sr Int fel, USPHS, 86. *Mem:* Am Soc Microbiol; Am Asn Immunol. *Res:* Pathogenic mechanisms in arena virus infection with special emphasis on virus replication and immune recognition. *Mailing Add:* Dept Biol Rensselaer Polytech Main Campus Troy NY 12180

PFEFFER, JANICE MARIE, CARDIOVASCULAR RESEARCH. *Educ:* Univ Okla, PhD, 77. *Prof Exp:* ASST PROF MED, BRIGHAM & WOMEN'S HOSP, HARVARD MED SCH. *Mailing Add:* Dept Med Brigham & Women's Hosp 75 Francis St Boston MA 02115

PFEFFER, JOHN T, b Ripley, Ohio, Oct 2, 35; m 58; c 2. SANITARY ENGINEERING. *Educ:* Univ Cincinnati, CE, 58, MS, 59; Univ Fla, PhD(sanit eng), 62. *Prof Exp:* Eng assoc, Univ Fla, 60-61; from asst prof to assoc prof sanit eng, Univ Kans, 62-67; assoc prof, 67-69, PROF SANIT ENG, UNIV ILL, URBANA, 69- *Mem:* Water Pollution Control Fedn; Am Soc Civil Engrs; Am Soc Microbiol. *Res:* Basic and applied research on biological processes for purification of wastewaters, for processing waste materials for energy recovery by methane fermentation; processing of hazardous and toxic wastes. *Mailing Add:* 3230 Newmark Lab Univ Ill 205 N Mathews Ave Urbana IL 61801-2397

PFEFFER, MARC ALAN, CARDIOLOGY. *Educ:* Univ Okla, MD & PhD, 72. *Prof Exp:* ASSOC PROF MED, BRIGHAM & WOMEN'S HOSP, HARVARD MED SCH. *Mailing Add:* 75 Francis St Boston MA 02115

PFEFFER, PHILIP ELLIOT, b New York, NY, Apr 8, 41; m 62; c 3. BIOPHYSICS, AGRICULTURE & FOOD CHEMISTRY. *Educ:* Hunter Col, City Univ NY, AB, 62; Rutgers Univ, MS, 64, PhD(chem), 66. *Prof Exp:* Res fel, Univ Chicago, 66-68; sr res chemist, Eastern Regional Res Ctr, 68-76, res leader milk components, 76-80; RES LEADER, SPECTROS LAB, 80-, LEAD SCIENTIST, PLANT & SOIL BIOPHYS LAB, USDA, 88- *Concurrent Pos:* Adj prof chem, Pa State Univ, Ogontz, 74-77; ed, J Carbohydrate Chem, 82-; vis scientist, Ctr Nuclear Study, Grenoble, France, 86, OECD fel, Oxford Univ, 89; fel, Agr Res Serv, 89. *Honors & Awards:* Distinguished Res Scientist Award, ERRC, 83. *Mem:* AAAS; Am Chem Soc; Sigma Xi; Soc Appl Spectros; Soc Magnetic Res. *Res:* Nuclear magnetic resonance spectroscopy of solids (CPMAS), polysaccharides and studies of molecular interactions; carbanion chemistry of carboxylic acids; 2H isotope effects in 13C nuclear magnetic resonance spectroscopy (deuterium induced differential isotope 13C nuclear magnetic resonance spectroscopy); in vivo nuclear magnetic resonance studies of plant tissues and symbiosis; 2D nuclear magnetic resonance studies of natural products of agriculture. *Mailing Add:* Plant & Soil Biophys Lab USDA Philadelphia PA 19118

PFEFFER, RICHARD LAWRENCE, b Brooklyn, NY, Nov 26, 30; m 53; c 4. DYNAMIC METEOROLOGY. *Educ:* City Col New York, BS, 52; Mass Inst Technol, MS, 54, PhD(meteorol), 57. *Prof Exp:* Res asst meteorol, Mass Inst Technol, 52-55; atmospheric physicist, Air Force Cambridge Res Ctr, 55-59; sr res scientist, Columbia Univ, 59-61, from lectr to asst prof geophys, 61-64; assoc prof, 64-67, PROF METEOROL & DIR GEOPHYS FLUID DYNAMICS INST, FLA STATE UNIV, 67- *Concurrent Pos:* Guest lectr, Mass Inst Technol, 54 & broadcast on tsunamis for The Voice of Am, 63; consult, N W Ayer Co, 62, Educ Testing Serv, Princeton, NJ, 63, Grolier Inc, 63 & Naval Res Lab, Washington, DC, 71-76; mem, Int Comn for Dynamical Meteorol, 72-76. *Mem:* Am Meteorol Soc. *Res:* Dynamics of atmospheric processes; momentum and energy exchanges in hurricanes, cyclones and the global atmospheric circulation; available potential energy; fluctuations of planetary atmospheric circulations; acoustic-gravity wave propagation; experimental and computer modeling of climatic variability. *Mailing Add:* Geophys Fluid Dynamics Inst Fla State Univ Tallahassee FL 32306

PFEFFER, ROBERT, b Vienna, Austria, Nov 26, 35; US citizen; m 60; c 2. CHEMICAL ENGINEERING. *Educ:* NY Univ, BChE, 56, MChE, 58, DEngSc(chem eng), 62. *Prof Exp:* From lectr to assoc prof, City Col New York, 57-70, prof, 71-80, chmn dept, 73-87, dean grad studies & res & dep provost, 87-88, HEBERT G KAYSER PROF CHEM ENG, CITY COL NEW YORK, 80-, PROVOST & VPRES ACAD AFFAIRS, 88- *Concurrent Pos:* Res assoc, Brookhaven Nat Lab, 63, 78; prin investr res grants, NSF, Environ Protection Agency, NIH, NASA, Office Naval Res, 63-; fac fel, NASA Lewis Res Ctr, 64 & 65; consult, NASA Lewis Res Ctr-Cleveland, 65-68, US Army Res-Durham, 70-76 & various indust co, 77-; vis prof, Imp Col, Univ London, 69-70; vis prof, Technion-Isreal Inst Technol, 76-77; sci adv, Int Fine Particle Res Inst, 80- *Honors & Awards:* Fulbright-Hayes Award, 76. *Mem:* Fel Am Inst Chem Engrs; Am Soc Eng Educ. *Res:* Low Reynolds-number hydrodynamics; heat and mass transfer with applications to biomedical problems; agglomeration in fluidized beds; filtration of aerosols by granular beds. *Mailing Add:* Dept Chem Eng City Col New York New York NY 10031

PFEFFER, WASHEK F, b Prague, Czech, Nov 14, 36; m 59. MATHEMATICS. *Educ:* Charles Univ, Prague, RNDr, 60; Univ Md, PhD(math anal), 66. *Prof Exp:* Res assoc math, Charles Univ, Prague, 59-60; res asst, Inst of Heat Technique, Prague, 60-63; res asst, Czech Acad Sci, 63-64; vis res asst, Polish Acad Sci, spring, 64; res asst, Royal Inst Technol, Stockholm, Sweden, fall, 64; asst prof, George Washington Univ, 65-66; asst prof, Univ Calif, Davis, 66-68; vis assoc prof, Univ Calif, Berkeley, 68-69; assoc prof, Univ Calif, Davis, 69-71; vis prof, Univ Ghana & dir Ghana Ctr, Univ Calif Educ Abroad Prog, 71-73; PROF MATH, UNIV CALIF, DAVIS, 73- *Concurrent Pos:* Vis prof, Royal Inst Technol, Stockholm, Sweden, 75. *Mem:* Am Math Soc; Swedish Math Soc. *Res:* Non-absolute integration in topological spaces; topology and topological measure theory. *Mailing Add:* Dept Math Univ Calif Davis CA 95616

PFEFFERKORN, ELMER ROY, JR, b Manitowoc, Wis, Dec 13, 31; m 64; c 3. MICROBIOLOGY, PARASITOLOGY. *Educ:* Lawrence Col, BA, 54; Oxford Univ, BA & MA, 56; Harvard Univ, PhD(bact), 60. *Prof Exp:* Instr bact & immunol, Harvard Med Sch, 62-63, assoc, 63-65, asst prof, 65-67; assoc prof, 67-70, PROF & CHMN MICROBIOL, DARTMOUTH MED SCH, 70- *Concurrent Pos:* Res fel bact & immunol, Harvard Med Sch, 60-62; adj prof biol, Dartmouth Col, 70-; mem, virol study sect, 70-74 & trop med parasitol study sect, NIH, 80- *Mem:* Am Soc Microbiol; Am Soc Biol Chemists; Am Soc Parasitologists; Soc Protozoologists. *Res:* Genetics and biochemistry of intracellular parasites. *Mailing Add:* Dept of Microbiol Dartmouth Med Sch Hanover NH 03755

PFEFFERKORN, HERMANN WILHELM, b Muenster, Ger, Sept 25, 40; m 69; c 1. PALEOBOTANY, STRATIGRAPHY. *Educ:* Univ Muenster, WGer, Dipl, 66, Drrernat(geol), 68. *Prof Exp:* Fel paleobot, Univ Ill, Champaign-Urbana, 68-69; asst geologist paleobot, Ill State Geol Surv, 69-71; res assoc paleobot, Univ Muenster, WGer, 71-73; vis lectr, geol, Univ Pa, 73-74, asst prof, 74-79, assoc prof, 79-; AT GEOL-PALEONT INST, UNIV HEIDELBERG, FEDERAL REPUBLIC GERMANY. *Mem:* Geol Soc Am; Bot Soc Am; Paleont Soc; Int Orgn Paleobot. *Res:* Carboniferous (Pennsylvanian) floras of North America with special reference to biostratigraphy and paleoecology; stratigraphic correlation between different coal basins and between North America and Europe. *Mailing Add:* Dept Geol Univ Pa Philadelphia PA 19104

PFEIFER, GERARD DAVID, b Chicago, Ill, Sept 5, 37; m 64; c 2. OCCUPATIONAL HEALTH, BIOCHEMISTRY. *Educ:* Univ Tulsa, BS, 64; Univ Louisville, PhD(biochem), 71. *Prof Exp:* Chemist, Ozark-Mahoning Co, 64-65; Nat Res Coun Fel & res chemist, Human Nutrit Res Div, Agr Res Serv, USDA, 71-73; asst prof chem, Cent Mo State Univ, 73-77; mem staff res & technol, 77-81, SUPVR-CORP OCCUP HEALTH, ARMCO, INC, 81- *Mem:* Am Chem Soc; AAAS; Am Indust Hyg Asn. *Res:* Industrial toxicology. *Mailing Add:* Res Ctr 703 Curtis St Middletown OH 45043

PFEIFER, RICHARD WALLACE, b Brooklyn, NY, Sept 29, 51; m 81; c 1. IMMUNOTOXICOLOGY, MOLECULAR TOXICOLOGY. *Educ:* Bucknell Univ, BS, 73; Univ Rochester, PhD(pharmacol), 79. *Prof Exp:* Res asst tumor immunobiol, Sloan-Kettering Inst Cancer Res, 73-74; NIH trainee cell molecular biol, dept pharmacol & toxicol, Sch Med & Dent, Univ Rochester, 74-79; postdoctoral fel toxicol, dept path, Chem Indust Inst Toxicol, 79-82; sr staff fel immunotoxicol, STB, Nat Inst Environ Health Sci, 82-84; ASST PROF TOXICOL, DEPT PHARMACOL & TOXICOL, SCH PHARM & PHARMACAL SCI, PURDUE UNIV, 84- *Mem:* AAAS; Int Soc Immunopharmacol; Sigma Xi; NY Acad Sci; Soc Toxicol. *Res:* Molecular mechanisms of chemical-induced immunotoxicity; cell-cell communication, tumor promoters and growth control regulation; the cytoskeleton as a subcellular target; quinones and sulfhydryl group biochemistry-inflammation. *Mailing Add:* Dept Pharmacol & Toxicol Sch Pharm & Pharmacal Sci Purdue Univ West Lafayette IN 47907

PFEIFFER, CARL CURT, pharmacology; deceased, see previous edition for last biography

PFEIFFER, CARL J, b Quincy, Ill, June 28, 37; m 62; c 2. PHYSIOLOGY, GASTROENTEROLOGY. *Educ:* Duke Univ, BA, 59; Southern Ill Univ, MA, 61, PhD(physiol), 64; Harvard Univ, MSHyg, 67. *Prof Exp:* Res asst cell physiol, Duke Univ, 58-59; res asst stress physiol, Ind Univ, 59-60; res asst pharmacol, Southern Ill Univ, 60-63; res scientist, Ames Res Ctr, NASA, Calif, 63-67; dir gastrointestinal res, Inst Gastroenterol, Presby-Univ Pa Med Ctr & asst prof pharmacol, Sch Med, Univ Pa, 67-71; assoc prof physiol, Fac Med, Mem Univ Nfld, 71-75, prof gastrointestinal physiol, 75-82; PROF VET MED, VA TECH, 82- *Concurrent Pos:* Guest prof, Kyoto Univ Fac Med, 77-78; Eleanor Roosevelt, Am Can Soc, Int Cancer Fel, 77. *Mem:* Am Soc Zool; Am Asn Hist Med; Marine Mammal Soc; Electron Micros Soc Am; Comp Gastroenterol Soc. *Res:* Physiology; toxicology; gastroenterology; electron microscopy. *Mailing Add:* Dept Biomed Sci Va Tech Blackburg VA 24061

PFEIFFER, CARROLL ATHEY, b Reno, Ohio, Apr 18, 06; m 37; c 1. ANATOMY. *Educ:* Marietta Col, AB, 29; Univ Iowa, MS, 31, PhD(zool), 35. *Hon Degrees:* ScD, Marietta Col, 57. *Prof Exp:* Asst zool, Univ Iowa, 29-35; res asst & instr primate biol, Sch Med, Yale Univ, 36-37, res asst & instr anat, 37-39, res asst & asst prof anat & endocrinol, 39-47, res assoc & asst prof histol, gross anat & endocrinol, 47-51; head dept, 54-66, prof, 51-75, EMER PROF ANAT, SCH MED, UNIV PR, SAN JUAN, 76- *Concurrent Pos:* Henry res fel med, New York Hosp & Med Col, Cornell Univ, 35-36. *Mem:* Am Soc Zoologists; Soc Exp Biol & Med; Endocrine Soc; Am Asn Cancer Res; Am Asn Anatomists. *Res:* Gonad-hypophyseal interrelationships; endocrine imbalance in relation to cancer; physiology of reproduction and of bone formation; relation of gonadotropin and sex hormones to cancer of the reproductive system. *Mailing Add:* c/o Virginia Pfeiffer 424 W 22nd St Apt 4 New York NY 10011

PFEIFFER, CURTIS DUDLEY, b Oshkosh, Wis, July 4, 49; m 68; c 2. SEPARATIONS SCIENCE, LIQUID CHROMATOGRAPHY. *Educ:* Univ Wis, Oshkosh, BS, 65. *Prof Exp:* ASSOC SCI ANAL CHEM, DOW CHEM CO, 66- *Honors & Awards:* V A Stenger Award, Dow Chem Co, 73, P A Traylor Award, 85. *Mem:* Sigma Xi; Am Soc Testing Mats; Am Chem Soc; AAAS. *Res:* Liquid chromatography; capillary HPLC, process scale separations and separations of optical isomers. *Mailing Add:* Dow Chem Co 1897 Bldg Midland MI 48674

PFEIFFER, DOUGLAS ROBERT, b Cedar Rapids, Iowa, June 3, 46; m 77; c 1. BIOCHEMISTRY, BIOPHYSICAL CHEMISTRY. *Educ:* Coe Col, BA, 68; Wayne State Univ, PhD(biochem), 73. *Prof Exp:* Asst, Wayne State Univ, 68-69, res asst biochem, 69-73, fel, 73; proj assoc, Enzyme Inst, Univ Wis, 73-76, asst scientist, 76-77; from asst prof to assoc prof, 77-85, PROF, HORMEL INST, UNIV MINN, 86- *Concurrent Pos:* Noeller Fel Chem, 71-72. *Mem:* Am Chem Soc; AAAS; NY Acad Sci. *Res:* Ionophore transport mechanisms; mitochondrial calcium metabolism; mitochondrial phospholipid metabolism; ischemic tissue metabolism; bioenergetics; membrane permeability. *Mailing Add:* 205 24th St SW Austin MN 55912

PFEIFFER, EGBERT WHEELER, b New York, NY, May 23, 15; m 49; c 3. VERTEBRATE ZOOLOGY. *Educ:* Cornell Univ, BA, 37; Univ BC, MA, 48; Univ Calif, PhD(zool), 54. *Prof Exp:* Vis prof biol, Col Idaho, 54-55; asst prof zool, Utah State Univ, 55-57; asst prof anat, Univ NDak, 57-59; from assoc prof to prof, 59-85, EMER PROF ZOOL, UNIV MONT, 85- *Concurrent Pos:* Prin investr, NIH & NSF grants, USPHS spec res fel, Western Reserve Univ, 65-66. *Mem:* AAAS; Am Soc Zool; Am Soc Mammal; Am Asn Anat. *Res:* Vertebrate reproduction; mammalian renal physiology and anatomy. *Mailing Add:* 855 Beverly Ave Missoula MT 59801

PFEIFFER, ERIC A, b Rauental, Ger, Sept 15, 35; US citizen; m 64; c 3. PSYCHIATRY, GERONTOLOGY. *Educ:* Washington Univ, AB, 56, MD, 60; Am Bd Psychiat & Neurol, dipl, 66. *Prof Exp:* Instr psychiat, Sch Med, Univ Rochester, 63-64; staff psychiatrist, USPHS Hosp, Lexington, Ky, 64-66; assoc, Med Ctr, Duke Univ, 66, asst prof, 67-69, assoc prof, 69-73, prof psychiat, 73-77, proj dir, Older Americans Resources & Serv Prog, 72-77, assoc dir, Ctr Study Aging & Human Develop, 74-77; dir, Davis Inst Care & Study Aging, Denver, 76-77; prof psychiat, Sch Med, Univ Colo, 76-78; PROF PSYCHIAT, COL MED, UNIV S FLA, 78-; CHIEF GERIAT PSYCHIAT, TAMPA VET ADMIN HOSP, 78- *Concurrent Pos:* Markle scholar acad med, 68-73; consult, Berea Col, 64-66; lectr, Law Sch, Duke Univ, 68-; dir, Suncoast Geront Ctr, Univ SFla, 70- *Mem:* Am Psychiat Asn; Geront Soc; Am Geriat Soc. *Res:* Psychiatric education; geriatric psychiatry; psychotherapy; behavior and adaptation in late life. *Mailing Add:* 12901 Bruce B Downs Blvd Tampa FL 33612

PFEIFFER, HEINZ GERHARD, b Pforzheim, Germany, Mar 31, 20; US citizen; m 48; c 2. ENERGY CONVERSION. *Educ:* Drew Univ, AB, 41; Syracuse Univ, AM, 44; Calif Inst Technol, PhD(chem), 49. *Prof Exp:* Asst, Am Platinum Works, NJ, 39-41; asst, Syracuse Univ, 41-42, instr, 42-43; asst, Calif Inst Technol, 43-44 & 47-48; res assoc, Gen Elec Co, 48-54, mgr dielectric studies, 54-64, div liaison scientist, 64-68, mgr educ technol, 68-71; dir, Ctr Study Sci & Soc, State Univ NY Albany, 71; mgr technol & energy assessment, Pa Power & Light Co, 72-79; CONSULT, 89- *Concurrent Pos:* Adj prof, Lehigh Univ, 73-77; chmn, thermal & mech prog comt, Elec Power Res Inst, 73-78; mem, bd dirs, Pa Sci & Eng Found, 75-79; chmn air & water qual tech adv comt, Pa Dept Environ Resources, 77-80; mem energy storage comt, Nat Acad Sci, 76-81; mem, environ comt, Nat Chamber Com, 84-89; prin vis scientist, Energy Res Ctr, Lehigh Univ. *Mem:* AAAS; fel Am Inst Chemists; emer mem, Am Chem Soc; sr mem Inst Elec & Electronics Engrs. *Res:* Diffraction; surface films; x-ray spectrography and electrochemistry, dielectrics; electronic properties of molecular compounds; educational technology; science and society; advanced energy systems; direct utilization of coal; use of reject heat; clean air legislation. *Mailing Add:* Barnes Lane Allentown PA 18103-1179

PFEIFFER, LOREN NEIL, b Waukesha, Wis, Aug 21, 39; m 64. EXPERIMENTAL PHYSICS. *Educ:* Univ Mich, BS, 61; Johns Hopkins Univ, PhD(physics), 67. *Prof Exp:* Res assoc physics, Johns Hopkins Univ, 67-68; MEM TECH STAFF, BELL TEL LABS, 68- *Mem:* Am Phys Soc. *Res:* Mossbauer effect; nuclei in solids; hyperfine interactions; gamma ray optics. *Mailing Add:* AT&T Bell Labs Inc Rm 1 C-445 Murray Hill NJ 07974

PFEIFFER, PAUL EDWIN, b Newark, Ohio, Sept 9, 17; m 43; c 4. MATHEMATICS. *Educ:* Rice Inst, BS, 38, MS, 48, PhD(math), 52; Southern Methodist Univ, BD, 43. *Prof Exp:* From instr to assoc prof elec eng, 47-59, chmn dept, 59-63, dean students, 65-69, chmn dept math sci, 74-75, PROF MATH SCI & ELEC ENG, RICE UNIV, 69- *Mem:* Am Math Soc; Math Asn Am; Inst Elec & Electronics Eng; Soc Indust & Appl Math; Inst Math Statist. *Res:* Applied probability and random processes. *Mailing Add:* Dept Math Sci Rice Univ Box 1892 Houston TX 77251

PFEIFFER, RAYMOND JOHN, b Trenton, NJ, Apr 29, 37; wid; c 1. ASTROPHYSICS, ASTRONOMY. *Educ:* Univ Mich, BSEd, 61; Temple Univ, MA, 68; Univ Pa, PhD(astron), 75. *Prof Exp:* from instr to assoc prof, 64-80, PROF PHYSICS & ASTRON, TRENTON STATE COL, 80- *Concurrent Pos:* Zaccheus Daniel fel, Univ PA, 72-73; Res assoc, Univ Pa Observ, 75-; IVE guest investr, 80. *Mem:* Am Astron Soc; Int Astron Union; Astron Soc of Pacific. *Res:* Measurement and analysis of linear & circular polarization in interacting binary stars; photometry and spectrophotometry of interacting binary stars; computer modeling of scattering in circumbinary material. *Mailing Add:* Dept Physics Trenton State Col Trenton NJ 08650-4700

PFEIFFER, STEVEN EUGENE, b Watertown, Wis, Aug 13, 40; m 65; c 3. CELL BIOLOGY, NEUROSCIENCES. *Educ:* Carleton Col, BA, 62; Wash Univ, PhD(molecular biol), 67. *Prof Exp:* Nat Inst Neurol Dis & Stroke fel biochem, Brandeis Univ, 67-69; from asst prof to assoc prof, 69-81, PROF MICROBIOL, HEALTH CTR, UNIV CONN, 81- *Concurrent Pos:* Neurosci res prog, Intensive Study Prog fel, 69; Merck Found fac develop award, 70; Max Planck Soc fel, 72; Josiah Macy Jr fac scholar award, 76-77; vis scientist sabbatical leave, Pasteur Inst, 76-77; Human Develop Study Sect, Basic Neurol Sci, 79-83; Fulbright fel, 83-84; NIH sr fel, 84. *Mem:* Am Asn Cell Biol; Soc Neurosci; Am Soc Neurochem. *Res:* Molecular cell biology of glial development. *Mailing Add:* Dept Microbiol Univ Conn Health Ctr Farmington CT 06030

PFENDER, EMIL, b Germany, May 25, 25; US citizen; m 54; c 3. MECHANICAL ENGINEERING, PLASMA TECHNOLOGY. *Educ:* Univ Stuttgart, Dipl, 53, Dr Ing(elec eng), 59. *Prof Exp:* Res assoc gaseous elec, Univ Stuttgart, 53-55, asst, 55-61, first asst, Inst High Temperature Res, 62-64; vis scientist arc technol, Aerospace Res Labs, Wright-Patterson AFB, Ohio, 61-62; assoc prof, 64-67, PROF, DEPT MECH ENG, UNIV MINN, 67- *Concurrent Pos:* Mem adv panel, NSF, 74-76; dir, Minn Metric Ctr, 74- *Honors & Awards:* Adams Mem Mem Award, Am Welding Soc, 66; US Sr Scientist Award, Humboldt Found, Fed Repub Ger, 78. *Mem:* Nat Acad Eng; Am Phys Soc; fel Am Soc Mech Engrs; Inst Elec & Electronic Engrs; fel Am Soc Mech Engrs. *Res:* Arc technology; plasma heat transfer; plasma chemistry and processing; energy conservation; electric arc technology. *Mailing Add:* 125 Dept Mech Eng Univ Minn 111 Church St SE Minneapolis MN 55455

PFENNIGWERTH, PAUL LEROY, b Ludlow, Ky, Apr 19, 29; m 55; c 3. MECHANICAL ENGINEERING. *Educ:* Univ Cincinnati, ME, 52; Univ Ky, MSME, 54; Univ Pittsburgh, PhD(mech eng), 63. *Prof Exp:* Instr mech eng, Univ Ky, 52-54; naval off, 54-57; fel engr, Pa, 57-68, mgr eng mech, 68-71, tech consult to US Naval Nuclear Power Sch, Mare Island Naval Shipyard, Vallejo, 71-74, mgr fuel syst performance, Lwbr Proj, 74-84, mgr shock & vibration anal, 84-89, CONSULT MECH, BETTIS ATOMIC POWER LAB, WESTINGHOUSE ELEC CORP, 89- *Mem:* Am Soc Mech Engrs. *Res:* Applied mechanics, including solid and structural mechanics, fluid flow and heat transfer; nuclear engineering including irradiation testing, radiation effects on materials and fuel element design; dynamics of nuclear reactor structures. *Mailing Add:* Bettis Atomic Power Lab Box 79 West Mifflin PA 15122-0079

PFISTER, DONALD HENRY, b Kenton, Ohio, Feb 17, 45; m 71. MYCOLOGY. *Educ:* Miami Univ, AB, 67; Cornell Univ, PhD(mycol), 71; Harvard, AM. *Prof Exp:* Asst prof biol & mycol, Univ PR, Mayaguez, 71-74; from asst prof biol & asst cur to assoc prof & assoc cur, 74-80, dir, 83-88, PROF & CUR, FARLOW REF LIBR & HERBARIUM, HARVARD UNIV, 80-, ASA GRAY PROF SYST BOT, 90- *Mem:* Mycol Soc Am; Sigma Xi. *Res:* Floristic and monographic studies of operculate Discomycetes; historical and bibliographical studies of mycological herbaria. *Mailing Add:* Farlow Herbarium Harvard Univ 20 Divinity Ave Cambridge MA 02138

PFISTER, PHILIP CARL, b New York, NY, Apr 12, 25; m 59; c 1. MECHANICAL ENGINEERING. *Educ:* City Col New York, BME, 47; Columbia Univ, MSME, 49; Ill Inst Technol, PhD(mech eng), 62. *Prof Exp:* Instr mech eng, City Col New York, 47-50 & WVa Univ, 50-52; asst prof, Univ Utah, 52-55; fel & asst prof, Ill Inst Technol, 55-67; head dept, 67-70, PROF MECH ENG, NDAK STATE UNIV, 67- *Concurrent Pos:* Vis prof, Kabul, Afghanistan, 64-66. *Mem:* Am Soc Mech Engr; Am Wind Energy Asn; Sigma Xi. *Res:* Vibrations; computer design; wind machines. *Mailing Add:* 1544 35th St S No 103 Fargo ND 58103-4535

PFISTER, RICHARD CHARLES, b Ypsilanti, Mich, Nov 27, 33; m 56; c 5. RADIOLOGY, URORADIOLOGY. *Educ:* Cent Mich Univ, BS, 71; Wayne State Col, MD, 62, Am Col Radiol, dipl radiol, 67. *Prof Exp:* Instr radiol, Harvard Med Sch, 69-71, from asst prof to assoc prof, Harvard Univ, 71-88; HEAD, GENITOURINARY RADIOL DIV, MASS GEN HOSP, 68- *Concurrent Pos:* Secy comt contrast media, Internal Comg, 71-75; prin investr, contrast media, Nat Inst Health, 74-76, E R Squibb, 73-77; vis prof to various schools in US & Eur, 74- *Mem:* Soc Uroradiol (pres-treas, 76-79, pres, 83-85); Radiol Soc N Am; Am Roentgen Ray Soc; Am Med Asn; NY Acad Sci. *Res:* Percutaneous procedures on the urinary tract; their introduction, development and application; adverse reactions to contrast media, types, pathogenesis and treatment; diagnostic imaging of urinary tract diseases. *Mailing Add:* Genitourinary Radiol Div Mass Gen Hosp Boston MA 02114

PFISTER, ROBERT M, b New York, NY, Feb 25, 33; m 54; c 4. MICROBIOLOGY. *Educ:* Syracuse Univ, AB, 57, MS, 60, PhD(microbiol), 64. *Prof Exp:* Res assoc microbiol, Res Corp, Syracuse Univ, 57-62; res assoc, Lamont Geol Observ, Columbia Univ, 64-66; from asst prof to prof microbiol, Ohio State Univ, 66-85, chmn dept, 73-85; CONSULT, 85- *Concurrent Pos:* NSF res grants, 65-67. *Mem:* Am Soc Microbiol; Electron Micros Soc Am. *Res:* Ultrastructure and function of bacterial organelles and microparticulate-microorganism interactions; microbial ecology. *Mailing Add:* Dept of Microbiol Ohio State Univ 484 W 12th St Columbus OH 43210

PFLANZER, RICHARD GARY, b Ashland, Wis, June 16, 40; m 64; c 2. MEDICAL PHYSIOLOGY. *Educ:* Ind Univ, Bloomington, AB, 64, PhD(physiol), 69. *Prof Exp:* From teaching asst to teaching assoc human physiol, Ind Univ, Bloomington, 63-69, asst prof physiol & anat, Ind Univ-Purdue Univ, Indianapolis, 69-75, assoc prof physiol, Sch Med, 75-, AT DEPT BIOL, IND UNIV MED CTR. *Concurrent Pos:* NASA-Univ Va fel biospace technol, Wallops Island, Va, 70; consult, John Wiley & Sons, Inc Publishers, 73- & Wm C Brown Co Publishers, 74- *Mem:* AAAS; Am Inst Biol Sci; Sigma Xi. *Res:* Comparative, environmental and adaptation physiology. *Mailing Add:* Dept Biol CA325 Ind Univ Med Ctr 925 W Michigan St Indianapolis IN 46202

PFLAUM, RONALD TRENDA, b Webster, Minn, June 21, 22; m 48; c 3. ANALYTICAL CHEMISTRY. *Educ:* St Olaf Col, BA, 48; Purdue Univ, MS, 51, PhD(chem), 53. *Prof Exp:* Asst chem, St Olaf Col, 48-49; asst, Purdue Univ, 49-51; from instr to assoc prof, 53-63, PROF CHEM, UNIV IOWA, 63- *Mem:* Am Chem Soc; Sigma Xi. *Res:* Spectrophotometric methods of analysis; analytical chemistry of coordination compounds; organic reagents in inorganic analysis. *Mailing Add:* Dept Chem State Univ Iowa Iowa City IA 52242

PFLIEGER, WILLIAM LEO, b Columbus, Ohio, Oct 26, 32; m 53; c 2. ICHTHYOLOGY, FISHERIES. *Educ:* Ohio State Univ, BS, 58, MS, 60; Univ Kans, PhD(zool), 69. *Prof Exp:* SR FISHERIES BIOLOGIST, MO DEPT CONSERV, 61- *Concurrent Pos:* Res assoc, Univ Mo-Columbia, 73- *Mem:* Am Soc Ichthyol & Herpet; Am Fisheries Soc; Sigma Xi. *Res:* Systematics, distribution and life history of North American freshwater fishes; protection of endangered species; classification, inventory and evaluation of stream habitats in Missouri. *Mailing Add:* Mo Dept Conserv 1110 College Ave Columbia MO 65201

PFLUG, GERALD RALPH, b Philadelphia, Pa, July 30, 41; m 65; c 2. PHARMACEUTICS. *Educ:* Philadelphia Col Pharm, BSc, 63, MSc, 65; State Univ NY Buffalo, PhD(pharmaceut), 69. *Prof Exp:* Sect head develop, Vicks Div Res & Develop, 69-74; dir pharmaceut develop, USV Pharmaceut, NY, 74-75; dir, 75-80, VPRES PROPRIETARY DRUG RES & DEVELOP, CARTER PRODS DIV, CARTER-WALLACE INC, 80- *Mem:* Am Pharmaceut Asn; Acad Pharmaceut Sci; Am Chem Soc. *Res:* Catalytic reactions involving penicillins; biopharmaceutics; drug interactions; kinetics. *Mailing Add:* Carter Prod Half Acre Rd Cranbury NJ 08512

PFLUG, IRVING JOHN, b Gibson Co, Ind, Sept 17, 23; div; c 4. FOOD SCIENCE, MICROBIOLOGY. *Educ:* Purdue Univ, BSA, 46, BSAE, 48; Univ Mass, MS, 50, PhD, 53. *Prof Exp:* Instr, Purdue Univ, 46-48; from asst prof to assoc prof, Univ Mass, 48-54; from assoc prof to prof, Mich State Univ, 54-67; PROF FOOD SCI & NUTRIT, SCH PUB HEALTH, UNIV MINN, MINNEAPOLIS, 67- *Concurrent Pos:* Indust consult. *Honors & Awards:* Kilmer Award. *Mem:* Am Soc Microbiol; fel Am Soc Heat, Refrig & Air Conditioning Engrs; fel Inst Food Technol; Soc Appl Bact; Int Inst Refrig; Sigma Xi. *Res:* Sterilization microbiology; sterilization of drugs; heating, cooling and thermal processing of food products; refrigerated storage of agricultural products; food engineering. *Mailing Add:* Environ Sterilization Lab Univ Minn 543 Shepherd Lab 100 Union St Minneapolis MN 55455

PFLUGER, CLARENCE EUGENE, b Coupland, Tex, Sept 1, 30; m 61; c 3. ANALYTICAL CHEMISTRY, CRYSTALLOGRAPHY. *Educ:* Univ Tex, BS, 51, PhD(analytical chem), 58. *Prof Exp:* Shift chemist, Dana Plant, E I du Pont de Nemours Co, Ind, 51-52, shift supvr chem control, Savannah River Plant, SC, 52-54; from instr to asst prof analytical chem, Univ Ill, Urbana, 59-66; asst prof, Univ Ga, 66-67; assoc prof, 67-76, PROF CHEM, SYRACUSE UNIV, 76- *Concurrent Pos:* Fulbright scholar, Darmstadt Tech Univ, 58-59; Fulbright lectr, Tech Univ Denmark, 65-66; vis prof chem, Tech Univ, Darmstadt, 78-79 & Univ Tex, Austin, 85; fac res prog partic, Navy-Am Soc Eng Educ, 83, 84. *Mem:* Am Chem Soc; Am Crystallog Asn. *Res:* Crystal structure determinations, primarily of metal chelate complexes; analytical instrumentation and methods development in the trace and ultratrace region; solid state photochemistry. *Mailing Add:* Dept Chem Pittsburgh State Univ Pittsburgh KS 66762

PFLUGFELDER, HALA, b Dec 3, 21; nat US; m 43; c 2. MATHEMATICS. *Educ:* Univ Gottingen, dipl, 47; Univ Freiburg, PhD, 49. *Prof Exp:* From instr to asst prof, 56-71, ASSOC PROF MATH, TEMPLE UNIV, 71- *Res:* Theory of loops. *Mailing Add:* Dept Math Temple Univ 42 Oak Forrest Pl Oakmont Santa Rosa CA 95409

PFLUKE, JOHN H, b Peoria, Ill, June 17, 31; m 57; c 5. GEOPHYSICS, SEISMOLOGY. *Educ:* St Louis Univ, BS, 53, MS, 61; Pa State Univ, PhD(geophys), 63. *Prof Exp:* Geophysicist, US Coast & Geod Surv, 59-60; sr seismologist, Teledyne, Inc, 63-65; chief anal sect, Earthquake Mechanism Lab, Inst Earth Sci, Environ Sci Serv Admin, 65-70, sr res geophysicist, Nat Oceanic & Atmospheric Admin, 70-74; res geophysicist, 74-81, mgr, External Res Prog, 81-87, RES GEOPHYSICIST, US GEOL SURV, 87- *Mem:* Seismol Soc Am; Am Geophys Union; Soc Explor Geophysicists. *Res:* Seismic field and model studies; deep well and seismic instrumentation; geophysical inverse problems; model studies of earth strain data; statistics of earthquakes. *Mailing Add:* 221 Kingsley Ave Palo Alto CA 94301

PFOHL, RONALD JOHN, b Baraboo, Wis, Aug 6, 37; m 67; c 1. DEVELOPMENTAL BIOLOGY. *Educ:* Wartburg Col, BA, 59; Mich State Univ, MS, 62, PhD(zool), 67. *Prof Exp:* USPHS-NIH fel, Res Unit Molecular Embryol, Univ Palermo, 67-69; asst prof, 69-76, ASSOC PROF ZOOL, MIAMI UNIV, 76- *Mem:* AAAS; Am Soc Zoologists; Soc Develop Biol; Am Chem Soc; Sigma Xi. *Res:* Assessment of toxic effects of environmental contaminants in drinking water on rat brain development based on analysis of biochemical, histological & behavioral parameters. *Mailing Add:* Dept Zool Miami Univ Oxford OH 45056

PFRANG, EDWARD OSCAR, b New Haven, Conn, Aug 9, 29; m 58; c 3. CIVIL ENGINEERING, STRUCTURAL ENGINEERING. *Educ:* Univ Conn, BS, 51; Yale Univ, ME, 52; Univ Ill, Urbana, PhD(struct eng), 61. *Prof Exp:* Struct designer, Gibbs & Hill Inc, NY, 52 & Singmaster & Bryer, NY, 52-53; construct officer, US Naval Base, Cuba, 53-55; proj mgt officer, Dist Pub Work Off, Calif, 55-56; gen field supt, Major Concrete Co Inc, New York, 56-57; asst prof civil eng, Univ Nev, Reno, 57-59; univ & Ford Found fel, Univ Ill, Urbana, 59-61; chief supvry civil engr, Nat Bur Standards, 66-70, actg prog mgr off housing tech, 70-72, prog mgr, 72, chief, 72-73, chief struct div, 73-83; EXEC DIR, AM SOC ENGRS, 83- *Concurrent Pos:* Chmn, US Panel Wind & Seismic Effect, 69-; chmn tech activ comt, Am Concrete Inst, 72- *Honors & Awards:* Raymond C Reese Struct Res Award, 71; ACI Wason Medal Award, Am Concrete Inst, 71; Spec Achievement Award, Dept Housing & Urban Develop, 72; Silver Medal, Dept Com, 81. *Mem:* Am Soc Civil Engrs; Am Soc Testing & Mat. *Mailing Add:* 345 E 47th St New York NY 10017

PFRIMMER, THEODORE ROSCOE, entomology, for more information see previous edition

PFROGNER, RAY LONG, b Meyersdale, Pa, Sept 30, 34; div; c 2. SOLID STATE PHYSICS, ENERGY CONVERSION. *Educ:* Juniata Col, BS, 60; Univ Del, MS, 64, PhD(physics), 71. *Prof Exp:* Geophysicist, Coast & Geod Surv, Dept Com, 61-62; ASSOC PROF PHYSICS, JUNIATA COL, 64- *Concurrent Pos:* Vis assoc prof, Inst Energy Conversion, Dept Eng, Univ Del, 75-76. *Mem:* Am Asn Physics Teachers; Sigma Xi. *Res:* Photovoltaic energy conversion; practical application of solar-thermal processes; deep-center luminescence. *Mailing Add:* 408 16th St Huntingdon PA 16652

PFUDERER, HELEN A, b Ames, Iowa, Mar 17, 39; m 59; c 2. ENVIRONMENTAL SCIENCE, NUTRITION. *Educ:* Iowa State Univ, BS, 61; Univ Tenn, MS, 69, MBA, 85. *Prof Exp:* Dir environ info, Ecol Sci Info Ctr, 74-78; head, dept energy & environ info, 78-79, tech asst, assoc lab dir, biomed & environ sci, 80-82, head, Prog Develop Off, Info Res & Anal, 84-86, DIR, ENVIRON INFO SYSTS, OAK RIDGE NAT LAB, 87- *Concurrent Pos:* Invitee, Numeric Data Adv Bd, Nat Acad Sci, 81-83; consult info progs, Space Nuclear Reactor Prog, 84. *Mem:* Am Nuclear Soc; Am Soc Info Sci. *Res:* Computerized information systems and newsletters on the environmental effects of energy technologies including the nuclear and fossil fuel cycles; marketing and long-range planning of research; computing, computerized systems and expert systems; research management and planning. *Mailing Add:* Oak Ridge Nat Lab PO Box 2008 Oak Ridge TN 37831-6050

PHADKE, KALINDI, IMMUNOLOGY, BIOCHEMISTRY. *Educ:* Univ Bombay, PhD(biochem), 62. *Prof Exp:* SR RES ASSOC BIOL, IND UNIV, 83- *Mailing Add:* 12730 Limberlost Dr Carmel IN 96032

PHADKE, MADHAV SHRIDHAR, b Bombay, India, Aug 28, 48; m 74; c 2. ENGINEERING, STATISTICS & QUALITY. *Educ:* Indian Inst Technol, Bombay, BTech, 69; Univ Rochester, MS, 71; Univ Wis-Madison, MS, 72, PhD(mech eng), 73. *Prof Exp:* Asst scientist eng statist, Math Res Ctr, 74-76, res assoc air pollution modeling & data anal, Dept of Statist, Univ Wis-Madison, 74-76; vis scientist, IBM T J Watson Res Ctr, 76-77; mem tech staff, Bell Tel Labs, 77-83, supv, AT&T Bell Labs, 83-91; PRES, PHADKE ASSOCS, INC, 91- *Honors & Awards:* Taquchi Award, 85. *Mem:* Am Statist Asn; Inst Elec & Electronic Engrs; Am Soc Mech Eng; Am Soc Qual Control. *Res:* Time series analysis; system identification; process control; mathematical modeling; air quality data analysis; quality control; experimental design; robust design; Taquchi methods. *Mailing Add:* 15 Fairfield Dr Tinton Falls NJ 07724-3113

PHAFF, HERMAN JAN, b Winschoten, Netherlands, May 30, 13; nat US; m 87. FOOD TECHNOLOGY, MICROBIOLOGY. *Educ:* Delft Univ Technol, Chem Engr, 38; Univ Calif, PhD(pectin chem), 43. *Prof Exp:* Asst prof food technol & asst microbiologist, Exp Sta, Univ Calif, Berkeley, 46-52, assoc prof food sci & assoc microbiologist, 52-58; chmn dept bacteriol, 70-75, prof food sci & technol & microbiologist, 58-83, prof bacteriol, 65-83, EMER PROF BACTERIOL, FOOD SCI & TECHNOL & MICROBIOLOGIST, UNIV CALIF, DAVIS, 83- *Concurrent Pos:* Ed, Yeast Newslett, 55-88; grants, NIH, 68-88, NSF, 81-91; hon mem Int Comn Yeast & Yeast-Like organisms, Int Union Microbiol Socs, 88-; lectr, Mycol Soc Am, 76; assoc ed, Int J Syst Bact, 80- *Honors & Awards:* James F Guymon Lectr, Am Soc Enology & Viticulture, 86; J Roger Porter Award, Am Soc Microbiol, 84; John L Etchells Mem Lect, NC State Univ, 86. *Mem:* Am Chem Soc; Am Soc Microbiol; Mycol Soc Am; Inst Food Technol; Am Acad Microbiol; Can Soc Microbiol; Soc Gen Microbiol, UK; Dutch Microbiol Soc. *Res:* Chemistry of pectins and pectic enzymes; biology, ecology and systematics of yeasts; prevention of microbiological spoilage of foods; hydrolytic enzymes; biochemistry of yeast cell walls; ecology and molecular taxonomy of yeasts. *Mailing Add:* Dept Food Sci & Technol Univ Calif Davis CA 95616

PHAIR, GEORGE, b Ballston, Va, 18. EARTH SCIENCES. *Educ:* Hamilton Col, BS, 40; Rutgers Univ, MS, 42; Princeton Univ, PhD(geol), 49. *Prof Exp:* Mineralogist & petrologist, US Geol Surv, Washington, DC, 49-88; RETIRED. *Concurrent Pos:* Rockefeller found fel, Princeton Univ, 49; comt mem, Earth Sci Div, Nat Res Coun, 59. *Mem:* Fel Mineral Soc Am. *Res:* Petrology of Colorado Rocky Mountains; miner element studies; geochemistry; uranium & thorium. *Mailing Add:* 14700 River Rd Potomac MD 20854

PHAIR, JOHN P, b Paris, France, July 17, 34; US citizen; m 58; c 2. MEDICINE. *Educ:* Yale Univ, BA, 56; Univ Cincinnati, MD, 60. *Prof Exp:* From intern to resident med, Yale New Haven Hosp, 60-65; from asst prof to prof med, Col Med, Univ Cincinnati, 67-76; CHIEF INFECTIOUS DIS, DEPT MED, NORTHWESTERN UNIV MED SCH, 76-, DIR COMPREHENSIVE AIDS CTR, 88- *Concurrent Pos:* USPHS trainee, Yale New Haven Hosp, 65-67. *Mem:* Am Soc Microbiol; Am Asn Immunologists; Am Fedn Clin Res; Infectious Dis Soc Am; Central Soc Clin Res. *Res:* Host defenses in infectious disease; acquired immunodeficiency syndrome. *Mailing Add:* Dept Med Northwestern Univ Sch Med Chicago IL 60611

PHALEN, WILLIAM EDMUND, b Nebr, Dec 18, 16; m 42; c 3. BIOCHEMISTRY. *Educ:* Univ Omaha, BA, 41. *Prof Exp:* Chemist, Cudahy Packing Co, 41-42 & 45-46, supvr meat res, 47-63, lab dir, Cudahy Labs, 63-68; Lab dir, Am Labs, Inc, 68-82; CONSULT, 82- *Concurrent Pos:* Mem comt food preserv, Am Meat Inst, 54-63, sci adv comt, 64-66. *Mem:* Am Oil Chem Soc; Inst Food Technol; Am Chem Soc. *Res:* Thermal processing of canned meats; curing and smoking of meat products; meat pigments; enzymology; conversion of animal tissues to pharmaceutical products. *Mailing Add:* 1339 S 93rd Ave Omaha NE 68124

PHAM, TUAN DUC, b Hanam, Vietnam, Dec 28, 38; US citizen; m 64; c 2. ANATOMY, CELL BIOLOGY. *Educ:* St Edward's Univ, BS, 63; Loyola Univ, MS, 67; Columbia Univ, PhD(anat), 75. *Prof Exp:* Staff assoc path, 73-80, asst prof anat & pharmacol, 80-85, ASST PROF ANAT & CELL BIOL, COL PHYSICIANS & SURGEONS, COLUMBIA UNIV, 86- *Concurrent Pos:* Kevin Doyle fel, NY Heart Asn, 75-77; sr investr, NY Heart Asn, 80-84. *Mem:* Am Asn Anatomists; NY Acad Sci; AAAS. *Res:* Electron microscopy; immunoelectron microscopy; role of targets in the development and survival of neurons. *Mailing Add:* 67 Rose Rd West New York NY 10994

PHAM-GIA, THU, b Ninh-Binh, Vietnam, July 9, 45; Can citizen; div; c 2. BAYESIAN ANALYSIS, RELIABILITY STUDIES. *Educ:* Univ Saigon, Lic Sci, 66; Univ Hawaii, MA, 69; Univ Toronto, PhD(probability), 72. *Prof Exp:* Fel math, Univ Toronto, 72-73; from asst prof to assoc prof probability & statist, Univ Moncton, 73-81; prog controller oper res, Pratt & Whitney Can Ltd, 81-82; mgr mgt sci, Bell Can, 82-83; PROF PROBABILITY & STATIST, UNIV MONCTON, 83- *Concurrent Pos:* Consult, Atlantic Co-op, Moncton, NB, Can, 78-81, Goodwin & Co, 77-80; vis statistician, Indust Mat Res Inst, Nat Res Coun, Montreal, Can, 81; lectr, Sch Admin Studies, Univ Quebec, Montreal, 82; group leader, Indust Statist Res Group, Montreal, 84- *Mem:* Fel Inst Statisticians; Can Statist Soc; Inst Elec & Electronic Engrs; sr mem Inst Indust Engrs. *Res:* Statistical methods used in industry; bayesian methods applied to work samplings and reliability; bayesian work sampling; software developed. *Mailing Add:* Fac Sci & Eng Univ Moncton Moncton NB E1A 3E9 Can

PHAN, CHON-TON, b Soctrang, S Viet-Nam, Apr 14, 30; m 81; c 2. PLANT PHYSIOLOGY, BIOCHEMISTRY. *Educ:* Univ Paris, Lic es Sci, 53, Dipl d'Etudes Superieures, 56, Dr es Sci, 61. *Prof Exp:* From jr searcher to sr searcher plant biol, Nat Ctr Sci Res, France, 56-68, master in res, 68; from asst prof to assoc prof plant physiol & biochem, Univ Alta, 68-74; PROF PLANT PHYSIOL, UNIV MONTREAL, 74 - *Mem:* Int Soc Hort Sci; Fr Soc Plant Physiol. *Res:* Plant metabolism as affected by the structure of the organ and by the physical and chemical changes in the environment; plant tissue culture. *Mailing Add:* Dept Biol Sci Univ Montreal CP 6128 Sta A Montreal PQ H3C 3J7 Can

PHAN, CONG LUAN, high voltage engineering, atmospheric glaciology; deceased, see previous edition for last biography

PHAN, KOK-WEE, b Segamat, Johore, Malaysia, Nov 27, 39; m 62; c 3. ALGEBRA. *Educ:* Univ Melbourne, BE, 62, BSc, 64; Monash Univ, Australia, PhD(math), 69. *Prof Exp:* ASSOC PROF MATH, UNIV NOTRE DAME, 68-, PROF MATH. *Mem:* Am Math Soc. *Res:* Theory of finite groups. *Mailing Add:* Dept of Math Univ Notre Dame Notre Dame IN 46556

PHAN, SEM HIN, b Jakarta, Indonesia, Sept 15, 49; m 76; c 2. IMMUNOPATHOLOGY, COLLAGEN BIOCHEMISTRY. *Educ:* Ind Univ, Bloomington, BSc, 71, PhD(chem), 75; Ind Univ, Indianapolis, MD, 76. *Prof Exp:* Resident path, Univ Conn Health Ctr, 76-78, fel immunopath, 78-80; asst prof, 80-86, ASSOC PROF PATH, UNIV MICH MED SCH, 86- *Concurrent Pos:* Prin invest, NIH grants, 82-; mem, Path A Study Sect, NIH, 83-87; estab investr, Am Heart Asn, 84-89. *Mem:* Am Asn Pathologists; NY Acad Sci; Am Chem Soc; Am Thoracic Soc; AAAS; Am Heart Asn; Soc Investigative Dermat; fel Col Am Pathologists. *Res:* Interstitial lung disease; lung injury and repair; role of cytokimes; arachidonate metabolites in fibrogenesis and regulation of collagen metabolism; renal and liver fibrosis; xanthine oxidase in oxidant induced injury. *Mailing Add:* Dept Path M0602 Univ Mich Med Sch Ann Arbor MI 48109

PHANEUF, RONALD ARTHUR, b Windsor, Ont, Jan 26, 47; m 80; c 2. ATOMIC PHYSICS. *Educ:* Univ Windsor, BSc, 69, MSc, 70 & PhD(physics), 73. *Prof Exp:* Postdoctoral res assoc fel, Joint Inst Lab Astrophysics, Univ Colo, 73-75; SR RES STAFF MEM, OAK RIDGE NAT LAB, 75- *Concurrent Pos:* Dir, Controlled Fusion Atomic Data Ctr, 84-; ed bd, J Phys & Chem Reference Data, 86-89; mem, Atomic, Molecular & Optical Physics Comt, Nat Res Coun, 90- *Mem:* Fel Am Phys Soc. *Res:* Atomic processes relevant to fusion research; inelastic collisions of highly charged ions with electrons and atoms. *Mailing Add:* Oak Ridge Nat Lab Bldg 6003 PO Box 2008 Oak Ridge TN 37831-6372

PHANG, JAMES MING, CELL ACTIVATION. *Educ:* Loma Linda Univ, MD, 63. *Prof Exp:* CHIEF ENDOCRINOL SEC, NAT CANCER INST, 65- *Res:* Regulatory mechanisms involving amino acids. *Mailing Add:* Metabol Br Bldg 10 Rm 3B40 Bethesda MD 20892

PHARES, ALAIN JOSEPH, b Beirut, Lebanon, Apr 20, 42; US citizen; m 68; c 3. STATISTICAL MECHANICS, MATHEMATICAL PHYSICS. *Educ:* St Joseph Univ, Beirut, BSCE, 64; Ecole Sup D'Electricite, Paris, BSEE, 66; Univ Paris, France, DSc, 71; Harvard Univ, PhD(physics), 73. *Prof Exp:* Assoc prof physics, Lebanese Univ, 73-75; res fel, Int Ctr Theoret Physics, Trieste, Italy, 74 & Harvard Univ, 75-76; from asst prof to assoc prof, 77-82, PROF PHYSICS & DIR SEC SCH SCI PROG, VILLANOVA UNIV, 82-, CHMN, PHYSICS DEPT, 81- *Concurrent Pos:* Vis prof, Univ Du Quebec a Trois, Rivieres, 75, Int Ctr Theoret Physics, Trieste, Italy, 79 & 85; vis asst prof physics, Univ Mont, Missoula, 76-77. *Mem:* Am Phys Soc; Sigma Xi. *Res:* Developed the Combinatorics Function Technique which gives the solutions of multidimensional linear partial difference equations; solution of the radial Shrodinger equation with a linear potential; problem of monomers and dimers on 2- and 3 dimensional lattices; high energy physics; author of physics textbooks; training high school science teachers. *Mailing Add:* Dept Physics Villanova Univ Villanova PA 19085

PHARES, CLEVELAND KIRK, b Many, La, Mar 7, 38; m 71; c 3. BIOCHEMISTRY, PARASITOLOGY. *Educ:* Univ SC, BS, 61, MS, 64; La State Univ, PhD(parasitol), 71. *Prof Exp:* Instr biol, Univ SC, 64-68; res assoc biochem, 71-72, asst prof, 72-77, ASSOC PROF BIOCHEM & MED MICROBIOL, COL MED, UNIV NEBR, 77- *Mem:* AAAS; Soc Exp Biol & Med; Am Soc Parasitologists; Sigma Xi. *Res:* Purification and characterization of the growth stimulating factor produced by the tapeworm, Spirometra mansonoides. *Mailing Add:* Dept Biochem Univ Nebr Col Med 42nd & Dewey Ave Omaha NE 68105

PHARES, RUSSELL EUGENE, JR, b Richmond, Ind, May 9, 37; m 72; c 6. PHYSICAL PHARMACY. *Educ:* Purdue Univ, BS, 59, MS, 60, PhD(pharm), 62. *Prof Exp:* Asst prof, Col Pharm, Univ Fla, 62-66; tech dir, Barnes-Hind Pharmaceut, 66-73; proj leader bioerodible polymers, ALZA Res, 73-76; dir biol res, Cooper Labs, Inc, 76-80, pres, 80-90; pres, Prototek, 90-91; PVT CONSULT, 91- *Mem:* Am Pharmaceut Asn. *Res:* Application of kinetics, mathematics, and statistics to drug delivery system design; ophthalmic and contact lens accessory solution development and contact lens design and fitting. *Mailing Add:* 115 Los Patios Los Gatos CA 95030

PHARIS, RICHARD PERSONS, b Indianapolis, Ind, Mar 19, 37; m 67. PLANT PHYSIOLOGY. *Educ:* Univ Wash, BS, 58; Duke Univ, MF, 59, DF(plant physiol), 61. *Prof Exp:* Plant physiologist, Forest Serv Exp Sta, USDA, Roseburg, Ore, 61-63; res fel plant physiol, Calif Inst Technol, 63-65; from asst prof to assoc prof, 65-72, PROF BOT, UNIV CALGARY, 72- *Concurrent Pos:* Mem, Grant Selection Comt, Nat Sci & Eng Res Coun Can; E W R Steacie mem fel, 70-73; assoc ed, Can J Bot, 80-; organizer, 13th Int Plant Growth Substance Asn Conf, Calgary, Alta, 88. *Honors & Awards:* Heaslip Award, 82. *Mem:* Fel AAAS; Am Soc Plant Physiol; Can Soc Plant Physiol (pres, 78-79); Int Plant Growth Substances Asn (vpres, 85-); fel Royal Soc Can; Can Inst Forestry. *Res:* Physiology of plant growth and development, especially flowering and apical dominance, gibberellin physiology and metabolism. *Mailing Add:* Dept Biol Univ Calgary Calgary AB T2N 1N4 Can

PHARO, RICHARD LEVERS, b Allentown, Pa, Jan 15, 36; m 59; c 3. BIOCHEMISTRY. *Educ:* Pa State Univ, BS, 57, MS, 59; Johns Hopkins Univ, ScD(biochem), 62. *Prof Exp:* Res investr, Geront Br, NIH, 62-67; admin dir & assoc scientist, Retina Found, 67-76, DIR RES ADMIN & SR SCIENTIST, EYE RES INST, 76- *Mem:* AAAS; Am Chem Soc; Friends of Eye Res Rehab & Treatment. *Res:* Enzymology, particularly enzyme mechanism in oxidative phosphorylation; biochemistry and physiology of the eye. *Mailing Add:* Six Ivy Circle Winchester MA 01890

PHARRISS, BRUCE BAILEY, b Springfield, Mo, Dec 12, 37; m 60; c 1. BIOCHEMISTRY, PHARMACOLOGY. *Educ:* Univ Mo, BA, 60, MA, 62, PhD(physiol, pharmacol), 66. *Prof Exp:* Res scientist, Upjohn Co, 66-70; prin scientist, Alza Corp, 70-80; VPRES SCI AFFAIRS, COLLAGEN CORP, 80- *Concurrent Pos:* Tech adv coord off, Fam Planning Prog, Mex; mem adv panel device classification, Food & Drug Admin. *Mem:* AAAS; Am Soc Pharmacol & Exp Therapeut; Soc Study Reproduction; Am Fertil Soc; Int Family Planning Res Asn; Sigma Xi. *Res:* Role of prostaglandins in mediating responses of trophic hormones and steroid hormone research in development of new forms of contraception. *Mailing Add:* Collagen Corp 2500 Faber Pl Palo Alto CA 94303

PHATAK, SHARAD CHINTAMAN, b Indore, India, Feb 28, 32; m 67; c 2. HORTICULTURE, PLANT PHYSIOLOGY. *Educ:* Agra Univ, BSc, 55; Indian Agr Res Inst, New Delhi, AIARI, 59; Mich State Univ, PhD(hort), 64. *Prof Exp:* Instr hort, Dept Agr, Madhya Pradesh, India, New Delhi, 59-60; regional agronomist, Indian Potash Supply Agency Ltd, 60-61; asst hort & veg crops, Mich State Univ, 61-64, fel & res assoc hort, 64-65; scientist div hort, Indian Agr Res Inst, New Delhi, 65-66; chief plant sci, Elanco Div, Eli Lilly & Co India Inc, 67-69; res scientist veg physiol & hort, Res Inst Ont, 69-75; assoc prof, 75-82, PROF HORT, COASTAL PLAIN EXP STA, COL AGR, UNIV GA, 82- *Mem:* Am Soc Hort Sci; Weed Sci Soc Am; Sigma Xi. *Res:* Use and physiology of herbicides and growth regulators; genetic basis for tolerance to herbicides; mechanism of herbicide action; physiology of flowering; precision seeding; plant population; vegetable production; chemigation. *Mailing Add:* Dept Hort Coastal Plain Exp Sta Ga PO Box 748 Tifton GA 31793

PHEASANT, RICHARD, b Brookline, Mass, May 3, 20; m 45; c 1. ORGANIC CHEMISTRY, RESEARCH ADMINISTRATION. *Educ:* Harvard Univ, SB, 40. *Prof Exp:* Chemist, Schering Corp, 41-49, pharmaceut specifications, 49-53, new prod coordr, 53-57; adminr, Res Inst Med & Chem,

57-59; dir pharmaceut develop, Int Div, Schering Corp, 59-60; proj coordr, Aeroprojects, Inc, 61-63, prin scientist, 64-68, mgr, Technidyne, Inc, 69-70; tech data mgr drug regulatory affairs, E R Squibb & Sons, Inc, 70-88; RETIRED. *Mem:* AAAS; Am Chem Soc. *Res:* Steroids; organic proximate analysis; fine particle technology; colloid and surface chemistry; gelation; aerospace and military applications of chemistry; pharmaceutical chemistry. *Mailing Add:* Five Stearns Rd East Brunswick NJ 08816-4167

PHEIFFER, CHESTER HARRY, optometry; deceased, see previous edition for last biography

PHELAN, EARL WALTER, b Rahway, NJ, Sept 24, 00; m 30; c 2. CHEMISTRY. *Educ:* Cornell Univ, BChem, 21, PhD(inorg chem), 28. *Hon Degrees:* LHD, Tusculum Col, 71. *Prof Exp:* Asst chem, Univ Wis, 21-23; instr, Ore State Col, 23-26; asst, Cornell Univ, 26-28; instr, Western Reserve Univ, 28-30; prof & head dept, Valdosta State Col, 30-52; liaison asst, Argonne Nat Lab, 52-65; prof, 65-71, EMER PROF CHEM, TUSCULUM COL, 71- *Concurrent Pos:* Tech aide, Nat Defense Res Comt, 43-46. *Mem:* AAAS; Am Chem Soc. *Res:* Oxidation of hydroxylamine; tests in inorganic chemistry; minerals in foods; chemical education. *Mailing Add:* 11211 Whisperwood Lane Rockville MD 20852-3634

PHELAN, JAMES FREDERICK, b Sedalia, Mo, Jan 13, 17; m 46; c 2. MATHEMATICS, ELECTRICAL ENGINEERING. *Educ:* US Naval Acad, BS, 40; Univ Ill, MA, 50, PhD(math), 64. *Prof Exp:* Res assoc elec eng, Univ Ill, 63-65; ASSOC PROF ELEC ENG TECHNOL, VA POLYTECH INST & STATE UNIV, 65- *Mem:* Inst Elec & Electronics Engrs. *Res:* Techniques for computer solutions of partial differential equations. *Mailing Add:* 503 Skyview Blacksburg VA 24060

PHELAN, JAMES JOSEPH, b San Francisco, Calif, Dec 28, 37; m 62; c 3. REAL-TIME COMPUTER SYSTEMS. *Educ:* Univ San Francisco, BS, 59; St Louis Univ, PhD(physics), 68. *Prof Exp:* Fel, Argonne Nat Lab, 67-68; res assoc, Rutherford High Energy Lab, Eng, 68-72; asst physicist, Argonne Nat Lab, 72-78; MEM TECH STAFF, AT&T BELL LABS, 78- *Res:* Design and performance analysis of real-time computer systems. *Mailing Add:* AT&T Bell Labs 200 N Naperville Rd Rm 2C418 Naperville IL 60566-7033

PHELAN, R(ICHARD) M(AGRUDER), b Moberly, Mo, Sept 20, 21; m 51; c 2. MECHANICAL ENGINEERING. *Educ:* Univ Mo, BS, 43; Cornell Univ, MME, 50. *Prof Exp:* From instr to prof mech eng, Cornell Univ, 47-88; RETIRED. *Mem:* AAAS; Am Soc Mech Engrs. *Res:* Feedback control systems. *Mailing Add:* Four Cornell Walk Ithaca NY 14850

PHELAN, ROBERT J, JR, b Pasadena, Calif, Oct 18, 33; m 60; c 3. OPTICAL PHYSICS. *Educ:* Calif Inst Technol, BS, 58; Univ Colo, PhD(physics), 62. *Prof Exp:* Staff physicist solid state device res, Lincoln Lab, Mass Inst Technol, 62-69; PROJ LEADER, NAT INST STANDARDS & TECHNOL, 69- *Concurrent Pos:* Adj prof, Univ Colo, 89- *Mem:* Am Phys Soc; Optical Soc Am; Inst Elec & Electronic Engrs. *Res:* Semiconductor lasers; optical detectors, integrated optics; optical measurement devices and systems. *Mailing Add:* Nat Inst Standards & Technol 814-02 325 Broadway Boulder CO 80303

PHELPS, ALLEN WARNER, b Washington, DC, Dec 1, 50. TOXICOLOGY, CELL CULTURE. *Educ:* Southwestern at Memphis, BS, 72; Univ Tenn, PhD(toxicol), 78. *Prof Exp:* Fel, Baylor Col Med, 77-80; res toxicologist, Shell Develop Co, 80-81; CELLULAR TOXICOLOGIST, STILLMEADOW, INC, 81- *Concurrent Pos:* Res assoc, Vet Admin, 77-; adj asst prof, Baylor Col Med, 80- *Mem:* Tissue Cult Asn; Environ Mutagen Soc; NY Acad Sci; AAAS; Sigma Xi. *Mailing Add:* 302 Cedarwood Dr Jamestown NC 27282

PHELPS, ARTHUR VAN RENSSELAER, b Dover, NH, July 20, 23; m 56; c 2. ELECTRON COLLISIONS, GASEOUS ELECTRONICS. *Educ:* Mass Inst Technol, ScD(physics), 51. *Prof Exp:* Consult physicist, Res Labs, Westinghouse Elec Corp, 51-70; sr res scientist, Nat Bur Standards, 70-88; RETIRED. *Concurrent Pos:* Fel, Joint Inst Lab Astrophysics, Univ Colo, Boulder, 70-88, chmn, 79-81 & fel adjoint, 88- *Honors & Awards:* W P Allis Prize, Am Phys Soc, 90. *Mem:* Am Phys Soc. *Res:* Electron and atomic collision processes involving low energy electrons, molecules, ions, metastable atoms and resonance radiation; laser processes and modeling. *Mailing Add:* 3405 Endicott Dr Boulder CO 80303

PHELPS, CHRISTOPHER PRINE, b Westfield, NJ, July 6, 43; m 69; c 2. NEUROENDOCRINOLOGY, NEUROSCIENCE. *Educ:* Lafayette Col, AB, 65; Rutgers Univ, PhD(endocrinol), 73. *Prof Exp:* Asst endocrinol, Grad Sch, Rutgers Univ, 65-72, fel, Bur Biol Res, 72-73; res anatomist, Sch Med, Univ Calif, Los Angeles, 73-75, fel neuroendocrinol, 75-76; asst prof, 76-81, ASSOC PROF ANAT, COL MED, UNIV SFLA, 81-, VCHMN DEPT, 87- *Concurrent Pos:* NIH proj dir, 78-81, 90-95, COSTE Prog, USPHS, 67-68. *Honors & Awards:* Nat Res Soc Award, NIH, 75-76. *Mem:* Am Asn Anatomists; Soc Neurosci; Endocrine Soc. *Res:* Neuroendocrinology; limbic brain regulation of endocrine function; neuroplasticity; recovery of brain function after damage, endocrine and behavioral. *Mailing Add:* Dept Anat Box 6 Med Ctr Univ SFla Tampa FL 33612

PHELPS, CREIGHTON HALSTEAD, b Haydenville, Ohio, Nov 12, 40. NEUROSCIENCE. *Educ:* Ohio Univ, BSc, 62; Univ Mich, Ann Arbor, MSc, 64, PhD(anat), 67. *Prof Exp:* Asst prof anat, Health Ctr, Univ Conn, 69-76; assoc prof anat, Sch Med, Wright State Univ, 76-85; prog dir, Neurobiol & Neuroplasticity, Nat Inst Aging, 85-89; SR VPRES, MED & SCI AFFAIRS, ALZHEIMER'S ASN, 89- *Concurrent Pos:* USPHS fel, Univ Col, Univ London, 67-69. *Mem:* AAAS; NY Acad Sci; Soc Neurosci; Geront Soc Am. *Res:* Ultrastructure and cytochemistry of central nervous tissue under normal conditions and during development, degeneration and regeneration. *Mailing Add:* Div Med & Sci Affairs Alzheimer's Asn 70 E Lake St Suite 600 Chicago IL 60601

PHELPS, DANIEL JAMES, b Jackson, Mich, July 15, 47; m 71. SOLID STATE PHYSICS. *Educ:* Rose Polytech Inst, BS, 69; Univ Ill, MS, 70, PhD(physics), 74. *Prof Exp:* RES PHYSICIST, KODAK RES LABS, EASTMAN KODAK CO, 74- *Mem:* Am Phys Soc; Inst Elec & Electronics Engrs. *Res:* Design and use of solid state imaging and light emitting devices. *Mailing Add:* 18 Kyle Dr Rochester NY 14626

PHELPS, DEAN G, b Grafton, NDak, Feb 7, 34; m 54; c 3. MATHEMATICS. *Educ:* Univ NDak, BS, 56; Univ Wis-Madison, MS, 59; Wash Univ, PhD(math), 68. *Prof Exp:* Proj engr, A C Spark Plug Div, Gen Motors Corp, 56-58; asst prof math, Univ NDak, 61-64; asst prof, Univ Fla, 68-71; chmn dept math & comput sci, 75-79, PROF MATH, 71-, VPRES, ADMIN & DEVELOP, LOCK HAVEN UNIV, 87- *Concurrent Pos:* Vis prof math, Trent Polytech, Nottingham, Eng, 79-80. *Mem:* Math Asn Am. *Res:* Complex analysis. *Mailing Add:* Six Magnolla Dr Lock Haven PA 17745

PHELPS, FREDERICK MARTIN, III, b Grand Rapids, Mich, June 11, 33; m 57; c 3. ATOMIC SPECTROSCOPY. *Educ:* Carleton Col, BA, 55; Univ Mich, MS, 58; Univ Alta, PhD(physics), 63. *Prof Exp:* Instr physics & math, Kalamazoo Col, 59-60; sessional lectr physics, Univ Alta, 62-64; res assoc, Univ Mich, 64-68, lectr, 66-68; assoc prof physics, Detroit Inst Technol, 67-69; head grating res lab, Bausch & Lomb Optical Co, 69-70; ASSOC PROF PHYSICS, CENT MICH UNIV, 70- *Concurrent Pos:* Assoc ed, J Optical Soc Am, 65-77; vis scientist, Am Inst Physics, 69, 70 & 72. *Mem:* Am Asn Physics Teachers; Optical Soc Am. *Res:* Precision spectroscopy, especially establishment of class A secondary standards of wavelength. *Mailing Add:* Dept Physics Cent Mich Univ Mt Pleasant MI 48859

PHELPS, FREDERICK MARTIN, IV, b Kalamazoo, Mich, Jan 7, 60; m 87; c 1. BIOMATHEMATICS. *Educ:* Kalamazoo Col, BS(math), 82, BS(physics), 82; Univ Utah, MS, 84, PhD(math), 89. *Prof Exp:* Postdoctoral fel math, Oxford Univ, UK, 89-90; DIR ACAD RES, CENT ASIAN FOUND, 90- *Res:* Mathematical biology. *Mailing Add:* 290 Cedar Dr Mt Pleasant MI 48858

PHELPS, GEORGE CLAYTON, b Lexington, Mo, Aug 21, 35; m 57; c 2. MATERIAL ANALYSIS, FORENSICS. *Educ:* Wichita State Univ, BS, 58. *Prof Exp:* Chemist, Cessna Aircraft Co, 56-58; first lieutenant, US Army Air Defense, 58-61; chemist, Nebr Testing Corp, 61-63, lab dir, 63-65, mgr, 65-67, vpres & owner, 67-86, mgr, 86-90; SR PROJ SCIENTIST, UNIV NEBR, OMAHA, 90- *Concurrent Pos:* Dir, Great Plains Aerial Surv, 69-72; instr, adult educ, Univ Nebr, Omaha, 71-72 , dept eng, 72-73. *Mem:* Am Chem Soc; fel Am Inst Chemists; Am Coun Independent Labs; Instrument Soc Am; Construct Specif Inst; Water Pollution Control Fedn. *Res:* Chemistry; environment; materials testing; engineering; product evaluation. *Mailing Add:* 4123 S 67th St Omaha NE 68117

PHELPS, HARRIETTE LONGACRE, b Exeter, NH, Mar 15, 36; div; c 2. PHYSIOLOGICAL ECOLOGY. *Educ:* Carleton Col, BA, 56; Mt Holyoke Col, MA, 58; Ohio State Univ, PhD(zool), 64. *Prof Exp:* Staff fel virol, Nat Cancer Inst, 65-68; assoc prof biol, Fed City Col, 68-78; assoc prof, 78-80, PROF BIOL, UNIV DC, 80- *Mem:* Fel AAAS; Estuarine Res Fedn. *Res:* Estuarine research including biogeochemistry of metals in sediments and benthics. *Mailing Add:* Dept Biol Univ DC Van Ness Campus Bldg 44 MB 4407 Washington DC 20008

PHELPS, JACK, b Guymon, Okla, Feb 15, 26; m 46; c 4. MATHEMATICS. *Educ:* Panhandle Agr & Mech Col, BS, 48; West State Col Colo, MA, 54; Okla State Univ, EdD(math ed), 63. *Prof Exp:* Teacher pub sch, Okla, 48-58, supt & teacher, 58-61; from asst prof to prof math, Northwestern State Col, Okla, 63-86; EMER PROF MATH, SOUTHWESTERN OKLA STATE UNIV, 86- *Concurrent Pos:* Math consult, State Dept Educ, 63-64, Alaska, 66; consult, Concord Col. *Mem:* Math Asn Am. *Res:* Education in mathematics. *Mailing Add:* 514 Maple Weatherford OK 73096

PHELPS, JAMES PARKHURST, b Medford, Mass, Feb 13, 24. REACTOR PHYSICS. *Educ:* Univ Maine, BS, 51; Mich State Univ, PhD(exchange reactions kinetics), 56. *Prof Exp:* Asst chemist, coating develop group, Ansco, Gen Aniline & Film Corp, NY, 51-52; teaching asst, Mich State Univ, 52-54; sr engr, Nuclear Div, Martin Co, Md, 56-57; asst chemist, Reactor Physics Div, Nuclear Eng Dept, Brookhaven Nat Lab, 57-60, from assoc chemist to chemist, 60-70; prof chg nuclear reactor, 70-75, CHMN DEPT NUCLEAR ENG, UNIV LOWELL, 75- *Concurrent Pos:* Lectr, continuing educ prog, Martin Co, 57; mem staff atoms at work exhibit, Reactor Exp Prog, AEC, Bangkok, Thailand, 62. *Mem:* Am Phys Soc; Am Nuclear Soc. *Res:* Anisotropic neutron diffusion in graphite; buckling, prompt lifetime, disadvantage factor and spectral index measurements in graphite-uranium; fast critical experiments; time characteristics and spectra of delayed neutrons; radiographer safety. *Mailing Add:* 64 Melendy Rd Hudson NH 03051

PHELPS, LEE BARRY, b Quirigua, Guatemala, May 14, 38; m 71; c 1. SURFACE MINING, DREDGING. *Educ:* Univ Idaho, BS, 66; Pa State Univ, MEng, 72, PhD(mining eng), 81. *Prof Exp:* Dredge supt, Int Mining Corp, 66-70; res asst, Pa State Univ, 70-72; mine supt, Aluminum Co Am, 72-76; proj entr, Dow Chem Co, 76-77; instr, 77-81, asst prof, 81-86, ASSOC PROF SURFACE MINING, PA STATE UNIV, 86- *Concurrent Pos:* Mem, Interstate Mining Compact Comn, 77-; consult, Dow Chem Co, 77-85; chmn, Surface Mining Comt, SME. *Mem:* Soc Mining Eng; Inst Mining & Metall Engrs. *Res:* Surface mining systems and design to include reclamation and environmental controls; dredging. *Mailing Add:* 125 Mineral Sci Bldg Pa State Univ University Park PA 16802

PHELPS, LEROY NASH, b Logan, Ohio, Feb 19, 30; m 62; c 2. MICROBIOLOGY, IMMUNOLOGY. *Educ:* Ohio State Univ, BSc, 55; Univ Southern Calif, PhD(immunol, bacteriophages), 64. *Prof Exp:* Assoc biologist, Eli Lilly & Co, 55-59; lectr bact, Univ SC, 63-64; staff fel bact genetics lab, Venereal Dis Res Lab, Commun Dis Ctr, Atlanta, Ga, 64-66; asst prof, 66-71, ASSOC PROF MICROBIOL, SAN DIEGO STATE UNIV, 71-

Concurrent Pos: Consult, San Diego Inst Path, 66- *Mem:* AAAS; Am Soc Microbiol; Brit Soc Gen Microbiol; Sigma Xi. *Res:* Antigens of bacteriophages and their hosts, including the effects of different hosts on specific bacteriophage antigens. *Mailing Add:* Dept of Microbiol San Diego State Univ San Diego CA 92182

PHELPS, MICHAEL EDWARD, b Cleveland, Ohio, Aug 24, 39; m 69; c 2. BIOPHYSICS, NEUROSCIENCES. *Educ:* Western Wash Univ, BA, 65; Wash Univ, PhD(nuclear chem), 70. *Prof Exp:* From asst prof to assoc prof radiation sci, Sch Med, Wash Univ, 71-75, from asst prof to assoc prof elec eng, 74-75; assoc prof radiol & biophys, Med Sch, Univ Pa, 75-76; PROF RADIOL & BIOPHYS, SCH MED, UNIV CALIF, LOS ANGELES, 76-, DIV CHIEF NUCLEAR MED, 80-, JENNIFER JONES SIMON PROF, 83- *Concurrent Pos:* Mem, Comput & Biomath Study Sect, NIH, 74-78; prin investr, Dept Energy, 81-; adv, Lab Cerebral Metab, 77-; chief, Lab Nuclear Med, Dept Energy, 84-; dir, Crump Inst, 89- *Honors & Awards:* Von Hevesy Prize, 78 & 82; Paul Aebersold Award, 83; Ernest O Lawrence Presidential Award, 84; Sarah L Poiley Award, 84; Rosenthal Award, 87; Landauer Award, 88; Ted Bloch Award, 89. *Mem:* Nuclear Med Soc; Neuroscience Soc; NY Acad Sci; Radiol Soc NAm. *Res:* Development of tomographic techniques for assaying local biochemical processes in the human subject atramatically; approach uses tracer kinetic techniques with biologically active compounds labeled with radioactive isotopes of carbon, nitrogen, oxygen and fluorine primarily and an analytical imaging technique referred to as positron emission tomography; brain development; neuronal plasticity and compensatory reorganization to brain injury, disease or surgery; metabolic basis of degenerative brain disease and coronary artery disease. *Mailing Add:* Radiol Sci B2-085 Chs Univ of Calif 405 Hilgard Ave Los Angeles CA 90024

PHELPS, PATRICIA C, b 1930. CYTOSKELETAL COMPONENTS, ELECTRON MICROSCOPY. *Educ:* Brown Univ, AB, 52. *Prof Exp:* INSTR, ELECTRON MICROSCOPY, SCH MED, UNIV MD, 75- *Mem:* Am Soc Cell Biol. *Res:* Toxic cell injury. *Mailing Add:* Dept Path Sch Med Univ Md 10 S Pine St Baltimore MD 21201

PHELPS, PAULDING, b Philadelphia, Pa, June 5, 33; c 2. RHEUMATOLOGY. *Educ:* Haverford Col, BA, 55, Col Physicians & Surgeons, MD, 60. *Prof Exp:* CHIEF RHEUMATOLOGY SECT, MAINE MED CTR, 78-, ATTEND PHYSICIAN, 71- *Concurrent Pos:* Attend physician, Mercy Hosp, 71- & Portland City Hosp, 72-; mem, Comt Rheumatology, Am Bd Internal Med, 76-82; chmn, Med Sci Comt, Maine Arthritis Found, 72-77; chmn, Comt Rheumatologic Pract, Am Rheumatism Asn, 84-86; mem, Nat Arthritis Adv Bd, 85- *Mem:* Am Rheumatism Asn (vpres, 85, pres, 87-88); fel Am Col Physicians; Am Bd Internal Med; Am Soc Int Med. *Mailing Add:* 51 Sewall St Portland ME 04102

PHELPS, PHARO A, b Biggs, Calif, Aug 24, 28; m 50; c 4. CIVIL ENGINEERING. *Educ:* US Naval Acad, BS, 50; Rensselaer Polytech Inst, BCE, 53; US Naval Postgrad Sch, PhD(physics), 63. *Prof Exp:* Antisubmarine warfare officer, USS Brinkley Bass, US Navy Civil Eng Corps, 50-52, spec asst civil eng, Dist Pub Works Off, 11th Naval Dist, 53-55, housing officer, Pub Works Ctr, Guam, 55-57, training officer, Bur Yards & Docks, 57-59, instr physics, US Naval Acad, 63-65, cmndg officer, US Naval Mobile Construct Battalion 8, 65-67, exec officer, Europ Div, Naval Facil Eng Command, London, 67-69, officer in chg construct, Naval Facil Eng Command Contracts, Madrid, 69-70, asst comdr res & develop, Naval Facil Eng Command, Washington, DC, 70-71, asst comdr design & eng, 71-74, dep comdr facil acquisition, 73-74; applns eng group mgr, Bechtel Corp, 74-76, mgr eng res & eng oper, Bechtel Nat, Inc, 76-78, mgr eastern opers, Int Bechtel Inc, Kuwait, 78-80, mgr eng & mat, Bechtel Nat Inc, San Francisco, 80-82, mgr defense projs & vpres, Bechtel Nat Inc, 83-85; prin, Phelps Assocs, Rheem Valley, Calif, 85-; MGR, MIDEASTERN OPERS, BECHTEL CORP. *Concurrent Pos:* Vis lectr, Univ Md, 64; assoc prof, Anne Arundel Community Col, 64-65; mem, Interagency Comt Excavation Technol, 71-72; proj mgr, Intecontinental Ballistic Missile Deep Basin Egress Syst, 83-84 & Very Large Floating Struct. *Mem:* Am Underground Space Asn; Am Phys Soc; Inst Shaft Boring Technol; Soc Am Mil Eng; Am Soc Civil Eng. *Res:* Underground construction; advanced civil engineering and construction techniques including robotics and automation; physics interdisciplinary topics in civil engineering; technical management. *Mailing Add:* Bechtel Corp Hydro Comm Facil Div PO Box 23930 Safat Kuwait

PHELPS, ROBERT RALPH, b San Bernardino, Calif, Mar 22, 26; m 55. MATHEMATICS. *Educ:* Univ Calif, Los Angeles, BA, 54; Univ Wash, PhD(math), 58. *Prof Exp:* Asst, Univ Calif, Los Angeles, 54-56; asst, Univ Wash, 56-57; mem, Inst Adv Study, 58-60, NSF fel, 59-60; lectr & asst res mathematician, Univ Calif, Berkeley, 60-62; from asst prof to assoc prof, 62-66, PROF MATH, UNIV WASH, 66- *Concurrent Pos:* Instr, Princeton Univ, 58-59. *Mem:* Am Math Soc; Math Asn Am. *Res:* Abstract functional analysis; convex sets in normed algebras, normed and topological vector spaces. *Mailing Add:* Dept Math Univ Wash GN-50 Seattle WA 98195

PHELPS, RONALD P, b Mobile, Ala, Oct 30, 47; m 75. FISHERIES. *Educ:* Auburn Univ, BS, 69, PhD(fisheries), 75. *Prof Exp:* Res asst, 70-75, ASST PROF FISHERIES, DEPT FISHERIES & ALLIED AQUACULT, AUBURN UNIV, 75- *Concurrent Pos:* Fel, Dept Fisheries & Allied Aquacult, Auburn Univ, 75. *Mem:* Am Fisheries Soc; Marine Technol Soc; Sigma Xi. *Res:* Chemical control if fish diseases and the related environmental effects; international aquaculture. *Mailing Add:* Auburn Univ Fisheries Bldg Auburn AL 36830

PHELPS, WILLIAM ROBERT, forest pathology; deceased, see previous edition for last biography

PHEMISTER, ROBERT DAVID, b Framingham, Mass, July 15, 36; m 60; c 3. VETERINARY PATHOLOGY. *Educ:* Cornell Univ, DVM, 60; Colo State Univ, PhD(path), 67. *Prof Exp:* Assoc res vet, Univ Calif, Davis, 60-61; staff scientist, Armed Forces Inst Path, 62-64; leader path sect, Collab Radiol Health Lab, Colo State Univ, 64-68, dep dir, 68-72, from assoc prof to prof path, 68-85, sci dir, 72-73, dir, Collab Radiol Health Lab, 73-76, assoc dean, 76-77, dean, Col Vet Med & Biomed Sci, 77-85, interim acad vpres, 82, interim pres, 83-84; DEAN & PROF PATH, COL VET MED, CORNELL UNIV, 85- *Concurrent Pos:* Vis res pathologist, Univ Calif, 74-75; consult, Miss State Univ, 77-81. *Mem:* AAAS; Am Vet Med Asn; Am Col Vet Path; Asn Am Vet Med Col (pres, 82-83). *Res:* Diseases of pre- and postnatal development; canine nephropathology; radiation effects. *Mailing Add:* Deans Off Cornell Univ Vet Med Ithaca NY 14853

PHIBBS, PAUL VESTER, JR, b Pulaski, Va, July 24, 42; m 68; c 2. MICROBIOLOGY. *Educ:* Bridgewater Col, BA, 64; Univ Ga, MS, 66, PhD(microbiol), 69. *Prof Exp:* Res fel microbial physiol, Univ Minn, Minneapolis, 69-70; from asst prof to prof microbiol, Med Col Va, Va Commonwealth Univ, 70-85; PROF & CHMN MICROBIOL & IMMUNOL, SCH MED, E CAROLINA UNIV, 86-, ADMIN DIR BIOTECHNOL PROG, 86- *Concurrent Pos:* NSF res grant, Va Commonwealth Univ, 71-, Nat Cystic Fibrosis Res Found grant, 75-78; vis prof microbiol, Med Ctr, Univ Ala, 73; dir microbiol grad prog, Va Commonwealth Univ, 75-76, leader microbiol physiol-genetics cluster, 75-83, dep chmn, 82-83; prog dir, NIH res training grant, 77-83; ed microbiol sect, Va J Sci, 77-79; vis prof genetics, Monash Univ, Melbourne, Australia, 84-85; Celanese Res Co res grant, 84- *Mem:* Am Soc Microbiol; Soc Exp Biol & Med. *Res:* Bacterial physiology and genetics with emphasis on the regulation of metabolic pathways and membrane transport sytems in Pseudomonas; genetics and mechanisms of exopolysaccharide synthesis by Pseudomonas. *Mailing Add:* Dept Microbiol & Immunol Sch Med E Carolina Univ Greenville NC 27858

PHIBBS, RODERIC H, b Chicago, Ill, Nov 3, 30; m 52; c 4. PEDIATRICS, NEONATOLOGY. *Educ:* Syracuse Univ, AB, 54; State Univ NY, MD, 58. *Prof Exp:* Intern rotating, Med Ctr, 58-59, from asst resident to chief resident pediat, 59-62, asst clin prof pediat, Cardiovasc Res Inst, 64-65, from asst prof to assoc prof, 65-80, PROF PEDIAT, MED CTR, UNIV CALIF, SAN FRANCISCO, 80-, ASSOC STAFF, CARDIOVASC RES INST, 68- *Concurrent Pos:* USPHS fel biophys, Univ Western Ont, 62-64; res fel physiol, Cardiovasc Res Inst, Med Ctr, Univ Calif, San Francisco, 64-65; vis prof, Oxford Univ, 74-75. *Mem:* Int Soc Hemorheology; Soc Pediat Res; Am Pediat Soc. *Mailing Add:* Dept Pediat Univ Calif Med Ctr San Francisco CA 94143

PHIFER, HAROLD EDWIN, physical chemistry, for more information see previous edition

PHIFER, LYLE HAMILTON, b Chester, SC, Nov 16, 27; m 83; c 4. ANALYTICAL CHEMISTRY. *Educ:* Wofford Col, BS, 48; Vanderbilt Univ, MS, 51, PhD(analytical chem), 53. *Prof Exp:* Asst dept med, Med Sch, Vanderbilt Univ, 51-53; res assoc, FMC Corp, 53-76; VPRES & TECH DIR, CHEM SERV, INC, 76- *Mem:* Am Chem Soc; Am Soc Test & Mat. *Res:* Ultraviolet spectroscopy kinetics; gas and liquid chromatography; x-ray diffraction and fluorescence. *Mailing Add:* Chem Serv Inc PO Box 3108 West Chester PA 19381-3108

PHILBIN, DANIEL MICHAEL, b Dunmore, Pa, June 2, 35; m 64; c 2. ANESTHESIOLOGY. *Educ:* Duquesne Univ, BS, 57; St Louis Univ, MD, 61; Am Bd Anesthesiol, dipl, 69. *Hon Degrees:* MA, Harvard Univ, 90. *Prof Exp:* Resident anesthesia, Columbia-Presby Med Ctr, 65-68, NIH res fel, Dept Anesthesia, Columbia Univ, 67-68, assoc, Columbia-Presby Med Ctr, 68-70, asst prof, Columbia Univ, 70; asst prof, 71-79, assoc prof, 79-88, DIR CARDIAC ANESTHESIA, 83-, PROF ANESTHESIA, HARVARD MED SCH, MASS GEN HOSP, 88- *Concurrent Pos:* Ed, Anesthesia Anal, 80-89; Fogarty Sr Int Fel, vis fel, Magdalen Col & vis res prof, Nuffield Dept Anaesthetics, Oxford Univ, 82-83. *Honors & Awards:* Res Award Prize, Am Soc Anesthesiologists, 69. *Mem:* Am Physiol Soc; fel Am Col Cardiol; Am Soc Anesthesiologists; Asn Cardiac Anesthesiologists (secy-treas, 75-77, vpres, 77-78, pres, 78-79); Int Anesthesia Res Soc. *Res:* Effect of anesthesia and open cardiac surgery on antidiurectic hormone and renal function; hormonal responses to anesthesia and surgery; effect of anesthesia on ventricular function. *Mailing Add:* Dept of Anesthesia Mass Gen Hosp 32 Fruit St Boston MA 02114

PHILBIN, JEFFREY STEPHEN, b Orange, NJ, Feb 1, 42; m 69; c 2. NUCLEAR ENGINEERING. *Educ:* Univ Notre Dame, BSME, 64; Northwestern Univ, Ill, MS, 69; Univ Ill, Urbana, PhD(nuclear eng), 71. *Prof Exp:* NUCLEAR ENGR, SANDIA LABS, 70- *Mem:* Am Nuclear Soc (pres, 75-76). *Res:* Pulse reactor design and development; pulse reactor operations and experiments; risks analysis, safeguards, space nuclear power. *Mailing Add:* Sandia Labs Div 6453 PO Box 5800 Bldg 6588 Rm 18 Albuquerque NM 87185-5800

PHILBRICK, CHARLES RUSSELL, b Jefferson City, Tenn, Sept 24, 40; m 60; c 2. ATMOSPHERIC & SOLID STATE PHYSICS. *Educ:* NC State Univ, BS, 62, MS, 64, PhD(physics), 66. *Prof Exp:* PROJ SCIENTIST, AIR FORCE CAMBRIDGE RES LAB, 66- *Concurrent Pos:* Mem working group II, Inter Union Comn on Solar-Terrestrial Physics; mem working group IV, panel IV A on struct of upper atmosphere, Comt on Space Res, Int Coun Sci Unions. *Honors & Awards:* Res & Develop Award, US Air Force. *Mem:* Am Phys Soc; Nat Asn Physics Teachers; Am Geophys Union; Sigma Xi. *Res:* Atmospheric composition studies in the mesosphere and thermosphere with rocket and satellite experiments; investigation of atmospheric structure, physical and chemical processes and models; ionospheric composition; mass spectrometry. *Mailing Add:* 425 Shadow Lane State College PA 16803

PHILBRICK, RALPH, b San Francisco, Calif, Jan 1, 34; div; c 3. BOTANY, BIOLOGICAL CONSULTING. *Educ:* Pomona Col, BA, 56; Univ Calif, Los Angeles, MA, 58; Cornell Univ, PhD(bot), 63. *Prof Exp:* Res assoc, Bailey Hortorium, Cornell Univ, 57-63; assoc bot, dept biol sci, Univ Calif, Santa Barbara, 63-64; biosystematist, Santa Barbara Bot Garden, 64-73, cur herbarium, 65-87, dir, 74-87; res assoc & lectr, Univ Calif, Santa Barbara, 64-82; BIOL CONSULT, 87- *Concurrent Pos:* Mem bd dirs, Int Camellia Soc, 63-66; mem, Santa Barbara County Planning Comn, 81-85. *Mem:* Sigma Xi. *Res:* Biosystematics of Pacific Coast cacti; taxonomy of Camellia cultivars; flora of the California Islands; biological planning. *Mailing Add:* 29 San Marcos Trout Club Santa Barbara CA 93105

PHILBRICK, SHAILER SHAW, b Columbia, Mo, May 11, 08; m 36; c 3. GEOLOGY. *Educ:* DePauw Univ, AB, 30; Johns Hopkins Univ, PhD(geol), 33. *Prof Exp:* Jr topographic engr, US Geol Surv, 34; jr soil expert, Soil Conserv Serv, USDA, 35; from jr geologist to sr geologist, Corps Engrs, US Dept Army, 35-49, prin geologist & div geologist, Ohio River Div, 49-56, head geologist & div geologist, 56-66; vis prof geol sci, 63-64, prof, 66-72, EMER PROF GEOL SCI, CORNELL UNIV, 73- *Concurrent Pos:* Lectr, Northwestern Univ, 60; consult, Corps Engrs, US Army, US Nuclear Regulatory Comn & various indust concerns; consult, 72-88. *Honors & Awards:* Claire P Holdredge Award, Asn Eng Geologists, 77; Distinguished Pract Award, Eng Geol Div, 85. *Mem:* Fel Geol Soc Am; hon mem Asn Eng Geologists. *Res:* Engineering geology; geology of nuclear power plant sites; Niagara Falls. *Mailing Add:* 117 Texas Lane Ithaca NY 14850

PHILIP, A G DAVIS, b New York, NY, Jan 9, 29; m 64. ASTRONOMY. *Educ:* Union Col, BS, 51; NMex State Univ, MS, 59; Case Inst Technol, PhD(astron), 64. *Prof Exp:* Asst prof astron, Univ NMex, 64-66; from asst prof to assoc prof astron, State Univ NY, Albany, 66-76; RES PROF, UNION COL, 76-; ADJ PROF, ASTRON, WESLEYEN UNIV, 82- *Concurrent Pos:* Astronomer, Dudley Observ, 66-81 & Van Vleck Observ, 82-; secy & treas, New York Astron Corp, 68-; vis prof, 72-73, vis fel, Yale Univ, 76; vis prof, La State Univ, 73, 76 & Acad Sci, Lithuania USSR, 73 & 76; secy comn 33, pres comn 30, Int Astron Union, 82-85, mem orgn comt & secy comn 30, Int Astronom Union, 73-; ed, Dudley Observ Report Series, 74; pres & treas, L Davis Press, Inc, 82- & Inst Space Observ, 86-; US Observer, Soviet Union's 6 meter reflector, 80; trustee, Found Astrophys Res, 85- *Honors & Awards:* Harlow Shapley Lectr, Am Astron Soc, 73. *Mem:* Am Phys Soc; Am Astron Soc; Royal Astron Soc; Sigma Xi; NY Acad Sci. *Res:* Photometry; spectroscopy; galactic structure; Mandelbrot Set. *Mailing Add:* Physics Dept Union Col Schenectady NY 12308

PHILIPP, HERBERT REYNOLD, b New York, NY, Nov 6, 28; m 67; c 4. EXPERIMENTAL SOLID STATE PHYSICS. *Educ:* Colgate Univ, AB, 50; Mass Inst Technol, SB, 50; Univ Mo, PhD(physics), 54. *Prof Exp:* PHYSICIST, RES & DEVELOP CTR, GEN ELEC CO, 54- *Mem:* Am Phys Soc. *Res:* Optical properties of crystalline and amorphous solids; physics of metal oxide varistor ceramic materials. *Mailing Add:* Res & Develop Ctr Gen Elec Co PO Box 8 Schenectady NY 12301

PHILIPP, MANFRED (HANS), b Rostock, Ger, Sept 30, 45; US citizen; m 75; c 2. ENZYMOLOGY. *Educ:* Mich Technol Univ, BS, 66; Northwestern Univ, PhD(biochem), 71. *Prof Exp:* Res assoc biochem, Inst Macromolecular Chem, Univ Freiburg, 71-75 & Sch of Hyg & Pub Health, Johns Hopkins Univ, 75-77; PROF BIOCHEM, LEHMAN COL, 77- & GRAD CTR, CITY UNIV NY, 79- *Concurrent Pos:* Predoctoral fel, NIH, 66-71; fel, Europ Molecular Biol Orgn, 72-75,; vis scientist, Weizmann Inst Sci, 74; vis prof, Univ Ulm, 83; vis scientist, Univ Munich, 84. *Mem:* Am Chem Soc; AAAS; Sigma Xi. *Res:* Chemical mechanism of serine hydrolases, enzyme models, abzymes; boron chemistry. *Mailing Add:* Bedford Park Blvd W Dept Chem Lehman Col City Univ NY Bronx NY 10468

PHILIPP, RONALD E, b Easton, Pa, Sept 1, 32; m 55; c 2. MECHANICAL ENGINEERING. *Educ:* Lafayette Col, BS, 54; Lehigh Univ, MS, 56; Columbia Univ, PhD(mech eng), 64. *Prof Exp:* Instr mech & elec eng, Lafayette Col, 54-56; res, Develop & Design, Ord Corps, US Army, 56-84; MGR, US GOLF ASN, RES & TEST CTR, 84- *Concurrent Pos:* Instr, US Mil Acad, West Point. *Mem:* Am Soc Mech Engrs; NY Acad Sci; Sigma Xi. *Res:* Kinematics synthesis of two degree of freedom linkages for function generation. *Mailing Add:* 40 Mockingbird Hackettstown NJ 07840

PHILIPP, WALTER V, b Vienna, Austria, Dec 14, 36; US citizen; m 84; c 4. MATHEMATICS. *Educ:* Univ Vienna, MS & PhD(math), 60, MS(physics), 60. *Prof Exp:* Asst math, Univ Vienna, 60-63; asst prof, Univ Mont, 63-64; Univ Ill, Urbana, 64-65; asst, Univ Vienna, 65-67, dozent, 67; from asst prof to assoc prof, Univ Ill, Urbana, 67-73, PROF MATH, UNIV ILL, 73-, PROF STATIST, DEPT STATIST, 88-, CHMN, 90-; AT DEPT MATH, UNIV ILL. *Concurrent Pos:* Vis prof, Univ NC, Chapel Hill, 72 & 88-89, Univ Vienna, 73, Mass Inst Technol, 80-82, Tufts Univ, 81, Univ Goettingen, 82 & 85, Imp Col, London, 85; assoc ed, Annals Probability, 76-81. *Mem:* fel Inst Math Statist; Int Statist Inst. *Res:* Limit theorems, strong approximation theorems and invariance principles for independent and weakly dependent Banach space valued random variables with applications to mathematical statistics. *Mailing Add:* Dept Statist 101 Illini Hall Univ Ill 725 S Wright St Champaign IL 61820

PHILIPPART, MICHEL PAUL, b Ixelles, Belgium, Aug 1, 35; m 58; c 2. NEUROLOGY, NEUROCHEMISTRY. *Educ:* Brussels Univ, BS, 56, MD, 60. *Prof Exp:* Head, Lab C, Inst Bunge, Belgium, 65, head lab neurol develop, 66-67; asst prof pediat & med, 67-69, assoc prof, 69-75, PROF PEDIAT, NEUROL & PSYCHIAT, UNIV CALIF, LOS ANGELES, 75- *Concurrent Pos:* Belg Am Educ Found fel, Johns Hopkins Univ, 62-64; Parkinson Dis Found sr fel, Inst Bunge, Belgium, 65-66; mem, Ment Retardation Ctr & Brain Res Inst, Los Angeles, 68- *Honors & Awards:* Travel award, Belg Govt, 63, Giannina Gaslini Award, 77. *Mem:* Soc Neurosci; Soc Pediat Res; Soc Exp Biol & Med; NY Acad Sci; Soc Neurochem. *Res:* Lipids, glycolipids; inborn errors of metabolism; hereditary degenerations of the nervous system; lysosomal functions; pediatric neurology. *Mailing Add:* Ment Retardation Inst 300 Westwood Plaza Los Angeles CA 90024

PHILIPPOU, ANDREAS NICOLAOU, b Nicosia, Cyprus, July 15, 44; m 84. FIBONACCI NUMBERS, ESTIMATION. *Educ:* Univ Athens, BSc, 67; Univ Wis, Madison, MSc, 70, PhD(statist), 72. *Prof Exp:* Teaching & res asst statist & probability, Univ Wis, Madison, 68-72; asst prof, Univ Tex at El Paso, 72-74; vis prof, Univ Patras, Greece, 74-78; asst prof, Am Univ Beirut, 78-79, assoc prof, 79-80; vice rector acad affairs & personnel, 83-86, chmn dept math, 84-86, PROF STATIST PROBABILITY & DECISION THEORY, UNIV PATRAS, GREECE, 80- *Concurrent Pos:* Prof & consult, Beirut Univ Col, 79-80; dep chmn, Hellenic Aerospace Indust, 81-82. *Mem:* Am Statist Asn; Greek Math Soc; Inst Math Statist; Math Asn Am; Fibonacci Asn. *Res:* Distributions of order k, Fibonacci polynomials of order k, success runs and system reliability; asymptotic properties of estimators and optimal hypothesis testing; odd-even games. *Mailing Add:* Dept Math Univ Patras Patras Greece

PHILIPS, BILLY ULYSES, b San Antonio, Tex, Oct 22, 46; m 75; c 1. PUBLIC HEALTH, SOCIAL EPIDEMIOLOGY. *Educ:* Oklahoma City Univ, BA, 69; Univ Okla, MPH, 71, PhD(public health), 74. *Prof Exp:* Asst dir prog eval, Nat Drug Educ Ctr, 72-73; eval specialist & instr, 73-74, assoc dir & asst prof continuing med educ, 74-76, dir, Off Spec Progs, 76-77, ASSOC PROF, UNIV TEX MED BR, GALVESTON, 78-, ASSOC DEAN ALLIED HEALTH SCI, 81- *Concurrent Pos:* Prin investr, Galveston Clin Cancer Educ Prog, Nat Cancer Inst, 79-, HCOP Prog Grant, Dept Health & Human Serv, 81-, prin investr, Behav Prescription for Smoking Cessation, 87-; grant reviewer, Am Heart Asn, 88-; comn on accreditation in educ, Am Phys Ther Asn, 88- *Mem:* Sigma Xi; Am Pub Health Asn; Asn Teachers Prev Med; Am Sociol Asn; Am Soc Allied Health Prof. *Res:* Community health; social epidemiology; health promotion and health behavior. *Mailing Add:* Univ Tex Med Br Sch Allied Health Sci Rm 4202 UTMB Zip Code J-28 Galveston TX 77550-2774

PHILIPS, JUDSON CHRISTOPHER, b Glen Ridge, NJ, Nov 15, 42; m 64. ORGANIC CHEMISTRY. *Educ:* Pa State Univ, BS, 64; Ohio State Univ, PhD(chem), 69. *Prof Exp:* Fel, Ohio State Univ, 69; from asst prof to assoc prof chem, Univ Detroit, 69-78; sr res scientist, 78-82, proj leader, 82-84, mgr, 84-86, ASST DIR, PFIZER CENT RES, 86- *Mem:* Am Chem Soc. *Res:* Synthesis and characterization of polymers; structural modification of biopolymers; organosulfur chemistry; enhanced oil recovery. *Mailing Add:* Pfizer Cent Res Eastern Point Rd Groton CT 06340

PHILIPS, LAURA ALMA, b San Francisco, Calif, Sept 4, 57. PHYSICAL CHEMISTRY. *Educ:* Williams Col, Williamstown, Mass, BA, 79; Univ Calif, Berkeley, PhD(chem), 85. *Prof Exp:* Postdoctoral fel chem, Univ Chicago, 85-87; ASST PROF CHEM, CORNELL UNIV, ITHACA, NY, 87- *Mem:* Am Chem Soc; Am Phys Soc. *Res:* High resolution infrared spectroscopy in molecular beams as a probe of the structure and dynamics of small organic molecules; role of intra- and inter-molecular interactions in chemical reactivity. *Mailing Add:* Dept Chem Baker Lab Ithaca NY 14853

PHILIPSEN, JUDITH LYNNE JOHNSON, b Casper, Wyo, Jan 25, 42; m 67; c 2. PHARMACEUTICAL PATENT LITERATURE. *Educ:* Lindenwood Col, Mo, BA, 64; Iowa State Univ, Ames, MS, 67. *Prof Exp:* Res assoc, Food & Nutrition Dept, Iowa State Univ, 66-67; assoc scientist, Parke Davis & Co, 67-77; sr assoc scientist, 77-89, patent searcher, 89-90, SR PATENT SEARCHER, WARNER-LAMBERT CO, DIV PARKE DAVIS CO, 90- *Mem:* Am Chem Soc. *Res:* Studies of chlorophyll and ribonucleic acids and enzymatic studies; chemotherapy of parasitic infections; cancer chemotherapy; novel heterocyclic ring systems. *Mailing Add:* Warner-Lambert Co 2800 Plymouth Rd Ann Arbor MI 48105

PHILIPSON, JOSEPH, b Chicago, Ill, Aug 30, 18; m 42; c 4. POLYMER CHEMISTRY. *Educ:* Univ Wis, BS, 40; Univ Minn, MS, 41; Univ Southern Calif, PhD, 44. *Prof Exp:* Process engr, NAm Aviation, Inc, Calif, 45-46; res chemist, Coast Paint & Chem Co, 46-47; develop chemist, Aerojet-Gen Corp, 47-52; chief chemist, Rocket Div, Grand Cent Rocket, 52-54; dir western div, Atlantic Res Corp, 55-61, dir European Opers, 61-63; CONSULT, 64- *Concurrent Pos:* Consult, US Navy, 68-, Am Hosp Supply, 74-, Hi-Shear Corp, 74-, Magic Mountain, 77-, Beckman Instruments, 78- & Bell Helmets, 81- *Mem:* Am Inst Aeronaut & Astronaut; Soc Plastics Eng; Soc Aerospace Mat & Process Eng; Asn Consult Chemists & Chem Engrs. *Res:* Propellants; plastics; rocketry; adhesives, polyurethane rubbers and sealants; new applications of polymers for coatings including corrosion resistant coatings and optical coating; applications of new polymers for optical use. *Mailing Add:* 2250 Kinclair Dr Pasadena CA 91107-1022

PHILIPSON, LLOYD LEWIS, b Utica, NY, June 19, 28; m 87. SAFETY ENGINEERING, RISK MANAGEMENT. *Educ:* Univ Calif, Los Angeles, BA, 50, MA, 51, PhD(math), 54. *Prof Exp:* Staff mem indust math several co, 51-58; sr assoc opers analysis, Planning Res Corp, 58-62; mgr opers anal dept, Hughes Aircraft Co, 62-66; mgr WCoast opers, Syst Sci Corp, Calif, 66-70; dir, Comput Sci Southeast Ctr, 70-72; prin, Planning Res Corp, 72-85; sr sci adv & tech dir, J H Wiggins Co, 79-86, consult, 75-88; CONSULT, 88- *Concurrent Pos:* Fel, Nat Bur Standards, 53-54; lectr & sr res assoc, Univ Southern Calif, 73-79. *Mem:* Syst Safety Soc; Opers Res Soc Am; Inst Elec & Electronics Engrs; Soc Risk Analysis. *Res:* Systems engineering; operations analysis; applied mathematics; differential equations; statistics; risk analysis. *Mailing Add:* 6364 W Chartres Dr Rancho Palos Verdes CA 90274

PHILLEO, ROBERT EUGENE, materials science, civil engineering; deceased, see previous edition for last biography

PHILLEY, JOHN CALVIN, b Indianola, Miss, Oct 17, 35; m 59; c 3. GEOLOGY. *Educ:* Millsaps Col, BS, 57; Univ Tenn, Knoxville, MS, 61; PhD(geol), 71. *Prof Exp:* From instr to assoc prof, 60-73, head dept phys sci, 74-86, PROF GEOL, MOREHEAD STATE UNIV, 73-, DEAN COL ARTS & SCI, 86 - *Concurrent Pos:* Geologist, US Geol Surv, 67-77; actg vpres acad affairs, Col Arts & Sci, 89- *Mem:* Am Asn Petrol Geol; Am Inst Prof Geologists. *Res:* Environmental carbonate stratigraphy; areal geologic mapping; paleozoic stratigraphy of southeastern United States. *Mailing Add:* 1001 Knapp Ave Morehead KY 40351

PHILLIES, GEORGE DAVID JOSEPH, b Buffalo, NY, July 23, 47. STATISTICAL MECHANICS, COMPLEX FLUIDS. *Educ:* Mass Inst Technol, SB(physics) & SB(biol), 69, SM, 71, DSc(physics), 73. *Prof Exp:* Res staff biophys, Harvard-Mass Inst Technol Health Sci & Technol Prog, 73-75; res chemist, Univ Calif, Los Angeles, 75-78; asst prof, dept chem, Univ Mich, Ann Arbor, 78-85; ASSOC PROF, DEPT PHYSICS, WORCESTER POLYTECH INST, MASS, 85- *Mem:* Am Phys Soc; Mat Res Soc. *Res:* Experimental and theoretical statistical mechanics, quasi-elastic light scattering, diffusion in nonideal solutions, Raman and infrared spectroscopy and membrane-active antibiotics. *Mailing Add:* Dept Physics Worcester Polytech Inst Worcester MA 01609

PHILLIP, MICHAEL J, genetics, for more information see previous edition

PHILLIPS, ALLAN ROBERT, b New York, NY, Oct 25, 14; m 64; c 3. ORNITHOLOGY. *Educ:* Cornell Univ, AB, 36 & PhD(ornith), 46; Univ Ariz, MS, 39. *Prof Exp:* Cur ornith, Mus Northern Ariz, 39-57; guest researcher, Inst Biol, Nat Univ Mex, 57-69, tech counsr, 69-71, prof ornith, Ornith Lab, 72-74; TAXON CONSULT & RES ASSOC, DENVER MUS NAT HIST, 77- *Concurrent Pos:* Asst prof, Univ Ariz, 49 & 58; res assoc, Del Mus Nat Hist, 71-77; mem, Int Ornith Comt, 70-; guest res assoc, Sch Biol Sci, Univ Nuevo León, 73-; vis prof, Univ Michoacana San Nicolás Hidalgo, 79 & 88; res assoc, Smithsonian Inst, 79-, San Diego Natural Hist Mus, 79 & Can Mus Natural Sci, 80- *Mem:* Wilson Ornith Soc; Western Field Ornithologists; Brit Ornithologists Club. *Res:* Birds of Mexico, Arizona, Central America and adjacent regions; taxonomy and migration of North and Central American birds, Tyrannidae, Loxia, oscines. *Mailing Add:* Reforma 825-A Col Chapultepec San Nicolas de los Garza Nuevo Leon Mexico

PHILLIPS, ALLEN THURMAN, b Vicksburg, Miss, Oct 30, 38; m 61; c 2. BIOCHEMISTRY, ENZYMOLOGY. *Educ:* La State Univ, BS, 60, MS, 61; Mich State Univ, PhD(biochem), 64. *Prof Exp:* Asst prof biochem, La State Univ, 64-66; assoc prof, 67-71, assoc head dept, 84-87, PROF, DEPT MOLECULAR & CELL BIOL, PA STATE UNIV, UNIVERSITY PARK, 71- *Concurrent Pos:* Ed, J Bact, 75-82; vis prof, Dept Biol Sci, Purdue Univ, 82-83; res career develop award, NIH, 72-77; mem, NIH MBC-1 Study Sect, 86-89; vis scientist, Lab Biochem Pharmacol, NIH, 73-74. *Mem:* AAAS; Am Chem Soc; Am Soc Microbiol; Am Soc Biol Chem; Int Soc Neurochemistry. *Res:* Enzymology; chemistry and metabolism of amino acids; metabolic regulatory mechanisms; neurobiochemistry. *Mailing Add:* Dept Molecular & Cell Biol Althouse Lab Pa State Univ University Park PA 16802

PHILLIPS, ALVAH H, b Buffalo, NY, Aug 21, 28; m 49; c 2. PHYSIOLOGICAL CHEMISTRY. *Educ:* Allegheny Col, BA, 49; Univ Buffalo, MA, 52; Johns Hopkins Univ, PhD(physiol chem), 57. *Prof Exp:* Res fel chem, Harvard Univ, 57-59; from instr to asst prof physiol chem, Sch Med, Johns Hopkins Univ, 59-70; assoc prof, 70- PROF BIOL AT UNIV CONN. *Mem:* Am Chem Soc. *Res:* Mechanisms of enzyme action; hormonal control of metabolism; role of lipids in biological energy transfer. *Mailing Add:* 18 Broad Oak Dr Ashford CT 06278

PHILLIPS, ALVIN B(URT), b Milwaukee, Wis, July 25, 20; m 45; c 2. CHEMICAL ENGINEERING. *Educ:* Univ Mo, BS, 42. *Prof Exp:* Chem engr, Tenn Valley Authority, 42-59, chief process eng br, 59-65, asst mgr agr & chem develop, Nat Fertilizer Develop Ctr, 65-82; RETIRED. *Mem:* Am Chem Soc. *Res:* Phosphate fertilizers; nitrogen fixation; carbonization of coal; aluminum alloys; alumnia from clay; wood sugar by hydrolysis; fertilizer technology. *Mailing Add:* Rte 7 Box 319 Florence AL 35630

PHILLIPS, ANTHONY GEORGE, b Barrow, Eng, Jan 30, 43; Can citizen. BIOPSYCHOLOGY, NEUROSCIENCE. *Educ:* Univ Western Ont, BA, 66, MA, 67, PhD(biopsychol), 70. *Prof Exp:* Res assoc neurophysiol, Fels Res Inst, 69 & biol, Calif Inst Technol, 69-70; from asst prof to assoc prof, 70-80, PROF PSYCHOL, UNIV BC, 80- *Concurrent Pos:* Vis prof, Oxford Univ, 75-76; mem psychol comt, Natural Sci & Eng Res Coun Can, 76-79, chmn, 78-79; mem behav sci comt, Med Res Coun Can, 78-85, chmn, 82-87; Killam Sr Res Scholar, Can Coun, 78 & sr fel, 85-; E W R Steacie fel, 79-80. *Mem:* Soc Neurosci; Can Col Neuropsychopharmacol; Can Asn Neurosci; fel Royal Soc Can. *Res:* Neural substrates of learning, memory and motivation; role of biogenic amines and neuropeptides in normal and pathological behavior. *Mailing Add:* Dept Psychol Univ BC Vancouver BC V6T 1W5 Can

PHILLIPS, ARTHUR PAGE, b Haverhill, Mass, Feb 14, 17; m 42, 50; c 2. ORGANIC CHEMISTRY. *Educ:* Tufts Col, BS, 38, MS, 39; NY Univ, PhD(org chem), 42. *Prof Exp:* Asst chem, NY Univ, 39-42; RES ORG CHEMIST, RES LABS, BURROUGHS WELLCOME & CO, INC, 42-, GROUP LEADER, 70- *Concurrent Pos:* Instr, Polytech Inst Brooklyn, 44-57. *Mem:* Am Chem Soc; The Chem Soc. *Res:* Synthetic and theoretical organic chemistry; medicinal chemistry; chemotherapy; heterocyclic compounds; mechanisms of reactions. *Mailing Add:* Burroughs Wellcome & Co Inc 3030 Cornwallis Rd Research Triangle Park NC 27709

PHILLIPS, ARTHUR WILLIAM, JR, b Claremont, NH, Sept 25, 15; m 50; c 2. MICROBIOLOGY, ENVIRONMENTAL SCIENCES. *Educ:* Univ Notre Dame, BS, 39, MS, 41; Mass Inst Technol, ScD, 47. *Prof Exp:* Asst, Univ Notre Dame, 36, asst bact, 39-41, res assoc, 43-45, prof bact & head bioeng, 49-53; prof microbiol, 54-, EMER PROF MICROBIOL, SYRACUSE UNIV. *Concurrent Pos:* Res assoc, Mass Inst Technol, 47-49. *Mem:* Am Soc Microbiol; Soc Gen Microbiol; Asn Gnotobiotics. *Res:* Microbiol interactions with phagocytic cells. *Mailing Add:* 115 Crawford Ave Syracuse NY 13224

PHILLIPS, BENJAMIN, b Galveston, Tex, Apr 1, 17; m 41; c 4. ORGANIC CHEMISTRY. *Educ:* Univ of the South, BS, 37; Johns Hopkins Univ, PhD(org chem), 41. *Prof Exp:* Res chemist, Chem Div, Union Carbide Corp, 41-55, asst dir res, 55-58, res consult, 58-60, assoc dir res, 60-61, dir res & develop, 61-64, mgr chem res develop dept, 64-65, mgr pharmaceut technol, 66-67, mgr corp res, 67-72, sr res fel, 72-78; CONSULT, 78- *Concurrent Pos:* Collabr south chem utilization res & develop, USDA, 56-66, mem chem utilization res & develop adv comt, 66-69. *Honors & Awards:* Chem Pioneer Award, Am Inst Chem, 67. *Mem:* Am Chem Soc; fel Am Inst Chem; Asn Consult Chemists & Chem Engrs. *Res:* Industrial organic chemistry; polymers; peracids; epoxides; high performance materials; chemical systems; chemurgy. *Mailing Add:* 2 Wesskum Wood Rd Riverside CT 06878

PHILLIPS, BOBBY MAL, b Morris Chapel, Tenn, Apr 17, 41; m 81. CHEMICAL ENGINEERING. *Educ:* Univ Tenn, BS, 62, MS, 63, PhD(chem eng), 68. *Prof Exp:* Res engr, 68-70, sr res chem engr, 70-77, res assoc, 77-86, SR RES ASSOC, EASTMAN CHEMICALS DIV, EASTMAN KODAK CO, 86- *Mem:* Am Inst Chem Engrs; Am Asn Textile Technol. *Res:* Rheology; fluid mechanics; fiber spinning and processing; futures. *Mailing Add:* Eastman Chem Div Fibers Res Lab PO Box 1972 Kingsport TN 37662

PHILLIPS, BRIAN ANTONY MORLEY, b Edinburgh, UK, Apr 10, 42; m 65; c 2. PHYSICAL GEOGRAPHY, CARTOGRAPHY. *Educ:* Univ Col Wales, BSc, 64, PhD(coastal geomorphol), 69. *Prof Exp:* ASSOC PROF GEOG & CHMN DEPT, LAKEHEAD UNIV, 67- *Concurrent Pos:* Nat Res Coun grant, Lakehead Univ, 69-72 & 78-81. *Mem:* Can Asn Geog; Am Asn Geog; Inst Brit Geog; fel Royal Geog Soc. *Res:* Present coastal processes and forms including beach stability and rock shore morphology; paleogeography of glacial lakes of Lake Superior basin; shoreline recreational potential and user perception. *Mailing Add:* Dept Geog Lakehead Univ Oliver Rd Thunder Bay ON P7B 5E1 Can

PHILLIPS, BRIAN ROSS, b Manchester, UK, Oct 6, 35; m 67. PHYSICAL CHEMISTRY. *Educ:* Univ Edinburgh, BS, 58, PhD(phys chem), 61. *Prof Exp:* Res scientist, Univ Copenhagen, 62-63; res scientist fibers, 64-82, staff asst to res & develop dir, 83-88, TECH PLANNING, DU PONT, 89- *Mem:* Am Chem Soc; Royal Soc Chem. *Res:* Chemistry and kinetics of solid state reactions; low pressure oxidation of single copper crystals; man made fibers; long range business and technology planning; research communication. *Mailing Add:* Fibers CRP Oak Run Du Pont PO Box 0701 Wilmington DE 19880-0701

PHILLIPS, BRUCE A, b Chicago, Ill, Aug 1, 38; m 63; c 2. MICROBIOLOGY, VIROLOGY. *Educ:* Univ Ill, BS, 60; Univ Mich, MS, 63, PhD(microbiol), 67. *Prof Exp:* From instr to assoc prof, 67-83, PROF MICROBIOL, SCH MED, UNIV PITTSBURGH, 85- *Concurrent Pos:* Res fel virol, Albert Einstein Col Med, 65-67; NIH grants, 68-70, 71-73, 74-78, 78-82, 84-87 & 87-88; sabbatical leave, Univ Cambridge, Eng, 79-80. *Mem:* AAAS; Am Soc Microbiol; NY Acad Sci; Am Soc Virol; Am Acad Microbiol. *Res:* Morphogenesis of polio virus. *Mailing Add:* Dept Molecular Genetics & Biochem Univ Pittsburgh Sch Med Rm E1240 Biomed Sci Tower Pittsburgh PA 15261

PHILLIPS, BRUCE EDWIN, b Olivet, Ill, May 7, 34; m 55; c 2. ORGANIC CHEMISTRY. *Educ:* Olivet Nazarene Col, AB, 56; Wash Univ, PhD(org chem), 61. *Prof Exp:* Chemist, Res Ctr, Gen Foods Corp, 61-62; asst prof chem, Eastern Nazarene Col, 62-68; chemist, Agr Res Serv, USDA, 68-70; asst prof, 70-80, ASSOC PROF CHEM, ST LOUIS COL PHARM, 80- *Concurrent Pos:* Res Corp res grant, 64; NSF res participation prog grant, 65-67. *Mem:* Am Chem Soc; AAAS; Am Asn Cols Pharm. *Res:* Stereochemistry of acetal condensations and synthetic organic chemistry; analysis of lipids from botanical sources. *Mailing Add:* St Louis Col Pharm 4588 Parkview Pl St Louis MO 63110

PHILLIPS, CARLETON JAFFREY, b Muskegon, Mich, Nov 17, 42; m 65; c 2. MAMMALOGY, ORAL BIOLOGY. *Educ:* Mich State Univ, BS, 64; Univ Kans, MA, 67, PhD(zool), 69. *Prof Exp:* Res assoc zool, Bernice P Bishop Mus, Hawaii, 64-65; res asst, State Biol Surv Kans, Univ Kans, 65-66; res assoc, Sch Med, Univ Md, Baltimore City, 66-67; res asst, Univ Kans, 67-69; aerospace biologist, Grumman Aerospace Corp, NY, 69-70; from asst prof to assoc prof, 70-80, PROF BIOL, HOFSTRA UNIV, 80-, NAT INST DENT RES FEL, 72-, DIR GRAD PROG BIOL & MEM UNIV RES COMT, 73- *Concurrent Pos:* Asst to pres, Synecology Corp, NY, 70-; chmn standing comt anat & physiol, Am Soc Mammal; res assoc, Carnegie Mus Natural Hist, 80-; assoc ed, J Mammal, 83-85, managing ed, spec publ, Am Soc Mammalogists, 86-; leader & partic, res expeds to various int locations. *Mem:* AAAS; Am Soc Mammal; fel Explor Club. *Res:* Distribution, systematics, ecology and evolution of mammals; organ system evolution; ultrastructure of salivary glands and digestive tract; evolution of cell types; digestive tract endocrine cells; neuropeptides; mtDNA in insular populations; international environmental problems; dentition. *Mailing Add:* Dept of Biol Hofstra Univ 1000 Fulton Ave Hempstead NY 11550

PHILLIPS, CAROL FENTON, b Bridgeport, Conn, Dec 17, 32; m 56; c 4. PEDIATRICS, INFECTIOUS DISEASES. *Educ:* NJ Col Women, BS, 54; Yale Univ, MD, 58. *Prof Exp:* From instr to assoc prof pediat, 68-76, asst dean admissions, 77-78, assoc dean, 78-83, ASSOC DEAN ACAD AFFAIRS, COL MED, UNIV VT, 85-, PROF PEDIAT, 76-, CHMN DEPT, 83- *Concurrent Pos:* Mem, Comt Infectious Dis, Am Acad Pediat, 85- *Mem:* Am Acad Pediat; Asn Med Sch Pediat Dept Chmn; Infectious Dis Soc; Ambulatory Pediat Asn. *Res:* Immunizations for children; influenza, rubella, respiratory syncytial virus and hemophilus influenza vaccines. *Mailing Add:* Dept Pediat Given Med Bldg Col Med Univ Vt Burlington VT 05405

PHILLIPS, CHANDLER ALLEN, b Los Angeles, Calif, Dec 21, 42. BIOMECHANICS, REHABILITATION. *Educ:* Stanford Univ, AB, 65; Univ Southern Calif, MD, 69. *Hon Degrees:* PhD, Univ Humanistic Studies, 85. *Prof Exp:* Intern med, Good Samaritan Hosp, Los Angeles, 69-70; gen med officer, 8th US Air Force Dispensary, Thailand, 70-71, res med officer, 6570th Aerospace Med Res Lab, US Air Force, 71-72; res physician biomed eng, Univ Dayton Res Inst, 72-74; from asst prof to prof eng & physiol, 75-88, PROF BIOMED & HUMAN FACTORS ENG, WRIGHT STATE UNIV, 89- *Concurrent Pos:* Chmn & mem var spec study sections, NIH, 84-; consult med, Stouder Mem Hosp, Troy, Ohio, 75-; Grantee, NIH, 78-83, NSF, 87-;

contractor, US Army, 80-84; US Vets Admin, 83-85; NASA, 85-86. *Honors & Awards:* Outstanding Eng Achievement Award, Nat Soc Prof Engrs, 84; Harry Rowe Mimno Award, Inst Elec & Electronics Engrs, 84. *Mem:* Inst Elec & Electronic Engrs; Am Heart Asn; Am Soc Biomechanics; Am Soc Eng Educ; Eng Med & Biol Soc; hon fel Am Acad Neurol & Orthop Surg, 85. *Res:* Muscle biomechanics, quantitative evaluation of cardiac and skeletal muscle; neuromuscular control systems utilizing technological sensory feedback; rehabilitation engineering (high-technology rehabilitation); biomedical engineering; design of medical devices. *Mailing Add:* Dept Biomed Eng Wright State Univ Dayton OH 45435

PHILLIPS, CHARLES W(ILLIAM), metal cutting; deceased, see previous edition for last biography

PHILLIPS, DAVID, b Marion, Ind, Oct 10, 36; m 58; c 3. PSYCHOPHYSIOLOGY. *Educ:* Wabash Col, BS, 58; Purdue Univ, MS, 60, PhD(olfaction), 62. *Prof Exp:* Cardiovasc grant, 62-63, from instr to asst prof, 63-67, assoc prof psychol, Med Sch, 67-78, PROF & CHIEF BIOSTATIST SECT, ORE HEALTH SCI UNIV, 78- *Concurrent Pos:* NIH fel, 63-65. *Mem:* Am Psychol Asn; Am Statist Asn; Psychonomic Asn; AAAS; Am Pub Health Asn. *Mailing Add:* Dept Pub Health & Prev Med L352 Ore Health Sci Univ Portland OR 97201

PHILLIPS, DAVID BERRY, b Highland Co, Ohio, May 30, 40. PHYSICAL CHEMISTRY. *Educ:* Miami Univ, Ohio, AB, 61; Univ Calif, Berkeley, PhD(phys chem), 65. *Prof Exp:* Asst prof, 65-75, ASSOC PROF CHEM, MIAMI UNIV, 75- *Mem:* Am Chem Soc; Am Phys Soc. *Res:* Electrical and magnetic properties of metals at high pressures. *Mailing Add:* Dept of Chem Miami Univ Oxford OH 45056

PHILLIPS, DAVID COLIN, b Rhosllanerchrugog, Wales, Sept 6, 40; m 68; c 2. CHEMISTRY. *Educ:* Univ Wales, BSc, 65, PhD(phys chem), 68. *Prof Exp:* Sr scientist chem, 68-76, mgr chem reaction dynamics, 76-80, MGR CHEM PHYSICS, WESTINGHOUSE RES LABS, 80- *Concurrent Pos:* Mem fac, Univ Wales, Aberystwyth, 65-68. *Honors & Awards:* Indust Res -100 Award, Indust Res Mag, 76. *Mem:* Am Chem Soc; Electrochem Soc; Royal Inst Chem. *Res:* Physical chemistry; gas kinetics; polymer chemistry; laser chemistry; electropolymerization; photopolymerization, particulate chemistry. *Mailing Add:* Greenbank Church St Rhos Wrexham Clwyd LL14 2BP England

PHILLIPS, DAVID LOWELL, b Rushville, Ind, May 21, 29; m 52; c 2. MATHEMATICS. *Educ:* Ball State Teachers Col, BS, 51; Purdue Univ, MS, 52, PhD(math), 57. *Prof Exp:* Asst math, Purdue Univ, 52-56; asst mathematician, 57-60, assoc mathematician, Argonne Nat Lab, 60-71; from asst prof to assoc prof, 71-81, PROF, UNIV SOUTHERN COLO, 81- *Mem:* Am Math Soc; Soc Indust & Appl Math; Math Asn Am; Sigma Xi. *Res:* Numerical analysis; matrix algebra; Hilbert space. *Mailing Add:* 4033 Hillside Dr Pueblo CO 81008-1749

PHILLIPS, DAVID MANN, b Cincinnati, Ohio, Apr 25, 38; m 64; c 4. CELL BIOLOGY. *Educ:* Northeastern Univ, BS, 61; Univ Chicago, PhD(zool), 66. *Prof Exp:* NIH fel, Harvard Med Sch, 66-68; asst prof biol, Wash Univ, 68-73; STAFF SCIENTIST POP COUN, ROCKEFELLER UNIV, 73- *Mem:* AAAS; Am Soc Cell Biol; Soc Study Reproduction. *Res:* Sperm development, structure and motility. *Mailing Add:* Pop Coun 1230 York Ave New York NY 10021

PHILLIPS, DAVID RICHARD, b Turlock, Calif, Feb 15, 42; m 61; c 2. BIOCHEMISTRY. *Educ:* Univ Calif, Los Angeles, BS, 64; Univ Southern Calif, PhD(biochem), 69. *Prof Exp:* Asst mem, St Jude Children's Res Hosp, 72-74; Roche Res Inst fel, Theodor Kocher Inst, Univ Bern, Switz, 74-75; assoc mem, St Jude Children's Res Hosp, Memphis, 75-80; MEM FAC, UNIV CALIF, SAN FRANCISCO, 80-, SR SCIENTIST, GLADSTONE RES FOUND CARDIOVASC DIS. *Concurrent Pos:* USPHS fel, St Jude Children's Res Hosp, 70-72; USPHS grant, 74-77, 77-80, 78-83; res career develop award, 75-80. *Mem:* Am Chem Soc; Am Soc Biol Chemists; Am Soc Hematol; Am Soc Cell Biol. *Res:* Membrane structure and function; mechanism of platelet aggregation. *Mailing Add:* Core Therapeut 256 E Grand Ave Suite 5180 South San Francisco CA 94080

PHILLIPS, DAVID T, b Charles City, Iowa, Oct 9, 38; m 63; c 2. PHYSICS. *Educ:* Iowa State Univ, BS, 61; Univ Calif, Berkeley, PhD, 66. *Prof Exp:* Physicist, Radiation Div, Naval Res Lab, DC, 60-61; asst prof physics, Univ Calif, Santa Barbara, 66-71, asst res scientist, 71-74; chief scientist, Sci Spectrum, Inc, Santa Barbara, 75-76, vpres, 77-78; lectr parapsychol, Univ Calif, Santa Barbara, 78 & mech eng, 79-80; sr staff engr, Renco Corp, Goleta, Calif, 80-85; pres, The Info Connection, 79-85; vpres, R & D Wyatt Technol, Santa Barbara, 86-87; PRES, GLENDAN CO, 71- *Concurrent Pos:* Technol staff, Gen Res Corp, Santa Barbara, Calif, 87-; lectr parapsychol, Univ Calif Exten, Santa Barbara, 75. *Mem:* Am Phys Soc; Parapsychol Asn. *Res:* Lasers; quantum optics; electronics; light scattering; instrumentation; bacterial bio-assay; parapsychology; instrumentation; electronic design; computer applications. *Mailing Add:* 5107 Calle Asilo Santa Barbara CA 93111

PHILLIPS, DAVID WILLIAM, b Chicago, Ill, Aug 17, 46; m 67, 81. MARINE BIOLOGY. *Educ:* Univ Wash, BS, 68; Stanford Univ, PhD(biol sci), 74. *Prof Exp:* Asst prof zool, Univ Calif, Davis, 74-81; CONSULT, 81- *Concurrent Pos:* Ed scientific J, The Villager, 85- *Mem:* Am Soc Zoologists; AAAS; Western Soc Naturalists. *Res:* Behavioral, ecological and physiological aspects of marine invertebrate prey-predator interactions; interactions involving mollusks and chemoreception. *Mailing Add:* 2410 Dakenshield Rd Davis CA 95616

PHILLIPS, DON IRWIN, b Kansas City, Mo, Nov 17, 45. SCIENCE POLICY, EDUCATION POLICY. *Educ:* Univ Mich, BS, 67; Harvard Univ, MA, 69, PhD(chem), 72. *Prof Exp:* Res fel chem, Harvard Univ, 73; staff assoc & proj dir sci educ, AAAS, 73-75; asst coordr aging & res admin, Ctr Aging, Univ Md, College Park, 74-75; educ policy prog fel & spec asst to assoc dir educ instnl res, Nat Inst Educ, 75-76; pub policy prog mgr, AAAS, 77-79; assoc dir, Sci & Pub Affairs, Duke Univ, Durham, 79-82, dir, NC Biotechnol Ctr, 82-83; EXEC DIR, GOV, RES ROUND TABLE, UNIV IND, 84- *Concurrent Pos:* Spec sci adv, Bd Sci & Technol, Off Gov, 79-82; consult, Off Environ Educ, 74, Fund for Improv PostSecondary Educ, 76-77, Nat Inst Educ, 78-80, NSF, 79-81, Nat Acad Aci, 83; treas, Intersoc Liason Comt, Environ, State Legis, 79-82, Comt Sci & Pub Affairs, Am Chem Soc, 79- & Sect X Comt, AAAS, 90- *Mem:* Am Chem Soc; AAAS; NY Acad Sci. *Res:* Federal and private research and development budgets and policies; relationship between academic science policy studies and government science policy; science in state and local governments. *Mailing Add:* 3733 Jocelyn St NW Washington DC 20015

PHILLIPS, DON T, b Bessemer, Ala, Feb 6, 42; m 66; c 2. OPERATIONS RESEARCH, COMPUTER INTEGRATED MANUFACTURING. *Educ:* Lamar State Col Technol, BS, 65; Univ Ark, Fayetteville, MS, 67, PhD(opers res, indust eng), 69. *Prof Exp:* Instr, Univ Ark, Fayetteville, 65-69; asst prof opers res & indust eng, Univ Tex, Austin, 69-71; assoc prof indust & systs eng, Purdue Univ, West Lafayette, 71-75; PROF INDUST ENG, TEX A&M UNIV, COLLEGE STATION, 75- *Honors & Awards:* David F Baker Distinguished Research Award, 89. *Mem:* Opers Res Soc Am; Am Inst Indust Engrs; Computer & Automated Systs Asn-Soc Mfg Eng. *Res:* Queueing theory; simulation analysis; nonlinear programming; manufacturing systems; network flow theory. *Mailing Add:* Dept Indust Eng Tex A&M Univ College Station TX 77843

PHILLIPS, DONALD ARTHUR, b Olean, NY, Apr 7, 45; m 68; c 1. PLANT PHYSIOLOGY. *Educ:* Duke Univ, BS, 67; Harvard Univ, PhD(biol), 71. *Prof Exp:* Asst prof life sci, Ind State Univ, 72-76; from asst prof to assoc prof, 76-81, chmn, Dept Agron & Range Sci, 87-88, PROF AGRON, UNIV CALIF, DAVIS, 81- *Concurrent Pos:* NSF fel, 71-72; Woodrow Wilson fel; Sr Fulbright fel. *Mem:* Am Soc Plant Physiol; fel Am Soc Agron; fel Crop Sci Soc Am; AAAS. *Res:* Symbiotic nitrogen fixation in legumes; plant-microbe interactions. *Mailing Add:* Dept Agron & Range Sci Univ Calif Davis CA 95616

PHILLIPS, DONALD DAVID, b Los Angeles, Calif, Apr 12, 26; m 50; c 4. ORGANIC CHEMISTRY. *Educ:* Univ Alta, BS, 49; Univ Calif, PhD(chem), 52. *Prof Exp:* Asst chem, Univ Calif, 49-51; from instr to asst prof chem, Cornell Univ, 52-58; chemist, Shell Develop Co, 58-62, supvr, 62-63; mgr cent res lab, Mobil Chem Co, 63-67, com develop, 67-71, mgr Edison Res & Develop Labs, 71-78, gen mgr, res & develop Mobil Chem Co, Edison, 78-87; RETIRED. *Concurrent Pos:* Alfred P Sloan fel, 55-58; chmn bd trustees, Plastics Inst Am, 72-73. *Mem:* Am Chem Soc; Sigma Xi; Indust Res Inst. *Res:* Petrochemicals; agricultural chemicals; polymer chemistry of natural products and polynuclear aromatic hydrocarbons. *Mailing Add:* 2486 Unicornio St Carlsbad CA 92009-5324

PHILLIPS, DONALD HERMAN, b Knoxville, Tenn, Mar 10, 41; m 59, 73; c 4. PHYSICS, PHYSICAL CHEMISTRY. *Educ:* Tenn Technol Univ, BS, 63; Va Polytech Inst, MS, 68, PhD(physics), 71. *Prof Exp:* SR RES SCIENTIST PHYSICS, NASA, 63- *Concurrent Pos:* Vis scholar, Appl Sci Div, Harvard Univ, 80-81; Floyd Thompson fel, 80-81. *Mem:* Am Phys Soc; Am Chem Soc. *Res:* Theoretical, atmospheric and surface chemistry; spectroscopy. *Mailing Add:* Langley Res Ctr NASA Hampton VA 23665

PHILLIPS, DONALD KENNEY, b Newark, Del, Apr 24, 31; m 71; c 2. ORGANIC CHEMISTRY. *Educ:* Univ Del, BS, 53; Ohio State Univ, PhD(org chem), 58. *Prof Exp:* Proj assoc chem, Univ Wis, 58-60; res chemist, Sterling-Winthrop Res Inst, 60-65, group leader, 65-68, sect head med chem, 68-79, head chem prog mgt, 79- 83, sect head anal chem, Sterling-Winthrop Res Inst, 83-89, proj resource mgr, 89-90, ASST RES DIR, ANAL SCI DEPT, STERLING RES GROUP, STERLING DRUG INC, 90- *Mem:* Am Chem Soc; Sigma Xi. *Res:* Medicinal chemistry; cardiovascular, lipid-lowering, anti-inflammatory and adrenergic agents; regulatory submissions; resource management. *Mailing Add:* Sterling Drug Inc 81 Columbia Turnpike Rensselaer NY 12144

PHILLIPS, DONALD LUNDAHL, b Wilmington, Del, July 15, 52; m 78; c 2. PLANT ECOLOGY. *Educ:* Mich State Univ, BS, 74; Utah State Univ, MS, 77, PhD(biol), 78. *Prof Exp:* Asst prof, Emory Univ, 78-83; biostatician, Ctr Environ Health, Ctr Dis Control, 83-88; RES BIOLOGIST, ENVIRON RES LAB, ENVIRON PROTECTION AGENCY, 88- *Concurrent Pos:* Co-prin investr, NSF grant, 80- & US Forest Serv grant, 80-; ecol consult, Haday Corp, 80 & Oak Ridge Nat Lab, 82. *Mem:* AAAS; Ecol Soc Am. *Res:* Evaluation of environmental contamination, human exposure and health effects; community ecology and mathematical ecology; ecological effects of man-made and natural disturbances; ecological succession; statistics in ecology and environmental health; global climate changes. *Mailing Add:* Environ Res Lab Environ Protection Agency Corvallis OR 97333

PHILLIPS, DOUGLAS J, b Franklin Co, Iowa, Dec 11, 31; m 56; c 3. PLANT PATHOLOGY. *Educ:* Colo State Univ, BS, 58, MS, 60, PhD(bot sci), 68. *Prof Exp:* Jr plant pathologist, Dept of Bot, Colo State Univ, 60-68; asst res plant path, Univ Calif, Berkeley, 68-70; RES PLANT PATHOLOGIST, AGR RES SERV, USDA, 70- *Mem:* Am Phytopathological Soc. *Res:* Post-harvest diseases of fruits and vegetables. *Mailing Add:* 2021 S Peach Fresno CA 93727

PHILLIPS, DWIGHT EDWARD, b Lewistown, Mont, Aug 16, 44; m 67; c 2. ANATOMY. *Educ:* Univ Mont, BA, 66; Tulane Univ, PhD(anat). *Prof Exp:* Asst prof anat, Sch Med, Univ SDak, 70-73; asst prof, 73-76; assoc prof human anat, 76-84, PROF ANAT, DEPT BIOL, MONT STATE UNIV, 84- *Mem:* Am Asn Anatomists; AAAS; Sigma Xi; Soc Neuroscience. *Res:* Development of the central nervous system; function of glial cells in mammalian central nervous system. *Mailing Add:* Dept of Biol Mont State Univ Bozeman MT 59715

PHILLIPS, E ALAN, b Boston, Mass, Oct 28, 37. OPTICS, NUCLEAR PHYSICS. *Educ:* Mass Inst Technol, BS, 57, PhD(physics), 61. *Prof Exp:* Instr physics, Mass Inst Technol, 61-62; from instr to asst prof, Princeton Univ, 62-69; physicist, Lawrence Berkeley Lab, 69-71; prin res scientist, Avco Everett Res Lab, Inc, 71-75; SCIENTIST, SCI APPLN INT CORP, 75- *Mem:* Am Phys Soc; Soc Photo-Optical Instrumentation Engrs; Sigma Xi; AAAS. *Res:* High energy lasers; diffraction; thermal blooming; atmospheric optics; nuclear moments. *Mailing Add:* 990 Leigh Mill Rd Great Falls VA 22066

PHILLIPS, EDWARD, b Roselle, NJ, Aug 21, 26; m 49; c 4. MOLECULAR PHYSICS. *Educ:* US Naval Postgrad Sch, BS, 62; Univ RI, MS, 65; Univ Southern Calif, PhD(physics), 75. *Prof Exp:* Served US Navy, 44-70; res asst, Univ Southern Calif, 71-74, res staff physicist, 75-77, asst prof physics, 77-80; staff mem, Litton Guidance & Control Systs Div, 81-91. *Res:* Study of the photodissociation process in simple atmospheric molecules through observation of fluorescence from photodissociation fragments. *Mailing Add:* 21820 Marylee St Apt 222 Woodland Hills CA 91367

PHILLIPS, EDWIN ALLEN, b Lowell, Fla, Mar 18, 15; m 42; c 2. PLANT ECOLOGY. *Educ:* Colgate Univ, AB, 37; Univ Mich, MA, 41, PhD(bot), 48. *Prof Exp:* Teacher high sch, NY, 37-39; instr bot, Colgate Univ, 46-48; from instr to prof bot, 48-80, chmn dept, 73-77, EMER PROF BOT, POMONA COL, 80- *Concurrent Pos:* Mem exped, Great Bear Lake, Northwest Territories, 48; vis prof ecol, Univ Mich Biol Sta, 55, 56, 58, 70 & 71; NSF sci fac fel, Oxford Univ, 61-62; consult, Agency Int Develop, India, 64-65, Indonesia, 72; consult, Biol Sci Curric Study, 61; NSF grant for physiol-ecol studies on Calif chaparral; Atomic Energy Comn grant for tracer studies in mosses, 68-; Schenck res fund for bot, 68-80; consult fundamental approaches sci teaching study, Univ Hawaii, 68-, contextual approaches to secondary educ, 83-; writer, comput progs vegetation data, 84-; lectr ecol, Nanjing Univ, People's Rep China. *Mem:* Fel AAAS; Bot Soc Am; Sigma Xi; Ecol Soc Am. *Res:* Physiological ecology; desert and chaparral; bryophytes; ecological and general biology textbooks; succession studies on Michigan forests; computer programming for vegetation studies using APL, a computer language; study of forests of Hill Tribes of northern Thailand. *Mailing Add:* Thille Bot Labs Pomona Col Claremont CA 91711

PHILLIPS, ESTHER RODLITZ, b Brooklyn, NY, May 27, 33; m 67. COMPUTER SCIENCE, MATHEMATICS. *Educ:* Brooklyn Col, BS, 55; NY Univ, MA, 56, PhD(math), 60. *Prof Exp:* Assoc prof math, NY Univ, 60-67; assoc prof, 67-78, PROF MATH, LEHMAN COL, 78- *Concurrent Pos:* Fulbright award, Rome, Italy, 59-60; managing ed, Historia Mathematica. *Mem:* Am Math Soc; Math Asn Am; Hist Sci Soc; Asn Comput Mach. *Res:* History of mathematics; modern theories of the integral; Russian mathematics; The Moscow School of the Theory of Functions. *Mailing Add:* 262 Central Park W Apt 8D New York NY 10024

PHILLIPS, FRED MELVILLE, b Bishop, Calif, May 12, 54; m 75; c 2. ISOTOPE HYDROLOGY, PALEOCLIMATOLOGY. *Educ:* Univ Calif, Santa Cruz, BA 76; Univ Ariz, MS, 79, PhD(hydrol), 81. *Prof Exp:* Asst prof, 81-86, ASSOC PROF HYDROL, NMEX INST MINING & TECHNOL, 86- *Honors & Awards:* F W Clarke Medal, Geochem Soc, 88. *Mem:* Am Geophys Union; Geol Soc Am; Sigma Xi; Geochem Soc; Am Quaternary Asn. *Res:* Isotope hydrology; applications of isotopic methods to dating of ground water and to paleoclimatic reconstruction. *Mailing Add:* Geosci Dept NMex Inst Mining & Technol Socorro NM 87801

PHILLIPS, GARY WILSON, b Golden City, Mo, June 11, 40; m 63; c 3. NUCLEAR SCIENCE, RADIATION PHYSICS. *Educ:* Mass Inst Technol, BS, 62; Univ Md, PhD(nuclear physics), 67. *Prof Exp:* Res assoc physics, Univ Wash, 66-69; res scientist assoc, Ctr Nuclear Studies, Univ Tex, Austin, 69-71; scientist, Geonuclear Dept, Teledyne Isotopes, Inc, 71-73; res physicist, Condensed Matter & Radiation Sci Div, 73-84, HEAD, RADIATION DETECTION SECT, RADIATION EFFECTS BR, CONDENSED MATTER & RADIATION DETECTION DIV, NAVAL RES LAB, 84- *Mem:* Am Phys Soc; Sigma Xi. *Res:* Nuclear radiation detection; nuclear materials; nuclear instrumentation and analysis; applied radiation physics. *Mailing Add:* Code 4616 Condensed Matter Radiation Sci Div Naval Res Lab Washington DC 20375-5000

PHILLIPS, GEORGE DOUGLAS, b Stoke-on-Trent, Eng, Dec 3, 28; m 55; c 3. VETERINARY PHYSIOLOGY. *Educ:* Univ Liverpool, BVSc & MRCVS, 52, BSc, 54, PhD, 68. *Prof Exp:* Res student ruminant physiol, Univ Liverpool & Rowett Inst, Aberdeen, Scotland, 54-56; vet res off physiol, EAfrican Vet Res Orgn, 56-60; assoc prof, 60-70, PROF PHYSIOL, UNIV MAN, 70-, DEPT HEAD, 80- *Mem:* Brit Vet Asn; Can Vet Med Asn; Can Soc Animal Sci; Sigma Xi. *Res:* Physiology of digestion and acid-base control in cattle and sheep; electrolyte metabolism in ruminants, including intestinal absorption and renal excretion. *Mailing Add:* CAB Int Centre Wallingford Oxon OX10 8DE England

PHILLIPS, GEORGE WYGANT, JR, b New York, NY, Aug 5, 29; m 58; c 2. PHYSICAL CHEMISTRY. *Educ:* Wesleyan Univ, BA, 51; Harvard Univ, PhD(chem), 57. *Prof Exp:* Asst, Ames Lab, Atomic Energy Comn, 51-53; res chemist, Air Res & Develop Command, US Air Force, 53; res scientist, Union Carbide Plastics Co, 57-62, licensing mgr, Union Carbide Int Co, 62-65, mgr tech rels group, Union Carbide Europa SA, Switz, 65-68, acct mgr, Chem & Plastics Div, 68-72, prod mgr, Chem & Plastics Div, 72-76, COORDR INT TRADE AFFAIRS, UNION CARBIDE CORP, 76- *Mem:* Sigma Xi; NY Acad Sci; Am Chem Soc. *Res:* Physical chemistry and structure of polymers; mechanism of polymerization; radiochemistry; patent licensing. *Mailing Add:* 94 Deer Hill Dr Ridgefield CT 06877

PHILLIPS, GERALD B, b Bethlehem, Pa, Mar 20, 25; m 70; c 2. MEDICINE. *Educ:* Princeton Univ, AB, 48; Harvard Med Sch, MD, 48; Am Bd Internal Med, dipl. *Prof Exp:* Intern med, Presby Hosp, New York, 48-50; res fel, Harvard Med Sch, 50-52, res assoc, 52-53; vis fel biochem, 54-56, from assoc to assoc prof, 56-73, PROF MED, COL PHYSICIANS & SURGEONS, COLUMBIA UNIV, 73- *Concurrent Pos:* Sr asst sergeon, Nat Inst Arthritis & Metab Dis, 52-54; Lederle Med Fac award, 63-66. *Mem:* AAAS; Soc Exp Biol & Med; Am Fedn Clin Res; Am Soc Clin Invest; Am Soc Biol Chemists. *Res:* Aging; atherosclerosis; hormones; nutrition; diabetes; liver disease; lipids; relationship of sex hormones to disorders of Western societies. *Mailing Add:* 428 W 59th St New York NY 10019

PHILLIPS, GERALD C, b Plainview, Tex, Feb 27, 22. PHYSICS. *Educ:* Rice Inst, AB, 44, AM, 46, PhD(physics), 49. *Prof Exp:* From instr to prof physics, Rice Univ, 49-88, chmn dept, 60-66, dir, Bonner Nuclear Labs, 60-87, EMER PROF PHYSICS, RICE UNIV, 88-; PRES & CHMN BD, COLUMBIA SCI INDUSTS, INC, 87- *Concurrent Pos:* Fel, Carnegie Inst, 50-51; Guggenheim fel, 57-58; NSF sr fel, 66-67; mem earth sci div, Nat Res Coun, 58-61; mem coun, Oak Ridge Asn Univs, 60-76; mem nuclear cross sect adv group, Atomic Energy Comn, 61-76; mem subcomt nuclear struct, Nat Acad Sci-Nat Res Coun, 70-76, nuclear sci, 68-; mem physics adv panel, NSF, 63-66; mem comt physics fac in cols, Am Inst Physics, 63-; sr vpres, Columbia Sci Industs, 68-; chmn sci comt, Univs Res Asn, 70-76; mem, policy bd, Meson Physics Facil, Los Alamos, 70-73 & chmn, users group, 71; chmn, Div Nuclear Physicists, Am Phys Soc, 71-72; mem energy res working group, Off Sci & Technol Policy, 77-78 & proj mgt group, Tex Accelerator Ctr, 83-87. *Mem:* Fel Am Phys Soc; Am Geophys Union; AAAS. *Res:* Nuclear and particle physics; geophysics. *Mailing Add:* Columbia Sci Industs Inc PO Box 203190 Austin TX 78720

PHILLIPS, GRACE BRIGGS, b Mobile, Ala, Apr 15, 23; m 51; c 3. MICROBIOLOGY. *Educ:* Univ Md, BS, 54; NY Univ, PhD, 65. *Prof Exp:* Bacteriologist, US Army Biol Labs, 44-63, asst dir indust health & safety, 63-66; microbiologist, Off of Chief, Commun Dis Ctr, USPHS, Tex, 66-69; dir, Becton-Dickinson Res Ctr, 69-76; vpres for sci & tech affairs, Health Indust Mfrs Asn, 76-84; dir qual assurance, Becton Dickinson & Co, 84-88; PRES, PETRA INC, 88- *Concurrent Pos:* Secy of Army res & study fel, 59-60. *Mem:* Fel Am Pub Health Asn; Am Soc Microbiol; fel Am Acad Microbiologists; Am Asn Contamination Control (past pres); Sigma Xi. *Res:* Microbiological safety and contamination control; microbiological containment system technology; design of containment facilities; industrial sterilization; medical device research; planetary quarantine and protection; air filtration and sterilization; gaseous decontaminants; germicidal ultraviolet radiation; animal cross-infection; microbial sampling of air and surfaces; industrial research and quality assurance. *Mailing Add:* Petra Inc PO Box 1054 St Michaels MD 21663

PHILLIPS, GREGORY CONRAD, b Covington, Ky, Sept 11, 54; m 75; c 2. PLANT CELL GENETICS, PLANT DEVELOPMENTAL BIOLOGY. *Educ:* Univ Ky, Lexington, BA, 75, PhD(crop sci), 81. *Prof Exp:* Asst prof, 81-85, ASSOC PROF PLANT CELL GENETICS, NMEX STATE UNIV, 85-, ASSOC DIR PLANT GENETIC ENG, 86- *Concurrent Pos:* Interim dir, Plant Genetic Eng Lab, NMex State Univ, 83-85; ed, Plant Cell Reports, 85- *Mem:* Am Soc Agron; Crop Sci Soc Am; Am Soc Hort Sci; Tissue Cult Asn Am; Int Asn Plant Tissue Cult. *Res:* Plant cellular totipotency mechanisms; applications of cell genetic approaches to crop plant improvement. *Mailing Add:* Dept Agron NMex State Univ Las Cruces NM 88003-0003

PHILLIPS, GUY FRANK, b Hartwell, Ga, May 24, 23; m 46; c 2. BEVERAGE RESEARCH. *Educ:* Emory Univ, AB, 48; Univ Tenn, MS, 49. *Prof Exp:* Chemist beverages, Nugrape Co, Atlanta, 50-57 & Royal Crown Cola, Columbus, Ga, 57-64; VPRES RES BEVERAGES, DR PEPPER CO, DALLAS, 64- *Concurrent Pos:* Dir, Inst Food Technologist, 78-81. *Mem:* Am Chem Soc; Am Water Works Asn; Soc Am Microbiologists; Soc Soft Drink Technol (pres, 81-82); Inst Food Technologists; Soc Am Qual Control. *Res:* Quality control in soft drinks. *Mailing Add:* 7538 Woodstone Lane Dallas TX 75248

PHILLIPS, HUGH JEFFERSON, physiology, pharmacology; deceased, see previous edition for last biography

PHILLIPS, JACOB ROBINSON, b Newport, Ark, June 24, 29; m 52; c 2. ENTOMOLOGY. *Educ:* Univ Ark, BSA, 52, MS, 61; La State Univ, PhD(entom), 65. *Prof Exp:* From asst prof to prof, 64-85, UNIV PROF ENTOM, UNIV ARK, FAYETTEVILLE, 85- *Honors & Awards:* Nat Award for Agr Excellence, 80; Mobay Cotton Achievement Award, 91. *Mem:* Entom Soc Am; Sigma Xi. *Res:* Insect biology and ecology; development of community-wide cotton insect management systems: an extension of integrated pest management to the inclusion of crop production systems that effect cotton insect population biology and dynamics. *Mailing Add:* Dept Entom A 320 Univ Ark Fayetteville AR 72701

PHILLIPS, JAMES A, b Johannesburg, SAfrica, May 17, 19; US citizen; m 48; c 3. PLASMA PHYSICS. *Educ:* Carleton Col, BA, 42; Univ Ill, MS, 43, PhD(physics), 49. *Prof Exp:* Tech asst, Tenn Eastman Corp, 44-45; asst, Univ Ill, 48-49; mem staff, 49-56, GROUP LEADER PHYSICS DIV, LOS ALAMOS NAT LAB, 56-, LAB FEL, 82- *Concurrent Pos:* Head physics sect, Int Atomic Energy Agency, Vienna, Austria, 75-79. *Mem:* Fel Am Phys Soc. *Mailing Add:* 48 Del Loma Escolar Los Alamos NM 87544

PHILLIPS, JAMES CHARLES, b New Orleans, La, Mar 9, 33. SEMICONDUCTORS, SUPERCONDUCTORS. *Educ:* Univ Chicago, BS, 53, PhD(physics), 56. *Prof Exp:* Mem tech staff physics, AT&T Bell Labs, 56-58; NSF fel physics, Univ Calif, 58-59, Cambridge Univ, 59-60, 62-63 & 66-67; from asst prof to prof physics, Univ Chicago, 60-68; MEM TECH SERV, AT&T BELL LABS, 69- *Honors & Awards:* Backley Prize, Am Phys Soc, 72. *Mem:* Nat Acad Sci. *Res:* Atomic and electronic structure of crystals, glasses and liquids; bonding and dielectric properties of semiconductors bonding mechanisms and properties of high temperature superconductors; author of three books. *Mailing Add:* AT&T Bell Labs Murray Hill NJ 07974-2070

PHILLIPS, JAMES M, b Fairfield, Ill, July 9, 34; m 54; c 2. PHYSICS. *Educ:* Cent Mo State Col, BS, 56; Univ Mo, Rolla, PhD(physics), 66. *Prof Exp:* Teacher high sch, Mo, 56-58; instr math, Univ Mo, Rolla, 58-61; asst prof, 65-69, assoc prof, 69-77, PROF PHYSICS, UNIV MO-KANSAS CITY, 77- *Mem:* Am Phys Soc; Am Asn Physics Teachers; Am Vacuum Soc. *Res:* Theory of liquids, especially structure of a cell-model liquid; surface physics, including theory of field ionization and binding energy of lamellar solids. *Mailing Add:* 2 W 54th St Kansas City MO 64112

PHILLIPS, JAMES W, b Terre Haute, Ind, Jan 12, 30; m 56; c 3. INSTRUMENT DESIGN. *Educ:* Rose-Hulman Inst Technol, BSME, 51; Univ Notre Dame, MSME, 64. *Prof Exp:* Proj engr, Bendix, South Bend, Ind, 51-60; chief engr, Robertshan Controls, Goshen, Ind, 60-66; VPRES ENGR, DWYER INSTRUMENTS INC, MICHIGAN CITY, IND, 66- *Mem:* Instrument Soc Am; Am Soc Testing & Mat. *Res:* Development of instruments used in heating, air conditioning and industrial control and monitoring systems; medical and general laboratory measurements of pressure and liquid flow; 25 US patents and 35 foreign patents. *Mailing Add:* 3239 Cleveland Ave Michigan City IN 46360

PHILLIPS, JAMES WOODWARD, b Washington, DC, Mar 8, 43; m 65; c 4. ENGINEERING MECHANICS. *Educ:* Cath Univ Am, BME, 64; Brown Univ, ScM, 66, PhD(eng), 69. *Prof Exp:* Res asst eng, Brown Univ, 65-69, res assoc, 69; from asst prof to assoc prof, 69-81, PROF THEORET & APPL MECH, UNIV ILL, URBANA, 81-, ASSOC HEAD, 85- *Concurrent Pos:* Ed, Mechanics, Am Acad Mech, 75-78; vis assoc prof mech eng, Univ Md, 78-79. *Honors & Awards:* Exp Tech Award, Soc Exp Mech, 87. *Mem:* Soc Exp Stress Anal; Am Acad Mech; Sigma Xi. *Res:* Theoretical and experimental stress pulse propagation in solids; stress analysis; computer graphics. *Mailing Add:* Dept Theoret & Appl Mech 216 Talbot Lab Univ of Ill 104 S Wright St Urbana IL 61801

PHILLIPS, JEAN ALLEN, b Knoxville, Tenn, Dec 10, 18; m 43. FOOD SCIENCE, NUTRITION. *Educ:* Univ Tenn, BSHE, 40, MS, 55; Purdue Univ, PhD(food sci), 66. *Prof Exp:* Teacher high sch, WVa, 40-42; instr foods & inst mgt, Univ Tenn, 54-58; instr foods & nutrit, Purdue Univ, 58-63, asst prof, 65-66; assoc prof, Univ Tenn, Martin, 66-69; from assoc prof to prof, 69-85, EMER PROF FOODS & NUTRIT, VA POLYTECH INST & STATE UNIV, 85- *Mem:* Am Home Econ Asn; Am Dietetic Asn; Inst Food Technol; Sigma Xi. *Res:* Cereal foods; lipids in flour products. *Mailing Add:* 508 South Gate Dr Blacksburg VA 24060

PHILLIPS, JERRY CLYDE, b McKenzie, Tenn, Sept 23, 35; m 59; c 3. RADIOLOGY. *Educ:* Memphis State Univ, BA, 57; Univ Tenn, Memphis, MD, 60. *Prof Exp:* From instr to assoc prof, 65-72, PROF RADIOL, COL MED, UNIV TENN, MEMPHIS, 72- *Mem:* Radiol Soc NAm; Am Col Radiol; AMA; Am Roentgen Ray Soc; Soc Gastrointestinal Radiol. *Res:* Diagnostic radiology, especially gastrointestinal radiology. *Mailing Add:* Univ Tenn Memphis 865 Jefferson Ave Memphis TN 38163

PHILLIPS, JOHN EDWARD, b Montreal, Que, Dec 20, 34; m 56; c 5. COMPARATIVE PHYSIOLOGY. *Educ:* Dalhousie Univ, BSc, 56, MSc, 57; Cambridge Univ, PhD(zool), 61. *Prof Exp:* Asst prof zool, Dalhousie Univ, 60-64; from asst prof to assoc prof, 64-71, PROF ZOOL, UNIV BC, 71- *Concurrent Pos:* Chmn animal biol grant selection comt, Nat Res Coun Can, 69-71; Killam sr fel, 80; mem, Natural Sci Eng Res Coun Can, 83-87. *Honors & Awards:* Killam Res Prize, 89-90. *Mem:* Am Physiol Soc; fel Royal Soc Can; Am Soc Zool; Can Soc Zool (pres, 78-79); Int Union Biol Sci. *Res:* Transport across biological membranes, particularly rectal absorption and renal function in insects and their hormonal control; phosphorus cycle in lakes. *Mailing Add:* Dept Zool Univ BC Vancouver BC V6T 2A9 Can

PHILLIPS, JOHN HOWELL, JR, b Fresno, Calif, Dec 19, 25; m 84; c 3. BIOCHEMISTRY, BACTERIOLOGY. *Educ:* Univ Calif, Berkeley, AB, 49, MA, 54, PhD(bact), 55. *Prof Exp:* Am Cancer Soc fel comp immunol, Hopkins Marine Sta, Stanford Univ, 55-56; Waksman Merch fel immunochem, Inst Microbiol, Rutgers Univ, 56-57; from instr to asst prof immunochemistry & microbiol, Univ Calif, Berkeley, 57-62; from asst prof to prof, 62-80, dir, 65-75, EMER PROF COMP BIOCHEM, HOPKINS MARINE STA, STANFORD UNIV, 80- *Concurrent Pos:* Res grants, NSF, 57-59, USPHS, 57-66 & Environ Protection Agency, 72-; mem pesticide adv comt, Calif Dept Agr, 70-; consult aquacult, Hawaiian Abalone Farms, 80-88. *Mem:* AAAS; NY Acad Sci. *Res:* Immunochemistry; comparative biochemistry; marine biology. *Mailing Add:* 73-1397 Hikimoe St Kailua-Kona HI 96740

PHILLIPS, JOHN HUNTER, JR, b Houston, Tex, Nov 2, 30; m 54; c 3. INTERNAL MEDICINE, CARDIOLOGY. *Educ:* Tulane Univ, BS, 52, MD, 55; Am Bd Internal Med, dipl cardiovasc dis. *Prof Exp:* Intern, Charity Hosp, New Orleans, 55-56; asst instr med, 56-57, from instr to assoc prof, 57-70, PROF MED, SCH MED, TULANE UNIV, 70-, CHIEF TULANE CARDIOL DIV, 74-, SIDNEY W & MARILYN S LASSEN PROF CARDIOVASC, 85- *Concurrent Pos:* Nat Heart Inst fel, 57-60; asst vis physician, Charity Hosp, New Orleans, 56-61, vis physician, 61-68, sr vis physician, 69-; consult, Lallie Kemp Charity Hosp, Independence, La, 56-63; clin asst, Vet Admin Hosp, New Orleans, 58-61, staff physician & chief cardiol sect, 61-; consult, Huey P Long Hosp, Alexandria, 60-62; fel, Coun Clin Cardiol, Am Heart Asn; asst ed, Am Heart J, 59-73; consult ed, Addison Wesley Publ Co, 78- *Mem:* AAAS; Am Fedn Clin Res; fel Am Col Physicians; fel Am Col Chest Physicians: fel Am Col Cardiol; Sigma Xi. *Res:* Peripheral vascular disease; geriatrics. *Mailing Add:* Dept Med Tulane Univ Sch Med New Orleans LA 70112

PHILLIPS, JOHN LYNCH, b Ft Belvoir, Va, Apr 15, 51; m 79; c 2. SPACE PLASMA PHYSICS, SOLAR PHYSICS. *Educ:* US Naval Acad, BS, 72; Univ Calif, Los Angeles, MS, 84, PhD(space physics), 87. *Prof Exp:* Oppenheimer fel, 87-89, STAFF MEM, LOS ALAMOS NAT LAB, 89- *Mem:* Am Geophys Union; Am Astron Soc. *Res:* Solar wind, its internal state and dynamic structure, and its interaction with planetary atmospheres, magnetospheres, and cometary bodies. *Mailing Add:* Los Alamos Nat Lab Mail Stop D438 Los Alamos NM 87545

PHILLIPS, JOHN PERROW, b Lynchburg, Va, June 17, 25; m 53; c 2. ANALYTICAL CHEMISTRY. *Educ:* Univ Va, BS, 45; Univ Ind, MA, 47, PhD(chem), 49. *Prof Exp:* From asst prof to assoc prof, 49-56, PROF CHEM, UNIV LOUISVILLE, 56- *Concurrent Pos:* NSF sci fac fel, 63-64. *Mem:* Am Chem Soc. *Res:* Substituted 8-quinolinols. *Mailing Add:* Dept of Chem Univ of Louisville Louisville KY 40292

PHILLIPS, JOHN R, b Philadelphia, Pa, May 3, 35. INORGANIC CHEMISTRY. *Educ:* Dartmouth Col, AB, 57; Harvard Univ, MA, 59, PhD(chem), 62. *Prof Exp:* Asst prof chem, Univ Ottawa, 62-67; asst prof, 67-73, ASSOC PROF CHEM, PURDUE UNIV, CALUMET CAMPUS, 73- *Mem:* Am Chem Soc; Royal Soc Chem. *Res:* Organometallic and coordination chemistry; fluorocarbon compounds of metals. *Mailing Add:* Dept Chem Purdue Univ Calumet Campus Hammond IN 46323

PHILLIPS, JOHN RICHARD, b Portland, Ore, Feb 17, 34; m 53; c 5. COMPUTER SCIENCES, APPLIED MATHEMATICS. *Educ:* Lewis & Clark Col, BS, 59; Ore State Univ, PhD(math), 66. *Prof Exp:* Instr math, Ore State Univ, 62-66; fel, Lawrence Livermore Lab, Calif, 66-67; asst prof comput sci, Univ Ill, Urbana-Champaign, 67-74; prog dir software eng, Div Math & Comput Sci, NSF, 74-76; ASSOC PROF COMPUT SCI & CHMN DEPT, OKLA STATE UNIV, STILLWATER, 76- *Mem:* Asn Comput Mach. *Res:* Computer languages; design of array languages; information structures; data structures; microcomputers. *Mailing Add:* Dept Sci & Sci Technol Dodge City Community Col 2501 N 14th Ave Dodge City KS 67801

PHILLIPS, JOHN RICHARD, b Albany, Calif, Jan 30, 34; m 57; c 3. THERMODYNAMICS, FLUID FLOW. *Educ:* Univ Calif, Berkeley, BS, 56; Yale Univ, MEng, 58, DEng, 60. *Prof Exp:* Chem engr, Stanford Res Inst, 60; res design & develop engr, Chevron Res Co, Calif, 62-66; from asst prof to assoc prof, 66-74, PROF ENG, HARVEY MUDD COL, 74-, DIR ENG CLIN, 77- *Concurrent Pos:* Pres, Claremont Eng Co, 73-; vis prof, Univ Edinburgh, 75, ESIEE Paris, 81 & Naval Postgrad Sch, 84-85; vis scientist, Southern Calif Edison Co, 80; vis scholar, Cambridge Univ, 81. *Mem:* Am Inst Chem Engrs. *Res:* Permeation of gases through metals; thermodynamics and separations; desalination; research management; energy. *Mailing Add:* Dept Eng Harvey Mudd Col Claremont CA 91711

PHILLIPS, JOHN SPENCER, b Baltimore, Md, May 22, 53; m 76. ELECTROANALYTICAL CHEMISTRY. *Educ:* Western Md Col, BA, 75; Purdue Univ, PhD(chem), 81. *Prof Exp:* Teaching asst chem, Purdue Univ, 75-78, res asst, 78-81; asst prof chem, James Madison Univ, 81-85; ASST PROF CHEM, UNIV WIS-RIVER FALLS, 85- *Mem:* Am Chem Soc. *Res:* Optimization in chemical analysis with emphasis in electroanalytical methods and flow injection analysis; innovations in the teaching of undergraduate chemistry. *Mailing Add:* Dept Chem Univ Wis River Falls WI 54022

PHILLIPS, JOSEPH D, b Woodbury, NJ, Sept 11, 38; m 61; c 3. GEOPHYSICS. *Educ:* Rutgers Univ, BA, 61; Princeton Univ, MSE, 63, MA, 64, PhD(geophys), 66. *Prof Exp:* Asst eng mgt, NJ Bell Tel Co, summer 61; res asst geol, Princeton Univ, 61-66; assoc scientist, Woods Hole Oceanog Inst, 66-77; prof, Mass Inst Technol, 77-78; RES SCI & LECTR, UNIV TEX, 78- *Concurrent Pos:* Consult, Princeton Appl Res Corp, Mobil Corp, Exxon Corp, Bell Labs, 65- & Tex Radioactive Waste Disposal Authority, 86-; NSF & Off Naval Res oceanog & geophys res grants, 66-; vis prof, Cambridge Univ, 74-75. *Mem:* AAAS; Am Geophys Union; Soc Explor Geophys. *Res:* Paleomagnetism; geomagnetism; marine geophysics and geology; seismology; nuclear waste disposal site evaluation; vertical seismic profiling. *Mailing Add:* Univ Tex Geol Sci Dept Inst Geophysics Austin TX 78759

PHILLIPS, JOY BURCHAM, b Decatur, Ill, Sept 7, 17; m 38. PHYSIOLOGY. *Educ:* Millikin Univ, AB, 39; Univ Ill, MA, 44; NY Univ, PhD(physiol), 54. *Prof Exp:* With biol dept, Queens Col, NY, 45-47; asst therapeut, NY Univ, 47-49; from asst prof to prof biol, Drew Univ, 52-86, chmn dept, 70-75; RETIRED. *Concurrent Pos:* NSF teaching fel, Columbia Univ, 59-60; Sigma Delta Epsilon grant, 61; Brown Hazen grant, Res Corp, 68-; mem res panel NSF Col Sci Improv Prog, 68 & 69. *Mem:* Sigma Xi; Am Soc Zool; Am Asn Anat; AAAS; NY Acad Sci. *Res:* Developmental endocrinology; histology. *Mailing Add:* Dept Biol Drew Univ 36 Madison Ave Madison NJ 07940

PHILLIPS, JULIA M, b Freeport, Ill, Aug, 17, 54. SOLID STATE PHYSICS. *Educ:* Col William & Mary, BS, 76; Yale Univ, PhD(appl physics), 81. *Prof Exp:* Mem tech staff physics, 81-88, SUPVR, THIN FILM RES GROUP, RES DIV, AT&T BELL LABS, 88- *Mem:* Am Phys Soc; Mat Res Soc (secy, 87-89). *Res:* Epitaxial growth; molecular beam epitaxy; ion beam analysis; rapid thermal processing; electrical and structural properties of epitaxial thin films; high temperature superconductors. *Mailing Add:* AT&T Bell Labs Rm 10-158 600 Mountain Ave Murray Hill NJ 07974-2070

PHILLIPS, KEITH L, b Broken Bow, Nebr, June 11, 37; m 60; c 2. MATHEMATICAL ANALYSIS, COMPUTER VISION. *Educ:* Univ Colo, BA, 59, MA, 61; Univ Wash, PhD(math), 64. *Prof Exp:* Instr math, Univ Wash, 64-65, NSF grant, 65; asst prof, Calif Inst Technol, 65-68; ASSOC PROF, NMEX STATE UNIV, 68- *Concurrent Pos:* NSF grant, Calif Inst Technol, 66-68; vis prof, Univ Colo, Boulder, 82-83; prin investr comput vision, Comput Res Lab, NMex State Univ, 83- *Mem:* Am Math Soc; Math Asn Am; Soc Indust & Appl Math; AAAS; Am Asn Artificial Intel; Inst Elec & Electronics Engrs. *Res:* Fourier analysis in locally compact groups, especially in locally compact fields; pattern theory and edge analysis in computer vision; theory of distribution and homogeneity in local fields and martingales; edge analysis by differential operators; shape theory by Fourier series and invariant theory. *Mailing Add:* Dept Math Univ Colo Colorado Springs CO 80933

PHILLIPS, KENNETH LLOYD, chemical engineering, bioengineering; deceased, see previous edition for last biography

PHILLIPS, LAWRENCE STONE, b Washington, DC, Sept 16, 41; m 71; c 2. ENDOCRINOLOGY, METABOLISM. *Educ:* Swarthmore Col, BA, 63; Harvard Univ, MD,67; Am Bd Internal Med, dipl, 72, cert endocrinol & metab, 73. *Prof Exp:* Intern & resident, Presby-St Luke's Hosp, Chicago, 67-69; staff assoc, Grady Mem Hosp, 69-71; fel metab, Sch Med, Wash Univ, 71-74; from asst prof to assoc prof, Northwestern Univ Med Sch, 74-83; PROF MED & DIR, DIV ENDOCRINOL & METAB, EMORY UNIV SCH MED, 83- *Concurrent Pos:* Mem lab prog, Ctr Dis Control, USPHS, 69-71; instr med, Sch Med, Emory Univ, 70-71; asst physician, Barnes Hosp, 72-74; adj attend physician, Northwestern Mem Hosp, 74-75, assoc attend physician, 75-81, attend physician, 81-83; attend physician, Emory Clin, 83- *Honors & Awards:* Elliott P Joslin Res & Develop Award, Am Diabetes Asn, 77. *Mem:* Endocrine Soc; Am Diabetes Asn; Am Fedn Clin Res; Am Soc Clin Investr. *Res:* Somatomedin; nutrition; growth. *Mailing Add:* Endocrinol Div Dept Med Emory Univ Sch Med 69 Butler St SE Atlanta GA 30303

PHILLIPS, LEE REVELL, b Salt Lake City, Utah, Apr 7, 53; m 73; c 3. FORENSIC CHEMISTRY. *Educ:* Brigham Young Univ, BS, 76, MS, 79, JD, 81. *Prof Exp:* Researcher, Phillips Petrol Co, 73-74; geophysics interpreter, Cities Serv Co, 75-76; DIR, ALPINE WEST LABS, 79-; RES ASSOC, BRIGHAM YOUNG UNIV, 81- *Concurrent Pos:* Forensic chemist, Cent & Southern Utah Law Enforcement Agencies, 81- *Mem:* Am Chem Soc; Sigma Xi; Soc Environ Geochem & Health. *Mailing Add:* 1515 E 1575 N Provo UT 84604

PHILLIPS, LEE VERN, b Checotah, Okla, Feb 19, 30; m 54; c 2. ORGANIC CHEMISTRY. *Educ:* Univ Mo, PhD(chem), 57. *Prof Exp:* Mem staff, Res Ctr, Spencer Chem Co, 57-66, group leader pesticide synthesis & process develop, Gulf Res & Develop Co, 66-67, sect supvr, pesticide synthesis sect, 67-69, div dir agr chem, 69-75; mgr chem res dept, Richmond Res Ctr, Stauffer Chem Co, 75-91; RETIRED. *Mem:* Am Chem Soc. *Res:* Synthetic organic chemistry. *Mailing Add:* 100 Quintas Lane Moraga CA 94556

PHILLIPS, LEO AUGUSTUS, molecular biology, biophysical chemistry; deceased, see previous edition for last biography

PHILLIPS, LEON A, b Mehoma, Ore, Sept 1, 23; m 48; c 2. MEDICINE, RADIOLOGY. *Educ:* Univ Wash, BS, 48; Yale Univ, MD, 52. *Prof Exp:* Instr radiol, Johns Hopkins Hosp, 56-59; asst prof, 60-64, ASSOC PROF RADIOL, SCH MED, UNIV WASH, 64- *Mem:* AMA; Am Col Radiol; Asn Univ Radiol. *Res:* Radiology utilization review. *Mailing Add:* 1959 NE Pacific St Seattle WA 98105

PHILLIPS, LYLE LLEWELLYN, b Long Beach, Calif, June 14, 23; m 53; c 2. CYTOGENETICS. *Educ:* Univ Redlands, BA, 50; Claremont Cols, MA, 51; Univ Wash, PhD(bot), 54. *Prof Exp:* Asst bot, Pomona Col, 51; asst, Univ Wash, 51-54; fel genetics, State Col Wash, 54-56; from asst prof to assoc prof, 56-65, PROF CROP SCI, NC STATE UNIV, 65-, PROF GENETICS, 70- *Mem:* Bot Soc Am; Genetics Soc Can. *Res:* Cytogenetics and evolution of Gossypium. *Mailing Add:* 77 Pompano Emerald Isle NC 28594

PHILLIPS, MARSHALL, b Yankton, SDak, Dec 1, 32; m 57; c 2. BIOCHEMISTRY, ORGANIC CHEMISTRY. *Educ:* Yankton Col, AB, 56; Univ SDak, MA, 59; Univ Kans, PhD(biochem), 64. *Prof Exp:* Nat Res Coun res assoc, Southern Regional Lab, 63-65, RES CHEMIST, NAT ANIMAL DIS LAB, USDA, 65-, OFF AGR BIOTECHNOL, 90- *Concurrent Pos:* Biotechnol policy, USDA. *Mem:* AAAS; Am Chem Soc. *Res:* Complex macromolecules; recombinant vaccines; immunomodulators; lipopolysaccharides. *Mailing Add:* Nat Animal Dis Lab PO Box 70 Ames IA 50010

PHILLIPS, MARVIN W, b Salem, Ind, Aug 9, 29; m 53; c 2. SOIL FERTILITY. *Educ:* Purdue Univ, BS, 53, MS, 58; Univ Minn, PhD(soil chem), 64. *Prof Exp:* From asst prof to assoc prof, 61-65, head dept, 71-91, PROF AGRON, PURDUE UNIV, WEST LAFAYETTE, 69- *Mem:* Fel Soil Sci Soc Am; fel Crop Sci Soc Am; Int Soil Sci Soc; fel Am Soc Agron; Soil Conserv Soc Am. *Res:* Soil fertility and management. *Mailing Add:* Dept Agron Purdue Univ West Lafayette IN 47907-1150

PHILLIPS, MELBA (NEWELL), b Hazleton, Ind, Feb 1, 1907. PHYSICS. *Educ:* Oakland City Col, Ind, AB, 26; Battle Creek Col, MA, 28; Univ Calif, Berkeley, PhD(physics), 33. *Hon Degrees:* DSc, Oakland City Col, Ind, 64. *Prof Exp:* Instr, Brooklyn Col, 34-44, asst prof physics, 44-52; assoc dir, Acad Year Inst, Washington Univ, St Louis, 57-62; prof, 62-72, EMER PROF PHYSICS, UNIV CHICAGO, 72- *Concurrent Pos:* Helen Huff res fel, Bryn Mawr Col, 35-36; Am Asn Univ Women fel, Inst Advan Study, 36-37; mem, Comn Col Physics, 62-68; mem gov bd, Am Inst Physics, 65-68 & 75-77; vis prof, State Univ NY Stony Brook, 72-75; actg exec officer, Am Asn Physics Teachers, 75-77. *Honors & Awards:* Oerstad Medal, Am Asn Physics Teachers, 74; Karl Taylor Compton Award, Am Inst Physics, 81; First Recipient, Melba Newell Phillips Award, Am Asn Physics Teachers, 82. *Mem:* Am Asn Physics Teachers (pres, 66-67); fel Am Phys Soc; AAAS. *Res:* Theory of complex spectra, theory of light nuclei; physics education; history of physics. *Mailing Add:* 351 W 24th St Apt 1F New York NY 10011

PHILLIPS, MELVILLE JAMES, b Newport, Eng, Dec 9, 30; Can citizen; m 56; c 2. PATHOLOGY. *Educ:* McGill Univ, MD, 56. *Prof Exp:* Lectr path, Univ Toronto, 61-62, asst prof, 63-67; assoc prof, McGill Univ, 67-70; assoc prof, 70-74, PROF PATH, BANTING INST, UNIV TORONTO, 74-, AT HOSP SICK CHILDREN. *Concurrent Pos:* Fel path, Royal Col Physicians & Surgeons Can, 61; Med Res Coun Can res grant, 62-; surg pathologist, Toronto Gen Hosp, 69- *Mem:* Am Soc Exp Path; Am Asn Path & Bact; Int Acad Path; fel Col Am Path; Can Asn Path. *Res:* Human and experimental liver disease, especially correlation of electron microscopic and biochemical changes in liver disease. *Mailing Add:* Hosp Sick Children 555 University Ave Toronto ON M5G 1X8 Can

PHILLIPS, MICHAEL CANAVAN, b London, Eng, Feb 23, 40; m 64; c 2. PHYSICAL BIOCHEMISTRY. *Educ:* Southampton Univ, BSc, 62, PhD(phys chem), 65. *Hon Degrees:* DSc, Southampton Univ, 76. *Prof Exp:* Fel theoret biol, State Univ NY Buffalo, 65-67; scientist & sect mgr biophys, Unilever Res Lab, Welwyn, Eng, 67-78; assoc prof, 78-80, PROF BIOCHEM, MED COL PA, 80- *Concurrent Pos:* Fel, Coun Arteriosclerosis, Am Heart Asn, 78- *Mem:* Am Chem Soc; fel Am Soc Biochem Mol Biol. *Res:* Interactions of serum lipoproteins with cells; lipid and sterol metabolism and their influence on atherosclerosis; lipid-protein interaction; interfacial phenomena in biological systems. *Mailing Add:* Dept Physiol & Biochem Med Col Pa Philadelphia PA 19129

PHILLIPS, MICHAEL IAN, b London, Eng, July 30, 38; US citizen; m 87. NEUROPHYSIOLOGY, NEUROENDOCRINOLOGY. *Educ:* Univ Exeter, BSc, 62; Univ Birmingham, MSc, 65, PhD(neuropharmacol), 67. *Hon Degrees:* DSc, Univ Birmingham, 86. *Prof Exp:* Tutor psychol, Univ Birmingham, 66-67; vis asst prof, Univ Mich, 67-69; res fel biol, Calif Inst Technol, 69-70; from asst prof to prof physiol & prof pharmacol, Univ Iowa, 70-80; PROF PHYSIOL, UNIV FLA, 80-, CHMN DEPT, 80- *Concurrent Pos:* Vis scientist, Brain Res Inst, Zurich, 75 & 78; Humboldt Found scholar & fel, Pharmacol Inst Heidelberg, WGer, 76-77; mem neurobiol rev panel, NSF, 77-80; fel, Coun Circulation, Am Heart Asn, 78, Coun High Blood Pressure, 81; vis investr, NIH, 79-80; mem, Exp Cardiovasc Res, NIH, 80-85, 89-93; ed, Int Brain Res Org News; dir Med Student Res Fels, Am Heart Asn, 86-89; pres, assoc chmn, Dept Physiol, 87-88; mem, Cardiovasc Study Sect, Am Heart Asn, 87-; chmn, Gordon Conf Angiotensin, 89; prog dir, Nat Science Found, 90; mem White House Off Sci Technol comt, Decade of the Brain, 90; ed, Regulatory Peptides, 90- *Honors & Awards:* Nat Inst Ment Health Career Develop Award, 73; Lucia Award for Cardiovasc Res, 89. *Mem:* Soc Neuroscience; Am Physiol Soc; Endocrine Soc; Int Brain Res Org. *Res:* Neuroendocrinology; site and mode of peptide angiotensin action in the brain; neurophysiology. *Mailing Add:* Dept Physiol Univ Fla Col Med Gainesville FL 32610

PHILLIPS, MILDRED E, b New York, NY, May 21, 28. PATHOLOGY. *Educ:* Hunter Col, BA, 46; Howard Univ, MD, 50. *Prof Exp:* Intern, King's County Hosp, Brooklyn, NY, 50-52; resident path, Mt Sinai Hosp, New York, 52-54; resident asst surg path, Presby Hosp, 54-55; instr path, State Univ NY Downstate Med Ctr, 55-56; from instr to asst prof, 57-65, assoc prof path, Med Ctr, NY Univ, 65-; AT State Univ NY, STONY BROOK. *Concurrent Pos:* Fel surg path, Presby Hosp, New York, 54-55; fel, Postgrad Med Sch, Univ London & London Hosp, 56-57. *Mem:* AMA; Col Am Path; Transplantation Soc; NY Acad Sci; Int Acad Path. *Res:* Tumor specific antigenicity and serum protein production by experimental and human tumors. *Mailing Add:* Dept Path State Univ NY Health Sci Ctr Stony Brook NY 11790

PHILLIPS, MONTE LEROY, b Valley City, NDak, Dec 12, 37; m 60; c 3. FORENSIC ENGINEERING, STRUCTURAL & GEOTECHNICAL ENGINEERING. *Educ:* Univ NDak, Bs, 59, MS, 61; Univ Ill, PhD(civil eng), 70. *Prof Exp:* Asst prof civil eng, Ohio Northern Univ, 62-63; instr gen eng, Univ Ill, 63-70; asst prof, 61-62, assoc prof, 70-89, PROF CIVIL ENG, UNIV NDAK, 89- *Concurrent Pos:* Proj engr, NDak State Hwy Dept, 56-61; construct mgr, Woerfel Corp, 71-73; consult, Eng Assocs, 73-80, pvt consult, engrs & architects, 81-; proj mgr, Nodak Contracting Corp, 81-91; pvt consult, engrs & architects, EAPC. *Honors & Awards:* Outstanding Serv Award, Am Soc Civil Engrs, 79. *Mem:* Am Soc Civil Engrs; Nat Acad Forensic Engrs; Nat Soc Prof Engrs; Am Soc Eng Educ. *Res:* Structural and geotechnical engineering; author of over 100 proprietary forensic engineering publications. *Mailing Add:* 820 N 25th St Grand Forks ND 58203

PHILLIPS, N CHRISTOPHER, US citizen. C-ALGEBRAS, K-THEORY. *Educ:* Univ Calif, Berkeley, AB, 78, PhD(math), 84. *Prof Exp:* Postdoctoral res fel math, Math Sci Res Inst, 84-85 & NSF, 85-88; vis asst prof math, Univ Calif, Los Angeles, 85-86 & 87-88; ASST PROF MATH, UNIV GA, 88- & UNIV ORE, 90- *Mem:* Am Math Soc. *Res:* Noncommutative topology; generalization of concepts, methods and results in topology to operator algebras; equivariant K-theory, noncommutative homtopy theory and ideas related to real rank. *Mailing Add:* Dept Math Univ Ore Eugene OR 97403-1222

PHILLIPS, NORMAN, METEOROLOGY. *Prof Exp:* At Nat Meteorol Ctr, Nat Weather Serv; RETIRED. *Mem:* Nat Acad Sci. *Mailing Add:* 18 Edward Lane Merrimack NH 03054

PHILLIPS, NORMAN EDGAR, b Detroit, Mich, Dec 20, 28; m 51; c 2. PHYSICAL CHEMISTRY. *Educ:* Univ BC, BA, 49, MA, 50; Univ Chicago, PhD(chem), 54. *Prof Exp:* Nat Res Coun fel chem, 54-55, from instr to assoc prof, 55-66, dean, Col Chem, 78-80, PROF CHEM, UNIV CALIF, BERKELEY, 66- *Concurrent Pos:* Sloan Found fel, 62-64; Guggenheim fel, 63-64; NSF sr fel, Tech Univ, Helsinki, 70-71; prin investr inorg mat, Res Div, 73-77. *Mem:* Am Chem Soc; Am Phys Soc; Sigma Xi. *Res:* Cryogenics; low temperature calorimetry. *Mailing Add:* Dept of Chem Univ of Calif Berkeley CA 94720

PHILLIPS, OWEN M, b Parramatta, Australia, Dec 30, 30; m 52; c 4. MECHANICS. *Educ:* Univ Sydney, BSc, 52; Cambridge Univ, PhD(appl math), 55. *Prof Exp:* Imp Chem Indust res fel theoret physics, Cambridge Univ, 55-57; from asst prof to assoc prof mech eng, 57-62, prof geophys mech, 62-68, chmn dept earth & planetary sci, 68-77, prof geophys 68-75, DECKER PROF SCI & ENG, JOHNS HOPKINS UNIV, 75- *Concurrent Pos:* Fel, St John's Col, Cambridge Univ, 57-60, asst dir res appl math, univ, 61-64; consult, Martin Co, 58, Westinghouse Elec Corp, 58-59; Hydronautics, Inc, 61- & Phila Elec Co, 65-66; mem coun mem & rev & goals comt, Nat Ctr Atmospheric Res, 64-68; assoc ed, J Fluid Mech, 64- *Honors & Awards:* Adams Prize, 65; Sverdrop Gold Medal, Am Meteorol Soc, 74. *Mem:* Geophys Union; fel Royal Soc; fel Am Meteorol Soc. *Res:* Fluid mechanics, particularly turbulence and its applications to meteorology; oceanography; theory of surface waves on fluids. *Mailing Add:* Dept of Earth & Planetary Sci Johns Hopkins Univ Baltimore MD 21218

PHILLIPS, PAUL J, b Carlisle, Eng, Nov 26, 42; m 67; c 2. POLYMER PHYSICS. *Educ:* Univ Liverpool, BSc, 65, PhD(phys chem), 68. *Prof Exp:* Fel polymer sci, Univ Mass, 68-70; sr res assoc polymer sci, Queen Mary Col, Univ London, 70-71, lectr mat sci, 71-75; asst prof chem eng, State Univ NY, Buffalo, 75-77; assoc prof, 77-80, prof mat sci, Univ Utah, 80-85; AT DEPT MAT SCI & ENG, UNIV TENN, 85- *Concurrent Pos:* Fiber Soc Lectr, 86-87. *Mem:* Am Chem Soc; Am Phys Soc; Soc Plastics Engrs; Sigma Xi. *Res:* Crystallization and morphology of polymers at atmospheric and elevated pressures; effects of morphology on the mechanical and electrical properties of polymers. *Mailing Add:* Univ Tenn Knoxville TN 37996-2200

PHILLIPS, PERRY EDWARD, b Los Angeles, Calif, Sept 22, 44; m 66; c 1. PLASMA PHYSICS. *Educ:* Univ Calif, Santa Barbara, BA, 66; Univ Tex, Austin, PhD(physics), 71. *Prof Exp:* Nuclear physicist, Los Alamos Sci Lab, 67; RES SCIENTIST PLASMA PHYSICS, FUSION RES CTR, UNIV TEX, AUSTIN, 71- *Mem:* Am Phys Soc; Inst Elec & Electronics Engrs; Sigma Xi. *Res:* Experimental plasma devices including microwave and laser scattering on toroidal devices; strong magneto-hydrodynamic shock waves propagating in magnetized plasma. *Mailing Add:* 4606 Creek Ridge Austin TX 78735

PHILLIPS, PETER CHARLES B, b Weymouth, Dorset, Eng, Mar 23, 48; NZ citizen; m 81; c 3. MULTIVARIATE ANALYSIS, ECONOMETRICS. *Educ:* Univ Auckland, NZ, BA, 69, MA, 71; Univ London, Eng, PhD(econ), 74. *Hon Degrees:* MA, Yale Univ, 79. *Prof Exp:* Jr lectr econ, Univ Auckland, 70-71; lectr, Univ Essex, UK, 72-76; prof econ & statist, Univ Birmingham, UK, 76-79; prof, 79-85, Stanley Resor prof econ & statist, 85-89, STERLING PROF ECON, YALE UNIV, 89- *Concurrent Pos:* External examr, Univ Kent, London, 76-83; vis scholar, Ecole Polytechnique, 77; prin investr, NSF res grants, 80-91; chmn, Winter Meetings of Econometric Soc, 83; fel, Guggenheim Found, 84; ed, Economet Theory, Cambridge Univ Press, 84-; vis univ prof, Monash Univ, Australia, 86; vis prof, Univ Auckland, 88; distinguished visitor, London Sch Econ, 89; commonwealth scholar, 71-72; J E Charles fel, 88. *Mem:* Fel Economet Soc; Am Statist Asn; Inst Math Statist; Math Asn Am; fel Japan Soc Promotion Sci. *Res:* Mathematical statistics, multivariate analysis, time series & nonlinear estimation; methodology, specification & estimation in dynamic economic models. *Mailing Add:* Cowles Found Res Econ PO Box 2125 Yale Sta New Haven CT 06520

PHILLIPS, PHILIP WIRTH, b Trinidad & Tobago, WI, June 28, 58; US citizen. CONDENSED MATTER THEORY. *Educ:* Walla Walla Col, BSc, 79; Univ Wash, PhD(phys chem), 82. *Prof Exp:* Res asst, Univ Wash, 79-81, Danforth-Compton predoctoral fel, 81-82; Miller postdoctoral fel, Univ Calif, Berkeley, 82-84; asst prof, 84-90, ASSOC PROF PHYS CHEM, MASS INST TECHNOL, 90- *Concurrent Pos:* Vis prof, Mich State Univ, 87; vis fel, Balliol Col, Oxford, 88. *Res:* Electron transport in condensed phases; hopping conduction in anisotropically-disordered solids; coupled transport in cytochrome oxidase; insulator/metal transitions in conducting polymers; transport in structurally-disordered solids. *Mailing Add:* Mass Inst Technol Rm 6-223 Cambridge MA 02139

PHILLIPS, RALPH SAUL, b Oakland, Calif, June 23, 13; m 42; c 1. PURE MATHEMATICS, APPLIED MATHEMATICS. *Educ:* Univ Calif, Los Angeles, AB, 35; Univ Mich, PhD(math), 39. *Prof Exp:* Rackham fel from Univ Mich, Inst Adv Study, 40; instr math, Univ Wash, 40-41; instr, Harvard Univ, 41-42; from staff mem to group leader, Radiation Lab, Mass Inst Technol, 42-46; asst prof math, NY Univ, 46-47; from assoc prof to prof, Univ Southern Calif, 47-58; prof, Univ Calif, Los Angeles, 58-60; prof, 60-83, EMER PROF MATH, STANFORD UNIV, 83- *Concurrent Pos:* Mem, Inst Adv Study, 49-50; res assoc, Yale Univ, 53-54; Guggenheim fel, 54-55; mem inst math sci, NY Univ, 58; vis prof, Aarhus Univ, 68; Guggenheim fel, 75; Nat Acad Sci exchange vis to USSR, 75; Robert Grimmett prof math, Stanford Univ. *Mem:* Am Math Soc; Am Acad Arts & Sci. *Res:* Functional analysis; partial differential equations; mathematical physics; scattering theory; number theory. *Mailing Add:* Dept of Math Stanford Univ Stanford CA 94305

PHILLIPS, RALPH W, b Farmland, Ind, Jan 12, 18; m 43; c 1. DENTAL MATERIALS. *Educ:* Ind Univ, BS, 40, MS, 55; Univ Ala, DSc, 62. *Prof Exp:* Asst, 40-41, from instr to prof, 41-62, asst dean res, 69-74, distinguished prof, assoc dean res & dir, Oral Health Res Inst, 74-87, RES PROF DENT MAT, SCH DENT, IND UNIV, INDIANAPOLIS, 62-, CHMN DEPT, 58- *Concurrent Pos:* Consult, Nat Inst Dent Res, US Pub Health Serv, Nat Acad Scis, Off Surg Gen, US Army, US Navy, US Air Force, Nat Inst Health, Vet Admin, Council Dent Mats, Instruments & Equip & Coun Dent Res, Am Dent Asn, Pankey Res Inst & Bio-Eng Res Progs, Rensselaer Polytech Inst; chmn, Dent Sect Opthalmic, Ear, Nose & Throat Panel, US Food & Drug Admin, Bio-Mats Res Adv Comt, Nat Inst Dent Res, Peer Rev Comts, Dent Health Res Inst & Coun Dent Mats, Instruments & Equip, Am Dent Asn; vis fac mem, Univ Calif & other dent schs. *Honors & Awards:* Gold Medal Award & The Mitch Nakayama Mem Award, Pierre Fauchard Acad; William J Gies Award, Am Col Dentists; Wilmer Souder Award, Int Asn Dent Res, 59; First Annual G V Black Distinguished Lect Award, Wash Univ; Hon Fel Award, Acad Gen Dent; Int Res Award, Acad Int Dent Studies; Hollenback Prize, Am Acad Oper Dent. *Mem:* Sigma Xi; Am Chem Soc; Am Asn Dent Schs; hon mem Am Dent Asn; Am Asn Dent Eds; Int Asn Dent Res; fel AAAS; Soc Bio-Mats; hon mem Am Acad Restorative Dent; hon mem Belg Dent Soc; fel Am Col Dent; fel Int Col Dent. *Res:* Application of fluorides in dentistry; physical properties of dental materials and tooth structure; clinical behavior of materials as related to their properties; author or coauthor of 14 publications. *Mailing Add:* Dept Dent Math DS112 Ind Univ Dent 1121 W Michigan St Indianapolis IN 46202

PHILLIPS, RICHARD ARLAN, b Detroit, Mich, May 30, 33; m 56; c 2. OPTICS, APPLIED PHYSICS. *Educ:* Univ Mich, Ann Arbor, BS, 55, MS, 62, PhD(physics), 66. *Prof Exp:* Syst analyst petrol reservoirs, Exxon Corp, Venezuela, 55-60; asst prof elec eng & physics, Univ Minn, 66-71; sr optical engr, Foster Grant Co, Inc, 71-72, mgr consumer prod res & develop, 72-85; PRES, INT POLARIZER, INC, 85- *Concurrent Pos:* US del, Int Standards Orgn, 74- *Mem:* Optical Soc Am. *Res:* Polarizers and polarized products, physical optics, lasers, nonlinear optics, color vision, dyeing of plastics; ophthalmic lens design; translation of new products from research laboratory into mass production; co-author. *Mailing Add:* 3 Betsy Ross Circle Acton MA 01720

PHILLIPS, RICHARD DEAN, b Sacramento, Calif, Sept 17, 29; m 50. ENVIRONMENTAL PHYSIOLOGY, RADIATION BIOLOGY. *Educ:* Univ Calif, Berkeley, BA, 58, PhD(physiol), 66. *Prof Exp:* Res physiologist, US Naval Radiol Defense Lab, Calif, 58-69; res & develop mgr, bioelectromagnetics prog, biol dept, Pac Northwest Labs, Battelle Mem Inst, 69-84; DIR, DEV & CELL TOX DIV, HEALTH EFFECTS RES LAB, US ENVIRON PROTECTION AGENCY, 84- *Concurrent Pos:* Mem, Nat Coun Rad Protection & Measurements Comt 53, Biol Effects RF Radiation & Comt 79, Biol Effects ELF Radiation; consult, WHO; ed-in-chief, Bioelectromagnetics, 83- *Mem:* Am Physiol Soc; NY Acad Sci; Inst Elec & Electronics Engrs; Power Eng Soc; Am Nat Standards Inst. *Res:* Physiological responses of mammals to electromagnetic radiations, drugs, toxic wastes, toxic chemicals, and environmental stresses; cardiovascular physiology, metabolism and thermoregulation. *Mailing Add:* 9610 N Ridgecrest Dr Spokane WA 99208

PHILLIPS, RICHARD E, JR, b US, Dec 3, 36; m 63; c 2. MATHEMATICS. *Educ:* Otterbein Col, BS, 61; Univ Kans, MA, 62, PhD(math), 65. *Prof Exp:* Asst prof math, State Univ NY, 62-63; asst prof, Wis State Univ, 65-66; asst prof, Univ Kans, 66-69; assoc prof, 69-74, PROF MATH, MICH STATE UNIV, 74- *Concurrent Pos:* Vis prof math, Univ Padua, Italy, 90. *Mem:* Am Math Soc; Sigma Xi. *Res:* Group theory. *Mailing Add:* Dept of Math Mich State Univ East Lansing MI 48824

PHILLIPS, RICHARD EDWARD, b Hammond, Ind, Oct 31, 30; m 65; c 3. NEUROSCIENCES, ORNITHOLOGY. *Educ:* Purdue Univ, BS, 52; Ore State Univ, MS, 54; Cornell Univ, PhD(physiol), 59. *Prof Exp:* Res assoc avian physiol, Cornell Univ, 59-62; asst prof zool, Va Polytech Inst, 62-64; from asst prof to assoc prof, 64-71, ECOL & BEHAV BIOL PROF ANIMAL SCI, UNIV MINN, ST PAUL, 71- *Concurrent Pos:* NSF res grants, 59-62 & 63-66; NIH res grants, 66-69, 72-74, 74-76, 76-79 & 78-81; vis prof, Ethopharmacol Lab, Univ Leiden, Neth, 78. *Mem:* AAAS; Am Ornith Union; Animal Behavior Soc; Sigma Xi; Soc Neuroscience. *Res:* Neural and endocrine mechanisms of behavior emphasizing affect, communication and reproduction. *Mailing Add:* Dept of Ecol & Behav Biol Univ Minn Minneapolis MN 55455

PHILLIPS, RICHARD HART, b Atlanta, Ga, June 23, 22; m 45; c 5. PSYCHIATRY. *Educ:* Univ NC, BS, 44; NY Univ, MD, 45; Am Bd Psychiat & Neurol, dipl, 54. *Prof Exp:* Intern med, US Naval Hosp, Camp LeJuene, NC, 45-46; resident, Harrisburg Hosp, Pa, 48; psychiat, Duke Univ Hosp, 49-50, chief resident, 50-51; ward psychiatrist, Vet Admin Hosp, Wilmington, Del, 52-53; from asst prof to assoc prof, 53-89, PROF PSYCHIAT, STATE UNIV NY UPSTATE MED CTR, 89- *Concurrent Pos:* Assoc psychiatrist, Crouse Irving Mem Hosp, Syracuse, 54-; consult psychiatrtist, Hutchings Psychiat Ctr, Syracuse, 55-; Peace Corps, 62-; lectr, Postgrad Prog, NY State Dept Ment Hyg, 58-60. *Mem:* Life fel Am Psychiat Asn; Acad Psychother. *Res:* Psychological aspects of environmental mastery, dreams and childrens games; creativity in art, music and literature; vocational rehabilitation of the handicapped. *Mailing Add:* Dept Psychiat State Univ NY Upstate Med Ctr Syracuse NY 13210

PHILLIPS, RICHARD LANG, b Saginaw, Mich, July 6, 34; m 56; c 3. COMPUTER GRAPHICS, AEROSPACE ENGINEERING. *Educ:* Univ Mich, BSE, 56, MSE, 57, PhD(elecarc behavior), 64. *Prof Exp:* Res engr, Space Technol Labs, Inc, 57-58 & Bendix Systs Div, Bendix Corp, 58-61; assoc res engr, 61-64, from asst prof to assoc prof aerospace eng, 65-74, PROF AEROSPACE ENG, UNIV MICH, ANN ARBOR, 74- *Concurrent Pos:* Consult, Clinton Motors Corp, 61-62; NSF scholar, Munich Tech Univ & Univ Liverpool, 64-65; vis staff mem, Los Alamos Sci Lab, 76-; consult, Tektronix, Inc, 78- *Mem:* Asn Comput Mach. *Res:* Data base management systems; computer aided design. *Mailing Add:* 1549 Mollie St Ypsilanti MI 48198

PHILLIPS, RICHARD LEE, horticulture, plant physiology, for more information see previous edition

PHILLIPS, RICHARD P, b Spokane, Wash, Mar 15, 28; m 85; c 6. EARTH SCIENCE, TEACHING. *Educ:* Stanford Univ, BS, 49, MS, 51 & 56; Univ Calif, San Diego, PhD(geophys), 64. *Prof Exp:* Explor geologist, Day Mines, Inc, Idaho, 50-54; chmn div math & natural sci, Univ San Diego, 58-61; res geophysicist marine phys lab, Scripps Inst, Univ Calif, 61-64; asst prof geol & geophys, San Diego State Col, 64-66; dir Natural Hist Mus, San Diego, Calif, 66-69; asst prof geol & geophys, San Diego State Col, 69-72; ASSOC PROF MARINE & ENVIRON STUDIES & COORDR ENVIRON STUDIES, UNIV SAN DIEGO, 72- *Concurrent Pos:* Mem working group 10 inter-union comn, Int Union Geodynamics & Geophys, 74-80; Fulbright scholar, 84-85. *Mem:* AAAS; Am Geophys Union; Geol Soc Am; Seismol Soc Am. *Res:* Geology and geophysics of continental margins; environmental geology of shallow marine and esturine environments. *Mailing Add:* Envrion Studies Univ San Diego San Diego CA 92110

PHILLIPS, ROBERT ALLAN, b St Louis, Mo, July 2, 37; m 59; c 3. IMMUNOLOGY, CANCER. *Educ:* Carleton Col, BA, 59; Wash Univ, PhD(molecular biol), 65. *Prof Exp:* Res staff, Biol Div, Ont Cancer Inst, 67-86; PROF, DEPT MED BIOPHYS, UNIV TORONTO, 76-, DIR HEMAT & ONCOL RES, PROF DEPT MED GENETICS, 86- *Concurrent Pos:* Nat Cancer Inst Can grant, 67-, mem grant panel, 74-79, mem res adv group, 77-79; fel, Ont Cancer Inst, 65-67; grant, Med Res Coun Can, 70-; chmn immunol grant panel, Ont Cancer Treat & Res Found, 75-76. *Mem:*

Transplantation Soc; Am Asn Immunol; Can Soc Immunol; Am Asn Cancer Res. *Res:* Regulation of early stages of lymphoid differentiation; molecular and cellular studies on the inherited form of retinoblastoma. *Mailing Add:* Hosp for Sick Children 555 University Ave Toronto ON M5G 1X8 Can

PHILLIPS, ROBERT BASS, JR, b Campbell County, Va, Aug 26, 32. MATHEMATICS. *Educ:* Lynchburg Col, BS, 54; Univ Va, MEd, 60, DEd, 69. *Prof Exp:* Teacher math, Altavista High Sch, 57-60; assoc prof, 61-74, PROF MATH, LYNCHBURG COL, 74- *Mem:* Math Asn Am. *Res:* Mathematics education. *Mailing Add:* Dept Math Lynchburg Col Lynchburg VA 24501

PHILLIPS, ROBERT EDWARD, b Evansville, Ind, Dec 30, 23; m 53; c 4. ORGANIC CHEMISTRY. *Educ:* Calif Inst Technol, BS, PhD(chem), 53. *Prof Exp:* Asst, Jr Org Chem Lab, Calif Inst Technol, 48-49, asst, Sr Org Chem Lab, 49-50, asst lectr, 51-52; chief chemist, Synthetic Lab, Calif Found Biochem Res, 52-56; res chemist, Aerojet-Gen Corp, Gen Tire & Rubber Co, Calif, 56-59; vpres, G K Turner Assocs, 59-71; consult, 72-74; vpres, Turner Designs, 74-90; RETIRED. *Mem:* Am Chem Soc. *Res:* Environmental and industrial applications of chemiluminescence; application of fluorometry; analytical instrumentation. *Mailing Add:* 2082 Sandlewood Ct Palo Alto CA 94303-3115

PHILLIPS, ROBERT GIBSON, b Los Angeles, Calif, May 29, 36; m 59; c 3. MATHEMATICS. *Educ:* Univ Calif, BA, 60, MA, 61, PhD(math), 68. *Prof Exp:* Asst prof math, Calif State Col, Los Angeles, 63-67 & Univ SC, Columbia, 67-75; assoc prof, 75-80, PROF MATH, UNIV SC, AIKEN, 80- *Mem:* Am Math Soc; Asn Symbolic Logic. *Res:* Non-standard model theory. *Mailing Add:* 545 Highland Park Terr Aiken SC 29801

PHILLIPS, ROBERT RHODES, b Norfolk, Va. SOCIAL BEHAVIOR, AGGRESSION. *Educ:* Old Dominion Univ, BS, 63; Univ Md, MS, 68, PhD(zool), 71. *Prof Exp:* NIMH fel res assoc marine ethol, Hawaii Inst Marine Biol, 71-72; res assoc zool, Okla State Univ, 72-73; from asst prof to assoc prof, 73-88, PROF & CHAIR BIOL, STATE UNIV COL ONEONTA, NY, 88- *Mem:* Animal Behav Soc; Ecol Soc Am; Int Soc Res Aggression. *Res:* Relationships between social organization and environment; behavioral mechanisms of spacing; behavioral ecology; aquatic predation. *Mailing Add:* Dept Biol State Univ Col Oneonta NY 13820

PHILLIPS, ROBERT WARD, b Peoria, Ill, Jan 21, 29; m 54; c 4. VETERINARY PHYSIOLOGY. *Educ:* Colo State Univ, BS, 59, DVM, 61; Univ Calif, PhD(physiol), 65. *Prof Exp:* NIH trainee, 61-64; from asst prof to prof animal sci, Colo State Univ, 64-80, prof, Dept Physiol, 71-91, prof, Dept Biophys, 80-91; payload specialist, NASA, Houston, 84-91; VISITING SR SCI, NASA HQ, WASHINGTON, DC, 91- *Mem:* AAAS; Am Phys Soc; Am Inst Nutrit; Am Soc Vet Physiol & Pharmacol; Am Vet Med Asn. *Res:* General interest whole animal biochemistry, particularly the interactions of carbohydrate-lipid metabolism and their control; metabolic adaptations to the environment; pathophysiology of diarrhea, altered transport and the individual's response to nutrient imbalance. *Mailing Add:* NASA Hq Div Life Sci 600 Independence Ave SW Washington DC 20546

PHILLIPS, ROGER GUY, DEVELOPMENTAL BIOLOGY, MOLECULAR GENETICS. *Educ:* Univ Calif, Santa Barbara, BA, 79. *Prof Exp:* RES ASST, UNIV TEX, AUSTIN, 80- *Res:* Early embryogenesis. *Mailing Add:* c/o Margaret Phillips Rte 2 Box 381 Bastrop TX 78603

PHILLIPS, ROGER WINSTON, b Bristol, Eng, June 14, 42; US citizen; m 67; c 3. SURFACE CHEMISTRY, THIN FILMS. *Educ:* Univ of Calif, Berkeley, AB, 64; Univ Calif, Davis, PhD(phys chem), 68. *Prof Exp:* Res chemist phys res, E I du Pont de Nemours & Co, Inc, 68-74; mem tech staff res surface chem, The Aerospace Corp, 74-78; research chemist, Optical Coating Lab, Inc, 78-89; PROD DEVELOP MGR, FLEX PROD INC, 89- *Mem:* Am Chem Soc; Sigma Xi. *Res:* Characterization of surfaces by spectroscopic methods, particularly by auger and x-ray photoelectron spectroscopies, scanning electron spectroscopy and by infrared reflection spectroscopy; product and process development of polymeric, magnetic and optical thin films for commercial applications. *Mailing Add:* 466 Jacqueline Dr Santa Rosa CA 95405

PHILLIPS, ROHAN HILARY, b Colombo, Sri Lanka; m; c 3. COMPUTER AIDED MANUFACTURING, ROBOTICS. *Educ:* Univ Sri Lanka, BSME, 70; Purdue Univ, MS, 75, PhD(indust eng), 78. *Prof Exp:* Mech engr, State Eng Corp, Sri Lanka, 70-71, Rice Mkt Bd, 71-73; instr indust eng, Purdue Univ, 75-78; asst prof mfg eng, Univ Ill, Chicago, 78-86; VPRES, KHAN, PHILLIPS & ASSOC, INC, 84- *Concurrent Pos:* Vpres, Khan, Phillips & Assoc, Inc, 84- *Honors & Awards:* Outstanding Young Mfg Eng Award, Soc Mfg Engrs, 84. *Mem:* Am Inst Indust Engrs; Soc Mfg Engrs; Am Soc Eng Educ. *Res:* Group technology applications in computer aided manufacturing; flexible manufacturing technology; integration of robotics; computer integrated manufacturing. *Mailing Add:* Khan, Phillips & Assoc, Inc 1140 Lake St, Suite 401 Oak Park IL 60301

PHILLIPS, RONALD CARL, b Carbondale, Ill, June 4, 32; c 3. MARINE BOTANY. *Educ:* Wheaton Col, BS, 54; Fla State Univ, MS, 56; Univ Wash, PhD, 72. *Prof Exp:* Biologist, Marine Lab, Fla State Bd Conserv, 57-61; from asst prof to prof biol, Seattle Pac Univ, 61-90; ENVIRON CONSULT, BEAK CONSULTS INC, 90- *Concurrent Pos:* Sigma Xi-Sci Res Soc Am grant-in-aid, 62-63; NSF Seagrass Ecosyst Study res grants, Int Decade Ocean Exp, 74-81 & US-E Asia Sci Agreement Prog, Philippines, 83-85; res consult, US Fish & Wildlife Serv, 63 & US Army Corps Engrs, 76-77; vis prof, Cath Univ, Nijmegen, Netherlands, 83,; NSF/USAID Seagrass Res, Philippines, 90-; vis scholar USSR, Nat Acad Sci. *Honors & Awards:* Burlington Northern Award. *Mem:* Am Inst Biol Sci; Ecol Soc Am. *Res:* Aquatic angiosperms; phycology. *Mailing Add:* Beak Consults Inc 12931 NE 126th Pl Kirkland WA 98034-7416

PHILLIPS, RONALD EDWARD, b Williamstown, Ky, Nov 30, 29; m 50; c 3. SOIL PHYSICS, AGRONOMY. *Educ:* Univ Ky, BS, 54, MS, 55; Iowa State Univ, PhD(soil physics), 59. *Prof Exp:* From asst prof to assoc prof soil physics, Univ Ark, 59-66; assoc prof, 66-69, PROF SOIL PHYSICS, UNIV KY, 69- *Honors & Awards:* Thomas Poe Cooper Award, Univ Ky Res Found. *Mem:* Fel Am Soc Agron; fel Soil Sci Soc Am. *Res:* Water and solute movement in soil; relationships of soil physical properties and plant growth; effects of no-tillage on soil properties and plant growth. *Mailing Add:* Dept Agron Univ Ky Lexington KY 40546-0091

PHILLIPS, RONALD LEWIS, b Huntington Co, Ind, Jan 1, 40; m 62; c 2. CYTOGENETICS, PLANT BREEDING. *Educ:* Purdue Univ, BS, 61, MS, 63; Univ Minn, PhD(genetics), 66. *Prof Exp:* NIH fel, Inst Gen Med Sci, Cornell Univ, 66-67; res assoc genetics, 67-68, asst prof, 68-72, assoc prof genetics, 72-76, PROF GENETICS & PLANT BREEDING, UNIV MINN, ST PAUL, 76- *Concurrent Pos:* Mem, Biol Stain Comn, 75-; assoc ed, Genetics, 78-81, Genome, 85-90; NSF, USDA & USAID adv grants panels; prog dir, USDA Res Grants Off, Washington, DC, 79; vis prof, Exp Inst Cereal Crops, Italy, 81 & Univ Guelph, Ont, Can, 83, People's Repub China, 86, Japan, 90; mem, Sci Adv Coun Plant Gene Expression Ctr, 86-; mem, Nat Plant Genetic Resource Bd, 88- *Mem:* Nat Acad Sci; fel AAAS; Genetics Soc Am; fel Am Soc Agron; Plant Molecular Biol Asn; Int Asn Plant Tissue Cell Cult; fel Crop Sci Soc Am. *Res:* Chromosome function and behavior; nucleolus organizers and RNA genes; male sterility; amino acid mutants; somatic cell genetics; recombination phenomena; applications of cytogenetics to plant breeding; plant genetic engineering; plant development; molecular genetics. *Mailing Add:* Dept of Agron & Plant Genetics Univ of Minn St Paul MN 55108

PHILLIPS, RUSSELL ALLAN, b Lakewood, Ohio, Feb 19, 35; m 56; c 3. PHYSICS. *Educ:* Iowa State Univ, BS, 61, PhD(physics), 67. *Prof Exp:* Asst physics, Iowa State Univ, 62-67, assoc, 67-68, asst physicist, Inst Atomic Res, 68-71; from asst prof to assoc prof, 71-78, PROF PHYSICS, DIV ARTS & SCI, MO SOUTHERN STATE COL, 78- *Mem:* Am Asn Physics Teachers; Sigma Xi. *Res:* Experimental study of electronic structure of metals. *Mailing Add:* 515 Grandview Dr Joplin MO 64801

PHILLIPS, RUSSELL C(OLE), b Bad Axe, Mich, Apr 28, 23; m 46; c 2. CHEMICAL ENGINEERING. *Educ:* Mich State Univ, BS, 44; Polytech Inst Brooklyn, MChE, 47. *Prof Exp:* Asst chem engr, Gen Foods Corp, 45-47; res engr, Huron Milling Co, 47-49; chem engr, Blaw-Knox Co, 49-51; sr chem engr, SRI Int, Menlo Park, 51-59, mgr res serv, Phys & Biol Sci Div, 59-61, mgr res & facil serv, 62-63, mgr chem eng, 64-76, dir, Chem Eng Lab, 76-83, sr consult, 87-88; PVT CONSULT, 88- *Concurrent Pos:* Gen mgr, Coal-Tec, 83-87; interim vpres, Res & Develop, Ralph Wilson Plastics Co, 90. *Mem:* Am Chem Soc; Am Inst Chem Engrs; Sigma Xi. *Res:* Synthetic fuels; pollution control; high temperature research; mass transfer. *Mailing Add:* 861 Garland Dr Palo Alto CA 94303

PHILLIPS, RUTH BROSI, b Providence, RI, June 21, 40; m 66; c 2. GENETICS. *Educ:* Swarthmore Col, BA, 62; Ind Univ, Bloomington, MA, 64; Univ Ill, Urbana, PhD(genetics), 67. *Prof Exp:* Res assoc zool, Univ Ill, Urbana, 67, cell biol traineeship, 67-68, asst prof zool, 68-70; lectr, 70-71, from asst prof to assoc prof zool, 71-89, PROF BIOL SCI, UNIV WIS-MILWAUKEE, 89- *Concurrent Pos:* Clin consult, Milwaukee Childrens Hosp, 72-; sr scientist, Ctr Great Lakes Studies, Univ Wis-Milwaukee, 82-; sr scientist, Nat Inst Environ Health Sci, Marine & Freshwater Biomed Core Ctr, Univ Wis-Milwaukee, 89- *Mem:* AAAS; Genetics Soc Am; Am Soc Human Genetics; Am Fisheries Soc; Am Soc Ichthyologists & Herpetologists; Soc Study Evolution. *Res:* Molecular evolutionary genetics; evolution of ribosomal DNA in salmonid fishes; application of molecular genetics techniques to phylogeny reconstruction and conservation biology of vertebrates. *Mailing Add:* Dept Biol Sci Univ Wis PO Box 413 Milwaukee WI 53201

PHILLIPS, S MICHAEL, b San Francisco, Calif, Oct 9, 40; m 82; c 2. IMMUNOLOGY. *Educ:* Univ Wis, BS, 62, MD, 66; Am Bd Internal Med, dipl, 74. *Prof Exp:* Fel immunol, Harvard Univ, 69-71, instr, 71-72; sr res assoc, Walter Reed Army Inst Res, 72-75; intern med, Univ Hosp, 68-69, resident & fel, 67-69, asst prof, 75-80, ASSOC PROF DEPT MED ALLERGY & IMMUNOL, UNIV PA, 80- *Mem:* Am Asn Immunologists; Am Fedn Clin Res; Am Soc Trop Med & Hyg. *Res:* Mechanisms of cellular immunity in diseases of human relevance. *Mailing Add:* Allergy & Immunol Sect Univ Pa Sch Med 519 Johnson Philadelphia PA 19104

PHILLIPS, SAMUEL C, aeronautical & astronautical engineering; deceased, see previous edition for last biography

PHILLIPS, SIDNEY FREDERICK, b Melbourne, Australia, Sept 4, 33; m 57; c 3. GASTROENTEROLOGY. *Educ:* Univ Melbourne, MB, BS, 56, MD, 60; FRACP, 60. *Prof Exp:* Resident med officer internal med, Royal Melbourne Hosp, Univ Melbourne, 57-60, asst sub-dean clin teaching, Clin Sch, 61-62; res asst physiol, 63-64, gastroenterol, 64-66, instr & consult internal med & gastroenterol, 66-69, asst prof internal med, Mayo Clin & Grad Sch Med, 69-72, assoc prof, 72-76, PROF INTERNAL MED, MAYO MED SCH, UNIV MINN, 76- *Concurrent Pos:* Royal Australian Col Physicians traveling scholar gastroenterol, Cent Middlesex Hosp, London, Eng, 62-63. *Mem:* AAAS; Am Gastroenterol Asn; Brit Soc Gastroenterol; Am Soc Clin Invest. *Res:* Gastrointestinal motility, absorption and secretion in health and disease. *Mailing Add:* Dept Med/Gastroent May Med Sch 200 First St SW Rochester MN 55902

PHILLIPS, STEPHEN LEE, b Monte Vista, Colo, Apr 28, 40; m 60; c 2. GENE REGULATION. *Educ:* Adams State Col, BA, 63; Pa State Univ, MS, 65, PhD(biophys), 68. *Prof Exp:* USPHS fel, Wash Univ, 68-70; asst prof, 70-76, ASSOC PROF BIOCHEM, SCH MED, UNIV PITTSBURGH, 76- *Concurrent Pos:* Land use planning. *Mem:* Am Soc Biol Chemists; Am Soc Microbiol. *Res:* Genetic regulation of basement membrane biogenesis; messenger RNA synthesis and processing in eucaryotic cells; mechanisms controlling cellular differentiation. *Mailing Add:* Sch Med Univ Pittsburgh 1256 W Biomed Sci Tower Pittsburgh PA 15261

PHILLIPS, STEVEN J, b Alliance, Nebr, Feb 10, 48; m 69; c 2. SOIL PHYSICS, ENGINEERING GEOLOGY. *Educ:* Western Wash Univ, BS, 70; Mont State Univ, MS, 73. *Prof Exp:* Res asst soil sci, Mont State Univ, 71-73; SR RES SCIENTIST SOIL SCI, BATTELLE PAC NORTHWEST LAB, WESTINGHOUSE-HANFORD CO, 73- *Concurrent Pos:* Mem, Nat Steering Comt, US Dept Energy, 76- *Mem:* Am Soc Agron; Soil Sci Soc Am; Int Soil Sci Soc. *Res:* Nuclear and hazardous waste management; soil chemistry. *Mailing Add:* Westinghouse-Hanford Co PO Box 1970 MSIN 84-14 Richland WA 99352

PHILLIPS, STEVEN JONES, b Atlantic City, NJ, Jan 2, 29; m 52; c 3. ANATOMY. *Educ:* Swarthmore Col, AB, 55; Hahnemann Med Col, MS, 58, MD, 60. *Prof Exp:* Intern, Philadelphia Gen Hosp, 60-61; from instr to assoc prof, 63-77, PROF ANAT, SCH MED, TEMPLE UNIV, 77- *Concurrent Pos:* USPHS trainee anat, Col Physicians & Surgeons, Columbia Univ, 61-63; NIH res grant, 67-69. *Mem:* AAAS; Am Asn Anatomists; Electron Micros Soc Am. *Res:* Fine structure of cardiovascular innervation cardiac and skeletal muscle. *Mailing Add:* Dept Anat Sch Med Temple Univ 3400 N Broad St Philadelphia PA 19140

PHILLIPS, TERENCE MARTYN, b Amersham, Eng, Aug 2, 46; m 75; c 1. IMMUNE REGULATORY MECHANISMS. *Educ:* London Univ, BSc, 73, MPhil, 75, PhD(immunochem), 77, DSc(immunol), 86. *Prof Exp:* Res technician immunol dept path, Westminster Hosp, 70-72; res asst, London Sch Hygiene & Trop Med, 72-75 & McGill Cancer Res Unit, 75-77; asst prof path & asst dir clin chem, 77-81, assoc prof med, 81-90, DIR IMMUNOGENETIC & IMMUNOCHEM LABS, GEORGE WASHINGTON UNIV MED CTR, 85-, PROF MED, 90- *Concurrent Pos:* Consult, Corning Glassware Co & Ctr Environ Health & Human Toxicol, 83- & Food & Drug Admin, 85-; reviewer, NSF, 84- & Nat Cancer Inst, NIH, 85- *Honors & Awards:* McDonald Prize, Inst Trop Med & Hygiene, 75. *Mem:* NY Acad Sci; AAAS; Electron Micros Soc Am; Nat Acad Biochem. *Res:* Regulatory mechanisms of immunological control; idiotypic regulation of antibody expression and the role of circulating immune complexes of lymphocyte reactivity. *Mailing Add:* Immunogenetics & Immunochem Lab 2300 Eye St NW Ross Hall Rm 413 Washington DC 20037

PHILLIPS, THEODORE LOCKE, b Philadelphia, Pa, June 4, 33; m 56; c 3. MEDICINE, RADIOBIOLOGY. *Educ:* Dickinson Col, BS, 55; Univ Pa, MD, 59. *Prof Exp:* Intern, Univ Hosps, Cleveland, Ohio, 59-60; resident therapeut radiol, 60-63, asst prof 65-68, assoc prof, 68-70, PROF RADIOL, MED CTR, UNIV CALIF, SAN FRANCISCO, 70-, CHMN, DIV RADIATION ONCOL, 73- *Concurrent Pos:* Consult, US Naval Radiol Defense Lab, 65-; head, Sect Radiation Oncol, Med Ctr, Univ Calif, San Francisco, 70-73. *Mem:* Radiation Res Soc; Radiol Soc NAm; Am Radium Soc; Am Soc Therapeut Radiol; Am Asn Cancer Res. *Res:* Radiation therapy; neutron radiobiology; normal tissue effects; experimental tumors; study of fractionation and pharmacologic effects on injury and recovery. *Mailing Add:* Dept Radiation Oncol L-75 Univ Calif San Francisco CA 94143

PHILLIPS, THOMAS GOULD, b London, Eng. RADIOASTRONOMY. *Educ:* Oxford Univ, BA, 61, MA & DPhil, 64. *Prof Exp:* Jr res fel physics, Jesus Col, Oxford, 63-67; res assoc, Stanford Univ, 65-66; res officer, Clarendon Lab, Oxford, 66-68; mem tech staff, Bell Lab, Murray Hill, NJ, 68-80; PROF PHYSICS, CALIF INST TECHNOL, 80- *Concurrent Pos:* Lectr, Magdalen Col, 67-68; univ reader, Queen Mary Col, London Univ, 74-75; Consult, Bell Labs, 80-; Mem, Comt Space Astron, Nat Acad Sci & Mgt Operations Working Group Space Astron, NASA, 81- *Mem:* Fel Am Phys Soc; Am Astron Soc; Int Astron Union. *Res:* Submillimeter-wave astronomy; development of heterodyne detection techniques including bolometer mixers and superconducting tunnel junctions; molecules (carbon monoxide) and atoms (carbon) in the interstellar medium to determine physical and chemical parameters of star forming clouds in the galaxy. *Mailing Add:* Dept Physics MS 320-47 Calif Inst Technol Pasadena CA 91125

PHILLIPS, THOMAS JAMES, b Ft Wayne, Ind, May 26, 58. COLLIDING BEAM PHYSICS. *Educ:* Stanford Univ, BS & MS, 80; Harvard Univ, MA, 82, PhD(physics), 86. *Prof Exp:* Res asst, Stanford Linear Accelerator Ctr, 76-80; res asst, Harvard Univ, 80-86, res assoc, 86-90; RES ASST PROF PHYSICS, DUKE UNIV, 90- *Mem:* Am Phys Soc; Am Sci Affil. *Res:* Measuring the W-Boson Mass; missing Et physics; construting a precision vertex detector; proton-decay and cosmic-ray experiments. *Mailing Add:* Fermilab MS318 PO Box 500 Batavia IL 60510

PHILLIPS, THOMAS JOSEPH, b Aug 29, 47; US citizen. CLIMATE RESEARCH, NUCLEAR PHYSICS. *Educ:* Kans State Univ, BS, 69; Univ Wis-Madison, MS, 71, PhD(meteorol), 79. *Prof Exp:* Postdoctoral scientist, Nat Ctr Atmospheric Res, 79-81; res assoc, Nat Res Coun, 81-83; assoc res scientist, Appl Res Corp, 83-86; sr analyst, Computer Sci Corp, 86-89; PHYSICIST, LAWRENCE LIVERMORE NAT LAB, 89- *Mem:* Am Meteorol Soc; Am Phys Soc; Am Geophys Soc. *Res:* Climate dynamics and computer modeling; atmosphere-ocean interaction; numerical/statistical algorithms. *Mailing Add:* Lawrence Livermore Nat Lab L-264 PO Box 808 Livermore CA 94550

PHILLIPS, THOMAS LEONARD, b Istanbul, Turkey, May 2, 24. ENGINEERING. *Educ:* Va Polytech Inst, BS, 47, MS, 48. *Hon Degrees:* DCS, Stonehill Col, 68; DS, Northeastern Univ, 68 & Lowell Technol Inst, 70; LLD, Gordon Col, 70; DBA, Boston Col, 74 & Babson Col, 81; Suffolk Univ, PhD, 86. *Prof Exp:* Var eng & mgt positions, Raytheon Co, 48-61, exec vpres, 61-64, pres, 64-75, chief oper officer, 64-68, dir, 62-91, chief exec officer, 68-91, chmn bd, 75-; RETIRED. *Concurrent Pos:* Dir, John Hancock Mutual Life Ins Co, State St Investment Corp, Knight-Ridder, Newspapers, Inc, Raytheon Co & Digital Equip Corp; trustee, Gordon Col & Northeastern Univ; mem corp, Joslin Diabetes Ctr & Boston Mus Sci. *Mem:* Nat Acad Eng. *Mailing Add:* Raytheon Co 141 Spring St Lexington MA 02173

PHILLIPS, TOM LEE, b Kingsport, Tenn, Dec 6, 31; m 67; c 4. PALEOBOTANY, PLANT MORPHOLOGY. *Educ:* Univ Tenn, BS, 53, BA, 57; Wash Univ, MA, 59, PhD(paleobot), 61. *Prof Exp:* From asst prof to assoc prof bot, 61-72, assoc head dept, 72-73, PROF BOT, UNIV ILL, URBANA, 72-; RES ASSOC, ILL GEOL SURV, 77-, PROF GEOL, 82- *Concurrent Pos:* Fel, John Simon Guggenheim Mem Found, 75-76; exchange scientist, Nat Acad Sci, USSR, 76; vis lectr, China Inst Mining, PRC, 82; head dept, plant biol, Ill Geol Surv, 84-88. *Mem:* AAAS; Bot Soc Am; Torrey Bot Club; Brit Paleont Asn; Geol Soc Am; Sigma Xi. *Res:* Study of plant evolution and morphology based on Paleozoic age fossil plants and analyses of Pennsylvanian age coal swamp floras and vegetation based on petrified peat; paleoecology. *Mailing Add:* Dept Plant Biol Univ of Ill Urbana IL 61801

PHILLIPS, TRAVIS J, b Moline, Ill, Dec 13, 19; m 45; c 4. PHYSICAL CHEMISTRY. *Educ:* Iowa State Teachers Col, BA, 41; Univ Iowa, MS, 48; Ohio State Univ, PhD(chem), 58. *Prof Exp:* Instr physics, Iowa State Teachers Col, 46-47; instr chem, Evansville Col, 48-51; engr, Westinghouse Elec Corp, 57-64; chmn natural sci, Robert Morris Jr Col, 64-66; asst prof 66-85, EMER PROF CHEM, PURDUE UNIV, CALUMET CAMPUS, 85- *Mem:* AAAS; Am Chem Soc; Am Crystallog Asn. *Res:* X-ray diffraction; solid state chemistry and physics. *Mailing Add:* 1750 S Flannery Rd Baton Rouge LA 70816

PHILLIPS, VERIL LEROY, b Denison, Tex, July 15, 43; m 64; c 1. ALGEBRA, COMPUTER SCIENCE. *Educ:* Univ Tulsa, BS, 65; Mich State Univ, MA, 72, PhD(math), 75. *Prof Exp:* Programmer & analyst eng, Amerada Petrol Corp, 64-66; instr physics, US Naval Nuclear Power Sch, 67-70; math teaching & consult, 74-77; from asst prof to prof, 78-87, PROF MATH & COMPUT SCI, SAN JOSE STATE UNIV, 87-, CHMN DEPT, 85- *Mem:* Math Asn Am; Am Math Soc; Asn Comput Mach; Am Asn Artificial Intel. *Res:* Theory of infinite groups, particularly the Cernikov p-groups; Computer Sciences. *Mailing Add:* Dept Math & Computer Sci San Jose State Univ San Jose CA 95192-0103

PHILLIPS, WENDELL FRANCIS, b Revere, Mass, May 1, 21; m 49; c 3. ANALYTICAL CHEMISTRY. *Educ:* Colby Col, AB, 48. *Prof Exp:* Control chemist, Berke Bros, 48-49; res chemist, Beech-Nut Packing Co, 49-51, chief chemist, food lab, 51-56, mgr res & develop, 56-60; staff chemist, 60-71, mgr labs-tech admin, 71-75, dir tech resources, 76-78, sr scientist, 78-80, sr res scientist, 80-85, PRIN SCIENTIST, CAMPBELL SOUP CO, 85- *Mem:* Am Chem Soc; Inst Food Technol. *Res:* Analysis of agricultural chemicals; research and development of food products. *Mailing Add:* 624 E Main St Morrestown NJ 08057

PHILLIPS, WILLIAM, solid state physics, material science, for more information see previous edition

PHILLIPS, WILLIAM BAARS, b Nashville, Tenn, Oct 18, 34; m 72. SOLID STATE PHYSICS. *Educ:* David Lipscomb Col, BA, 56; Vanderbilt Univ, MS, 59; Fla State Univ, PhD(physics), 67. *Prof Exp:* Instr physics, Murray State Univ, 59-60, asst prof, 60-62; from asst prof to assoc prof, Univ WFla, 67-73; dir sci & eng, Bd Regents, State Univ Syst Fla, 73-77; acad planning coordr, 77-79, assoc dir acad prof, Ariz Bd Regents, 79-; AT PHYSICS DEPT, PEPPERDINE UNIV. *Mem:* AAAS; Am Phys Soc; Am Asn Physics Teachers; Am Vacuum Soc. *Res:* Structure, surface morphology and electrical properties of very thin discontinuous vacuum-deposited metal films; high-vacuum techniques. *Mailing Add:* Dept Physics Pepperdine Univ 24255 Pac Coast Hwy Malibu CA 90265

PHILLIPS, WILLIAM DALE, b Kansas City, Mo, Oct 10, 25; m 50; c 2. BIOPHYSICS, PHYSICAL CHEMISTRY. *Educ:* Univ Kans, AB, 48; Mass Inst Technol, PhD(phys chem), 51. *Prof Exp:* Assoc dir basic sci, Cent Res Dept, 51-74, tech mgr mycoprotein venture, The Lord Rank Res Ctr, Eng, 74-76, asst dir res & develop, Plastics Prod & Resins Dept, E I du Pont de Nemours & Co, Inc, 76-78; Charles Allen Thomas prof chem & chmn dept, Sch Med, Wash Univ, 77-84, dir, Ctr Biotechnol, 82-84; sr vpres sci & technol, Mallinckrodt, 84-; ASSOC DIR OFF SCI & TECHNOL POLICY. *Mem:* Nat Acad Sci; AAAS; Am Chem Soc; Am Phys Soc; Am Acad Arts & Sci. *Res:* Molecular, electronic structure; proteins; magnetic resonance. *Mailing Add:* Off Sci Technol Policy Rm 5005 NEOB Washington DC 20504

PHILLIPS, WILLIAM DANIEL, b Wilkes-Barre, Pa, Nov 5, 48; m 70; c 2. LASER SPECTROSCOPY, QUANTUM OPTICS. *Educ:* Juniata Col, BS, 70; Mass Inst Technol, PhD(physics), 76. *Prof Exp:* Res assoc, Mass Inst Technol, 76-78; PHYSICIST, NAT BUR STANDARDS, 78- *Concurrent Pos:* Vis lectr, Col France, 87; vis prof, Ecole Normale, Supérieure, Paris, 89-90. *Honors & Awards:* S W Stratton Award, Nat Inst Standards & Technol, 87. *Mem:* Fel Am Phys Soc; Optical Soc Am. *Res:* Precision measurement of fundamental constants; laser-cooling & trapping of neutral atoms; physics of laser-cooling & laser-cooled atoms. *Mailing Add:* PHY B160 Nat Inst Standards & Technol Gaithersburg MD 20899

PHILLIPS, WILLIAM ERNEST JOHN, b Ottawa, Ont, July 31, 29. ANIMAL NUTRITION. *Educ:* McGill Univ, BSc, 51, MSc, 53; Univ Liverpool, PhD(biochem), 59. *Prof Exp:* Chemist, Sci Serv Div, Can Dept Agr, 53-59, res officer, Animal Res Inst, 59-64, chief biochem sect, 64-65; res scientist, 65-69, head pesticide sect, 69-73, RES SCIENTIST, NUTRIT RES DIV, HEALTH PROTECTION BR, DEPT NAT HEALTH & WELFARE, 73- *Concurrent Pos:* Mem subcomt, vitamin A, Nat Comt Animal Nutrit, 60. *Honors & Awards:* Borden Award Can, 71. *Mem:* Can Biochem Soc; Nutrit Soc Can; Prof Inst Pub Serv Can; Can Soc Animal Care; Am Inst Nutrit. *Res:* Vitamin metabolism; biochemical and nutritional aspects of the fat-soluble vitamins, particularly vitamin A; elucidating the ubiquinones; relationship of vitamin A to nonsaponifiable constituents; pesticide metabolism and toxicology. *Mailing Add:* 28 Gervin Crescent Ottawa ON K2G 0J8 Can

PHILLIPS, WILLIAM GEORGE, b Norristown, Pa, May 1, 29; m 55; c 2. ENTOMOLOGY. *Educ:* Pa State Univ, BA, 51; Univ Md, MS, 57, PhD(entom), 61. *Prof Exp:* Supvry entomologist, Dept Army, 63-67, entomologist, 67-70; entomologist, USDA, 70; supvry entomologist, Environ Protection Agency, 70-74, supvry ecologist, 74-85, supvry biologist, 85-90; RETIRED. *Concurrent Pos:* Substitute teaching, Sci, Frederick County, Md Pub Sch. *Mem:* Entom Soc Am; Am Mosquito Control Asn; Sigma Xi. *Res:* Dietary and fecundity studies associated with insect rearing and colonization; effects of temperature and methods of feeding on the survival and life span of insects; federal regulation pesticides. *Mailing Add:* 7188 W Sundown Ct Frederick MD 21702

PHILLIPS, WILLIAM H, b Port Sunlight, Eng, May 31, 18. AERONAUTICAL ENGINEERING. *Educ:* Mass Inst Technol, BAE, 39, MAE, 40. *Prof Exp:* Head aerospace mech div, NACA, 40-58; chief flight dynamics & control div, 58-79, DISTINGUISHED RES ASSOC, NASA, 79- *Honors & Awards:* Lawrence Sperry Award, Am Inst Aeronaut & Astronaut, 44, Mech & Controllers Flight Award, 88. *Mem:* Nat Acad Eng; Am Inst Aeronaut & Astronaut. *Mailing Add:* 310 Manteo Ave Hampton VA 23661

PHILLIPS, WILLIAM MAURICE, b Newton, Kans, Dec 4, 22; m 46; c 4. WEED CONTROL, CROP PRODUCTION. *Educ:* Kans State Univ, BS, 47, MS, 49. *Prof Exp:* Res agronomist weed control, US Dept Agr, 48-73; res agronomist, Kans State Univ, 73-76, head, Ft Hays Br, Kans Agr Exp Sta, 76-85; RETIRED. *Mem:* Weed Sci Soc Am; Coun Agr Sci & Technol. *Res:* Weed control in reduced tillage systems, weed control in sorghum and winter wheat, control of field bindweed. *Mailing Add:* 2013 Somerset Sq Manhattan KS 66502

PHILLIPS, WILLIAM REVELL, b Salt Lake City, Utah, Jan 9, 29; m 50; c 4. MINERALOGY, GEMOLOGY. *Educ:* Univ Utah, BS, 50, MS, 51, PhD(mineral), 54. *Prof Exp:* Teaching & res fels, Univ Utah, 49-50; res mineralogist & petrographer, Res Ctr, Kennecott Copper Corp, 54-56; asst prof geol, La Polytech Inst, 56-57; from asst prof to assoc prof, 57-66, chmn dept, 72-75, PROF GEOL, BRIGHAM YOUNG UNIV, 66- *Concurrent Pos:* Fulbright lectr, Univ Sind, Pakistan, 63-64 & Middle East Tech Univ, Ankara, 66-67; vis prof, Univ Waterloo, 71-72; vis res prof, Hacettepe Univ, Ankara, 75; asst dir, Fayum Archaeol Exped, Egypt, 81-91. *Mem:* Mineral Soc Am; Geol Soc Am. *Res:* Petrography; blueschist mineralogy of plate subduction zones; gemology. *Mailing Add:* Dept Geol Brigham Young Univ Provo UT 84601

PHILLIPS, WINFRED M(ARSHALL), b Richmond, Va, Oct 7, 40; div; c 2. FLUID MECHANICS, MECHANICAL ENGINEERING. *Educ:* VA Polytech Inst, BSME, 63; Univ Va, MAE, 66, DSc(aerospace eng), 68. *Prof Exp:* Res scientist, res labs, eng sci, Univ Va, 67-68; from asst prof to prof aerospace eng, Pa State Univ, 68-80, assoc dean res, Col Eng, 79-80; prof & head, Sch Mech Eng, Purdue Univ, W Lafayette, Ind, 80-88; DEAN & ASSOC VPRES, COL ENG, UNIV FLA, 88- *Concurrent Pos:* Res career develop award, US Pub Health Serv, NIH, 74-78; vpres educ, Am Soc Mech Engrs; NSF trainee, 63-67; bd dirs, Tokheim Corp, 85-, Accred Bd Eng & Technol, 89-92, Fla High Technol & Indust Coun, 90-94. *Mem:* Fel Am Soc Mech Engrs; Int Soc Biorheology; Am Soc Artificial Internal Organs (pres); fel AAAS; assoc fel Am Inst Aeronaut & Astronaut; fel Am Soc Eng Educ. *Res:* Fluid mechanics and gas dynamics; hemodynamics; blood rheology; cardiovascular dynamics; prosthetic devices; artificial heart and left ventricular assistance; bioengineering. *Mailing Add:* Col Eng Univ Fla Gainesville FL 32611

PHILLIPS, YORKE PETER, b Brooklyn, NY, Apr 1, 32; m 53; c 3. POLYMER CHEMISTRY. *Educ:* Polytech Inst Brooklyn, BChE, 53; Am Int Col, MBA, 59. *Prof Exp:* Res eng polymer processing, Monsanto Co, 56-60, res group leader polyvinyl chloride polymers, 60-68, mgr res polyvinyl chloride & polystyrene polymers, 68-77, -gr tech abs polymers, 78-81, mgr tech, Fome-Cor, 82-85; RETIRED. *Mem:* Am Chem Soc; Soc Plastics Engrs; Am Inst Chem Engrs. *Res:* Applications and processes for polystyrenes, polystyrene laminates and polyvinyl chloride polymers. *Mailing Add:* 235 Ames Rd Hampden MA 01036

PHILLIPSON, ELIOT ASHER, b Edmonton, Alta, Dec 21, 39; m 70; c 2. RESPIRATORY PHYSIOLOGY & DISEASE. *Educ:* Univ Alta, MD, 63, MSc, 65; FRCP(C), 68. *Prof Exp:* From asst prof to assoc prof, 71-80, PROF MED, UNIV TORONTO, 80-, DIR, RESPIRATORY DIV, 83-; CO-DIR, SLEEP RES LAB, QUEEN ELIZABETH HOSP, 78- *Concurrent Pos:* Res scholar, Med Res Coun Can, 71-76, mem Heart & Lung Rev Comt, 77-80; staff physician, Toronto Gen Hosp, 71-; vis prof, Inst Neurol, Univ London, 81-82. *Honors & Awards:* Cecile Lehman Mayer Res Award, Am Col Chest Physicians, 73. *Mem:* Am Physiol Soc; Am Soc Clin Invest; Am Thoracic Soc; Sleep Res Soc; Can Soc Clin Invest. *Res:* Regulation of respiration during wakefulness and sleep; respiratory disorders during sleep. *Mailing Add:* Dept Med Pulmonary Dis Univ Toronto Fac Med One Kings College Circle Toronto ON M5S 1A8 Can

PHILLIPSON, PAUL EDGAR, b Newark, NJ, May 22, 33; m 65. BIOPHYSICS, NONLINEAR DYNAMICS. *Educ:* Univ Chicago, BA, 53, MS, 56, PhD(physics), 62. *Prof Exp:* Res assoc physics, Univ Mich, 61-63; asst prof, 63-67, asst prof biophys, 65-71, assoc prof, 67-77, PROF PHYSICS, UNIV COLO, BOULDER, 77- *Concurrent Pos:* Alfred P Sloan Found fel, 65-67; Europ Molecular Biol Orgn res fel, Univ Rome, 72-74; vis prof, Univ of Vienna, 83. *Res:* Molecular quantum mechanics; electronic structure of molecules; theory of molecular force constants; molecular transition probabilities; mu-mesic molecules; macromolecular kinetics; statistical mechanics of liquids, nonlinear dynamics. *Mailing Add:* Dept Physics Univ Colo Boulder Campus Box 390 Boulder CO 80309

PHILLIPS-QUAGLIATA, JULIA MOLYNEUX, b London, Eng, July 22, 38; m 81. IMMUNOLOGY, ZOOLOGY. *Educ:* Univ London, BSc, 59; Univ Edinburgh, PhD(immunol), 63. *Prof Exp:* Sci officer, Med Res Coun Rheumatism Res Unit, Eng, 63-65; sci officer, Med Res Coun Nat Inst Med Res, Eng, 67-68; from asst prof to assoc prof, 70-89, PROF PATH, SCH MED, NY UNIV, 89- *Concurrent Pos:* Fel, Irvington House Inst, NY, 65-67; USPHS res grant, Sch Med, NY Univ, 69-, res career develop award, 73; Am Cancer Soc res grant, 77- *Mem:* Am Asn Immunologists; Am Soc Exp Path; Harvey Soc; Brit Soc Immunol; Soc Mucosal Immunol. *Res:* Cellular immunology; mucosal immunology. *Mailing Add:* Dept Path NY Univ Med Ctr New York NY 10016

PHILLIS, JOHN WHITFIELD, b Trinidad, WI, Apr 1, 36; m 69; c 3. NEUROPHYSIOLOGY, NEUROPHARMACOLOGY. *Educ:* Univ Sydney, BVSc, 58, DVSc, 76; Australian Nat Univ, PhD(neurophysiol), 61; Monash Univ, Australia, DSc(neurophysiol), 70. *Prof Exp:* Wellcome res fel, Agr Res Coun Inst Animal Physiol, 61-62; lectr physiol, Monash Univ, Australia, 63-66, sr lectr, 67-69; prof physiol & assoc dean med, Univ Man, 71-73; prof physiol & head dept, Univ Sask, 43-81; PROF PHYSIOL & CHMN, WAYNE STATE UNIV, 81- *Concurrent Pos:* Vis prof, Ind Univ, 69; Wellcome vis prof, Tulane Univ, 86; ed, Prog Neurobiology, 73-, Can J Physiol Pharmacol, 78-81, Int J Purine Pyrimid Res, 90- *Mem:* Brit Pharmacol Soc; Int Brain Res Orgn; Brit Physiol Soc; Am Physiol Soc; Soc Neuroscience. *Res:* Pharmacology and identification of synaptic transmitters in the central nervous system; role of purines in the modulation of central exutability; control of cerebral blood flow; prevention of stroke damage. *Mailing Add:* Dept Physiol Sch Med Wayne State Univ 540 E Canfield Detroit MI 48201

PHILLS, BOBBY RAY, b Shreveport, La, Sept 12, 45; m 68; c 2. HORTICULTURE, PLANT BREEDING. *Educ:* Southern Univ, Baton Rouge, La, BS, 68; La State Univ, MS, 72, PhD(hort), 75. *Prof Exp:* Res assoc veg crops, NY State Agr Exp Sta, Cornell Univ, 75-76; ASST PROF PLANT & SOIL SCI, TUSKEGEE INST, 76- *Concurrent Pos:* Rockefeller Found fel, Cornell Univ, 75-76. *Mem:* Am Genetics Asn; Am Soc Hort Sci; Sigma Xi. *Res:* Development and evaluation of tomatoes adapted to adverse environmental conditions coupled with pest resistance; development of sweet potato germplasm suitable for seedpiece propagation. *Mailing Add:* 13608 Cadiz Dr Baker LA 70714

PHILO, JOHN STERNER, b Washington, DC, Dec 28, 48; m 70; c 1. MAGNETIC SUSCEPTIBILITY, METALLOPROTEINS. *Educ:* Johns Hopkins Univ, BA, 69; Stanford Univ, PhD(physics), 77. *Prof Exp:* Sr physicist, Superconducting Technol Div, United Sci Corp, Mountain View, Calif, 77-78; postdoctoral fel, Dept Biochem & Biophys, 78-80, asst prof, 80-87, ASSOC PROF BIOPHYS, DEPT MOLECULAR & CELL BIOL, UNIV CONN, STORRS, 88- *Concurrent Pos:* Consult, Xenogen, Inc, Storrs, Conn, 82-84, Zeamer Systs Group, Hopkinton, Ma, 88- *Mem:* Biophys Soc; Am Phys Soc; AAAS; Asn Com Sci. *Res:* Structure and function of metalloproteins, as probed by the magnetic susceptibility of the metal ions and by rapid reaction kinetics; cooperative mechanism of hemoglobin. *Mailing Add:* Dept Molecular & Cell Biol Univ Conn Box U-125 Storrs CT 06268

PHILOGENE, BERNARD J R, b Beau-Bassin, Mauritius, May 4, 40; Can citizen; m 64; c 2. PEST MANAGEMENT, PESTICIDES. *Educ:* Univ Montreal, BSc, 64, McGill Univ, MSc, 66; Univ Wis-Madison, PhD(entom), 70. *Prof Exp:* Res officer, Can Forestry Serv, 66-71; asst prof entom, Univ BC, 71-74; from asst prof to assoc prof, 74-82, vice-dean, fac sci & eng, 82-85, actg dean, 85-86, PROF ENTOM, 82-, UNIV OTTAWA, DEAN FAC SCI, 86- *Concurrent Pos:* Consult, Can Int Develop Agency, 82- & Int Registry of Potentially Toxic Chemicals, 85-; mem, NSERC Strategic Grant Panel, 85-88. *Mem:* Fel Entom Soc Can; Can Pest Mgt Soc; French Can Asn Advan Sci; Entom Soc Am; Am Inst Biol Sci. *Res:* Influence of light on insect growth and development; plant-insect relationships; effects of phototoxic compounds; pest management in developing countries. *Mailing Add:* Dept Biol Univ Ottawa Ottawa ON K1N 6N5 Can

PHILOON, WALLACE C, JR, b Peking, China, Mar 19, 23; US citizen; m 54; c 2. CHEMICAL ENGINEERING. *Educ:* Bowdoin Col, BS, 45; Mass Inst Technol, MS, 47, ScD(chem eng), 50. *Prof Exp:* Proj engr, Mallinckrodt Chem Works, Mo, 50-57, asst mgr process develop, Uranium Div, 57-61, proj engr, 61-64; assoc prof chem eng, Univ Tulsa, 64-84; RETIRED. *Mem:* Am Inst Chem Engrs; Nat Asn Corrosion Engrs. *Res:* Corrosion, especially stress corrosion cracking. *Mailing Add:* 1210 E 30th Pl Tulsa OK 74114

PHILP, RICHARD BLAIN, b Guelph, Ont, Jan 19, 34; m 55, 87; c 4. PHARMACOLOGY. *Educ:* Univ Toronto, DVM, 57; Univ Western Ont, PhD(pharmacol), 64. *Prof Exp:* Can Defence Res Bd fel aviation med, 64-65; from asst prof to assoc prof, 65-73, chmn dept pharmacol & toxicol, 81-86, PROF PHARMACOL, FACULTIES MED, DENT & GRAD STUDIES, UNIV WESTERN ONT, 73- *Concurrent Pos:* Ont Heart Found res grant; Dept Nat Defence contract, US Office of Naval Res Contract. *Honors & Awards:* Stover-Link Award, Undersea Med Soc, 77. *Mem:* Undersea Med Soc; Aerospace Med Asn; Pharmacol Soc Can; NY Acad Sci; AAAS. *Res:* Biological factors influencing susceptibility to decompression sickness; platelet aggregation and thrombus formation; pharmacological effects of, and drug interactions with, inert gases at pressure. *Mailing Add:* Dept Pharmacol & Toxicol Med Sci Bldg Univ Western Ont London ON N6A 5C1 Can

PHILP, RICHARD PAUL, b Plymouth, Eng, Aug 26, 47; m 75; c 3. ORGANIC GEOCHEMISTRY. *Educ:* Univ Aberdeen, Scotland, BSc, 68; Univ Sydney, PhD(org chem), 72. *Prof Exp:* Fel org geochem, Univ Bristol, Eng, 72-73; fel, Univ Calif, Berkeley, 74-75, asst res chemist org geochemistry, 76-78; RES SCIENTIST, Commonwealth Sci & Indust Res Orgn, Sydney, Australia, 78-; AT SCH GEOL, UNIV OKLA. *Mem:* Royal Inst Chem London; AAAS; Sigma Xi. *Res:* Study of origin and method of formation of petroleum; crude oil correlation studies; coal and oil shale characterization and utilization. *Mailing Add:* Sch Geol Univ Okla Norman OK 73019

PHILP, ROBERT HERRON, JR, b Demorest, Ga, Sept 10, 34; m 57; c 4. ANALYTICAL CHEMISTRY. *Educ:* Wheaton Col, Ill, BS, 56; Emory Univ, PhD(anal chem), 62. *Prof Exp:* Res assoc chem, Univ Kans, 62-63; asst prof, 63-73, ASSOC PROF CHEM, UNIV SC, 73- *Mem:* Am Chem Soc; Sigma Xi. *Res:* Mechanisms of polarographic reductions of carbonyl compounds; techniques related to square wave polarography. *Mailing Add:* Dept Chem Univ SC Columbia SC 29208

PHILPOT, JOHN LEE, b Kansas City, Mo, May 7, 35; m 56, 82; c 3. PHYSICS. *Educ:* William Jewell Col, AB, 57, Univ Ark, MS, 61, PhD(physics), 65. *Prof Exp:* From asst prof to assoc prof, 62-74, PROF PHYSICS, WILLIAM JEWELL COL, 74- *Concurrent Pos:* Nat Defense Emergency Authorization Title IV fel, 59-62; consult, thin film lab, Bendix Corp, 65-66 & forensic physics. *Mem:* Am Asn Physics Teachers; Am Phys Soc. *Res:* Optics of thin films; solar heating; atomic and molecular collision phenomena. *Mailing Add:* Dept Physics William Jewell Col Liberty MO 64068

PHILPOTT, CHARLES WILLIAM, b Canadian, Tex, Jan 29, 32; m 60; c 2. CELL BIOLOGY. *Educ:* Tex Tech Col, BA, 57, MS, 58; Tulane Univ, PhD(zool), 62. *Prof Exp:* Fel cell biol, Harvard Univ, 62-64; from asst prof to prof, 64-69, PROF BIOL, RICE UNIV, 69- *Concurrent Pos:* Head master, Baker Residential Col, Rice Univ, 68- *Mem:* AAAS; Am Asn Anat; Am Soc Cell Biol; Am Soc Zool; Electron Micros Soc Am. *Res:* Comparative fine structure and cytochemistry of osmoregulatory tissues; ion transport and polyanions of the cell surface. *Mailing Add:* 2323 Gramercy St Houston TX 77030

PHILPOTT, DELBERT E, b Loyal, Wis, Sept 24, 23; m. MOLECULAR BIOLOGY, MEDICAL RESEARCH. *Educ:* Ind Univ, BA, 46, MS, 49; Boston Univ, PhD(cytol), 63. *Hon Degrees:* PhT, Mass Col Optom, 76. *Prof Exp:* Res asst electron micros, Ind Univ, 47-49; res assoc, Ultrastruct Lab, Univ Ill Med Ctr, 49-52; head electron micros lab, Inst Muscle Res & Marine Biol Lab, Woods Hole, 52-63; prof biochem, Med Sch, Univ Colo, 63-65; head & co-dir, Electron Micros Lab, Ultrastruct Lab, Inst Biomed Res, Mercy Hosp, Denver, 65-66; head ultrastruct lab, Ames Res Ctr, NASA, 66-90; RETIRED. *Concurrent Pos:* Res asst, Boston Univ, 60-63; mem bd dirs, Inst Biomed Res, Mercy Hosp, Denver, 71-75. *Honors & Awards:* NASA/USSR Soyu Bio Mission Award, 76. *Mem:* AAAS; Am Soc Cell Biol; Electron Micros Soc Am; Biophys Soc; Sigma Xi; NY Acad Sci. *Res:* Ultrastructure of retina; muscle and cartilage as changed by abnormal environments encountered in space. *Mailing Add:* 1602 Kamsack Dr Sunnyvale CA 94087

PHILPOTT, JANE, b Kansas City, Mo, June 12, 18. BOTANY. *Educ:* Harris Teachers Col, AB, 40; Univ Iowa, MS, 41, PhD(plant anat), 47. *Prof Exp:* Teacher pub schs, Mo, 41-44; asst bot, Univ Iowa, 44-47; instr biol, Univ Chicago, 47-51; from asst prof to assoc prof bot, 51-68, dean instr, 61-72, PROF BOT, DUKE UNIV, 68- *Concurrent Pos:* Am Asn Univ Women fel, Duke Univ, 58; consult, Encycl Brittanica Films, Inc, 59-60. *Mem:* AAAS; Bot Soc Am. *Res:* Ecological anatomy of foliage leaves; ecological plant anatomy of woody plants. *Mailing Add:* 131 Carol Woods Chapel Hill NC 27514

PHILPOTT, MICHAEL RONALD, b Bristol, Eng, Jan 2, 40; div; c 2. CHEMICAL PHYSICS, TRIBOLOGY. *Educ:* Univ Col, London, BSc, 61, PhD(chem), 64,. *Hon Degrees:* DSc, Univ Col, London,74. *Prof Exp:* Res assoc theoret sci, 65-67, from asst prof to assoc prof chem, Univ Ore, 67-73; res staff mem, 73-75, RES MGR PHYS SCI, RES DIV, INT BUS MACH CORP, 75- *Concurrent Pos:* Ramsay fel, Univ Col, London, 64-65; Alfred P Sloan fel, 70-72, Camille & Henry Dreyfus scholar, Univ Ore, 73. *Honors & Awards:* Fel Am Phys Soc. *Mem:* Am Phys Soc; Am Chem Soc; Electrochem Soc. *Res:* Tribology of magnetic recording; electronic processes in organic solids; surface polariton spectroscopy; in situ surface spectroscopy of metal electrodes. *Mailing Add:* Dept K33/801 IBM Res Div Almaden Res Ctr 650 Harry Rd San Jose CA 95120-6099

PHILPOTT, RICHARD JOHN, b London, Eng, Sept 4, 36; m 66. NUCLEAR PHYSICS. *Educ:* Oxford Univ, BA, 60, MA & DPhil(theoret physics), 63. *Prof Exp:* Res assoc nuclear theory, Univ Pittsburgh, 63-64, Univ Calif, Davis, 64-66 & Vanderbilt Univ, 66-68; asst prof, 68-74, assoc prof, 74-79, PROF PHYSICS, FLA STATE UNIV, 79- *Concurrent Pos:* Mellon fel, 63-64. *Mem:* Am Phys Soc; Sigma Xi. *Res:* Theory of nuclear structure and reactions; model calculations for nuclear continuum phenomena including many-body effects. *Mailing Add:* Dept of Physics Fla State Univ Tallahassee FL 32306

PHILPOTTS, ANTHONY ROBERT, b Bristol, Eng, July 27, 38; Can citizen; m 60; c 3. PETROLOGY, MINERALOGY. *Educ:* McGill Univ, BSc, 58, MSc, 60; Cambridge Univ, PhD(petrol), 63. *Prof Exp:* Asst prof petrol, McGill Univ, 63-70; assoc prof geol, 70-75, head, dept geol & geophys, 78-83, PROF GEOL, UNIV CONN, STORRS, 75- *Honors & Awards:* Hawley Award, Mineral Asn Can, 82. *Mem:* Mineral Soc Am; Mineral Asn Can; Brit Mineral Soc; Geol Soc Can; Geol Soc Am. *Res:* Liquid immiscibility in silicate melts. *Mailing Add:* Dept of Geol Univ of Conn Storrs CT 06269

PHINNEY, BERNARD ORRIN, b Superior, Wis, July 29, 17; m 51, 65; c 4. BOTANY. *Educ:* Univ Minn, BS, 40, PhD(bot), 46. *Hon Degrees:* DSc, Univ Bristol, 89. *Prof Exp:* Asst bot & plant physiol, Univ Minn, 40-46; inst fel, Calif Inst Technol, 46-48; from instr to prof bot, 47-73, prof biol, 73-88, EMER PROF BIOL, UNIV CALIF, LOS ANGELES, 88- *Concurrent Pos:* NSF sr fel, Genetics Inst, Copenhagen Univ, 59-60; US-Japan Coop Sci Prog NSF vis scientist, Int Christian Univ, Tokyo, 66-67; NSF vis prof, Dept Chem, Bristol Univ, UK, 73-74, 82-83; chair, Sect Plant Biol, Nat Acad Sci, 89-; res fel, Japan Soc Prom Sci, 91. *Honors & Awards:* IPGSA Res Award, 82; Stephen Hales Award, Am Soc Plant Physiol, 84; Charles Reid Barnes Award, 87. *Mem:* Nat Acad Sci; Bot Soc Am; Am Soc Plant Physiol (pres, 89); Genetics Soc Am; AAAS; Scand Soc Plant Physiol; hon foreign mem Japanese Soc Chem Regulation Plants. *Res:* Chemical genetics and molecular biology of gibberellins; plant growth; author 90 publications. *Mailing Add:* Dept Biol Univ Calif Los Angeles CA 90024

PHINNEY, GEORGE JAY, b Columbus, Ohio, July 29, 30; m 53; c 1. VERTEBRATE ECOLOGY. *Educ:* Ohio State Univ, BSc, 53, MSc, 56, PhD(zool), 67. *Prof Exp:* Asst instr zool, Ohio State Univ, 57-59, instr, 59-62; from asst prof to assoc prof, 62-80, PROF BIOL, OTTERBEIN COL, 80- *Mem:* Ecol Soc Am; Am Soc Limnol & Oceanog; Am Fisheries Soc; Am Soc Ichthyologists & Herpetologists. *Res:* Freshwater ecology; lentic and lotic communities. *Mailing Add:* Dept Life Sci Otterbein Col Westerville OH 43081

PHINNEY, HARRY KENYON, botany; deceased, see previous edition for last biography

PHINNEY, NANETTE, b Evanston, Ill, 1945. ACCELERATOR CONTROL SYSTEMS, HIGH ENERGY PHYSICS. *Educ:* Mich State Univ, BS, 66; State Univ NY Stony Brook, MA, 69, PhD(physics), 72. *Prof Exp:* Charge res physics, Ecole Polytech Paris, 72-74; vis scientist, Europ Orgn Nuclear Res, 74-75; res officer physics, Nuclear Physics Lab, Oxford Univ, 75-80; vis scientist, Europ Orgn Nuclear Res, 81; vis mathematician, 81-83, head, software eng, 84-86, ACCELERATOR PHYSICIST, SOFTWARE ENG, INSTRUMENTATION & CONTROL DEPT, STANFORD LINEAR ACCELERATOR CTR, 87- *Mem:* Am Phys Soc. *Res:* Accelerator physics; accelerator control systems; lepton pair production; high transverse momentum phenomena; experimental high energy physics. *Mailing Add:* Stanford Linear Accelerator Ctr BIN 66 Stanford CA 94309

PHINNEY, RALPH E(DWARD), b Cleveland, Ohio, Mar 3, 28. AERODYNAMICS. *Educ:* Univ Mich, BS, 50, MS, 51, PhD(aerodyn), 53. *Prof Exp:* Asst, Univ Mich, 51-53; res assoc, Johns Hopkins Univ, 53-54; res & develop scientist, Martin Marietta Corp, Md, 59-68; RES & DEVELOP SCIENTIST, NAVAL ORD LAB, SILVER SPRING, 68- *Mem:* AAAS; Am Inst Aeronaut & Astronaut. *Res:* Theoretical and experimental aerodynamics; supersonic flow; boundary layer flow; liquid jet breakup. *Mailing Add:* 220 W Lanvale St Baltimore MD 21217

PHINNEY, ROBERT A, b Rochester, NY, Oct 7, 36; m 59, 77; c 2. GEOPHYSICS. *Educ:* Mass Inst Technol, BS & MS, 59; Calif Inst Technol, PhD(geophys), 61. *Prof Exp:* Asst prof geophys, Calif Inst Technol, 61-63; assoc prof, 63-69, dept chair, 82-88, PROF GEOPHYS, PRINCETON UNIV, 69- *Concurrent Pos:* Chmn, Inc Res Insts Seismol, 86-88, Pres, 88-91. *Mem:* AAAS; Am Geophys Union; Soc Explor Geophys; Seismol Soc Am; Geol Soc Am; Sigma Xi. *Res:* Seismology; solid earth geophysics; planetary physics; theoretical geophysics; wave propagation. *Mailing Add:* 22 Lake Lane Princeton NJ 08540

PHINNEY, WILLIAM CHARLES, b South Portland, Maine, Nov 16, 30; m 53; c 4. GEOLOGY. *Educ:* Mass Inst Technol, SB, 53, SM, 56, PhD(geol), 59. *Prof Exp:* From asst prof to prof petrol, Univ Minn, Minneapolis, 59-71; geol br chief, 70-82, PLANETOLOGY BR CHIEF, JOHNSON SPACE CTR, NASA, 82- *Honors & Awards:* Except Sci Achievement Medal, NASA. *Mem:* AAAS; Am Geophys Union. *Res:* Phase relations in igneous and metamorphic rocks; lunar geology; early crustal evolution of the Earth. *Mailing Add:* Planetary Sci Br Sn2 NASA Johnson Space Ctr Houston TX 77058

PHIPPS, JAMES BIRD, b Birmingham, Eng, July 22, 34; m 67; c 3. SYSTEMATICS. *Educ:* Univ Birmingham, BSc, 56; Univ Western Ont, PhD(bot), 69. *Prof Exp:* Prof off, Br Bot & Plant Path, Ministry Agr, Govt Rhodesia & Nyasaland, Salisbury, Rhodesia & Nyasaland, 56-61; lectr bot, Dept Bot, Univ Western Ont, 61-64, from asst prof to assoc prof, 64-77, actg chmn bot dept, 79-80, chmn, 80-84, PROF BOT, DEPT PLANT SCI, UNIV WESTERN ONT, 77- *Concurrent Pos:* Mem panel plant systs, Study Fundamental Biol in Can-Privy Coun Sci Secretariat & Biol Coun Can, 68-70; mem, Pop Biol Grant Selection Comt, Nat Sci & Eng Res Coun Can, 79-81; dir, Sherwood Fox Arboretum, Univ Western Ont, 80- *Mem:* Can Bot Asn (pres, 81-82); Soc Syst Zool; Int Asn Plant Taxon; Bot Soc Am; Brit Ecol Soc; Int Orgn Plant Biosystems; Systs Asn; Am Soc Plant Taxonomists. *Res:* Plant systematics; numerical systematics; biosystematics; systematics and evolutionary biology of Crataegus; generic delimitation and evolution in Maloideae (Rosaceae). *Mailing Add:* Dept Plant Sci Univ Western Ont London ON N6A 5B7 Can

PHIPPS, PATRICK MICHAEL, b New Martinsville, WVa, Oct 19, 45; m 67; c 1. PLANT PATHOLOGY, MYCOLOGY. *Educ:* Fairmont State Col, BS, 70; Va Polytech Inst & State Univ, MS, 72; WVa Univ, PhD(plant path), 74. *Prof Exp:* Asst prof plant path, NC State Univ, 74-78; from asst prof to assoc prof, 78-89, PROF PLANT PATH, TIDEWATER RES CTR, VA POLYTECH INST & STATE UNIV, 89- *Mem:* Am Phytopathological Soc; Mycol Soc Am; Am Peanut Res Educ Soc. *Res:* Plant diseases caused by soilborne microorganisms; mycoparasitism; physiology of fungi. *Mailing Add:* Tidewater Res Ctr Holland Sta Suffolk VA 23437

PHIPPS, PETER BEVERLEY POWELL, b London, Eng, Aug 5, 36; m 62; c 1. SOLID STATE CHEMISTRY, CORROSION. *Educ:* Oxford Univ, BA, 61, DPhil(chem), 63. *Prof Exp:* Res assoc mat sci, Stanford Univ, 63-65 & Univ Southern Calif, 65-67; MEM RES STAFF, IBM RES LABS, 67- *Concurrent Pos:* Co-ed, Corrosion Chem, Am Chem Soc, 79. *Res:* Redox reactions; nuclear resonance; semiconductivity; photoconductivity; luminescence; defect chemistry; corrosion; wear. *Mailing Add:* IBM G20 14/2 5600 Cottle Rd San Jose CA 95193

PHIPPS, RICHARD L, b Coles Co, Ill, Apr 5, 35; m 55; c 2. PLANT ECOLOGY. *Educ:* Eastern Ill Univ, BSc, 58; Ohio State Univ, MSc, 60, PhD(bot), 65. *Prof Exp:* Res asst plant ecol, Ohio Agr Exp Sta, 58-64; RES BOTANIST, WATER RESOURCES DIV, US GEOL SURV, 64- *Concurrent Pos:* Prof lectr, George Washington Univ. *Mem:* Ecol Soc Am; Am Inst Biol Sci; Tree-Ring Soc. *Res:* Tree growth physiology; reconstruction of climatic and hydrologic information from tree rings; ecology and physiology of tree growth in relation to microclimate; computer simulation of forest vegetation dynamics. *Mailing Add:* 22 Foothill Rd Monticello IL 61856

PHIPPS, ROGER JOHN, RESPIRATION, EPITHELIAL TRANSPORT. *Educ:* London Univ, UK, PhD(physiol), 77. *Prof Exp:* Res assoc pulmonary physiol, Div Pulmonary Dis, Mt Sinai Med ctr, 81-88; STAFF SCIENTIST, PRODUCT DEVELOP, NORWICH EATON PHARMACEUT, 88- *Mailing Add:* Prod Develop Norwick Eaton Pharmaceut PO Box 191 Woods Corner NY 13815

PHISTER, MONTGOMERY, JR, b Calif, Feb 26, 26; m 49; c 3. COMPUTER PROGRAMMING. *Educ:* Stanford Univ, BS, 49, MS, 50; Cambridge Univ, PhD(physics, chem), 53. *Prof Exp:* Mem tech staff, Hughes Aircraft Co, 53-55 & Thompson-Ramo-Wooldridge, Inc, 55-60; vpres & dir, Scantlin Electronics, Inc, 60-66; vpres, Xerox Data Systs, 66-71; tech & sci res & writing, 71-; SYST CONSULT. *Concurrent Pos:* Lectr, Univ Calif, Los Angeles, 54-65, Harvard Univ, 74-75 & Univ Sydney, Australia, 75; dir eng, Thompson-Ramo-Wooldridge Prod Co, 58-60. *Mem:* Asn Comput Mach; fel Inst Elec & Electronic Engrs. *Res:* Economics of computers and data processing; software engineering; management in the information industry. *Mailing Add:* 414 Camino De Las Animas Santa Fe NM 87501

PHLEGER, CHARLES FREDERICK, b Northampton, Mass, June 30, 38; m 83; c 3. LIPIDS MARINE ANIMALS, AQUACULTURE. *Educ:* Stanford Univ, BA, 61; San Diego State Univ, MA, 64; Univ Calif San Diego, PhD(marine biol), 72. *Prof Exp:* Res asst, Scripps Inst Oceanog, Univ Calif, San Diego, 65-71; from lectr to assoc prof, 67-78, PROF OCEANOG, DEPT NATURAL SCI, SAN DIEGO STATE UNIV, 78- *Concurrent Pos:* Prin investr, NSF, 75, Calif Sea Grant Prog, San Diego State Univ Found, 76-83; vis res scientist, Lizard Island Res Sta, Australia, 84; vis fel, dept biochem, Univ Otago, Dunedin, NZ, 88; Fulbright Lectr, Univ WI Jamaica, 85-86. *Mem:* Sigma Xi; AAAS; World Maricult Soc. *Res:* Lipid biochemistry including cholesterol and hyperbaric oxygen in swimbladders of deep sea fishes; fish bone triglyceride storage; air breathing fish lung surfactant evolution; aquaculture of the purple-hinge rock scallop. *Mailing Add:* Dept Natural Sci San Diego State Univ San Diego CA 92182

PHLEGER, FRED B, b Kansas City, Kans, July 31, 09; m 33; c 2. OCEANOGRAPHY. *Educ:* Univ Southern Calif, AB, 31; Calif Inst Technol, MS, 32; Harvard Univ, PhD(geol), 36. *Prof Exp:* Asst paleont, Harvard Univ, 34-36; Sheldon traveling fel geol, 36-37; instr paleont, Amherst Col, 37-40, from asst prof to assoc prof, 40-49; vis assoc prof, 49-51, assoc prof, 51-57, chmn geol res div, 70-74, prof, 57-77, EMER PROF OCEANOG, SCRIPPS INST OCEANOG, UNIV CALIF, SAN DIEGO, 77- DIR MARINE FORAMINIFERA LAB, 49- *Concurrent Pos:* Fel, Cushman Found Foraminiferal Res, dir, 50-, pres, 56-57, 63; hon & distinguished sci investr, Inst Geol, Nat Univ Mex, 60; extraordinary investr, Nat Autonomous Univ Mex. *Honors & Awards:* Joseph A Cushman Award, Cushman Found Foraminiferal Res, 80. *Mem:* Geol Soc Am; Paleont Soc; Soc Econ Paleont & Mineral; Mex Geol Soc. *Res:* Ecology of Recent Foraminifera; coastal lagoon sedimentology. *Mailing Add:* Geol Res Div Scripps Inst Oceanog Univ of Calif at San Diego La Jolla CA 92093

PHOENIX, CHARLES HENRY, b Webster, Mass, Aug 18, 22; m 51; c 2. BEHAVIORAL PHYSIOLOGY, REPRODUCTIVE BIOLOGY. *Educ:* Univ Conn, BA, 45; Boston Univ, MA, 50, PhD(psychol), 54. *Prof Exp:* Asst prof psychol, Univ Kans, 54-57, res assoc anat, 59-63; assoc scientist, 63-64; assoc prof, 64-65, PROF MED PSYCHOL, ORE HEALTH SCI UNIV, 66-, SCIENTIST, ORE REGIONAL PRIMATE RES CTR, 65- *Concurrent Pos:* USPHS fel anat, Univ Kans, 57-59; vis prof, Sch Med, Univ Cincinnati, 61-62; vis scientist, Christ Hosp Inst Med Res, Cincinnati, 61-62. *Mem:* AAAS; Am Asn Anatomists; Am Psychol Asn; Sigma Xi. *Res:* Reproductive physiology and behavior; neuroendocrinology. *Mailing Add:* Ore Regional Primate Res Ctr 505 NW 185th Ave Beaverton OR 97006

PHOENIX, DAVID A, b Lompoc, Calif, June 25, 16; m 38; c 2. GEOLOGY, HYDROLOGY. *Educ:* Univ Calif, Berkeley, AB, 41; Stanford Univ, MS, 54. *Prof Exp:* Geologist, US Geol Surv, 41-62, res hydrologist, 62-72; CONSULT, GEOL-HYDROL, 73-, INT EXEC SERV CORP, 76- *Concurrent Pos:* From teaching asst to instr, Stanford Univ, 51-53; vis lectr, Chapman Col, 73-74, Calif State Univ, Fullerton, 81; field assoc, Int Exec Serv Corp, 83-; dir, Niguel Bot Preserve, 85-87. *Mem:* Fel Geol Soc Am; Am Inst Hydrol; Am Geophys Union; Soc Econ Geol; Sigma Xi; AAAS. *Res:* Geology of chromite, quicksilver, lead-zinc and uranium ore deposits; field geology; geophysical investigations; geology and permeability of host rocks, geochemistry of ground water; remote sensing applications and research in hydrology; regional water resource appraisals; origin, distribution and chemical characteristics of ground water; Numerous publications. *Mailing Add:* 450 Ruby St Laguna Beach CA 92651

PHOENIX, DONALD R, US citizen. BIOLOGY, ECOLOGY. *Educ:* Univ Pa, BA, 68, PhD(biol & ecol), 76. *Prof Exp:* Mem staff, Acad Natural Sci, Pa, 68-76; proj mgr, 76-77, dept mgr, 77-82, PRIN SCIENTIST & VPRES, WESTON, 82-, MGR SOUTHWEST REGION, 85- *Concurrent Pos:* Lectr limnol & ecol communities. *Res:* Large scale environmental impact assessment studies; impact of power-plant operations, acid mine drainage, toxin effluents, and thermal additions on rivers and other water bodies; fisheries population studies; trophic relationships between fish and their major food sources; large scale characterization and engineering feasibility studies in hazardous and radioactive waste management. *Mailing Add:* Weston Way Westchester PA 19380

PHONG, DUONG HONG, b Nam-Dinh, Vietnam, Aug 30, 53; US citizen. ANALYSIS, MATHEMATICAL PHYSICS. *Educ:* Princeton Univ, BA, 73, MA, 74, PhD(math), 77. *Prof Exp:* Instr math, Univ Chicago, 75-77; mem, Inst Advan Study, 77-78; asst prof, 78-81, ASSOC PROF MATH, COLUMBIA UNIV, 81- *Mem:* Am Math Soc. *Res:* Theory of partial differential equations; mathematical physics. *Mailing Add:* Dept Math Columbia Univ New York NY 10027

PHOTINOS, PANOS JOHN, b Greece, Dec 31, 48; m 79; c 3. LIQUID CRYSTALS, POROUS MEDIA. *Educ:* Nat Univ Athens, Greece, Bachelor, 71; Kent State Univ, PhD(physics), 75. *Prof Exp:* Vis prof physics, Kent State Univ, 76-79, res fel, Liquid Crystal Inst, 80-88; lectr, St Francis Xavier Univ, Can, 79-80; res prof chem, Univ Pittsburgh, 88-89; PROF PHYSICS, SOUTHERN ORE STATE COL, 89- *Concurrent Pos:* Consult, LaserOptics Assocs, 89- *Mem:* Am Chem Soc; Am Phys Soc; Int Liquid Crystal Soc; Optical Soc Am. *Res:* Optics, optic materials, and lasers; liquid crystals, and their electro-optical applications; applications of colloidal array filters; transport of trace contaminants and radioactive gases through porous adsorbers; properties and applications of micellar systems, microemulsions and colloids. *Mailing Add:* Dept Physics Southern Ore State Col Ashland OR 97520

PHUNG, DOAN LIEN, b Battrang, Vietnam, Jan 1, 40; US citizen; m 70; c 2. NUCLEAR SAFETY, NUCLEAR ENERGY ECONOMICS. *Educ:* Fla State Univ, BA, 61; Mass Inst Technol, MS, 63, ScD, 72. *Prof Exp:* Engr nuclear eng, Vietnam Atomic Energy Off, 64-66; consult nuclear safety, US Atomic Energy Comn, 72-73; consult engr nuclear eng, United Engr & Construct, 66-75; SR SCIENTIST ENERGY, INST ENERGY ANAL, OAK RIDGE ASSOC UNIV, 75- *Concurrent Pos:* Lectr, Univ Pa, 74. *Mem:* Sigma Xi; AAAS; NY Acad Sci; Am Nuclear Soc. *Res:* Nuclear plant design and safety; nuclear energy and alternatives; energy economics; cost analysis; industrial energy conservation. *Mailing Add:* Inst Energy Anal Orau PO Box 117 Oak Ridge TN 37830

PIACSEK, BELA EMERY, b Budapest, Hungary, Apr 17, 37; US citizen; m 62; c 4. REPRODUCTION. *Educ:* Univ Notre Dame, BS, 59, MS, 61; Mich State Univ, PhD(physiol), 66. *Prof Exp:* From asst prof to assoc prof biol, 68-79, asst prof physiol, Med Sch, 69-72, PROF BIOL, MARQUETTE UNIV, 79- *Concurrent Pos:* NIH res fel physiol, Med & Dent Sch, Harvard Univ, 66-68. *Mem:* Endocrine Soc; Am Physiol Soc; Soc Study Reproduction. *Res:* Maturation of the reproductive system; environmental and nutritional influences in reproduction; role of feedback mechanisms. *Mailing Add:* Dept Biol Marquette Univ Milwaukee WI 53233

PIALA, JOSEPH J, b Carrollville, Wis, Nov 1, 21; m 52; c 3. PHARMACOLOGY. *Educ:* Univ Wis, BS, 47; Univ Md, PhD(pharmacol), 51. *Prof Exp:* Sect head pharmacol, Squibb Inst Med Res, 50-59, sr res pharmacologist, 59-65, res assoc pharmacol, 65-67; dept head pharmacol, 67-77, sr res investr, 77-85, PRIN RES INVESTR, BRISTOL-MYERS PROD, 85- *Mem:* AAAS; Am Soc Pharmacol & Exp Therapeut; NY Acad Sci; Am Pharmaceut Asn. *Res:* Central nervous system pharmacology; diuretics; cardiovascular drugs; tranquilizers; muscle relaxants; general anesthetics; tuberculostatic agents; trypanosomiasis; analgesics; anti-inflammatory agents; antacids; skin and hair products. *Mailing Add:* 31 Beacon Hill Dr Metuchen NJ 08840

PIALET, JOSEPH WILLIAM, b Presto, Pa, June 6, 51. LUBRICANT & FUEL ADDITIVES, ELECTRORHEOLOGICAL FLUIDS. *Educ:* Carnegie-Mellon Univ, BS, 73, MS, 74. *Prof Exp:* RES CHEMIST, LUBRIZOL CORP, 74- *Mem:* Am Chem Soc. *Res:* Development of additives for lubricants and fuels to provide greater protection and fuel economy; development of electrorheological fluids; lubricants for adiabatic engines. *Mailing Add:* Lubrizol Corp 29400 Lakeland Blvd Wickliffe OH 44092-2298

PIAN, CARLSON CHAO-PING, b Beijing, China, Dec 31, 45; US citizen; m 69; c 3. MAGNETOFLUID DYNAMICS, ENERGY CONVERSION. *Educ:* Univ Mich, BSE, 68, MSE, 69, PhD(aerospace eng), 74. *Prof Exp:* Asst gas dynamics, Univ Mich, 68-74; vis scientist, Dept Elec Eng, Eindhoven Univ Tech, Neth, 74-75; res engr magneto-gas dynamics, Plasma Flow & Magnetohydrodynamics Sect, Lewis Res Ctr, NASA, 75-79; PRIN RES SCIENTIST, AVCO EVERETT RES LABS, 79-, PROG DIR, 88- *Concurrent Pos:* Fel, Eindhoven Univ Tech, Neth, 74-75; mem terrestrial energy systs tech comt, Am Inst Aeronaut & Astronaut, 82-83 & plasmadynamics & lasers tech comt, 83-84. *Honors & Awards:* Space Shuttle Flag Award, Am Inst Aeronaut & Astronaut, 84; Jean F Louis Award, Am Soc Mech Engrs, 88. *Mem:* Am Inst Aeronaut & Astronaut. *Res:* Magnetohydrodynamics; direct energy conversion; computational fluid dynamics. *Mailing Add:* 38 Kenmore St Newton MA 02159

PIAN, CHARLES HSUEH CHIEN, b Tientsin, China, June 21, 21; US citizen; m 45; c 3. ORGANIC CHEMISTRY. *Educ:* Fu Jen Univ, BS, 44; Northeastern Univ, MS, 64; State Univ NY, Albany, PhD(org chem), 69. *Prof Exp:* Asst engr, Ta-shin Paper & Pulp Factory, 45-46; chief engr & plant dir, Shinei Pharmaceut Factory, Taiwan Med & Supplies Co, 46-51; specialist, Bd Trustees Rehab Affairs, 51-52, Joint Comn Rural Reconstruct, 52-54, Joint Comt Med Supplies, 54-56 & Inst Nuclear Sci, Tsing Hua Univ, Taiwan, 56-58; chemist, Lexington Res Labs, Itek Corp, 59-62, sr chemist, 64-65, proj leader, 68-70; proj leader, 70-80, GROUP LEADER, NASHUA CORP, 81- *Mem:* Am Chem Soc; Soc Photog Sci & Eng. *Res:* Organic reaction mechanisms; organometallic chemistry; liquid transfer toner; photochemistry; dry toner; organic photoconductors; image formation studies. *Mailing Add:* 64 Fifer Lane Lexington MA 02173

PIAN, THEODORE H(SUEH) H(UANG), b Shanghai, China, Jan 18, 19; nat US; m 45; c 1. STRUCTURAL MECHANICS, AERONAUTICS. *Educ:* Tsing Hua Univ, China, BS, 40; Mass Inst Technol, MS, 44, ScD(aeronaut eng), 48. *Hon Degrees:* Dr, Beijing Univ, Beijing, China, 90. *Prof Exp:* Asst, 46-47, res assoc, 47-48, mem staff, Div Indust Coop, 48-52, from asst prof to assoc prof, 52-66, prof aeronaut eng, 66-89, EMER PROF AERONAUT ENG, MASS INST TECHNOL, 89- *Concurrent Pos:* Vis assoc, Cal Inst Technol, 65-66; assoc ed, Am Inst Aeronaut & Astronaut J, 73-75; vis prof, Univ Tokyo, 74, Tech Univ Berlin, 75 & Nat Tsing Hua Univ, Hsin Chu, Taiwan, 90; hon prof, Beijing Univ Aeronaut & Astronaut, 81-, Beijing Inst Technol, 82-, Southwestern Jiaotong Univ, 83-, Dalian Inst Technol, 85-, Changsha Railway Col, 86- & Huazhong Univ Sci & Technol, 90- *Honors & Awards:* Von Karman Mem Prize, TRE Corp, Beverly Hills, Calif, 74; Struct,

Struct Dynamics & Mat Award, Am Inst Aeronaut & Astronaut, 75. *Mem:* Nat Acad Eng; Am Soc Eng Educ; fel Am Inst Aeronaut & Astronaut; fel AAAS; hon mem Am Soc Mech Engrs. *Res:* Finite element methods; structural mechanics. *Mailing Add:* Dept Aeronaut & Astronaut Rm 33-413 Mass Inst of Technol Cambridge MA 02139

PIANETTA, PIERO ANTONIO, b Santa Rosa, Calif, Oct 19, 49; m 83; c 1. SURFACE PHYSICS, DEVICE TECHNOLOGY. *Educ:* Univ Santa Clara, BS, 71; Stanford Univ, MS, 74, PhD(appl physics), 77. *Prof Exp:* Mem tech staff eng, Hewlett-Packard Labs, Hewlett-Packard Co, 78-82; res assoc physics, 77-78, PROF ELEC ENG, STANFORD UNIV, 82-, ASSOC DIR, STANFORD SYNCHROTRON RADIATION LAB, 85- *Concurrent Pos:* Vis prof, Musashino Lab, Nippon Tel & Tel, 85. *Mem:* Am Vacuum Soc; Am Phys Soc; AAAS. *Res:* Development of synchrotron radiation techniques for the study of materials; use of photoelectron spectroscopy to study semiconductor surfaces and interfaces. *Mailing Add:* SLAC Bin 69 PO Box 4349 Stanford CA 94309

PIANFETTI, JOHN ANDREW, b Sawyerville, Ill, July 19, 07; m 40; c 2. ORGANIC CHEMISTRY. *Educ:* Univ Ill, BS, 30; Univ Wis, MA, 33; Purdue Univ, PhD(org chem), 41. *Prof Exp:* Asst phys chem, A O Smith Corp, Wis, 30-32; res chemist, Miner Labs, Chicago, 36-38; res chemist, Westvaco Chlorine Prod Corp, 40-48, Westvaco Chem Div, FMC Corp, 48-58 & sr res chemist, 59-68; CONSULT, 68- *Mem:* Am Chem Soc. *Res:* Diazonium salts; glycerol and its derivatives; chlorination and bromination of organic compounds; organic phosphates; resins and plasticizers; dichloro-diphenyl-trichloroethane; agricultural insecticides; war gases; nitrogen chemistry; high pressure synthesis; toxic agents; rocket fuels. *Mailing Add:* 5145 Russet Dr Charleston WV 25313

PIANKA, ERIC R, b Hilt, Calif, Jan 23, 39; div; c 2. ECOLOGY. *Educ:* Carleton Col, BA, 60; Univ Wash, PhD(ecol), 65. *Prof Exp:* NIH fel, Princeton Univ, 65-66, Univ Western Australia, 66-68 & Princeton Univ, 68; from asst prof to assoc prof, 68-76, PROF ZOOL, UNIV TEX, AUSTIN, 76- *Concurrent Pos:* Guggenheim fel, 78-79; vis prof, Univ Kansas, 78 & Univ Puerto Rico, 81; NSF grants & Nat Geog Soc grants; Denton A Cooley Centennial prof zool. *Mem:* Am Soc Nat; Am Soc Ichthyol & Herpet; Soc Study Evolution; Ecol Soc Am; fel AAAS. *Res:* Reproductive tactics; foraging tactics; thermoregulation, resource partitioning; species diversity and ecology of desert lizards; comparison of North American (Sonoran, Mojave & Great Basin), Western Australian (Great Victoria), and Southern African (Kalahari) deserts, especially environmental factors determining the number of coexisting species of lizards. *Mailing Add:* Dept Zool Univ Tex Austin Austin TX 78712-1064

PIANOTTI, ROLAND SALVATORE, b New York, NY, May 8, 30; m 58; c 1. MICROBIOLOGY, DENTAL RESEARCH. *Educ:* Iona Col, BS, 52; St Johns Univ, MS, 54. *Prof Exp:* Assoc scientist, Dept Microbiol, Warner-Lambert Res Inst, 57-66, scientist, 66-69, sr scientist, Dept Oral-Dent Sci, 69-73, sect head, Dept Oral-Dent Sci, 73-79, sr scientist, 79-86; RETIRED. *Mem:* Am Soc Microbiol; Int Asn Dent Res. *Res:* Metabolism of dental plaque; microbiol biochemistry. *Mailing Add:* 89 Field Crest Rd Parsippany NJ 07054

PIANTADOSI, CLAUDE, b Naples, Italy, Jan 4, 23; nat US; m 45; c 3. MEDICINAL CHEMISTRY. *Educ:* Brooklyn Col, BS, 49; Columbia Univ, MS, 52; Univ NC, PhD(biochem, pharmaceut chem), 56. *Prof Exp:* Asst prof pharmaceut chem, Butler Univ, 55-56; asst prof pharmaceut chem & biochem, Fordham Univ, 56-57; from asst prof to assoc prof, 57-63, chmn, Med Chem Div 71-75 & 83-90, dir grad studies, Sch Pharm, 71-75 & 77-87, PROF MED CHEM, SCH PHARM & PROF BIOCHEM, SCH MED, UNIV NC, CHAPEL HILL, 63- *Concurrent Pos:* Consult, Oak Ridge Inst Nuclear Studies, 63; vis scientist, Am Asn Cols Pharm, 66 & NC Acad Sci. *Mem:* AAAS; Am Chem Soc; Am Pharmaceut Asn; fel Am Inst Chemists; NY Acad Sci; Sigma Xi. *Res:* Lipid chemistry; chemistry of purine and pyrimidine; synthetic oriented medical chemistry and radioactive synthesis. *Mailing Add:* Col Pharm Univ NC Chapel Hill NC 27515

PIASECKI, FRANK NICHOLAS, b Philadelphia, Pa, Oct 24, 19; m 58; c 7. HELICOPTERS, HIGH-SPEED GROUND TRANSPORTATION. *Educ:* NY Univ, BS, 40. *Hon Degrees:* DSc, Pa Mil Col, 53, Alliance Col, 70; Dr Aeronaut Eng, NY Univ, 55. *Prof Exp:* Pres & chmn, Piasecki Helicopter Corp, 40-55; PRES & CHMN, PIASECKI AIRCRAFT CORP, 55- *Concurrent Pos:* Mem, Helicopter Coun, Aerospace Industs Asn, comt Innovation, NSF, citizen's adv comt transp qual, Dept Transp, indust adv coun, Dept Com Study Comn Innovation; dir, Crown Cork & Seal; trustee, Kosciuszko Found, NY. *Honors & Awards:* Lawrence Sperry Award, Inst Aeronaut Sci, 50; Tribute Appreciation, Nat Defense Transp Asn, 54; Leonardo Da Vinci Award, Navy Helicopter Asn, 74; Spirit of St Louis Award, Am Soc Mech Engrs, 83; Nat Medal Technol, 86. *Mem:* Soc Automotive Engrs; Nat Soc Prof Engrs; Soc Exp Test Pilots; Soc Mil Engrs; Am Soc Testing & Mat; Am Helicopter Soc; Am Inst Aeronaut & Astronaut; Soc Nondestructive Testing; Am Defense Preparedness Asn. *Res:* Built and flew the second successful helicopter in the United States; the world's first successful tandem rotor transport helicopter; numerous production designs built for the military; world's largest transport helicopter; developed and flew the Sea Bat, Aerial Geeps I and II, Ring-Tail shaft-driven compound helicopter, 16H-1, 16H1A and the Heli-Stat, the largest flying aircraft. *Mailing Add:* Tunbridge Rd Haverford PA 19041

PIASECKI, LEONARD R(ICHARD), b Michigan City, Ind, Nov 15, 23; m 54; c 5. CHEMICAL ENGINEERING. *Educ:* Purdue Univ, BS, 49. *Prof Exp:* Analytical chemist, US Steel Corp, 49-50; res chemist, Reynolds Metals Corp, 51-53; res engr, 53-56, eng group supvr, 56-60, chief solid propellant rockets sect, 60-63, dep div chief, 63-64, chief propulsion sect & Voyager propulsion, 64-65, mgr Voyager capsule syst, 66-68, MGR LONG RANGE PLANNING, OFF PLANS & PROGS, JET PROPULSION LAB, CALIF INST TECHNOL, 68- *Mem:* Am Inst Aeronaut & Astronaut; Am Chem Soc. *Res:* Voyager capsule. *Mailing Add:* 5702 Catherwood LaCanada Flintridge CA 91011

PIATAK, DAVID MICHAEL, b Simpson, Pa, Jan 25, 36; m 60; c 3. ORGANIC CHEMISTRY. *Educ:* Pa State Univ, BS, 57; Univ Maine, MS, 59, PhD(org chem), 62. *Prof Exp:* Staff scientist, Worcester Found Exp Biol, 61-66; asst prof chem, 66-72, ASSOC PROF CHEM, NORTHERN ILL UNIV, 72- *Concurrent Pos:* Sr res award, Fulbright-Hays Comn, 74-75; vis assoc prof chem, Univ Belgrade, 74-75; US-Israel Fulbright-Hays Comn travel grantee, 75; exchange participant, Nat Acad Sci, 78. *Mem:* Phytochemical Soc NAm; Am Chem Soc; Royal Soc Chem; Am Soc Pharmacog; Sigma Xi. *Res:* Isolation and identification of natural products; chemistry and biochemistry of steroids and related compounds; synthesis of potential antineoplastic agents; heterocyclic compounds; new organic synthetic methods. *Mailing Add:* 308 Augusta Ave DeKalb IL 60115

PIATIGORSKY, JORAM PAUL, b Elizabethtown, NY, Feb 4, 40. DEVELOPMENTAL BIOLOGY, BIOCHEMISTRY. *Educ:* Harvard Univ, AB, 62; Calif Inst Technol, PhD(develop biol), 67. *Prof Exp:* Sr staff fel develop biol, Nat Inst Child Health & Human Develop, 67-75, head, Sect Cellular Differentiation, Lab Molecular Genetics, 76-81; CHIEF, LAB MOLECULAR & DEVELOP BIOL, NAT EYE INST, 81- *Honors & Awards:* Friedenwald Award, 86; Alcon Award, 86. *Mem:* Am Soc Zool; Am Soc Cell Biol; Int Soc Embryol; Soc Develop Biol; NY Acad Sci; Sigma Xi. *Res:* Regulation of nucleic acid and protein synthesis in growing and differentiating cells, particularly in the ocular lens. *Mailing Add:* 8435 Persimmon Tree Rd Bethesda MD 20817

PIATKOWSKI, THOMAS FRANK, b Ann Arbor, Mich, Oct 24, 38; m 64; c 3. ELECTRICAL ENGINEERING. *Educ:* Univ Mich, BSE, 60, MSE, 61, PhD(elec eng), 63. *Prof Exp:* Asst res engr logic systs lab & asst lectr elec eng, Univ Mich, 62-63; asst prof & dir comput ctr, Tuskegee Inst, 64-66; res engr systs eng lab, Univ Mich, 66-67; assoc prof eng, Dartmouth Col, 67-72; staff systs analyst systs develop div, IBM Corp, 72-75; mgr commun archit, Burroughs Corp, 75-80; assoc prof elec eng, Iowa State Univ, 80-,; ASSOC PROF, DEPT COMPUT SCI, STATE UNIV NY, BINGHAMPTON. *Mem:* Inst Elec & Electronics Engrs; Asn Comput Mach. *Res:* Computer applications; engineering and education. *Mailing Add:* Dept Comput Sci SUNY Binghamton Binghamton NY 13901

PIAVIS, GEORGE WALTER, b Glen Lyon, Pa, May 4, 22; m 46; c 3. ZOOLOGY. *Educ:* Western Md Col, AB, 48, MEd, 52; Duke Univ, PhD(zool), 58. *Prof Exp:* Teacher high sch, Md, 50-53; asst instr comp anat embryol & zool, Duke Univ, 53-56; fisheries res biologist, Hammond Bay Lab, US Fish & Wildlife Serv, Mich, 55-58; from asst prof to assoc prof, 58-65, PROF ANAT, UNIV MD, BALTIMORE, 65- *Concurrent Pos:* Teacher high schs, Md, 56-58. *Honors & Awards:* Award, Bur Com Fisheries, US Fish & Wildlife Serv, 61. *Mem:* Fel AAAS; Am Soc Ichthyol & Herpet; Am Asn Anat; Am Soc Zool; NY Acad Sci; Sigma Xi. *Res:* Effects of antimetabolites on regeneration and embryological development in the sea lamprey. *Mailing Add:* 318 Mary Ave Westminster MD 21157

PICARD, DENNIS J, RESEARCH ADMINISTRATION. *Prof Exp:* CHMN & CHIEF EXEC OFFICER, RAYTHEON CO, 91- *Mem:* Nat Acad Eng. *Mailing Add:* Raytheon Co 141 Spring St Lexington MA 02173

PICARD, GASTON ARTHUR, food chain, analytical chemistry, for more information see previous edition

PICARD, M DANE, b Washburn, Mo, Aug 7, 27; m 58; c 4. SEDIMENTARY PETROLOGY. *Educ:* Univ Wyoming, BS, 50; Princeton Univ, AM, 62, PhD(geol), 63. *Prof Exp:* Field asst, Texaco Inc, 50; from jr geologist to dist stratigrapher, Shell Oil Co, 50-56; geologist, St Helens Petrol Corp, 56-57; dist mgr & resident geologist, Am Stratig Co, 57-60; asst instr, Princeton Univ, 60-61; NSF fel, 61-63; from assoc prof to prof geol, Univ Nebr, Lincoln, 63-68; PROF GEOL, UNIV UTAH, 68- *Concurrent Pos:* Instr night sch, Ft Lewis Agr & Mech Col, 57-59; NIH grants, 62-72; sr geologist, Pan Am Petrol Corp, 64; consult, Utah Geol Surv, 69-74; deleg, House of Deleg, Am Asn Petrol Geologists, 71-77, mem adv coun, 73-76; sr consult, Mountain Fuel Supply Co, 72-75. *Mem:* Fel AAAS; fel Geol Soc Am; Soc Econ Paleont & Mineral; Am Asn Petrol Geologists; Int Asn Sedimentol. *Res:* Ephemeral streams; stratigraphy; petroleum geology; paleomagnetism. *Mailing Add:* Dept of Geol & Geophys Univ of Utah Salt Lake City UT 84112

PICARD, RICHARD HENRY, b Springfield, Mass, Aug 26, 38; m 67; c 2. ATMOSPHERIC OPTICS, ATMOSPHERIC MODELS. *Educ:* Assumption Col, Mass, AB, 59; Boston Univ, MA, 62, PhD(physics), 68. *Prof Exp:* Res physicist, Air Force Cambridge Res Labs, 64-75; physicist, Rome Air Develop Ctr, 76-81, PHYSICIST, AIR FORCE GEOPHYSICS LAB, 81- *Concurrent Pos:* Vis indust prof, Tufts Univ, 84- *Mem:* Am Phys Soc; Sigma Xi; Optical Soc Am; Am Geophys Union. *Res:* Airglow and aurora, especially infrared emissions; radiative transfer; atmospheric gravity waves; upper atmospheric structure; laser and quantum optics. *Mailing Add:* Air Force Geophysics Lab Optical Environ Div (OPS) Hanscom AFB Bedford MA 01731

PICCHIONI, ALBERT LOUIS, b Klein, Mont, Aug 28, 21; m 53; c 4. PHARMACOLOGY. *Educ:* Univ Mont, BS, 43; Purdue Univ, MS, 50, PhD(pharmacol), 52. *Prof Exp:* Pharmacist supvr, Univ Mich Hosp, 46-48; res asst pharmacol, Purdue Univ, 48-52; actg dean, Col Pharm, Univ Ariz, 75-76, head dept, 70-81, prof pharmacol, 52-87, assoc dean, Col Pharm, 82-87; RETIRED. *Concurrent Pos:* Exec dir, Ariz Regional Poison Control Ctr, 79-83. *Honors & Awards:* Award, Am Asn Poison Control Ctr, 87. *Mem:* Am Soc Pharmacol & Exp Therapeut; Soc Toxicol; Acad Pharmaceut Sci; Sigma Xi. *Res:* Investigation of the role of the brain biogenic amines to susceptibility to seizures; application of pharmacokinetic principles in the management of acute poisoning. *Mailing Add:* Col Pharm Univ Ariz Tucson AZ 85721

PICCIANO, MARY FRANCES ANN, b Palmerton, Pa, Dec 19, 46; m 76; c 2. NUTRITION. *Educ:* St Francis Col, BS, 68; Pa State Univ, MS, 70, PhD(nutrit), 74. *Prof Exp:* From asst prof to prof nutrit, Sch Human Resources & Family Studies, Div Foods & Nutrit, Univ Ill, Urbana, 74-89, prof nutrit in med, Col Med Urbana-Champaign, 86-89; PROF NUTRIT, NUTRIT DEPT, COL HEALTH & HUMAN DEVELOP, PA STATE UNIV, 89- *Concurrent Pos:* Fac res award, Ill Home Economics Asn, 79, co-chmn res sect, 80-82; mem sci adv bd, Am Coun Sci & Health, 84-; vis scientist, USDA Vitamin & Nutrit Res Lab, Beltsville Human Nutrit Res Ctr, 84; mem exec comt, Int Soc Res Human Lactation & Milk, 84-; vchmn, Food & Nutrit Biochem Subdiv, Agr & Food Chem Div, Am Chem Soc, 86-88, chmn, 88-89; mem intervention rev comt, NIH/NCI Cancer Inst, 85-88; mem, Subcomt Nutrit During Lactation, Nat Acad Sci, 88-90, Comt Int Nutrit Prog, 89-92. *Honors & Awards:* Borden Award Achievements Infant Nutrit Res, 84; Lederle Award Human Nutrit for Distinguished Res Infant Nutrit, 87; Paul A Funk Recognition Award Achievements Maternal & Infant Nutrit, 88. *Mem:* Sigma Xi; AAAS; Am Inst Nutrit; Am Dietetic Asn; Am Soc Clin Nutrit; Am Chem Soc; Am Col Nutrit; Am Home Econ Asn; Am Pub Health Asn; Inst Food Technologists; Soc Nutrit Educ. *Res:* Nutrition during growth and development; maternal nutrition; infant nutrition; nutritional status; iron metabolism; flate metabolism; author of numerous publications. *Mailing Add:* Pa State Univ 126 B Henderson S University Park PA 16802-6597

PICCIOLO, GRACE LEE, b Baltimore, Md, Feb 20, 34; div; c 2. CELL PHYSIOLOGY, SPACE BIOLOGY. *Educ:* Univ Md, PhD(cell physiol), 64. *Prof Exp:* Teacher, Convent of Sacred Heart Sch, 56-58; life scientist, NASA-Goddard Space Flight Ctr, 64-77; mem staff regulatory clin microbiol prod, Bur Med Devices, 77-84, BIOMATERIALS RES, CTR DEVICES & RADIOL HEALTH, FOOD & DRUG ADMIN, 84- *Res:* Determination of methods of life detection using bioluminescence for extraterrestrial analysis of atmospheric and planetary surface samples; analysis of mutant strain of Astasia, including growth and cellular organelle experimentation; development of microbial detection methods for automation in clinical laboratory; quantitation of immunofluorescent microscopy; biomedical materials degradation. *Mailing Add:* 13007 Renfrew Circle Ft Washington MD 20744

PICCIONI, ORESTE, b Siena, Italy, Oct 24, 15; nat US; m 45; c 5. NUCLEAR PHYSICS. *Educ:* Ginnaiso-Liceo, Italy, Dipl, 29; Univ Rome, Dr(physics), 38. *Prof Exp:* Instr, Univ Rome, 35-38, asst, 38-44, prof electromagnetism, 44-46; res assoc, Mass Inst Technol, 46-48; scientist, Brookhaven Nat Lab, 48-60; PROF PHYSICS, UNIV CALIF, SAN DIEGO, 60- *Concurrent Pos:* Prin Investr, Dept Energy contract. *Res:* Cosmic rays; high energy physics; elementary particles; electronic circuits for nuclear researches; radiation; Einstein-Podolsky-Rosen paradox. *Mailing Add:* Dept Physics B-019 Univ Calif San Diego La Jolla CA 92093

PICCIOTTO, CHARLES EDWARD, b Buenos Aires, Arg, July 1, 42; US citizen; m 64; c 2. PARTICLE PHYSICS. *Educ:* Univ Calif, Berkeley, AB, 64; Univ Calif, Santa Barbara, MA, 66, PhD(physics), 68. *Prof Exp:* PROF PHYSICS, UNIV VICTORIA, BC, 68- *Mem:* Am Phys Soc. *Res:* Particle physics. *Mailing Add:* Dept Physics Univ Victoria Victoria BC V8W 2Y2 Can

PICCIRELLI, ROBERT ANTHONY, b New York, NY, Dec 9, 30; m 59; c 6. STATISTICAL & ENGINEERING PHYSICS. *Educ:* Cath Univ Am, AB, 52, MS, 53, PhD(physics), 56. *Prof Exp:* Res physicist, Nat Bur Standards, 57-58; PROF MECH ENG SCI & MEM RES INST ENG SCI, WAYNE STATE UNIV, 68- *Concurrent Pos:* NSF-Nat Res Coun fel, Nat Bur Standards, 56-58; Dept Com sci & technol fel, 65-66; consult, Available Energy, Inc, Detroit, Mich. *Mem:* AAAS; Am Phys Soc. *Res:* Combustion engineering; performance of alternate fuels; nonequilibrium statistical mechanics; generalizations of classical fluid dynamics; relaxation phenomena; nonlocal and nonlinear effects; propagation and scattering of sound and light from fluids. *Mailing Add:* 19969 Fairway Dr Grosse Pointe Shores MI 48236

PICCOLINI, RICHARD JOHN, b Los Angeles, Calif, Mar 9, 33; m 59; c 2. PHYSICAL CHEMISTRY, ORGANIC CHEMISTRY. *Educ:* Calif Inst Technol, BS, 55; Univ of Calif, Los Angeles, PhD(org chem), 60. *Prof Exp:* Asst chem, Calif Inst Technol, 54-55 & Univ Calif, Los Angeles, 56-59; sr chemist, 59-70, sr chemist chem div, 70-73, sr res assoc, 73-86, MGR, NEW TECHNOL, ROHM & HAAS CO, 86- *Mem:* Am Chem Soc; Royal Soc Chem; AAAS. *Res:* Chemistry in bridged ring systems; metal carbonyl chemistry; photochemistry as applied to polymers; free radical reactions and solution properties of polymers; polymers as petroleum modifiers; chemical process research; organometallic chemistry; catalysis. *Mailing Add:* 976 Washington Crossing Rd Newtown PA 18940

PICHA, KENNETH G(EORGE), b Chicago, Ill, July 24, 25; m 48; c 3. MECHANICAL ENGINEERING. *Educ:* Ga Inst Technol, BME, 46, MS, 48; Univ Minn, PhD(mech eng), 57. *Prof Exp:* Aeronaut res scientist, Nat Adv Comt Aeronaut, Ohio, 48-49; prof mech eng, Ga Inst Technol, 49-58, from assoc dir to dir, Sch Mech Eng, 60-66; prog dir eng sci, NSF, 58-60; dean, Sch Eng, Univ Mass, Amherst, 66-76, dir, Off Coord Energy Res & Educ, 77-80, prof mech eng, 80-82; exchange prof & consult to chancellor, Univ Puerto Rico, Mayaguez, 82; prof 83-89, EMER PROF MECH ENG, UNIV MASS, AMHERST, 89- *Concurrent Pos:* Consult, Orgn Europ Econ Coop, 60, NSF, 60-66 & Govt of Singapore, 70-72; chmn manpower develop comt, Environ Protection Agency, 71; chmn eng educ & accreditation comt, Eng Coun Prof Develop, 71-73; consult, Nat Acad Sci; dir, Off Univ Progs, US Energy Res & Develop Admin, 76-77; Fulbright fel, Univ Col Galway, Ireland, 86-87. *Mem:* Fel AAAS; fel Am Soc Mech Engrs; fel Am Soc Engr Educ. *Res:* Thermal sciences. *Mailing Add:* 56 Oak Knoll Amherst MA 01002

PICHAL, HENRI THOMAS, II, b London, UK, Feb 14, 23; US citizen; m 66; c 3. RF & MICROWAVE-COMMUNICATIONS ANTENNAE, NAVIGATION EQUIPMENT DESIGN. *Educ:* Southend Tech Col, UK, BS, 52; London Univ, MS, 55. *Prof Exp:* Sr staff, EK Cole Southend-on-Sea, Essex, UK, 52-57; sect head, Gen Dynamics, Rochester, NY, 57-59; prod mgr, ECI (E-Systs), St Petersburg, Fla, 59-64; task leader, Martin Marietta, Orlando, Fla, 64-65; sr staff, Honeywell Commun Develop Ctr, Fla, 65-70; eng mgr, John Fluke Corp, Everett, Wash, 70-73; specialist engr, Harris Corp, Melbourne, Fla, 73-75; CONSULT, PROF ENG CO, INC, KISSIMMEE, FLA, 75- *Concurrent Pos:* Sect head, Litton Res, Greenwich, Conn, 70; consult, several companies, 75- *Mem:* Am Phys Soc; sr mem Inst Environ Sci; sr mem Inst Elec & Electronic Engrs. *Res:* Noise reduction in electronic circuits, in low frequency and RF and microwave systems and components; increasing dynamic range in receivers and transmitters and processing electronic subsystems; reducing phase noise in oscillators and frequency synthesizers; granted nine patents; author of two publications. *Mailing Add:* 1922 N Easy St PO Box 1137 Kissimmee FL 34741-2115

PICHANICK, FRANCIS MARTIN, b Salisbury, Rhodesia, May 15, 36; div; c 1. ATOMIC PHYSICS. *Educ:* Univ Cape Town, MS, 58; Oxford Univ, PhD(physics), 61. *Prof Exp:* Instr physics, Yale Univ, 61-64, asst prof, 64-69; assoc prof, 69-86, PROF PHYSICS, UNIV MASS, AMHERST, 86- *Concurrent Pos:* Guest worker, Nat Bur Stand, 66-67; res collabr, Brookhaven Nat Lab, 69-; mem comt atoms & molecules, Nat Acad Sci, 71-; assoc ed, Phys Rev Lett, 74-; vis fel, Joint Inst Lab Astrophys, Boulder, Colo, 75-76. *Mem:* Fel Am Phys Soc. *Res:* Experimental investigations in basic atomic structure using radiofrequency spectroscopy and scattering techniques; electron-atom and electron-molecule scattering at low energies; nuclear moments; laser multiphoton excitation. *Mailing Add:* Dept Physics & Astron Univ of Mass Amherst MA 01003

PICK, JAMES RAYMOND, b Baltimore, Md, Mar 6, 36; c 1. LABORATORY ANIMAL SCIENCE. *Educ:* Univ Ga, DVM, 61. *Prof Exp:* Fel lab animal med, Bowman Gray Sch Med, 61-63; prof assoc, Inst Lab Animal Resources, Nat Acad Sci-Nat Res Coun, 63-64; instr comp path, 65-66, asst prof, 66-70, ASSOC PROF COMP PATH, SCH MED, UNIV NC, CHAPEL HILL, 70-, DIR, DIV LAB ANIMAL MED, 65-, PROF PATH, 81- *Mem:* Am Vet Med Asn; Am Asn Lab Animal Sci. *Res:* Infectious and metabolic diseases of animals analagous to those in humans; alcohol dependence in animals; fetal alcohol syndrome. *Mailing Add:* Div of Lab Animal Med Univ of NC Sch of Med Chapel Hill NC 27514

PICK, ROBERT ORVILLE, b Madison, Wis, Apr 13, 40; m 65; c 2. MEDICINAL CHEMISTRY, CLINICAL CHEMISTRY. *Educ:* Univ Wis, Madison, BS, 62; Univ Calif, Los Angeles, PhD(org chem), 67. *Prof Exp:* Actg instr chem, Univ Calif, Los Angeles, 67-68; chemist, Div Med Chem, Walter Reed Army Med Ctr, 68-71; chief lab serv & asst chief dept med res & develop, William Beaumont Gen Hosp, 71-74; chief, Chem Br, Acad Health Sci, Ft Sam Houston, 75-78; chief, Chem Handling & Data Anal Br, 78-80, chief, dept med chem, div exp therapeut, Walter Reed Army Inst Res, 80-84; USA Forensic Testing Lab, Wiesbaden 84-87; mem staff, Res & Develop Command, 87-90, LAB SCI CONSULT, OFF SURGEON GEN, US ARMY, 90- *Mem:* Am Chem Soc; Am Acad Forensic Sci. *Res:* Physical organic chemistry; mechanisms of reactions; structure-biological activity relationships; chemical information systems; drug development. *Mailing Add:* 9410 Erin Ave Walkersville MD 21793

PICK, ROY JAMES, b Vancouver, BC, Dec 28, 41; m 64; c 1. MECHANICAL ENGINEERING. *Educ:* Univ BC, BASc, 64; Univ London, MSc, 67, Imp Col, dipl, 67; Univ Waterloo, PhD(mech eng), 70. *Prof Exp:* Stress engr, Orenda Engines Ltd, 64-66; ASSOC PROF, UNIV WATERLOO, 68-, NAT RES COUN CAN FEL, 70- *Concurrent Pos:* Pres, R J Pick Ltd, 74-; consult, Welding Inst Can & Nova, Alberta Corp. *Res:* Nuclear pressure vessels; fracture analysis; finite element methods; high strain rate behavior; pipeline welding analysis. *Mailing Add:* Dept Mech Eng Univ Waterloo Waterloo ON N2L 3G1 Can

PICKANDS, JAMES, III, b Cleveland, Ohio, Sept 4, 31; m 61; c 2. STATISTICS. *Educ:* Yale Univ, BA, 54; Columbia Univ, PhD(math, statist), 65; Univ PA, MA, 73. *Prof Exp:* Mathematician-programmer, Ballistics Res Labs, Aberdeen Proving Ground, Md, 54-55; asst prof statist, Va Polytech Inst, 66-69; ASSOC PROF STATIST, UNIV PA, 69- *Concurrent Pos:* NSF grants, Va Polytech Inst, 68 & Univ Pa, 71. *Mem:* Am Statist Asn; Am Math Soc; Inst Math Statist; Sigma Xi. *Res:* Probability theory and stochastic processes; time series data analysis and interactive computing. *Mailing Add:* 253 Lawndale Ave King of Prussia PA 19406

PICKAR, DAVID, b New Brunswick, NJ, Mar 29, 48; m. CLINICAL NEUROSCIENCE. *Educ:* Rutgers Col, BA, 69; Yale Univ, MD, 73; Am Bd Psychiat & Neurol, dipl, 78. *Prof Exp:* Intern internal med, Univ Ky Med Ctr, 73-74; resident psychiat, Dept Psychiat, Yale Univ Sch Med, 74-77; clin assoc, Clin Neuropharmacol Br, NIMH, Bethesda, Md, 77-79, staff psychiatrist & chief, Unit Studies Drug Abuse, Biol Psychiat Br, 79-81, actg chief, Clin Neurosci Br, Intramural Res Prog, 88-90, CHIEF, SECT CLIN STUDIES, EXP THERAPEUT BR, INTRAMURAL RES PROG, NIMH, 82-, CHIEF, EXP THERAPEUT BR, 91- *Concurrent Pos:* Fel, NIMH Biol Scientist Training Prog; clin prof psychiat, Uniformed Serv Univ Health Sci, Bethesda, Md; fac, Wash Sch Psychiat, Washington, DC; chief resident, Clin Res Unit, Conn Mental Health Ctr, 76, 90-Day Inpatient Unit, Yale-New Haven Hosp, 77; ward adminr, 3-E Nursing Unit, Clin Neuropharmacol Br, NIMH, 78; lectr, Clin Psychopharmacol Prog, George Washington Univ Sch Med, 79-81, med student Psychopharmacol Elective, NIMH, 79-; med student preceptor & lectr, Psychiat Clerkship, Uniformed Serv Univ Health Sci, 82- *Mem:* Am Col Neuropsychopharmacol; NY Acad Sci. *Res:* Schizophrenia-pharmacotherapy and clinical course; mechanism of neuroleptic action; neurobiology of endogenous opioid peptides; relationship to behavior and psychiatric illness; affective disorders; phenomenology and catecholamine metabolism. *Mailing Add:* NIMH NIH Exp Therapeut Br Bldg 10 Rm 4N214 900 Rockville Pike Bethesda MD 20892

PICKARD, BARBARA GILLESPIE, b Charleston, WVa, Feb 24, 36; m 63; c 2. PLANT PHYSIOLOGY. *Educ:* Stanford Univ, BA, 58, MA, 59; Harvard Univ, PhD(biol), 63. *Prof Exp:* Fel biol, Harvard Univ, 63-64; NIH fel biol & res lab electronics, Mass Inst Technol, 64-66; mem staff, Wash Univ, 66-67, asst prof, 67-73, assoc prof biol, 73-, PROF BIOL, WASH UNIV. *Mem:* Am Soc Plant Physiol; Biophys Soc. *Res:* Biochemical and biophysical description of how plants transduce and respond to internally and externally generated mechanical and hormonal signals; emphasis on electrophysiological characterization of ion channels by patch clamp and molecular characterization of the channels as complexes connected both with force-focusing input polymers and with membrane proteins effecting output. *Mailing Add:* Dept of Biol Wash Univ St Louis MO 63130-4899

PICKARD, GEORGE LAWSON, b Cardiff, Wales, July 5, 13; m 38; c 2. OCEANOGRAPHY. *Educ:* Oxford Univ, BA, 35, DPhil(physics), 37, MA, 47. *Hon Degrees:* DMS, Royal Roads Mil Col, 80. *Prof Exp:* Sci officer, Royal Aircraft Estab, 37-42, prin sci officer, 42-47; from assoc prof to prof physics, 47-79, dir, Inst Oceanog, 58-79, EMER PROF PHYSICS, UNIV BC, 79- *Honors & Awards:* Tully Medal, Can Meteorol & Oceanog Soc, 87. *Mem:* AAAS; Am Soc Limnol & Oceanog; Am Geophys Union; fel Royal Soc Can. *Res:* Physical oceanography of inlets, estuaries and coral reef systems. *Mailing Add:* Inst of Oceanog Univ of BC Vancouver BC V6T 1W5 Can

PICKARD, MYRNA RAE, b Oct 10, 35; m 57; c 1. RURAL HEALTH NURSING. *Educ:* Tex Wesleyan Col, BS, 57, MEd, 64; Tex Woman's Univ, MS, 74; Nova Univ, Ft Lauderdale, EdD, 76. *Prof Exp:* Pub health nurse, Forest County Health Dept, Miss, 58-60; asst nurse admin, John Peter Smith Sch Nursing, 60-70, nurse adminr, 70-73; dean nursing, Syst Sch Nursing, Ft Worth, 71-76, DEAN NURSING, SCH NURSING, UNIV TEX, ARLINGTON, 76- *Concurrent Pos:* Mem bd dir, Mult Sclerosis, Tex, 78-85 & Nat Rural Health Asn, 86-; consult, Southwestern Col, Tex, Midwestern Univ, Tex & Cath Univ, Ponce, PR, 84 & 86; pres, Tex League Nursing, 86-89; treas, Nat Rural Health Asn, 90-91. *Mem:* Am Nurses Asn; Nat League Nursing. *Res:* Rural health and economics of health. *Mailing Add:* 8301 Anglin Dr Ft Worth TX 76140

PICKARD, PORTER LOUIS, JR, b Ft Worth, Tex, June 2, 22; m 44; c 4. ORGANIC CHEMISTRY, RESEARCH PERSONNEL ADMINISTRATION. *Educ:* NTex State Teachers Col, BA, 43, MA, 44; Univ Tex, PhD(org chem), 47. *Prof Exp:* Lab instr chem, Lamar Jr Col, 40-41 & NTex State Teachers Col, 41-44; from asst prof to assoc prof, Univ Okla, 47-56; sr res chemist, Celanese Chem Co Tech Ctr, 56-60, group leader, 60-66, res assoc, 66-67; asst to vpres res & develop, Union Camp Res & Develop Div, 67-87; RETIRED. *Concurrent Pos:* Proj leader, Res Inst, Univ Okla, 51-56. *Mem:* Tech Asn Pulp & Paper Indust; Am Chem Soc. *Res:* Proof of structure by synthesis; reactions of hindered nitriles; reactions of aliphatic hydrocarbons; technical recruiting; administration of research in pulp, paper and natural products. *Mailing Add:* 117 Ingleside Ave Pennington NJ 08534

PICKARD, WILLIAM FREEMAN, b Boston, Mass, Sept 16, 32; m 63; c 2. BIOPHYSICS, ELECTRICAL ENGINEERING. *Educ:* Boston Univ, AB, 54; Harvard Univ, MA, 55, PhD(appl physics), 62. *Prof Exp:* Assoc prof, 66-73, PROF ELEC ENG, WASHINGTON UNIV, 73- *Concurrent Pos:* Res fel electronics & lectr appl math, Harvard Univ, 62-63; res fel biol, Mass Inst Technol, 63-66. *Mem:* Am Phys Soc; Bioelectromagnetics Soc; fel Inst Elec & Electronic Engrs; Am Soc Plant Physiologists. *Res:* Energy and resource engineering; membrane electrobiology; biological effects of electromagnetic waves; intracellular transport. *Mailing Add:* Dept Elec Eng Washington Univ St Louis MO 63130

PICKART, STANLEY JOSEPH, b Norway, Iowa, May 12, 26; m 52; c 7. MAGNETISM, ADMINISTRATION. *Educ:* Univ Iowa, MA, 51; Univ Md, PhD(physics), 58. *Prof Exp:* Asst, Univ Iowa, 50-51; physicist, US Naval Ord Lab, 51-54, 56-74; chmn dept physics, 74-85, ACTG DIR RES, DEPT PHYSICS, UNIV RI, 87- *Concurrent Pos:* Guest physicist, Brookhaven Nat Lab, 55-67; temp res assoc, Atomic Energy Res Estab, Harwell, Eng, 65; guest worker, Nat Bur Standards, 68-74; Nat Acad Sci foreign exchange scientist to Romania, 71; prog officer, NSF, 78-79 & 85-86. *Mem:* Fel Am Phys Soc; Am Asn Physics Teachers; AAAS; Mat Res Soc. *Res:* Ferromagnetism and antiferromagnetism; metals and alloys; amorphous magnetism; neutron scattering. *Mailing Add:* Dept of Physics Univ of RI Kingston RI 02881

PICKENS, CHARLES GLENN, b Clinton, Okla, Sept 15, 36; m 58; c 3. MATHEMATICS. *Educ:* Cent State Col, Okla, BS, 58; Okla State Univ, MS, 59, EdD, 67. *Prof Exp:* From asst prof to assoc prof, 60-72, PROF MATH, KEARNEY STATE COL, 72-, CHMN DEPT, 80- *Concurrent Pos:* NSF Sci fac fel, 65-66. *Mem:* Math Asn Am; Nat Coun Teachers Math. *Res:* Number theory; analysis; linear algebra. *Mailing Add:* Dept of Math Kearney State Col Kearney NE 68847

PICKENS, PETER E, b Kuling, China, June 25, 28; US citizen; m 55; c 3. NEUROPHYSIOLOGY. *Educ:* Columbia Univ, AB, 53; Univ Calif, Los Angeles, PhD, 61. *Prof Exp:* Biologist, Lederle Labs, Am Cyanamid Co, 53-55; res scientist physiol, Inst Marine Sci, Univ Tex, 60-61; from asst prof to prof , 61-72, PROF ZOOL, UNIV ARIZ, 72- *Concurrent Pos:* NSF grant, 65-67. *Mem:* Fel AAAS; Am Soc Zool; Soc Neuroscience. *Res:* Neurophysiology in invertebrates. *Mailing Add:* Dept Cell Univ Ariz Col Med 1501 N Campbell Tucson AZ 85724

PICKER, HARVEY SHALOM, b Manila, Philippines, Sept 3, 42; US citizen; m 70; c 2. THEORETICAL PHYSICS. *Educ:* Mass Inst Technol, BS, 63, PhD(physics), 66. *Prof Exp:* Res physicist, Carnegie Inst Technol, 66-67; NSF fel nuclear theory, Princeton Univ, 67-69; res assoc, Univ Md, College Park, 69-70, Ctr Theoret Physics fel, 70-71; from asst prof to assoc prof, 71-85, PROF PHYSICS, TRINITY COL, CONN, 85- *Mem:* Am Phys Soc. *Res:* Few-body physics; applications of mathematical approximation theory to theoretical physics; atomic and solid-state physics; scattering theory. *Mailing Add:* Dept of Physics Trinity Col Hartford CT 06106

PICKERING, ED RICHARD, b Cincinnati, Ohio, Dec 15, 34; m 62; c 2. PLANT PHYSIOLOGY. *Educ:* Ohio State Univ, BS, 56, MS, 58; Univ Calif, Davis, PhD(bot), 64. *Prof Exp:* Res asst bot, Ohio State Univ, 57-58; asst, Univ Calif, Davis, 58-62; instr, Univ SFla, 62-63; asst, Univ Calif, Davis, 64; asst prof, Rutgers Univ, 64-66; assoc prof biol, Adrian Col, 66-74; instr hort, Springfield-Clark Co Joint Voc Sch, 74-77; assoc prof biol, Central State Univ, Ohio, 77-78. *Concurrent Pos:* NSF fac partic grant, Adrian Col, 69-71. *Mem:* Am Soc Plant Physiol; Bot Soc Am; Am Inst Biol Sci; Sigma Xi. *Res:* Translocation; plant morphogenesis. *Mailing Add:* 2399 Versailles Ct Springfield OH 45502-9114

PICKERING, FRANK E, AIRCRAFT ENGINEERING. *Prof Exp:* VPRES & GEN MGR, AIRCRAFT ENGINES ENG DIV, GEN ELEC CO, CINCINNATI, OHIO & LYNN, MASS, 90- *Mem:* Nat Acad Eng. *Mailing Add:* Gen Elec Co 1000 Western Ave Bldg 24501 Lynn MA 01910

PICKERING, HOWARD W, b Cleveland, Ohio, Dec 15, 35; m 63; c 4. CORROSION, SURFACE SCIENCE. *Educ:* Univ Cincinnati, BMetEng, 58; Ohio State Univ, MS, 59, PhD(metall eng, stress corrosion), 61. *Prof Exp:* Res metallurgist, Res Lab, US Steel Corp, Pa, 62-72; assoc prof, 72-76, prog chmn, 75-80, PROF METALL, PA STATE UNIV, UNIVERSITY PARK, 76- *Concurrent Pos:* Guest scientist, Max Planck Inst Phys Chem, Gottingen, WGer, 64-65; prin investr, Off Naval Res, 72-, NSF, 73-, Int Copper Asn, Ltd, 84-; Am ed, Corrosion Sci, 75-; mem adv panel, Corrosion Res Ctr, Univ Minn, 79-86; res prof, Inst Solid State Physics, Univ Tokyo, 82. *Honors & Awards:* Whitney Award, Nat Asn Corrosion Engrs, 85; H H Uhlig Award, Electrochem Soc. *Mem:* Electrochem Soc; Am Inst Mining, Metall & Petrol Engrs; Am Soc Metals; Nat Asn Corrosion Engrs; AAAS. *Res:* Corrosion and oxidation of metals; electroplating; hydrogen in metals; grain boundary diffusion and surface and grain boundary segregation in metals; atom probe field ion microscopy; surface science. *Mailing Add:* 326 Steidle Bldg Dept Mat Sci & Engr Pa State Univ University Park PA 16802

PICKERING, JERRY L, b Washington, Iowa, May 26, 42; m 63; c 3. SYSTEMATIC BOTANY. *Educ:* Iowa State Univ, BS, 64; Rutgers Univ, MS, 66, PhD(bot), 69. *Prof Exp:* Instr biol, Rutgers Univ, 69; asst prof, 69-74, assoc prof, 74-78, PROF BIOL, INDIANA UNIV, PA, 78- *Mem:* Am Inst Biol Sci; Am Soc Plant Taxon; Sigma Xi. *Res:* Comparative protein chemistry using plant materials and their value in systematics; seriological techniques; polyacrylamide gel electrophoresis; chemotaxonomy; flora of Western Pennsylvania. *Mailing Add:* Dept Biol Ind Univ Indiana PA 15705

PICKERING, MILES GILBERT, b Beaconsfield, Eng, Apr 8, 43; US citizen. CHEMICAL EDUCATION. *Educ:* Yale Univ, BS, 65; Univ Wash, MS, 66; State Univ NY Stony Brook, PhD(chem), 70. *Prof Exp:* Vis asst prof chem, Reed Col, 70-71; Petrol Res Fund fel, State Univ NY Stony Brook, 71-72, lectr & res assoc nuclear chem, 72-73; lectr chem, Columbia Univ, 73-76; LECTR WITH RANK OF ASSOC PROF, PRINCETON UNIV, 76- *Concurrent Pos:* Consult, State Univ NY Stony Brook, 73-76, Yale Univ, 77-78, Columbia Univ, 78-79, Harvard Univ, 80-81, Brown Univ, 82-83 & Bowdoin Col, 85. *Mem:* Sigma Xi. *Res:* Nuclear magnetic resonance applied to inorganic and biochemical systems; innovative teaching of laboratory courses and of physical chemistry. *Mailing Add:* Dept Chem Princeton Univ Princeton NJ 08540

PICKERING, RANARD JACKSON, b Goshen, Ind, Mar 24, 29; m 50; c 3. PRECIPITATION CHEMISTRY, WATER QUALITY. *Educ:* Ind Univ, AB, 51, MA, 52; Stanford Univ, PhD(geol), 61. *Prof Exp:* Geologist, NJ Zinc Co, 52-55; raw mat engr, Columbia Iron Mining Co, 55-58; geologist, Water Resources Div, Tenn, 61-65, hydrologist, Ohio, 65-67, assoc dist chief, Ohio, 67-70, asst chief, 70-72, chief, Qual of Water Br, 72-84, CHIEF, OFF ATMOSPHERIC DEPOSITION ANAL, US GEOL SURV, RESTON, VA, 84- *Concurrent Pos:* US Dept Interior Deposition Res, Interagency Sci Comt, Nat Acid Precipitation Assessment Prog. *Mem:* AAAS; Geochemical Soc; fel Geol Soc Am; Am Water Resources Asn; Sigma Xi. *Res:* Geochemistry of weathering processes; movement and fate of radionuclides in a fluvial environment; geochemistry and control of acid mine drainage; hydrologic processes affecting water quality; effect of atmospheric deposition on water quality. *Mailing Add:* 2321 N Richmond St Arlington VA 22207

PICKERING, RICHARD JOSEPH, b Medicine Hat, Alta, Aug 26, 34; m 59; c 2. PEDIATRICS, IMMUNOLOGY. *Educ:* Univ Sask, BA, 57, MD, 61; Am Acad Pediat, dipl, 74. *Prof Exp:* Asst prof pediat, Univ Minn, 69; assoc prof, Albany Med Col, 70-72; assoc prof microbiol & prof pediat, Dalhousie Univ, 72-73; res physician, NY State Kidney Dis Inst, 74-80; prof pediat & microbiol, 78-80, PROF MICROBIOL, IMMUNOL & PEDIAT, ALBANY COL, 80-; RES PHYSICIAN, NY STATE DEPT HEALTH, 80- *Concurrent Pos:* Queen Elizabeth II res fel, Univ Minn, 67-69; res physician, NY State Kidney Dis Inst, 70-72; prof pediat & microbiol, Albany Med Col, 74. *Mem:* AAAS; Am Soc Immunologists; Soc Exp Biol & Med; Soc Pediat Res. *Res:* Immunobiology of inflammation. *Mailing Add:* 47 Cross St Dundas ON L9H 2R5 Can

PICKERING, THOMAS G, b Gerrards Cross, Eng, May 5, 40; m 65; c 2. CARDIOVASCULAR MEDICINE. *Educ:* Cambridge Univ, BA, 62, MA, 68, MB, 66; Oxford Univ, PhD(med), 70- *Prof Exp:* Registr, Radcliffe Infirmary, 68-71; asst prof, Rockefeller Univ, 72-74; lectr, Oxford Univ, 74-76; assoc prof, 76-85, PROF, CORNELL UNIV MED SCH, 85- *Concurrent Pos:* Res fel, Oxford Univ, 68-71; assoc attend physician, NY Hop, 76-85, atend physician. 85- *Mem:* Intl Soc Hypertension; Am Soc Hypertension; Am Heart Asn; Brit Cardiac Soc; Soc Behav Med; Acad Behav Med Res. *Res:* Blood pressure measurement technique; renovascular hypertension; ambulatory blood pressure monitoring for the diagnosis of hypertension and evaluation of effects of stress on blood pressure. *Mailing Add:* Cardiovasc Ctr NY Hosp 520 E 70th St New York NY 10021

PICKERING, W(ILLIAM) H(AYWARD), b Wellington, NZ, Dec 24, 10; nat US; m 32; c 2. ELECTRICAL ENGINEERING, PHYSICS. *Educ:* Calif Inst Technol, BS, 32, MS, 33, PhD(physics), 36. *Hon Degrees:* DSc, Clark Univ, 66 & Occidental Col, 66. *Prof Exp:* Dir, Res Inst, Univ Petrol & Minerals, Saudi Arabia, 76-78; asst physics, Calif Inst Technol, 32-36, from instr to prof elec eng, 36-78, dir, Jet Propulsion Lab, 54-76, EMER PROF ELEC ENG, CALIF INST TECHNOL, 78-; PRES, PICKERING RES CORP, 78-; PRES, LIGNETICS, INC, 83- *Concurrent Pos:* Lectr, Univ Southern Calif, 38; mem sci adv bd, US Air Force, 45-48; mem, US Nat Comt Technol Panel Earth Satellite Prog, 56-59; mem, Army Sci Adv Panel, 63-65. *Honors & Awards:* Wyld Mem Award, Inst Aeronaut & Astronaut, 57, Louis W Hill Transp Award, Am Rocket Soc, 68; Space Flight Achievement Award, Nat Missile Indust Conf, 59; Distinguished Civilian Serv Medal, US Dept Army, 59; Columbus Gold Medal, 64; Crozier Gold Medal, 65; Goddard Mem Trophy, 65; Spirit of St Louis Medal, 65; Procter Prize, 65; Magellanic Premium, 66; Edison Medal, Inst Elec & Electronic Engrs, 72; Nat Medal Sci, 76; Fahrney Medal, Franklin Inst, 76. *Mem:* Nat Acad Sci; Nat Acad Eng; hon fel Inst Aeronaut & Astronaut (pres, 63); fel Inst Elec & Electronic Engrs; fel Am Acad Arts & Sci. *Res:* Unmanned lunar and planetary exploration. *Mailing Add:* Pickering Res Corp 1401 S Oak Knoll Ave Pasadena CA 91109

PICKETT, BILL WAYNE, b Cyril, Okla, Dec 14, 30; m 55; c 3. REPRODUCTIVE PHYSIOLOGY. *Educ:* Okla State Univ, BS, 52; Univ Mo, MS, 55, PhD(agr), 58. *Prof Exp:* From asst prof to assoc prof, Univ Conn, 57-67, res asst, 58-59; PROF PHYSIOL & BIOPHYS, COLO STATE UNIV, 67-, DIR ANIMAL REPROD LAB, COL VET MED & BIOMED SCI, 70- *Concurrent Pos:* Mem, Rockefeller Found Mex Agr Prog, 63. *Honors & Awards:* Physiol & Endocrinol Award, Am Soc Animal Sci, 78; Res Award, Nat Asn Animal Breeders, 80; L W Durrall Award, 81. *Mem:* Am Soc Animal Sci; AAAS; Soc Study Reproduction; Soc Study Fertil; Am Dairy Sci Asn. *Res:* Reproductive physiology of the male with special emphasis on preservation of spermatozoa of the equine and bovine species. *Mailing Add:* Animal Reprod Lab Dept Physiol Col Vet Med Colo State Univ Ft Collins CO 80523

PICKETT, CECIL BRUCE, b Canton, Ill, Oct 5, 45; m 67; c 2. CELL & MOLECULAR BIOLOGY. *Educ:* Calif State Univ, Hayward, BS, 71; Univ Calif, Los Angeles, PhD(biol), 76. *Prof Exp:* Fel cell biol, Univ Calif, Los Angeles, 76-78; SR RES BIOCHEMIST, MERCK SHARP & DOHME RES LABS, 78- *Concurrent Pos:* NSF scholar, 76-77; scholar, Univ Calif, 77-78; Macy scholar, Marine Biol Lab, Woods Hole, 78; vis asst prof anat, Col Med, Howard Univ, 79- *Mem:* Am Soc Cell Biol; AAAS. *Res:* Molecular basis of induction of glutathione transferance and cytochrome P-450 by xenobiotics. *Mailing Add:* 16711 Trans Canada Hwy Kirkland PQ H9H 3L1 Can

PICKETT, EDWARD ERNEST, b Greenfield, Ind, May 27, 20; m 50; c 3. ANALYTICAL CHEMISTRY. *Educ:* Purdue Univ, BS, 42; Ohio State Univ, PhD(org chem), 48. *Prof Exp:* Asst chem, Ohio State Univ, 42-44, Off Sci Res & Develop asst, Res Found, 44-48; from asst prof to assoc prof agr chem, 48-55, PROF AGR CHEM, UNIV MO-COLUMBIA, 55-, SUPVR SPECTROG LABS, 53- *Mem:* Am Chem Soc; Soc Appl Spectros; Sigma Xi. *Res:* Chemical spectroscopy; trace elements in biological materials; co-precipitation; atomic absorption analysis; role of lithium in living tissues. *Mailing Add:* Univ Mo 31 Chem Bldg Columbia MO 65211

PICKETT, GEORGE R, b El Paso, Tex, Jan 13, 25; m 48; c 2. GEOPHYSICS, ENGINEERING. *Educ:* Univ Okla, BA, 51, MS, 52; Colo Sch Mines, DSc(geophys eng), 55. *Prof Exp:* Exploitation engr, Shell Develop Co, 55-63; sr petrophys entr, Shell Oil Co, 63-64, staff petrophys engr, 64-66; asst prof, 66-69, PROF GEOPHYS, COLO SCH MINES, 69- *Concurrent Pos:* Consult short course teaching, 66-; head, Dept Petrol Eng, Colo Sch Mines, 69-71; distinguished lectr, Soc Explor Geophys, 81. *Mem:* Soc Explor Geophys; Am Geophys Union; Soc Petrol Eng; Soc Prof Well Log Analysts. *Res:* Formation evaluation in hydrocarbon exploration, particularly well log interpretation; training methods. *Mailing Add:* 2554 Devinney Ct Golden CO 80401

PICKETT, HERBERT MCWILLIAMS, b Baltimore, Md, Apr 2, 43; m 65; c 2. PHYSICAL CHEMISTRY. *Educ:* Williams Col, BA, 65; Univ Calif, Berkeley, PhD(chem), 70. *Prof Exp:* Fel chem, Harvard Univ, 70-72; Miller fel chem, Univ Calif, Berkeley, 72-73; asst prof chem, Univ Tex, Austin, 73-78; mem tech staff, 78-81, res scientist, 81-86, SR RES SCIENTIST, JET PROPULSION LAB, 86- *Concurrent Pos:* NIH fel, 70-71. *Mem:* Am Chem Soc; Am Phys Soc; Sigma Xi; Inst Elec & Electronics Engrs. *Res:* Microwave, millimeter wavelength and infrared spectroscopy; rotational energy transfer, molecular structure and conformational dynamics; astro-chemistry; millimeter and submillimeter wavelength radiometry. *Mailing Add:* 1112 Wiladonda Dr La Canada CA 91011

PICKETT, JACKSON BRITTAIN, b San Antonio, Tex, Apr 30, 43; m 67. NEUROLOGY, PHYSIOLOGY. *Educ:* Occidental Col, BA, 64; Yale Univ, MD, 68. *Prof Exp:* Intern, Grady Mem Hosp, Atlanta, 68-69; resident neurol, Univ Calif, San Francisco, 69-72; neurologist, Regional Hosp, US Air Force, 72-74; fel clin neurophysiology, Mayo Clin, Minn, 74-75; asst prof neurol, Univ Calif, San Francisco, 75-81; NEUROL SERV, VET ADMIN MED CTR, CHARLESTON, 81-; ASSOC PROF NEUROL, MED UNIV OF SC, 81- *Mem:* Am Acad Neurol; Am Asn Electromyography & Clin Neurophysiology. *Res:* Neuromuscular transmission. *Mailing Add:* Neurology 109 Bee St Charleston SC 29403

PICKETT, JAMES EDWARD, b Plainwell, Mich, May 10, 54; m 90. ORGANIC PHOTOCHEMISTRY. *Educ:* Kalamazoo Col, BA, 76; Yale Univ, MPhil, 78, PhD(org chem), 80. *Prof Exp:* STAFF CHEMIST, GEN ELEC CORP RES & DEVELOP, 80- *Mem:* Am Chem Soc; AAAS. *Res:* Mechanisms of polymer thermal and photo-degradation; accelerated testing; flame retardance; UV-cured coatings. *Mailing Add:* Gen Elec Corp Res & Develp Bldg K-1 PO Box Eight Schenectady NY 12301

PICKETT, JAMES M, b Pampa, Tex, Jan 20, 37; m 60; c 2. BIOMATHEMATICS. *Educ:* Rice Univ, BA, 60; Univ Tex, Austin, PhD(zool), 64. *Prof Exp:* Fel plant biol, Carnegie Inst Dept Plant Biol, 65-67; from asst prof to assoc prof, 67-78, head dept, 73-83, actg dir admin, 83-86, PROF BOT, 78-, PROF BIOL, 86- *Res:* Computer models and decision support software. *Mailing Add:* Dept of Biol Mont State Univ Bozeman MT 59717-0346

PICKETT, JOHN HAROLD, b Boston, Mass, June 21, 43; m 74; c 1. INFRARED & NMR SPECTROSCOPY. *Educ:* Tufts Univ, BS, 65; Purdue Univ, PhD(chem), 70. *Prof Exp:* ANAL CHEMIST, HOECHST CELANESE, 70- *Mem:* Am Chem Soc; Soc Appl Spectros. *Res:* Infrared spectroscopy; nuclear magnetic resonance spectroscopy. *Mailing Add:* Hoechst Celanese Box 32414 Charlotte NC 28232

PICKETT, LAWRENCE KIMBALL, b Baltimore, Md, Nov 10, 19; m 43; c 4. SURGERY. *Educ:* Yale Univ, BA, 41, MD, 44. *Prof Exp:* Clin asst prof surg, State Univ NY Upstate Med Ctr, 50-54, clin assoc prof, 54-64; PROF SURG & PEDIAT, SCH MED, YALE UNIV, 64-, ASSOC DEAN CLIN AFFAIRS, 73- *Concurrent Pos:* Chief staff & chmn med bd, Yale-New Haven Hosp, 73- *Mem:* Fel Am Acad Pediat; fel Am Col Surgeons. *Mailing Add:* 789 Howard Ave New Haven CT 06504

PICKETT, LEROY KENNETH, b Clay Center, Kans, May 8, 37; m 63; c 2. CONCEPT FEASIBILITY ANALYSIS. *Educ:* Kans State Univ, BS, 61; Univ Ill, Urbana, MS, 62; Purdue Univ, PhD(agr eng), 69. *Prof Exp:* Asst engr, Independence Works, Allis-Chalmers Mfg Co, 62-65; asst prof agr eng, Mich State Univ, 68-73; design engr, J I Case, Int Harvester, 73-75, proj engr, 75-81, prod engr, 81-82, SR PROJ ENGR, J I CASE, INT HARVESTER, 81- *Concurrent Pos:* Mem nominating comt, Am Soc Agr Engrs, 89-91. *Mem:* Am Soc Agr Engrs; Soc Automotive Engrs. *Res:* Development of combine harvesters including mechanism and structures for cutting, threshing, separating and conveying grain; application of fluid power; concept feasiblity analysis for agricultural and construction equipment; value engineering of parts & components. *Mailing Add:* 327 Eighth St Downer's Grove IL 60515

PICKETT, MORRIS JOHN, b Beloit, Kans, Dec 3, 15; m 46; c 4. BACTERIOLOGY. *Educ:* Stanford Univ, BS, 38, PhD(bact), 42. *Prof Exp:* Asst bact, Harvard Med Sch, 42-45, instr, 45-46; asst prof, Sch Med, Univ Louisville, 46-47; from asst prof to assoc prof, 47-57, PROF BACT, UNIV CALIF, LOS ANGELES, 57- *Concurrent Pos:* Consult bacteriologist, New Eng Deaconess Hosp, 44-45, Olive View Hosp, 63-71 & Food & Drug Admin, 71-81; USPHS spec res fel, Univ Queensland, 60; adv ed, Lab World, 61-80 & J Infectious Dis, 64-68; ed bd, J Clin Microbiol, 80- *Mem:* Am Soc Microbiol; fel Am Acad Microbiol; Brit Soc Gen Microbiol. *Res:* Medical and diagnostic bacteriology; taxonomy; immunology. *Mailing Add:* Dept Microbiol Univ of Calif Los Angeles CA 90024

PICKETT, PATRICIA BOOTH, cell biology; deceased, see previous edition for last biography

PICKETT, STEWARD T A, b Louisville, Ky, Nov 30, 50. ECOLOGY, BOTANY. *Educ:* Univ Ky, BS, 72; Univ Ill, PhD(bot), 77. *Prof Exp:* Asst prof, Rutgers Univ, 77-82, assoc prof bot & ecol, 82-87; assoc scientist, 87-90, SCIENTIST, INST ECOSYST STUDIES, 90- *Concurrent Pos:* Vis fac, Orgn Trop Studies, 77; vis scientist, Inst Ecosystem Studies, 85-86. *Mem:* Ecol Soc Am; Soc Study Evolution; Sigma Xi; AAAS; Brit Ecol Soc. *Res:* Plant population interaction and community organization; role of species adaptation in succession; successional theory. *Mailing Add:* Inst Ecosyst Studies NY Botanical Garden Box AB Millbrook NY 12545-0129

PICKETT, THOMAS ERNEST, b Griffin, Ga, June 4, 37; m 71; c 3. GEOLOGY. *Educ:* Duke Univ, BS, 59; Univ NC, MS, 62, PhD(geol), 65. *Prof Exp:* Geol supvr, Smithsonian Inst, 65-66; geologist, 66-70, sr geologist, 70-78, ASSOC DIR, DEL GEOL SURV, 78- *Concurrent Pos:* Asst prof, continuing educ div, Univ Del, 67-71, assoc prof, 71-; alt mem adv bd, Outer Continental Shelf Res Mgt, Dept Interior, 74-77. *Mem:* Fel Geol Soc Am; Hist Earth Sci Soc (treas, 91-). *Res:* Modern coastal sedimentary environments; sedimentary structures; coastal plain stratigraphy and sedimentation; trace fossils; history of geology. *Mailing Add:* Del Geol Surv Univ Del Newark DE 19716

PICKETT-HEAPS, JEREMY DAVID, b Bombay, India, June 5, 40; m 65; c 4. CELL BIOLOGY, PHYCOLOGY. *Educ:* Cambridge Univ, BS, 62, PhD(biochem), 65. *Prof Exp:* Res fel electron-micros, Australian Nat Univ, 65-69, fel, 69-70; from asst prof to assoc prof, 70-78, PROF MCD BIOL, UNIV COLO, BOULDER, 78- *Concurrent Pos:* NSF grant, 71-76; adv, NIH grant. *Honors & Awards:* Darbaker Prize, Am Bot Soc, 74. *Res:* Ultrastructure and function during cell division and differentiation in plant cells. *Mailing Add:* Sch Botany Univ Melbourne Parkville Victoria 3052 Australia

PICKHOLTZ, RAYMOND L, b New York, NY, Apr 12, 32; m 57; c 3. ELECTRONICS, COMMUNICATIONS. *Educ:* City Col New York, BS, 54, MS, 56; Polytech Inst Brooklyn, PhD(elec eng), 62. *Prof Exp:* Res engr, RCA Labs, 54-57; res specialist, ITT Labs, 57-61; from assoc prof to prof elec eng, Polytech Inst Brooklyn, 61-71; PROF & DEPT CHMN ELEC ENG & COMPUT SCI, GEORGE WASHINGTON UNIV, 72- *Concurrent Pos:* Adj assoc prof physics, Brooklyn Col, 63-71; prin investr, NASA space commun res grant, 65-; NSF res grant commun, 68-69; consult, IBM Corp, Fairchild Industs & Comput Sci Corp; vis prof, Inst Nat Reserches de la Sci, Que, 77-78, Univ Calif, 83, 89. *Honors & Awards:* RCA Labs Res Award; Centennial Medal, Inst Elec & Electronics Engrs, 84. *Mem:* Fel Inst Elec & Electronics Engrs; Math Asn Am; Asn Comput Mach; AAAS; Sigma Xi; fel AAAS; Cosmos Club. *Res:* Communication theory, including data transmission, information theory, detection and estimation theory and coding. *Mailing Add:* Sch Eng & Appl Sci George Washington Univ Washington DC 20052

PICKLE, LINDA WILLIAMS, b Hampton, Va, July 19, 48; m 84; c 1. BIOMATHEMATICS, PUBLIC HEALTH. *Educ:* Johns Hopkins Univ, BA, 74, PhD(biostatist), 77. *Prof Exp:* Comput programmer, Com Credit Comput Corp, 66-69; systs analyst, Greater Baltimore Med Ctr, 69-72; teaching asst biostatist, Johns Hopkins Univ, 74-77; staff fel, 77-84, HEALTH STATISTICIAN, NAT CANCER INST, NIH, 84- *Concurrent Pos:* Adj asst prof biostatist, Georgetown Univ, Washington, DC, 83- *Mem:* Biometric Soc; Am Statist Asn; Soc Epidemiol Res; Grad Women Sci; Soc Indust & Appl Math. *Res:* Develops and applies advanced statistical techniques and computer systems to the analysis of epidemiologic studies of cancer; biostatistical consultant; designs and supervises large questionnaire studies of cancer. *Mailing Add:* Environ Epidemiol Br Nat Cancer Inst NIH Bldg EPN Rm 430 Bethesda MD 20892

PICKRELL, JOHN A, b Decatur, Ill, May 8, 41; m 62; c 2. BIOCHEMISTRY, PHYSIOLOGY. *Educ:* Univ Ill, Urbana, BS, 63, DVM, 65, MS, 66, PhD(vet med sci), 68. *Prof Exp:* Inhalation Toxicol Res Inst, Lovelace Biomed & Environ Res Inc, 68-88. *Concurrent Pos:* Toxicologist, Dept Clin Sci, Kans State Univ, Manhattan. *Mem:* Am Vet Med Asn; Am Soc Vet Physiol & Pharmacol; Am Chem Soc; Fedn Am Soc Exp Biol. *Res:* Veterinary clinical pathology; porcine edema disease; response of pulmonary connective tissue in health and disease; clinical pathology; heart-lung disease; indoor air pollution. *Mailing Add:* Comp Toxicol Lab Dept Clin Sci Kans State Univ Manhattan KS 66506-5606

PICKRELL, KENNETH LEROY, surgery; deceased, see previous edition for last biography

PICKRELL, MARK M, b Englewood, NJ, Mar 24, 54; m; c 3. NONDESTRUCTIVE ASSAY, ELECTRONICS. *Educ:* Mass Inst Technol, BS, 76, PhD(physics), 83. *Prof Exp:* Res assoc plasma physics, Mass Inst Technol, 76-82; staff physicist plasma physics, 83-90, STAFF PHYSICIST NUCLEAR PHYSICS, LOS ALAMOS NAT LAB, 90- *Mem:* Am Phys Soc. *Res:* Investigating methods of non destructive assay of nuclear material for the verification of the nuclear non-proliferation treaty; nuclear spectroscopy and counting; plasma diagnostics and plasma physics on the reversed field pinch magnetic confinement device; instrumentation. *Mailing Add:* Mail Stop E540 Los Alamos Nat Lab Los Alamos NM 87544

PICKWELL, GEORGE VINCENT, b San Jose, Calif, Oct 20, 33. PHYSIOLOGY, MARINE BIOLOGY. *Educ:* San Jose State Col, BA, 57; Univ Calif, San Diego, PhD(marine biol), 64. *Prof Exp:* Res biologist, Scripps Inst, Univ Calif, 60-61, res asst marine biol, 61-64; Nat Acad Sci-Nat Res Coun resident res assoc oceanog, Navy Electronics Lab, 64-66; OCEANOGR, NAVAL OCEAN SYST CTR, 66-, HEAD, BIOTECHNOL LAB, 86- *Mem:* AAAS; Am Inst Biol Sci; Am Soc Zoologists; Sigma Xi. *Res:* Biological oceanography; venomous marine animals; sea snakes and their venom; physiology and behavior of deep-sea animals; diving physiology; sound scattering and gas production by marine fishes and invertebrates; biochemical stress factors in marine animals, marine biotechnology. *Mailing Add:* PO Box 80426 San Diego CA 92138

PICKWORTH, WALLACE BRUCE, b Rochester, NY, May 1, 46; m 74; c 1. PHARMACOLOGY. *Educ:* Albany Col Pharm, BS, 69; Univ Tenn, PhD(pharmacol), 74. *Prof Exp:* Staff fel, 74-78, PHARMACOLOGIST, ADDICTION RES CTR, NAT INST DRUG ABUSE, 78- *Mem:* Soc Neurosci; Sigma Xi. *Res:* Drug abuse; sleep; psychopharmacology; neuropharmacology. *Mailing Add:* 4418 Roland Ave Baltimore MD 21210

PICO, GUILLERMO, b Coamo, PR, Dec 9, 15; m 40; c 3. OPHTHALMOLOGY. *Educ:* Univ PR, BS, 36; Univ Md, MD, 40. *Prof Exp:* Resident ophthal, St Luke's Hosp, New York, 45-46; prof, 52-77, EMER PROF OPHTHAL & HEAD DEPT, SCH MED, UNIV PR, SAN JUAN, 77- *Concurrent Pos:* Attend & chief dept ophthal, Univ Hosp, PR & San Juan City Hosp, PR, 52-77; consult & dir training prof, Vet Hosp, San Juan, 60-77; mem, Vision Res Training Comt, Nat Eye Inst, 67-71. *Honors & Awards:* Physician of the Year Award, President's Comt for Crippled, 66; PR Med Asn Award, 66; First Am J Ophthal Lectr Award, Cong Pan-Am Asn Ophthal, 71. *Mem:* Am Acad Ophthal & Otolaryngol (3rd vpres, 70); PR Med Asn (pres, 57); AMA; Am Acad Ophthal; Asn Res Vision & Ophthal; Am Ophthal Soc; Pan Am Asn Ophthal (vpres, 83-87). *Res:* Diseases and surgery of the lacrimal system, cornea and cataract; hereditary and congenital anomalies of the eyes; pathogenesis; clinical manifestation and treatment of Pterygium. *Mailing Add:* 1475 Wilson Ave Santurce San Juan PR 00907

PICONE, J MICHAEL, b Galveston, Tex, Apr 12, 48; m 85; c 2. IONOSPHERIC MODELLING, MAGNETOHYDRODYNAMIC TURBULENCE. *Educ:* Rice Univ, BA, 70; Univ Tex, Austin, Ma, 72, PhD(physics), 74. *Prof Exp:* Res physicist, E I Du Pont de Nemours & Co, Inc, 74-75; tech staff, Anal Serv Inc, 75-79; res assoc, Nat Res Coun/Naval Res Lab, 79-80, RES PHYSICIST, NAVAL RES LAB, 80- *Mem:* Am Phys Soc; Am Astron Soc; Am Geophys Union; Int Neural Network Soc. *Res:* Develop detailed numerical simulation models and apply them to atmospheric physics and the study of turbulence; develop database systems for satellite remote sensing data and modelling. *Mailing Add:* Code 4143 Naval Res Lab Washington DC 20375-5000

PICOT, JULES JEAN CHARLES, b Edmundston, NB, July 23, 32; m 56; c 2. CHEMICAL ENGINEERING. *Educ:* NS Tech Col, BE, 55; Mass Inst Technol, SM, 57; Univ Minn, PhD(heat conduction), 66. *Prof Exp:* Control engr, Consol Bathurst Mills, 55-59; asst prof chem eng, Univ NB, 59-66; assoc prof, 66-70, chmn dept, 69-76, PROF CHEM ENG, UNIV NB, FREDERICTON, 70- *Concurrent Pos:* Mem grants selection comt, Nat Res Coun Can, 71-73 & 77-; dir, Can Soc Chem Engrs. *Mem:* Am Inst Chem Engrs; fel Can Inst Chem. *Res:* Transport phenomena, principally combined heat and momentum transfer; orientation effects in polymer liquids; atomization and dispersal of aerial sprays. *Mailing Add:* Dept Chem Eng Univ NB Box 4400 Fredericton NB E3B 5A3 Can

PICRAUX, SAMUEL THOMAS, b St Charles, Mo, Mar 3, 43; m 70; c 3. ELECTRONIC MATERIALS, ION-SOLID INTERACTIONS. *Educ:* Univ of Mo-Columbia, BS, 65; Calif Inst Technol, MS, 67, PhD(eng sci & physics), 69. *Prof Exp:* Tech staff mem, 69-72, supvr, Ion Solid Interactions Res, 72-86, MGR, SURFACE INTERFACE & ION BEAM RES DEPT, SANDIA NAT LABS, 86- *Concurrent Pos:* Fulbright fel, physics, Cambridge Univ, 65-66; vis prof, Aarhus Univ, Denmark, 75; mem, Int Adv Comt Ion Beam Anal Conf, 73-, Ion Beam Modification Mat, 73-, Nat Res Coun Adv Comt, Ion Implantation Appln, 78 & Mat Sci Panel, 85, Photonic, 86; lectr, NATO, 79, 81 & 83; ed, Int J Nuclear Instruments & Methods B, 83-90. *Honors & Awards:* Mat Sci Award, Dept Energy, 85; O E Lawrence Award, 90. *Mem:* Fel Am Phys Soc; Int Elec & Electronic Engrs; Electrochem Soc; Mat Res Soc; Metall Soc. *Res:* Ion-beam analysis techniques including ion backscattering, ion-induced nuclear reactions, ion channeling; ion implantation, surface modification, electron beam and laser pulsed annealing, hydrogen in solids, surface layer reactions, semiconductors, disorder studies, metals, molecular beam epitaxial growth and strained semiconductor structures. *Mailing Add:* Sandia Nat Labs Dept 1110 Albuquerque NM 87185

PICTON, HAROLD D, b Bowman, NDak, Oct 6, 32; m 60; c 3. ZOOLOGY. *Educ:* Mont State Col, BS, 54, MS, 59; Northwestern Univ, PhD(biol), 64. *Prof Exp:* Jr biologist, Mont Fish & Game Dept, 54-55, biologist, 57-58, proj biologist, 59-63; asst prof comp physiol, 63-69, assoc prof environ physiol, 69-77, ASSOC PROF WILDLIFE MGT, MONT STATE UNIV, 77- *Mem:* AAAS; Ecol Soc Am; Am Soc Mammalogists; Wildlife Soc; Sigma Xi. *Res:* Ecology of large mammals; comparative physiology; environmental management. *Mailing Add:* 3026 Candy Lane Bozeman MT 59715

PICUS, GERALD SHERMAN, b Madison, Wis, Jan 9, 26; m 52; c 3. SOLID STATE SCIENCE. *Educ:* Univ Chicago, BS, 47, MS, 49, PhD, 54. *Prof Exp:* Physicist, US Naval Res Lab, 53-59; mem tech staff, Semiconductor Div, 59-62, sr staff physicist, Res Lab, 62-66, sr scientist, 66-68, mgr, Chem Physics Dept, 68-82, DIR, TECH EDUC CTR, HUGHES AIRCRAFT CO, 82- *Concurrent Pos:* Vis prof appl physics, Calif Inst Technol, 76-77. *Mem:* AAAS; fel Am Phys Soc; sr mem Inst Elec & Electronics Engrs; Sigma Xi. *Res:* Film flow in liquid helium; infrared properties of semi-conductors and other nonmetallic solids; infrared detectors; photoemission in tunnel structures. *Mailing Add:* 22545 Marylee St Woodland Hills CA 91364

PIDGEON, REZIN E, JR, US citizen. ELECTRICAL ENGINEERING. *Educ:* Ga Inst Technol, BEE, 49, MSEE, 61. *Prof Exp:* Field engr mil electronic equip, RCA Serv Co, 50-52, engr exp airborne radar, RCA Labs, 53-55; engr reactor instrumentation, Oak Ridge Lab, 56-57; asst res engr, Eng Exp Sta, Ga Inst Technol, 57-58, res engr, 58-62; sr engr, 62-68, staff engr instrumentation, 68-72, prin engr, Telecommun Prod Line, 72-65, PRIN ENGR, COMMUN GROUP, SCI-ATLANTA, INC, 79- *Res:* Antenna instrumentation; electronic circuitry; cable television systems and equipment design. *Mailing Add:* Scientific Atlanta 4311 Communications Dr Norcross GA 30093

PIECH, KENNETH ROBERT, b Buffalo, NY. OPTICAL PHYSICS. *Educ:* Canisius Col, BS, 62; Cornell Univ, PhD(physics), 67. *Prof Exp:* Vpres, Scipar Inc, 79-85; STAFF SCIENTIST, CALSPAN CORP, 67-; PRES, ASPEN ANALYSIS, 85- *Honors & Awards:* Autometric Award, Am Soc Photogram, 73. *Mem:* Am Soc Photogram; Sigma Xi. *Res:* Application of photometry to imaging systems; optical properties of surfaces; atmospheric propagation; optical meteorology; information systems and information extraction. *Mailing Add:* 51 Hetzel Rd Williamsville NY 14221

PIECH, MARGARET ANN, b Bridgewater, NS, Apr 6, 42; m 65; c 2. MATHEMATICS. *Educ:* Mt Allison Univ, BA, 62; Cornell Univ, PhD(math), 67. *Prof Exp:* From asst prof to assoc prof, 67-78, PROF MATH, STATE UNIV NY BUFFALO, 78- *Concurrent Pos:* NSF res grants, 70-85, consult, 80-81; US Army res grants, 85-; consult, Aspen Anal, 86- *Mem:* Am Math Soc; Asn Comput Mach; Inst Elec & Electronic Engrs; Greater Yellowstone Coalition. *Res:* Functional analysis; differential equations; probability theory; fractal modeling. *Mailing Add:* Dept Math State Univ NY Diefendorf Hall Buffalo NY 14214

PIECZENIK, GEORGE, b Cuba, Dec, 19, 44; US citizen. MOLECULAR BIOLOGY. *Educ:* Harvard Univ, AB, 65; Univ Miami, MS, 67; NY Univ, PhD(biol), 72. *Prof Exp:* Vis scientist, MRC Lab Molecular Biol, Cambridge, Eng, 71-72; res assoc microbial genetics, Rockefeller Univ, 72-75; ASSOC PROF BIOCHEM, RUTGERS UNIV, NEW BRUNSWICK, 75- *Concurrent Pos:* Vis scientist, MRC Lab Molecular Biol, Cambridge, Eng, 75- *Honors & Awards:* Claude Fuess Award, 80. *Mem:* AAAS; NY Acad Sci; Roy Soc Chem. *Res:* Investigation on the origin of nucleotide sequences; the theory of genotypic selection; the genotype as a phenotype for syntactical and structural selection; development of non-palliogenie varient of HIV that is a treatment and vaccine for AIDS. *Mailing Add:* Dept Biochem Rutgers Univ New Brunswick NJ 08903

PIEHL, DONALD HERBERT, b Chicago, Ill, Jan 18, 39; m 65; c 2. TOBACCO & SMOKE CHEMISTRY, BEVERAGE TECHNOLOGY. *Educ:* Carthage Col, BS, 61; Univ Iowa, MS, 64, PhD(inorg chem), 66. *Prof Exp:* Res chemist, R J Reynolds Tobacco Co, 65-69, group leader res dept, 69-70, sect head phys chem res, 70-76, mgr, Chem Res Div, 76-80, dir appl res & develop, 80-83, dir tech serv, 83-85; VPRES, RES & DEVELOP HEUBLEIN, INC, 85-; VPRES, TECHNOL & BRAND DEVELOP, 89- *Mem:* Am Chem Soc; Sigma Xi; Inst Food Sci. *Res:* Tobacco and smoke chemistry; aerosol technology; heterogeneous catalysis; coordination chemistry; starch technology; wine and spirits technology. *Mailing Add:* Heublein, Inc 430 New Park Ave Hartford CT 06142-0778

PIEHL, FRANK JOHN, b Chicago, Ill, Oct 10, 26; m 55; c 1. ORGANIC CHEMISTRY. *Educ:* Univ Chicago, PhD(chem), 52. *Prof Exp:* Dir anal res, Amoco Corp, Naperville, Ill, 52-86, mgr anal serv, 80-86; RETIRED. *Res:* Analytical chemistry. *Mailing Add:* 1129 Mary Lane Naperville IL 60540

PIEKARSKI, KONSTANTY, material science, bioengineering; deceased, see previous edition for last biography

PIEL, CAROLYN F, b Birmingham, Ala, Oct 18, 18; m 51; c 4. PEDIATRICS. *Educ:* Agnes Scott Col, AB, 40; Emory Univ, MS, 43; Washington Univ, MD, 46. *Prof Exp:* from instr to asst prof, Sch Med, Stanford Univ, 51-57; from asst prof to prof, 59-89, EMER PROF PEDIAT, SCH MED, UNIV CALIF, SAN FRANCISCO, 89- *Concurrent Pos:* USPHS fel, Med Col, Cornell Univ, 49-51. *Mem:* AAAS; Soc Pediat Res; Soc Exp Biol & Med; Am Pediat Soc; Am Soc Nephrology; Am Bd Pediat (pres, 86). *Res:* Electron microscopy of normal kidney; experimental and clinical renal diseases; metabolism of parathormone and vitamin D. *Mailing Add:* Dept Pediat Univ Calif Sch Med San Francisco CA 94143

PIEL, GERARD, b Woodmere, NY, Mar 1, 15; m 38, 55; c 2. SCIENCE WRITING. *Educ:* Harvard Col, AB, 37. *Hon Degrees:* DSc, LittD, LLD, LHD from several univs. *Prof Exp:* Sci ed, Life Mag, 38-44; asst to pres, Henry J Kaiser Co, 45-46; publ mag, Sci Am, 47-84, pres, 46-84, chmn, 84-87, EMER, SCI AM, INC, 88- *Concurrent Pos:* Trustee, Phillips Acad, 69-85, Am Mus Natural Hist, Radcliffe Col, NY Bot Garden, NY Univ & Henry J Kaiser Family Found; chmn, Mayo Found & Found Child Develop, 76-85; mem bd overseers, Harvard Univ, 66-68 & 73-79 & Univ Calif-San Fransisco. *Honors & Awards:* George Polk Award, 61; Kalinga Prize, 62; Bradford Washburn Award, 66; Arches of Sci Award, 69. *Mem:* Inst Med-Nat Acad Sci; fel Am Acad Arts & Sci; Am Philos Soc; Sigma Xi; AAAS (pres, 85-86). *Mailing Add:* Sci Am 41 Madison Ave New York NY 10010

PIEL, KENNETH MARTIN, b Perry, Okla, Jan 19, 36; c 3. PALYNOLOGY, SEQUENCE STRATIGRAPHY. *Educ:* Univ Okla, BS, 57; Tulane Univ, MS, 65; Univ BC, PhD(bot), 69. *Prof Exp:* Asst palynologist, Shell Oil Co, 60-63; res palynologist, Union Oil Co, Calif, 67-73, sr res palynologist, 73-79, res assoc, 79-86, biostratigrapher, Northwest Europe, Unocal UK Ltd, 86-88; res assoc, 88-89, SR RES ASSOC, UNOCAL, CALIF, 89- *Mem:* Am Asn Stratig Palynologists (pres, 75-76, secy-treas, 82-86); Brit Micropaleont Soc; Int Asn Angiosperm Paleobot; Int Orgn Paleobot; Soc Econ Paleontologists & Mineralogists. *Res:* Dinoflagellate morphology and stratigraphy in the Jurassic and Cretaceous of Britain and the North Sea area; sequence stratigraphy; palynofacies; computer data handling. *Mailing Add:* Unocal Res Ctr PO Box 76 Brea CA 92621

PIEL, WILLIAM FREDERICK, b Indianapolis, Ind, Nov 14, 41. NUCLEAR PHYSICS. *Educ:* Ind Univ, BS, 64, MS & PhD(physics), 72. *Prof Exp:* Res assoc physics, Brookhaven Nat Lab, 72-75, asst physicist, 75-77; SR RES ASSOC, STATE UNIV NY, STONY BROOK, 77- *Mem:* Am Phys Soc. *Res:* Experimentation in the field of nuclear structure utilizing a tandem Van de Graaff; production of new nuclei far from beta stability; measurement of atomic masses utilizing a QDDD magnet facility; study of giant nuclear monopole resonances using a high-efficiency pair detector. *Mailing Add:* 460-10 E Old Town Rd Port Jefferson Station NY 11777

PIELET, HOWARD M, b Chicago, Ill, Nov 13, 42; m 66; c 2. PROCESS METALLURGY, CONTINUOUS CASTING. *Educ:* Mass Inst Technol, BS, 63, PhD(metall), 71; Columbia Univ, MS, 66. *Prof Exp:* From res engr to staff res engr, 71-86, SCIENTIST, STEELMAKING CASTING, INLAND STEEL, 86- *Concurrent Pos:* Bd rev metall trans, Am Inst Metall & Petrol Engrs, vchmn, 85 & chmn, 86; lectr phys chem steelmaking, Purdue Univ, 83. *Honors & Awards:* Extractive Metall Sci Award, Am Inst Mech Engrs, 86, John Chipman Award, 86. *Mem:* Am Inst Metall & Petrol Engrs. *Res:* Process metallurgy, ladle metallurgy, system dynamics, solidification, casting processes; tundish metallurgy, steelmaking, tundish addition of alloying elements to free-machining steels; prevention of nozzle blockage in calcium-treated steels. *Mailing Add:* Inland Steel 3001 E Columbus Dr East Chicago IN 46312

PIELKE, ROGER ALVIN, b Baltimore, Md, Oct 22, 46. METEOROLOGY. *Educ:* Towson State Col, BA, 68; Pa State Univ, MS, 69, PhD(meteorol), 73. *Prof Exp:* Res meteorologist, Exp Meteorol Lab, Nat Oceanic & Atmospheric Admin, 71-74; asst prof, Univ Va, 74-78, assoc prof meteorol, 78-81; assoc prof, 81-85, PROF ATMOSPHERIC SCI, COLO STATE UNIV, 85-; comnr, Colo Air Qual Control Comn, 83-89; PRES, ASTER, INC, 86- *Concurrent Pos:* Consulting meteorologist, 76-; prin investr grants & contracts, NSF, Environ Protection Agency, Nat Park Serv, Elec Power & Res Inst & Dept Energy, 74-; chief ed, Monthly Weather Rev, 81-85. *Honors & Awards:* Spec Achievement Award, Nat Oceanic & Atmospheric Admin, 74; LeRoy Meisinger Award, Am Meteorol Soc, 77. *Mem:* Fel Am Meteorol Soc; Air Pollution Control Assoc. *Res:* Study of regional and local weather phenomena, including air quality, using sophisticated mathematical simulation models; air pollution meteorology; mesoscale meteorology. *Mailing Add:* Dept Atmospheric Sci Cold St Univ Ft Collins CO 80523

PIELOU, DOUGLAS PATRICK, entomology, ecology, for more information see previous edition

PIELOU, EVELYN C, b Bognor, Eng, Feb 20, 24; Can citizen; m 44; c 3. ECOLOGY, BIOGEOGRAPHY. *Educ:* Univ London, BSc, 50, PhD(statist ecol), 62, DSc, 75. *Prof Exp:* Res off, Statist Res Serv, Can Dept Forestry, 63-65; res scientist, Can Dept Agr, 65-68; vis prof math ecol, NC State Univ, 68, Yale Univ, 69 & Univ Sydney, Australia, 75; prof biol, Queen's Univ, Ont, 69-71; Killam res prof, Dalhousie Univ, 71-74, prof biol, 74-83; vis res prof, Oil Sands Environ, Univ Lethbridge, Alta, 81-86; RETIRED. *Honors & Awards:* Lawson Medal, Can Bot Asn, 84; Eminent Ecologist Award, Ecol Soc Am, 86; Distinguished State Ecologist Award, Int Asn Ecol, 90. *Res:* Population and community ecology; statistical biogeography; statistical estimation and inference in ecology and biogeography; author of semi-popular books on ecology and biogeography. *Mailing Add:* S130-C17 RR 1 Denman Island BC V0R 1T0 Can

PIELOU, WILLIAM P, b Detroit, Mich, Mar 21, 22; m 46; c 2. NATURAL SCIENCE. *Educ:* Univ Mich, BS, 48, MS, 49, PhD(zool), 57. *Prof Exp:* Prof biol, Col Charleston, 49-50; asst prof, Alma Col, 51-53; instr zool, Mich State Univ, 53-54, from instr to asst prof, 54-63; asst prof, Ariz State Univ, 63-64; assoc prof, 64-80, PROF BIOL, FURMAN UNIV, 80- *Mem:* Am Ornith Union. *Res:* Ornithology; embryology; anatomy. *Mailing Add:* Dept Biol Furman Univ Poinsett Hwy Greenville SC 29613

PIENAAR, LEON VISSER, b Cape Town, SAfrica, June 13, 36; US citizen; m 59; c 3. FOREST BIOMETRY, POPULATION DYNAMICS. *Educ:* Univ Stellenbosch, MSc, 60; Univ Wash, PhD(forest biomet), 65. *Prof Exp:* Res scientist, Dept Forestry, Pretoria SAfrica, 59-62 & 65-67; res scientist, Fisheries Res Bd Can, 67-69; from asst prof to assoc prof forestry, Wash State Univ, 69-73; asst prof, 73-75, ASSOC PROF FORESTRY, UNIV GA, 75- *Concurrent Pos:* Mem biomet sect, Int Union Forest Res Orgns, 65-; consult, Forest Serv, USDA, 74- *Res:* Growth, yield and management of plantations; fish population dynamics. *Mailing Add:* 525 Pine Forest Dr Athens GA 30606

PIENKOWSKI, ROBERT LOUIS, b Cleveland, Ohio, Aug 22, 32; m 58; c 3. ENTOMOLOGY. *Educ:* Ohio State Univ, BS, 54; Univ Wis, MS, 58, PhD(entom), 62. *Prof Exp:* From asst prof to assoc prof, 61-72, PROF ENTOM, VA POLYTECH INST & STATE UNIV, 72- *Concurrent Pos:* Consult & teacher, Sri Lanka & Peoples Repub China. *Mem:* Entom Soc Am; Brit Ecol Soc. *Res:* Ecology, behavior and biological control of forage crop insects; biological control of weeds; ecology and control of stored product insects. *Mailing Add:* Dept Entom Va Polytech Inst & State Univ Blacksburg VA 24061-0319

PIENTA, NORBERT J, b Buffalo, NY, Oct 24, 52; m 80; c 3. PHOTOCHEMISTRY, MECHANISMS & REACTIVE INTERMEDIARIES. *Educ:* Univ Rochester, BS, 74; Univ NC, PhD(chem), 78. *Prof Exp:* Res asst chem, Univ Pittsburgh, 78-80; asst prof, 80-85, ASSOC PROF CHEM, UNIV ARK, 85- *Mem:* Am Chem Soc; AAAS; Int Photochem Soc. *Res:* Generation of radical ion pairs and radicals using photochemistry; solvent effects on reactive intermediate; use of molten organic salts as solvents for reaction; thermodynamics of radical ion formation and reactions; photochemistry of silica and alumina. *Mailing Add:* Dept Chem Univ Ark Fayetteville AR 72701

PIENTA, ROMAN JOSEPH, b Old Forge, Pa, Feb 28, 31; m 56; c 3. CARCINOGENESIS, TOXICOLOGY. *Educ:* Pa State Univ, BS, 54; Rutgers Univ, MS, 56, PhD(microbiol), 59. *Prof Exp:* Asst virol, Inst Microbiol, Rutgers Univ, 57-59, res assoc, 59-60, asst res specialist, 60-65; virologist, Univ Labs, Inc, NJ, 65; assoc prof microbiol & assoc microbiologist, Univ Tex M D Anderson Hosp & Tumor Inst, 65-69, actg chief microbiol unit, 67-69; sr scientist, Litton Bionetics, Inc, 69-70, actg dir dept exp oncol, 70-71, dir dept, 71-72, proj mgr in vitro carcinogenesis, 72-76, actg dir, 75-76, sect head & dep dir chem carcinogenesis prog, Frederick Cancer Res Ctr, 76-81; systs analyst & mem tech staff, Dept Toxic & Hazardous Mat Assessment & Control, 81-87, CIVIL SYSTS DIV, MITRE CORP, 88- *Concurrent Pos:* Res fel virol, Inst Microbiol, Rutgers Univ, 54-57, mem grad sch fac, 63-65. *Mem:* Am Soc Microbiol; Am Asn Cancer Res; NY Acad Sci; Environ Mutagen Soc; fel Am Acad Microbiol; Am Acad Clin Toxicol; Am Col Toxicol; Soc Toxicol. *Res:* Chemotherapy of viruses and tumors; virus-induced cancer; in vitro culture of human tumor and leukemia cells; in vitro chemical carcinogenesis; in vitro carcinogenesis bioassay model systems; genotoxicology; primate tumor viruses. *Mailing Add:* 14113 Flint Rock Rd Rockville MD 20853

PIEPER, DAVID ROBERT, REPRODUCTION, BIORHYTHMS. *Educ:* Wayne State Univ, PhD(physiol), 78. *Prof Exp:* DIR PHYSIOL, PROVIDENCE HOSP, 85- *Mailing Add:* Dept Physiol Providence Hosp 16001 Nine Mile Rd Southfield MI 48037

PIEPER, GEORGE FRANCIS, b Boston, Mass, Jan 1, 26; m 50; c 2. NUCLEAR PHYSICS & STRUCTURE, SPACE SCIENCES. *Educ:* Williams Col, BA, 46; Cornell Univ, MS, 49; Yale Univ, PhD(physics), 52. *Prof Exp:* Mem staff radiation lab, Mass Inst Technol, 44-45; instr physics, Williams Col, 46-47; assoc physicist, Tracerlab, Inc, 49-50; from instr to asst prof physics, Yale Univ, 52-60; sr physicist appl physics lab, Johns Hopkins Univ, 60-64; dep asst dir adv res, Goddard Space Flight Ctr, NASA, 64-65, dir sci, 65-83, assoc ctr dir, 84-86; PROF AEROSPACE ENG, US NAVAL ACAD, WASHINGTON, DC, 86- *Concurrent Pos:* Fel Dept Terrestrial Magnetism Carnegie Inst Technol, 56-57; consult to dir physics & astron prog, Off Space Sci, NASA Hq, 63-64; guest worker, Max Planck Inst Extraterrestrial Physics, 71-72; prof aerospace eng, Naval Space Command, US Naval Acad, 90-91. *Honors & Awards:* NASA Medal Outstanding Achievement, 69; NASA Medal Outstanding Leadership, 77. *Mem:* Fel Am Phys Soc; Am Geophys Union; Am Inst Aeronaut & Astronaut; AAAS; fel Am Astronaut Soc. *Res:* Experimental nuclear physics; particle measurements from satellites; administration of space science program. *Mailing Add:* 3155 Rolling Rd Edgewater MD 21037-2601

PIEPER, GUSTAV RENE, entomology, toxicology, for more information see previous edition

PIEPER, HEINZ PAUL, b Wuppertal-Barmen, Ger, Mar 24, 20; US citizen; m 45. CARDIOVASCULAR PHYSIOLOGY. *Educ:* Univ Munich, MD, 48. *Prof Exp:* Resident, Med Clin, Univ Munich, 49-50, res assoc & asst prof physiol, 50-57; from asst prof to assoc prof, 57-68, actg chmn dept, 73-74, prof physiol, 68-85, chmn dept, 74-85, PROF EMER PHYSIOL, COL MED, OHIO STATE UNIV, 85- *Concurrent Pos:* Res fel, Bremer Found, Youngstown, Ohio, 57-60; estab investr, Am Heart Asn, 62-67. *Mem:* Am Physiol Soc; Ger Physiol Soc; Sigma Xi. *Res:* Pressure flow relationships of the coronary system in intact anesthetized dogs; hemodynamic studies in intact dogs; aortic distensibility; instrumentation. *Mailing Add:* 2206 SE 36th St Cape Coral FL 33904

PIEPER, REX DELANE, b Idaho Falls, Idaho, Jan 18, 34; m 65; c 3. RANGE SCIENCE, PLANT ECOLOGY. *Educ:* Univ Idaho, BS, 56; Utah State Univ, MS, 58; Univ Calif, Berkeley, PhD(bot), 63. *Prof Exp:* From asst prof to assoc prof, 63-75, PROF RANGE SCI, NMEX STATE UNIV, 75- *Mem:* Fel AAAS; Sigma Xi; Soc Range Mgt; Am Soc Animal Sci. *Res:* Nutrient cycles in Alaska; vegetation studies in New Mexico; nutrition studies on desert ranges in Utah; ecosystem analysis of desert grassland. *Mailing Add:* Dept Animal & Range Sci NMex State Univ Las Cruces NM 88003

PIEPER, RICHARD EDWARD, b Whittier, Calif, Mar 22, 41. BIOLOGICAL OCEANOGRAPHY. *Educ:* Univ Calif, Santa Barbara, BA, 64, MA, 67; Univ BC, PhD(zool & oceanog), 71; Pepperdine Univ, MA, 87. *Prof Exp:* Asst prof biol & biol oceanographer, Allan Hancock Found, 71-76, RES SCIENTIST, HANCOCK INST MARINE STUDIES, UNIV SOUTHERN CALIF, 76-, ASSOC RES PROF, DEPT BIOL, 80- *Concurrent Pos:* Assoc curator ichthyol, Los Angeles County Mus Natural Hist, 73-; consult marine ecol, Tracor, Inc, San Diego, 75-; res grants, Off Naval Res & NSF; pres, Southern Calif Acad Sci, 83-84, Nat Asn Acad Sci, 88-89; intern counr, South Bay Free Clin, Manhattan Beach, Calif, 86- *Mem:* Oceanog Soc; Am Soc Limnol & Oceanog; AAAS. *Res:* Sound scattering by marine organisms; plankton and fish ecology; sampling gear evaluation and development. *Mailing Add:* Hancock Inst Marine Studies Univ Southern Calif Harbor Lab & Marine Facil 820 S Seaside Ave Terminal Island CA 90731

PIEPER, STEVEN CHARLES, b Oceanside, NY, Apr 25, 43; m 65; c 3. THEORETICAL NUCLEAR PHYSICS, MANY-BODY GROUND STATES SCATTERING CALCULATIONS. *Educ:* Univ Rochester, BS, 65; Univ Ill, Urbana, PhD(physics), 70. *Prof Exp:* Fel physics, Case Western Reserve Univ, 70-72; fel physics, 72-74, asst scientist physics, 74-77, SCIENTIST PHYSICS, ARGONNE NAT LAB, 77- *Mem:* Am Phys Soc. *Res:* The calculation of heavy-ion reactions by distorted-wave born approximation and coupled channel techniques; variational Monte-Carlo calculations for finite many-body systems; computer techniques for physics research. *Mailing Add:* Physics Div Argonne Nat Lab Argonne IL 60439

PIEPHO, ROBERT WALTER, b Chicago, Ill, July 31, 42; m 81; c 6. PHARMACOLOGY, NEUROCHEMISTRY. *Prof Exp:* Registered pharmacist, South Park Pharm, Walther Mem Hosp, 65-70; asst prof pharm, Univ Nebr Med Ctr, 70-74, assoc prof, 74-80, coordr clin educ, 75-80; PROF PHARM & ASSOC DEAN CLIN PROG, UNIV COLO MED CTR, SCH PHARM, 80-; PROF & DEAN, DEPT PHARMACOL, SCH PHARM, UNIV MO, KANS CITY. *Concurrent Pos:* Fel pharmacol, Sch Med, Loyola Univ, Chicago, 65-70; mem, Nat Adv Comt, Student Am Pharmaceut Asn, 71-74; regent, Am Col Clin Pharmacol, 83-88. *Mem:* Am Asn Col Pharm; Am Col Clin Pharmacol; Am Pharmaceut Asn. *Res:* Cardiovascular pharmacology; hypertension; investigation of central nervous system neurotransmitter candidates and their functional roles; neurochemical correlates of thermoregulation. *Mailing Add:* Dept Pharm Univ Missouri Sch Pharm, 5005 Rockhill Rd Kansas City MO 64110-2499

PIEPHO, SUSAN BRAND, b Pound Ridge, NY, Apr 28, 42; m 64. THEORETICAL CHEMISTRY, QUANTUM CHEMISTRY. *Educ:* Smith Col, BA, 64; Columbia Univ, MA, 65; Univ Va, PhD(phys chem), 70. *Prof Exp:* Teacher chem, math & phys sci, Riverdale Country Sch Girls, Bronx, NY, 65-66; teacher math, Burley High Sch, Charlottesville, Va, 66-67; res assoc, chem dept, Univ Va, 70-71, 73-75 & 77, Univ Southern Calif, 76; asst prof chem, Sweet Briar Col, 71-73; NATO fel, Phys Chem Lab, Oxford Univ, 75-76; from asst prof to assoc prof chem, Randolph-Macon Woman's Col, 77-81; assoc prof, 81-84, PROF CHEM, SWEET BRIAR COL, VA, 84- *Concurrent Pos:* Vis prof chem, Univ Va, 86-87. *Mem:* Am Chem Soc; Am Phys Soc; AAAS. *Res:* Theory of mixed-valence compounds; advanced group-theoretical techniques and their applications to spectroscopy; interpretation of electronic absorption and magnetic circular dichroism spectra. *Mailing Add:* Dept Chem Sweet Briar Col Sweet Briar VA 24595

PIEPMEIER, EDWARD HARMAN, b St Louis, Mo, June 6, 37; m 61; c 3. ANALYTICAL CHEMISTRY. *Educ:* Northwestern Univ, BS, 60; Univ Ill, PhD(chem), 66. *Prof Exp:* Asst prof, 66-73, assoc prof, 73-79, PROF CHEM, ORE STATE UNIV, 79- *Concurrent Pos:* Res fel, Delft Univ Technol, 73-74; fac fel, NASA, 79, 89 & 90, ASEE, 87. *Mem:* Am Chem Soc; Soc Appl Spectros; Am Asn Univ Prof; Sigma Xi. *Res:* Analytical applications of lasers; trace element analysis; instrumentation; plasmas for trace element analysis. *Mailing Add:* 3325 NW Garfield Ave Corvallis OR 97330

PIER, ALLAN CLARK, b Chicago, Ill, Jan 17, 28; m 50; c 3. MEDICAL MYCOLOGY, MYCOTOXICOLOGY. *Educ:* Univ Calif, Davis, BS, 51, DVM, 53, Univ Calif, Berkeley-Davis, PhD(comp path), 60. *Prof Exp:* Pvt pract vet, Arcata, Calif, 53-54; lectr & asst vet, Univ Calif, Davis, 54-61; proj leader mycotic dis, Nat Animal Dis Lab, Ames, Iowa, 61-64, chief, Bact & Mycol Res Lab, 64-81; prof vet microbiol & collabr, Iowa State Univ, 64-82; head microbiol & vet med & dir, Wyo State Vet Lab, 81-85, PROF VET SCI, UNIV WYO, 85- *Concurrent Pos:* Mem standard methods vet microbiol comt, Nat Acad Sci-Nat Res Coun, 69-78, mycotoxins & animal prod comt, 78-79 & protection against trichothecene mycotoxins comt, 82-83; ann sci lectr, Med Mycol Soc Am, 76; consult animal mycotoxicoses, NS, 85-; WHO expert panel, dermatophytosis immunization, 87- *Mem:* Am Col Vet Microbiol; Int Soc Human Animal Mycol; Am Vet Med Asn; Conf Res Workers Animal Dis. *Res:* Mycoses, actinomycoses, mycotoxicoses and mycotoxins immunosuppression in animals. *Mailing Add:* Box 3806 Laramie WY 82071

PIER, GERALD BRYAN, IMMUNOLOGY, MICROBIOLOGY. *Educ:* Univ Calif, Berkeley, PhD(microbiol), 76. *Prof Exp:* ASST PROF MED, SCH MED, HARVARD UNIV, 78- *Mailing Add:* Dept Med Harvard Sch Med Channing Lab 180 Longwood Ave Boston MA 02115

PIER, HAROLD WILLIAM, b Mt Jewett, Pa, Aug 9, 35; m 58; c 2. ORGANIC CHEMISTRY. *Educ:* Pa State Univ, BS, 57; Univ Del, MS, 59, PhD(org chem), 62. *Prof Exp:* Fel, Columbia Univ, 61-62; res assoc, Johns Hopkins Univ, 62-63; asst prof, 63-68, ASSOC PROF ORG CHEM, UTICA COL, 68- *Concurrent Pos:* Res Corp grant, 63. *Mem:* Am Chem Soc; Sigma Xi. *Res:* Molecular rearrangements; reaction mechanisms; nature of dissolved organic matter in water. *Mailing Add:* RD 2 Box 31 Barneveld NY 13304

PIER, STANLEY MORTON, b Brooklyn, NY; m 55. TOXICOLOGY. *Educ:* Brooklyn Col, BS, 48; Purdue Univ, MS, 49, PhD, 52. *Prof Exp:* Asst, Purdue Univ, 48-52; res chemist, Tex Co, 52-55, proj leader, 55-58; sect head, Cities Serv Res & Develop Co, 58-59, from div head to prof dir lab, 60-69; tech dir, Pace Co, 69-70, dir environ sci, 70-72; PROF OCCUP HEALTH & AEROSPACE MED, UNIV TEX, SCH PUB HEALTH, HOUSTON, 72- *Honors & Awards:* Achievement Award, Nat Aeronautics & Space Admin. *Mem:* AAAS; Am Indust Hygiene Asn; Am Pub Health Asn. *Res:* Industrial hygiene; occupational health; water and air pollution; toxicology. *Mailing Add:* 5326 Dora St Houston TX 77005

PIERARD, JEAN ARTHUR, b Liege, Belg, Jan 26, 34; Can citizen; m 58; c 5. VETERINARY ANATOMY, MAMMALOGY. *Educ:* Univ Montreal, DVM, 57; Cornell Univ, MSc, 63. *Prof Exp:* Asst chief diag lab, Que Dept Agr, 57-58; inspector, Can Dept Agr, 58-59; instr micros & gross anat, Que Vet Col, 59-61; asst, State Univ NY Vet Col, Cornell Univ, 61-63; secy fac, 67-73 & 78-80, head dept animal anat & physiol, 73-78, PROF MICROS & GROSS ANAT, FAC VET MED, UNIV MONTREAL, 63- *Concurrent Pos:* Pres, Québec Vet Asn, 90-; mem, Real Academia de Cieucias Vet, Madrid, Spain & Académie Royale de Médecine de Belgique. *Mem:* Am Asn Vet Anat; World Asn Vet Anatomists (secy gen, 75-83, vpres, 83-87, pres, 87-91); Can Asn Vet Anat (pres, 77-78); Can Vet Med Asn; World Asn Vet Educ (pres, 87-). *Res:* Pinnipeds and carnivores morphology; osteoarchaeology. *Mailing Add:* Dept Animal Anat & Physiol Fac Vet Med Univ Montreal Montreal PQ H3C 3J7 Can

PIERCE, ALAN KRAFT, b Houston, Tex, Sept 3, 31; m 58; c 2. INTERNAL MEDICINE. *Educ:* Baylor Univ, MD, 55. *Prof Exp:* Resident internal med, Parkland Mem Hosp, 58-61; from instr to assoc prof, 62-69, PROF MED, UNIV TEX SOUTHWESTERN MED SCH DALLAS, 69- *Concurrent Pos:* Res fel pulmonary dis, Univ Tex Southwestern Med Sch Dallas, 61-62, NIH res fel, 61-64. *Mem:* Am Fedn Clin Res; Am Thoracic Soc; Am Col Chest Physicians; fel Am Col Physicians; Am Soc Clin Invest; Am Asn Physicians. *Res:* Nosocomial pulmonary infections; pathogenesis of pneumonia. *Mailing Add:* Dept Med Univ Tex Southwestern Med Sch Dallas TX 75235

PIERCE, ALEXANDER WEBSTER, JR, b Nashville, Tenn, Nov 13, 31; wid; c 4. PEDIATRICS. *Educ:* Vanderbilt Univ, BA, 52, MD, 56; Am Bd Pediat, dipl, 62, recert, 81. *Prof Exp:* Instr pediat, Sch Med, Univ Okla, 61-63, from asst prof to assoc prof, 63-69; assoc prof, 69-75, dir pediat clin & dir ambulatory pediat, 69-78, PROF PEDIAT, UNIV TEX MED SCH, SAN ANTONIO, 78- *Concurrent Pos:* Asst dir outpatient clins, Children's Mem Hosp-Univ Okla Hosps, 65-67, dir med serv, Out-Patient Dept, 67-69; prof fam pract, Univ Tex Health Sci Ctr, 78-87. *Mem:* Am Acad Pediat; Ambulatory Pediat Asn. *Mailing Add:* Dept Pediat 7703 Floyd Curl Dr San Antonio TX 78284-7816

PIERCE, ALLAN DALE, b Clarinda, Iowa, Dec 18, 36; m 61; c 2. PHYSICAL ACOUSTICS, UNDERWATER SOUND. *Educ:* NMex State Univ, BS, 57; Mass Inst Technol, PhD(physics), 62. *Prof Exp:* Mem res staff physics, Rand Corp, 61-63; sr staff scientist, Avco Corp, 63-66; from asst prof to assoc prof mech eng, Mass Inst Technol, 66-73; prof, 73-76, REGENTS PROF MECH ENG, GA INST TECHNOL, 76- *Concurrent Pos:* Assoc ed, Acoust Soc Am, 73-79; vis prof, Max Planck Inst Aerodyn, 76-77; fac fel, US Dept Transp, 79-80. *Honors & Awards:* Sr US Scientist Award, Alexander von Humboldt Found, 76. *Mem:* Fel Acoust Soc Am; fel Am Soc Mech Engrs; Inst Elect Electron Engrs. *Res:* Theoretical and experimental studies of long range sound propagation in the atmosphere and water; study of diffraction, radiation and scattering of sound; laser generation of sound, shock waves, sonic booms, thermal acoustics. *Mailing Add:* Grad Prog in Acoustics Pa State Univ 157 Hammond Bldg University Park PA 16802

PIERCE, ARLEEN CECILIA, b New York, NY, Jan 29, 39. ORGANIC CHEMISTRY, BIOCHEMISTRY. *Educ:* Queens Col, NY, BS, 59; Univ Pa, PhD(chem), 62. *Prof Exp:* Sr res chemist, Cent Res Lab, Allied Chem Corp, 62-66; asst prof chem, Douglass Col, Rutgers Univ, 66-68; head dept, Unity Col, 68-72, dean fac, 70-72; head chemist, Northeast Labs, 72-74; CONSULT, 72- *Mem:* Sigma Xi; fel Am Inst Chemists; NY Acad Sci. *Res:* Aminolysis of esters; animal feed from waste products; organic fertilizers and their effectiveness for salt water farming; effect of light on natural products; vitamins, minerals and health; biochemical reactions to stress. *Mailing Add:* RFD 2 Lubec ME 04652

PIERCE, AUSTIN KEITH, solar physics, for more information see previous edition

PIERCE, CAMDEN BALLARD, b Shelbyville, Ky, July 17, 32; m 59; c 2. SOLID STATE PHYSICS. *Educ:* Univ Richmond, BS, 54; Univ Ill, MS, 56, PhD(exp physics), 60. *Prof Exp:* Staff mem, Sandia Lab, 60-66; coordr, Bronfman Sci Ctr, Williams Col, 74-77, actg chmn, Dept Physics & Astron, 77-79, assoc prof, 66-79, PROF PHYSICS, WILLIAMS COL, 66-, CHMN DEPT, 80- *Mem:* Am Phys Soc; Am Asn Physics Teachers; Sigma Xi. *Res:* Optical and electrical properties associated with point defects in crystalline solids; color centers; radiation damage; ionic conductivity; high pressure studies; diffusion. *Mailing Add:* Dept Physics Williams Col Williamstown MA 01267

PIERCE, CARL WILLIAM, b Buffalo, NY, Oct 2, 39; m 73. SOLUBLE SUPPRESSOR FACTORS, SCIENCE EDUCATION. *Educ:* Colgate Univ, AB, 62; Univ Chicago, MD & PhD(path), 66. *Prof Exp:* Messing prof & pathologist-in-chief, Jewish Hosp, 76-87, PROF PATH, WASH UNIV, ST LOUIS, 76- *Honors & Awards:* Parke Davis Award, Am Asn Pathologists, 79. *Mem:* Am Asn Immunologists; Am Asn Pathologists; NY Acad Sci; AAAS. *Res:* Regulatory T-cell subsets. *Mailing Add:* 216 S Kings Hwy CSB7 St Louis MO 63110

PIERCE, CAROL S, b Lockport, NY, Mar 4, 38. PATHOGENIC MECHANISMS. *Educ:* Mt Union Col, BS, 60; Univ Chicago, MS, 63; State Univ NY, Buffalo, MS, 78, PHD(microbiol), 81. *Prof Exp:* Res assoc, Argonne Cancer Res Hosp, Chicago, 63-66, Nat Jewish Hosp, Colo, 66-67, Dairy Prod Lab, USDA, 67-70 & Vet Admin Hosp, Buffalo, NY, 72-74; fel, Erie County Labs, Buffalo, 80-82; chmn dept med technol, 83-90, ASST PROF CLIN MICROBIOL, STATE UNIV NY, BUFFALO, 80-, CLIN ASSOC PROF, 86- *Concurrent Pos:* Dir grad studies, State Univ NY, Buffalo, 89- *Mem:* Am Soc Microbiol; AAAS. *Res:* Surface proteins of Pseudomonas cepocia; immunoassays. *Mailing Add:* Dept Med Technol State Univ NY 462 Grider St Buffalo NY 14215

PIERCE, CHESTER MIDDLEBROOK, b Glen Cove, NY, Mar 4, 27; m 49; c 2. PSYCHIATRY. *Educ:* Harvard Col, AB, 48; Harvard Med Sch, MD, 52; Am Bd Psychiat & Neurol, dipl, 58. *Hon Degrees:* ScD, Westfield Col, 77, Tufts Univ, 84. *Prof Exp:* Instr psychiat, Univ Cincinnati, 57-60; from asst prof to prof, Univ Okla, 60-68; PROF PSYCHIAT, HARVARD UNIV, 68- *Concurrent Pos:* Sr consult, Peace Corps, 65-69; Alfred North Whitehead fel, Harvard Univ, 68-69; mem adv br, Children's TV Workshop, 68-; mem & pres, Am Bd Psychiat & Neurol, 70-78; nat consult Surgeon Gen, US Air Force, 76-82; mem polar res bd, Nat Res Coun & mem working group human biol & med, Sci Comt Antarctic Res, 77-; counr, Am Asn Social Psychiat, 84-88; psychiatrist, Mass Gen Hosp & Mass Inst Technol. *Honors & Awards:* Pierce Peak in Antarctica, 68; Spec Recognition Award, Nat Med Asn, 71; Hon Fel, Australian & NZ Col Psychiatrists, 78; Solomon Carter Fuller Award, Am Psychiat Asn, 86. *Mem:* Inst of Med of Nat Acad Sci; Am Psychiat Asn; Am Orthopsychiat Asn (vpres, 76, pres, 83); Black Psychiatrists Am; Am Col Physicians; Am Asn Social Psychiat. *Res:* Extreme environments; media; racism; sports medicine; education. *Mailing Add:* Nichols House Appian Way Cambridge MA 02138

PIERCE, CYRIL MARVIN, b Boston, Mass, Feb 15, 39; m 62; c 2. METALLURGY. *Educ:* Mass Inst Technol, BS, 60, MS, 61; Ohio State Univ, PhD(metall eng), 66. *Prof Exp:* Chief struct metals develop, Air Force Mat Lab, 73-74, asst chief mfg technol, 74-76, chief, 76; dir mfg, Air Force F16 Prog Off, 76-77; mgr mfg technol lab, 77-79, PLANT MGR, AIRCRAFT ENGINE GROUP, GEN ELEC CO, MADISONVILLE, KY, 79- *Concurrent Pos:* US Air Force rep, Mat Adv Bd Metalworking, 64, Int Tech Coop Prog, 73-76 & Mfg Tech Adv Group, Dept Defense, 74-76; adj prof, Eng & Sci Inst Dayton, 71 & Univ Dayton, 74. *Honors & Awards:* Sci Achievement Award, US Air Force, 73. *Mem:* Sigma Xi; fel Am Soc Metals; Am Inst Metall Engrs. *Res:* Titanium alloys; superalloys; manufacturing technology. *Mailing Add:* 3747 Fallen Tree Lane Cincinnati OH 45236

PIERCE, DANIEL THORNTON, b Los Angeles, Calif, July 16, 40; c 2. SURFACE SCIENCE, SURFACE MAGNETISM. *Educ:* Stanford Univ, BS, 62, PhD(appl physics), 70; Wesleyan Univ, MA, 66. *Prof Exp:* Res assoc, Stanford Electronics Labs, Stanford Univ, 70-71; physicist, Swiss Fed Inst Technol, Zurich, 71-75; PHYSICIST, NAT INST STANDARDS & TECHNOL, 75- *Honors & Awards:* Indust Res 100 Award, 80, 85. *Mem:* Am Phys Soc; Am Vacuum Soc; Swiss Phys Soc. *Res:* Physics of spin polarized electrons; scattering; emission; sources and detectors; structural, electronic and magnetic properties of surfaces; photoemission. *Mailing Add:* Rm B206 Bldg 220 Nat Inst Standards & Technol Gaithersburg MD 20899

PIERCE, DONALD N(ORMAN), b Lincoln, Nebr, Oct 30, 21; m 43; c 2. SOLID MECHANICS, MATERIALS SCIENCE. *Educ:* Univ Nebr, BSc, 48, MSc, 54. *Prof Exp:* From instr to asst prof, 48-61, ASSOC PROF ENG MECH, UNIV NEBR, LINCOLN, 61- *Concurrent Pos:* Consult mech testing & stress analysis. *Mem:* Am Soc Mech Engrs; Am Soc Eng Educ; Am Acad Mech; Soc Exp Mech. *Res:* Dimensional analysis; solid mechanics; stresses in thin-wall shells; energy dissipation in concrete. *Mailing Add:* Dept Eng Mech Univ Nebr 219 Bancroft Hall Lincoln NE 68588-0347

PIERCE, DOROTHY HELEN, b Bronxville, NY. PHYSICAL CHEMISTRY, BIOCHEMISTRY. *Educ:* Bryn Mawr Col, BA, 76; Univ Pa, MS, 79, PhD(phy chem), 82. *Prof Exp:* Post-doctoral assoc phys chem, biochem & biophys, Univ Pa, 83-84; asst prof phys chem, org chem & biochem, Adelphi Univ, 84-85; asst prog org chem, phys chem & gen chem, Auburn Univ, 85-89; ASST PROG ADMINR, RES GRANTS, AM CHEM SOC, 89- *Concurrent Pos:* Part-time asst prof phys chem, Columbus Col, 86-87. *Mem:* Am Chem Soc; Sigma Xi. *Res:* Physical measurements in biochemistry; transport properties of membranes; kinetics; spectrophotometry; lasers and optics; diffraction. *Mailing Add:* c/o Res Grants Am Chem Soc 1155 16th St NW Washington DC 20036

PIERCE, EDWARD RONALD, b Chester, Pa, Mar 1, 37; m 63; c 2. GENETICS HEALTH SCIENCES, HUMAN & MEDICAL GENETICS. *Educ:* Univ Louisville, BA, 62, PhD(zool), 68; Johns Hopkins Univ, MPH, 70. *Prof Exp:* Asst prof zool, Ohio Univ, 67-68; fel, Div Med Genetics, Johns Hopkins Univ Hosp, 68-70; asst prof biol, American Univ, 70-73, assoc prof, 73-74; asst dean & assoc prof health studies, Sch Health Studies, Univ NH, 74-78; PROF, DIR DIV ALLIED HEALTH SCI & ASSOC DEAN SCH MED, IND UNIV, 78- *Concurrent Pos:* Asst prof, Div Med Genetics, Johns Hopkins Univ, 70-73. *Mem:* Genetics Soc Am; Am Pub Health Asn; Am Soc Allied Health Prof. *Res:* Cytogenetics, behavioral genetics; population genetics-human genetic counseling. *Mailing Add:* Four Wood Song Circle Newark DE 19711

PIERCE, ELLIOT STEARNS, b Attleboro, Mass, Apr 30, 22; m 46; c 3. ENERGY-RELATED CHEMISTRY. *Educ:* Yale Univ, BS, 43, MS, 48, PhD(org chem), 51. *Prof Exp:* Res chemist, Socony-Vacuum Oil Co, 43-44; instr org & phys-org chem, Univ Mass, 50-51; res chemist, Am Cyanamid Co, 51-54, group leader, 54-55, univ recruiter, 55-56, govt res liaison, 56-59; res adminr, US Air Force Off Sci Res, 59-61; chemist, US Atomic Energy Comn, Dept Energy, 61-66, dep asst dir res, 66-67, dep dir Div Nuclear Educ & Training, 67-70, dir, 70-73, asst dir, Div Phys Res, 73-75, asst dir basic energy sci, Energy Res & Develop Admin, 75-77, dir, Chem Sci Div, 77-86; CONSULT, 86- *Concurrent Pos:* Hon mem, Bd Chem Sci & Technol, Nat Res Coun. *Honors & Awards:* Hornbook Award, Montgomery County Educ Asn, 65. *Mem:* Fel AAAS; Am Chem Soc; Am Phys Soc; Sigma Xi; Am Inst Chemists. *Res:* Basic energy-related chemistry including coal chemistry, surface chemistry, photochemistry, combustion, photoelectron spectroscopy, molecular beam research, atomic physics, analytical chemistry, chemical and isotope separations, nuclear chemistry. *Mailing Add:* 10705 Brunswick Ave Kensington MD 20545

PIERCE, FELIX J(OHN), b Warren, RI, Nov 5, 32; m 55; c 3. FLUID MECHANICS. *Educ:* Univ RI, BS, 55; Cornell Univ, MS, 58, PhD(mech eng), 61. *Prof Exp:* From instr to asst prof mech eng, Cornell Univ, 58-66; assoc prof, 66-70, asst dean grad sch, 69-70, PROF MECH ENG, VA POLYTECH INST & STATE UNIV, 70- *Mem:* Am Soc Mech Engrs; Am Inst Aeronaut & Astronaut; Am Soc Eng Educ; Sigma Xi. *Res:* Three-dimensional turbulent flows; classical and statistical thermodynamics. *Mailing Add:* Dept of Mech Eng Col of Eng Va Polytech Inst & State Univ Blacksburg VA 24061

PIERCE, G(EORGE) ALVIN, b Philadelphia, Pa, Dec 22, 31; m 75; c 3. AEROELASTICITY, UNSTEADY AERODYNAMICS. *Educ:* Mass Inst Technol, BSc, 53; Ohio State Univ, PhD(aerospace eng), 66. *Prof Exp:* Jr engr, Nat Adv Comt Aeronaut, Langley Field, 51-52; mech engr, ADC Electronics Prod Corp, 53-54; dynamics engr, N Am Aviation, Inc, 56-58, sr engr, 58-61, res specialist aeroelasticity, 61-63; res assoc aerodyn, Ohio State Univ, 64-66; PROF AEROELASTICITY, GA INST TECHNOL, 66- *Concurrent Pos:* Consult various indust firms, 64-; Am Inst Aeronaut & Astronaut rep, Engr Coun Prof Develop, 70- *Mem:* Am Inst Aeronaut & Astronaut; Am Helicopter Soc; Sigma Xi; Am Soc Eng Educ. *Res:* Aeroelasticity and unsteady aerodynamics of fixed and rotary wing flight vehicles. *Mailing Add:* Sch of Aerospace Eng Ga Inst Technol Atlanta GA 30332

PIERCE, GORDON BARRY, b Westlock, Alta, July 21, 25; m 52; c 5. PATHOLOGY. *Educ:* Univ Alta, BSc, 49, MSc, 50, MD, 52. *Prof Exp:* Intern, Univ Hosp, Univ Alta, 52-53, lectr path, Univ, 53-55; asst prof, Sch Med, Univ Pittsburgh, 59-61; assoc prof, Univ Mich, Ann Arbor, 61-65, prof, 65-68; prof path & chmn dept, 68-82, ACS CENTENNIAL PROF, MED CTR, UNIV COLO, DENVER, 82- *Concurrent Pos:* Scaife fel, Sch Med, Univ Pittsburgh, 55-59; Markle scholar med sci, 59-63; Am Cancer Soc career prof, 64-; mem med scientist training comt, Nat Inst Gen Med Sci, 63-67; mem cell biol study sect, NIH, 68-72 & path study sect, 72-77. *Honors & Awards:* Guiteras Award, Am Urol Asn; Rous-Whipple Award, Am Asn Path. *Mem:* AAAS; Am Asn Path (pres, 77); Fedn Am Socs Exp Biol (pres, 78); Am Soc Cell Biol; Am Asn Cancer Res. *Res:* Oncology; aspects of developmental carcinomas; teratocarcinoma; biology of basement membranes. *Mailing Add:* Dept Path Univ Colo Med Sch 4200 E Ninth Ave Denver CO 80262

PIERCE, HARRY FREDERICK, b Baltimore, Md, Nov 29, 41; m 66; c 1. UNDERWATER ACOUSTICS, SONAR. *Educ:* Johns Hopkins Univ, BS, 67. *Prof Exp:* US Navy, 68-, officer acoustics, 68-72, physicist, 72-79, proj mgr, 79-82, sr proj mgr, 82-85, PROG MGR ACOUSTICS, DAVID TAYLOR RES CTR, US NAVY, 85- *Concurrent Pos:* Physicist, David Taylor Res Ctr, 72- *Mem:* Math Asn Am. *Res:* Underwater acoustics; sonar design and applications; signal processing. *Mailing Add:* Res & Develop David Taylor Res Ctr Bethesda MD 20084

PIERCE, JACK ROBERT, b Sturgis, Mich, Mar 18, 39; m 71; c 2. PARASITOLOGY. *Educ:* Western Mich Univ, BA, 61, MA, 63; Univ Tex, PhD(zool), 68. *Prof Exp:* Asst prof, 67-71, chmn dept, 76-83, PROF BIOL, AUSTIN COL, 71-, DIR, HEALTH SCIENCES, 87- *Concurrent Pos:* Dir, NSF student sci training prog environ qual; McGaw Chair Health Sci, 88- *Honors & Awards:* McGaw Chair of Health Sci, 88- *Mem:* AAAS; Nat Asn Advs Health Professions. *Res:* Marine parasitology. *Mailing Add:* Dept Biol Austin Col 900 N Grand Ave Sherman TX 75090

PIERCE, JACK VINCENT, b Kalamazoo, Mich, Feb 2, 19. BIOCHEMISTRY. *Educ:* Kalamazoo Col, BA, 40; Univ Mich, MA, 41; Columbia Univ, MA, 50; Univ Ill, PhD(biochem), 56. *Prof Exp:* Res assoc, Lederle Labs, Am Cyanamid Co, 47-52; res assoc chem, Univ Ill, 55; res chemist, Lederle Labs, Am Cyanamid Co, 56-57; chemist, Nat Heart Lung & Blood Inst, 57-85, vol biochem res, 85-89; RETIRED. *Mem:* AAAS; Am Chem Soc; NY Acad Sci; Am Soc Biochem & Molecular Biol; Sigma Xi. *Res:* Isolation and characterization of proteins, especially components of mammalian kinin system; methods for separating macromolecules. *Mailing Add:* 10508 Montrose Ave-204 Bethesda MD 20814

PIERCE, JACK WARREN, b Springfield, Ill, Jan 23, 27; m 52; c 3. GEOLOGY. *Educ:* Univ Ill, BS, 49, MS, 50; Univ Kans, PhD(geol), 64. *Prof Exp:* Geologist, Pure Oil Co, 50-56, dist geologist, 56-60; assoc prof geol, George Washington Univ, 63-65; CUR SEDIMENTOLOGY, US NAT MUS, SMITHSONIAN INST, 65- *Concurrent Pos:* Adj prof, George Washington Univ, 65- *Mem:* Am Asn Petrol Geol; Soc Econ Paleont & Mineral; Int Asn Sedimentol; Geol Soc Am; Marine Technol Soc; Sigma Xi. *Res:* Sedimentology; stratigraphy; marine geology. *Mailing Add:* MS 125-NHB Smithsonian Inst Washington DC 20560

PIERCE, JAMES BENJAMIN, b St Catherines, Ont, Aug 11, 39; m 64; c 2. ORGANIC CHEMISTRY. *Educ:* Univ Toronto, BSc, 62, MSc, 64, PhD(chem), 66. *Prof Exp:* post doctoral fel, Fla State Univ, 67; res scientist chem, Uniroyal Inc, 67-81, sect mgr org/anal, 81-87, sect mgr new prod res, 88-90, SECT MGR PROCESS & FORMULATIONS, UNIROYAL INC, 90- *Mem:* Fel Chem Inst Can; Am Chem Soc. *Res:* Organic chemistry particularly as applied to the synthesis of agricultural chemicals; process development; formulations. *Mailing Add:* Uniroyal Chem Inc World Headquarters Benson Rd Middlebury CT 06749

PIERCE, JAMES BRUCE, b Edmon, Pa, Feb 15, 22; m 43. ELECTROCHEMISTRY. *Educ:* Thiel Col, BS, 50; Case Inst Technol, MS, 55, PhD(chem), 58. *Prof Exp:* Chemist, Jamestown Paint & Varnish Co, 50-52; asst petrol chem, Case Inst Technol, 52-54, instr gen & org chem, 54-58; prof, Lowell Technol Inst, Univ Lowell, 58-75, prof org chem, Univ Lowell, 75-89; RETIRED. *Mem:* Am Chem Soc; Electrochem Soc. *Res:* Organic semiconductors; heterogeneous and stereospecific catalysis; organic reaction mechanisms; selective ion electrodes; organic electrochemistry; fuel chemistry. *Mailing Add:* 8090 NW 43rd Lane Ocala FL 32675

PIERCE, JAMES KENNETH, b Kansas City, Mo, Aug 31, 44; m 66; c 2. CHEMISTRY. *Educ:* William Jewell Col, AB, 66; Univ Kans, PhD(chem), 70. *Prof Exp:* RES CHEMIST, DOW CHEM CO, 70- *Mem:* Am Chem Soc; Sigma Xi. *Res:* Organic chemistry. *Mailing Add:* PO Box 1965 Midland MI 48641-1965

PIERCE, JAMES OTTO, II, b Memphis, Tenn, May 15, 37; m 59; c 2. ENVIRONMENTAL HEALTH, INDUSTRIAL HEALTH. *Educ:* Univ Ala, BS, 58; Univ Cincinnati, MS, 63, ScD(indust health), 64. *Prof Exp:* Res asst indust hyg, Kettering Lab, Cincinnati, Ohio, 64, asst prof, 65; from asst prof to assoc prof environ health, Univ Cincinnati, 66-69; assoc prof bioeng & community health & med pract & dir, Environ Trace Substances Ctr, Univ Mo-Columbia, 69-79; PROF & CHMN, SAFETY SCI DEPT & DIR NAT INST OCCUP SAFETY & HEALTH, REGION IX SOUTHERN CALIF EDUC RESOURCE CTR, INST SAFETY & SYSTS MGT, UNIV SOUTHERN CALIF, 79- *Concurrent Pos:* Mem panel chromium, Nat Acad Sci-Nat Res Coun; spec asst to asst sec, US Dept Labor, Occup Safety & Health Admin, 77-79. *Mem:* Am Chem Soc; Am Indust Hyg Asn; Am Conf Govt Indust Hygienists; Am Conf Govt Indust Hygienists; Am Soc Safety Engrs; Nat Safety Coun. *Mailing Add:* Inst Safety & Systs Mgt Bldg 108 Univ Southern Calif Univ Park Los Angeles CA 90089-0021

PIERCE, JOHN ALBERT, b Little Rock, Ark, Mar 10, 25; m 47, 84; c 3. INTERNAL MEDICINE. *Educ:* Univ Ark, MD, 48. *Prof Exp:* Intern med, USPHS Hosp, Galveston, Tex, 48-49, resident internal med, New Orleans, La, 51-54; from instr to prof med, Med Ctr, Univ Ark, 54-67; assoc prof, 67-72, PROF MED, SCH MED, WASH UNIV, 72-; HERMAN & SELMA SELDIN PROF PULMONARY MED, BARNES HOSP, 83- *Concurrent Pos:* Chief div pulmonary dis, Barnes Hosp, 67-85. *Mem:* Am Fedn Clin Res; fel Am Col Physicians. *Res:* Mechanics of pulmonary ventilation; chemistry of sclero proteins; alpha 1-antitrypsin. *Mailing Add:* Pulmonary Dis Wash Univ Sch Med 660 S Euclid Ave St Louis MO 63110

PIERCE, JOHN FRANK, b Mountain City, Tenn, Dec 27, 20; m 40; c 2. ELECTRICAL ENGINEERING. *Educ:* Univ Tenn, BS, 43; Univ Pittsburgh, MS, 46, PhD(elec eng), 53. *Prof Exp:* Engr, Res Labs, Westinghouse Elec Corp, 43-46; asst prof elec eng, Univ Pittsburgh, 46-50, asst prof clin sci & head instrumentation group, Sch Med, 51-53; engr in chg res, Wright Mach Co, 53-54; from assoc prof to prof elec eng, 54-64, distinguished serv prof, 64-76, head dept, 68-76, prof elec eng, Univ Tenn, Knoxville, 76-88; RETIRED. *Concurrent Pos:* Consult, Western State Psychiat Inst & Clin, 48-51, Am Inst Res, 52-53, Montefiore Inst Res, 53 & Instrumentation & Control Div, Oak Ridge Nat Lab, 59-69; chief engr, Pittsburgh Electronic Corp, 49-53; consult, ORTEC, Inc, 76- *Mem:* Sr mem Inst Elec & Electronics Engrs. *Res:* Design of radar equipment and electronic instruments; investigation of electro-shock; human engineering; automatic machinery; high-speed electonic circuits; nuclear electronics. *Mailing Add:* 137 N Oakhill Dr Lewisville TN 37777

PIERCE, JOHN GREGORY, b Cleveland, Ohio, Jan 6, 42. APPLIED MATHEMATICS. *Educ:* Case Western Reserve Univ, BS, 63, MS, 67, PhD(math), 69. *Prof Exp:* Off Naval Res fel, Courant Inst Math Sci, NY Univ, 69-70; asst prof math, Univ Southern Calif, 70-76; assoc prof, 76-80, PROF MATH, CALIF STATE UNIV, FULLERTON, 80- *Mem:* Soc Indust & Appl Math; Asn Comput Mach. *Res:* Optimization and control theory applied to biological problems; spline functions and the finite element method. *Mailing Add:* 1024 18th St No D Santa Monica CA 90403

PIERCE, JOHN GRISSIM, b San Jose, Calif, May 9, 20; m 49; c 4. BIOCHEMISTRY. *Educ:* Stanford Univ, AB, 41, AM, 42, PhD(biochem), 44. *Prof Exp:* Lab asst med biochem, Stanford Univ, 42, asst, Nutrit Proj, 44, actg instr biochem, 46-47; instr, Med Col, Cornell Univ, 48-49, asst prof, 49-52; asst prof, Sch Med, Univ Calif, Los Angeles, 52-55, assoc prof, Med Ctr, 55-61, vchmn, Dept Biol Chem, 63-79, prof biol chem, 61-84, chmn dept, 79-84, assoc dean, Sch Med, 84, EMER PROF BIOL CHEM, UNIV CALIF, LOS ANGELES, 84- *Concurrent Pos:* Fels, Am Chem Soc, Stanford Univ & Med Col, Cornell Univ, 46-47, Arthritis & Rheumatism Found, Cambridge Univ, 52-53 & Guggenheim Found, Mass Inst Technol, 60-61; Guggenheim Found fel, NIH, 76. *Honors & Awards:* Eli Lilly lectr award, Endocrine Soc Am, 71; Parke-Davis lectr award, Am Thyroid Asn, 84. *Mem:* AAAS; Am Chem Soc; Am Soc Biol Chem; Harvey Soc. *Res:* Pituitary hormones; chemistry of thyroid-stimulating and gonard-stimulating hormones. *Mailing Add:* PO Box 95 Cambria CA 93428

PIERCE, JOHN ROBINSON, b Des Moines, Iowa, Mar 27, 10; m 87; c 2. ELECTRICAL ENGINEERING, MUSICAL ACOUSTICS. *Educ:* Calif Inst Technol, BS, 33, MS, 34, PhD(elec eng), 36. *Hon Degrees:* DEng, Newark Col Eng, 51, Carnegie Inst Technol, 64, Univ Bologna, 74; DSc, Northwestern Univ, 61, Yale Univ, 63, Polytech Inst Brooklyn, 63, Columbia Univ, 65, Univ Nev, 70, Univ Southern Calif, 78; LLD, Univ Pa, 74. *Prof Exp:* Mem tech staff, Bell Tel Labs, Inc, 36-52, dir electronics res, 52-55, dir res elec commun, 55-58, dir res commun prin, 58-61, exec dir, 61-63, exec dir res commun prin & systs div, 62-65 & commun sci div, 65-71; prof eng, 71-80, EMER PROF ENG, CALIF INST TECHNOL, 80- *Concurrent Pos:* Ed, Inst Elec & Electronics Engrs, 54-55; coun, Nat Acad Sci, 71-74; Marconi int fel, 79; chief technologist, Jet Propulsion Lab, 81-82; vis prof music, musical acoust, Stanford Univ, 83- *Honors & Awards:* Liebman Mem Prize, Inst Elec & Electronics Engrs, 47; Ballantine Medal, Franklin Inst, 60; H H Arnold Trophy, Air Force Asn, 62; Gen Hoyt St Vandenberg Trophy, Arnold Air Soc, 63; Edison Medal, 63; Valdemar Poulsen Medal, 63; Med of Sci, 63; H T Cedergren Medal, 64; John Scott Award, City of Philadelphia 74; Marconi Award, 74; Medal of Honor, Inst Elec & Electronics Engrs, 75; Founders Award, Nat Acad Eng, 77. *Mem:* Nat Acad Sci; Nat Acad Eng; fel Am Acad Arts & Sci; fel Am Phys Soc; fel Acoust Soc Am. *Res:* Vacuum tubes; microwave oscillators and amplifiers; low voltage microwave reflex oscillator; high current electron guns; traveling-wave amplifiers; satellites. *Mailing Add:* CCRMA Music Stanford Univ Stanford CA 94305

PIERCE, JOHN THOMAS, b Coffeyville, Kans, Mar 15, 49; m 81. INDUSTRIAL HYGIENE, TOXICOLOGY. *Educ:* Northwestern Okla State Univ, BS, 72; Pittsburg State Univ, MS, 73; Univ Okla, MPH, 77, PhD(environ health), 78. *Prof Exp:* Adj asst prof chem, Okla City Univ, 76-78; med serv corps, US Navy, 78-81; assoc prof indust hyg, Univ NAla, 81-88; CENT MO STATE UNIV. *Concurrent Pos:* Adj asst prof chem, Old Dominion Univ, 80-84. *Mem:* AAAS; Am Acad Indust Hyg; Am Conf Govt Indust Hygienists; Am Chem Soc. *Res:* Environmental analytical chemistry, including biological monitoring of workers, heavy metals analysis, dermal absorption of toxicants and innovations such as immunoassay. *Mailing Add:* Humphries Bldg Rm 327 Cent Mo State Univ Warrensburg MO 64093

PIERCE, KEITH ROBERT, b Portland, Ore, Oct 15, 42; m 69; c 6. INTELLIGENT SYSTEMS, ALGEBRA. *Educ:* Carnegie Inst Technol, BS, 65; Univ Wis-Madison, MS, 68, PhD(math), 70. *Prof Exp:* From asst prof to assoc prof math, Univ Mo, Columbia, 70-80; from assoc prof to prof math sci, 80-86, PROF & HEAD DEPT COMPUT SCI, UNIV MINN, DULUTH, 86- *Mem:* Math Asn Am; Asn Comput Mach; Inst Elec & Electronics Engrs; Am Asn Artificial Intel. *Res:* Automated theorem proving; lattice ordered groups. *Mailing Add:* Dept Comput Sci Univ Minn Duluth MN 55812

PIERCE, KENNETH LEE, b Washington, DC, Oct 21, 37; m 60; c 3. QUATERNARY GEOLOGY, NEOTECTONICS. *Educ:* Stanford Univ, BS, 59; Yale Univ, PhD(geol), 64. *Prof Exp:* GEOLOGIST, US GEOL SURV, 63- *Concurrent Pos:* Panel mem, Quaternary Geol & Geomorphol Div, Geol Soc Am, 74-76 & 78-80; comnr, Am Comn Stratig Nomenclature, 75-78; prog chmn, Am Quaternary Asn Meeting, 85; vchmn & chmn, Quaternary Geol & Geomorphol Div, Geol Soc Am, 87-90; co-chmn, Surficial Processes Group, US Geol Surv, 90- *Honors & Awards:* Kirk Bryan Award, Geol Soc Am, 82; Meritorious Serv Award, US Dept Interior, 86. *Mem:* Geol Soc Am; Am Asn Quaternary Res; Am Geophys Union. *Res:* Glacial geology and paleoglaciology of northern Yellowstone Park; Quaternary dating in western United States, especially combined relative-age and numerical methods; pleistocene loess and gravel of southern Idaho; analysis of scarp degradation and slope processes; glacial geology, neotectonics and archeological geology of Jackson Hole; neotectonics of the Wyoming, Idaho and Montana areas; faulting, uplift, and volcanism associated with the Yellowstone hotspot. *Mailing Add:* US Geol Surv MS 913 Fed Ctr Denver CO 80225

PIERCE, KENNETH RAY, b Snyder, Tex, May 21, 34; m 56; c 2. VETERINARY PATHOLOGY, CLINICAL PATHOLOGY. *Educ:* Tex A&M Univ, DVM, 57, MS, 62, PhD(path), 65; Am Col Vet Path, Dipl, 64. *Prof Exp:* Instr vet anat, Tex A&M Univ, 57-59; practr vet med, San Angelo Vet Hosp, 59-61; from asst prof to prof, 61-78, prof & head vet pathobiol, 78-89, PROF VET PATHOBIOL, TEX A&M UNIV, 90- *Concurrent Pos:* NSF sci fac fel, 63-64; consult, Univ Tex M D Anderson Hosp & Tumor Inst, Houston; co-ed, J Vet Path, 71-80; comp pathologist, Inst Comp Med, Baylor Col Med & Tex A&M Univ, 76-79; guest scientist chem path, Nat Inst Environ Health Sci, 85; mem, conf Res Workers Animal Dis; lectr, USDA Food Safety Inspection Serv, 88- *Mem:* AAAS; Int Acad Path; Am Vet Med Asn; Am Col Vet Path (vpres, 83, pres, 84). *Res:* Role of growth factors and other oncogene products on the biological behavior of cancer in animals. *Mailing Add:* Dept Vet Pathobiol Tex A&M Univ College Station TX 77843-4463

PIERCE, LOUIS, physical chemistry; deceased, see previous edition for last biography

PIERCE, MADELENE EVANS, b Boston, Mass, Nov 7, 04. ECOLOGY. *Educ:* Radcliffe Col, AB, 26, AM, 27, PhD(zool), 33. *Prof Exp:* Instr zool, Smith Col, 27-29; instr, 31-38, from asst prof to prof, 38-70, EMER PROF ZOOL, VASSAR COL, 70- *Concurrent Pos:* Instr, Marine Biol Lab, Woods Hole, 43-52 & mem corp. *Mem:* Am Asn Univ Professors; Am Soc Limnol & Oceanog; Weed Sci Soc Am; Am Inst Biol Sci. *Res:* Aquatic ecology; effect of weedicides on pond fauna and flora; effect of heated effluent on river fauna and flora. *Mailing Add:* 104 Greer Crest Millbrook NY 12545

PIERCE, MARION ARMBRUSTER, b Folsomdale, NY, Mar 17, 09; m 47; c 2. PHYSICAL CHEMISTRY. *Educ:* Mt Holyoke Col, AB, 30; Bryn Mawr Col, AM, 32, PhD(phys chem), 34. *Prof Exp:* Asst chem, Barnard Col, Columbia Univ, 34-35, instr, 43-44, asst prof, 45-46; phys chemist, Res Lab, US Steel Corp, 35-43; sr res assoc, E I Du Pont de Nemours & Co, Inc, Arlington, NJ, 46-87; RETIRED. *Mem:* Am Chem Soc. *Res:* Chemical thermodynamics; electrochemistry; physico-chemical properties of metals; surface chemistry; high polymers; synthesis of hydantoins, amino acids. *Mailing Add:* 1706 N Bancroft Pkwy Wilmington DE 19806

PIERCE, MATTHEW LEE, b San Francisco, Calif, Nov 18, 52. ATOMIC SPECTROSCOPY. *Educ:* Univ San Francisco, BS, 74; Ariz State Univ, PhD(chem), 81. *Prof Exp:* Teaching assoc quantitative & instrumental analysis, Ariz State Univ, 76-78, res assoc, 78-81; Sr environ chemist, Environ Serv, Tex Div, 81-87, PROJ LEADER, ANALYTICAL RES, WESTERN DIV, DOW CHEM, 88- *Concurrent Pos:* Consult, SEM/TEC Labs, 77-80. *Mem:* Am Chem Soc; Soc Appl Spectros. *Res:* Rates, extents and mechanisms of adsorption on oxide surfaces as well as oxidation kinetics of various arsenic species. *Mailing Add:* PO Box 1398 Dow Chem Bldg 463 Pittsburg CA 94565

PIERCE, NATHANIEL FIELD, b Rudyard, Mich, July 27, 34; m 66; c 3. INFECTIOUS DISEASES. *Educ:* Univ Mich, MD, 58. *Prof Exp:* Instr med, Univ Louisville, 63-64 & Univ Southern Calif, 64-65; from instr to asst prof, 66-72, assoc prof, 72-79, PROF MED, SCH MED, JOHNS HOPKINS UNIV, 79- *Concurrent Pos:* Consult cholera, WHO, 71-74; mem, Cholera Adv Comt, Nat Inst Allergy & Infectious Dis, 71-73, Nat Inst Allergy & Infectious Dis res career develop award, 71; vis fel immunol, St Cross Col, Oxford Univ, 73-74; mem, Cholera Panel, US-Japan Coop Med Sci Prog, 72-76, chmn, 77-; mem adv comt health, biomed res & develop, Comn Int Rels, Nat Acad Sci. *Mem:* Infectious Dis Soc Am; Am Col Physicians; Am Soc Clin Invest; Am Fedn Clin Res; Am Soc Microbiol. *Res:* The mucosal immune system, especially that of the gut; improved means of immunizing against enteric infections. *Mailing Add:* Francis Scott Key Med Ctr Baltimore MD 21224

PIERCE, PERCY EVERETT, b Bayonne, NJ, Jan 16, 32; m 58; c 4. PHYSICAL CHEMISTRY. *Educ:* Case Western Univ, BS, 53; Yale Univ, MS, 56, PhD(chem), 58. *Prof Exp:* Asst chem, Case Western Reserve Univ, 53-55; asst chem, Yale Univ, 55-57; asst prof, Case Western Reserve Univ, 58-63; sect leader fundamental res, coatings & resins, Res Ctr, Glidden Co, 63-67, sr scientist, 67-69; scientist, 69-74, sr scientist, 74-78, MGR PHYS & ANAT RES, COATINGS & RESIN RES CTR, PPG INDUSTS, INC, 78- *Honors & Awards:* Mattiello lectr, Fedn Soc Coating Technol, 80. *Mem:* AAAS; Am Inst Chem; Am Chem Soc; Fedn Soc Coating Technol; Sigma Xi. *Res:* Rheology; physical chemistry of polymers and coatings. *Mailing Add:* 2178 Ramsey Rd Monroeville PA 15146

PIERCE, R(OBERT) DEAN, b Saginaw, Mich, Dec 7, 29; m 52; c 3. CHEMICAL ENGINEERING, PYROCHEMICAL PROCESSING. *Educ:* Univ Mich, BS, 51, MS, 52, PhD(chem eng), 55. *Prof Exp:* Sr engr, Atomic Energy Div, Babcock & Wilcox Co, 54-58; SR CHEM ENGR, ARGONNE NAT LAB, 58 - *Mem:* Am Inst Chem Engrs; Sigma Xi. *Res:* Liquid metal and liquid salt technology; nuclear reactor fuel reprocessing; actinide recovery. *Mailing Add:* Argonne Nat Lab Chem Tech Div Bldg 205 9700 S Cass Ave Argonne IL 60439-4837

PIERCE, RICHARD SCOTT, b Calif, Feb 26, 27; m 71; c 2. ALGEBRA. *Educ:* Calif Inst Technol, BS, 50, PhD(math), 52. *Prof Exp:* Fel math, Off Naval Res, Yale Univ, 52-53; Jewett res fel, Harvard Univ, 53-55; from asst prof to prof, Univ Wash, 55-70; prof, Univ Hawaii, 70-75; PROF MATH, UNIV ARIZ, 75- *Concurrent Pos:* NSF sr fel, 61-62. *Mem:* Am Math Soc; Math Asn Am; Asn Symbolic Logic. *Res:* Lattice and ring theory; Boolean algebras; Abelian groups. *Mailing Add:* Dept of Math Univ of Ariz Tucson AZ 85721

PIERCE, ROBERT CHARLES, b Newark, NJ, Mar 2, 47; m 70; c 3. PHYSICAL CHEMISTRY. *Educ:* Rutgers Univ, BA, 69; Cornell Univ, MS, 72, PhD(phys chem), 74. *Prof Exp:* Res assoc biol mass spectrometry, Brookhaven Nat Lab, 74-75; res chemist, Colgate-Palmolive, 75-78, sect head, 78-80, mgr chem res, 80-82, mgr, Fabric Care Prod Develop, 82-83, assoc dir, Hard Surface Care Prod Develop, 83-85, dir, Oral Care Prod Develop, 85-87, dir, Golbal Mkt Oral Car, 87-89, VPRES RES & DEVELOP, SKIN & HAIR CARE PRODUCTS, COLGATE-PALMOLIVE RES & DEVELOP, 89- *Mem:* Int Asn Dent Res; Am Oil Chemists Soc. *Res:* Gaseous ion chemistry; application of mass spectrometry to the structure elucidation of biologically important materials; physical chemistry of surface active agents; surface and colloid chemistry. *Mailing Add:* Colgate-Palmolive Co 909 River Rd Piscataway NJ 08854

PIERCE, ROBERT HENRY HORACE, JR, b Pittsburgh, Pa, July 18, 10; m 47; c 2. CHEMICAL ENGINEERING. *Educ:* Western Reserve Univ, BA, 31; Case Inst Technol, BS, 31; Ohio State Univ, MA, 37, PhD(phys chem), 40. *Prof Exp:* Phys chemist, Res Lab, US Steel Corp, NJ, 31-46; assoc dir, Cryogenic Lab, Res Found, Ohio State Univ, 46-49; res phys chemist, Ammonia Dept, 49-53, res supvr, Polychem Dept, 53-58, sr res chemist, 58-61, res assoc, Plastics Dept, E I du Pont de Nemours & Co, Inc, 61-75; RETIRED. *Concurrent Pos:* Res engr, Eng Exp Sta, Ohio State Univ, 36-39. *Mem:* Am Chem Soc. *Res:* Physical properties of metals; refractories; thermal expansion; thermal conductivity; pyrometry; stress analysis; state of gases; cryogenics; x-ray diffraction; polymers. *Mailing Add:* 1706 N Bancroft Pkwy Wilmington DE 19806

PIERCE, ROBERT WESLEY, b Atlanta, Ga, Apr 4, 45; m 69; c 2. MICROPALEONTOLOGY, BIOSTRATIGRAPHY. *Educ:* Univ Ala, BS, 67; La State Univ, MS, 69, PhD(paleont), 75. *Prof Exp:* Instr geol, La State Univ, 73-74; prof Auburn Univ, 74-77; res scientist paleont, 77-82, PROJ PALEONTOLOGIST, AMOCO PROD CO, AMOCO CORP, DENVER, 82- *Concurrent Pos:* Res assoc, Cambridge Univ, 70-71; instr geol, Tulsa Jr Col, 80. *Mem:* Soc Econ Paleontologists & Mineralogists; Sigma Xi; Paleont Res Inst. *Res:* Biostratigraphy of cenozoic calcareous nannoplankton; world-wide cenozoic stratigraphy; composite standard graphic correlations. *Mailing Add:* Dept Phys Sci Eastern NMex Univ Main Campus Portales NM 88130

PIERCE, ROBERT WILLIAM, b Des Moines, Iowa, Feb 26, 40; m 64; c 2. ELECTRON MICROSCOPY GEOLOGIC MATERIALS, COMPUTER APPLICATIONS IN STRATIGRAPHY. *Educ:* Monmouth Col, BA, 62; Univ Ill, MS, 67, PhD(geol), 69. *Prof Exp:* Instr geol, Univ Ill, 67-69; asst prof, 69-76, assoc prof geol, Univ Fla, 76-81; ASSOC PROF GEOL, EASTERN NMEX UNIV, 82-, DIR, ELECTRON MICROSCOPE FACIL, 85-, CHMN, DEPT PHYS SCI, 87- *Mem:* AAAS; Am Asn Petrol Geol; Geol Soc Am; Int Paleont Union; Soc Econ Paleont & Mineral. *Res:* Conodont biostratigraphy; electron microscopy; stratigraphy; field geology. *Mailing Add:* Dept Physical Scis & Geol Eastern NMex Univ Portales NM 88130

PIERCE, ROBERTA MARION, physical chemistry, organometallic chemistry, for more information see previous edition

PIERCE, ROGER J, b Des Moines, Iowa, Mar 19, 11. ELECTRICAL ENGINEERING. *Educ:* Iowa State Univ, BS, 32. *Prof Exp:* Dir radio technol, Collins, 51-54; pres, Hydrospace Systs, 64-71; PRES TRIAD CORP, 80- *Mem:* Fel Inst Elec & Electronics Engrs; Am Rocket Soc. *Mailing Add:* 900 Staub Ct NE Cedar Rapids IA 52402

PIERCE, RONALD CECIL, b Arnprior, Ont, Feb 25, 49; m 70; c 2. ENVIRONMENTAL SCIENCES. *Educ:* Univ Guelph, Ont, BSc, 70; York Univ, Downsview, Ont, PhD(chem), 75. *Prof Exp:* Res officer environ sci, Nat Res Coun Can, 75-85; WATER QUAL GUIDELINES OFFICER, ENVIRON CAN, 85- *Res:* Compilation and critical assessment of the scientific criteria required for the establishment of environmentally relevant standards pertaining to the aquatic environment. *Mailing Add:* 15 Shaw Ct Kanata Ottawa ON K2L 2L9 Can

PIERCE, RUSSELL DALE, b Iselin, Pa, July 17, 38; m 65; c 3. SEMICONDUCTORS, MAGNETISM. *Educ:* Carnegie Inst Technol, BS, 60, MS, 61, PhD(physics), 66. *Prof Exp:* Instr physics, Carnegie Inst Technol, 65-66; mem tech staff, 66-85, DISTINGUISHED MEM TECH STAFF, BELL LABS, 85- *Mem:* Sigma Xi. *Res:* Low temperature calorimetry of dilute magnetic solids; magnetic properties of solids, especially materials applicable to memory devices; instrumentation for magnetic bubble materials preparation and characterization; GaAs integrated circuit testing; silicon device modeling. *Mailing Add:* Rd 8 Box 32 Horseshoe Dr Sinking Spring PA 19608-9808

PIERCE, SIDNEY KENDRICK, b Holyoke, Mass, Sept 19, 44; m 74; c 2. COMPARATIVE PHYSIOLOGY. *Educ:* Univ Miami, BEd, 66; Fla State Univ, PhD(physiol), 70. *Prof Exp:* Asst prof, 70-73, assoc prof, 73-78, PROF ZOOL, UNIV MD, COLLEGE PARK, 78- *Concurrent Pos:* NSF res grants, 73-75, 75-77; NIH grants, 77- mem corp, Marine Biol Lab, 73-; assoc ed, J Exp Zool, 80-82; ed, Marine Biol, 81-84. *Mem:* AAAS; Am Soc Zool; Am Physiol Soc. *Res:* Physiological interactions of marine invertebrates with the environment and the control of cell membrane permeability. *Mailing Add:* Dept Zool Univ Md College Park MD 20742

PIERCE, TIMOTHY ELLIS, nuclear chemistry, medical technology, for more information see previous edition

PIERCE, WAYNE STANLEY, b Atascadero, Calif, Mar 8, 42; m 64; c 1. PLANT PHYSIOLOGY. *Educ:* Humboldt State Col, AB, 64; Wash State Univ, MS, 67, PhD(bot), 71. *Prof Exp:* From asst prof to assoc prof, 71-81, PROF BIOL, CALIF STATE UNIV, STANISLAUS, 81- *Concurrent Pos:* Fel, Univ Houston, 78; vis fac, Univ Md, 86-87. *Mem:* Am Soc Plant Physiol; Sigma Xi; Hydroponic Soc Am. *Res:* Mineral nutrition; cellular ion-transport mechanisms; electrophysiology. *Mailing Add:* Dept of Biol Sci Calif State Univ Stanislaus Turlock CA 95380

PIERCE, WILLIAM ARTHUR, JR, b Dayton, Ohio, Apr 11, 18; m 46; c 2. MICROBIOLOGY. *Educ:* Ohio Wesleyan Univ, BA, 41; Univ Wis, MS, 47, PhD(med microbiol), 49. *Prof Exp:* Instr microbiol, 49, from asst prof to assoc prof, 50-60, PROF MICROBIOL & IMMUNOL, SCH MED, TULANE UNIV, 61- *Mem:* Am Soc Microbiol; Am Acad Microbiol. *Res:* In vitro phagocytosis studies; antigens of Neisseria Gonorrhoeae. *Mailing Add:* 10720 Kinneil Rd New Orleans LA 70127

PIERCE, WILLIAM G, b Gettysburg, SDak, Sept 24, 04; m 30; c 3. STRUCTURAL GEOLOGY, EARTH SCIENCES. *Educ:* Univ SDak, AB, 27; Princeton Univ, MA, 29, PhD(geol), 31. *Prof Exp:* Field asst, SDak Geol Surv, 26; asst, Princeton Univ, 28-29; from geologist to staff geologist, Mo River Basin, 29-47, chief, Western Sect Fuels Br, 48-57 & Radioactive Waste Disposal & Salt Deposits, 57-60, res geologist, 61-74, EMER SCIENTIST, US GEOL SURV, 75- *Concurrent Pos:* NSF res grant tectonics of Italian Apennines & Swiss Jura, 63; adv lignite resources of Greece, Econ Coop Admin, 49; NSF & Nat Sci Coun Repub China res grant, Tectonics of Western Taiwan, 74-75. *Honors & Awards:* US Dept Interior Distinguished Serv Award, 65; Gold Plaque, Contrib Sci, Repub of China, 79. *Mem:* AAAS; fel Geol Soc Am; Am Asn Petrol Geologists. *Res:* Oil, gas and coal resources; structural geology; stratigraphy; radioactive waste disposal. *Mailing Add:* 14380 Manuella Rd Los Altos Hills CA 94022

PIERCE, WILLIAM H, b Washington, DC, July 10, 33; m 56; c 2. ELECTRICAL ENGINEERING. *Educ:* Harvard Univ, AB, 55; Stanford Univ, MS, 59, PhD(elec eng), 61. *Prof Exp:* From asst prof to assoc prof elec eng, Carnegie Inst Technol, 61-69; PROF ELEC ENG, UNIV LOUISVILLE, 69- *Mem:* Inst Elec & Electronics Engrs. *Res:* Communications; biomedical engineering and medical science. *Mailing Add:* 2036 Strathmore Blvd Louisville KY 40205

PIERCE, WILLIAM R, b Topeka, Kans, Aug 13, 15; m 42; c 2. FOREST MANAGEMENT. *Educ:* Wash Univ, BS, 40, PhD, 58; Yale Univ, MS, 47. *Prof Exp:* Dist ranger, US Forest Serv, 45-55; from assoc prof to prof forestry, Univ Mont, 55-81; RETIRED. *Mem:* AAAS; Soc Am Foresters; Sigma Xi; Nat Wildlife Fedn Nature Conservancy. *Res:* Forest resource management planning and inventory; statistics, computer modeling and programming. *Mailing Add:* 5801 28th Ave NW Gig Harbor WA 98335

PIERCE, WILLIAM SCHULER, b Wilkes Barre, Pa, Jan 12, 37; m 65; c 2. CARDIAC SURGERY. *Educ:* Lehigh Univ, BS, 58; Univ Pa, MD, 62. *Prof Exp:* Surg resident, Univ Pa Hosp, 62-69, sr cardiac surg resident, 69-70; from asst prof to assoc prof, 70-77, PROF THORACIC SURG, COL MED, PA STATE UNIV, 77- *Concurrent Pos:* USPHS grants, Nat Heart, Lung & Blood Inst, Bethesda, Md, 77-80, 78-82, 82-87 & 87-92; chmn, NIH Study Sect, Surg & Bioeng, 87- *Honors & Awards:* Becton-Dickinson Career Achievement Award, Asn Advan Med Instrumentation, 77; Clemson Award for Apple Res, 85. *Mem:* AMA; Am Col Surgeons; Soc Univ Surgeons; Am Surg Asn; Am Soc Artificial Internal Organs (secy-treas, 80-82, pres, 83-84). *Res:* Development of paracorporeal and implantable left ventricular assist devices and the artificial heart; mechanical cardiac valves. *Mailing Add:* Dept Surg Hershey Med Ctr 500 University Ave Hershey PA 17033

PIERCEY, MONTFORD F, b Meriden, Conn, July 25, 42; m 65; c 2. PHARMACOLOGY, NEUROBIOLOGY. *Educ:* Boston Univ, AB, 65, MS, 67; Yeshiva Univ, PhD(pharmacol), 72. *Prof Exp:* Fel, Albert Einstein Col Med, 72-74; res scientist, 74-82, SR RES SCIENTIST, THE UPJOHN CO, 82- *Mem:* Soc Neurosci; Am Soc Pharmacol & Exp Therapeut; Am Pain Soc; Int Narcotic Res Conf; AAAS. *Res:* Neurophysiology of motor control; neural control of respiration and circulation; analgesic drugs and their mechamism of action; neurotransmitter identification; anti-diarrheal mechanisms; pharmacological evaluation of central nervous system drugs and neuropeptides; computer imaging of brain function. *Mailing Add:* Upjohn Co Kalamazoo MI 49001

PIERCY, GEORGE ROBERT, b Vancouver, BC, May 27, 28; m 52; c 4. METALLURGY, SOLID STATE PHYSICS. *Educ:* Univ BC, BASc, 51, MASc, 52; Univ Birmingham, PhD(metall), 54. *Prof Exp:* Res scientist, Atomic Energy Can, 55-63; vis prof metall, Benares Hindu Univ, 64-65; res scientist, Atomic Energy Can, 65-69; PROF METALL & MAT SCI, MCMASTER UNIV, 69- *Res:* Deformation and irradiation damage in metals. *Mailing Add:* Dept Metall & Mat McMaster Univ 1280 Main St W Hamilton ON L8S 4L8 Can

PIERINGER, ARTHUR PAUL, b Weehawken, NJ, Oct 30, 24; m 49; c 5. PLANT BREEDING. *Educ:* Univ Ky, BSAgr, 51; Cornell Univ, PhD, 56. *Prof Exp:* Asst, Cornell Univ, 51-56; from asst horticulturist to assoc horticulturist, citrus exp sta, Univ Fla, 72-84; RETIRED. *Mem:* Sigma Xi. *Res:* Citrus variety improvement. *Mailing Add:* 524-B W Highland St Lakeland FL 33803

PIERINGER, RONALD ARTHUR, b Jersey City, NJ, Nov 23, 35; m 57; c 2. BIOCHEMISTRY. *Educ:* Lebanon Valley Col, BS, 57; Univ Wis, MS, 59, PhD(physiol chem), 61. *Prof Exp:* Res assoc biochem of lipids, Harvard Med Sch, 61-63; from instr to assoc prof, 63-74, PROF BIOCHEM, SCH MED, TEMPLE UNIV, 74- *Concurrent Pos:* NIH fel, 61-63, Nat Inst Neurol Dis & Stroke res career develop award, 71-76; res grants, NSF & Nat Inst Neurol Dis & Stroke. *Honors & Awards:* Javits Award. *Mem:* Am Soc Biol Chem; Am Soc Neurochem; Am Oil Chem Soc; Am Chem Soc; Am Soc Microbiol; Int Soc Neurochem. *Res:* Metabolism and function of lipids in bacteria and animals. *Mailing Add:* Dept of Biochem Temple Univ Sch of Med Philadelphia PA 19140

PIERMARINI, GASPER J, b Leominster, Mass, Apr 26, 33; m 60; c 2. PHYSICAL CHEMISTRY, CRYSTALLOGRAPHY. *Educ:* Boston Univ, AB, 55; Am Univ, PhD(phys chem), 71. *Prof Exp:* PHYS CHEMIST, NAT INST STANDARDS & TECHNOL, 58- *Concurrent Pos:* Adj prof chem, Am Univ, Washington, 76-80; Gesamt-Hochschule-Paderborn, Fed Repub Ger, 81-82. *Honors & Awards:* Spec Achievement Award, US Dept Com, 73, 74, Gold Medal Award, 74; Alexander von Humboldt Award, 81. *Mem:* AAAS; Am Chem Soc; Am Crystallog Asn; Sigma Xi. *Res:* Ceramics, ceramic science, ceramic processing and transformation toughening; crystal structures of inorganic compounds, especially determination of crystal structures of materials under high pressure; application of x-ray diffraction methods in solids under high pressure; high pressure measurement by ruby fluorescence method; diamond anvil cells; high energy materials, explosives; spectroscopy and mechanisms of thermal decomposition and structures of high energy materials at elevated pressures and temperatures. *Mailing Add:* Mat Sci & Eng Lab Nat Inst Standards & Technol Gaithersburg MD 20899

PIEROTTI, ROBERT AMADEO, b Newark, NJ, Nov 13, 31; m 59; c 2. PHYSICAL CHEMISTRY. *Educ:* Pomona Col, BA, 54; Univ Wash, PhD(chem), 58. *Prof Exp:* Asst chem, Pomona Col, 53-54 & Univ Wash, 54-58; instr, Univ Nev, 58-60; from asst prof to assoc prof, 60-68, PROF CHEM, GA INST TECHNOL, 68-, DIR, SCH CHEM, 82-, DEAN SCI, 89- *Concurrent Pos:* NATO sr fel, Univ Bristol, 68. *Honors & Awards:* Monie Ferst Res Award, Ga Tech Chapter Sigma Xi, 68. *Mem:* AAAS; Am Chem Soc; Am Phys Soc; Royal Soc Chem; Sigma Xi. *Res:* Adsorption; interaction of gases with solids; solutions. *Mailing Add:* Sch of Chem Ga Inst of Technol Atlanta GA 30332

PIERPONT, CORTLANDT GODWIN, b New York, NY, Jan 26, 42; m 63; c 2. INORGANIC CHEMISTRY. *Educ:* Columbia Univ, BS, 67; Brown Univ, PhD(chem), 71. *Prof Exp:* Asst prof, WVa Univ, 71-75; assoc prof, 75-81, PROF CHEM, UNIV COLO, 81-, CHMN, DEPT CHEM & BIOCHEM, 86- *Mem:* Am Crystallog Asn; Am Chem Soc; Royal Soc Chem. *Res:* Organo-transition metal chemistry; coordination complexes of transition metals; x-ray crystallography. *Mailing Add:* Dept Chem Univ Colo Campus Box 215 Boulder CO 80309

PIERRARD, JOHN MARTIN, b Chicago, Ill, Mar 26, 28; m 64; c 5. AIR POLLUTION, ENVIRONMENTAL SCIENCES. *Educ:* Ill Inst Technol, BS, 52; Tex A&M Univ, MS, 58; Univ Wash, PhD, 69. *Prof Exp:* Res assoc atmospheric chem, Cloud Physics Proj, Univ Chicago, 52-53; res assoc micrometeorol, Res Found, Tex A&M Univ, 57-58, res scientist, 58-59; assoc meteorologist, Armour Res Found, 59-61, res meteorologist, 61-62; res scientist, Nat Ctr Atmospheric Res, 62-66; atmospheric chemist, 69-72, eng assoc, 72-78, eng fel, 79-84, ENVIRON MGR, E I DU PONT DE NEMOURS & CO, INC, 84- *Honors & Awards:* Horning Mem Award, Soc Automotive Engrs. *Mem:* Sigma Xi; Air & Waste Mgt Asn. *Res:* Aerosol physics and chemistry; meteorological sensory systems; evaluation of social, econmic and environmental quality impacts of air pollution control strategies; air permitting and modeling; regulatory affairs consulting. *Mailing Add:* E I du Pont de Nemours & Co Inc Newark DE 19714-6090

PIERRE, DONALD ARTHUR, b Bloomington, Wis, July 2, 36; m 59; c 3. CONTROL SYSTEMS, OPTIMIZATION TECHNIQUES. *Educ:* Univ Ill, Urbana, BS, 58; Univ Southern Calif, MS, 60; Univ Wis, PhD(elec eng), 62. *Prof Exp:* Res asst, Ill State Geol Surv, 55-58; mem tech staff, Hughes Aircraft Co, 58-60; from asst prof to assoc prof, 62-65, head elec eng, 79-84, head comput sci, 80-84, PROF ELEC ENG, MONT STATE UNIV, 69- *Concurrent Pos:* Head syst group, Electronics Res Lab, Mont State Univ, 69-79; regist prof engr, State Mont, 75- *Honors & Awards:* Wiley Award Meritorious Res, Mont State Univ, 82. *Mem:* Fel Inst Elec & Electronics Engrs; Sigma Xi; Am Soc Eng Educ. *Res:* Design of control systems and in the development of optimization algorithms; computer control; author of two books and numerous technical pulications. *Mailing Add:* Dept Elec Eng Mont State Univ Bozeman MT 59717

PIERRE, LEON L, b Caracas, Venezuela, June 26, 22; m 65. BACTERIOLOGY, BIOCHEMISTRY. *Educ:* Dalhousie Univ, BS, 52; Fordham Univ, MS, 58, PhD(enzyme chem), 61. *Prof Exp:* Asst prof biol, Long Island Univ, 61-67; assoc prof, City Univ New York, 67-68; assoc prof biol, 68-69, ASSOC PROF LIFE SCI, NEW YORK INST TECHNOL, 69- *Mem:* Sigma Xi; Soc Exp Biol & Med; fel Royal Soc London. *Res:* Enzyme research into the function of the symbionts and fat bodies of the cockroach, Leucophaea maderae. *Mailing Add:* 4720 195th St Flushing NY 11358

PIERRE, ROBERT V, b Athens, Ohio, Aug 24, 28; div; c 4. INTERNAL MEDICINE. *Educ:* Univ Ohio, BS, 50; Northwestern Univ, MD, 54; Am Bd Internal Med, dipl, 63; cert hemat. *Prof Exp:* Intern, Chicago Wesley Mem Hosp, 54-55; resident internal med, Vet Admin Res Hosp, Chicago, 55-60, clin investr, 62-65, chief hemat sect, 65-67; asst prof med, 67-73, PROF LAB MED & INTERNAL MED, MAYO GRAD SCH MED, UNIV MINN, ROCHESTER, 73-, CONSULT LAB MED, MAYO CLIN, 67- *Concurrent Pos:* Fel hemat, Vet Admin Res Hosp, Chicago, 60-62; asst prof, Northwestern Univ, 65-67. *Mem:* Fel Am Col Physicians; Am Soc Hemat; Int Soc Hemat; Sigma Xi; Am Soc Clin Pathologists. *Res:* Cytogenetic studies in leukemia and preleukemia, automated hematology instruments. *Mailing Add:* Dept Path Mayo Clinic Hilton Bldg Rochester MN 55905

PIERREHUMBERT, RAYMOND T, b Passaic, NJ, May 15, 54; m 78. ATMOSPHERIC SCIENCES & OCEANOGRAPHY. *Educ:* Harvard Univ, BA, 75; Mass Inst Technol, PhD(meteorol), 80. *Prof Exp:* Asst prof, Dept Meteorol, Mass Inst Technol, Cambridge, 80-82; RES SCIENTIST, GEOPHYS FLUID DYNAMICS LAB, NAT OCEANIC & ATMOSPHERIC ADMIN, PRINCETON UNIV, 82- *Res:* Author of numerous articles in scientific journals. *Mailing Add:* 527 Morris Ave Elizabeth NJ 07208

PIERRET, ROBERT FRANCIS, b East Cleveland, Ohio, Aug 20, 40; m 65; c 3. SEMICONDUCTOR MATERIAL & DEVICE CHARACTERIZATION. *Educ:* Case Inst Technol, BS, 62; Univ Ill, Urbana, MS, 63, PhD(physics), 66. *Prof Exp:* Res assoc elec eng, Univ Ill, Urbana, 66-67, asst prof, 67-70; assoc prof, 70-77, PROF ELEC ENG, PURDUE UNIV, 77- *Concurrent Pos:* Consult ed, Addison-Wesley Publ Co. *Mem:* Inst Elec & Electronics Engrs. *Res:* Measurement of parameters characterizing solid-state materials and devices, with special emphasis on the development of measurement techniques and the interrogation of metal- insulator-semiconductor structures; author of four volumes in the Addison-Wesley Modular Series on solid state devices. *Mailing Add:* 1285 Elec Eng Bldg Purdue Univ West Lafayette IN 47907-1285

PIERRO, LOUIS JOHN, b Bristol, Pa, Sept 5, 31; m 55; c 2. BIOLOGY. *Educ:* St Joseph's Col, BS, 52; Marquette Univ, MS, 54; Brown Univ, PhD(biol), 57. *Prof Exp:* USPHS res fel, Calif Inst Technol, 57-58; asst prof biol, Wheeling Col, 58-60; assoc prof animal genetics, 60-66, actg dept head pathobiol, 80-82, PROF ANIMAL GENETICS, UNIV CONN, 66-, HEAD DEPT, 65- & ASSOC DIR, STORRS AGR EXP STA, 81- *Concurrent Pos:* Secy, Northeast Regional Asn Exp Sta Dirs, 82-83, vchmn, 83-84, chmn, 84-85; head, gen & cell biol sect, Biol Sci Group, Univ Conn, 84-85. *Mem:* AAAS; Genetics Soc Am; Am Genetics Asn; Am Soc Zoologists; Soc Develop Biol; Int Soc Develop Biol; Am Asn Advan Zoologists; Teratology Soc. *Res:* Developmental genetics; developmental abnormalities. *Mailing Add:* Univ Conn Box U-10 Storrs CT 06268

PIERSCHBACHER, MICHAEL DEAN, b Chariton, Iowa, Dec 19, 51. BIOMATERIALS, TISSUE REGENERATION. *Educ:* Northeast Mo State Univ, BS, 74; Univ Mo, PhD(biochem), 79. *Prof Exp:* Postdoctoral fel immunol, Scripps Clin & Res Found, 78-79; postdoctoral fel biochem, La Jolla Cancer Res Found, 79-81, res assoc, 82-84, asst staff scientist, 84-86, STAFF SCIENTIST, LA JOLLA CANCER RES FOUND, 86-; SR VPRES & SCI DIR, TELIOS PHARMACEUTICALS, INC, 87- *Mem:* Protein Soc; AAAS; Am Soc Biol Chemists. *Res:* Elucidating the mechanisms that cells use to interact with their extracellular surroundings and develop therapeutic ways to manipulate that interaction. *Mailing Add:* Telios Pharmaceut Inc 2909 Science Park Rd San Diego CA 92121

PIERSKALLA, WILLIAM P, b St Cloud, Minn, Oct 22, 34; m 53; c 3. OPERATIONS RESEARCH. *Educ:* Harvard Univ, AB, 56, MBA, 58; Univ Pittsburgh, MS, 62; Stanford Univ, MS & PhD(opers res), 65. *Hon Degrees:* MA, Univ Pa, 78. *Prof Exp:* From asst prof to assoc prof opers res, Case Western Reserve, 65-68; assoc prof & actg dir comput sci opers res ctr, Southern Methodist Univ, 68-70; from assoc prof to prof indust eng & mgt sci, Northwestern Univ, 70-78; exec dir, Leonard Davis Inst Health Econ & dir, Nat Health Care Mgt Ctr, Univ Pa, 78-83, prof, systs eng, Sch Engr & Appl Sci & chmn, Health Care Systs Dept, 81-90, dep dean, 83-89, PROF DECISION SCI DEPT & HEALTH CARE SYSTS DEPT, WHARTON SCH, UNIV PA, 78- *Concurrent Pos:* Ed, Opers Res J, 78-82. *Mem:* Soc Indust & Appl Math; Asn Comput Mach; Opers Res Soc Am (secy, 77-80, pres elect, 81-82, pres, 82-83); Inst Mgt Sci; Int Fed Oper Res Socs (pres, 89-91). *Res:* Inventory theory and mathematical programming; health care delivery. *Mailing Add:* Health Care Syst 206 Cpc 1Ce Univ Pa Philadelphia PA 19104

PIERSMA, BERNARD J, b Utica, NY, Mar 23, 38; m 66; c 3. PHYSICAL CHEMISTRY. *Educ:* Colgate Univ, BA, 59; St Lawrence Univ, MS, 61; Univ Pa, PhD(phys chem), 65. *Prof Exp:* Nat Acad Sci-Nat Res Coun res assoc electrochem, Naval Res Lab, 65-66; asst prof chem, Eastern Baptist Col, 66-

70, assoc prof, 70-71; PROF PHYS CHEM, HOUGHTON COL, 71- *Concurrent Pos:* sabbatical leave, univ resident res prof, Frank J Seiler Res Lab, US Air Force Acad, 81-82; fac res prof, UES, 86; fac res prof, UES, 86 & 90. *Mem:* Am Chem Soc; Electrochem Soc; Am Sci Affiliation. *Res:* Kinetics and mechanisms of anodic organic oxidation reactions; hydrogen and oxygen electrodes; study of the electrical double layer; fundamental electrochemistry of physiological electrodes; electrocatalysis; electrochemistry of room temperature molten salts, especially imidazolium chloride/aluminum chloride melts. *Mailing Add:* Dept Chem Houghton Col Houghton NY 14744

PIERSOL, ALLAN GERALD, b Pittsburgh, Pa, June 2, 30; m 58; c 3. MECHANICAL VIBRATIONS, STATISTICAL DATA ANALYSIS. *Educ:* Univ Ill, BS, 52; Univ Calif, Los Angeles, MS, 61. *Prof Exp:* Res engr, Douglas Aircraft Co, Inc, 52-59; tech staff, Ramo-Wooldridge Corp, 59-63; vpres, Measurement Analysis Corp, 63-71; prin scientist, Bolt Beranek & Newman, Inc, 71-85, sr scientist, Astron Res & Eng, 85-89; DEPT MECH ENGR, UNIV SOUTHERN CALIF, 68-; PRES, PIERSOL ENG CO, 89- *Concurrent Pos:* Lectr, Univ Calif, Los Angeles, 65-, Univ Southern Calif, 68- *Mem:* Am Soc Mech Engrs; Acoustical Soc Am; Inst Environ Sci. *Res:* Applications of random process theory to mechanical shock, vibration, and acoustic noise problems; author of four books on random data analysis and applications. *Mailing Add:* Piersol Eng Co 23021 Brenford St Woodland Hills CA 91364

PIERSON, BERNICE FRANCES, b Auburn, Nebr, Aug 17, 06. PROTOZOOLOGY. *Educ:* Case Western Reserve Univ, BA, 28; Johns Hopkins Univ, MA, 37, PhD(protozool), 41. *Prof Exp:* Asst zool, Schs Dent & Pharm, Univ Md, 28-31, 36-38; instr biol, Towson State Univ, Towson, Md, 34-35 & Nat Park Col, 41-42; biologist, Off Sci Res & Develop & Nat Defense Res Coun, Johns Hopkins Univ, 42-45; instr anat & physiol, Dept Nursing Educ, Univ Baltimore, 45-46; from instr to prof, 46-73, chmn dept, 66-69, EMER PROF BIOL, MONTGOMERY COL, 73- *Concurrent Pos:* Exchange prof, San Bernardino Valley Col, Calif, 53-54. *Mem:* AAAS; Soc Protozool; Am Soc Zool; Am Micros Soc; Sigma Xi. *Res:* Morphology and physiology of protozoa. *Mailing Add:* 19805 Bramble Bush Drive Gaithersburg MD 20879

PIERSON, BEVERLY KANDA, b Syracuse, NY, Jan 9, 44; m 66; c 1. MICROBIOLOGY. *Educ:* Oberlin Col, BA, 66; Univ Ore, MA, 69, PhD(biol), 73. *Prof Exp:* Asst prof biol, Oberlin Col, 74-75; from asst prof to assoc prof, 75-85, chmn dept, 83-86, PROF BIOL, UNIV PUGET SOUND, 85- *Concurrent Pos:* Vis scientist, Dept Biol, Univ Calif, Los Angeles, 81-82; mem fac, Planetary Biol & Microbiol Ecol, NASA, 82 & Marine Biol Lab, Woods Hole, Mass, 84; coun undergrad res, Univ Puget Sound, 89. *Mem:* Am Soc Microbiol; Sigma Xi; AAAS; Int Soc Study Origin of Life. *Res:* Physiology of photosynthetic bacteria; pigments and photosynthesis; ecology of microbial mats; extreme environments; early evolution; biology of Chloroflexaceae. *Mailing Add:* Dept of Biol Univ of Puget Sound Tacoma WA 98416

PIERSON, DAVID W, b Ottumwa, Iowa, Jan 16, 26; m 57; c 2. SCIENCE EDUCATION, CONSERVATION. *Educ:* Colo State Col, AB, 51; State Col Iowa, MA, 57; Univ Mo, EdD, 62. *Prof Exp:* Teacher high sch, Colo, 52-55 & Iowa, 56-57; instr appl sci, Monticello Col, 57-58; teacher high sch, Iowa, 58-59; part-time instr educ, Univ Mo, 60-62; assoc prof biol, Ft Hays Kans State Col, 62-88; RETIRED. *Concurrent Pos:* Dir, Kans Jr Acad Sci, 62-69; vis lectr, Scientist Prog, Jr Acad Sci Western Kans, 63-65. *Mem:* Fel AAAS; Am Asn Univ Professors; Am Nature Study Soc. *Res:* Development of teaching methods in soil and water conservation; mobility of Western Kansas science teachers and factors affecting movement; resource use education; wind erosion and soil bacteria dissemination; biological implications of energy production. *Mailing Add:* 2705 Woodrow Ct Hays KS 67401-1618

PIERSON, EDGAR FRANKLIN, b Fairfield, Iowa, Aug 31, 09; m 42; c 1. BIOLOGY. *Educ:* Iowa Wesleyan Col, BS, 33; State Univ Iowa, MS, 36, PhD(bot), 38. *Prof Exp:* From instr to prof biol, 38-42, 46-63, head dept, 47-63, dean grad prog, 63-69, prof, 69-80, EMER PROF BIOL, UNIV WIS-STEVENS POINT, 80- *Mem:* AAAS. *Res:* Plant taxonomy; limnology; plankton development in Lake Macbride; botany; general zoology; freshwater biology. *Mailing Add:* Tamarack Pl 84N 17147 W Menomonee Ave Menomonee Falls WI 53051

PIERSON, EDWARD S, b Syracuse, NY, June 27, 37; m 71; c 1. ENERGY CONVERSION, POWER SYSTEMS. *Educ:* Syracuse Univ, BS, 58; Mass Inst Technol, SM, 60, ScD, 64. *Prof Exp:* Asst prof elec eng, Mass Inst Technol, 64-66; assoc prof & assoc head eng, Univ Ill, Chicago, 66-75; prog mgr liquid-metal MHD, Argonne Nat Lab, 75-82; HEAD DEPT ENG, PURDUE UNIV CALUMET, 82- *Concurrent Pos:* Ford Found fel, Mass Inst Technol, 65-66; consult, Argonne Nat Lab, 70-; vis prof, Tech Univ, Berlin, 73; adj prof, Univ Ill, Chicago, 75-82; consult, Solmecs Corp, 82-, HMJ Corp, 83- *Mem:* Am Soc Eng Educ; Am Soc Mech Engrs; Inst Elec & Electronics Engrs. *Res:* Analysis and development of two-phase liquid-metal magnetohydrodynamic energy-conversion systems; analysis of induction generators for magnetohydrodynamic power systems and of new or novel energy-conversion systems. *Mailing Add:* Dept Eng Purdue Univ Calumet Hammond IN 46323-2094

PIERSON, ELLERY MERWIN, b Eugene, Ore, Mar 31, 35; m 58; c 2. STATISTICAL ANALYSIS. *Educ:* Portland State Col, BS, 57; Rutgers Univ, MEd, 65; Univ Pa, PhD(tech educ res), 75. *Prof Exp:* Res asst biochem, Col Physicians & Surgeons, Columbia Univ, 57-58; res asst physics, RCA Corp, Somerville, NJ, 58-60; welfare investr, Middlesex County, NJ, 60-61; res asst psychol, Educ Testing Serv, Princeton, NJ, 61-66; res psychologist, Franklin Inst Res Labs, Philadelphia, 66-67; res assoc educ, 67-70, res assoc design & analysis, 70-75, mgr statist analysis, 75-87, MGR MEASUREMENT RES, SCH DIST PHILADELPHIA, 88- *Mem:* Nat Coun Measurement Educ; Am Educ Res Asn. *Res:* Educational information management systems; semantic differential applications; optical mark scanning. *Mailing Add:* Off Res & Eval Rm 407 21st & The Parkway Philadelphia PA 19103

PIERSON, KEITH BERNARD, b San Francisco, Calif, Nov 19, 49. AQUATIC, BIOCHEMICAL. *Educ:* Calif Polytech State Univ, BSc, 72; Univ Wash, Seattle, PhD(fisheries), 83. *Prof Exp:* Lectr & instr fish classification & fisheries, Univ Wash, Seattle, 78-81, fisheries biologist, 81-84; postdoctoral fel, Univ Wis-Madison, 84-87, res assoc, 87-89; RES TOXICOLOGIST, E I DU PONT DE NEMOURS & CO, 89- *Mem:* Am Soc Zoologists; Am Fisheries Soc; Soc Environ Toxicol & Chem. *Res:* Toxicology of chemicals to aquatic organisms-effects on animals such as growth and reproduction and on biochemical parameters such as uptake, metabolism and fate; interactions of proteins and metals as related to structure-function. *Mailing Add:* 4610 Christiana Meadows Bear DE 19701

PIERSON, MERLE DEAN, b Mitchell, SDak, May 23, 42; m 62; c 3. FOOD MICROBIOLOGY, FOOD PROCESSING. *Educ:* Iowa State Univ, BS, 64; Univ Ill, MS, 69, PhD(food sci), 70. *Prof Exp:* Res chemist meat processing, George A Hormel & Co, 65-66; res asst food microbiol, Univ Ill, 66-68, Wright fel food sci, 68-70; from asst prof to assoc prof, 70-83, PROF FOOD SCI, VA POLYTECH INST & STATE UNIV, 83-, HEAD DEPT, 85- *Concurrent Pos:* Food indust consult, 70- *Mem:* Am Soc Microbiol; Soc Appl Bact; fel Inst Food Technologists; Int Asn Milk, Food & Environ Sanitarians; fel AAAS; Am Meat Sci Asn. *Res:* Food microbiology; physiology of foodborne microorganisms; heat-injury of microorganisms; foodborne infections and intoxications; thermal processing of foods; antimicrobial food additives; food processing sanitation. *Mailing Add:* Dept Food Sci & Technol Va Polytech Inst & State Univ Blacksburg VA 24061

PIERSON, RICHARD NORRIS, JR, b New York, NY, Sept 22, 29; m 54, 74; c 6. NUCLEAR MEDICINE. *Educ:* Princeton Univ, BA, 51; Columbia Univ, MD, 55. *Prof Exp:* Residency med, St Luke's Hosp, 55-61, assoc dir nuclear med, 61-65, dir nuclear med, 65-89, DIR BODY COMPOS UNIT, ST LUKE'S-ROOSEVELT HOSP CTR, 68- *Concurrent Pos:* Instr, Col Physicians & Surgeons, Columbia Univ, 62-67, from asst prof to assoc prof, 68-80, prof clin med, 81-; mem, House Delegates, AMA, 78-90; pres, Am Med Rev Res Ctr, 85-89. *Mem:* AAAS; Am Physiol Soc; Soc Nuclear Med; Alliance Continuing Med Educ (pres, 87-89). *Res:* Exploration of all major methods used for body composition research, special emphasis on physical methods: neutron activation; application of methods to acute and chronic illness and aging. *Mailing Add:* St Luke's-Roosevelt Hosp Ctr 425 W 113th St New York NY 10025

PIERSON, THOMAS CHARLES, b Trenton, NJ, Dec 4, 47; div; c 2. NON NEWTONIAN FLUID MECHANICS, SEDIMENTOLOGY. *Educ:* Middlebury Col, BA, 70; Univ Wash, MS, 72, PhD(geol), 77. *Prof Exp:* Petrol geologist, Texaco Inc, 72-73; fel, Forest Res Inst, NZ Forest Serv, 77-80; RES HYDROLOGIST, CASCADES VOLCANO OBSERV, US GEOL SURV, 80- *Mem:* Geol Soc Am; Am Geophys Union. *Res:* Dynamics and flow behavior of debris flows and the rheologic properties of mud-rock-water slurries; sedimentology of debris-flow deposits; assessment of hydrologic hazards at volcanoes. *Mailing Add:* US Geol Surv 5400 MacArthur Blvd Vancouver WA 98661

PIERSON, WILLARD JAMES, JR, b New York, NY, July 7, 22; m 54; c 3. METEOROLOGY, OCEANOGRAPHY. *Educ:* Univ Chicago, BS, 44; NY Univ, PhD(meteorol), 49. *Prof Exp:* Prof oceanog, NY Univ, 49-73; PROF, INST MARINE & ATMOSPHERIC SCI, CITY COL NEW YORK, 73- *Concurrent Pos:* Chmn, SASS team, SEASAT; mem, Sci Def Team, NASA Scatterometer, Sci Team ERS1; consult, losses ships at sea. *Honors & Awards:* Mil oceanog award, Oceanogr of Navy, 69. *Mem:* Fel Am Meteorol Soc; fel AAAS; Soc Naval Archit & Marine Eng; fel Am Geophys Union; sr mem Inst Elec & Electronic Engrs. *Res:* Ocean waves; ship motions; wave forecasting; satellite oceanography; study of radar data to measure waves and winds over the ocean; generation of capillary waves by wind and calculation of radar response to these waves. *Mailing Add:* IMAS City Col 138th St & Convent Ave New York NY 10031

PIERSON, WILLIAM GRANT, b Elizabeth, NJ, Jan 27, 33; m 58; c 2. ORGANIC CHEMISTRY, PHYSICAL CHEMISTRY. *Educ:* Seton Hall Univ, BA, 54, MS, 61, PhD(org chem), 68. *Prof Exp:* Res assoc, Ciba Pharmaceut Co, 58-67; res chemist, 68-70, sect head, oral prods, 70-78, mgr toiletries prod res, 78, dir res & develop, Far East Div, 78-81, dir res & develop, Oral Planning & Admin, 81-85, DIR RES & DEVELOP, HARD SURFACE CARE, COLGATE-PALMOLIVE CO, 85- *Mem:* Am Chem Soc; Int Asn Dent Res. *Res:* Exploration of electroorganic reaction mechanisms; organic synthesis; alkaloid structure elucidation; alkaloid synthesis and structural modification; pharmaceutical and oral health research. *Mailing Add:* 557 Hanson Ave Piscataway NJ 08854

PIERSON, WILLIAM R, aerospace medicine; deceased, see previous edition for last biography

PIERSON, WILLIAM R, b Charleston, WVa, Oct 21, 30; m 61; c 2. AIR POLLUTION, ATMOSPHERIC CHEMISTRY. *Educ:* Princeton Univ, BSE, 52; Mass Inst Technol, PhD, 59. *Prof Exp:* Res assoc, Enrico Fermi Inst Nuclear Studies, Univ Chicago, 59-62; mem sci res staff, Chem Dept, Ford Motor Co, 62-87; EXEC DIR, ENERGY & ENVIRON ENG CTR, DESERT RES INST, 87- *Concurrent Pos:* Lectr chem, Univ Mich, 68. *Mem:* AAAS; Am Chem Soc; Am Phys Soc; Sigma Xi; Air Pollution Control Asn; Am Asn Aerosol Res. *Res:* Nuclear level schemes and high-energy nuclear reactions; nuclear properties; atmospheric aerosols. *Mailing Add:* 6308 Meadow Creek Dr Reno NV 89509

PIERUCCI, MAURO, b Lucca, Italy, Jan 5, 42; US citizen; m 65; c 2. ACOUSTIC RADIATION, ENGINEERING ACOUSTICS. *Educ:* Polytech Inst New York, BS, 63, MS, 64, PhD(astronaut), 68. *Prof Exp:* Res assoc fluid dynamics, Polytech Inst New York, 65-68; prin engr acoust, Elec Boat Div Gen Dynamics, 68-79; PROF AERO ENG & ENG MECH, SAN DIEGO STATE UNIV, 79- *Concurrent Pos:* Adj prof, Univ Conn, 70-72 & Univ New Haven, 74-75; consult, Bolt Beranek & Newman, 80, Gen Atomic

Corp, 80-81, DATA Inc, 87-88 & Rohr, 88, GD EB div, 90; mem tech coun, Acoust Soc Am, 81-85, chmn eng acoust comn, 81-85, chmn tutorial comt, 88-91, tech chmn, 120th meeting, 90, chmn, book comt, 90- *Mem:* Sigma Xi; Am Acad Mech; assoc fel Am Inst Aeronaut & Astronaut; fel Acoust Soc Am. *Res:* Fluid structure interaction; acoustics and flow fields over compliant surfaces. *Mailing Add:* 2701 Summit Dr Escondido CA 92025

PIERUCCI, OLGA, b Crotone, Italy, Apr 17, 26. MOLECULAR BIOLOGY, PHYSICS. *Educ:* Univ Padua, Dipl, 43, DSc(physics), 48. *Prof Exp:* Fel physics, Univ Padua, 48-52; res asst, Indust Complex, Rome, 52-56; UNESCO vis prof, Gadjah Mada Univ, Jogjakarta, 56-58; Fulbright scholar; asst prof, Univ Buffalo, 59-60; radiation physicist, Grad Sch, Roswell Park Mem Inst, 60-63, sr cancer res scientist, 63-69, assoc cancer res scientist, 69-73, ASSOC RES PROF BIOL & CANCER RES SCIENTIST V, GRAD SCH, ROSWELL PARK MEM INST, 73- *Mem:* AAAS; Radiation Res Soc; Am Soc Microbiol; NY Acad Sci. *Res:* Control mechanisms in bacterial duplication; role of envelope in bacterial duplication. *Mailing Add:* Dept Cell-Molecular Biol Roswell Park Mem Inst, 666 Elm St Buffalo NY 14263

PIESCO, NICHOLAS PETER, b Havre DeGrace, Md, Sept 15, 46; m 70; c 2. TISSUE CULTURE, ULTRASTRUCTURE. *Educ:* Univ SFla, BA, 69, MA, 72; Univ Fla, MSA, 75, PhD(poultry sci & anat), 79. *Prof Exp:* Res fel, Sch Dent, Univ Conn, 79-81; ASST PROF ANAT & HISTOL, SCH DENT, UNIV PITTSBURGH, 81- *Mem:* Tissue Cult Asn; Sigma Xi. *Res:* Developing dentition; evolution, embryology, ultrastructure and histochemistry of dental hard tissues. *Mailing Add:* Dept Anat & Histol Sch Dent Med Univ Pittsburgh 345 Salk Hall Pittsburgh PA 15261

PIESKI, EDWIN THOMAS, b Dickson City, Pa, May 23, 24; m 60. POLYMER CHEMISTRY, PHYSICAL CHEMISTRY. *Educ:* Lehigh Univ, BS, 45, MS, 46, PhD(chem), 49. *Prof Exp:* Res chemist, Polychem Dept, 49-66, sr res chemist, 66-87, SR SCIENTIST, DUPONT POLYMERS, E I DU PONT DE NEMOURS & CO, 87- *Mem:* AAAS; Am Chem Soc; Sigma Xi; Soc Plastics Eng. *Res:* Physical chemistry of polymers; polyethylene; ethylene copolymers; polyvinyl alcohol. *Mailing Add:* 2301 B Inglewood Rd - Fairfax Wilmington DE 19803

PIETERS, CARLE M, b Ft Sill, Okla, Nov 11, 43; wid. PLANETARY SCIENCES. *Educ:* Antioch Col, BA, 66; Mass Inst Technol, BS, 71, MS, 72, PhD, 77. *Prof Exp:* Teacher math, Somerville High Sch, Mass, 66-67; teacher sci, Peace Corps, Sarawak, Malaysia, 67-69; staff scientist res, Planetary Astron Lab, Dept Earth & Planetary Sci, Mass Inst Technol, 72-75; space scientist, Johnson Space Ctr, NASA, 77-80; asst prof, 80-83, ASSOC PROF, DEPT GEOL SCI, BROWN UNIV, 83- *Mem:* Am Geophys Union (secy, 86-88); Am Astron Soc; AAAS; Meteoritical Soc. *Res:* Remote sensing of surface composition; planetary exploration; spectral reflectance properties of rocks and minerals. *Mailing Add:* Dept Geol Sci Brown Univ Providence RI 02912

PIETRA, GIUSEPPE G, b Piacenza, Italy, Dec 30, 30; m 57; c 2. PATHOLOGY. *Educ:* Univ Milan, MD, 55. *Prof Exp:* Resident path, Mass Gen Hosp, 60-62; asst, Univ Zurich, 62-63, prosector, 63-64; asst prof oncol, Chicago Med Sch, 64-65; asst pathologist, Michael Reese Hosp, Chicago, 65-69; from asst prof to assoc prof, 69-77, PROF PATH, MED SCH, UNIV PA, 77- *Concurrent Pos:* Res fel oncol, Chicago Med Sch, 57-60; clin asst prof, Univ Ill Col Med, 66-69; res assoc, Cardiovasc Inst, Michael Reese Hosp, Chicago, 68-69. *Mem:* Int Acad Pathologists; Am Thoracic Soc; Am Asn Pathologists. *Res:* Electron microscopy; developmental pathology; capillary permeability; cardiopulmonary pathology. *Mailing Add:* Dept Path Univ Pa Hosp Philadelphia PA 19104

PIETRAFACE, WILLIAM JOHN, b Scranton, Pa, Mar 23, 49; m 78. PLANT TISSUE CULTURE, BIOTECHNOLOGY. *Educ:* Pa State Univ, BS, 71; East Stroudsburg State Col, MS, 73; WVa Univ, PhD(biol), 79. *Prof Exp:* Grad asst biol, East Stroudsburg State Col, 71-73, instr biol & chem, 74-75, asst prof chem, 76; instr environ & man, Pa State Univ, Scranton, 75; from asst prof to assoc prof, 79-89, PROF BIOL, STATE UNIV NY, ONEONTA, 89- *Concurrent Pos:* Vis assoc prof, Cornell Univ, 86; adj fac mem, State Univ, Plattsburgh, 83- *Mem:* Am Soc Plant Physiologists; Sigma Xi; Int Plant Growth Substances Asn; Int Asn Plant Tissue Cult; Int Soc Plant Molecular Biol; AAAS. *Res:* Plant tissue culture methodologies; metabolism of the cytokinins during lettuce seed germination. *Mailing Add:* Biol Dept State Univ NY Oneonta NY 13820-4015

PIETRI, CHARLES EDWARD, b New York, NY, July 6, 30; c 3. ANALYTICAL CHEMISTRY, NUCLEAR CHEMISTRY. *Educ:* NY Univ, BA, 51. *Prof Exp:* Chemist, Oak Ridge Nat Lab, E I du Pont de Nemours & Co, 51-53 & Savannah River Lab, 53-56; res chemist, Curtiss-Wright Corp, 56-58; chief, Plutonium Chem Sect, USAEC, US Energy Res Develop Admin, 58-70, chief, Anal Chem Br, 70-75, asst dir oper, 75-77, asst dir oper, 77-83, sr scientist, 83-85, SCI ADMINR, US DEPT ENERGY, 85- *Concurrent Pos:* Dir, Planter Assocs (Consults), Chicago, Ill, 80-85; dir, Technol Assocs (Consults), 85- *Mem:* Am Chem Soc; fel Am Inst Chem; Am Nuclear Soc; Health Physics Soc; Int Standards Orgn; Inst Nuclear Mat Mgt. *Res:* Chemistry of plutonium; chemical and radiochemical standards; quality assurance in research and development; high radiation level chemical and radiochemical analyses; chemical separations; health physics; analytical methods development and evaluation; uranium chemistry; human resource management; computer information systems and technology transfer. *Mailing Add:* 5506 Grand Ave Western Springs IL 60558

PIETRO, WILLIAM JOSEPH, b Jersey City, NJ, Apr 23, 56. MOLECULAR ELECTRONICS. *Educ:* Polytech Inst Brooklyn, BS, 78; Univ Calif, Irvine, PhD(chem), 82. *Prof Exp:* ASST PROF INORG CHEM, UNIV WIS-MADISON, 85- *Mem:* Am Chem Soc; Am Phys Soc; Sigma Xi; AAAS. *Res:* Inorganic and physical chemistry; molecular electronic devices; polymeric electronic conductors; molecular machines. *Mailing Add:* 1158 Gammon Lane Madison WI 53719

PIETRUSEWSKY, MICHAEL, JR, b Boonville, NY, May 18, 44. PHYSICAL ANTHROPOLOGY. *Educ:* State Univ NY Buffalo, BA, 66; Univ Toronto, MA, 67, PhD(anthrop), 69. *Prof Exp:* From asst prof to assoc prof, 69-79, PROF ANTHROP, UNIV HAWAII, MANOA, 79-; RES ASSOC ANTHROP, B P BISHOP MUS, HAWAII, 70-; CONSULT FORENSIC ANTHROPOLOGY, 70- *Concurrent Pos:* Prin investr, Univ Hawaii res grants, 69-70, 71-72 & 74-79; B P Bishop Mus res grants, 69-70 & 71-72; Wenner-Gren Found grants, Bankok, Thailand, 70, Papua-New Guinea, 70-71, NZ, 84, 87 & Vietnam, 84; spec consult, Ford Found, 74, Nat Endowment Humanities, 81 & med examr off, City & County Honolulu, 87-; Nat Ctr Sci Res grant, Paris, 75; grants, German Acad Exchange Serv, 75 & 83, Australian Inst Aboriginal Studies, 76 & 79 & NSF, 76, contracts osteological res, 80-, res Vietnam, Nat Geog Soc, 85-86 & study Lapita remains, Univ Hawaii, 87; vis prof, Univ Toronto, 79, 80-81 & 82 & Col France, 83; vis cur, Australian Mus, Sydney, 83; vis scholar, Comt Scholarly Commun with People's Repub China, 88. *Mem:* Am Asn Phys Anthrop; Indo-Pac Prehist Asn; Am Acad Forensic Sci; Am Bd Forensic Anthrop. *Res:* Human biology of skeletal populations from the Pacific, Australia, Southeast Asia and East Asia, including paleodemography, skeletal and dental variation, paleopathology, and the application of multivariate procedures to cranial variation; forensic anthropology. *Mailing Add:* Dept of Anthrop Univ of Hawaii 2424 Maile Way Honolulu HI 96822

PIETRUSZA, EDWARD WALTER, organic chemistry, polymer chemistry; deceased, see previous edition for last biography

PIETRUSZKO, REGINA, b Hulicze, Poland, Feb 15, 29; UK citizen. BIOCHEMISTRY. *Educ:* Univ London, BS, 54, MS, 56, PhD(plant chem), 60. *Prof Exp:* Res asst lipid chem, Lister Inst Prev Med, London, 60-62; lectr enzymol, Royal Free Hosp Sch Med, London, 62-65; staff scientist biochem, Worcester Found Exp Biol, Mass, 65-67; docent, Nobel Inst, Karolinska Inst, Sweden, 67-70; asst prof, 70-74, assoc prof, 74-80, PROF BIOCHEM, RUTGERS UNIV, NEW BRUNSWICK, 80- *Concurrent Pos:* Res scientist develop award, Nat Inst Alcohol Abuse & Alcoholism, 78-89, res scientist award, 89-94. *Mem:* Am Soc Biol Chemists; Res Soc Alcoholism; Brit Biochem Soc; Am Chem Soc. *Res:* Structure and function relationships of isoenzymes of human aldehyde dehydrogenases; purification and characterization of isoenzymes. *Mailing Add:* Ctr of Alcohol Studies Rutgers Univ New Brunswick NJ 08903

PIETRZAK, LAWRENCE MICHAEL, b Hamtramck, Mich, June 27, 42; m 65; c 3. COMPUTER SIMULATION OF ENGINEERING & OPERATIONAL SYSTEMS, EXPERT SYSTEMS. *Educ:* Univ Detroit, BS, 65; Mass Inst Technol, MS, 66. *Prof Exp:* Res engr, Gen Res Corp, 67-74; DIR, PROTECTION TECHNOL SYSTS GROUP, MISSION RES CORP, 74- *Concurrent Pos:* NSF fel, 65-67; Am Inst Steel Construct fel, 66-67; Mem, Systs Synthesis panel, Advan Ballistic Defense Agency, 69 & 70, comt fire modeling, US Nat Inst Sci & Technol, 78-, comt fire suppression equip, Nat Fire Protection Asn, 79-83. *Mem:* AAAS; Am Soc Civil Engrs; Am Soc Indust Security. *Res:* Simulation modeling; systems analysis; operations research; computer aided analysis; design and expert systems techniques; physical security and fire safety systems; author of over 70 technical publications. *Mailing Add:* 6541 Camino Caseta Goleta CA 93117

PIETSCH, PAUL ANDREW, b New York, NY, Aug 8, 29; m 50; c 4. ANATOMY, REGENERATION. *Educ:* Syracuse Univ, AB, 54; Univ Pa, PhD(anat), 60. *Prof Exp:* Instr physiol, Sch Nursing, Univ Pa, 58; instr anat, Bowman Gray Sch Med, 59-61; asst prof, State Univ NY Buffalo, 61-64; sr res molecular biologist biochem, Dow Chem Co, 64-70; assoc prof anat, 70-78, chmn dept basic health sci, 77-82, PROF ANAT, SCH OPTOM, IND UNIV, BLOOMINGTON, 78-, ADJ PROF ANAT, SCH MED, MED SCI PROG, 79- *Honors & Awards:* Med J Award, AMA, 72. *Mem:* Am Asn Anatomists. *Res:* Regeneration; muscle differentiation; replication; biology of memory. *Mailing Add:* Sch of Optom Ind Univ Bloomington IN 47405

PIETSCH, THEODORE WELLS, b Royal Oak, Mich, Mar 6, 45; m 67; c 2. ICHTHYOLOGY, BIOSYSTEMATICS. *Educ:* Univ Mich, BA, 67; Univ Southern Calif, MS, 69, PhD(biol), 73. *Prof Exp:* Vis asst prof, Univ Southern Calif, 73; fel res, Mus Comparative Zool, Harvard Univ, 73-75; lectr, Dept Biol, Calif State Univ, Long Beach, 75-76, asst prof, 76-78; from asst prof to assoc prof, 75-76, 81-84, PROF, SCH FISHERIES, 84-, ADJ PROF DEPT ZOOL, UNIV WASH, 86- *Concurrent Pos:* Res assoc ichthyol, Natural Hist Mus Los Angeles County, 73-, Mus Comparative Zool, Harvard Univ, 75-; tutor biol, Harvard Univ, 74-75; ed, Bull Southern Calif Acad Sci, 76-78, Ichthy ed, Copeia, 86-; fel, Calif Acad Sci; fel, Linn Soc London; fel, Gilbert Soc Ichthy. *Mem:* Am Soc Ichthyologists & Herpetologists; AAAS; Brit Soc Hist Sci; Ichthy Soc Japan; Asn Syst Col; Soc Hist Nat Hist. *Res:* Biosystematics, geographic distribution, and the behavior and functional morphology of feeding in marine fishes, particularly euteleostean fishes. *Mailing Add:* Sch Fisheries Univ Wash Seattle WA 98195

PIETTE, LAWRENCE HECTOR, b Chicago, Ill, Jan 4, 32; m 57. BIOPHYSICS. *Educ:* Northwestern Univ, BS, 53, MS, 54; Stanford Univ, PhD, 57. *Prof Exp:* Res chemist, Varian Assocs, 56-65; prof biophys, Univ Hawaii, 65-84; prof biochem, 84-, DEAN GRAD SCH, 84-, ASSOC VPRES RES, UTAH STATE UNIV, 84- *Concurrent Pos:* Guggenheim fel, Inst Biophys & Biochem, Paris & Grenoble Nuclear Res Ctr, Grenoble, France, 71-72; consult, Varian Assocs & NAm Aviation, Inc; Exec dir, Cancer Ctr Hawaii, 74-84. *Mem:* Am Chem Soc; NY Acad Sci; Biophys Soc; Am Asn Univ Prof. *Res:* Application of nuclear magnetic resonance and electron paramagnetic resonance to the study of chemical kinetics, photochemistry and rapid biological reactions; studies of chemical carcinogenesis. *Mailing Add:* Utah State Univ Logan UT 84322

PIEZ, KARL ANTON, b Newton, Mass, Aug 30, 24; m 48; c 3. MOLECULAR BIOLOGY. *Educ:* Yale Univ, BS, 47; Northwestern Univ, PhD(biochem), 52. *Prof Exp:* Biochemist, Nat Inst Dent Res, 52-61, chief sect protein chem, 61-66, chief lab biochem, 66-82; dir res, Collagen Corp, 82-91,

CHIEF SCI ADV, CELTRIX LABS, 91- *Concurrent Pos:* Biotechnol res & develop. *Honors & Awards:* T Duckett Jones Mem Award, Helen Hay Whitney Found, 70. *Mem:* Am Soc Biochem & Molecular Biol; AAAS; Am Chem Soc. *Res:* Protein chemistry; biochemistry of connective tissues. *Mailing Add:* Celtrix Labs 2500 Faber Pl Palo Alto CA 94303

PIFKO, ALLAN BERT, b Bronx, NY, Dec 12, 38; m 60; c 1. APPLIED MECHANICS. *Educ:* NY Univ, BAE, 60, MAE, 61; Polytech Inst Brooklyn, PhD(appl mech), 74. *Prof Exp:* Res engr, 61-74, res scientist, 74-79, staff scientist, 79-81, SR STAFF SCIENTIST, GRUMMAN CORP TECHNOL CTR, 81- *Mem:* Am Soc Mech Engrs; Sigma Xi. *Res:* Applied mechanics; computer software and numerical techniques for nonlinear analysis of complex structures; author or coauthor of more than 20 publications. *Mailing Add:* Two George Ct Huntington NY 11746

PIGAGE, LEO C(HARLES), b Rochester, NY, Nov 22, 13. INDUSTRIAL ENGINEERING. *Educ:* Cornell Univ, ME, 36, MME, 38. *Prof Exp:* Instr, Duke Univ, 38-41; asst prof, Purdue Univ, 41-47; assoc prof mech eng, 47-52, EMER PROF INDUST ENG, UNIV ILL, URBANA, 52- *Mem:* Am Soc Mech Engrs; Am Soc Eng Educ; Am Inst Indust Engrs. *Res:* Industrial engineering, organization and management. *Mailing Add:* 206 Elmwood Dr Champaign IL 61821

PIGDEN, WALLACE JAMES, b Madoc, Ont, Jan 30, 20; m 49; c 3. RESEARCH MANAGEMENT. *Educ:* Ont Agr Col, Toronto, BSA, 48; Univ Alta, MSc, 50; Univ Sask, PhD(nutrit), 55. *Prof Exp:* Res officer, Exp Sta, Can Dept Agr, Sask, 50-53, Animal Husb Div, Exp Farms Serv, Ont, 53-59, chief, Nutrit Sect, Animal Res Inst, Res Br, 60-67, res coordr, Animal Sci Res Br, Ont, 67-78; PRES, FAW CONSULTS LTD, 79- *Concurrent Pos:* Secy, Nat Comt Animal Nutrit, Can, 58-64; Can Dept Agr fel, Nat Inst Res Dairying, Reading, Eng, 63-64; consult indust res assistance grant comt, Nat Res Coun Can, 63-66; res coordr, planning & evaluation, Agr Can, adv & consult, Can Govt res incentive & contract prog & Food & Agr Orgn United Nations, Can Int Develop Agency & Res Ctr, 67-78; pvt consult, 78-82. *Honors & Awards:* Borden Award, Nutrit Soc Can, 64; Cert Merit, Can Soc Animal Sci, 79; Queens Jubilee Medal, 77. *Mem:* Am Soc Animal Sci; Nutrit Soc Can; Agr Inst Can; Can Soc Animal Sci; Can Consult Agr Asn. *Res:* Forage utilization; methods of measuring herbage intake and animal production on pasture; in vivo, in vitro and chemical methods of evaluating forages; physical and chemical methods of increasing available energy content of forages. *Mailing Add:* 850 Norton Ave Ottawa ON K2B 5P6 Can

PIGFORD, THOMAS H(ARRINGTON), b Meridian, Miss, Apr 21, 22; m 48; c 2. CHEMICAL & NUCLEAR ENGINEERING. *Educ:* Ga Inst Technol, BS, 43; Mass Inst Technol, SM, 48, ScD(chem eng), 52. *Prof Exp:* Instr chem eng, Mass Inst Technol, 46-47, asst prof & dir sch eng pract, Oak Ridge, Tenn, 50-52, from asst prof to assoc prof nuclear & chem eng, 52-57; sr develop engr, Carbide & Carbon Chem Co, 52; chmn eng & asst dir lab, Gen Atomic Div, Gen Dynamics Corp, 57-59; chmn dept, 59-64, 74-79, 84-88, PROF NUCLEAR ENG, UNIV CALIF, BERKELEY, 59- *Concurrent Pos:* Consult, govt & various industs, 53-; mem, Nat Panel Atomic Safety & Licensing Bds; consult, Union Carbide & Carbon, 67- & US Geol Surv, 68-; vis prof, Kyoto Univ, Japan, 75 & Kuwait Univ, 76; mem, Am Phys Soc Study Group on Nuclear Fuel Cycles & Waste Mgt, 76-77; mem, Presidents Comn Accident of Three Mile Island, 79; mem bd radioactive mgt, Nat Acad Sci & Eng, 78-; mem lab adv bd, Oakridge Nat Lab; chmn adv bd, Inst Nuclear Power Opers; chmn, waste isolation syst panel, Nat Res Coun, 80-83; mem, performance assessment, Nat Rev Group, US Dept Energy, 84-85; chmn, Waste Isolation Systs Panel, Nat Res Coun, 80-83 & Safety Adv Comt, Gulf States Utilities Co, 80-; mem, Secy Energy's Expert Consults on the Chernobyl Accident, 86. *Honors & Awards:* Arthur H Compton Award, Am Nuclear Soc, 71; Robert E Wilson Award, Am Inst Chem Engrs, 80 & Serv Soc Award, 85; John Wesley Powell Award, US Geol Surv, 81. *Mem:* Nat Acad Eng; AAAS; fel Am Nuclear Soc; Am Inst Chem Engrs; Am Inst Mech Engrs. *Res:* Design analysis of nuclear reactors; nuclear power economics; safety analysis of nuclear reactors; environmental transport of radionuclides; environmental effects of electric power production; radioactive waste management; nuclear fuel cycles. *Mailing Add:* 1 Garden Dr Kensington CA 94708

PIGGOTT, MICHAEL R(ANTELL), b Cheadle Huime, Eng, Feb 13, 30; m 55; c 3. HIGH PERFORMANCE FIBRE COMPOSITES. *Educ:* Univ London, BSc, 51, Imp Col, MSC & dipl, 53, PhD(appl phys chem), 55. *Prof Exp:* Mem sci staff, Gen Elec Co, Eng, 55-56, proj engr, 56-58; res physicist, Bexford Ltd, 58-60 & Gillette Res Labs, 60-64; assoc res officer, Atomic Energy Can, Ltd, Ont, 64-68; assoc prof, 68-74, PROF CHEM ENG, UNIV TORONTO, 74- *Honors & Awards:* Delmonte Award, Soc Advan Mat & Process Eng. *Mem:* Soc Advan Mat & Process Eng; Asn Prof Engrs Ont; Am Soc Testing & Mat. *Res:* Surface physical and chemical interactions; mechanical properties of materials, with special emphasis on load bearing fibre composites. *Mailing Add:* Dept Chem Eng Univ Toronto Toronto ON M5S 1A4 Can

PIGIET, VINCENT P, b Yankers, NY, Nov 24, 43. BIOTECHNOLOGY DRUG DEVELOPMENT & MANUFACTURE. *Educ:* Syracuse Univ, BS, 65; Univ Calif, Berkeley, MS, 67, PhD(biochem), 71. *Prof Exp:* From asst prof to assoc prof biochem, Dept Biol, Johns Hopkins Univ, 73-83; assoc dir protein eng, Repligen Corp, 83-87; mgr protein develop, Ortho Pharmaceut/Biotechnol, NJ, 87-90; DIR TECH AFFAIRS, CHIRON CORP, EMERYVILLE, CALIF, 90- *Concurrent Pos:* NSF fel, Univ Calif, Berkeley, 67-71; Jane Coffin Childs fel, Karolinska Inst, Stockholm, 71-73; adj prof, Johns Hopkins Univ, 83-88; mem, Biotechnol Comt, Pharmaceut Mfrs Asn. *Mem:* Am Chem Soc; Pharmaceut Mfrs Asn; Am Soc Qual Control; Pharmaceut Drug Asn. *Res:* Direction and management of development of biotechnology drugs and oversight of the quality control of products; protein structure and function; imammalian cell growth. *Mailing Add:* Chiron Corp 4560 Horton St Emeryville CA 94611

PIGNATARO, AUGUSTUS, b Bronx, NY, Aug 12, 43; m 65; c 3. PHYSICS. *Educ:* Calif State Univ, Los Angeles, BS, 65; Calif Lutheran Col, MBA, 75. *Prof Exp:* Engr, Atlantic Res Corp, 65-67; SR PHYSICIST, NAVY PAC MISSILE TEST CTR, CALIF, 67- *Mem:* Naval Aviation Exec Inst. *Res:* Development, testing and evaluation of electro-optical systems for naval airborne applications; nonimaging infrared and ultraviolet sensors. *Mailing Add:* Pac Missile Test Ctr Code 4033 Point Mugu CA 93042-5000

PIGNATARO, LOUIS J(AMES), b Brooklyn, NY, Nov 30, 23; m 54; c 1. TRANSPORTATION ENGINEERING, PLANNING. *Educ:* Polytech Inst Brooklyn, BCE, 51; Columbia Univ, MS, 54; Graz Tech Univ, Dr Techn Sc, 61. *Prof Exp:* Fac, Polytech Inst New York, 51-85, prof civil eng, 67-70, dir Div transp planning, 70-82, head Dept transp, Planning & Eng, 75-82, dir, Transp Trainning & Res Ctr, 85-87; Kayser prof transp eng, assoc dir, Inst Transp Systs, City Col, 85-87; DISTINGUISHED PROF TRANSP ENG, DIR CTR TRANSP STUDIES & RES, NJ INST TECHNOL, 88- *Concurrent Pos:* Consult, var govt & pvt agencies. *Honors & Awards:* Distinguished Res Polytechnic Chpt, Sigma Xi, 75. *Mem:* Fel Am Soc Civil Engrs; Transp Res Forum; Nat Soc Prof Engrs; fel Inst Transp Engrs; Transp Res Bd. *Res:* Traffic engineering; transportation planning; transportation economics and policy. *Mailing Add:* Ctr Transp Studies & Res NJ Inst Technol Newark NJ 07102

PIGNOCCO, ARTHUR JOHN, b Jeanette, Pa, Apr 15, 29; m 59; c 2. PHYSICAL CHEMISTRY. *Educ:* Univ Pittsburgh, BS, 50; Carnegie Inst Technol, MS & PhD(chem), 54. *Prof Exp:* Technologist, US Steel Corp, 55-60, sr res technologist, 60-66, assoc res consult, 66-67, sect supvr, 68-80, res consult, 80-84; res mgr, Assoc Mat Inc, 85-88, consult, 88-90; RETIRED. *Mem:* Am Chem Soc. *Res:* Separations; mass spectroscopy; infrared and ultraviolet spectroscopy; auger and photoelectron spectroscopy; surface studies; low energy electron diffraction; steel cord-rubber interfaces, steel cord adhesion and corrosion characteristics. *Mailing Add:* 1030 Sourwood Ridge Greensboro GA 30642-4009

PIGNOLET, LOUIS H, b Orange, NJ, Mar 24, 43; m 69. METAL CLUSTERS, CATALYSIS. *Educ:* Lafayette Col, BS, 65; Princeton Univ, PhD(chem), 69. *Prof Exp:* Res assoc chem, Mass Inst Technol, 69-70; asst prof, 70-74, assoc prof, 74-81, PROF CHEM, UNIV MINN, MINNEAPOLIS, 81- *Concurrent Pos:* Res Corp grant, 71; NSF res grant, 73-84; consult, 3M Co. *Mem:* Am Chem Soc. *Res:* Synthesis of inorganic and organometallic compounds; metal cluster compounds of several metals; catalysis; design of alloy materials; x-ray crystallography; chemistry of gold and gold-transition metal clusters; metal hydrides. *Mailing Add:* Univ Minn Minneapolis 139 Smith Hall 207 Pleasant St SE Univ Minn Minneapolis MN 55455

PIGOTT, GEORGE M, b Vancouver, Wash, Oct 25, 28; m 48, 81; c 5. FOOD SCIENCE & FOOD ENGINEERING. *Educ:* Univ Wash, Seattle, BS, 50, MS, 53, PhD(chem & food sci), 62. *Prof Exp:* Chem engr, US Fish & Wildlife Serv, 48-51; field engr, Continental Can Co, 53-55; food engr, Nat Canners Asn, 55-57, consult eng, 57-63; lectr food sci, 63-65, from asst prof to assoc prof food eng, 65-72, PROF FOOD ENGR, UNIV WASH, 72-, DIR, INST FOOD SCI & TECHNOL, 89- *Concurrent Pos:* Consult food engr, Sea Resources Eng, Inc, 57- *Mem:* Inst Food Technol; Am Chem Soc; Am Soc Agr Engrs; Am Inst Chem Eng; Nat Soc Prof Eng; Am Inst Nutrit. *Res:* Application of scientific and engineering principles to processing of food products, especially sea foods. *Mailing Add:* Inst Food Sci & Technol HF-10 Univ Wash Seattle WA 98195

PIGOTT, MILES THOMAS, b Springfield, Ohio, Mar 23, 23; m 45; c 7. SPACE DUST. *Educ:* Miami Univ, AB, 47; Univ Ill, MA, 48; Pa State Univ, PhD(physics), 55. *Prof Exp:* Res assoc, Pa State Univ, 51-55, from asst prof to assoc prof, 55-65, head, Acoust Unit, 70-74, prof eng res, ord res lab, 65-, asst dir, appl res lab, 67-; AT SCI & MATH DEPT, PARKS COL. *Concurrent Pos:* Asst dir acad affairs, Pa State Univ, 74. *Mem:* Am Phys Soc; Sigma Xi. *Res:* General physics; space dust detection. *Mailing Add:* 1510 Doris Ave Cahokia IL 62206

PIGOZZI, DON, b Oakland, Calif, June 29, 35. MATHEMATICS. *Educ:* Univ Calif, BA, 58, PhD(math), 70. *Prof Exp:* Asst prof math, Indiana Univ, 69-71; PROF MATH, IOWA STATE UNIV, 71- *Mailing Add:* Dept Math Iowa State Univ Ames IA 50011

PIH, HUI, b Kunming, China, Feb 14, 22; US citizen; m 54; c 4. MECHANICAL ENGINEERING. *Educ:* Nat Inst Technol, China, BS, 45; Stanford Univ, MS, 49; Ill Inst Technol, PhD(mech), 53. *Prof Exp:* Supt, 53rd Arsenal, Kunming, China, 45-48; res asst, Ill Inst Technol, 51-53; proj engr, Res & Develop Lab, Int Harvester Co, 53-56; assoc engr, Int Bus Mach Corp, 56-57, staff engr, 57-59; assoc prof theoret & appl mech, Marquette Univ, 59-65; PROF ENG MECH, UNIV TENN, KNOXVILLE, 65- *Concurrent Pos:* Consult, Allis-Chalmers Mfg Corp, 61-62, A O Smith Corp, 62-65, Oak Ridge Nat Lab, 67-77, ACF Industs Corp, 71-80 & Oak Ridge Nat Lab, 87- *Honors & Awards:* M Hetényi Award, Soc Exp Stress Anal, 75. *Mem:* AAAS; Am Soc Mech Engrs; Am Acad Mech; Am Soc Eng Educ; Soc Mfg Engrs; Sigma Xi; Soc Exp Mech. *Res:* Solid mechanics; experimental stress analysis; photoelasticity; fracture mechanics. *Mailing Add:* Dept Eng Sci & Mech Univ Tenn Knoxville TN 37996-2030

PIHL, ROBERT O, b Milwaukee, Wis, Feb 2, 39; m 75; c 2. DRUG ABUSE. *Educ:* Lawrence Univ, BA, 61; Ariz State, MA, 65, PhD(psychol), 66. *Prof Exp:* assoc prof, 71-78, PROF PSYCHOL, MCGILL UNIV, 79-, PROF PSYCHIAT, 83- *Concurrent Pos:* Mem, Rev Comt, Health & Welfare Can, 74-78, & Prog Eval Comt, Med Res Coun Can, 75-; sci officer, Res Drug Abuse Prog, Med Res Coun, Health & Welfare, Can, 75-78; consult, Maritime Provinces Higher Educ Comt, 80 & Res Inst, Royal Victorian Hosp, 86; pres, Pihl-Ervin Inc, 81-; co-dir, Alcohol Studies, Douglas Hosp, Mc Gill Univ, 85- *Mem:* Fel Behav Sci Found; fel Can Psychol Asn; fel Am Psychol Asn. *Res:* Psychopharmacology; psychopathology; behavioral modification and history;

drugs, elements and psychological factors and aggressions; theory and direction of chemical psychology; identification and treatment of learning disabilities; alcoholism and the response to alcohol. *Mailing Add:* 225 Bedbrook Ave Montreal West PQ H4X 1S2 Can

PIIRMA, IRJA, b Tallinn, Estonia, Feb 4, 20; US citizen; m 43; c 2. POLYMER CHEMISTRY. *Educ:* Darmstadt Tech Univ, Dipl, 49; Univ Akron, MS, 57, PhD(polymer chem), 60. *Prof Exp:* Res chemist, Inst Polymer Sci, 52-60, from instr to asst prof chem, 63-76, assoc prof, 76-81, PROF POLYMER SCI, UNIV AKRON, 81-, RES ASSOC POLYMER CHEM, INST POLYMER SCI, 60- *Mem:* Am Chem Soc. *Res:* Polymerization kinetics in emulsion; properties of polymer colloids. *Mailing Add:* Inst of Polymer Sci Univ of Akron 302 E Buchtel Ave Akron OH 44325-3909

PIKAL, MICHAEL JON, b Henning, Minn, Aug 17, 39; m 63; c 5. PHYSICAL CHEMISTRY, PHARMACEUTICS. *Educ:* St John's Univ, Minn, BA, 61; Iowa State Univ, PhD(phys chem), 66. *Prof Exp:* Fel phys chem, Lawrence Radiation Lab, Univ Calif, 65-67; asst prof chem, Univ Tenn, Knoxville, 67-72; sr pharmaceut chemist, 72-76, res scientist, 76-81, SR RES SCIENTIST, ELI LILLY & CO, 81- *Concurrent Pos:* Adj prof, Dept Pharmaceut, Univ Mich & Univ Minn. *Honors & Awards:* Ebert Prize, Am Pharmaceut Asn, 77. *Mem:* Fel Am Asn Pharmaceut Scientists; Am Chem Soc. *Res:* Thermodynamic properties of pharmaceuticals, characterization of amorphous materials, decomposition of solids; freeze drying; iontophoresis. *Mailing Add:* Lilly Res Labs Eli Lilly & Co Lilly Corp Ctr Indianapolis IN 46285

PIKARSKY, MILTON, transportation; deceased, see previous edition for last biography

PIKE, ARTHUR CLAUSEN, meteorology; deceased, see previous edition for last biography

PIKE, CARL STEPHEN, b New York, NY, Apr 13, 45; m 69; c 2. PLANT PHYSIOLOGY. *Educ:* Yale Univ, BS, 66, MPhil, 67; Harvard Univ, PhD(biol), 72. *Prof Exp:* Chmn dept, 86-90, from asst prof to assoc prof, 71-86, PROF BIOL, FRANKLIN & MARSHALL COL, 86- *Concurrent Pos:* NSF Sci Fac Prof Develop Grant, 78-79; sr fel, Dept Plant Biol, Carnegie Inst Wash, 78-79; vis scholar, Dept Bot, Univ Texas-Austin, 85-86. *Mem:* AAAS; Am Soc Plant Physiol; Japanese Soc Plant Physiol. *Res:* Plant photomorphogenesis; control of plant function and development by phytochrome; role of membrane lipids in plant responses to environmental stresses; plant protein kinases. *Mailing Add:* Dept Biol Franklin & Marshall Col Box 3003 Lancaster PA 17604

PIKE, CHARLES P, b Feb 21, 41; US citizen; m 69; c 2. GEOPHYSICS. *Educ:* Boston Col, BS, 63; Northeastern Univ, MS, 69. *Prof Exp:* PHYSICIST SPACE PHYSICS, AIR FORCE GEOPHYSICS LAB, 67- *Mem:* Am Inst Aeronaut & Astronaut. *Res:* Ionospheric physics; space physics. *Mailing Add:* Air Force Geophysics Lab Space Physics Div Hanscom AFB MA 01731

PIKE, EILEEN HALSEY, b London, Eng, Apr 18, 18; US citizen. MEDICAL PARASITOLOGY, MEDICAL MICROBIOLOGY. *Educ:* Hunter Col, BA, 56; Tulane Univ, MS, 58; Columbia Univ, PhD, 63. *Prof Exp:* Res asst parasitol, Med Sch, Tulane Univ, 58-59; lab instr, Col Physicians & Surgeons, Columbia Univ, 59-63; from asst prof to assoc prof, 64-76, prof, 76-81, EMER PROF PARASITOL, NY MED COL, 81- *Concurrent Pos:* Health Res Coun of City of New York grant, 65-68. NIH fel trop med, Caribbean Prog, La State Univ, 64; consult parasitologist, Hackensack Hosp, 67-69. *Mem:* AAAS; Am Soc Parasitol; Am Soc Trop Med & Hyg; Royal Soc Trop Med & Hyg; Int Col Trop Med. *Res:* Host-parasite relationships; cellular and humoral responses of hosts to helminths and protozon. *Mailing Add:* Dept Microbiol & Immunol NY Med Col Sunshine Cottage Valhalla NY 10595-1690

PIKE, GORDON E, b Pittsburgh Pa, Jan 1, 42. ELECTRONIC MATERIALS. *Educ:* Carnegie-Mellon Univ, BS, 63; Univ Pittsburgh, PhD(physics), 69. *Prof Exp:* Mem staff, 69-85, SUPVR, ELECTRONIC PROPERTIES MATS DIV, SANDIA NAT LAB, 85- *Mem:* Mat Res Soc (pres, 86); Am Phys Soc; Am Ceramic Soc; AAAS. *Res:* Transport and microstructure relationships in composite materials; electronic properties of semiconductor grain boundaries; dielectric properties of insulators with defects; electrical properties of amorphous metal alloys; defect microstructures in silicon. *Mailing Add:* Div 1815 Sandia Nat Labs Albuquerque NM 87185

PIKE, J(OHN) G(IBSON), b Hamilton, Ont, Apr 12, 30; m 59; c 2. MECHANICAL ENGINEERING, FLUID DYNAMICS. *Educ:* Queen's Univ, Ont, BSc, 53; Univ Birmingham, MSc, 56; McGill Univ, PhD(mech eng), 63. *Prof Exp:* Lectr mech eng, Royal Mil Col, Ont, 53-54; engr, Bristol Aeroplane Co, 54-55; res asst fluid dynamics, McGill Univ, 56-57, 58-60, lectr mech eng, 57-58; from asst prof to assoc prof, 60-70, PROF MECH ENG, ROYAL MIL COL CAN, 70- *Concurrent Pos:* Can Defence Res Bd grant, 64- *Mem:* Brit Inst Mech Engrs. *Res:* Thin liquid film flow; two-dimensional incompressible flow. *Mailing Add:* 150 Macdonnell Kingston ON K7L 4B8 Can

PIKE, JOHN NAZARIAN, b Boston, Mass, Feb 13, 29; m 57; c 2. OPTICAL PHYSICS. *Educ:* Princeton Univ, AB, 51; Univ Rochester, PhD(physics, optics), 58. *Prof Exp:* physicist, Tarrytown Tech Ctr, Union Carbide Corp, 56-85; PRES, J J PIKE & CO, INC, INDUST OPTICS CONSULT, 85- *Mem:* Optical Soc Am; Soc Photo-optical Instrumentation Engrs. *Res:* Optical instrumentation; remote sensing methods; particulate scattering; radiative and conductive heat transfer in fibrous and porous materials; photosensitive materials systems; solar energy conversion; optical analytical methods; position sensing and metrology. *Mailing Add:* J J Pike & Co Inc PO Box 186 Pleasantville NY 10570-0186

PIKE, JULIAN M, b Eugene, Ore, Mar 14, 30; m 51; c 3. ATMOSPHERIC PHYSICS. *Educ:* Cascade Col, AB, 53; Oregon State Univ, MA, 56, PhD(physics), 58. *Prof Exp:* Res assoc meteorol, Ore State Univ, 57-58; prof physics, Ashbury Col, 58-66; scientist, 66-80, STAFF PHYSICIST, NAT CTR ATMOSPHERIC RES, 80- *Mem:* Am Meteorol Soc. *Res:* Meteorological instrumentation. *Mailing Add:* 1705 Upland Ave Boulder CO 80304

PIKE, KEITH SCHADE, b Ogden, Utah, Jan 20, 47; m 68; c 4. AGRICULTURAL ENTOMOLOGY. *Educ:* Utah State Univ, BS, 71; Univ Wyo, MS, 73, PhD(entom), 74. *Prof Exp:* Res assoc entom, Univ Nebr, 74-76; from asst entomologist to assoc entomologist, 76-89, ENTOMOLOGIST, WASH STATE UNIV, 89- *Concurrent Pos:* Asst entomologist, res grants, Wash Wheat Comn, 76-; Wash Mint Comn, 77-, Cent Wash Res Assoc, 77-, IR-4, USDA, 77- & var chem indust grants, 78- *Mem:* Entom Soc Am; Am Registry Prof Entomologists. *Res:* Resistance of plants to insects; insects affecting cereal grains, corn and mint; aphid barley yellow dwarf virus studies; Russian wheat-aphid; biological control; foreign exploration for insect pest natural enemies. *Mailing Add:* Irrigated Agr Res & Exten Ctr Wash State Univ Prosser WA 99350-9687

PIKE, LEROY, b Duncan, SC, Dec 25, 28; m 51; c 5. GAS TRANSMISSION COMPOSITION AND PROPERTIES OF PACKAGING FILMS, IRRIDATION DOSIMETRY AND EFFECT ON POLYMERS. *Educ:* Wofford Col, AB, 54; Clemson Univ, MS, 59; Univ Va, PhD(anal chem), 64. *Prof Exp:* Teacher high sch, SC, 54-56; lab asst, Clemson Univ, 55-58, assoc res chemist, 58-59; instr chem, Va Mil Inst, 59-61; grad asst, Univ Va, 61-63; asst prof, Appalachian State Univ, 63-65; sect head, Anal Serv, Cryovac Div, W R Grace & Co, 65-73, mgr, 73-90; CONSULT, 90- *Concurrent Pos:* Instr food packaging mat properties, Clemson Univ. *Mem:* Am Chem Soc; Am Soc Testing Mat. *Res:* Absorption and infrared spectroscopy, polymer properties, identification and characterization; analytical methods development; irradiation dosimetry; gas transmission rate measurement of polymeric packaging films. *Mailing Add:* 3021 Clark Rd Inman SC 29349-9764

PIKE, LINDA JOY, b Philadelphia, Pa, Jan 21, 53; m 81; c 2. GROWTH CONTROL, PROTEIN PHOSPHORYLATION. *Educ:* Univ Del, BS, 75; Duke Univ, PhD(biochem), 80. *Prof Exp:* Fel biochem, Univ Wash, 80-84; ASST PROF BIOL CHEM, WASH UNIV, 84- *Concurrent Pos:* Assoc investr, Howard Hughes Med Inst, 83- *Mem:* Am Soc Biochem & Molecular Biol; Am Soc Cell Biol. *Res:* Control of cell growth by mitogens, especially the role of protein phosphorylation and phosphatidyl inositol metabolism. *Mailing Add:* Dept Biol Chem Wash Univ Med Sch Box 8045 St Louis MO 63110

PIKE, LOY DEAN, b Howard Co, Tex, June 4, 40. MICROBIOLOGY. *Educ:* Univ Tex, BA, 67, MA, 70, PhD(microbiol), 73. *Prof Exp:* Lectr, Tex A&M Univ, 73-74; res assoc, Ind Univ, Bloomington, 74-76; ASST PROF MICROBIOL, IND UNIV, SOUTH BEND, 76- *Mem:* Am Soc Microbiol; Sigma Xi. *Res:* Control of glycerol catabolism in the purple nonsulfur bacteria. *Mailing Add:* Dept Biol Ind Univ South Bend 1700 Mishawaka Ave PO Box 7111 South Bend IN 46634

PIKE, MARILYN CECILE, CHEMOTAXIS. *Educ:* Duke Univ, MD, 84. *Prof Exp:* Resisent internal med, Sch Med, Harvard Univ, 84-88; ASST PROF, ARTHRITIS UNIT, MASS GEN HOSP, 88- *Mailing Add:* 2817 Laurel Hill Rd Ann Arbor MI 48103

PIKE, RALPH EDWIN, b Rochester, NY, May 26, 15; m 41; c 2. ORGANIC COATINGS, CORROSION. *Educ:* Univ Rochester, BS, 37, MS, 39. *Prof Exp:* Asst chem, Pa State Col, 37-38 & Univ Rochester, 38-39; chemist, E I du Pont de Nemours & Co, 39-42, chem engr, Wabash River Ord Works, 42-45, chemist, Pa, 45-48, res supvr, 48-52, sr develop supvr, Color Styling, Automotive Finishes & Develop, Data Processing, 52-64, from asst mgr to mgr, Marshall Develop Lab, Indust & Tech Serv, 64-69, mkt mgr wood finishes, Finishes Mkt Div, Fabrics & Finishes Dept, 69-70, mgr new prod develop & mkt, 70-76, res fel, 76-78; CONSULT, 78- *Mem:* Inter-Soc Color Coun (pres, 64-66); Am Chem Soc: Asn Finishing Processes (pres, 76-77); Soc Mfg Engrs. *Res:* Synthetic enamels; process engineering; automotive color styling; administration; color technology; organic finishes. *Mailing Add:* 280 Crum Creek Rd Media PA 19063

PIKE, RALPH W, b Tampa, Fla, Nov 10, 35; m 58; c 2. CHEMICAL ENGINEERING. *Educ:* Ga Inst Technol, BChE, 58, PhD(chem eng), 62. *Prof Exp:* Res engr, Swift & Co, Fla, 57-58 & Exxon Res & Eng Co, Tex, 62-64; from asst prof to assoc prof, 64-72, PROF CHEM ENG, LA STATE UNIV, BATON ROUGE, 72-, ASSOC VCHANCELLOR RES, 74-, DIR, LA MINERAL INST, 80- *Concurrent Pos:* Numerous res grants from fed, state and pvt orgn; vis scholar, Stanford Univ, 71. *Mem:* Fel AmInst Chem Engrs; Am Chem Soc; Am Soc Eng Educ; AAAS; Am Inst Aeronaut & Astronaut; Sigma Xi. *Res:* Reaction engineering; transport phenomena; systems engineering optimization; research management. *Mailing Add:* 6053 Hibiscus Dr Baton Rouge LA 70808

PIKE, RICHARD JOSEPH, b Nantucket, Mass, June 28, 37; m 86; c 2. TERRAIN ANALYSIS, GEOMORPHOLOGY. *Educ:* Tufts Univ, BS, 59; Clark Univ, MA, 63; Univ Mich, Ann Arbor, PhD(geol), 68. *Prof Exp:* Geographer, US Army Natick Labs, Mass, 62-63; geologist, US Geol Surv, 68-87; GEOLOGIST, BR W REG GEOL, 87- *Concurrent Pos:* Assoc ed, Proc Lunar & Planet Sci Conf, 78 & 80; prin investr, NASA, 78-84; vis lectr, Hydrol Hazards Summer Sch, Perugia Italy 88 - 89. *Honors & Awards:* Apollo 11 Medallion, Marshall Space Flight Ctr, NASA, 71. *Mem:* AAAS; Geol Soc Am; Am Geophys Union. *Res:* Computer mapping from digital data; quantitative physiography; numerical landform taxonomy; morphometry of impact craters and volcanoes; terrestrial and planetary geomorphology. *Mailing Add:* Stop 975 US Geol Surv 345 Middlefield Rd Menlo Park CA 94025

PIKE, ROBERT MERRETT, b Hiram, Maine, Apr 5, 06; m 32; c 3. BACTERIOLOGY. *Educ:* Brown Univ, AB, 28, MA, 30, PhD(bact), 32; Am Bd Med Microbiol, dipl. *Prof Exp:* Asst biol, Brown Univ, 28-29, demonstr bact, 29-31; bacteriologist, Otsego County Lab & M I Bassett Hosp, NY, 32-43; from asst prof to prof, 43-74, EMER PROF BACT, UNIV TEX SOUTHWESTERN MED SCH DALLAS, 74- *Mem:* Am Soc Microbiol; assoc Soc Exp Biol & Med; Am Asn Immunol; fel Am Acad Microbiol. *Res:* Phagocytosis; virulence and antigenic structure; streptococci; laboratory infections; rheumatoid arthritis; leptospirosis; antibody activity. *Mailing Add:* 5815 Elderwood Dr Dallas TX 75230

PIKE, RONALD MARSTON, b Calais, Maine, Aug 16, 25; m 52; c 2. ORGANIC CHEMISTRY. *Educ:* Univ NH, BS, 49, MS, 50; Mass Inst Technol, PhD(org chem), 53. *Prof Exp:* Res chemist, Linde Air Prod Co Div, Union Carbide Corp, 53-57; prof chem, Lowell Tech Inst, 57-65; PROF CHEM, MERRIMACK COL, 65- *Concurrent Pos:* Vis Charles Weston Pichard prof chem, Bowdoin Col, Brunswick, Maine, 80-81, 84; vis assoc, Comt Prof Training, Am Chem Soc, 86-; vis prof, US Mil Acad, West Point, NY, 90-91; screening comt, Stanley Kipping Award, 78-81, chmn, 81. *Honors & Awards:* Charles Dana Found Award, 86; Chem Health & Safety Award, Am Chem Soc, 87, James Flack Norris Award, 88; John A Timm Award, Northeast Asn Chem Teachers, 87; Catalyst Award, Chem Mfgs Asn, 90. *Mem:* Am Chem Soc; Sigma Xi. *Res:* Rearrangement of benzyl, phenyl and benzyl o-tolyl ethers; synthesis of cyclo-octatetraene derivatives; silicone polymers; organofunctional silanes; organogermanes; silicon and germanium coordination compounds; development of organic microscale chemistry laboratory. *Mailing Add:* Dept of Chem Merrimack Col North Andover MA 01845

PIKE, ROSCOE ADAMS, b Calais, Maine, Aug 16, 25; m 51; c 5. ORGANIC CHEMISTRY. *Educ:* Univ NH, BS, 49, MS, 50; Mass Inst Technol, PhD(chem), 53. *Prof Exp:* Res chemist, Silicones Div, Union Carbide Corp, 53-60, res supvr, 60-61; abrasives div, Norton Co, Mass, 61-64, sr res engr, 64-66, res assoc, 66-68; sr res scientist, United Aircraft Res Labs, 68-80, sr mat scientist, 80-85, MGR POLYMER SCI, UNITED TECHNOL RES CTR, 85- *Mem:* Am Chem Soc. *Res:* Rearrangement of triphenyl methyl o-cresol ethers; sterochemistry of Hofmann exhaustive methylation and amine oxide decomposition reactions; silicone monomers and polymers; aluminum oxide, silicon carbides and diamond resinoid abrasive products; grinding fluids and lubricants; high temperature organic resins and adhesives; structural composites. *Mailing Add:* United Technol Res Ctr Silver Lane East Hartford CT 06108

PIKE, RUTH LILLIAN, b New York, NY, Apr 5, 16. NUTRITION. *Educ:* Hunter Col, AB, 36; Columbia Univ, MA, 37; Univ Chicago, PhD(nutrit), 50. *Prof Exp:* From instr to prof, 43-77, EMER PROF NUTRIT SCI, PA STATE UNIV, UNIVERSITY PARK, 77- *Honors & Awards:* Borden Award, 67. *Mem:* Fel AAAS; Fel Am Inst Nutrit, 87; NY Acad Sci; Soc Exp Biol & Med; Sigma Xi. *Res:* Nutrition during pregnancy; sodium homeostasis and blood pressure; written three advance textbooks on nutrition science. *Mailing Add:* 1275 Penfield Rd State College PA 16801

PIKLER, GEORGE MAURICE, b Quito, Ecuador, June 4, 42. ONCOLOGY, HEMATOLOGY. *Educ:* Cent Univ Ecuador, MD, 68. *Prof Exp:* DIR ONCOL PROG, HILLCREST MED CTR, TULSA, 78-; CLIN ASSOC PROF, DEPT MED, UNIV OKLA, 78- *Concurrent Pos:* Pvt pract, hematol & oncol, 78- *Mem:* AMA; Am Soc Internal Med; Am Soc Clin Oncol; Am Soc Cell Biol; Sigma Xi; AAAS; Am Clin Res; NY Acad Sci; Int Soc Prev Oncol. *Mailing Add:* Cancer Specialists Inc 1145 S Utica Suite 806 Tulsa OK 74104

PIKUS, ANITA, b New York, NY, June 13, 38; m 59; c 1. CLINICAL RESEARCH IN HEARING LOSS, GENETICS & HEARING LOSS. *Educ:* Hunter Col, BA, 60; Temple Univ, MA, 65. *Prof Exp:* Teacher, Montgomery County, Pa, Pub Sch, 60-62, hearing consult, 62-63; asst prof audiol, Dept Otorhinol, Temple Univ Med Sch, 63-75, field supvr, Audiol Practicum, Temple Univ Col Liberal Arts, Dept Speech & Hearing, 63-75; supvr, Clin Audiol, Dept Hearing & Speech Sci, Univ Md, 76-78; CHIEF CLIN AUDIOL, NAT INST DEAFNESS & OTHER COMMUN DIS, NIH, BETHESDA, MD, 78- *Concurrent Pos:* Clin audiologist, Temple Univ Health Sci Ctr, 63-75, dir, Infant Screening Prog, 67-75, chief audiologist, Hereditary Hearing Loss Clin, 71-73; audiol consult, Elwyn Inst, 67-75; proj dir, Am Speech-Lang-Hearing Asn, 75-76; consult, Technol Advan Prog, Univ Md, 85, Off Personnel Mgt, USAF, 89-, Audiol Found, 90- *Mem:* Fel Am Acad Audiol; Am Speech Lang & Hearing Asn; Am Auditory Soc. *Res:* Genetics; audiology syndrome delineation. *Mailing Add:* Nat Inst Deafness & Commun Dis NIH Clin Audiol Unit Bldg 10/5C306 9000 Rockville Pike Bethesda MD 20892

PIKUS, IRWIN MARK, b Philadelphia, Pa, Apr 21, 36; m 59; c 1. PHYSICS, LAW. *Educ:* Drexel Univ, BS, 58; Univ Pa, MS, 60; Temple Univ, PhD(physics), 66, JD, 72. *Prof Exp:* Engr physics, Advan Airborn Systs Div, RCA Corp, 58-62; staff scientist, Phys Res Lab, Budd Co, 62-65; sr scientist environ, Space Sci Labs, Gen Elec Co, 66-73; atty & consult, pvt pract, 73-75; dep dir, Off Technol Policy, US Dept State, 75-79; dir, Div Planning & Policy anal, Nat Sci Found, 79-84, sr adv, Int Sci & Technol Policy, 84-87; DIR, OFF FOREIGN AVAILABILITY TECHNOL, US DEPT COM, 87- *Concurrent Pos:* Vchmn, Philadelphia Air Pollution Control Bd, 73-75; mem, Mayor's Sci & Technol Adv Comt, 73-75. *Mem:* AAAS; Am Inst for Aeronaut & Astronaut; Am Bar Asn; Am Phys Soc; Int Inst Space Law; Int Acad Astronaut; Fed Bar Asn. *Res:* Space sciences; science policy; US foreign policy; advanced technology; communications; materials; metals; re-entry physics; environmental science; nuclear effects; radio propagation. *Mailing Add:* Ofc of Foreign Availability, SB 097 14th Constitution Ave Washington DC 20230

PILACHOWSKI, CATHERINE ANDERSON, b Sacramento, Calif, Aug 20, 49. COMPOSITIONS OF STARS, STAR CLUSTERS. *Educ:* Harvey Mudd Col, BS, 71; Univ Hawaii, MS, 73, PhD(astron), 75. *Prof Exp:* res assoc astron, Univ Wash, 75-79; asst support scientist, 79-82, assoc astron, 82-86, ASTRON WITH TENURE, KITT PEAK NAT OBSERV, 86- *Concurrent Pos:* Fel, Am Asn Univ Women, 74-75; mem, Comt Status Women, Am Astron Soc, 80-82, chair, 81-82, Shapley lectr, 82-91, mem publ bd, 89-91, chair, 90-93; mem bd dirs, Astron Soc Pac, 80-86; mem, Organizing Comt, Comn 37, Int Astron Union, 82-88, Organizing Comt, Comn 29, 88-91, vpres, Comn 37, 88-91; mem, IUE Peer Rev Panel, NASA, 84, 85, IPAC Proposal Rev Bd (IRAS), 87, Starlab mission, Spectrograph Subcomt, 82-84, & Calibration Subcomt, 83-84; mem, Adv Comt Astron Sci, NSF, 89-91, Telescope Allocation Comt, MMT, 82-86. *Mem:* Am Astron Soc; AAAS; Astron Soc Pac; Int Astron Union. *Res:* Analysis of the composition of stellar atmospheres, with particular interest in stellar evolution and nucleosynthesis. *Mailing Add:* Kitt Peak Nat Observ Nat Optical Astron Obs 950 N Cherry Ave PO Box 26732 Tucson AZ 85726-6732

PILANT, WALTER L, b Los Angeles, Calif, May 15, 31; m 54; c 5. GEOPHYSICS. *Educ:* Calif Inst Technol, BS, 53; Univ Calif, Los Angeles, MS, 56, PhD(physics), 60. *Prof Exp:* Asst res geophysicist, Univ Calif, Los Angeles, 55-62; res fel geol, Calif Inst Technol, 62-63; ASSOC PROF GEOPHYS, UNIV PITTSBURGH, 63- *Mem:* Inst Elec & Electronics Engrs; Geol Soc Am; Am Geophys Union; Soc Explor Geophysicists. *Res:* Elastic wave propagation; seismology; solid earth geophysics. *Mailing Add:* Dept Geol & Planetary Sci Univ Pittsburgh Pittsburgh PA 15260

PILAR, FRANK LOUIS, b Verdigre, Nebr, Aug 28, 27; m 54; c 5. PHYSICAL CHEMISTRY. *Educ:* Univ Nebr, BS, 51, MS, 53; Univ Cincinnati, PhD(chem), 57. *Prof Exp:* Jr chemist, Standard Oil Co(Ind), 52, asst chemist, 54-55; asst, Univ Wis, 53-54; from asst prof to assoc prof, 57-68, PROF CHEM, UNIV NH, 68-, CHMN, 82- *Concurrent Pos:* Vis scientist, Math Inst, Oxford Univ, 63-64; vis prof, Univ Wis, 70-71; proj specialist, Ningxia Univ, People's Repub China, 87. *Mem:* Am Chem Soc; Am Phys Soc; Sigma Xi. *Res:* Quantum chemistry. *Mailing Add:* Parsons Hall Univ NH Durham NH 03824

PILAR, GUILLERMO ROMAN, b Buenos Aires, Arg, Oct 31, 33; div; c 3. NEUROPHYSIOLOGY. *Educ:* Univ Buenos Aires, MD, 54. *Prof Exp:* Teaching asst physiol, Univ Buenos Aires, 50-54, res assoc med, Inst Med Invest, 57-59; res assoc physiol, Univ Utah, 61-64; vis sr lectr, Monash Univ, Australia, 65; from asst res prof to assoc res prof, Univ Utah, 66-71; head physiol sect, 73-85, PROF BIOL, UNIV CONN, 71-, CHMN, DEPT PHYSIOL & NEUROBIOL, 85- *Concurrent Pos:* Squibb Found fel, 56-57; fel, Nat Coun Sci Invest Arg, 59-61; Guggenheim Found fel, 61-63; vis fel, Australian Nat Univ, 64-65; prin investr, Nat Heart Inst grant, 66- & NSF grant, 84-; USPHS res career develop award, 67-71; vis assoc prof, Univ Calif, Los Angeles, 71, vis prof, 72; mem, molecular & cellular neurobiol panel, NSF, 84-; NIH, NINCOS prog, proj B, 86-90; res vis fel, Div Biol, Calif Inst Technol, 90. *Honors & Awards:* J Javits Neurosci Investr Award, 85. *Mem:* Soc Neurosci; NY Acad Sci; Am Physiol Soc. *Res:* Nerve cell interactions and communications, influences and changes during development, use and disuse; environment of the nerve cells. *Mailing Add:* Physiol & Neurobiol Dept Box 42 Univ Conn 75 N Eagleville Rd Storrs CT 06269

PILAT, MICHAEL JOSEPH, b Longview, Wash, Feb 19, 38. ENVIRONMENTAL ENGINEERING. *Educ:* Univ Wash, BSChE, 60, MSChE, 63, PhD(civil eng), 67. *Prof Exp:* Engr, Boeing Co, 61-66; from asst prof to assoc prof, 67-78, PROF CIVIL ENG, UNIV WASH, 78- *Mem:* Air & Waste Mgt Asn; Am Inst Chem Engrs; Sigma Xi; Am Chem Soc; Am Asn Aerosol Res; Tech Asn Pulp & Paper Indust; Am Indust Hyg Asn. *Res:* Air pollution control technology; air pollutant source testing; plume opacity and atmospheric visibility; particulate control with wet scrubbers and electrostatic precipitators; particle sizing with cascade impactors. *Mailing Add:* Dept Civil Eng Univ Wash Seattle WA 98195

PILATI, CHARLES FRANCIS, b Wheeling, WVa, Aug 4, 45; m 82. CARDIOVASCULAR. *Educ:* Kent State Univ, BS, 69, PhD(physiol), 79. *Prof Exp:* Asst instr, Kent State Univ, 73-75, fel, 75-79; fel, Northeastern Ohio Univ, 79-81, from instr to asst prof, 81-91, ASSOC PROF PHYSIOL, COL MED, NORTHEASTERN OHIO UNIV, 91- *Concurrent Pos:* Instr, Kent State Univ, 84- *Mem:* Microcirculatory Soc. *Res:* Cardiac research; interstitial fluid balance. *Mailing Add:* Prog Physiol Col Med Northeastern Ohio Univ Rootstown OH 44272

PILCH, SUSAN MARIE, b Washington, DC, Aug 23, 54; m 78. NUTRITION. *Educ:* Univ Va, BA, 76; Cornell Univ, PhD(nutrit), 82. *Prof Exp:* Ed asst, Am J Clin Nutrit, 81, asst ed, 81-82; assoc staff scientist, Life Sci Res Off, Fedn Am Socs Exp Biol, 83, staff scientist, 84-86, sr staff scientist, 87-90; DEP DIR, DIV NUTRIT RES COORD, NIH, 90- *Concurrent Pos:* Consult, Health Officers Asn Calif, 87-89. *Mem:* Am Inst Nutrit; AAAS; NY Acad Sci; Asn Women Sci. *Res:* National nutrition monitoring; nutrition and health; nutrition and public policy. *Mailing Add:* Div Nutrit Res Coord NIH Bldg 31 Rm 4B63 Bethesda MD 20892

PILCHARD, EDWIN IVAN, b Urbana, Ill, Dec 14, 25; m 49; c 3. VETERINARY VIROLOGY, VETERINARY IMMUNOLOGY. *Educ:* Mich State Univ, DVM, 47; Univ Ill, MS, 59, PhD(immunol), 64. *Prof Exp:* Vet, 48-51, 53-57; pathologist, Ill Dept Agr, 57-58, dir vet path, Diag & Res Labs, 58-64; asst prof vet virol & immunol, Col Vet Med, Univ Ill, 64-69; prin vet, Coop State Res Serv, 69-77, SR STAFF VET, VET SERV, USDA, 77- *Concurrent Pos:* Res asst, Col Vet Med, Univ Ill, 62-63; assoc mem, Ill Ctr Zoonoses Res, 66-69; NIH res grant, 66-69; prin investr proj livestock reproduction res, NSF Coop Prog Spain, 73-75; asst dir, Agr Exp Sta, Cornell Univ, 74-75. *Mem:* Am Vet Med Asn; Conf Res Workers Animal Dis; US Am Health Asn; Sigma Xi. *Res:* Infectious diseases; immunology with emphasis in immunopathology and cellular immunity. *Mailing Add:* 10510 Royal Rd Silver Spring MD 20903

PILCHER, BENJAMIN LEE, b Corpus Christi, Tex, Feb 25, 38; m 65; c 2. BOTANY. *Educ:* Tex Technol Col, BS, 61, MS, 63; Univ NMex, PhD(biol), 69. *Prof Exp:* Instr biol, Univ Tex, Arlington, 63-66; assoc prof, 69-79, head dept, 70-74, PROF, MCMURRY UNIV, 69- *Concurrent Pos:* Dir honors prog, McMurry Col, 86- *Mem:* Cactus & Succulent Soc; Sigma Xi. *Res:* Developmental aspects of plants on both an anatomical and physiological basis; cactus seedling anatomy and development. *Mailing Add:* 3234 Beltway S Abilene TX 79606

PILCHER, CARL BERNARD, b Brooklyn, NY, Apr 7, 47. PLANETARY SCIENCES. *Educ:* Polytech Inst Brooklyn, BS, 68; Mass Inst Technol, PhD(chem), 73. *Prof Exp:* Asst astronomer, 73-77, asst prof, 77-79, ASSOC PROF, DEPT PHYSICS & ASTRON, INST ASTRON, UNIV HAWAII, HONOLULU, 79- *Mem:* Am Astron Soc; AAAS; Astron Soc Pac; Int Astron Union. *Res:* Study of atmospheres, surfaces and magnetospheres of solar system objects, particularly those in outer solar system. *Mailing Add:* 4316 Ellicott St Washington DC 20016

PILCHER, JAMES ERIC, b Toronto, Ont, Apr 23, 42; m 70; c 3. ELEMENTARY PARTICLE PHYSICS. *Educ:* Univ Toronto, BASc, 64, MSc, 66; Princeton Univ, PhD(physics), 68. *Prof Exp:* Res assoc, Princeton Univ, 68-69; vis scientist, Europ Orgn Nuclear Res, Geneva, Switz, 69-70; asst prof, Harvard Univ, 70-72; asst prof to assoc prof, 72-79, PROF ELEM PARTICLE PHYSICS, ENRICO FERMI INST, UNIV CHICAGO, 79- *Concurrent Pos:* Sci assoc, Europ Orgn Nuclear Res, Geneva, Switz, 79-80. *Mem:* fel Am Phys Soc. *Res:* Experimental weak interactions including charge conjugation-parity violation and high energy neutrino scattering; proton-proton scattering at ultra-high energies; lepton pair production in strong interactions; high energy e+ e- annihilations. *Mailing Add:* Enrico Fermi Inst Univ of Chicago 5640 S Ellis Ave Chicago IL 60637

PILCHER, VALTER ENNIS, b Savannah, Ga, Nov 7, 25; m 50; c 2. PHYSICS. *Educ:* Emory Univ, Ab, 48, MS, 49; NC State Col, PhD, 55. *Prof Exp:* Instr physics, Armstrong State Col, Ga, 49-50 & NC State Col, 50-53; jr res assoc, Brookhaven Nat Lab, 53, res assoc, 54-55; from asst prof to assoc prof physics, Union Col, NY, 56-72, prof, 72-86, chmn dept, 78-80; RETIRED. *Concurrent Pos:* Guest scientist, Swed Atomic Energy Co, 62-63; Fulbright lectr, Haile Selassie Univ, 68-69. *Mem:* Am Phys Soc; Am Asn Physics Teachers. *Res:* Neutron physics; nuclear spectroscopy. *Mailing Add:* 171 Wendell Ave Schenectady NY 12308

PILE, PHILIP H, b Paris, Ark, July 18, 46; m 70; c 3. HYPERNUCLEAR PHYSICS. *Educ:* Univ Ark, BS, 69; Ind Univ, PhD(physics), 78. *Prof Exp:* Res assoc, Cyclotron Facil, Ind Univ, 78-79; res physicist, Carnegie Mellon Univ, 79-81; asst physicist, Brookhaven Nat Lab, 81-83, assoc physicist, 83-85, physicist, 85-89, HEAD, EXP PLANNING & SUPPORT DIV, BROOKHAVEN NAT LAB, 89- *Mem:* Am Phys Soc; NY Acad Sci. *Res:* Intermediate energy nuclear physics; hypernuclear and Dibaryon physics using pion and kaon beams; design of high rate drift chamber systems; design of separated kaon beam lines; rare kaon decays. *Mailing Add:* Physics Dept Brookhaven Nat Lab Bldg 911B Upton NY 11973

PILEGGI, VINCENT JOSEPH, b Philadelphia, Pa, May 10, 28; m 57; c 3. CLINICAL CHEMISTRY. *Educ:* Drexel Inst, BS, 50; Univ Wis, MS, 53, PhD(biochem), 54. *Prof Exp:* Chief biochem, Letterman Army Hosp, San Francisco, Calif, 55-57; asst prof, Univ Pa, 57-59; chief, Iodine Div, Bio-Sci Labs, 59-65 & Spec Projs Prods, 65-66, dir res, 66-77; dir, 77-82, VPRES SCI AFFAIRS, NAT HEALTH LABS, LA JOLLA, CALIF, 82- *Mem:* Am Thyroid Asn; Am Asn Clin Chem. *Res:* Clinical biochemistry, especially thyroid; radioimmunoassay of hormones. *Mailing Add:* 1057 Via Mil Cumbres Solana Beach CA 92075

PILET, STANFORD CHRISTIAN, b Ft Benning, Ga, Nov 6, 31; m 57; c 2. OPERATIONS RESEARCH, ASTRODYNAMICS. *Educ:* US Mil Acad, BS, 54; Univ Cincinnati, MS, 60, PhD(dynamical astron), 63. *Prof Exp:* Res specialist, Orbital Flight Mech, Aero-Space Div, Boeing Co, Seattle, 63-65, RES MGR, MIL OPERS SYSTS ANALYSIS, BOEING AEROSPACE, 85- *Mem:* Fel AAAS; Sigma Xi. *Res:* Long term prediction of orbital motion; orbital motion of artificial satellites effected by various perturbations; analysis and evaluation of military systems and policies. *Mailing Add:* Boeing Co PO Box 3999 Seattle WA 98124-2499

PILGER, REX HERBERT, JR, b North Platte, Nebr, June 8, 48; m 73; c 5. TECTONOPHYSICS, GEOPHYSICS. *Educ:* Univ Nebr, BS, 70, MS, 72; Univ Southern Calif, PhD(geol), 76. *Prof Exp:* Geol field asst, US Geol Surv, 75-76; from asst prof to assoc prof, 76-85, assoc chmn, 85-87, PROF GEOL & GEOPHYS, LA STATE UNIV, 85- *Concurrent Pos:* Geophysicist, Naval Ocean Res & Develop Activ, 78-82. *Mem:* Geol Soc Am; Soc Explor Geophysicists; Am Geophys Union. *Res:* Reflection seismology; plate tectonics of western United States, South America and Gulf of Mexico; hotspots; geophysics of sedimentary basins and continental margins. *Mailing Add:* Dept Geol La State Univ Baton Rouge LA 70803

PILGER, RICHARD CHRISTIAN, JR, b Hartford, Conn, June 13, 32; m 57; c 2. PHYSICAL CHEMISTRY, INORGANIC CHEMISTRY. *Educ:* Univ Notre Dame, BS, 54; Univ Calif, Berkeley, PhD(chem), 57. *Prof Exp:* Asst, Univ Calif, Berkeley, 55-56; asst prof chem, Wash Univ, 57-59; Univ Notre Dame, 59-64; assoc prof, 64-71, PROF CHEM & PHYSICS ST MARY'S COL, IND, 71- *Res:* Nuclear spectroscopy; coordination chemistry. *Mailing Add:* Dept Chem & Physics St Marys Col Notre Dame IN 46556

PILGERAM, LAURENCE OSCAR, b Great Falls, Mont, June 23, 24; m 51; c 2. BIOCHEMISTRY, PHYSIOLOGY. *Educ:* Univ Calif, BA, 49, PhD(biochem), 53. *Prof Exp:* Asst, Inst Exp Biol, Univ Calif, 45-47, asst, Div Biochem, 51-52; instr physiol, Col Med, Univ Ill, 54-55; asst prof biochem, Dept Chem, Stanford Univ, 55-57; dir arteriosclerosis res lab, Sch Med, Univ Minn & St Barnabas Hosp Res Found, 57-65; dir, Arteriosclerosis Res Lab, Calif, 65-71; dir, Coagulation Lab & assoc dir, Cerebrovascular Res Ctr, Baylor Col Med, 71-75; DIR CELL CULT LAB, CALIF, 75- *Concurrent Pos:* Life Insurance Med Res Fund fel physiol chem, Sch Med, Univ Calif, 52-54. *Honors & Awards:* Ciba Found Award, London, 58; Karl Thomae Award, Ger, 73. *Mem:* Am Soc Biochem & Molecular Biol. *Res:* Blood coagulation, thrombosis-stroke & coronary; cellular control mechanisms. *Mailing Add:* PO Box 1583 Goleta Sta Santa Barbara CA 93116

PILGRIM, DONALD, b Sioux City, Iowa, May 31, 29; m 56; c 5. MATHEMATICS. *Educ:* Morningside Col, AB, 51; Univ Iowa, MS, 52; Univ Wis, PhD(math), 63. *Prof Exp:* Mathematician, Gen Labs, US Rubber Co, 52-53; teacher high sch, Iowa, 54-55; instr math, Keokuk Community Col, 55-56; PROF MATH, LUTHER COL, IOWA, 56- *Concurrent Pos:* Treas, Iowa Acad Sci, 75- *Mem:* Am Math Soc. *Res:* Engel conditions on groups; groups of exponent four; additive theory of prime numbers. *Mailing Add:* Luther Col Decorah IA 52101

PILGRIM, HYMAN IRA, b New York, NY, Feb 10, 25; m 47, 68; c 8. CANCER, GENETICS. *Educ:* Univ Calif, AB, 50, MA, 53, PhD, 56. *Prof Exp:* Asst zool, Univ Calif, 52-53; from instr to asst prof anat, Sch Med, Univ Buffalo, 56-61; asst res prof anat & surg, Col Med, Univ Utah, 61-68, assoc res prof anat, 68-72; vis reader & actg head dept anat, Univ Ibadan, 76-77; RETIRED. *Mem:* Am Asn Cancer Res. *Res:* Experimental oncology; mouse genetics and pathology; genesis of adrenal and reticuloendothelial tumors; metastasis and kinetics of tumor growth; gnotobiotics; genetics. *Mailing Add:* PO Box 1082 Laytonville CA 95454

PILITT, PATRICIA ANN, b Washington, DC, Dec 8, 42; m 68; c 2. PLANT & INSECT PARASITIC NEMATODES. *Educ:* Univ Md, BS, 65. *Prof Exp:* Res asst plant nematol, Nematol Lab, 66-75, SUPPORT SCIENTIST PARASITOL, BIOSYST PARASITOL LAB, AGR RES SERV, USDA, BELTSVILLE, MD, 76- *Concurrent Pos:* Librn, Helminth Soc Wash, 80-84, asst ed, 84-88, vpres, 85-86, pres, 86-87, archivist/librn, 91. *Mem:* Soc Nematologists. *Res:* Preparation of guides or keys to nematode parasites; morphological characters of nematode parasites of domestic animals. *Mailing Add:* Animal Parasitol Inst ARS USDA Biosyst Parasitol Lab BARC-East Bldg 1180 Beltsville MD 20705

PILKERTON, A RAYMOND, b Washington, DC, Mar 27, 35; m 66; c 5. OPHTHALMOLOGY, VITREORETINAL SURGERY. *Educ:* Georgetown Univ, BS, 56, MD, 60. *Prof Exp:* Intern, Univ Pittsburgh, 60-61; resident ophthal, Med Ctr, Georgetown Univ, 61-64; fel, Wills Eye Hosp, Philadelphia, 64-65; from instr to asst prof, 65-73, dir retina serv, 65-85, assoc prof, 73-81, CLIN PROF OPHTHAL, MED CTR, GEORGETOWN UNIV, 81- *Concurrent Pos:* Nat Inst Neurol Dis & Blindness spec fel retinal surg, Wills Eye Hosp, Philadelphia, 64-65; chief ophthal serv, Vet Admin Hosp, Washington, DC, 65- *Mem:* Am Col Surg; Am Acad Ophthal & Otolaryngol; Asn Res Vision & Ophthal; Am Med Asn. *Res:* Retina adhesion characteristics; vitreous fibraplasia; retinopathy of prematurity. *Mailing Add:* 5454 Wisconsin Ave Suite 1540 Chevy Chase MD 20815

PILKEY, ORRIN H, JR, b New York, NY, Sept 19, 34; m 56; c 5. MARINE GEOLOGY. *Educ:* Wash State Univ, BS, 57; Mont State Univ, MS, 59; Fla State Univ, PhD(geol), 62. *Prof Exp:* Res assoc & asst prof, Marine Inst, Univ GA, 62-65; from assoc prof to prof, 67-83, JAMES B DUKE PROF GEOL, DUKE UNIV, 83- *Concurrent Pos:* Vis prof, Univ PR, 71-72 & US Geol Surv, Woods Hole, Mass, 75-76; ed, J Sedimentary Petrol, 78-83; co-ed, Duke Univ Press series; pres, NC Acad Sci, 83-84, Soc Econ Paleontologists & Mineralogists, 85-86. *Honors & Awards:* Francis Shepard Medal, 87. *Mem:* Am Asn Petrol Geol; Geol Soc Am; Int Asn Sedimentol; Soc Econ Paleontologists & Mineralogists (pres, 85-86); AAAS. *Res:* Geological oceanography; applied coastal geology; shoreline conservation. *Mailing Add:* Dept Geol Duke Univ Durham NC 27706

PILKEY, WALTER DAVID, b Chicago, Ill, Oct 28, 36; m 65; c 2. APPLIED MECHANICS, STRUCTURAL ENGINEERING. *Educ:* Wash State Univ, BA, 58; Purdue Univ, MS, 60; Pa State Univ, PhD(eng mech), 62. *Prof Exp:* Scientist mech, Ill Inst Technol Res Inst, 62-69; PROF MECH ENG, UNIV VA, 69- *Concurrent Pos:* Vis prof, Kabul Univ, Afghanistan, 63-64; adj assoc prof, Ill Inst Technol, 64-69; vis res scientist, Ruhr Univ, Germany, 75-76; sr lectr, Fulbright-Hays, Soviet Union, 78; chair prof, Naval Sea Systs Command, Monterey, 83-84; fel, Japan Phys Soc, 85; hon prof, Zhejiang Univ, China. *Honors & Awards:* Sr von Humboldt Prize, 90. *Mem:* Am Soc Mech Engrs; Am Soc Civil Engrs; Am Acad Mech; Am Soc Eng Educ. *Res:* Numerical methods of solid mechanics; optimization of mechanical systems; crashworthy vehicle design; balancing of rotating shafts; transient response of structural members. *Mailing Add:* Dept Mech & Aerospace Eng Univ Va Charlottesville VA 22901

PILKIEWICZ, FRANK GEORGE, b Jersey City, NJ, June 21, 46; m 69; c 2. CHEMISTRY. *Educ:* St Peters Col, BS, 68; Rutgers Univ, MS, 73, PhD(chem), 74. *Prof Exp:* Teaching asst chem, Rutgers Univ, 69-71, res asst, 71-74; res assoc fel, Columbia Univ, 74-77; prod mgr, Waters Assoc, 77-79; sr res investr, Squibb Inst Med Res, 79-86; dir anal res & develop, 86-89, VPRES RES & DEVELOP, LIPOSOME CO, 89- *Mem:* Am Chem Soc; Am Asn Off Anal Chemists. *Res:* Direct the research and development efforts of a pharmaceutical company dedicated toward the goal of producing liposomal drug formulation for use in such areas as cancer and infectious disease. *Mailing Add:* Liposome Co Forrestral Ctr One Res Way Princeton NJ 08540

PILKINGTON, HAROLD DEAN, geology, petrology, for more information see previous edition

PILKINGTON, LOU ANN, b Wichita, Kans, Apr 2, 24. PHYSIOLOGY. *Educ:* Univ Okla, BS, 48, MS, 59, PhD(physiol), 61. *Prof Exp:* NIH fel & instr physiol, Sch Med, Univ Okla, 62-63; fel, 63-65, from instr to asst prof physiol, Med Col, Cornell Univ, 66-72; dir pulmonary function & Work Phsyiol Lab, Goldwater Mem Hosp, New York Univ Med Ctr, 72-82; RETIRED. *Concurrent Pos:* NIH career develop award, 65-70. *Mem:* AAAS; Am Physiol Soc; Harvey Soc; Sigma Xi. *Res:* Respiratory physiology; work physiology. intact functioning kidney; renal blood flow. *Mailing Add:* 315 E 72 Apt 5A New York NY 10021

PILKINGTON, THEO C(LYDE), b Durham, NC, June 23, 35; m 57; c 2. ELECTRICAL ENGINEERING. *Educ:* NC State Univ, BEE, 58; Duke Univ, MS, 60, PhD(elec eng), 63. *Prof Exp:* From instr to assoc prof elec eng, 61-69, chmn dept, biomed eng, 67-78, PROF BIOMED & ELEC ENG, DUKE UNIV, 69-; DIR, CARD TECHNOL, NSF/ERC, 87- *Concurrent Pos:* Ed, CRC Critical Rev Bioeng, 75-79; ed, Transactions on Biomed Eng, Inst Elec & Electronic Engrs, 79-84. *Honors & Awards:* Centennial Medal, Inst Elec & Electronic Engrs. *Mem:* Am Phys Soc; fel Inst Elec & Electronic Engrs; Biomed Eng Soc. *Res:* Biomedical engineering; field theory; numerical analysis. *Mailing Add:* Dept of Biomed Eng Duke Univ Durham NC 27706

PILKIS, SIMON J, b Chicago, Ill, Dec 30, 42; m; c 1. PHYSIOLOGY. *Educ:* Univ Chicago, Ill, BS, 64, MD, 69, PhD(physiol), 69. *Prof Exp:* Postdoctoral fel, NIH, Dept Physiol, Univ Sussex, Eng, 69-70; vis investr, Howard Hughes Med Inst, Dept Physiol, Vanderbilt Univ Sch Med, Nashville, Tenn, 71-72, Howard Hughes investr, 72-78, from asst prof to assoc prof physiol, 72-82, prof molecular physiol & biophys, 82-86; PROF & CHMN PHYSIOL & BIOPHYS, SCH MED, HEALTH SCI CTR, STATE UNIV NY, STONY BROOK, 86- *Concurrent Pos:* Career investr, Howard Hughes Med Inst, Dept Physiol, Vanderbilt Univ, Nashville, Tenn, 72-79; consult, Lederle Labs, Am Cyanimide Corp, Pearl River, NJ, Sandoz Pharmaceut Co, East Hanover, NJ, McNeil Pharmaceut, Ft Washington, Pa & Gensia Labs, La Jolla, Calif; mem, Cellular Physiol Study Sect, NSF, 80-82, Biochem Study Sect, 90-; mem, Physiol Chem Study Sect, NIH, 82-86, ad hoc mem, Biochem Study Sect, Metab Study Sect, Endocrinol Study Sect; grant reviewer, NY Downstate Affil, Am Diabetes Asn; reviewer, fel appln, NY State Coun Diabetes; prin investr, NIH grant, 88-93, NSF grant, 90-93; res fel, Juv Diabetes Found, 89-91; mem, Burger Ctr Peer Rev Comt, 91. *Mem:* Am Soc Biochem & Molecular Biol; Am Diabetes Asn; NY Acad Sci. *Res:* Molecular basis for regulation of glycolytic/gluconeogenic flux in mammalian liver; author of various publications. *Mailing Add:* Dept Physiol & Biophys State Univ NY Stony Brook NY 11794-8661

PILLA, ARTHUR ANTHONY, b New York, NY, Aug 12, 36. BIOELECTROCHEMISTRY, BIOELECTRICITY. *Educ:* St Joseph's Col, BS, 56; Univ Pa, MS, 59; Univ Paris, PhD, 64. *Prof Exp:* Sr scientist, US Army Res Labs, Ft Monmouth, 66-69; dir, Bioelectrochem Lab, 69-75, PROF BIOELECTROCHEM, DEPT APPL CHEM, COLUMBIA UNIV, 72- *Concurrent Pos:* Ed, J Electrochem Soc, 74-80, Bioelectrochem & Bioenergetics, 75- & Electrochimica Acta, 75- *Mem:* Electrochem Soc; Int Bioelectrochem Soc; Am Chem Soc; AAAS. *Res:* Bioelectric phenomena in tissue growth and repair via electrochemical kinetics at cell surfaces; electrochemical relaxation techniques for cell membrane impedance; in vivo applications of electromagnetic modulation of bone and other tissue repair. *Mailing Add:* 133 Heights Rd Ridgewood NJ 07450

PILLAI, PADMANABHA S, b Nov 13, 31; Indian citizen. SOLID STATE & POLYMER PHYSICS. *Educ:* Univ Kerala, BSc, 50; Banaras Hindu Univ, MSc, 53; WVa Univ, PhD(physics), 66. *Prof Exp:* Lectr physics, Sree Narayan Col, India, 53-61 & Annamalai Univ, Madras, 59-61; prof, M D T Hindu Col, India, 61-63; NSF grant, Kans State Univ, 63-64; Atomic Energy Comn fel, WVa Univ, 65-66; SR RES PHYSICIST POLYMER PHYSICS, GOODYEAR TIRE & RUBBER CO, 66- *Mem:* Am Phys Soc; Am Chem Soc. *Mailing Add:* 509 Fairwood Dr Tallmadge OH 44278

PILLAI, THANKAPPAN A K, b Chavara, India, Jan 21, 50; m 86; c 1. ELASTIC WAVE SCATTERING, NON-DESTRUCTIVE EVALUATION. *Educ:* Univ Kerala, India, BS, 70, MS, 72; Univ Louisville, PhD(physics), 80. *Prof Exp:* CSIR fel physics, Univ Calicut, India, 72-75; lectr, Univ Kerala, India, 75-76; teaching asst, Univ Louisville, 76-80; res assoc eng mech, Ohio State Univ, 80-83; asst prof, 83-87, ASSOC PROF PHYSICS, UNIV WIS-LA CROSSE, 87- *Concurrent Pos:* Fac res assoc, Argonne Nat Lab, Ill, 84- *Mem:* Am Phys Soc; Am Geophys Union; Am soc Nondestructive Testing. *Res:* Acoustic and elastic wave scattering problems with applications to non-destructive evaluation; composite materials; physical oceanography. *Mailing Add:* Dept Physics Univ Wis La Crosse WI 54601

PILLAR, WALTER OSCAR, b Philadelphia, Pa, Apr 14, 40; m 62; c 2. POLYMER CHEMISTRY, FOREST PRODUCTS. *Educ:* Pa State Univ, BS, 62; Univ Mich, PhD(wood technol), 68. *Prof Exp:* Lab chemist, Perkins Glue Co, 62-63; res scientist adhesives, Koppers Co, 66-70, res scientist expendable polystyrene, 70-72, sr res scientist, 72-74; sr res scientist, ARCO/Polymers, Inc, 74-77, prin scientist, Expandable Polystrene Thermoplastic Elastomers, 77-81; SR ENGR, ADVAN TECHNOL, GE AEROSPACE, 81- *Concurrent Pos:* Dir, Gateway Sch Dist, 72-79; tech dir, AP/SP Mfg Technol Prog. *Mem:* Am Chem Soc; Sigma Xi. *Res:* Printed circuit assemblies; plastic foams; wood products; adhesives; phenolic resins; flame retardance in plastics. *Mailing Add:* GE Aerospace MD 150 Utica NY 13503

PILLARD, RICHARD COLESTOCK, b Springfield, Ohio, Oct 11, 33; div; c 3. PSYCHIATRY. *Educ:* Antioch Col, BA, 55; Univ Rochester, MD, 59. *Prof Exp:* Asst psychiatrist, Boston State Hosp, 63-67; from instr to assoc prof, 66-84, head, basic studies unit, Psychopharmacol Lab, 68-78, PROF PSYCHIAT, SCH MED, BOSTON UNIV, 84-, DIR, FAMILY STUDIES LAB, 78- *Concurrent Pos:* Pvt Pract, 63-; USPHS res scientist, Develop Award, 67-72; dir, Res Training Prog, Sch Med, Boston Univ, 70-77; med dir, Dr Solomon Carter, Fuller Med Health Ctr, 85- *Honors & Awards:* Res Scientist Devel Award, NIMH, 67; Hugo Biegel Award, 88. *Mem:* Am Psychiat Asn; Am Col Neuropsychopharmacol; Am Psychopathalogic Asn; Int Acad Sex Res; Soc Sci Study Sex. *Res:* Psychopharmacology; homosexuality; familial studies of homosexual orientation; effects of prenatal exposure to sex hormones. *Mailing Add:* 80 East Concord St Boston MA 02118

PILLAY, DATHATHRY TRICHINOPOLY NATRAJ, b Hyderabad, India, Dec 16, 31; m 65; c 2. PLANT PHYSIOLOGY. *Educ:* Osmania Med Col, BScAg, 53; Cornell Univ MS, 59, PhD(pomol, plant physiol), 61. *Prof Exp:* Hort asst, Dept Agr, Hyderabad, India, 54-58; jr horticulturist tree fruit sta, Wash State Univ, 62-63; from asst prof to assoc prof, 63-72, PROF PLANT PHYSIOL, UNIV WINDSOR, 72- *Concurrent Pos:* Grants, Nat Res Coun Can, 63-72 & Ministry Univ Affairs Can, 66-67. *Mem:* AAAS; Am Soc Hort Sci; Am Soc Plant Physiol; Can Bot Soc; Can Soc Plant Physiol. *Res:* Physiological and biochemical effects of plant growth regulating chemicals, with special emphasis on nucleic acid metabolism during plant growth and development. *Mailing Add:* Dept Biol Sci Univ Windsor Windsor ON N9B 3P4 Can

PILLAY, K K SIVASANKARA, b Puliyoor, India, Jan 28, 35; m 64; c 1. RADIOANALYTICAL CHEMISTRY, NUCLEAR ENGINEERING. *Educ:* Cent Col, Bangalore, India, BSc, 55; Univ Mysore, MSc, 56; Pa State Univ, PhD(chem), 65. *Prof Exp:* Lectr chem, Univ Mysore, 56-60; asst, Pa State Univ, 60-65; resident res assoc nuclear chem, Argonne Nat Lab, 65-66; res scientist, Western NY Nuclear Res Ctr, Inc, 66-67, sr res scientist, 67-71; res assoc nuclear eng, Pa State Univ, 71-74, asst prof, 75-77, assoc prof, 77-81; STAFF SCIENTIST, LOS ALAMOS NAT LAB, 81- *Concurrent Pos:* Consult radionuclear appln criminal invests, Pa State Police, 71-74, nuclear process chem & waste mgt, US Dept Energy, 76-81; Govt India Scholar, 52-56; Kopper Chem fel, 61-63, Energy Res & Develop Admin fel, 76. *Mem:* AAAS; Am Chem Soc; fel Am Nuclear Soc; fel Am Inst Chem; NY Acad Sci; Health Physics Soc; Inst Nuclear Mat Mgr; Am Soc Testing & Mat. *Res:* Applications of neutron activation analysis in physical and life sciences; reactor fuel burn up analysis; industrial applications of nuclear analytical methods in process and quality control; environmental pollution studies and nuclear forensic applications; nuclear process chemistry; radioactive waste management; nuclear nonproliferation and safeguards; arms control and disarmament. *Mailing Add:* Los Alamos Nat Lab Safeguards Systs Group Mail Stop E-541 Los Alamos NM 87545

PILLER, HERBERT, b Hartmanitz, Bohemia, May 18, 26; m 56; c 4. SOLID STATE PHYSICS. *Educ:* Univ Vienna, PhD(physics), 55. *Prof Exp:* Physicist, ZW Labor, Siemens AG, 55-60; SUPVRY RES PHYSICIST, NAVAL WEAPONS CTR LAB, CORONA & MICHELSON LAB, CHINA LAKE, CALIF, 60-; ASSOC PROF PHYSICS, LA STATE UNIV, BATON ROUGE, 69- *Mem:* Am Phys Soc; Am Asn Univ Prof. *Res:* Optical properties of solids; semiconductor physics. *Mailing Add:* 12467 Sherbrook Ave Baton Rouge LA 70815

PILLIAR, ROBERT MATHEWS, b Beamsville, Ont, Dec 14, 39; m 78; c 2. BIOMATERIALS, METALLURGY. *Educ:* Univ Toronto, BASc, 61; Univ Leeds, PhD(metall), 65. *Prof Exp:* Univ grant metall, McMaster Univ, 65-67; res engr, Int Nickel, Inc, 67-68; res scientist, Ont Res Found, 68-78; PROF, FAC DENT, UNIV TORONTO, 78-, PROF METALL & MAT SCI, 80- *Concurrent Pos:* Adj prof metall & mat sci, Univ Toronto, 77-78; adj prof, Mech Eng Dept, Univ Waterloo, 76-77; vis fel, Dental Biomats, Liverpool Univ, 84-85. *Honors & Awards:* Clemson Award for Appl Biomat Res, 89. *Mem:* Am Soc Testing & Mat; Can Soc Biomats (secy-treas, 77-78 & 80-82, pres, 82-84); Soc Biomats; Int Asn Dent Res; Orthop Res Soc. *Res:* Powder metallurgy; interface studies; composite materials; mechanical properties; biomaterials use in surgical implant design; biomaterials degradation due to in vivo or in vitro exposure; biomaterials effect on body tissues, specifically strain-related bone remode. *Mailing Add:* Fac of Dent Univ Toronto 124 Edward St Toronto ON M5G 1G6 Can

PILLINGER, WILLIAM LEWIS, applied physics, for more information see previous edition

PILLION, DENNIS JOSEPH, b Hartford, Conn, Aug 15, 50; m 75. BIOCHEMISTRY, IMMUNOLOGY. *Educ:* Univ Hartford, BA, 72; Med Col Ga, PhD(cell & molecular biol), 76. *Prof Exp:* NIH fel insulin action, 76-79, ASSOC PROF PHARMACOL, BROWN UNIV, 79- *Concurrent Pos:* Assoc scientist, Vision Sci Res Ctr & Cystic Fibrosis Res Ctr, Univ Ala, Birmingham. *Honors & Awards:* Excellence in Res Award, Sigma Xi, 76; Baladimos Award, Am Diabetes Asn, 78; Louis N Katz award, Am Heart Asn. *Mem:* Sigma Xi; AAAS; NY Acad Sci; Immunopharmacol Soc; Am Diabetes Asn. *Res:* Insulin action; isolation of transport proteins; immunoregulation of membrane function; microvascular tissue response to insulin; renal response to insulin. *Mailing Add:* 1100 Regent Dr Birmingham AL 35226

PILLMORE, CHARLES LEE, b Boulder, Colo, Apr 7, 30; m 54; c 3. COAL GEOLOGY, GEOLOGIC PHOTOGRAMMETRY. *Educ:* Univ Colo, BA, 52, MS, 54. *Prof Exp:* Res geologist, 54-75, supvry geologist, 75-80, RES GEOLOGIST, US GEOL SURV, 80- *Concurrent Pos:* Pres, Colo Sci Soc, 83-84. *Mem:* Geol Soc Am; Am Soc Photogram & Remote Sensing. *Res:* Geology and coal resources of Raton coal field in New Mexico; the Tertiary/Cretaceous boundary in continental rocks; geologic photogrammetry; application of photogrammetry to geology; development of a computer-supported photogrammetric mapping system for geologic studies; development of close-range photogrammetric mapping system based on Kern DSR-11 analytical plotter for shaft and tunnel geologic mapping at Yucca Mountain nuclear waste repository, Nevada. *Mailing Add:* US Geol Surv Fed Ctr Box 25046-MS 913 Denver CO 80225

PILLOFF, HERSCHEL SYDNEY, b Uhrichsville, Ohio, Aug 15, 40; m 68; c 3. MOLECULAR PHYSICS, LASERS. *Educ:* Mass Inst Technol, BS, 61; Cornell Univ, PhD(phys chem), 66. *Prof Exp:* USPHS fel molecular beams, Univ Toronto, 67-69; res chemist, Naval Res Lab, 69-74; PHYSICIST, OFF NAVAL RES, 74- *Mem:* Optic Soc Am. *Res:* Laser and optical physics. *Mailing Add:* Code 1112LO Off Naval Res 800 N Quincy St Arlington VA 22217

PILLSBURY, DALE RONALD, b Chico, Calif, Aug 10, 40; c 2. PHYSICAL OCEANOGRAPHY. *Educ:* Chico State Col, BS, 61; Univ Calif, Davis, MA, 64; Ore State Univ, PhD(oceanog), 71. *Prof Exp:* Asst, Univ Calif, Davis, 61-64; res asst oceanog, 64-67, RES ASSOC OCEANOG, ORE STATE UNIV, 67- *Mem:* Am Geophys Union; Am Meteorol Soc; Marine Technol Soc. *Res:* Coastal upwelling; deep ocean current measurements. *Mailing Add:* Col Oceanog Admin Bldg 104 Ore State Univ Corvallis OR 97331-5503

PILLSBURY, HAROLD C, b Baltimore, Md, Oct 18, 22; m 47; c 3. BIOCHEMISTRY. *Educ:* Loyola Col, Md, BS, 50. *Prof Exp:* Res chemist, Food & Drug Admin, 58-66; RES CHEMIST, FED TRADE COMN, 66-, DEPT SURG, UNIV NC. *Mem:* Fel Am Inst Chem. *Mailing Add:* Dept Surg Div Ent Univ NC 610 Burnett Womack Bldg CB No 7070 Chapel Hill NC 27599-7070

PILON, JEAN-GUY, b Ste Therese, Que, Can, Apr 7, 31; m 60; c 2. ENTOMOLOGY. *Educ:* Univ Montreal, BA, 53, BSc, 56; McGill Univ, MSc, 60; Yale Univ, PhD(bioclimat), 65. *Prof Exp:* Res officer entom, Can Dept Forestry, 56-65; asst prof, 65-67, assoc prof, 67-73, PROF ENTOM, UNIV MONTREAL, 73- *Mem:* AAAS; Entom Soc Can; Entom Soc Am. *Mailing Add:* Dept Biol Sci Univ Montreal CP 6128 Sta A Montreal PQ H3C 3J7 Can

PILOT, CHRISTOPHER H, b Ont, Can, Sept 3, 53; US citizen. SUPERGRAVITY, SUPERSYMETRY. *Educ:* Boston Univ, BA, 74; Tech Univ WGer, Dipl, 78, PhD(physics), 81. *Prof Exp:* Researcher, Max-Planck Inst Physics, Munich, WGer, 81-82, Mc Gill Univ, Montreal, 82-85; asst prof, Dept Physics, Univ Tulsa, 86-89; RES ASSOC, APPL INST RES, BOSTON, MA, 90- *Concurrent Pos:* Vis asst prof, Okla State Univ, 85-86. *Mem:* Am Phys Soc; Int Asn Particle Physicists; Am Asn Univ Prof. *Res:* Elementary particles; gravitation; unified field theory. *Mailing Add:* 233 Massachusetts Apt 215 Arlington MA 02174

PILSON, MICHAEL EDWARD QUINTON, b Ottawa, Ont, Oct 25, 33; m 57; c 2. MARINE CHEMISTRY. *Educ:* Bishop's Univ, BSc, 54; McGill Univ, MSc, 59; Univ Calif, PhD(marine biol), 64. *Prof Exp:* Asst res biologist, Inst Comp Biol, Zool Soc San Diego, 63-66; from asst prof to assoc prof, 67-78, PROF OCEANOG, GRAD SCH, UNIV RI, 78- *Concurrent Pos:* Dir, Marine Ecosysts Res Lab, 76- *Mem:* AAAS; Am Soc Mammal; Am Soc Limnol & Oceanog; Am Geophys Union. *Res:* Marine chemistry; milk of marine mammals; chemical ecology of coral reefs; chemistry of marine sediments; ecology of marine microcosms; estuarine chemistry; experimental biogeochemistry in large mesocosms simulating the coastal environment; fates and effects of interesting substances. *Mailing Add:* Grad Sch of Oceanog Univ of RI Narragansett RI 02882

PILTCH, MARTIN STANLEY, b Brooklyn, NY, Aug 11, 39; m 69; c 3. LASERS. *Educ:* Columbia Univ, AB, 60; Polytech Inst Brooklyn, MS, 68, PhD(electrophys), 71; Univ NMex, MBA, 80. *Prof Exp:* Staff mem lasers, TRG Control Data Corp, 60-66; GROUP LEADER LASERS, LOS ALAMOS SCI LAB, UNIV CALIF, 72- *Concurrent Pos:* NSF fel, 66-70; fel, Max Planck Inst Biophys Chem, 71-72; adj asst prof physics, Univ NMex, 75-; guest scientist, Max Planck Inst, 81. *Mem:* Inst Elec & Electronic Engrs; Am Phys Soc. *Res:* Atomic and molecular physics as applied to high power ultraviolet and infrared lasers and to nonlinear optical processes. *Mailing Add:* Los Alamos Nat Lab Univ Calif Box 1663 MS564 Los Alamos NM 87545

PILZ, CLIFFORD G, b Chicago, Ill, Apr 20, 21; m 57; c 1. INTERNAL MEDICINE, CARDIOLOGY. *Educ:* Univ Ill, Chicago, BS, 44, MD, 45. *Prof Exp:* Intern med, Cook County Hosp, 45-46, resident, 49; resident, Hines Vet Admin Hosp, 49-52; physician, Vet Admin Hosp, Iowa, 52-53; asst to chief med serv, 53-58, asst chief med, 58-71, actg chief, 69-71, CHIEF MED, VET ADMIN W SIDE HOSP, 71- *Concurrent Pos:* Fel path, Cook County Hosp, 48; clin assoc prof, Chicago Med Sch, 55-60, assoc prof, 60-66; assoc prof, Univ Ill Col Med, 66-68, prof, 68- *Mem:* Fel Am Col Physicians; fel Am Col Cardiol. *Mailing Add:* Vet Admin W Side Hosp PO Box 8195 Chicago IL 60612

PIMBLEY, GEORGE HERBERT, JR, b Cleveland, Ohio, Mar 11, 22; m 51; c 3. MATHEMATICS. *Educ:* Western Reserve Univ, AB, 43; Univ Calif, Los Angeles, MA, 49; NY Univ, PhD(math), 57. *Prof Exp:* Mathematician, US Naval Electronics Lab, Calif, 47-50; MEM STAFF, LOS ALAMOS SCI LAB, 50- *Concurrent Pos:* Asst, Inst Math Sci, NY Univ, 53-55; Ger Res Asn travel grant; vis lectr, Cologne Univ, 74; prog dir, Appl Math & Statist, NSF, 77-78. *Mem:* Am Math Soc. *Res:* Differential, integral and nonlinear functional equations; nonlinear functional analysis; neutron transport theory; mathematical problems in physics; biophysical problems. *Mailing Add:* Pajarito Acres 145 Monte Rey Dr S Los Alamos NM 87544

PIMBLEY, WALTER THORNTON, b Cleveland Heights, Ohio, July 12, 30; m 56; c 2. PHYSICS. *Educ:* Kent State Univ, AB, 52; Pa State Univ, MS, 55, PhD(physics, math), 59. *Prof Exp:* Instr physics, Pa State Univ, 58-59; staff physicist, IBM Corp, 59-61, adv physicist, 61-67, SR PHYSICIST, IBM CORP, 67- *Mem:* Am Phys Soc. *Res:* Surface studies, including emission of electrons and adsorption phenomena; propagation of electromagnetic waves; fluid dynamics. *Mailing Add:* 540 Torrance Vestal NY 13850

PIMBLOTT, SIMON MARTIN, b Derby, UK, Jan 8, 62. RADIATION CHEMISTRY, STOCHASTIC PROCESSES. *Educ:* Oxford Univ, UK, BA, 85, DPhil, 88. *Hon Degrees:* MA, Oxford Univ, UK, 88. *Prof Exp:* Res assoc, 88-90, ASST PROF SPEC, RADIATION LAB, UNIV NOTRE DAME, 90- *Mem:* Am Chem Soc; AAAS; Radiation Res Soc; Sigma Xi. *Res:* Description of nonhomogeneous systems such as those found in radiation chemistry and photochemistry; transfer of energy from ionizing particles to the medium; fragmentation and thermalization of the species produced and the short period of diffusion and reaction that follows. *Mailing Add:* Radiation Lab Univ Notre Dame Notre Dame IN 46556

PIMENTEL, DAVID, b Fresno, Calif, May 24, 25; m 49; c 3. ECOLOGY, ENTOMOLOGY. *Educ:* Univ Mass, BS, 48; Cornell Univ, PhD(entom), 51. *Prof Exp:* Asst entom, Cornell Univ, 48-51; chief trop res lab, USPHS, San Juan, PR, 51-53, proj leader tech develop lab, Savannah, Ga, 54-55; from asst prof to prof insect ecol & head dept entom & limnol, State Univ NY Col Agr, Cornell Univ, 55-69, prof insect ecol, 69-76, PROF INSECT ECOL & AGR SCI, CORNELL UNIV, 76- *Concurrent Pos:* Fel, Univ Chicago, 54-55; Orgn Europ Econ Coop fel, Oxford Univ, 61; NSF comput scholar, Mass Inst Technol, 61; invited mem panel environ pollution, President's Sci Adv Coun, 64-66; chmn biol & renewable resources panel, Life Sci Comt, Nat Acad Sci, 66-67, mem ad hoc comt environ aspects foreign assistance progs, 70, mem panel on water in man's life in India, 70, mem comt agr & the environ, 70- & co-chmn panel innovative mosquito control, Off Foreign Secy, 72-; mem coun, Nat Inst Environ Health Sci, 67-68; chmn training comt, Int Biol Prog, 67-70; mem comt pesticides, Dept Health, Educ & Welfare, 69; US del, UNESCO Conf Univ Governance & Role of Stud, 69; ecol consult environ qual, Off Sci & Technol, Exec Off of the Pres, 69-70; chmn biol panel, Advan Training Proj, NSF, 70; chmn panel environ impact herbicides, Environ Protection Agency, 72-; chmn, Panel Econ & Environ Aspects Pest Mgt in Cent Am, Nat Acad Sci, 73-77, comt food & food prod, 74-76, comt world food health & pop, 74-75; consult, Food & Agr Org UN, 74; chmn, Bd Sci & Technol Int Develop, Off Foreign Secy, Nat Acad Sci, 75-79; chmn, Nat Adv Coun Environ Educ, Off Educ, HEW, 75-78; pesticide adv coun, Environ Protection Agency, 75-78; comn int rel, Nat Acad Sci, 75-79; chair biomass panel, Energy Res Adv Bd, US Dept Energy, 79-; mem genetics panel & chair land degradation panel, Off Technol Assessment, 80-; mem res adv comt, USAID, 79-; chair, Environ Studies Bd, Nat Acad Sci, 81-82. *Mem:* AAAS; Ecol Soc Am; Soc Study Evolution; Entom Soc Am; Am Soc Zool; Sigma Xi. *Res:* Ecology and genetics of insect-plant, parasite-host and predator-prey population systems; environmental resource management and pollution, energy and land resources in the food system; ecosystems management and pest control. *Mailing Add:* Dept Entom & Limnol Cornell Univ Ithaca NY 14850

PIMENTEL, GEORGE CLAUDE, chemistry; deceased, see previous edition for last biography

PINA, EDUARDO ISIDORIO, b New Bedford, Mass, Oct 2, 31; m 53; c 4. MATHEMATICAL STATISTICS, OPERATIONS RESEARCH. *Educ:* Univ Mass, BS, 53, MA, 54. *Prof Exp:* Instr math, Univ Mass, 53-54; mathematician, Boeing Co, 54-55, lead engr, Systs Criteria Group, 55-56, acting supvr, Appl Math Res Group, 56-58, supvr, 58-59, chief, Anal Unit, 59-63, mgr, Tech Support, 63-65, Prog Appln, 65-66 & Opers & Comput Tech Sect, 66-67, dir, Opers Res-Mgt Sci, Commercial Airplane Div, 67-71, dir sales technol, 71-78, dir sales strategy anal & comput, 78-81, dir int airline anal, 81-83, dir airline anal, 83-85, dir sales & mkt comput, Boeing Commercial Airline Co, 85-87; dir, United Way Support, 87-89; RETIRED. *Concurrent Pos:* Mem sch bd, Highline Sch Dist 401, 87-, pres, 91- *Mem:* Sigma Xi. *Res:* Operations research; modeling; simulation; mathematical programming; airline passenger flight selector process considering departure/arrival time, airline cost, airplane services and preferences; aircraft routing and scheduling; competitive analysis of airlines and resulting market and profit impacts. *Mailing Add:* PO Box 98085 Des Moines WA 98198

PIÑA, ENRIQUE, b Mexico City, Mex, Oct 21, 36; m 61; c 4. BIOCHEMISTRY, ENDOCRINOLOGY. *Educ:* Nat Univ Mex, BSci, 53, MD, 61, PhD(biochem), 69. *Prof Exp:* Asst prof, 61-65, PROF BIOCHEM, SCH MED, NAT UNIV MEX, 66- *Mem:* Sigma Xi; Am Soc Biochem & Molecular Biol. *Res:* Metabolic action of ethanol, nucleosides and non steroidal anti-flammatory drugs on isolated hepatocytes. *Mailing Add:* A Postal 70159 Mexico City 04510 Mexico

PINAJIAN, JOHN JOSEPH, b Clinton, NJ, Oct 31, 21; m 49; c 1. PHARMACEUTICAL CHEMISTRY, NUCLEAR PHYSICS. *Educ:* Rutgers Univ, BSc, 49; Purdue Univ, MSc, 54, PhD(pharmaceut chem), 55. *Prof Exp:* Res assoc radiochem, US AEC, Agr Res Prog, Univ Tenn, 55, prof & res scientist, 55-56; chemist 56-60, group leader appl physics, Electronuclear Res Div, 60-62, group leader neutron & cyclotron prod, Isotopes Div, 62-65, develop specialist, Isotopes Div, 65-75, RES STAFF MEM, EMPLOYEE RELATIONS DIV, OAK RIDGE NAT LAB, 75- *Concurrent Pos:* Sci expert, Off Atomic Energy for Peace, Thailand, 66-67; consult, Cyclotron & Isotopes Labs, N V Philips-Duphar, Petten, N Holland, 68-69, tech mgr, 69; USAEC sci rep, US Consulate, Bombay, India, 72-73; adj prof, Vanderbilt Univ, 70- *Mem:* Fel AAAS; Am Chem Soc; Am Nuclear Soc; Am Phys Soc; Sigma Xi. *Res:* Production of radioisotopes in nuclear reactors and cyclotrons; internal conversion coefficient measurements; nuclear decay schemes; nuclear chemistry. *Mailing Add:* 161 LaSalle Rd Oak Ridge TN 37830-8547

PINCH, HARRY LOUIS, b Toronto, Ont, July 7, 29; nat US; m 55; c 2. PHYSICAL INORGANIC CHEMISTRY. *Educ:* City Col New York, BS, 51; Pa State Univ, PhD(chem), 55. *Prof Exp:* Res chemist, Mineral Beneficiation Lab, Sch Mines, Columbia Univ, 55-57; RES CHEMIST, DAVID SARNOFF RES CTR, 57- *Mem:* Am Chem Soc; Am Vacuum Soc, Sigma Xi. *Res:* Inorganic preparative chemistry; thin films; vacuum technics and sputtering. *Mailing Add:* David Sarnoff Res Ctr Princeton NJ 08543-5300

PINCHAK, ALFRED CYRIL, b Cleveland, Ohio, Aug 5, 35; m 71; c 6. FLUID SCIENCES, ANESTHESIOLOGY. *Educ:* Case Inst Technol, BSEE, 57; Purdue Univ, MSE, 59; Calif Inst Technol, PhD(mech eng & physics), 63; Case Western Reserve Univ, MD, 73, MS, 90. *Prof Exp:* Res scientist thermomech, Aerospace Res Labs, Off Aerospace Res, 63-66; res assoc fluid sci, Case Western Res Univ, 66-68, sr res assoc, 68-76, ASST PROF ANESTHESIOL, CAST WESTERN RESERVE UNIV, 76-, RES ENG-PHYSICIAN ANESTHESIOLOGIST, METRO HEALTH SYSTS. *Concurrent Pos:* Prin investr, Analytical Res Assocs, Ohio, 67-; dir, Nat Avalanche Observ, Juneau Icefield, Alaska, 68-; consult, Forestry Sci Lab, US Forest Serv, 70-; sr res fel, Am Heart Asn, 72-73. *Mem:* Inst Elec & Electronics Engrs; Am Soc Mech Engrs; Am Soc Anesthesiologists; Int Anesthesia Res Soc; Am Statist Asn. *Res:* Glaciology; supra-glacial meltwater streams; seismic detection of avalanches; two-phase flows; air-sea interaction; internal vortex flows; hydraulic transients; cardiovascular-pulmonary physiology; intragravel water flow; esophageal accelerometry; physiological measurements with accelerometers; flow measurement with self heated thermistors. *Mailing Add:* Metro Health Systs 3395 Scranton Rd Cleveland OH 44109

PINCK, ROBERT LLOYD, b Passaic, NJ, Aug 20, 20; m 49; c 3. RADIOLOGY. *Educ:* Washington & Lee Univ, AB, 42; Duke Univ, MD, 46; Am Bd Radiol, dipl, 53. *Prof Exp:* Clin assoc prof, 46-80, CLIN PROF RADIOL, STATE UNIV NY DOWNSTATE MED CTR, 80-; DIR DEPT RADIOL, LONG ISLAND COL HOSP, BROOKLYN, 46-, PRES MED BD, 72- *Concurrent Pos:* Asst attend radiologist, Roosevelt Hosp, New York, 53-54, assoc attend, 54-56. *Mem:* AMA; fel Am Col Radiol; fel Am Col Physicians; Am Roentgen Ray Soc; Radiol Soc NAm. *Res:* Diagnostic and therapeutic radiology. *Mailing Add:* 340 Henry St Brooklyn NY 11201

PINCKAERS, B(ALTHASAR) HUBERT, b Heerlen, Netherlands, Mar 21, 24; US citizen; m 47; c 4. ELECTRICAL ENGINEERING. *Educ:* HTS Eng Col, 44; Univ Minn, Minneapolis, MS, 60. *Prof Exp:* Civilian tech employee, US Army Signal Corps, Ger, 45-47; test engr, Honeywell Inc, 48-50, design engr, 50-57, prin develop engr, 57-66, staff engr, 66-71, design supvr, 71-77, mgr, Electronics Eng Residential Div, 77-78, chief engr advan technol residential group, 78-81; PRES & PRIN CONSULT, PINCKAERS ENG INC, 81- *Res:* Solid state electronics and control systems; holder of 80 US patents. *Mailing Add:* 7411 W 114th St Bloomington MN 55438

PINCKARD, ROBERT NEAL, b Chicago, Ill, Apr 16, 41; c 2. IMMUNOPATHOLOGY. *Educ:* Univ Kans, BA, 63; Univ Edinburgh, PhD(microbiol), 67. *Prof Exp:* From asst prof to assoc prof microbiol, Col Med, Univ Ariz, 68-76; PROF PATH, UNIV TEX HEALTH SCI CTR, 76-, PROF MED, 78- *Concurrent Pos:* Fel immunol, Scripps Clin & Res Found, 67-68; Nat Heart & Lung Inst grants, 85-91. *Mem:* Am Asn Path; Am Fedn Clin Res; Am Asn Immunol; Brit Soc Immunol; Col Int Allergologium; Reticuloendothelial Soc. *Res:* Immunopathology acute allergic and inflammatory reactions; structure and function of platelet activating factor; mechanisms for complement activation during ischemia and its role in tissue injury. *Mailing Add:* Dept Path Univ Tex Health Sci Ctr 7703 Floyd Curl Dr San Antonio TX 78284

PINCKNEY, ROBERT L, b Sept 26, 23; US citizen; m 45; c 4. AUTOMATED MANUFACTURING, PROGRAM MANAGEMENT RESEARCH & DEVELOPMENT. *Educ:* Iowa State Univ, BSE, 51. *Prof Exp:* Mgr mat technol, Wichita Div, Boeing Co, 51-64, Helicopter Div, 63-67, prog mgr advan rotors, 67-72, PROG MGR RES & DEVELOP, HELICOPTER DIV, BOEING CO, 72- *Concurrent Pos:* Sampe fel, Soc Aerospace Mat & Process Eng, 89; Boeing tech fel, Boeing Co, 90. *Honors & Awards:* Jud Hall Award, Soc Mfg Engrs, 89. *Mem:* Soc Advan Mat & Process Eng; Soc Mfg Engrs; Am Helicopter Soc. *Res:* Design, development, and manufacture of high performance composite helicopter components including glass, boron and carbon fiber motor systems; metal matrix transmission components; jet & engine thrust augmentation fuels and additives; Radan transparent structures; low abserrable aircraft materials and structures. *Mailing Add:* PO Box 16858 P38-48 Philadelphia PA 19142

PINCOCK, RICHARD EARL, b Ogden, Utah, Sept 14, 35; c 3. ORGANIC CHEMISTRY. *Educ:* Univ Utah, BS, 56; Harvard Univ, AM, 57, PhD(chem), 59. *Prof Exp:* Res fel chem, Calif Inst Technol, 59-60; instr, 60-62, from asst prof to assoc prof, 62-70, PROF CHEM, UNIV BC, 70- *Res:* Organic reactions and stereochemistry. *Mailing Add:* Dept Chem Univ BC Vancouver BC V6T 1Y6 Can

PINCUS, GEORGE, b Havana, Cuba, July 5, 35; US citizen; m 58; c 3. STRUCTURAL ENGINEERING. *Educ:* Ga Inst Technol, BCE, 59, MS, 60; Cornell Univ, PhD(struct), 63; Univ Houston, MBA, 74. *Prof Exp:* From res asst to res assoc struct eng, Cornell Univ, 61-63; from asst prof to assoc prof, Univ Ky, 63-67; chief-of-party, Grad Eng Prog, Brazil, 67-68, chmn dept, 75-79, prof civil eng, Univ Houston, 69-87; DEAN ENG, NJ INST TECHNOL, 87- *Concurrent Pos:* Consult, Architects Registr Bd, Ky, 63-64 & Engrs Registr Bd, 63-; fallout shelter instr & analyst, Off Civil Defense, 65-; consult, petrochem firms, 69-; syst design fel, NASA-AM Soc Eng Educ, 69, aeronaut & space res fel, 70; fel physiol with mod instrumentation, Baylor Col Med, 71; registered prof eng, Tex, Ky, Fla, Calif. *Honors & Awards:* D V Terrel Award, Am Soc Civil Engrs, 65. *Mem:* Am Soc Civil Engrs; Am Soc Testing & Mat; Am Soc Eng Educ; Am Concrete Inst. *Res:* Structural performance of wood members; tension strength of concrete; wood-concrete beams; strength of pipeline joints; fatigue strength of thin shells; wind effects on structures; behavior of biomaterials; foundations. *Mailing Add:* NJ Inst Technol Info Tech Bldg Central & Lock St Newark NJ 07107

PINCUS, HOWARD JONAH, b New York, NY, June 24, 22; m 53; c 2. ROCK PROPERTY TESTING, UNDERGROUND CONSTRUCTION. *Educ:* City Col New York, BS, 42; Columbia Univ, AM, 48, PhD(geol), 49. *Prof Exp:* Instr geol, Ohio State Univ, 49-51, from asst prof to prof, 51-69, chmn dept, 60-65; NSF sr postdoctoral fel, Appl Physics Lab, US Bur Mines, 62, res mgr, Twin Cities Mining Res Ctr, 67-68; prof dept civil eng, Univ Wis-Milwaukee, 69-72, prof dept geol & geol sci, 72-87, dean Col Letters & Sci, 87, EMER PROF, UNIV WIS-MILWAUKEE, 87- *Concurrent Pos:* Consult, eng geol & geotech eng, 55-; mem, US Nat Comt Tunneling Technol, Nat Res Coun, 72-74, Comt Rock Mech, 75-77, 80-89, comt chair, 85-87, chair, panel int activities, 87-89, study dir rock mech res req resource recovery, construction & earthquake hazard mitigation, 79-81; mem adv bd, Milwaukee Water Pollution Abatement Prog, 80-83; dir task force, US Nuclear Regulatory Comn, 82-86; chair US Nat Comt Int Asn Eng Geol, Nat Res Coun, 87-90; mech review panel, Yucca Mt Proj, Sandia Nat Lab, 89-; chair steering comt, Interlab Testing Prog, Inst Standards Res, Am Soc Test & Mats, 89- *Honors & Awards:* Frank W Reinhart Award, Am Soc Test & Mat, 87, Merit Award, 90. *Mem:* Fel AAAS; fel Geol Soc Am; Am Geophys Union; Soc Mining Engrs; Am Soc Test & Mat; Int Soc Rock Mechs; Sigma Xi; Am Inst Prof Geologists; Nat Soc Prof Engrs; Int Asn Eng Geologists; Am Asn Univ Prof. *Res:* Rock mechanics; engineering and geotechnical engineering; statistical analysis; optical data processing; fabric analysis as applied to deformation mechanics of rocks; tectonics. *Mailing Add:* PO Box 27598 San Diego CA 92198-1598

PINCUS, IRVING, b Brooklyn, NY, July 3, 18; m 43; c 4. MATERIALS ENGINEERING. *Educ:* City Col New York, BS, 38; George Washington Univ, MS, 47; Pa State Univ, PhD(fuel tech), 50; Rensselaer Polytech Univ, MS, 60. *Prof Exp:* Chemist, US Naval Powder Factory, 41-43, 45-48; res asst, Pa State Univ, 48-50; sect leader, Org By-prod, Great Lakes Carbon Corp, 50-56; coal chem, Curtiss-Wright Corp, 56-57; sr develop engr, Carbon Prod, Gen Elec Co, 57-60; prod mgr, Res Div, Raytheon Co, 60-63; chief mat eng & develop, Baltimore Div, Martin Co, 63-66; asst to pres, IIT Res Inst, 66-84; CONSULT, 85- *Mem:* Am Chem Soc; Am Inst Aeronaut & Astronaut; Sigma Xi; Am Defense Prepardness Asn; Am Arbitration Asn. *Res:* High temperature materials; graphites; carbons; cokes; advanced metals; composites; ceramics; plastics; fuels; tars; pitches; chemicals; management engineering. *Mailing Add:* 1099 Linda Lane Glencoe IL 60022

PINCUS, JACK HOWARD, b New York, NY, Jan 4, 39; m 66. BUSINESS DEVELOPMENT, TECHNOLOGY TRANSFER. *Educ:* NY Univ, AB, 59; Columbia Univ, PhD(biochem), 66. *Prof Exp:* Res asst biochem, Res Found Ment Hyg, New York, 62-66; res fel med, Mass Gen Hosp, Boston, 66-69; spec fel, Lab Immunol, Nat Inst Allergy & Infectious Dis, Bethesda, Md, 69-70, sr staff fel, Lab Microbiol & Immunol, Nat Inst Dent Res, 70-73; res chemist, VA Res Hosp, Chicago Ill, 73-75; asst prof surgery & biochem, Med Sch, Northwestern Univ, Chicago, 73-75; sr immunochemist, 75-77; assoc dir, Biomed Res Dept, Stanford Res Inst Int, 77-81; dir res & develop, Becton Dickinson Immunodiag, 81-84; dir mkt, Appl DNA Syst, Pittsburg, Pa, 84-86; VPRES ECON DEVELOP, MICH BIOTECHNOL INST, LANSING, 86- *Mem:* AAAS; Am Chem Soc; Am Inst Chem; Am Asn Immunol; Licensing Exec Soc; Technol Transfer Soc. *Res:* Research and technology management; technology transfer. *Mailing Add:* Mich Biotechnol Inst PO Box 27609 Lansing MI 48909

PINCUS, JONATHAN HENRY, b Brooklyn, NY, May 4, 35; m 61, 83; c 6. NEUROLOGY. *Educ:* Amherst Col, AB, 56; Columbia Univ, MD, 60. *Prof Exp:* From instr to prof neurol, Yale Univ, 64-87; CHMN NEUROL, GEORGETOWN UNIV, 87- *Concurrent Pos:* Clin fel neurol, Yale Univ, 61-64; co-auth, Behav Neurol, 78. *Res:* Role of thiamine in nervous system; mechanism of action of diphenylhydantoin; treament of Parkinsons; violence. *Mailing Add:* Dept Neurol Georgetown Univ 3800 Reservoir Ave NW Washington DC 20007

PINCUS, PHILIP A, b New York, NY, May 4, 36; m 59; c 2. THEORETICAL SOLID STATE PHYSICS. *Educ:* Univ Calif, Berkeley, AB, 57, PhD(physics), 61. *Prof Exp:* NSF fel, Saclay Nuclear Res Ctr, France, 61-62; from asst prof to assoc prof, Univ Calif, Los Angeles, 61-68, prof physics, 68-82; scientific adv, 82-85, PROF MAT & PHYSICS, UNIV CALIF, SANTA BARBARA, 85- *Concurrent Pos:* Alfred P Sloan Found fel, 64-67; chmn dept physics, Univ Calif, Los Angeles, 71-74. *Mem:* Am Phys Soc. *Res:* Magnetic properties of solids; molecular crystals. *Mailing Add:* Dept Mat Univ Calif Santa Barbara CA 93106

PINCUS, SETH HENRY, b New York, NY, July 25, 48; m 72; c 2. RHEUMATOLOGY. *Educ:* Middlebury Col, BA, 69; NY Univ, MD, 73. *Prof Exp:* Investr immunol, Immunol Br, Nat Cancer Inst, 75-78; ASSOC PROF PEDIAT, SCH MED, UNIV UTAH, 79- *Mem:* Am Acad Pediat; Am Rheumatol Asn; Am Asn Immunologists. *Res:* Antibody variable region genes; use of these genes, derived from monoclonal antibodies to produce genetically engineered antibodies; relationship between autoimmune disease and antibody variable genes. *Mailing Add:* Dept Pediatrics Univ Utah Med Ctr Salt Lake City UT 84132

PINCUS, THEODORE, RHEUMATOLOGY, IMMUNOLOGY. *Educ:* Columbia Univ, AB, 61; Harvard Univ, MD, 66. *Prof Exp:* Assoc, Sloan-Kettering Inst, 73-75; asst prof med-immunol, Stanford Univ Sch Med, 75-76; adj assoc prof med-rheumatology, Sch Med, Univ Pa, 76-80; PROF MED & MICROBIOL, SCH MED, VANDERBILT UNIV, 80- *Concurrent Pos:* Prof, Wistar Inst, Philadelphia, 76-80; dir, Clin Immunol Lab, Stanford Univ Hosp, 75-76. *Mem:* Fel Am Col Physicians; Am Rheumatism Asn; Am Soc Microbiol. *Res:* Morbidity and mortility of rheumatoid arthritis; host variables in chronic diseases; host genetic control of virus infection; psychological and economic consequences of chronic disease; socioeconomic status and chronic disease. *Mailing Add:* Div Rheumat & Immunol B-3219 MCN Vanderbilt Univ Nashville TN 37232

PINCZUK, ARON, b San Martin, Buenos Aires, Argentina, Feb 15, 39; m 62; c 2. OPTICAL PROPERTIES, SEMICONDUCTOR PHYSICS. *Educ:* Univ Buenos Aires, BS, 62; Univ Pa, PhD(physics), 69. *Prof Exp:* Asst prof, physics dept, Univ Pa 69-70; vis scientist, Nat Res Council, Argentina, 71-75, Max Planck Inst, 76 & IBM, 77; TECHNICAL, AT&T - BELL LABS, 78- *Honors & Awards:* Fel, Am Physical Soc. *Mem:* AAAS; Am Physical Soc. *Res:* Physics of condensed matter systems; advanced optical techniques; free carriers; semiconductors quantum wells; heterojunctions using inelastic light scattering; magneto-optics; optical properties of solids. *Mailing Add:* Rm 1D 433 AT&T Bell Labs Murray Hill NJ 07974

PINDER, ALBERT REGINALD, b Sheffield, Eng, June 10, 20; m 47; c 2. ORGANIC CHEMISTRY. *Educ:* Univ Sheffield, BSc, 41, PhD(chem), 48, DSc(chem), 63; Oxford Univ, DPhil(chem), 50. *Prof Exp:* Fel, Oxford Univ, 50-53; lectr org chem, Queen's Univ, Belfast, 53-56; sr lectr, Univ Wales, 56-65; prof org chem, 66-87, DISTINGUISHED EMER PROF ORG CHEM, CLEMSON UNIV, 87- *Concurrent Pos:* Mem, Chem Defence Res Dept, Brit Govt, 41-45. *Mem:* Fel Royal Soc Chem; Am Chem Soc. *Res:* Natural product chemistry, including isolation, synthesis, stereochemistry, and application of modern physical methods; organometallic chemistry. *Mailing Add:* 111 Lincoln Rd Edgewater FL 32032

PINDER, GEORGE FRANCIS, b Windsor, Can, Feb 6, 42; m 63; c 2. GROUNDWATER HYDROLOGY, NUMERICAL MATHEMATICS. *Educ:* Univ Western Ont, BSc, 65; Univ Ill, PhD(geol), 68. *Prof Exp:* Res hydrologist, US Geol Surv, 68-72; prof & dept chair civil eng, Princeton Univ,

72-89; DEAN ENG & MATH, UNIV VT, 89- *Concurrent Pos:* Pres, Hydrol Sect, Am Geophys Union, 88-90. *Honors & Awards:* Horton Award, Am Geophys Union, 69; O E Meinzer Award, Geol Soc Am, 75. *Mem:* Am Geophys Union; Am Soc Civil Engrs; Am Soc Petrol Engrs; Soc Indust & Appl Math. *Res:* Development and application of numerical models for groundwater flow and transport of contaminants. *Mailing Add:* Seven Bishop Rd Shelburne VT 05482

PINDER, JOHN EDGAR, III, b Baltimore, MD, Oct 19, 44. ECOLOGY, BIOSTATISTICS. *Educ:* Towson State Univ, BS, 67; Univ Ga, MS, 71, PhD(zool), 77. *Prof Exp:* Res assoc, 77-81, ASST RES ECOL, SAVANNAH RIVER ECOL LAB, UNIV GA, 81- *Mem:* Ecol Soc Am; Am Statist Asn. *Res:* Structure and function of plant communities; succession; radioecology of transuranic elements; biostatistics. *Mailing Add:* Savannah River Ecol Lab Drawer E Aiken SC 29801

PINDER, KENNETH LYLE, b Chaplin, Sask, May 20, 29; m 54; c 3. CHEMICAL ENGINEERING. *Educ:* McGill Univ, BEng, 51, MEng, 52; Univ Birmingham, PhD(chem eng), 54. *Prof Exp:* Engr, Shell Oil Co, 55-56; res engr, Pulp & Paper Res Inst Can, 56-59 & Dow Chem Can Ltd, 59-63; assoc prof chem eng, 63-71, PROF CHEM ENG, UNIV BC, 71-, ASST TO HEAD DEPT, 78- *Concurrent Pos:* Nat Res Coun Can res grant, 63-; consult & dir, Multifibre Process Ltd, 76- *Mem:* Chem Inst Can. *Res:* Time dependent rheology; direct contact boiling and condensing; gas hydrate formation; desalination; pollution; biomedical engineering; computer simulation. *Mailing Add:* Dept Chem Eng Univ BC 2075 Wesbrook Pl Vancouver BC V6T 1W5 Can

PINDERA, JERZY TADEUSZ, b Czchow, Poland, Dec 4, 14; Can citizen; m 49; c 2. ADVANCED EXPERIMENTAL MECHANICS. *Educ:* Warsaw Tech Univ, BEng, 36, MEng, 47; Lodz Tech Univ, MEng, 47; Polish Acad Sci, Dr Appl Sci, 59; Cracoco Tech Univ, DSc, 62. *Prof Exp:* Asst tech serv, Polish Air Lines, Lot, 47; lab dir, Aeronaut Inst, Warsaw, 47-52; Inst Metallog, 52-54; asst prof exp mech, Inst Fund Technol Res, Polish Acad Sci, 54-56, dep prof & lab dir, Dept Civil Eng, 56-59; lab dir, Inst Bldg Technol, 59-62; vis prof exp mech, Mich State Univ, 63-65; prof, 65-83, adj prof & dir, Fac Eng, Inst Exp Mech, 83-86, EMER PROF EXP MECH, UNIV WATERLOO, 87- *Concurrent Pos:* Consult, 50-; Nat Res Coun Can res grants, 65-; vis prof, Ruhr-Universitt Bochum, 72-73, Univ Poitiers, 73 & 75, Univ Braunschweig, 78, Nat Super Sch, Toulouse & Univ Stuttgart, 79; hon adv prof, Chongging Univ, 88-, Shanghai Col Archit Eng, 88-; mem, Int Sci Comt & Teaching Staff, Optical Ctr, Nuoro, Italy. *Honors & Awards:* M M Frocht Award, Soc Exp Mech, 78; G H Dugan Medal, Can Soc Mech Eng, 86. *Mem:* Fel Soc Exp Mech; Ger Soc Appl Math & Mech; NY Acad Sci; Am Soc Mech Eng; Soc Eng Sci; fel Can Soc Mech Eng. *Res:* Perception and modeling reality; theoretical foundation of experimental research; stress analysis; stress and flow binefringence; contact stresses; fracture mechanics; composite structures; new methods of stress analysis, iodynesi, strain gradient coupling thermoelastic coupling; paradigm in science and technology; theories of modeling reality, observations and measurement; stress and flow birefringence; fracture mechanics; composite structures; theory and techniques of isodynes; theory and techniques of stress gradient methods. *Mailing Add:* 310 Grant Crescent Waterloo ON N2K 2A2 Can

PINDOK, MARIE THERESA, b Chicago, Ill, Jan 20, 41. CARDIOVASCULAR PHYSIOLOGY. *Educ:* Loyola Univ, Chicago, BS, 63; Chicago Med Sch-Univ Health Sci, PhD(physiol), 78. *Prof Exp:* Res asst physiol, Stritch Sch Med, Loyola Univ, 63-70; res & teaching asst, 70-78, instr, 78-83, asst prof physiol, Chicago Med Sch-Univ Health Sci, 83-85; CO-DIR, BIOTECHNOL RESEARCH ASSOCS, 86- *Mem:* Am Physiol Soc; AAAS; Sigma Xi. *Res:* Hemodynamic and pharmacological intervention on cardiovascular dynamics with emphasis on changes in mycardial lipid metabolism. *Mailing Add:* 2775 Awayffaring Lane Lisle IL 60532

PINDZOLA, MICHAEL STUART, b Hartford, Conn, July 12, 48; m; c 2. THEORETICAL PHYSICS. *Educ:* Univ South, BA, 70; Univ Va, PhD(physics), 75. *Prof Exp:* PROF, PHYSICS DEPT, AUBURN UNIV, 77- *Concurrent Pos:* Consult, Oak Ridge Nat Lab, 80-, Lawrence Livermore Nat Lab, 85- *Mem:* Am Phys Soc; Brit Inst Physics. *Res:* Theoretical atomic, molecular, and optical physics. *Mailing Add:* Physics Dept Auburn Univ Auburn AL 36849-5311

PINE, JEROME, b New York, NY, Apr 14, 28; c 2. PHYSICS. *Educ:* Princeton Univ, AB, 49, Cornell Univ, MS, 52, PhD, 56. *Prof Exp:* From instr to asst prof physics, Stanford Univ, 56-63; assoc prof, 63-67, PROF PHYSICS, CALIF INST TECHNOL, 67- *Mem:* Soc Neurosci. *Res:* High energy experimental physics. *Mailing Add:* Dept Physics Calif Inst Technol Mail Sta 256-48 Pasadena CA 91125

PINE, LEO, b Tucson, Ariz, Feb 13, 22; m 43; c 6. BIOCHEMISTRY, MICROBIOLOGY. *Educ:* Univ Ariz, BS, 43; Univ Wis, MS, 48; Univ Calif, Berkeley, PhD, 52. *Prof Exp:* Med bacteriologist, Nat Microbiol Inst, 52-56, biochemist, Nat Inst Allergy & Infectious Dis, 56-57; asst prof & sr res fel microbiol, Sch Med, Duke Univ, 57-62; res prof mycol lab, Pasteur Inst, Paris, 62-63; biochemist Mycol Unit, Nat Ctr Dis Control, USPHS, 63-65, chief Res & Develop Unit, Biol Reagents Sect, 65-81, chemist, Bacterial Dis Div, 81-90; RETIRED. *Concurrent Pos:* Mem, Int Comt Taxon & Anaerobic Actinomycetes, 62- *Mem:* Fel AAAS; fel Am Soc Microbiol; Int Soc Human & Animal Mycol; Am Chem Soc. *Res:* Growth, physiology and metabolism of pathogenic bacteria and fungi. *Mailing Add:* Prods Develop Br Ctr for Dis Control Atlanta GA 30333

PINE, LLOYD A, b Emporia, Kans, June 5, 33; m 67; c 2. ORGANIC CHEMISTRY. *Educ:* Univ Kans, BA, 55, PhD(chem), 62. *Prof Exp:* Chemist, Esso Res Lab, La, 62-65, sr chemist, 65-67, group head, Enjay Chem Plant, 67-69; sr res chemist, 69-74, SR RES ASSOC EXXON RES LABS, 74- *Mem:* Fel Am Inst Chem. *Res:* Heterogeneous catalysis; petroleum resins; nitrogen heterocyclic compounds. *Mailing Add:* 5858 Berkshire Baton Rouge LA 70806

PINE, MARTIN J, b Forest Hills, NY, Nov 6, 27; m 54; c 3. BIOCHEMISTRY, ONCOLOGY. *Educ:* Cornell Univ, BS, 48; Univ Wis, MS, 50; Ind Univ, PhD(bact), 52. *Prof Exp:* Res assoc, Roswell Park Mem Inst, 54-56, sr cancer res scientist, 56-67, assoc cancer res scientist, 67-90; RETIRED. *Concurrent Pos:* NIH fel, Univ Calif, 52-53 & Yale Univ, 53-54. *Mem:* AAAS; Am Asn Cancer Res; Am Soc Biol Chem. *Res:* Oncology; pharmacology. *Mailing Add:* Dept Exp Therapeut Roswell Park Mem Inst 660 N Elm St Buffalo NY 14263

PINE, MICHAEL B, QUALITY ASSURANCE, DECISION THEORY. *Educ:* Harvard Univ, MD, 66. *Prof Exp:* ASSOC PROF MED, SCH MED, UNIV CINCINNATI, 81- *Concurrent Pos:* Prog dir, Med Dist 13, MEDIPRO Pilot, Vet Admin Med Ctr, 85- *Mailing Add:* Michael Pine & Assocs Inc 5050 S Lake Shore Dr Chicago IL 60615

PINE, STANLEY H, b Los Angeles, Calif, June 27, 35; m 57; c 3. ORGANIC CHEMISTRY. *Educ:* Univ Calif, Los Angeles, BS, 57, PhD(carbanion stereochem), 63. *Prof Exp:* Fel, Harvard Univ, 63-64; PROF CHEM, CALIF STATE UNIV, LOS ANGELES, 64- *Concurrent Pos:* Vis prof, Univ Strasbourg, France, 70-71, 76 & Calif Inst Technol, 78-79 & 86-88; vis sr chemist, Occidental Res Corp, 81. *Honors & Awards:* Chem Health & Safety Award, CHAS Div, Am Chem Soc, 90. *Mem:* Am Chem Soc. *Res:* Organic reaction mechanisms; organometallic synthesis. *Mailing Add:* Dept Chem Calif State Univ 5151 State Univ Dr Los Angeles CA 90032

PINEAULT, SERGE RENE, b Rimouski, Que, July 7, 47; m 76; c 2. ASTROPHYSICS, RADIO ASTRONOMY. *Educ:* Univ Laval, BSc, 70; Univ Toronto, MSc, 71, PhD(astron), 75. *Prof Exp:* Researcher astron, Inst Astron, Cambridge Univ, 76-77; lectr astron, Univ BC, 77-80; res assoc, Dominion Radio Astrophys Observ, BC, 80-82; NSERC res assoc, 82-87, ASSOC PROF, UNIV LAVAL, 87- *Concurrent Pos:* Nat Res Coun Can fel, 76-77. *Mem:* Can Astron Soc; Int Astron Union; Am Astron Soc; Can Inst Theoret Astron. *Res:* Theoretical astrophysics: compact objects, black holes, pulsars, active galactic nuclei; plasma astrophysics; radio astronomy: radio source spectra, galactic recombination lines, supernova remnants. *Mailing Add:* Dept Physics Univ Laval Ste Foy PQ G1K 7P4 Can

PINERO, GERALD JOSEPH, b New Orleans, La, Feb 26, 43; m 69; c 2. HISTOLOGY, CYTOLOGY. *Educ:* La State Univ, New Orleans, BS, 64; La State Univ, Baton Rouge, PhD(zool), 70. *Prof Exp:* from asst prof to assoc prof, 71-80, PROF HISTOL, DENT BR, HEALTH SCI CTR, UNIV TEX, HOUSTON, 80-, MEM FAC GRAD SCH BIOMED SCI, 81- *Concurrent Pos:* Rockefeller Pop Coun fel, La State Univ, 70-71. *Mem:* AAAS; Am Asn Dent Schs; Am Asn Anatomists; Tissue Cult Soc; Am Asn Dent Res. *Res:* Phagocytosis and intracellular digestion; connective tissue biology; fine structure; tissue culture. *Mailing Add:* 8302 Burning Hills Houston TX 77071

PINES, ALEXANDER, b Tel Aviv, Israel, June 22, 45; US citizen. SPECTROSCOPY & NUCLEAR MAGNETIC RESONANCE, MATERIALS SCIENCE. *Educ:* Hebrew Univ Jerusalem, BSc, 67; Mass Inst Technol, PhD(chem physics), 72. *Prof Exp:* From asst prof to assoc prof, 72-80, PROF CHEM, UNIV CALIF BERKELEY, 80-, FAC SR SCIENTIST NUCLEAR MAGNETIC RESONANCE, MATS & CHEMS SCI DIV, LAWRENCE BERKELEY LAB, 75- *Concurrent Pos:* Alfred P Sloan fel, 74-78; Camille & Henry Dreyfus teacher & scholar, 76-80; Richard Merton guest prof, Ger Sci Found, 77; distinguished vis lectr, Univ Tex, 78; chmn, Gordon Conf Magnetic Resonance, 85; permanent guest prof, E China Normal Univ, Shanghai, 85-; Prof Joliot-Curie, Phys & Chem Grad Sch, Paris, 87; mem, acad comt, Lab Magnetic Resonance & Atomic & Molecular Physics, Acad Sinica, 87-, ampere group, NSF High Magnetic Field Panel & sci adv bd. *Honors & Awards:* Vaughan Mem lectr, Calif Inst Technol, 81; Louis Strait Award, 82; Alexander von Humboldt Sr Scientist Award, 83; Baekeland Medal, Am Chem Soc, 85; Dept Energy Award, 83 & 87; Bourke lectr, Royal Soc Chem, 88; Ernest O Lawrence Award, Dept Energy, 88; Hinshelwood lectr, Oxford Univ, 89; Wolf Prize in Chem, 91. *Mem:* Nat Acad Sci; Royal Soc Chem; Int Soc Magnetic Resonance; AAAS; Am Chem Soc; fel Am Phys Soc. *Res:* Understanding of coherence and guantum phase helping in the development and application of modern nuclear magnetic resonance spectrascopy; quantum phase and topology in optics and magnetic resonance, two-dimensional and multiple-quantum spectroscopy, zero field NMR and SQUID detectors, high resolution solid state NMR, iterative maps on rotation group, relaxation and many-body dynamics and magnetic resonance imaging. *Mailing Add:* Dept Chem Univ Calif Berkeley Berkeley CA 94720

PINES, DAVID, b Kansas City, Mo, June 8, 24; m 48; c 2. CONDENSED MATTER THEORY, THEORETICAL ASTROPHYSICS. *Educ:* Univ Calif, AB, 44; Princeton Univ, MA, 48, PhD(physics), 50. *Prof Exp:* Instr physics, Univ Pa, 50-52; res asst prof, Univ Ill, 52-55; asst prof, Princeton Univ, 55-58; mem, Inst Adv Study, NJ, 58-59; dir, Ctr Advan Study, 67-70, PROF PHYSICS & ELEC ENG, UNIV ILL, URBANA-CHAMPAIGN, 59-, CTR ADVAN STUDY PROF PHYSICS & ELEC COMPUTER ENG, 90-; VPRES, SANTA FE INST, 85-, EXTERNAL PROF, 89- *Concurrent Pos:* NSF sr fel, Inst Theoret Physics, Copenhagen & Univ Paris, 57-58; Guggenheim fel & prof, Univ Paris, 62-63; co-chmn, Joint Soviet-US Symp Condensed Matter Theory, 68-79; vpres, Aspen Ctr Physics, 68-80; mem, Coun Biol Human Affairs, Salk Inst, 69-73; Guggenheim fel & Nordita prof, Copenhagen Univ, 70; Lorentz prof, Univ Leiden, 71; chmn, Bd Int Sci Exchange, Nat Res Coun, Nat Acad Sci, mem, Comt Int Rels & Comt Scholarly Commun Peoples Repub China, 74-79; Sherman Fairchild distinguished scholar, Calif Inst Technol, 77-78; mem, theory adv comt, Los Alamos Nat Lab, 77-85, chmn, 80-85; mem, Space Sci Bd, Nat Acad Sci, Nat Res Ctr, 78-81; mem, Ctr Adv Study, Univ Ill, Urbana, 79-; mem adv bd, Inst Theoret Physics, Univ Calif, Santa Barbara, 79-81; Gordon Godfrey prof, Univ New SWales, Austrailia, 85; B T Matthias vis scholar, Los Alamos, 86; prof, Col France, 89. *Honors & Awards:* Fritz London Mem Lectr, Duke Univ, 72; Guilio Racah Mem Lectr, Hebrew Univ, 74; Marchon Lectr, Univ Newcastle-upon-Tyne, 76; Eugene Feenberg Mem Lectr, Wash Univ, 82;

Eastman Kodak Distinguished Lectr, Univ Rochester, 83; Friemann Prize, 83; Dirac Medal, 85; Emil Warburg Lectr, Univ Bayreuth, 85; Eugene Feenberg Medal, 85. *Mem:* Nat Acad Sci; fel Am Astron Soc; fel Am Acad Arts & Sci; fel Am Physics Soc; fel AAAS; Am Philos Soc. *Res:* Theoretical studies of neutron stars; compact x-ray sources; quantum liquids. *Mailing Add:* Dept Elec Eng Univ Ill 1110 W Green St Urbana IL 61801

PINES, HERMAN, b Lodz, Poland, Jan 17, 02; nat US; m 27; c 1. ORGANIC CHEMISTRY. *Educ:* Univ Lyon, ChEng, 27; Univ Chicago, PhD(org chem), 35. *Hon Degrees:* DSc, Claude Bernard Univ, Lyon, France, 83. *Prof Exp:* From asst dir to dir, Ipatieff Lab, 41-70, from asst prof to assoc prof chem, 41-53, Ipatieff res prof, 53-70, IPATIEFF RES EMER PROF CHEM, NORTHWESTERN UNIV, EVANSTON, 70- *Concurrent Pos:* Res chemist, Universal Oil Prods Co, 30-45, coord explor res, 45-53; chmn, Gordon Conf Catalysis, 60; co-ed, Advances in Catalysis, 62-; vis prof, Bar-Ilan Univ Israel, 71, 72-77, Weizmann Inst Sci, Israel, 72-77, Fed Univ Rio de Janeiro, 73, Univ Wis, Milwaukee, 76 & Univ Calif, Berkeley, 85. *Honors & Awards:* Fritzsche Award, Am Chem Soc, 56 & Midwest Award, 63; Petrol Chem Award, Am Chem Soc, 81; Eugene J Houdry Award, Catalysis Soc, 81; Chem Pioneer Award, Am Inst Chemists, 82; E V Murphree Award, Am Chem Soc, 83. *Mem:* Am Chem Soc; Am Inst Chem; Catalysis Soc. *Res:* Petroleum; catalysis; terpenes; isomerization; hydrogen transfer; hydrogenolysis; aromatization; dehydration; base-catalyzed reaction of hydrocarbons; study of intrinsic acidity and catalytic activity of transition metals. *Mailing Add:* Dept Chem Northwestern Univ Evanston IL 60201

PINES, KERMIT L, b Brooklyn, NY, Oct 18, 16; m 40; c 3. INTERNAL MEDICINE. *Educ:* Columbia Univ, AB, 37, MD, 42, DMedSci, 47; Am Bd Internal Med, dipl, 50. *Prof Exp:* from asst to assoc, 48-58, from asst prof to prof clin med, 58-81, EMER PROF CLIN MED, COL PHYSICIANS & SURGEONS, COLUMBIA UNIV, 81- *Concurrent Pos:* Consult med, Presby Hosp, NY, 81- *Mem:* Am Diabetes Asn; Harvey Soc; fel Am Col Physicians. *Res:* Metabolic diseases. *Mailing Add:* 161 Ft Washington Ave New York NY 10032

PINES, SEEMON H, b Portland, Maine, Jan 3, 26; m 49; c 2. ORGANIC CHEMISTRY. *Educ:* Lehigh Univ, BS, 48; Univ Ill, MS, 49, PhD(org chem), 51. *Prof Exp:* Sr chemist, Merck & Co, 51-58, sect leader, Merck Sharp & Dohme Res Labs, 58-69, sr res fel, 69-75, sr investr, 75-77, SR DIR, MERCK & CO, 77-, EXEC DIR, 79- *Mem:* Am Chem Soc. *Res:* Process research and development. *Mailing Add:* 24 Candlewood Dr Murray Hill NJ 07974

PINGALI, KESHAV KUMAR, b Vijayawada, India, Feb 22, 57; m 88; c 1. PARALLELIZING COMPILERS, DATAFLOW ARCHITECTURE. *Educ:* Indian Inst Technol, Kanpur, BTech, 78; Mass Inst Technol, SM, 83, ScD(elec eng & computer sci), 86. *Prof Exp:* ASST PROF COMPUTER SCI, CORNELL UNIV, 86- *Concurrent Pos:* Fac develop award, Int Bus Mach, 86; consult, Hewlett-Packard Corp, 88-, Digital Equip Corp, 89-90 & Int Bus Mach, 90-; NSF presidential young investr, 89. *Honors & Awards:* President's Gold Medal, India, 78. *Mem:* Asn Comput Mach; Inst Elec & Electronics Engrs; Sigma Xi. *Res:* Software systems for parallel computers; programming languages for parallel machines, as well as parallelizing compiler technology. *Mailing Add:* Dept Computer Sci Cornell Univ 4152 Upson Hall Ithaca NY 14853

PINGLETON, SUSAN KASPER, b Elsworth, Kans, Sept 27, 46; m 77; c 1. PULMONARY DISEASE. *Educ:* Univ Kans, BA, 68, Med Sch, MD, 72. *Prof Exp:* Asst prof med, Univ Mo-Kansas City Med Sch, 77-79, assoc prof, 79-, head, Div Pulmonary Med, Truman Med Ctr, 77-; AT MED DEPT, COL HEALTH, UNIV KANS. *Concurrent Pos:* Prin investr, Pulmonary Academic Award, NIH, 77-82; Pulmonary Academic Award, NIH, 77-85; Res Day Award, Univ Kans, 67-68; NIH training grant, NIH, 75-77. *Mem:* Am Col Physicians; Am Col Chest Physicians; Am Thoracic Soc; Am Fedn Clin Res; Soc Critical Care Med. *Res:* Intensive care pulmonary disease; complications of respiratory failure, such as gastrointestinal hemorrhage and pulmonary emboli; effect of various treatment modalities on gastrointestinal hemorrhage in respiratory failure patients; complications of bronchodilators in respiratory failure secondary to spinal cord trauma. *Mailing Add:* Dept Med Univ Kans Sch Med Med Ctr Kansas City KS 66103

PINGS, C(ORNELIUS) J(OHN), b Conrad, Mont, Mar 15, 29; m 60; c 3. CHEMICAL ENGINEERING, CHEMISTRY. *Educ:* Calif Inst Technol, BS, 51, MS, 52, PhD(chem eng), 55. *Prof Exp:* Reservoir engr, Shell Oil Co, 52-53; from instr to asst prof chem eng, Stanford Univ, 55-59; from assoc prof to prof chem eng, Calif Inst Technol, 59-81, exec officer, 69-73, vprovost & dean grad studies, 70-81; PROVOST & SR VPRES ACAD AFFAIRS & PROF CHEM ENG, UNIV SOUTHERN CALIF, 81- *Concurrent Pos:* Co-ed, Phys & Chem Liquids, 67-78; ed, Chem Eng Commun, 72-77; consult adv bd, US Naval Weapons Ctr & Bd Dirs, Asn Univs Res Astron, 72-79; dir, Nat Comn Res, Washington, DC, 78-80, Hughes Aircraft Co, 84-85; chmn, Comt Sci, Eng & Pub Policy, 88- *Honors & Awards:* Lectureship Award, Am Soc Eng Educ, 69; Prof Progress Award, Am Inst Chem Engrs, 69, Tech Achievement Award, 72. *Mem:* Nat Acad Eng; fel Am Inst Chem Engrs; Am Chem Soc; Am Phys Soc; Am Soc Eng Educ. *Res:* Statistical mechanics; liquid state physics; applied chemical thermodynamics. *Mailing Add:* Provost Admin 101 Univ Southern Calif Los Angeles CA 90089-4019

PINHEIRO, MARILYN LAYS, b Brockton, Mass, Apr 18, 24; m 46; c 2. NEUROSCIENCES, AUDIOLOGY. *Educ:* Boston Univ, BA, 45; Western Reserve Univ, MA, 57, Case Western Reserve Univ, PhD(exp audiol), 69. *Prof Exp:* Dir clin audiol, Rio de Janeiro, 58-65; from asst prof to prof neurosci & dir Hearing & Speech Serv, Med Col Ohio, 71-80; res asst & instr, Case Western Reserve Univ, 69-70. *Concurrent Pos:* NIH fel, Case Western Reserve Univ, 69-71; NIH grant, 75-77; audiologist & speech therapist, Inst Neurol, Rio de Janeiro, 57-60; consult audiol, Cleveland Hearing & Speech Ctr, 70-71; adj prof, Grad Sch, Bowling Green State Univ, 71-73; adj asst prof commun, Univ Toledo, 73-80, adj assoc prof, Grad Sch, 73-80, clin supv, Clinical Audiol, 89-; consult, Family Learning Ctr, Toledo, 73-74. *Mem:* Am Speech & Hearing Asn; Soc Neurosci; Acoust Soc Am; Int Neuropsychology Soc. *Res:* Auditory perception and auditory dysfunction in central nervous system pathology. *Mailing Add:* 3021 Frampton Dr Toledo OH 43614

PINK, DAVID ANTHONY HERBERT, b St Lucia, WI, Nov 1, 36; Brit & Can citizen. THEORETICAL SOLID STATE PHYSICS, BIOPHYSICS. *Educ:* St Francis Xavier Univ, BSc, 61; Univ BC, PhD(theoret physics), 64. *Prof Exp:* Nat Res Coun fel, Dept Theoret Physics, Oxford Univ, 65-67; asst prof physics, 67-70, from actg chmn to chmn dept, 70-74, assoc prof, 70-77, PROF PHYSICS, ST FRANCIS XAVIER UNIV, 77- *Concurrent Pos:* Vis fac mem physics, Univ Guelph, 74-75; vis researcher, Max Planck Inst Solid State Res, Stuttgart, Ger, 75; Humboldt Fel, Inst Neurolbiol Kernforchungsanlage, Julich, Ger, 77-78. *Mem:* Biophys Soc; Can Asn Physicists; Am Phys Soc. *Res:* Theoretical studies of magnetically ordered systems with emphasis on surface excitations and related phenomena; cooperative effects in model and bio-membranes. *Mailing Add:* Dept Physics St Francis Xavier Univ Antigonish NS B2G 1C0 Can

PINKAVA, DONALD JOHN, b Cleveland, Ohio, Aug 29, 33; m 76; c 1. BOTANY. *Educ:* Ohio State Univ, BSc, 55, MS, 61, PhD(bot), 64. *Prof Exp:* From asst prof to assoc prof, 64-78, PROF BOT, ARIZ STATE UNIV, 78-DIR, HERBARIUM, 64- *Mem:* Am Soc Plant Taxon; Int Asn Plant Taxon. *Res:* Biosystematic studies of Compositae and Opuntia; floristic studies in Arizona, Northern Mexico, Southwestern United States; cytogenetics. *Mailing Add:* Dept Bot Ariz State Univ Tempe AZ 85287-1601

PINKEL, B(ENJAMIN), b Gloversville, NY, Mar 31, 09; m 40; c 2. ELECTRICAL ENGINEERING. *Educ:* Univ Pa, BS, 30. *Prof Exp:* Head Engine Anal Sect, Nat Adv Comt Aeronaut, 38-42, chief, Thermodyn Res Div, 42-45, Fuels & Thermodyn Res Div, 45-50 & Mat & Thermodyn Res Div, 50-56; assoc head Aero-Astronaut Dept, Rand Corp, 56-68, sr staff, 68-72; sr scientist, Bolt Beranek & Newman Inc, 72-74; consult engr, 74-78; RETIRED. *Concurrent Pos:* Served 17 govt & tech soc adv comts. *Mem:* Am Soc Mech Engrs; Am Nuclear Soc; fel Am Inst Aeronaut & Astronaut; Planetary Soc; AAAS. *Res:* Heat transfer; thermodynamics; propulsion systems; nuclear reactor analysis; author of 70 technical papers and one book. *Mailing Add:* 726 Adelaide Pl Santa Monica CA 90402

PINKEL, DONALD PAUL, b Buffalo, NY, Sept 7, 26; m 49; c 9. PEDIATRICS. *Educ:* Canisius Col, BS, 47; Univ Buffalo, MD, 51. *Prof Exp:* Intern pediat, Children's Hosp Buffalo, NY, 51-52, resident, 52-53, chief resident, 53-54; chief pediat, Roswell Park Mem Inst, 56-61; prof pediat & prev med, Univ Tenn, Memphis, 61-73; chmn dept, Med Col Wis, 74-78, prof pediat, 74-78, dir Midwest Children's Cancer Ctr, Milwaukee Children's Hosp, 78; chmn pediat, City of Hope Nat Med Ctr, 78-; CHMN, DEPT PEDIAT, TEMPLE UNIV. *Concurrent Pos:* Mead Johnson resident fel, Children's Hosp Buffalo, NY, 53-54; res fel, Children's Cancer Res Found, Boston, 55-56; med dir, St Jude Children's Res Hosp, Memphis, 61-73; assoc dir, Wis Clin Cancer Ctr, Madison, 74-78; pediatrician-in-chief, Milwaukee Children's Hosp, 74-78. *Mem:* Am Asn Cancer Res; Soc Pediat Res; Soc Exp Biol & Med; Am Pediat Soc. *Res:* Childhood cancer; leukemia virology; cancer chemotherapy; infectious diseases; child nutrition. *Mailing Add:* St Christophers Hosp for Children Fifth & Lehigh Ave Philadelphia PA 19133

PINKEL, ROBERT, b Cleveland, Ohio, Mar 30, 46; c 2. PHYSICAL OCEANOGRAPHY. *Educ:* Univ Mich, BA, 68; Univ Calif, San Diego, MA, 69, PhD(phys oceanog), 74. *Prof Exp:* Fel phys oceanog, 74-75, asst res oceanogr, 75-81, from asst prof to assoc prof, 81-87, PROF OCEANOG, MARINE PHYS LAB, UNIV CALIF, SAN DIEGO, 87- *Mem:* Am Geophys Union; fel Acoust Soc Am; AAAS. *Res:* Upper ocean studies of internal waves, air-sea interactions and mixing processes. *Mailing Add:* Scripps Inst of Oceanog Marine Phys Lab San Diego CA 92152

PINKERSON, ALAN LEE, GERIATRICS, CARDIOLOGY. *Educ:* Boston Univ, MD, 56. *Prof Exp:* Actg assoc dir, Nat Inst Aging, NIH, 85-88; RETIRED. *Mailing Add:* 4330 Klingle St NW Washington DC 20016

PINKERTON, A ALAN, b Watlington, UK, Aug 6, 43; m; c 1. X-RAY STRUCTURE DETERMINATION, LANTHANIDE INDUCED SHIFTS. *Educ:* Royal Inst Chem, UK, Grad RIC, 66; Univ Alta, Can, PhD(chem), 71. *Prof Exp:* Teaching fel chem, Univ Sussex, UK, 71-72; prin asst, Univ Lausanne, Switz, 72-74; teaching asst, Univ Nice, France, 74-76; prin asst crystallog, Univ Lausanne, 76-78, asst prof chem, 78-84; ASSOC PROF CHEM, UNIV TOLEDO, OHIO, 84- *Mem:* Am chem Soc; Royal Soc Chem; Am Crystallog Asn; Sigma Xi. *Res:* Chemistry of lanthanides, actinides and early transition metals; x-ray crystallography; nuclear magnetic resonance of diamagnetics and paramagnetics. *Mailing Add:* Dept Chem Univ Toledo Toledo OH 43606

PINKERTON, FRANK HENRY, b Nashville, Tenn, Sept 4, 45; m 67. CHEMISTRY. *Educ:* Vanderbilt Univ, BA, 67; Univ Southern Miss, PhD(chem), 71. *Prof Exp:* NASA fel, Univ Southern Miss, 71-72; asst prof chem, William Carey Col, 71-77; dept staff, 77-80, ASSOC PROF CHEM, CARSON-NEWMAN COL, 80- *Mem:* Am Chem Soc. *Res:* Organometallic chemistry; organosilicon chemistry; physiologically active organosilicon compounds; polymeric intermediates; thermally stable polymers; heterocyclic chemistry. *Mailing Add:* 918 Lakewood Dr Jefferson City TN 37760

PINKERTON, JOHN EDWARD, b Cortland, NY, Mar 2, 39; m 61. EXPERIMENTAL NUCLEAR PHYSICS, ELECTRONICS ENGINEERING. *Educ:* Geneva Col, BS; Univ Wis-Milwaukee, MS, 65; Univ SC, PhD(nuclear spectroscopy), 73. *Prof Exp:* Teacher, Pine Richland Schs, 60-61 & Audubon High Sch, 61-65; from asst prof to assoc prof physics, 65-79, PROF ELEC ENG, GENEVA COL, 79-, DIR ELEC ENG, 85- *Concurrent Pos:* Teaching asst, Univ SC, 69-71; pres, SPL Systems, 74-; fac physicist, US Bur Mines, Theoret Support Group, Bruceton Res Ctr, 76-; dir, Geneva Col Ctr for Technol Develop, 89- *Honors & Awards:* IR-100 Award, 85. *Mem:* Inst Elec & Electronic Engrs; Am Asn Physics Teachers; Optical Soc Am; Am Phys Soc. *Res:* Low energy nuclear physics, including associated solid state detectors and electronic instrumentation; design and development of digital systems for mini and micro computer experiment control and data acquisition; coal and rock dust deposition research. *Mailing Add:* Dept Elec Engr Geneva Va Col Col Ave Beaver Fall PA 15010

PINKERTON, PETER HARVEY, b Glasgow, Scotland, Feb 14, 34; m 66; c 2. HEMATOLOGY. *Educ:* Univ Glasgow, MB & ChB, 58, MD, 69; FRCP(E); FRCPath; FRCP(C). *Prof Exp:* Lectr haemat, Univ Glasgow, 63-65; asst res prof med, State Univ NY Buffalo, 65-67; PROF PATH, UNIV TORONTO, 67-; DIR LAB HAEMAT, SUNNYBROOK MED CTR, TORONTO, 67- *Concurrent Pos:* Med Res Coun Can grant, Univ Toronto, 68-75. *Honors & Awards:* Jr Sci Award, Can Asn Path, 69. *Mem:* Fel Am Col Physicians; Brit Soc Haemat; Am Soc Hemat; Can Hemat Soc. *Res:* Iron absorption; animal models of human blood diseases; cell surface markers on malignant cells, cancer cytogenetics. *Mailing Add:* Sunnybrook Med Ctr Toronto ON M4N 3M5 Can

PINKHAM, CHESTER ALLEN, III, b Ft Wayne, Ind, Aug 2, 36; m 59; c 2. PHYSICAL CHEMISTRY. *Educ:* Univ Ind, AB, 58; Purdue Univ, MS, 63; Rensselaer Polytech Inst, PhD(chem), 67. *Prof Exp:* Teacher high sch, 61-64; instr chem, Hudson Valley Community Col, 64-65; from asst prof to assoc prof, 67-77, PROF, TRI-STATE UNIV, 77-, CHMN SCI DEPT, 88- *Concurrent Pos:* Abstractor, Chem Abstracts, 59-73; dir, NE Ind Regional Sci Fair, 70-87. *Mem:* Fel Am Inst Chem; NY Acad Sci; Ind Acad Sci; Am Chem Soc; Nat Sci Teachers Asn; Soc Col Sci Teachers. *Res:* chemical education. *Mailing Add:* Dept Chem Tri-State Univ Angola IN 46703

PINKHAM, HENRY CHARLES, b New York City, NY, Sept 26, 48; m; c 1. ALGEBRAIC GEOMETRY. *Educ:* Harvard Col, BA, 70; Univ Paris, DEA, 71; Harvard Univ, PhD(math), 74. *Prof Exp:* Asst prof, 74-79, ASSOC PROF MATH, COLUMBIA UNIV, 79-, PROF MATH. *Res:* Algebraic geometry: deformations of singularities, moduli problems. *Mailing Add:* Dept Math Columbia Univ New York NY 10027

PINKSTAFF, CARLIN ADAM, b Louisville, Ill, June 10, 34; m 58; c 1. HISTOLOGY, HISTOCHEMISTRY. *Educ:* Eastern Ill Univ, BS, 60; Emory Univ, PhD(anat), 64. *Prof Exp:* From instr to asst prof anat, Dent Sch, Univ Ore, 64-67; asst prof, 67-70, assoc prof, 70-81, PROF ANAT, MED SCH, WEST VA UNIV, 81- *Concurrent Pos:* External examr anat, Univ Ibadan, 73; vis scientist, Yerkes Regional Primate Res Ctr, 73; vis prof, Semmelweis Univ, Budapest, Hungary, 82, St George's Univ Sch Med, Grenada, 79, 90 & 91. *Mem:* Am Asn Dent Res; NY Acad Sci; Int Asn Dent Res; Am Asn Anat; Histochem Soc; Am Soc Mammalogists. *Res:* Cytology, histology and histochemistry of salivary glands in normal and diabetic animals. *Mailing Add:* Dept Anat WVa Univ Med Ctr Morgantown WV 26506

PINKSTON, EARL ROLAND, physics; deceased, see previous edition for last biography

PINKSTON, JOHN TURNER, b Meridian, Miss, Mar 5, 15; m 40; c 2. PHYSICAL CHEMISTRY. *Educ:* Miss State Col, BS, 36; Ind Univ, PhD(phys chem), 40. *Prof Exp:* Res chemist, Universal Oil Prod Co, Ill, 40-44; sr chemist, Manhattan Dist, NY, 44-45; chemist, Harshaw Chem Co, Ohio, 45-50, asst dir, Tech Develop, 50-52; process consult, Catalytic Construct Co, 52-55, mgr process, 55-59; vpres & process consult, United Engrs & Constructors, Inc, Philadelphia, 59-66, vpres chem, 66-81; RETIRED. *Concurrent Pos:* With AEC, 44-46, 50- *Mem:* Am Chem Soc. *Res:* Catalytic conversions of hydrocarbons; fuel gas technology; uranium technology; chemical process engineering. *Mailing Add:* 60 Forest Lane Swarthmore PA 19081

PINKSTON, JOHN TURNER, III, b Chicago, Ill, Nov 28, 42; m; c 2. ELECTRICAL ENGINEERING. *Educ:* Princeton Univ, BSE, 64; Mass Inst Technol, PhD(elec eng), 67. *Prof Exp:* Elec engr, Nat Security Agency, 67-69; asst prof elec eng, Univ Md, 69-70; mem res staff, Nat Security Agency, 70-83; VPRES & CHIEF SCIENTIST, MICROELECTRONICS & COMPUT TECHNOL CORP, 83- *Mem:* Inst Elec & Electronic Engrs; Sigma Xi. *Res:* Information and statistical communication theory; error correcting codes; signal processing by computer; architecture of computer systems. *Mailing Add:* Microelectronics & Comput Technol Corp 3500 W Balcones Ctr Dr Austin TX 78759-6509

PINKSTON, MARGARET FOUNTAIN, b Macon, Ga, Jan 27, 19; m 46; c 3. BIOCHEMISTRY. *Educ:* Brooklyn Col, BA, 71; City Univ NY, PhD(biochem), 76. *Prof Exp:* ASST PROF BIOCHEM, MARY BALDWIN COL, 76- *Mem:* Am Chem Soc; Biophys Soc; Am Oil Chemists Soc; Sigma Xi. *Res:* Study of possible relationship between plasma lipoproteins and delayed hypersensitivity reactions to certain foods characterized by a high percentage of saturated triglycerides; interaction between model proteins and deoxyribonucleic acids. *Mailing Add:* 16 Church St Staunton VA 24401

PINKSTON, WILLIAM THOMAS, b Albany, Ga, Jan 19, 31; m 57; c 5. NUCLEAR PHYSICS, THEORETICAL PHYSICS. *Educ:* Cath Univ, AB, 52, MS, 55, PhD, 57. *Prof Exp:* Physicist, US Naval Ord Lab, Md, 55-57; instr physics, Princeton Univ, 57-59; from asst prof to assoc prof, 59-69, PROF PHYSICS, VANDERBILT UNIV, 69-, CHAIR DEPT, 69-76 & 85- *Concurrent Pos:* Sr fel, Bartol Res Found, Pa, 63-64; consult, Oak Ridge Nat Lab; Sr US Scientist Award, Alexander von Humboldt Soc. *Mem:* AAAS; fel Am Phys Soc; Am Asn Physics Teachers; Fedn Am Sci; Am Meteorol Soc. *Res:* Theory of nuclear structure and reactions; heavy-ion collisions. *Mailing Add:* Dept Physics Vanderbilt Univ Nashville TN 37235

PINKUS, A(LBIN) G(EORGE), physical organic chemistry, stereochemistry, for more information see previous edition

PINKUS, HERMANN (KARL BENNO), b Berlin, Ger, Nov 18, 05; nat US; m 35; c 1. DERMATOLOGY, ONCOLOGY. *Educ:* Univ Berlin, MD, 30; Univ Mich, MS, 35. *Prof Exp:* Asst dermat, Breslau Univ, 30-33; from instr to prof, 44-76, assoc path, 46-76, from actg chmn to chmn dept, 58-73, EMER PROF DERMAT, COL MED, WAYNE STATE UNIV, 76- *Concurrent Pos:* Res fel dermatopath, Wayne County Gen Hosp, Mich, 36-38; res assoc, Detroit Inst Cancer Res, 51-62. *Mem:* AAAS; Soc Invest Dermat (vpres, 57-58, pres, 58-59); Am Dermat Asn (vpres, 77-78); Am Asn Cancer Res; Am Acad Dermat (vpres, 71-72); Sigma Xi. *Res:* Anatomy and pathology of the skin, especially oncology. *Mailing Add:* Box 360 Monroe MI 48161

PINKUS, JACK LEON, b Syracuse, NY, July 17, 30; m 61. ORGANIC CHEMISTRY. *Educ:* Syracuse Univ, BS, 52; Univ Southern Calif, PhD(chem), 56. *Prof Exp:* Fulbright scholar, Groningen, 56-58; res assoc biochem, Upstate Med Ctr, State Univ NY, 58; instr & res assoc chem, Univ Pittsburgh, 58-60, asst prof, 60-61; asst prof, Wheeling Col, 61-63; NASA fel, Univ Pittsburgh, 63-65, asst res prof obstet & gynec, Sch Med, Univ Pittsburgh, 65-67; asst res prof obstet & gynec, Sch Med, Boston Univ, 67-71, assoc res prof, 71-75; PRES, BOSTON BIOCHEM CORP, 76-; CLIN ASSOC, MASS REHAB HOSP, 79- *Concurrent Pos:* Vis assoc prof, Dept Chem, Clark Univ, 75-76. *Mem:* Am Chem Soc; Royal Soc Chem; Royal Neth Chem Soc. *Res:* Heterocyclic compounds; reaction mechanisms; organic synthesis; steroid hormones. *Mailing Add:* 75 Cleveland Rd Wellesley MA 02181

PINNAS, JACOB LOUIS, b Newark, NJ, Jan 31, 40; m 64; c 3. ALLERGY, IMMUNOLOGY. *Educ:* Rutgers Univ, AB, 61; Univ Chicago, MD & MS, 65. *Prof Exp:* Resident med, State Univ NY, Syracuse, 65-66 & 68-70; med epidemiologist, Ctr Dis Control, 66-68; fel allergy & immunol, Scripps Clin & Res Found, 70-73; DIR ALLERGY SERV, SCH MED, UNIV ARIZ, 73-, ASSOC PROF INTERNAL MED, 79- *Concurrent Pos:* Mem attend staff, Univ Hosp, Univ Ariz Health Sci Ctr, 73-, dir allergy serv, 73-79; consult & dir allergy clin, Vet Admin Hosp, Tucson, 73- *Mem:* Am Asn Immunologists; Am Fedn Clin Res; fel Am Acad Allergy; fel Am Col Physicians. *Res:* Hypersensitivity reactions in respiratory tissues and skin; insect allergens and aeroallergens. *Mailing Add:* Dept Int Med Univ Ariz Sch Med 1501 N Campbell Tucson AZ 85724

PINNAVAIA, THOMAS J, b Buffalo, NY, Feb 16, 38; m 59; c 2. INORGANIC CHEMISTRY. *Educ:* State Univ NY Buffalo, BA, 62; Cornell Univ, PhD(inorg chem), 67. *Prof Exp:* From asst prof to assoc prof, 66-76, PROF CHEM, MICH STATE UNIV, 76- *Concurrent Pos:* Dir, Ctr Fundamental Mat Res, Mich State Univ, 89- *Honors & Awards:* George W Brindley Lectr, Clay Mineral Soc, 91. *Mem:* Clay Mineral Soc (pres, 90-91); AAAS; Am Chem Soc; Mat Res Soc. *Res:* Molecular and particle engineering of lamellar compounds for applications in heterogeneous catalysis, composite design, ceramics, electronic devices and related areas of materials chemistry; intercalation chemistry of layered silicate clays and complex layered oxides; phase transfer catalysis; surface coordination chemistry and organometallic chemistry; solid acids and bases. *Mailing Add:* 5901 Sleepy Hollow Rd East Lansing MI 48823

PINNELL, CHARLES, b Midland, Tex, Mar 16, 29; m 50; c 2. CIVIL ENGINEERING. *Educ:* Tex Tech Col, BS, 52; Purdue Univ, MS, 58; Tex A&M Univ, PhD, 64. *Prof Exp:* Asst prof civil eng, Tex Tech Col, 54-57; asst, Purdue Univ, 57-58; asst prof & asst res engr, Tex A&M Univ, 58-64, assoc prof, 64-70, prof civil eng, 70-89, assoc dean acad affairs, 64-89, asst dir, Tex Transp Inst, 76-89, assoc dep chancellor eng, 80-89; RETIRED. *Concurrent Pos:* Mem, Hwy Res Bd, Nat Acad Sci-Nat Res Coun, 58-; pres, PAWA Inc, 71-76. *Honors & Awards:* Award, Nat Acad Sci-Nat Res Coun, 59; Award, Inst Traffic Eng, 60. *Mem:* Am Soc Civil Engrs; Inst Traffic Eng. *Res:* Transportation; traffic control; freeway operations; administration; management science; computer utilization. *Mailing Add:* 1205 Munson College Station TX 77841

PINNELL, ROBERT PEYTON, b Fresno, Calif, Dec 5, 38; m 62. ORGANIC CHEMISTRY. *Educ:* Fresno State Col, BS, 60; Univ Kans, PhD(chem), 64. *Prof Exp:* Robert A Welch Found fel, 64-66; from asst prof to assoc prof inorg chem, 66-78, PROF CHEM, CLAREMONT COLS, 78-, CHMN, JOINT SCI DEPT, 74- *Concurrent Pos:* Vis Assoc Prof, Calif Inst Technol, 73-74. *Mem:* Am Chem Soc; AAAS; Sigma Xi. *Res:* Chemistry of organometallic and organometalloidal compounds, phosphorus compounds. *Mailing Add:* Joint Sci Dept Claremont Cols Claremont CA 91711

PINNEY, EDMUND JOY, b Seattle, Wash, Aug 19, 17; m 45; c 2. MATHEMATICAL ANALYSIS. *Educ:* Calif Inst Technol, BS, 39, PhD(math), 42. *Prof Exp:* Res assoc, Mass Inst Technol, 42-43; res analyst, Consol Vultee Aircraft Co, Calif, 43-45; instr math, Ore State Col, 45-46; dir, Off Naval Res Contract, 53-70; lectr, 46-47, from asst prof to assoc prof, 47-59, PROF MATH, UNIV CALIF, BERKELEY, 59- *Concurrent Pos:* Civilian with Off Sci Res & Develop, 44. *Mem:* Fel AAAS; Am Phys Soc; Am Math Soc. *Res:* Electromagnetic theory; integral equations; hydrodynamics; mechanics; electricity; nonlinear mechanics; calculus of variations in abstract spaces; plasma theory. *Mailing Add:* Dept Math Univ Calif Berkeley CA 94720

PINNICK, HARRY THOMAS, b Manhattan, Kans, Dec 18, 21; m 49; c 3. SOLID STATE PHYSICS. *Educ:* Southwestern Col, Kans, BA, 43; Univ Buffalo, PhD, 55. *Prof Exp:* Res physicist, Union Carbide Metals Co, 54-61; asst prof physics, Colo Sch Mines, 61-64; ASSOC PROF PHYSICS, UNIV AKRON, 64- *Concurrent Pos:* Vis staff mem, Los Alamos Sci Lab. *Mem:* Am Phys Soc; Sigma Xi. *Res:* Electron transport properties of metallic solid solutions; positron annihilation in transuranic elements; inelastic electron tunneling spectroscopy. *Mailing Add:* Dept of Physics Univ of Akron Akron OH 44304

PINNINGTON, ERIC HENRY, b London, Eng, Aug 19, 38; m 68; c 3. ATOMIC SPECTROSCOPY, ASTROPHYSICS. *Educ:* Univ London, BSc, 59, PhD(physics), 62; Imp Col London, ARCS, 59, dipl, 62. *Prof Exp:* Nat Res Coun Can fel, McMaster Univ, 62-64; Alexander von Humboldt Stiftung res fel, Max Planck Inst Physics & Astrophys, Munich, Ger, 64-65; from asst prof to assoc prof, 65-75, PROF PHYSICS, UNIV ALTA, 75- *Concurrent Pos:* Alexander von Humboldt res fel, Nuclear Res Estab, Julich, Ger, 71-72 & Bielefeld Univ, 76; sr vis fel, Oxford Univ, 80. *Mem:* Optical Soc Am; Can Asn Physicists (chmn, Atomic Molecular Physics Div, 86-88); NY Acad Sci. *Res:* Analysis of atomic spectra; measurement of atomic transition probabilities and fine structure effects using the beam-foil technique; investigations of new light sources for atomic spectra; laser excitation of fast ions. *Mailing Add:* Dept Physics Univ Alta Edmonton AB T6G 2J1 Can

PINNOW, KENNETH ELMER, b Villa Park, Ill, Nov 30, 28; m 58; c 3. METALLURGY. *Educ:* Mich Col Mining & Technol, BS, 51; Pa State Univ, MS, 55, PhD(metall), 61. *Prof Exp:* Jr metallurgist, Gen Motors Res, Mich, 51-53; instr metall, Pa State Univ, 58-60; sr res scientist, 61-78, mgr process res, 78-81, TECH DIR STAINLESS, TOOL & VALVE STEELS, CRUCIBLE RES, CRUCIBLE MAT CORP, 83- *Mem:* Nat Asn Corrosion Engrs; fel Am Soc Metals. *Res:* Physical and process metallurgy of stainless steels, tool steels, and other high alloy steels, especially studies of product properties including mechanical properties, formability, corrosion, hot and cold working, new alloys; continuous casting. *Mailing Add:* Crucible Res Crucible Mat Corp PO Box 88 Pittsburgh PA 15230

PINO, LEWIS NICHOLAS, b Niagara Falls, NY, June 1, 24; m 47; c 5. RESEARCH ADMINISTRATION. *Educ:* Univ Buffalo, AB, 47, PhD(chem), 50. *Prof Exp:* Res assoc, Univ Buffalo, 49-50; from instr to assoc prof chem, Alleghany Col, 50-56; from asst dean to assoc dean, Colo Col, 56-59; mem insts staff, NSF, 59-61, prog dir undergrad sci educ, 61-66; asst to chancellor, 66-67, dir res serv & dean spring & summer sessions, 67-71, dir res serv, 71-83, actg dean, Grad Sch, 81-82, prof, 71-86, EMER PROF CHEM, OAKLAND UNIV, 87- *Concurrent Pos:* Trustee, Cranbrook Inst Sci, 66-78; mem nat comn undergrad educ & educ of teachers, US Off Educ, 71-74. *Mem:* Am Chem Soc; Sigma Xi. *Res:* Heterocyclic chemistry; educational administration; higher education and public policy. *Mailing Add:* 2911 Heidelberg Rochester MI 48309

PINO, RICHARD M, b Brooklyn, NY, Mar 10, 50. ANESTHESIOLOGY, CYTOCHEMISTRY. *Educ:* Villanova Univ, BS, 72; Rutgers Univ, MS, 74; Hahnemann Med Col, PhD(anat), 78; La State Univ Sch Med, MD, 90. *Prof Exp:* Fel cell biol, Sch Med, Wayne State Univ, 78-80; from asst prof to assoc prof anat, La State Univ Med Ctr, 84-86, intern transitional prog, 90-91; RESIDENT ANESTHESIOL, MASS GEN HOSP, 91- *Mem:* Asn Military Surgeons US; Am Med Asn; Histochem Soc. *Res:* Structure and function of vascular endothelia in the eye, liver, and bone marrow; structure of Kupffer cells and retinal pigment epithelium; cell biology; cytochemistry; electron microscopy. *Mailing Add:* Dept Anaesthesia Mass Gen Hosp Boston MA 02114

PINSCHMIDT, MARY WARREN, b Washington, DC, Aug 10, 34; m 57; c 2. COMPARATIVE PHYSIOLOGY, EVOLUTION. *Educ:* Western Md Col, AB, 56; Duke Univ, MA, 61; Med Col Va, PhD(physiol), 73. *Prof Exp:* Instr, 61-67, asst prof, 68-75, assoc prof biol, 75-80, PROF BIOL SCI, MARY WASHINGTON COL, 81- *Concurrent Pos:* Assoc dean, Grad & Extended Progs. *Mem:* Am Soc Zool; Am Inst Biol Sci; AAAS. *Res:* Biological clocks; history of biology; aging. *Mailing Add:* Dept of Biol COMBS-108 Mary Washington Col Fredericksburg VA 22401

PINSCHMIDT, ROBERT KRANTZ, JR, b Los Angeles, Calif, July 25, 45; m 68; c 3. PHYSICAL ORGANIC CHEMISTRY, POLYMER CHEMISTRY. *Educ:* Wabash Col, BA, 67; Univ Ore, PhD(org chem), 71. *Prof Exp:* Vis asst prof phys org chem, Tech Univ Wroclaw, Poland, 71-73; independent fel, Dept Chem, Univ Ore, 73-74; sr res chemist ion exchange resins, Fluid Process Res, Rohm & Haas Co, Pa, 74-76; res chemist, environ & nitration chem, Indust Chem Res & Develop, 76-80, prin res chemist, polymer chem & technol, 80-81, sr prin res chemist, 81-82, group leader, 82-85, RES ASSOC, CORP SCI & TECHNOL CTR, AIR PROD & CHEM INC, 85- *Mem:* Am Chem Soc. *Res:* New monomer synthesis, water soluble polymers, emulsion and suspension; new reactions and molecular rearrangements; ion chromatography and nuclear magnetic resonance spectroscopy; waste abatement techniques. *Mailing Add:* Sci Ctr Air Prod & Chem Inc 7201 Hamilton Blvd Allentown PA 18195-1501

PINSCHMIDT, WILLIAM CONRAD, JR, b Richmond, Va, Oct 30, 26; m 57; c 2. ZOOLOGY. *Educ:* Mt Union Col, BS, 50; Ohio State Univ, MS, 52; Duke Univ, PhD(zool), 63. *Prof Exp:* Asst zool, Ohio State Univ, 50-52, asst instr, 51-52; instr, Mary Wash Col, 52-59, from asst prof to prof, 59-88, chmn dept biol, 68-70, EMER PROF BIOL, MARY WASH COL, 88- *Concurrent Pos:* Instr, Duke Univ, 56-57, asst to dir, Marine Lab, 59, investr, 64-; trustee, Va Chap Nature Conservancy, 71-75 & consult property acquisitions, 71-74. *Mem:* Atlantic Estuarine Res Soc; Estuarine Res Fedn. *Res:* Marine invertebrate zoology; crab larval development and ecological distribution; animal behavior and physiology; ecology. *Mailing Add:* Eight Nelson St Fredericksburg VA 22405

PINSKI, GABRIEL, b New York, NY, Sept 9, 37; m 60; c 3. PSYCHIATRIC EPIDEMIOLOGY, PSYCHOPHARMACOLOGY. *Educ:* Columbia Univ, AB, 57; Univ Rochester, PhD(physics), 64; Temple Univ, MS, 80; Univ Autonoma de Cd Juarez, MD, 82. *Prof Exp:* Instr physics, Syracuse Univ, 63-65; asst prof physics, Drexel Univ, 65-70; sr opers res analyst, Sun Oil Co, 70-73; res adv, Comput Horizons, Inc, 73-77; sr res scientist, Inst Sci Info, 77-80; resident internal med, Mercy Cath Med Ctr, Darby, Pa, 82-83; res psychiat, Thomas Jefferson Univ, Phila, Pa, 83-86; STAFF PSYCHIATRIST, COMMUNITY COUN, 86-; PSYCHIATRIST, UNIV CITY COUN CTR, 88- *Concurrent Pos:* Prin investr, NSF grant, 80-81. *Mem:* Sigma Xi; Opers Res Soc Am; Am Phys Soc; Am Psychiat Asn; Am Med Asn. *Res:* Statistical applications to medical problems. *Mailing Add:* 411 Witley Rd Wynnewood PA 19096

PINSKY, CARL, b Montreal, Que, May 15, 28; m 55; c 2. NEUROPHARMACOLOGY, NEUROPHYSIOLOGY. *Educ:* Sir George Williams Univ, BSc, 55; McGill Univ, MSc, 57, PhD(neurophysiol), 61. *Prof Exp:* From lectr to assoc prof, 58-78, PROF PHARMACOL & THERAPEUT, UNIV MAN, 78- *Concurrent Pos:* Operating & major equip grants, Med Res Coun Can, 62-79, Med Res Coun Can Scholar, 63-67; chmn, summer scholar comt, 74-75 & biomed grant rev comt, Non-Med Use of Drugs Directorate, Health & Welfare Can, 76-78. *Mem:* Pharmacol Soc Can; Am Soc for Pharmacol & Exp Therapeut. *Res:* Endogenous opiate-mimetic substances, their physiological mechanisms and role in opiate narcotic dependency; cholinergic mechanisms at the synaptic, membrane and molecular levels; cholinergic interactions with opiate narcotic drugs. *Mailing Add:* Dept Pharmacol & Therapeut Fac Med Univ Man 770 Bannatyne Ave Winnipeg MB R3T 2N2 Can

PINSKY, CARL MUNI, b Philadelphia, Pa, May 7, 38; m 84. ONCOLOGY, IMMUNOLOGY. *Educ:* Univ Pa, AB, 60; Jefferson Med Col, MD, 64. *Prof Exp:* From intern to resident internal med, Med Ctr, Univ Ky, 64-66; resident med & Am Cancer Soc clin fel, Mem Hosp Cancer & Allied Dis, 66-67, USPHS clin res trainee, 67-70; instr med, Cornell Univ, 70-75, from asst to assoc prof, 75-85; asst mem, Sloan-Kettering Inst Cancer Res, 74-85; chief, Biol Resources Br, Biol Response Modifiers Prog, Nat Cancer Inst, DCT & NIH, 85-88; VPRES, MED AFFAIRS, IMRE CORP, 88- *Concurrent Pos:* Res fel tumor immunol, Sloan-Kettering Inst Cancer Res, 67-70, res assoc, 70-74; fel med, Med Col, Cornell Univ, 67-70; clin asst physician med oncol, Mem Hosp Cancer & Allied Dis, 70-74, asst attend physician, 74-79 & assoc attend physician, 79-85. *Mem:* AAAS; Am Fedn Clin Res; Am Asn Cancer Res; Am Soc Clin Oncol; Soc Biol Therapy. *Res:* Evaluation of immunocompetence; development of immunodiagnostic techniques; exploration of immunotherapy in patients with cancer; evaluation and development of new biological response modifiers and immunoabsorption columns for preclinical and clinical trials. *Mailing Add:* Immune Medics 1500 Mt Bethel Rd Warren NJ 07060

PINSKY, LEONARD, b Montreal, Que, July 2, 35; m 60; c 4. GENETICS. *Educ:* McGill Univ, BSc, 56, MD, CM, 60; FRCPS(C), 67. *Prof Exp:* INVESTR SOMATIC CELL GENETICS, LADY DAVIS INST MED RES, JEWISH GEN HOSP, 67-, DIR DIV MED GENETICS, 67- *Concurrent Pos:* Med Res Coun Can operating grants, 67-; asst prof, McGill Univ, 68-, assoc prof, 73-; prof pediat, Ctr Human Genetics, 79- *Mem:* Am Soc Human Genetics; Soc Pediat Res; Can Col Med Geneticists (pres, 78-80). *Res:* Androgen resistance; disorders of sexual development; nosology of malformation syndromes. *Mailing Add:* Cell Genetics Lab Lady Davis Inst Jewish Gen Hosp Montreal PQ H3T 1E2 Can

PINSKY, MARK A, b Philadelphia, Pa, July 15, 40; m 63; c 3. MATHEMATICS. *Educ:* Antioch Col, BA, 62; Mass Inst Technol, PhD(math), 66. *Prof Exp:* Instr math, Stanford Univ, 66-68; asst prof, 68-72, assoc prof, 72-77, PROF, NORTHWESTERN UNIV, 77- *Concurrent Pos:* Vis prof, Univ Paris, 72-73. *Mem:* Am Math Soc; Inst Math Statist; Math Asn Am. *Res:* Stochastic differential equations; Markov processes; limit theorems in probability; asymptotic analysis of kinetic models. *Mailing Add:* Dept Math Northwestern Univ Evanston IL 60208

PINSLEY, EDWARD ALLAN, b New York, NY, June 18, 27; m 60; c 3. LASERS, ENGINEERING MANAGEMENT. *Educ:* City Col New York, BME, 49; Mass Inst Technol, MS, 50. *Prof Exp:* Res engr res labs, United Aircraft Corp, Pratt & Whitney Aircraft group, 55-59, supvr Turbomach Group, 55-59, supvr Elec Propulsion Group, 59-64, chief advan technol, 64-67,prog mgr advan laser technol, 67-68, prog mgr lasers, 68-73, chief laser eng, 73-79, chief tech develop, 79-80, dep dir tech & res, Govt Prod Div, United Technol Corp, 81-87; ASST DIR RES & TECHNOL, KAMAN AEROSPACE CORP, 87- *Honors & Awards:* George Mead Medal Eng Achievement, United Aircraft Corp, 67. *Mem:* Am Phys Soc; Am Inst Aeronaut & Astronaut; Inst Elec & Electronics Engrs. *Res:* Laser systems and components. *Mailing Add:* Kaman Aerospace Corp PO Box 2 Bloomfield CT 06002

PINSON, ELLIOT N, b New York, NY, Mar 18, 35; m 56; c 2. COMPUTER SCIENCE, ELECTRICAL ENGINEERING. *Educ:* Princeton Univ, BS, 56; Mass Inst Technol, SM, 57; Calif Inst Technol, PhD(elec eng), 61. *Prof Exp:* Mem tech staff, 61-65, supvr info processing res, 65-68, head, Comput Systs Res Dept, 68-77, dir, Bus Commun Systs Lab, 77-80, DIR, ADVAN SOFTWARE TECHNOL LAB, BELL LABS, 80- *Concurrent Pos:* Vis MacKay lectr, Univ Calif, Berkeley, 69-70. *Mem:* AAAS; Asn Comput Mach; Sigma Xi. *Res:* Computer science; graphical data processing; speech analysis and synthesis; operating systems design, modeling and evaluation; computer communications networks; private branch telephone exchange and office communications systems. *Mailing Add:* 581 Phillip Lane Watchung NJ 07060

PINSON, ELLIS REX, JR, b Wichita, Kans, Oct 23, 25; m 54; c 3. BIOCHEMICAL PHARMACOLOGY. *Educ:* Univ South, BSc, 48; Univ Rochester, PhD(org chem), 51. *Prof Exp:* Res chemist, Pfizer Inc, 51-59, proj leader, 59-61, sect mgr, 61-65, asst dir pharmacol res, 65-67, dir, 67-71, dir res, 71-72, vpres mech prod res & develop, 72-81, exec vpres, Pfizer Cent Res, 81-86; CONSULT, 86- *Mem:* Fel NY Acad Sci; Am Chem Soc; Am Soc Pharmacol & Exp Therapeut; AAAS. *Res:* Biochemical pharmacology; research management; toxicology. *Mailing Add:* 66 Braman Rd Waterford CT 06385

PINSON, ERNEST ALEXANDER, BIOLOGY. *Educ:* Univ Rochester, PhD(physiol), 39; Univ Calif, Berkeley, PhD(physics), 48. *Prof Exp:* Commandant, Air Force Inst Technol, 67-73; RETIRED. *Mailing Add:* 4917 Ravenswood Dr San Antonio TX 78227

PINSON, JAMES WESLEY, b Baton Rouge, La, Apr 10, 37; m 61; c 4. PHYSICAL CHEMISTRY & RADIATION CHEMISTRY. *Educ:* William Carey Col, BS, 59; Univ Miss, MSCS, 64, PhD(chem), 67. *Prof Exp:* Bookkeeper, W R Aldrich & Co, 59-60; head sci dept high sch, Miss, 60-63; from asst prof to assoc prof, 67-74, PROF CHEM, UNIV SOUTHERN MISS, 74-, ASST TO DEAN COL SCI, 71- *Concurrent Pos:* AEC-US Army mat loan grant, 68-70; Res Corp res grant, AEC nuclear educ & training grant & Miss State Bd Inst Higher Learning res grant, 69-70; Oak Ridge Assoc Univs-Atomic Energy Comn self-serv contract, Oak Ridge Nat Lab, 65-, consult, 67-; Sigma Xi grant, 70-71; Miss State Bd Inst Higher Learning grants, 70-72; NASA Earth Resources Lab grant, Air Pollution, Univ Miss, 71; mem, NASA Remote Sensing Study, Miss, 72- *Mem:* AAAS; Am Chem Soc; Am Phys Soc; Am Soc Mass Spectrometry; NAm Thermal Anal Soc; Sigma Xi. *Res:* Airborne radionuclides; remote sensing; air pollution; vibrational spectroscopy; cobalt 60 gamma ray radiolysis of boron, silicon, germanium and tin hydrides in the gas phase; characteriaztion of solid and liquid polymeric radiolytic products using mass spectrometric techniques. *Mailing Add:* 1008 SE Circle Hattiesburg MS 39402

PINSON, JOHN C(ARVER), b Lubbock, Tex, July 8, 31; m 52; c 1. SYSTEMS ENGINEERING, ELECTRICAL ENGINEERING. *Educ:* Tex Tech Col, BS, 52; Mass Inst Technol, SM, 54, ScD(elec eng), 57. *Prof Exp:* Sr res engr, 57-58, res specialist, 58-59, supvr systs eng, 59-60, group chief, 60-62, sect mgr, 62-63, asst chief engr, Systs Div, 63-65, CHIEF SCIENTIST, AUTONETICS STRATEGIC SYSTS DIV, ROCKWELL, INT, 65- *Mem:* Sigma Xi. *Res:* Systems engineering, mechanization and accuracy analysis on inertial guidance systems for marine, aircraft, missile and space vehicles. *Mailing Add:* 5069 Crescent Dr Anaheim CA 92807

PINSON, REX, JR, b Wichita, Kans, Oct 23, 25; m 54; c 3. BIOCHEMICAL PHARMACOLOGY. *Educ:* Univ of the South, BSc, 48; Univ Rochester, PhD(chem), 51. *Prof Exp:* Res chemist, Chas Pfizer & Co, Inc, 51-59, proj leader, 59-61, sect mgr, 61-65, asst dir pharmacol res, 65-67, dir, 67-71, dir res, 71-72, vpres med prod res & develop, 72-80, exec vpres, Pfizer Cent Res, Inc, 80-86; CONSULT, 86- *Mem:* AAAS; Am Soc Pharmacol & Exp Therapeut; Am Chem Soc; fel NY Acad Sci; Int Soc Biochem Pharmacol. *Res:* Physiological disposition of drugs; lipid and carbohydrate metabolism; research administration; toxicology. *Mailing Add:* Pfizer Inc Cent Res Eastern Point Rd Groton CT 06340

PINSON, WILLIAM HAMET, JR, b Atlanta, Ga, Sept 6, 19; m 42; c 3. GEOLOGY. *Educ:* Emory Univ, BA, 48, MS, 49; Mass Inst Technol, PhD, 51. *Prof Exp:* Res fel geol & instr astron, Harvard Univ, 51-53; from res assoc to assoc prof geol, 53-, EMER PROF GEOL, MASS INST TECHNOL. *Mem:* Geol Soc Am; Am Astron Soc; Mineral Soc Am; Meteoritical Soc. *Res:* Distribution and abundances of chemical elements; age determinations of rocks and meteorites; origin of meteorites, tektites; meteorite craters. *Mailing Add:* Dept Geol & Geophys 54 Bldg 1712 Rm 1528 Mass Inst Technol 77 Massachusetts Ave Cambridge MA 02139

PINTAR, M(ILAN) MIK, b Celje, Yugoslavia, Jan 17, 34; m 74; c 3. NUCLEAR MAGNETIC RESONANCE. *Educ:* Univ Ljubljana, BS, 59, MS, 64, PhD(physics), 66. *Prof Exp:* Res fel physics, Jozef Stefan Inst, Ljubljana, 59-66; fel, McMaster Univ, 66-67; from asst prof to assoc prof, 67-75, PROF, UNIV WATERLOO, 75- *Concurrent Pos:* Chmn, Waterloo Nuclear Magnetic Resonance Summer Sch, 69-89; dir, Nuclear Magnetic Resonance Ctr, Univ Waterloo, 85-90. *Honors & Awards:* B Kidric Nat Award, Slovenian Acad Sci & Arts, 61. *Mem:* Can Asn Physicists; Am Phys Soc; Ampere Soc; Int Soc Magnetic Resonances (secy gen, 78-83); Soc Magnetic Resonance Med. *Res:* Nuclear spin relaxation; atomic group tunneling in solids; nuclear magnetic resonance of disordered solids; nuclear magnetic resonance in medicine. *Mailing Add:* Dept Physics Univ Waterloo Waterloo ON N2L 3G1 Can

PINTER, CHARLES CLAUDE, b Budapest, Hungary, Mar 5, 32; US citizen; m 64; c 4. MATHEMATICS. *Educ:* Columbia Univ, BSc, 56; Univ Paris, PhD(math), 64. *Prof Exp:* Lectr math, European Div, Univ Md, 61-64; assoc prof, Northern Mich, 64-65; assoc prof, 65-72, PROF MATH, BUCKNELL UNIV, 72- *Concurrent Pos:* Fulbright grants, 66-67 & 78; NSF res prog col teachers res grant, 69-70, fac fel, 71-72. *Res:* Algebraic logic; universal algebra. *Mailing Add:* 232 S Second St Lewisburg PA 17837

PINTER, GABRIEL GEORGE, b Bekes, Hungary, June 23, 25; nat US; m 84; c 2. MEDICINE, PHYSIOLOGY. *Educ:* Univ Budapest, MD, 51. *Prof Exp:* From asst to asst prof physiol, Med Univ Budapest, 47-56; res assoc, Inst Exp Med Res, Oslo, Norway, 57; res assoc, Univ Tenn, 57-58, asst prof, 59-61; from asst prof to assoc prof, 61-70, PROF PHYSIOL, SCH MED, UNIV MD, BALTIMORE CITY, 70- *Concurrent Pos:* Vis prof, Univ Uppsala, Sweden, 72-73; hon sr res fel, King's Col, London, UK, 87. *Honors & Awards:* Alexander von Humboldt Prize, WGer, 80. *Mem:* Am Physiol Soc; Scand Physiol Soc. *Res:* Renal physiology; hematology; capillary permeability. *Mailing Add:* Dept Physiol Univ Md Baltimore MD 21201

PINTER, PAUL JAMES, JR, b Abington, Pa, Oct 27, 46. ENTOMOLOGY, BIOMETEOROLOGY. *Educ:* Ariz State Univ, BS, 67, MS, 69, PhD(zool), 76. *Prof Exp:* Res assoc entom, Ariz State Univ, 72-75; RES ENTOMOLOGIST, US WATER CONSERV LAB, 77- *Concurrent Pos:* Africa travel grant, Sigma Xi, 74. *Mem:* Entom Soc Am; Sigma Xi. *Res:* Remote sensing applications to agricultural resource managemtn; microclimatic influences on insect development and survival. *Mailing Add:* US Water Conserv Lab 4331 E Broadway Phoenix AZ 85040

PINTO, FRANK G, b New York, NY, Jan 2, 34; m 59; c 2. ORGANIC CHEMISTRY. *Educ:* Fordham Univ, BS, 55; Seton Hall Univ, MS, 64, PhD(org chem), 66. *Prof Exp:* Chemist, Semet-Solvay Div, Allied Chem Corp, 55-57; chemist, Bound Brook, Am Cyanamid Co, Wayne, 59-64, res chemist, 64-68, group leader org pigments, 68-69, dyes mfg, chief chemist rubber chem & specialty elastomers mfg, 72-77, prod mgr rubber chem & chem intermediates mfg, Bound Brook, 77-79, mkt mgr elastomers, 79-84; BUS MGR, CHEM PROD LATIN AM GROUP, 84- *Mem:* Am Chem Soc. *Res:* Synthesis of antioxidants, ultraviolet and infrared absorbers; process development of organic pigments; process and new product development of dyes; rubber chemicals and specialty elastomers; decomposition studies of mixed carboxylic and carbonic anhydrides. *Mailing Add:* 565 Emerald Trail Martinsville NJ 08836

PINTO, JOHN DARWIN, b Chicago, Ill, Dec 10, 40; m 63; c 2. ENTOMOLOGY, SYSTEMATICS. *Educ:* Humboldt State Col, BA, 63; Univ Ill, Urbana, PhD(entom), 68. *Prof Exp:* Asst prof biol, Calif State Polytech Col, San Luis Obispo, 68-70; NSF grant, 72-74, asst prof entom, 70-77, ASST ENTOMOLOGIST, UNIV CALIF, RIVERSIDE, 70-, ASSOC PROF ENTOM, 77-, PROF ENTOM. *Mem:* Am Entom Soc; Soc Syst Zool; Soc Study Evolution; Entom Soc Can. *Res:* Taxonomy, ethology and evolutionary studies of Meloidae. *Mailing Add:* Dept Entom Univ Calif Riverside CA 92521

PINTO, JOHN GILBERT, b Gubgull, India, Mar 7, 40; US citizen; m 72; c 21. CARDIAC MECHANICS, EDUCATIONAL ADMINISTRATION. *Educ:* Mysore Univ, India, BE, 64; Univ Toronto, MASc, 67; Univ Calif, San Diego, PhD(eng sci), 73. *Prof Exp:* Jr lectr mech eng, BDT Col, Mysore Univ, India, 63-64, asst lectr, 64-65; instr mach design, Univ Toronto, 65-68; postdoctoral fel bioeng, Univ Calif, San Diego, 72-74, lectr, 74-76, prin scientist & lectr, 76-79; PROF & CHMN MECH ENG & DIR BIODYNAMICS LAB, SAN DIEGO STATE UNIV, 79- *Concurrent Pos:* Consult, Bioeng Group, NOSC, San Diego, 79-87; co-researcher, Gait Lab, Children's Hosp, San Diego, 80-81; mem, Prog Proj, NIH, 86 & Panel Rev Bioeng Res Aid Handicapped, 87; chmn, Panel Young Investr Awards, NSF, 89; tech reviewer, Appl Mech Rev, J Biomech Eng, 77-85. *Honors & Awards:* Lamport Award, Int Soc Biorheology, 78; Ralph R Teetor Award, Soc Automotive Engrs, 81. *Mem:* Am Soc Mech Engrs; Am Heart Asn. *Res:* Study of cardiac-cardiovascular mechanics for the purpose of understanding and defining cardiac performance; development of invasive-non-invasive diagnostic techniques. *Mailing Add:* 18646 Aceituno St San Diego CA 92128

PINTO, LAWRENCE HENRY, b Paterson, NJ, Feb 9, 43. NEUROPHYSIOLOGY. *Educ:* Villanova Univ, BEE, 64; Northwestern Univ, Evanston, MS, 67, PhD(physiol), 70. *Prof Exp:* From asst prof to prof biol, Purdue Univ, 70-86; DEPT NEUROBIOL, NORTHWESTERN UNIV, 86- *Mem:* AAAS; Asn Res Vision & Opthal; Biophys Soc; Soc Neurosci. *Res:* Genetics of visual system; mechanism for generation of potentials in vertebrate retinal neurons. *Mailing Add:* Neurobiol Dept Northwestern Univ Hogan Hall Evanston IL 60201

PINTO DA SILVA, PEDRO GONCALVES, MEMBRANE & CELL BIOLOGY, ELECTRON MICROSCOPY. *Educ:* Univ Lisbon, Portugal, PhD(agr eng), 62; Univ Calif, Berkeley, PhD(plant physiol), 71. *Prof Exp:* CHIEF, MEMBRANE BIOL SECT, NAT CANCER INST, NIH, 75- *Res:* Cytochemistry. *Mailing Add:* Membrane Biol Sect Math Biol Dept Nat Cancer Inst NIH Bldg 538 Rm 104 FCRDC Frederick MD 21701

PINZKA, CHARLES FREDERICK, b New York, NY, Feb 5, 18; m 51; c 2. MATHEMATICS. *Educ:* Rutgers Univ, BS, 48; Univ Cincinnati, PhD(math), 62. *Prof Exp:* Res asst physics, Am Cyanamid Corp, 37-43; instr math, Xavier Univ, Ohio, 49-51; biometrician, E I du Pont de Nemours & Co, 51; math test specialist, Educ Testing Serv, 51-53, prog dir, 53-56; instr math, Xavier Univ, Ohio, 56-57; instr, 57-61, NSF sci fac fel, 61-62, asst prof, 62-70, ASSOC PROF MATH, UNIV CINCINNATI, 71- *Concurrent Pos:* Sr lectr, Univ Queensland, 68. *Mem:* Am Math Soc; Australian Math Soc; Math Asn Am. *Res:* Combinatorics; differential equations; probability; mathematics test construction; number theory. *Mailing Add:* Dept Math Univ Cincinnati Cincinnati OH 45221

PIOCH, RICHARD PAUL, b South Haven, Mich, Sept 5, 22; m 51; c 5. MEDICINAL CHEMISTRY. *Educ:* Mich State Col, BS, 48; Pa State Col, MS, 50, PhD(org chem), 52. *Prof Exp:* From org chemist to sr org chemist, Eli Lilly & Co, 52-65, res scientist, 66-72, res assoc, 73-87; RETIRED. *Mem:* Am Chem Soc. *Res:* Synthesis of compounds affecting the human central nervous system or gastrointestinal system; structure activity relationships among compounds affecting the central nervous system or gastrointestinal system; synthesis of heterocyclic compounds. *Mailing Add:* 3750 Briarwood Dr Indianapolis IN 46240

PIOMELLI, SERGIO, b Naples, Italy, Mar 29, 31; m 56; c 2. HEMATOLOGY, PEDIATRICS. *Educ:* Univ Naples, MD, 54. *Prof Exp:* Asst med, Harvard Med Sch, 58-60; asst prof, Med Sch, Univ Naples, 60-61; investr, Ital Nat Res Coun, Univ Rome, 61-63; assoc hemat, Mt Sinai Hosp, New York, 63-64; from asst prof to assoc prof pediat, Sch Med, NY Univ, 64-71, dir pediat hemat, Med Ctr, 64-79, prof, 71-79; PROF PEDIAT & DIR PEDIAT HEMATOL, COL PHYSICIANS & SURGEONS, COLUMBIA UNIV, 79- *Concurrent Pos:* NIH grant & City of New York Health Res Coun career scientist, Bellevue Hosp, NY Univ, 64-; assoc attend physician, Univ & Bellevue Hosps, New York, 64-71, attend physician, 72-; vis prof, Med Fac, Rotterdam, Holland, 68-69. *Mem:* Am Soc Pediat Res; Am Soc Hemat; Am Soc Human Genetics; Am Soc Clin Invest; NY Acad Sci. *Res:* Red cell metabolism; lead poisoning; porphyrin metabolism; biochemical genetics; childhood cancer chemotherapy. *Mailing Add:* Dept Pediat Columbia Univ 360 W 168th St New York NY 10032

PION, LAWRENCE V, b Providence, RI, July 11, 26; m 49; c 8. ANIMAL BEHAVIOR. *Educ:* Providence Col, BS, 50; Univ NH, MS, 52; Pa State Univ, PhD(animal behav), 70. *Prof Exp:* Instr, 51-64, actg head dept biol, 54-63, assoc prof, 64-71, chmn dept, 72-77, PROF BIOL, ST FRANCIS COL, PA, 71- *Mem:* Animal Behav Soc; AAAS. *Res:* Animal aggression. *Mailing Add:* Dept Biol St Francis Col Loretto PA 15940

PIONKE, HARRY BERNHARD, b Brooklyn, NY. SOIL CHEMISTRY. *Educ:* Univ Wis-Madison, BS, 63, MS, 66, PhD(soils), 67. *Prof Exp:* Asst prof soils, Univ Wis-Madison, 67-68; soil scientist, 68-74, RES LEADER, NORTHEAST WATERSHED RES CTR, AGR RES SERV, USDA, 74- *Mem:* Soil Sci Soc Am; Am Chem Soc. *Res:* Water chemistry of runoff and groundwaters related to geology and pollution potentials. *Mailing Add:* 111 Res Bldg A University Park PA 16802

PIORE, EMANUEL RUBEN, b Wilno, Russia, July 19, 08; US citizen; m 31; c 3. PHYSICS. *Educ:* Univ Wis, BA, 30, PhD(physics), 35. *Hon Degrees:* ScD, Union Col, 62, Univ Wis, 66; PhD, Bar Ilan Univ. *Prof Exp:* Asst instr physics, Univ Wis, 30-35; physicist, Electronic Res Lab, Radio Corp Am Mfg co, 35-38; engr in charge TV lab, Columbia Broadcasting Syst, 38-42; sr physicist, Bur Ships, Dept Navy, 42-44, head, Electronics Br, Off Naval Res, 46-47, dir, Phys Sci Div, 47-48; dir, Res Lab Electronics, Mass Inst Tech, 48-49; dep nat sci, Off Naval Res, 49-51, chief scientist, 51-55; vpres res & dir, Avco Corp, 55-56; dir res, IBM Corp, 56-61, vpres res & eng, 60-63, vpres & group exec, 63-65, vpres & chief scientist, 65-72, mem bd dirs, 62-73; RETIRED. *Concurrent Pos:* Trustee, Woods Hole Oceanog Inst, 56-64, 65-;

mem, Naval Res Adv Comt, 56-65; mem, President's Sci Adv Comt, 59-62, Nat Sci Bd, 61- & Bd Higher Educ, New York; bd mem, NY City Hall Sci, 60- & Resources for New Future, 70-73; mem corp, Polytech Inst Brooklyn, 62-; trustee, Sloan-Kettering Inst Cancer Res, 64-; mem bd dirs, Sci Res Assocs; adj prof, Rockefeller Univ, 74- *Honors & Awards:* Gold Medal, Indust Res Inst; Kaplun Int Prize, Hebrew Univ, 75. *Mem:* Nat Acad Sci (treas, 66-); Nat Acad Eng; Am Philos Soc; fel Inst Elec & Electronic Engrs; Am Acad Arts & Sci. *Res:* Electronics; thermionic, photoelectric and secondary emission; composite surfaces; metallic and solid state. *Mailing Add:* 115 Central Park W New York NY 10023

PIORE, NORA, b New York, NY, Nov 28, 12; m 31; c 3. HEALTH ECONOMICS, HEALTH POLICY. *Educ:* Univ Wis, AB, 33, MA, 34. *Prof Exp:* Economist, Comt on Labor & Pub Welfare, US Senate, 50-53; prog analyst, Interdept Comt on Low Incomes, NY, 56-57; spec asst to comnr, Dept Health, New York, 62-68; vis scientist & dir, Hosp Out-Patient Studies Prog, Asn for Aid to Crippled Children NY, 68-71; prof, Health Admin Columbia Sch Pub Health, Columbia Univ, 72-82, assoc dir, Ctr Community Health Systs, 72-86; CONSULT, HEALTH SERV RES, 86- *Concurrent Pos:* Adj prof urban studies, Hunter Col, 62-71; consult, Cunegie Corp, R W Johnson Found, HEW, 65-; vis fel, Ctr for Health Care Studies, United Hosp Found, 78-; prog health admin & econ, Sch Pub Health, Columbia Univ, 72-82; app mem, Nat Health Adv Coun, USPHS, HEW, 64-69, NY State Hosp Rev & Planning Coun, 75-85 & US Nat Comn Vital & Health Statist, 78-; mem, Nat Coun, Alan Guttmacher Inst, 74- & Tech Bd, Milbank Mem Fund, 76-78. *Mem:* Inst Med-Nat Acad Sci; assoc fel NY Acad Med; fel Am Pub Health Asn; Indust Relation Res Asn; Am Econ Asn. *Res:* Data systems and analytic tools for urban health planning, resource allocation and monitoring; health needs and services for urban children; changing role of hospitals in health delivery system. *Mailing Add:* 115 Cent Park W New York NY 10023

PIORKO, ADAM M, b Chorow, Poland, Sept 21, 47; Can citizen; m 76; c 2. ORGANOMETALLIC, ORGANIC SYNTHESIS. *Educ:* Silesian Tech Univ, Gliwice, Poland, BSc Eng, 71, MSc Chem, 71; Silesian Univ, Katowice, Poland, PhD, 79. *Prof Exp:* Teacher & res assoc, chem, Inst Chem, Silesian Univ, Katowice, Poland, 72-79, asst prof, 79-81; res assoc chem, Dept Chem Univ Saskatchewan, Can, 84-90; ASST PROF, DEPT CHEM, ST MARY'S UNIV, HALIFAX, NS, CAN, 90- *Concurrent Pos:* Engr, develop unit, Factory Paints & Enamels, Poland, 71-72; Postdoctoral, Dept Chem, Univ Saskatchewan, Can, 81-84; lectr, Dept Chem, Univ Saskatchewan, Can, 88-90. *Res:* Chemistry of biologically active compounds, synthesis and properties of functionalized arenes and heterocycles; synthetic organic chemistry; use of organometallic compounds in synthesis of heterocycles. *Mailing Add:* Dept Chem St Mary's Univ Halifax NS B3H 3C3 Can

PIOTROWICZ, STEPHEN R, b Cleveland, Ohio, Feb 20, 45. CHEMICAL OCEANOGRAPHY. *Educ:* Purdue Univ, BS, 67; Univ RI, MS, 72, PhD(oceanog), 77. *Prof Exp:* Res asst chem oceanog, Sch Oceanog, Univ RI, 70-72, teaching asst, 72-73; res assoc, Nat Acad Sci, 73; instr, Univ RI, 74-75 & res asst, Sch Oceanog, 75-76; prin scientist environ res, Energy Resources Co, Inc, Cambridge, 76-79; RES OCEANOGRAPHER CHEM, ATLANTIC OCEANOG & METEOROL LABS & NAT OCEANIC & ATMOSPHERIC ADMIN, US DEPT COM, 79- *Concurrent Pos:* Adj asst prof oceanog, Rosenstiel Sch Marine Atmospheric Sci, Univ Miami, 82- *Mem:* Am Chem Soc; Am Geophys Union; Geochem Soc; Am Meteorol Soc; AAAS; Am Soc Limnol Oceanog. *Res:* Marine chemistry, especially the biogeochemistry of trace elements in the marine environment; chemical fractionation at the air-sea interface; atmospheric chemistry primarily the cycles of trace elements and gases between the ocean, land and atmosphere. *Mailing Add:* Nat Oceanic & Atmospheric Admin 4301 Rickenbacker Causeway Miami FL 33149

PIOTROWSKI, GEORGE, b Koenigsberg, Ger, Jan 4, 42; c 4. MECHANICAL ENGINEERING. *Educ:* Mass Inst Technol, BSME, 64, MSME, 65; Case Western Reserve Univ, PhD(biomed eng), 75. *Prof Exp:* From asst prof to assoc prof, Mech Eng & Orthop Surg, Univ Fla, 69-90; PARTNER, DESIGN/ANALYSIS SERV CO, 83- *Concurrent Pos:* Consult, Am Acad Orthop Surg, 72-86 & Food & Drug Admin, Dept Health & Human Serv, 76-83. *Mem:* Am Soc Mech Engrs; Am Soc Eng Educ; Am Soc Testing & Mat; Vet Orthop Soc; Soc Automotive Engrs; Asn Advan Automotive Med. *Res:* Biomechanics and machine design. *Mailing Add:* 4001-37 NW 43rd St Gainesville FL 32606

PIOTROWSKI, JOSEPH MARTIN, b New Haven, Conn, July 1, 36; m 61; c 2. PETROLOGY, MINERALOGY. *Educ:* Univ Conn, BA, 61; Univ Western Ont, PhD(igneous petrol), 66. *Prof Exp:* From asst prof to assoc prof petrol & mineral, 66-74, PROF BIOL, SOUTHERN CONN STATE COL, 74-, PROF GEOL, 80- *Concurrent Pos:* Res grants, Geol Soc Am & Sigma Xi, 65; Ont grad fel, 65-66. *Mem:* Mineral Soc Am. *Res:* Experimental studies pertinent to genesis of alkaline under saturated rocks and lithium bearing pegmatites; experimental igneous petrology. *Mailing Add:* Dept Earth Sci Southern Conn State Univ 501 Crescent St New Haven CT 06515

PIOTROWSKI, ZBIGNIEW, b Kościan, Poland, Jan 26, 53; div; c 2. ANALYSIS & FUNCTIONAL ANALYSIS. *Educ:* Univ Wroclaw, BS, 74, MS, 76, PhD(math), 79. *Prof Exp:* From jr to asst prof math, Univ Wroclaw, Poland, 76-81; vis asst prof, Cleveland State Univ, 81, John Carroll Univ, 81-82, Auburn Univ, 82-84; asst prof, 84-88, ASSOC PROF MATH, YOUNGSTOWN STATE UNIV, 88- *Concurrent Pos:* Referee, Colloquium Mathematicum, 79-, Proc Am Math Soc, 86-, Int J Math & Math Sci, 88-, Mathematica Slovaca, 89-; reviewer, Math Rev, 80-, Zentralblatt fur Mathematik, 88-; translator, Am Math Soc, 85- *Mem:* Polish Math Soc; Am Math Soc; Sci Res Soc; Math Asn Am. *Res:* General topology and real functions; Blumberg and Baire spaces; quasi-continuity of functions; separate versus joint continuity problems; closed graph theorem; product topologies and continuous functions on products of spaces. *Mailing Add:* Dept Math & Computer Sci Youngstown State Univ Youngstown OH 44555

PIOUS, DONALD A, b Bridgeport, Conn, Feb 12, 30; m 54; c 2. MEDICINE. *Educ:* Univ Pa, AB, 52, MD, 56; Am Bd Pediat, dipl, 61. *Prof Exp:* Intern, Cincinnati Gen Hosp, 56-57; intern pediat, Med Ctr, Yale Univ, 57-58, asst resident, 58-59; from asst prof to assoc prof, 64-76, PROF PEDIAT, SCH MED & ADJ PROF GENETICS, UNIV WASH, 76- *Concurrent Pos:* Fel biol, Univ Calif, San Diego, 61-64; NIH res career develop award, 66 & 71; mem attend staff, Univ Wash Hosp, 64- & Children's Orthopedic Hosp, 71; mem, Mammalian Genetics Study Sect, NIH, 77-81; vis prof, Col de France, 78; mem, Genetic Basis Dis Study Sect, NIH, 83- *Mem:* Am Soc Human Genetics; Am Asn Immunol; Soc Pediat Res; Am Soc Cell Biol. *Res:* Somatic cell genetics; immunogenetics; regulation of gene expression. *Mailing Add:* Dept Pediat RD-20 Univ Wash Genetics Seattle WA 98195

PIPA, RUDOLPH LOUIS, b East Canaan, Conn, July 3, 30; m 53; c 2. ENTOMOLOGY. *Educ:* Cent Conn State Col, BS, 52; Univ Conn, MS, 55; Univ Minn, PhD(entom), 59. *Prof Exp:* NSF fel, Albert Einstein Col Med, 59-60; from asst prof to assoc prof, 60-72, PROF ENTOM, UNIV CALIF, BERKELEY, 72- *Concurrent Pos:* NIH postdoctoral fel, Cambridge Univ, Eng, 72-73. *Mem:* AAAS; Am Soc Zoologists; Sigma Xi. *Res:* Structure and histochemistry of the insect nervous system; endocrines and insect metamorphosis; hormones and insect neurometamorphosis; insect neurocytology. *Mailing Add:* Dept Entom Univ Calif Berkeley CA 94720

PIPBERGER, HUBERT V, b Camberg, Germany, May 29, 20; m 55. MEDICINE. *Educ:* Bad Godesberg Col, Ger, BA, 38; Univ Bonn, MD, 51. *Hon Degrees:* MD, Univ Mainz, Ger, 82. *Prof Exp:* Chief lab clin electrophysiol, Univ Hosp, Univ Zurich, 53-55; from instr to assoc prof med, Sch Med, Georgetown Univ, 56-71; chief, cardiovasc data processing, Vet Admin Res Ctr, 57-81; PROF MED & CLIN ENG, SCH MED, GEORGE WASHINGTON UNIV, 71- *Concurrent Pos:* Res fel, Los Angeles County Heart Asn, Calif, 55-56; Life Ins Med Res Found res fel, 57-59. *Honors & Awards:* William S Middleton Award, 61; Einthoven Medal, 74. *Mem:* AAAS; Am Col Cardiol; Am Heart Asn. *Res:* Electrocardiography; vectorcardiography; electronic data processing. *Mailing Add:* Dept Computer Med George Washington Univ 2300 K St NW Washington DC 20037

PIPENBERG, KENNETH JAMES, b Racine, Wis, Sept 20, 20; m 64; c 3. ANALYTICAL CHEMISTRY. *Educ:* Wheaton Col, Ill, BS, 43; Univ Ill, PhD(anal chem), 46. *Prof Exp:* Lab asst, Wheaton Col, 41-43; asst chem, Univ Ill, 43-45 & Penicillin Res Proj, Off Sci Res & Develop, 45-46; res assoc biol, Mass Inst Technol, 46-50; res chemist, Jackson Lab, 50-53, Petrol Lab, 53-58 & Eastern Lab, 58-61, sr res chemist, Repauno Develop Lab, 61-70 & Eastern Lab, 70-72, sr res chemist, Polymer Intermediates Dept, 72-77, SR RES CHEMIST, PETROCHEM DEPT, EXP STA, E I DU PONT DE NEMOURS & CO, 78- *Concurrent Pos:* Asst prof, King's Col, Del, 52-54, assoc prof, 54-55. *Mem:* AAAS; Am Chem Soc. *Res:* X-ray diffraction; starch; spectroscopy; combustion; air pollution; x-ray spectroscopy; trace element analysis. *Mailing Add:* North Rd Bernard VT 05031

PIPER, DAVID ZINK, b Lexington, Ky, Mar 12, 35; c 4. OCEANOGRAPHY. *Educ:* Univ Ky, BS, 60; Syracuse Univ, MS, 63; Scripps Inst, Calif, PhD(oceanog), 69. *Prof Exp:* Asst prof oceanog, Univ Wash, 68-75; OCEANOGR, US GEOL SURV, 75- *Mem:* AAAS; Am Geophys Union; Geochem Soc; Geol Soc Am. *Res:* Geochemistry of sedimentary phosphate deposits, their minor element and stable isotopic composition; geochemical relationships between seawater and marine sediments, with particular interest in the transition elements and rare earth elements. *Mailing Add:* 3180 Wash St San Francisco CA 94115

PIPER, DOUGLAS EDWARD, b Cobourg, Ont, July 6, 23; nat US. ORGANIC CHEMISTRY. *Educ:* Univ Man, BSc, 46; Univ Toronto, PhD(chem), 49. *Prof Exp:* Res chemist, Eastman Kodak Co, 50-58, asst div head, 58-61, assoc div head, 61-68, div head res labs, 68-; RETIRED. *Mem:* Am Chem Soc; Soc Photog Sci & Eng. *Res:* Photographic chemistry. *Mailing Add:* 30 Woodbury Way Fairport NY 14450-2475

PIPER, EDGAR L, b Sacramento, Calif, Apr 23, 37; m 59; c 1. ANIMAL PHYSIOLOGY, REPRODUCTIVE PHYSIOLOGY. *Educ:* Univ Nev, Reno, BS, 60, MS, 61; Utah State Univ, PhD(animal physiol), 67. *Prof Exp:* Res asst animal sci, Utah State Univ, 64-67; from asst prof to assoc prof, 67-83, PROF ANIMAL SCI, UNIV ARK, FAYETTEVILLE, 83- *Mem:* Am Soc Animal Sci. *Res:* physiology mechanisms involved in fescue toxicosis. *Mailing Add:* Animal Sci Ctr Univ of Ark Fayetteville AR 72701

PIPER, ERVIN L, b Cleveland, Ohio, Mar 31, 23; m 49; c 2. LUBRICATION ENGINEERING. *Educ:* Case Inst Technol, BS, 49; Univ Ill, MS, 51. *Prof Exp:* Sr group leader, carbon prod div, Union Carbide Corp, 51-81; TECHNOL MGR, DYLON INDUSTS, INC, 81- *Mem:* Soc Tribiologists & Lubrication Engrs. *Res:* Artificial graphite technology; plastic flow of graphite at elevated temperature. *Mailing Add:* 430 Cranston Dr Berea OH 44017

PIPER, JAMES ROBERT, b Tallassee, Ala, Jan 14, 33; m 57; c 3. ORGANIC CHEMISTRY. *Educ:* Auburn Univ, BS, 55, PhD(org chem), 60. *Prof Exp:* sr chemist, 59-82, head, Drug Synthesis Sect, 82-90, HEAD, BIO-ORGANIC CHEM DIV, KETTERING-MEYER LAB, SOUTHERN RES INST, 90- *Mem:* Am Chem Soc; Am Asn Cancer Res; Sigma Xi. *Res:* Synthesis of organic compounds; indole derivatives; radioprotective agents; antimalarial agents; folate antagonists. *Mailing Add:* Kettering-Meyer Lab Southern Res Inst 2000 Ninth Ave S PO Box 55305 Birmingham AL 35255-5305

PIPER, JAMES UNDERHILL, b Flint, Mich, July 30, 37; m 59; c 2. ORGANIC CHEMISTRY. *Educ:* Mass Inst Technol, BS, 59; Emory Univ, MS, 61, PhD(org chem), 63. *Prof Exp:* Asst prof chem, New Haven Col, 63-66; from asst prof to assoc prof, 66-78, chmn dept, 73-79, PROF CHEM, SIMMONS COL, 79- *Concurrent Pos:* Res staff chemist, Yale Univ, 63-66; res assoc, Mass Inst Technol, 66-67 & 72-73; vis fel, Worcester Found Exp Biol, 79-80. *Mem:* Am Chem Soc; AAAS; Royal Soc Chem. *Res:* Cycloaddition reactions; intramolecular interactions in heterocyclic compounds; chemistry of small ring compounds; penicillin chemistry. *Mailing Add:* 19 Mill Rd Harvard MA 01451

PIPER, JOHN, b London, Eng, Nov 25, 34; US citizen; m 58; c 3. PHYSICAL CHEMISTRY, METALLURGY. *Educ:* Trinity Col, BS, 56; Mass Inst Technol, PhD(phys chem), 60. *Prof Exp:* Phys chemist, Union Carbide Res Inst, 60-65, proj leader inorg mat, 65-67, mgr technol, 67-82, dir Technol, 82-87; VPRES TECHNOL KEMER ELECTRONICS CORP, 87- *Mem:* Electrochem Soc; Inst Elec & Electronics Engrs; Am Ceramics Soc. *Res:* Electronic components; high performance materials; thin metal films; inorganic fibers; superconductivity; transition metals, alloys and compounds; solid state diffusion; electronic band structure; physical measurements; far infrared spectrometry; ceramics; non-ferrous metallurgy. *Mailing Add:* Kemet Electronics Corp PO Box 5928 Greenville SC 29606

PIPER, RICHARD CARL, b Cleveland, Ohio, Mar 10, 32; m 56; c 2. TOXICOLOGY, VETERINARY PATHOLOGY. *Educ:* Ohio State Univ, DVM, 56, MSc, 57, PhD(vet path), 60; Am Col Vet Path, dipl, 60; Am Bd Toxicol, cert, 80. *Prof Exp:* Instr vet path, Ohio State Univ, 56-57, res asst, 57-58, instr, 58-60; assoc prof, Kans State Univ, 61-64; prof vet path, Wash State Univ, 64-75; res head, 75-85, ASSOC DIR PATH & TOXICOL, UPJOHN CO, 85- *Mem:* AAAS; Am Vet Med Asn; Am Col Vet Path; Soc Toxicol; Int Acad Path. *Res:* Toxicology and pathology of drugs and chemicals; general comparative pathology; mercurialism. *Mailing Add:* 5252 Birchwood Kalamazoo MI 49009

PIPER, ROGER D, b Oberlin, Ohio, Apr 28, 28; m 56; c 2. PHYSICAL CHEMISTRY. *Educ:* Oberlin Col, BA, 49; Emory Univ, MS, 51. *Prof Exp:* Chemist, 51-62, proj supvr, 62-66, res chemist, 66-69, Res Assoc, Mallinckrodt Chem Works, 69-; RETIRED. *Mem:* Am Chem Soc. *Res:* Process development in uranium production technology; fused salt electrochemistry; inorganic precipitation processes. *Mailing Add:* 908 Oge Ave St Louis MO 63131

PIPER, WALTER NELSON, b Ravenna, Ohio, Dec 28, 40; m 65; c 3. BIOCHEMISTRY, ENDOCRINOLOGY. *Educ:* Kent State Univ, BS, 62; Ohio State Univ, MS, 64; Purdue Univ, PhD(pharmacol), 69. *Prof Exp:* Biochemist, Warren-Teed Pharmaceut, Ohio, 64-65; res biochemist, Dow Chem Co, Mich, 69-72; res assoc pharmacol, Col Med, Univ Iowa, 72-74; asst prof, Col Med, Univ Okla, 74-75; asst prof, Univ Calif, San Francisco, 75-79; from assoc prof to prof pharmacol, Med Ctr, Univ Nebr, 79-86; PROF & DIR TOXICOL PROG, SCH PUB HEALTH, PROF PHARMACOL, MED CTR, PROF REPRODUCTIVE SCI PROG, UNIV MICH, 87- *Mem:* AAAS; Am Chem Soc; Sigma Xi; Soc Toxicol; Am Soc Pharmacol & Exp Therapeut; Am Soc Biochem & Molecular Biol; Am Pub Health Asn; NY Acad Sci; Soc Study Reproduction. *Res:* Biochemical pharmacology; toxicology; drug and toxicant regulation of the heme synthesis and steroid hormone biosynthesis; chemically induced and inherited porphyric diseases in liver bone marrow and endocrine organs; mechanisms of induction of drug hormone and toxicant metabolizing enzymes; endocrine/reproduction pharmacology toxicology. *Mailing Add:* Toxicol Prog 6108 SPH-II Univ Mich 109 S Observatory St Ann Arbor MI 48109-2029

PIPER, WILLIAM WEIDMAN, b Columbus, Ohio, July 1, 25; m 50; c 2. SOLID STATE PHYSICS. *Educ:* Columbia Univ, BS, 46; Ohio State Univ, PhD(physics), 50. *Prof Exp:* PHYSICIST, GEN ELEC RES & DEVELOP CTR, 50- *Mem:* Am Phys Soc; Inst Elec & Electronics Engrs; Electrochem Soc; Mat Res Soc. *Res:* Luminescence; electrical and optical properties of semiconductors and amorphous silicon. *Mailing Add:* Gen Elec Res & Develop Corp K1W B301 PO Box 8 Schenectady NY 12301

PIPERNO, ELLIOT, b New York, NY, Dec 30, 38; wid; c 2. PHARMACOLOGY, OTHER MEDICAL. *Educ:* Long Island Univ, BS, 61; Univ Mich, MS, 65; Mich State Univ, DVM, 68. *Prof Exp:* Instr pharmacol, Mich State Univ, 65-68; sr res toxicologist, Merck Inst Therapeut Res, 68-72; group leader, McNeil Labs, McNeil Pharmaceut Inc, 72-78, prin scientist toxicol, 78-79, res fel, Toxicol Res Unit, 79-84, mgr Drug Experience Reporting Sect, 84-87, mgr med info, 87-90; CONSULT, 90- *Concurrent Pos:* Wild game biologist, Mich Dept Conserv, 65-68; adj assoc prof pharmacol & toxicol, Sch Vet Med, Univ Pa, 79- *Honors & Awards:* Philip B Hofman Award, 80. *Mem:* Am Vet Med Asn; Am Acad Clin Toxicol; Am Acad Vet Pharmacol & Therapeut; Europ Soc Toxicol. *Res:* Comparative pharmacology; toxicology; mechanisms of drug-induced toxicity. *Mailing Add:* Piperno Adv Group PO Box 1058 Ft Washington PA 19034

PIPES, GAYLE WOODY, bionucleonics, chemistry, for more information see previous edition

PIPES, PAUL BRUCE, b Ft Worth, Tex, July 22, 41; m 66; c 2. LOW TEMPERATURE PHYSICS. *Educ:* Rice Univ, BA, 63; Stanford Univ, MS, 64, PhD(physics), 70. *Prof Exp:* NATO fel, Kamerlingh Onnes Lab, Leiden, Holland, 69-70; res instr physics, La State Univ, 70-72; from asst prof to assoc prof, 72-84, PROF PHYSICS & ASSOC DEAN SCI & DEAN GRAD STUDIES, DARTMOUTH COL, 84- *Concurrent Pos:* Alfred P Sloan res fel, 76-80. *Mem:* Am Phys Soc; Sigma Xi. *Res:* Low temperature physics. *Mailing Add:* Dept Physics & Astron Dartmouth Col Hanover NH 03755

PIPES, ROBERT BYRON, b Shreveport, La, Aug 14, 41; m 64; c 2. MECHANICAL ENGINEERING. *Educ:* La Tech Univ, BS, 64; Princeton Univ, MA & MSE, 69; Univ Tex, PhD(eng), 72. *Prof Exp:* Sr structures eng, Gen Dynamics, 69-72; asst prof mech eng, Drexel Univ, 72-74; assoc prof, 74-80, dir Ctr Composite Mat, 78-85, PROF MECH ENG, UNIV DEL, 80-, DEAN ENG, 85-, ROBERT L SPENCER PROF ENG, 88- *Concurrent Pos:* Prin investr, 15 Grants from Nat Aeronaut & Space Admin, Air Force Mat Lab, Naval Air Systs, Air Force Off, Off Naval Res & Gen Motors Corp; auth/consult, NASA/Dept Defense Composites Design Guide, 77-; mem Nat Mat Adv Bd, 78-, OTA Adv Panel & Mat Panel, Nat Eng Res Bd, US Army Res Off, Aeronautics & Space Eng Bd. *Honors & Awards:* Gustus L Larson Award, 83; Nat Acad Eng, 87. *Mem:* Am Soc Testing Mat; Soc Exp Stress Anal; Am Soc Mech Engrs; Nat Res Coun; Soc Mfg Eng; Am Soc Composites; Nat Mat Adv Bd; Soc Advan Mat Process Eng. *Res:* Experimental and analytical research in composite materials, manufacturing science, mechanics, characterization, finite element methods, computer aided design, structure-property relations, technical management. *Mailing Add:* Col Eng Univ Del Newark DE 19716

PIPES, WESLEY O, b Dallas, Tex, Jan 28, 32; div; c 4. ENVIRONMENTAL ENGINEERING, AQUATIC ECOLOGY. *Educ:* NTex State Univ, BS, 53, MS, 55; Northwestern Univ, Ill, PhD, 59. *Prof Exp:* Res engr, Univ Calif, 55-57; lectr civil eng, Northwestern Univ, Ill, 58-59, from asst prof to prof, 59-74, prof biol sci, 69-74; Betz prof ecol, prof biol sci & prof environ eng & sci, 75-83, head civil eng, 83-88, PROF CIVIL ENG, DREXEL UNIV, 83- *Concurrent Pos:* Consult, Commonwealth Edison Co, Pa Power & Light, US Environ Protection Agency & SKF Lab Co. *Honors & Awards:* Bd Dirs Award, Asn Environ Eng Professors, 75. *Mem:* Am Soc Microbiologists; Am Water Works Asn; Am Soc Civil Engrs; Water Pollution Control Fedn. *Res:* Water and waste treatment; biology of polluted waters; drinking water microbiology; microbiological monitoring and environmental measurements; biodegradation of hazardous wastes. *Mailing Add:* Dept Civil Eng Drexel Univ Philadelphia PA 19104

PIPKIN, ALLEN COMPERE, b Mena, Ark, May 21, 31; m 56; c 3. APPLIED MATHEMATICS. *Educ:* Mass Inst Technol, ScB, 52; Brown Univ, PhD(appl math), 59. *Prof Exp:* Res assoc fluid dynamics, Inst Fluid Dynamics & Appl Math, Univ Md, 58-60; from asst prof to assoc prof, 60-66, PROF APPL MATH, BROWN UNIV, 66- *Concurrent Pos:* Guggenheim fel, Univ Nottingham, 68; Sci Res Coun sr vis fel, Nottingham, 78 & 82. *Mem:* Soc Rheol; Soc Nat Philos. *Res:* Constitutive equations; viscoelastic fluids; fiber-reinforced composites. *Mailing Add:* Div Appl Math Brown Univ Providence RI 02912

PIPKIN, BERNARD WALLACE, b Los Angeles, Calif, Dec 5, 27; m 57; c 3. ENVIRONMENTAL GEOLOGY. *Educ:* Univ Southern Calif, BA, 53, MA, 56; Univ Ariz, PhD(geol), 65. *Prof Exp:* Consult, Thomas Clements Assoc, 60-69; adj prof geol & asst dean, Col Lett, Arts & Sci, 69-86, PROF GEOL, UNIV SOUTHERN CALIF, 86- *Concurrent Pos:* Consult, Los Angeles Sch Sci Adv Bd, 72- & Planning Comn, City Palos Verdes Estates, 75; panelist, Nat Res Coun Postdoctoral Assoc Prog, 74-; expert witness, State of Calif; coastal comnr, Arthur D Little & Co & TRW; pres, Western Sect, Nat Asn Geol Teachers, 83-84 & 86-87. *Mem:* Geol Soc Am; Am Asn Petrol Geologists; Asn Eng Geologists; Nat Asn Geol Teachers (pres, 86-87); Sigma Xi. *Res:* Geologic hazards in the urban environment, particularly landslides, seismic hazards and coastal erosion. *Mailing Add:* Box 2391 Rolling Hills Estates CA 90274

PIPKIN, FRANCIS MARION, b Marianna, Ark, Nov 27, 25; m 58; c 2. EXPERIMENTAL ATOMIC PHYSICS. *Educ:* Univ Iowa, BA, 50; Princeton Univ, MA, 53, PhD, 54. *Hon Degrees:* MA, Harvard Univ, 60. *Prof Exp:* Soc Fellows jr fel, Harvard Univ, 54-57, from asst prof to prof physics, 57-74, assoc dean, Fac Arts & Sci, 74-77, chmn physics dept, 85-88, BAIRD PROF SCI, HARVARD UNIV, 76- *Concurrent Pos:* Sloan Found fel, 59-61. *Mem:* Fel Am Phys Soc; AAAS. *Res:* Atomic beams; elementary particle physics; experimental high energy physics. *Mailing Add:* 10 Kilburn Rd Belmont MA 02178

PIPPEN, RICHARD WAYNE, b Villa Grove, Ill, Sept 8, 35; m 61; c 1. SYSTEMATIC BOTANY, PLANT ECOLOGY. *Educ:* Eastern Ill Univ, BS, 57; Univ Mich, MA, 59, PhD(bot), 63. *Prof Exp:* From asst prof to assoc prof, 63-78, PROF BIOL, WESTERN MICH UNIV, 78- *Concurrent Pos:* Dir trop studies prog, Assoc Univs Int Educ, 74-76; actg chairperson biol dept, Western Mich Univ, 75-76, chairperson, 77- *Mem:* Am Soc Plant Taxon; Asn Tropical Biol; Sigma Xi. *Res:* Reproductive biology of flowering plants; ecology of rare and endangered plants; systematics and ecology of cacaliod complex of composites; ecology of wetlands in southern Michigan. *Mailing Add:* Dept Biol Sci Western Mich Univ Kalamazoo MI 49008-5050

PIPPERT, RAYMOND ELMER, b Lawrence, Kans, May 18, 38; m 58; c 2. GRAPH THEORY. *Educ:* Univ Kans, BA, 59, PhD(math), 65. *Prof Exp:* From asst prof to assoc prof, 65-73, actg dept chmn, 79-81, PROF MATH, PURDUE UNIV, FT WAYNE, 73-; prin investr res contract, Off Naval Res, 86-89; vis prof, 90-91, MATH COORDR, INST TECHNOL MARA, MALAYSIA, 91- *Concurrent Pos:* Sabbatical, Western Mich Univ, 71-72; vis prof, Univ Paris, 88. *Mem:* Am Math Soc; Math Asn Am; Sigma Xi; Soc Indust Appl Math. *Res:* Graph theory and combinatorics, especially characterization and enumeration problems, trees and networks. *Mailing Add:* Dept Math Purdue Univ 2101 Coliseum Blvd E Ft Wayne IN 46805

PIQUETTE, JEAN CONRAD, b St Augustine, Fla, Feb 15, 50; m 72; c 2. PANEL-TEST MEASUREMENT METHODOLOGY, REVERBERATION-LIMITED MEASUREMENT METHODOLOGY. *Educ:* Rutgers Univ, BA, 72; Stevens Inst Technol, MS, 74, PhD(physics), 83. *Prof Exp:* RES PHYSICIST, UNDERWATER SOUND REF DETACHMENT, NAVAL RES LAB, 81- *Concurrent Pos:* Adj instr, Valencia Community Col, 82- *Mem:* Acoust Soc Am. *Res:* New measurement methods to evaluate passive acoustical materials used in underwater applications; new measurement methods for the accurate calibration of acoustic sources and receivers used in underwater applications; symbolic methods for automatic computer evaluation of indefinite integrals containing special functions. *Mailing Add:* Naval Res Lab Code 5983 PO Box 568337 Orlando FL 32856-8337

PIRANI, CONRAD LEVI, b Pisa, Italy, July 29, 14; nat US; m 48, 55; c 3. PATHOLOGY. *Educ:* Univ Milan, MD, 38. *Prof Exp:* Asst, Inst Path Anat, Univ Milan, 38-39; intern, Mother Cabrini Mem Hosp, Chicago, 40-41 & Columbus Mem Hosp, 41-42; resident, Michael Reese Hosp, 42-45; from instr to prof path, Univ Ill Col Med, 45-71; prof, Pritzker Sch Med, Univ Chicago, 71-72; prof, 72-84, EMER PROF PATH, COL PHYSICIANS & SURGEONS, COLUMBIA UNIV, 84- *Concurrent Pos:* Chmn dept path, Michael Reese Hosp & Med Ctr, 65-72; consult, WSide & Hines Vet Hosps,

65-72; attend pathologist, Presby Hosp, 72-84, consult, 85- *Honors & Awards:* John Peters Award, Am Soc Nephrology, 87. *Mem:* AAAS; Am Soc Clin Path; Am Soc Exp Path; Am Soc Nephrology; Int Soc Nephrology. *Res:* Pathology of renal diseases; rheumatic diseases. *Mailing Add:* Col Physicians & Surgeons Columbia Univ New York NY 10032

PIRANIAN, GEORGE, b Thalwil, Switz, May 2, 14; nat US; m 41; c 5. MATHEMATICS, COMPLEX ANALYSIS. *Educ:* Utah State Univ, BS, 36, MS, 38; Rice Inst, MA, 41, PhD(math), 43. *Prof Exp:* Instr math, Rice Inst, 43; from instr to prof, 45-84, EMER PROF MATH, UNIV MICH, ANN ARBOR, 84- *Concurrent Pos:* Ed, Mich Math J, 54-75. *Mem:* Am Math Soc; Math Asn Am. *Res:* Functions of a complex variable; prime ends; power series; boundary behavior; conformal maps. *Mailing Add:* Dept Math Univ Mich Ann Arbor MI 48109

PIRCH, JAMES HERMAN, b Henrietta, Mo, July 10, 37; m 59; c 2. NEUROPHARMACOLOGY, ELECTROPHYSIOLOGY. *Educ:* William Jewell Col, AB, 59; Univ Kans, PhD(pharmacol), 66. *Prof Exp:* Asst prof pharmacol, Bowman Gray Sch Med, 67-69; from asst prof to assoc prof, Univ Tex Med Br Galveston, 69-76; assoc prof, 76-79, prof pharmacol, Sch Med, 79-87, PROF PHARMACOL, HEALTH SCI CTR, TEX TECH UNIV, 89- *Concurrent Pos:* Fel pharmacol, Dartmouth Med Sch, 65-67; USPHS fel, 66-67; prof & chmn, Pharmacol, Kirksville Col Osteopathic Med, 87-89. *Mem:* AAAS; Am Soc Pharmacol & Exp Therapeut; Soc Neurosci; Soc Exp Biol & Med. *Res:* Central mechanisms of drug induced behavioral changes; electrophysiological and biochemical correlates; neural mechanisms of conditioning; neurotransmitters and conditioning-related neuronal activity. *Mailing Add:* Dept Pharmacol Tex Tech Univ Sch Med Lubbock TX 79430

PIRES, RENATO GUEDES, b Sao Paulo, Brazil, June 7, 61; m 88. SEMICONDUCTOR DEVICES, MOS DEVICES. *Educ:* Inst Mavá Technol, BS, 83; Univ Sao Paulo, MS, 87; Univ Vt, mat sci, 87- *Prof Exp:* Lectr computer sci, Inst Mavá Technol, 84-87. *Mem:* Inst Elec & Electronic Engrs. *Res:* Semiconductor physical electronics; electrical properties of materials; cryoelectronics; characterization of MOS capacities with polycide gate and low-voltage low-temperature operation of mosfetis. *Mailing Add:* 20 Ethan Allen Ave No 62 Colchester VT 05446

PIRIE, ROBERT GORDON, b Dundas, Ont, June 19, 36; m 76; c 2. SEDIMENTOLOGY, PETROLEUM GEOLOGY. *Educ:* Indiana Univ, BS, 58, MS, 61, PhD(geol, sedimentology), 63. *Prof Exp:* From asst prof to assoc prof, 63-75, dept head, 69-70, prof sedimentology, geol oceanog, Univ Wis, 75-78; res geologist, Schlumberger-Doll Res, 78-85, INTERPRETATION DEVELOP, SCHLUMBERGER WELL SERV, 85- *Concurrent Pos:* Grants, NSF, 64-66, Water Resources Ctr, 65-66, AEC, PR Nuclear Ctr, 66-70, Coastal Eng Res Ctr, 68-69, NDEA, 68, sea grant, 75 & US Corps Engrs, Dept Natural Resources Wis, 75. *Honors & Awards:* Jules Braunstein Mem Award, Am Asn Petrol Geologists, 90. *Mem:* Am Asn Petrol Geologists; AAAS; Am Soc Limnol & Oceanog; Soc Econ Paleont & Mineral; Int Asn Sedimentologists. *Res:* Marine and freshwater sedimentology; oceanography; analogs of modern and ancient depositional systems; dynamics of sediment-water interface phenomena, especially fine sediment-trace element chemistry and their variations in nearshore environments. *Mailing Add:* Schlumberger Well Serv 5000 Gulf Freeway PO Box 2175 Houston TX 77252-2175

PIRIE, WALTER RONALD, b Toronto, Ont, Dec 31, 34; US citizen; div; c 2. MATHEMATICAL STATISTICS, APPLIED STATISTICS. *Educ:* Temple Univ, BA, 61; Univ Toronto, MA, 64; Fla State Univ, NIH fel & PhD(math statist), 70. *Prof Exp:* Electronics engr, RCA Corp, 58-61, res scientist physics, 61; res assoc, Am Can Co, 64-65; res assoc physics & math, Univ Pa, 65-66; comput consult, Fla State Univ, 66-67; asst prof, 70-80, ASSOC PROF STATIST, VA POLYTECH INST & STATE UNIV, 80- *Concurrent Pos:* Instr electronics, Tech Inst, Temple Univ, 58-61; Woodrow Wilson Fel, 61. *Honors & Awards:* Woodrow Wilson Fel, 61. *Mem:* Am Statist Asn; Inst Math Statist; Statist Sci Asn Can; Nat Speleol Soc. *Res:* Nonparametric statistics, theory and applications; theory of statistical inference; limit theory of statistics. *Mailing Add:* Dept of Statistics Va Polytech Inst & State Univ Blacksburg VA 24061

PIRINGER, ALBERT ALOYSIUS, JR, b St Paul, Minn, Apr 24, 21; m 44; c 4. HORTICULTURE. *Educ:* Univ Minn, BS, 47, 48; Univ Md, PhD(hort), 53. *Prof Exp:* Asst hort, Univ Minn, 47-48, Univ Md, 48-52; asst horticulturist, Hort Crops Res Br, Plant Indust Sta, USDA, 52-53, assoc plant physiologist, 53-54; asst prof hort, Univ Minn, 54-56; plant physiologist, Crops Res Div, USDA, 56-62, asst chief, Fruit & Nut Crops Br, 62-68, asst dir, US Nat Arboretum, 68-72, asst area dir, Chesapeake-Potomac Area, Northeast Region, 72-79, supvry horticulturist & dir, Hort Sci Inst, Agr Res Serv, 79-86; RETIRED. *Honors & Awards:* Vaughn Award, Am Soc Hort Sci, 56, 57. *Mem:* Am Soc Plant Physiol; Am Soc Hort Sci. *Res:* Photoperiodism and light effects on plant growth and development; physiology of horticultural plants; photoperiodism in tropical plants; fruit and nut crop culture; administration of varied agricultural research. *Mailing Add:* 4409 Beechwood Rd University Park MD 20782

PIRKLE, EARL C, b Morgan Co, Ga, Jan 8, 22; m 42; c 3. ECONOMIC GEOLOGY. *Educ:* Emory Univ, AB, 43, MS, 47; Univ Cincinnati, PhD(geol), 56. *Prof Exp:* Prod coordr & res crystallogr, Pan Electronics Labs, Inc, 42-45; instr geol, Univ Tenn, 47-50; from asst prof to assoc prof, 50-63, chmn dept, 72-79, dir Phys Sci, 79-82, PROF PHYS SCI & GEOL, UNIV FLA, 63-, ASSOC CHMN GEOL DEPT, 83- *Concurrent Pos:* Vis instr, Univ Cincinnati, 55-56; consult geologist, 55- *Mem:* Geol Soc Am; Am Asn Petrol Geol; Soc Econ Geol; Am Inst Mining, Metall & Petrol Engrs, Inc. *Res:* Economic mineral deposits; physiography and sedimentation; origin of heavy mineral deposits and phosphate deposits. *Mailing Add:* Dept Geol Univ Fla Gainesville FL 32611

PIRKLE, FREDRIC LEE, b Atlanta, Ga, Dec 23, 49; m 70; c 3. ECONOMIC GEOLOGY, GEOMATHEMATICS. *Educ:* Fla State Univ, BS, 72; Univ Fla, MS, 74; Pa State Univ, PhD(geol), 77. *Prof Exp:* Lab & plant technician, E I Du Pont de Nemours & Co, Inc, 72, geologist & proj leader, 73, geologist, 74 & 75; teaching asst phys geol, Univ Fla, 72-74; teaching asst hist geol, Pa State Univ, 74-76; field geologist engr, Franklin Co, 77-78; geophysics engr III, Bendix Field Eng Corp, 78-79, res geoscientist, 79-80, advan res geoscientist, 80-81; sr geologist, Conoco, Inc, 81, staff geologist, 81-84; sr geologist, 84-85, RES ASSOC GEOL, E I DU PONT DE NEMOURS & CO, INC, 85- *Concurrent Pos:* Consult geologist, E I Du Pont de Nemours & Co, Inc, 73-76. *Mem:* Soc Mining Metall & Explor Am; Inst Mining Metall & Petrol Engrs; Int Asn Math Geol; Math Geologists US; Am Asn Petrol Geologists. *Res:* Modeling mineral sand deposits: modeling host environments for heavy-mineral deposits, determining transportation characteristics of and mechanisms for concentrating heavy minerals into commercial concentrations and quantities; evaluation of heavy-mineral deposits. *Mailing Add:* E I Du Pont de Nemours & Co PO Box 753 Starke FL 32091

PIRKLE, HUBERT CHAILLE, b Indianapolis, Ind, Feb 18, 24; m 78; c 2. BIOCHEMISTRY. *Educ:* Ind Univ, MD, 49. *Prof Exp:* Intern, Univ Chicago, 49-50, resident path, 52-54, asst, 54-55; asst prof, Sch Med, Univ Louisville, 55-61, assoc prof, 61-70; assoc prof, 70-86, vchmn, 78-86, actg chmn, 86-88, PROF PATH, COL MED, UNIV CALIF, IRVINE, 86-, CHIEF, DIV CHEM PATH, 81- *Concurrent Pos:* Markle scholar, 60-65; NIH spec res fel, Karolinska Inst, Sweden, 64-65; vis investr, Karolinska Inst, Sweden, 64-65; mem path & anat fel rev comt, NIH; prin investr, NIH grants; ed, Thrombosis Res, 89; chmn subcomt, Nomenclature Exogenous Hemostatic Factors, Int Soc Thrombosis & Haemostasis, 87- *Mem:* Am Asn Path; Am Chem Soc; Am Soc Exp Path; Int Soc Thrombosis & Haemostasis; Am Heart Asn. *Res:* Biochemistry of blood coagulation; cardiovascular disease. *Mailing Add:* Dept Path Univ Calif Irvine CA 92717

PIRKLE, WILLIAM ARTHUR, b Atlanta, Ga, May 11, 45; m 68; c 2. GEOMORPHOLOGY. *Educ:* Emory Univ, BS, 67; Univ NC, Chapel Hill, MS, 70, PhD(geol), 72. *Prof Exp:* From asst prof to assoc prof, 72-82, chmn, div natural sci, 84-86, PROF GEOL, UNIV SC, AIKEN, 82-, DEAN, COL MATH & NATURAL HEALTH SCI, 86- *Concurrent Pos:* Geologist, Rosario Resources Inc, 69; prof geologist, SCGeol Surv, 73-85. *Mem:* Geol Soc Am; Am Inst Mining Metall & Petrol Engrs; Sigma Xi. *Res:* Physiography and mineral resources, especially heavy mineral occurrences, of the Atlantic-Gulf Coastal Plain; physiography, sedimentation and structure of slate belt rocks of the Appalachian Piedmont. *Mailing Add:* Col Math Natural & Health Sci Univ SC 171 Univ Pkwy Aiken SC 29801

PIRKLE, WILLIAM H, b Shreveport, La, May 2, 34; m 56; c 4. ORGANIC CHEMISTRY. *Educ:* Univ Calif, Berkeley, BS, 59; Univ Rochester, PhD(chem), 63. *Prof Exp:* NSF fel, Harvard Univ, 63-64; asst prof, 64-69, ASSOC PROF CHEM, UNIV ILL, URBANA, 69- *Concurrent Pos:* Alfred P Sloan fel, 71-72. *Mem:* Am Chem Soc. *Mailing Add:* 161 Roger Adam Lab Univ Ill 1209 Calif Urbana IL 61801

PIRLOT, PAUL, zoology; deceased, see previous edition for last biography

PIRNOT, THOMAS LEONARD, b Scranton, Pa, Sept 3, 43; m 68; c 4. ALGEBRA, COMPUTER GRAPHICS. *Educ:* Wilkes Col, BA, 65; Pa State Univ, Univ Park, PhD(math), 70. *Prof Exp:* PROF MATH & COMPUTER SCI, KUTZTOWN UNIV, 70- *Mem:* Math Asn Am; Am Math Soc; Asn Comp Mach. *Res:* Algebraic theory of semigroups; graphics software; simulation software. *Mailing Add:* Dept of Math & Comput Sci Kutztown Univ Kutztown PA 19530

PIROFSKY, BERNARD, b New York, NY, Mar 27, 26; m 53; c 3. IMMUNOHEMATOLOGY, CLINICAL IMMUNOLOGY. *Educ:* NY Univ, AB, 46, MD, 50. *Prof Exp:* Intern med, Bellevue Hosp, New York, 50-51, resident, 51-52 & 54-55; dir, Pac Red Cross Blood Ctr, 56-58; chief immunol, allergy, rheumatology, Portland Vet Admin Hosp, 80-83; from instr to assoc prof, 58-66, head div immunol allergy & rheumatology, 65-85, PROF MED, ORE HEALTH SCI UNIV, 66-, PROF MICROBIOL & IMMUNOL, 72- *Concurrent Pos:* Am Cancer Soc res fel hemat, Med Sch, Univ Ore, 55-56; Commonwealth fel, 66-67; consult immunohemat, Pac Red Cross Blood Ctr, 58-; mem med adv bd, Leukemia Soc, 65-70; vis prof immunol, Nat Inst Nutrit, Mexico City, 66-67; vis prof, Nat Res Coun SAfrica, 77, vis res prof, Nat Acad Sci, Hiroshima, Japan, 78-79; asst dean res, Ore Health Sci Univ, 85- *Honors & Awards:* Emily Cooley Mem Award, Am Asn Blood Banks, 72. *Mem:* Fel Am Col Physicians; Am Soc Hemat; Int Soc Hemat; Int Soc Blood Transfusion; NY Acad Sci; Am Asn Immunol. *Res:* Influence of immunologic events in development of human disease; autoimmunity; homotransplantation; malignancies. *Mailing Add:* Ore Health Sci Univ 3181 SW Sam Jackson Rd Portland OR 97201

PIRONE, LOUIS ANTHONY, b Laredo, Tex, Apr 28, 45; m 70; c 2. MEDICAL DEVICE DEVELOPMENT, WOUND HEALING MODEL DEVELOPMENT. *Educ:* Univ Maine, BS, 75. *Prof Exp:* Res asst, Res Dept, Maine Med Ctr, 69-85; asst res investr, ConvaTec Div, E R Squibb, 85-87, res investr, 88-90, SR RES INVESTR, CONVATEC, BRISTOL MYERS SQUIBB, 90- *Mem:* Soc Invest Dermat. *Res:* Wound care materials and products that will be used in the treatment and cure of chronic, non-healing wounds and burns; novel, wound healing products and device development; development of quantitative wound healing models. *Mailing Add:* 45 Morgan Ave Yardley PA 19067

PIRONE, THOMAS PASCAL, b Ithaca, NY, Jan 3, 36; m 61; c 2. PLANT PATHOLOGY, PLANT VIROLOGY. *Educ:* Cornell Univ, BS, 57; Univ Wis, PhD(plant path), 60. *Prof Exp:* From asst prof to assoc prof, La State Univ, 60-67; assoc prof, 67-71, chmn, 78-86, PROF PLANT PATH, UNIV KY, 71- *Concurrent Pos:* Fulbright sr res fel & vis scientist, Rothamsted Exp Sta, 74-75; sr ed, Phytopath, 77-78; mem, recombinant DNA adv comt, NIH,

84-87. *Honors & Awards:* Ruth Allen Award, Am Phytopath Soc, 89. *Mem:* AAAS; fel Am Phytopath Soc; Am Soc Virol. *Res:* Virology; insect transmission of plant viruses; characterization of viral genes and gene products which regulate virus spread. *Mailing Add:* Dept Plant Path Univ Ky Lexington KY 40546

PIROOZ, PERRY PARVIZ, b Ahwaz, Iran, June 2, 28; US citizen; m 54; c 6. GLASS SCIENCE & TECHNOLOGY, INORGANIC CHEMISTRY. *Educ:* Univ Toledo, BS, 54, MS, 62. *Prof Exp:* Anal chemist, 55-59, glass res & develop engr, 59-69, PROJ MGR RES & DEVELOP, OWENS ILL, INC, 69- *Mem:* Am Chem Soc; Am Ceramic Soc; Brit Soc Glass Technol; Sigma Xi. *Res:* Compositions, physical and chemical properties, structure and manufacture of glass. *Mailing Add:* Owens Ill Tech Ctr One Seagate Toledo OH 43666

PIROUE, PIERRE ADRIEN, b Switz, Sept 18, 31. PHYSICS. *Educ:* Univ Geneva, Lic es sc, 53, PhD(physics), 58. *Prof Exp:* Asst exp physics, Univ Geneva, 53-55, asst theoret physics, 55-56; asst physics, Princeton Univ, 57-60; res fel, European Orgn Nuclear Res, Geneva, Switz, 60-61; from instr to assoc prof, 61-70, PROF PHYSICS, PRINCETON UNIV, 70- *Mem:* Am Phys Soc. *Res:* High energy physics; electronics. *Mailing Add:* Dept Physics Princeton Univ Princeton NJ 08540

PIRRAGLIA, JOSEPH A, b Providence, RI, July 15, 28; m 65; c 3. ATMOSPHERIC DYNAMICS. *Educ:* Univ RI, BS, 59; NY Univ, MS, 61; Polytech Inst Brooklyn, PhD(electrophysics), 71. *Prof Exp:* Mem tech staff, Bell Telephone Labs, Inc, 59-65; ASTROPHYSICIST, GODDARD SPACE FLIGHT CTR, NASA, 71- *Mem:* Am Math Soc; Am Phys Soc; NY Acad Sci. *Res:* Theoretical studies of the dynamics and energy balance of planetary atmospheres. *Mailing Add:* 5617 Gulf Stream Row Columbia MD 21044

PIRRI, ANTHONY NICHOLAS, b Providence, RI, Apr 13, 43; m 68; c 2. APPLIED PHYSICS, ENGINEERING SCIENCE. *Educ:* Boston Univ, BS, 64; Mass Inst Technol, MS, 66; Brown Univ, PhD(eng), 70. *Prof Exp:* Res asst aeronaut & astronaut, Mass Inst Technol, 64-66; prin res scientist aerophysics, Avco Everett Res Lab, 69-73; mgr laser appln, 73-80, vpres defense progs, 80-88, SR VPRES CORP DEVELOP, PHYS SCI, INC, 88- *Mem:* Am Inst Aeronaut & Astronaut; Am Inst Physics. *Res:* High temperature gas dynamics and fluid mechanics; radiation gas dynamics and high power laser effects upon materials; laser produced plasmas; turbulent flows and classical fluid mechanics. *Mailing Add:* 20 New Eng Bus Ctr Andover MA 01810

PIRRUNG, MICHAEL CRAIG, b Cincinnati, Ohio, July 31, 55. ORGANIC CHEMISTRY, BIOCHEMISTRY. *Educ:* Univ Tex, Austin, BA, 75; Univ Calif, Berkeley, PhD(org chem), 80. *Prof Exp:* asst prof chem, Stanford Univ, 81-89; sr scientist, Affymax, 89; ASSOC PROF CHEM, DUKE UNIV, 89- *Concurrent Pos:* Presidential Young Investr, 85-90. *Mem:* Am Chem Soc. *Res:* Organic chemistry; organic synthesis; photochemistry; synthetic methodology; bio-organic chemistry; biosynthesis; plant growth hormones. *Mailing Add:* Dept Chem Duke Univ Durham NC 27706

PIRTLE, EUGENE CLAUDE, b Wichita Falls, Tex, Nov 17, 21; m 44; c 5. VETERINARY MICROBIOLOGY. *Educ:* Univ Colo, BA, 47; Univ Iowa, MS, 50, PhD(bact), 52. *Prof Exp:* From asst prof to assoc prof microbiol, Med Sch, Univ SDak, 52-60; dir, Virus Lab, Hawaii State Dept Health, 60-61; res microbiologist virol invests, Nat Animal Dis Ctr, 61-81; VIS ASSOC PROF, DEPT MICROBIOL, IOWA STATE UNIV, 88- *Concurrent Pos:* Grants, USPHS, 54-56, AMA, 58, SDak Game Dept, 58-60, Nat Renderers Asn, 90 & Upjohn Co, 90. *Mem:* Sigma Xi; Am Soc Virol. *Res:* Virus replication; growth of cells in culture; viral nucleic acids, immune responses to viral diseases of animals. *Mailing Add:* Dept Vet Microbiol & Prev Med Col Vet Med Iowa State Univ Ames IA 50011

PIRTLE, ROBERT M, b Memphis, Tenn, Sept 25, 45; m 68. NUCLEIC ACIDS, BIOCHEMISTRY. *Educ:* Ga Inst Technol, BS, 67; Univ Louisville, PhD(chem), 73. *Prof Exp:* Asst prof, 80-87, ASSOC PROF BIOCHEM, TEX COL OSTEOP MED, UNIV N TEX, 87- *Concurrent Pos:* Res assoc, State Univ NY, Stony Brook, 73-80. *Res:* Human transfer RNA gene organization and expression; structure and expression of eukaryotic genes. *Mailing Add:* 731 Londonderry Apt 132 Denton TX 76205

PISACANE, VINCENT L, b Philadelphia, Pa, Apr 6, 33; m; c 1. ENGINEERING PHYSICS. *Educ:* Drexel Univ, BS, 55; Mich State Univ, MS, 57, PhD(appl mech/physics), 62. *Prof Exp:* Head space dept, 85-90, ASST DIR RES & EXPLOR DEVELOP, 90- *Concurrent Pos:* Instr, Johns Hopkins Univ, 68- *Mem:* Am Geophys Union; Am Inst Aeronaut & Astronaut; Inst Elect & Electronics Engrs. *Res:* Classical and celestral mechanics and geophysics; technical management of satellite borne instrumentation systems. *Mailing Add:* Appl Physics Lab Johns Hopkins Univ Johns Hopkins Rd Laurel MD 21818

PISANO, DANIEL JOSEPH, JR, b Mt Vernon, NY, Oct 30, 46; m 69; c 1. NUCLEAR PHYSICS. *Educ:* Columbia Univ, BA, 68; Yale Univ, MPhil, 73, PhD(physics), 73. *Prof Exp:* Res staff physicist, Wright Nuclear Struct Lab, Yale Univ, 73-74; res assoc nuclear physics, Brookhaven Nat Lab, 74-77; mem staff, FMI Med Inc, 77-80; mem staff, Oakbrook Div, 80-, MEM STAFF, RIDGEFIELD DIV, PERKIN ELMER CORP. *Mem:* Am Phys Soc; Am Asn Physics Teachers. *Res:* Heavy-ion induced nuclear reactions and the atomic physics of highly-stripped heavy ions. *Mailing Add:* Perkin Elmer 761 Main Ave Norwalk CT 06859

PISANO, JOSEPH CARMEN, b Wilkes-Barre, Pa, July 19, 41; m 63; c 3. CELL PHYSIOLOGY, IMMUNOBIOLOGY. *Educ:* King's Col (Pa), BS, 63; Univ Notre Dame, PhD(microbiol), 67. *Prof Exp:* Instr physiol, Med Units, Univ Tenn, Memphis, 67-68; from asst prof to assoc prof physiol, 68-90, DIR, FINANCIAL AID OFF, SCH MED, TULANE UNIV, 90- *Mem:* AAAS; Am Soc Microbiol; Am Physiol Soc; Reticuloendothelial Soc. *Res:* Physiology of phagocytic cells; cellular involvement in the immune response. *Mailing Add:* Financial Aid Dept Tulane Univ Sch Med 1430 Tulane Ave New Orleans LA 70112

PISANO, MICHAEL A, b Brooklyn, NY, Mar 31, 23; m 48; c 4. MICROBIOLOGY. *Educ:* Brooklyn Col, BA, 46; St John's Univ, NY, MS, 49; Mich State Univ, PhD(bact, biochem), 53. *Prof Exp:* Bacteriologist, Mich State Dept Health, 51-53; microbiologist, Res Div, S B Penick & Co, 53-55; chmn dept biol, 81-83, PROF MICROBIOL, GRAD SCH, ST JOHN'S UNIV NY, 55- *Concurrent Pos:* Res assoc, Mt Sinai Hosp, New York, 56-; consult, Merck & Co, Inc 57 & E R Squibb & Sons, 60; res contract, US Naval Appl Sci Lab, 63-; consult, Miles Chem Co, 63-65 & Macrosonics Corp, 65-; fac res award, St John's Univ NY, 66; res fel, Dept Health, Educ & Welfare, 68-69; consult, Amstar Corp, 72-, Deltown Chemurgic, 79; pres bd dir, Am Soc Microbiol Found, 76, chmn, Div Fermentation Microbiol, 78; chmn, microbiol sect, NY Acad Sci, 80-82. *Mem:* Am Soc Microbiol; Soc Indust Microbiol (pres, 72-73); NY Acad Sci; Brit Soc Gen Microbiol; Mycol Soc Am. *Res:* Fermentation, antibiotics, vitamins; microbiological nutrition; cell morphology; sterilization procedures; microbial chemiluminescence; microbial lipids; microbial enzymes; interest in antibiotics, enzymes, microbial lipids; antibiotic synthesis by actinomycetes from marine and estuarine sources. *Mailing Add:* Dept Biol Sci St John's Univ Jamaica NY 11439

PISCHEL, KEN DONALD, b Glendale, Calif, Aug 11, 50; m 80. RHEUMATOLOGY. *Educ:* Calif Inst Technol, BS, 72; Wash Univ, St Louis, MD & PhD(immunol), 79. *Prof Exp:* Teaching fel, 82-85, ASST PROF RHEUMATOLOGY, DIV RHEUMATOLOGY, UNIV CALIF, SAN DIEGO, 85- *Concurrent Pos:* Attend physician, Med Ctr, Univ Calif San Diego, 85- *Honors & Awards:* Arthritis Investr Award, Arthritis Found, 85; First Investr Award, NIH, 87. *Mem:* Am Asn Immunologists; Am Rheumatism Asn; Am Fedn Clin Res; AAAS; Am Col Physicians. *Res:* Lymphocyte cell surface adhesion proteins, normal immune response and mechanism of autoimmune diseases; identification of surface proteins recognized by autoimmune antibodies; delineating structural variation and functional properties of lymphocyte cell surface proteins. *Mailing Add:* Dept Med Div Rheumatol H- 811-G Univ Calif SD Med Ctr 225 Dickinson St San Diego CA 92103

PISCIOTTA, ANTHONY VITO, b New York, NY, Mar 3, 21; m 51; c 3. INTERNAL MEDICINE. *Educ:* Fordham Univ, BS, 41; Marquette Univ, MD, 44, MS, 52; Am Bd Internal Med, dipl. *Prof Exp:* Intern, Med Ctr, Jersey City, NJ, 44-45; resident path, Fordham Hosp, New York, 47-48; resident internal med, Milwaukee County Gen Hosp, 48-51; instr internal med, Sch Med, Tufts Univ, 51-52; from instr to assoc prof, 52-66, PROF MED, MED COL WIS, 66-, ROBERT A UIHLEIN JR PROF HEMAT RES, 83- *Concurrent Pos:* Mem attend staff & dir blood res lab, Milwaukee County Gen Hosp, 52-; consult, St Mary's, St Joseph's, Deaconess, Columbia, St Luke's & Lutheran Hosps; vchmn, Radiation Effects Res Found, Hiroshima, Japan, 81-83. *Honors & Awards:* Encaenia Award, Fordham Univ, 56. *Mem:* AAAS; AMA; Am Fedn Clin Res; fel Am Col Physicians; Am Soc Exp Path. *Res:* Mechanisms of autoimmunization; drug induced bone marrow damage; hematology. *Mailing Add:* 8700 W Wisconsin Ave Milwaukee WI 53226

PISCITELLI, JOSEPH, b Reading, Pa, June 28, 34; m; c 2. PHYSIOLOGY. *Educ:* Cath Univ Am, AB, 59, MS, 61, PhD(physiol), 68. *Prof Exp:* Instr biol, Allentown Col, 66-68; asst prof, Philadelphia Community Col, 68-69 & Philadelphia Col Osteop, 69-72; PROF BIOL, KUTZTOWN STATE COL, 72- *Mem:* Sigma Xi; Nat Asn Biol Teachers. *Res:* Effects of ultrasound and 6-chloropurine on the growth of Krebs-2 ascites cells; protein metabolism and amino-acid composition of regenerating Stentor coeruleus. *Mailing Add:* 435 W Main St Kutztown PA 19530

PISERCHIO, ROBERT J, b Pueblo, Colo, Feb 12, 34; m 63. PHYSICS. *Educ:* Univ Ariz, BS, 58, MS, 62, PhD(physics), 63. *Prof Exp:* Asst physics, Univ Ariz, 58-63; res asst prof & actg asst prof, Univ Wash, Seattle, 63-66, from asst prof to assoc prof, 66-74, PROF PHYSICS, SAN DIEGO STATE UNIV, 74- *Res:* High energy physics; cosmic rays. *Mailing Add:* Dept Physics 5300 Campanile Dr San Diego CA 92182

PISTER, KARL S(TARK), b Stockton, Calif, June 27, 25; m 50; c 6. APPLIED MECHANICS, SYSTEMS ENGINEERING. *Educ:* Univ Calif, BS, 45; MS, 48; Univ Ill, PhD(eng mech), 52. *Prof Exp:* Instr civil eng, Univ Calif, 47-49; instr theoret & appl mech, Univ Ill, 49-52; from asst prof to assoc prof civil eng, Univ Calif, 52-62, dean, Col Eng, 80-90, Roy W Carlson prof eng, 85-90, PROF ENG SCI, UNIV CALIF, BERKELEY, 62- *Concurrent Pos:* Fulbright lectr & researcher, Ireland, 65-66; sr Fulbright scholar, Univ Stuttgart, 73-74, Richard Merton vis prof, 78. *Honors & Awards:* Wason Medal Res, Am Concrete Inst, 60; Bandix Award for Minorities in Eng, Am Soc Engr Educ, 88. *Mem:* Nat Acad Eng; Am Soc Mech Engrs; Earthquake Eng Res Inst; Soc Eng Sci; fel Am Acad Mechanics. *Res:* Structural and continuum mechanics; system science; optimal design; earthquake engineering. *Mailing Add:* Dept Civil Eng Univ Calif Berkeley CA 94720

PISTOLE, THOMAS GORDON, b Detroit, Mich, Sept 17, 42; m 65; c 2. COMPARATIVE IMMUNOLOGY, HOST DEFENSE MECHANISMS. *Educ:* Wayne State Univ, PhB, 64, MS, 66; Univ Utah, PhD(microbiol), 69. *Prof Exp:* Nat Res Coun res assoc microbiol, immunol br, US Army Biol Ctr, Ft Detrick, 69-70; res assoc dept microbiol, Univ Minn, Minneapolis, 70-71; from asst prof to assoc prof, 71-83, PROF & CHMN DEPT MICROBIOL, UNIV NH, 83- *Concurrent Pos:* Vis scientist, Weizmann Inst Sci, Rehovot, Israel, 79; vis prof, Univ Edinburgh, Scotland, 86. *Mem:* Am Asn Immunologists; Am Soc Zoologists; AAAS; Am Soc Microbiol; Int Soc Develop & Comp Immunol. *Res:* Invertebrate defense mechanisms and mammalian counterparts; microbial adherence in pathogenicity; macrophage biology. *Mailing Add:* Dept Microbiol Univ NH Durham NH 03824-3544

PI-SUNYER, F XAVIER, b Barcelona, Spain, Dec 3, 33; US citizen; m 61; c 3. ENDOCRINOLOGY, NUTRITION. *Educ:* Oberlin Col, BA, 55; Columbia Univ, MD, 59; Harvard Univ, MPH, 63. *Prof Exp:* Intern & asst resident med, St Luke's Hosp, New York, 59-61; jr registr, St Bartholomew's Hosp & Med Sch, London, Eng, 61-62; from instr to assoc prof, 65-85, PROF

CLIN MED, COL PHYSICIANS & SURGEONS, COLUMBIA UNIV, 85- *Concurrent Pos:* USPHS res fel, Sch Pub Health, Harvard Univ, 63-64, fel, Thorndike Mem Lab, 64-65; Norman Jolliffe fel nutrit, Inst Human Nutrit, Columbia Univ, 66-67; fel, NY Heart Asn, 67-68, sr investr, 68-72; dir, Div Endocrinol, St Luke's Roosevelt Hosp, 78-88; assoc dir med, St Luke's Roosevelt Hosp, 83-86, dir, Div Endocrinol, Diabetes & Nutrit, 88-, Obesity Res Ctr, 88-; mem, Nutrit Study Sect, NIH, 83-86, C Study Sect, Nat Inst Diabetes & Digestive & Kidney Dis; vis physician, Rockefeller Univ, 85-; mem bd dirs, Am Diabetes Asn, 85- *Honors & Awards:* Fogarty Int Fel, 79-80. *Mem:* AAAS; Am Diabetes Asn (pres-elect, 91-92); Am Inst Nutrit; Am Soc Clin Nutrit (pres, 89-90); Am Fedn Clin Res; Endocrine Soc; NY Acad Med; bd dirs Am Diabetes Asn, 85-; dir, Am Bd Nutrit. *Res:* Hormonal control of carbohydrate metabolism; obesity; diabetes mellitus; food intake regulation. *Mailing Add:* Dept of Med St Luke's Hosp Ctr 114th & Amsterdam Ave New York NY 10025

PISZKIEWICZ, DENNIS, b Chicago, Ill, Dec 21, 41. BIOCHEMISTRY. *Educ:* Loyola Univ Chicago, BS, 63; San Diego State Col, MS, 65; Univ Calif, Santa Barbara, PhD(chem), 68. *Prof Exp:* Res assoc biol chem, Sch Med, Univ Calif, Los Angeles, 68-71; asst prof biol chem, Univ Calif, Irvine-Calif Col Med, 71-78; assoc prof chem, Duquesne Univ, 78-83; RES MGR, HYLAND DIV, BAXTER HEALTHCARE CORP, 84- *Mem:* AAAS; Am Soc Biol Chem; Am Chem Soc. *Res:* Structure and function of proteins; protein chemistry; plasma proteins. *Mailing Add:* 385 Calliope St Laguna Beach CA 92651

PITA, EDWARD GERALD, b New York, NY, Jan 22, 25; m 55; c 2. MECHANICAL ENGINEERING. *Educ:* Purdue Univ, BS, 44; Univ Mich, Ann Arbor, MA, 49; Columbia Univ, MS, 56; Univ Md, College Park, PhD(mech eng), 69. *Prof Exp:* Proj engr, Worthington Corp, 53-56; chief engr, Panero-Weidlinger-Salvadori, Engrs, 56-62; assoc prof mech eng, Manhattan Col, 63-71; ASSOC PROF ENVIRON TECHNOL, NEW YORK CITY TECH COL, 71- *Mem:* Am Soc Heating, Refrig & Air Conditioning Engrs; NY Acad Sci; Sigma Xi. *Res:* Heat transfer, thermal pollution, air conditioning and refrigeration. *Mailing Add:* Access Ctr New York Tech Col 300 Jay St Rm M4111 Brooklyn NY 11201

PITA, JULIO C, PROTEOGLYCANS, AGGREGATION. *Educ:* Univ Havana, Cuba, PhD(phys chem), 41. *Prof Exp:* RES PROF, UNIV MIAMI, 78- *Mailing Add:* Vet Admin Hosp & Univ Miami Med Sch PO Box 016960 Miami FL 33101

PITAS, ROBERT E, ATHEROSCLEROSIS, LIPOPROTEIN METABOLISM. *Educ:* Univ Conn, PhD(nutrit & biochem), 76. *Prof Exp:* ASSOC PROF PATH, UNIV CALIF, SAN FRANCISCO, 84-, SR SCIENTIST, GLADSTONE FOUND LAB CARDIOVASC DIS, 79- *Mailing Add:* Cardiovasc Dis Univ Calif Gladstone Found Lab PO Box 40608 San Francisco CA 94140

PITCAIRN, DONALD M, b Portland, Ore, Mar 23, 21; m 52; c 3. INTERNAL MEDICINE. *Educ:* Univ Ore, BA, 44, MD, 45. *Prof Exp:* Intern, Med Sch, Univ Ore, 45-46, resident thoracic surg, 48-49, instr physiol, 49-52, resident internal med, 52-53; asst med, Peter Bent Brigham Hosp, Boston, 53-55; from instr to prof, Med Sch, Univ Ore, 55-68, head div chest dis, 59-67; chief physician educ br, Div Physician Manpower, Bur Health Professions Educ & Manpower Training, 68-71, spec asst to dir, Fogarty Int Ctr Advan Study Health Sci, 71-80, actg chief, cancer centers br, Nat Cancer Inst, NI,, 80-84; RETIRED. *Concurrent Pos:* Life Ins Med Res Fund fel med, Peter Bent Brigham Hosp, Boston, 54-55. *Res:* Bioassay of adrenal medullary hormones; oxyhemoglobin dissociation characteristics of old and young human erythrocytes and hemoglobin; pulmonary physiology and function; cardiopulmonary diseases; medical education. *Mailing Add:* 24 Orchard Way N Rockville MD 20854

PITCHER, ARTHUR EVERETT, b Hanover, NH, July 18, 12; m 36, 73; c 2. MATHEMATICS. *Educ:* Case Western Reserve Univ, AB, 32; Harvard Univ, MA, 33, PhD(math), 35. *Hon Degrees:* DSc, Case Western Reserve Univ, 57. *Prof Exp:* Instr math, Harvard Univ, 34-35; asst, Inst Advan Study, 35-36; Benjamin Pierce instr, Harvard Univ, 36-38; from instr to assoc prof, Lehigh Univ, 38-48, prof math, 48-78, chmn dept math & astron, 60-78, CONSULT TO PRES, LEHIGH UNIV, 78- *Concurrent Pos:* Mem Inst Advan Study, 45-46 & 47-50; consult, US Army & US Air Force, 47-54; Guggenheim Mem Found fel, 52-53. *Honors & Awards:* Distinguished Serv Award, Math Asn Am. *Mem:* AAAS; Am Math Soc (assoc secy, 59-66, secy, 68-88); Soc Indust & Appl Math; Math Asn Am. *Res:* Critical point theory; calculus of variations; exterior and terminal ballistics. *Mailing Add:* Dept Math Bldg 14 Lehigh Univ Bethlehem PA 18015

PITCHER, DEPUYSTER GILBERT, b Bay Shore, NY, Mar 2, 28; m 52; c 3. REACTOR INSTRUMENTATION. *Prof Exp:* Engr, 54-61, group leader reactor instrumentation nuclear sci, Assoc Univs Inc, 61-87; RETIRED. *Mem:* Inst Elec & Electronics Engrs; Am Nuclear Soc. *Res:* Safety and control instrumentation for research reactors. *Mailing Add:* 1024 Farrington Dr Knoxville TN 37923

PITCHER, ERIC JOHN, b St John's, Nfld, Aug 19, 46; US citizen; m 73; c 2. CLIMATE DYNAMICS. *Educ:* Mem Univ Nfld, BSc, 68; McGill Univ, MSc, 70; Univ Mich, PhD(atmospheric sci), 74. *Prof Exp:* Fel, Nat Ctr Atmospheric Res, 74-75; res assoc, Nat Res Coun, 75-76; from asst prof to assoc prof meteorol, Univ Miami, 76-89; DIR, UNIV PROGS, CRAY RES, INC, 88- *Concurrent Pos:* Vis scientist, Nat Ctr Atmospheric Res, 79-80, 81 & 82. *Mem:* Am Meteorol Soc; Can Meteorol & Oceanog Soc; Sigma Xi. *Res:* Modeling of the general circulation of the atmosphere; climate dynamics; geophysical fluid dynamics. *Mailing Add:* Cray Res Inc 655F Lone Oak Dr Eagan MN 55121

PITCHER, TOM STEPHEN, b Wenatchee, Wash, Apr 17, 26; m 50; c 3. MATHEMATICS. *Educ:* Univ Wash, BA, 49; Mass Inst Technol, PhD(math), 53. *Prof Exp:* Staff mem math, Lincoln Lab, Mass Inst Technol, 53-65; prof, Univ Southern Calif, 65-70; PROF MATH, UNIV HAWAII, 70- *Concurrent Pos:* Consult, Jet Propulsion Lab, 65-68, TRW Corp, 68-69 & Rand Corp, 69-70. *Mem:* Am Math Soc. *Res:* Probability theory and applications to engineering; statistics; number theory; complex variables. *Mailing Add:* 3456 Keanu St Honolulu HI 96816

PITCHER, WAYNE HAROLD, JR, b St Louis, Mo, Jan 5, 44; m 70; c 2. CHEMICAL ENGINEERING, BIOENGINEERING. *Educ:* Calif Inst Technol, BS, 66; Mass Inst Technol, SM, 68, ScD(chem eng), 72. *Prof Exp:* Sr chem engr bio-mat res & develop, Corning Glass Works, 72-74, sr res chem engr, 74-76, eng supvr biotech develop, 80-81, mgr biotech develop, 81-; AT GENENCOR CORP. *Mem:* Am Inst Chem Engrs; Am Chem Soc. *Res:* Development of immobilized biological systems for industrial applications. *Mailing Add:* 319 Castilian Way San Mateo CA 94402

PITCHFORD, ARMIN CLOYST, b Mt Home, Ark, May 12, 22; m 44; c 2. PETROLEUM CHEMISTRY. *Educ:* Univ Ark, BS, 47. *Prof Exp:* Chemist, Phillips Petrol Co, 48-49; pvt consult, 49-50; chemist, Kans, 50-52, res chemist, Okla, 52-56, group leader, Rocket Fuels Div, Tex, 56-57, sr res chemist, Res Div Chem Labs, 57-69, PROJ MGR, PHILLIPS PETROL CO, PHILLIPS RES CTR, OKLA, 69- *Concurrent Pos:* Mem, Asphalt Inst Mat Comt & Int Tech Comt, Phillips Petrol Co. *Mem:* Am Chem Soc. *Res:* Emulsification, surfactants, rheology and composition of residual fractions and methods for enhancing durability, performance and ecological utilization of residual products as construction materials; processing, evaluation and analysis of foreign, domestic and synthetic crude oils derived from oil shale and coal; asphalt composition; demetalization processes. *Mailing Add:* 4912 SE Princeton Dr Bartlesville OK 74006

PITCHFORD, LEANNE CAROLYN, b Sedalia, Mo, Feb 9, 50. EXPERIMENTAL ATOMIC PHYSICS. *Educ:* ETex State Univ, BS, 70; Univ Tex, Dallas, MS, 73, PhD(physics), 76. *Prof Exp:* Physicist, Ctr Nuclear Res, France, 76-77; res assoc, Joint Inst Lab Astrophys, Univ Colo, 77-80; PHYSICIST, SANDIA NAT LAB, 80- *Mem:* Am Phys Soc. *Res:* Computer modeling of high pressure gas lasers including radiation fields and pumping chemistry; calculation of electron energy distribution; functions in gaseons electronics; stopping power and opaut; calculations in high temperature, high density plasmas. *Mailing Add:* Centre de Physique Atomic de Toulouse Univ Paul Sabatier 118 Rte de Narbonne Toulouse Cedex 31062 France

PITCOCK, JAMES ALLISON, b Little Rock, Ark, Sept 13, 29; m 54; c 2. SURGICAL PATHOLOGY, HYPERTENSION. *Educ:* Mass Inst Technol, BS, 51; Washington Univ, St Louis, MD, 55; Am Bd Path, dipl, 62. *Prof Exp:* Asst pathologist, Barnes Hosp, St Louis, 61-62, St Vincent's Hosp, Little Rock, 62-63, Baptist Mem Hosp, Memphis, 64-75; asst dir, 75-87, DIR HOSP LABS PATH, BAPTIST MEM HOSP, MEMPHIS, 87-; CLIN PROF, MED SCH, UNIV TENN, 81- *Concurrent Pos:* Clin asst prof path, Med Sch, Univ Ark, 63; from clin asst prof to clin assoc prof, Med Sch, Univ Tenn, 65-81; actg chmn path, Med Ctr, Univ Tenn, 86-89. *Mem:* Sigma Xi; Am Asn Pathologists; Int Acad Path; Coun High Blood Pressure Res; Int Soc Hypertension; AMA. *Res:* The mechanisms and pathogenesis of hypertension with particular emphasis on anti-hypertensive function of the kidney. *Mailing Add:* Dept Path Baptist Mem Hosp 899 Madison Ave Memphis TN 38146

PITELKA, DOROTHY RIGGS, zoology, for more information see previous edition

PITELKA, FRANK ALOIS, b Chicago, Ill, Mar 27, 16; m 43; c 3. ZOOLOGY. *Educ:* Univ Ill AB, 39; Univ Calif, PhD(zool), 46. *Prof Exp:* Asst, Univ Calif, 39-44, lectr, 44-46, from instr to assoc prof, 46-58, chmn dept, 63-66 & 69-71, PROF ZOOL, UNIV CALIF, BERKELEY, 58-; RES ECOLOGIST, MUS VERT ZOOL, 67-, ASSOC DIR, 82- *Concurrent Pos:* Field researcher, Calif, Ore, Ariz & Nev, 40-51, Mex, 46-50, Alaska, 51-80; from asst cur to cur birds, Mus Vert Zool, 45-63; Guggenheim fel, 49-50; NSF sr fel, Oxford Univ & field researcher, England & Norway, 57-58; mem adv panel environ biol, NSF, 59-62; panel biol & med sci, Comt Polar Res, Nat Acad Sci, 60-65; ed, Ecol J, 62-64; res prof, Miller Inst, 65-66; mem, Int Biol Prog Subcomt Terrestrial Prod, Nat Acad Sci, 65-67, tundra biome dir, 67-69; chmn exec comt, Miller Inst, 67-70; mem US nat comm, UNESCO, 70-72; fel, Ctr Advan Study Behav Sci, Stanford Univ, 71; mem, Adv Panel Polar Progs, NSF, 78-80. *Honors & Awards:* Mercer Award, Ecol Soc Am, 53; Brewster Award, Am Ornith Union, 80. *Mem:* Fel AAAS; Ecol Soc Am; fel Am Ornith Union; hon mem Cooper Ornith Soc; fel Arctic Inst NAm; Brit Ecol Soc; Am Soc Naturalists; fel Animal Behav Soc. *Res:* Animal ecology; population and behavior studies of birds and mammals; arctic biology. *Mailing Add:* Mus Vert Zool Univ Calif Berkeley CA 94720

PITELKA, LOUIS FRANK, b Berkeley, Calif, Mar 28, 47; m 69; c 2. PLANT ECOLOGY. *Educ:* Univ Calif, Davis, BS, 69; Stanford Univ, PhD(biol), 74. *Prof Exp:* From asst prof to assoc prof biol, Bates Col, 74-84, chmn, 82-84; prog dir, Pop Biol & Physiol Ecol Prog, NSF, 83-84; proj mgr, 84-89, SR PROJ MGR, ELEC POWER RES INST, 89- *Concurrent Pos:* Prin investr res grants, NSF, 80-84; mem, Integration Team, Forest Response Prog, Nat Acid Precipitation Assessment Prog, Tech Liason Comt, 84-87; prog chair, Ecol Soc Am, 88-90. *Mem:* Ecol Soc Am (treas, 90-); Am Inst Biol Sci; Int Soc Plant Pop Biologists; Brit Ecol Soc; AAAS. *Res:* Evolutionary ecology of plants, including demography, life history patterns and physiological ecology; ecological effects of air pollutants and climate change on plants and plant communities. *Mailing Add:* Elec Power Res Inst PO Box 10412 Palo Alto CA 94303

PITESKY, ISADORE, b Chicago, Ill, Apr 18, 18; m 69. BIOCHEMISTRY, IMMUNOLOGY. *Educ:* Univ Chicago, SB, 40; Univ Ill, BS, 48, MS & MD, 51. *Prof Exp:* Chemist, Univ Chicago, 40-42 & Northern Regional Res Lab, Bur Agr & Indust Chem, USDA, 42-46; asst pharmacol, Univ Ill Col Med,

48-51; intern, resident internal med & in chg renal clin, Vet Admin Hosp, Long Beach, Calif, 51-54; DIR, IMMUNO-CHEM LABS, 65-; PRES & CHMN BD, CREATIVE SCI EQUIP CORP, LONG BEACH, 71- *Concurrent Pos:* Attend staff, Mem Hosp, Long Beach, Calif; pres, Creative Sci Equip Corp, 71- *Mem:* AAAS; Am Acad Allergy; Am Chem Soc; Soc Exp Biol & Med; AMA; Am Inst Chemists; Am Asn Clin Chemists. *Res:* Allergy; biochemistry of allergens. *Mailing Add:* Immuno-Chem Labs 3711 Long Beach Blvd Long Beach CA 90807

PITHA, JOHN JOSEPH, b New York, NY, Mar 19, 20; wid; c 4. INORGANIC CHEMISTRY. *Educ:* Polytech Inst Brooklyn, BS, 41, MS, 43, PhD(inorg chem), 46. *Prof Exp:* Asst anal chem, Polytech Inst Brooklyn, 41-43, res assoc inorg phosphors, 43-45; from instr to asst prof gen & inorg chem, Mich State Univ, 45-52; chemist, Gen Elec Co, Lenox, 52-66, mgr thyrite & ceramics res & develop, 66-75, sr ceramic engr, 75-81; sr ceramic engr, Westinghouse Elec, 81-85; CONSULT, 85. *Mem:* Am Chem Soc; Am Ceramic Soc; Sigma Xi. *Res:* Reactions at elevated temperatures; semiconductors; design of materials; non-linear resistors; ceramics; devices; metal oxide varistors. *Mailing Add:* 3913 Sage Court Bloomington IN 47401

PITHA, JOSEF, b Tabor, Czech, Oct 28, 33; US citizen. MEDICINAL CHEMISTRY. *Educ:* Czech Acad Sci, PhD(chem), 60. *Prof Exp:* Res fel, Inst Org Chem & Biochem, Czech, 59-65; fel Nat Res Coun, Can, 65-66; res assoc, Johns Hopkins Univ, 66-68; RES CHEMIST, NIH GERONT RES CTR, BALTIMORE CITY HOSPS, 68-, CHIEF SECT MACROMOLECULAR CHEM, 80- *Honors & Awards:* Inventors Award, US Dept Comm. *Mem:* Am Chem Soc. *Res:* Affinity probes for hormonal receptors, cyclodesctrins, solubilizers for drugs. *Mailing Add:* 417 Anglesea St Baltimore MD 21224

PITHA-ROWE, PAULA MARIE, b Prague, Czech, Nov 7, 37; US citizen; m 61, 81; c 2. BIOCHEMISTRY, MOLECULAR BIOLOGY. *Educ:* Prague Tech Univ, BS & MS, 60; Czech Acad Sci, PhD(biochem), 64. *Prof Exp:* Instr radiol sci, Sch Hyg & Pub Health, Johns Hopkins Univ, 66-68; assoc res scientist, Dept Path, Sch Med, NY Univ, 68-69; asst researcher, Inst Phys-Chem Biol, Paris, 69; asst prof med, 71-75, from asst prof to assoc prof microbiol, 72-85, asst prof oncol, 73-75, assoc prof med & oncol, 75-85, PROF ONCOL, MOLECULAR BIOL & GENETICS, JOHNS HOPKINS UNIV, 85- *Concurrent Pos:* Res fel, Inst Org Chem & Biochem, 64-65; jr res fel, Nat Res Coun Can, 65-66; fel, Nat Ctr Res, France, 69; Europ Molecular Biol Orgn travel award, 69; fel, Dept Med, Sch Med, Johns Hopkins Univ, 69-71; Eleanor Roosevelt cancer fel, 79 & 85. *Honors & Awards:* Prize, Czech Acad Sci, 65. *Mem:* Biophys Soc; Am Soc Microbiol; Am Asn Cancer Res; fel Leukemia Soc. *Res:* Interferon-molecular mechanism of induction and action; regulation of gene expression; RNA tumor viruses; HIV. *Mailing Add:* 5503 Huntley Sq Baltimore MD 21210

PITKIN, EDWARD THADDEUS, b Putnam, Conn, Dec 14, 30; m 53; c 2. SOLAR ENERGY, MODELING DYNAMIC SYSTEMS. *Educ:* Univ Conn, BS, 52; Princeton Univ, MS, 53; Univ Calif, Los Angeles, PhD(astrodyn), 64. *Prof Exp:* Proj engr, Astro Div, Marquardt Corp, 56-58, mgr space propulsion dept, 58-61; consult astronaut, 61-64; assoc prof aerospace eng, 64-70, prof, 70-89, EMER PROF AEROSPACE & MECH ENG, UNIV CONN, 89-; ENG CONSULT, 89- *Concurrent Pos:* Vis prof eng, Univ Canterbury, 70-71, 78 & 85; consult eng, 65- *Mem:* Assoc fel Am Inst Aeronaut & Astronaut; Solar Energy Soc. *Res:* Propulsion; fluid dynamics. *Mailing Add:* Dept of Mech Eng Univ Conn 191 Auditorium Rd Storrs CT 06268

PITKIN, ROY MACBETH, b Anthon, Iowa, May 24, 34; m 57; c 4. OBSTETRICS & GYNECOLOGY. *Educ:* Univ Iowa, BA, 56, MD, 59; Am Bd Obstet & Gynec, dipl, 67, cert maternal-fetal med, 74; FRCOG, 86. *Prof Exp:* Intern, King County Hosp, Seattle, 59-60; resident obstet & gynec, Univ Iowa Hosps, 60-63; asst prof, Univ Ill, 65-68; from assoc prof to prof obstet & gynec, Col Med, Univ Iowa, 66-77, chmn dept, 77-87; PROF OBSTET & GYNEC & CHMN DEPT, SCH MED, UNIV CALIF, LOS ANGELES, 87- *Concurrent Pos:* Chmn, Comt Nutrit, Am Col Obstet & Gynec, 70-74, Comt Nutrit Mother & Pre-sch Child, Food & Nutrit Bd, Nat Acad Sci, Nat Res Coun, 74-81; mem, Comt Dietary Allowances, Food & Nutrit Bd, Nat Acad Sci, Nat Res Coun, 75-80, Human Embryol & Develop Study Sect, NIH, 80-84 & Eval Comt, Am Bd Obstet & Gynec, 83-; ed, Yr Bk Obstet & Gynec, 75-86, ed-in-chief, Clin Obstet & Gynec, 78-, assoc ed, Obstet & Gynec, 85-86, ed, 86- *Honors & Awards:* Joseph B Goldberger Award, AMA, 82. *Mem:* Nat Acad Sci; Am Col Obstet & Gynec; Soc Gynec Invest (pres, 85-86); Perinatal Res Soc; Soc Exp Biol & Med; Am Gynec & Obstet Soc; AMA; Soc Perinatal Obstetricians (pres, 78-79); Am Fedn Clin Res. *Res:* Maternal and fetal physiology; author of numerous technical publications. *Mailing Add:* Univ Calif Sch Med 10833 Le Conte Ave Los Angeles CA 90024

PITKIN, RUTHANNE B, b Springfield, Mass, Oct 10, 44; m 66; c 2. INTRODUCTORY BIOLOGY EDUCATION, PHYSIOLOGY OF RED-SPOTTED NEWT. *Educ:* Univ Mass, Amherst, BS, 66, PhD(zool), 78; Univ Wash, MS, 68. *Prof Exp:* Instr biol, Smith Col, 77-78, lab teaching assoc, 78-80; asst prof, Allegheny Col, 80-87; ASSOC PROF BIOL, SHIPPENSBURG UNIV, 87- *Mem:* Sigma Xi; Asn Biol Lab Educ; Am Soc Zoologists; Soc Study Amphibians & Reptiles; Nat Asn Biol Teachers; Nat Sci Teachers Asn. *Res:* How the adult red-spotted newt has adapted to survive its environmental challenges; role of the anatomic nervous system in the control of heart rate; diving bradycardia; seasonal patterns in blood parameters and the effects of anemia on oxygen consumption. *Mailing Add:* Biol Dept Shippensburg Univ Shippensburg PA 17257

PITKOW, HOWARD SPENCER, b Philadelphia, Pa, May 21, 41; m 69; c 2. REPRODUCTIVE ENDOCRINOLOGY, ENVIRONMENTAL TOXICOLOGY. *Educ:* Univ Pa, BA, 62, MS, 63; Rutgers Univ, New Brunswick, PhD(reproductive physiol), 71. *Prof Exp:* Instr histol & embryol, 65-66, asst prof anat, 67-68, asst prof physiol & biochem, 69-71, assoc prof physiol, 72-77, chmn dept physiol sci, 74-84, PROF PHYSIOL, PA COL PODIATRIC MED, 78- *Concurrent Pos:* Lectr, Dept Biol & guest lectr, Dept Allied Health Studies, Community Col Philadelphia, 72-75 & lectr Dept Sci, Bucks County Community Col, 78, 79 & 81; lectr, Dept Biol & guest lectr, Dept Allied Health Studies, Community Col Philadelphia, 72-75 & lectr, Dept Sci, Bucks County Community Col, 78, 79 & 81; vis assoc prof, Dept Obstet & Gynec, Hahnemann Med Col, Philadelphia, 80-84; adj prof, Dept Biol, Drexel Univ, 80; res prof, Dept Physiol & Biophysics, Health Sci Ctr, Temple Univ, Philadelphia, 81-86; fel, Am Podiatry Asn, 66 & 68; travel grant, 27th Cong Int Sci Physiol, Paris, France, Argonne Nat Labs, 78 & 79; comt res grants, Pa Acad Sci, 80-82; session chmn, Pa Acad Sci, 80 & 82, Nat Asn Biol Teachers, 81 & 83; vis scientist, Inter Am Univ Puerto Rico, Fedn Am Soc Exp Biol, 84; grantee, 7th Int Cong Endocrin, Quebec City, Can, 84; mem educ comt, Am Physiol Soc, 85-88. *Mem:* Fedn Am Soc Exp Biol; assoc Am Physiol Soc; fel Am Podiatry Asn; Am Asn Univ Professors; Nat Asn Biol Teachers. *Res:* Effects of dimenthylbenzanthracene, a polycylic paramogenic hydrocarbon, on pregnancy, lactation and neonatal development; teratology; effects of anabolic steroids on growth and wound healing of lower extremity muscles in normal and diabetics. *Mailing Add:* Dept of Physiol Sci Eighth & Race Sts Philadelphia PA 19107

PITLICK, FRANCES ANN, b Pasadena, Calif, Feb 24, 40. BIOCHEMISTRY, CELL BIOLOGY. *Educ:* Univ Calif, Berkeley, AB, 61; Univ Wash, PhD(biochem), 68. *Prof Exp:* From instr to asst prof med, Sch Med, Yale Univ, 70-77, sr res assoc, 77-78; health scientist adminr, devices & technol br, 78-83, ASSOC DIR, DIV BLOOD DIS & RESOURCES, NAT HEART, LUNG & BLOOD INST, NIH, 83-, DIR, DIV EXTRAMURAL AFFAIRS, 86- *Concurrent Pos:* Conn Heart Asn fel, Sch Med, Yale Univ, 68-70. *Mem:* Am Soc Biol Chemists; Am Soc Hemat; Am Fedn Clin Res; AAAS. *Res:* Enzymology of blood coagulation, specifically mechanism of initiation by cell-surface tissue factor. *Mailing Add:* Am Assoc Pathologists & Univs Assoc for Res & Educ in Pathol 9650 Rockville Pike Bethesda MD 20814

PITMAN, GARY BOYD, entomology, for more information see previous edition

PITNER, SAMUEL ELLIS, b Knoxville, Tenn, Mar 8, 32; m 55; c 3. NEUROLOGY, FORENSIC NEUROLOGY. *Educ:* Univ Tenn, Knoxville, AB, 53; Univ Tenn, Memphis, MD, 56. *Prof Exp:* From instr to assoc prof, 64-75, asst dir, Cerebral Vascular Clin Res Ctr, 68-72, prof neurol, Col Med, Univ Tenn, Memphis, 75-78; PROF & CHMN NEUROL, SCH MED, WRIGHT STATE UNIV, 78- *Concurrent Pos:* Fel clin neurol, Tulane Univ, 61-63; Nat Inst Neurol Dis & Blindness fel, Univ Tenn, Memphis, 64; consult, City of Memphis Hosps, Baptist Mem Hosp, Memphis, Memphis Vet Admin Hosp & St Jude Children's Res Hosp, 64-78, dir clin for neuromuscular dis, 66-78 & multiple sclerosis clin, 79-88; mem staff, Miami Valley, Kettering, Good Samaritan, St Elizabeth's & Children's Hosp, Dayton, 78-; consult, US Air Force Wright Patterson Med Ctr & Dayton Vet Admin Ctr, 78- mem coun cerebrovascular dis, Am Heart Asn, 69- *Mem:* AAAS; fel Am Acad Neurol; fel Am Col Physicians. *Res:* Cerebrovascular diseases and neuromuscular disorders. *Mailing Add:* Dept of Neurol Box 927 Dayton OH 45401-0927

PITOT, HENRY C, III, b New York, NY, May 12, 30; m 54; c 8. TOXICOLOGY, MOLECULAR BIOLOGY. *Educ:* Va Mil Inst, BS, 51; Tulane Univ, MD, 55, PhD, 59; Am Bd Path, dipl, 60. *Prof Exp:* Instr path, Tulane Univ, 55-59; from asst prof to assoc prof, 60-66, chmn dept path, 68-71, actg dean, 71-73, PROF ONCOL & PATH, SCH MED, UNIV WIS-MADISON, 66-, DIR McCARDLE LAB CANCER RES, 73- *Concurrent Pos:* Israel Mem fel, 57-60; fel oncol, Sch Med, Univ Wis-Madison, 59-60; Lederle med fac award, 63-65; Nat Cancer Inst res career develop award, 65-68; Lucy Wortham James res award, Soc Surg Oncol, 81. *Honors & Awards:* Borden Award, 55; Parke Davis Award Exp Path, 68; Noble Found Res Recgonition Award, 83; Esther Langa Award cancer res, Univ Chicago, 84. *Mem:* AAAS; Am Chem Soc; Am Asn Path; Am Soc Biol Chem; NY Acad Sci; Am Soc Cell Biol. *Res:* Biochemical mechanisms underlying the morphological changes in disease; biochemical and genetic pathology; oncology; regulatory mechanisms. *Mailing Add:* McArdle Lab Sch Med Univ Wis Madison WI 53706

PITRAT, CHARLES WILLIAM, b Kansas City, Mo, June 1, 28; m 63; c 3. PALEONTOLOGY, STRATIGRAPHY. *Educ:* Univ Kans, AB, 49; Univ Wis, MS, 51, PhD(geol), 53. *Prof Exp:* Asst, Univ Wis, 49-53, proj assoc, 53-55; asst prof geol, NMex Inst Mining & Technol, 55-56; from asst prof to assoc prof, Univ Kans, 56-64; assoc prof, 64-78, PROF GEOL, UNIV MASS, AMHERST, 78- *Concurrent Pos:* Co-ed, J Paleontol, 81-86. *Mem:* Geol Soc Am; Paleont Soc; Soc Econ Paleont & Mineral. *Res:* Devonian stratigraphy and paleontology, particularly brachiopods and corals; brachiopod systematics. *Mailing Add:* Dept Geol & Geog Univ Mass Amherst Campus Amherst MA 01003

PITRE, HENRY NOLLE, JR, b Opelousas, La, Oct 23, 37; m 62; c 2. ENTOMOLOGY, PLANT PATHOLOGY. *Educ:* Univ Southwest La, BS, 60; La State Univ, MS, 62; Univ Wis, PhD(entom), 65. *Prof Exp:* PROF ENTOM, MISS STATE UNIV, 65- *Mem:* Entom Soc Am; Phytopath Soc Am. *Res:* Basic and applied aspects of field crop entomology, particularly dealing with insect biology, ecology and behavior related to pest management, and insect transmission of plant diseases. *Mailing Add:* PO Drawer EM Mississippi State MS 39762

PITT, BERTRAM, b Kew Gardens, NY, Apr 27, 32; m 65; c 3. HEART FAILURE, MYOCARDIAL INFARCTION. *Educ:* Cornell Univ, BA, 53; Univ Basel, Switz, MD, 59. *Prof Exp:* Intern, Beth Israel Hosp, NY, 59-60, resident med, Boston, 60-63; res assoc & captain, Walter Reed Army Res Inst, 63-66; fel cardiol, Johns Hopkins Hosp, 66-67; from instr to prof med, Johns Hopkins Univ Sch Med, 67-76; PROF MED, UNIV MICH, 76- *Mem:* Am Soc Clin Invest; Am Asn Physicians; Am Heart Asn; Am Col Cardiol; Am Physiol Soc; Asn Univ Cardiologists. *Res:* Develop new strategies for the diagnosis and treatment of ischemic heart disease and congestive heart failure. *Mailing Add:* 24 Ridgeway Ann Arbor MI 48104

PITT, BRUCE R, b Brooklyn, NY, Mar 13, 49. PULMONARY PHARMACOLOGY, PULMONARY PHYSIOLOGY. *Educ:* Johns Hopkins Univ, PhD(environ med), 77. *Prof Exp:* Asst prof, 81-85, ASSOC PROF ANESTHESIOL, SCH MED, YALE UNIV, 85- *Mem:* Am Physiol Soc; Am Thoracic Soc. *Mailing Add:* Dept Pharmacol Pittsburgh Univ Sch Med Pittsburgh PA 15261

PITT, CHARLES H, b Fremont, Wis, Aug 9, 29; m 56; c 5. METALLURGY, SURFACE CHEMISTRY. *Educ:* Univ Wis, BS, 51; Univ Utah, PhD(metall), 59. *Prof Exp:* Process engr, Gen Elec Co, 51-53; asst res prof metall, Univ Utah, 59-60; NSF fel, Cambridge Univ, 60-61; from asst prof to assoc prof, 61-71, PROF METALL & METALL ENG, UNIV UTAH, 71- *Mem:* Am Soc Metals; Am Inst Mining, Metall & Petrol Engrs; AAAS; Nat Asn Corrosion Engrs; Am Soc Testing & Mat. *Res:* Corrosion and wear of grinding balls using electrochemical techniques and electron microscopy of dispersion hardened alloys; surface reactions. *Mailing Add:* 416 Browning Bldg Dept Metall Univ Utah Salt Lake City UT 84112

PITT, COLIN GEOFFREY, b London, Eng, Sept 15, 35; m 60; c 2. ORGANOMETALLIC CHEMISTRY, POLYMER CHEMISTRY. *Educ:* Univ London, BSc, 56, PhD(silicon chem), 59. *Prof Exp:* Chemist, Midland Silicones, Eng, 59-60; fel, Fla State Univ, 60-62 & Mass Inst Technol, 62-63; chemist, asst dir, Chem & Life Sci Div, Res Triangle Inst, 71-75, dir, Phys Sci, 75-87; DIR, PHARM DRUG DELIVERY, AMGEN, 87- *Concurrent Pos:* Adj Duke Univ, 69- *Mem:* Am Chem Soc; AAAS. *Res:* Synthetic and physical chemistry of polymers and organometallic compounds; medicinal chemistry. *Mailing Add:* Amgen 1840 Dehavilland Dr Thousand Oaks CA 91320-1789

PITT, DONALD ALFRED, b Glen Ridge, NJ, June 19, 26; m 48; c 2. PHYSICAL CHEMISTRY. *Educ:* Princeton Univ, MA, 51, PhD(chem), 57. *Prof Exp:* Sr chemist, Cent Res Dept, Minn Mining & Mfg Co, 51-63, supvr chem physics res, 63-67, sr res specialist, Corp Res Labs, 67-83, staff scientist, 83-85, STAFF SCIENTIST, SCI STA SERVS, CORP RES LABS, MINN MINING & MFG CO, 85- *Mem:* Am Chem Soc; Am Phys Soc; Optical Soc Am; Soc Photog Scientists & Engrs; Fine Particle Soc. *Res:* Thermochemistry and thermodynamics; surface chemistry and adhesion; rheology; electrochemistry; dielectric relaxation phenomena; applied mathematics; optical scattering; submicron particle size distributions; physics of photographic emulsions. *Mailing Add:* 8277 Shadow Pine Way Sarasota FL 34238-5518

PITT, JANE, b Frankfurt, Germany, Aug 25, 38; US citizen; m 62, 86; c 2. MICROBIOLOGY, IMMUNOLOGY. *Educ:* Radcliffe Col, BA, 60; Harvard Univ, MD, 64. *Prof Exp:* From intern to resident pediat, Children's Hosp Med Ctr, 64-66; fel infectious dis, Med Sch, Tufts Univ, 66-67 & Harvard Med Sch, 67-70; asst prof pediat, State Univ NY Downstate Med Sch, 70-71; instr med, 71-73, asst prof, 73-77, ASSOC PROF PEDIAT, MED SCH, COLUMBIA UNIV, 77- *Concurrent Pos:* NIH spec fel, 71-73; prin investr, res grants, Food & Drug Admin, HEW, 75-77, NIH, HEW, 75-82 & Nat Found, 76-79. *Mem:* Soc Pediat Res; Am Fedn Clin Res; Harvey Soc; fel Infectious Dis Soc. *Res:* Bacterial virulence; development of host defense in the neonate; phagocytic function; pediatric acquired immunodeficiency. *Mailing Add:* Dept of Pediat Columbia Univ 630 W 168th St New York NY 10032

PITT, LOREN DALLAS, b Chewela, Wash, Sept 18, 39; m 61; c 2. MATHEMATICS. *Educ:* Univ Idaho, BS, 61; Cath Univ Am, MA, 64; Princeton Univ, MA, 66, PhD(math), 67. *Prof Exp:* Res assoc math, Rockefeller Univ, 67-68, asst prof, 68-70; from asst prof to assoc prof, 70-78, chmn math dept, 80-83, PROF MATH, UNIV VA, 78- *Concurrent Pos:* Assoc ed, Multivariate Analysis J; ed, Progress in Probability. *Mem:* Am Math Soc; Inst Math Statist; Math Asn Am. *Res:* Probability; stochastic processes; mathematics education; random fields. *Mailing Add:* Dept Math Math Astron Bldg 204 Univ Va Charlottesville VA 22903-3199

PITT, WOODROW WILSON, JR, b Rocky Mount, NC, Aug 14, 35; m 58; c 3. CHEMICAL ENGINEERING. *Educ:* Univ SC, BS, 57; Univ Tenn, Knoxville, MS, 66, PhD(chem eng), 69. *Prof Exp:* Develop engr, Union Carbide Nuclear Div, 60-70, group leader, 70-76, mgr biotechnol & environ progs, Advan Technol Sect, 76-81, HEAD, ENG DEVELOP SECT, CHEM TECHNOL DIV, OAK RIDGE NAT LAB, 81- *Honors & Awards:* IR-100 Award, 71 & 80. *Mem:* Am Inst Chem Engrs; Am Chem Soc; Nat Soc Prof Engrs; Am Nuclear Soc; Water Pollution Control Fedn; NY Acad Sci. *Res:* Analytical liquid chromatography; analytical instrumentation for complex biochemical mixtures, such as physiological fluids and polluted waters; biochemical separations and processes; bioreactors; environmental control technology; nuclear fuel cycle; hazardous waste disposal. *Mailing Add:* Dept Nuclear Eng Tex A&M Univ 129 Zachry Collage Station TX 77843

PITTEL, STUART, b Brooklyn, NY, Apr 13, 44; m; c 1. THEORETICAL NUCLEAR PHYSICS. *Educ:* Rensselaer Polytech Inst, BS, 64; Univ Minn, PhD(physics), 68. *Prof Exp:* Res assoc physics, Bartol Res Found, Franklin Inst, 68-70, Univ Pittsburgh, 71-73; vis asst prof physics, Univ Colo, 70-71; asst prof, 73-78, assoc prof, 78-88, PROF PHYSICS, BARTOL RES FOUND, UNIV DEL, 88- *Mem:* Am Phys Soc. *Res:* The development of microscopic theories of nuclear structure; relating properties of nuclear many-body systems to properties of their constituent nucleons; the development of a unified theory of nuclear collective motion. *Mailing Add:* Bartol Res Found Univ Del Newark DE 19716

PITTENDRIGH, COLIN STEPHENSON, b Whitley Bay, Eng, Oct 13, 18; nat US; m 43; c 2. BIOLOGY. *Educ:* Univ Durham, BSc, 40; Imp Col, Trinidad, AICTA, 42; Columbia Univ, PhD(zool), 50. *Hon Degrees:* DSc, Univ Newcastle-upon-Tyne, Eng. *Prof Exp:* Biologist int health div, Rockefeller Found, 42-45; adv malaria, Ministry Health & Educ, Brazil, 45; from asst prof to prof biol, Princeton Univ, 47-62, class of 1877 prof, 62-69, dean grad sch, 65-69; prof, Stanford Univ, 69-70, Bing prof human biol, 70-76, Miller prof biol & dir, Hopkins Marine Sta, 76-84; RETIRED. *Concurrent Pos:* Guggenheim fel, 59-60; vis prof, Rockefeller Inst, 60; adv malaria, Brit Colonial Off, 58; trustee, Rocky Mt Biol Lab, Colo; mem comt oceanog, Nat Acad Sci, 58-59; mem space sci bd, 60-; mem comt biotins & biol facil panel, NSF, 58-59; adv comt biol, NASA, 60-; Alexander von Homboldt Prize, 87. *Honors & Awards:* Phillipps Lectr, Haverford Col, 57; Hopkins lectr, Stanford Univ, 57; Stearns lectr, Albert Einstein Col Med & Yeshiva Univ, 60; Van Uxem lectr, Princeton Univ, 67; Alexander Von Humboldt Prize, Alexander Von Humboldt Found, W Ger, 87. *Mem:* Nat Acad Sci; AAAS(vpres & chmn sect F); Am Soc Nat(pres, 67); Am Soc Zool; Soc Study Evolution; Am Philos Soc; Am Acad Arts & Sci. *Res:* Evolution; comparative and cellular physiology of daily rhythms. *Mailing Add:* PO Box 343 Sonita AZ 85637

PITTENGER, ARTHUR O, b Indianapolis, Ind, Oct 24, 36; m 65; c 3. MATHEMATICS. *Educ:* Stanford Univ, BS, 58, MS, 60, PhD(math), 67. *Prof Exp:* Inter-Univ Comt Travel Grants grant, Moscow State Univ, 67-68; res assoc math, Rockefeller Univ, 68-70; asst prof, Univ Mich, Ann Arbor, 70-72; assoc prof, 72-85, PROF MATH, UNIV MD BALTIMORE COUNTY, 85-, DEAN ARTS & SCI, 88- *Mem:* Am Math Soc; fel Inst Math Statist. *Res:* Probability theory. *Mailing Add:* Dept Math Univ Md Baltimore County Baltimore MD 21228

PITTENGER, ROBERT CARLTON, microbiology, for more information see previous edition

PITTENGER, THAD HECKLE, JR, b Omaha, Nebr, May 28, 21; m 43; c 3. GENETICS. *Educ:* Univ Nebr, BS, 47, PhD(genetics), 51. *Prof Exp:* AEC fel genetics, Calif Inst Technol, 51-53; geneticist, Biol Div, Oak Ridge Nat Lab, 53-57; asst prof biol, Marquette Univ, 57-59; PROF GENETICS, AGR EXP STA, KANS STATE UNIV, 59- *Mem:* AAAS; Genetics Soc Am; Am Soc Nat. *Res:* Genetics of Neurospora. *Mailing Add:* Div Biol Ackert Hall Bldg Kans State Univ Manhattan KS 66506

PITTER, RICHARD LEON, b Whittier, Calif, Apr 4, 47; div; c 2. CLOUD PHYSICS, WEATHER MODIFICATION. *Educ:* Univ Calif, Los Angeles, AB, 69, MS, 70, PhD(meteorol), 73. *Prof Exp:* Instr, 73-74, asst prof environ technol, Ore Grad Ctr Study & Res, 74-77; asst prof meteorol, Univ Md, 77-81; ASSOC RES PROF PHYSICS, DESERT RES INST, UNIV NEV, RENO, 81- *Concurrent Pos:* Consult, Lawrence Livermore Nat Lab, 74-83, Comput Sci Corp, 78-79 & Mitre Corp, 78-81. *Mem:* Am Meteorol Soc; Am Geophys Union; Royal Meteorol Soc; AAAS; Weather Modification Asn. *Res:* Riming and scavenging by ice crystals; electrical nature of snowflakes; ice crystal formation and growth in the atmosphere; interactions of cloud microphysics and chemistry. *Mailing Add:* Desert Res Inst PO Box 60220 Reno NV 89506-0220

PITTILLO, JACK DANIEL, b Hendersonville, NC, Oct 25, 38; m 66; c 2. PLANT ECOLOGY. *Educ:* Berea Col, AB, 61; Univ Ky, MS, 63; Univ Ga, PhD(bot), 66. *Prof Exp:* From asst prof to assoc prof, 66-77, PROF BIOL, WESTERN CAROLINA UNIV, 77- *Concurrent Pos:* Consult, Natural Landmarks Prog, Nat Park Serv, 75, Blue Ridge Pkwy, 85, NC Dept Natural Resources & Econ Develop, 85. *Honors & Awards:* Co-recipient Ann Res Award, Asn Southeastern Biol, 85; O Max Gardner Award Nominee, 87. *Mem:* Ecol Soc Am; Bot Soc Am. *Res:* Cladrastis kentukea; fallout radionuclides; biogeochemical cycling; granitic outcrop studies; autecology; endangered and threatened plants; phytogeography of the Balsam and Craggy Mountains, North Carolina; vascular plants of Coweeta Ecological Reserve; vegetational history of Southern Appalachians and Yunnan, China. *Mailing Add:* Dept of Biol Western Carolina Univ Cullowhee NC 28723-4073

PITTMAN, CHARLES U, JR, b Rahway, NJ, Oct 26, 39; m 65; c 3. ORGANIC CHEMISTRY, CHEMICAL ENGINEERING. *Educ:* Lafayette Col, BS, 61; Pa State Univ, PhD(chem), 64. *Prof Exp:* Fel chem, Eastern Res Lab, Dow Chem Co, 64-65; fel, Case Western Reserve Univ, 65; from asst prof to prof org chem, Univ Ala, 67-77, res prof 77-81; PROF INDUST CHEM & CATALYSIS & RES DIR, INDUST CHEM RES CTR, MISS STATE UNIV, 83- *Concurrent Pos:* Vis lectr, Natick Army Lab, Lowell Tech Inst & Koppers Co Res Labs, Pa, Univ Mass, Univ Strasbourg & Univ Liege; mem vis fac, Oakwood Col 66; consult, 83- *Mem:* Am Chem Soc. *Res:* Carbonium ion chemistry; organic reactions in strong acids; organometallic polymers; propellant combustion mechanisms and catalysis; application of quantum mechanical calculations to organic structures; homogeneous catalysis and attaching homogeneous catalysts to polymers; polymer anchored photocatalysts and metal cluster catalysis; process organic chemistry; lithographic polymers, polyolefins. *Mailing Add:* PO Drawer CH Miss State Univ Mississippi State MS 39762

PITTMAN, CHATTY ROGER, b Blakely, Ga; m 60; c 2. MATHEMATICS. *Educ:* N Ga Col, BS, 59; Univ Ga, MA, 62, PhD (math), 65; Ga Tech, MS, 78. *Prof Exp:* From asst prof to assoc prof, 65-76, PROF MATH, WEST GA COL, 76-, CHMN DEPT, 70- *Concurrent Pos:* Actg head dept math, W Ga Col, 69-70. *Mem:* Am Math Soc; Math Asn Am. *Res:* Paracompact and ordered spaces; general and point-set topology. *Mailing Add:* Dept of Math West Ga Col Carrollton GA 30118

PITTMAN, EDWARD D, b Dublin, Tex, Feb 17, 30; m 55; c 2. PETROLOGY, SEDIMENTOLOGY. *Educ:* Univ Calif, Los Angeles, BA, 56, MA, 58, PhD(geol), 62. *Prof Exp:* Geologist, Pan Am Petrol Corp, 62-65; sr res scientist, Amoco Prod Co, 66-71, staff res scientist, 71-73, res group supvr, 73-79, special res assoc, 79-85, sr res assoc, 85-89; VIS PROF, UNIV TULSA, 89- *Concurrent Pos:* Chmn geol res comt, Am Petrol Inst, 73-77; distinguished lectr, Am Asn Petrol Geologists, 78-79; assoc ed, J Sedimentary Petrol, 77-79 & Am Asn Petrol Geologists, 77-; adj prof, Univ Tulsa, 79-; vpres, Soc Econ Paleont & Mineral, 88-89. *Honors & Awards:* Distinguished Serv Award, Am Asn Petrol Geologists, 91. *Mem:* Am Asn Petrol Geologists; Soc Econ Paleont & mineral. *Res:* Sedimentary petrology; carbonate petrology; sandstone diagenesis. *Mailing Add:* Geosci Dept Univ Tulsa 600 S College Tulsa OK 74104

PITTMAN, FRED ESTES, b Cleveland, Miss, June 29, 32; m 61; c 3. GASTROENTEROLOGY, MEDICAL RESEARCH. *Educ:* Yale Univ, BA, 55; Columbia Univ, MD, 59; Univ Birmingham, Eng, PhD(med sci), 66. *Prof Exp:* Asst med, Sch Med, Cornell Univ, 59-61; vis fel, Sch Med, Columbia Univ, 61-63; sr registr, Sch Med, Univ Birmingham, 63-66; asst prof, Sch Med, Tulane Univ, 66-69; assoc prof, 69-75, PROF MED, MED UNIV SC, 75- *Concurrent Pos:* Trainee, NIH, 61-63; fel, Helen Hay Whitney Found, 63-66; res fel, Cancer Res Inst, Nat Ctr Sci Res, France, 66; fel, Roche Found, Switz, 76. *Mem:* Am Fedn Clin Res; Am Gastroenterol Asn; Am Soc Cell Biol; Am Soc Trop Med & Hyg; AAAS. *Res:* Tissue toxicity of entamoeba histolytica using whole and cell fractions of axenically cultured organisms; mechanism of action of certain antibiotics in the production of colonic mucosal injury; electron microscopy of intestinal mucosa; geographic-tropical medicine. *Mailing Add:* Dept Med Med Univ SC 171 Ashley Ave Charleston SC 29425

PITTMAN, G(EORGE) F(RANK), JR, b Pittsburgh, Pa, June 9, 27; m 49; c 2. ELECTRICAL ENGINEERING. *Educ:* Carnegie Inst Technol, BSEE, 49, MSEE, 50, DSc(elec eng), 52. *Prof Exp:* Lab asst elec eng, Carnegie Inst Technol, 49-50, res engr, 50-52; engr, Mat Eng Dept, 52-54, supv engr, 54-55, sect mgr, New Prod Labs, 55-59, mgr elec equip develop dept, 59-62, mgr power conversion eng, Indust Electronics Div, 62; mgr elec eng, Molecular Electronics Div, Westinghouse Elec Corp, 63-66, mgr eng, 66-69, sr projs mgr, Prod Transition Lab, 69-76, mgr spec progs, Westinghouse Design & Develop Ctr, 77-79, mgr opers, Westinghouse Prod & Qual Ctr, Pittsburgh, 79-; CONSULT. *Mem:* Inst Elec & Electronics Engrs. *Res:* Energy storage systems; power electronics; development and application of solid state devices. *Mailing Add:* 1830 Foxcroft Lane Allison Park PA 15101

PITTMAN, JAMES ALLEN, JR, b Orlando, Fla, Apr 12, 27; m 55; c 2. INTERNAL MEDICINE, ENDOCRINOLOGY. *Educ:* Davidson Col, BS, 48; Harvard Med Sch, MD, 52; Am Bd Internal Med, dipl, 59; recert, 74. *Hon Degrees:* DSc, Davidson Col, 80 & Univ Ala, 83. *Prof Exp:* Intern & resident med, Mass Gen Hosp, Boston, 52-54; clin assoc, Endocrinol Br, Nat Cancer Inst, 54-56; from instr to assoc prof med, Med Ctr, Univ Ala, Birmingham, 56-64, prof med & dir, Endocrinol & Metab Div, 62-71, co-chmn dept med, 69-71; prof med, Sch Med, Georgetown Univ, 71-73; EXEC DEAN, UNIV ALA SCH MED SYST, 73- *Concurrent Pos:* Fel, Harvard Med Sch, 53-54; instr, Sch Med, George Washington Univ, 55-56; chief radioisotope serv, Birmingham Vet Admin Hosp, 58-71; asst chief med dir res & educ in med, US Vet Admin, 71-73. *Honors & Awards:* Flexner Award, Asn Am Med Col, 90. *Mem:* Inst Med-Nat Acad Sci; Endocrine Soc; Soc Nuclear Med; Am Col Physicians; Am Thyroid Asn; Am Physiol Soc; Am Soc Pharmacol Exp Therapeut. *Res:* Control of thyroid function. *Mailing Add:* 5 Ridge Dr Birmingham AL 35213

PITTMAN, KENNETH ARTHUR, b Baltimore, Md, May 7, 37. DRUG METABOLISM, TOXICOLOGY. *Educ:* Univ Md, BS, 60, PhD(biochem), 66. *Prof Exp:* Res chemist, Dairy Cattle Res Br, Animal Health Res Div, USDA, 62-67; res biologist, Sterling-Winthrop Res Inst, 67-73; asst res prof environ toxicol, Inst Human & Comp Toxicol, Albany Med Col, 73-76; assoc dir drug metab, 76-82, EXEC DIR METAB PHARMACOKINETICS, PHARMACEUT RES INST, BRISTOL-MYERS SQUIBB, 82- *Mem:* Am Chem Soc; Am Soc Pharmacol & Exp Therapeut. *Res:* Metabolism, pharmacokinetics, biopharmaceutics and mechanism of toxic action of drugs, pesticides and food additives. *Mailing Add:* 8184 Bluffview Dr Manlius NY 13104

PITTMAN, MARGARET, b Prairie Grove, Ark, Jan 20, 01. PATHOGENIC BACTERIA, BACTERIAL VACCINES. *Educ:* Hendrix Col, AB, 23; Univ Chicago, MS, 26, PhD(bact), 29; Dipl, Am Bd Microbiol, 62. *Hon Degrees:* LLD, Hendrix Col, 54. *Prof Exp:* Prin acad, Galloway Woman's Col, 23-25; asst scientist, Rockefeller Inst, 28-34; asst bacteriologist, NY State Dept Health, 34-36; bacteriologist, NIH, 36-71, chief lab bact prod, Div Biol Stand, 58-71, guest worker, 71-72; GUEST WORKER & CONSULT, CTR BIOL EVAL & RES, FOOD & DRUG ADMIN, 72- *Concurrent Pos:* Consult at large, WHO, 58, 59, 62, 69, & 71-73; mem, US Pharmacopeia Panels, 66-75 & 70-73; guest lectr, Howard Univ, Washington DC, 67-70, Lederle Lab, 85; guest scientist, NIH, Bilthoven, Neth, 74, Univ Glasgow, Scotland, 77; consult, Connaught Labs, Ont, Can 74 & 75, State Inst for Serum & Vaccine, Razi, Teheran, Iran, Dept Microbiol, Glasgow, Scotland, 78, 79 & 84; emer distinguished scientist, Soc Exp Biol & Med, 86; mem, numerous adv comt for grants contracts. *Honors & Awards:* Superior Serv Award, Dept Health, Educ & Welfare, 63 & Distinguished Serv Award, 68; Fed Woman's Award, 70; Cert Appreciation, Ctr Dis Control, US Dept Health & Human Serv, Pub Health Serv, 86; Citation, Infectious Dis Soc, 87; Alice Evans Award, Am Soc Microbiol, 90. *Mem:* AAAS; hon mem Am Soc Microbiol; Am Acad Microbiol; Soc Exp Biol & Med; Int Asn Biol Standard; hon fel Am Acad Pediat. *Res:* Haemophilus, pneumococcus; meningococcus; respiratory infections, meningitis and conjuctivitis; standardization of biological products; pertussis, typhoid and cholera vaccines; tetanus toxoid; Bordetella pertussis systematics; nature of pathophysiology of pertussis effected by pertussis toxin; cause of rare severe vaccine reactions; role of the adrenergic receptor system; protective antibodies; vaccine standards. *Mailing Add:* 3133 Connecticut Ave NW Apt 912 Washington DC 20008

PITTMAN, MELVIN AMOS, b Chester, SC, Aug 4, 05; m 43; c 3. OPTICS. *Educ:* The Citadel, BS, 25; Univ SC, MS, 29; Johns Hopkins Univ, PhD(physics), 36. *Prof Exp:* Instr physics, Univ Md, 29-36, asst prof & head dept, Baltimore br, 36-37, prof physics & head dept, Madison Univ, 37-42; radar engr, Signal Corps, US Army 42, inst elect engr, US Naval Acad, 42-46, prof physics & head dept, Madison Univ, 46-55, Col Of William & Mary 55-60, dir of inst, 66-67; dean, 67-74, EMER DEAN SCH SCI, OLD DOMINION UNIV, 74- *Mem:* Am Phys Soc; assoc Optical Soc Am. *Res:* Infrared dispersion and absorption; polarization in infrared; effects of radiation, laser, laser optics. *Mailing Add:* 3449 Waverly Dock Rd Jacksonville FL 32223

PITTMAN, QUENTIN J, b Lethbridge, Alta, Sept 23, 52; c 3. NEUROSCIENCES, AUTONOMIC NERVOUS SYSTEMS. *Educ:* Univ Lethbridge, BSc, 72; Univ Calgary, PhD(med physiol), 76. *Prof Exp:* Fel neurol, Montreal Gen Hosp, 76-78 & Arthur V Davis Ctr Behav Neurobiol, Salk Inst, La Jolla, Calif, 78-80; from asst prof to assoc prof pharmacol, Univ Calgary, 80-84, assoc prof med physiol, 84-88, asst dean med sci, 89-90, PROF MED PHYSIOL, UNIV CALGARY, 88- *Concurrent Pos:* Mem, Grant Rev Comt Neuroregulatory Mechanisms, Med Res Coun Can, 82-87 & Maj Equip Grants Comt, Alta Heritage Found, 85-86; scientist, Med Res Coun Can, 85-89; spec reviewer, Neurobiol Study Sect, NIH, 86; mem, sci rev comt, Can Heart Found, 86-89; vis lectr, Int Brain Res Orgn, 87; scientist, Alta Heritage Found, 90. *Honors & Awards:* JAF Stevenson Award, Can Physiol Soc, 85; Med Res Scholar, Alta Heritage Found, 85. *Mem:* Am Physiol Soc; Can Physiol Soc; Soc Neurosci; AAAS. *Res:* Central nervous system control of pituitary function; physiological anatomy of brain limbic structures; electrophysiological and pharmacological properties of peptide neurons and peptidergic neuronal transmission; central nervous system control of blood pressure, thermoregulation, fever; febrile convulsions. *Mailing Add:* Dept Med Physiol Univ Calgary 3300 Hosp Dr NW Calgary AB T2N 4N1 Can

PITTMAN, RAY CALVIN, BIOMEDICAL RESEARCH, ATHEROSCLEROSIS. *Educ:* Univ Calif, PhD(biochem), 75. *Prof Exp:* RES BIOCHEMIST, DEPT MED, UNIV CALIF, SAN DIEGO, 69- *Res:* Lipoprotein metabolism. *Mailing Add:* Dept Med Univ Calif San Diego La Jolla CA 92093

PITTMAN, ROBERT PRESTON, b Lubbock, Tex, May 26, 40. CARDIOVASCULAR PHYSIOLOGY. *Educ:* Tex Tech Univ, BA, 63, MS, 66; Univ Houston, PhD(physiol), 70. *Prof Exp:* ASSOC PROF PHYSIOL, MICH STATE UNIV, 71- *Concurrent Pos:* NIH fel, Mich State Univ, 70-71. *Mem:* AAAS; Am Soc Zool; Sigma Xi. *Res:* Hemorrhagic shock; hepatic and gastrointestinal blood flow. *Mailing Add:* Off VProvost Mich State Univ Health Fee Hall East Lansing MI 48824

PITTMAN, ROLAND NATHAN, b Waco, Tex, Feb 13, 44; m 68; c 2. CARDIOVASCULAR PHYSIOLOGY. *Educ:* Mass Inst Technol, SB, 66; State Univ NY Stony Brook, MA, 68, PhD(physics), 71. *Prof Exp:* Fel physiol, Univ Va, 71-74; from asst prof to assoc prof, 74-87, PROF PHYSIOL, MED COL VA, 87- *Mem:* Europ Soc Microcirculation; Am Physiol Soc; Biophys Soc; Microcirculatory Soc (secy, 87-); Sigma Xi; Int Soc Oxygen Tranport Tissue. *Res:* The study of oxygen transport in the microcirculation and the role of oxygen in local regulation of blood flow. *Mailing Add:* Med Col Va Box 551 MCV Sta Richmond VA 23298-0551

PITTS, BARRY JAMES ROGER, b UK, Sept 27, 40. CARDIOVASCULAR BIOCHEMISTRY. *Educ:* Dublin Univ, BA, 64, PhD(biochem), 69. *Prof Exp:* Fel pharmacol, Med Col, Cornell Univ, 69-72; from instr to asst prof cell biophys, Baylor Col Med, 72-77, asst prof med & biochem, 77-84; HEAD, CARDIOVASC BIOCHEM SECT, PHARMACEUT RES DIV, SCHERING-PLOUGH CORP, BLOOMFIELD, NJ, 84- *Mem:* Am Soc Biol Chem; Am Soc Pharmacol & Exp Therapeut; Biophys Soc; NY Acad Sci; Int Heart Res Soc. *Res:* Ion transport systems in the heart and vascular smooth muscles and their role in regulation of contractions; role in excitation, contraction coupling and in myocardial ischemia; physiology. *Mailing Add:* 1295 Denmark Rd Plainfield NJ 07062

PITTS, CHARLES W, b Corinth, Miss, Dec 11, 33; m 56; c 4. MEDICAL ENTOMOLOGY, INSECT PHYSIOLOGY. *Educ:* Miss State Univ, BS, 60; Kans State Univ, MS, 62, PhD(entom), 65. *Prof Exp:* From instr to prof entom, Kans State Univ, 62-78; PROF ENTOM & HEAD DEPT, PA STATE UNIV, 78- *Mem:* AAAS; Entom Soc Am. *Res:* Chemical control of insects attacking man and animal; insect attractants and sensory physiology and behavior; scanning electron microscopy. *Mailing Add:* Dept Entom Pa State Univ University Park PA 16802

PITTS, DAVID EUGENE, b Oklahoma City, Okla, July 29, 39; m 63; c 2. IMAGE ANALYSIS, REMOTE SENSING. *Educ:* Univ Okla, BS, 61, MS, 64, DrEng, 71. *Prof Exp:* Res meteorologist, Johnson Space Ctr, NASA, 63-70, chief, Remote Sensing Sect, 70-74, mgr, Accuracy Assessment Subsyst, 74-79 & Agristars Field Res, 79-81, chief, Radiation Characterization, 81-83 & Biospheric Sci Sect, 83-84, head, Image Analysis Lab, 84-89, MGR, FLIGHT SCI SUPPORT OFF, NASA, 89- *Mem:* AAAS; Am Meteorol Soc; Am Soc Photogram & Remote Sensing; Inst Elec & Electronic Engrs. *Res:* Image analysis of photo and television for space shuttle launch, landing and on-orbit; support for space shuttle astronauts photography of the earth; image analysis of high speed photography of projectile impacts; image analysis of medical ultrasound; planetary atmospheres; remote sensing; holder of one US patent. *Mailing Add:* 16011 Stonehaven Dr Houston TX 77059

PITTS, DONALD GRAVES, b Perry Twp, Ark, Apr 11, 26; m 46; c 3. OPTOMETRY, PHYSIOLOGICAL OPTICS. *Educ:* Southern Col Optom, OD, 50; Ind Univ, MS, 59, PhD(physiol optics), 64. *Prof Exp:* Clin optometrist, Biomed Sci Corps, US Air Force, 51-57, mem vision res staff, Aerospace Med Res Lab, Wright-Patterson Air Force Base, Ohio, 59-62, mem fac, Sch Aerospace Med, 62-69, mem optics sect, 64-69; assoc dean, 73-76, PROF OPTOM, COL OPTOM, UNIV HOUSTON, 69- *Concurrent Pos:* Mem Armed Forces vision comt, Nat Res Coun; mem panel ophthal devices, Food & Drug Admin, 74-77; mem adv coun, AAAS, 74-76. *Mem:* Fel AAAS; Am Asn Univ Prof; Optical Soc Am; Am Optom Asn; fel Am Acad Optom. *Res:* Visual neurophysiology; aerospace vision effects; effects of radiant energy on the visual system; color vision. *Mailing Add:* Col Optom Univ Houston Houston TX 77004

PITTS, DONALD JAMES, b San Diego, Calif, Dec 30, 45. AGRICULTURAL ENGINEERING. *Educ:* San Diego State Univ, BA, 70, MA, 76; Univ Ark, MS, 83, PhD(eng), 86. *Prof Exp:* Asst prof agr eng, Southwest Fla Res & Educ Ctr, Univ Fla, 87-90; asst prof agr eng, Northeast

Res & Exten Ctr, 86-87, DIR & HEAD, AGR DEPT, SOUTHEAST RES & EXTEN CTR, UNIV ARK, MONTICELLO, 91- *Mem:* Am Soc Agr Engrs; Irrig Asn Am; Water Well Asn Am. *Mailing Add:* Southeast Res & Exten Ctr Univ Ark PO Box 3508 Monticello AR 71655

PITTS, DONALD ROSS, b Anniston, Ala, Sept 21, 29; m 51; c 3. MECHANICAL ENGINEERING. *Educ:* Auburn Univ, BME, 51; Ga Inst Technol, MSME, 60, PhD(mech eng), 68. *Prof Exp:* Develop engr, Goodyear Tire & Rubber Co, Ohio, 53-55; aircraft design engr, Lockheed Aircraft Corp, Ga, 55-58, sr aircraft res engr, Res & Develop Lab, 59-67; asst plant engr, H W Lay Co Div, Frito-Lay Corp, 58-59; from assoc prof to prof heat transfer &fluid mech, Tenn Technol Univ, 67-71; Prof Mech Eng, Clemson Univ, 78-82; DEPT MECH & AEROSPACE ENG, UNIV TENN,82- *Mem:* Am Soc Mech Engrs; Am Soc Eng Educ; Am Inst Aeronaut & Astronaut. *Res:* Transient film and nucleate boiling; free convection heat transfer; liquid droplet interaction with external boundary layer flow; fluidized bed heat transfer. *Mailing Add:* Dept Mech & Aerospace Eng Univ Tenn 414 Dougherty Hall Knoxville TN 37996

PITTS, GERALD NELSON, b Brownwood, Tex, June 14, 43; m 64; c 1. COMPUTER SCIENCES. *Educ:* Tex A&M Univ, BA, 66, MS, 67, PhD(comput sci), 71. *Prof Exp:* Teaching asst math, Tex A&M Univ, 66-71; asst prof comput sci, Univ Southwestern La, 71-72; assoc prof & head dept, Central Tex Col, 72-74; assoc prof, Tex Tech Univ, 74-75; PROF COMPUT SCI, MISS STATE UNIV, 75-; AT DEPT COMPUT & INFO SCI, TRINITY UNIV. *Concurrent Pos:* Adj prof comput sci, Fla Inst Technol, 72-73. *Mem:* Sigma Xi; Asn Comput Mach; Data Processing Mgt Asn. *Mailing Add:* 1601 Avenue D Brownwood TX 78284

PITTS, GROVER CLEVELAND, b Richmond, Va, Apr 4, 18; m 47; c 3. PHYSIOLOGY. *Educ:* Univ Richmond, BA, 39; Harvard Univ, AM, 40, PhD(physiol), 43. *Prof Exp:* Lab asst anat, Harvard Univ, 40-41, asst invert physiol, 42-43, asst, Fatigue Lab, 42-44; investr physiol, Naval Med Res Inst, 44-49; asst prof, 50-58, assoc prof, 58-79, PROF PHYSIOL, SCH MED, UNIV VA, 79- *Concurrent Pos:* Field studies Alaska, Us Air Force, 53-54; experimenter, Univ Wisc exped SAmer, 59 & Soviet Cosmos 1129 mission, 70; prin investr rat study, Biosatellite Proj, 65-70; mem working group 5, panel gravitational physiol, Comt Space Res, 77-79; corresp gravitational physiol, Int Union Physiol Sci, 85- *Honors & Awards:* Laudatory cert, Inst Biomed Probs, Ministry Pub Health, USSR, 79; Cosmos Achievement Award, NASA, 81. *Mem:* Fel AAAS; Am Physiol Soc; Am Soc Zool. *Res:* Physiologic effects of gravitation, exercise; regulation of body composition. *Mailing Add:* Dept Physiol Univ Va Sch Med Charlottesville VA 22908

PITTS, JAMES NINDE, JR, b Salt Lake City, Utah, Jan 10, 21; m 45, 76; c 3. PHOTOCHEMICAL AIR POLLUTION & AIRBORNE TOXICS. *Educ:* Univ Calif, Los Angeles, BS, 45, PhD(chem), 49. *Hon Degrees:* MA, Oxford Univ, 65. *Prof Exp:* Asst, Nat Defense Res Comt, Northwestern Univ, 42-45, res assoc spec projs div, US Army Serv Forces, 45-46, from instr to asst prof chem, 49-54; assoc prof phys chem, 54-59, PROF CHEM, UNIV CALIF, RIVERSIDE, 59-, DIR, STATEWIDE AIR POLLUTION RES CTR, 70-, PROJ CLEAN AIR, 71- *Concurrent Pos:* Guggenheim fel, 60-61; chmn dept chem, Univ Calif, Riverside, 61-63, fac res lectr, 65-66, dir dry lands res inst, 70-71; acad adv to chancellor, Univ Calif, Santa Cruz, 62-63; vis res fel, Merton Col, Oxford Univ, 65; mem Inst Geophys & Planetary Physics, 68-; Sigma Xi-Sci Res Soc Am lectr, 72-73; consult, USPHS & mem environ sci & eng study sect air pollution; mem adv coun protective comt, Chem Corps US Army; adv bd mil personnel supplies comt textile functional finishing, Nat Acad Sci-Nat Res Coun; co-chmn, 2nd Int Union Pure & Appl Chem Int Symp Photochem, Netherlands; adj fac mem, Off Manpower Develop, Environ Protection Agency, mem air pollution chem adv comt, nat air qual adv comt, 75-76; tech & sci adv comt, Calif Air Resources Bd & Comt Biol Effects of Atmos Pollutants; chmn panel polycyclic org mat comt kinetics of chem reactions, Nat Acad Sci; mem, Nat Air Conserv Comn, Am Lung Asn, 75-78; mem sci adv comt, High Altitude Pollution Prog, Fed Aviation Admin, US Dept Transp, 78-80 & Acid Deposition, Calif State Assembly, 83-; co-ed, Advan Environ Sci & Technol & Advan Photochem. *Honors & Awards:* Dreyfus Lectr, Hope Col, Mich, 78; Frank A Chambers Award, Air Pollution Control Asn, 82. *Mem:* Air Pollution Control Asn; Am Chem Soc; Inter-Am Photochem Soc; Sigma Xi; Environ Mutagen Soc; fel AAAS. *Res:* Fundamental processes in photochemistry and photooxidations and their application to atmospheric chemistry; photochemical smog, acid rain, airborne toxics and mutagenic and/or carcinogenic pollutants. *Mailing Add:* PO Box 409 Fawnskin CA 92333

PITTS, JOHN ROLAND, b Dallas, Tex, Jan 16, 48; m 79; c 3. SURFACE PHYSICS, THIN FILMS. *Educ:* NMex Inst Mining & Technol, BS, 70; Ore Grad Ctr, MS, 72; Denver Univ, PhD(physics), 85. *Prof Exp:* Eng duty officer, US Navy, 72-75; Teaching asst physics, Univ Colo, Boulder, 75-77, res asst 77-78; staff physicist, 78-84, SR PHYSICIST, SOLAR ENERGY RES INST, 84- *Concurrent Pos:* vis fel, Fritz-Haber-Institut der Max-Plauck-Gesellschaft, Berlin, 85-86; mem bd dirs, Rocky Mountain Chap Am Vacuum Soc, 88- *Mem:* Am Phys Soc; Am Vacuum Soc; Mat Res Soc. *Res:* Surfaces, interfaces and thin films used in energy, optical and electronic materials; surface analytical tools, such as x-ray photoemission spectroscopy, auger electron spectroscopy, ion scattering spectrometry and secondary ion mass spectrometry; surface modification using photon, electron, and ion beams; rapid thermal processing; metallurgical coatings; chemical vapor deposition. *Mailing Add:* Solar Energy Res Inst 1617 Cole Blvd Golden CO 80401

PITTS, JON T, b St Marcos, Tex, Jan 10, 48. GEOMETRIC MEASURE THEORY. *Educ:* Princeton Univ, PhD(math), 74. *Prof Exp:* PROF MATH, TEX A&M UNIV, 80- *Mem:* Am Math Soc. *Mailing Add:* Dept Math Tex A&M Univ College Station TX 77843

PITTS, MALCOLM JOHN, b Paris, France, May 13, 50; US citizen; m 76; c 2. CHEMICAL ENHANCED OIL RECOVERY, TWO PHASE FLOW IN POROUS MEDIA. *Educ:* Univ Colo, Boulder, BA, 72; Purdue Univ, MSc, 74; Georgetown Univ, PhD(chem), 79. *Prof Exp:* Res assoc, Purdue Univ, 72-74 & Georgetown Univ, 75-79; chemist & div supvr, Hazen Res, 74-75; postdoctoral res assoc, Nat Jewish Hosp & Res Ctr, 79-80; DIR, PETROL TECHNOL, SURTEK, INC, 80- *Concurrent Pos:* Mem, Tech Info Comt, Soc Petrol Engrs, 86-89, co-chmn, Joint Can Inst Mining-Soc Petrol Engrs Tech Meeting, 89-90; partic, Workshop DNAPL removal hazardous waste sites, Environ Protection Agency, 91. *Mem:* Fel Am Inst Chemists; Soc Petrol Engrs; Am Chem Soc; Nat Well Water Asn; Can Inst Mining & Metall. *Res:* Application of chemical systems to economically remove oil from porous media. *Mailing Add:* Surtek Inc 1511 Washington Ave Golden CO 80401

PITTS, ROBERT GARY, b Auburn, Ala, Aug 17, 45; m 83; c 2. DENTAL RESEARCH, MICROBIAL PHYSIOLOGY. *Educ:* Auburn Univ, BS, 67, MS, 69; La State Univ, Baton Rouge, PhD, 71; Rutgers Univ Law Sch, JD, 82. *Prof Exp:* Res assoc microbiol & NIH fel, Inst Dent Res, Univ Ala, Birmingham, 71-72; scientist, Warner-Lambert Co, 72-76, sr scientist, 76-77, mgr chem & biol res, Personal Prod Div, 77-78, dir res serv, Personal Prod Div, 78-80, dir biol res, Consumer Prod Group, 80-83; mgr appl res, 83-84, mgr biochem, microbiol & immunol, Unilever NV, 84-85; vpres biol res, Schering-Plough, 85-86; VPRES RES & DEVELOP, ORAL B LABS DIV, GILLETTE, 86- *Concurrent Pos:* Pvt law pract, NJ, 83-85; outside dir, US Biothane Inc, 85- *Mem:* Am Soc Microbiol; Am Chem Soc; Sigma Xi; Int Asn Dent Res; Am Bar Asn; NY Acad Sci. *Res:* Periodontitis; oral malodor; acne. *Mailing Add:* Oral-B Labs Inc One Lagoon Dr Redwood City CA 94065

PITTS, THOMAS GRIFFIN, b Clinton, SC, Aug, 19, 35; m 61; c 2. COMPUTER SOFTWARE, EXPERIMENTAL NUCLEAR PHYSICS. *Educ:* Presby Col SC, BS, 57. *Prof Exp:* Mathematician atomic energy div, Babcock & Wilcox, Co, 57- 58, physicist, 59-76, sr res engr, Lynchburg Res Ctr, Res Develop Div, 77-91; RETIRED. *Res:* Computer programming and evaluation; nuclear instrumentation design and testing. *Mailing Add:* 1804 Parkland Dr Lynchburg VA 24503-2421

PITTS, WANNA DENE, b Tonkawa, Okla, Jan 24, 32; m 47; c 2. POLLINATION ECOLOGY, FIRE ECOLOGY. *Educ:* San Jose State Col, BS, 70; San Jose State Univ, MS, 72; Univ Calif, Davis, PhD(ecol), 76. *Prof Exp:* Lectr, 72-76, asst prof, 76-80, ASSOC PROF BIOL SCI, SAN JOSE STATE UNIV, 80-, ASSOC DEAN, SCH SCI, 78- *Concurrent Pos:* Lectr biol sci, San Jose City Col, 71-73, Gavilan Commun Col, 72-73. *Mem:* AAAS; Am Bot Soc; Ecol Soc Am. *Res:* Biology of cupressus abramsiana; plant ecology. *Mailing Add:* Dept Biol Sci San Jose State Univ San Jose CA 95192

PITTZ, EUGENE P, b Albany, NY, Sept 17, 39; m 62; c 2. TOXICOLOGY, PHYSICAL BIOCHEMISTRY. *Educ:* Siena Col, BS, 62; State Univ NY, Buffalo, PhD(biophysics), 70. *Prof Exp:* Fel biochem, Brandeis Univ, 70-74; res asst prof toxicol, Albany Med Col, 74-78; sr res scientist dermato-toxicol, Lever Brothers Co, 78-82, Warner Lambert, 82-84, Sandoz, 84-86, SCHERING PLOUGH, 86- *Mem:* AAAS; Am Chem Soc; Sigma Xi; Soc Cosmetic Chemists; Soc Invest Dermatol. *Res:* Efficiency and safety evaluation of products and new technologies; dermatoxicology; claim support research; carcinogenesis; systemic toxicology; physical biochemsitry; respiratory, gastroenterology and dermatologic research and product development. *Mailing Add:* 2255 E Glenalden Dr Germantown TN 38138

PITZER, KENNETH SANBORN, b Pomona, Calif, Jan 6, 14; m 35; c 3. PHYSICAL CHEMISTRY. *Educ:* Calif Inst Technol, BS, 35; Univ Calif, PhD(chem), 37. *Hon Degrees:* ScD, Wesleyan Col, 62; LLD, Univ Calif, 63, Mills Col, 69. *Prof Exp:* Asst chem, Univ Calif, Berkeley, 35-36, from instr to prof, 37-61, asst dean col letters & sci, 47-48, dean col chem, 51-60; prof chem & pres, Rice Univ, 61-68, Stanford Univ, 68-70; PROF CHEM, UNIV CALIF, BERKELEY, 71- *Concurrent Pos:* Tech dir, Md Res Lab, Off Sci Res & Develop, 43-44; assoc dir proj, Am Petrol Inst, 47-52, dir proj, 50, 52-58; dir div res, US AEC, DC, 49-51, mem gen adv comt, 58-65, chmn, 60-62; Guggenheim fel, 51; chmn sect chem, Nat Acad Sci, 59-62, coun, 64-68 & 73-76; trustee, Rand Corp, 62-72, Carnegie Found Advan Teaching, 66-70 & Pitzer Col, 66-; dir, Fed Reserve Bank Dallas, 65-68; mem, President's Sci Adv Comt, 65-68; chmn coun presidents, Univ Res Asn, 66-67, mem bd dirs, Owens-Ill, Inc, 66-86. *Honors & Awards:* Awards, Am Chem Soc, 43 & 50, G N Lewis Medal, 65, Priestly Medal, 69 & Gibbs Medal, 76; Clayton Prize, Brit Inst Mech Eng, 58; Priestley Mem Award, Dickinson Col, 63; Nat Medal of Sci, 75; Gold Medal, Am Inst Chem, 76; Centenary Lectr, Chem Soc (Gt Brit), 78; Robert A Welch Award Chem, 84; Rossini Lect, Int Union Pure & Appl Chem. *Mem:* Nat Acad Sci; Am Chem Soc; Am Philos Soc; Am Nuclear Soc; fel Am Phys Soc; Faraday Soc; fel Am Inst Chem; Geochem Soc. *Res:* Chemical thermodynamics; quantum theory and statistical mechanics applied to chemistry; molecular spectroscopy. *Mailing Add:* Dept Chem Univ Calif Berkeley CA 94720

PITZER, RUSSELL MOSHER, b Berkeley, Calif, May 10, 38; m 59; c 3. THEORETICAL CHEMISTRY. *Educ:* Calif Inst Technol, BS, 59; Harvard Univ, AM, 61, PhD(chem physics), 63. *Prof Exp:* Mem div sponsored res staff, Mass Inst Technol, 63; res instr chem, Calif Inst Technol, 63-66, asst prof, 66-68; assoc prof, 68-79, actg assoc dir, Ohio Supercomputer Ctr, 86-87, PROF CHEM, OHIO STATE UNIV, 79-, CHMN DEPT, 89- *Concurrent Pos:* Consult, Battelle Mem Inst, 75-77, Lawrence Berkeley Lab, 78-87 & Lawrence Livermore Nat Lab, 80- *Mem:* Am Phys Soc; Am Chem Soc. *Res:* Treatment of polyatomic molecules by quantum mechanics; use of symmetry and relativity in the computation of molecular electronic structure. *Mailing Add:* Dept of Chem Ohio State Univ Columbus OH 43210

PIVER, M STEVEN, b Washington, DC, Sept 29, 34; m 58; c 3. GYNECOLOGIC ONCOLOGY. *Educ:* Gettysburg Col, BS, 57; Temple Univ Sch Med, MD, 61. *Prof Exp:* Postgrad residency, Pa Hosp, 67; fel gynec oncol, Md Anderson Hosp & Tumor Inst, 70; dir gynec oncol, Univ NC Sch Med, 70-71; assoc chief gynec oncol, 72-84, CHIEF GYNEC ONCOL,

ROSWELL PARK MEM INST, 85-; CLIN PROF GYNEC, STATE UNIV NY BUFFALO, 74- *Mem:* Am Soc Clin Oncol; Soc Gynec Oncol; Soc Surg Oncol; Am Radium Soc; fel Am Col Obstetricians & Gynecologists; fel Am Col Surgeons. *Res:* Radiation sensitizers to improve cure rates in cancer using radiation therapy; genetic link in ovarian cancer. *Mailing Add:* Roswell Park Mem Inst 666 Elm St Buffalo NY 14263

PIVONKA, WILLIAM, b Albert, Kans, Mar 31, 30; m 59; c 1. ORGANIC CHEMISTRY. *Educ:* St Benedict's Col, BS, 51, Dow Chem Co fel & PhD(chem), 59. *Prof Exp:* Shift chemist, Hercules Powder Co, 51-52; asst chem, Univ Kans, 54-59; asst prof org chem & chmn dept, 59-63, from assoc prof to prof, 63-72, DEAN SCH ARTS & SCI, PARKS COL, 72-, PROF CHEM, 76- *Mem:* Am Chem Soc; Am Inst Chem; Am Inst Chem Eng. *Res:* Organophosphorus and fluorocarbon chemistry; reaction mechanisms. *Mailing Add:* Dept of Chem Parks Col Parkville MO 64152

PIVORUN, EDWARD BRONI, b Nashua, NH, Nov 5, 46; m 71. COMPARATIVE PHYSIOLOGY, ZOOLOGY. *Educ:* Tufts Univ, BS, 68; Univ Minn, PhD(zool), 74. *Prof Exp:* Asst prof, 74-80, ASSOC PROF ZOOL, CLEMSON UNIV, 80- *Mem:* Am Soc Zoologists; Ecol Soc Am; Sigma Xi; Int Hibernation Soc. *Res:* Physiology of hibernation; radiotelemetry of body temperature and heart rate patterns of mammals during hibernation; effects of blood-borne hibernation factors on homeothermic tissue; effects of opiods and peptides on daily torpor; influence of the pineal gland on hibernation and daily torpor. *Mailing Add:* Dept of Zool Clemson Univ Clemson SC 29631

PIWONI, MARVIN DENNIS, b Stanley, Wis, Mar 9, 47; m 74; c 1. ENVIRONMENTAL CHEMISTRY, ENVIRONMENTAL POLLUTANT TRANSPORT. *Educ:* Univ Wis, BS, 70, MS, 75; Univ Tex, Dallas, PhD(environ sci), 75. *Prof Exp:* Asst prof environ sci, Univ Tex, Dallas, 75-80; res phys scientist, RS Kerr Environ Res Lab, 80-87, LAB MGR, HAZARDOUS WASTE RES CTR, STATE ILL, 87- *Concurrent Pos:* Parts coordr, Standard Methods Comt, Water Pollution Control Fedn, 89- *Res:* Organic pollutant behavior in the subsurface and in ground water; sorption of organic pollutants; analytical methods development. *Mailing Add:* Hazardous Waste Res Ctr One E Hazelwood Dr Champaign IL 61820

PIWONKA, THOMAS SQUIRE, b Cleveland, Ohio, Jan 8, 37; m 61; c 3. METALLURGY. *Educ:* Case Inst Technol, BS, 59; Mass Inst Technol, MS, 60, DSc, 63. *Prof Exp:* Sr engr, Gen Motors Corp, 63-66; sr metallurgist, TRW Metals Div, 66-70, prin engr, Mat Technol Lab, 70-73, tech dir, Kelsey-Hayes Invest Casting Orgn, 73-74, mgr casting res, TRW Mat Technol, 74-80, MGR, MAT RES DEPT, TRW INC, 80- *Concurrent Pos:* Mem comt high strength steel & titanium castings, Nat Mat Adv Bd, 72. *Mem:* Am Inst Mining, Metall & Petrol Engrs; Am Soc Metals. *Res:* Solidification; titanium casting; directional solidification; microporosity in solidification; vacuum melting; manufacturing systems. *Mailing Add:* Box 870204 Tuscaloosa AL 35487

PIXLEY, ALDEN F, b Pasadena, Calif, Feb 27, 28; m 56; c 3. MATHEMATICS. *Educ:* Univ Calif, Berkeley, AB, 49, MA, 50, PhD(math), 61. *Prof Exp:* Spec rep to petrol indust, IBM Corp, 55-58; asst prof math, Col Notre Dame, 58-60 & San Francisco State Col, 60-62; from asst prof to assoc prof, 62-72, PROF MATH, HARVEY MUDD COL, 72- *Mem:* Am Math Soc; Math Asn Am. *Res:* Universal algebra and equational logic. *Mailing Add:* Dept Math Harvey Mudd Col Claremont CA 91711

PIXLEY, CARL PRESTON, b Omaha, Nebr, Nov 3, 42; m 68. TOPOLOGY. *Educ:* Univ Omaha, BA, 66; Rutgers Univ, MS, 68; State Univ NY Binghamton, PhD(math), 73. *Prof Exp:* From instr to asst prof math, Univ Tex, Austin, 74-78; ASSOC PROF MATH, SOUTHWEST TEX STATE UNIV, 78- *Mem:* Am Math Soc; Asn Symbolic Logic. *Res:* Topological selection theory and geometrical topology. *Mailing Add:* 1016 W Hopkins St San Marcos TX 78666

PIXLEY, EMILY CHANDLER, b Knoxville, Tenn, Aug 19, 04; m 31; c 3. MATHEMATICS. *Educ:* Randolph-Macon Woman's Col, AB, 26; Univ Chicago, SM, 27, PhD, 31. *Prof Exp:* From instr to prof & head dept math, St Xavier Col, 27-36; res mathematician, US Govt, 34-35; from instr to asst prof math, Wayne State Univ, 36-48; from asst prof to prof, 48-73, EMER PROF MATH, UNIV DETROIT, 73- *Mem:* Am Math Soc; Math Asn Am; Sigma Xi. *Res:* Modern abstract algebra; theory numbers; analytical and additive number theory. *Mailing Add:* 17383 Garfield Detroit MI 48240

PIZER, LEWIS IVAN, b New York, NY, Sept 14, 32; m 57; c 3. MICROBIOLOGY, BIOCHEMISTRY. *Educ:* Univ NZ, BS, 54; Univ Calif, Berkeley, PhD(biochem), 58. *Prof Exp:* Donner fel, 58-59; NIH fel, 59-60; from asst prof to assoc prof microbiol, Univ Pa, 61-77, prof, 77-80; PROF MICROBIOL & CHMN, DEPT MICROBIOL & IMMUNOL, SCH MED, UNIV COLO HEALTH SCI CTR, 80- *Mem:* Am Soc Biol Chem; Am Soc Microbiol. *Res:* Metabolic control mechanisms and the physiology of bacteriophage infected cells. *Mailing Add:* Dept Microbiol & Immunol Univ Colo Health Sci Ctr 4200 E Ninth Ave Denver CO 80262

PIZER, RICHARD DAVID, b Washington, DC, Mar 15, 44; m 89; c 2. INORGANIC CHEMISTRY. *Educ:* Johns Hopkins Univ, BA, 65, Brandeis Univ, PhD(chem), 69. *Prof Exp:* Asst prof chem, State Univ NY Col Potsdam, 69-71; asst prof, NY Univ, 71-73; PROF CHEM, BROOKLYN COL, CITY UNIV NEW YORK, 73- *Concurrent Pos:* Vis fel, Res Sch Chem, Australian Nat Univ, 79-80; exec officer, PhD prog chem, City Univ NY, 90- *Mem:* Sigma Xi; Am Chem Soc. *Res:* Study of fast reactions in aqueous solution studied by relaxation and stopped-flow methods; complexation reactions of boron acids; the dynamics of cryptate formation; metal ion inclusion complexes. *Mailing Add:* Dept Chem Brooklyn Col City Univ New York Brooklyn NY 11210

PIZER, STEPHEN M, b Boston, Mass, Oct 4, 41; m 64; c 2. GRAPHICS & IMAGING PROCESSING, VISUAL PERCEPTION. *Educ:* Brown Univ, BA & BSc, 63; Harvard Univ, AM, 64, PhD(appl math comput sci), 67. *Prof Exp:* Res fel radiol med, Mass Gen Hosp, 67-75; PROF COMPUT SCI, UNIV NC, 67- *Concurrent Pos:* Vis res fel, Univ Col Hosp Med Sch, London, 73-74; adj prof, Dept Radiol, Univ NC, 82; vis prof, Rijks Univ, Utrech, Netherlands, 83-84; prin investr, NIH prog proj, NIH grants, NATO grant, 83-; assoc ed, Inst Elec & Electronics Engrs, Trans Med Imaging, 88. *Mem:* Am Asn Univ Prof; Asn Comput Mach; Inst Elec & Electronics Engrs; Asn Univ Radiologists. *Res:* Medical Image Display: image processing, computer vision, computer graphics, visual perception, observer performance. *Mailing Add:* Dept Comp Sci Univ NC Sitterson Hall Campus Box 3175 Chapel Hill NC 27599

PIZIAK, ROBERT, b Hadley, Mass, Jan 17, 43; m 66. MATHEMATICS. *Educ:* Amherst Col, BA, 64; Univ Mass, MA, 66, PhD(math), 69. *Prof Exp:* Asst prof math, Univ Mass, Amherst, 69-70 & Univ Fla, 70-73; from asst prof to assoc prof math, Centre Col, KY, 73-77; lectr, 81-89, ASST PROF, BAYLOR UNIV, 89- *Mem:* Am Math Soc; Math Asn Am. *Res:* Hilbert space geometry; logic of quantum mechanics. *Mailing Add:* 100 Ottoway Dr Temple TX 76501

PIZIAK, VERONICA KELLY, b Oak Bluffs, Mass, May 31, 42; m 66. ENDOCRINOLOGY, METABOLISM. *Educ:* Univ Mass, BS, 63, MS, 65, PhD(biochem), 70; Univ Ky, MD, 76; Am Bd Internal Med, cert, 79 & cert endocrinol & metab, 81. *Prof Exp:* Fel biochem, Univ Fla, 70-73; resident internal med, Akron Gen Med Ctr, 76-79; fel endocrinol, Univ Cincinnati, 79-81; SR STAFF PHYSICIAN ENDOCRINOL, 81-, DIR, DIABETES SECT, SCOTT & WHITE CLIN, 85-; ASSOC PROF, ENDOCRINOL, TEX A&M UNIV, 81- *Mem:* AMA; fel Am Col Physicians; Am Diabetes Asn; Endocrinol Soc; Soc Med Anthropol. *Res:* Effect of acute and chronic disease and antineoplastic therapy on endocrine organ functions; osteoporosis; diabetes education; obesity. *Mailing Add:* Scott & White Clin 2401 S 31st St Temple TX 76501

PIZZARELLO, DONALD JOSEPH, b Mt Vernon, NY, Aug 14, 33; m 61; c 4. RADIOBIOLOGY. *Educ:* Fordham Univ, AB, 55, MS, 57, PhD(exp embryol), 59. *Prof Exp:* AEC res assoc, Argonne Nat Lab, 59-60; from instr to assoc prof radiol, Bowman Gray Sch Med, 60-70; assoc prof, 70-73, PROF RADIOL, SCH MED, NY UNIV, 74- *Concurrent Pos:* Nat Cancer Inst res grants, Nat Ctr Radiation Health, 61 & 64; examr, Am Bd Radiol, 68- *Mem:* AAAS; Radiation Res Soc. *Mailing Add:* Dept Radiol NY Univ Med Ctr 550 First Ave New York NY 10016

PIZZICA, PHILIP ANDREW, b Chicago, Ill, Sept 22, 45; m 75. NUCLEAR ENGINEERING. *Educ:* Univ Ill, BS, 68, MA, 73. *Prof Exp:* Nuclear engr, Argonne Nat Lab, 67-71; systs analyst math, United Aircraft Res Lab, 73-74; NUCLEAR ENGR, ARGONNE NAT LAB, 74- *Res:* Fast breeder reactor safety research; building computer models to calculate accident conditions in a liquid metal fast breeder reactor; fuel cycle and core design studies. *Mailing Add:* 29W331 Staffeldt Dr Naperville IL 60564

PIZZIMENTI, JOHN JOSEPH, b New York, NY, Oct 12, 46; m 74; c 1. ENVIRONMENTAL IMPACTS & ASPECTS. *Educ:* Calif State Univ, Northbridge, AB, 69; Univ Kans, Lawrence, PhD(ecol), 74. *Prof Exp:* Asst cur mammals, Field Mus Natural Hist, Chicago, 74-77; asst prof biol sci, Univ Ill, Chicago, 77-81; SR SCIENTIST ENVIRON SCI, HARZA ENG CO, 81-, REGIONAL MGR, 89- *Concurrent Pos:* Adj prof biol, Comt Evol Biol, Univ Chicago, 75-77. *Mem:* Am Fisheries Soc. *Res:* Environmental planning and analysis of land and water resource projects; dams and reservoirs; abandoned and active mine lands; hydroelectric and energy generation; environmental rehabilitation. *Mailing Add:* Harza Eng Co 11675 SW 66th Portland OR 97223

PIZZINI, LOUIS CELESTE, b Black Eagle, Mont, June 7, 32; m 55; c 3. ORGANIC CHEMISTRY. *Educ:* Col Great Falls, BS, 54; Notre Dame, PhD(org chem), 59. *Prof Exp:* Res chemist, Wyandotte Chem Corp, 59-64 & Allied Chem Corp, 64; sr res chemist, 65-69, res assoc, 69-74, mgr, Urethane Applns Res & Develop, Tech Serv, 74-77, dir, Organic Res & Develop, 77-80, dir, Indust Chem Res & Develop, 80-82; dir, Chem Res & Develop & Urethane Res & Develop, dir, Polymers Res & Develop, BASF Corp, 83-89; PUB, WRITER & PHOTOG, CELESTIAL CREATIONS, 89- *Mem:* Am Chem Soc; Sci Res Soc Am. *Res:* Reactions of aromatic nitro compounds, surface active agents, detergents, alkylene oxides, polyalkylene oxides, polyols, polyethers and polyurethane polymers; polymer chemistry; microcellular, flexible and rigid polyurethane foams; automotive coolants; flame resistant polyurethane foams; corporate restructuring. *Mailing Add:* 9792 Hawthorn Glen Dr Grosse Ile MI 48138-1686

PIZZO, JOSEPH FRANCIS, b Houston, Tex; m 60; c 5. PHYSICS. *Educ:* Univ St Thomas, BA, 61; Univ Fla, PhD(physics), 64. *Prof Exp:* Head dept physics, 80-86, PROF PHYSICS, LAMAR UNIV, 64- *Mem:* Am Asn Physics Teachers. *Res:* Interactive physics demonstrations. *Mailing Add:* Dept Physics Lamar Univ Beaumont TX 77710

PIZZO, PHILIP A, b New York, NY, Dec 6, 44; m; c 2. PEDIATRICS. *Educ:* Fordham Univ, BA, 66; Univ Rochester, MD, 70; Am Bd Pediat, dipl, 75, cert hemat/oncol, 76. *Prof Exp:* Intern, Children's Hosp Med Ctr, Boston, 70-71, jr asst resident, 71-72, sr resident, 72-73; teaching fel, Harvard Med Sch, 72-73; clin assoc, Pediat Oncol Br, Nat Cancer Inst, 73-75, investr, 75-76, sr investr, 76-80, HEAD, INFECTIOUS DIS SECT, PEDIAT BR, NAT CANCER INST, BETHESDA, MD, 80-, CHIEF PEDIAT, 82-; PROF PEDIAT, UNIFORMED SERV UNIV HEALTH SCI, F EDWARD HEBERT SCH MED, BETHESDA, MD, 87- *Concurrent Pos:* Mem, Clin Res Subpanel, Nat Cancer Inst, 78-81, Infect Comt Clin Ctr, NIH, 78-, Transfusion Comt, 84-87, Pediat Core Comt, Pediat AIDS Clin Trials Group, 88-, Secy Task Force, HIV Infection in Children, 88, sci adv bd, Children's Hospice Int, 88-, sci adv bd, AIDS Prog, Nat Inst Allergy & Infectious Dis,

88-89, Pediat Sect, Am Fedn AIDS Res, 89-, Clin Res Subcomt AIDS Res Adv Comt, 90- *Honors & Awards:* Myron Karon Mem Lectr, 86; Melissa Anne Krinsky Mem Lectr, 89. *Mem:* Am Asn Clin Res; AAAS; Am Fedn Clin Res; Am Soc Clin Invest; Am Soc Clin Oncol; Am Soc Hemat; Am Soc Microbiol; Am Soc Pediat Hemat/Oncol; fel Am Acad Pediat; Soc Pediat Res. *Res:* Diagnosis, management and prevention of infections in the cancer compromised host; childhood cancer; pediatric AIDS; role of immunoregulatory agents and biological modifiers in bolstering host defenses following immunosuppressive therapy in both in vivo and in vitro model systems. *Mailing Add:* Nat Cancer Inst Pediat Br Bldg 10 Rm 13N240 Bethesda MD 20892

PIZZO, SALVATORE VINCENT, b Philadelphia, Pa, June 22, 44; m 88. EXPERIMENTAL PATHOLOGY. *Educ:* St Joseph's Col, BS, 66; Duke Univ, PhD(biochem), 72, MD, 73. *Prof Exp:* From asst prof to assoc prof, 76-85, PROF PATH, MED CTR, DUKE UNIV, 85-, DIR, MED SCIENTIST TRAINING PROG, 86- & HEAD CELL BIOL & BIOCHEM PROG, DUKE COMPREHENSIVE CANCER CTR, 88- *Concurrent Pos:* Mem, Am Asn Pathologists Prog Comt, 85-89; mem, Prog Proj Review Comt B, Nat Heart, Lung & Blood Inst, 86-88, chmn, 88-90; mem long range planning comt, Am Asn Pathologists, 90- *Mem:* Am Chem Soc; Am Asn Pathologists; Am Soc Biol Chemists; Int Soc Thrombosis & Haemostasis; NY Acad Sci; Am Soc Clin Invest; Asn Univ Pathologists. *Res:* Protease control by the plasma protease inhibitors; cell receptors for protease inhibitor-protease complexes; regulation of cells by the complexes; protease regulation of tumor kill. *Mailing Add:* Dept Path Duke Med Ctr Box 3712 Durham NC 27710

PIZZOLATO, THOMPSON DEMETRIO, b New Orleans, La, Nov 8, 43; m 71; c 2. PLANT ANATOMY. *Educ:* La State Univ, BS, 65, MS, 68; Miami Univ, PhD(bot), 74. *Prof Exp:* Fel, Tex Tech Univ, 74-75, N Cent Forest Exp Sta, 75-76 & Harvard Univ, 76-77; asst prof, 77-82, ASSOC PROF, UNIV DEL, 82- *Concurrent Pos:* Collabr, Dept Bot, Smithsonian Inst, Washington, DC; co-cur, Phillips Herbarium, Dover, Del. *Mem:* Bot Soc Am; Int Asn Wood Anatomists; Torrey Bot Club. *Res:* Anatomical studies on the vascular systems of the florets of grasses. *Mailing Add:* Dept Plant & Soil Sci Univ Del Newark DE 19717-1303

PLA, GWENDOLYN W, b North Little Rock, Ark, Nov, 7, 39. NUTRITION, BIOAVAILABILITY OF TRACE MINERALS. *Educ:* Univ Md, PhD(nutrit sci), 78. *Prof Exp:* ASSOC PROF NUTRIT, COL DENT, HOWARD UNIV, 82- *Mem:* Am Inst Nutrit; Am Dietetic Asn; Soc Nutrit Educ. *Res:* Special iron. *Mailing Add:* 1509 Upshur St NW Washington DC 20011

PLAA, GABRIEL LEON, b San Francisco, Calif, May 15, 30; m 51; c 8. PHARMACOLOGY. *Educ:* Univ Calif, BS, 52, MS, 56, PhD(pharmacol), 58, Am Bd Toxicol, Dipl, 81. *Prof Exp:* Asst pharmacologist, Univ Calif, 54-58; from instr to asst prof, Tulane Univ, 58-62; from asst prof to assoc prof, Univ Iowa, 62-68; chmn dept, Univ Montreal, 68-80, vdean, Fac Grad Studies, 79-82, vdean, Fac Med, 82-89, PROF PHARMACOL, UNIV MONTREAL, 68- *Concurrent Pos:* Asst toxicologist, San Francisco Coroner's Off, Calif, 54-58; mem toxicol study sect, NIH, 65-69, mem pharmacol-toxicol prog comt, 71-74; mem, Select Comt Generally Recognized As Safe Substances, Fedn Am Socs Exp Biol, 72-82; mem comt scholars, Med Res Coun, 69-75; ed, Toxicol & Appl Pharmacol, 72-80. *Honors & Awards:* Thienes Mem Award, Am Acad Clin Toxicol, 77; Lehman Award, Soc Toxicol, 81; Educ Award, Soc Toxicol, 87. *Mem:* Am Soc Pharmacol & Exp Therapeut; Soc Toxicol; Soc Exp Biol & Med; Am Acad Forensic Sci; Soc Toxicol Can; Pharmacol Soc Can. *Res:* Forensic and clinical toxicology; liver injury; effects of drugs on liver function. *Mailing Add:* Dept Pharmacol Univ Montreal Fac of Med Montreal PQ H3C 3J7 Can

PLACE, JANET DOBBINS, b Decatur, Ill, Sept 14, 53; m 80; c 1. CLINICAL CHEMISTRY, PROCESS DEVELOPMENT. *Educ:* Univ Mo, St Louis, BA, 74; Univ Kans, PhD(biochem), 79. *Prof Exp:* SR ASSOC RES SCIENTIST, SR APPLN SCIENTIST, MILES INC, DIAGNOSTICS DIV, 79- *Mem:* Am Chem Soc; Am Asn Clin Chem; Am Soc Biol Chemists. *Res:* Non-isotopic immunoassays for monitoring therapeutic drug levels in serum; monoclonal antibody production and characterization; liquid and dry phase immunoassays for monitoring therapeutic drug levels in serum. *Mailing Add:* Technicon Corp Rte 1145 3400 Middlebury St Middletown VA 22645

PLACE, RALPH L, b Chicago, Ill, 36; m 62; c 2. INTELLIGENT SYSTEMS. *Educ:* Ball State Univ, BS, 58; Univ Ky, MS, 62, PhD(physics), 69. *Prof Exp:* Instr physics, Northwest Ctr, Univ Ky, 61-63 & Ball State Univ, 63-65; res asst, Univ Ky, 65-68; from asst prof to assoc prof physics, 68-75, admin asst to head dept physics, 72-82, PROF COMPUTER SCI & PHYSICS, BALL STATE UNIV, 75- *Concurrent Pos:* Cottrell res grant, Ball State Univ, 71-72; consult, Ball Res Corp, 71-72; res grant, NSF, 89-90. *Mem:* Asn Comput Mach; Am Phys Soc; Inst Elec & Electronic Engrs; Am Asn Artificial Intel. *Res:* Computer vision, robotics, and hardware systems. *Mailing Add:* Dept Computer Sci Ball State Univ Muncie IN 47306

PLACE, ROBERT DANIEL, b St Johns, Mich, May 10, 41; m 64; c 2. INORGANIC CHEMISTRY, PHYSICAL CHEMISTRY. *Educ:* Albion Col, AB, 63; Univ Calif, Berkeley, PhD(chem), 67. *Prof Exp:* asst prof, 67-80, ASSOC PROF CHEM, OTTERBEIN COL, 80- *Mem:* AAAS; Am Chem Soc. *Res:* Analysis of trace organic molecules in water and air; acid rain transport from major utilities. *Mailing Add:* Dept Chem Otterbein Col Westerville OH 43081-2006

PLACE, THOMAS ALAN, b Leeds, Eng, Apr 6, 38; Can citizen; m 64. METALLURGY, MATERIALS SCIENCE. *Educ:* Univ Nottingham, BSc, 61; McMaster Univ, MEng, 66; Univ BC, PhD(metall), 69. *Prof Exp:* Res metallurgist, Rolls-Royce Advan Mat Res Lab, Derby, Eng, 61-63; lectr phys metall, Teesside Polytech, Middlesbrough, 63-64; teaching fel mat sci, Univ BC, 69-70; assoc prof mech eng, 70-75, PROF MECH ENG, UNIV IDAHO, 75- *Concurrent Pos:* Vis prof, Univ Auckland, New Zealand, 76-77. *Mem:* Brit Inst Metall; Am Soc Metals; Am Soc Testing & Mat. *Res:* Mechanical properties and fracture mechanics of metals, particleboard and pesticide-damaged eggshells. *Mailing Add:* 1034 S Harding St Moscow ID 83843

PLACE, VIRGIL ALAN, b Crown Point, Ind, Oct 24, 24; m 82; c 3. CLINICAL PHARMACOLOGY, INTERNAL MEDICINE. *Educ:* Ind Univ, AB, 44; Johns Hopkins Univ, MD, 48. *Prof Exp:* Intern med & biochem, Presby Hosp, Chicago, 48-51; asst to staff med, Mayo Clin, 55; pvt pract, Calif, 55-58; assoc dir clin pharmacol, Lederle Labs Div, Am Cyanamid Co, 58-59, dir, 59-66; assoc med dir, Syntex Res, 66-67, dir clin pharmacol, 67-68; pres & founder, Pharmaceut Res Int, 68-69; vpres & med dir, 69-84, SR DIR MED & REGULATORY AFFAIRS, ALZA CORP. 84- *Concurrent Pos:* Res fel, Presby Hosp, Chicago, 48-51; fel, Mayo Grad Sch Med, Univ Minn, 53-55; instr, Univ Ill, Chicago, 50; attend physician, City Hosp, County Hosp & Community Hosp, Modesto, Calif, 55-58, Bellevue Hosp, Cornell Div, New York, 59-61 & Bergen Pines Hosp, Paramus, NJ, 62-66. *Mem:* AAAS; Am Col Physicians; Am Diabetes Asn; Am Fedn Clin Res. *Res:* Human studies of therapeutic effects of drugs related to their chemical structure, physical form and methods of delivery. *Mailing Add:* 950 Page Mill Rd Palo Alto CA 94304

PLACIOUS, ROBERT CHARLES, b Rochester, NY, Jan 20, 23; m 54; c 5. RADIATION PHYSICS. *Educ:* Univ Rochester, BS, 50; Univ Iowa, MS, 53; Cath Univ, PhD(physics), 65. *Prof Exp:* Jr physicist, 53-56, physicist, 56-66, SR PHYSICIST, NAT BUR STAND, 67- *Mem:* Am Phys Soc; Am Asn Physics Teachers; NY Acad Sc; Am Soc Nondestructive Testing. *Res:* Electron physics; x-ray interactions; electron interactions using electron accelerators in the energy range of several kilo-electron volts to 4 mega-electron volts; radiation effects studies; Bremsstrahlung spectroscopy; optical spectroscopy of fluorescent materials; radiographic imaging and industrial radiographic standards. *Mailing Add:* 13340 Bondy Way Gaithersburg MD 20878

PLAFKER, GEORGE, b Upland, Pa, Mar 6, 29; m 49; c 3. GEOLOGY. *Educ:* Brooklyn Col, BS, 49; Univ Calif, MS, 56; Stanford Univ, PhD(geol), 71. *Prof Exp:* Eng geologist, Sacramento Dist, US Corps Engrs, 49-50; geologist, Mil Geol Br, US Geol Surv, 51- 52, Alaskan Geol Br, 52-56, Guatemala Calif Petrol Co Ltd Div, Standard Oil Co Calif, 56-59 & Bolivia Calif Petrol Co Ltd Div, 59-62; RES GEOLOGIST, ALASKAN GEOL BR, US GEOL SURV, 62- *Honors & Awards:* Harry Oscar Wood Award Seismol, 67; Distinguished Serv Award, US Dept Interior, 79. *Mem:* AAAS; Geol Soc Am; Am Geophys Union; Asn Eng Geol; Sigma Xi. *Res:* Geologic earthquake hazards in Alaska; tectonic deformation and geologic effects related to major earthquakes; geology, geophysics and petroleum potential of the Gulf of Alaska margin and contiguous offshore areas; deep crustal structure of Alaska. *Mailing Add:* 10558 Creston Dr Los Altos CA 94022

PLAGEMANN, PETER GUENTER WILHELM, b Magdeburg, Ger, Oct 24, 28; US citizen; m 56; c 3. VIROLOGY, BIOCHEMISTRY. *Educ:* Univ Toronto, BSA, 59, MSA, 60; Case Western Reserve Univ, PhD(microbiol), 65. *Prof Exp:* From asst prof to assoc prof, 65-74, PROF MICROBIOL, MED SCH, UNIV MINN, MINNEAPOLIS, 74- *Mem:* AAAS; Am Soc Microbiol; Am Soc Biol Chem; Am Soc Cell Biol. *Res:* Biochemistry of virus replication and of cultured cells. *Mailing Add:* Dept of Microbiol Univ of Minn Med Sch Minneapolis MN 55455

PLAGER, JOHN EVERETT, b New York, NY, Mar 2, 27; m 50; c 3. BIOCHEMISTRY, INTERNAL MEDICINE. *Educ:* Tufts Univ, MS, 49; Univ Utah, PhD(biochem, physiol), 53, MD, 57. *Prof Exp:* Instr internal med, Univ Rochester, 59-60, sr instr, 60-61; res dir, Med Found Buffalo, NY, 61-66; ASSOC CHIEF MED, ROSWELL PARK MEM INST, 66- *Concurrent Pos:* USPHS fel, 59-60; assoc res prof med, State Univ NY Buffalo, 66- *Mem:* Am Soc Clin Oncol; Am Asn Cancer Res; fel Am Col Physicians; Endocrine Soc; Am Physiol Soc. *Res:* Pituitary-adrenal function; steroid biochemistry and enzymology; steroid-protein interactions; cancer medicine; pharmacology of anti-cancer drugs. *Mailing Add:* 666 Elm St Buffalo NY 14203

PLAGGE, JAMES CLARENCE, b Barrington, Ill, Apr 5, 11; m 38, 72; c 3. ANATOMY. *Educ:* Univ Chicago, SB, 37, PhD(zool), 40. *Prof Exp:* Asst zool, Univ Chicago, 37-40; from instr to asst prof gross anat, Univ Md, 40-43; from assoc to prof anat, Col Med, Univ Ill Med Ctr, 43-80, asst dean int activities, 72-80, assoc dir, Off Spec Progs, 74-80. *Concurrent Pos:* Med ed adv, Int Coop Admin, Saigon, Vietnam, 58-60; coordr, Univ Ill-Chiang Mai Univ proj, 62, chief party, Thailand, 64-68; Univ Ill fac liaison officer, Med Opportunities Prog, 69-; mem, Int Health Survey Team, Nicaragua, 72; consult anat, Fac Med, Univ Azarabadegan, Tabriz, Iran, 77. *Mem:* Soc Exp Biol & Med; Am Asn Anat. *Res:* Physiology of the endocrines; sex organs; irradiation; liver diseases. *Mailing Add:* 2248 Kent St Okemos MI 48864

PLAISTED, ROBERT LEROY, b Hornell, NY, Jan 1, 29; m 51; c 4. PLANT BREEDING. *Educ:* Cornell Univ, BS, 50; Iowa State Col, MS & PhD(hort), 56. *Prof Exp:* From asst prof to assoc prof, 56-65, head dept, 65-80, PROF PLANT BREEDING & BIOMETRY, CORNELL UNIV, 65- *Honors & Awards:* Hon Life Mem, Potato Asn Am, 84. *Mem:* Potato Asn Am; Crop Sci Soc Am. *Res:* Methods of breeding potatoes and development of new varieties. *Mailing Add:* Dept of Plant Breeding & Biomet Cornell Univ Ithaca NY 14853

PLAIT, ALAN OSCAR, b Chicago, Ill, Aug 14, 26; m 53; c 4. ELECTRONICS ENGINEERING, MATHEMATICS. *Educ:* Ill Inst Technol, BS, 51 & 57; Va Polytech Inst & State Univ, MS, 76. *Prof Exp:* Engr electronics, Admiral Corp, 52-55; assoc engr, Armour Res Found, 55-59; sr engr, Magnavox Co, 59-62; mgr reliability eng, Melpar, Inc, 62-67; mgr reliability eng & opers res, Comput Sci Corp, 67-75; mgr reliability assurance, Amecom-Litton Systs, 75-77; TECH DIR, MANTECH INT CORP, 77- *Concurrent Pos:* Instr, Ill Inst Technol, 56-59, Grad Sch, USDA, 63- & Univ Conn, 73-75; adj prof, George Washington Univ, 70 & Va Polytech Inst & State Univ, 73-75, 81. *Mem:* Fel Am Soc for Qual Control; sr mem Inst Elec & Electronics Engrs; Nat Soc Prof Engrs. *Res:* Development of mathematical basis for reliability studies; application of reliability principles in the improvement of electronic systems; exploration of methods of failure recurrence control in minimizing system failures. *Mailing Add:* Mantech Adv Systs Int Box 1588 Vint Hill Farms Sta Warrenton VA 22186

PLAKKE, RONALD KEITH, b Holland, Mich, July 21, 35; m 58; c 2. VERTEBRATE ANATOMY, PHYSIOLOGY. *Educ:* Univ Northern Colo, AB, 62; Univ Mont, PhD(zool), 66. *Prof Exp:* From asst prof to assoc prof, 66-74, PROF ZOOL & WOMEN'S STUDIES, UNIV NORTHERN COLO, 74-, CHAIRPERSON, DEPT BIOL SCI, 80- *Mem:* AAAS; Am Soc Mammal; Am Soc Zool; Nat Asn Biol Teachers; Sigma Xi. *Res:* Mammalian renal anatomy and physiology; descriptive histology and histochemistry. *Mailing Add:* Dept of Biol Sci Univ of Northern Colo Greeley CO 80639

PLAMBECK, JAMES ALAN, b Chicago, Ill, Sept 16, 38; m 70; c 4. ANALYTICAL CHEMISTRY, ELECTROCHEMISTRY. *Educ:* Carleton Col, BS, 60; Univ Ill, Urbana, MSc, 62, PhD(chem), 65. *Prof Exp:* Res assoc chem eng, Argonne Nat Lab, 64-65; asst prof, 65-70; ASSOC PROF CHEM, UNIV ALTA, 70- *Concurrent Pos:* Assoc mem comm I.3 electrochem, Int Union Pure & Appl Chem, 71-76, 80-85; consult electrochem; instr comput prog lang. *Mem:* Am Chem Soc; Electrochem Soc; Chem Inst Can. *Res:* Electroanalytical methods; electrochemistry of fused salts; electrochemical thermodynamics; environmental analytical chemistry; electrochemistry of antitumor antibiotics; computer interfacing of analytical instrumentation. *Mailing Add:* Dept Chem Univ Alta Edmonton AB T6G 2G7 Can

PLAMBECK, LOUIS, JR, b Moline, Ill, May 15, 12; m 40; c 1. ORGANIC CHEMISTRY. *Educ:* Univ Ill, BS, 35; Pa State Col, MS, 37, PhD(org chem), 39. *Prof Exp:* Asst chem, Pa State Col, 35-37; chemist, Chas Lennig & Co, 37; chemist, Exp Sta, E I du Pont de Nemours & Co, Inc, 39-77; CONSULT, 77- *Honors & Awards:* Kosar Mem Award, Soc Photog Sci & Eng, 70; Robert F Reed Technol Medal, Graphic Arts Tech Found, 88. *Mem:* Am Chem Soc. *Res:* Photopolymerization; photographic chemistry; coatings for plastics; biopolymers. *Mailing Add:* 107 Hoiland Dr Wilmington DE 19803-3227

PLAMONDON, JOSEPH EDWARD, b Dubuque, Iowa, Mar 3, 41; m 66; c 1. ADHESIVES TECHNOLOGY. *Educ:* Loras Col, BS, 63; Univ Calif, Berkeley, MS, 65; Univ Calif, Davis, PhD(chem), 69. *Prof Exp:* RES CHEMIST SYNTHETIC FIBERS, ROHM AND HAAS CO, 69- *Mem:* AAAS; Am Chem Soc. *Res:* Hydroboration; organometallic chemistry; polymer synthesis and characterization; synthetic fibers; coated fabrics; pressure sensitive adhesives; textile chemistry; coated fabrics; automotive trim adhesives; speciality monomers; radiation care; government regulations; toxic substances control act. *Mailing Add:* 55 Mill Creek Rd Southampton PA 18966

PLAMTHOTTAM, SEBASTIAN S, b Kerala, India, Jan 22, 51; m 85. MATERIALS SCIENCE. *Educ:* Univ Kerala, India, BSc, 72; Univ Cochin, India, BTech, 75; Univ Akron, Ohio, PhD(polymer sci), 81. *Prof Exp:* RES ASSOC, RES CTR, AVERY INT, PASADENA, CALIF, 80- *Mem:* Plastics & Rubber Inst; Am Chem Soc. *Res:* Polymer synthesis and radiation (ultraviolet and electron beam); curing of adhesives and coatings. *Mailing Add:* 1496 Moonridge Ct Upland CA 91786

PLANE, ROBERT ALLEN, b Evansville, Ind, Sept 30, 27; m 50; c 4. BIOINORGANIC CHEMISTRY, SPECTROSCOPY. *Educ:* Evansville Col, AB, 48; Univ Chicago, SM, 49, PhD(chem), 51. *Hon Degrees:* DSc, Univ Evansville, 68 & Clarkson Univ, 85, LHD, St Lawrence Univ, 86, LLD, Hobart & William Smith Col, 90. *Prof Exp:* Chemist, Oak Ridge Nat Lab, 51-52; from instr to prof chem, Cornell Univ, 52-74, chmn dept, 67-70, provost, 69-73; prof chem & pres, 74-85, EMER PRES, CLARKSON COL TECHNOL, 85- *Concurrent Pos:* NIH spec fel, Nobel Inst, 60; Oxford Univ, 61; consult, Procter & Gamble Co, 65-; vis scientist, Univ Calif, 69; dir, NY State Agr Exp Sta, 86-90. *Mem:* Am Chem Soc; Am Soc Enol & Viticulture; fel AAAS. *Res:* Metal ions in biological systems; fast reactions in solution; Raman spectroscopy. *Mailing Add:* RD 2 Box 273 Ovid NY 14521

PLANK, CHARLES ANDREWS, b Charlotte, NC, Oct 12, 28; m 50; c 3. CHEMICAL ENGINEERING. *Educ:* NC State Univ, BChE, 49, MS, 51, PhD(chem eng), 57. *Prof Exp:* Res asst liquid extraction, NC State Univ, 51-52, instr chem eng, 52-57; assoc prof, Univ Louisville, 57-63; spec proj engr, Olin Mathieson Co, Ky, 63 & 66; assoc prof chem eng, 64-67, dir interdisciplinary prog eng, 71-73, chmn dept, 73-79, PROF CHEM ENG, UNIV LOUISVILLE, 67- *Concurrent Pos:* Consult, Matt Corcoran, Inc, 58-, Tube Turns Plastics Co, 59-60, Am Air Filter Co, 64- & Ro Tech Corp, 77-; vis prof, Washington Univ, St Louis, 79. *Mem:* Am Inst Chem Engrs. *Res:* Mass transfer and thermodynamics; distillation and absorption; sorption and transport properties of polymers. *Mailing Add:* Dept of Chem Eng Univ of Louisville Speed Sci Sch Louisville KY 40292

PLANK, DON ALLEN, b Memphis, Tenn, Aug 8, 39. ORGANIC CHEMISTRY, PHOTOCHEMISTRY. *Educ:* Carleton Col, BA, 61; Purdue Univ, PhD(chem), 65. *Prof Exp:* NIH fel, Iowa State Univ, 65-67; res chemist, Esso Res & Eng Co, 67-74, staff chemist, Exxon Chem Co, 74-79, SR STAFF CHEMIST, EXXON CHEM AM, 79- *Mem:* Am Chem Soc. *Res:* Polymer stabilization. *Mailing Add:* 1 Thunderbird Circle Baytown TX 77520

PLANK, DONALD LEROY, b Manchester, Conn, Aug 20, 37; m 61; c 3. MATHEMATICS. *Educ:* Trinity Col, Conn, BS, 60; Yale Univ, MA, 63; Univ Rochester, PhD(math), 67. *Prof Exp:* Asst prof math, Case Western Reserve Univ, 66-69, asst prof, Amherst Col, 69-71; prof & chmn, Dept Math, Univ Bridgeport, 83-85; assoc prof math, 71-82, chmn nat sci & math, 85-88, dean nat sci & math, 88-90, PROF MATH, STOCKTON STATE COL, 82- *Mem:* Am Math Soc; Math Asn Am. *Res:* Lattice-ordered rings; rings of functions; topology. *Mailing Add:* Math Prog Stockton State Col Pomona NJ 08240

PLANK, STEPHEN J, b Rochester, NY, May 8, 27; m 56; c 3. PUBLIC HEALTH, POPULATION BIOLOGY. *Educ:* Univ Chicago, PhB, 48; Univ Calif, AB, 51, MD, 55; Harvard Univ, MPH, 61, DrPH, 64. *Prof Exp:* Intern, Gorgas Hosp, Balboa, CZ, 55-56; staff physician, Coco Solo Hosp, Cristobal, 56-60; asst prof pop studies, Sch Pub Health, Harvard Univ, 64-73, lectr, 69-73; mem staff health sci, Rockefeller Found, Bahia, Brazil, 73-78; mem staff Planning Off, exec pres, 78-79, HEALTH OFFICER, SHASTA COUNTY, CALIF, 80- *Concurrent Pos:* Asst surg, Peter Bent Brigham Hosp, Boston, 62-; temp adv, Pan-Am Health Orgn, 65-68; mem adv coun, Pathfinder Fund, 65-68; consult, Latin Am Prog, Ford Found, 66-67, WHO, 70 & World Bank, 77. *Mem:* Am Pub Health Asn. *Res:* Immunological control of fertility; relationships between health and population change; international health. *Mailing Add:* 2650 Hosp Lane Redding CA 96001

PLANKEY, FRANCIS WILLIAM, JR, b Malden, Mass, Feb 28, 45; m 68; c 2. ANALYTICAL CHEMISTRY. *Educ:* Univ Mass, BS, 71; Univ Fla, PhD(chem), 74. *Prof Exp:* ASST PROF CHEM, UNIV PITTSBURGH, 74- *Mem:* Am Chem Soc; Soc Appl Spectros; Optical Soc Am. *Res:* Analytical atomic spectroscopy; trace metal analysis; chemical instrumentation development and the application of computers to chemical instrumentation; forensic applications of analytical chemistry. *Mailing Add:* Purvis Systs Inc 470 Streets Run Rd Pittsburgh PA 15236

PLANO, RICHARD JAMES, b Merrill, Wis, Apr 15, 29; m 56; c 2. PHYSICS. *Educ:* Univ Chicago, PhD(physics), 56. *Prof Exp:* From instr to asst prof physics, Columbia Univ, 56-60; from assoc prof to prof physics, 60-80, PROF II, RUTGERS UNIV, NEW BRUNSWICK, 80- *Mem:* Am Phys Soc; Am Asn Physics Teachers. *Res:* Elementary particle physics. *Mailing Add:* Dept of Physics Rutgers Univ New Brunswick NJ 08903

PLANT, B J, b Smith Falls, Ont, Aug 2, 33; m 56; c 4. ELECTRICAL ENGINEERING. *Educ:* Mass Inst Technol, PhD(elec eng), 65. *Prof Exp:* Lectr elec eng, 65-66, asst prof, 66-67, prof & head dept, 67-72, dean grad studies & res, 72-84, CHMN GRAD STUDIES & RES DIV & PROF ELEC ENG, ROYAL MIL COL CAN, 80-, PRIN & DIR STUDIES, 84- *Concurrent Pos:* Defense Res Bd Can grant, 67-84; sabbatical with Indust, Thompson CSF, Paris, France; pres, Can Soc Elec & Computer Eng, 89. *Honors & Awards:* Centennial Medal, Inst Elec & Electronics Engrs, 84. *Mem:* AAAS; Inst Elec & Electronic Engrs; fel Eng Inst Can. *Res:* Automatic control; optimization techniques and digital computer control; sonar and radar signal processing. *Mailing Add:* Royal Mil Col of Can Kingston ON K7K 5L0 Can

PLANT, RICHARD E, b Oct 10, 47; m; c 2. BIOMATHEMATICS, INTELLIGENT SYSTEMS. *Educ:* Cornell Univ, PhD(math). *Prof Exp:* MEM FAC, DEPT AGRON, UNIV CALIF, DAVIS. *Mem:* Am Asn Artificial Intel. *Res:* Application of computer science and mathematics in agriculture; agricultural engineering. *Mailing Add:* Dept Agron Univ Calif Davis Davis CA 95616-8515

PLANT, WILLIAM J, b Burlington, Vt, Mar 29, 26; m 55; c 7. ORGANIC POLYMER CHEMISTRY. *Educ:* Trinity Col, Conn, BS, 48; Univ Vt, MS, 50; Univ Tex, PhD(chem), 55. *Prof Exp:* Sr chemist, Gen Dynamics Corp, 55; res chemist, Celanese Corp Am , NJ, 55-59; ROBERTSON FEL, CARNEGIE-MELLON UNIV, 59- *Concurrent Pos:* Sr fel Carnegie-Mellon Inst Res, 59- *Mem:* Am Chem Soc; Soc Plastics Eng. *Res:* Organometallics; monomer synthesis; synthetic polymers, resins and coatings. *Mailing Add:* 1702 President Dr Glenshaw PA 15116-2149

PLANT, WILLIAM JAMES, b Wichita, Kans, Sept 18, 44; div; c 2. AIR-SEA INTERACTIONS, MICROWAVE REMOTE SENSING. *Educ:* Kans State Univ, BS, 66; Purdue Univ, MS, 68, PhD(physics), 72. *Prof Exp:* Nat Res Coun assoc physics, Naval Res Lab, 71-73, career appointment, 73-88; SR SCIENTIST, WOODS HOLE OCEANOG INST, 88- *Mem:* Int Union Radio Sci; Am Geophys Union; Sigma Xi. *Res:* Electromagnetic scattering from water waves. *Mailing Add:* Woods Hole Oceanog Inst Woods Hole MA 02543

PLANTS, HELEN LESTER, b Desloge, Mo, Mar 9, 25; m 50; c 3. ENGINEERING EDUCATION. *Educ:* Univ Mo-Columbia, BSCE, 45; WVa Univ, MSCE, 53. *Prof Exp:* Jr engr reservoir develop, Atlantic Refining Co, 45; engr map drafting, Upham Eng Co, 45-46; struct consult, Harry E Graham, Consult Engr, 46-47; instr mech, WVa Univ, 47-56, from asst prof to prof mech, 56-82; chmn, Civil & Archit Eng Technol, Purdue, Ft Wayne, 82-90; RETIRED. *Concurrent Pos:* US Off Educ res grant, WVa Univ, 67-68; mem acad adv bd, US Naval Acad, 76-80; adv, Nat Eng Ctr Phillipines, 80; sr lectr, Mech & Aeronaut Eng, Kingston Polytech, London & Educ Adv, Brighton Polytech, 82-83; develop, Int Exchange Eng Students, Gt Brit, Austria, various other Europ countries, Africa, Spac & Asia, 89- *Honors & Awards:* Helen Plants Award, Inst Elec & Electronic Engrs & Am Soc Eng Educ, 79. *Mem:* Am Educ Res Asn; Nat Soc Prof Engrs; Am Soc Eng Educ (vpres, 75-76); Am Soc Mech Engrs; Am Soc Civil Engrs. *Res:* Development and evaluation of educational systems for teaching engineering subjects; programmed instruction applied to technical education; behavioral psychology in education. *Mailing Add:* 2536 Thompson Ave Ft Wayne IN 46807

PLAPINGER, ROBERT EDWIN, b New York NY, Feb 23, 23; m 54; c 3. ORGANIC CHEMISTRY. *Educ:* City Col New York, BS, 44; Univ Md, PhD(chem), 51. *Prof Exp:* Org chemist, Montrose Chem Co, 43-44; asst, Univ Md, 46-49; res chemist, Chem Corps Med Labs, US Army Chem Ctr, 51-57; sr res chemist, Niagara Chem Div, Food Mach & Chem Corp, 57-58; sr res chemist, Chem Warfare Labs, US Army Chem Ctr, 58-62; res assoc Sinai Hosp Baltimore, 62-74, head div surg res, 70-72; res chemist, US Customs Lab, 74-75; res chemist, Bur Med Devices, Food & Drug Admin, 75-86; CONSULT, 87- *Concurrent Pos:* Consult, Biochem Res Div, Sinai Hosp Baltimore, 74- *Honors & Awards:* Dept of the Army Performance Award, 61. *Mem:* AAAS; Am Chem Soc; Royal Soc Chem; NY Acad Sci; Am Asn Clin Chemists. *Res:* Synthetic organic chemistry; chemistry of organic insecticides; organophosphorus compounds; peptides; chemical warfare agents; molecular rearrangements; cancer chemotherapy; enzyme histochemistry; clinical chemistry. *Mailing Add:* 825 Center St Unit 6D Jupiter FL 33458-4114

PLAPP, BRYCE VERNON, b DeKalb, Ill, Sept 11, 39; m 62; c 2. BIOCHEMISTRY. *Educ:* Mich State Univ, BS, 61; Univ Calif, Berkeley, PhD(enzym), 66. *Prof Exp:* Am Cancer Soc fel, Inst Org Chem, Frankfurt, Ger, 66-68; res assoc, Rockefeller Univ, 68-70; from asst prof to assoc prof, 70-79, PROF BIOCHEM, UNIV IOWA, 79- *Concurrent Pos:* Merck Found grant fac develop, 70; res sci develop award, 75; vis scientist, Swed Univ Agr Sci, Uppsala, 76-77; mem, NIH Study Sect, 78-82. *Mem:* Am Chem Soc; Am Soc Biol Chemists; Sigma Xi. *Res:* Structure and function of alcohol dehydrogenases; protein chemistry; enzyme mechanisms; active site directed reagents; alcohol metabolism. *Mailing Add:* Dept of Biochem Univ of Iowa Iowa City IA 52242

PLAPP, FREDERICK VAUGHN, b Kansas City, Mo, Jan 27, 49; m 75; c 2. CLINICAL PATHOLOGY, BLOOD BANKING. *Educ:* Univ Kans, BA, 71, PhD(path), 74, MD, 75. *Prof Exp:* Intern path, Univ Chicago, 75-76; resident, Univ Kans Med Ctr, 76-78, from asst prof to assoc prof, 78-82, assoc dean med, 81-82; asst med dir, Blood Bank Community Blood Ctr Greater Kansas City, 82-87; CLIN PATHOLOGIST, ST LUKES HOSP, 87-, MED DIR LAB MED & CHMN PATH, 91- *Concurrent Pos:* Consult, Immucor, Inc, 83-, Osborn Labs, 89- *Mem:* Am Asn Pathologists; Am Asn Blood Banks; AMA; Am Asn Clin Chem; Int Soc Blood Transfusion; Am Soc Microbiol. *Res:* Structure and function of red blood cell antigens, particularly the Rh blood group system; new solid phase red cell adherence method for blood grouping. *Mailing Add:* Dept Path St Lukes Hosp 44th & Wornall Rd Kansas City MO 64111

PLAPP, JOHN E(LMER), b El Paso, Tex, Feb 24, 29. MECHANICAL ENGINEERING. *Educ:* Rice Inst, BS, 50; Calif Inst Technol, MS, 51, PhD(mech eng), 57. *Prof Exp:* From asst prof to assoc prof mech eng, Rice Univ, 55-68; assoc prof Eng Sci, Trinity Univ, Tex, 68-89; CONSULT, 89- *Concurrent Pos:* USAID vis instr, Kabul Univ, Afghanistan, 63. *Mem:* Am Soc Mech Engrs; Am Inst Aeronaut & Astronaut. *Res:* Convective heat transfer; density currents; applied mathematics; fluid mechanics; thermodynamics. *Mailing Add:* 326 E Rosewood Apt 5 San Antonio TX 78212

PLASIL, FRANZ, b Prague, Czech, May 17, 39; m 64, 80; c 2. NUCLEAR PHYSICS. *Educ:* London Univ, BSc, 60; Univ Calif, Berkeley, PhD(chem), 64. *Prof Exp:* Chemist, Lawrence Radiation Lab, Calif, 64-65; res assoc nuclear chem, Brookhaven Nat Lab, 65-67; physicist, 67-79, mem sr res staff & group leader, 79-86, SECT HEAD, PHYS DIV, OAK RIDGE NAT LAB, 86- *Concurrent Pos:* Maitre de Recherche, Institut de Physique Nucleaire, Orsay, France, 74-75. *Honors & Awards:* Alexander von Humboldt Award, 84. *Mem:* Fel Am Phys Soc; Nuclear Div Am Chem Soc. *Res:* Nuclear fission; heavy ion reactions; nucleus-nucleus collisons at ultra relativistic energies; elementary particle physics. *Mailing Add:* Oak Ridge Nat Lab PO Box 2008 Oak Ridge TN 37831-6372

PLASS, GILBERT NORMAN, b Toronto, Ont, Mar 22, 20; US citizen; m 62; c 2. PLANETARY ATMOSPHERES. *Educ:* Harvard Univ, BS, 41; Princeton Univ, PhD(physics), 47. *Prof Exp:* Assoc physicist metall lab, Univ Chicago, 42-45; from instr to assoc prof physics, Johns Hopkins Univ, 46-55; staff scientist, Lockheed Aircraft Corp, 55-56; adv res staff, Aeronutronic Div, Ford Motor Co, 56-60, mgr theoret physics dept res lab, 60-63; prof atmospheric & space sci, Southwestern Ctr Advan Studies, 63-68; head dept, 68-77, PROF PHYSICS, TEX A&M UNIV, 68- *Concurrent Pos:* Vis assoc prof, Northwestern Univ, 49 & Mich State Univ, 54-55, consult ed, Infared Physics, 60- *Mem:* Am Phys Soc; Optical Soc Am; Am Meteorol Soc; Am Asn Physics Teachers; AAAS. *Res:* Infrared absorption and emission; spectroscopy; electromagnetic and gravitational action at a distance; electrostatic electron lenses; fission and neutron physics; planetary atmospheres. *Mailing Add:* Dept Physics Tex A&M Univ College Station TX 77843

PLASS, HAROLD J(OHN), JR, b Merrill, Wis, Dec 1, 22; m 46; c 5. ENGINEERING MECHANICS. *Educ:* Univ Wis, BS, 44, MS, 48; Stanford Univ, PhD(eng mech), 50. *Prof Exp:* From asst prof to prof eng mech, Univ Tex, Austin, 50-67; chmn dept mech eng, 67-76, PROF MECH ENG, UNIV MIAMI, 67- *Concurrent Pos:* Res engr, Defense Res Lab, Tex, 50-67. *Mem:* Am Soc Eng Educ; Am Soc Mech Engrs; Sigma Xi. *Res:* Structural dynamics. *Mailing Add:* Dept Mech Eng Univ Miami PO Box 248294 Coral Gables FL 33124

PLASSMANN, ELIZABETH HEBB, b Boston, Mass, Oct 4, 28; m 55; c 4. NUCLEAR PHYSICS. *Educ:* Bryn Mawr Col, AB, 50; Ind Univ, MS, 51, PhD, 55. *Prof Exp:* Asst prof physics, Bradley Univ, 55; mem staff, 55-60, asst group leader, 60-70, group leader, 70-81, asst div leader, 81-82, assoc div leader, 82-87, PROG MGR, LOS ALAMOS NAT LAB, UNIV CALIF, 83- *Mem:* Am Phys Soc; Am Nuclear Soc. *Res:* Nuclear components of atomic and thermonuclear weapons. *Mailing Add:* 103 Chamisa Los Alamos NM 87544

PLASSMANN, EUGENE ADOLPH, b New York, NY, June 7, 21; m 55; c 4. NUCLEAR PHYSICS. *Educ:* Hastings Col, AB, 49; Ind Univ, MS, 52, PhD(physics), 55. *Prof Exp:* Instr, Univ Ky, 55; asst group leader, Los Alamos Nat Lab, Univ Calif, 55-89; RETIRED. *Concurrent Pos:* Temp fac mem, Tex A&M Univ, 70-74; consult, 90- *Honors & Awards:* Award of Excellence, Off Military Applications, Dept Energy, 88. *Mem:* Am Nuclear Soc; Sigma Xi; Am Phys Soc; Health Physics Soc. *Res:* Critical assemblies of fissionable materials; theoretical investigations of nuclear reactor systems; measurement and calculation of radiation effects; beta ray spectroscopy. *Mailing Add:* 103 Chamisa Los Alamos NM 87544

PLATAU, GERARD OSCAR, b Potsdam, Ger, June 29, 26; nat US; m 61; c 1. INFORMATION SCIENCE, PATENT DOCUMENTATION. *Educ:* Brooklyn Col, BA, 46; Purdue Univ, MS, 48, PhD(org chem), 50. *Prof Exp:* Asst, Purdue Univ, 46-50; from asst ed to sr assoc ed, Chem Abstract Serv, 50-61, head, Assignment & Abstracting Dept, 61-71, asst managing ed, 71-79, sr adv edit opers, Chem Abstract Serv, 79-90; CONSULT, 90- *Concurrent Pos:* Consult, NSF, 74-80; dir & parliamentarian, Am Soc Info Sci, 76-; CHMN, CO data '90, 90. *Honors & Awards:* Watson Davis Award, Am Soc Info Sci, 80. *Mem:* Am Chem Soc; Am Soc Info Sci; hon fel Nat Fed Abstr & Info Serv. *Res:* Information storage and retrieval; abstracting and indexing; on-line data bases; synthesis of sympathomimetic drugs; synthetic sweetening agents. *Mailing Add:* 1686 Arlingate Dr S Columbus OH 43220

PLATE, CHARLES ALFRED, b Butte, Mont, Sept 25, 41; m 64; c 4. BIOCHEMISTRY, MICROBIOLOGY. *Educ:* Jamestown Col, BA, 63; Duke Univ Med Ctr, PhD(biochem), 70. *Prof Exp:* Am Cancer Soc postdoctoral fel, Mass Inst Technol, 70-72, res assoc, 72-78; asst prof microbiol, Northwestern Univ Med Sch, 78-85; sr scientist cancer diagnostics, 85-91, SR SCIENTIST INFECTIOUS DIS, ABBOTT LABS, 91- *Concurrent Pos:* Vis prof, Sect Med Oncol, Rush Presby St Lukes Med Ctr, 85-87. *Mem:* Am Soc Biochem & Molecular Biol; Am Soc Microbiol. *Res:* Prototype in vitro diagnostic assay systems. *Mailing Add:* D93L AP20 Abbott Labs Abbott Park IL 60064

PLATE, HENRY, b Jersey City, NJ, Oct 12, 25; m 58; c 2. SOILS & SOIL SCIENCE. *Educ:* Univ Maine, BS, 48, MS, 49. *Prof Exp:* Agronomist fertilizer res div, Eastern States Farmers' Exchange, Inc, 49-62, head, 62-64; mgr tech serv, Agway Inc, 64-66, mgr blend plants, 66-67, mgr agron servs, Fertilizer Div, 67-85, sr agronomist, Res Dept, 85, mgr fertilizer develop, Agway Inc, 86-88; AGR CONSULT, 88- *Mem:* Am Soc Agron. *Res:* Plant nutrition; fertilizer technology and usage. *Mailing Add:* 8330 Craine Dr Manlius NY 13104

PLATE, JANET MARGARET, b Minot, NDak, Nov 27, 43; m 64; c 4. MOLECULAR CLONING. *Educ:* Jamestown Col, BA, 64; Duke Univ, PhD(immunol), 70. *Prof Exp:* Asst immunol, Mass Gen Hosp, 72-78, assoc, Harvard Med Sch, 70-77, asst prof immunol, Harvard Univ Sch Pub Health, 77-78; assoc prof, 78-89, PROF MED & IMMUNOL, RUSH-PRESBY ST LUKE'S MED CTR, 89- *Concurrent Pos:* Am Cancer Soc fel, Mass Gen Hosp, Harvard Med Sch, 70-72, res fel transplantation immunol, 70-71; mem immunobiol study sect, Div Res Grants, NIH, 79-83, Cancer Ctr Support Rev Comt, Nat Cancer Inst, 90-, med adv bd, Nat Cancer Cytol Ctr, NY, 82-; mem, res rev comt, Am Cancer Soc, 78-88 & med adv comt, Leukemia Res Found, 84-88. *Mem:* Transplantation Soc; Am Asn Immunol; Am Asn Clin Histocompatibility Testing; Am Asn Cancer Res. *Res:* Immunobiology; cell-cell interactions; differentiation; molecular biology of IL-1 receptors; immunogenetics; lymphokines, cytokines. *Mailing Add:* Sect Med Oncol Rush-Presby St Luke's Med Ctr 1753 W Congress Pkwy Chicago IL 60612

PLATEK, RICHARD ALAN, b Brooklyn, NY, Sept 27, 40; m 60; c 2. MATHEMATICAL LOGIC. *Educ:* Mass Inst Technol, BS, 61; Stanford Univ, PhD(math), 65. *Prof Exp:* NATO fel, 65-66, C L E Moore instr math, Mass Inst Technol, 66-67; asst prof, 67-70, ASSOC PROF MATH, CORNELL UNIV, 70- *Res:* Axiomatic set theory; theory of recursive functions. *Mailing Add:* Odyssey Res Assoc 301A Harris B Oates Dr Ithaca NY 14850

PLATFORD, ROBERT FREDERICK, b Unity, Sask, Can, June 29, 33; m 61; c 4. PHYSICAL CHEMISTRY. *Educ:* Univ BC, BA, 55, MSc, 56; PhD(radiation chem), Univ Sask, 58. *Prof Exp:* Sci officer, Suffield Exp Sta, Alta, 58-61; scientist, Atlantic Oceanog Group, Bedford Inst Oceanog, 61-67; RES SCIENTIST, CAN CTR INLAND WATERS, 67- *Concurrent Pos:* Cominco fel, 58. *Res:* Electrochemistry; thermodynamics; surface chemistry; radiochemistry. *Mailing Add:* RR 2 2215 Centre Rd Campbellville ON L0P 1B0 Can

PLATNER, EDWARD D, b Los Angeles, Calif, Apr 1, 34; c 2. PHYSICS. *Educ:* Walla Walla Col, BA, 56; Univ Wash, MS, 60, PhD(physics), 61. *Prof Exp:* PHYSICIST, BROOKHAVEN NAT LAB, 64- *Mem:* Inst Elec & Electronics Engrs. *Res:* High energy particle experimental physics. *Mailing Add:* Dept of Physics 510A Brookhaven Nat Lab Upton NY 11973

PLATNER, WESLEY STANLEY, b Newark, NJ, Sept 26, 15; m 42; c 1. MAMMALIAN PHYSIOLOGY. *Educ:* Philadelphia Col Pharm & Sci, BS, 39; Univ Pa, MS, 42; Univ Mo, PhD(mammalian physiol), 48. *Prof Exp:* Aquatic biologist, US Dept Interior, Washington, DC, 42-47; instr physiol & pharmacol, 46-48, from asst prof to prof physiol, 49-81, chmn physiol area, 77-80, EMER PROF PHYSIOL, SCH MED, UNIV MO-COLUMBIA, 81- *Concurrent Pos:* Res asst & fel dairy husb, Univ Mo, 48-55; NIH grant physiol, Univ Mo, 57-69, USPHS grant, 59-72; NSF travel fel, 21st Int Physiol Cong, Buenos Aires, Arg, 59; res investr, Space Sci Ctr, Univ Mo, 65-69; vis res prof & NIH fel, Univ Calif, Santa Barbara, 71-72. *Mem:* Fel AAAS; Am Physiol Soc; Soc Exp Biol & Med; Am Inst Biol Sci; Am Oil Chem Soc; Sigma Xi. *Res:* Adaptive mechanisms, especially hormonal, nutritional, and environmental stress factors which affect trace minerals, carbohydrates, and lipids at the tissue and cellular levels of physiology; environmental physiology. *Mailing Add:* Dept of Physiol Univ of Mo Sch of Med Columbia MO 65212

PLATNICK, NORMAN I, b Bluefield, WVa, Dec 30, 51; m 70; c 1. ARACHNOLOGY, SYSTEMATICS. *Educ:* Concord Col, BS, 68; Mich State Univ, MS, 70; Harvard Univ, PhD(biol), 73. *Prof Exp:* Asst cur, 73-77, assoc cur, 77-82, CUR ENTOM, AM MUSEUM NATURAL HIST, 82- *Concurrent Pos:* Ed, Cladistics, 84-86; adj prof, City Univ New York, 78-, Cornell Univ, 88- *Mem:* Am Arachnological Soc; Soc Syst Zool; British Arachnological Soc. *Res:* Systematics and biogeography of spiders; theory of systematics and biogeography. *Mailing Add:* Dept Entom Am Museum Natural Hist New York NY 10024

PLATO, CHRIS C, b Nicosia, Cyprus, Oct 10, 31; US citizen; m 61; c 3. GERONTOLOGY, EPIDEMIOLOGY. *Educ:* Univ Ga, BSc, 55; Iowa State Univ, MSc, 56; Univ Mich, MSc, 60, PhD(growth & develop), 76. *Prof Exp:* Geneticist human, Nat Inst Neurol Dis & Stroke, 62-67 & Nat Inst Child

Health & Human Develop, 67-72, GENETICIST AGING, NAT INST AGING, 72-, SR RES GENETICIST, NIH. *Concurrent Pos:* Adj prof, Dept Med Genetics, Univ SAla Med Sch, 76-; consult, Ctr Human Growth & Develop, Univ Mich, 75-, Ctr Demog & Pop Genetics, Univ Tex, Houston, 78-; adj sr scientist biol anthrop, Penn State Univ, 85- *Mem:* Geront Soc; fel Int Soc Twin Studies; Am Dermatoglyphics Asn (pres, 75-78); Am Soc Human Genetics; fel Am Col Epidemiol. *Res:* Osteoarthritis, bone loss, amyotrophic lateral sclerosis, dermatoglyphics, genetics of aging, population genetics. *Mailing Add:* Geront Res Ctr Nat Inst Aging NIH 4940 Eastern Ave Baltimore MD 21224

PLATO, PHILLIP ALEXANDER, b Nashville, Tenn, Nov 16, 43; m 64; c 2. RADIOLOGICAL HEALTH, PUBLIC HEALTH. *Educ:* Univ Miami, BS, 65, MS, 66; Iowa State Univ, PhD(sanitary eng), 68. *Prof Exp:* From asst prof to assoc prof, 69-78, PROF RADIOL HEALTH, UNIV MICH, 78- *Concurrent Pos:* Consult to various pvt & pub orgn, 69- *Mem:* Health Physics Soc; Am Pub Health Asn; Am Nat Standards Inst. *Res:* Radiation dosimetry, including calibration of radiation sources and movement of radionuclides from sources to people; design and operation of environmental monitoring programs around nuclear facilities. *Mailing Add:* 2819 SE 22nd Pl Cape Coral FL 33904

PLATSOUCAS, CHRIS DIMITRIOS, b Athens, Greece, Apr 17, 51; m 85. MOLECULAR IMMUNOLOGY. *Educ:* Univ Patras, BS, 73; Mass Inst Technol, PhD(biochem & immunol), 78. *Prof Exp:* From res fel to asst mem immunol, Mem Sloan-Kettering Cancer Ctr, 78-85; assoc prof, 85-89, PROF IMMUNOL, M D ANDERSON CANCER CTR, UNIV TEX, 89- *Concurrent Pos:* Nat res serv award, NIH, 78-79; asst prof immunol, Med Col, Cornell Univ, 81-85; head lab & biol response modifiers, Mem Sloan-Kettering Cancer Ctr, 81-85; prin investr, Am Cancer Soc, 80-, NIH grants, 82-, NSF, 82-84 & Advan Technol Prog, Tex, 88-90; biotech consult & mem, NIH Study Sect, 82 & 89; patentee, US Patent Off, 89 & 91; dep chmn immunol, M D Anderson Cancer Ctr, Univ Tex, 89-, Ashbel Smith prof, 91- *Mem:* Am Asn Immunologists; Am Soc Hemat; Am Asn Biochem & Molecular Biol; Am Asn Pathologists; Am Asn Cancer Res; Am Fed Clin Res. *Res:* Molecular and cellular immunology of human T cells; T-cell antigen receptors; tumor-infiltrating lymphocytes in malignant melanoma and ovarian carcinoma; lymphoproliferative disorders; immunoregulatory factors; T-T cell hydrids; tumor-specific antigens in malignant melanoma. *Mailing Add:* M D Anderson Cancer Ctr 1515 Holcombe Box 178 Houston TX 77030

PLATT, ALAN EDWARD, b London, Eng, Dec 30, 36; m 65; c 2. PHYSICAL ORGANIC CHEMISTRY. *Educ:* Univ London, BSc, 58, PhD(chem), 61. *Prof Exp:* Res assoc chem, Univ Mich, 61-62; res fel, Calif Inst Technol, 62-63; res chemist, 63-68, group leader, 68-75, from assoc scientist to sr assoc scientist, 75-84, RES SCIENTIST, DOW CHEM USA, 84- *Mem:* Sigma Xi; Asn Comput Sci. *Res:* Polymerization kinetics; polymer characterization and processes. *Mailing Add:* 4402 Andre Midland MI 48640

PLATT, AUSTIN PICKARD, b Evanston, Ill, Oct 29, 37; m 63; c 2. ECOLOGY, POPULATION BIOLOGY. *Educ:* Williams Col, BA, 59; Univ Mass, MA, 63, PhD(zool), 65. *Prof Exp:* Instr zool, Univ RI, 65-66; asst prof biol, Wesleyan Univ, 66-69; asst prof, 69-71, ASSOC PROF BIOL SCI, UNIV MD BALTIMORE COUNTY, 71- *Concurrent Pos:* Univ RI grant, 65-66; NSF grant, 67-69; ed, J Lepid Soc, 78-80. *Mem:* Am Inst Biol Sci; Soc Study Evolution; Royal Entomol Soc London; Lepid Soc. *Res:* Calculation of small mammal home ranges; population fluctuations of small mammals; insect diapause; genetics and evolution of Nearctic Limenitis butterflies. *Mailing Add:* Dept Biol Sci Univ Md Baltimore County Catonsville MD 21228

PLATT, DAVID, immunology, antibiotics, for more information see previous edition

PLATT, DWIGHT RICH, b Chicago, Ill, Aug 4, 31; m 56; c 2. HERPETOLOGY. *Educ:* Bethel Col, BS, 52; Univ Kans, MA, 54, PhD(zool), 66. *Prof Exp:* Asst instr zool, Univ Kans, 52-54; ed tech village develop, Am Friends Serv Comt, Barpali, India, 54-57; from instr to assoc prof, 57-69, PROF BIOL, BETHEL COL, 69- *Concurrent Pos:* Res asst, Univ Kans, 59-60; vis prof Sambalpur Univ, India, 70-71. *Honors & Awards:* Chevron Nat Conserv Award, 90. *Mem:* Ecol Soc Am; Am Soc Ichthyol & Herpet; Am Ornith Union; Wilson Ornith Soc; Soc Study Amphibians & Reptiles; Herpetologists League. *Res:* Vertebrate population ecology; population dynamics of snakes as related to changes in their prey populations; prairie ecology. *Mailing Add:* Dept Biol Bethel Col North Newton KS 67117

PLATT, JAMES EARL, b Greensburg, Ind, June 19, 44; m 65; c 1. COMPARATIVE ENDOCRINOLOGY. *Educ:* Ind Univ, BA, 66, MAT, 67; Univ Colo, PhD(biol), 74. *Prof Exp:* ASST PROF BIOL SCI, UNIV DENVER, 74- *Mem:* Am Soc Zoologists. *Res:* Role of prolactin in larval amphibian growth and development; prolactin-thyrozine interaction in amphibians and other lower vertebrates; amphibian metamorphosis. *Mailing Add:* Dept Biol Sci Univ Denver University Park Denver CO 80208

PLATT, JOHN RADER, b Jacksonville, Fla, June 29, 18; m 41, 79; c 2. ENVIRONMENTAL SCIENCES. *Educ:* Northwestern Univ, BS, 36, MS, 37; Univ Mich, PhD(physics), 41. *Hon Degrees:* Dr psychol, Utah State Univ, 74. *Prof Exp:* Rockefeller Found Proj fel & res assoc, Univ Minn, 41-43; instr physics, Northwestern Univ, 43-45, res assoc Nat Defense Res Comt Prog, 44-45; from asst prof to prof physics, Univ Chicago, 46-65; res scientist, Ment Health Res Inst, Univ Mich, Ann Arbor, 65-77; vis prof, Univ Calif, Santa Barbara, 77-79; lectr, Sch Med, Boston Univ, 79-81; vis lectr, Harvard Univ, 81-83; RETIRED. *Concurrent Pos:* Guggenheim fel, 51-52; fel, Ctr Adv Study, Stanford, 72-73. *Honors & Awards:* USPHS Career Award, 64; Regent's lectr, Univ Calif, Santa Barbara, 76. *Mem:* AAAS; Biophys Soc; Am Acad Polit & Soc Sci. *Res:* Theory of chemical spectra; interpretation of light absorption by biological systems; physics of perception; social change and global problems; future global problems. *Mailing Add:* 14 Concord Ave Apt 624 Cambridge MA 02138

PLATT, JOSEPH BEAVEN, b Portland, Ore, Aug 12, 15; m 46; c 2. PHYSICS. *Educ:* Univ Rochester, BA, 37; Cornell Univ, PhD(exp physics), 42. *Hon Degrees:* LLD, Univ Southern Calif, Claremont McKenna Col; DSc, Harvey Mudd Col. *Prof Exp:* Asst physics, Cornell Univ, 37-41; instr, Univ Rochester, 41-43; staff mem & sect chief radiation lab, Mass Inst Technol, 43-46; from asst prof to assoc prof physics, Univ Rochester, 46-49; chief physics br, Res Div, AEC, 49-51; from assoc prof to prof physics & assoc chmn dept, Univ Rochester, 51-56; pres, Harvey Mudd Col, 56-76, Claremont Univ Ctr, 76-81; SR PROF, HARVEY MUDD COL, 81- *Concurrent Pos:* Consult, Nat Defense Res Comt, 41-45; adv comt sci educ, NSF, 65-70 & 73-74, vchmn, 67-68 & 73, chmn, 69-70 & 74; mem mine adv comt, Nat Acad Sci-Nat Res Coun, 55-61, mem comt sci in UNESCO, 60-62, comt int orgns & progs, 62-64, chmn, Subcomt Sino-Am Sci Coop, 64-78; Gov comt study med aid & health, Calif, 59-60; trustee, Anal Serv, Inc, 59-89, chmn, 62-89; sci adv, US deleg, UNESCO Gen Conf, Paris, 60, alternate deleg, 62; mem panel int sci, President's Sci Adv Comt, 61-64, int tech coop & assistance panel, 66-67, comput in educ panel, 66-68; trustee, Thacher Sch, 62-74; chmn select comt on master plan for higher educ, Coord Coun Higher Educ, Calif, 71-73; study comt on NIH, 64-65; bd trustees, Carnegie Found Advan Teaching, 70-76; bd trustees, Aerospace Corp, 71-85; trustee, China Found Prom Educ & Cult, 66-; mem, Carnegie Coun Policy Studies in Higher Educ, 75-80; dir, Bell & Howell Co, 78-88, Jacobs Eng Group, 78-86, Am Mutual Fund, 81-88, Sigma Res, Inc, 82-87, DeVry, Inc, 84-87 & Automobile Club Southern Calif, 71-89. *Mem:* Fel Am Phys Soc; sr mem Inst Elec & Electronic Engrs. *Res:* X-ray spectroscopy; radar systems engineering; nuclear and meson physics. *Mailing Add:* Dept Physics Harvey Mudd Col Claremont CA 91711

PLATT, KENNETH ALLAN, medicine; deceased, see previous edition for last biography

PLATT, LAWRENCE DAVID, b Detroit, Mich, Oct 19, 47; c 2. OBSTETRICS & GYNECOLOGY. *Educ:* Wayne State Univ, BS, 68, MD, 72; Am Bd Obstet & Gynec, dipl, 79. *Prof Exp:* Resident obstet & gynec, Sinai Hosp Detroit, 73-76; clin res fel obstet & gynec, 76-78, from asst prof to assoc prof, 78-86, PROF OBSTET & GYNEC, LOS ANGELES CITY, UNIV SOUTHERN CALIF MED CTR, 86- *Concurrent Pos:* Clin instr obstet & gynec, Univ Southern Calif Sch Med, 76- 78; attend staff, Calif Hosp, Los Angeles, 80-, Childrens Hosp & Cedar Sinai Hosp, 81-; consult staff, Huntington Mem Hosp, 81-; courtesy staff, St Johns Hosp, 81-; mem sci rev comt, Soc Gynec Invest, 83-, Am Inst Ultrasound Med, 82-, Am Col Obstetricians & Gynecologists, 84-, Soc Perinatal Obstetricians, 85-, Am J Cardiac Imaging, 87-; prin investr, Los Angeles County, Univ Southern Calif, 78, Pfizer & Co, 80-82, ADR Corp, 77, Xerox Med Syst, 80-82, Smith Kline Inst, 78-79. *Mem:* Fel Am Col Obstetricians & Gynecologists; Soc Obstet & Gynec Ultrasound (secy-treas, 78-80); fel Am Inst Ultrasound Med (vpres, 86); Soc Obstet Anesthesia & Perinatology; Soc Gynec Invest; Soc Perinatal Obstetricians. *Res:* Ultrasound in obstetrics and gynecology; biophysical assessment of fetal condition; Rh disease; prenatal diagnosis. *Mailing Add:* Womens Hosp LAC/USC Med Ctr 1240 N Mission Rd No 5K9 Los Angeles CA 90033

PLATT, LUCIAN B, b Syracuse, NY, Apr 9, 31; m 57; c 2. STRUCTURAL GEOLOGY. *Educ:* Yale Univ, BS, 53, MS, 57, PhD(geol), 60. *Prof Exp:* Fel, Princeton Univ, 60-61; asst prof geol, Villanova Univ, 61-64; assoc prof, George Washington Univ, 64-70; chmn dept, 71-75, assoc prof, 70-79, PROF GEOL, BRYN MAWR COL, 79- *Concurrent Pos:* Consult, Pa Geol Surv, 61-64; part time geologist, US Geol Surv, 65-85. *Mem:* Geol Soc Am; Am Geophys Union; Am Asn Petrol Geol. *Res:* Structure and stratigraphy in complex shale areas. *Mailing Add:* Dept of Geol Bryn Mawr Col Bryn Mawr PA 19010

PLATT, MARVIN STANLEY, PEDIATRICS, FORENSICS. *Educ:* Univ Md, MD, 56; Univ Akron, JD, 77. *Prof Exp:* STAFF PATHOLOGIST, CHILDREN'S HOSP MED CTR, 73- *Mailing Add:* Children's Hosp Med Ctr 281 Locust St Akron OH 44308

PLATT, MILTON M, b Bayonne, NJ, Mar 7, 21; m 43; c 3. MATERIALS ENGINEERING, TEXTILE MECHANICS. *Educ:* Mass Inst Technol, BS, 42, ScD(struct eng), 46. *Prof Exp:* Res technician, Mass Inst Technol, 42-46; sr res assoc, Fabric Res Labs, Inc, 46-47, asst dir res, 47-52, assoc dir res, 52-74, vpres, 53-74, treas, 70-74; dir, FRL Div, Albany Int Res Co, Albany Int Corp, 74-78, sr dir, 78-; RETIRED. *Concurrent Pos:* Lectr, Mass Inst Technol, 53-; consult, Off Qm Corps; mem comt textiles, Nat Acad Sci-Nat Res Coun; mem, Textile Res Inst. *Honors & Awards:* Fiber Soc Award, 61; Harold DeWitt Smith Mem Medal, Am Soc Testing & Mat, 67; Space Shuttle Orbiter Award, Rockwell Int Corp, 78; Edward R Schwarz Award, Am Soc Mech Engrs, 83; Smith Medal, Brit Textile Inst, 83. *Mem:* Fiber Soc (vpres, 68-69, pres, 70-71); Soc Rheology; fel Brit Textile Inst. *Res:* Theory of structures; mechanics of textile materials; analysis of flat plates subject to transverse distributed loads; relation between mechanical properties and structural geometry of textile structures. *Mailing Add:* c/o John Gray 777 West St Mansfield MA 02048

PLATT, ROBERT SWANTON, JR, b Chicago, Ill, Jan 2, 25; m 51; c 2. PLANT PHYSIOLOGY. *Educ:* Wash Univ, St Louis, AB, 49; Harvard Univ, AM, 51, PhD(biol), 54. *Prof Exp:* Fulbright fel, State Agr Univ, Wageningen, 54-55; instr, 56-57, asst dean col biol sci, 66-68, ASST PROF BOT, OHIO STATE UNIV, 57-66 & 68- *Concurrent Pos:* Danforth assoc, 70- *Mem:* AAAS; Am Soc Plant Physiol; Bot Soc Am; Sigma Xi. *Res:* Metabolism of plant growth substances and light-growth reactions. *Mailing Add:* 2006 Coventry rd Ohio State Univ Columbus OH 43212

PLATT, RONALD DEAN, b Council Bluffs, Iowa, May 21, 42; m 63; c 2. MATHEMATICAL STATISTICS. *Educ:* Univ Northern Iowa, BA, 63; Univ Iowa, MS, 66, PhD(statist), 71. *Prof Exp:* Instr math, Univ Iowa, 65-66, programmer, Measurement Res Ctr, 66-67; instr math, Grinnell Col, 67-69; instr math & sr programmer, Measurement Res Ctr, Univ Iowa, 69-70, dir statist serv, 70-71; assoc prof statist, Northwest Mo State Univ, 71-74; biostatistician, Miles Labs Inc, 75-78; mgr prod develop & prog, Cap Comput Ctr, 78-82; STATIST CONSULT, 82- *Mem:* Am Statist Asn. *Res:* Probability theory; clinical studies and management of toxicity data. *Mailing Add:* 11708 W 108th Shawnee Mission KS 66210

PLATT, THOMAS BOYNE, b Evanston, Ill, Sept 30, 27; m 54, 63; c 5. BIOCHEMISTRY. *Educ:* Iowa State Col, BS, 50; NC State Col, MS, 52; Univ Wis, PhD(bact), 56. *Prof Exp:* ASSOC DIR, BRISTOL-MYERS PHARMACEUT RES INST, 56- *Mem:* Am Soc Microbiol; Brit Soc Gen Microbiol. *Res:* Analytical microbiology; chemotherapy; fermentation technology; analytical chemistry. *Mailing Add:* Bristol-Myers Squibb Pharmaceut Res Inst New Brunswick NJ 08903

PLATT, THOMAS REID, b Youngstown, Ohio, June 20, 49; m 73; c 2. PARASITOLOGY, SYSTEMATICS & EVOLUTION. *Educ:* Hiram Col, BA, 71; Bowling Green State Univ, MS, 73; Univ Alta, PhD(zool), 78. *Prof Exp:* Asst prof biol, Univ Richmond, 78-85; ASST PROF BIOL, ST MARY'S COL, 86-; 219-291-6439. *Mem:* Am Soc Parasitologists; Soc Syst Zoologists; Am Microscopical Soc. *Res:* Systematics and biogeography of helminth parasites of turtles, particularly the blood-flukes, (Drgenea: Spirorchidae). *Mailing Add:* Dept Biol Saint Mary's Col Notre Dame IN 46556

PLATT, WILLIAM JOSHUA, III, b Gainesville, Fla, Nov 21, 42; m 68. ECOLOGY, VERTEBRATE BIOLOGY. *Educ:* Univ Fla, BS, 64; Cornell Univ, PhD(ecol), 71. *Prof Exp:* Asst prof zool, Univ Iowa, 69-77; asst prof biol sci, Univ Ill, 77-; Tall Timbers Res Sta; PROF, DEPT BOT, LA STATE UNIV. *Concurrent Pos:* Army Corps Engrs Environ Resources Study, Univ Iowa-Iowa State Univ, 72. *Mem:* AAAS; Am Inst Biol Sci; Ecol Soc Am; Am Soc Mammal. *Res:* Predator-prey relationships in prairie ecosystems; effects of predation upon ecosystems structure and dynamics; perturbations upon ecosystems stability; biology of shrews. *Mailing Add:* Dept Bot La State Univ 502 Life Sci Bldg Baton Rouge LA 70803-1705

PLATT, WILLIAM RADY, b Baltimore, Md, July 25, 15; m 49, 79; c 3. HEMATOPATHOLOGY. *Educ:* Univ Md, BS, 36, MD, 40. *Prof Exp:* Resident path, Sch Med, Emory Univ, 41-44; instr, Yale Univ, 44-45, Washington Univ, 45-46, Univ Louisville, 46-48, Univ Penn, 48-52; assoc prof, Washington Univ, 52-76; prof, Southwestern Med Sch, Dallas, Tex, 76-77; LECTR PATHOL, SCH MED, JOHNS HOPKINS UNIV, 78- *Concurrent Pos:* Vis prof pathol, Sch Med, Chinese Univ, Hong Kong, 85-86; chmn & ed-in-chief, Path Update J, 82-87. *Mem:* Am Soc Hemat; Am Soc Clin Path; Am Asn Blood Banks; Col Am Pathologists; Am Col Physicians. *Res:* Laboratory medicine; hematology and hematopathology; pathology; anatomy; effects of chemotherapy and radiotherapy on solid and hematopoietic tissue. *Mailing Add:* 12 Hamlet Hill Rd Baltimore MD 21210

PLATT ALOIA, KATHRYN ANN, ENVIRONMENTAL STRESS, SENESCENCE. *Educ:* Univ Calif, PhD(bot), 80. *Prof Exp:* RES ASSOC, DEPT BOT & PLANT SCI, UNIV CALIF, RIVERSIDE, 69- *Mailing Add:* Dept Bot & Plant Sci Univ Calif Riverside CA 92521

PLATTER, SANFORD, b New York, NY, Jan 16, 39; m 60; c 2. COMPUTER PERIPHERAL, PNEUMATICS. *Educ:* Rutgers Univ, BSME, 60; Syracuse Univ, MSME, 65. *Prof Exp:* Staff engr, Int Bus Mach, 60-70; dir technol, STC, 70-76, Aspen Peripherals, 82-85 & Exabyte, 85-86; vpres, 76-82, PRES, PAD, 86- *Mem:* Am Soc Mech Engrs; Inst Elec & Electronic Engrs. *Res:* Development of computer tape drive; tape path technology; produced technology and basic design for most of the major high performance computer tape drives; servos. *Mailing Add:* 421 Kelly Rd E Boulder CO 80302

PLATTS, DENNIS ROBERT, b Kittaning, Pa, Mar 22, 33; m 54; c 4. GLASS TECHNOLOGY, CERAMICS. *Educ:* Alfred Univ, BS, 61, PhD(ceramic sci), 65. *Prof Exp:* Res engr, Glass & Chem Prod Div, Ford Motor Co, 64-66; res scientist, Linden Labs, Inc, 66-68 & Ceramic Finishing Co, 68-72; RES ENG ASSOC, FORD MOTOR CO, 72- *Mem:* Am Ceramic Soc. *Res:* Glass strength and defects; chemically strengthened glass. *Mailing Add:* 15000 Commerce Dr N Dearborn MI 48120

PLATTS, WILLIAM SIDNEY, b Burley, Idaho, Apr 18, 28; m 51; c 2. FISHERIES. *Educ:* Idaho State Univ, BS, 55; Utah State Univ, MS, 57, PhD(fishery mgt), 74. *Prof Exp:* Regional biologist & enforcement supt, Idaho Fish & Game, 61-64; zone biologist, 67-74, RES BIOLOGIST, US FOREST SERV, 74- *Mem:* Am Fisheries Soc; Nature Conservancy; Nat Wildlife Fedn. *Res:* Aquatic documentation, methodology and classification; land use impacts. *Mailing Add:* Don Chapman Consults Inc 3653 Rickenbacker Suite 200 Boise ID 83705

PLATUS, DANIEL HERSCHEL, b Los Angeles, Calif, Jan 19, 32; m 69; c 2. ENGINEERING. *Educ:* Univ Calif, Los Angeles, BS, 54, PhD(eng), 61. *Prof Exp:* Sr nuclear engr, Holmes & Narver, Inc, 58; sr res engr, Atomics Int Div, NAm Aviation, 61-62; proj scientist, ARA, Inc, 62-65; SR SCIENTIST, AEROPHYSICS RES LAB, AEROSPACE CORP, 65- *Mem:* Assoc fel Am Soc Aeronaut & Astronaut; Sigma Xi. *Res:* Applied mechanics; flight dynamics; solid mechanics; nuclear reactor technology. *Mailing Add:* 29816 Grandpoint Lane Rancho Palos Verdes CA 90274

PLATZ, JAMES ERNEST, b Bloomington, Ind, Sept 4, 43; m 69; c 1. EVOLUTIONARY BIOLOGY. *Educ:* Tex Tech Univ, BS, 67, MS, 70; Ariz State Univ, PhD(zool), 74. *Prof Exp:* From asst prof to assoc prof, 74-85, PROF BIOL, CREIGHTON UNIV, 85- *Concurrent Pos:* Vis assoc prof, Dept Zoology, Univ Tex, Austin, 86-87. *Mem:* AAAS; Soc Syst Zool; Soc Study Evolution; Am Soc Ichthyol & Herpetol. *Res:* Vertebrate evolutionary biology involving the study of the dynamics of the speciation process. *Mailing Add:* Dept Biol Creighton Univ Omaha NE 68178

PLATZ, MATTHEW S, b New York, NY, July 22, 51; m 73; c 3. PHOTOCHEMISTRY. *Educ:* State Univ NY-Albany, BSc, 73; Yale Univ, PhD(chem), 77. *Prof Exp:* PROF CHEM, OHIO STATE UNIV, 78- *Mem:* Am Chem Soc; Am Soc Photobiol; Intra Am Photochem Soc; AAAS; Europ Photochem Asn; Am Asn Univ Prof. *Res:* The interaction of light & organic compounds; the characterization of the high energy species formed on photoactivation. *Mailing Add:* Chem Dept Ohio State Univ 140 W 18th Ave Columbus OH 43210

PLATZ, ROBERT DOLE, ONCOLOGY, CELL BIOLOGY. *Educ:* Univ Mich, PhD, 72. *Prof Exp:* SR SCIENTIST & PROG MGR, DYNAMAC CORP, 80- *Mailing Add:* Dynamac Corp 7940 Edgewood Farm Rd Frederick MD 21701

PLATZER, EDWARD GEORGE, b Vancouver, BC, Oct 3, 38; m 62; c 1. PARASITOLOGY, NUTRITION. *Educ:* Univ BC, BSc, 61, MSc, 64; Univ Mass, PhD(zool), 68. *Prof Exp:* Guest investr biochem malaria, Rockefeller Univ, 68-71; asst prof nematol, 71-77, assoc prof nematol biol, 77-82, PROF NEMATOL BIOL, UNIV CALIF, RIVERSIDE, 82- *Concurrent Pos:* USPHS fel, 68-70; mem trop med & parasitol study sect, NIH, 82-85. *Mem:* Am Soc Parasitol; AAAS; Soc Nematol. *Res:* Life history, biochemistry and physiology of nematodes; nematodes as agents for insect control. *Mailing Add:* Dept of Nematol Univ of Calif Riverside CA 92521

PLATZER, MAXIMILIAN FRANZ, b Vienna, Austria, June 26, 33; US citizen; m 58; c 3. FLUID MECHANICS, AEROSPACE ENGINEERING. *Educ:* Vienna Tech Univ, Dipl Ing, 57, Dr Tech Sci(appl mech), 64. *Prof Exp:* Aerospace engr, Marshall Space Flight Ctr, NASA, 60-66; aerospace scientist, Lockheed-Ga Co, 66-70; from assoc prof to prof aeronaut, Naval Postgrad Sch, 70-76, chmn dept aeronaut, 78-88; CONSULT. *Honors & Awards:* NASA Incentive Award, Marshall Space Flight Ctr, 65. *Mem:* Assoc fel Am Inst Aeronaut & Astronaut; Am Soc Mech Engrs; NY Acad Sci; Sigma Xi. *Res:* Fluid and flight mechanics; propulsion aerodynamics. *Mailing Add:* Dept of Aeronaut Naval Postgrad Sch Monterey CA 93940

PLATZMAN, GEORGE WILLIAM, b Chicago, Ill, Apr 19, 20; m 45. METEOROLOGY, OCEANOGRAPHY. *Educ:* Univ Chicago, BS, 40, PhD, 48; Univ Ariz, MS, 41. *Prof Exp:* Instr eng, Sci & Mgt Defense Training Prog, Univ Chicago, 41, instr meteorol, 42-45; jr sci aide, USDA, Ill, 41-42; instr, Univ Ore, 45; engr hydrol, US Eng Dept, 45-46; res assoc, 46-49, secy-counsr dept meteorol, 53-59, head phys sci sect, col, 59-60, chmn dept, 71-74, from asst prof to prof, 49-90, EMER PROF GEOPHYS SCI, UNIV CHICAGO, 90- *Concurrent Pos:* Ed, J Meteorol, Am Meteorol Soc, 48-50; consult, Comput Proj, Inst Advan Study, 50-53; Guggenheim Fel, Imp Col, Univ London, 67-68; Green Scholar, Inst Geophys & Planetary Physics, Univ Calif, San Diego, 75; sr postdoctorate fel, Nat Ctr Atmospheric Res, 76. *Honors & Awards:* Meisinger Award, Am Meteorol Soc, 66, Ed Award, 73. *Mem:* Fel Am Meteorol Soc; fel Am Geophys Union; fel AAAS. *Res:* Circulation and wave theory of atmosphere and ocean; dynamical prediction; storm surges; tides. *Mailing Add:* Dept Geophys Sci Univ Chicago 5734 S Ellis Ave Chicago IL 60637-1434

PLATZMAN, PHILIP M, b Brooklyn, NY, May 1, 35; m 56; c 2. SOLID STATE PHYSICS, PLASMA PHYSICS. *Educ:* Mass Inst Technol, BS, 56; Calif Inst Technol, PhD(physics), 60. *Prof Exp:* Mem tech staff, Hughes Aircraft Co, 56-60; res physicist, 60-68, DEPT HEAD, AT&T BELL LABS, INC, 68- *Concurrent Pos:* Asst, Calif Inst Technol, 59-60; vis prof, Univ Calif, Berkeley, 68, ISSP Tokyo, 78, Hebrew Univ, 79 & KFA Julich, Ger, 80; adj prof, Univ Calif, San Diego, 83- *Mem:* Fel Am Phys Soc; NY Acad Sci. *Res:* Nuclear structure corrections to the hyper-fine structure in hydrogen; mesonic origins of the two nucleon L-S potential; microwave breakdown of gases; polarons; wave propagation in plasmas; plasma effects in metals; light scattering in solids; inelastic x-ray scattering from electrons in solids, liquids and gases; properties of 2-d electron gas at the surface of liquid helium and in semi-conductor structures; quantum hall effect interactions of positrons with solids; tunnelling phenomenon. *Mailing Add:* AT & T Bell Labs Inc, ID 427 PO Box 261 Murray Hill NJ 07974

PLAUSH, ALBERT CHARLES, b Cleveland, Ohio; m 83. PHYSICAL CHEMISTRY. *Educ:* Adelbert Col, BA, 57; Case Western Reserve Univ, MS, 58, PhD(chem), 65. *Prof Exp:* NIH fel biochem, Case Western Reserve Univ, 65-67; from asst prof to assoc prof, 67-85, PROF CHEM, SAGINAW VALLEY STATE COL, 85-, CHMN DEPT, 82- *Concurrent Pos:* Vis scholar, Univ Mich, 74-75. *Mem:* Sigma Xi; Am Chem Soc; AAAS. *Res:* Nuclear magnetic studies of biological molecules. *Mailing Add:* Sagina Valley State Univ 2250 Pierce Rd University Center MI 48710

PLAUT, ANDREW GEORGE, b Leipzig, Ger, Feb 19, 37; US citizen; m 65; c 2. GASTROENTEROLOGY, IMMUNOLOGY. *Educ:* Ohio State Univ, BS, 58; Tufts Univ, MD, 62. *Prof Exp:* NIH res fel gastroenterol, Mem Hosp Cancer & Allied Dis & Second Med Div, Bellevue Hosp, 64-65; Buswell fel med, State Univ NY Buffalo, 68-73, asst prof med, 71-73; NIH spec fel infectious dis, 65-66, assoc prof, 73-77, PROF MED, SCH MED, TUFTS UNIV, 77-, PHYSICIAN, TUFTS-NEW ENG MED CTR HOSP, 73-; DIR, DIGESTIVE DIS RES CTR, 88- *Concurrent Pos:* Mem, Gen Med Study Sect, Div Res Grants, NIH, 73- & Sci Adv Bd, Ileitis & Colitis Found; Nat Inst Allergy & Infectious Dis res career develop award, 72; consult, Nat Inst Dent Res, Navy Med Res Lab & Walker Reed Army Inst Res. *Mem:* Am Asn Immunologists; Am Gastroenterol Asn; Am Soc Microbiol; AAAS; Royal Soc Trop Med; Am Soc Clin Invest; Am Asn Physicians. *Res:* Immunology of mucosal surfaces, mainly secretory immunoglobulin A; interactions of microbial proteolytic enzymes with mucosal immunoglobulin A antibody; pathogenesis of dental caries, gonorrhea, inflammatory bowel disease; intestinal microflora; AIDS. *Mailing Add:* Dept Med Tufts New Eng Med Ctr 750 Washington St Boston MA 02111

PLAUT, GERHARD WOLFGANG EUGEN, b Frankfurt, Germany, Jan 9, 21; nat US; m 50; c 3. BIOCHEMISTRY. *Educ:* Iowa State Col, BS, 43; Univ Wis, MS, 49, PhD(biochem), 51. *Prof Exp:* Develop biochemist, E R Squibb & Sons, 43-47; asst biochem, Univ Wis, 47-49, asst prof, Inst Enzyme Res, 51-54; assoc prof biochem, NY Univ, 54-58; res prof med, Univ Utah, 58-66, assoc prof biol chem, 58-62, prof biol chem & head biochem lab study hereditary & metab disorders, 62-66; prof biochem & chmn dept, Rutgers Med Sch, 66-70; prof biochem & chmn dept, 70-89, EMER PROF BIOCHEM, SCH MED, TEMPLE UNIV, 90- *Concurrent Pos:* Alumni Res Found fel, Univ Wis, 51; USPHS sr res fel, 59-62, res career award, 62-66; estab investr, Am Heart Asn, 53-58; consult to var nat orgns & asns, 60-88. *Honors & Awards:* Swern Res Award, 83; Am Swiss Found Lectr, 66. *Mem:* Am Soc Biol Chemists; Am Inst Nutrit; Am Chem Soc; Brit Biochem Soc; Sigma Xi. *Res:* Intermediary metabolism; enzyme chemistry; vitamins; metabolic regulation. *Mailing Add:* Dept Biochem Temple Univ Sch Med Philadelphia PA 19140

PLAUT, MARSHALL, b Baltimore, Md, Apr 14, 44; m 82; c 2. ALLERGY, IMMUNOLOGY. *Educ:* Johns Hopkins Univ, MD, 67. *Prof Exp:* ASSOC PROF MED, SCH MED, JOHNS HOPKINS UNIV, 82- *Mem:* Am Asn Immunologists; fel Am Acad Allergy & Immunol; Am Fedn Clin Res. *Res:* Cellular interactions in allergic inflammatory disease: the roles of lymphocytes, mast cells, and basophils and cytokines. *Mailing Add:* Dept Med Johns Hopkins Asthma & Allergy Ctr 301 Bayview Blvd Baltimore MD 21224

PLAUT, WALTER (SIGMUND), b Darmstadt, Ger, Nov 21, 23; nat US; m 53; c 4. CELL BIOLOGY. *Educ:* Rutgers Univ, BSc, 49; Univ Wis, MS, 50, PhD(genetics), 52. *Prof Exp:* NIH fel, 52-54; asst res zoologist, Univ Calif, 54-56; asst prof bot, 56-60, assoc prof zool, chmn dept, 78-86, PROF ZOOL, UNIV WIS-MADISON, 65- *Concurrent Pos:* Vis prof, Agrarian Univ, Lima, Peru, 70 & Fac Med, Ribeirao, Preto, Brazil, 74. *Mem:* Genetics Soc Am; Am Soc Cell Biol. *Res:* Chromosome structure and function; light microscopy. *Mailing Add:* Dept Zool Univ Wis 1117 W Johnsten St Zool Res Bldg Madison WI 53706

PLAUTZ, DONALD MELVIN, b Milwaukee, Wis, Dec 26, 42. FORENSIC SCIENCE. *Educ:* Elmhurst Col, BA, 65; SDak State Univ, PhD(chem), 71. *Prof Exp:* Asst prof chem, Madonna Col, 71-72; lab adv, Advan Med & Res Corp, 72; lab scientist forensic chem, Sci Lab, Mich Dept State Police, 72-78; asst lab dir, 78-82, LAB DIR, BUR FORENSIC SCI, ILL DEPT STATE POLICE, 82- *Concurrent Pos:* Comnr, Community Comn on Drug Abuse, 75-78. *Mem:* Am Chem Soc; Int Asn Arson Investrs; Am Soc Crime Lab Dirs. *Res:* Forensic chemistry. *Mailing Add:* Div Forensic Serv & ID 2873 S 25th Ave Broadview IL 60153

PLAVEC, MIREK JOSEF, b Sedlcany, Czech, Oct 7, 25; US citizen; m 50; c 2. ASTROPHYSICS. *Educ:* Charles Univ, Prague, RNDr(astron), 49; Charles Univ, Prague & Czech Acad Sci, PhD(astron), 55, DSc(astron), 68. *Prof Exp:* Prin sci off astrophys, Astron Inst, Ondrejov, Czech, 54-67, chmn dept stellar astrophys, 68-70; vis prof astron, Ohio State Univ, 70; chmn dept, 75-77, PROF ASTRON, UNIV CALIF, LOS ANGELES, 71- *Concurrent Pos:* Int Astron Union grant, Dom Astrophys Observ, Victoria, BC, 61 & 65; assoc prof astron, Charles Univ, Prague, 68-70; pres comn eclipsing binaries, Int Astron Union, 70-73; NSF res grant, Univ Calif, Los Angeles, 71- & NASA res grants, Univ Calif, Los Angeles, 78- *Mem:* Int Astron Union; Am Astron Soc. *Res:* Evolution of close binary stars; mass transfer in binary stars; stars with extended atmospheres; physical characteristics of eclipsing binaries; circumstellar matter; satellite ultraviolet stellar spectra. *Mailing Add:* Dept Astron Univ Calif Los Angeles CA 90024-1562

PLAXICO, JAMES SAMUEL, b Chester, SC, Oct 2, 24; m 45; c 4. AGRICULTURAL ECONOMICS. *Educ:* Clemson Col, BS, 45, MS, 48; Univ Minn, PhD(agr econ), 53. *Prof Exp:* From asst prof to prof agr econ, Va Polytech Inst, 47-55; head dept, 61-77, PROF AGR ECON, OKLA STATE UNIV, 55- *Concurrent Pos:* Ford Found prog adv, 68-69; state exec dir, Agr Stabilization & Conserv Serv, 79-81. *Mem:* Am Econ Asn; Am Agr Econ Asn. *Res:* Production economics; agricultural finance policy. *Mailing Add:* 1123 W Will Rogers Dr Stillwater OK 74075

PLAYER, MARY ANNE, b Oak Park, Ill, June 6, 20. PLANT PHYSIOLOGY. *Educ:* Northwestern Univ, BS, 43, MS, 45, PhD(plant physiol), 48. *Prof Exp:* Res asst, Surg Dept, Univ Chicago, 48-49; asst prof biol, Elmhurst Col, 54-56; fac mem, 56-64, assoc prof, 64-72, PROF BIOL, WRIGHT COL, 72- *Concurrent Pos:* Lectr, Roosevelt Univ, 56-80; mem deleg, People to People Citizen Ambassador Prog, China, 88. *Mem:* Fel AAAS; Bot Soc Am; Am Soc Plant Physiol; Sigma Xi. *Res:* Effects of plant hormones on transpiration; plant metabolism and growth. *Mailing Add:* 1600 Hinman Ave Apt 4-L Evanston IL 60201

PLAYTER, ROBERT FRANKLIN, b Perth, W Australia, Oct 2, 34; US citizen; m 55; c 3. VETERINARY SURGERY, VETERINARY OPHTHALMOLOGY. *Educ:* Kans State Univ, BS, 59, DVM, 61; Tex A&M Univ, MS, 74; Am Col Vet Ophthalmologists, dipl, 76. *Prof Exp:* Vet practr vet med & surg, Hays Vet Hosp, 61-70; from asst prof to assoc prof, 70-76, chief small animal clin, 76-77, PROF & HEAD DEPT SMALL ANIMAL MED & SURG, COL VET MED, 77-, ASSOC DEAN, CLIN & OUTREACH PROG, 90- *Mem:* Am Vet Med Asn; m Asn Vet Clinicians; Asn Am Vet Med Col; Am Soc Vet Ophthal; Am Col Vet Ophthal. *Mailing Add:* Dept Vet Med Tex A&M Univ College Station TX 77843

PLAZEK, DONALD JOHN, b Milwaukee, Wis, Jan 12, 31; m 55; c 7. MATERIALS SCIENCE ENGINEERING. *Educ:* Univ Wis, BS, 53, PhD, 57. *Prof Exp:* Res assoc phys chem, Univ Wis, 57-58; fel fundamental res, Mellon Inst, 58-67; assoc prof, 67-75, assoc chmn dept, 71-74, PROF MAT SCI & ENG, UNIV PITTSBURGH, 75- *Concurrent Pos:* Sr vis res fel, Univ Glasgow, Scotland, 76-77, Japanese Soc Promotion Sci fel, 88. *Mem:* Am Chem Soc; Soc Rheology; fel Am Phys Soc; NAm Thermal Analysis Soc; Mat Res Soc. *Res:* Rheology; polymer chemistry and physics. *Mailing Add:* Mat Sci & Engr Univ Pittsburgh Pittsburgh PA 15261

PLEASANT, JAMES CARROLL, b Greenville, NC, Jan 9, 36; m 57; c 3. MATHEMATICS, COMPUTER SCIENCE. *Educ:* ECarolina Col, BS, 58, MA, 60; Univ SC, PhD(math), 65. *Prof Exp:* Teacher high sch, NC, 58-60; asst prof math, ECarolina Col, 60-61, assoc prof, 64-66; from asst prof to assoc prof math, 66-85, PROF COMPUT SCI, EAST TENN STATE UNIV, 85- *Mem:* Math Asn Am; Am Math Soc; Asn Comput Mach. *Res:* Ring and module theory; numerical analysis; applied mathematics; theory of computation. *Mailing Add:* 2221 Nantucket Dr Johnson City TN 37601

PLEASANTS, ELSIE W, b Cincinnati, Ohio; m 47; c 3. INORGANIC CHEMISTRY, ORGANIC CHEMISTRY. *Educ:* Univ Cincinnati, ChE, 38; Columbia Univ, MA, 43, PhD(chem), 46. *Prof Exp:* Res worker biochem, Inst Divi Thomae, Cincinnati, 38-50; res worker skin cancer, NY Skin & Cancer Clin, 40-46; res worker enzymes, St Vincent Hosp, NY, 46-50; res scientist, St Anthony's Guild, 50-67; adj asst prof chem & lectr, 67-71, asst prof nutrit & food sci, Hunter Col, City Univ New York, 80-85; asst prof chem, Grad Sch Nursing, NY Med Col, 71-73; adj asst prof chem, Dent Sch, Fairleigh Dickinson Univ, 73-74; from adj asst prof to asst prof, Queensborough Community Col, City Univ New York, 74-76; mem chem fac, Montclair State Col, 76-78; asst prof nutrit, Queens Col, City Univ NY, 78-79; asst prof chem, Seton Hall Univ, 79-80; RETIRED. *Concurrent Pos:* Adj asst prof & researcher, Prog Nutrit & Food Sci, Sch Health Sci, Hunter Col, 85- *Mem:* AAAS; Sigma Xi; NY Acad Sci. *Res:* Skin cancer; antibiotics from soil organisms; human pigmentation; effect of various diets on rat weight; study of dietary variables in cadmium exposure of rats. *Mailing Add:* 27 Linden Terr Leonia NJ 07605

PLEASANTS, JULIAN RANDOLPH, b Palmetto, Fla, May 1, 18; m 48; c 7. ANIMAL NUTRITION, MICROBIAL ECOLOGY. *Educ:* Univ Notre Dame, BS, 39, MA, 50, PhD(microbiol), 66. *Prof Exp:* Asst biochemist, Lobund Lab, Univ Notre Dame 46-53, chief mammal rearing, 53-60, res assoc biochem, 60-66, assoc res scientist, 66-68, asst prof microbiol, 68-73, assoc prof microbiol, Lobund Lab, 73-83; RETIRED. *Concurrent Pos:* Consult, Am Biogenetic Scis Inc, 88- *Mem:* Asn Gnotobiotics. *Res:* Nutrition and physiology of germfree animals fed chemically defined diets from birth. *Mailing Add:* 52631 Gumwood Rd Granger IN 46530

PLEASURE, DAVID, b Dannemora, NY, Aug 23, 38. NEUROLOGY, PEDIATRICS. *Educ:* Yale Univ, BA, 59; Columbia Univ, MD, 63. *Prof Exp:* Intern, Mary Fletcher Hosp, Univ Vt, 63-64; resident neurol, Neurol Inst, NY, 64-66; clin assoc neurol, NIH, Bethesda, Md, 66-68, asst neurologist, 68-69; PROF NEUROL, UNIV PA, 69-, PROF PEDIAT & ORTHOP SURG, 77-; DIR NEUROL RES, 74-, VCHMN NEUROL, UNIV PA, 87- *Concurrent Pos:* Assoc, Muscle & Nerve, 84- & Ann Neurol, 85-; mem, Neurobiol Grad Group, Univ Pa, 85-; chmn, prog comt, Am Acad Neurol, 85-87; dir prog, NIH Neuromuscular Ctr, Univ Pa, 86- *Mem:* Am Neurol Asn; Soc Neurosci; Am Soc Neurochem; Am Acad Neurol. *Res:* Biology of neuroglia; pathogenesis of denyelinative diseases of the central and peripheral nervous system. *Mailing Add:* Children's Hosp Philadelphia PA 19104

PLEBUCH, RICHARD KARL, b Longview, Wash, Sept 7, 35; m 56; c 4. NUCLEAR SCIENCE, NUCLEAR ENGINEERING. *Educ:* Univ Wash, BS, 57; Univ Mich, MS, 58; Mass Inst Technol, ScD(nuclear eng), 63. *Prof Exp:* Proj officer, US Army Chem Corps Bd, 59-60; sect head nuclear systs anal, TRW Defense Systs Group, 63-68, mgr nuclear anal dept, 68-70, mgr nuclear survivability dept, 70-83, mgr hardness & survivability lab, 83-89, PROJ MGR, TRW SYST INTEGRATION GROUP, REDONDO BEACH, 89- *Mem:* Am Nuclear Soc; Am Inst Aeronaut & Astronaut. *Res:* Nuclear rocket engine system analysis; nuclear gauging systems; weapon environments and effects; weapon system survivability; nuclear safeguards; security systems analysis. *Mailing Add:* 3602 W Estates Lane No 121 Rancho Palos Verdes CA 90274

PLEDGER, RICHARD ALFRED, b Mineola, NY, July 3, 32; m 55; c 2. MICROBIOLOGY. *Educ:* Rensselaer Polytech Inst, BS, 54; Rutgers Univ, PhD(microbiol), 57. *Prof Exp:* Res biochemist, Plum Island Animal Dis Lab, USDA, 57-58, sr res biochemist, 58-63, prin res biochemist, 63-64; virologist, Hazleton Labs, Va, 64-66, head microbiol-virol sect, 66-69, dir biosci dept, 69-71; scientist-adminr, Nat Cancer Inst, 71-77; HEALTH SCIENTIST-ADMINR, NAT INST ARTHRITIS, DIABETES, DIGESTIVE & KIDNEY DIS, 77- *Res:* Mechanism of growth and reproduction of microbes, especially viral infection and replication within living tissue; chemical carcinogenesis; animal model development of major forms of human cancer. *Mailing Add:* NIH Div Extramural Activ 5333 Westbard Ave Bethesda MD 20892

PLEIN, ELMER MICHAEL, b Dubuque, Iowa, Nov 21, 06; m 32, 52. PHARMACY. *Educ:* Univ Colo, BS & PhC, 29, MS, 31, PhD(chem), 36. *Prof Exp:* Instr pharm, Univ Colo, 29-38; from instr to prof pharm, Univ Wash, 38-77, coordr pharmaceut servs & dir drug serv dept, Hosp Pharm Educ, 58-77; RETIRED. *Mem:* Am Chem Soc; Am Soc Hosp Pharmacists; Am Pharmaceut Asn; Am Asn Col Pharm. *Res:* Therapeutic efficiency of drug products; clinical and hospital pharmacy; pharmacokinetics and pharmacodynamics in the geriatric patient. *Mailing Add:* Col Pharm Univ Wash Seattle WA 98195

PLEIN, JOY BICKMORE, b Nov 10, 25; US citizen; m 52. PHARMACY. *Educ:* Idaho State Univ, BSPharm, 47; Univ Wash, MSPharm, 52, PhD(pharm), 56. *Prof Exp:* Pharmacist, Heinz Apothecary, Salt Lake City, Utah, 47-49; lectr pharm & pharmacol, Seattle Pac Col, 54-59 & 61-65; asst to chmn comt pharm & pharmaceut, Am Soc Hosp Pharmacists, 59-62, consult, Dept Sci Serv, 62-63, asst ed, Am Hosp Formulary Serv, 64-66; lectr, 66-72, assoc prof, 72-74, PROF PHARM, SCH PHARM, UNIV WASH, 74- *Concurrent Pos:* Mem continuing educ subcomt, Wash-Alaska Regional Med Prog, 69-71; mem ref panel, Am Hosp Formulary Serv, 69-76 & OTC Rev Panel Dentifrices & Dent Care Agents, Food & Drug Admin, 73-78; coordr, Interdisciplinary Nursing Home Prog, Univ Washington, Foss Home, Seattle. *Mem:* Am Soc Hosp Pharm; Am Pharmaceut Asn; Am Asn Cols Pharm; Geront Soc Am. *Res:* Clinical pharmacy; educational methods and materials; effects of aging on drug therapy; long term care. *Mailing Add:* Sch Pharm Univ Wash Seattle WA 98195

PLEMMONS, ROBERT JAMES, b Old Fort, NC, Dec 18, 39; m 63; c 2. MATHEMATICS, COMPUTER SCIENCES. *Educ:* Wake Forest Univ, BS, 61; Auburn Univ, MS, 62, PhD(math), 65. *Prof Exp:* Res mathematician, Nat Security Agency, 65-66; assoc prof math, Univ Miss, 66-67; assoc prof math & comput sci, Univ Tenn, Knoxville, 67-74, prof, 67-81; PROF COMPUT SCI & MATH, NC STATE UNIV, RALEIGH, 81- *Concurrent Pos:* Consult, Oak Ridge Nat Lab. *Mem:* Soc Indust & Appl Math; Asn Comput Mach; Math Asn Am; Am Math Soc. *Res:* Numerical analysis; computer science. *Mailing Add:* Dept Math & Comput Sci Wake Forest Univ Winston-Salem NC 27109

PLEMONS, TERRY D, b Seminole, Okla, Feb 19, 41; m 70; c 3. ACOUSTICS, SYSTEMS ANALYSIS. *Educ:* Univ Tex, Austin, BS, 63, PhD(physics), 71. *Prof Exp:* Peace Corps vol teacher, Am Peace Corps, India, 64-66; res scientist, underwater acoust, Appl Res Lab, Univ Tex, Austin, 66-75; physicist, Appl Physics Lab & lectr elec eng, Univ Wash, Seattle, 75-78; ENGR SCIENTIST, TRACOR, INC, 78- *Concurrent Pos:* Mem, US Naval Inst. *Mem:* Sigma Xi; Acoust Soc Am; Int Test & Eval Asn. *Res:* Underwater acoustics; signal processing; systems analysis; sonar development. *Mailing Add:* Tracor Inc 6500 Tracor Lane Austin TX 78725

PLENDL, HANS SIEGFRIED, b Berlin, Germany, June 12, 27; nat US; m 57; c 4. MEDIUM-ENERGY & APPLIED NUCLEAR PHYSICS. *Educ:* Harvard Univ, BA, 52; Yale Univ, MS, 54, PhD(physics), 58. *Prof Exp:* Asst physics, Yale Univ, 52-54, asst nuclear physics, 54-56; from asst prof to assoc prof, 56-77, PROF PHYSICS, FLA STATE UNIV, 77- *Concurrent Pos:* Res partic, Oak Ridge Nat Lab, 57-68; vis scientist, Nuclear Res Ctr, Karlsruhe & Max Planck Inst Nuclear Physics, Heidelberg, 62-63; consult, Sch X-ray Technol, Tallahassee Mem Hosp, Fla, 64-72; mem bd dirs, Recon, Inc, 65-68; mem users adv comt, Space Radiation Effects Lab, Va, 69-78; vis staff mem, Los Alamos Nat Lab, 69-; consult, LCE Ltd, Worthington, Minn & CTI, Tallahassee, Fla, 80-86; guest sr physicist, Brookhaven Nat Lab, 82-; guest prof, Tech Univ, Munich, Germany, 85-86; guest physicist, Nuclear Res Ctr, Juelich, Ger, 90-; Fulbright travel award, 63. *Mem:* Am Phys Soc; Europ Phys Soc; Sigma Xi. *Res:* Nuclear structure; reactions of hadrons with nuclei; applied nuclear physics. *Mailing Add:* Dept of Physics Fla State Univ Tallahassee FL 32306-3016

PLESCIA, OTTO JOHN, b Utica, NY, Apr, 12, 21; m 49; c 4. IMMUNOCHEMISTRY, IMMUNOPHARMACOLOGY. *Educ:* Hamilton Col BS, 43; Cornell Univ, PhD(chem), 47. *Prof Exp:* Instr math, Hamilton Col, 43-44; res chemist, M W Kellogg Co, 47-49; res assoc chem, Univ Wis, 50-51; res assoc immunochem, Col Physicians & Surgeons, Columbia Univ, 51-55; assoc prof, 55-60, PROF IMMMUNOCHEM, WAKSMAN INST MICROBIOL, RUTGERS STATE UNIV NJ, 60- *Concurrent Pos:* Lectr, Brooklyn Col, 49; de pont fel, Univ Wis, 49-50; assoc mem, Comn Immunization, Dept Defense, 57-62; vis prof, Univ Puerto Rico, 65-68; vis investr, Salk Inst, 76-77. *Mem:* Am Asn Immunologists; AAAS; Am Soc Microbiol; Am Chem Soc. *Res:* Role of prostaglandins in the physiology and pathophysiology of immmune responses; host-tumor interactions; pathogenesis and regulation of antoimmune diseases. *Mailing Add:* Waksman Inst Microbiol Rutgers State Univ NJ PO Box 759 Piscataway NJ 08854

PLESS, IRWIN ABRAHAM, b New York, NY, Mar 11, 25; div; c 3. PHYSICS. *Educ:* Univ Chicago, SB, 50, SM, 51 & 53, PhD(physics), 55. *Prof Exp:* From instr to assoc prof, 56-64, PROF PHYSICS, MASS INST TECHNOL, 64- *Mem:* Sigma Xi; Am Phys Soc; Am Asn Physics Teachers. *Res:* Elementary particle physics. *Mailing Add:* 290 Massachusetts Ave Cambridge MA 02139-4307

PLESS, VERA STEPEN, b Chicago, Ill, Mar 5, 31; c 3. MATHEMATICS. *Educ:* Univ Chicago, PhB, 49, MS, 52; Northwestern Univ, PhD(math), 57. *Prof Exp:* Lectr, Boston Univ, 57-61; res mathematician, Air Force Cambridge Res Labs, 61-66; mathematician, Argonne Nat Lab, 67; res mathematician, Air Force Cambridge Res Labs, 68-72; res assoc elec eng, Mass Inst Technol, 72-75; PROF MATH & COMPUT SCI, UNIV ILL, CHICAGO CIRCLE, 75- *Concurrent Pos:* Ed, J Combinatorial Theory Ser A, 82-; vis prof, Cal Inst Technol, 85-86, Univ Wis, Madison, 87 & Inst Math & Applns, Univ Minn, 88; bd gov info theory group, Inst Electronics & Elec Eng, 85-88. *Honors & Awards:* Marcus O'Day Award, Air Force Cambridge Res Lab, 69; Patricia Keyes Glass Award, US Air Force, 71. *Mem:* Am Math Soc; Math Asn Am; Asn Women Math. *Res:* Error correcting codes; combinatorial analysis. *Mailing Add:* Dept of Math Col Lib Arts & Sci Univ Ill Chicago Circle Box 4348 Chicago IL 60680

PLESSET, MILTON SPINOZA, b Pittsburgh, Pa, Feb 7, 08; m 34; c 4. PHYSICS. *Educ:* Univ Pittsburgh, BS, 29, MS, 30; Yale Univ, PhD(physics), 32. *Prof Exp:* Nat Res Coun fel, Calif Inst Technol, 32-33, inst fel, 40-42; Nat Res Coun fel, Inst Theoret Physics, Copenhagen Univ, 33-34 & Belgium-Am Educ Found, 34-35; instr physics, Univ Rochester, 35-40; head anal group, Douglas Aircraft Co, Calif, 42-45; theoret physics consult, US Naval Ord Testing Sta, 46-47; assoc prof appl mech, 47-51, prof eng sci, 51-78, EMER PROF ENG SCI, CALIF INST TECHNOL, 78- *Concurrent Pos:* Mem, Sci Adv Bd, US Air Force, 51-53 & Maritime Res Adv Comt, 58-60; mem mech adv panel, Nat Bur Standards, 59-61, chmn exec comt, Fluid Eng Div, 70-72 & Fluid Dynamics Div, 72-74, mem, Adv Comt on Reactor Safeguards, 75-; consult, Rand Corp, 74-76. *Mem:* Nat Acad Eng; fel Am Soc Mech Eng; fel Am Phys Soc; Am Nuclear Soc. *Res:* Theoretical physics; fluid dynamics; two-phase flow and cavitation; radiation effects; theory of two phase flow. *Mailing Add:* Dept of Eng Sci Calif Inst of Technol Pasadena CA 91125

PLESSY, BOAKE LUCIEN, b New Orleans, La, Sept 12, 38; m 59; c 3. PHYSICAL CHEMISTRY, POLYMER CHEMISTRY. *Educ:* Dillard Univ, BA, 59; Adelphi Univ, PhD(chem), 74. *Prof Exp:* Staff chemist, Pfizer Inc, 62-73; asst prof, 73-80, ASSOC PROF CHEM, DILLARD UNIV, 80-, CHMN DIV NATURAL SCI. *Concurrent Pos:* Woodrow Wilson fel, 59. *Mem:* Soc Complex Carbohydrates. *Res:* Hydration of proteoglycans from bovine cornea; polymer dimensions in binary solvent systems. *Mailing Add:* 5741 Congress Dr New Orleans LA 70126

PLETCHER, RICHARD H, b Elkhart, Ind, May 21, 35; m 57; c 3. MECHANICAL ENGINEERING. *Educ:* Purdue Univ, BSME, 57; Cornell Univ, MS, 62, PhD(mech eng), 66. *Prof Exp:* Instr mech eng, Cornell Univ, 61-64; sr res engr, Propulsion Sect, United Aircraft Res Labs, 66-67; assoc prof, 67-76, PROF MECH ENG, IOWA STATE UNIV, 76-, ASSOC MGR, COMPUTATIONAL FLUID DYNAMICS CTR, 84- *Mem:* Am Soc Mech Engrs; Am Inst Aeronaut & Astronaut; fel Am Soc Mech Engr. *Res:* Heat transfer and fluid mechanics of two-phase flows; analysis of turbulent flows; computational fluid mechanics and heat transfer. *Mailing Add:* Dept Mech Eng Iowa State Univ Ames IA 50011

PLETCHER, WAYNE ALBERT, b Crooksville, Ohio, Aug 24, 42; m 68; c 2. ORGANIC & POLYMER CHEMISTRY. *Educ:* Ohio Univ, BS, 66; Univ Mich, MS, 67, PhD(chem), 71. *Prof Exp:* Teaching fel & lectr org chem, Univ Mich, 66-70, fel azide res, 71; fel natural prod res, Fla State Univ, 71; res specialist & proj coordr, CRL, 71-76, supvr Gen Adhesives, Adhesives, Coatings & Sealers Div, 76-78, mgr Res Lab, 78-80, mgr Indust Prod Lab, 80-83, lab mgr, Indust, Aerodyn & Construct Prod Lab, Adhesives, Coatings & Sealers Div, 83-85, TECH DIR, PROF PROD LAB, MEMORY TECH GROUP, 3M CO, 86- *Mem:* Am Chem Soc; Adhesive & Sealant Coun. *Res:* Basic structure-property relationship in polymer chemistry, particularly the investigation of new adhesive systems, pressure sensitive, thermoplastic, reactive liquids and crosslinked and new binder systems for data-video memory storage products involving the synthesis of polymers having unique physical properties. *Mailing Add:* 2050 Loren Rd St Paul MN 55113-5402

PLETSCH, DONALD JAMES, b Lake City, Minn, May 8, 12; div; c 3. ENTOMOLOGY, MALARIOLOGY. *Educ:* Hamline Univ, BS, 32; Univ Minn, MS, 36, PhD(entom), 42. *Prof Exp:* Asst exp sta, Univ Minn, 33-37; asst state entomologist, Mont, 37; from asst prof to assoc prof entom, Mont State Col, 37-47, from asst entomologist to assoc entomologist, Exp Sta, 37-47; tech instr, Med Admin Corps, Officers Candidate Sch, 43-44; cmndg officer, Malaria Surv Unit, US Army & Sanitary Corps, 44-46; biologist, Sci & Tech Div, Econ & Sci Sect, Supreme Comdr Allied Powers, Japan, 47-50; biologist, Div Int Health, USPHS, 50-52; leader malaria team, WHO, Formosa, 52-55 & Pan Am Sanit Bur, Mex, 56-63; entomologist, AID, Addis Ababa, 64-66; dir, Cent Am Malaria Res Sta, USPHS, El Salvador, 67-69; MALARIA CONSULT, AID, 70- *Concurrent Pos:* Med entom consult, Dept Health, Govt Indonesia, 77-78; Consult, Pan-Am Health Orgn & WHO, Guyana, SAm, 81. *Honors & Awards:* Haile Selassie I Gold Medal, Ethiopia, 66. *Mem:* Am Soc Trop Med & Hyg; Sigma Xi; Am Pub Health Asn; Soc Vector Ecol; Am Mosquito Control Asn. *Res:* Insect ecology; medical entomology; research administration. *Mailing Add:* 8620-238 NW 13th St Gainesville FL 32606

PLETT, EDELBERT GREGORY, b Bassano, Alta, Jan 17, 39; m 66; c 2. MECHANICAL ENGINEERING. *Educ:* Univ BC, BASc, 62; Mass Inst Technol, SM, 64, ScD(mech eng), 66. *Prof Exp:* Asst prof eng, Carleton Univ, 66-70, assoc prof, 70; mem res staff, Princeton Univ, 70-76, res engr, 76; assoc prof, 76-79, PROF, CARLETON UNIV, 79- *Concurrent Pos:* Consult, Comput Devices of Can, 69-70, Dashwood Indust Ltd & Energy North Int Ltd, 78-; pres, Asecor Ltd, 78- *Mem:* Combustion Inst; assoc fel Am Inst Aeronaut & Astronaut; Am Soc Heating Refrig & Air Conditioning Engrs. *Res:* Combustion noise; coal gasification; coal gas combustion; solar energy; combustion; thermal comfort. *Mailing Add:* Dept Mech Eng Carleton Univ Ottawa ON K1S 5B6 Can

PLETTA, DAN HENRY, b South Bend, Ind, Dec 31, 03; m 31; c 2. MECHANICS. *Educ:* Univ Ill, BS, 27, CE, 38; Univ Wis, MS, 31. *Prof Exp:* Instr mech, Univ Wis, 27-30; asst prof civil eng, Univ SDak, 30-32; from asst prof to prof appl mech, 32-42, prof & head dept eng mech, 46-70, univ distinguished prof, 69-72, EMER UNIV DISTINGUISHED PROF ENG MECH, VA POLYTECH INST, 72- *Concurrent Pos:* Pvt engr, 22-30; instr & asst dir mech, US Mil Acad, 44-46; mem bd dirs, Va Polytech Inst, 48-53; consult var govt, pvt & acad insts, 48-79. *Honors & Awards:* Western Elec Fund Award Excellence in Teaching, Am Soc Eng Educ, 68. *Mem:* Fel Am Soc Eng Educ; hon mem Am Soc Civil Engrs; Nat Soc Prof Engrs; Am Concrete Inst; fel Am Acad Mech; Sigma Xi; Soc Exp Mech. *Res:* Engineering materials; stress analysis; structural mechanics; engineering professionalism and ethics. *Mailing Add:* 15 Dogwood Circle Blacksburg VA 24060-6234

PLEWA, MICHAEL JACOB, b Chicago, Ill, June 5, 47; m 78. GENETIC TOXICOLOGY. *Educ:* Loyola Univ, Chicago, BS, 69; Ill State Univ, MS, 71, PhD(genetics), 74. *Prof Exp:* Res assoc, Univ Ill, 74-76, asst res geneticist, 76-78, from asst prof to assoc prof, 78-86, PROF GENETICS, UNIV ILL, 86- *Concurrent Pos:* Prin investr, US Environ Protection Agency, 76-82, Nat Inst Environ Health Sci, 78-83 & Joyce Found, 81-83; mem, Gene-Tox Prog, US Environ Protection Agency, 78-80. *Mem:* Environ Mutagen Soc; AAAS; Maize Genetics Coop; Sigma Xi. *Res:* Activiation of promutagens to mutagens by plant systems; comparison of the mutation induction kinetics between somatic and germinal cells after acute or chronic mutagen exposure regimens. *Mailing Add:* Environ Res Lab Inst Environ Studies Univ Ill 1005 W Western Ave Urbana IL 61801

PLIMMER, JACK REYNOLDS, b Liverpool, Eng, July 18, 27; m 49; c 2. ORGANIC CHEMISTRY. *Educ:* Trinity Col, Dublin, BA, 52; Univ Edinburgh, PhD(org chem), 55. *Prof Exp:* Asst chem, Univ Edinburgh, 52-55; Eli Lilly res fel, Univ WIndies, 55-56, res fel natural prod chem, 59-62, lectr, 62-64; scientist, Med Res Coun, Univ Exeter, 56-59; vis scientist, NIH, Md, 64-66; res chemist, Plant Sci Res Div, 66-75, CHIEF ORG CHEM SYNTHESIS LAB, AGR ENVIRON QUAL INST, AGR RES SERV, USDA, 75- *Concurrent Pos:* Hon res asst, Univ Col, London, 62. *Mem:* Am Chem Soc; Royal Soc Chem; Asn Off Anal Chemists; Sigma Xi. *Res:* Chemistry and synthesis of natural products; photochemistry of pesticides; pesticide degradation in soils; identification, synthesis and application of chemicals affecting insect behavior. *Mailing Add:* PO Box 25 Clarksville MD 21029

PLIMMER, JACK REYNOLDS, b Liverpool, Eng, July 18, 27; US citizen; m 49; c 2. PESTICIDE ANALYSIS, ENVIRONMENTAL FATE OF PESTICIDES. *Educ:* Univ Dublin, Ireland, BA, 52; Univ Edinburgh, Scotland, PhD(org chem), 55. *Prof Exp:* Lectr phys chem, Univ WI, Kingston, Jamaica, 59-64; vis scientist, NIH, Bethesda, Md, 64-66; res chemist, Pesticide Degradation Lab, Agr Res Serv, USDA, Beltsville, Md, 66-74, lab chief, Org Chem Synthesis Lab, 74-83, res leader, Environ Chem Lab, 86-90; sect head, Food & Agr Orgn & Int Atomic Energy Agency, Vienna, Austria, 83-86; SR MGR, ABC LABS, COLUMBIA, MO, 91- *Concurrent Pos:* Corresp Collab Int Pesticides Anal Coun, 77-; chmn, Div Pesticide Chem, Am Chem Soc, 80; vchmn, Plant Chem Working Group, Int Orgn Chem Develop, 88-; ed, Pesticide Sci J, 91- *Honors & Awards:* Burdick & Jackson Int Award in Pesticide Chem, Am Chem Soc, 83. *Mem:* Am Chem Soc; Royal Chem Soc; Am Asn Off Analytical Chemists; AAAS. *Res:* Environmental fate of pesticides, their transport, photochemistry, microbial transformations, analysis; lepidopteran pheromones; organic chemistry of natural products. *Mailing Add:* ABC Labs PO Box 1097 Columbia MO 65205

PLIMPTON, RODNEY F, JR, b Poughkeepsie, NY, May 16, 37; m 58; c 4. ANIMAL SCIENCE, MEAT SCIENCE. *Educ:* Rutgers Univ, BS, 59; Ohio State Univ, MS, 61, PhD(animal sci), 65. *Prof Exp:* From instr to assoc prof, 65-74, PROF ANIMAL SCI, OHIO STATE UNIV, 74- *Honors & Awards:* Nat Independent Meat Packers Asn-Wilbur LaRoe Mem Award Meat Res, 63. *Mem:* Am Soc Animal Sci; Am Meat Sci Asn; Inst Food Technol. *Res:* Hormone effect on carcass development and evaluation in swine; protein and fat studies in sausage processing; mechanical muscle treatment in cured meat processing. *Mailing Add:* Dept Animal Sci Ohio State Univ Columbus OH 43210

PLINT, COLIN ARNOLD, b Liverpool, Eng, Jan 16, 26; m 47; c 4. PHYSICS. *Educ:* Oxford Univ, BA, 50; Univ Toronto, PhD(physics), 53. *Prof Exp:* From asst prof to prof, Univ Okla, 53-68, chmn dept, 59-63; prof physics & chmn dept, 68-69, dean fac arts & sci, 69-75, dean div math & sci, 75-76, PROF PHYSICS, BROCK UNIV, 68- *Mem:* Can Asn Physicists. *Res:* Solid state physics--Raman and Brillouin spectroscopy; thermal physics. *Mailing Add:* Brock Univ Brock Univ Merrittville Hwy St Catharines ON L2S 3A1 Can

PLISETSKAYA, ERIKA MICHAEL, b Leningrad, USSR, Dec 8, 29; US citizen. COMPARATIVE ENDOCRINOLOGY. *Educ:* State Univ Leningrad, BS, 52, PhD(physiol), 58; Inst Physiol, Acad Sci USSR, DSc, 72. *Prof Exp:* Researcher physiol, Sechenov Inst Evolutionary Physiol & Biochem, Acad Sci USSR, Leningrad, 57-67, sr researcher, Comp Endocrinol, 67-74, head lab, 74-79; res scientist, Dept Zool, 80-89, RES SCIENTIST COMP ENDOCRINOL, SCH FISHERIES, UNIV WASH, 89- *Concurrent Pos:* Consult, Sechenov Inst Evolutionary Physiol & Biochem, Acad Sci USSR, Leningrad, 65-79, head, grad dept, 67-72, lectr, 67-79, ed, 75-76. *Mem:* Am Soc Zoologists; fel AAAS; Europ Soc Comp Endocrinologists; Am Fisheries Soc. *Res:* Structure and role of pancreatic hormones in regulation of metabolism in lower vertebrates (cyclostomes and fishes) and in some invertebrates (mollusks); endocrinology. *Mailing Add:* 2310 NE 48th No 734 Seattle WA 98105-4138

PLISKIN, WILLIAM AARON, b Akron, Ohio, Aug 9, 20; m 44; c 3. SEMICONDUCTORS, CHEMICAL PHYSICS. *Educ:* Kent State Univ, BSEd, 41; Univ Ohio, MS, 43; Ohio State Univ, PhD(physics), 49. *Prof Exp:* Asst, Ohio State Univ, 47-48; res physicist, Texaco Res Ctr, Texaco, Inc, 49-59; staff physicist, IBM Corp, 59-60, adv physicist, 60-63, mgr spec studies oxides & insulators dept, 64-70, sr physicist, 63-79, mgr thin film insulators, IBM Components, Hopewell Junction, 70-79, sr tech staff mem, 79-90; CONSULT, 90- *Honors & Awards:* Corp Invention Award, IBM Corp, 66; Electronics Div Award, Electrochem Soc, 73; Corp Contribution Award, IBM Corp, 79; Annual Award, Am Chem Soc, 64. *Mem:* Am Phys Soc; Am Chem Soc; Electrochem Soc; fel Inst Elec & Electronic Engrs; AAAS. *Res:* Infrared spectroscopy; chemisorption; surface physics; semiconductor surfaces; passivation; oxidation; thin film measurements; thin insulating films; semiconductor device processing. *Mailing Add:* 31 Greenvale Farms Rd Poughkeepsie NY 12603

PLOCK, RICHARD JAMES, b Mineola, NY, Apr 22, 31. CHEMICAL PHYSICS, COMPUTATIONAL PHYSICS. *Educ:* Polytech Inst Brooklyn, BS, 52; Yale Univ, MS, 54, PhD(chem), 57. *Prof Exp:* Res physicist, Lawrence Radiation Lab, 56-62 & Goddard Space Flight Ctr, NASA, Md, 62-64; asst prof physics, Univ Redlands, 64-68; consult theoret chem, Lockheed Propulsion Co, 66-78; consult specialist, Systs Consults, Inc, 79-81; scientist, Sci Applns Inc, 81-90; CONSULT, 90- *Mem:* AAAS; Am Phys Soc; Am Inst Chem; Am Chem Soc. *Res:* Computational chemical physics. *Mailing Add:* PO Box 26876 San Diego CA 92196-0876

PLOCKE, DONALD J, b Ansonia, Conn, May 5, 29. BIOPHYSICAL CHEMISTRY. *Educ:* Yale Univ, BS, 50; Boston Col, MA, 56; Mass Inst Technol, PhD(biophys), 61. *Prof Exp:* Res fel biol chem, Harvard Univ, 64; asst prof, 66-71, chmn dept, 71-80, ASSOC PROF BIOL, BOSTON COL, 71- *Concurrent Pos:* Lectr biol chem, Harvard Med Sch, 74. *Mem:* Sigma Xi; Biophys Soc; AAAS. *Res:* Role of metal ions in cell cycle and cell surface interactions; metalloenzymes; spectroscopy; chemistry of macromolecules. *Mailing Add:* Dept of Biol Boston Col Chestnut Hill MA 02167

PLODINEC, MATTHEW JOHN, b Kansas City, Mo, Mar 29, 46; m 68; c 2. PROCESS CONTROL. *Educ:* Franklin & Marshall Col, BA, 68; Univ Fla, PhD(phys chem), 74. *Prof Exp:* Res chemist photochem, E I du Pont de Nemours & Co, Inc, Wilmington, Del, 74-75, res chemist glass & cement, 75-80, res supvr, 80-86, res assoc, Glass Technol Group, Aiken, SC, 86-89; SR ADV SCIENTIST, WESTINGHOUSE SAVANNAH RIVER CO, AIKEN, SC, 89- *Mem:* Am Chem Soc; Am Ceramic Soc; Mat Res Soc. *Res:* Encapsulation of nuclear waste in durable form; study of interactions of different waste components with encapsulating material. *Mailing Add:* Savannah River Lab Westinghouse Savannah River Co Aiken SC 29802

PLONA, THOMAS JOSEPH, b Hartford, Conn, Sept 5, 48; m 69; c 2. ACOUSTICS OF POROUS MEDIA, SONIC WELL LOGGING. *Educ:* Providence Col, BS, 70; Georgetown Univ, MS, 73, PhD(physics), 75. *Prof Exp:* SR RES SCIENTIST, SCHLUMBERGER-DOLL RES, 76- *Mem:* Acoust Soc Am; Am Geophys Union; Inst Elec & Electronic Engrs. *Res:* Acoustics, ultrasonics and fluid flow in sedimentary rocks, specializing in novel experimental techniques; porous media; tool development for oil well logging. *Mailing Add:* 32 Maplewood Dr New Milford CT 06776

PLONSEY, ROBERT, b New York, NY, July 17, 24; m 48; c 1. BIOMEDICAL ENGINEERING. *Educ:* Cooper Union, BEE, 43; NY Univ, MEE, 48; Univ Calif, PhD(elec eng), 56. *Prof Exp:* Asst antenna lab, Univ Calif, 53-55, actg asst prof elec eng, 55-57; prof, Case Western Reserve Univ, 57-68, chmn biomed engr, 76-80, dir bioeng group, 62-68, prof biomed eng, 68-83, dir biomed eng training prog, 70-83; PROF BIOMED ENG, DUKE UNIV, 83- & HUDSON PROF ENG, 90- *Concurrent Pos:* Mem biomed eng fel rev comt, NIH, 66-70, eng in med & biol training comt, 71-73; mem, Alliance Eng Med & Biol Coun, 72-74, treas, 76-78; from vpres to pres, Group Eng in Med & Biol, Inst Elec & Electronics Engrs, 71-74, vpres, 91-, chmn ad hoc comt cert, 75- 77; pres, Am Bd Clin Eng, 75, mem comn trustees, 75-; vis prof & biomed engr, Duke Univ, 80-81; sr fel, NIH, 80-81; fel comt, Inst Elec & Electronic Engrs, 82-85. *Honors & Awards:* Morlock Award, Inst Elec & Electronics Engrs, 79; Centennial Medal, Inst Elec & Electronic Engrs, 84; Nat Acad Eng, 86; Alza Distinguished Lectr, Biomed Eng Soc, 88. *Mem:* Fel AAAS; fel Inst Elec & Electronics Engrs; Am Soc Eng Educ (pres, Biomed Eng Div, 82-83); Biomed Eng Soc (from pres-elect to pres, 80-82). *Res:* Electrophysiology with special emphasis on electrocardiography and application of electromagnetic field theory to volume conductor problems. *Mailing Add:* Dept Biomed Eng Duke Univ-136 Eng Durham NC 27706

PLONSKER, LARRY, b Brooklyn, NY, Aug 20, 34; m 62; c 1. ORGANIC CHEMISTRY, PETROLEUM PRODUCTS. *Educ:* City Col NY, BS, 56; Fla State Univ, PhD(org chem), 60. *Prof Exp:* Res chemist, 60-67, res supvr, 67-74, com develop assoc, Baton Rouge, 74-76, prod mgr, 76-77, dir com develop, Europe, 77-81, mgr petrol chem com develop, 81-84, mgr spec studies com develop, 84-88, RES & DEVELOP ADV, ETHYL CORP, 88- *Mem:* Am Chem Soc; Technol Transfer Soc. *Res:* Small ring compounds; organo-metallics; phosphorus chemistry; agricultural chemicals; aromatic amines and phenols; fuels and lubricants. *Mailing Add:* 7600 GSRI Rd Ethyl Corp Baton Rouge LA 70820

PLONSKY, ANDREW WALTER, b Pa, Dec 19, 12; m 39; c 3. ELECTRICITY, PHYSICS. *Educ:* Univ Scranton, BS, 40; Mass Inst Technol, BS, 43, MS, 45. *Prof Exp:* Instr, Mass Inst Technol, 43-45; asst engr, New Haven RR, 45-47; PROF ENG, UNIV SCRANTON, 47- *Mem:* Math Asn Am; Inst Elec & Electronics Engrs; Asn Comput Mach; Nat Soc Prof Engrs. *Res:* Electric fields and circuits; electromagnetic fields; circuits and systems analysis; electrical controls and signal amplification; software systems; engineering mechanics. *Mailing Add:* Dept Comput Sci Univ Scranton Scranton PA 18510

PLONUS, MARTIN, b Trumpininken, Lithuania, Dec 21, 33; US citizen; m 65; c 4. ELECTRICAL ENGINEERING. *Educ:* Univ Ill, BS, 56, MS, 57; Univ Mich, PhD(elec eng), 61. *Prof Exp:* From asst prof to assoc prof elec eng, 61-69, PROF ELEC ENG, NORTHWESTERN UNIV, 69-, DIR, INTEGRATED MICROELECTRONIC SYSTS. *Concurrent Pos:* Mem bd dirs, Nat Electronics Conf, 63-; US Air Force contrOff Sci Res contract, 64-65; NSF grant, 66-67 & 75-76; assoc ed, Transactions on Antennas & Propagation, 68-72; mem US comn VI, Int Union Radio Sci; dir, Grad Prog, EECS, Northwestern Univ. *Honors & Awards:* Best Paper Award, Transactions on Antennas & Propagation, 71. *Mem:* Fel Inst Elec & Electronics Engrs. *Res:* Electromagnetic theory; scattering and diffraction; antennas and propagation; propagation through turbulent media. *Mailing Add:* Dept Elec Eng & Comput Sci Northwestern Univ 2145 Sheridan Rd Rm 2696 Evanston IL 60208

PLOPPER, CHARLES GEORGE, b Berkeley, Calif, June 16, 44; m 69. RESPIRATORY SYSTEM, PULMONARY TOXICOLOGY. *Educ:* Univ Calif-Davis, AB, 67, PhD(anat), 72. *Prof Exp:* Chief electron micros, US Army Med Res & Nutrit Lab, 72-73; chief electron micros, Letterman Army Inst Res, 74-75; asst prof anat, Sch Med, Univ Hawaii, 75-77; assoc prof cell biol, Sch Med, Univ Kuwait, 77-78; sr staff fel cell biol, Nat Inst Environ Health Sci, 78-79; from asst prof to assoc prof, 79-86, chair dept, 84-88, PROF ANAT, SCH VET MED, UNIV CALIF, DAVIS, 86- *Concurrent Pos:* Vis scientist, Calif Primate Res Ctr, 74-75, Nat Cancer Inst, 86-87; mem study sect, NIH, 86-90; mem res subomt, Am Lung Asn, Calif, 86-92; vis prof, Boston Univ Sch Med, 85 & Kans State Univ Sch Vet Med, 88. *Mem:* Am Soc Cell Biol; AAAS; Am Thoracic Soc; Am Asn Anatomists; Am Asn Pathologists; Histochem Soc. *Res:* air pollution. *Mailing Add:* Sch Vet Med Dept Anat Univ Calif Davis CA 95616

PLOTH, DAVID W, b Carroll, Iowa, June 22, 41. INTERNAL MEDICINE, RENAL PHYSIOLOGY. *Educ:* Univ Iowa, MD, 67. *Prof Exp:* From assoc prof to prof med, Sch Med, Univ Ala, 81-87; DIR, DIV NEPHROLOGY MUSG, 87- *Mem:* Am Col Physicians; Am Physiol Soc. *Mailing Add:* Dept Med Med Univ SC 171 Ashley Ave Charleston SC 29425-2220

PLOTKA, EDWARD DENNIS, b Utica, NY, Oct 10, 38; m 66. ENDOCRINOLOGY, PHYSIOLOGY. *Educ:* Del Valley Col, BS, 60; Ore State Univ, MS, 63; Purdue Univ, PhD(physiol), 66. *Prof Exp:* Asst prof, Purdue Univ, 66-67; asst res prof, Univ Ga, 67-69; RES SCIENTIST, MARSHFIELD MED FOUND, 69- *Mem:* Endocrine Soc; Am Soc Animal Sci; Am Fertil Soc; Soc Exp Biol & Med; Am Physiol Soc; Soc Study Reproduction; Soc Conserv Biol; NY Acad Sci; AAAS. *Res:* Biochemical and physiological studies in the metabolism of reproductive hormones with special reference to steroids. *Mailing Add:* Marshfield Med Found 1000 N Oak Ave Marshfield WI 54449

PLOTKE, FREDERICK, b Kattowitz, Ger, Nov 19, 09; nat US; m 39; c 1. MEDICINE. *Educ:* Univ Frankfurt, MD, 35; Vanderbilt Univ, MPH, 42; Am Bd Prev Med, dipl, 49. *Prof Exp:* Pub health physician, State Dept Pub Welfare, Ill, 39-43; venereal dis control officer, Chicago Bd Health, 47-50; chief, USPHS, State Dept Pub Welfare, Ill, 51-65; asst clin prof, 55-58, ASSOC CLIN PROF PREV MED, STRITCH SCH MED, LOYOLA UNIV CHICAGO, 58-; MED REP COMMUN DIS CTR, USPHS, STATE DEPT PUB WELFARE, ILL, 65- *Concurrent Pos:* Consult, State Civil Defense Agency, Ill, 55-; lectr, Sch Hosp Admin, Northwestern Univ, 58- *Mem:* AMA; fel Am Pub Health Asn; Asn Teachers Prev Med; Am Col Prev Med. *Res:* Communicable and venereal disease control; epidemiology and mental health. *Mailing Add:* 777 N Michigan Ave Chicago IL 60611

PLOTKIN, ALLEN, b New York, NY, May 4, 42; m 66; c 2. AERODYNAMICS, INCOMPRESSIBLE FLUID MECHANICS. *Educ:* Columbia Univ, BS, 63, MS, 64; Stanford Univ, PhD(eng mech), 68. *Prof Exp:* From asst prof to prof aerospace eng, Univ Md, 68-85; chmn, 85-90, PROF AEROSPACE ENG, SAN DIEGO STATE UNIV, 85- *Concurrent Pos:* Vis assoc eng sci, Caltech, 75-76; assoc ed, Am Inst Aeronaut & Astronaut J, 86-91. *Honors & Awards:* Young Engr-Scientist Award, Am Inst Aeronaut & Astronaut, Nat Capital Sect, 76. *Mem:* Assoc fel Am Inst Aeronaut & Astronaut; Am Soc Mech Engrs; Soc Naval Architects & Marine Engrs; Am Soc Eng Educ. *Res:* Application of analytical and computational techniques to the study of the low speed aerodynamics of airfoils and wings and to laminar viscous flows with separation. *Mailing Add:* 17364 St Andrews Dr Poway CA 92064

PLOTKIN, EUGENE ISAAK, b Leningrad, USSR, May 8, 32; m 56; c 2. INFORMATION SCIENCE & SYSTEMS. *Educ:* Inst Commun Eng, USSR, BSc & MSc, 55, PhD(signal processing), 63. *Prof Exp:* Assoc prof commun, Inst Commun Eng, 63-73; assoc prof commun networks, Ben Gurion Univ, Israel, 73-80; prof commun signal processing, Pa State Univ, 80-82; prof, Ben-Gurion Univ, Israel, 82-85; PROF SIGNAL PROCESSING, CONCORDIA UNIV, MONTREAL, CAN, 86- *Concurrent Pos:* Assoc prof commun networks, Tel-Aviv Univ, 78-80. *Mem:* NY Acad Sci; Asn Engrs & Architects; sr mem Inst Elec & Electronic Engrs. *Res:* Problems of signal detection and parameter estimation; adaptive and non-linear filtering for suppression of non-gaussian interferences; phase space approach to speech signal recognition; signal processing. *Mailing Add:* Dept Elec Eng Concordia Univ 1455 De Maisonneuve Blvd W Montreal PQ H3G 1M8 Can

PLOTKIN, HENRY H, b New York, NY, 1927; m 53; c 2. SPACE INSTRUMENTS FOR EARTH OBSERVATIONS, ROBOTICS IN SPACE ASSEMBLY & MAINTENANCE. *Educ:* City Col NY, BS, 46; NY Univ, MS, 48; Mass Inst Technol, PhD(physics), 57. *Prof Exp:* Physicist, Westinghouse Res Labs, 48-50; asst br head, US Army Signal Corps Res & Develop Lab, 54-60; br head, 60-73, div chief, 74-81, ASST DIR ENG, GODDARD SPACE FLIGHT CTR, NASA, 81- *Mem:* Am Phys Soc. *Res:* Laser applications; space ranging; optical communications; atomic frequency control and standards; space remote sensing; robotics in space; space instrumentation for astronomy and earth sciences. *Mailing Add:* NASA Goddard Space Flight Ctr Code 700 Greenbelt MD 20771

PLOTKIN, JACOB MANUEL, b Bloomfield, Conn, Mar 10, 41; m 66. MATHEMATICS. *Educ:* Yale Univ, BA, 63; Cornell Univ, PhD(math), 68. *Prof Exp:* assoc prof, 68-80, PROF MATH, MICH STATE UNIV, 80- *Mem:* Am Math Soc; Asn Symbolic Logic; Math Asn. *Res:* Mathematical logic and set theory. *Mailing Add:* Dept Math Mich State Univ East Lansing MI 48824-1027

PLOTKIN, MARTIN, b Brooklyn, NY, July 22, 22; c 1. ELECTRONICS. *Educ:* City Col New York, BEE, 43; Polytech Inst Brooklyn, MEE, 51. *Prof Exp:* Electronic engr, Gunfire Control, Bendix Aviation Corp, 43-44; engr high energy accelerator electronics, Brookhaven Nat Lab, group leader, 56-65, chief elec engr, Accelerator Dept, 65-76, sr engr Isabelle Proj, 77-84; CONSULT, 86- *Concurrent Pos:* Chmn steering comt, Trans Med Imaging , Inst Elec & Electronics Engrs, 81-84 & 86-90. *Mem:* Fel Inst Elec & Electronics Engrs; Nuclear & Plasma Sci Soc (pres, 77-78). *Res:* Radio frequency and electronics for alternating gradient synchrotron; high energy accelerator electronics; ferromagnetic ferrites. *Mailing Add:* 117 Clover Dr Massapequa Park NY 11762

PLOTKIN, STANLEY ALAN, b New York, NY, May 12, 32; m 56, 79; c 2. VIROLOGY, PEDIATRICS. *Educ:* NY Univ, BA, 52; State Univ NY Downstate Med Ctr, MD, 56; Am Bd Pediat, dipl, 65. *Hon Degrees:* MA, Univ Pa, 74. *Prof Exp:* Intern med, Cleveland Metrop Gen Hosp, 56-57; from instr to assoc prof, 59-73, PROF PEDIAT, SCH MED, UNIV PA, 74- *Concurrent Pos:* Joseph P Kennedy Jr Found scholar, 64-; assoc mem, Wistar Inst Anat & Biol, 60-73, prof, 74-; resident, Children's Hosp Philadelphia, 61-63, assoc physician, 65-73, sr physician, 74-, pres med staff, 84-86; registr, Hosp Sick Children, London, Eng, 62-63. *Honors & Awards:* Bruce Award, Am Col Phys. *Mem:* AAAS; Pediat Infectious Dis Soc; Soc Pediat Res; Infectious Dis Soc Am; Am Epidemiol Soc. *Res:* Rubella; poliomyelitis; cytomegalovirus; vaccination against viral diseases; infectious diseases; antivirals. *Mailing Add:* Children's Hosp 34th St & Civic Ctr Blvd Philadelphia PA 19104

PLOTNICK, GARY DAVID, b Baltimore, Md, Nov 23, 41; m 67; c 2. CARDIOLOGY. *Educ:* Johns Hopkins Univ, BA, 62; Univ Md, MD, 66. *Prof Exp:* Resident med, Sch Med, Univ Md, 66-68 & 70-71, chief resident & instr, 71-72, from asst prof to assoc prof med, 74-89; res fel cardiol, Sch Med, Johns Hopkins Univ, 72-74, from instr to asst prof med, 74-88, ASSOC PROF MED, SCH MED, JOHNS HOPKINS UNIV, 88-; PROF MED, SCH MED, UNIV MD, 89- *Concurrent Pos:* Mem, Nat Coop Study Group Unstable Angina, NIH, 72-80 & Coun Clin Cardiol, Am Heart Asn, 78-; dir, Cardiac Graphics Lab, Vet Admin Med Ctr, 74-, Cardiol Educ, 83- & co-dir, Cardiol Res, 83-; distinguished prof, Am Col Angiology, 81; chmn, Cardiovasc Sect, Eastern Sect, Am Fedn Clin Res, 81; asst dean student affairs, Univ Md, 75-, dir echocardiography, Univ Md Hosp, 87- *Mem:* Fel Am Heart Asn; fel Am Col Cardiol; fel Am Col Physicians; Am Fedn Clin Res. *Res:* Ischemic heart disease, particularly unstable angina pectoris, hypertensive heart disease and left ventricular diastolic function. *Mailing Add:* Univ Md Hosp C3S17 22 S Greene St Baltimore MD 21201

PLOTNIKOFF, NICHOLAS PETER, b Aug 13, 27; m 59; c 2. PHARMACOLOGY. *Educ:* Univ Calif, AB, 49, MS, 51; Univ Tex, PhD(pharmacol), 56. *Prof Exp:* Pharmacologist, Merck Inst Therapeut Res, 57-59 & Stanford Res Inst, 59-62; group leader neuropharmacol, Abbott Labs, 62-68, sect head neuropharmacol, 68-76; staff mem, Trans-Neuro, Inc, 76-80; prof pharmacol, Dept Pharmacol, Oral Roberts Med Sch, Tulsa, Okla, 80-87; PROF PHARMACOL, COL PHARM & COL MED, UNIV ILL, 87- *Mem:* AAAS; Am Soc Pharmacol & Exp Therapeut; NY Acad Sci; Sigma Xi. *Res:* Neuropharmacology; psychopharmacology; pharmacodynamics; drug metabolism. *Mailing Add:* Dept MC Col Pharm Univ Ill 833 S Wood St Chicago IL 60612

PLOTZ, CHARLES M, b New York, NY, Dec 6, 21; m; c 3. INTERNAL MEDICINE, RHEUMATOLOGY. *Educ:* Columbia Col, BA, 41; Long Island Col Med, MD, 44; Columbia Univ, ScD(med), 51. *Prof Exp:* Intern internal med, New Haven Hosp, 44-45; resident, Kings County Hosp, 45-46 & Maimonides Hosp, 48-49; chief arthritis clin, Mt Sinai Hosp, 55-65; DIR CONTINUING EDUC, STATE UNIV NY DOWNSTATE MED CTR, 66-, PROF MED, 67-, CHMN DEPT FAMILY PRACT, 71- *Concurrent Pos:* USPHS res fel, Col Physicians & Surgeons, Columbia Univ, 49-50; attend physician, Kings County Hosp Ctr, Brooklyn, Long Island Col Hosp; asst attend physician, Mt Sinai Hosp; asst physician, Columbia-Presby Med Ctr; chief arthritis clin, Kings County Hosp, Brooklyn, 50, Long Island Col Hosp, 65-80 & State Univ Hosp, Brooklyn, 67-80; attend physician, Brooklyn State Hosp; consult physician, Peninsula Gen Hosp; vis consult, Jewish Gen Hosp, Can & Avicenna Hosp & Wazir Akbar Hosp, Afghanistan, 65; consult, Bur Hearings & Appeals, Soc Security Admin, 67- & Bur Dis Prev & Environ Control, Nat Ctr Chronic Dis Control, Diabetes & Arthritis Control Prog, USPHS, 67-; chief sci consult sect immunol, Vet Admin, 72-76; mem bd gov, Arthritis Found, 64-, secy-treas & mem var comts, 66-; mem med adv coun, Iran Found, 65-; chmn comt educ, NY Acad Med, 76- *Honors & Awards:* Distinguished Serv Award, Arthritis Found, 77 & 83. *Mem:* Am Rheumatism Asn (secy-treas, 66-70); Am Acad Family Physicians; Int League Against Rheumatism (treas, 81-89). *Res:* Latex fixation reaction for rheumatoid arthritis. *Mailing Add:* Dept Med State Univ NY Downstate Med Ctr Brooklyn NY 11203

PLOTZ, PAUL HUNTER, b New York, NY, Oct 19, 37; m 63; c 2. IMMUNOLOGY, RHEUMATOLOGY. *Educ:* Harvard Univ, AB, 58, MD, 63. *Prof Exp:* From intern to resident, Beth Israel Hosp, Boston, 63-65; clin assoc arthritis, NIH, 65-68; Helen Hay Whitney fel immunol, Nat Inst for Med Res, London, 68-70; SR INVESTR IMMUNOL & RHEUMATOLOGY, NIH, 70-, CHIEF, CONNECTIVE TISSUE DIS SECT, ARTHRITIS & RHEUMATISM BR, 84- *Concurrent Pos:* Mem, Am Rheumatism Asn, Ctr Grants Comt, 76- *Honors & Awards:* Rheumatology Soc France Prize, 81; Philip Hench Award, 84. *Mem:* Am Asn Immunologists; fel Am Col Physicians; Am Rheumatism Asn; Am Soc Clin Invest. *Res:* Pathogenesis and therapy of connective tissue diseases. *Mailing Add:* Arthritis & Rheumatism Br 9N 244 Bldg 10 NIAMS-NIH Bethesda MD 20892

PLOTZ, RICHARD DOUGLAS, b Brooklyn, NY, Aug 15, 48; m 71; c 2. CLINICAL PATHOLOGY, PERINATAL PATHOLOGY. *Educ:* Harvard Univ, AB, 71; Univ Pittsburgh, MD, 77. *Prof Exp:* Resident, 77-81, instr, 82-85, ASST PROF PATH, BROWN UNIV, 85-; STAFF PATHOLOGIST, WOMEN & INFANTS HOSP RI, 82- *Mem:* Fel Col Am Pathologists; Soc Pediat Path; Am Asn Pathologists. *Res:* Perinatal clinical pathology; blood banking and areas of clinical chemistry associated with the evaluation of fetal lung maturity; applications of morphometry in perinatal pathology. *Mailing Add:* 104 11th St Providence RI 02906

PLOUGHE, WILLIAM D, b Fort Wayne, Ind, Mar 30, 29; m 52; c 3. NUCLEAR PHYSICS, NUCLEAR SCATTERING. *Educ:* Ind Univ, BS, 51, MS, 53; Purdue Univ, PhD(physics), 61. *Prof Exp:* Instr physics, Univ Fla, 53-56; res assoc, 61-62, asst prof, 62-66, ASSOC PROF PHYSICS, OHIO STATE UNIV, 66- *Concurrent Pos:* Physicist, US AEC, 66-68. *Mem:* Am Phys Soc; Am Asn Physics Teachers. *Res:* Experimental nuclear physics; nuclear reactions and spectroscopy; nuclear scattering of alpha particles and lithium ion at medium energies; optical model analysis; computers in physics education. *Mailing Add:* 211 Sinsbury Dr N Worthington OH 43085

PLOURDE, J ROSAIRE, b Cap-de-la-Madeleine, Que, July 20, 23; m 45; c 2. PHARMACY. *Educ:* Univ Montreal, BPharm, 47, MPharm, 53, DPharm, 60. *Prof Exp:* From asst prof to assoc prof chem anal, 48-74, PROF PHARMACEUT ANALYSIS, UNIV MONTREAL, 74- *Concurrent Pos:* Res fel alkaloids, Univ RI, 60-61; co-ed, Int Pharmaceut Abstr. 14. *Mem:* Am Chem Soc; Am Soc Pharmacog; Can Pharmaceut Asn; Acad Pharmaceut Sci; Am Pharm Asn. *Res:* Phytochemical investigations on Xanthium strumarium, Arctium minus, Eupatorium purpureum and pleiocarpa mutica; analytical investigation on tranquilizers and anti-depressants; qualitative, quantitative and instrumental analysis of pharmaceuticals; quantitative methods of separation and determination of pharmaceuticals; biotransformation of steroids and antibiotics by the spores of microorganisms; ion pair extraction of alkaloids. *Mailing Add:* Fac Pharm Univ Montreal 2900 Blvd Edouard Montetit Montreal PQ H3C 3J7 Can

PLOVNICK, ROSS HARRIS, b Charlotte, NC, May 4, 42; div; c 2. CHEMISTRY. *Educ:* Northeastern Univ, BA, 64; Brown Univ, PhD(chem), 69. *Prof Exp:* Res mgr crystal growing facility, Mat Sci Ctr, Cornell Univ, 68-72; sr chemist, 72-75, res specialist, 75-79, supvr new prod, 79-81, prog mgr, 81-87, GLOBAL PROF SERV MGR, 3M CO, 87- *Mem:* Am Chem Soc. *Res:* Solid state inorganic chemistry; materials preparation, characterization, crystal growth. *Mailing Add:* 3M Co 3M Ctr Bldg 201-4N-01 St Paul MN 55144-1000

PLOWMAN, RONALD DEAN, b Smithfield, Utah, Aug 25, 28; m 51; c 5. DAIRY HUSBANDRY, RESEARCH ADMINISTRATION. *Educ:* Utah State Univ, BS, 51; Univ Minn, MS, 55, PhD(dairy cattle breeding), 56. *Prof Exp:* Herd mgr, Far Hills Farms, Ore, 52-53; county agr agent, Weber County, Utah, 53-54; asst, Univ Minn, 54-56; dairy res scientist, Dairy Cattle Res Br, Agr Res Serv, USDA, 57-63, leader genetics & mgt invests, 63-66, leader, dairy herd improv invests, 66-68, chief, 68-72, area dir, Idaho, Mont & Utah, 72-75; MEM FAC, UTAH STATE UNIV, 77-, HEAD DEPT ANIMAL, DAIRY & VET SCI, 84- *Concurrent Pos:* Mem fac, Dept Dairy Sci, Univ Md, College Park, 72- *Mem:* Am Dairy Sci Asn. *Res:* Dairy cattle breeding, especially various types of mating systems; dairy cattle management; milk quality and influencing factors; nutrition; physiology. *Mailing Add:* USDA ARSOA Rm 300A 14th & Independence Ave SW Washington DC 20250

PLOWMAN, TIMOTHY, systematic botany, ethnobotany; deceased, see previous edition for last biography

PLOWS, WILLIAM HERBERT, b Newark, NJ, Mar 22, 35; m 62; c 2. REENTRY VEHICLE FRATRICIDE, COMPUTER MODELING. *Educ:* St Lawrence Univ, BS, 56; Brooklyn Col, MA, 60; Columbia Univ, MS, 62; Univ Calif, Davis, PhD(appl sci), 71. *Prof Exp:* Instr math, St Lawrence Univ, Canton, NY, 57-61; computational physicist, Lawrence Livermore Lab, 63-73; sr scientist, 73-88, CONSULT, SCI APPLICATIONS INT CORP, 88- *Mem:* Soc Indust & Appl Math. *Res:* Nuclear weapon effects; reentry vehicle fratricide and related targeting constraints, collateral damage, fallout, computer models of the above; shock waves in soil/rock from buried nuclear bursts; Bénard convection; geophysical fluid dynamics models. *Mailing Add:* 2406 Briarwood Dr Boulder CO 80303

PLUCKNETT, WILLIAM KENNEDY, b De Witt, Nebr, Dec 20, 16; m 42; c 3. CHEMISTRY, CHEMICAL KINETICS. *Educ:* Nebr State Col, Peru, AB, 37; Iowa State Univ, PhD(phys chem), 42. *Prof Exp:* Asst chem, Iowa State Col, 37-42, asst prof, 47-51; technologist, Shell Oil Co, Calif, 42-46; instr chem, Univ Calif, 47; assoc chemist, Ames Lab, AEC, 47-51; assoc prof & dir high temp lab, Fordham Univ, 51-53; assoc prof chem, 53-83, dir gen chem, 75-83, EMER PROF CHEM, UNIV KY, 83- *Mem:* Am Chem Soc. *Res:* Inorganic hydrides; hafnium-zirconium; quantum chemistry; thermodynamics; polarizability of the halogen atoms in a series of organic halides; chemical kinetics; dielectric properties. *Mailing Add:* 850 NE County Line Rd Sadieville KY 40370

PLUE, ARNOLD FREDERICK, b Saugerties, NY, Mar 21, 17; m 38; c 3. ORGANIC CHEMISTRY. *Educ:* Siena Col, BS, 50; Rensselaer Polytech Inst, PhD(chem), 60. *Prof Exp:* Lab technician, GAF Corp, 34-42, prod supvr, 42-48, res & develop chemist, 48-55, sr develop chemist, 55-63, tech assoc, 63-74, mgr new prod introd, 74-75, mgr process develop, 75-78; RETIRED. *Mem:* Am Chem Soc; fel Am Inst Chem. *Res:* Dyestuffs; intermediates; optical bleaches; ultraviolet absorbers; nitrogen heterocycles. *Mailing Add:* 9 Parkview Dr Woodland Park Rensselaer NY 12144

PLUM, FRED, b Atlantic City, NJ, Jan 10, 24; m 90; c 3. NEUROSCIENCE. *Educ:* Dartmouth Col, BA, 44; Cornell Univ, MD, 47. *Hon Degrees:* MD, Karolinska Inst, Stockholm, 87; DSc, Long Island Col, 90. *Prof Exp:* Instr med, Cornell Univ, 50-53; from asst prof to prof, Univ Wash, 53-63; TITZELL PROF NEUROL & CHMN DEPT, MED COL, CORNELL UNIV, 63-; NEUROLOGIST-IN-CHIEF, NEW YORK HOSP, 63- *Concurrent Pos:* Mem grad training comt, Nat Inst Neurol Dis & Stroke, 59-63, study sect, 64-68, Nat Adv Coun, 77-81 & 84-86; pres, McKnight Endowment Fund, Neurosci, 86-89; ed, Annals Neurol, 76; pres, Soc Cerebral Blood Flow Metabolism, 89-91. *Honors & Awards:* Jacoby Award, Am Neurol Asn, 84 & Inst Med-Nat Acad Scis, 85. *Mem:* Am Neurol Asn (pres, 76-77); Am Acad Neurol; Am Soc Clin Invest; Soc Neurosci. *Res:* Cerebral energy metabolism and blood flow; experimental epilepsy; mechanisms neural cell death. *Mailing Add:* Dept Neurol Med Col Cornell Univ 1300 York Ave New York NY 10021

PLUMB, JOHN ALFRED, b Waynesboro, Va, Nov 29, 33; m 60; c 2. FISHERIES, MICROBIOLOGY. *Educ:* Bridgewater Col, BA, 60; Southern Ill Univ, MS, 63; Auburn Univ, PhD(fisheries), 72. *Prof Exp:* Hatchery mgr, US Fish & Wildlife Serv, 62-66, hatchery biologist, 66-69; res assoc fisheries, 69-72, from asst prof to assoc prof, 72-85, PROF FISHERIES, AUBURN UNIV, 85- *Concurrent Pos:* Pres, Fish Health Sect, Auburn Univ, 78. *Mem:* Am Fisheries Soc; Am Fisheries Asn; Sigma Xi. *Res:* Bacterial and viral diseases and immunology of fish; the effects of environmental stress on fish diseases; control of bacterial fish pathogens. *Mailing Add:* Dept Fish Univ Auburn Auburn AL 36849-5419

PLUMB, JOHN LAVERNE, b Harlan, Iowa, Dec 24, 33; m 63; c 2. LIGHT SOURCES, GAS DISCHARGES. *Educ:* Iowa State Univ, BS, 55; Univ Minn, MS, 63; NY Univ, PhD(elec eng), 67. *Prof Exp:* Test engr, Honeywell, Inc, 55, eval engr, 57-60; teaching asst elec eng, Univ Minn, 61-62; res asst, NY Univ, 64-67; asst prof, Univ Conn, 67-73; design engr, Transitron Inc, 73-75; pvt consult, MCE Res, 75-78; ADV RES & DEVELOP ENGR, GTE SYLVANIA, INC, 78- *Mem:* Inst Elec & Electronics Engrs. *Res:* High field effects in insulating thin films; thin film electroluminescent devices; electric breakdown in insulating thin films; dielectric films; electroluminescent displays; electric discharge lamps; high pressure sodium lamps. *Mailing Add:* GTE Sylvania Inc Sylvania Lighting Ctr Danvers MA 01923

PLUMB, ROBERT CHARLES, b Springfield, Mass, Jan 24, 26; m 73; c 6. PHYSICAL CHEMISTRY. *Educ:* Clark Univ, AB, 49; Brown Univ, PhD(chem), 53. *Prof Exp:* Researcher chelation chem, Bersworth Chem Co, 47-49; asst chief phys chem, Aluminum Co Am, 52-58; from asst prof to assoc prof, 58-64, head dept, 64-76, PROF CHEM, WORCESTER POLYTECH INST, 64- *Concurrent Pos:* NSF fel, Cambridge Univ, 56-57; res prof, Brown Univ, 90- *Mem:* Am Chem Soc; Sigma Xi; New Eng Asn Chem Teachers; Int Asn Colloid & Surface Scientists. *Res:* Surface chemistry; metal oxidation; metal complex ions in solution; demonstrations for the teaching of statistical mechanics; chemistry of Mars. *Mailing Add:* Dept Chem Worcester Polytech Inst 100 Institute Rd Worcester MA 01609

PLUMLEE, KARL WARREN, b Richland, Wash, Sept 30, 49; m 70; c 3. PETROLEUM CATALYSTS, ENVIRONMENTAL HAZARDOUS EMISSIONS. *Educ:* North Cent Col, BA 71; Northwestern Univ, MS, 72, PhD(chem), 75. *Prof Exp:* Res chemist, Coaliquefaction Labs, Exxon Res & Eng Co, 75-77, staff chemist, 77-78, sr staff chemist, 78-79, proj dir staff, Exxon Donor Solvent Process, 79-81, Sect Head, Coal Liquefaction Labs, 81-83, sect head, Shale Res Lab, 83-85, sect head, exploratory res, Exxon Res & Eng Co, 85-87; DIR, CATALYSIS RES, WASH RES CTR, W R GRACE CO, 87- *Mem:* Am Chem Soc; Am Inst Chem Engr. *Res:* Petroleum catalysts, automobile exhaust catalysts, and stationary emission control catalysts and processes for all types of hazardous environmental emissions. *Mailing Add:* 7379 Rt 32 Columbia MD 21043

PLUMLEE, MILLARD P, JR, b Celina, Tenn, July 14, 21; m 49; c 5. NUTRITION. *Educ:* Tenn Tech Univ, BS, 44; Purdue Univ, MS, 48, PhD(nutrit), 53. *Prof Exp:* Asst nutrit, Univ Tenn, 48-51; res fel, 51-52, from instr to assoc prof, 52-62, PROF NUTRIT, PURDUE UNIV, LAFAYETTE, 62- *Mem:* Am Soc Animal Sci. *Res:* Mineral requirements and metabolism in farm animals. *Mailing Add:* Dept of Animal Sci Purdue Univ West Lafayette IN 47907

PLUMMER, ALBERT J, b Somerville, Mass, Apr 16, 08; m 40; c 2. PHARMACOLOGY. *Educ:* Boston Univ, AB, 29, AM, 30, PhD(biol chem), 35, MD, 49. *Prof Exp:* Asst, Mass Mem Hosp, 30-32; from instr to asst prof pharmacol, Sch Med, Boston Univ, 35-49; sr pharmacologist, Ciba Pharmaceut Co, 49-51, dir pharmacol res, 51-57, dir macrobiol res, 58-66, dir biol res, 66-67, exec dir biol res, 67-73, CONSULT BIOL RES, CIBA-GEIGY CORP, 73- *Concurrent Pos:* Lectr, Sch Med, Boston Univ, 49-; mem, Coun High Blood Pressure Res, Am Heart Asn. *Mem:* Soc Pharmacol & Exp Therapeut; Am Soc Clin Pharmacol & Therapeut; Am Col Neuropsychopharmacol. *Res:* Cardiovascular, gastrointestinal and central nervous system drugs; metabolism of drugs; diuretics. *Mailing Add:* 13 Harding Terr Morristown NJ 07960

PLUMMER, BENJAMIN FRANK, b Burlington Junction, Mo, Feb 29, 36; m 62; c 4. ORGANIC CHEMISTRY. *Educ:* Iowa State Univ, BS, 58; Ohio State Univ, PhD(org chem), 62. *Prof Exp:* Res asst, Ga Inst Technol, 62-63; asst prof chem, SDak State Univ, 63-67; assoc prof, 67-74, chmn chem dept, 76-80, PROF CHEM, TRINITY UNIV, TEX, 74- *Concurrent Pos:* Robert A Welch Found grant; dir, NSF Undergrad Res Partic Proj, 72 & 75, Petrol Res Fund, NSF-RUI; Camille & Henry Dreyfus Scholar, 91-93. *Mem:* AAAS; Am Chem Soc; NY Acad Sci; Sigma Xi; InterAm Photochem Soc; Coun Undergrad Res. *Res:* Photochemistry; organic synthesis; molecular spectroscopy; luminescence; polycyclic aromatic hydrocarbons. *Mailing Add:* Dept Chem Trinity Univ 715 Stadium Dr San Antonio TX 78212

PLUMMER, CHARLES CARLTON, b Mexico City, Mex, Apr 21, 37; US citizen; m 72. PETROLOGY. *Educ:* Dartmouth Col, AB, 59; Univ Wash, MS, 64, PhD(geol), 69. *Prof Exp:* Glaciologist-geologist, Inst Polar Studies, Ohio State Univ, 64-66; instr geol, Olympic Col, 67-69; geologist, Antarctic Proj, US Geol Surv, 69-70; from asst prof to assoc prof, 75-80, chmn dept, 75-78, PROF GEOL, CALIF STATE UNIV, SACRAMENTO, 80- *Concurrent Pos:* Leader, Australian, NZ & US geol invest Daniels Range, Antarctica, 81-82. *Honors & Awards:* Antarctic Serv Medal, US Govt, 66, 70 & 82. *Mem:* AAAS; Geol Soc Am; Nat Asn Geol Teachers; Am Alpine Club. *Res:* Metamorphic petrology; geology of alpine and polar regions; writer of physical geology textbooks. *Mailing Add:* Dept of Geol Calif State Univ Sacramento CA 95819-6043

PLUMMER, E WARD, b Astoria, Ore, Oct 30, 40; m 61; c 2. SOLID STATE PHYSICS. *Educ:* Lewis & Clark Col, BA, 62; Cornell Univ, PhD(physics), 67. *Prof Exp:* Nat Res Coun assoc, Nat Bur Standards, 67-69, physicist, 69-73; assoc prof, 73-77, PROF PHYSICS, UNIV PA, 77- *Honors & Awards:* Wayne Nottingham Prize, 68. *Mem:* Am Phys Soc. *Res:* Properties of solid surfaces; electron emission spectroscopies; atomic and molecular spectroscopy and electron optics. *Mailing Add:* 17 Penarth Rd Bala-Cynwyd PA 19004

PLUMMER, GAYTHER LYNN, b Marion Co, Ind, Jan 27, 25; m 50. PLANT ECOLOGY, CLIMATOLOGY. *Educ:* Butler Univ, BS, 48; Kans State Col, MS, 50; Purdue Univ, PhD(ecol), 54. *Prof Exp:* Asst, Kans State Col, 48-50; instr biol, Knox Col, 50-51; asst, Purdue Univ, 51-53; asst prof, Antioch Col, 54-55; from asst prof to assoc prof, 55-67, PROF BOT & ENVIRON DESIGN, UNIV GA, 67-; STATE CLIMATOLOGIST, GA, 78- *Concurrent Pos:* Naturalist, Ind State Parks, 50-51 & Oak Ridge Nat Lab, 58- 60; ed Ga J Sci, 78-85. *Mem:* Fel AAAS; Ecol Soc Am; Am Asn State Climatologists; Sigma Xi. *Res:* Bioclimatology; environmental biology; remote sensing. *Mailing Add:* Inst Natural Res Univ Ga Athens GA 30602

PLUMMER, JAMES WALTER, b Idaho Springs, Colo, Jan 29, 20; m 48; c 4. ELECTRONICS, ASTRONAUTICS. *Educ:* Univ Calif, Berkeley, BS, 42; Univ Md, MS, 53. *Prof Exp:* Vpres & asst gen mgr space syst, Lockheed Missiles & Space Co Inc, 65-68, vpres & asst gen mgr res & develop, 68-69, vpres & gen mgr space syst, 69-73; undersecy Air Force, USAF, 73-76; exec pres, Lockheed Missiles & Space Co Inc, 76-83; RETIRED. *Concurrent Pos:* Mem, Space Appl Bd, Nat Acad Eng, 78-83; bd trustees, Aerospace Corp, Los Angeles, Calif, 83-85, chmn bd, 85- *Mem:* Nat Acad Eng; fel Am Inst Aeronaut & Astronaut; fel Am Astronaut Soc. *Mailing Add:* PO Box 910 Jackson OR 97530

PLUMMER, MARK ALAN, b Wichita, Kans, July 30, 36; m 67; c 2. CHEMICAL ENGINEERING. *Educ:* Colo Sch Mines, BS, 58, MS, 63, PhD(chem eng), 72. *Prof Exp:* Process engr, Skelly Oil Co, 58-60; advan res engr, 62-80, SR RES ENGR, MARATHON OIL CO, 80- *Concurrent Pos:* Mem hydrocarbon compressibility & thermal expansion comt, Am Petrol Inst. *Mem:* Am Inst Chem Engrs. *Res:* petrochemical manufacturing, separation techniques and heavy oils upgrading. *Mailing Add:* Marathon Oil Co PO Box 269 Littleton CO 80160

PLUMMER, MICHAEL DAVID, b Akron, Ohio, Aug 31, 37; m 68; c 2. MATHEMATICS, GRAPH THEORY. *Educ:* Wabash Col, BA, 59; Univ Mich, MS, 61, PhD(math), 66. *Prof Exp:* Instr math, Yale Univ, 66-68; asst prof comput sci, City Col New York, 68-70; assoc prof, 70-78, PROF MATH, VANDERBILT UNIV, 78- *Concurrent Pos:* Vis res mathematician, CIMAS, Univ Mex, Mex City, 74, Math Inst, Hungarian Acad Sci, Budapest, 75-76 & 80; vis sr assoc, Univ Melbourne, 79; vis prof, Rheinische Friedrich-Wilhelms-Univ Bonn, 80 & 81, Math Inst Hungarian Acad Sci, Budapest, 83, Univ Paris-Sud, Orsay, France, 90; William Evans vis fel, Univ Otago, Dunedin, NZ, 86. *Mem:* AAAS; Am Math Soc; Math Asn Am; Soc Indust & Appl Math. *Res:* Theory of graphs. *Mailing Add:* Dept Math Vanderbilt Univ Nashville TN 37235

PLUMMER, MICHAEL V(AN), b Tuscumbia, Ala, Mar 9, 45; m 67; c 2. HERPETOLOGY, ECOLOGY. *Educ:* Harding Col, BS, 67; Utah State Univ, MS, 69; Univ Kans, PhD(biol), 76. *Prof Exp:* Instr, 71-72, asst prof, Harding Col, 76-81, assoc prof biol, 81-85, PROF BIOL, HARDING UNIV, 85-, CHMN, DEPT BIOL, 86-, COONS ENDOWED CHAIR BIOMED SCI, 90- *Concurrent Pos:* Index ed, Herpetologica, 80-85; adj prof biol, Memphis State Univ, 81-83, 89-; vis scientist, Savannah River Ecol Lab, 89; consult, Arkansas Natural Heritage Comn, 82, US Fish & Wildlife Serv, 87. *Mem:* Am Inst Biol Scis; Am Soc Ichthyologists & Herpetologists; Soc Study Amphibians & Reptiles; Herpetologists' League. *Res:* Physiological and population ecology of reptiles (especially turtles and snakes). *Mailing Add:* Dept Biol Harding Univ Searcy AR 72143

PLUMMER, OTHO RAYMOND, b Beaumont, Tex, Sept 27, 38; m 65. APPLIED MATHEMATICS, MATHEMATICAL PHYSICS. *Educ:* Univ Tex, BS, 61, MA, 63, PhD(physics), 66. *Prof Exp:* Res assoc physics & spec instr chem, Univ Tex, 66; asst prof physics, Univ Ark, Fayetteville, 66-68; asst prof, 68-71, ASSOC PROF MATH, UNIV MO-ROLLA, 71- *Concurrent Pos:* Actg chmn math, Univ Mo-Rolla, 78-79. *Mem:* Am Phys Soc; Soc Indust & Appl Math. *Res:* Group theory; operator theory; quantum theory; applications. *Mailing Add:* 2600 Johnson Dr Columbia MO 65203

PLUMMER, PATRICIA LYNNE MOORE, b Tyler, Tex; m 65; c 2. THEORETICAL CHEMICAL PHYSICS. *Educ:* Tex Christian Univ, BA, 60; Univ Tex, PhD(theoret chem), 64. *Prof Exp:* Res assoc theoret chem, Univ Tex, 64-66, instr chem, 65-66; res assoc theoret chem, Univ Ark, Fayetteville, 66-68; res assoc grad ctr cloud physics res, 68-73, asst prof physics & sr investr, 73-77, assoc prof physics & assoc dir, cloud phys res ctr, Univ Mo-Rolla, 77-85, PROF PHYSICS, ASTRON & CHEM, UNIV MO-COLUMBIA, 85- *Concurrent Pos:* Res grants, Off Naval Res, 69-71, NSF, 72-74, 74-80, 81-85, 87-88 & NASA, 75-81; chmn, Int Symp Physics & Chem Ice, 82, mem, Sci Adv Comt, 82-; fac fel, Air Force, 88. *Honors & Awards:* Award Molecular Model Heterogeneous Nucleation, Atmospheric Sci Sect, NSF, 74. *Mem:* Am Chem Soc; Am Phys Soc; Am Geophys Union; Sigma Xi. *Res:* Ab initio calculations on small molecules; potential energy surfaces; applications of group algebra to quantum mechanical systems; molecular theory for nucleation of water droplets and ices; calculated molecular properties of small clusters; stability and structure of atmospheric ions; molecular dynamic modeling of vapor clusters and processes on surfaces. *Mailing Add:* 314 Physics Univ Mo Columbia MO 65211

PLUMMER, THOMAS H, JR, b Cleveland, Ohio, Aug 25, 33; m 60; c 4. BIOCHEMISTRY. *Educ:* Cornell Univ, BS, 55; Purdue Univ, MS, 57, PhD(plant physiol), 60. *Prof Exp:* Res assoc protein struct, Brookhaven Nat Lab, 60-63; ASSOC RES SCIENTIST, DIV LABS & RES, NY STATE HEALTH DEPT, 63- *Mem:* AAAS; Am Chem Soc; Am Soc Biol Chem; Soc Complex Carbohydrates. *Res:* Glycoprotein isolation and structure, particularly protein in pancreatic secretions and blood plasma; endoglycosidase isolation and mechanism. *Mailing Add:* Div Labs & Res Empire State Hall NY State Dept Health Albany NY 12201

PLUMMER, WILLIAM ALLAN, b Wilkes-Barre, Pa, Jan 20, 27; m 57, 78; c 2. PHYSICAL CHEMISTRY. *Educ:* Wilkes Col, BS, 50; Univ Pittsburgh, PhD(chem), 56. *Prof Exp:* Sr chemist, Tech Staff Div, Corning Glass Works, 56-60, res chemist, 60-76, sr res scientist, 76-80, res assoc, 80-87; RETIRED. *Mem:* AAAS; Am Chem Soc; Am Ceramic Soc. *Res:* Thermal expansion, specific heat, thermal diffusivity and conductivity; thermal properties of glasses and ceramics; glass viscosity. *Mailing Add:* Ten E Fox Lane Rd Painted Post NY 14870

PLUMMER, WILLIAM TORSCH, b Baltimore, Md, Mar 25, 39; m 61; c 2. OPTICAL DESIGN, OPTICAL ENGINEERING. *Educ:* Johns Hopkins Univ, AB, 60, PhD(physics), 65. *Prof Exp:* Res asst, Johns Hopkins Univ, 65; asst prof astron, Univ Mass, Amherst, 67-69; scientist optics, Polaroid Corp, 69-70, sr scientist, 71-72, assoc fel, 73-75, sr mgr, 78-80, ENG FEL OPTICS, POLAROID CORP, 76-, DIR OPTICAL ENG, 80- *Concurrent Pos:* Consult, Muffoletto Optical Co, Baltimore, 63-69; dir, Nat Speleological Soc, 66-68; vis indust prof, ElectroOptics Tech Ctr, Tufts Univ, 84-88. *Honors & Awards:* Richardson Medal, Optical Soc Am, 80. *Mem:* Optical Soc Am; Nat Speleological Soc. *Res:* Optical design and development and technical research and development management, directed toward high-volume precision manufacture of image-forming systems for consumer, medical and industrial use; aspheric plastic optics and diffraction optics; precision mechanical design. *Mailing Add:* Polaroid Corp 38 Henry St 2A Cambridge MA 02139

PLUMTREE, A(LAN), b Scunthorpe, Eng, Aug 1, 36; m 59; c 2. PHYSICAL & MECHANICAL METALLURGY, MATERIALS ENGINEERING. *Educ:* Univ Nottingham, BSc, 60, PhD(metall), 63. *Prof Exp:* Metallurgist, Appleby-Frodingham Steel Works Ltd, Eng, 54-57; lectr, Nottingham & Dist Tech Col, 60-63; fel, Univ Toronto, 63, asst prof phys metall, 63-65; from asst prof to assoc prof, Univ Waterloo, 65-77, assoc chmn dept, 81-84, assoc dean eng, 84-87, PROF MECH ENG, UNIV WATERLOO, 68- *Concurrent Pos:* Vis prof, Univ Paris, France, 79, Tech Univ Braunschweig, Ger, 87. *Mem:* Am Soc Metals; fel Brit Inst Metall. *Res:* Fatigue behaviour of metals and polymers; fracture mechanisms in materials; metal deformation and forgeability. *Mailing Add:* Dept Mech Eng Univ Waterloo Waterloo ON N2L 3G1 Can

PLUNKETT, ROBERT, b New York, NY, Mar 15, 19; m 46; c 3. MECHANICS. *Educ:* Mass Inst Technol, SB, 39, ScD, 48. *Prof Exp:* Asst elec eng, Mass Inst Technol, 39-41 from instr to asst prof mech eng, 41-48; asst prof, Rice Inst, 48-51; vibration engr, Gen Elec Co, 51-56; consult engr, 56-60; PROF MECH, UNIV MINN, MINNEAPOLIS, 60- *Mem:* Nat Acad Eng; hon mem Am Soc Mech Engrs; Acoust Soc Am; Am Soc Eng Educ; Soc Exp Mech. *Res:* Vibration analysis; mechanics of materials; applied mechanics. *Mailing Add:* Dept Aeronaut & Eng Mech Univ Minn Minneapolis MN 55455

PLUNKETT, ROBERT LEE, b Lynchburg, Va, Mar 18, 22; m 43, 79; c 1. TOPOLOGY, MATHEMATICAL ANALYSIS. *Educ:* Randolph-Macon Col, BS, 43; Univ Chicago, MS, 49; Univ Va, PhD(math), 53. *Prof Exp:* Asst prof math, Vanderbilt Univ, 53-54; from asst prof to prof, Fla State Univ, 54-63; mem tech staff, Northrop Space Lab, Ala, 63-64; chmn dept, 66-73, prof, 63-87, EMER PROF MATH, UNIV ALA, 87- *Concurrent Pos:* Radar specialist, USN Reserve, 43-46; via assoc prof math, Univ Va, 57. *Mem:* Math Asn Am. *Mailing Add:* eight N Pinehurst Tuscaloosa AL 35401

PLUNKETT, WILLIAM KINGSBURY, b Boston, Mass, May 4, 43; m 65; c 1. BIOCHEMISTRY, BIOCHEMICAL PHARMACOLOGY. *Educ:* Springfield Col, BS, 65; Univ Mass, PhD(biochem), 70. *Prof Exp:* Fel biochem, Univ Pa, 70-71; res assoc, Univ Colo Med Ctr, 71-75; from asst prof to assoc prof, 75-86, PROF MED & PHARMACOL, UNIV TEX M D ANDERSON CANCER CTR, 86- *Concurrent Pos:* From asst prof to assoc prof, Grad Sch Biomed Sci, Univ Tex, Houston, 76-86, prof biochem, 86-; Hubert L & Oliver Stringer prof med oncol, 89. *Honors & Awards:* Serv to Mankind Award, Leukemia Soc Am, 89. *Mem:* Am Asn Cancer Res; Am Soc Cell Biol; AAAS; Am Soc Clin Oncologists; Am Soc Pharm Exp Therapeuts; Am Soc Hemat. *Res:* Biochemical basis for drug effectiveness and molecular basis of drug action. *Mailing Add:* Dept Med Oncol Univ Tex MD Anderson Cancer Ctr Houston TX 77030

PLUSCEC, JOSIP, b Gornji Milanovac, Yugoslavia, June 6, 28; US citizen; m 61; c 2. ORGANIC CHEMISTRY. *Educ:* Univ Zagreb, BSc, 53, PhD(org chem), 60. *Prof Exp:* Chemist, Inst Invest Drugs, Zagreb, Yugoslavia, 55-57; asst prof org synthesis, Inst Org Chem, Sch Pharmacy, Univ Zagreb, 57-60 & Inst Chem, Sch Med, 60-62; res assoc biochem synthesis, Dept Bot, Univ Chicago, 65-66; sr res scientist, Dept Biochem Pharmacol, 66-67 & Dept Org Chem, 67-86, sr res investr, 86-88, sect head, 86-88, mgr, 88, DEPT HEAD, BRISTOL-MYERS SQUIBB CO, 88- *Concurrent Pos:* Nat Res Coun Can Div Pure Chem fel, 62-64. *Mem:* AAAS; Am Chem Soc; Ger Chem Soc; Croatian Chem Soc. *Res:* Investigations in biochemistry of porphyrins; synthetic work in lipid, porphyrin, peptide chemistry and antibiotics; new antibiotics isolation; analytical chemistry, GC, HPLC, TGA-methods develop. *Mailing Add:* Bristol-Myers Squibb Co One Squibb Dr PO Box 191 New Brunswick NJ 08903

PLUTH, DONALD JOHN, b Emmet Co, Iowa, Feb 5, 36; m 60; c 3. SOIL SCIENCE. *Educ:* Univ Minn, Minneapolis, BS, 58, MS, 62, PhD(soil sci), 65. *Prof Exp:* Res assoc soil sci, Univ Minn, 65-67; ASSOC PROF, UNIV ALTA, 67- *Mem:* Am Soc Agron; Sigma Xi. *Res:* Soil-vegetation relationships; site evaluation; nutrient cycling in forest ecosystems. *Mailing Add:* Dept Soil Sci Edmonton AB T6G 2E3 Can

PLUTH, JOSEPH JOHN, b Cook, Minn, Feb 18, 43; m 73; c 3. STRUCTURAL CHEMISTRY. *Educ:* Bemidji State Col, BA, 65; Univ Wash, PhD(chem), 71. *Prof Exp:* Res assoc geophys sci, 71-72, res scientist chem & geophys sci, 71-80, SR RES ASSOC GEOPHYS SCI, UNIV CHICAGO, 81- *Mem:* Am Crystallog Asn; Sigma Xi; Am Chem Soc; Am Inst Physics. *Res:* Studies of the crystal structures of zeolite molecular sieves and related materials and their relationship to observed physical chemical and catalytic properties; structural studies using synchrotron and neutron radiation sources. *Mailing Add:* Dept Geophys Sci Univ Chicago 5734 S Ellis Ave Chicago IL 60637

PLUTZER, MARTIN DAVID, b New York, NY, Apr 27, 44; m 72; c 1. PSYCHIATRY. *Educ:* Alfred Univ, BA, 64; Univ Pittsburgh, MD, 68; Am Bd Psychiat & Neurol, dipl, 76. *Prof Exp:* Asst prof, 74-80, ASSOC PROF PSYCHIAT, MED COL PA, 80- *Concurrent Pos:* Ward chief, inpatient psychiat unit, Eastern Pa Psychiat Inst, 74-; assoc dir undergrad educ psychiat, Med Col Pa, 77-80, dir, 80- *Mem:* Assoc mem Am Psychoanal Asn. *Mailing Add:* Dept of Psychiat 3300 Henry Ave Philadelphia PA 19129

PLUZNIK, DOV HERBERT, b Vienna, Austria, Nov 7, 35; US citizen; m 61; c 2. LYMPHOKINES, CYTOKINES. *Educ:* Weizllmann Inst Sci, Israel, PhD (cell biol), 66. *Prof Exp:* Prof & chmn, Dept Life Sci, Bar-Ilan Univ, Israel, 76; vis prof, George Washington Univ, DC, 80; VIS SCIENTIST, NIH, 82- *Concurrent Pos:* Assoc ed, Exp Hemat J, 88- *Mem:* Am Soc Exp Hemat; Am Asn Immunologist. *Res:* Hematopoietic growth factor, regulation and function. *Mailing Add:* Cytokine Biol Ctr Biol Eval & Res FDA Bldg 29A Rm 3B19 8800 Rockville Pike Bethesda MD 20892

PLYBON, BENJAMIN FRANCIS, b Huntington, WVa, Feb 12, 30; m 62; c 4. MATHEMATICAL PHYSICS, OTHER COMPUTER SCIENCES. *Educ:* Marshall Univ, BS, 57; Carnegie-Mellon Univ, MS, 59; Ohio State Univ, PhD(math), 68. *Prof Exp:* Electronics technologist, Cubic Corp, 56; from instr to asst prof math, Marshall Univ, 59-64; instr, Ohio Wesleyan Univ, 64-66; asst prof, 66-74, ASSOC PROF MATH, MIAMI UNIV, 74- *Mem:* Soc Indust & Appl Math; Math Asn Am; Sigma Xi. *Res:* Variational principles and invariance groups for physical systems; conservation laws for physical systems; applied mathematics; mechanics. *Mailing Add:* Dept Math Miami Univ Oxford OH 45056

PLYMALE, CHARLES E, b Gallipolis, Ohio, May 21, 35; m 56; c 3. PLASTICS ENGINEERING, CHEMICAL ENGINEERING. *Educ:* Marietta Col, BS, 58; C ase Inst Technol, BSChE, 58; Princeton Univ, MSE, 59. *Prof Exp:* Proj engr, Owens-Ill Tech Ctr, 59-61, group leader plastics, 61-

63, sect chief, 63-66, dir develop, Forest Prod Div, 66-73, vpres & tech dir, 73-75, VPRES & PROD MGR, BOX OPERS, OWENS-ILL TECH CTR, 75- *Mem:* Am Chem Soc; Am Inst Chem Engrs; Soc Plastics Engrs; Nat Soc Prof Engrs. *Res:* Diaelectrophoresis; plastics engineering and applications in packaging industry; blow molding; film extrusion; injection molding; rheology; corrugated box plant automation. *Mailing Add:* Avant Garde Computing Inc 8000 Commerce Pkwy Mt Laurel NJ 08054

PLYMALE, DONALD LEE, b Huntington, WVa, June 12, 34; m 59; c 2. INORGANIC CHEMISTRY. *Educ:* Marshall Univ, BS, 57, MS, 62; Ga Inst Technol, PhD(inorg chem), 66. *Prof Exp:* Spectroscopist, Int Nickel Co, 59-62; sr res chemist, Monsanto Res Corp, 66-69; from asst prof to assoc prof chem, Roanoke Col, 71-75; PROG COORDR, US DEPT ENERGY, 76- *Mem:* AAAS; Am Chem Soc. *Res:* Transition metal complexes. *Mailing Add:* 1114 Santa Ana Ave SE Albuquerque NM 87123

PLYMALE, EDWARD LEWIS, b Kenova, WVa, Sept 20, 14; m 48; c 2. BOTANY. *Educ:* Marshall Univ, AB, 35; Univ Ky, MS, 39; Univ Iowa, PhD(bot), 42. *Prof Exp:* Lab instr, 35-37, from asst prof to prof, 46-74, EMER PROF BOT, MARSHALL UNIV, 74- *Mem:* AAAS; Bot Soc Am. *Res:* Veins of mesomorphic leaves; taxonomy of vascular plants of Southern West Virginia; plant morphology and taxonomy. *Mailing Add:* 661 Buffalo Rd Huntington WV 25704-9641

PLYMALE, HARRY HAMBLETON, b Huntington, WVa, July 13, 28; m 78; c 4. ZOOLOGY, VETERINARY MEDICINE. *Educ:* Mich State Univ, BS, 54, DVM, 56. *Prof Exp:* Instr vet sci, Calif Polytech Col, San Luis Obispo, 58-60; ASSOC PROF BIOL, SAN DIEGO STATE UNIV, 62- *Mem:* Asn Mil Surgeons US. *Mailing Add:* Dept Biol San Diego State Univ San Diego CA 92182

PNEUMAN, GERALD W, b Gary, Ind, Aug 4, 31; m 57; c 3. ASTROPHYSICS. *Educ:* Purdue Univ, BS, 58, MS, 59, PhD(magnetohydrodyn), 63. *Prof Exp:* Instr aerodyn, Purdue Univ, 59-63; staff scientist, AC Electronics Defense Res Labs, 63-68; staff scientist, 68-81, SR SCIENTIST, HIGH ALTITUDE OBSERV, 81- *Concurrent Pos:* Lectr, Dept Astrophysics, Univ Colo, Boulder. *Honors & Awards:* Nat Ctr Atmospheric Res Pub Award, 68 & 75. *Mem:* Am Astron Soc; Am Phys Soc; Int Astron Union. *Res:* Influence of magnetic fields on phenomena occuring in the solar corona and interplanetary space; solar and interplanetary physics. *Mailing Add:* 550 Roe Rd Paradise CA 95969

PO, HENRY N, b San Fernando, Philippines, Oct 4, 37; m 66; c 2. PHYSICAL & INORGANIC CHEMISTRY, ELECTROCHEMISTRY. *Educ:* Mapua Inst Technol, BS, 60; Univ Wis-Madison, MS, 62; Univ Calif, Davis, PhD(phys inorg chem), 67. *Prof Exp:* Res engr, Allis-Chalmers Mfg Co, 62-63; res fel chem, Brookhaven Nat Lab, 67-68; from asst prof to assoc prof, 68-76, PROF CHEM, CALIF STATE UNIV, LONG BEACH, 76- *Mem:* Am Chem Soc; Sigma Xi. *Res:* Mechanism of hexa-aquometal ions redox reaction, ligand substitution and coordination chemistry; cryogenic electronic spectral of binuclear complexes of transition metal; electronic spectroscopy of inorganic complexes; energy conversion system; solar energy research; electrochemical inorganic system; bio-inorganic complexes; electron-transfer reactions involving hyper-valent transitions metal ions and macrocyclic complexes. *Mailing Add:* Dept of Chem Calif State Univ Long Beach CA 90840

POAG, CLAUDE WYLIE, b Deland, Fla, Aug 12, 37; m 62; c 3. MICROPALEONTOLOGY, FORAMINIFERA. *Educ:* Fla State Univ, BS, 59; La State Univ, MS, 63; Tulane Univ, PhD(paleont), 71. *Prof Exp:* Micropaleontologist, Chevron Oil Co, 62-70; asst prof geol oceanog, Tex A&M Univ, 70-74; GEOLOGIST, US GEOL SURV, 74- *Concurrent Pos:* Asst field officer, Nat Sci Found, 73-74; co-chief scientist, Deep Sea Drilling Proj, 81- *Mem:* Am Asn Petrol Geologists; Soc Econ Paleontologists & Mineralogists; Paleont Soc; Paleont Res Inst; Am Quaternary Asn; Sigma Xi. *Res:* Microfossil analyses leading to intepretation of ecology, paleoecology, biostratigraphy, paleobiogeography, paleoceanography and depositional history of triassic to quaternary sodimentary rocks along the Atlantic and Gulf margins of the United States. *Mailing Add:* 14 Sandpiper Circle East Falmouth MA 02536

POAGE, SCOTT T(ABOR), b Waco, Tex, Dec 5, 31. INDUSTRIAL ENGINEERING, OPERATIONS RESEARCH. *Educ:* Tex Tech Col, BS, 53; Agr & Mech Col, Tex, MS, 57; Okla State Univ, PhD(indust eng), 62. *Prof Exp:* Prod control engr, Phillips Petrol Co, 53; instr indust eng, Agr & Mech Col, Tex, 57-59; asst prof, Okla State Univ, 59-61; from assoc prof to prof, Arlington State Col, 61-68, head dept, 61-68; chmn dept, 68-77, PROF INDUST ENG, UNIV HOUSTON, 68- *Concurrent Pos:* Consult, Am Mfg Co, Tex & Tex Instruments, Inc, 59, Am Airlines, 60-62, Ling-Temco-Vought, 63, US Govt AID, India, 66 & Grocers' Supply Inc, 76. *Mem:* Fel Am Inst Indust Engrs (vpres, 65-67); Am Soc Eng Educ; Am Soc Qual Control; Opers Res Soc Am. *Res:* Operations research in production control, especially applications of queuing theory; statistical quality control; organizational and management theory. *Mailing Add:* Poage Land & Cattle 600 Edgewood Waco TX 76708

POATE, JOHN MILO, b Devonport, Eng, Nov 12, 40; US citizen; m 65; c 2. PHYSICS, MATERIALS SCIENCE. *Educ:* Univ Melbourne, BSc, 62, MSc, 65; Australian Nat Univ, PhD(nuclear physics), 67. *Prof Exp:* Res fel nuclear physics, Atomic Energy Res Estab, Harwell, Eng, 67-71; mem tech staff physics, 71-82, DEPT HEAD, AT&T BELL LABS, 82- *Concurrent Pos:* Chair, Div Mat Physics, Am Phys Soc. *Mem:* Fel Am Phys Soc; Electrochemical Soc; Mat Res Soc (pres). *Res:* Surface, near-surface and thin film properties of solids using energetic ion beams for analysis; material modification by implantation. *Mailing Add:* Bell Labs Rm 1E-338 600 Mountain Ave PO Box 636 Murray Hill NJ 07974-0636

POBER, JORDAN S, IMMUNOLOGY, BIOCHEMISTRY. *Educ:* Yale Univ, MD & PhD(molecular biol, physics & chem), 77. *Prof Exp:* ASSOC PROF PATH, BRIGHAM WOMEN'S HOSP, HARVARD UNIV, 81- *Mailing Add:* Dept Pathol Harvard Med Sch Brigham & Women's Hosp 75 Francis St Boston MA 02115

POBER, KENNETH WILLIAM, b Chicago, Ill, Jan 3, 40; m 65; c 2. OILFIELD CHEMISTRY, ORGANIC CHEMISTRY. *Educ:* Univ Colo, BS, 62; Univ Idaho, PhD(org chem), 67. *Prof Exp:* Fel, Boston Univ, 67-69; sr chemist process res & develop, Texaco Inc, 69-72; sr develop chemist photo prod, 3M Co, 72-75; sr res chemist fuels, Occidental Res Corp, 75-77; res scientist prod develop, NL Baroid Petrol Serv, 77-80, group leader, completion & drilling fluids develop, 80-82; STAFF SCIENTIST OIL PROD RES, CONOCO INC, 82- *Mem:* Am Chem Soc; Sigma Xi; Soc Petrol Engrs. *Res:* Completion/workover fluids; drilling fluids and oilfield chemicals; resource recovery; synthetic fuels; lignite and coal conversion; photopolymer systems development; homogeneous catalysis; organic photochemistry. *Mailing Add:* Conoco Inc PO Box 1267 Ponca City OK 74603

POBER, ZALMON, b Philadelphia, Pa, July 31, 39; m 69. MAMMALIAN PHYSIOLOGY. *Educ:* Drexel Univ, BS, 62; Thomas Jefferson Univ, MS, 65, PhD(physiol), 68; Western New Eng Col, MBA, 84. *Prof Exp:* Physiologist, Food Labs, US Army Natick Labs, 68-76; asst prof, 76-81, ASSOC PROF, MASS COL PHARM, SPRINGFIELD, 81- *Concurrent Pos:* Lectr, W New Eng Col, 83-; consult, Continuing Prof Educ Serv, 84- *Mem:* Am Physiol Soc; NY Acad Sci; Sigma Xi. *Res:* Mechanisms by which the intestine controls the body's response to food; intestinal regulation of food intake; stimulus and pathways for the intestinal phase of gastric secretion; compatibility of biomaterials. *Mailing Add:* 5 Applewood Circle Easthampton MA 01027-1309

POBERESKIN, MEYER, b Philadelphia, Pa, Sept 13, 16; m 46; c 2. PHYSICAL CHEMISTRY. *Educ:* City Col New York, BS, 37; Columbia Univ, MA, 39. *Prof Exp:* Chemist, Fales Chem Co, 39 & Mt Sinai Hosp, New York, 39-41; assoc chemist, Cent Concrete Lab, 41-42; chemist, Brush Develop Co Div, Clevite Corp, 46; assoc chemist, Argonne Nat Lab, 46-48; proj leader nuclear chem, Vitro Corp Am, 48-52; div consult, Battelle Mem Inst, 52-59; res assoc, M & C Nuclear, Inc, 59-61; sect mgr, Nuclear Div, Martin Co, 61-63; FEL, BATTELLE MEM INST, 63- *Mem:* Fel AAAS; Am Nuclear Soc; Am Chem Soc; fel Am Inst Chem. *Res:* Nuclear and radiochemistry; nuclear fuels; applications of radioisotopes; radiation processes; energy conversion devices; nuclear waste isolation. *Mailing Add:* 271 S Cassingham Rd Columbus OH 43209-1804

POBINER, BONNIE FAY, b Plainfield, NJ, Mar 15, 59. ANTIBIOTICS. *Educ:* Col William & Mary, BS, 80; Univ Va, Charlottesville, PhD (pharmacol), 85. *Prof Exp:* Sr res pharmacologist, Opthalmic Res, 85-87 & Int Prod Develop, 87-88, MGR, INT PROD DEVELOP, HOECHST-ROUSSEL PHARMACEUTICALS, INC, 88- *Mem:* Inflammation Res Asn; Am Soc Microbiol; AAAS. *Res:* Preclinical liaison and coordination of pharmacological data from research and development of compounds originating Hoechst AG (Frankfurt), Roussel Uclaf (Paris), Behringwerke AG (Marburg, Germany) to be developed on an international level; compilation and writing of preclinical pharmacology reports for Food and Drug Administration Submission. *Mailing Add:* Hoechst-Roussel Pharmaceuticals, Inc Rte 202-206 PO Box 2500 Somerville NJ 08876

POBINER, HARVEY, b New York, NY, Mar 2, 27; m 54; c 2. ANALYTICAL CHEMISTRY, PHYSICAL CHEMISTRY. *Educ:* Brooklyn Col, BA, 48, MA, 56. *Prof Exp:* Chemist, Sperry Gyroscope Co, NY, 48-54; res chemist, Gen Aniline & Film Corp, NJ, 54-57 & Esso Res & Eng Co, 57-63; staff scientist, Gen Precision Aerospace Res Ctr, 63-64; supvr anal chem, Am Can Co, 64-74, mgr org chem & anal chem, 74-83; MGR, ANALYTICAL CHEM, HYDROCARBON RES, INC, 83- *Mem:* Am Chem Soc. *Res:* Instrumental methods of organic analysis; absorption and emission spectroscopy; chromatography; rare earth analysis; thin film characterization; radiation-curable coatings; organic separations; catalyst characterization; chromatography of petroleum distillates. *Mailing Add:* 29 Taylor Rd RD 4 Princeton NJ 08540-9522

POCCIA, DOMINIC LOUIS, b Utica, NY, Aug 8, 45; c 1. FERTILIZATION, SPERMATOGENESIS. *Educ:* Union Col, BS, 67; Harvard Univ, AM, 68, PhD(biochem & molecular biol), 71. *Hon Degrees:* AM, Amherst Col, 87. *Prof Exp:* Asst prof chem, Wellesley Col, 71-72; postdoctoral fel, Univ Calif, Berkeley, 72-74; asst prof cell & develop biol, State Univ NY, Stony Brook, 74-78; from asst prof to assoc prof biol, 78-87, chair Biol Dept, 83-85, PROF BIOL, AMHERST COL, 87-; ADJ PROF MOLECULAR CELL BIOL, UNIV MASS, AMHERST, 84- *Concurrent Pos:* Postdoctoral fel, Harvard Univ, 71-72; prin investr, NIH, 75-86 & 89-91, NSF, 90-93; fel, Nat Acad Sci interacad exchange, 85; vis scholar, Univ Calif, Berkeley, 86-87; vis prof, Stanford Univ, 88. *Mem:* Sigma Xi; Am Soc Cell Biol; Soc Develop Biol; AAAS. *Res:* Chromatin structure and histose biochemistry of sperm nuclei; male germ line differentiation in the sea urchin; pronuclear development following fertilization. *Mailing Add:* Dept Biol Amherst Col Amherst MA 01002

POCHAN, JOHN MICHAEL, b Kittanning, Pa, Apr 8, 42; m 76; c 4. POLYMER PHYSICS, PHYSICAL CHEMISTRY. *Educ:* Rensselaer Polytech Inst, BChE, 64; Univ Ill, Urbana, MS, 68, PhD(phys chem), 69. *Prof Exp:* Sr scientist polymer physics, Rochester Corp Res Ctr, Xerox Corp, 69-81; scientist adhesion res, Chem Res Lab, Eastman Kodak, 81-84; dir corp res, Johnson Wax, Racine, Wis, 84-87; LAB HEAD, POLYMER SCI, EASTMAN KODAK, 87- *Concurrent Pos:* Vis prof, Chem Dept, Univ Ill, 78. *Mem:* NY Acad Sci; Am Chem Soc; Am Phys Soc; Soc Prof Educ. *Res:* Liquid crystals, physical and optical properties, rheooptic phenomena; polymers, dielectric, electrical and triboelectric properties, morphology, photoconduction and transport, oxidative stability; high resolution microwave spectroscopy; adhesion chemistry; surface physics and chemistry. *Mailing Add:* Five Glen Valley Dr Penfield NY 14526-9765

POCHI, PETER E, medicine, dermatology, for more information see previous edition

POCHOP, LARRY OTTO, b Winner, SDak, Sept 12, 40; m 65. AGRICULTURAL ENGINEERING. *Educ:* SDak State Univ, BS, 62; Univ Mo-Columbia, MS, 64, PhD(agr eng), 67. *Prof Exp:* Instr agr eng, Univ Mo-Columbia, 63-67; from asst prof to assoc prof, 67-77, PROF AGR ENG, UNIV WYO, 77- *Mem:* Am Soc Agr Engrs. *Res:* Agricultural animal and plant environment. *Mailing Add:* Dept Agr Eng Univ Wyo Box 3354 Laramie WY 82071

POCHRON, SHARON, BIOCHEMISTRY, MOLECULAR BIOLOGY. *Educ:* Boston Univ, PhD(biochem), 78. *Prof Exp:* ASSOC DIR, CELL GENETICS LAB, CATHOLIC MED CTR BROOKLYN & QUEENS, 81- *Mailing Add:* Seven Martha St Seymour CT 06483

POCIUS, ALPHONSUS VYTAUTAS, b Emsdetten, Germany, May 20, 48; US citizen; m 70; c 2. POLYMER CHEMISTRY. *Educ:* Knox Col, BA, 70; Univ Ill, Urbana, PhD(phys chem), 74. *Prof Exp:* Sr res chemist, Chem Resources Div/3M, 74-75, sr res chemist, Cent Res Div-3M, 75-79, res specialist, Adhesives, Coatings & Sealers, Div/3M, 79-83, sr res specialist, Adhesives, Coatings & Sealers Div-3M, 83-87, sr res specialist, Speciality Film Div-3M, 87-89, div Scientist, Film Tech Ctr-3M, 89-90, DIV SCIENTIST, ADHESIVE TECHNOL CTR-3M, 3M CO, 90- *Honors & Awards:* Delmonte Award, Soc Advan Mat & Process Eng. *Mem:* Adhesion Soc; Am Soc Testing & Mat. *Res:* Surface chemistry, corrosion and adhesion science as well as adhesive polymer characterization and adhesive and primer development. *Mailing Add:* 445 Highpoint Curve Maplewood MN 55119

POCKER, YESHAYAU, b Kishineff, Romania, Oct 10, 28; m 50; c 2. PHYSICAL ORGANIC CHEMISTRY, BIOCHEMISTRY. *Educ:* Hebrew Univ, Israel, MSc, 49; Univ London, PhD(phys chem), 53, DSc(phys & org chem), 60. *Prof Exp:* Res assoc, Weizmann Inst, 49-50; from asst lectr to lectr chem, Univ Col London, 52-61, recognized teacher, Senate House, 59, exam sci Russian, 60; PROF CHEM, UNIV WASH, 61- *Concurrent Pos:* Vpres, Chem & Phys Soc, London, 59-60; vis assoc prof, Ind Univ, 60-61; nat speaker, Am Chem Soc, 70, 74 & 84; bicentennial lectr, Mont State Univ, 76; horizons chem lectr, Univ NC, 77; chmn, Pauling Award Comt, 78; cong lectr, Int Union Pure & Appl Chem, 80, Int Conf Bioinorg Chem, 83; guest lectr, Acad Sinica, Shanghai, China, 82, Univ Kyoto, Nagoya, Tokyo & Hong Kong, 84; conf lectr, Chem Inst Can, 83, Royal Soc Chem, 85; plenary lectr, NY Acad Sci Conf, 83, Int Workshop Zinc Enzymes, 85; prog consult, NIH, 84; cong lectr, Biomed Res Alcoholism, Helsinki, Finland, 86, Biomed Gerontol, Hamburg, WGer, 87; plenary lectr, Enzym Carbonyl Metab, Espoo, Finland, 86, NATO Advan Study Inst, Italy, 89; conf lectr, Genetic Physico-Chem Approaches to Biol Catalysis, Florence, Italy, 86, Royal Soc Chem, Sackville, NB, Can, 87; adv bd, Inorg Chim Acta, Bioinorg Chem, 82-89; bd rev ed, Science, Wash, DC, 84-; Edward A Doisy vis prof, Sch Med, St Louis Univ, 90; Consiglio Nazionale delle Richerche Plenory Lect, Univ Bari, Italy, 89. *Honors & Awards:* Plaque Award, Am Chem Soc, 70, 74 & 84; Int Union Pure & Appl Chem Award, 80. *Mem:* Am Chem Soc; fel NY Acad Sci; Am Soc Biol Chem; The Chem Soc; Royal Soc Chem; Int Soc Biomed Res Alcoholism; Int Soc Biomed Gerontol; Protein Soc. *Res:* Kinetics and mechanisms of organic reactions; molecular rearrangements; mechanisms of chemical and enzymatic catalysis; metalloenzymes; respiration; inhibitory pollutants; hydration and hydrolysis; very fast reactions; high pressure effects; kinetic isotope effects; biophysical chemistry; storage and release of solar energy; regulation of alcohol metabolism; hormones and receptors; plant biochemistry; photosynthesis. *Mailing Add:* 3515 NE 42nd Seattle WA 98105

POCOCK, STANLEY ALBERT JOHN, b London, Eng, Dec 12, 28; Can citizen; m 55. GEOLOGY, PALYNOLOGY. *Educ:* Univ Col, Univ London, BSc, 50, PhD(palynol), 64. *Prof Exp:* Exp officer, Geol Surv & Mus, London, Eng, 52-56; res geologist, Imp Oil Ltd, 56-70, geol res specialist, 70-75, sr geol res specialist, res & tech serv dept, 75-80; res adv, res & tech serv dept, Esso Resources Can, 81-85; founding consult, Stanley A J Pocock, Biostratigraphic Consults Ltd, 85-90; RETIRED. *Mem:* Can Bot Asn; Int Asn Plant Taxon; Am Asn Stratig Palynologists. *Res:* Recent palynology and palynology of Jurassic and Cretaceous sediments of Canada and worldwide; qualitative and quantitative aspects of the visual identification of the organic components of sediments. *Mailing Add:* RR No 1 Creston BC V0B 1G0 Can

PODAS, WILLIAM M(ORRIS), b Minneapolis, Minn, June 15, 16; m 42; c 2. CHEMICAL ENGINEERING. *Educ:* Univ Minn, BChE, 38. *Prof Exp:* Chem engr, Econ Lab, Inc, 39-42, Chem Warfare Serv, Edgewood Arsenal, 42 & Rocky Mountain Arsenal, Colo, 42-43; ord engr, Ord Res Ctr, Aberdeen Proving Ground, Md, 43-45; chem engr, 45-47, asst dir res & develop div, 47-60, vpres & dir res & develop, 60-73, sr vpres, Eco Lab, Inc, 73-82; RETIRED. *Mem:* Am Chem Soc; Commercial Develop Asn; Am Inst Chem Engrs; AAAS; Am Pub Health Asn. *Res:* Detergents and sanitation; ecology; washing equipment and controls; chemical processing; water treatment. *Mailing Add:* 1021 Sibley Mem Hwy #306 St Paul MN 55118

PODDAR, SYAMAL K, b Calcutta, India, Jan 14, 45, US citizen; m 71; c 1. PRODUCT UPGRADING, SPECIALTY CHEMICALS & FERTILIZERS. *Educ:* Jadavpur Univ, Calcutta, BCHE Hons, 65, MCHE, 67; Univ Pa, PhD(chem eng), 74. *Prof Exp:* Lectr chem eng, Jadavpur Univ, Calcutta, 68-71; res assoc polymer rheology, Sch Chem Eng, Univ Pa, 74-76; engr, Exxon Res & Eng, 76-77, res engr, 77-80, sr engr, 80-83, staff engr, 83-86; pres, Chementerprise Int, 87-89; TECH-PROJ COORDR & STUDY MGR, BECHTEL CORP, 90- *Concurrent Pos:* Tech teacher trainee, Jadavpur Univ, Calcutta, 65-67; mem, Chem Eng Prod Eval Panel, Am Inst Chem Engrs, 80-84; adj fac mem, Univ Houston, Clearlake, 88- & Downtown, 89- *Honors & Awards:* G R Mem Award-Gold Medal, Inst Engrs, India, 71 & 72. *Mem:* Am Inst Chem Engrs. *Res:* Direct-indirect liquefaction of coal to produce liquid fuels; assessment of technologies and cost-economics of commercial plant; two United States patents. *Mailing Add:* 16434 Larkfield Houston TX 77059

PODGERS, ALEXANDER ROBERT, b Beverly, Mass, May 19, 46; m 74; c 3. CREATIVE ACHIEVEMENT IN ENGINEERING, PRODUCT DESIGN. *Educ:* Northeastern Univ, BSME, 70; Gannon Univ, MBA, 74. *Prof Exp:* Design engr, BF Goodrich Footwear Co, 66-70; design engr, Swanson Erie Corp 70-72; proj engr, Reed Mfg Co, 72-75; sr prod engr, Smith Meter Sys, 75-77; proj design engr, Copes-Vulcan, Inc, 77-80; eng mgr, Emco Wheaton Inc, 80-90; MGR, SPECIALTY PROD, SNAP-TITE, INC, 90- *Concurrent Pos:* Pipeline Valves Comn, Am Petrol Inst, 78-80; Mem, Methanol Fuel Task Force, Calif Energy Comn, 87-; mem, Fuel Supply Sys Comn, Soc Automotive Engrs, 86-; consult expert witness various oil co & petrol equip mfrs, 84- *Mem:* Am Soc Mech Engrs; sr mem, Soc Mfr Engrs; Am Petrol Inst; Soc Automotive Engrs. *Res:* Research, development and application of fluid handling equipment for fuel and chemical distribution which enhances safety and controls environmental pollution; developed microprocessor gasoline dispenser, process control valve, service station equipment, tank truck equipment, automation machinery and sub-sea hydraulic coupling; 5 US patents. *Mailing Add:* 231 Elk Creek Rd Lake City PA 16423

PODGORSAK, ERVIN B, b Vienna, Austria, Sept 28, 43; Yugoslavian & Can citizen; m 65; c 2. MEDICAL PHYSICS, RADIATION PHYSICS. *Educ:* Univ Ljubljana, Yugoslavia, dipl ing, 68; Univ Wis-Madison, MS, 70, PhD(physics), 73. *Prof Exp:* Fel med biophys, Univ Toronto, 73-74; med physicist, Ont Cancer Inst, 74; sr physicist radiother, 75-78, DIR DEPT MED PHYSICS, MONTREAL GEN HOSP, 78- *Concurrent Pos:* Asst prof, Fac Med, McGill Univ, 75-79, assoc prof, 79-85, prof, 85-; assoc scientist, Royal Victoria Hosp, 75-, Sir Mortimer B Davis Jewish Gen Hosp, 78- *Mem:* Am Asn Physicists Med; Can Asn Physicists; fel Can Col Physicists Med; Can Col Phys Med (pres, 87-89). *Res:* Solid state radiation dosimetry; high energy equipment used in cancer therapy; radiosurgery. *Mailing Add:* Dept Radiation Oncol McGill Univ Mont Gen Hosp 1650 Ave des Cedres Montreal PQ H3G 1A4 Can

PODILA, GOPI KRISHNA, b Orissa, Sept 14, 57; Indian citizen; m 89. PLANT BIOTECHNOLOGY, MOLECULAR GENETICS OF BACTERIA & FUNGI. *Educ:* Nagarjuna Univ Andhra Pradesh, India, BSc, 77, MSc, 80; La State Univ, MS, 83; Ind State Univ, PhD(microbiol), 86. *Prof Exp:* Res asst plant path, La State Univ, 81-83; teaching fel microbiol, Ind State Univ, 83-86; res fel fungal molecular biol, Ohio State Univ, 87-90; ASST PROF MOLECULAR BIOL, MICH TECH UNIV, 90- *Res:* Molecular biology of plant-microbe interactions; biotechnological applications of both beneficial and pathogenic interactions; utilization of microbes for bioremediation of water and soil pollution and forest productivity. *Mailing Add:* Dept Biol Sci Mich Tech Univ Houghton MI 49931

PODIO, AUGUSTO L, b Bogota, Colombia, Aug 14, 40; US citizen; m 65; c 1. PETROLEUM ENGINEERING. *Educ:* Univ Tex, Austin, BS, 63, MS, 65, PhD(petrol eng), 68; Univ of the Andes, Colombia, BS, 63. *Prof Exp:* Asst prof petrol eng, 68-69, res scientist petrol eng, 69-80, ASSOC PROF PETROL ENG, UNIV TEX, AUSTIN, 77- *Mem:* Am Inst Mining, Metall & Petrol Engrs; Soc Explor Geophys. *Res:* Drilling; exploration; rock mechanics; seismic exploration; wave propagation; well-logging; ultrasonics. *Mailing Add:* Dept Petrol & Mining Eng Univ Tex Austin TX 78712

PODLAS, THOMAS JOSEPH, b East Orange, NJ, Oct 15, 40; m 68; c 2. CHEMISTRY. *Educ:* Seton Hall Univ, BS, 62, MS, 64, PhD(phys chem), 68. *Prof Exp:* Res chemist, 68-74, SR RES CHEMIST, HERCULES INC, 74- *Mem:* Am Chem Soc. *Res:* Use of natural and synthetic polymers in foods, pharmaceutical and household products; use of cellulosic polymers for enhanced petroleum recovery; use of cellulose related polymers in building materials; chemical modification of cellulose polymers. *Mailing Add:* Nine Paddington Ct Hockessin DE 19707

PODLECKIS, EDWARD VIDAS, b Elizabeth, NJ, Sept 27, 56. PLANT VIROLOGY, DECIDUOUS FRUIT PATHOLOGY. *Educ:* Rutgers Univ, BS, 79; Univ Del, MS, 82; Univ Md, PhD(plant virol), 86. *Prof Exp:* Res asst, dept plant sci, Univ Del, 79-81; teaching asst bot, Univ Md, 81-82, res asst, 82-85; RES FEL, DEPT PLANT PATH, COOK COL, RUTGERS UNIV, 85- *Mem:* Am Phytopath Soc; AAAS; Sigma Xi; NY Acad Sci. *Res:* Isolation, characterization and identification of unknown etiological agents, especially viruses, from deciduous fruits; detection of deciduous fruit and other plant viruses by electron microscopy. *Mailing Add:* Dept Plant Path Cook Col Rutgers Univ PO Box 231 New Brunswick NJ 08903

PODLESKI, THOMAS ROGER, b Bloomington, Ill, May 16, 34; div; c 2. BIOCHEMISTRY, MOLECULAR BIOLOGY. *Educ:* Univ NMex, BS, 56; Univ Rochester, MS, 60; Columbia Univ, PhD(physiol), 66. *Prof Exp:* Res assoc neurobiol, Columbia Univ, 66-68; USPHS fel, Pasteur Inst, Paris, 68, Gen Del Nat Ministry Educ fel, 69, NSF fel, 69-70; sr researcher, Nat Ctr Sci Res, France, 70; assoc prof, 70-80, chmn dept, 79-82, PROF NEUROBIOL & BEHAV, CORNELL UNIV,80- *Mem:* Am Gen Physiologists; Am Soc Cell Biol; AAAS; Soc Neurosci. *Res:* Molecular events associated with the formation of the neuromuscular synapse and the regulation of the synthesis of acetylcholine receptors in CNS; neurobiology. *Mailing Add:* Neurobiol & Behavior Dept W139 Mudd Hall Cornell Univ Ithaca NY 14853

PODOLSKY, RICHARD JAMES, b Chicago, Ill, Aug 20, 23; c 2. PHYSIOLOGY, BIOPHYSICS. *Educ:* Univ Chicago, BS, 46, PhD(biophys), 52. *Prof Exp:* Biophysicist, Naval Med Res Inst, 56-62, CHIEF SECT CELLULAR PHYSICS, LAB PHYS BIOL, NAT INST ARTHRITIS, DIABETES, DIGESTIVE & KIDNEY DIS, 62-, CHIEF LAB PHYS BIOL, 74- *Concurrent Pos:* Nat Found Infantile Paralysis res fel, Naval Med Res Inst, Md, 53-54; USPHS spec fel, Univ Col, Univ London, 55-56; mem physiol study sect, NIH, 65-69; vis prof, Univ Calif, Santa Cruz, 69; mem US nat comt, Int Union Pure & Appl Biophys, 75-81. *Mem:* Biophys Soc; Am Physiol Soc; Soc Gen Physiol. *Res:* Muscle physiology; membrane transport. *Mailing Add:* Lab Phys Biol Nat Inst Arthritis Musculoskeletal & Skin Dis Rm 114 Bethesda MD 20014

PODOSEK, FRANK A, b Ludlow, Mass, Nov 26, 41; m 84; c 2. MASS SPECTROMETRY, METEORITICS. *Educ:* Harvard Univ, BA, 64; Univ Calif, Berkeley, PhD(physics), 69. *Prof Exp:* Asst physics, Univ Calif, Berkeley, 64-69; Kellogg Lab fel, Calif Inst Technol, 69-74; from asst prof to assoc prof, 73- 83, PROF EARTH SCI, DEPT EARTH & PLANETARY SCI, WASH UNIV, ST LOUIS, 83- *Concurrent Pos:* Vis assoc prof, Univ Tokyo, 79-80; vis prof, Univ Cambridge, 82-83. *Mem:* AAAS; Am Phys Soc; Am Geophys Union; Meteoritical Soc; Geol Soc Am. *Res:* Rare gas and thermal emission mass spectroscopy of meteorites and lunar samples; extinct radioactivities. *Mailing Add:* Dept of Earth & Planetary Sci Wash Univ St Louis MO 63130-4899

PODOWSKI, MICHAEL ZBIGNIEW, b Warsaw, Poland, May 15, 40; m 66; c 2. SYSTEM STABILITY ANALYSIS. *Educ:* Warsaw Tech Univ, Poland, MSc, 65, PhD(nuclear eng), 72; Univ Warsaw, MSc, 70. *Prof Exp:* From asst prof to assoc prof nuclear eng, Warsaw Tech Univ, Poland, 65-790; vis assoc prof, Ore State Univ, Corvallis, 79-80; vis assoc prof, 80-82, assoc prof, 82-88, PROF NUCLEAR ENG & ENG PHYSICS, RENSSELAER POLYTECH INST, TROY, NY, 88- *Concurrent Pos:* Consult, Westinghouse Elec Corp, 80, 84 & 85, Yankee Atomic Elec Co, 82, Philadelphia Elec Co, 84; prin investr, US Nuclear Regulatory Comn, 80-83, Oak Ridge Nat Lab, 81-, Elec Power Res Inst, 82-83, EMPIRE STATE ELEC ENERGY RES CORP, 83-, TECH EXPERT, INT ATOMIC ENERGY AGENCY, 88- *Mem:* Sigma Xi; NY Acad Sci; Am Nuclear Soc; Am Inst Chem Engrs; Am Soc Mech Engrs. *Res:* Development of advanced methods for system stability analysis; two-phase flow instabilities with applications in nuclear reactor technology; modelling of core meltdown phenomena for analysis of severe reactor accidents; multidimensional effects in two-phase flows and boiling heat transfer. *Mailing Add:* Dept Nuclear Eng & Eng Physics Rensselaer Polytech Inst Tibbits Ave Troy NY 12180-3590

PODREBARAC, EUGENE GEORGE, b Kansas City, Kans, Nov 12, 29; m 60; c 2. MEDICINAL CHEMISTRY, QUALITY ASSURANCE. *Educ:* Rockhurst Col, BS, 52; Univ Kans, MS, 57, PhD(chem), 60. *Prof Exp:* Res chemist, US Army, 52-54; asst chem, Univ Kans, 55-56 & 57-59; assoc chemist, Midwest Res Inst, 56-57; sr chemist, 60-79, mgr qual assurance, 79-90, MGR QUAL ASSURANCE & SAFETY, MIDWEST RES INST, 90- *Concurrent Pos:* Consult, Good Lab Practice/QA, 86- *Mem:* Am Chem Soc; Sigma Xi; Soc Qual Assurance. *Res:* Synthetic organic chemistry; organometallics; pharmaceutical chemistry; design and synthesis of compounds for cancer chemotherapy; immobilized enzymes. *Mailing Add:* Midwest Res Inst 425 Volker Blvd Kansas City MO 64110-9990

PODSHADLEY, ARLON GEORGE, b Farmersville, Ill, Jan 5, 28; m 49; c 5. RESTORATIVE DENTISTRY. *Educ:* Springfield Col, AA, 47; St Louis Univ, DDS, 51, MS, 64. *Prof Exp:* Captain, pract dent, US Air Force, 51-53; dentist, Jacksonville, Ill, 53-62; asst prof dent, Univ Ky, 65-69; PROF, RESTORATIVE DENT & CHMN DEPT, UNIV LOUISVILLE, 69- *Concurrent Pos:* NIH fel, 62-65. *Mem:* Am Asn Dent Schs; Acad Oper Dent. *Res:* Clinical procedures in restorative dentistry; preventive dentistry evaluation methods; accuracy of various dental materials and technical procedures. *Mailing Add:* 305 Rannoch Ct Louisville KY 40243

PODSKALNY, JUDITH MARY, ENDOCRINOLOGY, BIOCHEMISTRY. *Educ:* George Washington Univ, PhD(genetics), 81. *Prof Exp:* BIOLOGIST, NAT INST DIABETES, DIGESTIVE & KIDNEY DIS, NIH, 74- *Mailing Add:* Nat Inst Diabetes Digestive & Kidney Br NIH Bldg 10 Rm 8S 243 Bethesda MD 20892

PODUSKA, JOHN W, SR, b Memphis, Tenn, Dec 30, 37. SCIENCE ADMINISTRATION. *Educ:* Mass Inst Technol, BS & MS, 60, ScD, 62. *Prof Exp:* Dir, Honeywell Info Syst, 70-72; vpres res & develop, Prime Comput Inc, 72-79; chmn, pres & chief exec officer, Apollo Comput Inc, 80-85; chmn & chief exec officer, Stellar Comput Inc, 86-89; PRES & CHIEF EXEC OFFICER, STARDENT COMPUTERS INC, 89- *Mem:* Nat Acad Eng; fel Inst Elec & Electronics Engrs. *Mailing Add:* Stardent Computers Inc 521 Virginia Rd Concord MA 01742

PODUSLO, SHIRLEY ELLEN, b Richeyville, Pa. NEUROCHEMISTRY. *Educ:* Ohio State Univ, BS; Johns Hopkins Univ, MA & PhD. *Prof Exp:* ASSOC PROF NEUROL & NEUROCHEM, SCH MED, JOHNS HOPKINS UNIV, 76- *Mem:* Am Soc Neurochem; Int Soc Neurochem; Soc Neurosci; Am Soc Cell Biol; Sigma Xi; Soc Biol Chem. *Res:* Characterization of purified oligodendroglia and their plasma membranes; gene regulation of their surface glycoproteins; comparison of their properties to myelin and cells from normal human brain and from demyelinating brain and to cell lines derived from glial tumors. *Mailing Add:* 4623 Fourth St Lubbock TX 79416-4725

PODZIMEK, JOSEF, b Brandys NL, Czech, Mar 24, 23; m 50; c 2. CLOUD PHYSICS, METEOROLOGY. *Educ:* Charles Univ, Prague, BS & MS, 49, MS, 52, PhD(physics & meteorol), 59. *Prof Exp:* Res asst aerodyn, Flight Res Inst, Prague, 49-53; dir & head dept cloud physics, Inst Atmospheric Physics, Czech Acad Sci, 54-69; res assoc, Atmospheric Sci Res Ctr, State Univ NY, Albany, 69-71; SR INVESTR, GRAD CTR CLOUD PHYSICS RES & PROF MECH ENG, UNIV MO-ROLLA, 71- *Concurrent Pos:* From asst prof to assoc prof, Charles Univ, Prague, 62-69, lectr, 62-66, fel, 66-69. *Mem:* Am Geophys Union; Am Meteorol Soc; WGer Soc Aerosol Res; Am Asn Aerosol Res; Sigma Xi. *Res:* Cloud and aerosol physics; aerodynamics of precipitation elements; tropospheric and stratospheric nuclei; formation of precipitation elements. *Mailing Add:* Cloud & Aerosol Sci Lab Univ of Mo Rolla MO 65401

POE, ANTHONY JOHN, b Portsmouth, Eng, July 10, 29; UK & Can citizen. INORGANIC REACTION MECHANISMS. *Educ:* Oxford Univ, BA, 50, BSc, 51; Univ London, PhD(inorg chem) & DIC, 61 DSc(inorg chem), 71; Cambridge Univ, MA, 76 & ScD(inorg chem), 90. *Prof Exp:* Sci officer, Atomic Energy Res Estab, Eng, 51-55; asst lectr inorg chem, Imp Col, Univ London, 55-57, lectr, 57-66, sr lectr, 66-70; PROF CHEM, ERINDALE COL, UNIV TORONTO, 70- *Concurrent Pos:* Vis lectr, Northwestern Univ, 61-62; vis prof, State Univ NY Buffalo, 68-69; fel, St John's Col, Cambridge Univ, Eng, 76-77. *Mem:* Am Chem Soc; fel Chem Inst Can; fel Royal Soc Chem. *Res:* Kinetics and mechanisms of thermal and photochemical reactions, especially of metal carbonyls, metal carbonyl clusters, and their organic derivatives. *Mailing Add:* Dept Chem Erindale Col Univ Toronto Mississauga ON L5L 1C6 Can

POE, DONALD PATRICK, b DuQuoin, Ill, Aug 19, 48. ANALYTICAL CHEMISTRY. *Educ:* Southern Ill Univ, BS, 70; Iowa State Univ, MS, 72, PhD(analytical chem), 74. *Prof Exp:* asst prof, 74-80, ASSOC PROF ANALYTICAL CHEM, UNIV MINN, DULUTH, 80- *Concurrent Pos:* Vis assoc prof, Georgetown Univ, 87-88. *Honors & Awards:* NSF Res Opportunity Award, 87. *Mem:* Am Chem Soc; Sigma Xi. *Res:* Oxidation-reduction properties of metal complexes; spectrophotometric and electrochemical studies of mixed ligand complexes; liquid chromatography with electrochemical detection; chromatographic theory. *Mailing Add:* Dept of Chem Univ of Minn Duluth MN 55812

POE, MARTIN, b St Louis, Mo, Sept 26, 42; m 65; c 1. BIOPHYSICS, PHYSICAL BIOCHEMISTRY. *Educ:* Mass Inst Technol, BS, 64; Univ Pa, PhD(biophys), 69. *Prof Exp:* Fel biophys, E I du Pont de Nemours & Co, Inc, 69-70; sr res biophysicist, 70-74, res fel, 74-79, sr res fel biophysics, 79-84, SR INVESTR, MERCK INST THERAPEUT RES, 84- *Concurrent Pos:* Byron Riegel lectr, Northwestern Univ, 79. *Mem:* Biophys Soc; Am Soc Biol Chemists; Am Chem Soc; Am Soc Hypertension. *Res:* Physical chemistry of macromolecules, particularly dihydrofolate reductase, renin, and lymphocyte proteases; biophysical chemistry, particularly rational design of drugs; high resolution nuclear magnetic resonance; enzymology. *Mailing Add:* PO Box 2000 Merck Sharp & Dohme Res Labs Rahway NJ 07065

POE, RICHARD D, b Commerce, Tex, Dec 13, 31; m 55; c 3. ANALYTICAL CHEMISTRY, ORGANIC CHEMISTRY. *Educ:* ETex State Col, BS, 57, MS, 58; Tex A&M Univ, PhD(anal chem), 64. *Prof Exp:* From instr to asst prof chem, Tarleton State Col, 58-61; head anal chem, Alcon Labs Inc, 64-65, dir chem res, 65-68, head anal develop, Sci & Technol Div, 68-72, mgr, Chem Serv, 72-74, mgr ophthalmic res, 74-78, assoc dir ophthal res & develop, 78-80, dir chem, 80-83, DIR PROD TECH, ALCON LABS INC, 84- *Mem:* Am Chem Soc; fel Am Inst Chem. *Res:* Analytical chemistry, especially organic and pharmaceutical applications; organic chemistry, especially medicinal applications. *Mailing Add:* Alcon Labs Inc PO Box 6600 Ft Worth TX 76115-0600

POE, ROBERT HILLEARY, b Cincinnati, Ohio, Apr 29, 34; m 63; c 3. PULMONARY DISEASE. *Educ:* Univ Cincinnati, BS, 55, MD, 59. *Prof Exp:* ASSOC PROF MED, PULMONARY DIS, SCH MED & DENT, UNIV ROCHESTER, 74-; DIR PULMONARY, HIGHLAND HOSP ROCHESTER, 74- *Mem:* Am Fedn Clin Res; Fel Am Col Chest Physicians; Am Thoracic Soc; fel Am Col Physicians. *Res:* Assessment of procedures used in pulmonary medicine; factors influencing reactive airway disease. *Mailing Add:* Highland Hosp South Ave & Bellevue Dr Rochester NY 14620

POE, SIDNEY LAMARR, b Leesville, La, Sept 29, 42; m 66; c 2. ENTOMOLOGY. *Educ:* Northwestern State Univ, BS, 64, MS, 66; Univ Mo-Columbia, PhD(entom), 70. *Prof Exp:* From asst prof to prof entom, Univ Fla, 69-79; PROF & HEAD, DEPT ENTOM, VA POLYTECH INST & STATE UNIV, 79- *Mem:* Entom Soc Am. *Res:* Acarology; biology and ecology of saprophytic and phytophagous mites; host plant resistance and integrated management of agronomic insect and mite pests; biological control of arthropods; zoology. *Mailing Add:* 1555 Beasley Rd Jackson MS 39206

POEHLEIN, GARY WAYNE, b Tell City, Ind, Oct 17, 36; m 58; c 4. CHEMICAL ENGINEERING, POLYMER CHEMISTRY. *Educ:* Purdue Univ, BS, 58, MS, 63, PhD(chem eng), 66. *Prof Exp:* Design engr, Procter & Gamble Co, 58-61; from asst prof to prof chem eng, Lehigh Univ, 65-78; prof chem eng & dir Sch Chem Eng, 78-86, assoc vpres grad studies & res, 86-89, VPRES INTERDISCIPLINARY PROGS, GA INST TECHNOL, 89- *Concurrent Pos:* Indust consult, 66-; bd dir, Flexible Product Co, Marietta, Ga, 81- *Honors & Awards:* MacPruitt Award, Coun Chem Res, 89. *Mem:* Am Inst Chem Engrs; Am Chem Soc; Am Soc Eng Educ. *Res:* Polymerization kinetics; latex technology; applied rheology. *Mailing Add:* Cent Res Blod Ga Inst Technol Atlanta GA 30332-0370

POEHLER, THEODORE O, b Baltimore, Md, Oct 20, 35; m 61; c 2. SOLID STATE PHYSICS, LASERS. *Educ:* Johns Hopkins Univ, BS, 56, MS, 58, DEng(solid state physics), 61. *Prof Exp:* Instr, Johns Hopkins Univ, 58-60, from res staff asst to res staff assoc, 59-62, solid state physicist, 63-69, prin staff physicist, 69-74, head, Quantum Electronics Group, Appl Physics Lab, 74-83, dir, Milton S Eisenhower Res Ctr, 83-89, ASSOC DEAN, WHITING SCH ENG, JOHNS HOPKINS UNIV, 89- *Concurrent Pos:* Consult, Appl Physics lab, Johns Hopkins Univ, 62, William S Parsons vis prof, 72-73. *Honors & Awards:* Nat Capital Award Eng, 71. *Mem:* AAAS; Am Phys Soc; Inst Elec & Electronics Engrs; Am Chem Soc; Mat Res Soc. *Res:* Semiconductors; thin films; optical masers; crystal growth techniques; solid state plasmas; millimeter waves; cyclotron resonance; band structure; quantum electronics; infrared and chemical lasers; organic conductors. *Mailing Add:* Whiting Sch Eng Johns Hopkins Univ 34th & Charles St Baltimore MD 21218

POEHLMAN, JOHN MILTON, b Macon, Mo, May 9, 10; m 36; c 2. PLANT BREEDING. *Educ:* Univ Mo, BS, 31, PhD(bot), 36. *Prof Exp:* From instr to assoc prof, 34-50, prof agron, 50-80, EMER PROF AGRON, UNIV MO-COLUMBIA, 80- *Concurrent Pos:* Res adv, Orissa Univ, India, 63-65. *Honors & Awards:* Thomas Jefferson Award, Univ Mo, 78. *Mem:* Fel Am Soc Agron; fel Crop Sci Soc Am; fel Indian Soc Genetics & Plant Breeding. *Res:* Breeding wheat, oats, barley, rice and mungbeans; international agricultural development. *Mailing Add:* Dept of Agron Univ of Mo Columbia MO 65211

POEL, ROBERT HERMAN, b Kalamazoo, Mich, Feb 10, 41; m 65; c 2. SCIENCE EDUCATION. *Educ:* Kalamazoo Col, BA, 62; Western Mich Univ, MA, 64, PhD(sci educ), 70. *Prof Exp:* Teacher pub sch, Mich, 63-66; asst prof, 70-74, assoc prof natural sci, 74-82, coordr grad sci educ, 78-88, PROF NATURAL SCI, WESTERN MICH UNIV, 82-, DIR CTR SCI EDUC, 88- *Concurrent Pos:* Consult Ministry Educ, Repub Liberia, WAfrica. *Mem:* Nat Asn Res Sci Teaching; Am Asn Physics Teachers; Sch Sci & Math Asn. *Res:* Development of critical thinking skills within the context of science; problem solving and decision making. *Mailing Add:* Ctr Sci Educ A-428 Ellsworth Hall Western Mich Univ Kalamazoo MI 49008

POEL, RUSSELL J, organic chemistry; deceased, see previous edition for last biography

POENIE, MARTIN FRANCIS, b Downey, Calif, Nov 4, 51; m 85. CELL BIOLOGY, CELL SIGNALING. *Educ:* Calif State Univ Northridge, BA, 73; Calif State Univ, Fullerton, MA, 79; Stanford Univ, PhD (biol), 86. *Prof Exp:* Postdoctoral, Univ Calif, Berkeley, 84-87; ASST PROF ZOOL, UNIV TEX, AUSTIN, 87- *Concurrent Pos:* NSF presidential young investr, 87. *Mem:* AAAS. *Res:* Intracellular signals that regulate cell responses to external stimuli; calcium regulation T cell activation; directionality of cytotoxic T lymphocytes. *Mailing Add:* Dept Zool 141 Patterson Labs Univ Tex Austin TX 78712

POENITZ, WOFGANG P, b Erfurt, Ger, Apr 24, 38; m 64; c 1. PHYSICS. *Educ:* Univ Gottingen, vordiplon, 59; Karlsruhe Tech Univ, dipl, 62, Dr rer nat(physics), 66. *Prof Exp:* Physicist, Nuclear Res Ctr, Karlsruhe Tech Univ, 62-67; from assoc physicist to physicist, 67-87, PHYSICIST, ARGONNE NAT LAB, 87-- *Mem:* Am Phys Soc. *Res:* Neutron physics; nuclear structure; nuclear reactor. *Mailing Add:* 329 E Mink Creek Rd Pocatello ID 83204

POEPPEL, ROGER BRIAN, b Staten Island, NY, July 6, 41; m 67; c 3. MATERIALS SCIENCE, CERAMICS. *Educ:* Cornell Univ, BEngPhys, 64, PhD(mat sci), 69. *Prof Exp:* Instr mat sci, Cornell Univ, 68-69; asst metallurgist, 69-72, group leader ceramics, 76-86, metallurgist, 72-89, SECT MGR CERAMICS, ARGONNE NAT LAB, 86-, SR SCIENTIST, 89- *Concurrent Pos:* Instr dept gen surg, Rush Med Col, 71-78. *Mem:* Am Ceramic Soc; Sigma Xi. *Res:* Ceramic development and testing for new energy applications, including storage batteries, fuel cells, magnetohydrodynamic power generation; refractories for coal gasification; structural ceramics; nuclear reactor fuel reprocessing, fabrication, testing and safety analysis; solid breeder blanket materials for fusion reactors; high temperature ceramic superconductors superconductivity. *Mailing Add:* Argonne Nat Lab 9700 S Cass Ave Argonne IL 60439

POEPPELMEIER, KENNETH REINHARD, b St Louis, Mo, Oct 6, 49; m 78; c 3. SOLID-STATE CHEMISTRY, CATALYSIS. *Educ:* Univ Mo, Columbia, BS, 71; Iowa State Univ, PhD(chem), 78. *Prof Exp:* Instr chem, Peace Corps, Samoa Col, 71-74; res chemist, Exxon Res & Eng Co, 78-80, sr chemist, 80-81, staff chemist, 81-84, sr staff chemist, 84; assoc prof, 84-88, PROF CHEM, NORTHWESTERN UNIV, 88- *Concurrent Pos:* Chmn, Solid State Div Inorg Chem, Am Chem Soc, 88-89; assoc dir, Sci & Technol Ctr Superconductivity, NSF, 88- *Mem:* Am Chem Soc; Catalysis Soc; AAAS; Sigma Xi. *Res:* Heterogeneous catalysis and ceramics with focus on the fundamental aspects and reactivity of solids and new pathways for synthesis of new and technically important solid-state compounds. *Mailing Add:* Chem Dept Northwestern Univ 2145 Sheridan Rd Evanston IL 60208

POESCHEL, GORDON PAUL, b Milwaukee, Wis, Nov 21, 42; m 66; c 2. PARASITOLOGY. *Educ:* Univ Wis-Madison, BS, 65, MS, 66, PhD(vet sci), 68. *Prof Exp:* Res parasitologist, Walter Reed Army Inst Res, US Army, 68-69, clin lab off, 6th Convalescent Ctr, Vietnam, 69-70; group leader anthelmintic chemother, Am Cyanamid Co, Princeton, 70-73, group leader parasitol discovery, 74, mgr prod develop & agr, Cyanamid Latin Am-Asia, 74-76, mkt mgr animal prod, Cyanamid Americas Far East Div, Wayne, 76-78, mgr, Agr Div, Cyanamid Australia, Ltd, 79-80, mkt mgr, Amdro Fire Ant Insecticide, Agr Div, Am Cyanamid Co, Wayne, 80-81; mgr mkt servs, Agr Chem Div, 83-90, EASTERN REGION SALES MGR, ANIMAL HEALTH DEPT, MOBAY CORP, 82-, VPRES MARKETING, 90- *Mem:* Nat Agr-Mkt Asn. *Res:* Parasitology. *Mailing Add:* Agr Chem D Mobay Corp Box 4913 Hawthorn Rd Kansas City MO 64120

POESE, LESTER E, b St Charles, Mo, Apr 1, 13; m 37; c 2. CHEMICAL ENGINEERING, ENGINEERING. *Educ:* Mo Sch Mines & Metall, BS, 34. *Prof Exp:* Control chemist, Titanium Div, Nat Lead Co, Mo, 34-35 & NJ, 35-36, process foreman, 36-38, process engr, 38-42; area head, Merck & Co, Inc, 42-55; owner, Eastern Penn Oil Co, 55-65; GEN ENGR, US ARMY TECH DETACHMENT, US ARMY ARMAMENT RES & DEVELOP COMMAND, INDIAN HEAD, MD, 65- *Mem:* Am Chem Soc; Am Inst Chem Engrs; Nat Soc Prof Engrs. *Mailing Add:* 934 Portsmouth Circle Maryville TN 37801-6714

POET, RAYMOND B, b Port Chester, NY, Feb 12, 20; m 43; c 2. BIOCHEMISTRY. *Educ:* City Col New York, BS, 42. *Prof Exp:* Jr chemist, NY Univ Res Div, Goldwater Mem Hosp, NY, 45-50; res asst anal chem, Squibb Inst Med Res, 50-57, sr res chemist, 58-68, res assoc, 68-69, res fel, 69-74, sect head, 74-82; RETIRED. *Mem:* Am Chem Soc. *Res:* Development of analytical methods for synthetic and natural products. *Mailing Add:* 1122 SE 4th St Crystal River FL 32629

POETHIG, RICHARD SCOTT, b Buffalo, NY, July 17, 53; m 89. DEVELOPMENTAL GENETICS. *Educ:* Col Wooster, BA, 74; Yale Univ, MPhil, 81, PhD(biol), 81. *Prof Exp:* Fel, Stanford Univ, 81; fel, Univ Mo, Columbia, 81-83; asst prof, 83-89, ASSOC PROF BIOL, DEPT BIOL, UNIV PA, 89- *Honors & Awards:* Nickenson Prize, Exp Biol, Yale Univ, 81. *Mem:* Soc Develop Biol; Genetics Soc Am. *Res:* Genetic and developmental analysis of plant morphogenesis. *Mailing Add:* Biol Dept Univ Penn Philadelphia PA 19104-6018

POETTMANN, FRED H(EINZ), b Moers, Ger, Dec 20, 19; US citizen; m 46; c 2. CHEMICAL ENGINEERING, PETROLEUM ENGINEERING. *Educ:* Case Inst Technol, BS, 42; Univ Mich, MS & ScD(chem eng), 46. *Prof Exp:* Chemist, Lubrizol Corp, Ohio, 42-43; mgr prod res sect, Phillips Petrol Co, 46-55; supvr engr, Res Dept, Marathon Oil Co, 55-61, mgr eng & chem dept, 61-63, assoc res dir, 63-72, res adv to vpres res & chem, 72-73, mgr, com develop div, 72-83; PROF PETROL ENG, COLORADO SCH MINES, 83- *Concurrent Pos:* Spec lectr, Exten Div, Okla State Univ, 47-55; lectr, Ente Nazional Idrocarburi, Milan, Italy, 68; mem res comt, Interstate Oil Compact Comn, 67-; mem adv panel, US Off Technol Assessment, 76-77; textbk ed, Soc Petrol Engrs, 83- *Honors & Awards:* Lester C Uren Award, Soc Petrol Engrs, 66, John Franklin Carll Award, 71; Enhanced Oil Recovery Pioneer, Dept Energy & Soc Petrol Engrs, 84; Halliburton Educ Award, Colo Sch Mines, 86; Everette Lee Degolya Metal, Soc Petrol Engrs, 90. *Mem:* Nat Acad Eng; Am Chem Soc; Am Inst Chem Engrs; hon mem Am Inst Mining, Metall & Petrol Engrs; hon mem Soc Petrol Engrs. *Res:* Reservoir mechanics; natural gas engineering; phase behavior of hydrocarbons under pressure; multiphase-fluid flow; oil recovery processes. *Mailing Add:* 47 Eagle Dr Littleton CO 80123

POFFENBARGER, PHILLIP LYNN, b Lafayette, Ind, Oct 13, 37; m 57; c 6. INTERNAL MEDICINE, BIOCHEMISTRY. *Educ:* Ind Univ, Bloomington, AB, 59; Ind Univ, Indianapolis, MS & MD, 63. *Prof Exp:* Intern med, Univ Wash Hosp, 63-64, resident, Affiliated Hosps, 64-65; asst prof med, 70-73, asst prof biochem, 71-73, assoc prof, 73-76, PROF MED, HUMAN BIOL CHEM & GENETICS, UNIV TEX MED BR GALVESTON, 76-, DIR, CLIN RES CTR & DIV ENDOCRINOL & METAB, 75-; FEL AM COL PHYSICIANS, DIABETES & INTERNAL MED CLIN. *Concurrent Pos:* NIH fel, Univ Wash, 65-66; Am Diabetes Asn res & develop award, Harvard Med Sch, 68-69, NIH spec fel, 69-70. *Mem:* Am Soc Clin Investr; Am Fedn Clin Res; Endocrine Soc; Am Diabetes Asn. *Res:* Diabetes mellitus, insulin biosynthesis; insulin-like proteins in human blood. *Mailing Add:* Diabetes & Internal Med Clinic 450 Medical Center Blvd Suite 205 Webster TX 77598

POGANY, GILBERT CLAUDE, b Brussels, Belg, July 4, 32; US citizen; m 62; c 2. EXPERIMENTAL EMBRYOLOGY. *Educ:* Belmont Col, BS, 62; Tulane Univ La, MS, 64, PhD(biol), 66. *Prof Exp:* Res asst streptococci, Med Sch, Vanderbilt Univ, 61-62; asst prof embryol, Emory & Henry Col, 66-69; asst prof embryol, 69-74, asst prof biol, 74-80, ASSOC PROF BIOL, NORTHERN ARIZ UNIV, 80- *Concurrent Pos:* Biol Res Ctr fel, 67-68; Sigma Xi grant, 70. *Mem:* Am Soc Zool; Soc Develop Biol. *Res:* Effects of radiation on embryonic development. *Mailing Add:* Northern Ariz Univ Box 5640 Flagstaff AZ 86011

POGELL, BURTON M, b Baltimore, Md, Mar 3, 28; m 87; c 1. BIOCHEMISTRY, MICROBIOLOGY. *Educ:* Johns Hopkins Univ, BA, 48; Univ Wis, PhD(physiol chem), 52. *Prof Exp:* Asst prof biochem, Wilmer Ophthal Inst, Johns Hopkins Univ, 55-58; asst prof microbiol, Sch Med, Vanderbilt Univ, 60-64; assoc prof biochem, Albany Med Col, 64-69; prof microbiol, Sch Med, St Louis Univ, 69-80; PROF BIOMED CHEM, SCH PHARM, 80-, RES PROF, CTR AGR BIOTECHNOL, UNIV MD, BALTIMORE, 90- *Concurrent Pos:* NIH spec trainee, Biochem Res Inst, Buenos Aires, 58-59, Carlsburg Lab, Copenhagen, 59-60 & career develop award, 60-69. *Mem:* Am Soc Biol Chem; Am Soc Microbiol. *Res:* Enzymology and intermediary metabolism; control and regulation of fructose-1, 6-diphosphatase activity and its role in glyconeogenesis; biosynthesis of puromycin; control of differentiation in actinomycetes and chick embryo liver. *Mailing Add:* Dept Biomedicinal Chem Sch Pharmacy Univ MD 20 N Pine St Baltimore MD 21201

POGGENBURG, JOHN KENNETH, JR, b New York, NY, Jan 18, 35; m 63; c 5. NUCLEAR CHEMISTRY, RADIO PHARMACEUTICALS. *Educ:* Col Holy Cross, BS, 56; Univ Calif, Berkeley, PhD(chem), 66. *Prof Exp:* Res asst nuclear chem, Univ Calif, Berkeley, 59-65; res staff neutron prod, 65-73, group leader, Biomed Radioisotopes Group, Oak Ridge Nat Lab, 73-77, sect head biomed effects & instrumentation, 77-78; mgr nuclear prod res & develop, Med Prod Div, Union Carbide Corp, 78-81; dir res & develop, Analytab Prod Div, Ayerst Labs, 81-84; dir, radiopharmaceut, 84-90, DIR THERAPEUT PROJ MGT, HYBRITECH, INC, 90- *Concurrent Pos:* Exten instr, Univ Tenn, 67-70. *Mem:* Soc Nuclear Med; Am Chem Soc; AAAS; Sigma Xi; NY Acad Sci. *Res:* Nuclear spectroscopy; nuclear structure; theoretical alpha decay rates of heavy elements; radioisotope production, including nuclear reactions and chemical process development; radiopharmaceuticals; diagnostic and therapeutic applications immunology. *Mailing Add:* Hybritech Inc PO Box 269006 San Diego CA 92126-9006

POGGIO, ANDREW JOHN, b New York, NY, Nov 3, 41; c 1. MICROWAVE THEORY & TECHNIQUES. *Educ:* Cooper Union Sch Eng, BEE, 63; Univ Ill, MS, 64, PhD, 69. *Prof Exp:* Res specialist electrosci, MB Assocs, 69-71; res engr, Cornell Aeronaut Lab, 71-73; proj engr, Diag Systs, 73-77, group leader, 77-79, nuclear safeguard prog leader, 79-83, proj engr microwave diag, 83-85, group leader, 85-87; DEP DIV LEADER, ENG RES DIV, LAWRENCE LIVERMORE NAT LAB, UNIV CALIF, 87- *Concurrent Pos:* Scientific ed, Energy & Technol Rev & Res Monthly, 88-89. *Mem:* Inst Elec & Electronics Engrs; Sigma Xi; Inst Elec & Electronic Engrs Antennas & Propagation Soc; Inst Elec & Electronic Engrs Microwave Theory & Techniques Soc; Comn B Int Union Radio Sci. *Res:* Electromagnetic theory, antennas, scattering, propagation; microwave theory and techniques, acoustics; numerical analysis; systems analysis. *Mailing Add:* Lawrence Livermore Nat Lab Univ of Calif Livermore CA 94550

POGGIO, GIAN FRANCO, b Genoa, Italy, Sept 21, 27; US citizen. NEUROLOGY. *Educ:* Univ Genoa, MD, 51. *Prof Exp:* Asst neurol & psychiat, Univ Genoa, 52-54; from asst prof to prof physiol, 60-85, PROF NEUROSCI, SCH MED, JOHNS HOPKINS UNIV, 80- *Concurrent Pos:* Fel neurol surg, Sch Med, Johns Hopkins Univ, 54-56, fel physiol, 57-60. *Res:* Anatomy and physiology of the mammalian visual system; organization of the thalamus and cerebral cortex. *Mailing Add:* Dept Neurosci Johns Hopkins Sch Med Baltimore MD 21205

POGO, A OSCAR, b Buenos Aires, Arg, Aug 28, 27; m 55; c 2. CELL BIOLOGY. *Educ:* Univ Buenos Aires, MD, DMSci, 59. *Prof Exp:* Res assoc, Inst Anat & Embryol, Fac Med Sci, Univ Buenos Aires, 55-59; asst prof, Inst Cell Biol, Fac Med Sci, Univ Cordoba, Arg, 61-64; res assoc, Rockefeller Univ, 64-66, asst prof, 66-67; investr, 67-74, HEAD, LAB CELL BIOL, NY BLOOD CTR, 67-, SR INVESTR, 74- *Concurrent Pos:* Fel, Guggenheim Found, 59-61; career develop award, NIH, 66; prin investr, NIH & NSF grants, 67- *Mem:* Am Soc Biol Chemists; Am Chem Soc; AAAS; Am Soc Cell Biol. *Res:* Gene structure and functions; molecular biology of blood group genes; expression of blood group genes in nonerythroid cells; molecular basis of malaria resistance; genetic predisposition to disease; structure and functions of the cell nucleus; transcription and process of primary transcripts. *Mailing Add:* Dept Cell Biol Lindsley F Kimball Res Inst NY Blood Ctr 310 E 67th St New York NY 10021

POGO, BEATRIZ G T, b Buenos Aires, Arg, Dec 24, 32; US citizen; m 56; c 2. ONCOGENES, GENE EXPRESSION & REGULATION. *Educ:* Liceo Nacional No 1, Buenos Aires, BS, 50; Sch Med, Buenos Aires Univ, MD, 56, DMSc, 61. *Prof Exp:* Instr cell biol, Inst Anat & Embryol, Sch Med, Buenos Aires Univ, 57-59; res fel viral oncol, Sloan Kettering Inst, NY, 59-60; res fel cell biol, Rockefeller Univ, 60-61, res assoc, 64-67; asst prof, Inst Cell Biol, Sch Med, Cordoba Univ, 62-64; asst, Pub Health Res Inst, NY, 67-69, assoc, 69-73, assoc mem, 73-78; actg dir, Ctr Exp Cell Biol, 87-89, PROF CELL BIOL & MICROBIOL, MT SINAI SCH MED, 78-, PROF NEOPL DIS, 89- *Concurrent Pos:* Res career award, Nat Res Coun Arg, 62-64; mem, Grad Sch Biomed Sci, Mt Sinai Sch Med, City Univ NY, 80-; Damon Runyon fel, 64-65. *Mem:* Am Soc Cell Biol; Am Asn Cancer Res; Am Soc Microbiol; fel NY Acad Sci; Am Soc Virol; AAAS; Sigma Xi. *Res:* Molecular mechanisms of replication and expression of cytolytic and oncogenic viruses; identifying the sites and the products of the viral genome responsible for the effects on cells; cell biology; virology; viral oncology. *Mailing Add:* Mt Sinai Sch Med One Gustave Levy Pl New York NY 10029

POGORSKI, LOUIS AUGUST, b Lwow, Poland, Nov 7, 22; Can citizen; m 55; c 2. CHEMICAL ENGINEERING, EARTH SCIENCES. *Educ:* Polish Univ Col, London, dipl eng, 51; Univ Toronto, MASc, 53, PhD(chem eng), 58. *Prof Exp:* Chem engr, Barvue Mines, Que, 51-52; res chemist, Imp Oil Esso Res, Ont, 52; demonstr chem eng, Univ Toronto, 52-55; res engr, Air Prod, Inc, Pa, 55-57; sr process & proj engr, Chem Construct Corp, NY, 57-60; PRES & RES DIR, CHEM PROJS LTD, 60-; PRES & RES DIR, HELIUM SURVEYS INC, 75- *Concurrent Pos:* Consult, US & Can industs, 61- *Res:* Geochemical exploration for oil, natural gases, minerals; stable isotope ratio determination; sampling techniques; trace determination; pollution detection and control; cryogenic processes; chemical plant design methods; separation and ultrapurification of gases and liquids; plasma techniques. *Mailing Add:* 339 Maple Leaf Dr Toronto ON M6L 1P4 Can

POGORZELSKI, HENRY ANDREW, b Harrison, NJ, Sept 26, 22; wid; c 1. THEORETICAL COMPUTER SCIENCE. *Educ:* Princeton Univ, MA, 68; City Univ, New York, PhD(math), 69. *Prof Exp:* Sci ed, Math Rev, Am Math Soc, RI, 53-64; res asst math, Inst Advan Study, 64-65; lectr, City Col, New York, 65-66; res asst, Inst Advan Study, 66-67; lectr, Fordham Univ, 67-69; assoc prof, 69-74, PROF MATH, UNIV MAINE, ORONO, 74-; PRES, RES INST SEMIOLOGICAL MATH. *Mem:* Inst Advan Study. *Res:* Recursive functions and number theory; mathematical logic. *Mailing Add:* Res Inst for Semiological Math 383 College Ave Orono ME 04473

POGORZELSKI, RONALD JAMES, b Detroit, Mich, Mar 25, 44; m 72; c 2. ELECTROMAGNETIC THEORY, PHYSICS. *Educ:* Wayne State Univ, BS, 64, MS, 65; Calif Inst Technol, PhD(elec eng), 70. *Prof Exp:* Mem tech staff, Hughes Aircraft Co, 66; asst prof eng & appl sci, Univ Calif, Los Angeles, 69-73; assoc prof elec eng, Univ Miss, 73-77; sr staff eng, TRW, 77-90; DIR, ENG RES GROUP, GEN RES CORP, 90- *Concurrent Pos:* Consult, Sci & Technol Assocs, Inc, 70-71, Assoc Sci Advisers, 72-73 & Univ Colo, 76; instr, Univ Southern Calif, 81-85, adj prof, 85-89; assoc ed, Inst Elec & Electronic Engrs Trans Antennas & Propagation, 80-86, ed, 86-89. *Honors & Awards:* R W P King Award, Inst Elec & Electronic Engrs, 79. *Mem:* sr mem, Inst Elec & Electronic Engrs; Am Phys Soc; Int Union Radio Sci; Optical Soc Am; Sigma Xi; AAAS; Appl Computational Electromagnetics Soc. *Res:* Electromagnetic theory. *Mailing Add:* 1133 Nirvana Rd Santa Barbara CA 93101-4946

POGUE, JOHN PARKER, b Maysville, Ky, May 21, 25; m 50; c 6. VETERINARY MEDICINE. *Educ:* Miami Univ, AB, 49; Auburn Univ, DVM, 58. *Prof Exp:* Gen pract, 58-67; res vet, Commercial Solvents Corp, 68-73; clin vet, 74, MGR VET SERV, NORWICH EATON PHARMACEUT, 75- *Mem:* Am Vet Med Asn; Am Asn Swine Practrs; Indust Vet Asn; Am Asn Lab Animal Sci; Am Soc Lab Animal Practrs. *Res:* The effects of dantrolene sodium, a peripheral muscle relaxant, on muscle spasm resulting from trauma to the long bones. *Mailing Add:* 52 Randall Ave Norwich NY 13815

POGUE, RANDALL F, b Joplin, Mo, Apr 9, 62. CATALYSIS-ZEOLITE & METAL OXIDES, SYNTHESIS & UTILIZATION OF AROMATIC MONOMERS. *Educ:* Drury Col, Springfield, Mo, BA, 84; Okla Univ, MS, 86, PhD(phys chem), 88. *Prof Exp:* SR RES CHEMIST, HYDROCARBONS RES, DOW CHEM CO, 88- *Mem:* Am Chem Soc. *Res:* Cutting edge technology for the upgrade of low value hydrocarbon feedstocks for the production of aromatic monomers; catalyst development; micro-reactor design; analytical analysis. *Mailing Add:* 15546 Pensgate Houston TX 77062

POGUE, RICHARD EWERT, b Loma Linda, Calif, Jan 7, 30; m 53; c 3. MEDICAL STATISTICS, COMPUTER SCIENCE. *Educ:* Univ Minn, BA, 52, MS, 64, PhD(biostatist), 66. *Prof Exp:* Asst prof biomet, Col Med Sci, Univ Minn, 66-68; assoc prof community med, 68-73, assoc prof family pract & health syst & info sci, 73-78, assoc dir syst & comput serv, 73-80; PROF HEALTH SYST & INFO SCI, DEPT FAMILY PRACT, MED COL GA, 78- *Mem:* Asn Comput Mach; Biomet Soc; Asn Develop Comput Based Instrnl Systs; Soc Comput Med. *Res:* Computers in medicine and education; education in computer medicine. *Mailing Add:* 3527 Interlachen Lane Augusta GA 30907

POHL, DOUGLAS GEORGE, b Washington, DC, July 10, 44; m 67; c 3. BIOLOGICAL DIELECTROPHORESIS, CELLULAR ROTATION. *Educ:* John Hopkins Univ, BA, 66; Pa State Univ, PhD(chem), 74. *Prof Exp:* Asst prof chem, Univ Toledo, 74-75, Goucher Col, 75; post doc fel chem, John Hopkins Univ, 75-76; from asst prof to assoc prof, 76-86, PROF CHEM, GA COL, 86-; PRES & RES DIR, POHL CANCER RES LAB, INC, 86- *Concurrent Pos:* Dir, Ga Jr Acad Sci, 87-88. *Mem:* Am Chem Soc; Am Asn Advan Sci; Sigma Xi. *Res:* Examination of electrical properties of living organisms and polymers; dielectrophoresis, cellular rotation, cellular sorting and electrofusion of cells. *Mailing Add:* Dept Chem Ga Col Box 082 Milledgeville GA 31061

POHL, LANCE RUDY, DRUG METABOLISM TOXICITY, IMMUNOLOGY. *Educ:* Univ Calif, PhD(med chem), 74. *Prof Exp:* SECT CHIEF, NAT HEART, LUNG & BLOOD INST, NIH, 80- *Res:* Medicinal chemistry. *Mailing Add:* Nat Heart Blood & Lung Inst NIH Bldg 10 Rm 8N115 Bethesda MD 20892

POHL, RICHARD WALTER, b Milwaukee, Wis, May 21, 16; m 41; c 3. BOTANY. *Educ:* Marquette Univ, BS, 39; Univ Pa, PhD(bot), 47. *Prof Exp:* Asst instr bot, Univ Pa, 39-42; range conservationist, Soil Conserv Serv, USDA, Tex, 42 & 44-45; from asst prof to prof, 47-75, DISTINGUISHED PROF BOT, IOWA STATE UNIV, 75- *Concurrent Pos:* Ford Found fel, Univ Calif, 52-53; res assoc, Field Mus & Milwaukee Pub Mus; Hon Cur, Nat Mus Costa Rica; distinguished prof, Iowa State Univ, 75; Fulbright Res Scholar, Costa Rica, 82 & 89. *Mem:* Bot Soc Am; Am Soc Plant Taxon (treas, 59-65, pres, 73). *Res:* Taxonomy of seed plants; Central American Gramineae, grassland vegetation. *Mailing Add:* Dept Bot Iowa State Univ Ames IA 50011

POHL, ROBERT O, b Gottingen, Germany, Dec 17, 29; nat US; m 61; c 3. SOLID STATE PHYSICS. *Educ:* Univ Erlangen, MS, 55, PhD(physics), 57. *Prof Exp:* Asst physics, Univ Erlangen, 57-58; res assoc, 58-60, from asst prof to assoc prof, 60-68, PROF PHYSICS, CORNELL UNIV, 68- *Concurrent Pos:* Vis prof, Aachen Tech Univ, 64 & Stuttgart Tech Univ, 66-67; Guggenheim fel, 73-74; Alexander von Humboldt Sr Scientist fel, 80. *Honors & Awards:* Oliver E Buckley Prize, Am Phy Soc, 85. *Mem:* Fel Am Phys Soc; fel AAAS. *Res:* Low temperature solid state physics; lattice dynamics. *Mailing Add:* Dept of Physics Cornell Univ Ithaca NY 14853-2501

POHL, VICTORIA MARY, b Haubstadt, Ind, Mar 29, 30. ALGEBRA, GEOMETRY. *Educ:* St Benedict Col, Ind, BS, 54; Univ Notre Dame, MS, 60; Univ Iowa, PhD(math), 69. *Prof Exp:* Teacher, St Theresa Elem Sch, 50-55; teacher math, Mater Dei High Sch, 55-59 & 60-62, chmn dept, 60-62; from instr to asst prof, St Benedict Col, Ind, 62-66 & 69-70, chmn dept, 65-66 & 69-70; from asst prof to assoc prof math, Ind State Univ, Evansville, 73-83; prof, 83-, CHMN DEPT MATH, UNIV SOUTHERN IND, 87- *Concurrent Pos:* Consult math, Dubois County Sch Syst, 69-70; vis prof, Ohio State Univ, 83-84. *Mem:* Math Asn Am; Nat Coun Teachers Math. *Res:* Enrichment materials for mathematics teachers in middle school through high school, through publications correlating geometry and art. *Mailing Add:* Dept Math & Sci Univ Southern Ind Evansville IN 47712

POHL, WILLIAM FRANCIS, mathematics; deceased, see previous edition for last biography

POHLAND, ALBERT, b Latrobe, Pa, Jan 23, 19; m 44; c 2. ORGANIC CHEMISTRY. *Educ:* St Vincent Col, BS, 40; Pa State Col, MS, 41, PhD(org chem), 44. *Prof Exp:* Asst war gases, Nat Defense Res Comt Proj, Univ Ill, 43-44; sr org chemist, Res Labs, 44-57, res assoc, 57-58, dept head, Div Org Chem, 58-64, res adv, 64-69, res consult, 69-80, LILLY RES FEL, ELI LILLY & CO, 80- *Mem:* Am Chem Soc. *Res:* War gases; structure-activity relationships of synthetic organic medicinals, especially analgesics; drug metabolism; structure of natural products. *Mailing Add:* 12306 Shadetree Lane Laurel MD 20708

POHLAND, FREDERICK G, b Oconomowoc, Wis, May 3, 31; m 66; c 1. SANITARY ENGINEERING, CIVIL ENGINEERING. *Educ:* Valparaiso Univ, BS, 53; Purdue Univ, MS, 58, PhD(sanit eng), 61. *Prof Exp:* From asst prof to assoc prof, 61-71, PROF CIVIL ENG, GA INST TECHNOL, 71- *Concurrent Pos:* Vis scholar, Univ Mich, 67-68; vis prof, Delft Univ Technol, Neth, 76-77. *Honors & Awards:* Harrison Prescott Eddy Medal, Water Pollution Control Fedn, 65, Charles Alvin Emerson Medal, 83. *Mem:* AAAS; Am Water Works Asn; Am Chem Soc; Water Pollution Control Fed; Int Asn Water Pollution Res & Control. *Res:* Environmental engineering operations and processes; water and waste chemistry and microbiology; solid and hazardous waste management; environmental impact monitoring and assessment. *Mailing Add:* Dept Civil Eng Univ Pittsburgh Pittsburgh PA 15261

POHLAND, HERMANN W, b Karlsruhe, Germany, Nov 17, 34; m 62; c 2. ORGANIC CHEMISTRY. *Educ:* Univ Bonn, Vordiplom, 57, Hauptdiplom, 59, Dr rer nat(synthesis of isothiazoles), 61. *Prof Exp:* Lectr org analysis, Univ Bonn, 60-61, Karl Duisberg fel, 61; fel, Univ Toronto, 61-62; res chemist, Jackson Lab, 62-68, sr res chemist, 68-69, res supvr, 69-71, sr supvr opers, dyes & chem, Org Chem Dept, 71-74, staff asst mfg dyes, 75-77, sr tech specialist, 77-79, tech consult, 79-84, US SALES MGR, PERMASEP REVERSE OSMOSIS SYSTS, E I DU PONT DE NEMOURS & CO, INC, 84- *Mem:* Am Chem Soc; Soc German Chem; Int Desalination Asn; Nat Water Supply & Improvement Asn; Am Water Works Asn. *Res:* Synthesis of hetero-aromatic compounds and phospholipids; preparation and characterization of fluoropolymers; synthesis and application of dyes and textile chemicals; manufacture of dyes and intermediates; process for preventing biological fouling in membrane devices. *Mailing Add:* 21 Orchard View Chadds Ford PA 19317-9238

POHLE, FREDERICK V, b Buena Vista, Va, Feb 21, 19; m 49; c 3. APPLIED MATHEMATICS. *Educ:* Cooper Union Inst Technol, BCE, 40; NY Univ, MS, 43, PhD(math), 50. *Prof Exp:* Instr mech, Eng Sch, NY Univ, 41-43, res asst math, Courant Inst, 46-48, instr, NY Univ, 48-50; asst prof mech, Polytech Inst Brooklyn, 50-53, from assoc prof to prof, 53-60; PROF MATH, ADELPHI UNIV, 60- *Concurrent Pos:* Consult, Knolls Atomic Power Lab, Gen Elec Co, 51-54; US Nat Bur Standards, 55; chmn dept math, Adelphi Univ, 62-72; US Air Force Off Sci Res grant, 64-66. *Mem:* Am Soc Eng Educ; Am Math Soc; Math Asn Am; Am Inst Aeronaut & Astronaut; Am Soc Mech Eng; Sigma Xi. *Res:* Theory of elasticity and of shells; vibration theory; nonlinear mechanics; celestial mechanics. *Mailing Add:* 61 Kilburn Rd Garden City NY 11530

POHLMANN, HANS PETER, b Waldenburg, Germany, Jan 30, 33; m 61; c 4. INORGANIC CHEMISTRY, PHYSICAL CHEMISTRY. *Educ:* Hanover Tech Univ, BS, 56, MS, 59, PhD(inorg chem), 61. *Prof Exp:* Asst prep inorg chem, Hanover Tech Univ, 62; asst proj chemist, Am Oil Co, Whiting, 62-63, proj chemist, 63-66; sr proj chemist, 67-69, group leader, 69-75, div dir, 75-79, MGR RES & DEVELOP, AMOCO CHEM CORP, 79- *Mem:* Am Chem Soc; Soc German Chem. *Res:* Liquid-liquid extraction for isolation and purification of rare earths, zirconium, hafnium; catalysis; high-pressure hydrogenations; liquid phase oxidations; reaction kinetics; petroleum technology; process development. *Mailing Add:* One S 275 S Cantigny Dr Winfield IL 60190-1704

POHLMANN, JUERGEN LOTHAR WOLFGANG, b Berlin-Lichtenberg, Ger, Feb 2, 34; US citizen; m 61; c 3. SOLID STATE CHEMISTRY. *Educ:* Univ Tuebingen, Ger, BS, 56, MS, 59, PhD(org & phys chem), 61. *Prof Exp:* Res chemist, US Naval Propellant Plant, Indian Head, Md, 61-64; res chemist, US Army Engrs Res & Develop Labs, 64-66, res chemist, US Army Night Vision & Electro-Optics Lab, Ft Belvoir, Va, 66-90; asst exec secy, US Army Sci Bd, Pentagon, 90-91; PROG MGR, STRATEGIC DEFENSE INITIATIVE ORGN, PENTAGON, 91- *Concurrent Pos:* OSD, O Dept Defense Res & Engr & Off Dir Res & Lab Mgt, 88-89. *Mem:* Am Chem Soc; Ger Chem Soc. *Res:* Preparation of metalorganic compounds; synthesis of compounds exhibiting mesophases; liquid crystals; influence of impurities and molecular structure on mesomorphic behavior; preparation of solid state materials and infrared detector materials; investigation of organic materials with nonlinear optical characteristics; surface passivation; research policy and manpower; optics and laser technology; nonlinear optical materials. *Mailing Add:* 8407 Camden St Alexandria VA 22308

POHLO, ROSS, b Chicago, Ill, Jan 31, 31; m 58; c 2. INVERTEBRATE ZOOLOGY. *Educ:* Univ Ill, BS, 54; Univ Ark, MS, 57; Univ Chicago, PhD(paleozool), 61. *Prof Exp:* Fel marine biol, Kristinebergs Zool Sta, Sweden, 61-62; Scripps Inst, Calif, 62-63; from asst prof to assoc prof, 63-74, PROF BIOL, CALIF STATE UNIV, NORTHRIDGE, 74- *Mem:* AAAS; Am Soc Limnol & Oceanog; Am Soc Zool. *Res:* Ecology and evolution of the bivalved mollusca. *Mailing Add:* Dept of Biol Calif State Univ Northridge CA 91330

POHM, A(RTHUR) V(INCENT), b Olmsted Falls, Ohio, Jan 11, 27; m 52; c 3. ELECTRICAL ENGINEERING. *Educ:* Fenn Col, BEE & BES, 50; Iowa State Col, MS, 53, PhD(physics), 54. *Prof Exp:* Lab instr, Fenn Col, 49-50; asst, Ames Lab, AEC, 50-54; res physicist, Remington Rand Univac Div, Sperry Rand Corp, 54-55, proj physicist, 55-56, res supvr, 57-58; assoc prof elec eng, 58-62, PROF ELEC ENG, IOWA STATE UNIV, 62- *Concurrent Pos:* Lectr, Univ Mich, 56. *Mem:* Fel Inst Elec & Electronics Engrs; Sigma Xi. *Res:* Thin films and solid electrical devices. *Mailing Add:* 1704 Woodhaven Circle Ames IA 50010

POHORECKY, LARISSA ALEXANDRA, b Cholm, Ukraine, Jan 16, 42; US citizen. NEUROPHARMACOLOGY, ALCOHOLISM. *Educ:* Univ Ill, Chicago, BS, 63; Univ Chicago, PhD(pharmacol), 67. *Prof Exp:* Res assoc neuropharmacol, Mass Inst Technol, 67-71; asst prof neuropharmacol, Rockefeller Univ, 71-78; prof neuropharmacol, Rutgers Univ, 79-86; CONSULT, 86- *Concurrent Pos:* USPHS fel, Mass Inst Technol, 67-71 & Rockefeller Univ, 71-78; mem alcohol res rev comt, Nat Inst Alcoholism & Alcohol Abuse, 77-79; adj assoc prof, Rockefeller Univ, 78-83. *Mem:* Res Soc Alcoholism; Sigma Xi; Am Soc Pharmacol & Exp Therapeut; AAAS; Int Soc Alcoholism. *Res:* Influence of ethanol, hormonal and psychological factors on noradrenergic neurons and endocrine organs; psychopharmacology; interaction of ethanol and stress; prenatal stress. *Mailing Add:* Ctr for Alcohol Studies Rutgers Univ New Brunswick NJ 08903

POIANI, EILEEN LOUISE, b Newark, NJ, Dec 17, 43. MATHEMATICS, HIGHER EDUCATION PLANNING AND RESEARCH. *Educ:* Douglass Col, AB, 65; Rutgers Univ, MS, 67, PhD(math), 71. *Prof Exp:* From instr to assoc prof 67, PROF MATH ST PETER'S COL , NJ, 80-, ASST TO PRES FOR PLANNING, 76- *Concurrent Pos:* Residence counsr, Rutgers Univ, 66-67; Danforth Assoc; dir mid states self-study, St Peter's Col, 74-76 & 81-83; founding dir, Women and Math Nat Prog, Math Asn Am, 75-81, consult, 81-; speaker, Int Cong Math Educ, Berkeley, 80-, US Deleg, Budapest, 88; vis lectr, Math Asn Am; chair, US Comn Math Instr, Nat Res Coun, 85-90; pres, Pi Mu Epsilon, Nat Math Hon Soc, 87-90, past pres, 90- *Mem:* Asn Women in Math; Math Asn Am; Am Math Asn. *Res:* Real analysis; special functions; Fourier analysis; participation of women in mathematics; math avoidance; higher education planning; NSF reviewer. *Mailing Add:* Dept Math St Peter's Col Jersey City NJ 07306

POILLON, WILLIAM NEVILLE, MOLECULAR GENETICS. *Educ:* Columbia Univ, PhD(chem), 62. *Prof Exp:* ASST PROF PEDIAT & CHILD HEALTH & RES BIOCHEMIST, CTR SICKLE CELL DIS, HOWARD UNIV, 79. *Mailing Add:* Ctr Sickle Cell Dis 2121 Georgia Ave Washington DC 20059

POINAR, GEORGE O, JR, b Spokane, Wash, Apr 25, 36; m 62; c 3. NEMATOLOGY, PALEONTOLOGY. *Educ:* Cornell Univ, BS, 58, MS, 60, PhD(entom), 62. *Prof Exp:* Res asst entom, Cornell Univ, 58-62; asst entomologist, Univ Calif, Riverside, 62-63; nematologist, Rothamsted Exp Sta, 63-64; INSECT PATHOLOGIST & LECTR, DEPT ENTOM SCI, UNIV CALIF, BERKELEY, 77- *Concurrent Pos:* NIH fel, 63-64; consult, Nutrilite Corp, 67-68; mem, South Pac Comn, 68; Nat Acad Sci partic exchange prog, Soviet Union, 69, Rumania,74; vis prof, Univ Amsterdam, 71, Nat Mus Natural Hist, Paris, 72 & Lab Nematol, Antibes, 78-79; consult, W Africa Onchocerciasis prog, 81, 82 & 84, Mex, 82, Ecuador, 84, New Zealand, Australia, 88. *Mem:* Soc Nematol; Entom Soc Am; Soc Invert Path; Soc European Nematol. *Res:* Taxonomy, morphology, biology and host-parasite relationship of invertebrate-parasitic nematodes; fossil invertebrates and invertebrates in amber. *Mailing Add:* Div of Entom Univ of Calif Berkeley CA 94720

POINCELOT, RAYMOND PAUL, JR, b New Haven, Conn, June 10, 44; div; c 3. SUSTAINABLE AGRICULTURAL. *Educ:* Southern Conn State Univ, BA, 66; Case Western Reserve Univ, PhD(biochem), 70. *Prof Exp:* Asst agr scientist, 70-74, assoc agr scientist, Conn Agr Exp Sta, 74-77; from asst prof to assoc prof biol, 77-90, chmn dept, 79-85, UNIV SAFETY OFFICER, FAIRFIELD UNIV, 88-, PROF BIOL, 90- *Concurrent Pos:* Plant sci writer, 74-; compost consult, 79-; sr ed, Food Products Press; ed, J Sustainable Agr, 89, J Home & Consumer Hort; Yale Mellon Vis Fac Fel, 90-91. *Honors & Awards:* Cert of Achievement, Livestock Environ Sci Comt of Great Plains States, 74. *Mem:* Am Soc Hort Sci; Garden Writers Asn America; Nat Asn Sci Writers; Am Soc Plant Physiol; Sigma Xi; Coun Agr Sci & Technol; Agr, Food & Human Values Soc; Inst Alternative Agr. *Res:* Horticultural uses of seaweed extract; composting; succulent propagation; author of over 60 publications. *Mailing Add:* Dept of Biol Fairfield Univ Fairfield CT 06430

POINDEXTER, EDWARD HAVILAND, b Lansing, Mich, Dec 15, 30; m 59; c 2. MATERIALS SCIENCE. *Educ:* Univ Mich, BS, 52, MS, 53, PhD(mineral), 56. *Prof Exp:* Res assoc, Univ Mich, 55-57; res physicist, Calif Res Corp, Standard Oil Co Calif, 57-62; leader, Magnetic Resonance Team, US Army Inst Explor Res, 62-69, chief, Magnetism & Molecular Elec Res Area, 69-71, chief, Elec Mat Res Tech Area, 71-73, chief, Device Physics & Anal Br, 73-87, LAB FEL, US ARMY ELEC TECHNOL & DEVICES LAB, 87- *Mem:* Am Chem Soc; Am Phys Soc; Inst Elec & Electron Engrs; Mineral Soc Am; Electrochem Soc; Mat Res Soc. *Res:* Crystal optics; magnetic resonance; transient complexes; free radical solutions; asphaltene colloids; high pressure techniques; semiconductor defects, oxide traps and interface states; radiation damage in polymer resists. *Mailing Add:* Three Rosedale Terr Holmdel NJ 07733

POINDEXTER, GRAHAM STUART, b Louisville, Ky, Dec 18, 48; m 68; c 2. ORGANIC CHEMISTRY. *Educ:* Univ Louisville, BS, 71; Univ NC, PhD(org chem), 75. *Prof Exp:* Res assoc, Colo State Univ, 75-76, Nat Res Serv fel, 76-77; sr res chemist, Dow Chem USA, 77-80; mem staff, chem res dept, Mead Johnson & Co, 80-89; MEM STAFF, PHARM RES, BRISTOL MYERS SQUIBB CO, 89- *Mem:* Am Chem Soc; AAAS; Int Soc Heterocyclic Chem. *Res:* Heterocyclic chemistry; synthetic methods; asymmetric synthesis; organic photochemistry. *Mailing Add:* Bristol Myers Squibb Co 5 Research Pkwy Wallingford CT 06492

POINDEXTER, JEANNE STOVE, b Lamoine Twp, Ill, Oct 5, 36; m 64; c 2. PROKARYOTE DEVELOPMENT, MICROBIAL ECOLOGY. *Educ:* Ind Univ, Bloomington, AB, 58; Univ Calif, Berkeley, MA, 61, PhD(bacteriol), 63. *Prof Exp:* Fel, Hopkins Marine Sta, Stanford Univ, 63; vis lectr bacteriol, Med Sch, Univ Edinburgh, 64; asst prof microbiol, Ind Univ, Bloomington, 64-67; assoc sci, Sch Med, NY Univ, 69-71; assoc prof biol, Medgar Evers Col, City Univ New York, 71-75 & Marymount Manhattan Col, 75-76; assoc microbiol, 76-84, ASSOC MEM, PUB HEALTH RES INST, NEW YORK CITY, 84- *Concurrent Pos:* Adj asst prof biol, Hunter Col, City Univ New York, 71; chmn, Nat Sci Math, Marymount Manhattan Col, 75-76; resident instr, Marine Biol Lab, 78-84; res assoc prof, Sch Med, NY Univ, 81-; adj assoc prof biol, Washington Sq Col, NY Univ, 83- *Mem:* AAAS; Am Soc Microbiol; Am Inst Biol Sci. *Res:* Physiology, ultrastructure, cellular differentiation of stalked bacteria (Caulobacteraceae); bacterial anatomy; prokaryote development; microbial ecology. *Mailing Add:* Sci Div Long Island Univ Brooklyn Campus University Plaza Brooklyn NY 11201

POINTER, RICHARD HAMILTON, b Covington, Ga, June 4, 44; m 66; c 3. ENDOCRINOLOGY. *Educ:* Morehouse Col, BS, 68; Brown Univ, MS, 73, PhD(physiol chem), 75. *Prof Exp:* Res technician physiol, Emory Univ, 69-70; res assoc physiol, Sch Med, Vanderbilt Univ, 75-77; instr, Harvard Med Sch & Res Assoc, Howard Hughes Med Inst, 77-80; asst prof, 80-87, ASSOC PROF, DEPT BIOCHEM, COL MED, HOWARD UNIV, WASHINGTON, DC, 80- *Concurrent Pos:* Fel, Diabetes Prog, Nat Inst Arthritis Metab & Digestive Dis, NIH, 75-78. *Mem:* AAAS; Sigma Xi; Am Diabetes Asn; Am Physiol Soc. *Res:* The mechanisms of hormone action at the cellular and molecular levels, specifically the mode of insulin action. *Mailing Add:* Dept Biochem Howard Univ Med Col Washington DC 20059

POIRIER, CHARLES PHILIP, b Wyandotte, Mich, Feb 28, 37; m 63; c 4. NUCLEAR PHYSICS. *Educ:* Univ Detroit, BS, 59; Ind Univ, Bloomington, MS, 62, PhD(physics), 65. *Prof Exp:* res physicist, Aerospace Res Labs, US Air Force, 64-80; MEM STAFF, GEOPHYS TECH CORP, 80- *Mem:* Am Phys Soc. *Res:* Nuclear spectroscopy using charged particle, neutron and electromagnetic interactions; application of computers to problems in nuclear physics, both experimental and theoretical. *Mailing Add:* 4434 Rustic Trail Boulder CO 80301

POIRIER, FRANK EUGENE, b Paterson, NJ, Aug 7, 40; m 63; c 2. BIOLOGICAL ANTHROPOLOGY. *Educ:* Paterson State Col, BA, 62; Univ Ore, MA, 64, PhD(anthrop), 67. *Prof Exp:* Asst prof psychiat & anthrop, Univ Fla, 67-68; asst prof anthrop, Ohio State Univ, 68-70; assoc prof, 72-74, PROF ANTHROP, OHIO STATE UNIV, 74- *Concurrent Pos:* NSF, NIMH,

Pacific Cultural Found grants, 65-82, Fulbright, 87, Nat Geog Soc, 82, Explorers Club, 82 & 87; res prof, Academia Sinica, 78. *Honors & Awards:* Outstanding Young Scientist, Sigma Xi, 74. *Mem:* AAAS; Int Primatol Soc; Am Asn Physical Anthrop; Am Anthrop Asn; Am Primatological Soc; Sigma Xi. *Res:* Primate ethology; paleoanthropology; primate socialization; ethnology of West Indies, India, China and Middle East; behavioral anthropology. *Mailing Add:* Dept of Anthrop 124 W 17 Ave Ohio State Univ Columbus OH 43210

POIRIER, GARY RAYMOND, b Bay City, Mich, Apr 10, 38; m 63; c 5. REPRODUCTIVE BIOLOGY. *Educ:* Mich State Univ, BS, 61, MS, 64, PhD(zool), 70. *Prof Exp:* Res asst chemother, Lederle Labs, 64-65; res asst immunochem, Hyland Labs, 65-67; asst prof biol, 70-74, ASSOC PROF BIOL, UNIV ALA, BIRMINGHAM, 74- *Mem:* AAAS; Am Soc Zool; Sigma Xi. *Res:* Gametogenesis, fertilization and early development. *Mailing Add:* Biol-Col Gen Studies Univ of Ala Birmingham AL 35294

POIRIER, JACQUES CHARLES, b Mehun sur Yevre, France, Jan 3, 27; US citizen; m 51; c 3. THEORETICAL CHEMISTRY, PHYSICAL CHEMISTRY. *Educ:* Univ Chicago, PhB, 47, SB, 48, SM, 50, PhD, 52. *Prof Exp:* NSF fel, Yale Univ, 52-53; Corning Glass Works Found fel, Univ Calif, Berkeley, 53-55; from asst prof to assoc prof 55-67, prof, 67-88, EMER PROF CHEM, DUKE UNIV, 88- *Concurrent Pos:* Sloan fel, 59-63; vis assoc prof chem, Ind Univ, 61; vis prof appl sci, Univ Calif, Davis, 84-85; Mem, Comt concerned Scientist. *Honors & Awards:* Sigma Xi. *Mem:* Am Phys Soc; Am Chem Soc. *Res:* Statistical mechanical theory of fluids; liquid crystals. *Mailing Add:* Dept of Chem Duke Univ Durham NC 27706

POIRIER, JOHN ANTHONY, b Lewistown, Mont, May 15, 32; m 61; c 5. PHYSICS AND ASTROPHYSICS. *Educ:* Univ Notre Dame, BS, 54; Stanford Univ, MS, 57, PhD(physics), 59. *Prof Exp:* Mem staff, Lawrence Berkeley Lab & lectr, Univ Calif, Berkeley, 59-63, NSF fel, European Orgn Nuclear Res, Switz, 63-64; assoc prof physics, 64-72, PROF PHYSICS, UNIV NOTRE DAME, 72- *Concurrent Pos:* Consult, NASA, 62; prog assoc, Elementary Particle Physics, Nat Sci Found, 77-78. *Mem:* Am Phys Soc; Sigma Xi; Am Astron Soc. *Res:* Ultra high energy cosmic rays and experimental research in the field of elementary particle interactions. *Mailing Add:* Dept Physics Univ Notre Dame Notre Dame IN 46556

POIRIER, LIONEL ALBERT, b Providence, RI, Aug 4, 37; div; c 3. CANCER. *Educ:* Providence Col, BS, 59; Univ Wis-Madison, PhD(oncol), 65. *Prof Exp:* Fel oncol, McArdle Lab, Univ Wis-Madison, 65-67; investr cancer, Montreal Cancer Inst, Notre Dame Hosp, 67-71; chemist, Carcinogen Screening Sect, Nat Cancer Inst, 71-73 & Carcinogen Metab & Toxicol Br, 73-76, head Nutrit & Metab Sect, 76-87, DIR DIV COMP TOXICOL, NAT CANCER CTR, NAT CTR TOXICOL RES, 88- *Concurrent Pos:* Adj asst prof biochem, Univ Montreal, 68-71; consult, Health Protection Br, Can Govt, 72-75 & Food & Drug Admin, 73-87; mem, Chem Sel Working Group, Nat Cancer Inst, 74- *Mem:* Am Asn Cancer Res; Am Inst Nutrit; AAAS. *Res:* Chemical carcinogenesis; methyl metabolism in carcinogenesis; nutrition and cancer. *Mailing Add:* Div Comp Toxicol Nat Ctr Toxicol Res NCTR Dr Jefferson AR 72079-9502

POIRIER, LOUIS, b Montreal, Que, Dec 30, 18; m 47; c 4. NEUROBIOLOGY, EXPERIMENTAL NEUROLOGY. *Educ:* Col Jean-de-Brebeuf, BA, 41; Univ Montreal, MD; Univ Mich, MSc, 49, PhD(neuroanat), 50. *Prof Exp:* Asst prof histol, Univ Montreal, 50-55, assoc prof, 55-58, prof neurol, 58-65; prof exp neurol, Fac Med, Laval Univ, 65-85, chmn dept, 70-78; RETIRED. *Concurrent Pos:* Mem, Med Res Coun Can, 75-81; officer, Order of Can, 75. *Mem:* Am Asn Anat; AAAS; Can Asn Anatomists; Am Phys Soc; Soc Neurosci. *Res:* Neuroanatomy, neurochemistry, neuropharmacology, neurophysiology and histophathology of the brain; motor and psychomotor disorders; scientific translation of German, Italian and Spanish texts. *Mailing Add:* CP310 RR 1 Montpellier Quebec PQ J0V 1M0 Can

POIRIER, MIRIAM CHRISTINE MOHRHOFF, b Detroit, Mich, Apr 30, 40; div; c 3. ONCOLOGY & BIOCHEMISTRY, MICROBIOLOGY. *Educ:* Marygrove Col, BSc, 62; Univ Wis-Madison, MSc, 64; Cath Univ Am, PhD(microbiol), 77. *Prof Exp:* Res asst oncol, Univ Wis, 62-64, lab asst biochem, 64-65, lab asst oncol, 65-67; lab asst, 71-78, RES CHEMIST ONCOL, NAT CANCER INST, NIH, 78- *Mem:* AAAS; Am Asn Cancer Res; Grad Women Sci; Environ Mutagen Soc. *Res:* Interactions of chemical carcinogens with DNA; carcinogen-DNA adduct antisera used for quantitation (immunoassays) and localization (immunohistochemistry) of fenitomale adduct quantities in tissues of humans and animal models. *Mailing Add:* Rm 3B25 Bldg 37 NIH Bethesda MD 20892

POIRIER, ROBERT VICTOR, b San Francisco, Calif, Oct 27, 39; m 67; c 2. CHEMICAL ENGINEERING. *Educ:* Stanford Univ, BS, 65; Univ Minn, Minneapolis, PhD(chem eng), 70. *Prof Exp:* Res engr, Fluorocarbons Div, Plastics Dept, E I du Pont de Nemours & Co, Inc, 70-76; res supvr, 76-77, mgr finance & planning, 77-78, develop eng supvr, Plastics Div, 78-80, TECH SUPERINTENDENT, FILMS DIV, MOBIL CHEM CO, 80- *Mem:* Am Inst Chem Engrs; Soc Plastics Engrs. *Res:* Gas phase kinetics; film extrusion and materials development. *Mailing Add:* Mobil Chem Co Tech Ctr Box 798 Macedon NY 14502

POIRIER, VICTOR L, b Lowell, Mass, July 3, 41; m 61; c 2. BUSINESS ADMINISTRATION, POLYMER ENGINEERING. *Educ:* Northeastern Univ, BME, 70, MBA, 81. *Prof Exp:* Prog mgr, Thermo Electron Corp, 69-73, sr engr, 73-79, group mgr, 79-83; exec vpres, 88-90, PRES, THERMO CARDIOSYSTS, 90-; SR VPRES, THERMEDICS INC, 84- *Concurrent Pos:* Asst prof, Tufts Univ Med, 80- *Mem:* Am Soc Artificial Internal Organs; Int Soc Artificial Organs; NY Acad Sci. *Res:* Direct energy conversion systems; artificial hearts and related technology; biomaterials and medical products; author of over 100 publications; holder of 14 patents. *Mailing Add:* Thermedics Inc 470 Wildwood St Woburn MA 01888-1799

POIRRIER, MICHAEL ANTHONY, b Edgard, La, Oct 2, 42; m 73; c 2. FRESHWATER SPONGE TAXONOMY. *Educ:* Univ New Orleans, BS, 63; La State Univ, MS, 65, PhD(zool), 69. *Prof Exp:* Instr zool, La State Univ, 68-69; from asst prof to assoc prof, 69-79, chmn dept biol sci, 79-86, PROF BIOL, UNIV NEW ORLEANS, 79- *Concurrent Pos:* Sci writer & ed, US Army Corps Engrs, New Orleans Dist, 71-72; vis prof, Gulf Coast Res Lab, 70-73. *Mem:* Estuarine Res Fedn; Nat Shell Fisheries Asn; Sigma Xi; Am Asn Univ Profs; Am Micros Soc. *Res:* Invertebrate zoology and aquatic ecology; systematics, ecology and physiology of freshwater sponges; epifaunal invertebrates as indicators of environmental quality in the Lake Pontchartrain estuary; production of soft-shelled blue crabs, Callinectes sapidus. *Mailing Add:* Dept Biol Sci Univ New Orleans Lake Front New Orleans LA 70148

POISNER, ALAN MARK, b Kansas City, Mo, Oct 15, 34; m 62; c 2. PHARMACOLOGY, CELL PHYSIOLOGY. *Educ:* Calif Inst Technol, BS, 56; Univ Kans, MD, 60. *Prof Exp:* Intern, Univ Ill Res Hosp, 60-61; asst prof pharmacol, Albert Einstein Col Med, 64-68; assoc prof, 68-73, PROF PHARMACOL, UNIV KANS MED CTR, KANSAS CITY, 73- *Concurrent Pos:* Fel pharmacol, Albert Einstein Col Med, 61-64; res career develop award, 65-68; vis scientist, Nat Inst Med Res, London, 63; vis scientist, Nat Inst Env Health Sci, NC, 87-88; vis med res prof, Duke Univ, Durham, 87-88. *Mem:* Am Soc Pharmacol & Exp Therapeut; Endocrine Soc; Neurosci Soc. *Res:* Physiology and pharmacology of secretory phenomena including release of hormones and neurotransmitters; renin and endocrinology of the placenta. *Mailing Add:* Dept Pharmacol Univ Kans Med Ctr Kansas City KS 66103

POJASEK, ROBERT B, US citizen. HAZARDOUS MATERIALS MANAGEMENT. *Educ:* Rutgers Univ, BS; Univ Mass, PhD(chem). *Prof Exp:* Consult, New Eng Res Inc, Worchester, Mass, 71-73; lab dir & sr staff chemist, Lawler, Matusky & Skelly Engrs, Pearl River, NY, 73-75; lab dir & sr environ chemist, JBF Sci Corp, Wilmington, Mass, 75-77; lab dir & hazardous waste consult, ERCO, Cambridge, Mass, 77-80; vpres, Weston, 80-; AT CHAS T MAIN, INC. *Concurrent Pos:* Consult, Exec Off Pres & Off Sci & Technol Policy; mem, Hazardous Waste Adv Comt; Consult, Regional Off Europe, WHO. *Mem:* Am Chem Soc; Sigma Xi; Am Soc Testing & Mat. *Res:* Environmental monitoring. *Mailing Add:* 57 Jaconet Cambridge MA 02139

POJETA, JOHN, JR, b New York, NY, Sept 9, 35; m 57; c 2. INVERTEBRATE PALEONTOLOGY, TAXONOMY. *Educ:* Capital Univ, BS, 57; Univ Cincinnati, MS, 61, PhD(paleont, zool), 63. *Prof Exp:* PALEONTOLOGIST, PALEONT & STRATIG BR, US GEOL SURV, 63- *Concurrent Pos:* Assoc prof lectr, George Washington Univ. *Mem:* Fel AAAS; Paleont Soc; fel Geol Soc Am. *Res:* Taxonomy, morphology, ontogeny, variation, phylogeny and ecology of early Paleozoic pelecypod faunas, especially comparison of these forms to recent pelecypod faunas and molluscan evolution. *Mailing Add:* US Geol Surv Nat Ctr MS 970 Reston VA 22092

POKORNY, ALEX DANIEL, b Taylor, Tex, Oct 18, 18; m 48; c 4. PSYCHIATRY. *Educ:* Univ Tex, BA, 39, MD, 42. *Prof Exp:* Intern, Hermann Hosp, Houston, Tex, 43; resident psychiat, McKinney Vet Admin Hosp, 46-47; resident, Topeka Vet Admin Hosp, 47-49; from instr to assoc prof, 49-63, clin prof, 63-67, PROF PSYCHIAT, BAYLOR COL MED, 67-, VCHMN DEPT, 72- *Concurrent Pos:* Menninger Found fel, Topeka Vet Admin Hosp, 47-49; asst chief psychiat & neurol serv, Vet Admin Hosp, Houston, 49-55, chief, 55-73; actg chmn dept psychiat, Baylor Col Med, 68-72. *Mem:* Fel AAAS; AMA; fel Am Psychiat Asn; fel Am Col Psychiat; Soc Biol Psychiat. *Res:* Suicidal behavior and violence; psychiatric classification; psychopharmacology; addictions and alcoholism. *Mailing Add:* 813 Atwell Bellaire TX 77401

POKORNY, FRANKLIN ALBERT, b Chicago, Ill, Feb 7, 30; m 57; c 2. HORTICULTURE. *Educ:* Univ Ill, BS, 53, MS, 56, PhD(hort), 59. *Prof Exp:* Asst prof, 59-66, ASSOC PROF HORT, UNIV GA, 66- *Mem:* Int Soc Hort Sci; Am Soc Hort Sci. *Res:* Plant nutrition; plant propagation; photoperiod; soils. *Mailing Add:* Dept of Hort Univ of Ga Athens GA 30602

POKORNY, GEROLD E(RWIN), b Czech, May 25, 28; US citizen; m 58; c 3. PHYSICS, ELECTRONICS. *Educ:* Vienna Tech Univ, MS, 51, PhD(tech sci), 53. *Prof Exp:* Physicist, US Army Signal Corps Labs, 53-56, sect chief microwave tubes, 56-60; sr scientist, 60-63, eng mgr display devices, 63-64, assoc dir res, 64-66, mgr res dept, 66-70, div vpres, 68-84, mgr electro optics dept, 70-76, facility mgr, 76-84, PRES, LITTON ELECTRON DEVICES DIV, 84-, CORP VPRES LITTON INDUST & GROUP EXEC, ELECTRONIC DEVICES & MATS, ARIZ, 86- *Mem:* Sr mem Inst Elec & Electronics Engrs; Electron Devices Soc. *Res:* Electrooptics; night vision and infrared technology; solid state physics; electrooptic image detection and display; microwave physics and technology; solid state microwave devices; high power microwave generation; electron beam optics. *Mailing Add:* 1046 E Buena Vista Dr Tempe AZ 85284

POKORNY, JOEL, b Brooklyn, NY, July 29, 40; m 65; c 2. PSYCHOPHYSIOLOGY. *Educ:* Middlebury Col, AB, 62; Columbia Univ, PhD(psychol), 67. *Prof Exp:* Res assoc ophthal, Univ Chicago, 66-69, asst prof ophthal, 69-72, asst prof psychol, 71-72, assoc prof ophthal & psychol, 72-78, assoc prof ophthal & behav sci, 78, PROF OPHTHAL VISUAL SCI & PSYCHOL, UNIV CHICAGO, 78- *Concurrent Pos:* Mem, Int Res Group Colour Vision Deficiencies. *Honors & Awards:* Tillyer Award, Optical Soc Am. *Mem:* Asn Res Vision & Ophthal; Optical Soc Am; Sigma Xi. *Res:* Mechanisms of normal and anomalous color vision in humans; theories of color vision; spatial and temporal factors in vision. *Mailing Add:* 5431 S Blackstone Ave Chicago IL 60615

POKORNY, KATHRYN STEIN, b New York, NY, Mar 22, 44. PROTOZOOLOGY, CELL BIOLOGY. *Educ:* Hunter Col, BA, 63; Columbia Univ, MA, 69, PhD(biol), 71. *Prof Exp:* Electron microscopist, Osborn Labs Marine Sci, New York Aquarium, 72-80; MEM STAFF, DEPT

OPHTHAL, MT SINAI SCH MED, NEW YORK, 80- *Concurrent Pos:* Biol consult, Columbia Univ Press, 71-74; adj asst prof biol, Baruch Col, City Univ NY, 75, 76. *Mem:* NY Acad Sci; Sigma Xi; Soc Protozool; Soc Cell Biol. *Res:* Ultrastructure and morphogenesis of cells and cellular organelles, particularly in marine parasitic protozoa, with emphasis on problems involved with host-parasite relationships and secretion of extracellular materials. *Mailing Add:* Dept Opthal Mt Sinai Med Ctr One Gustave L Levy Pl New York NY 10029

POKOSKI, JOHN LEONARD, b St Louis, Mo, July 19, 37; m 62; c 4. ELECTRICAL ENGINEERING. *Educ:* St Louis Univ, BS, 59; Ariz State Univ, MSE, 65; Mont State Univ, PhD(elec eng), 67. *Prof Exp:* Assoc engr comput res & develop, Int Bus Mach Corp, 59-63; asst elec eng, Ariz State Univ, 63-65 & Mont State Univ, 65-67; assoc prof, 67-80, PROF ELEC ENG, UNIV NH, 80- *Mem:* Am Soc Eng Educ; Instrument Soc Am; Inst Elec & Electronics Engrs. *Res:* Digital computer design; switching theory and logic design; applications of control theory; sampled-data control systems. *Mailing Add:* Dept Elec Eng & Comput Eng Univ NH Kingsbury Hall Durham NH 03824

POKRANT, MARVIN ARTHUR, b Chicago, Ill, June 12, 43; m 67. STATISTICAL MECHANICS. *Educ:* Ill Inst Technol, BS, 65; Univ Fla, PhD(physics), 70. *Prof Exp:* Asst engr, Controls Co of Am, 65; postdoctoral assoc, Physics Dept, Univ Fla, 70-76; sr systs analyst, Mantech Int Corp, 76-78, proj mgr, 78-82, tech dir, 82-84; MEM RES STAFF, CTR NAVAL ANALYSIS, 84- *Concurrent Pos:* Rep, Comdr in Chief, Pac Fleet, 86-87; comdr, Third Fleet, 87-89, comdr in chief, Pac Fleet, 89-90, comdr, Seventh Fleet, 90-91, comdr, US Naval Forces Cent Command, 90-91. *Mem:* Opers Res Soc Am; Am Phys Soc; Am Asn Physics Teachers. *Res:* Military operations research; many-body problem; structure of quantum fluids. *Mailing Add:* Ctr Naval Anal PO Box 16268 Alexandria VA 22302

POKRAS, HAROLD HERBERT, b Chicago, Ill, Sept 3, 18; m 43; c 2. CHEMISTRY. *Educ:* Univ Calif, BA, Los Angeles, 40, MA, 41; Ill Inst Technol, PhD(org chem), 44. *Prof Exp:* Jr instr chem, Ill Inst Technol, 41-43; res chemist, Emulsol Corp, 44 & Quincy Labs, 46; res assoc, Lever Bros, Co, 46-56; chemist, Louis Milani Foods, Inc, 56-65; assoc prof, 65-77, PROF CHEM, LOS ANGELES CITY COL, 65-, CHMN DEPT, 77- *Concurrent Pos:* Jr instr chem, Ill Inst Technol, 46. *Mem:* Am Chem Soc; Inst Food Technol. *Res:* Organic synthesis; surfactants; kinetic measurement; emulsion technology; products development of toilet goods; cosmetics; dentifrices and food products; food technology and heat processing; teaching of chemistry. *Mailing Add:* 15157A Magnolia Blvd Sherman Oaks CA 91403-1777

POLACK, JOSEPH A(LBERT), b New Orleans, La, Sept 11, 20; m 43; c 2. CHEMICAL ENGINEERING. *Educ:* Tulane Univ, BE, 41; Mass Inst Technol, MS, 43, ScD(chem eng), 48. *Prof Exp:* Res assoc, Chem Warfare Serv, Mass Inst Technol, 43-45, instr chem eng, 46, asst, 47-48; chem engr, Esso Res Labs, Humble Oil & Refining Co, 48-54, sect head, 54-57, asst dir, 57-61, asst to mgr res, 61-63, admin asst, Esso Res Labs, 63-66, asst head, tech div, Baton Rouge Refinery, 66, dir, 66-70; prof chem eng & head dept, 70-76, dir, Audubon Sugar Inst, La State Univ, 76-87; CONSULT, 87- *Honors & Awards:* Charles Coates Award, Am Inst Chem Engrs & Am Chem Soc, 82. *Mem:* Fel Am Inst Chem Engrs; Am Soc Sugar Cane Technologists; Sigma Xi. *Res:* Chemical process development; pilot plants; research and technical administration; cane sugar processing. *Mailing Add:* 4332 Claycut Rd Baton Rouge LA 70806

POLAHAR, ANDREW FRANCIS, b Somerville, NJ, June 24, 50; m 74; c 2. ENVIRONMENTAL ENGINEERING. *Educ:* Univ Louisville, BChE, 73. *Prof Exp:* Chem engr, Process Improv, Holston Defense Corp, 73-76, Prototype & Pilot Plant Projs, 77-79, Eng Spec Projs, 80-82 & Eng Environ Group, 83-85, environ coordr, 85-88, SR CHEM ENG, PROJ MGT, HOLSTON DEFENSE CORP, 88- *Concurrent Pos:* Adv, Jr Achievement, 83-85. *Mem:* Am Inst Chem Engrs; Nat Soc Prof Engrs; Water Pollution Control Fedn; Air & Waste Mgt Asn; Am Mgt Asn. *Mailing Add:* 4800 Cork Lane Kingsport TN 37664

POLAK, ARNOLD, b Michalovce, Czech, Aug 7, 27; US citizen. FLUID MECHANICS. *Educ:* Prague Tech Univ, Dipl Ing, 53; Univ Cincinnati, MS, 61, PhD(aerospace eng), 66. *Prof Exp:* Asst prof eng math, Col Transp, Prague, Czech, 53-57; asst prof prod planning, Am Ball Bearing Co, NY, 58; instr math & eng mech, Centralia Col, 58-59; instr math & mech, Univ Cincinnati, 59-62, res assoc gas dynamics, Dept Aerospace Eng, 62-66, asst prof fluid dynamics, 66-67; res aerospace engr appl aerodyn, US Naval Ord Lab, Md, 67-69; assoc prof, 69-76, PROF FLUID DYNAMICS, DEPT AEROSPACE ENG & APPL MECH, UNIV CINCINNATI, 76- *Concurrent Pos:* Nat Acad Sci fel, 67. *Mem:* Am Inst Aeronaut & Astronaut; Am Soc Eng Educ. *Res:* High speed gas dynamics; applied aerodynamics; separated flows. *Mailing Add:* Dept Aerospace Eng Univ Cincinnati Cincinnati OH 45221

POLAK, ELIJAH, b Bialystok, Poland, Aug 11, 31; m 61; c 2. ELECTRICAL ENGINEERING. *Educ:* Univ Melbourne, BEE, 57; Univ Calif, Berkeley, MSEE, 59, PhD(elec eng), 61. *Prof Exp:* Instrument engr, Imp Chem Industs, Ltd, 57-58; teaching asst elec eng, 58-59, teaching assoc, 59-61, from asst prof to assoc prof, 61-69, PROF ELEC ENG, UNIV CALIF, BERKELEY, 69 - *Concurrent Pos:* Guggenheim fel, Inst Blaise Pascal, Paris, France, 68-69. *Mem:* Fel Inst Elec & Electronics Engrs; Soc Indust & Appl Math. *Res:* Optimization theory; optimization of computer-aided-design of engineering systems; modern systems theory. *Mailing Add:* Dept Elec Eng & Comput Sci Univ Calif Berkeley CA 94720

POLAK, JOEL ALLAN, b New Haven, Conn, Sept 17, 37; m 60; c 2. PHYSICS, UNDERWATER ACOUSTICS. *Educ:* Yale Univ, BS, 59, MS, 60, PhD(theoret physics), 65. *Prof Exp:* Res assoc nuclear physics, Yale Univ, 65-67; mem tech staff, 67-71, SUPVR OCEAN ACOUST & SYSTS ANALYSIS, BELL LABS, 71- *Mem:* Am Phys Soc; Am Math Soc; Math Asn Am; Sigma Xi. *Res:* Ocean acoustics; influence of rheological properties of bottom on sound propagation; undersea communications and sonar systems. *Mailing Add:* 22 Porter Place New Providence NJ 07974

POLAKOSKI, KENNETH LEO, b Fond du Lac, Wis, Sept 6, 44. BIOCHEMISTRY, ANDROLOGY. *Educ:* Wis State Univ, BS, 66; Univ Ga, MS, 71, PhD(biochem), 72. *Prof Exp:* Res tech biochem, Marshfield Clin Found, 66-68; res assoc biochem, Univ Ga, 72-73; res asst prof, 73-77, asst prof, 77-79, assoc prof, 79-84, PROF OBSTET & GYNEC, WASHINGTON UNIV, 84- *Concurrent Pos:* Prin investr, 77-87, NIH Res Grant, 75-85 & NIH Biomed Res Grant, 77-92. *Mem:* Am Chem Soc; AAAS; Am Fertil Soc; Soc Dynamic Invest. *Res:* Biochemical mechanisms of sperm penetration of ovum; mechanisms of sperm motility; regulatory mechanisms of proteolytic enzymes, proteinase; proacrosin, acyosin, sperminogen sperm in reproduction tissues of mammals and plants, which are required for fertilization; semen analysis and male infertility; drug toxicology and male reproduction. *Mailing Add:* Dept of Obstet & Gynec Washington Univ Sch of Med St Louis MO 63110

POLAN, CARL E, b Blandville, WVa, Sept 14, 31; m 56; c 4. DAIRY NUTRITION, BIOCHEMISTRY. *Educ:* Univ WVa, BS, 53, MS, 60; NC State Univ, PhD(animal nutrit), 64. *Prof Exp:* Farm supt, Potomac State Col, WVa, 56-58; res asst dairy sci & exten dairyman, Univ WVa, 58-60; res asst nutrit, NC State Univ, 60-63; res fel biochem, Univ Minn, St Paul, 63-65; from asst prof to assoc prof, 65-73, PROF NUTRIT, VA POLYTECH INST & STATE UNIV, 73- *Mem:* AAAS; Am Soc Animal Sci; Am Dairy Sci Asn; Am Inst Nutrit. *Res:* Nutrition and metabolism of amino acids and lipids; general nutritional considerations of large animals, particularly ruminants. *Mailing Add:* Dept Dairy Sci Va Polytech Inst & State Univ Blacksburg VA 24061-6999

POLAND, ALAN P, b Baltimore, Md, June 5, 40; m 75; c 1. PHARMACOLOGY, BIOCHEMISTRY. *Educ:* Univ Rochester, MS & MD, 66. *Prof Exp:* Fel drug metab, Rockefeller Univ, 69-71; asst prof pharmacol, Sch Med, Univ Rochester, 71-77; from asst prof to assoc prof, 77-85, PROF ONCOL, MCARDLE LAB CANCER RES, UNIV WIS-MADISON, 85- *Concurrent Pos:* NIH res career develop award, 75-; mem comt pentachlorophenol & its contaminants, Environ Protection Agency, 77-78; Burroughs-Wellcome scholar toxicol, 81-86. *Honors & Awards:* John Jacob Abel Award, Am Soc Pharmacol & Exp Therapeut, 76. *Mem:* Am Soc Pharmacol & Exp Therapeut; AAAS; Am Soc Cancer Res; Fedn Am Soc Exp Biol; Soc Toxicol. *Res:* Toxicity of halogenated aromatic hydrocarbons; gene expression of aryl hydrocarbon hydroxylase activity by 2,3,7,8-tetrachlorodibenzo-p-dioxin. *Mailing Add:* McArdle Lab Cancer Res Univ Wis Madison WI 53706

POLAND, ARTHUR I, b Asbury Park, NJ, Mar 30, 43; m 64. ASTROPHYSICS, SOLAR PHYSICS. *Educ:* Univ Mass, Amherst, BS, 64; Ind Univ, Bloomington, PhD(astrophys), 69. *Prof Exp:* Vis scientist, High Altitude Observ, Nat Ctr Atmospheric Res, 69-71, scientist, 71-80; ASTROPHYSICIST, GODDARD SPACE FLIGHT CTR, NASA, 80- *Mem:* Am Astron Soc; Int Astron Union. *Res:* Solar research on the corona, solar prominences and flares. *Mailing Add:* Code 682 Goddard Space Flight Ctr NASA Greenbelt MD 20771

POLAND, JACK DEAN, b Isabel, Kans, Jan 30, 30; m 52; c 2. MEDICINE. *Educ:* Harding Col, BS, 52; Univ Kans, MD, 59; Univ Mich, MPH, 65; Am Bd Prev Med, cert. *Prof Exp:* Intern, USPHS Hosp, Norfolk, Va, 59-60; officer, Commun Dis Ctr, USPHS, 60-62, epidemiologist, Respiratory & Enteric Virus Dis Unit, 65, chief neurotropic virus dis unit, 65-67, actg chief virol, 65-68, chief virus dis sect, 67-68, chief zoonoses sect, Ecol Invest Prog, 68-72, asst chief, Ft Collins Br, 72-74, chief epidemiologist, 74-81, asst dir med sci, 59-80, dir med sci, Vector Borne Div, Bur Labs, Ctr Dis Control, USPHS, 80-87; AFFIL PROF MICROBIOL, COLO STATE UNIV, 68- *Concurrent Pos:* Instr, Sch Med, Univ Kans, 66-68. *Mem:* AAAS; Am Epidemiol Soc; Am Soc Trop Med & Hyg. *Res:* Epidemiology of diseases of animals transmissible to man; primarily plaque and arthropod borne viral diseases. *Mailing Add:* 1704 Concord Dr Ft Collins CO 80526

POLAND, JAMES LEROY, b Washington, Pa, Nov 20, 40; m 66; c 2. PHYSIOLOGY. *Educ:* Waynesburg Col, BS, 62; WVa Univ, MS, 64, PhD(physiol), 67. *Prof Exp:* asst prof, 67-80, ASSOC PROF PHYSIOL, MED COL VA, 80- *Concurrent Pos:* Consult, Vet Admin Hosp, Richmond, Va. *Mem:* Am Physiol Soc; Soc Exp Biol & Med; Sigma Xi; Am Col Sports Med. *Res:* Functional integrity of skeletal muscle grafts. *Mailing Add:* Dept Physiol Med Col Va Richmond VA 23298

POLAND, JOHN C, b Toronto, Ont, Apr 25, 39; m 62; c 2. ALGEBRA. *Educ:* McGill Univ, BSc, 60, MSc, 63, PhD(group theory), 66. *Prof Exp:* Lectr math, McGill Univ, 61-66; Nat Res Coun Can fel, Australian Nat Univ, 66-67; asst prof, 67-74, ASSOC PROF MATH, CARLETON UNIV, 74- *Mem:* Am Math Soc; Can Math Cong. *Res:* Group theory. *Mailing Add:* Dept Math & Stats Carleton Univ Ottawa ON K1S 5B6 Can

POLAND, JOSEPH FAIRFIELD, hydrogeology; deceased, see previous edition for last biography

POLAND, RUSSELL E, SLEEP PSYCHOLOGY, PSYCHOPHARMACOLOGY. *Educ:* Univ Calif, Los Angeles, PhD(pharmacol), 79. *Prof Exp:* ASSOC RES PROF PHARMACOL & PSYCHOL, HARBOR MED CTR, UNIV CALIF, LOS ANGELES, 78- *Mailing Add:* Dept Psychol Div Biol Psych Harbor Med Ctr Univ Calif Los Angeles Torrance CA 90509

POLANER, JEROME L(ESTER), b Newark, NJ, Sept 13, 15; m 53; c 3. MECHANICAL ENGINEERING. *Educ:* Newark Col Eng, BS, 38; Stevens Inst Technol, MS, 51. *Prof Exp:* Engr, Newark Gear Cutting Mach Co, 37-38; instr mech eng, Newark Col Eng, 38-42; engr, Robins Conveying Belt Co, 42; from asst prof to prof, 42-83, assoc chmn dept, 52-66, consult dept, 83-85, EMER PROF MECH ENG, NJ INST TECHNOL, 83- *Concurrent Pos:* Consult engr, Am Leather Belting Assocs, 47-49; consult engr, 44-; Adj lectr, Rutgers Univ, Newark, 52-67. *Mem:* AAAS; Am Soc Mech Engrs; Am Soc Eng Educ; NY Acad Sci; Nat Soc Prof Engrs. *Res:* Environmental control; industrial and automotive safety devices; accident reconstruction; product design. *Mailing Add:* 30 Tiffany Dr Livingston NJ 07039

POLANSKY, MARILYN MACARTHUR, FOODS & NUTRITION. *Educ:* Cornell Univ, MS, 52. *Prof Exp:* CHEMIST, BELTSVILLE HUMAN NUTRIT RES CTR, SCI & EDUC ADMIN, USDA, 52- *Mailing Add:* Beltsville Human Nutrit Res Ctr Vitamins & Minerals Lab Bldg 307 Rm 218 Beltsville MD 20705

POLANYI, JOHN CHARLES, b Berlin, Ger, Jan 23, 29; m 58; c 2. PHYSICAL CHEMISTRY. *Educ:* Univ Manchester, BSc, 49, MSc, 50, PhD(chem), 52, DSc, 65. *Hon Degrees:* DSc, Univ Waterloo, 70, Mem Univ Nfld, 76, McMaster Univ, 77, Carleton Univ, 81, Harvard Univ, 82, Renssalaer, 84, Brock Univ, 84, Univ Lethbridge, 87, Univ Victoria, 87, Univ Ottawa, 87, Univ Sherbrooke, 87, Laval Univ, 87, Univ Manchester, 88, York Univ, 88; LLD, Trent Univ, 77, Dalhousie Univ, 83, St Francis Xavier, 84. *Prof Exp:* Res fel chem, Nat Res Coun Ottawa, 52-54; asst, Princeton Univ, 54-56; lectr, 56-57, from asst prof to assoc prof, 57-62, PROF CHEM, UNIV TORONTO, 62- *Concurrent Pos:* Sloan fel, 59-63; lectr numerous US & Can univs & insts, 65-; Guggenheim fel, 70 & 79; chmn, Can Pugwash Comt. *Honors & Awards:* Nobel Prize in Chem, 86; Steacie Prize Natural Sci, 65; Centennial Medal, The Chem Soc, 65, Award Kinetics, 71; Noranda Award, Chem Inst Can, 67, Medal, 76; Mack Award, Ohio State Univ, 69; Companion, Order of Can, 79; Henry Marshall Tory Medal, Royal Soc Can, 77; Remsen Award, Am Chem Soc, 78; Wolf Prize, 82; Marlow Medal, Faraday Soc, 63; Isaac Walton Killam Mem Prize, 88. *Mem:* Foreign assoc Nat Acad Sci; Fel Royal Soc Can; hon foreign mem Am Acad Arts & Sci; Royal Soc Chem; fel Royal Soc; Pontifical Acad Sci; fel Royal Soc Edinburgh. *Res:* Reaction kinetics; photochemistry; chemiluminescence; energy exchange; surface chemistry. *Mailing Add:* Dept Chem Univ Toronto Toronto ON M5S 1A1 Can

POLATNICK, JEROME, b New York, NY, Oct 4, 22; m 48; c 3. BIOCHEMISTRY. *Educ:* City Col New York, BS, 43; Columbia Univ, PhD(biochem), 54. *Prof Exp:* Res chemist, Schenley Res Inst, 43-47; res chemist, NY Bot Gardens, 48-50; asst biochem, Columbia Univ, 50-54; prin investr, Manhattan Eye & Ear Hosp, 54-57; res biochemist, Plum Island Animal Dis Lab, USDA, 57-85; CONSULT, 85- *Mem:* Am Chem Soc; Am Soc Microbiol. *Res:* Cell culture; virology; molecular biology of virus-host cell interactions; radioisotope safety programs. *Mailing Add:* 1230 Crittens Lane Southold NY 11971-1914

POLAVARAPU, PRASAD LEELA, b Gudlavalle ru, Andhra Pradesh, India; m 81; c 2. VIBRATIONAL SPECTROSCOPY. *Educ:* Andhra Univ, BSc, 70; Birla Inst Technol & Sci, MSc, 72; Indian Inst Technol, PhD(chem), 76. *Prof Exp:* Res assoc chem, Univ Toledo, 77-78 & Syracuse Univ, 78-80; res ass, Syracuse Univ, 78-80; asst prof, 80-86, ASSOC PROF CHEM, VANDERBILT UNIV, 86- *Mem:* Am Chem Soc; AAAS; Soc Appl Spectros. *Res:* Vibrational circular dichroism; raman optical activity; molecular vibrations; vibrational absorption intensities. *Mailing Add:* Dept Chem Vanderbilt Univ Nashville TN 37235

POLCYN, DANIEL STEPHEN, b Montello, Wis, Sept 17, 33; m 69; c 2. ANALYTICAL CHEMISTRY. *Educ:* Wis State Col, Oshkosh, BS, 55; Univ Wis, PhD(chem), 65. *Prof Exp:* Teaching asst physics, Univ Wis, 55-57, res asst chem, 59-65, fel, 65; sr electrochemist, Gould, Inc, Minn, 65-70; res assoc, Univ Wis-Madison, 71-73, lectr, 73-75; from asst prof to assoc prof chem, Univ Wis, Oshkosh, 75-80; res dir, Pope Sci, Inc, 80-82, tech dir, Pro-Tech, 83, Rolite Chem, 84-85 & Consult, 85-87; SR CHEMIST, MERCURY MARINE DIV, BRUNSWICK CORP, 87- *Concurrent Pos:* Res assoc, chem dept, Univ Wis-Madison, 73-75, vis assoc prof, 79. *Mem:* AAAS; Am Chem Soc; Electrochem Soc; Sigma Xi; ASTM. *Res:* Electrode kinetics; electroanalytical chemistry; trace analysis; air pollution; environmental chemistry; coatings. *Mailing Add:* N 81 W 18359 Tours Dr Menomonee Falls WI 53051

POLEJES, J(ACOB) D, b Ellenville, NY, June 5, 34; m 57. CHEMICAL ENGINEERING. *Educ:* Polytech Inst Brooklyn, BChE, 55; Univ Wis, MS, 56, PhD(chem eng), 59. *Prof Exp:* Res engr, Film Dept, NY, 59-69, SR RES ENGR, TEXTILE FIBERS DEPT, E I DU PONT DE NEMOURS & CO, INC, 69- *Mem:* Am Chem Soc. *Mailing Add:* 2538 Eaton Rd Wilmington DE 19810-3566

POLEN, PERCY B, agricultural chemistry, for more information see previous edition

POLESTAK, WALTER JOHN S, b New York, NY, Oct 27, 26; m 59; c 3. PHYSICAL CHEMISTRY. *Educ:* Manhattan Col, BS, 50; Tex A&M Univ, MS, 55; Univ Pa, PhD(phys chem), 60. *Prof Exp:* Chemist, Fisher Sci Co, 51-53; chemist, Fritzsche Bros, Inc, 54; asst instr chem, Tex A&M Univ, 54-55; asst instr, Univ Pa, 55-59; res chemist fibers, E I du Pont de Nemours & Co, 59-62; res chemist, Celanese Res Co, 62-68; sect leader basic group, ITT-Rayonier, Inc, 68-71; group leader resins & polymers, Process Chem Div, Diamond Shamrock Chem Co, 71-75, lab mgr resins & polymers, 75-80; GROUP LEADER, RESINS & POLYMERS, GEORGIA KAOLIN RES, 80- *Mem:* Am Inst Chem; NY Acad Sci; Am Chem Soc; Sigma Xi; Soc Plastics Engrs; Am Ceramic Soc; Fedn of Socs for Coatings Technol. *Res:* Kaolin, natural silicas and related minerals research for paint, plastics, ceramics and rubber applications. *Mailing Add:* 36 Beekman Rd Summit NJ 07901-1721

POLET, HERMAN, b Ghent, Belg, Aug 15, 30; m 65. CELL BIOLOGY. *Educ:* Univ Ghent, MD, 56. *Prof Exp:* Part-time free collabr pharmacol, Univ Ghent, 54-58; asst resident path, Children's Med Ctr, 59-60; jr & sr asst resident, Peter Bent Brigham Hosp, 60-62; med officer, Walter Reed Army Inst Res, 64-68; asst prof path, Harvard Med Sch, 69-78; assoc prof, 78-90, PROF PATH, COL MED, UNIV ILL, CHICAGO, 90- *Concurrent Pos:* Res fel cancer chemother, Children's Cancer Res Found, Boston, 58-59; res fel microbiol, Sch Pub Health, Harvard Univ, 62-64. *Mem:* Am Soc Cell Biol; Am Asn Path; Soc Exp Biol Med. *Res:* Cellular pharmacology and immunology; transplantation; growth control of mammalian cells. *Mailing Add:* Dept Path Col Med Univ Ill 1853 West Polk St Rm 446 CMW M/C 847 Chicago IL 60612

POLGAR, GEORGE, b Gyongyos, Hungary, Apr 5, 19; US citizen; m 45, 71; c 3. PEDIATRICS, PHYSIOLOGY. *Educ:* Univ Szeged, MD, 43. *Prof Exp:* Resident pediat, City Hosp, Szt Istvan, Budapest, 45-50; assoc pediat lung dis, State Sanitarium, 50-56; from asst prof to assoc prof pediat & physiol, Sch Med, Univ Pa, 67-74; chief div respiratory dis, Children's Hosp, Philadelphia, 71-74; prof pediat & assoc physiol, Wayne State Univ, 74-85; chief Respiratory Dis Div & dir Cystic Fibrosis Ctr, Children's Hosp Mich, 74-85; ADJ PROF PEDIAT, SCH MED, UNIV PA, 85- *Concurrent Pos:* Res fel pediat & physiol, Children's Hosp, Philadelphia, Univ Pa, 57-61; USPHS res fel, 59-60, spec res fel, 61, res career develop award, 62-; mem care-comt, Nat Cystic Fibrosis Res Found, 62-65; vis prof, Dept Pediat, Univ Geneva, 70-71; admin chmn, Philadelphia Pediat Pulmonary Prog, 72-74; mem exec comt, Assoc Pediat Pulmonary Centers, 73-; mem govt rels comt, Cystic Fibrosis Found, 74-75; ed, Pediat Pulmonology, 84- *Mem:* Am Physiol Soc; Am Acad Pediat; Soc Pediat Res; Am Thoracic Soc. *Res:* Respiratory diseases in children; pulmonary physiology in infants and children. *Mailing Add:* Univ PA PO Box 1118 King of Prussia PA 19406

POLGE, ROBERT J, b Anduze, France, Mar 11, 28; m 65; c 2. ELECTRICAL ENGINEERING. *Educ:* Montpellier Univ, Lic es sc, 50; Advan Sch Elec, Paris, EE, 52; Carnegie Inst Technol, MS, 61, PhD(elec eng, control), 63. *Prof Exp:* Engr, Andre Citroen Co, Paris, 53-56; head electronic lab, Sciaky Co, Paris, 56-59; head tech dept, Thompson-CSF Co, Paris, 59-60; asst elec eng, Carnegie Inst Technol, 60-63; assoc prof elec eng, 63-70, chmn dept, 81-85, PROF ELEC ENG, UNIV ALA, 70- *Concurrent Pos:* Consult, French Air Force. *Mem:* Inst Elec & Electronics Engrs; French Elec Soc; Sigma Xi. *Res:* Radar; communication; digital processing. *Mailing Add:* Dept Elec Univ Ala Huntsville Huntsville AL 35899

POLGLASE, WILLIAM JAMES, b Vancouver, BC, May 31, 17; m 60; c 1. CHEMISTRY. *Educ:* Univ BC, BA, 43, MA, 44; Ohio State Univ, PhD(chem), 48. *Prof Exp:* Asst, Univ BC, 43-44; chemist, Imp Oil Co, Ont, 44-45; asst, Res Found, Ohio State Univ, 45-47; res instr, Sch Med, Univ Utah, 48-51; res chemist, Rayonier, Inc, Wash, 51-52; assoc prof, 52-62, actg head dept, 74-77, PROF BIOCHEM, FAC MED, UNIV BC, 62-, HEAD DEPT, 77- *Mem:* Am Chem Soc; Am Soc Microbiol; Am Soc Biol Chemists. *Res:* Microbial chemistry and metabolism; regulation; antibiotics. *Mailing Add:* 1402 9500 Erickson Burnaby BC V3J 1M8 Can

POLHEMUS, JOHN THOMAS, b Ames, Iowa, Sept 11, 29; m 55; c 2. BIOENGINEERING & BIOMEDICAL ENGINEERING, ENTOMOLOGY. *Educ:* Iowa State Univ, BSEE, 56; Univ Colo, PhD(syst biol), 77. *Prof Exp:* Res engr, Calif Res Corp, Standard Oil Calif, 56-60; sr staff engr, Martin Marietta Corp, 60-85, sr staff scientist, 85-86; chief engr, Computer Technol Assocs Inc, 86-90; RETIRED. *Concurrent Pos:* Res assoc, Univ Colo Mus, 63-, Am Mus Natural Hist, 81-, Calif Acad Sci, 86-, B P Bishop Mus, 87-, Smithsonian Inst, 89-; adj prof, Univ Southern Ill, 88- *Mem:* Inst Elec & Electronic Engrs; Sigma Xi; Entom Soc Am; Soc Syst Zool. *Res:* Bioengineering research on non-invasive cardiovascular measurements; application of space technology to public needs; zoogeography; entomological research on aquatic Hemiptera, insect physiology, biogeography; published approximately 120 books and articles in refere ed jours. *Mailing Add:* 3115 S York Englewood CO 80110

POLHEMUS, NEIL W, b Fountain Hill, Pa, Apr 29, 51. STATISTICAL GRAPHICS, RISK ANALYSIS. *Educ:* Princeton Univ, BSE, 73, PhD(civil eng), 76. *Prof Exp:* Asst prof, Sch Bus Admin, Univ NC at Chapel Hill, 76-78; ASST PROF DEPT CIVIL ENG, PRINCETON UNIV, 78- *Concurrent Pos:* Mem, Trans Res Bd Quality Assurance, 81-; Consult, Statist Graphics Corp, 82-; vis scholar, Univ Calif, Berkeley, 82. *Mem:* Am Statist Asn; Inst Mgt Sci; Operations Res Soc Am. *Res:* Application of statistical methods to engineering systems, including stochastic modeling, time series analysis, collision risk estimation, and statistical graphics. *Mailing Add:* Statistical Graphics Corp Five Independence Way Princeton Corp Ctr Princeton NJ 08540

POLI, CORRADO, b Troy, NY, Aug 9, 35; m 60; c 3. MECHANICAL ENGINEERING. *Educ:* Rensselaer Polytech Inst, BS, 57, MS, 58; Ohio State Univ, PhD(eng mech), 65. *Prof Exp:* Aerospace eng, Air Force Res & Develop Command, Wright-Patterson AFB, 58-65, asst prof aeronaut eng, Air Force Inst Technol, 65-67; assoc prof mech eng, 67-71, head dept, 78-81, PROF MECH ENG, UNIV MASS, AMHERST, 72- *Mem:* Soc Plastic Engrs; Am Soc Mech Engrs. *Res:* Design for manufacture. *Mailing Add:* Dept of Mech Eng Univ of Mass Amherst MA 01003

POLI, RINALDO, b Barga, Italy, Aug 17, 56; m 85; c 2. STRUCTURE & BONDING, INORGANIC REACTION MECHANISMS. *Educ:* Univ Pisa, Italy, LAUREA, 81; Scuola Normale Superiore, Pisa, PhD(chem), 85. *Hon Degrees:* Dottore di Ricerca, Ital Ministry Educ, 88. *Prof Exp:* Res assoc inorg chem, Tex A&M Univ, 85-87; ASST PROF INORG CHEM, UNIV MD, 87- *Concurrent Pos:* Distinguished new fac, Camille & Henry Dreyfus Found, 87; NSF presidential young investr, 90. *Mem:* Am Chem Soc; Italian Chem Soc. *Res:* Structure and bonding of inorganic and organometallic compounds; metal-metal bonds and clusters; odd-electron organometallic reactivity; mechanistic studies of inorganic reactions. *Mailing Add:* Dept Chem & Biochem Univ Md College Park MD 20742

POLIAK, AARON, b Buenos Aires, Arg, Apr 30, 25; m 51; c 3. OBSTETRICS & GYNECOLOGY. *Educ:* Univ Buenos Aires, MD, 49, PhD(med), 55; Am Bd Obstet & Gynec, dipl, 69 & recertified obstet & gynec, 86- *Prof Exp:* Docent gynec, Med Sch, Univ Buenos Aires, 63-68; from asst prof to assoc prof, 72-76, PROF OBSTET, ALBERT EINSTEIN COL MED, 76-; DIR, GYNEC DIV, JACOBI HOSP, 78-; CHIEF GYNEC, BRONX MUNICIPAL HOSP CTR, 84- *Concurrent Pos:* Fel obstet & gynec, Sch Med, Johns Hopkins Univ, 66-67, instr, 67-68; dir, Dept Obstet & Gynec, Lincoln Hosp, 72-77; dir gynec serv, Bronx Munic Hosp Ctr, 78-79, 84-, dir, Dept Obstet & Gynec, 79-8. *Mem:* Fel Am Col Surg; fel Am Col Obstet & Gynec; Am Fertil Soc. *Res:* Infertility, ovarian post-menopausal function; ovarian inervation; urinary stress incontinence. *Mailing Add:* Dept Obstet & Gynec Albert Einstein Col of Med Bronx NY 10461

POLICANSKY, DAVID J, b Cape Town, SAfrica, Dec 9, 44; US citizen; m. ENVIRONMENTAL SCIENCES, EDUCATION. *Educ:* Stanford Univ, BA, 67; Univ Ore, MS, 68, PhD(biol), 72. *Prof Exp:* Asst prof biol, Univ Mass, 73-81; res assoc, Harvard Univ, 81-83; STAFF OFFICER, NAT RES COUN, 83- *Concurrent Pos:* Vis asst prof biol, Harvard Univ, 81-82. *Mem:* Am Soc Naturalists; Am Fisheries Soc. *Res:* Ecology, evolution of life history patterns; sex changes; fishery biology. *Mailing Add:* Nat Res Coun GF 354 2101 Constitution Ave, NW Washington DC 20418

POLIFERNO, MARIO JOSEPH, b Derby, Conn, Feb 21, 30. MATHEMATICS. *Educ:* Yale Univ, BA, 52, MA, 54, PhD(math), 58. *Prof Exp:* Asst instr math, Yale Univ, 54-57; instr, Williams Col, 57-58; from instr to asst prof, 58-63, ASSOC PROF MATH, TRINITY COL, CONN, 63- *Mem:* Am Math Soc; Math Asn Am; Asn Symbolic Logic. *Res:* Symbolic logic; foundations of mathematics; set theory; point-set topology. *Mailing Add:* 442 New Haven Ave Derby CT 06418

POLIK, WILLIAM FREDERICK, b Washington, DC, Dec 30, 60; m 88. LASER SPECTROSCOPY, CHEMICAL REACTION DYNAMICS. *Educ:* Dartmouth Col, BA, 82; Univ Calif, Berkeley, PhD(chem), 88. *Prof Exp:* Grad student instr chem, Univ Calif, Berkeley, 82-85; ASST PROF CHEM, HOPE COL, 88- *Concurrent Pos:* Consult, Coherent, Inc, 86-88; presidential young investr award, NSF, 91. *Mem:* Coun Undergrad Res; Am Chem Soc; Am Phys Soc; AAAS. *Res:* Highly excited molecular states; gas phase chemical reaction theories; energy transfer; chaotic systems; high-resolution laser spectroscopy; molecular beams; computer interfacing; numerical algorithms. *Mailing Add:* Dept Chem Hope Col Holland MI 49423-3698

POLIMENI, ALBERT D, b Canandaigua, NY, Mar 14, 38. MATHEMATICS. *Educ:* Univ Buffalo, BA, 60, MA, 62; Mich State Univ, PhD(math), 65. *Prof Exp:* Lectr math, Univ Mich, Ann Arbor, 65-67; asst prof, Syracuse Univ, 67-70; assoc prof, 70-81, PROF MATH, STATE UNIV NY COL FREDONIA, 81- *Mem:* Am Math Soc; Math Asn Am. *Res:* Finite group theory. *Mailing Add:* Dept Comput Sci State Univ NY Col Fredonia NY 14063

POLIMENI, PHILIP INIZIATO, b New York, NY, May 23, 34; m 60; c 2. MYOCARDIAL ELECTROLYTE DISTRIBUTIONS, POLYMER DRAG REDUCTION. *Educ:* City Col New York, BSc, 57; State Univ NY, PhD(physiol), 69. *Prof Exp:* Biochemist, State Univ NY, 59-64, teaching asst physiol, 66-68; instr, Pritzker Sch Med, Univ Chicago, 69-73, res assoc & asst prof, 73-78; ASSOC PROF FAC MED, UNIV MAN, 78- *Mem:* Int Soc Heart Res; Biophys Soc; Am Physiol Soc; Can Physiol Soc. *Res:* Polymer drag reduction in vivo; pharmacological application of drag-reducing polymers to heart disease; cardiac pharmacology; study of electrolyte and water distributions in the healthy and diseased myocardium. *Mailing Add:* Dept Pharmacol & Therapeut Univ Man Winnipeg MB R3E 0W3 Can

POLIN, DONALD, b Arlington, Mass, Dec 7, 25; m 54; c 3. AVIAN PHYSIOLOGY, AVIAN NUTRITION. *Educ:* US Merchant Marine Acad, BS, 50; Rutgers Univ, BS, 51, PhD(physiol), 55. *Prof Exp:* Asst avian physiol, Rutgers Univ, 51-55; res fel, Merck Inst Therapeut Res, NJ, 55-67; unit leader biochem, Norwich Pharmacal Co, NY, 67-69; from assoc prof to prof poultry sci, Mich State Univ, 69-90; RETIRED. *Concurrent Pos:* Mem subcomt on effect of environ on nutrient requirements & subcomt on nutrient requirements of poultry, Nat Acad Sci-Nat Res Coun; chief exec officer, Avi-Sci, Inc, 89- *Honors & Awards:* Am Feed Mfg Award, Poultry Sci Asn, 84. *Mem:* Soc Exp Biol & Med; World's Poultry Sci Asn; Soc Toxicol; Am Inst Nutrit; Sigma Xi; Am Fedn Avicult. *Res:* Avian biochemistry; nutritional physiology; tissue residues; energy utilization; metabolism and mode of action of drugs; regulation of feed intake; obesity and fatty liver-hemorrhagic syndrome; nutritional toxicology of farm animals; nutrition of pet (exotic) birds. *Mailing Add:* Animal Sci Dept Mich State Univ East Lansing MI 48823

POLING, BRUCE EARL, b Columbus, Ohio, Oct 8, 44; m 70; c 2. CHEMICAL ENGINEERING, CHEMISTRY. *Educ:* Ohio State Univ, BChE & MSc, 67; Univ Ill, PhD(chem eng), 71. *Prof Exp:* ASSOC PROF CHEM ENG, UNIV MO-ROLLA, 71- *Concurrent Pos:* Consult, Bartlesville Energy Res Ctr, 76; researcher, NSF, 76-78. *Mem:* Sigma Xi; Am Inst Chem Engrs. *Res:* Storage of thermal energy by means of reversible, liquid phase chemical reactions; kinetics of chemical reactions between diazo esters and olefins. *Mailing Add:* Dept Chem Eng Univ Mo Rolla MO 65401

POLING, CLYDE EDWARD, b Darke Co, Ohio, Dec 15, 14; m 40; c 4. NUTRITION. *Educ:* Ohio Wesleyan Univ, BA, 36; Syracuse Univ, MS, 39; Case Western Reserve Univ, PhD(biochem), 41. *Prof Exp:* Asst chem, Syracuse Univ, 37-39; jr chemist, City Dept Pub Health & Welfare, Cleveland, Ohio, 41-42; res chemist, Nutrit Div, Res & Develop Ctr, Swift & Co, 42-75, chief nutritionist & res biochemist, Biochem & Nutrit Div, 70-72, res nutritionist, 72-80; RETIRED. *Concurrent Pos:* Exec secy, comt vet serv farm animals, Nat Res Coun, 45-50. *Mem:* Emer mem Am Chem Soc; Am Oil Chemists' Soc; emer mem Asn Vitamin Chemists; emer mem Am Inst Nutrit; Inst Food Technologists; Sigma Xi. *Res:* Nutritional values and interrelationships of proteins, fats, vitamins, and carbohydrates; physiological effects of insecticides; nutritional composition and labeling of food products. *Mailing Add:* 1903 Maple Ave Downers Grove IL 60515-4409

POLING, CRAIG, b Vienna, Austria, Oct 2, 53; m; c 2. ACOUSTICS, SONAR. *Educ:* Col William & Mary, BS, 76, MA, 78; Carnegie-Mellon Univ, MS, 79. *Prof Exp:* Mathematician, Naval Avionics Ctr, 82-83; sr engr, Honeywell, Inc, 83-88, fel, 88-89, sr fel, 89-90; SR FEL, ALLIANT TECH SYSTS, 90- *Mem:* Math Asn Am; Am Math Soc; Soc Indust & Appl Math. *Res:* Acoustics and sonar. *Mailing Add:* 18980 Pheasant Circle Eden Prairie MN 55346

POLING, GEORGE WESLEY, b Lloydminster, Alta, June 3, 35; m 58; c 5. METALLURGICAL ENGINEERING. *Educ:* Univ Alta, BS, 57, MSc, 61, PhD(metall eng), 63. *Prof Exp:* Trainee, Roan Antelope Copper Mines, North Rhodesia, 57-58; sr metall engr, Texaco Res Ctr, 63-68; from assoc prof to prof mineral eng, 68-78, PROF & HEAD, DEPT MINING & MINERAL PROCESS ENG, UNIV BC, 78- *Concurrent Pos:* Distinguished lectr, Can Inst Mining & Metall. *Mem:* Fel Can Inst Mining & Metall; fel Brit Inst Mining & Metall; Am Inst Mining, Metall & Petrol Engrs; Nat Asn Corrosion Engrs; Can Mineral Processors. *Res:* Flotation-mineral beneficiation; bulk materials transportation; recovery of placer gold; infrared spectroscopic study of adsorbed molecules and reactions at solid surfaces; surface chemistry of corrosion and corrosion inhibition processes; environmental protection in mining. *Mailing Add:* 4050 W 36th Vancouver BC V6N 2S1 Can

POLING, STEPHEN MICHAEL, b Long Beach, Calif, Sept 29, 46. PLANT BIOCHEMISTRY. *Educ:* Pomona Col, BA, 68; Univ Wash, MS, 70. *Prof Exp:* Res chemist citrus, 72-78, RES CHEMIST, FRUIT & VEGETABLE CHEM LAB, AGR RES SERV, USDA, 78- *Mem:* Am Chem Soc; AAAS. *Res:* Use of bioregulators to improve citrus quality and to increase the rubber content of guayule. *Mailing Add:* 263 S Chester Ave Pasadena CA 91106-3108

POLINSKY, RONALD JOHN, b Reading, Pa, Dec 6, 48; m 70; c 3. NEUROLOGY, CLINICAL NEUROPHARMACOLOGY. *Educ:* Mass Inst Technol, BS, 70; Dartmouth Med Sch, MD, 73. *Prof Exp:* Intern, med, Cleveland Metrop Gen Hosp, 73-74; resident, neurol, Johns Hopkins Hosp, 74-77; clin assoc, neuropharmacol, NIMH, 77-81; asst prof, neurol, Uniformed Servs Univ Health Sci, 81-82; staff neurologist, NIMH, 82-84; SECT CHIEF, CLIN NEUROPHARMACOL, NAT INST NEUROL DISORDERS & STROKE, 84- *Concurrent Pos:* Med officer, US Pub Health Serv, 77-81; vis prof, neurol, Univ Vt Med Ctr, 85. *Mem:* Fel Am Acad Neurol; Clin Autonomic Res Soc; Am Neurol Asn. *Res:* Investigation and treatment of patients with disorders of the autonomic nervous system; development of methods for assessing neurotransmitter/neuropeptide metabolism and function in man, studies of familial and genetic forms of Alzheimer's disease. *Mailing Add:* Nat Inst Health Bldg 10 5N-262 9000 Rockville Pike Bethesda MD 20892

POLIS, GARY ALLAN, b Los Angeles, Calif, Aug 28, 46; m; c 1. ECOLOGY, ZOOLOGY. *Educ:* Loyola Univ Los Angeles, BS, 69; Univ Calif, Riverside, MS, 75, PhD(biol), 77. *Prof Exp:* Asst prof zool, Dept Zool, Ore State Univ, 77-79; asst prof, 79-85, ASSOC PROF ECOL, DEPT BIOL, VANDERBILT UNIV, 85- *Concurrent Pos:* High sch sci teacher. *Mem:* Ecol Soc Am; Am Soc Naturalists; Am Arachnological Asn; Sigma Xi; Brit Ecol Soc. *Res:* Evolution and dynamics of population self-regulatory systems; analysis of patterns and determinants of community structure; dynamics and patterns of predation; biology of Scorpionida, Araneae and predaceous insects; desert biology. *Mailing Add:* Dept Biol Vanderbilt Univ Nashville TN 37285

POLIS, MICHAEL PHILIP, b New York, NY, Oct 24, 43; m 66; c 3. AUTOMATIC CONTROL, OPTIMIZATION. *Educ:* Univ Fla, BSEE, 66; Purdue Univ, MSEE, 68, PhD(elec eng), 72. *Prof Exp:* Engr, Monsanto Corp, 66; grad instr elec eng, Purdue Univ, 66-71; res assoc, Polytech Sch, Montreal, 72-73, from asst prof to prof control, 73-86; prog dir, NSF, 83-87; PROF & CHAIR ELEC ENG, WAYNE STATE UNIV, 87- *Concurrent Pos:* Consult, Metrop Transit Bur, Montreal, Que, 73-77, Transp Res & Develop Ctr, Montreal, Can, 79-80, Sidbec-Dosco, Montreal, Que, Can, 79-81, Mich Bell, Detroit, 89-91; vis researcher, Laboratoire d'Automatique et d'Analyse des Sys, France, 78; mem I E grant selection comt, Natural Sci & Eng Res Coun, Can, 89-; vpres mem activ, Inst Elec & Electronic Engrs, 90-91. *Mem:* Inst Elec & Electronic Engrs; Sigma Xi. *Res:* Identification and control of distributed parameter systems, robust control, computer-aided design and transportation systems. *Mailing Add:* Dept Elec & Computer Eng Wayne State Univ Detroit MI 48202

POLITZER, HUGH DAVID, US citizen; m 75; c 2. PHYSICS. *Educ:* Univ Mich, BS, 69; Harvard Univ, PhD(physics), 74. *Prof Exp:* Jr fel, Harvard Soc Fels, 74-77; assoc prof physics, 77-79, exec officer physics, 86-88, PROF THEORET PHYSICS, CALIF INST TECHNOL, 79- *Honors & Awards:* Sakurai Award, Am Physics Soc, 86. *Res:* Theoretical physics. *Mailing Add:* Dept Physics 405-47 Calif Inst Technol Pasadena CA 91125

POLITZER, PETER ANDREW, b Prague, Czech, Dec 12, 37; US citizen; div; c 3. THEORETICAL CHEMISTRY. *Educ:* Case Western Reserve Univ, BA, 60, MS, 61, PhD(chem), 64. *Prof Exp:* Lectr chem, Case Western Reserve Univ, 64; res assoc, Ind Univ, 64-66; from asst prof to assoc prof, 66-74, PROF CHEM, UNIV NEW ORLEANS, 74- *Concurrent Pos:* Vis fel, Johns Hopkins Univ, 73-74, res scientist, 74. *Honors & Awards:* Distinguished Prof Chem, Univ New Orleans. *Mem:* Am Chem Soc; Royal Soc Chem; Int Soc Quantum Biol; Catalysis Soc; Am Phys Soc; Sigma Xi. *Res:* Distribution of electrons in molecules; properties of chemical bonds; chemical reactivity; applications of theoretical chemistry to biological processes; molecular electrostatic potentials; relationship of energy to electronic density; chemical carcinogenesis. *Mailing Add:* Dept Chem Univ New Orleans New Orleans LA 70148

POLIVKA, RAYMOND PETER, b Chicago, Ill, July 18, 29; m 54; c 2. COMPUTER SCIENCES EDUCATION. *Educ:* North Cent Col, Ill, BA, 51; Univ Ill, MS, 53, PhD(math), 58. *Prof Exp:* Asst math, Univ Ill, 52-53, asst comput lab, 53-55 & 57-58; eng programmer, Int Bus Mach Corp, 58-65; assoc prof math, North Cent Col, Ill, 65-66; programmer, Int Bus Mach Corp, 66-71, staff instr, 71-75, ADV INSTR, IBM CORP, 76- *Concurrent Pos:* Adj prof comput sci, Union Col, 77- *Mem:* Am Math Soc; Math Asn Am; Asn Comput Mach. *Res:* System programming; engineering analysis; programming educators and specializing in programming languages. *Mailing Add:* 60 Timberline Dr Poughkeepsie NY 12603

POLK, C(HARLES), b Vienna, Austria, Jan 15, 20; nat US; m 46; c 2. APPLIED ELECTROMAGNETICS, BIOELECTROMAGNETICS. *Educ:* Washington Univ, BSEE, 48; Univ Pa, MS, 53, PhD(elec eng), 56. *Prof Exp:* Engr, Victor Div, Radio Corp Am, 48-52, mem tech staff, David Sarnoff Res Ctr Labs, 57-59; res assoc, Moore Sch Elec Eng, Univ Pa, 52-57; prof elec eng & chmn dept, Univ RI, 59-75; head elec sci & anal sect, Eng Div, NSF, 75-76, actg dir, Eng Div, 76-77; chmn dept, 77-79, PROF ELEC ENG, UNIV RI, 77- *Concurrent Pos:* Consult, Philco Corp & Am Electronics Labs, 55-56, Labs, Radio Corp Am, 59 US Navy, 71-72, Tex Instruments Co, 86, Electrobiol Inc, 86-; adj prof, Grad Div, Drexel Inst Technol, 57-59; chmn, Providence Sect, Inst Elec & Electronics Engrs, 64-65; vis prof, Stanford Univ, 68-69, Univ Wis, Madison, 83-84; vpres Bioelectromagnetics Soc, 87-88, pres, 88-89; distinguished lect, Inst Elec & Electronic Engrs-Engn Med & Biol Soc, 91. *Honors & Awards:* Super Accomplishment Award, NSF, 77. *Mem:* AAAS; fel Inst Elec & Electronics Engrs; Bioelec Repair & Growth Soc; Am Soc Eng Educ; Int Sci Radio Union; Bioelectromagnetics Soc. *Res:* Biological effects of electromagnetic fields, particularly low frequency effects at cellular level; applied electromagnetic theory; geomagnetism and geoelectricity; upper atmosphere and ionospheric physics; extremely low frequency noise; antennas and radio propagation. *Mailing Add:* Dept Elec Eng Univ RI Kingston RI 02881

POLK, CONRAD JOSEPH, b Chicago, Ill, Nov 9, 39; m 63. EQUIPMENT FAILURE ANALYSIS, MATERIALS ENGINEERING. *Educ:* Loyola Univ, BS, 61; Univ Toledo, MS, 63; Ill Inst Technol, PhD(mat eng), 68. *Prof Exp:* Instr metall eng, Ill Inst Technol, 66-68; from res metallurgist to sr res metallurgist, 68-72, assoc engr, 72-75, leader metall eng group, 73-77, sr assoc engr, 75-82, leader failure anal group, 78-82, MGR MAT ENG SECT, MOBIL RES & DEVELOP CORP, 82-; leader task group on amine corrosion (RP 945), 86-89, CHMN, SUB COMT CORROSION & MAT, AM PETROL INST, 89- *Concurrent Pos:* Leader Corrosion Res Panel, Am Petrol Inst, 81-89. *Mem:* Am Soc Metals; Nat Asn Corrosion Engrs; Am Petrol Inst. *Res:* Mechanical behavior of materials for refinery and chemical processing equipment under various temperature, pressure and stream composition conditions; lubrication and rolling contact bearing technology; corrosion control. *Mailing Add:* Mobil Res & Develop Corp Box 1026 Princeton NJ 08543-1026

POLK, HIRAM CAREY, JR, b Jackson, Miss, Mar 23, 36; m 56; c 2. SURGERY. *Educ:* Millsaps Col, BS, 56; Harvard Med Sch, MD, 60. *Prof Exp:* Instr surg, Wash Univ, 64-65; from asst prof to assoc prof, Univ Miami, 65-71; PROF SURG & CHMN DEPT, UNIV LOUISVILLE, 71- *Concurrent Pos:* Clin trainee cancer control, USPHS, 64-65; dir tumor clin, Jackson Mem Hosp, Miami, Fla, 69-71; res assoc path, Lister Inst Prev Med, 69; mem consult staff, Vet Admin Hosp, 71-; mem attend staff, Univ Hosp & Norton-Children's Hosp, 71-; dir, Am Bd Surg. *Honors & Awards:* Seale Harris Award, Southern Med Asn. *Mem:* Asn Acad Surg; Soc Univ Surg; AMA; Am Surg Asn; Soc Surg Alimentary Tract; Am Soc Surg Oncol. *Res:* Surgical infection; cancer; burns. *Mailing Add:* Dept of Surg Univ Louisville Louisville KY 40292

POLK, MALCOLM BENNY, b Chicago, Ill, Feb 2, 38. NUCLEAR MAGNETIC RESONANCE SPECTROSCOPY, INFRARED SPECTROSCOPY. *Educ:* Univ Ill, BS, 60; Univ Pa, PhD(chem), 64. *Prof Exp:* Asst instr chem, Univ Pa, 60-64; fel, Univ Calif, Davis, 64-65; res chemist, E I DuPont Co, 65-67 & 68-72; assoc prof chem, Prairie View Col, 67-68; res chemist, US Dept Interior, 72-73; prof org & polymer chem, Atlanta Univ, 73-85; ASSOC PROF, SCH TEXTILE ENG, GA INST TECHNOL, 85- *Mem:* Am Chem Soc. *Res:* Synthesis and characterization of liquid crystalline polymers. *Mailing Add:* GA Inst Technol Sch Textile Eng Atlanta GA 30332

POLKING, JOHN C, b Breda, Iowa, June 6, 34. MATHEMATICS. *Educ:* Univ Notre Dame, BS, 56; Univ Chicago, SM, 61, PhD(math), 66. *Prof Exp:* Instr math, Univ Chicago, 65-66; Off Naval Res res assoc, Brandeis Univ, 66-67, lectr, 67-68; from asst prof to assoc prof, 68-78, chmn dept, 79-82, PROF MATH, RICE UNIV, 78- *Concurrent Pos:* vis prof, Univ Utah, 78; dir, math Sci Div, NSF, 84-87. *Mem:* Am Math Soc; Math Asn Am; AAAS. *Res:* Partial differential equations; several complex variables. *Mailing Add:* Dept Math Rice Univ Box 1892 Houston TX 77251

POLKOWSKI, LAWRENCE B(ENJAMIN), b Rockford, Ill, Feb 22, 29; m 51; c 3. SANITARY ENGINEERING. *Educ:* Univ Ill, BS, 50; Univ Wis, MS, 51, PhD(sanit eng), 58. *Prof Exp:* Asst, Univ Wis, 50-51, instr civil eng, 54-55, proj assoc, 55-57, from asst prof to assoc prof civil eng, 57-61; prof sanit eng, Univ Iowa, 61-64; fel, Ford Found, 60, assoc prof civil eng, 64, PROF CIVIL ENG, UNIV WIS-MADISON, 65- *Concurrent Pos:* Consult, Procter & Gamble, Co, 60-, S B Foot Tanning Co, 65-, Am Concrete Pipe Asn, 68-, Campbell Soup Co, 70, Libby, McNeil Libby, 71 & Lake to Lake Creameries, 72-; consult partner, Polkowski, Boyle & Assocs, 66-; comnr, Madison Metrop Sewerage Comn, 71- *Honors & Awards:* Harrison Prescott Eddy Medal, Water Pollution Control Fedn, 71. *Mem:* AAAS; Am Soc Eng Educ; Am Soc Civil Engrs; Am Water Works Asn; Water Pollution Control Fedn. *Res:* Biological treatment of wastes; design and development of waste treatment processes; combined treatment of industrial and municipal water treatment; solids-liquid separation; filtration wastewater effluents; farm waste disposal. *Mailing Add:* Dept Civil Eng 2205 Eng Bldg Univ Wis Madison 1415 Johnson Dr Madison WI 53706

POLL, JACOBUS DANIEL, b Deventer, Neth, Mar 28, 30; m 58; c 4. THEORETICAL PHYSICS. *Educ:* Univ Leiden, Drs, 56; Univ Toronto, PhD(theoret physics), 60. *Prof Exp:* Lectr physics, Univ Toronto, 60-61, from asst prof to assoc prof, 61-70; PROF PHYSICS, UNIV GUELPH, 70- *Concurrent Pos:* Vis prof, Univ Calif, Berkeley, 68-69, Univ Nijmegen, Neth, 75-76, Univ Victoria, Can, 84-85. *Mem:* Am Phys Soc; Can Asn Physicists. *Res:* Theoretical molecular physics and statistical mechanics; infrared spectra; molecular distribution functions and interactions; molecular structure; astrophysics. *Mailing Add:* Dept Physics Univ Guelph Guelph ON N1G 2W1 Can

POLLACK, BERNARD LEONARD, b Detroit, Mich, Jan 18, 20; m 49; c 2. PLANT BREEDING. *Educ:* Pa State Univ, BS, 49, MS, 51, PhD(hort), 53. *Prof Exp:* Asst plant breeding, Pa State Univ, 49-52, from instr to asst prof, 52-60; veg exten specialist, 60-85, chmn dept Hort & Forestry, Cook Col, 77-85, EMER PROF, RUTGERS UNIV, NEW BRUNSWICK, 85- *Concurrent Pos:* consult, Int Agr, WAfrica, Puerto Rico & Jamaica. *Mem:* Am Soc Hort Sci. *Res:* Breeding early, disease resistant, fresh market tomatoes; disease resistance in eggplant; tomato genetics; plasticulture; degradable plastic mulches; eggplant and pepper genetics; trickle irrigation research. *Mailing Add:* 7595 Dehesa Ct Carlsbad CA 92009

POLLACK, EDWARD, b New York, NY, Apr 28, 31; m 55; c 3. ATOMIC PHYSICS. *Educ:* City Col New York, BS, 52; NY Univ, MS, 54, PhD(physics), 63. *Prof Exp:* Asst physics, NY Univ, 52-54; lectr, City Col New York, 56-60; instr, NY Univ, 60-63; from asst prof to assoc prof, 63-74, PROF PHYSICS, UNIV CONN, 74- *Concurrent Pos:* Visiting physicist, Univ Paris, Orsay & NATO res grant, 74 & Service de Physique Atomique-Saclay, France, 81; vis prof, Yale Univ, 87, res affil, 87-89. *Mem:* Am Phys Soc. *Res:* Polarizabilities of the alkalis and metastable argon; differential cross sections for kilo electron volt region atom-atom scattering; thermal energy atom-atom collisions; inelastic ion-molecule collisions; energy loss scaling in atom-molecule collisions; electron capture processes; rydberg states of helium; two photon spectroscopy of hydrogen. *Mailing Add:* Dept Physics Univ Conn Storrs CT 06269-3046

POLLACK, EMANUEL DAVIS, b Chicago, Ill, Mar 29, 42; m 77; c 2. ZOOLOGY, EMBRYOLOGY. *Educ:* Roosevelt Univ, BS, 64; Univ Iowa, MS, 67, PhD(zool), 70; Cell Sci Ctr, dipl, 74; NATO Advan Study Inst, 71. *Prof Exp:* Teaching asst zool & embryol, res asst zool & instr embryol, Univ Iowa, 65-70; postdoctoral NIMH trainee & res fel, Albert Einstein Col Med, 70-72; res scientist neuroembryol, 72-79, PROG DIR, INST STUDY OF DEVELOP DISABILITIES, UNIV ILL, 79-, ASST PROF ANAT, COL MED & ASSOC PROF BIOL SCI, 75-,. *Concurrent Pos:* Consult, Ill Environ Protection Agency, 73-80 & Mus of Sci & Indust, 85-; chmn, Comt Neurosci, Univ Ill, 83-, res assoc prof develop disabilities, 78- *Mem:* Am Soc Cell Biol; AAAS; Soc Neurosci; NY Acad Sci; Sigma Xi; Soc Develop Biol; Am Soc Zoologists. *Res:* Developmental neurobiology and neurophysiology; development of neuro-neuronal and neuromuscular relationships; neural tissue culture; neural ontogeny in amphibian metamorphosis; behavioral ontogeny; neuronal growth factors. *Mailing Add:* 1640 W Roosevelt Rd Chicago IL 60608

POLLACK, GERALD H, b New York, NY, May 20, 40; div; c 3. PHYSIOLOGY, BIOMEDICAL ENGINEERING. *Educ:* Polytech Inst Brooklyn, BS, 61; Univ Pa, PhD(biomed eng), 68. *Prof Exp:* From asst prof to assoc prof, 68-77, prof anesthesiol & bioeng, 77-82, PROF BIOENG, UNIV WASH, 82- *Concurrent Pos:* Estab investr, Am Heart Asn, 74-79; mem bd dir, Biomed Eng Soc, 77-80; mem, Basic Sci Coun, Am Heart Asn, 79- *Honors & Awards:* Kulka Award. *Mem:* Am Physiol Soc; Biophys Soc; Cardiac Muscle Soc; Inst Elec & Electronic Engrs; Biomed Eng Soc. *Res:* Molecular mechanisms of contraction in muscle; mechanisms of cardiac function. *Mailing Add:* Div Bioengineering WD-12 Univ Wash Seattle WA 98195

POLLACK, GERALD LESLIE, b Brooklyn, NY, July 8, 33; m 58; c 4. PHYSICS LIQUIDS & SOLIDS, BIOPHYSICS. *Educ:* Brooklyn Col, BS, 54; Calif Inst Technol, MS, 57, PhD(physics), 62. *Prof Exp:* Asst math, Calif Inst Technol, 55-57, asst physics, 57-61; solid state physicist inorg mat div, Nat Bur Standards, 61-65; assoc prof, 65-69, PROF PHYSICS, MICH STATE UNIV, 69- *Concurrent Pos:* Consult, Nat Bur Standards, 65-71, Nuclear Regulatory Comn, 79-84. *Mem:* Fel Am Phys Soc; Am Asn Physics Teachers. *Res:* Liquid helium physics; rare-gas solids; molecular solids; crystal growth; surface physics; diffusion in biological systems; liquid helium films; solubility of gases in liquids. *Mailing Add:* Dept Physics & Astron Mich State Univ East Lansing MI 48824-1116

POLLACK, HENRY NATHAN, b Omaha, Nebr, July 13, 36; m 63; c 2. GEOPHYSICS, GEOTHERMICS. *Educ:* Cornell Univ, AB, 58; Univ Nebr, MS, 60; Univ Mich, PhD(geophys), 63. *Prof Exp:* Res fel geophys, Harvard Univ, 63-64; from asst prof to assoc prof, 64-74, PROF GEOPHYS, UNIV MICH, ANN ARBOR, 74- *Concurrent Pos:* Sr lectr, Univ Zambia, 70-71; res scientist, Univ Durham & Newcastle upon Tyne, UK, 77-78; assoc dean res, Univ Mich, 82-85; vis prof, Univ Western Ont, Can, 85-86. *Mem:* fel AAAS; Am Geophys Union; fel Geol Soc Am. *Res:* State, composition, dynamics and evolution of the earth's interior; heat flow and geothermics. *Mailing Add:* Dept Geol Sci Univ of Mich Ann Arbor MI 48109

POLLACK, HERBERT, medicine, biomedical engineering; deceased, see previous edition for last biography

POLLACK, IRWIN W, b Philadelphia, Pa, Aug 14, 27; m 57; c 3. PSYCHIATRY, REHABILITATION. *Educ:* Temple Univ, BA, 50; Columbia Univ, MA, 51; Univ Vt, MD, 56. *Prof Exp:* Rotating intern, Grad Hosp, Univ Pa, 56-57; res asst psychiat, Henry Phipps Psychiat Clin, Johns Hopkins Univ, 57-60, chief psychiat resident, 60-61, adminr clin & psychiatrist, Psychiat Liaison Serv, 61-64, asst prof psychiat, 64-68; assoc prof, 68-70, dir, Community Mental Health Ctr, 68-77, chmn dept, 70-77, PROF PSYCHIAT & PROF NEUROL, ROBERT WOOD JOHNSON MED SCH, UNIV MED & DENT NJ, 70- *Concurrent Pos:* Psychiatrist-in-chief, Sinai Hosp, Baltimore, 64-68; consult, Middlesex Gen Univ Hosp, New Brunswick & Muhlenberg Hosp, NJ, 80-; mem, Middlesex Ment Health Bd, 73-77. *Mem:* Fel Am Psychiat Asn; Am Psychosom Soc; Am Cong Rehab Med. *Res:* Relationship of spatial and temporal orientation to personality and to psychopathology; muscle activity and emotions; rehabilitation of cognitive dys function. *Mailing Add:* Dept Neurol Ctr Cognitive Rehab Robert Wood Johnson Med Sch Univ Med & Dent NJ Piscataway NJ 08854

POLLACK, J DENNIS, b Brooklyn, NY, Apr 3, 31; m 62. MICROBIOLOGY, MEDICAL MYCOLOGY. *Educ:* Univ Conn, BA, 58, MS, 64, PhD(bact), 66. *Prof Exp:* Res assoc microbiol, Hadassah Med Sch, Hebrew Univ Jerusalem, 66-67; from instr to assoc prof med microbiol, 66-77, from asst prof to assoc prof pediat, 68-78, from co-dir to dir phase III med curric, 76-80, PROF MED MICROBIOL, COL MED, OHIO STATE UNIV, 77-, PROF PEDIAT, 78- *Concurrent Pos:* NIH fel, Hadassah Med Sch, Hebrew Univ Jerusalem, 66-67; chmn sci adv bd, Nat Reyes Syndrome Found, 76-, ed-in-chief, J Nat Reyes Syndrome Found, 79- *Mem:* Am Soc Microbiol; Int Orgn Mycoplasmology; Med Mycol Soc Am; Int Soc Human & Animal Mycol; Sigma Xi. *Res:* Medical microbiology; biochemistry; Reye's Syndrome; medical administration. *Mailing Add:* Dept Med Microbiol & Immunol Ohio State Univ Col of Med Columbus OH 43210

POLLACK, JAMES BARNEY, b New York, NY, July 9, 38. PLANETARY SCIENCE, CLIMATOLOGY. *Educ:* Princeton Univ, AB, 60; Univ Calif, Berkeley, MA, 62; Harvard Univ, PhD(astron), 65. *Prof Exp:* Res physicist planets, Smithsonian Astrophys Observ, 65-68; sr res assoc, Cornell Univ, 68-70; RES SCIENTIST SPACE, NASA AMES RES CTR, 70- *Concurrent Pos:* Team mem, Mariner 9 Imaging Exp, 67-73 & Viking Lander Imaging Exp, 75-; inter-disciplinary scientist, Venus Pioneer Mission, 72- & Galileo Jupiter Mission, 78-, voyager imaging team, Saturn, 80-90, Mars Observer, 86, Cassini (Saturn), 90-, partic scientist, Mars '94, 90-; mem exec coun, Div Planetary Sci, 76-78; chief scientist, Ames Climate Off, 78-84; mem, Space Sci Adv Comt, NASA, 78-82; actg probe proj scientist, Galileo Jupiter Mission, 85-86; Sherman Fairchild Distinguished Scholar, Calif Tech, 90. *Honors & Awards:* Medal for Exceptional Sci Achievement, NASA, 76 & 79; Space Sci Award, Am Inst Aeronaut & Astronaut; Leo Szilard Res Award, Am Phys Soc, 85; Kuiper Prize, Div Planetary Sci, 89. *Mem:* fel Am Astron Soc; fel Am Geophys Union; fel AAAS. *Res:* Planetary atmospheres; origin and evolution of the solar system; climate. *Mailing Add:* Space Sci Div NASA Ames Res Ctr Moffett Field CA 94035

POLLACK, JEROME MARVIN, b Chicago, Ill, Apr 16, 26; m 52; c 4. GEOLOGY. *Educ:* Univ Okla, BS, 49, MS, 51, PhD(geol), 59. *Prof Exp:* Asst geol, Univ Okla, 49-51; geologist, Humble Oil & Refining Co, 51-55; instr, Univ Okla, 55-58; asst prof, Okla City Univ, 59; asst prof, Harpur Col, 59-61; assoc prof & chmn dept, Univ NH, 61-65; prof, Univ RI, 65-71; prof, 71-76, vpres acad affairs, 71-74, pres, 74-83, distinguished univ prof, Fairleigh Dickinson Univ, 83-85; RETIRED. *Concurrent Pos:* Dean, Col Arts & Sci, Univ RI, 65-68 & 69-71, vpres acad affairs, 68-69; Am Asn Petrol Geol grant-in-aid, 58. *Mem:* Geol Soc Am; Soc Econ Paleont & Mineral; Am Asn Petrol Geol; Nat Asn Geol Teachers. *Res:* Sedimentology; statistical problems in geology; petroleum geology; science education; invertebrate paleontology. *Mailing Add:* Box 380 Anacortes WA 98221-0380

POLLACK, LOUIS, b New York, NY, Nov 4, 20; m 45; c 3. COMMUNICATION ENGINEERING. *Educ:* City Col New York, BEE, 53. *Prof Exp:* Engr, ITT Fed Labs, 43-47, sr proj engr, 47-54, dept head commun, 54-55, exec engr, 55-60, assoc lab dir, 60-62, lab dir, 62-66, dir transmission systs oper, 66-67; mgr, Radio Frequency Transmission Lab, Commun Satellite Corp Labs, 67-73, dir, Technol Div, 73-74, asst tech dir, 74-78, exec dir, 78-80, vpres, COMSAT World Systs Div, 80-84; OWNER, LOUIS POLLACK ASSOCS, 84- *Concurrent Pos:* Nat Acad Sci deleg, XIIIth gen assembly, Int Union Radio Sci, 60; mem US deleg, Int Radio Consult Comt study group IV, Int Telecommun Union, 62, mem study group I & II. *Mem:* Fel Inst Elec & Electronics Engrs; assoc fel Am Inst Aeronaut & Astronaut; sr mem Nat Soc Prof Engrs; Sigma Xi. *Res:* Communications equipment design; microwave; solid state devices; spacecraft structures and power; transmission system engineering; communications satellite systems. *Mailing Add:* 15321 Delphinium Ln Rockville MD 20853-1725

POLLACK, LOUIS RUBIN, b Chicago, Ill, Nov 3, 19; m 42. RADIOCHEMISTRY, POLYMER STABILITY. *Educ:* Univ Calif, Los Angeles, AB, 40. *Prof Exp:* Chemist, Mare Island Naval Shipyard, US Dept Navy, Vallejo, Calif, 41-49, supvry chemist, 49-61, chief chemist, 61-72; DIR, CHEM CONSULTS, 72-; CONSULT NUCLEAR ENGR, 76- *Honors & Awards:* Petersen Award, Am Chem Soc, 84. *Mem:* AAAS; Am Chem Soc; Am Soc Metals; Fedn Am Scientists; fel Am Inst Chemists. *Res:* Nuclear reactor and analytical chemistry; radiochemistry; electrochemical power sources; materials compatability; marine corrosion; oxidative degradation of elastomens. *Mailing Add:* 401 Santa Clara Ave No 223 Oakland CA 94610

POLLACK, MAXWELL AARON, b New York, NY, June 11, 15; m 39; c 1. ORGANIC CHEMISTRY. *Educ:* City Col New York, BS, 34; Northwestern Univ, MS, 35, PhD(chem), 37. *Prof Exp:* Asst chem, Northwestern Univ, 34-37; res chemist & supvr org res, Pittsburgh Plate Glass Co, 37-40; asst prof chem & res chemist, Univ Tex, 40-42; chem dir, E F Drew & Co, Inc, NJ, 42-47; chem consult, 47-64; pres & chmn bd, Van Dyk Res Corp, 64-83; RETIRED. *Concurrent Pos:* Pres, Parco Chems, Inc & Garden State Chem Co, 60-64. *Mem:* AAAS; Am Chem Soc; Geront Soc Am; NY Acad Sci; Am Aging Asn. *Res:* Plastics and plasticizers; fats and oils; synthetic organic chemicals; cancer metabolism; B vitamins; allylic rearrangements; gerontology; photo-imaging processes; nutrition and aging. *Mailing Add:* 121 Glenbrook Rd Morris Plains NJ 07950

POLLACK, RALPH MARTIN, b Boston, Mass, May 25, 43; m 64; c 2. BIO-ORGANIC CHEMISTRY, ENZYMOLOGY. *Educ:* Brown Univ, ScBChem, 65; Univ Calif, Berkeley, PhD(chem), 68. *Prof Exp:* NIH fel chem, Northwestern Univ, 68-70; from asst prof to assoc prof, 70-81, PROF CHEM, UNIV MD, BALTIMORE COUNTY, 81-; SR STAFF SCIENTIST, CTR ADVAN RES BIOTECHNOLOGY, 87- *Concurrent Pos:* Vis prof, Univ Montpellier, France, 78-79. *Mem:* AAAS; Am Chem Soc; Sigma Xi. *Res:* Mechanisms of organic reactions; mechanisms of enzyme reactions; organic models for enzymes. *Mailing Add:* Dept of Chem Univ of Md Baltimore County Baltimore MD 21228-5329

POLLACK, RICHARD, b Brooklyn, NY, Jan 25, 35; m 57; c 2. MATHEMATICS. *Educ:* Brooklyn Col, BA, 56; NY Univ, PhD(math), 62. *Prof Exp:* From asst to assoc prof, 62-82, PROF MATH, NY UNIV, 82- *Mem:* Am Math Soc; Math Asn Am. *Res:* Combinatorics and finite maths; geometry; discrete and computational geometry. *Mailing Add:* Dept Math NY Univ New York NY 10012

POLLACK, ROBERT ELLIOT, b Brooklyn, NY, Sept 2, 40; m 61; c 1. CELL BIOLOGY, ONCOLOGY. *Educ:* Columbia Univ, BA, 61; Brandeis Univ, PhD(biol), 66. *Prof Exp:* Asst prof path, Med Ctr, NY Univ, 68-70; Nat Cancer Inst spec fel, Weizmann Inst Sci, 70-71; sr staff scientist, Cold Spring Harbor Lab Quant Biol, 71-75; from assoc prof to prof microbiol, Sch Basic Health Sci, State Univ NY Stony Brook, 75-78; dean, Columbia Col, 82-89, PROF BIOL SCI, COLUMBIA UNIV, 78- *Concurrent Pos:* Nat Cancer Inst fel med ctr, NY Univ, 66-68, Nat Cancer Inst & Am Cancer Soc res grants, 68-75; NSF res grant, 75-78; NIH res grants, 78-; adj prof molecular pharmacol, Albert Einstein Col Med, 78- *Honors & Awards:* McGregory Lectr, Colgate Univ, 78; Alexander Hamilton Medal, 90. *Mem:* AAAS; Am Soc Microbiol; NY Acad Sci; Am Soc Cell Biol. *Res:* Separate growth controls of normal cells; linkages among them that give rise to a healthy tissue; perturbations in such growth controls give rise to cancer; muscle-proteins in non-muscle cells; nude mice; detection in vitro of pre-malignant phenotypes. *Mailing Add:* 813 Fairchild Hall Columbia Univ New York NY 10027

POLLACK, ROBERT LEON, b Philadelphia, Pa, Apr 29, 26; m 52; c 2. BIOCHEMISTRY. *Educ:* Philadelphia Col Pharm, BS, 48 & 49, MS, 50; Univ Tenn, PhD(biochem), 54. *Prof Exp:* Instr chem, Philadelphia Col Pharm, 48-49, asst bact, 49-50; instr bact, Univ Tenn, 52-54, res assoc, Cancer Res Lab, 54; res scientist, Eastern Utilization Res Lab, USDA, 54-62; from assoc prof to prof biochem, Sch Dent, Temple Univ, 61-86, chmn dept, 62-86, dir, Nutrit Health Ctr, 75-86, prof biochem, Sch Med, 86-88, EMER PROF BIOCHEM, SCH MED, TEMPLE UNIV, 88- *Concurrent Pos:* Instr eve col, Drexel Inst, 57-62. *Mem:* AAAS; Am Inst Chem; Int Asn Dent Res; Am Chem Soc. *Res:* Human nutrition. *Mailing Add:* 8442 Chippewa Rd Philadelphia PA 19128-1206

POLLACK, SIDNEY SOLOMON, b New York, NY, Sept 24, 29; wid; c 3. CRYSTALLOGRAPHY. *Educ:* Mich State Univ, BS, 51, MS, 53; Univ Wis, PhD(soil sci), 56. *Prof Exp:* Fel x-ray diffraction, Mellon Inst, 56-76; MEM STAFF, DEPT ENERGY, 76- *Mem:* Am Chem Soc; Mineral Soc Am; Am Crystallog Asn. *Res:* 520 Zeolites, reactivity of pyrites; catalysts; coal; low crystallinity carbons; small-angle x-ray scattering and crystallography. *Mailing Add:* Pittsburgh Energy Technol Ctr Dept Energy Box 10940 Pittsburgh PA 15236

POLLACK, SOLOMON R, b Philadelphia, Pa, May 7, 34; m 55; c 3. BIOENGINEERING, MATERIALS SCIENCE. *Educ:* Univ Pa, AB, 55, MS, 57, PhD(physics), 61. *Prof Exp:* Physicist, Univac Div, Sperry Rand Corp, 60-64; from asst prof to assoc prof metall, 64-77, PROF METALL, UNIV PA, 77-, PROF BIOENG, 77- *Concurrent Pos:* Consult, Univac Div, Sperry Rand Corp, 64-; vis scientist prog physics, Am Inst Physics, 64-; vpres, Cara Corp, 69-72, pres, 72-87, chmn, 87-; chair bioeng, Univ Pa, 77-81, assoc dean eng, 81-86. *Honors & Awards:* Lindback Award, Soc Biomat; Galvani Medal. *Mem:* Orthop Res Soc; Soc Biomat; Bioelectric Repair & Growth Soc; AMA; Inst Elec & Electronic Engrs Eng Med & Biol Sci. *Res:* Electrical effects in bone; biomaterials. *Mailing Add:* Dept Bioeng Univ Pa Philadelphia PA 19104

POLLACK, SYLVIA BYRNE, b Ithaca, NY, Oct 18, 40; c 2. IMMUNOLOGY, LYMPHOPOIESIS. *Educ:* Syracuse Univ, BA, 62; Univ Pa, PhD(develop biol), 67. *Prof Exp:* Instr anat, Woman's Med Col Pa, 67-68; res assoc, 68-73, res asst prof microbiol, 73-77, res assoc prof microbiol & immunol, 77-81, res assoc prof biol structure, 81-85, RES PROF BIOL STRUCTURE, UNIV WASH, 85- *Concurrent Pos:* Asst mem, Fred Hutchinson Cancer Res Ctr, 75-79, assoc mem, 79-81. *Mem:* Am Asn Immunologists; Soc Develop Biol; Reticuloendothelial Soc; Am Asn Cancer Res. *Res:* Analysis of production and activity of natural killer cells; long term bone marrow cultures; role of natural killer cells in pregnancy. *Mailing Add:* Biol Structure SM-20 Univ Washington Seattle WA 98195

POLLACK, WILLIAM, b London, Eng, Feb 26, 26; m 54; c 2. IMMUNOLOGY, IMMUNOCHEMISTRY. *Educ:* Univ London, BSc, 48, MSc, 50; Rutgers Univ, PhD(zool), 64. *Prof Exp:* Sci officer hemat & serol, St Georges Hosp Med Sch, London, Eng, 48-64; tech head clin lab, Royal Columbian Hosp, Vancouver, BC, 64-66; res fel immunol & immunochem, Ortho Res Found, 63-81, sr scientist, 66-69, vpres & dir res diag, 69-75, vpres & dir res diag, Ortho Res Inst Med Sci, 75-81; VPRES & DIR RES & DEVELOP, PURDUE FREDERICK CO, 81- *Concurrent Pos:* Assoc clin prof path, Col Physicians & Surgeons, Columbia Univ, 68; adj assoc prof immunol, Col Med & Dent NJ, Rutgers Med Sch, 74. *Honors & Awards:* Karl Landsteiner Award, 69. *Mem:* AAAS: fel NY Acad Sci; Inst Soc Blood Transfusion; Int Soc Hemat. *Res:* Physical chemistry of immune reactions; immunochemistry in diagnostic research; prevention of immune induction as applied to Rhesus disease; immunology of cancer and tissue transplantation. *Mailing Add:* 2660 Acuna Ct Carlsbad CA 92008-0530

POLLAK, BARTH, b Chicago, Ill, Aug 14, 28; m 54; c 2. QUADRATIC FORMS. *Educ:* Ill Inst Technol, BS, 50, MS, 51; Princeton Univ, PhD(math), 57. *Prof Exp:* Instr math, Ill Inst Technol, 56-58; asst prof, Syracuse Univ, 58-63; assoc prof, 63-67, PROF MATH, UNIV NOTRE DAME, 67- *Concurrent Pos:* Tech staff mem, Inst Defense Anal, 60-62. *Mem:* Am Math Soc; Math Asn Am. *Res:* Algebra and the theory of numbers. *Mailing Add:* Dept of Math Univ of Notre Dame Notre Dame IN 46556

POLLAK, EDWARD, b New York, NY, June 29, 32. MATHEMATICAL STATISTICS. *Educ:* Cornell Univ, BS, 54; NC State Col, MS, 56; Columbia Univ, PhD(math statist), 64. *Prof Exp:* Instr math, Lehigh Univ, 63-64; from asst prof to assoc prof, 64-72, PROF STATIST & GENETICS, STATIST

LAB, IOWA STATE UNIV, 72- *Mem:* AAAS; fel Am Statist Asn; Inst Math Statist; Biomet Soc; Am Inst Biol Sci; Genetics Soc of Am; Am Soc of Naturalists. *Res:* Population genetics; theory of branching processes. *Mailing Add:* Rm 111C Snedecor Hall Iowa State Univ Ames IA 50011

POLLAK, EMIL JOHN, b Portland, Maine, June 17, 47; m 72. ANIMAL BREEDING. *Educ:* Cornell Univ, BS, 69; Iowa State Univ, MS, 74, PhD(animal breeding), 75. *Prof Exp:* Res asst animal breeding, Iowa State Univ, 69-75; ASST PROF ANIMAL SCI, UNIV CALIF, DAVIS, 75- *Mem:* Am Dairy Sci Asn; Am Soc Animal Sci. *Res:* Genetic improvement in beef cattle; estimation of genetic parameters and the application of selection techniques to the improvement in efficiency and production of beef cattle in the United States. *Mailing Add:* Cornell Univ Main Campus 149 Morrison Hall Animal Sci Ithaca NY 14853

POLLAK, FRED HUGO, b Vienna, Austria, May 3, 35; US citizen; m 60; c 1. SOLID STATE PHYSICS. *Educ:* Franklin & Marshall Col, BS, 57; Univ Chicago, MS, 59, PhD(physics), 65. *Prof Exp:* From instr to assoc prof physics, Brown Univ, 64-72; prof physics, Belfer Grad Sch Sci, Yeshiva Univ, 72-80; Broeklundian prof physics, 83-88, mem fac dept physics, Brooklyn Col, NY, 80-, DISTINGUISHED PROF PHYSICS, 88- *Concurrent Pos:* Vis scientist & group proj leader, Francis Bitter Nat Magnet Lab, Mass Inst Technol, 69-72; assoc dir & head, Mat Group, NY State Ctr Advan Technol Telecommun, Polytech Univ, Brooklyn, NY, 86-; dir, Semiconductor Inst Brooklyn Col, City Univ New York, 88- *Mem:* Am Phys Soc; Am Asn Physics Teachers; Soc Photo-Optical Instrumentation Engrs; Electrochem Soc. *Res:* Optical and electronic properties of solids; modulation spectroscopy. *Mailing Add:* Dept Physics Brooklyn Col Bedford Ave & Ave H New York NY 11210

POLLAK, HENRY OTTO, b Vienna, Austria, Dec 13, 27; nat US; m 49; c 2. MATHEMATICS, EDUCATION. *Educ:* Yale Univ, BA, 47; Harvard Univ, MA, 48, PhD(math), 51. *Hon Degrees:* DSc, Rose-Hulman Inst Technol, 64, Monmouth Col, NJ, 75, Bowdoin Col, 77, Tech Univ Eindhoven, 81 & Montclair State Col, 84. *Prof Exp:* Res mathematician, Bell Tel Labs, Inc, 51-83; asst vpres, Bell Commun Res, Inc, 83-86; RETIRED. *Concurrent Pos:* Vis prof, Teachers Col, Columbia Univ, 87- *Mem:* Am Math Soc; Math Asn Am (pres, 75-76). *Res:* Function theory; discrete mathematics; mathematics education. *Mailing Add:* 40 Edgewood Rd Summit NJ 07901

POLLAK, KURT, b Vienna, Austria, Dec 27, 33; US citizen; m 61; c 2. ORGANIC CHEMISTRY. *Educ:* Harvard Univ, AB, 54; Mass Inst Technol, PhD(org chem), 60. *Prof Exp:* Chemist, Rayonier, Inc, 60-62; chemist, Collab Res Inc, 62-65; chemist, Esso Res & Eng Co, 65-73, chemist, Exxon Res & Eng Co, Linden, 73-80; CONSULT & EXEC RECRUITER, CHEM & PETROCHEM INDUST, 80- *Mem:* Soc Automotive Eng; Am Chem Soc. *Res:* Amino acid and peptide synthesis; cellulose modifications; cancer chemotherapy; heterocyclics; lubricant additives. *Mailing Add:* 622 Maye St Westfield NJ 07090-2423

POLLAK, MICHAEL, b Ostrava, Czech, Sept 1, 26; m 64; c 2. SEMICONDUCTORS. *Educ:* Israel Inst Technol, BSc, 53, DiplIng, 54; Univ Pittsburgh, MS, 57, PhD(physics), 58. *Prof Exp:* Res engr electronics, Israeli Inst Defense, 53-54; asst physics & electronics, Israel Inst Technol, 54-55; physicist, Westinghouse Res Labs, Pa, 58-59; mem tech staff, Bell Tel Labs, 59-60; sr physicist, Westinghouse Res Labs, 60-66; assoc prof biophys, 66-69, PROF PHYSICS, UNIV CALIF, RIVERSIDE, 69- *Concurrent Pos:* Vis scientist quantum chem group, Univ Uppsala, 64-65; vis lectr physics, Okla State Univ, 66; vis prof, Israel Inst Technol, 70 & 77, Univ Wales, 72, Univ Calif, Los Angeles, 75 & Univ Cambridge, 77. *Mem:* AAAS. *Res:* Transport properties in disordered systems. *Mailing Add:* Dept of Physics Univ of Calif Riverside CA 92501

POLLAK, VICTOR EUGENE, b Johannesburg, SAfrica, Sept 7, 26; m 56; c 4. INTERNAL MEDICINE, NEPHROLOGY. *Educ:* Univ Witwatersrand, BA, 44, MB, BCh, 50; FRCP(E), 65. *Prof Exp:* Asst med, Univ Ill Col Med, 54-56, instr, 56-57, res assoc, 57-59, from res asst prof to res assoc prof, 59-65, from clin assoc prof to clin prof, 65-70; prof med, Pritzker Sch Med, Univ Chicago, 70-73; PROF MED, COL MED, UNIV CINCINNATI, 73- *Concurrent Pos:* Attend physician, Res & Educ Hosp, Chicago, 56-73; res assoc, Hektoen Inst, Cook County Hosp, 59-65; asst attend physician, Presby-St Luke's Hosp, 59-65; estab investr, Am Heart Asn, 59-64; NIH grants, 60-85; USPHS res career develop award, 64-65; dir renal div, Michael Reese Hosp & Med Ctr, 65-72; dir div nephrol, Univ Cincinnati, 73-88, attend physician, Univ Cincinnati & Vet Admin Hosps, 73- *Mem:* Am Soc Nephrology; Am Soc Clin Invest; Cent Soc Clin Res; Am Heart Asn; fel Am Col Physicians; Sigma Xi. *Res:* Renal disease-immunopathogenesis and role of coagulation; informatics and computerization of medicine. *Mailing Add:* 400 Rawson Woods Lane Cincinnati OH 45220

POLLAK, VICTOR LOUIS, b Vienna, Austria, Mar 25, 30; US citizen; m 57; c 1. PHYSICS, SCIENCE EDUCATION. *Educ:* Case Western Reserve Univ, BSc, 52; Wash Univ, St Louis, PhD(physics), 60. *Prof Exp:* Sr physicist, Schlumberger Corp, 60-62; asst prof physics, Okla State Univ, 62-68; assoc prof, 68-73, chmn dept, 68-76, PROF PHYSICS, UNIV NC, CHARLOTTE, 73- *Concurrent Pos:* Lectr, Univ Houston, 61-62; vis asst prof, Case Western Reserve Univ, 67-68. *Mem:* Am Phys Soc; Am Asn Physics Teachers. *Res:* Magnetic resonance; liquid state; chemical physics. *Mailing Add:* Dept Physics Univ NC Charlotte NC 28223

POLLAK, VIKTOR A, b Vienna, Austria, Mar 17, 17; m 66. BIOMEDICAL ENGINEERING, INFORMATION SCIENCE. *Educ:* Prague Tech Univ, Engr, 46, Dr Tech Sci(info theory), 53, CSc, 66; Univ Sask, PhD, 72, DSc, 82. *Prof Exp:* Designer, Tesla Tel Works, Prague, 47-48; head dept indust commun, Res Inst Telecommun, Prague, 48-56; med electronics, Res Inst Med Eng, Prague, 56-58; res scientist, Inst Fuel Res, Prague, 58-66; vis assoc prof electronics, Tech Univ Denmark, 66-68; assoc prof elec eng & secy div biomed eng, 68-72, prof elec eng, 70-84, chmn biomed eng div, 72-83, EMER PROF, UNIV SASK, 83- *Concurrent Pos:* Consult, var insts, Govt of China, 54-55; external lectr info theory, Prague Tech Univ, 55-65; independent res scientist, Czech Acad Sci, 56; vis prof biomed eng, Fed Univ Rio de Janeiro, 74-75, exchange prof, 75-76; exchange vis, Hokkaido Univ, 79, Tokyo Univ, 81 & Aachen, Germany, 86; vis prof, Univ Paris, 83, Tang-Min Med Col, Taiwan, 84 & Tokyo Denki Univ, 90. *Mem:* Sr mem Inst Elec & Electronics Engrs; Can Med & Biol Eng Soc. *Res:* Pattern recognition in medicine; electronic aids for the handicapped; general systems theory applied to biological, social and economic problems; applied biophysics; optical quantitative assessment of thin media chromatograms and electropherograms; theoretical neurophysiology. *Mailing Add:* Div Biomed Eng Univ Sask Saskatoon SK S7N 0W0 Can

POLLARA, BERNARD, b Chicago, Ill, July 28, 27; m 58; c 5. PEDIATRICS, IMMUNOLOGY. *Educ:* Northwestern Univ, PhB, 51, MS, 54; Univ Minn, MD, 60, PhD, 63. *Prof Exp:* Clin chemist, St Luke's Hosp, Chicago, 48; res technician, Argonne Nat Lab, 51-52; asst biochem, Northwestern Univ, 52-53; asst physiol chem, Univ Minn, 54-56, asst scientist pediat, 57-60; intern, USPHS, Seattle, Wash, 60-61; from asst prof to assoc prof pediat & biochem, Med Sch, Univ Minn, Minneapolis, 64-69; PROF PEDIAT, ALBANY MED COL, 69-, CHMN DEPT, 79- *Concurrent Pos:* Consult, St Mary's Hosp, 56-59; Arthritis & Rheumatism Found res fel, 61-65; dir, Kidney Dis Inst, NY State Dept Health, 69-79; vpres, res affairs, Albany Med Ctr, 86-89. *Mem:* AAAS; Am Pediat Soc; Am Asn Immunol; Am Soc Cell Biol; Soc Exp Biol & Med. *Res:* Immunoglobulin structure; clinical investigation of immunodeficiency disease; transplantation antigens; phylogenetic studies of immunity; pediatric AIDS. *Mailing Add:* Dept Pediat Albany Med Col Albany NY 12208

POLLARD, ARTHUR JOSEPH, b San Francisco, Calif, Aug 14, 56; m 81. POPULATION BIOLOGY, ECOLOGICAL GENETICS. *Educ:* Duke Univ, BS, 77; Cambridge Univ, PhD(bot), 81. *Prof Exp:* Asst prof bot,Okla State, 81-86, assoc prof biol, 86-88; ASSOC PROF BIOL, FURMAN UNIV, 88- *Mem:* Ecol Soc Am; Brit Ecol Soc; Am Soc Plant Taxonomists; Bot Soc Am; Int Soc Plant Pop Biol. *Res:* Study of variation within plant species, and its ecological and evolutionary implications; variation in plant defense mechanisms, and the consequences for plant-herbivore interactions; systematics and herbarium curation. *Mailing Add:* Dept Biol Furman Univ Greenville SC 29613

POLLARD, CHARLES OSCAR, JR, b Henderson, Tex, Sept 6, 37; m 60; c 1. GEOCHEMISTRY, MINERALOGY. *Educ:* Rice Univ, BA, 60; Fla State Univ, PhD(geol), 67. *Prof Exp:* Asst prof phys geol, mineral & geochem, 67-72, ASSOC PROF EARTH & ATMOSPHERIC SCI, GA INST TECHNOL, 72- *Concurrent Pos:* Grants, Am Chem Soc-Petrol Res Fund, 68-72; grant, US Army Res Off-Durham, 69-72; grants, Off Water Resources & Technol, 74-76, US Dept Interior, Bur Mines, 83, Ga Dept Transp, 89-91. *Mem:* Mineral Soc Am; Clay Minerals Soc; Am Geophys Union; Sigma Xi. *Res:* Correlation of physical properties with mineral structure and composition; growth mechanisms of natural crystals; geochemical classification of groundwaters; x-ray diffraction; scanning electron microscopy; petrology; petrography. *Mailing Add:* Sch Earth & Atmospheric Sci Ga Inst of Technol Atlanta GA 30332-0340

POLLARD, DOUGLAS FREDERICK WILLIAM, b Morden, Eng, Feb 3, 40; m 78; c 3. PLANT PHYSIOLOGY. *Educ:* Univ Wales, BSc, 62, PhD(bot), 66. *Prof Exp:* Sci officer, Wildfowl Trust, Slimbridge, Eng, 65-67; res scientist tree physiol, Dept Environ, 67-81, sr policy analyst, Dept Agr, 82-88, RES SCIENTIST TREE PHYSIOL, FORESTRY CAN, 88- *Concurrent Pos:* Vis scientist, Can Sci Deleg, People's Repub China, 75. *Honors & Awards:* E B Eddy Distinguished Lectr, 87. *Mem:* Can Inst Forestry. *Res:* Forest productivity; growth physiology of genetic variation of conifers; physiology of flowering; seed certification; ecological interpretation; global carbon dioxide issue. *Mailing Add:* Pac Forest Ctr Victoria BC V8Z 1M5 Can

POLLARD, HARVEY BRUCE, b San Antonio, Tex, May 26, 43; m 66; c 2. BIOCHEMISTRY, NEUROBIOLOGY. *Educ:* Rice Univ, BA, 64; Univ Chicago, MS & MD, 69, PhD(biochem), 73. *Prof Exp:* Res assoc, Lab Chem Biol, NIH, 69-71; med officer, Lab Molecular Biophys, Oxford Univ, 71-72; SR INVESTR NEUROBIOL, REPROD RES BR, NIH, 72-, SR INVESTR, NEUROENDOCRINOL-CLIN HEMAT BR, NIADDK, 76-, CHIEF, SECT CELL BIOL & BIOCHEM, 80-, CHIEF, LAB CELL BIOL & GENETICS, 81- *Concurrent Pos:* Mem, Corp Marine Biol Lab, Woods Hole, Mass, 75- *Mem:* Biophys Soc; Soc Neurosci; Am Soc Pharmacol & Exp Therapeut; Soc Develop Biol; Am Soc Cell Biol. *Res:* Regulation of neurotransmitter and hormone release from cells and secretory vesicles; communication between cells; membrane structure and function; endocrinology; cell biology. *Mailing Add:* Bldg 4 Rm 310 NIH Bethesda MD 20892

POLLARD, JAMES EDWARD, b Little Rock, Ark, Apr 13, 43; m 67; c 1. PHYSIOLOGY, HORTICULTURE. *Educ:* Duke Univ, AB, 65; Univ Fla, PhD(hort), 69. *Prof Exp:* Res assoc plant physiol, Univ Fla, 69-70; asst prof, 70-77, ASSOC PROF PLANT SCI, AGR EXP STA, UNIV NH, 77- *Mem:* Am Soc Hort Sci; NY Acad Sci; Am Soc Plant Physiol. *Res:* Tree physiology including production of tree fruit crops, with emphasis on stress physiology; post-harvest physiology of fruit crops. *Mailing Add:* PO Box 14192 Torrance CA 90503

POLLARD, JEFFREY WILLIAM, b Rochford, Eng, Jan 23, 50. MOLECULAR GENETICS, REPRODUCTIVE ENDOCRINOLOGY. *Educ:* Sheffield Univ, BS, 71; London Univ, PhD(biochem), 75. *Prof Exp:* Nat Cancer Inst Can fel, Ont Cancer Inst & Dept Med Biophys, Univ Toronto, 71-79; lectr, Queen Elizabeth Col, Univ London, Eng, 79-85, lectr, dept biochem, Kings Col, 85-88; ASSOC PROF, ALBERT EINSTEIN COL MED, 88- *Concurrent Pos:* Prin investr, MRC group human genetic dis, Kings Col, Univ London, 83-88; NATO travel fel, dept microbiol & immunol,

Albert Einstein Col Med, 84-88. *Honors & Awards:* Darwin Lectr, Brit Asn Advan Sci, 89. *Mem:* Brit Biochem Soc; Brit Soc Study Fertil; Am Soc Microbiol. *Res:* Utilization of molecular genetic approaches to examine the control of protein synthesis and cell proliferationin eukaryotic cells; regulation of protein synthesis by female sex steroids; evolutionary biology; growth factors and steroid hormones. *Mailing Add:* Dept Develop Biol & Cancer Albert Einstein Col Med 1300 Morris Park Ave Bronx NY 10461

POLLARD, JOHN HENRY, b Bristol, Eng, Nov 17, 33; m 66; c 4. ELECTRONICS. *Educ:* Bristol Univ, Eng, BSc Hons, 57; Aberdeen Univ, Scotland, PhD(natural philos), 61. *Prof Exp:* Asst lectr natural philos, Aberdeen Univ, Scotland, 60-62; res fel physics, Bartol Res Found, Pa, 62-68; SUPVRY PHYSICIST, CTR FOR NIGHT VISION & ELECTRO-OPTICS, US ARMY, VA, 68- *Concurrent Pos:* Assoc mem, Adv Group Electron Devices, 79-; chmn, IRIS Mat Specialty Group, 88- *Mem:* Sigma Xi. *Res:* Photoconductive and photovoltaic detectors; charge coupled devices; heterojuction devices for infrared sensors; photoemission from negative affinity surfaces; auger and angular photoemission spectroscopy. *Mailing Add:* 3610 Surrey Dr Alexandria VA 22309

POLLARD, MORRIS, b Hartford, Conn, May 24, 16; m 38; c 3. MEDICAL MICROBIOLOGY. *Educ:* Ohio State Univ, DVM, 38; Va Polytech Inst, MSc, 39; Univ Calif, Berkeley, PhD, 50. *Hon Degrees:* DSc, Miami Univ, Ohio, 81. *Prof Exp:* Asst prof prev med & dir virus lab, Med Br, Univ Tex, 46-61; prof biol, Univ Notre Dame, 61-66, chmn dept, 66-81, prof microbiol, 66-81, DIR LOBUND LAB, UNIV NOTRE DAME, 61-, EMER PROF MICROBIOL, 81-, COLEMAN PROF LIFE SCI, 85- *Concurrent Pos:* Nat Found Infantile Paralysis fel, 49; McLaughlin fac fel, Cambridge Univ, 56; consult, Brooke Army Med Ctr, 46-50, M D Anderson Hosp, 54-61, US Dept Health, Educ & Welfare, 65-70 & Inst Lab Animal Resources, Nat Acad Sci, 65-69; mem microbiol adv panel, Off Naval Res, 66-70, chmn, 68-70; mem colo-rectal cancer comt, Nat Cancer Inst, 73-79, chmn tumor immunol comt, 77-80; external examr, Med Res Coun, Australia, 75-; vis prof, Fed Univ Rio de Janeiro, Brazil, 76; corresp mem, Nat Acad Sci, Brazil; Raine Found Prof med microbiol, Univ Western Australia; consult, Chinese Acad Med Sci, 81-, hon prof, 82-; mem prog rev panel, Argonne Nat Lab, 79-85, chmn, 83-85; guest prof, Katholieke Univ, Leuven, Belgium, 81; ed, Perspectives in Virol, 56- *Mem:* AAAS; Soc Biol & Med; Asn Pathologists; fel Am Acad Microbiol; Asn Gnotobiotics (pres, 77); Am Soc Microbiol; Am Soc Virol; Am Asn Lab Animal Sci; Int Asn Gnatobiol (pres, 87-88). *Res:* Virology; comparative pathology. *Mailing Add:* Lobund Lab Univ Notre Dame Notre Dame IN 46556

POLLARD, RICHARD BYRD, INFECTIOUS DISEASES, INTERNAL MEDICINE. *Educ:* Univ Md, MD, 70. *Prof Exp:* ASSOC PROF INTERNAL MED, UNIV TEX, GALVESTON, 78- *Res:* Virology. *Mailing Add:* Dept Internal Med Univ Tex Med Br Galveston TX 77550

POLLARD, ROBERT EUGENE, b Taunton, Mass, Nov 13, 24; m 47; c 4. POLYMER CHEMISTRY. *Educ:* Brown Univ, BS, 47. *Prof Exp:* Res chemist, vinyl chloride polymerization, 47-55, res group leader vinyl applns, 55-61, res specialist polyethylene applns, 62-67, res specialist vinyl applns, 67-70, res specialist reinforced thermoplastics, 70-77, SR RES SPECIALIST ENG THERMOPLASTICS, MONSANTO CO, 77- *Mem:* Soc Plastics Engrs; Am Chem Soc. *Res:* Polymer structure and properties; injection molding and extrusion; rheology of polymers; compounding, reinforcement of thermoplastics; testing of plastics. *Mailing Add:* 8631 Scenic Hills Dr Pensacola FL 32514

POLLARD, THOMAS DEAN, b Pasadena, Calif, July 7, 42; m 64; c 2. CELL BIOLOGY, MEDICINE. *Educ:* Pomona Col, BA, 64; Harvard Med Sch, MD, 68. *Prof Exp:* Intern med, Mass Gen Hosp, Boston, 68-69; staff assoc biochem, Nat Heart & Lung Inst, 69-72; from asst prof to assoc prof anat, Harvard Med Sch, 72-77; BAYARD HALSTED PROF ANAT & DIR DEPT CELL BIOL & ANAT, MED SCH, JOHNS HOPKINS UNIV, 77- *Concurrent Pos:* Assoc ed, Ann Rev Biophys, 82-, J Cell Biol, 84-91; trustee, Marine Biol Lab, Woods Hole, Mass, 90-, Comn on Life Sci of the Nat Res Coun, 90- *Mem:* Soc Gen Physiol; Am Soc Cell Biol (pres, 88); Biophys Soc; Am Soc Biochem & Molecular Biol; Am Acad Arts & Sci. *Res:* Biochemical and ultrastructural investigation of the molecular mechanism of cell motility. *Mailing Add:* Dept Cell Biol & Anat Med Sch Johns Hopkins Univ 725 N Wolfe St Baltimore MD 21205

POLLARD, WILLIAM BLAKE, b Tuskegee, Ala, Dec 4, 50. SEMICONDUCTOR PHYSICS. *Educ:* Fisk Univ, BA, 73; Mass Inst Technol, PhD(physics), 79. *Prof Exp:* Res assoc fel, Mass Inst Technol, 78-79; asst prof res & teaching, Atlanta Univ, 79-80; asst prof, NC State Univ, 80-81; dep dir admin, Inst Res, Develop & Eng, AMAF Ind, Inc, 81-82; asst prof, NC State Univ, 81-; Physics Dept, Morehouse Col, NC State Univ, 88; MEM FAC, PHYSICS DEPT, VALDOSTA STATE COL, 88- *Concurrent Pos:* Adj asst prof, Atlanta Univ, 81- *Mem:* Am Phys Soc; Nat Soc Black Physicists. *Res:* Electronic and optical properties of bulk amorphous semiconductors and alloys; bonding structure and electronic properties of semiconductor-insulator interfaces. *Mailing Add:* Dept Physics Valdosta State Col Valdosta GA 31698-0001

POLLART, DALE FLAVIAN, b Holly, Colo, Jan 7, 32; m 54; c 5. PETROLEUM CONVERSION, PETROLEUM PRODUCTS. *Educ:* Regis Col, Colo, BS, 52; Northwestern Univ, PhD(chem), 56. *Prof Exp:* Res chemist, Plastics Div, Union Carbide Corp, NJ, 56-60, proj scientist, Occidental Chem, 60-63, group leader, 63-67, group leader res & develop, Tech Ctr, WVa, 67-69, asst dir res & develop, 69-74, prod mgr urethane intermediates, 74-76, dir res & develop, NY, 76-79, bus mgr, 79-81; vpres technol, Occidental Chem, 81-82; dir, Res & Environ Affairs, 82-88, GEN MGR, RES & DEVELOP DEPT, TEXACO INC, 88- *Mem:* Am Chem Soc; Sigma Xi; AAAS; Chemists Club NY. *Res:* Mechanism of organic reactions; organometallic catalysts; high temperature reactions; polymer synthesis; polyurethanes; petroleum products. *Mailing Add:* Texaco Res Ctr PO Box 509 Beacon NY 12508

POLLATSEK, HARRIET SUZANNE, b Detroit, Mich, May 2, 42; m 64; c 2. MATHEMATICS, GROUP THEORY. *Educ:* Univ Mich, Ann Arbor, BA, 63, MA, 64, PhD(math), 67. *Prof Exp:* Instr math, Western Mich Univ, 67; asst prof, Univ Toledo, 68-69 & Univ Mass, 69-70; from asst prof to assoc prof math, 70-80, dean studies, 77-80, PROF MATH, MT HOLYOKE COL, 80- *Concurrent Pos:* NSF grantee, 70-71 & 87-89. *Mem:* Am Math Soc; Math Asn Am; Asn Women Math; Fedn Am Scientists; Sigma Xi. *Res:* Finite groups, especially the classical linear groups. *Mailing Add:* Two Campbell Ct Amherst MA 01002

POLLAY, MICHAEL, b Boston, Mass, Mar 25, 31; m 57; c 3. NEUROSURGERY, NEUROPHYSIOLOGY. *Educ:* Univ Wis, BS, 52, MD, 55; Univ Colo, MS, 61; Am Bd Neurol Surg, dipl, 65. *Prof Exp:* Clin asst surg, Univ Colo, 62-63; from instr to asst prof neurosurg & neurobiol sci, Sch Med, Univ NMex, 63-68, assoc prof, 68-71, prof neurosurg & physiol, 71-76; chief neurosurg sect, neurol B study sect, 74-78, PROF NEUROSURG, SCH MED, UNIV OKLA, 76- *Concurrent Pos:* NIH spec fel physiol, Univ Col, Univ London, 61-62, res grants, Univ NMex, 64-68; consult, Albuquerque Vet Admin Hosp, 63-, Bernalillo County Med Ctr & Soc Security Comn. *Mem:* AAAS; Am Physiol Soc; Am Asn Neurol Surg; NY Acad Sci; Soc Neurosci. *Res:* Blood-brain barrier; central nervous system transport; blood-cerebrospinal fluid barrier and acid base balance. *Mailing Add:* Dept Neurol Surg Univ Okla Sch Med Oklahoma City OK 73190

POLLEY, EDWARD HERMAN, b Chicago, Ill, Sept 20, 23; m 53; c 2. NEUROSCIENCES. *Educ:* DePauw Univ, AB, 47; Univ St Louis, MS, 49, PhD(anat), 51. *Prof Exp:* USPHS fel, Lab Neurophysiol, Wash Univ, 51-53; instr anat, Hahnemann Med Col, 53-59; chief neurol br, Exp Med Dept, Med Res Lab, Army Edgewood Arsenal, 59-70; PROF ANAT & RES PROF OPHTHAL, UNIV ILL COL MED, 70- *Concurrent Pos:* From asst prof to assoc prof, Sch Med, Univ Md, 64-70; mem comns vision & bioacoust, Nat Res Coun. *Mem:* AAAS; Soc Neurosci; Am Asn Anatomists; Asn Res Vision & Ophthal; Sigma Xi. *Res:* Anatomy of autonomic nervous system; anatomy and physiology of visual system; plasticity in nervous system; retinal anatomy and organization; physiology. *Mailing Add:* Dept Anat & Cell Biol Univ Ill Med Sch 808 S Wood St Chicago IL 60612

POLLEY, JOHN RICHARD, b Toronto, Ont, Dec 29, 17; m 45; c 2. BIOCHEMISTRY. *Educ:* Univ Western Ont, BA, 42, MA, 43; McGill Univ, PhD(chem), 46. *Prof Exp:* Fel, Life Ins Med Res Found, 47-49; biochemist, Lab Hyg, Can Dept Nat Health & Welfare, 49-74, dir, Bur Virol, Lab Ctr Dis Control, 74-80; REGISTR, CAN COL MICROBIOLOGISTS, 81- *Mem:* Am Chem Soc; Can Soc Microbiol; Can Physiol Soc; Am Nuclear Soc. *Res:* Virology; development of diagnostic antigens and vaccines. *Mailing Add:* 1502-2001 Carling Ave Ottawa ON K2A 3W5 Can

POLLEY, LOWELL DAVID, b Columbus, Ohio, Dec 31, 48; m 78; c 1. GENETICS, PLANT PHYSIOLOGY. *Educ:* Miami Univ, Ohio, BA, 70; Yale Univ, PhD(biol), 74. *Prof Exp:* Fel, Univ Utah, 75-76; res assoc, Wash Univ, 76-78; asst prof, 78-84, ASSOC PROF BIOL, WABASH COL, 84- *Concurrent Pos:* Res grant, NSF, 77-79, Res Corp, 80-82 & 83-85. *Mem:* AAAS; Am Soc Plant Physiologists; Genetics Soc Am. *Res:* Genetic control of processes essential to plants; regulation of ion transport; isolation of mutants defective in ion transport. *Mailing Add:* Dept Biol Wabash Col Crawfordsville IN 47933

POLLEY, MARGARET J, b Barry, Wales, Mar 18, 33. COMPLEMENT IN THROMBOSIS. *Educ:* Univ London, Eng, PhD(med), 64. *Prof Exp:* ASSOC PROF MED, MED COL, CORNELL UNIV, 72- *Mem:* Am Asn Immunologists; Am Soc Exp Path; NY Acad Sci; Harvey Soc. *Mailing Add:* Pharmingen 11555 Sorrento Valley Rd Suite E San Diego CA 92121

POLLIN, JACK MURPH, b Lawton, Okla, Apr 26, 22; m 44; c 2. MATHEMATICS, SYSTEMS ENGINEERING. *Educ:* US Mil Acad, BS, 44; Univ Pa, MS, 49; Rensselaer Polytech Inst, MS, 57; George Washington Univ, MA, 64; Univ Ariz, PhD(systs eng), 69. *Prof Exp:* Brigadier gen, US Army, 44-85; instr math, US Mil Acad, 49-52, from asst prof to prof, 57-85, dep dir commun & electronics, Artillery & Missile Ctr, 64-65, dep head dept, Acad, 68-74, head dept, 74-85; RETIRED. *Mem:* Soc Indust & Appl Math; Math Asn Am. *Res:* General systems. *Mailing Add:* 6840 N Nanini Dr Tucson AZ 85704

POLLIN, WILLIAM, b Philadelphia, Pa, May 13, 22; m 51; c 2. PSYCHIATRY, PSYCHOANALYSIS. *Educ:* Brooklyn Col, BA, 47; Columbia Univ, MD, 52. *Prof Exp:* Tutor & lectr biol, Brooklyn Col, 47-51; fel psychiat, Sch Med, Univ Pittsburgh, 53-56; res psychiatrist, NIMH, 56-68, chief, Sect Psychiat, Lab Clin Sci, 58-63, chief sect twin & sibling studies, Adult Psychiat Br, NIMH, 63-72, coordr res progs, Div Narcotic Addiction & Drug Abuse, 73-74, dir, Div Res, 74-79, DIR, NAT INST DRUG ABUSE, 79- *Concurrent Pos:* Resident & teaching fel, Western Psychiat Inst & Clin Med Ctr, 53-56; mem, Int Post-Doctoral Fel Comt, 62-66; chmn, Surgeon Gen Med Rev Bd, 69-72; temp adv, WHO Sci Group Biol Res Schizophrenia, 69; res consult, Thudicum Res Lab, Galesburg State Hosp, Ill State Dept Ment Health, 71-72; res coordr, US-USSR Joint Schizophrenia Res Proj, 72-73; chief clin res, NIMH Task Force, 72-74; res dir, White House Spec Action Off Drug Abuse Prevention, 74-75; mem adv comt res, Ill Dept Mental Health & Develop Disabilities, 73-79; Alcohol, Drug Abuse & Mental Health Admin rep & Nat Inst Drug Abuse mem to NIH Clin Trials Discussion Group, 75-76; mem res & eval subcomt, Cabinet Comt Drug Abuse Prevention, Treatment & Rehab, 76-80; chmn, Sect Drug Dependency & Alcoholism, World Psychiat Asn, 80- *Mem:* Soc Life Hist Res Psychpath; assoc mem Am Psychoanal Asn; Soc Neurosci; Am Col Neuropsychopharmacol; Behav Genetics Asn. *Res:* Schizophrenia; psychotherapy; application of psychiatric knowledge to international relationships; monozygotic twins discordant for schizophrenia; conflict resolution; personality development; drug abuse. *Mailing Add:* 7720 Sebago Rd Bethesda MD 20817

POLLITZER, ERNEST LEO, petroleum chemistry, surface chemistry, for more information see previous edition

POLLITZER, WILLIAM SPROTT, b Charleston, SC, May 6, 23; m 55; c 2. PHYSICAL ANTHROPOLOGY, GENETICS. *Educ:* Emory Univ, AB, 44, MA, 47; Columbia Univ, PhD(human variation), 57. *Prof Exp:* Instr biol, Armstrong Col, Ga, 47-49; instr anthrop, Hunter Col, 54 & 56; from instr to assoc prof, 57-73, PROF ANAT, UNIV NC, CHAPEL HILL, 73-, ADJ PROF ANTHROP, 76- *Concurrent Pos:* Ed, Am J Phys Anthrop, 70-77. *Mem:* Am Soc Human Genetics; Am Asn Phys Anthrop (vpres, 77-79, pres, 79-80); Human Biol Coun (pres, 85-87); Soc Study Social Biol; Am Asn Anatomists; Am Anthrop Asn. *Res:* Human evolution; populations; human genetics. *Mailing Add:* Dept Cell Biol & Anat Univ of NC Chapel Hill NC 27599-7090

POLLNOW, GILBERT FREDERICK, b Oshkosh, Wis, Jan 17, 25; c 2. PHYSICAL CHEMISTRY. *Educ:* Wis State Univ, Oshkosh, BS, 50; Univ Iowa, MS, 51, PhD(phys chem), 54. *Prof Exp:* Res chemist, Dow-Corning Corp, 54-58; res scientist, Allis-Chalmers Mfg Co, 58-61; from asst prof to assoc prof chem, 61-66, chmn dept, 65-68, PROF CHEM, UNIV WIS-OSHKOSH, 66- *Mem:* Am Chem Soc. *Res:* Macromolecular physical chemistry; computer automation of experiments and processes; application of energy conversion and cybernetic principles to social networks and deviant behavior. *Mailing Add:* Dept of Chem Univ of Wis Oshkosh WI 54901

POLLOCK, BRUCE MCFARLAND, b Rochester, NY, Mar 23, 26; m 47; c 2. PLANT PHYSIOLOGY. *Educ:* Cornell Univ, BS, 47; Univ Rochester, PhD(plant physiol), 50. *Prof Exp:* NIH res fel, Carlsberg Lab, Copenhagen, 50-51; res assoc, Conn Col, 51-53; from asst prof to assoc prof biol sci, Univ Del, 53-59; leader veg seed invests, Veg & Ornamentals Res Br, Crops Res Div, Agr Res Serv, USDA, Colo, 59-71; owner, Taylor Mt Enterprises, 71-76; attendant prof, 73-80, prof environ pop & organismic biol, 73-80, dir Health Sci Adv, Univ Colo, Boulder, 73-80; OWNER, SCI MEDIATION SERV, 80- *Concurrent Pos:* Res assoc, Mich State Univ-AEC Plant Res Lab, Mich State Univ, 69-70. *Honors & Awards:* Asgrow Award, Am Soc Hort Sci, 70. *Mem:* Am Soc Hort Sci; AAAS. *Res:* Physiology of the rest period in buds and seeds; seed and germination physiology; seed production; ethics in science and agricultural research. *Mailing Add:* 500 Butte St Suite 28 Boulder CO 80301-2236

POLLOCK, D(ANIEL) D(AVID), b Philadelphia, Pa, Mar 28, 18; m 42; c 1. SOLID STATE PHYSICS, PHYSICAL METALLURGY. *Educ:* Temple Univ, BS, 47; Lehigh Univ, MS, 48, PhD, 61. *Prof Exp:* Asst supt, Fogel Refrig Co, 36-39 & Metall Labs, Inc, 39-46; res metallurgist, Naval Air Mat Ctr, 46-47; res metallurgist, Leeds & Northrup Co, 48-62, chief alloy group res & develop dept, 57-62; chief metals-ceramics mat res, Douglas Aircraft Co, 62-66; prof, 66-85, EMER PROF METALL ENG, STATE UNIV NY BUFFALO, 85- *Mem:* AAAS; Am Soc Metals; Am Soc Testing & Mat; Am Phys Soc; Am Inst Mining, Metall & Petrol Engrs; Sigma Xi. *Res:* Aqueous and high temperature corrosion of nonferrous alloys; electrical properties of alloys; thermoelectricity; electrical resistivity; temperature coefficient of resistance; Hall coefficient; semiconductors; insulators. *Mailing Add:* 126 Robinhill Rd Williamsville NY 14221

POLLOCK, EDWARD G, b Glen Lyon, Pa, July 9, 31; m 57; c 3. DEVELOPMENTAL BIOLOGY, CELL BIOLOGY. *Educ:* Wilkes Col, AB, 53; Univ Va, MS, 57; Univ Calif, Berkeley, PhD(bot), 62. *Prof Exp:* Chmn dept biol, 64-65, PROF BIOL, CALIF STATE UNIV, NORTHRIDGE, 61-, DIR LAB ELECTRON MICROS, 71- *Concurrent Pos:* NSF instnl awards, 63-67, in serv sch-col sci teaching grant, 63-64, sci fac fel, 68-69; consult, Los Angeles Sch Dist, 63 & 64; Lalor Found res fel reprod physiol, 66; Sigma Xi res award, 66; Calif Biochem Res Found award, 67 & 69; researcher, Ger Marine Biol Labs, Helgoland, Ger, 68; Swiss Fed Inst Technol, 68-70; Europ Molecular Biol Orgn fel, 69-70; researcher, Hopkins Marine Lab, Stanford Univ. *Mem:* AAAS; Soc Develop Biol; Am Soc Cell Biol; NY Acad Sci; Int Phycol Soc; Sigma Xi. *Res:* Biochemical and ultrastructural study of cell differentiation, with emphasis on membrane characteristics, in Fucus distichus; membrane studies on sarcoma 180 cells via biochemistry and freeze-fracture. *Mailing Add:* Dept of Biol Calif State Univ Northridge CA 91330

POLLOCK, FRANKLIN, b New York, NY, Sept 1, 26; m 53; c 2. PHYSICS. *Educ:* Columbia Univ, AB, 47, AM, 49, PhD(physics), 52. *Prof Exp:* Res scientist, Hudson Labs, Columbia Univ, 51-54, Cyclotron Lab, 54-56; from asst prof to assoc prof, 56-67, PROF PHYSICS, STEVENS INST TECHNOL, 67- *Mem:* Am Phys Soc. *Res:* Low temperature physics. *Mailing Add:* Dept Physics Stevens Inst Technol Castle Point Hoboken NJ 07030

POLLOCK, G DONALD, b Brampton, Ont, July 7, 28; m 58; c 2. MEDICINE. *Educ:* Butler Univ, BSc, 72. *Prof Exp:* Asst pharmacologist, res labs, Bathesda Hosp, 64-66, Proctor & Gamble, 66-67; assoc pharmacologist, Eli Lilly Res Labs, 67-71, pharmacologist, 71-82; pharmacologist, 73-82, ASSOC SR PHARMACOLOGIST, 82- *Mem:* Am Heart Asn. *Res:* Co-author of 28 publications, technical on 11 publications; dobutamine iv inotrope, aprindine antiarrhythmic, butopamine iv inotrope; isomazole oral inotrope, indolidan 24 hour oral inotrope, pailine calorigenic; dobutamine shock. *Mailing Add:* Eli Lilly & Co Lilly Res Lab Lilly Corp Ctr Indianapolis IN 46285

POLLOCK, GEORGE HOWARD, b Chicago, Ill, June 19, 23; m 46; c 5. PSYCHIATRY. *Educ:* Univ Ill, BS, 44, MD, 45, MS, 48, PhD(physiol), 52. *Prof Exp:* Intern, Cook County Hosp, Chicago, 45-46; resident psychiat, Neuropsychiat Inst, Univ Ill, 48-50, from asst prof to assoc prof, Univ Ill Col Med, 52-62, clin prof, 62-72; PROF PSYCHIAT, NORTHWESTERN UNIV, CHICAGO, 72- *Concurrent Pos:* Mem attend staff, Ill Mason Hosp, 54-90; res assoc, Inst Psychoanal, 53-56, mem staff, 56-60, asst dean & dir res, 60-71, Pres, 71-89; Found Fund Res Psychiat fel, 60-65; pres, Ctr Psychosocial Studies, 72-90; pres, Ill Psychiat Asn, 73-74, Am Psychiat Asn, 87-88. *Mem:* Soc Exp Biol & Med; Am Electroencephalog Soc; Am Psychosom Soc; Soc Biol Psychiat; Am Psychiat Asn (treas, 80-86, pres, 87-88). *Res:* Psychosomatic correlations; neuropharmacological and neurophysiological studies; psychoanalytic investigations; psychoanalysis; psychosomatic medicine. *Mailing Add:* 5759 S Dorchester Ave Chicago IL 60637-1726

POLLOCK, HERBERT CHERMSIDE, b Staunton, Va, 1913; m 42, 80; c 4. PHYSICS. *Educ:* Univ Va, BA, 33; Oxford Univ, PhD(physics), 37. *Prof Exp:* Res assoc res lab, Gen Elec Co, 37-43; dir, Eglin Field Sta, Radio Res Lab, Harvard Univ, 43-44; physicist, Res & Develop Ctr, Gen Elec Co, 44-78, consult, 78-84; RETIRED. *Concurrent Pos:* Mem staff, Manhattan Proj, Radiation Lab, Univ Calif, 44-45; mem adv bd Ctr Advan Studies, Univ Va, 70-76; pres bd trustees, 76-80, Dudley Observ, trustee, 72-; trustee, Alumni Patent Found, Univ Va, 78-84. *Mem:* Fel Am Phys Soc. *Res:* Electron physics; plasma physics. *Mailing Add:* 456 Quinnipiac Ave North Haven CT 06473

POLLOCK, JOHN ARCHIE, b Kenmore, NY, Jan 10, 57. NEURODEVELOPMENT, CELL BIOLOGY. *Educ:* Syracuse Univ, BS, 78, MS, 83, PhD(biophys), 84. *Prof Exp:* Sr res fel, Calif Inst Technol, 84-89; ASST PROF NEUROBIOL, CARNEGIE MELLON UNIV, 89- *Concurrent Pos:* Vis res scholar, high-voltage electron micros, Univ Colo, Boulder, 88-89 & IVEM, Univ Calif, San Diego, 91- *Mem:* Am Soc Cell Biol; Asn Res Vision & Ophthal; Biophys Soc; Genetics Soc Am; Soc Neurosci; NY Acad Sci. *Mailing Add:* Dept Biol Sci Carnegie Mellon Univ 4400 Fifth Ave Pittsburgh PA 15213

POLLOCK, JOHN JOSEPH, b Kulpmont, Pa, Feb 6, 33; m 60; c 3. TOXICOLOGY, PHARMACOLOGY. *Educ:* Mt St Mary's Col, Md, BS, 54; Univ Man, MSc, 61, PhD(pharmacol), 65; Am Bd Toxicol, dipl. *Prof Exp:* Jr pharmacologist, Smith Kline & French Labs, 56-59, head toxicol unit, 65-68; sr toxicologist, Wyeth Labs, Pa, 68-72; ASST VPRES, DRUG SAFETY EVAL, WYETH-AYERST RES, 72- *Concurrent Pos:* Vis lectr, Univ Pa, 66-72; mem adv coun animal health sci, State Univ Canton, NY, 75-84; mem adv coun regional continuing educ, State Univ Plattsburgh, NY, 75-85. *Mem:* AAAS; Soc Toxicol; Can Asn Res Toxicol; Soc Exp Biol & Med; European Soc Toxicol; Expert Toxicologue. *Res:* Safety evaluation in animals of drugs which have potential human clinical use. *Mailing Add:* Drug Safety Eval Wyeth-Ayerst Res Chazy NY 12921

POLLOCK, L(YLE) W(ILLIAM), b Bowman, NDak, June 14, 18; m 44; c 4. CHEMICAL ENGINEERING. *Educ:* Univ NDak, BS, 40; Univ Wash, MS, 42. *Prof Exp:* Process design engr, 43-45, design supvr, 48-56, mgr math eng, 56-58, mgr planning & correlation, 58-66, mgr advan eng sci, 66-69, staff engr, 69-74, ENVIRON ENGR, PHILLIPS PETROL CO, 74- *Mem:* Am Inst Chem Engrs. *Res:* Process development; environmental engineering. *Mailing Add:* 1416 Melmart Dr Bartlesville OK 74006

POLLOCK, LELAND WELLS, b Boston, Mass, Apr 3, 43; m 69; c 2. ECOLOGY, INVERTEBRATE ZOOLOGY. *Educ:* Bates Col, BS, 64; Univ NH, MS, 66, PhD(zool), 69. *Prof Exp:* Fed Water Qual Admin fel syst-ecol, Wellcome Marine Lab, Robin Hood's Bay, Yorkshire, Eng, 69-70; fel, Marine Biol Lab, Woods Hole, Mass, 70-72; from asst prof to assoc prof, 72-85, PROF ZOOL, DREW UNIV, 85- *Mem:* AAAS; Asn Meiobenthologists; Am Micros Soc. *Res:* Ecology of marine beaches and biology of Tardigrada; role of environmental variables in shaping meiofaunal distribution. *Mailing Add:* Dept of Biol Drew Univ Madison NJ 07940

POLLOCK, MICHAEL L, b Los Angeles, Calif, June 24, 36; m 63, 79. EXERCISE PHYSIOLOGY. *Educ:* Univ Ariz, BS, 58; Univ Ill, MS, 61, PhD(phys educ, physiol), 67. *Prof Exp:* Instr phys educ, Univ Ill, 61-67, asst supvr, Phys Fitness Res Lab, 64-67; dir, Phys Fitness Res Lab, Wake Forest Univ, 67-73, mem grad fac phys educ, 69-73, assoc prof, 71-73, assoc in med, Bowman Gray Sch Med, 71-73; dir res, Inst Aerobics Res, 73-77; assoc prof, 77-80, prof med, Sch Med, Univ Wis-Milwaukee, 80-; at dept med, Univ Wis-Madison; dir cardiac rehab, Mt Sinai Med Ctr, 77-; AT FLORIDA GYM, UNIV FLA. *Concurrent Pos:* Consult & symp staff mem, President's Coun Phys Fitness & Sport, 72-; mem, Am Heart Asn; fel, Epidemiol Coun, Am Heart Asn. *Mem:* Am Asn Health, Phys Educ & Recreation; fel Am Col Sports Med; fel Am Col Cardiol. *Res:* Sports medicine; adult fitness; cardiac rehabilitation. *Mailing Add:* Florida Gym Univ Florida Rm 27 Gainesville FL 32610

POLLOCK, ROBERT ELWOOD, b Regina, Sask, Mar 2, 36; m 59; c 4. NUCLEAR PHYSICS, ACCELERATOR DESIGN. *Educ:* Univ Man, BSc, 57; Princeton Univ, MA, 59, PhD(physics), 63. *Prof Exp:* Instr physics, Princeton Univ, 61-63; Nat Res Coun Can fel nuclear physics, Atomic Energy Res Estab, Harwell, Eng, 63-64; asst prof physics, Princeton Univ, 64-69, res physicist, 69-70; from actg dir to dir, Cyclotron Facil, 72-79, from assoc prof to prof, 70-84, DISTINGUISHED PROF PHYSICS, IND UNIV, BLOOMINGTON, 85- *Concurrent Pos:* Mem, nuclear sci adv comt, 77-80. *Honors & Awards:* Humboldt Sr US Scientist Award, 85. *Mem:* Fel Am Phys Soc; Can Asn Physicists. *Res:* Intermediate energy nuclear reactions; accelerator development. *Mailing Add:* Dept Physics Ind Univ Swain Hall W Bloomington IN 47405

POLLOCK, STEPHEN M, b New York, NY, Feb 15, 36; m 62; c 3. OPERATIONS RESEARCH. *Educ:* Cornell Univ, BEngPhys, 58; Mass Inst Technol, SM, 60, PhD(physics, opers res), 64. *Prof Exp:* Mem tech staff, Arthur D Little, Inc, 64-65; from asst prof to assoc prof opers res, US Naval Postgrad Sch, 65-69; assoc prof, 69-74, chmn, 80-89, PROF INDUST & OPERS ENG & RES SCIENTIST, INST PUB POLICY STUDIES, UNIV MICH, ANN ARBOR, 74- *Concurrent Pos:* Consult to over thirty co & govt agencies; NSF lectr & Opers Res Soc Am vis lectr, 68-, NSF adv panels, 74- *Honors & Awards:* Stephen S Attwood Award. *Mem:* AAAS; Sigma Xi; Opers Res Soc Am (pres, 86-87); Inst Mgt Sci. *Res:* Stochastic systems analysis, reliability and Queueing networks; search and detection theory; sequential decision theory; social systems modeling. *Mailing Add:* Dept Indust and Oper Eng Univ Mich Ann Arbor MI 48109-2117

POLLOK, NICHOLAS LEWIS, III, b Danville, Va, May 26, 26; m 51; c 3. MICROBIOLOGY. *Educ:* Univ Va, AB, 51; Univ Md, Baltimore, PhD(microbiol), 72. *Prof Exp:* Microbiologist, US Army Biol Labs, Ft Detrick, Md, 52-72; USPHS grant, Sch Dent, Univ Md, 72-73;

MICROBIOLOGIST & DIR, EXTRAMURAL SCI ADV STAFF, CTR BIOLOGICS, FOOD & DRUG ADMIN, 73- *Mem:* AAAS; Am Soc Microbiol; Sigma Xi; Am Pub Health Asn. *Res:* Airborne mixed-infections; viral and rickettsial pathogenesis and immunology; airborne contamination in dentistry; nosocomial infections; airborne contamination control; aerogenic vaccination and respiratory diseases; aerobiology. *Mailing Add:* Ctr Biologics, Food & Drug Admin Nicholson Lane Res Ctr 5516 Nicholson Lane Rm 110 Kensington MD 20895

POLLYCOVE, MYRON, b Nogales, Ariz, 1921; m 41; c 4. PATHOLOGY. *Educ:* Calif Inst Technol, BS, 42; Univ Calif, MD, 50; Am Bd Path, cert path; Am Bd Nuclear Med , cert nuclear med. *Prof Exp:* Intern, Harvard Med Serv, Boston City Hosp, 50-51; resident med, Boston VA Hosp; lectr, Med Sch, Tufts Univ, 53-55; DIR, DEPT NUCLEAR MED, SAN FRANCISCO GEN HOSP. *Concurrent Pos:* Teaching fel, Med Sch, Tufts Univ, 53-55; from resident to sr resident, Med Serv, Boston Vet Admin Hosp, 53-55; dir, Donner-Highland Radioisotopes Unit, Highland Alameda County Hosp, 55-61; resident physician, Cowell Mem Hosp, 55-; attend physician, Vet Admin Hosp, San Francisco, 57-, Martinez Vets Admin Hosp, Oak Knoll US Naval Hosp, Oakland, Letterman Gen US Army Hosp, San Francisco, Travis AFB Hosp, Kaiser Found Hosp, Oakland & Univ Calif Cowell Mem Hosp, Berkeley; dir, Clin Labs & chief, Nuclear Med Div, San Francisco Gen Hosp, Calif, 64-81, chief, Nuclear Med Serv, 81-; mem, Am Bd Nuclear Med, 77-83, Nuclear Med Exam Comt, Bd Registry, Am Soc Clin Pathologists, 79-85. *Mem:* Am Soc Clin Pathologists; Acad Clin Lab Physicians & Sci; Soc Nuclear Med; Am Soc Hemat; Am Fedn Clin Res; Am Col Nuclear Physicians (pres, 88-89); Am Col Pathologists; AMA; CMA. *Res:* Laboratory and nuclear medicine; iron kinetics and metabolism; functional kinetics in hematology; amino acid; vitamin B-12 and folic acid intermediary metabolism. *Mailing Add:* Dept Nuclear Med San Francisco Gen Hosp 22nd & Potrero St San Francisco CA 94110

POLMANTEER, KEITH EARL, b Midland, Mich, Mar 19, 23; m 46; c 2. PHYSICS, CHEMISTRY. *Educ:* Cent Mich Univ, BS 48; Case Western Reserve Univ, MS, 50. *Prof Exp:* Technician, Dow Chem Co, 41-42; lectr & asst physics, Case Western Reserve Univ, 48-49; physicist silicone rubber, Dow Corning Corp, 49-84, supvr res, 52-65, mgr elastomer develop, 65-66, mgr polymer physics, 66-69, scientist, 69-80, res scientist, 80-84; CONSULT, KEP ENTERPRISES, 84- *Honors & Awards:* I R 100 Award, 82. *Mem:* Am Chem Soc; Am Phys Soc. *Res:* Polymer chemistry; organo-silicon polymers as related to elastomers; physics and physical chemistry of polymers and elastomers; synthesis of new elastomeric polymers. *Mailing Add:* Consult KEP Enterprises 904 Balfour Midland MI 48640

POLMAR, STEPHEN HOLLAND, c 2. IMMUNOLOGY, IMMUNOPHARMACOLOGY. *Educ:* Case Western Univ, PhD(genetics), 66, MD, 67. *Prof Exp:* PROF PEDIAT, IMMUNOL & MICROBIOL, SCH MED, WASH UNIV, 82- *Concurrent Pos:* Med adv bd, Immune Deficiency Found, 81-; mem, immunol sci study sect, NIH, 83-87. *Mem:* Int Asn Immunopharmacol; Am Asn Immunologist; Am Acad Allergy & Immunol. *Res:* Cellular immunity, immunodeficiency diseases, immunopharmacology; cellular and molecular basis of human immunodeficiency diseases; relation of immunologic responses in man using antacoids as models. *Mailing Add:* Sch Med Wash Univ 400 S Kingshighway Blvd St Louis MO 63110

POLNASZEK, CARL FRANCIS, b Nanticoke, Pa, Jan 6, 45; m 82. BIOPHYSICAL CHEMISTRY. *Educ:* Wilkes Col, BS, 66; Cornell Univ, PhD(phys chem), 76; Univ Minn, Comput Sci, 85. *Prof Exp:* asst res officer, Nat Res Coun Can, 73-78; res chemist, Vet Admin Med Ctr, Minn, 78-82; res assoc, Univ Minn, 82-85; SR TECH PROG & ANALYST, TECH COMPUT, FMC CO, MINN, 85- *Mem:* Am Chem Soc. *Res:* Theoretical and computational modeling of the molecular dynamics and organization of biological and model systems as studied by means of electron spin resonance and nuclear magnetic resonance spectroscopies. *Mailing Add:* Tech Comput T-025 FMC Corp 4800 E River Rd Minneapolis MN 55421-1498

POLO, SANTIAGO RAMOS, b Salamanca, Spain, Dec 15, 22; m 62; c 4. MOLECULAR PHYSICS, SOLID STATE PHYSICS. *Educ:* Univ Madrid, MS, 45, PhD, 49. *Prof Exp:* Res assoc phys chem, Higher Coun Sci Res, Spain, 49-50; res assoc, Ill Inst Technol, 50-51; res assoc, Harvard Univ, 51-55; res assoc physics, Nat Res Coun Can, 55-58; mem tech staff, RCA Labs, 58-65; PROF PHYSICS, PA STATE UNIV, 65- *Honors & Awards:* Award, Spanish Royal Acad Sci. *Mem:* Am Phys Soc. *Res:* Molecular structure and spectroscopy. *Mailing Add:* Davey Bldg Dept of Physics Pa State Univ University Park PA 16802

POLONIS, DOUGLAS HUGH, b North Vancouver, BC, Sept 2, 28; nat US; m 53; c 4. ENGINEERING PHYSICS, ENGINEERING MECHANICS. *Educ:* Univ BC, BASc, 51, PhD(metall), 55; Univ Toronto, MASc, 53. *Prof Exp:* Metall engr, Steel Co Can, Ltd, 51-52; from asst prof to assoc prof, 55-62, chmn, dept mining, metall & ceramic eng, 69-71 & 73-82, PROF METALL ENG, UNIV WASH, 62- *Mem:* Fel Am Soc Metals; Am Inst Mining, Metall & Petrol Engrs; Sigma Xi. *Res:* Materials science; physical metallurgy; phase transformations in solids; reactive metal technology; mechanical behavior and microstructure of materials. *Mailing Add:* Dept Mat Sci Univ Wash Seattle WA 98195

POLONSKY, IVAN PAUL, b Brooklyn, NY, Aug 23, 29; m 53; c 3. MATHEMATICS, COMPUTER SCIENCE. *Educ:* NY Univ, BA, 49, MS, 52, PhD(math), 57. *Prof Exp:* Instr math, NY Univ, 51-53; instr, Queens Col, 53-60; MEM TECH STAFF COMPUT PROGRAMMING, BELL TEL LABS, INC, 60- *Concurrent Pos:* Assoc, Grad Fac, Rutgers Univ, 72- *Mem:* Am Math Soc; Math Asn Am; Asn Comput Mach. Res; Complex function theory; programming languages; applications of computers to nonnumeric problems. *Mailing Add:* 151 Harding Rd Red Bank NJ 07701

POLOSA, CANIO, b Rome, Italy, Oct 17, 28; m 60; c 3. MEDICINE, PHYSIOLOGY. *Educ:* Univ Rome, MD, 53; McGill Univ, PhD(physiol), 66. *Prof Exp:* Asst med, Inst Med Path, Univ Rome, 55-57 & Inst Clin Med, 57-59; from lectr to assoc prof, 63-77, PROF PHYSIOL, MCGILL UNIV, 77- *Concurrent Pos:* Nat Res Coun Italy fel, 57-58; USPHS trainee cardiovasc res, 59-62. *Mem:* Can Physiol Soc; Am Physiol Soc; Soc Neurosci. *Res:* Cardio-pulmonary physio-pathology; physiology of nervous control of circulation; neurophysiology; physiology of sympathetic neurons. *Mailing Add:* Dept Physiol McIntyre Med Sci Bldg McGill Univ 3655 Drummond St Montreal PQ H3G 1Y6 Can

POLOUJADOFF, MICHEL E, b Paris, France, Apr 2, 32; m 64; c 2. ELECTRIC POWER, ELECTRIC MAGNETISM. *Educ:* Ecole Supérieure d'Electricité, Ingénieur dipl; Harvard Univ, MSc; Paris Univ, Docteur és Sci (physics). *Hon Degrees:* Dr, Univ Liège, 83, Budapest Univ, 89. *Prof Exp:* Prof elec eng, Inst Polytech Grenoble, 61-85; PROF ELEC ENG, UNIV PIERRE ET MARIE CURIE, 85- *Concurrent Pos:* Vis prof, McMaster Univ, Ont, Can, 83-84; prof, Ecole Centrale de Paris, 85-; mem, numerous sci & steering comts int meetings. *Honors & Awards:* Nikola Tesla Award, Inst Elec & Electronics Engrs, 91. *Mem:* Fel Inst Elec & Electronics Engrs; fel NY Acad Sci. *Res:* Generation and utilization of electric power, including electromagnetism, design and optimization; power electronics; electrical lines; rotating squirrel cage and linear induction machines. *Mailing Add:* U PM Curie, Electrotech Tour 22-12-El, 4 Place Jussieu, Cedex 05 Paris 75252 France

POLOWCZYK, CARL JOHN, b New York, NY, Oct 3, 33; m 53; c 3. PHOTOCHEMISTRY, ELECTROCHEMISTRY. *Educ:* City Col New York, BS, 58; NY Univ, MS, 64, PhD(chem), 65. *Prof Exp:* Chief chemist, NY Testing Lab, 51-55; chief chemist, Mich Res Div, Sun Chem, 55-58 & Faberge, Inc, 58-60; from instr to assoc prof, 60-74, head dept, 70-78, PROF CHEM, BRONX COMMUNITY COL, 74-, DEAN ACAD AFFAIRS, 78- *Concurrent Pos:* Adj asst prof, NY Univ, 65-; chmn, Metrop Leadership Prog, NY Univ, 70-; consult & prin investr, Minority Biomed Res Support Prog. *Honors & Awards:* Mfg Chemists Asn Medalist, 79. *Mem:* Am Chem Soc; AAAS. *Res:* Photochemistry and electrochemistry of organic compounds; detection and estimation of trace organic and inorganic materials; resource management, educational research and science education. *Mailing Add:* Dean Acad Affairs Bronx Community Col Univ Ave & W 181 St Bronx NY 10453

POLOWY, HENRY, b Elberfeld-Wuppertal, Ger, Mar 10, 18; nat US; m 55; c 6. APPLIED MATHEMATICS. *Educ:* NY Univ, AB, 48, MS, 49, PhD(math), 57. *Prof Exp:* Engr, Brewster Aeronaut Corp, 40-43; teacher high sch, NY, 50-51; engr, Gen Motors Corp, 51-52; teacher high sch, NJ, 52; asst prof math, Stevens Inst Technol, 52-67; head dept math, Lincoln Univ, 67-75, prof math, 67-82; RETIRED. *Concurrent Pos:* Instr, Am Fedn Labor, 50-51; lectr, Int Tel & Tel, NJ, 57-58 & 63-64; lectr, NSF, 60-61, dir undergrad res math, 62-67; adj prof math, Columbia Col, 84-87. *Mem:* Math Asn Am; Am Math Soc; Inst Math Statist; Am Soc Qual Control; Sigma Xi. *Res:* Real and complex analysis; probability; game theory and decision theory; mathematical statistics; stability-catastrophe theory. *Mailing Add:* Rte 2 New Bloomfield MO 65063

POLSON, CHARLES DAVID ALLEN, b Alpena, Mich, Dec 18, 47; m 70; c 1. MOLECULAR BIOLOGY. *Educ:* Cent Mich Univ, BS, 70, MS, 75; Fla Inst Technol, PhD(biol), 79. *Prof Exp:* From instr to asst prof, 78-87, ASSOC PROF BIOL, FLA INST TECHNOL, 87-, ASSOC DEPT HEAD UNDERGRAD PROGS, 89- *Concurrent Pos:* Fulbright fel, Dept Biochem, Univ Zimbabwe, 87. *Mem:* Am Asn Univ Prof; Sigma Xi. *Res:* Application of biotechnology techniques to undergraduate education; electrophoretic separation of DNA molecules. *Mailing Add:* Dept Biol Sci Fla Inst Technol Melbourne FL 32901-6988

POLSON, JAMES BERNARD, b Kansas City, Mo, Apr 29, 38; m 59; c 2. PHARMACOLOGY. *Educ:* Univ Mo-Columbia, AB, 61, MS, 66, PhD(pharmacol), 68. *Prof Exp:* Fel pharmacol, Sch Med, Univ Mo-Columbia, 68-69; trainee pharmacol & toxicol, Med Sch, Univ Minn, Minneapolis, 69-71; from asst dean to assoc dean res grad affairs, 82-88, from asst prof to assoc prof, 71-83, PROF PHARMACOL, COL MED, UNIV SFLA, 83- *Mem:* Am Soc Pharmacol & Exp Therapeut. *Res:* Biochemical mechanisms of drug action. *Mailing Add:* Dept Pharmacol Univ S Fla Col of Med Tampa FL 33612

POLSON, WILLIAM JERRY, b Sulphur Springs, Tex, July 14, 43; m 66; c 2. NUCLEAR PHYSICS. *Educ:* ETex State Univ, BS, 64; Stephen F Austin State Univ, MS, 66; Auburn Univ, PhD(physics), 69. *Prof Exp:* From asst prof to assoc prof, 69-80, PROF PHYSICS, SOUTHEASTERN OKLA STATE UNIV, 80- *Mem:* Am Phys Soc. *Res:* Theoretical nuclear physics; visual astronomy. *Mailing Add:* Dept Physics Southeastern Okla State Univ Durant OK 74701

POLT, SARAH STEPHENS, b Leavittsburg, Ohio, Aug 22, 26; m 64; c 2. CLINICAL PATHOLOGY. *Educ:* Western Reserve Univ, BA, 48, MS, 50; Woman's Med Col Pa, MD, 57. *Prof Exp:* Assoc path, Duke Univ Med Ctr, 62-64; pathologist, Vet Admin Hosp, Birmingham, Ala, 65-67; asst prof clin path & dir serol clin labs, 67-71, assoc prof, dept path, 71-84, PROF, DEPT PATH, UNIV ALA HOSPS, BIRMINGHAM, 84- *Mem:* Am Soc Clin Path; Am Soc Microbiol. *Res:* Diagnosis of infectious disease by immunological techniques. *Mailing Add:* Dept Path Univ Ala Hosp Birmingham AL 35233

POLTORAK, ANDREW STEPHEN, b Somerville, NJ, Jan 11, 38; m 62; c 3. OPERATIONS RESEARCH, PHYSICS. *Educ:* Univ Notre Dame, BS, 60; Univ Mich, Ann Arbor, MS, 61, PhD(physics), 66. *Prof Exp:* Opers res analyst mil systs, Conductron Corp, 67-68; opers res analyst mil systs, Naval Ord Lab, Naval Surface Weapons Ctr, 68-73, br head syst anal, 74-76; analyst, Nuclear Regulatory Comn, 76-77; DIV MGR, SCI APPL, INC, 77- *Mem:* Inst Elec & Electronics Engrs; Opers Res Soc Am. *Res:* Systems analysis of advanced military systems, determining feasibility, requirements and effectiveness of proposed systems with introductory dates of five to thirty years. *Mailing Add:* 2932 Beaverwood Ln Wheaton MD 20906

POLUHOWICH, JOHN JACOB, b Bridgeport, Conn, Mar 25, 39; m 62; c 4. PHYSIOLOGICAL ECOLOGY, NUTRITION. *Educ:* Univ Bridgeport, BA, 62, MS, 82; Iowa State Univ, MS, 66, Univ Conn, PhD(zool), 69. *Prof Exp:* from asst prof to prof biol, Univ Bridgeport, 69-87, dir, Inst Anguilliform Res & Maricult, 74-87; asst prof, 87-90, ASSOC PROF, WTEX STATE UNIV, 90- *Concurrent Pos:* Mem, Conn Coastal Zone Mgt Comt, 71-72; environ consult, Oley-Pavia, Stamford, Conn. *Mem:* Sigma Xi; World Maricult Soc; AAAS. *Res:* Physiology of catadromous and anadromous fishes; mariculture; control of morphogenesis. *Mailing Add:* Dept Biol & Geosci WTex State Univ Box 808 Canyon TX 79016

POLVE, JAMES HERSCHAL, b Kenilworth, Utah, Feb 7, 21; m 48; c 3. AERONAUTICAL ENGINEERING. *Educ:* Univ Utah, BS, 48, ME, 54; Princeton Univ, MS, 51; Univ Ariz, PhD(aeronaut eng), 66. *Prof Exp:* Instr & chief opers, Exp Flight Test Pilot Sch, US Air Force, 51-55, from asst prof to assoc prof math, Air Force Acad, 55-58, assoc prof aeronaut, 58-60, dir flight test eng, Air Force Flight Test Ctr, Edwards AFB, 63-66, staff officer, Fed Aviation Admin, 66-67, chief Seattle Off at Boeing Co, 67-69; PROF MECH ENG, BRIGHAM YOUNG UNIV, 69- *Mem:* Am Soc Eng Educ; assoc fel, Am Inst Aeronaut & Astronaut; Am Soc Mech Eng. *Res:* Stability and control; flight testing; mechanical engineering. *Mailing Add:* Box 303 Ferron UT 84523

POLYAKOV, ALEXANDER, b Moscow, USSR, Sept 27, 45; m 81; c 2. QUANTUM GRAVITY, HIGH ENERGY PHYSICS. *Educ:* Landau Inst, PhD(physics), 69. *Prof Exp:* Prof physics, Landau Inst, 69-89; PROF PHYSICS, PRINCETON UNIV, 89- *Honors & Awards:* D Heineman Prize, Am Phys Soc, 85; P Dirac Medal, Int Ctr Theoret Physics, 86. *Mem:* Acad Arts & Sci; Acad Sci USSR. *Res:* Quantum field theory; statistical mechanics and quantum gravity; string theory and gauge theories; critical phenomena; turbulence. *Mailing Add:* 93 MacLean Circle Princeton NJ 08540

POLYCHRONAKOS, VENETIOS ALEXANDER, b Thessaloniki, Greece, July 3, 45; m 74; c 2. ELEMENTARY PARTICLE PHYSICS. *Educ:* Univ Thessaloniki, BSc, 69; Univ Notre Dame, PhD(physics), 76. *Prof Exp:* Res assoc, Fermi Nat Accelerator Lab, 76-79; PHYSICIST, BROOKHAVEN NAT LAB, 79- *Res:* Experimental patricle physics; lepton production in high energy proton-proton collisions; relativistic heavy ion interactions; search for rare kaon decays; semiconductor detector development. *Mailing Add:* 85 Mahogany Rd Rocky Point NY 11778

POLYCHRONOPOULOS, CONSTANTINE DIMITRIUS, b Patras, Greece, Apr 5, 58; m 84; c 2. SUPERCOMPUTER COMPILERS, PARALLEL & HIGH-PERFORMANCE COMPUTER ARCHITECTURES. *Educ:* Univ Athens, Greece, dipl math & computer sci, 80; Vanderbilt Univ, MS, 82; Univ Ill, Urbana-Champaign, PhD(computer sci), 86. *Prof Exp:* Res asst, 82-86, SR SOFTWARE ENGR, CTR SUPERCOMPUT RES & DEVELOP & ASST PROF, DEPT ELEC & COMPUTER ENG, UNIV ILL, URBANA- CHAMPAIGN, 86- *Concurrent Pos:* Consult scientist, Computer Technol Inst, Univ Patras, 87-; consult, Tera Computer Co, 88-; vis prof, Univ Rennes, France, 88; NSF presidential young investr, 89. *Mem:* Asn Comput Mach; Inst Elec & Electronic Engrs. *Res:* Parallel processing; design of parallel computer systems; theory and development of parallelizing compilers; operating systems for multiprocessors; scheduling and synchronization of parallel computations. *Mailing Add:* Ctr Supercomput Res & Develop Univ Ill Urbana IL 61801

POLYE, WILLIAM RONALD, b New York, NY, Dec 12, 14; m 41; c 3. PHYSICS. *Educ:* St Lawrence Univ, BS, 37; Univ Notre Dame, MS, 39. *Prof Exp:* Asst physics, Univ Notre Dame, 37-40; physicist, 40-65, sr scientist, 65-77, CONSULT APPL PHYSICS, BENDIX CORP, 77- *Concurrent Pos:* pvt tech consult. *Mem:* Am Phys Soc; Am Vacuum Soc. *Res:* Atmospheric physics; electro-optical and solid state devices; microelectronics; vacuum deposition; sensor research; infra red. *Mailing Add:* 154 Lake End Rd RR Two Newfoundland NJ 07435

POLZER, WILFRED L, b Cameron, Tex, Apr 10, 31; m 58; c 6. GEOCHEMISTRY, ENVIRONMENTAL SCIENCES. *Educ:* Tex A&M Univ, BS, 53, MS, 56; Mich State Univ, PhD(soil sci), 60. *Prof Exp:* Res chemist, US Geol Surv, 60-67; geochemist, US AEC, US Energy Res & Develop Admin, 67-71, environ scientist, 71-76; SOIL CHEMIST, LOS ALAMOS NAT LAB, 76- *Mem:* Am Soc Agron. *Res:* Modeling of contaminants in unsaturated soil zones; behavior of plutonium and americium in soil-water environments; radioactive waste and soil interactions; geochemical controls of water composition. *Mailing Add:* 7 La Rosa Court Los Alamos NM 87544

POLZIN, DAVID J, b Chicago, Ill, Nov 10, 51; m 86; c 1. NEPHROLOGY, INTERNAL MEDICINE. *Educ:* Univ Ill, DVM, 75; Univ Minn, PhD(vet med), 81. *Prof Exp:* Intern vet med, Col Vet Med, Univ Ga, 75-76; resident, Col Vet Med, Univ Minn, 76-81, res assoc, 81-84, asst prof, 84-88, ASSOC PROF VET MED, COL VET MED, UNIV MINN, 88- *Concurrent Pos:* Secy-treas, Am Soc Vet Nephrol & Urol, 81- *Mem:* Am Vet Med Asn; Am Soc Nephrol; Int Soc Nephrol; Am Soc Vet Nephrol & Urol; Am Col Vet Internal Med; Am Animal Hosp Asn. *Res:* Pathophysiology, diagnosis and treatment of chronic renal failure with emphasis on medical means of limiting progression of chronic renal failure, particularly using nutrition therapy. *Mailing Add:* Dept Small Animal Clin Sci Col Vet Med Univ Minn 1352 Boyd Ave Rm C-325 St Paul MN 55108

POMERANCE, CARL, b Joplin, Mo, Nov 24, 44; m 71; c 2. MATHEMATICS, NUMBER THEORY. *Educ:* Brown Univ, BA, 66; Harvard Univ, MA, 70, PhD(math), 72. *Prof Exp:* from asst prof to prof, 72-82, RES PROF MATH, UNIV GA, 84- *Concurrent Pos:* Vis assoc prof math, Univ Ill, 78-79; Vis prof math, Univ Limoges, 84; Vis mem, Tech Staff, Bell Commun Res, 84-85. *Honors & Awards:* Chauvenet Prize, Math Asn Am, 85. *Mem:* Am Math Soc; Math Asn Am; Soc Indust & Appl Math. *Res:* Number theory including number theoretic algorithms. *Mailing Add:* Dept Math Univ Ga Athens GA 30602

POMERANCE, HERBERT S(OLOMON), b Chicago, Ill, July 24, 17; m 48; c 1. PHYSICS. *Educ:* Univ Chicago, SB, 37, PhD(physics), 50. *Prof Exp:* Asst chemist metall lab, Univ Chicago, 42-43; assoc chemist, Clinton Labs, Oak Ridge, 43-46; physicist, Oak Ridge Nat Lab, 46-64; ed, Int Atomic Energy Agency, 64-66; physicist, Oak Ridge Nat Lab, 66-82; RETIRED. *Mem:* AAAS; Am Asn Physics Teachers; Am Nuclear Soc. *Res:* Molecular spectra; spectrochemistry; neutron absorption cross sections; nuclear physics; information services. *Mailing Add:* 104 Ulena Lane Oak Ridge TN 37830-5237

POMERANTZ, IRWIN HERMAN, b Floral Park, NY, July 18, 34. ENVIRONMENTAL CHEMISTRY, QUALITY ASSURANCE OF ENVIRONMENTAL MEASUREMENTS. *Educ:* Antioch Col, BS, 57; Univ Colo, PhD(org chem), 62. *Prof Exp:* Steroid Biochem Training Prog fel, Worcester Found Exp Biol, 61-62; asst prof chem, Bethany Col, WVa, 62-64; res assoc, Antioch Col, 64-65, asst prof, 65-66; res chemist, Food & Drug Admin, 66-70, supvry res chemist, 70-77; qual assurance officer, Off Drinking Water, 81-87; adminr phys sci, US Environ Protection Agency, 77-89; CONSULT, 89- *Concurrent Pos:* Adj prof chem, Am Univ, DC, 81-82. *Mem:* Am Chem Soc; NY Acad Sci; Asn Off Analytical Chemists. *Res:* Pesticide chemistry; industrial chemicals; analytical procedures for trace analysis; organic chemicals in drinking water; chemistry of chlorinated aromatic compounds. *Mailing Add:* 10813 Inwood Ave Silver Spring MD 20902

POMERANTZ, MARTIN, b New York, NY, May 3, 39; m 61; c 3. ORGANIC CHEMISTRY, POLYMER & COAL CHEMISTRY. *Educ:* City Col New York, BS, 59; Yale Univ, MS, 61, PhD(chem), 64. *Prof Exp:* NSF fel chem, Univ Wis, 63-64; asst prof, Case Western Reserve Univ, 64-69; from assoc prof to prof, Belfer Grad Sch Sci, Yeshiva Univ, 69-76, chmn dept, 71-72, 73-76; PROF, DEPT CHEM, UNIV TEX, ARLINGTON, 76- *Concurrent Pos:* Alfred P Sloan Found res fel, 71-76; vis assoc prof, dept chem, Univ Wis-Madison, 72; vis prof, dept chem, Columbia Univ, 70-75, Ben-Gurion Univ, Negev, Beer Sheva, Israel, 81 & 85; NSF postdoctoral fel, 63-64, grad fel, 62-63; Woodrow Wilson fel, 59-60; Leeds & Northrup fel, 60-62. *Mem:* Am Chem Soc; Royal Soc Chem; Sigma Xi. *Res:* Organic chemical reaction mechanisms; synthesis and study of phosphorus-nitrogen compounds; polymer chemistry; NMR spectroscopy; study of reactive intermediates; study of microcyclic compounds; conducting polymers; coal chemistry. *Mailing Add:* Dept Chem Univ Tex Box 19065 Arlington TX 76019-0065

POMERANTZ, MARTIN ARTHUR, b Brooklyn, NY, Dec 17, 16; m 41; c 2. PHYSICS. *Educ:* Syracuse Univ, AB, 37; Univ Pa, MS, 38; Temple Univ, PhD, 51. *Hon Degrees:* Fil Dr, Univ Uppsala, 67; ScD, Swarthmore Col, 73. *Prof Exp:* Asst, 38-41, res fel, 41-43, physicist, 43-59, dir, 59-85, pres, 85-87, EMER PROF, BARTOL RES FOUND, FRANKLIN INST, UNIV DEL, 87- *Concurrent Pos:* Fulbright scholar & vis prof, Aligarh Muslim Univ India, 52-53; vis prof, Swarthmore Col; ed, J, vpres & exec vpres, Franklin Inst, 85-87; prof, Thomas Jefferson Univ, Bartol prof; mem comt polar res, Geophys Res Bd & Space Sci Bd & Comt Solar Terrestrial Res, Nat Acad Sci; vpres comt & chmn US comt, Int Years of Quiet Sun & vpres int geophys comt, Int Coun Sci Unions; mem & leader numerous sci expeds; vis prof, Orgn Am States, 73, Tokyo Univ, 83, Potchefstroom Univ, 87. *Honors & Awards:* Centennial Medal, Syracuse Univ, 70. *Mem:* Fel AAAS; fel Am Phys Soc; fel Am Geophys Union; Am Astron Soc. *Res:* Cosmic rays; secondary electron emission; semiconductors; electronic physics; solid state; radiation effects; geophysics; solar terrestrial physics; astrophysics; astronomy. *Mailing Add:* Bartol Found Univ Del Newark DE 19711

POMERANTZ, MELVIN, b Brooklyn, NY, May 10, 32. PHYSICS. *Educ:* Polytech Inst Brooklyn, BS, 53; Univ Calif, Berkeley, MS, 55, PhD(physics), 59. *Prof Exp:* RES STAFF MEM, THOMAS J WATSON RES CTR, IBM CORP, 60- *Concurrent Pos:* Vis lectr, Univ Calif, Berkeley, 69-70. *Mem:* Am Phys Soc; fel NY Acad Sci; Sigma Xi. *Res:* Microwave frequency ultrasonics; properties of semiconductors; phonons in solids; properties of organic film; magnetic films. *Mailing Add:* IBM Res Ctr 20-132 Box 218 Yorktown Heights NY 10598

POMERANTZ, SEYMOUR HERBERT, b Houston, Tex, Oct 4, 28; m 53; c 4. BIOCHEMISTRY. *Educ:* Rice Inst, BA, 48; Univ Tex, PhD(org chem), 52. *Prof Exp:* Asst chem, Univ Ill, 52-53; from sr instr to asst prof biochem, Western Reserve Univ, 53-61; asst prof pharmacol, St Louis Univ, 61-63; from asst prof to assoc prof, 63-69, PROF BIOCHEM, SCH MED, UNIV MD, BALTIMORE, 69- *Concurrent Pos:* Vis prof, Weizmann Inst Sci, 72 & 79, Ben-Gurion Univ Negev, 85-86; Fogarty Fel, Weizmann Inst Sci, 76; Fulbright Fel, Ben-Gurion Univ Negev, 85-86. *Mem:* Am Soc Biol Chem; Am Chem Soc; Endocrine Soc. *Res:* Tyrosinase; oocyte maturation; hydroxylations. *Mailing Add:* Dept of Biochem Univ of Md Sch of Med Baltimore MD 21201

POMERANZ, BRUCE HERBERT, b Montreal, Que, July 24, 37; m 57; c 1. NEUROPHYSIOLOGY, NEUROSCIENCES. *Educ:* McGill Univ, BSc, MD & CM, 61; Harvard Univ, PhD(physiol), 67. *Hon Degrees:* DSc, Sri Lanka Open Univ. *Prof Exp:* NIH fel, Harvard Med Sch, 62-66; asst prof biol, Mass Inst Technol, 66-68; from asst prof to assoc prof, 68-80, PROF ZOOL, UNIV TORONTO, 80-, PROF PHYSIOL, 82- *Honors & Awards:* Dag Hammarskjold Prize, 86. *Mem:* AAAS; Am Physiol Soc; Soc Neurosci. *Res:* Physiology of endorphins such as, electrophysiology, pharmacology and behaviour; acupuncture and endorphins; nerve regeneration; wound healing. *Mailing Add:* Dept Zool Univ Toronto Toronto ON M5S 1A1 Can

POMERANZ, JANET BELLCOURT, b Albany, NY, Jan 15, 30. MATHEMATICS, STATISTICS. *Educ:* Col St Rose, BA, 56; Cath Univ, MS, 59, PhD(math), 62. *Prof Exp:* Teacher, St Patrick's High Sch, 53; teacher, Sacred Heart Sch, 53-54; teacher, St Vincent's Sch, 54-56; teacher, St Anthony's High Sch, 56-57; assoc prof math & chmn dept, Col St Rose, 62-68; assoc prof, State Univ NY Col Oneonta, 68-71; assoc prof, 71-78, PROF MATH, STATE UNIV NY MARITIME COL, FT SCHUYLER, 78-

Concurrent Pos: NSF res participation for col teachers grant, Univ Okla, 64-65. *Mem:* Math Asn Am. *Res:* Functional analysis; abstract algebra; theory of linear operators on H-algebras; matrices over the quaternions; combinatorial problems in the theory of relations. *Mailing Add:* 23 Eastland Dr Glen Cove NY 11542

POMERANZ, YESHAJAHU, b Poland, Nov 28, 22; m 48; c 2. BIOCHEMISTRY, CEREAL CHEMISTRY. *Educ:* Israel Inst Technol, BSc, 48, Chem E, 49; Kans State Univ, PhD(cereal chem), 62. *Prof Exp:* Chief chemist, Israel Govt, 49-54, head cent food lab, 55-59; UN Food & Agr Orgn res fel cereal chem, Brit Milling Industs Res Asn, 54-55; res chemist, USDA, 62-69, chemist in charge barley & malt lab, 69-73, dir, US Grain Mkt Res Ctr, 73-86; RES PROF, WASH STATE UNIV, 86- *Concurrent Pos:* ed, Adv Cereal Sci Technol; prof, Kans State Univ, 63-69; prof, Univ Wis-Madison; ed-in-chief, Cereal Chem. *Honors & Awards:* Osborne Medal, Am Asn Cereal Chem; Harvey W Wiley Award, Asn Off Anal Chemists; Agr Food Chem Award, Am Chem Soc; Sr US Scientist Award, Von Humbold; Newmann Medal. *Mem:* Fel AAAS; Am Chem Soc; Am Asn Cereal Chem; fel Inst Food Technol; Asn Off Anal Chemists. *Res:* Chemical composition and functional properties of cereal grains; food analysis, nutritional value of cereals; high-protein foods; food rheology; microbiology of stored grain products. *Mailing Add:* Dept Food Sci & Human Nutrit Washington State Univ Pullman WA 99164-6376

POMERENE, JAMES HERBERT, b Yonkers, NY, June 22, 20; m 44; c 3. COMPUTER ARCHITECTURE. *Educ:* Northwestern Univ, BS, 42. *Prof Exp:* Design engr radar, Hazeltine Corp, Long Island, NY, 42-46; staff engr comput, Inst Advan Study, Princeton, NJ, 46-51, chief engr, 51-56; sr engr comput design, IBM Corp, Poughkeepsie, NY, 56-67, sr staff mem corp tech comt, Armonk, NY, 67-76, IBM FEL COMPUT ARCHIT, IBM TJ WATSON RES CTR, 76- *Mem:* Nat Acad Eng; AAAS; fel Inst Elec & Electronic Engrs; Sigma Xi. *Res:* High speed computer organizations, highly concurrent processors. *Mailing Add:* 403 N Bedford Rd Chappaqua NY 10514

POMERENING, JAMES ALBERT, b New London, Wis, May 3, 29; m 58; c 2. SOIL CLASSIFICATION. *Educ:* Univ Wis, BS, 51; Cornell Univ, MS, 56; Ore State Univ, PhD(soil classification), 61. *Prof Exp:* Asst soils, Cornell Univ, 51-53 & 56; res fel soil classification, Ore State Univ, 56-60; asst prof, Univ Md, 60-65, from asst prof to assoc prof plant & soil sci, 65-73, assoc dean, Sch Agr, 74-75, PROF PLANT & SOIL SCI, CALIF STATE POLYTECH UNIV, POMONA, 73- *Concurrent Pos:* Soil scientist, US Dept Interior, Bur Land Mgt, 71-88. *Mem:* Soil Sci Soc Am; Am Soc Agron. *Res:* Soil genesis, morphology, classification and survey. *Mailing Add:* Dept Plant/Soil Sci Calif State Polytech Pomona CA 91768-4043

POMERLEAU, OVIDE F, b Waterville, Maine, June 4, 40. BEHAVIOR THERAPY, GENERAL PSYCHOTHERAPY. *Educ:* Bowdoin Col, AB, 62; Columbia Univ, MS, 65, PhD(exp psychol), 69. *Prof Exp:* Asst prof psychiat, Temple Univ Med Sch, 69-72; from asst prof to assoc prof psychol, Univ Pa Sch Med, 72-79; prof, psychiat, Univ Conn Sch Med, 79-85; PROF PSYCHOL, UNIV MICH SCH MED, 85-, DIR, BEHAV MED PROG, 85- *Concurrent Pos:* Dir res, Temple Univ Unit, Philadelphia State Hosp, 69-71; dir, Univ Conn Behav Med Prog, Vet Admin Med Ctr, Newington, Conn, 79-85, chief psychol, 80-85; tech consult, Surgeon Gen's Adv Comt on Health Consequences of Smokeless Tobacco, 85-86; tech consult, Am Psychiat Asn Comt for DSM-III-R, 86- *Mem:* Sigma Xi; AAAS; Am Psychol Asn; NY Acad Med; Behav Pharmacol Soc; Behav Ther & Res Soc. *Res:* Behavior therapy and general psychotherapy; self-management disorders; author of numerous articles. *Mailing Add:* 2101 Devonshire Ann Arbor MI 48104

POMEROY, BENJAMIN SHERWOOD, b St Paul, Minn, Apr 24, 11; m 38; c 4. VETERINARY MICROBIOLOGY. *Educ:* Iowa State Univ, DVM, 33; Cornell Univ, MS, 34; Univ Minn, PhD(bact), 44. *Prof Exp:* From asst vet to vet, 34-38, instr, 38-41, from asst prof to assoc prof, 41-48, prof vet microbiol & pub health & head dept, 48-73, assoc dean col vet med, 70-74, prof vet microbiol, 73-81, actg dean, 79, EMER PROF, UNIV MINN, ST PAUL, 81- *Concurrent Pos:* Nat res coun, Nat Acad Sci, 67-72; adv, Animal & Plant Health Insp Serv, USDA, 72-75 & 83-84, Nat Poultry Improv Plan, USDA, 50-72 & 78-84. *Honors & Awards:* Animal Health Award, USDA, 86- *Mem:* US Animal Health Asn; Am Vet Med Asn; Poultry Sci Asn; Conf Res Workers Animal Dis. *Res:* Salmonella infections and respiratory diseases of poultry; mycoplasma infections and enteric diseases of turkeys. *Mailing Add:* 1443 Raymond Ave St Paul MN 55108

POMEROY, GORDON ASHBY, chemical computer simulations, for more information see previous edition

POMEROY, JOHN S, b Bethlehem, Pa, May 7, 29; m 57; c 4. GEOLOGY. *Educ:* Lehigh Univ, BA, 51. *Prof Exp:* Geologist, NAm Geol Bibliog Unit, US Geol Surv, 53-55, geologist, Mineral Deposits Br, 55, geologist, Alaska Geol Br, Photogeol Sect, 56-64, geologist, Br Regional Geol Ky, 64-68, supvry geologist, Wash Tech Reports Unit, 68-70, geologist, Br Atlantic Environ Geol, Beltsville, Md, 70-75, geologist, Br Eastern Regional Geol, Reston, Va, 75-89; RETIRED. *Honors & Awards:* Super Serv Award, Dept Interior, 89. *Res:* Photogeologic reconnaissance of sedimentary, igneous and metamorphic terrains; reconnaissance geologic mapping of southeastern Alaska and the Colorado Plateau; quadrangle geologic mapping in Kentucky and Massachusetts; applications of geology to environmental studies; landslide susceptibility studies in western Pennsylvania, southeastern Ohio and Maryland. *Mailing Add:* 1568 Katella Way Escondido CA 92027-3638

POMEROY, LAWRENCE RICHARDS, b Sayre, Pa, June 2, 25; m 52; c 2. ECOLOGY. *Educ:* Univ Mich, BS, 47, MS, 48; Rutgers Univ, PhD(zool), 51. *Prof Exp:* Res assoc, Bur Biol Res, Rutgers Univ, 51-52, asst res specialist, NJ Oyster Res Lab, 52-54; from asst prof to assoc prof zool, Marine Inst, 54-60, assoc prof, Univ, 60-65, prof zool, 66-80, ALUMNI FOUND DISTINGUISHED PROF ZOOL, UNIV GA, 80- *Concurrent Pos:* Assoc prog dir, environ biol, NSF, 66-67. *Honors & Awards:* Hutchinson Medal, Am Soc Limnol Oceanog, 87; Huntsman Medal, Bedford Inst Oceanog, Can, 89. *Mem:* AAAS; Am Soc Limnol & Oceanog (pres, 85-86); Am Geophys Union; Oceanog Soc. *Res:* Microbiol food web of the ocean; energetics and cycles of elements in marine and aquatic ecosystems. *Mailing Add:* Inst of Ecol Univ of Ga Athens GA 30602

POMEROY, RICHARD JAMES, b Prince George, BC, Oct 23, 37; m 64; c 2. VEHICLE DISPLAY TECHNOLOGY. *Educ:* Univ BC, BASc, 61, MASc, 63; Cambridge Univ, PhD(appl mech), 68. *Prof Exp:* Fel stess analysis, Univ Col, Swansea, Wales, 68-69; sr engr, random data analysis, Comput Devices Co, 69-75; mgr, transp systs, Glenayre Electronics Ltd, 75-85; vpres, DSL Dynamic Sci Ltd, 85-90; PRES, SEAMOUNT TECHNOLOGIES INC, 90- *Honors & Awards:* James Clayton Prize, Inst Mech Engrs, 69. *Mem:* Inst Elec & Electronics Engrs; Inst Mech Engr UK; Can Soc Mech Eng. *Res:* Application of microcomputer and data communications technology to freight railroad operations; computerized railroad traffic management systems, vehicle information displays and automatic location devices. *Mailing Add:* Seamount Technologies Inc 901-2280 Bellevue Ave West Vancouver BC V7V 1C6 Can

POMEROY, ROLAND KENNETH, b London, Eng, Oct 20, 45; Can citizen; m 82; c 1. INORGANIC CHEMISTRY, ORGANOMETALLIC CHEMISTRY. *Educ:* Imperial Col Sci & Technol, London, Eng, BSc, 67; Univ Alta, PhD(chem), 71. *Prof Exp:* Fel chem, Univ BC, 72-74; res assoc, Univ Alta, 74-75; asst prof, Univ Sask, 75-76; from asst prof, to assoc prof, 76-89, PROF CHEM, SIMON FRASER UNIV, 89- *Mem:* Royal Soc Chem; Am Chem Soc; Can Inst Chem. *Res:* Synthesis and characterization of organometallic compounds, especially those involving derivatives of iron, ruthenium and osmium carbonyls. *Mailing Add:* Dept Chem Simon Fraser Univ Burnaby BC V5A 1S6 Can

POMES, ADRIAN FRANCIS, b Hattiesburg, Miss, Apr 18, 20; m 47; c 6. CHEMISTRY, CARBOHYDRATES. *Educ:* Loyola Univ, La, BS, 40; Tulane Univ, MS, 47. *Prof Exp:* Jr chemist, Breaux Bridge Sugar Coop, 40; jr chemist, Southern Regional Res Lab, USDA, 41-47; res chemist, Drackett Co, 47-49; fel, Mellon Inst, 49-57; chemist & zein & fermentation specialist, Moffett Tech Ctr, Corn Prod, Div CPC NAm, CPC Int, 57-75, sr appln spec, 75-82; RETIRED. *Concurrent Pos:* Mem bd dirs, Omni Res Found, 63-68, comdr reserve police, Downers Grove, Ill, 68-79, Red Cross First Aid Inst, 54-82; Vol, VITA, 89- *Mem:* Master Brewers Asn Am. *Res:* Applications of vegetable proteins in paper and textiles; brewing and food applications of corn syrup, starch and dextrose; zein applications. *Mailing Add:* RR 1 Box 82A Custer OK 73639

POMIAN, RONALD J, b Philadelphia, Pa, Feb 26, 57. PHYSICS, SOLID STATE PHYSICS. *Educ:* Pa State Univ, BS, 79, MS, 82. *Prof Exp:* RES DEVELOP, AIL, DIV OF EATON CORP, 82- *Mem:* Inst Elec & Electronics Engrs; Am Phys Soc. *Mailing Add:* 172 Oakfield Ave Dix Hills NY 11746

POMILLA, FRANK R, b Brooklyn, NY, Oct 1, 26; m 53; c 4. ATOMIC PHYSICS, SCIENCE EDUCATION. *Educ:* Fordham Univ, BS, 48, MS, 49, PhD(atomic scattering), 63. *Prof Exp:* Instr math, St John's Univ, NY, 49-51, instr physics, 51-53, from asst prof to assoc prof, 53-64, actg chmn dept, 51-52 & 60; res physicist res dept, Grumman Aircraft Eng Corp, 64-67; assoc prof physics, 67-70, chmn dept natural sci, 72-73, coordr physics & phys sci, 70-72, NSF-Alfred P Sloan Found dir educ res projs, 69-72, PROF PHYSICS, YORK COL, NY, 71-, DIR SCI & MATH CTR, 86- *Concurrent Pos:* NSF fac fel, 60-61; assoc prof, 68-71, prof physics, doctoral fac, City Univ NY, 71-; proj dir, SST progs, NSF, 73-76, 79 & 81, CCSS progs, 69-72 & CAUSE prog, 77-80, 81-84, Minority Inst Sci Prog grants, US Dept Educ, 82-85, NSF teacher enhancement grant, 87-90. *Mem:* Inst Elec & Electronics Eng; Am Phys Soc; Am Asn Physics Teachers; Sigma Xi. *Res:* Atomic scattering; lunar and planetary atmospheres. *Mailing Add:* Dept Physics York Col Jamaica NY 11451

POMMERSHEIM, JAMES MARTIN, b Pittsburgh, Pa, Dec 5, 37; m 62; c 2. CHEMICAL ENGINEERING. *Educ:* Univ Pittsburgh, BS, 60, MS, 62, PhD(chem eng), 69. *Prof Exp:* From asst prof to assoc prof, 65-76, PROF CHEM ENG, BUCKNELL UNIV, 76- *Mem:* Am Inst Chem Engrs. *Res:* Transport phenomena; modeling; reaction analysis; quasi-steady-state processes; cement hydration. *Mailing Add:* Dept Chem Eng Bucknell Univ Lewisburg PA 17837

POMMERVILLE, JEFFREY CARL, b Santa Barbara, Calif, Mar 18, 48; m 73. CELL BIOLOGY, MICROBIOLOGY. *Educ:* Univ Calif, Santa Barbara, BA, 71, MA, 73, PhD(cell biol), 76. *Prof Exp:* Postdoctoral fel, Univ Ga, 76-81; asst prof teaching & res, Tex A&M Univ, 81-89; PROF, GLENDALE COL, 89- *Concurrent Pos:* Fel, Monsanto Chem Co, 83; teacher/scholar award, Tex A&M Univ, 88; prof, Glendale Community Col, 89-; adj fac, Ariz State Univ W, 90- *Mem:* Am Soc Cell Biol; Am Soc Microbiol; AAAS. *Res:* Cell communication in fungi; currently writing textbook on cell biology; author numerous articles. *Mailing Add:* Biol Dept Glendale Col 6000 W Olive Ave Glendale AZ 85302

POMONIS, JAMES GEORGE, b Santa Fe, NMex, Aug 11, 32; m 56; c 5. ORGANIC CHEMISTRY. *Educ:* Univ Denver, BS, 55, MS, 61. *Prof Exp:* Res asst chem, Denver Res Inst, 57-58; res chemist, Lasdon Found Res Inst Chemother, 59-63; res chemist, Geigy Chem Corp, 64-65; RES CHEMIST METAB & RADIATION RES LAB, USDA, 65- *Concurrent Pos:* Vis scientist & consult, Entomol Res Dept Biol, Demokritos, Nuclear Res Ctr, Athens, Greece, 81. *Mem:* Am Chem Soc; Sigma Xi. *Res:* Synthesis of steroids and heterocyclic compounds; isolation and identification of natural products from insects; synthesis and mass spectrometry of branched and straight chain hydrocarbons, thiophenes, and fatty acid esters. *Mailing Add:* Biosci Res Lab Agr Res SVC-USDA Fargo ND 58105

POMRANING, GERALD C, b Oshkosh, Wis, Feb 23, 36; m 61; c 2. REACTOR PHYSICS, TRANSPORT THEORY. *Educ:* Univ Wis, BS, 57; Delft Technol Univ, cert, 58; Mass Inst Technol, PhD(nuclear eng), 62. *Prof Exp:* Physicist & group leader, Vallecitos Atomic Lab, Gen Elec Co, 62-64; physicist & group leader, Gen Atomic Div, Gen Dynamics Corp, 64-66, staff mem & physicist, 66-68, group leader, 68-69; staff scientist & asst to pres & vpres, Sci Applns Int Corp, 69-76; PROF, SCH ENG & APPLIED SCI, UNIV CALIF, LOS ANGELES, 76- *Concurrent Pos:* Vis scientist, Brookhaven Nat Lab, 66; vis staff mem, Los Alamos Sci Lab, 68; vis lectr, Univ Calif, Los Angeles, 73-74; vis scientist, Nuclear Res Ct, Karlsruhe, W Germany, 82-83, 89; vis prof, Univ Bologna, Italy, 90. *Honors & Awards:* Mark Mills Award, Am Nuclear Soc, 63. *Mem:* Fel AAAS; fel Am Nuclear Soc; fel Am Phys Soc; Soc Indust & Appl Math; Math Asn Am; Am Math Soc. *Res:* Transport phenomena; reactor physics; radiative transfer; radiation hydrodynamics; applied mathematics. *Mailing Add:* Univ Calif 38-137H Eng IV 405 Hilgard Ave Los Angeles CA 90024-1597

PONARAS, ANTHONY A, b New York, NY, May 12, 47. ORGANIC CHEMISTRY. *Educ:* Columbia Col, NY, AB, 67; Columbia Univ, PhD(chem), 72. *Prof Exp:* Fel chem, Swiss Fed Inst Technol, 73-75 & Univ Calif, Berkeley, 75-77; asst prof, Univ Md, 77-81; ASSOC PROF, CATH UNIV, 81- *Res:* Synthetic organic chemistry; stereospecific synthesis; pericyclic reactions. *Mailing Add:* Chem Dept Cath Univ Washington DC 20064

PONCHA, RUSTOM PESTONJI, b Bombay, India; US citzen. INORGANIC CHEMISTRY, PHYSICAL CHEMISTRY. *Educ:* Univ Bombay, BSc, 47, MSc, 49; Northeastern Univ, PhD(chem), 68. *Prof Exp:* Res chemist, Amzel Ltd, Ind, 50-55 & Bhabha Atomic Res Ctr, 55-62; teaching asst chem, Northeastern Univ, 62-67; res chemist, 67-84, SR RES CHEMIST, ALLIED-SIGNAL CORP, 84- *Mem:* Am Chem Soc. *Res:* Nuclear chemistry; foam fractionation; separation of isotopes; ion exchange; development of new alkali products and processes; water treatment research; metallization of plastics; printed circuits research; organic and inorganic fluorine compounds. *Mailing Add:* Allied Signal Corp 20 Peabody St Buffalo NY 14210

POND, DAVID MARTIN, b Washington, DC, Sept 14, 42; m 67; c 3. ORGANIC CHEMISTRY, PHOTOCHEMISTRY. *Educ:* Col William & Mary, BS, 64; Univ SC, PhD(org chem), 68. *Prof Exp:* Prof NJ Turro fel, Columbia Univ, 68-70; sr res chemist, Tenn Eastman Co, 70-77, res assoc, 77-83, supt, 83-85, staff asst, 86, mgr, 86-87, DIR, TENN EASTMAN CO, 87- *Mem:* Am Chem Soc. *Res:* Organic synthesis; synthetic photochemistry; homogeneous catacysis. *Mailing Add:* 1333 Dupont Kingsport TN 37664

POND, GEORGE STEPHEN, b Victoria, BC, Aug 16, 40; m 68; c 3. PHYSICAL OCEANOGRAPHY. *Educ:* Univ BC, BSc, 62, PhD(phys oceanog), 65. *Prof Exp:* Nat Res Coun Can res fel phys oceanog, Nat Inst Oceanog, Eng, 65-66 & exchange scientist, Inst Atmospheric Physics, Acad Sci, USSR, 67; from asst prof to assoc prof phys oceanog, Ore State Univ, 67-71; assoc prof, 71-77, PROF PHYSICS & OCEANOG, UNIV BC, 77- *Mem:* Am Geophys Soc; Can Meteorol & Oceanog Soc. *Res:* Coastal circulation; air-sea interaction; turbulence and turbulent fluxes in the air-sea boundary layer. *Mailing Add:* Dept Oceanog Univ of BC Vancouver BC V6T 1W5 Can

POND, JUDSON SAMUEL, b Minneapolis, Minn, July 15, 22; m 59. INORGANIC CHEMISTRY. *Educ:* Univ Minn, BMechE, 43, BBA, 55, PhD(inorg chem), 64. *Prof Exp:* Asst, Univ Minn, 57-61, instr, 62-63; instr, Univ Mont, 63-64, asst prof, 64-66; res assoc inorg chem, Univ Chicago, 66-67; assoc prof, Lane Community Col, 67-68; sr instr inorg chem, Univ Ore, 68-77; exec asst to chmn, Dept Biol Sci, Univ Pittsburgh, 77-84; ASST DIR, VOLLUM INST ADVAN BIOMED RES, ORE HEALTH SCI UNIV, PORTLAND, ORE, 84- *Mem:* Am Chem Soc; Sigma Xi. *Res:* Sulfur-amine chemistry; inorganic reaction mechanisms. *Mailing Add:* 1470 Forest Dr SW Beaverton OR 97007

POND, THOMAS ALEXANDER, b Los Angeles, Calif, Dec 4, 24; m 58; c 2. NUCLEAR PHYSICS. *Educ:* Princeton Univ, AB, 47, AM, 49, PhD(physics), 53. *Prof Exp:* Instr physics, Princeton Univ, 51-53; from asst prof to assoc prof, Wash Univ, 53-62; chmn dept, 62-68, actg pres, 70, 75 & 78, prof physics, 62-82, exec vpres, 67-80, EMER PROF PHYSICS, STATE UNIV NY, STONY BROOK, 82-; actg pres, 90, PROF PHYSICS, EXEC VPRES & CHIEF ACAD OFFICER, RUTGERS UNIV, NJ, 82- *Mem:* Am Phys Soc. *Res:* Positron processes; beta-decay. *Mailing Add:* Old Queens Bldg New Brunswick NJ 08903

POND, WILSON GIDEON, b Minneapolis, Minn, Feb 16, 30; m 53; c 1. GROWTH & DEVELOPMENT, PRENATAL NUTRITION. *Educ:* Univ Minn, BS, 52; Okla State Univ, MS, 54, PhD(animal nutrit), 57. *Prof Exp:* From asst prof to prof animal sci, Cornell Univ, 57-78; res animal scientist, US Meat Animal Res Ctr, 78-90, RES ANIMAL SCIENTIST, CHILDREN'S NUTRIT RES CTR, AGR RES SERV, USDA, 90- *Concurrent Pos:* Adj prof, Div Nutrit Sci, Cornell Univ, 63-78, Dept Animal Sci, Univ Nebr, 78-90 & Dept Pediat, Baylor Col Med, 90-; vis prof, Danish Agr Res Lab, Copenhagen, 63-64; vis scientist, Ctr Int Agr Tropics, 70; ed-in-chief, J Animal Sci, 76-78. *Honors & Awards:* Am Feed Mfg Award, Am Soc Animal Sci, 69, Gustav Bohstedt Trace Mineral Nutrit Award, 79 & Morrison Award, 85. *Mem:* Hon fel Am Soc Animal Sci (vpres, 80, pres, 81); Am Inst Nutrit; Soc Exp Biol Med; NY Acad Sci; fel AAAS. *Res:* Effects of maternal nutrition on fetal and postnatal growth and development; role of dietary lipids and genetic background on cholesterol metabolism; use of the pig as an animal model for human infant nutrition and metabolism. *Mailing Add:* Nutrit Res Ctr USDA Agr Res Serv 1100 Bates St Houston TX 77030-2600

PONDER, BILLY WAYNE, b Chatham, La, Nov 4, 33; m 58; c 2. ORGANIC CHEMISTRY. *Educ:* La Polytech Inst, BS, 56; Iowa State Univ, MS, 59, PhD(org chem), 60. *Prof Exp:* Sr res chemist, Aerospace Res Labs, Wright-Patterson AFB, Ohio, 61-64; asst prof, 64-69, ASSOC PROF CHEM, UNIV ALA, 69-, ASSOC CHMN DEPT, 81- *Concurrent Pos:* Res chemist, Jefferson Chem Co, 60-61; US Army res grant, 65-69; NASA res grant, 67-71; chmn, Health Careers Off, 74-; Exxon Educ Found grant, 75. *Mem:* Am Chem Soc; Royal Soc Chem; Sigma Xi (pres, 73). *Res:* New synthetic methods; stereochemistry; organic reaction mechanisms. *Mailing Add:* Dept Chem Univ Ala Tuscaloosa AL 35487

PONDER, FELIX, JR, b Quitman, Ga, Oct 7, 46; m 69; c 2. PLANT SCIENCE, SOIL SCIENCE. *Educ:* Ft Valley State Col, BS, 69; Tuskegee Inst, MS, 71; Southern Ill Univ, PhD(bot), 78. *Prof Exp:* Soil conservationist, Soil Conserv Serv, USDA, 69; res asst soil sci, Tuskegee Inst, 69-71; soil scientist, Forest Serv, Nat Forest Syst, 71-75, RES SOIL SCIENTIST, N CENT FOREST EXP STA, USDA, 75- *Mem:* Am Soc Agron; Soil Sci Soc Am; Soc Am Foresters; Sigma Xi. *Res:* Growing black walnut faster on old field sites; assessing the contributions of mycorrhizal fungi; fertilization; physical, biological, and chemical soil mechanisms associated with root and shoot growth and dieback in relation to regeneration; below ground competition between roots, other flora, fauna, and the impacts of naturally occuring plant produced chemical inhibitors to determine their impact on tree growth. *Mailing Add:* N Cent Forest Exp Sta 1-26 Aug Bldg Univ Mo Columbia MO 65211

PONDER, HERMAN, b Light, Ark, Jan 31, 28; m 47; c 2. MINERALS ENGINEERING, MINERALOGY. *Educ:* Univ Mo, AB, 55, PhD(geol), 59. *Prof Exp:* Asst, Univ Mo, 55-56, asst instr, 56; geologist, A P Green Fire Brick Co, 57; geologist & mineralogist, Shell Develop Co, 58; res engr & mineralogist, A P Green Fire Brick Co, Mo, 59-61, group leader, 61-62, lab mgr, 62-63; sr proj engr, 63-66, mgr mining div, 66, dir res, 66-70, pres, Colo Sch Mines Res Inst, 70-85; PRES, ATI EXPLOR, 85-; VPRES & DIR, NORTHERN COPPER CO, 85- *Concurrent Pos:* Geologist, US Geol Surv, 55-57; Dir & chmn bd, Golden State Bank. *Mem:* Am Inst Mining, Metall & Petrol Eng; Am Inst Prof Geol. *Res:* Mineral exploration and beneficiation; mining engineering. *Mailing Add:* 1919 Mt Zion Dr Golden CO 80401

PONDROM, LEE GIRARD, b Dallas, Tex, Dec 26, 33; m 61. EXPERIMENTAL ELEMENTARY PARTICLE PHYSICS. *Educ:* Southern Methodist Univ, BS, 53; Univ Chicago, MS, 56, PhD(physics), 58. *Prof Exp:* Instr physics, Columbia Univ, 60-63; assoc prof, 63-69, PROF PHYSICS, UNIV WIS-MADISON, 69- *Concurrent Pos:* Guggenheim fel, 71-72; secy, Users Exec Comt, Nat Accelerator Lab, Ill; mem, Univs Res Asn, 73-76 & 82-85; mem, High Energy Adv Comt, Brookhaven Nat Lab, 74-76; fel, Japan Soc Prom Sci, 81; mem, High Energy Phys Adv Panel, Us Dept Energy, 81-84 & 86-88 & Phys Adv Comt, Nat Sci Found, 81-; chmn, High Energy Phys Rev Comt, Argonne Univ Asn, 80-81 & Phys Adv Comt, Fermi Nat Accelerator Lab, 81-82; assoc ed, Phys Rev Lett, 84-86; chmn, rev comt physics, Lawrence Berkeley Lab, 83-85; chmn, comput coord comt, Fermilab, 85-87; mem, Int Comt Future Accelerators, 84-89; chmn, Div Particles & Fields, Am Phys Soc, 87; chmn, Users Orgn, SSC, 87-89. *Mem:* Fel Am Phys Soc. *Res:* Experimental studies in elementary particle physics using electronic techniques; general instrumentation problems in elementary particle physics. *Mailing Add:* Dept Physics Univ Wis Madison WI 53706

PONELEIT, CHARLES GUSTAV, b Collinsville, Ill, Oct 14, 40; m 63; c 2. PLANT GENETICS, PLANT BREEDING. *Educ:* Univ Ill, Champaign, BS, 62, MS, 64; Purdue Univ, Lafayette, PhD(plant genetics), 68. *Prof Exp:* From asst prof to assoc prof, 67-83, PROF PLANT BREEDING & GENETICS, UNIV KY, 83- *Mem:* Am Soc Agron; Crop Sci Soc Am; Sigma Xi. *Res:* Biochemical and physiological genetics of corn; corn breeding. *Mailing Add:* Dept of Agron Univ of Ky Lexington KY 40546

PONESSA, JOSEPH THOMAS, b New York, NY, Sept 10, 41; m 65; c 3. ENERGY CONSERVATION, DESIGN CONSTRUCTION. *Educ:* St Francis Col, BS, 62; Univ NDak, MS, 65; Loyola Univ Chicago, PhD(physiol), 69. *Prof Exp:* NIH fel, Hahnemann Med Col, 69-71, asst prof physiol, 69-81; ASSOC PROF HOUSING & ENERGY CONSERVATION, COOP EXTEN, COOK COL RUTGERS UNIV, 81- *Concurrent Pos:* Foreign Exchange res fel, Semmelweis Med Univ, Budapest, 73. *Mem:* Sigma Xi; Asn Housing Educators. *Res:* Cardiovascular control; cerebral blood flow; autonomic nervous system control. *Mailing Add:* 215 Locust St Moorestown NJ 08057

PONG, SCHWE FANG, b Hsin-Chu, Taiwan, June 4, 36; m 64; c 3. PHARMACOLOGY. *Educ:* Nat Taiwan Univ, BS, 62; Univ Tenn, MS, 66; Univ Miss, PhD(pharmacol), 70. *Prof Exp:* Res specialist neurochem & neuropharmacol, Tex Res Inst Ment Sci, 70-71; NIMH fel neurobiol, Inst Psychiat Res, Sch Med, Ind Univ, Indianapolis, 71-73, res assoc, 73-74; sr res pharmacologist, 74-78, res assoc, 78-83, GROUP LEADER, NORWICH EATON PHARMACEUT, 83- *Mem:* Am Soc Pharmacol Exp Therapeut; Am Soc Med Chem; Am Pharmaceut Asn; Am Soc Neurochem. *Res:* Evaluation and mechanism of efficacy and toxicity study of CNS, muscle relaxant, cardiac and other drugs; drug metabolism. *Mailing Add:* Norwich Eaton Pharmaceut Norwich NY 13815-0191

PONG, TING-CHUEN, b Hong Kong, June 27, 57. COMPUTER VISION, ARTIFICIAL INTELLIGENCE. *Educ:* Univ Wis-Eau Claire, BS, 78; Va Polytechnic Inst & State Univ, MS, 81, PhD(computer sci), 84. *Prof Exp:* Asst prof, 84-90, ASSOC PROF COMPUTER SCI, UNIV MINN, 90- *Concurrent Pos:* Assoc ed, Pattern Recognition J, 91- *Honors & Awards:* First Place Winner, Ann Pattern Recognition Soc Award, 90. *Mem:* Inst Elec & Electronic Engrs; Inst Elec & Electronic Engrs Computer Soc; Asn Comput Mach. *Res:* Field of computer vision care; image segmentations; stereo and motion analysis; shape-from-x; object recognition; design and analysis of parallel algorithms for pattern recognition and image analysis. *Mailing Add:* 4-192 EE-C Sci Bldg 200 Union St SE Minneapolis MN 55455

PONG, WILLIAM, b Cincinnati, Ohio, Dec 26, 27; c 4. PHYSICS OF SENSORS FOR BIOMEDICAL APPLICATIONS. *Educ:* Univ Cincinnati, PhD(physics), 54. *Prof Exp:* Res physicist, Electronics Div, Baldwin Co, 54-58; asst prof physics, Xavier Univ, Ohio, 58-60; assoc prof, 60-68, PROF PHYSICS, UNIV HAWAII, 68- *Mem:* Am Phys Soc. *Res:* Nonlinear acoustics; UV photoemission from solids and materials; physics of sensors for biomedical applications. *Mailing Add:* Dept Physics & Astron Univ Hawaii Honolulu HI 96822

PONGRATZ, MORRIS BERNARD, b O'Neill, Nebr, Apr 10, 42; m 67; c 2. SPACE PHYSICS. *Educ:* Creighton Univ, BS, 64; Univ Md, College Park, PhD(space physics), 72. *Prof Exp:* Res assoc space physics, Inst Fluid Dynamics, Univ Md, 72-73; sci specialist neutron physics, Los Alamos Div, EG&G Co, 73-75; STAFF MEM SPACE PHYSICS, LOS ALAMOS NAT LAB, 75- *Mem:* Am Geophys Union; AAAS. *Res:* Active experiments in the magnetosphere and study of the magnetospheric electric and magnetic fields using high explosive shaped charge barium plasma injections; plasma physics associated with such injections. *Mailing Add:* 1438 Canada Los Alamos NM 87544

PONKA, PREMYSL, b Prague, Czech, Sept 21, 41; Can citizen; m 73; c 2. HEMATOLOGY, PATHOPHYSIOLOGY. *Educ:* Charles Univ, Prague, Czech, MD, 64, PhD(physiol & pathophysiol), 69. *Prof Exp:* Demonstr pathophysiol, Fac Med, Charles Univ, Prague, 61-64, lectr, 64-68, asst prof, 68-79; assoc prof physiol, 79-87, assoc prof med, 83-87, PROF PHYSIOL & MED, MCGILL UNIV, MONTREAL, CAN, 87- *Concurrent Pos:* Prin investr, Med Res Coun Can, 80-, external reviewer, 88-; staff investr, Lady Davis Inst, Sir Mortimer B Davis-Jewish Gen Hosp, Montreal, 84-; vis prof, Lab Molecular Genetics, Univ Paris, 87-88; ad hoc mem, NIH Hemat Study Sect, 88, mem, 90-; mem adv bd, Can Porphyria Found, 88-; ed, Iron Transp & Storage, 90. *Mem:* Am Soc Hemat; NY Acad Sci. *Res:* Cellular iron metabolism; erythroid differentiation; heme synthesis regulation; regulation of ferrochelatase and transferrin receptor gene expression; role of inflammatory lymphokines in ferritin synthesis; role of iron in cell proliferation; iron chelation. *Mailing Add:* Lady Davis Inst Jewish Gen Hosp 3755 Cote Ste Catherine Rd Montreal PQ H3T 1E2

PONNAMPERUMA, CYRIL ANDREW, b Galle, Ceylon, Oct 16, 23; m 55; c 1. BIO-ORGANIC CHEMISTRY. *Educ:* Univ Madras, BA, 48; Univ London, BSc, 59; Univ Calif, Berkeley, PhD(chem), 62. *Hon Degrees:* DSc, Univ Sri Lanka, 78. *Prof Exp:* Res asst, Lawrence Radiation Lab, Univ Calif, Berkeley, 60-62; scientist, Ames Res Ctr, NASA, 63-65, chief chem evolution br, 65-71; PROF CHEM & DIR LAB CHEM EVOLUTION, UNIV MD, COLLEGE PARK, 71- *Concurrent Pos:* Prin investr org anal, Apollo Prog, 67; distinguished vis prof, Indian Atomic Energy Comn, 67; fac sci, Univ Sorbonne, 69; adj prof, Boston Col, 69; dir UNESCO Inst Early Evolution, Ceylon, 70; ed, J Molecular Evolution, 70-72; ed, Origins Life, 73-; distinguished prof, Univ Md, 78; Foreign mem, Indian Nat Sci Acad, 78. *Mem:* AAAS; Am Chem Soc; Royal Soc Chem; fel Royal Inst Chem; Int Soc Study Origin Life. *Res:* Chemical evolution exobiology; geochemistry; space sciences. *Mailing Add:* Dept of Chem Univ of Md College Park MD 20742

PONNAPALLI, RAMACHANDRAMURTY, b Hyderabad, India, Mar 15, 33. MATHEMATICS, STATISTICS. *Educ:* Andhra Univ, India, BA, 53, MSc, 55; Univ Calif, Berkeley, PhD(statist), 64. *Prof Exp:* Lectr statist, Andhra Univ, India, 55-60; asst, Univ Calif, Berkeley, 60-64; asst prof, Case Inst Technol, 64-65; reader, Osmania Univ, India, 65-69, head dept, 66-69; asst prof math, Univ Alta, 69-76, Univ West Ont, 76-78; asst prof math, Univ Windsor, 78-82; assoc prof, State Col Technol, Univ NY, 83-88; SR STATISTICIAN, BIOMETRICS DIV, FOOD & DRUG ADMIN, ROCKVILLE, MD, 88- *Concurrent Pos:* Vis prof, Dept Statist, Univ Calif, Riverside, 82-83. *Mem:* Inst Math Statist; Can Math Cong; Indian Statist Inst; Indian Sci Cong Asn. *Mailing Add:* 8101 Brucar St Gaithersburg MD 20877

PONOMAREV, PAUL, b Czestochowa, Poland, Oct 5, 44; US citizen; div; c 2. MATHEMATICS. *Educ:* Princeton Univ, AB, 66; Yale Univ, PhD(math), 70. *Prof Exp:* Asst prof math, Johns Hopkins Univ, 70-74; asst prof, 75-77, ASSOC PROF, OHIO STATE UNIV, 77- *Concurrent Pos:* Vis mem, Inst Advan Study, 74-75; vis prof, Goettingen Univ, 75; mem, Math Sci Res Inst, Berkeley, 87. *Mem:* Am Math Soc. *Res:* Number theory; algebra. *Mailing Add:* Dept Math Ohio State Univ 231 W 18th Ave Columbus OH 43210-1174

PONS, STANLEY, US citizen. SPECTROELECTROCHEMISTRY, ELECTROCHEMICAL KINETICS. *Educ:* Wake Forest Univ, BS, 65; Univ Southampton, PhD(chem), 79. *Prof Exp:* Asst prof, Oakland Univ, 78-80; asst prof chem, Univ Alberta, 80- DEPT CHEM, OAKLAND, DEPT CHEM, 88- *Mem:* Int Soc Electrochem; Electrochem Soc; Am Chem Soc; Can Inst Chem. *Res:* Spectroelectrochemistry; electrochemical kinetics; simulation of diffusion processes; electrochemistry of metallporphyrins; electroorganic synthesis; infrared structure of the electrode-electrolyte intertake. *Mailing Add:* Dept Chem Univ Utah Salt Lake City UT 84112

PONSETI, IGNACIO VIVES, b Balearic Islands, Spain, June 3, 14; nat; m 61; c 1. MEDICINE. *Educ:* Univ Barcelona, BA, 30, MD, 36. *Hon Degrees:* Dr, Univ Barcelona, 84. *Prof Exp:* Instr, 44-46. asst, 46-48, from asst prof to assoc prof, 48- 57, PROF ORTHOP SURG, UNIV IOWA, 57- *Mem:* Am Acad Orthop Surgeons; Am Col Surgeons; Am Acad Cerebral Palsy; Am Orthop Asn; Orthop Res Soc (past pres). *Res:* Congenital and developmental skeletal deformities; factors influencing the growth of epiphyseal plates of bone; congenital anomalies of the skeleton. *Mailing Add:* Dept Orthop Surg Univ Hosp Iowa City IA 52242

PONTARELLI, DOMENIC JOSEPH, b Philadelphia, Pa, May 26, 12; m 47; c 2. OBSTETRICS & GYNECOLOGY. *Educ:* Hahnemann Med Col, MD, 40. *Prof Exp:* From instr to assoc prof, 47-66, PROF OBSTET & GYNEC, HAHNEMANN MED COL & HOSP, 66- *Concurrent Pos:* Asst attend obstet & gynec, Philadelphia Gen Hosp, 53-64, chief serv, 64-66; chief obstet & gynec, Jeanes Hosp, Fox Chase. *Mem:* Fel Am Col Surgeons; fel Am Col Obstetricians & Gynecologists; Asn Profs Gynec & Obstet; AMA. *Mailing Add:* Dept Obstet-Gynec Hahnemann Univ Sch Med Broad & Vine St Philadelphia PA 19102

PONTAU, ARTHUR E, b Philadelphia, Pa, July 27, 51. PLASMA MATERIALS INTERACTIONS, ION BEAM ANALYSIS. *Educ:* Univ Calif, San Diego, BA, 73; Univ Ill, MS, 75, PhD(physics), 78. *Prof Exp:* PHYSICIST, SANDIA NAT LABS, 78- *Mem:* Am Phys Soc; AAAS. *Res:* Analysis of materials appropriate for use in in-vessel components for magnetic confinement fusion reactors; characterization of plasma edge parameters. *Mailing Add:* Div 8347 Sandia Labs Box 969 Livermore CA 94551

PONTE, JOSEPH G, JR, b New Bedford, Mass, Aug 9, 25; m 56; c 3. FOOD SCIENCE. *Educ:* Northwestern Univ, BA, 56; Univ Minn, St Paul, MS, 58. *Prof Exp:* Baking technologist, Am Inst Baking, 54-56; res chemist, ITT Continental Baking Co, 58-65, sr res chemist, 65-70, res supvr, 70-74, mgr res serv, 74-75; PROF GRAIN SCI & INDUST, KANS STATE UNIV, 75- *Concurrent Pos:* Chmn, Comt on Cereal Prod, Nat Res Coun, 74-78, Milling & Baking Div, Am Asn Cereal Chemists, 89-90. *Mem:* Fel Am Asn Cereal Chem (treas, 70-72, secy, 75-77); Sigma Xi; Am Soc Bakery Eng; Inst Food Technol. *Res:* Cereal chemistry; food technology; baking technology; nutrition food products. *Mailing Add:* Dept Grain Sci & Indust Shellenberger Hall Kans State Univ Manhattan KS 66506-2201

PONTER, ANTHONY BARRIE, b Staffordshire, Eng, Sept 27, 33; m 60; c 2. CHEMICAL ENGINEERING. *Educ:* Birmingham Univ, BSc, 56, DSc, 86; Univ Manchester Inst Sci Technol, MSc, 62; Manchester Univ, PhD, 66. *Prof Exp:* Mem staff mech eng, Rubery Owens Orgn, 56-57; tech officer & plant mgr, Pilkington-Sullivan Works, 57-59; lectr, Inst Sci & Technol, Manchester Univ, 60-69; assoc prof, Univ NB, 69-72; prof & dir, Swiss Fed Inst Technol, 72-77; vis prof, Univ Aston, Eng, 77-78; dept head & prof, Mich Technol Univ, 78-83; ASST TO PRES, PROF CHEM ENGR, 84- *Mem:* fel Inst Chem Engrs UK; fel Chem Inst Can; fel Am Inst Chem Engrs; Ver Deut Ingenieure. *Res:* Distillation; absorption; condensation; wetting; surface phenomena; coalescence; nucleate boiling; diffusion; liquid flow in packed columns; boiling acoustics; liquid films; author of approximately 125 papers. *Mailing Add:* Chem Eng Dept 1983 E 24th St Cleveland State Univ Cleveland OH 44115

PONTICELLO, GERALD S, b Rochester, NY, Mar 31, 39; m 67; c 2. ORGANIC CHEMISTRY. *Educ:* Univ Toronto, BS, 61; Univ Rochester, PhD(org chem), 69. *Prof Exp:* Prod chemist, Eastman Kodak Co, 61-62; fel with Dr Henry Sobel, 69-70; SR RES CHEMIST, MERCK & CO, INC, WEST POINT, 70- *Mem:* Am Chem Soc. *Res:* Medicinal chemistry; testing of organic molecules in health related fields. *Mailing Add:* 2045 Spring Valley Rd Lansdale PA 19446-5111

PONTICELLO, IGNAZIO SALVATORE, b Rochester, NY, Mar 31, 39; m 67; c 2. ORGANIC CHEMISTRY. *Educ:* Univ Toronto, BSc, 61; Univ Rochester, PhD(org chem), 69. *Prof Exp:* SR RES CHEMIST, RES LABS, CHEM DIV, EASTMAN KODAK CO, 69- *Mem:* Am Chem Soc. *Res:* Synthesis and polymerization of new monomers for evaluation in photographic processes. *Mailing Add:* 21 Copper Woods Pittsford NY 14534

PONTINEN, RICHARD ERNEST, b Eveleth, Minn, Sept 21, 33; m 56; c 2. PHYSICS. *Educ:* Hamline Univ, BS, 55; Univ Minn, Minneapolis, PhD(solid state physics), 62. *Prof Exp:* From asst prof to assoc prof, 61-68, PROF PHYSICS, HAMLINE UNIV, 68-, CHMN DEPT, 71- *Mem:* Am Phys Soc; Am Asn Physics Teachers. *Res:* Liquids using lasers; low temperature solid state physics. *Mailing Add:* Dept of Physics Hamline Univ St Paul MN 55104

PONTIUS, DIETER J J, b Zurich, Switz, July 31, 14; US citizen; m 51. ORGANIC CHEMISTRY, ANALYTICAL CHEMISTRY. *Educ:* Univ Frankfurt, MSc, 47, PhD(chem physics & physiol chem), 50. *Prof Exp:* Ger Scientist's Res Asn scholar, Steroid Res, Endocrine Res Unit, Trier, Ger, 51-54; Univ Hosp, Hamburg, 54-55; Montreal Gen Hosp, Que, 56; asst chem, Clark Univ, 57-59; res scientist, Neuro-Endocrine Res Unit, Willowbrook State Sch, NY, 59-60; res chemist, Fairmount Chem Co, 60-69; PVT RES, 69- *Mem:* Am Chem Soc; Am Inst Chemists. *Res:* Organic synthesis; steroid biochemistry; absorption spectroscopy; development of new color reactions. *Mailing Add:* 165 E 60th St New York NY 10022

PONTIUS, DUANE HENRY, b Urbana, Ill, Feb 16, 39; m 60; c 2. SOLID STATE PHYSICS. *Educ:* Auburn Univ, BS, 62, MS, 69, PhD(physics), 72. *Prof Exp:* Fel, Auburn Univ, 72-73; asst prof physics, Univ WFla, 73-75; RES PHYSICIST, SOUTHERN RES INST, 75- *Mem:* Am Asn Physics Teachers. *Res:* Electrical charging of fine particles in aerosol suspension. *Mailing Add:* 324 Thompson Dr Gardendale AL 35071

PONTIUS, E(UGENE) C(AMERON), b Napoleon, Ohio, Mar 18, 15; m 41; c 2. CHEMICAL ENGINEERING. *Educ:* Ohio State Univ, BChE, 39, MS, 49, PhD(chem eng), 50. *Prof Exp:* Chem engr res & develop, Va Smelting Co, 39-42; asst chem eng, Ohio State Univ, 46-50; res engr, Textile Fibers Dept, E I Du Pont de Nemours & Co, Inc, 50-56, process design engr, Eng Dept, 56-69, sr process engr, Design Div, 69-70, sr design consult, Eng Dept, Design Div, 70-80; RETIRED. *Mem:* Am Chem Soc; Instrument Soc Am; fel Am Inst Chemists; Am Inst Chem Engrs; NY Acad Sci. *Res:* Process design and development; plant design; materials and science engineering. *Mailing Add:* Possum Hollow Rd No R Newark DE 19711

PONTIUS, STEVEN KENT, b Auburn, Ind, Feb 5, 45; m 71; c 2. DIFFUSION, ECONOMIC DEVELOPMENT. *Educ:* Ind Univ, AB, 67, AM, 71; Univ Minn, PhD(geog), 77. *Prof Exp:* Res fel geog, Kasetsart Univ, Thailand, 74-76; teaching fel, Univ Minn, 76-77; asst prof, 77-83, ASSOC PROF GEOG, RADFORD UNIV, 83-, CHMN DEPT, 78- *Mem:* Asn Am Geographers; Thailand Geog Soc. *Res:* Economic development in the Third World; diffusion process of agricultural technology. *Mailing Add:* Dept Geog Radford Univ Radford VA 24142

PONTOPPIDAN, HENNING, b Copenhagen, Denmark, July 24, 25; US citizen; m 49; c 3. ANESTHESIOLOGY, RESPIRATORY PHYSIOLOGY. *Educ:* Copenhagen Univ, MD, 52. *Prof Exp:* From instr to asst prof, 59-70, assoc prof, 70-75, PROF ANESTHESIA, HARVARD MED SCH, 75-; chief, Respiratory Unit, 61-85, ANESTHETIST, MASS GEN HOSP, 67- *Mem:* Am Soc Anesthesiol; Am Thoracic Soc; Soc Critical Care Med; Am Col Chest Physicians. *Res:* Acute respiratory failure; physiology of artificial ventilation. *Mailing Add:* Dept Anesthesia Mass Gen Hosp Boston MA 02114

PONTRELLI, GENE J, b Bronx, NY, Dec 25, 33; div; c 3. TEXTILE PHYSICS. *Educ:* City Col New York, BS, 58; Univ Wis, PhD(chem), 63. *Prof Exp:* Teaching asst chem, Univ Wis, 59-60; res phys chemist, Eng Physics Lab, E I du Pont de Nemours & Co, 62-65; res chemist, Textile Res Lab, 65-66, sr res chemist, 66, res supvr, 66-69; tech mgr, Fibers Div, Allied Chem Corp, 70-73; res assoc, 76-81, res fel, Textile Fibers Dept, 81-85, SR RES FEL ELECTRONICS, E I DU PONT DE NEMOURS & CO, INC, 85- *Concurrent Pos:* Woodrow Wilson fel, Esso fel & Procter & Gamble fel. *Honors & Awards:* Medal, Am Inst Chemists. *Mem:* Am Chem Soc; Am Phys Soc; Fiber Soc. *Res:* Mechanism of moisture transport in clothing as a function of fiber, yarn and fabric construction; technology forecasting for synthetic fibers; advanced composites; electro-optics. *Mailing Add:* Chestnut Run Lab E I Du Pont de Nemours & Co Inc Wilmington DE 19898

PONZIO, NICHOLAS MICHAEL, US citizen. IMMUNOLOGY. *Educ:* Seton Hall Univ, BA, 68, MS, 70; State Univ NY Brooklyn, PhD(immunol), 73. *Prof Exp:* Fel, Albert Einstein Col Med, 73-74; fel, Sch Med, NY Univ, 74-76; asst prof, Med Sch, Northwestern Univ, 76-80, assoc prof, 80-81; assoc prof, 81-86, PROF, SCH MED, UNIV MED & DENT NJ, 86- *Concurrent Pos:* Res career develop award, 79-84. *Mem:* AAAS; Sigma Xi; Am Asn Immunologists; Am Asn Pathologists; Am Asn Cancer Res; NY Acad Sci. *Res:* In vivo and in vitro analysis of lymphocyte responses to mitogens, alloantigens, tumor antigens and hapten-carrier conjugates; immunoglobulin allotype suppression; lymphocyte migration and localization in vivo. *Mailing Add:* 140 Harrison Av Westfield NJ 07090

PONZO, PETER JAMES, b Toronto, Ont, Nov 6, 34; m 60; c 1. APPLIED MATHEMATICS. *Educ:* Univ Toronto, BASc, 57, MA, 59; Univ Ill, PhD(elec eng), 64. *Prof Exp:* Lectr math, Univ Waterloo, 57-58 & 59-60; res assoc elec eng, Univ Ill, 63-64; asst prof, 64-66, assoc prof, 66-76, PROF MATH, UNIV WATERLOO, 76- *Res:* Nonlinear differential equations. *Mailing Add:* Dept of Appl Math Univ of Waterloo Waterloo ON N2L 3G1 Can

POOCH, UDO WALTER, b Berlin, WGer, Apr 20, 43; m 64; c 2. THEORETICAL PHYSICS, COMPUTER SCIENCE. *Educ:* Univ Calif, Los Angeles, BS, 65; Univ Notre Dame, PhD(theoret physics), 69. *Prof Exp:* Asst prof indust eng, Texas A&M Univ, 69-74, mem Computer Sci Consult Group, 70-72, computer consult, Tex Transp Inst, 71-73, assoc prof indust eng, 74-80, PROF, COMPUTER SCI, TEXAS A&M UNIV, 80-, MEM GRAD FAC, 69-; prof, 79-88, PROF COMPUT SCI & MEM GRAD FAC, TEX A&M UNIV, 80- *Concurrent Pos:* Pres, Micro Systs Technol Inc, College Station, 78- & Independent Computer Consults Inc, 73-; partner, Appl Comput, Bryan, 73-; mem, USAF Sci Adv Bd, 86-87; consult various co; mem, Res & Develop Coun Advert Performance, 86-87; nat lectr, Asn Comput Mach, 75 & IBM, 81. *Mem:* Am Phys Soc; Asn Comput Mach; Simulation Coun; Soc Indust & Appl Math; Opers Res Soc Am; Am Statist Asn; sr mem Inst Elec & Electronic Engrs Computer Soc; Sigma Xi. *Res:* Softwave systems for mini-maxi computers; interactive computer graphics; information storage-retrieval systems; natural language query systems; artificial intelligence; teleprocessing applications; data structures; pattern recognition; mathematical linguistics; time sharing; author of 13 books and 29 publications. *Mailing Add:* Dept Comput Sci Tex A&M Univ H R Bright Bldg College Station TX 77843-3112

POOCHIKIAN, GUIRAGOS K, b Beirut, Lebanon, 45; US citizen; m 75; c 2. PHARMACEUTICAL CHEMISTRY, MEDICINAL CHEMISTRY. *Educ:* Am Univ Beirut, BS, 70; Ohio State Univ, PhD(pharmaceut & med chem), 74. *Prof Exp:* Dir qual control lab, Scierlabs, Sci & Pharmaceut Res Labs, 74-76; formulation chemist, Anal & Prod Develop Sect, Pharmaceut Resources Br, Nat Cancer Inst, NIH, 76-80; CHEMIST, CTR DRUG EVALUATION & RES, FOOD & DRUG ADMIN, 80- *Concurrent Pos:* Res asst, Col Pharmacy, Ohio State Univ, 70-71, res assoc, 71-74. *Mem:* Am Chem Soc; Sigma Xi; NY Acad Sci; AAAS. *Res:* Product development of delivery systems; study of new formulation approaches, evaluation of dosage form problems, and development of new analytical methods pertaining to the formulation studies; structure activity relationships; manufacture, control and stability of pharmaceutical dosage forms. *Mailing Add:* 7812 Rydal Terr Rockville MD 20855-2058

POODRY, CLIFTON ARTHUR, b Buffalo, NY, July 31, 43; m 84; c 1. DEVELOPMENTAL BIOLOGY. *Educ:* State Univ NY, Buffalo, BA, 65, MA, 68; Western Reserve Univ, PhD(biol), 71. *Prof Exp:* Res assoc genetics, Univ BC, 71-72; from asst prof to assoc prof biol, Univ Calif, Santa Cruz, 72-83, NSF Develop Biol Prog dir, 82-84, chmn, Biol Dept, 86-89, actg dean natural sci, 89-90, PROF BIOL, UNIV CALIF, SANTA CRUZ, 83- *Concurrent Pos:* Mem, Gen Res Adv Comt, NIH, 77-81, Marc Rev Comt, 86-90; adv mem & fac, Headlands Indian Health Careers Prog, Univ Okla, 77-; adv mem, NSF Develop Biol Prog, 80-82, adv coun mem, 81-82. *Mem:* Genetics Soc Am; Am Soc Zoologists; Sigma Xi; Soc Develop Biol; Am Indian Sci & Eng Soc; AAAS; Soc Advan Chicanos & Native Am Sci. *Res:* Developmental genetics; cell surface proteins and morphogenesis. *Mailing Add:* Biol Dept Univ Calif Santa Cruz CA 95064

POOL, EDWIN LEWIS, b Eagle Grove, Iowa, Mar 2, 21; m 49; c 3. ANALYTICAL CHEMISTRY, ENVIRONMENTAL CHEMISTRY. *Educ:* Iowa State Univ, BS, 43, MS, 49, PhD(biophys chem), 52. *Prof Exp:* Asst prof chem, DePauw Univ, 53-54; from instr to assoc prof, 54-70, PROF CHEM, MIDDLEBURY COL, 71- *Mem:* Am Chem Soc; Sigma Xi. *Res:* Environmental analyses; analytical chemistry; colorimetric analysis; enzyme production by microorganisms. *Mailing Add:* Rogers Rd RD 4 Box 2052 Middlebury VT 05753

POOL, JAMES C T, b Wellsville, Kans, Feb 27, 37; m 56; c 2. COMPUTATIONAL SCIENCE. *Educ:* Univ Kans, AB, 59; Univ Iowa, PhD(physics), 63. *Prof Exp:* Res scientist, Sci Lab, Ford Motor Co, 63-65; NSF fel, Brandeis Univ, 64-65; asst mathematician, Appl Math Div, Argonne Nat Lab, 66-68; asst prof math, Univ Mass, 68-69; asst div dir, Argonne Nat Lab, 69-70, actg div dir, 70-71, assoc div dir res, Appl Math Div, 71-78; prog dir appl math sci, Basic Energy Sci, Dept Energy, Washington, DC, 78-80; dir, Math & Info Sci Div, Off Naval Res, Washington, DC, 80-82; exec vpres, Numerical Algorithms Group, Inc, Downers Grove, Ill, 82-88; PROF & DEPT HEAD, DEPT MATH & COMPUTER SCI, DREXEL UNIV, PHILADELPHIA, PA, 88- *Concurrent Pos:* Fel, Inst Theoret Physics, Univ Hamburg, Ger, 66-67; Swiss Nat Fund fel, Inst Theoret Physics, Univ Geneva, 67; chmn working group numerical software, Int Fed Info Processing, 74-78. *Mem:* AAAS; Soc Indust & Appl Math; Inst Math & its Appln; Asn Comput Mach; Inst Elec & Electronic Engrs Computer Soc; Am Math Soc. *Res:* Software for numerical computation. *Mailing Add:* Math & Computer Sci 206 Korman Ctr Drexel Univ Philadelphia PA 19104-2884

POOL, KARL HALLMAN, b Norristown, Pa, Aug 3, 39; m 61; c 2. ELECTROCHEMICAL CORROSION. *Educ:* Calif Inst Tech, BS, 61; Univ Wash, PhD(analytical chem), 65. *Prof Exp:* From instr to assoc prof anal chem, Wash State Univ, 55-81; SR RES SCIENTIST CORROSION, BATTELLE MEM INST, 81- *Concurrent Pos:* Scientist, Nat Bur Standards, 67; res assoc, State Univ NY, Buffalo, 72; vis prof, Univ NC, 77-78; consult, Battelle Mem Inst, 80-81. *Mem:* Am Chem Soc; Sigma Xi. *Res:* Electrochemical aspects of corrosion; laboratory management and methods development in aqueous inorganic chemical analysis. *Mailing Add:* Battelle Northwest 314 Bldg 300 Area PO Box 999 Richland WA 99352

POOL, MONTE J, b Toledo, Ohio, Nov 18, 34; m 52; c 3. METALLURGY, THERMODYNAMICS. *Educ:* Univ Cincinnati, BMetE, 58; Ohio State Univ, MS, 59, PhD(metall), 61. *Prof Exp:* From asst prof to assoc prof metall, Univ Denver, 61-68, res metallurgist, Denver Res Inst, 61-68; assoc prof mat sci & metall eng, 68-72, PROF METALL ENG, UNIV CINCINNATI, 72- *Concurrent Pos:* Vis sr scientist, Max Planck Inst Metall, Stuttgart, 76-77. *Mem:* AAAS; Am Soc Metals; Am Inst Mining, Metall & Petrol Engrs; Am Soc Eng Educ. *Res:* Alloy thermochemistry, especially binary systems; dilute metallic liquid solutions using solution calorimetry. *Mailing Add:* Dept Mat Sci & Eng Univ Cincinnati Cincinnati OH 45221-0012

POOL, PETER EDWARD, b Chicago, Ill, Dec 30, 36; m 87; c 1. CARDIOLOGY, CARDIOVASCULAR PHARMACOLOGY. *Educ:* Yale Univ, BA, 58; NY Univ, MD, 62. *Prof Exp:* Res assoc cardiol, Cardiol Br, Nat Heart Inst, 64-68; dir cardiol, Scripps Mem Hosp-Encinitas, 71-82, chief staff, 79-80; CLIN PROF MED, CARDIOL, SCH MED, UNIV CALIF, SAN DIEGO, 68-; DIR RES, N COUNTY CARDIOL RES LAB, 71- *Concurrent Pos:* Counr, Western Sect, Am Fedn Clin Res, 69-72; lectr, Off Continuing Med Educ, Univ Calif, San Diego, 77-; pres, Calif Affil & fel, Coun Clin Cardiol, Basic Sci Coun, Am Heart Asn, 80-81; consult, Pharmaceut Indust, 81- & areas of drug development. *Honors & Awards:* Res Career Develop Award, USPHS, 69; Silver Serv Medallion, Calif Affil, Am Heart Asn, 81. *Mem:* Fel Am Col Cardiol; Am Heart Asn; Am Fedn Clin Res; Am Physiol Soc; Am Soc Pharmacol & Exp Therapeut; Western Soc Clin Res. *Res:* Clinical trials in cardiovascular pharmacology, including hypertension, angina, arrhythmias, heart failure and lipid abnormalities; analysis of problems in the area of drug treatment of heart disease. *Mailing Add:* 1441 Cooper Point Circle Reno NV 89509

POOL, ROBERT MORRIS, b Sacramento, Calif, Feb 22, 40; m; c 3. VITICULTURE, PLANT PHYSIOLOGY. *Educ:* Univ Calif, Davis, BS, 62, MS, 69; Cornell Univ, PhD(pomol), 74. *Prof Exp:* Staff res assoc viticult, Univ Calif, Davis, 62-71; res asst pomol, Cornell Univ, 71-74; from asst prof to assoc prof viticult, 74-87, PROF VITICULT, NY STATE AGR EXP STA, 87- *Concurrent Pos:* Chmn adv comt, Nat Grape Germplasm Commodity, 74-; assoc ed, Am Soc Hort Sci, 76- *Mem:* Am Soc Hort Sci; Am Soc Enology; Sigma Xi. *Res:* Cultural practices in vineyards in relation to wine quality; growth regulators; grapevine growth and development. *Mailing Add:* Dept Pomol & Viticult NY State Agr Exp Sta Geneva NY 14456

POOL, ROY RANSOM, b Raleigh, NC, Oct 8, 34; m 67; c 1. VETERINARY PATHOLOGY, RADIOBIOLOGY. *Educ:* Duke Univ, BS, 57; Okla State Univ, BS, 62, DVM, 64; Univ Calif, Davis, PhD(comp path), 67. *Prof Exp:* Asst prof vet anat, Okla State Univ, 67-69; asst prof & asst res pathologist, 69-75, ASSOC PROF VET PATH & ASSOC RES PATHOLOGIST, SCH VET MED, UNIV CALIF, DAVIS, 75- *Mem:* Am Vet Med Asn; Vet Cancer Soc; Sigma Xi. *Res:* Bone pathology; radionuclide toxicity. *Mailing Add:* Dept Vet Path Univ Calif Davis CA 95616

POOL, THOMAS BURGESS, b Houston, Tex, Oct 10, 48; m 81. CELL BIOLOGY, DEVELOPMENTAL BIOLOGY. *Educ:* Sam Houston Univ, BS, 70, MA, 73; Univ Va, PhD(biol), 76. *Prof Exp:* Res assoc, Univ Tex Health Sci Ctr, San Antonio, 76-78, asst prof anat, 78-86. *Honors & Awards:* Outstanding Res Award, Sigma Xi, 73. *Mem:* Am Soc Cell Biol; Am Asn Anatom. *Res:* Ionic regulation of mitosis in cancer and development; hormone-directed cell differentiation; cell cycle regulation. *Mailing Add:* 11418 Whisper Breeze San Antonio TX 78230

POOL, WILLIAM ROBERT, b Ft Lauderdale, Fla, May 14, 37; m 58; c 4. TOXICOLOGY, PHARMACOLOGY. *Educ:* Fla State Univ, BS, 60, MS, 63, PhD(physiol), 66, Am Bd Toxicol, dipl, 80. *Prof Exp:* Teacher, high sch, Fla, 60-63; res assoc, Inst Molecular Biol, Fla State Univ, 66-67; sr pharmacologist, 67-70, res group chief, 71-73, sect head, 73-75, assoc dir toxicol, Hoffmann-La Roche, Inc, 75-76; dri toxicol, G D Searle & Co, 76-84; dir toxicol, Pfizer Labs, France, 84-88; DIR SAFETY ASSESSMENT MED RES DIV, AM CYANAMID, 88- *Mem:* AAAS; NY Acad Sci; Am Soc Pharmacol & Exp Therapeut; Europ Soc Toxicol; Soc Toxicol; Am Asn Pharmaceut Scientists. *Res:* Drug safety evaluation. *Mailing Add:* Med Res Div Am Cyanamid Co Lederle Pearl River NY 10965

POOL, WINFORD H, JR, b Atlanta, Ga, Oct 2, 26; m 50; c 5. RADIOLOGY. *Educ:* Mercer Univ, BS, 49; Med Col Ga, MD, 52. *Prof Exp:* Assoc prof, 64-68, PROF RADIOL, MED COL GA, 68-, CHMN DEPT, 74-, CHIEF DIAG RADIOL, 70- *Mem:* Fel Am Col Radiol; Radiol Soc NAm; AMA; Am Roentgen Ray Soc; Asn Univ Radiologists. *Mailing Add:* Med Col Ga Augusta GA 30901

POOLE, ANDREW E, b Eng, Aug 4, 35; m 62; c 3. HUMAN GENETICS, PEDODONTICS. *Educ:* London Hosp, BDS, 60; Royal Col Surgeons, LDS, 60; Univ Rochester, MS, 67, PhD(genetics, anat), 71. *Prof Exp:* Instr pedodont, London Hosp Med Sch, 61-62; from asst prof to assoc prof, 70-85, DIR, CRANIOFACIAL DISORDERS TEAM, HEALTH CTR, UNIV CONN, 82-, PROF PEDIAT DENT, 85- *Mem:* Am Soc Human Genetics; Int Asn Dent Res; AAAS; Sigma Xi; Soc Craniofacial Genetics. *Res:* Biochemical human genetics; genetic polymorphisms; inborn error of metabolism; birth defects; dental caries prevention; anthropology and dental health of the Aleuts; craniofacial aspects of growth hormone deficiency; Prader Willi Syndrome and Turners Syndrome; genetic aspects of cleft lip and palate. *Mailing Add:* Dept Pediat Dent Univ Conn Health Ctr Farmington CT 06032

POOLE, ANTHONY ROBIN, b Southend-on-sea, Eng, Oct 1, 39; m 64; c 3. CELL BIOLOGY, IMMUNOLOGY. *Educ:* Reading Univ, BSc, 61, PhD(biochem, physiol), 69, DSc, 85. *Hon Degrees:* DSc, 85. *Prof Exp:* Res scientist microbiol, Unigate, 61-63; res scientist biochem, Marie Curie Mem Found, 63-69 & Royal Free Hosp Med Sch, London, 69-70; res scientist cell biol, Strangeways Res Lab, Cambridge Univ, 70-77, scientist, dept path, 75-77; DIR, JOINT DIS LAB, SHRINERS HOSP CRIPPLED CHILDREN, 77-; PROF EXP SURG & ASSOC MEM IMMUNOL, MICROBIOL & MED, MCGILL UNIV, 81- *Concurrent Pos:* Res assoc ed, J Bone Foot Surg Am, 77-90; mem res grants panel, Can Arthritis Soc, 78-83; mem, Manpower Develop Comt, Arthritis Soc Can, 83-84; mem, orthop & musculoskeletal study sect, Div Res Grants, NIH, 86-90; mem, res grants comt clin invest, Med Res Coun Can, 87-89; chmn, organizing & prog comt, First Combined Meeting Orthop Res Socs of US, Japan & Can, 91. *Mem:* Am Soc Cell Biol; Biochem Soc; Orthopaedic Res Soc; Am Soc Immunol; Am Rheumatism Asn; Brit Corrective Tissue Soc; Histochem Soc; Can Orthop Res Soc (pres, 88-89); Can Rheumatism Asn. *Res:* Immunology and biochemistry of cartilage and arthritis; particularly proteoglycans, structure, type II collagen degradation and autoimmunity; calcification of cartilage. *Mailing Add:* Joint Dis Lab Shriners Hosp 1529 Cedar Ave Montreal PQ H3G 1A6 Can

POOLE, CHARLES PATTON, JR, b Panama City, Panama, June 7, 27; US citizen; m 53; c 5. PHYSICS. *Educ:* Fordham Univ, BS, 50, MS, 52; Univ Md, PhD(physics), 58. *Prof Exp:* Physicist, Westinghouse Elec Corp, 52-53; physicist, Gulf Res & Develop Co, Pa, 58-64; assoc prof, 64-66, PROF PHYSICS, UNIV SC, 66- *Concurrent Pos:* Ed, J Magnetic Resonance Rev. *Honors & Awards:* Jesse Beams Award, Am Phys Soc, 80. *Mem:* Fel Am Phys Soc; Philos Sci Asn; Int Soc Magnetic Resonance (treas); Groupment Ampere. *Res:* Solid state physics; magnetic resonance; Clifford algebras; chemical physics; super conductivity. *Mailing Add:* Dept of Physics Univ of SC Columbia SC 29208

POOLE, COLIN FRANK, b Huyton, Lancashire, UK, Apr 7, 50; m 86. CHROMATOGRAPHY, LIQUID ORGANIC SALTS. *Educ:* Univ Leeds, Eng, BSc, 71; Univ Bristol, Eng, MSc, 72; Univ Keele, PhD(chem), 75. *Prof Exp:* Postdoctoral, Univ Aslon, Birmingham, Gt Brit, 75-76; postdoctoral, Univ Ghent, Belgium, 76-77; res assoc, Univ Houston, 77-80; assoc prof, 84-90, PROF ANAL CHEM, WAYNE STATE UNIV, 90- *Concurrent Pos:* Regional ed, J Planar Chromatog. *Honors & Awards:* Tswett Medal, 85; Jubilee Medal, 91. *Mem:* Am Chem Soc; Chromatog Soc. *Res:* Separation of complex mixtures using chromatographic techniques, the chemistry of liquid organic salts and mass spectrometry; gas and supercritical fluid; liquid and thin-layer chromatography. *Mailing Add:* Rm 171 Chem Wayne State Univ Detroit MI 48202

POOLE, DONALD RAY, b Pontiac, Mich, Jan 8, 32; m 54. ROCKET PROPELLANT CHEMISTRY. *Educ:* Cent Mich Col Educ, BS, 54; Univ Maine, MS, 56; Univ Ore, PhD, 61. *Prof Exp:* Asst chem, Univ Maine, 54-56; asst anal chem, Univ Ore, 56-57; chemist, US Naval Ord Test Sta, 60-66; sr res chemist, 66-71, dir chem technol, 71-76, SR STAFF SCIENTIST, ROCKET RES CO, OLIN DEFENSE SYSTS, 76- *Mem:* Am Chem Soc; Sigma Xi. *Res:* Combustion and chemistry of rocket propellants; catalysts; hydrazine chemistry; chemical processing. *Mailing Add:* Box 525 Woodinville WA 98072

POOLE, DORIS THEODORE, b Greenville, SC, Mar 9, 23; m 47; c 2. BIOCHEMISTRY, PHARMACOLOGY. *Educ:* Winthrop Col, BS, 44; Univ NC, PhD(biochem), 62. *Prof Exp:* Chem analyst, Standard Oil Development Co, 44-46; asst prof zool, Madison Col, 58-59; from instr to asst prof, 62-71, assoc prof, 71-88, EMER PROF PHARMACOL, SCH MED, UNIV NC, CHAPEL HILL, 88- *Mem:* Am Soc Pharmacol & Exp Therapeut. *Res:* Determination of intracellular pH of various tissues from the distribution of the weak acid 5, 5-dimethyl-2, 4-oxazolidinedione; effects produced by various metabolic inhibitors and environmental changes on intracellular pH; ion transport. *Mailing Add:* Dept Pharmacol Univ NC Sch Med 2108 Ridgefield Dr Chapel Hill NC 27514

POOLE, GEORGE DOUGLAS, b Miami, Fla, Nov 30, 42; m 63. MATHEMATICS. *Educ:* Kans State Teachers Col, BSEduc, 64; Colo State Univ, MS, 66; Tex Tech Univ, PhD(math), 72. *Prof Exp:* Instr math, Washburn Univ, Topeka, 66-67 & Tex Tech Univ, 67-68; asst prof, 71-76, ASSOC PROF MATH, EMPORIA STATE UNIV, 76- *Mem:* Math Asn Am; Am Math Soc; Sigma Xi. *Res:* Abelian group theory; numerical linear algebra; matrix theory. *Mailing Add:* 762 Dunlap Rd Box 247 Kingsport TN 37663-9533

POOLE, H K, b Hagerstown, Md, Apr 16, 31; m 57; c 2. AIR POLLUTION, ENVIRONMENTAL BIOLOGY. *Educ:* Gettysburg Col, AB, 53; Univ Md, MS, 57. *Prof Exp:* Asst zool, Univ Md, 53-56; agr res scientist & geneticist, Agr Res Serv, USDA, 56-67, res zoologist, Bee Res Lab, 67-74; dep air qual control officer, Pima County Air Qual Control Dist, Pima County Health Dept, 74-80; environ coordr, Davis Monthan AFB, Ariz, 80-88; RETIRED. *Mem:* Air Pollution Control Asn; Am Inst Biol Sci. *Res:* Parthenogenesis; cytology and cytogenetics of birds and honey bees; reproductive biology of honey bees; biological effects of air pollution; environmental compliance and planning. *Mailing Add:* 4620 N Avenida Pocacca Tucson AZ 85749-9524

POOLE, JOHN ANTHONY, b New York, NY, Apr 19, 32; m 60; c 2. PHYSICAL CHEMISTRY. *Educ:* St John's Univ, NY, BS, 54; Univ Alta, PhD(chem), 62. *Prof Exp:* Fel chem, Univ Alta, 62 & Univ Rochester, 63-64; asst prof, 64-69, ASSOC PROF CHEM, TEMPLE UNIV, 69- *Mem:* Am Asn Univ Prof; Inter-Am Photochem Soc. *Res:* Photochemistry; spectroscopy. *Mailing Add:* Dept Chem Temple Univ Philadelphia PA 19122

POOLE, JOHN TERRY, b Washington, DC, Oct 4, 37; m 60; c 2. MATHEMATICS. *Educ:* Univ NC, BS, 59; Univ Md, MA, 62, PhD(math), 65. *Prof Exp:* Asst prof math, Fla State Univ, 64-68; assoc prof, Clemson Univ, 68-69; PROF MATH, FURMAN UNIV, 69- *Mem:* Math Asn Am. *Res:* Complex analysis; geometric function theory. *Mailing Add:* Dept Math Furman Univ Greenville SC 29613

POOLE, JOHN WILLIAM, b Nanticoke, Pa, Nov 28, 31; m 54; c 1. PHARMACEUTICAL CHEMISTRY. *Educ:* Philadelphia Col Pharm, BS, 54; Temple Univ, MS, 56; Univ Wis, PhD(phys pharm), 59. *Prof Exp:* Instr chem, Temple Univ, 54-59; res pharmaceut chemist, McNeil Labs, Inc, 59-65; mgr, Explor Pharm Sect, Wyeth Labs Inc, 65-81; mgr drug delivery/exp concepts, 81-87, DIR, PROD DEVELOP/DRUG DELIVERY SYSTS, MCNEIL CONSUMER PROD CO, 86- *Mem:* Am Chem Soc; Am Pharmaceut Asn; NY Acad Sci; Am Asn Pharmaceut Scientists. *Res:* Physiochemical studies of drug substances as a means of developing efficient dosage forms; biopharmaceutics; drug delivery systems. *Mailing Add:* 3002 Arch Rd Norristown PA 19401

POOLE, RICHARD TURK, JR, b Memphis, Tenn, June 16, 31; m 55; c 4. ORNAMENTAL HORTICULTURE. *Educ:* Principia Col, BS, 53; Southwestern La Inst, BS, 59; Univ Fla, MSA, 61, PhD(bot), 65. *Prof Exp:* Res assoc ornamental hort, Univ Fla, 61-65, asst prof, 65-66; asst prof, Univ Hawaii, 66-68; from asst prof to assoc prof, 68-75, PROF ORNAMENTAL HORT, UNIV FLA, 75-, PLANT PHYSIOLOGIST, AGR RES CTR, APOPKA, FLA, 68- *Mem:* Am Soc Hort Sci. *Res:* Production of tropical indoor foliage plants. *Mailing Add:* AREC 2807 Binion Rd Apopka FL 32703

POOLE, ROBERT WAYNE, b Oakland, Calif, Nov 11, 44; m 69, 89. ECOLOGY, STATISTICS. *Educ:* Cornell Univ, BS, 66, PhD, 69. *Prof Exp:* Asst Prof, Div Biomed Sci, Brown Univ, 74-80; ENTOMOLOGIST, SYSTEMATIC ENTOMOLOGY LAB, AGR RES SERV, 80- *Res:* Systematics of moths of the family Noctuidae. *Mailing Add:* Systematic Entomology Lab c/o Nat Museum Nat Hist Wash DC 20560

POOLE, RONALD JOHN, b Leicester, Eng, Mar 16, 36; m 63, 90; c 1. PLANT PHYSIOLOGY, BIOPHYSICS. *Educ:* Univ Birmingham, BSc, 57, PhD(plant physiol), 60. *Prof Exp:* Cabot fel, 60-62; NATO fel, 62-63; asst lectr biophys, Univ Edinburgh, 63-64, lectr, 64-65; asst prof bot, 65-71, assoc prof, 71-86, PROF BIOL, MCGILL UNIV, 86- *Concurrent Pos:* Assoc ed, Botanica Acta, 88-91. *Mem:* Am Soc Plant Physiol. *Res:* Membrane transport in plants; transport in membrane vesicles from plasma membrane and tonoplast; purification and reconstitution of transport systems; role of membranes and transport in stress tolerance of plants. *Mailing Add:* Dept Biol McGill Univ 1205 Docteur Penfield Ave Montreal PQ H3A 1B1 Can

POOLE, WILLIAM HOPE, b Kimberley, BC, Can, Feb 20, 27; m 54; c 2. REGIONAL GEOLOGY, RESEARCH MANAGEMENT. *Educ:* Univ BC, BASc, 49; Princeton Univ, PhD(geol), 56. *Prof Exp:* head, Appalachian Sect, 67-79, head, Correlation & Standards Subdiv, 74-79, spec proj, 79-82, field geologist, 52-82, COORDR, FED-PROV MINERAL DEVELOP AGREEMENTS, GEOL SURV CAN, 82- *Honors & Awards:* Medal of Merit, Alberta Soc Petrol Geologists, 69; W J Wright Award, CIM New Brunswick Br, 87. *Mem:* Fel Geol Soc Am; fel Geol Asn Can. *Res:* Field geology; structure; igneous and metamorphic petrology; stratigraphy; economic geology. *Mailing Add:* Geol Surv Can 601 Booth St Ottawa ON K1A 0E8 Can

POOLE, WILLIAM KENNETH, b Cheatham Co, Tenn, Apr 9, 39; c 3. STATISTICS. *Educ:* Austin Peay State Univ, BS, 61; Univ NC, MPH, 63, PhD(statist), 68. *Prof Exp:* Statistician, 67-75, dir, 76-82, VPRES, STATIST METHODOLOGY & ANAL CTR, RESEARCH TRIANGLE INST, 83- *Concurrent Pos:* Mem, Am Heart Asn. *Mem:* Am Statist Asn; Soc Clin Trials. *Res:* Statistical methodology and applications; multivariate analysis; stochastic processes. *Mailing Add:* Research Triangle Inst PO Box 12194 Research Triangle Park NC 27709

POOLER, FRANCIS, JR, b Waltham, Mass, Mar 30, 26; m 52; c 4. METEOROLOGY. *Educ:* Mass Inst Technol, BS, 49; Pa State Univ, PhD, 72. *Prof Exp:* Meteorologist, US Weather Bur, 50-52 & Wallace E Howell Assocs, 52; meteorologist, US Weather Bur, 53-55, Robert A Taft Sanit Eng Ctr, 55-69; RES METEOROLOGIST, NAT ENVIRON RES CTR, ENVIRON PROTECTION AGENCY, 69- *Mem:* Am Meteorol Soc; Geophys Union. *Res:* Atmospheric turbulence and diffusion; transport of pollutants in the atmosphere; micrometeorology. *Mailing Add:* 2131 Meares Rd Chapel Hill NC 27514

POOLER, JOHN PRESTON, b Waltham, Mass, June 10, 35; m 65; c 2. PHOTOBIOLOGY, PHOTOSENSITIZATION. *Educ:* Brown Univ, AB, 62; Duke Univ, PhD(physiol), 67. *Prof Exp:* Training prog, Duke Univ, 67-69; asst prof, 69-74, ASSOC PROF PHYSIOL, EMORY UNIV, 74- *Mem:* Soc Photobiol; Biophys Soc; Am Asn Univ Profs; AAAS. *Res:* Photosensitization; membrane biophysics; phototoxicity at cell membrane level. *Mailing Add:* Dept Physiol Emory Univ Atlanta GA 30322

POOLEY, ALAN SETZLER, b Madison, Wis, Dec 1, 38; m 65; c 3. SCANNING ELECTRON MICROSCOPY. *Educ:* Univ Wis-Madison, BS, 61, MS, 64, PhD(zool), 68. *Prof Exp:* Fel, Imp Cancer Res Fund, London, 68-70; scientist, Searle Res Labs, 70-72; res assoc, Univ Conn, 72-74; ASSOC MICROSCOPIST, PEABODY MUS, YALE UNIV, 74- *Mem:* Electron Micros Soc Am; Sigma Xi. *Res:* Scanning electron microscopy; energy dispersive x-ray analysis; mineralized biological structures. *Mailing Add:* Peabody Mus Yale Univ New Haven CT 06520

POON, BING TOY, b San Francisco, Calif, Feb 26, 24. CHEMISTRY, RESEARCH ADMINISTRATION. *Educ:* Univ Calif, BS, 48; Univ Colo, MS, 50, PhD(org chem), 51. *Prof Exp:* Res asst, Univ Colo, 48-51; res chemist, Solvay Process Div, Allied Chem & Dye Corp, 51-52, Nitrogen Div, 52-55, sr res chemist, Allied Chem Corp, 55-65; ORG CHEMIST, DIV EXP THERAPEUT, WALTER REED ARMY INST RES, 65- *Mem:* AAAS; Am Soc Mass Spectrometry; Am Chem Soc; Sigma Xi. *Res:* Diene synthesis; chlorinations; industrial organic chemicals and nitrogen derivatives; process development; applications research; polymers and resins. *Mailing Add:* Walter Reed Army Inst Res-ET Washington DC 20307-5100

POON, CALVIN P C, b China, Nov 8, 35; US citizen; m 63; c 2. WATER POLLUTION CONTROL, WATER RESOURCES. *Educ:* Nat Taiwan Univ, BS, 58; Univ Mo, MS, 60; Univ Ill, PhD(sanit eng), 64. *Prof Exp:* Design engr wastewater treatment, J Stephen Watkins Consult Eng, 60; proj engr water treatment, Gannet Fleming Corddry & Carpenter Eng Inc, Pa, 63-65; from asst prof to assoc prof, 65-75, PROF POLLUTION CONTROL, UNIV RI, 75-, DIR WATER RESOURCES CTR, 82- *Concurrent Pos:* Mem, Nat Sci Adv Comt, US Coast Guard, 72-74; consult, US Environ Protection Agency & var other co; UN adv & consult, Pollution Mgt Prog, Nat Taiwan Univ & Rep China govt; vis prof, Univ Edinburgh, Scotland; Fulbright Sr Scholar, Egypt. *Mem:* Water Pollution Control Fedn; Am Soc Civil Engrs; Am Asn Environ Eng Prof; Inst Chinese-Am Engrs. *Res:* Industrial wastewater treatment processes; biological treatment, water quality control and advanced treatment technology. *Mailing Add:* Dept Civil Eng Univ RI Kingston RI 02881

POON, CHI-SANG, b Hong Kong. MATHEMATICAL MODELING, CONTROL THEORY. *Educ:* Univ Hong Kong, BSc, 75, MPhil, 77; Univ Calif Los Angeles, PhD(eng), 81. *Prof Exp:* From asst prof to assoc prof elec eng, NDak State Univ, 81-89; PRIN RES SCIENTIST, MASS INST TECHNOL, 89- *Concurrent Pos:* Vis assoc prof biomed eng, Mass Inst Technol, 88-89; consult, NIH, 89 & NSF, 90. *Honors & Awards:* Harold Lamport Award, Biomed Eng Soc, 83. *Mem:* Inst Elec & Electronics Engrs; Biomed Eng Soc; AAAS; Neural Networks Soc; Am Physiol Soc. *Res:* Neural mechanisms underlying the control of breathing during muscular exercise; elucidation of the optimal and adaptive behaviors in the maintenance of respiratory homeostasis. *Mailing Add:* Biomed Eng Ctr Rm 20A-126 Mass Inst Technol Cambridge MA 02139

POON, TING-CHUNG, Hong Kong, Mar 19, 55; m 80; c 2. OPTICAL IMAGE PROCESSING, PHYSICAL OPTICS. *Educ:* Univ Iowa, BS, 77, MS, 79, PhD(elec eng), 82. *Prof Exp:* Vis asst prof elec eng, Univ Iowa, 82-83; asst prof, 83-89, ASSOC PROF ELEC ENG, VA POLYTECH INST & STATE UNIV, 89-, DIR, OPTICAL IMAGE PROCESSING LAB, 88- *Mem:* Soc Photo-Optical Instrumentation Engrs; Optical Soc Am; sr mem Inst Elec & Electronic Engrs. *Res:* Acousto-optics; hybrid (optical and electronic) image processing; optical scanning holography; pattern recognition; vision systems. *Mailing Add:* 515 Cranwell Circle Blacksburg VA 24060

POOR, HAROLD VINCENT, b Columbus, Ga, Oct 2, 51; m 73; c 2. ELECTRICAL ENGINEERING. *Educ:* Auburn Univ, BEE, 72, MS, 74; Princeton Univ, MA, 76, PhD(elec eng), 77. *Prof Exp:* From asst prof & res asst prof to assoc prof & res assoc prof, Coord Sci Lab, Univ Ill, Urbana-Champaign, 77-84, prof elec eng & res prof, 84-90; PROF ELEC ENG, PRINCETON UNIV, 90- *Concurrent Pos:* Consult var indust & govt orgn; assoc ed, Trans Automatic Control, Inst Elec & Electronics Engrs, 81-82, Trans Info Theory, 82-85 & Maths of Control Signals & Systs, 87-; acad visitor, Imperial Col, London Univ, 85; vis prof, Univ Newcastle, Australia, 87. *Mem:* Fel Inst Elec & Electronic Engrs; Sigma Xi. *Res:* Statistical signal processing; signal detection and estimation. *Mailing Add:* Dept Elec Eng Princeton Univ Princeton NJ 08544

POORE, AUBREY BONNER, b LaGrange, Ga, Feb 4, 45; m 66; c 2. APPLIED MATHEMATICS, OPTIMIZATION. *Educ:* Ga Inst Technol, BS, 67, MS, 69; Calif Inst Technol, PhD(appl math), 73. *Prof Exp:* Asst prof math, State Univ NY, Buffalo, 72-74; asst prof, 74-77, assoc prof, 77-82, PROF MATH, COLO STATE UNIV, 82-, PROF ELEC ENG, 89- *Concurrent Pos:* Assoc prof math, Math Res Ctr, Univ Wisc, 79-80. *Mem:* Soc Indust & Appl Math; Soc Natural Philos. *Res:* Differential equations; applied mathematics; numerical analysis; computational optimization and control. *Mailing Add:* Dept Math Colo State Univ Ft Collins CO 80523

POORE, EMERY RAY VAUGHN, b Mumfordville, Ky, July 30, 37; m 59; c 2. NUCLEAR PHYSICS. *Educ:* Duke Univ, BS, 59; Univ NC, PhD(physics), 66. *Prof Exp:* From res assoc to asst prof nuclear physics, Duke Univ, 66-70; asst dir, Comput Res Ctr, Univ SFla, 70-72; MEM STAFF, LOS ALAMOS SCI LAB, 72- *Res:* Nuclear spectroscopy with direct reactions. *Mailing Add:* Los Alamos Sci Lab MP-6 MS-H852 Los Alamos NM 87544

POORE, JESSE H, JR, b Louisville, Ky. COMPUTER SCIENCE. *Educ:* Ga Inst Technol, PhD(comput sci), 70. *Prof Exp:* Assoc Prof Math & Dir Comput Ctr, Fla State Univ, 71-, Assoc Dean Grad Studies & Res, 80-; AT DEPT INFO TECHNOL, GA INST TECHNOL. *Concurrent Pos:* Proj mgr comput-based educ, NSF, 74-75; vis prof, Univ Simon Bolivar; pres, Exec Loan Prog, President's Reorgn Prog. *Mem:* Asn Comput Mach. *Mailing Add:* Dept Comput Sci Univ Tenn 105 Ayres Hall Knoxville TN 37996

POORMAN, LAWRENCE EUGENE, b Ft Wayne, Ind, July 27, 26; m 51; c 2. PHYSICS. *Educ:* Ball State Univ, BS, 50; Ind Univ, MS, 53, EdD(sci educ), 67; Purdue Univ, MS, 64. *Prof Exp:* Teacher, Salem Pub Schs, Ind, 50-51, Jamestown Schs, 51-52, Edinburg Schs, 52-54, Columbus Schs, 54-60; instr physics, Lab Sch, Ind Univ, Bloomington, 60-66; res assoc proj physics, Harvard Univ, 66-67; from asst prof to prof physics, Ind State Univ, 67-91; RETIRED. *Concurrent Pos:* Field consult, Proj Physics, Harvard Univ, 64-66, 67-69. *Mem:* Am Asn Physics Teacher; Nat Educ Asn; Nat Sci Teachers Asn; Sch Sci & Math Asn. *Res:* Computer assisted instruction in physics; multi-media systems approach in physics teaching; programmed instruction used in physics; curricular development in physical science courses. *Mailing Add:* 51 Lakeview Dr Terre Haute IN 47803

POOROOSHASB, HORMOZD BAHMAN, b Tehran, Feb 23, 36; m 55; c 2. FOUNDATION ENGINEERING. *Educ:* Univ Manchester, BSc, 58; Cambridge Univ, PhD(soil sci), 61. *Prof Exp:* Fel soil mech, Emmanuel Col, 61-63; from asst prof civil eng to prof, Univ Waterloo, 63-69; dir, Iranian Geotech Inst, 69-71; consult engr geotech eng, Kantab Consult Engrs, 71-79; PROF CIVIL ENG, CONCORDIA UNIV, 79- *Concurrent Pos:* Vis prof, Danish Geotech Inst, 69-70; pres, Iranian Geotech Soc, 71-73; sr vis scientist, McGill Univ, 81- *Res:* Constitutive relationships of granular media soil structure interaction; numerical techniques applied to geotechnical engineering; tailings deposits. *Mailing Add:* Dept Civil Eng Concordia Univ Sir G Williams 1455 DeMaisonneuve Montreal PQ H3G 1M8 Can

POORVIN, DAVID WALTER, b Brooklyn, NY, Aug 25, 46; m 70; c 3. CLINICAL RESEARCH. *Educ:* Hunter Col, BA, 67; Rutgers Univ, PhD(physiol), 72. *Prof Exp:* From res asst to res assoc physiol, Rutgers Univ, 67-73; res physiologist, 73-75, group leader pharmacol, Lederle Labs, 75-77; asst dir new drug develop, Pfizer Pharmaceut, 77-79, assoc dir, 79-81; assoc med dir, 81-82, sr assoc dir, 83-86, DIR, SCHERING-PLOUGH, 86- *Mem:* NY Acad Sci; Am Heart Asn; Sigma Xi. *Res:* Clinical research management of cardiovascular drugs. *Mailing Add:* Schering-Plough Galloping Hill Rd Kenilworth NJ 07033

POOS, GEORGE IRELAND, b Sandusky, Ohio, Apr 15, 23; m 76; c 3. ORGANIC CHEMISTRY, RESEARCH ADMINISTRATION. *Educ:* Va Mil Inst, BS, 47; George Wash Univ, BS, 47; Univ Ill, PhD(chem), 50. *Prof Exp:* Res chemist, Merck & Co, Inc, 50-56; sr chemist, McNeil Pharmaceut, 56-57, asst dir chem res, 57-65, assoc dir chem res, 65-66, dir chem res, 66-67, exec dir res, 67-71, vpres sci affairs, 71-79, vpres planning & corp develop, 79-83; vpres, Int Johnson & Johnson, 83-85; RETIRED. *Concurrent Pos:* Consult, 85- *Honors & Awards:* Am Inst Chemists Medal; Johnson Medal. *Mem:* Am Chem Soc; AAAS; Sigma Xi; NAm Soc Corp Planning; Soc Chemists & Indust. *Res:* Total synthesis of adrenal steroids; steroid chemistry; heterocycles; bicyclic compounds; stereochemistry; medicinal chemistry. *Mailing Add:* 150 Camp Hill Rd Ft Washington PA 19034

POOTJES, CHRISTINE FREDRICKA, b Montclair, NJ, Sept 19, 27. MICROBIOLOGY. *Educ:* Rutgers Univ, BS, 49, PhD(microbiol), 64; Purdue Univ, MS, 55. *Prof Exp:* Lab asst microbiol, Wallace & Tiernan, Inc, NJ, 49-52; lab asst, Schering Corp, 52-53; res asst biochem, Fels Res Inst, Antioch Col, 54-55; res assoc microbiol, Rutgers Univ, 55-62; from asst prof to assoc prof microbiol, Col Sci, Pa State Univ, University Park, 68-91; RETIRED. *Mem:* Am Soc Microbiol. *Res:* Bacterial viruses, especially those associated with the facultative autotroph Hydrogenomonas facilis; lysogeny in Agrobacterium tumefaciens. *Mailing Add:* 646 Franklin St State College PA 16803

POOVAIAH, BACHETTIRA WTHAPPA, b Coorg, India, May 10, 43; m 70; c 2. PLANT PHYSIOLOGY, MOLECULAR BIOLOGY. *Educ:* Karnataka Univ, India, BS, 64; Utah State Univ, MS, 68, PhD(plant physiol), 70. *Prof Exp:* Lectr agr chem, Agr Col, Dharwar, 64-65; fel, Mich State Univ, 70-71 & Purdue Univ, 71-75; from asst prof to assoc prof, 75-82, PROF HORT, WASH STATE UNIV, 82- *Mem:* Am Soc Plant Physiol; Am Soc Hort Sci; Sigma Xi; AAAS; Japanese Soc Plant Physiologists; Fedn Europ Soc Plant Physiologists. *Res:* Signal transduction in plants: calcium messenger system and auxin-regulated gene expression. *Mailing Add:* Dept Hort/Landscape Archit Wash State Univ Pullman WA 99164-6414

POP, EMIL, b Tirgu Mures, Romania, Aug 12, 39; m 64; c 1. SYNTHESES & STUDY OF BRAIN-SPECIFIC CHEMICAL DELIVERY SYSTEMS. *Educ:* Babes-Bolyai Univ, Romania, BS & MS, 61; Inst Chem, PhD(org chem), 73. *Prof Exp:* Chemist, Chem Pharmaceut Res Inst, Romania, 62-65, res scientist, 65-78, sr res scientist & group leader, 78-83; postdoctoral res assoc, Univ Fla, Gainesville, 83-86; sr res scientist, 86-87, group leader, 87-88, ASSOC DIR CHEM DEVELOP, PHARMATECH, ALACHUA, 89- *Concurrent Pos:* Vis res chemist, Rugjer Boskovic Inst, Yugoslavia, 71-72. *Honors & Awards:* N Teclu Award for Chem, Romanian Acad Sci, Bucharest, 80. *Mem:* Am Chem Soc; AAAS; Am Asn Pharmaceut Scientists; fel Am Inst Chemists; NY Acad Sci; Int Union Pure & Appl Chem; Int Soc Quantum Biol & Pharmacol. *Res:* Syntheses and pharmacological study of novel drugs, drug analogs and site-specific chemical drug delivery systems; pharmaceutical chemistry; theoretical, molecular orbital, studies of organic molecules and reaction mechanisms. *Mailing Add:* 810 SW 51st Way Gainesville FL 32607

POPE, BARBARA L, b Kelowna, BC, Feb 3, 45; m 65. ADHESION MOLECULES, IMMUNE STIMULATION. *Educ:* Univ BC, BSc, 73, PhD(microbiol & immunol), 78. *Prof Exp:* Res asst, Vancouver Gen Hosp, 68-74; teaching asst immunol, Univ BC, 74-78; fel, NIH, 78-80; from asst prof to assoc prof pharmacol & immunol, Dalhousie Univ, 80-89; RES MGR, R W JOHNSON PHARMACEUT RES INST, 89- *Concurrent Pos:* Vis prof med microbiol & immunol, Univ SFla, 86-87; assoc prof microbiol, Dalhousie Univ, 88-89. *Mem:* Am Asn Immunol; Can Soc Immunologists; Pharmacol Soc Can; Int Soc Immunopharmacol. *Res:* Establishment of technologies which lead to the discovery and development of small molecular weight compounds which will act as antagonists of adhesion molecules or as immunostimulants; development of novel anti-inflammatory, immunosuppressive or immunostimulatory drugs. *Mailing Add:* R W Johnson Pharmaceut Res Inst 19 Green Belt Dr Don Mills ON M3C 1L9

POPE, BERNARD G, b Caerleon, Wales, Oct 7, 43; m 67; c 2. LEPTON PAIR PRODUCTION & TRANSVERSE MOMENTUM PHENOMENA IN HADRONIC INTERACTIONS. *Educ:* Univ Birmingham, UK, BS, 65; Columbia Univ, MA, 67, PhD(physics), 71. *Prof Exp:* Res fel, NP Div, European Orgn Nuclear Res, 71-73, staff physicist, EP Div, 73-77; asst prof physics, Princeton Univ, 77-82; PROF PHYSICS, MICH STATE UNIV, EAST LANSING, 83- *Concurrent Pos:* ISR coordinator, European Orgn Nuclear Res, 76; mem, High Energy Adv Comt, Brookhaven Nat Lab, 79-82; mem, Stanford Linear Accelerator Ctr, 82-83; secy & treas, Exp Prog Adv Comt, Am Physical Soc, 86-88. *Mem:* Am Inst Physics; Am Physical Soc; European Physical Soc. *Res:* High energy physics, with particular emphasis on Lepton Pair production and high transverse momentum phenomena in Hadronic interactions. *Mailing Add:* Dept Physics Michigan State Univ East Lansing MI 48824-1113

POPE, BILL JORDAN, b Salt Lake City, Utah, Sept 12, 22; m 43; c 4. CHEMICAL ENGINEERING. *Educ:* Univ Utah, BA, 47; Univ Wash, MS, 49, PhD, 59. *Prof Exp:* Proj chem engr, Utah Oil Refining Co, 51-58; from assoc prof to prof, 58-78, chmn dept, 66-70, EMER PROF CHEM ENG, BRIGHAM YOUNG UNIV, 78- *Concurrent Pos:* Lectr grad sch, Univ Utah, 53-54; prof & actg pres, Abadan Inst Technol, Iran, 59-62; consult, Hercules Co, 63-66 & Keysor Chem Co, 64-66; vpres, Megadiamond Corp, 66-69, exec vpres, 70-73, pres & chmn bd, 73-78; pres, US Synthetic Corp, 78-87, Chief exec officer & chmn bd, 87- *Mem:* Am Inst Chem Engrs; Am Chem Soc; Nat Soc Prof Engrs; Am Soc Eng Educ. *Res:* Chemical kinetics; heat transfer; process design; ultra high pressure synthesis of diamond and cubic boron nitride; sintering of diamond and other materials at high pressure. *Mailing Add:* US Synthetic Corp 744 S 100th E Provo UT 84601-1617

POPE, CHARLES EDWARD, II, b Cleveland, Ohio, Sept 18, 31; m 58; c 2. GASTROENTEROLOGY. *Educ:* Cornell Univ, BA, 53; Western Reserve Univ, MD, 57. *Prof Exp:* Trainee gastroenterol, Royal Free Hosp, London, 61-62; res fel, Mass Mem Hosp & Boston Hosp, 61-64; clin investr, Vet Admin Hosp, 64-67; asst prof, 66-70, ASSOC PROF MED, UNIV WASH, 70-; CHIEF GASTROENTEROL, VET ADMIN HOSP, 66- *Mem:* AMA; Am Fedn Clin Res; Am Gastroenterol Asn. *Mailing Add:* 4435 Beacon Ave Seattle WA 98108

POPE, DANIEL LORING, b Buffalo, NY, Mar 4, 31; m 52; c 3. ENGINEERING MECHANICS. *Educ:* Cornell Univ, BCE, 53, PhD(mech), 61. *Prof Exp:* Struct engr, Pittsburgh-Des Moines Steel Co, 53; instr eng mech, Cornell Univ, 56-59; mem tech staff, Continuum Mech Group, Bell Tel Labs, Inc, 60-63, supvr, 63-69, supvr phys design group, 69-72, supvr below ground installation group, 72-83, SUPVR, CHESTER OPERS, 83- *Mem:* Soc Cable Television Engrs. *Res:* Nonlinear problems; soil mechanics; dynamics of machines. *Mailing Add:* AT&T Bell Labs 50 North Rd Chester NJ 07930

POPE, DANIEL TOWNSEND, b Edisto Island, SC, Dec 24, 13; wid; c 1. HORTICULTURE. *Educ:* Clemson Univ, BS, 35; La State Univ, MS, 37; Cornell Univ, PhD, 52. *Prof Exp:* Asst prof agr, Berry Col, 37-40; asst, Cornell Univ, 40-42; res assoc prof hort, NC State Col, 47-50; asst agr, Cornell Univ, 50-52; res assoc prof, NC State Univ, 52-62, prof hort sci, 62-76; RETIRED. *Mem:* Fel Am Soc Hort Sci. *Res:* Breeding new varieties of sweet potatoes. *Mailing Add:* Box 126 Edisto Island SC 29438

POPE, DAVID PETER, b Waukesha, Wis, July 31, 39; m 65. METALLURGY, MATERIALS SCIENCE. *Educ:* Univ Wis, BS, 61; Calif Inst Technol, MS, 62, PhD(mat sci), 67. *Prof Exp:* Res fel mat sci, Calif Inst Technol, 67-68; from asst prof to assoc prof, 68-82, assoc dean undergrad educ, Sch Eng & Appl Sci, 84-88, PROF METALL & MAT SCI, UNIV PA, 82-, CHAIR, DEPT MECH ENG, 88- *Mem:* Am Inst Mining, Metall & Petrol Engrs; fel Am Soc Metals Int; Am Soc Eng Educ; Mat Res Soc; AAAS. *Res:* Plastic deformation of metals; mechanical properties of ordered alloys; high temperature deformation and fracture of metals; properties of metal-ceramic interfaces. *Mailing Add:* Dept Mech Eng Univ Pa Philadelphia PA 19104-6315

POPE, JOHN KEYLER, b Cincinnati, Ohio, July 27, 31; m 54; c 2. PALEONTOLOGY, GEOLOGY. *Educ:* Harvard Univ, AB, 53; Univ Mich, MS, 56; Univ Cincinnati, PhD(geol), 66. *Prof Exp:* From instr to assoc prof, 60-76, PROF GEOL, MIAMI UNIV, 76-, DIR GRAD STUDIES, 80- *Concurrent Pos:* Vis asst prof, Univ Cincinnati, 65; mem, Paleont Res Inst. *Mem:* Paleont Soc; Brit Paleont Asn. *Res:* Ordovician brachiopods, faunas and stratigraphy; Cambrian and Ordovician echinoderms; zoological nomenclature; techniques of paleontological preparation. *Mailing Add:* Dept of Geol Miami Univ Oxford OH 45056

POPE, JOSEPH, US citizen. NOISE CONTROL, DIGITAL SIGNAL ANALYSIS. *Educ:* Mass Inst Technol, SB, 72; Stanford Univ, MS, 73, PhD(mech eng), 78. *Prof Exp:* Res asst, Dept Mech Eng, Stanford Univ, 73-77; sr res engr, Res Labs, Gen Motors, 77-81, staff res engr, 81-82; appln engr, Bruel & Kjaer Instruments, Inc, 82-89, sr tech staff, 89-90; PRIN CONSULT, POPE ENG CO, 90- *Mem:* Acoust Soc Am; Am Soc Mech Engrs; Soc Automotive Engrs. *Res:* Digital signal analysis; measurement techniques in acoustics; noise control. *Mailing Add:* Pope Eng Co PO Box 236 Newton Centre MA 02159-0002

POPE, LARRY DEBBS, b Corpus Christi, Tex, Dec 22, 40; m 63; c 3. VIBROACOUSTICS. *Educ:* Tex Tech Col, BSME, 63; Southern Methodist Univ, MSME, 66; Univ Houston, PhD(mech eng), 70. *Prof Exp:* Design engr, Collins Radio Co, Tex, 63-66; prin engr, Houston Aerospace Systs Div, Lockheed Electronics Co, 66-68; sr scientist, Bolt Beranek & Newman, Inc, Los Angeles, Calif, 70-82; PRIN, L D POPE & ASSOCS, 82- *Concurrent Pos:* Assoc prof mech eng, Tex A&M Univ, 77-78; NASA Langley Res Ctr, 83-84; consult, Lockheed Calif Co, 84-87, Boeing Com Airplane 86-90, General Dynamics Space Systs Div, 90- *Mem:* Acoust Soc Am. *Res:* Theoretical vibroacoustics; sound transmission and radiation; random vibration; turbulent boundary layers; statistical energy methods; aircraft interior noise. *Mailing Add:* L D Pope & Assocs 1776 Woodstead Ct Suite 111 The Woodlands TX 77380

POPE, LARRY ELMER, b Vernal, Utah, Jan 6, 41; m 63; c 5. SURFACE METALLURGY, TRIBOLOGY. *Educ:* Univ Utah, BA, 65, PhD(metall), 69. *Prof Exp:* Vis asst prof metall, Univ Utah, 69-70; mem tech staff, Sandia Labs, 70-73; mgr res projs, Fansteel, Inc, 73-74, mgr grad res, 74-79; mem tech staff process & surface metall, 79-83, surface & interface technol, 83-89, SUPVR, THIN FILM VACUUM & BRAZING, SANDIA NAT LABS, 89- *Honors & Awards:* Mat Sci Res Award, US Dept Energy, 85. *Mem:* Am Soc for Metals; Sigma Xi; Am Welding Soc. *Res:* Effect of surface preparation, ion implantation, and surface alloying on friction coefficients, electrical contact resistance and wear of unlubricated material couples; solid film lubrication; precious metal electrical contacts; surface characterization; carbide cermets; thin film deposition. *Mailing Add:* 9000 Galaxia Way NE Albuquerque NM 87111

POPE, MALCOLM H, b London, UK, Feb 11, 41; Brit & US citizen; m 71; c 3. ERGONOMICS WORKPLACE, BIOMECHANICS KNEE. *Educ:* Southall Col, London, HND, 62; Univ Bridgeport, Conn, MS, 69; Univ Vt, PhD(biomech), 72; Gothenburg Univ, Gotaborg, Sweden, Dr Med Sc, 90. *Prof Exp:* Assoc prof mech eng, dir, Orthop Res Lab & mem, Grad Fac, 72-75, PROF ORTHOP SURG & PROF MECH ENG, ORTHOP SURG BIOMECH, UNIV VT, 76-, DIR ORTHOP RES & CO-DIR, REHAB ENG CTR, 83- *Concurrent Pos:* Reviewer, NATO, NIH, Vet Admin & NSF, 82-86; mem, NIH Orthop Study Sect, 82-86 & rev bd, Nat Inst Handicapped Res-Nat Inst on Disability & Rehab Res, 84 & 85-87; vis prof, Dept Orthop Surg, Sahlgrenska Inst, Gothenburg, Sweden, 84-88. *Honors & Awards:* O'Donoghue Award, Am Orthop Soc Sports Med, 83; O'Connor Award, Am Arthroscopy Asn, 89, Sicot Maurice E Muller Award, 90. *Mem:* Fel Ergonomics Soc; Orthop Res Soc; Am Acad Orthop Surgeons; fel Inst Mech Engrs; Am So Biomech. *Res:* Biomechanics of lumbar and thoracic spine; vibration on human spine; mechanical analysis of low back pain; low back pain etiology, diagnosis & prevention; dynamic properties of soft tissues; relationship between ski injuries & ski equipment; strain behavior of knee ligaments, ligament reconstruction, in vivo anterior cruciate ligament strains; strain of the intervertebral disc. *Mailing Add:* Dept Orthop & Rehab Univ Vt C417 Given Bldg Burlington VT 05405-0068

POPE, MARTIN, b New York, NY, Aug 22, 18; m 46; c 2. SOLID STATE PHYSICS. *Educ:* City Col New York, BS, 39; Polytech Inst Brooklyn, PhD(chem), 51. *Prof Exp:* Res chemist, Balco Res Lab, 46-47, asst tech dir, 51-56; res assoc prof physics, 56-66, assoc prof chem, 66-68, assoc dir solid state & radiation lab, 68-83, PROF CHEM & DIR SOLID STATE & RADIATION LAB, NY UNIV, 83- *Mem:* Fel AAAS; fel Am Phys Soc; Am Chem Soc; fel NY Acad Sci. *Res:* Photoconductivity and semiconductivity in organic materials; electrical properties of thin metallic films; ion exchange kinetics; exciton dynamics in organic crystals. *Mailing Add:* 1005 E Fourth St Brooklyn NY 11230

POPE, MARY E, MATERIALS TESTING SOFTWARE. *Educ:* Stanford Univ, BS, 76; Mass Inst Technol, 78. *Prof Exp:* Res asst, dept mech eng, Mass Inst Technol, 76-78; res engr, Gen Motors Res Labs, 78-82; ENGR, SCI PROG, INSTRON CORP, 83- *Concurrent Pos:* Consult, 82-83. *Honors & Awards:* Ralph H Isbrandt Automotive Safety Eng Award, Soc Automotive Engrs, 82. *Mem:* Sigma Xi. *Res:* Development of engineering applications software for real time control of laboratory instruments; materials testing software. *Mailing Add:* PO Box 236 Newton Centre MA 02159

POPE, MICHAEL THOR, b Exeter, Eng, Apr 14, 33; m 57; c 2. INORGANIC CHEMISTRY. *Educ:* Oxford Univ, BA, 54, DPhil(chem), 57, MA, 58. *Prof Exp:* Monsanto fel inorg chem, Boston Univ, 57-58, AEC fel, 58-59; res chemist, Laporte Chem Ltd, Eng, 59-61, jr sect leader inorg chem, 61-62; from asst prof to assoc prof, 62-73, PROF INORG CHEM, GEORGETOWN UNIV, 73-, DEPT CHAIR, 90- *Concurrent Pos:* Petrol Res Fund int award fel & vis prof, Tech Univ Vienna, 70-71; vis scientist, Oxford Univ, 78; assoc prof, Univ Pierre & Marie Curie, Paris & guest prof, Free Univ Berlin, 79; vis prof, Northeast Norm Univ, Changchun, China, 85, Univ Umeå, Sweden, 89, Univ Bielefeld, Ger, 89-90. *Honors & Awards:* Sr US Scientist Award, Alexander Von Humboldt Found, 89. *Mem:* Am Chem Soc; Royal Soc Chem. *Res:* Structures and reactions of heteropoly oxoanions of transition metals and their organic derivatives; mixed-valence chemistry; catalysis; electron and nuclear magnetic resonance; electrochemistry; crystallography; activation of dioxygen. *Mailing Add:* Dept Chem Georgetown Univ Washington DC 20057

POPE, NOEL KYNASTON, b Christchurch, NZ, July 31, 18; m 46; c 5. KINETIC EQUATIONS, OPTIMIZATION OF STRUCTURES. *Educ:* Univ NZ, BSc, 40, MSc, 42; Univ Edinburgh, PhD(theoret physics), 48. *Prof Exp:* Lectr demonstr physics, Univ Victoria Wellington, 43-45; fel, Nat Res Coun Can, 49-50; asst res officer theoret physics, Chalk River Labs, Atomic Energy Can, Ltd, 51-53, assoc res officer theoret physics, 56-61; lectr physics, Univ Canterbury, 53-55; prof, 61-83, head dept, 67-82, EMER PROF MATH, ROYAL MIL COL CAN, 83- *Concurrent Pos:* Consult, Solid State Div, Harshaw Chem Co, 63-67 & Struct Eng Group, Royal Mil Col, Can, 83-88; vis prof, Mass Inst Technol, 80-81. *Mem:* Am Phys Soc; Can Asn Physicists. *Res:* Theory of kinetic equations; optimization of structures; inelastic scattering of thermal neutrons by fluids and solids; lattice dynamics; positron annihilation in solids. *Mailing Add:* 15 Nottingham Pl Kingston ON K7M 7H1 Can

POPE, PAUL TERRELL, b Ft Worth, Tex, Mar 28, 42; m 64; c 3. STATISTICS, MATHEMATICS. *Educ:* Abilene Christian Col, BA, 64; Univ Tex, MA, 66; Southern Methodist Univ, PhD(statist), 69. *Prof Exp:* Asst prof math, Univ Tulsa, 69-74; RES MATHEMATICIAN, CITIES SERV OIL CO, 74- *Mem:* Am Statist Asn. *Res:* Linear models; simulation. *Mailing Add:* 3128 W 58th Ct Tulsa OK 74107

POPE, ROBERT EUGENE, veterinary public health, epidemiology, for more information see previous edition

POPE, WENDELL LAVON, b Arco, Idaho, Jan 16, 28; m 50; c 9. COMPUTER ANALYSIS. *Educ:* Utah State Univ, BS, 56; Stanford Univ, MS, 58; Univ Wis, MS, 68. *Prof Exp:* Math analyst, Missiles & Space Div, Lockheed Aircraft Corp, 56-59; from asst to assoc prof math, 59-77, dir comput ctr, 69-77, PROF COMPUT SCI, UTAH STATE UNIV, 77- *Mem:* Asn Comput Mach. *Res:* Automatic software generation. *Mailing Add:* Dept Comput Sci Utah State Univ Logan UT 84322-4205

POPEJOY, WILLIAM DEAN, b Norfolk, Va, Feb 22, 25; m 57; c 1. MATHEMATICS. *Educ:* Ill State Univ, BS, 49, MS, 50; Colo State Col, EdD, 59. *Prof Exp:* Teacher high schs, Ill, 50-53; instr math, Ill State Univ, 53-58; from instr to assoc prof, 58-67, PROF MATH, UNIV NORTHERN COLO, 67- *Concurrent Pos:* Consult, 58-; fel, NY Univ, 70-71. *Res:* Mathematical education. *Mailing Add:* 2531 17th Avenue Ct Greeley CO 80631

POPEL, ALEKSANDER S, b Moscow, USSR, Oct 8, 45; citizen US; m 66; c 1. FLUID MECHANICS, MATHEMATICAL MODELING. *Educ:* Moscow Univ, USSR, MS, 67, PhD(fluid mech), 72. *Prof Exp:* Res scientist fluid mech, Inst Mech, Moscow State Univ, 70-75; asst prof mech eng, Tulane Univ, 76; res assoc prof, Depts Chem Eng, Aerospace & Mech Eng & Physiol, Univ Ariz, 76-80; assoc prof mech eng, Univ Houston, 80-84; assoc prof, 84-88, PROF DEPT BIOMED ENG, JOHNS HOPKINS UNIV, 88- *Mem:* Am Soc Mech Engrs; Microcirculatory Soc; Am Phys Soc; Am Heart Asn; Am Physiol Soc; Biomed Eng Soc. *Res:* Mechanics of blood flow; oxygen transport to tissue; mathematical modeling of biological processes; biomechanics; injury biomechanics. *Mailing Add:* Dept Biomed Eng Sch Med Johns Hopkins Univ Baltimore MD 21205

POPELAR, CARL H(ARRY), b Dundee, Mich, Feb 5, 38; m 61; c 3. ENGINEERING MECHANICS. *Educ:* Mich State Univ, BS, 60, MS, 61; Univ Mich, PhD(eng mech), 65. *Prof Exp:* Engr, Martin Col, 61-62; instr eng mech, Univ Mich, 64-66; from asst prof to assoc prof, 66-72, PROF ENG MECH, OHIO STATE UNIV, 72- *Mem:* Am Acad Mech; Am Soc Mech Engrs. *Res:* Fracture mechanics; dynamic stability of thin-walled members and shells; elastic stability; structural optimization. *Mailing Add:* Dept Eng Mech Ohio State Univ 155 W Woodruff Ave Columbus OH 43210

POPENOE, EDWIN ALONZO, b Topeka, Kans, Aug 3, 22; m 50; c 2. BIOCHEMISTRY. *Educ:* Pomona Col, BA, 47; Univ Wis, MS, 48, PhD(biochem), 50. *Prof Exp:* Res assoc biochem, Med Col, Cornell Univ, 50-54; head, Biochem Div, 70-78, res biochemist, 54-86, EMER RES BIOCHEMIST, MED DEPT, BROOKHAVEN NAT LAB, 86- *Concurrent Pos:* NIH res fel, Med Col, Cornell Univ, 50-52; consult, Off Sci & Technol, 64. *Mem:* Am Chem Soc; Am Soc Biochem & Molecular Biol; AAAS; Sigma Xi. *Res:* Chemistry of proteins, peptides and mucoproteins; biosynthesis of DNA. *Mailing Add:* Med Dept Brookhaven Nat Lab Upton NY 11973

POPENOE, HUGH, b Tela, Honduras, Aug 28, 29; US citizen. SOIL SCIENCE. *Educ:* Univ Calif, BS, 51; Univ Fla, PhD(soils), 60. *Prof Exp:* Asst soil scientist, Econ Coop Admin, Thailand, 51-52; res assoc, Univ Fla, 54-59, interim instr, 59-60, from asst prof to assoc prof soils, 60-71, asst to provost, Inst Food & Agr Sci, 65-66, actg dir, Ctr Aquatic Sci, 69-81, dir, State Univ Syst Fla Sea Grant Prog, 71-81, PROF SOILS, UNIV FLA, 71-, DIR CTR TROP AGR, 65-, DIR INT PROG AGR, 66- *Concurrent Pos:* Consult, USDA, Nicaragua; Soc Sci Res Coun, Latin Am; Rockefeller Found, Columbia Univ; US Agency Int Develop, Panama, El Salvador, Honduras, Costa Rica, Jamaica & Vietnam; Univ Hawaii; For Area Fel Prog, 69-72; mem bd dirs, Orgn Trop Studies, 64-67; vchmn, Gulf Univ Res; mem bd trustees, Pan Am Sch Agr, Honduras, 70, chmn, 71-80; Latin Am Scholar Prog, Am Univ, 71-74; vis lectr trop pub health, Harvard Univ, 75-; mem int sci & educ coun, Nat Asn State Univ & Land-Grant Cols & USDA, 74-81 & int affairs comt, Nat Asn State Univs & Land-Grant Cols, 74-; chmn, Sea Grant Asn, 76-77; mem bd dirs, League Int Food Educ, 76-80; mem, Joint Res Comt, Bd Int Food & Agr Develop, Title XII, 77-87 & chmn, Adv Comt Technol Innovations, Nat Acad Sci, 77-82; mem bd, Sci Technol Int Develop, Nat Acad Sci, 79-82; mem bd, Int Found Sci, 84-87. *Mem:* Fel AAAS; fel Am Geog Soc; fel Int Soc Soil Sci; Soil Sci Soc Am; Int Soc Trop Ecol; Sigma Xi; Am Water Buffalo Asn (pres, 88). *Res:* Development of tropical agriculture; tropical ecology. *Mailing Add:* Int Prog Agr 3028 McCarty Hall Univ Fla Gainesville FL 32611

POPENOE, JOHN, horticulture, for more information see previous edition

POPHAM, RICHARD ALLEN, botany; deceased, see previous edition for last biography

POPJAK, GEORGE JOSEPH, b Kiskundorozsma, Hungary, May 5, 14; m 41. BIOCHEMISTRY. *Educ:* Univ Szeged, MD, 38, FRIC, 55; Univ London, DSc(biochem), 61. *Prof Exp:* Demonstr path, St Thomas' Hosp Med Sch, Univ London, 41-47; res scientist, Nat Inst Med Res, London, 47-53; dir exp radiopath res unit, Med Res Coun, Hammersmith Hosp, 53-62; co-dir chem enzymol lab, Shell Res Ltd, Sittingbourne, 62-68; prof, 68-84, EMER PROF BIOCHEM, UNIV CALIF, LOS ANGELES, 84- *Honors & Awards:* Stouffer Found Prize, 67; Ciba Medal, Brit Biochem Soc, 66; Davy Medal, Royal Soc, 68; Lipid Chem Award, Am Oil Chemists Soc, 77. *Mem:* AAAS; fel Royal Soc; Brit Biochem Soc; Am Acad Arts & Sci; hon mem Am Soc Biol Chemists. *Res:* Biosynthesis of lipids, particularly that of cholesterol; enzyme stereochemistry and regulation. *Mailing Add:* 511 Cashmere Terrace Los Angeles CA 90049

POPKIN, A(LEXANDER) H, b New York, NY, Nov 7, 13; c 2. ORGANIC CHEMISTRY. *Educ:* Brooklyn Col, BS, 34; Pa State Univ, MS, 35, PhD(org chem), 39. *Prof Exp:* Asst chem, Pa State Univ, 35-36, asst org chem, 37-38, instr, 38-39; res chemist, Sun Chem Corp, NY, 39-42, group leader, 42-45; dir res, Maltbie Chem Co, NJ, 45-46; res chemist, 46-58, res assoc, 58-68, CONSULT, EXXON RES & ENG CO, 69- *Concurrent Pos:* Pres, A H Popkin Assocs, 78- *Mem:* Am Chem Soc; Sigma Xi; Soc Automotive Engrs; Am Inst Chemists. *Res:* Chemical market studies, petroleum additives; gasolines; automotive and aviation lubricants; improved uses of petroleum products; petrochemicals; new energy sources; patent searches and analyses. *Mailing Add:* 534 Academy St Maplewood NJ 07040

POPKIN, MICHAEL KENNETH, b Trenton, NJ, Dec 31, 43; m 71; c 3. PSYCHOSOMATIC MEDICINE. *Educ:* Princeton Univ, BA, 65; Univ Chicago Med Sch, MD, 69. *Prof Exp:* Intern, NY Univ-Bellevue Hosp, 69-70; resident, Mass Gen Hosp, 70-73; from asst prof to assoc prof, 75-83, PROF PSYCHIAT & MED, UNIV MINN MED SCH, 83- *Concurrent Pos:* Consult, Brevard County Mental Health Ctr, 73-75; dir, Consultation Psychiat Serv, Univ Minn Hosps, 75-; mem, Alcohol, Drug Abuse & Ment Health Assoc Epidemiol & Serv Res Comt, 85-; adv ed, Psychosomatics, 83-, assoc ed, Psychiatric Med, 84-; consult, Behav Sci Task Force, Diabetes Complication & Cont Trial, 82- *Mem:* Am Psychopathol Asn; Am Psychosomatic Soc; Am Psychiat Asn; Am Col Sports Med; Asn Acad Psychiat; fel Am Psychiat Assoc; fel Am Ortopsychiat Assoc (dir, 85-88); Acad Psychosomatic Med. *Res:* Studies in consultation psychiatry directed at increasing clinical effectiveness and revising practices; depressive disorder in the medically ill; psychiatric epidemiology in medical surgical patients. *Mailing Add:* Box 345 Mayo Bldg Univ Hosps Minneapolis MN 55455

POPLAWSKY, ALEX JAMES, b Scranton, Pa, July 12, 48; m 72; c 2. PHYSIOLOGICAL PSYCHOLOGY. *Educ:* Univ Scranton, BS, 70; Ohio Univ, MS, 72, PhD(exp psychol), 74. *Prof Exp:* From asst prof to assoc prof, 74-82, PROF PSYCHOL, BLOOMSBURG UNIV PA, 82- *Concurrent Pos:* Vis assoc prof, State Univ NY, Binghamton, 81-82. *Mem:* Am Psychol Asn; Sigma Xi; Psychonomic Soc. *Res:* Effects of septal lesions on learning and motivation; recovery of function. *Mailing Add:* Dept Psychol Bloomsburg Univ Bloomsburg PA 17815

POPLE, JOHN ANTHONY, b Somerset, Eng, Oct 31, 25; m 52; c 4. THEORETICAL CHEMISTRY. *Educ:* Cambridge Univ, BA, 46, MA, 50, PhD(math), 51. *Prof Exp:* Res fel math, Trinity Col, Cambridge Univ, 51-54, lectr, 54-58; supt basic Physics Div, Mat Phys Lab, Teddington, Eng, 58-61, 62-64; Ford vis prof chem, Carnegie Inst Technol, 61-62, Carnegie Prof chem physics, 64-74, adj fel chem, Mellon Inst, 64-76, JOHN CHRISTIAN WARNER PROF NATURAL SCI, CARNEGIE-MELLON UNIV, 74- *Honors & Awards:* Irving Langmuir Award, Am Chem Soc, 70, Harrison Howe Award, 71; Marlow Medal, Faraday Soc, 58; Gilbert Newton Lewis Award, 72; Morley Award, 76; Pittsburgh Award, 75; Pauling Award, 77; Sr Scientist Award, Alexander von Humboldt Found, 81; G Willard Wheland Award, Univ Chicago, 81; Evans Award, Ohio State Univ, 82. *Mem:* Foreign assoc Nat Acad Sci; Am Phys Soc; fel Am Acad Arts & Sci; Royal Soc Chem; fel Royal Soc; fel AAAS. *Res:* Application of quantum mechanics to the structure of molecules and theories of physical properties. *Mailing Add:* Dept Chem Carnegie-Mellon Univ 4400 Fifth Ave Pittsburgh PA 15213-3876

POPLI, SHANKAR D, pharmaceutics, for more information see previous edition

POPOV, ALEXANDER IVAN, b Vladivostok, Russia, Mar 22, 21; nat US; m 50; c 1. ANALYTICAL CHEMISTRY, INORGANIC CHEMISTRY. *Educ:* Aurora Univ, China, BS, 44; Univ Iowa, MS, 48, PhD(phys chem), 50. *Prof Exp:* Instr chem, Aurora Univ, 44-46; res assoc, Iowa Univ, 49-50, from instr to assoc prof anal chem, 50-60; prof chem & dean col lib arts & sci, Northern Ill Univ, 60-62; chmn dept, 62-69, PROF CHEM, MICH STATE UNIV, 62- *Concurrent Pos:* Brotherton res prof, Univ of Leeds, 84. *Mem:* AAAS; Am Chem Soc. *Res:* Spectroscopic studies; structure of electrolyte solutions; non-aqueous solvents; complex compounds of alkali metal ions. *Mailing Add:* Dept of Chem Mich State Univ East Lansing MI 48824

POPOV, E(GOR) P(AUL), b Kiev, Russia, Feb 19, 13; nat US; m 39; c 2. STRUCTURAL ENGINEERING. *Educ:* Univ Calif, BS, 33; Mass Inst Technol, MS, 34; Stanford Univ, PhD(civil eng), 46. *Prof Exp:* Struct designer, Calif, 35-39; asst prod engr, Southwestern Portland Cement Co, 39-42; mach designer, Goodyear Tire & Rubber Co, 42-43; design engr, Aerojet Corp, 43-45; from asst prof to prof, 46-83, chmn, Div Struct Eng & Mech & dir, Struct Eng Lab, 57-60, Miller res prof, Miller Inst Basic Res Sci, 67-68, EMER PROF CIVIL ENG, UNIV CALIF, BERKELEY, 83- *Honors & Awards:* Hetenyi Award, Soc Exp Stress Anal, 69; T R Higgins Award, Am Inst Steel Construct, 71; E E Howard Award, Am Soc Civil Engrs, 76, J James R Croes Medal, 79 & 82, Nathan M Newmark Medal, 81, Struct Div Res Award, 84 & Norman Medal, 87, Theodore Von Karman Medal, 89. *Mem:* Nat Acad Eng; hon mem Int Asn Shell Struct (vpres, 70-83); Am Concrete Inst; Am Soc Eng Sci; Am Soc Eng Educ; hon mem Am Soc Civil Engrs, 86. *Res:* Structural mechanics; limit design; shell theory; earthquake-resistant design. *Mailing Add:* Davis Hall Univ Calif Berkeley CA 94720

POPOVIC, VOJIN, b Belgrade, Yugoslavia, Sept 18, 22; m 53; c 1. PHYSIOLOGY. *Educ:* Univ Belgrade, dipl, 49, PhD, 51. *Prof Exp:* Asst prof physiol, Univ Belgrade, 49-53, prof, 53-56, chmn dept, 54-56; res fel, Nat Ctr Sci Res, France, 56-57; res assoc, Univ Rochester, 57-58; Nat Res Coun Can fel, 58-60; assoc prof physiol, Univ Houston, 60-61; assoc prof, 61-66, PROF PHYSIOL, SCH MED, EMORY UNIV, 66- *Mem:* Am Physiol Soc; Soc Exp Biol & Med; Int Soc Chronobiol; Aerospace Med Asn; fel Royal Soc Health. *Res:* cardiovascular research; space research. *Mailing Add:* Dept Physiol Emory Univ Sch Med Atlanta GA 30322

POPOVIC, ZORAN, b Belgrade, Yugoslavia, Dec 7, 41; m 71; c 2. SOLID STATE PHYSICS, SEMICONDUCTORS. *Educ:* Univ Belgrade, dipl eng, 65, MSc, 69; McMaster Univ, PhD(mat sci), 74. *Prof Exp:* Asst, Fac Elec Eng, Univ Belgrade, 66-71; mem sci staff, 74-80, SR SCIENTIST ORG SEMICONDUCTORS, XEROX RES CTR CAN LTD, 80- *Mem:* Am Phys Soc; Asn Prof Engrs, Ont. *Res:* Photovoltaic energy conversion; photovoltaic phenomena in organic semiconductors. *Mailing Add:* 3349 Sawmill Valley Dr Mississauga ON L5L 2Z8 Can

POPOVICH, FRANK, b Czech, Nov 2, 23; Can citizen; m 53; c 4. DENTISTRY. *Educ:* Univ Toronto, DDS, 51, MScD, 53; FRCD(C), 67. *Prof Exp:* Orthodontist, 53, dir, 55-60, co-dir, 62-65, assoc prof orthod, 69-74, PROF ORTHOD, FAC DENT, UNIV TORONTO, 74-, DIR & CUR, BURLINGTON GROWTH STUDY, 65- *Mem:* Can Dent Asn; Can Asn Orthod; Int Asn Dent Res; Human Biol Coun. *Res:* Facial growth and development; orthodontic treatment; differential diagnosis. *Mailing Add:* Eight Golf Crest Toronto ON M9A 1L1 Can

POPOVICH, M(ILOSH), b Pittsburg, Calif, Sept 3, 17; m 39; c 1. MECHANICAL ENGINEERING. *Educ:* Ore State Col, BS, 39, MS, 41. *Prof Exp:* Asst prof mech eng, Ore State Col, 45-46; engr, Union Oil Co, Calif, 46-47; from asst prof to assoc prof mech eng, 47-50, chmn dept mech eng, 50-54, asst dean eng, 54-59, dean admin, 59-76, vpres admin, 76-79, PROF, ORE STATE UNIV, 50-, VPRES EMER, 79- *Mem:* Fel Am Soc Mech Engrs. *Res:* Fuels and lubricants. *Mailing Add:* 1390 NW 14 St Corvallis OR 97330

POPOVICH, ROBERT PETER, b Sheboygan, Wis, Jan 9, 39; div; c 4. BIOMEDICAL ENGINEERING, CHEMICAL ENGINEERING. *Educ:* Univ Wis, BS, 63; Univ Wash, MS, 68, PhD(chem eng), 70. *Prof Exp:* Chem engr, Pac Northwest Labs, Battelle Mem Inst, 63-65; teaching asst chem eng, Univ Wash, 65-70, asst prof, 70-72; from asst prof to prof, 70-83, E P SHOCH PROF CHEM & BIOCHEM ENG, UNIV TEX, 83- *Concurrent Pos:* Biomed & chem eng consult to var pvt, pub & govt agencies, 72-; pres, Hemotherapy Instruments Inc, 74-; co-dir, Moncriet-Popovich Res Inst Inc, 77-; biomed eng consult, Baxter-Travenol Labs Inc, 77- *Honors & Awards:* Dialysis Pioneering Award, Nat Kidney Found, 83. *Mem:* Am Soc Artificial Internal Organs; Am Inst Chem Engrs; Nat Soc Prof Engrs; Am Soc Nephrol; Europ Dialysis & Transplant Asn; Int Soc Peritoneal Dialysis. *Res:* Industrial and biomedical applications of membrane systems; development of artificial internal organs; transport phenomena and biomedical instrumentation; continuous ambulatory peritoneal dialysis. *Mailing Add:* ENS-617 Dept Chem Eng Univ Tex Austin TX 78712

POPOVICS, SANDOR, b Budapest, Hungary, Dec 24, 21; US citizen; m 61; c 2. CONCRETE TECHNOLOGY, OPTIMIZATION TECHNIQUES. *Educ:* Polytech Univ Budapest, BS, 44; Hungarian Acad Sci, MS, 56; Purdue Univ, PhD(civil eng), 61. *Prof Exp:* Res engr mat, Metrop Lab, 44-48; mgr mat, Inst Bldg Sci, 49-56; prof civil eng, Auburn Univ, 59-67; prof civil eng, Northern Ariz Univ, 67-76; prof civil eng, King Abdulaziz Univ, Saudi Arabia, 77-78; SAMUEL S BAXTER PROF CIVIL ENG, DEPT CIVIL ENG, DREXEL UNIV, 78- *Concurrent Pos:* Consult, Hungarian Bur Standards, 48-56 & NASA, 69; mem, Nat Highway Safety Adv Comn, 72-73; prin investr, Ala Dept Transp, 60-67 & Ariz Dept Transp, 72-76, Air Force Sci Res. *Mem:* Fel Am Concrete Inst; fel Am Soc Civil Engrs; Transp Res Bd; Am Soc Testing & Mat. *Res:* Mathematical modeling of materials properties; mechanical properties of concrete and concrete making materials; deformability and fracture of composite materials; linear and non-linear optimization; applied mathematics and statistics and highway safety. *Mailing Add:* Dept Civil Eng Drexel Univ Philadelphia PA 19104

POPP, CARL JOHN, b Chicago, Ill, Apr 3, 41; m 65; c 3. ACID RAIN, TRACE MATERIALS IN THE ENVIRONMENT. *Educ:* Colo State Univ, BS, 62; Southern Ill Univ, Carbondale, MA, 65; Univ Utah, PhD(inorg chem), 68. *Prof Exp:* Instr chem, Southern Ill Univ, 65; fel, Univ Utah, 68-69; from asst prof to assoc prof, 69-72, PROF CHEM, NMEX INST MINING & TECHNOL, 82-, VPRES ACAD AFFAIRS, 83- *Mem:* AAAS; Am Chem Soc; Am Geophys Union; Sigma Xi. *Res:* Acid rain in the western United States; natural (lightning) and anthropogenic sources; effects studies; inorganic aspects of water chemistry. *Mailing Add:* Dept of Chem NMex Inst of Mining & Technol Socorro NM 87801

POPP, FRANK DONALD, b New York, NY, Dec 25, 32; m 55, 77; c 2. REISSERT COMPOUNDS. *Educ:* Rensselaer Polytech Inst, BS, 54; Univ Kans, PhD(chem), 57. *Prof Exp:* Asst, Univ Kans, 54-57; jr res chemist, Univ Calif, 57-58; res assoc, Univ Mich, 58-59; asst prof chem, Univ Miami, 59-62; from asst prof to prof chem, Clarkson Col Technol, 62-76; prof med, chmn dept chem, 76-82, PROF CHEM, UNIV MO-KANSAS CITY, 82- *Concurrent Pos:* USPHS fel, 59 & 64-65. *Mem:* AAAS; Am Chem Soc; Am Asn Cancer Res; NY Acad Sci; Royal Soc Chem; Am Inst Chem; Int Soc Heterocyclic Chem; Sigma Xi. *Res:* Synthesis and reactions of nitrogen heterocyclic compounds; medicinal chemistry. *Mailing Add:* Dept of Chem Univ Missouri-Kansas City Kansas City MO 64110-2499

POPP, GERHARD, b Obernzenn, Ger, Apr 10, 40; c 2. INORGANIC CHEMISTRY. *Educ:* Univ Würzburg, MS, 65, PhD(chem), 67. *Prof Exp:* Res asst chem, Phys Chem Inst, Univ Würzburg, 65-67; fel inorg chem, Univ Calif, Riverside, 67-69; SR RES CHEMIST, RES LABS, EASTMAN KODAK CO, 69- *Concurrent Pos:* Dir, Color Negative Technol Div, Prl, EK Co. *Res:* Electroorganic chemistry, particularly electro-organic synthesis; transition metal chemistry. *Mailing Add:* 1116 Webster Rd Webster NY 14850

POPP, JAMES ALAN, b Salem, Ohio, Mar 13, 45; m 66; c 1. CHEMICAL CARCINOGENESIS, ULTRASTRUCTURAL PATHOLOGY. *Educ:* Ohio State Univ, DVM, 68; Univ Calif, PhD(comp path), 72. *Prof Exp:* Fel biochem path, Fels Res Inst, Temple Univ Sch Med, 72-74; asst prof path, Univ Fla, 74-78; scientist path, 78-84, DEPT HEAD EXP PATH & TOXICOL, CHEM INDUST INST TOXICOL, 84-, VPRES, 89- *Concurrent Pos:* Adj asst prof, Col Med, Univ NC, 82-, Col Med, Duke Univ, 86- & NC State Univ, 89- *Mem:* Am Asn Cancer Res; Soc Toxicol; Am Col Vet Pathologists; Am Asn Pathologists; Am Vet Med Asn; Electron Micros Soc Am. *Res:* Elucidate the mechanisms of hepatocarcinogenicity of chemicals in rodents; mechanisms of promotion of non-genotoxic carcinogens. *Mailing Add:* Chem Indust Inst Toxicol PO Box 12137 Res Triangle Pk NC 27709

POPP, RAYMOND ARTHUR, b Northport, Mich, Nov 23, 30; m 54; c 4. EMBRYOLOGY, GENETICS. *Educ:* Univ Mich, BS, 52, MA, 54, PhD PhD(zool), 57. *Prof Exp:* Res assoc, Biol Div, 57-58, biologist, 58-65, SR STAFF BIOLOGIST, OAK RIDGE NAT LAB, 65- *Concurrent Pos:* Mem fac, Univ Tenn Biomed Grad Sch, 66-; prin investr, Genetics Training Grant, 77-88; dir, Univ Tenn Biomed Grad Sch, 86- *Mem:* AAAS; Sigma Xi; Am Genetics Asn; Genetics Soc Am; Environ Mutagen Soc; Am Soc Biolchem. *Res:* Experimental embryology; transplantation and cellular differentiation of hematopoietic tissues in mice; genetics of mouse hemoglobins; protein structure and function; murine retroviruses and gene therapy in experimental animals. *Mailing Add:* 9901 Emory Rd Knoxville TN 37931

POPPE, CARL HUGO, b Chicago, Ill, Nov 23, 36; m 57; c 4. NUCLEAR REACTIONS, NUCLEAR SCATTERING. *Educ:* DePauw Univ, BA, 57; Univ Wis, MS, 59, PhD(physics), 62. *Prof Exp:* From asst prof to assoc prof physics, Univ Minn, Minneapolis, 62-76; assoc div leader, exp physics div, 79-83, div leader, exp physics div, 83-84, prompt diagnostics prog leader & div leader, 84-86, MEM STAFF, LAWRENCE LIVERMORE NAT LAB, UNIV CALIF, 76-, NUCLEAR CHEM DIV LEADER, 86- *Concurrent Pos:* x-ray calibration & standards facil proj dir, 85-88. *Mem:* Fel Am Phys Soc; Sigma Xi. *Res:* Nuclear physics; polarized ion sources; synchrotron radiation beam lines. *Mailing Add:* Lawrence Livermore Lab Univ Calif PO Box 808 Livermore CA 94550

POPPE, WASSILY, b Riga, Latvia, Nov 10, 18; US citizen; m 42; c 1. PHYSICAL CHEMISTRY. *Educ:* Univ Tubingen, BS, 47; Univ Stuttgart, MS, 49; Univ Pittsburgh, PhD(phys chem), 66. *Prof Exp:* Chemist, Dr Hans Kittel Chem Lab, Ger, 49-50; develop chemist, Karl Worwag Lack & Farbenfabrik, 50-51; prod mgr paint, Pinturas Iris, Venezuela, 51-53; lab supvr paint chem, Pinturas Tucan affiliated PPG Industs, Inc, Venezuela, 53-54, tech dir, 54-57, plant mgr paint prod, 57-59; res chemist, Springfield Res Ctr, PPG Industs, Inc, 59-64, res asst phys chem, Univ Pittsburgh, 64-66; group leader surface chem, Avisun Corp, 66-68; group leader polypropylene prod res, Amoco Chem, Naperville, 68-76, res supvr, Plastic Prod Div, 76-78, res assoc, 78-85; RETIRED. *Concurrent Pos:* consult, 85- *Mem:* Am Chem Soc; fel Am Inst Chemists. *Res:* Surface chemistry; nonequilibrium thermodynamics; color research of plastic products; adhesion of coatings and metals to plastic surfaces. *Mailing Add:* 105 N Main St Lombard IL 60148

POPPELBAUM, WOLFGANG JOHANNES, b Frankfurt, Ger, Aug 28, 24. ELECTRICAL ENGINEERING, COMPUTER SCIENCE. *Educ:* Univ Lausanne, MS, 52, PhD(physics), 54. *Prof Exp:* Swiss fel, Univ Lausanne, 54-55; res assoc elec eng, 55-56, from asst prof to assoc prof, 56-63, prof elec eng & comput sci, Univ Ill, Urbana-Champaign, 63-, DIR, INFO ENG LAB. *Concurrent Pos:* Consult, Elliott Automation, 60- & Los Alamos Sci Lab, 71- *Mem:* Fel Inst Elec & Electronics Engrs; Swiss Phys Soc; Am Phys Soc; Soc Info Display. *Res:* Semiconductor physics; circuit design and research; electro optics and hybrid digital-analog circuitry; display devices with bandwidth compression; probabilistic and deterministic averaging processors. *Mailing Add:* 2007 S Anderson Urbana IL 61801

POPPELE, RICHARD E, b Irvington, NJ, Mar 6, 36; m 59; c 4. NEUROPHYSIOLOGY. *Educ:* Tufts Univ, BSEE, 58; Univ Minn, PhD(physiol), 65. *Prof Exp:* Res assoc neurophysiol, Univ Minn, 65; Nat Inst Neurol Dis & Stroke fel, Univ Pisa, 65-67; from asst prof to assoc prof, 67-75, prof 86-88, HEAD PHYSIOL DEPT, UNIV MINN, MINNEAPOLIS, 75- *Concurrent Pos:* Dir Neurophysiol Lab, 76-; prin investr, NIH grant, 80-, mem NIH study sect, 87-; vis prof, CNR, Milano, italy, 89-90. *Mem:* Int Brain Res Orgn; Soc Neurosci; NY Acad Sci; Biophys Soc: Am Physiol Soc; Sigma Xi. *Res:* Sensory motor integration in the central nervous system; computational approach to analysis of neuronal coding. *Mailing Add:* Dept Physiol 6-255 Millard Hall 435 Delaware St SE Univ of Minn Minneapolis MN 55455

POPPENDIEK, HEINZ FRANK, b Altona, Ger, Nov 8, 19; nat US; m 43; c 3. PHYSICS. *Educ:* Univ Calif, BS, 42, MS, 44, PhD(mech eng), 49. *Prof Exp:* Res engr, Univ Calif, 42-46, lectr & res engr, Univ Calif, Los Angeles, 46-49, asst prof & res engr, 49-50; group leader, Oak Ridge Nat Lab, 50-52, sect chief, 52-56; mem staff, Gen Atomic Div, Gen Dynamics Corp, 56-58, staff specialist, Convair Div, 58-60; dir appl res, Geophys Corp, Am, 60-61; PRES, GEOSCI LTD, 61- *Concurrent Pos:* Lectr, Univ Mich, 53 & Oak Ridge Sch Reactor Tech, 52-56; consult, Los Alamos Sci Lab, 56- *Honors & Awards:* Award, Am Geophys Union, 53. *Mem:* Am Nuclear Soc; Sigma Xi. *Res:* Applied physics and mathematics; heat, mass and momentum transfer; nuclear reactor physics; elasticity; micro-meteorology and atmospheric turbulence; selenography; physical properties; sea water conversion; electromagnetic pumping of blood. *Mailing Add:* 7834 Esterel Dr La Jolla CA 92037

POPPENSIEK, GEORGE CHARLES, b New York, NY, June 18, 18; m 43; c 2. VETERINARY MEDICINE, MICROBIOLOGY. *Educ:* Univ Pa, VMD, 42; Cornell Univ, MS, 51; Am Bd Microbiol, dipl; Am Col Vet Microbiol, dipl. *Prof Exp:* Asst instr med, Sch Vet Med, Univ Pa, 42-43; asst prof vet sci, Univ Md, 43-44; dept head, Lederle Labs Div, Am Cyanamid Co, 44-49; dir diag lab, State Univ NY Vet Col, Cornell Univ, 49-51, res assoc vet virus res inst, 51-54, actg prof bact, 53-54; vet, Plum Island Dis Res Ctr, Agr Res Serv, USDA, 55-56, supvry vet, Immunol Invests, 56-59; prof microbiol & dean, Col Vet Med, 59-74, James Law Prof Comp Med, NY State Col Vet Med, Cornell Univ, 74-88; res prof, Dept Prev Med, Med Col, Upstate Med Ctr, State Univ NY Syracuse, 74-88; RETIRED. *Concurrent Pos:* Guest prof, Univ Bern, Switz, 74-75; chmn bd dirs, Cornell Vet, Inc, 76-86; external examr, Col Vet Med, Miss State Univ, 78-86; external adv vet med, Mount

Hope Med Complex, Univ West Indies, Trinidad/Tobago, 78-85. *Honors & Awards:* XII Int Vet Cong Prize, Am Vet Med Asn, 77. *Mem:* Fel AAAS; US Animal Health Asn; Am Vet Med Asn; Am Col Vet Microbiol; fel Am Acad Microbiol; hon mem, Polish Soc Vet Med, Polish Acad Sci; hon mem, Nat Acad Agron & Vet Med, Argentina. *Res:* Virus diseases of domestic animals, especially food-producing animals. *Mailing Add:* 122 E Remington Rd Ithaca NY 14850

POPPER, ARTHUR N, b New York, May 9, 43; m 68; c 2. AUDITORY PHYSIOLOGY. *Educ:* NY Univ, BA, 64; City Univ NY, PhD(biol), 69. *Prof Exp:* From asst prof to assoc prof zool, Univ Hawaii, 69-78; from assoc prof to prof anat & cell biol, Georgetown Univ Sch Med, 78-87; PROF & CHAIR ZOOL, UNIV MD, 87- *Concurrent Pos:* Assoc zoologist, Lab Sensory Sci, Univ Hawaii, 72-78; postdoctoral scholar, Kresge Hearing Inst, Univ Mich, 75-76; res career develop award, NIH, 78-83; adj sr scientist, Mote Marine Lab, Sarasota, Fla, 79- *Mem:* Fel AAAS; Soc Neurosci; Asn Res Otolaryngol; Acoust Soc Am; Am Soc Zoologists; Int Soc Neuroethology. *Res:* Sound detection and processing mechanisms in vertebrate animals; hearing and acoustic mechanisms of fishes. *Mailing Add:* Dept Zool Univ Md College Park MD 20742

POPPER, DANIEL MAGNES, b Oakland, Calif, Aug 11, 13; m 40; c 1. CLOSE BINARY STARS. *Educ:* Univ Calif, AB, 34, PhD(astron), 38. *Prof Exp:* Kellogg fel, Lick Observ, Univ Calif, 38-39; asst astron, McDonald Observ, Chicago-Tex, 39-40, instr, 40-42; instr, Yerkes Observ, Univ Chicago, 42-43; physicist, Radiation Lab, Univ Calif, 43-45; instr astron, Yerkes Observ, Univ Chicago, 45-46, asst prof, 46-47; from asst prof to assoc prof, 47-55, chmn dept, 51-57, 59-63, prof astron, 55-78, PROF EMER & RES ASTRON, UNIV CALIF, LOS ANGELES, 78- *Concurrent Pos:* Guest investr, Mt Wilson & Palomar Observs, 49-75; NSF sr fel, 64-65. *Honors & Awards:* Karl Schwarzschild Lectr, Astronomische Gesellschaft, 84. *Mem:* Am Astron Soc; Astron Soc Pac. *Res:* Stellar spectroscopy; radial velocities of stars; spectroscopic and eclipsing binaries. *Mailing Add:* Dept of Astron Univ of Calif Los Angeles CA 90024

POPPER, ROBERT DAVID, b Budapest, Hungary, Dec 10, 27; m 56; c 7. CONSULTING COMPUTER SCIENCES, MANAGEMENT SCIENCES. *Educ:* George Washington Univ, BA, 57, MA, 61. *Prof Exp:* Dir oper res, SCM, 65-70; dir, Sandoz Corp, 70-72; sr vpres & chief planning officer, US Trust Co, 72-82; CHIEF EXEC OFFICER, NEW DIMENSIONS CONSULT INC, 82- *Concurrent Pos:* Scientist, HSR & Dunlap & Assoc, 59-60; sr vpres, MIS J Walter Thompson, 81-82; pres, info prod group, Squibb Corp Res & Develop, 82-87. *Mem:* Math Asn Am; Am Statist Asn; Am Econ Asn; Asn Comput Mach; Am Arbitration Asn; Opers Res Soc Am. *Res:* Quantitive decision making; strategic planning; math and computer sciences; co-author book on the effects on nuclear explosions. *Mailing Add:* 539 Kingston Rd Princeton NJ 08540

POPPER, THOMAS LESLIE, b Budapest, Hungary, May 5, 33; US citizen; m 62; c 2. ORGANIC CHEMISTRY. *Educ:* Eotvos Lorand Univ, Budapest, dipl chem, 56; Mass Inst Technol, PhD(org chem), 62. *Prof Exp:* Res chemist, Wander Pharmaceut Co, Budapest, 56; asst chemist, Schering Corp, NJ, 57-58; Nat Inst Gen Med Sci fel, Biochem Inst, Med Sch, Vienna Univ, 62-63; sr chemist, Rohm & Haas Co, Pa, 63-65; sr res scientist, 65-69, sect head, 69-70, mgr natural prod res, 70-74, ASSOC DIR CHEM RES, SCHERING CORP, 74- *Mem:* AAAS; Am Chem Soc; Royal Soc Chem; NY Acad Sci. *Res:* Isolation, characterization and synthesis of natural products; compounds of biological activity for pharmaceutical uses, particularly on steroidal field; radiochemistry; medicinal chemistry (allergy and inflammation). *Mailing Add:* Infectious Dis & Tumor Biol Chem Schering Corp 60 Orange St Bloomfield NJ 07003

POPPLEWELL, JAMES MALCOLM, b UK, June 2, 42; US citizen; m 66; c 2. METALLURGY, CORROSION. *Educ:* Univ Leeds, BSc, 64, PhD(metall), 68. *Prof Exp:* Eng specialist metall, 68-69, supvr corrosion group, 69-74, chief chem metallurgist, 74-76, assoc dir chem metall & eng, 76-81, DIR PROD RES, OLIN CORP, 81- *Mem:* Nat Asn Corrosion Engrs; Electrochem Soc; Am Soc Testing & Mat. *Res:* Corrosion and surface studies; physical and chemical metallurgy; electron optics; glass sealing; materials usage in solar energy, chemical plants and power generation. *Mailing Add:* Olin Corp 120 Long Ridge Rd Stamford CT 06904-1355

POPPOFF, ILIA GEORGE, b San Diego, Calif, Apr 9, 24; m 44; c 3. AERONOMY. *Educ:* Whittier Col, BA, 47. *Prof Exp:* Physicist, US Naval Radiol Defense Lab, 43-53; assoc physicist, Stanford Res Inst, 53-56, sr physicist, 56-63, chmn dept atmospheric sci, 63-67; chief atmospheres & astrophysics br, Space Sci Div, NASA Ames Res Ctr, 67-70, actg chief space sci div, 68-69, chief earth sci applns off, 71-74, asst chief space div, 70-79, chief stratospheric proj off, 74-79; pres, Flick Point Assocs, Inc, 79-82; SCI WRITER, 82- *Concurrent Pos:* Ed adv bd, Geophys Surv, 72-85; mem, Tahoe Area Land Acquisition Comn, 82-83 & Calif Lay Adv Planning Comn of the Tahoe Regional Planning Agency; pres, Tahoe Resource Conserv Dist, 82-87; chmn, Lahontan Regional Water Qual Control Bd, 85. *Mem:* AAAS; Am Geophys Union. *Res:* Chemistry and physics of the stratosphere. *Mailing Add:* 2909 17-Mile Dr Pebble Beach CA 93953

POPPY, WILLARD JOSEPH, b New London, Wis, June 23, 07; m 34; c 2. PHYSICS. *Educ:* Univ Wis, Oshkosh, EdB, 30; Univ Iowa, MS, 33, PhD(physics), 34. *Prof Exp:* Assoc prof physics & head dept, Cleveland State Univ, 34-42, prof & head dept, 43-49; lectr, Univ Iowa, 42-43; from assoc prof to prof, 49-75, EMER PROF PHYSICS, UNIV NORTHERN IOWA, 75- *Concurrent Pos:* Pres, Iowa Acad Sci, 76-77. *Mem:* AAAS; Am Asn Physics Teachers. *Res:* Zinc crystals; electrical resistivity of zinc crystals. *Mailing Add:* 2117 Clay St Cedar Falls IA 50613

POPS, HORACE, b Queens, NY, July 20, 36; m 58; c 3. METALLURGY. *Educ:* Rensselaer Polytech Inst, BMetE, 57; Lehigh Univ, MMetE, 58; Univ Pittsburgh, DSc(metall), 62. *Prof Exp:* Jr fel ferrous alloys, Mellon Inst, 58-63, fel & head ferrous alloy group, 62-63, fel metal physics, 63-65, sr fel & head copper alloys, 65-67; head res & develop metall, 67-72, DIR METALS LAB, ESSEX INT, 72- *Concurrent Pos:* Mem adj staff, Univ Pittsburgh, 67-69. *Honors & Awards:* Medal Award, Wire Asn, 76 & 80. *Mem:* Am Soc Metals; Am Inst Mining, Metall & Petrol Engrs. *Res:* Physical metallurgy; phase transformations; cryogenics; x-ray diffraction; dispersion hardening; copper alloys; martensite. *Mailing Add:* 10610 Pine Mills Rd Ft Wayne IN 46825

POP-STOJANOVIC, ZORAN RISTA, b Belgrade, Yugoslavia, May 30, 35; US citizen; m 66. MATHEMATICS. *Educ:* Univ Belgrade, BS, 58, PhD(math), 64. *Prof Exp:* Instr math, Univ Belgrade, 59-64, asst prof, 64-65; from asst prof to assoc prof, 65-72, PROF MATH, UNIV FLA, 72-, ASSOC CHMN MATH. *Concurrent Pos:* Army Res Off-Durham fel, Univ Fla, 68-70. *Mem:* Am Math Soc. *Res:* Probability theory; random processes; laws of large numbers; Martingales; stochastic integration; spectral theory of random fields. *Mailing Add:* Dept Math Univ Fla Gainesville FL 32611

PORANSKI, CHESTER F, JR, b Brooklyn, NY, Mar 13, 37; m 59; c 4. COMPOSITES, RUBBER. *Educ:* Univ San Francisco, BS, 59; St Joseph's Univ, Pa, MS, 61; Univ Md, PhD(phys chem), 73. *Prof Exp:* Qual control & applns chemist, Prod Res & Chem Corp, 59-61; RES CHEMIST, NAVAL RES LAB, 63- *Mem:* Am Chem Soc; Sigma Xi. *Res:* Basic and applied research for the characterization of the processes occuring during thermal and radiation induced degradation of materials; characterization of polymers by spectroscopic and other techniques; reinforced elastomeric composites. *Mailing Add:* Code 6120 Naval Res Lab Washington DC 20375-5000

PORCELLA, DONALD BURKE, b Modesto, Calif, Oct 2, 37; m 61; c 3. ENVIRONMENTAL SCIENCE. *Educ:* Univ Calif, Berkeley, AB, 59, MA, 61, PhD(environ health sci), 67. *Prof Exp:* Asst specialist, USPHS, 61-63; asst specialist, Univ Calif, Berkeley, 63-65; Fulbright fel, Norweg Inst Water Res, 67-68; asst res zool, Univ Calif, Berkeley & Lake Tahoe Area Coun, 68-70; asst prof, Utah State Univ, 70-72, assoc prof biol, 72-78, prof, Civil Environ Eng, 78-79, assoc dir, Utah Water Res Lab, 77-79; prin scientist, Tetra Tech Inc, 79-84; PROJ MGR, ELEC POWER RES INST, 84- *Concurrent Pos:* Consult, Bechtel Corp, 70-71, Procter & Gamble, 71-72, Intermountain Consults & Planners, 75-79, Kennecott Copper Corp, 76-80 & PUD, South Lake Tahoe, 77-81; head div environ eng, IPA, US Environ Protect Agency, 76-77. *Mem:* AAAS; Am Soc Limnol & Oceanog; Water Pollution Control Fedn; Sigma Xi; Int Asn Water Pollution Res. *Res:* Eutrophication; nutrients and productivity; radiological health and radioecology; resource utilization and environmental management; water quality and pollution control; minimum flow requirements in streams; stream productivity; ecosystem modeling; hazardous wastes and toxicity; bioassay-soils, fresh and salt water; acidic deposition in lakes; metal biogeochemistry; genetic ecology. *Mailing Add:* 1034 Lindsay Ct Lafayette CA 94549

PORCELLO, LEONARD J(OSEPH), b New York, NY, Mar 1, 34; m 62; c 2. ELECTRICAL ENGINEERING, PHYSICS. *Educ:* Cornell Univ, BA, 55; Univ Mich, MS, 57, MSE, 59, PhD(elec eng), 63. *Prof Exp:* Res asst radar & optics, Inst Sci & Technol, Univ Mich, Ann Arbor, 55-57, grad res asst, 57-58, instr elec eng, Univ, 58-61, res engr, Inst, 61-69, asst head radar & optics lab, 63-68, head, 68-72, from assoc prof to prof elec eng, Univ, 69-72, assoc dir, Willow Run Labs, 70-72; dir radar & optics div & vpres, Environ Res Inst Mich, 73-76; asst vpres & div mgr, 76-79, vpres, Sci Appl, Inc, 79-85, corp vpres, 85-87, SR VPRES, SCI APPL INT CORP, 87-, MGR, DEFENSE SYSTS GROUP, 86- *Concurrent Pos:* Consult, Radar & Optics Lab, Inst Sci & Technol, Univ Mich, 58-61; lectr, Eng Summer Conf Prog, Univ Mich, 63-70, course co-chmn, 66-70; adj prof elec eng, Univ Mich, 73-76; spec adv, Div Adv Group, Aeronaut Systs Div, US Air Force, 73-84; trustee, Environ Res Inst Mich, 75. *Mem:* Fel Inst Elec & Electronics Engrs; Optical Soc Am; AAAS; Sigma Xi. *Res:* Radar technology and applications; synthetic aperture radar; coherent optics technology; optical data processing; radio wave propagation through the atmosphere; remote sensing. *Mailing Add:* 13190 E Camino La Cebadilla Tucson AZ 85749-9324

PORCH, WILLIAM MORGAN, b Athens, Ohio, Nov 8, 44; m 78; c 1. ATMOSPHERIC PHYSICS, GEOPHYSICS. *Educ:* Univ Utah, BA, 66; Univ Wash, MS, 68, PhD(geophys), 71. *Prof Exp:* Optical engr, Boeing Aircraft Co, 66-68; res assoc geophys, Univ Wash, 68-72; physicist atmospheric sci, Lawrence Livermore Lab, 72-87; ATMOSPHERIC PHYSICIST, LOS ALAMOS NAT LAB, 87- *Concurrent Pos:* Fel, Univ Wash, 71-72; lectr atmospheric sci, Univ Calif, Davis, 73. *Mem:* Am Meteorol Soc; Am Geophys Union; Optical Soc Am. *Res:* Physical and chemical properties of atmospheric aerosols; remote sensing of transport and diffusion of atmospheric pollutants in complex terrain and wind energy prospecting. *Mailing Add:* O-466 Los Alamos Nat Lab Los Alamos NM 87545

PORCHER, RICHARD DWIGHT, b Charleston, SC, Jan 1, 39; m 69; c 2. BOTANY, TAXONOMY. *Educ:* Col Charleston, BS, 62; Univ SC, MS, 66, PhD(biol), 74. *Prof Exp:* Teacher sci, N Charleston High Sch, 62-63; lab asst, Univ SC, 64-67; from instr to asst prof, Voorhees Col, 69-70; asst prof, 70-78, ASSOC PROF BIOL, THE CITADEL, 78- *Concurrent Pos:* Dir & cur, The Citadel Herbarium, 71-; contractor, Belle W Baruch Found grant, 76-80, Nat Audubon Soc grant, 77-80 & US Forest Serv grant, 78-80. *Mem:* Sigma Xi; Bot Soc Am. *Res:* Taxonomy and ecology of vascular flora of Coastal Plain of South Carolina; endangered and threatened species. *Mailing Add:* Dept of Biol The Citadel Charleston SC 29409

PORE, ROBERT SCOTT, b Toledo, Ohio, Feb 26, 38; m 65; c 1. MEDICAL MYCOLOGY, MICROBIOLOGY. *Educ:* Univ Tex, Austin, BA, 60; Tex A&M Univ, MS, 62; Univ Calif, Los Angeles, PhD, 65; Am Bd Med Microbiol, dipl, 76. *Prof Exp:* NIH fel med mycol, Univ Okla, 65-67, asst prof microbiol, 67-68; from asst prof to assoc prof, 68-83, PROF MICROBIOL, MED SCH, WVA UNIV, 84- *Mem:* Med Mycol Soc of the Americas; Int Soc

Human & Animal Mycol; Am Soc Microbiol; Mycol Soc Am. *Res:* Pathobiology and systematics of fungi and algae; all aspects of the biology of Prototheca species; microbial biopolymers; antibiotic suseptibility tests. *Mailing Add:* Dept Microbiol WVa Univ Md Morgantown WV 26506

PORETZ, RONALD DAVID, b New York, NY, 40; m 65; c 1. BIOCHEMISTRY, IMMUNOCHEMISTRY. *Educ:* Hartwick Col, BA, 61; Long Island Univ, MS, 63; State Univ NY Buffalo, PhD(biochem), 68. *Prof Exp:* Nat Heart Inst fel, Lister Inst Prev Med, London, 68-70; asst prof biochem, Med Ctr, Univ Kans, 70-73; from asst prof to assoc prof, 73-80, PROF BIOCHEM, NELSON BIOL LAB, RUTGERS UNIV, 80- *Mem:* AAAS; Sigma Xi; Am Soc Molecular Biol & Biochem; Am Chem Soc; Am Asn Immunol. *Res:* Properties characteristics and biosynthesis of glycoproteins. *Mailing Add:* Dept Molecular Biol & Biochem Nelson Biol Lab Rutgers Univ PO Box 1059 Piscataway NJ 08855-1059

PORIES, WALTER J, b Munich, Ger, Jan 18, 30; US citizen; m 51; c 6. SURGERY, NUTRITION. *Educ:* Univ Rochester, MD, 55; Wesleyan Univ, BA, 73; Am Bd Surg & Bd Thoracic Surg, dipl, 63. *Prof Exp:* Chief surg & obstet, Rosieres AFB Hosp, US Air Force, Toul, France, 56-58, resident gen & thoracic surg, Strong Mem Hosp, Rochester, NY, 58-62, instr, 60-62; surgeon, Wright-Patterson AFB Hosp, 62-64, chief gen & thoracic surg, 64-67; asst prof surg & assoc surgeon, Sch Med & Dent, Univ Rochester, 67-69; prof surg & assoc dir dept, Sch Med, Case Western Reserve Univ, 69-77; chief dept surg, Cleveland Metrop Gen Hosp, 69-77; PROF & CHMN DEPT SURG, SCH MED, EAST CAROLINA UNIV, 77-; CHIEF SURG, PITT CO MEM HOSP, 77- *Concurrent Pos:* Atomic Energy Comn fel radiation biol, Univ Rochester, 58-59; dir intern & residency training prog, Wright-Patterson AFB Hosp, 65-67; dir, Cancer Ctr, Northeast Ohio, 72-74; chief of staff, Cuyahoga County Hosp, 72-74. *Mem:* Am Surg Asn; Soc Univ Surgeons; Vascular Soc; Am Col Surgeons. *Res:* Metabolism of trace metals; wound healing; problems in pulmonary embolism and vascular surgery; obesity; cancer. *Mailing Add:* Dept of Surg East Carolina Univ Sch Med Greenville NC 27858

PORILE, NORBERT THOMAS, b Vienna, Austria, May 18, 32; nat US; m 57; c 1. NUCLEAR CHEMISTRY. *Educ:* Univ Chicago, AB, 52, SM, 54, PhD(chem), 57. *Prof Exp:* Res assoc chem, Brookhaven Nat Lab, 57-59, from assoc chemist to chemist, 59-64; assoc prof, 65-69, PROF CHEM, PURDUE UNIV, LAFAYETTE, 69- *Concurrent Pos:* Fel, Soc Prom Sci Japan, Univ Tokyo Inst Nuclear Study & Kyoto, 61; vis prof, McGill Univ, 63-65; Guggenheim Mem fel, Inst Nuclear Physics, France, 71-72; Alexander von Humboldt sr scientist award, Marburg, 82; prog adv comt, Los Alamos Meson Physics Facil, Los Alamos Nat Lab, 85-89; comt nuclear & radiochemistry, Nat Res Coun, 85-90; ed, Monogr on Radiochemistry of the Elements & Radiochem Techniques, Nat Res Coun, 86-90. *Mem:* Am Chem Soc; Am Phys Soc. *Res:* Nuclear reactions and fission; high and intermediate energy nuclear reactions; collider physics. *Mailing Add:* Dept of Chem Purdue Univ West Lafayette IN 47907

PORKOLAB, MIKLOS, b Budapest, Hungary, Mar 24, 39; US citizen. PLASMA PHYSICS. *Educ:* Univ BC, BASc, 63; Stanford Univ, MS, 64, PhD(appl physics), 67. *Prof Exp:* Res asst, Plasma Physics, Microwave Lab, Stanford Univ, 64-67; res assoc, Princeton Univ, 67-68, mem res staff, 68-71, res physicist, 71-75, sr res physicist, Plasma Physics Labs, 75-77, from assoc prof to prof astrophys sci, 73-77; PROF PHYSICS, MASS INST TECHNOL, 77- *Concurrent Pos:* Consult, Bell Labs, 70-77, Plasma Physics Labs, Princeton Univ, 77-81, Lawrence Livermore Nat Lab, 78- & United Technol, 78-80; vis prof, Univ Colo, Boulder, 75; vis scientist, Max Planck Inst, WGer, 75-76, 88; Alexander von Humboldt US sr scientist award, 75; consult, Gen Atomics, 87-, Spire Corp, 87-89; ed, Physics Letters, 91-; US deleg, Int Union Pure & Appl Physics, 90- *Honors & Awards:* Excellence in Plasma Physics Res Award, Am Phys Soc, 84. *Mem:* Fel Am Phys Soc. *Res:* Experimental and theoretical plasma physics; plasma waves; instabilities; nonlinear phenomena; radio frequency heating of thermonuclear plasmas; plasma confinement experiments. *Mailing Add:* Dept Physics 36-213 Mass Inst Technol Cambridge MA 02139

PORSCHE, JULES D(OWNES), b London, Eng, July 7, 09; US citizen; m 35; c 3. CHEMISTRY. *Educ:* Univ Chicago, BS, 30, PhD(org chem), 33. *Prof Exp:* Res chemist, Munic Tuberc Sanit, Chicago, 33-35; res chemist, Armour & Co, 35-40, head biol sect, 40-44, asst dir chem res & develop dept, 44-50, dir res & develop chem & by-prod dept, 50-52, mgr cent res dept, 52-56, staff asst to vpres, 58-61; PRES, JULES D PORSCHE & ASSOCS, 61- *Mem:* AAAS; Am Chem Soc; fel Am Inst Chem; Am Oil Chem Soc; Inst Food Technologists. *Res:* Hydrolytic instability of carbon to carbon and carbon to nitrogen bonds; silicosis chemistry; fractionation of plasma and plasma proteins; chemistry of fatty and amino acids; chemical market research; research administration. *Mailing Add:* Beacon Hill 2400 S Finley Rd E317 Lombard IL 60148-4870

PORSCHING, THOMAS AUGUST, b Pittsburgh, Pa, Jan 25, 36; m 81; c 1. MATHEMATICS. *Educ:* Carnegie Inst Technol, BS, 57, MS, 58, PhD(math), 64. *Prof Exp:* Staff mathematician, Bettis Atomic Power Lab, Westinghouse Elec Corp, 60-68, fel mathematician, 68-71, adv mathematician, 71-73; assoc prof, 73-79, PROF MATH, UNIV PITTSBURGH, 79- *Concurrent Pos:* Lectr, Carnegie Inst Technol, 65-71; consult, Westinghouse Elec Corp, 73-88, Inel, 88-, NASA, 89- & EPRI, 89- *Mem:* Math Asn Am; Soc Indust & Appl Math. *Res:* Numerical analysis; differential equations; computational fluid dynamics. *Mailing Add:* Dept Math Univ Pittsburgh Pittsburgh PA 15260

PORT, CURTIS DEWITT, b Cheyenne, Wyo, Aug 25, 30; m 57; c 2. LABORATORY ANIMAL MEDICINE, VETERINARY PATHOLOGY. *Educ:* Univ Calif, Davis, BS, 54, DVM, 56; Northwestern Univ, MS, 67; Univ Chicago, PhD(pathol), 72. *Prof Exp:* Dir, animal care, Northwestern Univ, 64-67; vet med res investr, Hines Vets Hosp, 67-70; sr veterinary pathologist, Ill Inst Technol Res Inst, 70-74, sci adv, 74-77; actg dir, Lab Animal Med, Northwestern Univ, 78-80, assoc prof & assoc dir, Path & Animal Resources, 77-82; pathologist, 82-87, DIR, LAB ANIMAL RESOURCES, G D SEARLE & CO, 87- *Concurrent Pos:* Instr pathol, Ill Col Podiat Med, 75-77; consult pathologist, Travenol Labs, 76-, Lincoln Park Zoo, Chicago, 78- & Standard Oil Corp, 78-; lectr, Safety Carcinogenesis Lab, Nat Cancer Inst, 77-81; adj prof path, Univ Ill Med Sch, Chicago, 89- *Mem:* AAAS; Soc Toxicol Pathologists; Am Asn Lab Animal Sci. *Res:* Lung pathology; lung cancer; laboratory animal science; laboratory animal pathology. *Mailing Add:* 1448 Lois Ave Park Ridge IL 60068

PORT, SIDNEY C, b Chicago, Ill, Nov 27, 35; m 57; c 3. MATHEMATICS. *Educ:* Northwestern Univ, AB, 57, MS, 58, PhD(math), 62. *Prof Exp:* Mathematician, Rand Corp, 62-66; lectr, 65-66, assoc prof, 66-69, PROF MATH, UNIV CALIF, LOS ANGELES, 69- *Mem:* Am Math Soc; fel Inst Math Statist. *Res:* Mathematical probability theory. *Mailing Add:* Dept Math Univ Calif Los Angeles CA 90024

PORTA, EDUARDO ANGEL, b Buenos Aires, Arg, Dec 12, 24; nat; m 55; c 3. EXPERIMENTAL PATHOLOGY. *Educ:* Univ Buenos Aires, BSc, 43, MD, 52. *Prof Exp:* Asst pathologist, Children Hosp, Buenos Aires, Arg, 52-57; asst resident, Sch Med, Wash Univ, 57-58, lectr, 57-60, instr path & Life Ins Found Res fel, 58-60; chmn dept path, Inst Biol & Exp Med, Buenos Aires, 60-61; res assoc path, Albany Med Col, 61-63; res assoc, Res Inst, Hosp Sick Children, 63-65, asst scientist, 65-68, assoc scientist, 68-70; prof path & chmn dept, Univ Buenos Aires, 70-71; PROF PATH, UNIV HAWAII, MANOA, 71- *Honors & Awards:* Price Award, Arg Soc Pediat, 56-57. *Mem:* Am Soc Exp Path; Am Asn Path & Bact; Am Inst Nutrit; Nutrit Soc Can; Int Acad Path. *Res:* Nutritional hepatic injuries; electron microscopy of hepatic diseases; ceroid pigment formation; acute and chronic alcoholism. *Mailing Add:* Dept of Path Univ of Hawaii at Manoa Honolulu HI 96822

PORTA, HORACIO A, b Cordoba, Arg, July 1, 39; m 64; c 3. MATHEMATICS. *Educ:* Univ Buenos Aires, Mat Lic, 60; NY Univ, PhD(math), 66. *Prof Exp:* Instr math, NY Univ, 65-66 & Univ Ill, 66-67; prof, Univ La Plata, 67-69; from asst prof to assoc prof, 69-80, PROF MATH, UNIV ILL, URBANA, 80- *Concurrent Pos:* Prof, Univ Buenos Aires, 73-74; vis prof, Inst Univ Dini, Florence, Italy, 76 & Univ SBolivar, Venezuela, 77. *Honors & Awards:* Math Prize, Odol Found, 69. *Res:* Analysis; operator theory. *Mailing Add:* Univ Ill 1409 W Green St Urbana IL 61801-2875

PORTE, DANIEL, JR, b New York, NY, Aug 13, 31; m 51; c 3. MEDICINE, ENDOCRINOLOGY. *Educ:* Brown Univ, AB, 53; Univ Chicago, MD, 57; Am Bd Internal Med, dipl, 64. *Prof Exp:* Asst prof med & clin investr, 65-70, assoc prof, 70-73, PROF MED, UNIV WASH, 73- *Concurrent Pos:* Assoc chief of staff, Vet Admin Med Ctr, Seattle, 70- *Honors & Awards:* Lilly Award, Am Diabetes Asn. *Mem:* Am Fedn Clin Res; Am Diabetes Asn (pres, 86-87); Am Soc Clin Invest; Asn Am Physicians; fel Am Col Physicians. *Res:* Endocrinology; mechanism of insulin secretion in man; effects of catecholamines; mechanism for lipemia in man; effects of diet on serum lipid levels; pathophysiology of diabetes; diabetic neuropathy. *Mailing Add:* Dept Med Univ Wash Sch Med Vet Admin Med Ctr 1660 S Columbian Way Bldg 13 Rm 113 Seattle WA 98108

PORTELANCE, VINCENT DAMIEN, b Woonsocket, RI, June 5, 23; Can citizen. MICROBIOLOGY. *Educ:* Univ Montreal, BSc, 46, MSc, 47, PhD(org chem), 50. *Prof Exp:* Govt France fel, 50-51; asst prof org chem, Univ Montreal, 51-52, res assoc, Microbial Biochem, Inst Microbiol & Hyg, 52-55, res officer, 55-57, head lab & exec res secy, 57-75, lectr bact physiol, Fac Med, 57-87; dir bact, Inst Armand-Frappier, 75-88; RETIRED. *Concurrent Pos:* Res prof fac med, Univ Montreal, 69-87; prof, Univ Quebec, 74-88. *Mem:* Am Soc Microbiol; Am Chem Soc; Soc Nuclear Med; sr mem Chem Inst Can; Soc Can Microbiol (secy-treas, 61-64, 2nd vpres, 68, pres, 75-76). *Res:* Chemistry of natural products; structure determination and synthesis of alkaloids; chemistry and biochemistry of mycobacteria. *Mailing Add:* 130 Clermont L Rap Laval PQ H7N 2Z7 Can

PORTER, A DUANE, b Detroit, Mich, Dec 31, 36; m 60; c 2. MATHEMATICS. *Educ:* Mich State Univ, BS, 60, MS, 61; Univ Colo, Boulder, PhD(math), 64. *Prof Exp:* From asst prof to assoc prof, Univ Wyo, 64-69, actg head, Math Dept, 78-79, dir, Sci-Math Teaching Ctr, 79-83, PROF MATH, UNIV WYO, 69- *Concurrent Pos:* Vis prof, Clemson Univ & Humboldt State Univ, 78. *Mem:* Am Math Soc; Math Asn Am; Nat Coun Teachers Math; Nat Sci Teachers Asn. *Res:* Matrix theory; combinatorial problems; finite fields. *Mailing Add:* Dept Math Univ Wyo Laramie WY 82071

PORTER, ALAN LESLIE, b Jersey City, NJ, June 22, 45; m 68; c 3. TECHNOLOGY ASSESSMENT, INDUSTRIAL ENGINEERING. *Educ:* Calif Inst Technol, BS, 67; Univ Calif, Los Angeles, MA, 68, PhD(eng psychol), 72. *Prof Exp:* Res assoc, Prog in Social Mgt Technol, Univ Wash, 72-74; from asst prof to assoc prof, 75-86, PROF INDUST & SYSTS ENG, GA INST TECHNOL, 86-, DIR, TECHNOL POLICY & ASSESSMENT CTR, 89- *Concurrent Pos:* Co-prin investr, NSF grants, 74-75, 81-83, 82-85, 84-86, 89-91, Fund for Improv Post Sec Educ grant, 77-78 & Dept Labor, 82-84; prin investr, Dept Transp grant, 77-79, NSF grant, 78-81, 82-83 & Off Technol Assessment, 84-85; exec dir, Int Asn Impact Assessment, 87-89. *Mem:* AAAS; Am Soc Eng Educ; Int Asn Impact Assessment (secy, 81-87). *Res:* Public policy analysis for technology intensive issues; quasi-experimental statistical designs; social studies of science. *Mailing Add:* Sch of Indust & Systs Eng Ga Inst of Technol Atlanta GA 30332

PORTER, BEVERLY FEARN, b Elizabeth, NJ, Aug 11, 35; m 61. RESEARCH ADMINISTRATION, SCIENCE POLICY. *Educ:* Antioch Col, BA, 58; NY Univ, MA, 64. *Prof Exp:* Lectr, Univ Pa, 64-67 & City Col NY, 67-69; res asst hist & sociol sci, 69-71, dept dir, 72-75, DIR & MGR STATIST, AM INST PHYSICS, 76- *Concurrent Pos:* Consult, Physics Surv Comt, Nat Acad Sci, 70-71 & 84-85; mem bd dirs, Comn Professions Sci & Technol, 88 & Adv Comt, Am Chem Soc, 89-; adv, NSF. *Mem:* AAAS. *Res:* Supply, utilization and demand for physicists and related scientists and engineers; demography of science; women in science; survey research methods. *Mailing Add:* Am Inst Physics 335 E 45th St New York NY 10017

PORTER, CHARLES WARREN, b Patterson, La, Aug 16, 25; m 49; c 4. ZOOLOGY, PHYSIOLOGY. *Educ:* Fisk Univ, AB, 49; Univ Mich, AM, 51, PhD, 70. *Prof Exp:* Prof biol, Miss Indust Col, 50-51; asst prof, San Antonio Jr Col, 51-52; asst prof, Southern Univ, 52-55; res asst genetics, Univ Mich, 56-57, res assoc surg, 57-60; chief technician, Mich Sci Co, 60-62; asst prof, 62-70, ASSOC PROF BIOL, SAN JOSE STATE COL, 70- *Concurrent Pos:* Nat Found fel, 56-57; Ford Found fel, Univ Mich, 69-70, NSF fel, 70; researcher reproduction & early development, MBL, Woods Hole, Mass, 78, 79, 81, 87. *Res:* Steroid hormones and their effects on metabolism in certain target organs, particularly the estrogens and adrenal cortical hormones. *Mailing Add:* 611 Astor Pl Dr New Iberia LA 70560

PORTER, CHASTAIN KENDALL, dentistry, oral pathology, for more information see previous edition

PORTER, CLARENCE A, b McAlester, Okla, Mar 19, 39; div; c 2. PARASITOLOGY. *Educ:* Portland State Col, BS, 62; Ore State Univ, MS, 64, PhD(parasitol), 66. *Prof Exp:* Asst prof biol, Portland State Univ, 66-71, assoc prof biol & exec asst to pres, 71-72; asst v-provost acad affairs & adj assoc prof zool, Univ NH, 72-76; assoc vchancellor acad affairs, State Univ Minn, 77-79; EXEC DIR, PHYLLIS WHEATLEY COMMUNITY CTR, MINNEAPOLIS, 79- *Mem:* AAAS; Am Soc Parasitol; Sigma Xi. *Res:* Molluscan taxonomy of Northwest pulmonates; parasite-host relationships. *Mailing Add:* 7186 Lasting Light Way Columbia MD 21045

PORTER, CLARK ALFRED, b Chico, Calif, Apr 3, 25; m 50; c 2. PLANT GROWTH REGULATORS, HERBICIDES. *Educ:* Chico State Col, AB, 48; Ore State Col, PhD(plant path), 53. *Prof Exp:* Assoc plant pathologist, Boyce Thompson Inst Plant Res, 52-59, plant pathologist, 59-62; scientist, 62-70, sci fel, 70-79, SR SCI FEL, MONSANTO CO, 79- *Mem:* Am Soc Plant Physiol; Am Phytopath Soc; Am Chem Soc; Phytochem Soc NAm; Plant Growth Regulators Soc Am; Weed Sci Soc Am. *Res:* Biochemistry and physiology of virus-diseased plants; chemotherapy and its biochemical basis; metabolism of pesticidal chemicals; plant growth regulators; crop yield physiology: photosynthesis, transport, nitrogen, metabolism and drought; physiology/biochemistry of herbicidal action. *Mailing Add:* 406 Parkwood Ave Kirkwood MO 63122

PORTER, CLYDE L, JR, b Williston, SC, Sept 24, 32; m 57; c 2. ECOLOGY. *Educ:* Univ SC, BS, 57, MS, 59; Univ Okla, PhD(bot), 63. *Prof Exp:* Instr, Univ Okla, 62; asst prof biol, Univ Miami, 63-68; assoc prof, 68-80, PROF BIOL, PURDUE UNIV, NORTH CENT CAMPUS, 80-, CHMN BIOL & CHEM SECT. *Concurrent Pos:* NSF grant, Univ Miami, 64-66. *Mem:* Ecol Soc Am. *Res:* Methods in determining trophic relationships; effects of various pollutants on community structure and metabolism. *Mailing Add:* Dept Biol & Chem Purdue Univ North Cent Campus Westville IN 46391

PORTER, CURT CULWELL, b Weatherford, Tex, Sept 14, 14; m 37; c 1. BIOCHEMISTRY. *Educ:* Agr & Mech Col, Tex, BS, 36; Johns Hopkins Univ, PhD(biochem), 41. *Prof Exp:* Asst org chem, Johns Hopkins Univ, 41-45; res chemist, Merck Inst Therapeut Res, Merck & Co Inc, 45-58, dir physiol chem, 58-71, sr scientist, 71-81; RETIRED. *Mem:* Am Soc Biol Chem; Am Soc Pharmacol. *Res:* Biochemical pharmacology. *Mailing Add:* 711 Fitzwatertown Rd Glenside PA 19038

PORTER, DANIEL MORRIS, b Newton Grove, NC, July 6, 36; m 59; c 2. PHYTOPATHOLOGY. *Educ:* Atlantic Christian Col, BS, 60; NC State Univ, MS, 63, PhD(plant path), 66. *Prof Exp:* PLANT PATHOLOGIST, SOUTH ATLANTIC AREA, TIDEWATER RES CTR, AGR RES SERV, USDA, 66-, RES & LOCATION LEADER, 81- *Mem:* Am Phytopath Soc; fel Am Peanut Res & Educ Soc. *Res:* Diseases of the peanut; ecology of soil-borne peanut diseases; peanut pod mycoflora. *Mailing Add:* USDA Agr Res Serv Tidewater Res Ctr Suffolk VA 23437

PORTER, DARRELL DEAN, b Fennimore, Wis, May 21, 38; m 64; c 3. MINING ENGINEERING. *Educ:* Wis State Col & Inst Technol, BS, 60; Colo Sch Mines, MS, 62; Univ Minn, PhD(mining eng), 71. *Prof Exp:* Res engr, 62-76, TECH SPECIALIST, EXPLOSIVES PROD DIV, E I DU PONT DE NEMOURS & CO, INC, 76- *Mem:* Am Inst Mining, Metall & Petrol Engrs. *Res:* Rock mechanics pertaining mainly to science of rock fragmentation and blasting mechanics; design and application of commercial explosives. *Mailing Add:* 5438 S Iola Way Englewood CO 80111

PORTER, DAVID, b Bronxville, NY, Oct 20, 41; m 65; c 2. BOTANY. *Educ:* Yale Univ, BS, 63; Univ Wash, 66-67, PhD(bot), 67. *Prof Exp:* Fel, Woods Hole Oceanog Inst, 67-68; NIH fel, Harvard Univ, 68-69; asst prof, 69-74, ASSOC PROF BOT, UNIV GA, 74- *Concurrent Pos:* NIH grant, 72-74. *Mem:* Mycol Soc Am; Soc Protozool; Bot Soc Am; Am Soc Cell Biol; Sigma Xi. *Res:* Marine mycology; ultrastructural cytology and cell motility. *Mailing Add:* Dept of Bot Univ of Ga Athens GA 30602

PORTER, DAVID DIXON, b Columbus, Ohio, Apr 21, 35; m 66; c 1. VIROLOGY, PATHOLOGY. *Educ:* Univ Pittsburgh, BS, 57, MD, 61. *Prof Exp:* Intern, Med Ctr, Duke Univ, 61-62; NIH & San Diego County United Fund fel, Scripps Clin & Res Found, Calif, 62-65; NIH fel Wistar Inst Anat & Biol, 65-67; asst prof virol, Baylor Col Med, 67-69; from asst prof to assoc prof, 69-75, PROF PATH, SCH MED, UNIV CALIF, LOS ANGELES, 75- *Mem:* NY Acad Sci; Am Asn Immunol; Am Soc Microbiol; Reticuloendothelial Soc; Am Asn Pathologists. *Res:* Persistent virus infections; viral immunology and pathology. *Mailing Add:* Dept Path Univ of Calif Sch Med Los Angeles CA 90024-1732

PORTER, DUNCAN MACNAIR, b Kelseyville, Calif, Apr 20, 37; m 66; c 4. SYSTEMATIC BOTANY, PHYTOGEOGRAPHY. *Educ:* Stanford Univ, AB, 59, AM, 61; Harvard Univ, PhD(biol), 67. *Prof Exp:* Asst, Calif Acad Sci, San Francisco, 66-67; asst prof biol, Univ San Francisco, 67-68; asst botanist & cur flora panama, Mo Bot Garden, 68-72; ed-in-chief, Flora NAm Prog, Smithsonian Inst, 72-73; assoc prog dir syst biol, NSF, 73-75; assoc prof, 75-84, PROF BOT, VA POLYTECH INST & STATE UNIV, 84- *Concurrent Pos:* Asst prof, Washington Univ, 68-73; res assoc, Smithsonian Inst, 75-; vis Erskine fel, Univ Canterbury, Christchurch, NZ, 86. *Mem:* Fel AAAS; Am Soc Plant Taxonomists (secy, 72-73); Hist Sci Soc; Sigma Xi. *Res:* Systematics of the families Burseraceae, Rutaceae, and Zygophyllaceae; phytogeography of tropical America, especially the Galapagos Islands; Charles Darwin's botanical work. *Mailing Add:* Dept Biol Va Polytech Inst & State Univ Blacksburg VA 24061

PORTER, FREDERIC EDWIN, b Cleveland, Ohio, Dec 9, 22; wid; c 4. BACTERIOLOGY. *Educ:* Ohio State Univ, BSc, 48, MSc, 50, PhD(bact), 52. *Prof Exp:* Asst instr bact, Ohio State Univ, 51-52; proj leader, Battelle Mem Inst, 52-59; res bacteriologist, Northrup King & Co, 59-77, proj mgr, 77-82, dir agron res, 82-83, dir, Res Serv Div, 83-84; RETIRED. *Res:* Seed processing technology; control seed germination; microbial interactions; legume inoculation. *Mailing Add:* 9255 W 23rd St Minneapolis MN 55426

PORTER, FREDERICK STANLEY, JR, b Baltimore, Md, Sept 18, 26; m 51; c 2. PEDIATRICS. *Educ:* Princeton Univ, BSE, 48; Johns Hopkins Univ, MD, 52. *Prof Exp:* Intern pediat, Johns Hopkins Hosp, 52-54, asst resident, 54-55; fel pediat hemat, Children's Med Ctr, Boston, 55-58; asst prof, Sch Med, Univ Ark, 58-64; from assoc prof to prof pediat hemat, Med Ctr, Duke Univ, 64-75; vpres med affairs, 75-89, Children's Hosp of the King's Daughters, Norfolk, 75-87; prof & chmn, Dept Pediat, 75-89, PROF PEDIAT, EASTERN VA MED SCH, 89- *Res:* Pediatric hematology. *Mailing Add:* Dept Pediat Eastern Va Med Sch Box 1980 Norfolk VA 23501

PORTER, GARY DEAN, b Alliance, Nebr, June 3, 42; m 64; c 2. MIRROR CONFINEMENT PHYSICS, PLASMA CONFINEMENT MODELING. *Educ:* Pa State Univ, MS, 66, PhD(nuclear eng), 68. *Prof Exp:* PHYSICIST, LAWRENCE LIVERMORE LAB, 68- *Mem:* Am Physics Soc. *Res:* Magnetic confinement of controlled fusion plasmas. *Mailing Add:* L-637 Lawrence Livermore Lab PO Box 5511 Livermore CA 94550

PORTER, GEORGE A, b Medford, Ore, Dec 22, 31; c 4. MEDICINE. *Educ:* Ore State Univ, BS, 53; Univ Ore, MS & MD, 57. *Prof Exp:* Instr med, Univ Ore, 61-62; res assoc, Cardiovasc Res Inst, San Francisco, 62-64; from asst prof to assoc prof, 64-73, chief nephrology sect, 70-73, head div nephrology, 73-77, PROF MED, MED SCH, UNIV ORE, 73-, CHMN MED, UNIV ORE HEALTH SCI CTR, 77- *Concurrent Pos:* Ore Heart Asn Howard Irwin fel cardiol, Med Sch, Univ Ore, 60-62; NIH career develop award, 65-70; mem exec comt, Coun Kidney in Cardiovasc Dis, Am Heart Asn, 71-, chmn coun kidney in cardiovasc dis, 77-79, pres sci councils, 80-; ed-in-chief, Am J Kidney Disease, 80- *Mem:* Endocrine Soc; Am Fedn Clin Res; Am Heart Asn; Am Physiol Soc; Am Soc Nephrology. *Res:* Regulation of sodium transport; drug induced nephrotoxicity. *Mailing Add:* Dept Med Ore Health Sci Univ Sch Med 3181 SW Sam Jackson Portland OR 97201

PORTER, GERALD BASSETT, b Eng, Apr 5, 26; Can citizen; m 50; c 2. ANALYTICAL CHEMISTRY, INORGANIC CHEMISTRY. *Educ:* Univ Calif, BS, 50; Univ Southern Calif, PhD(chem), 55. *Prof Exp:* Res chemist, Calif Res Corp, Standard Oil Co Calif, 50-51; Shell fel, Univ Rochester, 54-56; from instr to prof, Univ BC 56-81, hon prof chem, 81; pres, Enchem Develop Co, 81-87; RETIRED. *Concurrent Pos:* Vis prof, Univ Frankfurt, 62-63, Univ Bologna, 70-71 & Univ Victoria, 77-78. *Mem:* Sigma Xi; Int-Am Photochem Soc; fel Chem Inst Can. *Res:* Photochemistry; inorganic complexes; coordination compounds; solar energy conversion; emission spectroscopy; energy conversion; energy conservation. *Mailing Add:* 754-1515 W Second Ave Vancouver BC V6J 5C5 Can

PORTER, GERALD JOSEPH, b Elizabeth, NJ, Feb 27, 37; m 60; c 3. MATHEMATICS. *Educ:* Princeton Univ, AB, 58; Cornell Univ, PhD(math), 63. *Hon Degrees:* MA, Univ Pa, 71. *Prof Exp:* Instr math, Mass Inst Technol, 63-65; from asst prof to assoc prof math, Univ Pa, 65-75, undergrad chmn dept, 71-73, assoc dean comput, Sch Arts & Sci, 81-91, PROF MATH, UNIV PA, 75- *Concurrent Pos:* Off Naval Res res grant, Brandeis Univ, 65-66; vis, Inst Adv Studies, 69-70; bd gov, Math Asn Am, 80-83 & 86-, chmn comt Comput Math Educ, 83-87, Investment Comt, 86-, Audit & Budget Comt, 86-, Finance Comt, 86- *Mem:* Am Math Soc; Asn Comput Mach; Math Asn Am; AAAS; Asn Women Math. *Res:* Algebraic topology; homotopy theory, especially homotopy operations, higher order Whitehead products; computer graphics; use of computers in undergraduate mathematics instruction. *Mailing Add:* Dept of Math Univ Pa 209 S 33rd St Philadelphia PA 19104-6395

PORTER, GILBERT HARRIS, b Northampton, Mass, Nov 25, 25; m 50; c 2. ANIMAL NUTRITION. *Educ:* Univ Mass, BS, 49; Univ Conn, MS, 51; Pa State Univ, PhD(dairy prod, biochem), 56. *Prof Exp:* Asst dairy husb, Univ Conn, 49-51; asst prof, Univ Mass, 51-52; instr, Pa State Univ, 54-56; dairy & livestock specialist, Beacon Milling Co, 56-61, dir dairy & livestock res, 61-63, dir res & develop, 64; dir nutrit & tech sales, Specialty Div, Albers Milling Co, 64-65; dir animal nutrit & health dept, Div, Allied Chem Corp, 65-69; dir dairy & livestock mgt, 69-70, dir res, 70-72, VPRES RES & DEVELOP, AGWAY INC, 72- *Mem:* Am Soc Animal Sci; Am Dairy Sci Asn; Agr Res Inst (treas, 76). *Res:* Utilization of alfalfa hay and alfalfa silage by the mature dairy cow and the dairy calf; feeding high-producing cows to obtain maximum net profits; intensive, semi-environmentally controlled veal production; ruminant utilization of non-protein nitrogen. *Mailing Add:* 4881 Candy Ln Manlius NY 13104

PORTER, HARDIN KIBBE, b Kingsville, Tex, Aug 19, 17; m 45; c 4. ORGANIC CHEMISTRY. *Educ:* Univ Tex, BS, 40; Ga Inst Technol, MS, 42; Duke Univ, PhD(chem), 52. *Prof Exp:* Chemist, Merck & Co, Inc, 46-48; asst, Duke Univ, 50-51; chemist, 51-58, PROCESS SUPVR, E I DU PONT DE NEMOURS & CO, 58- *Mem:* Fel Am Inst Chem; Am Chem Soc. *Res:* Antimalarial quinoline compounds; fungicidal phenolic compounds and effect of structure; amino acids; development and organic research; synthetic chemistry. *Mailing Add:* 2005 Kynwyd Rd Wilmington DE 19810

PORTER, HAYDEN SAMUEL, JR, b Cincinnati, Ohio, June 2, 45; m 67; c 2. APPLIED PHYSICS, AERONOMY. *Educ:* Univ Cincinnati, BS, 67, PhD(physics), 73. *Prof Exp:* Instr & res assoc, Univ Fla, 73-75, asst res scientist physics, 75-76; sr mem tech staff, Comput Sci Corp, 76-79; from asst prof to assoc prof, 79-88, DANIEL DISTINGUISHED PROF, DEPT COMPUTER SCI, FURMAN UNIV, 89-, DEPT CHMN, 86- *Concurrent Pos:* Consult, Comput Sci Corp, 79-80; prin investr, NASA, 80-; rev, jour Geophys Res, 86- *Mem:* Am Phys Soc; Am Geophys Union; Sigma Xi; Asn Comput Mach; Inst Elec & Electronic Engrs. *Res:* Computer simulation, computer vision, real-time computation, fluid dynamics, and the interaction of photons and electrons with matter. *Mailing Add:* Dept Comput Sci Furman Univ Greenville SC 29613

PORTER, HERSCHEL DONOVAN, b Silverton, Ohio, Aug 9, 24; m 64; c 1. ORGANIC CHEMISTRY. *Educ:* Wilmington Col, BS, 44; Univ Ill, PhD(org chem), 47. *Prof Exp:* Asst chem, Univ Ill, 44-47; res chemist, 47-70, RES SCIENTIST, ELI LILLY & CO, 70- *Concurrent Pos:* Res chemist, Monsanto Chem Co, Ala, 46. *Mem:* Am Chem Soc. *Res:* Steric hindrance; addition of Grignard reagents to an olefinic hydrocarbon; Willgerodt reaction; esters of geometrically isomeric vinyl alcohols; herbicides; anthelmintics. *Mailing Add:* 5911 Central Indianapolis IN 46220-2511

PORTER, JACK R, b Oklahoma City, Okla, Mar 31, 38; m 59; c 2. TOPOLOGY. *Educ:* Univ Okla, BS, 60; NMex State Univ, MS, 64, PhD(math), 66. *Prof Exp:* From asst prof to assoc prof, 66-77, PROF MATH, UNIV KANS, 77- *Mem:* Am Math Soc; Math Asn Am. *Res:* Hausdorff extensions (especially H-closed extensions) and absolutes of Hausdorff spaces; minimal topological spaces. *Mailing Add:* Dept Math Univ Kans Lawrence KS 66045-2142

PORTER, JAMES ARMER, JR, b Fredonia, Kans, Oct 8, 22; m 43; c 3. PRIMATOLOGY, VETERINARY PARASITOLOGY. *Educ:* Kans State Univ, DVM, 44; Tulane Univ, MPH, 61; Univ Ill, Urbana, PhD(parasitol), 65; Col Lab Animal Med, dipl, 71. *Prof Exp:* Vet, US War Food Admin, 44; pvt practice, 45-47, 49-60; vet, Foot & Mouth Dis Eradication, Bur Animal Industs, 47-49; parasitologist, Univ Miami, 62; vet ext serv, Univ Ill, Urbana, 62-63; vet parasitologist, Gorgas Mem Lab, 65-68; vet lab animal med, Vet Admin, 68-70; vet primatologist, & dir S Am primates 70-85, DIR PRIMATE SERV, 85-; VETERINARIAN, 85- *Concurrent Pos:* Consult vet, Monkey Jungle, 80-; lab animal vet, Fla Intern Univ, 80-, Barry Univ, 89-90; US AID, Univ Wis, Univ Riogrande do Sul, Porto Alegre, Brazil, 70-72. *Mem:* Am Soc Lab Animal Pract; Am Soc Parasitol; Am Col Lab Animal Med; Am Vet Med Asn; Am Asn Lab Animal Sci. *Res:* Developed non-human primate as model for studies on human malaria; Author of several books. *Mailing Add:* Primate Services PO Box 970768 Miami FL 33197

PORTER, JAMES COLEGROVE, b Nov 5, 37; m 66. ACOUSTICS. *Educ:* Purdue Univ, BS, 59, PhD(physics), 64. *Prof Exp:* Res assoc physics, Univ Mich, 64, instr, 64-67; asst prof, 67-70, ASSOC PROF PHYSICS, EASTERN MICH UNIV, 70- *Mem:* Am Phys Soc; Acoust Soc Am; Audio Engr Soc. *Res:* Acoustics via finite element and boundary element methods. *Mailing Add:* Dept Physics Eastern Mich Univ 303 Strong Bldg Ypsilanti MI 48197

PORTER, JAMES FRANKLIN, b Coshocton, Ohio; m 63; c 1. MATHEMATICS, TOPOLOGY. *Educ:* Ohio State Univ, BS, 57, MS, 62; Syracuse Univ, PhD(math), 71. *Prof Exp:* From instr to asst prof math, Hiram Col, 63-66; from asst prof to assoc prof, Univ Ark, Fayetteville, 71-83; chmn, math dept, 83-89, PROF, UNIV MISS, 83- *Concurrent Pos:* Sr vis res assoc, Calif Inst Technol, 81. *Mem:* Am Math Soc; Math Asn Am. *Res:* Riesz spaces and representations; analysis of positive operations. *Mailing Add:* Dept Math Univ Miss University MS 38677

PORTER, JAMES W, b Tiffin, Ohio, Oct 5, 46; m 72; c 1. MARINE BIOLOGY, MARINE ECOLOGY. *Educ:* Yale Col, BS, 69; Yale Univ, PhD(biol), 73. *Prof Exp:* Predoctoral fel, Smithsonian Trop Res Inst, Panama, 71-72; asst prof natural resources, Univ Mich, 73-77; biologist, Environ Protection Agency, Gulf Breeze Lab, Fla, 80; assoc prof, 77-85, PROF ZOOL, UNIV GA, 85-, CUR INVERT, MUS NATURAL HIST, 85- *Concurrent Pos:* Prin investr, NSF, Environ Protection Agency, Nat Oceanic & Atmospheric Asn, NPS grants, 73-; vis lectr, Marine Biol Lab, Woods Hole, 74-; ed, Ecol & Ecol Monographs & Ecol Soc Am, 75-80. *Mem:* Am Soc Limnol & Oceanogr; Ecol Soc Am; fel AAAS. *Res:* Marine ecology; coral reef biology; biodiversity; ecological physiology. *Mailing Add:* Zool Dept Univ Ga Athens GA 30602

PORTER, JOHN CHARLES, b Paducah, Tex, Mar 1, 25; m 52. PHYSIOLOGY. *Educ:* Baylor Univ, BA, 49; Tex Technol Col, MA, 50; Iowa State Col, PhD(zool, physiol), 52. *Prof Exp:* Asst physiol, Iowa State Col, 50-52; res instr endocrinol, Duke Univ, 52-53; from instr to assoc prof, 53-63, PROF PHYSIOL, UNIV TEX HEALTH SCI CTR DALLAS, 63- *Mem:* AAAS; Endocrine Soc. *Res:* Metabolic function of androgens; functional relationship of the hypothalamus and the pituitary. *Mailing Add:* Dept Obstet-Gynec Univ Tex Southwestern Med Ctr 5323 Harry Hines Blvd Dallas TX 75235-9032

PORTER, JOHN E(DWARD), b Charleston, WVa, July 6, 21. ELECTRONICS ENGINEERING, ASTRONOMY. *Educ:* Ore State Col, BS, 46. *Prof Exp:* Instr physics, Northrop Aeronaut Inst, 47-55, PROF ELECTRONIC ENG & TECHNOL, NORTHROP UNIV, 55- *Res:* Atomic physics; celestial mechanics; circuit theory; field theory. *Mailing Add:* Dept Electronic Eng Northrop Univ W 5800 Arbor Vitae St Los Angeles CA 90045

PORTER, JOHN J, b East Orange, NJ, Feb 15, 32; m 59; c 3. PHYSICAL ORGANIC CHEMISTRY. *Educ:* Ga Inst Technol, BS, 56, PhD(org chem), 61. *Prof Exp:* Res chemist, Am Cyanamid Corp, 60-62; asst prof, 62-67, assoc prof, 67-76, PROF TEXTILE CHEM, CLEMSON UNIV, 76- *Concurrent Pos:* Mem, Nat Comt Textile Educ, Am Asn Textile Chemists & Colorists, 67-; consult, Fed Water Pollution Control Admin, US Dept Interior, 68-, US Dept Interior grant, 69-70. *Mem:* Am Chem Soc; Am Inst Chem Engrs; Am Asn Textile Chemists & Colorists. *Res:* Water pollution and waste treatment; membrane recovery of chemicals, water and energy from wastewater; photochemistry of dye degradation; organometallic polymers, dyeing thermodynamics; cellulose chemistry. *Mailing Add:* Dept of Textiles Clemson Univ Clemson SC 29631

PORTER, JOHN ROBERT, b Conroe, Tex, May 25, 40; m 62. MATHEMATICS, MATHEMATICAL PHYSICS. *Educ:* Univ Tex, BS, 62, PhD(physics), 64. *Prof Exp:* Res assoc low energy nuclear physics, Tex Nuclear Corp, 59-62; res assoc gen relativity, Southwest Inst Adv Studies, 63-64; vis lectr, King's Col, Univ London, 65-67; vis prof, Free Univ Brussels, 66; asst prof, 67-74, ASSOC PROF MATH, UNIV PITTSBURGH & RES ASSOC, CTR PHILOS SCI, 74- *Mem:* AAAS; Am Math Soc; Sigma Xi; Am Phys Soc. *Res:* Application of modern methods of mathematics to the problems of mathematical physics, particularly to the general theory of relativity; differential topology. *Mailing Add:* Dept Math Univ Pittsburgh Main Campus 922 William Pitt Union Pittsburgh PA 15260

PORTER, JOHN WILLARD, b Mukwonago, Wis, June 12, 15; m 41; c 4. BIOCHEMISTRY, GENETICS. *Educ:* Univ Wis, BS, 38, PhD(biochem), 42. *Prof Exp:* Fel, Dept Agr Chem, Purdue Univ, 42-44; asst chemist nutrit, Med Nutrit Lab, Chicago, 45; assoc chemist & asst prof agr chem, Purdue Univ, 45-49; sr scientist biol sect, Hanford Atomic Works, Gen Elec Co, Richland, Wash, 49-54; Nat Heart Inst fel, Enzyme Inst, Univ Wis, 54-56; asst chief, Radioisotope Serv, 56-64, CHIEF, LIPID METAB LAB, VET ADMIN HOSP, MADISON, 64-; PROF PHYSIOL CHEM, UNIV WIS-MADISON, 64- *Concurrent Pos:* From asst prof to assoc prof, Univ Wis-Madison, 56-64. *Mem:* AAAS; Am Inst Chem; Am Chem Soc; Am Soc Biol Chem; Am Inst Biol Sci. *Res:* Enzymatic synthesis of fatty acids, cholesterol and carotenoids; isolation and identification of intermediates; purification of enzymes; study of control mechanisms, synthesis of enzymes; mechanism of plaque formation in atherosclerosis. *Mailing Add:* 1710 Baker Ave Madison WI 53705

PORTER, JOHN WILLIAM, b Longview, Tex, Aug 24, 37. FLUID MECHANICS. *Educ:* Rice Univ, BA, 59; Calif Inst Technol, MS, 60, PhD(eng sci), 63. *Prof Exp:* Asst prof aerospace eng, WVa Univ, 63-64; asst prof, 64-69, ASSOC PROF AEROSPACE ENG & ENG MECH, UNIV TEX, AUSTIN, 69- *Concurrent Pos:* Consult, Res & Eng Support Div, Inst Defense Anal, 64- *Mem:* Am Inst Aeronaut & Astronaut. *Res:* High-temperature gas dynamics; jet propulsion; combustion; solid-gas interface interactions. *Mailing Add:* Dept Aerospace Studies Univ Tex Austin TX 78712

PORTER, JOHNNY RAY, b Bernice, La, Sept 18, 44; m 68; c 1. PHYSIOLOGY, ENDOCRINOLOGY. *Educ:* Western Ky Univ, BS, 66; Northeast La Univ, MS, 68; La State Univ, PhD(physiol), 73. *Prof Exp:* Asst biol, Northeast La Univ, 66-68; grad asst physiol, 70-73, from instr to asst prof, 73-83, ASSOC PROF PHYSIOL, LA STATE UNIV MED CTR, 83- *Mem:* Am Physiol Soc; Soc Exp Biol Med; AAAS; Sigma Xi. *Res:* Hypothalamic-pituitary adrenalaxis; pituitary biosynthesis of rat ACTH; pineal interaction with hypothalamic-pituitary-adrenal axis; brain uptake of melanin; brain gut peptides: possible interaction with hypothalamic-pituitary-adrenal axis; involvement of pituitary-adrenal in obesity and diabetes. *Mailing Add:* Dept Physiol La State Univ Med Ctr New Orleans LA 70119

PORTER, KEITH ROBERTS, b Yarmouth, NS, June 11, 12; nat US; m 38; c 1. CELL BIOLOGY. *Educ:* Acadia Univ, BSc, 34; Harvard Univ, MA, 35, PhD(biol), 38. *Hon Degrees:* DSc, Acadia Univ, 64; LLD, Queen's Univ, Ont, 66; DSc, Med Col Ohio, Toledo & Univ Toronto, 78, Rockefeller Univ, 80, Univ Colo, 82. *Prof Exp:* Nat Res Coun fel zool, Princeton Univ, 38-39; asst, Rockefeller Univ, 39-45, assoc, 45-50, assoc mem, 50-56, mem & prof, 56-61; prof biol, Harvard Univ, 61-70, chmn dept, 65-67; prof molecular, cellular & develop biol & chmn dept, Univ Colo, Boulder, 68-74, distinguished prof, 78, prof biol, 74-84; prof & chmn, Dept Biol Sci, Univ Md, 84-88; RETIRED. *Concurrent Pos:* Co-ed, Protoplasma, 64-; ed, J Biophys & Biochem Cytol; dir, Marine Biol Lab, Woods Hole, Mass, 75-76; organizer, First Int Cong Cell Biol, 86. *Honors & Awards:* Warren Triennial Award, 62; Passano & Gardner Found Awards, 64; co-recipient, Louisa Gross Horwitz Prize, 70; co-recipient, Paul-Ehrlich-Ludwig-Darmstaedter Prize, 71; co-recipient, Dickson Prize, 71; Electron Micros Soc Am Award, 75; Nat Medal Sci Award, 77; Waterford Biomedical Award, Scripps Clinic, 79; Henry Gray Award, Am Asn Anat, 81; E B Wilson Award, Am Soc Cell Biol, 81. *Mem:* Nat Acad Sci; Am Soc Zool; Tissue Cult Asn (pres, 78-80); Am Soc Cell Biol (pres, 61 & 78); Electron Micros Soc Am (pres, 62 & 90). *Res:* Experimental embryology; tissue culture; electron microscopy; cell fine structure. *Mailing Add:* Dept Biol Sci Leidy Bldg Univ Pa 34th & Spruce St Philadelphia PA 19104

PORTER, KENNETH RAYMOND, b Laramie, Wyo, Oct 20, 31; m 82; c 3. VERTEBRATE ECOLOGY, SPECIATION & EVOLUTION. *Educ:* Univ Wyo, BS, 53; Ore State Col, MS, 59; Univ Tex, PhD(zool), 62. *Prof Exp:* From asst prof to assoc prof ecol, Univ Denver, 62-74; sr ecologist & proj mgr, 74-80, PARTNER & CONSULT, DAMES & MOORE ENVIRON CONSULT, 80- *Concurrent Pos:* NSF res grants, 62-64 & 65-66; adj assoc prof zool, Univ Denver, 73- *Mem:* Ecol Soc Am. *Res:* Vertebrate evolution and speciation; ecology and biogeography; interdisciplinary environmental impact assessments; ecology and systematics of Eutamias; physiological ecology of amphibians. *Mailing Add:* 6681 Village Rd Parker CO 80134

PORTER, LAWRENCE DELPINO, b Los Angeles, Calif, Oct 5, 32; m 64; c 2. GEOPHYSICS, APPLIED MATHEMATICS. *Educ:* Stanford Univ, BS, 55; Univ Calif, Los Angeles, MS, 57, PhD(physics), 62. *Prof Exp:* Res geophysicist, Univ Calif, Los Angeles, 55-61; Alexander von Humboldt Found res fel theoret physics, Aachen Tech Univ, 61-63; res assoc, Courant Inst Math Sci, NY Univ, 63-64; asst prof aerospace eng, Polytech Inst Brooklyn, 64-68; physicist, Lawrence Radiation Lab, 68-71; sci specialist,

EG&G, Inc, 71-72; vis assoc prof geol, Northern Ill Univ, 72-73; Nat Res Coun-NASA sr resident res assoc, Jet Propulsion Lab, 73-75, mem tech staff, 75-76; seismologist, Calif Div Mines & Geol, 76-81; lead analyst, Sohio Petrol Co, 81-84; sr analyst, Systs Control Technol Inc, 84-90; SR ANALYST, GEOSPECTRA INC, 90- *Concurrent Pos:* Lectr geol, Calif State Univ, Los Angeles, 75-76 & Calif State Univ, Northridge, 76. *Mem:* Am Geophys Union; Royal Astron Soc; Soc Indust & Appl Math; Am Math Soc; Seismol Soc Am; Earthquake Eng Res Inst. *Res:* Geophysics, geodesy and seismology; use of applied mathematics in data analysis; statistics; engineering studies of earthquakes and explosions; wave propagation; diffraction theory. *Mailing Add:* PO Box 564 Alamo CA 94507

PORTER, LEE ALBERT, b Cleveland, Ohio, July 24, 34; m 68; c 2. NATURAL PRODUCTS CHEMISTRY. *Educ:* Univ Calif, Los Angeles, BSc, 57, PhD(org chem), 64. *Prof Exp:* Instr chem, Univ Calif, Los Angeles, 63; res assoc med chem, Sch Med, Univ Louisville, 64-65; Nat Inst Arthritis & Metab Dis fel chem inst, Univ Vienna, 65-66; res assoc org chem, Univ Minn, Minneapolis, 66-67; career res chemist, Lab Exp Metab Dis, Vet Admin Hosp, Long Beach, Calif, 67-71; asst prof, 71-78, ASSOC PROF CHEM, JACKSONVILLE UNIV, 78-, DIR MILLAR WILSON LAB CHEM RES, 71- *Mem:* Am Chem Soc; AAAS; NY Acad Sci. *Res:* Isolation, structure, chemistry, biosynthesis and biological action of substances of plant and marine origin; structure-activity relations; organometallic chemistry; chemical taxonomy. *Mailing Add:* 5039 Timuguana Rd No Ten Jacksonville FL 32210

PORTER, LEO EARLE, b Pueblo, Colo, Feb 3, 39; m 59; c 2. CHEMICAL MICROSCOPY, MATERIALS SCIENCE. *Educ:* Univ Colo, BA, 63. *Prof Exp:* SR RES MICROSCOPIST, GOODYEAR TIRE & RUBBER CO, 63- *Honors & Awards:* Best Paper Award, Am Chem Soc, 69. *Mem:* Am Chem Soc. *Res:* The microscopy of rubber chemicals, their identification and characterization in trace amounts; failure analysis; modes and interactions of components in composite structures. *Mailing Add:* Goodyear Tire & Rubber Co 1144 E Market St Akron OH 44316-0001

PORTER, LEONARD EDGAR, b New Limerick, Maine, Nov 19, 34; m 58; c 2. NUCLEAR PHYSICS. *Educ:* Miami Univ, AB, 56; Univ Wis-Madison, MS, 61, PhD(physics), 65. *Prof Exp:* Res assoc res in particle physics, Univ Calif, Riverside, 65-66, lectr physics & asst res physicist, 66-67; from asst prof to assoc prof, Univ Mont, 67-77, prof physics & astron, 77-90, chmn dept, 76-90; DIR, RADIATION SAFETY OFF & PROF PHYSICS, WASH STATE UNIV, 90- *Concurrent Pos:* Mem adj fac, Marine Corps Command & Staff Col, 74-80; dir, Mont Sci Fair, 74-85. *Mem:* Am Phys Soc; Am Asn Physics Teachers; Sigma Xi. *Res:* Nucleon polarization in reactions and scatterings; hypernuclei; stopping power of matter for charged projectiles. *Mailing Add:* Dept Physics Wash State Univ Pullman WA 99164-2814

PORTER, LEW F(ORSTER), b Madison, Wis, July 18, 18; m 52; c 3. PHYSICAL METALLURGY. *Educ:* Univ Wis, BS, 40, MS, 50, PhD(metall eng), 55. *Prof Exp:* Chief chemist, Ind Harbor Works, Am Steel Foundries, 40-44; res metallurgist, Chain Belt Co, Wis, 46-49; instr metall eng & proj asst, Univ Wis, 49-54; res technologist, Res Lab, US Steel Corp, 55-62, sect supvr, 62-74, res consult phys metall, 74-76, sr res consult phys metall, Res Lab, 76-83; RETIRED. *Honors & Awards:* Howe Medal, Am Soc Metals, 53 & 66. *Mem:* Am Soc Testing & Mat; fel Am Soc Metals; Am Inst Mining, Metall & Petrol Engrs; The Metals Soc. *Res:* Basic research in physical metallurgy of steel; vitreous enameling of gray cast iron; effect of stress on transformations in steel; radiation effects in steel; alloy-steel development. *Mailing Add:* 1234 Northwestern Dr Monroeville PA 15146

PORTER, LYNN K, b Helper, Utah, June 26, 29; m 54; c 4. SOIL CHEMISTRY. *Educ:* Utah State Univ, BS, 53, MS, 54; NC State Col, PhD(soil chem), 57. *Prof Exp:* Asst agron, Utah State Univ, 53-54 & NC State Col, 54-56; soil scientist, 57-74, SUPVRY SOIL SCIENTIST, AGR RES SERV, USDA, 74- *Mem:* Am Soc Agron; fel Soil Sci Soc Am; Sigma Xi. *Res:* Chemistry and biochemistry and the efficiency of nitrogen in the soil-water-plant system; investigations of nitrogen pollution and mechanisms of nitrogen loss from soils. *Mailing Add:* 1320 Lory St Ft Collins CO 80521

PORTER, MARCELLUS CLAY, b Louisville, Ky, May 14, 38; m 62; c 2. CHEMICAL ENGINEERING. *Educ:* Mass Inst Technol, BS, 60, MS, 62, ScD(chem eng), 64. *Prof Exp:* Group leader mat sci, Avco Space Systs Div, 64-69; prod develop mgr, Amicon Corp, Lexington, Mass, 69-70, mgr indust separations div, 70-72; VPRES RES & DEVELOP & FOUNDER, NUCLEPORE CORP, 72- *Concurrent Pos:* Tech consult, Indust Membrane Technol, Ctr Prof Advan. *Mem:* Am Inst Chem Engrs; Am Chem Soc. *Res:* Membrane separations, particularly microfiltration, ultrafiltration and reverse osmosis. *Mailing Add:* 3449 Byron Ct Pleasanton CA 94588

PORTER, NED ALLEN, b Marion, Ohio, May 10, 43; m 64; c 2. ORGANIC CHEMISTRY. *Educ:* Princeton Univ, BSChE, 65; Harvard Univ, PhD(chem), 70. *Prof Exp:* Asst prof, 69-74, assoc prof, 74-79, PROF CHEM, DUKE UNIV, 79- *Mem:* Royal Soc Chem; Am Chem Soc. *Res:* Physical organic chemistry; mechanisms of photochemical and free radical reactions; concerted reactions. *Mailing Add:* Dept Chem Duke Univ Durham NC 27706-8001

PORTER, OWEN ARCHUEL, b Tuckerman, Ark, Dec 21, 43. AGRONOMY. *Educ:* Agr, Mech & Norm Col, Ark, BS, 66; NDak State Univ, MS, 70, PhD(soils), 75. *Prof Exp:* RES AGRONOMIST, UNIV ARK, PINE BLUFF, 72- *Mem:* Am Soc Agron; Sigma Xi; Am Soybean Asn. *Res:* Soybean production, fertility and nutrition. *Mailing Add:* Dept Agr Univ Ark Pine Bluff AR 71601

PORTER, RAYMOND P, b Worcester, Mass, Nov 1, 30; m 57; c 3. PHYSICAL CHEMISTRY. *Educ:* Worcester Polytech Inst, BS, 53; Univ Rochester, PhD(phys chem), 58. *Prof Exp:* Res chemist, E I du Pont de Nemours & Co, 58-61; fel, Mellon Inst, 61-65; chemist, Space & Missile Div, Gen Elec Co, Pa, 65-68, res chemist, Res & Develop Ctr, NY, 68-72; RES CHEMIST, ACUSHNET CO, 72- *Mem:* Sigma Xi. *Res:* Photochemistry of gases; mechanical behavior of polymers; heterogeneous catalysis; high temperature kinetics; friction, lubrication and wear of rubber. *Mailing Add:* 1034 S Broadway St Akron OH 44311

PORTER, RICHARD A, b Edwardsville, Ill, Feb 25, 33; m 63; c 1. PHYSICAL CHEMISTRY. *Educ:* Northwestern Univ, BS, 54; Univ Calif, Los Angeles, PhD(chem), 59. *Prof Exp:* Physicist, Lawrence Radiation Lab, Univ Calif, Berkeley, 58-61; vis res chemist, Univ Calif, San Diego, 61-62; asst prof chem, Univ Idaho, 62-70, assoc prof, 70-80; mem fac, dept chem, Univ Wis, 80-; AT CHEM DEPT, UNIV LOUISVILLE. *Concurrent Pos:* Consult, Rand Corp, 63- *Mem:* Am Chem Soc. *Res:* Quantum chemistry; Mossbauer effect; surface adsorption. *Mailing Add:* Dept Chem Univ Louisville Louisville KY 40292-0001

PORTER, RICHARD D, b Livingston, NJ, Dec 28, 44. GEOMETRY, TOPOLOGY. *Educ:* Lafayette Univ, BA, 67; Yale Univ, PhD(math), 71. *Prof Exp:* Instr math, Brown Univ, 71-74; PROF MATH, NORTHEASTERN UNIV, 75- *Mem:* Am Math Soc. *Mailing Add:* Dept Math Northeastern Univ Boston MA 02115

PORTER, RICHARD DEE, b Harrisville, Utah, July 11, 23; m 52; c 2. WILDLIFE ECOLOGY. *Educ:* Univ Utah, BS, 50, MS, 52; Tex A&M Univ, PhD(wildlife mgt), 62. *Prof Exp:* Assoc ecologist & chief ornith sect, Dugway Ecol Res Proj, Univ Utah, 52-55; asst prof biol, Tex Western Col, 59-60; from asst prof to assoc prof, Wis State Univ-Whitewater, 61-65; asst prof, NMex Inst Mining & Technol, 65-67; pollution biologist, Patuxent Wildlife Res Ctr, US Fish & Wildlife Serv, 67-73, res biologist, Denver Wildlife Res Ctr, 73-80; RETIRED. *Mem:* Am Ornith Union; Cooper Ornith Soc; Wilson Ornith Soc; Raptor Res Found; Am Soc Mammal. *Res:* Ecology, taxonomy and distribution of birds and mammals, especially habitat requirements and study of population; reproduction and ecology of raptors; effects of environmental pollutants on wildlife. *Mailing Add:* 325 N 300 West RR #1 Mapleton UT 84664

PORTER, RICHARD FRANCIS, b Fargo, NDak, Feb 8, 28; m 55; c 2. PHYSICAL INORGANIC CHEMISTRY. *Educ:* Marquette Univ, BS, 51; Univ Calif, PhD(chem), 54. *Prof Exp:* Res assoc physics, Univ Chicago, 54-55; from instr to asst prof chem, 55-59, assoc prof, 60-64, PROF CHEM, CORNELL UNIV, 64- *Concurrent Pos:* Sloan fel, 60-64; Guggenheim fel, Nat Res Coun Can, 64; vis prof, Univ Fla, 64; NATO sr fel, Univ Freiburg, 70-71; vis collabr, Brookhaven Nat Lab, 78-79. *Mem:* Am Chem Soc; Am Soc Mass Spectrometry. *Res:* Mass spectrometric, spectroscopic and electron diffraction studies of gaseous systems at high temperatures; thermodynamic studies of vaporization reactions; high temperature boron chemistry; photochemistry of boron compounds; chemical ionization studies of inorganic compounds. *Mailing Add:* Dept Chem Cornell Univ Baker Lab Ithaca NY 14850

PORTER, RICHARD NEEDHAM, b Texarkana, Ark, Oct 22, 32; m 60; c 2. PHYSICAL CHEMISTRY, THEORETICAL CHEMISTRY. *Educ:* Tex A&M Univ, BS, 54; Univ Ill, PhD(phys chem), 60. *Prof Exp:* Instr & res assoc, Univ Ill, 60-61; mem theoret chem staff, IBM Watson Lab, Columbia Univ, 61-62; from asst prof to prof chem, Univ Ark, 62-69; assoc prof, 69-70, PROF CHEM, STATE UNIV NY STONY BROOK, 70- *Concurrent Pos:* Vis fel, Joint Inst Lab Astrophys, Univ Colo, 68-69; Alfred P Sloan fel, 68-70; consult, Los Alamos Sci Lab, 72-77; vis prof chem, Harvard Univ, 73, 80 & 87; actg dean phys sci & math, State Univ NY Stony Brook, 90. *Honors & Awards:* Alexander von Humboldt Sr US Scientist Award, 78. *Mem:* AAAS; fel Am Inst Chemists; Am Chem Soc; Am Phys Soc; Sigma Xi. *Res:* Chemical dynamics; theoretical studies of reactive collisions; hot atom chemistry; potential energy surfaces for collision complexes; quantum theory of molecules; nonadiabatic molecular dynamics; field theory of molecules and photons. *Mailing Add:* Dept Chem State Univ NY Stony Brook NY 11790

PORTER, RICHARD W(ILLIAM), b Salina, Kans, Mar 24, 13; m 46; c 3. SYSTEMS ENGINEERING, AEROSPACE ENGINEERING. *Educ:* Univ Kans, BS, 34; Yale Univ, PhD(elec eng), 37. *Hon Degrees:* ScD, Yale Univ, 47. *Prof Exp:* Assoc prof elec eng, New Haven Jr Col, 36-37; engr, Gen Elec Co, 37-50, mgr guided missiles dept, 50-55, consult corp eng staff, 55-70, mgr sci & tech affairs, Aerospace Group, 70-76. *Concurrent Pos:* Mem tech eval group, Guided Missiles Comt Res & Develop Bd, US Dept Defense, 47-50, adv comt spec capabilities, 55-56; mem sci adv bd, US Air Force, 48-51 & 60-78; chmn panel earth satellites, US Comt for Int Geophys Year, Nat Acad Sci, 56-58, mem, Space Sci Bd, 58 & chmn int rels comt, 59-70; sci adv, Geophys Inst, Univ Alaska, 61-69; mem planetary quarantine adv panel, NASA-Am Inst Biol Sci, 70-75; aerospace, energy & environ consult, 76- *Honors & Awards:* Except Civilian Serv Award, US Air Force, 75. *Mem:* Fel Am Inst Aeronaut & Astronaut; sr mem Inst Elec & Electronics Engrs; Am Geophys Union; NY Acad Sci; Am Rocket Soc (pres, 55). *Res:* Industrial and military control equipment; servomechanisms; radar; aircraft and space technology; space research; meteorology; environmental pollution monitoring and control. *Mailing Add:* R W Porter Consult PO Box 28 Riverside CT 06878

PORTER, RICK ANTHONY, latex synthesis, pressure sensitive adhesives, for more information see previous edition

PORTER, ROBERT WILLIAM, b Pontiac, Mich, Jan 19, 38; m. MECHANICAL & AEROSPACE ENGINEERING. *Educ:* Univ Ill, BSME, 61; Northwestern Univ, MS, 63, PhD(mech eng), 66. *Prof Exp:* Res asst plasma physics, Northwestern Univ, 62; engr missiles div, Bendix Corp, 62-63; res asst magnetogasdynamics, Northwestern Univ, 64-65 & 66; assoc prof, 66-80, PROF MECH ENG, ILL INST TECHNOL, 80- *Concurrent Pos:* Consult, Dir Am Power Conf. *Honors & Awards:* Chanute Medal, Western Soc Engrs, 76. *Mem:* Am Soc Mech Engrs. *Res:* Heat transfer; electrical power generation; cogeneration; engineering economics. *Mailing Add:* Dept Mech & Aerospace Eng Ill Inst Technol Chicago IL 60616

PORTER, ROBERT WILLIS, b San Diego, Calif, Dec 18, 26; m 68; c 1. NEUROSURGERY, NEUROPHYSIOLOGY. *Educ:* Northwestern Univ, BS, 47, MS, 48, MB, 50, MD, 51, PhD(anat), 52; Am Bd Neurol Surg, dipl, 58. *Prof Exp:* Asst anat, Northwestern Univ, 49-50; intern, Los Angeles County Hosp, 50-51; from lectr anat to assoc clin prof surg, Univ Calif, Los Angeles, 51-69; PROF NEUROSURG, UNIV CALIF, IRVINE, 69- *Concurrent Pos:* USPHS fel, Univ Calif, Los Angeles, 51-52; from resident to staff neurosurgeon, Vet Admin Hosp, 52-55, chief neurosurgeon, 58-, assoc chief staff res, 59-; NIH spec fel surg neurol, Univ Edinburgh, 63-64. *Mem:* Am Asn Neurol Surgeons; Am Physiol Soc; Am Asn Anatomists; Am Col Surgeons; Am Acad Neurol Surg. *Res:* Experimental neurosurgery; neuroendocrinology; psychosomatic interrelationships. *Mailing Add:* Univ Calif 5901 E Seventh St Long Beach CA 90822

PORTER, ROGER J, b Pittsburgh, Pa, April 4, 42; m 68; c 2. EPILEPSY, PHARMACOLOGY. *Educ:* Eckerd Col, BS, 64; Duke Univ, MD, 68. *Prof Exp:* Med intern, Univ Calif, San Diego, 68-69; staff assoc, Epilepsy Br, NIH, 69-71; resident neurol, Univ Calif, San Francisco, 71-74; sr staff assoc, Epilepsy Br, 74-78, asst chief, 78-79, actg chief, 79-80, chief, Epilepsy Br, 80-84, chief, Med Neurol Br, 84-87, DEP DIR, NAT INST NEUROL & COMMUN DISORDERS & STROKE, NIH, 87- *Concurrent Pos:* Consult & lectr, Nat Naval Med Ctr, Bethesda, Md, 79-; clin prof neurol & adj prof pharmacol, Uniform Serv Univ Health Sci, 80- *Mem:* Am Acad Neurol; Am Epilepsy Soc; Soc Neurosci; Am Soc Clin Pharmacol & Therapeut; Am Neurol Asn; Am EEG Soc; hon mem Am Soc Neurol Invest. *Res:* Epilepsy and the development of new antiepileptic drugs. *Mailing Add:* NIH Bldg 31 Rm 8A-52 Bethesda MD 20892

PORTER, ROGER STEPHEN, b Windom, Minn, June 2, 28; m 53, 68; c 4. POLYMER CHEMISTRY. *Educ:* Univ Calif, Los Angeles, BS, 50; Univ Wash, PhD(chem), 56. *Prof Exp:* From res chemist to sr res assoc, Chevron Res Co, Stand Oil Co, Calif, 56-66; assoc prof, 66-70, head dept, 66-76, co-dir, Mat Res Lab, 73-85, PROF POLYMER SCI & ENG, UNIV MASS, AMHERST, 70- *Concurrent Pos:* Ed, Polymer Eng & Sci, Polymer Composites; lectr, USSR, Brazil, & Rumanian Acad Sci; mem adv panels chem & mat res, NSF. *Honors & Awards:* Bingham Medal, Soc Rheol; Gold Medal, Soc Plastics Engrs; Coatings & Plastics Award, Am Chem Soc; Mettler Prize, NAm Thermal Anal Soc. *Mem:* Am Chem Soc; Soc Rheol; fel Am Phys Soc; Brit Soc Rheol; Plastics Inst Am. *Res:* Thermodynamic measurements; polymer physics; rheology; liquid crystals. *Mailing Add:* 220 Rolling Ridge Rd Amherst MA 01003

PORTER, RONALD DEAN, b Elmira, NY, Dec 20, 45. MOLECULAR GENETICS. *Educ:* Cornell Univ, BS, 67; Duke Univ, MA, 69, PhD(biochem & genetics), 76. *Prof Exp:* Fel assoc, Radiobiol Lab, Yale Univ, 76-78; asst prof, 78-84, ASSOC PROF MICROBIOL & GENETICS, PA STATE UNIV, 84- *Concurrent Pos:* Prin investr, Pub Health Serv, 79-, NSF, 85- *Mem:* Am Soc Microbiol; Biophhys Soc. *Res:* Studies of recombination mechanism and kinetics in E coli and the development of genetic systems in blue-green algae; genetic regulation of photosynthetic pigments; mutation mechanism. *Mailing Add:* Dept Molecular Biol Pa State Univ Main Campus University Park PA 16802

PORTER, SANFORD DEE, b Salt Lake City, Utah, Dec 7, 53; m 78; c 4. ECOLOGY & BEHAVIOR OF ANTS. *Educ:* Brigham Young Univ, BS, 78, MS, 80; Fla State Univ, PhD(biol sci), 84. *Prof Exp:* Res assoc fire ants, Fla State Univ, 84-86 & Univ Tex, 86-89; RES ENTOMOLOGIST, AGR RES SERV, USDA, MAVERL, 89- *Mem:* Entom Soc Am; Int Union Study Social Insects (pres, 91-92). *Res:* Ecology and behavior of ants, especially fire ants. *Mailing Add:* USDA, ARS-MAVERL 1600 SW 23rd Dr, PO Box 14565 Gainesville FL 32604

PORTER, SPENCER KELLOGG, b Hartford, Conn, Aug 7, 37; m 64; c 1. PHYSICAL CHEMISTRY, INORGANIC CHEMISTRY. *Educ:* Grinnell Col, AB, 59; DePauw Univ, MA, 64; Iowa State Univ, PhD(phys chem), 68. *Prof Exp:* Asst prof phys chem, Concordia Col, 68-71; fel, Ames Lab, Iowa State Univ, 71-72; asst prof chem, physics & math, St Meinrad Col, 72-74; fel phys chem, Ames Lab, Iowa State Univ, 74-75; from asst prof to assoc prof inorg chem, 75-85, chmn dept, 78-81, PROF CHEM, CAPITAL UNIV, 85-, CHMN DEPT, 84- *Concurrent Pos:* Northwest Col & Univ Asn for Sci fel, Pac Northwest Div, Battelle Mem Inst, 77-79; NASA-Am Soc Elec Engrs fel, Stanford Univ, 81 & 83; sabbatical leave, chem dept, Ariz State Univ, 81-82. *Mem:* Am Chem Soc; Royal Soc Chem; Am Asn Univ Prof. *Res:* Heterogeneous catalysis especially with the rare earths; surface science especially surface composition of alloys; single crystal x-ray crystallography; hystoresis in solid-state reactions. *Mailing Add:* Dept Chem Capital Univ E Main St Columbus OH 43209

PORTER, STEPHEN CUMMINGS, b Santa Barbara, Calif, Apr 18, 34; m 59; c 3. QUATERNARY GEOLOGY. *Educ:* Yale Univ, BS, 55, MS, 58, PhD(geol), 62. *Prof Exp:* From asst prof to assoc prof geol, 62-71, assoc dir, 67-72, PROF GEOL SCI, UNIV WASH, 71-, DIR, QUATERNARY RES CTR, 80- *Concurrent Pos:* NSF res grants, 64-83; Fulbright-Hays sr res fel, NZ, 73-74; ed, Quaternary Res, 76-; assoc ed, Radiocarbon, 77-; vis fel, Clare Hall, Univ Cambridge, 80-81; mem bd, Earth Sci, NAS, 83-85; exchange prof, Univ Bergen, Norway, 88. *Honors & Awards:* Benjamin Sullivan Prize, Yale Univ, 62; Guest Prof, Academia Sinica (China), 87. *Mem:* AAAS; fel Geol Soc Am; Int Glaciol Soc; fel Arctic Inst NAm; Am Quaternary Asn; Int Union Quaternary Res. *Res:* Stratigraphy and chronology of glaciated alpine regions; volcanic stratigraphy and tephrochronology; Quaternary snowlines. *Mailing Add:* Quarternary Res Ctr Univ of Wash Seattle WA 98195

PORTER, SYDNEY W, JR, b Baltimore, Md, June 27, 32; m 61; c 1. HEALTH PHYSICS, PHYSICAL CHEMISTRY. *Educ:* St John's Col, Md, BA, 54; Am Bd Health Physics, cert, 66. *Prof Exp:* Health physics coordr power reactors, Elec Boat Div, Gen Dynamics Corp, 58-63; head radiol safety dept health physics, Armed Forces Radiobiol Res Inst, 63-69; vpres, Radiation Mgt Corp, 69-74; PRES HEALTH PHYSICS, PORTER CONSULT INC, 74- *Concurrent Pos:* Consult, Univ Pa, Dept of the Navy & Tellurometer Inc; mem, Army Inspector General's Reactor Inspection Team; health physics consult, Philadelphia Elec Co, 68. *Mem:* Am Asn Physicists in Med; Am Chem Soc; Am Indust Hyg Asn; Am Nuclear Soc; Health Physics Soc; Am Asn Radon Scientists & Technologists (pres, 91-93). *Res:* Radiological environmental monitoring; emergency planning. *Mailing Add:* Porter Consults Inc 125 Argyle Rd Ardmore PA 19003-3201

PORTER, TERENCE LEE, b New Britain, Conn, Aug 31, 35; m 64; c 2. OPTICAL PHYSICS. *Educ:* Mass Inst Technol, SB, 57; Univ Calif, Berkeley, PhD(physics), 61. *Prof Exp:* Assoc, Nat Bur Stand, 61-63, physicist, 63-65; from asst prog dir to assoc prog dir, Div Grad Educ Sci, 65-73, prog mgr, Div Higher Educ Sci, 73-76, prog dir, Div Sci Manpower Improvement, 76-77, dep div dir, Sci Educ Resources Improv, 77-82, head, fel section, 82-84, DIV DIR, RES CAREER DEVELOP, NSF, 84- *Mem:* Optical Soc Am. *Res:* Atomic and molecular physics; optical spectroscopy; science education. *Mailing Add:* Div Res Career Develop Nat Sci Found Washington DC 20550

PORTER, THOMAS WAYNE, b Bowling Green, Ohio, Aug 8, 11; m 35; c 2. INVERTEBRATE ZOOLOGY. *Educ:* Bowling Green State Univ, AB, 35, BS, 36; Univ Mich, MA, 40; Univ Kans, PhD(entom), 50. *Prof Exp:* Teacher high sch, Ohio, 36-42 & Mich, 42-44; instr biol, Univ Kans, 46-48; asst prof entom & biol, Iowa State Col, 48-50; asst prof invert zool, 50-59, prof & asst dir, Kellogg Gull Lake Biol Sta, 59-83, EMER PROF ZOOL, MICH STATE UNIV, 83- *Mem:* AAAS; Am Inst Biol Sci; Am Micros Soc (treas, 56-74, pres elect, 75, pres, 76); Entom Soc Am; Wilson Ornith Soc. *Res:* Taxonomy, life cycles and ecology of arthropods, aquatic and semiaquatic Hemiptera; Hebridae of the world; systematics of phylopoda. *Mailing Add:* Dept of Zool East Lansing MI 48824

PORTER, VERNON RAY, b Huntington Park, Calif, Sept 9, 35; m 59; c 2. INORGANIC CHEMISTRY. *Educ:* NTex State Univ, BS, 58; Ga Inst Technol, MS, 61; Pa State Univ, PhD(solid state sci), 65. *Prof Exp:* Mem tech staff, Corp Res Lab, Tex Instruments Inc, 65-79, sr mem tech staff semiconductor process res, 80-90; CONSULT, V R PORTER & ASSOC, 91- *Mem:* Soc Photo-Optical Engrs; Am Chem Soc; Coun Radiation Measurements. *Res:* Precision instrumentation; laser lithography; semiconductor processing; chemical reactions in cool plasmas; electrochemistry of group IV elements; laser microchemistry. *Mailing Add:* 450 N Maxwell Creek Rd Plano TX 75094

PORTER, WARREN PAUL, b Madison, Wis, Jan 26, 39; m 61; c 2. TOXICOLOGY. *Educ:* Univ Wis-Madison, BS, 61; Univ Calif, Los Angeles, MA, 63, PhD(physiol ecol), 66. *Prof Exp:* NIH res assoc, Wash Univ & Mo Bot Garden, 66-68; from asst prof to assoc prof, 68-74, PROF ZOOL, UNIV WIS-MADISON, 74- *Concurrent Pos:* Guggenheim fel, 79. *Mem:* Fel AAAS; Ecol Soc Am; Am Soc Ichthyologists & Herpetologists. *Res:* Animal-physical environment interactions; system modeling of large and small ecosystems; multiple low level stress effects. *Mailing Add:* Dept Zool Univ Wis 1117 W Johnson St Madison WI 53706

PORTER, WAYNE MELVIN, plant breeding, for more information see previous edition

PORTER, WILBUR ARTHUR, b Dallas, Tex, Mar 29, 41; m 63; c 2. MATERIALS SCIENCE. *Educ:* NTex State Univ, BS, 63, MS, 64; Tex A&M Univ, PhD(interdisciplinary eng), 70. *Prof Exp:* Mem tech staff, Semiconductor Res & Develop Labs, Tex Instruments Inc, 66-68; from asst prof to assoc prof, Tex A&M Univ, 68-76, dir, Inst Solid State Electronics, 72-79, asst dir, Tex Eng Exp Sta, Tex A&M Univ Syst, 79-80, dir, Tex Eng Exp Sta, 80-85, PROF ELEC ENG, TEX A&M UNIV, 76- *Concurrent Pos:* Consult, Electro Sci Industs, 75- & Tex Instruments Inc, 76; pres, Houston Area Res Ctr, 85-; mem bd regents, Univ Portland; bd mem, Woodlands Venture Capital, Dallas Venture Corp & Keplinger Co. *Mem:* Am Phys Soc; sr mem Inst Elec & Electronics Engrs (pres, 81-); Electro-Chem Soc. *Res:* Solid state device development; integrated circuit fabrication and materials processing; manufacturing systems development. *Mailing Add:* Houston Area Res Ctr 4802 Research Forest Dr The Woodlands TX 77381

PORTER, WILLIAM A, b South Haven, Mich; c 3. ELECTRICAL ENGINEERING. *Educ:* Mich Tech, 57; Univ Mich, MS, 58, PhD(elec eng), 61. *Prof Exp:* Res asst elec eng, Univ Mich, Ann Arbor, 59-61, assoc res engr, 61-62, from asst prof to prof elec eng, 61-77, res engr, 64-77; PROF ELEC ENG & CHMN DEPT, LA STATE UNIV, BATON ROUGE, 77- *Mem:* Soc Indust & Appl Math; Inst Elec & Electronics Engrs. *Res:* Automatic control and system theory; pattern recognition; applications of functional analysis to system engineering problems. *Mailing Add:* Dept of Elec Eng La State Univ Baton Rouge LA 70803

PORTER, WILLIAM FRANK, b Cedar Falls, Iowa, Apr 27, 51; m 73; c 2. HABITAT & POPULATION MODELING. *Educ:* Univ Northern Iowa, BA, 73; Univ Minn, MS, 76, PhD(ecol & behav biol), 79. *Prof Exp:* Teaching asst biol & ecol, Univ Minn, 73-75, NIH fel, 75-77; from asst prof to assoc prof, 78-88, PROF WILDLIFE BIOL, COL ENVIRON SCI & FORESTRY, STATE UNIV NY, SYRACUSE, 88-; DIR, ADIRONDACK ECOL CTR, 80- *Mem:* AAAS; Wildlife Soc; Ecol Soc Am. *Res:* Population dynamics of large vertebrates and their interaction with vegetation habitat; ecological linkages; application of knowledge to management of wildlife resources. *Mailing Add:* Illick Hall Col Environ Sci & Forestry State Univ NY Syracuse NY 13210

PORTER, WILLIAM HUDSON, b Wilson, NC, Mar 12, 40; m 63; c 2. CLINICAL CHEMISTRY & TOXICOLOGY. *Educ:* The Citadel, BS, 62; Med Univ SC, MS, 66; Vanderbilt Univ, PhD(biochem), 71. *Prof Exp:* Develop chemist, Derring Milliken Inc, 62-64; instr gen chem, Baptist Col SC, 65-66; res fel biochem, Fla State Univ, 70-72; res asst prof biochem, Mem Res Ctr, Univ Tenn, 72-74; asst prof clin chem, dept lab med, Med Univ SC, 74-78; assoc prof, 78-86, PROF CLIN CHEM, DEPT PATH, UNIV KY, 86-

Concurrent Pos: Secy-treas & pres, Am Bd Clin Chem, 86- *Mem:* Am Asn Clin Chem; Am Soc Biol Chemists; Am Chem Soc; AAAS. *Res:* Development of new analytical procedures and their application to diagnosis and the management and understanding of human disease; appropriate use of drug therapy; assessment and management of drug intoxication. *Mailing Add:* Dept Path Univ Ky Med Ctr Lexington KY 40536

PORTER, WILLIAM L, b Philadelphia, Pa, Nov 3, 17. BIOCHEMISTRY. *Educ:* Oberlin Col, AB, 38; Univ Chicago, MS, 52; Harvard Univ, PhD, 64. *Prof Exp:* Meteorologist, 52-64, RES CHEMIST, SOLDIER SCI DIRECTORATE, US ARMY NATICK RES, DEVELOP & ENG CTR, MASS, 64- *Mem:* AAAS; Sigma Xi; Am Oil Chem Soc. *Res:* Autoxidative degradation of membrane lipids; structural requirements for action of plant auxins; mechanism of action of plant auxins; effect of storage temperature on food degradation; oxidation products of lipids and alpha-tocopherol in membranes; relative effectiveness of antioxidants on membranes of freeze-dried and whole tissue foods; fluorescence detection of autoxidation; Maillard browning of sugars and ascorbic acid. *Mailing Add:* 10 Oak Ridge Ave Natick MA 01760

PORTER, WILLIAM SAMUEL, b Niagara Falls, NY, May 5, 30; m 55; c 1. THEORETICAL PHYSICS. *Educ:* Univ Buffalo, BA, 52; Yale Univ, MS, 53, PhD(physics), 58. *Prof Exp:* Mem staff physics, Los Alamos Sci Lab, Univ Calif, 57-59; from asst prof to assoc prof, Bucknell Univ, 59-66; assoc prof, 66-69, PROF PHYSICS, SOUTHERN CONN STATE COL, 69- *Concurrent Pos:* NSF grant, 61-66. *Mem:* AAAS; Am Asn Physics Teachers; Am Phys Soc; Sigma Xi. *Res:* Nuclear reaction theory; plasma physics; planetary constitutions. *Mailing Add:* Dept Physics Southern Conn State Univ New Haven CT 06515

PORTERFIELD, IRA DEWARD, b Greenville, WVa, July 8, 20; c 4. DAIRY HUSBANDRY. *Educ:* Univ Md, BS, 44; WVa Univ, MS, 50; Univ Minn, PhD, 56. *Prof Exp:* Asst exten dairyman, WVa Univ, 46-51, assoc prof dairy husb, 51-54; asst, Univ Minn, 54-55; prof dairy sci & head dept, WVa Univ, 56-62; head dept, NC State Univ, 62-76, prof animal sci, 76; RETIRED. *Mem:* Fel AAAS; Am Soc Animal Sci; Am Dairy Sci Asn. *Res:* Breeding and physiology of dairy cattle. *Mailing Add:* Rte 12 Box 175P Statesville NC 28677

PORTERFIELD, JAY G, b Holton, Kans, July 1, 21; m 57; c 1. AGRICULTURAL ENGINEERING. *Educ:* Iowa State Univ, BS, 47, MS, 50. *Prof Exp:* From instr to asst prof agr eng, Iowa State Univ, 47-51; from assoc prof to prof, 52-82, head dept agr eng, 74-78, EMER PROF AGR ENG, OKLA STATE UNIV, 82- *Concurrent Pos:* Serv engr, Iowa Ford Tractor Co, 49; consult, Brazil, 58; res reviewer, USDA SEA-CR, Washington, DC, 78-79. *Mem:* Am Soc Agr Engrs; Sigma Xi. *Res:* Agricultural power and field machinery. *Mailing Add:* Mayberry Dr Apt 1004 Tahlequah OK 74464

PORTERFIELD, SUSAN PAYNE, b Plainfield, NJ, Sept 6, 43; m 64; c 2. ENDOCRINOLOGY, PERINATAL PHYSIOLOGY. *Educ:* Ohio State Univ, BS, 65, MS, 67, PhD(physiol), 73. *Prof Exp:* Instr biol, Ohio Northern Univ, 69-73; asst prof, Ga Inst Technol, 73-76; asst prof, 76-79, ASSOC PROF PHYSIOL, MED COL GA, 79- *Mem:* Am Physiol Soc; Soc Exp Biol & Med; Endocrine Soc; Am Thyroid Asn. *Res:* Perinatal endocrinology; perinatal thyroid function; effects of maternal thyroid disturbances on fetal development; thyroid hormones and fetal brain development. *Mailing Add:* Dept Physiol Med Col Ga Augusta GA 30912

PORTERFIELD, WILLIAM WENDELL, b Winchester, Va, Aug 24, 36; m 57; c 2. INORGANIC CHEMISTRY. *Educ:* Univ NC, BS, 57, PhD(inorg chem), 62; Calif Inst Technol, MS, 60. *Prof Exp:* Sr res chemist, Allegany Ballistics Lab, 62-64; from asst prof to assoc prof, 64-68, PROF CHEM, HAMPDEN-SYDNEY COL, 68-, CHARLES SCOTT VENABLE PROF CHEM, 89- *Mem:* Am Chem Soc; Royal Soc Chem. *Res:* Theory of cluster structures; coordination chemistry; molten salt coordination studies. *Mailing Add:* Dept Chem Hampden-Sydney Col Hampden-Sydney VA 23943

PORTEUS, JAMES OLIVER, b Wilmington, Del, Oct 25, 29; m 53; c 3. SURFACE PHYSICS. *Educ:* Univ Del, BS, 51; Cornell Univ, PhD(physics), 58. *Prof Exp:* Asst x-ray spectros, Cornell Univ, 52-58, res assoc, 58-59; res physicist, 59-69, HEAD QUANTUM SURFACE DYNAMICS BR, NAVAL WEAPONS CTR, 69- *Mem:* Am Phys Soc; Am Vacuum Soc; Sigma Xi; Soc Photo-Optical Instrumentation Engrs. *Res:* Low energy electron diffraction and spectroscopy; physical optics; laser damage; thermal modeling. *Mailing Add:* Code 3817 Naval Weapons Ctr China Lake CA 93555

PORTH, CAROL MATTSON, b Wis, Feb 15, 30. PATHOPHYSIOLOGY. *Educ:* Univ Wis-Madison, BS, 54; Marquette Univ, MS, 68; Med Col Wis, PhD(physiol), 80. *Prof Exp:* Staff nurse, Wis & Ind, 54-68; instr nursing, 68-71, asst prof, 71-74, ASSOC PROF NURSING PATHOPHYSIOL, UNIV WIS, MILWAUKEE, 74- *Concurrent Pos:* Adj asst prof, dept physiol, Med Col Wis, 80-; fel physiol, Med Col Wis, 83-84. *Mem:* Coun Nurse Researchers; Am Phys Soc; Sigma Xi. *Res:* Effect of aging and gender on autonomic control of the circulation. *Mailing Add:* 1901 E Elmdale Ct Shorewood WI 53211

PORTIS, ALAN MARK, b Chicago, Ill, July 17, 26; m 48; c 4. SOLID STATE PHYSICS. *Educ:* Univ Chicago, PhB, 48; Univ Calif, AB, 49, PhD(physics), 53. *Prof Exp:* Asst physics, Univ Calif, 49-51; asst prof, Univ Pittsburgh, 53-56; From asst prof to assoc prof physics, 56-63, dir, Lawrence Hall Sci, 69-72, Univ Ombudsman, 81-83, assoc dean eng, 83-87, PROF PHYSICS, UNIV CALIF, BERKELEY, 63- *Honors & Awards:* Millikan Award, Am Asn Physics Teachers, 66. *Mem:* Fel Am Phys Soc; fel AAAS; Am Asn Physics Teachers. *Res:* Nuclear and electron spin resonace; microwave magnetoconductivity; high temperature superconductivity; muon spin rotation spectroscopy. *Mailing Add:* Dept Physics Univ Calif Berkeley CA 94720

PORTIS, ARCHIE RAY, JR, b Winston-Salem, NC, Aug 17, 49; m 75; c 2. PHOTOSYNTHESIS, CARBON METABOLISM. *Educ:* Duke Univ, BS, 71; Cornell Univ, PhD(biochem), 76. *Prof Exp:* Res assoc, Univ of Munich, 75-77, Roswell Mem Inst, 77-78 & Univ Calif, San Francisco, 78; from asst prof to assoc prof physiol, 78-88, PROF PLANT PHYSIOL, UNIV ILL, 88-; PLANT PHYSIOLOGIST, AGR RES SERV, USDA, 78- *Mem:* Am Soc Plant Physiologists; AAAS; Am Soc Biochem & Molecular Biol. *Res:* Identification of rate limiting steps in photosynthesis and regulatory aspects of the process. *Mailing Add:* Dept Agron Univ Ill USDA Ars Urbana IL 61801

PORTIS, JOHN L, b Los Angeles, Calif, July 1, 43. VIROLOGY. *Educ:* Univ Calif, Los Angeles, MD, 71. *Prof Exp:* RES IMMUNOLOGIST, ROCKY MOUNTAIN LABS, INC, NIAID, NIH, 76- *Mem:* Am Asn Immunologists. *Mailing Add:* Lab Persistent Viral Dis NIH-NIAID Rocky Mountain Labs Hamilton MT 59840

PORTLOCK, DAVID EDWARD, b Fall River, Mass, Dec 5, 44; m; c 1. MEDICINAL & PHARMACEUTICAL CHEMISTRY. *Educ:* Southeastern Mass Univ, BS, 66; Cent Mich Univ, MS, 68; Va Polytech Inst & State Univ, PhD(chem), 72. *Prof Exp:* Fel alkaloids, Univ NH, 71-72 & Sheehan Inst Res, 72-74; proj dir narcotics, Sharps Assocs, 74-77; MEM STAFF, NORWICH PHARMACEUT INC, 77- *Concurrent Pos:* post doc fel, Univ NH, 73; group leader, Org Chem, 78- *Mem:* Sigma Xi; Am Chem Soc. *Res:* Design and synthesis of peptide, carbohydrates and alkaloid. *Mailing Add:* Norwich Eaton Pharmaceut Inc Norwich NY 13815

PORTMAN, DONALD JAMES, b Amherst, Ohio, Apr 25, 22; div; c 3. MICROMETEOROLOGY. *Educ:* Univ Mich, BS, 46; Johns Hopkins Univ, PhD(phys sci), 55. *Prof Exp:* Meteorologist, Univ Akron, 47-49; res staff asst, Johns Hopkins Univ, 49-54; physicist, Beckman & Whitley, Inc, 54-56; assoc res meteorologist, Res Inst, 56-62, lectr meteorol, 57-62, PROF METEOROL, UNIV MICH, ANN ARBOR, 62- *Concurrent Pos:* Consult, Durant Insulated Pipe Co, 55-56, qm res & eng comn, US Dept Army, 58-59, Bendix Aviation Corp, 59, Inst Defense Analysis, 65-66, Detroit Edison Co, 70-, Consumers Power Co, 71- & Murray & Trettel, Inc, 72-; Am Meteorol Soc vis scientist, NSF, 59-60; vis res prof, Univ Calif, 60. *Mem:* AAAS; Am Meteorol Soc; Am Geophys Union; Royal Meteorol Soc. *Res:* Turbulent trasfer of heat, mass and momentum; optical propagation in turbulence; air-sea interaction; turbulent diffusion of air pollutants; formation and dissipation of steam fogs. *Mailing Add:* Dept Atmospheric Sci Univ Mich Main Campus Ann Arbor MI 48109-2143

PORTMAN, OSCAR WILLIAM, b Denison, Tex, July 20, 24; m 48; c 3. MEDICINE. *Educ:* US Mil Acad, BS, 45; Harvard Med Sch, MD, 54. *Prof Exp:* Life Ins Med Res Fund fel nutrit, Sch Pub Health, Harvard Univ, 54-56, from assoc to asst prof nutrit, 56-64; SR SCIENTIST, ORE REGIONAL PRIMATE RES CTR, 64-; PROF BIOCHEM, MED SCH, UNIV ORE, 65- *Concurrent Pos:* Estab investr, Am Heart Asn, 59-64, mem coun atherosclerosis. *Mem:* Am Soc Clin Nutrit; Am Inst Nutrit. *Res:* Nutritional factors; cardiovascular disease; gallstone disease; metabolism of phospholipids and lipoproteins. *Mailing Add:* 13250 Thama Rd Lake Oswego OR 97034

PORTMANN, GLENN ARTHUR, b Canton, Ohio, Aug 13, 31; m 60; c 4. PHARMACEUTICS. *Educ:* Ohio State Univ, BS, 54, MS, 55, PhD(pharmaceut chem), 57. *Prof Exp:* Res pharmacist, Chas Pfizer & Co, 57-60 & White Labs, 60-63; RES PHARMACIST, STERLING RES GROUP, 63- *Mem:* Am Asn Pharmaceut Scientists. *Res:* Mechanism and rates of drug decomposition and interaction; dissolution, bioavailability and disposition of drugs; preformulation physical and chemical studies; analytical method development. *Mailing Add:* Sterling Res Group Rensselaer NY 12144

PORTMANN, WALTER ODDO, b Canton, Ohio, Oct 5, 29; m 56; c 3. MATHEMATICS. *Educ:* Kent State Univ, BS, 52; Case Inst Technol, MS, 54, PhD(math), 57. *Prof Exp:* Instr math, Case Inst Technol, 56-57, asst prof, 58-59; aeronaut res scientist, NASA, 57-58; assoc prof math, Ariz State Univ, 59-65; prof math & chmn dept, Wilson Col, 65-85; RETIRED. *Mem:* Sigma Xi; Math Asn Am. *Res:* Matrix theory; analytic functions of matrices or hyper-complex variables. *Mailing Add:* 146 Harvest Lane Chambersburg PA 17201

PORTNER, ALLEN, b Brooklyn, NY, Jan 19, 34; m 63; c 2. VIROLOGY, MOLECULAR BIOLOGY. *Educ:* Brooklyn Col, BA, 57; Long Island Univ, MS, 64; Univ Kans, PhD(microbiol), 68. *Prof Exp:* Sr res aide, Sloan-Kettering Inst Cancer Res, 59-63; fel virol, St Jude's Children's Res Hosp, 68-70; ASSOC PROF MICROBIOL, MED UNIV TENN CTR HEALTH SCI, MEMPHIS, 70-, MEM DEPT VIROL & MOLECULAR BIOL, ST JUDE'S CHILDREN'S RES HOSP, 70- *Mem:* Am Soc Microbiol; Am Soc Virol; Sigma Xi. *Res:* Virus replication, pathogenesis and immunity. *Mailing Add:* St Jude's Children's Res Hosp 332 N Lauderdale Memphis TN 38101

PORTNOFF, MICHAEL RODNEY, b Newark, NJ, July 1, 49. SIGNAL PROCESSING. *Educ:* Mass Inst Technol, SB, 73, SM, 73, EE, 73, ScD, 78. *Prof Exp:* Coop student, Bell Tel Labs, 69-71; res asst, Res Lab Electronics, Mass Inst Technol, 71-78, teaching asst, Dept Elec Eng & Comput Sci, 71-78, res assoc, Res Lab Electronics, 78-79; proj engr, Eng Res Div, 79-83, group leader, Signal & Image Process Res Group, 83-88, SR STAFF RES ENGR, LAWRENCE LIVERMORE NAT LAB, 88- *Honors & Awards:* Browder J Thompson Mem Prize, Inst Elec & Electronic Engrs, 77, Acoust, Speech & Signal Processing Soc Award, 80. *Mem:* Inst Elec & Electronic Engrs. *Res:* Theory of digital signal processing and its application to speech, image, radar and seismic signal processing. *Mailing Add:* Lawrence Livermore Nat Lab L-156 Univ Calif PO Box 808 Livermore CA 94550

PORTNOY, BERNARD, b Brooklyn, NY, Jan 20, 29; m 50; c 2. EPIDEMIOLOGY, PEDIATRICS. *Educ:* Syracuse Univ, BA, 50, MA, 51; State Univ NY, MD, 55. *Prof Exp:* From asst to instr psychol, Grad Sch, Syracuse Univ, 50-51; from instr to asst prof pediat, State Univ NY Upstate Med Ctr, 60-61; from asst prof pediat to assoc prof pediat & community med & pub health, 61-72, PROF PEDIAT & COMMUNITY & FAMILY MED, SCH MED, UNIV SOUTHERN CALIF, 72-; DEP REGIONAL DIR, CENT HEALTH SERV REGION, LOS ANGELES COUNTY DEPT HEALTH SERV, 74- *Concurrent Pos:* Dep dir infectious dis lab, Los Angeles County-Univ Southern Calif Med Ctr, 61-64, dir lab, 64-70, asst med dir, 70-74; USPHS res career develop award, 64; Nat Inst Allergy & Infectious Dis res grants. *Mem:* Am Epidemiol Asn; fel Am Pub Health Asn; Infectious Dis Soc Am; NY Acad Sci. *Res:* Psychology; infectious diseases; clinical virology; environmental studies. *Mailing Add:* Los Angeles County Univ Southern Calif Med Ctr 1129 N State St Los Angeles CA 90033

PORTNOY, DEBBI, b New York, NY, Jan 6, 62; m 82; c 2. BIOCHEMISTRY. *Educ:* Manl Sch Med Asst, MA, 80. *Prof Exp:* CLIN RES ASSOC, SOC PSYCHIAT RES INST, 85- *Mailing Add:* SPRI c/o Debbi Portnoy 150 E 69th St New York NY 10021

PORTNOY, ESTHER, b Dearborn, Mich, Jan 4, 45. GEOMETRY. *Educ:* Stanford Univ, BS, 66, MS, 68, PhD(math), 69. *Prof Exp:* Instr math, Lowell State Col, 70-73; VIS LECTR MATH, UNIV ILL, URBANA-CHAMPAIGN, 74-, ASST PROF, 81- *Mem:* Am Math Soc; Asn Women in Math. *Res:* Hyperbolic geometry, especially global analysis of hypersurfaces; Gauss-Bonnet theory. *Mailing Add:* Dept Math Univ Ill 1409 W Green St Urbana IL 61801

PORTNOY, NORMAN ABBYE, b New Orleans, La, Oct 26, 44; m 66; c 2. ORGANIC CHEMISTRY, CELLULOSE CHEMISTRY. *Educ:* La State Univ, New Orleans, BS, 66; NDak State Univ, PhD(chem), 70. *Prof Exp:* Fel organophosphorus chem, Newcomb Col, Tulane Univ, 70-71; Nat Res Coun res assoc cellulose chem, Southern Regional Res Ctr, USDA, 71-73; res chemist, Eastern Res Div, ITT Rayonier, Inc, 73-77, group leader, 78-81; sr res chemist, GAF Corp, 81-84; SR RES CHEMIST, A E STALEY MFG CO, 84- *Mem:* Am Chem Soc; Tech Asn Pulp & Paper Indust. *Res:* Novel methods of cellulosic fiber and film production; chemical modifications of cellulose for end product improvement; photochemically initiated processes for cellulose modification; flame retardance of cellulosic fibers and textiles; starch derivatives and chemistry. *Mailing Add:* A E Staley Manuf 220 E Eldorado St Decatur IL 62525

PORTNOY, STEPHEN LANE, b Kankakee, Ill, Dec 2, 42; m 65; c 2. MATHEMATICAL STATISTICS, APPLIED STATISTICS. *Educ:* Mass Inst Technol, BS, 64; Stanford Univ, MS, 66, PhD(statist), 69. *Prof Exp:* Asst prof statist, Harvard Univ, 69-74; assoc prof, 74-81, PROF STATIST, UNIV ILL, URBANA-CHAMPAIGN, 81- *Mem:* Inst Math Statist; Am Statist Asn. *Res:* Mathematical statistics especially asymptotics and robust estimation. *Mailing Add:* Dept Statist Univ Ill 725 S Wright St Champaign IL 61820

PORTNOY, WILLIAM M, b Chicago, Ill, Oct 28, 30; m 56; c 2. ELECTRICAL ENGINEERING, PHYSICS. *Educ:* Univ Ill, BSc & MSc, 52, PhD(physics), 59. *Prof Exp:* Mem tech staff, Semiconductor Div, Hughes Aircraft Co, 59-61; mem tech staff, Semiconductor-components Div, Tex Instruments, Inc, 61-67; from assoc prof to prof elec eng, 67-73, prof biomed eng, Health Sci Ctr, 73-85, PROF ELEC ENG, TEX TEX UNIV, 75-, PROF PHYSICS, 85- *Concurrent Pos:* Resident res associateship, NASA Manned Spacecraft Ctr, 68; Nat Heart Inst trainee, Baylor Col Med, 69, adj assoc prof, 69-73; Fulbright prof elec eng, Univ Warwick, Coventry, UK, 75-76; ed newsletter, Inst Elec & Electronic Engrs; ed, Inst Elec & Electronic Engrs Circuits & Systs Mag; treas, Power Electronics Coun; consult, Hughes Res Lab, 72, NOM Corp, 74-76, Los Alamos Nat Lab, 80-81, ERADCOM, US Army, 81- & Westinghouse Elec Corp, 84-, Lawrence Nat Lab, 86-, IAP & GRC, 90-; assoc ed, Inst Elec & Electronic Engrs Trans Power Electronics. *Honors & Awards:* Western Elec Fund Award, Am Soc Eng Educ, 80. *Mem:* Sigma Xi; sr mem & fel Inst Elec & Electronic Engrs; Am Phys Soc; Am Soc Eng Educ. *Res:* Semiconductor device and integrated circuit physics and technology; biomedical instrumentation; power semiconductor switches. *Mailing Add:* Dept Elec Eng Tex Tech Univ Lubbock TX 79409

PORTOGHESE, PHILIP S, b Brooklyn, NY, June 4, 31; m 60; c 3. MEDICINAL CHEMISTRY. *Educ:* Columbia Univ, BS, 53, MS, 58; Univ Wis, PhD(pharmaceut chem), 61. *Hon Degrees:* Hon Doctorate, Univ Catania, Italy, 86. *Prof Exp:* From asst prof to assoc prof, 61-69, dir grad studies, 74-86, PROF MED CHEM, COL PHARMACOL, UNIV MINN, MINNEAPOLIS, 69-, PROF PHARMCOL, MED SCH, 87- *Concurrent Pos:* NIH res grants, 62-; consult, NIMH, 71-72, Med Chem B Study Sect, NIH, 72-76; ed-in-chief, J Med Chem, Am Chem Soc, 72-; Gustavus Pfeiffer fel, 73-74; Presidents Biomed Res Panel, 75; Watkins vis prof, Wichita State Univ, 86. *Honors & Awards:* McPike lectr, Univ Kans, 68; Res Achievement Award, Acad Pharmaceut Sci, 90; Ernest E Just lectr, Howard Univ, 75; Byron Riegel lectr, Northwestern Univ, 80; Res Achiev Award, Acad Pharmaceut Sci, 80; Ernest H Volwiler Award, Am Found Pharmaceut Educ,84; Grollman lectr, Univ Md, 85; Res Achievement Med Chem, Am Asn Pharmaceut Scientists, 90; Med Chem Award, Am Chem Soc, 90, Edward E Smissman Award, 91. *Mem:* Fel Acad Pharmaceut Sci; fel AAAS; Am Chem Soc; Am Pharmaceut Asn; Am Soc Pharmacol & Exp Therapeut; Am Asn Col Pharm; Am Asn Pharmaceut Scientists; Am Soc Neurosci; Sigma Xi. *Res:* Stereochemical factors in drug action, including design and synthesis of drugs and configurational and conformational analysis of medicinal agents; chemistry and pharmacology of opioids. *Mailing Add:* Col Pharm Univ Minn Minneapolis MN 55455

PORTWOOD, LUCILE MITCHELL, b Ft Worth, Tex, Apr 8, 13. BACTERIOLOGY. *Educ:* Mary Hardin-Baylor Col, BA, 33; Okla State Univ, MS, 34; Mich State Univ, PhD(bact), 44. *Prof Exp:* Sr bacteriologist, Globe Labs, Tex, 36-40; asst, Bur Animal Indust, USDA, Mich State Col, 41-42; bacteriologist, Mich Dept Pub Health, 42-47, biochemist, 47-80; RETIRED. *Mem:* Am Chem Soc; Am Soc Microbiol; Am Pub Health Asn. *Res:* Antibiotics; vaccines; antitoxin and antiserum production; purification and testing. *Mailing Add:* PO Box 76 Okemos MI 48805-0076

PORTZ, HERBERT LESTER, b Waukesha, Wis, July 8, 21; m 44; c 3. TURFGRASS MANAGEMENT, PLANT PHYSIOLOGY. *Educ:* Univ Wis, BS, 48, MS, 52; Univ Ill, PhD(agron), 54. *Prof Exp:* Instr high sch, Wis, 48-52; asst crop prod, Univ Ill, 52-54; from asst prof to prof, 54-86, EMER PROF AGRON, SOUTHERN ILL UNIV, CARBONDALE, 86- *Concurrent Pos:* Asst dean, Sch Agr, Southern Ill Univ, 57-67; int agriculturalist, US AID, Nepal, 67-69 & Food & Agr Orgn, UN, Brazil, 72-74. *Mem:* Am Soc Agron; Am Forage & Grassland Coun. *Res:* Forage crop production; turfgrass mgt; international agricultural development; zoysiagrass seed production and physiology. *Mailing Add:* RR 2 Box 164 Makanda IL 62958

PORVAZNIK, MARTIN, b Hammond, Ind, Oct 30, 47; m 71; c 3. EXPERIMENTAL HEMATOLOGY, PERIODONTICS. *Educ:* St John's Univ, Collegeville, Minn, BA, 69; Univ Minn, Minneapolis, MS, 75, PhD(biol), 76. *Prof Exp:* Fel, Univ Calif, Irvine, 76; prin investr, Armed Forces Radiobiol Res Inst, Bethesda, Md, 76-80; prin investr, Naval Dent Res Inst, Great Lakes, Ill, 80-82; mem staff, Naval Hosp, Great Lakes, Ill, 82-84; mem staff, Naval Toxicol Lab, Wright-Patterson Air Force Base, Ohio, 84-88; mem staff, Naval Drug Screening Lab, Great Lakes, Ill, 88-91; EXEC OFFICER, NAVY DRUG SCREENING LAB, NAVAL AIR STA, JACKSONVILLE, FLA, 91- *Mem:* Am Soc Forensic Toxicol. *Res:* Localization of organotins in cell membranes using electron microscopic techniques; analysis of combustion products formed from lubricating oil. *Mailing Add:* 2734 Flynn Cove Rd Jacksonville FL 32223

PORZEL, FRANCIS BERNARD, b Chicago, Ill, July 28, 13; m 40; c 7. EXPLOSION SAFETY, UNIFIED DYNAMICS. *Educ:* Univ Idaho, BS, 40; Princeton Univ, MS, 48. *Prof Exp:* Chemist, Bunker Hill & Sullivan Smetting, 40-41; Lt Colonel, US Army, 41-48; group leader, Los Alamos Nat Lab, 48-54; sr sci adv, Armour Res Found, Ill Inst Technol, 54-61; sr staff, Inst Defense Anal, 61-68; sr scientist, Naval Surface Weapons Ctr, 68-83; RETIRED. *Concurrent Pos:* Prog & proj dir, Nuclear Tests, US Atomic Energy, 51-58; official dele, UN Conf Peaceful Uses Atomic Energy, 58; mem, Panel Ocean Eng, Nat Acad Sci, 64-65; consult, US Atomic Energy Agencies, 54-59; consult, NASA, 84-86. *Honors & Awards:* Air Power Award, US Air Force Asn, 59. *Res:* Definitive theory and tests on nuclear blast and thermal phenomena; reactor containment techniques against internal explosion; comprehensive technology base for improved safety in handling-storage of explosives; "unidynamic" model for physical and social behavior. *Mailing Add:* 500 Hillwood Ave Falls Church VA 22042

PORZIO, MICHAEL ANTHONY, b East Orange, NJ, Apr 5, 42; m 71. FOOD CHEMISTRY. *Educ:* Rutgers Univ, BS, 64; Univ Hawaii, PhD(chem), 69. *Prof Exp:* Mem staff fish protein, Dept Food Sci & Human Nutrit, Univ Hawaii, 68-70; consult oil pollution, Dept Trans, State Hawaii, 71-72; res assoc muscle chem, Dept Food Sci, Mich State Univ, 73-75; SR RES CHEMIST, GEN MILLS INC, 76- *Mem:* Am Chem Soc; AAAS; Inst Food Technologists. *Res:* Thermoanalysis of carbohydrates; food carbohydrates and fibres; new techniques of protein analyses; structure and composition of contractile proteins; hydrolyses mechanisms of 3- and 4-member substituted ring ethers. *Mailing Add:* PO Box 743 Cockeysville MD 21030

POSAMENTIER, ALFRED S, b New York, NY, Oct 18, 42; c 2. MATHEMATICS EDUCATION. *Educ:* City Univ New York, AB, 64, MA, 66; Fordham Univ, PhD(math educ), 73. *Prof Exp:* From instr to assoc prof, 70-80, PROF MATH EDUC, CITY COL, CITY UNIV NEW YORK, 81-, ASSOC DEAN, SCH EDUC, 86- *Concurrent Pos:* Educ consult, var schs, 71-; dir, Sch Educ, Educ Mat Ctr, City Univ New York, 72-78, Math Develop Prog Sec Sch Teachers, NSF, 78-82, Select Prog Sci & Eng, 78-, UK Initiatives Prog, 83-, Ctr Sci & Math Educ, 86- & Global Educ Telecommun, 87-; chmn, Dept Sec & Continuing Educ, City Univ New York, 80-86, coordr, Northeast Resource Ctr Sci & Eng & Sci Lect Prog, 81-; lectr, Univ Linz, Austria, 87; vis prof, Tech Univ Berlin, Fed Repub Ger, 89; vis lectr, Univ Warsaw, Poland, 89; hon fel, Polytech SBank, London, 89; Fulbright scholar lectr res, Univ Vienna, Austria, 90-91. *Mem:* Math Asn Am; Nat Coun Teachers Math; Nat Coun Supervisors Math. *Res:* Mathematics curriculum for the pre-college audience; author and co-author of numerous publications. *Mailing Add:* Sch Educ City Col City Univ Convent Ave & 138th St New York NY 10031

POSCHEL, BRUNO PAUL HENRY, b Brooklyn, NY, June 6, 29; m 56; c 2. PSYCHOPHARMACOLOGY. *Educ:* Roosevelt Univ, BS, 51; Univ Ill, PhD(psychol), 56. *Prof Exp:* NIMH res fel, Univ Ill, 56-57; asst prof psychol, Wayne State Univ, 57-59; assoc res pharmacologist, 59, sect dir psychopharmacol, 59-81, SR RES FEL, PARKE, DAVIS & CO, 81- *Mem:* Am Soc Pharmacol & Exp Therapeut; Am Psychol Asn; Neurosci Soc. *Res:* Brain-behavior relations; psychopathology. *Mailing Add:* 2923 Sunnywood Dr Ann Arbor MI 48103

POSEN, GERALD, b Toronto, Ont, Sept 23, 35; m 63, 87; c 3. INTERNAL MEDICINE, NEPHROLOGY. *Educ:* Univ Western Ont, MD, 61; FRCP(C), 67. *Prof Exp:* Fel med, Sch Med, Johns Hopkins Univ, 65-66; res asst nephrology, Montreal Gen Hosp, 66-67; Chief, Div Nephrology, Ottawa Civic Hosp, 74-89; lectr, 67-70, asst prof, 70-77, ASSOC PROF MED, UNIV OTTAWA, 77- *Concurrent Pos:* Dir, Can Renal Failure Register. *Mem:* Am Soc Nephrology; Am Soc Artificial Internal Organs; Can Soc Nephrology; Int Soc Nephrology. *Res:* Fluoride metabolism and bone disease in renal failure and hemodialysis; immunological integrity in renal failure; dialysis and transplantation; immunological studies in renal failure. *Mailing Add:* 1053 St Carling Ave Ottawa ON K1Y 4E9 Can

POSEY, C(HESLEY) J(OHNSTON), b Mankato, Minn, June 12, 06; m 40; c 2. HYDRAULICS, STRUCTURAL ENGINEERING. *Educ:* Univ Kans, BS & BSCE, 26, CE, 33; Univ Ill, MS, 27. *Prof Exp:* Struct draftsman, Am Bridge Co & Modjeski & Chase, Pa, 28-29; from instr to asst prof mech & hydraul, Univ Iowa, 29-39, asst prof struct eng, 37-39, in-chg mat test lab, 37-50, assoc prof hydraul & struct eng, 40-46, res engr, 42-50, prof civil eng & head dept, 50-62, consult engr, Inst Hydraul Res, 50-62; prof civil eng, 62-76, EMER PROF CIVIL ENG, UNIV CONN, STORRS, 76- *Concurrent Pos:* Dir, Rocky Mountain Hydraul Lab, 45-; vol teaching & res. *Mem:* Am Soc Civil Engrs; Am Geophys Union. *Res:* Open channel flow; reinforced concrete structures; engineering education. *Mailing Add:* Box U-37 Univ Conn Storrs CT 06268

POSEY, DANIEL EARL, b Corpus Christi, Tex, Apr 9, 47; m 68; c 2. ANALYTICAL CHEMISTRY. *Educ:* Univ Houston, BS, 70. *Prof Exp:* Lab tech, Getty Oil Co, 68-69; lab mgr, Inst Res, Inc, 69-79, Convertors Div, Am Hosp Supply Corp, 79-84; vpres & lab dir, Inst Res-Austin, 84-86; Independant consult, 86-88; CHEM QUALITY CONTROL SUPV, ADV MICRO DEVICES, 88- *Mem:* Am Chem Soc; fel Am Inst Chemists; Am Soc Qual Control. *Res:* Design and conduct testing programs to identify and remedy failure mechanisms for aerospace, biomedical, electronics and petroleum products and processes; spectrochemical techniques. *Mailing Add:* 11902 Rustle Ln Austin TX 78750

POSEY, ELDON EUGENE, b Oneida, Tenn; m 43; c 2. MATHEMATICS. *Educ:* Eastern Tenn State Col, BS, 47; Univ Tenn, MA, 49, PhD(math), 54. *Prof Exp:* Instr math, Univ Tenn, 52-54; asst prof, WVa Univ, 54-59; from assoc prof to prof, Va Polytech Inst & State Univ, 59-64; head dept, 64-80, PROF MATH, UNIV NC, GREENSBORO, 64- *Mem:* Am Math Soc; Math Asn Am. *Res:* Theory of Knotts; free groups; imbedding of cells in space. *Mailing Add:* 4311 Dogwood Dr Greensboro NC 27410

POSEY, FRANZ ADRIAN, b Jackson, Miss, Jan 10, 30; m 52; c 3. ELECTROCHEMISTRY, PHYSICAL CHEMISTRY. *Educ:* Millsaps Col, BS, 51; Univ Chicago, MS, 52, PhD(chem), 55. *Prof Exp:* Chemist, Oak Ridge Nat Lab, 54-61, group leader chem div, 61-84; RETIRED. *Honors & Awards:* Young Author's Award, Electrochem Soc, 60; IR-100 Award, Indust Res Mag, 83. *Res:* Theoretical electrochemistry and electrochemical kinetics; electrochemistry of corrosion and porous electrodes; electroanalysis. *Mailing Add:* 2408 Perch Cove Gautier MS 39553-6710

POSEY, JOE WESLEY, acoustics, for more information see previous edition

POSEY, ROBERT GILES, b Plainfield, NJ, Sep 18, 47; m 69; c 2. POLYMER CHEMISTRY. *Educ:* Furman Univ, BS, 79, MS, 71; Univ Fla, PhD(chem), 75. *Prof Exp:* Fel chem, Univ Maine, Orono, 75-76; res chemist, Celanese Plastics Co, 76-79; SR STAFF CHEMIST, AM HOECHST CORP, 79- *Concurrent Pos:* Instr polymer chem, Greenville Tech Col, 80- *Mem:* Am Chem Soc. *Res:* Synthetic routes to methal enecyclopropene and spiropentene; diels-alder reactions of halocyclopropenes; organometallic complexes of cyclopropenes; polyester chemistry. *Mailing Add:* 307 Pebble Creek Dr Taylors SC 29687-9272

POSHUSTA, RONALD D, b Calmar, Iowa, June 28, 35; m 70; c 4. CHEMICAL PHYSICS. *Educ:* Iowa State Univ, BS, 57; Univ Tex, PhD(chem), 63. *Prof Exp:* Instr physics, Univ Tex, 63-65; NSF fel, Univ Keele, 65-66; from asst prof to assoc prof, 66-76, PROF CHEM & PHYSICS, WASH STATE UNIV, 76- *Mem:* Am Phys Soc; Am Chem Soc; Sigma Xi. *Res:* Energy and properties of small molecules such as H-3 positive, H-5 positive; permutation group theory; evaluation of matrix elements; lower bounds; properties of solids. *Mailing Add:* Dept of Chem Wash State Univ Pullman WA 99163

POSKA, F(ORREST) L(YNN), chemical engineering, for more information see previous edition

POSKANZER, ARTHUR M, b New York, NY, June 28, 31; m 54; c 3. NUCLEAR PHYSICS, NUCLEAR CHEMISTRY. *Educ:* Harvard Col, AB, 53; Columbia Univ, MA, 54; Mass Inst Technol, PhD(phys chem), 57. *Prof Exp:* Chemist, Brookhaven Nat Lab, 57-66; staff scientist, 66-68, STAFF SR SCIENTIST, LAWRENCE BERKELEY LAB, 68- *Concurrent Pos:* Chmn Gordon Conf Nuclear Chem, 70; John Simon Guggenheim Mem fel, Orsay, France, 70-71; NATO sr fel, 75; mem, Panel Future Nuclear Sci, Nat Res Coun, 76; chmn, Div Nuclear Chem & Technol, Am Chem Soc, 77; sci dir of Bevalac, 78-79; sci assoc, Europ Orgn Nuclear Res, Geneva, Switz, 79-80; Alexander von Humboldt sr US scientist award, 86. *Honors & Awards:* Nuclear Chem Award, Am Chem Soc, 80. *Mem:* Fel Am Phys Soc; Am Chem Soc. *Res:* Relativistic nuclear collisions; high energy nuclear reactions; isotopes far from stability. *Mailing Add:* Lawrence Berkeley Lab Berkeley CA 94720

POSKANZER, DAVID CHARLES, neurology, preventive medicine, for more information see previous edition

POSKOZIM, PAUL STANLEY, b Chicago, Ill, Dec 15, 40; m 70; c 4. INORGANIC CHEMISTRY. *Educ:* Loyola Univ, Ill, BS, 61; Northwestern Univ, PhD(chem), 67. *Prof Exp:* Asst prof chem, Sam Houston State Col, 66-68; from asst prof to assoc prof, 68-74, assoc dean, Grad Col, 87-89, PROF CHEM, NORTHEASTERN UNIV, 74-,. *Concurrent Pos:* Fel, Univ Calif, Berkeley, 74. *Mem:* Am Chem Soc; Royal Soc Chem. *Res:* Organogermanium chemistry; divalent germanium chemistry; germanium-cobalt, metal-metal bonds; cobalt-cyanide complex chemistry. *Mailing Add:* Dept Chem Northeastern Ill Univ Chicago IL 60625

POSLER, GERRY LYNN, b Cainsville, Mo, July 24, 42; m 63; c 3. AGRONOMY, FORAGE UTILIZATION. *Educ:* Univ Mo, BS, 64, MS, 66; Iowa State Univ, PhD(crop breeding), 69. *Prof Exp:* Res asst, Univ Mo, 64-66; res assoc, Iowa State Univ, 66-69; asst prof agr, Western Ill Univ, 69-74; assoc prof, 74-80, PROF AGRON, KANS STATE UNIV, 80-, HEAD DEPT, 90- *Honors & Awards:* Resident Inst Award, Am Soc Agron. *Mem:* Am Forage & Grassland Coun; fel Am Soc Agron; Crop Sci Soc Am; Nat Asn Col & Teachers Agr; Coun Agr Sci & Technol. *Res:* Forage utilization; forage systems planning. *Mailing Add:* Dept of Agron Kans State Univ Manhattan KS 66506

POSLUSNY, JERROLD NEAL, b Chicago, Ill, Oct 19, 44; m 67; c 2. ORGANIC CHEMISTRY, ELECTROCHEMISTRY. *Educ:* Stevens Inst Technol, BS, 66, MS, 68, PhD(org chem), 72. *Prof Exp:* Fel synthetic org chem, Natural Prod Res Ctr, Univ NB, 71-73; SR RES CHEMIST, EASTMAN KODAK CO, 73- *Mem:* Am Chem Soc; Sigma Xi. *Res:* Electroorganic chemistry, particularly electroorganic synthesis. *Mailing Add:* 118 Burrows Hills Dr Rochester NY 14625

POSLUSZNY, USHER, b Hof, WGer, July 20, 47; Can citizen. PLANT MORPHOLOGY. *Educ:* McGill Univ, BSc, 69, PhD(bot), 75. *Prof Exp:* Fel bot, Cabot Found, Harvard Univ, 75-77; ASSOC PROF BOT, UNIV GUELPH, ONT, 77- *Mem:* Can Bot Asn; Bot Soc Am; Sigma Xi; Am Inst Biol Sci; Linnean Soc. *Res:* Floral developmental study of freshwater and marine monocotyledons using both organographic and histological data; morphological and ampelographic studies of wild and cultivated grape plants. *Mailing Add:* Dept Bot Univ Guelph Guelph ON N1G 2W1 Can

POSNER, AARON SIDNEY, b Newark, NJ, Nov 10, 20; m 44; c 2. PHYSICAL CHEMISTRY, BIOCHEMISTRY. *Educ:* Rutgers Univ, BS, 41; Polytech Inst Brooklyn, MS, 49; Univ Liege, PhD(chem), 54. *Prof Exp:* Res chemist, P J Schweitzer, Inc, 41-42; res chemist, Johnson & Johnson, 46-47; chem crystallogr, Squier Signal Lab, 47-50; res assoc, Am Dent Asn Res Div, Nat Bur Standards, 50-61; chem crystallogr, NIH, 61-63; assoc prof ultrastruct biochem, Med Col, Cornell Univ, 63-69, assoc dir res, Hosp Spec Surg, 63-84, prof biochem, Med Col, 69-87, dir res opers, Hosp Spec Surg, 84-87, EMER SCIENTIST, HOSP SPEC SURG & EMER PROF, MED COL, CORNELL UNIV, 87- *Concurrent Pos:* Chem crystallogr, Inst Exp Therapeut, Univ Liege, 53-54; Claude Bernard guest prof, Univ Montreal, 62; adj prof, Columbia Univ, 87-; assoc fac, Empire State Col, State Univ NY, 87- *Honors & Awards:* Kappa Delta Award, Am Acad Orthop Surgeons, 77; Award for Basic Res in Biol Mineralization, Int Asn Dent Res, 78. *Mem:* AAAS; Am Chem Soc; Am Crystallog Asn; Biophys Soc; Am Dent Asn; Am Soc Bone & Mineral Res. *Res:* Crystal chemistry of bone and tooth mineral and allied phosphates; crystal chemistry of biological materials; ultrastructural biochemistry. *Mailing Add:* Two Longview Dr Scarsdale NY 10583

POSNER, EDWARD CHARLES, b Brooklyn, NY, Aug 10, 33; m 56; c 2. COMMUNICATIONS, SPACE SYSTEMS. *Educ:* Univ Chicago, BA, 52, MS, 53, PhD(math), 57. *Prof Exp:* Res asst, Univ Chicago, 56; mem tech staff, Bell Tel Labs, NY, 56-57; instr math, Univ Wis, 57-60, Wis Alumni Res Found fel, 58-60; asst prof, Harvey Mudd Col, 60-62; res group supvr info processing group, Calif Inst Technol, 61-67, dep sect mgr commun systs res sect, 67-69, res & develop mgr telecommun div, 69-73, mgr data processing & mgt sci, 73-78, mgr planning telecommunication & data acquisition off, 78-81, CHIEF TELECOMMUNICATION & DATA ACQUISITION TECHNOLOGIST, JET PROPULSION LAB, CALIF INST TECHNOL, 81- *Concurrent Pos:* Vis prof elec eng, Calif Inst Technol, 78-; consult, Inst Defense Anals, Commun Res Div, 83- *Mem:* Math Asn Am; Soc Indust & Appl Math; Am Math Soc; fel Inst Elec & Electronics Engrs; Int Neural Network Soc; Am Inst Aeronaut & Astronaut; AAAS. *Res:* Digital networks; space communication systems; information theory; neural networks; acoustic signal processing. *Mailing Add:* 1460 Rose Villa Pasadena CA 91106

POSNER, GARY HERBERT, b New York, NY, June 2, 43; m 65; c 2. ORGANIC CHEMISTRY. *Educ:* Brandeis Univ, BA, 65; Harvard Univ, MA, 65, PhD(org chem), 68. *Prof Exp:* Res asst, Univ Calif, Berkeley, 68-69; from asst prof to prof, 69-84, SCOWE PROF CHEM, JOHNS HOPKINS UNIV, 84- *Concurrent Pos:* Johns Hopkins Univ fac res grant, 69-70; Res Corp grant, 70-71; NIH res grant, 71-74, 75-79, 80-88 & 90-95; NSF res grant, 72-74, 75-78, 81-84, 85-89; Petrol Res Fund res grant, 79-81, 82-86, 87-89 & 90-92; res grant, US Army contract, 80-81 & 81-84; Nat Inst Environ Health Sci res grant, 80-86. *Honors & Awards:* Sr Fulbright lectr, Univ Paris, 75-76. *Mem:* AAAS; Am Chem Soc; The Chem Soc. *Res:* New synthetic methods; synthesis of natural products; organic reactions at solid surfaces; organometallic chemistry; environmental carcinogenesis. *Mailing Add:* Dept of Chem Johns Hopkins Univ Baltimore MD 21218-2685

POSNER, GERALD SEYMOUR, b New York, NY, Sept 30, 27; m 51, 78; c 3. OCEANOGRAPHY. *Educ:* City Col New York, BS, 49; Univ Miami, MS, 51; Yale Univ, PhD(zool), 56. *Prof Exp:* Oceanogr, Yale exped, SAm, 53; instr biol, Wesleyan Univ, 54-55; instr & investr marine biol, Inst Fisheries Res, Univ NC, 55-56, from asst prof to assoc prof, 56-61; from asst prof to prof biol, City Col NY, 61-91, actg dir, Inst Oceanog, 71-74, dir, Ctr Educ Exp & Develop, 71-75, EMER PROF BIOL, CITY COL, NY, 91- *Concurrent Pos:* Vis sr res assoc, Lamont-Doherty Geol Observ, Columbia Univ; mem, Mayor's Oceanog Adv Comt, 69-75; intern & consult to Nat Oceanic & Atmospheric Admin, Dept Com, 77-79; consult, NOAA, Sea Grant, Food & Agr Org, US Agency Int Develop, 80-85; coord, Foreign Reinbursable Asst Prog, NOAA, 80-85. *Mem:* Am Soc Limnol & Oceanog; Am Inst Biol Sci; AAAS; Sigma Xi. *Res:* Plankton dynamics and hydrography of estuarine and near shore waters; fisheries oceanography, development and management in Morocco, West Africa and Marshall Islands. *Mailing Add:* 9955 Westview Dr 225 Coral Springs FL 33076

POSNER, HERBERT BERNARD, b Brooklyn, NY, Apr 16, 30; m 58; c 2. PLANT PHYSIOLOGY. *Educ:* Brooklyn Col, BS, 53, MA, 58; Yale Univ, PhD(bot), 62. *Prof Exp:* Res assoc radiobot, Brookhaven Nat Lab, 62-64; asst prof biol, 64-67, ASSOC PROF BIOL, STATE UNIV NY BINGHAMTON, 67- *Concurrent Pos:* NSF grants, 65-73; vis prof, Weizmann Inst, 71-72. *Mem:* Am Soc Plant Physiol; AAAS; Sigma Xi. *Res:* Physiology of plant growth and development. *Mailing Add:* Dept Biol Sci State Univ NY Binghamton NY 13902

POSNER, HERBERT S, b New York, NY, Aug 30, 31; m 53; c 3. BIOCHEMISTRY, PHARMACOLOGY. *Educ:* City Col New York, BS, 53; Purdue Univ, MS, 55; George Washington Univ, PhD(biochem), 58; Univ NC, Chapel Hill, MRP, 90. *Prof Exp:* Biochemist, Nat Heart Inst, 54-58, Clin Neuropharmacol Res Ctr, NIMH, 58-65, lab biochem Nat Inst Dent Res, 65-67, head sect growth & develop, 67-71, pharmacologist, Nat Inst Environ Health Sci, 71-86; RETIRED. *Concurrent Pos:* Fel genetics, Sch Med, Stanford Univ, 63-64; assoc clin prof, Duke Univ, 68-78. *Mem:* AAAS; Am Chem Soc; NY Acad Sci. *Res:* Chemical metabolism; pharmacologic effects of chemicals and metabolites; mechanisms of action; causes of individual differences; abnormal growth and development; chemical and physical environmental health interactions; structure-activity relationships. *Mailing Add:* 307 N Estes Dr Chapel Hill NC 27514-2738

POSNER, JEROME B, b Cincinnati, Ohio, Mar 20, 32; m 54; c 3. NEUROLOGY. *Educ:* Univ Wash, BS, 51, MD, 55; Am Bd Psychiat & Neurol, dipl, 62. *Prof Exp:* Intern med, King County Hosp, Seattle, 55-56; asst resident neurol, affiliated hosps, Univ Wash, 56-59; instr med, Sch Med, Univ Louisville, 59-61; res instr biochem & neurol, Univ Wash, 61-63; from asst prof to assoc prof, 63-70, PROF NEUROL, MED COL, CORNELL UNIV, 70- *Concurrent Pos:* Fel neurol, Univ Wash, 58-59, NIH spec fel biochem, 61-63; attend neurologist, King County Hosp, Seattle, 62-63; from asst attend neurologist to assoc attend neurologist, Mem Hosp Cancer & Allied Dis, 63-70, attend neurologist, 70-, chief neurol serv, 67-75; assoc, Sloan-Kettering Inst Cancer Res, 67-76; mem, Cotzias Lab Neuro-Oncol Sloan Kettering Inst Cancer Res, 77- & mem Neurol B Study Sect, NIH, 72-76; mem med adv bd, Burke Rehab Ctr, 73-; adj prof & vis physician, Rockefeller Univ & Hosp, 73-75; mem med adv bd, Asn Brain Tumor Res; chmn dept neurol, Mem Hosp Cancer & Allied Dis, 75-87, 89- *Honors & Awards:* George C Cotzias Chair in Neuro-Oncol; Farber Award, Brain Tumor Res, 83; Javits Neurosci Investr Award, NINCDS, 88. *Mem:* Am Asn Cancer Res; Am Acad Neurol; Soc Neurosci; Am Electroencephalog Soc; Am Neurol Asn. *Res:* Neuro-oncology with a special interest of the immunology of paraneoplastic syndromes. *Mailing Add:* Mem Sloan-Kettering Cancer Ctr 1275 York Ave New York NY 10021

POSNER, MARTIN, b Pasadena, Calif, Jan 26, 35. NUCLEAR PHYSICS, ATOMIC PHYSICS. *Educ:* Univ Calif, Los Angeles, BA, 56; Princeton Univ, PhD(physics), 61. *Prof Exp:* NSF fel nuclear physics, Weizmann Inst, 61-62, NATO fel, 62-63; res assoc & lectr physics, Yale Univ, 63-67; ASST PROF PHYSICS, UNIV MASS, BOSTON, 67- *Concurrent Pos:* Am Cancer Soc scholar's grant, 69-71. *Mem:* Am Phys Soc. *Res:* Experimental nuclear and atomic physics; radioactive atomic beams; nuclear reactions and spectroscopy; production of high intensity polarized electron source; cancer research and endocrinology using hormone dependent cell lines in tissue culture. *Mailing Add:* Dept Physics Univ Mass Boston Harbor Campus Boston MA 02125

POSNER, MORTON JACOB, b Toronto, Ont, Jan 9, 42; m 65; c 5. OPERATIONS RESEARCH. *Educ:* Univ Toronto, BASc, 63, PhD(indust eng), 67. *Prof Exp:* Asst prof indust eng, 67- 73, assoc prof, 73-84, PROF, INDUST ENG, UNIV TORONTO, 85- *Concurrent Pos:* Consult, Bur Mgt Consult, Govt Can, 72-73; Ford Found & Am Soc Eng Educ fel, 72-73. *Mem:* Opers Res Soc Am; Can Opers Res Soc. *Res:* Stochastic models in operations research, primarily queueing theory with applications in networks and communications systems. *Mailing Add:* Dept of Indust Eng Univ of Toronto Toronto ON M5S 1A4 Can

POSNER, PHILIP, b Brooklyn, NY, Nov 14, 44. PHYSIOLOGY. *Educ:* Wagner Col, BS, 63; State Univ NY Downstate, PhD, 72. *Prof Exp:* Instr surg, State Univ NY, 64-67; NIH fel, 67-70; from asst prof to assoc prof, 72-85, PROF PHYSIOL, UNIV FLA, 85- *Honors & Awards:* Fogarty Sr Int Fel, Oxford, 87. *Mem:* Am Physiol Soc; Soc Gen Physiologists; Sigma Xi. *Res:* Cardiac cell physiology; pharmacology. *Mailing Add:* 60 NW 44th Terr Gainesville FL 32607

POSS, HOWARD LIONEL, b Brooklyn, NY, Oct 25, 25; m 63; c 1. PHOTOELECTRIC ASTRONOMY. *Educ:* Harvard Univ, BS, 44; Yale Univ, MS, 45; Mass Inst Technol, PhD(physics), 48. *Prof Exp:* Staff mem radiation lab, Mass Inst Technol, 45, res asst physics, 45-48; assoc scientist, Brookhaven Nat Lab, 48-51; res physicist, Hudson Labs, Columbia Univ, 51-53, asst dir, 53-56; sr physicist, Res & Adv Develop Div, Avco Corp, 56-59; mem tech staff, RCA Labs, 59-61; assoc prof, 61-88, SR PROF PHYSICS & ASTRON, TEMPLE UNIV, 88- *Concurrent Pos:* Vis scientist, Am Asn Physics Teachers & Am Inst Physics, 71-72; lectr, Mid-Atlantic Region, Sigma Xi, 79-87. *Mem:* Am Phys Soc; Am Astron Soc; Sigma Xi. *Res:* Photoelectric observations of occultations, eclipses, solar oscillations; history of science and technology. *Mailing Add:* 1246 Knox Dr Yardley PA 19067-4424

POSS, RICHARD LEON, b Aurora, Ill, Aug 21, 44; m 67; c 2. MATHEMATICS. *Educ:* St Procopius Col, BS, 66; Univ Notre Dame, MS, 69, PhD(math), 70. *Prof Exp:* Asst prof, 70-77, ASSOC PROF MATH, ST NORBERT COL, 77- *Mem:* Math Asn Am; Am Math Soc. *Res:* Set theory-strong axiom systems, general principles of induction. *Mailing Add:* Dept of Math St Norbert Col De Pere WI 54115-2099

POSS, STANLEY M, b Nanticoke, Pa, July 10, 29; m 55; c 3. ENVIRONMENTAL SCIENCES. *Educ:* St Bonaventure Univ, BS, 71; Syracuse Univ, MBA, 73. *Prof Exp:* Factory engr lamp phosphors, GTE Prod Corp, 56-62, sr engr chem pilot plt, 62-66, proj engr, 66-70, proj engr spec proj, 70-71, div environ engr, 71-73, proj engr plt servs, 73-81, engr-in- charge environ & plt servs, 81-85, div environ specialist, 85-88; CONSULT ENVIRON, NORTHEAST PA CONSULTS, 89- *Mem:* Am Chem Soc; Am Inst Chem Eng. *Res:* Provide environmental technical support including preparing permit applications, interpreting regulations for applicability, monitor control, develop and implement regulatory compliance; liason with regulatory agencies; participate in litigation; design training programs; holder of several patents. *Mailing Add:* 302 Wilmot Dr Towanda PA 18848

POSSANZA, GENUS JOHN, b Jessup, Pa, Dec 21, 37; m 70. PHARMACOLOGY. *Educ:* Univ Scranton, BS, 59; Univ Tenn, Memphis, MS, 62, PhD(pharmacol), 64. *Prof Exp:* Fel, Worcester Found Exp Biol, 64-66; pharmacologist, Ayerst Labs, Am Home Prod, 66-67; pharmacologist, Pharma-Res Can, Ltd, 67-78; PHARMACOLOGIST, BOEHRINGER INGELHEIM LTD, 78- *Mem:* Pharmacol Soc Can. *Res:* Effects of nonsteroidal and immunosuppressant agents on inflammatory reactions. *Mailing Add:* Dept Pharmacol Boehringer Ingelheim Pharmaceut Inc 174 Briar Ridge Rd Ridgefield CT 06877

POSSIDENTE, BERNARD PHILIP, JR, b New Haven, Conn, Apr 17, 54; m 82; c 1. FUNCTIONAL ANALYSIS OF MAMMALIAN CIRCADIAN SYSTEMS, UNDERGRADUATE BIOLOGY TEACHING. *Educ:* Wesleyan Univ, BA, 76; Univ Iowa, PhD(genetics), 81. *Prof Exp:* NSF postdoctoral fel psychol, Fla State Univ, 81-82; postdoctoral zool, Univ Iowa, 82-83; asst prof, 83-89, ASSOC PROF BIOL, SKIDMORE COL, 89- *Concurrent Pos:* Chair, Biol Dept, Skidmore Col, 90- *Mem:* AAAS; Behavior Genetics Asn; Soc Study Biol Rhythms. *Res:* Quantitative genetic analysis of mammalian biological clock function; olfactory bulbectomized rodents as a chronobiological model for agitated depression; quantitative genetic analysis of homeotic gene expression. *Mailing Add:* Biol Dept Skidmore Col Saratoga Springs NY 12866

POSSIN, GEORGE EDWARD, b Fond du Lac, Wis, Oct 23, 41. EXPERIMENTAL SOLID STATE PHYSICS. *Educ:* Univ Wis-Madison, BS, 63; Stanford Univ, MS, 65, PhD(physics), 69. *Prof Exp:* PHYSICIST, GEN ELEC CORP RES & DEVELOP, 70- *Concurrent Pos:* NSF fel, 63-69. *Mem:* Am Phys Soc. *Res:* Solid state physics; semiconductors, radiation effects in solids, information recording; amorphous silicon and flat panel displays. *Mailing Add:* 2361 Algonquin Rd Schenectady NY 12309

POSSLEY, LEROY HENRY, b St Anna, Wis, May 22, 28; m 57; c 3. PHARMACEUTICAL CHEMISTRY. *Educ:* St Ambrose Col, BS, 50; Univ Ill, MS, 54. *Prof Exp:* Biochemist, Vet Admin, 55-60; sr biochemist, Anar-Stone Lab, Inc, 60-64, chief chemist, 64-69, dir qual control, 69-75, dir drug regulatory affairs, 75-80; dir regulatory affairs, Am Critical Care, Am Hosp Supply Corp, 80-87; RETIRED. *Mem:* Drug Info Asn; Am Pharmaceut Asn; Pharmaceut Mfrs Asn. *Res:* Pharmaceutical analysis; neurochemistry; new drug evaluation; drug stability. *Mailing Add:* 641 Sterling Lane Prospect Heights IL 60085

POSSMAYER, FRED, b Montreal, Que, Apr 24, 39; m 69; c 2. BIOCHEMISTRY, NEUROBIOLOGY. *Educ:* Univ Toronto, BSc, 61; Univ Western Ont, PhD(biochem), 65. *Prof Exp:* Med Res Coun Can fels, Physiol Chem Inst, Univ Cologne, 66- 67 & Biochem Lab, State Univ Utrecht, 67-68; res assoc, Air Pollution Res Ctr, Univ Calif, Riverside, 68-71; from asst prof to assoc prof, 71-81, PROF BIOCHEM, OBSTET, GYNEC & BIOCHEM, UNIV WESTERN ONT, 81- *Concurrent Pos:* Med Res Coun Can scholar, Univ Western Ont, 72-78; Genetics Lab, Oxford, 84-85. *Honors & Awards:* Soc Obstetricians & Cynaecologists Award, 73; Sr Res Scientist Award, NATO, 87. *Mem:* Am Soc Biol Chemists; Biochem Soc; Can Biochem Soc; Perinertol Soc. *Res:* Control of lipid metabolism; pulmonary surfactant production. *Mailing Add:* Dept Obstet & Gynec & Biochem Univ Western Ont London ON N6A 5A5 Can

POST, BENJAMIN, b New York, NY, July 23, 11; m 39; c 3. CRYSTALLOGRAPHY. *Educ:* City Col New York, BS, 30; Polytech Inst Brooklyn, MS, 46, PhD(chem), 49. *Prof Exp:* From res assoc to prof physics & chem, Polytech Inst NY, 60-88; RETIRED. *Mem:* Am Chem Soc; fel Am Phys Soc; Am Crystallog Asn (treas, 62-65, pres, 66); Ital Asn Crystallog. *Res:* Crystal structure analysis; dynamical diffraction effects; x-ray instrumentation; powder methods. *Mailing Add:* 1111 Beacon St Brookline MA 02146

POST, BOYD WALLACE, b Glouster, Ohio, Oct 5, 28; m 52; c 4. ECOLOGY, PLANT PHYSIOLOGY. *Educ:* Ohio Univ, BS, 50; Duke Univ, MF, 58, DF, 62. *Prof Exp:* Asst ranger, Ohio Div Forestry, 53-55; from asst prof forestry & asst forester to assoc prof forestry & assoc forester, Univ Vt, 59-69; asst dep adminr, 81-82, natural resources coodr, 82-83, FOREST BIOLOGIST, COOP STATE RES SERV, USDA, 69- *Concurrent Pos:* Asst dir, Hawaii Inst Trop Agr & Human Resources, Univ Hawaii, 83-84. *Mem:* Fel AAAS; fel Soc Am Foresters; Sigma Xi; Nitrogen Fixing Tree Asn; Am Forestry Asn; Int Soc Trop Foresters. *Res:* Forest ecology and physiology; administration of forestry; remote sensing; agricultural meteorology; wildlife and fish research; research administration. *Mailing Add:* Coop State Res Serv US Dept Agr Washington DC 20250-2200

POST, DANIEL, b Brooklyn, NY, Apr 5, 29; m 48; c 3. ENGINEERING MECHANICS. *Educ:* Univ Ill, BS, 50, MS, 51, PhD(theoret & appl mech), 57. *Prof Exp:* Physicist, US Naval Res Lab, 51-55; instr appl mech, Univ Ill, 55-57; sr res engr, Res Labs, United Aircraft Corp, 57-60; assoc prof, Rensselaer Polytech Inst, 60-63; staff consult, Vishay Intertech, Inc, 62-78; PROF ENG SCI & MECH, VA POLYTECH INST & STATE UNIV, 78- *Concurrent Pos:* Adj prof, Rensselaer Polytech Inst, 64-78; vis assoc prof, Col Physicians & Surgeons, Columbia Univ, 65; vis sr res fel, Univ Sheffield,

65-66; mem, US-India Exchange Prog, NSF, 75; sr ed, Phytopath, 77-78. *Honors & Awards:* W M Murray lectr, 71. *Mem:* Fel Soc Exp Stress Anal; Am Soc Mech Engrs; Sigma Xi. *Res:* Experimental mechanics, particularly by optical methods and moire interferometry; optical interferometry. *Mailing Add:* Dept Eng Sci & Mech Va Polytech Inst Blacksburg VA 24061-0219

POST, DOUGLAS MANNERS, b Elko, Nev, May 31, 20; m 50; c 3. BOTANY. *Educ:* Univ Wash, BS, 48, MA, 50; Univ Calif, PhD(bot), 56. *Prof Exp:* Instr bot, Univ Ill, 55-57, asst prof, 57-61; assoc prof, San Francisco State Univ, 61-68, prof biol, 68-83; RETIRED. *Mem:* Bot Soc Am; Int Soc Plant Morphol. *Res:* Plant anatomy; systematics of Gentianaceae; cytology. *Mailing Add:* 9527 Crystal Lake Dr 1600 Hollway NAD 467 Woodinville WA 98072-9501

POST, DOUGLASS EDMUND, b Gulfport, Miss, Mar 16, 45; m 69; c 2. PLASMA PHYSICS. *Educ:* Southwestern at Memphis, BS, 67; Stanford Univ, MS, 68, PhD(physics), 75. *Prof Exp:* Physicist, Lawrence Livermore Lab, 68-71; PHYSICIST PLASMA PHYSICS LAB, PRINCETON UNIV, 74- *Concurrent Pos:* Consult, Lawrence Livermore Lab, 71-72 & 74-; actg instr physics, Stanford Univ, 74. *Mem:* Fel Am Phys Soc; Am Vacuum Soc; Am Nuclear Soc; Europ Phys Soc. *Res:* Computational modeling of plasma physics experiments, with emphasis on plasma physics, atomic physics and numerical techniques; plasma diagnostics; Tokamak design. *Mailing Add:* Plasma Physics Lab Princeton Univ PO Box 451 Princeton NJ 08543

POST, EDWIN VAN HORN, economic geology, geochemistry, for more information see previous edition

POST, ELROY WAYNE, b Raymond, Minn, July 18, 43; m 67. INORGANIC CHEMISTRY, ORGANOMETALLIC CHEMISTRY. *Educ:* Dordt Col, AB, 65; Kans State Univ, PhD(inorg chem), 69. *Prof Exp:* Asst prof, 69-77, ASSOC PROF CHEM, UNIV WIS-OSHKOSH, 77- *Concurrent Pos:* NSF res grant, Univ Wis-Oshkosh, 70-72. *Mem:* Am Chem Soc; Am Sci Affiliation. *Res:* Synthesis of organometallic compounds; transition metal carbonyl carbenes; amine activated metalation of ferrocene; boron halide reactions with organometallic compounds; photo chemistry of metal carbonyls. *Mailing Add:* Dept Chem Univ Wis Oshkosh WI 54901

POST, FREDERICK JUST, b Berkeley, Calif, Feb 20, 29; m 58; c 3. MICROBIAL ECOLOGY, POLLUTION BIOLOGY. *Educ:* Univ Calif, BS, 52; Mich State Univ, MS, 53, PhD(microbiol), 58. *Prof Exp:* Asst, Mich State Univ, 53, 55-58; from instr to asst prof pub health microbiol, Univ Calif, Los Angeles, 58-65; assoc prof pub health microbiol, 65-76, PROF BIOL, UTAH STATE UNIV, 76- *Mem:* AAAS; Am Soc Microbiol; Sigma Xi. *Res:* Water microbiology; food poisoning organisms; ecology of microorganisms in natural environments; halophilic microorganisms; biotechnology. *Mailing Add:* Dept Biol Utah State Univ Logan UT 84322-5500

POST, GEORGE, b Rapid City, SDak, May 12, 18; m 43; c 2. FISH PATHOLOGY, WILDLIFE PATHOLOGY. *Educ:* Univ Wyo, BS, 47, MS, 48; Utah State Univ, PhD(fish path), 63. *Prof Exp:* Parasitologist, Wyo State Vet Lab, Laramie, 47-48; biologist, Wyo Game & Fish Res Lab, 48-52, lab dir, 56-60; fish pathologist & nutritionist, Utah Dept Fish & Game, 60-64; asst leader, Colo Coop Fishery Unit, 64-66; from assoc prof to prof fishery & wildlife biol, Colo State Univ, 66-84, from assoc prof to prof microbiol, 76-84; RETIRED. *Concurrent Pos:* Auth. *Mem:* Wildlife Dis Asn; Am Fisheries Soc; Sigma Xi. *Res:* Diseases, parasites and nutrition of fishes and wildlife; effects of energy development on fish and wildlife. *Mailing Add:* 1803 Paseo Venado Rio Rico AZ 85621

POST, HOWARD WILLIAM, b Syracuse, NY, 96; wid. RUBBER & SILICONES, CHEMICAL WARFARE. *Educ:* Syracuse Univ, BS, 19, MS, 21; Johns Hopkins Univ, PhD(org chem), 27. *Prof Exp:* Fac mem organic chem, State Univ NY, Buffalo, 27-67; RETIRED. *Concurrent Pos:* Consult, US Army, 78. *Mem:* Am Chem Soc; Am Inst Chemists; Am Sci Affil. *Res:* Organic silicon chemistry. *Mailing Add:* 2235 Millersport Hwy c/o Methodist Home Getzville NY 14068

POST, IRVING GILBERT, b New York, NY, Feb 11, 37; m 65; c 3. SOLID STATE ELECTRONICS. *Educ:* Lafayette Col, BS(elec eng) & BS(physics), 58; Ohio State Univ, MS, 59, PhD(physics), 62. *Prof Exp:* Res asst, Ohio State Univ, 58-62; distinguished mem tech staff, AT&T Bell Labs, 63-90; CONSULT, 90- *Mem:* AAAS; Inst Elec & Electronics Engrs. *Res:* Integrated circuit design; semiconductor device development and applications; electron spin resonance in solids. *Mailing Add:* 112 Harvey Ave Reading PA 19606

POST, J(AMES) L(EWIS), b South Bend, Ind, Sept 25, 29; m 67. CIVIL & MINING ENGINEERING. *Educ:* NMex Inst Mining & Technol, BS, 51; Univ Ariz, MS, 63, PhD(civil eng), 66. *Prof Exp:* Mining engr, United Park City Mines Co, 54-56; mining engr, US Bur Mines, 57-58; civil engr asst, County of Los Angeles Rd Dept, 59-61; teaching assistantships, Univ Ariz, 63-66; assoc res engr, E H Wang Civil Eng Res Facility, Univ NMex, 66-68; assoc prof, 68-77, PROF CIVIL ENG, CALIF STATE UNIV, SACRAMENTO, 77- *Mem:* Fel AAAS; Am Soc Civil Engrs; Am Inst Mining, Metall, & Petrol Engrs; Clay Minerals Soc. *Res:* Physical-chemical nature of soils including mineralogy and soil stabilization for construction purposes; applied x-ray diffraction procedures; microstructure investigations; geological engineering. *Mailing Add:* Dept Civil Eng Calif State Univ 6000 J St Sacramento CA 95819

POST, JOHN E, b Sussex, NJ, Mar 2, 26; m 52; c 4. VETERINARY PATHOLOGY, VETERINARY MICROBIOLOGY. *Educ:* Rutgers Univ, BS, 51; Cornell Univ, DVM, 58, PhD(vet path), 68. *Prof Exp:* Vet gen pract, Middlebury, Vt, 58-61; sr res asst path & virol, Dept Path, Univ Conn, 62-79, prof & dir diag testing, 79-91. *Mem:* Am Vet Med Asn; NY Acad Sci; US Animal Health Asn; Am Asn Vet Lab Diagnosticians. *Res:* Tumor virology; retroviruses; lyme disease. *Mailing Add:* Dept Path Univ Conn Storrs CT 06268

POST, JOSEPH, b New York, NY, Mar 6, 13; m 42; c 2. MEDICINE. *Educ:* City Col New York, BS, 32; Univ Chicago, MD, 37; Columbia Univ, MedScD, 41. *Prof Exp:* Asst liver physiol, Dept Surg, Univ Chicago, 36-37; intern, Michael Reese Hosp, Chicago, 37-38; from asst to instr med, Col Physicians & Surgeons, Columbia Univ, 38-47, resident, Res Div Chronic Dis, Univ, 38-42; from asst prof to prof, 46-91, EMER PROF CLIN MED, SCH MED, NY UNIV, 91- *Concurrent Pos:* Attend consult, Vet Admin Hosp, New York, 46-54; consult gastroenterologist, US Naval Hosp, NY, 47-56; res fel, Goldwater Mem Hosp, 48, attend physician, 80-88; assoc physician, Lenox Hill Hosp, New York, 47-64, attend physician, 64-91; assoc attend physician, Hosp, NY Univ, 52-71, attend physician, 71-91. *Honors & Awards:* Plaque Award, NY Acad Med, 82. *Mem:* Am Gastroenterol Asn; Radiation Res Soc; Am Asn Cancer Res; fel Am Col Physicians; fel NY Acad Sci; Soc for Cell Biol; Am Soc Clin Oncol; Cellkinetics Soc; NY Acad Med (pres, 79, 80). *Res:* Physiology and pathology of liver; nutrition and metabolism; cell replication kinetics; tumor cell replication; radiation biology. *Mailing Add:* 29 Washington Square W New York NY 10011

POST, M(AURICE) DEAN, b Topeka, Kans, Sept 16, 16; m 47; c 3. ELECTRONICS. *Educ:* Univ Colo, BSEE, 38. *Prof Exp:* Radio engr, Sta KOBH, SDak, 38-39; audio engr, Shirley Savoy Hotel, 39-40; instr commun & radar, Army Air Force Tech Training Command, 40-44; tech aide, Off Sci Res & Develop, Eng, 44-45; chief electronics intel, Asst Chief of Staff, Pentagon, 46-51; staff asst res & develop, Asst Secy Defense, 51-56; spec asst to dept dir, Nat Security Agency, DC, 56-58; planning adminr defense electronics prod, Radio Corp Am, NJ, 58-60, adminr electronic data processing custom appln, 60-61; Washington mgr, Emerson Elec Mfg Co, Mo, 61-62; mkt mgr, Babcock Electronics Corp, 62-63; consult, US Off Emergency Planning, 63-64; vpres, Computronics E, Washington, 64-65; mem staff, US Army Automatic Data Field Systs Command, Ft Belvoir, 65-67; mem staff, Defense Intel Agency, Pentagon, 67-77; RETIRED. *Concurrent Pos:* Adv, Nat Security Agency, 60-61. *Mem:* Sr mem Inst Elec & Electronic Engrs. *Res:* Radio-frequency oscillation; class C and high fidelity amplifiers. *Mailing Add:* 704 Cedar Ave Fairhope AL 36532

POST, MADISON JOHN, b Detroit, Mich, Oct 4, 46; m 83; c 1. ATMOSPHERIC PHYSICS, OPTICS. *Educ:* Univ Ill, Champaign, BS(elec eng) & BS(math), 69; Univ Colo, Boulder, MS, 75; Univ Ariz, Tucson, MS, 83, PhD(optical sci), 85. *Prof Exp:* Physicist, Hach Co, 86-87; elec engr, Environ Sci Serv Admin, US Dept Com, 69-71, Wave Propagation Lab, 71-75, PHYSICIST, WAVE PROPAGATION LAB, US DEPT COM, 75-85, 88- *Concurrent Pos:* Asst Sta Sci leader, Byrd Sta, Antarctica, 70. *Mem:* Optical Soc Am; Sigma Xi; SPIE. *Res:* Atmospheric propagation and scattering of electromagnetic energy, with application to remote sensing by laser probing; development of pulsed, coherent Doppler lidar for remote sensing of atmospheric winds. *Mailing Add:* Nat Oceanic & Atmospheric Admin Wave Propagation Lab R/E/WP2, 325 Broadway Boulder CO 80303

POST, RICHARD FREEMAN, b Pomona, Calif, Nov 14, 18; m 46; c 3. PHYSICS. *Educ:* Pomona Col, BA, 40; Stanford Univ, PhD(physics), 51. *Hon Degrees:* ScD, Pomona Col, 59. *Prof Exp:* From asst to instr physics, Pomona Col, 40-42; physicist underwater sound, Naval Res Lab, 42-46; res assoc physics, Stanford Univ, 47-51; res group leader controlled thermonuclear res, 51-74, dep assoc dir, magnetic fusion physics, 74-87, SR SCIENTIST, MAGNETIC FUSION PHYSICS, 87-, PROF IN RESIDENCE ENG & APPL SCI, LAWRENCE LIVERMORE NAT LAB, 63- *Concurrent Pos:* Mem, NASA Adv Comt, Nuclear Energy Systs, 60-62; mem Physics Surv Comt, Nat Acad Sci, 64-65 & 84-85; vis comt, Dept Nuclear Eng, Mass Inst Technol, 65-71 & Plasma Inst, 82-; mem, adv comt to Air Force Systs Command, 66. *Honors & Awards:* Robert Henry Thurston Award, Am Soc Mech Engr, 63; Am Acad Achievement Golden Plate Award, Dallas, 67; Distinguished Assoc Award, US Energy Res & Develop Admin, 77; Outstanding Achievement Award, Am Nuclear Soc, 78; James Clark Maxwell Prize Award, Am Phys Soc, 78; Distinguished Career Award, Fusion Power Assoc, 87. *Mem:* Fel Am Phys Soc; fel Am Nuclear Soc. *Res:* Electron physics; traveling wave electron linear accelerators; controlled fusion; scintillation counter resolving time; high temperature plasmas; energy storage. *Mailing Add:* Lawrence Livermore Nat Lab PO Box 808 MS L-644 Livermore CA 94550

POST, RICHARD S, PLASMA PHYSICS, ENGINEERING PHYSICS. *Educ:* Univ Calif, Berkeley, BS, 66; Columbia Univ, PhD(plasma physics), 73. *Prof Exp:* Instr physics & teacher training, Peace Corps, Nat Univ, Columbia, SAm, 66-68; asst prof, nuclear eng, Univ Wis, 73-81, assoc prof, 81-87; SR SCIENTIST & HEAD ADV CONCEPTS DIV, MASS INST TECHNOL PLASMA FUSION CTR, 87- *Concurrent Pos:* Vis physicist, Naval Res Lab, 75; vis physicist, Phaedrus Proj, 77-79; mem, Attached Payloads Panel, Nat Aeron Space Admin; various publ, 67-87. *Mem:* Fel Am Phys Soc. *Res:* Plasma-wall interactions and surface physics; radio frequency heating, diagnostics and low frequency stability; plasma transport in the toroidal multipole; high beta plasmas and collective transport of collisionality; fluctuation spectra in the dense plasma focus with carbon scattering and extended Bremsstrahlung measurements; sloshing ions and axisymmetric confinement. *Mailing Add:* Appl Sci & Technol Inc 40 Allisten St Cambridge MA 02139

POST, ROBERT ELLIOTT, b Paterson, NJ, Mar 17, 24; m 53; c 2. ELECTROCHEMISTRY. *Educ:* Yale Univ, BE, 44; Univ Wis, MS, 48; Univ Ill, PhD(phys chem), 57; Cleveland State Univ, JD, 66. *Prof Exp:* Engr indust rayon div, Midland-Ross Corp, 49-51, chemist, 55-61; sr res chemist, Tee-Pak, Inc, 61-62; chemist, Lewis Res Ctr, NASA, Cleveland, 62-81, engr, 81-; RETIRED. *Res:* Electrochemically related aspects of catalysis; mass transport and analytical techniques. *Mailing Add:* 23935 E Oakland Bay Village OH 44140

POST, ROBERT LICKELY, b Philadelphia, Pa, Nov 4, 20; m 47. BIOCHEMISTRY. *Educ:* Harvard Univ, AB, 42, MD, 45. *Prof Exp:* Instr physiol, Sch Med, Univ Pa, 46-48; from instr to prof molecular physiol & biophysics, Sch Med, Vanderbilt Univ, 48-91; VIS PROF PHYSIOL, SCH MED, UNIV PA, 91- *Concurrent Pos:* Biomed Sci study sect, NIH, 83-87. *Honors & Awards:* Cole Award, Biophys Soc, 83. *Mem:* Am Physiol Soc; Biophys Soc; Soc Gen Physiol; Am Soc Biol Chem. *Res:* Adenosine triphosphate dependent transport of sodium and potassium ions across cell membranes. *Mailing Add:* Dept Physiol Univ Pa Sch Med Philadelphia PA 19104-6085

POST, ROBERT M, b New Haven, Conn, Sept 16, 42; m 66; c 2. PSYCHOPHARMACOLOGY. *Educ:* Yale Univ, BA, 64; Univ Pa, MD, 68. *Prof Exp:* Clin assoc, Lab Clin Sci, Sect Psychiat, 70-72, res fel, 72-73, chief, 3-West Clin Res Unit, 73-77, Sect Psychobiol, 77-82, ACTG CHIEF, BIOL PSYCHIAT BR, NAT INST MENTAL HEALTH, 82- *Concurrent Pos:* Intern neurol & pediat, Albert Einstein Sch Med, Jacobi Hosp, New York, 68-69; resident psychiat, Mass Gen Hosp, 69-70; A E Bennett Neuropsychiat Res Awarc, Soc Biol Psychiat, 73; psychiat residency, Dept Psychiat & Behav Sci Prog, George Washington Univ, 73-75. *Mem:* Am Col Neuropsychopharmacol; Am Psychiat Asn; Soc Biol Psychiat; Soc Neurosci. *Res:* Interactions between psychological and biological phenomena in patients with manic-depressive and anxiety disorders and in laboratory animal models of psychopathology. *Mailing Add:* NIH Nat Inst Mental Health Biol Psychiat Br Bldg 10 Rm 3N212 Bethesda MD 20892

POST, ROY G, b Asherton, Tex, June 24, 23; m 46; c 4. PHYSICAL INORGANIC CHEMISTRY. *Educ:* Univ Tex, BS, 44, PhD(chem), 52. *Prof Exp:* Assoc engr, Argonne Nat Lab, 44-49; chemist, Magnolia Petrol Co, 51-52; sr engr, Gen Elec Co, 52-58; sect head, Tex Instruments, Inc, 58-61; dir spec prof educ, Col Eng, 74-76, prof nuclear eng, 61-88, EMER PROF, UNIV ARIZ, 88- *Concurrent Pos:* Ed, Nuclear Technol, 69-88. *Mem:* Fel Am Nuclear Soc; AAAS; Sigma Xi. *Res:* Solvent extraction; surface chemistry; nucleonics; water desalting; nuclear fuel cycles; nuclear wastes. *Mailing Add:* Dept Nuclear & Energy Eng Univ of Ariz Tucson AZ 85721

POSTE, GEORGE HENRY, b Polegate, Suxxex, Eng, Apr 30, 44; m 68; c 1. CELL BIOLOGY, MOLECULAR GENETICS. *Educ:* Univ Bristol, UK, DVM, 66, PhD(virol), 69; Royal Col Vet Surgeons, FRCVS, 87; Royal Col Path, FRC, 89. *Hon Degrees:* DSc, Univ Bristol, UK, 87. *Prof Exp:* Lectr virol, Univ London, Eng, 69-72; assoc prof path, Roswell Park Mem Inst, 72-75; prof path, State Univ, NY, 75-80; PROF PATH, UNIV PA MED SCH, 80-; vpres res & develop, Smith Kline Beckman, 83-86, vpres pharmacol res, 80-83, vpres worldwide res & preclin develop, 87- 88, pres res & develop, 88-89, pres res & develop technologies, 89-90, VCHMN & EXEC VPRES, RES & DEVELOP, SMITHKLINE BEECHAM PHARMACEUTICALS, 90- *Concurrent Pos:* Ed, Cell Surface Rev, 75-, Cancer Matastasis Rev, 84- & Advan Drug Delivery Rev, 85-; mem, path bd sect, NIH, 79-83; mem, Univ Fedn Space Res, 84-85; prof, Anderson Hosp & Tumor Inst, Univ Tex, 84-; bd dirs, Nat Asn Biomed Res, 85-86; res & develop steering comt, Pharmaceut Mfg Asn; mem bd, Life Sci Res Found. *Mem:* Am Asn Cancer Res; Am Soc Cell Biol; Path Soc Gt Brit & Ireland; Fedn Am Soc Exp Biol Med; AAAS. *Res:* Mechanisms of cancer metastasis; drug delivery systems; colorectal cancer; lymphokine biology; macrophages amd host defense. *Mailing Add:* SmithKline Beecham Pharmaceuticals L328 PO Box 1539 King of Prussia PA 19406-0939

POSTELNEK, WILLIAM, b Peoria, Ill, Jan 22, 18; m 45; c 3. ORGANIC CHEMISTRY, MATERIALS SCIENCE. *Educ:* Ill Inst Technol, BS, 42; Ohio State Univ, MSc, 54; Univ Chicago, MBA, 59. *Prof Exp:* Res chemist, L R Kerns Co, 46-48 & Dawe's Mfg Co, 48-50; proj officer polymer res sect, Air Force Mat Lab, US Air Force, 51-53, chief polymer res sect, 54-58, chief mat div, Europ Off, Off Aerospace Res, 59-62, chief mat appln div, Air Force Mat Lab, 62-67; consult, Gen Elec Co, Pa, 67-70; dir mat & processes lab, Kearfott Div, Singer Co, 70-80; CONSULT, MAT & PROCESS ENG, 80- *Honors & Awards:* Legion of Merit, 59. *Mem:* Am Chem Soc; Am Inst Chem; Am Soc Test & Mat; Soc Advan Mat & Process Eng. *Res:* Organic fluorine and organometallic compounds; polymer and high temperature chemistry; combustion; propulsion; aerospace materials. *Mailing Add:* 4725 Cove Circle 907 St Petersburg FL 33708

POSTEN, HARRY OWEN, b Middletown, NY, Feb 6, 28; m 57; c 3. MATHEMATICAL STATISTICS. *Educ:* Cent Conn State Col, BS, 56; Kans State Univ, MS, 58; Va Polytech Inst, PhD, 60. *Prof Exp:* Mem res staff, IBM Res Ctr, 60-62; asst prof math, Univ RI, 62-63; PROF MATH STATIST, UNIV CONN, 63- *Mem:* Fel Am Statist Asn; Sigma Xi. *Res:* Statistical robustness; statistical computing. *Mailing Add:* Dept Statist Univ Conn Storrs CT 06269

POSTIC, BOSKO, b Novi Sad, Yugoslavia, Feb 9, 31; m 67; c 2. MEDICINE, INFECTIOUS DISEASES. *Educ:* Univ Zagreb, MD, 55; Harvard Univ, MPH, 62; Univ Pittsburgh, DSc(microbiol), 65; Am Bd Internal Med, cert, 74 & 80, cert infectious disease, 76. *Prof Exp:* Intern med, Mt Auburn Hosp, Cambridge, Mass, 57-58; asst resident, Lemuel Shattuck Hosp, Boston, 58-59; resident med bact, Boston City Hosp, 59-60, res fel, Thorndike Mem Lab, 60-61; trainee microbiol, Sch Pub Health, Harvard Univ, 61-62; trainee, Grad Sch Pub Health, Univ Pittsburgh, 62-65, from asst prof to assoc prof epidemiol, Grad Sch Pub Health, 66-78, assoc prof med, Sch Med, 72-78; chief infectious dis sect, Vet Admin Hosp, Pittsburgh, 75-78, chief med serv, 76-78; PROF MED, SCH MED, UNIV SC, COLUMBIA, SC, 78-, CHIEF MED SERV, VET ADMIN HOSP, 78- *Concurrent Pos:* Am Cancer Soc fel, Sch Med, Univ Pittsburgh, 70-71, WHO fel, 70-72, asst prof med, 71-72, Health Res & Serv Found fel, 72-73; fel, Res Dept, Vet Admin, 76-78. *Mem:* Am Soc Microbiol; fel Am Col Physicians; fel Infectious Dis Soc Am. *Res:* Medical virology; interferon; epidemiology; infectious diseases. *Mailing Add:* Dept Int Med Richland Med Park No 506 Columbia SC 29203

POSTL, ANTON, b Graz, Austria, June 16, 16; nat US; m 42; c 3. ORGANIC CHEMISTRY. *Educ:* Univ Hawaii, BS, 40, MS, 42; Ore State Col, PhD(gen sci), 55. *Prof Exp:* Staff, Mid-Pac Inst, 41-43; instr chem, Univ Hawaii, Honolulu, 43-45; from instr to prof chem, 47-81, chmn div sci & math, 55-72, EMER PROF CHEM, DIV NATURAL SCI & MATH, WESTERN ORE STATE COL, MONMOUTH, 82- *Concurrent Pos:* Chmn, Ore Sect Am Chem Soc, 63-64. *Mem:* Fel AAAS; Am Chem Soc; Am Asn Univ Prof. *Res:* Colloid chemistry of soils; history of science. *Mailing Add:* 1971 Manor View Lane NW Salem OR 97304-4400

POSTLE, DONALD SLOAN, b Columbus, Ohio, July 18, 22; m 47; c 4. VETERINARY SCIENCE. *Educ:* Ohio State Univ, DVM, 50; Univ Wis-Madison, MS, 67. *Prof Exp:* Pvt pract, 50-62; from instr to asst prof vet sci, Univ Wis-Madison, 62-69, proj assoc mastitis res & prin investr, USPHS res grant, 62-67; ASSOC PROF VET SCI, NY STATE COL VET MED, 69- *Concurrent Pos:* Dir, NY State Mastitis Control Prog, 72. *Mem:* Am Vet Med Asn; US Animal Health Asn. *Res:* Mastitis research and control; bovine mastitis. *Mailing Add:* NY State Col Vet Med Cornell Univ C 117 Schurman Hall Ithaca NY 14853-6401

POSTLETHWAIT, JOHN HARVEY, b Kittery, Maine, July 15, 44; m 64; c 2. DEVELOPMENTAL BIOLOGY, MOLECULAR GENETICS. *Educ:* Purdue Univ, BA, 66; Case Western Reserve Univ, PhD(biol), 70. *Prof Exp:* Fel, Harvard Univ, 70-71; asst prof, 71-77, assoc prof, 77-81, PROF BIOL, UNIV ORE, 81- *Concurrent Pos:* Fulbright fel, Inst Molecular Biol, Austria, 77-78; NSF/Nat Ctr Sci Res fel, Inst Molecular Genetics, Strasbourg, France, 82-83; fel Am Acad Sci; vis scientist, Develop Biol Univ, Imp Cancer Res Fund, Oxford, Gt Brit. *Mem:* AAAS; Genetics Soc Am; Soc Develop Biol. *Res:* Molecular genetics of development in Drosophila melanogaster; insect endocrinology; molecular mechanisms of insect immunity. *Mailing Add:* Dept Biol Univ Ore Eugene OR 97403

POSTLETHWAIT, RAYMOND WOODROW, b New Martinsville, WVa, Oct 9, 13; m 37; c 4. SURGERY. *Educ:* WVa Univ, BS, 35; Duke Univ, MD, 37; Am Bd Surg, dipl. *Prof Exp:* Intern, Duke Hosp, NC, 37-39; house officer surg, Henry Ford Hosp, Mich, 39-40, Palmerton Hosp, Pa, 40-41 & Duke Hosp, NC, 45-47; instr, Bowman Gray Sch Med, 48-49; asst prof, Med Col SC, 49-55; from assoc prof to prof surg, Sch Med, Duke Univ, 55-85, emer prof, 85; RETIRED. *Concurrent Pos:* Chief surg serv, Vet Admin Hosp, Durham, 55-65, chief of surg, 65-85. *Mem:* Soc Univ Surgeons; AMA; Am Col Surgeons. *Res:* Surgery of the esophagus. *Mailing Add:* 1413 Pinecrest Rd Durham NC 27705

POSTLETHWAIT, SAMUEL NOEL, b Wileysville, WVa, Apr 16, 18; m 41; c 2. BOTANY. *Educ:* Fairmont State Col, AB, 40; WVa Univ, MS, 47; State Univ Iowa, PhD, 49. *Hon Degrees:* PhD, Doane Col, 62. *Prof Exp:* Teacher pub schs, WVa, 40-41; instr bot & biol, Univ Iowa, 48-49; from asst prof to prof, 49-84, EMER PROF BOT & BIOL, PURDUE UNIV, W LAFAYETTE, 84- *Concurrent Pos:* Nat Sci Found fac fel, Manchester Univ, 57-58; Fulbright fel, Macquarie Univ, Australia, 68; Lilly Endowment fac open fel award, 79; vis prof, Windsor Univ, Can. *Mem:* Bot Soc Am; Am Inst Biol Sci; AAAS; Sigma Xi; Am Genetic Asn. *Res:* Plant morphology; science education. *Mailing Add:* 3180 Soldiers Home Rd West Lafayette IN 47906

POSTLETHWAITE, ARNOLD EUGENE, CONNECTIVE TISSUE DISEASES. *Educ:* Cornell Univ, MD, 66. *Prof Exp:* DIR, DIV CONNECTIVE TISSUE DIS, UNIV TENN, MEMPHIS, 73- *Res:* Collagen biochemistry; cellular immunology. *Mailing Add:* 635 Bethany Rd Eads TN 38028-9723

POSTMA, HERMAN, b Wilmington, NC, Mar 29, 33; m 60; c 2. PHYSICS. *Educ:* Duke Univ, BS, 55; Harvard Univ, MA, 58, PhD(physics), 59. *Prof Exp:* Physicist, Oak Ridge Nat Lab, 59-63 & Inst Plasma Physics, Neth, 63-64; physicist, 64-67, dir thermonuclear div, 67-73, DIR, OAK RIDGE NAT LAB, 74- *Concurrent Pos:* Consult, Lab Laser Energetics, Rochester Univ, 74-; bd adv, Univ Tenn Bus Sch; bd adv, NC Energy Inst. *Mem:* Fel AAAS; fel Am Phys Soc; Am Nuclear Soc. *Res:* Nuclear, atomic and plasma physics; controlled fusion feasibility. *Mailing Add:* 104 Berea Rd Oak Ridge TN 37830

POSTMAN, ROBERT DEREK, b Jersey City, NJ, July 13, 41; m 65; c 3. MATHEMATICS, COMPUTER EDUCATION. *Educ:* Kean Col, NJ, BA, 66; Columbia Univ, MA, 67, EdD(math), 71. *Prof Exp:* Asst prof math, Hunter Col, City Univ New York, 71-76; dir, Computer Workshops, 81-86, dean grad educ progs, 87, PROF MATH & EDUC, MERCY COL, DOBBS FERRY, NY, 76-, DIR, INST GIFTED CHILDREN, 79- *Concurrent Pos:* Math consult, Columbia Univ-Royal Afghan Univ Coop Proj Math, 72-73 & Harcourt, Brace & Jovanovich Inc, 75-; auth, numerous texts & comput progs, 74-; fac mem, Dept Math, Statist & Comput, Educ Teachers Col, Columbia Univ, 76-; math consult, Sch Dists & State Educ Depts, 79-; consult, Psychol Corp & CBS Publ, 84-; dir, Comput Workshops, 81- *Mem:* Am Math Soc; Math Asn Am. *Res:* Microcomputers in education; writing computer programs and mathematical texts. *Mailing Add:* Dept Math Mercy Col Dobbs Ferry NY 10522

POSTMUS, CLARENCE, JR, b Grand Rapids, Mich, Dec 12, 27; m 52; c 3. INORGANIC CHEMISTRY. *Educ:* Calvin Col, AB, 50; Univ Wis, PhD(chem), 54. *Prof Exp:* Assoc chemist, Argonne Nat Lab, 54-69; from asst prof to assoc prof, 69-78, PROF CHEM, NORTH PARK COL, 73- *Mem:* Soc Appl Spectros; Am Chem Soc. *Res:* Solution chemistry; far infrared. *Mailing Add:* 5032 Woodland Western Springs IL 60558

POSTON, FREDDIE LEE, JR, b Jacksonville, Fla, Nov 19, 46; m 67; c 2. ENTOMOLOGY. *Educ:* WTex State Univ, BS, 71; Iowa State Univ, MS, 73, PhD(entom), 75. *Prof Exp:* Res assoc entom, Iowa State Univ, 73-75; from asst prof to assoc prof, Kans State Univ, 75-84, assoc dir exten & prof entom, 84-87; ASSOC DEAN, DIR EXTEN & PROF ENTOM, WASH STATE UNIV, 87- *Mem:* Entom Soc Am. *Res:* Crop management and insect damage/crop loss relationship. *Mailing Add:* Coop Exten Wash State Univ Pullman WA 99164-6230

POSTON, HUGH ARTHUR, b Canton, NC, May 5, 29; m 54; c 3. REPRODUCTIVE PHYSIOLOGY, ANIMAL NUTRITION. *Educ:* Berea Col, BS, 51; NC State Univ, MS, 54, PhD(reproductive physiol), 62. *Prof Exp:* Mgr dairy animal res, NC State Cent Dairy Res Sta, 56-58; res physiologist, Tunison Lab Fish Nutrit, Fish & Wildlife Serv, US Dept Interior, 61-90; ADJ ASSOC PROF, COL VET MED, CORNELL UNIV, 78- *Mem:* Am Fisheries Soc; Am Inst Nutrit; Am Inst Fishery Res Biologists; World Aquaculture Soc. *Res:* Mammalian nutrition and physiology; fish nutrition, physiology and endocrinology. *Mailing Add:* 2644 Sugar Bush Lane Dryden NY 13053

POSTON, JOHN MICHAEL, b Kalispell, Mont, Oct 16, 35; m 67; c 3. BIOCHEMISTRY. *Educ:* Mont State Col, BS, 58; Univ NDak, Grand Forks, MS, 60, PhD(biochem), 70. *Prof Exp:* Res fel, Biochem Dept, Univ NDak, 60-61; chemist, Nat Heart Inst, NIH, 61-69; RES CHEMIST, ENZYME SECT, LAB BIOCHEM, NAT HEART, LUNG & BLOOD INST, NIH, 70- *Mem:* Am Chem Soc; Am Soc Biol Chemists; Am Soc Microbiol; Sigma Xi. *Res:* Metabolism of branched-chain amino acids, especially with regard to the involvement of cobalamins, in man, animals, and bacteria; in-born errors of metabolism, oxidation of amino acids and proteins. *Mailing Add:* Bldg 3 Rm 216 NIH Bethesda MD 20892

POSTON, JOHN WARE, b Sparta, Tenn, July 8, 37; m 58; c 3. HEALTH PHYSICS, NUCLEAR ENGINEERING. *Educ:* Lynchburg Col, BS, 58; Ga Inst Technol, MS, 69, PhD(nuclear eng), 71. *Prof Exp:* Exp reactor physicist, Critical Exp Lab, Babcock & Wilcox Co, 57-64; health physicist, Oak Ridge Nat Lab, 64-74, chief, Med Physics Internal Dosimetry Sect, Health Physics Div, 74-77; assoc prof, Sch Nuclear Eng, Ga Inst Technol, Atlanta, 77-84; PROF NUCLEAR ENG, TEX A&M UNIV, 85-, HEAD, DEPT NUCLEAR ENG, 88- *Concurrent Pos:* Consult, Int Atomic Energy Agency, Austria, 65-70; lectr & consult, Inst Atomic Energy, Brazil, 71; Halliburton prof, 87-88. *Honors & Awards:* Landauer Mem lectr, 90. *Mem:* AAAS; Soc Nuclear Med; Int Radiation Protection Asn; Health Physics Soc (secy, 74-76, pres, 86-87); Am Nuclear Soc; Sigma Xi; Soc Nuclear Med; Nat Coun Radiation Protection & Measurements. *Res:* Neutron and gamma ray dosimetry; nuclear accident dosimetry; internal dosimetry; dosimetry for medical physics purposes. *Mailing Add:* Dept Nuclear Eng Tex A&M Univ College Station TX 77843-3133

POSTOW, ELLIOT, b New York, NY, June 21, 40; m 70; c 2. BIOPHYSICS. *Educ:* City Col New York, BS, 61; Mich State Univ, PhD(biophys), 68. *Prof Exp:* Biophysicist, Off Naval Res, Navy Dept, Naval Med Res & Develop Command, 68-72, prog mgr, Navy Bur Med & Surg, 72-74, assoc sci dir, 74-87; health scientist adminr, 87-90, CHIEF SPEC REV SECT, DIV RES GRANTS, NIH, 91- *Concurrent Pos:* Mem, Am Nat Standards Inst Comt C95, 76-90; ed, Bioelectromagnetics, 78-83; mem, Comt Man & Radiation, Inst Elec & Electronic Engrs, 82-90. *Mem:* AAAS; Bioelectromagnetics Soc (pres, 85-86). *Res:* Biological effects and medical uses of electromagnetic radiation; bioenergetics; radiation; research administration. *Mailing Add:* Div Res Grants Nat Inst Health Bethesda MD 20892

POSVIC, HARVEY WALTER, b Cicero, Ill, Apr 22, 21. ORGANIC CHEMISTRY. *Educ:* Univ Chicago, BS, 40; Carleton Col, MA, 42; Univ Wis, PhD(org chem), 46. *Prof Exp:* Asst chem, Carleton Col, 40-42; asst, Univ Wis, 42-44, instr, 44-45; DuPont fel, Cornell Univ, 46-47, instr chem, 47-52; from asst prof to assoc prof, 52-83, EMER PROF CHEM, LOYOLA UNIV CHICAGO, 83- *Mem:* Am Chem Soc. *Res:* Stereochemistry; physical organic chemistry; reaction mechanisms; natural products; drugs. *Mailing Add:* 1571 Popko Circle W Mercer WI 54547-9625

POSWILLO, D E, b Gisborne, NZ, Jan 1, 27; m; c 4. SURGERY. *Educ:* Univ Otago, BDS, 48, DDS, 62, DSc, 75; Univ Zurich, MDhc, 83; MRCPath. *Hon Degrees:* Dr, Univ Zurich, 83. *Prof Exp:* Oral surgeon pvt pract, Christchurch, 53-68; prof teratol, Dept Dent Sci & Res Estab, Royal Col Surgeons Eng, 69-77; sr hon oral & maxillofacial surgeon, Royal Adelaide Hosp, Queen Elizabeth Hosp, Adelaide Children's Hosp & Modbury Hosp, 77-79; prof oral surg, Sch Dent Surg, Royal Dent Hosp, London, 79-83; PROF & HEAD, DEPT ORAL & MAXILLOFACIAL SURG, UNITED MED & DENT SCHS, GUY'S & ST THOMAS' HOSPS, 83-; HON CONSULT, GUY'S & LEWISHAM HOSPS, 83- *Concurrent Pos:* Assoc ed, Int J Orthod, 62-68; Nuffield Dominion travelling fel, Univ Zurich, 67-68; Hunterian prof, Royal Col Surgeons, Eng, 67-68 & 75-76; consult oral surgeon, Oral Surg Unit, Queen Victoria Hosp, Sussex, 69-77; num vis prof foreign & US, 70-88; counr, Sect Odont, Royal Soc Med, 72-76, hon sr secy, 74-76, vpres, 80-85, pres, 89-90; prof & chmn, Dept Oral Path & Oral Surg, Univ Adelaide, 77-79, assoc dean, Fac Dent, 78-79; hon consult oral surgeon, St George's Hosp, London, 79-83, St James' Hosp, 80-83; consult adv craniofacial anomalies to dir, Dept Health & Human Serv, NIH, 80- *Honors & Awards:* John Tomes Prize, Royal Col Surgeons, Eng, 68, Colyer Gold Medal, 90; Maurice Down Prize for Surg Res, London, 73; Kay-Kilner Prize, Brit Asn Plastic Surgeons, 75; Richardson Lectr in Surg, Harvard Med Sch & Mass Gen Hosp, 81; Award for Excellence, Am Oral & Maxillofacial Surg Asn, 82; Lawson Tait Lectr, Univ Birmingham, 86; Friel Mem Lectr, Europ Orthod Soc, 87; Sarnat Lectr, Univ Col, Los Angeles, 89; William Guy Lectr, Royal Col Surgeons, Edinburgh, 90. *Mem:* Foreign assoc Inst Med-Nat Acad Sci; fel Royal Soc Med; Int Asn Oral & Maxillofacial Surgeons; Int Fedn Dent; Royal Inst Biol; Royal Col Pathologists. *Res:* Dental science; oral surgery. *Mailing Add:* United Med & Dent Schs Guy's & St Thomas Hosps Univ London Lambeth Palace Rd Guys Tower Fl 24 London SE1 7EH England

POTASH, MILTON, b New York, NY, Nov 23, 24; m 49; c 3. ECOLOGY, LIMNOLOGY. *Educ:* Univ Louisville, AB, 47; Ind Univ, MA, 50; Cornell Univ, PhD(zool), 53. *Prof Exp:* Asst zool, Cornell Univ, 49-51; from instr to assoc prof, 51-67, PROF ZOOL, UNIV VT, 67- *Concurrent Pos:* Ed, Water Resources Bull, 81-86. *Mem:* Ecol Soc Am; Am Soc Limnol & Oceanog; Am Water Resources Asn; Int Soc Limnol; Int Asn Great Lakes Res; N Am Lakes Mgt Soc; fel Am Water Resources Asn. *Res:* Water quality; limnology of Lake Champlain. *Mailing Add:* Dept Zool Univ Vt Burlington VT 05405

POTASHNER, STEVEN JAY, b Lowell, Mass, July 29, 45; m 66; c 2. NEUROCHEMISTRY & NEUROANATOMY. *Educ:* McGill Univ, BSc, 66, PhD(biochem), 71. *Prof Exp:* Fel, Physiol Lab, Univ Cambridge, 70-73; res assoc, dept res anesthesia & asst prof, dept physiol, McGill Univ, 73-78; asst prof, 78-85, ASSOC PROF, DEPT ANAT, UNIV CONN HEALTH CTR, 85- *Mem:* Int Soc Neurochem; Soc Neurosci. *Res:* Chemical basis of information processing in the normal and diseased nervous system. *Mailing Add:* Dept Anat Univ Conn Health Ctr Farmington CT 06032

POTCHEN, E JAMES, b Queens Co, NY, Dec 2, 32; m 56; c 4. RADIOLOGY, NUCLEAR MEDICINE. *Educ:* Mich State Univ, BS, 54 JD, 84; Wayne State Univ, MD, 58; Am Bd Radiol, dipl, 65; Am Bd Nuclear Med, dipl, 71; Mass Inst Technol, Sloan Sch Mgt, MS, 73. *Prof Exp:* Chief res radiologist, Peter Bent Brigham Hosp, 64, jr assoc, 65; asst radiol, Harvard Med Sch, 65; dir nuclear med, Mallinckrodt Inst, Wash Univ, 66-73, from asst prof to prof, Sch Med, 70-73; prof radiol & dean mgt resources, Sch Med, Johns Hopkins Univ, 73-75; PROF RADIOL & CHMN DEPT, MICH STATE UNIV, 75-, PROF MGT, COL BUS, 85-; PROF PHYSIOL, COL OF HUMAN MED, 88- *Concurrent Pos:* Teaching fel radiol, Harvard Med Sch, 64-66; Nat Acad Sci-Nat Res Coun advan fel acad radiol, James Picker Found, 65-66 & scholar radiol res, 67-68; radiologist, Barnes Hosp, St Louis, 66-71 & Johns Hopkins Hosp, 73-75; consult, Nat Heart Inst, 67-73, Nat Inst Gen Med Sci, 68-, Nat Heart & Lung Inst, 71-, Vet Admin Cent Off Prof Serv Div, 71-73, Bur Drugs, Food & Drug Admin, 72-, Health Systs Group, Westinghouse Corp, 74- & Gen Bus Develop, IBM Corp, 74-; examr, Am Bd Radiol, 68-; mem comt radiol, Nat Acad Sci-Nat Res Coun, 71-75; mem coun stroke & thrombosis, Am Heart Asn, 71-; Sloan fel, Mass Inst Technol, 72-73, vis lectr, 73-74; prof philos, Lyman Briggs Col, 85-87. *Honors & Awards:* John J Lankin Award, Basic Med Res. *Mem:* Am Soc Clin Invest; Am Fedn Clin Res; Soc Nuclear Med (pres, 75-76); fel Am Col Chest Physicians; fel Am Col Radiol; fel Am Col Legal Med. *Res:* Social and economic implications of transfer technology in medicine; diagnostic decision making; efficacy of diagnostic procedure; reconstructive imaging using transmitted photons, emitted protons and ultrasound. *Mailing Add:* Dept Radiol B220 Clin Ctr Mich State Univ East Lansing MI 48824

POTEAT, WILLIAM LOUIS, b Brooklyn, NY, May 14, 44; m 68; c 1. ANATOMY. *Educ:* Wake Forest Univ, BS, 66, PhD(anat), 71. *Prof Exp:* From instr to asst prof anat, Med Col Va, 71-75; ASSOC PROF ANAT, SCH MED, UNIV SC, 75- *Concurrent Pos:* Mem fac, dept anat, Va Med Col, 80- *Res:* Uterine glycogen metabolism; blastocyst implantation; pertussis vaccine and experimental allergic encephalomyelitis; uterotrophic action of clomiphene citrate. *Mailing Add:* Dept Anat Univ SC Sch Med Columbia SC 29208

POTEMPA, LAWRENCE ALBERT, b Chicago, Ill, May 6, 51; m 80; c 1. IMMUNOCHEMISTRY, IMMUNOPATHOLOGY. *Educ:* Bradley Univ, BS, 73; Northwestern Univ, PhD(biochem), 77. *Prof Exp:* Instr biochem, Ill Col Podiatric Med, 75-77; fel immunol, 77-79, ASST PROF IMMUNOL & MICROBIOL & ASST SCIENTIST, RUSH PRESBY-ST LUKE'S MED CTR, 79-; ASST PROF, RUSH UNIV, 79-, DIR GRAD PROG, DIV IMMUNOL, 85- *Concurrent Pos:* Assoc course dir, Rush Med Col, 80-84. *Mem:* Am Asn Immunologists; Am Chem Soc; Sigma Xi. *Res:* Recognition and activation mechanisms of the immune response in acute and chronic inflammation; immune-complex disease; amyloid disease; C-reactive protein. *Mailing Add:* Immtech Int Inc 906 Univ Pl Evanston IL 60201

POTEMRA, THOMAS ANDREW, b Cleveland, Ohio, Oct 23, 38; m 62; c 3. PLASMA PHYSICS, ATMOSPHERIC CHEMISTRY & PHYSICS. *Educ:* Case Inst Technol, BS, 60; NY Univ, MEE, 62; Stanford Univ, PhD(elec eng), 66. *Prof Exp:* Mem tech staff, Bell Tel Labs, 60-62; res asst, Radiosci Lab, Stanford Univ, 62-65; sr policy analyst, Off Sci & Technol Policy, Exec Off of the Pres, 84-85, SUPVR SPACE PHYSICS GROUP, APPL PHYSICS LAB, JOHNS HOPKINS UNIV, 65-, FAC MEM, EVE COL, 68- *Concurrent Pos:* Mem, Comt Nasen Drift Sta, Nat Acad Sci, 75-76, Comt Eval NSF Arctic Res Progs, 76-77, Comt Solar-Terrestrial Res, 82-85 & Comt Space & Solar Physics, Nat Acad Sci, 86-89; assoc ed, J Geophs Res, 77-81; vis mem fac, US Naval Acad, Annapolis, 84-85; chmn, Space Physics Mgt Opers Working Group, NASA, 87. *Mem:* Am Geophys Union. *Res:* Auroral and magnetospheric phenomena; ionospheric dynamics; solar-terrestrial relationships. *Mailing Add:* Appl Physics Lab Johns Hopkins Univ Laurel MD 20723

POTENZA, JOSEPH ANTHONY, b New York, NY, Nov 13, 41; m 64; c 2. PHYSICAL CHEMISTRY, INORGANIC CHEMISTRY. *Educ:* Polytech Inst Brooklyn, BS, 62; Harvard Univ, PhD(chem), 67. *Prof Exp:* From asst prof to assoc prof, 68-77, dir, Sch Chem, 77-80, chmn, dept chem & dir, grad prog chem, 80-84, PROF CHEM, RUTGERS UNIV, NEW BRUNSWICK, 77-, ASSOC PROVOST, ACAD AFFAIRS SCI, 90- *Concurrent Pos:* Alfred P Sloan Found fel, 71-73; Alexander von Humboldt Found sr US scientist award for res & teaching, 74; guest prof, Univ Münster, WGermany, 74-75; vis prof, Jilin Univ, Changchun, China, 83. *Mem:* Am Chem Soc; Am Crystallog Asn; Am Phys Soc. *Res:* X-ray crystallography; molecular structure and dynamics; magnetic resonance. *Mailing Add:* Dept Chem Rutgers Univ New Brunswick NJ 08903

POTH, EDGAR J, surgery; deceased, see previous edition for last biography

POTH, JAMES EDWARD, b Galion, Ohio, May 19, 33; m 60; c 3. PHYSICS EDUCATION, SPORTS PHYSICS. *Educ:* Miami Univ, BS, 55, MA, 60; Yale Univ, MS, 62, PhD(nuclear physics), 66. *Prof Exp:* Res physicist, Yale Univ, 66; from asst prof to assoc prof, 66-70, PROF PHYSICS, MIAMI UNIV, 76- *Concurrent Pos:* NASA-Am Soc Eng Educ fel, 67-68; vis res physicist, Yale Univ, 74; coun, Am Asn Physics Teachers, 87- *Mem:* Am Phys Soc; Am Asn Physics Teachers; Sigma Xi. *Res:* Low energy nuclear physics; physics education; sports physics. *Mailing Add:* Dept of Physics Miami Univ Oxford OH 45056

POTHOVEN, MARVIN ARLO, b Pella, Iowa, Apr 15, 46; m 69; c 2. ANIMAL NUTRITION. *Educ:* Cent Col, Iowa, BA, 68; Iowa State Univ, MS, 72, PhD(animal nutrit), 74. *Prof Exp:* Researcher qual control, 74-78, coordr, lab serv & qual assurance, 78-85, COORDR, FORMULATIONS & QUAL ASSURANCE, MOORMAN MFG CO, 85- *Mem:* Sigma Xi; Am Soc Animal Sci; Am Dairy Sci Asn. *Res:* Maintaining and improving the product quality of livestock feed through testing and revising present and future manufacturing techniques and developing new quality tests. *Mailing Add:* Moorman Mfg Co 1000 N 30 St Quincy IL 62305-3115

POTKAY, STEPHEN, b New Bedford, Mass, July 15, 37; m 71; c 2. LABORATORY ANIMAL MEDICINE. *Educ:* Univ Pa, VMD, 62. *Prof Exp:* Assoc vet, Salem Vet Hosp, Mass, 62-63; vet officer lab animal prod, med & surg, 63-78, asst chief, 78-80, actg chief, 80-84, DEPT CHIEF, VET RESOURCES BR, DIV RES SERV, NIH, 84- *Concurrent Pos:* Assoc ed, Lab Animal Sci, 75-; adj prof, Va-Md Regional Col Vet Med, 83- *Mem:* Soc Trop Vet Med; Am Col Lab Animal Med; AAAS; Am Asn Lab Animal Sci; Am Soc Vet Ethology; Asn Mil Surgeons US. *Res:* Diseases of laboratory and exotic animals; experimental surgery; management of laboratory animal resources. *Mailing Add:* NIH Div Res Serv Vet Resources Bldg 14G Rm 102 Bethesda MD 20892

POTMESIL, MILAM, b Prague, Czech, Sept 22, 26; US citizen; m 51, 74; c 3. ONCOLOGY, HEMATOLOGY. *Educ:* Charles Univ, Prague, MD, 51, PhD(exp hematol), 67. *Prof Exp:* Clin hematologist, Cent Mil Hosp, Prague, 60-68; vis scientist, Wash Sq Col, 68-70, res scientist, 70-75, sr res scientist, 75-77, adj assoc prof biol, Fac Arts & Sci, 77-78, ASSOC PROF RADIOL, MED SCH & DIR EXP THER RADIOBIOL UNIT, CANCER CTR, NY UNIV, 78- *Concurrent Pos:* Co-prin investr, Nat Cancer Inst, NIH res grants, 71-75, 75-78 & 78-81; prin investr, Am Cancer Soc res grant, 79-82 & Inst res grant, NY Univ, 78. *Mem:* Am Soc Hematol; Am Soc Cell Biol; Cell Kinetic Soc; Radiation Res Soc; Soc Exp Biol & Med; Am Asn Cancer Res. *Res:* Clinical and experimental hematology; experimental oncology; cell and tissue kinetics; drug and radiation effects; predictive and screening systems for therapy of neoplasias. *Mailing Add:* 345 E 80th St New York NY 10021

POTNIS, VASANT RAGHUNATH, b Lashkar, India, Mar 10, 28; m 57. PHYSICS. *Educ:* Agra Univ, BSc, 48, MSc, 52, PhD(nuclear physics), 59. *Prof Exp:* Teacher high sch, Lashkar, India, 49-50; sci asst, Phys Res Lab, Ahmedabad, 52-54; physicist, Bartol Res Found, Pa, 57-60; sci pool officer, Aligarh Muslim Univ, India, 61-62; asst prof physics, Kans State Univ, 63-68; assoc prof, 68-72, PROF PHYSICS, MICH TECHNOL UNIV, 72- *Mem:* AAAS; fel Am Phys Soc; Am Asn Physics Teachers; Sigma Xi. *Res:* Cosmic ray time variations and their relation with solar activity; nuclear physics; radioactivity; energies and intensities of gamma and beta rays; study of nuclear energy levels and their properties. *Mailing Add:* Dept of Physics Mich Technol Univ Houghton MI 49931

POTOCKI, KENNETH ANTHONY, b Chicago, Ill, Oct 8, 40; m; c 3. APPLIED PHYSICS, TECHNICAL MANAGEMENT. *Educ:* Loyola Univ Chicago, BS, 62; Ind Univ, Bloomington, MS, 65, PhD(physics), 68. *Prof Exp:* Physicist analysis, US Army Missile Command, 68-70; sr staff physicist, 70-77, nagiv group supvr systs anal, 77-81, SATELLITE PROG MGR, APPL PHYSICS LAB, JOHNS HOPKINS UNIV, 81- *Concurrent Pos:* Instr appl physics, Johns Hopkins Univ, 74- *Mem:* Am Phys Soc; Inst Navig; Am Geophys Union. *Res:* Ocean acoustics; navigation; nuclear physics; satellite systems engineering. *Mailing Add:* Appl Physics Lab Johns Hopkins Univ Johns Hopkins Rd Laurel MD 20723-6099

POTOCZNY, HENRY BASIL, b Philadelphia, Pa, Mar 18, 44; c 3. MATHEMATICS. *Educ:* La Salle Col, BA, 65; Univ Ky, MA, 67, PhD(math), 69. *Prof Exp:* Prof, 69-75, assoc prof math, Univ Dayton, 75- AT DEPT MATH, AIR FORCE INST TECHNOL. *Mem:* Am Math Soc; Math Asn Am; Asn Comput Mach. *Res:* Point set topology. *Mailing Add:* ENC Air Force Inst Technol Dayton OH 45433

POTRAFKE, EARL MARK, b Wellsville, NY, Oct 25, 30. ENVIRONMENTAL CONSULTANT. *Educ:* Alfred Univ, BA, 52; Pa State Univ, MS, 56; Univ Tex, PhD(inorg chem), 60. *Prof Exp:* Res chemist, Res & Develop Div, Pioneering Res Sect, E I du Pont de Nemours & Co, 60-64, Petrol Chem ResSect, 64-65 & Freon Prod Div, 65-69, proj leader, Permasep Prod Div, 69-75, Environ Control coordr, Chem & Pigments 75-85; RETIRED. *Mem:* Am Chem Soc; Sigma Xi. *Res:* Reactions in liquid ammonia; coordination chemistry of transitional metal complexes; phosphine-metal salt complexes; inorganic photochemistry; ceramics; fluorocarbon chemistry; chemistry of water and waste water. *Mailing Add:* 1314 Newcomb Rd Green Acres Wilmington DE 19803

POTTASCH, STUART ROBERT, b New York, NY, Jan 16, 32; m 56; c 3. ASTROPHYSICS. *Educ:* Cornell Univ, BEngPhys, 54; Harvard Univ, MA, 57; Univ Colo, PhD(astrophys), 58. *Prof Exp:* Astrophysicist, Nat Bur Stand, Colo, 57-59, Observ of Paris & Inst Astrophys, 59-60, Princeton Univ & Inst Advan Study, 60-62 & Ind Univ, 62-63; PROF ASTROPHYS, KAPTEYN ASTRON INST 63- *Concurrent Pos:* Ed-in-chief, Astron & Astrophys, 69-76, Lett ed, 76- *Honors & Awards:* Gold Medal, Int Astron Union, Astron Soc Mex. *Mem:* Int Astron Union. *Res:* Theory of stellar atmospheres; solar physics; interstellar medium; novae outburst; infrared astronomy; planetary nebulae. *Mailing Add:* Kapteyn Astron Inst PO Box 800 Groningen 9700 AV Netherlands

POTTER, ALLAN G, b Frankfort, Kans, May 9, 30; m 52; c 3. ELECTRICAL & BIOMEDICAL ENGINEERING. *Educ:* Kans State Univ, BSEE, 55; Iowa State Univ, MSEE, 59, PhD(elec eng), 66. *Prof Exp:* Elec engr, Minneapolis-Honeywell Regulator Co, 52-53; assoc elec engr, Magnavox Corp, 55-56; from instr to assoc prof, 56-76, PROF ELEC ENG, IOWA STATE UNIV, 77- *Concurrent Pos:* Co-prin investr, HEW res grant, 60-63; consult, Bendix Corp, 66-68; prin investr, Voc Rehab Admin res grant, 68- *Mem:* Assoc Inst Elec & Electronics Engrs; Am Soc Eng Educ. *Res:* Application of engineering principles in the development of myoelectric control systems for upper extremity braces used by quadraplegic patients. *Mailing Add:* Iowa State Univ Sci & Technol 231 Coover Hall Ames IA 50010

POTTER, ANDREW ELWIN, JR, b St Petersburg, Fla, Nov 29, 26; m 51; c 3. PLANETARY ASTRONOMY, ORBITAL DEBRIS. *Educ:* Univ Fla, BS, 48; Univ Wis, PhD(chem), 53. *Prof Exp:* Res scientist, Lewis Flight Propulsion Sect, NASA, Cleveland, Ohio, 53-57, head, Combustion Sect, 57-61; chief, Energy Conversion Br, Solar Energy, Lewis Res Ctr, Cleveland, Ohio, 61-67, res scientist upper atmosphere, 67-70, chief, Res Br Earth Observations, 70-76, mgr, shuttle environ effects, Environ Effects Off, 76-81, CHIEF, SPACE SCI BR, ORBITAL DEBRIS, JOHNSON SPACE CTR, NASA, HOUSTON, TEX, 81- *Concurrent Pos:* Postdoctoral study, Univ Col, London, 60. *Honors & Awards:* Exceptional Serv Medal, Nat Aeronaut & Space Admin, 81 & Exceptional Sci Achievement Medal, 90. *Mem:* Am Astron Soc; Am Inst Aeronaut & Astronaut. *Res:* Pioneering research on flame quenching; invention of opposed jet combustion measurement technique; early development of thin-film and radiation-resistant solar cells; environmental impact statement for the space shuttle program; radar and optical measurements of orbital debris; discovery of the sodium and potassium metal vapor atmospheres of the moon and planet mercury. *Mailing Add:* 1018 Woodbank Dr Seabrook TX 77586

POTTER, DAVID DICKINSON, b Chicago, Ill, Dec 22, 30; m 52; c 4. NEUROPHYSIOLOGY. *Educ:* Swarthmore Col, BA, 52; Harvard Univ, PhD, 56. *Prof Exp:* Hon res asst biophys, Univ Col, Univ London, 56-58; fel neurophysiol, Med Sch, Johns Hopkins Univ, 58-59; from instr to assoc prof, 59-69, PROF NEUROBIOL, HARVARD MED SCH, 69-, CHMN DEPT. *Res:* Development, physiology and chemistry of synaptic transmission. *Mailing Add:* Dept Neurobiol Harvard Med Sch 25 Shattuck St Boston MA 02115

POTTER, DAVID EDWARD, b Tyler, Tex, June 30, 37; m 59; c 4. PHARMACOLOGY. *Educ:* Tex Tech Univ, AB, 60; Univ Kans, PhD(pharmacol), 69. *Prof Exp:* Res asst neuropharmacol, Parke, Davis & Co, 60-64; jr scientist, Alcon Labs, Inc, 64-65; from instr to assoc prof pharmacol, Univ Tex Med Br, Galveston, 69-77; prof pharmacol/ophthal, Sch Med, Tex Technol Univ, Calif, Irvine, 77-85; ADJ PROF & DIR, BIOL SCI ALLERGAN, UNIV CALIF, IRVINE, 85- *Mem:* Am Soc Pharmacol & Exp Therapeut; Soc Exp Biol & Med; Endocrine Soc; Sigma Xi; Asn Res Vision & Ophthal; Int Cong Eye Res. *Res:* Autonomic pharmacology; polypeptides; drug-receptor interaction; metabolic effects of catecholamines and alcohol; ocular pharmacology. *Mailing Add:* Ctr for Biotechnol 4000 Research Forest Dr Woodlands TX 77381

POTTER, DAVID EDWARD, b Tyler, Tex, June 30, 37; m 81; c 4. OCULAR PHARMACOLOGY, NEUROENDOCRINOLOGY. *Educ:* Tex Tech Univ, BA, 60; Univ Kansas Med Sch, PhD(pharmacol), 69. *Prof Exp:* Assoc prof pharmacol, Med Br, Univ Tex, 68-76; prof pharmacol, Health Sci Ctr, Tex Tech Univ, 76-85; dir biol sci, Allergan Inc, 85-87; vpres & dir res, Houston Biotech Inc, 87-90; PROF & CHMN PHARMACOL, MOREHOUSE SCH MED, 90- *Concurrent Pos:* Adj prof, Univ Calif, Irvine, 85-87, Baylor Col Med, 87-90. *Mem:* Asn Res Vision & Ophthal; Am Soc Pharmacol & Exp therapeut; Endocrine Soc; AAAS. *Res:* Neuroendocrinology of the eye and pharmacologic approaches to the therapy of glaucoma. *Mailing Add:* Dept Pharmacol Morehouse Sch Med Atlanta GA 30310-1495

POTTER, DAVID ERIC, b San Jose, Calif, Sept 12, 49; m 74; c 1. SOFTWARE DEBUGGING, REAL-TIME SYSTEMS. *Educ:* Univ Calif, Berkeley, BA, 71, MA, 74. *Prof Exp:* Training instr, Intel Corp, 78-79, mkt mgr, 79-80; consult, 80-83; PRES, CONCURRENT SCI, 83- *Res:* Design and implementation of interactive, source-level software debuggers for microprocessor-based embedded computer systems; software debugging of real-time systems. *Mailing Add:* 1191 Tolo Trail Moscow ID 83843

POTTER, DAVID SAMUEL, b Seattle, Wash, Jan 16, 25. ACOUSTICS. *Educ:* Yale Univ, BS, 45; Univ Wash, PhD(physics), 51. *Prof Exp:* Asst dir, Appl Physics Lab, Univ Wash, 55-60; head, Sea Opers, Defense Res Lab, Gen Motors, 60-66, dir, Santa Barbara Opers, 66-69, dir res & develop, Delco Electronics Div, 69-70, chief engr, 70-73, dir res, Detroit Diesel Allison Div, 73; asst secy res & develop, Navy, Dept Defense, US Govt, 73-74, under secy Navy, 74-76; vpres, environ actg staff, Gen Motors Corp, 76-78, vpres pub affairs group, 78-83, vpres & group exec power prods & defense opers, 83-85; RETIRED. *Mem:* Nat Acad Eng; Soc Automotive Engrs; Am Phys Soc; Acoust Soc Am; Am Inst Aeronaut & Astronaut. *Res:* Underwater acoustics with primary emphasis on absorption and scattering phenomenon. *Mailing Add:* 877 Lilac Dr Santa Barbara CA 93108-1438

POTTER, DONALD B, b Utica, NY, June 4, 23; m 45; c 6. FIELD GEOLOGY. *Educ:* Williams Col, AB, 47; Brown Univ, MS, 49; Calif Inst Technol, PhD, 54. *Prof Exp:* Asst, Brown Univ, 47-49; asst, Calif Inst Technol, 49-50 & 52-54; from asst prof to prof geol, Hamilton Col, 54-67, chair, Dept Geol, 73-87; RETIRED. *Concurrent Pos:* Geologist, US Geol Surv, 50-53, 71-72 & 76-79; prin investr, NASA, 71-75; vis scientist, Los Alamos Nat Lab, 81-82. *Res:* High alumina metamorphic rocks; stratigraphy and structure of central Taconic region; seismic refraction studies of drumlin cores; photo interpretation of Martian surface; flow direction indicators in ash flows. *Mailing Add:* Reservoir Rd One Box 459 Clinton NY 13323

POTTER, DONALD IRWIN, b Detroit, Mich, May 28, 41; m 61; c 1. ELECTRON MICROSCOPY, ION IMPLANTATION. *Educ:* Wayne State Univ, BPh, 64, BS, 66; Univ Ill, Urbana, MS, 67, PhD(metall eng), 70. *Prof Exp:* Asst prof mat sci, Union Col, NY, 70-74; metallurgist, Mat Sci Div, Argonne Nat Lab, 74-79; PROF METALL, UNIV CONN, 79- *Concurrent Pos:* NSF initiation and continuing grants & Res Corp Frederick Cottrell award; Alcoa Found Award. *Honors & Awards:* NSF Creative Investr Award,

83 and 88. *Mem:* Am Inst Mining, Metall & Petrol Engrs. *Res:* Phase transformations in metals, especially interstitial elements in refractory metals; gas-metal equilibria; internal friction; electron microscopy and microchemical analysis; ion irradiation damage and implantation; phase stability and precipitate growth and dissolution during irradiation; surface modification of metals and alloys. *Mailing Add:* Inst Mat Sci U-136 Univ Conn Storrs CT 06269-3136

POTTER, DOUGLAS MARION, b Mineola, NY, Feb 22, 45. ELEMENTARY PARTICLE PHYSICS, DETECTOR DEVELOPMENT. *Educ:* Mass Inst Technol, BS, 67, MS, 69, PhD(physics), 72. *Prof Exp:* Fel physics, Northeastern Univ, 72-77; asst prof, dept physics, Rutgers Univ, 77-; AT DEPT PHYSICS, CARNEGIE MELLON UNIV. *Mem:* Am Phys Soc. *Res:* Polarization phenomena in high energy photon physics; large momentum transfer elastic scattering. *Mailing Add:* Dept Physics Carnegie Mellon Univ Pittsburgh PA 15213

POTTER, ELIZABETH VAUGHAN, b Detroit, Mich, Jan 27, 14; m 37; c 3. MEDICINE. *Educ:* Univ Chicago, BS, 35, MD, 39. *Prof Exp:* Intern, Swed Covenant Hosp, Chicago, 39-40; resident, Chicago Contagious Dis Hosp, 40-41; Am Heart Asn fel, 60-61, from instr to assoc, 61-68, from asst prof to assoc prof, 68-84, EMER PROF MED, NORTHWESTERN UNIV, CHICAGO, 84- *Concurrent Pos:* Mem coun cardiovasc-renal dis & coun thrombosis, Am Heart Asn; mem expert adv panel bact dis, WHO. *Mem:* Int Soc Nephrology; Am Soc Nephrology; Infectious Dis Soc Am. *Res:* Immunological and coagulation associated aspects of renal disease, particularly of post streptococcal glomerulonephritis and acute rheumatic fever. *Mailing Add:* Dept Med-Nephrol Northwestern Univ Med Sch 303 E Chicago Ave Chicago IL 60611

POTTER, FRANK ELWOOD, food science, dairy chemistry, for more information see previous edition

POTTER, FRANK WALTER, JR, b Worcester, Mass, June 27, 42; m 68; c 2. PALEOBOTANY, PALEOECOLOGY. *Educ:* Pa State Univ, BS, 65, MS, 70; Ind Univ, PhD(paleobot), 75. *Prof Exp:* Instr biol, Pa State Univ, Beaver Campus, 66-67 & Ball State Univ, 74-75; res assoc paleobot, Ind Univ, 75-76; ASST PROF BIOL, FT HAYS STATE UNIV, 76- *Mem:* Bot Soc Am; Ecol Soc Am; Sigma Xi; Int Asn Angiosperm Paleobotanists; Am Asn Stratig Palynologists. *Res:* Evolution and phylogeny of early Angiosperms; paleoecology and terrestrial ecosystem evolution during the Cretaceous and early Tertiary. *Mailing Add:* Dept Biol Sci Ft Hays State Univ Hays KS 67601

POTTER, GEORGE HENRY, b Bolton Landing, NY, Mar 28, 32; m 53; c 4. INSTRUMENTAL ANALYSIS. *Educ:* Clarkson Col, BS, 53; Rensselaer Polytech Inst, PhD(org chem), 58. *Prof Exp:* Proj scientist, Union Carbide Corp, 58-71; teacher chem & physics, Cardinal McClosky High Sch, 71-76; from instr to assoc prof chem, 76-86, chmn, Dept Math, Sci & Technol, 81-88, PROF CHEM, SCHENECTADY COUNTY COMMUNITY COL, 86- *Concurrent Pos:* Adj prof, WVa State Col, 63-65; prin investr, NSF, 90- *Mem:* Am Chem Soc; Soc Plastics Engrs. *Res:* Plasticizers; vinyl resins; new polymer synthesis; plant processes; quality control; hydrogenation of amines. *Mailing Add:* RD 1 Box 107 Duanesburg NY 12056

POTTER, GERALD LEE, b Klamath Falls, Ore, Jan 24, 45. CLIMATOLOGY. *Educ:* Univ Calif, Los Angeles, BA, 67, MA, 70, PhD(geog), 75. *Prof Exp:* Teaching & res asst geog, Univ Calif, Los Angeles, 66-71; instr, Calif State Univ, Hayward, 71-72; GEOGR & CLIMATOLOGIST, LAWRENCE LIVERMORE LAB, UNIV CALIF, 72- *Res:* Investigation of anthropogenic induced and natural climate variations using numerical modeling techniques; comparison of various models sensitivities. *Mailing Add:* Lawrence Livermore Lab PO Box 808 Livermore CA 94550

POTTER, GILBERT DAVID, b Calgary, Alta, May 23, 24; m 55; c 3. PHYSIOLOGY, PHARMACOLOGY. *Educ:* Univ BC, MA, 50, PhD(physiol), 56. *Prof Exp:* From jr res physiologist to asst res physiologist, Univ Calif, Berkeley, 56-59; res pharmacologist & asst lab dir, Miles Labs, Inc, 59-63; res physiologist & group leader biomed div, Lawrence Livermore Lab, Univ Calif, 63-73; biologist & br chief, Environ Monitoring Systs Lab, US Environ Protection Agency, Las Vegas, Nev, 73-81, res radiobiologist, 81-86; RETIRED. *Concurrent Pos:* Adj prof biol, Univ Nev, Las Vegas, 76-79. *Mem:* AAAS; Am Physiol Soc; Health Physics Soc. *Res:* Endocrinology; trace element metabolism; radiobiology; health physics. *Mailing Add:* 3208 Mason Ave Las Vegas NV 89102-1937

POTTER, HOWARD SPENCER, plant pathology, entomology, for more information see previous edition

POTTER, JAMES D, b Waterbury, Conn, Sept 26, 44; m; c 2. PHARMACOLOGY. *Educ:* Univ Conn, PhD(biochem), 70. *Prof Exp:* Head, sect contractile proteins, Univ Cincinnati, 78-83, prof pharmacol, 81-83; PROF PHARMACOL & CHMN DEPT, UNIV MIAMI, 83- *Concurrent Pos:* Merit Award, NIH, 89-99; chmn, Am Med Sch Pharmacol. *Mem:* AAAS; Am Heart Asn; Am Chem Soc; Am Soc Biol Chem; Am Soc Pharmacol & Exp Therapeut; Biophys Soc; Cardiac Muscle Soc; Sigma Xi; Int Soc Heart Res. *Res:* Calcium regulation of skeletal and cardiac muscle contraction. *Mailing Add:* Dept Molecular & Cellular Pharmacol Univ Miami PO Box 016189 Miami FL 33101

POTTER, JAMES GREGOR, b Manhattan, Kans, Apr 2, 07; m 41; c 2. PHYSICS. *Educ:* Princeton Univ, BS, 28; NY Univ, MS, 31; Yale Univ, PhD(physics), 39. *Prof Exp:* Jr physicist, US Naval Res Lab, 30-31; asst physics, Mass Inst Technol, 31-32; instr math, Armour Inst Technol, 35-36, physics, 36-39, asst prof, 39-40; prof physics & head dept, adminr gen eng, SDak Sch Mines & Technol, 40-44; mem tech staff, Bell Tel Labs, Inc, NY, 44-45; prof physics & head dept, Tex A&M Univ, 45-66, prof & asst dean student resources, 66-67; head dept, 67-74, prof, 67-87, EMER PROF PHYSICS, FLA INST TECHNOL, 87- *Concurrent Pos:* Assoc prog dir advan sci educ, NSF, 59-60. *Mem:* AAAS; Am Phys Soc; Am Soc Eng Educ; Am Asn Physics Teachers. *Res:* Electronic work functions and emission; cathode sputtering; education in physics. *Mailing Add:* Dept Physics & Space Sci Fla Inst Technol Melbourne FL 32901-6988

POTTER, JAMES MARTIN, b Peoria, Ill, Aug 15, 41; m 64; c 5. PHYSICS, ELECTRONICS. *Educ:* Univ Ill, Urbana, BS, 64, MS, 70, PhD(physics), 75. *Prof Exp:* Staff mem physics, 64-68 & 70-80, SECT LEADER ACCELERATOR STRUCTURE DEVELOP, LOS ALAMOS NAT LAB, 80- *Concurrent Pos:* Consult accelerator structures & microwave radio frequency power. *Mem:* Sigma Xi. *Res:* Accelerator structures; parity violation in p-nucleon scattering; radio frequency quadruple accelerator; analysis of periodic systems. *Mailing Add:* 2245 47th St Los Alamos NM 87544

POTTER, JANE HUNTINGTON, b Chicago, Ill, Feb 21, 21; m 42; c 3. ZOOLOGY. *Educ:* Univ Chicago, SB, 42, MS, 48, PhD(zool), 49. *Prof Exp:* Asst prof zool, Am Univ Beirut, 49-52; from instr to asst prof, Univ Md, College Park, 62-70, assoc prof zool, 70-; RETIRED. *Res:* Factors affecting sexual isolation in sympatic species of Drosophila. *Mailing Add:* 4804 Wellington Dr Chevy Chase MD 20015

POTTER, JOHN CLARKSON, b Chicago, Ill, May 15, 21; m 46; c 2. RADIOCHEMISTRY, ANALYTICAL CHEMISTRY. *Educ:* Univ Wash, BS, 43; Duke Univ, PhD(chem), 50. *Prof Exp:* Asst prof chem, NC State Col, 50-51; fel, Mellon Inst, 51-52; res chemist, Koppers Co, Inc, 52-54; RES CHEMIST, SHELL DEVELOP CO DIV, SHELL OIL CO, 54- *Mem:* Am Chem Soc; Sigma Xi. *Res:* Chemical analysis; radioactive tracers; metabolic fate of pesticides in biological systems; mathematical models in biology. *Mailing Add:* 649 Geer Ct Modesto CA 95354

POTTER, JOHN F, b New York, NY, July 26, 25; m 55; c 3. SURGERY, ONCOLOGY. *Educ:* Georgetown Univ, MD, 49; Am Bd Surg, dipl, 58. *Prof Exp:* Resident surg, Hosp, Georgetown Univ, 50-56; sr investr, Nat Cancer Inst, 57-60; from asst prof to assoc prof, Vincent T Lombardi Cancer Res Ctr, Georgetown Univ, 59-72, prof surg, med sch & dir, 72-85, head div oncol surg, 60-85; CONSULT, 85- *Concurrent Pos:* Mem cancer clin invest res comt, Nat Cancer Inst, 69; consult, Clin Ctr, NIH & Vet Admin Hosp, Washington, DC; hon prof, Universidad Cayetano Heredia Peruana, Lima, Peru. *Mem:* Fel Am Col Surgeons; Soc Surg Oncol; Asn Am Cancer Inst. *Res:* Cancer surgery and cancer biology. *Mailing Add:* Dept Surg Georgetown Univ Med Sch Washington DC 20007

POTTER, JOHN FRED, atmospheric science, for more information see previous edition

POTTER, JOHN LEITH, b Metz, Mo, Feb 5, 23; m 57; c 3. AEROSPACE & MECHANICAL ENGINEERING. *Educ:* Univ Ala, BS, 44, MS, 49; Vanderbilt Univ, PhD(mech eng), 74, MS, 76. *Prof Exp:* Engr, Curtis-Wright Corp, 44-47; from instr to asst prof aeronaut eng, Univ Ala, 47-51; res engr, Naval Ord Lab, 51-52; chief flight & aerodynamics lab, Ord Missile Labs, 52-56; mgr res & aerophysics brs, ARO, Inc, 56-71, chief aerospace div, 71-73, dep dir, von Karman Facil, 73-77, dep tech dir, 77-81; sr staff scientist, Sverdrup Technol, Inc, 81-83; RES PROF MECH ENG, VANDERBILT UNIV, 84-; CHMN, PATHFINDER ASSOC, 88- *Concurrent Pos:* Aerodynamicist, NAm Aviation, Inc, 49; prof, Univ Tenn, 56-; hon adj prof, Univ Ala, 65-; USSR Acad lectr, Russia, 67; mem, Tech Comt on Fluid Dynamics, 70-72; assoc ed, Am Inst Aeronaut & Astronaut J, 70-73, mem publ comt, 70-79, assoc ed, Am Inst Aeronaut & Astronaut Progress Series, 81-85; adj prof, Vanderbilt Univ, 81-84, res prof, 84-; consult to indust, NASA & USAF; mem, Nat Res Coun Comt, 87-88; distinguished eng fel, Univ Ala, 88; mem, Eng Accreditation Comn, 85-90. *Honors & Awards:* Gen H H Arnold Award, Am Inst Aeronaut & Astronaut, 64. *Mem:* Sigma Xi; fel Am Inst Aeronaut & Astronaut. *Res:* Aerodynamics and gas dynamics; stressing boundary layers; rarefied flows; development of experimental facilities and simulation; engineering management. *Mailing Add:* Pathfinder Assoc 200 Sheffield Pl Nashville TN 37215-3235

POTTER, LAWRENCE MERLE, b Sabattus, Maine, Dec 28, 24; m 50; c 4. POULTRY NUTRITION. *Educ:* Univ Maine, BS, 51; Univ NH, MS, 53; Univ Conn, PhD(animal nutrit), 58. *Prof Exp:* Asst, Univ NH, 51-53; res asst, Univ Conn, 53-55, asst instr, 55-58, res assoc, 58-60; from assoc prof to prof, 60-89, EMER PROF POULTRY SCI, VA POLYTECH INST & STATE UNIV, 90- *Concurrent Pos:* sect ed, J Poultry Sci, 87-89. *Honors & Awards:* Res award, Nat Turkey Fedn, 80; Poultry Res Award, Am Feed Mfr Asn, 82. *Mem:* Poultry Sci Asn; Animal Nutrit Res Coun; Sigma Xi; Biometric Soc. *Res:* Determination of metabolizable energy values of feed ingredients for poultry; statistical analysis of experimental data; turkey nutrition; amino acid requirements of poultry. *Mailing Add:* Dept Poultry Husbandry Va Polytech Inst & State Univ Blacksburg VA 24061

POTTER, LINCOLN TRUSLOW, b Chicago, Ill, Mar 26, 33; m 56; c 4. NEUROBIOLOGY, NEUROSCIENCES. *Educ:* Swarthmore Col, AB, 55; Yale Univ, MD, 59. *Prof Exp:* From intern to resident med, Peter Bent Brigham Hosp, Boston, 59-61; res assoc, NIMH, 61-63; from instr to asst prof pharmacol, Harvard Med Sch, 63-70; sr res assoc biophys, Univ Col, Univ London, 66-72; PROF PHARMACOL, PHYSIOL & BIOPHYS, MED SCH, UNIV MIAMI, 72- *Concurrent Pos:* USPHS spec fel, Harvard Med Sch, 63-66, Markle Found scholar, 65-70; training in Gestalt, NLP, Ericksonian & family ther tech, 79-88; chmn, Neurol Scis Study Sect, NIH, 85-88. *Mem:* Am Soc Pharmacol & Exp Therapeut; Brit Physiol Soc; Neurosci Soc. *Res:* Neurobiology of synaptic transmission; receptor mechanisms. *Mailing Add:* Dept Molecular & Cellular Pharmacol Med Sch Univ Miami PO Box 016189 Miami FL 33101

POTTER, LOREN DAVID, b Fargo, NDak, June 23, 18; m 41; c 4. BOTANY. *Educ:* NDak Col, BS, 40; Oberlin Col, MA, 46; Univ Minn, PhD(bot), 48. *Prof Exp:* Field asst, NDak State Game & Fish Dept, 41-42; mem staff fiber & textile res, Goodyear Tire & Rubber Co, Ohio, 43-45; from asst prof to assoc prof bot, NDak Agr Col, 48-58; chmn dept, 58-72, prof, 58-88, EMER PROF, BIOL, UNIV NMEX, 88- *Mem:* AAAS; Ecol Soc Am; Soc Range Mgt. *Res:* Postglacial forest vegetation by pollen analysis; life history of grasses; thermosetting resin for strengthening cord; ecological plant geography; reclamation science. *Mailing Add:* Dept Biol Univ NMex Albuquerque NM 87131

POTTER, MEREDITH WOODS, mathematics, computer science, for more information see previous edition

POTTER, MERLE C(LARENCE), b Grand Rapids, Mich, Oct 13, 36; m 57; c 4. MECHANICAL ENGINEERING, FLUID MECHANICS. *Educ:* Mich Tech, BS, 58, MS, 61; Univ Mich, MS, 64, PhD(eng mech), 65. *Prof Exp:* Instr eng mech, Mich Tech, 58-61 & Univ Mich, 63-65; from asst prof to assoc prof, 65-74, PROF MECH ENG, MICH STATE UNIV, 74- *Mem:* Am Soc Mech Engrs; Am Soc Eng Educ. *Res:* Experimental transition studies in channel flow; numerical solution of internally seperated flows; numerical and experimental studies of air pipes. *Mailing Add:* Dept Mech Eng Mich State Univ East Lansing MI 48824

POTTER, MICHAEL, b East Orange, NJ, Feb 27, 24; m; c 2. GENETICS. *Educ:* Princeton Univ, AB, 45; Univ Va, MD, 49. *Prof Exp:* Res asst, dept microbiol, Med Sch, Univ Va, 52-54; biologist, 54-70, head immunochem sect, Lab Cell Biol, 70-82, CHIEF, LAB GENETICS, NAT CANCER CTR, NIH, BETHESDA, MD, 82- *Concurrent Pos:* Vis prof, Weizman Inst Sci, Rehovoth, Israel, 72; chmn, Workshop Multiple Myeloma, Int Union Against Cancer, Geneva, 74; adj prof, Dept Zool & Biochem, Univ Md, 76-; Albert Lasker med res award, 84. *Honors & Awards:* Meritorious Serv Award, USPHS, 69, Distinguished Serv Medal, 81; Reilly Lectr, Univ Notre Dame, 76; Phillip Levine Lectr, Rockefeller Univ, 78; R E Dyer Lectr, NIH, 82; Paul-Ehrlich & Ludwig-Darmstaedter Prize, 83. *Mem:* Nat Acad Sci; Am Asn Cancer Res; Am Asn Immunologists. *Mailing Add:* 9820 Parkwood Dr Bethesda MD 20014

POTTER, NEIL H, b Lancaster, Pa, Oct 14, 38; m 62; c 4. ORGANIC CHEMISTRY. *Educ:* Franklin & Marshall Col, BS, 60; Middlebury Col, MS, 62; Pa State Univ, PhD, 66. *Prof Exp:* From asst prof to assoc prof, 66-81, PROF CHEM, SUSQUEHANNA UNIV, 81- *Concurrent Pos:* Consult, Pa Dept Hwy, 67- *Mem:* Am Chem Soc. *Res:* Oxidation of organic compounds by bromine and chromic acid; effect of paraffin wax on bituminous concrete. *Mailing Add:* Dept of Chem Susquehanna Univ Selinsgrove PA 17870

POTTER, NOEL, JR, b Burlington, Vt, Jan 24, 40; m 89. GEOLOGY. *Educ:* Franklin & Marshall Col, AB, 61; Darmouth Col, MA, 63; Univ Minn, Minneapolis, PhD(geol), 69. *Prof Exp:* From teaching asst to teaching assoc geol, Univ Minn, 63-68, instr, 68-69; from asst prof to assoc prof, 69-88, chmn, 77-81, PROF GEOL, DICKINSON COL, 88- *Concurrent Pos:* Instr, Col St Thomas, 67-68. *Mem:* AAAS; Geol Soc Am; Glaciol Soc; Am Geophys Union; Am Quaternary Asn. *Res:* Geomorphology; glacial geology; Antarctic glacial history; rock glaciers; effects of snow avalanches; periglacial features. *Mailing Add:* Dept Geol Dickinson Col Carlisle PA 17013

POTTER, NOEL MARSHALL, b Machias, Maine, May 11, 45; m 69; c 2. CLASSICAL INORGANIC ANALYSIS, ATOMIC SPECTROSCOPY. *Educ:* Worcester Polytech Inst, BS, 67; Cornell Univ, MS, 69, PhD(anal chem), 71. *Prof Exp:* SR STAFF RES SCIENTIST, GEN MOTORS RES LABS, 71- *Mem:* Am Chem Soc. *Res:* Classical inorganic analytical chemistry, including geological materials, catalyst materials and noble metals; atomic absorption spectroscopy; atomic emission; petroleum products. *Mailing Add:* 7007 Grenadier Ct Utica MI 48087

POTTER, NORMAN D, b Portland, Ore, Sept 22, 28; m 57; c 2. POLYMERS & ADHESIVE COATING COMPOSITES. *Educ:* Willamette Univ, BS, 52; Ore State Univ, MS, 55, PhD(phys chem), 62. *Prof Exp:* Res engr, Atomics Int Div, NAm Aviation, Inc, 54-57; teaching fel, Ore State Univ, 57-60; res asst, Ore State Univ, 60-62; scientist, Aeronutronic Div, Ford Aerospace & Commun Corp, Newport Beach, 62-90; SPEC SCIENTIST, LORAL AERONUTRONIC, NEWPORT BEACH, 90- *Mem:* Sigma Xi; Am Chem Soc; AAAS. *Res:* Division consultant in polymers; adhesives; coatings; composites; spectrometric techniques; air pollution monitoring instrumentation research. *Mailing Add:* Loral Aerospace Corp Loral Aeronutronic Newport Beach CA 92658

POTTER, NORMAN N, b New York, NY, Oct 10, 26; m 50; c 2. FOOD SCIENCE, NUTRITION. *Educ:* Cornell Univ, BS, 50; Iowa State Col, MS, 51, PhD(dairy bact), 53. *Prof Exp:* Asst bact viruses, Iowa State Col, 50-53; asst head cereal chem div, Fleischmann Labs, Stand Brands, Inc, 53-60; mgr food tech, Am Mach & Foundry Co, 60-66; assoc prof food sci, 66-73, PROF FOOD SCI, CORNELL UNIV, 73- *Concurrent Pos:* Consult to food indust, 66- & US Food & Drug Admin, 81-; vis scientist, Univ PR, 72, FDA, 81; chmn, Gordon Res Conf Food & Nutrit, 80; vis prof, Univ Roading, Eng, 87. *Honors & Awards:* William V Cruess Award, Inst Food Technol, 85. *Mem:* Inst Food Technol; Int Asn Milk Food & Environ Sanitarians; AAAS. *Res:* Food processing; nutrition, food safety; food microbiology. *Mailing Add:* Dept Food Sci Stocking Hall Cornell Univ Ithaca NY 14853

POTTER, PAUL EDWIN, b Springfield, Ohio, Aug 30, 25. GEOLOGY. *Educ:* Univ Chicago, PhB, 49, MS, 50, PhD(geol), 52; Univ Ill, MS, 59. *Prof Exp:* From asst geologist to assoc geologist, Ill State Geol Surv, 52-61; Guggenheim fel, Johns Hopkins Univ, 61-62; from assoc prof to prof geol, Ind Univ, Bloomington, 63-71; MEM FAC GEOL, UNIV CINCINNATI, 71- *Concurrent Pos:* NSF sr fel, Univ Ill, 57-58. *Res:* Sedimentary petrology; sedimentation; applications of statistics to geological problems; megasedimentology (basin analysis), big rivers and geological history; South America. *Mailing Add:* Dept Geol Univ Cincinnati Cincinnati OH 45221

POTTER, RALPH MILES, b Mt Vernon, Wash, Aug 21, 27; m 57; c 1. PHYSICAL CHEMISTRY. *Educ:* State Col Wash, BS, 48; Purdue Univ, MS, 50, PhD(phys chem), 53. *Prof Exp:* Res phys chemist, Lamp Div, Gen Elec Co, 52-87; RES ASSOC, UNIV TORONTO, 91- *Mem:* Am Chem Soc. *Res:* Physical chemistry of phosphor preparation; nature of luminescent centers in sulfide type phosphors; optical and electrical properties of wide band gap semiconductors as related to injection electroluminescence; growth of single-crystal alpha-silicon carbon; tin oxide thin films; chemical photoflash lamps; measurements and thermal modeling of metal halide arc lamps. *Mailing Add:* 2618 Brainard Rd Pepper Pike OH 44124

POTTER, RICHARD C(ARTER), b Ekalaka, Mont, May 19, 19; m 48; c 6. MECHANICAL ENGINEERING. *Educ:* Purdue Univ, BS, 40, MS, 47, PhD(eng), 50. *Prof Exp:* Test engr, Crane Co, Ill, 40-41, asst res engr, 46; instr mech eng, Purdue Univ, 47-48; prof mech & assoc dean, Kansas State Univ, 49-59; mem res staff, Gen Atomic Div, Gen Dynamics Corp, 59-60; head tech staff develop, Ramo Wooldridge Div, Thompson Ramo Wooldridge, Inc, 60, mgr prof placement & develop, Space Tech Labs, Inc, 60-63; prof & dir res, Inst Indust Res, Univ Louisville, 63-65; pres, Northrop Inst Technol, 66-67; dir educ serv, Calif State Univ, 67-69, prof mech eng & dean, Sch Eng, 69-83; HYGEIA INC, 83- *Mem:* Am Soc Mech Engrs; Am Soc Eng Educ. *Res:* Heat transfer; fluid flow; thermodynamics; power. *Mailing Add:* 209 N Hillcrest Blvd Inglewood CA 90301

POTTER, RICHARD LYLE, b Regina, Sask, June 5, 26; nat US; m 65. ANIMAL PHYSIOLOGY. *Educ:* Reed Col, BA, 50; Wash State Univ, MS, 52; Univ Rochester, PhD(biochem), 57. *Prof Exp:* Instr gen biol, State Univ NY Teachers Col, Geneseo, 56-58; res fel neurochem, Biol Div, Calif Inst Technol, 58-61; from asst prof to assoc prof, 61-78, PROF BIOL, CALIF STATE UNIV, NORTHRIDGE, 78- *Mem:* Sigma Xi; Am Soc Zoologists. *Res:* Amino acids, respiration and cell counts of bullfrog brain regions; extra- and intra-cellular volumes of hypo- and hyper-osmotic toad brain. *Mailing Add:* Dept of Biol Calif State Univ Northridge CA 91330

POTTER, RICHARD R(ALPH), b Lawrence, Kans, May 9, 26; m 50; c 1. ELECTRONICS ENGINEERING. *Educ:* Univ Kans, MS, 50. *Prof Exp:* Jr electronic engr, Proving Ground, US Dept Navy, 50-51, electronic engr, 52, electronic scientist, 53-54, supvry gen engr, 55, head ballistic measurement div, 56-58, head eval br, Electromagnetic Hazards Div, 59-64, asst dir weapons develop & eval lab, Naval Weapons Lab, 64-68, asst head advan systs dept, 68-74; vpres, LP Weedon Hauling Inc, 80-87; RETIRED. *Concurrent Pos:* Instr exten serv, Univ Va, 53; assoc, George Washington Univ, 59. *Mem:* Inst Elec & Electronics Engrs. *Res:* Hazards of electromagnetic radiation to ordnance; range instrumentation for ordnance tests; electronic countermeasures. *Mailing Add:* Rte 1 Box 992 King George VA 22485

POTTER, ROBERT JOSEPH, b New York, NY, Oct 29, 32; c 3. TELECOMMUNICATIONS, DATA PROCESSING. *Educ:* Lafayette Col, BS, 54; Univ Rochester, MA, 57, PhD(optics), 60. *Prof Exp:* Asst, Univ Rochester, 54-60; optical sci & pattern recognition mgr, T J Watson Res Ctr, Int Bus Mach Corp, 60-65; info tech mgr, Res Labs, Xerox Corp, 65-66, vpres & mgr, Advan Eng Dept, 67-68, Develop Eng, 68-69, vpres & gen mgr, New Ventures Div, 69-70, Spec Prod & Systs Div, Calif, 70-71, vpres, Info Technol Group, NY, 71-75, pres, Off Systs Div, 75-78; sr vpres & chief tech officer, Int Harvester Co, 78-82; pres, R J Potter Co, 82-84; group vpres, Northern Telecom Inc, 85-87; pres & chief exec officer, Datapoint Corp, 87-89; PRES, R J POTTER CO, 89- *Concurrent Pos:* Mem, US Nat Comn on Optics, 65-68; mem, Nat Acad Sci-Int Comn on Optics, Paris, 66; mem bd dirs, Xerox Ctr Health Care Res, Baylor Sch Med; mem comt telecommun, Nat Acad Eng; mem, President's Task Force, 67; mem bd dirs, Inst Technol, Southern Methodist Univ, 73-85, Molex, Inc, 81, First City Bank of Dallas, 80-82; trustee, Ill Inst Technol, 78. *Honors & Awards:* Outstanding Tech Contrib Award, IBM Corp, 64; Soc Mfg Engrs Distinguished Engrs Award, 81. *Mem:* Fel Optical Soc Am; Am Phys Soc. *Res:* Information processing, optical and electronic image and character processing; technologies for image reproduction, particularly xerography, electrography, and photoelectrophoresis; optical pattern recognition; fiber optics; radiometry; optical instrumentation and scanning techniques; word processing; communications; technologies for transportation; agricultural equipment. *Mailing Add:* R J Potter Co 5624 Covehaven Dallas TX 75252

POTTER, ROSARIO H YAP, b Manila, Philippines, Aug 21, 28; m 64; c 1. GENETICS, DENTAL RESEARCH. *Educ:* Univ of the East, Manila, DMD, 52; Univ Ore, MSD, 63; Ind Univ, MS, 67. *Prof Exp:* Fel, Eastman Dent Ctr, NY, 59-61; res assoc, Univ Ore, 61-63; USPHS training grant med genetics, Sch Med, 63-66, from asst prof to assoc prof, 67-80, PROF DENT RES, SCH DENT, IND UNIV, INDIANAPOLIS, 80- *Concurrent Pos:* USPHS res grant awards, 80-; Am Fund Dent Health grantee, 82-; mem, Spec Study Sect, NIH, 83-; dir, Biostatist, Biomet & Acad Comput, 84- *Honors & Awards:* First Place Award, Philippine Nat Dent Bd Exam, 55; WHO Lectr, Beijing, China, 83. *Mem:* Am Soc Human Genetics; Int Asn Dent Res; Am Asn Dent Schs; Am Asn Dent Res; Behav Scientists Dent Res; Sigma Xi. *Res:* Biometrical genetics of dento-facial variables; experimental designs in dental research; dental education research. *Mailing Add:* Gral Health Res Inst Ind Univ 415 Lansing St Indianapolis IN 46202

POTTER, THOMAS FRANKLIN, b New York, NY, July 1, 41. MATHEMATICS. *Educ:* Yale Univ, BA, 62; Univ Calif, Berkeley, MA, 64, PhD(math), 70. *Prof Exp:* Asst prof, 68-77, ASSOC PROF MATH, FISK UNIV, 77-; SR SYSTS ANALYST, TELCO RES CORP. *Mem:* Am Math Soc. *Res:* Algebraic groups; group representations; functional analysis. *Mailing Add:* 2004 20th Ave S Nashville TN 37212

POTTER, VAN RENSSELAER, b Day Co, SDak, Aug 27, 11; m 35; c 3. BIOCHEMISTRY. *Educ:* SDak State Col, BS, 33; Univ Wis, MS, 36, PhD(biochem), 38. *Hon Degrees:* ScD, SDak State Col, 59. *Prof Exp:* Asst chem, Exp Sta, SDak State Col, 30-35; asst biochem, Univ Wis, 36-38; Nat Res Coun fel, Stockholm, Sweden, 38-39; Rockefeller traveling fel, Sheffield,

Eng & Chicago, 39-40; Bowman res fel cancer res, Univ Wis-Madison, 40-42, from asst prof to prof, 42-80, asst dir, McArdle Lab, 58-72, Hilldale prof, 80-82, EMER PROF ONCOL, UNIV WIS-MADISON, 82- *Concurrent Pos:* Vis prof, Lima, 52-53; mem comt environ physiol, Nat Res Coun, 65-67. *Honors & Awards:* Paul-Lewis Award, Am Chem Soc, 47; Award, Bertner Found, 61; Clowes Medal, Am Asn Cancer Res, 64; Noble Fedn Award, 79; Bristol-Myers Award, 81; Medal of Honor, Am Cancer Soc, 86. *Mem:* Nat Acad Sci; Am Soc Biol Chem; hon mem Am Asn Cancer Res (pres, 74-75); Am Soc Cell Biol (pres, 64-65); Am Acad Arts & Sci; fel AAAS. *Res:* Enzymes of normal and cancer tissue; control mechanisms in intermediary metabolism; enzyme inhibition in metabolic sequences related to cancer chemotherapy; ecological needs for human survival. *Mailing Add:* McArdle Lab Univ Wis Madison WI 53706

POTTHOFF, RICHARD FREDERICK, b Champaign, Ill, Mar 17, 32; m 63; c 2. APPLIED STATISTICS. *Educ:* Swarthmore Col, BA, 53; Univ NC, Chapel Hill, PhD(math statist), 59. *Prof Exp:* Mem staff, Blue Bell, Inc, 58-61; res assoc statist, Univ NC, Chapel Hill, 61-65; opers res analyst, Burlington Industs Inc, 65-89; adj assoc prof, Sch Bus & Econ, NC Agr & Tech State Univ, Greensboro, 89-90; SR RES SCIENTIST, CTR DEMOGRAPHIC STUDIES, DUKE UNIV, 90- *Concurrent Pos:* Consult, Educ Testing Serv; mem, comt E-11 Statist Methods, Am Soc Testing & Mat. *Mem:* Am Statist Asn; Inst Math Statist; Biomet Soc. *Res:* Various uses of theoretical and applied statistics in the social sciences. *Mailing Add:* 803 Winview Dr Greensboro NC 27410

POTTLE, CHRISTOPHER, b New Haven, Conn, Feb 14, 32; m 61; c 3. ELECTRICAL ENGINEERING, COMPUTER SCIENCE. *Educ:* Yale Univ, BE, 53; Univ Ill, MS, 58, PhD(elec eng), 62. *Prof Exp:* Asst proj engr, Sperry Gyroscope Co, 53-54; instr elec eng, Univ Ill, 59-60, res assoc, 60-62, asst prof, 62; asst prof elec eng & comput sci, 62-66, assoc prof, 66-80, PROF ELEC ENG, CORNELL UNIV, 80- *Concurrent Pos:* Fulbright lectr, Univ Erlangen, 66-67; consult math sci dept, IBM Corp, 70-71; resident prog prof, EUSED, Gen Elec Co, 77-78; vis prof, Dept Elec & Comput Eng, Carnegie-Mellon Univ, 83-84; assoc dean comput, Col Eng, Cornell Univ, 86-90. *Mem:* Asn Comput Mach; Inst Elec & Electronics Engrs. *Res:* Application of digital computers to electrical system theory and design. *Mailing Add:* 384 Eng & Theory Ctr Sch Elec Eng Cornell Univ Ithaca NY 14853

POTTMEYER, JUDITH ANN, b Pittsburgh, Pa, Nov 17, 54; m 86. PHYSIOLOGICAL PLANT ECOLOGY. *Educ:* Clarion State Col, BSEd, 75; Wash State Univ, PhD(bot), 84. *Prof Exp:* ASST PROF BIOL, LYCOMING COL, 84- *Mem:* Am Soc Plant Physiologists; Ecol Soc Am; Bot Soc Am. *Res:* Physiological plant ecology; the effects of volcanic ash on the water relations and growth of plants; the effects of flooding on plants. *Mailing Add:* Dept Biol Lycoming Col Williamsport PA 17701

POTTS, ALBERT MINTZ, b Baltimore, Md, June 8, 14; m 38; c 3. BIOCHEMISTRY. *Educ:* Johns Hopkins Univ, AB, 34; Univ Chicago, PhD(biochem), 38; Western Reserve Univ, MD, 48. *Prof Exp:* Asst, Univ Chicago, 38-42, res assoc & instr biochem, Nat Defense Res Coun Proj, 42-44, biochemist, Metall Lab, 44-45; sr instr biochem in ophthal, Sch Med, Western Reserve Univ, 48-50, from asst prof to assoc prof ophthal res, 51-59; prof ophthal & dir res ophthal, Sch Med, Univ Chicago, 59-75; PROF OPHTHAL & CHMN DEPT, UNIV LOUISVILLE, 75- *Mem:* AAAS; Am Chem Soc; Asn Res Vision & Ophthal. *Res:* Ocular physiology and pharmacology; chromatographic adsorption; chemical endocrinology; poisons and cellular metabolism; radiobiology; chemical ophthalmology. *Mailing Add:* 1501 N Campbell Ave Tucson AZ 85724

POTTS, BYRON C, b Springfield, Ohio, Oct 2, 30; m 52, 76; c 4. ELECTRICAL & ELECTRONICS ENGINEERING. *Educ:* Ohio State Univ, BEE & MS, 59, PhD(elec eng), 63. *Prof Exp:* Tech asst, Ohio State Univ, 55-59, res assoc, Antenna Lab, 59-63, assoc supvr lab & asst prof elec eng, 63-64; mem tech staff, Rand Corp, Santa Monica, 64-72; mem tech staff, R&D Assocs, Marina Del Rey, Calif, 72-76, Munich, WGermany, 76-79, Marina Del Rey, Calif, 79-85; mem tech staff, Rand Corp, Santa Monica, Calif, 85-86; SR SCIENTIST, R & D ASSOCS, MUNICH, WGER, 86- *Concurrent Pos:* Coldwell Scholar, 56, Sq D Scholar, 57, Mershon Scholar, 58; instr, Eng Exten, Univ Calif, Los Angeles, 65-66. *Mem:* Inst Elec & Electronics Engrs; Prof Soc Antenna & Propagation; Prof Soc Aerospace Electronics; Prof Soc Commun; Sigma Xi; Asn Old Crows. *Res:* Military electronics including such specialties as command and control, communications, electronic warfare, radar and effects of nuclear bursts. *Mailing Add:* 8352 Chase Ave Westchester CA 90045

POTTS, DONALD CAMERON, b Edinburgh, Scotland, Apr 4, 42; m 70; c 2. POPULATION BIOLOGY, CORAL REEFS. *Educ:* Univ Queensland, Australia, BSc, 63, Hons, 65; Univ Calif, Santa Barbara, PhD(biol), 72. *Prof Exp:* Res assoc, US Antartic Res Prog, 66; lectr biol, Bishops Univ, Que, 71-72 & Flinders Univ, SAustralia, 72-73; res fel, Australian Nat Univ, 74-78; asst prof, 78-85, ASSOC PROF BIOL, UNIV CALIF, SANTA CRUZ, 85- *Concurrent Pos:* Vis fel, Australian Nat Univ, 79-85; vis scientist, Australian Inst Marine Sci, 82- & State Univ NY, Stony Brook, 85-86; mem, Int Asn Biol Oceanogr Coral Reef Comt, 86- *Mem:* Am Soc Limnol & Oceanog; Australian Coral Reef Soc (counr, 74-77); Australian Marine Sci Asn (counr, 75-77); Ecol Soc Am; Int Soc Reef Studies; Soc Study Evolution. *Res:* Ecology, genetics and evolution of clonal organisms; ecology, paleoecology and biogeography of corals and coral reefs. *Mailing Add:* Inst Marine Sci Univ Calif Santa Cruz CA 95064

POTTS, DONALD HARRY, b Seattle, Wash, Dec 15, 21; m 62; c 4. MATHEMATICS. *Educ:* Calif Inst Technol, BS, 43, PhD(math), 47. *Prof Exp:* Asst math, Calif Inst Technol, 43-46; from instr to asst prof, Northwestern Univ, 46-51; mathematician, US Navy Electronics Lab, 51-58; assoc prof math, Long Beach State Col, 58-61; lectr, Univ Calif, Santa Barbara, 61-64; res assoc, Univ Calif, Berkeley, 64-65; PROF MATH, CALIF STATE UNIV, NORTHRIDGE, 65- *Concurrent Pos:* NSF sci fac fel, 64-65. *Mem:* AAAS; Am Math Soc; Math Asn Am; Asn Symbolic Logic. *Res:* Diophantine equations; real variables; wave propagation; mathematical logic; foundations of mathematics; universal algebra. *Mailing Add:* Dept Math Calif State Univ Northridge CA 91330

POTTS, GORDON OLIVER, b Wheeling, WVa, Oct 10, 24; m 43; c 2. ENDOCRINOLOGY. *Educ:* Bethany Col, WVa, BS, 48; WVa Univ, MS, 49; Univ Cincinnati, PhD(zool), 52. *Prof Exp:* Res chemist, Hilton Davis Co, 51-54; assoc res endocrinologist, 54-59, res endocrinologist, 59-64, group leader. 64-69, sect head, 69-73, dept head, 73-76, vpres res admin, 76-83, EXEC VPRES DRUG DEVELOP, STERLING-WINTHROP RES INST, 83- *Mem:* AAAS; Endocrine Soc; NY Acad Sci; Soc Study Reproduction; Sigma Xi. *Res:* Pituitary, gonodal and adrenal hormones; physiology of reproduction, anti-hormones; hormonal control of metabolism. *Mailing Add:* Box 602 Greentrees North Chatham NY 12132

POTTS, HOWARD CALVIN, plant breeding; deceased, see previous edition for last biography

POTTS, JAMES EDWARD, b Alexandria, La, Oct 28, 18; m 44; c 2. PHYSICAL & RADIATION CHEMISTRY. *Educ:* La Col, BS, 39; La State Univ, MS, 41, PhD(phys chem), 48. *Prof Exp:* Res chemist, Dept Chem Eng, Tenn Valley Authority, 42-46; res chemist, Bakelite Co Div, Union Carbide Corp, NJ, 47-52, group leader, 52-58 & Union Carbide Plastics Co, 58-70, sr res scientist chem & plastics, Res & Develop Lab, 70-80, sr res scientist, Coatings Mat Div, 80-82, consult, 82-84; RETIRED. *Concurrent Pos:* Contract Res NASA, 65-71. *Mem:* Am Chem Soc; Soc Mach Engrs. *Res:* Radiation chemistry of polymers; photodegradable plastics; biodegradable plastics; polymer synthesis; biomedical and agricultural plastics; polymeric coatings for concrete. *Mailing Add:* 1619 Treehouse Circle 118 Sarasota FL 34231

POTTS, JOHN CALVIN, b St Louis, Mo, June 23, 06; m 41. INORGANIC & PHYSICAL CHEMISTRY. *Educ:* Univ Calif, BS, 30, MS, 32, PhD(phys chem), 35. *Prof Exp:* With Food & Drug Admin, USDA, 36-37; res engr, Owens-Ill Glass Co, 37-42; phys chemist, Res Div, Int Minerals & Chem Corp, NMex, 42-44, Calif Res Corp, 45-48 & Los Alamos Sci Lab, 48-56; asst to Comnr W F Libby, Atomic Energy Comn, 56-59; asst to assoc dir, Lawrence Radiation Lab, Univ Calif, 59-61; mem staff, Aerojet-Gen Corp Div, Gen Tire & Rubber Co, 61-64; from asst prof to prof chem, Univ of the Pac, 65-75; RETIRED. *Mem:* AAAS; Am Chem Soc; Am Phys Soc. *Res:* Kinetics of gaseous reactions; glass melting process; phase equilibria in salt solutions; atomic energy; chemical education. *Mailing Add:* 324 Indio Dr Pismo Beach CA 93449

POTTS, JOHN EARL, b Lake City, Minn, Nov 25, 44; m 71. SOLID STATE PHYSICS. *Educ:* Marquette Univ, BS, 66; Northwestern Univ, Evanston, PhD(physics), 72. *Prof Exp:* Lectr physics, Ariz State Univ, 71-73; asst prof, Univ Mich, Dearborn, 73-77, assoc prof physics, 77-; RETIRED. *Concurrent Pos:* Vis scientist, Argonne Nat Lab, 80-81. *Mem:* Am Phys Soc; Am Asn Physics Teachers. *Res:* Experimental solid state physics; studies of elementary excitations in solids using laser light scattering techniques. *Mailing Add:* 43834 Brandywine Canton MI 48187

POTTS, JOHN THOMAS, JR, b Philadelphia, Pa, Jan 19, 32; m 61; c 3. ENDOCRINOLOGY, INTERNAL MEDICINE. *Educ:* La Salle Col, BA, 53; Univ Pa, MD, 57. *Prof Exp:* From intern to asst resident med, Mass Gen Hosp, Boston, 57-59; resident, Nat Heart Inst, 59-60, res fel med, 60-63, sr res staff, 63-66, head sect polypeptide hormones, 66-68; from asst prof to assoc prof med, 68-75, prof, 75-81, JACKSON PROF CLIN MED, HARVARD MED SCH, 81-; PHYSICIAN IN CHIEF, MASS GEN HOSP, 81- *Concurrent Pos:* Chief endocrine unit, Mass Gen Hosp, 68-81. *Honors & Awards:* Ernest Oppenheimer Award & Andre Lichwit Prize, Endocrine Soc, 68. *Mem:* Am Soc Biol Chem; Endocrine Soc; Asn Am Physicians; Am Fedn Clin Res; Am Soc Clin Invest. *Res:* Polypeptide hormones; parathroid hormone; calcitonin; polypeptide structure and synthesis; hormone radioimmunoassay; vitamin D; biosynthesis, secretion and metabolism of hormones; mode of action of hormones; structure and function of genes for peptide hormones. *Mailing Add:* Med Serv Mass Gen Hosp Fruit St Boston MA 02114

POTTS, KEVIN T, b Sydney, Australia, Oct 26, 28; m 59; c 4. ORGANIC CHEMISTRY. *Educ:* Univ Sydney, BSc, 50, MSc, 51; Oxford Univ, DPhil(org chem), 54; DSc, Oxford Univ, 73. *Hon Degrees:* DSc, Oxford Univ, 73. *Prof Exp:* Demonstr chem, Univ Sydney, 50; res asst org chem, Oxford Univ, 51-54; scientist, Med Res Coun, Eng, 54-56; res asst org chem, Harvard Univ, 56-58; lectr, Univ Adelaide, 58-61; assoc prof chem, Univ Louisville, 61-65; assoc prof, 65-66, chmn dept, 73-80, PROF CHEM, RENSSELAER POLYTECH INST, 66- *Concurrent Pos:* Res grants, Nat Cancer Inst, Nat Heart & Lung Inst, NSF, US Army Med Res & Develop Command & Dept Energy, ACS-PRF. *Mem:* Am Chem Soc; Royal Soc Chem; Int Soc Heterocyclic Chem. *Res:* Synthetic organic chemistry. *Mailing Add:* Dept Chem Cogswell Lab Rensselaer Polytech Inst Troy NY 12181

POTTS, LAWRENCE WALTER, b Buffalo, NY, Sept 22, 45; m 70; c 3. ELECTROANALYTICAL CHEMISTRY, TRACE METALS ANALYSIS. *Educ:* Oberlin Col, AB, 67; Univ Minn, PhD(chem), 72. *Prof Exp:* PROF ANALYTICAL & INORG CHEM, GUSTAVUS ADOLPHUS COL, 72- *Concurrent Pos:* Vis assoc prof, Oberlin Col, 81-82; vis prof, Univ Minn, 88 & 90-91. *Mem:* Am Chem Soc; Sigma Xi; AAAS. *Res:* Chemisorption of polymers on metal and metal oxide surfaces; infrared spectroscopy of surfaces. *Mailing Add:* Dept Chem Gustavus Adolphus Col St Peter MN 56082

POTTS, MALCOLM, b Durham, Eng, Jan 8, 35; m 83; c 4. FAMILY PLANNING. *Educ:* Cambridge Univ, MA, 60, MB, BChir, 62, PhD(embryol), 65. *Prof Exp:* Dir med studies, Sidney Sussex Col, Cambridge Univ, 64-67; med dir, Int Planned Parenthood Fedn, London, Eng, 68-78;

PRES, FAMILY HEALTH INT, NC, 78- *Concurrent Pos:* Consult, World Bank, 71 & UK Ministry Overseas Develop, 78; Chmn Bd, Clin Res Int, 87. *Honors & Awards:* Hugh Moore Prize for Int Family Planning, 73. *Mem:* Int Union Sci Study Pop; Am Fertil Soc. *Res:* Family planning; clinical trials of long acting steroids, intrauterine devices and spermicides; maternal mortality; lactation. *Mailing Add:* Dept Biochem VPI & State Univ Engel Hall W Campus Dr Blacksburg VA 24061

POTTS, MARK JOHN, b St Louis, Mo. GEOPHYSICS, GEOCHEMISTRY. *Educ:* Wash Univ, BS, 66, PhD(geochem), 71. *Prof Exp:* Res radiochemist uranium purification, Mallinckrodt Chem Corp, 66; asst prof neutron activation anal, Geochem Inst, Univ Gottingen, 71-72; asst prof geochem, Dept Earth Sci, Wash Univ, 72-73, res assoc environ geochem, Ctr Biol of Natural Systs, 73; staff geoscientist uranium explor, Airborne Geophys Serv, 73-79, STAFF SCIENTIST, SEISMIC RES DEPT, TEX INSTRUMENTS INC, 79- *Mem:* Asn Explor Geochemists; Geol Soc Am. *Res:* Application of geochemistry, statistics and data processing techniques to the analysis of geophysical and geochemical data; hardware and software development for geophysical systems; software systems; technical management. *Mailing Add:* 6931 Cliffbrook Dallas TX 75240-7907

POTTS, MELVIN LESTER, b Lewellen, Nebr, July 19, 21; m 46; c 2. PHYSICAL INORGANIC CHEMISTRY. *Educ:* Kans State Col, Pittsburg, BSc, 54, MSc, 57; Colo State Univ, PhD(phys chem), 63. *Prof Exp:* Teacher high sch, Kans, 55-56; instr chem & physics, Chanute Jr Col, 56-57; instr physics, Kans State Col Pittsburg, 57-59; instr chem, Colo State Univ, 62-63; PROF CHEM, KANS STATE COL PITTSBURG, 63-, CHMN DEPT, 76- *Mem:* Am Chem Soc. *Res:* Surface chemistry. *Mailing Add:* 1707 S Walnut Pittsburg KS 66762-5751

POTTS, RICHARD ALLEN, b Massillon, Ohio, Jan 2, 40; m 66; c 1. INORGANIC CHEMISTRY. *Educ:* Hiram Col, BA, 62; Northwestern Univ, PhD(inorg chem), 66. *Prof Exp:* AEC res assoc, Ames Lab, Iowa State Univ, 65-66; from asst prof to assoc prof chem, 66-73, PROF CHEM, UNIV MICH-DEARBORN, 73- *Mem:* Am Chem Soc; AAAS. *Res:* Compounds containing metal-to-metal bonds; coordination chemistry of mercury and gold. *Mailing Add:* Dept of Chem Univ of Mich Dearborn MI 48128

POTTS, RICHARD CARMECHIAL, agronomy, for more information see previous edition

POTTSEPP, L(EMBIT), b Tartu, Estonia, Feb 24, 29; nat US; m 66. AERONAUTICAL ENGINEERING. *Educ:* Univ Ill, BS, 53, MS, 54, PhD(aeronaut eng), 59. *Prof Exp:* Design specialist, Douglas Aircraft Co, 59-62, chief, Flight Mech Res Sect, McDonnell Douglas Astronaut Co, 62-65, chief, Appl Math Br, 65-76; PRES, ENG RES ASSOCS, 77- *Concurrent Pos:* Res fel appl math, Harvard Univ, 63-64. *Mem:* Am Inst Aeronaut & Astronaut. *Res:* Optimal control theory; fluid mechanics; analytical mechanics. *Mailing Add:* 5642 Highgate Terr Irvine CA 92715

POTVIN, ALFRED RAOUL, b Worcester, Mass, Feb 5, 42; m 65. BIOMEDICAL ENGINEERING. *Educ:* Worcester Polytech Inst, BSEE, 64; Stanford Univ, MSEE, 65; Univ Mich, MS(bioeng) & MS(psychol), 70, PhD(bioeng), 71. *Prof Exp:* Asst prof elec eng, Univ Tex, Arlington, 66-68; clin asst prof biophys, Univ Tex Southwestern Med Sch, Dallas, 67-70; assoc prof biomed eng, Univ Tex, Arlington, 71-76, prof elec eng & chmn dept, 76-84; dir, Med Instrument Syst Res Div, 84-90, DIR TECHNOL ASSESSMENT & PROJ MGT, MED DEVICES & DIAG DIV, LILLY RES LAB, 90- *Concurrent Pos:* NSF res grant, 72-74; consult, Merck, Sharpe & Dohme, 72-84 & NASA, 72-77; NASA life scientist, 74-75; consult, US Food & Drug Admin, 77-84, Vet Admin Rehab Eng Res & Develop Serv, 79-84; mem bd dir, Rocky Mountain Bioeng Symp Inc, 80-84; pres, Eng Med & Biol Soc, Inst Elec & Electronic Engrs, 83-85; NIH fel, 68; pres, Alliance Eng Med & Biol (vpres nat affairs, 87-88, pres, 89-92). *Honors & Awards:* Centennial Medal, 84. *Mem:* fel Inst Elec & Electronics Engrs; Alliance Eng Med & Biol (vpres nat affairs, 87-88); Sigma Xi. *Res:* Biomedical engineering; biosensors; cardiovascular diagnostics and therapeutics. *Mailing Add:* Med Devices & Diag Div Lilly Corp Ctr 2053 Indianapolis IN 46285

POTVIN, PIERRE, b Quebec, Que, Jan 5, 32; m 63; c 2. PHYSIOLOGY. *Educ:* Laval Univ, BA, 50, MD, 55; Univ Toronto, PhD(physiol), 62. *Prof Exp:* From asst prof to assoc prof, Laval Univ, 60-69, secy, 75- 86, univ exec coun, 76-80, assoc dean, 77-86, PROF PHYSIOL, FAC MED, LAVAL UNIV, 69-, DEAN, 86- *Mem:* Can Physiol Soc; hon mem Royal Col Physicians Can. *Res:* Hepatic physiology, especially hepatic circulation. *Mailing Add:* Dept Physiol Laval Univ Fac Med Quebec G1K 7P4 PQ G1K 7P4 Can

POTWOROWSKI, EDOUARD FRANÇOIS, b Lyon, France, May 15, 40; Can citizen; m 69; c 2. IMMUNOLOGY. *Educ:* Loyola Col Montreal, BA, 62; Univ Montreal, MSc, 64; Monash Univ, Australia, PhD(immunol), 67. *Prof Exp:* Sr res asst, 67-70, head tissue antigens lab, Inst Microbiol & Hyg, 70-75; mem res staff, 75-80, chmn, 86-90, PROF, IMMUNOL RES CTR, INST ARMAND-FRAPPIER, 80- *Concurrent Pos:* Med Res Coun Can & Nat Cancer Inst Can grants, 67-79; assoc, Dept Med, McGill Univ & McGill Cancer Ctr; Eleanor Roosevelt Int Cancer fel, 82-83. *Mem:* Can Soc Immunol (vpres, 85-86, pres, 87-89); Transplantation Soc; Sigma Xi; Am Asn Immunologists. *Res:* Role of thymic microenvironment in T cell maturation and in leukemogenesis, in vitro thymic culture. *Mailing Add:* Immunol Res Ctr Inst Armand-Frappier Laval PQ H7N 4Z3 Can

POTYRAJ, PAUL ANTHONY, b Baltimore, Md, Jan 15, 60; m 82; c 3. MICROELECTRONICS. *Educ:* Johns Hopkins Univ, BA, 82; Carnegie-Mellon Univ, MS, 84, PhD(physics), 87. *Prof Exp:* Teaching asst, Dept Physics, Carnegie-Mellon Univ, 82-83, res asst, Dept Elec & Computer Eng, 83-87; SR ENGR, WESTINGHOUSE ELEC CORP, 87- *Mem:* Inst Elec & Electronic Engrs; Electrochem Soc; Am Physics Soc. *Res:* Design, development and analysis of silicon semiconductor devices, particularly polycrystalline-silicon-emitter transistors and high-power high-frequency (microwave) transistors. *Mailing Add:* 1315 North Point Rd Baltimore MD 21222-1420

POTZICK, JAMES EDWARD, b Cincinnati, Ohio, Apr 20, 41. PHYSICS, PHYSICAL INSTRUMENTATION. *Educ:* Xavier Univ, BS, 63, MS, 65. *Prof Exp:* Physicist, Eng Mech Sect, 66-70, physicist, Fluid Eng Div, 70-85, PHYSICIST, PRECISION ENG DIV, NAT INST STANDARDS & TECH, 85- *Mem:* Am Phys Soc; Am Soc Precision Eng; SPIE. *Res:* Mechanical and electronic instrumentation for fluid dynamics; acoustic flowmeter for dynamic measurement of mass flowrate of hot gases such as automotive exhaust; precision dimensional metrology; optical submicron linewidth meterology; automated measurements. *Mailing Add:* Tech A347 Nat Inst Standards & Tech Gaithersburg MD 20899

POU, JACK WENDELL, b Shereveport, La, Sept 30, 23. OTOLARYNGOLOGY. *Educ:* Tulane Univ, BS, 43, MD, 46; Am Bd Otolaryngol, dipl. *Prof Exp:* Ear, nose & throat consult, Schumpert Hosp, Physicians & Surgeons Hosp & Doctors Hosp, 51; CHIEF EAR, NOSE & THROAT SERV, CONFEDERATE MEM MED CTR, 62-; PROF OTOLARYNGOL & CHMN DEPT, SCH MED, LA STATE UNIV, SHREVEPORT, 70- *Concurrent Pos:* Chief med adv, Deaf Oral Dept, Caddo Parish, La, 52-; mem bd dirs, Caddo Found Except Children, 52-; ear, nose & throat consult, Barksdale AFB Hosp, 62 & Vet Admin Hosp, Shreveport, 68; mem adv admis bd, Speech Correction Ctr, 62-; chief otolaryngol, Confederate Mem Med Ctr, 63. *Mem:* AMA; Am Acad Ophthal & Otolaryngol; Am Laryngol, Rhinol & Otol Soc. *Res:* Feasibility of decompression of the facial nerve. *Mailing Add:* Schumpert Med Ctr 2121 Line Ave Shreveport LA 71104

POU, WENDELL MORSE, b Laurel, Miss, Oct 11, 37. PHYSICS. *Educ:* Millsaps Col, BS, 59; Vanderbilt Univ, MS, 62, PhD(physics), 69. *Prof Exp:* Asst prof, 66-74, ASSOC PROF PHYSICS & ASTRON, ST CLOUD STATE UNIV, 74-, PROF PHYSICS, ASTRON & ENG SCI. *Mem:* Am Phys Soc; Am Asn Physics Teachers. *Res:* Radiation chemistry; electron spectroscopy. *Mailing Add:* Dept Physics & Astron St Cloud State Univ St Cloud MN 56301

POUCHER, JOHN SCOTT, b Evanston, Ill, Apr 10, 45; m 69; c 2. SYSTEMS DESIGN. *Educ:* Mass Inst Technol, BS, 67, PhD(physics), 71. *Prof Exp:* Instr, Mass Inst Technol, 71-74; asst prof physics, Vanderbilt Univ, 74-80; DISTINGUISHED MEM TECH STAFF, BELL LABS, 81- *Concurrent Pos:* Vis fel, Lab Nuclear Studies, Cornell Univ, 77 & 78-79; Univ Res Coun fel, Vanderbilt Univ, 79. *Mem:* Am Phys Soc; AAAS; Fedn Am Scientists; NY Acad Sci; Sigma Xi; Inst Elec & Electronics Engrs; Am Inst Aeronaut & Astronaut. *Res:* Systems planning and design especially for government computer and communications applications; experimental high energy physics: hadron structure and interactions, deep inelastic electron scattering and electron- positron collisions at high energies. *Mailing Add:* AT&T Bell Labs Crawfords Corner Rd Rm 2G-231 Holmdel NJ 07733-1988

POUCHER, MELLOR PROCTOR, b London, Eng, Jan 14, 29; m 57; c 3. CIVIL ENGINEERING. *Educ:* Univ London, BSc, 49, MSc, 51. *Prof Exp:* Assoc prof eng sci, 54-67, PROF ENG SCI, UNIV WESTERN ONT, 67-, CHMN, DEPT CIVIL, 60- *Mem:* Can Soc Civil Engrs; Am Concrete Inst. *Res:* Concrete technology; structural analysis and design. *Mailing Add:* 69 Bloomfield Dr London ON N6G 1P4 Can

POUCHER, WILLIAM B, b Auburn, Ala, Dec 9, 48. OBJECT ORIENTED PROGRAMMING, COMPUTER SCIENCE EDUCATIONAL NETWORKS. *Educ:* Auburn Univ, BS, 71, MS, 72, PhD(combinatorics math comput), 75. *Prof Exp:* Assoc prof, 83-90, PROF COMPUTER SCI, BAYLOR UNIV, 90- *Concurrent Pos:* Consult, Computer Network Applications, 77-; reviewer, Comput Reviews, 82-; mem coun, Asn Comput Mach, 86-90, dir student competition, 88-89 & Int Scholastic Prog Contest, 89- *Honors & Awards:* Outstanding Contrib Award, Asn Comput Mach, 91. *Mem:* Asn Comput Mach. *Mailing Add:* Dept Computer Sci Baylor Univ PO Box 97356 Waco TX 76798-7356

POUGH, FREDERICK HARVEY, b Brooklyn, NY, June 26, 06; m 38; c 2. MINERALOGY. *Educ:* Harvard Univ, SB, 28, PhD(mineral), 35; Wash Univ, MS, 32. *Prof Exp:* Instr crystallog, Harvard Univ, 32-34; asst cur mineral, Am Mus Natural Hist, 35-41, actg cur, 41-43, cur phys geol & mineral, 43-52; consult mineralogist, 53-64; dir, Santa Barbara Mus Natural Hist, 65-66; CONSULT MINERALOGIST, 66- *Concurrent Pos:* Am Fedn Mineral Socs scholar, 66. *Honors & Awards:* Derby Medal, Brazil, 48; Haneman Educ Award, 88; Carnegie Mineral Award, 89. *Mem:* Fel Mineral Soc Am; fel Geol Soc Am; Mineral Soc Gt Brit & Ireland; Gemmol Asn Gt Brit. *Res:* Morphological crystallography; mineral paragenesis; gemmology; vulcanism; synthetic crystal growth. *Mailing Add:* PO Box 7004 Reno NV 89510-7004

POUGH, FREDERICK HARVEY, b New York, NY, Jan 13, 42; m 67; c 1. ENVIRONMENTAL PHYSIOLOGY, HERPETOLOGY. *Educ:* Amherst Col, BA, 64; Univ Calif, Los Angeles, MA, 66, PhD(zool), 68. *Prof Exp:* From asst prof to assoc prof, 69-82, PROF HERPET, NY STATE COL AGR & LIFE SCI, CORNELL UNIV, 83- *Concurrent Pos:* Cur, Herpet Collection, 69-79. *Res:* Biology of amphibians and reptiles. *Mailing Add:* Ecol/Syst Cornell Univ E211 Corson Hall Ithaca NY 14853-2701

POUGH, RICHARD HOOPER, b New York, NY, Apr 19, 04; m 37; c 2. ORNITHOLOGY, ECOLOGY. *Educ:* Mass Inst Technol, SB, 26. *Hon Degrees:* LLD, Haverford Col, 70. *Prof Exp:* Chem engr, Southern Acid & Sulphur Co, Tex, 27-28; mech engr, Fulton Iron Works, Mo, 29-31; retail exec, MacCallam, Inc, Pa, 31-36; res assoc & mem tech staff, Nat Audubon Soc, NY, 36-48; cur & chmn dept conserv & gen ecol, Am Mus Natural Hist, 48-56; pres, Natural Area Coun, Inc, 57-82; RETIRED. *Concurrent Pos:* Gov & past pres, Nature Conserv; trustee, Nat Parks & Conserv Asn; Natural Sci Youth Found, Thorne Ecol Inst & Sapelo Island Res Found; dir, Scenic Hudson Preservation Conf; dir & past pres, Open Space Inst; adv coun, Trust for Pub Lands; dir & past pres, Defenders of Wildlife; chmn, Am The Beautiful Fund. *Mem:* Fel AAAS. *Res:* Conservation; plant-animal distribution; preservation of biotic communities in a natural condition. *Mailing Add:* 33 Highbrook Ave Pelham NY 10803-1713

POUKEY, JAMES W, b Indianapolis, Ind, Feb 17, 40. PHYSICS, BEAM PHYSICS. *Educ:* Univ Wisconsin, BS, 61, MS, 63, PhD(appl physics), 66. *Prof Exp:* Fel, Univ Wis, 66-67; RES PHYSICIST, SANDIA NAT LAB, 67- *Mem:* Am Phys Soc. *Mailing Add:* 1717 Buffalo Dancer Trail NE Albuquerque NM 87112

POULARIKAS, ALEXANDER D, b Desylla, Greece, Sept 1, 33; m 62. PLASMA PHYSICS, ELECTROMAGNETICS. *Educ:* Univ Ark, BS, 60, MS, 63, PhD(plasmas), 66. *Prof Exp:* From asst prof to assoc prof elec eng, Univ RI, 65-76, prof, 76-; CHMN & PROF ENG DEPT, UNIV DENVER. *Concurrent Pos:* NASA Fac fel, 68 & 71; vis scientist, Mass Inst Technol, 71-72; consult, Naval Underwater Systs Ctr, Newport, RI, 71-73. *Mem:* Optical Soc Am; Soc Photo-Optical Instrumentation Engrs; Inst Elec & Electronics Engrs. *Res:* Nuclear reactions; magnetohydrodynamics power generation; plasma interactions with magnetic fields; plasma waves; optics. *Mailing Add:* Dept Elec Comput Eng Univ Ala Huntsville AL 35899

POULIK, MIROSLAV DAVE, b Brno, Czech, June 6, 23; US citizen; m 50; c 2. IMMUNOLOGY, IMMUNOCHEMISTRY. *Educ:* I Real Gym, Brno, BA, 42; Univ Toronto, MD, 60. *Prof Exp:* Res asst hyg & prev med, Univ Toronto, 51-56; res assoc, Child Res Ctr, Children's Hosp, Detroit, 56-60; guest investr, Rockefeller Inst, 60-61; asst dir res, Am Red Cross, Washington, DC, 61-62; from asst prof to prof pediat, 62-72, coordr med genetics, Sch Med, 65-68, PROF IMMUNOL MICROBIOL, WAYNE STATE UNIV, 72-; CHIEF IMMUNOL, WILLIAM BEAUMONT HOSP, 71- *Concurrent Pos:* Sr res assoc, Child Res Ctr, Children's Hosp, Detroit, 62-71; temp adv, Immunol Lab, WHO, Lausanne, 65-68, mem Expert Adv Comt & mem WHO, 68-73; consult, Vet Admin Hosp, Dearborn, 69- *Mem:* Am Asn Immunol; Am Soc Human Genetics; NY Acad Sci; Soc Exp Biol & Med; Can Soc Immunol. *Res:* Structure and function of antibodies; deficiency diseases; serum and urinary proteins in health and disease; cancer cell immunology; cyto fluorometry. *Mailing Add:* Dept Immunopathol William Beaumont Hosp 3601 W 13 Mile Rd Royal Oak MI 48072

POULIKAKOS, DIMOS, b Athens, Greece, Aug 29, 55; m 83; c 1. FLUID MECHANICS & HEAT TRANSFER, MASS TRANSFER. *Educ:* Nat Tech Univ Athens, Greece, BS, 78; Univ Colo, MS, 80, PhD(mech eng), 83. *Prof Exp:* Lectr & res assoc heat transfer, Univ Colorado, Boulder, 82-83; asst prof mech eng, 83-87, ASSOC PROF MECH ENG, UNIV ILL, CHICAGO, 87- *Concurrent Pos:* Dir grad study, Mech Eng Dept, Univ Ill, Chicago, 87- *Honors & Awards:* Presidential Young Investr Award. *Mem:* Am Soc Mech Engrs; Soc Automotive Engrs. *Res:* Fundamental contributions in natural and forced convection, heat transfer in porous media, alloy solidification, boiling, and cooling of electronic equipment; published over 70 articles in technical journals. *Mailing Add:* Dept Mech Eng Univ Ill PO Box 4348 Chicago IL 60680

POULOS, DENNIS A, b Brooklyn, NY, Jan 18, 32. NEUROPHYSIOLOGY. *Educ:* Univ Wis, BS, 60, MS, 63, PhD(physiol), 66. *Prof Exp:* Nat Inst Neurol Dis & Blindness fel, Univ Wis, 66-67; from asst res prof to assoc res prof neurosurg & from asst prof to assoc prof physiol, 67-74, res prof neurosurg & prof physiol, 74-78, PROF ANAT, ALBANY MED COL, 78-, ASST DEAN ACAD AFFAIRS, 84- *Mem:* AAAS; Am Physiol Soc; Int Brain Res Orgn; Sigma Xi; Soc Neurosci; Am Asn Anat. *Res:* Sensory physiology, especially peripheral and central neural mechanisms of thermal sensation; physiological psychology. *Mailing Add:* Dept of Anat Albany Med Col Albany NY 12208

POULOS, NICHOLAS A, b Maywood, Ill, Dec 20, 26. ORGANIC CHEMISTRY. *Educ:* Univ Ill, BS, 48; Northwestern Univ, PhD(org chem), 57. *Prof Exp:* Lab asst, Northwestern Univ, 49, 51, asst, 50, 52; mem staff, Olin Mathieson Chem Corp, 53-61; task engr, Wright-Patterson AFB, 62-64; proj supvr, Aeroproj, Inc, 64; assoc prof chem, Union Col, Ky, 65-70; INSTR ORG CHEM, TRITON COL, 70- *Res:* Chemistry. *Mailing Add:* 800 Sherman St Melrose Park IL 60160

POULOSE, PATHICKAL K, b Kerala, India, Jan 22, 39; m 71; c 2. MATERIALS SCIENCE ENGINEERING. *Educ:* Indian Inst Sci, BS, 65, MS, 67; Univ Conn, PhD(mat sci), 73. *Prof Exp:* Res assoc, George Washington Univ, 73-74, res scientist mat eng, 74-77, asst res prof mat sci & eng, 77-86. *Mem:* Am Soc Metals. *Res:* Determination of fracture characteristics of brittle and semibrittle materials; fatigue crack growth and life studies in structural alloys; effect of microstructure on mechanical properties of titanium alloys. *Mailing Add:* 11306 Attingham Lane Glenn Dale MD 20769

POULSEN, BOYD JOSEPH, b Tetonia, Idaho, Aug 30, 33; m 53; c 2. PHARMACY. *Educ:* Idaho State Col, BS, 56; Univ Wis, PhD(pharm), 63. *Prof Exp:* Sr chemist, Pharmaceut Prod Develop, Ayerst Labs, Inc, 63-65; staff researcher, Syntex Corp, 65-69, assoc dir, 69-77, dir, Inst Pharmaceut Sci, 77-79, VPRES, SYNTEX RES DIV, 79- *Concurrent Pos:* Adj prof, Univ of Pac, 74- *Mem:* Fel Acad Pharmaceut Sci; Am Pharmaceut Asn; NY Acad Sci; AAAS; Sigma Xi. *Res:* Pharmaceutics; surface chemistry; percutaneous absorption of drugs. *Mailing Add:* Syntex Res R1-138 3401 Hillview Ave Palo Alto CA 94304

POULSEN, LAWRENCE LEROY, b Salmon, Idaho, Nov 27, 33; m 57; c 6. BIOCHEMISTRY, BIOCHEMICAL PHARMACOLOGY. *Educ:* Univ Calif, Riverside, BA, 65, PhD(biochem), 69. *Prof Exp:* Phys sci technician chem, USDA, 57-65; fel biol, Tex A&M Univ, 69-70; res assoc, Univ Tex, Austin, 71-72, USPHS fel, 72-74, res scientist assoc, Clayton Found Biochem Inst, 75-81; RES ASSOC, CLAYTON FOUND BIOCHEM INST, 81- *Concurrent Pos:* Consult, Alcon Labs, Inc, 81. *Mem:* Am Soc Biochem & Molecular Biol. *Res:* Enzymology and control of drug metabolism; interactions of enzymes and environmental contaminants and carcinogens in the control of growth and metabolism; mechanism of plant and mammalian growth control. *Mailing Add:* Clayton Found Biochem Inst Univ of Tex Austin TX 78712

POULSON, DONALD FREDERICK, b Idaho Falls, Idaho, Oct 5, 10; m 34; c 2. GENETICS. *Educ:* Calif Inst Technol, BS, 33, PhD(genetics), 36. *Hon Degrees:* MA, Yale Univ, 55. *Prof Exp:* Res asst embryol, Carnegie Inst, 36-37; from instr to prof biol, 37-81, chmn dept, 62-65, fel 54-81, EMER PROF BIOL & EMER FEL, CALHOUN COL, YALE UNIV, 81- *Concurrent Pos:* Gosney fel, Calif Inst Technol, 49; res collabr, dept biol, Brookhaven Nat Lab, 51-55; Fulbright sr res scholar, Commonwealth Sci & Indust Res Orgn, Canberra, Australia, 57-58, 66-67; Guggenheim Mem Found fel, 57-58; Japan Soc Promotion Sci fel, 79. *Mem:* AAAS; Am Soc Naturalists (treas, 51-53); Am Soc Zoologists; Genetics Soc Am; Soc Develop Biol. *Res:* Cytology; genetics of Drosophila; embryology of insects; physiology of development; hereditary infections. *Mailing Add:* 96 Greenhill Rd Orange CT 06477

POULSON, RICHARD EDWIN, b Detroit, Mich, Sept 26, 28; m 54; c 4. PHYSICAL CHEMISTRY. *Educ:* Mich State Univ, BS, 53, PhD(phys chem), 59; Univ Calif, Berkeley, MS, 57. *Prof Exp:* Chemist, Mich Dept Agr, 54-55; res asst, Lawrence Radiation Lab, Univ Calif, 56; res chemist, Beckman Instruments Inc, 59-63; res chemist, US Dept Energy, 63-74, proj leader, 74-76, mgr, div environ sci, Laramie Energy Technol Ctr, 76-83; SR RES SCIENTIST, WESTERN RES INST, UNIV WYO RES CORP, 83- *Concurrent Pos:* Chmn, E-19 Comt Chromatography, Am Soc Testing & Mat. *Mem:* Am Chem Soc; Am Soc Testing & Mat. *Res:* Environmental chemistry of pollutants in aqueous media; development of physiochemical process models through chromatographic analogies; detection methods development. *Mailing Add:* Univ Sta Box 3136 Laramie WY 82071

POULTER, CHARLES DALE, b Monroe, La, Aug 29, 42; m 64; c 2. CHEMISTRY. *Educ:* La State Univ, Baton Rouge, BS, 64; Univ Calif, Berkeley, PhD(org chem), 67. *Prof Exp:* NIH fel, Univ Calif, Los Angeles, from asst prof to assoc prof, 69-78, PROF CHEM, UNIV UTAH, 78- *Concurrent Pos:* Alfred P Sloan Found fel & NIH res develop award, 75. *Mem:* AAAS; Am Chem Soc; Royal Soc Chem; Am Asn Biol Chemists. *Res:* Bioorganic chemistry; biosynthesis of terpenes; applications of nuclear magnetic resonance techniques to polynucleotides. *Mailing Add:* Dept of Chem Univ of Utah Salt Lake City UT 84112-1102

POULTER, DOLORES IRMA, b Mt Pleasant, Iowa, Mar 22, 41; m 84. ECOLOGY, PLANT MORPHOLOGY. *Educ:* Iowa Wesleyan Col, BA, 63; Iowa State Univ, MS, 65, PhD(bot), 69. *Prof Exp:* From instr to assoc prof, 65-72, dept head, 73-83, chmn nat sci div, 80-83, PROF BIOL, IOWA WESLEYAN COL, 72-, COORD BIOL, 84- *Mem:* Sigma Xi; Nature Conservancy. *Res:* Distribution patterns of Eastern Red Cedar in Iowa. *Mailing Add:* 301 S Van Buren Mt Pleasant IA 52641

POULTER, HOWARD C, b Chicago, Ill, Mar 16, 25; m 49, 69; c 6. ELECTRICAL ENGINEERING. *Educ:* Ill Inst Technol, BS, 46, MS, 47; Stanford Univ, PhD(elec eng), 55. *Prof Exp:* Res assoc microwave tubes, Stanford Univ, 52-54; engr, Hewlett-Packard Co, 54-56, proj supvr instrumentation, 56-58, engr mgr, Microwave Div, 59-64, sr staff engr, 65-66, sect mgr anal instruments, 66-70; INSTR ELECTRONICS, MATH & PHYSICS, PERALTA JR COL DIST, 70- *Concurrent Pos:* Sci fac professional develop grant, NSF, 81-82. *Res:* Electrical measurement techniques to the development of instrumentation for chemical and biological areas; microwave instrumentation. *Mailing Add:* 4375 Bridgeview Dr Oakland CA 94602

POULTNEY, SHERMAN KING, b Leominster, Mass, Mar 18, 37; div; c 1. APPLIED & OPTICAL PHYSICS. *Educ:* Worcester Polytech Inst, BS, 58; Princeton Univ, MA, 60, PhD(physics), 62. *Prof Exp:* Res assoc physics, Univ Md, College Park, 64-66, asst prof, 66-73; res assoc prof, Old Dominion Univ, 73-75; sr staff scientist, Perkin-Elmer Corp, 75-89; SR STAFF SCIENTIST, HUGHES DANBURY, 89- *Concurrent Pos:* Consult, Naval Ord Sta, Indian Head, Md, 71-73. *Mem:* Am Phys Soc; Optical Soc Am. *Res:* Active and passive remote sensing of atmosphere and earth resources; upper atmosphere physics; molecular infrared spectroscopy, imaging interferometers; satellite and lunar laser ranging; single photon detection, timing and imaging; electro-optical system design and engineering. *Mailing Add:* Seven Stebbins Close Ridgefield CT 06877

POULTON, BRUCE R, b Yonkers, NY, Mar 7, 27; m 50; c 4. ANIMAL SCIENCE. *Educ:* Rutgers Univ, BS, 50, MS, 52, PhD, 56. *Prof Exp:* Asst animal sci, Rutgers Univ, 50-51, from instr to asst prof, 51-56; from assoc prof to prof, Univ Maine, Orono, 56-75, chmn dept, 58-66, dean, Col Life Sci & Agr, 68-71, dir, Life Sci Exp Sta, 68-72, vpres, Res & Pub Serv, 72-75; chancellor, Univ Syst NH, 75-, dean, Sch Lifelong Learning, 77-; CHANCELLOR, NC STATE UNIV. *Concurrent Pos:* Am Coun Educ fel acad admin & vis prof, Mich State Univ, 66-67; dir, Univ Maine, Bangor, 67-68. *Mem:* Am Soc Animal Sci; Am Dairy Sci Asn; Am Inst Nutrit; Brit Nutrit Soc; Sigma Xi. *Res:* Rumen nutrition; role of the endocrine system in certain physiological disorders; parathyroid physiology and effect of soil fertility on nutritive value of forages; author and co-author of over 32 publications. *Mailing Add:* NC State Univ Box 7401 Raleigh NC 27695

POULTON, CHARLES EDGAR, b Oakley, Idaho, Aug 2, 17; m 39; c 4. ECOLOGY, RANGE SCIENCE & MANAGEMENT. *Educ:* Univ Idaho, BS, 39, MS, 48; Wash State Univ, PhD(ecol), 55. *Prof Exp:* Forest adminr & dist forest ranger, US Forest Serv, 37-39, 40-47, field asst, Inter-Mt Forest & Range Exp Sta, 39-40; instr range mgt, Univ Idaho, 47-49; assoc prof & organizer, Range Mgt Prog, Ore State Univ, 49-59, prof range ecol, 59-74, dir, Environ Remote Sensing Appln Lab, 72-74, head range mgt, 49-70; dir, Rangeland & Resource Ecol Div, Earth Satellite Corp, 74-75; sr officer range & pastures, Food & Agr Orgn UN, Rome, Italy, 76-78; training officer remote sensing applns, Ames Res Ctr, NASA, 79-81; REMOTE SENSING & RESOURCE ANALYST, 71- *Concurrent Pos:* Asst prof, Mont State Col, 46-47; consult, Southern Forest Exp Sta, US Forest Serv, 58-59 & Ore State Land Bd, 65-70; post doctoral studies, Univ Calif, Berkeley, 66-67; remote sensing & resource analyst, 71-; consult range sci & mgt, remote sensing appl renewable natural resources mgt & bus develop, coord resource develop & mgt, 78-; natural resources curric design, Asian Inst Technol, Bangkok, Thailand, 86-87. *Mem:* Fel Soc Range Mgt. *Res:* Rangeland resources improvement and management planning; plant ecology-soils; environmental impact analysis; application of remote sensing technology in solving renewable natural resource problems. *Mailing Add:* PO Box 2081 Gresham OR 97030-0601

POUND, ROBERT VIVIAN, b Ridgeway, Ont, May 16, 19; nat US; m 41; c 1. PHYSICS. *Educ:* Univ Buffalo, BA, 41. *Hon Degrees:* AM, Harvard Univ, 50. *Prof Exp:* Res physicist, Submarine Signal Co, Mass, 41-42; mem staff, Radiation Lab, Mass Inst Technol, 42-46; Soc of Fels jr fel, 45-48, from asst prof to prof physics, 48-68, chmn dept, 68-72, Mallinckrodt prof, 68-69, dir physics labs, 75- 83, EMER PROF PHYSICS, HARVARD UNIV, 89- *Concurrent Pos:* Fulbright scholar, Clarendon Lab, Oxford Univ, 51; Guggenheim fel, 57-58 & 72-73; vis prof, Col France, 73; assoc etranger, Acad Sci, 78; vis fel, Joint Inst Lab Astrophys, Univ Colo, 79-80 & Merton Col, Oxford, 80; vis Zernike prof, Groningen, Neth, 82; vis sr scientist, Brookhaven Nat Lab, 86-87; vis prof, Univ Fla, 87; trustee, Assoc Univ Inc, 76- *Honors & Awards:* Thompson Mem Award, Inst Radio Engrs, 48; Fulbright lectr, Univ Paris, 58; Eddington Medal, Royal Astron Soc, 65; Nat Medal Sci, 90. *Mem:* Nat Acad Sci; fel Am Phys Soc; fel AAAS; fel Am Acad Arts & Sci. *Res:* Nuclear moments by radio-frequency spectroscopy; nuclear physics; microwaves and radar; experimental relativity. *Mailing Add:* Lyman Lab of Physics Harvard Univ Cambridge MA 02138

POUNDER, ELTON ROY, b Montreal, Que, Jan 10, 16; wid; c 2. ICE PHYSICS, PHYSICAL OCEANOGRAPHY. *Educ:* McGill Univ, BSc, 34, PhD(physics), 37. *Prof Exp:* Engr, Can Bell Tel Co, 37-39; from asst prof to prof, 45-82, EMER PROF PHYSICS, MCGILL UNIV, 82- *Mem:* Am Phys Soc; Am Geophys Union; Can Asn Physicists (secy, 56-59, pres, 61-62); fel Royal Soc Can. *Mailing Add:* Dept Physics McGill Univ Ernest Rutherford Bldg 3600 University St Montreal PQ H3A 2T8 Can

POUNDSTONE, WILLIAM N, b Morgantown, WVa, Aug 12, 25; c 3. COAL MINING EQUIPMENT METHOD. *Educ:* WVa Univ, BSE, 49. *Hon Degrees:* DSc, WVa Univ, 81. *Prof Exp:* Exec vpres, Consolidation Coal Co, 82; dir, Standard Haven, 83-89; RETIRED. *Honors & Awards:* Perry Nicolls Medal, 79; Erskin Pamsay Medal, 80; Distinguished Serv Award, Nat Coal Asn, 81. *Mem:* Nat Acad Eng. *Mailing Add:* 11730 Turtle Beach Rd North Palm Beach FL 33408

POUPARD, JAMES ARTHUR, b Philadelphia, Pa, Jan 23, 43; m 62; c 3. INFECTIOUS DISEASES, CLINICAL MICROBIOLOGY. *Educ:* Temple Univ, BA, 72; Thomas Jefferson Univ, MS, 74; Univ Pa, PhD(interdisciplinary), 82. *Prof Exp:* Virol technologist, Smith Kline & French Labs, 62-67; toxicologist, Nat Drug Co, 67-68; microbiol supvr, Univ Pa, 68-74; dir, Bryn Mawr Asn, 74-86; ASST PROF PATH, MED COL PA, 86-; DIR, SMITHKLINE BEECHAM LABS, 90- *Concurrent Pos:* Consult, Educ Commun Inc, 74-90, Beecham Labs, 84-90 & Searle Labs, 88-90; dir, Ctr Hist Microbiol, 89- *Mem:* Am Soc Microbiol; Am Acad Microbiol; AAAS; Hist Sci Soc; Nat Asn Sci Teachers. *Res:* Automated systems for antimicrobial susceptibility testing; infections in comprised hosts; resistance mechanisms of microrganisms; history of biological warfare; history of microbiology; normal human flora. *Mailing Add:* 3612 Earlham St Philadelphia PA 19129

POURCHO, ROBERTA GRACE, b Toledo, Ohio, Mar 6, 34; m 55; c 3. ANATOMY, CELL BIOLOGY. *Educ:* Univ Mich, BS, 54; Wayne State Univ, PhD(anat), 72. *Prof Exp:* From asst prof to assoc prof, 73-88, PROF ANAT, SCH MED, WAYNE STATE UNIV, 88- *Mem:* AAAS; Asn Res Vision & Ophthal; Soc Neurosci; Am Asn Anatomists. *Res:* Identification of neuronal subpopulations in the mammalian retina which employ specific neurotransmitters and their synaptic relationships; electron cytochemistry. *Mailing Add:* Dept Anat Sch Med Wayne State Univ 540 E Canfield Ave Detroit MI 48201

POUR-EL, AKIVA, b Tel Aviv, Israel, Sept 26, 25; US citizen; m 55; c 1. RENEWABLE RESOURCE UTILIZATION. *Educ:* Phila Col Pharm, BS, 52; Univ Calif, Berkeley, PhD(biochem), 60. *Prof Exp:* Analytical chemist, US Vitamin Corp, 55-56; anal develop chemist, Calif Spray Chem Corp, 56-57; NIH fel, Pa State Univ, 60-61, fel biophys, 61-62; biochemist, Vet Admin Hosp, Phila, 63-64; biochemist, T L Daniels Res Ctr, Minn, 64-69, mgr biochem sect, Res Dept, Archer Daniels Midland Co, Decatur, Ill, 69-75, consult, 75-77; dir, Peaco, 78-80; consult, 80-90; RETIRED. *Concurrent Pos:* Vis prof, Fed Univ Rio De Janeiro, 78; consult, 81-; vchmn, Am Chem Soc, 80, chmn elect, 81, chmn, 82. *Mem:* Am Chem Soc (secy-treas, 77-78); Inst Food Technol; Am Asn Cereal Chemists; fel Am Inst Chem; NY Acad Sci; AAAS. *Res:* Physical chemistry of nucleic acids; rotatory dispersion of macromolecules; protein modifications by chemical and biological means; functional properties of foods and ingredients; analytical chemistry of natural products; financial management. *Mailing Add:* 1389 Keston St St Paul MN 55108

POUR-EL, MARIAN BOYKAN, b New York, NY; m 55; c 1. MATHEMATICS. *Educ:* Hunter Col, AB, 49; Harvard Univ, AM, 51, PhD(math), 58. *Prof Exp:* From asst prof to assoc prof math, Pa State Univ, 58-64; assoc prof, 64-68, PROF MATH, UNIV MINN, MINNEAPOLIS, 68- *Concurrent Pos:* Mem, Inst Advan Study, 62-64; Nat Acad Sci grant, Int Cong Mathematicians, Moscow, USSR, 66; vis prof, Univ Bristol, England, 69-70; invited lectr, Int Congresses, Logic & Comput Sci, Hungary, 67, Eng, 71, Czech, 73, Ger, 83, Japan, 85, China, 87. *Mem:* Am Math Soc; Math Asn Am; AAAS; Asn Symbolic Logic. *Res:* Computable and noncomputable aspects of mathematics and physical theory; ideas from mathematical logic (recursion theory), mathematical analysis (both classical and functional analyses) and physical theory are used in the work. *Mailing Add:* Sch Math Univ Minn Minneapolis MN 55455

POURING, ANDREW A, b New York, NY, Feb 12, 32; m 57; c 4. PHYSICS, GAS DYNAMICS. *Educ:* Rensselaer Polytech Inst, BME, 54, MME, 59; Yale Univ, PhD, 63. *Prof Exp:* Instr mech eng, Rensselaer Polytech Inst, 56-59; lectr, Yale Univ, 62-63, res fel, 63-64; assoc prof, US Naval Acad, 64-70, prof aerospace eng, 70-; CHMN, SONEY RES INC. *Concurrent Pos:* Consult, Trident Eng Assocs. *Mem:* Am Phys Soc; Am Soc Mech Engrs; Sigma Xi. *Res:* Steady and non-steady fluid dynamics phenomena, especially thermal choking; nucleation and phase change; relaxation effects; multi-phase flows; simulation of gas kinetic systems. *Mailing Add:* Soney Res Inc 23 Hudson St Annapolis MD 21401

POUSSAINT, ALVIN FRANCIS, b East Harlem, NY, May 15, 34; div; c 1. PSYCHIATRY. *Educ:* Columbia Col, BA, 56; Cornell Univ, MD, 60; Univ Calif, Los Angeles, MS, 64; Am Bd Psychiat & Neurol, cert, 70. *Prof Exp:* Intern, Ctr for Health Sci, 60-61, resident, 61-64, chief resident psychiat, Neuropsychiat Inst, Univ Calif, Los Angeles, 64-65; sr clin instr psychiat, Tufts Univ, 65-66, asst prof prev med & psychiat, 67-69; dean students, 75-78, ASSOC DEAN STUDENTS & ASSOC PROF PSYCHIAT, HARVARD MED SCH, 69- *Concurrent Pos:* Southern field dir, Med Comt for Human Rights, Miss, 65-66; dir psychiat, Columbia Point Health Ctr, Boston, 68-69; mem bd trustees, Nat Asn Afro-Am Artists, 68-, Wesleyan Col, 68-69 & Oper People United To Serve Humanity, 71-85; assoc psychiatrist, Mass Mental Health Ctr, 69-78; fac sponsor, Student Health Orgn, 69 & Student Health League, Boston, 69-73; consult & adv, Dept HEW, 69-73; chmn of bd, Solomon Fuller Inst, 72-81; deleg, Official Med Deleg to People's Repub China, 75; health consult, Cong Black Caucus, 76-; affil mem, Am Acad Child Psychiat. *Mem:* Nat Med Asn; fel Am Psychiat Asn; Am Acad Child Psychiat; fel Am Orthopsychiat Asn; fel AAAS. *Mailing Add:* Judge Baker Guid Ctr 295 Longwood Ave Boston MA 02115

POUSSART, DENIS, b France, Oct 3, 40; Can citizen; m 64; c 3. BIOENGINEERING & BIOMEDICAL ENGINEERING, ELECTRONICS ENGINEERING. *Educ:* Laval Univ, BSc, 63; Mass Inst Technol, MSc, 65, PhD(elec eng), 68. *Prof Exp:* Chmn dept, 88-91, PROF ELEC ENG, LAVAL UNIV, 68-, COORDR INTELLIGENT SENSING, INST ROBOTICS & INTELLIGENT SYSTS, 90- *Mem:* Inst Elec & Electronic Engrs. *Res:* Bioelectric signals; membrane biophysics; processing in nervous system; biomedical instrumentation; computer vision; imaging; 3-dimensional shape analysis; hardware systems; biophysics. *Mailing Add:* Dept Elec Eng Laval Univ Quebec City PQ G1K 7P4 Can

POUST, ROLLAND IRVIN, b Warren, Pa, June 19, 43; m 66; c 2. PHARMACEUTICS. *Educ:* Univ Pittsburgh, BS, 66, MS, 68; Purdue Univ, PhD(pharm), 71. *Prof Exp:* From asst prof to assoc prof pharmaceut, Sch Pharm, Univ Pittsburgh, 71-79; SECT HEAD, PHARMACEUT RES & DEVELOP LABS, BURROUGHS WELLCOME CO, 79- *Mem:* Acad Pharmaceut Sci; NY Acad Sci; Am Pharmaceut Asn; Controlled Release Soc; Sigma Xi; Am Asn Pharmaceut Scientists. *Res:* Transdermal delivery systems; preformulation research; drug stability. *Mailing Add:* 1200 Kingsbrook Rd Greenville NC 27858

POUTSIAKA, JOHN WILLIAM, b Newark, NJ, Feb 6, 25; m 47; c 2. PHYSIOLOGY. *Educ:* Seton Hall Univ, BS, 48; Fordham Univ, MS, 49, PhD(physiol), 54. *Prof Exp:* Lab instr biol, Fordham Univ, 48-50; sr res scientist, 50-60, head dept toxicol, 60-68, dir toxicol dept, 68-72, assoc dir res, 72-76, V PRES DRUG REGULATION & CONTROL, SQUIBB INST MED RES, 76- *Mem:* Am Soc Pharmacol & Exp Therapeut; Soc Toxicol; NY Acad Sci. *Res:* Pharmacology and toxicology of new drugs; renal physiology; body fluids and electrolytes. *Mailing Add:* 56 Woodcrest Dr Morristown NJ 07961

POUTSMA, MARVIN LLOYD, b Grand Rapids, Mich, Aug 7, 37; m 68; c 2. CHEMISTRY, CATALYSIS. *Educ:* Calvin Col, AB, 58; Univ Ill, Urbana, PhD(org chem), 62. *Prof Exp:* Staff scientist, Corp Res Lab, Union Carbide Corp, 61-64, group leader, 64-68, sr res scientist, 68-73, sr group leader, 73-78; group leader, 78-80, sect head, 80-84, DIR, CHEM DIV, OAK RIDGE NAT LAB, 84- *Mem:* Am Chem Soc; AAAS; NY Acad Sci. *Res:* Coal chemistry; free-radical chemistry; catalysis; research administration. *Mailing Add:* Oak Ridge Nat Lab PO Box 2008 Oak Ridge TN 37831-6182

POVAR, MORRIS LEON, b New York, NY, Feb 19, 20; m 45; c 2. VETERINARY MEDICINE, PRIMATOLOGY. *Educ:* State Univ NY Col Agr, BS, 41; Cornell Univ, DVM, 44; Brown Univ, MA, 69. *Prof Exp:* Assoc pathologist, Rutgers Univ, 44-45; pathologist, Kimber Farms, 45-46; veterinarian, 46-52, 54-67; assoc prof, Brown Univ, 68-73, dir animal care facility & dir animal health, 68-85, prof psychol & med sci, 73-85,; RETIRED. *Concurrent Pos:* Consult ed, Lab Primate Newsletter, Brown Univ, 62-; lectr, Sch Dent Med, Harvard Univ, 71-85; consult, Northeastern Univ, 71-82 & Health Ctr, Univ Conn, 72-85. *Mem:* AAAS; Am Soc Lab Animal Practitioners (vpres, 74-75, pres, 75-76); Am Vet Med Asn; Am Asn Lab Animal Sci. *Res:* Nonhuman primate husbandry and behavior; dental implants in experimental animals. *Mailing Add:* 21365 Cypress Hammock Dr Boca Raton FL 33428-1980

POVINELLI, LOUIS A, b New York, NY, June 10, 31; m 58; c 7. MECHANICAL & AERONAUTICAL ENGINEERING. *Educ:* Univ Detroit, BME, 54; Univ Ky, MSME, 56; Northwestern Univ, PhD(mech eng), 59. *Prof Exp:* Rocket design engr, Bell Aircraft Corp, 51-56; asst, Northwestern Univ, 56-58; abstractor, Comt Fire Res, Nat Acad Sci-Nat Res Coun, 57-76; aeronaut res scientist, NASA, 60-81, sect head turbine aerodynamics, 81-84, chief, Computational Appln Br, Lewis Res Ctr, 84-85, DEP CHIEF, INTERNAL FLUID MECH DIV, NASA, 85- & PROG DIR, INST COMPUTATIONAL MECH PROPULSION, 85- *Concurrent Pos:* Instr, Univ Ky, 55-56 & Siebel Inst Technol, 56-57; res engr, Armour Res Found, Ill Inst Technol, 57-59; Fiat fel & Fulbright grant, Turin Polytech Inst, 59-60; consult, Off Sci & Technol, Exec Off of the President, 63 & Dept of Defense Working Group, 65-72; lectr, Lewis Advan Study Prog, NASA, 66-72; consult NASA study team, 72. *Honors & Awards:* Except Sci

Achievement Medal, NASA, 83; Gene Zara Award, 89; Collier Trophey, 88. *Mem:* Am Inst Aeronaut & Astronaut; Am Soc Mech Engrs. *Res:* Turbine aerodynamics; computational fluid mechanics; jet propulsion; gas dynamics; turbulence; solid propellants; supersonic combustion. *Mailing Add:* NASA Lewis Res Ctr 21000 Brookpark Rd MS5-7 Cleveland OH 44135

POVZHITKOV, MOYSEY MICHAEL, b Kamenka, Ukraine, USSR, Mar 16, 28; m 56; c 3. CARDIOVASCULAR PHYSIOLOGY & PHARMACOLOGY. *Educ:* Kiev Med Inst, USSR, MD, 49; Inst Physiol, Kiev, USSR, PhD(physiol), 63, DSc(med), 70. *Hon Degrees:* FACC, Am Col Cardiol, 83. *Prof Exp:* Sr sci researcher cardiol, Inst Physiol, USSR, 63-75; dir, dept cardiol, Cardiol Clin, USSR, 75-76; dir, dept cardiol, Inst Physiol & Path Effects of High Altitude, USSR, 76-78; res assoc, Cedars-Sinai Med Ctr, Los Angeles, 80-81; PRIN SCIENTIST PHARMACOL, SYNTEX RES, SYNTEX CORP, 81- *Mem:* Am Col Cardiol. *Res:* Medicine: cardiology; cardiovascular research; regulation of cardiac output; physiology and pathophysiology of coronary circulation; cardiovascular pharmacology. *Mailing Add:* 1190 N Clark Ave Mountain View CA 94040

POWANDA, MICHAEL CHRISTOPHER, b Jersey City, NJ, Jan 15, 42. INFECTIOUS DISEASES, TRAUMA. *Educ:* St Peter's Col, BS, 63; Univ Miami, PhD(biochem), 68. *Prof Exp:* Biochemist, US Army Med Res Inst Infectious Dis, 68-77; chief, biochem br, US Army Inst Surg Res, 78-82; chief Cutaneous Hazards Div, Letterman Army Inst Res, 82-88; COFOUNDER & PRIN, CONCEPT DEVELOP & COMMERCIALIZATION CORP, 88- *Concurrent Pos:* Lectr biochem, Hood Col, Frederick, Md, 75-77; sponsored vis scientist, Nat Acad Sci, Poland, 78, Czech, 81. *Mem:* Am Soc Biotechnol; Am Soc Cell Biol; Am Soc Microbiol; Am Physiol Soc; Am Fedn Clin Res; Am Inst Nutrit. *Res:* Alterations in metabolism and physiology during injury, inflammation and infection; mode of induction and regulation and role of cytokines in host defense against infection and repair of injury; metabolism and physiology during injury, infection and inflammation; biomedical product development; project management. *Mailing Add:* Concept Develop & Commercialization Corp 435 Marin Ave Mill Valley CA 94941

POWDERS, VERNON NEIL, b Woodward, Okla, June 14, 41; m 67; c 2. PARASITOLOGY. *Educ:* Northwestern State Col, BS, 63; Okla State Univ, MS, 64; Univ Tenn, Knoxville, PhD(parasitol), 67. *Prof Exp:* Asst prof biol, Ga Southwestern Col, 67-73, assoc prof, 73-80; mem fac, dept entomol, Univ Tenn, 80-; asst prof biol, 80-86, ASSOC PROF BIOL, NORTHWESTERN OKLA STATE UNIV, 86- *Mem:* Am Soc Parasitol; Am Soc Ichthyol & Herpet; Soc Study Amphibia & Reptiles. *Res:* Systematics of Orthoptera and parasitism in salamanders; reptilian and avian parasites; biology of amphibians. *Mailing Add:* Dept Biol Northwestern Okla State Univ Alva OK 73717

POWE, RALPH ELWARD, b Tylertown, Miss, July 27, 44; m 62; c 3. MECHANICAL ENGINEERING, APPLIED PHYSICS. *Educ:* Miss State Univ, BS, 67, MS, 68; Mont State Univ, PhD(mech eng), 70. *Prof Exp:* Trainee mech eng, NASA, 62-65; res asst, Miss State Univ, 65-68, lab instr physics, 66-67, instr mech eng, 68; res asst, Mont State Univ, 68-70, teaching asst civil eng, 69-70, asst prof mech eng, 70-74; assoc prof, 74-78, assoc dean eng, 79-80, assoc vpres res, 80-86, PROF MECH ENG, MISS STATE UNIV, 78-, VPRES RES, 86 - *Concurrent Pos:* AEC contract, Mont State Univ, 71-72, NSF grant, 72-74; US Dept Energy contract, 76-79; NSF Exp Prog Stimulated Competitive Res Grant, 89- *Honors & Awards:* Ralph R Teetor Award, Soc Automotive Engrs, 71. *Mem:* AAAS; fel Am Soc Mech Engrs; Soc Automotive Engrs; Nat Soc Prof Engrs; Am Soc Eng Educ; Nat Council Univ Res Admin. *Res:* Heat transfer; natural convection; fluid mechanics; magnetohydrodynamics power generation. *Mailing Add:* PO Box 6343 Miss State Univ Mississippi State MS 39762

POWELL, ALAN, b Buxton, Eng, Feb 17, 28; US citizen; m 56. ENGINEERING, PHYSICS. *Educ:* Loughborough Col, DLC, 48 & Hons, 49; Univ London, BSc Hons, 49; Univ Southampton, PhD(eng), 53. *Hon Degrees:* DTech, Loughborough Univ Technol, 80. *Prof Exp:* Tech asst aeronaut eng, Percival Aircraft Co, Eng, 49-51; asst aeronaut, Univ Southampton, 51-53, lectr, 53-56; res fel aeronaut, Calif Inst Technol, 56-57; from assoc prof to prof eng, Univ Calif, Los Angeles, 57-64, head aerosonics lab, 57-64; assoc tech dir & head acoust & vibration lab, David W Taylor Model Basin, 64-65, tech dir, 66-67; tech dir, David W Taylor Naval Ship Res & Develop Ctr, 67-85; PROF MECH ENG, UNIV HOUSTON, 86- *Concurrent Pos:* Consult, Gen Elec Co, 56, Douglas Aircraft Co, 56-64, Ramo-Wooldridge Corp, 57, Indust Acoust, Inc, 58 & Rocketdyne Div, NAm Aircraft Co, 60; mem noise suppression comt, Brit Ministry of Supply, 52-56, mem exec comt, 54-56; mem, Aeronaut Res Coun, London, 53-56; mem comt hearing, bioacoust & biomech, Nat Acad Sci-Nat Res Coun, 62-, mem exec coun, 63-65, chmn exec coun, 65-66, mem comt undersea warfare panel on submarine noise measurement, 65-; mem comt shock & vibration, Am Standards Asn; initial mem & dir, Inst Noise Control Engrs, 74-77; mem US-Japan Coop Prog, Natural Resources, 80-86, adv, 86-; adv coun, Int Towing Tank Conf, 81-85, sci comt, Int Union Theoret & Appl Mech, 83-85; chmn, Int Conf Comput Aided Design, Manufacture & Oper in Marine & Offshore Indust, 87-; mem, Naval Studies Bd, Nat Acad Sci-Nat Res Coun, 89- *Honors & Awards:* Baden-Powell Prize, 48 & Wright Prize, 53, Royal Aeronaut Soc; Biennial Award, Acoust Soc Am, 62; Aeroacoust Award, Am Inst Aeronaut & Astronaut, 80; Capt Robert Dexter Conrad Gold Medal for Sci Achievement, Secy Navy, 84; Rayleigh Lectr, Am Soc Mech Engrs, 88. *Mem:* Sr mem Inst Elec & Electronics Engrs; fel Royal Aeronaut Soc; fel Brit Inst Mech Engrs; assoc fel Am Inst Aeronaut & Astronaut; fel Acoust Soc Am (pres-elect, 78-90, vpres-elect, 80-81, vpres, 80-82, pres, 90-91); Am Soc Naval Engrs; Inst Noise Control Eng (vpres, 82-85); Am Soc Mech Engrs. *Res:* Basic and applied research in aero-acoustics; noise control in aircraft and ships; engineering design, especially aircraft; research direction; engineering and science education. *Mailing Add:* Dept Mech Eng Univ Houston Houston TX 77204-4792

POWELL, ALBERT E, b Larimer, Pa, Nov 8, 19; m 43; c 4. AGRICULTURAL ENGINEERING. *Educ:* Pa State Univ, BSc, 41; Purdue Univ, MSc, 43; Iowa State Univ, PhD(agr & mech eng), 60. *Prof Exp:* Prof consult engr, 48-49; agr engr, Midwest Plan Serv, 49-50; agr engr, Douglas Fir Plywood Asn, 50-56, regional mgr, 56-57, agr engr, 57-58, sr agr engr, 58-59; from asst prof to prof, 60-86, EMER PROF AGR ENG, WASH STATE UNIV, 86- *Concurrent Pos:* Consult agr & archit engr, 80-, Powell Eng Assocs, 86- *Mem:* Am Soc Agr Engrs; Nat Soc Prof Engrs; Am Acad Environ Engrs. *Res:* Connections between wood members; adhesives; structural analysis; grain storage; solar energy applications. *Mailing Add:* Dept of Agr Eng Wash State Univ Pullman WA 99164-6120

POWELL, ARNET L, b Worcester, Mass, Feb 4, 15; m 48; c 2. PHYSICAL CHEMISTRY. *Educ:* Worcester Polytech Inst, BS, 38, MS, 40; Clark Univ, PhD(chem), 56. *Prof Exp:* Asst physics, Worcester Polytech Inst, 38-40; phys chemist, Dow Chem Co, 40-44; sr res engr, Sylvania Elec Co, 44-47; chemist, US Off Naval Res, Boston Br, 47-54, asst chief scientist, 54-59, dep dir sci, 67-77, chief scientist, 59-79, dir sci, 77-79; CONSULT, 79- *Concurrent Pos:* Guest researcher, Mass Inst Technol, 56- & Oxford Univ, 64. *Mem:* AAAS; Am Chem Soc; NY Acad Sci. *Res:* Determination of halogens in organic compounds; indicators; electroanalytical methods; corrosion; Friedel-Crafts reactions; fluorescence; glow discharge; electrical conductivity in nonaqueous solvents; mechanism of Cannizzaro reaction; halogenation kinetics; correlation of solvent effects. *Mailing Add:* 65 Woodridge Rd Wayland MA 01778

POWELL, BENJAMIN NEFF, b Montclair, NJ, Oct 28, 41; m 64, 89; c 1. PETROLOGY, ELECTRON MICROSCOPY. *Educ:* Amherst Col, BA, 64; Columbia Univ, MA, 66, PhD(geol), 69. *Prof Exp:* Res assoc geol, Smithsonian Astrophys Observ & Harvard Col Observ, 69-70; asst prof, Rice Univ, 70-76, lectr geol, 76-80; sr res geologist res & develop, 80-84, SR RES GEOLOGIST EXPLOR & PROD, PHILLIPS PETROL CO, 84- *Concurrent Pos:* Prin investr, NASA, 71-76 & NSF, 73 & 76-79. *Honors & Awards:* Nininger Meteorite Prize, Ariz State Univ, 68. *Mem:* Am Asn Petrol Geologists; Geol Soc Am; Mineral Soc Am; Sigma Xi. *Res:* Mineralogy and petrology of lunar samples and meteorites; igneous petrology; geochemistry; terrestrial igneous petrology; ore deposits; sandstone diagenesis; chalk compaction. *Mailing Add:* 2390 Mountain Dr Bartlesville OK 74003

POWELL, BERNARD LAWRENCE, b Kansas City, Kans, June 7, 45; m 67; c 3. STEROID HORMONE ENDOCRINOLOGY. *Educ:* Rockhurst Col, BA, 67; Univ Okla, PhD(biochem), 75. *Prof Exp:* NIH trainee, Res Inst Hosp Joint Dis, Mt Sinai Sch Med, 75-77; res assoc, Dept Med & Oncol, Univ Tex Health Sci Ctr, 77-81; SR RES ASSOC, DEPT MED & INFECTIOUS DIS, AUDIE MURPHY VET ADMIN HOSP & UNIV TEX HEALTH SCI CTR, 81- *Mem:* AAAS. *Res:* Steroid and polypeptide hormone receptor in cancer and fungal diseases, in particular the isolation and physicochemical characterization of such receptors; radioisotope technology. *Mailing Add:* Dept Phys Sci Univ Tex San Antonio 6700 N Fm 1604 W San Antonio TX 78285

POWELL, BOBBY EARL, b Moultrie, Ga, Jan 15, 41; m 65; c 2. SOLID STATE PHYSICS, ASTRONOMY. *Educ:* Ga Inst Technol, BS, 63; Clemson Univ, MS, 65, PhD(physics), 67. *Prof Exp:* From asst prof to assoc prof, 67-75, PROF PHYSICS, WEST GA COL, 75-, DIR, COL OBSERV, 79- *Mem:* Am Phys Soc; Am Asn Physics Teachers; Sigma Xi. *Res:* Measurement of elastic constants of crystals from finite deformations; photometry of eclipsing binary stars; x-ray diffraction studies. *Mailing Add:* Dept Physics West Ga Col Carrollton GA 30118

POWELL, BRIAN M, b Wigan, Eng, Sept 16, 38; Can citizen. DYNAMICS OF MOLECULAR SOLIDS. *Educ:* Univ London, BSc, 60, PhD(physics), 64. *Prof Exp:* RES OFFICER, CHALK RIVER NUCLEAR LABS, ATOMIC ENERGY CAN, LTD, 64- *Mem:* Can Asn Physicists; Am Phys Soc. *Res:* Properties of condensed matter using the techniques of elastic and inelastic scattering of thermal neutrons; neutron powder diffraction. *Mailing Add:* Neutron & Solid State Phys Branch Atomic Energy Can Ltd Chalk River ON K0J 1J0 Can

POWELL, BRUCE ALLAN, b Waterbury, Conn, Aug 12, 41; m 64; c 3. OPERATIONS RESEARCH. *Educ:* Denison Univ, BS, 63; Case Western Reserve Univ, MS, 65, PhD(opers res), 67. *Prof Exp:* Sr mathematician, Westinghouse Res & Develop Ctr, 67-73, fel mathematician, 73-77, adv mathematician, 77-89; PRIN RES ENGR, OTIS ELEVATOR CO, 89- *Mem:* Opers Res Soc Am; Soc Comput Simulation. *Res:* Inventory systems; queuing theory; practical application of mathematical models; optimization; development of elevator control strategies; computer simulation of manufacturing plants. *Mailing Add:* Otis Elevator Co Five Farm Springs Farmington CT 06032

POWELL, BURWELL FREDERICK, b Washington, DC, June 9, 33; m 54; c 2. ORGANIC CHEMISTRY. *Educ:* Univ Md, BS, 56; Polytech Inst Brooklyn, MS, 66, PhD(chem), 70. *Prof Exp:* CHEMIST, MERCK SHARP & DOHME RES LABS, MERCK & CO, 56- *Mem:* Am Chem Soc; The Chem Soc; AAAS; Sigma Xi. *Res:* Synthetic organic chemistry. *Mailing Add:* 251 N 7th Ave Manville NJ 08835

POWELL, CEDRIC JOHN, b Australia, Feb 19, 35; US citizen; m 58; c 3. SURFACE SCIENCE. *Educ:* Univ Western Australia, BSc Hons, 56, PhD(physics), 62. *Prof Exp:* Res physicist, Imp Col, London, 60-62; physicist, 62-78, CHIEF SURFACE SCI DIV, NAT BUR STANDARDS, 78- *Concurrent Pos:* Gen chmn, Phys Electronics Conf, 80-84; chmn, Comt E-42 Surface Anal, Am Soc Testing & Mat, 80-85, Surface Sci Div, Am Vacuum Soc, 83-84 & Appl Surface Sci Div, 85-86, Surface Chem Anal Working Party, Versailles Proj Advan Mat & Standards, 84-87; mem bd trustees, Gordon Res Conf, 82-88 & chmn bd trustees, 85-86. *Honors & Awards:* Silver Medal, US Dept Com, 83, Gold Medal, 86. *Mem:* AAAS; fel Am Phys Soc; Am Chem Soc; fel Brit Inst Physics; Am Soc Testing & Mat; Am Vacuum Soc. *Res:* Electron spectroscopy of solids, surfaces and atoms; surface science; surface analysis. *Mailing Add:* Nat Inst Standards & Technol Nat Bur of Standards Gaithersburg MD 20899

POWELL, CHARLES CARLETON, JR, b Massillon, Ohio, May 11, 42; m 64; c 2. PLANT PATHOLOGY. *Educ:* Ohio State Univ, BSc, 64; Univ Calif, Berkeley, PhD(plant path), 69. *Prof Exp:* Res scientist, Agr Res Serv, US Dept Agr, 69-70; from asst prof to assoc prof, 70-82, PROF PLANT PATH, OHIO STATE UNIV, 83- *Mem:* Am Phytopath Soc. *Res:* Diseases of ornamental plants; flower crops; woody ornamentals; landscape plantings; research and extension appointment. *Mailing Add:* Dept Plant Path Ohio State Univ Main Campus 201 Kottman Hall Columbus OH 43210

POWELL, CLINTON COBB, b Hartford, Conn, Mar 9, 18; m 44; c 3. RADIOLOGY. *Educ:* Mass Inst Technol, SB, 40; Boston Univ, MD, 44; Am Bd Radiol, dipl, 54. *Prof Exp:* Intern, US Marine Hosp, Boston, 44-45; asst surgeon, NIH, 46-47, from sr asst surgeon to sr surgeon, 47-56, med dir, 56-64, exec secy radiation & surg study sect, 58-59, asst chief res grants rev br, 59-60, dep dir div res grants, 60-61, assoc dir extramural progs, Nat Inst Allergy & Infectious Dis, 61-62, dir, Nat Inst Gen Med Sci, 62-64; assoc coord & actg coordr, Med & Health Sci, 64-66, coordr, 66-70, spec asst to pres health affairs, Systwide Admin, 70-79, EMER SPEC ASST TO PRES HEALTH AFFAIRS, UNIV CALIF, 79- *Concurrent Pos:* Instr, Univ Pa, 52-54; chief radiol health med prog, USPHS, 56-58; mem comn Radiol units, Standards & Protection, Am Col Radiol, 57-, chmn comt radiation exposure of women, 65-72. *Mem:* Radiation Res Soc; Am Col Radiol. *Res:* Radiotherapy; radiation protection; medical administration. *Mailing Add:* 542 Tahos Rd Orinda CA 94563

POWELL, DAVID LEE, b Bucyrus, Ohio, July 27, 36; m 59; c 2. PHYSICAL CHEMISTRY. *Educ:* Oberlin Col, AB, 58; Univ Wis, PhD(phys chem), 62. *Prof Exp:* Fel molecular spectros, Univ Minn, 62-64; from asst prof to assoc prof, 64-73, chmn dept, 77-81, BENJAMIN S BROWN PROF CHEM, COL WOOSTER, 73- *Concurrent Pos:* Res leave, Univ Oslo, 71-72, 76-77, 81-82 & 86-87. *Mem:* Am Chem Soc; Sigma Xi. *Res:* Vibrational spectroscopy; conformational behavior of organic compounds as studied by infrared and Raman spectroscopy. *Mailing Add:* Dept Chem Col Wooster Wooster OH 44691

POWELL, DON WATSON, b Gadsden, Ala, Aug 29, 38; m 60; c 3. GASTROENTEROLOGY. *Educ:* Auburn Univ, BS, 60; Med Col Ala, MD, 63; Am Bd Internal Med, dipl, 72, dipl gastroenterol, 79. *Prof Exp:* From intern to jr asst resident med, Peter Bent Brigham Hosp, Boston, 63-65; assoc investr gastroenterol, Walter Reed Army Inst Res, 65-68; asst resident & teaching asst internal med, Yale-New Haven Hosp, 68-69, USPHS spec fel, Physiol, Yale Univ, 69-71; asst prof, Sch Med, Univ NC, Chapel Hill, 71-74, assoc prof med, 74-78, prof med & chair Div Digestive Dis & Nutrit, 78-91, adj prof physiol, Sch Vet Med, 86-91, assoc chmn dept med, 89-91; RANDALL PROF INTERNAL MED, PROF PHYSIOL & BIOPHYSICS, CHAIR DEPT INTERNAL MED, UNIV TEX MED BR, GALVESTON, 91- *Concurrent Pos:* USPHS res grants, 71-81, Nat Inst Arthritis, Metab & Digestive Dis res career develop award, 73-78; NIH res career develop award, 73-78; NIH merit award, 87-92, gen med (gastroentrol & nutrit) study sect, 85-88, NIH; consult, diarrhocal dis control prog, WHO, 80-85; mem gen med study sect, NIH, 85-88; mem, MKSAP VI gastroenterol subcomt, Am Col Physicians; NIADDK Dig Dis Res Ctr, 85-92. *Mem:* AAAS; Am Fedn Clin Res; Am Gastroentrol Asn (vpres); Biophys Soc; Am Physiol Soc; Am Soc Clin Invest; Am Soc Internal Med. *Res:* Transport of water and electrolytes by gastrointestinal epithelia; intestinal absorption and secretion; pathophysiology of diarrheal diseases; neuro-immunophysiology of intestinal electrolyte transport. *Mailing Add:* Dept Internal Med Univ Tex Med Br 4-108 Old John Sealy Hosp R-E 67 Galveston TX 77550

POWELL, DONALD ASHMORE, b Spartanburg, SC, Oct 29, 38; m 60; c 4. NEUROBIOLOGY, BIOBEHAVIOR. *Educ:* Univ SC, BS, 60, MA, 62; Fla State Univ, PhD(psychol), 67. *Prof Exp:* USPHS res fel, dept psychol, Univ Miami, Fla, 67-69; DIR, NEUROSCI LAB, VET ADMIN MED CTR, COLUMBIA, SC, 69-; ASSOC PROF, DEPT NEUROPSYCHIAT & BEHAV SCI, SCH MED, UNIV SC, COLUMBIA, 79- *Concurrent Pos:* Adj asst prof, dept psychol, Univ Miami, Coral Gables, Fla, 67-69; from adj asst prof to prof, dept psychol Univ SC, 69-; coordr res & develop, Vet Admin Med Ctr, Columbia, SC, 76-81. *Mem:* Am Psychol Asn; Soc Neurosci; Psychonomic Soc; Soc Psychophysiol Res. *Res:* Central nervous system control of behavior; mediation of the emotional and behavioral components of learning by different neural substrates. *Mailing Add:* Neurosci Lab 151A Vet Admin Med Ctr Columbia SC 29201

POWELL, EDWARD GORDON, b Washington, DC, Apr 5, 46; m 77; c 4. EXPERT SYSTEMS, NEUROSCIENCES. *Educ:* Univ Md, BS, 68, MS, 72. *Prof Exp:* Res physicist, US Naval Ordnance Sta, 68-73, US Naval Ordnance Lab, 73-74, US Naval Surface Weapons Ctr, 74-81, proj mgr artificial intelligence, 81-83; lead scientist, Mitre Corp, 83-89; LEAD SCIENTIST, ACCOTECH, 89- *Concurrent Pos:* Guest worker, Nat Bur Standards, 76-79; adj prof, Va Polytech Inst, 86-88. *Mem:* Am Astron Soc; Am Asn Artificial Intel; Asn Comput Mach; Europ Asn Theoret Comput Sci; AAAS. *Res:* Biological neural network architectures. *Mailing Add:* 1126 Apple Valley Rd Accokeek MD 20607-9605

POWELL, ERVIN WILLIAM, b Niles, Ohio, Nov 8, 22; m 45; c 3. NEUROANATOMY. *Educ:* Youngstown Univ, AB, 48; Western Reserve Univ, MS, 50, PhD(zool), 53. *Prof Exp:* Lectr anat & physiol, Western Reserve Univ, 51-53; from instr to asst prof anat, Sch Med, Creighton Univ, 54-60; assoc prof, Med Ctr, Univ Miss, 60-66; prof, 66-87, EMER PROF ANAT, SCH MED, UNIV ARK, LITTLE ROCK, 87- *Mem:* AAAS; Am Asn Anatomists; Soc Neurosci; Soc Chronobiol; Sigma Xi. *Res:* Neurophysiology; limbic system; neuropathology. *Mailing Add:* 9905 Satterfield Dr Little Rock AR 72205

POWELL, FRANCIS X, b Hoxie, Ark, Sept 23, 29; m 53; c 1. CHEMISTRY. *Educ:* Univ Calif, Berkeley, BS, 56; Univ Md, PhD(chem), 62. *Prof Exp:* Res engr, Jet Propulsion Lab, Calif Inst Technol, 56-57; fel chem, Univ Md, 62; asst prof, 62-68, ASSOC PROF CHEM, CATH UNIV AM, 68- *Concurrent Pos:* Consult, Nat Bur Standards, 63-70, Am Instrument Co, 64-70, Naval Res Lab, 70- & Goddard Space Flight Ctr, 71- *Mem:* Am Phys Soc; Am Chem Soc. *Res:* Spectroscopy of free radicals; electrical discharges through gases. *Mailing Add:* 13205 Locksley Lane Silver Spring MD 20904

POWELL, FRANK LUDWIG, JR, b Pasadena, Calif, Aug 10, 52; m 85; c 2. RESPIRATORY PHYSIOLOGY, COMPARATIVE PHYSIOLOGY. *Educ:* Univ Calif, Irvine, BS, 74, Davis, MS, 78, PhD(syst physiol), 78. *Prof Exp:* Res asst, Univ Calif, Davis, 74-75, asst instr, 75-76,; DAAD fel, Max-Planck Inst Exp Med, WGer, 76-77; assoc, systs physiol, Univ Calif, Davis, 77-78; res fel, 78-80, asst prof med, 80-86, res scientist, White Mountain Res Sta, 85-89, ASSOC PROF MED, UNIV CALIF, SAN DIEGO, 86- *Concurrent Pos:* Accurate alcohol breath testing contract, Meridian Med Corp, 88; PI mechanisms of chemosensitivity, NIH grant, 90-95; sci adv bd, Quantum Group Inc, 90- *Mem:* Am Physiol Soc; Am Thoracic Soc; AAAS. *Res:* Control of breathing, especially carotid body physiology; alcohol breath testing, physiological (O2 and CO2) and inert gas measurement; high altitude acclimation and intrapulmonary carbon dioxide sensitivity; comparative respiratory physiology, especially structure- function relationships in avian and reptilian lungs. *Mailing Add:* Div Physiol 0623A Dept Med Univ Calif San Diego Sch Med La Jolla CA 92093-0623

POWELL, G(EORGE) M(ATTHEWS), III, b Montgomery, Ala, Mar 29, 10; m 37; c 2. CHEMICAL ENGINEERING. *Educ:* Columbia Univ, AB, 31, BS, 32, ChE, 33. *Prof Exp:* Researcher, Carbide & Carbon Chem Co, 33-44, tech head res div, 44-55, asst dir res, Union Carbide Chem Co, 55-60, res supt plastics, 60-61, supt polymer chem div, 61-64, assoc dir res & develop dept, Chem Div, Union Carbide Corp, 64-69, res consult coatings mat, 69; consult coatings technol, 69-77; RETIRED. *Honors & Awards:* Hyatt Award, 50. *Mem:* Am Chem Soc; Soc Rheology; Am Inst Chem Engrs. *Res:* Vinyl resin technology; surface coatings; organosols; plastisols; water-soluble polymers, paper chemicals and resins; solvents, intermediates and polymer latexes for industrial and trade paints. *Mailing Add:* 2114 Glenfield Terr South Charleston WV 25303-3007

POWELL, GARY LEE, b Fullerton, Calif, Jan 24, 41; m 65; c 2. BIOCHEMISTRY. *Educ:* Univ Calif, Los Angeles, BS, 62; Purdue Univ, PhD(chem), 67. *Prof Exp:* From asst prof to assoc prof, 67-79, PROF BIOL SCI, CLEMSON UNIV, 79- *Concurrent Pos:* NIH trainee biol chem, Sch Med, Wash Univ, 67-69; vis prof, Dept Chem, Univ, Ore, 75-76; guest prof, Max-Planck Soc, WGermany, 83-84; sr Fulbright fel, 83-84; basic sci coun, Am Heart Asn, chmn, Res Comt, SC Affil, Am Heart Asn; sr prof/res scholar, Fulbright Comn, 83-84; guest prof, Max-Planck Inst Biophys Chem, Max-Planck Soc, 86; coop exchange scientist, NSF USA/Fed Repub Ger, 86-87. *Mem:* Am Chem Soc; Am Soc Biol Chemists; Biophys Soc; Am Heart Asn. *Res:* Control of lipid metabolism; lipid-protein interactions in biological membranes; cardiolipin structure and function; unsaturated fatty acid biosynthesis in plants. *Mailing Add:* Dept of Biol Sci Clemson Univ Clemson SC 29634-1903

POWELL, GEORGE LOUIS, b Wilmington, NC, Oct 17, 40; m 64; c 1. PHYSICAL CHEMISTRY. *Educ:* Presby Col, BS, 63; Univ NC, Chapel Hill, PhD(phys chem), 67. *Prof Exp:* Chemist, 67-78, SECT LEADER PHYS CHEM, OAK RIDGE Y-12 PLANT, NUCLEAR DIV, UNION CARBIDE CORP, 69- *Mem:* Sigma Xi; Am Chem Soc. *Res:* Gas phase kinetics, including gas-solid interactions and energy transfer reactions; metal-hydrogen phase relationships; hydrogen embrittlement; trace analysis of gases in solids; mass- and optical-kinetic spectroscopy; surface chemistry of lithium compounds. *Mailing Add:* 298 East Dr Oak Ridge TN 37830

POWELL, GEORGE WYTHE, b Aiken, SC, Oct 26, 36; m 58; c 1. ANIMAL NUTRITION. *Educ:* Clemson Univ, BS, 58; Univ Ga, MS, 64, PhD(animal nutrit), 67. *Prof Exp:* Res assoc dairy sci, Univ Fla, 67-68; PROF BIOL, ABRAHAM BALDWIN AGR COL, 68-, CHMN SCI & MATH DIV, 74- *Mem:* Sigma Xi. *Res:* Trace mineral metabolism in ruminants. *Mailing Add:* Abraham Baldwin Col Tifton GA 31794

POWELL, GORDON W, b Providence, RI, Mar, 10, 28; m 55, 76; c 4. FAILURE ANALYSIS. *Educ:* Mass Inst Technol, BS, 51, MS, 52, ScD(phys metall), 55. *Prof Exp:* Res metallurgist, Nuclear Metals, Inc, 55-57; asst prof phys metall, Univ Wis, 57-58; PROF METALL ENG, OHIO STATE UNIV, 58- *Concurrent Pos:* Consult, Monsanto Res Corp, 66-; consult to var law & ins firms. *Mem:* Am Powder Metall Inst; Am Soc Metals; Nat Asn Corrosion Engrs; Int Metall Soc. *Res:* Powder Metallurgy, specifically with high-temperature aluminum alloys. *Mailing Add:* Mat Sci & Eng Dept 143 Fontana Labs 116 W 19th Ave Columbus OH 43210

POWELL, GRAHAM REGINALD, b Hampton, Eng, Mar 19, 35; Can citizen; m 60; c 2. TREE DEVELOPMENT & REPRODUCTION. *Educ:* Univ Edinburgh, BSc, 56, Hons, 57, PhD(tree develop), 71; Univ NB, Fredericton, MSc, 61. *Prof Exp:* Lectr, 61-64, from asst prof to assoc prof, 64-77, PROF FORESTRY, UNIV NB, FREDERICTON, 77- *Mem:* Can Inst Forestry; Can Bot Asn; Can Tree Improv Asn. *Res:* Tree crown development, morphogenesis of buds, shoots, cones; interrelations between reproductive and vegetative forms of growth in tree species; seed production; seed germination; seedling development; litter fall in forest stands. *Mailing Add:* Dept Forest Resources Bag #44555 Fredericton NB E3B 6C2 Can

POWELL, HAROLD, b Jacksonboro, SC, Dec 17, 32. SPEECH PATHOLOGY, AUDIOLOGY. *Educ:* SC State Col, BA, 61; Pa State Univ, MS, 63, PhD(speech path), 66. *Prof Exp:* Assoc prof speech correction, 65-69, PROF SPEECH PATH, SC STATE COL, 69-, KIRKLAND W GREEN PROF, 68- *Concurrent Pos:* Mem, SC State Bd Examrs in Speech Path & Audiol, 73-78, chmn, 77-78; mem adv comt, Deaf-Blind in SC, 74-79; mem, Am Bd Examrs in Speech Path & Audiol, 75-78, secy, 76-78; mem adv comt, Off Progs for Handicapped, SC Dept Educ, 75-78; mem, SC Gov's Comt Employment of Handicapped, 77-85; mem bd dirs, Nat Black Asn Speech, Lang & Hearing, 78-85. *Mem:* Fel Am Speech, Lang & Hearing Asn. *Res:* Communicative problems of the aging in rural areas. *Mailing Add:* Box 1596 SC State Col Orangeburg SC 29117

POWELL, HARRY DOUGLAS, b Wallace, NC, Oct 14, 37; m 63; c 1. SOLID STATE PHYSICS. *Educ:* Davidson Col, BS, 60; Clemson Univ, MS, 62, PhD(physics), 65. *Prof Exp:* Assoc prof, 65-70, PROF PHYSICS, E TENN STATE UNIV, 70- *Mem:* Am Phys Soc; Am Asn Physics Teachers; Sigma Xi. *Res:* Methods of crystal growth; electron spin resonance of impurities in crystalline solids. *Mailing Add:* E Tenn State Univ Et Su box 22060A Johnson City TN 37601

POWELL, HOWARD B, b Benton, Ky, May 2, 33; m 52; c 3. INORGANIC CHEMISTRY. *Educ:* Murray State Col, BS, 55; Univ Tex, MA, 61, PhD(inorg chem), 63. *Prof Exp:* Control chemist, Nat Carbide Co, Ky, 53-57; teacher, Benton Independent Schs, 57-59; asst, Univ Tex, 60-61; asst prof chem, Univ Miami, 63-69; assoc prof, 69-74, PROF CHEM, EASTERN KY UNIV, 74- *Concurrent Pos:* Vis prof, Dept Chem, Univ Utah, 76-77. *Mem:* Am Chem Soc; Sigma Xi; NAm Thermal Anal Soc. *Res:* Investigation of the structure and reactivity of coordination compounds containing the ligand pyridine 1-oxide, particularly the thermal behavior of these compounds. *Mailing Add:* Dept Chem Eastern Ky Univ Richmond KY 40475

POWELL, HUGH N, b Birmingham, Ala, June 16, 22; m 50; c 3. CHEMICAL ENGINEERING. *Educ:* Ga Inst Technol, BS, 43; Univ Del, PhD(chem eng), 50. *Prof Exp:* Res engr combustion and thermodynamics, Aircraft Gas Turbine Div, Gen Elec Co, Ohio, 51-57; assoc prof, Cornell, 57-61; PROF MECH ENG, UNIV WIS-MADISON, 61- *Concurrent Pos:* Consult, Gen Elec Co, 57-61. *Mem:* Am Inst Aeronaut & Astronaut. *Res:* Properties of high temperature gases; shock tube technique; plasma dynamics; spectroscopy. *Mailing Add:* Dept of Mech Eng Univ of Wis Madison WI 53706

POWELL, JACK EDWARD, b Avon, Ill, June 11, 21; m 44; c 4. PHYSICAL & INORGANIC CHEMISTRY. *Educ:* Monmouth Col, BS, 43; Iowa State Univ, PhD(chem), 52. *Prof Exp:* Asst, 43-46, jr scientist, 46-52, from asst prof to assoc prof, 52-63, PROF CHEM, AMES LAB, IOWA STATE UNIV, 63- *Mem:* Sigma Xi; Am Chem Soc; Am Asn Univ Profs. *Res:* Ion exchange; rare earth chemistry; chelate formation constants; chelate structures; synthesis; isotope separations; ion exclusion techniques. *Mailing Add:* 1023 Grand Ave Ames IA 50000

POWELL, JAMES DANIEL, b Paducah, Tex, Sept 13, 34; m 58; c 3. GEOLOGY. *Educ:* Tex Tech Col, BS, 56, MS, 58; Univ Tex, PhD(geol), 61. *Prof Exp:* Res geologist, Res Div, Continental Oil Co, 61-63; asst prof geol, WTex State Univ, 63-64 & Univ Tex, Arlington, 64-70; asst prof, Univ Idaho, 70-74, affil prof geol, US Geol Surv, 74-76; chief res geologist, Union Carbide Corp, 76-79; CONSULT GEOLOGIST, POWELL & ASSOCS, 79- *Mem:* Am Asn Petrol Geologists; Paleont Soc; Soc Econ Paleont & Mineral. *Res:* Mesozoic Ammonoidea; stratigraphy in North and Central America. *Mailing Add:* Powell & Assoc PO Box 11478 Aspen CO 81612-9560

POWELL, JAMES HENRY, b Columbus, Ohio, May 21, 26. MATHEMATICAL STATISTICS. *Educ:* Mich State Univ, BA, 49, MA, 51, PhD(math statist), 54. *Prof Exp:* Instr math, Univ Detroit, 54-55; from asst prof to assoc prof, 55-60, head dept, 60-66, assoc dean sci & arts, 66-69, assoc chmn dept, 73-77, PROF MATH, WESTERN MICH UNIV, 60-, CHMN DEPT, 77- *Mem:* Inst Math Statist; Am Statist Asn; Am Math Soc; Math Asn Am. *Res:* Probability sociometric investigations. *Mailing Add:* 417 W South St Kalamazoo MI 49007

POWELL, JAMES LAWRENCE, b Berea, Ky, July 17, 36; m 59, 83; c 2. GEOCHEMISTRY, PETROLOGY. *Educ:* Berea Col, AB, 58; Mass Inst Technol, PhD(geochem), 62. *Hon Degrees:* DSc, Oberlin Col, 83. *Prof Exp:* Asst prof geol, Oberlin Col, 62-71, chmn dept, 65-74, prof, 71-83, provost col, 74-83, actg pres, 81-83; PRES, FRANKLIN MARSHALL COL, 83- *Mem:* AAAS; fel Geol Soc Am. *Res:* Application of isotope geochemistry to problems of petrology; use of radiogenic isotopes as natural tracers. *Mailing Add:* 6230 SE 36th Ave Portland OR 97207

POWELL, JAMES R, JR, b Rochester, Pa, June 3, 32; m 65. NUCLEAR ENGINEERING. *Educ:* Carnegie Inst Technol, BS, 53; Mass Inst Technol, ScD(nuclear eng), 57. *Prof Exp:* HEAD, FUSION TECHNOL GROUP, BROOKHAVEN NAT LAB, 56- *Mem:* AAAS; Am Phys Soc; Am Geophys Union; Am Inst Aeronaut & Astronaut; Am Nuclear Soc. *Res:* Fusion and fission technology and cryogenic applications. *Mailing Add:* Brookhaven Nat Lab Bldg 701 143rd Level Upton NY 11973

POWELL, JEANNE ADELE, b Los Angeles, Calif, Jan 18, 33. DEVELOPMENTAL BIOLOGY. *Educ:* Brown Univ, AB, 54; Bryn Mawr Col, MA, 59, PhD(develop biol), 67. *Prof Exp:* Res asst biol, Brown Univ, 54-55; teacher, Baldwin Sch, 55-64; instr biol, Bryn Mawr Col, 66-67; from asst prof to assoc prof biol, 67-79, prof biol, 79-85, SOPHIA SMITH PROF BIOL SCI, SMITH COL, 85- *Concurrent Pos:* Assoc investr, NIH fel, Smith Col, 68-73; res fel, Carnegie Inst Wash, 72; vis scientist, Inst Nat de la Sante et de la Recherche Medicale, Paris, 79; adj vis prof, Univ Bourdeaux II, France. *Mem:* AAAS; Soc Develop Biol; Am Soc Cell Biologists; Sigma Xi. *Res:* Tissue and organ culture of muscle and nerves. *Mailing Add:* Dept of Biol Sci Smith Col Northampton MA 01063

POWELL, JEFF, b Syracuse, NY, Nov 16, 39; m 63; c 1. RANGE SCIENCE. *Educ:* Southeastern La Col, BS, 61; Ore State Univ, BS, 64; Tex Tech Col, MS, 66; Colo State Univ, PhD(range sci), 68. *Prof Exp:* Asst prof range mgt, Humboldt State Col, 68-70; from asst prof to assoc prof, 70-81, prof range mgt, Okla State Univ, 81-82; PROF RANGE SCI, UNIV WYO, 82- *Mem:* Fel Soc Range Mgt; Am Soc Animal Sci; Wildlife Soc; Am Soc Agron; Sigma Xi. *Res:* Range nutrition; management and nutritive value of range plants; ranch recreation. *Mailing Add:* Dept Range Mgt Univ Wyo Laramie WY 82071

POWELL, JERREL B, b Knowles, Okla, Aug 15, 30; m 52; c 4. GENETICS, AGRONOMY. *Educ:* Okla State Univ, BS, 53, MS, 58; Wash State Univ, PhD(agron), 63. *Prof Exp:* Chief, Field Crops Lab, 77-80, asst dir, Beltsville Agr Res Ctr, 80-84, asst dir, S Atlantic Area, 84-87, RES GENETICIST, AGR RES SERV, USDA, 63-, DIR NAT SOIL LAB, 87- *Mem:* Am Soc Agron; Crop Sci Soc Am; Am Genetic Asn; Sigma Xi. *Res:* Breeding and cytogenetics of annual and perennial grasses; mutation breeding of turf grasses; cropping system research; controlled traffic research. *Mailing Add:* RR 1 Box 81-A Council Hill OK 74428

POWELL, JERRY ALAN, b Glendale, Calif, May 23, 33; m 56, 77; c 3. ECOLOGY. *Educ:* Univ Calif, Berkeley, BS, 55, PhD(entom), 61. *Prof Exp:* Asst entomologist, 61-67, assoc prof entom, Univ, 69-73, assoc entomologist, 67-73, ENTOMOLOGIST, EXP STA, UNIV CALIF, BERKELEY, 73-, PROF ENTOM, UNIV, 73- *Concurrent Pos:* Ed, J Lepidop Soc, 64-69; chief investr, NSF grant, 65-70 & 85-88; vis res fel, Smithsonian Inst, 70-71; vis res assoc, Commonwealth Sci & Indust Res Orgn, Canberra, Australia, 80-81. *Honors & Awards:* Karl Jordan Medal, Lepidopterists' Soc, 82. *Mem:* Pac Coast Entom Soc (pres, 64); Soc Syst Zool (pres elect, 88, pres, 89); Lepidopterists' Soc (vpres, 85-86, pres, 87-88); Entom Soc Am. *Res:* Comparative biology; systematics and phylogeny of Lepidoptera, particularly the Tortricoidea, Ethmiidae; behavior of solitary wasps; prolonged diapause. *Mailing Add:* Dept Entom & Sci Univ Calif Berkeley CA 94720

POWELL, JOHN DAVID, electromagnetic propulsion, for more information see previous edition

POWELL, JOHN LEONARD, b Portland, Ore, May 12, 19; m 43; c 2. PHYSICS. *Educ:* Reed Col, BA, 43; Univ Wis, PhD(physics), 48. *Prof Exp:* Res assoc, Radiation Lab, Mass Inst Technol, 43-45; asst physics, Univ Wis, 45-48, from asst prof to assoc prof, 49-54; res assoc, Inst Nuclear Studies, Univ Chicago, 48-49; lab dir, Theoret Group, Missile Systs Div, Lockheed Aircraft Corp, 54-55; assoc prof, 55-57, head dept, 60-66, PROF PHYSICS, UNIV ORE, 57- *Mem:* Fel Am Phys Soc; Am Asn Physics Teachers. *Res:* Quantum electrodynamics; nuclear scattering; particle accelerators; neutron transport theory. *Mailing Add:* 355 W 27 Pl Eugene OR 97405

POWELL, JUSTIN CHRISTOPHER, b Chester, Pa, Oct 24, 43; m 64; c 2. MICELLAR CATALYSIS, PETROLEUM PRODUCT ADDITIVES. *Educ:* Carnegie-Mellon Univ, BS, 65; Univ Del, PhD(phys org chem), 69. *Prof Exp:* Asst prof, Exten Div, Univ Del, 67-68; sr res chemist, Beacon Labs, Texaco, Inc, 69-80; sect chief, Off Toxic Substances, US Environ Protection Agency, 80-89, spec asst compliance & enforcement to dir of the Off Toxic Substances, 89; SR SCIENTIST, LAW FIRM KELLER & HECKMAN, 89- *Honors & Awards:* Bronze Medal, US Environ Protection Agency, 84. *Mem:* Am Chem Soc. *Res:* Scientific review regulatory compliance; chemical data bases; structure-activity relationships; categorical nomenclature; chemical analog identification, chemical use projections and chemical identification to fulfill use requirements; chemical identity and composition characterizations; chemical regulation teaching. *Mailing Add:* 5196 Lewisham Rd Fairfax VA 22030

POWELL, KENNETH GRANT, b Euclid, Ohio, July 3, 60; m 91. FLUID DYNAMICS, COMPUTATIONAL FLUID DYNAMICS. *Educ:* Mass Inst Technol, SB(math) & SB(aerospace eng), 82, SM, 84, ScD(aerospace eng), 87. *Prof Exp:* ASST PROF AEROSPACE ENG, UNIV MICH, 87- *Concurrent Pos:* NSF presidential young investr, 88; vis scientist, NASA Lewis Res Ctr, 89, NASA Ames Res Ctr, 90 & 91. *Mem:* Sigma Xi; AAAS; Am Inst Aeronaut & Astronaut; Soc Indust & Appl Math. *Res:* Scientific computing, specializing in computational fluid dynamics; study of, and algorithm development for, compressible fluid flow. *Mailing Add:* Dept Aerospace Eng Univ Mich Ann Arbor MI 48109-2140

POWELL, LESLIE CHARLES, b Beaumont, Tex, Dec 13, 27; m 52, 83; c 4. OBSTETRICS & GYNECOLOGY. *Educ:* Southern Methodist Univ, BS, 48; Johns Hopkins Univ, MD, 52; Am Bd Obstet & Gynec, dipl, 61. *Prof Exp:* Rotating intern, 52-53, resident, 53-55, from instr to assoc prof, 57-68, PROF OBSTET & GYNEC, UNIV TEX MED BR, GALVESTON, 68- *Concurrent Pos:* Ed consult, Tex Med, 64-; USPHS grant cancer screening, 67-71; consult, Tex State Dept Ment Health & Retardation, 70-; Fulbright-Hayes lectr & vis prof obstet & gynec, Union of Burma, 75-76; DATTA comt, AMA, 83- *Mem:* Fel Am Col Obstetricians & Gynecologists; AMA; Int Soc Study Vulvar Dis; Soc Gynec Surgeons. *Res:* Isoimmunization in pregnancy; hormones, effects and side effects; gynecological malignancy; treatment of sexual dysfunction; fertility control. *Mailing Add:* Dept Obstet & Gynec Univ Tex Med Br Galveston TX 77550

POWELL, LOYD EARL, JR, b Ravenwood, Mo, Oct 17, 28; m 56; c 3. POMOLOGY, PLANT PHYSIOLOGY. *Educ:* Univ Mo, BS, 51; Ohio State Univ, MS, 52; Cornell Univ, PhD, 55. *Prof Exp:* Asst prof, NY State Agr Exp Sta, Geneva, 55-61; assoc prof, 61-74, PROF POMOL, NY STATE COL AGR & LIFE SCI, CORNELL UNIV, 74- *Concurrent Pos:* Res assoc, Yale Univ, 62-63; NATO vis prof, Univ Pisa, Italy, 75; chmn, Local Prog Comt Ann Meeting, NE Sect Am Soc Hort Sci, 77, pres, 79-80; prog chmn, Fourth Int Symp Growth Regulators Fruit Prod, Cornell Univ, 81; chmn, NE Sect Am Soc Plant Physiologists, 82; pres, Working Group Growth Regulators Fruit Prod, Int Soc Hort Sci, 81-85. *Honors & Awards:* J H Gourley Award, 75. *Mem:* Am Soc Plant Physiologists; Scandinavian Soc Plant Physiol; Int Plant Growth Substances Asn; fel Am Soc Hort Sci; Int Soc Hort Sci. *Res:* Plant hormones; dormancy; fruit development; growth and development of woody plants; hormone physiology of dormancy in apple trees. *Mailing Add:* Dept Pomol Cornell Univ Ithaca NY 14853

POWELL, MICHAEL A, b Coleman, Tex, Aug 26, 37; m 56; c 2. BOTANY. *Educ:* Sul Ross State Univ, BS & MA, 60; Univ Tex, PhD(bot), 63. *Prof Exp:* Assoc prof, 63-68, chmn dept, 79, PROF BIOL, SUL ROSS STATE UNIV, 68-, DEPT CHAIR, 78- *Concurrent Pos:* Sigma Xi res grant, 64; NSF res grants, 64-78. *Mem:* AAAS; Am Soc Plant Taxon. *Res:* Botanical systematics; cytotaxonomy of Compositae. *Mailing Add:* Dept Biol Sul Ross State Univ Alpine TX 79830

POWELL, MICHAEL ROBERT, b Detroit, Mich, Nov 23, 41; m 64; c 4. PHYSIOLOGY, BIOPHYSICS. *Educ:* Mich State Univ, BS, 63, MS, 66, PhD(biophys), 69. *Prof Exp:* Res biophysicist, Res Inst, Union Carbide Corp, 69-75; dir hyperbaric lab, Inst Appl Physiol & Med, 75-77; mem staff, Inst Flugmedizin, 77-80; dir, hyperbaric & biophys dept, Inst Appl Physiol & Med, 80- 89; HEAD ENVIRON PHYSIOL, SPACE BIOMED RES INST, NASA - JOHNSON SPACE CTR, HOUSTON, TEX, 90- *Mem:* Am Physiol Soc; Am Chem Soc; Biophys Soc; Undersea Hyperbaric Med Soc; Aerospace Med Asn. *Res:* Hyperbaric/hypobaric physiology; bioelectromagnetics; charge transport in biomacromolecules; pathophysiology of decompression sideness; oxygen toxicity; Hyperbaric medicine. *Mailing Add:* Mail Code SD5 NASA Johnson Space Ctr Houston TX 77058

POWELL, NATHANIEL THOMAS, b Halifax, Va, July 7, 28; m 51; c 3. PLANT PATHOLOGY. *Educ:* Va Polytech Inst & State Univ, BS, 50; NC State Univ, MSc, 56, PhD(plant path), 58. *Prof Exp:* From res asst prof to assoc prof plant path, NC State Univ, 58-67, prof plant path & genetics, 67-87; RETIRED. *Mem:* Am Phytopath Soc. *Res:* Disease complexes in plants; genetics and nature of disease resistance in plants; host-parasite relationships; physiology of parasitism. *Mailing Add:* 114 Merwin Rd Raleigh NC 27606

POWELL, NOBLE R, b New Kensington, Pa, Apr 19, 30; m 53; c 3. ELECTRICAL ENGINEERING. *Educ:* Columbia Univ, BSEE, 58; Syracuse Univ, PhD(elec eng), 68. *Prof Exp:* Consult engr, Signal Processing, Instrumentation, 76, mgr, Integrated Functional Processors, 76-79, mgr, Architecture & Processor Devint, 79-81, PRIN STAFF ENGR, ELECTRONICS LAB, GEN ELEC, 81- *Concurrent Pos:* Adj prof very-large-scale integration, Syracuse Univ, 79 & Cornell Univ, 80- *Honors & Awards:* Indust Res-100 Award, Indust Res Corp, 75. *Mem:* Sigma Xi. *Res:* Critical issues pertaining to large scale electronic integration of modular functions; digital filters; graphics functional generators; matrix array processors. *Mailing Add:* 4896 Westview Dr Syracuse NY 13215

POWELL, NORBORNE BERKELEY, b Montgomery, Ala, July 24, 14; m 39; c 2. UROLOGY. *Educ:* Baylor Univ, MD, 38, Am Bd Urology, dipl, 45. *Prof Exp:* Asst urol, Sch Med, Tulane Univ, 40-42; assoc prof, 43-65, PROF CLIN UROL, BAYLOR COL MED, 65-; CHIEF UROL, TWELVE OAKS HOSP, HOUSTON, 66- *Concurrent Pos:* Trustee, Baylor Med Found, 49-; chief of staff, Twelve Oaks Hosp, 74; pvt pract; chmn bd trustees, Twelve Oaks Med Ctr. *Honors & Awards:* Prize, Am Urol Asn, 51. *Mem:* Am Urol Asn; Am Fertil Soc; Biol Photog Asn; fel Am Col Surgeons; fel Int Col Surgeons; fel Can Urol Asn. *Res:* Clinical research on urological problems. *Mailing Add:* Twelve Oaks Hosp 4200 Westhimer Suite 288 Houston TX 77027

POWELL, R(AY) B(EDENKAPP), b Buffalo, NY, June 13, 20; m 48; c 3. MECHANICAL ENGINEERING. *Educ:* Univ Mich, BSME, 43. *Prof Exp:* Tech asst to supt area plant, Manhattan Eng Dist, Res Labs, Union Carbide Carbon Corp, 43-44; mech engr & leader design & maintenance group, Los Alamos Sci Lab, 44-45, admin asst leader chem & metall div, 45-46, asst to personnel dir, 47; personnel mgr, Sandia Nat Labs, 47-49, mgr, Employ & Personnel Dept, 49-54, supt, Personnel & Pub Rels, 54-59, asst vpres admin, 59, vpres, 59-85; CONSULT, 85- *Mailing Add:* 805 Pueblo Solano Rd NW Albuquerque NM 87107

POWELL, RALPH ROBERT, b Beech Grove, Ind, June 23, 36; m 65; c 4. PHYSICAL CHEMISTRY. *Educ:* Marian Col, Ind, BS, 58; Purdue Univ, PhD(chem), 65. *Prof Exp:* Instr chem, Purdue Univ, 64-65, asst prof, 65-66; asst prof, 66-76, ASSOC PROF CHEM, EASTERN MICH UNIV, 76- *Mem:* AAAS; Am Chem Soc; Sigma Xi. *Res:* Application of quantum theory to chemical systems using computer programs; development of the use of computers for instructional purposes both in the laboratory and in the classroom. *Mailing Add:* 2887 Dalton Ann Arbor MI 48108

POWELL, REX LYNN, b Lansing, Mich, Nov 19, 42; m 64; c 2. ANIMAL BREEDING. *Educ:* Mich State Univ, BS, 64, MS, 69; Iowa State Univ, PhD(animal breeding), 72. *Prof Exp:* RES GENETICIST, ANIMAL IMPROV PROG LAB, LIVESTOCK & POULTRY SCI INST, AGR RES SERV, USDA, 72- *Concurrent Pos:* Mem, comt standards for estimating yield and other performance traits, Coord Group Nat Coop Dairy Herd Improv Prog, 75-80, Res Comt Policy Bd, 80-83; NAm rep, comt conversions of sire evaluations, Int Bull Eval Serv, 82-; adv, Genetic Advan Comt, Holstein Asn, 86-87 & 90-91. *Mem:* Am Dairy Sci Asn; Dairy Shrine; Sigma Xi. *Res:* Developing and analyzing methods for conversions of genetic evaluations of dairy cattle across countries; examination of genotype-environment interaction and estimation of genetic trends. *Mailing Add:* Bldg 263 Agr Res Ctr Agr Res Serv USDA Beltsville MD 20705-2350

POWELL, RICHARD ANTHONY, b Buffalo, NY, Apr 11, 17; m 43; c 5. DENTISTRY. *Educ:* Syracuse Univ, AB, 39; Univ Buffalo, DDS, 49. *Prof Exp:* Clin instr oper dent, State Univ NY, Buffalo, 49-53, from asst prof to assoc prof oper dent & endodontics, 53-61, asst dean clin opers & student affairs, 65-68, prof oper dent & endodontics, 61-, assoc dean clin opers & student affairs, 68, dir clins, 71-; RETIRED. *Concurrent Pos:* Consult, Vet Admin Hosp, Buffalo, NY, 65-; chmn coun sects, Am Asn Dent Schs. *Honors & Awards:* Jarvie-Burrhardt Award, Dent Soc NY. *Mem:* Am Dent Asn; fel Am Col Dent; fel Int Col Dentists. *Res:* Operative dentistry; endodontics; silver amalgam restorations. *Mailing Add:* 161 Southwood Dr Kenmore NY 14223

POWELL, RICHARD CINCLAIR, b South Bend, Ind, July 28, 29; m 54; c 2. INTERNAL MEDICINE, ENDOCRINOLOGY. *Educ:* DePauw Univ, AB, 51; Northwestern Univ, MD, 55; Am Bd Internal Med, dipl, 63. *Prof Exp:* From instr to assoc prof, 61-72, PROF MED, SCH MED, IND UNIV, INDIANAPOLIS, 72- *Concurrent Pos:* Attend physician, Vet Admin Hosp, Indianapolis, 63-; assoc staff, med sect, Marion County Gen Hosp, Ind, 65- *Mem:* AMA; fel Am Col Physicians; Endocrine Soc; Am Diabetes Asn; Am Heart Asn. *Res:* Thyroid physiology; lipid metabolism. *Mailing Add:* Dept of Med Ind Univ Med Ctr Indianapolis IN 46202-5124

POWELL, RICHARD CONGER, b Lincoln, Nebr, Dec 20, 39; m 62; c 2. SOLID STATE PHYSICS. *Educ:* US Naval Acad, BS, 62; Ariz State Univ, MS, 64, PhD(physics), 67. *Prof Exp:* Res scientist, Air Force Cambridge Res Labs, 64-68; res scientist, Sandia Labs, NMex, 68-71; assoc prof, 71-73, PROF PHYSICS, OKLA STATE UNIV, 73- *Mem:* Fel Am Phys Soc; Sigma Xi; fel Optical Soc Am. *Res:* Optical spectra of solids; solid state lasers. *Mailing Add:* Ctr Laser Res Okla State Univ Stillwater OK 74078

POWELL, RICHARD GRANT, b Avon, Ill, Oct 29, 38; m 60; c 2. NATURAL PRODUCTS CHEMISTRY. *Educ:* Western Ill Univ, BS, 61, MSEd, 63. *Prof Exp:* RES CHEMIST, NORTHERN REGIONAL RES CTR, AGR RES SERV, USDA, 63- *Concurrent Pos:* Res fel, Univ St Andrews, 66-67. *Mem:* Am Chem Soc; Phytochem Soc NAm; Soc Econ Bot; Am Soc Pharmacog. *Res:* Alkaloids; lipids; antitumor compounds; natural products. *Mailing Add:* Nat Ctr Agr Utilization Res 1815 N University St Peoria IL 61604

POWELL, RICHARD JAMES, b Muskogee, Okla, Mar 15, 31; m 54; c 5. POLYMER CHEMISTRY. *Educ:* Okla State Univ, BS, 53. *Prof Exp:* Chem engr, 53-65, sr res engr, 65-72, res assoc, 72-78, res fel, 78-86, DEPT FEL, E I DU PONT DE NEMOURS & CO, INC, 86- *Mem:* Am Chem Soc. *Res:* Product and process development on Surlyn ionomer resins; all product end uses, including molding, film, sheeting and extrusion coating. *Mailing Add:* 4217 Meeks Dr Orange TX 77630-1514

POWELL, ROBERT DELAFIELD, plant physiology; deceased, see previous edition for last biography

POWELL, ROBERT ELLIS, b Lansing, Mich, Mar 16, 36; m 58; c 3. MATHEMATICS. *Educ:* Mich State Univ, BA, 58, MA, 59; Lehigh Univ, PhD(math), 66. *Prof Exp:* Assoc res engr, Boeing Co, Wash, 59-61; instr math, Highline Col, 61-63; instr, Lehigh Univ, 64-66; asst prof, Univ Kans, 66-69; assoc prof, 69-74, assoc dean, 74-75, dean, Grad Col, 75-76, PROF MATH, KENT STATE UNIV, 74-, DEAN, GRAD COL, 80- *Concurrent Pos:* Vis res asst prof, Univ Ky, 67-68. *Mem:* Math Asn Am. *Res:* Examination of summability and related approximation procedures in complex analysis. *Mailing Add:* Kent State Univ Grad Col Kent OH 44242

POWELL, ROBERT LEE, b Chicago, Ill, Mar 11, 28; div; c 4. ORTHOPEDICS, THERMOMETRY. *Educ:* Univ Colo, BS, 50, MA, 51; Cambridge Univ, PhD, 66. *Prof Exp:* Proj leader, sect chief & consult, Cryogenics Div, Nat Bur Standards, 51-61, 63-76; ORTHOP PHYSICIANS ASST, UNIV COLO HEALTH SCI CTR, 81- *Concurrent Pos:* Consult, 76- *Mem:* Am Phys Soc; Am Soc Test & Mat; fel Brit Inst Physics; Am Soc Orthop Phys Asst. *Res:* Low temperature transport properties of solids; thermal and electrical conductivity of pure metals and dilute alloys; thermometry; applications of superconductivity; basic science of orthopedics. *Mailing Add:* 2795 Stanford Ave Boulder CO 80303

POWELL, ROBERT W, JR, b Mobile, Ala, Nov 11, 29; m 58; c 4. BOTANY. *Educ:* Memphis State Univ, BS, 51; Univ Houston, MS, 53; Duke Univ, PhD(bot), 60. *Prof Exp:* Asst prof biol, Memphis State Univ, 59-61; asst prof bot, Humboldt State Col, 61-63; PROF BIOL & CHMN DEPT, CONVERSE COL, 63- *Mem:* Sigma Xi; Am Inst Biol Sci. *Res:* Plant taxonomy; ecology; local flora; horticulture; plant propagation. *Mailing Add:* Dept Biol Converse Col Spartanburg SC 29301

POWELL, ROBIN DALE, b Indianapolis, Ind, Apr 19, 34; m 57; c 3. MEDICINE. *Educ:* Univ Chicago, MD, 57. *Prof Exp:* Intern med, Minneapolis Gen Hosp, 57-58; resident internal med, Univ Chicago, 58-61, from instr to asst prof med, 63-69; assoc prof, Col Med, Univ Iowa, 69-72, prof internal med, 72-78; PROF INTERNAL MED, SCH MED & ASSOC DEAN ACAD AFFAIRS, NORTHWESTERN UNIV, 78- *Concurrent Pos:* Mem comn malaria, Armed Forces Epidemiol Bd, 64-72; consult, Surgeon Gen, US Dept Army, 64-; mem expert adv panel malaria, WHO, 65-; assoc chief of staff res & educ, Iowa City Vet Admin Hosp, 70- *Mem:* AAAS; Am Heart Asn; AMA; Am Fedn Clin Res; Am Soc Trop Med & Hyg. *Res:* Internal medicine; human malaria; erythrocyte biochemistry. *Mailing Add:* Nat Bd Med Examiners 3930 Chestnut St Philadelphia PA 19104

POWELL, ROGER ALLEN, b Joliet, Ill, Jan 24, 49; m 72; c 1. BEHAVIORAL ECOLOGY, MAMMALOGY. *Educ:* Carleton Col, BA, 71; Univ Chicago, PhD(biol), 77. *Prof Exp:* Lectr natural resources, Univ Mich, 77; lectr biol, Univ Chicago, 78; asst prof, 79-85, ASSOC PROF, DEPTS ZOOL & FORESTRY, NC STATE UNIV, 85- *Concurrent Pos:* Coop res asst, US Fisheries & Wildlife Serv 71 & US Forest Serv, 73-76; vis asst prof behav ecol, Dept Ecol, Ethol & Evolution, Univ Ill, 78-79, prof zool, Dept Zool, Oulu Univ, Oulu, Finland, 82; vis assoc prof, Dept Zool & Physiol, Univ Wyo. *Mem:* Am Soc Mammalogists; Animal Behav Soc; British Ecol Soc; Ecol Soc Am; Am Soc Naturalists; AAAS; Int Soc Behav Ecol. *Res:* Ecology with special interests in theory, predation, vertebrate predators and mammals; broad experience with carnivora, especially Mustelidae. *Mailing Add:* Depts Zool & Forestry NC State Univ Raleigh NC 27695-7617

POWELL, RONALD ALLAN, b Brooklyn, NY, Oct 18, 46; m 75. SOLID STATE PHYSICS. *Educ:* Union Col, BS, 67; Stanford Univ, MS, 69, PhD(physics), 73. *Prof Exp:* Teaching & res asst physics, Stanford Univ, 68-73, res assoc physics, 74-76, sr res assoc elec eng, 76-79; sr staff scientist, Varian Assoc, 78-85, assoc dir 85-87, DIR MAT & EQUIP LAB, VARIAN RES CTR, PALO ALTO, CA, 87- *Mem:* Am Phys Soc; Am Vacuum Soc; Am Asn Physics Teachers. *Res:* Advanced processing techniques for integrated circuit technology of silicon and III-V compound semiconductors; plasma etching; ion implantation; photon and electron beam processing of semiconductors; chemical vapor deposition. *Mailing Add:* 427 Crest Dr Redwood City CA 94062

POWELL, SHARON KAY, b Bethpage, NY, Sept 12, 61. PROTEIN TRAFFICKING, ADHESION MOLECULES. *Educ:* Cornell Univ, BS, 83; Univ Calif, Berkeley, PhD(cell biol), 88. *Prof Exp:* Teaching asst microbiol, Univ Calif, Berkeley, 84 & 86, grad res fel, 83-88; POSTDOCTORAL FEL, CORNELL UNIV MED COL, 88- *Concurrent Pos:* Res asst, Cornell Univ, 81-82 & 83, teaching asst microbiol, 83. *Mem:* Am Soc Cell Biol. *Res:* Polarized epithelia sort proteins and lipids to their apical and basolateral plasma membrane domains; glycosyl phosphatidyl inositol anchors as a targeting signal for the apical surface of many polarized epithelia; define apical and basolateral targeting signals and proteins which they interact with. *Mailing Add:* Cell Biol & Anat Dept Cornell Univ Med Col 1300 York Ave New York NY 10021

POWELL, SMITH THOMPSON, III, b Kirksville, Mo, Jan 8, 40; m 78; c 7. PHYSICS, ASTRONOMY. *Educ:* Berea Col, BA, 61; Univ Mich, Ann Arbor, MS, 63, PhD(physics), 70. *Prof Exp:* Asst prof, 70-78, ASSOC PROF PHYSICS, BEREA COL, 78-, PROF, 85- *Mem:* AAAS; Am Asn Physics Teachers; Am Phys Soc. *Res:* Elementary particle physics; astronomy; optics; demography; computer sciences, planetarium director. *Mailing Add:* CPO 1630 Berea KY 40404

POWELL, THOMAS MABREY, b Oakland, Calif, Aug 6, 42; c 2. LIMNOLOGY, ESTUARINE PROCESSES. *Educ:* Univ Calif, Berkeley, AB, 64, PhD(physics), 70. *Prof Exp:* Researcher ecol, Inst Ecol, 70-71, lectr physics, dept physics, 71, from asst prof to assoc prof, 71-85, PROF ENVIRON STUDIES, DIV ENVIRON STUDIES, UNIV CALIF, DAVIS, 85- *Concurrent Pos:* Prin investr, Inst Ecol, Univ Calif, Davis, for NSF & NASA projs; vis sr researcher, Lawrence Berkeley Lab, Univ Calif, Berkeley, 77-78; vis sr prof, Ctr Math Biol, Math Inst, Univ Oxford, Eng, 86. *Mem:* AAAS; Sigma Xi; Am Soc Limnol Oceanog; Am Geophys Union; Soc Indust Appl Math. *Res:* Impact of physical processes like currents, waves, and mixing on the ecology of plankton in lakes, estuaries, and the coastal ocean; measurements of physical and biological quantities as well as mathematical models. *Mailing Add:* Div Environ Studies Univ Calif Davis Davis CA 95616

POWELL, THOMAS SHAW, b Hahira, Ga, Aug 28, 46; m 72; c 1. POULTRY NUTRITION, MANAGEMENT. *Educ:* Univ Fla, BS, 68, MS, 71; NC State Univ, PhD(animal Sci), 74. *Prof Exp:* Poultry nutritionist, Cent Soya Res, 74-79; nutritionist, Crystal Farms, Inc, 79-80; nutritinsist, Carnation Co, 80-87; nutritioinst, Moorman Mfg, 87-91; NUTRITIONIST, CENT SOYA, 91- *Mem:* Poultry Sci Asn. *Res:* Nutritional requirements of broiler breeders and turkeys; nutritional requirements of broilers and management factors affecting leghorn pullets. *Mailing Add:* 115 Brandywine Lane Decatur IN 46733

POWELL, WARREN HOWARD, b North Collins, NY, Jan 1, 34; m 55, 84; c 3. ORGANIC CHEMISTRY. *Educ:* Antioch Col, BS, 55; Ohio State Univ, PhD(chem), 59. *Prof Exp:* Asst, Kettering Found, Antioch Col, 52-53; jr technician, Battelle Mem Inst, 53-54; asst, Kettering Found, Antioch Col, 54-55; asst, Ohio State Univ, 55-57, asst instr, 57-58, instr, 58; res chemist org chem dept, E I du Pont de Nemours & Co, 59-64; asst ed, 64-67, assoc ed, 67-69, sr assoc ed, 69-73, SR ED, CHEM ABSTRACTS SERV, 73- *Concurrent Pos:* Mem comt nomenclature inorg chem div chem & chem technol, Nat Acad Sci-Nat Res Coun, 68-72; assoc mem comn nomenclature inorg chem, Int Union Pure & Appl Chem, 69-75, titular mem, 75-, secy, 79-, assoc mem comn nomenclature org chem, 75-79, titular mem & secy, 79-, assoc mem interdiv comt nomenclature & symbols, 83- *Mem:* Am Chem Soc. *Res:* Heterocyclic compounds, especially acridines and benzothiazoles; aromatic fluorine compounds; biphenyl derivatives; chemical nomenclature: organic, inorganic, biochemical, & macromolecular. *Mailing Add:* Chem Abstracts Serv PO Box 3012 2540 Olentangy River Rd Columbus OH 43210-0012

POWELL, WILLIAM ALLAN, b Wallace, NC, May 28, 21; m 41; c 3. ANALYTICAL CHEMISTRY. *Educ:* Wake Forest Col, BS, 42; Duke Univ, PhD(chem), 53. *Prof Exp:* Asst chief chemist, Carolina Aluminum Co, 42-46; indust hyg chemist, Aluminum Co Am, 46-48; instr chem, Wake Forest Col, 48-49 & Duke Univ, 49-51; from asst prof to prof, Univ Richmond, 52-86, chmn dept, 59-82; CONSULT, PHILLIP MORRIS, 57- *Mem:* Am Chem Soc. *Res:* Instrumental methods of analysis. *Mailing Add:* Dept of Chem Univ of Richmond Richmond VA 23173

POWELL, WILLIAM JOHN, JR, b Pittsburgh, Pa, Sept 18, 35. CARDIOVASCULAR PHYSIOLOGY. *Educ:* Columbia Univ, MD, 61. *Prof Exp:* Asst prof, 71-74, ASSOC PROF MED, HARVARD UNIV, 74-; VPRES CLIN RES, MERCK SHARP & DOHME, 86- *Concurrent Pos:* Assoc physician, Mass Gen Hosp, 78-87, consult med, 87- *Mem:* Am Fedn Clin Res; Am Physiol Soc; Am Soc Pharmacol & Exp Therapeut; Am Heart Asn; fel Am Soc Clin Invest. *Mailing Add:* Merck Sharp & Dohme Ten Century Pkwy West Point PA 19486

POWELL, WILLIAM MORTON, b Halifax, Va, May 12, 30; m 53; c 4. PLANT PATHOLOGY. *Educ:* Va Polytech, BS, 53; NC State Univ, MS, 57, PhD(plant path), 60. *Prof Exp:* From asst prof to assoc prof, 60-73, PROF PLANT PATH, UNIV GA, 73- *Mem:* Am Phytopath Soc; Soc Nematol. *Res:* Host-parasite relations of plant parasitic nematodes; nematode-fungus interactions in plant disease. *Mailing Add:* Dept Plant Path & Genetics Univ Ga Four Towers Blvd Athens GA 30602

POWELL, WILLIAM ST JOHN, b Dublin, Ireland, Dec 6, 45; Can & Irish citizen; m 68; c 3. PROSTAGLANDINS & LEUKOTRIENES, LIPID PEROXIDATION. *Educ:* Univ Sask, Saskatoon, BA, 67; Dalhousie Univ, PhD(chem), 72. *Prof Exp:* Fel, dept physiol chem, Karolinska Inst, Stockholm, 72-75; asst prof, 76-82, assoc prof, 82-88, PROF DEPT MED, MCGILL UNIV, 88- *Mem:* Am Soc Biochem & Molecular Biol; Endocrine Soc; Can Biochem Soc; Am Oil Chemists Soc. *Res:* Biosynthesis and metabolism of prostaglandins and leukotrienes; role of lipid peroxides and eicosanoids in atherosclerosis; metabolism of leukotrienes by leukocytes; role of prostaglandins and leukotrienes in asthma. *Mailing Add:* Meakins-Christie Labs McGill Univ 3626 rue St Urbain Montreal PQ H2X 2P2 Can

POWELSON, ELIZABETH EUGENIE, b New York, NY, Jan 30, 24. GENETICS. *Educ:* Oberlin Col, AB, 46; Wellesley Col, MA, 48; Ind Univ, PhD(zool), 57. *Prof Exp:* Instr biol, Hood Col, 48-52; from instr to assoc prof, 57-71, chmn dept, 64-73, prof, 71-73, GREENAWALT PROF BIOL, WITTENBERG UNIV, 73- *Concurrent Pos:* Fel, Johns Hopkins Med Sch, 73-75. *Mem:* AAAS; Am Soc Human Genetics; Sigma Xi. *Res:* The relationship of chromosomal abnormalities to birth defects in humans and its implication for genetic counseling. *Mailing Add:* Box 720 Springfield OH 45501

POWELSON, ROBERT LORAN, b Salt Lake City, Utah, Sept 23, 29; m 48, 73; c 4. PLANT PATHOLOGY. *Educ:* Utah State Univ, BS, 51, MS, 56; Ore State Col, PhD(plant path), 59. *Prof Exp:* From asst plant pathologist to assoc plant pathologist, 59-74, prof plant path, 74-, EMER PROF BOT & PLANT PATH, ORE STATE UNIV. *Mem:* Am Phytopath Soc. *Res:* Ecology of soil-borne plant pathogens; epidemiology of plant diseases. *Mailing Add:* Dept Bot Ore State Univ Corvallis OR 97331

POWER, DENNIS MICHAEL, b Pasadena, Calif, Feb 18, 41; m 65, 85; c 3. EVOLUTIONARY BIOLOGY, ZOOGEOGRAPHY. *Educ:* Occidental Col, BA, 62, MA, 64; Univ Kans, PhD(zool), 67. *Prof Exp:* From asst cur to assoc cur ornith, Royal Ont Mus, 67-72; DIR, SANTA BARBARA MUS NATURAL HIST, 72- *Concurrent Pos:* Nat Res Coun Can res grant, 68-; Dept Univ Affairs Ont res grant, 68-72 & NSF grants, 73-74; asst prof, Univ Toronto, 68-72; assoc res zoologist, Univ Calif, Santa Barbara, 72- *Mem:* AAAS; Soc Syst Zool; Soc Study Evolution; Am Ornith Union; Cooper Ornith Soc; Soc Am Naturalists. *Res:* Statistical analysis in systematic and evolutionary biology; evolutionary and ecological studies of birds and bird communities. *Mailing Add:* Santa Barbara Mus Natural Hist 2559 Puesta del Sol Rd Santa Barbara CA 93105

POWER, GEOFFREY, b Accrington, Eng, Sept 21, 33; Can citizen; m 56; c 2. FISH BIOLOGY. *Educ:* Univ Durham, BSc, 54; McGill Univ, PhD(zool), 59. *Prof Exp:* From asst prof to assoc prof, 57-71, PROF BIOL, UNIV WATERLOO, 71- *Concurrent Pos:* Res prof, Univ Laval, 74-78. *Mem:* Am Fisheries Soc; Can Soc Zoologists; Brit Freshwater Biol Asn; fel Arctic Inst NAm. *Res:* Arctic and subarctic anadromous and freshwater fisheries. *Mailing Add:* Dept Biol Sci Univ Waterloo Waterloo ON N2L 3G1 Can

POWER, GORDON G, b Baltimore, Md, July 12, 35; m 60; c 3. PHYSIOLOGY, INTERNAL MEDICINE. *Educ:* Swarthmore Col, BA, 57; Univ Pa, MD, 61. *Prof Exp:* Intern, Philadelphia Gen Hosp, 61-62; resident internal med, Hosp, Univ Va, 62-63; Nat Heart Inst fel pulmonary physiol, Univ Pa, 63-65; resident, Philadelphia Gen Hosp, 65-66; from asst prof to assoc prof, 69-75, PROF OBSTET & GYNEC, SCH MED, LOMA LINDA UNIV, 75- *Concurrent Pos:* Nat Inst Child Health career develop award, NIH, 69-74. *Mem:* Am Fedn Clin Res; Am Physiol Soc; Soc Gynec Invest; Perinatal Res Soc; Sigma Xi. *Res:* Pulmonary physiology; gas exchanges; diffusion, fetal and placental physiology; blood flow regulation; blood gases; control of circulation. *Mailing Add:* Dept Obstet & Gynec Loma Linda Univ Loma Linda CA 92354

POWER, HARRY WALDO, III, b Conrad, Mont, Jan 24, 45; m 83; c 1. EVOLUTIONARY BIOLOGY, SOCIOBIOLOGY. *Educ:* Univ Mont, BA, 67; Univ Mich, MA, 69, PhD(zool), 74. *Prof Exp:* Vis asst prof, Syracuse Univ, 74-76; asst prof, 76-81, ASSOC PROF BIOL SCI, RUTGERS UNIV, 81- *Concurrent Pos:* NSF grant, 76-79, 84-86. *Honors & Awards:* Roberts Award, Cooper Ornith Soc, 73; Res Award, NAm Bluebird Soc, 87. *Mem:* Ecol Soc Am; Am Ornithologists Union; Cooper Ornith Soc; Animal Behav Soc; Wilson Ornith Soc; Am Soc Naturalist. *Res:* Evolution of social behavior and clutch sizes. *Mailing Add:* Dept Biol Sci Rutgers Univ PO Box 1059 Piscataway NJ 08855-1059

POWER, JAMES FRANCIS, b Saybrook, Ill, Sept 18, 29; m 58; c 9. SOIL MANAGEMENT, WATER QUALITY. *Educ:* Univ Ill, BS, 51, MS, 52; Mich State Univ, PhD, 54. *Prof Exp:* RES SOIL SCIENTIST, US DEPT AGR, 55- *Concurrent Pos:* Vis staff mem, Cornell Univ, 66-67; fel, Grad Facil, Univ Nebr; Food & Agr Orgn consult, India, 88. *Mem:* Soil Sci Soc Am; fel Am Inst Chemists; Int Soil Sci Soc; Am Soc Agron; Soil & Water Conserv Soc. *Res:* Nitrogen and phosphorus transformations as affected by soil management; reclamation of surface mined land; nutrient cycling in rangeland; nitrates in groundwater. *Mailing Add:* USDA Agr Res Serv 122 Keim Hall, UNL Lincoln NE 68583

POWER, JOAN F, b Montreal, Que, May 22, 58. PHOTOCHEMISTRY, PHOTOTHERMAL MATERIALS EVALUATION. *Educ:* Concordia Univ, BSc, 81, PhD(chem), 86. *Prof Exp:* Res assoc appl physics, dept mech eng, Univ Toronto, 86-87; ASST PROF CHEM, MCGILL UNIV, 87- *Mem:* Sigma Xi. *Res:* Photoacoustic & photothermal sciences including microvolume photothermal measurement, & thermooptic theory; thermal wave imaging & non-destructive evaluation using thermal wave techniques; development of photothermal instrumentation. *Mailing Add:* Dept Chem McGill Univ 801 Sherbrooke St W Montreal PQ H3A 2K6 Can

POWER, RICHARD B, biology, for more information see previous edition

POWER, WALTER ROBERT, b Seattle, Wash, Nov 7, 24; m 60; c 2. GEOLOGY. *Educ:* Univ Wash, BS, 50; Johns Hopkins Univ, PhD, 59. *Prof Exp:* Geologist, US Geol Surv, 50-54; asst prof geol, Univ Ga, 57-61; chief geologist, Ga Marble Co, 61-67; PROF GEOL, GA STATE UNIV, 67- *Concurrent Pos:* Consult geologist indust minerals. *Mem:* AAAS; Geol Soc Am; Am Inst Mining, Metall & Petrol Engrs; Soc Econ Geol; Sigma Xi. *Res:* Geology of industrial minerals; economic geology. *Mailing Add:* 3474 Havalyn Lane Doraville GA 30340

POWERS, CHARLES F, limnology, for more information see previous edition

POWERS, DALE ROBERT, b Yankton, SDak, Mar 13, 49; m 74; c 1. PHYSICAL CHEMISTRY, CHEMICAL ENGINEERING. *Educ:* Iowa State Univ, BS, 70; Calif Inst Technol, PhD(phys chem), 75. *Prof Exp:* Consult chem anal, Jet Propulsion Labs, 73-74; SR CHEMIST, CORNING GLASS WORKS, 74- *Mem:* Am Chem Soc. *Res:* Process research on optical waveguide preparation; kinetics of hydrocarbon reactions; kinetics of organometalic compounds; flame science and burner design. *Mailing Add:* 112 Weston Lane Painted Post NY 14870-9356

POWERS, DANA AUBURN, b Ironton, Mo, July 16, 48; m 74. INORGANIC CHEMISTRY, HIGH TEMPERATURE CHEMISTRY. *Educ:* Calif Inst Technol, BS, 70, PhD(inorg chem), 74. *Prof Exp:* MEM TECH STAFF MAT SCI, SANDIA LABS, 75- *Mem:* Am Chem Soc. *Res:* Experimental investigations with the consequences of nuclear reactor core-meltdown accidents; kinetics of metallothermic reactions. *Mailing Add:* 7964 Sarton Way NE Albuquerque NM 87109-3128

POWERS, DANIEL D, b Wichita, Kans, Jan 23, 35; m 60; c 2. ORGANIC CHEMISTRY. *Educ:* Univ Wichita, BS, 56, MS, 58; Univ Kans, PhD(chem), 66. *Prof Exp:* From asst prof to prof chem, Sterling Col, 63-83, chmn dept, 71-83; res assoc, Dept Biochem, Univ NC, Chapel Hill, 84-89; RES ASSOC, DEPT CHEM ENG, NC STATE UNIV, RALEIGH, 89- *Concurrent Pos:* Vis prof, Biochem Dept, Kans State Univ, 78-79. *Mem:* Am Chem Soc. *Res:* Blood coagulation; protein chemistry. *Mailing Add:* 301 Hickory Dr Chapel Hill NC 27514

POWERS, DARDEN, b Holly Springs, Miss, Nov 15, 32; m 57; c 4. ATOMIC & MOLECULAR PHYSICS. *Educ:* Univ Okla, BS, 55; Calif Inst Technol, MS, 57, PhD(physics), 61. *Prof Exp:* From asst prof to assoc prof, 61-68, actg dean grad studies & res, 82, PROF PHYSICS, BAYLOR UNIV, 68-, ASSOC DEAN GRAD STUDIES & RES, 84-; DIR, INST BIOMED STUDIES, 88- *Concurrent Pos:* Head, Van de Graaff Accelerator Lab, Baylor Univ; NSF res grants 64-74 & Robert A Welch Found, 68-88; mem, Int Comn on Radiation Units Stopping Power Comt. *Mem:* Am Phys Soc; Am Asn Physics Teachers. *Res:* Atomic & molecular structure effects on energy loss of alpha particles in matter; Auser electron studies induced by heavy ions; author of 43 publications. *Mailing Add:* Dept Physics Baylor Univ Waco TX 76703

POWERS, DAVID LEUSCH, b Abington, Pa, Feb 17, 39; m 66; c 3. GRAPH & MATRIX THEORY. *Educ:* Carnegie Inst Technol, BS, 60, MS, 61; Univ Pittsburgh, PhD(math), 66. *Prof Exp:* Prof math, Tech Univ Santa Maria, Chile, 65-67; asst prof, 67-71, assoc prof, 71-87, PROF MATH, CLARKSON UNIV, 87- *Concurrent Pos:* Fulbright lectr, State Tech Univ, Chile, 71-72; Orgn Am States res fel, Tech Univ Santa Maria, Chile, 79-80. *Mem:* Soc Indust & Appl Math; Am Math Soc; Sigma Xi; Math Asn Am. *Mailing Add:* Dept Math Clarkson Univ Potsdam NY 13676

POWERS, DENNIS A, b Detroit, Mich, May 4, 38; m 63; c 3. MOLECULAR & PHYSIOLOGICAL ECOLOGY. *Educ:* Ottawa Univ, BA, 63; Univ Kans, PhD(biol), 70. *Prof Exp:* NIH trainee biol, Univ Kans, 67-70; AEC fel biochem, Argonne Nat Labs, 70-71; NSF fel biol, Woods Hole Marine Biol Labs & State Univ NY, Stony Brook, 71-72; from asst prof to prof biol, John Hopkins Univ, 72-88, chmn biol & dir, McCollum-Pratt Inst, 83-88; DIR, HOPKINS MARINE STA, JOHNS HOPKINS, 88-; HAROLD MILLER PROF BIOLSCI, STAN FORD UNIV, CALIF, 88- *Concurrent Pos:* NSF fel, Univ Hawaii, 67; physiol teaching staff, Woods Hole Marine Biol Labs, 72-76; consult, UN Food & Agr Orgn Comt Fish Genetics, 80; ed, Physiol Zool & Biol Oceanog, 82-; consult marine biotechnol. *Mem:* Am Soc Biophysics; Sigma Xi; NY Acad Sci; Genetics Soc Am; Am Soc Zoologists; Am Soc Biol Chemists; Am Chem Soc; Protein Soc; Soc Gen Physiologists; Evolution Soc; Protein Soc. *Res:* Multivariant-interdisciplinary approach to resolve the molecular mechanisms that marine animals use to adapt to environmental stress; genetic engineering of fish; cloning of fish genes; protein chemistry. *Mailing Add:* 1903 W Rogers Ave Baltimore MD 21209

POWERS, DONALD HOWARD, JR, b Providence, RI, Feb 23, 30; m 50; c 3. INORGANIC CHEMISTRY. *Educ:* Amherst Col, Mass, AB, 51; Brown Univ, Providence, RI, PhD(org chem), 55. *Prof Exp:* Fel, Univ Rochester, 54-55; chemist, Owens-Corning Fiberglas Corp, 55-59; sr res assoc, Fabric Res Labs, Inc, 59-71; consult, Fram Corp, 72-77; SR RES CHEMIST, CHOMERICS, INC, 77- *Mem:* Am Chem Soc; Am Electroplaters Soc; Sigma Xi. *Res:* Mechanism of organic reactions; adhesion to glass fibers; wool processing; cotton finishing, fiber spinning; glass fiber manufacuture; crosslinking of thermoplastic resins; electroless plating of fine particles. *Mailing Add:* Chomerics 77 Dragon Ct Woburn MA 01888

POWERS, EDMUND MAURICE, b Tarrytown, NY; m 62; c 2. FOOD MICROBIOLOGY. *Educ:* Univ NH, BA, 59; Univ Md, MS, 64. *Prof Exp:* Bacteriologist, Norwich Pharmacal Co, 59-62; life scientist space biol, Goddard Space Flight Ctr, NASA, 64-68; RES FOOD MICROBIOLOGIST, US ARMY NATICK RES DEVELOP & ENG CTR, 68- *Concurrent Pos:* Mem intersoc, Agency Comt Compendium Methods Microbiol Exam Foods, 70-80. *Honors & Awards:* Dirs Silver Pin Award Res, 79. *Mem:* Am Soc Microbiol; Int Asn Milk Food & Environ Sanitarians; Nat Registry Microbiologists. *Res:* Recovery of microbes from food; rapid methods for microbiological analysis of foods; sanitation; water purification; one patent. *Mailing Add:* SSD US Army Natick Res Develop & Eng Ctr Natick MA 01760

POWERS, EDWARD JAMES, b Pittsfield, Mass, Nov 27, 36; m 67; c 2. MATERIALS SCIENCE, APPLICATIONS DEVELOPMENT. *Educ:* Siena Col, BS, 61; Fla State Univ, PhD(org chem), 69. *Prof Exp:* Res chemist, Celanese Res Co, Summit, NJ, 69-72, sr res chemist, 74-76; sr develop engr, Fiber Industs Inc, Charlotte, NC, 76-80; develop assoc, Celanese Plastics & Specialties Co, Louisville, Ky, 80-83; RES ASSOC, HOECHEST CELANESE, STRATEGIC BUS DEVELOP, CHARLOTTE, NC, 83- *Mem:* Am Chem Soc; AAAS; fel Am Inst Chemists. *Res:* Specialty product development; application development; product safety; coatings applications development (flexible packaging); adhesion; fiber finish development; structure/property relationships of polymer forms. *Mailing Add:* 3501 Haverstick Pl Charlotte NC 28226

POWERS, EDWARD JOSEPH, JR, b Winchester, Mass, Nov 29, 35; m 59; c 4. ELECTRICAL ENGINEERING, PHYSICS. *Educ:* Tufts Univ, BS, 57; Mass Inst Technol, MS, 59; Stanford Univ, PhD(elec eng), 65. *Prof Exp:* Res asst microwave circuits & antennas, Mass Inst Technol, 57-59; scientist, Lockheed Missiles & Space Co, 59-65; res asst plasmas, Stanford Univ, 65; from asst prof to assoc prof, 65-73, prof elec eng, 73-81, N GAFFORD PROF ELEC ENG, UNIV TEX, AUSTIN, 81-, DIR ELECTRONICS RES CTR, 77, CHMN DEPT ELEC ENG, 81- *Concurrent Pos:* Mem honor coop prog, Lockheed Aircraft Corp, Stanford Univ, 59-65. *Mem:* Am Phys Soc; Inst Elec & Electronics Engrs; Int Sci Radio Union; Sigma Xi. *Res:* Digital time series analysis; plasma diagnostics; laser applications; electromagnetics. *Mailing Add:* 8703 Mountainwood Circle Austin TX 78759

POWERS, EDWARD LAWRENCE, b Columbia, SC, Dec 30, 15; m 39; c 7. RADIATION BIOLOGY. *Educ:* Col Charleston, BSc, 38; Johns Hopkins Univ, PhD(zool), 41. *Hon Degrees:* LHD, Col Charleston, 74. *Prof Exp:* From instr to asst prof biol, Univ Notre Dame, 41-46; from assoc biologist to sr biologist, Argonne Nat Lab, 46-65, assoc dir div biol & med res, 49-62; prof zool & dir lab radiation biol, Univ Tex, Austin, 65-87, dir Ctr Fast Kinetics Res, 74-86, TS Painter Centennial Prof Genetics, 85-87, TS PAINTER CENTENNIAL EMER PROF GENETICS, UNIV TEX, AUSTIN, 87-; PROF IN RESIDENCE, COL CHARLESTON, 87- *Concurrent Pos:* Guggenheim fel, Eng, 58-59; Douglas Lea mem lectr, Univ Leeds, 61; mem, Radiation Study Sect, NIH, 70-74. *Mem:* Fel AAAS; Genetics Soc Am; Am Soc Microbiol; Soc Protozool; Radiation Res Soc (treas, secy, secy-treas, 55-63, vpres, 63-64, pres, 64-65). *Res:* Action of genetic determiners in cells; ultrastructure of cell inclusions; genetic effects of atomic transmutations in cells; fast kinetic techniques applied to biology; roles of water in radiation damage in living cells; general radiation biophysics; primary physico-chemical effects, especially oxygen, in radiation biology. *Mailing Add:* Grice Marine Biol Lab Col Charleston 205 Ft Johnson Charleston SC 29412

POWERS, GARY JAMES, b Highland Park, Mich, Sept 18, 45; m 66; c 3. CHEMICAL ENGINEERING. *Educ:* Univ Mich, BSChE, 67; Univ Wis, PhD(chem eng), 71. *Prof Exp:* Engr, Dow Chem Co, 68-70; asst prof chem eng, Mass Inst Technol, 71-74; assoc prof, 74-75, PROF CHEM ENG, CARNEGIE-MELLON UNIV, 75- *Concurrent Pos:* Dir, Design Res Ctr; pres, Design Sci, Inc, 77- *Mem:* AAAS; Am Inst Chem Engrs; Am Chem Soc. *Res:* Computer-aided synthesis of chemical processing systems; separation science; risk assessment and fault free analysis. *Mailing Add:* Witherow Rd Sewickley PA 15143

POWERS, HARRY ROBERT, JR, b Suffolk, Va, July 25, 23; m 47; c 2. PLANT PATHOLOGY. *Educ:* NC State Col, BS, 49, PhD(plant path), 53; Duke Univ, MF, 50. *Prof Exp:* Pathologist, Agr Res Serv, 53-59, Southeastern Forest Exp Sta, 59-68, prin plant pathologist, 68-80, CHIEF PLANT PATHOLOGIST, FORESTRY SCI LAB, US DEPT AGR, 80- *Mem:* Am Phytopath Soc; Soc Am Foresters. *Res:* Tree diseases; wheat mildew and rust; genetics. *Mailing Add:* Forestry Sci Lab Carlton St Athens GA 30601

POWERS, J BRADLEY, b Framingham, Mass, Dec 16, 37; m 66; c 2. BEHAVIORAL NEUROSCIENCE. *Educ:* Harvard Col, AB, 59; Brown Univ, MA, 63; Univ Calif, Berkeley, PhD(physiol psychol), 70. *Prof Exp:* Asst res scientist, Univ Mich, 74-77; sr fel, Univ Wash, 78-79; RES ASSOC PROF, VANDERBILT UNIV, 80- *Mem:* Soc Neurosci; Animal Behav Soc; Sigma Xi. *Res:* Elucidating neuroendocrine and environmental mechanisms regulating mammalian social behavior, especially reproduction. *Mailing Add:* Vanderbilt Univ A&S Psychol Bldg Nashville TN 37240

POWERS, JAMES CECIL, b Highland Park, Mich, Dec 13, 37; m 66; c 2. BIOORGANIC CHEMISTRY. *Educ:* Wayne State Univ, BS, 59; Mass Inst Technol, PhD(chem), 63. *Prof Exp:* Asst prof chem, Univ Calif, Los Angeles, 63-67; NIH spec fel, Univ Wash, 67-69, actg asst prof chem, 69-70; from asst prof to prof chem, 70-87, REGENTS PROF CHEM, GA INST TECHNOL, 87- *Concurrent Pos:* Consult, Merck, Inc, 75-87. *Mem:* Am Chem Soc; AAAS; Am Soc Biol Chemists. *Res:* Enzyme inhibitors; enzyme mechanism; drug design; proteases; chemical defense; blood coagulation. *Mailing Add:* Dept Chem Ga Inst Technol Atlanta GA 30332

POWERS, JAMES MATTHEW, b Cleveland, Ohio, Sept 15, 43; c 2. NEUROPATHOLOGY, PATHOLOGY. *Educ:* Manhattan Col, BS, 65; Med Univ SC, MD, 69. *Prof Exp:* From asst prof to prof path, Med Sch SC, 73-88; PROF PATH, COLUMBIA UNIV, NY, 88-, VCHMN DEPT PATH, 89- *Concurrent Pos:* Dir electron micros, Vet Admin Hosp, SC, 73-76. *Honors & Awards:* Moore Award, Am Asn Neuropathologists, 75, 76, 77 & 81. *Mem:* Am Asn Neuropathologists (asst secy-treas, 76-81); Am Asn Pathologists; Int Acad Path; NY Acad Sci. *Res:* Adreno-leukodystrophy; diseases of myelin; dementia, especially Alzheimer's senile and presenile. *Mailing Add:* Dept Pathol-Neuropath Columbia Univ 630 W 168th St New York NY 10032

POWERS, JEAN D, b Hamilton, Ohio, Dec 24, 30; m 75; c 2. BIOMETRICS, BIOSTATISTICS. *Educ:* Ohio State Univ, BS, 50, MS, 51, PhD(biostatist), 68. *Prof Exp:* Statistician, Ohio Dept Health, 51-52; instr, St Mary's High Sch, 63-64; assoc prof math, Capital Univ, 68-73; asst prof, 73-80, ASSOC PROF BIOSTATIST, OHIO STATE UNIV, 80- *Concurrent Pos:* Vis asst prof, Ohio State Univ, 68-73; statistician res found cancer grant, Ohio State Univ, 78-80; consult, Abbott Labs, Am Hoechst, Schering Corp, 82- *Mem:* Am Statist Asn; Am Acad Pharmacologists & Therapeut; Am Soc Vet Physiologists & Pharmacologists; Sigma Xi. *Res:* Estimating pharmacokinetic parameters. *Mailing Add:* Vet Clin Sci Ohio State Univ 1935 Coffey Rd Columbus OH 43210

POWERS, JOHN CLANCEY, JR, b Billings, Mont, Dec 27, 31; m 64; c 2. ORGANIC CHEMISTRY, BIOCHEMISTRY. *Educ:* Yale Univ, BS, 53; Harvard Univ, AM, 54, PhD(chem), 58. *Prof Exp:* Fel chem, Bryn Mawr Col, 58-59; chemist, Metal Hydrides, Inc, 59-61; res chemist, Int Bus Mach Corp, 61-67; asst prof chem, Hunter Col, 67-70; assoc prof, 70-80, PROF CHEM,

PACE UNIV, 81- *Mem:* Am Chem Soc; Sigma Xi. *Res:* Reaction mechanisms; photochemistry; electrooptic effects in dyes and polymers; chemistry of biologically important macromolecules. *Mailing Add:* Dept Chem Pace Univ New York NY 10038

POWERS, JOHN E(DWARD), b Wilkensburg, Pa, Oct 12, 27; m 51; c 3. CHEMICAL ENGINEERING. *Educ:* Univ Mich, BS, 51; Univ Calif, Berkeley, PhD(chem eng), 54. *Prof Exp:* Assoc chem eng, Univ Calif, Berkeley, 53-54; res engr, Shell Develop Co, 54-56; asst prof chem eng, Univ Okla, 56-58, assoc prof & chmn dept, 58-61, prof, 61-63; PROF CHEM METALL ENG, UNIV MICH, ANN ARBOR, 63- *Concurrent Pos:* Mem at large, Gordon Res Confs Coun, 59-62, chmn conf & purification, 60; NSF sr fel, Univ Erlangen, 62-63. *Mem:* AAAS; Am Inst Chem Engrs; Inst Chem Eng, PR. *Res:* Separation processes; crystallization; thermodynamics; biomedical applications of chemical engineering. *Mailing Add:* Castle Valley SR Box 2310 Moab UT 84532

POWERS, JOHN JOSEPH, b Pittsfield, Mass, Feb 3, 18; m 45; c 4. SENSORY ANALYSIS. *Educ:* Univ Mass, BS, 40, PhD(food technol), 45. *Prof Exp:* Instr food technol, Univ Mass, 42-46; asst prof, Exp Sta, Ohio State Univ, 46-47; assoc prof, 47-52, head dept, 52-67, prof food technol, 52-79, William Terrell distinguished prof agr, 79-88, EMER WILLIAM TERRELL DISTINGUISHED PROF, UNIV GA, 88- *Concurrent Pos:* Mem, Toxicol Study Sect, USPHS, 64-67, Environ Biol & Chem Study Sect, 67-70; sr Fulbright-Hays fel, 72 & 86; mem, Nat Acad Sci Exchange Visit, Bulgaria, 72, Yugoslavia & Hungary, 74 & Poland, 77, US-Spain, 83; Am deleg to ISO meetings, 83, 87, 89, 90. *Honors & Awards:* Res Award, Soc Med Friends Wine, 72; Nicholas Appert Award, 84; Wm V Cruess Award, 82; Merit Award, Am Soc Testing & Mat, 84. *Mem:* Am Chem Soc; fel Am Soc Testing & Mat; fel Inst food technol (pres-elect, 85-86, pres, 86-87, treas, 90-91). *Res:* Correlation of sensory patterns with chemical composition; plant compounds; food industries methods; acidification of foods; flavor chemistry; application of multivariate statistical procedures to analysis of multidimensional profile of foods and other materials possessing sensory properties. *Mailing Add:* Dept of Food Sci Univ of Ga Athens GA 30602

POWERS, JOHN MICHAEL, b Ft Wayne, Ind, Apr 16, 46; m 68; c 2. DENTAL MATERIALS. *Educ:* Univ Mich, Ann Arbor, BSCh, 67, PhD(dent mat, mech eng), 72. *Prof Exp:* Res asst, Sch Dent, Univ Mich, Ann Arbor, 65-67; instr health sci, Washtenaw Community Col, 68-69; teaching fel, 69-72, from asst prof to assoc prof, 72-79, PROF DENT MAT, SCH DENT, UNIV MICH, ANN ARBOR, 79- *Mem:* Int Asn Dent Res; Soc Biomat; Acad Dent Mat. *Res:* Physical and mechanical properties of materials as applied to dentistry; thermal analysis of dental materials; friction and wear of dental materials; color and optical properties of dental materials. *Mailing Add:* Oral Biomat Dent Br Univ Tex Health Sci Ctr 6516 John Freeman Houston TX 77030

POWERS, JOHN ORIN, b Mt Olive, Ill, Mar 6, 22; m 48; c 2. AERODYNAMICS. *Educ:* Univ Mich, BS(aeronaut eng) & BS(mech eng), 43; Cath Univ, MS, 51; Univ Md, PhD(aeronaut eng), 65. *Prof Exp:* Aeronaut engr, David Taylor Model Basin, Dept Navy, 48-50, aeronaut engr, Bur of Aeronaut, 50-51; aeronaut engr, Proj Hermes, Gen Elec Co, 51-53, Guided Missiles Dept, 53-55 & Missile & Ord Dept, 55-57; aeronaut engr, US Naval Ord Lab, 57-66; chief tech support staff, Off Noise Abatement, 66-69, actg dir, 69-71, chief environ scientist, off environ & energy, fed aviation admin, 71-87; RETIRED. *Concurrent Pos:* Mem sonic boom res panel, Interagency Noise Prog, 68- *Honors & Awards:* Distinguished Career Service Award Fed Aviation Admin, 87. *Mem:* Soc Automotive Engrs; Am Inst Aeronaut & Astronaut. *Res:* Boundary layer stability; stability and control characteristics of aircraft and missiles; aircraft noise emissions; sonic boom. *Mailing Add:* 13700 Carlisle Silver Spring MD 20904

POWERS, JOHN PATRICK, b Winchester, Mass, Dec 28, 43; m 67; c 2. ELECTRICAL ENGINEERING. *Educ:* Tufts Univ, BS, 65; Stanford Univ, MS, 66; Univ Calif, Santa Barbara, PhD(elec eng), 70. *Prof Exp:* Res asst elec eng, Univ Calif, Santa Barbara, 67-70; asst prof, 70-76, ASSOC PROF ELEC ENG, NAVAL POSTGRAD SCH, 76- *Concurrent Pos:* Vis exchange scientist, Univ Paris VI, 74-75. *Mem:* Inst Elec & Electronics Engrs; Acoust Soc Am; Optical Soc Am; Sigma Xi. *Res:* Acoustic imaging systems; electro optics. *Mailing Add:* US Naval Postgrad Sch Code 62 Po Monterey CA 93943

POWERS, JOSEPH, b Fall River, Mass, June 4, 31; m 57; c 2. MATERIALS SCIENCE, ADVANCED COMPOSITES PROCESSING. *Educ:* Univ Mass, BS, 53, MS, 59, PhD(phys chem), 61. *Prof Exp:* Off Naval Res fel, 61; proj leader polymer physics, Nat Bur Standards, 61-65; proj leader phys chem, Am Cyanamid Co, Stamford, 65-69, group leader, 69-72; Instr, Univ Bridgeport, 72-73; sr anal engr, 73-77, res engr, Fuel Cell Opers, SR MAT ENGR, SIKORSKY AIRCRAFT, UNITED TECHNOLOGIES CORP, 82- *Concurrent Pos:* Nat Acad Sci-Nat Res Coun fel, 61-62; managing ed, Macromolecular Ed, Chemtracts. *Mem:* Am Phys Soc; Am Chem Soc; AAAS; Soc Plastics Engrs; Soc Adman Mat & Processing Eng; Sigma Xi. *Res:* Processing and chemical characterization of fiber reinforced plastics; physical properties of polymers, polymer stability in severe environments, thermal, chemical and electrical; use of polymers in fuel cell applications; states of aggregation in polymeric systems. *Mailing Add:* 26 Spring St Riverside CT 06878

POWERS, JOSEPH EDWARD, b Gustine, Calif, Feb 19, 49; m 73. FISHERIES MANAGEMENT, OPERATIONS RESEARCH. *Educ:* Univ Calif, Davis, AB, 71; Calif State Univ, Humboldt, MS, 73; Va Polytech Inst & State Univ, PhD(fish & wildlife), 75. *Prof Exp:* Systs analyst fisheries, Dept Fisheries & Wildlife Sci, Va Polytech Inst & State Univ, 75; opers res analyst fisheries, Southwest Fisheries Ctr, Nat Marine Fisheries serv, 75-80; MEM STAFF, NAT MARINE FISHERIES SERV, SOUTHEAST FISHERIES CTR, 80- *Mem:* Am Fisheries Soc; Soc Comput Simulation; Opers Res Soc. *Res:* Application of operations research technology to the management and planning of environmental and ecological systems. *Mailing Add:* Nat Marine Fisheries Serv 75 Virginia Beach Dr Miami FL 33149

POWERS, JOSEPH ROBERT, b Tillamook, Ore, July 31, 48; m 75; c 2. BIOCHEMISTRY OF FOODS PROTEINS, ENZYMOLOGY. *Educ:* Ore State Univ, BS, 70; Wash State Univ, MS, 73; Univ Calif, Davis, PhD(biochem), 77. *Prof Exp:* Asst food scientist, Irrigated Agr Res Ext Ctr, 76-80, asst prof, 80-83, ASSOC PROF FOOD SCI, DEPT FOOD SCI & TECHNOL, WASH STATE UNIV, 83- *Concurrent Pos:* Prin investr, Pac NW Regional Comn, 77-80. *Mem:* Inst Food Technologists; Am Soc Enologists; Am Asn Cereal Chemists; Sigma Xi. *Res:* Biochemistry of food proteins including structure, function relations, biological activity; enzymes of foods especially plant products in relation to food quality specifically related to fiber (lignin) biosynthesis. *Mailing Add:* Dept Food Sci Wash State Univ Pullman WA 99164-6330

POWERS, K(ERNS) H(ARRINGTON), b Waco, Tex, Apr 15, 25; m 52; c 2. COMMUNICATIONS. *Educ:* Univ Tex, BS & MS, 51; Mass Inst Technol, ScD(elec eng), 56. *Prof Exp:* Chief engr, Radio Sta KTXN, Tex, 49-51; mem tech staff, RCA Labs, 51-53; res asst, Mass Inst Technol, 55-56; mem tech staff, RCA Labs, 56-60, tech dir new systs & spec projs, 60-66, dir commun res lab, 66-77, STAFF V PRES COMMUN RES, RCA CORP, 77- *Concurrent Pos:* Adj prof elec eng, Newark Col Eng, 57-60. *Mem:* Fel Inst Elec & Electronics Engrs; fel Soc Motion Picture & Television Engrs. *Mailing Add:* David Sarnoff Res Ctr RCA Corp Princeton NJ 08540

POWERS, KENDALL GARDNER, b Rockland, Maine, May 15, 30; m 60; c 1. VETERINARY PARASITOLOGY. *Educ:* Univ Wis, BS, 56, MS, 58, PhD(vet parasitol-bacteriol), 61. *Prof Exp:* Instr vet sci, Univ Wis, 60-61; head parasitol res, Schering Corp, 61-63; res parasitologist, Nat Inst Allergy & Infectious Dis, NIH, 63-75; head parasitol res, Bur Vet Med, Fed Drug Admin, 75-82; HEALTH SCIENTIST ADMINR, NAT CANCER INST, 82-, DIV RES GRANTS, NIH. *Concurrent Pos:* Captain, Med Serv Corps, US Naval Reserve. *Mem:* Am Soc Trop Med & Hyg; Am Soc Parasitol; Royal Soc Trop Med & Hyg; Sigma Xi. *Res:* Chemotherapy and immunology of parasitic diseases. *Mailing Add:* 6311 Alcott Rd Bethesda MD 20817

POWERS, LARRY JAMES, b Dillsboro, Ind, July 11, 44; m 68; c 2. MEDICINAL CHEMISTRY. *Educ:* Purdue Univ, Lafayette, BS, 66; Univ Kans, PhD(med chem), 69. *Prof Exp:* Asst prof, Med Units, Univ Tenn, Memphis, 69-73, assoc prof, 73-75; sr res chemist, Diamond Shamrock Corp, 75-77, res assoc, 77-79, group leader, 80-85; MGR, RICERCA INC, 85- *Mem:* Am Chem Soc; AAAS; NY Acad Sci. *Res:* Herbicide and fungicide synthesis; contract services in agricultural research/chemistry/fermentation. *Mailing Add:* 5820 Chapel Rd Madison OH 44057-1754

POWERS, LINDA SUE, b Pittsburgh, Pa, Feb 8, 48. SYNCHROTRON RADIATION, MOLECULAR STRUCTURE & ELECTRONICS. *Educ:* Va Polytech Inst & State Univ, BS, 70; Harvard Univ, MA, 72, PhD, 76. *Prof Exp:* Mem tech staff, AT&T Bell Labs, 76-88; PROF CHEM & BIOCHEM, UTAH STATE UNIV, 88- *Concurrent Pos:* Adj prof biophys, Univ Pa, 78-; mem exec bd, Div Biol Physics, Am Phys Soc, 79-83, chairperson, 86-87; vis fel, Princeton Univ, 81 & 86; organizer, VII Int Biophys Cong, 81; mem coun, Am Phys Soc, 88-92. *Honors & Awards:* First US Bioenergetics Award, Biophys Soc, 82. *Mem:* Fel Am Inst Chemists; fel Am Phys Soc; Biophys Soc; Am Soc Biochem & Molecular Biol; Am Chem Soc; Protein Soc; Soc Appl Spectros; NY Acad Sci; AAAS. *Res:* Physical studies of biological molecules to investigate structure-function relationship; development and application of synchrotron radiation and optical spectroscopy to metalloenzymes and proteins; structural kinetics; enzyme mechanics; molecular electronics. *Mailing Add:* Ctr Bio Catalysis Sci & Technol Utah State Univ Logan UT 84322-4630

POWERS, LOUIS JOHN, b St Louis, Mo, Oct 8, 12; m 35; c 3. MECHANICAL ENGINEERING. *Educ:* Tex Tech Col, BS, 39; Univ Tex, MS, 50. *Prof Exp:* Mech engr, Gen Petrol Corp, 39-42; assoc prof mech eng, 42-52, PROF MECH ENG, TEX TECH UNIV, 52- *Mem:* Am Soc Eng Educ; Am Soc Mech Engrs; Nat Soc Prof Engrs. *Res:* Structural vibrations. *Mailing Add:* Dept Mech Eng Tex Tech Univ Lubbock TX 79409

POWERS, MARCELINA VENUS, b Abra, Philippines, Mar 22, 27; m 60; c 1. TOXICOLOGY. *Educ:* Univ Philippines, DVM, 51; Univ Wis, MS, 57. *Prof Exp:* Instr biol sci, Araneta Univ, Philippines, 51-55; res asst vet sci, Univ Wis, 55-57, staff vet & toxicologist, Wis Alumni Res Found, 57-59, head dept biol, 59-63; res coordr, Hazelton Labs, 63-69, asst dir drugs & indust chem dept, 69-70, dir dept, 70-72, proj mgr, NCI/NIH ,72-78; staff toxicologist, 78-84, HEALTH SCIENTIST ADMINR, DRG/NIH, 84- *Mem:* Am Vet Med Asn; Am Soc Trop Med & Hyg; Soc Toxicol; Am Col Vet Parasitol. *Res:* Toxicology of drugs, industrial and agricultural chemicals, chemical carcinogenesis; environmental carcinogenesis, nutrition and cancer. *Mailing Add:* 6311 Alcott Rd Bethesda MD 20817

POWERS, MICHAEL JEROME, b Freeport, Ill, May 5, 41; m 64; c 3. SYSTEMS ANALYSIS & DESIGN. *Educ:* Carthage Col, BA, 63; Ind Univ, Bloomington, MA, 65, PhD(math), 68. *Prof Exp:* Asst prof math, Ind State Univ, 67-68 & Northern Ill Univ, 68-74; mem staff, De Kalb Agresearch, 74-76; ASSOC PROF & CHMN APPL COMPUT SCI DEPT, ILL STATE UNIV, 76- *Mem:* Asn Comput Mach; Data Processing Mgt Asn; Asn Educ Data Systs; Inst Elec & Electronic Engrs; Sigma Xi. *Res:* System development and design; database processing. *Mailing Add:* Appl Comput Sci Ill State Univ Normal IL 61761

POWERS, RICHARD JAMES, b Chicago, Ill, Jan 20, 40. PHYSICS. *Educ:* Ill Inst Technol, BS, 61; Univ Chicago, MS, 62, PhD(physics), 67. *Prof Exp:* Res assoc physics, Univ Chicago, 67-68; asst prof, Va Polytech Inst & State Univ, 68-72; sr res fel, 73-76, from res assoc to sr res assoc physics, Calif Inst Techno, 76-82; mem sci, staff, Arete Assoc, 85-86; VIS SCIENTIST, UNIV BASEL, 87- *Concurrent Pos:* Vis scientist, CEN-Saclay, 72-73 & 90, Swiss Fed Inst Technol, 79-81 & 82, Univ Zürich, 80 & 83, Nuclear Res Ctr, Karlsruhe, Fed Rep Ger, 84-85, Col William & Mary, 89. *Mem:* Am Phys Soc. *Res:* Nuclear charge and matter distribution using x-rays from exotic atoms; weak interactions of leptons; oceanography; internal waves; remote sensing; synthetic aperture radar. *Mailing Add:* Inst Physics Klingelbergstrasse 82 Basel CH-4056 Switzerland

POWERS, ROBERT D, b Scotts Hill, Tenn, Sept 19, 33; m 62; c 1. PATHOLOGY, LABORATORY ANIMAL MEDICINE. *Educ:* Univ Tenn, BS, 57, PhD(exp path), 67; DVM, Auburn Univ, 62. *Prof Exp:* NIH training fel, 63-67; asst prof path & assoc dir, Vivarium, Med Units, Univ Tenn, 67-70; PROF PATH & PARASITOL, AUBURN UNIV, 70- *Concurrent Pos:* Am Cancer Soc study grant, 64-65. *Mem:* AAAS; Am Vet Med Asn; Am Asn Zool Vets; Am Asn Lab Animal Sci; Am Soc Lab Animal Practrs. *Res:* Veterinary pathology; experimental and comparative pathology; zoo animal pathology. *Mailing Add:* Dept Path Auburn Univ Auburn AL 36830

POWERS, ROBERT FIELD, b Los Angeles, Calif; c 2. FOREST SOILS, PLANT NUTRITION. *Educ:* Humboldt State Univ, BS, 63; Univ Calif, Berkeley, PhD, 81. *Prof Exp:* Res asst forest ecol, Humboldt State Univ, 64-65; PRIN SILVICULTURIST, TIMBER MGT RES, PAC SOUTHWEST FOREST & RANGE EXP STA, FOREST SERV, USDA, 66- *Concurrent Pos:* Vchmn, Calif Forest Fertil Coun, 74-81, chmn, 81-82; mem, Regionwide Forest Soil Fertil Steering Comt, Forest Serv, USDA, 76-, Working Group Soil & Site Factors, Int Union Forest Res Orgn, 76- & Planning & Selection Comt, 7th & 8th NAm Forest Soils Conf, 88; guest lectr, Univ Calif, Humboldt State Univ, N Ariz State Univ, Wash State Univ, Univ Melbourne, 78; assoc ed, J Am Soil Sci Soc, 81-88; chmn, Calif Forest Soils Coun, 81-83, Forest & Range Soils Div, Soil Sci Soc Am, 89, Nat Long-Term Productivity Tech Comt, 89- & PSW Comt Scientists; mem adv comt, 7th NAm Forest Biol Workshop, 82, mem selection, 87- 88, co ed, 88-89; mem & chmn, Agr Natural Res Adv Comt, Shasta Col, 84 & 87-90. *Mem:* AAAS; Int Energy Asn; Int Union Forest Res Orgn; Soc Am Foresters; Soil Sci Soc Am. *Res:* Nutrient requirements of forest trees; properties of forest ecosystems affecting productivity; effects of forest practices on soil fertility and forest productivity; author of 60 publications. *Mailing Add:* Pac Southwest Exp Sta 2400 Washington Ave Redding CA 96001

POWERS, ROBERT S(INCLAIR), JR, b Dallas, Tex, Apr 7, 34; m 62; c 2. PHYSICAL CHEMISTRY, ELECTRICAL ENGINEERING. *Educ:* Southern Methodist Univ, BS, 55; Univ Wis, PhD(phys chem), 60. *Prof Exp:* Nat Bur Standards-Nat Res Coun fel plasma physics, Boulder Labs, Nat Bur Standards, 60-61, chemist, 61-65, gen phys scientist, 65-70; spec asst urban telecommun, Off Telecommun, US Dept Commerce, 70-75; sr engr, Cable TV Bur, Fed Commun Comn, 75-79, rep, off Sci & Technol, 79-83, chief scientist, 83-85; ADV ENGR, MCI TELECOMMUNICATIONS CORP, 85- *Concurrent Pos:* Chair, US comt commun & info policy, Inst Elec & Electronic Engrs. *Honors & Awards:* Achievement Award, US Dept Com, 73 & Fed Commun Comn, 80. *Mem:* AAAS; Inst Elec & Electronic Engrs. *Res:* Urban communications systems analysis; communications technology and standards; communications security. *Mailing Add:* MCI Telecom 1077/107 2400 N Glenville Dr Richardson TX 75083

POWERS, ROBERT WILLIAM, b Peoria, Ill, Feb 3, 22; m 50; c 1. PHYSICAL CHEMISTRY. *Educ:* Bradley Univ, BS, 43; Univ Ill, PhD(phys chem), 46. *Prof Exp:* Chemist, Exp Sta, E I du Pont de Nemours & Co, 46-47; res assoc, Res Found, Ohio State Univ, 47-51; phys chemist, Res & Develop, Gen Elec Corp, 51-87; RETIRED. *Mem:* Am Ceramic Soc; Am Chem Soc; Electrochem Soc; Am Phys Soc. *Res:* Polymerization kinetics; thermal conductivity measurements at low temperatures; mechanical relaxation in metals; anodic film formation; alkaline zinc electrochemistry; fabrication properties and degradation of beta-alumina. *Mailing Add:* 1507 Kingston Ave Schenectady NY 12308-1507

POWERS, THOMAS E, b Cedarville, Ohio, Dec 29, 25; m 48, 75; c 3. VETERINARY PHARMACOLOGY. *Educ:* Ohio State Univ, DVM, 53, MSc, 54, PhD(physiol, pharmacol), 60. *Prof Exp:* Instr bact, Ohio State Univ, 54-55, from instr to assoc prof, 55-66, chmn dept, 70-84, mem grad fac, 60-89, prof vet physiol & pharmacol, 66-89, prof clin pharmacol, 66-89, CHMN & EMER PROF VET PHYSIOL, PHARMACOL & CLIN PHARMACOL, OHIO STATE UNIV, 90- *Honors & Awards:* Am Acad Vet Pharmacol & Therapeut Award, 83. *Mem:* Am Col Vet Pharmacol & Therapeut (pres, 81-83); AAAS; Am Vet Med Asn; Am Asn Vet Physiologists & Pharmacologists (pres, 75); Am Col Vet Clin Pharmacol (pres, 90-). *Res:* Veterinary pharmacodynamics; renal physiology; comparative pharmacology and physiology; antimicrobial therapy; animal disease modeling; clinical pharmacology. *Mailing Add:* Dept Vet Pharmacol Ohio State Univ 1900 Coffey Rd Columbus OH 43210

POWERS, WENDELL HOLMES, b Richford, Vt, Mar 30, 15; m 42; c 2. BIOCHEMISTRY. *Educ:* Middlebury Col, BS, 37; Univ NH, MS, 39; Columbia Univ, PhD, 43. *Prof Exp:* Instr chem, Univ NH, 38-39; asst, Columbia Univ, 39-42; from instr to prof, 42-81, prof, EMER PROF CHEM, WAYNE STATE UNIV, 81- *Concurrent Pos:* Asst dir, Kresge-Hooker Sci Libr Assocs, 43-49, actg dir, 49-50, assoc dir, 50-56, exec secy, 56-69; ed, Record Chem Prog, 52-71. *Mem:* Sigma Xi; AAAS; Am Chem Soc. *Res:* Synthetic organic chemistry involving preparation of various aromatic ethers; respiratory enzymes and biological oxidations. *Mailing Add:* 24600 Greater Mock No 218 St Clair Shores MI 48080-1321

POWERS, WILLIAM ALLEN, III, b Baltimore, Md. STATISTICS. *Educ:* Univ Richmond, BS, 66; Univ Conn, MS, 70, PhD(statist), 72. *Prof Exp:* Lectr math, Univ Bridgeport, 70-71; lectr, 71-72, asst prof, 72-77, assoc prof math, Univ NC Greensboro, 77-84, GROUP MGR, EDUC DIV, SAS INST, CARY, NC, 84- *Mem:* Am Statist Asn; Math Asn Am. *Res:* Applied statistical methods; unbalanced A nova; robustness of statistical procedures. *Mailing Add:* PO Box 4145 Cary NC 27519

POWERS, WILLIAM FRANCIS, b Philadelphia, Pa, Dec 11, 40; m 63; c 2. ELECTRONICS ENGINEERING, MECHANICAL ENGINEERING. *Educ:* Univ Fla, BS, 63; Univ Tex, Austin, MS, 66, PhD(eng mech), 68. *Prof Exp:* Mathematician, Marshall Space Flight Ctr, NASA, 63-64; from asst prof to assoc prof aerospace eng, Univ Mich, Ann Arbor, 68-75, prof aerospace eng, 76-80; mgr, Control Systs Dept, Ford Motor Co, 80-86, dir, Powertrain Syst & Res, 86-87, dir prod & mfr systs, 87-89, PROG MGR, SPECIALTY CAR PROGS, FORD MOTOR CO, 89- *Concurrent Pos:* Consult, Johnson Space Ctr, NASA, 71-79; pres, Am Automatic Control Coun, 87-89; chmn, NSF Adv Comt Elec Comm & Systs Eng, 88-89. *Mem:* Am Inst Aeronaut & Astronaut; Soc Automotive Engrs; Am Soc Mech Engrs; Inst Elec & Electronic Engrs. *Res:* Optimization of dynamic systems; guidance and control; automotive control systems; automotive systems engineering. *Mailing Add:* 2032 Greenview Ann Arbor MI 48103

POWERS, WILLIAM JOHN, b Northampton, Mass, April 28, 49; m 83; c 2. CEREBRAL BLOOD FLOW & METABOLISM, POSITRON EMISSION. *Educ:* Dartmouth Col, AB, 71; Cornell Univ, MD, 75. *Prof Exp:* Resident internal med, Duke Univ Med Ctr, 75-77; resident neurol, Univ Calif, San Francisco, 77-79, chief resident, 79-80; from instr to asst prof neurol & radiol, 80-87, ASSOC PROF NEUROL & RADIOL, WASH UNIV SCH MED, 87- *Concurrent Pos:* Attend neurologist, Barnes Hosp, 80-; neurologist-in-chief, Jewish Hosp St Louis, 86; mem Stroke Coun, Am Heart Asn, exec comt, 87-91. *Mem:* Fel Am Col Physicians; Am Acad Neurol; Int Soc Cerebral Blood Flow & Metab; Am Neurol Asn. *Res:* Cerebral blood flow and metabolism. *Mailing Add:* Dept Neurol Jewish Hosp St Louis 216 S Kingshwy St Louis MO 63110

POWERS LEE, SUSAN GLENN, ENZYMOLOGY, PROTEIN CHEMISTRY. *Educ:* Univ Calif, Berkeley, PhD(biochem), 75. *Prof Exp:* ASST PROF BIOCHEM, NORTHEASTERN UNIV, 83- *Mailing Add:* Dept Biol Northeastern Univ Boston MA 02115

POWIS, GARTH, b West Bromwich, Eng, June 12, 46; m 67; c 2. PHARMACOLOGY, BIOCHEMISTRY. *Educ:* Univ Birmingham, BSc, 67; Oxford Univ, DPhil(biochem, physiol), 70. *Prof Exp:* Univ lectr pharmacol, Glasgow Univ, 70-77; CONSULT ONCOL & PROF PHARMACOL, MAYO CLIN, 77- *Mem:* Soc Toxicol; Am Asn Cancer Res; Am Soc Pharmacol & Exp Therapeut. *Res:* Pharmacology of anticancer agents. *Mailing Add:* Dept Pharmacol Mayo Clin 200 First St SW Rochester MN 55905

POWITZ, ROBERT W, b New York, NY, Aug 29, 42; c 2. INSTITUTIONAL ENVIRONMENTAL HEALTH & SAFETY-HOSPITALS & PRISONS, RISK ASSESSMENT-PROPERTY TRANSFER-LIABILITY ISSUES. *Educ:* State Univ NY, Cobleskill, AAS, 62; Univ Ga, BS, 64; Univ Minn, MPH, 74, PhD(environ health), 78. *Prof Exp:* Sanitarian, NJ Dept Health, 65-69; sr sanitarian, NJ Inst & Agencies, 69-74; health officer, Ringwood, NJ, 74-75; epidemiologist, Mercy Health Care Corp, 78-80; DIR, ENVIRON HEALTH & SAFETY, CHEM ENG, WAYNE STATE UNIV, 80- *Concurrent Pos:* Pres, Biosafety Systs, Inc, 81-88; prin, R W Powitz & Assocs, PC, 88- *Mem:* Nat Environ Health Asn. *Res:* Curriculum development in environmental health; hazardous materials; water purification; biofilm assessment in high purity water systems; risk assessment of infections vs accidents in hospitals. *Mailing Add:* Wayne State Univ 625 Mullet Detroit MI 48226-2324

POWLES, PERCIVAL MOUNT, b Takada, Japan, Jan 16, 30; Can citizen; m 60; c 3. LARVAL-REPRODUCTION. *Educ:* McGill Univ, BA, 51; Univ Western Ont, BSc, 55, MSc, 57; McGill Univ, PhD(marine zool), 64. *Prof Exp:* Scientist pleuronectid res, Fisheries Res Bd, Can, NB, 57-68; assoc prof, 68-74, chmn dept, 75-76, PROF BIOL, TRENT UNIV, 74- *Concurrent Pos:* Res assoc, Plymouth, Eng, 78 & Nat Marine Fisheries Serv Lab, Beaufort, NC, 83; pres, Can Comt Fishery Res, 84-85; assoc ed, Am Fisheries Soc Trans, 90-92. *Mem:* AAAS; Am Fisheries Soc; Can Soc Zoologists. *Res:* Fish biology; age and growth; larval fish ecology. *Mailing Add:* Dept Biol Trent Univ Peterborough ON K9J 7B8 Can

POWLES, WILLIAM EARNEST, b Matsumoto, Japan, Sept 6, 19; Can citizen; m 43; c 2. PSYCHIATRY. *Educ:* McGill Univ, BA, 40, MD, CM, 43, dipl, 52. *Prof Exp:* Clin instr psychiat, Univ BC, 52-58; asst prof psychiat & indust med, Univ Cincinnati, 58-66; assoc prof, 66-72, PROF PSYCHIAT, QUEEN'S UNIV, ONT, 72- *Concurrent Pos:* Asst clin dir, Prov Ment Health Serv, BC, 52-56; sr specialist, Ment Health Ctr, 57-58; attend psychiatrist, Kingston Gen Hosp, 66-; consult, Ont Hosp, 66- *Mem:* Fel Am Group Psychother Asn; fel Royal Col Physicians & Surgeons Can; fel Am Psychiat Asn; Can Psychiat Asn; Can Med Asn. *Res:* Clinical psychiatry; group psychotherapy and milieu therapy; medical education; health manpower studies; psychosomatic medicine. *Mailing Add:* Dept Psychiat Queen's Univ Kingston ON K7L 3N6 Can

POWLEY, GEORGE R(EINHOLD), b New London, Ohio, Mar 7, 16; m 40; c 4. ELECTRICAL ENGINEERING. *Educ:* Va Polytech Inst, BS, 38, MS, 39. *Prof Exp:* Induction motor design engr, Electrodyn Works, Elec Boat Co, 39; test & control engr, Gen Elec Co, 39-41, control engr, 45-49; from assoc prof to prof elec eng, Va Polytech Inst & State Univ, 49-64, head dept, 58-64, Westinghouse prof, 64-68, prof, 68-81, chmn elec eng technol, 74-81; RETIRED. *Concurrent Pos:* Vis prof elec eng, Va Mil Inst, 81, 83 & 84. *Honors & Awards:* Wine Award, Va Polytech Inst, 57. *Mem:* Inst Elec & Electronics Engrs. *Res:* Automatic controls; rotating machinery; transformers. *Mailing Add:* 1401 Hillcrest Dr Blacksburg VA 24060

POWNALL, HENRY JOSEPH, b Lancaster, Pa, Dec 9, 42; m 71. BIOCHEMISTRY. *Educ:* Elizabethtown Col, BS, 65; Wilkes Col, MS, 67; Northeastern Univ, PhD(chem), 70. *Prof Exp:* NSF grant, Univ Houston, 71; NIH grant, 72-73, asst prof, 73-79, ASSOC PROF EXP MED, BAYLOR COL MED, 79- *Concurrent Pos:* Estab investr, Am Heart Asn. *Mem:* Am Chem Soc; Sigma Xi. *Res:* Investigation of biological systems by physical chemical methods; principally molecular spectroscopy; model aromatic hydrocarbons. *Mailing Add:* Dept Med Baylor Col Med 6565 Fannin MS A601 Houston TX 77030

POWNALL, MALCOLM WILMOR, b Coatesville, Pa, Jan 6, 33; m 61; c 4. MATHEMATICS. *Educ:* Princeton Univ, AB, 54; Univ Pa, AM, 57, PhD(math), 60. *Prof Exp:* Instr, 59-61, from asst prof to assoc prof, 61-71, PROF MATH, COLGATE UNIV, 71- *Concurrent Pos:* Assoc dir comn undergrad prog, Math Asn Am, 66-67, dir, 67-68. *Mem:* Math Asn Am. *Res:* Graph theory; mathematical education. *Mailing Add:* Dept of Math Colgate Univ Hamilton NY 13346

POWRIE, WILLIAM DUNCAN, b Toronto, Ont, Nov 1, 26; m 55; c 4. FOOD SCIENCE. *Educ:* Univ Toronto, BA, 49, MA, 51; Univ Mass, PhD(food tech), 55. *Prof Exp:* Sr food technologist, Can Dept Agr, 55-56; asst prof food technol, Mich State Univ, 56-59; from asst prof to assoc prof, Univ Wis-Madison, 59-69; PROF FOOD SCI & CHMN DEPT, UNIV BC, 69- *Honors & Awards:* William J Eva Award, Can Inst Food Sci & Technologists. *Mem:* Fel Chem Inst Can; fel, Inst Food Technologists; fel Can Inst Food Sci & Technol; Can Soc Nutrit Sci; Sigma Xi. *Res:* Denaturation of proteins; egg yolk proteins; steroids; freezing of foods; mutagens in food; modified atmosphere packaging of food. *Mailing Add:* 505 Ventura Crescent North Vancouver BC V7N 3G8 Can

POWSNER, EDWARD R, b New York, NY, Mar 17, 26; c 4. PATHOLOGY, NUCLEAR MEDICINE. *Educ:* Mass Inst Technol, SB, 48, SM, 49; Yale Univ, MD, 53; Wayne State Univ, MS, 57; Am Bd Internal Med, dipl, 60; Am Bd Path, dipl & cert clin path, 63, cert anat path, 78; Am Bd Nuclear Med, 72. *Prof Exp:* Asst instr med, Sch Med, Wayne State Univ, 54-56, fels hemat, Sch Med & Detroit Receiving Hosp, 56-58, from instr med to prof path, Sch Med, 59-78; PROF PATH, SCH MED, MICH STATE UNIV, 78- *Mem:* Am Soc Clin Path; Am Soc Hemat; Soc Nuclear Med; Int Acad Path; Col Am Path. *Mailing Add:* Nuclear Med St John Hosp Detroit MI 48236

POYDOCK, MARY EYMARD, b Skyesville, Pa, Dec 3, 10. EXPERIMENTAL MEDICINE. *Educ:* Mercyhurst Col, BA, 43; Univ Pittsburgh, MA, 46; St Thomas Inst, PhD(biol & exp med), 65. *Prof Exp:* Elem sch teacher, Erie Diocese, Pa, 35-41; high sch teacher, Pittsburgh Diocese, 41-47; from asst prof to prof biol, 47-75, EMER PROF BIOL & DIR RES, MERCYHURST COL, 75- *Mem:* NY Acad Sci; Sigma Xi; Am Asn Cancer Res. *Res:* Preparation and synthesis of the compound Mercytamin; effect of vitamins C and B12 on normal and neoplastic tissues in vitro and in vivo; effect of Cobalt compounds and vitamin C as antitumor agents. *Mailing Add:* Mercyhurst Col 501 E 38th st Erie PA 16546

POYER, JOE LEE, b Tulsa, Okla, Dec 29, 31. BIOCHEMISTRY. *Educ:* Univ Okla, BS, 54, PhD(biochem), 69. *Prof Exp:* Res assoc, 59-64, fel, 69-72, staff scientist, 72-75, asst mem biochem, 75-81, SR RES SCIENTIST, OKLA MED RES FOUND, 81- *Res:* The mechanism of enzymic formation of activated oxygen species and free radicals and the effects on biological systems. *Mailing Add:* Okla Med Res Found 825 NE 13th St Oklahoma City OK 73104

POYNTER, JAMES WILLIAM, b Winchester, Ky, Mar 30, 09; m 38; c 1. PHYSICAL CHEMISTRY. *Educ:* Univ Ky, BS, 30; Mass Inst Technol, MS, 32. *Prof Exp:* Lab asst, Res Labs, Armco Steel Corp, 33-36; metallurgist, Mat Lab, US Air Force, 37-52, asst chief, Metall & Ceramics Br, Aeronaut Res Lab, 53-61, phys metallurgist, Metals & Ceramics Div, Air Force Mat Lab, 61-72; RETIRED. *Concurrent Pos:* Abstr, Chem Abstracts Serv, 36-80. *Mem:* Am Soc Metals. *Res:* Alloy development; kinetics of heat treatment of alloys; research administration. *Mailing Add:* 4279 Catalpa Dr Dayton OH 45405

POYNTER, ROBERT LOUIS, b St Louis, Mo, Feb 25, 26; m 60; c 5. MOLECULAR SPECTROSCOPY. *Educ:* Univ Ill, Urbana, BSc, 50; Univ Iowa, MS, 52, PhD(chem), 54. *Prof Exp:* Res assoc physics, Ohio State Univ, 53-54; res assoc chem, Columbia Univ, 54-56; res engr, 56-60, from scientist to sr scientist, 60-69, MEM TECH STAFF, JET PROPULSION LAB, CALIF INST TECHNOL, 69- *Mem:* AAAS; Am Phys Soc; Am Chem Soc; Am Inst Chemists; Sigma Xi. *Res:* Microwave spectroscopy; earth and planetary spectra; interstellar microwave spectra. *Mailing Add:* 2541 N Marengo Altadena CA 91001

POYNTON, JOSEPH PATRICK, b Chicago, Ill, Aug 28, 34; m 71; c 1. LOW SPEED VISUAL FLOW AERODYNAMICS, CARGO HANDLING. *Educ:* Univ Notre Dame, BSAE, 56, MSAE, 58. *Prof Exp:* Assoc engr, Convair, 56; res fel visual flow aerodyn, Univ Notre Dame, 56-58; aerospace engr, Douglas Aircraft Co, 58-59, Honeywell Corp, 59-64, Foreign Technol Div, USAF, 64-65, Lockheed Corp, 65-66, Aviation Systs Command, US Army, 66-85; AEROSPACE CONSULT, AM ELECTRO LABS, INC, 90- *Concurrent Pos:* Mem, Nat Tech Support Comt, 76-78; substitute teacher math, Spec Sch Dist St Louis County, 86- *Honors & Awards:* Cert Appreciation, Am Inst Aeronaut & Astronaut, 90. *Mem:* Am Inst Aeronaut & Astronaut; emer mem Sigma Xi. *Res:* Boundary layer vortex transition striation lines; magnus force lift reversal; cruciform fin autorotation; nike zeus; SR-71 applications; Apollo flight controls; Polaris; aircraft survivability countermeasures; ballistic and crashworthy fuel systems. *Mailing Add:* 9624 Yorkshire Estates Dr Crestwood MO 63126

POYTON, HERBERT GUY, b London, Eng, Nov 24, 11; m 38. RADIOLOGY. *Educ:* Univ London, LDS, 34, HDD(Edin), 49, FDS(Eng), 51, FDS(Edin), 72; FRCD(C), 69. *Prof Exp:* Lectr dent radiol, Inst Dent Surg, Univ Loncon, 54-59; prof, Fac Dent, Univ Toronto, 59-80, emer prof radiol, 80-88; RETIRED. *Concurrent Pos:* Consult, Hosp Sick Children, Toronto & Ont Crippled Children's Ctr; dent radiologist, Toronto Gen Hosp. *Mem:* Am Acad Dent Radiol; Int Asn Dent Res; Can Acad Oral Radiol (founding pres); Brit Soc Dent Radiol. *Res:* Dental radiology. *Mailing Add:* Five Elm Ave Apt 411 Toronto ON M4W 1N1 Can

POYTON, ROBERT OLIVER, b Providence, RI, Sept 3, 44; c 2. MOLECULAR BIOLOGY, MICROBIOLOGY. *Educ:* Brown Univ, AB, 66; Univ Calif, Berkeley, PhD(microbiol), 71. *Prof Exp:* NIH fel molecular biol, Cornell Univ, 71-73; asst prof microbiol, Health Ctr, Univ Conn, 73-76, assoc prof, 76-; PROF, DEPT MOLECULAR, CELL & DEVELOP BIOL, UNIV COLO. *Mem:* Am Soc Microbiol; Am Soc Cell Biol; AAAS. *Res:* Membrane biochemistry; mitochondrial biogenesis. *Mailing Add:* Dept Molecular Cell & Develop Biol Univ Colo Boulder CO 80309

POZIOMEK, EDWARD JOHN, b Albany, NY, June 15, 33; m 54; c 4. SCIENCE ADMINISTRATION, RESEARCH MANAGEMENT. *Educ:* Rensselaer Polytech Inst, BS, 54; Univ Del, MS, 60, PhD(chem), 61. *Prof Exp:* Chemist, Durez Plastics Div, Hooker Electrochem Co, NY, 54-55; asst chem, Rensselaer Polytech, 55-56; chemist, Protective Develop Div, Army Chem Ctr, Md, 58-62, res chemist, Res Labs, Chem & Res Develop Ctr, 62-71, chief, Mat & Eval Sect, Develop & Eng Directorate, 71-74, physical scientist, 74-75, dep dir, Chem Lab, Edwood Arsenal, 75-77, asst chief, 77-79, chief scientist, 79-80, chief, Res Div, Chen Systs Lab, 81, dir res, 81-84; consult, 84-89; SR SCIENTIST, ENVIRON RES CTR, 89- *Concurrent Pos:* Secy Army res & study fel, State Univ NY Stony Brook, 62-63; Fulbright res scholar, Univ Leiden, 65-66; instr, Div Univ Exten, Univ Del, 66-67; fel, Educ Pub Mgt, Mass Inst Technol, 75-76 & Ctr Advan Study, 75-76; Fed Exec Inst fel, 77; adj prof chem, Drexel Univ, 83-88; prof sci, Aberdeen Grad Ctr, Fla Inst Technol, 84-88. *Mem:* AAAS; Am Chem Soc; Sigma Xi; Am Defense Preparedness Asn. *Res:* Heterocyclic chemistry; chemotherapy; olfaction; microchemistry; free radicals; organonitrogen compounds; photochemistry; detection; pollution control; air filtration; gas chromatography; charcoal; sorption; fluorescence; multidisciplianry approaches to scientific research. *Mailing Add:* Environ Res Ctr Univ Nev Las Vegas NV 89154-4009

POZNANSKI, ANDREW K, b Czestochowa, Poland, Oct 11, 31; US citizen; m 57; c 2. RADIOLOGY. *Educ:* McGill Univ, BSc, 52, MD & CM, 56. *Prof Exp:* Intern, Montreal Gen Hosp, 56-57; from resident radiol to radiologist, Henry Ford Hosp, 57-68; from assoc prof to prof radiol, Med Sch, Univ Mich, Ann Arbor, 68-79; RADIOLOGIST IN CHIEF, CHILDREN'S MEM HOSP, 79-; PROF RADIOL, NORTHWESTERN UNIV, 79- *Concurrent Pos:* Mem, Nat Coun Radiation Protection. Int Comn Radiation Protection. *Honors & Awards:* Hon Mem, Europ Soc Pediat Radiol, Can Asn Radiol. *Mem:* Radiol Soc NAm; Am Col Radiol; Soc Pediat Radiol (past pres); Am Roentgen Ray Soc; Asn Univ Radiologists; hon mem Europ Soc Pediat Radiol & Can Asn Radiol. *Res:* Congenital abnormalities of hands and feet; skeletal maturation; radiation protection; growth and development; physical foundation of radiology; Rheumatic diseases. *Mailing Add:* Children's Mem Hosp Dept of Radiol 2300 Children's Plaza Chicago IL 60614

POZNANSKY, MARK JOAB, b Montreal, Que, Apr 25, 46. PHYSIOLOGY, BIOPHYSICS. *Educ:* McGill Univ, BSc, 67, PhD(physiol), 70. *Prof Exp:* Fel biophys, Harvard Med Sch, 70-71, instr, 71-73; sr researcher physiol, Col France, 73-74; lectr biophys, Harvard Med Sch, 74-75; assoc prof, 75-82, PROF PHYSIOL, UNIV ALTA, 82-, ASSOC DEAN MED, 84- *Res:* Structure and function studies of biological and model membrane systems; enzyme replacement therapy for treatment of genetic and metabolic diseases; lipid metabolism and atherosclerosis. *Mailing Add:* Dept Physiol Univ Alta Edmonton AB T6G 2E1 Can

POZOS, ROBERT STEVEN, b Ventura, Calif, Dec 28, 42; m 66; c 3. NEUROPHYSIOLOGY, BIOPHYSICS. *Educ:* St Southern Ill Univ, MA, 67, PhD(physiol, biophys), 69. *Prof Exp:* Fel clin physiol & instr physiol & biophys, Univ Tenn, Memphis, 69-71; asst prof physiol, 71-75, ASSOC PROF PHYSIOL, SCH MED, UNIV MINN, DULUTH, 75- *Mem:* AAAS; Am Inst Biol Sci; Inst Elec & Electronics Eng; Am Acad Neurol; NY Acad Sci; Soc Neurosci. *Res:* Tremor, shivering, electromyogram, rigidity, Parkinson's disease and computer application to these problems; catecholamine levels of central nervous system; motor control diseases; nerve tissue culture and electrical recordings. *Mailing Add:* NHRC PO Box 85122 San Diego CA 92186

PRABHAKAR, BELLUR SUBBANNA, b Tiptur, Karnataka, Apr 19, 51; US citizen; m 80; c 2. VIROLOGY, AUTOIMMUNITY. *Educ:* Univ Mysore, India, BSc, 70, MSc, 73; Johns Hopkins Univ, PhD(viral immunol), 80. *Prof Exp:* Fel microbiol, Univ Pa, 79-80; staff fel, 80-83, sr staff fel viral immunol, 83-87, RES MICROBIOLOGIST, NIH, 87- *Mem:* Am Asn Immunologists; Am Soc Virologists; NY Acad Sci; AAAS. *Res:* Immunological perturbation caused by viral infections and how it may cause immunopathology in the host; role of viruses in triggering autoimmunity; anti-idiotypic antibodies as probes of autoimmune disorders. *Mailing Add:* NIH NIDR Bldg 30 Rm 122 Bethesda MD 20892

PRABHAKAR, JAGDISH CHANDRA, b Malikwal, Pakistan, Sept 14, 25; US citizen; m 50; c 2. ELECTRICAL ENGINEERING & EDUCATION. *Educ:* Panjab Univ, India, BS, 46, MS, 48; Ill Inst Technol, MS, 64; Southern Methodist Univ, PhD(elec eng), 69. *Prof Exp:* Sr engr, All India Radio, New Delhi, 49-62; design engr, Oak Mfg Co, Ill, 64-66; sr engr, Tex Instruments Inc, 66-69; asst prof elec eng, Tex Tech Univ, 70-72, assoc prof, 72-77; PROF, CALIF STATE UNIV, NORTHRIDGE, 78- *Concurrent Pos:* Tex State grant, 71-72. *Honors & Awards:* Res Award, US Air Force, 69. *Mem:* Sr mem Inst Elec & Electronics Engrs; Int Soc Hybrid Microelectronics. *Res:* Stochastic processes and optional control; homomorphic filters for communication systems; information and coding theories. *Mailing Add:* Elec Eng Dept Calif State Univ Northridge CA 91330

PRABHAKARA, CUDDAPAH, b India, Feb 8, 34; US citizen; m 62; c 3. ATMOSPHERIC SCIENCES, PHYSICS. *Educ:* Univ Madras, BSc, 54; Andhra Univ, India, MSc, 57; NY Univ, PhD(meteorol), 64. *Prof Exp:* Res asst, Tata Inst Fundamental Res, Bombay, India, 57-58; res asst, NY Univ, 58-62; res assoc, Nat Acad Sci-Nat Res Coun, 63-66; aerospace technologist, 66-77, SR RES METEOROLOGIST, GODDARD SPACE FLIGHT CTR, NASA, 77- *Honors & Awards:* Except Performance Award, Goddard Space

Flight Ctr, NASA, 77. *Mem:* Am Meteorol Soc; Am Geophys Union. *Res:* Satellite meteorology; remote sensing; climate modelling; radiative transfer; stratospheric ozone; stratospheric climatology. *Mailing Add:* NASA Goddard Space Fl Ctr Mail Code 613 Greenbelt MD 20771

PRABHU, NARAHARI UMANATH, b Calicut, India, Apr 25, 24; m 51; c 2. MATHEMATICS, STATISTICS. *Educ:* Univ Madras, BA, 46; Univ Bombay, MA, 50; Univ Manchester, MSc, 57. *Prof Exp:* Lectr math, Baroda Col, Bombay, 46-47 & Victoria Jubilee Tech Inst, 47-48; lectr math & statist, Gauhati Univ, India, 50-52; reader statist, Karnatak Univ, India, 52-61 & Univ Western Australia, 61-64; assoc prof, Mich State Univ, 64-65; assoc prof, 65-67, PROF OPERS RES, CORNELL UNIV, 67- *Concurrent Pos:* Prin ed, 73-79, chief ed, Stochastic Process Applin, Amsterdam, 80-84; chief ed, Queueing Systems, Basel, 86- *Mem:* Inst Math Statist; Am Math Soc; Opers Res Soc Am; Opers Res Soc India; Int Statist Inst. *Res:* Probability and stochastic processes with special reference to applications. *Mailing Add:* Dept Opers Res Cornell Univ Ithaca NY 14853-4902

PRABHU, VASANT K, b Kumta, India, May 13, 39; US citizen; m 66; c 1. COMMUNICATIONS, TRANSMISSION. *Educ:* Karnatak Univ, BSc, 58; Indian Inst Sci, 61, BE; Mass Inst Technol, SM, 63, ScD(elec eng), 66. *Prof Exp:* Mem tech staff, AT&T Bell Labs, 66-86, dir tech staff, 86-91; PROF, UNIV TEX, ARLINGTON, 91- *Honors & Awards:* Centennial Medal, Inst Elec & Electronic Engrs, 85. *Mem:* Fel Inst Elec & Electronics Engrs. *Res:* Analog and digital communications systems; communication theory; network theory and optical communications systems; propagation effects on microwave radio and cellular systems. *Mailing Add:* Univ Tex Arlington PO Box 1019 Arlington TX 76019-0019

PRABHU, VENKATRAY G, b Shirali, India, Mar 15, 30; m 57; c 2. PHARMACOLOGY. *Educ:* Univ Bombay, BS, 53 & 55, MS, 58; Loyola Univ, Ill, PhD(pharmacol), 62. *Prof Exp:* Sr pharmacologist, Arnar-Stone Labs, Ill, 62-63; assoc dir pharmacol, Sarabhai Chem Res Inst, India, 64-67; from instr to assoc prof, 67-74, actg chmn dept, 71-74, PROF PHARMACOL & CHMN DEPT, CHICAGO COL OSTEOP MED, 74- *Mem:* Am Soc Pharmacol & Exp Therapeut. *Mailing Add:* Dept Pharmacol Chicago Col Osteo Med 515 31st St Downers Grove IL 60515

PRABHU, VILAS ANANDRAO, b Bombay, India, Oct 11, 48; US citizen; m 75; c 3. MEDICINAL CHEMISTRY, PHARMACOLOGY. *Educ:* Univ Bombay, India, BS, 70; Idaho State Univ, MS, 73; Univ Tex, Austin, PhD(pharmaceut chem), 77. *Prof Exp:* Teaching asst pharm, Sch Pharm, Idaho State Univ, 71-73; teaching asst pharmaceut chem, Sch Pharm, Univ Tex, Austin, 73-77; asst prof, Sch Pharm, Wash State Univ, 77-80; ASSOC PROF PHARMACEUT CHEM, SCH PHARM, SOUTHWESTERN OKLA STATE UNIV, 80- *Concurrent Pos:* Reviewer, US Pharmacopeia XIX, 70-75. *Mem:* Sigma Xi; Am Pharmaceut Asn; Am Asn Cols Pharm; Acad Pharmaceut Sci; Am Chem Soc. *Res:* Synthesis and structure activity relationships of potential medicinal agents with special emphasis on agents affecting central nervous system and the autonomic nervous system; effects of chemical substances on the consumption and-or preference for alcohol in animals. *Mailing Add:* Dept Pharm Southwestern Okla State Univ 100 Campus Dr Weatherford OK 73096

PRABHUDESAI, MUKUND M, b Goa, India, Mar 17, 42; US citizen; m 72; c 1. NUCLEAR MEDICINE, ELECTRON MICROSCOPY. *Educ:* G S Med Col, Bombay, India, MD, 66. *Prof Exp:* From asst pathologist to assoc pathologist, Fordham Hosp, Bronx, NY, 73-76; assoc dir clin path, Lincoln Med & Mental Health Ctr, Bronx, NY, 76, dep dir path, 77-79; CHIEF LAB SERV, VA MED CTR, DANVILLE, ILL & VA OUTPATIENT CLIN, PEORIA, ILL, 79- *Concurrent Pos:* Clin asst prof lab med, Albert Einstein Col Med, 77-80; clin assoc prof path & med, Col Med Univ Ill, Urbana, 79-83; lab inspector, CAP Lab Inspection, Accreditation Dept, 81-; chair ADP, Geriatric Res, Educ & Clin Ctr, Med Dist 15, VHSRA, Wash, DC, 82-88; sen, Col Med, Univ Ill, Chicago, 82-83; coordr res & develop, VA Med Ctr, Danville, Ill, 83-, dir electron micros, 87-; Ill State deleg, House Deleg, Col Am Pathologists, 88-93; bd dirs, Ill Soc Pathologists, 91. *Honors & Awards:* Recognition Award, Am Cancer Soc, 84. *Mem:* Fel Col Am Pathologists; Am Asn Physician Execs. *Res:* Effects of alcohol on organ systems to include biochemical and ultrastructural studies; new methods and developments in the clinical laboratory; dietary intervention in cholesterol reduction; structural basis of coronary artery disease; health services research including medical management and cost containment. *Mailing Add:* VA Med Ctr 1900 E Main St Danville IL 61832

PRABULOS, JOSEPH J, JR, b Cincinnati, Ohio, Nov 21, 38; m 63; c 2. NUCLEAR & CHEMICAL ENGINEERING. *Educ:* Purdue Univ, BS, MS, 64, PhD(nuclear eng), 66. *Prof Exp:* Staff mem nuclear eng, Los Alamos Sci Lab, Univ 66-67; staff physicist, Fast Breeder Reactor Group, Utility Div, Combustion Eng Inc, 67-75, nuclear engr, Advan Develop Dept, 75-87, Advan Nuclear Systs Dept, 88-89; NUCLEAR ENGR, FLUID SYSTS ENG, ABB, INC, 90- *Mem:* Am Nuclear Soc. *Res:* Development of fast reactor nuclear design computing programs; nuclear design of fast breeder reactors; thermal hydraulic analysis. *Mailing Add:* ASEA Brown Boveri Inc 1000 Prospect Hill Rd Windsor CT 06095

PRADDAUDE, HERNAN CAMILO, b Rosario, Arg, Oct 18, 32; m 57; c 3. SOLID STATE PHYSICS. *Educ:* Nat Univ Litoral, CEng, 57; Mass Inst Technol, PhD(physics), 64. *Prof Exp:* Instr physics, Fac Math, Nat Univ Litoral, 54-59; res collabr accelerator dept, Brookhaven Nat Lab, 59-60; sr scientist, 63-77, ASST LEADER, NAT MAGNET LAB, MASS INST TECHNOL, 77- *Mem:* Fel Am Phys Soc; sr mem Inst Elec & Electronics Engrs; Sigma Xi. *Res:* Theory of magnetism; magnetic measurements; resonance and transport phenomena in solid state. *Mailing Add:* 11 Athens Dr Saugus MA 01906

PRADOS, JOHN W(ILLIAM), b Spring Hill, Tenn, Oct 12, 29; m 51; c 3. CHEMICAL ENGINEERING. *Educ:* Univ Miss, BS, 51; Univ Tenn, MS, 54, PhD(chem eng), 57. *Prof Exp:* Asst, Univ Tenn, Knoxville, 53-55, from instr to prof chem eng, 55-89, assoc dean eng, 69-71, dean admin & rec, 71-73, actg chancellor, Knoxville Campus, 73, Martin Campus, 79, vpres acad affairs, 73-81, vpres acad affairs & res, 81-88, EMER VPRES & UNIV PROF, UNIV TENN, KNOXVILLE, 89- *Concurrent Pos:* Consult, Oak Ridge Nat Lab, 57-73 & Ford Found Res, 65-66; mem, Eng Accreditation Comn, Accreditation Bd Eng & Technol, Inc, 78-86, chmn, 84-85, secy, 89-90, pres-elect, 90-91; consult eng educ, SC Comn Higher Educ, 81, 86, Ark Dept Higher Educ, 81, 88, La State Univ-Shreveport, 82, 83, Idaho Bd Educ, 84, Univ Mo Syst, 84; dir, Am Inst Chem Engrs, 75-77; mem, Comn Cols, Southern Asn Cols & Schs, 87-; consult, eng educ to Univ Ala, 86, Clemson Univ, 87, King Saud Univ, 87, Kan bd regents, 88, Tex Higher Educ Coord Bd, 89, Ohio Bd Regents & Kuwait Univ, 90. *Mem:* Am Chem Soc; Am Soc Eng Educ; fel Am Inst Chem Engrs; Sigma Xi (pres, 83-84, treas, 90-); fel Am Inst Chemists. *Res:* Transfer and rate processes; chemical reactor design; mathematical analysis of chemical and nuclear systems; engineering education. *Mailing Add:* 419 Dougherty Eng Bldg Univ Tenn Knoxville TN 37996-2200

PRADOS, RONALD ANTHONY, b New Orleans, La, July 9, 46; m 71; c 3. PHYSICAL CHEMISTRY, INORGANIC CHEMISTRY. *Educ:* Univ New Orleans, BS, 68, PhD(chem), 71. *Prof Exp:* Teacher chem, Univ Va, 71-72; res, Georgetown Univ, 72-74; res chemist, E I Du Pont de Nemours & Co, Inc, 74-77, sr res chemist, 77, res supvr chem, 77-79, mgr process eng, 79-80, site tech supt, 80-84, mgr prod develop, 85-90, LAB DIR, E I DU PONT DE NEMOURS & CO, INC, PA, 90- *Mem:* Sigma Xi; Am Chem Soc. *Res:* Silver halide and photo polymer photographic systems; electroplating theory and processes; oxidation-reduction mechanisms. *Mailing Add:* 112 Hillcrest Dr Sayre PA 18840

PRAG, ARTHUR BARRY, b Portland, Ore, Apr 14, 38; m 86; c 1. ATMOSPHERIC PHYSICS, IONOSPHERIC PHYSICS. *Educ:* Univ Portland, BS, 59; Univ Wash, MS, 62, PhD(physics), 64. *Prof Exp:* Mem tech staff, 64-72, staff scientist, 72-80, RES SCIENTIST, AEROSPACE CORP, 80- *Mem:* NY Acad Sci; AAAS; Am Phys Soc; Am Geophys Union; Sigma Xi. *Res:* Aeronomy; upper atmospheric composition; particle precipitation; satellite instrumentation; upper atmospheric density; ionospheric structure. *Mailing Add:* Space Sci Lab M2/255 The Aerospace Corp PO Box 92957 Los Angeles CA 90009-2957

PRAGER, DENIS JULES, b Dayton, Ohio, Nov 7, 38. SCIENCE POLICY. *Educ:* Univ Cincinnati, BEE, 62; Stanford Univ, PhD(physiol), 68. *Prof Exp:* Investr med instrumentation, Lab Tech Develop, Nat Heart Inst, Md, 62-69; chief contraceptive develop br, Ctr Pop Res, Nat Inst Child Health & Human Develop, 69-71; dir pop study ctr, Battelle Mem Inst, 71-77; sr policy analyst, White House, 77-79, assoc dir, Off Sci & Technol Policy, exec off pres, 79-83; dep dir, 83-89, DIR, HEALTH PROG, MACARTHUR FOUND, 89- *Concurrent Pos:* Affil assoc prof, Sch Pub Health, Ctr Bioeng, Univ Wash, 72-83; exec dir, govt-univ-indust res roundtable, Nat Acad Sci, 83. *Mem:* AAAS. *Res:* Science policy, science administration; biomedical and health sciences. *Mailing Add:* 503 N Wells #2C Chicago IL 60610

PRAGER, JAN CLEMENT, b Cincinnati, Ohio, Mar 17, 34; m 59; c 2. MARINE MICROBIOLOGY. *Educ:* Univ Cincinnati, BS, 54, MSc, 56; NY Univ, PhD(microbiol), 61. *Prof Exp:* Res assoc biol, Hasksins Labs, Inc, NY, 58-61; chief microbiologist, Sandy Hook Marine Lab, US Fish & Wildlife Serv, 61-66, estab & directed Plankton Biol prog, 61-65; unit leader, Fed Water Pollution Control Admin, 66-69, prof coordr plankton environ studies, Nat Marine Water Qual Lab, 69-77; chief tech assistance br, Environ Res Lab, 77-80, RES BIOL, US ENVIRON PROTECTION AGENCY, 80- *Concurrent Pos:* Lectr, Queens Col, NY, 60-61; adv, Philipp Co, NJ, 65-66. *Mem:* AAAS; Soc Protozool; Am Soc Microbiol; Am Soc Limnol & Oceanog; Phycol Soc Am; NY Acad Sci. *Res:* Marine phytoplankton physiology, nutrition, biochemistry, morphology and environmental requirements of coastal and estuarine water column plankton; man's effects on growth of phytoplankton species in the coastal marine environment expert systems; expert systems. *Mailing Add:* 140 Shadow Farm Way Wakefield RI 02879-3632

PRAGER, JULIANNE HELLER, b Boston, Mass, June 5, 24; m 48. ORGANIC CHEMISTRY. *Educ:* Brown Univ, ScB, 46; Cornell Univ, PhD(org chem), 53. *Prof Exp:* Mem staff cent res, 52-73, exec dir corp tech planning & coord, 73-89, COMMUNITY SERV EXEC, MINN MINING & MFG CO, 89- *Concurrent Pos:* Trustee, Brown Univ, 78-82. *Honors & Awards:* Spurgeon Award, 80; Am Chem Soc Award, 86. *Mem:* Am Chem Soc; Indust Res Inst. *Res:* Information handling; polymer chemistry; fluorine chemistry. *Mailing Add:* 3320 N Dunlap St St Paul MN 55112

PRAGER, MARTIN, b Brooklyn, May 23, 39; div; c 2. PHYSICAL METALLURGY, MATERIALS SCIENCE. *Educ:* Cornell Univ, BChE, 61; M MetE, 62; Univ Calif, Los Angeles, PhD(mat eng), 69. *Prof Exp:* Sr res engr, Rocketdyne Div, NAm Rockwell Corp, 61-68; supvr metall develop, Copper Develop Asn, 69-72, supvr tech serv, 72-74; CONSULT, 74- *Concurrent Pos:* exec dir, Metal Properties Coun, 78-; tech dir, Welding Res Coun, 85-, Pressure Vessel Res Comt, 85- *Honors & Awards:* Davis Silver Medal, Am Welding Soc. *Mem:* Am Soc Testing & Mat; Am Soc Metals; Am Welding Soc; Am Mining Metall & Petrol Engrs; Nat Asn Corrosion Engrs. *Res:* Environmental embrittlement; fracture; welding; weld cracking; precipitation hardenable nickel base alloys; high pressure technology; thermodynamics; parametic analysis; corrosion resistance; creep damage; copper and copper alloys. *Mailing Add:* 125 E 87th St New York NY 10128

PRAGER, MORTON DAVID, b Dallas, Tex, Dec 12, 27; m 51; c 4. BIOCHEMISTRY. *Educ:* Univ Tex, BA, 47; Purdue Univ, MS, 49, PhD(chem), 51. *Prof Exp:* Res chemist, Org Chem Res Div, B F Goodrich Co, 51-53; consult chemist, Dallas Labs, 53-54; res chemist, Wadley Res Inst

& Blood Bank, 54-67, asst dir res, 63-67; assoc prof, 67-72, PROF BIOCHEM, UNIV TEX HEALTH SCI CTR, DALLAS, 72-, PROF SURG, 67- *Concurrent Pos:* From asst prof to prof chem, Grad Res Inst, Baylor Univ, 54-68, prof microbiol, 64-68; consult, Vet Admin Hosp, Temple, 60-68; assoc ed, Cancer Res, 79-88. *Mem:* Am Chem Soc; Am Soc Hemat; Am Asn Immunol; Int Soc Hemat; Am Asn Cancer Res; AAAS. *Res:* Biochemistry and immunology of cancer; biochemical studies of the traumatized patient; interaction of tissue with biomaterials. *Mailing Add:* Dept Surg Univ Tex Health Sci Ctr Dallas TX 75235-9031

PRAGER, STEPHEN, b Darmstadt, Ger, July 20, 28; nat US; m 48. PHYSICAL CHEMISTRY. *Educ:* Brown Univ, BSc, 47; Cornell Univ, PhD(phys chem), 51. *Prof Exp:* Res assoc phys chem, Cornell Univ, 50-51; Jewett fel, Univ Utah, 51-52; from asst prof to assoc prof, 52-62, PROF PHYS CHEM, UNIV MINN, MINNEAPOLIS, 62- *Concurrent Pos:* Consult, Union Carbide Corp, Tenn, 54-; Fulbright grant & Guggenheim fel, Univ Brussels, 58-59; Fulbright lectr & Guggenheim fel, Univ Erlangen, 66-67; assoc ed, J Phys Chem, 70-79. *Mem:* Am Chem Soc; fel Am Phys Soc; Soc Rheol. *Res:* Diffusion; rheology; statistical and quantum mechanics; hydrodynamics; polymers. *Mailing Add:* Dept Chem Univ Minn Minneapolis MN 55455-0100

PRAGER, STEWART CHARLES, b New York, NY, Oct 21, 48; m 72; c 3. PLASMA STABILITY. *Educ:* Queens Col, BA, 70; Columbia Univ, BS, 70, PhD(plasma physics), 75. *Prof Exp:* Sr scientist, Gen Atomic Co, 75-77; from asst prof to assoc prof, 77-85, PROF PHYSICS, UNIV WIS, MADISON, 85- *Mem:* Fel Am Phys Soc. *Res:* Physics of magnetically confined plasmas; micro and macrostability of plasmas using the tokamak, octupole and reversed-field-pinch as experimental devices. *Mailing Add:* Dept Physics Univ Wis Madison WI 53706

PRAHL, HELMUT FERDINAND, organic chemistry, chemical engineering, for more information see previous edition

PRAHL, JOSEPH MARKEL, b Beverly, Mass, Mar 30, 43; m 77; c 2. FLUID MECHANICS, HEAT TRANSFER. *Educ:* Harvard Col, BA, 63; Harvard Univ, PhD(mech eng), 68. *Prof Exp:* From asst prof to assoc prof, 68-85, PROF ENG, CASE WESTERN RESERVE UNIV, 85- *Concurrent Pos:* Lectr eng, Harvard Col, 68, lectr & res assoc, Div Appl Sci, Harvard Univ, 74-75. *Mem:* Sigma Xi; Am Soc Mech Engrs; Am Inst Aeronaut & Astronaut; Am Soc Eng Educ. *Res:* Bouyancy driven flows; boundary layer stability with the effects of heat transfer; tribology, including friction and wear; parched elastohydrodynamic lubrication; mechanics of droplet formation from jets and sheets. *Mailing Add:* 16143 Cleviden Rd East Cleveland OH 44112

PRAHLAD, KADABA V, b Tumkure, India, Oct 18, 26; US citizen; c 3. DEVELOPMENTAL BIOLOGY. *Educ:* Univ Mysore, BSc, 48, MSc, 49; Univ Mo, PhD(zool), 59. *Prof Exp:* Fel reprod physiol, Univ Wis, 59-60; guest investr, Cent Drug Res Inst, India, 60-62; res assoc biol, Wabash Col, 63-65, vis asst prof, 65-66; assoc prof, 66-70, PROF DEVELOP BIOL, NORTHERN ILL UNIV, 70- *Mem:* AAAS; Soc Develop Biol; Am Soc Zool. *Res:* Influence of endocrines on developmental processes; influence of pesticides on development. *Mailing Add:* Dept of Biol Sci Northern Ill Univ De Kalb IL 60115

PRAIRIE, MICHAEL L, b Philadelphia, Pa, June 11, 50; m 75; c 3. RESEARCH ADMINISTRATION, INDUSTRIAL & MANUFACTURING ENGINEERING. *Educ:* Rutgers Univ, BA, 72. *Prof Exp:* Methods engr, Boeing Vertol, 76-77; sr indust engr, 747 Div, Boeing, 77-79, Martin Marrieta Aerospace, 79-82; staff engr, 82-83, eng supvr, 83, PROJ MGR, TEXTRON AEROSTRUCTURES, 84- *Mem:* Soc Mfg Engrs; Inst Indust Engrs; Nat Mgt Asn. *Res:* Manufacturing technology for large aircraft composite wing structures. *Mailing Add:* Textron Aerostructures PO Box 210 Nashville TN 37202

PRAIRIE, RICHARD LANE, b Chicago, Ill, Apr 25, 34; m 63; c 2. BIOCHEMISTRY. *Educ:* Univ Chicago, BA, 56, BS, 57, PhD(biochem), 61. *Prof Exp:* Res assoc biochem, Pub Health Res Inst New York, 61-63; from instr to asst prof, Univ Ill, Urbana, 63-67; asst prof, 67-74, ADJ ASST PROF BIOCHEM, COL MED, UNIV CINCINNATI, 74- *Concurrent Pos:* NSF fel, 61-63. *Mem:* Am Chem Soc; AAAS. *Res:* Mechanism and specificity of mammalian dehydrogenases; application of computer techniques to biochemistry. *Mailing Add:* UCCC Mail Loc 149 2900 Reading Rd Cincinnati OH 45221

PRAIS, MICHAEL GENE, b Detroit, Mich. CHEMISTRY. *Educ:* Univ Chicago, BS, 72; Univ Calif, San Diego, MS, 74, PhD(chem), 81. *Prof Exp:* Instr chem & math, San Diego City Col, 75-80; post doctoral chem & physics, Univ Calif, San Diego, 81; asst prof chem, 81-86, DIR ACAD COMPUT FACIL, ROOSEVELT UNIV, 84- , ASSOC PROF CHEM, 86-, CHMN CHEM, 87- *Concurrent Pos:* Vis res asst chem, Univ Calif, San Diego, 82- *Mem:* Am Chem Soc; Asn Comput Mach. *Res:* Numerical analysis of thermodynamic properties of crystal surfaces; laboratory data acquisition and analysis. *Mailing Add:* Dept Chem Roosevelt Univ 430 S Michigan Ave Chicago IL 60605

PRAISSMAN, MELVIN, b Philadelphia, Pa, Aug 23, 40; m 63; c 1. BIOCHEMISTRY, PHYSICAL CHEMISTRY. *Educ:* Univ Pittsburgh, BS, 62; Univ Ariz, PhD(chem), 67. *Prof Exp:* Res assoc biochem, Purdue Univ, 67-68; RES BIOCHEMIST, DEPT MED, MEADOWBROOK HOSP, 68-; ASST PROF MED, MED SCH, STATE UNIV NY STONY BROOK, 72- *Concurrent Pos:* NIH fel, 67-; assoc, Dept Physiol, Mt Sinai Sch Med, 72- *Mem:* AAAS; Am Chem Soc. *Res:* Membrane phenomena with emphasis on ion and water movement across biologic membranes; molecular factors in gastrin release and action; gastric physiology. *Mailing Add:* Nassau County Med Ctr 2201 Hempstead Turnpike East Meadow NY 11554-1146

PRAKASAM, TATA B S, b Vizianagram, AP, India, Jan 5, 36; US citizen; m 53; c 3. ENVIRONMENTAL PROBLEMS OF DEVELOPING COUNTRIES, TEACHING ENVIRONMENTAL ENGINEERING. *Educ:* MR Col, India, BS, 53; Nagpur Univ, India, MS, 55; Rutgers Univ, PhD(environ sci), 66. *Prof Exp:* Asst res officer, All India Inst Hyg & Pub Health, 55-59; jr sci officer, Cent Pub Health Eng Res Inst, India, 59-62; res asst, Rutgers Univ, NJ, 62-66; res assoc, Cornell Univ, Ithaca, NY, 66-70, sr res assoc & lectr, 70-74; proj mgr, 74-90, COORDR TECH SERV, ENVIRON ENG & SCI, METROP WATER RECLAMATION DIST GREATER CHICAGO, 90- *Concurrent Pos:* Mem comt, Water Pollution Control Fedn, standard methods, 72-80, res, 76-80, prog, 91; adj prof, dept environ eng, Ill Inst Technol, 74-; panel mem, Bostid, Nat Acad Sci, 74-76, adv, 78-84, consult, 85. *Mem:* Water Pollution Control Fedn; Air & Waste Mgt Asn; Int Asn Water Pollution Res & Control; Sigma Xi. *Res:* Sludge conditioning and dewatering, odor control activated sludge process performance optimization, nutrient control and anaerobic digestion; author of 80 publications and reports. *Mailing Add:* 7014 Richmond Ave Darien IL 60559

PRAKASH, ANAND, b India, Jan 1, 15; m 56; c 2. GROUNDWATER, HYDRAULIC STRUCTURES & HYDROLOGY. *Educ:* Univ Roorkee, India, BS, 57, MS, 69; Colo State Univ, PhD(civil eng), 74. *Prof Exp:* Asst engr, Irrigation Dept, Univ Roorkee, India, 58-69; res asst civil eng, Colo State Univ, 72-74; eng supvr water resources, Sargent & Lundy Engrs, Chicago, 74-77, hydrol, Bechtel Inc, San Francisco, 77-79; CHIEF WATER RESOURCES ENG, DAMES & MOORE, DENVER, 79- *Concurrent Pos:* Lectr civil eng, Univ Roorkee, India, 66-69, exec engr, 66-71. *Mem:* Am Soc Civil Engrs; Am Water Resources Asn; Asn Hydraulic Res; Am Geophys Union; Am Nuclear Soc; Soc Am Mil Engrs. *Res:* Hydrology; groundwater; hydraulic structures; hydraulics; water resources engineering. *Mailing Add:* Dames & Moore 1550 NW Hwy Parkridge IL 60068

PRAKASH, LOUISE, b Lyon, France, Apr 11, 43; US citizen; m 65. MOLECULAR GENETICS. *Educ:* Bryn Mawr Col, BA, 63; Wash Univ, MA, 65; Univ Chicago, PhD(microbiol), 70. *Prof Exp:* Fel, 70-72, from asst prof to assoc prof radiation biol & biophys, 72-86, PROF BIOPHYS, UNIV ROCHESTER, 86- *Mem:* Biophys Soc; Environ Mutagen Soc; Genetics Soc Am; Am Soc Microbiol. *Res:* Repair of damaged DNA; mechanisms of mutagenesis and recombination in yeast. *Mailing Add:* Radiation Biol & Biophys Univ Rochester Rochester NY 14642

PRAKASH, SATYA, b Pilkhuwa, India, July 8, 38; m 65. GENETICS. *Educ:* Meerut Col Agra Univ, BSc, 56; Vet Col, Mhow, BVScAH, 60; Indian Vet Res Inst, Izatnagar, MVSc, 62; Wash Univ, PhD(zool), 66. *Prof Exp:* NIH trainee, Univ Chicago, 66-67, res assoc zool, 68-69; asst prof, 69-74, assoc prof, 74-80, PROF BIOL, UNIV ROCHESTER, 80- *Mem:* AAAS; Genetics Soc Am. *Res:* Mechanisms of DNA repair and recombination in eukaryotes. *Mailing Add:* Dept of Biol Univ of Rochester Rochester NY 14627

PRAKASH, SHAMSHER, b Panjab, India, Jan 3, 33; m. GEOTECHNICAL ENGINEERING, EARTHQUAKE ENGINEERING. *Educ:* Univ Roorkee, BE, 54; Univ Ill, Urbana, MS, 61, PhD(civil eng), 62. *Prof Exp:* Asst engr, Panjab Pub Works Dept, India, 54-57; lectr civil eng, Univ Roorkee, 57-62, reader, 62-66, prof civil eng & earthquake eng, 66-78; assoc prof, 78-80, PROF CIVIL ENG, UNIV MO, ROLLA, 80- *Honors & Awards:* Khosla Res Award & Gold Medal, 78; Instrumentation Prize, Indian Geotech Soc, 80; Res & Develop Prize, Sci & Technol, Fedn Indian Chamber Com & Indust, 84. *Mem:* Fel Am Soc Civil Engrs; fel Brit Inst Civil Engrs; fel Indian Inst Engrs; fel Indian Geotech Soc (pres, 71-75); Indian Soc Earthquake Technol (secy, 69-71); Sigma Xi; Am Soc Testing & Mat; Earthquake Eng Res Inst; Indian Soc Tech Educ. *Res:* Effect of earthquakes on pile foundations; seismic stability of retaining walls; liquefaction of soils; machine foundation design and analysis and dynamic soil constants; soil dynamics; geotechnical earthquake engineering; author and co-author of three books. *Mailing Add:* 308 Dept of Civil Eng Univ of Mo Rolla MO 65401-0249

PRAKASH, SURYA G K, b Bangalore, India, Oct 7, 53; m 81; c 2. HYDROCARBON CHEMISTRY, SYNTHETIC METHODOLOGY. *Educ:* Bangalore Univ, BSc Hons, 72; Indian Inst Technol, MSc, 74; Univ Southern Calif, PhD(chem), 78. *Prof Exp:* From res asst prof to res assoc prof, 81-90, ASSOC PROF CHEM, UNIV SOUTHERN CALIF, 90- *Concurrent Pos:* Consult, Avery Int, 82-89, Global Geochem Corp, 84-88, Ultrasystems Inc, 90- & Chemsys, 90- *Honors & Awards:* Res Excellence Award, Hydrocarbon Res Inst, 84. *Mem:* Am Chem Soc. *Res:* Synthetic, mechanistic, physical-organic and superacid chemistry; application of nuclear magnetic resonance spectroscopy. *Mailing Add:* Dept Chem Univ Southern Calif Los Angeles CA 90089-1661

PRALL, BRUCE RANDALL, b Bremen, Kans, Mar 22, 42; m; c 3. DEVELOPMENT OF CHEMICAL DEMONSTRATIONS ON OVERHEAD PROJECTORS. *Educ:* Kans State Univ, Emporia, BS, 63, MS, 65; Univ Nev, Reno, PhD(chem), 71. *Prof Exp:* Postdoctoral fel phys chem, Univ Wis-Milwaukee, 70-71; PROF PHYS CHEM, MARIAN COL, 71- *Concurrent Pos:* Vis prof, Univ Nev, Reno, 81 & 83. *Mem:* Am Chem Soc. *Res:* Interfacial phenomena in the presence of non-uniform electric fields. *Mailing Add:* 45 S National Ave Fond Du Lac WI 54935

PRAMER, DAVID, b Mt Vernon, NY, Mar 25, 23; m 50; c 2. MICROBIAL ECOLOGY. *Educ:* Rutgers Univ, BSc, 48, PhD(microbiol), 52. *Prof Exp:* Vis investr, Imp Chem Indust, Ltd, Eng, 52-54; from asst prof to assoc prof, Rutgers Univ, 54-61, chmn dept biochem & microbiol, 65-69, dir biol sci, 69-73, dir univ res, 73-80, dir, Waksman Inst Microbiol, 80-88, PROF MICROBIOL, RUTGERS UNIV, NEW BRUNSWICK, 61-, ASSOC VPRES RES, 88- *Concurrent Pos:* Fulbright-Hays res scholar & lectr; mem, Int Comn Microbial Ecol, 71- *Mem:* AAAS; Am Soc Microbiol; Brit Soc Gen Microbiol; Sigma Xi. *Res:* Microbial ecology; nature and activity of microorganisms in aquatic and terrestrial environments; biological control; biochemical bases of microbial interrelationships; chemical transformations mediated by microorganisms. *Mailing Add:* 377 Hoes Lane Piscataway NJ 08854

PRANCE, GHILLEAN T, b Brandeston, Eng, July 13, 37; m 61; c 2. BOTANY. *Educ:* Oxford Univ, BA, 60, DPhil(bot), 63, MA, 65. *Hon Degrees:* Dr, Goteborg Univ, Sweden, 83. *Prof Exp:* Res fel trop bot, 63-66, assoc cur, 66-68, B A krukoff Cur Amazonian Bot, 68-75, DIR BOT, BOT GARDEN NY, 75-, FROM VPRES BOT RES TO SR VPRES RES, NEWYORK. *Concurrent Pos:* NSF study grants, 66-67, 68-70, 72-75, 75-76, 76-78, 79-80 81-85, 85-88; adj prof, Herbert H Lehman Col, City Univ New York, 69-; ed adv, Torrey Bot Club, 70-; dir grad prog, Instituto Nacional de Pesquisas da Amazonia, Brazil, 73-75. *Mem:* Am Soc Plant Taxon (pres, 84-85); Asn Trop Biol (pres, 79-80); Am Inst Biol Sci; Acad /Brasilera de Ciencias. *Res:* Taxonomy of tropical plant families, especially Chrysobalanaceae, Dichapetalaceae, Caryocaraceae and Lecythidaceae; floristic studies of the Amazon basin; ethnobotany. *Mailing Add:* Royal Bot Gardens Q Richmond Surrey TW9 3AB England

PRANE, JOSEPH W(ILLIAM), b New York, NY, June 18, 23; m 45; c 2. COATINGS, ADHESIVES. *Educ:* City Col New York, BChE, 43; Columbia Univ, MS, 48. *Prof Exp:* Metallurgist, Bendix Aviation Corp, 43; asst, SAM Labs, Columbia Univ, 43-44, Manhattan Proj, 44-46; head oil & resin develop labs, Nat Lead Co, 46-61, tech dir, Pecora, 61-64; tech serv & develop mgr specialty chem div, Celanese Corp Am, 64-65; consult, Skeist Labs, NJ, 65-68; INDUST CONSULT, 69- *Concurrent Pos:* Adj prof, Dept Chem Eng, Polytech Inst New York; contrib ed, Adhesives Age, Polymer News. *Mem:* Am Chem Soc; Am Soc Testing & Mat; Am Inst Chem; Asn Consult Chemists & Chem Engrs; Soc Mfg Engrs; Fedn Soc Coatings Technol. *Res:* Organic coatings; synthetic resins; emulsion polymers; adhesives; sealants; statistical methods. *Mailing Add:* 213 Church Rd Elkins Park PA 19117

PRANGE, ARTHUR JERGEN, JR, b Grand Rapids, Mich, Sept 19, 26; c 4. PSYCHIATRY. *Educ:* Univ Mich, BS, 47, MD, 50. *Prof Exp:* Intern, Wayne County Gen Hosp, Eloise, Mich, 50-51; resident anesthesiologist, Detroit Receiving Hosp, 51-52; resident, Mem Hosp, 54-57, from instr to assoc prof, 57-68, PROF PSYCHIAT, SCH MED, UNIV NC, CHAPEL HILL, 68- *Concurrent Pos:* Consult, Vet Admin Hosp, Fayetteville, NC, 60-74; NIH grant, 61-64, NIMH career develop awards, 61-69 & career scientist award, 69-; consult, Dorothea Dix Hosp, Raleigh, 63- & Cherry State Hosp, Goldsboro, 68-70; mem med adv coun, NC State Bd Ment Health, 70- *Honors & Awards:* Eugene A Hargrove Award, NC Found Ment Health Res, 85. *Mem:* Fel Am Psychiat Asn; Psychosom Soc; Am Psychiat Asn; Am Col Neuropsychopharmacol; Int Col Neuropsychopharmacol. *Res:* Biological aspects of mental depression and its nosology; relationsip of hormones to catecholamine metabolism in depression; role of thyroid status in drug response and in psychological processes. *Mailing Add:* Dept of Psychiat Univ of NC Sch of Med Chapel Hill NC 27599-7160

PRANGE, HENRY DAVIES, b Chicago, Ill, Oct 28, 42; c 2. ANIMAL PHYSIOLOGY. *Educ:* Duke Univ, BA, 64, MA, 67, PhD, 70. *Prof Exp:* Asst prof zool, Univ Fla, 70-75; asst prof, 76-79, ASSOC PROF PHYSIOL, SCH MED, IND UNIV, 79- *Mem:* Am Physiol Soc; AAAS. *Res:* Energetics of animal locomotion; comparative respiratory physiology; skeletal allometry; temperature regulation. *Mailing Add:* Med Sci Prog Ind Univ Neyers Hall Bloomington IN 47405

PRANGE, RICHARD E, b Ohio, Sept 23, 32; m. PHYSICS. *Educ:* Univ Chicago, MS, 55, PhD(physics), 57. *Prof Exp:* Instr physics, Univ Pa, 57-59; fel, Inst Theoret Physics, Copenhagen, Denmark, 59-60; staff mem, Inst Advan Study, 60-61; from asst prof to assoc prof, 61-68, PROF PHYSICS & ASTRON, UNIV MD, COLLEGE PARK, 68- *Concurrent Pos:* NSF fel, 59-61. *Mem:* Am Phys Soc. *Res:* Theoretical, many-particle, and solid state physics. *Mailing Add:* Dept Physics & Astron Univ Md College Park MD 20742

PRANGER, WALTER, b Chicago, Ill, May 2, 34. COMPLEX ANALYSIS. *Educ:* Ill Inst, PhD(math), 64. *Prof Exp:* PROF MATH DEPAUL UNIV, 64- *Concurrent Pos:* Res grant, Nat Sci Found, 72. *Mailing Add:* Dept Math DePaul Univ Chicago IL 60614

PRAPAS, ARISTOTLE G, b Athens, Greece, Feb 21, 22; US citizen. ORGANIC POLYMER CHEMISTRY, PETROLEUM CHEMISTRY. *Educ:* Athens Tech Univ, ChemE, 45; Northwestern Univ, PhD(chem), 56. *Prof Exp:* Sr res chemist, Davison Div, W R Grace & Co, 55-57, Res Div, 57-59; sr res chemist, Mobil Chem Co, 59-60, sr res chemist, 61-86; RETIRED. *Mem:* Am Chem Soc; Sigma Xi. *Res:* Product and process research and development; development of polymerization processes, and of styrenic, acrylic, polyester and polyolefin resins development of supported catalysts; synthesis of specialty chemicals, and additives for plastics and petroleum; fast curing protective coatings. *Mailing Add:* 1144 Blueberry Ct Edison NJ 08817-2609

PRASAD, ANANDA S, b Buxar, India, Jan 1, 28; m 52; c 4. INTERNAL MEDICINE, HEMATOLOGY. *Educ:* Patna Univ, BSc, 46, MB, BS, 51; Univ Minn, Minneapolis, PhD(internal med), 57. *Prof Exp:* Instr med, Hosp, Univ Minn, 57-58; vis assoc prof, Fac Med, Univ Shiraz, 58-60, vis prof & chmn dept, 60; asst prof med & nutrit, Vanderbilt Univ, 61-63; assoc prof, 63-68, chief hemat, 63-84, PROF MED, SCH MED, WAYNE STATE UNIV, 68-, DIR RES, 84- *Concurrent Pos:* Assoc, Nemazee Hosp, Shiraz, Iran, 58-60; hon prof, Univ Shiraz, 60-; head nutrit proj, US Naval Med Res Unit 3, Cairo, UAR, 61-63; mem subcomt trace elements, food & nutrit bd, Nat Acad Sci-Nat Res Coun, 65; ed, Am J Hemat, J Trace Elements Exp Med. *Honors & Awards:* Goldberger Award, 75; Am Col Nutrit Award, 76; Robert H Herman Award, 84. *Mem:* Am Physiol Soc; Am Fedn Clin Res; Am Soc Hemat; Am Inst Nutrit; fel Am Col Physicians; fel AAAS; Am Soc Clin Invest; Asn Am Physicians. *Res:* Trace metal metabolism; dysproteinemias and various hematological disorders; author of numerous articles and books. *Mailing Add:* 4710 Cove Rd Orchard Lake MI 48033

PRASAD, ARUN, m; c 2. PRECIOUS & NON-PRECIOUS ALLOY DEVELOPMENT, REACTIVE METALS & REFRACTORY MATERIALS FOR CASTING. *Educ:* Banaras H Univ, BS, 67, MS, 69; Polytechnic Inst NY, PhD(phys metall), 76. *Prof Exp:* Chief metallurgist, Rx Jeneric Gold Co, 76-79; dir res & develop, Jeneric Industs Inc, 79-82, vpres, 82-87; SR VPRES TECHNOL, JENERIC/PENTRON INC, 87- *Concurrent Pos:* Adj assoc prof, Univ Med & Dent, NJ, 89-91; prin investr, NIH-SBIR grant, Jeneric Pentron Inc, 90. *Mem:* Am Ceramic Soc; Am Soc Metall; Am Inst Mech Engrs; Int Asn Dent Res. *Res:* New biocompatible titanium alloys and compatible porcelains to be used for porcelain-fused to-metal dental restorations. *Mailing Add:* Jeneric Pentron Inc 53 N Plains Indust Rd Wallingford CT 06492

PRASAD, BIREN, June 30, 49; m; c 3. MECHANICAL & AEROSPACE ENGINEERING. *Educ:* Bihar Col Eng, Patna Univ, BS, 69; Indian Inst Technol, India, MS, 71; Stanford Univ, DEng, 75; Ill Inst Technol, PhD(mech & aerospace eng), 77. *Prof Exp:* Res engr, Stanford Univ & Failure Anal Assocs, Calif, 74-75; res asst, Ill Inst Technol, 76-77; sr res engr, Am Asn Railroad, Tech Ctr, Ill, 78-80; res scientist, Sci Res Lab, Ford Motor Co, 80-85; strategic planning & tech develop, Electronic Data Systs, Gen Motors Corp, Troy, Mich, 85-86, mgr applns develop group, 86-87, eng prod des, 87-88, spec proj mgr, Artificial Intel Serv, 88-89, SR ENG CONSULT, ELECTRONIC DATA SYSTS, GEN MOTORS, CORP, TROY, MICH, 89- *Concurrent Pos:* Adj prof mech & aerospace eng, WVa Univ; consult, US & foreign cos. *Honors & Awards:* Creative Develop Tech Innovation, NASA, 81; Survey Paper Citation Award, AIAA, 82; Commemorative Medal Honor, Am Biographical Inst, 87. *Mem:* Assoc fel, Am Inst Aeronaut & Astronaut, 77-; sr mem Am Soc Mech Eng, 78; sr mem Am Soc Civil Eng, 80; sr mem Soc Automotive Eng, 81; sr mem Am Asn Artificial Intell, 88; fel Int Soc Prod Enhancement (pres, 88-). *Res:* Structural optimization in vehicle design and applications; author of over 70 publications. *Mailing Add:* Gen Motors Electronic Data Systs 750 Tower Dr Troy MI 48007-7019

PRASAD, CHANDAN, b Chain Patti, India, Jan 1, 41; m 74; c 3. NEUROPEPTIDES, ENDOCRINOLOGY. *Educ:* Uttar Pradesh Agr Univ, India, BSc, 64, MSc, 66; La State Univ, PhD(microbiol biochem), 70. *Prof Exp:* Sr staff fel, NIH, 74-78; from asst prof to assoc prof med & biochem, 78-86, PROF MED, LA STATE UNIV MED CTR, 86-, CHIEF NEUROSCIENCES, 89- *Mem:* Soc Neurosci; Brit Brain Res Asn; European Brain & Behav Soc; Am Soc Biol Chemists; Endocrine Soc; Int Soc Neurochem; NY Acad Sci; AAAS; Sigma Xi. *Res:* Biochemistry, pharmacology and endocrinology of brain peptides; nutrition and behavior; obesity; aging. *Mailing Add:* Dept of Med La State Univ-Med 1542 Tulane Ave New Orleans LA 70112

PRASAD, KAILASH, b Peshaur, India, Apr 2, 30; m 78; c 2. PHARMACOLOGY, PHYSIOLOGY. *Educ:* Patna Univ, BSc, 52; Univ Bihar, MB, BS, 57; Univ Delhi, MD, 61; Univ Alta, PhD(pharmacol), 67. *Prof Exp:* Surgeon obstet & gynec, Darbhanga Med Col & Hosp, India, 57; demonstr pharmacol, Gandhi Med Col & Hosp, 58; from demonstr to lectr, Lady Hardinge Med Col & Hosp, 58-64; teaching asst, Univ Alta, 64-65, from asst prof to assoc prof pharmacol, 67-73; assoc prof, 73-77, actg head physiol, 87-88, PROF PHYSIOL, COL MED, UNIV SSK, 77- *Concurrent Pos:* Can Heart Found fel, 65-69; Alta Heart Found & Med Res Coun Can res grants, 68-; vpres, Am Col Angiol, 86- & Int Col Angiol, 91- *Mem:* Can Cardiovasc Soc; Int Col Angiol; Pharmacol Soc Can; Am Soc Clin Pharmacol & Therapeut; NY Acad Sci; fel Am Col Angiol; fel Am Col Cardiol; Physiol Soc Can; fel Royal Col Physicians & Surgeons. *Res:* mechanism of cardiac arrhythmias and congestive heart failure; cardiovascular pharmacology; electrophysiology of guinea pig and human heart in relation to the mechanism of drug action; development of noninvasive diagnostic techniques; free oxygen radicals and heart disease; protection of ischemic myocardium; mechanism of artherosclerosis. *Mailing Add:* Dept Physiol Univ Sask Col Med Saskatoon SK S7H 0W0 Can

PRASAD, KEDAR N, b Barhiya, India, Jan 8, 35. RADIOBIOLOGY, ONCOLOGY. *Educ:* Univ Bihar, BSc, 55, MSc, 57; Univ Iowa, PhD(radiation biol), 63. *Prof Exp:* Asst scientist radiation biol, Brookhaven Nat Lab, 63-66; instr, Univ Rochester, 66-67; asst prof, 68-73, assoc prof, 74-79, PROF RADIATION BIOL, COL MED, UNIV COLO HEALTH SCI CTR, DENVER, 80- *Mem:* Radiation Res Soc; Soc Exp Biol & Med; Am Asn Cancer Res; Am Soc Neurochem; Am Soc Pharmacol & Exp Therapeut. *Res:* Differentiation of neuroblastoma cells in culture and cyclic nucleotides; vitamins and cancers. *Mailing Add:* Dept Radiol Univ Colo Health Sci Ctr 4200 E Ninth Ave Denver CO 80262

PRASAD, MAREHALLI GOPALAN, b Karnataka State, India, July 22, 50; m 76; c 2. ENGINEERING ACOUSTICS, NOISE CONTROL. *Educ:* Univ Col Eng, Bangalore, India, BE, 71; Indian Inst Technol, Madras, India, MS, 74; Purdue Univ, PhD(mech eng), 80. *Prof Exp:* Lectr mech eng, BMS Col Eng, Bangalore, India, 71-72; res asst mech, Indian Inst Technol, Madras, India, 72-74; aeronaut engr design, Hindustan Aeronaut Ltd, Bangalore, India, 74-77; grad res asst acoust res, Herrick Labs, Purdue Univ, 77-80; from asst prof to assoc prof, 80-90, PROF RES, ACOUST & MECH ENG, STEVENS INST TECHNOL, 90- *Concurrent Pos:* Grad instr, Dept Civil Eng, Purdue Univ, 78-80; chmn tech comt, Am Soc Mech Engrs, 83-87, assoc ed, Appl Mch Rev; dir, Noise & Vibration Control Lab, Stevens Inst Technol, 84-; res consult, AT&T Bell Labs, NJ, 86 & IBM Corp, NY, 88; expert noise control, UN Indust Develop Org, Vienna, 89- *Mem:* Inst Noise Control Eng (vpres, 87-); Am Soc Mech Engrs; Acoust Soc Am; Am Soc Eng Educ. *Res:* Acoustical modeling of sources in ducts including acoustical source-load interactions; boundary element method and acoustic intensity; established noise and vibration at Stevens Institute of Technology. *Mailing Add:* Dept Mech Eng Stevens Inst Technol Hoboken NJ 07030

PRASAD, NARESH, b Auria, Bihar, India, Dec 11, 39; m 65; c 3. RADIOBIOLOGY, CYTOGENETICS. *Educ:* Patna Univ, BSc, 59, MSc, 61; NDak State Univ, PhD(bot), 67. *Prof Exp:* Lectr bot, Ranchi Univ, India, 61-63; USPHS fel, Baylor Col Med, 67-69; asst prof radiobiol, 69-74; assoc dir & consult, Vet Admin Hosp, Houston, 74-78; assoc prof, 74-82, PROF RADIOL, BAYLOR COL MED, 82-; DIR RADIOBIOL RES, VET AFFAIRS MED CTR, HOUSTON, 78- *Concurrent Pos:* Asst dir radiobiol res & consult, Vet Admin Hosp, Houston, 69-74; ed-in-chief, Radiation & Immunol Series, CRC Press, Inc, Boca Raton, Fla. *Mem:* AAAS; Radiation Res Soc; Am Asn Cancer Res; Am Inst Ultrasound Med; Sigma Xi; Soc Magnetic Resonance Med; Am Col Med Physics. *Res:* Chromosomes; radiation effects; lymphocytes; lung cancer; ultrasound; nuclear magnetic resonance. *Mailing Add:* Dept Radiol Baylor Col Med Houston TX 77030

PRASAD, PARAS NATH, b Bihar, India, June 4, 46. LASER CHEMISTRY, SOLID STATE SPECTROSCOPY. *Educ:* Bihar Univ, BSc, 64, MSc, 66; Univ Philadelphia, PhD(chem), 71. *Prof Exp:* Lectr chem, Bhagalpur Univ, India, 67; fel, Univ Mich, 71-74; from asst prof to assoc prof, 74-82, PROF CHEM, STATE UNIV NY, BUFFALO, 82- *Concurrent Pos:* Prin investr, Petrol Res Found, 83-86, Air Force Off Sci Res, 84-85 & 85-87 & NSF, 84-87; consult, Foster-Miller, Inc, 86. *Mem:* Am Chem Soc. *Res:* Picosecond optical double resonance of molecular solids at low temperatures; phonon-assisted spectral-diffusion in disordered solids; dynamics of chemical transformations in organic solids and monolayer films; time-resolved studies of vibrational energy redistribution; design, ultrastructure and dynamics of thin films of electroactive polymers; phonon echo in organic solids. *Mailing Add:* Dept Chem State Univ NY 57 Acheson Hall Buffalo NY 14214

PRASAD, RAGHUBIR (RAJ), b Allahabad, India, July 5, 36; m 66; c 3. PLANT PHYSIOLOGY & FORESTRY. *Educ:* Univ Allahabad, BSA, 54, MS, 56; Oxford Univ, PhD(plant physiol), 61. *Prof Exp:* AEC fel, Univ Calif, Davis, 61-63; res biochemist, Univ Calif, Berkeley, 63-65; assoc prof agron, Univ Ibadan, 65-67; prof assoc bot, Univ Man, 67-68; res scientist, Fed Dept Agr, 68-70, Chem Control Res Inst, Ottawa, 70-78 & Forest Pest Mgt Inst, Sault Ste Marie, Ont, 78-88; RES SCIENTIST, PAC FOREST RES CTR, 88- *Concurrent Pos:* Consult, Food & Agr Orgn, Kenya, 68-72. *Mem:* Am Soc Plant Physiol; Weed Sci Soc Am; Am Chem Soc; Can Soc Plant Physiol; Soc Environ Toxicol Chem. *Res:* Pesticide physiology and biochemistry; fate of pesticides in aquatic and forest environment; pesticide pollution; systemic pesticides in forests; weed science; herbicides for forest regeneration; mode of action of pesticides (herbicides) in plants; physiological and biochemical changes induced by pesticides in plants; translocation and metabolism of pesticides in forest plants; the influence of droplet size, volume rate and dosage rate of herbicides on their efficacy. *Mailing Add:* Pac Forest Res Ctr 506 E Burnside Rd Victoria BC V8Z 1M5 Can

PRASAD, RAMESHWAR, b Sayadpur, India, July 3, 36; m 61; c 3. IMMUNOLOGY, IMMUNOCHEMISTRY. *Educ:* Allahabad Univ, India, PhD(chem), 59, DSc, 67. *Prof Exp:* Reader chem, Patna Univ, 68-70; head res & develop immunol, 72-79, CHIEF RADIOIMMUNOASSAY, UNIV ILL MED CTR, 79-, ASSOC PROF PATH. *Mem:* Fel Indian Nat Acad Sci; Fel Nat Acad Clin Biochem; Fel Assoc Clin Sci. *Res:* Study of immunoglobulin levels in cerebrospinal fluid of patients with central nervous system diseases; purification, characterization and clinical evaluation of organ specific antigens. *Mailing Add:* Dept Path Univ Ill Med Ctr Box 6998 Chicago IL 60680

PRASAD, RUPI, b Ranchi, India. MEDICAL GENETICS, ENZYMOLOGY. *Educ:* Ranchi Univ, BS, 62, MS, 64; NDak State Univ, PhD(bot), 68. *Prof Exp:* Fel med genetics, Univ Tex M D Anderson Hosp & Tumor Inst, 69-71; asst prof genetics, Dept Obstet & Gynec, Baylor Col Med, 72-79; RES CHEMIST, CHAMPION CHEM, INC, 80- *Concurrent Pos:* Biochem geneticist, Vet Admin Hosp, Houston, 74- *Mem:* Sigma Xi; Am Chem Soc; Nat Asn Corrosion Eng. *Mailing Add:* Champion Chem Inc PO Box 450499 Houston TX 77245-0499

PRASAD, S E, ULTRASONICS, MATERIALS ENGINEERING. *Educ:* Andhra Univ, India, BSc, 69, MSc, 72, PhD(eng physics), 76. *Prof Exp:* Systs engr, Nat Aeronaut Lab, 77; res assoc, Queen's Univ, 78-79; physicist, Ont Res Found, 80; scientist, Almax Industs Ltd, 80-83; opers mgr, 83-85, PRES, SENSOR/BM HIGH TECH INC, 85- *Mem:* Am Ceramic Soc. *Res:* Piezoelectric materials, transducers and systems. *Mailing Add:* B M Hi-Tech PO Box 97 Collingwood ON L9Y 3Z4 Can

PRASAD, SHEO SHANKER, physics, for more information see previous edition

PRASAD, SURESH, b Chapra, India, Jan 26, 37; m 60; c 3. POULTRY SCIENCE, LABORATORY ANIMAL SCIENCE. *Educ:* Bihar Vet Col, India, DVM, 59, Univ Mo-Columbia, MS, 64, PhD(poultry sci), 67. *Prof Exp:* Animal husb exten supvr, Animal Husb Dept, Govt Bihar, India, 59-60, key village inspector, 60-63; dir res & qual control, Mountaire Poultry, Mountaire Corp, 66-68, dir res & diag serv, 68-72; vet, Biol Vet Serv, 73-75, MGR BIOL VET SERV, MERCK, SHARP & DOHME, MERCK & CO, INC, 75- *Mem:* Am Vet Med Asn; Am Asn Avian Path; Poultry Sci Asn. *Res:* Poultry and laboratory animal care administration. *Mailing Add:* 1520 Tennis Circle Lansdale PA 19446

PRASANNA, HULLAHALLI RANGASWAMY, b Mysore, Karnataka, India, Dec 1, 46; US citizen; m 76; c 2. HORMONES, CARCINOGENESIS. *Educ:* Univ Bangalore, India, BSc, 65, MSc, 67; Univ Delhi, India, MSc, 71, PhD(med biochem), 75. *Prof Exp:* Postdoctoral fel cell & molecular biol, State Univ NY, Buffalo, 75-78; res assoc, St Jude Childrens Hosp, Memphis, 78-80 & Memphis State Univ, 80-81; res asst prof toxicol, Col Pharm, Univ Tenn, 82-83; sr res assoc, Fels Res Inst, Temple Univ Med, 83-87; vis scientist, Nat Ctr Toxicol Res, Food & Drug Admin, Jefferson, Ariz, 87-90, RES CHEMIST, CTR DRUG EVAL RES, FOOD & DRUG ADMIN, ROCKVILLE, MD, 90- *Mem:* Am Soc Biochem & Molecular Biol; Am Asn Cancer Res; AAAS; Sigma Xi. *Res:* Chemistry of toxins and carcinogens, their mode of action in various species; toxicology and carcinogenises; chemoprevention of cancer; anti cancer drugs; nucleic acids and hormones. *Mailing Add:* Div Clin Pharmacol Eval Res Ctr Drug Food & Drug Admin 5600 Fishers Lane Rockville MD 20857

PRASANNA KUMAR, V K, b Mysore, India, Aug 10, 56. PARALLEL PROCESSING, PARALLEL ARCHITECTURES. *Educ:* Bagalore Univ, BS, 76; Indian Inst Sci, MS, 78; Pa State Univ, PhD(comput sci), 83. *Prof Exp:* Sr engr, Processor Systs Ltd, India, 78-79; asst prof, 83-89, ASSOC PROF COMPUT ENG, UNIV SOUTHERN CALIF, 89- *Concurrent Pos:* Prin investr, NSF, Off Naval Res, Air Force Off Sci Rese; prog chair, Inst Elec & Electronic Engrs Int Parallel Processing Symp, 91; IBM fel; subj area ed, J Parallel & Distrib Comput. *Mem:* Inst Elec & Electronics Engrs. *Res:* Computer architecture; parallel processing; VLSI systems; computational aspects of image processing and computer vision. *Mailing Add:* Dept EE Systs Univ Southern Calif Sal 344 Los Angeles CA 90089-0781

PRASHAD, NAGINDRA, MICROBIOLOGY. *Educ:* Univ Houston, PhD(microbiol), 70. *Prof Exp:* ASST PROF BIOCHEM, HEALTH SCI CTR, UNIV TEX, HOUSTON, 81- *Res:* Reversal of malignancy; gene regulation of camp; dependent protein kinases. *Mailing Add:* Grad Sch Biomed Sci Univ Tex Health Sci Ctr Box 20334 Astromdome Sta Houston TX 77025

PRASHAR, PAUL D, b Lahore, India, Sept 10, 30; US citizen; m 66. HORTICULTURE, PLANT BREEDING. *Educ:* Punjab Univ, BSc, 52; Univ Minn, MS, 55; Univ Mo, PhD(hort), 60. *Prof Exp:* Asst hort, Univ Mo, 57-60; from instr to assoc prof, 60-77, PROF HORT, SDAK STATE UNIV, 78- *Mem:* Am Soc Hort Sci; Int Soc Hort Sci. *Res:* Tomato breeding for early maturity; Septoria resistance; pepper breeding to develop early bell pepper for home gardeners. *Mailing Add:* Dept Plant Sci SDak State Univ PO Box 2207-A Brookings SD 57007-0996

PRASK, HENRY JOSEPH, b Detroit, Mich, Sept 26, 36; m 68; c 2. SOLID STATE PHYSICS. *Educ:* Univ Notre Dame, BS, 58, PhD(nuclear physics), 63. *Prof Exp:* Consult, Feltman Res Lab, 64-65, solid state physicist, Explosives Div, Feltman Res Lab, Picatinny Arsenal, 65-80, solid state physicist, Energetics & Warheads Div, US Army Armament R, D & E Ctr, Dover, NJ, 80-89; RES PHYSICIST, REACTOR RADIATION DIV, NAT INST STANDARDS & TECHNOL, GAITHERSBURG, MD, 89- *Concurrent Pos:* Guest scientist, Reactor Radiation Div, Nat Bur Standards, Gaithersburg, MD, 71- *Mem:* Am Phys Soc; Am Soc Nondestructive Testing; Am Soc Metals. *Res:* Lattice dynamics and crystallographic studies by means of thermal neutron scattering with particular emphasis on energetic materials and metallurgical samples. *Mailing Add:* Reactor Radiation Div Nat Inst Standards & Technol Gaithersburg MD 20899

PRASUHN, ALAN LEE, b Columbus, Ohio, Feb 19, 38; m 59; c 3. FLUID MECHANICS, HYDRAULICS. *Educ:* Ohio State Univ, BCE, 61; State Univ Iowa, MS, 63; Univ Conn, PhD(fluid mech), 68. *Prof Exp:* Res specialist mixing stratified flows, Univ Conn, 65-68; from asst prof to prof civil eng, Calif State Univ, Sacramento, 68-78; prof civil eng, SDak State Univ, 78-90; CHMN CIVIL ENG, LAWRENCE TECH UNIV, 90- *Concurrent Pos:* Hon res fel, Birmingham Univ, 77; Nat Sci Found travel grant, 77; dir, Am Soc Civil Engrs, 85-88. *Mem:* Am Soc Civil Engrs; Am Soc Eng Educ; Int Asn Hydraul Res. *Res:* Sediment transport, determination of turbulence field flows over sand beds using hot film anemometers; various model studies; scour at bridges; computer modeling. *Mailing Add:* Dept Civil Eng Lawrence Technol Univ 21000 W Ten Mile Rd Southfield MI 48075-1058

PRATER, ARTHUR NICKOLAUS, b Driscoll, NDak, Oct 22, 09; m 36; c 1. NUTRITION, FOOD SCIENCE. *Educ:* Univ Calif, Los Angeles, AB, 32; Calif Inst Technol, MS, 33, PhD(chem), 35. *Prof Exp:* Asst org chem, Calif Inst Technol, 33-35, res asst & res fel bio-org chem, 38-41; res chemist, Continental Oil Co, Okla, 35-36; refining technologist, Shell Oil Co, Calif, 36-38; assoc chemist, Western Regional Res Lab, Bur Agr & Indust Chem, USDA, 41-45; dir, Gentry Div Consol Foods Corp, 45-54, vpres, tech dir & prod mgr, 54-55, exec vpres, 55-56, pres, 56-64; CONSULT, 64- *Concurrent Pos:* Past pres, Res & Develop Assocs Food & Container Inst & League Int Food Educ; mem, Assocs Calif Inst Technol; mem mgt comt, Int Food Info Serv, 67- *Mem:* Fel AAAS; Am Chem Soc; fel Inst Food Technologists (treas, 58-70); fel Am Inst Chemists. *Res:* Bio-organic chemistry; technical assistance to legal counsel; food development and processing; food dehydration; food science and technology. *Mailing Add:* 17400 Weddington St Encino CA 91316-2562

PRATER, C(HARLES) D(WIGHT), b Sylacauga, Ala, Jan 2, 17; m 38; c 2. CHEMICAL ENGINEERING, BIOPHYSICS. *Educ:* Auburn Univ, BS, 40; Univ Pa, PhD(biophys), 51. *Prof Exp:* Physicist, Bartol Res Found, 42-46; res asst biophys, Johnson Found, Pa, 46-51; sr res physicist, Res Dept, Paulsboro Lab, Cent Res Div, Socony Mobil Oil Co, Inc, 51-55, group leader, 55-57, res assoc, 57-62, sr res assoc, Res Dept, Princeton Lab, Mobil Oil Corp, 62-67, mgr process res & develop, Process Res & Tech Serv Div, Res Dept, Mobil Res & Develop Corp, 67-77, sr scientist, 77-80, CONSULT, MOBIL RES & DEVELOP CORP, 82- *Concurrent Pos:* Adj prof chem eng, Calif Tech Inst, 82- *Mem:* Nat Acad Eng; AAAS; Am Chem Soc; fel Am Inst Chem Engrs; Sigma Xi. *Res:* Thermionic emission; magnetron; bacteriophage; catalysis; solid state physics; chemical kinetics; heat and mass transport; coal conversion and gasification; brains in computers. *Mailing Add:* 1500 E Karen Ave Apt 89 Las Vegas NV 89109

PRATER, JOHN D, b Dalroy, Alta, June 24, 17; US citizen; m 84; c 4. METALLURGY. *Educ:* Mont Sch Mines, BS, 39; Univ Idaho, MS, 40. *Prof Exp:* Chemist, Anaconda Co, 40-42; metallurgist, US Bur Mines, 42-52; sr scientist, 52-65, sect head hydrometall, Western Mining Div, Kennecott Copper Co, 65-77, mgr spec proj, Kennecott Minerals Co, 77-83; RETIRED. *Res:* Chemical metallurgy; extraction and purification of nonferrous metals. *Mailing Add:* 2860 E 3185 S Salt Lake City UT 84109

PRATER, JOHN THOMAS, b Belfonte, Pa, Apr 18, 51; m 80; c 2. MATERIALS SCIENCE ENGINEERING, METALLURGY & PHYSICAL METALLURGICAL ENGINEERING. *Educ:* Middlebury Col, BA, 73; Univ Pa, PhD(mat sci), 78. *Prof Exp:* Res scientist, Mat Dept, Battelle Pac Northwest Lab, 78-84, sr res scientist, 84-88, group leader, 86-88; MAT ENGR & BR CHIEF, MAT SCI DIV, USA RES OFF, 88- *Concurrent Pos:* Secy, NC Chap-Mat Res Soc, 89-90, pres, 90-91; adj assoc prof, Dept Mat Sci & Eng, NC State Univ, 89-; adv bd mem, Nat Ctr Integrated Photonic Res, Univ SC, 90- *Honors & Awards:* Jacquet-Lucas Award, Am Soc Metals & Int Metallog Soc, 84. *Mem:* Mat Res Soc; Am Vacuum Soc. *Res:* Nucleation and growth of diamond films; magnetic behavior of dilute spin systems; ceramic coatings for engine applications; physical properties and oxidation behavior of nuclear fuel rod reactor materials above 1500 degrees centigate; electronic, optical and magnetic materials research oversight. *Mailing Add:* 2113 New Hope Dr Chapel Hill NC 27514

PRATER, T(HOMAS) A(LLEN), b Can, Mar 5, 20; nat US; m 48; c 3. PHYSICAL METALLURGY. *Educ:* Mont Sch Mines, BS, 41; Pa State Col, MS, 42, PhD(metall), 50. *Prof Exp:* Asst metall, Pa State Col, 43-44 & 45-48, from instr to asst prof, 48-51; asst, Carnegie Inst Technol, 44-45; metall engr, Carboloy Dept, Gen Elec Co, 51-52, res assoc metall, Res Lab, 52-60, mat engr, 60-61, mgr metals processing, Res & Develop Ctr, 61-69, metallurgist, Res & Develop Ctr, 69-86; RETIRED. *Mem:* Am Soc Metals; Am Inst Mining, Metall & Petrol Engrs; Am Soc Testing & Mat. *Res:* High temperature materials development. *Mailing Add:* 2120 Morrow Ave Niskayuna NY 12309

PRATHER, ELBERT CHARLTON, b Jasper, Fla, Mar 13, 30; m 54; c 2. PREVENTIVE MEDICINE, PUBLIC HEALTH. *Educ:* Univ Fla, BS, 52, MS, 54; Bowman Gray Med Sch, MD, 59; Univ NC, MPH, 63. *Prof Exp:* Mem staff, Fla State Div Health, Jacksonville, 52-62; mem staff, Div Health, Hillsborough County Health Dept, Tampa, Fla, 62-63; state epidemiologist, 64-70, chief, Bur Prev Dis, 70-74, state health officer, 74-79, 86-88, health prog supvr, 74-85, supvr, div health, 80-88, State epidemiologist, Fla Dept Health & Rehab Serv, 86-88; RETIRED. *Res:* Epidemiology of disease and health. *Mailing Add:* 2816 Terry Rd Tallahassee FL 32312

PRATHER, MARY ELIZABETH STURKIE, b Auburn, Ala, Dec 2, 29; wid; c 1. NUTRITION. *Educ:* Auburn Univ, BS, 51, MS, 55; Iowa State Univ, PhD(body composition), 63. *Prof Exp:* Res asst human nutrit, Ala Agr Exp Sta, Auburn Univ, 52-55, from instr to asst prof foods & nutrit, 55-63, assoc prof nutrit, 63-67; PROF FOOD, NUTRIT & INST ADMIN & HEAD DEPT, COL HUMAN ECOL, UNIV MD, COLLEGE PARK, 67- *Mem:* AAAS; Am Dietetic Asn; Am Home Econ Asn; Am Inst Nutrit; Soc Nutrit Educ; Sigma Xi. *Res:* Dietary factors affecting blood lipids; body composition and obesity; nutrition of the elderly. *Mailing Add:* Dept Human Nutrit & Food Systs Univ Md College Park MD 20742

PRATHER, MICHAEL JOHN, b Pittsburgh, Pa, Aug 31, 47; m 74; c 2. ATMOSPHERIC SCIENCES, PLANETARY ATMOSPHERES. *Educ:* Yale Univ, BS, 69 PhD, (astron), 76; Oxford Univ, BA, 71. *Prof Exp:* RES FEL ATMOSPHERIC CHEM, HARVARD UNIV, 75- *Mem:* Sigma Xi. *Mailing Add:* NASA/GISS 2880 Broadway New York NY 10025

PRATHER, THOMAS LEIGH, b Hastings, Nebr, Sept 12, 36. GEOLOGY. *Educ:* Carleton Col, BA, 59; Univ Colo, MS, 61, PhD(geol), 64. *Prof Exp:* Geologist, US Geol Surv, 64-65; from asst prof to assoc prof, 65-75, PROF GEOL, WESTERN STATE COL COLO, 75- *Mem:* Geol Soc Am. *Res:* Structural geology and stratigraphy of the Elk Mountains of Colorado. *Mailing Add:* Dept Sci Western State Col Gunnison CO 81230

PRATLEY, JAMES NICHOLAS, b Eastland, Tex, Sept 20, 28. PHYSIOLOGY. *Educ:* Oberlin Col, AB, 49; Univ Tex, MA, 51, PhD(physiol), 57. *Prof Exp:* From asst prof to assoc prof, 57-66, prof biol sci, San Jose State Univ, 66-; AT BIOL SCI DEPT, CALIF ST UNIV. *Mem:* AAAS; Am Soc Cell Biol; Am Soc Zoologists. *Res:* Cell physiology; bioelectric phenomena and permeability properties of plant and animal tissue; cellular correlation and biological oxidation. *Mailing Add:* Dept Biol Sci Calif St Univ San Jose CA 95192

PRATS, FRANCISCO, b Guadalajara, Spain, June 3, 22; m 60; c 2. THEORETICAL PHYSICS. *Educ:* Univ Madrid, BS, 46; Univ Md, PhD(physics), 58. *Prof Exp:* Res fel physics, Univ Birmingham, 58-60; res assoc, Univ Md, 60-61; physicist, Nat Bur Standards, 61-65; from asst prof to assoc prof, 65-70, chmn dept, 80-87, PROF PHYSICS, GEORGE WASH UNIV, 70- *Mem:* Am Phys Soc; Sigma Xi. *Res:* Theoretical nuclear physics; photonuclear processes on light nuclei at intermediate energy. *Mailing Add:* Dept Physics George Washington Univ Washington DC 20052

PRATS, MICHAEL, b Tampa, Fla, Dec 18, 25; m 51; c 4. RESERVOIR ENGINEERING, HEAT & MASS TRANSFER POROUS MEDIA. *Educ:* Univ Tex, Austin, BS, 49, MA, 51. *Prof Exp:* Jr physicist, Shell Develop Co, 50-52, physicist, 52-63, res assoc, 63-66, sr res assoc, 66-72 & 80-89, consult res engr, 72- 80; RETIRED. *Concurrent Pos:* Consult, numerous insts; vis prof, petrol eng dept, Univ Tex, Austin, 89. *Honors & Awards:* Lester C Uren Award, Soc Petrol Engrs, 74 & Thermal Recovery Distinguished Achievement Award, 91. *Mem:* Nat Acad Eng; hon mem Soc Petrol Eng; Can Inst Mining; Venezuela Soc Petrol Engrs; Argentina Inst Petrol; hon mem Am Inst Mining & Metall Engrs. *Res:* Enhanced oil recovery methods, with emphasis on thermal methods; general reservoir engineering, with emphasis on water flooding; production enhancement resulting from hydraulic fracturing; pressure well testing; compaction and subsidence. *Mailing Add:* 2834 Bellefontaine Houston TX 77025-1610

PRATT, ARNOLD WARBURTON, b Binghamton, NY, Nov 24, 20; m. BIOPHYSICS. *Educ:* Univ Rochester, MD, 46. *Prof Exp:* Intern & fel, New York Hosp, 46-47; assoc pub health & prev med, Med Col, Cornell Univ, 47-48; officer, USPHS, NIH, 48-66, dir, Div Comput Res & Technol, 66-85; RETIRED. *Res:* Biomedical applications of computers. *Mailing Add:* 4104 Dewmar Ct Kensington MD 20895

PRATT, ARTHUR JOHN, b Norwich, NY, May 3, 05; m 31; c 3. VEGETABLE CROPS. *Educ:* Cornell Univ, BS, 26, PhD(veg crops), 33. *Prof Exp:* Asst veg crops, Cornell Univ, 29-32, from exten instr to exten assoc prof, 32-48, prof, 48-62; consult, Tasmania & Victoria Depts Agr, 63-64; prof veg crops, Univ Liberia, 64-65; USAID contracting veg specialist, Jamaica Sch Agr, 65-67; secy-treas & gen mgr, Perry City Farms, Inc, 68-74; CONSULT, 74- *Concurrent Pos:* Vchmn, Nat Jr Veg Growers Found, 35-58; dir, Eastern Coop League, 40-42; asst horticulturist, Veg Crops Lab, Fla, 45-46; vpres, Consumers Coop, NY, 46-47, pres, 47-49, 54-57, 73-74 & 77-78; researcher, Maple Leaf Farms, Calif, 53; consult, Exp Sta, Alaska, 59; mem adv comt, Nat Jr Veg Growers; vis scholar irrigation res, Univ Ariz, 79-80; vis prof, Dept Veg Crops, Cornell Univ, 80- *Mem:* Potato Asn Am; Am Soc Hort Sci. *Res:* Irrigation of vegetable crops; vegetable production; potato breeding; practical farm management. *Mailing Add:* Maple Grove Place Ithaca NY 14850

PRATT, CHARLES BENTON, b Madison, NC, Aug 7, 30; m 58; c 5. PEDIATRICS, CANCER. *Educ:* Univ NC, Chapel Hill, BA, 51; Univ Md, Baltimore City, MD, 55. *Prof Exp:* Intern, Hosps, Med Col Va, 55-56; resident pediat, Babies Hosp, Columbia-Presby Med Ctr, 58-60, fel cancer chemother, 60-61; Nat Cancer Inst spec fel, 65-66, from asst mem to assoc mem, 67-75, MEM, ST JUDE CHILDREN'S RES HOSP, 75- *Mem:* Am Acad Pediat; Am Asn Cancer Res; Am Soc Clin Oncol; Am Pediat Soc. *Res:* Clinical pediatric cancer chemotherapy; pharmacology of oncolytic agents. *Mailing Add:* St Jude Children's Res Hosp 332 N Lauderdale Memphis TN 38101

PRATT, CHARLES WALTER, b Kansas City, Mo, Feb 5, 44; m 68. MICROBIOLOGY EDUCATION, GENE REGULATION. *Educ:* Ore State Univ, BA, 66; Univ Wash, Seattle, PhD(genetics), 71. *Prof Exp:* NIH res fel, Inst Molecular Biol, Univ Ore, 71-73; res assoc, biol dept, Mass Inst Technol, 73-74, 75-76 & 78-79, instr exp biol, 74-75 & 77; ASST PROF MICROBIOL & RES MICROBIOLOGIST, UNIV ILL, URBANA-CHAMPAIGN, 79-, ASST HEAD DEPT MICROBIOL, 83- *Mem:* Am Soc Microbiol; Genetics Soc Am; AAAS. *Res:* Biological regulation of metabolic activities in methanogenic bacteria. *Mailing Add:* Microbiol Dept 131 Burrill Hall Univ Ill 407 S Goodwin Ave Urbana IL 61801

PRATT, DAN EDWIN, b High Point, NC, Feb 7, 24; m 59; c 1. FOOD CHEMISTRY, NUTRITION. *Educ:* Univ Ga, BSA, 50, MS, 52; Fla State Univ, PhD(food sci), 63. *Prof Exp:* Teaching fel food technol, Univ Ga, 50-52, from instr to asst prof, 52-61; Nuclear Sci fel food sci, Fla State Univ, 61-62, res asst, 62-63; from asst prof to assoc prof foods & nutrit, Univ Wis-Madison, 63-69; assoc prof, 69-79, PROF FOODS & NUTRIT, PURDUE UNIV, WEST LAFAYETTE, 78- *Concurrent Pos:* Deleg, Int Cong Food Scientists, Tokyo, 78 & Dublin, Ireland, 84; vis res scientist, Natick Res & Develop Command, 80; lectr, Harvard Univ, 80; res scientist, Am Oil Chemists Soc, Cannes, France, 85. *Mem:* Inst Food Technologists; fel Am Inst Chemists; NY Acad Sci; Sigma Xi. *Res:* Thermal processing; food color stability; lipid oxidation; pesticides; fruit and vegetable processing; methods of food analyses; enzymatic and non-enzymatic browning of fruits and vegetables; author of over 100 publications. *Mailing Add:* Dept Foods & Nutrit Purdue Univ Main Campus West Lafayette IN 47907

PRATT, DARRELL BRADFORD, b Millinocket, Maine, Oct 22, 20; m 43; c 1. BACTERIOLOGY. *Educ:* Univ Maine, BS, 42; Purdue Univ, MS, 45; Harvard Med Sch, PhD, 51. *Prof Exp:* Instr bact & immunol, Harvard Med Sch, 51-52; Hite fel, Univ Tex, 52-53; from asst prof to prof bact, Univ Fla, 53-66; prof biol, Univ Houston, 66-67; prof, 67-85, chmn dept, 67-79, EMER PROF MICROBIOL & ZOOL, UNIV MAINE, ORONO, 85- *Concurrent Pos:* Vis prof, Univ Leeds, 56-57. *Mem:* Sigma Xi; Am Acad Microbiologists; Am Soc Microbiologists. *Res:* Bacterial nutrition; marine microbiology. *Mailing Add:* Eight Glenwood St Orono ME 04473

PRATT, DAVID, b Ithaca, NY, June 7, 32; m 54; c 4. GENETICS, VIROLOGY. *Educ:* Cornell Univ, BS, 54; Univ Calif, Davis, PhD(genetics), 58. *Prof Exp:* Virol trainee, Virus Lab, Univ Calif, Berkeley, 59-60; NIH fel microbiol, Univ Copenhagen, 60-61; Helen Hay Whitney fel biophys, Univ Geneva, 61-62; from asst prof to assoc prof bact, Univ Wis-Madison, 62-70; assoc prof, 70-74, PROF BACT, UNIV CALIF, DAVIS, 74- *Res:* Plant somatic cell genetics. *Mailing Add:* Dept Microbiol Univ Calif Davis CA 95616

PRATT, DAVID MARIOTTI, b Williamstown, Mass, Feb 18, 18; m 41; c 2. BIOLOGICAL OCEANOGRAPHY. *Educ:* Williams Col, BA, 39; Harvard Univ, MA, 41, PhD(biol), 43. *Prof Exp:* Instr biol, Harvard Univ, 46-49; from asst prof to assoc prof marine biol, 49-60, prof, 60-80, EMER PROF OCEANOG, UNIV RI, 80- *Concurrent Pos:* Res assoc, Oceanog Inst, Woods Hole, 46-48, instr, Comn Exten Courses, 48-49. *Res:* Animal behavior, especially predator-prey relationships and chemoreception in mollusks; social development and relationships in giraffe; swans. *Mailing Add:* 115 Little Rest Rd Kingston RI 02881

PRATT, DAVID R, b Dallas, Tex, Oct 20, 29; m 51; c 4. PHYSIOLOGY, GENETICS. *Educ:* Southwest Tex State Col, BS, 51, MEd, 54; Mich State Univ, PhD(animal breeding), 60. *Prof Exp:* Assoc prof animal husb, Panhandle Agr & Mech Col, 58-60; assoc prof, 61-72, PROF BIOL, TEX A&I UNIV, 72- *Mem:* Am Soc Animal Sci. *Res:* Animal nutrition; reproductive physiology of Whitetail deer; population dynamics in kangaroo rats, respiratory, circulatory functions related to injection of venom into rats; inbreeding in laboratory rats. *Mailing Add:* Dept Biol Tex A&I Univ Santa Gertrudis Kingsville TX 78363

PRATT, DAVID TERRY, b Shelley, Idaho, Sept 14, 34; m 56; c 3. MECHANICAL ENGINEERING. *Educ:* Univ Wash, BSc, 56; Univ Calif, Berkeley, MSc, 62, PhD(mech eng), 68. *Prof Exp:* Instr eng, US Naval Acad, 57-60, asst prof, 61-64; lectr mech eng, Univ Calif, Berkeley, 64-65, asst res engr, 66-68; from asst prof to prof, Wash State Univ, 68-76, asst dean col eng, 70-73; prof mech eng & adj prof chem eng, Univ Utah, 76-78; prof mech eng

& chmn dept, Univ Mich, Ann Arbor, 78-; DEPT MECH ENG, UNIV WASH. *Concurrent Pos:* Fulbright-Hays sr res fel, Imp Col, Univ London, 74-75; res dir supercomput, Aerojet Propulsion Res Inst, Sacramento, CA, 86-87; NSF Sci fac fel, 65-66; David Pierpont Gardner fac fel, 78. *Mem:* Combustion Inst; Am Inst Aeronaut & Astronaut; Am Soc Mech Engrs; Am Soc Eng Educ. *Res:* Aerothermochemistry; prediction and measurement of emissions from steady-flow combustion; mathematical modelling of combustion phenomena; hypersonic propulsion by stabilized detonation waves. *Mailing Add:* Dept Mech Eng FU-10 Univ Wash Seattle WA 98195

PRATT, DAVID W, b Providence, RI, Sept 14, 37; m 61; c 2. CHEMICAL PHYSICS, PHYSICAL CHEMISTRY. *Educ:* Princeton Univ, AB, 59; Univ Calif, Berkeley, PhD(chem), 67. *Prof Exp:* Teaching asst chem, Univ Calif, Berkeley, 62-63, res asst, Lawrence Radiation Lab, 63-67; NIH fel chem physics, Univ Calif, Santa Barbara, 67-68; NIH fel, 68, from asst prof to assoc prof, 68-78, PROF CHEM, UNIV PITTSBURGH, 78- *Concurrent Pos:* Merck Found fac develop award, 69; vis prof & Fulbright fel, Univ Leiden, 79; Guggengeim fel, 85; guest scholar & Japan Soc Prom Sci fel, Kyoto Univ, 88. *Mem:* AAAS; fel Am Phys Soc; Am Chem Soc. *Res:* Molecular spectroscopy; laser physics; magnetic resonance and optical spectroscopy; application to problems in molecular structure and dynamics; radiationless transitions and light-initiated chemical reactions; patterns of energy levels in large molecules; chaotic behaviors and structures. *Mailing Add:* Dept Chem Univ Pittsburgh Pittsburgh PA 15260

PRATT, DIANE MCMAHON, PARASITOLOGY, MEMBRANE BIOCHEMISTRY. *Educ:* Harvard Univ, PhD(immunol), 78. *Prof Exp:* Asst prof immunol & parasitol, Harvard Univ, 81-85; ASSOC PROF IMMUNOL & PARASITOL, SCH MED, YALE UNIV, 85- *Mailing Add:* Dept Epidemiol & Pub Health Yale Univ Sch Med 60 College St New Haven CT 06510

PRATT, DOUGLAS CHARLES, b Minneapolis, Minn, Mar 31, 31; m 51; c 7. PHOTOPHYSIOLOGY. *Educ:* Univ Minn, BS, 52, MA, 59, PhD, 60. *Prof Exp:* Asst natural sci, Univ Minn, 55-56, res assoc photosynthesis & nitrogen fixation, 59-60, res fel phys chem, 60-62; asst prof, Carleton Col, 62-66; assoc prof, 66-70, head dept, 74-85, actg dean, Col Biol Sci, 84-86, PROF BOT, UNIV MINN, ST PAUL, 70- *Mem:* AAAS; Am Soc Plant Physiologists; Am Soc Photobiol; Int Peat Soc; Am Inst Biol Sci. *Res:* Photophysiology; bioenergy. *Mailing Add:* Dept Plant Biol Univ Minn 220 Bio Sci Ctr St Paul MN 55108

PRATT, ELIZABETH ANN, b Orange, NJ, Jan 2, 33; m 58; c 4. MEMBRANE PROTEINS. *Educ:* Oberlin Col, BA, 54; Univ Chicago, MS, 57, PhD(microbiol), 61. *Prof Exp:* Res asst, Biochem Dept, Stanford Univ, 59-61, fel, 61-63; asst prof biol sci, Univ Pittsburgh, 66-73, res assoc, 73-80; RES ASSOC BIOL SCI, CARNEGIE MELLON UNIV, 80- *Mem:* Am Soc Microbiol; Am Inst Biol Sci. *Res:* Membrane proteins and their interaction with lipids; labeling with fluorine for nuclear magnetic resonance spectroscopy; recombinant DNA; oligonucleotide-directed mutagenesis. *Mailing Add:* Biol Sci Dept Carnegie-Mellon Univ Pittsburgh PA 15213

PRATT, GEORGE L(EWIS), b Fargo, NDak, Jan 31, 26; m 55; c 2. AGRICULTURAL ENGINEERING. *Educ:* NDak State Univ, BS, 50; Kans State Univ, MS, 51; Okla State Univ, PhD, 67. *Prof Exp:* Asst prof agr eng, NDak State Univ, 51; salesman, 52; from asst prof to assoc prof, 53-67, PROF AGR ENG, NDAK STATE UNIV, 67-, DEPT CHMN, 77- *Mem:* Am Soc Agr Engrs; Sigma Xi. *Res:* Agricultural buildings design; control of pollution from agricultural sources; nonpetroleum fuels for tractors. *Mailing Add:* Dept of Agr Eng NDak State Univ Fargo ND 58102

PRATT, GEORGE WOODMAN, JR, b Boston, Mass, Aug 13, 27; m 48; c 2. THEORETICAL PHYSICS. *Educ:* Mass Inst Technol, BS, 49, PhD(physics), 52. *Prof Exp:* Instr physics, Boston Univ, 51-52; mem staff, Lincoln Lab, 52-60, vis lectr, Inst, 60, assoc prof elec eng, 60-65, consult, Lincoln Lab, 61-77, PROF ELEC ENG, MASS INST TECHNOL, 65- *Concurrent Pos:* Vis asst prof, Brandeis Univ, 57-58; vis prof, Dartmouth Col, 68; consult, Monsanto Co, Kennecott Copper Corp & Ford Motor Co; adj prof vet med, Tufts Univ, 79- *Mem:* Fel Am Phys Soc; fel Phys Soc Japan. *Res:* Solid state physics; quantum electronics; magnetism; optoelectronic devices; biological effects of laser radiation; biomedical engineering; gait analysis. *Mailing Add:* Glezen Lane Wayland MA 01778

PRATT, HARLAN KELLEY, b Berkeley, Calif, Mar 18, 14; m 39; c 2. PLANT PHYSIOLOGY, HORTICULTURE. *Educ:* Univ Calif, Los Angeles, BS, 39, PhD(plant physiol), 44. *Prof Exp:* Res fel biol, Calif Inst Technol, 44-46; from instr veg crops & jr plant physiologist to assoc prof & assoc plant physiologist, 46-63, prof, 63-81, EMER PROF VEG CROPS & EMER PLANT PHYSIOLOGIST, AGR EXP STA, UNIV CALIF, DAVIS, 81-; CONSULT, 81- *Concurrent Pos:* Fulbright res scholar food res, Australia, 56; USPHS spec fel, UK, 63-64; scientist, Plant Dis Div, Dept Sci & Indust Res, NZ, 70-71; vis prof & NSF-SEED grant, Dept Hort, Univ Philippines, Los Banos, 77-78; vol hort consult, IESC & VOCA, Guatemala, Costa Rica, Panama, Indonesia. *Honors & Awards:* Campbell Award for Outstanding Veg Res, Am Inst Biol Sci, 69. *Mem:* Bot Soc Am; Am Soc Hort Sci; Am Soc Plant Physiologists. *Res:* Postharvest physiology of fruits and vegetables; handling, transportation and storage; senescence; role of ethylene in plant physiology. *Mailing Add:* Dept of Veg Crops Mann Lab Univ of Calif Davis CA 95616

PRATT, HARRY DAVIS, b North Adams, Mass, Apr 13, 15; m 44, 52; c 3. ENTOMOLOGY. *Educ:* Mass State Col, BS, 36, MS, 38; Univ Minn, PhD(entom, zool), 41. *Prof Exp:* Asst entom, Mass State Col, 36-38; jr biologist, Upper Miss River Malaria Surv, 40-41; asst entomologist & dist entomologist, Malaria Control, USPHS, PR, 41-42, assoc entomologist, 42-43, asst sanitarian, 43-46, chief med entom lab, Communicable Dis Ctr, Ga, 46-53, chief insect & rodent control training sect, 53-64, chief training sect, Aedes Aegypti Eradication Br, 64-68, chief insect & rodent control br, Environ Control Admin, 69-72; RETIRED. *Concurrent Pos:* Spec consult, Econ Coop Admin, Vietnam, 50, Pan Am Health Orgn Guatemala, 57 & Jamaica, 58, WHO, Switz, 66 & SPac Comn, 65; dir training, Stephenson Chem Co, College Park, Ga, 74- *Honors & Awards:* Gorgas Medal, Asn Mil Surgeons US, 64; Distinguished Serv Medal, US Dept Health, Educ & Welfare, 71; Commendation Medal, USPHS. *Mem:* Entom Soc Am; Am Mosquito Control Asn (pres, 67); Sigma Xi; hon mem Nat Pest Control Asn. *Res:* Medical entomology; vector control; mosquito taxonomy and control; sucking louse taxonomy and control. *Mailing Add:* 879 Glen Arden Way NE Atlanta GA 30306

PRATT, HERBERT T, b Eden, NC, Jan 19, 26; m 48; c 4. FIBER CHEMISTRY, HISTORY OF CHEMISTRY. *Educ:* Tri-State Univ, BS, 45; Goddard Col, MA, 88. *Prof Exp:* Develop engr, King-Seeley Corp, Ann Arbor, Mich, 45-46; chemist, Fieldcrest Mills, Eden, NC, 46-47, head anal & appl chem, 48-51; customer serv rep, Textile Fibers Dept, E I duPont Co, 52-62, res engr, 63-67, tech rep, 68-71, tech specialist, 72-75, sr tech specialist, 76-80, tech mkt assoc, 80-85; RETIRED. *Concurrent Pos:* Adj prof, NC State Univ, 74-79. *Honors & Awards:* Am Dyestuff Reporter Award, Am Asn Textile Chemists & Colorists; Frank W Reinhart Award, Am Soc Testing & Mats, 82. *Mem:* Am Chem Soc; Am Asn Textile Chemists & Colorists; Am Soc Testing & Mat; Hist Sci Soc; Mus Am Textile Hist. *Res:* Physical and chemical properties of fibers; textile processing and manufacture; textile dyeing and finishing; quality management systems; chemical literature before 1900; history of chemistry. *Mailing Add:* 23 Colesbery Dr New Castle DE 19720

PRATT, LEE HERBERT, b Oakland, Calif, Dec 3, 42; m 63; c 2. PLANT PHYSIOLOGY, MOLECULAR BIOLOGY. *Educ:* Stanford Univ, BA, 63, MA, 64; Ore State Univ, PhD(bot), 67. *Prof Exp:* NSF fel, Univ Calif, San Diego, 67-68, USPHS fel, 68; from asst prof to assoc prof biol, Vanderbilt Univ, 69-74, prof, 74-79; prof, 79-86, RES PROF BOT, UNIV GA, ATHENS, 86- *Concurrent Pos:* NSF res grants, Vanderbilt Univ, 70-81; res assoc, Inst Molecular Genetics, Univ Freiburg, WGer, 75-76; counr, Am Soc Photobiol, 78-85; assoc ed, J Photochem & Photobiol, 78-81; invited prof, Univ Geneva, Switz, 82-86, NSF Cell Biol panel, 84-88; res grants, NSF, DOE, Univ Ga, 79-; prog dir, Plant Growth & Develop Panel, USDA Nat Res Initiative Competitive Grants Prog. *Honors & Awards:* Charles Albert Shull Award, Am Soc Plant Physiologists, 81. *Mem:* Fel AAAS; Am Soc Plant Physiologists; Bot Soc Am; Am Soc Photobiol; Int Soc Plant Molecular Biol. *Res:* Photobiology; plant photomorphogenesis; photochemistry, biochemistry, immunochemistry and molecular biology of phytochrome; immunocytochemistry; monoclonal antibody technology, development and application. *Mailing Add:* Dept Bot Univ Ga Athens GA 30602

PRATT, MELANIE M, b Wash, DC, 52. CELL BIOLOGY. *Educ:* Brandeis Univ, PhD(biol), 79. *Prof Exp:* ASSOC PROF ANAT & CELL BIOL, SCH MED, UNIV MIAMI, 83- *Concurrent Pos:* Res assoc, Dept Anat, Harvard Med Sch, 82-83. *Mem:* Am Soc Cell Biol; NY Acad Sci. *Res:* Microtubule-based cell motility; cytoplasmic dynein ATPase; sea urchin develop. *Mailing Add:* Dept Anat & Cell Biol Univ Miami Sch Med PO Box 016960 Miami FL 33101

PRATT, NEAL EDWIN, b Cincinnati, Ohio, July 19, 37; m 60; c 1. PHYSICAL THERAPY. *Educ:* Wheaton Col, BS, 59; Univ Pa, cert phys therapy, 60; Temple Univ, PhD(anat), 67. *Prof Exp:* From asst prof to prof anat, Sch Med, Temple Univ, 67-82; prof anat, Sch Med, Mercer Univ, 82-85, assoc dean student affairs, 85-88; PROF ORTHOP SURG/ANAT, HAHNEMANN UNIV, 88- *Honors & Awards:* Christian R & Mary F Lindbach Found Award, 74. *Mem:* Am Asn Anatomists; Am Phys Ther Asn. *Res:* Ultrastructure of irradiated rat parotid gland; mitochondrial evaluation in the hypertrophied and failing mammalian heart; carpal tunnel biomechanics. *Mailing Add:* Dept Orthop Surg & Rehab MS 502 Hahnemann Univ 201 N 15th St Philadelphia PA 19102-1192

PRATT, PARKER FROST, b Virden, NMex, Nov 21, 18; m 45; c 4. SOIL CHEMISTRY, WATER CHEMISTRY. *Educ:* Utah State Univ, BS, 47, MS, 48; Iowa State Univ, PhD, 50. *Prof Exp:* Asst, Utah State Univ, 47-48; from asst to instr, Iowa State Univ, 48-50; asst prof agron, Ohio State Univ, 50-55; asst chemist, Dept Soils & Plant Nutrit, 55-58, assoc chemist, 58-64, assoc prof, 60-64, chmn dept soil sci & agr eng, 65-75, CHEMIST, DEPT SOIL SCI & AGR ENG, UNIV CALIF, RIVERSIDE, 64-, PROF, 65-, CHMN DEPT, 80- *Honors & Awards:* Agron Serv Award, Am Soc Agron, 75. *Mem:* Soil Sci Soc Am (pres, 78). *Res:* Criteria for water quality for use in agriculture; effect of agricultural production on water quality; crop production and animal waste disposal in relation to nitrates in drainage waters. *Mailing Add:* Dept Soil & Environ Sci Univ Calif Riverside CA 92521

PRATT, PHILIP CHASE, b Livermore Falls, Maine, Oct 19, 20; m 45; c 2. PULMONARY PATHOLOGY, AUTOPSY EPIDEMIOLOGY. *Educ:* Bowdoin Col, AB, 41; Johns Hopkins Univ, MD, 44; Am Bd Path, dipl, 58. *Prof Exp:* Instr path, Johns Hopkins Hosp, 44-46; pathologist, Saranac Lab, NY, 46-52, asst dir, 52-55; chief lab, Ohio Tuberc Hosp, Columbus, 55-66; assoc prof, 66-71, PROF PATH, MED CTR, DUKE UNIV, 71- *Concurrent Pos:* Consult, Raybrook State Tuberc Hosp, NY, 52-54, Nat Inst Occup Safety & Health, 70- & Nat Inst Environ Health Sci, 71-; from clin instr to assoc prof path, Col Med, Ohio State Univ, 55-66; chief lab serv, Vet Admin Hosp, Durham, 71- *Mem:* AAAS; Int Acad Path; Am Thoracic Soc; Am Col Chest Physicians; Am Asn Pathologists; Royal Soc Health. *Res:* Pulmonary pathology, especially pneumoconiosis, tuberculosis and emphysema in man and experimental animals; pathology of adult respiratory distress syndrome; autopsy epidemiology. *Mailing Add:* Dept Path Duke Univ Med Ctr Durham NC 27710

PRATT, RICHARD HOUGHTON, b New York, NY, May 5, 34; m 58; c 4. THEORETICAL PHYSICS. *Educ:* Univ Chicago, AB, 52, SM, 55, PhD(physics), 59. *Prof Exp:* Res assoc physics, Stanford Univ, 59-61, asst prof, 61-64; assoc prof, 64-69, PROF PHYSICS, UNIV PITTSBURGH, 69-;

prog dir atomic physics, NSF, 87-89. *Concurrent Pos:* Consult, Lawrence Livermore Lab; prin investr, Nat Sci Found; acad dean, Semester-at-Sea, Fall, 84, admin dean, Spring, 90. *Mem:* Fel Am Phys Soc; fel AAAS; Int Radiation Phys (gen secy). *Res:* Electron-photon interactions in atomic and nuclear fields, bremsstrahlung, photoeffect, photon and electron scattering, quantum electrodynamics; atomic processes at high temperature and pressure. *Mailing Add:* Dept Physics Univ Pittsburgh Pittsburgh PA 15260

PRATT, RICHARD J, b Chicago, Ill, May 17, 27; m 55; c 2. ORGANIC CHEMISTRY. *Educ:* Univ Ill, BS, 49; Wayne State Univ, MS, 52, PhD(chem), 54. *Prof Exp:* Res assoc, Ben May Lab Cancer Res, Chicago, 54-56; asst prof chem, Colo Sch Mines, 56-58; res chemist, A E Staley Mfg Co, 58-62; proj leader new prod, Allis Chalmers Mfg Co, 62-63; head resins modifications lab, Sinclair Res Inc, 63-69; res assoc, Charles Bruning Div, Addressograph-Multigraph Corp, 69-71, sr res chemist, Graphics Res & Develop Ctr, 71-75; supplies develop mgr, Multigraphics Develop Ctr, 75-77; GROUP LEADER PAPER PIGMENTS RES, ENGELHARD MINERALS & CHEM DIV, 77- *Concurrent Pos:* Colo Sch Mines Res Found grant. *Mem:* Am Chem Soc. *Res:* Polymer modification; paper and textile chemistry; coatings; water soluble films and paints; enamels; flocculants; dielectrics; dispersants; photoconductors; polymeric corrosion inhibitors, lubricants, antiwear and antiscalants; electrographics; product development and applications. *Mailing Add:* PO Box 705 Campton NH 03223

PRATT, STEPHEN TURNHAM, b Minneapolis, Minn, June 13, 55. MOLECULAR PHOTOIONIZATION. *Educ:* Bennington Col, BA, 77; Yale Univ, MS, 79, MPhil, 80, PhD(chem), 82. *Prof Exp:* Res fel, 82-84, asst chemist, 84-85, CHEMIST, ARGONNE NAT LAB, 85- *Mem:* Am Phys Soc; Am Chem Soc. *Res:* Chemical physics; molecular photoionization, multiphoton ionization and photoionization from excited states; unimolecular reaction dynamics; spectroscopy of Van der Waals clusters; optical-optical double resonance spectroscopy. *Mailing Add:* Argonne Nat Lab Bldg 203 C-141 Argonne IL 60439

PRATT, TERRENCE WENDALL, b Minneapolis, Minn, Mar 9, 40; div; c 3. PROGRAMMING LANGUAGES. *Educ:* Univ Tex, BA, 61, MA, 63, PhD(math), 65. *Prof Exp:* Asst prof math & eng, Mich State Univ, 65-66; from asst prof to assoc prof comput sci, Univ Tex, Austin,66-77; from assoc prof to prof, 77-86, RES PROF COMPUT SCI, UNIV VA, CHARLOTTESVILLE, 86- *Mem:* Asn Comput Mach; Math Asn Am; Inst Elec & Electronics Engrs; Comput Soc. *Res:* Programming languages; software engineering. *Mailing Add:* Univ Va Thornton Hall Charlottesville VA 22901

PRATT, THOMAS HERRING, JR, physical chemistry, for more information see previous edition

PRATT, VAUGHAN RONALD, b Melbourne, Australia, Apr 12, 44; m 69; c 2. CONCURRENCY THEORY, DIGITAL TYPOGRAPHY. *Educ:* Sydney Univ, BSc, 67, MSc, 70; Stanford Univ, PhD(comput sci), 72. *Prof Exp:* From asst to assoc prof comput sci, Mass Inst Technol, 72-82; PROF COMPUT SCI, STANFORD UNIV, 81-; PRES, TRIANGLE CONCEPTS INC, 88- *Concurrent Pos:* Consult, Honeywell Info Systs, 72, Off Naval Res, 74, IBM, 72-74, 78-79, VLSI Systs Inc, 81-82, Sun Microsysts Inc, 82-88; head res, Sun Microsysts Inc, 83-85. *Mem:* Asn Comput Mach; Asn Symbolic Logic; Am Math Soc; Math Asn Am. *Res:* Programming language foundations, applying algebraic methods to concurrency theory; digital typography, applying techniques; computational geometry; image processing; computer vision. *Mailing Add:* 2215 Old Page Mill Rd Palo Alto CA 94304

PRATT, WALDEN PENFIELD, b Columbus, Ohio, Mar 22, 28; m 57; c 3. GEOLOGY. *Educ:* Univ Rochester, AB, 48; Stanford Univ, MS, 56, PhD(geol), 64. *Prof Exp:* Geologist, 49-55, res geologist, 55-89, EMER SCIENTIST, US GEOL SURV, 89- *Concurrent Pos:* Explor geologist, Pac Coast Borax Co, Arg, 55; ed, newsletter, Soc Econ Geol, 86-91. *Honors & Awards:* Meritorious Serv Award, US Dept Interior, 73. *Mem:* Fel Geol Soc Am; fel Soc Econ Geol; Am Asn Petrol Geologists. *Res:* Geology of borate deposits; areal geology of southwestern New Mexico, southwestern Colorado and southeast Missouri; regional mineral-resource appraisal; exploration models for Mississippi Valley-type deposits. *Mailing Add:* US Geol Surv, MS 905 25046 Fed Ctr Denver CO 80225

PRATT, WILLIAM WINSTON, b Columbus, Ohio, Aug 9, 21; m 47; c 4. NUCLEAR PHYSICS. *Educ:* Univ Rochester, BS, 43; Iowa State Univ, PhD, 50. *Prof Exp:* Assoc physicist, Brookhaven Nat Lab, 50-52; res assoc, Univ Iowa, 52-54; from asst prof to assoc prof, 54-63, PROF PHYSICS, PA STATE UNIV, UNIVERSITY PARK, 63- *Mem:* Am Phys Soc. *Res:* Nuclear decay schemes. *Mailing Add:* 1176 Smithfield Ave State College PA 19124

PRATT-THOMAS, HAROLD RAWLING, b Barnsley, Eng, June 9, 13; US Citizen; m 41; c 4. PATHOLOGY. *Educ:* Davidson Col, AB, 34; Med Col SC, MD, 38. *Hon Degrees:* LLD, Col Charleston, 64; DSc, Davidson Col, 71. *Prof Exp:* From instr to assoc prof path, 40-51, chmn dept path & dean sch med, 60-62, pres, 62, PROF PATH, MED UNIV SC, 52- *Concurrent Pos:* Consult, Charleston Vet Admin Hosp, 54. *Honors & Awards:* Am Cancer Soc Award, 58. *Mem:* Am Asn Path & Bact; Am Col Physicians. *Res:* General anatomic and surgical pathology with emphasis on cancer; etiologic and environmental factors in carcinoma of cervix, lung and stomach; toxoplasmosis; ovarian tumors. *Mailing Add:* Dept Path Med Univ SC 171 Ashley Ave Charleston SC 29425

PRAUSNITZ, JOHN MICHAEL, b Berlin, Ger, Jan 7, 28; nat US; m 56; c 2. CHEMICAL ENGINEERING. *Educ:* Cornell Univ, BChE, 50; Univ Rochester, MS, 51; Princeton Univ, PhD(chem eng), 55. *Hon Degrees:* DrEng, Univ Aquila, Italy, 83; DrEng, Tech Univ Berlin, 89. *Prof Exp:* Instr chem eng, Princeton Univ, 53-55; from asst prof to assoc prof, 55-63, PROF CHEM ENG, UNIV CALIF, BERKELEY, 63- *Concurrent Pos:* Guggenheim fels, 62 & 73; res prof, Miller Inst Basic Sci, Univ Calif, Berkeley, 65-66 & 78-79; consult to several cryogenic, petrol & petrochem industs; vis prof, Univ Karlsruhe, 73 & Tech Univ, Berlin, 76, 81; fel Inst Advan Study, Berlin, Ger, 85; fac res lectr, Univ Calif, Berkeley, 81. *Honors & Awards:* Honor Scroll, Am Chem Soc, 60, E V Murphree Award, 79; Colburn Award, Am Inst Chem Engrs, 62, Walker Award, 67; Lacey lectr, Calif Inst Technol, 69; Kelly lectr, Purdue Univ, 69; Wohl lectr, Univ Del, 73; Humboldt Sr Scientist Award, 76; G N Lewis lectr, Univ Calif, Berkeley, 78; Centennial lectr, La State Univ, 80; Wilhelm lectr, Princeton Univ, 80; Lindsay lectr, Tex A&M Univ, 83; McCabe lectr, NC State Univ, 84; Phillips Lectr, Okla State Univ, 86; Carl von Linde Mem Medal, Ger Soc Cryogenics; Solvay Prize, 90. *Mem:* Nat Acad Sci; Nat Acad Eng; fel Am Inst Chem Engrs; Am Chem Soc; AAAS; Bunsen Soc; Ger Soc Engrs; fel Am Acad Arts & Sci. *Res:* Thermodynamic properties and intermolecular forces in fluid mixtures; applications of molecular physics to fluid-phase equilibria as required for chemical engineering design. *Mailing Add:* Dept Chem Eng Univ Calif Berkeley CA 94720

PRAWEL, SHERWOOD PETER, JR, b Buffalo, NY, Jan 17, 32; m 54; c 5. CIVIL ENGINEERING. *Educ:* Ga Inst Technol, BCE, 53, MSCE, 59; Univ Waterloo, PhD(civil eng), 71. *Prof Exp:* Asst prof, 58-62, ASSOC PROF CIVIL ENG, STATE UNIV NY, BUFFALO, 62- *Concurrent Pos:* Reviewer var jour, Am Soc Civil Engrs, Am Concrete Inst & NSF; mem, Comt Rehab, Am Concrete Inst; mem, Comt Educ, Masonry Soc. *Mem:* Am Soc Civil Engrs; Am Concrete Inst; Masonry Soc; Earthquake Eng Res Inst. *Res:* Structural mechanics; numerical analysis; earthquake engineering; seismic renovation; masonry behavior. *Mailing Add:* Civil Eng 233 Ketter Hall Univ Buffalo Buffalo NY 14260

PRAY, DONALD GEORGE, b Troy, NY, Jan 19, 28; m 50; c 4. FRACTURE MECHANICS, SERVICE LIFE ASSESSMENT. *Educ:* Tex Christian Univ, BA, 55; Southern Methodist Univ, MSME, 79. *Prof Exp:* Jr engr, Radio Sonic Corp, NY, 48-49; sr engr, Gen Dynamics/Ft Worth, 53-62; eng specialist, Astronaut Div, Ling-Temco-Vought Aerospace Corp, 62-65; group supvr aerospace physics, Space Div, Chrysler Corp, La, 65-67; sr engr aerospace technol, Gen Dynamics Corp, 67-74, sr struct engr, fatigue & fracture anal, 74-84; sr engr specialist, JVX Airframe Struct, 84-85; group engr, V-22 Struct Life Assurance, Bell Helicopter Textron, 85-89; CONTRACT ENGR, DALLAS, TEX, 89- *Concurrent Pos:* US deleg, 2nd Cong, Int Comn Acoust, 56; instr, Tex Christian Univ, 58-59; consult, Ft Worth, Tex, 61-62. *Mem:* Acoust Soc Am; Am Soc Mech Engrs; Nat Soc Prof Engrs. *Res:* Jet noise; random vibration; aeroelasticity; noise control; acoustic emission; structural service-life assessment of tilt-rotor aircraft. *Mailing Add:* 3628 Wedgway Dr Ft Worth TX 76133

PRAY, LLOYD CHARLES, b Chicago, Ill, June 25, 19; m 46; c 4. GEOLOGY. *Educ:* Carleton Col, BA, 41; Calif Inst Technol, MS, 43, PhD(geol), 52. *Prof Exp:* From asst to assoc prof geol, Calif Inst Technol, 49-56; sr res geologist, Marathon Oil Co, 56-62, res assoc, 62-68; prof, 68-88, EMER PROF GEOL, UNIV WIS-MADISON, 89- *Concurrent Pos:* Geologist, Magnolia Petrol Co, 42 & Mineral Deposits Br, US Geol Surv, 43-44 & 46-56; vis prof, Univ Tex, 64; mem coun educ, Am Geol Inst, 64-66, mem comt educ, 66-68, house delegs, 71-72; vis prof, Univ Colo, 67 & Univ Miami, 71; mem adv panel earth sci, NSF, 73-76; distinguished lectr, Am Asn Petrol Geol, 86-87; res assoc, Univ Calif-Santa Cruz, 87. *Honors & Awards:* George C Matson Trophy, Am Asn Petrol Geologists, 67; Am Diabetes Soc Award, 68. *Mem:* Fel AAAS; Am Inst Prof Geologists; Geol Soc Am; hon mem Soc Econ Paleontologists & Mineralogists (secy-treas, 61-63, vpres, 66, pres, 69); Am Asn Petrol Geologists; Sigma Xi. *Res:* Geology of sedimentary carbonates; classification; diagenesis; facies relationships; porosity; petroleum reservoirs; reef facies; allochthonous basin carbonates; stratigraphy of New Mexico and West Texas; rare earth deposits in California; geological education. *Mailing Add:* Dept Geol & Geophys Univ Wis 1215 W Dayton St Madison WI 53706

PRAY, RALPH EMERSON, b Troy, NY, May 12, 26; m 59; c 4. METALLURGICAL ENGINEERING, EARTH SCIENCE. *Educ:* Univ Alaska, BS, 61; Colo Sch Mines, DSc(metall eng), 66. *Prof Exp:* Partner, Santa Fe Lead & Zinc Co, NMex, 47-53; owner, Socorro Assay Lab, NMex, 55-57; assayer & engr-in-chg, Alaska Div Mines & Minerals, 57-61; asst mgr res, Universal Atlas Cement Div, US Steel, 65-66; res metallurgist, Inland Steel Co, 66-67; proj mgr, Jacobs Eng Co, 67-68; PRES, MINERAL RES LABS, 68- *Concurrent Pos:* Instr, Ketchikan Community Col, Univ Alaska, 58-60; lectr, Calumet Campus, Purdue Univ, 66-67; vpres, Wilbur Foote Plastics, Pasadena, 68-72; panelist, Nat Mining Seminar, Barstow Col, Calif, 69-70; dir, Bagdad-Chase Inc, Pasadena-Los Angeles, 69-72; pres, Keystone Canyon Mining Co, Inc, Pasadena, 72-76, US Western Mines, Pasadena, 73-; guest lectr, Calif State Tech Univ, Pomona, 77-80; guest ed, Calif Mining J, 78-80; prime contractor, US Dept Defense, 89. *Mem:* Can Inst Mining & Metall Engrs; Am Inst Mining, Metall & Petrol Engrs; fel Geol Mining & Metall Soc India; Am Chem Soc; S African Inst Mining & Metall; SAfrica Geol Soc. *Res:* Metallic mineral exploration, acquisition, mining, and extraction; analytical chemistry. *Mailing Add:* 805 S Shamrock Ave Monrovia CA 91016-3651

PRAY, THOMAS RICHARD, b Modesto, Calif, May 5, 23. BOTANY. *Educ:* Univ Calif, BA, 47, PhD(bot), 53. *Prof Exp:* Asst bot, Univ Calif, 50-53; from instr to prof, 53-85, EMER PROF BOT, UNIV SOUTHERN CALIF, 85- *Concurrent Pos:* Res assoc, Dept Plant Biol, Carnegie Inst, 60-61; vis assoc prof, Univ Calif, 64 & Calif State Col, Palos Verdes, 66. *Mem:* Am Fern Soc; Bot Soc Am; Int Soc Plant Morphologists. *Res:* Anatomy and organogenesis of seed plants and ferns; comparative anatomy and ontogeny of foliar venation. *Mailing Add:* 4500 Palos Verdes Dr E Rancho Palos Verdes CA 90274

PRAZMA, JIRI, OTOLARYNGOLOGY. *Educ:* Charles Univ, Prague, Czech, MD, 60. *Prof Exp:* ASST PROF EAR, NOSE, & THROAT, SCH MED, UNIV NC, CHAPEL HILL, 69- *Res:* Blood flow in the inner ear; noise effects in the inner ear. *Mailing Add:* Dept Surgery Univ NC Sch Med Chapel Hill NC 27514

PREBLE, OLIVIA TOBY, b Oak Park, Ill, June 3, 47; m 67; c 1. VIROLOGY. *Educ:* Univ Ill, Urbana, BS, 68; Univ Pittsburgh, PhD(virol), 73. *Prof Exp:* Res assoc, 73-74, instr virol, 74-80, RES INSTR, DEPT MICROBIOL, SCH MED, UNIV PITTSBURGH, 80-; AT DEPT PATHOL, UNIFORMED SERV UNIV OF HEALTH. *Mem:* Am Soc Microbiol; AAAS. *Res:* In vitro models of the regulation and maintenance of persistent virus infections. *Mailing Add:* Prog & Proj Rev Br NIH Nat Inst Allergy & Infectious Dis Westwood Bldg Rm 3A-10 533 Westband Ave Bethesda MD 20816

PREBLUD, STEPHEN ROBERT, pediatric infectious diseases; deceased, see previous edition for last biography

PREBLUDA, HARRY JACOB, biochemistry, organic chemistry; deceased, see previous edition for last biography

PRECKSHOT, G(EORGE) W(ILLIAM), b Collinsville, Ill, Nov 18, 18; m 42; c 4. CHEMICAL ENGINEERING, FLUID DYNAMICS. *Educ:* Univ Ill, BS, 40; Univ Mich, MS, 41, PhD(chem eng), 51. *Prof Exp:* Res assoc, Univ Mich, 46-48; from asst prof to assoc prof chem eng, Univ Minn, Minneapolis, 48-64; prof & chmn dept, 64-85, EMER PROF CHEM ENG, UNIV MO-COLUMBIA, 85- *Concurrent Pos:* Consult, Minneapolis-Honeywell Regulator Co, 52, Gen Mills, 55-56, Minn Mining & Mfg Co, 59 & Mo Rolling Mills Corp, 69-; Guggenheim fel, Univ Edinburgh, 57-58. *Honors & Awards:* Res Award, Fatty Acid Producers & Am Oil Chemists Soc, 56. *Mem:* AAAS; Am Chem Soc; Am Inst Chem Engrs. *Res:* Thermodynamics; diffusion; heat transfer; biochemical engineering. *Mailing Add:* 1101 Parkridge Dr Columbia MO 65203

PRECOPIO, FRANK MARIO, b Providence, RI, Mar 12, 25; m 56; c 3. ORGANIC CHEMISTRY. *Educ:* Brown Univ, ScB, 48; Yale Univ, PhD(chem), 52. *Prof Exp:* Res assoc, Res Lab, Gen Elec Co, 51-55, mgr chem & insulation, Direct Current Motor & Generator Dept, 55-61, mgr res, Develop & Eng, Wire & Cable Dept, 61-66; corp tech dir, 66-68, vpres, 68-83, exec vpres & gen mgr, Amchem Prod, Inc, Ambler, 83-85; exec vpres & chief oper officer, Henkel Corp, 85-87, exec vpres-tech, 87-90; RETIRED. *Honors & Awards:* Mordica Mem Award, Wire Asn Int. *Mem:* Am Chem Soc; fel Am Inst Chemists. *Res:* High temperature polymers; plastics; elastomers; electrical insulation; wire and cable; chemical conversion coatings for metals; herbicides; plant growth regulators. *Mailing Add:* 310 Powder Horn Rd Ft Washington PA 19034

PREDECKI, PAUL K, b Warsaw, Poland, May 29, 38; US citizen; m 69. MATERIALS SCIENCE. *Educ:* Univ Witwatersrand, BSc, 58; Mass Inst Technol, MS, 61, PhD(metall), 64. *Prof Exp:* Res metallurgist, E I du Pont de Nemours & Co, Inc, 64-66; asst prof mat sci, 66-69, ASSOC PROF METALL & MAT SCI, UNIV DENVER, 69-; AT DEPT ENG, UNIV DENVER. *Concurrent Pos:* Consult, Coors Porcelain Co, Colo. *Mem:* Am Ceram Soc; Am Crystallog Asn; Am Inst Mining, Metall & Petrol Engrs. *Res:* Application of x-ray diffraction to materials problems: x-ray stress measurement in ceramics, composites and films; biomaterials. *Mailing Add:* Dept of Eng Univ of Denver Denver CO 80208

PREECE, SHERMAN JOY, JR, b Salt Lake City, Utah, May 2, 23. PLANT TAXONOMY. *Educ:* Univ Utah, BA, 45, MS, 50; Wash State Univ, PhD(bot), 56. *Prof Exp:* Asst, Univ Utah, 47-50; asst, Wash State Univ, 50-54, actg instr bot, 54-56; from instr to assoc prof, 56-68, PROF BOT, UNIV MONT, 68-, CHMN DEPT, 66-, CUR HERBARIUM, 77- *Concurrent Pos:* Actg cur herbarium, Wash State Univ, 54-55. *Mem:* Bot Soc Am; Am Soc Plant Taxonomists; Int Asn Plant Taxonomists. *Res:* Botany; cytotaxonomy. *Mailing Add:* 200 Pine Needle Lane Big Fork MT 59911

PREEDOM, BARRY MASON, b Stamford, Conn, Dec 31, 40; m 63; c 2. NUCLEAR PHYSICS. *Educ:* Spring Hill Col, BS, 62; Univ Tenn, MS, 64, PhD(nuclear physics), 67. *Prof Exp:* Res assoc nuclear physics, Cyclotron Lab, Mich State Univ, 67-70; asst prof, 70-73, assoc prof, 73-76, PROF PHYSICS, UNIV SC, 76- *Concurrent Pos:* Sabbatical, Swiss Inst Nuclear Res, 76-77. *Mem:* Am Phys Soc. *Res:* Meson-nucleus reactions; including elastic and inelastic scattering and meson absorption and production; nuclear structure. *Mailing Add:* Dept of Physics Univ of SC Columbia SC 29208

PREEDY, JOHN ROBERT KNOWLTON, b Leeds, Eng, Feb 20, 18; nat US; m 66. MEDICINE. *Educ:* Cambridge Univ, MB, BCh, 42, MA, 45, MD, 56; FRCP. *Prof Exp:* House surgeon, Postgrad Med Sch, London, 42, jr med registr, 46; from med registr to sr med registr, London Hosp, Univ London, 46-52; from instr to sr lectr, 52-57; from assoc prof to prof, 57-87, EMER PROF MED, SCH MED, EMORY UNIV, 87- *Concurrent Pos:* Vis fel, Col Physicians & Surgeons, Columbia Univ, 53-54. *Mem:* Am Soc Clin Invest; Endocrine Soc; Am Fedn Clin Res. *Res:* Clinical endocrinology; methods for chemical estimation of estrogens in body fluids; estrogen metabolism in the human in normal and disease states. *Mailing Add:* Nine Steam Gun Pl Hilton Head Island SC 29928

PREER, JAMES RANDOLPH, b Monahans, Tex, May 22, 44; m 67; c 2. ENVIRONMENTAL CHEMISTRY, WATER QUALITY. *Educ:* Swarthmore Col, AB, 65; Calif Inst Technol, PhD(chem), 70. *Prof Exp:* Woodrow Wilson teaching intern & asst prof chem, Fed City Col, 69-72, assoc prof chem, 72-73, assoc prof interdisciplinary sci, 73-78; prof interdisciplinary sci, 78-80, actg chair environ sci dept, 86-89, PROF ENVIRON SCI, UNIV DC, 80-, COORDR, INTEGRATED SCI COURSES, 74- *Concurrent Pos:* Fel, Mass Inst Technol, 76-77; proj leader, Agr Exp Sta, Univ DC, 78-, coordr, Water Qual Prog, 80-82. *Mem:* Am Chem Soc; Soil Sci Soc Am; Water Pollution Control Fedn. *Res:* Environmental contamination of urban gardens, soil and sediment with heavy metals; chemistry of heavy metals in soil and sludge; chemical analysis of water and wastewater. *Mailing Add:* 2005 Klingle Rd NW Washington DC 20010

PREER, JOHN RANDOLPH, JR, b Ocala, Fla, Apr 4, 18; m 41; c 2. ZOOLOGY, MOLECULAR BIOLOGY. *Educ:* Univ Fla, BS, 39; Ind Univ, PhD(zool), 47. *Prof Exp:* Assoc zool, Univ Pa, 47-50, from assoc prof to prof, 50-68, chmn grad group, 58-62, admis officer, Grad Sch Arts & Sci, 61-62; chmn dept biol, 77-79, prof zool, 68-88, EMER PROF BIOL, IND UNIV, BLOOMINGTON, 88- *Concurrent Pos:* Guggenheim fel, 76-77. *Mem:* Nat Acad Sci; Genetics Soc Am; Am Soc Naturalists; Am Soc Zoologists; Soc Protozool (pres, 86-). *Res:* Genetics of microorganisms; cytoplasmic inheritance in paramecium; genetics of proteins. *Mailing Add:* Dept Biol Ind Univ Bloomington IN 47405

PREGGER, FRED TITUS, b Paterson, NJ, May 14, 24; m 53; c 2. PHYSICS, SCIENCE & ENERGY EDUCATION. *Educ:* Montclair State Col, BA, 48, MA, 50; Columbia Univ, EdD(sci educ), 56. *Prof Exp:* High sch teacher, NJ, 48-55; from asst prof to assoc prof sci, 55-62, prof physics, 62- 88, EMER PROF PHYSICS, TRENTON STATE COL, 89- *Concurrent Pos:* Dir, NSF Inst Physics, Elem Sch Sci, Radioisotope Tech Energy, 70-73 & 75-81; mem, Pub Serv Elec & Gas Co, res adv coun, 88-92, educ comt, 83-; lectr astron, Georgian Court Col, 88- *Mem:* Am Asn Physics Teachers; Nat Sci Teachers Asn. *Res:* Teaching of science at the secondary school and college levels; finding and updating physics lecture demonstrations which have been lost. *Mailing Add:* 42 Harbourton Woodsville Rd Pennington NJ 08534

PREHN, RICHMOND TALBOT, b New York, NY, Dec 8, 22; m 46; c 3. EXPERIMENTAL PATHOLOGY. *Educ:* Long Island Col Med, MD, 47. *Prof Exp:* Intern, Philadelphia Naval Hosp, 47-48; fel, Nat Cancer Inst, 48-50, mem staff, 50-56, res physician, USPHS Hosp, Seattle, 56-58; from asst prof to prof path, Sch Med, Univ Wash, 58-66; prof path, Sch Med, Univ Pa & sr mem, Inst Cancer Res, 66-76; dir, Jackson Lab, 76-81; sci dir, Inst Med Res, San Jose, 81-86; RETIRED. *Concurrent Pos:* Walker-Ames prof, Univ Wash, 78; mem bd dirs, Am Asn Cancer Res, 80-83. *Honors & Awards:* Langer-Tiplitz Prize Cancer Res, 64; Ludvig Hektoen Mem lectr, Inst Med Chicago, 73; Award Cancer Immunol, Cancer Res Inst, 75; Rous-Whipple Award Path, 77. *Mem:* Am Soc Cancer Res; Am Soc Exp Path. *Res:* Tissue transplantation; immunopathology; oncology. *Mailing Add:* Dept Pathol Univ Wash Med Sch Pacific Ave Seattle WA 98195

PREIKSCHAT, EKKEHARD, b Insterburg, Ger, Apr 25, 43; US citizen; m 68; c 3. ELECTRICAL ENGINEERING, PHYSICS. *Educ:* Univ Wash, BS, 64, MS, 65; Univ Birmingham, PhD(physics), 68. *Prof Exp:* Res asst nuclear physics, Univ Wash, 64-65; res assoc physics, Univ Birmingham, 65-68, fel, 68-69; vpres eng, F P Res Lab, Inc, 70-79, vpres eng, Eur-Control M&D USA, Inc, 79-83; PRES, LASER SENSOR TECHNOL, INC, 83- *Concurrent Pos:* Res assoc, Univ Wash, 69-72. *Mem:* Tech Asn of Pulp & Paper Indust; Inst Elec & Electronics Engrs; Am Phys Soc. *Res:* Development of instrumentation for pulp and paper industry; development and optimization of moisture measurement in various mineral and organic materials. *Mailing Add:* 9048 NE 41st Bellevue WA 98004

PREIKSCHAT, F(RITZ) K(ARL), b Finkengrund, Ger, Sept 11, 10; m 37; c 2. ELECTRICAL ENGINEERING. *Educ:* Sch Eng, Oldenburg, BSEE, 34, PE, 72. *Prof Exp:* Develop engr, Siemens, Berlin, Ger, 36-39; dir lab group radar & infrared develop, Gema, Berlin, 40-45; dir lab radar & radio guid systs, NII 88, Moscow, Russia, 46-52; dir lab teletypewriter develop, Telefon Bau und Normal Zeit Co, Frankfurt, 53-57; sr staff mem, Appl Physics Lab, Johns Hopkins Univ, 57-59; res specialist radio guid commun, Boeing Co, 59-62; prin engr, Seattle Develop Lab, Honeywell Inc, 62-64; develop engr, Laucks Labs, 64-66; res specialist, Space Div, Boeing Co, 66-71; dir res & develop, F P Res Lab, 71-79; DIR RES & DEVELOP, LASER SENSOR TECHNOL, 80- *Mem:* Sr mem Inst Elec & Electronics Engrs. *Res:* Radar equipment research and development; infrared target seekers; missile guidance systems; teletypewriter equipment; blind landing systems; industrial control equipment; particle counters. *Mailing Add:* Laser Sensor Technol Inc Box 3912 14926 NE 31st Circle Redmond WA 98052

PREISER, STANLEY, b New York, NY, Nov 25, 27; wid; c 1. NUMERICAL ANALYSIS, COMPUTATION. *Educ:* City Col New York, BS, 49; NY Univ, MS, 50, PhD, 58. *Prof Exp:* Mathematician, Nuclear Develop Corp Am, 50-60, mgr math sect, United Nuclear Corp, 60-65; from asst prof to assoc prof, Polytech Inst NY, 65-69, dir, Comput Sci Div, 75-76, dean, Westchester Ctr, 80-82, PROF MATH & COMPUTER SCI, POLYTECH UNIV, 69- *Concurrent Pos:* Consult scientist, United Nuclear Corp, 65-66; consult, Radioptics, Inc, IBM, Inc, 66-69, State Univ NY Downstate Med Ctr, 70 & Bridgeport Eng Inst, 71-75, Software Systs Technol, 82- *Mem:* AAAS; Am Math Soc; Soc Indust & Appl Math; Math Asn Am; Asn Comput Mach; Am Asn Artificial Intel. *Res:* Numerical analysis; theory of computation; mathematics of computation; computer sciences; software engineering. *Mailing Add:* Dept of EE/CS New York NY 11201

PREISLER, HARVEY D, b Brooklyn, NY, Feb 5, 41; m 63; c 3. ONCOLOGY, CELL BIOLOGY. *Educ:* Brooklyn Col, BA, 61; Univ Rochester, MD, 65. *Prof Exp:* Res assoc, Leukemia Serv, Nat Cancer Inst Med Br, 67-69; fel med, Columbia Presby Hosp, 69-71; asst prof med, Mt Sinai Sch Med, 71-74; clinician II, 74-75, ASSOC CHIEF DEPT MED A, ROSWELL PARK MEM INST, 75- *Mem:* Am Asn Cancer Res; Am Soc Clin Oncol. *Res:* Regulation of cell proliferation and differentiation; anti-cancer chemotherapy. *Mailing Add:* Dept Med A Roswell Pk Mem Inst 666 Elm St Buffalo NY 14263

PREISLER, JOSEPH J(OHN), b New York, NY, May 17, 19; m 42; c 2. METALLURGY, CHEMICAL ENGINEERING. *Educ:* City Col New York, BChE, 40. *Prof Exp:* Metallurgist, Sperry Rand Corp, 40-43, sr methods engr, 43-44, methods eng supvr, 44-45, standards mat engr, 45-51, sr mat engr, 52-53, head sect mat & processes, 53-59, dept head, Mat Labs, 59-70, mgr eng dept, Sperry Gyroscope Co Div, 70-; RETIRED. *Concurrent Pos:* Instr, Polytech Inst Brooklyn, 44; mem conf elec insulation, Nat Acad Sci-Nat Res Coun, 56-60. *Mem:* Fel Am Inst Chemists; Am Soc Metals; Am Soc Testing & Mat; Soc Automotive Engrs; Am Foundrymen's Soc. *Res:*

Materials engineering; electric and magnetic alloys; ferrous and nonferrous alloys; electrochemical and organic finishes; microelectronic device processing; environmental testings; electrical testing. *Mailing Add:* 36 Donald East Williston NY 11596-2407

PREISMAN, ALBERT, b New York, NY, Feb 8, 01; m 34; c 2. ELECTRICAL ENGINEERING. *Educ:* Columbia Univ, AB, 22, EE, 24. *Prof Exp:* Sr eng, AT&T, 42-43; eng vpres, Capital Radio Engr Inst, 43-59; assoc prof, Catholic Univ Am, 58-69; elec engr, Naval Ordinance Lab, 60-74; RETIRED. *Honors & Awards:* Inst Radio Eng, fel, 53. *Mem:* Sigma Xi; Inst Audio Engrs; Acoustical Soc Am. *Mailing Add:* 1921 Merrifields Dr Silver Springs MD 20906

PREISS, BENJAMIN, b Hamburg, Ger, Sept 16, 33; Israeli & Can citizen. BIOLOGICAL CHEMISTRY. *Educ:* Hebrew Univ Jerusalem, MSc, 57, PhD(biochem), 62. *Prof Exp:* Fel biochem, Harvard Univ, 62-64; res assoc, Yale Univ, 64-66; vis prof, Nat Univ Mex, 67-69; res fel, McGill Univ, 69-70; asst prof, 70-77, ASSOC PROF BIOCHEM, UNIV SHERBROOKE, 78- *Mem:* Am Chem Soc; Can Biochem Soc; Sigma Xi; NY Acad Sci. *Res:* Regulation of cholesterol biosynthesis. *Mailing Add:* Dept Biochem Univ Sherbrooke Fac Med Sherbrooke PQ J1H 5M4 Can

PREISS, DONALD MERLE, b Minn, Jan 5, 27; m 49; c 2. ORGANIC CHEMISTRY. *Educ:* Willamette Univ, BS, 49; Univ Del, MS, 50, PhD(org chem), 52. *Prof Exp:* Asst, Univ Del, 49-51; res technologist, Shell Oil Co, 52-59, res chemist, Shell Develop Co, 59-62; MGR MAT LAB, IBM CORP, 62-; AT DEPT CHEM ENG, NC STATE UNIV. *Concurrent Pos:* Adj prof, NC State Univ. *Mem:* Am Chem Soc; Soc Plastics Eng. *Res:* Elastomers, plastics, high temperature greases, lubricating oil additives, synthetic rubber; extending oils and plasticizers for synthetic rubber; reaction mechanisms; engineering uses of materials. *Mailing Add:* Dept Mat Eng NC State Univ Box 7907 Raleigh NC 27609-7907

PREISS, FREDERICK JOHN, b White Plains, NY, Oct 25, 42; m 64; c 2. INSECTICIDE RESEARCH, MEDICAL ENTOMOLOGY. *Educ:* Unvi Del, BS, 65, MS, 67; Unvi Md, PhD(entom), 69. *Prof Exp:* Captain, US Army Med Serv Corp, 69-71; entomologist, 71-73, mgr tech serv, 73-82, MGR RES & DEVELOP, MCLAUGHLIN GORNLEY KING CO, 82-, DIR CORP RES. *Mem:* Entom Soc Am; Am Mosquito Control Asn; Sigma Xi; Am Soc Testing & Mat. *Res:* Development of pesticide for all non-agricultural products; pyrethroids, pyrethrum; synergists, repellents and insect growth regulators. *Mailing Add:* McLaughlin Gornley King Co 8810 Tenth Ave N Minneapolis MN 55427

PREISS, IVOR LOUIS, b New York, NY, Mar 24, 33; m 56; c 5. NUCLEAR CHEMISTRY. *Educ:* Rensselaer Polytech Inst, BS, 55; Univ Ark, MS, 57, PhD(chem), 60. *Prof Exp:* Res asst nuclear chem, Univ Ark, 56-58; res assoc physics, Yale Univ, 60-65, res assoc chem, 65-66, lectr, 60-65, lectr, Foreign Students' Inst, 63-65, asst dir, Heavy Ion Accelerator Labs, 64-66; assoc prof chem, 66-73, PROF PHYS CHEM, RENSSELAER POLYTECH INST, 73- *Concurrent Pos:* Mem comt nuclear educ & employment, Nat Res Coun-Nat Acad Sci, 69-71; chmn, Interdisciplinary Ctr Nuclear Studies, Rensselaer Polytech Inst, 70-; vis prof, State Univ NY Albany, 70-; resident staff nuclear med, Albany Med Ctr Hosp, 73-; consult, Div Spec Progs, NIH, NY Power Authority, Nat Inst Occup Safety & Health, Nuclear Indust, US-Israel Binational Progs, NSF Binational Progs, & Comt Nuclear & Radiochem, Nat Acad Sci, 88; panel mem, spec progs, NIH, Environ Toxicol, NIEHS; consult, radiation detection monitoring technol; expert, radiation detection & monitoring, trace element anal & radioisotope identification & application. *Mem:* AAAS; Am Chem Soc; Am Phys Soc; Sigma Xi; NY Acad Sci. *Res:* Nuclear and atomic spectroscopy; trace element analysis; heavy ion and neutron reactions; new isotopes. *Mailing Add:* Cogswel Lab Rensselaer Polytech Inst Troy NY 12181

PREISS, JACK, b Brooklyn, NY, June 2, 32; m 59; c 3. BIOCHEMISTRY. *Educ:* City Col New York, BS, 53; Duke Univ, PhD(biochem), 57. *Prof Exp:* Asst, Duke Univ, 53-56; Am Cancer Soc fel microbiol, Washington Univ, 58-59; Am Cancer Soc fel biochem, Stanford Univ, 59-60; biochemist, Nat Inst Arthritis & Metab Dis, 60-62; from asst prof to prof biochem & biophys, Univ Calif, Davis, 62-85, chmn dept, 71-74 & 77-81; chmn dept, 85-89, PROF BIOCHEM, MICH STATE UNIV, 85- *Concurrent Pos:* Consult, Physiol Chem Study Sect, NIH, 67-72 & Metab Biol Study Sect, NSF, 72-75; Guggenheim mem fel, 69-70; pub health sr postdoc fel, 71; Camille & Henry Dreyfus distinguished scholar, Calif State Univ, Los Angeles, 83; Pub Health Serv res grant merit award, 86; Nat Sci Coun lectureship, Repub China, 88. *Honors & Awards:* Pfizer Prize, Am Chem Soc, 71; Fulbright Scholar, 69-70; Alexander von Humboldt Stiftung Sr US Scientist Award, 84; Alsberg-Schoch Mem Lectr Award, Am Asn Cereal Chemists. *Mem:* Am Soc Biol Chem; Am Soc Plant Physiologists; Am Soc Microbiol; NY Acad Sci; Brit Biochem Soc; Am Chem Soc; Am Asn Cereal Chemists. *Res:* Carbohydrate metabolism; regulation of glycogen and starch synthesis; enzymology; molecular biology. *Mailing Add:* Dept of Biochem Mich State Univ East Lansing MI 48824

PREJEAN, JOE DAVID, b Pampa, Tex, Feb 9, 40; m 78; c 2. PHYSIOLOGY, BIOCHEMISTRY. *Educ:* Stephen F Austin State Col, BS, 63; E Tex State Univ, MA, 65; Tex A&M Univ, PhD(physiol), 69. *Prof Exp:* Sr biologist, Southern Res Inst, 69-72, head, Chem Carcinogenesis Div, 72-84, assoc dir chemother toxicol dept, 83-89, head, Toxicol Carcinogenesis Div, 84-89, DIR, TOXICOL RES, SOUTHERN RES INST, 89-,. *Concurrent Pos:* Fel, Tex A&M Univ, 69. *Mem:* NY Acad Sci; Am Col Toxicol; Sigma Xi; Am Asn Cancer Res. *Res:* Effects of intra-arterial hydrogen peroxide on isotope localization and wound healing; effects of chronic cobalt 60 irradiation on longevity and fertility; cancer chemotherapy; carcinogenesis; toxicology. *Mailing Add:* Southern Res Inst 2000 Ninth Ave S PO Box 55305 Birmingham AL 35255

PREKOPA, ANDRAS, b Nyiregyhaza, Hungary, Sept 11, 29; m 62; c 2. OPTIMIZATION OF STOCHASTIC SYSTEMS, OPTIMAL ENGINEERING DESIGN. *Educ:* Univ Debrecen, BA, 49, MA, 52; Univ Budapest, PhD(math), 60. *Prof Exp:* From asst prof to assoc prof probability & statist, Univ Budapest, 56-68; prof math, Tech Univ Budapest, 68-83, prof opers res EOTVOS, 83-85; PROF OPERS RES, RUTGERS UNIV, 85- *Concurrent Pos:* Prof opers res, Univ Budapest, 83; head oper res group, Math Inst, Hungarian Acad Sci, 58-70, head dept oper res, Comput & Automation Inst, 70-75, head dept appl math, 77-85; head math group, Cent Statist Bur Hungary, 59-65; chmn comt stochastic prog, Math Prog Soc. *Honors & Awards:* Benedikt Prize, Hungarian Acad Sci, 79. *Mem:* Int Statist Inst; fel Econometric Soc; Math Prog Soc; Hungarian Acad Sci; foreign mem Nat Acad Eng Mex; NY Acad Sci. *Res:* Mathematical model building; computerized solution of the problems of stochastic systems & their applications to engineering design, in particular to power systems & water resource problems; further applications to inventory control & manufacturing. *Mailing Add:* Rutgers Ctr Opers Res Rutgers Univ PO Box 5062 New Brunswick NJ 08903-5062

PRELAS, MARK ANTONIO, b Pueblo, Colo, July 2, 53; m 79; c 2. LASERS, FUSION ENGINEERING. *Educ:* Colo State Univ, BS, 75; Univ Ill, MS, 76, PhD(nuclear eng), 79. *Prof Exp:* Vpres, Nuclear-Pumped Laser Corp, 81-86; from asst prof to assoc prof nuclear eng, 79-89, Ketcham Res Chair, 88-90, RES SCIENTIST, UNIV MO, 80-, DIR FUSION ENERGY RES LAB, 80-, H O CROFT PROF, 90- *Concurrent Pos:* Prin investr, NSF res grants, 80, 82-89, Dept of Energy, 87-91; fel, Workshop Fuels Inorg Resources, Gas Res Inst, 81; vis scientist, Idaho Nat Engr Lab, 87, consult; sr res scientist Columbia Res Instruments Corp, 88-, consult, 88-91. *Honors & Awards:* Presidential Young Investr Award, NSF, 84. *Mem:* Am Nuclear Soc; Am Phys Soc; Inst Elec & Electronics Engrs; Am Soc Eng Educ. *Res:* Laser physics; plasma chemistry; plasma physics; gaseous electronics; direct energy conversion; spectroscopy. *Mailing Add:* Nuclear Eng Dept Univ Mo Columbia MO 65211

PRELL, GEORGE D, b Syracuse, NY, Sept, 13, 51; m 85. NEUROCHEMISTRY. *Educ:* Univ Ottawa, PhD(med), 84. *Prof Exp:* Instr, 85-86, ASST PROF, MT SINAI MED CTR, 86- *Mem:* Soc Neurosci; Asn Off Anal Chemists; Am Soc Pharmacol & Exp Therapeut; Histamine Neurosci Res Group; Am Soc Mass Spectrometry. *Res:* Pharmacology and neurochemistry of histamine and imidazoleacetic acid. *Mailing Add:* Dept Pharm Mt Sinai Med Ctr 1 Gustave Levy Pl New York NY 10029-6574

PRELL, WARREN LEE, b Oakpark, Ill, June 18, 43; m 67; c 2. MARINE GEOLOGY, PALEOCEANOGRAPHY. *Educ:* Hanover Col, BA, 66; Columbia Univ, PhD(marine geol), 74. *Prof Exp:* Dir geol & oceanog projs, NY State Energy & Res Develop Authority, 73-75; asst prof oceanog & paleoceanog, 75-80, ASSOC PROF MARINE GEOL, BROWN UNIV, 80- *Concurrent Pos:* Actg dir, Climate: Long-range Invest, Mapping & Prediction Proj, 78-81; consult, Ocean Sci Div, NSF, 80-82; cochief scientist, Leg 68, Deep Sea Drilling Proj, 79. *Mem:* AAAS; Am Geophys Union; Geol Soc Am; Soc Econ Paleontologists & Mineralogists. *Res:* Evolution of climate as recorded in deep-sea sediments; history of monsoonal variations in the Indian Ocean; marine geology of the Caribbean; reconstruction of past oceanic circulation. *Mailing Add:* Dept Geol Brown Univ Providence RI 02912

PRELOG, VLADIMIR, b Sarajevo, Yugoslavia, July 23, 06; m 33; c 1. ORGANIC CHEMISTRY. *Educ:* Czech Inst Technol, DChem, 29. *Hon Degrees:* Dhc, Univ Zagreb, 54, Univ Paris, 63, Univ Cambridge, 69. *Prof Exp:* From lectr to assoc prof chem, Univ Zagreb, 35-41; from instr to prof, Inst Technol, Zurich, 42-76; RETIRED. *Honors & Awards:* Nobel Prize, Swed Acad Sci, 75; Roger Adams Award, Am Chem Soc, 69. *Mem:* Foreign assoc Nat Acad Sci; hon mem Am Acad Arts & Sci; Foreign mem Royal Soc London; Foreign mem Acad Sci USSR; Am Philos Soc. *Res:* Author of approximately 400 scientific papers mainly about structure of natural compounds and stereochemistry. *Mailing Add:* Chem Dept Eidgenossische Technische Hochschule, ETH-Zentrum 8092 Zurich Switzerland

PREM, KONALD ARTHUR, b St Cloud, Minn, Nov 6, 20; m 47; c 3. GYNECOLOGIC ONCOLOGY. *Educ:* Univ Minn, BS, 47, MB, 50, MD, 51; Am Bd Obstet & Gynec, dipl, 58, cert gynec oncol, 74. *Prof Exp:* Med fel, 51-54, Am Cancer Soc clin fel, 53-54, from instr to assoc prof, 55-70, dir gynec oncol, 70-83, head obstet & gynec, 76-84, PROF OBSTET & GYNEC, SCH MED, UNIV MINN, 70- *Concurrent Pos:* Mem bd dirs, Nat Comn Human Life, Reproduction & Rhythm, 66- *Mem:* Am Col Obstet & Gynec; Soc Gynec Oncol; Soc Pelvic Surg; Am Gynec & Obstet Soc; Soc Gynec Surgeons. *Res:* Natural family planning; gynecologic oncology; stress incontinence. *Mailing Add:* Box 395 Univ Hosps Minneapolis MN 55455

PREMACHANDRA, BHARTUR N, b Bangalore, India, May 17, 30; US citizen; m 63; c 2. ENDOCRINOLOGY, IMMUNOLOGY. *Educ:* Univ Mysore, BSc, 50; Univ Bombay, MSc, 55; Univ Mo-Columbia, PhD(endocrinol), 58; FRIC. *Hon Degrees:* DSc, Univ Bombay, 81. *Prof Exp:* Res assoc & asst prof endocrinol, Univ Mo-Columbia, 58-60; res assoc, Jewish Hosp, St Louis, 60-62; ENDOCRINOLOGIST, VET ADMIN HOSP, ST LOUIS, 62-, RES PHYSIOLOGIST & BIOCHEMIST. *Concurrent Pos:* Fel endocrinol, Univ Mo-Columbia, 58-59; res prof, Wash Univ, 67; consult, Nat Sci Found, 67. *Mem:* Royal Soc Health; Endocrine Soc; Am Physiol Soc; Am Thyroid Asn; Soc Exp Biol & Med. *Res:* Autoimmune diseases, diabetes, thyroid physiology and thyroid hormone-protein interaction; immunoendocrinology; author and coauthor of over 100 publications. *Mailing Add:* Vet Admin Hosp St Louis MO 63125

PREMANAND, VISVANATHA, b Cannanore, India, Oct 2, 29; US citizen; m 63. MICROELECTRONICS. *Educ:* Univ Madras, MA, 50; Univ Paris, PhD(physics), 62. *Prof Exp:* Res asst microwaves, Tata Inst Fundamental Res, India, 51-57; sci officer electron micros, Bhabha Atomic Res Ctr, 62-67; prin scientist, Appl Sci Div, Litton Indust, 67-68; res assoc electron micros, Sch Metall, Univ Minn, 68-69; mgr process develop, Control Data Corp,

69-74; sr engr spec solid state sensor develop, Emerson Elec, 75-76; consult microelectronics, 76-80; eng mgr, Dahlberg Electronics, 80-81; mgr, Hybrid Microelectronics, Resistance Technol Inc, 81-85; MGR, METAX ASSOC, 85- *Res:* Electron microscopy; physical metallurgy. *Mailing Add:* 10 W 107th St Minneapolis MN 55420-5502

PREMUZIC, EUGENE T, b Zagreb, Yugoslavia, Feb 2, 29; US citizen; m 52; c 1. SYNTHETIC ORGANIC CHEMISTRY. *Educ:* Univ Birmingham, BSc, 54, MSc, 55; Univ Sussex Sch Molecular Sci, D Phil(chem natural prod), 67. *Prof Exp:* Scientist, NY Ocean Sci Lab, 72-76; consult, 76-77, sr res assoc, 77-81, CHEMIST, BROOKHAVEN NAT LAB, 81-; PROF CHEM, LONG ISLAND UNIV, 75- *Concurrent Pos:* Sr scientist, World Life Res Inst, 71-72; adj prof, Fordham Univ, 73-74. *Mem:* Am Chem Soc; sr mem Can Inst Chem; fel Chem Soc UK; NY Acad Sci. *Res:* Chemistry of primitive organisms: the role of primitive organisms in the process of chemical transformation, chemical communication and transport; organic bio-geochemistry: applied biochemistry; use of biotechnology in energy related areas. *Mailing Add:* Dept Appl Sci Brookhaven Nat Lab Upton NY 11973

PREND, JOSEPH, b Slawkow, Poland, Apr 18, 20; nat US; m 49; c 3. HORTICULTURE. *Educ:* Munich Tech Univ, MS, 48, ScD, 50. *Prof Exp:* Plant breeder, F H Woodruff & Sons, Inc, 52-58; sr res horticulturist, Crops Res Dept, H J Heinz Co, 58-67, chief res horticulturist, Agr Res Dept, 67-75, assoc mgr agr res, 75-81, mgr agr res, 81-85; RES & DEVELOP CONSULT, BREEDING & DEVELOP SEED PROD, 86- *Mem:* AAAS; Am Soc Hort Sci; Am Phytopathological Soc. *Res:* Breeding and development of tomatoes and cucumber varieties; introducing disease resistance; laboratory and processing evaluation of varieties; corn breeding. *Mailing Add:* 2531 Sheridan Way Stockton CA 95207

PRENDERGAST, FRANKLIN G, b Kingston, Jamaica, Mar 7, 45. PROTEIN BIOCHEMISTRY & BIOPHYSICS. *Educ:* Univ WI, MD, 68; Univ Minn, PhD(biochem), 76. *Prof Exp:* PROF BIOCHEM & MOLECULAR BIOL & CHMN DEPT, MAYO FOUND, 76- *Mem:* Am Chem Soc; Biophys Soc; Am Soc Photobiol; Am Soc Biol Chemists; NY Acad Sci; AAAS. *Mailing Add:* Pharmacol Dept Mayo Clin Rochester MN 55901

PRENDERGAST, KEVIN HENRY, b Brooklyn, NY, July 9, 29; m 65; c 1. ASTRONOMY. *Educ:* Columbia Univ, AB, 50, PhD, 54. *Prof Exp:* Asst astron, Columbia Univ, 50-54, Adams res fel, 54-56; asst prof astrophys, Yerkes Observ, Univ Chicago, 56-66; PROF ASTRON, COLUMBIA UNIV, 66- *Mem:* Am Astron Soc; Am Phys Soc. *Res:* Hydromagnetics; theoretical astrophysics. *Mailing Add:* Dept Astron Columbia Univ Main Div New York NY 10027

PRENDERGAST, ROBERT ANTHONY, b Brooklyn, NY, Nov 6, 31. IMMUNOLOGY, PATHOLOGY. *Educ:* Columbia Univ, BA, 53; Boston Univ, MD, 57. *Prof Exp:* Intern med, Cornell Div, Bellevue Hosp, 57-58; resident path, Mallory Inst, Boston City Hosp, 58-59; resident, Mem-Sloan Kettering Hosp, 59-61; vis physician, Rockefeller Univ, 63-65; asst prof, 65-70, ASSOC PROF OPHTHAL & PATH, SCH MED, JOHNS HOPKINS UNIV, 70- *Concurrent Pos:* USPHS fel, Rockefeller Univ, 63-65; prof, Res Prev Blindness, Inc, 71- *Mem:* Am Asn Immunol; Am Soc Exp Path; Transplantation Soc; Reticuloendothelial Soc. *Res:* Delayed hypersensitivity and cellular immunology; ontogeny of the immune response; transplantation immunology; immunology of neoplasia; viral immunopathology; immunopathology of ocular inflammatory diseases. *Mailing Add:* Johns Hopkins Univ Sch of Med Baltimore MD 21205

PRENER, ROBERT, b Brooklyn, NY, Feb 27, 39; m 58. MATHEMATICS EDUCATION. *Educ:* Cornell Univ, AB, 60; Columbia Univ, MA, 63; Polytech Inst Brooklyn, PhD(math), 69. *Prof Exp:* Lectr math, Cornell Univ, 59-60; asst, Columbia Univ, 60-61; from instr to assoc prof, 61-84, PROF MATH, LONG ISLAND UNIV, 84- *Concurrent Pos:* Math adv, Rockland Proj Sch, 69-70 & Skunk Hollow High Sch, NY, 70- *Res:* Group theory; secondary mathematics and science teaching methods; elementary arithmetic and mathematics learning. *Mailing Add:* Dept Math Long Island Univ Brooklyn NY 11201

PRENGLE, H(ERMAN) WILLIAM, JR, b Connelsville, Pa, Nov 6, 19; m 41; c 3. CHEMICAL ENGINEERING. *Educ:* Carnegie Mellon Univ, BS, 41, MS, 47, DSc(chem eng), 49. *Prof Exp:* Jr res engr, Linde Air Prod Co, NY, 41; from sr engr to sr technologist, Shell Oil Co, Tex, 49-52; assoc prof, 52-59, assoc dean, 81-85, PROF CHEM ENG, UNIV HOUSTON, 59- *Concurrent Pos:* Consult, Houston Res Inst, Inc, 52-85, mem bd dirs, 63-80; res grants, NSF, 56-59 & 62-66, Dept Interior Off Saline Water, 64-66, Environ Protection Agency, 68-71, US Dept Energy, 72-; vis scholar, Cambridge Univ, 71-72, 88 & 90; prin investr, 36 res grants & contracts. *Mem:* Fel Am Inst Chemists; Am Inst Chem Engrs; Am Chem Soc; Royal Soc Chem (London). *Res:* Thermodynamics and structure of liquids and solutions; molecular spectroscopy; kinetics of chemical reactions in liquid and vapor phases; remote of pollutants; environmental science and engineering. *Mailing Add:* Dept Chem Eng Univ Houston Houston TX 77204-4792

PRENSKY, WOLF, b Koenigsberg, Ger, Dec 26, 30; US citizen; m 66. GENETICS, CELL BIOLOGY. *Educ:* Cornell Univ, BS, 53; Univ Ill, MS, 57, PhD(genetics, plant breeding), 61. *Prof Exp:* NIH fel biol, Brookhaven Nat Lab, 61-63, res assoc biol, 63-64; from instr to assoc prof physiol & genetics, Sch Med, Tufts Univ, 65-73; AT SLOAN-KETTERING INST CANCER RES, 73- *Mem:* AAAS; Genetics Soc Am; Am Soc Cell Biol; Biomet Soc; Sigma Xi. *Mailing Add:* 28-10 High St Fairlawn NJ 07410

PRENTICE, GEOFFREY ALLAN, b Cleveland, Ohio, Oct 15, 46; m 68; c 2. ELECTROCHEMICAL ENGINEERING. *Educ:* Ohio State Univ, BS & MS, 69; Univ Calif, Berkeley, PhD(chem eng), 81. *Prof Exp:* Engr, Goodyear Tire & Rubber Co, 69-71, Sweden, 71-73, chief chemist, Zaire, 73-75; asst prof, 81-88, ASSOC PROF CHEM ENG, JOHNS HOPKINS UNIV, 88- *Mem:* Electrochem Soc; Am Inst Chem Engrs. *Res:* Modeling electrochemical cells and calculating current distributions especially when the electrode shape changes with time; electrode kinetics, passivation. *Mailing Add:* Dept Chem Eng Johns Hopkins Univ 34th & Charles St Baltimore MD 21218

PRENTICE, JACK L, b Santa Barbara, Calif, Sept 25, 31; m 51; c 3. PHYSICAL CHEMISTRY, HIGH TEMPERATURE CHEMISTRY. *Educ:* Calif State Polytech Col, BS, 59. *Prof Exp:* res chemist, Naval Weapons Ctr, 59-74; Consult, Polymer Abrasion Pads, construction indust, 79-80; CONSULT & PRACTR, SMALL ARMS INTERN BALLISTICS, 81- *Res:* Metal combustion; combustion chemistry; internal ballistics of small arms ammunition. *Mailing Add:* PO Box 639 Athena OR 97813

PRENTICE, JAMES DOUGLAS, b Auckland, NZ, Oct 31, 30; Can citizen; m 60; c 2. PHYSICS. *Educ:* McGill Univ, BSc, 51, MSc, 53; Univ Glasgow, PhD(physics), 59. *Prof Exp:* Lectr, 58-59, from asst prof to assoc prof, 59-70, PROF PHYSICS, UNIV TORONTO, 70- *Concurrent Pos:* Sr sci officer, Rutherford High Energy Lab, 63-64. *Mem:* Can Asn Physicists; Am Phys Soc. *Res:* High energy particle physics; high energy Hadron interactions and Hadron spectroscopy. *Mailing Add:* Dept Physics Univ Toronto Toronto ON M5S 1A1 Can

PRENTICE, NEVILLE, b Longbenton, Eng, Feb 10, 20; US citizen; m 59; c 2. AGRICULTURAL BIOCHEMISTRY. *Educ:* Univ Man, BSc, 50; Univ Minn, MS, 55, PhD(biochem), 56. *Prof Exp:* RES CHEMIST, USDA CEREAL CROPS RES UNIT, 56- *Concurrent Pos:* Assoc ed, J Am Asn Cereal Chemists, 73-76 & J Am Soc Brewing Chemists, 80- *Mem:* Am Asn Cereal Chemists; Sigma Xi; Am Soc Brewing Chemists. *Res:* Microbiological metabolites toxic to animals or plants; enzyme systems involved in germination of cereals. *Mailing Add:* 5803 Spurwood Ct Colorado Springs CO 80918

PRENTICE, NORMAN MACDONALD, b Yonkers, NY, Feb 25, 25. CLINICAL PSYCHOLOGY, CHILD PSYCHOLOGY. *Educ:* Princeton Univ, AB, 49; Harvard Univ, MA, 52, PhD(clin psychol), 56; Am Bd Prof Psychol, dipl, 62. *Prof Exp:* Res fel, Child Psychiat Unit, Mass Gen Hosp, Boston, 53-55; NIMH fel, Judge Baker Guid Ctr, Boston, 55-57, staff psychologist to coordr training, 58-65; chief, adolescent sect psychologist, dept psychol, Children's Hosp, Boston, 57-58; assoc prof psychol, 65-68, dir, Clin Psychol Training Prog, 74-76, PROF PSYCHOL & EDUC PSYCHOL, UNIV TEX, AUSTIN, 68- *Mem:* Fel Am Psychol Asn; fel Am Orthopsychiat Asn (vpres, 79-80); fel Soc Personality Assessment. *Res:* Etiology and remediation of psychopathological conditions in children, adolescents, young adults and their families (especially delinquency and psychogenic learning disabilities); socioaffective development in normal childhood (especially humor and fantasy). *Mailing Add:* Dept Psychol Univ Tex Austin TX 78712

PRENTICE, ROSS L, b Ont, Can, Oct 16, 46; m; c 2. BIOSTATISTICS. *Educ:* Univ Waterloo, BS, 67, MS, 68, PhD(statist), 70. *Prof Exp:* Asst prof statist, Univ Waterloo, 70-74; assoc mem, 74-82, SR BIOSTATIST, FRED HUTCHINSON CANCER RES CTR, 74-, DIR, DIV PUB HEALTH SCI, 83- *Concurrent Pos:* Vis asst prof statist, State Univ NY, Buffalo, 71-72; consult, Lung Cancer Study Group, Vet Admin, 72-74; assoc prof biostatist, Univ Wash, 74-77, prof, 77-; dir, Statist Ctr, Northwest Oncol Group, 76-78; biostatistician, Radiation Effects Res Found, Hiroshima, Japan, 80-81; assoc ed, Am J Epidemiol, 86-; Rosenstatt prof, Univ Toronto Med Sch, 88-89; mem, Data Monitoring Bd, Physicians Health Study, 90- *Honors & Awards:* Benz Lectr, Univ Iowa, 90; Greenberg Distinguished Lectr, Univ NC, 90. *Mem:* Inst Med-Nat Acad Sci; fel Am Statist Asn. *Res:* Author or co-author of over 140 publications. *Mailing Add:* Fred Hutchinson Cancer Res Ctr 1124 Columbia St Seattle WA 98104

PRENTICE, WILBERT NEIL, b Nagpur, India, Nov 19, 23; nat US; m 55; c 2. NUMERICAL ANALYSIS. *Educ:* Middlebury Col, AB, 44; Brown Univ, AM, 50; Syracuse Univ, PhD(math), 59. *Prof Exp:* Instr math, Clarkson Tech Univ, 49-51; asst instr, Syracuse Univ, 52-57; from instr to assoc prof, 57-74, dir, Denison Comput Ctr, 64-71, PROF MATH, DENISON UNIV, 74- *Concurrent Pos:* Vis fel, Ohio State Univ, 72. *Mem:* Math Asn Am; Asn Comput Mach. *Res:* Computer programming and its use in liberal arts education. *Mailing Add:* 117 Locust Pl Granville OH 43023

PRENTISS, WILLIAM CASE, b Baltimore, Md, Sept 5, 24; m 45; c 2. LEATHER CHEMISTRY. *Educ:* Amherst Col, AB, 47; Brown Univ, PhD(phys chem), 51. *Prof Exp:* Chemist, Res & Develop, 50-57, head lab, 57-62, sales develop rep, 62-64, sr res chemist, 64-66, head lab, 66-70, leader tech serv group, Leather Chem & Finishes, 70-76, proj leader, leather & coated fabrics chem res, 76-80, sect mgr leather chem res, 80-85, MGR LEATHER CHEM RES, ROHM AND HAAS CO, 85- *Mem:* Am Chem Soc; Am Leather Chemists Asn (pres, 84-); Am Soc Testing & Mat. *Res:* Waterborn polymers; surfactants; acrylic monomers and polymers; leather chemicals. *Mailing Add:* 260 Mathews Ave Doylestown PA 18901

PRENZLOW, CARL FREDERICK, b Long Beach, Calif, Apr 11, 30; m 67; c 2. PHYSICAL CHEMISTRY. *Educ:* Univ Colo, BA, 51; Univ Wash, PhD(phys chem), 57. *Prof Exp:* Res chemist, Dow Chem Co, 56-58; asst adminr chem sci div, Off Ord Res, US Dept Army, 58-60; res chemist, Nat Bur Standards, 60-64; chemist, Allied Chem Co, 64-65; PROF CHEM, CALIF STATE UNIV, FULLERTON, 65-, CHMN DEPT, 71- *Mem:* Am Chem Soc. *Res:* Physical adsorption; organic reaction mechanisms and synthesis; surface chemistry and related fields. *Mailing Add:* 1065 N Granada Orange CA 92669-1227

PREPARATA, FRANCO PAOLO, b Reggio Emilia, Italy, Dec 29, 35; m 64; c 2. COMPUTER SCIENCE, ENGINEERING. *Educ:* Univ Rome, Dr Ing, 59, Libera docenza, 69. *Prof Exp:* Fel, Nat Inst Appl Calculation, Rome, 59-60; tech mgr comput, Univac Italia, Rome, 60-63; sr engr, Selenia, SpA, Rome, 63-65; res assoc elec eng, Univ Ill, Urbana, 65-66, from asst prof to prof elec & computer eng, 66-90; AN WANG PROF COMPUTER SCI, BROWN

UNIV, PROVIDENCE, 91- *Mem:* Fel Inst Elec & Electronics Engrs; Asn Comput Mach; Europ Asn Theoret Comput Sci. *Res:* Theory of computation, design and analysis of algorithms, computational geometry, digital systems, information and coding theory. *Mailing Add:* Dept Computer Sci Brown Univ Providence RI 02906

PREPAS, ELLIE E, b Hamilton, Ont, July 31, 47; c 1. LIMNOLOGY, ECOLOGICAL MODELLING. *Educ:* Univ Waterloo, BA, 71; York Univ, MA, 74; Univ Toronto, PhD(limnol), 80. *Prof Exp:* Lectr biomet, McGill Univ, 77; from asst prof to assoc prof, 79-90, PROF ZOOL, UNIV ALTA, 90- *Concurrent Pos:* Dir, Meanook Biol Res Sta, 83-; actg dir, Environ Studies & Res Ctr, 90- *Mem:* Int Soc Limnol; Soc Can Limnologists; Am Soc Limnol & Oceanog; Am Fisheries Soc; NAm Lake Mgt Soc. *Res:* Limnology of eutropic lakes and rivers; nutrient (phosphorus and nitrogen) cycling in lakes (including saline) and streams; lake restoration; climate change and eutrophie lakes; whole lake modelling; saline lakes. *Mailing Add:* Zool Dept Univ Alta Bio Sci Bldg Edmonton AB T6G 2E9 Can

PREPOST, RICHARD, b New York, NY, Mar 20, 35. ELEMENTARY PARTICLE PHYSICS. *Educ:* Rensselaer Polytech Inst, BS, 56; Columbia Univ, PhD(physics), 61. *Prof Exp:* Res assoc physics, Yale Univ, 61-63; asst prof, Stanford Univ, 63-67; assoc prof, 67-73, PROF PHYSICS, UNIV WIS-MADISON, 73- *Honors & Awards:* Fel, Am Phys Soc. *Mem:* Am Phys Soc. *Res:* Particle and muon physics; electron-positron colliding beams; strong and electromagnetic interactions. *Mailing Add:* Dept Physics Univ Wis 1150 University Ave Madison WI 53706

PRERAU, DAVID STEWART, b New York, NY; c 1. EXPERT SYSTEMS, ARTIFICIAL INTELLIGENCE. *Educ:* City Col NY, BE, 64; Mass Inst Technol, MS, 66, PhD(computer sci), 70. *Prof Exp:* Prog mgr, US Dept Transp, 70-81; PRIN MEM TECH STAFF, GTE LABS, 81- *Concurrent Pos:* Lectr expert syst technol, numerous univs, prof socs, conferences & corp. *Honors & Awards:* Leslie H Warner Tech Achievement Award. *Mem:* Am Asn Artificial Intel; Asn Comput Mach; Inst Elec & Electronics Engrs Computer Soc; Sigma Xi. *Res:* Lead development of major expert systems with high corporate impact; conceive many important, widely-used techniques for building expert systems; author of one book on expert systems. *Mailing Add:* 187 South St Chestnut Hill MA 02167

PRESBY, HERMAN M, b Jersey City, NJ, Sept 25, 41; m; c 4. PHYSICS. *Educ:* Yeshiva Univ, BA, 62, PhD(physics), 66. *Prof Exp:* Res scientist plasma physics, Columbia Univ, 66-67; asst prof physics, Belfer Grad Sch, Yeshiva Univ, 67-72; DISTINGUISHED MEM TECH STAFF OPTICAL FIBERS, BELL LABS, 72- *Mem:* Optical Soc Am. *Res:* Optical fiber waveguides. *Mailing Add:* Bell Labs HOH Rm 231 Crawford Hill Holmdel NJ 07733

PRESCH, WILLIAM FREDERICK, b Columbus, Ohio, Aug 27, 42; m 63; c 2. HERPETOLOGY. *Educ:* San Diego State Univ, BS, 65, MA, 67; Univ Southern Calif, PhD(biol), 70. *Prof Exp:* Asst prof biol, Univ Southern Calif & marine biol, Santa Catalina Marine Sta, 70-71; asst prof zool, Univ Calif, Berkeley, 71-73; asst prof, 73-76, ASSOC PROF ZOOL, CALIF STATE UNIV, FULLERTON, 76- *Concurrent Pos:* Orgn Trop Studies fel, Univ Costa Rica, 67. *Mem:* AAAS; Am Soc Ichthyol & Herpet; Am Soc Zoologists; Soc Study Amphibians & Reptiles; Am Soc Naturalists; Sigma Xi. *Res:* Comparative osteology; systematics and evolution of lizards; functional anatomy and mechanics in lower vertebrates; biogeography of reptiles. *Mailing Add:* Dept Biol Sci Calif State Univ Fullerton CA 92634

PRESCOTT, BASIL OSBORNE, b Matehuala, Mex, Nov 23, 11; US citizen; m 41; c 2. GEOLOGY. *Educ:* Colo Sch Mines, GE, 35. *Prof Exp:* Seismol computer, Shell Oil Co, 35-36, asst seismologist, 36-41, seismologist, 41-47, geologist, 47-49, spec probs geologist, 49-52, sr geologist, 52-56, explor petrophysicist, 51-56; explor geologist & physicist, Shell Develop Co, 56-71, res assoc instrumentation, 57-71; CONSULT INSTRUMENTATION, 71- *Res:* Macropaleontology; structural geology; petroleum hydrodynamics; inorganic and organic geochemistry; analog computator design; chromatography of hydrocarbon gases; special instrument design. *Mailing Add:* 3618 Elmridge Houston TX 77025

PRESCOTT, BENJAMIN, b Fall River, Mass, Feb 7, 07; m 32; c 2. BIOCHEMISTRY. *Educ:* Univ Chicago, BS, 30; Georgetown Univ, PhD(biochem), 41. *Prof Exp:* Asst metab in Bright's dis, Peter Bent Brigham Hosp, Boston, 33-34; asst chem, Franklin Union, 34-35; immunochemist, Johns Hopkins Hosp, 35-38; immunochemist, NIH, 38-79; RETIRED. *Mem:* Hon mem Am Soc Microbiol; emer mem Am Chem Soc; NY Acad Sci. *Res:* Biochemical and serological studies of bacterial products; chemistry of antigenic and immunogenic fractions from mycoplasma, pneumococcal and Streptococci. *Mailing Add:* 16424 Felice Dr San Diego CA 92128-2804

PRESCOTT, CHARLES YOUNG, b Ponca City, Okla, Dec 14, 38; m 64; c 1. ELEMENTARY PARTICLE PHYSICS. *Educ:* Rice Univ, BA, 61; Calif Inst Technol, PhD(physics), 66. *Prof Exp:* Res assoc physics, Calif Inst Technol, 66-70; asst prof, Univ Calif, Santa Cruz, 70-71; mem staff, 71-80, assoc prof physics, 80-84, PROF PHYSICS, STANFORD LINEAR ACCELERATOR CTR, 84-, ASSOC DIR, RES DIV, 86- *Honors & Awards:* Wolfgang KH Panofsky Prize, Am Physiol Soc, 88. *Res:* Photoproduction; electroproduction reactions; weak interactions; electron-positron annihilation. *Mailing Add:* Stanford Linear Accelerator Ctr PO Box 4349 Stanford CA 94309

PRESCOTT, DAVID JULIUS, b Philadelphia, Pa, Oct 8, 39; m 64; c 3. PROTEIN CHEMISTRY, NEUROCHEMISTRY. *Educ:* Univ Pa, BA, 61, PhD(biochem), 67. *Prof Exp:* Instr biochem, Univ Pa, 67-68; asst chief radioisotopes res, Vet Admin, 67-69; fel biol chem, Sch Med, Washington Univ, 68-70; asst prof, 70-77, ASSOC PROF BIOL, BRYN MAWR COL, 77- *Concurrent Pos:* Vis asst prof neurobiol, Harvard Sch Med, 74-75. *Honors & Awards:* Dreyfus Found Award, 74. *Mem:* AAAS; Am Chem Soc; Soc Neurosci; Sigma Xi. *Res:* Enzymology of fatty acid biosynthesis; chemical synthesis of proteins; structure and function of acyl carrier protein; lepidopteran cholinergic development. *Mailing Add:* Dept Biol Bryn Mawr Col Bryn Mawr PA 19010

PRESCOTT, DAVID MARSHALL, b Clearwater, Fla, Aug 3, 26; m 69; c 3. CELL BIOLOGY, GENETICS. *Educ:* Wesleyan Univ, BA, 50; Univ Calif, Berkeley, PhD(zool), 54. *Prof Exp:* Res assoc biol, Univ Copenhagen, 51-52; Am Cancer Soc fel, Carlsberg Lab, Copenhagen, 54-55; asst prof anat, Univ Calif, Los Angeles, 55-59; biologist, Oak Ridge Nat Lab, 59-63; prof & chmn, Dept Anat, Med Sch, 63-66, prof, Inst for Develop Biol, 66-68, chmn, 74-75, prof, 68-80, DISTINGUISHED PROF, DEPT MOLECULAR, CELLULAR & DEVELOP BIOL, UNIV COLO, 80- *Concurrent Pos:* Humboldt Found Sr US Scientist Award, 79-80. *Mem:* Nat Acad Sci; fel Am Acad Arts & Sci; Am Soc Cell Biol (pres, 65-66). *Res:* Factors regulating initiation of synthesis of DNA during cellular reproduction; mechanisms of chromosome replication and function; structure of the chromosome; biology of cancer. *Mailing Add:* Dept Molecular Cellular & Develop Biol Univ Colo Boulder CO 80309

PRESCOTT, GERALD H, b Wendell, Idaho, June 16, 37; div; c 3. MEDICAL GENETICS, ORAL MEDICINE. *Educ:* Col Idaho, BS, 59; Wash Univ, DMD, 64; Ind Univ, Indianapolis, MS, 66. *Prof Exp:* Fel med genetics, Ind Univ, Indianapolis, 66-68; prof med genetics & perinatal med, Dir Antenatal Diag Clin, Med Ctr, Univ Ore, 68-81; DIR PRENATAL DIAG PROG, EMANUEL HOSP & ASSOC DIR, ORE MED GENETICS & BIRTH DEFECTS CTR, 82-; ASST STATE SURGEON, US ARMY NAT GUARD, ORE, 88- *Concurrent Pos:* Consult, Good Samaritan Hosp, St Vincents Hosp & Providence Hosp. *Honors & Awards:* Down's Syndrome Serv Award. *Mem:* AAAS; Genetics Soc Am; Am Acad Oral Path; Sigma Xi; Am Soc Human Genetics; Am Soc Mil Surgeons; Am Soc Craniofacial Genetics. *Res:* Bioethics of genetic counseling; amniotic fluid studies; craniofacial anomalies; prenatal diagnosis. *Mailing Add:* 2801 N Gantenbein Ave Portland OR 97227

PRESCOTT, GLENN CARLETON, JR, b Northampton, Mass, Feb 23, 23; m 50; c 6. HYDROLOGY. *Educ:* Brown Univ, AB, 44, MS, 50; Harvard Univ, MPA, 54. *Prof Exp:* Geologist, State Geol Surv, Kans, 47-48; geologist, Ground Water Br, US Geol Surv, 48-57, geologist in chg, 57-66, hydrologist, 66-80; RETIRED. *Mem:* Geol Soc Am; Am Geophys Union. *Res:* Glacial geology; ground-water hydrology. *Mailing Add:* Four Highland Ave Camden ME 04843

PRESCOTT, HENRY EMIL, JR, b Philadelphia, Pa, July 1, 36; m 65; c 1. FOOD SCIENCE, BOTANY. *Educ:* Drexel Inst Technol, BS, 59; Univ Pa, PhD(bot), 66. *Prof Exp:* Asst instr biol, Rutgers Univ, 65-66; chemist, 66-67, sr chemist, 67-75, proj specialist, 75-78, TECH SPECIALIST, TECH CTR, GEN FOODS USA, 78- *Mem:* Bot Soc Am; Mycol Soc Am; Sigma Xi; Microbeam Anal Soc. *Res:* Preservation methods; freezing phenomena; techniques for microscopy; food structure; food processing. *Mailing Add:* Tech Ctr Gen Foods USA Tarrytown NY 10591

PRESCOTT, JOHN HERNAGE, b Corona, Calif, March 16, 35. SCIENCE & RESEARCH ADMINISTRATION. *Educ:* Univ Calif, Los Angeles, BA, 57. *Prof Exp:* Vpres & curator, Marineland Pac, 57-70, gen mgr, 70-72; EXEC DIR, NEW ENG AQUARIUM, 72- *Concurrent Pos:* Chmn & mem comt sci adv, Marine Mammal Comn, 77-80; mem, Marine Fisheries Adv Comt, 88- *Mem:* Soc Marine Mammalogy; Am Asn Zool Parks & Aquarium; AAAS; Am Fisheries Soc; Nat Inst Fisheries Res Biol. *Res:* Cetacean behavior, biology and status, principally the North Atlantic whale and harbor porpoise; pelagic ecosystems. *Mailing Add:* New Eng Aquarium Ctr Wharf Boston MA 02110

PRESCOTT, JOHN MACK, b San Marcos, Tex, Jan 22, 21; m 46; c 2. BIOCHEMISTRY. *Educ:* Southwest Tex State Col, BS, 41; Tex A&M Univ, MS, 49; Univ Wis, PhD(biochem), 52. *Prof Exp:* Instr chem, Tex A&M Univ, 46-48, asst biochem, 48-49; asst, Univ Wis, 49-51; asst, Biochem Inst, Univ Tex, 51-52; from asst prof to assoc prof, 52-59, head dept, 69, dean col sci, 70-77, prof biochem & biophys, 59-81, vpres acad affairs, 77-81, prof med biochem, 81-85, dir inst occup & environ med, 81-87, EMER PROF, TEX A&M UNIV, 85- *Concurrent Pos:* Vis prof, Ctr Biomed & Boiphys Sci & Med, Harvard Med Sch, 82. *Mem:* Am Chem Soc; Am Soc Biol Chemists; Soc Exp Biol & Med. *Res:* Proteolytic enzymes; microbial metabolism. *Mailing Add:* 31 Forest Dr College Station TX 77840

PRESCOTT, JON MICHAEL, b Amarillo, Tex, June 17, 39; m 73; c 3. PLANT PATHOLOGY, PLANT BREEDING. *Educ:* Okla State Univ, BS, 65, MS, 67; Univ Minn, PhD(plant path), 70. *Prof Exp:* Res asst plant path, Okla State Univ, 65-67; res fel, Univ Minn, 67-70; asst plant pathologist, Rockefeller Found, 70-71; cereal pathologist, Int Maize & Wheat Improv Ctr, Cimmyt, 71-75, regional cereal pathologist, 72-87; DIR, TECH SERV, INT DIV, DEKALB PLANT GENETICS, 87- *Mem:* Am Phytopath Soc; Am Soc Agron; Crop Sci Soc Am; Indian Phytopath Soc; Turkish Phytopath Soc; Sigma Xi. *Res:* Surveillance, epidemiology, virulence analysis, host-pathogen interactions of cereal diseases, especially of wheat and barley; disease resistance breeding; international testing/evaluation maize, sorghum, sunflower, soybean and alfalfa. *Mailing Add:* Tech Serv Int Div DeKalb-Pfizer Genetics 3100 Sycamore Rd DeKalb IL 60115

PRESCOTT, LANSING M, b Tulsa, Okla, Sept 29, 41; m 64; c 2. MICROBIOLOGY, BIOCHEMISTRY. *Educ:* Rice Univ, BA, 63, MA, 64; Brandeis Univ, PhD(biochem), 69. *Prof Exp:* From asst prof to assoc prof, 69-82, chmn dept, 72-75 & 85-90, PROF BIOL, AUGUSTANA COL, 82- *Concurrent Pos:* Adj fac mem, Dept Microbiol, Univ Ga, 80- *Mem:* Am Chem Soc; Am Soc Microbiol; Sigma Xi; Electron Micros Soc Am; AAAS. *Res:* Toxicology and microbial ecology; morphology of bacteria and diatoms; microbial physiology; aspartate carbamoyl transferase kinetics and regulation; author of one textbook. *Mailing Add:* Dept Biol Augustana Col Sioux Falls SD 57197

PRESCOTT, PAUL ITHEL, b Sanford, Maine, Dec 7, 31; m 55; c 2. CALCIUM CARBONATE & KAOLIN CHEMYTECHNOLOGY, FILLER & EXTENDER PIGMENTS. *Educ:* Univ Maine, BS, 58, MS, 60. *Prof Exp:* Chemist silicones, Res & Develop Ctr, Gen Elec Co, 60-62, develop chemist, Insulating Mat Dept, 62-65, mgr mat anal & testing, Wire & Cable Dept, 65-69; supvr plastics res, Freeport Minerals Co, 69-74, asst res dir clay, 74-79, mgr mkt develop, Freeport Kaolin, 79-85, consult, 85; mgr new bus develop, ECC Am, 85-86, mkt mgr, 86-88; TECH MGR, ECC INT, 89- *Mem:* Sigma Xi; Am Chem Soc; Soc Plastics Engrs; Fedn Socs Coating Technol; Nat Stone Asn; Tech Asn Pulp & Paper Industs. *Res:* Surface modification of pigments and fillers to improve their compatability on polymeric systems; development of kaolin clays for paint, rubber, plastics, and paper industries; research and development on calcium carbonate products. *Mailing Add:* ECC Int PO Box 330 Sylacauga AL 35150

PRESCOTT, STEPHEN M, b Bryan, Tex, Feb 22, 48; m 69; c 2. LIPID METABOLISM, CELL TO CELL INTERACTIONS. *Educ:* Texas A&M Univ, BS, 70; Baylor Col Med, MD, 73. *Prof Exp:* Res fel biochem, Washington Univ Sch Med, 79-82; asst prof, 82-86, ASSOC PROF INTERNAL MED & BIOCHEM, UNIV UTAH, 86- *Concurrent Pos:* Ed bd, J Biol Chem, 86-; estab investr, Am Heart Asn, 87- *Mem:* Am Soc Molecular Biol; Am Soc Cell Biol; Am Soc Clin Invest. *Res:* Regulation of the enzymes that control synthesis and degradation of physiologically active lipids such as prostaglandins, leukotrienes and platelet-activating factor. *Mailing Add:* Dept Internal Med & Biochem Univ Utah Sch Med CVRTI Bldg 100 Salt Lake City UT 84112

PRESES, JACK MICHAEL, b Amityville, NY, Oct 11, 47. PHYSICAL CHEMISTRY. *Educ:* Hofstra Univ, BS, 69; Columbia Univ, PhD(chem), 75. *Prof Exp:* Res asst chem, Columbia Univ, 69-75; res assoc, 75-78, assoc chemist, 78-82, CHEMIST, BROOKHAVEN NAT LAB, 82- *Concurrent Pos:* Mem staff, Div Chem Sci, US Dept Energy, 78-80. *Mem:* Am Chem Soc; Am Phys Soc. *Res:* Gas-phase energy transfer in polyatomic molecules; infrared laser induced chemistry; applications of synchrotron radiation to photochemistry and photophysics. *Mailing Add:* Dept of Chem 555A Brookhaven Nat Lab Upton NY 11973

PRESKILL, JOHN P, b Highland Park, Ill, Jan 19, 53; m 75; c 2. THEORETICAL PHYSICS. *Educ:* Princeton Univ, AB, 75; Harvard Univ, AM, 76, PhD(physics), 80. *Prof Exp:* Jr fel, Harvard Soc Fels, 80-81; from asst prof to assoc prof physics, Harvard Univ, 81-83; assoc prof, 83-90, PROF PHYSICS, CALIF INST TECHNOL, 90- *Concurrent Pos:* A P Sloan fel, 82-86; Presidential Young Investr, NSF, 84-89. *Mem:* Am Phys Soc. *Res:* Theory of elementary particles; cosmology. *Mailing Add:* Dept Physics Calif Inst Technol Pasadena CA 91125

PRESLEY, BOBBY JOE, b Poplar Bluffs, Mo, July 27, 35; div; c 2. CHEMICAL OCEANOGRAPHY. *Educ:* Okla State Univ, BS, 57; WVa Univ, MA, 65; Univ Calif, Los Angeles, PhD(geol), 69. *Prof Exp:* Comput, Geophys Serv, Inc, 57-59; teacher, Sacramento City Unified Schs, 60-61 & 63-64 & Marysville Union High Sch, 61-63; fel, Univ Calif, Los Angeles, 69-70; from asst prof to assoc prof, 70-80, PROF OCEANOG TEX A&M UNIV, 81- *Concurrent Pos:* Prog mgr, NSF, 74-75; res chemist, Environ Protection Agency, 81-82. *Mem:* Am Geophys Union; Geochem Soc. *Res:* Pore water chemistry; diagenesis of sediments; stable isotope geochemistry; trace elements; heavy metal pollution. *Mailing Add:* Dept Oceanog Tex A&M Univ College Station TX 77843

PRESLEY, CECIL TRAVIS, b Marshall, Ark, Nov 3, 41. PHYSICAL CHEMISTRY. *Educ:* Ark Polytech Col, BS, 63; Rice Univ, PhD(chem), 68; Univ Denver, JD, 78. *Prof Exp:* RES SCIENTIST, DENVER RES CTR, MARATHON OIL CO, 67- *Mem:* AAAS; Am Chem Soc; Sigma Xi; Soc Petrol Engrs. *Res:* Areas of x-ray crystallography; light scattering, small angle x-ray scattering; ultracentrifugation; micro-emulsions; water-soluble polymers. *Mailing Add:* Denver Res Ctr Marathon Oil Co Littleton CO 80120

PRESLOCK, JAMES PETER, b Newark, NJ. ENDOCRINOLOGY, REPRODUCTIVE PHYSIOLOGY. *Educ:* Univ Scranton, BS, 64; Univ Md, College Park, PhD(endocrinol), 69. *Prof Exp:* Fel steroid biochem, Worcester Found, 69-70; asst prof, 70-80, ASSOC PROF OBSTET, GYNEC & REPRODUCTIVE SCI, UNIV TEX MED SCH, HOUSTON, 80- *Mem:* AAAS; Soc Study Reproduction; Endocrine Soc. *Res:* Neuroendocrine regulation of luteinizing hormone, follicle-stimulating hormone and prolactin secretion in male rats; metabolic pathways in the biosynthesis of testicular steroids in hamsters. *Mailing Add:* Dept Obstet Gynec & Reproductive Sci Univ Tex Med Sch 6431 Fannin Suite 3204 Houston TX 77030

PRESNALL, DEAN C, b Cedar City, Utah, Nov 6, 35; m 64; c 2. GEOCHEMISTRY, PETROLOGY. *Educ:* Pa State Univ, BS, 57, PhD(geochem), 63; Calif Inst Technol, MS, 59. *Prof Exp:* Fel, Carnegie Inst Geophys Lab, 63-67; asst prof geosci, Southwest Ctr Advan Studies, 67-69; from asst prof to assoc prof, 69-78, PROF GEOSCI, UNIV TEX, DALLAS, 78- *Mem:* Fel Geol Soc Am; Am Geophys Union; fel Mineral Soc Am; Geochem Soc. *Res:* Experimental and theoretical igneous petrology; phase equilibrium studies of silicate systems at high temperatures and pressures; petrogenesis of granites and oceanic volcanic rocks; chemical evolution of the earth's mantle and crust. *Mailing Add:* Prog in Geosci PO Box 830688 Richardson TX 75083-0688

PRESNELL, RONALD I, b Boulder, Colo, Feb 12, 33; m 57; c 2. ELECTRONICS ENGINEERING. *Educ:* Stanford Univ, BS, 54, MS, 55, Engr, 59. *Prof Exp:* Jr res engr, 54-56, res engr, 56-64, SR RES ENGR ELECTRONICS, SRI INT, 64- *Mem:* Inst Elec & Electronics Engrs. *Res:* Radar systems; upper atmospheric physics; nuclear detonation effects on radar systems. *Mailing Add:* ESI Inc 495 Java Dr Sunnyvale CA 94088

PRESS, FRANK, b Brooklyn, NY, Dec 4, 24; m 46; c 2. GEOPHYSICS. *Educ:* City Col New York, BS, 44; Columbia Univ, MA, 46, PhD(geophys), 49. *Hon Degrees:* LLD, Univ Notre Dame, 73; DSc, City Univ New York, 74. *Prof Exp:* From asst to assoc prof physics, Columbia Univ, 45-55; prof geophys, Calif Inst Technol, 55-65, dir seismol lab, 57-65; dir, Off Sci & Technol Policy, Exec Off Pres, Washington, DC, 77-81; inst prof, Mass Inst Technol, 81; PRES, NAT ACAD SCI, 81- *Concurrent Pos:* Consult, US Navy, 56-57, US Geol Surv, 57-59, Dept State & Dept Defense, 58-62, President's Asst for Sci & Technol, 59-60 & 64-, NASA, 60-62 & 65-, AID, 62-63 & Arms Control & Disarmament Agency, 62-; mem, Gov Adv Coun Atomic Activities, Calif, 59; mem, President's Sci Adv Comt, 61-64; mem bd adv, Nat Ctr Earthquake Res, US Geol Surv, 66- Mem glaciol & seismol panel, Int Geophys Year, 55-59, comt polar res, 57-, interdisciplinary res panel, 58-, seismol working group, Upper Mantle Proj, 64-, int geophys comt, Int Coun Sci Unions, 59-, panel solid earth probs, Geophys Res Bd, Nat Acad Sci, 61 & mem coun, 75-78; chmn earthquake prediction panel, Off Sci & Technol, 65-; mem planetology subcomt, NASA, 66- & Nat Sci Bd, NSF, 71- Mem, UNESCO Tech Assistance Mission, 53; US deleg, Nuclear Test Ban Conf, Geneva, 59-61, Moscow, 63 & UN Conf Sci & Technol Underdeveloped Nations, 63; prof geophys & head dept earth & planetary sci, Mass Inst Technol, 65-; mem, Comt Anticipated Advan Sci & Tech, 74-76; mem, lunar and planetary missions bd, NASA; partic, bilateral sci agreement, Peoples Repub China and USSR; mem, US deleg Nuclear Test Ban Negotiations, Geneva, Moscow; bd trustees, Sloan Found & Rockefeller Univ; co-ed, Physics & Chem Earth. *Honors & Awards:* Arthur L Day Medal, Geol Soc Am; Bowie Medal, Am Geophys Union; Ewing Medal, Soc Explor Geophysicists, 82; Gold Medal, Royal Astron Soc, 72. *Mem:* Nat Acad Sci; Am Acad Arts & Sci; Geol Soc Am; Seismol Soc Am (secy, 57-58, vpres, 59-61, pres, 62); Am Geophys Union (pres, 74-76); Am Philos Soc; Am French Acad Sci; Am Asn Univ Professors. *Res:* Planetary interiors; crustal and mantle structure; regional and submarine geophysics; seismology, including earthquake mechanism and elastic wave propagation. *Mailing Add:* Nat Acad Sci 2101 Constitution Ave NW Washington DC 20418

PRESS, JEFFERY BRUCE, b Rochester, NY, May 24, 47; m 76; c 2. MEDICINAL CHEMISTRY. *Educ:* Bucknell Univ, BS, 69; Ohio State Univ, PhD(org chem), 73. *Prof Exp:* Fel org chem, Harvard Univ, 73-74, NIH fel, 74-75; res chemist, Lederle Lab, Pearl River, New York, 75-77, sr res chemist, 77-81, group leader, 81-83; res mgr, med chem, Ortho Pharm Corp, Raritan, NJ, 83-89; ASST DIR, MED CHEM, R W JOHNSON PHARMACEUT RES INST, SPRING HOUSE, PA. *Concurrent Pos:* NIH postdoctoral fel, 74-75; managing ed, Chemtracts Org Chem. *Mem:* Am Chem Soc; Sigma Xi; Int Soc Heterocyclic Chem; Int Union Pure & Appl Chem. *Res:* Heterocyclic chemistry; natural products; reactive intermediates; chemistry of strained ring systems; cardiovascular drugs, thiophene compounds; CNS drugs. *Mailing Add:* R W Johnson Pharmaceut Res Inst Spring House PA 19477

PRESS, LINDA SEGHERS, b Jacksonville, Fla, Dec 2, 50; m 76; c 2. ORGANIC CHEMISTRY. *Educ:* Ohio State Univ, BSc, 72, PhD(org chem), 77. *Prof Exp:* Res assoc org chem, Int Paper Co, 77-81, mgr biochem res, 81-84; dir biotechnol, FMC Corp, 84-86; CONSULT BIOTECHNOL, 86- *Res:* Biochemical processes in plants; plant tissue culture; technology evaluation; market analysis; contracts and licensing; proposal preparation. *Mailing Add:* 716 Buckley Circle Penllyn PA 19422

PRESS, NEWTOL, b New York, NY, Nov 27, 30; m 51; c 3. CELL BIOLOGY. *Educ:* NY Univ, BA, 51; Univ Iowa, MS, 55, PhD(zool), 56. *Prof Exp:* Asst zool, Univ Iowa, 52-56; from instr to assoc prof, 56-60, chmn dept, 64-66, PROF ZOOL, UNIV WIS-MILWAUKEE, 77- *Res:* Histology; photoreception; embryology; history of science; theoretical biology; aging; bioweapons. *Mailing Add:* Dept Biol Sci Univ of Wis PO Box 413 Milwaukee WI 53201

PRESS, S JAMES, b New York, NY, Feb 4, 31; m 51; c 3. MATHEMATICAL STATISTICS. *Educ:* NY Univ, BA, 50; Univ Southern Calif, MS, 55; Stanford Univ, PhD(statist), 64. *Prof Exp:* Physicist, Brookhaven Nat Lab, 49-50; res analyst, Northrop Aircraft Corp, 51-54; res engr, Douglas Aircraft Co, Inc, Long Beach, Calif, 54-57, design specialist, Santa Monica, 58-61; lectr statist, Univ Calif, Los Angeles, 64-66, assoc prof, Grad Sch Bus, Univ Chicago, 66-74; prof statist, Univ BC, 74-77; PROF & CHMN DEPT STATIST, UNIV CALIF, RIVERSIDE, 77- *Concurrent Pos:* Res statistician, Rand Corp, 64-66, consult, 66-; consult, Real Estate Res Corp, Ill, 66, probs mgr educ in India, AID, 67, John Wiley & Sons & McGraw-Hill Bk Co; vis prof, Univ Col, Univ London, 70-71, London Sch Econ & Polit Sci, 70-71 & Yale Univ, 72-73; assoc ed, J Am Statist Asn & J Economet; prin investr, Biometric Proj, Citrus Res Ctr & Agr Exp Sta. *Mem:* Fel Inst Math Statist; fel Am Statist Asn; Economet Soc; Inst Mgt Sci; fel Royal Statist Soc. *Res:* Multivariate statistical analysis; Bayesian statistics; econometrics; probability distribution theory; operations research. *Mailing Add:* Dept Statist Univ Calif Riverside CA 92521

PRESS, WILLIAM HENRY, b New York, NY, May 23, 48; div; c 1. ASTROPHYSICS, COMPUTER SCIENCE. *Educ:* Harvard Univ, AB, 69; Calif Inst Technol, MS, 71, PhD(physics), 72. *Prof Exp:* Richard Chace Tolman res fel theoret physics, Calif Inst Technol, 72-73, asst prof, 73-74; asst prof physics, Princeton Univ, 74-76; chmn dept astron, 82-85, PROF ASTRON & PHYSICS, HARVARD UNIV, 76- *Concurrent Pos:* Consult, Lawrence Livermore Lab, 73-, Mitre Corp, 77- & Los Alamos Nat Lab, 84-; Alfred P Sloan Found res fel, 74-78; mem adv comt on physics, NSF, 78-81, adv bd, Inst Theoret Physics, 84-, chmn, 85-86; mem ad hoc comt innovation, NASA, 79 & comt on role of theory in space sci, 80; mem, Astron Surv Comt Panel on Theoret Astrophys, Nat Acad Sci/Nat Res Coun, 79-80, Advocacy Panel on Physics of the Sun, 80-81 & Astron & Astrophys Surv Comt, 89-91; chmn, Subcomt Comput Facil for Theoret Physics, NSF, 80-81; assoc ed, Ann Physics, 84-90; mem, Defense Sci Bd, 85-90; mem prog comt, Alfred P Sloan Found, 85-91 & Princeton Univ Physics Adv Coun, 86-90; trustree Inst Defense Analysis, 88-; mem, Sci Adv Comt, Packard Fel Sci & Engineering, 88-; mem, Sci Policy Comt, Superconducting Super Collider, 89- *Honors &*

Awards: Warner Prize, Am Astron Soc, 81. *Mem:* Am Phys Soc; Am Astron Soc; Int Soc on Gen Relativity & Gravitation; Int Astron Union; Asn Comput Mach. *Res:* Relativistic astrophysics; theoretical astrophysics; cosmology; galaxy formation; general relativity; numerical methods. *Mailing Add:* Harvard-Smithsonian Ctr Astrophys 60 Garden St Cambridge MA 02138

PRESSBURG, BERNARD S(AMUEL), b Alexandria, La, Jan 23, 18; m 47; c 2. CHEMICAL ENGINEERING. *Educ:* La State Univ, BS, 37, MS, 39, PhD(chem eng), 41. *Prof Exp:* Asst prof, La State Univ, Baton Rouge 41-42, from asst prof to assoc prof, 46-55, actg dean, 64-65, prof chem eng, 55-83, assoc dean, 65-80; RETIRED. *Concurrent Pos:* Consult, 46- *Mem:* Am Chem Soc; Am Soc Eng Educ; Am Inst Chem Engrs. *Res:* Unit operations; fluid flow; heat and mass transfer. *Mailing Add:* 5937 Clematis Dr Baton Rouge LA 70808-8802

PRESSER, BRUCE DOUGLAS, b Bellmawr, NJ; m 59. ENTOMOLOGY. *Educ:* Temple Univ, BA, 49, MA, 51; Pa State Univ, PhD(entom), 55. *Prof Exp:* Prof biol, Belmont Col, 57-61; ASSOC PROF BIOL, SUSQUEHANNA UNIV, 61- *Mem:* Entom Soc Am; Am Soc Zoologists; Sigma Xi. *Res:* Insect embryology and morphology. *Mailing Add:* Five Linda Lane Selinsgrove PA 17870

PRESSER, CARY, b Brooklyn, NY, June 20, 52; m 77; c 2. COMBUSTION SCIENCE & TECHNOLOGY, ENERGY UTILIZATION & CONSERVATION. *Educ:* Polytech Inst Brooklyn, BSc, 74, MSc, 76; Technion, Israel Inst Technol, DSc, 80. *Prof Exp:* Teaching fel, Polytech Inst Brooklyn, 74-75; teaching instr, Technion, Israel Inst Technol, 75-80, sr res asst, 80; RES ENGR, NAT INST STANDARDS & TECHNOL, 80- *Concurrent Pos:* Grad fel, Lady Davis Fel Trust, 75-76; mem, K-6 Comn Heat Transfer Energy Systs, Am Soc Mech Engrs, 86-, Propellants & Combustion Tech Comn, Am Inst Aeronaut & Astronaut, 87-89, E29-04 Comn Liquid Particle Measurements, Am Soc Testing & Mat, 91. *Mem:* Assoc fel Am Inst Aeronaut & Astronaut; Am Soc Mech Engrs; Am Inst Chem Engrs; Combustion Inst; NY Acad Sci; Sigma Xi. *Res:* Spray combustion; swirl flows; numerical modelling of internal reaction flows; flame stabilization and gas emission monitoring of combustors; laser diagnostics of high temperature and reacting flow fields. *Mailing Add:* Nat Inst Standards & Technol Bldg 221 Rm B312 Gaithersburg MD 20899

PRESSER, LEON, b Matanzas, Cuba, Jan 23, 40; US citizen; m 69; c 2. COMPUTER SCIENCE. *Educ:* Univ Ill, Champaign, BS, 61; Univ Southern Calif, MS, 64; Univ Calif, Los Angeles, PhD(eng, comput sci), 68. *Prof Exp:* Design engr, Comput Measurements Co, 61-62; mem tech staff, Comput Div, Bendix Corp, 62-63; systs engr, Gen Instrument Corp, 63-64; asst comput, Univ Calif, Los Angeles, 64-68, asst prof comput sci, 68-69; asst prof elec eng & comput sci, Univ Calif, Santa Barbara, 69-77; PRES, SOFTOOL CORP, 77- *Concurrent Pos:* Grant, Univ Calif, Santa Barbara, 69-76, NSF grant, 72-74. *Mem:* Asn Comput Mach; Inst Elec & Electronics Engrs. *Res:* Software development and management. *Mailing Add:* Softool Corp 340 S Kellogg Ave Goleta CA 93017

PRESSEY, RUSSELL, b Man, Can, Sept 5, 35; US citizen; m 62; c 4. BIOCHEMISTRY, PLANT PHYSIOLOGY. *Educ:* Univ Man, BSc, 58, MSc, 59; Iowa State Univ, PhD(biochem), 62. *Prof Exp:* Res chemist, Gen Mills, Inc, Minn, 62-64 & USDA, Minn, 64-69; RES CHEMIST, RICHARD B RUSSELL AGR RES CTR, 69- *Mem:* Am Soc Hort Sci; Am Soc Plant Physiologists. *Res:* Plant biochemistry; cell wall polysaccharides and the enzymes involved in their changes; regulation of plant metabolism by natural enzyme inhibitors; biochemistry of fruit ripening. *Mailing Add:* 410 Greencrest Dr Athens GA 30605

PRESSMAN, ADA IRENE, b Shelby Co, Ohio, March 3, 27; m 69. TECHNICAL MANAGEMENT. *Educ:* Ohio State Univ, BME, 50; Golden Gate Univ, MBA, 74. *Prof Exp:* Engr, Bailey Meter Co, Cleveland, Ohio, 50-55; engr, Bechtel Power Corp, Los Angeles, 55-74, chief engr, 74-79, eng mgr, 79-87; RETIRED. *Honors & Awards:* E G Bailey Award, Instr Soc Am, 85. *Mem:* Am Nuclear Soc; Instr Soc Am (vpres 73-78); Soc Women Engrs (pres 79-80). *Mailing Add:* 1301 S Alantic Blvd Monterey Park CA 91754

PRESSMAN, BERTON CHARLES, b Brooklyn, NY, Sept 25, 26; m 56; c 3. BIOCHEMISTRY, BIOPHYSICS. *Educ:* Univ Wis, BS, 48, MS, 50, PhD(biochem), 53. *Prof Exp:* USPHS fel, Cath Univ Louvain, 53 & Wenner-Gren Inst, Sweden, 54; proj assoc, Enzyme Inst, Univ Wis, 55-58; biochemist, Radioisotope Serv, Vet Admin Hosp, Kansas City, Mo, 59-61; from asst prof to assoc prof phys biochem, Univ Pa, 61-69; PROF PHARMACOL, PHYSIOL & BIOPHYS, MED SCH, UNIV MIAMI, 70- *Concurrent Pos:* Asst prof, Univ Kans, 59-61; USPHS career develop award, 64-71. *Mem:* Am Soc Biol Chemists; Biophys Soc; Am Soc Pharmacol & Exp Therapeut. *Res:* Molecular basis of action and biological effects of transport-inducing antibiotics, known as ionophores; mechanism of ion transport in mitochondria and other membrane systems; cardiovascular effects of ionophores. *Mailing Add:* Dept Pharmacol Univ Miami Med Sch Box 016189 Miami FL 33101

PRESSMAN, IRWIN SAMUEL, b Port Arthur, Ont, Aug 17, 39; m 66; c 3. MATHEMATICS, TOPOLOGY. *Educ:* Univ Man, BSc, 60; Cornell Univ, PhD(math), 65; Univ Toronto, MBA, 84. *Prof Exp:* Res asst math, Swiss Fed Inst Technol, 65-67; asst prof, Ohio State Univ, 67-71; ASSOC PROF MATH, CARLETON UNIV, 71- *Concurrent Pos:* Consult, Dept Finance, Health & Welfare, Govt Can; Woodrow Wilson fel; Nat Res Coun fel; Sr Indust Res Fel, Bell Northern Res. *Mem:* Can Oper Res Soc; Am Math Soc; Inst Elec & Electronics Engrs. *Res:* Operations research; grain transportation; computer aided learning systems; linear programming; telecommunication; office automation. *Mailing Add:* Dept Math & Statist Carleton Univ Colonel By Dr Ottawa ON K1S 5B6 Can

PRESSMAN, NORMAN JULES, b New York, NY, Sept 30, 48; m 71. BIOMEDICAL ENGINEERING. *Educ:* Columbia Univ, BS, 70; Univ Pa, MS, 72, PhD(biomed eng), 76. *Prof Exp:* Biomed engr, Lawrence Livermore Lab, Univ Calif, 73-76; HEAD, QUANT CYTOPATH LABS, JOHNS HOPKINS MED INSTS & ASST PROF PATH, MED SCH, UNIV, 76- *Concurrent Pos:* Consult, NIH, 77-78 & US Food & Drug Admin, 78-; chmn, Int Coun on Automated & Quant Cytol, Int Acad Cytol, 77- *Mem:* Inst Elec & Electronics Engrs; Biomed Eng Soc; Int Acad Cytol; Am Soc Cytol; Asn for Advan Med Instrumentation; Sigma Xi. *Res:* Quantitative and automated cytopathology; medical image processing and analysis; flow system analysis and sorting of biological cells; clinical laboratory computerized information systems. *Mailing Add:* Dept Cent Res & Develop E I du Pont de Nemours & Co Exp Sta Bldg 328 Wilmington DE 19898

PRESSMAN, RALPH, bacteriology, for more information see previous edition

PREST, VICTOR KENT, b Edmonton, Alta, Apr 2, 13; m 42; c 2. EARTH SCIENCE. *Educ:* Univ Man, BSc, 35, MSc, 36; Univ Toronto, PhD(geol), 41. *Prof Exp:* Asst geol, Univ Toronto, 36-39, lectr, 39-40; geologist, Res Lab, Int Nickel Co, 41-42 & Ont Dept Mines, 45-50; geologist, Geol Surv Can, 50-78; RETIRED. *Concurrent Pos:* Mus asst, Royal Ont Mus, 37-39; geologist, Ont Dept Mines, 37-41; lieutenant, Royal Can Naval Volunteer Res, 42-45; consult, 78. *Honors & Awards:* W A Johnston Medal, 87; Can Quaternary Asn; Kirk Bryan Award, Geol Soc Am, 90. *Mem:* Royal Soc Can; Geol Asn Can; Int Quaternary Asn. *Res:* Quaternary geology. *Mailing Add:* 405-1465 Baseline Rd Ottawa ON K2C 3L9 Can

PREST, WILLIAM MARCHANT, JR, b Boston, Mass, Nov 4, 41; m 69; c 2. POLYMER PHYSICS. *Educ:* Union Col, BS, 63; Univ Pa, MS, 65; Univ Mass, MS, 69, PhD(polymer sci & eng), 72. *Prof Exp:* Physicist, res technol prog, Polymer Physics Br, Gen Elec Res & Develop Ctr, Schenectady, 66-68; assoc scientist polymer sci, 71-72, scientist, 72-77, sr scientist, 77-81, prin scientist, mat sci lab, 81-83, prin scientist & mgr physics mat, 83-85, PRIN SCIENTIST & MGR, POLYMER SCI GROUP, WEBSTER RES CTR, XEROX CORP, 85- *Concurrent Pos:* Herman Mark fel, Plastics Inst Am, 70. *Mem:* Fel Am Phys Soc; Soc Rheol; Am Chem Soc; NAm Thermal Analysis Soc; Soc Plastics Engrs. *Res:* Thermodynamic and spectroscopic properties of single and multi component glasses; compatibility of polymer blends; orientation processes in amorphous polymers; polymorphism of semi-crystalline polymers; rheology of polymer melts; Fourier transform infrared spectroscopy; viscoelastic properties of polymer melts and rheology of polymer blends; optical properties, internal stresses and structure of solvent cast polymer films. *Mailing Add:* 616 Lake Rd Webster NY 14580-1520

PRESTAYKO, ARCHIE WILLIAM, b Dauphin, Man, July 27, 41; m 69; c 2. BIOCHEMISTRY. *Educ:* Colo Col, BA, 62; Univ Tenn, Knoxville, PhD(radiation biol), 66. *Prof Exp:* Am Cancer Soc fel, Dept Pharmacol, Baylor Col Med, Houston, Tex, 67-68, res asst prof pharmacol, 69-71, asst prof, 72-75, adj assoc prof, 76-88; assoc dir/dir res & develop, Cancer Diagnostics, Bristol Labs, Bristol Myers Corp, 76-81, dir, Bristol-Baylor Lab Molecular Pharmacol, Baylor Col Med, 80-81; vpres, Worldwide Res & Develop, Smith Kline & French Labs, 81-85, vpres corp res & develop, Smith Kline Beckman Corp, 86-88; vpres res & develop, Lyphomed/Fujisawa USA, Rosemont, Ill, 89-90; exec vpres, 90, DIR, MEDCO RES INC, LOS ANGELES, CALIF, 90-, PRES & CHIEF EXEC OFFICER, 91- *Mem:* Am Asn Cancer Res; AAAS. *Res:* Molecular and clinical pharmacology of anticancer drugs. *Mailing Add:* Medco Res Inc 8733 Beverly Blvd Suite 404 Los Angeles CA 90048

PRESTEGARD, JAMES HAROLD, b Minneapolis, Minn, Jan 27, 44; m 66; c 2. NUCLEAR MAGNETIC RESONANCE, MOLECULAR MODELING. *Educ:* Univ Minn, BS, 66; Calif Inst Technol, PhD(chem), 71. *Prof Exp:* From asst prof to assoc prof, 70-84, PROF CHEM, YALE UNIV, 84- *Concurrent Pos:* Vis assoc prof, Univ Utah, 76. *Mem:* Am Chem Soc; Biophys Soc. *Res:* Development of nuclear magnetic resonance methods for study of biological systems; major systems include biological membranes; lipids, carbohydrates and proteins associated with membranes. *Mailing Add:* Dept Chem Yale Univ New Haven CT 06511

PRESTEMON, DEAN R, b Waukon, Iowa, Oct 4, 34; m 56; c 6. FOREST PRODUCTS. *Educ:* Iowa State Univ, BS, 56; Univ Minn, St Paul, MS, 57; Univ Calif, Berkeley, PhD(forestry), 66. *Prof Exp:* Res technician wood technol, Douglas Fir Plywood Asn, Wash, 58; tech rep, Nat Lumber Mfrs Asn, Calif, 58-62; asst specialist, Forest Prod Lab, Univ Calif, 62-65; from asst to assoc prof, 65-76, PROF WOOD TECHNOL, IOWA STATE UNIV, 76- *Mem:* Forest Prod Res Soc; Soc Am Foresters. *Res:* Glues and gluing; log grade-lumber yield; wood quality-growth relations; use and marketing of forest products in residential construction; mechanical processing; wood mechanics; wood quality evaluation. *Mailing Add:* Dept of Forestry Iowa State Univ Ames IA 50011

PRESTON, ERIC MILES, b Cortland, NY. ECOLOGY. *Educ:* State Univ NY Binghamton, BA, 66; Univ Hawaii, MS, 68, PhD(zool), 71. *Prof Exp:* NIH trainee, Univ Minn, 71-72; asst prof marine ecol, Univ PR, Rio Piedras, 72-75; proj leader, Mont Coal-Fired Power Plant Proj, 75-80, proj leader, Nat Crop Loss Assessment Network, 81-84, dep nat prog mgr, acid deposition effects forests, 84-85, MGR, WETLANDS RES PROG, US ENVIRON PROTECTION AGENCY, 86- *Mem:* Ecol Soc Am; Sigma Xi; Soc Wetland Scientists. *Res:* Population and community dynamics; ecological and agronomic effects of air pollution; wetlands. *Mailing Add:* Corvallis Environ Res Lab US Environ Protection Agency Corvallis OR 97333

PRESTON, FLOYD W, b Albuquerque, NMex, Feb 11, 23; m 45; c 4. PETROLEUM ENGINEERING. *Educ:* Calif Inst Technol, BS, 44; Univ Mich, MS, 48; Pa State Univ, PhD(petrol eng), 57. *Prof Exp:* Res chemist, Nat Defense Res Comt Proj, Calif Inst Technol, 44-45; res engr, Eng Exp Sta, Univ Mich, 45-46; res engr, Calif Res Corp, Standard Oil Co Calif, 48-50; res assoc, Pa State Univ, 50-55; from asst prof to assoc prof, 57-65, asst chmn, Dept Chem & Petrol Eng, 66-76, chmn, Dept Chem & Petrol Eng, 74-79 PROF PETROL ENG, UNIV KANS, 65-,. *Concurrent Pos:* Consult, Ministry Mines & Hydrol, Venezuela, 59-61. *Mem:* AAAS; Am Chem Soc; Am Inst Chem Engrs; Soc Petrol Engrs. *Res:* Petroleum production; digital computers. *Mailing Add:* 832 Sunset Dr Lawrence KS 66044

PRESTON, FREDERICK WILLARD, b Chicago, Ill, June 27, 12; m 42, 61; c 3. SURGERY. *Educ:* Yale Univ, BA, 35; Northwestern Univ, MD, 40, MS, 42; Univ Minn, MS, 47. *Prof Exp:* Fel surg, Mayo Clin, 42-48; clin asst, Col Med, Univ Ill, 48-49; from instr to prof surg, Med Sch, Northwestern Univ, Chicago, 60-76; chmn, Dept Surg, Santa Gen Hosp, 75-78. *Concurrent Pos:* Assoc surgeon, Cook County Hosp, Chicago, Ill, 48-49; attend surgeon, Vet Admin Hosp, Hines, 50-53; assoc attend surgeon, Chicago Wesley Mem Hosp, 50-75; chief surg serv, Vet Admin Res Hosp, Chicago, 53-68; sr attend surgeon, Henrotin Hosp, Chicago & Skokie Valley Community Hosp, Skokie, 64-75. *Mem:* Am Asn Cancer Res; Am Surg Asn; fel Am Col Surgeons; Int Soc Surg; Soc Surg Alimentary Tract; AMA. *Res:* Cancer; antibiotics; cirrhosis of liver; trauma; decompression of the portal veneons system; effects of hormones on breast cancer; chemotherapy for visceral cancer. *Mailing Add:* 755 Via Airosa Santa Barbara CA 93110

PRESTON, GEORGE W, III, b Los Angeles, Calif, Aug 25, 30; m 52; c 4. ASTRONOMY. *Educ:* Yale Univ, BS, 52; Univ Calif, Berkeley, PhD(astron), 59. *Prof Exp:* Carnegie fel, Mt Wilson Observ, Carnegie Inst Wash, 59-60; asst astronr, Lick Observ, Univ Calif, Santa Cruz, 60-64, assoc astronr, 64-68; STAFF MEM, MT WILSON & LAS CAMPANAS OBSERV, CARNEGIE INST WASH, 68- *Concurrent Pos:* Dir, Mt Wilson & Las Campanas Observ, 80-86. *Honors & Awards:* Robert J Trumpler Lectr, Astron Soc Pac, 63; Helen B Warner Prize, Am Astron Soc, 65. *Mem:* Nat Acad Sci; Am Astron Soc; Int Astron Union. *Res:* Stellar spectroscopy; magnetic stars; galactic structure. *Mailing Add:* Carnegie Inst Wash Mt Wilson & Las Campanas Observ 813 Santa Barbara St Pasadena CA 91101

PRESTON, GERALD COWLES, mathematics, for more information see previous edition

PRESTON, GLENN WETHERBY, b Welch, WVa, Mar 30, 22; m 44; c 3. COMPUTER SCIENCES, ELECTRONICS. *Educ:* Trinity Col, BS, 47; Yale Univ, MS, 48. *Prof Exp:* Mathematician systs anal, Goodyear Aircraft Corp, 50-51; math consult electronic physics, Philco Corp, 51-56; vpres eng, Gen Atronics Corp, 56-62; pres, Preston Assocs Inc, 62-66; MEM TECH STAFF MICROELECTRONICS, INST DEFENSE ANALYSIS, 66- *Concurrent Pos:* Noyes-Clark Fel, Yale Univ, 48; consult, Rand Corp, 60-62, Nat Acad Sci, 67-68 & Air Traffic Control Adv Comt, 69. *Mem:* Inst Elec & Electronics Engrs. *Res:* Integrated circuit architecture; algorithmic analysis; logic synthesis. *Mailing Add:* PO Box 1298 Rancho Santa Fe CA 92067-1298

PRESTON, JACK, b Birmingham, Ala, Aug 7, 31; m 60; c 2. ORGANIC CHEMISTRY, POLYMER CHEMISTRY. *Educ:* Howard Col, BS, 52; Univ Ala, MS, 54, PhD, 57. *Prof Exp:* Res chemist, Chemstrand Res Ctr, Inc, Durham, 57-63, sr res chemist, 63-65, assoc scientist, 65-66, scientist, 66-70, sci fel, 70-74, sr sci fel, 74-81; sr Monsanto fel, Monsanto Textiles Co, Pensacola, Fla, 81-85; SR RES SCIENTIST, RES TRIANGLE INST, RESEARCH TRIANGLE PARK, 87- *Concurrent Pos:* Res assoc, Duke Univ, 77-87, adj prof, 87-; adj prof, Univ NC, Chapel Hill, 88-, NC State Univ, 89- *Mem:* AAAS; Am Chem Soc; NY Acad Sci; fel Am Inst Chemists; Fiber Soc. *Res:* Preparation and characterization of linear condensation polymers, including polyaromatic heterocycles, amides, amide-hydrazides and amide-heterocycles; thermally stable fibers and films; high strength and high modulus fibers from organic polymers; evaluation of fibers, especially under extreme environments; liquid crystalline polymers and molecular composites. *Mailing Add:* 4914 Rembert Dr Raleigh NC 27612-6238

PRESTON, JAMES BENSON, b Nelsonville, Ohio, Feb 4, 26; m 47; c 3. PHYSIOLOGY. *Educ:* Univ Cincinnati, MD, 52. *Prof Exp:* Intern, Res & Educ Hosps & instr pharmacol, Col Med, Univ Ill, 52-54; from instr to assoc prof, 54-60, PROF PHYSIOL & CHMN DEPT, STATE UNIV NY HEALTH SCI CTR, 60- *Concurrent Pos:* USPHS sr res fel, 58-60; mem physiol training comt, Nat Inst Gen Med Sci, 71-73; mem study sect, NIH, 63-67, phys test comt, Nat Bd Med Exam, 72-76, multiple sclerosis subcomt, Nat Adv Comt, 73-74, exec bd, Am Asn Med Col, 77-80; chmn, phys & molecular biol eval comt, 68, mem comt, Am Phys Soc, 68-71, chairperson, WCBR, 87-88; pres, Army Control Prog Directives, 73-74; med lab sci rev comt, Nat Inst Gen Med Sci, 76-77; dir, State Univ NY Res Found, 88. *Mem:* Soc Gen Physiol; Am Physiol Soc; Asn Chmn Depts Physiol (pres, 73-74); Soc Neurosci. *Res:* Physiology of the nervous system; primate motor control systems with emphasis on cerebral cortex influences on spinal motor systems. *Mailing Add:* Dept Physiol State Univ NY Health Sci Ctr 750 E Adams St Syracuse NY 13210

PRESTON, JAMES FAULKNER, III, b Boston, Mass, July 4, 39; m 66; c 2. BIOCHEMISTRY, MICROBIOLOGY. *Educ:* Colgate Univ, BA, 61; Ohio State Univ, MS, 64; Univ Minn, PhD(biochem), 67. *Prof Exp:* NIH fel microbiol, Yale Univ, 67, Am Cancer Soc fel, 67-69; from asst prof to assoc prof, 69-81, PROF MICROBIOL, UNIV FLA, 81- *Mem:* AAAS; Am Chem Soc; Am Soc Microbiol. *Res:* Biosynthesis and function of peptides of fungi; antibody and lectin medrated drug targeting; somatic cell hybridization and plant development; enzymology of alginate and lectin degradation. *Mailing Add:* Dept Microbiol Univ Fla 1053 McCarty Hall Gainesville FL 32601

PRESTON, KEITH FONCELL, b Chase Terrace, Eng, Apr 3, 38; Can citizen; m 62; c 3. CHEMICAL PHYSICS, PHYSICAL CHEMISTRY. *Educ:* Univ Cambridge, BA, 59, MA, 62, PhD(chem), 63. *Prof Exp:* Fel photochem, 63-65, SR RES OFFICER KINETICS, NAT RES COUN CAN, 65- *Mem:* Chem Inst Can. *Res:* Electron spin resonance of inorganic free radicals; kinetics of gas-phase reactions; photochemistry; combustion. *Mailing Add:* Steacie Inst Molecular Sci Montreal Rd Ottawa ON K1A 0R9 Can

PRESTON, KENDALL, JR, b Boston, Mass, Oct 22, 27; m 52; c 1. COMPUTER ENGINEERING, HEALTH SCIENCE. *Educ:* Harvard Col, AB, 50; Harvard Univ, SM, 52. *Prof Exp:* Engr hydraul, United Shoe Mach Corp, 51; mem tech staff electronics, Bell Tel Labs, 52-60; mem sci staff electro-optics, Perkin-Elmer Corp, 60-74; PROF ELEC ENG, CARNEGIE-MELLON UNIV, 74-; PROF RADIATION HEALTH, GRAD SCH PUB HEALTH, UNIV PITTSBURGH, 77- *Concurrent Pos:* Ed adv, Biocharacterist, 70-73; consult, Perkin-Elmer Corp, 74- & Bausch & Lomb Ind, 76-77; assoc ed, Pattern Recognition, 75-; mem, Tech Audit Bd, 76-; gen mgr, Kensal Consult, 81-; pres, Path Imaging Corp, 86- *Honors & Awards:* Henry Warden Cary Prize; Centennial Plaque, Inst Elec Electronics Engs, 84. *Mem:* Inst Elec & Electronics Engrs; NY Acad Sci; Biomed Eng Soc. *Res:* Computer science and engineering with application to coherent radiation systems, health science and pattern recognition. *Mailing Add:* Dept Elec Eng Carnegie-Mellon Univ 5000 Forbes Ave Pittsburgh PA 15213

PRESTON, MELVIN ALEXANDER, b Toronto, Ont, May 28, 21; m 47, 66; c 2. THEORETICAL NUCLEAR PHYSICS, HADRON STRUCTURE. *Educ:* Univ Toronto, BA, 42, MA, 46; Univ Birmingham, PhD, 49. *Hon Degrees:* DSc, McMaster Univ, 83. *Prof Exp:* Asst lectr math physics, Univ Birmingham, 47-49; asst prof physics, Univ Toronto, 49-53; assoc prof physics, McMaster Univ, 53-59, dean grad studies, 65-71, prof physics, 59-77, prof appl math, 67-77, chmn dept appl math, 75-77; acad vpres, 77-82, prof physics, Univ Sask, 77-86; EMER PROF PHYSICS, MCMASTER UNIV, 86- *Concurrent Pos:* Nuffield fel, 57; Nat Res Coun Can sr fel, 63-64; chmn, Ont Coun Grad Studies, 70-71; pres, Can Asn Grad Schs, 71; exec vchmn adv comt acad planning, Coun Ont Univs, 71-75. *Honors & Awards:* Centennial Medal of Can, 67. *Mem:* Fel AAAS; fel Am Phys Soc; Can Asn Physicists (treas, 52-58); fel Royal Soc Can; Brit Inst Physics. *Res:* Theoretical nuclear physics, especially nuclear structure, nuclear forces and hadron structure. *Mailing Add:* Dept Physics McMaster Univ Hamilton ON L8S 4M1 Can

PRESTON, RICHARD SWAIN, b Natick, Mass, Feb 4, 25; m 54; c 2. MOSSBAUER SPECTROSCOPY, HIGH ENERGY EXPERIMENTATION. *Educ:* Wesleyan Univ, BA, 49, MA, 50; Yale Univ, PhD(physics), 54. *Prof Exp:* Assoc dir, Geochronometric Lab, Yale Univ, 54-55; physicist, Argonne Nat Lab, Ill, 55-72; PROF PHYSICS, NORTHERN ILL UNIV, 70-, CHAIR, PHYSICS DEPT, 86- *Concurrent Pos:* Vis scientist, UK Atomic Energy Res Estab, Harwell, Eng, 65-66; vis prof, Univ Saarlandes, WGer, 77-78 & 81; Fulbright scholar, Univ Fed Minas Gerais, Brazil, 83. *Mem:* Am Phys Soc; Sigma Xi. *Res:* Mossbauer studies of dilute alloys and amorphous metals; decay of B mesons and baryons. *Mailing Add:* Physics Dept Northern Ill Univ De Kalb IL 60115

PRESTON, ROBERT ARTHUR, b New York, NY, June 29, 44; m 70; c 1. RADIO ASTRONOMY. *Educ:* Cornell Univ, BS, 66, MS, 67; Mass Inst Technol, PhD(astronaut), 72. *Prof Exp:* Sr scientist, Lockheed Res Lab, 72-73; scientist, 73-76, SUPVR, ASTRON MEASUREMENTS GROUP, JET PROPULSION LAB, 76- *Mem:* Am Astron Soc; Int Union Radio Sci. *Res:* Investigations of quasars and galaxies; tracking of interplanetary spacecraft; planetary dynamics. *Mailing Add:* Jet Propulsion Lab 238- 700 4800 Oak Grove Dr Pasadena CA 91109

PRESTON, ROBERT JULIAN, b London, Eng, June 5, 42; m 80; c 3. CYTOGENETICS. *Educ:* Cambridge Univ, BA, 63, MA, 66; Reading Univ, PhD(radiation biol), 70. *Prof Exp:* Staff mem biophys, Radiol Unit, Med Res Coun, Harwell, Eng, 63-70; SECT HEAD HUMAN GENETICS, BIOL DIV, OAK RIDGE NAT LAB, 70- *Concurrent Pos:* Adj prof, Biomed Grad Sch, Univ Tenn, Knoxville, 70-; ed, Mutation Res Letters, 80-89. *Mem:* Radiation Res Soc; Environ Mutagen Soc (pres, 89-90); Am Soc Human Genetics. *Res:* Assessment of radiation and chemical genetic and carcinogenic hazards to man; mechanisms of chromosome aberration induction by radiation and chemicals; molecular characterization of mutagens in human cells. *Mailing Add:* 12 Mona Lane Oak Ridge TN 37831

PRESTON, ROBERT LESLIE, b Stevens Point, Wis, Apr 24, 42; m 65; c 4. CELL PHYSIOLOGY, MEMBRANE PHYSIOLOGY. *Educ:* Univ Minn, BA, 66; Univ Calif, Irvine, PhD(biol), 70. *Prof Exp:* Res asst, Dept Develop & Cell Biol, Univ Calif, Irvine, 66-70; NIH fel membrane physiol, Med Sch, Yale Univ, 70-73, res assoc, 73-74; from asst prof to assoc prof, 74-88, PROF PHYSIOL, ILL STATE UNIV, 88- *Concurrent Pos:* Lectr gastrointestinal physiol, Physicians Asn Prog, Med Sch, Yale Univ, 72; vis prof, Univ Ill, Urbana, 78-79, Physiol Lab, Univ Cambridge, Eng, 81-82 & 84, Univ Oxford, Eng, 86, 88 & Univ Hawaii, 87; prin investr & mem, bd trustees, Mt Desert Island Biol Lab, Salsbury Cove, Maine, 85- *Mem:* Am Soc Zoologists; AAAS; Biochem Soc; Soc Gen Physiologists; Sigma Xi; NY Acad Sci. *Res:* Membrane transport of nonelectrolytes; comparative physiology and transport in marine organisms; energetics of ion coupled cotransport systems; amino acid transport by red blood cells; effects of heavy metals on membrane transport. *Mailing Add:* Dept Biol Sci Ill State Univ Normal IL 61761

PRESTON, RODNEY LEROY, b Denver, Colo, Jan 11, 31; m 50; c 3. ANIMAL NUTRITION & PHYSIOLOGY. *Educ:* Colo State Univ, BS, 53; Iowa State Univ, MS, 55, PhD(animal nutrit), 57. *Prof Exp:* From asst prof to prof animal husb, Univ Mo-Columbia, 57-69; prof animal sci, Ohio Agr Res & Develop Ctr, 69-75; prof animal sci & dept chmn, Wash State Univ, 75-82; THORNTON DISTINGUISHED PROF, DEPT ANIMAL SCI, TEX TECH UNIV, 82- *Concurrent Pos:* NIH spec fel, Lab Vet Biochem, State Univ Utrecht, The Netherlands, 64-65. *Mem:* Am Soc Animal Sci; Am Inst Nutrit; Soc Exp Biol & Med. *Res:* Nutrition, biochemistry, physiology, growth and body composition of animals. *Mailing Add:* Dept Animal Sci Tex Tech Univ Lubbock TX 79409-2141

PRESTON, STEPHEN BOYLAN, b Burnwell, WVa, May 5, 19; m 43. FOREST PRODUCTS. *Educ:* Colo Agr & Mech Col, BS, 46, MF, 47; Yale Univ, MF, 48, DF, 51. *Prof Exp:* Forester, Tex Longleaf Lumber Co, 47; res asst & asst instr wood technol, Yale Univ, 49-50; from instr to assoc prof wood technol, 50-55, res consult & wood technologist, Res Inst, 51-65, chmn dept wood technol, 54-65, actg dean sch natural resources, 69-70, chmn dept resource planning & conserv, 70-71, prof wood technol, 55-59, PROF NATURAL RESOURCES, UNIV MICH, ANN ARBOR, 59-, ASSOC DEAN SCH NATURAL RESOURCES, 71- *Concurrent Pos:* Exec dir, Orgn Trop Studies, Costa Rica, 65-69, vpres, 70-72, pres, 72-74. *Honors & Awards:* Wood Mag Award, 50. *Mem:* Soc Am Foresters; Forest Prod Res Soc; Soc Wood Sci & Technol (pres, 61-62 & 63-64); Int Soc Trop Foresters. *Res:* Physical and mechanical properties of veneer laminates drying of wood; tropical wood; tropical forestry. *Mailing Add:* 2681 Appleway Ann Arbor MI 48104

PRESTON, THOMAS ALEXANDER, b Hampton-on-Thames, Eng, Aug 13, 27; UK & Can citizen; m 54, 74; c 5. AGRICULTURAL ENGINEERING, PRODUCTION ENGINEERING. *Educ:* Univ Cambridge, BA, 51, MA, 56. *Prof Exp:* Mgr feeds div, UNGA Nairobi, Kenya, 53-57; mgr agr div, Prod Eng plc, Egham, 57-63; mem efficiency surv, Sierra Leone govt, 60-61; assoc prof, 63-72, PROF AGR ENG, UNIV ALTA, 72-; PROJ ERGONOMICS TROPICS, INT AGR CTR NETH STATE AGR, UNIV WAGENINGEN. *Concurrent Pos:* Consult orgn & methods, Tea Res Found Cent Africa, 64-69; fel, Nat Col Agr Eng, UK, 70-71; coordr, Fac Agr, Univ Nigeria, 72-74; ed, Rural Work Sci Abstr, Commonwealth Agr Bur, 78; mem, Irish Geneal Res Soc Coun; coordr, fac agr, Univ Nigeria, 72-74; ed, Biotechnol Trop Crop Improvement, Int Crops Res Inst Semi-Arid Tropics, 88; chmn, Small Farmers. *Mem:* Fel Brit Inst Mgt Services; Am Soc Agr Engrs; Agr Inst Can. *Res:* Ergonomics; work design; productivity methods; chronocyclegraphy; work study; network analysis and operations research; rural medicine. *Mailing Add:* 30 Russell Dr Christchurch Dorset BH23 3PA England

PRESTON, WILLIAM BURTON, b Penticton, BC, Mar 6, 37; m 63; c 3. HERPETOLOGY, ECOLOGY. *Educ:* Univ BC, BSc, 61, MSc, 64; Univ Okla, PhD(zool), 70. *Prof Exp:* Cur lower vert & invert, 70-76, CUR HERPETOLOGY & ICHTHYOLOGY, MANITOBA MUS MAN & NATURE, 76- *Mem:* Orthopterists' Soc; Soc Study Amphibians & Reptiles; Lepidopterists Soc; Entom Soc Can; Coleopterists' Soc; Am Arachnological Soc. *Res:* Ethoecology; ecology, behavior and distribution of amphibians and reptiles in Manitoba; ecology and distribution of insects and arachnids. *Mailing Add:* Man Mus Man & Nature 190 Rupert Ave Winnipeg MB R3B 0N2 Can

PRESTON, WILLIAM M, physics, nuclear physics; deceased, see previous edition for last biography

PRESTON-THOMAS, HUGH, b Eng, Dec 26, 23. EXPERIMENTAL PHYSICS. *Educ:* Bristol Univ, BSc, 45, PhD(physics), 50. *Prof Exp:* Res physicist, Div Physics, Nat Res Coun Can, 51-89; RETIRED. *Res:* Advances in both fundamental and practical metrology of the basic units of measurement. *Mailing Add:* 1109 Blasdell Ave Ottawa ON K1K 0C1 Can

PRESTWICH, GLENN DOWNES, b Canal Zone, Panama, Nov 29, 48; US citizen. ORGANIC & BIOCHEMISTRY, ENTOMOLOGY. *Educ:* Calif Inst Technol, BS, 70; Stanford Univ, PhD(chem), 74. *Prof Exp:* Res scientist, Int Ctr Insect Physiol & Ecol, Kenya, 74-75; NIH fel, Cornell Univ, 76-77; from asst prof to assoc prof, 77-84, PROF CHEM, STATE UNIV NY, STONY BROOK, 84- *Concurrent Pos:* Alfred P Sloan res fel, 81-83; Camille & Henry Dreyfus teacher scholar grant, 81- *Mem:* Am Chem Soc; AAAS; Am Entom Soc; Int Union Study Social Insects. *Res:* Insect juvenile hormone and steroid metabolism; synthetic organic and natural products chemistry; pheromone biochemistry; inhibitors of cholesterol metabolism; tritium NMR; receptors for inositol polyphosphates. *Mailing Add:* 56 Farm Rd Head of the Harbour NY 11780

PRESTWICH, KENNETH NEAL, b Inglewood, Calif, Mar 8, 49; m 89. PHYSIOLOGICAL ECOLOGY, COMPARATIVE PHYSIOLOGY. *Educ:* Davidson Col, BS, 71; Univ Fla, MS, 75, PhD(zool), 82. *Prof Exp:* Postdoctoral fel cardiac electrophysiol, Dept Physiol, Univ Fla, 82-83; asst prof physiol & anat, Dept Biol, Swarthmore Col, 83-84; asst prof physiol & invert zool, 84-89, ASSOC PROF PHYSIOL & BEHAV ECOL, DEPT BIOL, HOLY CROSS COL, 89- *Mem:* Am Soc Zoologists; Am Inst Biol Sci; Am Arachnological Soc; Sigma Xi. *Res:* Physiology of locomotion in spiders; energetics of locomotion; energetics and physiology of acoustic communication. *Mailing Add:* Dept Biol Holy Cross Col Worcester MA 01610

PRESTWIDGE, KATHLEEN JOYCE, b New York, NY, Jan 7, 27. BIOLOGY, OTHER MEDICAL & HEALTH SCIENCES. *Educ:* Hunter Col, BA, 49; Brooklyn Col, MA, 57; St John's Univ, NY, PhD(biol), 70. *Prof Exp:* Teacher, Bd Educ, City of New York, 56-59; prof biol & med lab technol, Bronx Community Col, City Univ NY, 59-78, sci columnist, 82-85, resident scholar, 87-88; RETIRED. *Concurrent Pos:* Pres, bd dir Sr Ctr, Hudson Community Ctr, Bronx. *Mem:* AAAS; Am Inst Biol Sci; fel NY Acad Sci; Nat Tech Asn; Am Asn Univ Women. *Res:* Medical laboratory technology; human physiology, health; delivery of community health services; career training, curriculum development and biological education. *Mailing Add:* 162-01 77th Rd Flushing NY 11366

PRESTWOOD, ANNIE KATHERINE, b Lenoir, NC, July 4, 35. VETERINARY PARASITOLOGY, WILDLIFE DISEASES. *Educ:* NC State Univ, BS, 60; Univ Ga, DVM, 62, MS, 64, PhD(parasitol), 68. *Prof Exp:* Fel, Southeastern Coop Wildlife Dis Study, 64-66, res assoc, 66-76, assoc prof, 76-81, PROF PARASITOL, COL VET MED, UNIV GA, 81- *Concurrent Pos:* Pres, Animal Dis Res Workers in Southern States, 88-89. *Mem:* Wildlife Dis Asn (secy, 77-80); Am Soc Parasitologists; Am Vet Med Asn; Wildlife Soc, (pres, 83-85); Can Soc Zoologists. *Res:* Helminth and Protozoan diseases of domestic and wild animals, particularly the lungworm fauna of these animals; tissue coccidians of domestic livestock. *Mailing Add:* Dept Parasitol Univ Ga Athens GA 30602

PRESZLER, ALAN MELVIN, b Missoula, Mont, July 11, 39; m 66; c 1. SPACE PHYSICS. *Educ:* Univ Wash, BA, 63; Univ Calif, Riverside, MS, 72, PhD(physics), 73. *Prof Exp:* Res assoc physics, Univ Calif, Riverside, 69-73; res assoc, Univ NH, 73-75; res physicist, Univ Calif, Riverside, 75-77; asst prof, Loma Linda Univ, 77-80; physicist, Appl Automation, Inc, 80-83; sr res physicist, Phillips Petrol Co, 83-88; SR SCI SPECIALIST, IDAHO NAT ENG LAB, EG & G IDAHO, INC, 88- *Concurrent Pos:* Instr, Riverside City Col, 75- & Med Ctr, Loma Linda Univ; tech mem, Conf Disarmament, Geneva. *Mem:* Am Asn Advan Med Instr. *Res:* Interpretation of atmospheric neutron data obtained from a large area double-scattering detector and from neutron spectrometers flown on Skylab and the Apollo-Soyuz test project; pulmonary system data interpretation through measurements of gas composition, airway flow, pressure and temperature; development of elemental analyses for process control. *Mailing Add:* EG&G Idaho PO Box 1625 MS 3411 Idaho Falls ID 83415-3411

PRETI, GEORGE, b Brooklyn, NY, Oct 7, 44; m 73. ORGANIC CHEMISTRY, MASS SPECTROMETRY. *Educ:* Polytech Inst Brooklyn, BS, 66; Mass Inst Technol, PhD(org chem), 71. *Prof Exp:* Res assoc, 71-73, MEM, MONELL CHEM SENSES CTR & ADJ ASSOC PROF OBSTET & GYNEC, UNIV PA, 78- *Mem:* Am Chem Soc. *Res:* Use of gas chromatography and mass spectrometry to profile the small, organic constituents of human female reproductive tract secretions for metabolites diagnostic of the fertile period and/or pathological conditions. *Mailing Add:* Monell Chem Senses Ctr 3500 Market St Philadelphia PA 19104-3308

PRETKA, JOHN E, b Lawrence, Mass, May 21, 19; m 43; c 2. TEXTILE FIBERS, ORGANIC CHEMISTRY. *Educ:* Tufts Col, BS, 42; Columbia Univ, MA, 47; NY Univ, PhD(chem), 50. *Prof Exp:* Org res chemist, Evans Assocs, 42-44, Stauffer Chem Co, 46, Calco Div, Am Cyanamid Co, 50-55; org res chemist, textile fibers dept, E I Du Pont de Nemours & Co, Inc, 55-60, sr res chemist, 60-82, textile consult, 82-84; RETIRED. *Concurrent Pos:* DuPont fel, NY Univ; consult, Delaware Sr Consult, 84-86; chmn, Res Comt Static Elec Textiles, Am Asn Texile Chemists & Colorists, 60-65. *Mem:* Am Chem Soc; Am Asn Textile Chemists & Colorists; Sigma Xi. *Res:* Applied textile end use research; textile fibers; new product development; static electricity; finishing; fabric uniformity; physical tests; organic synthesis; optical bleaches; dyes; pigments; heterocycles; organic intermediates; agricultural chemicals; resin finishes. *Mailing Add:* 2617 Turnstone Dr Wilmington DE 19808-1638

PRETLOW, THOMAS GARRETT, II, b Warrenton, Va, Dec 11, 39; m 63; c 3. PATHOLOGY. *Educ:* Oberlin Col, AB, 60; Univ Rochester, MD, 65. *Prof Exp:* From intern internal med to resident path, Univ Hosps, Univ Wis, 65-67; res assoc cell biol sect, Viral Etiol Br, Nat Cancer Inst, 67-69; asst prof path, Rutgers Med Sch, 69-71; assoc prof, Med Ctr, Univ Ala, Birmingham, 71-74, asst prof biophys, 71-76, dir, Div Cancer Res, 76-78, prof path, 74-, assoc prof biochem, 76-; PROF, DEPT PATH, CASE WESTERN RESERVE UNIV. *Concurrent Pos:* USPHS res career develop award, 73-78; mem path B study sect, NIH, 76-80; ed, Cell Biophys. *Mem:* Am Asn Cancer Res; Am Soc Biol Chem; Am Asn Path; Int Acad Path; Am Asn Immunol. *Res:* Neoplastic diseases; cellular physiology; mechanism of drug action; isolation of cell types in vitro. *Mailing Add:* Dept Path Case Western Reserve Univ Sch Med Cleveland OH 44106

PRETTY, KENNETH MCALPINE, b Wilkesport, Ont, June 19, 29. SOIL SCIENCE, CROP SCIENCE. *Educ:* Univ Toronto, BSA, 51; Mich State Univ, MS, 55, PhD(soil sci), 58. *Prof Exp:* Soils fieldman, Soil & Crop Improv Asn, 54; from instr to asst prof soil sci, Mich State Univ, 57-59; Can dir indust agron, 59-67, VPRES INDUST AGRON & COORDR FOREIGN PROGS, AM POTASH INST, 67-; PRES, POTASH & PHOSPHATE INST CAN, 71- *Concurrent Pos:* Hon prof, Nanjing Inst Soil Sci, Academia Sinica, China. *Mem:* AAAS; Am Soc Agron; Soil Sci Soc Am; Crop Sci Soc Am; Am Chem Soc. *Res:* Nutrient availability in soils; factors affecting plant uptake; components of yield; sources of plant nutrients; distribution and function of nutrients in plants. *Mailing Add:* 311 Sixth Ave N Suite 2602 Saskatchewan SK S7K 7A9 Can

PRETZER, C ANDREW, b Scranton, Pa, Sept 10, 28; c 8. STRUCTURAL ENGINEERING, CIVIL ENGINEERING. *Educ:* Univ Mich, BSE, 50, MS, 51; Mass Inst Technol, PhD(civil eng), 63. *Prof Exp:* Instr civil eng, Mass Inst Technol, 60-63; engr struct eng, Le Messurier Assoc, Inc, 63-66; assoc prof civil eng, Northeastern Univ, 66-72; PRES STRUCT ENG, C A PRETZER ASSOCS INC, 72- *Concurrent Pos:* Consult civil eng, Douglas G Peterson & Assoc Inc, 72- *Mem:* Am Concrete Inst; Am Soc Civil Engrs; Prestressed Concrete Inst. *Mailing Add:* East Rd North Scituate RI 02857

PRETZER, DONAVON DONALD, image processing, for more information see previous edition

PREUL, HERBERT C, b Berger, Mo, Jan 11, 26; m 50; c 1. CIVIL & ENVIRONMENTAL ENGINEERING. *Educ:* Univ Iowa, BS, 50; Univ Minn, Minneapolis, MS, 55, PhD(civil eng), 64. *Prof Exp:* Civil engr, US Bur Reclamation, Colo & Calif, 50-52; supvry civil engr, US Corps Engrs, Minn & Nebr, 52-60; teaching asst, res fel & civil eng, Univ Minn, Minneapolis, 60-64; assoc prof, 64-75, PROF CIVIL & ENVIRON ENG, COL ENG, UNIV CINCINNATI, 75- *Concurrent Pos:* Consult sanit & hydraul eng for var orgn, 55-; res grants, Off Water Resources Res, Univ Cincinnati, 65-67, USPHS, 68-70 & Environ Protection Agency, 69-82; eng educ consult, USAID & NSF for water resources & sanit eng in India, 66, 68 & 70; int consult sanit eng for WHO, World Bank , var foreign govt, US Agency Int Develop. *Mem:* Am Soc Civil Engrs; Am Soc Eng Educ; Water Pollution Control Fedn; Am Water Works Asn; Am Water Resources Asn. *Res:* Water resources and sanitary engineering; water pollution control; waste water treatment; groundwater contamination. *Mailing Add:* Dept Civil & Environ Eng Col Eng Univ Cincinnati Cincinnati OH 45221

PREUS, MARILYN IONE, b Edmonton, Alta, Oct 20, 44; m 73. HUMAN GENETICS. *Educ:* Univ Alta, BS, 67; McGill Univ, MS, 71, PhD(biol), 75. *Prof Exp:* Assoc prof human genetics, Mem Univ Nfld, Fac Med & Discipline Pediat, 82-85; res scientist, McGill Univ-Montreal Children's Hosp, Res Inst, 75-82, asst prof, Dept Pediat & Ctr Human Genetics, 80-82, affil prof biol, 80-82, ASST PROF, DEPT PEDIAT & CTR HUMAN GENETICS, McGILL UNIV, 85- *Concurrent Pos:* Fac lectr, Dept of Pediat, McGill Res Inst, 78-81. *Mem:* Teratology Soc; Am Soc Human Genetics; fel Can Col Med Geneticists. *Res:* Numerical classification of dysmorphic syndromes by phenotypic resemblance; health care and the law. *Mailing Add:* Dept Biol Ctr Human Genetics 1205 Rue Docteur Penfield Montreal PQ H3A 1B1 Can

PREUS, MARTIN WILLIAM, b Chicago, Ill, Feb 26, 54; m 76. OIL FIELD CHEMICALS. *Educ:* Calif State Polytech Univ, BS, 75; Univ Calif, San Diego, MS, 78, PhD(org chem), 79. *Prof Exp:* Res chemist, Rohm and Haas Co, 80-81; RES CHEMIST, MAGNA CORP, 81- *Mem:* Am Chem Soc. *Res:* Biocides, sulfide scavengers, corrosion inhibitors and emulsion breakers; herbicides for aquatic weed controll; abrasion resistant coated transparent plastic sheet; synthesis of biologically active natural products. *Mailing Add:* 419 Indiana St Anaheim CA 92805-3621

PREUSCH, CHARLES D, b Elmhurst, NY, June 19, 17; m 42; c 3. METALLURGICAL ENGINEERING. *Educ:* Columbia Univ, BS, 39, MS, 40. *Prof Exp:* Chief metallurgist, Crucible Steel Co, 55-60, asst dir res & develop, 60-64, dir planning & develop, 64-69, dir advan planning & magnetics, 69-72, sr consult engr, 72-81, CONSULT PRIN ENGR, CRUCIBLE, INC, COLT INDUSTS, 81- *Honors & Awards:* Gary Medal, Am Iron & Steel Inst, 61. *Mem:* Am Inst Mining, Metall & Petrol Engrs; Am Soc Metals; Asn Iron & Steel Engrs. *Res:* Specialty steels melting; steel processing; casting; vacuum metallurgy; materials systems; product design; manufacturing technology; powder metallurgy. *Mailing Add:* 555 Greenhurst Dr Pittsburgh PA 15243

PREUSCH, PETER CHARLES, b Orange, NJ, May 13, 53; m 87; c 3. ENZYMOLOGY, MEDICINAL CHEMISTRY. *Educ:* Pa State Univ, BS, 74; Cornell Univ, PhD(biochem), 79. *Prof Exp:* Postdoctoral fel biochem, Univ Wis, 79-83; asst prof chem, Univ Akron, 83-90; SCI REV ADMINR, NIH, 90- *Mem:* Am Chem Soc; Am Soc Biochem & Molecular Biol; AAAS. *Res:* Enzymology; vitamin K; anticoagulants; flavins. *Mailing Add:* Div Res Grants Westwood Bldg 2A17 NIH Bethesda MD 20892

PREUSS, ALBERT F, b Gleason, Wis, Sept 17, 26; m 49; c 2. CHEMISTRY. *Educ:* Univ Wis, BS, 49, PhD(chem), 53. *Prof Exp:* Asst anal chem, Univ Wis, 47-52; res chemist, Rohm and Haas Co, Pa, 52-63, lab head, 63-68; assoc dir res, Ionac Chem Co, NJ, 68-69, dir res, 69-71, vpres res & develop, 71-72; pres, Aldex Co, Inc, 72-75, PRES, ALDEX CHEM CO, LTD, 75- *Mem:* Am Chem Soc; Soc Automotive Engrs; Soc Plastics Engrs. *Res:* Ion exchange technology; chemistry of rhenium; hydrometallurgy of uranium; solvent extraction; nonaqueous chemistry; oil additives. *Mailing Add:* Seven Martindale Rd Shelburne VT 05482-7283

PREVATT, RUBERT WALDEMAR, b Seville, Fla, May 15, 25; m 53; c 2. SOIL CHEMISTRY, HORTICULTURE. *Educ:* Univ Fla, BSA, 48, MSA, 51, PhD, 59. *Prof Exp:* Lab asst in charge soil testing, Univ Fla, 48-51; soil chemist, Dr P Phillips Co, 51-54; asst, Cornell Univ, 54-56; asst, Univ Fla, 56-59; sr res agronomist, Int Minerals & Chem Corp, Ill, 59-64, supvr Fla Agr Res, 64-70; PROF CITRUS & SOILS, FLA SOUTHERN COL, 71-, DIR 80-, JOHN V & RUTH A TYNDALL CITRUS CHAIR, 81- *Concurrent Pos:* Consult hort, 71- *Mem:* Am Soc Agron; Soil Sci Soc Am; Am Hort Soc. *Res:* Soil fertility; citrus; vegetables and ornamentals; roses; chrysanthemums; agronomic crops; fertilizers; growth regulators. *Mailing Add:* 2705 Collins Ave Lakeland FL 33802

PREVEC, LUDVIK ANTHONY, b Kirkland Lake, Ont, Aug 19, 36; m 62; c 3. BIOLOGY. *Educ:* Univ Toronto, BA, 59, MA, 62, PhD(biophys), 65. *Prof Exp:* Res assoc virol, Wistar Inst Anat & Biol, 65-67; asst prof, 67-71, assoc prof, 71-79, PROF BIOL, MCMASTER UNIV, 79- *Concurrent Pos:* Res fel, Inst Virol, Glasgow, Scotland, 73-74; assoc ed, J Can Soc Cell Biol, 71-73 & J Virol, 76-78; vis prof, Univ Ottawa Med Sch. *Mem:* Can Soc Cell Biol; Can Soc Biochem; Am Soc Microbiol. *Res:* Biological and biochemical aspects of virus replication. *Mailing Add:* Dept Biol Sci McMaster Univ Hamilton ON L8S 4L8 Can

PREVIC, EDWARD PAUL, b Export, Pa, June 29, 31; m 81; c 3. MICROBIAL PHYSIOLOGY. *Educ:* Carnegie Inst Technol, BS, 53; Univ Ill, PhD(biochem), 63. *Prof Exp:* Exp coatings formulator, Thompson & Co, 54-55; org res chemist, Res & Develop Div, Consol Coal Co, 55-60; fel microbiol, Col Med, Tufts Univ, 63-64; Nat Inst Allergy & Infectious Dis fel, Inst Microbiol, Copenhagen, Denmark, 64-65; from instr to assoc prof, 65-76, ASSOC PROF MICROBIOL & CELL SCI, INST FOOD & AGR SCI, UNIV FLA, 76- *Mem:* Am Soc Microbiol; Am Inst Chemists. *Res:* Endospore-forming bacterial pathogen of root-knot nematodes; mechanisms and control of cell division processes in bacteria and other microorganisms; biochemistry and enzymology of cell division and other growth related phenomena; microbial biochemistry and physiology. *Mailing Add:* Dept Microbiol & Cell Sci Univ Fla 1059 McCarthy Hall Gainesville FL 32611

PREVITE, JOSEPH JAMES, b Lawrence, Mass, Jan 25, 36; m 59; c 5. PHYSIOLOGY, MICROBIOLOGY. *Educ:* Merrimack Col, AB, 56; Boston Col, MS, 59; Bryn Mawr Col, PhD(physiol), 62. *Prof Exp:* Instr biol, Villanova Univ, 58-62; asst prof zool, Smith Col, 62-63; res physiologist-microbiologist, US Army Navy Natick Labs, 63-68 & Exp Path Lab, US Army Res Inst Environ Med, 68-70; PROF PHYSIOL & IMMUNOL, FRAMINGHAM STATE COL, 70- *Concurrent Pos:* Consult, Div Allied Health Manpower, USPHS, US Army Res Off. *Mem:* AAAS; Am Physiol Soc; fel Am Acad Microbiol. *Res:* Host-parasite relations; endotoxins; environmental physiology and disease; food microbiology; radiation microbiology; salmonellae; Cl botulinum toxins; staphylococci; food borne infections and intoxications. *Mailing Add:* Dept Biol Framingham State Col A 519 Framingham MA 01701

PREVORSEK, DUSAN CIRIL, b Ljubljana, Yugoslavia, Feb 14, 22; US citizen; m 63; c 1. POLYMER PHYSICS, POLYMER CHEMISTRY. *Educ:* Univ Ljubljana, BS, 50, PhD(chem), 56. *Prof Exp:* Fel physics, Univ Paris, 56-57, res assoc, 57-58; proj leader polymer physics, Goodyear Tire & Rubber Co, 58-61; fel, Textile Res Inst, 61-62, prin scientist, 62-65; sr res assoc, 65-77, sr res scientist, Chem Res Ctr, Allied Chem Corp, 77-80, MGR POLYMER SCI, CORP TECHNOL, ALLIED SIGNAL INC, 80- *Concurrent Pos:* Lectr, Fiber Soc, 70; chmn, Gordon Conf Fiber Sci, 76; ed-in-chief, Int J Polymeric Mat. *Honors & Awards:* Harold DeWitt Smith Award, Am Soc Testing & Mat, 75; IR-1000 Award, 85; Del Monte Excellence Award, Soc Advan Mat & Process Eng, 89. *Mem:* Am Chem Soc; Am Phys Soc; Soc Rheol; Fiber Soc; NY Acad Sci. *Res:* Structure of complex organic molecules; theoretical and experimental viscoelasticity, diffusion and fracture; textile and tire mechanics; synthesis and properties of ordered copolymers; polymer compatibility; interpenetrating networks; composites; adhesion. *Mailing Add:* PO Box 1021R Allied Signal Inc Morristown NJ 07960

PREVOST, JEAN HERVE, b Fez, Morocco, Sept 10, 47; French citizen; m 75; c 3. GEOTECHNICAL ENGINEERING, NUMERICAL METHODS. *Educ:* ETP, France, 71; Stanford Univ, MS, 72, PhD(soil mech), 74. *Prof Exp:* Teaching asst civil eng, Stanford Univ, 72-74; res fel soil mech, Norwegian Geotech Inst, Oslo, 74-76; lectr civil eng, Calif Inst Technol, 76-78; from asst prof to assoc prof, 78-86, PROF CIVIL ENG, PRINCETON UNIV, 86-, CHMN CIVIL ENG & OR, 89- *Concurrent Pos:* Fulbright fel, 71-74; Royal Norwegian Coun Sci & Indust Res fel, 74-76. *Mem:* Am Soc Civil Engrs; Soc Eng Sci; Int Soc Soil Mech; Am Acad Mech. *Res:* Dynamics; nonlinear continuum mechanics; mixture theories; finite element methods; constitutive theories; soil mechanics; centrifuge soil testing; author or coauthor of over 60 publications. *Mailing Add:* Dept Civil Eng & OR Princeton Univ Princeton NJ 08544

PREWITT, CHARLES THOMPSON, b Lexington, Ky, Mar 3, 33; m 58; c 1. EARTH SCIENCE. *Educ:* Mass Inst Technol, SB, 55, SM, 60, PhD(crystallog), 62. *Prof Exp:* Crystallogr, Cent Res Dept, E I du Pont de Nemours & Co, Inc, 62-69; from assoc prof to prof crystallog, 69-75, chmn dept, 77-80, prof earth sci & mat sci, State Univ NY Stony Brook, 75-86; DIR GEOPHYS LAB, CARNEGIE INST WASH, 86- *Concurrent Pos:* Churchill Found fel, 75; vis prof, Monash Univ, Australia, 76, Univ Copenhagen, 82, Ariz State Univ, 83, Univ Tsukuba, Japan, 83; ed, Physics & Chem of Minerals, 77-85. *Mem:* Am Crystallog Asn; Mineral Soc Am (pres, 83-84); Am Geophys Union; Geol Soc Am; Mat Res Soc. *Res:* Solid state geochemistry and geophysics; mineralogy, crystallography and chemistry of silicates and sulfides. *Mailing Add:* Geophys Lab 5251 Broad Branch Rd NW Washington DC 20015-1305

PREWITT, JUDITH MARTHA SHIMANSKY, b Brooklyn, NY, Oct 16, 35; m 56; c 1. MATHEMATICS, COMPUTER SCIENCE. *Educ:* Swarthmore Col, BA, 57; Univ Pa, MA, 59; Univ Uppsala, PhD(comput sci & numerical anal), 78. *Prof Exp:* Mathematician, Burroughs Corp, 56-58, Auerbach Corp, 61-62; instr math & radiol, Grad Sch Med, Univ Pa, 62-71; mathematician, NIH, 71-74, res mathematician, 74-83; vis prof, Dept Elec Eng & Comput Eng, Ohio Univ, 83-84; supvr & mem tech sta, AT&T Bell Labs, 84-86; PRES & CHIEF SCI, PAXX GROUP, INC, 86- *Concurrent Pos:* Instr math, Univ Pa, 60-63; mem comt cytol automation, Nat Cancer Inst, 72-77, diag radiol adv comt, 73-77 & diag res adv comt, 77-; consult & vis scientist, Uppsala Univ & Hosp, 73-; assoc ed, Comput Graphics & Image Processing, 74-; co-chmn, NSF Proj Ultrasonic Tissue Characterization, 76-; pres, Medimatics, 86-. *Honors & Awards:* Sustained High Performance Qual Award, NIH, 73. *Mem:* Soc Indust & Appl Math; Biomed Eng Soc; fel Inst Elec & Electronics Engrs; Math Sci Asn; Asn Comput Mach. *Res:* Machine intelligence (sensors, pattern recognition and decision-making) applied to quantitative characterization of biological images; use of computers in cancer detection, diagnosis and therapy optimization. *Mailing Add:* Paxx Group Inc 14300 Poplar Hill Rd Germantown MD 20874

PREWITT, LARRY R, b Harlingen, Tex, Aug 19, 45; m 71; c 3. INGREDIENT, FORMULATION. *Educ:* Univ Mo, BS, 67, MS, 68; Univ Ky, PhD(ruminant nutrit), 72. *Prof Exp:* Asst prof dairy, Mich State Univ, 72-75 & Univ Ky, 75-76; mgr, Appl Res, Ralston Purina Co, 76-80, dir, 80-84, Chow Res, 84-86 & Int Res, 86-88, VPRES & DIR INT RES, RALSTON PURINA INT, 88- *Mem:* Am Dairy Sci Asn; Am Soc Animal Sci; Am Asn Cereal Chemists. *Res:* Product development for animals, pets, and human ready-to-eat cereal products; procedural implementation; technical service training of technical, marketing and sales staff; formulation ingredient evaluation; product-ingredient standards, quality control, development and supervision of institutional research projects. *Mailing Add:* Ralston Purina Int Checkerboard Sq-IRS St Louis MO 63164

PREWITT, RUSSELL LAWRENCE, JR, b St Louis, Mo, June 19, 43; m 67; c 3. CARDIOVASCULAR PHYSIOLOGY. *Educ:* St Louis Univ, AB, 65; Univ Mo, PhD(physiol), 71. *Prof Exp:* Res assoc physiol, Col Med, Univ Ariz, 71-74; asst prof physiol, Ctr Health Sci, Univ Tenn, 74-81; assoc prof physiol, La State Univ Med Ctr, 81-85; assoc prof, 85-87, PROF PHYSIOL, EASTERN VA MED SCH, 87- *Concurrent Pos:* Res grants, Tenn Heart Asn, 75; NIH grants, 76-, res career develop award, 82-87; fel, Coun High Blood Pressure Res, Am Heart Asn, 83. *Mem:* Microcirculatory Soc; Am Physiol Soc; Int Soc Hypertension. *Res:* Control of blood flow in the microcirculation and microvascular alterations in hypertension. *Mailing Add:* Dept Physiol Easter Va Med Sch PO Box 1980 Norfolk VA 23501

PREZANT, ROBERT STEVEN, b Brooklyn, NY, Aug 29, 51; m 74; c 2. MALACOLOGY, FUNCTIONAL MICROSTRUCTURE. *Educ:* Adelphi Univ, BA, 73; Northeastern Univ, MS, 76; Univ Del, PhD(marine sci), 81. *Prof Exp:* Interim instr marine sci, Col Marine Studies, Univ Del, 79-80; from asst prof to assoc prof, Univ Southern Miss, 81-87; IND UNIV PA, 87- *Concurrent Pos:* Publ ed, Am Malacol Union, 82-, ed-in-chief, Am Malacol Bull, 82-; affil ed, Malacol Rev, 83-; ed, Perspectives Malacol, 85 & Entrainment of Larval Oysters, 85-86; managing ed, Int Corbicula Symp,

85-86. *Mem:* Am Malacol Union; Nat Shellfisheries Asn; Am Soc Zoologists; AAAS. *Res:* Environmental influence on biomineralization of shell microstructure; systematics of bivalve molluscs; antipredation mechanisms of marine invertebrates; reproductive ecology of Antarctic bivalves. *Mailing Add:* Dept Biol Ind Univ Pa Indiana PA 15705-1090

PREZELIN, BARBARA BERNTSEN, b Portland, Ore, Apr 13, 48; m 72. MARINE PHYCOLOGY, PHOTOSYNTHESIS. *Educ:* Univ Ore, BS, 70; Scripps Inst Oceanog, Univ Calif, San Diego, PhD(marine biol), 75. *Prof Exp:* NSF res partic phycol, Univ Ore, 69, teaching asst biol, 68-70; NIH trainee marine bot, Marine Biol Lab, Woods Hole, 70; NIH trainee marine biol, Scripps Inst Oceanog, Univ Calif, San Diego, 70-75, res asst, 75; NSF res assoc, 75-77, ASST PROF, DEPT BIOL SCI, MARINE SCI INST, UNIV CALIF, SANTA BARBARA, 77- *Concurrent Pos:* Lectr, NATO, 80. *Mem:* Am Inst Biol Sci; Am Soc Photobiol; Phycol Soc Am; Am Soc Limnol & Oceanog; AAAS. *Res:* Studies of mechanisms regulating algal photosynthesis, primary productivity and growth with special interest in light-regulation, diel periodicity and biological clocks and photonutritive interactions of the photosynthetic apparatus of red tide species of dinoflagellates; study of mechanism of circadian rhythmicity in photosynthetic capacity. *Mailing Add:* Dept Biol Sci Univ Calif Santa Barbara CA 93106

PRIBBLE, MARY JO, b Macfarlan, WVa, Jan 20, 30. ANALYTICAL CHEMISTRY. *Educ:* Maryville Col, BA, 52; Duke Univ, MA, 57; La State Univ, Baton Rouge, PhD(chem), 70. *Prof Exp:* Instr chem, Davis & Elkins Col, 54-55 & Marietta Col, 57-59; documentation chemist, Ethyl Corp, 59-61; instr, Marshall Univ, 61-63; from asst prof to assoc prof chem, Limestone Col, 63-70, prof & chmn, Div Sci & Math, 70-77, chmn dept, 69-70; assoc prof, 77-84, PROF CHEM, GLENVILLE STATE COL, 84- *Concurrent Pos:* NSF sci fac fel, La State Univ, Baton Rouge, 68-69. *Mem:* AAAS; Am Chem Soc; Sigma Xi. *Res:* Coordination compounds; preparation, interpretation of electronic and vibrational spectra, magnetic properties, unusual oxidation states; chemical education. *Mailing Add:* Dept Chem Glenville State Col Glenville WV 26351

PRIBIL, STEPHEN, b Pressbourg, Czech, Oct 27, 19; US citizen; m 59; c 2. THERMODYNAMICS, LOW TEMPERATURE PHYSICS. *Educ:* Slovak Tech Univ, DiplIng, 45, DrIngSc(appl physics), 50. *Prof Exp:* Scientist, SKODA, Czech, 45-48; sr scientist, A G Dynamit Nobel, 48-51; prof mech eng, Sch Eng, Pressbourg, 52-58; scientist, Div Natural Sci, Inst Appl Physics, Czech Acad Sci, 58-62; res consult thermodyn & heat transfer, Worthington Corp, NY, 62-63; PROF PHYSICS, STATE UNIV NY COL BROCKPORT, 64- *Concurrent Pos:* Sci adv, Czech Govt, Ministry Health, Mining Indust & Heavy Indust, 54-62. *Mem:* Am Asn Physics Teachers; Am Phys Soc; Am Soc Heat, Refrig & Air-Conditioning; Royal Soc Health; Int Inst Refrig. *Res:* Phase-equilibria of cryogens; solidification of gases and mixtures of gases; cryoacoustics; dielectric properties of cryogens; solid state biophysics. *Mailing Add:* Dept Physics State Univ NY Col Brockport Brockport NY 14420

PRIBOR, DONALD B, b Detroit, Mich, Dec 4, 32; m 61; c 4. CELL PHYSIOLOGY, CRYOBIOLOGY. *Educ:* St Louis Univ, BS, 54; Cath Univ Am, MS, 61, PhD(physiol), 64. *Prof Exp:* Consult, Inst Lab Med, Perth Amboy Gen Hosp, 68-70. *Concurrent Pos:* Res assoc, Inst Med Res, Toledo Hosp, Ohio, 68-70. *Mem:* Soc Gen Physiol; Nat Asn Biol Teachers; Am Asn Higher Educ. *Res:* Integration of science and human values; prevention of psychosomatic illness; psychobiology; human sexuality; development of human personality. *Mailing Add:* Dept Biol Univ Toledo 2801 W Bancroft Toledo OH 43606

PRIBOR, HUGO C, b Detroit, Mich, June 12, 28; m 55; c 3. CLINICAL PATHOLOGY, LABORATORY MANAGEMENT. *Educ:* St Mary Col, BS, 49; St Louis Univ, MS, 51, PhD(anat & pathol), 54, MD, 55. *Prof Exp:* Intern, Providence Hosp, 55-56; resident, NIH, 56-59; field investr, Gastric Cytol Res, Bowman-Gray Sch Med, 59-60; assoc pathologist & dir, Clin Lab, Bon Secours Hosp, 60-63; dir, labs & pathologist, Samaritan Hosp Assoc, 63-64; dir, labs & chief pathologist, Perth Amboy Gen Hosp, 64-73; chmn & chief exec officer, Ctr Lab Med, Inc, 68-77; exec med dir, MDS Health Group, Inc, 78-80; MED DIR, SMITHKLINE BEECHAM CLIN LABS, 81-; PATHOLOGIST, ASSOC PATHOLOGISTS, PC, 81- *Concurrent Pos:* Consult pathologist, Med Examr Off, Middlesex Co, 64-73; prof biochem eng, Col Eng, 71-75, Rutgers State Univ, 80-82; vpres med affairs, Med Serv Group, Damon Corp, 77-78; clin prof pathol, Sch Med, Vanderbilt Univ, 81- *Honors & Awards:* Silver Award, Am Soc Clin Pathologists & Col Am Pathologists, 68. *Mem:* AMA; Am Soc Clin Pathologists; Col Am Pathologists; Am Path Found; Int Acad Path. *Res:* Applications of computers to laboratory medicine; expert systems (artificial intelligence) techniques in medicine; pathology consultation system, the laboratory consultant. *Mailing Add:* Smithkline Beecham Clin Labs 2545 Park Plaza Nashville TN 37203

PRIBRAM, JOHN KARL, b Chicago, Ill, Feb 1, 41; m 63; c 2. OPTICAL PROPERTIES. *Educ:* Middlebury Col, AB, 62; Wesleyan Univ, MA, 65; Univ Mass, Amherst, PhD(physics), 73. *Prof Exp:* From instr to assoc prof, 70-89, PROF PHYSICS, BATES COL, 89-, CHMN, DEPT PHYSICS & ASTRON, 85- *Concurrent Pos:* Mellon fel, 77-78; vis asst prof physics, Univ Ill, Urbana, 79-81; vis assoc prof physics, Dartmouth Col, 87-88. *Mem:* Am Phys Soc; Am Asn Physics Teachers; History Sci Soc; Sigma Xi. *Res:* Optical properties of semiconductors and insulators; superlattices. *Mailing Add:* Dept Physics & Astron Bates Col Lewiston ME 04240

PRIBRAM, KARL HARRY, b Vienna, Austria, Feb 25, 19; nat US; m 40, 60; c 5. NEUROPSYCHOLOGY. *Educ:* Univ Chicago, SB, 39, MD, 41; Am Bd Neurol Surg, dipl, 48. *Prof Exp:* Asst resident neurol surg, Chicago Mem Hosp, 42-43; resident, St Luke's Hosp, Chicago, 43-45; instr neurol surg, Univ Tenn, 45-46; neurosurgeon & neurophysiologist, Yerkes Labs Primate Biol, Fla, 46-48; res asst prof physiol & psychol, Sch Med, Yale Univ, 48-51, lectr, 51-58; fel, Ctr Advan Study Behav Sci, 58-59; from assoc prof to prof psychiat & psychol & NIMH res prof, Stanford Univ, 59-89, emer prof, 89-; JAMES P & ANNA KING UNIV PROF, EMMINENT SCHOLAR COMMONWEALTH VA, DEPT PSYCHOL, CTR BRAIN RES, RADFORD UNIV, 89- *Concurrent Pos:* Asst, Univ Ill, 43-45; vis lectr, Mass Inst Technol, 54, Clark Univ, 56 & Harvard Univ, 56; vis prof, Trinity Col, Conn, 57 & Menninger Found, 73-75. *Honors & Awards:* Phillips Haverford Col, 61; Spencer lectr, Univ Chicago, 73. *Mem:* Am Psychol Asn; Psychonom Soc; Soc Exp Psychologists; Int Soc Res on Agression; Sigma Xi. *Res:* Neurophysiology. *Mailing Add:* Dept Psychol Ctr Brain Res-Radford Univ Radford VA 24142

PRIBULA, ALAN JOSEPH, b Irvington, NJ, Jan 18, 48; m 83. INORGANIC CHEMISTRY. *Educ:* Bucknell Univ, BS, 69; Univ Ill, Urbana, PhD(inorg chem), 74. *Prof Exp:* Res assoc, Princeton Univ, 74-75, instr, 75-76; instr 76-80, asst prof, 80-90, ASSOC PROF CHEM, TOWSON STATE UNIV, 90- *Mem:* Am Chem Soc; Sigma Xi; Am Asn Univ Prof. *Res:* Transition-metal compounds; complexes of unusual sulfur and phosphorus ligands; metal complexes of anticancer agents; thermodynamics and mechanism of metal-ligand bond formation and cleavage. *Mailing Add:* Dept Chem Towson State Univ Baltimore MD 21204

PRIBUSH, ROBERT A, b Elizabeth, NJ, Sept 27, 46; m 68, 81; c 3. ANALYTICAL CHEMISTRY, INORGANIC CHEMISTRY. *Educ:* Univ Del, BS, 68; Univ Mass, PhD(chem), 72. *Prof Exp:* Res fel chem, Univ Southern Calif, Los Angeles, 72-74; PROF CHEM, BUTLER UNIV, INDIANAPOLIS, IND, 74- *Concurrent Pos:* Consult, Wolf Tech Serv, Indianapolis, 77- *Mem:* Fel Am Inst Chemists; Am Chem Soc; AAAS. *Res:* Acid deposition studies; emissions scavenging by rain, fog, dew and foliage; chemistry of fires and explosions. *Mailing Add:* Dept Chem Butler Univ Indianapolis IN 46208

PRICE, ALAN ROGER, b Pontiac, Mich, Jan 15, 42; m 62; c 4. BIOCHEMISTRY. *Educ:* Fla State Univ, BS, 64; Univ Minn, PhD(biochem), 68. *Prof Exp:* NIH fel, Med Sch, Univ Mich, Ann Arbor, 68-69, asst prof, 70-75, asst dean, 78-81, asst vpres, 80-83, assoc prof biol chem, 75-87, assoc vpres res, 83-87; HEALTH SCIENTIST ADMINR, GENETICS PROG, NAT INST AGING, NIH, 87- *Honors & Awards:* Rumsey Mem Res Award, Fla State Univ, 68; Sigma Xi Meritorious Res Award, Mich State Univ, 69. *Mem:* AAAS; Am Soc Microbiol; Am Soc Biol Chemists; fel Am Coun Educ. *Res:* Enzymology of nucleotide and nucleic acid biosynthesis; biochemistry of bacterial virus infections; biosynthesis and function of unusual nucleosides in deoxyribonucleic acids. *Mailing Add:* Dept Health & Human Serv NIH Pub Health Serv AIDS Unit Bldg 31 Rm 5859 Bethesda MD 20892

PRICE, ALSON K, b Putney, WVa, Feb 14, 38. ORGANIC CHEMISTRY, CATALYSIS. *Educ:* Morris Harvey Col, BS, 60; Yale Univ, PhD(org chem), 64. *Prof Exp:* Res chemist polymers, Chem Div, Union Carbide Corp, 60; res chemist org fluorine chem, Div Gen Chem, 63-70, res coordr, 70-71, mgr nutrit res, 71-72, mgr spec projs, 72-74, mgr catalysis, 74-79, mgr org chem, 79-80, MGR CATALYST & PROCESS RES, CORP RES CTR, ALLIED CORP, 80- *Mem:* Am Chem Soc. *Res:* Inorganic chemistry; homogeneous and heterogeneous catalysis; coal chemistry. *Mailing Add:* Box 90A North Rd Chester NJ 07930

PRICE, ALVIN AUDIS, b Lingleville, Tex, Oct 8, 17; m 43; c 2. VETERINARY MEDICINE. *Educ:* Tex A&M Univ, BS, 40, DVM, 49, MS, 56. *Prof Exp:* Prod mgr, Lockhart Creamery, Tex, 40-42; from instr to asst prof vet anat, Tex A&M Univ, 49-57, dean, Col Vet Med, 57-73, prof physiol & dir biomed sci, 73-88, emer dean, Col Vet Med, 88-; RETIRED. *Concurrent Pos:* Consult, Intercol Exchange Prog, Int Coop Admin, 59, Univ Puerto Rico & Miss State Univ; nat civilian consult, Surgeon Gen, US Air Force, 66; mem adv comt, Nat Health Resources & Selective Serv Syst. *Mem:* Am Vet Med Asn; Am Asn Vet Anat (pres, 57); Asn Am Vet Med Cols (secy, 63-66, pres, 67); Am Asn Vet Clinicians; US Animal Health Asn; Am Asn Vet Physiologists & Pharmacologists. *Res:* Reproductive physiology; embryology. *Mailing Add:* 1203 Walton Dr College Station TX 77840

PRICE, BOBBY EARL, b Henderson, Tex, Nov 21, 37; m 58; c 2. CIVIL ENGINEERING. *Educ:* Arlington State Col, BS, 62; Okla State Univ, MS, 63; Univ Tex, Austin, PhD(civil eng, water resources), 67. *Prof Exp:* Civil engr aide, Drainage Div, City of Dallas, Tex Pub Works Dept, 57-60, civil engr, 60-62, civil engr, City of Austin, 64; grad asst water resources, Okla State Univ, 62-64; res engr, Univ Tex, Austin, 66-67; PROF CIVIL ENG, DIR ENG UNDERGRAD STUDIES & DIR WATER RESOURCES CTR, LA TECH UNIV, 67- *Mem:* Am Soc Civil Engrs; Am Water Works Asn; Am Soc Eng Educ; Nat Soc Prof Engrs. *Res:* Water resources; culverts set on steep grades; computer applications to water resources. *Mailing Add:* Dept Civil Eng La Tech Univ Ruston LA 71272-0046

PRICE, BYRON FREDERICK, b Pittsburgh, Pa, May 13, 36; m 63; c 1. ANALYTICAL BIOCHEMISTRY. *Educ:* Pa State Univ, BS, 58; Univ Mo, MS, 64. *Prof Exp:* Res asst anal chem, Chem Lab, Mo Exp Sta, 61-64, res instr anal biochem, 64-65; jr res assoc, New Prod Div, Am Tobacco Co, 65-66, res assoc, 66-68, asst mgr, Anal Res Div, 68-71, asst mgr res, New Prod Div, 71-81, res mgr, Dept Res & Develop, 81-90, mgr, Res & Tech Serv, Dept Res & Qual Assurance, 90-91, RES DIR, DEPT RES & QUAL ASSURANCE, AM TOBACCO CO, 91- *Mem:* Am Chem Soc. *Res:* Development of analytical methods and basic design parameters for tobacco products, smoking article and other consumer products. *Mailing Add:* Am Tobacco Co PO Box 899 Hopewell VA 23860

PRICE, C A, b Long Beach, Calif, Feb 16, 27; m 51; c 4. PLANT MOLECULAR BIOLOGY, PLASTIDS. *Educ:* Harvard Univ, PhD(physiol), 52. *Prof Exp:* Assoc prof, 59-65, PROF PLANT BIOCHEM, RUTGERS UNIV, 65- *Concurrent Pos:* Vis prof, Princeton Univ, 76; vis prof, Woods Hole Oceanographic Inst, 62-84; vis prof, Japan Soc Promotion Sci, 79; vis prof, Univ Ariz, 85-86. *Res:* Protein Synthesis. *Mailing Add:* Waksman Inst Microbiol Rutgers Univ Piscataway NJ 08855-0759

PRICE, CHARLES MORTON, b Chicago, Ill, Apr 12, 26; m 51, 87; c 3. MATHEMATICS. *Educ:* Univ Chicago, BPh & BS, 44, MS, 47, PhD(math), 50. *Prof Exp:* Instr math, Ill Inst Technol, 46-53; supvr design anal, NAm Aviation, Inc, 53-61; head, Satellite Navig Dept, Aerospace Corp, 61-89; RETIRED. *Concurrent Pos:* Sr mathematician, Inst Air Weapons Res, 47-53. *Mem:* Am Math Soc; Soc Indust & Appl Math; Am Inst Aeronaut & Astronaut. *Res:* Orbit determination and prediction; atmosphere and gravitational potential of the earth; optimal filtering theory; matrix theory; numerical and operations analysis; technical management. *Mailing Add:* 340 Via El Chico Redondo Beach CA 90277

PRICE, CHARLES R(ONALD), b Pocahontas, Va, Sept 17, 32; m 61; c 2. CHEMICAL ENGINEERING. *Educ:* Va Polytech Inst, BS, 53, MS, 56, PhD(chem eng), 58. *Prof Exp:* Asst plant mgr, Southern Cotton Oil Co, Ga, 53-56; design engr, Union Carbide Chem Co Div, Union Carbide Corp, 56-58; res engr, E I du Pont de Nemours & Co, 58-62, tech investr econ eval, 62-63; sr res engr, J M Huber Corp, 63-71, tech dir, Clay Div, Res Dept, 71-85; RETIRED. *Mem:* Am Chem Soc; Am Inst Chem Engrs; Can Pulp & Paper Asn; Am Ceramic Soc; Tech Asn Pulp & Paper Indust. *Res:* Characterization of molecular sieves; processing of highly viscous liquids; cellulosic and plastic films development; economic evaluations; carbon black process and product development; research and development activity with clay products and processes. *Mailing Add:* 4655 Savage Hills Dr Macon GA 31210

PRICE, DAVID ALAN, b New York, NY, July 16, 48; m 74; c 4. NEUROPEPTIDE CHEMISTRY & FUNCTION. *Educ:* Cooper Union, BS, 70; Fla State Univ, PhD(molecular biophys), 77. *Prof Exp:* Res assoc, Fla State Univ, 75-81; res assoc, 81-84, asst res scientist, 84-89, ASSOC RES SCIENTIST, WHITNEY LAB, UNIV FLA, 89- *Concurrent Pos:* Vis lectr neuropeptides & neurosecretion, St Andrews Univ, Scotland, 82. *Mem:* Soc Neurosci; Am Chem Soc. *Res:* Characterization of neuropeptides, particularly those found in invertebrate groups; peptide sequencing; radioimmunoassay and purification. *Mailing Add:* Whitney Lab Univ Fla 9505 Ocean Shore Blvd St Augustine FL 32086-8623

PRICE, DAVID C, b Toronto, Ont, July 3, 34; div; c 2. NUCLEAR MEDICINE. *Educ:* Univ Toronto, MD, 58; FRCP(C), 64; ABNM, 72. *Prof Exp:* Res assoc, Med Res Ctr, Brookhaven Nat Lab, 60-62; Med Res Coun Can fel, Donner Lab, Univ Calif, Berkeley, 65-66; scientist, Lawrence Radiation Lab, 67; from asst prof to assoc prof 68-80, PROF RADIOL & MED, SCH MED, UNIV CALIF, SAN FRANCISCO, 80- *Concurrent Pos:* Bd dir, Am Col Nuclear Physicians, 82; pres, Northern Calif Chapter, & chmn, Western Regional Steering Comt, 83-84, chmn, Credentials & Mem Comt, Soc Nuclear Med, 85; pres, Calif Chapter & chmn bd dir, Am Col Nuclear Physicians, 88-89. *Mem:* Royal Col Physicians Can, 64; AAAS; Am Soc Hemat. *Res:* Hematologic applications of radioisotopes; iron kinetics; vitamin B-12 metabolism; isotopic imaging devices; whole body counting; experimental hematology, particularly radiation effects on bone marrow; radioimmunodiagnosis and therapy of tumors. *Mailing Add:* Nuclear Med Box 0252 Univ Calif Med Sch 455 Sci Bldg San Francisco CA 94143-0252

PRICE, DAVID CECIL LONG, b London, Eng, Jan 17, 40; m 89; c 2. CONDENSED MATTER PHYSICS. *Educ:* Cambridge Univ, BA, 61, PhD(physics), 66. *Prof Exp:* Res assoc physics, Brookhaven Nat Lab, 66-68; from asst physicist to assoc physicist, 68-74, dir, Solid State Sci Div, 74-79, dir, Intense Pulsed Neutron Source Prog, 79-81, SR PHYSICIST, ARGONNE NAT LAB, 74- *Mem:* Fel Am Phys Soc; AAAS. *Res:* Structure and dynamics of glasses and liquids; neutron scattering. *Mailing Add:* Mat Sci Div Argonne Nat Lab Argonne IL 60439

PRICE, DAVID EDGAR, b San Diego, Calif, July 5, 14; m 36; c 2. PUBLIC HEALTH ADMINISTRATION. *Educ:* Univ Calif, AB, 36, MA, 37, MD, 40; Johns Hopkins Univ, MPH, 45, DrPH, 46; Am Bd Prev Med, dipl, 50. *Prof Exp:* Med intern, USPHS, DC, 40-41, epidemiologist, San Diego City & County Health Dept, 41-42, staff physician, Venereal Dis Ctr, Ark, 42-44, fever therapy trainee & clin dir, 44; Rockefeller Found fel, Johns Hopkins Univ, 44-46; asst to chief, Res Grants Div, NIH, 46-47, chief cancer res grants br, Nat Cancer Inst, 47-48, chief div res grants & fels, NIH, 48-50, assoc dir, 50-52, asst surgeon gen, 52-57, dep chief bur med serv, 57, chief bur state serv, 57-60, dep dir, NIH, 60-62, dep surgeon gen, USPHS, 62-65; consult, Ford Found, India, 65-67; dir planning, Med Insts, 67-75, prof pub health admin, 67-80, dir res progs, 75-80, EMER PROF HEALTH SERV ADMIN & DIV PUB HEALTH ADMIN, JOHNS HOPKINS UNIV, 80- *Res:* Experimental and pituitary endocrinology; population; family planning. *Mailing Add:* 1405-B Ocean Pines Berlin MD 21811-9138

PRICE, DAVID THOMAS, b Chicago, Ill, Nov 21, 43; m 69. ALGEBRA. *Educ:* Calif Inst Technol, BS, 65; Univ Chicago, MS, 67, PhD(math), 71. *Prof Exp:* Asst prof, 69-74, assoc prof, 74-80, PROF MATH, WHEATON COL, 80- *Concurrent Pos:* Consult software, Western Elec Corp, 80-83 & consult financial models, Chicago Res & Trading, 83- *Mem:* Math Asn Am; Am Math Soc; Asn Comput Mach; Soc Indust & Appl Math. *Res:* Nilpotent rings, finite groups. *Mailing Add:* 611 E Prairie Ave Wheaton IL 60187-3824

PRICE, DONALD RAY, b Rockville, Ind, July 20, 39; m 63; c 4. ENGINEERING, AGRICULTURE. *Educ:* Purdue Univ, BS, 61, PhD(agr eng), 71; Cornell Univ, MS, 63. *Prof Exp:* Test engr hydraul, Int Harvester Co, 60-61; exten specialist agr eng, Cornell Univ, 62-63, asst prof, 63-69, assoc prof, 69-77, prof & dir energy progs, 77-80; from asst dean to assoc dean, 80-83, dean grad studies & res, 83, PROF, UNIV FLA, 80-, VPRES RES, 84-, PRES, RES FOUND, 87- *Concurrent Pos:* NY Farm Electrification res grant, Cornell Univ, 64-72; proj leader, NY Farm Electrification Coun, 64-72; owner, Donormac Corp, 68-72; Northeast rep, Nat Food & Energy Coun, 75-; prog mgr, US Dept of Energy, 77-78; consult, President Carter's Reorgn Proj, 78- *Mem:* Am Soc Agr Engrs; Illum Eng Soc; Nat Coun Res Admin; Soc Res Admin. *Res:* Systems engineering; environmental quality; electric power and processing; environmental control; thermal pollution. *Mailing Add:* Grinter Hall Univ Fla Gainesville FL 32611

PRICE, DONNA, b Baltimore, Md, Oct 23, 13. PHYSICAL CHEMISTRY, EXPLOSIVES. *Educ:* Goucher Col, AB, 34; Cornell Univ, PhD(phys chem), 37. *Hon Degrees:* DSc, Goucher Col, 74. *Prof Exp:* Instr chem, Rockford Col, 38-40; Berliner fel, Radcliffe Col, 40-41; fel, Univ Chicago, 42; res chemist, Hercules Powder Co, Del, 42-48; sr scientist, Naval Surface Warfare Ctr, White Oak, 49-71, actg chief phys chem div, 61-62, sr res scientist, explosives div, 71-80; retired annuitant, 80-86; sr res chemist, Advan Technol & Res, Inc, 86-90; RETIRED. *Concurrent Pos:* Consult, Am Marietta Co, 55 & Spec Proj Off & Armed Serv Explosives Safety Bd, 61-80; lectr explosives, Brit Ministry of Supply, 59; mem ad hoc comts on safety, US Air Force; mem, Fourth Detonation Symp Comn, 65, Fifth, 70; mem propellant hazard assessment panel, Strategic Systs Proj Off, Navy, Dept of Defense, 72-86. *Honors & Awards:* Fliedner Trophy Award, US Bur Naval Weapons, 62. *Mem:* Am Chem Soc; Combustion Inst; Am Phys Soc. *Res:* Characterization of explosives; thermodynamics of dense gases and of explosives; sensitivity of explosives and propellants. *Mailing Add:* 13801 York Rd Cockeysville MD 21030

PRICE, EDWARD HECTOR, b Baltimore, Md, Nov 25, 20; m 45; c 3. ORGANIC CHEMISTRY. *Educ:* Univ Md, BS, 42, PhD(org chem), 49. *Prof Exp:* Res chemist, EI du Pont de Nemours & Co, Inc, 49-57, supvr plants technol, 57-59, sr supvr res & develop, 59-66, lab mgr, 66-70, res mgr, 70-77, res mfg mgr, Plastics Prod Div, Plastics Dept, 78-85; TECH CONSULT, 85- *Concurrent Pos:* Int negotiations Japan & USSR. *Mem:* Electrochem Soc; Am Chem Soc; Sigma Xi. *Res:* Podophyllotoxin; polymer and fluorochemical chemistry and electrochemical membranes; process product development; research management. *Mailing Add:* 726 Isaac Taylor Dr W Chester PA 19380

PRICE, EDWARD WARREN, b Detroit, Mich, Dec 6, 20; m 51; c 3. COMBUSTION, ROCKET PROPULSION. *Educ:* Univ Calif, Los Angeles, BA, 48. *Prof Exp:* Ord engr rocket propulsion, Calif Inst Technol, 41-44 & US Naval Ord Test Sta, 44-46, physicist gas dynamics, 48-56, head gas dynamics br, 56-57, head aerothermochem group, 57-60, head, Aerothermochem Div, 60-74; prof aerospace eng, 67-68 & 74-85, REGENT'S PROF, AEROSPACE ENG, GA INST TECHNOL, 85- *Concurrent Pos:* Lectr, Univ Conn, 58-64; chmn combustion subcomt, Joint Army-Navy-NASA-Air Force Interagency Propulsion Comt, 64-72; dir, Tech Am Inst Aeronaut & Astronaut, 66, vpres, 67-68; mem, Nat Res Coun Panel Eval Space Shuttle Booster Redesign Prog, 86-89. *Honors & Awards:* Dryden Res Award, Am Inst Aeronaut & Astronaut, 67, G Edward Pendray Award, 71, R H Goddard Award, 76; Super Civilian Serv Medal, USN, 74; Cert Recognition, Joint Army-Navy-NASA-Air Force, 85; Pub Serv Award, NASA, 88, Astronauts Silver Snoopy Award, 89. *Mem:* Fel Am Inst Aeronaut & Astronaut; Int Combustion Inst; AAAS. *Res:* High temperature gas dynamics; combustion and rocket propulsion; solid rocket propellant combustion; combustor instability. *Mailing Add:* Sch Aerospace Eng Ga Inst Technol Atlanta GA 30332-1501

PRICE, ELTON, b Pitt Co, NC, Feb 5, 33; m 55; c 3. PHYSICAL CHEMISTRY. *Educ:* Howard Univ, BS, 54; Yale Univ, MS, 56; Boston Univ, PhD(chem), 61. *Prof Exp:* Lab asst chem, Yale Univ, 54-56; fel, Pa State Univ, 60-62; mem tech staff, Bell Tel Labs, 62-64; from asst prof to assoc prof, 64-71, PROF CHEM, HOWARD UNIV, 71- *Mem:* Am Chem Soc; Sigma Xi. *Res:* Chemical kinetics; application of nuclear magnetic resonance techniques to study relaxation phenomena of molecules in the liquid and solid states. *Mailing Add:* Dept of Chem Howard Univ Washington DC 20001

PRICE, FRANK DUBOIS, dielectric materials, engineering, for more information see previous edition

PRICE, FREDERICK WILLIAM, b London, Eng, Mar 4, 32. MOLECULAR BIOLOGY, BIOCHEMISTRY. *Educ:* Univ Bristol, BSc, 53; Univ London, PhD(biochem), 63. *Prof Exp:* Chemist, Int Chem Co, Eng, 56-57; res assoc enzyme assay, Brit Empire Cancer Campaign Unit, Dept Zool, King's Col, Univ London, 57-62; res assoc, Dept Biochem Pharmacol, State Univ NY Buffalo, 63-64; cancer res scientist, Viral Oncol Dept, Roswell Park Mem Inst, 64-67; assoc prof biol, 67-74, NSF grant, 72-73, PROF BIOL, STATE UNIV NY COL BUFFALO, 74- *Concurrent Pos:* Res assoc mycology, Buffalo Mus Sci, 81-84. *Mem:* AAAS; Brit Astron Asn; Nat Asn Biol Teachers; Sigma Xi. *Res:* Reactions of carbohydrates with cuprammonium compounds; spectrophotometric methods of enzyme assay; toxohormone; precipitation reactions in biocolloids; methods of biochemical analysis; thin layer chromatography of plant lipids. *Mailing Add:* Dept Biol SUNY 1300 Elmwood Ave Buffalo NY 14222

PRICE, GLENN ALBERT, b Minn, Feb 9, 23; m 50; c 4. ACADEMIC ADMINISTRATION, REACTOR PHYSICS. *Educ:* Univ Ky, BS, 46; Univ Ill, MS, 48, PhD(physics), 52. *Prof Exp:* Assoc physicist, Brookhaven Nat Lab, 52-58, physicist, 58-85, head, Off Acad Rels, 73-85, mem staff, 52-85; training sect, IAEA Safeguards, 87-89; RETIRED. *Concurrent Pos:* Vis scientist, Atomic Energy Res Estab-Harwell, 58-59; consult, India AEC, 67. *Mem:* Am Phys Soc; Am Nuclear Soc. *Res:* Photoneutrons; reactor physics. *Mailing Add:* 92 Pondview Rd Weare NH 03281-9802

PRICE, GRIFFITH BALEY, b Brookhaven, Miss, Mar 14, 05; m 40; c 6. MATHEMATICS. *Educ:* Miss Col, AB, 25; Harvard Univ, MA, 28, PhD(math), 32. *Hon Degrees:* LLD, Miss Col, 62. *Prof Exp:* Instr math, Miss Col, 25-26 & 29-30, Union Univ, NY, 32-33, Univ Rochester, 33-36 & Brown Univ, 36-37; from asst prof to prof, 37-75, chmn dept, 51-70, EMER PROF MATH, UNIV KANS, 75- *Concurrent Pos:* Opers anal, Hq Eighth Air Force, Eng, 43-45; Guggenheim fel, 46-47; ed bull, Am Math Soc, 50-57; chmn comt regional develop math, Nat Res Coun, 52-54; vis prof, Calif Inst Technol, 59-60; exec secy, Conf Bd Math Sci, Washington, DC, 60-62; mem, US Nat Comn, UNESCO, 61-66; trustee, Argonne Univs Asn, 65-69; vis prof, Univ Western Australia, Perth, 75; consult, Orgn Am States, 78. *Honors & Awards:* Math Asn Am Award, 70. *Mem:* Am Math Soc; Nat Coun Teachers Math; Math Asn Am (vpres, 55-56, pres, 57-58); AAAS; NY Acad Sci; Sigma Xi. *Res:* Analysis: functions of several variables, multicomplex spaces and functions. *Mailing Add:* Dept Math Univ Kans Lawrence KS 66045-2142

PRICE, HAROLD ANTHONY, b Glenwood, Minn, May 27, 19; m 52; c 1. ORGANIC CHEMISTRY. *Educ:* Univ Ill, BS, 45; Ind Univ, AM, 46; Purdue Univ, PhD(chem), 50. *Prof Exp:* Asst prof res, Mich State Univ, 50-57; res chemist, Dow Chem Co, 57-63; lab mgr, Allied Paper Corp, 64-65; chief, Environ Epidemiol Lab, Mich Dept Pub Health, 65-85; RETIRED. *Mem:* Am Chem Soc. *Res:* Polyolefins; pesticide residue analysis and industrial chemical residue analysis in biological systems. *Mailing Add:* 3406 Aragon Dr Lansing MI 48906-3505

PRICE, HAROLD JAMES, b Bremerton, Wash, Oct 7, 43; m 66; c 2. PLANT EVOLUTION & CYTOGENETICS. *Educ:* Western Wash State Col, BA, 65; Brigham Young Univ, MS, 67; Univ Calif, Davis, PhD(genetics), 70. *Prof Exp:* Res assoc, Brookhaven Nat Lab, 70-72; asst prof biol, Fla Technol Univ, 72-74; assoc prof, 75-84, PROF GENETICS, TEX A&M UNIV, 84- *Concurrent Pos:* Vchmn, Genetics Sect, Bot Soc Am, 85-87, chmn, 87-89; fac senator, Tex A&M Univ, 87; mem coun, Bot Soc Am, 88-89. *Mem:* Genetics Soc Am; Bot Soc Am; AAAS. *Res:* Plant evolution and cytogenetics, with emphasis on the interactions of DNA content and genotype on plant development and evolution; plant genome organization; cotton genetics. *Mailing Add:* Dept Soil & Crop Sci Tex A&M Univ College Station TX 77843

PRICE, HAROLD M, b Chicago, Ill, Aug 24, 31; m 54; c 4. PATHOLOGY. *Educ:* Univ Southern Calif, BA, 53, MD, 57; Am Bd Path, dipl; Univ Pittsburgh, MPH, 73. *Prof Exp:* Intern, Univ Hosps, Cleveland, 57-58; resident path, Western Reserve Univ, 58-61; USPHS fel, Brain Res Inst, Univ Calif, Los Angeles, 61-62; pathologist, Armed Forces Inst Path, 62-64; resident clin path, Walter Reed Hosp, 64-65; pathologist, Armed Forces Inst Path, 65-67; assoc prof path, Sch Med, Univ Pittsburgh, 67-78; pathologist, Magee-Womens Hosp, 67-78; dir anatomic path, 78-87, DIR LAB, VALLEY MED CTR, FRESNO, 87-; ASSOC CLIN PROF PATH, SCH MED, UNIV CALIF, SAN FRANCISCO, 78- *Mem:* AAAS; fel Soc Clin Path; Am Asn Pathologists; fel Col Am Pathologists; NY Acad Sci. *Res:* Disorders affecting skeletal, cardiac and smooth muscle; electron microscopy. *Mailing Add:* Dept of Path Valley Med Ctr Fresno CA 93702

PRICE, HARRY JAMES, b Reading, Pa, June 1, 41; m 64; c 4. INORGANIC CHEMISTRY. *Educ:* Univ Rochester, BS, 63; Stanford Univ, PhD(inorg chem), 67. *Prof Exp:* RES ASSOC, RES LABS, EASTMAN KODAK CO, 66- *Mem:* Am Chem Soc; Soc Photog Sci & Eng. *Res:* Mechanisms of inorganic electron transfer reactions; photographic processing chemistry; silver halide surface chemistry. *Mailing Add:* 1629 Bridgeboro Dr Webster NY 14580

PRICE, HARVEY SIMON, applied mathematics, for more information see previous edition

PRICE, HENRY LOCHER, b Philadelphia, Pa, Oct 21, 22; m 53; c 2. PHARMACOLOGY, ANESTHESIOLOGY. *Educ:* Swarthmore Col, AB, 45; Univ Pa, MD, 46; Am Bd Anesthesiol, dipl, 53. *Prof Exp:* Chief oper room sect, Tripler Gen Hosp, 49-50; from instr to res assoc anesthesiol, Sch Med, Univ Pa, 50-52; Nat Res Coun fel med sci, Harvard Med Sch, 52-53; from asst prof to prof anesthesiol, Sch Med, Univ Pa, 53-70; prof, Sch Med, Temple Univ, 70-71; chmn dept, 71-78, prof, 71-87, EMER PROF ANESTHESIOL, HAHNEMANN MED COL, 87- *Concurrent Pos:* Mem comt anesthesia, Nat Res Coun, 59-66; NIH fel, Univ Calif, San Francisco, 60-61; prin investr, USPHS Grants; mem clin res training comn, Nat Inst Gen Med Sci, 63-65 & 66-67, mem nat anesthesiol training comn, 66-67; consult, US Naval & Vet Admin Hosps, Philadelphia. *Honors & Awards:* Huffeld Mem Lectr, Copenhagen, 72. *Mem:* Am Soc Clin Invest; Sigma Xi; Am Physiol Soc; Am Soc Pharmacol; Am Soc Anesthesiol. *Res:* Applied physiology; pharmacology of anesthetic agents. *Mailing Add:* 510 Lynmere Rd Bryn Mawr PA 19010

PRICE, HOWARD CHARLES, b South Gibson, Pa, Feb 26, 42; m 67; c 2. BIOCHEMISTRY, ORGANIC CHEMISTRY. *Educ:* Dickinson Col, BS, 63; State Univ NY Binghamton, PhD(org chem), 71. *Prof Exp:* NIH fel, Dept Neurol, Albert Einstein Col Med, Yeshiva Univ, 70-71; from asst prof to assoc prof chem, Marshall Univ, 71-81; mem staff, 81-84, mgr & group mgr res, 84-89, DIR, RES LABS & ADVAN TECHNOL, ZIMMER INC, 89- *Mem:* Am Chem Soc; AAAS; Sigma Xi; Soc Plastic Engrs; Soc Advan Mat & Process Eng; Am Soc Testing & Mat. *Res:* Synthesis, chiroptical phenomena and biochemistry of amino acid analogues; gangliosides; neurochemistry; medical polymers and devices; engineering composites; bioresorbable materials; non-destructive test methods. *Mailing Add:* Zimmer Inc PO Box 708 Warsaw IN 46581

PRICE, HUGH CRISWELL, b Newark, Ohio, Nov 1, 39; m 65; c 2. HORTICULTURE. *Educ:* Ohio State Univ, BS, 61; Univ Del, MS, 66; Mich State Univ, PhD(hort), 69. *Prof Exp:* Biologist herbicide res, Am Cyanamid Co, 69-71; asst prof, 71-74, assoc prof, 74-80, PROF HORT, MICH STATE UNIV, 80- *Mem:* Am Soc Hort Sci; Sigma Xi. *Res:* Growth and development of vegetable crops. *Mailing Add:* Dept Hort NYS Agr Exp Sta Geneva NY 14456

PRICE, JAMES CLARENCE, b London, Ark, Nov 20, 32; m 61; c 4. PHARMACY. *Educ:* Univ Ark, BS, 53; Univ Utah, MS, 57; Univ RI, PhD(pharm), 63. *Prof Exp:* From instr to assoc prof pharm, Univ RI, 61-68; ASSOC PROF PHARM, UNIV GA, 68- *Mem:* Am Pharmaceut Asn; Acad Pharmaceut Sci; Sigma Xi. *Res:* Microencapsulation; tablet technology; suspension-emulsion stability. *Mailing Add:* Col Pharm Univ Ga Athens GA 30601

PRICE, JAMES F, b Columbia, Tenn, Apr 13, 32; m 59; c 2. FOOD SCIENCE. *Educ:* Univ Tenn, BS, 54; Mich State Univ, MS, 56, PhD(animal sci), 60. *Prof Exp:* Specialist livestock & meat mkt, USDA, 56; asst prof animal sci, Auburn Univ, 60-62; asst prof food sci, 63-77, PROF FOOD SCI & HUMAN NUTRIT, MICH STATE UNIV, 77- *Mem:* Am Soc Animal Sci; Am Meat Sci Asn; Inst Food Technol; Sigma Xi. *Res:* Body composition of meat animals; post mortem changes in muscle; meat processing, preservation and microbiology-technologies; improve utilization and value of undesirable fish. *Mailing Add:* Dept Food Sci Mich State Univ Campus East Lansing MI 48823

PRICE, JAMES FELIX, b Hartselle, Ala, June 17, 47; m 68; c 2. ENTOMOLOGY. *Educ:* Univ Fla, BS, 72, MS, 74; Clemson Univ, PhD(entom), 77. *Prof Exp:* Asst prof, 78-83, ASSOC PROF ENTOM, UNIV FLA, 83- *Mem:* Entom Soc Am. *Res:* Biological, cultural and chemical control of insects and mites; floricultural entomolgy; ornamental entomology. *Mailing Add:* Gulf Coast Res & Educ Ctr 5007 60th St E Bradenton FL 34203

PRICE, JAMES FRANKLIN, b Washington, DC, Sept 9, 48; div; c 3. PHYSICAL OCEANOGRAPHY. *Educ:* Okla, BS, 70; Univ Miami, MS, 72, PhD(oceanog), 77. *Prof Exp:* RES ASSOC, GRAD SCH OCEANOG, UNIV RI, 77-; AT WOODS HOLE OCEANOGRAPHIC INST. *Res:* Dynamics of upper ocean; dynamics of oceanic eddies. *Mailing Add:* Woods Hole Oceanographic Inst Woods Hole MA 02543

PRICE, JAMES GORDON, b Brush, Colo, June 20, 26; m 49; c 4. FAMILY PRACTICE. *Educ:* Univ Colo, BA, 47, MD, 51; Am Bd Family Pract, dipl, cert, 70, re-cert, 76 & 82. *Prof Exp:* Intern, Denver Gen Hosp, 51-52; assoc clin instr med, Sch Med, Univ Colo, 67-79; assoc prof, 80-82, prof family pract & chmn dept, 82-90, DEAN, SCH MED, UNIV KANS MED CTR, 90- *Concurrent Pos:* Pvt pract gen med, Brush, Colo, 52-78; trustee, Family Health Found & Univ Colo Med Develop Fund. *Mem:* Inst Med-Nat Acad Sci; fel Am Acad Family Physicians (pres-elect, 72, pres, 73); Am Bd Family Pract (pres, 79); Am Acad Family Pract (pres, 73). *Mailing Add:* Univ Kans Med Ctr 39th & Rainbow Blvd Kansas City KS 66103

PRICE, JOEL MCCLENDON, SMOOTH MUSCLE PHYSIOLOGY, MECHANICS. *Educ:* Univ Calif, San Diego, PhD(bioeng), 76. *Prof Exp:* ASSOC PROF HEART & MUSCLE PHYSIOL, UNIV S FLA, 82- *Res:* Hypertension. *Mailing Add:* Col Med Univ SFla 12901 N 30th St Tampa FL 33612

PRICE, JOHN AVNER, b Ogdensburg, NY, July 13, 32; m 53; c 3. ORGANIC CHEMISTRY, CHEMICAL ENGINEERING. *Educ:* St Lawrence Univ, BS, 53; Univ Rochester, PhD(org chem), 57. *Prof Exp:* Res chemist elastomers, E I du Pont de Nemours & Co, Inc, 56-58; res chemist, 58-70, res assoc & group head new prod, 70-76, sect head info technol, Exxon Res & Eng Co, 76-81; vpres corp planning, Augsbury Orgn, 81-82; Prof, Union County Col, 82-84; MGR, ASHLAND CHEM CO, 84- *Concurrent Pos:* NSF grant, 77-78. *Mem:* Am Chem Soc; Am Inst Chem Engrs. *Res:* Management, administration and the innovation of information technology; mini-computers, distributed processing, micrographics, information networking. *Mailing Add:* 5501 Loch More Ct E Dublin OH 43017-9403

PRICE, JOHN C, b New Haven, Conn, Feb 17, 59. LOW TEMPERATURE PHYSICS. *Educ:* Yale Univ, BS, 80; Stanford Univ, PhD(physics), 86. *Prof Exp:* Lab asst, Dept Physics, Yale Univ, 77-80; res asst physics, Stanford Linear Accelerator Ctr, Stanford Univ, 81-85, res assoc, Dept Physics, 86-89; ASST PROF PHYSICS, UNIV COLO, 89- *Concurrent Pos:* NSF presidential young investr & Off Naval Res young investr, 90-; sci & eng fel, David & Lucille Packard Found, 90; Alfred P Sloan Found fel, 91. *Mem:* Am Phys Soc. *Res:* Solid state physics; low temperature physics; low temperature electronics; mesoscopic physics; nanostructure physics. *Mailing Add:* Dept Physics Campus Box 390 Univ of Colorado, Boulder Boulder CO 80302

PRICE, JOHN CHARLES, b Chicago, Ill, Feb 9, 37; m 62; c 2. REMOTE SENSING, METEOROLOGY. *Educ:* Calif Inst Technol, BS, 59; Univ Calif, Berkeley, PhD(plasma physics), 66. *Prof Exp:* Aerospace scientist, Theoret Studies Br, NASA Goddard Space Flight Ctr, 66-71 & Lab Meteorol & Earth Sci, 71-80; PHYSICIST, REMOTE SENSING RES LAB, BELTSVILLE AGR RES CTR, USDA, 80- *Concurrent Pos:* Proj scientist, Heat Capacity Mapping Mission, 76-80. *Mem:* Am Phys Soc; Am Geophys Union; Am Soc Photo Eng. *Res:* Plasma kinetic theory; fluid mechanics; atmospheric radiative transfer; remote sensing. *Mailing Add:* 707 Orchard Way Silver Springs MD 20904

PRICE, JOHN DAVID EWART, b Kingswood, Eng, Apr 26, 27; Can citizen; m 58; c 5. INTERNAL MEDICINE. *Educ:* McGill Univ, BSc, 50, MD, CM, 54; FRCPS(C), 61. *Prof Exp:* Can Life Ins fel, 57-58, from instr to assoc prof, 61-73, PROF MED, UNIV BC, 73- *Concurrent Pos:* Mem attend staff internal med, Vancouver Gen Hosp, 61- *Mem:* Fel Am Col Physicians; Am Soc Artificial Internal Organs; Can Soc Clin Invest; Can Soc Nephrology (pres); Am Soc Nephrology. *Res:* Renal disease; physiology, natural history and the treatment of medical diseases of the kidney; design, application and biochemistry of artificial kidneys. *Mailing Add:* 4303 W Tenth Ave Vancouver BC V6R 2H6 Can

PRICE, JONATHAN GREENWAY, b Danville, Pa, Feb 1, 50; m 72; c 2. GEOLOGY APPLIED ECONOMIC & ENGINEERING, GEOCHEMISTRY. *Educ:* Lehigh Univ, BA, 72; Univ Calif, Berkeley, MS, 75, PhD(geol), 77. *Prof Exp:* Geologist, Anaconda Co, 74-75 & US Steel Corp, 77-81; res scientist, Bur Econ Geol, Univ Tex, Austin, 81-88; DIR & STATE GEOLOGIST, NEV BUR MINES & GEOL, UNIV NEV, RENO, 88- *Concurrent Pos:* Adj asst prof geol, Bucknell Univ, 77-78. *Mem:* AAAS; Am Geophys Union; fel Geol Soc Am; Mineral Soc Am; Soc Econ Geologists; Soc Mining Metall & Explor-Am Inst Mech Engrs. *Res:* Geology and geochemistry of ore deposits, igneous petrology and geochemistry applied to environmental issues. *Mailing Add:* Nev Bur Mines & Geol Univ Nev MS 178 Reno NV 89557-0088

PRICE, JOSEPH EARL, b Denver, Colo, Sept 23, 30; m 63; c 2. PHYSICS, MICROCOMPUTER USAGE. *Educ:* Colo Col, BS, 52; Rice Univ, MS, 54, PhD(physics), 56. *Prof Exp:* Asst prof physics, Lamar State Col, 56-59; from asst prof to assoc prof, 59-67, chmn dept, 66-72 & 77-83, PROF PHYSICS, IDAHO STATE UNIV, 67- *Concurrent Pos:* Westinghouse Res Lab, 56 & Idaho Nat Engr Lab, 60, 68 & 70. *Mem:* Am Phys Soc; Sigma Xi; Am Asn Physics Teachers. *Res:* Use of computers in teaching situations; nuclear physics; programming microcomputers in assembler language for teaching and interfacing situations. *Mailing Add:* Dept of Physics Idaho State Univ PO Box 8106 Pocatello ID 83209-0009

PRICE, JOSEPH LEVERING, b Ala, Oct 17, 42; m 67; c 3. ANATOMY, NEUROSCIENCE. *Educ:* Univ South, BA, 63; Oxford Univ, BA, 66, DPhil(anat), 69. *Prof Exp:* From instr to asst prof anat, 69-76, assoc prof, 76-83, PROF ANAT & NEUROBIOL, SCH MED, WASH UNIV, 83- *Concurrent Pos:* USPHS res grant, 70-; assoc dir, Alzheimer's Dis Res Ctr, 85- *Honors & Awards:* C J Herrick Award Neuroanatomy, 73; Javits Neurosci Investr Award, 87. *Mem:* Am Asn Anat; Soc Neurosci. *Res:* Neuroanatomy; structure, organization and development of the cerebral cortex, with special reference to the limbic system. *Mailing Add:* Dept Anat & Neurobiol Wash Univ Sch Med 660 S Euclid Ave St Louis MO 63110

PRICE, KEITH ROBINSON, b Oxnard, Calif, Nov 24, 30; m 58; c 4. PLANT ECOLOGY, RADIOECOLOGY. *Educ:* Univ Calif, Davis, BS, 58; Wash State Univ, PhD(bot), 65. *Prof Exp:* Asst prof bot, Univ Nev, Reno, 64-68; mgr soil relationships sect, Pac Northwest Labs, Battelle Mem Inst, 68-71, sr res scientist, 71-74; staff environ scientist, Atlantic Richfield Hanford Co, 74-78; SR RES SCIENTIST, PAC NORTHWEST LAB, BATTELLE MEM INST, 78- *Mem:* AAAS; Am Inst Biol Sci; Ecol Soc Am; Health Physics Soc; Sigma Xi. *Res:* Behavior of radionuclides important to radioactive waste management in plants and soils; environmental management; environmental surveillance of radioactive and nonradioactive waste materials. *Mailing Add:* 715 Lynnwood Loop Richland WA 99352

PRICE, KENNETH ELBERT, b Cumberland, Md, Aug 12, 26; c 2. MICROBIOLOGY, IMMUNOLOGY. *Educ:* Univ Md, BS, 50, MS, 52, PhD(bact), 54. *Prof Exp:* Head dept bact, Agr Res & Develop Ctr, Chas Pfizer & Co, Inc, Ind, 54-60; coordr Cancer Res Prog, Bristol-Meyers Co, NY, 60-65, asst dir Microbiol Res, 65-70, dir, 70-76, assoc dir res & develop, 76-81, dir pre-clin res & develop, 81-85; consult, 85- 90; RETIRED. *Concurrent Pos:* Chmn, Gordon Res Conf Med Chem, 77; mem ed bd, Antimicrobiol Agents & Chemotherapy. *Mem:* AAAS; Am Soc Microbiol; NY Acad Sci; Infectious Dis Soc Am; Sigma Xi; Am Chem Soc. *Res:* Discovery, characterization and determination of mode of action and mechanisms of resistance of antibiotics and the chemotherapeutic and chemoprophylatic agents used for treatment of malignant and infectious diseases of animals and man; mechanisms of resistance to chemotherapeutic agents. *Mailing Add:* 1713 Ebb Dr Wilmington NC 28409-9572

PRICE, KENNETH HUGH, b Greenville, Tex. MATHEMATICS. *Educ:* Univ Tex, Austin, BA, 63, PhD(math), 70. *Prof Exp:* Asst prof, 70-77, ASSOC PROF MATH, STEPHEN F AUSTIN STATE UNIV, 77- *Concurrent Pos:* Software engr, 80-81; consult, 82-86. *Mem:* AAAS; Am Math Soc; Math Asn Am; Sigma Xi. *Res:* Approximation theory; functional analysis. *Mailing Add:* Dept Math Stephen F Austin State Univ Nacogdoches TX 75962

PRICE, KENT SPARKS, JR, b Trenton, NJ, Oct 2, 36; m 58; c 2. MARINE BIOLOGY. *Educ:* Univ Md, BS, 59; Univ Del, MS, 61, PhD(zool), 64. *Prof Exp:* Biol aid insecticide toxicity, Patuxent Refuge, US Dept Interior, Md, 59; asst prof biol, Old Dom Col, 64-67; asst prof biol sci & dir field sta, Marine Labs, 67-70, dir, Lewes Field Sta, 67-72, asst dean, Col Marine Studies, 70-72, assoc prof biol sci, 70-80, dir marine adv serv, 71-75, assoc dean, Col Marine Studies & dir, Lewes Marine Studies Ctr, 72-79, ASSOC PROF MARINE STUDIES, UNIV DEL, 72- *Concurrent Pos:* Assoc prog dir fisheries & aquaculture, Off Sea Grant, 79-81. *Res:* Ecology of estuaries; fisheries; water regulation and reproduction in Elasmobrachs; mariculture. *Mailing Add:* Univ Del 700 Pilot Town Rd Lewes DE 19958

PRICE, LAWRENCE EDWARD, b Altadena, Calif, Jul 25, 43; m 68; c 2. NUCLEON DECAY. *Educ:* Pomona Col, BA, 65; Harvard Univ, MA, 66, PhD(physics), 70. *Prof Exp:* Res assoc, physics dept, Columbia Univ, 70-72, asst prof physics, 72-78; from asst to physicist, 78-88, ASSOC DIV DIR, ARGONNE NAT LAB, 86-, SR PHYSICIST, 88- *Concurrent Pos:* Mem, Fastbus Review Comt, 80, Int Adv Comt, Gen Detector Res & Develop, SSC, 87-; chmn, Rev Comt, HEPNET, 87-88. *Mem:* Fel Am Phys Soc. *Res:* Physics of elementary particles; nucleon decay; study of E+E- collisions; hadronic elastic scattering; ssc detector. *Mailing Add:* Argonne Nat Lab 9700 Cass Ave Bldg 362 Argonne IL 60439

PRICE, LAWRENCE HOWARD, b Detroit, Mich, July 26, 52; m 82; c 2. BIOLOGICAL PSYCHIATRY, PSYCHOPHARMACOLOGY. *Educ:* Univ Mich, BS, 74, MD, 78. *Prof Exp:* Instr, 82-83, asst prof, 83-89, ASSOC PROF PSYCHIAT, YALE UNIV, 89- *Concurrent Pos:* Intern, Norwalk Hosp, Yale Univ, 79, resident, 82; consult, Conn Valley Hosp, 83- *Mem:* Soc Neurosci; Am Col Neuropsychopharm. *Res:* Depression and neuro disorders; etiology; refractory depression; obsessive compulsive disorders. *Mailing Add:* Clin Neurosci Res Unit Conn Mental Health Ctr Ribicoff Res Facil 34 Pk St New Haven CT 06508

PRICE, LEONARD, b New Orleans, La, Nov 3, 33. ORGANIC CHEMISTRY. *Educ:* Xavier Univ, La, BS, 57; Univ Notre Dame, PhD(chem), 62. *Prof Exp:* From asst prof to assoc prof, 61-63, PROF CHEM, XAVIER UNIV, LA, 63-, CHMN DEPT, 70- *Concurrent Pos:* Am Chem Soc-Petrol Res Fund fel fundamental res undergrad level, 64-66; NSF fel undergrad res participation proj, 64-65. *Res:* Synthesis of acetylenic amino alcohols and amino ethers. *Mailing Add:* Dept Chem Xavier Univ La Palmetto & Pine Sts New Orleans LA 70125

PRICE, MARTIN BURTON, b New York, NY, June 6, 28; m 48; c 3. ORGANIC CHEMISTRY. *Educ:* Syracuse Univ, BA, 48, MS, 50; Univ Del, PhD, 56. *Prof Exp:* Res chemist, Pa Indust Chem Co, 50-57, group leader, 57-60; res assoc, Celanese Res Co, Celanese Corp, 60, group leader, 60-63, sect head, 63-71, mgr phys chem & prod technol, 71-73; dir res & develop, Reliance Universal Inc, 73-85, vpres, 85-89; VPRES, AKZO COATINGS INC, 89- *Mem:* AAAS; Am Chem Soc; Sigma Xi. *Res:* Polymers, preparation and characterization; polyacetals; polyolefins; chemical coatings for wood and metal industries; gas chromatography; ultraviolet light reactions. *Mailing Add:* Reliance Universal Inc 4730 Crittenden Dr Louisville KY 40233

PRICE, MARY VAUGHAN, b Bethesda, Md, July 19, 49; m. COMMUNITY ECOLOGY, EVOLUTIONARY ECOLOGY. *Educ:* Vassar Col, AB, 71; Univ Ariz, PhD(zool), 76. *Prof Exp:* Teaching asst biol, dept ecol & evolutionary biol, Univ Ariz, 71-76; Teaching fel biol, Dalhousie Univ, 76-77; assoc instr, Univ Utah, 77-79; from asst prof to assoc prof, 79-90, PROF BIOL, UNIV CALIF, RIVERSIDE, 90- *Concurrent Pos:* Trustee, Rocky Mountain Biol Lab, 81- *Mem:* Ecol Soc Am; Am Soc Mammalogists; Soc Study Evolution; Soc Conserv Biol; Am Soc Naturalists. *Res:* Factors determining the structure of natural communities; desert rodents; plant-pollinator interactions; evolution of plant breeding systems. *Mailing Add:* Dept Biol Univ Calif Riverside CA 92521-0427

PRICE, MAUREEN G, b Rochester, NY, June 26, 51; m 86; c 1. CELL BIOLOGY. *Educ:* Univ Pa, PhD(anat), 80. *Prof Exp:* Res fel, Col Inst Technol, 80-82; asst research physiol, Univ Calif, Los Angeles, 82-85; RES ASSOC MOLECULAR BIOL, SCRIPPS CLIN & RES FOUND, 85- *Concurrent Pos:* Fel, NIH, 81-83,; sr investr, Am Heart Asn, 83-84. *Honors & Awards:* Lambisch Award, 84 & 85. *Mem:* AAAS; Sigma Xi; Am Soc Cell Biol; Am Asn Anat. *Res:* Structure and function relationships, cytoskeleton, mitochondria. *Mailing Add:* Res Inst Scripps 10666 N Torrey Pines Rd La Jolla CA 92037

PRICE, MICHAEL GLENDON, b Washington, DC, July 27, 47; m 82; c 1. ELECTRICAL ENGINEERING. *Educ:* Univ Va, BS & ME, 75, PhD(elec eng), 78. *Prof Exp:* Res engr, Res Labs for Eng Sci, Univ Va, 72-77; anal engr controls & dynamic systs, United Technol Res Ctr, 77-78; sr scientist, Amecom Div, Litton Systs, Inc, 78-85; mgr, Syst Eng & Develop Corp, 85-86; CEO, SYSTEKA, INC, 86- *Concurrent Pos:* Design engr, Dept Pharmacol, Univ Va, 75-77. *Mem:* Inst Elec & Electronics Engrs; Sigma Xi. *Res:* Modelling and simulation of dynamic physical systems; analysis and synthesis of automatic control systems; acousto/optic wideband receiver development; optical signal processing. *Mailing Add:* 9525 Worrell Ave Lamham MD 20706-4032

PRICE, PAUL ARMS, b Inglewood, Calif, Apr 17, 42; m 78; c 3. BIOCHEMISTRY. *Educ:* Pomona Col, BA, 64; Rockefeller Univ, PhD(biochem), 68. *Prof Exp:* From asst prof to assoc prof, 68-80, PROF BIOL, UNIV CALIF, SAN DIEGO, 81- *Mem:* AAAS; Am Soc Biol Chem; Am Chem Soc; Am Soc Bone & Mineral Res. *Res:* Protein structure and function; chemical modification of proteins; biochemistry of the gamma carboxyglutamic acid containing bone protein and of other mineralized tissue proteins; biochemical and physiological mechanisms of action for vitamin K and vitamin D. *Mailing Add:* Dept Biol Univ Calif at San Diego La Jolla CA 92093

PRICE, PAUL BUFORD, JR, b Memphis, Tenn, Nov 8, 32; m 58; c 4. EXPERIMENTAL PHYSICS. *Educ:* Davidson Col, BS, 54; Univ Va, MS, 56, PhD(physics), 58. *Hon Degrees:* ScD, Davidson Col, 73. *Prof Exp:* Fulbright fel, H H Wills Physics Lab, Bristol Univ, 58-59; NSF fel, Cavendish Lab, Cambridge Univ, 59-60; physicist, Res Lab, Gen Elec Co, 60-65, physicist, Res & Develop Ctr, 65-69; Adolf C & Mary Sprague Miller res prof, 72-73, dir, Space Sci Lab, 79-85, PROF PHYSICS, UNIV CALIF, BERKELEY, 69-, CHMN PHYSICS DEPT, 87-, McADAMS CHAIR PHYSICS, 90- *Concurrent Pos:* Adj prof metall dept, Rensselaer Polytech Inst, 61; vis scientist, Lawrence Radiation Lab, 63; Fulbright sr vis prof, Tata Inst Fundamental Res, India, 65-66; consult, NASA, Lunar Sample Analysis Planning Team, 71-72; mem space sci bd, Nat Acad Sci & mem comt space astron & astrophys, 73-76; Guggenheim fel, 76-77; mem bd dir, Terradex Corp, 81-86; vis prof, Univ Rome, 83; sci assoc, CERN, 84; chmn class 1, phys & math sci, Nat Acad Sci, 88- *Honors & Awards:* Co-recipient, Am Nuclear Soc Award, 64 & Ernest O Lawrence Award, AEC, 71; Exceptional Sci Achievement Medal, Nat Aeronaut & Space Admin, 73. *Mem:* Nat Acad Sci; fel Am Phys Soc; fel Am Geophys Union; Am Astron Soc. *Res:* Cosmic rays; solar system physics; nuclear tracks and radiation effects in solids; relativistic heavy ion physics; high energy interactions; experimental cosmology. *Mailing Add:* Dept Physics Univ Calif Berkeley CA 94720

PRICE, PAUL JAY, cell biology, for more information see previous edition

PRICE, PETER J, b London, Eng, July 29, 24; m 56; c 2. THEORETICAL PHYSICS. *Educ:* Oxford Univ, BA, 48; Cambridge Univ, PhD, 51. *Prof Exp:* Asst lectr, Univ Southampton, 46-47; Off Naval Res, res asst, Duke Univ, 51-52; mem, Inst Advan Study, 52-53; MEM RES STAFF, IBM CORP, 53- *Mem:* Fel Am Phys Soc. *Res:* Solid state physics. *Mailing Add:* Watson Res Ctr IBM Corp PO Box 218 Yorktown Heights NY 10598

PRICE, PETER MICHAEL, b Rockville Center, NY, Aug 7, 46; m 71; c 1. MOLECULAR BIOLOGY. *Educ:* Cornell Univ, BS, 68; NY Univ, New York, PhD(genetics), 73. *Prof Exp:* Teaching fel pediat, Mt Sinai Sch Med, 71- 73, instr, 73-75, assoc, 75-77, sr instr, 77-83, asst prof biochem, 83-88, ASST PROF PEDIAT, MT SINAI SCH MED, 79-, ASSOC PROF BIOCHEM, 88- *Concurrent Pos:* Vis scientist, Univ Calif, San Francisco, 79. *Mem:* Am Soc Biol Chemists. *Res:* Hepatitis B virus replication and its relationship to host genome integration as a mechanism of induction of heptocellular carcinoma; using bacterial and eukaryotic expression vectors for production of hepatitis protein. *Mailing Add:* Mt Sinai Sch Med One Gustav Levy Pl New York NY 10029

PRICE, PETER WILFRID, b London, Eng, Apr 17, 38; m 68; c 2. ECOLOGY, ENTOMOLOGY. *Educ:* Univ Wales, BSc, 62; Univ NB, MSc, 64; Cornell Univ, PhD(ecol), 70. *Prof Exp:* Res officer entom, Can Dept Fisheries & Forestry, 64-70, res scientist, 70-71; from asst prof to assoc prof entom, Univ Ill, Urbana, 71-79; assoc prof biol, 80-85, PROF BIOL, NORTHERN ARIZ UNIV, 85- *Concurrent Pos:* Asst prof, Ill Agr Exp Sta, 74-79; asst entomologist, Ill Natural Hist Surv, 74-79; Guggenheim fel, 77-78; NSF panelist, pop biol, 79-81, ecol, 91-93. *Honors & Awards:* Burlington Found Award, 89. *Mem:* Ecol Soc Am; Brit Ecol Soc; Entom Soc Am; Entom Soc Can; Soc Study Evolution; Sigma Xi; fel AAAS. *Res:* Three trophic level interactions; plant and insect interactions; community organization of insects. *Mailing Add:* Dept Biol Sci Box 5640 Northern Ariz Univ Flagstaff AZ 86011-5640

PRICE, RALPH LORIN, b Arco, Idaho, Sept 8, 39; m 64; c 6. FOOD SCIENCE, BIOCHEMISTRY. *Educ:* Brigham Young Univ, BA, 65; Purdue Univ, PhD(food sci), 69. *Prof Exp:* Asst prof food sci, 69-74, ASSOC PROF FOOD SCI, UNIV ARIZ, 74- *Concurrent Pos:* Int adv food technol, Univ Ariz-Fed Univ Ceara, Brazil, 70-74, Fed Univ Rio de Janeiro, 78, 83; Fulbright fel, Esc Catolica Portuguesa, 88. *Mem:* Inst Food Technol. *Res:* Detoxification and utilization of aflatoxin-contaminated foods and feeds; mycotoxins. *Mailing Add:* Dept Nutrit & Food Sci Univ Ariz Tucson AZ 85721

PRICE, RAYMOND ALEX, b Winnipeg, Man, Mar 25, 33; m 56; c 3. STRUCTURAL GEOLOGY, TECTONICS. *Educ:* Univ Man, BSc, 55; Princeton Univ, AM, 57, PhD(geol), 58. *Prof Exp:* Geologist, Geol Surv Can, 58-68; assoc prof geol, Queen's Univ, Ont, 68-70, head, Dept Geol Sci, 72-77, prof geol, 70-81; DIR, GEN GEOL SURV, CAN, 81- *Concurrent Pos:* Pres, Inter-Union Comn Lithosphere, Int Coun Sci-Unions; assoc ed, Can Bull Petrol Geol & Geol Soc Am, NAm ed, Tectonics; mem, Can Nat Comt Int Geol Correlation Prog & Int Union Geol Sci; foreign secy, Can Geosci Coun; Killam sr res fel, 78-80. *Honors & Awards:* Robert J W Douglas Medal, Can Soc Petrol Geol, 83; Sir William Logan Medal, Geol Asn Can, 84. *Mem:* Fel Geol Soc Am; fel Geol Asn Can; Am Asn Petrol Geologists; fel Royal Soc Can; Am Geophys Union. *Res:* Regional geology of southern Canadian Rockies; tectonic evolution of North American Cordillera; nature and significance of variations in tectonic style; mechanics of large-scale thrust faulting. *Mailing Add:* Dept Geol Sci Queens Univ Kinston ON K7L 3N6 Can

PRICE, RICHARD GRAYDON, b Bartlesville, Okla, Nov 8, 33; m 61; c 2. ENTOMOLOGY. *Educ:* Okla State Univ, BS, 55, MS, 59, PhD(entom), 63. *Prof Exp:* Technologist, Shell Chem Co, NY, 63-64 & SC, 64-65; from asst prof to assoc prof, Okla State Univ, 66-78, prof entom, 78-; RETIRED. *Mem:* Entom Soc Am; Am Registry Prof Entomologist. *Res:* Evaluation of insecticidal, ecological and cultural controls of cotton insects; biology and control of arthropod pests of the home. *Mailing Add:* Rte 1 Box 194 Stillwater OK 74074

PRICE, RICHARD HENRY, b New York, NY, Mar 1, 43. THEORETICAL PHYSICS, ASTROPHYSICS. *Educ:* Cornell Univ, BEng, 65; Calif Inst Technol, PhD(physics), 71. *Prof Exp:* Fel, Calif Inst Technol, 71; asst prof, 71-75, ASSOC PROF PHYSICS, UNIV UTAH, 75- *Concurrent Pos:* Lectr, Fulbright Comn, Nat Seminar Physics, 70 & Inst Cultural Integration, Colombia, 70. *Mem:* Am Phys Soc. *Res:* Relativistic astrophysics; black holes; gravitational waves; cosmology; x-ray astronomy. *Mailing Add:* Dept Physics Univ Utah 201 James Fletcher Bldg Salt Lake City UT 84112

PRICE, RICHARD MARCUS, b Colorado Springs, Colo, Jan 18, 40; m 68; c 2. ASTRONOMY, PHYSICS EDUCATION. *Educ:* Colo State Univ, BS, 61; Australian Nat Univ, PhD(astron), 66. *Prof Exp:* Vis res officer radiophysics, Radiophysics Lab, Commonwealth Sci & Indust Res Orgn, 61-65; from asst prof to assoc prof physics, Mass Inst Technol, 67-75, staff mem, Res Lab Electronics, 67-75, staff mem, Educ Res Ctr, 69-73; radio spectrum mgr, NSF, 75-77, head, Astron Res Sect, 76-79; prof & chmn physics & astron, 79-85, dean grad studies, 85-88, PROF PHYSICS, UNIV NMEX, 88- *Concurrent Pos:* Vis scientist, Max Planck Inst Radioastron, 70-71; consult, Off Int Progs, NSF, 72-73 & Mass Inst Technol-Del Inst Technol Educ Asn, 72-75; fel, Res Lab Electronics, Mass Inst Technol, 70-71; assoc scientist, Nat Radio Astron Observ, 74-75; Harlow Shapley vis lectr, Am Astron Soc, 74-; NSF rep, Interdept Radio Adv Comt, Off Telecommun, 75-77; mem, Comt on Radio Frequences, Nat Acad Sci, 83-, chmn, 85; mem, US Nat Comt, Union Radio Sci Int, 75-77, 90-; vis scientist, Nat Radio Astron Observ, 90, Australia Telescope, Commonwealth Sci, Indust & Res Orgn, 90- *Mem:* Inst Astron Union; Am Astron Soc; NY Acad Sci; Am Asn Physics Teachers; Int Union Radio Sci. *Res:* Galactic structure; radiometer systems; development of undergraduate physics laboratory programs; radio galaxies; nuclei of galaxies; galactic structure; develop of physics and astronomy programs; research administration; electromagnetic measurements; electromagnetism. *Mailing Add:* Dept Physics & Astron Univ NMex Albuquerque NM 87131-1156

PRICE, ROBERT, b West Chester, Pa, July 7, 29; m 58; c 3. TELECOMMUNICATIONS, DIGITAL SIGNAL PROCESSING. *Educ:* Princeton Univ, AB, 50; Mass Inst Technol, ScD(elec eng), 53. *Prof Exp:* Asst, Res Lab Electronics, Mass Inst Technol, 51-53; Fulbright fel radio astron, Commonwealth Sci & Indust Res Orgn, Sydney, Australia, 53-54; staff mem, Lincoln Lab, Mass Inst Technol, 54-62, 63-65; lectr elec eng, Univ Calif, Berkeley, 62-63; head radar res, Sperry Res Ctr, 65-67, head systs studies, 67-70, mgr data transmission dept, 70-77, staff consult commun sci, 77-83; chief scientist, Govt Systs Div, M-A-COM, 83-87; consult, 87-88; CONSULT SCIENTIST, RES DIV, RAYTHEON CO, 88- *Concurrent Pos:* Dir, Northeast Electronics Res & Eng Meeting, 68-70; mem adv coun, Dept Elec Eng & Comput Sci, Princeton Univ, 71-77 & Franklin Inst. *Mem:* Nat Acad Eng; fel Inst Elec & Electronics Engrs; Sigma Xi. *Res:* Statistical communication and noise theory; signal design and processing with data, sensor and magnetic recording applications; techniques innovation, systems engineering and project consultations in advanced electronics, including spread spectrum communications. *Mailing Add:* 80 Hill St Lexington MA 02173

PRICE, ROBERT ALLEN, b Chicago, Ill, Sept 4, 34. PEDIATRICS, PATHOLOGY. *Educ:* Univ Ill, Urbana, BS, 57, DVM, 59; Univ Tenn, Memphis, MD, 65. *Prof Exp:* Resident path, Univ Tenn & John Gaston Hosp, Memphis, 66-67; resident, Baptist Mem Hosp, Memphis, 68-69; chief resident, Children's Hosp Med Ctr, Boston, 69-70; asst mem, St Jude Children's Res Hosp, 71-74, assoc mem path, 74-89; PVT PRACT, METHODIST HOSP, 89- *Concurrent Pos:* Res fel path, Harvard Med Sch, Children's Hosp Med Ctr, Boston & New Eng Regional Primate Res Ctr, Boston, 70; asst clin prof path, Ctr Health Sci, Univ Tenn, 73- *Mem:* Am Asn Cancer Res. *Res:* Pathology of childhood cancer; central nervous system diseases in children with cancer; comparative pathology. *Mailing Add:* Methodist Hosp 400 Tickle Rd Dyersburg TN 38024-3182

PRICE, ROBERT HAROLD, b Lansing, Mich, Apr 9, 49. PLASMA PHYSICS. *Educ:* Mass Inst Technol, BS, 71, PhD(physics), 77. *Prof Exp:* Res asst high energy particle physics, Univ Md, 66, phys chem, 67, atomic & low temperature physics, Res Lab Electronics, Mass Inst Technol, 68, plasma physics, 69-71; res asst physicist, Lawrence Livermore Nat Lab, 71; mem staff laser plasma, Res Lab Electronics, Mass Inst Technol, 71-76; STAFF PHYSICIST PLASMA PHYSICS & X-RAY OPTICS, LAWRENCE LIVERMORE NAT LAB, 76- *Concurrent Pos:* Hertz fel, Fannie & John Hertz Found, Mass Inst Technol, 71-75. *Mem:* Am Phys Soc. *Res:* Development of high resolution x-ray optics and high speed imaging devices to study the dynamics of laser fusion targets using the images produced to study plasma physics under ICF conditions. *Mailing Add:* PO Box 520 Santa Fe NM 87504-0520

PRICE, ROGER DEFORREST, b Lawrence, Kans, Aug 4, 29; m 52; c 3. ENTOMOLOGY. *Educ:* Univ Kans, AB, 51, MA, 53, PhD(entom), 55. *Prof Exp:* From instr to assoc prof, 55-70, PROF ENTOM, UNIV MINN, ST PAUL, 70- *Mem:* Entom Soc Am. *Res:* Insect taxonomy, especially chewing lice. *Mailing Add:* Dept Entom 219 Hodson Hall Univ Minn 180 Folwell Ave St Paul MN 55108

PRICE, SAMUEL, b Burley, Idaho, May 14, 23; m 53; c 3. GENETICS. *Educ:* Utah State Agr Col, BS, 49; Univ Calif, PhD(genetics), 52. *Prof Exp:* Asst bot, Utah State Agr Col, 47-49; asst genetics, Univ Calif, 49-52; asst prof agron, Univ Hawaii, 52-53; geneticist, Crops Res Div, Agr Res Serv, USDA, Md, 53-67; grants assoc, Div Res Grants, NIH, 67-68, health sci adminr, Nat Inst Environ Health Sci, NIH, 68-69, health sci adminr, Nat Eye Inst, 69-70, chief, Sci Progs Br, Nat Eye Inst, 70-71, health sci adminr, Nat Cancer Inst, 71-72, chief Nat Organ Site Progs Br, Nat Cancer Inst, 72-80; RES COORDR, UNIV MD, 80- *Mem:* Am Genetic Asn (treas, 63, pres, 72); AAAS. *Res:* Cytogenetics. *Mailing Add:* PO Box 299 St Inigoes MD 20684

PRICE, STEVEN, b New York, NY, May 13, 37; m 58; c 2. CELL & SENSORY PHYIOLOGY. *Educ:* Adelphi Col, AB, 58; Princeton Univ, MA, 60, PhD(biol), 61. *Prof Exp:* Nat Cancer Inst fel chem, Fla State Univ, 61-63; from sr res biologist to res group leader biochem, Boston Lab, Monsanto Res Corp, 63-66; from asst prof to assoc prof, 66-75, PROF PHYSIOL, MED COL VA, 75- *Concurrent Pos:* Nat Inst Dent Res res career develop award, 72-76. *Mem:* Am Physiol Soc; Asn Chemoreception Sci; Europ Chemoreception Res Orgn; Soc Gen Physiologists. *Res:* Receptor mechanisms in taste and smell, including stimulus recognition and transduction; insulin secretion. *Mailing Add:* Dept Physiol Med Col Va Richmond VA 23219

PRICE, THOMAS R, b Hampton, Va, July 31, 34; m 58; c 2. STROKE. *Educ:* Univ Va, BS, 56, MD, 60. *Prof Exp:* Internship med, Cincinnati Gen Hosp, 60-61; captain med, US Air Force Med Corps, 61-63; resident neurol, Univ Va Hosp, 63-66, instr, 66-67; from instr to assoc prof, 67-78, PROF NEUROL, SCH MED, UNIV MD, 78- *Concurrent Pos:* Prin investr, Stroke Ctr, Sch Med, Univ Md, 80-, prof epidemiol, 90-; mem, Epidemiol & Dis Control Study Sect, 88-; chmn, Bylaws Comt, Am Acad Neurol & Stroke Subcomt, Am Heart Asn. *Mem:* Am Acad Neurol; Am Heart Asn; Am Neurol Asn; AMA. *Res:* Transient ischemic attacks; stroke diagnosis and treatment; depression of mood after stroke; dementia and pseudodementia after stroke; Rocky Mountain Spotted Fever predictors of patient outcome. *Mailing Add:* Dept Neurol Univ Md Hosp 22 S Greene St Baltimore MD 21201

PRICE, WALTER VAN VRANKEN, b Schenectady, NY, Dec 10, 96; m 21; c 2. FOOD SCIENCE & TECHNOLOGY. *Educ:* Cornell Univ, BS, 20, MS, 21, PhD(agr), 25. *Prof Exp:* Asst dairy indust, Cornell Univ, 19-21, instr, 21-22; plant mgr, Hygeia Ice Cream Co, Elmira, NY, 22-23; from instr to asst prof dairy indust, Cornell Univ, 23-29; prof, 29-64, EMER PROF DAIRY INDUST, UNIV WIS, MADISON, 64- *Concurrent Pos:* Mgr, Sanit Milk & Ice Cream Co, Ithaca, NY, 24-25; consult, Marshall Dairy Lab, Marshall Products Div, Miles Lab, 41-; adv, Refrig Res Found, 55-64. *Honors & Awards:* Borden Award, Am Dairy Sci Asn, 50 & Paul-Lewis Award, 59. *Mem:* Sigma Xi; fel AAAS; Am Dairy Sci Asn (pres, 53); Am Chem Soc. *Res:* Production, improvement, quality control and development of milk products. *Mailing Add:* 1605 Linden Dr Madison WI 53706

PRICE, WILLIAM ALRICH, JR, b Chester, Pa, Jan 5, 24; m 47; c 2. PHARMACOLOGY. *Educ:* Pa State Univ, BS, 48, MS, 50; Rutgers Univ, PhD(biochem), 51. *Prof Exp:* Res chemist, E I du Pont de Nemours & Co, Inc, 51-62, sr res biochemist, 62-85, res assoc pharmacol, 85; RETIRED. *Mem:* Am Chem Soc; NY Acad Sci. *Res:* Animal nutrition; chemistry of natural products; neuropharmacology; cardiovascular pharmacology. *Mailing Add:* Clermont Seven Clermont Rd Wilmington DE 19803

PRICE, WILLIAM CHARLES, b Fort Worth, Tex, Aug 26, 30; m 52; c 2. ENGINEERING GEOLOGIST, ENGINEERING GEOPHYSICS. *Educ:* Univ Tex, Austin, BS, 58. *Prof Exp:* Engr, electro-mech, Missile Systs, Martin-Marietta Co, 60-64; systs engr & geologist, Lockheed Aircraft, 64-70; eng geologist, Dept Main Rds, Australia, 70-74; consult geologist, Geotechnics, 74-78; eng geologist, Bur Municipal Solid Waste Mgt, 78-81, PROG CHIEF, HYDROL/GEOTECH SECT, BUR RADIATION CONTROL, TEX DEPT HEALTH, 81- *Mem:* Am Inst Prof Geologists; Asn Eng Geologists; Geol Soc Am; Nat Waterwell Asn. *Res:* Developed a simplified earth resistivity sounding array for monitoring subsoils for contaminant migration from waste disposal facilities as a suppliment to groundwater monitoring. *Mailing Add:* 8909 Briardale Austin TX 78758

PRICER, WILBUR DAVID, b Des Moines, Iowa, July 22, 35; m 64; c 4. ELECTRONICS ENGINEERING. *Educ:* Middlebury Col, BA, 59; Mass Inst Technol, BS, 59, MS, 59. *Prof Exp:* Eng, 59-70, sr eng, 70-83, SR MEM TECH STAFF, IBM, 83- *Concurrent Pos:* ed, Inst Elec & ELectronics J Solid State Circuits, 83-86; pres, Solid State Circuits Coun; adj prof, Elec Eng Dept, Univ Vt, 84- *Mem:* Fel Inst Elec & Electronics Engrs; Sigma Xi. *Res:* Integrated solid state circuits, digital circuit design, process integration and fabrication, systems architecture, electronic packaging. *Mailing Add:* 58 Van Patten Pkwy Burlington VT 05401

PRICHARD, GEORGE EDWARDS, b Corbin, Ky, July 29, 16; m 42; c 2. GEOLOGY. *Educ:* Univ Ky, BS, 40. *Prof Exp:* Asst geol, Univ Denver, 40-41; asst comput, Seismog Serv Corp, Okla, 46-47; geologist, Org Fuels Br, US Geol Surv, 48-66, Br Regional Geol Southern Rocky Mt, 66-69 & Br Rocky Mt Environ Geol, 69-72, Br Cent Environ Geol, 72-81, Br Cent Regional Geol, 81; RETIRED. *Mem:* Am Asn Petrol Geologists. *Res:* Mineral fuels; sedimentology; stratigraphy; structure. *Mailing Add:* 3665 W Floyd Ave Denver CO 80236

PRICHARD, ROBERT WILLIAMS, b Jersey City, NJ, May 30, 23; m 46; c 2. PATHOLOGY. *Educ:* George Washington Univ, MD, 47. *Prof Exp:* Asst instr path, Univ Pa, 50-51; from instr to assoc prof, 51-55, PROF PATH, BOWMAN GRAY SCH MED, 55-, CHMN DEPT, 73- *Concurrent Pos:* Adv med educ, US Opers Mission, Thailand, 55-57. *Mem:* Am Soc Clin Path; Col Am Path; Am Asn Hist Med; Am Asn Path; AMA; Sigma Xi. *Res:* Atherosclerosis; diseases of the hemolytopoietic organs; medical history. *Mailing Add:* Dept of Path Bowman Gray Sch of Med Winston-Salem NC 27103

PRIDDY, STEWART BEAUREGARD, b Lumberton, NC, Mar 31, 40; m 63; c 2. MATHEMATICS. *Educ:* Univ NC, BS, 61, MA, 63; Mass Inst Technol, PhD(math), 68. *Prof Exp:* From asst prof to assoc prof, 68-78, PROF MATH, NORTHWESTERN UNIV, EVANSTON, 78- *Mem:* Am Math Soc. *Res:* Algebraic topology; homotopy theory; infinite loop space theory; simplicial homotopy theory; homological algebra. *Mailing Add:* Northwestern Univ Evanston IL 60208

PRIDE, DOUGLAS ELBRIDGE, b Eau Claire, Wis, July 18, 42; m 66; c 2. ECONOMIC GEOLOGY, EXPLORATION GEOCHEMISTRY. *Educ:* Univ Wis-Madison, BS, 64, MS, 66; Univ Ill, Urbana, PhD(geol), 69. *Prof Exp:* Asst prof geol, Univ Ill, Urbana, 69-73; ASST PROF GEOL, OHIO STATE UNIV, 73- *Mem:* Geol Soc Am; Am Inst Mining, Metall & Petrol Engrs; Sigma Xi. *Res:* Geologic field mapping, microscopic studies and researches in exploration geochemistry, coordinated to locate metallic mineral deposits, in particular those associated with deep-seated and near-surface intrusive igneous activity. *Mailing Add:* 7656 Summerwood Dr West Worthington OH 43235

PRIDMORE-BROWN, DAVID CLIFFORD, b Mexico City, Mex, Sept 1, 28; m 58; c 2. PHYSICS, COMPUTER SIMULATION. *Educ:* Principia Col, BS, 49; Mass Inst Technol, SM, 51, PhD(physics), 54. *Prof Exp:* Researcher acoustics, Mass Inst Technol, 54-56; Imp Chem Industs fel, Univ Manchester, 56-57; asst prof mech eng, Mass Inst Technol, 57-62; STAFF SCIENTIST, AEROSPACE CORP, 62- *Mem:* Am Phys Soc. *Res:* Acoustic, magnetohydrodynamic and plasma wave motion; radiation problems; computational physics; mathematical modelling; orbital mechanics. *Mailing Add:* Space Sci Lab M2/269 Aerospace Corp Box 92957 Los Angeles CA 90009-2957

PRIELIPP, ROBERT WALTER, b Wausau, Wis, Mar 11, 36. MATHEMATICS. *Educ:* Univ Wis, Stevens Point, BS, 58; Univ Ill, Urbana, MS, 60; Univ Wis-Madison, PhD(math, math educ), 67. *Prof Exp:* Teacher high sch, Wis, 58-59; instr math, Univ Wis, Stevens Point, 60-62; proj asst math educ, Univ Wis-Madison, 62-63 & Res & Develop Ctr, 66-67; asst prof, 67-69, assoc prof, 69-79, PROF MATH, UNIV WIS-OSHKOSH, 79- *Concurrent Pos:* Co-ed, Prob Dept, Sch Sci & Math Asn, 76-86, SSMArrt Newsletter, 86; chair elect, Wis Sect, Math Asn Am, 89-90, chair, 90-91. *Honors & Awards:* George G Mallinson Distinguished Serv Award, Sch Sci & Math Asn, 89. *Mem:* Math Asn Am; Sch Sci & Math Asn; Nat Coun Teachers Math. *Res:* Algebra; elementary number theory; mathematics education. *Mailing Add:* Dept Math Univ Wis Oshkosh WI 54901

PRIEMER, ROLAND, b Oct 28, 43; m; c 4. ELECTRICAL ENGINEERING. *Educ:* Ill Inst Technol, BS, 65, PhD(elec eng). *Prof Exp:* Asst prof, 69-77, ASSOC PROF EECS, UNIV ILL, CHICAGO 77- *Concurrent Pos:* Consult, Argonne Nat Labs, 76-82, Motorola, Inc, 83-87, Bally Mfg, 86- & numerous other co. *Mem:* Inst Elec & Electronic Engrs. *Res:* Digital signal processing; microprocessor based design. *Mailing Add:* EECS Dept MC 154 Univ Ill PO Box 4348 Chicago IL 60680

PRIER, JAMES EDDY, b Richmond Co, NY, July 20, 24; m 46; c 4. PUBLIC HEALTH. *Educ:* Cornell Univ, DVM, 46; Univ Ill, MS, 49, PhD(path), 53; Southland Univ, JD, 82. *Prof Exp:* Bacteriologist, Lederle Labs, Am Cyanamid Co, NY, 46-47; exten vet, Univ Ill, 47, res instr path, 47-50; head dept bact & vet sci, Univ Wyo, 50-53; asst prof microbiol, Col Med, State Univ NY Upstate Med Ctr, 53-56; dir biol develop labs, Merck Sharpe & Dohme, Inc, 56-60; assoc prof microbiol, Sch Vet Med, Univ Pa, 60-64; from asst dir to dir div labs, Pa Dept Health, 64-72, dir bur labs, 72-76; DIR, CENTRE SQ VET CLIN, ARDMORE ANIMAL HOSP, 78-; DIR, COMP MED LAB & ADJ PROF, MICROBIOL, PHILADELPHIA COL OSTEOPATH, 78- *Concurrent Pos:* Vis assoc, Sch Vet Med, Univ Pa, 57-60; prof microbiol, Sch Med, Temple Univ, 64-76; dir labs, Henry R Landis State Hosp, 71-73; co-dir, Penndel Med Lab, Ardmore, 76-82. *Mem:* Am Soc Microbiol; fel Am Pub Health Asn; Am Vet Med Asn; US Animal Health Asn; Royal Soc Health. *Res:* Virology, especially host-parasite relationships; epidemiology; infectious diseases of domestic animals; viruses of man and animals. *Mailing Add:* 1030 DeKalb Pike Centre Square PA 19422

PRIESING, CHARLES PAUL, b Boston, Mass, June 26, 29; m 55; c 6. SANITARY & ENVIRONMENTAL ENGINEERING, TECHNICAL MANAGEMENT. *Educ:* Merrimack Col, BS, 52; Mass Inst Technol, PhD(org chem), 57; Fairleigh Dickenson Univ, MBA, 80. *Prof Exp:* Res chemist, Dow Chem Co, Mich, 57-64, proj leader, 64-65; chief treatment & control res progs, Robert S Kerr Water Res Ctr, Fed Water Pollution Control Admin, 65-68; from supvr environ eng to mgr environ eng, Am Cyanamid Co, NJ, 68-74, dir, Environ Protection Dept & corp dir environ affairs, 74-84; PRES & PRIN CONSULT, ARALIE, INC, 84- *Concurrent Pos:* Environ liaison chem indust, USSR, 75, 78-79. *Mem:* Am Chem Soc; Water Pollution Control Fedn; Am Water Works Asn. *Res:* Vitamin syntheses; functional monomers and polymers; polymer association; flocculation and solid-liquid separations; polyelectrolytes; water and waste treatment; product and process technology; environmental quality and pollution prevention, treatment and control; toxic and hazardous waste handling, storage and disposal; corporate environmental programs management and administration; environmental control facilities engineering design, construction and operation; environmental regulatory affairs management, auditing and compliance. *Mailing Add:* Aralie Inc 8 William St Pequannock NJ 07440

PRIEST, DAVID GERARD, b Tampa, Fla, Oct 16, 38; m 57; c 2. BIOCHEMISTRY, ENZYMOLOGY. *Educ:* Univ S Fla, BA, 64; Fla State Univ, PhD(chem), 69. *Prof Exp:* Chemist, Eastman Kodak, 69-70; assoc, 70-72, asst prof, 72-75, assoc prof, 75-81, PROF BIOCHEM, MED UNIV SC, 81- *Mem:* Am Chem Soc; Am Soc Biol Chemists; Am Asn Cancer Res. *Res:* Studies of the regulation of folate metabolism in eucaryotic cells. *Mailing Add:* Dept of Biochem Med Univ SC 171 Ashley Ave Charleston SC 29425

PRIEST, HOMER FARNUM, b Nelson, NH, June 14, 16; m 41. CHEMISTRY. *Educ:* Univ NH, BS, 38; Williams Col, MA, 40; Mass Inst Technol, PhD(chem), 48. *Prof Exp:* Asst instr, Univ NH, 34-37; asst chem, Williams Col, 38-40; instr, Columbia Univ, 40-41, div & sect head, Manhattan Proj, Div War Res, 41-44; works chemist, Carbide & Carbon Chem Corp, Tenn, 44-46, dir chem res, 46; tech adv to chief tech command, Decontamination Br, Radiol Warfare Div, US Army Chem Ctr, 48-50, consult, 51; asst group leader, Solid State Transistor Group, Lincoln Lab, Mass Inst Technol, 51-54; solid state res group, Baird Assocs, 54-57; supvry phys chemist, Mat Res Lab, Ord Mat Res Off, Watertown Arsenal, US Army, 57-63, dir mat res lab, Mat Res Agency, Watertown, 63-72, chief mat sci div, Army Mat & Mech Res Ctr, 72-80; RETIRED. *Concurrent Pos:* Exec asst to pres, High Voltage Eng Corp, 50-51. *Mem:* Am Chem Soc. *Res:* Uranium chemistry; solid state semiconductors; pure materials; fluorides; fluorine. *Mailing Add:* 180 Turtle Lake Ct, Apt 104 Naples FL 33942

PRIEST, JEAN LANE HIRSCH, b Chicago, Ill, Apr 5, 28; m 58; c 3. PEDIATRICS, MEDICAL GENETICS. *Educ:* Univ Chicago, PhB, 47, BS, 50, MD, 53; Am Bd Pediat, dipl, 58. *Prof Exp:* Intern, King County Hosp Syst, Seattle, Wash, 53-54; resident pediat, Univ Chicago Clins, 54-56; clin instr pediat, Univ Ill, 57-58; staff pediatrician & res cytogeneticist, Ranier Sch, Buckley, Wash, 58-62; instr pediat, Med Ctr, Univ Colo, 63-65, asst prof pediat & path, 65-71; assoc prof pediat & asst prof path, Sch Med, Emory Univ, 71-78, prof pediat, 71-90; LAB DIR, SHODAIR HOSP, HELENA, MONT, 90- *Concurrent Pos:* Fel, Nat Found for Infantile Paralysis, 57 & Chicago Heart Asn, 57-58; physician, Infant Welfare Soc, Chicago, 57-58 & Seattle-King County Dept Pub Health, 58-61; fel genetics, Univ Wash, 58-62, clin instr pediat, 60-62; vis mem staff, Dept Zool, Univ St Andrews, Scotland, 69-70; vis prof, Univ Auckland, NZ, 80-81 & Univ Würzburg, Ger, 87-88. *Mem:* Am Soc Cell Biol; Am Asn Pathologists; Am Soc Human Genetics; Am Dermatoglyphics Soc; Tissue Cult Asn; Sigma Xi. *Res:* Cytogenetics of human chromosome abnormalities; differentiated function in cultured cells. *Mailing Add:* Cytogenetics Lab Shodair Hosp 840 Helena Ave PO Box 5539 Helena MT 59604

PRIEST, JOSEPH ROGER, b Highland, Ohio, Oct 14, 29; m 59; c 4. NUCLEAR PHYSICS. *Educ:* Wilmington Col, BS, 51; Miami Univ, MS, 56; Purdue Univ, PhD(physics), 60. *Prof Exp:* Asst, Purdue Univ, 56-60; physicist, Int Bus Mach Res Lab, 60-62; from asst prof to assoc prof, 62-70, PROF PHYSICS, MIAMI UNIV, 70- *Mem:* Am Assoc Phys Teachers; Sigma Xi. *Res:* Low energy nuclear reactions; radioisotope techniques. *Mailing Add:* Dept Physics Miami Univ 35 Culler Oxford OH 45056

PRIEST, MATTHEW A, b Rochester, NY, Feb 21, 49; m 80; c 2. ANALYTICAL & PHARMACEUTICAL CHEMISTRY. *Educ:* Univ Notre Dame, BS, 71, Johns Hopkins Univ, PhD(org chem), 76. *Prof Exp:* Res assoc, dept chem, Univ Pittsburgh, 76-77; sr res chemist, Ash Stevens Inc, 77-86; SR DEVELOP CHEMIST, PFIZER INC, 86- *Mem:* Am Chem Soc; Am Soc Brewing Chemists. *Res:* Development of new bittering compounds for the brewing industry; chemical processing of substances derived from extracts of hops. *Mailing Add:* Pfizer Inc 4215 N Port Washington Milwaukee WI 53212

PRIEST, MELVILLE S(TANTON), b Cassvile, Mo, Oct 16, 12; m 41, 83. HYDRAULICS. *Educ:* Univ Mo, BS, 35; Univ Colo, MS, 43; Univ Mich, PhD(civil eng), 54. *Prof Exp:* Jr engr, US Corps Eng, Ill, 37-39; jr engr, US Bur Reclamation, Colo, 39-40, asst engr, 40-41; from instr to assoc prof mech, Cornell Univ, 41-55; prof hydraul, Auburn Univ, 55-65, head dept civil eng, 58-65; exec dir, Water Resources Res Inst, Miss State Univ, 65-77; CONSULT, 77- *Concurrent Pos:* Adv, UN Tech Asst Admin to Egypt, 56, 57, 60. *Mem:* Am Soc Civil Engrs; Int Asn Hydraul Res; Am Water Resources Asn. *Res:* Hydraulic engineering. *Mailing Add:* PO Box 541 Starkville MS 39759

PRIEST, ROBERT EUGENE, b Anacortes, Wash, Jan 6, 26; m 58; c 3. EXPERIMENTAL PATHOLOGY, CELL BIOLOGY. *Educ:* Reed Col, BA, 50; Univ Chicago, MD, 54. *Prof Exp:* From instr to res asst prof path, Univ Wash, 57-62; from asst prof to prof, Univ Colo, 62-72; PROF PATH, EMORY UNIV, 71- *Concurrent Pos:* Clin investr, Vet Admin Hosp, Seattle, 60-62; vis scientist, Univ St Andrews, Scotland, 69-70; fel coun arteriosclerosis, Am Heart Asn; Fogarty sr int fel, Univ Auckland, New Zealand, 80-81. *Mem:* AAAS; Am Asn Pathologists; Am Soc Cell Biol; Int Acad Path. *Res:* Connective tissue biochemistry and disease; cell culture; differentiation. *Mailing Add:* 904 Barten Woods Rd NE Atlanta GA 30307

PRIEUR, DAVID JOHN, b Flint, Mich, June 18, 42; m 66; c 2. GENETIC DISEASES OF ANIMALS, INHERITED ANIMAL MODELS. *Educ:* Mich State Univ, BS, 64, DVM, 66, MS, 67; Wash State Univ, PhD(path), 71. *Prof Exp:* Upjohn fel path, Mich State Univ, 66-67; res fel path, Wash State Univ, 67-71; sr staff fel, NIH, 71-74; from asst prof to assoc prof, 74-86, PROF PATH, WASH STATE UNIV, 86-, CHMN, DEPT VET

MICROBIOL & PATH, 88- *Concurrent Pos:* ed, animal models sect, Am J Med Genetics, 83-; assoc prog genetics & cell biol, Wash State Univ, 83- *Honors & Awards:* Ralston Purina Small Animal Res Award; Res career Development Award, NIH, 75. *Mem:* Am Asn Pathologists; Int Acad Path; Am Soc Human Genetics; Am Genetic Asn; Soc Exp Biol & Med. *Res:* Indentification, characterization and study of genetic diseases of animals that are similar to human genetic diseases; application of information derived from animal models to understanding of human diseases. *Mailing Add:* Dept Vet Microbiol & Path Wash State Univ Pullman WA 99164-7040

PRIEVE, DENNIS CHARLES, b Portage, Wis, Sept 1, 47. COLLOID SCIENCE. *Educ:* Univ Fla, BSChE, 70; Univ Del, MChE, 73, PhD(chem eng), 75. *Prof Exp:* Res engr chem eng, Westvaco Corp, 70; res fel, Univ Del, 70-73, teaching assoc, 73-74; from asst prof to assoc prof, 75-80, PROF CHEM ENG, CARNEGIE-MELLON UNIV, 83- *Concurrent Pos:* Chmn Procter & Gamble Fel Award Comt, Div Colloid & Surface Sci, Am Chem Soc, 83-84; Adv Bd J Colloid Interface Sci, 82-84; vis prof Chem eng, Princeton Univ, 84-85; chmn fac, Col Eng, Carnegie Mellon Univ, 86-87, vchmn, 85-86; chmn, Area 1c (Interfacial Phenomena) Nat Prog Comt Amer Inst Chem Engrs, 87-90, vchmn, 84-87. *Honors & Awards:* George Tallman Ladd Award, Carnegie-Mellon Univ, 77. *Mem:* Am Inst Chem Engrs; Am Chem Soc; Sigma Xi; AAAS. *Res:* Chemically induced locomotion of colloidal particles; measurement of surface forces in liquids; total-internal-reflection microscopy; electrokinetic phenomena. *Mailing Add:* Dept Chem Eng Carnegie Mellon Univ Pittsburgh PA 15213

PRIGODICH, RICHARD VICTOR, b St Albans, NY, Dec 8, 51; m 75; c 3. PHYSICAL BIOCHEMISTRY, PHYSICAL CHEMISTRY. *Educ:* Lake Forest Col, BA, 74; Wesleyan Univ, PhD(chem), 82. *Prof Exp:* Teacher, sign lang, Perkins Sch Blind, 74-75; teacher, chem, Marvelwood Sch, 75-76; anal chemist, Chandler-Evans Corp, 76-78; NIH teaching fel, Yale Univ, 82-85; ASST PROF CHEM, TRINITY COL, 85- *Mem:* Am Chem Soc; AAAS. *Res:* Equilibria; association of metal ions with organophosphorus compounds, proteins and nucleic acids; association of proteins with nucleic acids. *Mailing Add:* Dept Chem Trinity Col Hartford CT 06106

PRIGOGINE, ILYA, b Moscow, Russia, Jan 25, 17; Belg citizen; m 61; c 2. THERMODYNAMICS. *Educ:* Free Univ of Brussells, Lic Chem Sci, 39, Lic Phys Sci, 39, Dr Chem Sci, 41. *Hon Degrees:* Dr Honoris Causa from 16 US & foreign univs. *Prof Exp:* DIR, INT INSTS PHYSICS & CHEM, SOLVAY, 59-; PROF PHYSICS & CHEM ENG, UNIV TEX, AUSTIN, 67-, REGENTAL PROF, 77-, ASHBEL SMITH PROF, 84-, DIR, ILYA PRIGOGINE CTR STUDIES STATIST MECH, 67- *Concurrent Pos:* Asst lectr, Free Univ Brussels, 47, prof extraordinaire, 50, prof ordinaire, 51, hon prof, 87-; vis prof, Dept Chem, Enrico Fermi Inst Nuclear Studies & Study Metals, Univ Chicago, 61-66; RGK Centennial fel, Inst Construct Capitalism, Univ Tex, Austin, 85-; chmn, Reactor's Adv Comt, UN Univ, Tokyo, 86-; vpres, Codest Comt Stimulation Sci & Technol, 86; hon prof, Benares Hindu Univ, Varasani, India, 88; distinguished lectr, NSF, 89. *Honors & Awards:* Nobel Prize Chem, Royal Acad Sweden, 77; Van Laar Prize, Chem Soc Belg, 47; A De Potter Prize, Royal Acad Belg, 49, A Wetrems Prize, 50; Swante Arrenius Gold Medal, Royal Acad Sci Sweden, 76; Rumford Gold Medal, Royal Soc London, 76; Commander in the Order of Arts and Letters, France, 84; Award for Sci Achievement, Int Found Artificial Intel, 90; Medal Imp Order Rising Sun, Emperor Japan, 91; Homer Smith Mem Lectr, 91. *Mem:* Foreign assoc Nat Acad Sci; hon foreign mem Am Acad Arts & Sci; fel NY Acad Sci; hon mem Chem Soc Poland; Int Acad Philos Sci; hon foreign fel Am Chem Soc; fel World Acad Arts & Sci; hon mem Int Acad Hist Sci. *Res:* Author of numerous articles on physical chemistry, published internationally. *Mailing Add:* Ctr Statist Mech Univ Tex Austin TX 78712

PRIGOT, MELVIN, b Boston, Mass, Nov 16, 20; m 43; c 2. ORGANIC CHEMISTRY, ANALYTICAL CHEMISTRY. *Educ:* Boston Univ, AB, 43; Univ Fla, MS, 48, PhD(org chem), 51. *Prof Exp:* Chemist, Monsanto Chem Co, 51-52, Pennwalt Corp, 52-54, ICI-Am, 55-60 & Gen Aniline & Film Corp, 60-61; CHEMIST, H KOHNSTAMM & CO, INC, 61- *Mem:* Am Chem Soc; AAAS. *Res:* Process development; food, drug and cosmetic dyes; textile dyes. *Mailing Add:* 411 Wynnewood Rd Pelham Manor NY 10803

PRIKRY, KAREL LIBOR, b Chvalkovice, Czech, Apr 28, 44. MATHEMATICS. *Educ:* Univ Warsaw, Magister of Math, 65; Univ Calif, Berkeley, PhD(math), 68. *Prof Exp:* Res asst math, Univ Calif, Berkeley, 65-68; asst prof, Univ Wis-Madison, 68-72; assoc prof, 72-80, PROF, UNIV MINN, MINNEAPOLIS, 80- *Concurrent Pos:* Vis asst prof, Univ Calif, Los Angeles, 71-72. *Mem:* Asn Symbolic Logic. *Res:* Theory of sets; combinatorial properties of families of sets; ultrafilters; consistence and independence proofs. *Mailing Add:* Univ Minn Minneapolis MN 55455

PRILL, ARNOLD L, b New York, NY. METALLURGY. *Educ:* Mass Inst Technol, BS, 62, MS, 63; Lehigh Univ, PhD(metall), 66. *Prof Exp:* Res metallurgist, Int Nickel Co, 63-66; chief metallurgist, Sintercast Div, Chromalloy Am Corp, 66-68, sect mgr, Res & Develop Div, 68-69, dir eng, Chromalloy Div, 69-70, vpres eng & opers, 70-77, exec vpres, 77-; RETIRED. *Mem:* Mfg Engrs; Am Powder Metall Inst. *Res:* Powder metallurgy, particularly cermets; superalloy joining and forming; ternary diffusion. *Mailing Add:* 13269 Meadowside Dr Dallas TX 75240

PRIM, ROBERT CLAY, b Sweetwater, Tex, Sept 25, 21; m 42; c 3. MATHEMATICS. *Educ:* Univ Tex, BSEE, 41; Princeton Univ, PhD(math), 49. *Prof Exp:* Asst appl math, Univ Tex, 40-41; engr, Gen Elec Co, NY, 41-44; engr & mathematician, US Naval Ordn Lab, 44-49; res assoc dynamics, Princeton Univ, 48-49; res mathematician, 49-58, dir math & mech res Bell Tel Labs, Inc, 58-61; spec asst to Dir Defense Res & Eng, 61-63; vpres res, Sandia Corp, 63-64; vpres, Data Systs Div & Defense Systs Group, 64-66, Litton Industs Inc, Calif, 66-69; exec res assoc, Bell Labs Inc, 69-71, exec dir res, 71-80; RETIRED. *Concurrent Pos:* Mem adv comt eng, NSF, 72, mem adv comt res, 73-74; mem vis comt eng & appl physics, Harvard Univ, 73-78. *Mem:* Am Math Soc; Math Asn Am; Asn Comput Mach; AAAS; Am Sci Affiliation; Sigma Xi. *Res:* Research and development management; management science; systems analysis; systems engineering. *Mailing Add:* 17421 Tam O'Shanter Dr Poway CA 92064

PRIMACK, JOEL ROBERT, b Santa Barbara, Calif, July 14, 45; m 77; c 1. THEORETICAL PHYSICS, ELEMENTARY PARTICLE PHYSICS. *Educ:* Princeton Univ, AB, 66; Stanford Univ, PhD(physics), 70. *Prof Exp:* Harvard Soc Fellows jr fel physics, Harvard Univ, 70-73; from asst prof to assoc prof, 73-83, PROF PHYSICS, UNIV CALIF, SANTA CRUZ, 83- *Concurrent Pos:* Sloan Found res fel, 74-78; vis prof, Higher Normal Sch, Paris, 78, Stanford Linear Accelerator Ctr, 84, Ctr Particle Astrophysics, Univ Calif Berkeley, 89-90 & Inst Advan Studies, Hebrew Univ, 90; mem, citizens adv comt, NSF, 76-78, Fed Am Scientist Nat Coun, 70-74, comt A, Am Asn Univ Prof, 83-86; consult, Three Mile Island Spec Inquiry Group, Nuclear Regulatory Comn, 79-80; dir, elem particle physics, 3D Theoret Advan Study Inst, 86; mem, steering comt, Univ Calif Inst Global Conflict & Coop, 86-, Fed Am Scientist & Comt Soviet Scientists Joint Disarmament Proj, 87-; A P Sloan Found Res fel, 74-78. *Honors & Awards:* Award, Am Phys Soc Forum on Physics & Soc, 77. *Mem:* AAAS; fel Am Phys Soc; Am Astron Soc. *Res:* Theoretical physics and astrophysics, especially gauge theory, dark matter and cosmology; national science and technology policy, especially nuclear power, arms control and space policy. *Mailing Add:* Dept Physics Univ Calif Santa Cruz CA 95064

PRIMACK, MARSHALL PHILIP, b Louisville, Ky, Aug 30, 41; m 67; c 2. ENDOCRINOLOGY. *Educ:* Univ Louisville, BA, 61; Johns Hopkins Univ, MD, 65. *Prof Exp:* From intern to resident internal med, Bellevue Hosp, New York, 65-68; fel endocrinol, 68-71, assoc, 71-72, instr, 72-73, ASST PROF MED, COL PHYSICIANS & SURGEONS, COLUMBIA UNIV, 73- *Mem:* Endocrine Soc; Am Thyroid Asn. *Res:* The mechanism of action of thyroid hormones; emphasis on the way these hormones increase the rate of metabolism. *Mailing Add:* Presby Hosp 101 Central Park W New York NY 10023

PRIMACK, RICHARD BART, b Boston, Mass, Jan, 15, 50; m 83; c 2. PLANT ECOLOGY, TROPICAL ECOLOGY. *Educ:* Harvard Univ, BA, 72; Duke Univ, PhD(bot), 76. *Prof Exp:* Postdoctoral ecol, Univ Canterbury, 76-78; postdoctoral trop ecol, Harvard Univ, 80-81, assoc botanist, Arnold Arboretum, 86-89; from asst prof to assoc prof, 78-90, PROF PLANT ECOL, BOSTON UNIV, 90- *Concurrent Pos:* Panel mem, Pop Biol, NSF, 83 & Man & Biosphere Prog, Trop Ecosysts, 90-91; chair, Ecol Sect, Bot Soc Am, 84 & Res Awards Comt, New Eng Bot Club, 87-91; consult forestry eval, Australian Govt, 89; corresp, Tropinet Newsletter, Asn Trop Biol, 89-; prin investr, NSF, 86- & State Dept, 90-; mem, Comt Develop Nations, Ecol Soc Am, 90-91. *Mem:* Bot Soc Am; New Eng Bot Club; Ecol Soc Am. *Res:* Plant population biology; factors affecting plant distribution, particularly rare species in New England; tropical forest ecology and biological diversity in Sarawak, Malaysia and Karnataka, India; poisonous plants; pollination ecology. *Mailing Add:* Biol Dept Boston Univ Five Cummington St Boston MA 02215

PRIMAK, WILLIAM L, b New York, NY, June 4, 17; wid; c 3. RADIATION DAMAGE, INTERFEROMETRY & POLARIMETRY. *Educ:* City Col New York, BS, 37; Polytech Inst Brooklyn, MS, 43, PhD(phys chem), 46. *Prof Exp:* Jr physicist, Queens Col NY, 38-42; res fel, Polytech Inst Brooklyn, 42-43, res assoc, 43-46; assoc chemist, Argonne Nat Lab, 46-60, sr chemist, 60-84, guest scientist, 85-87; CONSULT, WOLFE LOEB & CO, 85- *Mem:* Emer mem Am Chem Soc; emer mem Am Phys Soc; Sigma Xi. *Res:* Radiation effects in solids; dimensional stability, dimensional changes and properties of silica and silicate glasses; kinetics of damaging and thermal annealing, stored energy; interferometry, polarimetry, stress-strain measurements; surface effects; dimensional and surface changes; compacted states of vitreous silica. *Mailing Add:* Wolfe Loeb & Co 735 S Quincy St Hinsdale IL 60521-4346

PRIMAKOFF, PAUL, b New York, NY, March 14, 44. BIOCHEMISTRY. *Educ:* Haverford Col, BA, 66; Stanford Univ, PhD(biochem), 73. *Prof Exp:* Res assoc physiol, Harvard Med Sch, 77-81; ASST PROF PHYSIOL, MED SCH, UNIV CONN, 81- *Mem:* Am Soc Cell Biol; AAAS. *Res:* Localization of molecules in domains in membranes; localized surface antigens of the sperm cell; immunological approaches to contraception. *Mailing Add:* Dept Physiol Univ Conn Sch Med Farmington CT 06032

PRIMIANO, FRANK PAUL, JR, b Cleveland, Ohio, July 5, 39. BIOMEDICAL ENGINEERING. *Educ:* Drexel Univ, BSME, 62, MSBME, 63; Case Western Reserve Univ, PhD(biomed eng), 71. *Prof Exp:* Engr, Gen Elec Co, 63-64; res engr, Technol, Inc, 64-65; asst prof biochem eng, Case Western Reserve Univ, 71-; VPRES, TECHNIDYNE INC. *Concurrent Pos:* Pulmonary consult, Cleveland Vet Admin Hosp. *Mem:* AAAS; Am Soc Mech Eng; Soc Automotive Eng; Inst Elec & Electronics Eng. *Res:* Biomedical engineering, modeling and control of physiological systems; respiratory physiology; biomechanics. *Mailing Add:* 5027 Corliss Dr Cleveland OH 44124

PRINCE, ALAN THEODORE, b Toronto, Ont, Feb 15, 15; m 42; c 2. CHEMISTRY, GEOLOGY. *Educ:* Univ Toronto, BA, 37, MA, 38; Univ Chicago, PhD(physico-chem petrol), 41. *Prof Exp:* Jr res chemist, Div Chem, Nat Res Coun Can, 40-42; res chemist & petrogr, Can Refractories Ltd, Que, 43-45; lectr petrol, Univ Man, 45-46; head, Ceramic Chem Sect, Mines Br, Dept Mines & Tech Surv, Can, 46-50, head, Phys & Crystal Chem Sect, 50-55, sr res officer, Mineral Dressing & Process Metall Div, 55-59, chief, Mineral Sci Div, 59-65, dir, Water Res Br, 65-67, dir, Inland Waters Br, 67-70, Dept Energy, Mines & Resources, dir gen, Inland Waters Br, Environ Can, 70-72, asst dep minister planning & eval, Dept Energy, Mines & Resources, 72-75, pres, Atomic Energy Control Bd, 75-78; RETIRED. *Concurrent Pos:* Chmn Can-Saskatchewan-Nelson Basin Bd, Fraser River Flood Control Bd, Okenagan Basin Planning Bd, 69-72, Great Lakes Water Qual Bd, 70-72. *Mem:* Fel Mineral Soc Am; Mineral Asn Can; Can Inst Mining & Metall; fel Chem Inst Can. *Res:* Science and technology of earth-forming materials, minerals, rocks, water and technological materials, metals and ceramics. *Mailing Add:* 5445 Riverside Dr Box 106 Manotick ON K0A 2N0 Can

PRINCE, ALFRED M, b Berlin, Ger, Dec 16, 28; US citizen; m 62; c 2. VIROLOGY, PATHOLOGY. *Educ:* Yale Univ, BA, 49; Columbia Univ, MA, 51; Western Reserve Univ, MD, 55. *Hon Degrees:* MD, Univ Tampere, Finland. *Prof Exp:* Intern path, Sch Med, Yale Univ, 55-56, res fel virol, 56-59; assoc mem virol, Wistar Inst Anat & Biol, 62-63; asst prof path, Sch Med, Yale Univ, 63-65; SR INVESTR VIROL, NEW YORK BLOOD CTR, 65- *Concurrent Pos:* NIH res grant, 62-; clin assoc prof path, Sch Med, Cornell Univ, 65-; career investr, New York City Res Coun, 66-74. *Honors & Awards:* Landsteiner Award, Am Asn Blood Banks, 75. *Mem:* Am Asn Immunologists. *Res:* Hepatitis viruses; chronic liver disease; AIDS virus; onchocerca volvulus. *Mailing Add:* Stonehill Rd Pound Ridge NY 10576

PRINCE, CHARLES WILLIAM, b Florence, Ala. BIOCHEMISTRY, CELL DIFFERENTIATION. *Educ:* Univ North Ala, BS, 73; Univ Ala, Birmingham, PhD(biochem), 81. *Prof Exp:* Asst biologist, Southern Res Inst, 74-76; predoctoral fel, 76-81, teaching fel, 81-84, res instr, 84-85, res asst prof biochem, 85-87, ASST PROF, NUTRIT SCI, UNIV ALA, BIRMINGHAM, 87- *Concurrent Pos:* Assoc scientist, Comprehensive Cancer Ctr, Univ Ala, Birmingham, 85- & investr, Inst Dent Res, 85-87. *Mem:* AAAS; Am Soc Bone & Mineral Res; Soc Complex Carbohydrates; NY Acad Sci. *Res:* Alterations of extracellular matrix components due to nutritional or hormonal effects; role of extracellular matrix components in function of bones and teeth; effects of extracellular matrix microenvironment on hematopoiesis. *Mailing Add:* Dept Nutrit Sci Rm 328 Volker Hall Univ Ala Birmingham UAB Birmingham AL 35294

PRINCE, DAVID ALLAN, b Newark, NJ, July 14, 32; m 56; c 3. NEUROLOGY, NEUROPHYSIOLOGY. *Educ:* Univ Vt, BS, 52; Univ Pa, MD, 56. *Prof Exp:* From asst prof to assoc prof med, 63-71, actg chmn dept neurol, 70-71, PROF MED & CHMN DEPT NEUROL, SCH MED, STANFORD UNIV, 71- *Concurrent Pos:* NIH spec fel neurophysiol, Sch Med, Stanford Univ, 62-63, Nat Inst Neurol Dis & Stroke res & training grant neurol, 69-73, Epilepsy Found Am grant, 71-72; Guggenheim fel, 71-72. *Mem:* Am Neurol Asn; Am Acad Neurol; Asn Res Nerv & Ment Dis; Am Epilepsy Soc (secy, 70-72, pres 73-74); Soc Neurosci; Sigma Xi. *Res:* Neurophysiology-synaptic potentials and impulse generation in experimental epilepsy; effects of convulsant agents on electrical activity of the brain. *Mailing Add:* Dept of Neurol Stanford Univ Med Ctr Stanford CA 94305

PRINCE, DENVER LEE, b Beaton, Ark, Oct 16, 32; m 51; c 3. PHYSICS. *Educ:* Henderson State Col, BS, 54; Univ Utah, MScEd, 58; Okla State Univ, EdD(higher educ in physics), 65. *Prof Exp:* Teacher high sch, Ark, 54-59; asst prof, 59-65, PROF PHYSICS & CHMN DEPT, UNIV CENT ARK, 65- *Concurrent Pos:* NSF res grant, 66-; consult pub schs, Ark, 67-; Intermediate Sci Curric Study; teacher, Phys Sci for Nonsci Student Trial Ctr. *Mem:* Am Asn Physics Teachers; Nat Sci Teachers Asn. *Res:* Science education. *Mailing Add:* Dept of Physics Univ of Cent Ark Conway AR 72032

PRINCE, EDWARD, b Schenectady, NY, Nov 29, 28; m 54; c 3. SOLID STATE PHYSICS. *Educ:* Harvard Univ, AB, 49; Cambridge Univ, PhD(physics), 52. *Prof Exp:* Mem tech staff, Bell Tel Labs, Inc, 52-58; physicist, US Naval Res Lab, Washington, DC, 58-68; RES PHYSICIST, NAT INST STANDARDS & TECHNOL, 68- *Concurrent Pos:* Co-ed, J Appl Crystallog, 86- *Mem:* AAAS; Am Phys Soc; Am Crystallog Asn. *Res:* Structure of crystals; thermal motion; experimental automation; neutron diffraction data analysis methods. *Mailing Add:* Reactor Radiation Div Nat Inst Standards & Technol Gaithersburg MD 20899

PRINCE, ERIC D, b New York, NY, Oct 10, 46. FISHERIES. *Educ:* Univ Pac, BS, 69; Humboldt State Univ, MS, 72; Va Polytech Inst & State Univ, PhD(fish & wildlife sci), 77. *Prof Exp:* Fishery biologist, US Fish & Wildlife Serv, Southeast Reservoir Invests, 77-80; FISHERY RES BIOLOGIST, NAT MARINE FISHERIES SERV, SOUTHEAST FISHERIES CTR, MIAMI LAB, 80- *Concurrent Pos:* Adj asst prof, Rosenstiel Sch Marine & Atmospheric Sci, Univ Miami, 81-83, adj assoc prof, 84-; assoc ed, NAm J Fisheries Mgt, 81-83, ed, 81-; referee for Trans Am Fisheries Soc 78-, Progressive Culturist, 79-, J Marine Sci, 80-, Fishery Bull, 80-& Can J Fisheries Mgt, 81-, coordr ICCAT Enhanced Res Prog for Bullfish, 87or Bullfish, 87. *Mem:* Am Fisheries Soc; Am Inst Fishery Res Biologists; Coastal Soc; Sigma Xi. *Res:* Biological effects of artificial reefs in aquatic environments; tracking movements of fishes using ultrasonic telemetry; trophic dynamics of marine and freshwater habitats; determining entrainment mortality of ichthyoplankton passing through power plants; age and growth of oceanic pelagic fishes; marking & tagging fish. *Mailing Add:* 75 Virginia Beach Dr Miami FL 33149-1099

PRINCE, GREGORY ANTONE, VIRAL IMMUNOLOGY, VACCINE DEVELOPMENT. *Educ:* Univ Calif, Los Angeles, DDS, 73, PhD(path), 75. *Prof Exp:* EXPERT, NAT INST ALLERGY & INFECTIOUS DIS, NIH, 75- *Mailing Add:* NIH Bldg 7 Rm 301 Bethesda MD 20892

PRINCE, HAROLD HOOPES, b Keokuk, Iowa, Mar 29, 41; m 62; c 3. BEHAVIORAL ECOLOGY. *Educ:* Iowa State Univ, BSc, 62; Univ NB, MSc, 65; Va Polytech Inst, PhD(wildlife biol), 69. *Prof Exp:* from asst prof to assoc prof, 68-78, PROF & ASST CHMN FISHERIES & WILDLIFE, MICH STATE UNIV, 78- *Concurrent Pos:* Vis Scholar, Delta Waterfowl Res Sta, Delta, Man, 75; Dansforth Assoc, 75. *Mem:* Wilson Ornith Soc; Wildlife Soc; Am Ornith Union; Cooper Ornith Soc. *Res:* Effects of genetic and environmental parameters on reproduction, survival and behavior of game birds, especially waterfowl; interaction of biocides with avian populations. *Mailing Add:* Dept Fisheries & Wildlife Mich State Univ East Lansing MI 48824

PRINCE, HARRY E, b Spartanburg, SC, Sept 2, 53. CLINICAL IMMUNOLOGY. *Educ:* Univ NC, PhD(bacteriol & immunol), 79. *Prof Exp:* Res fel, Med Immunol Lab, Univ Calif, Los Angeles, 81-83; HEAD, DIV CELLULAR IMMUNOL, AM RED CROSS BLOOD SERV, LOS ANGELES, LA, 83- *Mem:* Am Asn Immunologists; Am Asn Blood Banks; Am Soc Microbiol; Clin Immunol Soc; Asn Med Lab Immunologists. *Res:* Regulation of T lymphocyte activation and the role played by cell- surface activation antigens; immune dysfunction in Human Immunodeficiency Virus Infection. *Mailing Add:* Cell Immunol Lab Am Red Cross Blood Serv 1130 S Vermont Ave Los Angeles CA 90006

PRINCE, HERBERT N, b New York, NY, Aug 8, 29; m 55; c 3. MICROBIOLOGY, TOXICOLOGY. *Educ:* NY Univ, AB, 50; Univ Conn, PhD(bact), 56. *Prof Exp:* Bacteriologist, New York City Health Dept, 50; head microbiol res & qual control, Wallace & Tiernan, Inc, NJ, 56-60; sr virologist, Hoffmann-La Roche, Inc, 60-63, asst to dir, 63-66, asst dir chemother, 67-70; PRES, GIBRALTAR BIOL LABS, INC, 70- *Concurrent Pos:* Adj prof, Fairleigh Dickinson Univ, 57-72; vis prof, Seton Hall Univ, 70; adj prof, Rutgers Col, 79-83; consult infectious dis, St Michaels Med Ctr, Newark, NJ, 72-; vis scientist, UMDNJ, Newark, 86- *Mem:* Am Soc Microbiol; Soc Exp Biol & Med; Soc Indust Microbiol; fel Am Acad Microbiol; NY Acad Sci. *Res:* Industrial and public health microbiology; chemotherapy and experimental toxicology. *Mailing Add:* 21 Sylvan Way Caldwell NJ 07006

PRINCE, JACK, b Bronx, NY, Feb 12, 38; m 57; c 4. PHYSICS. *Educ:* Yeshiva Univ, BA, 59; NY Univ, MS, 63, PhD(physics), 67. *Prof Exp:* NY State Regents teaching fel physics, NY Univ, 59-61; instr, Bronx Community Col, 60-63 & NY Univ, 63-64; from asst prof to assoc prof, 64-71, chmn Physics, 71-80, PROF, BRONX COMMUNITY COL, 71-, CHMN, 86- *Concurrent Pos:* Dir, Nuclear Med Tech, Bronx Community, 83- *Mem:* Am Asn Physics Teachers; Asn Orthodox Jewish Scientists. *Res:* Theoretical solid state physics; many-body problem; differential and integral variational methods. *Mailing Add:* Dept of Physics Bronx Commun Col Bronx NY 10453

PRINCE, JAMES T, b Winchester, Tenn, June 29, 20; m 47; c 2. VIROLOGY, MEDICAL MICROBIOLOGY. *Educ:* Univ Minn, Minneapolis, BS, 53, MS, 61. *Prof Exp:* From instr to asst prof, med sch, Univ Minn, Minneapolis, 58-68, assoc prof microbiol, 68-, dir, MS degree prog med microbiol, 74-; RETIRED. *Mem:* Am Soc Microbiol. *Res:* Virology, including Shope papilloma to carcinoma sequence; medical diagnostic microbiology. *Mailing Add:* Dept of Microbiol Univ of Minn Med Sch Minneapolis MN 55455

PRINCE, JOHN LUTHER, III, b Austin, Tex, Nov 13, 41; m 60; c 4. ELECTRONIC PACKAGING STRUCTURES. *Educ:* Southern Methodist Univ, BSEE, 65; NC State Univ, MEE, 68, PhD(elec eng), 69. *Prof Exp:* Res engr, Res Triangle Inst, 68-70; mem tech staff, Tex Instruments, Inc, 70-75; prof elec eng, Clemson Univ, 75-80; dir, Intermedics, Inc, 80-83; PROF ELEC ENG, UNIV ARIZ, 83- *Mem:* Fel Inst Elec & Electronics Engrs. *Res:* Electrical and thermal characteristics of electronic packaging structures. *Mailing Add:* 7542 N San Lorenzo Tucson AZ 85704

PRINCE, M(ORRIS) DAVID, b Greensboro, NC; m 47; c 4. COMPUTER GRAPHICS, COMPUTER AIDED DESIGN. *Educ:* Ga Inst Technol, BS, 46, MS, 49. *Prof Exp:* Instr math, Ga Inst Technol, 47-51, res engr, Eng Exp Sta, 51-56; sr staff mem, ITT Labs, 56-59; res & develop engr, Lockheed-Ga Co, 59-64, assoc dir res, 64-68, res lab dir, 68-70, dir res progs, 70-71, sr res & develop engr, 71-75, staff specialist, 75-83, sr staff specialist, 83-90; CONSULT, 90- *Concurrent Pos:* Consult, NSF, 72-73; pres, First Am Systs, Inc, 72-75; prof, Sch Info & Comput Sci, Ga Inst Technol; vis lectr, Inst Elec & Electronics Engrs Comput Soc, 73. *Mem:* Fel Inst Elec & Electronics Engrs. *Res:* Artificial intelligence and expert systems for use in aircraft cockpits and other intelligent systems; computer graphics for aircraft cockpit displays; graphics for the engineer's work station and for management information systems; research and engineering management technology transfer and utilization. *Mailing Add:* 3132 Frontenac Ct Atlanta GA 30319

PRINCE, MACK J(AMES), b Vienna, Austria, Sept 28, 19; nat US; m 54; c 3. ELECTRICAL ENGINEERING. *Educ:* Worcester Polytech Inst, BS, 49; Univ RI, MS, 54. *Prof Exp:* Asst physics, 49-54, from instr to asst prof elec eng, 55-61, assoc prof, 61-85, EMER PROF UNIV RI, 85- *Res:* Circuit theory; instrumentation. *Mailing Add:* 876 Usquepaugh Rd West Kingston RI 02892

PRINCE, MARTIN IRWIN, b New York, NY, Sept 19, 37; m 60; c 3. ORGANOMETALLIC CHEMISTRY. *Educ:* NY Univ, BA, 58, MS & PhD(org chem), 62. *Prof Exp:* Assoc res scientist, Res Div, NY Univ, 62-64, res scientist, 64-67; PRES, SYNTHATRON CORP, 67- *Mem:* Am Inst Chemists; Am Chem Soc; NY Acad Sci; AAAS; Sigma Xi. *Res:* Organic synthesis; high pressure reactions; radiation chemistry; organometallic synthesis; preparation of new polymer systems; polymer coatings; metal hydrides; fiber optic chemicals gases. *Mailing Add:* Synthatron Corp PO Box 2600 30 Two Bridges Rd Suite 330 Fairfield NJ 07004

PRINCE, MORTON BRONENBERG, b Philadelphia, Pa, Apr 1, 24; m 47; c 1. SEMICONDUCTORS. *Educ:* Temple Univ, AB, 47; Mass Inst Technol, PhD, 51. *Prof Exp:* Mem tech staff, Bell Tel Labs, 51-56; dir res & develop, Semiconductor Div, Hoffman Electronics Corp, 56-59, vpres & gen mgr, 60-61, Electro-Optical Systs, Inc, Calif, 61-69; pres, SSR Instruments Co, 70-74; gen mgr & vpres, Meret, Inc, 74-75; br chief, Energy Res & Develop Admin, 75-77, DEPT OF ENERGY, 77- *Honors & Awards:* Marconi Premium, Inst Radio Engrs, Gt Brit, 59. *Mem:* Am Phys Soc; Inst Elec & Electronics Engrs; Int Solar Energy Soc. *Res:* Semiconductor device development; semiconductor materials; photovoltaic devices; solar energy; scientific instrumentation. *Mailing Add:* 2700 Virginia Ave NW Washington DC 20037

PRINCE, ROBERT HARRY, b Slough, Eng, Jan 26, 42; m 64; c 2. SURFACE PHYSICS, SPACECRAFT MATERIALS. *Educ:* Univ Toronto, BASc, 63, MASc, 64, PhD(aerophysics), 68. *Prof Exp:* Nat Res Coun Can fel, Univ Col, London, 68-69; res assoc aerophysics, Inst Aerospace Studies, Univ Toronto, 69-70; from asst prof to assoc prof, 70-81, dir, Grad Prog Physics,

85-89, PROF PHYSICS, YORK UNIV, 81-, CHMN, DEPT PHYSICS, 89- *Concurrent Pos:* Consult, Ont Res Found, 66-67, Martin Marietta Inc, 70- & Commun Res Ctr, Ottawa, 71-73, Nuffield fel, Univ Cambridge, 78; vis scholar phys chem, Univ Cambridge, 83-84; vis prof physics, Penn State Univ, 90-91. *Mem:* Can Asn Physicists. *Res:* Particle-surface interactions; surface science. *Mailing Add:* Dept Physics York Univ North York ON M3J 1P3 Can

PRINCE, ROGER CHARLES, b London, Eng, June 14, 50. REDOX CHEMISTRY, PHOTOSYNTHESIS. *Educ:* Univ York, Eng, BA, 71; Univ Bristol, Eng, PhD(biochem), 74. *Prof Exp:* Res asst prof biochem & biophysics, Univ Pa, 80-83; SR STAFF BIOCHEMIST, CORP RES LABS, EXXON RES & ENG, 83-, GROUP HEAD MOLECULAR BIOL, 85- *Concurrent Pos:* Adj assoc prof, dept biochem & biophysics, Univ Pa, 83- *Mem:* Am Soc Biol Chemists; Biophys Soc; Int Soc Toxicol. *Res:* Biological oxidation-reduction chemistry, both in enzymes that perform redox chemistry on their substrates and in membrane systems, particularly photosynthetic ones, where redox energy is conserved in forms more suitable for chemical manipulation. *Mailing Add:* Exxon Res & Eng Annandale NJ 08801

PRINCE, TERRY JAMISON, b Elwood, Ind, Mar 18, 49; m 78. ANIMAL NUTRITION. *Educ:* Purdue Univ, BSc, 71; Univ Ky, PhD(monogastric nutrit), 76. *Prof Exp:* ASST PROF SWINE NUTRIT & PROD, AUBURN UNIV, 76- *Mem:* Am Soc Animal Sci; Animal Nutrit Res Coun. *Res:* Swine nutrition and production. *Mailing Add:* Carl S Akey Inc CS 5002 Lewisburg OH 45338

PRINCEN, HENRICUS MATTHEUS, b Eindhoven, Neth, Apr 15, 37; div; c 2. SURFACE SCIENCE, COLLOID SCIENCE. *Educ:* State Univ Utrecht, BSc, 56, Drs, 60, PhD(colloid chem), 65. *Prof Exp:* Res assoc, Res Lab, N V Philips' Gloeilampenfabrieken, Eindhoven, 60; res scientist, Lever Bros Co, Edgewater, 65-73, mgr phys chem, 73-80; res assoc, Exxon Res & Eng Co, 80-86; PRIN SCIENTIST, GENERAL FOODS CORP, 87- *Concurrent Pos:* Vis lectr/scientist, Col de France, Paris, 86-87. *Mem:* Am Chem Soc. *Res:* Surface chemistry, especially detergency, capillarity, wetting, thin liquid films, foams, emulsions and rheology. *Mailing Add:* Mobil Res & Develop Corp Cent Res Lab Box 1025 Princeton NJ 08543-1025

PRINCEN, LAMBERTUS HENRICUS, b Eindhoven, Neth, Aug 31, 30; US citizen; m 55; c 1. CHEMISTRY, AGRICULTURE. *Educ:* State Univ Utrecht, BS, 52, Drs, 55, DSc(colloid chem), 59. *Prof Exp:* Res assoc, Univ Southern Calif, 55-58; res scientist, Nat Defense Res Coun, Neth, 58-60 & Nat Flaxseed Processors Asn, 60-62; proj leader linseed oil paints, Northern Regional Res Ctr, USDA, 62-75, chief, Hort & Spec Crops Lab, 75-82, assoc ctr dir, 82-84, dir, 84-90; ED, J AM OIL CHEMIST SOC, 91- *Concurrent Pos:* Lectr, Univ Southern Calif, 56-58; lectr, Bradley Univ, 63-73, adj prof, 91- *Mem:* Am Chem Soc; Am Oil Chemists' Soc; Sigma Xi. *Res:* Colloid chemistry; light scattering of detergent solutions, aerosols and aerosol filtration; pigment interactions in emulsion paint systems; emulsion technology; paint degradation; particle size distributions in colloidal systems; electron microscopy; agricultural chemistry; synthetic modification of natural products; new crops development; natural toxicants; renewable resources. *Mailing Add:* 677 E High Point Terr Peoria IL 61614

PRINDLE, BRYCE, b Brule, Wis, Oct 23, 09; m 33; c 4. MICROBIOLOGY. *Educ:* Mass Inst Technol, SB, 31, PhD(microbiol), 35. *Prof Exp:* Textile Found sr res fel, Mass Inst Technol, 35-37; from instr to asst prof bact, Iowa State Col, 37-43; field supvr dehydration, Qm Corp, US Dept Army, 43-45; biologist, Plymouth Cordage Co, 45-60; instr, 60-76, prof sci, Babson Col, 63-76; vis investr, Woods Hole Oceanog Inst & instr biol, Bridgewater State Col, 76-78; CONSULT, 78- *Mem:* AAAS; Am Soc Microbiol; Soc Indust Microbiol; Inst Food Technol. *Res:* Biological oceanography; marine deterioration; preservation of materials. *Mailing Add:* 15 Flintlock Way Yarmouth Port MA 02675

PRINDLE, RICHARD ALAN, b Mansfield, Ohio, Dec 28, 25; m 79; c 2. PUBLIC HEALTH, EPIDEMIOLOGY. *Educ:* Harvard Univ, MD, 48, MPH, 54; Am Bd Prev Med & Pub Health, dipl, Nat Bd Med Exam, dipl. *Prof Exp:* Intern, Columbia-Presby Med Ctr, 49-50; asst chief poliomyelitis invests sect, USPHS, 51-53, tech consult, US Opers Mission to Haiti, 54-57, epidemiologist, Air Pollution Med Prog, 57-58, chief, 58-60, dep chief div air pollution, 60-63, chief div pub health methods, 63-66, asst surgeon gen & chief bur state serv, 66, dir bur dis prev & environ control, 67-68; fed exec fel, Brookings Inst, 68-69; chief div family health, Pan-Am Health Orgn, 70-77; dir, Thomas Jefferson Health Dist, 77-87; PROF INTERNAL MED, UNIV VA & ASSOC DIR, CTR PREV DIS & INJURY, 87- *Concurrent Pos:* Res assoc, Harvard Univ, 54-56, vis lectr, 56-60 & 63-65; vis lectr, Univ Pa, 60-65; adj prof pop sci & int health, Sch Pub Health, Univ Ill, 72-78; clin prof med, Univ Va, 78-87, prof epidemiol, 81-87, dir, Occup Health Prog, 87- *Mem:* Fel AAAS; fel Am Col Prev Med; Am Pub Health Asn; Asn Teachers Prev Med; Am Occup Med Asn. *Res:* Epidemiology; public health administration; population dynamics and family planning; demography; environmental health. *Mailing Add:* Univ Va 630 Kearsarge Circle Charlottesville VA 22901-7836

PRINDLE, WILLIAM ROSCOE, b San Francisco, Calif, Dec 19, 26; m 47; c 2. GLASS TECHNOLOGY. *Educ:* Univ Calif, Berkeley, BS, 48, MS, 50; Mass Inst Technol, ScD(ceramics), 55. *Prof Exp:* Asst tech dir, Hazel-Atlas Glass Co, 54-56, mgr res, 57-59, gen mgr res & develop, 59-62; mgr mat res, Am Optical Co, 62-65, vpres res, 71-76; dir res, Ferro Corp, 66-67, vpres res, 67-71; exec dir, Nat Mat Adv Bd, Nat Res Coun, 76-80; dir, Admin & Tech Serv, Corning Inc, 81-85, dir, Mat Res, 85-87, assoc dir, Res, Develop & Eng, 87-90, DIV VPRES & ASSOC DIR, TECHNOL GROUP, CORNING INC, 90- *Concurrent Pos:* Mem, coun, IntComn Glass, 67-70 & steering comt, 83-91, US rep, 83-91, vpres, 84-85, pres, 85-88; mem, Eval Panel Inorg Mat Div Nat Bur Standards, Nat Res Coun, 70-75, chmn, 75-76 & Nat Mat Adv Bd, 73-76; pres, XII Int Cong Glass, 80; distinguished visitor, NY State Col Ceramics, Alfred Univ, 84. *Honors & Awards:* John F McMahon Eng Lectr, 83; Friedberg Mem Lectr, Nat Inst Ceramic Engrs, 90. *Mem:* Nat Acad Eng; fel Soc Glass Technol; fel Am Soc Metals Int; Minerals Metals & Mat Soc; Am Soc Testing & Mat; Nat Inst Ceramic Engrs; fel Am Ceramic Soc (treas, 71, vpres, 72, pres, 80-81); Acad Ceramics. *Res:* Direction of industrial research and development of glass, ceramics and metals. *Mailing Add:* Corning Inc Sullivan Park FR-02-10 Corning NY 14831

PRINE, GORDON MADISON, b Valdosta, Ga, Feb 16, 28; m 57; c 2. AGRONOMY. *Educ:* Univ Ga, BSA, 52, MS, 54; Ohio State Univ, PhD, 57. *Prof Exp:* Asst agronomist, Ga Coastal Exp Sta, 54-55; asst agronomist, 58-65, assoc agronomist, 65-77, AGRONOMIST, AGR EXP STA, UNIV FLA, 77- *Concurrent Pos:* Consult, Guyana,69-71, El Salvador, 71-72,78, Mexico, 72, Bolivia, 79 &Equador, 86. *Mem:* Am Soc Agron; Crop Sci Soc Am. *Res:* Field crop ecology, with emphasis on new crops and introductions; low light intensities on crop production; multiple cropping; selection of superior reseeding annual rye grasses, perennial peanuts, lupines, fababeans, crotalarias, napier grasses and pigeon peas. *Mailing Add:* Dept of Agron Univ of Fla, 304 Newell Hall Gainesville FL 32611

PRINEAS, JOHN WILLIAM, b Junee, Australia, May 30, 35; m 60; c 3. NEUROLOGY, NEUROPATHOLOGY. *Educ:* Univ Sydney, MB & BS, 58. *Prof Exp:* House physician, Royal Prince Alfred Hosp, Sydney, 58; sr house physician cardiol, Glasgow Royal Infirmary, Scotland, 60-61; house physician neurol, Nat Hosps Nervous Dis, London, 61-62, sr registr, 63-64; first asst neurol, Royal Victoria Infirmary, Eng, 65-67; res fel, Albert Einstein Col Med, 67-69; sr res fel, Univ Sydney, 69-72, sr lectr med, 72-74; PROF NEUROSCI, UNIV MED & DENT NJ, 74-, PROF PATH, 81- *Concurrent Pos:* Hon neurologist, Royal Prince Alfred Hosp, Sydney, 69-74; attending neurologist, Vet Admin Hosp, East Orange, 76-86; vis prof path, Albert Einstein Col Med, 77-86. *Honors & Awards:* Weil Award, Am Asn Neuropathologists, 69 & 75, Moore Award, 76. *Mem:* Am Asn Neuropathologists; Australian Asn Neurologists. *Res:* Myelin pathology; multiple sclerosis. *Mailing Add:* Dept of Neurol Vet Admin Hosp Tremont Ave East Orange NJ 07019

PRINEAS, RONALD JAMES, b NSW, Australia, Sept 19, 37; m 61; c 4. CARDIOVASCULAR DISEASES. *Educ:* Univ Sydney, MB & BS, 61; Univ London, PhD(epidemiol), 69. *Prof Exp:* Jr resident med, Prince Henry Hosp, Australia, 61; sr resident, Royal Perth Hosp, 62; sr house officer cardiol, Glasgow Royal Infirmary, 63, registr, 64; res fel epidemiol, Sch Hyg & Trop Med, London, 65-67, lectr, 68; from second asst to third asst, Dept Med, Royal Melbourne Hosp, Univ Melbourne, 69-72; assoc prof, Sch Pub Health, 73-78, assoc prof, Sch Med, 74-80, prof, Sch Pub Health, 79-88, prof Sch Med, 81-88, PROF CHMN EPIDEMIOL, UNIV MIAMI, 88- SCH MED, 81- *Concurrent Pos:* Fel, Coun Epidemiol, Am Heart Asn, 73- *Honors & Awards:* Shared Spec Lasker Award, 80. *Mem:* Am Col Cardiol; Royal Col Physicians Edinburgh; Royal Col Physicians London; Am Heart Asn; Am Col Epidemiol. *Res:* Epidemiology of cardiovascular diseases; hypertension; sudden death; population electrocardiography; preventive cardiology; methodology of cardiovascular epidemiology. *Mailing Add:* Dept Epidemiol R669 Univ Miami Sch Med 1600 NW Tenth Ave Miami FL 33101

PRING, DARYL ROGER, b Morris, Minn, June 6, 42; m 63; c 3. PLANT PATHOLOGY. *Educ:* NDak State Univ, BS, 64, MS, 66, PhD(plant path), 67. *Prof Exp:* Fel plant path, Univ Nebr, Lincoln, 68-71; res plant pathologist, 71-77, ASSOC PLANT PATHOLOGIST, PLANT VIRUS LAB, AGR RES SERV, USDA, FLA, 77-; ASSOC PROF PLANT PATH, UNIV FLA, 78- *Mem:* Am Phytopath Soc. *Res:* Plant virology; nucleic acid replication; nature of cytoplasmic inheritance; gene-cytoplasm interactions. *Mailing Add:* Plant Virus Lab Univ Fla Gainesville FL 32611

PRINGLE, DOROTHY JUTTON, b Evanston, Ill, Feb 20, 19; m 61. NUTRITION. *Educ:* Univ Ill, BS, 40; Univ Wis, MS, 51, PhD(nutrit biochem), 56. *Prof Exp:* Dietetic intern, Univ Hosp, Univ Mich, 40-41; staff & consult dietitian, St Luke's Hosp, Chicago, Ill, 42-49; asst prof foods & nutrit, 51-53, prof nutrit sci, 56-85, EMER PROF, UNIV WIS-MADISON, 85- *Mem:* Am Dietetic Asn. *Res:* Socioeconomic influences on food habits and nutritional adequacy of minority families; evaluation of nutrition education methods. *Mailing Add:* 2230 Monroe St Madison WI 53711

PRINGLE, JAMES SCOTT, b Danvers, Mass, Aug 14, 37. BOTANY. *Educ:* Dartmouth Col, AB, 58; Univ NH, MS, 60; Univ Tenn, PhD(bot), 63. *Prof Exp:* Asst, Univ Tenn, 60-62; TAXONOMIST, ROYAL BOT GARDENS, 63-, ASST DIR SCI, 85-; ASSOC PROF BIOL, MCMASTER UNIV, 85- *Concurrent Pos:* Spec lectr, McMaster Univ, 74-85. *Mem:* Am Soc Plant Taxon; Soc Study Evolution; Bot Soc Am; Can Bot Asn; Int Asn Plant Taxon. *Res:* Taxonomy and evolutionary biology of vascular plants, especially Gentianaceae; taxonomy and breeding of Syringa; history of Canadian floristics. *Mailing Add:* Royal Bot Gardens Box 399 Hamilton ON L8N 3H8 Can

PRINGLE, ORAN A(LLAN), b Lawrence, Kans, Sept 14, 23; m 47; c 4. MECHANICAL ENGINEERING. *Educ:* Univ Kans, BS, 47; Univ Wis, MS, 48, PhD(mech eng), 67. *Prof Exp:* Design engr, Black & Veatch, Consult Engrs, 48-50; from asst prof to assoc prof mech eng, 50-68, PROF MECH & AEROSPACE ENG, UNIV MO-COLUMBIA, 68- *Mem:* Am Soc Mech Engrs; Am Soc Eng Educ. *Res:* Dynamics of machinery; nonlinear vibrations; fatigue of metals and polymers; automatic control; machine elements. *Mailing Add:* Dept of Mech & Aerospace Eng Univ of Mo Columbia MO 65211

PRINGLE, WALLACE C, JR, b Hartford, Conn, Oct 23, 41; m 63; c 3. PHYSICAL CHEMISTRY. *Educ:* Middlebury Col, AB, 63; Mass Inst Technol, PhD(chem), 66. *Prof Exp:* Res assoc atomic physics, Nat Bur Stand, 66-67; asst prof, 68-74, ASSOC PROF CHEM, WESLEYAN UNIV, 74- *Concurrent Pos:* Nat Res Coun fel, 66-67. *Res:* Molecular spectroscopy and structure of small molecules; collision induced for infrared spectroscopy; hazardous waste management. *Mailing Add:* Dept of Chem Wesleyan Univ Middletown CT 06457

PRINS, RUDOLPH, b Grand Rapids, Mich, Dec 21, 35; m 58; c 4. AQUATIC ECOLOGY, ZOOLOGY. *Educ:* Calvin Col, BA, 58; Western Mich Univ, MA, 62; Univ Louisville, PhD(zool), 65. *Prof Exp:* Teacher high sch, Mich, 58-60 & 61-62; asst zool, Univ Louisville, 62-65; asst prof entom & zool, Clemson Univ, 65-68; assoc prof, 68-74, PROF BIOL, WESTERN KY UNIV, 74- *Concurrent Pos:* Guest prof, Univ Australia, Valdivia, Chile, 84; fel Orgn Am States; sabbatical leave, Inst Zool, Univ Australia. *Mem:* NAm Benthological Soc; Am Soc Limnol & Oceanog; Asn SE Biologists. *Res:* Ecology and taxonomy of freshwater invertebrates; aquatic biology. *Mailing Add:* Dept Biol Western Ky Univ Bowling Green KY 42101

PRINTZ, MORTON PHILIP, b Portsmouth, Va, Sept 24, 36; m 59; c 2. PHARMACOLOGY. *Educ:* Univ Pittsburgh, BS, 58, PhD(biophys), 63. *Prof Exp:* USPHS fel, 62-64; res assoc biochem, Dartmouth Med Sch, 64-65; asst prof, Rockefeller Univ, 65-72; asst prof med, 72-77, ASSOC PROF MED, UNIV CALIF, SAN DIEGO, 78- *Mem:* Am Soc Biol Chemists; Am Soc Pharmacol & Exp Therapeut. *Res:* Endocrine and neuroendocrine mechanisms of blood pressure regulation; reninangiotensin system; protaglandin biosynthesis in blood vessels and relation to cardiocirculatory diseases. *Mailing Add:* Dept Pharm Univ Calif San Diego M-036 La Jolla CA 92093

PRINZ, DIANNE KASNIC, b Conway, Pa, Sept 29, 38; m 60. PHYSICS. *Educ:* Univ Pittsburgh, BS, 60; Johns Hopkins Univ, PhD(physics), 67. *Prof Exp:* Res scientist, Space Sci Div, Naval Res Lab, 67-68; E O Hulburt fel physics & astron, Univ Md, College Park, 68-71; RES SCIENTIST, SPACE SCI DIV, NAVAL RES LAB, 71- *Concurrent Pos:* NSF grants, 68-71. *Mem:* Am Geophys Union. *Res:* Infrared spectroscopy of atmospheric gases; ultraviolet spectroscopy of solar and atmospheric gases. *Mailing Add:* Naval Res Lab Code 4142 Washington DC 20375

PRINZ, GARY A, b Cleveland, Ohio, Oct 2, 38; m 60. SOLID STATE PHYSICS. *Educ:* Univ Pittsburgh, BS, 60; Johns Hopkins Univ, PhD(physics), 66. *Prof Exp:* Res asst, Johns Hopkins Univ, 66-67; RES PHYSICIST, US NAVAL RES LAB, 67- *Mem:* Am Phys Soc; Sigma Xi. *Res:* Magnetism of rare earth ions in insulators. *Mailing Add:* Naval Res Lab Code 4635 Washington DC 20375

PRINZ, MARTIN, b New York, NY, Sept 5, 31; m 61, 84; c 3. PETROLOGY, MINERALOGY. *Educ:* City Col New York, BS, 53; Ind Univ, MA, 57; Columbia Univ, PhD(geol), 61. *Prof Exp:* Lectr geol, Brooklyn Col, 60-61; from instr to assoc prof, Tufts Univ, 61-69; sr res scientist, Inst Meteoritics, Univ NMex, 69-76; CUR DEPT OF MINERAL SCI, AM MUS NATURAL HIST, 76- *Mem:* Fel Geol Soc Am; fel Mineral Soc Am; Int Asn Geochem & Cosmochem (asst secy, 73-76); fel Meteoritical Soc; AAAS; Am Geophys Union. *Res:* Igneous petrology; lunar petrology; mafic and ultramafic rocks; oceanic rocks; meteorites; volcanology; rock forming minerals; electron microprobe analysis. *Mailing Add:* Dept Mineral Sci Am Mus Natural Hist New York NY 10024-5192

PRINZ, PATRICIA N, b Huntington, WVa, July 17, 42; m 63; c 1. NEUROSCIENCES, GERONTOLOGY. *Educ:* Duke Univ, BS, 63; Stanford Univ, PhD(pharmacol), 69. *Prof Exp:* Fel neurosci, Med Ctr, Duke Univ, 69-71 & Ctr for Aging, 71-74, asst prof, Dept Physiol, 74-76; asst prof, 76-81, assoc prof, 81-86, PROF DEPT PSYCHIAT, UNIV WASH, 86- *Concurrent Pos:* Preceptor, NIH Res Training Prog & sr fel sleep res, Ctr for Study Aging, Duke Univ, 74-76; Dir sleep & aging prog, Seattle-Tacoma Vet Admin Hosp & Univ Wash, 76-; res grants, NIMH, 78-, NC Alcoholism Res Authority, 74-76, Univ Wash, 76-78, Nat Inst Aging, 76-80 & Vet Admin Merit Rev, 78-; res career scientist, Vet Admin Med Ctr. *Honors & Awards:* Merit Award, NIMH. *Mem:* Geront Soc; Neurosci Soc; Sleep Res Soc. *Res:* Neurobiology of aging, including sleep/wakefulness patterns, neuroendocrine rhythms, electroencephalographic signal analysis by computer; correlation of neurobiological changes with neuroanatomical change in normal aging and in Alzheimer's disease. *Mailing Add:* Dept Psychiat Univ Wash Seattle WA 98195

PRIOLA, DONALD VICTOR, b Chicago, Ill, Sept 8, 38; m 60; c 4. PHYSIOLOGY, PHARMACOLOGY. *Educ:* Loyola Univ Chicago, BS, 60, PhD(physiol), 66. *Prof Exp:* Res assoc physiol, Stritch Sch Med, Loyola Univ Chicago, 65-66; from instr to assoc prof pharmacol, 66-70, assoc prof physiol, 70-75, asst chmn dept, 71-78, PROF PHYSIOL, SCH MED, UNIV NMEX, 75-, CHMN DEPT, 79- *Concurrent Pos:* NIH res grant, 66-, NIH career develop award, 70-75; prin investr, USPHS Grant, 67-; responsible investr, Life Ins Med Res Fund grant, 68-71; NMex Heart Asn grant, 70-71; mem cardiovasc/pulmonary study sect, Nat Heart, Lung & Blood Inst, 80-84; ed, J Pharmacol & Exp Therapeut, 80- *Mem:* Am Physiol Soc; Am Heart Asn; Fedn Am Socs Exp Biol; Am Soc Pharmacol & Exp Therapeut; Soc Exp Biol & Med; Sigma Xi. *Res:* Nervous control of the heart and vascular system; autonomic nervous system especially related to the pharmacology of drugs affecting the cardiovascular system. *Mailing Add:* Dept of Physiol Univ of NMex Sch of Med Albuquerque NM 87131

PRIOR, DAVID JAMES, US citizen. NEUROBIOLOGY. *Educ:* Olivet Col, AB, 65; Cent Mich Univ, MS, 68; Univ Va, PhD(physiol), 72. *Prof Exp:* Fel neurobiol & behav, Princeton Univ, 72-73; asst prof, 73-78, ASSOC PROF BIOL SCI, UNIV KY, 78-; AT DEPT ZOOL,. *Mem:* Soc Neurosci; Am Soc Zool; Sigma Xi; Am Soc Physiol; AAAS. *Res:* Use of certain molluscan nervous systems to study aspects of cyclical motor output, including initiation, maintenance, modulation, and termination of the rhythm. *Mailing Add:* Biol Sci Univ Ky Lexington KY 40506

PRIOR, JEAN CUTLER, b Akron, Ohio, Nov 5, 40; m 64. GEOLOGY, GEOMORPHOLOGY. *Educ:* Purdue Univ, BA, 62; Univ Ill, MS, 64. *Prof Exp:* Res asst, Coal Petrog, Ill State Geol Surv, 62-64; RES GEOLOGIST, IOWA GEOL SURV, 65- *Concurrent Pos:* Adv bd, Iowa State Preserves, 83-89; ed, Iowa Geol Mag, 84- *Mem:* Geol Soc Am; Asn Am Geogr; Asn Earth Sci Ed. *Res:* Quaternary geology; environmental geology; history of geology; water resources; Iowa land forms. *Mailing Add:* Geol Surv Bur Iowa DNR 123 N Capitol Iowa City IA 52242

PRIOR, JOHN ALAN, b Columbus, Ohio, Apr 17, 13; m 36; c 3. MEDICINE. *Educ:* Ohio State Univ, BS, 35, MD, 38. *Prof Exp:* Instr med, Univ Cincinnati, 43-44; from instr to assoc prof, 44-51, from assoc dean to dean col med, 63-72, PROF MED, OHIO STATE UNIV, 51- *Concurrent Pos:* Consult, USPHS, 46-51, Ohio State Dept Health, 48-51, Vet Admin Hosp & US Air Force Hosp, Dayton, 60-; mem bd dirs, Ohio State Univ Res Found & Willson Children's Ctr. *Mem:* Am Col Chest Physicians; Am Thoracic Soc; fel Am Col Physicians. *Res:* Pulmonary and fungus diseases; physical diagnosis. *Mailing Add:* Dunham Hosp 566 W Tenth Ave Columbus OH 43210

PRIOR, JOHN THOMPSON, b St Albans, Vt, Oct 8, 17; m 48; c 6. PATHOLOGY. *Educ:* Univ Vt, BS, 39, MD, 43. *Prof Exp:* Lab dir, Crouse Irving Hosp, 50-71; PROF PATH, COL MED, STATE UNIV NY UPSTATE MED CTR, 54-, ASSOC LAB DIR, UNIV HOSP, 50- *Concurrent Pos:* Attend pathologist, Vet Admin Hosp, 53-72. *Mem:* Am Asn Path & Bact; Am Fedn Clin Res. *Res:* Pathologic anatomy; cardiovascular and pulmonary disease. *Mailing Add:* 4615 Pewter Lane Manlius NY 13104

PRIOR, MICHAEL HERBERT, b Riverside, Calif, July 14, 39; m 63; c 2. PHYSICS. *Educ:* Pomona Col, BA, 61; Univ Calif, Berkeley, PhD(physics), 67. *Prof Exp:* Res physicist, Lawrence Berkeley Lab, Univ Calif, 67-68 & Roman Cath Univ Nijmegen, 68-69; STAFF SCIENTIST, LAWRENCE BERKELEY LAB, UNIV CALIF, 69- *Honors & Awards:* Fel, Am Phys Soc. *Mem:* Am Phys Soc. *Res:* Atomic beam magnetic resonance; atomic hyperfine structure; lifetimes of metastable ions; magnetic resonance spectroscopy of stored ions; optical pumping; laser fluorescence spectroscopy; ion-neutral collision studies. *Mailing Add:* Lawrence Berkeley Lab Univ Calif Berkeley CA 94720

PRIOR, PAUL VERDAYNE, b Cedar Rapids, Iowa, July 2, 21; m 43; c 2. PLANT MORPHOLOGY, PLANT PHYSIOLOGY. *Educ:* Univ Iowa, BA, 46, MS, 47, PhD, 54. *Prof Exp:* Instr biol, Bethany Col Kans, 47-48; assoc prof, Northern Ill Univ, 48-56; prof, Tex Tech Univ, 56-70; PROF BIOL & CHMN DEPT, UNIV NEBR, OMAHA, 70- *Mem:* AAAS; Am Inst Biol Sci; Bot Soc Am; Am Bryol & Lichenological Soc. *Res:* Stem apex; bryophyte ecology and physiology. *Mailing Add:* 10907 Elsalido Austin TX 78750

PRIOR, RICHARD MARION, b Dunedin, Fla, July 17, 42; m 64; c 2. NUCLEAR MEDICINE. *Educ:* Univ Fla, BS, 64, MS, 65, PhD(physics), 68. *Prof Exp:* Res assoc physics, Univ Notre Dame, 68-70; asst prof physics, 70-74, asst prof nuclear med, Sch Med, 72-74, assoc prof physics & assoc prof nuclear med, 74-81, PROF PHYSICS, SCH MED, UNIV ARK, LITTLE ROCK, 79- *Mem:* Am Phys Soc; Am Asn Physics Teachers; Sigma Xi. *Res:* Effects of fast neutrons on mammalian cells; neutron dosimetry; nuclear reactions and scattering using polarized beams. *Mailing Add:* Dept Phusics WGa Col Carollton GA 30118

PRIOR, RONALD L, b McCook, Nebr, Mar 9, 45. ANIMAL SCIENCE & NUTRITION. *Educ:* Univ Nebr, BS, 67; Cornell Univ, PhD(nutrit), 71. *Prof Exp:* NIH postdoctoral trainee, NY State Vet Col, Cornell Univ, 71-73; res chemist & asst prof, Meat Animal Res Ctr, USDA, Univ Nebr, Clay Ctr, 73-83; pres, Sunshine Audio, Inc & gen partner R&M Enterprises, 83-87; SCI PROG OFFICER & NUTRITIONIST, USDA HUMAN NUTRIT RES CTR AGING, TUFTS UNIV, 87- *Concurrent Pos:* Lectr, Animal Nutrit Lab, Cornell Univ, 68 & 69 & Univ Nebr, 78-81; chmn, Ruminant Nutrit Session, Am Soc Animal Sci, 73, Ruminant Nutrit Workshop, US Meat Animal Res Ctr, 77 & 22nd Ann Ruminant Nutrit Conf, Fed Am Soc Exp Biol & Med, 81; coop scientist, Spec Foreign Currency Res Progs, Univ Alexandria, Egypt & Warsaw Agr Univ, Poland. *Mem:* Sigma Xi; Am Soc Animal Sci; AAAS; Am Inst Nutrit. *Res:* Author of more than 70 technical publications. *Mailing Add:* USDA Human Nutrit Res Ctr Tufts Univ 711 Washington St Rm 919 Boston MA 02111

PRIORE, ROGER L, b Buffalo, NY, Apr 21, 38; m 60; c 3. BIOMETRICS & BIOSTATISTICS, COMPUTER SCIENCE. *Educ:* State Univ NY, Buffalo, BA, 60, MS, 62; Johns Hopkins Univ, ScD(biostatist), 65. *Prof Exp:* Sr cancer res scientist, 65-67, assoc cancer res scientist, 67-69, prin cancer res scientist, 69-74, dir Comput Ctr, 69-83, dir Comput Sci, 74-83, DIR BIOMATH, ROSWELL PARK MEM INST, 83-; RES PROF & DIR GRAD STUDIES BIOMET, STATE UNIV NY, BUFFALO, 80- *Concurrent Pos:* From asst res prof to assoc res prof, State Univ NY, Buffalo, 66-69; group statistician, Cancer Clin Trial Coop Groups; res prof, Niagara Univ, 68- *Mem:* Am Statist Asn; Biomet Soc; Asn Comput Mach; Soc Epidemiol Res; AAAS; Am Joint Comt Cancer. *Res:* Biostatistics and computer science methodology; statistical design and analysis of cancer research projects. *Mailing Add:* 666 Elm St Buffalo NY 14263

PRIORESCHI, PLINIO, b Florence, Italy, May 20, 30; US citizen; m 67; c 2. PHARMACOLOGY, HISTORY OF MEDICINE. *Educ:* Univ Pavia, MD, 54; Univ Montreal, PhD(exp med), 61. *Prof Exp:* Res assoc exp med, Univ Montreal, 61-63; asst prof pharmacol, Queen's Univ, Ont, 63-67; assoc prof, 67-73, PROF PHARMACOL, CREIGHTON UNIV, 73-, ASST PROF MED, 77- *Concurrent Pos:* Chief, Creighton Geriat Serv, Douglas County Hosp, Omaha, Nebr, 71-78; attend physician, Vet Admin Hosp, Omaha. *Mem:* Am Soc Pharmacol & Exp Therapeut; Can Physiol Soc; Pharmacol Soc Can. *Res:* History of Medicine. *Mailing Add:* Dept Pharmacol Sch Med Creighton Univ Omaha NE 68131

PRIPSTEIN, MORRIS, b Montreal, Que, Nov 24, 35; m 58; c 2. PHYSICS. *Educ:* McGill Univ, BEng, 57; Univ Calif, Berkeley, PhD(physics), 62. *Prof Exp:* Asst physics, Univ Calif, Berkeley, 57-62; res physicist, Col France, 62-63; res asst prof elem particle physics, Univ Ill, 63-65; res physicist, 65-73, GROUP LEADER, LAWRENCE BERKELEY LAB, UNIV CALIF, 73- *Concurrent Pos:* Atomic Energy Commissariat France Louis de Broglie fel, 62-63. *Mem:* Am Phys Soc. *Res:* Experimental elementary particle physics. *Mailing Add:* Univ Calif Lawrence Berkeley Lab Berkeley CA 94720

PRISBREY, KEITH A, b St George, Utah, June 23, 45; m 69; c 4. METALLURGY, MINERAL PROCESSING. *Educ:* Univ Utah, BS, 69, PhD(metall), 76; Stanford Univ, MS, 71. *Prof Exp:* Res officer minerals process, Julius Kruttschwitt Res Ctr, Univ Queensland, 72-73; res engr, Amax Res, 75-76; ASST PROF METALL, UNIV IDAHO, 76- *Concurrent Pos:* Res grants, Univ Idaho, 77-78, NSF, 78-80. *Mem:* Am Inst Metall Eng; Sigma Xi. *Res:* Comminution; control of mineral processing plants; surface chemistry of minerals; hydrometallurgy. *Mailing Add:* Dept Mining & Metall Univ Idaho Moscow ID 83843

PRITCHARD, AUSTIN WYATT, b Loma Linda, Calif, Mar 8, 29; m 49, 85; c 5. COMPARATIVE PHYSIOLOGY. *Educ:* Stanford Univ, AB, 48, MA, 49; Univ Hawaii, PhD(zool), 53. *Prof Exp:* From instr to assoc prof, 53-65, PROF ZOOL, ORE STATE UNIV, 65- *Mem:* Fel AAAS; Am Soc Zoologists; Am Physiol Soc; Marine Biol Asn UK. *Res:* Body fluid regulation in aquatic animals; respiratory, circulatory and metabolic adaptations of intertidal animals. *Mailing Add:* Dept of Zool Ore State Univ Corvallis OR 97331

PRITCHARD, DALTON H, b Crystal Springs, Miss, Sept 1, 21. COLOR TELEVISION SYSTEMS & CIRCUITS. *Educ:* Miss State Univ, BSEE, 43. *Prof Exp:* Tech trainer, US Army Signals Corp, 43-46; mem tech staff, RCA Labs, 46-75, fel tech staff, 75-87; RETIRED. *Honors & Awards:* Vladimir Zworykin Award, Inst Elec & Electronics Engrs, 77; Edward Rhein Prize, Edward Rhein Found, 80; David Sarnoff Award, RCA Labs, 81. *Mem:* Nat Acad Eng; fel Inst Elec & Electronics Engrs; fel Soc Info Display. *Mailing Add:* Three Bent Tree Lane Hilton Head SC 29926

PRITCHARD, DAVID EDWARD, b New York, NY, Oct 15, 41; m 67; c 2. ATOMIC & MOLECULAR PHYSICS. *Educ:* Calif Inst Technol, BS, 62; Harvard Univ, PhD(physics), 68. *Prof Exp:* Mem staff, Div Sponsored Res, 68, from instr to assoc prof, 68-80, PROF PHYSICS, MASS INST TECHNOL, 80- *Concurrent Pos:* Dir, Interbuild Corp, 69-73. *Honors & Awards:* Broida Prize, 91. *Mem:* Fel Am Phys Soc; Optical Soc Am; AAAS. *Res:* Atomic collisions; inter-atomic forces; dye laser spectroscopy; trapping atoms and ions; atom interferometers; precision measurements. *Mailing Add:* 88 Washington Ave Cambridge MA 02140

PRITCHARD, DAVID GRAHAM, b Montreal, Que, Oct 2, 45; m 75; c 2. CARBOHYDRATE CHEMISTRY, MICROBIOLOGY. *Educ:* Univ Sask, BA, 66, Hons, 67; Univ Calif, Los Angeles, PhD(biochem), 75. *Prof Exp:* Jr res scientist biochem, Dept Immunol, Nat Med Ctr, City of Hope, 73-75, asst res scientist, 75-76; res assoc, Dept Microbiol, 76-81, res asst prof, 81-85, ASST PROF PEDIAT, UNIV ALA, BIRMINGHAM, 84-, RES ASSOC PROF MICROBIOL, 85- *Mem:* Am Asn Immunologists; Am Soc Mass Spectrometry; Am Soc Microbiol; Soc Complex Carbohydrates. *Res:* Chemical and immunochemical charaterization of bacterial polysaccharide antigens, particularly streptococci. *Mailing Add:* Dept Microbiol Univ Ala Univ Sta Birmingham AL 35294

PRITCHARD, DONALD WILLIAM, b Santa Ana, Calif, Oct 20, 22; m 43; c 5. OCEANOGRAPHY. *Educ:* Univ Calif, Los Angeles, AB, 46; Scripps Inst, Univ Calif, San Diego, MS, 48, PhD(oceanog), 51. *Hon Degrees:* DS, Col William & Mary, 85. *Prof Exp:* Chg current anal sect, Scripps Inst, Univ Calif, 46-47; oceanogr, Navy Electronics Lab, Calif, 47-49; from assoc dir to dir, Chesapeake Bay Inst, Johns Hopkins Univ, 49-74, prof oceanog, 58-77, prof earth & planetary sci, 78-80; assoc dir res, Marine Sci Res Ctr, 80-86, ASSOC DEAN, STATE UNIV NY, STONY BROOK, 87- *Concurrent Pos:* Consult, Comt Hydraul & Waterways Exp Sta, Corps Engrs, US Dept Army, 54-; mem, Bd Rev Dept Natural Resources, Md, 59-78; mem radiation adv bd, Health Dept, 60-75; mem comt oceanog, Nat Acad Sci, 61-68, chmn panel radioactivity in ocean, 62-66; mem, Air Pollution Control Coun, Md, 66-73; mem adv comt isotopes & radiation develop, AEC, 66-68; mem steering comt & environ qual comt, Gov Sci Adv Coun, State of Md, 71-78. *Mem:* AAAS; Am Meteorol Soc; Am Soc Limnol & Oceanog (vpres, 55-56); fel Am Geophys Union. *Res:* Dynamics of estuarine circulation and mixing; inshore and coastal oceanography; turbulent diffusion of natural waters. *Mailing Add:* Chesapeake Bay Inst Johns Hopkins Univ 4800 Atwell Rd Shadyside MD 20764

PRITCHARD, ERNEST THACKERAY, b Ottawa, Ont, Mar 29, 28. BIOCHEMISTRY. *Educ:* Carleton Univ, BSc, 52; McGill Univ, MSc, 54; Univ Western Ont, PhD(biochem), 57. *Prof Exp:* Asst prof biochem, Univ Alta, 57-61; Nat Multiple Schlerosis Soc fel, McLean Hosp, Harvard Univ, 61-62; assoc prof, 62-70, PROF BIOCHEM & ORAL BIOL, UNIV MAN, 70- *Mem:* Am Soc Neurochem; Int Soc Neurochem; Can Biochem Soc; Nutrit Soc Can; Brit Biochem Soc. *Res:* Lipid metabolism of the nervous tissue during growth and development; secretion mechanisms of salivary glands and placental tissues. *Mailing Add:* Dept Dent Careers Univ Man Winnipeg MB R3T 2N2 Can

PRITCHARD, G IAN, b Alcove, Que, Nov 28, 29; m 56; c 4. FISHERIES. *Educ:* McGill Univ, BSc, 52; Univ Vt, MS, 54; NC State Univ, PhD(nutrit), 59. *Prof Exp:* Res scientist, Animal Res Inst, Can Dept Agr, 58-64; assoc ed & sci adv to chmn, Fisheries Res Bd, Can, 65-71; aquacult coordr, Fisheries Serv, Environ Can, 71-76, assoc dir fisheries res, 76-80, dir aquacult & resource develop, Fisheries & Oceans, Can, 80-88; DIR, MOUNTAIN AQUACULTURE, WESTERN CAROLINA UNIV, 88- *Concurrent Pos:* Consult int develop, Turkey, 75, Latin Am, 77-88, Philippines, 81 & Kenya, 83; mem governing bd, Ewart Col, Toronto, 85-88. *Mem:* Am Fisheries Soc; Nutrit Soc Can; Agr Inst Can. *Res:* Aquaculture; applied nutrition, fish health, strategic planning. *Mailing Add:* Natural Sci Bldg Western Carolina Univ Cullowhee NC 28723

PRITCHARD, GLYN O, b Bangor, Wales, Sept 26, 31; m 60; c 3. PHYSICAL CHEMISTRY. *Educ:* Univ Manchester, BSc, 52, PhD, 55. *Prof Exp:* Nat Res Coun Can fel phys chem, 55-57; phys chemist, Defence Res Bd Can, Royal Mil Col Can, 57-58; from asst prof to assoc prof, 58-69, PROF CHEM, UNIV CALIF, SANTA BARBARA, 69- *Mem:* Am Chem Soc. *Res:* Fluorine chemistry; kinetics; photochemistry. *Mailing Add:* Dept of Chem Univ of Calif Santa Barbara CA 93106

PRITCHARD, GORDON, b Burton-on-Trent, Eng, Feb 9, 39; m 63; c 2. ENTOMOLOGY. *Educ:* Univ London, BSc, 60; Univ Alta, PhD(entom), 63. *Prof Exp:* Res scientist, Commonwealth Sci & Indust Res Orgn, Australia, 63-66; sessional instr entom, Univ Alta, 66-67; from spec instr to assoc prof, 67-70, PROF BIOL, UNIV CALGARY, 76- *Mem:* Can Soc Zoologists; Entom Soc Can; Royal Entom Soc London; Sigma Xi; Int Soc Odontol. *Res:* Structure, behavior and population ecology of insects. *Mailing Add:* Dept Biol Univ Calgary Calgary AB T2N 1N4 Can

PRITCHARD, HAYDEN N, b Bangor, Pa, Feb 13, 33; m 62; c 2. BOTANY, HISTOCHEMISTRY. *Educ:* Princeton Univ, AB, 55; Lehigh Univ, MS, 60, PhD(biol), 63. *Prof Exp:* Instr biol, Cedar Crest Col, 59-60; res asst, Lehigh Univ, 59-62; res fel path, Univ Fla, 62-63, asst prof biol, 63-64; asst prof, 64-70, ASSOC PROF BIOL, LEHIGH UNIV, 70- *Concurrent Pos:* Co-investr, Am Cancer Soc res grant, 66-67. *Res:* Botanical histochemistry and botanical growth and development, especially botanical embryology. *Mailing Add:* Dept of Biol Lehigh Univ Bethlehem PA 18015

PRITCHARD, HUW OWEN, b Bangor, Wales, July 23, 28; m 56; c 2. PHYSICAL & THEORET CHEMISTRY. *Educ:* Univ Manchester, BSc, 48, MSc, 49, PhD(chem), 51, DSc, 64. *Prof Exp:* From asst lectr to lectr, Univ Manchester, 51-65; PROF CHEM, YORK UNIV, 65- *Concurrent Pos:* Vis res fel, Calif Inst Technol, 57-58. *Mem:* Royal Soc Chem; fel Royal Soc Can. *Res:* Thermochemistry; kinetics, photochemistry and thermodynamics of simple systems; theory of unimolecular reactions and of relaxation processes in gaseous systems; theory of thermal explosions. *Mailing Add:* Dept of Chem York Univ Downsview ON M3J 1P3 Can

PRITCHARD, JACK ARTHUR, b Painesville, Ohio, July 25, 21; m 45; c 3. OBSTETRICS & GYNECOLOGY. *Educ:* Ohio Northern Univ, BS, 42; Western Reserve Univ, MD, 46. *Prof Exp:* Intern, Univ Hosps, Western Reserve Univ, 46-47, fel pharmacol, Univ, 47-48, resident obstet & gynec, Hosps, 50-54, Oglebay fel, Univ, 52-54, asst prof, 54-55; PROF OBSTET & GYNEC & CHMN DEPT, UNIV TEX HEALTH SCI CTR DALLAS, 55- *Concurrent Pos:* Chief obstet & gynec serv, Parkland Mem Hosp, Dallas, 55-; consult, Vet Admin Hosp, St Paul Hosp, Baylor Hosp & Methodist Hosp, Dallas, William Beaumont Army Hosp, El Paso, Brooke Army Hosp, Ft Sam Houston, US Air Force Hosp, Lackland AFB & Carswell AFB, Ft Worth; mem obstet & gynec test comt, Nat Bd Med Examr, 56-60; mem human embryol & develop study sect, NIH, 60-64. *Mem:* Soc Gynec Invest; Am Asn Obstet & Gynec; Am Col Obstet & Gynec. *Res:* Blood cell production and destruction; blood coagulation. *Mailing Add:* Parkland Mem Hosp 5323 Harry Hines Blvd Dallas TX 75235

PRITCHARD, JAMES EDWARD, b Swaledale, Iowa, Jan 28, 22; m 42; c 3. CHEMISTRY. *Educ:* Univ Iowa, BA, 42, MS, 43; Purdue Univ, PhD(pharmaceut chem), 48. *Prof Exp:* Asst chem, Univ Iowa, 42-43; res chemist, Phillips Petrol Co, Okla, 43-46; asst chem, Purdue Univ, 46-47; res chemist, Phillips Petrol Co, 48-59, mgr polyolefins br, Res Div, 59-74, consult, Mgt Serv, 73-74, dir develop planning, 74-85; RETIRED. *Mem:* Am Chem Soc. *Res:* Surface chemistry; halogenation of organic compounds; synthetic rubber emulsion polymers; study of surface lipids of skin; polyolefins. *Mailing Add:* 2710 Cherokee Hills Dr Bartlesville OK 74006-4215

PRITCHARD, JOHN B, b Buffalo, NY, July 31, 43; m; c 3. RENAL PHYSIOLOGY, TRANSPORT. *Educ:* Oberlin Col, AB, 65; Harvard Univ, PhD(physiol), 70. *Prof Exp:* NSF fel, State Univ NY Upstate Med Ctr, 70-71; NIH fel, Mt Desert Island Biol Lab, 71-72; head, Marine Lab, 79-84, RES PHYSIOLOGIST, NAT INST ENVIRON HEALTH SCI, 76-, HEAD, COMP MEMBRANE PHARMACOL SECT, 84- *Concurrent Pos:* Asst prof, 72-76, assoc prof physiol, Med Univ SC, 76-; adj assoc prof pharmacol, Sch Med, Univ Fla, 79-84 & physiol, Duke Univ Sch Med, 84- *Honors & Awards:* Kinter Mem Lectr, Mt Desert Island Biol Lab, 86. *Mem:* Am Physiol Soc; Am Soc Zoologists; Am Soc Pharmacol & Exp Therapeut. *Res:* Interaction of environmental contaminants with physiological systems, particularly renal function and transport. *Mailing Add:* Lab Pharmacol Nat Inst Environ Health Sci PO Box 12233 Res Triangle Park NC 27709

PRITCHARD, JOHN PAUL, JR, b Washington, Pa, June 23, 29; m 57; c 2. ELECTRICAL ENGINEERING, APPLIED MATHEMATICS. *Educ:* Univ Okla, BA, 50, MS, 55; Iowa State Univ, PhD(elec eng), 60; Univ Dallas, MBA, 78. *Prof Exp:* Comput programmer, Boeing Airplane Co, Kans, 55-56, group head appl math serv group, 57-60; mem tech staff, Corp Res & Eng Div, Tex Instruments, Inc, 60, sect head adv systs res, 61-64, br mgr info components res, 65-71; div staff engr, Electronic Eng Div, Martin Marietta Corp, 71-74; proj engr, 75-80, STAFF ENGR, AIRCRAFT SYSTS GROUP, E-SYSTS, INC, 80- *Mem:* Inst Elec & Electronics Engrs. *Res:* Intrusion detection systems and devices; electronic warfare systems; optical counter-measures; thin film technology; data processing theory; real-time multiprocessor signal acquisition and analysis systems. *Mailing Add:* 5404 Stonewall Greenville TX 75401

PRITCHARD, MARY (LOUISE) HANSON, b Lincoln, Nebr, July 16, 24; m 56. PARASITOLOGY. *Educ:* Univ Nebr, BS, 46, MA, 49. *Prof Exp:* Assoc cur, State Mus, 48-57, 68-72, res assoc parasitol, 59-68, from asst prof to assoc prof, 68-80, PROF ZOOL, UNIV NEBR, LINCOLN, 80-, CUR PARASITOLOGY, HAROLD W MANTER LAB, UNIV NEBR STATE MUS, 72- *Concurrent Pos:* Mem, Coun on Standards, Asn Systs Collections, 73-79, Coun on Biol Nomenclature, 81-85 & ecol adv comt, Sci Adv Bd, US Environ Protection Agency, 74-78. *Mem:* Fel AAAS; Soc Syst Zool; Am Soc Parasitol; Am Micros Soc; Sigma Xi. *Res:* Taxonomy, geographical distribution and host-parasite relationships of digenetic trematodes. *Mailing Add:* Div Parasitol Harold W Manter Lab Univ Nebr State Mus Lincoln NE 68588-0514

PRITCHARD, PARMELY HERBERT, b Evanston, Ill, Aug 25, 41; m 64; c 2. MICROBIAL ECOLOGY. *Educ:* Univ RI, BS, 63; Cornell Univ, MS, 66, PhD(microbiol), 69. *Prof Exp:* Fel, Woods Hole Oceanog Inst, 69-71; assoc prof microbiol, State Univ NY Col Brockport, 71-78; ENVIRON SCIENTIST, US ENVIRON PROTECTION AGENCY, 78- *Mem:* Soc Indust Microbiol; Am Soc Microbiol. *Res:* Microbial degradation of organic pollutants; oil degradation; physiology of autochthonous bacterial populations in aquatic environments; continuous culture in microbial ecology; fate and transport of toxic organic pollutants in estuarine environments. *Mailing Add:* 1250 Tall Pine Tr Gulf Breeze FL 32561

PRITCHARD, ROBERT LESLIE, b Irvington, NJ, Sept 8, 24; m 48. APPLIED PHYSICS. *Educ:* Brown Univ, BSc, 46; Harvard Univ, MSc, 47, PhD(acoust), 50. *Prof Exp:* Tech asst, Bell Tel Labs, Inc, 42-43; res asst, Acoust Res Lab, Harvard Univ, 47-50; res assoc, Res Lab, Gen Elec Co, 50-57; from develop engr to staff dir eng, Semiconductor Div, Tex Instruments, Inc, 57-63; dir eng, Motorola, Inc, 63-64; prof elec eng, Stanford Univ, 64-70; mem res & develop staff, Gen Elec Co, 70-; COMP DIR TECH, GEN INSTRUMENT CORP. *Concurrent Pos:* Tech adv, US Nat Comt, Int Electrotech Comn Semiconductor Devices. *Mem:* Acoust Soc Am; fel Inst Elec & Electronics Engrs. *Res:* Acoustics, sound radiation and reception problems in technical acoustics; transistor, from electric-circuit point of view; integrated circuits. *Mailing Add:* 509 E 77th St New York NY 10021

PRITCHARD, WENTON MAURICE, b Wilson, NC, Nov 8, 31. SOLID STATE PHYSICS. *Educ:* NC State Univ, BNE, 54, MSNE, 55; Ga Inst Technol, PhD(physics), 60. *Prof Exp:* Nuclear engr, Newport News Shipbuilding & Dry Dock Co, 55-57, sr lab analyst, 60-62; PROF PHYSICS, OLD DOMINION UNIV, 62-, PROF GEOPHYS SCI, 78- *Concurrent Pos:* Tech asst expert, Int Atomic Energy Agency, 62-63; NASA grant, Old Dominion Univ, 65-68; tech asst expert, UNESCO, 68-69. *Mem:* AAAS; Am Phys Soc; Am Nuclear Soc. *Res:* Radiation effects on solids; physical properties of metals; nuclear reactor theory. *Mailing Add:* 601 Pembrook Ave-Apt 614 Norfolk VA 23507-2051

PRITCHARD, WILBUR L(OUIS), b New York, NY, May 31, 23; m 49; c 3. ELECTRICAL ENGINEERING. *Educ:* City Col New York, BSEE, 43. *Prof Exp:* Jr engr, Philco Corp, Pa, 42-46; sr commun engr, Raytheon Co, Mass, 46-49, engr, Microwave Sect, 49-54, mgr microwave-transmitter br, 54-57, dept mgr surface radar, 58-59, mgr eng, surface radar & navig opers, 59-60, dir eng, Europe, 60-62; group dir, Commun Satellite Systs, Aerospace Corp, 62-67; asst vpres & dir, Comsat Labs, Commun Satellite Corp, 67-72, vpres & dir, 72-73; pres, Fairchild Space & Electronics Co, 73-74; pres, Satellite Systs Eng, Inc, 74-87, Direct Broadcast Satellite Corp, 80-86; chmn bd, Satellite Systs Eng, Inc, 86-89, pres, SSE Telecom, 86-89; PRES, W L PRITCHARD & CO, 89- *Concurrent Pos:* Chmn panel broadcast satellites, Nat Acad Eng, 67-68; adv comt, Space Appln, NASA, 84 & Space & Earth Sci, 84-; adj prof, Polytechnic Univ, NY, 85-; prof lectr, George Washington Univ, 87- *Honors & Awards:* Aerospace Commun Award, Am Inst Aeronaut & Astronaut, 72; Lloyd V Berkner Space Utilization Award, Am Astronaut Soc, 83. *Mem:* Fel Am Inst Aeronaut & Astronaut; fel Inst Elec & Electronics Engrs; Int Acad Aeronaut & Astronaut; Soc Satellite Prof. *Res:* Communications satellites; microwave systems. *Mailing Add:* 7315 Winconsin Ave Bethesda MD 20814

PRITCHARD, WILLIAM ROY, b Portage, Wis, Nov 15, 24; c 4. VETERINARY MEDICINE. *Educ:* Kans State Univ, DVM, 46; Univ Minn, PhD, 53; Univ Ind, JD, 57. *Hon Degrees:* DSc, Kans State Univ, 70, Purdue Univ, 77 & Tufts Univ, 88. *Prof Exp:* Asst prof vet sci, Univ Wis, 46-49; assoc prof vet med & head div, Univ Minn, 49-53; prof & dir cattle dis res, Purdue Univ, 53-57; prof vet sci & head dept, Univ Fla, 57-61; prof & assoc dir, Vet Med Res Inst, Iowa State Univ, 61-62; assoc dir, Agr Exp Sta, 62-72, prof vet med & dean, Sch Vet Med, 62-82, PROF & CO-DIR, PEW NAT VET EDUC PROG, UNIV CALIF, DAVIS, 87- *Concurrent Pos:* Mem bd consults agr sci, Rockefeller Found, 62-66; med consult, US Air Force, 63-64; mem panel world food supply, President's Sci Adv Comt, 66-67, panel biol & med, 69-71, construct sch vet med comt, USPHS, 67-70, comt vet med res & educ, Nat Acad Sci, 68-72 & adv comt res to USDA, 68-71; mem int task force, African Livestock Ctr, 71; chmn, Int Exec Comt Estab Int Lab Res Animal Dis, 72-73; mem, Joint Res Comt, Bd Int Food & Agr Develop, USAID, 77-85; mem bd dirs, Int Lab Res An Dis, 82-88. *Honors & Awards:* Vet Cong Prize, Am Vet Med Asn, 88. *Mem:* Am Vet Med Asn; Am Pub Health Asn; US Animal Health Asn; AAAS. *Res:* International development; veterinary medical law. *Mailing Add:* Sch of Vet Med Univ of Calif Davis CA 95616

PRITCHETT, ERVIN GARRISON, b Hamlet, NC, Mar 6, 20; m 42; c 1. ORGANIC CHEMISTRY. *Educ:* The Citadel, BS, 41; Johns Hopkins Univ, MA, 48, PhD(chem), 51. *Prof Exp:* Asst prof chem, Washington & Jefferson Col, 50-51; res assoc, US Indust Chem Co, 51-74, sr res assoc, 74-85; RETIRED. *Mem:* Am Chem Soc. *Res:* Organic synthesis; organometallics; polyolefins; vinyl polymers; polyblends. *Mailing Add:* 10597 Swindon Ct Cincinnati OH 45241

PRITCHETT, JOHN FRANKLYN, b Evanston, Ill, June 30, 42; m 64; c 2. ENDOCRINOLOGY. *Educ:* Auburn Univ, BS, 65, MS, 68; Iowa State Univ, PhD(physiol), 73. *Prof Exp:* Asst prof biol, Lambuth Col, 67-70; instr physiol, Iowa State Univ, 71-72; assoc prof, 73-81, PROF PHYSIOL & DEPT HEAD ZOOL-WILDLIFE, AUBURN UNIV, 82- *Mem:* Sigma Xi; AAAS; Soc Exp Biol & Med. *Res:* Interaction of aging and/or environmental stressors with mammalian endocrine systems. *Mailing Add:* Dept of Zool & Wildlife Auburn Univ Auburn AL 36849

PRITCHETT, P(HILIP) W(ALTER), b Austin, Tex, July 31, 28; m 54; c 4. PETROCHEMICALS, HANDLING GRANULAR SOLIDS. *Educ:* Univ Tex, BS, 49, MS, 51; Univ Del, PhD(chem eng), 60. *Prof Exp:* Process engr, Ethyl Corp, 50-55; sr engr, Esso Res Labs, 58-66, res specialist, Esso Res & Eng Co, 66-70, eng assoc, Plastics Technol Div, Exxon Chem Co, 70-82; ASSOC PROF CHEM & NATURAL GAS ENG, TEX A&I UNIV, 83- *Mem:* Am Inst Chem Engrs. *Res:* Process development and design in polymerization of olefins, polymer extrusion, flow of granular solids, conveying, reaction kinetics, absorption with reaction, phase equilibrium, petrochemical processes and economic analysis. *Mailing Add:* Dept Chem Eng Tex A&I Univ Kingsville TX 78363

PRITCHETT, PHILIP LENTNER, b Chicago, Ill, Jan 29, 44. COMPUTATIONAL PLASMA PHYSICS, MAGNETOSPHERIC PHYSICS. *Educ:* Oberlin Col, BA, 65; Stanford Univ, MS, 66, PhD(physics), 70. *Prof Exp:* Fel High Energy physics, Ger Electron Synchrotron; res assoc, high energy physics, Northwestern Univ, 71-73, vis asst prof physics, 73-75; NSF fel, 75-76, asst res physicist, 76-81, assoc res physicist, 81-86, RES PHYSICIST PLASMA PHYSICS, UNIV CALIF, LOS ANGELES, 86- *Concurrent Pos:* Prin invest, US-France Coop Res Magnetospheric Physics, 85-88; adj prof physics, Univ Calif, Los Angeles, 87- *Mem:* Am Physical Soc; Am Geophys Union; Sigma Xi. *Res:* Computational plasma physics with applications to fundamental kinetic and fluid processes and instabilities in magnetospheric and laboratory plasmas. *Mailing Add:* Dept Physics Univ Calif Los Angeles CA 90024-1547

PRITCHETT, RANDALL FLOYD, animal virology; deceased, see previous edition for last biography

PRITCHETT, THOMAS RONALD, b Colorado City, Tex, Sept 2, 25; m 48; c 3. METALLURGY, PHYSICAL CHEMISTRY. *Educ:* Univ Tex, BS, 48, MA, 49, PhD(chem & chem eng), 51. *Prof Exp:* Res scientist corrosion, Defense Res Lab, US Navy, 48-51; res chemist, Monsanto Chem Co, 51-52; res engr & head corrosion, Kaiser Aluminum & Chem Corp, 52-60, tech mgr & lab mgr, 60-68, vpres res & develop, 68-89; RETIRED. *Concurrent Pos:* Instr chem, Univ Tex, 48-50; dir, Metal Properties Coun, 75-; chmn, Aluminum Asn Tech Comt, 81-, Aluminum Asn Acad Comt, 83-86; adv bd, Dept Engr, Univ Calif, Berkeley, 82- *Mem:* Fel Am Soc Metals; Am Inst Mining, Metall & Petrol Engrs; Nat Asn Corrosion Engrs; Am Chem Soc; Sigma Xi; Electrochem Soc. *Res:* Solidification phenomena of molten aluminum; corrosion and inhibition of metals; surface chemistry and physics of metals; aluminum metallurgy. *Mailing Add:* 1430 Laurenita Way Alamo CA 94507

PRITCHETT, WILLIAM CARR, b Austin, Tex, Aug 1, 25; m 51; c 4. SEISMIC DATA ACQUISITION, SHEAR WAVE SEISMOLOGY. *Educ:* Univ Colo, BS, 46; Univ Tex, MS, 48. *Prof Exp:* Dir well logging & uranium explor, Res & Develop, Arco, 63-69, res assoc shear wave seismic plate tectonics, 69-80, dir, seismic acquisition, Frontler Dist, Arco, Atlantic Richfield, 80-82, mgr, seismic acquisition, data processing & storage, Regional Explor Serv, Arco Explor Co, 82-85; CONSULT SEISMOL, WILLIAM CARR PRITCHETT, INC, 86- *Mem:* Soc Explor Geophysicists. *Res:* Author of dominant book on seismic data acquisition, shear wave seismic, exploration implications of plate tectonics and high pressure shales; well logging tools and techniques, mostly nuclear; electrical techniques for exploration. *Mailing Add:* 1109 Janwood Circle Plano TX 75075

PRITCHETT, WILLIAM HENRY, b Davidson, NC, Jan 16, 30; m 53; c 2. DEVELOPMENTAL BIOLOGY. *Educ:* Davidson Col, BS, 51; Univ Va, PhD, 67. *Prof Exp:* From jr buyer to supvr, Vick Chem Co, 54-57; instr biol, Augusta Mil Acad, 57-58 & Woodberry Forest Sch, 58-63; asst prof, 67-70, assoc prof, 70-73, PROF BIOL, FAIRMONT COL, 73- *Concurrent Pos:* Pres fac Senate, Fairmont Col, 72-74. *Mem:* AAAS. *Res:* Regeneration in vertebrates; factors affecting the rate of regeneration. *Mailing Add:* Dept Biol Fairmont State Col Fairmont WV 26554

PRITCHETT, WILLIAM LAWRENCE, b Lavaca, Ark, Sept 17, 18; m 49; c 4. FOREST SOILS. *Educ:* Univ Ark, BSA, 41; Iowa State Col, MS, 47, PhD(soils), 50. *Prof Exp:* Agriculturist, United Fruit Co, Guatemala, 41; res assoc, Iowa State Col, 46-48 & 49-50; asst soil chemist, Univ Fla, 48-49, prof soil sci & forest soil scientist, 53-85; RETIRED. *Concurrent Pos:* Foreman, United Fruit Co, Guatemala, 41; res agronomist, USDA & dir agr exp sta, Serv Agr Interam, Bolivia, 50-53; Fulbright lectr, Univ WI, Trinidad, 64; consult, Costa Rican Ministry of Agr, 66 & coordr coop res, Forest Fertilization proj, 67-85; Bullard fel, Harvard Univ, 73-74; consult pulp & paper co, Fla, 55, Bahamas, 62, Venezuela, 75, French Guiana, 76, Ala, 77 & UN Develop Prog/Food & Agr Orgn, Burma, 81; USAID, Cameroon, 83-85. *Mem:* AAAS; fel Am Soc Agron; fel Soil Sci Soc Am; Sigma Xi. *Res:* Soil management of forests; improvement of methods and interpretation of results of soil tests; fertilizer evaluation; forest fertilization and tree nutrition. *Mailing Add:* Rte 2 Box 1204 Palatka FL 32177

PRITHAM, GORDON HERMAN, b Freeport, Maine, Apr 7, 07; m 32; c 4. PHYSIOLOGICAL CHEMISTRY. *Educ:* Pa State Col, BS, 30, MS, 31, PhD(physiol chem), 36; Nat Registry Clin Chem, cert. *Prof Exp:* From assoc prof to prof chem, Univ Scranton, 33-47; from assoc prof to prof physiol chem, Pa State Univ, 47-67; prof biochem, 67-75, assoc dean basic sci, 71-73, EMER PROF OSTEOP MED, KANSAS CITY COL, 73- *Concurrent Pos:* Consult, Mo State Dept Mental Health, 73-74; affil fac, Univ Mo, Kans City, 80-81. *Honors & Awards:* Benjamin Rush Award, 59; Am Asn Ment Deficiency, 81. *Mem:* AAAS; Am Chem Soc; Am Asn Clin Chem; Am Asn Ment Deficiency; NY Acad Sci; Sigma Xi. *Res:* Metabolism of fungi; methods of blood analysis; biosynthesis of cholesterol; biochemistry of mental retardation; blood chemistry in relation to disease; survey of college courses in biochemistry; preparation and editing of multiple-choice exams, in chemistry. *Mailing Add:* 113 Chipola Ave Apt 509 De Land FL 32720-7783

PRITSKER, A(DRIAN) ALAN B(ERYL), b Philadelphia, Pa, Feb 5, 33; m 56; c 4. INDUSTRIAL ENGINEERING. *Educ:* Columbia Univ, BSEE, 55, MSIE, 56; Ohio State Univ, PhD(stochastic processes), 61. *Prof Exp:* Dir systs anal, Battelle Mem Inst, 56-62; prof indust eng, Ariz State Univ, 62-69 & Va Polytech Inst, 69-70; prof indust eng, Purdue Univ, West Lafayette, 70-81; pres, 73-86, CHMN, PRITSKER CORP, 73- *Concurrent Pos:* Consult, Battelle Mem Inst, 62-63, Rand Corp, 63- & Motorola, Inc, 65-68. *Honors*

& Awards: H B Maynard Innovation Award, Am Inst Indust Engrs, 78. *Mem:* Nat Acad Eng; Opers Res Soc Am; Inst Mgt Sci; fel Am Inst Indust Engrs. *Res:* Simulation; large-scale systems; developer of combined simulation techniques; developer of network modeling languages. *Mailing Add:* Pritsker Corp PO Box 2413 West Lafayette IN 47906

PRITZKER, ROBERT A, b Chicago, Ill, June 30, 26. INDUSTRIAL ENGINEERING. *Educ:* Ill Inst Technol, BSIE, 46. *Hon Degrees:* LHD, Ill Inst Technol. *Prof Exp:* Pres, Colson Co, 53; PRES & CHIEF EXEC OFFICER, MARMON GROUP INC, 53- *Concurrent Pos:* Dir, Hyatt Corp, Union Tank Car Co & S&W Berisford PLC; chmn bd trustees, Ill Inst Technol; lectr, Univ Chicago; dir, Marmon Group Inc. *Mem:* Nat Acad Eng. *Res:* Author of various publications. *Mailing Add:* Off Pres Marmon Group Inc 225 W Washington St Chicago IL 60606

PRITZLAFF, AUGUST H(ENRY), JR, b Culver, Ind, Aug 25, 24; m 52; c 2. CHEMICAL ENGINEERING. *Educ:* Northwestern Univ, PhD(chem eng), 52. *Prof Exp:* Chem engr res dept, Standard Oil Co Ind, 51-60; eng res dept, Swift & Co, 60-67; Commonwealth Assocs, 67-70; res engr, Beatrice Foods Res Ctr, Chicago, 70-89; RETIRED. *Mem:* Am Soc Naval Engrs; Am Inst Chem Engrs; Inst Food Technol. *Res:* Fluidization; unit operations; food processing mechanization. *Mailing Add:* PO Box 338 Princeton IL 61356

PRIVAL, MICHAEL JOSEPH, b New York, NY, Aug 22, 45; m; c 2. MUTAGENESIS. *Educ:* Columbia Col, BA, 66; Mass Inst Technol, PhD(microbiol), 72. *Prof Exp:* Proj dir, Ctr Sci Pub Interest, 72; microbiologist, US Environ Protection Agency, 73-76; RES MICROBIOLOGIST, US FOOD & DRUG ADMIN, 76- *Concurrent Pos:* Mem, Comt Diet, Nutrit & Cancer, Nat Acad Sci, 80-83. *Mem:* Environ Mutagen Soc; Am Soc Microbiol; Sigma Xi; AAAS. *Res:* Methodologies used for testing chemicals for mutagenicity in bacteria, particularly chemicals that are false negatives in standard test procedures, such as nitrosamines and azo dyes. *Mailing Add:* Genetic Toxicol Br Food & Drug Admin Washington DC 20204

PRIVITERA, CARMELO ANTHONY, b Buffalo, NY, May 12, 23; m 48; c 6. BIOLOGY. *Educ:* Canisius Col, BS, 47; Univ Notre Dame, MS, 49; St Louis Univ, PhD(biol), 57. *Prof Exp:* Instr zool, Col St Thomas, 48-54, asst prof, 56-59; from asst prof to assoc prof biol, St Louis Univ, 59-65; assoc prof, 65-72, vchmn dept, 65-68, PROF BIOL, STATE UNIV NY BUFFALO, 73- *Mem:* Sigma Xi; Am Soc Zoologists; Am Physiol Soc; Am Soc Cryobiol; Ecol Soc Am. *Res:* Comparative physiology; depressed metabolism; temperature adaptation; environmental physiology; physiological adaptation of cells and organisms to cold, metals and aging. *Mailing Add:* Dept Biol Univ NY Buffalo 659 Hochstetter Amherst NY 14260

PRIVITERA, PHILIP JOSEPH, b New York, NY, Aug 29, 38; m 61; c 3. PHARMACOLOGY. *Educ:* St John's Univ, NY, BS, 59; Albany Med Col, PhD(pharmacol), 66. *Prof Exp:* Fel pharmacol, Col Med, Univ Cincinnati, 66-68, from instr pharmacol to asst prof exp med & pharmacol, 68-72; assoc prof, 72-79, PROF PHARMACOL, COL MED, MED UNIV SC, 79- *Mem:* Sigma Xi; Am Soc Pharmacol & Exp Therapeut; NY Acad Sci; Am Fedn Clin Res. *Res:* Cardiovascular neuro-pharmacology; reinin-angiotensin system; central nervous system control of blood pressure and renin release. *Mailing Add:* Dept Pharmacol Med Univ SC 171 Ashley Ave Charleston SC 29425

PRIVMAN, VLADIMIR, b Lvov, Ukraine, USSR, Jan 2, 55; Israel citizen; m 80; c 3. PHASE TRANSITIONS THEORY, SURFACE POLYMER CRITICAL PHENOMENA. *Educ:* Technion, Haifa, Israel, BSc, 75, MSc, 79, PhD(physics), 82. *Prof Exp:* Res assoc, physics, Cornell Univ, 82-84; res fel, physics, Calif Inst Technol, 84-85; ASSOC PROF, PHYSICS, CLARKSON UNIV, 85- *Mem:* Am Phys Soc. *Res:* Theoretical solid state and statistical physics, phase transitions, finite size effects, polymer scaling, surface critical phenomena. *Mailing Add:* Physics Dept Clarkson Univ Potsdam NY 13676

PRIVOTT, WILBUR JOSEPH, JR, b Chowan County, NC, Oct 8, 38; m 60. TECHNICAL MANAGEMENT, COMMERCIAL DEVELOPMENT. *Educ:* NC State Univ, BS, 61, PhD(chem eng), 65. *Prof Exp:* Res chem engr, Chemstrand Res Ctr, Inc Div, Monsanto Co, 64-68, group leader, New Enterprise Div, 68-71, prod develop mgr, 71-73, dir commercial prods div, 73-78, com dir food ingredients, 78-90; PRES, NOVUS INT, INC, 90- *Mem:* Am Inst Chem Engrs. *Res:* Basic chemical engineering phenomena of heat transfer, momentum transfer, mass transfer, process control and reaction kinetics with analog and digital computer applications. *Mailing Add:* 16118 Wilson Manor Dr Chesterfield MO 63005

PRIZER, CHARLES J(OHN), b Lake Forest, Ill, Apr 24, 24; m 44; c 4. CHEMICAL ENGINEERING. *Educ:* Univ Ill, BS, 44; Drexel Inst, MS, 56. *Prof Exp:* Chem engr, Tenn Eastman Corp, Div, 44-45, co, 45-46; chem engr, Brooklyn Naval Shipyard, 46-47 & Edwal Labs, Inc, 47-51; chem engr, 51-54, group leader process develop, 54-57, head, Petrol Chem Res Lab, 57-63, supvr petrol chem res, 63-65, asst dir res, 65-68, mgr petrol chem dept, 68-70, PROD MGR, CHEM DIV, ROHM & HAAS CO, 70-, VPRES & GEN MGR ENG & ENVIRON CONTROL, 72-, ASST GEN MGR, INT DIV, 74-, CORP BUS GROUP DIR, INDUST CHEM, POLYMERS, RESINS & MONOMERS, 75-, REGIONAL DIR N AM, 78-, GROUP VPRES, CORP OPERS, 83- *Concurrent Pos:* Lectr, Drexel Inst, 56-59. *Mem:* Am Inst Chem Engrs; Am Chem Soc. *Res:* Process research and development; pilot plant design and operation; process economics. *Mailing Add:* 614 Chestnut St Moorestown NJ 08057-2004

PROAKIS, ANTHONY GEORGE, b Chios, Greece, May 26, 40; US citizen; m 63; c 3. PHARMACOLOGY. *Educ:* WVa Univ, BS, 62; Purdue Univ, PhD(pharmacol), 72. *Prof Exp:* Mgr, Medco Pharm Inc, 62-66, vpres, 66-67; gen mgr, Guardian Pharm Inc, 67-69; sr res pharmacologist, A H Robbins Co, 72-77, res assoc, 77-80, mgr cardiovasc pharmacol, 80-90; CARDIOVASC PHARMACOLOGIST, US FOOD & DRUG ADMIN, 90- *Mem:* AAAS; Sigma Xi; Soc Exp Biol Med; Am Heart Asn; Am Soc Pharmacol & Exp Therapeut; Am Col Cin Pharmacol. *Res:* Cardiovascular and autonomic pharmacology; mechanism of action of antihypentensive, antianginal and cardiac antiarrhythmic agents. *Mailing Add:* 2200 Normandstone Dr Midlothian VA 23113

PROAKIS, JOHN GEORGE, b Chios, Greece, June 10, 35; US citizen; m 68. ELECTRICAL ENGINEERING. *Educ:* Univ Cincinnati, BS, 59; Mass Inst Technol, MS, 61; Harvard Univ, PhD(eng), 67. *Prof Exp:* Staff mem elec eng, Mass Inst Technol, 59-63; res asst, Harvard Univ, 63-66; staff mem, Gen Tel & Electronics, 67-69; PROF ELEC ENG, NORTHEASTERN UNIV, 69- *Concurrent Pos:* Consult, Gen Tel & Electronics, 69-71 & 74-, Arthur D Little, Inc, 70-71, Stein Assocs, Inc, 70-74, EG&G, Inc, 71-72 & Digital Equip Corp, 77-78; NSF grant, Northeastern Univ, 71-78. *Mem:* Inst Elec & Electronics Engrs. *Res:* Communication systems; adaptive filtering and equalization; digital signal processing; data transmission; computer optimization techniques; radar detection and estimation. *Mailing Add:* 256 Emerson Rd Lexington MA 02173

PROBER, DANIEL ETHAN, b Schenectady, NY, Oct 16, 48; m 70. LOW TEMPERATURE PHYSICS, SOLID STATE PHYSICS. *Educ:* Brandeis Univ, AB, 70; Harvard Univ, MA, 71, PhD(physics), 75. *Prof Exp:* Asst prof eng & appl sci, 75-80, ASSOC PROF, APPL PHYSICS, YALE UNIV, 80- *Concurrent Pos:* Fulbright fel, Tel Aviv Univ, 85. *Mem:* Am Phys Soc; Inst Elec & Electronics Engrs; Am Vacuum Soc. *Res:* Superconductivity, especially research on superconducting quantum devices and materials; microstructure science and microlithography. *Mailing Add:* Yale Univ Becten Ctr 15 Prospect St Box 2157 Yale Sta New Haven CT 06520

PROBER, RICHARD, b Chicago, Ill, May 30, 37; m 59; c 3. ENVIRONMENTAL ENGINEERING, CHEMICAL ENGINEERING. *Educ:* Ill Inst Technol, BS, 57; Univ Wis, MS, 58, PhD(chem eng), 62. *Prof Exp:* Engr, Shell Develop Co, 61-65; chem engr, Aerochem Res Labs, Inc, NJ, 65-68; sr res scientist, Permutit Res Labs, NJ, 68-70; from asst prof to assoc prof chem eng, Case Western Reserve Univ, 70-77; sr engr & secy-treas, GMP Assoc Inc, 77-81; ENVIRON-CHEM ENGR, HAVENS & EMERSON INC, CLEVELAND, 81- *Concurrent Pos:* Adj prof chem eng, Case Western Reserve Univ, 77-79. *Honors & Awards:* EPRI Award, Int Ozone Inst, 77. *Mem:* Water Pollution Control Fedn; Am Inst Chem Engrs; Am Chem Soc. *Res:* Water resources; water pollution control; membrane technology; ion exchange; industrial waste treatment. *Mailing Add:* 17610 Van Aken Blvd Cleveland OH 44120

PROBST, ALBERT HENRY, b Lawrenceburg, Ind, Jan 25, 12; m 39; c 2. AGRONOMY. *Educ:* Purdue Univ, BSA, 36, MS, 38, PhD(plant breeding, genetics), 48. *Prof Exp:* From jr agronomist to assoc agronomist in chg soybean invests in Ind, Bur Plant Indust, Soils & Agr Eng, USDA, Exp Sta, 36-50, res agronomist, Crops Res Div & prof agron, Univ, 50-70, EMER PROF AGRON, PURDUE UNIV, LAFAYETTE, 70- *Mem:* Fel Am Soc Agron; Am Soybean Asn. *Res:* Breeding new soybean varieties; genetics of soybeans; cultural studies; inheritance of leaf abscission and other characters in soybeans. *Mailing Add:* 418 Evergreen St West Lafayette IN 47906

PROBST, GERALD WILLIAM, b Morris, Minn, May 4, 22; m 46; c 4. BIOCHEMISTRY. *Educ:* Col St Thomas, BS, 43; Univ Minn, MS, 49, PhD, 50. *Prof Exp:* Asst biochem, Univ Minn, 46-50; from biochemist to sr biochemist, Eli Lilly & Co, 50-61, res scientist, 61-63, dept head insulin testing, 63-65, asst head agr anal develop, 65-67, head agr biochem, 67-69, dir regulatory serv, 69-82, dir agr spec projs, 82-85; RETIRED. *Concurrent Pos:* Dir serving as liaison with state and fed regulatory agencies, Food & Drug Admin & Environ Protection Agency, Elanco Regulatory Serv. *Mem:* AAAS; Am Chem Soc; NY Acad Sci. *Res:* Antibiotics; cancer; immunochemistry; pesticides; analytical biochemistry; animal drugs. *Mailing Add:* 7205 Kingswood Circle Indianapolis IN 46256-2914

PROBSTEIN, RONALD F(ILMORE), b Bronx, NY, Mar 11, 28; m 50; c 1. FLUID MECHANICS. *Educ:* NY Univ, BME, 48; Princeton Univ, MSE, 50, AM, 51, PhD(aeronaut eng), 52. *Hon Degrees:* AM, Brown Univ, 57. *Prof Exp:* Asst, Eng Res Div, NY Univ, 46-48, instr eng mech, 47-48; asst aeronaut eng, Princeton Univ, 48-52, res assoc, 52-53, asst prof, 53-54; from asst prof to prof eng & appl math, Brown Univ, 54-62; prof mech eng, 62-89, FORD PROF ENG, MASS INST TECHNOL, 89- *Concurrent Pos:* Consult ed, Plenum Pub Corp, 67-; sr partner, Water Purification Assocs, 74-81; counr, Am Acad Arts Sci, 75-79; comr, Comn Engr & Tech Systs, Nat Res Coun, 80-83; chmn Water Gen Corp, 82-83; sr corp tech adv, Foster-Miller Inc. 83-; ed-in-chief, Physiocochem Hydrodynamics, 88-90; Guggenheim fel, Mass Inst Technol, 60-61. *Honors & Awards:* Freeman Award, Am Soc Mech Engrs, 71. *Mem:* Nat Acad Eng; fel Am Acad Arts Sci; fel AAAS; fel Am Phys Soc; fel Am Soc Mech Engrs; Am Inst Chem Engrs. *Res:* Physiochemical hydrodynamics; environmental control technologies. *Mailing Add:* Dept Mech Eng Rm 3-246 Mass Inst Technol Cambridge MA 02139

PROCACCINI, DONALD J, developmental physiology, systematics; deceased, see previous edition for last biography

PROCCI, WARREN R, b Staten Island, NY, Jan 19, 47; m 72. PSYCHIATRY. *Educ:* Wagner Col, BS, 68; Univ Wis, MD, 72. *Prof Exp:* From asst prof to assoc prof psychiat, 75-82; ASSOC PROF, PSYCHIAT DEPT, UNIV CALIF, LOS ANGELES, 82- *Concurrent Pos:* Fel, Sch Med, Univ Southern Calif, 74-75. *Mem:* Am Psychoanal Asn. *Res:* Sexual dysfunction in end stage renal disease; residency education and medical student education in psychiatry; psychological problems in end stage renal disease. *Mailing Add:* 181 N Oak Knoll Ave Pasadena CA 91101

PROCHAZKA, SVANTE, b Prague, Czech, May 19, 28; m 51; c 2. CERAMICS. *Educ:* Prague Tech Univ, EngChem, 51, PhD(chem), 55; Mass Inst Technol, SM, 68. *Prof Exp:* Asst prof ceramics, Prague Tech Univ, 54-58; head res & develop ceramics, PAL, Tabor, Czech, 58-65; sr res chemist, VUEK, Czech, 69-70; STAFF SCIENTIST CERAMICS, CORP RES & DEVELOP, GEN ELEC CO, 71- *Mem:* Am Ceramic Soc. *Res:* Sintering covalent solids; silicon carbide ceramics; high temperature structural materials; multilayer ceramics. *Mailing Add:* 23 Midline Rd Ballston Lake NY 12019

PROCK, ALFRED, b Baltimore, Md, Aug 11, 30; m 53; c 2. PHYSICAL CHEMISTRY. *Educ:* Johns Hopkins Univ, BE, 51, MA, 53, PhD(chem), 55. *Prof Exp:* NSF fel chem, Harvard Univ, 55-56; res assoc, Cornell Univ, 56-61; from asst prof to assoc prof, 61-72, PROF CHEM, BOSTON UNIV, 72- *Mem:* Am Chem Soc; Am Phys Soc. *Res:* electrochemistry of organmetallic systems; Atomic and molecular electronic energy transfer to surfaces. *Mailing Add:* Dept Chem Boston Univ 675 Commonwealth Ave Boston MA 02215-1401

PROCKNOW, DONALD EUGENE, b Madison, SDak, May 27, 23; m 53; c 2. ENGINEERING ADMINISTRATION. *Educ:* Univ Wisc, BS, 47. *Hon Degrees:* SD Sch Mines & Technol, DrSci(eng), 47; Clarkson Univ, Dr Sci, 84. *Prof Exp:* Gen mgr eng, Western Elec Co, 63-64, gen mgr servs, 64-65, vpres servs, 65-68, vpres personnel, 69, exec vpres, 69-71, pres & CEO, 71-83; vchmn, AT&T Technols, 84-86; RETIRED. *Concurrent Pos:* Trustee, Clarkson Univ, 74-84, emer trustee, 84-; trustee, Drew Univ, 84-, Logistics Mgt Inst, 87- *Honors & Awards:* Chairmans Award, Am Assoc Eng Soc. *Mem:* Nat Acad Eng. *Mailing Add:* 18 Saw Mill Rd Saddle River NJ 07458

PROCKOP, DARWIN J, b Palmerton, Pa, Aug 31, 29; m 61; c 2. BIOCHEMISTRY, MOLECULAR BIOLOGY. *Educ:* Haverford Col, BA, 51; Oxford Univ, MA, 53; Univ Pa, MD, 56; George Washington Univ, PhD(biochem), 61. *Hon Degrees:* DSc, Univ Oulu, Finland. *Prof Exp:* Intern med, New York Hosp-Cornell Med Ctr, 56-57; res fel pharmacol, Nat Heart Inst, 57-58, res investr biochem, 58-61; from assoc biochem to prof biochem & med, Univ Pa, 61-72; prof biochem & chmn dept, Rutgers Med Sch, 72-86; PROF BIOCHEM, CHMN DEPT & DIR, JEFFERSON INST MOLECULAR MED, THOMAS JEFFERSON UNIV, 86- *Mem:* Nat Acad Sci; Am Soc Biol Chemists; Am Soc Clin Invest; Asn Am Physicians; fel AAAS; Am Chem Soc. *Res:* Synthesis, catabolism, gene expression and genetic diseases of collagen and other components of connective tissue. *Mailing Add:* Jefferson Inst Molecular Med Thomas Jefferson Univ 11th & Walnut Sts Philadelphia PA 19107

PROCKOP, LEON D, b Palmerton, Pa, Mar 28, 34; div; c 2. NEUROLOGY. *Educ:* Princeton Univ, AB, 55; Univ Pa, MD, 59. *Prof Exp:* Intern, Lenox Hill Hosp & Hunterdon Med Ctr, 59-60; res asst, NIH, 60-62; assoc physiol, George Washington Univ, 62; from resident to chief resident neurol, Columbia-Presby Med Ctr, 62-65, asst neurologist & independent investr, 65-67; from asst prof to assoc prof neurol, Univ Pa, 67-73, assoc in neurol, Univ Hosp, 67-73; PROF MED & CHIEF NEUROL SECT, COL MED, UNIV S FLA, 73- *Concurrent Pos:* Asst, NY Neurol Inst, 62-64; from asst to instr neurol, Columbia Univ, 65-67; Nat Inst Neurol Dis & Blindness fels, 65-67; chief, Penn Neurol Serv, Philadelphia Gen Hosp, 67-73; neurol consult, Tampa Vet Admin Hosp, Tampa Gen Hosp & St Joseph Hosp, Tampa, Fla, 73-; mem coun stroke, Am Heart Asn. *Mem:* AAAS; AMA; Am Neurol Asn; Am Soc Pharmacol & Exp Therapeut; Am Acad Neurol. *Res:* Rate and mechanisms of transfer of various substances between blood and cerebrospinal fluid; pathophysiology of hypoglycorrhachia in meningitis; pathology of cerebral edema following sustained hyperglycemia; biochemical correlates of spinal cord injury; peripheral neuropathy secondary to organic solvents. *Mailing Add:* Dept Neurol Univ SFla Col Med 12901 N 30th St Box 55 Tampa FL 33612

PROCTER, ALAN ROBERT, b London, Eng, Nov 13, 39; m 65; c 2. PULP CHEMISTRY, PAPER CHEMISTRY. *Educ:* Univ Edinburgh, BSc, 62, PhD(chem), 65. *Prof Exp:* Otto Maass fel, McGill Univ, 65-66; sr res chemist, 66-73, sect head, 71-78, mgr, 78-85, DIR, PULP & PAPER RES, MACMILLAN BLOEDEL RES LTD, 85- *Mem:* Tech Asn Pulp & Paper Indust; Can Pulp & Paper Asn; Royal Soc Chem; Royal Inst Chem; assoc mem Am Chem Soc. *Res:* Wood chemistry, chemical pulping, wood fiber properties, pulp and paper technology; carbohydrates, lignin, cellulose chemistry; polymers. *Mailing Add:* 8075 Pasco Rd W Vancouver BC V7W 2T5 Can

PROCTOR, ALAN RAY, b Edgecombe Co, NC, June 22, 45; m 64; c 2. GENE EXPRESSION, DRUG DISCOVERY. *Educ:* Univ NC, Chapel Hill, AB, 67; NC State Univ, MS, 69, PhD(genetics), 71. *Prof Exp:* Am Cancer Soc fel microbiol, Scripps Clin & Res Found, 71-73; res staff scientist, Fermentation Res & Develop Dept, 73-80, DIR, MOLECULAR GENETICS & PROTEIN CHEM DEPT, PFIZER CENT RES, PFIZER, INC, 80- *Mem:* Am Soc Microbiol; AAAS; Sigma Xi. *Res:* Technical management of DNA research; cloning and in vitro alteration of gene structure and expression; regulation gene expression in eucaryotic organisms. *Mailing Add:* 12 Mainsail Dr Noank CT 06340

PROCTOR, CHARLES DARNELL, b St Louis, Mo, Sept 10, 22; m 50; c 4. NEUROPHARMACOLOGY, TOXICOLOGY. *Educ:* Fisk Univ, BA, 43, MA, 46; Loyola Univ, Chicago, PhD(pharmacol & toxicol), 50. *Prof Exp:* Sr toxicologist, Cook County, Ill, 46-50; from instr to assoc prof pharmacol, Med Sch, Loyola Univ, Chicago, 50-62; dir, lab for chem pharmacol, McCrone Res Inst, Chicago, 62-64; prof, Meharry Med Col, 64-78, chmn dept, 68-82, Bent Scholar prof, 78-82; PROF PHARMACOL & ASSOC DEAN, SCH MED, MERCER UNIV, 82- *Concurrent Pos:* Sr toxicol consult, Ill Bd Health, Chicago, 62-64; HEW consult, Off of the Pres of the US, 67-68; mem, Nat Inst Environ Health Sci Coun, 80-84; assoc ed, Drug Develop Res, 83-; consult, Staff Col, NIMH, 84- *Honors & Awards:* Frontiers of Am Award, Frontiers Int, 59; Meritorious Serv Award, NIH, 85. *Mem:* Am Soc for Pharmacol & Exp Therapeut; Soc Toxicol; Am Chem Soc; Am Inst Chemists; AAAS; Nat Inst Sci. *Res:* Biological correlates to schizophrenic reaction; development process involved in opiate and barbiturate addiction; actions of addictive and abused drugs. *Mailing Add:* Mercer Univ Sch Med 1550 College St Macon GA 31207

PROCTOR, CHARLES LAFAYETTE, b Collinsville, Okla, Feb 28, 23. INDUSTRIAL & SYSTEMS ENGINEERING. *Educ:* Okla State Univ, BSME, 51, PhD(indust & systs eng), 63; Purdue Univ, MSME, 55. *Prof Exp:* Engr, Tinker AFB, 46-49; asst mech eng, Okla State Univ, 49-51; aeronaut design engr, Beech Aircraft Corp, Kans, 51-53; instr gen & mech eng, Purdue Univ, 53-55; missile thermodynamicist, McDonnell Aircraft Corp, Mo, 55-56; asst prof mech eng, Univ Toronto, 56-63; dir indust eng, Univ Okla, 64-66; resident dir grad eng educ syst & prof indust & systs eng, Univ Fla, 66-72; prof & head dept, Univ Windsor, 72-75; chmn dept indust eng, Western Mich Univ, 75-77, prof, 75-80; prof & head dept eng & sci mgt, Univ Alaska, 80-84; prof mech eng, King Saud Univ, Saudi Arabia, 84-88; CONSULT, 88- *Concurrent Pos:* Ford grant, Mass Inst Technol, 65; NASA fac fel, Jet Propulsion Lab, Calif Inst Technol, 69; mem bd trustees, Embry Riddle Aeronaut Univ, 68-; res grants, NASA, 70 & Nat Res Coun Can, 74; prof, Univ Windsor, 75-76; mem grants & award selection comt, Nat Res Coun Can, 75-78; fac fel, US Air Force, Wright-Patterson AFB, 77; consult, Alaska Archit & Eng, 80-81, Alaska Mech, 81-82, City of Delta Junction, Alaska, 82-83, Jahn-Wayne Construct, 83-84 & Saudi Arabia Ctr Sci & Technol, 84-85. *Mem:* Am Soc Mech Engrs; Am Soc Qual Control; Am Inst Indust Engrs; Opers Res Soc Am; Inst Elec & Electronics Engrs. *Res:* The study of operating systems, including mathematical modeling of system reliablity, risk assessment and production research; construction management. *Mailing Add:* 7931 Asaro Rd Hobe Sound FL 33455

PROCTOR, CHARLES LAFAYETTE, II, b Crawfordsville, Ind, Nov 21, 54; m 76; c 1. COMBUSTION, HAZARDOUS WASTE INCINERATION. *Educ:* Purdue Univ, BSc(ME), 76, MS, 79, PhD(mech eng), 81. *Prof Exp:* Asst prof, 81-86, dir res, Combustion Lab, 81-88, ASSOC PROF MECH ENG, UNIV FLA, 86- *Concurrent Pos:* Prin investr, US Air Force, 82-, Allied Bendix Corp, 85-; consult, Prod Liability & Accident Reconstruction, 82- , Allied Bendix Corp, 86-; mem, fac intern prog, Am Soc Testing & Mat. *Mem:* Am Soc Mech Eng; Am Soc Testing & Mat; Air Polution Control Asn; Am Inst Aeronaut & Astronaut; Am Soc Eng Educ; Combustion Inst; Nat Fire Protection Asn. *Res:* Combustion research investigating the interaction of fluids mechanics and chemistry using advanced optical diagnostic techniques; emphasis on soot formation and turbulence; mechanics; fire protection and fire safety; hazardous waste disposal; combustion process; fluid mechanics. *Mailing Add:* Dept Mech Eng 336 MEB Univ Fla Gainesville FL 32611

PROCTOR, CHARLES MAHAN, b Belva, WVa, Oct 4, 17; c 1. BIOSCIENCES. *Educ:* Berea Col, AB, 39; Univ Utah, MS, 49; Tex A&M Univ, PhD(oceanog), 58. *Prof Exp:* Asst chem, Univ Ky, 40-41; chemist, Test Plant, Pa RR, 41-42; res assoc, Manhattan Proj, Univ Rochester, 46-47; asst & biochemist, Univ Utah, 47-51; in chg chem test sect, Dugway Proving Ground, Dept Army, 51-52; asst & actg asst prof oceanog, Tex A&M Univ, 52-57; in chg chem studies germicides for water treatment, Robert A Taft Sanit Eng Ctr, USPHS, 57-59; sr res specialist, 59-61, Boeing Aerospace Co, 59-61, sr bioscientist, 61-76; sr assoc, Envirodyne Engrs, Inc, 77-78; CONSULT, 79- *Mem:* Fel AAAS; Am Inst Biol Sci; fel Am Inst Chemists. *Res:* Biological systems; marine resources; human engineering; environmental science. *Mailing Add:* 2540 122nd Ave SE Bellevue WA 98005

PROCTOR, CLARKE WAYNE, b Houston, Tex, Feb 27, 42; m. MATHEMATICS. *Educ:* Univ Houston, BS, 65, MS, 66, PhD(math), 69. *Prof Exp:* Instr math, Univ Houston, 68-69; asst prof, Memphis State Univ, 69-70; from asst prof to assoc prof, 70-85, PROF MATH, STEPHEN F AUSTIN STATE UNIV, 85- *Mem:* Am Math Soc; Math Asn Am. *Res:* Topology. *Mailing Add:* Dept Math & Statist Stephen F Austin State Univ Box 13040 Sfa Sta Nacogdoches TX 75962

PROCTOR, DAVID GEORGE, b Youngstown, Ohio, Jan 20, 28; m 49; c 6. PHYSICS EDUCATION, AUTOMATIC TESTING OF MATERIALS. *Educ:* Case Inst Technol, BS, 52, MS, 55, PhD(physics), 58. *Prof Exp:* Asst physics, Case Inst Technol, 52-57; physicist, Phillips Petrol Co, 57-60; assoc prof physics & chmn dept, 60-66, chmn sci div, 66-86, PROF PHYSICS, BALDWIN-WALLACE COL, 66- *Concurrent Pos:* Consult, Union Carbide Corp, 61- & USAID, India, 66. *Mem:* Am Asn Physics Teachers; Acoust Soc Am; Am Phys Soc; Catgut Acoust Soc. *Res:* Reactor physics; nuclear spectroscopy; photonuclear reactions; mechanical properties of solids; high current arc physics. *Mailing Add:* 485 Wyleswood Dr Baldwin- Wallace Col Berea OH 44017

PROCTOR, DONALD FREDERICK, b Red Bank, NJ, Apr 19, 13; m 37; c 2. OTOLARYNGOLOGY, ENVIRONMENTAL MEDICINE. *Educ:* Johns Hopkins Univ, AB, 33, MD, 37; Am Bd Otolaryngol, dipl. *Prof Exp:* Assoc prof laryngol & otol, 46-51, prof anesthesiol, 51-55, asst prof physiol, 55-65, assoc prof laryngol & otol, 58-74, PROF LARYNGOL & OTOL, SCH MED, JOHNS HOPKINS UNIV, PROF LARYNGOL & OTOL, SCH MED, 74-, EMER PROF, ENVIRON HEALTH SCI, OTOLARYNGOL & ANESTHESIOL, JOHN HOPKINS UNIV, 78- *Concurrent Pos:* Prof environ health sci, Sch Hyg & Pub Health, 65-77. *Mem:* Am Physiol Soc; Am Thoracic Soc; fel Am Col Surg; Am Acad Ophthal & Otolaryngol; Am Broncho-Esophagol Asn. *Res:* Conservation of hearing in children; pathogenesis of the common cold; significance of the pneumotachogram; mechanics of breathing; measurement of intrapleural pressure; anesthesia and otolaryngology; the pulmonary circulation; airborne disease, air pollution and air hygiene; physiology of singing; upper airway physiology; air pollution. *Mailing Add:* Sch Hyg & Pub Health Johns Hopkins Univ Baltimore MD 21205

PROCTOR, IVAN D, b Bloomfield, Iowa, Nov 15, 39; m 66; c 2. NUCLEAR PHYSICS. *Educ:* Iowa State Univ, BS, 66; Mich State Univ, MS, 69, PhD(physics), 72. *Prof Exp:* PHYSICIST NUCLEAR PHYSICS, LAWRENCE LIVERMORE LAB, 72- *Mem:* Am Phys Soc. *Res:* Charged particle spectroscopy; accelerator technology. *Mailing Add:* Lawrence Livermore Lab L-397 Livermore CA 94550

PROCTOR, J(OHN) F(RANCIS), b Newark, NJ, Oct 21, 28; m 84; c 4. CHEMICAL ENGINEERING. *Educ:* Cooper Union, BChE, 51. *Prof Exp:* Engr, Dana Plant, Atomic Energy Div, 51-54, supvr tech studies group, 54-57, engr, Savannah River Lab, SC, 57-64, sr engr, 64-68, sr res supvr, 68-70, process supt separations areas, Atomic Energy Div, 70-73, process mgr, 73-81, sr tech consult, Atomic Energy Div, Petrol Chem Dept, E I Du Pont de Nemours & Co, Inc, 81-83, PROCESS FEL, SAVANNAH RIVER LAB, ATOMIC ENERGY DIV, DUPONT CO, 83- *Concurrent Pos:* Tech consult, nuclear fac, Savannah River Lab. *Mem:* Am Inst Chem Engrs; Am Nuclear Soc. *Res:* Administrator and manager of process design for nuclear reactor, fuel fabrication, fuel reprocessing, radioactive waste storage, and heavy water facilities of the Atomic Energy Division. *Mailing Add:* 106 Englewood Dr Aiken SC 29803

PROCTOR, JERRY FRANKLIN, b Westminster, Tex, Jan 4, 32; m 71; c 4. NUTRITION. *Educ:* Tex A&M Univ, BS, 55; Cornell Univ, MS, 58, PhD(animal nutrit), 60. *Prof Exp:* Lab analyst, Nat Lead Co, Tex, 53-54; asst nutrit, Cornell Univ, 55-60; res nutritionist, Res & Develop Div, Nat Dairy Prod Corp, 60-68, mgr, Nutrit Res Lab, Kraft Inc, 68-80, dep managing dir, 80-82, DIR TECHNOL FORECASTING, TINAMENOR, SUBSID KRAFT INC, 83- *Concurrent Pos:* Mem res comt, Nat Asn Margarine Mfrs, 61-; mem sci adv comt, Spec Dairy Indust Bd, 62-; mem food additives comt, Fatty Acid Producers' Coun, 64- *Mem:* AAAS; Am Nutrit Soc; Animal Nutrit Res Coun; Am Soc Animal Sci; Inst Food Technologists; Agr Res Inst. *Res:* Nutritional muscular dystrophy; antibiotic response in animals; nutrition deficiency diseases; toxicity testing; bio-assay. *Mailing Add:* 18937 W Rose Ave Mundelein IL 60060

PROCTOR, JOHN THOMAS ARTHUR, b Armagh, Ireland, Dec 19, 40; m 68. HORTICULTURE, PLANT PHYSIOLOGY. *Educ:* Univ Reading, BSc, 64; Cornell Univ, MS, 67, PhD(pomol), 70; Inst Biol, Eng, MIBiol, 69. *Prof Exp:* Res scientist, 69-71, actg chief res scientist, 71-72, actg dir, 74-75, res scientist, Simcoe Sta, Hort Res Inst Ont, 72-78; ASSOC PROF, 78-84, PROF, UNIV GUELPH, 84- *Concurrent Pos:* Vis res fel, Long Ashton Res Sta, Univ Bristol, Eng, 75; vis scientist, Agr Can, Kentville, 82. *Honors & Awards:* STARK Award, 76; VPres Hedrick Awards, 84, 87. *Mem:* Fel Am Soc Hort Sci; Inst Hort; Am Pomol Soc (pres, 86-88); Can Soc Hort Sci (pres, 87-88); fel Int of Hort. *Res:* Plant growth and development; microclimatology of crop production; ginseng production. *Mailing Add:* Dept of Hort Sci Univ Guelph Guelph ON N1G 2W1 Can

PROCTOR, JULIAN WARRILOW, b Cheltenham, UK, Mar 20, 42. CANCER, IMMUNOLOGY. *Educ:* Univ London, MBBS, 67, PhD(cancer), 73. *Prof Exp:* Res assoc path, Med Sch, Mem Univ Nfld, 72-73; lectr, McGill Univ, 73-76, sr scientist cancer res, 73-76; assoc biologist cancer res, Allegheny Gen Hosp, 76-79, sr biologist, 76-79, chief immunobiol, 80-83; staff therapist, Med Sch, Johns Hopkins Univ, 84-87, asst prof oncol & radiol, 84-87; PARTNER, TRIANGLE RADIATION, 87- *Concurrent Pos:* Gordon Jacobs fel, Royal Marsden Hosp, UK, 69-72; adj asst prof, Univ Pittsburgh, 76-; sr fel radiother radiation oncol, Allegheny Gen Hosp, 76-78, assoc staff radiation therapist, 79-83; NIH fel, 78. *Mem:* Can Cancer Soc; Can Soc Oncol; Reticuloendothelial Soc; Am Soc Theoret Radiol. *Res:* Tumor immunology; biology and immunology of metastatic cancer; function of macrophages; immunotherapy of metastatic animal tumors; human melanoma; epidemiology endocrinology of malignant melanoma; breast cancer. *Mailing Add:* 176 Valley Rd Wexford PA 15090

PROCTOR, PAUL DEAN, b Salt Lake City, Utah, Nov 24, 18; m 45; c 4. ECONOMIC GEOLOGY, STRUCTURAL GEOLOGY. *Educ:* Univ Utah, BA, 42; Cornell Univ, MA, 43; Ind Univ, PhD(geol), 49. *Prof Exp:* From jr geologist to geologist, US Geol Surv, 43-53; assoc prof, Ind Univ, 50-52; asst supvr raw mat explor, US Steel Western States, 53-57; prof & chmn dept, Mo Sch Mines, Univ Mo, Rolla, 57-64, dean, Sch Sci, 64-70, prof geol, 64-84; from asst prof to assoc prof, Brigham Young Univ, 47-50, assoc prof, 52-53, PROF GEOL, BRIGHAM YOUNG UNIV, 84- *Concurrent Pos:* Deleg, Int Geol Cong, 56, 60 & 68; mem, Int Geol Field Inst, Italy, 64; UNESCO geol specialist, Middle Eastern Tech Univ, Ankara, 66-67; consult, McDonnell Aircraft, Miss River Fuel, US Steel, Rohm & Haas & USAID, Vietnam; scientist econ develop, Turkey, NSF, 73-74 & 75; spec lectr, Belem, Para, Brazil, 73-75; consult, Phillips Petrol Co, 74-; consult Egypt, Smithsonian Inst, 80. *Mem:* AAAS; fel Geol Soc Am; Am Inst Mining, Metall & Petrol Eng; Soc Econ Geologists. *Res:* Ore genesis, localization and evaluation; trace elements; chemistry of water supplies. *Mailing Add:* 2949 Apache Way Provo UT 84604

PROCTOR, RICHARD JAMES, b Los Angeles, Calif, Aug 2, 31; m 55; c 3. GEOLOGICAL HAZARDS, EXPERT WITNESS. *Educ:* Calif State Univ, Los Angeles, BA, 54; Univ Calif Los Angeles, MA, 58. *Prof Exp:* Chief geologist, Metrop Water Dist S Calif, 58-80; PRES, R J PROCTOR INC, 80- *Concurrent Pos:* Mem, US Nat Comt Tunneling Technol, 72-74; vis prof, Calif Inst Technol, 75-78; sr consult, Wagner, Hohns, Inglis Inc, 86-; mem, US Comt Large Dams. *Honors & Awards:* EB Burwell Mem Award, Geol Soc Am, 72. *Mem:* Asn Eng Geologists (pres, 79); Am Inst Prof Geologists (pres, 89); Am Geol Inst (secy-treas, 79-83); mem Eartquake Eng Res Inst; Brit Tunneling Soc; Am Arbit Asn. *Mailing Add:* 327 Fairview Ave Arcadia CA 91007

PROCTOR, STANLEY IRVING, JR, b Belleville, Ill, Dec, 23, 36; m 60; c 2. TECHNICAL MANAGEMENT. *Educ:* Wash Univ, BSChE, 57, MS, 62, DSc, 72. *Prof Exp:* Engr simulation, Monsanto Co, 59-63, sr engr chem reaction, 63-70, eng specialist & supvr, 70-71, eng supt tech computations, 71-73, mgr eng, 73-76, gen supt mfg, 76-78, mgr, Proc Tech Plant Opers, Monsanto Co, 78-79, dir eng technol, 79-83, appl sci res & develop, 84-85 & Biotechnol Proj, 85-86, DIR, ENG TECHNOL, MONSANTO CO, 86- *Concurrent Pos:* Bd Dirs, Nat Consortium Grad Degrees for Minorities Eng Inc, 81-84, Heat Transfer Res, Inc, 88-, Nat Found Hist Chem, 88-, Accreditation Bd Eng & Technol, 90-; adv comt, Dept Chem Eng, Okla State Univ, 82-85, Mo Scholars Acad, 84-88; adv coun, Dept Chem Eng, Tex A&M Univ, 88-; adv bd, Dept Chem Eng, Tex Tech Univ, 89-; corp bd visitors, Univ Mo-Rolla, 91- *Honors & Awards:* R L Jacks Award, Am Inst Chem Engrs, 86, T J Hamilton Award, 88. *Mem:* Fel Am Inst Chem Engrs (vpres, 86, pres, 87); Nat Soc Prof Engrs; Am Chem Soc; Am Soc Eng Educ. *Res:* Director of an organization of technical specialists in various engineering disciplines; areas of technical specialty include reaction engineering and process simulation. *Mailing Add:* 50 High Valley Dr Chesterfield MO 63017

PROCTOR, THOMAS GILMER, b Winston-Salem, NC, Oct 4, 33; m 57. DIFFERENTIAL EQUATIONS. *Educ:* NC State Univ, BNE, 56, MS, 62, PhD(math), 64. *Prof Exp:* Nuclear engr, Ingalls Shipbldg Corp, Miss, 56-60; instr math, NC State Univ, 64-65; from asst prof to assoc prof, 65-74, PROF MATH, CLEMSON UNIV, 74- *Mem:* Math Asn Am; Soc Indust & Appl Math. *Res:* Linear and nonlinear ordinary differential equations; functional differential equations; dynamical systems. *Mailing Add:* Dept of Math Sci Clemson Univ Clemson SC 29634-1907

PROCTOR, VERNON WILLARD, b Valentine, Nebr, Apr 25, 27; m 64; c 2. PHYCOLOGY. *Educ:* Univ Mo, AB, 50, AM, 51, PhD(bot), 55. *Prof Exp:* From asst prof to assoc prof, 56-63, PROF BIOL, TEX TECH UNIV, 63- *Mem:* Bot Soc Am; Phycol Soc Am. *Res:* Systematics, phylogeny and biogeography of Charophytes. *Mailing Add:* Dept of Biol Tex Tech Univ Lubbock TX 79409

PROCTOR, W(ILLIAM) J(EFFERSON), JR, b Dublin, Ga, Sept 21, 25. CERAMICS, KAOLIN PROCESSING. *Educ:* Univ Ala, BS, 49. *Prof Exp:* Chemist, Southern Clays Inc, 49-56, ceramist, 57-63; ceramist, 63-76, res chemist, Freeport Kaolin Co, Inc, 76-85; anal chemist, Engelhard Corp, 85-87, sr process eng tech, 87-90; RETIRED. *Mem:* Am Ceramic Soc; Nat Inst Ceramic Eng. *Res:* Minerals processing; paper, Cracking Catalyst. *Mailing Add:* Box 411 Gordon GA 31031

PRODANY, NICHOLAS W, b Boston, Mass; m 68; c 2. ENVIRONMENTAL ENGINEERING. *Educ:* Tufts Univ, BS, 58; Univ Mich, PhD(chem eng), 67. *Prof Exp:* Engr, E I du Pont, 58-62; prof chem eng, Univ Mich, 67-70 & Northeastern Univ, 70-73; mgr, Gillette Co, 73-89; ENVIRON ENGR, US ENVIRON PROTECTION AGENCY, 89- *Concurrent Pos:* Adj prof, Univ Mass, Boston, 85-89. *Mem:* Am Inst Chem Engrs; Sigma Xi; AAAS. *Mailing Add:* Two Selwyn Rd Belmont MA 02178

PRODELL, ALBERT GERALD, b Newark, NJ, Feb 28, 25; m 58; c 2. PHYSICS. *Educ:* Columbia Univ, AB, 47, MA, 48, PhD(physics), 54. *Prof Exp:* Asst physics, Columbia Univ, 47-50, lectr, 50-51, lectr, Barnard Col, 51-52, from instr to asst prof, 52-56, res assoc, 56-59; PHYSICIST, BROOKHAVEN NAT LAB, 59- *Mem:* Am Phys Soc. *Res:* Atomic and molecular beams; high energy nuclear physics. *Mailing Add:* Box 322 Wading River NY 11792

PRODY, GERRY ANN, b Minneapolis, Minn, June 24, 52. RNA AUTO CATALYSIS, VIRUS-SATELLITE SYSTEMS. *Educ:* Univ Calif, Davis, BS, 74, PhD(biochem), 81. *Prof Exp:* Instr chem, Calif State Univ, Sacramento, 79-82, Sacramento City Col, 80-83; asst prof nursing, Calif State Univ Consortium, 81-83; asst prof biochem, Univ Calif, Davis, 82-84; asst prof, 84-88, ASSOC PROF CHEM, WESTERN WASH UNIV, 88- *Concurrent Pos:* Vis plant pathologist, Univ Calif, Davis, 85. *Mem:* AAAS; Sigma Xi. *Res:* Tobacco ringspot virus and its associated satellite; sequencing the genomic RNAs and infecting plant cell protoplasts with virus and satellite by electroporation. *Mailing Add:* Dept Chem Western Wash Univ MS 9058 Bellingham WA 98225-9058

PROEBSTING, EDWARD LOUIS, JR, b Calif, Mar 2, 26; m 47; c 3. HORTICULTURE. *Educ:* Univ Calif, BS, 48; Mich State Col, PhD(hort), 51. *Prof Exp:* Asst horticulturist, Irrig Exp Sta, 51-59, assoc horticulturist, 59-63, HORTICULTURIST, IRRIG AGR RES & EXTEN CTR, WASH STATE UNIV, 63 - *Concurrent Pos:* Vis prof, Cornell Univ, 66, Hokkaido Univ, 78, Victoria Australia Dept Agr, 86; vpres, Res Div, Am Soc Hort Sci, 78-79, pres, 83-84, chmn bd dirs, 84-85. *Honors & Awards:* Joseph Gourley Award, Am Soc Hort Sci, 55, Charles Woodbury Award, 58; Nat Food Processors Asn Award, 81; Stark Award, 84 & 85. *Mem:* AAAS; fel Am Soc Hort Sci (pres, 83-84). *Res:* Pomology; orchard management; cold hardiness; fruit tree water relations; sweet cherry culture including growth regulators and fruit quality. *Mailing Add:* Irrig Agr Res & Exten Ctr Wash State Univ Prosser WA 99350-0030

PROEBSTING, WILLIAM MARTIN, b Lansing, Mich, Feb 4, 51. PLANT PHYSIOLOGY. *Educ:* Univ Wash, BS, 73; Cornell Univ, PhD(plant physiol), 78. *Prof Exp:* Plant Physiologist, Sci & Educ Admin, USDA, 77-79; PLANT PHYSIOLOGIST, DEPT HORT, ORE STATE UNIV, 80- *Mem:* Am Soc Plant Physiologists; Bot Soc Am; Am Soc Hort Sci; Scand Soc Plant Physiol. *Res:* Physiological genetics of senescence and flowering. *Mailing Add:* Dept Hort Ore State Univ Corvallis OR 97331

PROENZA, LUIS MARIANO, b Mex, Dec 22, 44; US citizen; m 83. RETINAL PHYSIOLOGY, ION HOMEOSTASIS. *Educ:* Emory Univ, BA, 65; Ohio State Univ, MA, 66; Univ Minn, Minneapolis, PhD(neurosci), 71. *Prof Exp:* Res fel neurosci, Univ Minn, 68-71; asst prof neurobiol, dept psychol, Univ Ga, 71-77, from asst prof to assoc prof neurobiol, 77-84, prof neurobiol, dept zool, 84-87, asst to pres, 84-86; VCHANCELLOR FOR RES & DEAN OF GRAD SCH, UNIV ALASKA, FAIRBANKS, 87- *Concurrent Pos:* USPHS fel, 65-66; grantee, NIH-Nat Eye Inst, 73-, res career develop award, 76-83; coordr fac neurobiol, Univ Ga, 75-; study dir, comt vision, Nat

Acad Sci, 77-79, mem, 79-82; mem, social issues comt, Soc Neurosci, 78-81; res scientist, Marine Biol Lab, Woods Hole, 82; Am Coun Educ fel, 83-84. *Mem:* Soc Neurosci; Asn Res Vision & Ophthal; AAAS; Sigma Xi. *Res:* Physiology of vertebrate retina with an emphasis on neuro-glial interactions, ion homeostasis and mechanisms related to generation of the electroretinogram. *Mailing Add:* Chief Planning Officer Univ Alaska Fairbanks AK 99770

PROFFIT, WILLIAM R, b Harnett Co, NC, Apr 19, 36; c 3. ORTHODONTICS, PHYSIOLOGY. *Educ:* Univ NC, BS, 56, DDS, 59; Med Col Va, PhD(physiol), 62; Univ Wash, MSD, 63. *Hon Degrees:* FDS, Royal Col of Surgeons, London, 90. *Prof Exp:* Nat Inst Dent Res clin investr, 63-65; from asst prof to prof orthod, Univ Ky, 65-73, chmn dept, 65-73; prof orthod & chmn dept pediat dent, Univ Fla, 73-75; PROF ORTHOD & CHMN DEPT, SCH DENT, UNIV NC, CHAPEL HILL, 75- *Mem:* Am Dent Asn; Am Asn Orthod; Am Speech & Hearing Asn; Int Asn Dent Res; Int Soc Cranio-Facial Biol. *Res:* Functional influences on facial growth; bone growth in vivo; physiology of swallowing. *Mailing Add:* Dept Orthod Univ NC Sch of Dent Chapel Hill NC 27514

PROFFITT, MAX ROWLAND, b Knoxville, Tenn, Oct 27, 40. VIROLOGY, IMMUNOLOGY. *Educ:* Univ Tenn, Knoxville, BS, 64, MS, 66, PhD(exp path), 70. *Prof Exp:* Res fel med, Mass Gen Hosp, 70-73; res assoc, Harvard Med Sch, 73-74, instr med, 74-76; STAFF SCIENTIST, CLEVELAND CLIN FOUND, 76-; ADJ ASSOC PROF BIOL, CASE WESTERN RESERVE UNIV, 76- *Concurrent Pos:* Gen res support grant, Mass Gen Hosp, 70-71, Damon Runyon Mem Fund fel, 71-73; asst microbiol, Mass Gen Hosp, 73-76; Charles A King Trust fel, 73-74; Leukemia Soc Am spec fel, 74-76; NIH res grant, 74-; scholar, Leukemia Soc Am, 77- *Res:* Viral and tumor immunology; effects of oncogenic and non-oncogenic viruses on immunological functions of the host; host-virus relationships in oncogenesis and in other diseases. *Mailing Add:* Dept Microbiol Cleveland Clin Found 9500 Euclid Ave Cleveland OH 44106

PROFFITT, THOMAS JEFFERSON, JR, b Columbia, Va, Dec 23, 32; m 56; c 2. ORGANIC CHEMISTRY. *Educ:* Va Polytech Inst, BS, 55. *Prof Exp:* Chemist, E I du Pont de Nemours & Co, Inc, 55-56 & 58-62, res chemist, 62-72, sr res chemist, 72-88, RES ASSOC, E I DU PONT DE NEMOURS & CO, INC, 88- *Mem:* Am Chem Soc. *Res:* Formulation of synthetic textile fiber lubricating and antistatic compositions; surface chemistry and emulsion technology. *Mailing Add:* 2211 Sparre Dr Kinston NC 28501

PROFIO, A(MEDEUS) EDWARD, b New Castle, Pa, Apr 18, 31; m 54; c 3. NUCLEAR ENGINEERING. *Educ:* Mass Inst Technol, SB, 53, PhD(nuclear eng), 63. *Prof Exp:* Scientist, Westinghouse Bettis Atomic Power Lab, 53-55; res assoc nuclear eng, Mass Inst Technol, 57-63, asst prof, 63-64; mem res & develop staff, Gen Atomic Div, Gen Dynamics Corp, 64-69; assoc prof nuclear eng, 69-74, PROF NUCLEAR ENG, UNIV CALIF, SANTA BARBARA, 74- *Concurrent Pos:* Fel eng, 63-64. *Mem:* Am Nuclear Soc. *Res:* Bioengineering, including applications of lasers to medicine; nuclear instrumentation. *Mailing Add:* Dept Chem Univ Calif Santa Barbara CA 93106

PROGELHOF, RICHARD CARL, b Orange, NJ, Aug 13, 36; m 63, 90; c 7. MECHANICAL ENGINEERING. *Educ:* Newark Col Eng, BSME, 58; Stanford Univ, MS, 59; Lehigh Univ, PhD(mech eng), 62. *Prof Exp:* Asst prof mech eng, Lehigh Univ, 62-64; res scientist heat transfer, Am Standard Res, 64-67; from assoc prof to prof, Newark Col Eng, 67-72; prof mech eng, NJ Inst Technol, 72-86; PROF & CHMN MECH ENG, UNIV SC, 86- *Concurrent Pos:* Dir, NJ Inst Technol, CIM Ctr. *Mem:* Am Soc Mech Engrs; Soc Plastics Engrs; Plastics & Rubber Inst; Am Soc Eng Educ; Plastics Inst Am. *Res:* Heat transfer, fluid mechanics, plastics processing and plastic properties. *Mailing Add:* 455 Running Fox Rd W Columbia SC 29223-2918

PROHAMMER, FREDERICK GEORGE, b New York, NY, Mar 26, 22; m 51; c 3. PHYSICS. *Educ:* City Col New York, BME, 43. *Prof Exp:* Mem staff mech eng, Radiation Lab, Mass Inst Technol. 43-46; asst instr physics, Univ Pa, 46-50; physicist, Oak Ridge Nat Lab, 50-55; assoc physicist, 55-80, PHYSICIST, ARGONNE NAT LAB, 80- *Concurrent Pos:* Vis prof, Northwestern Univ, 58-59; Kans State, 64, Ill Benedictine Col, 70. *Mem:* Am Phys Soc; Am Nuclear Soc; AAAS. *Res:* Low energy nuclear physics; reactor physics; reactor safety computer analysis. *Mailing Add:* Argonne Nat Lab Bldg 208 Argonne IL 60439

PROHASKA, CHARLES ANTON, b Oak Park, Ill, Mar 21, 20. NUCLEAR CHEMISTRY. *Educ:* Mass Inst Technol, BS, 42; Wesleyan Univ, MA, 49; Univ Calif, PhD(chem), 51. *Prof Exp:* Asst, Univ Calif, 48-49, chemist radiation lab, 49-51; sr chemist, E I du Pont de Nemours & Co, Inc, 51-78; RETIRED. *Mem:* Am Chem Soc; Sigma Xi. *Res:* Computer programming for chemical analyses; analytical research. *Mailing Add:* RD 2 Box 470 Dingman's Ferry PA 18328

PROHASKA, JOSEPH ROBERT, b Chicago, Ill, June 5, 45; m 68; c 3. BIOCHEMISTRY, NUTRITION. *Educ:* Univ Wis-Oshkosh, BS, 68; Mich State Univ, PhD(biochem), 74. *Prof Exp:* Res assoc nutrit sci, Univ Wis-Madison, 74-77; asst prof, 77-82, assoc prof, 82-88, PROF BIOCHEM, UNIV MINN, DULUTH, 88- *Concurrent Pos:* Vis prof, Univ Mo-Columbia, 90. *Mem:* Am Inst Nutrit; Int Asn Bioinorganic Scientists; Sigma Xi. *Res:* Trace element metabolism. *Mailing Add:* Dept Biochem & Molecular Biol Univ Minn Sch Med Duluth MN 55812

PROHOFSKY, EARL WILLIAM, b St Paul, Minn, Feb 8, 35; m 60; c 4. MOLECULAR BIOLOGY. *Educ:* Univ Minn, BS, 57; Cornell Univ, PhD(physics), 63. *Prof Exp:* Asst physics, Cornell Univ, 57-62, res assoc, 62-63; mem staff, Sperry Rand Res Ctr, Mass, 63-66; assoc prof physics, 66-73, PROF PHYSICS, PURDUE UNIV, 73- *Mem:* Fel Am Phys Soc; Biophys Soc. *Res:* Collective effects; second sound; ultrasonics; thermal conductivity; magnetism; fluorescence; lattice dynamics of nucleic acids. *Mailing Add:* Dept Physics Purdue Univ Lafayette IN 47907

PROKAI, BELA, b Eger, Hungary, Feb 19, 37; m 62; c 2. ORGANOMETALLIC CHEMISTRY. *Educ:* Univ Leicester, BSc, 60; Univ Manchester, MSc, 62, PhD(chem), 64. *Prof Exp:* Res fel chem, Mass Inst Technol, 64-66; proj scientist, 66-73, res scientist, 73-75, group leader, 75-79, assoc dir res & develop, 79-85, BUS DIR, UNION CARBIDE CORP, 85- *Mem:* Am Chem Soc; NY Acad Sci; Royal Chem Soc. *Res:* Investigation of the preparation, mechanism and properties of organometallic derivatives. *Mailing Add:* Union Carbide Corp Old Ridgebury Rd Danbury CT 06817

PROKIPCAK, JOSEPH MICHAEL, b Niagara Falls, Ont, Oct 22, 37; m 60; c 3. ORGANIC CHEMISTRY. *Educ:* Univ Windsor, BSc, 59, PhD(chem), 64. *Prof Exp:* Vis res assoc org chem, Ohio State Univ, 64-66; asst prof, 66-71, ASSOC PROF ORG CHEM, UNIV GUELPH, 71- *Mem:* Am Chem Soc; Chem Inst Can. *Res:* Physical organic-mechanisms. *Mailing Add:* Dept Chem Univ Guelph Guelph ON N1G 2W1 Can

PROKOP, JAN STUART, b Cleveland, Ohio, June 23, 34; m 57; c 3. COMPUTER SCIENCE. *Educ:* US Naval Acad, BS, 56; Naval Postgrad Sch, MS, 64; Univ NC, PhD(comput sci). *Prof Exp:* US Navy, 56-76, instr marine eng, US Naval Acad, 58-60, logistical adv to Venezuelan Navy, US Military Group Venezuela, 60-63, fel, Naval Postgrad Sch, 63-64, chief, Data Systs Support Off, US Navy, 64-69, asst dep comptroller, Dept Defense, Off Secy Defense, 69-72, dir, US Navy Comput Selection Off, 72-76; dir, Automatic Data Processing Mgt, US Dept Com, 76-79; assoc commr, US Social Security Admin, 79-80; SR VPRES, DATA RESOURCES INC, 80- *Concurrent Pos:* Fel, Univ NC, 66-69. *Mem:* Asn Comput Mach. *Res:* Graphical displays; statistical inventory control; simulation; natural and languages. *Mailing Add:* 1517 San Gabriel Way San Jose CA 95125-9343

PROKOP, MICHAEL JOSEPH, b Leigh, Nebr, May 6, 44; m 66; c 3. RUMINANT NUTRITION, ANIMAL SCIENCE. *Educ:* Univ Nebr, BS, 66, MS, 71, PhD(nutrit), 76. *Prof Exp:* Lab supvr nutrit, Univ Nebr, 67-73, dir nutrit lab, 73-74, staff res assoc animal sci, 74-76, asst prof, 76-77; ASST PROF ANIMAL SCI, IMPERIAL VALLEY FIELD STA, UNIV CALIF, 77- *Mem:* Am Soc Animal Sci; Am Inst Biol Scientists; Coun Agr Sci & Technol; Sigma Xi. *Res:* Improvement of beef cattle production with emphasis on feedlot nutrition; increased utilization of by-product feeds; reduced stress in management systems; relationship of environment and animal performance. *Mailing Add:* Imperial Valley Field Sta 1004 E Holton Rd El Centro CA 92243

PROKSCH, GARY J, b La Crosse, Wis, May 25, 44; m 64; c 2. BIOCHEMISTRY, CLINICAL CHEMISTRY. *Educ:* Univ Ariz, BS, 65; Univ Iowa, MS, 68, PhD(biochem), 70; Nat Registry Clin Chem, cert. *Prof Exp:* Asst prof clin path, Med Ctr, Ind Univ, Indianapolis, 70-80, assoc prof path, 80-84; vpres & owner, Anal Control Systs, 84-87; PRES & OWNER, CREATIVE LAB PRODS, INDIANAPOLIS, IND, 87- *Mem:* Am Chem Soc; Am Asn Clin Chemists. *Res:* Metabolism and regulatory role of human protease inhibitors; clinical chemistry methodolgy; development of optically clear lyophilized human serum base standards for clinical laboratory; isolated lipoproteins and lipid controls. *Mailing Add:* 7764 Hoover Rd Indianapolis IN 46260

PROMERSBERGER, WILLIAM J, b Littlefork, Minn, May 28, 12; m 38; c 2. AGRICULTURAL ENGINEERING. *Educ:* Univ Minn, BS, 35; Kans State Univ, MS, 41. *Prof Exp:* Instr agr eng, Univ Minn, 35-38; asst prof, NDak Col, 38-40; asst, Kans State Univ, 40-41; asst prof agr eng, 41-45, prof, 46-77, chmn dept, 41-77, EMER PROF AGR ENG, N DAK STATE UNIV, 77- *Concurrent Pos:* Vis prof, Univ Col, Dublin, 65-66. *Honors & Awards:* Massey-Ferguson Gold Medal, Am Soc Agr Engrs, 79. *Mem:* Fel Am Soc Agr Engrs; Am Soc Eng Educ. *Res:* Farm power and machinery; farm buildings. *Mailing Add:* 55-18th Ave N Fargo ND 58102

PROMISEL, NATHAN E, b Malden, Mass, June 20, 08; m 31; c 2. METALLURGICAL ENGINEERING, ELECTROCHEMISTRY. *Educ:* Mass Inst Technol, BS, 29, MS, 30; Yale Univ, Doctoral Work, 33. *Hon Degrees:* DEng, Mich Technol Univ, 78. *Prof Exp:* Res electrochemist & asst tech dir, Int Silver Co, 30-40; consult, 40-41; chief mat scientist, Bur Aeronaut, US Dept Navy, 41-59, chief mat scientist & dir mat div, Bur Naval Weapons, 59-66, mat adminr, US Dept Navy & chief mat engr, Naval Air Systs Command, 66; exec dir nat mat adv bd, 66-74, EMER EXEC DIR, NAT MAT ADV BD, NAT ACAD SCI, 74-; INT CONSULT, 74- *Concurrent Pos:* Mem & chmn struct & mat panel, NATO Adv Group, 59-71; US rep mat adv group, Orgn Econ Coop & Develop, 67-70; mem adv bds, Navy Labs, Univs, Govt & Nat Labs; chmn, US Working Group on Electromet in US/USSR Sci Exchange Prog, 73-77; mem, Cong Off Technol Assessment, Mat Adv Comn, 74-80 & Nat Mat Adv Bd, Nat Acad Sci, 77-80; consult, Oak Ridge Nat Lab, 75-78; chmn, Soc Automotive Engrs, 59-72. *Honors & Awards:* Carnegie lectr, 59; Sauver lectr, 60; Burgess Award, 61; Gillet lectr, 64; Burgess lectr, 67; Nat Captial Eng Year Award, Coun Eng & Archit Socs, 74; Decennial Award, Fedn Mat Socs, 82. *Mem:* Nat Acad Eng; hon mem Am Inst Mining, Metall & Petrol Engrs; fel & hon mem Am Soc Mat Int (pres, 71-72); Fedn Mat Socs (pres, 72-73); hon mem Am Soc Testing & Mat; fel Brit Inst Metals; fel Soc Advan Mats & Process Eng; hon mem Soc Automotive Engrs. *Res:* Aeronautical, space, missile and underwater vehicle materials sciences and processes, including disciplines of chemistry, electrochemistry, metallurgy; behavioristic phenomena of materials in earth, space and ocean environments; national materials policy, resource availability; author of over 70 related articles and books. *Mailing Add:* 12519 Davan Dr Silver Spring MD 20904

PROMNITZ, LAWRENCE CHARLES, b Berwyn, Ill, Nov 21, 44; m 78; c 1. FORESTRY, STATISTICS. *Educ:* Southern Ill Univ, BS, 67; Iowa State Univ, MS, 70, PhD(forest biomet), 72. *Prof Exp:* Assoc prof forestry & statist, Iowa State Univ, 71-77; res forester biomet, Crown Zellerbach Com Ltd, 77-81, BUS PLANNING, CROWN FOREST ENDUST, 82- *Mem:* Am Statist Asn. *Res:* Forest biometry; quantitative biology; dynamics of forest populations. *Mailing Add:* Fletcher Challenge Can Ltd Coast Wood Prods Div 6th Floor 815 W Hastings St PO Box 9502 Vancouver BC V6B 5K9 Can

PRONI, JOHN ROBERTO, b Hempstead, NY, June 5, 42; m 65; c 3. UNDERWATER ACOUSTICS, OCEANOGRAPHY. *Educ:* Univ Miami, BS, 63, MS, 65; NC State Univ, PhD(physics), 70. *Prof Exp:* Mem tech staff underwater acoust, Bell Tel Labs, 69-72; PHYSICIST & OCEANOGR, ATLANTIC OCEANOG & METEOROL LABS, NAT OCEANIC & ATMOSPHERIC ADMIN, 72- *Concurrent Pos:* Tech adv, Off Naval Res, 72- *Mem:* Am Geophys Union; Acoust Soc Am. *Res:* Environmental acoustics; acoustical studies of internal waves, sewage dumping, dredging operations, sewage outfalls and oceanic microstructure; artificial seawater. *Mailing Add:* 10020 SW 102 Ave Rd Miami FL 33176

PRONKO, PETER PAUL, b Peckville, Pa, Mar 29, 38; m 67; c 2. PHYSICS, MATERIALS SCIENCE. *Educ:* Univ Scranton, BS, 60; Univ Pittsburgh, MS, 62; Univ Alta, PhD(physics), 66. *Prof Exp:* Asst prof physics, Univ Scranton, 66-68; res assoc mat res, McMaster Univ, 68-72; res assoc physics, Argonne Nat Lab, 72-74, asst metallurgist physics, 74-75, physicist solid state physics, 75-80; sr scientist, 80-82, chief scientist, 82-84, DIR MATS RES DIV, UNIVERSAL ENERGY SYSTS, 84- *Mem:* Am Phys Soc; Mat Res Soc; Inst Elec & Electronics Engrs; Am Mgt Asn. *Res:* Ion beam interactions in solids; laser solid interactions; metallurgy and materials science of ion implanted solids; advanced surface treatment of materials; semiconductor wafer engineering. *Mailing Add:* Universal Energy Systs 4401 Dayton-Xenia Rd Dayton OH 45432

PRONOVE, PACITA, pediatrics, for more information see previous edition

PROOPS, WILLIAM ROBERT, b Brooklyn, NY, Dec 5, 29; m 52, 75; c 4. ORGANIC CHEMISTRY. *Educ:* Polytech Inst Brooklyn, BS, 49; Columbia Univ, AM, 51, PhD(chem), 53. *Prof Exp:* Res chemist, Union Carbide Chem Co, 52-71; res & develop mgr, 71-77, VPRES RES & DEVELOP, ISOCYANATE PROD DIV, WITCO CHEM CORP, 77- *Mem:* Am Chem Soc. *Res:* Synthetic organic and polymer chemistry. *Mailing Add:* WITCO Chem Corp 3230 Broofield St Houston TX 77045-6610

PROPES, ERNEST (A), b Jacksonville, Ala, Feb 5, 09; m 33; c 4. ALGEBRA. *Educ:* Univ Ala, BS, 37, MA, 39. *Prof Exp:* Instr high sch, Ala, 36-37; instr math, Univ Ala, 39-42; asst prof, Birmingham-Southern Col, 43-44; instr, Univ Ill, 45-48; assoc prof, 48-74, prof math, 74-80, RES PROF MATH, LLANO ESTACADO CTR ADVAN PROF STUDIES & RES, EASTERN NMEX UNIV, 80-, MEM RES INST, 48- *Concurrent Pos:* Consult, Sandia Corp. *Mem:* Am Math Soc; Math Asn Am. *Res:* Groups, rings, fields in modern algebra. *Mailing Add:* 1325 W 17th Lane Portales NM 88130

PROPHET, CARL WRIGHT, b Anthony, Kans, Sept 20, 29; m 51; c 3. AQUATIC ECOLOGY. *Educ:* Kans State Teachers Col, BS, 55, MS, 57; Univ Okla, PhD(zool), 62. *Prof Exp:* Jr high sch teacher, 49-51; from instr to assoc prof, 56-58, PROF BIOL, EMPORIA STATE UNIV, 68- *Mem:* Am Soc Limnol & Oceanog; NAm Benthological Soc. *Res:* Limnology of artificial impoundments; invertebrate ecology; impacts of planktivory on zooplankton community structure. *Mailing Add:* Div of Biol Sci Emporia State Univ Emporia KS 66801

PROPP, JACOB HENRY, b Ft Morgan, Colo, Nov 28, 41; m 63; c 1. ANALYTICAL CHEMISTRY. *Educ:* Univ Colo, BA, 63; Univ Minn, MS, 65; Univ Ill, PhD, 68. *Prof Exp:* asst prof, 68-77, ASSOC PROF ANALYTICAL CHEM, UNIV WIS-OSHKOSH, 77- *Concurrent Pos:* Dir, Med Technol Prog, Univ Wis, Oshkosh, 75- *Mem:* Sigma Xi; Am Soc Med Technol; Am Chem Soc. *Res:* Electrochemical methods of trace analysis; environmental analysis of pollutants. *Mailing Add:* 1936 Hazel St Oshkosh WI 54901

PROPPE, DUANE W, b Walla Walla, Wash, Jan 13, 45; m 62, 67; c 3. CARDIOVASCULAR PHYSIOLOGY. *Educ:* Univ Wash, Seattle, PhD(physiol & biophys), 74. *Prof Exp:* ASSOC PROF PHYSIOL, UNIV TEX HEALTH SCI CTR, SAN ANTONIO, 88- *Concurrent Pos:* Prin investr, NIH Grants, 77- *Mem:* Am Physiol Soc; Sigma Xi. *Res:* Investigates mechanisms of cardiovascular system control during environmental heating and/or altered body fluid balance. *Mailing Add:* Dept Physiol Univ Tex Health Sci Ctr 7703 Floyd Curl Dr San Antonio TX 78284-7756

PROPST, CATHERINE LAMB, b Charlotte, NC, Mar 10, 46. MOLECULAR BIOLOGY. *Educ:* Vanderbilt Univ, BA, 67; Yale Univ, MPhil, 70, PhD(microbiol), 73. *Prof Exp:* Instr genetics & microbiol, Miss State Col Women, 67-68; asst prof chem, Quinnipiac Col, 73-74; head microbiol, GTE, Inc, 74-77; mgr microbiol & inspection qual assurance, Abbott Labs, 77-78, mgr biol develop & advan technol, 78-80, head, Microbiol & Molecular Biol, 80; vpres res & develop, diagnostics, Ayerst Labs, 80-83; vpres res & develop worldwide, Flow Gen Inc, 83-85; PRES & CHIEF EXEC OFFICER, AFFIL SCI INC, 85- *Concurrent Pos:* Vis genetics prof, Univ Ill, Chicago, 89- *Mem:* Am Soc Microbiol; Soc Indust Microbiol; Nat Wildlife Fedn; Sigma Xi. *Res:* New diagnostic and pharmaceutical products research and development; applied research in recombinant DNA, interferon, cell growth and aging. *Mailing Add:* 24794 W Nippersink Rd Round Lake IL 60073

PROPST, FRANKLIN MOORE, b Ohatchee, Ala, Mar 30, 35; m 60; c 1. PHYSICS. *Educ:* Ala Polytech Inst, BS, 57; Univ Ill, Urbana, MS, 59, PhD(physics), 62. *Prof Exp:* Res assoc, 62-63, asst prof, 63-70, assoc prof physics, coordr sci lab, 70-77, PROF COMPUT BASED EDUC RES LAB & ASSOC PLANNING UNIV ILL, URBANA, 78- *Concurrent Pos:* Ed, J Vacuum Sci & Technol, 66- *Mem:* Am Phys Soc; Am Vacuum Soc. *Res:* Surface physics measurements, specifically studies of interactions of electrons, ions and neutral particles with solid surfaces; vacuum technology; solid state. *Mailing Add:* 501 W Michigan Urbana IL 61801

PROSCHAN, FRANK, b New York, Apr 7, 21; m 52; c 2. STATISTICS. *Educ:* City Col New York, BS, 41; George Wash Univ, MA, 48; Stanford Univ, PhD, 59. *Prof Exp:* Lab asst cement testing, Nat Bur Standards, 41-42, mathematician, 51-52; geod computer, US Geol Surv, 42-46; res analyst, US Army Security Agency, 46-49; statistician, AEC, 49-51 & Sylvania Elec Prod Inc, 52-60; mem staff, Sci Res Lab, Boeing Co, 60-70; PROF STATIST, FLA STATE UNIV, 70- *Concurrent Pos:* Consult, US Army Materiel Command, 70- & Aberdeen Proving Ground, 72-73. *Honors & Awards:* Wilks Medal, Am Statist Asn; Elected Mem, Int Statist Inst. *Mem:* Int Statist Inst; fel Am Statist Asn; fel Inst Math Statist. *Res:* Mathematical inequalities; reliability theory using probability and statistics. *Mailing Add:* Dept of Statist Fla State Univ Tallahassee FL 32306

PROSE, PHILIP H, pathology; deceased, see previous edition for last biography

PROSEN, HARRY, b Saskatoon, Sask, June 27, 30; m 54; c 4. PSYCHIATRY. *Educ:* Univ Man, MD, 55, MSc, 57; FRCP(C). *Prof Exp:* Am Fund Psychiat bursary, 58-60; assoc prof psychiat, 65-71, co-dir grad training prog, 65-70, chmn postgrad educ comt psychiat, 70-75, actg head dept psychiat, 71-72, PROF PSYCHIAT, UNIV MAN, 71- HEAD DEPT PSYCHIAT, 75-; CHIEF PSYCHIAT, HEALTH SCI CTR, WINNIPEG, 75- *Concurrent Pos:* Consult, Deer Lodge Vet Hosp, Winnipeg, 59-66; dir psychiat out-patient dept, Winnipeg Gen Hosp, 59-68. *Mem:* Fel AAAS; fel Am Psychiat Asn; Can Psychiat Asn; fel Am Col Psychiatrists; fel Royal Col Psychiatrists Eng. *Res:* Life cycle; psychotherapy; administration. *Mailing Add:* Dept Psychiat Med Col Wis 9455 Watertown Plank Milwaukee WI 53226

PROSKAUER, ERIC S, b Frankfurt, Ger, Mar 19, 03; US citizen; m 31. PHYSICAL CHEMISTRY. *Educ:* Univ Leipzig, PhD(phys chem), 33. *Hon Degrees:* DSc, Polytechnic, New York, 84. *Prof Exp:* Ed, Akademische Verlagsgesellschaft, Ger, 25-37; secy-treas, Nordeman Publ Co, NY, 37-40; chmn, Intersci Publ, Inc, 40-61; vpres & mgr, Wiley-Intersci Div, 61-70, sr vpres, 70-73, dir, 70-81, EMER DIR, JOHN WILEY & SONS, INC, 81- *Concurrent Pos:* Consult, 73-87. *Mem:* AAAS; Am Chem Soc; Am Phys Soc; fel NY Acad Sci. *Res:* Publishing. *Mailing Add:* 220 Central Park S New York NY 10019

PROSKE, JOSEPH WALTER, b DuBois, Pa, Oct 11, 36; m 66; c 2. CERAMICS, METALLURGY. *Educ:* Pa State Univ, BS, 58, MS, 60, PhD(metall), 64. *Prof Exp:* Res engr, Carbon Co, 64-68, asst dir magnetic res, 68-69, dir, 69-71, sr res engr, Magnet Div, 71-73, SR RES ENGR, ELECTRONIC DIV, STACKPOLE CORP, 73- *Mem:* Am Ceramic Soc; Am Soc Metals; Am Inst Mining, Metall & Petrol Engrs; Inst Elec & Electronics Engrs; Sigma Xi. *Res:* Infiltration in metal-ceramic systems; properties and microstructure of magnetic ceramics; hard and soft ferrite materials development and characterization. *Mailing Add:* Stonehedge Meadows Ridgeway PA 15853

PROSKUROWSKI, WLODZIMIERZ, b Warsaw, Poland, Jan 19, 36; Swed citizen; m 74; c 1. NUMERICAL ANALYSIS, SCIENTIFIC COMPUTING. *Educ:* Warsaw Univ, MSc, 69; Royal Inst Technol, Stockholm, Sweden, PhD(numerical anal), 76. *Prof Exp:* Res assoc appl math, Lawrence Berkeley Lab, 76-78; asst prof, 78-84, ASSOC PROF MATH, UNIV SOUTHERN CALIF, LOS ANGELES, 84- *Concurrent Pos:* Chair Los Angeles chap, spec interest group numerical math, Asn Comput Mach, 81-84; chair Southern Calif chap, Soc Indust & Appl Math, 84-86. *Res:* Numerical methods for partial differential equations; scientific computing and mathematical software; large scale problems; sequential and parallel partial differential equation solvers. *Mailing Add:* Math Dept DRB 155 Univ Southern Calif Los Angeles CA 90089-1113

PROSKY, LEON, b New York, NY, Aug 2, 33; m 73; c 2. NUTRITIONAL BIOCHEMISTRY. *Educ:* Brooklyn Col, BS, 54; Rutgers Univ, MS, 55, PhD(biochem), 58. *Prof Exp:* Fel pharmacol, Sch Med, Washington Univ, 60-62; res assoc biochem, Albert Einstein Med Col, 63-65; sect chief biosynthesis, 65-67, asst br chief div nutrit, 67-74, prog mgr nutrit, 74-79, chief, proteins, amino acids & carbohydrates sect, 71-79, DEP CHIEF, EXP NUTRIT BR, CFSAN FOOD & DRUG ADMIN, 79- *Concurrent Pos:* Adj prof biochem, George Washington Univ Med Ctr, 80- *Honors & Awards:* Award of Merit, FDA, 82; Asn Referee of the Year Award, Asn Off Anal Chemists, 86. *Mem:* Am Inst Nutrit; fel AAAS; Am Chem Soc; fel Am Inst Chemists; Soc Exp Biol & Med. *Res:* Carbohydrate metabolism; protein synthesis and metabolism; vitamin-enzyme interactions; nutritional status and drug efficacy; nucleic acids; nutrition and development; brain biochemistry. *Mailing Add:* Div Nutrit Food & Drug Admin 200 C St SW Washington DC 20204

PROSNITZ, DONALD, US citizen. PHYSICS. *Educ:* Yale Univ, BSc, 70; Mass Inst Technol, PhD(physics), 75. *Prof Exp:* Asst prof eng, Yale Univ, 75-77; PHYSICIST, LAWRENCE LIVERMORE LAB, 77- *Mem:* Sigma Xi; AAAS. *Mailing Add:* 1910 Whitecliff Ct Walnut Creek CA 94596

PROSS, HUGH FREDERICK, b Wales, UK, Nov 8, 42; m 65; c 3. TUMOR BIOLOGY. *Educ:* Queen's Univ, BA, 63, MSc & MD, 68, PhD(microbiol & immunol), 72. *Prof Exp:* Intern med, Med Ctr, Duke Univ, 68-69; fel tumor biol, Karolinska Inst Stockholm, 72-74; from asst prof to assoc prof radiation oncol, Queen's Univ, 74-83, assoc prof microbiol & immunol, 79-83, head, Microbiol & Immunol, 86-90, PROF ONCOL, MICROBIOL & IMMUNOL, QUEEN'S UNIV, 83-, ASSOC DEAN, UNDERGRAD MED EDUC, 90- *Concurrent Pos:* Res assoc, Ont Cancer Treatment & Res Found, 74-86. *Mem:* Can Soc Immunol; Am Asn Immunologists; NY Acad Sci; Am Asn Med Cols. *Res:* Immunology and tumor biology; human natural killer cells. *Mailing Add:* Microbiol & Immunol Queens Univ Kingston ON K7L 3N6 Can

PROSSER, CLIFFORD LADD, b Avon, NY, May 12, 07; m 34; c 3. COMPARATIVE PHYSIOLOGY. *Educ:* Univ Rochester, AB, 29; Johns Hopkins Univ, PhD(zool), 32. *Hon Degrees:* DSc, Clark Univ, 75. *Prof Exp:* Asst zool, Johns Hopkins Univ, 29-32; Parker fel physiol, Harvard Med Sch & Cambridge Univ, 32-34; res assoc, Clark Univ, 34-35, from asst prof to assoc prof physiol, 35-39; from asst prof to assoc prof zool, 39-49, prof physiol, 49-75, head dept, 60-70, EMER PROF PHYSIOL, UNIV ILL, URBANA, 75- *Concurrent Pos:* Mem physiol staff, Marine Biol Lab, Woods Hole, 35-41, trustee, 50-; biologist & assoc sect chief metall lab, Univ Chicago, 43-46; Hopkins lectr, Stanford Univ, 54; Carnegie prof, Univ Hawaii, 57; Guggenheim fel, 63-64; vis prof, Monash Univ Australia, 71-72. Consult, Nat Defense Res Comt Proj, 42-43 & NSF, 51-; mem, Macy Conf Nerve Impulse; Maytag prof zool, Ariz State Univ, 78. *Mem:* Nat Acad Sci; fel Am Acad Arts & Sci; AAAS (pres zool sci sect, 65); Am Soc Zoologists (pres, 61); Am Physiol Soc (pres, 69-70); Soc Gen Physiologists. *Res:* Nerve and muscle physiology; comparative animal physiology; physiological adaptation; smooth muscle; editor and author Comparative Animal Physiology, 4 editions. *Mailing Add:* Dept Physiol Univ Ill Urbana IL 61801

PROSSER, FRANCIS WARE, JR, b Wichita, Kans, June 30, 27; m 52; c 3. NUCLEAR PHYSICS. *Educ:* Univ Kans, BS, 50, MS, 54, PhD(physics), 55. *Prof Exp:* Res assoc physics, Rice Inst, 55-57; from asst prof to assoc prof, 57-67, PROF PHYSICS, UNIV KANS, 67- *Concurrent Pos:* Nat Res Coun-Off Aerospace Res sr assoc, Aerospace Res Labs, Wright-Patterson AFB, 69-70; fac res partic, Argonne Nat Lab, 75-76, vis scientist, 77, 78, 85, 86 & 87. *Mem:* Fel Am Phys Soc; Am Asn Physics Teachers; Am Asn Univ Professors; NY Acad Sci. *Res:* Nuclear structure; heavy-ion reactions; gamma ray spectroscopy; angular correlations. *Mailing Add:* Dept Physics Univ Kans Lawrence KS 66045

PROSSER, FRANKLIN PIERCE, b Atlanta, Ga, July 4, 35; m 60; c 2. COMPUTER SCIENCE. *Educ:* Ga Inst Technol, BS, 56, MS, 58; Pa State Univ, PhD(chem), 61. *Prof Exp:* NSF fel, 61-63, lectr, Comput Ctr, 63-67, asst dir, 67-71, assoc prof, 69-77, chmn dept, 71-77, 87-, assoc dir, Ctr, 71-80, PROF COMPUT SCI, IND UNIV, BLOOMINGTON, 77-; VPRES, LOGIC DESIGN, INC, 82- *Mem:* Asn Comput Mach; Inst Elec & Electronics Engrs. *Res:* Digital hardware design; programming systems and languages; operating systems; computer science education. *Mailing Add:* Dept of Comput Sci Ind Univ Bloomington IN 47405

PROSSER, REESE TREGO, b Minneapolis, Minn, May 18, 27; m 52; c 4. MATHEMATICS. *Educ:* Harvard Univ, AB, 49; Univ Calif, PhD(math), 55. *Prof Exp:* Mathematician, Lawrence Radiation Lab, Univ Calif, 54-55; res instr math, Duke Univ, 55-56; Moore instr, Mass Inst Technol, 56-58, mem staff, Lincoln Lab, 58-62, group leader, 62-66; assoc prof math, 66-68, vchmn dept, 67-70, chmn sci div, 71-72, PROF MATH, DARTMOUTH COL, 68- *Concurrent Pos:* Ed, Duke Math J, 64-72. *Mem:* Am Math Soc; Am Phys Soc; Asn Comput Mach; Soc Indust & Appl Math; sr mem Inst Elec & Electronics Engrs. *Res:* Functional analysis. *Mailing Add:* Dept Math & Comput Sci Dartmouth Col Hanover NH 03755

PROSSER, ROBERT M, b Philadelphia, Pa, Apr 3, 23. ORGANIC CHEMISTRY. *Educ:* Yale Univ, BS, 43. *Prof Exp:* Supvr prod, E I Du Pont De Nemours & Co, Inc, 47-49, engr, 49-52, sr supvr develop, 52-55, supt prod, 55-58, supt develop, 58-65, gen supt, 65-67, mgr res & develop, 67-70, mgr new ventures, 70-72, prod mgr neoprene res & develop, 72-81; RETIRED. *Mem:* Am Chem Soc; AAAS. *Res:* Product and process development on dyes and intermediates; neoprene, isocyanates; urethanes; fluoroelastomers; rubber chemicals. *Mailing Add:* 2205 W 11th St Wilmington DE 19805-2603

PROSSER, THOMAS JOHN, b Chicago, Ill, Feb 17, 31; m 56; c 2. ORGANIC CHEMISTRY. *Educ:* Ill Inst Technol, BS, 53; Univ Notre Dame, PhD(org chem), 57. *Prof Exp:* Res chemist, 56-73, SR RES CHEMIST, HERCULES INC, 73- *Mem:* Am Chem Soc. *Res:* Synthetic organic chemistry; chemistry of rosin and rosin derivatives; synthetic resins; polymerization, copolymerization, polymer modification and characterization; cycloaddition reactions. *Mailing Add:* Five Galaxy Dr Newark DE 19711

PROSTAK, ARNOLD S, b Brooklyn, NY, Apr 8, 29; m 71; c 2. CHEMICAL INSTRUMENTATION. *Educ:* Col William & Mary, BS, 50; Johns Hopkins Univ, MA, 55; Univ Mich, Ann Arbor, MS, 60, PhD(chem), 69. *Prof Exp:* Chemist, Edgewood Arsenal, US Army, 50-58; res assoc infrared technol, Willow Run Labs, Univ Mich 58-60; sr res & develop engr, Aerospace Systs Div, Bendix Corp, 60-66 & 68-71; develop engr, 72-80, STAFF PROJ ENGR, VEHICLE EMISSION LAB, GEN MOTORS PROVING GROUND, MILFORD, 80- *Mem:* AAAS; Am Chem Soc; Soc Automotive Engrs. *Res:* Detection and measurement of trace materials in air using chemical and infrared techniques; infrared technology; gas chromatography; electrochemical and optical instrumentation. *Mailing Add:* 1707 Harding Ann Arbor MI 48104-4541

PROTHERO, DONALD ROSS, b Glendale, Calif, Feb 21, 54. VERTEBRATE PALEONTOLOGY, MAGNETOSTRATIGRAPHY. *Educ:* Univ Calif, Riverside, BA, 76; Columbia Univ, MA, 78, MPhil, 79, PhD(geol sci), 82. *Prof Exp:* Instr geol, Vassar Col, 79-81; asst prof geol, Knox Col, 82-85; ASST PROF GEOL, OCCIDENTAL COL, 85- *Honors & Awards:* Guggenheim Fel, 88-89. *Mem:* Sigma Xi; Soc Vert Paleont; Am Geophys Union; Soc Syst Zool; Paleont Soc; Geol Soc Am. *Res:* Climatic and faunal changes during the late Eocene and Oligocene, as recorded in the land mammals and marine record; magnetostratigraphic correlation of Tertiary rocks; patterns of evolution and systematics of fossil mammals, especially rhinos, camels and horses; petrography and provenance of chert artifacts from the Northeastern United States. *Mailing Add:* Dept Geol Occidental Col Los Angeles CA 90041

PROTHERO, JOHN W, b Toronto, Ont, Feb 5, 32; m 55; c 3. BIOPHYSICS, BIOLOGICAL STRUCTURE. *Educ:* Univ Western Ont, BSc, 56, PhD(biophys), 60. *Prof Exp:* Med Res Coun Can fel molecular biol, Royal Inst, London, Eng, 61-62 & Lab Molecular Biol, Cambridge Univ, 62-64; res assoc biophys, Mass Inst Technol, 65; asst prof, 65-69, ASSOC PROF BIOL STRUCT, UNIV WASH, 70- *Mem:* Can Asn Physicists; Am Soc Zoologists; Inst Elec & Electronic Engrs. *Res:* Morphometrics; morphogenesis; models; scaling. *Mailing Add:* Dept Biol Struct SM-20 Univ Wash Seattle WA 98195

PROTHEROE, WILLIAM MANSEL, b Wales, UK, Oct 16, 25; nat US; m 49; c 4. ASTRONOMY. *Educ:* Ohio State Univ, BSc, 50, PhD(astron), 55. *Prof Exp:* Res assoc astron, Ohio State Univ, 53-55; from asst prof to prof, Univ Pa, 55-67, vdean grad sch, 60-66; assoc dean grad sch, 67-70, PROF ASTRON, OHIO STATE UNIV, 67- *Concurrent Pos:* Consult, Franklin Inst, 58-66; Guggenheim fel, 66-67. *Mem:* AAAS; Am Astron Soc; Int Astron Union; Royal Astron Soc. *Res:* Photoelectric photometry; eclipsing binary stars. *Mailing Add:* Dept Astron Ohio State Univ Columbus OH 43210

PROTHRO, JOHNNIE W, b Atlanta, Ga, Feb 26, 22; m 49, 64; c 1. FOODS, NUTRITION. *Educ:* Spelman Col, BS, 41; Columbia Univ, MS, 46; Univ Chicago, PhD, 52. *Prof Exp:* High sch instr, 41-45; instr chem, Southern Univ, 47-48; from asst prof to prof, Tuskegee Inst & res assoc, Carver Found, 52-63; assoc prof, Univ Conn, 63-68; prof nutrit & res, Tuskegee Inst, 68-72; nutrit adv, Cent Dis Control, Dept Health, Educ & Welfare, Ga, 72-73; prof nutrit, Tuskegee Inst, 73-75; prof nutrit, Emory Univ, 75-79; prof nutrit & dietetics, Ga State Univ, 79-89; RETIRED. *Concurrent Pos:* NIH spec fel, Univ Calif, Los Angeles, 58-59. *Mem:* Am Inst Nutrit; Inst Food Technologists. *Res:* Physiological availability and patterns of amino acids; presence of nonantiscorbutic reducing substances in baked foods; protein utilization in adolescent male. *Mailing Add:* 919 Falcon Dr SW Atlanta GA 30311

PROTOPOPESCU, SERBAN DAN, b Yacuma-Beni, Bolivia, Oct 25, 45; m 72. EXPERIMENTAL HIGH ENERGY PHYSICS. *Educ:* Princeton Univ, BA, 68; Univ Calif, Berkeley, PhD(physics), 72. *Prof Exp:* Res asst high energy physics, Lawrence Berkeley Lab, 69-72; from asst physicist to assoc physicist, 72-76, PHYSICIST, BROOKHAVEN NAT LAB, 77- *Concurrent Pos:* Vis scientist, Lab de l'Accelerator Lineaire, Univ Paris-Sud, Orsay, 76-77; assoc prof, Univ NY, Stony Brook. *Mem:* Am Phys Soc; AAAS; NY Acad Sci. *Res:* Study of production and decay of meson resonances; partial wave analysis; high energy pp collisions. *Mailing Add:* Dept Physics Bldg 510A Brookhaven Nat Lab Upton NY 11973

PROTTER, MURRAY HAROLD, b New York, NY, Feb 13, 18; m 45; c 2. MATHEMATICS. *Educ:* Univ Mich, AB, 37, AM, 38; Brown Univ, PhD(math), 46. *Prof Exp:* Instr math, Ore State Col, 38-40; appl mathematician, David Taylor Model Basin, US Dept Navy, Washington, DC, 42-43 & Chance Vought Aircraft, Conn, 43-45; instr math, Brown Univ, 45-47; asst prof, Syracuse Univ, 47-51; mem, Inst Advan Study, Princeton, 51-53; assoc prof, 53-58, chmn dept, 62-65, PROF MATH, UNIV CALIF, BERKELEY, 58- *Mem:* Am Math Soc; Math Asn Am; Soc Indust & Appl Math. *Res:* Partial differential equations; fluid dynamics. *Mailing Add:* Univ Calif Berkeley CA 94720

PROTTER, PHILIP ELLIOTT, b Syracuse, NY, Aug 19, 49; m. PURE MATHEMATICS. *Educ:* Yale Univ, BA, 71; Univ Calif, San Diego, MA, 73, PhD(math), 75. *Prof Exp:* Asst prof math, Duke Univ, 75-78; ast prof math & Statist, 78-81, assoc prof math, 81-88, PROF MATH, PURDUE UNIV, 88- *Concurrent Pos:* NSF fel, Nat Ctr Sci Res, France, 79-80; mem, Inst Advan Study, 77-78; assoc prof, Univ Rennes, France, 81-82; vis prof, Univ Wisc, 85-86 & Univ Roven, 87-88. *Mem:* Am Math Soc; Inst Math Statist; AAAS; Soc Indust & Appl Math. *Res:* Probability theory, in particular stochastic integration and stochastic differential equations. *Mailing Add:* Dept Math Purdue Univ West Lafayette IN 47907-1395

PROTZ, RICHARD, b Willowbrook, Sask, Aug 8, 34; m 59; c 7. SOIL SCIENCE. *Educ:* Univ Sask, BSA, 56, MSc, 60; Iowa State Univ, PhD(soils), 65. *Prof Exp:* Soil surveyor, Univ Sask, 56-60; res assoc soils, Iowa State Univ, 60-65; res officer, Can Dept Agr, Guelph, 65-66; from asst prof to assoc prof, 66-74, PROF SOIL SCI, UNIV GUELPH, 74- *Concurrent Pos:* Consult soil sci, Malaysia, Egypt, Saudi Arabia, Nigeria, Indonesia, Jamaica & Kenya. *Mem:* Soil Sci Soc Am; Am Soc Agr Sci; Can Soc Soil Sci; Agr Inst Can; Int Soil Sci Soc; Am Asn Adv Sci. *Res:* Soil genesis; morphology and classification; remote sensing; soil micromorphology; image analysis. *Mailing Add:* Land Resource Sci Univ of Guelph Guelph ON N1G 2W1 Can

PROUDFIT, CAROL MARIE, b Cincinnati, Ohio, Nov 22, 37; div; c 1. PHARMACOLOGY, REPRODUCTIVE PHYSIOLOGY. *Educ:* Univ Cincinnati, BS, 59; Univ Kans, Kansas City, PhD(med physiol), 71; Rosary Col, River Forest, IL, MBA, 86. *Prof Exp:* Trainee reprod physiol, Univ Ill, 71-74; sr scientist endocrine pharmacol, 76-88, ASST DIV DIR & DRUG EVALUATIONS PROJ MGR, AMA, 88- *Concurrent Pos:* Lectr physiol, Triton Col, 74 & Northwestern Univ, 75; lectr pharmacol, Med Ctr, Univ Ill, 74- *Mem:* Am Fertil Soc; Am Women Sci; Am Soc Clin Pharmacol & Therapeut; Drug Info Asn; Am Diabetes Assoc. *Res:* Endocrine pharmacology. *Mailing Add:* Div Drugs & Toxicol AMA 515 N State St Chicago IL 60610

PROUDFIT, HERBERT K(ERR), b Mar 6, 40; c 1. NEUROPHARMACOLOGY. *Educ:* Univ Kans, PhD(pharmacol), 71. *Prof Exp:* PROF PHARMACOL, COL MED, UNIV ILL, 73- *Mem:* Am Soc Pharmacol Exp Therapeut. *Res:* Opiates; pain perception. *Mailing Add:* Col Med Univ Ill 901 S Wolcott St Chicago IL 60680

PROUDFOOT, BERNADETTE AGNES, b Jersey City, NJ, June 22, 21. BIOLOGY. *Educ:* Caldwell Col Women, BA, 52; St Thomas Inst, MS, 59, PhD(biol, exp med), 72. *Prof Exp:* Lectr, 55-56, from instr to assoc prof, 56-72, chmn dept, 72-83, PROF BIOL, CALDWELL COL, 72- *Mem:* NY Acad Sci. *Res:* Microrespiration studies on neoplasms and normal tissues; offsetting toxicity of neoplastic drugs. *Mailing Add:* Dept Biol Caldwell Col Caldwell NJ 07006

PROUGH, RUSSELL ALLEN, b Twin Falls, Idaho, Nov 5, 43; m 65; c 2. BIOCHEMISTRY. *Educ:* Col Idaho, BSc, 65; Ore State Univ, PhD(biochem), 69. *Prof Exp:* Vet Admin fel, Vet Admin Hosp, Kansas City, Mo, 69-72; from instr to prof biochem, Univ Tex Health Sci Ctr, Dallas, 72-86; PROF & CHMN, BIOCHEM, UNIV LOUISVILLE SCH OF MED, 86- *Mem:* Am Soc Biol Chemists; Am Soc Pharmacol & Exp Therapeut; Am Asn Cancer Res; Sigma Xi. *Res:* Hydrazine and azo metabolism; regulation of monooxygenesis and drug/carcinogen metobolizing enzymes. *Mailing Add:* Dept of Biochem Univ Louisville Sch Med Louisville KY 40292

PROULX, PIERRE R, b Montreal, Que, June 27, 38; m 61; c 3. BIOCHEMISTRY. *Educ:* McGill Univ, BSc, 59, MSc, 60, PhD(biochem), 62. *Prof Exp:* Lectr biochem, McGill Univ, 62-63; from lectr to assoc prof, 63-73, PROF BIOCHEM, UNIV OTTAWA, 73- *Concurrent Pos:* Med Res Coun Can fel, 66-67 & scholar, 67-72. *Mem:* Can Biochem Soc. *Res:* Metabolism of phospholipids and effects of lipid-soluble vitamins on tumor cells; tumor cell differentiation. *Mailing Add:* Dept of Biochem Univ of Ottawa Ottawa ON K1H 8M5 Can

PROUT, CURTIS, b Swampscott, Mass, Oct 13, 15. COLLEGE STUDENT HEALTH, MEDICINE. *Educ:* Harvard Col, BA, 37; Harvard Med Sch, MD, 41; Am Bd Internal Med, dipl, 48. *Prof Exp:* Intern path, Boston City Hosp, 41; intern med, Peter Brent Brigham Hosp, 42-43; med resident, Johns Hopkins Hosp, 43-44; med res fel, Mass Gen Hosp, 44-45; asst med, Peter Brent Brigham Hosp, 47-48, jr assoc, 49-54, assoc, 54-60, sr assoc, 60-82, sr physician, Brigham & Women's Hosp, 82; RETIRED. *Concurrent Pos:* Assoc clin prof med, Harvard Med Sch, 72-85, asst dean student affairs, 84-88, lectr med, 85; chief med & dir med educ, USPHS, Brighton, Mass, 80; bd dirs, Med Found, Boston, Mass, 81, Nat Comn Correctional Health Care, 82, chmn-elect, 89. *Mem:* Inst Med-Nat Acad Sci; AMA; fel Am Col Physicians; Am Col Health Asn (pres, 65-66); Am Clin & Climat Asn. *Res:* Author and co-author of various medical subjects; college health; longevity of athletes; illnesses peculiar to prison inmates. *Mailing Add:* 319 Longwood Ave Boston MA 02115

PROUT, FRANKLIN SINCLAIR, b Washington, DC, Jan 10, 20; m 55; c 7. ORGANIC CHEMISTRY. *Educ:* Univ Colo, AB, 41; Vanderbilt Univ, BS, 43, PhD(chem), 47. *Prof Exp:* Asst, Off Sci Res & Develop, Vanderbilt Univ, 43-45; res assoc, Northwestern Univ, 47-48; from instr to assoc prof, 48-80, PROF CHEM, DEPAUL UNIV, 80- *Mem:* Am Chem Soc. *Res:* Organic syntheses of aliphatic, optically active compounds. *Mailing Add:* 339 S Crescent Ave Park Ridge IL 60068-4109

PROUT, GEORGE RUSSELL, JR, b Boston, Mass, July 23, 24; m 50; c 2. UROLOGY. *Educ:* Albany Med Col, MD, 47. *Hon Degrees:* MA, Harvard Univ, 70; DSc, Albany Med Col, 90. *Prof Exp:* Instr surg, New York Hosp, 52-54; asst clinician urol, Sloan-Kettering Inst, 54-58; asst prof urol & chmn dept, Sch Med, Univ Miami, 57-60; from assoc prof to prof, Med Col, Va, 60-69, chmn dept, 60-69; prof, 69-89, EMER PROF SURG, HARVARD MED SCH, 89-; Chief urol serv, 69-89, CONSULT UROL, MASS GEN HOSP, 89- *Concurrent Pos:* Spec fel, Mem Ctr Cancer & Allied Dis, 54-55, asst attend, 56-57; asst clinician, James Ewing Hosp, 56-57; consult, McGuire Vet Admin Hosp, Richmond, Va, 61-69 & Crippled Children's Hosp, 63-69; chmn study group adjuvants in surg treatment bladder cancer, Nat Cancer Inst, 65-; chmn, Nat Bladder Cancer Collab Group A, NIH, 73-86. *Mem:* Am Urol Asn; fel Am Col Surg; Soc Pelvic Surg; Am Asn Genitourinary Surg; Soc Peidat Urol. *Res:* Urological neoplasms. *Mailing Add:* Mass Gen Hosp 32 Fruit St Boston MA 02114

PROUT, JAMES HAROLD, b Oak Park, Ill, Dec 23, 27; m 54; c 1. APPLIED PHYSICS, ACOUSTICS. *Educ:* Purdue Univ, BS, 52; Univ Mich, MS, 58. *Prof Exp:* Res asst physics, Willow Run Labs, Univ Mich, 53-57, res assoc, 57-61; asst prof, Pa State Univ, Univ Park, 61-67, assoc prof eng res, 61-91; RETIRED. *Concurrent Pos:* Consult mem, Standards Writing Group, 53-57. *Mem:* Acoust Soc Am. *Res:* Air and waterborne acoustic signal analysis and data acquisition systems; design and testing of ear protectors, noise monitoring systems and speech communication systems; audiological electronics; ultrasonic characterization of materials; co-author of two books. *Mailing Add:* Appl Res Lab Pa State Univ PO Box 30 State College PA 16804

PROUT, THADDEUS EDMUND, b Owings, Md, Dec 8, 23; m 50; c 3. MEDICINE. *Educ:* Harvard Med Sch, MD, 48; Am Bd Internal Med, dipl, 56. *Prof Exp:* Asst med, Harvard Med Sch, 49-50; asst, Sch Med, Boston Univ, 53-54; sr fel, 55-58, asst prof, 58-64, assoc prof med, Sch Med, Johns Hopkins Univ, 64-; CHIEF MED, GREATER BALTIMORE MED CTR, 64-; DIR, MD METAB INST, 87- *Concurrent Pos:* Resident, Boston Vet Admin Hosp, 53-54; dir diabetes training prof & assoc physician in chg, Endocrine Diabetes Clin, Johns Hopkins Hosp, 58-, physician, 58-; consult, Food & Drug Admin & Fed Trade Comn, diabetes, thyroid dis & gen metab probs. *Mem:* Endocrine Soc; assoc Royal Soc Med; Am Diabetes Asn; fel Am Col Physicians; Am Thyroid Asn. *Res:* Clincal trials on therapeutic agents. *Mailing Add:* 6565 N Charles St 411 Pavilion GBMC Baltimore MD 21204

PROUT, TIMOTHY, b Watertown, Conn, June 14, 23; m 50; c 2. GENETICS. *Educ:* Hobart Col, BA, 48; Columbia Univ, MA & PhD, 54. *Prof Exp:* From asst prof to prof zool, Univ Calif, Riverside, 54-74, PROF GENETICS, UNIV CALIF, DAVIS, 74- *Concurrent Pos:* Guest prof, Genetics Inst, Aarhus Univ, 67-68 & 74-75. *Mem:* AAAS; Soc Study Evolution; Am Soc Human Genetics; Genetics Soc Am; Am Soc Naturalists. *Res:* Population genetics in Drosophila. *Mailing Add:* 339 S Crescent Parkridge IL 60068

PROUTY, CHILTON E, b Tuscaloosa, Ala, Sept 8, 14. TECTONICS, PALEONTOLOGY. *Educ:* Univ NC, BS, 36; Univ Mo-Rolla, MS, 38; Columbia Univ, PhD(geol), 44. *Prof Exp:* Mem staff, US Geol Survey, 44-46; mem staff, Univ Pittsburgh, 46-47; prof, 57-85, chmn, 57-69, EMER PROF GEOL, MICH STATE UNIV, 85- *Concurrent Pos:* Ed, Am Inst Prof Geologists, 77 & 78. *Mem:* Paleont Soc; Nat Asn Geol Teachers (pres, 58); fel Geol Soc Am; Am Asn Petrol Geologist; Soc Econ Petrol & Mineral; Am Inst Prof Geologists. *Res:* Structural, mechanic Michigan Basin. *Mailing Add:* Dept Geol & Sci Mich State Univ Natural Sci Bldg Rm 206 East Lansing MI 48824

PROUTY, RICHARD METCALF, b Milford, Mass, May 20, 24; m 51; c 3. ANALYTICAL CHEMISTRY. *Educ:* Univ Mass, BS, 51; Univ Conn, MS, 57; Univ Minn, PhD(animal husb), 63. *Prof Exp:* Res fel, Sch Pub Health, Univ Minn, 61-63; res chemist, Environ Chem Sect, Patuxent Wildlife Res Ctr, US Dept Interior, 63-; RETIRED. *Mem:* Am Chem Soc. *Res:* Identification and measurement of organochlorine compounds accumulating in the tissues of wild birds. *Mailing Add:* 12602 Memory Lane Bowie MD 20715

PROVDER, THEODORE, b Brooklyn, NY, Dec 26, 39; m 61; c 3. PHYSICAL CHEMISTRY. *Educ:* Univ Miami, BS, 61; Univ Wis-Madison, PhD(phys chem), 66. *Prof Exp:* Sr res chemist, Monsanto Co, Mo, 65-70; tech mgr fundamental & comput sci dept coatings res, 70-77; SR SCIENTIST POLYMER RES & COMPUT SCI GROUP, POLYMER & COATINGS RES DEPT, GLIDDEN COATINGS & RESINS DIV, SCM CORP, STRONGSVILLE, 79- *Concurrent Pos:* Mem exec comt, Div Org Coatings & Plastics, 81-82, adv bd, Advan in Chem & Symp series, Am Chem Soc, 81-84. *Honors & Awards:* SCM Corp Sci & Tech Award, 77; ACM Glidden Coatings & Resins Div Award for Tech Excellence, 81. *Mem:* Am Soc Testing & Mat; fel Am Inst Chemists; Am Chem Soc; Sigma Xi. *Res:* Solution property and solid state characterization; mechanical properties and rheological studies of polymer, colloids and coatings, and new coatings concepts; use of computerization in polymer science and laboratory automation and instrument and process control. *Mailing Add:* Glidden Co 16651 Sprauge Rd Strongsville OH 44136-1757

PROVENCHER, GERALD MARTIN, b Detroit, Mich, Jan 16, 37; m 60; c 3. COMBUSTION CHEMISTRY. *Educ:* Wayne State Univ, BS, 62, MS, 66; Univ Windsor, Ont, PhD(chem), 72; Bellevue Col, BA, 85. *Prof Exp:* Instr chem, Highland Park Community Col, 66-72; teaching fel, Ctr Res Exp Space Sci, York Univ, Toronto, 72-73 & Acadia Univ, Wolfeville, NS, 73-74; from asst prof to assoc prof, Univ Southern Maine, 75-80; res scientist chem, 80-83, RES DIR SCIENTIFIC RES, HNG-INTERNORTH, 83- *Concurrent Pos:* Vis prof, Dalhousie Univ, Halifax, NS, 79-80 & Brookhaven Nat Lab, 80-; fel, NATO-Italian govt, 76. *Mem:* Am Chem Soc. *Res:* Dynamics of gas phase chemical reactions, laser induced fluorescence, non-linear (multi-photon) induced spectroscopy and the use of lasers and electro-optics as diagnostics for chemical events. *Mailing Add:* 4415 Pine St Omaha NE 68105-2418

PROVENCIO, JESUS ROBERTO, b Juarez, Mex, Feb 24, 25; US citizen; m 49; c 6. COMMUNICATIONS, PHYSICS. *Educ:* Tex Col Mines, BS, 48; Tex Western Col, MS, 65; Univ Chihuahua, DUniv, 72. *Prof Exp:* Instr physics, Univ Denver, 48-51; asst prof math & elec eng, Fed Schs, Mex, 51-53; systs scientist commun, Motorola, Inc, 53-63; asst prof math & physics, 63-73, ASST PROF MATH, UNIV TEX, EL PASO, 73- *Concurrent Pos:* Consult, Dept Health, Educ & Welfare, 70-, mem, Nat Adv Comt Educ, 70-72, Nat Task Force Res Phys Sci, 72-73; consult, US Off Educ, 70- & Nat Ctr Vol Action; systs analyst, Dept Housing & Urban Develop; mem, Nat Ctr Latin Am Studies Asn. *Mem:* Soc Advan Educ (pres, 69-70); Mex Phys Soc; Am Inst Physics. *Res:* Information-communications theory, especially communications among persons of different cultural backgrounds; propagation and decay of shocks in air. *Mailing Add:* Univ Tex El Paso TX 79968

PROVENZA, DOMINIC VINCENT, b Baltimore, Md, Nov 27, 17; m 64; c 2. ZOOLOGY. *Educ:* Univ Md, BS, 39, MS, 41, PhD(zool), 52. *Prof Exp:* From instr to assoc prof biol, Loyola Col, Md, 47-55; assoc prof anat, Univ Col, 55-56, assoc prof, Dent Sch, 55-57, actg head dept, 57-61; prof anat, Dent Sch, Univ Md, Baltimore, 57-, head dept, 61-, asst dean biol sci, 81-; RETIRED. *Concurrent Pos:* Prof biol & chmn dept sci, Eve Sch, Loyola Col Md, 47-55; NIH spec fel, 62-63; ed, J Baltimore Col Dent Surg, 65- *Mem:* Am Asn Anatomists; NY Acad Sci; Int Asn Dent Res; fel Am Col Dentists. *Res:* Histophysiology of mammalian microcirculation; intercellular substances of periodontium; physiology of spermatozoa; feeding mechanisms of Coccidae; physical-chemical ecology; electron microscopy dentino-genesis; adhesive restorative materials; ultrastructure of accessory boring organ of gastropods; formed intercellular components of oral mucosa; histo-cytochemistry of odontogenesis; histology and ultrastructure of miniature swine pulp. *Mailing Add:* 4000 N Charles St Univ of Md Sch of Dent Baltimore MD 21218

PROVERB, ROBERT JOSEPH, b Chelsea, Mass, Dec 20, 46; m 68; c 2. PAPER CHEMISTRY, COMPUTER APPLICATIONS. *Educ:* Northeastern Univ, BA, 69, PhD(org chem), 73. *Prof Exp:* Fel org chem, Univ Alta, 73-75, Rice Univ, 75-76 & Univ Calif, Los Angeles, 77; GROUP LEADER, AM CYANAMID CO, 77- *Mem:* Am Chem Soc; Sigma Xi. *Res:* Synthetic organic chemistry; organic reaction mechanisms; synthetic polymer and paper chemistry; application of polymer and organic chemistry to paper chemistry with emphasis on synthesis and mechanisms through employment of physical organic and molecular modeling techniques; isocyanate chemistry. *Mailing Add:* Am Cyanamid Co Chem Res Div 1937 W Main St Stamford CT 06904-0060

PROVINE, ROBERT RAYMOND, b Tulsa, Okla, May 11, 43. NEUROEMBRYOLOGY, NEUROETHOLOGY. *Educ:* Okla State Univ, BS, 65; Washington Univ, St Louis, PhD(neuroembryol), 71. *Prof Exp:* Res assoc biol & ophthal to res asst prof psychol, Washington Univ, St Louis, 71-74; PROF PSYCHOL, UNIV MD, BALTIMORE COUNTY, 74- *Concurrent Pos:* Prin investr, NIH. *Mem:* Sigma Xi; Soc Neurosci; Psychonomic Soc; Animal Behav Soc; Int Soc Develop Psychobiol; Int Soc Human Ethology; Int Soc Neuroethology. *Res:* Prenatal and postnatal development of the nervous system; behavior in man and animals; human ethology. *Mailing Add:* Dept of Psychol Univ Md Baltimore County Baltimore MD 21228

PROVOST, ERNEST EDMUND, b Needham Heights, Mass, June 19, 21; m 47; c 2. ZOOLOGY. *Educ:* Purdue Univ, BS, 50; Wash State Univ, MS, 53, PhD(zool), 58. *Prof Exp:* From asst to instr zool, Wash State Univ, 50-58; asst prof, Univ Ill, 58-60; assoc prof wildlife mgr & zool, 60-74, MEM STAFF, EXP STA, UNIV GA, 60-, PROF WILDLIFE & ZOOL, 74- *Mem:* AAAS; Wildlife Soc; Am Soc Mammalogists. *Res:* Wildlife biology; vertebrate zoology; mammalian reproduction and ecology of feral populations. *Mailing Add:* Dept Zool Univ of Ga Athens GA 30602

PROVOST, PHILIP JOSEPH, b Bristol, Conn, May 28, 35; m 62; c 2. MICROBIOLOGY, VIROLOGY. *Educ:* Univ Conn, AB, 57; Univ Md, MS, 59, PhD(microbiol), 61. *Prof Exp:* Res fel microbiol, Sch Med, Univ Md, 61-63 & McCollum-Pratt Inst, Johns Hopkins Univ, 63-64; asst prof microbiol, Fac Med, Dalhousie Univ, 64-66; res virologist, Lederle Labs, Am Cyanamid Co, 66-69, head dept biol res & develop, 69-70; res fel virus & cell biol res, 70-76, sr res fel, 76-80, SR INVESTR, MERCK INST THERAPEUT RES, MERCK & CO, INC, 80- *Mem:* AAAS; Am Soc Microbiol; NY Acad Sci; Can Soc Microbiologists; Sigma Xi; Am Soc Virol. *Res:* Viral hepatitis; mechanisms of microbial pathogenesis; immunity to infectious diseases; viral vaccines. *Mailing Add:* 1680 Jacks Circle Lansdale PA 19446

PROVOST, RONALD HAROLD, b Burlington, Vt, May 25, 42; m 68; c 2. PHYSICAL CHEMISTRY. *Educ:* St Michael's Col, Vt, AB, 64; Univ Vt, PhD(phys chem), 68. *Prof Exp:* Prof chem & chmn dept, 68-79, ACAD VPRES, ST MICHAEL'S COL, 79- *Concurrent Pos:* Environ consult. *Mem:* Am Chem Soc; Sigma Xi; AAAS. *Res:* Thermodynamics characterization of transition metal systems. *Mailing Add:* St Michael's Col Colchester VT 05439

PROWSE, STEPHEN JAMES, AUTOIMMUNITY, PARASITE IMMUNOLOGY. *Educ:* Univ Adelaide, Australia, PhD(immunol), 79. *Prof Exp:* ASST PROF MICROBIOL & IMMUNOL, UNIV COLO HEALTH SCI CTR, 82- *Mailing Add:* Div Animal Health CS RIO Private Bag 1 Parkville Victoria 3052 Australia

PRUCHA, JOHN JAMES, b River Falls, Wis, Sept 22, 24; m 48; c 10. STRUCTURAL GEOLOGY. *Educ:* Univ Wis, PhB, 45, MPh, 46; Princeton Univ, MA, 48, PhD(geol), 50. *Prof Exp:* From instr to asst prof geol, Rutgers Univ, 48-51; sr geologist, NY State Geol Surv, 51-56; res geologist, Shell Develop Co, 56-63; chmn dept geol, Syracuse Univ, 63-70 & 89-90, dean col arts & sci, 70-72, vchancellor acad affairs, 72-85, prof, 63-90, EMER PROF GEOL, SYRACUSE UNIV, 90- *Mem:* Fel AAAS; fel Geol Soc Am; Am Asn Petrol Geologists. *Res:* Precambrian geology; structural geology; regional tectonics; salt deformation in folding; basement tectonics. *Mailing Add:* Dept Geol Heroy Geol Lab, Syracuse Univ Syracuse NY 13244-1070

PRUCKMAYR, GERFRIED, b Vienna, Austria, Feb 1, 33; m 63. ORGANIC CHEMISTRY. *Educ:* Univ Vienna, PhD(org chem), 61. *Prof Exp:* Res assoc, Case Inst Technol, 61-62; res chemist, 62-71, staff scientist, 71-76, RES ASSOC, E I DU PONT DE NEMOURS & CO, INC, 76- *Mem:* Am Chem Soc. *Res:* Synthesis of pharmacologically active heterocyclic compounds; heterocyclic polymers; reaction mechanisms; ring-opening polymerizations. *Mailing Add:* Chestnut Run Plaza 709 PO Box 80 Wilmington DE 19880-0709

PRUDENCE, ROBERT THOMAS, b San Francisco, Calif, Nov 19, 44; m 68; c 1. ORGANIC POLYMER CHEMISTRY. *Educ:* John Carroll Univ, BS, 66; Purdue Univ, PhD(org chem), 71. *Prof Exp:* SR RES CHEMIST, GOODYEAR RES LABS, GOODYEAR TIRE & RUBBER CO, 70- *Mem:* Am Chem Soc. *Res:* Anionic polymerization techniques used to develop thermoplastic block polymers and diene homopolymers. *Mailing Add:* 2125 Forest Oak Dr Akron OH 44312-2229

PRUDER, GARY DAVID, b Detroit, Mich, June 23, 32; m 57; c 2. BIOENGINEERING, AQUACULTURE. *Educ:* Gen Motors Inst, BME, 56; Univ Del, MBA, 67, PhD PhD(marine studies), 78. *Prof Exp:* Sr exp engr, Hydra-matic Div, Gen Motors Corp, 52-62; head applns res, Elkton Div, Thiokol Chem Corp, 62-66; gen mgr, Technidyne Div, Aeroprojs Inc, 66-69; dir eng & mfg, Chem Data Systs, 69-71; prin investr controlled environ sys develop, 71- AT DEPT MARINE STUDIES, UNIV DEL. *Concurrent Pos:* Consult maricult, Kahuku Seafood Plantation, Hawaii, 77-78. *Mem:* World Maricult Soc; AAAS. *Res:* Intensive controlled mariculture; mass production of algae; solar ponds; waste treatment and recycle; rearing bivalve molluscs. *Mailing Add:* 2936 Alphonse Pl Honolulu HI 96816

PRUD'HOMME, JACQUES, b Montreal, July 15, 40; m 63; c 3. POLYMER SCIENCE. *Educ:* Univ Montreal, PhD, 67. *Prof Exp:* Nat Res Coun Can fel, 67-69, from asst prof to assoc prof, 69-80, PROF CHEM, UNIV MONTREAL, 80- *Mem:* Chem Inst Can. *Res:* Polymer chemistry; synthesis, characterization and properties of block copolymers. *Mailing Add:* Dept Chem Univ Montreal PO Box 6128 Sta A Montreal PQ H3C 3J7 Can

PRUD'HOMME, ROBERT EMERY, b Ste-Martine, Que, July 12, 46. POLYMER SCIENCE. *Educ:* Univ Montreal, BSc, 69, MSc, 70; Univ Mass, Amherst, PhD(chem), 73. *Prof Exp:* Res scientist pulp & paper, Paprican, 73-75; from asst prof to assoc prof, 75-84, PROF CHEM, LAVAL UNIV, 84- *Concurrent Pos:* E W R Steacie Mem fel, Nat Sci & Eng Res Coun Can, 85-87. *Mem:* Am Chem Soc; Inst Chim Can. *Res:* The solid state of polymers, including crystallization, morphology and mechanical properties. *Mailing Add:* Dept Chem Laval Univ Quebec City PQ G1K 7P4 Can

PRUD'HOMME, ROBERT KRAFFT, b Sacramento, Calif, Jan 28, 48; m 73; c 3. CHEMICAL ENGINEERING, RHEOLOGY. *Educ:* Stanford Univ, BS, 69; Univ Wis, PhD(chem eng), 78. *Prof Exp:* Environ engr, US Army Armaments Command, 71-72; res asst, Dept Chem Eng, Univ Wis, 73-78; ASST PROF CHEM ENG, PRINCETON UNIV, 78-, DEPT REP, 78- *Concurrent Pos:* Mem comt on undergrad educ, Univ Wis, 75-77; consult, Mobil Oil Co, 78- *Mem:* Sigma Xi; Am Inst Chem Engrs; Am Soc Rheol. *Res:* Transport processes; fluid mechanics of polymeric materials; polymerization reactions; environmental science. *Mailing Add:* Dept of Chem Eng Princeton Univ Princeton NJ 08544

PRUDHON, ROLLAND A, JR, b Princeton, NJ, June 16, 42; m 65; c 3. PARASITOLOGY. *Educ:* Maryville Col, BS, 65; Univ Tenn, Knoxville, MS, 68, PhD(zool), 72. *Prof Exp:* Asst prof natural sci, 72-80, chmn dept, 73-80, ASSOC PROF BIOL & HEAD DEPT NATURAL SCI, SHELBY STATE COMMUNITY COL, 80- *Concurrent Pos:* Mem, Audio-tutorial Cong. *Mem:* Am Soc Parasitologists. *Mailing Add:* Dept Nat Sci Shelby State Community Col PO Box 40568 Memphis TN 38174

PRUESS, KENNETH PAUL, b Troy, Ind, June 21, 32; m 64; c 2. ENTOMOLOGY. *Educ:* Purdue Univ, BS, 54; Ohio State Univ, MS, 55, PhD(entom), 57. *Prof Exp:* From asst prof to assoc prof, 57-75, PROF ENTOM, UNIV NEBR-LINCOLN, 75- *Mem:* Entom Soc Am; NAm Benth Soc. *Res:* Aquatic entomology; applied ecology; molecular systs. *Mailing Add:* Dept Entom Univ Nebr Lincoln NE 68583

PRUESS, STEVEN ARTHUR, b Cedar Rapids, Iowa, June 8, 44; m 85. NUMERICAL ANALYSIS. *Educ:* Iowa State Univ, BS, 66; Purdue Univ, MS, 68, PhD(comput sci), 70. *Prof Exp:* From asst prof to assoc prof, PROF MATH, UNIV NMEX, 82- *Concurrent Pos:* Sabbatical leave, Univ BC, 80-81; summer res, Australian Nat Univ, 82, Argonne Nat Lab, 85. *Mem:* Soc Indust & Appl Math. *Res:* Solutions of differential equations; eigenvalue problems; approximation theory. *Mailing Add:* 12298 W Connecticut Lakewood CO 80228

PRUESS, STEVEN ARTHUR, b Cedar Rapids, Iowa, June 8, 44; m 85. NUMERICAL ANALYSIS, APPROXIMATION THEORY. *Educ:* Iowa State Univ, BS, 66; Purdue Univ, MS, 68, PhD(computer sci), 70. *Prof Exp:* From asst prof to prof math, Univ NMex, 70-86; PROF MATH & COMPUTER SCI, COLO SCH MINES, 86- *Concurrent Pos:* Vis assoc prof, Univ BC, 80-81; vis lectr, Australian Nat Univ, 82; vis prof, Mex Nat Univ, 87. *Mem:* Soc Indust & Appl Math. *Res:* Mathematical software; numerical methods for eigenvalue problems and two point boundary value problems; curve and surface fitting; estimating fractal dimensions. *Mailing Add:* Math & Computer Sci Dept Colo Sch Mines Golden CO 80401

PRUETT, CHARLES DAVID, b Durham, NC. NUMERICAL METHODS, SUPERCOMPUTING. *Educ:* Va Polytech Inst & State Univ, BS, 70; Univ Richmond, Va, MEd, 74; Univ Va, MS, 80; Univ Ariz, PhD(appl math), 86. *Prof Exp:* Teacher math, Henrico County Schs, Va, 73-76; engr, Kentron (LIV), Hampton, Va, 76-78; eng analyst, Anal Mech Assocs, Hampton, Va, 80-82; asst prof math, Va Commonwealth, Univ Richmond, 86-89; RES ASSOC, NAT RES COUN, LANGLEY RES CTR, NASA, 89- *Concurrent Pos:* Prin investr, Numerical Experiments in High-Speed Transition, Nat Aerodyn Simulator Porj, Theoret Flow Physics Br, Fluid Mech Div, Langley Res Ctr, NASA co investigators, T A Zary & G Erlebacher, 91-92. *Mem:* Soc Indust & Appl Math. *Res:* Direct numerical simulation of stability and transition in high-speed boundary layer flow, by means of spectral and high-order accurate numerical methods. *Mailing Add:* 327 Merrimac Trail No 16C Williamsburg VA 23185

PRUETT, ESTHER D R, b June 25, 20; div. DIABETES, EXERCISE PHYSIOLOGY. *Educ:* Wellesley Col, BA, 42; Bryn Mawr Col, MA, 48; Univ Oslo, Norway, DPhil, 71. *Prof Exp:* Res investr, Inst Work Physiol, Oslo, Norway, 65-75; RETIRED. *Mem:* AAAS; Am Physiol Soc; Europ Soc Study Diabetes; Nordic Physiol Soc; Nordis Soc Diabetes Res. *Res:* Exercise and the endocrine system in humans; metabolism. *Mailing Add:* 80 Palmers Mill Rd Media PA 19063

PRUETT, JACK KENNETH, b Kinston, Ala, Sept 22, 32; m 56; c 3. PHARMACOLOGY. *Educ:* Auburn Univ, BS, 60, MS, 63; Med Col SC, PhD(pharmacol), 67. *Prof Exp:* From instr to assoc prof pharmacol, Med Univ SC, 66-77; assoc prof anesthesiol, 77-82, ASSOC PROF PHARMACOL, MED COL GA, 80-, PROF ANESTHESIOL, 82-, DIR ANESTHESIA RES, 82- *Mem:* Am Soc Pharmacol & Exp Therapeut; Am Soc Anesthesiologists; Am Heart Asn; Soc Cardiovasc Anesthesiologists; Int Anesthesia Res Soc. *Res:* Cardiovascular pharmacology and cardiac electrophysiology. *Mailing Add:* Dept Anesthesiol Med Col Ga Augusta GA 30912-2277

PRUETT, JOHN ROBERT, b Plankinton, SDak, Mar 8, 23; m 60; c 1. PHYSICS, COMPUTER SCIENCE. *Educ:* Ind Univ, PhD(physics), 49. *Prof Exp:* Asst physics, Ind Univ, 42, asst, Manhattan Proj, 42-44; jr physicist, Appl Physics Lab, Johns Hopkins Univ, 44 & Naval Res Lab, 45; asst physics, Ind Univ, 45-49; from asst prof to prof physics, 49-89, chmn dept, 78-89, EMER PROF PHYSICS & COMPUTER SCI, BRYN MAWR COL, 89- *Concurrent Pos:* US State Dept rep, Int Conf Orgn Econ Coop & Develop, Milan, Italy, 63 & Uppsala, Sweden, 64; owner, Pruett Assco, Inc. *Mem:* Am Phys Soc; Am Asn Physics Teachers. *Res:* Beta and gamma ray spectroscopy; beta decay theory; nuclear resonance fluorescence; nuclear structure; hardware and software design for direct computer controlled experimentation in the sciences. *Mailing Add:* 823 Old Gulph Rd Bryn Mawr PA 19010

PRUETT, PATRICIA ONDERDONK, b Chicago, Ill, June 11, 30; m 60; c 1. CELL PHYSIOLOGY, BIOCHEMISTRY. *Educ:* Bryn Mawr Col, AB, 52, MA, 61, PhD(biol), 65, PhD(biochem). *Prof Exp:* Teaching asst biol, 58-60, asst dean, 60-67, actg dean, 68-69, 76-77 & 77-78, ASSOC DEAN & LECTR BIOL, BRYN MAWR COL, 70- *Concurrent Pos:* Owner, Pruett Assoc, Inc. *Mem:* AAAS; Am Soc Cell Biologists; Soc Protozoologists. *Res:* Computer applications in the biological sciences, molecular modeling of phospholipids; membrane transport; anion accumulation; transport mechanisms; orthophosphate fluxes across the membrane of a ciliated protozoan. *Mailing Add:* 823 Old Gulph Rd Bryn Mawr PA 19010

PRUETT, ROY L, b Union City, Tenn, Nov 26, 24; m 61; c 2. TRANSITION METAL CHEMISTRY. *Educ:* Murray State Univ, BS, 44; Univ Tenn, MS, 48, PhD(org chem), 51. *Prof Exp:* Corp res fel, Union Carbide Corp, 48-79; sr res assoc, Exxon Res & Eng Co, 79-81, sci adv, 81-83, chief scientist, Exxon Chem Co, 83-87; RETIRED. *Mem:* Am Chem Soc; Am Inst Chemists. *Res:* Homogeneous catalysis with transition metal complexes; rhodium-catalyzed hydroformylation reaction; synthesis and reaction of metal cluster compounds. *Mailing Add:* 8431 Willow Oak Lane Harrisburg NC 28075

PRUETT-JONES, STEPHEN GLEN, b Richmond, Calif, Nov 17, 52; m 79. POPULATION BIOLOGY, EVOLUTIONARY BIOLOGY. *Educ:* Univ Calif, Davis, BS, 76; Brigham Young Univ, MS, 81; Univ Calif, Berkeley, PhD(zool), 85. *Prof Exp:* Postdoctoral res assoc, Univ Calif, San Diego, 85-88; ASST PROF ECOL & EVOLUTION, UNIV CHICAGO, 88-; RES ASSOC, FIELD MUS NAT HIST, 88- *Concurrent Pos:* Vis fel, Australian Nat Univ, 88- *Res:* Evolutionary biology; population biology; behavioral ecology; sexual selection; evolution of social behavior; animal communication; host-parasite interactions. *Mailing Add:* Dept Ecol & Evolution Univ Chicago Chicago IL 60637

PRUGH, JOHN DREW, b Greybull, Wyo, Dec 18, 32; m 54; c 3. MEDICINAL CHEMISTRY. *Educ:* Univ Wyo, BS, 58; Univ Wash, PhD, 64. *Prof Exp:* NIH fel chem, Yale Univ, 64-66; sr res chemist, 66-80, RES FEL, MERCK SHARP & DOHME RES LABS, 80- *Mem:* Am Chem Soc. *Res:* Furan chemistry; mechanism of acetal oxidation; dibenzocycloheptene chemistry; heterocycles; carbohydrates. *Mailing Add:* Dept Med Chem Merck Sharp & Dohme West Point PA 19486

PRUGOVECKI, EDUARD, b Craiova, Romania, Mar 19, 37; Can citizen. MATHEMATICAL PHYSICS. *Educ:* Univ Zagreb, dipl physics, 59; Princeton Univ, PhD(theoret physics), 64. *Prof Exp:* Prof assoc theoret physics, Princeton Univ, 64-65; lectr, Univ Alta, 66-67; from asst prof to assoc prof, 67-74, PROF APPL MATH, UNIV TORONTO, 75- *Concurrent Pos:* Fel, Theoret Physics Inst, Univ Alta, 65-67; Nat Res Coun Can grant, Univ Toronto, 68- *Res:* Scattering theory; foundations of quantum mechanics; quantum field theory; functional analysis; group theory; quantum geometry and gravity; quantum geometrics, aimed at a consistant unification of general relativity and quantum theory, and at producing a formulation of quantum gravity in terms of Hilbert superfibre bundles, whose gauge group would incorporate the Poincaré. *Mailing Add:* Dept Math Univ Toronto Toronto ON M5S 1A1 Can

PRUITT, ALBERT WESLEY, b Anderson, SC, Jan 1, 40; m 61; c 1. PHARMACOLOGY. *Educ:* Emory Univ, BA, 61, MD, 65. *Prof Exp:* Resident pediat, Grady Mem Hosp, 65-68, chief pediat, 81-82; fel, 70-72, asst prof, 72-77, assoc prof pediat, 75-81, ASSOC PROF PHARMACOL, EMORY UNIV, 77-; PROF & CHMN PEDIAT, MED COL GA, 82- *Mem:* Soc Pediat Res; Am Soc Pharmacol & Exp Therapeut; Am Soc Clin Pharmacol & Therapeut; Am Acad Pediat; Am Fedn Clin Res. *Res:* Age-related changes in metabolism of drugs and responses to drugs. *Mailing Add:* Dept Pediat Med Col Ga Augusta GA 30912

PRUITT, BASIL ARTHUR, JR, b Nyack, NY, Aug 21, 30; m 54; c 3. CARE OF BURN PATIENTS, SURGICAL CRITICAL CARE. *Educ:* Harvard Col, AB, 52; Tufts Univ, MD, 57. *Hon Degrees:* Int Hon Prof Surg, Third Mil Med Col, People's Liberation Army, Chongqing, People's Repub China, 89. *Prof Exp:* Surgeon, 24th Evac Hosp, Vietnam, 67; C prof serv surg, 12th Evac Hosp, Vietnam, 67-68; gen surgeon, Surg Res Unit, US Army, 64-65, C Burn Study Br surg, 65-66, C Clin Div, 66-67, C trauma study, Walter Reed Army Inst Res, Vietnam, 68, COMDR & DIR SURG, INST SURG RES, US ARMY, 68- *Concurrent Pos:* Clin prof surg, Health Sci Ctr, Univ Tex, San Antonio, 75-; consult surg, San Antonio Chest Hosp, 76-; prof surg, US Univ Health Sci, Bethesda, 78-; mem, Surg, Anesthesiol & Trauma Study Sect, NIH, 78-82, Burns Adv Bd, Shriners Hosp Crippled Children, 85- & Vet Admin Merit Rev Bd Surg, 90-; dir, Am Bd Surg, 82-88; consult, Kiwanis Pediat Trauma Inst, Boston, 83- *Honors & Awards:* Curtis P Artz Mem Award, Am Trauma Soc, 82; Harvey Stuart Allen Distinguished Serv Award, Am Burn Asn, 84; A B Wallace Mem Lectr, Brit Burn Asn, 84; Baron Dominque Larrey Award Surg Excellence, Uniformed Serv Univ Health Sci, 85; Churchill/Excelsior Surg Soc Lectr, Am Col Surgeons, 88. *Mem:* Am Burn Asn (pres, 75-76); Am Asn Surg Trauma (pres, 82-83); Surg Infection Soc (pres, 85-86); Am Surg Asn; Am Trauma Soc (pres elect, 90-); Int Soc Burn Injuries (pres elect, 90-). *Res:* Post-injury physiologic changes such as hypovolemic shock and fluid resuscitation, respiratory insufficiency, infection and sepsis, immunosuppression, and metabolic alterations. *Mailing Add:* Inst Surg Res US Army Ft Sam Houston San Antonio TX 78234-5012

PRUITT, KENNETH M, b Winston-Salem, NC, Oct 3, 33; c 4. BIOCHEMISTRY. *Educ:* Univ NC, BS, 56; Pa State Univ, BS, 57; Brown Univ, PhD(phys chem), 65. *Hon Degrees:* Dr Odontol, Univ Umea, Sweden, 88. *Prof Exp:* Chief chemist, Diabetic Res Lab, Birmingham Vet Admin Hosp, 64-65; investr, Lab Molecular Biol, Univ Ala, Birmingham, 65-68, scientist, Inst Dent Res, 67-77, from asst prof to assoc prof biochem, 64-78, assoc prof biomath, 78-80, asst vpres res, 79-85, dir, Div Molecular Biol, 80-81, PROF BIOCHEM & PROF BIOMATH, SCH MED, UNIV ALA, BIRMINGHAM, 79-, ASSOC VPRES RES, 85-; SR SCIENTIST, CYSTIC FIBROSIS RES CTR, 81- *Concurrent Pos:* Assoc prof chem, Univ Col, Birmingham, Ala, 68-71, 73-77; vis scientist pharmacol, Hershey Med Ctr, Pa State Univ, 70-71; vis scientist, Fac Odontol, 71-72, 73, Gothenburg vis prof, Lab Oral Biochem, 76; vis prof, fac Odontol, Univ Umea, Umea, Sweden, 77. 71-72. Am Soc Biol Chemists; vis prof, Lab Oral Biochem, Univ Umea, 78 & 79, Dept Physics, Univ Linkoping, Sweden, 79; chmn, Saliva Res Physiol Session, Int Asn Dent Res, Japan, 80; ad hoc mem, Oral Biol & Med Study Sect, NIH, 78-79; external rev, NSF, 80-81, exten rev, Cell & Physiol Biosci Sect, 81; invited lectr, Int Res Course, Univ Umea, Sweden, 81, Int Seminar Umea, Sweden, 87; state proj dir, Exp Prog, Nat Chmn, Coalition EPSCOR States, 88-90; Stimulate Comp Res, NSF, 88- *Mem:* AAAS; Am Soc Biol Chem; Int Asn Dent Res; Am Asn Dent Res; Am Chem Soc; Am Meteorol Soc; Am Phys Soc. *Res:* The peroxidase system and other non-immunoglobulin defense mechanisms; defense mechanisms on mucosal surfaces. *Mailing Add:* Univ Ala Birmingham Sta Birmingham AL 35294

PRUITT, NANCY L, b Philadelphia, Pa, Feb 24, 53; m; c 1. ENZYME KINETICS, LIPID METABOLISM. *Educ:* Ariz State Univ, PhD(zool), 83. *Prof Exp:* ASSOC PROF BIOL, COLGATE UNIV, 83- *Mem:* Am Soc Zoologists; Am Physiol Soc. *Res:* Lipids and enzymes of thermally-acclimated poikilothermic animals. *Mailing Add:* Dept Biol Colgate Univ Hamilton NY 13346

PRUITT, ROGER ARTHUR, b Kansas City, Mo, July 10, 36; m 60; c 2. PHYSICS. *Educ:* Univ Kans, BS, 59; Colo State Univ, MS, 67, PhD(physics), 69. *Prof Exp:* High sch teacher, Kans, 59-62; pub sch teacher, Colo, 62-64; instr physics, Colo State Univ, 69-70; from asst prof to assoc prof, 70-76, PROF PHYSICS, FT HAYS KANS STATE COL, 76- *Mem:* Am Asn Physics Teachers; Am Phys Soc; Sigma Xi. *Res:* Mossbauer spectroscopy in intermallic compounds. *Mailing Add:* Dept Physics Ft Hays Kans State Univ Hays KS 67601

PRUITT, WILLIAM EDWIN, b Oklahoma City, Okla, May 11, 34; m 56; c 3. PROBABILITY. *Educ:* Okla State Univ, BS, 55, MS, 56; Stanford Univ, PhD(statist), 60. *Prof Exp:* From asst prof to assoc prof, 60-70, PROF MATH, UNIV MINN, MINNEAPOLIS, 70- *Concurrent Pos:* Vis prof, Cornell Univ, 77-78; assoc ed, Annals Probability, 72-78. *Mem:* Am Math Soc; Inst Math Statist; Math Asn Am. *Res:* Probability theory. *Mailing Add:* Univ Minn Minneapolis MN 55455

PRUITT, WILLIAM O, JR, b Sept 1, 22; m 51; c 2. BOREAL & WINTER ECOLOGY. *Educ:* Univ Md, BSc, 47; Univ Mich, MA, 48, PhD(zool), 52. *Prof Exp:* Res biologist, Arctic Aeromed Lab, 53-56; contract biologist, Coop Caribou Invest, Can Wildlife Serv, 57-58; assoc prof biol, Univ Alaska, 59-62; vis assoc prof, Univ Okla, 63-65; assoc prof biol, Mem Univ, Nfld, 65-69; assoc prof, 69-70, PROF ZOOL, UNIV MAN, 70- *Mem:* Am Soc Mammalogists; fel Arctic Inst NAm; Can Soc Zoologists; Sigma Xi; fel Explorers Club. *Res:* Ecology of northern mammals; winter ecology; snow ecology; boreal ecology. *Mailing Add:* Dept Zool Univ Man Winnipeg MB R3T 2N2 Can

PRUNTY, LYLE DELMAR, b Charles City, Iowa, Oct 8, 45; m 75; c 2. SOIL PHYSICS. *Educ:* Iowa State Univ, BS, 67, MS, 69, PhD(soil physics), 78. *Prof Exp:* Develop engr, Western Elec Co, 70-75; res assoc, Dept Agron, Iowa State Univ, 76-78; asst prof soils sci, NDak State Univ, 78-83; ASSOC PROF SOIL SCI, NDAK STATE UNIV, 83- *Mem:* Am Soc Agron; Sigma Xi. *Res:* Coal surface mine reclamation; soil crusting; water use by sunflowers; irrigation return flow. *Mailing Add:* 2246 Carriage Ave Richland WA 99352-1804

PRUSACZYK, JOSEPH EDWARD, b Cleveland, Ohio, 1944; m 67; c 1. PHYSICAL CHEMISTRY. *Educ:* Univ Dayton, BS, 66; Case Western Reserve Univ, PhD(phys chem), 71. *Prof Exp:* Interim asst prof chem, Univ Fla, 70-71, fel radiation chem, 71-72; res assoc chem, Univ Md, College Park, 72-73 & Fla State Univ, 73-74; supvr, Owens Corning Fiberglas Corp, 74-86; scientist, 87-90, SUPVR, GEORGIA-PACIFIC RESINS, INC, 90- *Mem:* Am Chem Soc; Am Soc Mass Spectrometry. *Res:* Mass spectrometry; high temperature chemistry; thermodynamics and material properties; polymer chemistry; thermosets; phenol-formaldehyde resins; synthesis; thermoplastics; polymer characterization. *Mailing Add:* 1928 Londondale Pkwy Newark OH 43055-1600

PRUSAS, ZENON C, b Butenai, Lithuania, June 5, 21; US citizen; m 52; c 1. PULP & PAPER TECHNOLOGY. *Educ:* Univ Vilnius, Lithuania, MF, 43; Univ Munich, PhD(econ), 48; Syracuse Univ, MS, 55. *Prof Exp:* Res scientist pulp & paper, 55-63, sr res engr, 63-66, sr consult, 67-78, res fel pulp & paper, Mead Corp, 79-85, PRIN RES SCIENTIST, 85- *Mem:* Sigma Xi; Tech Asn Pulp & Paper Indust. *Res:* Solving of technical problems in pulp and paper mills; high yield pulping of hardwoods. *Mailing Add:* Seven Tecumseh Dr Chillicothe OH 45601

PRUSCH, ROBERT DANIEL, b Lackawanna, NY, July 13, 39; m 69. INVERTEBRATE PHYSIOLOGY. *Educ:* Univ Portland, BSc, 64; Syracuse Univ, PhD(physiol), 69. *Prof Exp:* USPHS fel, Case Western Reserve Univ, 69-71; asst prof physiol, Brown Univ, 71-78; assoc prof biol, RI Col, 78-; CHMN, DEPT BIOL, GONZAGA UNIV. *Mem:* Am Physiol Soc; Soc Exp Biol; Soc Gen Physiologists; Am Soc Zoologists. *Res:* Physiology of inorganic ion and water relationships in invertebrate cellular and epithelial systems. *Mailing Add:* Dept Biol Gonzaga Univ ES02 Boone Ave Spokane WA 99205

PRUSINER, STANLEY BEN, b Des Moines, Iowa, May 28, 42; m 70; c 1. NEUROLOGY. *Educ:* Univ Pa, AB, 64, MD, 68. *Prof Exp:* Asst prof neurol, Univ Calif, San Francisco, 74-80; ASST PROF VIROL, UNIV CALIF, BERKELEY, 79-; ASSOC PROF NEUROL, UNIV CALIF, SAN FRANCISCO, 80-, LECTR, BIOCHEM & BIOPHYS, 76- *Concurrent Pos:* Investr, Howard Hughes Med Inst, Univ Calif, San Francisco, 76-81; Res Career Develop Award, NIH, 75-76. *Mem:* Am Soc Biol Chemists; Am Soc Clin Invest; Am Asn Immunologists; Am Chem Soc; Am Acad Neurol. *Res:* Molecular structure of unusual slow infectious agents (prions) that cause degenerative diseases in humans and animals. *Mailing Add:* Dept Neurol Univ Calif Med Ctr Rm 781 HSE San Francisco CA 94143-0518

PRUSLIN, FRED HOWARD, b New York, NY, Oct 24, 51; m 85. AIDS RESEARCH. *Educ:* Yeshiva Univ, BA, 73; Cornell Univ, PhD(microbiol), 78. *Prof Exp:* Postdoctoral fel, Cornell Univ Med Col, 78-84; asst prof, 84-90, SR RES ASSOC, ROCKEFELLER UNIV, 90- *Concurrent Pos:* Consult, Inst Human Genetics & Biochem, Geneva, Switz, 89- *Mem:* Sigma Xi; Am Soc Cell Biol. *Res:* Investigating the role of naturally-occurring antibodies to sperm proteins in the etiology and pathogenesis of AIDS. *Mailing Add:* 166 E 61st St Apt 7J New York NY 10021

PRUSOFF, WILLIAM HERMAN, b New York, NY, June 25, 20; m 48; c 2. BIOCHEMISTRY, PHARMACOLOGY. *Educ:* Univ Miami, BS, 41; Columbia Univ, MS, 47, PhD(chem), 49. *Prof Exp:* From res assoc to sr instr pharmacol res, Western Reserve Univ, 49-53; from res assoc to assoc prof, 53-66, prof, 66-90, EMER PROF PHARMACOL & SR RES SCIENTIST, SCH MED, YALE UNIV, 90- *Honors & Awards:* Exp Therapeut Award, Am Soc Pharmacol & Exp Therapeut, 82; Int Soc Antiviral Res, 88. *Mem:* Am Chem Soc; Am Soc Biol Chemists; Am Asn Cancer Res; Am Soc Pharmacol & Exp Therapeut; Am Soc Microbiol; Am Soc Virol; Soc Chinese Bioscientists Am. *Res:* Nucleic acids; protein isolation; cancer chemotherapy;

biopharmacology; viral chemotherapy; radiation; Design and synthesis of compounds for antiviral and anticancer activity and elucidation of the molecular basis for the biological activity for those compounds that are active. *Mailing Add:* Dept Pharmacol Yale Univ Sch Med New Haven CT 06510

PRUSS, REBECCA M, b Minneapolis, Minn, Sept 5, 50. NEUROSCIENCES. *Educ:* Ill Inst Technol, Chicago, BS, 72; Univ Calif, Los Angeles, PhD(biol chem), 77. *Prof Exp:* Lab res assoc, Dept Renal Res, Michael Reese Hosp, Chicago, Ill, 72-73; postdoctoral fel, Dept Biol Chem, Univ Calif, Los Angeles, 77 & Dept Zool, Univ Col London, Eng, 78-80; staff res assoc, Diabetes Br, Nat Inst Arthritis, Metab & Digestive Dis, NIH, Bethesda, Md, 80; res assoc, Pharmacol & Toxicol Prog, Nat Inst Gen Med Sci, Sect Neuroendocrinol, Lab Clin Sci, NIMH Bethesda, Md, 81-82, staff fel, Lab Cell Biol, 82 & sr staff fel, 83-86; sr res biochemist III, 86-88, res assoc, 88, RES SCIENTIST, MARION MERRELL DOW RES INST, CINCINNATI, OHIO, 86- *Concurrent Pos:* Jane Coffin Childs Mem Fund Med Res postdoctoral fel, 78-80; Juv Diabetes Found postdoctoral fel, 80; Pharmacol Res Assoc Training Prog staff fel, 80-82; adj assoc prof, Dept Anat & Cell Biol, Col Med, Univ Cincinnati, Ohio, 88, lectr, Molecular Biol Cell Grad Prog, 88- *Mem:* AAAS; NY Acad Sci; Am Soc Cell Biol; Am Soc Neurochem. *Res:* Regulation of intracellular calcium and other second messenger systems; excitatory amino acids; ligand-gated and voltage-operated ion channels; neural development and degeneration; neuropeptide synthesis, storage and secretion; developmental biology; growth regulation; immunology; author of various publications. *Mailing Add:* Div Biochem Marion Merrell Dow Res Inst 2110 E Galbraith Rd Cincinnati OH 45215-6300

PRUSS, STANLEY MCQUAIDE, b Detroit, Mich, Sept 24, 43; m. EXPERIMENTAL HIGH ENERGY PHYSICS, ACCELERATOR TECHNOLOGY. *Educ:* Univ Mich, Ann Arbor, BS, 65, MS, 66, PhD(physics), 69. *Prof Exp:* Assoc physicist, Nat Accelerator Lab, 69-72, assoc physicist, 72-81, APPL SCIENTIST, FERMI NAT LAB, 81- *Mem:* Am Phys Soc; AAAS. *Res:* Experimental particle physics-strong interactions. *Mailing Add:* Cross-Gallery Fermi Nat Lab PO Box 500 Batavia IL 60510

PRUSS, THADDEUS P, b Baltimore, Md, Jan 13, 34; m 63; c 2. PHARMACOLOGY. *Educ:* Univ Md, BS, 56, MS, 59; Georgetown Univ, PhD(pharmacol), 62. *Prof Exp:* Sect head cardiovasc pharmacol, Cutter Labs, Inc, 62-66; from sect head cardiovasc pharmacol to actg dir pharmacol, McNeil Labs, Inc, 66-69, dir pharmacol res, 69-80; dir biol res & develop, Revlon Health Care Group, 80-86; DIR SCI EVAL, RORER GROUP, 86- *Mem:* Am Soc Pharmacol & Exp Therapeut; Am Soc Biochem Pharmacol; NY Acad Sci; Soc Exp Biol & Med; Am Heart Asn; Sigma Xi. *Res:* Cardiovascular research in development of drugs for treatment of hypertension, arrhythmias of the heart, coronary artery disease and pulmonary vascular disease; evaluation in licensing and acquisitions. *Mailing Add:* PO Box 642 Gwynedd Valley PA 19437-9999

PRUSSIN, STANLEY GERALD, b Bridgeport, Conn, Nov 20, 39. NUCLEAR CHEMISTRY, RADIOCHEMISTRY. *Educ:* Mass Inst Technol, BS, 60; Univ Mich, MS, 62, PhD(chem), 64. *Prof Exp:* Nuclear chemist, Lawrence Radiation Lab, 64-66, from asst prof to assoc prof nuclear eng, 66-77, PROF NUCLEAR ENG, UNIV CALIF, BERKELEY, 77- *Concurrent Pos:* NSF fel, 64-65; Lawrence Radiation Lab fel, 65-66; consult, Lawrence Radiation Lab, 68-; NATO fel, 72. *Mem:* Am Chem Soc; Am Phys Soc; NY Acad Sci. *Res:* Nuclear spectroscopy; activation analysis. *Mailing Add:* Dept Nuclear Eng Univ Calif Berkeley CA 94720

PRUSSING, JOHN E(DWARD), b Oak Park, Ill, Aug 19, 40; m 65; c 3. ASTRONAUTICS, CONTROL ENGINEERING. *Educ:* Mass Inst Technol, SB, 62, SM, 63, ScD(aeronaut & astronaut), 67. *Prof Exp:* Res asst aeronaut & astronaut, Mass Inst Technol, 63-67; res engr aerospace eng, Univ Calif, San Diego, 67-69; from asst prof to assoc prof, 69-81, asst dean, 76-77, PROF AERONAUT ENG, UNIV ILL, URBANA-CHAMPAIGN, 81- *Concurrent Pos:* Chmn, Astrodynamics Tech Comt, Am Inst Aeronaut & Astronaut, 82; researcher, NASA & US Army Res Off; aerospace consult; assoc ed, Am Inst Aeronaut & Astronaut J Guidance, Control & Dynamics, 90-93. *Mem:* Assoc fel Am Inst Aeronaut & Astronaut; sr mem Am Astronaut Soc; Sigma Xi. *Res:* Orbital mechanics; dynamics, control theory; optimal spacecraft trajectories. *Mailing Add:* Univ Ill 101 TB 104 S Mathews Ave Urbana IL 61801

PRUTKIN, LAWRENCE, b Brooklyn, NY, Dec 6, 35; m 70; c 2. ANATOMY. *Educ:* Brooklyn Col, BS, 57, MA, 60; NY Univ, PhD(basic med sci), 64. *Prof Exp:* From instr to asst prof, 64-70, ASSOC PROF ANAT, MED CTR, NY UNIV, 70- *Concurrent Pos:* USPHS fel, 64-65; vis scientist, Mass Inst Technol, 78-79. *Mem:* Fel NY Acad Sci; Am Asn Anatomists; Am Asn Cancer Res; Am Asn Clin Anatomists; AAAS. *Res:* Electron microscopy of skin tumors; electron microscopic autoradiography; cancer chemotherapy; evaluation of new dermatologic drugs. *Mailing Add:* Dept Cell Biol NY Univ Med Ctr New York NY 10016

PRUZANSKI, WALDEMAR, b Warsaw, Poland, Oct 23, 28; Can citizen; m 63; c 2. MEDICINE, IMMUNOLOGY. *Educ:* Hebrew Univ, Israel, MD, 56; FRCP(C), 68; FACP, 72; FACR, 87. *Prof Exp:* Chief physician, Asaf Harofe Hosp, Tel Aviv Univ, 64-65; res assoc immunol, Inst Cancer Res, Columbia Univ, 65-66; from asst prof to assoc prof, 68-77, PROF MED, UNIV TORONTO, 77-, DIR IMMUNOL, DIAG & RES CTR, 68-, SR STAFF MED, WELLESLEY HOSP, 68-, HEAD DIV IMMUNOL, 75- *Concurrent Pos:* Mem, Inst Med Sci, Univ Toronto, 72-, mem Dept Immunol, 75-, dir inflammation res group, 88- *Mem:* Am Asn Immunol; Am Rheumatism Asn; Am Fedn Clin Res; Can Soc Immunol; fel Am Col Physicians. *Res:* Immunoglobulins in human diseases; metabolism and biological function of lysozyme; humoral and cellular defense mechanisms; rheumatology; arachidonic acid cascade. *Mailing Add:* Wellesley Hosp Toronto ON M4Y 1J3 Can

PRUZANSKY, JACOB JULIUS, b Brooklyn, NY, June 20, 21; m 51; c 3. BIOCHEMISTRY. *Educ:* Brooklyn Col, BA, 41; Iowa State Univ, PhD(chem), 49. *Prof Exp:* Dir biochem res lab, Red Star Yeast Co, 49-51; instr biochem, Western Reserve Univ, 51-54; res assoc, Univ Pittsburgh, 54-56; from asst prof to assoc prof, 56-74, PROF MICROBIOL, MED SCH, NORTHWESTERN UNIV, CHICAGO, 74-, ASSOC DIR ALLERGY RES LAB, 56- *Res:* Immunochemistry. *Mailing Add:* 11945 143rd St N No 7322 Largo FL 34644

PRY, ROBERT HENRY, b Dormont, Pa, Dec 28, 23; m 49; c 5. TECHNOLOGY TRANSFER, GOVERNMENT TECHNOLOGY. *Educ:* Tex Col Arts & Indust, BA, 47; Rice Inst, MA, 49, PhD(physics), 51. *Prof Exp:* Res & develop ctr, Gen Elec Co, 51-76, res & develop mgr, Mat Sci & Eng, 73-74, elec sci & eng, 74-76; vpres res & develop, Combustion Eng, Inc, 76-77; vchmn technol, Gould Inc, 77-83; found pres, Va Ctr Innovative Technol, 84-86; dir, 87-90, EXEC CONSULT, INT INST FOR APPL SYST ANALYSIS, VIENNA, AUSTRIA, 90- *Concurrent Pos:* mem bd dir, Am Mgt Asn, 78-82; mem, Energy Res Adv Bd, Dept of Energy, 80-84; mem & chmn adv bd, Nat Bur Standards, 82-87; adj prof, Mass Inst Technol, 84. *Mem:* Fel AAAS; fel Am Soc Metals; Am Phys Soc; Am Inst Mining Metall & Petrol Engrs; Inst Elec & Electronics Engrs. *Res:* Electrical, magnetic and structural properties of solids; physical metallurgy; technological forecasting and assessment; research application processes; technology policy; management. *Mailing Add:* 11171 Rich Meadow Dr Great Falls VA 22066

PRYCHODKO, WILLIAM WASYL, b Mlyny, Ukraine, Feb 21, 22; nat US; m 49; c 1. ECOLOGY. *Educ:* Univ Munich, PhD(zool), 49. *Prof Exp:* Sr res asst, 51-56, res assoc, 56-63, ASSOC INVESTR, DETROIT INST CANCER RES, 63-; PROF BIOL, WAYNE STATE UNIV, 74- *Concurrent Pos:* Assoc investr, Inst Human Biol, Ann Arbor, 51-55; res assoc biol, Wayne State Univ, 58-60, from asst prof to assoc prof, 60-74. *Mem:* Am Soc Mammalogists; Am Asn Lab Animal Sci; Asn Trop Biol; Ecol Soc Am; Soc Study Evolution. *Res:* Geographic variations of morphological and physiological features in small mammals; responses of small mammals to isolation and aggregation; environmental physiology; biogeography. *Mailing Add:* 67 Oakdale Blvd Pleasant Ridge MI 48069

PRYDE, EVERETT HILTON, b Chicago, Ill, Feb 18, 18; m 59; c 2. LIPID CHEMISTRY, INDUSTRIAL ORGANIC CHEMISTRY. *Educ:* Amherst Col, AB, 39; Univ Mich, AM, 40; Univ Wis, PhD(chem), 48. *Prof Exp:* Res chemist synthetic resin & adhesives, Chem Div, Borden Co, 41-42; sr chemist, Control Lab, Butadiene Div, Koppers Co, 43-45; chemist, Res Div, Electrochem Dept, E I du Pont de Nemours & Co, Inc, 48-57; prin chemist, Northern Regional Res Ctr, 57-61, leader explor org reactions res, 62-76, res coordr, Oilseed Crops Lab, 77-79, LEADER EXPLOR ORG REACTIONS RES, REGIONAL RES CTR, 79- *Mem:* AAAS; Am Chem Soc; Am Oil Chemists Soc; Inst Food Technologists; Am Soc Agr Engrs. *Res:* Polymers; aliphatic acids; aldehydes, amines; ozonization; plasticizers; utilization of lipids and fatty acids; hybrid fuels. *Mailing Add:* 234 E Tower Lane Peoria IL 61614

PRYER, NANCY KATHRYN, b San Francisco, Calif, Apr 26, 59. CYTOSKELETON, PROTEIN SECRETION. *Educ:* Univ Calif, Davis, BS, 82; Univ NC, PhD(biol), 89. *Prof Exp:* POSTDOCTORAL RES FEL, UNIV CALIF, BERKELEY, 89- *Mem:* Am Soc Cell Biol. *Res:* Cytoskeletal structure and assembly dynamics; mechanism of vehicle formation and transport in secretory pathway; interactions between cytoskeleton and secretory pathway. *Mailing Add:* Div Biochem & Molecular Biol 401 Barker Hall Univ Calif Berkeley CA 94720

PRYOR, C(ABELL) NICHOLAS, JR, b Washington, DC, Feb 26, 38; m 57; c 2. ELECTRICAL ENGINEERING. *Educ:* Mass Inst Technol, BS & MS, 60; Univ Md, PhD(elec eng), 66. *Prof Exp:* Chief signal processing div, US Naval Ord Lab, 60-76; tech dir hq, Newport Lab, Naval Underwater Systs Ctr, 77-82; sr scientist, dep chief Naval Opers for Submarine Warfare, 82-83; CONSULT, 83- *Concurrent Pos:* Instr, Mass Inst Technol, 60-; lectr, Catholic Univ, 62-; Univ Md, 66- *Mem:* Inst Elec & Electronics Engrs; Sigma Xi. *Res:* Development of digital signal processing techniques; digital computers; communication theory. *Mailing Add:* 3715 Prosperity Ave Fairfax VA 22031-3324

PRYOR, DAVID BRAM, b Charleston, SC, Oct 18, 51; m 78; c 3. ISCHEMIC HEART DISEASE. *Educ:* Univ Mich Med Sch, MD, 76. *Prof Exp:* Res assoc med, 81, assoc med, 82-83, dir clin epidemiol & biostatist, 84-89, asst prof, 83-89, PROF MED, DUKE UNIV MED CTR, 89. *Concurrent Pos:* Dir cardiol consult serv, Duke Univ Med Ctr, 85-; consult, NIH, spec study sect, 83-, Health Serv Res & Develop Serv, 84-, Nat Ctr Health Serv Res & Health Care Technol, 85-; mem Coun Clin Cardiol, Am Heart Asn. *Mem:* Fel Am Col Cardiol; Am Col Physicians; Am Fedn Clin Res; Am Asn Med Syst & Informatics; fel Am Col Med Informatics. *Res:* Ischemic heart disease - diagnosis, prognosis and clinical research methodology. *Mailing Add:* Duke Univ Med Ctr Box 3531 Durham NC 27710

PRYOR, GORDON ROY, b Warroad, Minn, June 14, 35; m 60; c 3. PLANT NUTRITION, PLANT GENETICS. *Educ:* St Cloud State Univ, BS, 60; Univ Minn, MS, 70, PhD(agron), 76. *Prof Exp:* Scientist, dept agron, Univ Minn, 62-66, sr scientist, dept soil sci, 66-77; RES MGR, CONKLIN CO, INC, 77- *Mem:* Am Soc Agron; Soil Sci Soc Am; Crop Sci Soc; Plant Growth Regulation Soc Am. *Res:* Genetic control of element uptake and translocation; radioisotope assimilation; plant nutrition; fertilizer use and efficiency; spray adjuvants; influence of chemical/biological seed treatments on seed germination and seedling vigor; hybrid seedcorn performance. *Mailing Add:* Conklin Co Inc 889 Valley Park Dr S Shakopee MN 55379

PRYOR, JOSEPH EHRMAN, b Melber, Ky, Mar 19, 18; m 46; c 3. PHYSICAL CHEMISTRY. *Educ:* Harding Col, BA & BS, 37; La State Univ, MA, 39, PhD(phys chem), 43. *Prof Exp:* Instr math, La State Univ, 42-44; PROF CHEM & PHYSICS & DEAN, HARDING COL, 44-, VPRES

ACAD AFFAIRS, 74- *Mem:* AAAS; Am Chem Soc; Am Asn Physics Teachers; Am Sci Affil. *Res:* Electrokinetics; zeta potentials; polygons of maximum area inscribed in an ellipse. *Mailing Add:* Dept Chem Harding Univ Box 773 Searcy AR 72143-5500

PRYOR, MARILYN ANN ZIRK, b Harrisonburg, Va, Mar 9, 36. CELL PHYSIOLOGY, COMPARATIVE PHYSIOLOGY. *Educ:* Madison Col, Va, BS, 56; Univ Tenn, MS, 58, PhD(physiol), 61. *Prof Exp:* Instr cell physiol & biochem, Bryn Mawr Col, 61-62; res assoc renal physiol, 62-64, dean studies, 74-77, from asst prof to assoc prof, 70-77, PROF BIOL SCI, MT HOLYOKE COL, 77- *Concurrent Pos:* NSF fac fel, Inst Arctic & Alpine Res, Univ Colo, 70-71. *Mem:* Am Soc Zoologists; AAAS; Sigma Xi. *Res:* Influence of melatonin on thermogenesis in brown adipose tissue; biochemical mechanisms of environmental adaptation. *Mailing Add:* Dept of Biol Sci Mt Holyoke Col South Hadley MA 01075

PRYOR, MICHAEL J, b Leek, Eng, Mar 31, 25; nat US; m 51; c 3. METALLURGY. *Educ:* Cambridge Univ, BA, 46, MA & PhD(metallic corrosion), 49. *Prof Exp:* Fel phys chem, Nat Res Coun Can, 49-51, asst res officer, 51-53; head, Basic Physics Br, Kaiser Aluminum & Chem Corp, 53-57; chief, Metall Sect, 57-63, assoc dir, 63-71, dir, 71-80, vpres, Metals Res Labs, Olin Corp, 80-85; RETIRED. *Mem:* Hon mem Electrochem Soc; Am Inst Mining, Metall & Petrol Engrs. *Res:* Chemistry and physics of metals. *Mailing Add:* 98 Maplevale Dr WB Woodbridge CT 06525

PRYOR, RICHARD J, b Pittsburgh, Pa, Aug 25, 40; m 65; c 1. NUCLEAR PHYSICS. *Educ:* Pa State Univ, BS, 65; Univ Pittsburgh, PhD(nuclear physics), 70. *Prof Exp:* Res physicist, Savannah River Lab, E I Du Pont de Nemours & Co, Inc, 70-72, sr res supvr, 72-76; mem staff, Los Alamos Sci Lab, 76-84; RES MGR, SAVANNAH RIVER LAB, 84- *Concurrent Pos:* Lectr, Univ SC, 71-76. *Mem:* Am Phys Soc; Am Nuclear Soc. *Res:* Direct transport solutions of reactor problems, numerical methods and analysis; fluid mechanics. *Mailing Add:* Dept 6410 Sandia Nat Lab Albuquerque NM 87185

PRYOR, WAYNE ARTHUR, b Bellvue, Pa, May 6, 28; m 48. GEOLOGY. *Educ:* Centenary Col, BS, 52; Univ Ill, MS, 54; Rutgers Univ, PhD(geol), 59. *Prof Exp:* Sr asst geol, Centenary Col, 50-52; asst, Univ Ill, 52-53; asst groundwater, Ill Geol Surv, 53-56; asst geol, Rutgers Univ, 56-57; asst groundwater, Ill Geol Surv, 57-59; geologist, Gulf Res & Develop Co, 59-66; assoc prof geol, 66-71, PROF GEOL & FEL, GRAD SCH, UNIV CINCINNATI, 71- *Mem:* Geol Soc Am; Am Asn Petrol Geologists; Sigma Xi. *Res:* Sedimentary petrology and environmental stratigraphy; groundwater geology and micropaleontology. *Mailing Add:* Dept of Geol Univ of Cincinnati Cincinnati OH 45221

PRYOR, WILLIAM AUSTIN, b St Louis, Mo, Mar 10, 29. ORGANIC CHEMISTRY, BIOLOGICAL CHEMISTRY. *Educ:* Univ Chicago, PhB, 48, BS, 51; Univ Calif, PhD(chem), 54. *Prof Exp:* Res chemist, Calif Res Corp, 54-60; asst prof chem, Purdue Univ, 60-63; from assoc prof to prof, 63-72, THOMAS & DAVID BOYD PROF CHEM & BIOCHEM & DIR BIODYNAMICS INST, LA STATE UNIV, BATON ROUGE, 72- *Concurrent Pos:* Spec lectr, Univ Calif, Berkeley, 55-60; lectr, Fla State Univ, 67; instr, Off Continuing Educ, Washington Univ, 68, San Francisco State Univ, 68, Wright State Univ, 68 & Eastman Kodak Labs, NY, 70; Guggenheim fel, Univ Calif, Los Angeles, 70-71; La State Univ Found distinguished fac fel, 70-71; NIH spec fel, Berkeley, 71; vis scientist, Lab Chem Biodyn, Calif, 71; indust consult; vis prof, Univ Calif, Davis, 78, Univ Calif, San Diego, 78 & Duke Univ, 79; ed-in-chief, Free Radicals in Biol, 76-84, co-ed-chief, Free Radiol Biol & Med, 84- *Honors & Awards:* Du Pont Lectr, Univ SC, 68; Award, Am Chem Soc, 75, Petrol Chem Award, 80, Southern Chemist Medal, 84; Merit Award, Nat Inst Health, 86; Harper Award, Am Coun Advan Med, 87; Charles Coates Award, Am Chem Soc & Am Inst Chem Engrs, 89. *Mem:* Am Chem Soc; Royal Soc Chem; fel Geront Soc; Radiation Res Soc; Pan-Am Med Asn; fel AAAS; Am Aging Asn; Am Asn Cancer Res; fel Am Col Nutrit; fel Am Inst Chem; Am Soc Biochem & Molecular Biol; Am Soc Photobiol; Chem Soc London; Soc Toxicol; Soc Free Radiol Res; Sigma Xi; Radiation Res Soc; Oxygen Soc; NY Acad Sci; Lipid Peroxidation Soc. *Res:* Kinetics and mechanisms of organic reactions; free radical chemistry; vinyl polymerization; sulfur chemistry; chemistry of aging; chemistry of smog; destructive reactions of free radicals in biological systems; toxicity of smoke, soot and tar. *Mailing Add:* Biodynamics Inst 711 Choppin Baton Rouge LA 70803-1800

PRYSE-PHILLIPS, WILLIAM EDWARD MAIBEN, b London, Eng, Dec 13, 37; Brit & Can citizen; m 77; c 3. MEDICINE, NEUROLOGY. *Educ:* Univ London, MB, 61; MD, 68, MRCP, 68; FRCP(C), 71, MRCPsychiat, 72, FRCP, 81. *Prof Exp:* Cullip sr res fel psychiat, Univ Birmingham, 64-66; Med Res Coun fel biochem, 68; assoc prof, 72-80, PROF NEUROL, MEM UNIV NFLD, 80- *Concurrent Pos:* Attend mem staff, Gen Hosp, St John's, Nfld, 72-; ed, Nfld Med Asn J, 76-81; univ orator, Mem Univ; vis scholar, Green Col, Oxford, 89-90. *Mem:* Can Epilepsy Asn; Multiple Sclerosis Soc Can; Can Neurol Soc; Am Asn Electrodiagnostic Med. *Res:* Migraine, multiple sclerosis, myotonic dystrophy and Alzheimer disease. *Mailing Add:* Dept Med Neurol Mem Univ St John's NF A1B 3V6 Can

PRYSTOWSKY, HARRY, b Charleston, SC, May 18, 25; m 51; c 3. MEDICINE. *Educ:* The Citadel, BS, 44; Med Col SC, MD, 48; Am Bd Obstet & Gynec, dipl, 59. *Prof Exp:* Intern, Johns Hopkins Hosp, 48-49; res fel, Cincinnati Gen Hosp, 49-50; asst resident, Johns Hopkins Hosp, 50-51 & 53-54, resident & instr, 54-55; res fel, Yale Univ, 55-56; asst prof obstet & gynec, Sch Med, Johns Hopkins Univ, 56-58; prof obstet & gynec & head dept, Col Med, Univ Fla, 58-73; PROF OBSTET & GYNEC, DEAN & PROVOST, COL MED, M S HERSHEY MED CTR, PA STATE UNIV, 73- *Concurrent Pos:* Consult, Jacksonville Naval Hosp; nat consult, US Air Force; mem bd dirs, Am Bd Obstet & Gynec. *Mem:* Am Fertil Soc; Am Col Surg; Am Gynec Soc; Am Col Obstet & Gynec; Am Fedn Clin Res. *Res:* Delivery of female health care. *Mailing Add:* Pa State Univ 500 University Dr Hershey PA 17033

PRYWES, MOSHE, MEDICAL EDUCATION. *Prof Exp:* CHMN, CTR MED EDUC, BEN GOURION UNIV NEGEV. *Mem:* Inst Med-Nat Acad Sci. *Mailing Add:* Ctr Med Educ Ben Gourion Univ Negev PO Box 653 Beersheba 84105 Israel

PRYWES, NOAH S(HMARYA), b Warsaw, Poland, Nov 28, 25; nat US; m 53; c 3. COMPUTER SCIENCE, ENGINEERING. *Educ:* Israel Inst Technol, BSc, 49; Carnegie Inst Technol, MSc, 51; Harvard Univ, PhD(appl physics), 54. *Prof Exp:* Asst, Carnegie Inst Technol, 50-52; develop engr, Radio Corp Am, 54-55; dept mgr, Univac Div, Sperry Rand Corp, 56-58; from asst prof to assoc prof elec eng, 58-68, PROF COMPUT SCI, UNIV PA, 68- *Concurrent Pos:* Consult, Industry & US Gov; pres, Comput Command & Control Co, 63-; adj prof, Shanghai Jiao Teng Univ. *Mem:* Fel Inst Elec & Electronics Engrs; Sigma Xi. *Res:* Design and application of computers, computing devices; automatic programming; information retrieval. *Mailing Add:* 122 Moore Sch of Elec Eng Univ of Pa Philadelphia PA 19104

PRZIREMBEL, CHRISTIAN E G, b Brunn, Ger, Mar 30, 42; US citizen; m 63; c 2. SEPARATED FLOWS, GAS DYNAMICS. *Educ:* Rutgers Univ, BS, 63, MS, 64, PhD(mech eng), 67. *Prof Exp:* From asst prof to prof mech eng, Rutgers Univ, 67-81, assoc dean, Eng Acad Affairs, 77-81; PROF MECH ENG & HEAD DEPT, CLEMSON UNIV, 81- *Concurrent Pos:* Consult, Singer-Gen Precision, Kearfolt Div, 71-72, L B Foster Co, 75, Flo-Tron Inc, 77-80 & Porzio & Bromberg, 79-80; fac fel, US Air Force & Am Soc Eng Educ fac res prog, 76. *Honors & Awards:* Gold Medal, Pi Tau Sigma, 73. *Mem:* Fel Am Soc Mech Engrs; assoc fel Am Inst Aeronaut & Astronaut; fel Am Soc Eng Educ; Sigma Xi; fel Am Asn Adv Sci. *Res:* Fluid mechanics and gas dynamics; subsonic and supersonic near-wakes; vehicle aerodynamics; aerothermodynamics; slender body aerodynamics. *Mailing Add:* Dept Mech Eng Clemson Univ Clemson SC 29634-0921

PRZYBYLOWICZ, EDWIN P, b Detroit, Mich, June 29, 33; m 54; c 11. ANALYTICAL CHEMISTRY. *Educ:* Univ Mich, BS, 53; Mass Inst Technol, PhD(chem), 56. *Prof Exp:* Res chemist, Dow Chem Co, 54; res chemist, 56-69, head gen anal lab, 59, sr staff res labs, 60-68, asst div head, Methods Res & Tech Div, 69-72, asst head emulsion res div, 72-74, tech asst to dir res labs, 74, dir photog prog develop, 75-77, asst dir res labs, 77-85, DIR RES & SR VPRES, RES LABS, EASTMAN KODAK CO, 85- *Concurrent Pos:* Acad assignment, Nat Bur Standards & Mass Inst Technol, 68-69; mem, Nat Res Coun Eval Panel Anal Chem, 74-77, chmn, 78-81; cor assoc rep, Am Chem Soc. *Mem:* Nat Acad Eng; Am Chem Soc; AAAS; Indust Res Inst. *Res:* Electroanalytical methods of analysis; trace analysis; chromatography; photographic chemistry; neutron activation analysis. *Mailing Add:* 1219 Crown Point Dr Webster NY 14580

PRZYBYLSKA, MARIA, b Warsaw, Poland, Mar 2, 23; nat Can; m 47; c 2. X-RAY CRYSTALLOGRAPHY. *Educ:* Univ Glasgow, BSc, 46, PhD(x-ray crystallog), 49. *Prof Exp:* Fel, Div Physics, 50-52, asst res officer, Div Appl Chem, 52-54 & Div Pure Chem, 54-56, assoc res officer, 56-60; sr res officer, Biochem Lab, 60-69, sr res officer, 69-72, sr res officer, div biol sci, Nat Res Coun, Can, 72-88; RETIRED. *Mem:* Am Crystallog Asn. *Res:* Determination of molecular and crystal structures of organic compounds and proteins. *Mailing Add:* 14 Wick Cresent Gloucester ON K1J 7H2 Can

PRZYBYLSKI, RONALD J, b Chicago, Ill, Jan 5, 36; m 59, 90; c 3. ZOOLOGY, CELL BIOLOGY. *Educ:* DePaul Univ, BS, 57, MS, 59; Univ Chicago, PhD(zool), 63. *Prof Exp:* From sr instr to asst prof, 65-72, ASSOC PROF ANAT, SCH MED, CASE WESTERN RESERVE UNIV, 72- *Concurrent Pos:* Estab investr, Am Heart Asn, 72-77; fel, Coun Clin Cardiol, Am Heart Asn. *Mem:* Am Soc Cell Biol; Am Soc Zool; Soc Develop Biol; Am Asn Anatomists; Int Soc Develop Biol; Am Soc Physiol. *Res:* Cell surface proteins and glycoproteins cell physiology of developing cardiac and skeletal muscle cells; experimental cardiac hypertrophy; autoradiography techniques; insulin and growth factor action in developing skeletal, cardiac and smooth muscle; cellular endocrinology. *Mailing Add:* Dept of Anat Sch of Med Case Western Reserve Univ Cleveland OH 44106

PRZYBYSZ, JOHN XAVIER, b Steubenville, Ohio, July 22, 50; m 73; c 5. DIGITAL ELECTRONICS. *Educ:* Mich State Univ, BS, 71; Univ Ill, Urbana, MS, 74, PhD(physics), 76. *Prof Exp:* Instr physics, Wayne State Univ, 76-78; SR ENGR APPL SCI, WESTINGHOUSE RES CTR, 78- *Mem:* Am Phys Soc; Inst Elec & Electronic Engrs; Sigma Xi. *Res:* Cryogenic electronics; Josephson junctions, high electron mobility transistors (HEMT); high speed digital integrated circuit design and fabrication. *Mailing Add:* 682 Twin Oak Dr Pittsburgh PA 15235

PRZYBYTEK, JAMES THEODORE, b Chicago, Ill, July 27, 45; m 70; c 2. ANALYTICAL CHEMISTRY, ORGANIC CHEMISTRY. *Educ:* Univ Ill, Chicago Circle, BS, 67, MS, 69, PhD(chem), 72. *Prof Exp:* Chemist, Nalco Chem Co, 72-73, sr chemist, 73-75, group leader, 75-77; qual control mgr org chem, 77-78, lab dir, 79-91, TECH DIR, BURDICK & JACKSON DIV, 91- *Mem:* Am Chem Soc; Soc Appl Spectros; Asn Off Anal Chemists. *Res:* Chromatography; spectroscopy; photochemistry; solvent purification and contamination control. *Mailing Add:* Burdick & Jackson Baxter Diag Inc 1953 S Harvey St Muskegon MI 49442

PSAROUTHAKIS, JOHN, b Chania, Crete, Greece, June 29, 32; US citizen; m 59; c 2. MECHANICAL ENGINEERING. *Educ:* Mass Inst Technol, BSc, 57, MSc, 62; Univ Md, PhD(mech eng), 65; Carnegie-Mellon Univ, cert exec mgt, 68. *Prof Exp:* Engr, Boston Edison Co, 57-58; tech, Mgr Isotope Dept, Thermo Electron Corp, 58-62; sr staff scientist & mgr thermionic res, Martin-Marietta Co, 62-66; dir physics, controls & elec res, Allis-Chalmers Co, 66-68, dir, Technol Ctr, 68-70; vpres corp planning & technol, Masco Co, 70-73; group vpres, Int Opers, 73-77; pres, 77-81, CHMN & PRES, J P INDUSTS, INC, ANN ARBOR, MICH, 81- *Concurrent Pos:* Consult, Lab Electronics, Inc, 62; lectr, Chalmers Tech Univ, Sweden, 63 & 65 & Grad Sch Indust Admin, Carnegie-Mellon Univ, 69-71. *Mem:* Am Mgt Asn; Am Soc Metals. *Res:* Energy conversion; surface phenomena; controls; metal forming;

thermionic energy conversion, adsorption and emission phenomena of alkaline, and alkaline earths on high temperature metal surfaces, high temperature heat transfer, and hydroextrusion of metals and plastics. *Mailing Add:* 3820 Gulf Blvd No P3 St Petersburg FL 33706

PSCHEIDT, GORDON ROBERT, b Madison, Wis, Nov 4, 28; div; c 4. BIOCHEMISTRY. *Educ:* Univ Wis, BS, 50, MS, 53, PhD(physiol chem), 57. *Prof Exp:* Res assoc, Sch Med, Univ Wash, 57-58; med res assoc, Galesburg State Hosp, 58-68; counr chem & eng, Personnel Inc, 68-74; res assoc psychiat, Univ Chicago, 74-81; RETIRED. *Res:* Intermediary metabolism of the brain; enzyme inhibitor-neurohormone relationships; relationship of blood enzymes to mental illness; clinical research; theory of mental illness. *Mailing Add:* 527 N Barstow St Eau Claire WI 54701

PSCHUNDER, RALPH J(OSEF), b Marburg, Austria, June 20, 20; US citizen; m 43; c 4. MECHANICAL ENGINEERING. *Educ:* Vienna Tech Univ, MME, 41, PhD(mech eng), 65. *Prof Exp:* Res engr, Viennese Locomotive Works, Austria, 41-47; lectr mech eng, Inst Technol, Vienna, 48-54, consult, Fed Testing Lab, 51-54; res engr, NY Air Brake Co, 54-59; prin design engr, RCA Corp, 59-87; CONSULT, 87- *Honors & Awards:* David Sarnoff Award, RCA Corp, 74. *Res:* Structural dynamics of radars; stress analysis of antennas and radomes; cable structures; guyed towers; radar foundations; structural dynamics finite element analysis; ship structures. *Mailing Add:* 25 Valley View Terr Moorestown NJ 08057

PSIODA, JOSEPH ADAM, b Pittsburgh, Pa, Apr 22, 49; m 71. MATERIALS SCIENCE. *Educ:* Carnegie-Mellon Univ, BS, 71, MS, 73, PhD(mat sci), 77. *Prof Exp:* Mfg mgt enr, Gen Elec Co, 71-72; asst instr mat sci, Carnegie-Mellon Univ, 74-75; sr res metallurgist, Linde Div, Union Carbide Corp, 76-77, St Joe Minerals Corp, 77-78; proj scientist, 78-79, RES SCIENTIST, UNION CARBIDE CORP, 81- *Mem:* Am Soc Metals; Am Inst Mining, Metall & Petrol Engrs Metall Soc; Am Soc Testing & Mat; Electrochem Soc; Nat Asn Corrosion Engrs. *Res:* Effects of microstructure on the properties of structural alloys and materials; role of microstructure in the corrosion resistance of alloys with particular emphasis of applying a multidisciplinary approach; electrochemical and surface science techniques. *Mailing Add:* 312 Bear Ridge Road Pleasantville NY 10570

PSUTY, NORBERT PHILLIP, b Hamtramck, Mich, June 13, 37; m 59; c 3. COASTAL GEOMORPHOLOGY, COASTAL ECOLOGY. *Educ:* Wayne State Univ, BS, 59; Miami Univ, Ohio, MS, 60; La State Univ, PhD(geog), 66. *Prof Exp:* Instr geol & geog, Univ Miami, Fla, 64-65; asst prof geol, Univ Wis, Madison, 65-69; assoc prof, 69-73, PROF GEOG & GEOL, RUTGERS UNIV, 73-, DIR CTR COASTAL & ENVIRON STUDIES, 76-, DISTINGUISHED PROF, 79- *Concurrent Pos:* Mem, Comn Coastal Environ Int Geog Union, 77-, vchmn, 88-, US mem, Pan Am Inst Geog & Hist Orgn Am States, 78-82, Mid & NAtlantic Outer Continental Shelf Tech Working Group, Bur Land Mgt, 79-83; chair, Marine Geog Spec Group, Asn Am Geog, 76-78 & 81-82; mem exec bd, Comn Coastal Environ Int Geog Union, 84-, mem, Barrier Islands Comt, Nat Park Serv Dept Interior, 84-86; assoc dir, Inst Marine & Coastal Sci, 88- *Mem:* Coastal Soc (pres, 80-82); AAAS; Asn Am Geographers. *Res:* Coastal geomorphology, incorporating both the natural processes and coastal management based on scientific data and theory; barrier island and coastal dune systems. *Mailing Add:* Ctr Coastal Environ Studies Rutgers Univ Busch Campus New Brunswick NJ 08903

PTACEK, ANTON D, b Chicago, Ill, Nov 17, 33; m 60; c 2. GEOLOGY. *Educ:* Univ Wis, BS, 59; Univ Wash, MS, 62, PhD(geol), 65. *Prof Exp:* Instr geol, Olympic Col, 64-65; asst prof, 65-70, ASSOC PROF GEOL, SAN DIEGO STATE UNIV, 70- *Mem:* AAAS; Geol Soc Am; Geochem Soc. *Res:* Volcanology and the distribution of trace elements in various volcanic tectonic settings. *Mailing Add:* Dept Geol San Diego State Univ San Diego CA 92182

PTAK, ROGER LEON, b Wyandotte, Mich, Sept 17, 38; m 63; c 3. ASTROPHYSICS, ASTRONOMY. *Educ:* Univ Detroit, BS, 60; Cornell Univ, MS, 64, PhD(physics), 66. *Prof Exp:* Asst prof physics, De Pauw Univ, 66-68; from asst prof to assoc prof physics, 68-77, PROF PHYSICS, BOWLING GREEN STATE UNIV, 77- *Mem:* Am Astron Soc; Sigma Xi. *Res:* Ultraviolet spectroscopy of Seyfert galaxies; interpretation of the broad emission lines observed in Seyfert galaxies and quasars and understanding the nature of these objects. *Mailing Add:* Physics & Astron Bowling Green State Univ Bowling Green OH 43403

PTASHNE, MARK, b Chicago, Ill, June 5, 40. MOLECULAR BIOLOGY. *Educ:* Reed Col, BA, 61; Harvard Univ, PhD(molecular biol), 68. *Prof Exp:* Soc Fels jr fel, 65-68, lectr, 68-71, chmn dept, 80-83, PROF BIOCHEM & MOLECULAR BIOL, HARVARD UNIV, 71- *Concurrent Pos:* Guggenheim fel, 73-74. *Honors & Awards:* Harvey Lectr, 75; Eli Lilly Award in Biol Chem, 75; Charles-Leopold Mayer Prize, Acad Sci, Inst France, 77; US Steel Found Award in Molecular Biol, 79; Gairdner Found Int Award, 85; Louisa Gross Horwitz Prize, Columbia Univ, 85; Feodor Lynen Lectr, 88; George Drummond Mem Lectr, 89; Sloan Prize, Gen Motors Cancer Res Found, 90. *Mem:* Nat Acad Sci; fel Am Acad Arts & Sci; fel NY Acad Sci; Fedn Am Scientists. *Res:* Gene control by repressors; author of numerous publications. *Mailing Add:* Dept Biochem & Molecular Biol Harvard Univ 7 Divinity Ave Cambridge MA 02138

PUAR, MOHINDAR S, b Ghurial, India, Jan 3, 35; m 66; c 1. PHYSICAL ORGANIC CHEMISTRY, ANALYTICAL CHEMISTRY. *Educ:* Punjab Univ, India, BSc, 56, MSc, 58; Boston Univ, PhD(phys org chem), 66. *Prof Exp:* Fel, Brandeis Univ, 65-67; sr res chemist, Olin Mathieson Chem Corp, 67-69; sr res investr anal res & develop, Squibb Inst Med Res, 69-78; RES GROUP LEADER NMR PHYS ORG CHEM, SCHERING CORP, 78- *Mem:* Am Chem Soc. *Res:* Equilibrium and kinetics; reaction mechanisms; spectroscopic methods; nuclear magnetic resonance spectroscopy; drugs. *Mailing Add:* 94 Fairview Dr S Basking Ridge NJ 07920

PUBOLS, BENJAMIN HENRY, JR, b Washington, DC, Sept 1, 31; m 64; c 1. NEUROSCIENCES. *Educ:* Univ Va, BA, 53; Univ Wis, MS, 55, PhD(psychol), 57. *Prof Exp:* Asst prof psychol, Utah State Univ, 57-58; asst prof, Univ Miami, 58-61; Nat Inst Neurol Dis & Blindness fel neurophysiol, Univ Wis, 61-64; res assoc animal behav, Hershey Med Ctr, Pa State Univ, 64-67, from assoc prof zool & asst prof anat, 66-70, assoc prof anat, 64-82; SR SCI NEUROL SCI INST, GOOD SAMARITAN HOSP & MED CTR, 82- *Mem:* Am Asn Anatomists; Neurosci Soc. *Res:* Neural electrophysiology and functional neuroanatomy of mammalian somatic sensory systems. *Mailing Add:* Neurol Sci Inst Good Samaritan Hosp 1120 NW 20th Ave Portland OR 97209

PUBOLS, LILLIAN MENGES, b Columbus, Ohio, Oct 13, 39; m 64; c 1. NEUROPHYSIOLOGY. *Educ:* Rutgers Univ, AB, 61; Univ Wis, MA, 64, PhD(psychol), 66. *Prof Exp:* Proj assoc, Animal Behav Res Lab, Pa State Univ, 66-67, res assoc, Hershey Med Ctr, 67-71, asst prof anat, 71-74; asst prof anat, Med Col Pa, 74-77, assoc prof, 77; AT NEUROL SCI INST, GOOD SAMARITAN HOSP & MED CTR, 77- *Concurrent Pos:* Prin investr, NIH grant, 80-83; Fogarty Sr Int fel, 80-81; mem bd sci counrs, Nat Inst Dent Res, 80-84. *Mem:* Am Asn Anat; Neurosci Soc. *Res:* Anatomy and physiology of the mammalian somatic sensory nervous system. *Mailing Add:* Neurol Sci Inst Good Samaritan Hosp & Med Ctr 1120 NW 20th Ave Portland OR 97209

PUBOLS, MERTON HAROLD, b Portland, Ore, Sept 2, 31; m 53; c 4. NUTRITIONAL BIOCHEMISTRY. *Educ:* Lewis & Clark Col, BA, 53; Purdue Univ, MS, 60, PhD(biochem), 62. *Prof Exp:* Chemist, Ore State Bd Health, 53-55; analyst, State Chemist, Purdue Univ, 55-61, res asst biochem, 61-62; asst prof, 66-70, ASST SCIENTIST, DEPT ANIMAL SCI, WASH STATE UNIV, 62-, ASSOC PROF, 70- *Concurrent Pos:* NIH spec fel; guest assoc scientist, Brookhaven Nat Lab, 68-69. *Mem:* Am Chem Soc. *Res:* Naturally occuring toxicants and enzyme inhibitors; polypeptide hormones. *Mailing Add:* Dept of Animal Sci Clark Hall Wash State Univ Pullman WA 99164-6320

PUCCI, PAUL F(RANCIS), b Haledon, NJ, Dec 16, 23; m 47; c 2. MECHANICAL ENGINEERING. *Educ:* Purdue Univ, BSME, 49, MSME, 50; Stanford Univ, PhD(mech eng), 55. *Prof Exp:* Sr res engr, Ford Motor Co, 53-55; sr thermodyn engr, Convair Div, Gen Corp, 55-56; assoc prof, 56-66, PROF MECH ENG, NAVAL POSTGRAD SCH, 66- *Mem:* Am Soc Mech Engrs. *Res:* Fluid mechanics; heat transfer; gas turbines. *Mailing Add:* Dept Mech Eng Naval Postgrad Sch Monterey CA 93940

PUCEL, ROBERT A(LBIN), b Ely, Minn, Dec 27, 26; m 52; c 5. ELECTRICAL ENGINEERING. *Educ:* Mass Inst Technol, MS, 51, DSc(elec eng, network synthesis), 55. *Prof Exp:* Staff mem, Microwave Tube Noise Group, 55-65, staff mem, Solid State Physics Group & proj mgr, Microwave Semiconductor Prog, 65-70, consult microwave solid state devices, 70-74, CONSULT SCIENTIST, RES DIV, RAYTHEON CO, LEXINGTON, 74- *Concurrent Pos:* Nat lectr, Microwave Theory & Technol Soc, 80-; ed, Microwave Intergrated Circuits, Inst Elec & Electronics Engrs, 85; lectr, Monolithic Microwave Intergrated Circuits. *Honors & Awards:* Microwave Prize, Microwave Theory & Tech Soc, 77, Microwave Career Award, 90. *Mem:* Fel Inst Elec & Electronic Engrs; Microwave Theory & Tech Soc; Electron Devices Soc. *Res:* Novel semiconductor concepts for microwave applications; feasibility studies of integrated microwave circuits using novel materials and device designs; network synthesis theory; solid state devices and microwave devices research; gallium arsenic monolithic microwave circuits. *Mailing Add:* 427 South St Needham MA 02192

PUCHTLER, HOLDE, b Kleinlosnitz, Ger, Jan 1, 20; nat US. HISTOCHEMISTRY, PATHOLOGY. *Educ:* Univ Cologne, MD, 51. *Prof Exp:* Res assoc path, Univ Cologne, 51-55; Damon Runyon Mem Fund res fel, 55-58; Nat Cancer Inst Can res fel, 58-59; from res assoc to assoc res prof, 59-68, PROF PATH & DIR, HISTOCHEM RES UNIT, MED COL GA, 68- *Mem:* Fel Am Inst Chem; fel Royal Micros Soc; Anatomische Gesellschaft; Histochem Soc; Ger Histochem Soc; Royal Soc Chem. *Res:* Chemical basis of staining technics; application of principles of textile and leather dyeing to histology; development of staining, polarization and fluorescence microscopic methods; problems of quantitative staining; infrared fluorescence microscopy. *Mailing Add:* Dept Path Med Col of Ga Augusta GA 30912-3600

PUCK, MARY HILL, b Takoma Park, Md, Oct 29, 19; m 46; c 3. HUMAN GENETICS, NEUROSCIENCES. *Educ:* Judson Col, BA, 40; Univ Chicago, MA, 46. *Prof Exp:* Psychiat social worker child welfare, NMex State Dept, Sante Fe, 45-47; psychiat social worker adolescents, Scholar & Guid Asn, Chicago, 47-48; psychiat social worker diag, 63-75, res assoc chromosome abnormalities, Dept Biochem, Biophys, Genetics & Psychiat, Med Ctr, Univ Colo, 75-85; CLIN RES ASSOC, DEPT PSYCHIAT, UNIV COLO HEALTH SCI CTR, DENVER, 85- *Concurrent Pos:* Res fel, Eleanor Roosevelt Inst Cancer Res, Denver, 76-; clin res assoc, Dept Psychiat, Univ Colo Health Sci Ctr, Denver, 85- *Mem:* Am Soc Human Genetics; AAAS; Nat Asn Social Workers. *Res:* Application of genetic and cell biology concepts to study of human developmental diseases; assessment of biological and cultural needs in specific populations; study of child and adolescent development; molecular genetics in clinical neurological disorders. *Mailing Add:* Ten S Albion St Denver CO 80222

PUCK, THEODORE THOMAS, b Chicago, Ill, Sept 24, 16; m 46; c 3. BIOPHYSICS, GENETICS. *Educ:* Univ Chicago, BS, 37, PhD(phys chem), 40. *Prof Exp:* Res assoc med, Univ Chicago, 41-45, asst prof med & biochem, 45-47; Am Cancer Soc sr fel, Calif Inst Technol, 47-48; prof biophys & chmn dept, 48-67, RES PROF BIOCHEM, BIOPHYS & GENETICS, UNIV COLO MED CTR, DENVER, 67-, PROF MED, 81-, DISTINGUISHED PROF, 87- *Concurrent Pos:* Mem, Comn Airborne Infections, Army Epidemiol Bd, Off Surgeon Gen, 44-46; mem, Comn Physicians for Future, Macy Found; vis prof, Univ Ind, 60, Rockefeller Univ, 60, Stanford Univ, 78, Univ Tex, 80, Baylor Sch Med, 82, Yale Univ, 84 & Claremon Cols, 86; dir,

Eleanor Roosevelt Inst Cancer Res, 62-; distinguished res prof, Am Cancer Soc, 66; lectrs, var asns, orgn & univs, 75-; Heritage Found scholar, 84; distinguished prof, Univ Colo, 87; mem, Steering Comt, Given Inst Path, Coun Nat Inst Arthritis & Metab Dis, NIH, sci be, Santa Fe Inst, 87 & Terrapin, Diagnostics, 88, Joint Sci Comt, Claremont Cols, 89; vis fel, Los Alamos Nat Lab, 87-; mem, Adv Comt Radiol Health, Food & Drug Admin, 87- *Honors & Awards:* Lasker Award, Am Pub Health Asn, 58; Heidelberger Lectr, Col Physicians & Surgeons, 58; Borden Award, Asn Am Med Cols, 59; Squibb Mem Lectr, 59; Harvey Soc Lectr, 59; Karl F Muenzinger Lectureship, 63; Voice of Am Lectr, 72; Louisa Gross Horwitz Award, 73; Flow Lectr, Birmingham, Eng, 73; Gordon Wilson Medal, Am Clin & Climat Asn, 77; Edmund Hall Mem Lectr, Univ Louisville, 77; Ann Award, Environ Mutagen Soc, 81; E B Wilson Medal, Am Soc Cell Biol, 84; Bonfils-Stanton Award Sci, 84; Hon Award, Tissue Cult Asn, 87. *Mem:* Nat Acad Sci; Inst Med-Nat Acad Sci; Am Soc Cell Biol; Am Soc Human Genetics; fel Am Acad Arts & Sci; Res Prof Am Cancer Soc; Sigma Xi; Radiation Res Soc; Am Asn Immunologists; AAAS. *Res:* Somatic cell genetics and genetic biochemistry; mammalian development; cancer. *Mailing Add:* E Roosevelt Inst Cancer Res Med Ctr 1899 Gaylord St Denver CO 80208

PUCKETT, ALLEN EMERSON, b Springfield, Ohio, July 25, 19; m 40, 63; c 5. AERONAUTICAL ENGINEERING. *Educ:* Harvard Univ, BS, 39, MS, 41; Calif Inst Technol, PhD(aeronaut eng), 49. *Hon Degrees:* DSc, Salem Col, 77. *Prof Exp:* Consult aerodyn, Calif Inst Technol, 41-45, lectr aeronaut & chief wind tunnel sect, Jet Propulsion Lab, 45-49; from dept head to exec vpres & asst gen mgr, Hughes Aircraft Co, 49-76, pres, 77-78, chmn & chief exec officer, 78-, EMER CHMN, HUGHES AIRCRAFT CO, 87- *Concurrent Pos:* Consult various indust & govt; mem, US Mil Tech Mission, 45, US Mil Intel Mission, 47 & guided missile comt, Res & Develop Bd, past chmn subcomt high speed aerodyn & mem comt aerodyn, Nat Adv Comt Aeronaut, 53-55, mem subcomt automatic stabilization & control, NASA, 56-59, chmn res adv comt on control guid & navig, 59-64; mem group, Adv Panel Aeronaut, Off Asst Secy Defense, Res & Eng, chmn mutual weapons develop prog, 56-61; mem exec comt & full bd, Defense Sci Bd, 62-64, vchmn, 62-64; mem army sci adv panel, Dept Army, 65-69; chmn, Aerospace Industs Asn, 79 & mem, sr adv bd; chmn, Eng Res Bd & Space Applications Bd, Nat Acad Eng, mem, Indust Adv Bd. *Honors & Awards:* Lawrence Sperry Award, Inst Aeronaut Sci, 49; Lloyd V Berkner Award, Am Astron Soc, 74; Frederik Philips Award, Inst Elec & Electronics Engrs, 83; Int Award, Britian Inst Production Engrs, 83; Medal of Honor, Electroni; Medal of Honor, Electronic Indust Asn, 84; Williard F Rockwell Jr Medal, Int Technol Inst, 84; Nat Technol Medal, Pres Reagan, 85. *Mem:* Nat Acad Eng; Aerospace Indust Asn; hon fel Am Inst Aeronaut & Astronaut; Int Acad Astronaut. *Res:* Supersonic aerodynamics; supersonic airfoil theory; guided missiles; co-author of one book. *Mailing Add:* Hughes Aircraft Co PO Box 45066 Los Angeles CA 90045-0066

PUCKETT, HOYLE BROOKS, b Jesup, Ga, Oct 15, 25; m 45; c 3. AGRICULTURAL ENGINEERING. *Educ:* Univ Ga, BS, 48; Mich State Univ, MS, 49. *Prof Exp:* Res agr engr, 49-59, invest leader, 59-72, res leader, Agr Res Serv, 72-85, CONSULT AGR ENGR, USDA, 85- *Mem:* Inst Elec & Electronics Engrs; Am Soc Agr Engrs; Sigma Xi. *Res:* Development of controls and equipment for farmstead materials handling and processing operations. *Mailing Add:* RR 2 Box 181 Urbana IL 61801

PUCKETT, HUGH, b Louisville, Ky, Aug 9, 29; m 55; c 1. INVERTEBRATE ENDOCRINOLOGY. *Educ:* Western Ky State Col, BS, 54; Univ Va, MA, 59, PhD(biol), 64. *Prof Exp:* Asst prof biol, Norfolk Div, Col William & Mary, 59-61 & Univ Tenn, 62-64; asst prof, Western Ky Univ, 64-69, prof biol, 69-88; RETIRED. *Mem:* Sigma Xi. *Res:* Sexual endocrinology of the crustaceans; function and development of the androgenic gland of the crayfish; vertebrate developmental endocrinology. *Mailing Add:* 729 Morehead Way Bowling Green KY 42104-3006

PUCKETT, RUSSELL ELWOOD, b Ewing, Ky, Mar 28, 29; m 49; c 5. ELECTRICAL ENGINEERING. *Educ:* Univ Ky, BS, 56, MS, 59. *Prof Exp:* Engr, Dept of Defense, 56-57; asst prof elec eng, Col Eng, Univ Ky, 57-66, res engr, dept physics, 67-68, assoc dir eng res, 68-77; ASSOC PROF ENG TECHNOL, TEX A&M UNIV, 77- *Concurrent Pos:* Mem, spec comt electronic res, Hwy Res Bd, Nat Acad Sci-Nat Res Coun, 64-69 & comt traffic control devices, 69-70; assoc prof, Lexington Tech Inst, Univ Ky, 71-75; tech writer, Arabian-Am Oil Co, Houston, Tex, 85. *Mem:* Am Soc Eng Educ; Inst Elec & Electronics Engrs. *Res:* Administration and management of engineering research; electronic instrumentation with emphasis on development and design utilizing components in novel applications. *Mailing Add:* Eng Technol Dept Texas A&M Univ Col Station TX 77843-3367

PUCKETTE, STEPHEN ELLIOTT, b Ridgewood, NJ, Oct 18, 27; m 57; c 5. MATHEMATICS. *Educ:* Univ of the South, BS, 49; Yale Univ, MS, 50, MA, 51, PhD(math), 57. *Prof Exp:* Asst, Yale Univ, 53-56; from asst prof to assoc prof math, Univ of the South, 56-66; assoc prof, Univ Ky, 66-69; dean col arts & sci, 69-79, PROF, UNIV OF THE SOUTH, 79- *Concurrent Pos:* NSF fac fel, 64-65; Fulbright-Hays prof, Univ Abidjan, Ivory Coast, 79-80. *Mem:* Am Math Soc; Math Soc France; Ger Math Soc. *Res:* Functional analysis. *Mailing Add:* Univ of the South Sewanee TN 37375

PUCKNAT, JOHN GODFREY, b New York, NY, Feb 22, 31; m 60; c 1. ORGANIC CHEMISTRY. *Educ:* City Col New York, BS, 57; NY Univ, MS, 62, PhD(org chem), 66. *Prof Exp:* Res chemist, Cent Res Labs, Interchem Corp, 57-63; fel, Israel Inst Technol, 66-67; fel, Org Chem Inst, Univ Zurich, 67-68; sr res chemist, Argus Chem Corp, Brooklyn, 69-70; res assoc, Inmont Corp, NJ, 70-85, BASF INMONT DIV, 85-91; SUN CHEMICAL CORP, 91- *Mem:* AAAS; Am Chem Soc; NY Acad Sci. *Res:* Organic synthesis and reaction mechanisms; thiophene and natural product chemistry; dyes and pigments; polymer research and development, complaint and all-aqueous coatings and ink systems. *Mailing Add:* 565 West End Ave New York NY 10024

PUCKORIUS, PAUL RONALD, b Chicago, Ill, Apr 7, 30; m 53; c 3. CORROSION CONTROL, WATER TREATMENT. *Educ:* NCent Col, Naperville, Ill, BA, 53. *Prof Exp:* Mkt mgr & prod mgt, Water Treat Serv Co, Nalco Chem Co, Chicago, 53-69; exec vpres, Water Treat Serv Co, Zimmite Corp, Cleveland, 69-75; PRES, PUCKORIUS & ASSOCS, INC, 75- *Mem:* Nat Asn Corrosion Engrs; Am Soc Testing Mat; Am Petrol Inst; Nat Petrol Refineries Asn. *Res:* Cooling water systems; scale occurrence and control; microbiological corrosion and control. *Mailing Add:* 2338 Hearth Dr No 27 Evergreen CO 80439

PUDDEPHATT, RICHARD JOHN, b Aylesbury, Eng, Oct 12, 43; m 69; c 2. ORGANOMETALLIC CHEMISTRY, INORGANIC CHEMISTRY. *Educ:* Univ London, BSc, 65, PhD(chem), 68. *Prof Exp:* Teaching fel, Univ Western Ont, 68-70; lectr, Univ Liverpool, 70-77, sr lectr, 77-78; PROF CHEM, UNIV WESTERN ONT, 78- *Honors & Awards:* Alcan Award, 85; Florence Bucke Sci Prize, 85. *Mem:* Royal Soc Chem; Can Inst Chem; Am Chem Soc. *Res:* Reactivity and mechanism in organometallic chemistry, particularly of the elements platinum, gold and lead. *Mailing Add:* Dept Chem Univ Western Ont London ON N6A 5B7 Can

PUDDINGTON, IRA E(DWIN), b Clifton, NB, Can, Jan 8, 11; m 36; c 1. COLLOID CHEMISTRY. *Educ:* Mt Allison Univ, BSc, 33; McGill Univ, MSc, 36, PhD(chem), 38. *Hon Degrees:* DSc, Mt Allison Univ, 67, Carleton Univ, 75 & Mem Univ, 77. *Prof Exp:* Instr chem, Mt Allison Univ, 33-34; lectr, McGill Univ, 36-37 & Sir George Williams Univ, 37-38; res officer, Chem Div, 38-52, dir Appl Chem Div, 52-69, dir, Chem Div, 69-74, CONSULT, NAT RES COUN CAN, 74- *Concurrent Pos:* Lectr, dept chem, Univ Ottawa, 53-57; adj prof, Carleton Univ, 74- *Honors & Awards:* Montreal Medal, Chem Inst Can, 71; R S Jane Award, Can Soc Chem Eng, 75; Pioneer Award, Am Inst Chemists, 84. *Mem:* Chem Inst Can, (pres, 67-68); Royal Soc Can; Am Chem Soc; Am Inst Chemists; NY Acad Sci. *Res:* Behavior of small particles; particle enlargement; slimes separation. *Mailing Add:* 2324 Alta Vista Dr Ottawa ON K1H 7M7 Can

PUDDINGTON, LYNN, b Pleasanton, Calif, July 24, 56. CELL BIOLOGY. *Educ:* Iowa State Univ, Ames, BS, 78; Wake Forest Univ, Winston-Salem, PhD(microbiol & immunol), 84. *Prof Exp:* Teaching asst, Iowa State Univ, 77 & Bowman Gray Sch Med, Wake Forest Univ, 82-84; NIH postdoctoral fel, Salk Inst Biol Studies, 84-86; res scientist hairgrowth, 88-90, RES SCIENTIST CELL BIOL, UPJOHN CO, 90- *Concurrent Pos:* Vis scientist cardiovasc dis, Upjohn Co, 86-88; mem, Coun Basic Sci, Am Heart Asn, 89- *Mem:* Am Soc Microbiol; Am Soc Cell Biol; Am Heart Asn. *Res:* Hairgrowth research; cell biology; author of more than 20 technical publications. *Mailing Add:* Cell Biol Res Upjohn Co 301 Henrietta St 7239-25-11 Kalamazoo MI 49001

PUDELKIEWICZ, WALTER JOSEPH, b Oil City, Pa, Feb 3, 23; m 60; c 2. BIOCHEMISTRY. *Educ:* Pa State Univ, BS, 51, MS, 53, PhD, 55. *Prof Exp:* ASSOC PROF NUTRIT SCI, UNIV CONN, 55- *Mem:* AAAS; Am Chem Soc; Am Inst Nutrit. *Res:* Fat soluble vitamins; chlorinated hydrocarbon toxicants. *Mailing Add:* 80 Navratil Rd West Willington CT 06279

PUEPPKE, STEVEN GLENN, b Fargo, NDak, Aug 1, 50. PHYTOPATHOLOGY, PLANT BIOCHEMISTRY. *Educ:* Mich State Univ, BSc, 71; Cornell Univ, PhD(phytopath), 75. *Prof Exp:* Sr res assoc, C F Kettering Res Lab, 75-76; asst prof biol, Univ Mo-St Louis, 76-79; asst prof plant path, Univ Fla, 79-; AT DEPT PLANT PATH, UNIV MO, COLUMBIA. *Mem:* Am Phytopath Soc; Am Soc Plant Physiologists; Bot Soc Am; Am Soc Microbiol; Sigma Xi. *Res:* Biochemistry of plant-microorganism interactions; recognition of bacteria by plants; phytoalexins; lectins. *Mailing Add:* Dept Plant Path Univ Mo Columbia 108 Waters Hall Columbia MO 65211

PUERCKHAUER, GERHARD WILHELM RICHARD, b Regensburg, Ger, Apr 28, 30. ORGANIC CHEMISTRY. *Educ:* Purdue Univ, PhD, 56; Univ Minn, MBA, 64. *Prof Exp:* Fel org fluorine compounds, Purdue Univ, 56; sr res chemist, 57-64, group leader, 64-65, develop specialist, 66, int tech mgr, 66-70, DEVELOP MGR, MINN MINING & MFG CO, 70- *Mem:* Am Chem Soc; Sigma Xi. *Res:* Production of organic and organometallic compounds, research and development; copying products; non-impact printing papers; data base management and systems analysis. *Mailing Add:* Rte 1 Box 559 Roan Mountain TN 37687-9730

PUESCHEL, RUDOLF FRANZ, b Tetschen, Czech, Apr 9, 34; US citizen; m 63; c 3. ATMOSPHERIC PHYSICS, ATMOSPHERIC CHEMISTRY. *Educ:* Univ Frankfurt, BS, 58; Univ Giessen, MS, 62; Univ Wash, PhD(civil eng), 69. *Prof Exp:* Assoc res physicist, Calif Inst Technol, 62-64; res physicist, Univ Wash, 64-66; Nat Res Coun res assoc aerosol sci, Environ Res Labs, 69-70, dir, Mauna Loa Observ, 70-72, SUPV PHYSICIST, ENVIRON RES LABS, NAT OCEANIC & ATMOSPHERIC ADMIN, 72- *Concurrent Pos:* Consult, Spec Meteorol Appln Dept, World Meteorol Orgn, Geneva, Switz, 74-; adj prof physics, Denver Univ, 77- *Mem:* Ger Phys Soc; Am Geophys Union; AAAS; Am Meteorol Soc; Sigma Xi. *Res:* Effects of aerosols on cloud nucleation and on radiative transfer. *Mailing Add:* NASA Ames Res Ctr MS 245-4 Moffett Field CA 94035

PUESCHEL, SIEGFRIED M, b Waldenburg, Ger, July 28, 31; US citizen; m 62; c 4. PEDIATRICS. *Educ:* Braunschweig Col, BS, 55; Univ Düsseldorf, MD, 60; Harvard Univ, MPH, 67; Univ RI, PhD, 85. *Prof Exp:* Instr pediat, Harvard Med Sch, 67-74, asst prof, 74-75; from asst prof to assoc prof, 75-85, PROF PEDIAT, PROG MED, BROWN UNIV, 85-, DIR CHILD DEVELOP CTR, RI HOSP, 75- *Concurrent Pos:* Dir Down syndrome prog, Children's Hosp Med Ctr, 70-75 & PKU Clin, 72-75. *Mem:* Am Acad Pediat; Soc Human Genetics. *Res:* Phenylketonuria, particularly maternal phenylketonuria and amino acid transfer across placenta; amino acid metabolism, Down syndrome, etiology, epidemiology and counseling. *Mailing Add:* Child Develop Ctr RI Hosp 593 Eddy St Providence RI 02902

PUETT, J DAVID, b Morganton, NC, Mar 23, 39; m 61; c 2. BIOCHEMISTRY, ENDOCRINOLOGY. *Educ:* NC State Univ, BS, 61; Univ NC, Chapel Hill, MS, 65, PhD(biochem), 69. *Prof Exp:* Scientist biopolymers, Chemstrand Res Ctr Inc, Monsanto Co, 62-67; from asst prof to prof biochem, Vanderbilt Univ, 69-83; PROF BIOCHEM & DIR REPSCEND LABS, UNIV MIAMI, 83- *Concurrent Pos:* USPHS res grant, Vanderbilt Univ, 72-83 & Univ Miami, 83- *Honors & Awards:* Camille & Henry Dreyfus Found Award, 71. *Mem:* Biophys Soc; Am Chem Soc; Am Soc Biol Chemists; Endocrine Soc; AAAS. *Res:* Protein structure-function; polypeptide and glycoprotein hormone structure, mechanism-of-action and metabolism; physical biochemistry; opiate peptides; site-directed mutagenesis; hormones and cancer. *Mailing Add:* Dept Biochem Vanderbilt Univ Nashville TN 37232

PUFF, ROBERT DAVID, b St Louis, Mo, Sept 30, 33; m 54; c 4. THEORETICAL PHYSICS. *Educ:* Washington Univ, BS, 54; Harvard Univ, PhD(physics), 60. *Prof Exp:* Asst physics res, Lincoln Lab, Mass Inst Technol, 59-60; res assoc, Univ Ill, Urbana, 60-62; from asst to assoc prof, 62-74, PROF PHYSICS, UNIV WASH, 74- *Mem:* Am Phys Soc. *Res:* Nuclear and solid state many-body theory. *Mailing Add:* Dept Physics Univ Wash Seattle WA 98195

PUFFER, JOHN H, b Buffalo, NY, July 4, 41; m 61; c 4. PETROLOGY, GEOCHEMISTRY. *Educ:* Mich State Univ, BS, 63, MS, 65; Stanford Univ, PhD(geol), 69. *Prof Exp:* Asst prof geol, Eastern N Mex Univ, 68-70; from asst prof to assoc prof, 70-82, PROF GEOL, RUTGERS UNIV, NEWARK, 82- *Mem:* Geol Soc Am. *Res:* Granite petrology; geochemistry of basaltic rocks. *Mailing Add:* 34 Pine Grove Rd Berkeley Heights NJ 07922

PUGH, CLAUD ERVIN, b Asheboro, NC, Oct 6, 39; m 67; c 2. INELASTIC SOLID MECHANICS. *Educ:* NC State Univ, BS, 61, MS, 64, PhD(eng mech), 68. *Prof Exp:* Instr eng mech, NC State Univ, 65-68; researcher, 68-79, PROG MGR DESIGN TECHNOL, OAK RIDGE NAT LAB, 79- *Concurrent Pos:* Instr mech eng, Univ Tenn, 69. *Mem:* Am Soc Mech Engrs; Soc Exp Stress Analysis; Sigma Xi. *Res:* Structural design technology for high temperature components; emphasis on elastic-plastic and creep behavior of alloys and structures; constitutive equation development; creep-fatigue failure criteria; confirmatory structural testing. *Mailing Add:* 9724 Franklin Hill Blvd Knoxville TN 37922-3331

PUGH, DAVID MILTON, b Philadelphia, Pa, July 13, 29; m 55; c 3. INTERNAL MEDICINE, CARDIOLOGY. *Educ:* Univ Rochester, BA, 51; Yale Univ, MD, 58. *Prof Exp:* Intern, US Naval Hosp, Bethesda, MD, 58-59; resident, Univ Wash, 59-61, NIH fel, 61-64; from instr to assoc prof, 64-76, PROF CARDIOL, SCH MED, UNIV KANS MED CTR, KANSAS CITY, 76- *Concurrent Pos:* Consult, Vet Admin Hosp, Kansas City, 64-, St Margaret's Hosp & Providence Hosp, Kansas City, 70-, Vet Admin Hosp, Leavenworth, Kans, 70- *Mem:* Am Soc Clin Res; AMA; fel Am Col Physicians; fel Am Col Cardiol; fel Am Col Chest Physicians; fel Clin Cardiol; Am Med Asn. *Res:* Cardiovascular physiology with particular reference to left ventricular function. *Mailing Add:* Dept Med E116 Univ Kans Col Med 39th St & Rainbow Blvd Kansas City KS 66103

PUGH, E(DISON) NEVILLE, b Glamorgan, Gt Brit, May 5, 35, US citizen; m 56; c 2. METALLURGY. *Educ:* Univ Wales, BSc, 56, PhD(metall), 59. *Prof Exp:* Sci officer, Defence Standards Labs, NSWBr, Australian Dept Supply, 59-63; scientist, Res Inst Advan Studies, Martin Marietta Corp, Md, 63-65, staff scientist, 65-66, sr res scientist, 66-69; assoc prof eng & appl sci, George Washington Univ, 69-70; assoc prof metall & mining eng, Univ Ill, Urbana, 70-73, prof, 73-80; mem staff, 80-85, CHIEF, METALL DIV, NAT INST STANDARDS & TECHNOL, 85- *Honors & Awards:* Willis Rodney Whitney Award, Nat Asn Corrosion Engrs, 84. *Mem:* Metall Soc; Am Inst Mining, Metall & Petrol Engrs; fel Am Soc Metals; Nat Asn Corrosion Engrs; Am Soc Testing & Mat. *Res:* Fracture of crystalline solids, principally the influence of environments on fracture; corrosion of metals. *Mailing Add:* 9541 Ash Hollow Pl Gaithersburg MD 20879

PUGH, EMERSON WILLIAM, b Pasadena, Calif, May 1, 29; m 58; c 3. SOLID STATE ELECTRONICS. *Educ:* Carnegie Inst Technol, BSc, 51, PhD(physics), 56. *Prof Exp:* Asst prof physics, Carnegie Inst Technol, 56-57; staff physicist, Res Lab, 57-61, mgr metals physics group, 58-61, vis scientist, Zurich Lab, 61-62, sr engr components div, 62-65, dir oper memory, 65-66, dir tech planning, Res Div, 66-68, spec asst to vpres & chief scientist, 68-71, consult to dir res, 71-75, mgr explor magnetics, Res Div, 75-80, RES STAFF MEM, RES DIV, IBM CORP, 81- *Mem:* Fel AAAS; fel Am Phys Soc; fel Inst Elec & Electronics Engrs (pres, 89); Sigma Xi. *Res:* Data processing devices and systems; magnetic materials and device physics; history of data processing systems. *Mailing Add:* Brandon Dr Mt Kisco NY 10549

PUGH, HOWEL GRIFFITH, b Abercarn, Eng, Oct 13, 33; m 60; c 2. NUCLEAR PHYSICS, HIGH ENERGY PHYSICS. *Educ:* Cambridge Univ, BA, 55, PhD(nuclear physics), 61. *Prof Exp:* Fel nuclear physics, Atomic Energy Res Estab, Eng, 59-62; physicist, Lawrence Radiation Lab, Univ Calif, 62-65; assoc prof, Univ Md, College Park, 65-70, prof physics & astron, 70-77; head, Nuclear Sci Sect, NSF, 77-79; sci dir, Bevalac, 79-86, STAFF SR SCIENTIST, LAWRENCE BERKELEY LAB, UNIV CALIF, 79- *Concurrent Pos:* Prog dir intermediate energy physics, NSF, Washington, DC, 75-77. *Mem:* Fel Am Phys Soc. *Res:* Nuclear spectroscopy; quasi-free scattering; nuclear reactions at medium energies; ultra-relativistic proton-proton and nucleus-nucleus interactions; relativistic heavy ion collisions. *Mailing Add:* Nuclear Sci Div 70A-3307 Lawrence Berkeley Lab Berkeley CA 94720

PUGH, JEAN ELIZABETH, b Newport News, Va, Apr 25, 28. INVERTEBRATE ZOOLOGY. *Educ:* Madison Col, BS, 50; Univ Va, MA, 54, PhD(invert histol), 61. *Prof Exp:* From instr to assoc prof biol, Old Dominion Col, 54-65; assoc prof, 65-66, chmn dept, 66-74, PROF BIOL, CHRISTOPHER NEWPORT COL, 66- *Mem:* AAAS; Am Micros Soc; Sigma Xi. *Res:* Invertebrate histology; hindgut of the decapod crustaceans. *Mailing Add:* Rte 4 Box 882 Hayes VA 23072

PUGH, JOHN W(ILLIAM), b Pekin, Ill, July 4, 23; m 56; c 1. METALLURGY, REFRACTORY METALS. *Educ:* Univ Ill, BS, 48; Rensselaer Polytech Inst, MMetE, 54. *Prof Exp:* Student engr, 48-49, metallurgist, Res Lab, 49-56, assoc in metall, Lamp Div, 56-59, mgr metall studies & alloy develop, Lamp Metals & Components Dept, 59-63, mgr metall res, 63-68, mgr res & develop, 68-70, res adv, Gen Elec Lighting Res & Technol Serv Oper, 70-88, SR CONSULT METALLURGIST, GEN ELEC CO, 88- *Concurrent Pos:* Chmn, Refractory Metals comt, Metall Soc of Am Inst Mining, Metall & Petrol Engrs, 80-82. *Mem:* Am Soc Metals; Am Inst Mining, Metall & Petrol Engrs. *Res:* High temperature alloys; refractory metals; arc melting; powder metallurgy; mechanical properties; lighting systems development; lamp engineering. *Mailing Add:* Gen Elec Lighting Res & Tech Serv Oper Nela Park Cleveland OH 44112

PUGH, ROBERT E, b Kelowna, BC, Apr 29, 33; m 61, 68; c 1. THEORETICAL PHYSICS. *Educ:* Univ BC, BA, 54, MA, 55; Univ Iowa, PhD(physics), 63. *Prof Exp:* Sci officer, Defense Res Bd, 58-59; lectr math, Queen's Univ, Ont, 59-60; instr physics, Univ Iowa, 60-62, res assoc, 62-63; from asst prof to prof, 63-72, assoc dean arts & sci, 78-81, PROF PHYSICS, UNIV TORONTO, 72- *Concurrent Pos:* Secy-treas, Inst Particle Physics, 73-86. *Mem:* Am Phys Soc. *Res:* Quantum field theory; elementary particle physics. *Mailing Add:* Dept of Physics Univ Toronto Toronto ON M5S 1A7 Can

PUGLIA, CHARLES DAVID, b Paterson, NJ, Sept 5, 41; m 67; c 1. PHARMACOLOGY, ENVIRONMENTAL PHYSIOLOGY. *Educ:* St Johns Univ, NY, BS, 63; Rutgers Univ, MS, 65; Temple Univ, PhD(pharmacol), 71. *Prof Exp:* Res assoc environ physiol, Inst Environ Med, Univ Pa, 71-73, assoc pharmacol, Sch Med, 73-77; ASST PROF PHARMACOL, MED COL, PA, 77- *Mem:* AAAS; Undersea Med Soc; Sigma Xi. *Res:* Mechanisms of oxygen and oxidant toxicity; endogenous antioxidants; thermal homeostasis in extreme environments. *Mailing Add:* 14 Winslow Way Dresher PA 19025

PUGLIA, CHARLES RAYMOND, b Oak Park, Ill, Aug 3, 27; m 54; c 3. INVERTEBRATE ZOOLOGY. *Educ:* Elmhurst Col, BS, 52; Univ Ill, MS, 53, PhD(zool), 59. *Prof Exp:* Instr biol, Elmhurst Col, 53-54; from assoc prof to prof, Wells Col, 59-70; dean, Div Natural Sci, New Eng Col, 70-74, dean admin, 74-76, prof biol, 76-85; var positions, 85-91, DIV DIR, DIV TEACHER PREP & ENHANCEMENT, NSF, 91- *Concurrent Pos:* NSF sci fac fel, 64-65. *Mem:* Am Inst Biol Sci; Am Micros Soc; Soc Syst Zool; Sigma Xi; AAAS. *Res:* Tardigrade taxonomy and ecology. *Mailing Add:* NSF 1800 G St NW Washington DC 20550

PUGLIELLI, VINCENT GEORGE, b Newton, Mass, Feb 12, 43; m 68; c 2. SOLID STATE PHYSICS, ELECTRICAL ENGINEERING. *Educ:* Boston Col, BS, 64; Univ Mass, Amherst, MS, 66, PhD(physics), 70. *Prof Exp:* Res asst high energy physics, Brookhaven Nat Lab, 65; res asst physics, Univ Mass, Amherst, 68-70; SR SCIENTIST PROG MGR ELECTROMAGNETICS, NAVAL SURFACE WEAPONS CTR, 70- *Concurrent Pos:* Vis prof physics, US Naval Acad, 80-81. *Mem:* Inst Elec & Electronics Engrs (pres, 77). *Res:* Semiconductor devices; electromagnetic compatibility; microwaves and radar systems. *Mailing Add:* Battelle Memorial Inst 505 King Ave 4-133 Columbus OH 43201

PUGLIESE, MICHAEL, b Jan 10, 27; US citizen; m 54; c 4. ORGANIC CHEMISTRY. *Educ:* Univ Bologna, PhD(org chem), 51. *Prof Exp:* Develop chemist, Interchem Corp, 55-57; res org chemist, S B Penick & Co, 57-59; res org chemist, 59-72, MGR RES & DEVELOP PRINTING INK VEHICLES, RES DEPT, J M HUBER CORP, EDISON, 72-, TECH DIR PRINTING INK DIV, 84- *Mem:* Am Chem Soc. *Res:* Development of printing ink vehicles with particular interest in letterpress, lithographic, and flexographic inks. *Mailing Add:* J M Huber Corp Raritan Indust Ctr Pershing Ave Edison NJ 08837

PUGMIRE, RONALD J, b Shelley, Idaho, Jan 6, 37; m 65; c 2. PHYSICAL CHEMISTRY. *Educ:* Idaho State Univ, BS, 59; Univ Utah, PhD(phys chem), 66. *Prof Exp:* Fel, 66-68, asst to vpres res, 70-73, asst vpres res, 73-78, sr res chemist, 68-80, PROF FUELS ENG, UNIV UTAH, 80-, ASSOC VPRES RES, 78- *Mem:* Am Chem Soc. *Res:* Nuclear magnetic resonance. *Mailing Add:* Univ Utah 210 Park Bldg Salt Lake City UT 84112-1102

PUGSLEY, JAMES H(ARWOOD), b Berea, Ky, Mar 28, 36; m 57; c 2. ELECTRICAL ENGINEERING. *Educ:* Oberlin Col, AB, 56; Univ Ill, MS, 58, PhD(elec eng), 63. *Prof Exp:* ASSOC PROF ELEC ENG, UNIV MD, COLLEGE PARK, 63- *Mem:* Inst Elec & Electronics Engrs; Asn Comput Mach; Am Soc Eng Educ; Sigma Xi. *Res:* Switching circuit theory and digital computer design; use of computers in electrical engineering problems. *Mailing Add:* 11003 Childs St Silver Spring MD 20901

PUHALLA, JOHN EDWARD, b Scranton, Pa, Mar 30, 39. MICROBIAL GENETICS. *Educ:* Pa State Univ, BS, 61; Cornell Univ, PhD(genetics), 66. *Prof Exp:* Asst geneticist, Conn Agr Exp Sta, 65-70; res geneticist, Nat Cotton Path Lab, USDA, 71-80; MEM FAC, DEPT PLANT PATH, UNIV CALIF, 80- *Concurrent Pos:* Fac mem, Grad Sch, Tex A&M Univ, 75- *Mem:* Am Phytopath Soc; Genetics Soc Am. *Res:* Inheritance of mating type in fungi; genetics of pathogenicity in plant parasitic fungi; parasexual recombination. *Mailing Add:* PO Box 7583 Berkeley CA 94707

PUHL, RICHARD JAMES, b Eau Claire, Wis, Oct 12, 44; m 67; c 1. METABOLISM OF XENOBIOTIC COMPOUNDS. *Educ:* Univ Wis-Eau Claire, BS, 66; Univ Minn, Minneapolis, PhD(org chem), 71. *Prof Exp:* Res chemist, Uniroyal Chem Div, Uniroyal, Inc, 71-74; res chemist, Chemagro Agr Div, 74-81, sr res chemist, Mobay Chem Corp, 81-84; mgr, Metab Dept, 84-91, ASSOC DIR METAB & DISPOSITION, HAZLETON LABS, 91- *Mem:* Am Chem Soc; Am Asn Pharmaceut Scientists; NY Acad Sci; AAAS; Int Soc Study Xenobiotics. *Res:* Metabolic and environmental fate of agricultural chemicals; disposition of pharmaceutical compounds in mammals. *Mailing Add:* Hazleton Lab 3301 Kinsman Blvd Madison WI 53704

PUHVEL, SIRJE MADLI, b Tallinn, Estonia, July 26, 39; US citizen; m 60; c 3. MEDICAL MICROBIOLOGY. *Educ:* Univ Calif, Los Angeles, BA, 61, PhD(med microbiol, immunol), 63. *Prof Exp:* NIH fel dermat, 63-65, asst res immunologist, 65-72, assoc res immunologist, 72-73, adj assoc prof, 73-78, ADJ PROF, DEPT MED, UNIV CALIF, LOS ANGELES, 78- *Mem:* Am Soc Microbiol; Soc Invest Dermat. *Res:* Microbiology and immunology of skin diseases, particularly acne vulgaris; microbiology of normal skin; biochemistry of keratinization; mechanism of action by TCDD in skin. *Mailing Add:* Dept Med Div Dermat Univ Calif Ctr Health Sci Los Angeles CA 90024

PUI, CHING-HON, b Hong Kong, Aug 20, 51. PEDIATRIC HEMATOLOGY-ONCOLOGY. *Educ:* Nat Taiwan Univ, 76. *Prof Exp:* Res pediat, St Judes Children's Hosp, 77-79, fel pediat hemat-oncol, 79-81, res assoc hemat-oncol, 81-82, asst mem, 82- 86, assoc mem hemat-oncol, 86-90; assoc prof, 86-90, PROF PEDIAT, UNIV TENN SCH MED, 90-; MEM HEMAT-ONCOL & PATH, ST JUDE CHILDREN'S RES HOSP, 90- *Concurrent Pos:* prin investr, Pediat Oncol Group; mem ad hoc spec rev comt, Nat Cancer Inst. *Mem:* Soc Pediat Res; Am Soc Hemat; Am Soc Clin Oncol; Am Asn Cancer Res; Am Acad Pediat; Pediat Oncol Group. *Res:* Biology and treatment of childhood leukemia and lymphand. *Mailing Add:* St Jude Children's Res Hosp 332 N Lauderdale PO Box 318 Memphis TN 38101

PUKKILA, PATRICIA JEAN, b Woodbury, NJ, Sept 28, 48; m 74; c 1. MOLECULAR GENETICS, DEVELOPMENTAL GENETICS. *Educ:* Univ Wis-Madison, BS, 70; Yale Univ, Phd(biol), 75. *Prof Exp:* Postdoctoral fel genetics, Nat Inst Med Res, London, 75-77 & Harvard Univ, 77-79; asst prof, 79-86, ASSOC PROF BIOL, UNIV NC, CHAPEL HILL, 86- *Concurrent Pos:* Mem, Curric Genetics, Univ NC, 79- & Cell Biol Adv Panel, NSF, 88-; prin investr, NIH & NSF, 80-; assoc ed, Genetics, 90- *Mem:* Genetics Soc Am; Am Soc Cell Biol. *Res:* Molecular mechanisms that underlie chromosome behavior during meiosis; development of new model system, Coprinus cinereus, to analyze relationships between genetic exchange and chromosome pairing. *Mailing Add:* Dept Biol CB No 3280 Univ NC Chapel Hill NC 27599-3280

PULAY, PETER, b Veszprem, Hungary, Sept 20, 41; m 70; c 2. QUANTUM CHEMISTRY. *Educ:* Eotvos Univ, Budapest, dipl, 63; Univ Stuttgart, MD, 70. *Prof Exp:* Res assoc inorg chem, Hungarian Acad Sci, 63-66, res fel, 70-74; assoc prof chem, Eotvos Univ, 75-81; sr res fel, Univ Tex, Austin, 81-82; prof, 82-83, DISTINGUISHED PROF CHEM, UNIV ARK, FAYETTEVILLE, 83- *Concurrent Pos:* Vis scientist, Univ Tex, Austin, 76, vis prof, 80; vis assoc prof, Univ Calif, Berkeley, 76. *Honors & Awards:* Hungarian Acad Sci award, 79; Medal, Int Acad Quantum Molecular Sci, 82. *Mem:* Am Chem Soc; World Asn Theoret Organic Chemists; Sigma Xi; Int Acad Quantum Molecular Sci. *Res:* Improved methods for calculating molecular wave functions and potential surfaces; application of quantum chemical techniques for the study of molecular structure and reactivity, including geometries, vibrational spectra, reaction barriers, and nuclear magnetic resonance chemical shifts. *Mailing Add:* Dept Chem Univ Ark Fayetteville AR 72101

PULESTON, HARRY SAMUEL, b Wellington, Colo, July 6, 17; m 41; c 3. CHEMISTRY. *Educ:* Colo State Univ, BS, 41; Iowa State Col, MS, 42; Univ Colo, PhD(pharmaceut chem), 50. *Prof Exp:* Asst state chemist, State Dept Agr, Iowa, 42; asst res fel, Mellon Inst, 42-43; res chemist, Wander Co, 43-46; asst instr chem, Univ Colo, 46-47, asst, Atomic Energy Proj, 49-50; asst prof chem, Colo State Univ, 50-55; assoc prof, Univ Wyo, 55-56; from asst prof to assoc prof, 56-64, PROF CHEM, COLO STATE UNIV, 64- *Mem:* Am Chem Soc. *Res:* Bacterial and plant irradiation and its counteraction; radiochemistry; biochemistry. *Mailing Add:* Dept of Chem Colo State Univ Ft Collins CO 80523

PULIDO, MIGUEL L, weed science, plant pathology; deceased, see previous edition for last biography

PULIGANDLA, VISWANADHAM, b Vijayawada, Andhra, India, Jan 12, 38; m 62; c 2. PHYSICAL CHEMISTRY, INORGANIC CHEMISTRY. *Educ:* Sri Venkateswara Univ, India, BSc, 57; Univ Saugor, MSc, 59; Univ Toledo, PhD(chem), 74. *Prof Exp:* Sr teacher chem, Andhra Educ Soc Higher Secondary Sch, New Delhi, 60-61; sr res asst, Indian Inst Astrophys & Indian Meteorol Dept, 61-68; asst prof chem, Ohio Dominican Col, 74-78; asst res chemist magnetohydrodyn & energy, Dept Chem, Mont State Univ, 78-79; adv engr mats & process eng, Rochester, Minn, 79-85, ADV ENGR & MGR, MAT & PROCESS ENG, PROD ANALYSIS DEPTS, SYSTS TECHNOL DIV, IBM CORP, AUSTIN, TEX, 85-, ADV SCIENTIST, ASSEMBLY PROCESS DEVELOP ELECTRONIC CARD ASSEMBLY & TEST DEVELOP CTR, 85- *Mem:* Am Chem Soc; Astron Soc India; fel Indian Chem Soc; Sigma Xi; NY Acad Sci; Nat Asn Corrosion Engrs; fel Am Inst Chemists. *Res:* Process development in Interconnection Technology and reliability assessments of second level electronic packaging; materials engineering and analytical laboratories management; materials and process engineering, contamination control technology, corrosion research; analytical chemistry; high temperature chemistry of organic materials, specifically synthesis, characterization, vaporization, thermodynamics and phase studies; stellar physics, specifically studies of eclipsing and intrinsic variable stars. *Mailing Add:* 2147 Surrender Ave Austin TX 78728

PULITO, ALDO MARTIN, b West Cromwell, Conn, Aug 5, 20; m 48; c 3. CHEMISTRY. *Educ:* Trinity Col, Conn, BS, 42; Va Polytech Inst, BS, 44; Univ Conn, PhD(chem), 50. *Prof Exp:* Pilot plant operator, Columbia Univ, 44-45; plant operator, Fercleve Corp, Tenn, 45; scientist, Los Alamos Sci Lab, 45-46; asst chem, Univ Conn, 47-50; chemist & treas, Ames Labs, 50-70; ASST PROF CHEM, FAIRFIELD UNIV, 70- *Concurrent Pos:* Lectr, Fairfield Univ, 53-70. *Mem:* Am Chem Soc. *Res:* Preparation of amines and diamines; organic chemistry; oximes and antioxidants. *Mailing Add:* 502 Ferry Rd Orange CT 06477-2507

PULLAN, GEORGE THOMAS, b Tadcaster, Eng, Feb 1, 29; m 53; c 5. RESEARCH ADMINISTRATION. *Educ:* Cambridge Univ, BA, 49, MA, 53, PhD(physics), 52. *Prof Exp:* Fel physics, Nat Res Coun Can, Toronto, 52-54; sr sci officer, Nat Phys Lab, Dept Sci & Indust Res, Eng, 54-56; defence sci serv officer, Can Armament Res & Develop Estab, 56-69; planning staff, Defence Res Bd Can, 69-74; dir gen res & develop, Plans & Prog, Dept Nat Defense, Can, 74-76, dir technol appln land, 76-84; RETIRED. *Mailing Add:* 1693 Featherston Dr Ottawa ON K1H 6P3 Can

PULLAN, HARRY, b Shipley, Eng, Nov 12, 28; Can citizen; m 53; c 4. PHYSICS. *Educ:* Oxford Univ, BA, 50, MA & DPhil(physics), 53. *Prof Exp:* Fel physics, Nat Res Coun Can, 54-56; group head infrared physics, Res Labs, RCA Victor Co Ltd, 56-64, dir solid state physics, 64-66; dept dir physics, Ont Res Found, 66-77, dir dept appl physics, 77-; assoc dir eng sci div, Ont Res Found, 84; CONSULT, MCINTOSH COMPUTERS. *Mem:* Can Asn Physicists; sr mem Inst Elec & Electronics Engrs. *Res:* Properties of semiconductor materials and their application in cooled infrared detectors and detector systems for specific applications requiring the highest achievable performance. *Mailing Add:* 1859 Lincoln Green Close Mississauga ON L5K 1C4 Can

PULLARKAT, RAJU KRISHNAN, b North Parur, India, Aug 13, 39; m 69; c 2. BIOCHEMISTRY. *Educ:* Univ Kerala, BS, 58, MS, 60; Tex A&M Univ, PhD(biochem), 66. *Prof Exp:* Asst oils & fats, Regional Res Labs, Hyderabad, India, 60-62; asst biochem, Tex A&M Univ, 62-65, technician II, 65-66, res assoc, 66-68, res scientist, 68-73; sr res scientist, 73-74, assoc res scientist, 74-75, HEAD, LIPID STORAGE DIS LAB, INST BASIC RES DEVELOP DISABILITIES, 75- *Mem:* Am Soc Neurochem; Int Soc Neurochem; Am Soc Biochem & Molecular Biol. *Res:* Lipid metabolism in mental retardation; glycoproteins. *Mailing Add:* Inst Basic Res Develop Disabilities Dept Neurochem 1050 Forest Hill Rd Staten Island NY 10314-6399

PULLEN, BAILEY PRICE, b Town Creek, Ala, June 21, 40; m 67; c 2. CHEMICAL PHYSICS. *Educ:* Univ Tenn, BS, 63, MS, 67, PhD(chem), 70. *Prof Exp:* Chemist, Spec Training Div, Oak Ridge Inst Nuclear Studies, 63-66; radiation safety off, Univ Tenn, 67-68; AEC fel, Oak Ridge Nat Lab, 68-70; PROF CHEM, SOUTHEASTERN LA UNIV, 70- *Concurrent Pos:* Vis prof, Oak Ridge Nat Lab, 70-74, consult, Health Physics Div, 75-76; consult, Gamma Indust, 78-84. *Res:* Synchrotron and laser spectroscopy. *Mailing Add:* Dept Chem & Physics Southeastern La Univ Hammond LA 70401

PULLEN, DAVID JOHN, b London, Eng, June 28, 36; m 60; c 4. NUCLEAR PHYSICS. *Educ:* Univ London, BSc, 58; Oxford Univ, DPhil(nuclear physics), 63. *Prof Exp:* Res assoc nuclear physics, Mass Inst Technol, 63-64, instr, 64-65; asst prof, Univ Pa, 65-70; assoc prof, 70-74, PROF PHYSICS & APPL PHYSICS, UNIV LOWELL, 74- *Concurrent Pos:* Royal Comn Exhibition of 1851 res fel, Oxford Univ, 61-65; vis scientist, Niels Bohr Inst, 69 & Los Alamos Sci Lab, 70. *Mem:* Am Phys Soc; Sigma Xi; Am Asn Physics Teachers. *Res:* Low energy nuclear physics, especially reaction mechanisms and nuclear structure employing high resolution experimental techniques; neutron physics; time-of-flight. *Mailing Add:* Dept Physics & Appl Physics Univ Lowell Lowell MA 01854

PULLEN, EDWIN WESLEY, b Flushing, NY, June 2, 23; m 46; c 4. ANATOMY. *Educ:* Colgate Univ, AB, 43; Univ Mass, MS, 48; Univ Va, PhD(biol), 53. *Prof Exp:* Instr biol, 50-51, from instr to assoc prof, 51-73, PROF ANAT, SCH MED, UNIV VA, 73-, ASSOC DEAN, 74- *Mem:* AAAS; Sigma Xi; Am Asn Anat; Asn Am Med Cols. *Res:* Invertebrate histology; mesenchyme of rhabdocoels; microscopy of circulation in the amphibian kidney; human surgical anatomy. *Mailing Add:* Dept of Anat Univ of Va Sch of Med Charlottesville VA 22908

PULLEN, KEATS A(BBOTT), JR, b Onawa, Iowa, Nov 12, 16; m 45; c 5. ELECTRICAL ENGINEERING, CIRCUITS ANALYSIS. *Educ:* Calif Inst Technol, BS, 39; Johns Hopkins Univ, EngD(elec eng), 46. *Prof Exp:* Teacher eng sci & mgt, War Training Prog, Johns Hopkins Univ, 41-44; electronic engr, Liberty Motors & Eng Corp, Md, 43-45; res engr, Columbia Mach Works, NY, 45-46; instr elec eng, Pratt Inst, 46-47; electronic res engr, Army Mat Systs Anal Activ, Aberdeen Proving Ground, 46-90; CONSULT, ASI, ABERDEEN, MD. *Concurrent Pos:* Adj prof, Drexel Inst Technol; mem exten fac, Univ Del. *Honors & Awards:* Marconi Mem Award, Vet Wireless Operators Asn. *Mem:* Fel Inst Elec & Electronics Engrs. *Res:* Tube and transistor circuits; circuit theory; information theory and engineering; electron devices, antennas and propagation; communication systems. *Mailing Add:* 2807 Jerusalem Rd Kingsville MD 21087-1098

PULLEN, MILTON WILLIAM, JR, b Waterville, NY, Nov 5, 14; m 40; c 2. GEOLOGY. *Educ:* Colgate Univ, BA, 38; Syracuse Univ, MS, 40; Univ Ill, PhD(geol), 50. *Prof Exp:* Instr gen geol, Syracuse Univ, 38-40 & Univ Ill, 40-42; asst geologist, Coal Div, State Geol Surv, Ill, 42-46, geologist, Div Ground Water Geol & Geophys Explor, 46-50, geologist & head, Div Geophys Res, 50-53; geologist & geophysicist, C A Bays & Assocs, Inc, 53-57; assoc prof, 57-79, EMER ASSOC PROF GEOSCI, PURDUE UNIV, WEST LAFAYETTE, 79- *Concurrent Pos:* Consult, C A Bays & Assocs, Inc, 47-53. *Mem:* Geol Soc Am; Am Asn Petrol Geologists; Am Inst Mining, Metall & Petrol Engrs. *Res:* Solution mining; subsurface geology; waste disposal. *Mailing Add:* 618 Eastview Rd Largo FL 33540

PULLEN, THOMAS MARION, b Elmodel, Ga, June 17, 19; m 40; c 1. BOTANY. *Educ:* Univ Ga, BSA, 40, MEd, 58, PhD(bot), 63. *Prof Exp:* Pub sch teacher, Ga, 40-44; teacher, Mt Berry Sch Boys, 46-47; high sch teacher, Ga, 50-57, chmn sci dept, 58-59; teaching asst bot, Univ Ga, 60-63; from asst prof to assoc prof, 63-66, PROF BIOL, UNIV MISS, 66- *Mem:* AAAS; Am Soc Plant Taxon; Int Asn Plant Taxon; Am Inst Biol Sci; Sigma Xi. *Res:* Taxonomy and biosystematics of the vascular flora of the southeastern United States. *Mailing Add:* Box 1655 Vicksburg MS 39180

PULLIAM, H RONALD, b Miami Beach, Fla, Sept 7, 45; m 69; c 2. BEHAVIORAL AND POPULATION ECOLOGY. *Educ:* Univ Georgia, BS, 68; Duke Univ, PhD(biol), 70. *Prof Exp:* Res fel, Univ Chicago, 70-71; from asst prof to assoc prof biol, Univ Ariz, 71-78; biologist, Mus Northern Ariz, 78-80; assoc prof biol, State Univ NY, Albany, 80-84; PROF ZOOL, UNIV GA, 84-,DIR, INST ECOL, 87-; dir, Inst Ecol, Univ Ga, 87- *Concurrent Pos:* Vis scientist, Sch Biol Sci, Sussex Univ, 77-78; ed, Ecology, Ecol Soc Am, 81-; prin investr, NSF grant, 74- *Honors & Awards:* Ecol Soc Am (vpres, 85-86, pres-elect, 90- 91, pres, 91-92). *Mem:* Ecol Soc Am; Animal Behav Soc. *Res:* Ecology and behavior of vertebrates including social behavior and feeding behavior of birds; theoretical projects concerning the evolution of learning and cultural evolution. *Mailing Add:* Inst Ecol Univ Ga Athens GA 30605

PULLIAM, JAMES AUGUSTUS, neuroanatomy, radiation biology; deceased, see previous edition for last biography

PULLIG, TILLMAN R(UPERT), b Ashland, La, Dec 13, 22. ORGANIC CHEMISTRY. *Educ:* Northwest State Col, BS, 43; La State Univ, BS, 47, MS, 49, PhD(chem), 52. *Prof Exp:* Asst chem, La State, 50-52; chemist, Texaco Inc, 52-57, group leader, 57-60, asst to dir res, 60-61, sr proj chemist, 61-65, chief process eng, 65-69, technologist, 69-83; RETIRED. *Mem:* AAAS; Am Chem Soc; Sigma Xi. *Res:* Process development and utilization of petroleum products. *Mailing Add:* PO Box 180 Jasper TX 75951

PULLING, NATHANIEL H(OSLER), b Boston, Mass, Jan 10, 20; m 55. AUTOMOTIVE SAFETY. *Educ:* Brown Univ, AB, 42; Harvard Univ, PhD, 51. *Prof Exp:* Res fel, Harvard Univ, 51-53; develop engr, Gen Elec Co, 53-61, develop engr bus & prod planning, 61-67; proj dir automotive safety, Liberty Mutual Ins Co, 67-89; RETIRED. *Concurrent Pos:* Adj prof mech eng, Worcester Polytech Inst, 73-; mem, Transp Res Bd, Nat Acad Sci-Nat Res Coun. *Mem:* Soc Automotive Engrs; Am Soc Mech Engrs; Human Factors Soc. *Res:* Directing automotive safety research in driving impairments; driving performance measurement; accident avoidance techniques; crash injury protection of vehicles and car occupants. *Mailing Add:* 11 Manito Rd PO Box 608 East Orleans MA 02643-0608

PULLING, RONALD W, RESEARCH ADMINISTRATION. *Prof Exp:* PVT CONSULT, 91- *Mem:* Nat Acad Eng. *Mailing Add:* Ronald W Pulling Assoc PO Box 9526 Alexandria VA 22304

PULLMAN, IRA, b New York, NY, May 31, 21; m 44, 56; c 3. PHYSICS. *Educ:* Columbia Univ, BS, 41, ChE, 42; Univ Ill, MS, 49, PhD(physics), 53. *Prof Exp:* Jr engr, Tenn Valley Authority, 42-44; engr, Publicker Com Alcohol Co, 44-45; asst, SAM Lab, Columbia Univ, 45-46 & Argonne Nat Lab, 46; asst physics, Univ Ill, 46-50; scientist, Nuclear Develop Assoc, 52-55; asst prof biophys, Sloan-Kettering Div, Cornell Univ, 55-72; asst prof radiol & biochem, NY Med Col, 72-78; res assoc prof, 78-81, ADJ ASSOC PROF, DENT CTR, NY UNIV, 81-, ADJ ASSOC PROF, DEPT BIOL, 84- *Concurrent Pos:* Consult, Genetics Comt, Nat Acad Sci, 55; from asst to assoc, Sloan-Kettering Inst Cancer Res, 55-72. *Mem:* Fel NY Acad Sci; Am Asn Phys Teachers. *Res:* Electron spin resonance spectroscopy of free radicals; radiation physics; computational chemistry. *Mailing Add:* 140 Grand Blvd Scarsdale NY 10583

PULLMAN, MAYNARD EDWARD, b Chicago, Ill, Oct 26, 27; m 48; c 3. BIOCHEMISTRY. *Educ:* Univ Ill, BS, 48, MS, 50; Johns Hopkins Univ, PhD(biochem), 53. *Prof Exp:* Res assoc pediat, Johns Hopkins Hosp, 53-54; from asst to assoc mem, 54-65, mem dept biochem, 65-76, CHIEF DEPT BIOCHEM, PUB HEALTH RES INST OF CITY OF NEW YORK, INC, 76-; RES PROF, NY UNIV SCH MED, 76- *Concurrent Pos:* Mem biochem study sect, NIH, 69-73. *Mem:* AAAS; Am Soc Biol Chemists; Am Chem Soc; fel NY Acad Sci. *Res:* Oxidative phosphorylation; biosynthetic processes in mitochondria. *Mailing Add:* 338 Archer St Freeport NY 11520-4233

PULLMAN, NORMAN J, b New York, NY, Mar 31, 31; m 53; c 2. MATHEMATICS. *Educ:* NY Univ, BA, 56; Harvard Univ, MA, 57; Syracuse Univ, PhD(math), 62. *Prof Exp:* Instr math, Syracuse Univ, 60-62; lectr, McGill Univ, 62-63, asst prof, 63-66; assoc prof, 66-71, PROF MATH, QUEEN'S UNIV, ONT, 71- *Concurrent Pos:* Nat Res Coun Can fel, 65-66; Nat Sci & Eng Res Coun Can operating grant, 67-; NY Acad Sci Acad scholar, 71. *Mem:* Am Math Soc; Can Math Soc; Soc Indust & Appl Math; Sigma Xi; Combinatorial Math Soc Australia; NY Acad Sci; fel Inst Combinatorics & its Applications. *Res:* Matrix theory; combinatorics. *Mailing Add:* Dept Math Queen's Univ Kingston ON K7L 3N6 Can

PULLMAN, THEODORE NEIL, b New York, NY, Sept 30, 18; m 48. INTERNAL MEDICINE, NEPHROLOGY. *Educ:* Harvard Univ, BS, 38; Columbia Univ, MD, 43. *Prof Exp:* Intern med, New Haven Hosp, Conn, 43-44; asst resident med, Sch Med, Univ Chicago, 47-49, from instr to prof, 50-73; prof med, 73-87, EMER PROF, MED, DEPT MED, SCH MED, NORTHWESTERN UNIV, CHICAGO, 87- *Concurrent Pos:* Asst, Yale Univ, 43-44; attend physician, Albert Merritt Billings Mem Hosp, 51-73 & Northwestern Mem Hosp, 73-; chmn, subcomt physiol, Int Comt Nomenclature & Nosology Renal Dis, 66-75; cardiovasc syst res eval comt, Dept Med & Surg Res Serv, Vet Admin Cent Off, 69-71; spec fel, USPHS, 72-73; assoc chief staff, Vet Admin Lakeside Hosp, 73-81; comt on animal care & experimentation, Am Physiol Soc, 84-87. *Mem:* Soc Exp Biol & Med; Am Soc Clin Invest; Am Physiol Soc; Am Heart Asn; Am Col Physicians; Int Soc Nephrol. *Res:* Renal physiology; fluid and electrolyte metabolism; cardiovascular and endocrine physiology. *Mailing Add:* Dept Med Northwestern Univ Sch of Med 303 E Chicago Ave Chicago IL 60611

PULLUKAT, THOMAS JOSEPH, b India, Apr 28, 41; m 70; c 2. CHEMISTRY. *Educ:* Univ Kerala, BSc, 61; Purdue Univ, Lafayette, PhD(inorg chem), 68. *Prof Exp:* Res scientist, 67-71, group leader anal group, 71-78, asst mgr, 78-80, mgr, 80-85, SR RES SCIENTIST, CHEMPLEX CO, 71-; MGR, NORCHEM, INC, 85- *Mem:* Am Chem Soc. *Res:* Heterogeneous catalyst; olefin polymerization; organo-silicon chemistry; general organometallic chemistry; material science; polyolefin product development. *Mailing Add:* Quantum Chem Co USI Div 3100 Gulf Rd Rolling Meadows IL 60008-4070

PULS, ROBERT W, b Washington, DC, Sept 16, 50; m 76; c 3. SOIL CHEMISTRY, SOIL MINERALOGY. *Educ:* Univ Wis Madison, BS, 78; Univ Wash, MS, 79; Univ Ariz, PhD(soil & water sci), 86. *Prof Exp:* Forest soil specialist, Dept Natural Resources, State Wash, 79-80; sr soil scientist, Soil & Land Use Technol Inc, 80-82; res assoc, Univ Ariz, 82-86; sr res scientist, Westinghouse Hanford Co, 86-87; SR RES SOL SCIENTIST, US ENVIRON PROTECTION AGENCY, R S KERR LAB, 87- *Mem:* Soil Sci Soc Am; Am Geophys Union; Asn Ground Water Scientist & Engrs. *Res:* Metal, organic and mineral interactions in subsurface; environments for prediction of transport and fate of contamitants. *Mailing Add:* Rte 1 Box 194 Ada OK 74820

PULSIFER, ALLEN HUNTINGTON, b Johnstown, NY, Sept 18, 37; m 62; c 4. CHEMICAL ENGINEERING. *Educ:* Dartmouth Col, BA, 58; Mass Inst Technol, MS, 60, ChE, 61; Syracuse Univ, PhD(chem eng), 65. *Prof Exp:* From asst prof to prof chem eng, Iowa State Univ, 65-85; PROF & HEAD, DIV SCI ENG TECHNOL, BEHREND COL, PA STATE UNIV, 85- *Concurrent Pos:* Vis prof, Prairie View Agr & Mech Col, 70-71; Fulbright travel grant, 75; Humboldt Found res fel, 75-76; vis scientist, Bergbau-Forshung, W Ger, 75-76. *Mem:* Am Inst Chem Engrs; Am Chem Soc; Am Soc Eng Educ. *Res:* Fluidized bed reactors; coal gasification; solid-gas reactions. *Mailing Add:* Div Sci Eng Technol Behrend Col Pa State Univ Station Rd Erie PA 16563

PULVARI, CHARLES F, b Karlsbad, Austria-Hungary, July 19, 07; nat US; m 37; c 3. ELECTRICAL ENGINEERING. *Educ:* Royal Hungarian Univ Tech Sci, DrEng, 29. *Prof Exp:* Res engr lab, Tel Factory Co, Ltd, Hungary, 29-33; tech adv, Hungarian Radio & Commun Co, 33-35; chief engr, Hungarian Film Co, 35-41, tech mgr, 41-45; owner, Pulvari Electrophys Lab, 43-49; prof elec eng, George Washington Univ, 49-51; mem staff, 51-53, res prof, 53-56, prof elec eng & head solid state physics lab, 56-72, EMER PROF ELEC ENG, CATH UNIV AM, 72- *Concurrent Pos:* Prin investr, Mil Res & Instrumentation Projs, Hungary; consult, various insts, Hungary, 33-43; consult engr, Scophony, Ltd, London, 37; lectr, Hungarian Univ Tech Sci, 43 & Wayne State Univ, 53; pres, Electrocristal Corp, DC. *Honors & Awards:* IR 100 Award, 63. *Mem:* Fel Inst Elec & Electronics Engrs; fel NY Acad Sci; Sigma Xi (pres, 66); fel Am Ceramic Soc. *Res:* Electrostatic recording and reproducing; electric switch; electrical condensers; cathode ray tube; transpolarizer; ferrielectricity in crystals; solid state engineering; crystal growth; solar to electrical energy conversion research. *Mailing Add:* 4600 Duke St No 1620 Alexandria VA 22304-2515

PUMMER, WALTER JOHN, polymer chemistry, for more information see previous edition

PUMO, DOROTHY ELLEN, b Charleston, WVa, Oct 27, 51. MITOCHONDRIAL DNA ANALYSIS, CYTOGENETICS. *Educ:* Univ Mich, Ann Arbor, BS, 73, MS, 74, PhD(biol), 76. *Prof Exp:* Teaching asst biol & cell biol, Univ Mich, Ann Arbor, 74-75, res asst, 75-76; res assoc, Univ Colo, Boulder, 76-78, Univ Vt, Burlington, 78-79 & 81; asst prof, 81-87, ASSOC PROF BIOL, HOFSTRA UNIV, 87- *Concurrent Pos:* Trainee, NIH Training Grant, Univ Vt, 81. *Mem:* AAAS; Am Soc Cell Biol; Am Soc Mammalogists. *Res:* The application of mitochondrial DNA analysis to zoogeography and population dispersal in mammals. *Mailing Add:* Biol Dept Hofstra Univ Hempstead NY 11550

PUMPER, ROBERT WILLIAM, b Clinton, Iowa, Sept 12, 21; m 51; c 1. VIROLOGY, MEDICAL MICROBIOLOGY. *Educ:* Univ Iowa, BA, 51, MSc, 53, PhD, 55; Am Bd Med Microbiol, dipl & cert pub health & med lab virol. *Prof Exp:* Mem staff, Virus & Rickettsial Div, US Army Chem Res & Develop Labs, Ft Detrick, Md, 55-56; asst prof microbiol, Hahnemann Med Col, 56-57; from asst prof to assoc prof, 57-61, PROF MICROBIOL, UNIV ILL COL MED, 61- *Concurrent Pos:* Consult, St James Hosp. *Honors & Awards:* Raymond B Allen, Med Lectr Award, 70, 74, 76, 79 & 85. *Mem:* AAAS; Sigma Xi; Tissue Cult Asn. *Res:* Effect of viruses on in vitro tissue cell growth; growth of tissue in defined media and its use for disease diagnosis. *Mailing Add:* Dept Microbiol Univ Ill Col Med PO Box 6998 M/C 790 Chicago IL 60680

PUN, PATTLE PAK-TOE, b Hong Kong, China, Sept 30, 46; US citizen; m 70; c 2. MOLECULAR GENETICS, GENETIC ENGINEERING. *Educ:* San Diego State Univ, BS, 69; State Univ NY, Buffalo, MA, 72, PhD(biol), 74; Wheaton Grad Sch, MA, 85. *Prof Exp:* Res asst enzym, State Univ NY, Buffalo, 69, teaching asst biol, 69-72, res asst bact, 72-73; from asst prof to assoc prof, 73-87, PROF BIOL, WHEATON COL, 88- *Concurrent Pos:* Res assoc immunol, Argonne Nat Lab, 74-76; vis microbiologist, Med Ctr, Univ Ill, 80-81 & Northern Ill Univ, Dekalb, 85-88; vis scientist, Am Critical Care Inc, Ill, 84; book rev consult, William C Brown Publ Co, Dubuque, Iowa, 84-; res grants, Res Corp, N.S.F.. *Mem:* Am Soc Microbiol; NY Acad Sci; fel Am Sci Affil. *Res:* Bacterial physiology and nucleic acid enzymology; immunochemistry and immunogenetics; gene cloning in Bacillus subtilis and Bacillus megaterium. *Mailing Add:* Dept Biol Wheaton Col Wheaton IL 60187

PUNCH, JAMES DARRELL, b Conover, NC, Dec 31, 36; m 59; c 2. MICROBIOLOGY, BIOCHEMISTRY. *Educ:* NC State Univ, BS, 59; Univ Minn, MS, 63, PhD(microbiol), 65. *Prof Exp:* Asst prof microbiol, Med Col Va, 67-73; HEAD CULT DEVELOP & GENETICS, UPJOHN CO, 73- *Concurrent Pos:* Consult, A H Robins Pharmaceut Co, Richmond, Va, 68-73. *Mem:* Am Soc Microbiol; Am Inst Biol Sci; NY Acad Sci; Am Chem Soc. *Res:* R factors, bacterial plasmids and molecular genetics; molecular biology; DNA and RNA replication; protein synthesis; cell permeability. *Mailing Add:* The Upjohn Co Kalamazoo MI 49001

PUNCH, JERRY L, b Newton, NC; c 4. AUDIOLOGY. *Educ:* Wake Forest Univ, BA, 65; Vanderbilt Univ, MS, 67; Northwestern Univ, PhD(audiol), 72. *Prof Exp:* Asst prof commun dis , Univ Miss, 71-73; asst prof, Memphis State Univ, 73-75; res assoc audiol, Univ Md, 75-80; dir Res Div, Am Speech-Lang-Hearing Asn, 80-84; chief audiol, Med Ctr, Ind Univ, 84-87; sr audiologist, Nicolet Instrument Corp, 87-90; ASSOC PROF, DEPT AUDIOL & SPEECH SCI, MICH STATE UNIV, 90- *Concurrent Pos:* Proj dir, Vet Admin Contract, Univ Md, 76-; instr audiol, Towson State Univ, 77- *Mem:* Am Speech-Lang-Hearing Asn; Acoust Soc Am; Am Auditory Soc. *Res:* Psychoacoustics; diagnostic and rehabilitative aspects of hearing disorders; hearing aid electroacoustic measurements and hearing aid evaluation methods. *Mailing Add:* Mich State Univ 378 Commun Arts & Sci Bldg East Lansing MI 48824

PUNDERSON, JOHN OLIVER, b Northfield, Minn, Oct 28, 18; m 47; c 3. MATERIALS SCIENCE, POLYMER SCIENCE. *Educ:* Univ Chicago, BS, 40; Univ Minn, PhD(chem), 50. *Prof Exp:* Res chemist, Columbia Chem Div, Pittsburgh Plate Glass Co, 41-44; res chemist, Plastics Dept, E I du Pont de Nemours & Co, Inc, 50-56, res supvr, 56-61, sr res chemist, 61-67, res assoc, Plastics Prods & Resins Dept, 67-80, res assoc, Polymer Prods Dept, 80-82; PRES, PUNDERSON CONSULT, INC, MENDOTA HEIGHTS, MN, 83- *Concurrent Pos:* Mem, Int Standards Orgn comts on fire standards, Int Electrotech Comn; mem adv bd, J Fire Sci. *Mem:* Am Soc Testing & Mat; Inst Elec & Electronics Engrs; Am Chem Soc; Soc Plastics Indust; Nat Fire Protection Asn. *Res:* Polymer chemistry; fluoropolymer resins-properties and applications; electrical insulations for wire and cable; fire safety tests and standards, smoke and toxicity of combustion products. *Mailing Add:* 934 S Highview Circle Mendota Heights MN 55118-3686

PUNDSACK, ARNOLD L, b Albany, Minn, Apr 9, 38; m 61; c 4. PHYSICS, MATHEMATICS. *Educ:* St John's Univ, Minn, BS, 60. *Prof Exp:* Physicist, Nat Security Agency, 60-63; from assoc scientist to scientist mat sci, Xerox Res Ctr Can, 63-83, sr scientist, Novel Imaging Processes XDM, 83-87, prin scientist, coating technol, 87-89, MGR, COATING & SUBSTRATES TECHNOLOGIES, XEROX RES CTR CAN, 89- *Honors & Awards:* Excellence in Sci & Technol, Xerox, 87. *Mem:* Soc Photog Scientists & Engrs; Am Inst Chem Engrs. *Res:* Thin films, including nucleation and growth; electron microscopy and materials science; electrography and novel imaging systems; polymer coating processes including solvent extrusion techniques. *Mailing Add:* Xerox Res Ctr Can Ltd 2660 Speakman Dr Mississauga ON L5K 2L1 Can

PUNDSACK, FREDERICK LEIGH, b Pinckneyville, Ill, Sept 14, 25; m 48; c 5. INORGANIC CHEMISTRY. *Educ:* Univ Ill, BS, 49, PhD(chem), 52. *Prof Exp:* Sr res chemist, Johns-Manville Res Ctr, 52-60, sect chief, 60-61, mgr cent res, 62-63, dir res, 63-69, vpres res & develop, 69-71, sr vpres res, Technol & Corp Bus Develop, Johns-Manville Corp, 75-76, sr vpres res & develop group, 76-78, exec vpres-opers, 78-79, pres & chief oper officer, 79-82; CHMN & PRES, SUSQUEHANNA CORP, 84- *Concurrent Pos:* Mem bd dirs, Indust Res Inst, 76-79. *Mem:* AAAS; Am Chem Soc; Sigma Xi. *Res:* Chemistry of asbestos, silicates; surface chemistry; reinforced plastics. *Mailing Add:* 8 Sedgwick Dr Englewood CO 80110

PUNG, OSCAR J, b Pontiac, Mich, Feb 27, 51. PARASITOLOGY, EXPERIMENTAL BIOLOGY. *Educ:* Oakland Univ, BA, 73; Seton Hall Univ, MS, 81; Univ NC, PhD(med parasitol), 84. *Prof Exp:* Sci teacher, US Peace Corps, Repub Zaire, Africa, 75-77; teaching asst, Dept Biol, Seton Hall Univ, 79-81; microbiologist, Nat Inst Environ Health Sci, 82-85; postdoctoral res assoc, Dept Biol, Wake Forest Univ, 85-87; RES & DEVELOP MGR, ENVIRON DIAG, INC, 87- *Mem:* Am Soc Parasitologists. *Res:* Environmental health sciences; author of 8 technical publications. *Mailing Add:* Environ Diag Inc PO Box 908 2990 Anthony Rd Burlington NC 27215

PUNNETT, HOPE HANDLER, b New York, NY, Jan 24, 27; m 50; c 3. CYTOGENETICS. *Educ:* Smith Col, AB, 48; Yale Univ, MS, 49, PhD(bot & microbiol), 55. *Prof Exp:* Asst bot, Univ Ill, 50-52; asst lectr genetics, Univ Rochester, 59-63, res instr obstet & gynec, Sch Med, 62-63; from asst prof to assoc prof, 63-74, PROF PEDIAT, SCH MED, TEMPLE UNIV, 74-; DIR GENETICS LAB, ST CHRISTOPHER'S HOSP FOR CHILDREN, 63-; ASSOC-BIOSCI STAFF PEDIAT, ALBERT EINSTEIN MED CTR, 79- *Concurrent Pos:* NIH fel, Johns Hopkins Hosp, 61, sr fel, Fel Res Inst Cancer & Molecular Biol, 89-90; NIH grant meiotic & mitotic chromosomes in man, 63-68 & NIH grant cystic fibrosis & other inherited disorders, 70-72; study leave, Univ Col, Univ London, 69; bd dirs, Planned Parenthood Southeastern Pa, 72-78; NIH grant genetic constitutions predispose cancer, 76-78; cystic fibrosis fedn grant, carrier detection, NIH; genetics adv comt, Pa Dept Health, 78-, chmn, 80-85, mem, Am Bd Med Genetics, 79-83; pres, Middle Atlantic Human Genetics Network, 85-88; study leave, Weizmann Inst, Rehovot, Israel, 85-86. *Mem:* Soc Pediat Res; AAAS; Genetics Soc Am; Am Soc Human Genetics (secy, 77-80); Sigma Xi; Am Genetic Asn. *Res:* Human genetics; cytogenetics; relationships of chromosome abnormalities; congenital malformations and cancer; molecular cytogenetics. *Mailing Add:* St Christopher's Hosp Children Erie Ave at Front St Philadelphia PA 19134-1095

PUNNETT, THOMAS R, b Buffalo, NY, May 25, 26; m 50; c 3. PLANT BIOCHEMISTRY. *Educ:* Yale Univ, BS, 50; Univ Ill, PhD(physico-chem, biol), 54. *Prof Exp:* Nat Found Infantile Paralysis fel, Sch Biochem, Cambridge Univ, 54-56; from instr to asst prof biol, Univ Rochester, 56-63; from asst prof to assoc prof, 63-69, PROF BIOL, TEMPLE UNIV, 69- *Mem:* Am Soc Plant Physiologists; Am Soc Biol Chemists; Am Soc Photobiol. *Res:* Photosynthesis; control of chloroplast biochemistry and fine structure. *Mailing Add:* Dept Biol Temple Univ Broad & Montgomery Sts Philadelphia PA 19122

PUNTENNEY, DEE GREGORY, b Muncie, Ind, Feb 21, 48; m 71. PHYSICS, RADIOLOGICAL HEALTH. *Educ:* Taylor Univ, BA, 70; Purdue Univ, MS, 72, PhD(health physics), 73. *Prof Exp:* Asst prof physics, Asbury Col, 73-79, assoc prof, 79-; AT POINT LOMA NAZARENE COL. *Mem:* Am Phys Soc; Am Asn Physics Teachers; Sigma Xi. *Res:* Microwave dosimetry; microwave bio-effects in amphibians and reptiles. *Mailing Add:* Point Loma Nazarene Col 3900 Lomaland Dr San Diego CA 92106

PUNWANI, DHARAM VIR, b Multan, India, Aug 23, 42; US citizen; m 70; c 2. SYNTHETIC FUELS, BIOTECHNOLOGY. *Educ:* Indian Inst Technol, Bombay, BS, 65; Ill Inst Technol, Chicago, MS, 67; Loyola Univ, Chicago, MBA, 74. *Prof Exp:* Chem engr, 67-74, supvr, 74-76, mgr, 76-77, asst dir, 77-86, ASSOC DIR PROCESS RES, INST GAS TECHNOL, 86- *Mem:* Am Inst Chem Engrs. *Res:* Coal gasification; peat gasification; oil shale processing; high-temperature and high-pressure fluidization; high pressure solids transport; energy and environmental biotechnology. *Mailing Add:* 427 Prairie Knolls Dr Naperville IL 60565

PUNWAR, JALAMSINH K, b Santrampur, India, Mar 25, 23; US citizen; m 58; c 2. CHEMISTRY. *Educ:* Univ Bombay, BSc, 48; Mich State Univ, MS, 50; Univ Wis, PhD(soil chem), 54. *Prof Exp:* Chemist, Pennsalt Int Corp, 53-54; anal res chemist, Wis Alumni Res Found, 55-69; anal res chemist & group leader, Warf Inst, Inc, 69-74, asst head, Food & Drug Dept, 74-77, RES SCIENTIST, RES & DEVELOP DEPT, RALTECH SCI SERV, INC, 77- *Mem:* Am Chem Soc; Am Oil Chemists' Soc; Sigma Xi. *Res:* Analytical method research in fats and oils; gas chromatography of fats, oils and pesticides; polarography and chemical analysis; analytical method development for determination of total cholesterol in multicomponent foods; nitrosation reaction rates inhibition and determination of N-nitrosomines. *Mailing Add:* 614 Orchard Dr Madison WI 53711-1323

PURANDARE, YESHWANT K, b Poona, India, Sept 19, 34; US citizen; m 64; c 4. CLINICAL CHEMISTRY, PHYSIOLOGICAL CHEMISTRY. *Educ:* Univ Bombay, India, BSc(Hons), 57; Univ Poona, India, MSc, 60; Fordham Univ, MS, 68; NY Univ, PhD(biomech), 82; Spartan Health Sci Univ, St Lucia, MD, 85. *Prof Exp:* Chemist, Sardesai Bros, India, 60-61; clin chemist, Brunswick Hosp Ctr, Amityville, NY, 65-82; from instr to assoc prof, 64-74, PROF CHEM, COL TECHNOL, STATE UNIV NY, FARMINGDALE, 74-, DEPT CHMN, 84- *Concurrent Pos:* Consult, Life Industs, Inc, 87- *Mem:* Am Chem Soc; Am Soc Clin Chem; Am Asn Bioanal; Nat Acad Clin Biochem. *Res:* Chemistry of drying oils; chemistry of vision; abused drugs. *Mailing Add:* 39 Walnut Ave East Farmingdale NY 11735

PURBO, ONNO WÍDODO, b Bandung, Indonesia, Aug 17, 62; m 89; c 1. SILICON-ON-INSULATOR DEVICES, PACKET RADIO COMPUTER NETWORK. *Educ:* Inst Technol, Bandung, Indonesia, Ir, 87; McMaster Univ, Can, MEng, 90. *Prof Exp:* ASST PROF ELEC ENG, INST TECHNOL, BANDUNG, INDONESIA, 87-; RES STAFF ELEC ENG, INT UNIV CTR MICROELECTRONIC, INDONESIA, 87- *Concurrent Pos:* Founder & pres, Computer Div, Inst Technol, Bandung, 85-86; proj mgr, Packet Radio Network, Indonesian Amateur Radio Orgn, Bandung, 86-87; prin investr, Indonesian Packet Radio Computer Network, 91- *Mem:* Inst Elec & Electronics Engrs. *Res:* Fabrication and characterization of polysilicon emitter transistors and Sci-x Gex devices on silicon-on-insulator wafers; development of low cost wide area computer network based on packet radio network technology for use in developing country such as Indonesia. *Mailing Add:* Dept Elec Eng Univ Waterloo Waterloo ON N2L 3G1 Can

PURBRICK, ROBERT LAMBURN, b Salem, Ore, Dec 5, 19; m 53; c 3. PHYSICS. *Educ:* Willamette Univ, AB, 42; Univ Wis, MA, 44, PhD(physics), 47. *Prof Exp:* Asst physics, Univ Wis, 42-44; jr physicist, Argonne Nat Lab, Ill, 44-46; asst physics, Alumni Res Found, 46-47; from asst prof to prof physics, Willamette Univ, 47-85; RETIRED. *Mem:* AAAS; Am Phys Soc; Am Asn Physics Teachers; Inst Elec & Electronics Engrs; Optical Soc Am; Sigma Xi. *Res:* Molecular spectroscopy; nuclear physics; electronics and nuclear counter work. *Mailing Add:* 255 W Vista Ave S Salem OR 97302

PURCELL, ALEXANDER HOLMES, III, b Summit, Miss, Oct 12, 42; m 70; c 2. ENTOMOLOGY, PLANT PATHOLOGY. *Educ:* US Air Force Acad, BS, 64; Univ Calif, Davis, PhD(entom), 74. *Prof Exp:* Asst prof, 74-81, assoc prof entom, 81-88, PROF ENTOM, UNIV CALIF, BERKELEY, 88- *Concurrent Pos:* Consult, UN Food & Agr Orgn, 81, Crop Genetics Int, 87-89. *Mem:* Entom Soc Am; Am Phytopath Soc. *Res:* Insect transmission of plant pathogens, expecially Pierce's disease of grapevines, X-disease and aster yellows and the role of insects in the epidemiology of plant diseases; microbial ecology of the homeptera. *Mailing Add:* Dept Entom Sci Univ of Calif Berkeley CA 94720

PURCELL, EDWARD MILLS, b Taylorville, Ill, Aug 30, 12; m 37; c 2. PHYSICS. *Educ:* Purdue Univ, BSEE, 33; Harvard Univ, AM, 35, PhD(physics), 38. *Hon Degrees:* DEng, Purdue Univ, 53. *Prof Exp:* From instr to assoc prof, 38-49, Soc of Fels sr fel, 50-71, prof, 49-80, Gerhard Gade Univ Prof, 60-80, EMER PROF PHYSICS, HARVARD UNIV, 80- *Concurrent Pos:* Mem staff, Radiation Lab, Mass Inst Technol, 41-46; mem sci adv bd, US Air Force, 53-57 & President's Sci Adv Comt, 57-60 & 62-66. *Honors & Awards:* Nobel Prize in Physics, 52; Oersted Medal, Am Asn Physics Teachers, 68; Nat Medal Sci, NSF, 78. *Mem:* Nat Acad Sci; Am Acad Arts & Sci; fel Am Phys Soc; Am Philos Soc. *Res:* Nuclear magnetism; radio astronomy; biophysics; astrophysics. *Mailing Add:* Lyman Lab Harvard Univ Cambridge MA 02138

PURCELL, EVERETT WAYNE, b Omaha, Nebr, May 18, 24; m 47; c 3. MATHEMATICS, AEROSPACE ENGINEERING. *Educ:* Univ Nebr, BS, 47, MS, 49; Univ Southern Calif, MS, 60. *Prof Exp:* Instr math & physics, Univ Wyo, 49-50; prog analyst, Douglas Aircraft Co, Inc, 55-59; res engr, Ford Aerospace Corp, Ford Motor Co, 59-62, sect supvr, 62-64, sr eng specialist, 65-90; SR ENG SPECIALIST, LORAL AEROSPACE CORP, 90- *Concurrent Pos:* Instr, Biola Col, 61-64; mem, Creation-Sci Res Ctr; bd mem, Bible Sci Asn. *Mem:* Am Inst Aeronaut & Astronaut; Am Sci Affiliation; Creation Res Soc. *Res:* Digital and analog computing; numerical analysis and methods; redundant structural analysis; aircraft stability and dynamics; missile trajectory calculations and missile dynamics and stability; Kalman filters; statistical studies; radar data processing. *Mailing Add:* 18451 Saugus Ave Santa Ana CA 92705

PURCELL, JAMES EUGENE, b Henry Co, Va, Sept 25, 36; div; c 3. PHYSICS. *Educ:* Berea Col, BA, 59; Case Inst Technol, PhD(physics), 66. *Prof Exp:* Res assoc nuclear theory, Fla State Univ, 66-68; asst prof physics, Univ Fla, 68-70; asst prof, 70-74, ASSOC PROF PHYSICS, GA STATE UNIV, 74- *Mem:* Am Phys Soc; Am Asn Physics Teachers; AAAS. *Res:* Theory of nuclear structure and nuclear reactions; scattering of electrons by atoms. *Mailing Add:* Dept Physics Ga State Univ Atlanta GA 30303

PURCELL, JOSEPH CARROLL, b Cleveland, Va, Feb 5, 21; m 55. AGRICULTURAL ECONOMICS. *Educ:* Va Polytech Inst, BA, 48, MS, 51; Iowa State Univ, PhD(agr econ), 56. *Prof Exp:* Agr economist, USDA, 51-52; from asst prof to assoc prof, 53-60, head dept, 71-88, PROF AGR ECON, UNIV GA, 60- *Concurrent Pos:* Econ adv panelist, Milk Indust Found, 68; prin agr economist, Coop State Res Serv, USDA, 75-76; mem, Econ Adv Bd, Dairyman, Inc, 70-; assoc dir, Agr Res Planning & Eval, Sci & Educ Admin, USDA, 79-81. *Honors & Awards:* Res Award, Sears Found, 62. *Mem:* Am Agr Econ Asn; Int Asn Agr Econ; Europ Asn Agr Econ. *Res:* Planning and execution of research in the economics of commercial agriculture production and marketing; monitoring of the domestic market by product, space and time detail, international trade, food policy. *Mailing Add:* Dept Agr Econ Univ Ga Ga Sta Griffin GA 30212-1797

PURCELL, KEITH FREDERICK, b St Louis, Mo, Sept 12, 39; c 2. INORGANIC CHEMISTRY. *Educ:* Cent Col, AB, 61; Univ Ill, PhD(chem), 65. *Prof Exp:* Asst prof chem, Wake Forest Col, 65-67; asst prof, 67-70, assoc prof, 70-78, PROF CHEM, KANS STATE UNIV, 78- *Concurrent Pos:* Fulbright fel; NATO sr scientist. *Mem:* Am Chem Soc. *Res:* Synthesis of and electronic structures of transition metal complexes; electronic structure calculations; Mossbauer spectroscopy; solid state electron transfer; conductance in molecular solids. *Mailing Add:* Willard Hall Kans State Univ Manhattan KS 66502

PURCELL, KENNETH, b New York, NY, Oct 21, 28; m 49, 87; c 2. CLINICAL PSYCHOLOGY. *Educ:* Univ Nebr, BA, 49, PhD, 53. *Prof Exp:* Asst prof psychol, Univ Ky, 56-58; dir state training prog clin psychol, Commonwealth of Ky, 56-58; dir behav sci div, Children's Asthma Res Inst & Hosp, 59-68; prof psychol & dir grad training prog clin psychol, Univ Mass, Amherst, 68-70, chmn dept psychol, 69-70; chmn dept psychol. 70-76, dean Col Arts & Sci, 76-84, PROF PSYCHOL, UNIV DENVER, 84- *Mem:* Fel AAAS; fel Soc Res Child Develop; fel Am Psychol Asn. *Res:* Personality theory and function; child development; psychosomatic research. *Mailing Add:* Dept Psychol Univ Denver Denver CO 80208

PURCELL, ROBERT HARRY, b Keokuk, Iowa, Dec 19, 35; m 61; c 2. VIROLOGY, EPIDEMIOLOGY. *Educ:* Okla State Univ, BA, 57; Baylor Univ, MS, 60; Duke Univ, MD, 62. *Prof Exp:* Intern pediat, Duke Hosp, 62-63; officer, Epidemic Intel Serv, Commun Dis Ctr, 63-65; sr surgeon, 65-69, med officer, 69-72, med dir, 72-74, HEAD HEPATITIS VIRUS SECT, LAB INFECTIOUS DIS, NAT INST ALLERGY & INFECTIOUS DIS, USPHS, NIH, 74- *Honors & Awards:* Gorgas Medal, 77; Squibb Award, Infectious Dis Soc Am, 80. *Mem:* Nat Acad Sci; NY Acad Sci; Soc Epidemiol Res; Infectious Dis Soc Am; Am Epidemiol Soc; AAAS; Am Soc Virol; Am Soc Microbiol; Asn Am Physicians; Am Col Epidemiol; Am Soc Clin Invest; Soc Exp Biol & Med. *Res:* Viral hepatitis. *Mailing Add:* NIH Lab of Infect Dis Bldg 7, Rm 202, 9000 Rockville Pike Bethesda MD 20892

PURCELL, WILLIAM PAUL, b Tulsa, Okla, Dec 11, 35; m 57; c 3. DRUG DESIGN. *Educ:* Ind Univ, AB, 57; Princeton Univ, AM, 59, PhD(phys chem), 60. *Prof Exp:* Chief inorg res chemist, Buckman Labs, 60-63; instr physics & chem, Christian Bros Col, Tenn, 61-63; from assoc prof to prof med chem, Col Pharm, 63-69, chmn dept molecular & quantum biol, 69-74, PROF MOLECULAR & QUANTUM BIOL, COL PHARM, CTR HEALTH SCI, UNIV TENN, MEMPHIS, 69-, DIR DIV DRUG DESIGN, 74-, PROF, DEPT PHARMACOL & COL COMMUNITY & ALLIED HEALTH PROFS, 75-, PROF, DEPT MED CHEM, 82- *Concurrent Pos:* Co-prin investr, NSF grant, 64-67; prin investr, US Army Med Res & Develop Command res contract, 65-71, Cotton Prod Inst contract, 69-, FMC Corp, 71- & Diamond Shamrock Corp, 72-; grants, Eli Lilly & Co, 68- & Smith, Kline & French Found, 69-; co-prin investr, Nat Inst Dent Res grant, 69-; vis prof, Univ Lausanne, Switz, 75-77. *Mem:* Int Soc Quantum Biol (secy, 71-); Am Chem Soc; Royal Soc Chem; Acad Pharmaceut Sci; Sigma Xi. *Res:* Dipole moments; molecular orbital calculations; partition chromatography of molecules of biological and biochemical interest; application of regression analyses to structure-activity data; quantitative structure-activity relationships. *Mailing Add:* PO Box 41777 Memphis TN 38174-1777

PURCHASE, EARL RALPH, b Oceana Co, Mich, July 2, 19; m 41; c 2. POLLUTION CHEMISTRY. *Educ:* Hope Col, AB, 40; Univ Vt, MS, 43; Ohio State Univ, PhD(org chem), 48. *Prof Exp:* Instr chem, Univ Vt, 40-41; assoc chemist, Clinton Labs, Tenn, 44-45; res chemist, E I du Pont de Nemours & Co, Inc, 48-50 group supvr, 50-62, sr supvr int dept, 62-67, lycra prod supt, 67-70, Neth, process specialist, 71-81; RETIRED. *Mem:* Am Chem Soc. *Res:* Halogenated hydrocarbons; cellulose and polymer chemistry; wastewater treatment and air pollution control; hazardous material control and disposal. *Mailing Add:* 12 A Moratuck Manor Rd Littleton NC 27850

PURCHASE, HARVEY GRAHAM, b Livingstone, NRhodesia, Aug 8, 36; US citizen; m 63; c 2. VIROLOGY. *Educ:* Univ Witwatersrand, BSc, 55; Univ Pretoria, BVSc, 59; Univ London, MRCVS, 61; Mich State Univ, MS, 65, PhD, 70. *Prof Exp:* Pvt pract, SAfrica, 60-61 & Eng, 61; vet med off, Regional Poultry Res Lab, USDA, Mich, 61-74; staff scientist, 74-78, actg chief livestock & vet sci, Nat Prog Staff, 78-83, SPEC SCI ADV TO DIR, AGR RES SERV, USDA, 83- *Concurrent Pos:* Res exchange, Leukosis Unit, Houghton Poultry Res Sta, Eng, 65-66; asst prof microbiol, Mich State Univ, 70-74. *Honors & Awards:* Sir Arnold Theiler Medal, Agricura Prize & Imp Chem Ltd Prize, 59; CPC Int Award, Poultry Sci Asn, 71; Arthur S Flemming Award, 72; Tom Newman Mem Award, Poultry Stock Asn Gt Brit, 73. *Mem:* Am Vet Med Asn; Poultry Sci Asn; Am Asn Avian Path; Brit Vet Asn; US Animal Health Asn. *Res:* Immunology, virology and pathology of oncogenic viruses, mainly of the lymphoid leukosis-sarcoma group and Marek's disease of poultry and all diseases of poultry; biotechnology; research management. *Mailing Add:* Col Vet Med Miss State Univ Mississippi State MS 39762

PURCIFULL, DAN ELWOOD, b Woodland, Calif, July 1, 35; m 66; c 2. PLANT VIROLOGY. *Educ:* Univ Calif, Davis, BS, 57, MS, 59, PhD(plant path), 64. *Prof Exp:* Res asst plant path, Univ Calif, Davis, 59-64; from asst prof to assoc prof, 64-75, PROF PLANT PATH, UNIV FLA, 75- *Honors & Awards:* Lee M Hutchins Award, Am Phytopath Soc, 81. *Mem:* AAAS; Am Phytopath Soc; Sigma Xi; NY Acad Sci; Am Soc Virol. *Res:* Biological, serological and cytological properties of plant viruses; etiology and control of plant virus diseases. *Mailing Add:* 3106 NW 1st Ave Gainesville FL 32607

PURDIE, NEIL, b Castle Douglas, Scotland, Sept 16, 35; wid; c 3. INORGANIC CHEMISTRY, PHYSICAL CHEMISTRY. *Educ:* Univ Glasgow, BSc, 58, PhD(chem), 61. *Prof Exp:* Asst lectr chem, Univ Glasgow, 60-62; res assoc, Brookhaven Nat Lab, 62-63; Turner & Newall res fel inorg chem, Univ Glasgow, 63-65; from asst prof to assoc prof, 65-75, PROF CHEM, OKLA STATE UNIV, 75- *Concurrent Pos:* Vis assoc prof chem, Univ Utah, 73-74. *Mem:* Am Chem Soc; Royal Soc Chem. *Res:* Chemical reaction kinetics using relaxation methods; analysis of complex mixtures for dangerous drugs using circular dichroism; thermodynamics of metal complexes and polybasic acids in solution by potentiometry, calorimetry, spectroscopy and conductance. *Mailing Add:* Dept of Chem Okla State Univ Stillwater OK 74078

PURDOM, JAMES FRANCIS WHITEHURST, b Atlanta, Ga, Oct 5, 43; m 64; c 2. SATELLITE METROLOGY. *Educ:* Transylvania Col, AB, 65; St Louis Univ, MS, 68; Colo State Univ, PhD(atmospheric sci), 86. *Prof Exp:* CHIEF, MESOSCALE METEOROL, NAT ENVIRON SATELLITE DATA & INFO SERV, NAT OCEANIC & ATMOSPHERIC ADMIN, DEPT COM, 80- *Concurrent Pos:* Mem, adv coun, Comt Int Ref Atmosphere, 82-, Comn Satellite Meteorol & Oceanog, Am Meteorol, Soc 90-93 & Comn Dynamic Meteorol, Int Union Geophys & Geod Mesoscale Working Group; adv, Nat Res Coun. *Mem:* Am Meteorol Soc; Int Union Geophys & Geod. *Res:* Satellite metrology; mesoscale metrology; development and evolution of deep convection; mesoscale processes using satellite data; use of satellite data for severe and tornadic storm identification. *Mailing Add:* NOAA-NESDIS-RAMM Br Colo State Univ Ft Collins CO 80523

PURDOM, PAUL W, JR, b Atlanta, Ga, Apr 5, 40; m 65; c 3. ANALYSIS OF ALGORITHMS. *Educ:* Calif Inst Technol, BS, 61, MS, 62, PhD(physics), 66. *Prof Exp:* From asst prof to assoc prof comput sci, Univ Wis-Madison, 65-71; chmn dept, 78-82, assoc prof, 72-82, PROF COMPUT SCI, IND UNIV, BLOOMINGTON, 82- *Concurrent Pos:* Mem tech staff, Bell Tel Labs, 70-71; assoc ed, Computer Surveys. *Mem:* Asn Comput Mach; AAAS; Soc Indust & Appl Math. *Res:* Analysis of algorithms with an emphasis on the average time required for the satisfiability problem. *Mailing Add:* Dept Comput Sci Ind Univ Bloomington IN 47401

PURDOM, RAY CALDWELL, b Lebanon, Ky, Sept 8, 43; m 65; c 3. PHYSICS. *Educ:* Duke Univ, BS, 65; Purdue Univ, MS, 67, PhD(physics), 70. *Prof Exp:* From asst prof to assoc prof physics, 70-80, chmn dept physics, math & computer sci, 82-89, PROF PHYSICS, KY WESLEYAN COL, 80-, ACAD DEAN, 88- *Concurrent Pos:* Res assoc, Dept Radiol, Col Med, Univ Cincinnati, 77-78. *Mem:* Am Phys Soc; Am Asn Physics Teachers. *Res:* Phonon-phonon interactions in solids; ultrasonic attenuation in solids at low temperatures; medical physics. *Mailing Add:* Dept Physics, Math & Computer Sci Ky Wesleyan Col Owensboro KY 42301

PURDOM, WILLIAM BERLIN, b De Kalb, Ill, Apr 6, 34; m 60. ECONOMIC GEOLOGY. *Educ:* Univ Ky, BS, 56; Univ Ariz, PhD(econ geol), 60. *Prof Exp:* Asst prof geol, Univ Ore, 60-62, mineral examr, US Bur Land Mgt, 62-63; asst prof geol, Univ Ore, 63-64; from asst prof to assoc prof geol, 64-74, PROF GEOL & GEN SCI, SOUTHERN ORE COL, 74-, CHMN DEPT GEOL, 77- *Mem:* AAAS; Mineral Soc Am; Geol Soc Am; Nat Asn Geol Teachers; Am Inst Prof Geologists; Sigma Xi. *Res:* Mineralogy. *Mailing Add:* Dept Geol Southern Ore State Col Ashland OR 97520

PURDON, JAMES RALPH, JR, b Akron, Ohio, Feb 9, 33; m 54; c 5. POLYMER CHEMISTRY. *Educ:* Univ Akron, BS & MS, 55, PhD(polymer chem), 61. *Prof Exp:* Chemist colloid appln, 57-63, sect head phys & chem anal, 63-66, mgr prod develop rubbers & chem, 66-74, mgr chem develop serv, 74-75, mgr chem mat & processes develop, 75-79, mgr latex & coatings develop, 79-80, mgr emulsion polymers & resins develop, 80-82, mgr latex & coatings res & develop, 80-82, mgr prod appln & customer serv, 82-84, MGR CUSTOMER APPLN & TESTING SERV, GOODYEAR TIRE & RUBBER CO, 84- *Honors & Awards:* Rubber Age Award for Rubber Technol, Palmerton Publ Co, 61. *Mem:* Am Chem Soc. *Res:* Applications research and development; customer technical service; physical and analytical testing. *Mailing Add:* 3167 Englewood Dr Stow OH 44224-3849

PURDUE, JACK OLEN, b McLoud, Okla, Mar 2, 13; c 3. CHEMISTRY. *Educ:* Okla Baptist Univ, BS, 34; Univ Okla, MS, 39, PhD(chem), 48. *Prof Exp:* Instr physics, Okla Baptist Univ, 38-43; dir ground sch, Regional Sch, Civil Aeronaut Admin, 43-44; from assoc prof to prof, 46-64, distinguished serv prof, 64-80, EMER DISTINGUISHED SERV PROF CHEM, OKLA BAPTIST UNIV, 80- *Mem:* Am Chem Soc; Math Asn Am. *Res:* Colloids; some factors in the stability of emulsions; preparation and properties of clay systems; modification of carbon surfaces to enhance anion adsorption. *Mailing Add:* 2000 N Market Shawnee OK 74801

PURDUE, PETER, b Dublin, Ireland, Sept 18, 43; m 71. MATHEMATICS, STATISTICS. *Educ:* Univ London, BS, 67; Purdue Univ, Lafayette, MS, 69, PhD(statist), 72. *Prof Exp:* From asst prof to prof statist, Univ Ky, 71-86; PROF & CHMN OPERS RES, NAVAL POSTGRAD SCH, 86- *Concurrent Pos:* Consult, Oak Ridge Nat Lab; prog dir, NSF, 84-86. *Mem:* Fel Inst Math Statist; Am Statist Asn; Opers Res Soc Am; Int Statist Inst. *Res:* Queueing theory; stochastic processes; stochastic models in biology. *Mailing Add:* Dept Opers Res Naval Postgrad Sch Monterey CA 93943

PURDY, ALAN HARRIS, b Mt Clemens, Mich, Dec 13, 23; m; c 4. BIOMEDICAL ENGINEERING, OCCUPATIONAL HEALTH. *Educ:* Univ Miami, BS, 54; Univ Calif, Los Angeles, MS, 67; Univ Mo-Columbia, PhD(biomed eng), 70. *Prof Exp:* Proj engr acoust, Arvin Industs, 54-56 & AC Spark Plug Co, Mich, 56-61; asst prof eng, Calif State Polytech Univ, 61-67; assoc dir biomed eng, Univ Mo, 67-71; dep assoc dir, Nat Inst Occup Safety & Health, 71-76, biomed engr, 76-80; asst dir, Fla Inst Oceanog, 81-83; scientist, Biomed Eng, Nat Inst Occup Safety & Health, 83-86; PRES, ALPHA BETA RES & DEVELOP CORP, 87- *Concurrent Pos:* Consult acoust, Smithy Muffler Co, 61-63; Nat Heart Inst spec fel, 63-67; consult biomed, Statham Instruments, Inc, 66-67; consult fac, Sch Med, Tex Tech Univ, 71-74; mem adv bd, Fla Inst Technol, 76-; adj prof, marine sci, Univ SFla, 81-83; proj engr, Gen Motors Corp, 56-60. *Mem:* Asn Advan Med Instrumentation; Biomed Eng Soc; Oceanic Soc; Undersea Med Soc; Acoust Soc Am; Int Oceanog Found; Am Inst Physics. *Res:* Occupational safety and health; man-machine interface; man undersea; solar power; whole body vibration; acoustics. *Mailing Add:* Alpha Beta Res & Develop Corp 5228 SW Fifth Pl Cape Coral FL 33914

PURDY, D(AVID) C(ARL), b Bethlehem, Pa, Jan 17, 29; m 52; c 4. MECHANICAL & NUCLEAR ENGINEERING. *Educ:* Webb Inst, BS, 50. *Prof Exp:* Naval archit, Preliminary Design Div, Bur Ships, Navy Dept, 50-51, marine engr, 51-55; proj engr, Reactor Develop Dept, Chrysler Corp, 56-57; sr engr, Atomic Energy Div, Babcock & Wilcox Co, 57-59, eng supvr, 59-61, prof engr, 61-66; supvry engr, Gibbs & Hill, Inc, 66-72, asst chief nuclear engr, 72-73, chief mech engr, 73-79, chief nuclear engr, 79-80, mgr advan technol, 80-88; RETIRED. *Concurrent Pos:* Adj prof nuclear & mech eng, Polytech Inst NY, 78- *Mem:* Am Soc Mech Engrs; Int Solar Energy Soc. *Res:* Design and construction of nuclear power plants and alternate energy systems. *Mailing Add:* Eight Manetto Hill Rd Huntington NY 11743

PURDY, DAVID LAWRENCE, b New York, NY, Sept 18, 28; m 51; c 4. BIOMEDICAL ENGINEERING. *Educ:* Cornell Univ, Mech Engr, 51. *Prof Exp:* Prof engr, Gen Elec Co, 53-55, mgr eng prog, Eng Serv Div, 55-57, space propulsion engr, Space Power Oper, 57-60, systs engr, Missile & Space Div, 60-64; mgr, Nuclear Mat & Equip Corp, 64-72; pres, Coratomic, Inc, 72-86; PRES, BIOCONTROL TECHNOL, INC, 86- *Concurrent Pos:* Dir radioisotope powered artificial heart contract, Nat Heart & Lung Inst; mgr radioisotope powered cardiac pacemaker contract, AEC. *Mem:* Fel Am Soc Mech Engrs; Am Inst Aeronaut & Astronaut; Asn Advan Med Instrumentation; Am Diabetes Asn. *Res:* Application of atomic energy, mechanical and electrical engineering and electronics to implantable prosthetics such as the artificial heart; radioisotope powered cardiac pacemaker; glucose detector. *Mailing Add:* Biocontrol Technol Inc 300 Indian Springs Rd Indiana PA 15701

PURDY, GARY RUSH, b Edmonton, Alta, Oct 8, 36; m 61; c 4. PHYSICAL METALLURGY. *Educ:* Univ Alta, BS, 57, MS, 59; McMaster Univ, PhD(metall), 62. *Prof Exp:* Res assoc metall, Univ Alta, 59-60; fel, McMaster Univ, 62-63, from asst prof to assoc prof metall & metall eng, 63-71, assoc dean grad studies, 71-74, chmn, 78-86, actg dean eng, 87-88, PROF MAT SCI & ENG, MCMASTER UNIV, 71-, DEAN ENG, 90- *Concurrent Pos:* Vis prof, Cent Elec Res Labs, Eng, 69-70, Royal Inst Technol, Stockholm, 76-77 & LPTCM, Grenoble, 83-84; hon prof, Beijing Univ of Iron & Steel, 83, Chongging Univ, 85; C D Howe Mem Found Fel, 69. *Honors & Awards:* Can Metal Physics Medal, 90. *Mem:* Can Inst Mining & Metall; Am Inst Mining, Metall & Petrol Engrs; Am Soc Metals. *Res:* Thermodynamic properties of alloys; diffusion; phase transformations in solids; solidification of alloys. *Mailing Add:* Dept Mat Sci & Eng McMaster Univ Hamilton ON L8S 4L8 Can

PURDY, LAURENCE HENRY, b Miami, Ariz, Sept 28, 26; m 48; c 4. PLANT PATHOLOGY. *Educ:* San Diego State Col, BS, 49; Univ Calif, Davis, PhD(plant path), 54. *Prof Exp:* Res asst, Univ Calif, 52-53; pathologist, Agr Res Serv, USDA, 53-67; chmn dept, 67-79, PROF PLANT PATH, INST FOOD & AGR SCI, UNIV FLA, 67- *Concurrent Pos:* Consult, Am Cocoa Res Inst. *Mem:* Fel Am Phytopath Soc (treas, 70-76, vpres, 78, pres elect, 79, pres, 80). *Res:* Biology of rust and smut of sugar cane and other diseases of sugar cane; epidemiology and control of diseases of cocoa (Theobroma cacao); tropical agriculture and international cooperative research programs. *Mailing Add:* Dept Plant Path Univ Fla Gainesville FL 32611

PURDY, ROBERT H, b Bloomfield, NJ, Feb 26, 30. BIOCHEMISTRY. *Educ:* Yale Univ, BS, 51; Harvard Univ, PhD(biochem), 60. *Prof Exp:* USPHS fel, 59-61; Milton Found fel, 61-62; Lalor Found fel, 62-63; res assoc biol chem, Harvard Med Sch & Boston City Hosp, 62-64; FOUND SCIENTIST, SOUTHWEST FOUND BIOMED RES, 65- *Mem:* Am Chem Soc; Endocrine Soc; Am Soc Neurochem. *Res:* Anxiolytic steroids and GABAergic activity; steroid metabolism. *Mailing Add:* Southwest Found Res & Educ PO Box 28147 San Antonio TX 78284

PURDY, WILLIAM CROSSLEY, b Brooklyn, NY, Sept 14, 30; m 53; c 3. ANALYTICAL CHEMISTRY. *Educ:* Amherst Col, BA, 51; Mass Inst Technol, PhD(chem), 55. *Prof Exp:* Instr chem, Univ Conn, 55-58; from asst prof to prof, Univ Md, College Park, 58-76, head Anal Chem Div, 68-76; prof chem & assoc in med, 76-86, SIR WILLIAM MACDONALD PROF CHEM & ASSOC VPRIN ACAD, MCGILL UNIV, MONTREAL, 86- *Concurrent Pos:* Consult, US Dept Army, 58-76; vis prof inst nutrit sci, Univ Giessen, Ger, 65-66. *Honors & Awards:* Fisher Lectr Award, Chem Inst Can, 82; Ames Award, Outstanding Contrib Res, Am Asn Clin Chem, 86; Kodak Can Award, Outstanding Contrib Res Award, Can Soc Clin Chemists, 90. *Mem:* Am Chem Soc; fel Nat Acad Clin Biochem; Am Asn Clin Chem; fel Chem Inst Can; fel Royal Soc Chem UK; Can Soc Clin Chemists; fel Can Acad Clin Biochemists; Asn Clin Scientists; Soc Electroanal Chem. *Res:* Electroanalytical chemistry; electrode studies in aqueous and nonaqueous media; clinical analyses; application of electroanalytical techniques to biochemistry; clinical chemistry; lipid analyses. *Mailing Add:* Dept Chem McGill Univ Montreal PQ H3A 2K6 Can

PURE, ELLEN, B-CELL ACTIVATION, LYMPHOKINES. *Educ:* Univ Tex, PhD(med sci), 81. *Prof Exp:* ASST PROF IMMUNOL, ROCKEFELLER UNIV, 84- *Mailing Add:* Dept Cell Physiol & Immunol Rockefeller Univ 1230 York Ave New York NY 10021

PURI, KEWAL KRISHAN, b Sialkot, India, Oct 1, 33; m 67; c 3. MATHEMATICS, FLUID DYNAMICS. *Educ:* Delhi Univ, BA, 53, MA, 55; NY Univ, MS, 65, PhD(math), 67. *Prof Exp:* Sci asst, Defense Sci & Res Orgn, Delhi, India, 57-63; asst prof appl math, NY Univ, 67-70; assoc prof, Ala State Univ, 70-76; assoc prof, 77-81, PROF MATH, UNIV MAINE, 81- *Mem:* Soc Indust & Appl Math; Am Geophys Union; Int Asn Hydraul Res. *Res:* Viscous and inviscid flows in channels. *Mailing Add:* Dept Math 326 Neville Hall Univ Maine Orono ME 04469

PURI, NARINDRA NATH, b New Delhi, India, Nov 30, 33; US citizen; m 62; c 1. ENGINEERING, MATHEMATICS. *Educ:* Indian Inst Tech, Kharagpur, BSEE, 55; Univ Wis, MS, 58; Univ Pa, PhD(elec eng), 60. *Prof Exp:* From asst prof to assoc prof elec eng, Drexel Inst, 60-63; assoc prof, Univ Pa, 63-64; mgr guid & control systs, Missile & Space Div, Gen Elec Co, 64-68; prof elec eng, Rutgers Univ, New Brunswick, 68-; AT ELEC ENG DEPT, RUTGERS COL. *Concurrent Pos:* Consult, Honeywell Spec Syst Div, Pa, 62-63, Gen Elec Co, 63-64 & Astro Div, RCA, 69- *Mem:* Inst Elec & Electronics Engrs. *Res:* Optimal and adaptive control systems; design of control systems for artificial heart including computer simulation; research in guidance and control of aerospace vehicles; communication satellites; mathematical modeling of multi-element antennas. *Mailing Add:* Dept Elec Eng Rutgers Univ New Brunswick NJ 08903

PURI, OM PARKASH, b Sialkot, India, Apr 27, 35; m 63; c 3. SOLID STATE PHYSICS. *Educ:* Punjab Univ, India, BA, 55; Univ Saugar, MS, 58, PhD(physics), 61. *Prof Exp:* Assoc prof, 61-64, PROF PHYSICS, CLARK COL, 64-, ASSOC DEAN & SPEC ASST TO PRES, 89- *Concurrent Pos:* Assoc prof physics, State Univ NY, 63-64; dir, Coop Gen Sci Proj, Clark Col & NSF Prog, 66-; consult, US Off Educ; chmn Dept Physics, Clark Col, 61-, chmn, Div Nat Sci & Math, 77-, dir, Nigerian Manpower Proj, 78- *Mem:* Am Phys Soc; Am Soc Metals; Am Inst Mining, Metall & Petrol Engrs; Am Asn Physics Teachers. *Res:* X-ray diffraction; dielectrics; fiber glass; wax electrets; stresses in metals; magnetic materials; polarization; energy storage. *Mailing Add:* Dept Physics Clark Atlanta Univ 240 Chestnut St SW Atlanta GA 30314

PURI, PRATAP, b Lahore, India, Mar 15, 38; m 68; c 2. APPLIED MATHEMATICS, CONTINUUM MECHANICS. *Educ:* Panjab Univ, India, BA Hons, 57; Delhi Univ, MA, 59; Indian Inst Technol, Kharagpur, MTech, 60, PhD(appl math), 65. *Prof Exp:* Assoc lectr math, Indian Inst Technol, Bombay, 62-63, lectr, 63-68; from asst prof to assoc prof, 68-76, PROF MATH, UNIV NEW ORLEANS, 76- *Concurrent Pos:* Polish Acad Sci fel, Inst Fundamental Tech Probs, Warsaw, Poland, 66. *Mem:* Am Acad Mech; Calcutta Math Soc; Soc Indust & Appl Math. *Res:* Fluid mechanics, solid mechanics. *Mailing Add:* Dept Math Univ New Orleans New Orleans LA 70122

PURI, PREM SINGH, statistics, operations research; deceased, see previous edition for last biography

PURI, SURENDRA KUMAR, b New Delhi, India, Oct 26, 40; m 68; c 4. PHARMACOLOGY, PHARMACEUTICAL SCIENCES. *Educ:* Univ Saugar, BPharm, 66; Univ RI, MS, 71, PhD(pharmacol), 74. *Prof Exp:* Teaching asst biochem pharm, Col Pharm, Univ RI, 70-73; res pharmacologist, Sch Med, Boston Univ, 73-74, asst prof pharm, 74-76; sr res biochemist, 76-78, asst dir, 78-81, ASSOC DIR CLIN PHARM, HOECHST-ROUSSEL PHARMACEUTS INC, 81- *Concurrent Pos:* Adj asst prof pharmacol, Sch Med, Boston Univ, 76- *Mem:* Am Soc Pharmacol & Exp Therapeut; Soc Neurosci; Am Pharmacent Asn; Am Soc Clin Pharmacol & Therapeut. *Res:* Behavioral and biochemical approaches to study drug action and disease states; pharmacokinetic, metabolism and toxicological aspects of drugs; neuropharmacological aspects of aging. *Mailing Add:* Hoechst-Roussel Pharmaceuts Inc Rte 202-206 N Somerville NJ 08876

PURI, YESH PAUL, b Sangrur, India, Sept 20, 29; m 71; c 3. PLANT BREEDING, GENETICS. *Educ:* Punjab Univ, India, BSc, 49; Benares Hindu Univ, BSc, 54; Ore State Univ, MS, 58, PhD(plant breeding, genetics), 60. *Prof Exp:* Asst farm crops, Ore State Univ, 55-56, fiber crops, 56-59, jr agronomist, 59-61; asst prof agron, Univ Baghdad, 61; sr res technician, 62-65, specialist & supt res & admin, Tulelake Field Sta, 65-83, specialist & mgr, res & mgt, 84-87, SPECIALIST BARLEY BREEDING & GENETICS, AGRON RES STA, UNIV CALIF, DAVIS, 88- *Mem:* AAAS; Am Soc Agron; Crop Sci Soc Am; Soil Sci Soc Am; Am Genetic Asn; NY Acad Sci. *Res:* Breeding, genetics, pollen physiology and other physiological aspects of cereal, oil and forage crops. *Mailing Add:* Dept of Agron & Range Sci Univ of Calif Davis CA 95616

PURICH, DANIEL LEE, BIOCHEMISTRY. *Educ:* Slippery Rock State Col, BA, 67; Iowa State Univ, PhD(biochem & biophys), 73. *Prof Exp:* Teaching asst, Dept Biochem & Biophys, Iowa State Univ, 67-69; from asst prof to prof chem, Univ Calif, Santa Barbara, 73-84; PROF & CHMN, DEPT BIOCHEM & MOLECULAR BIOL, COL MED, UNIV FLA, 84- *Concurrent Pos:* Lectr, numerous univs, 72-91; NIH res career develop award, 77-82; Alfred

P Sloan fel chem, 78-80; mem, Biochem Study Sect, NIH, 82-85. *Mem:* Am Soc Biol Chemists; Am Chem Soc; NY Acad Sci; Biochem Soc. *Res:* Microtubule cytoskeleton self-assembly; cell processes depending on microtubules; enzyme catalysis and regulation; biophysics and enzymology in urolithiasis; author of numerous technical publications. *Mailing Add:* Dept Biochem & Molecular Biol Univ Fla Col Med J-245 Gainesville FL 32610-0245

PURISCH, STEVEN DONALD, b New York, NY, Mar 25, 45; div; c 1. TOPOLOGY. *Educ:* Carnegie-Mellon Univ, BS, 67, PhD(math), 73; Univ Wis-Madison, MA, 69. *Prof Exp:* Asst prof math, Rust Col, 73-74; lectr, Univ Ibadan, Nigeria, 74-78; vis asst prof, State Univ NY, Oswego, 78-79, Lehigh Univ, 79-81; ASST PROF MATH, LAFAYETTE COL, 81- *Concurrent Pos:* Vis lectr, Univ Toronto, 78. *Mem:* Am Math Soc. *Res:* General topology particularly total orderabiltiy and suborderability; categorical topology. *Mailing Add:* Tenn Tech Univ Box 5054 Cookeville TN 38505

PURKERSON, MABEL LOUISE, b Goldville, SC, Apr 3, 31. MEDICAL EDUCATOR, NEPHROLOGY. *Educ:* Erskine Col, AB, 51; Med Univ SC, MD, 56. *Prof Exp:* Intern, St Louis City Hosp, Mo, 56-57; resident pediat, St Louis Children's Hosp, 57-60, fel pediat metab, 60-62; from instr to asst prof pediat, Sch Med, Wash Univ, 62-76, from instr to assoc prof med, 66-89, asst dean curric, 76-78, ASSOC DEAN, SCH MED, WASH UNIV, 78-, PROF MED, 89- *Concurrent Pos:* Vis asst prof anat & NIH spec fel, Dept Anat, Col Physicians & Surgeons, Columbia Univ, NY, 71-72; asst physician, Barnes Hosp, St Louis; chief nephrol, Med Serv, Wash Univ, John Cochran Vet Admin Med Ctr, 73-76, res grantee, 73-78; mem merit rev bd, Nephrol, Vet Admin Cent Off, 77-80; pres, Women Nephrology, 88-90. *Mem:* Sigma Xi; Am Soc Nephrol; Int Soc Nephrol; Am Physiol Soc; Am Soc Renal Biochem & Metab; Int Soc Renal Nutrit & Metab. *Res:* Renal physiology, notably pathophysiology of kidney; pathophysiology of acute and chronic renal failure utilizing experimental models of renal disease; mechanisms underlying renal reabsorption post-release of obstruction; role of factors which may ameliorate progression of renal failure-anticoagulant drugs, hormones, dietary manipulation. *Mailing Add:* Renal Div Dept Internal Med Box 8126 Wash Univ Sch Med 660 Euclid Ave St Louis MO 63110

PURKHISER, E DALE, b Reynolds, Ind, Aug 8, 31; m 60; c 3. ANIMAL NUTRITION, BIOCHEMISTRY. *Educ:* Purdue Univ, Lafayette, BS, 57, PhD(animal nutrit), 62; Mich State Univ, MS, 58. *Prof Exp:* Asst prof animal sci, Rutgers Univ, 61-63 & Univ Ky, 63-67; SWINE SPECIALIST, AGR EXTEN OFF, MICH STATE UNIV, 67- *Concurrent Pos:* Mem, Swine Exten Adv Comt, USDA & Southern Regional Swine Exten Comt, USDA, 63-67. *Mem:* Am Soc Animal Sci; Am Chem Soc. *Res:* Swine nutrition; monogastric nutrition; physiology; reproductive physiology; animal breeding. *Mailing Add:* Agr Exten Off Mich State Univ Cassopolis MI 49031

PURKIS, IAN EDWARD, b London, Eng, July 22, 25; m 47; c 2. ANESTHESIOLOGY. *Educ:* Univ London, MB, BS, 53. *Prof Exp:* Registr anesthesia, St Thomas Hosp, London, Eng, 55-57; from asst prof to assoc prof, 58-74, PROF ANESTHESIA, DALHOUSIE UNIV, 74- *Concurrent Pos:* Clin fel, Royal Victoria Hosp, Montreal, Que, 57-58; chmn, Respiratory Curric Comt, 75-, Dalhousie Univ, asst dir, Continuing Med Educ, 76- *Mem:* Can Anesthetists Soc (pres, 66-67); Acad Anesthesiol. *Res:* Pharmacological action of drugs in anesthesia; medical education. *Mailing Add:* Dept Anesthesia Dalhousie Univ Halifax NS B3H 3J5 Can

PURKO, JOHN, b Telacze, Ukraine, Feb 29, 29; Can citizen; m 52; c 4. DEVELOPMENTAL BIOLOGY. *Educ:* Univ Western Ont, BA, 52, MSc, 58, PhD(biochem), 63. *Prof Exp:* Instr biol, St Francis Xavier Univ, 53-56; res asst med res, Univ Western Ont, 58-59; Nat Cancer Inst Can fels, Chester Beatty Res Inst, Univ London, 63-65 & Cancer Res Lab, Univ Western Ont, 65-66; from asst prof to assoc prof develop biol, Univ Western Ont, 66-77, assoc prof zool, 68-; RETIRED. *Mem:* Can Biochem Soc; Can Soc Cell Biologists. *Res:* Regulation and role of macromolecular, nucleic acid and protein synthesis in early teleost development; integrity and role of chorion in teleost development; biological action of alkylating agents in embryogenesis. *Mailing Add:* Dept Zool Univ Western Ont London ON N6A 3K7 Can

PURL, O(LIVER) THOMAS, b East St Louis, Ill, June 5, 24; m 48; c 2. ELECTRICAL ENGINEERING. *Educ:* Univ Ill, BS, 48 & 51, MS, 52, PhD(elec eng), 55. *Prof Exp:* Res engr, Collins Radio Co, Iowa, 48-49; asst phys electronics, Univ Ill, 50-54, res assoc, 54-55; mem tech staff, Res Labs, Hughes Aircraft Co, 55-57, sr mem tech staff & head power tube sect, Microwave Tube Dept, 57-58; mem tech staff, Watkins-Johnson Co, 58-66, vpres & mgr electron devices group, 67-78, vpres & mgr systs group, 78-81, vpres shareowner rels & planning coord, 81-86; RETIRED. *Mem:* Am Meteorol Soc; Sigma Xi; fel Inst Elec & Electronics Engrs. *Res:* Microwave electron devices, particularly microwave tubes. *Mailing Add:* 466 La Mesa Court Portola Valley CA 94028

PURNELL, DALLAS MICHAEL, b Tacoma, Wash, July 9, 39; m 72; c 2. EXPERIMENTAL PATHOLOGY, CANCER. *Educ:* Univ Puget Sound, BS, 63; Idaho State Univ, MS, 65; Univ Wash, PhD(exp path), 71. *Prof Exp:* Res asst, Pa State Univ, 72-73, instr, 73-75, asst prof path, Hershey Med Ctr, 75-77; assoc prof path, Sch Med, Univ Md, Baltimore, 77-87; RETIRED. *Concurrent Pos:* Fel path, Hershey Med Ctr, Pa State Univ, 71-72. *Mem:* Nat Acad Sci; AAAS; Int Soc Human & Animal Mycol; Am Asn Pathologists; Am Asn Cancer Res. *Res:* Mammary gland biology; mammary neoplasia and preneoplasia; fungal pathogenicity and genetics; cancer biology. *Mailing Add:* 22136 11th Ave S Seattle WA 98198

PURO, DONALD GEORGE, b Elmira, NY, Nov 2, 47; m 75; c 4. OPHTHALMOLOGY, NEUROBIOLOGY. *Educ:* Univ Pa, BA, 69; Univ Rochester, MD, 74, PhD(physiol), 75; Am Bd Ophthal, Dipl, 81. *Prof Exp:* Intern path, Univ Rochester, 74-75; res assoc pharmacol, NIH, 75-77; resident ophthal, Bascom Palmer Eye Inst, Univ Miami, 77-80; med off ophthal, Nat Eye Inst, NIH, 80-85; assoc prof, 85-90, PROF, UNIV MICH, 90- *Mem:* Soc Neurosci; Asn Res Vision & Ophthal. *Res:* Physiology and pathobiology of retinal cells; cell biology and culture. *Mailing Add:* Dept Ophthal Univ Mich Sch Med 1000 Wall St Ann Arbor MI 48105

PUROHIT, MILIND VASANT, b Delhi, India, Aug 16, 57; m 86; c 1. HEAVY QUARK PHYSICS, STRUCTURE FUNCTION OF QUARKS & GLUONS. *Educ:* Indian Inst Technol, Delhi, MS, 78; Calif Inst Technol, Pasadena. PhD(physics), 83. *Prof Exp:* Res assoc physics, Fermilab, 83-86, Wilson fel, 86-88; ASST PROF PHYSICS, PRINCETON UNIV, 88- *Honors & Awards:* Outstanding Jr Investr Award, US Dept Energy, 89. *Mem:* Am Phys Soc. *Res:* Nucleon structure functions using neutrino scattering; gluon structure function using the photoproduction of charm; charm decay. *Mailing Add:* Physics Dept Princeton Univ Princeton NJ 08544-0708

PURPLE, RICHARD L, b Cooperstown, NY, Oct 1, 36; m 60; c 4. PHYSIOLOGY, NEUROPHYSIOLOGY. *Educ:* Hamilton Col, AB, 58; Rockefeller Univ, PhD(life sci), 64. *Prof Exp:* From instr to assoc prof, 64-76, PROF PHYSIOL, SCH MED, UNIV MINN, MINNEAPOLIS, 76-, JOINT PROF PHYSIOL & OPHTHAL, 78-, DIR ELECTRORETINOGRAPHY LAB, DEPT OPHTHAL, 75- *Concurrent Pos:* Lectr, Gen Exten Div, Univ Calif, Berkeley, 67; consult, reading comt screening grad applns fel support, Danforth Found, 67-68. *Mem:* AAAS; Am Physiol Soc; Soc Neurosci. *Res:* Neurophysiology of sensory systems; integrative mechanisms of single cells in the visual system; systems analysis of sensory pathways. *Mailing Add:* Neurophys 339 Millard Hall Univ Minn Minneapolis MN 55455

PURPURA, DOMINICK PAUL, b New York, NY, Apr 2, 27; m 48; c 4. NEUROPHYSIOLOGY. *Educ:* Columbia Univ, AB, 49; Harvard Med Sch, MD, 53. *Prof Exp:* Intern, Presby Hosp, New York, 53-54; asst resident neurol, Neurol Inst, New York, 54-55; spec res fel, Nat Inst Neurol Dis & Blindness, USPHS, 55-57, res career develop fel, 61-65; from asst prof to assoc prof neurol surg, Col Physicians & Surgeons, Columbia Univ, 56-65, assoc prof anat & neurol, 65-66; dean, Sch Med, assoc vpres med affairs & prof neurobiol & neurol, Stanford Univ, Calif, 82-84; prof anat & chmn dept, Albert Einstein Col Med, 67-74, dir training prog neurol & behav sci, 69-82, prof neurosci & chmn dept, 74-82, PROF NEUROSCI & DEAN, COL MED, ALBERT EINSTEIN COL MED, 84-, VPRES MED AFFAIRS, YESHIVA UNIV, 87- *Concurrent Pos:* Bowen-Brooks scholar, NY Acad Med, 53-54; scholar, Sister Elizabeth Kenny Found, 59-61; mem, Neurophysiol Panel, Int Brain Res Orgn, Unesco, 61-; mem neurol A study sect, NIH, 66-70, chmn, 75-79, chmn, Basic Sci Talk Force Subcomt, 70-74; sect ed neurophysiol, Am J Physiol & J Appl Physiol, 66-70; sci dir, Rose F Kennedy Ctr Res in Mental Retardation & Human Develop, Bronx, NY, 69-72, dir, 72-82; counr, Soc Neurosci, 70-72, chmn, Govt & Pub Affairs Comt, 88-91; managing ed, Brain Res, 70-, ed-in-chief, 75-; mem neurobiol adv panel, NSF, 74-75; mem, Panel Biomed & Behav Sci Training, Nat Res Coun-Nat Acad Sci, 75-76; ed-in-chief, Brain Res Rev, 75-, Develop Brain Res, 81-, Molecular Brain Res, 85-; chmn, Int Cong Comt, Int Brain Res Orgn/World Fedn Neuroscientists, 83-; mem nat adv comt, Infant Health & Develop Prog, Robert Wood Johnson Found, 85-89, chmn, Nat Adv Comt for Minority Med Fac Develop Prog, 89; mem, Bd Mental Health & Behav Med, Inst Med-Nat Acad Sci, 86, Comt to Study Res on Child & Adolescent Mental Disorders, 88- 89; mem, Comn Life Sci & Inst Med Comt Use of Lab Animals in Biomed & Behav Res, Nat Res Coun, 86-87, Neurosci & Physiol 1 Panel, Hughes Doctoral Fel Prog Biol Sci, 89-; mem, Nat Adv Mental Health Coun, Alcohol Drug Abuse & Mental Health Admin, Dept Health & Human Serv, 89-92. *Honors & Awards:* Cressy-Morrison Award in Natural Sci, NY Acad Sci, 61; Lucy Moses Prize in Basic Neurol, 64; James Arthur Lectr on Evolution of the Brain, Am Mus of Natural Hist, 67; William G Lennox Lectr, Am Epilepsy Soc, 71, William G Lennox Award, 73; Martin E Rehfuss Lectr, Thomas Jefferson Univ, 83; Nat Med Res Award, Nat Health Coun, 88; Herbert H Jasper Award, Am EEG Soc, 89. *Mem:* Nat Acad Sci; Inst Med-Nat Acad Sci; Am Acad Neurol; Am Asn Anatomists; fel AAAS; Am Epilepsy Soc (pres, 77); Asn Res Nervous & Ment Dis (vpres, 77); Int Brain Res Orgn (pres, 86-92); fel NY Acad Sci; Soc Neurosci (pres, 83. *Res:* Neuroanatomy; neuropharmacology; functions of cerebral cortex and related structures. *Mailing Add:* Dept Neurosci Albert Einstein Col Med Bronx NY 10461

PURRINGTON, ROBERT DANIEL, b Alamosa, Colo, Apr 11, 36; m 59; c 4. THEORETICAL NUCLEAR PHYSICS. *Educ:* Tex A&M Univ, BS, 58, MS, 63, PhD(physics), 66. *Prof Exp:* From asst prof to assoc prof, 66-74, chmn dept, 79-85, PROF PHYSICS, TULANE UNIV LA, 75- *Concurrent Pos:* Grants, Dept Energy, NSF, NEH & Navoceano. *Mem:* Sigma Xi; Am Phys Soc. *Res:* Study of few-nucleon systems; excitations of vibrational nuclei; structure of medium-mass transition nuclei; ocean acoustics; history and philosophy of science; archaeoastronomy. *Mailing Add:* Dept Physics Tulane Univ New Orleans LA 70118

PURRINGTON, SUZANNE T, b New York, NY, Apr 30, 38; m 63; c 3. ORGANIC CHEMISTRY. *Educ:* Wheaton Col, Mass, AB, 60; Radcliffe Col, AM, 62; Harvard Univ, PhD(chem), 63. *Prof Exp:* Fel & instr chem, Duke Univ, 63-65; from asst prof to assoc prof chem, Shaw Univ, 65-68; assoc prof chem, NY Inst Technol, 68-70; chmn chem, Peace Col, 72-77; asst prof, 78-85, ASSOC PROF CHEM, NC STATE, 85- *Concurrent Pos:* Petrol Res Fund grant, 65-67 & 88-90; Res Corp grant, 83; consult, 69- *Mem:* Am Chem Soc. *Res:* Carbenes and fluorine chemistry. *Mailing Add:* Dept Chem NC State Univ Raleigh NC 27695-8204

PURSEL, VERNON GEORGE, b Yerington, Nev, Nov 11, 36; m 76; c 2. PHYSIOLOGY, ANIMAL SCIENCE. *Educ:* Univ Nev, BS, 58; Univ Minn, MS, 61, PhD(dairy husb), 65. *Prof Exp:* Instr dairy sci, Univ Minn, 64-65, asst prof, 65-67; RES PHYSIOLOGIST, REPRODUCTION LAB, AGR RES, USDA, 67- *Mem:* Soc Study Reproduction; Am Soc Animal Sci; Int Embryo Transfer Soc; Coun Agr Sci & Technol. *Res:* Gene transfer in swine; embryo transfer in swine; control of estrus and ovulation; freezing of swine spermatozoa. *Mailing Add:* Reproduction Lab Agr Res USDA Beltsville MD 20705

PURSELL, LYLE EUGENE, b Paola, Kans, Apr 23, 26; m 50; c 4. MATHEMATICAL ANALYSIS. *Educ:* Purdue Univ, BS, 49, MS, 50, PhD(math), 52. *Prof Exp:* Instr math, Ohio State Univ, 52-55; from asst prof to assoc prof math, Grinnell Col, 55-67; assoc prof, 67-68, prof, 68-88, EMER PROF MATH, UNIV MO-ROLLA, 88- *Concurrent Pos:* Vis asst prof math, Purdue Univ, 58-59; vis fel, Dartmouth Col, 64. *Mem:* Am Math Soc; Math Asn Am. *Res:* Rings of real functions and their transformation groups; non-Archimedean ordered fields. *Mailing Add:* One Hyer Ct Rte 4 Box 51 Rolla MO 65401-0249

PURSELL, MARY HELEN, b Philadelphia, Pa, July 5, 39; m 71. GENETICS. *Educ:* Glassboro State Col, BA, 63, MA, 66; Pa State Univ, PhD(genetics), 75. *Prof Exp:* Teacher high sch, NJ, 63-65; asst prof biol, Glassboro State Col, 66-68; assoc prof biol, 75-, PROF BIOL & ACTG VPRES ACAD & STUDENT AFFAIRS, LOCK HAVEN UNIV. *Res:* DNA evolution in lower vascular plants. *Mailing Add:* Dept Biol Sci Lock Haven Univ Lock Haven PA 17745

PURSELL, RONALD A, b Pa, Dec 7, 30; m 71. BOTANY. *Educ:* Pa State Univ, BS, 52, MS, 54; Fla State Univ, PhD, 57. *Prof Exp:* Actg asst prof biol, Univ Colo, 57-58; instr bot, Univ Tenn, 58-59; from instr to assoc prof, 59-77, PROF BOT, SCH FORESTRY, PA STATE UNIV, 77- *Mem:* Am Bryol & Lichenological Soc (vpres, 73-75, pres, 75-77); Am Soc Plant Taxon; Int Asn Plant Taxon; Bot Soc Am; Sigma Xi. *Res:* Taxonomy and geography of bryophytes. *Mailing Add:* 208 Mueller Lab Pa State Univ University Park PA 16802

PURSER, FRED O, b Greenville, NC, June 6, 31; m 55; c 4. NUCLEAR PHYSICS, PARTICLE PHYSICS. *Educ:* US Naval Acad, BS, 53; Duke Univ, PhD(nuclear physics), 66. *Prof Exp:* RES ASSOC, DUKE UNIV, 66-, ASST TO DIR, TRIANGLE UNIVS NUCLEAR LAB & PROJ ENGR, CYCLO-GRAAFF PROJ, 68- *Mem:* Am Phys Soc. *Res:* Nuclear structure; nuclear fission; particle accelerator development. *Mailing Add:* 527 Paul Pl Los Alamos NM 87544

PURSER, PAUL EMIL, b Amite, La, Dec 9, 18; m 39; c 2. SYSTEMS ENGINEERING. *Educ:* La State Univ, BS, 39. *Prof Exp:* Res engr, Nat Adv Comt Aeronaut, Langley Res Ctr, 39-58; spec asst to dir, Manned Spacecraft Ctr, NASA, 58-70; spec asst to pres, Univ Houston, 68-69; INDEPENDENT CONSULT, 70- *Concurrent Pos:* Vis lectr manned spacecraft eng, La State Univ, 63-64, Univ Houston, 64, Rice Univ, 64-65 & Univ Tex, Austin, 65; consult, Nat Acad Sci-Nat Acad Eng-Nat Res Coun Marine Bd, 71-82 & Cardiol Div, Stanford Sch Med, 72-78. *Mem:* Am Inst Aeronaut & Astronaut; Am Rocket Soc; Am Soc Oceanog; Marine Technol Soc. *Res:* Stability and control of aircraft; development of high-temperature research techniques and use of rocket models for aerodynamic research in free flight; early development of US manned space program; application of systems engineering techniques to biomedical instruments, offshore oil and gas safety, maritime safety and casualty investigation, oil-spill prevention and response, and maritime salvage and diving. *Mailing Add:* 8950 Shoreview Lane Humble TX 77346

PURSEY, DEREK LINDSAY, b Glasgow, Scotland, Oct 22, 27; US citizen; m 62; c 1. THEORY OF GROUPS APPLIED TO PHYSICS, SUPERSYMMETRIC QUANTUM MECHANICS. *Educ:* Univ Glasgow, BSc, 48, PhD(physics), 52. *Prof Exp:* Lect math physics, King's Col, Univ London, 51-54, Univ Edinburgh, 54-60 & Univ Glasgow, 61-64; PROF PHYSICS, IOWA STATE UNIV, 65- *Concurrent Pos:* Vis physicist, Univ Theoret Physic, Copenhagen, 57; res physicist, Univ Calif, Los Angeles, 59-60, vis prof, 79-80. *Mem:* Sigma Xi; fel Am Phys Soc; AAAS; fel Royal Soc Edinburgh. *Res:* Theory of groups applied to physics, quantum mechanics, nuclear physics and elementary particle physics; general quantum mechanics; theory of isospectral Hamiltonians and supersymmetric quantum mechanics. *Mailing Add:* Dept Physics Iowa State Univ Ames IA 50011

PURSGLOVE, LAURENCE ALBERT, b Pa, July 29, 24; m 44; c 5. PHARMACEUTICAL CHEMISTRY, CHEMICAL ENGINEERING. *Educ:* Carnegie Inst Technol, BS, 44, MS, 47, ScD(org chem), 49. *Prof Exp:* Jr chem engr, Minn Mining & Mfg Co, 44-46; consult, Food Mach Corp, 47-48; asst prof chem, WVa Univ, 49-53; chemist, Dow Chem Co, 53-59; instr chem, Bay City Jr Col, 59-61; asst prof chem, Delta Col, 61-63; SCI WRITING, COMPUT PROG, TRANSLATING & CONSULT, 64- *Concurrent Pos:* Chemist, US Bur Mines, 51-53. *Mem:* Am Chem Soc; Sigma Xi. *Res:* Heterocyclic synthesis; metabolite antagonists; programed self-instruction; educational technology; computer programing. *Mailing Add:* PO Box 4418 Santa Clara CA 95056-4418

PURSLEY, MICHAEL BADER, b Winchester, Ind, Aug 10, 45; m 68; c 1. COMMUNICATIONS SYSTEMS, INFORMATION THEORY. *Educ:* Purdue Univ, BS, 67, MS, 68; Univ Southern Calif, PhD(elec eng), 74. *Prof Exp:* Engr, Northrop Corp, 68; staff engr commun, Hughes Aircraft Co, 68-74; from asst prof to assoc prof elec eng, 74-80, PROF ELEC ENG, UNIV ILL, URBANA, 80- *Concurrent Pos:* Actg asst prof syst sci, Univ Calif, Los Angeles, 74, vis prof elec eng, 85; prin investr, NSF grants, 76-80, 76-78 & 78-80 & Army Res Off grant; consult, Army Res Off, 77-78, & 83-86, Int Tel & Tel, 78-; assoc mem, Ctr Advan Study, Univ Ill, 80-81. *Honors & Awards:* Centennial Medal, Inst Elec & Electronics Engrs, 84. *Mem:* Fel Inst Elec & Electronics Engrs; Inst Math Statist. *Res:* Communication systems, information theory and error-control coding, spread spectrum communications, and communication networks. *Mailing Add:* Coord Sci Lab Univ of Ill 1101 W Springfield Ave Urbana IL 61801-3082

PURTILO, DAVID THEODORE, PATHOLOGY. *Educ:* Univ Minn, BA, 61; Univ NDak, MS, 63; Northwestern Univ, MD, 67; Am Bd Path, dipl, 72. *Prof Exp:* From asst prof to prof path, Med Sch, Univ Mass, 73-81; prof & chmn, Dept Path & Lab Med, 81-85, PROF & CHMN, DEPT PATH & MICROBIOL, MED CTR, UNIV NEBR, 85- *Concurrent Pos:* Res pathologist, St Vincent Hosp, 73-74; clin pathologist, Worcester City Hosp, 75-76; adj prof nursing, 78-81; teaching fac hemat, Am Soc Clin Pathologists, 78-; prof pediat, Med Sch, Univ Mass, 78-81; vis scientist, Karolinska Inst, Stockholm, Sweden, 79-80; prof pediat, Dept Pediat, Med Ctr, Univ Nebr, 81-, grad fac fel, 81-; numerous res grants, 75-96; biosci lectr, State Univ NY Health Sci Ctr, 89, Univ Kans Med Ctr, 90. *Honors & Awards:* David Golan Mem Lectr, NShore Univ Hosp, 81; Dale Darst Mem Lectr, Med Ctr, Univ Iowa, 87. *Mem:* Am Soc Clin Pathologists; Int Acad Pathologists; Am Soc Microbiol; Am Fedn Clin Res; Am Asn Immunologists; Pediat Path Soc; AAAS; Col Am Pathologists; Int Asn Comp Res Leukemia & Related Dis; AMA. *Res:* Virus; determination of antibody-dependent cellular cytotoxicity of human blood mononuclear cells. *Mailing Add:* Dept Path & Microbiol Univ Nebr Med Ctr 600 S 42nd St Omaha NE 68198-3135

PURTON, CHRISTOPHER ROGER, b Toronto, Ont, Sept 10, 38; m 61, 69; c 3. ASTRONOMY. *Educ:* Univ Toronto, BA, 60, MA, 62; Cambridge Univ, PhD(astron), 66. *Prof Exp:* Asst res officer astron, Nat Res Coun Can, 66-68; from asst prof to prof physics, York Univ, 68-81; SR RES OFFICER, NAT RES COUN CAN, 81- *Mem:* Royal Astron Soc Can; Royal Astron Soc; Am Astron Soc; Can Astron Soc; Astron Soc Pac. *Res:* Radio astronomy. *Mailing Add:* Dominion Radio Astrophys Obs Box 248 Penticton BC V2A 6J9 Can

PURVES, DALE, b Philadelphia, Pa, Mar 11, 38; m 68; c 2. NEUROBIOLOGY. *Educ:* Yale Univ, BA, 60; Harvard Univ, MD, 64. *Prof Exp:* From asst prof to prof physiol, Med Sch, Wash Univ, 73-85, prof neurobiol, 85-90; GEORGE BARTH GELLER PROF RES IN NEUROBIOL & CHMN, DEPT NEUROBIOL, DUKE UNIV, DURHAM, NC, 90- *Concurrent Pos:* Assoc ed, J Neurocytol, 77-83, J Neurosci, 80-87, ed-in-chief, 88-; Jacob K Javits Neurosci Investr Award, Nat Inst Neurol Commun Dis & Stroke, 84, Nat Inst Neurol Dis & Stroke, 90; ann distinguished lectr neurosci, State Univ NY, 87, Univ Edinburgh, 90. *Honors & Awards:* Mathilde Solowey Award Neurosci, 79; Alexander Forbes Lectr, Grass Found, 83; Viktor Hamburger Lectr, Wash Univ, 84; Clinton Woolsey Ann Lectr, Univ Wis, 86; Grass Lectr, Soc Neurosci, 90; John H Krantz Jr Mem Lectr, Univ Md, 91. *Mem:* Nat Acad Sci; AAAS; Soc Neurosci; Soc Gen Physiologists; Soc Develop Biol. *Res:* Formation and maintenance of synaptic connections. *Mailing Add:* Dept Neurobiol Duke Univ Med Ctr Durham NC 27710

PURVES, WILLIAM KIRKWOOD, b Sacramento, Calif, Oct 28, 34; m 59; c 1. PLANT PHYSIOLOGY. *Educ:* Calif Inst Technol, BS, 56; Yale Univ, MS, 57, PhD(bot), 59. *Prof Exp:* NSF fel, Univ Tübingen, Ger, 59-60; Nat Cancer Inst fel, Univ Calif, Los Angeles, 60-61; asst prof bot, Univ Calif, Santa Barbara, 61-65, assoc prof biochem, 65-70, prof biol, 70-73, chmn dept biol sci, 72-73; prof biol & head Biol Sci Group, Univ Conn, 73-77; chmn dept computer sci, 85-90, STUART MUDD PROF BIOL, HARVEY MUDD COL, 77-, CHMN DEPT BIOL, 85- *Concurrent Pos:* Res collabr, Brookhaven Nat Lab, 65, 66; NSF sr fel, Univ London, 67, Harvard Univ, 68; adj prof plant physiol, Univ Calif, Riverside, 79-85; vis fel, Yale Univ, 83-84. *Mem:* Fel AAAS; Sigma Xi; Am Soc Plant Physiol; Am Inst Biol Sci. *Res:* Chemical and physical regulation of plant cell expansion; enzymology. *Mailing Add:* Dept Biol Harvey Mudd Col Claremont CA 91711

PURVIS, COLBERT THAXTON, b Odum, Ga, May 31, 20; m 41; c 2. MATHEMATICS. *Educ:* Ga Teachers Col, BS, 41; ETex State Teachers Col, MS, 48; Peabody Col, PhD(math), 57. *Prof Exp:* From instr to asst prof math, Ga Inst Technol, 46-60; prof math, Calif State Univ, Hayward, 60-83, emer prof, 83-; RETIRED. *Concurrent Pos:* Dir & teacher, NSF Insts Teachers Math. *Mailing Add:* Dept of Math & Comput Sci Calif State Univ Hayward CA 94542

PURVIS, GEORGE ALLEN, b Las Animas, Colo, Mar 30, 33; m 59; c 4. FOOD SCIENCE, NUTRITION. *Educ:* Colo State Univ, BS, 54, MS, 62; Mich State Univ, East Lansing, PhD(food sci, nutrit), 69. *Prof Exp:* Mgt trainee, Safeway Stores, Inc, 54-58; state chemist, Colo Dept Agr, 58-60; res assoc, Colo Exp Sta, Ft Collins, 60-62 & Ohio Exp Sta, Columbus, 62-63; lab mgr, 63-69, res mgr nutrit, 69-71, dir nutrit, 77-80, vpres nutrit sci, 80-85, VPRES RES & DEVELOP, GERBER PROD CO, 85- *Mem:* Inst Food Technologists; NY Acad Sci; Am Inst Nutrit; Soc Nutrit Educ; Sigma Xi. *Res:* Food and nutrition related to infants, research and management; evaluation of infant nutrition status in the United States; evaluation and design of foods for infants. *Mailing Add:* Res Dept Gerber Prod Co Fremont MI 49412

PURVIS, JOHN L, b Montclair, NJ, Mar 3, 26; m 64. BIOCHEMISTRY. *Educ:* McGill Univ, BSc, 52, MSc, 54, PhD, 56. *Prof Exp:* Res fel, Univ Amsterdam, 56-58 & Brandeis Univ, 59-61; assoc prof, 61-69, chmn dept, 66-77, PROF BIOCHEM, UNIV RI, 69- *Mem:* Am Chem Soc. *Res:* Pyridine nucleotides; oxidative phosphorylation; steroid biochemistry. *Mailing Add:* Dept of Biochem Univ of RI Kingston RI 02881

PURVIS, JOHN THOMAS, b Meridan, Conn, May 17, 51; m 73; c 2. COMPUTERIZATION OF RCRA SYSTEMS, DATABASE COMPUTER APPLICATIONS. *Educ:* Univ New Haven, BS, 79. *Prof Exp:* Pilot plant supvr, Corp Res Ctr, Uniroyal Inc, 72-80; plant supt, Solvent Recovery Serv, 80-84, tech & opers mgr, 85-88; plant engr, Laticrete Int, 84-85; TECH MGR, ECOLOTEC, INC, 88- *Concurrent Pos:* Hazard analyst, Sara Lerc, Montgomery Greene County, 87- *Res:* ethylene propylene synthetic rubber; Polymers, chemicals and synthetic oils. *Mailing Add:* Ecolotec Inc PO Box 175 Dayton OH 45404

PURVIS, MERTON BROWN, b Dubuque, Iowa, Feb 16, 23; m 44; c 7. MECHANICAL ENGINEERING, APPLIED PHYSICS. *Educ:* Iowa State Univ, BS, 44, MS, 49; Pa State Univ, PhD(mech eng), 54. *Prof Exp:* Engr, Hardsocg Pneumatic Tool Co, Iowa, 46-48; instr mech eng, Iowa State Univ, 48-49; from instr to asst prof, Pa State Univ, 49-55; mem tech staff appl optics, Bell Tell Labs, Inc, 55-58, supvr appl optics & magnetics, 58-69, head, Dept Electromech Components, 69-84, HEAD, TECHNOL TRANSFER DEPT, AT&T BELL LABS, 84- *Concurrent Pos:* Chmn comt P5.7, Am Nat Standards Inst. *Mem:* AAAS; Am Soc Mech Engrs. *Res:* Design and development on optical and photographic systems for computer memory, on switching matrices for both voice and radio frequencies, on relays, switches and manual apparatus for the Bell System. *Mailing Add:* 1425 Merrimac Trail Garland TX 75043

PUSCH, ALLEN LEWIS, b Richmond, Va, July 4, 34; m 55; c 3. CLINICAL PATHOLOGY. *Educ:* Johns Hopkins Univ, BA, 56, MD, 60; Am Bd Path, dipl & cert anat & clin path, 68. *Prof Exp:* From intern to asst resident path, Johns Hopkins Hosp, 60-64; resident clin path, State Univ NY Upstate Med Ctr, 64-65, asst prof path, 67-71, assoc prof, 71-, dir med tech progs, 78-; AT DEPT PATH, RUSH UNIV. *Concurrent Pos:* Fel path, Johns Hopkins Univ, 60-64. *Mem:* AAAS; Am Soc Clin Path; Am Soc Microbiol; Am Soc Cytol. *Res:* Congenital heart disease; diseases of skeletal muscle with emphasis on muscular dystrophy; infectious diseases; resistant transfer factor in bacteria; immunoglobulins in man. *Mailing Add:* Dept Path Rush Univ 600 S Paulina St Chicago IL 60612

PUSEY, ANNE E, b Oxford, UK, Mar 23, 49; m 77; c 2. PRIMATOLOGY, BEHAVIORAL ECOLOGY. *Educ:* Oxford Univ, BA, 70; Stanford Univ, PhD(ethology), 78. *Prof Exp:* Res assoc biol, Univ Sussex, 77-80 & Univ Chicago, 80-83; vis asst prof ecol, ethology & evolution, Univ Ill, Urbana, 83; asst prof, 83-89, ASSOC PROF BEHAV, DEPT EEB, UNIV MINN, 89- *Concurrent Pos:* Res scientist, Serengeti Wildlife Res Inst, 78-; vis scientist, Dept Zool, Oxford Univ, 90-91; vis fel, St Hugh's Col, Oxford, 90-91; John Simon Guggenheim mem fel, 90-91. *Mem:* Animal Behav Soc; Am Soc Primatology; Int Soc Primatology; Am Soc Zool. *Res:* Behavioral ecology of mammals; dispersal patterns; mating systems; behavioral development; field research on primates and lions. *Mailing Add:* Dept EEB Univ Minn 318 Church St SE Minneapolis MN 55455

PUSEY, P LAWRENCE, b Kankakee, Ill, June 15, 52; m 78; c 3. FRUIT PATHOLOGY, MYCOLOGY. *Educ:* Ball State Univ, BS, 74; Kent State Univ, MS, 76; Ohio State Univ, PhD(plant path), 80. *Prof Exp:* Grad teaching asst biol & bot, Kent State Univ, 74-76; grad res assoc, Ohio State Univ, 76-80; plant pathologist, Appalachian Fruit Res Sta, 80-82, RES PLANT PATHOLOGIST, SOUTHEASTERN FRUIT & TREE NUT RES LAB, AGR RES SERV, USDA, 82- *Concurrent Pos:* Chmn, postharvest path & mycotoxicol comt, Am Phytopath Soc, 87-88; speaker, Int Mycol Cong, Tokyo, 83, Int Cong Plant Protection, Manilla, Philippines, 87, Int Cong Plant Path, Kyoto, Japan, 88 & Soc Chem Indust, London, UK, 89. *Mem:* Am Phytopath Soc; Decidious Tree Fruit Dis Workers. *Res:* Etiology and epidemiology of deciduous fruit diseases and development of non-chemical approaches to disease control; biological control of postharvest fruit decay with microbial antagonists. *Mailing Add:* S E Fruit & Tree Nut Res Lab USDA PO Box 87 Byron GA 31008

PUSHKAR, PAUL, b Winnipeg, Man, Dec 14, 36; m 60. GEOLOGY, GEOCHEMISTRY. *Educ:* Univ Man, BSc, 59; Univ Calif, San Diego, PhD(earth sci), 66. *Prof Exp:* Asst prof geol, Algoma Col, 67-68; asst prof to assoc prof, 68-78, PROF GEOL, WRIGHT STATE UNIV, 78- *Mem:* Geol Soc Am; AAAS; Geochem Soc; Sigma Xi; Tobacco Root Geol Asn. *Res:* Geochemistry of strontium isotopes; petrology of igneous rocks. *Mailing Add:* Dept Geol Wright State Univ Col Glenn Hwy Dayton OH 45431

PUSKI, GABOR, b Budapest, Hungary, Jan 2, 38; US citizen; m 62; c 3. FOOD SCIENCE, BIOCHEMISTRY. *Educ:* Univ Toronto, BA, 63; Univ Mass, Amherst, PhD(food sci & technol), 66. *Prof Exp:* Chemist, Food & Drug Directorate, Can, 63; res scientist, Carnation Res Labs, Calif, 66-70; sr res scientist-group leader food sci res, Food Res Lab, Cent Soya, 70-74, group leader explor protein res, 74-78; group leader, protein lab, Kraft, Inc, 78-83; SR RES SCIENTIST, BRISTOL-MYERS, USPNG, 83- *Mem:* Am Chem Soc; Inst Food Technol; Am Asn Cereal Chem. *Res:* Process development for vegetable protein products; development of adult and infant nutritional beverages. *Mailing Add:* 2400 Lloyd Expressway No R-2 Evansville IN 47721

PUSKIN, JEROME SANFORD, b Akron, Ohio, Oct 15, 42; m 69. BIOPHYSICS. *Educ:* Johns Hopkins Univ, BA, 64; Harvard Univ, MA, 66, PhD(physics), 70. *Prof Exp:* NIH fel, Univ Rochester, 70-74, asst prof biophys, 74-82; biophysicist, US Nuclear Regulatory Comn, 82-85; CHIEF, BIOEFFECTS ANALYSIS BR, OFF RADIATION PROGS, US ENVIRON PROTECTION AGENCY, 85- *Mem:* Soc Risk Analysis; Health Physics Soc. *Res:* Mitochondrial ion transport; ion binding to membranes; electron paramagnetic resonance; radiation risk assessment. *Mailing Add:* 11103 Old Coach Rd Potomac MD 20854

PUSTER, RICHARD LEE, b Richmond, Va, Apr 29, 40; m 76; c 1. COMBUSTION RESEARCH, CERAMIC COMPOSITES. *Educ:* Va Polytech Inst & State Univ, BS, 63. *Prof Exp:* AEROSPACE ENGR, LANGLEY RES CTR, NASA, 63- *Mem:* Am Inst Aeronaut & Astronaut. *Res:* Testing of tactical and strategic missiles and development and use of wind tunnels for testing; space shuttles thermal protection systems; advanced high temperature inorganic composites for missiles and hypersonic craft; combustion device development. *Mailing Add:* 525 Stockton St Hampton VA 23669

PUSZKIN, ELENA GETNER, PLATELETS, HEMOSTASIS. *Educ:* Univ Tucuman, Arg, PhD(biochem), 67. *Prof Exp:* Assoc prof med, 74-89, ASSOC PROF PATH, ALBERT EINSTEIN COL MED, 89- *Concurrent Pos:* Appl mem NIH, Nat Health Blood Inst training prog, 87. *Mem:* Fel Am Heart Asn; NY Acad Sci; Am Path Soc; Soc Biol Chem; Int Soc Thrombosis & Hemat; Am Hemat Soc; Am Soc Exp Biol. *Mailing Add:* Dept Path Monpefiore Med Ctr 111 E 210 St Bronx NY 10467

PUSZKIN, SAUL, b Tucuman, Arg, Feb 3, 38; US citizen; m 62; c 2. CELL BIOLOGY, NEUROBIOLOGY. *Educ:* Univ Tucuman, BS, 60, MS, 63, PhD(biochem), 68. *Prof Exp:* Asst neurol, Col Physicians & Surgeons, Columbia Univ, 68-70, res assoc neurochem & neurol, 70-72; res asst prof, 72-77, res assoc prof, 77-80, PROF PATH, MT SINAI SCH MED, 80- *Concurrent Pos:* Estab investr path, Am Heart Asn, 74- *Mem:* Am Soc Biol Chemists; Am Soc Neurochem; Am Heart Asn; Biophys Soc. *Res:* Protein chemistry and function of contractile systems from muscle and nonmuscle tissues. *Mailing Add:* Dept Path Mt Sinai Sch Med 100th St & Fifth Ave New York NY 10029

PUTALA, EUGENE CHARLES, b Turnersfalls, Mass, Oct 7, 22; m 61; c 2. BOTANY. *Educ:* Univ Mass, BS, 50, MS, 52; Univ Calif, PhD, 67. *Prof Exp:* From instr to asst prof bot, Univ Mass, 50-68; prof bot, Kirkland Col, 68-78, chmn div sci, 68-70; prof, 78-89, EMER PROF BOT, HAMILTON COL, 89- *Res:* Plant morphology and anatomy; plant microtechnique and ultrastructure. *Mailing Add:* RR 4 Box 424 Clinton NY 13323

PUTHOFF, HAROLD EDWARD, b Chicago, Ill, June 20, 36; m 74; c 5. QUANTUM PHYSICS. *Educ:* Univ Fla, BEE, 58, MSE, 60; Stanford Univ, PhD(elec eng), 67. *Prof Exp:* Res engr comput design, Gen Eng Lab, Gen Elec Co, Schenectady, NY, 58; Sperry res fel microwave eng, Univ Fla, 58-59; res engr microwave tube eng, Sperry Electronic Tube Div, Gainesville, Fla, 59-60; res engr comput, Nat Security Agency, Ft Meade, Md, 60-63; res assoc quantum electronics & lasers, Hansen Labs, Stanford Univ, 63-71; sr res engr extrasensory perception, SRI Int, 71-85; SR FEL PHYSICS, INST ADVAN STUDIES AUSTIN, TEX, 85- *Concurrent Pos:* Lectr elec eng, Stanford Univ, 69-70. *Mem:* AAAS; Inst Elec & Electronics Engrs; Am Phys Soc; Sigma Xi; Parapsychol Asn; Sci Explor Soc. *Res:* Laser research and development; quantum electronics; electron beam and microwave tube research; parapsychological phenomena; vacuum zero-point energy studies; general relativity; cosmology. *Mailing Add:* 511 Konstanty Circle Austin TX 78746-6435

PUTMAN, DONALD LEE, b Cuero, Tex, Apr 26, 44; m 69; c 3. MICROBIOLOGY. *Educ:* Lamar Univ, BS, 66; George Washington Univ, PhD(microbiol), 76. *Prof Exp:* Teaching fel microbiol, George Washington Univ, 71-74; sr lab technician immunol, 74-76, investr immunol, 76-79, GENETIC TOXICOLOGIST, MICROBIOL ASSOC, 79- *Mem:* Environ Mutagen Soc; Am Col Toxicol; Genetic Toxicol Asn; Soc Toxicol. *Res:* Chemical mutagenesis; chemical carcinogenesis; immunology; cellular biology. *Mailing Add:* Microbial Assoc 9900 Blackwell Rd Rockville MD 20850

PUTMAN, EDISON WALKER, b Omaha, Nebr, Oct 31, 16; m 44; c 2. PLANT BIOCHEMISTRY. *Educ:* Univ Calif, Berkeley, AB, 42, PhD(plant physiol), 52. *Prof Exp:* From jr res biochemist to asst res biochemist, Univ Calif, Berkeley, 52-59; assoc plant physiologist, Exp Sta, Univ Hawaii, 59-86; RETIRED. *Mem:* Am Chem Soc. *Res:* Carbohydrate metabolism in plants; structural analysis of carbohydrates; biochemical preparations of C-14 labeled compounds; intermediary metabolism in mineral deficient plant tissue. *Mailing Add:* 1910 Ala Moana Blvd No 22A Honolulu HI 96815

PUTMAN, GEORGE WENDELL, b Schenectady, NY, Dec 11, 29; m 57; c 2. ENVIRONMENTAL GEOLOGY, ESP HAZARDOUS WASTE INVESTIGATION & REMEDIATION. *Educ:* Union Col, NY, BS, 51; Pa State Univ, MS, 58, PhD(geol, geochem), 61. *Prof Exp:* Assoc geochemist, Calif Div Mines & Geol, 61-66; chmn dept geol sci, 72-73, ASSOC PROF GEOL & GEOCHEM, STATE UNIV NY ALBANY, 66-; PRES, TERRAN RES, INC, 88- *Concurrent Pos:* Consult, geol, hydrogeol & hydrochem, 80- *Honors & Awards:* Fel Geol Soc Am. *Mem:* Geol Soc Am; Am Geophys Union. *Res:* Environmental hazard evaluation; chemical remediation methods; co-holder 2 US patents; hydrochemistry. *Mailing Add:* Dept Geol Sci State Univ NY Albany NY 12222

PUTMAN, THOMAS HAROLD, b Pittsburgh, Pa, Nov 22, 30; m 56; c 3. CONTROL SYSTEMS DESIGN, POWER ELECTRONICS. *Educ:* Union Col, BS, 52; Mass Inst Technol, SM, 54, ScD(elec eng), 58. *Prof Exp:* MGR ELECTROMECH TECHNOL & CONSULT ENGR, RES & DEVELOP CTR, RES LABS, WESTINGHOUSE ELEC CORP, 58- *Mem:* Inst Elec & Electronics Engrs; Soc Naval Archit & Marine Engrs. *Res:* Application of disciplines of field theory, circuit theory, mechanics, hydraulics, heat transfer to design and development of all types of electromechanical equipment, control systems, electric drives, static var generators, power systems, magnetic circuits and transformers. *Mailing Add:* 354 Stoneledge Dr Pittsburgh PA 15235

PUTNAM, ABBOTT (ALLEN), b Wellsboro, Pa, Nov 24, 20; m 42; c 2. ENGINEERING. *Educ:* Cornell Univ, BME, 42. *Prof Exp:* Jr engr, Dravo Corp, Pa, 42-43; instr, Cornell Univ, 43-44; res engr, Battelle Mem Inst, 46-52, asst div chief, 53-58, div consult, 59-60, staff mech engr, 60-65, fel, 65-76, sr res engr, 76-85; RETIRED. *Concurrent Pos:* Mech engr, Nat Adv Comt Aeronaut, Langley Field, Va. *Mem:* Am Soc Mech Engrs; Am Inst Aeronaut & Astronaut; Combustion Inst; Brit Inst Energy. *Res:* Gas dynamics of combustion systems, including flame holding, spreading and turbulence; flame driven oscillations; fluid flow modeling; combustion roar; pulse combustion. *Mailing Add:* 471 Village Dr Columbus OH 43214

PUTNAM, ALAN R, b Keene, NH, Apr 3, 39; m 70; c 3. WEED SCIENCE, HORTICULTURE. *Educ:* Univ NH, BS, 61; Mich State Univ, MS, 63, PhD(plant physiol), 66. *Prof Exp:* Res asst hort, 61-66, from asst prof to assoc prof, 66-74, PROF WEED SCI, MICH STATE UNIV, 74- *Concurrent Pos:* Vis prof, Cornell Univ, 73, Univ Calif, Davis, 82. *Honors & Awards:* Sterling Hendrick's Mem Lectr. *Mem:* Weed Sci Soc Am; Ctr Agr Sci & Technol; Soc Am Hort Sci; Sigma Xi; AAAS. *Res:* Weed biology, ecology and weed crop interaction; allelopathy and plant competition; herbicide action and fate in agroecosystems. *Mailing Add:* Dept Hort Mich State Univ East Lansing MI 48823

PUTNAM, ALFRED LUNT, b Dunkirk, NY, Mar 10, 16; m 66. MATHEMATICS. *Educ:* Hamilton Col, BS, 38; Harvard Univ, AM, 39, PhD(math), 42. *Prof Exp:* Instr math, Yale Univ, 42-45; from asst prof to assoc prof, 45-86, EMER PROF MATH, UNIV CHICAGO, 86- *Concurrent Pos:* Fulbright fel, Western Australia, 60; NSF sci faculty fel, 66-67. *Mem:* AAAS; Am Math Soc; Math Asn Am. *Res:* Origins and development of abstract algebra. *Mailing Add:* Dept of Math Univ of Chicago Chicago IL 60637

PUTNAM, CALVIN RICHARD, b Baltimore, Md, May 25, 24; m 52; c 3. MATHEMATICS. *Educ:* Johns Hopkins Univ, AB, 44, MA, 46, PhD(math), 48. *Prof Exp:* Instr math, Johns Hopkins Univ, 48-50; mem, Inst Adv Study, NJ, 50-51; from asst prof to assoc prof, 51-59, PROF MATH, PURDUE UNIV, WEST LAFAYETTE, 59- *Mem:* Am Math Soc; Math Asn Am. *Res:* Singular boundary value problems; differential equations; operator theory; hyponormal operators. *Mailing Add:* Dept of Math Purdue Univ West Lafayette IN 47907

PUTNAM, CHARLES E, b Cleburne, Tex, July 22, 41; m 66; c 3. MEDICINE. *Educ:* Univ Tex, BA, 63, MD, 67. *Prof Exp:* From asst prof to assoc prof radiol & internal med, Yale Univ Sch Med, 73-77; chmn, Radiol Dept, Duke Univ Med Ctr, 77-85, chief staff med, 82-85, vchancellor health affairs, 85-86, dean, Sch Med, 86-87, vprovost res & develop, 86-90, JAMES B DUKE PROF RADIOL & PROF MED, DUKE UNIV MED CTR, 83-, EXEC VPRES ADMIN, 90- *Concurrent Pos:* Chief chest radiol, Yale Univ Sch Med, 74-77, clin dir diag radiol, 76-77; ed in chief, Invest Radiol, 84-, co-ed, J Respiratory Dis, 84- *Mem:* Nat Acad Sci; fel Am Col Physicians; Am Col Radiol; fel Am Col Chest Physicians. *Res:* Pulmonary imaging, thoracic neoplasm & computed tomography. *Mailing Add:* Duke Univ 203 Allen Bldg Durham NC 27706

PUTNAM, FRANK WILLIAM, b New Britain, Conn, Aug 3, 17; m 42; c 2. BIOCHEMISTRY, IMMUNOLOGY. *Educ:* Wesleyan Univ, BA, 39, MA, 40; Univ Minn, PhD(biochem), 42; Cambridge Univ, MA, 73. *Hon Degrees:* MA, Cambridge Univ, 73. *Prof Exp:* Instr & res assoc biochem, Sch Med, Duke Univ, 42-46; biochemist, Res & Develop Dept, US Chem Corps, Md, 46; from asst prof to assoc prof biochem, Sch Med, Univ Chicago, 47-55; prof & head dept, Col Med, Univ Fla, 55-65; prof biol & dir div biol sci, Ind Univ, Bloomington, 65-69, prof molecular biol & biochem, 69-74, distinguished prof, 74-88, EMER PROF, IND UNIV, BLOOMINGTON, 88- *Concurrent Pos:* Markle scholar; Lasdon fel, Cambridge Univ, 52-53; ed, Arch Biochem & Biophys, 54-59 & Proc Fedn Am Socs Exp Biol, 58-63; specialist educr, US Dept State, 56; mem comt med biol, NSF, chmn adv comn instnl rels, Nat Adv Gen Med Sci Coun; chmn biophys study sect, NIH; ed sci, Biomed News, 69-73 & Immunochem, 72-75; Guggenheim fel, 70 & overseas fel, Churchhill Col, 70; bd visitors, Duke Med Ctr, 70-75; overseas fel, Churchill Col, Cambridge Univ, 72-; chmn, Assembly of Life Sci, Nat Res Coun- Nat Acad Sci, 77-81; mem, bd trustees, Argonne Univ Asn, 81-83; bd dir, Radiation Effects Res Found, Hiroshima, Japan, 82-87; mem, Univ Chicago bd gov for Argonne Nat Labs, 83-89; Eastman Kodak Sr Lectureship Award, 82. *Mem:* Nat Acad Sci; Am Asn Immunologists; fel Am Soc Biol Chemists (secy, 58-63); Fedn Am Socs Exp Biol; Am Acad Arts & Sci; fel NY Acad Sci. *Res:* Denaturation of proteins; isolation, physical chemistry and structure of plasma proteins, enzymes, toxins and viruses; isotopic study of protein synthesis and virus reproduction; immunoglobulins; albumin genetics. *Mailing Add:* Dept Biol Ind Univ 206 Jordan Hall Bloomington IN 47405

PUTNAM, G(ARTH) L(OUIS), b Reardan, Wash, June 11, 13; m 38; c 3. CHEMICAL ENGINEERING. *Educ:* Univ Wash, Seattle, BS, 35, MS, 37; Columbia Univ, PhD(chem eng), 42. *Prof Exp:* Asst to Dr C G Fink, electrochem develop, Columbia Univ, 38-42; res engr, Carborundum Co, 42-46; assoc prof chem eng, Ore State Col, 46-47; res assoc, Univ Wash, Seattle, 47-51; CONSULT ENGR, 51- *Mem:* AAAS; Am Chem Soc. *Res:* Industrial and applied electrochemistry; prevention of scale and corrosion in heating systems; chemistry of gold; occupational health hazards. *Mailing Add:* PO Box 7022 Seattle WA 98133-2022

PUTNAM, HUGH D, b Carrington, NDak, Feb 12, 28; m 50; c 3. MICROBIOLOGY. *Educ:* Univ Minn, BA, 53, MS, 56, PhD(pub health, biol), 63. *Prof Exp:* Asst soil microbiol, Univ Minn, 52-54, asst microbiol, 54-55, instr sanit biol, 56-60, res fel microbiol, 60-63; from asst prof to prof environ eng sci, Univ Fla, 63-74; VPRES, WATER & AIR RES, INC, 74- *Concurrent Pos:* USPHS res grants, 64-68; comt mem, Assembly Life Sci, Nat Res Coun. *Mem:* Am Soc Limnol & Oceanog; Water Pollution Control Fedn; Am Soc Microbiol; Ecol Soc Am. *Res:* Aquatic microbiology, especially marine and fresh water phytoplankton and transformations of organic matter in sediments; lake eutrophication. *Mailing Add:* 3747 SW 56th Rd Gainesville FL 32608

PUTNAM, JEREMIAH (JERRY) L, b Ft Worth, Tex, Dec 29, 39; m 59; c 2. ELECTRON MICROSCOPY. *Educ:* Tex A&M Univ, BS, 65, MS, 67, PhD(zool), 70. *Prof Exp:* Asst prof biol, Holyoke Community Col, 70-73; from asst prof to assoc prof, 73-88, PROF BIOL, DAVIDSON COL, 88- *Mem:* Am Soc Zoologists; Am Soc Ichthyologists & Herpetologists. *Res:* Comparative anatomy, histology and electron microscopy of the vertebrate heart with emphasis on the amphibians. *Mailing Add:* Dept Biol Davidson Col Davidson NC 28036

PUTNAM, LOREN SMITH, b Morrison, Ill, Oct 24, 13; m 40; c 1. ORNITHOLOGY. *Educ:* Murray State Col, BS, 35; Ohio State Univ, MS, 41, PhD(zool), 47. *Prof Exp:* Asst ornith, F T Stone Lab, 41-42; civilian instr, US Dept Navy, 43-45; assoc prof zool & entom, Ohio State Univ, 45-61, dir, F T Stone Lab, 56-73, prof zool, 61-, EMER PROF ZOOL, OHIO STATE UNIV. *Mem:* Wilson Ornith Soc; Am Ornith Union. *Res:* Biology of birds; bird behavior. *Mailing Add:* 580 Morning St Worthington OH 43085

PUTNAM, PAUL A, b Springfield, Vt, July 12, 30; m 56; c 4. ANIMAL NUTRITION, PHYSIOLOGY. *Educ:* Univ Vt, BS, 52; Wash State Univ, MS, 54; Cornell Univ, PhD(animal nutrit), 57. *Prof Exp:* Res animal husbandman, Animal Sci Res Div, 57-64, leader nutrit & mgt invest, 64-68, chief beef cattle res br, 68-72, asst dir, 72-80, dir, Beltsville Agr Res Ctr, 80-83, dir, cent plains area, 83-87, ASSOC DIR, MIDSOUTH AREA, USDA, 87- *Mem:* Am Dairy Sci Asn; AAAS; Coun Agr Sci & Technol; fel Am Soc Animal Sci. *Res:* Ruminant nutrition, specifically feeding, management and physiology including physical form salivary secretion, feeding behavior, digestibility, digestibility indicators and noncompetitive feedstuffs. *Mailing Add:* 10532 Edgemont Dr Hyattsville MD 20783

PUTNAM, STEARNS TYLER, b Springfield, Vt, July 30, 17; m 40; c 3. PAPER CHEMISTRY, POLYMER CHEMISTRY. *Educ:* Brown Univ, ScB, 38; Harvard Univ, AM, 40, PhD(org chem), 42. *Prof Exp:* Asst org chem, Harvard Univ, 38-41, Nat Defense Res Comt, 41-42; res chemist, Hercules Inc, 42-46, res supvr, 46-71, res assoc, 71-79, sr res assoc, 79-82; RETIRED. *Mem:* Am Chem Soc; assoc Tech Asn Pulp & Paper Indust. *Res:* Carcinogenic hydrocarbons; terpenes and derivatives; rosin derivatives; fatty acid derivatives; paper chemicals; adhesives; polymers. *Mailing Add:* Coffee Run Condominium 614 Loveville Rd Unit E-4D Hockessin DE 19707

PUTNAM, THOMAS MILTON, b Oakland, Calif, Jan 3, 45; m 70; c 1. VECTOR SUPERCOMPUTING, TECHNICAL MANAGEMENT. *Educ:* Univ Colo, BA, 66; Univ Wis-Madison, MS, 68. *Prof Exp:* Syst programmer time-sharing, Honeywell Inc, 68-72, supvr fed syst, Med Syst Ctr, 73-74, mgr res & develop, 74-77; assoc sr analyst info retrieval syst, Cindas, Purdue Univ, 77-79; mgr user serv, 79-87, ASST DIR, PURDUE UNIV COMPUT CTR, 87- *Mem:* Asn Comput Mach. *Res:* Operating systems design; computer user education; computer networks. *Mailing Add:* Math Sci Bldg Purdue Univ Comput Ctr West Lafayette IN 47907

PUTNAM, THOMAS MILTON, JR, b Oakland, Calif, Feb 26, 22; m 43; c 2. SYSTEM SAFETY. *Educ:* Univ Calif, PhD(physics), 51. *Prof Exp:* Supvr 60 inch cyclotron, Radiation Lab, Univ Calif, Berkeley, 44-52; mem staff, Physics Div, Los Alamos Nat Lab, 52-58, group leader, Sherwood Eng Group, 59-64, group leader, Accelerator Controls & Instrumentation Group, Los Alamos Meson Physics Facil, 65-72, asst div leader, medium energy physics div & lampf safety officer, 72-82; RETIRED. *Mem:* Am Phys Soc; Health Physics Soc; Systs Safety Soc. *Res:* Accelerator development and operational safety; nuclear physics. *Mailing Add:* PO Box 6293 Carmel CA 93921

PUTNEY, BLAKE FUQUA, b Farmville, Va, July 16, 23; m 50; c 3. PHARMACY. *Educ:* Med Col Va, BS, 47; Univ Minn, PhD(pharmaceut chem), 52. *Prof Exp:* Asst pharm, Med Col Va, 47-48, Univ Minn, 48-50; from asst prof to assoc prof, Rutgers Univ, 52-67; asst dean, 70-80, prof & chmn dept, 67-88, assoc dean, 81-88, EMER PROF PHARM, COL PHARM, MED UNIV SC, 88- *Concurrent Pos:* Consult pharm, Prof Exam Serv, Am Pub Health Asn, 60-67. *Mem:* Am Pharmaceut Asn; Am Assoc Col Pharm. *Res:* Synthesis of plant alkaloids; steroid analysis. *Mailing Add:* 479 Wade Hampton Dr Charleston SC 29412

PUTNEY, FLOYD JOHNSON, b Easton, Md, Jan 30, 10; m 41, 57; c 5. MEDICINE, SURGERY. *Educ:* Furman Univ, BS, 30; Jefferson Med Col, MD, 34. *Prof Exp:* PROF OTOLARYNGOL, MED UNIV SC, 67- *Concurrent Pos:* Ed, Am Laryngol Asn, 64-71. *Honors & Awards:* Newcomb Award, Am Laryngol Asn. *Mem:* Am Laryngol Asn (pres, 72); Am Laryngol, Rhinol & Otol Soc (vpres, 60); Am Broncho-Esophagol Asn (secy, 53-60, pres, 60); Am Col Surg; Am Acad Ophthal & Otolaryngol. *Res:* Diseases of the larynx, bronchi and esophagus. *Mailing Add:* 705 Knotty Pine Rd Charleston SC 29412

PUTNEY, JAMES WILEY, JR, b Farmville, Va, Mar 8, 46. PHARMACOLOGY. *Educ:* Univ Va, BA, 68; Med Col Va, PhD(pharmacol), 72. *Prof Exp:* asst prof pharmacol, Sch Med, Wayne State Univ, 74-78, assoc prof, 78-80; ASSOC PROF PHARMACOL, MED COL VA, VA COMMONWEALTH UNIV, 80- *Mem:* AAAS; Biophys Soc. *Res:* Calcium, cyclic nucleotides and coupling mechanisms, specifically excitation-contraction and stimulus-secretion coupling. *Mailing Add:* Box 12233 Research Triangle Park NC 27709

PUTNEY, SCOTT DAVID, US citizen. VACCINE DEVELOPMENT, GENE EXPRESSION. *Educ:* Univ Calif, Irvine, BA(chem) & BS(biol), 76; Mass Inst Technol, PhD(biochem), 81. *Prof Exp:* Teaching asst chem, Mass Inst Technol, 76-77; Am Heart Asn fel, Mass Inst Technol, 81, Am Cancer Soc fel, 81-83; prin investr, 82-85, DIR BIOPROCESS ENG, REPLIGEN CORP, 85- *Mem:* AAAS; Am Chem Soc; Am Soc Microbiol. *Res:* Development of new methods to design and express genes for proteins for medical and industrial uses; development of an Acquired Immune Deficiency Syndrome vaccine. *Mailing Add:* Repligen Corp One Kendall Sq Bldg 700 Cambridge MA 02139

PUTT, ERIC DOUGLAS, b Sask, Can, Aug 27, 15; m 57; c 5. PLANT BREEDING. *Educ:* Univ Sask, BSA, 38, MSc, 40; Univ Minn, PhD(genetics), 50. *Prof Exp:* Agronomist, Co-op Veg Oil Ltd, 47-52; res officer plant breeding, Can Dept Agr, 52-65, res scientist, plant breeding, 65-66, dir res mgt, 66-78; RETIRED. *Concurrent Pos:* Plant breeder, FAO, Chile, 59-60 & Can Int Develop Agency, India, 74-75; fel, Agr Inst, Can, 82. *Honors & Awards:* V S Pustovoit Award, Int Sunflower Asn, 80. *Mem:* Am Soc Agron; Crop Sci Soc Am; Agr Inst Can. *Res:* Breeding, genetics and pathology of sunflowers; management of research in oil crops and horticultural crops. *Mailing Add:* Box 544 Creston BC V0B 1G0 Can

PUTT, JOHN WARD, b Huntingdon, Pa, Feb 11, 24; m 67; c 1. SPACE PROPULSION. *Educ:* Juniata Col, BS, 45. *Prof Exp:* Res chemist fuels, Tide Water Assoc Oil Co, 45-52; chief chemist mat & processes, Walter Kidde & Co Inc, 52-63; mem tech staff, Hughes Aircraft Co, 63-66, head satellite propulsion, 66-74, mgr propulsion dept, 74-77, mgr propulsion systs lab, 84-86; RETIRED. *Concurrent Pos:* Consult, Thiokol Corp, 86- *Mem:* Assoc fel Am Inst Aeronaut & Astronaut. *Mailing Add:* 8031 Bleriot Ave Los Angeles CA 90045

PUTTASWAMAIAH, BANNIKUPPE M, b Bannikuppe, India, June 24, 32; m 58; c 2. PURE MATHEMATICS. *Educ:* Univ Mysore, BSc, 55, MSc, 57; Univ Toronto, MA, 61, PhD(algebra), 63. *Prof Exp:* Lectr math, Univ Mysore, 56-63; from asst prof to assoc prof, 63-84, PROF & CHMN, DEPT MATH & STATIST, CARLETON UNIV, 84- *Mem:* Am Math Soc; Math Asn Am; Can Math Soc; Indian Math Soc. *Res:* Representation theory of finite groups; group theory; application of modular representation theory. *Mailing Add:* Dept Math Carleton Univ Ottawa ON K1S 5B6 Can

PUTTER, IRVING, b New York, NY, Dec 27, 23; m 47; c 2. BIOCHEMISTRY, ENTOMOLOGY. *Educ:* City Col New York, BS, 45. *Prof Exp:* Sr chemist, 45-67, head sect biophys & pharmacol, 67-69, sr res fel, Basic Res Dept, 69-75, dir natural prod isolation dept, 80-82, sr tech adv, Explor Biol Res, Merck Sharpe & Dohme Res Labs, 82-84; CONSULTANT, 84- *Mem:* Am Chem Soc; NY Acad Sci; AAAS; Am Soc Microbiol; Entom Soc Am. *Res:* Isolation of natural products; ion exchange purifications; chromatography; antibiotics; steroids; vitamins; amino acid process development and factory break in of new processes; counter current extraction; insecticides; pesticides; agricultural chemicals. *Mailing Add:* 1089 Eastbrook Rd Martinsville NJ 08836

PUTTERMAN, GERALD JOSEPH, b Norwalk, Conn, July 2, 37; m 66; c 2. BIOCHEMISTRY, METABOLISM. *Educ:* Johns Hopkins Univ, AB, 59; Yale Univ, PhD(biochem), 65. *Prof Exp:* Fel biochem, Yale Univ, 65-66; biochemist, Abbott Labs, 66-69; assoc scientist, Papanicolaou Cancer Res Inst, 69-71; res supvr biochem, Gillette Res Inst, 71-73, sr res assoc, 73-76; scientist, Frederick Cancer Res Ctr, 76-82; sr res assoc, Med Sch, Yale Univ, 82-83, res scientist ophthal & visual sci, 83-87; progs officer, Sigma Xi, 88-89; SR GROUP LEADER, UNIROYAL CHEM CO, 89- *Concurrent Pos:* Adj asst prof, Sch Med, Univ Miami, 69-71. *Mem:* Am Chem Soc; Sigma Xi; AAAS; Am Inst Chemists; NY Acad Sci; Int Soc Study Xenobiotics. *Res:* Isolation and purification of proteins and peptides; amino acid sequencing and chemical modification of peptides and proteins; hormonal regulation of intraocular pressure; metabolism of agrochemicals. *Mailing Add:* 18 Wakefield St Hamden CT 06517

PUTTERMAN, SETH JAY, b Brooklyn, NY. FLUID MECHANICS. *Educ:* Calif Inst Technol, BS, 66; Rockefeller Univ, PhD(physics), 70. *Prof Exp:* Asst prof, 70-75, assoc prof, 75-81, PROF PHYSICS, UNIV CALIF, LOS ANGELES, 81- *Concurrent Pos:* Lectr, Cent Invest, Mexico, 72 & Scottish Univ, 74; Researcher, Philips Res Lab, 69 & 74, Ultrasonics Lab, Univ Paris, 74, Cent Nat Res Sci, Grenoble, 74, Inst Lorentz, 78, Univ New Castle upon Tyne, 81 & 85, Physics Dept, Univ Nanjing, 84; Alfred P Sloan res fel 72-74; sr fel, Sci & Eng Res Coun, 81 & 85. *Res:* Non-linear fluid mechanics, in and out of equilibrium in the presence of thermal and quantum fluctuations with applications to superfluids, mode coupling and the formation localized structures. *Mailing Add:* Dept Physics Univ Calif Los Angeles CA 90024

PUTTLER, BENJAMIN, b Bronx, NY, Dec 21, 30; m 61; c 2. ENTOMOLOGY. *Educ:* Univ Calif, Berkeley, BS, 55. *Prof Exp:* Sr lab technician, Citrus Exp Sta, Univ Calif, Riverside, 54-55; entomologist, Central Plains Area, Agr Res, USDA, 55-89; RETIRED. *Concurrent Pos:* Exten assoc entom, Univ Mo, Columbia, 89- *Mem:* AAAS; Entom Soc Am; Am Entom Soc; Int Orgn Biol Control. *Res:* Biological control of insects; insect parasite biology; host-parasite interactions, particularly on encapsulation. *Mailing Add:* Dept Entom Univ Mo Columbia MO 65211

PUTTLITZ, DONALD HERBERT, b Kingston, NY, Apr 21, 38; m 69; c 3. MEDICAL MICROBIOLOGY. *Educ:* State Univ NY Col New Paltz, BS, 59; State Univ NY Albany, MS, 61; Cornell Univ, PhD(bact), 65; Am Bd Med Microbiol, dipl, 72. *Prof Exp:* ASSOC MICROBIOLOGIST, BETH ISRAEL MED CTR, 67-; INSTR PATH, MT SINAI SCH MED, 68- *Concurrent Pos:* USPHS trainee clin microbiol, Col Physicians & Surgeons, Columbia Univ, 65-67. *Mem:* Am Soc Microbiol; Am Pub Health Asn. *Res:* Clinical microbiology; microbial nutrition. *Mailing Add:* 116 Horace Harding Blvd Great Neck NY 11020

PUTTLITZ, KARL JOSEPH, b Kingston, NY, Aug 4, 41; m 67; c 4. HARDWARE SYSTEMS. *Educ:* Mich State Univ, BS, 65, MS, 67, PhD(metall & mat sci), 71. *Prof Exp:* Chem technician, 61-65, assoc engr, 65-67, sr assoc engr, 67-71, staff engr, 71-72, adv engr, 72-84, SR ENGR, IBM CORP, 84- *Concurrent Pos:* Consult, 74- *Mem:* Am Soc Mat Int; Int Soc Hybrid Microelectronics; Sigma Xi; NY Acad Sci. *Res:* Develop new methods, materials and replacement techniques for computer interconnections, primarily device-to-substrate and substrate-to-card which satisfy both functionality and reliability requirements of cost-performance and high-end technology machines; ceramics engineering; analytical chemistry. *Mailing Add:* B-330-81A E Fishkill Facil IBM Corp Hopewell Jct NY 12533

PUTZ, FRANCIS EDWARD, b Summit, NJ, Mar 23, 52. BOTANY. *Educ:* Univ Wis, BSc, 73; Cornell Univ, PhD(ecol), 82. *Prof Exp:* Ecologist, Forest Res Inst, Malaysia, 73-76; ASSOC PROF ECOL, DEPT BOT, UNIV FLA, 83- *Concurrent Pos:* Adj prof, dept forestry, Univ Fla, 85- & Latin Am Studies, 85-; postdoctoral fel, NATO, 87. *Mem:* Ecol Soc Am; Bot Soc Am; Asn Trop Biologists; Int Soc Trop Foresters. *Res:* Experimental community ecology; plant population biology; ecological anatomy; physiological ecology; management and dynamics of tropical forests. *Mailing Add:* Dept Bot Univ Fla Bartram Hall Gainesville FL 32611

PUTZ, GERARD JOSEPH, b Philadelphia, Pa, May 23, 43; m 83; c 3. SCIENCE ADMINISTRATION. *Educ:* St Norbert Col, Wis, BS, 67; Northwestern Univ, MS, 70 PhD(chem), 73. *Prof Exp:* Instr physics, Bishop Newmann High Sch, Philadelphia, 67-68; fel chem, Univ Minn, Duluth, 72-73; lectr, Univ Colo, Denver, 73-74; asst prof, Barat Col, 74-79; mgr, Anal Lab Serv, Northview Labs, Ill, 79-81; mgr, Qual Dept, In Vitro Diag, New Eng Nuclear Corp, 81-84; mgr qual assurance, Ames Div, Miles Labs Inc, Fla, 84-86, dir qual assurance, Baxter Dade Div, 86-90; DIR QUAL ASSURANCE, PORTON DIAG, 90- *Mem:* Am Chem Soc; Am Soc Qual Control. *Res:* Stereochemistry and mechanistic partitioning during solvolytic reactions; quality control and assurance of in vitro diagnostic products. *Mailing Add:* 3853 Blackwood St Newbury Park CA 91320-4901

PUTZIG, DONALD EDWARD, b Rochester, NY, Mar 10, 43; m 65; c 1. ORGANIC CHEMISTRY. *Educ:* Univ Rochester, BS, 65; Cornell Univ, PhD(org chem), 69. *Prof Exp:* Res chemist, 69-72, res supvr, 73-81, RES FEL, E I DU PONT DE NEMOURS & CO INC, DEEPWATER, NJ, 82- *Mem:* Am Chem Soc. *Res:* Synthesis and use applications of organic titanates, antioxidants and dye stuffs. *Mailing Add:* 11 Forge Rd Drummond Hill Newark DE 19711

PUYAU, FRANCIS A, b New Orleans, La, Dec 1, 28; m 51; c 5. MEDICINE. *Educ:* Univ Notre Dame, BS, 48; La State Univ, New Orleans, MD, 52. *Prof Exp:* From instr to asst prof pediat, Sch Med, La State Univ, 57-61; asst prof, Vanderbilt Univ, 61-68; clin assoc prof, La State Univ Med Ctr, 68-71, prof radiol & pediat & head dept radiol, 71-74; PROF PEDIAT, RADIOL & MED, SCH MED, TULANE UNIV, 74- *Concurrent Pos:* NIH fel, Charity Hosp New Orleans, 69-70; asst radiologist, Charity Hosp, New Orleans, 69-, vis physician peidat & sr vis physician radiol, 71. *Mem:* Fel Am Col Cardiol; fel Am Col Radiol; Am Roentgen Ray Soc. *Res:* Cardiac radiology; heat metabolism; angiography; interventional radiology. *Mailing Add:* 458 Shady Lake Pkwy Baton Rouge LA 70810

PUYEAR, DONALD E(MPSON), b Cape Girardeau, Mo, Aug 21, 32; m 57; c 4. CHEMICAL ENGINEERING. *Educ:* Univ Mo, BS, 54; MS, 48; Va Polytech Inst, PhD(chem eng), 65. *Prof Exp:* Jr engr, E I du Pont de Nemours & Co, 54-55; asst prof chem eng, Va Polytech Inst, 58-64, prof eng & dir, Clifton Forge-Covington Div, 64-67; pres, Dabney S Lancaster Community Col, 67-69; pres, Va Highlands Community Col, 69-74; PRES, CENT VA COMMUNITY COL, 74- *Res:* Mass transfer; extraction; college administration. *Mailing Add:* 11708 Dewberry Lane Chester VA 23831

PUZANTIAN, VAHE ROPEN, psychiatry; deceased, see previous edition for last biography

PUZISS, MILTON, b Philadelphia, Pa, Feb 17, 20; m 51; c 3. MICROBIOLOGY. *Educ:* Kans State Col, BS, 48; Univ Wis, MS, 49; Univ Southern Calif, PhD(bact), 56. *Prof Exp:* Asst, Univ Wis, 48-49; res bacteriologist, Biol Labs, US Dept Army, Ft Detrick, Md, 51-53; asst, Univ Southern Calif, 53-55; med res bacteriologist, Chem Res & Develop Labs, US Dept Army, Ft Detrick, 55-62, microbiologist, Biol Labs, 62-68; asst chief, Virol & Rickettsial Br, 68, actg chief, 68-70, chief, Bact & Virol Br, Extramural Progs, Nat Inst Allergy & Infectious Dis, NIH, 73-86; RETIRED. *Concurrent Pos:* Secy Army Res & Study Fel, Karolinska Inst, Stockholm, Sweden, 63-64; sr asst scientist, USPHSR, 56-80. *Honors & Awards:* US Army Res & Develop Award, 62; NIH Dirs Award, 76. *Mem:* AAAS; Am Soc Microbiol; Infectious Dis Soc; Sigma Xi. *Res:* Microbial physiology; toxin formation; biphasic culture techniques; vaccine development; fermentation and mass culture methods; infectious diseases and immunology; health administration. *Mailing Add:* Five Watchwater Ct Rockville MD 20850

PYBURN, WILLIAM F, b Corsicana, Tex, Mar 22, 27; m 49; c 1. VERTEBRATE ZOOLOGY. *Educ:* Univ Tex, BA, 51, MA, 53, PhD, 56. *Prof Exp:* From asst prof to assoc prof, 56-70, prof, 70-82, EMER PROF BIOL, UNIV TEX, ARLINGTON, 82- *Mem:* Am Soc Ichthyologists & Herpetologists; Soc Study Amphibians & Reptiles. *Res:* Amphibian behavior-ecology; behavior and systematics of microhylid and phyllomedusine anurans. *Mailing Add:* 3808 Lake Ridge Rd Arlington TX 76016

PYE, EARL LOUIS, b Merino, Colo, Aug 18, 26; m 49, 87; c 6. ELECTROCHEMISTRY, CORROSION ENGINEERING. *Educ:* Chico State Col, AB, 58; Univ Calif, Davis, MS, 61; La State Univ, PhD(physical chem), 66. *Prof Exp:* Asst prof chem, Calif State Polytech Univ, 61-65; assoc prof, Parsons Col, 66-67; chmn dept, 67-69, dean grad studies & res, 78-83, assoc prof, 67-68, PROF CHEM, CALIF STATE POLYTECH UNIV, 68-; PRES, CCS CONTROL SYST, 81- *Concurrent Pos:* Eng & corrosion consult. *Mem:* Am Chem Soc; Nat Asn Corrosion Eng; Sigma Xi. *Res:* Electrode kinetics; electrochemical techniques of corrosion rate determinations and control; surface and corrosion chemistry; underground corrosion problems; computer interfaced systems for instrument and process control and data acquisition. *Mailing Add:* Dept Chem Calif State Polytech Univ Pomona CA 91768

PYE, EDGAR GEORGE, b Toronto, Ont, May 18, 25; m 53; c 2. ECONOMIC GEOLOGY. *Educ:* Univ Toronto, BASc, 47, MASc, 49, PhD(geol), 54. *Prof Exp:* Field geologist, 48-52, from resident geologist to chief geologist, 52-72, dir, Ont Geol Survey, Ont Div Mines, Ministry Natural Resources, 72-83; RETIRED. *Mem:* Geol Asn Can; Geol Soc Can; Am Inst Mining Engrs; Can Inst Mining & Metall. *Res:* Stratigraphic, petrologic and tectonic evolution of the Canadian Shield and the temporal and spatial association of mineral deposits with rock units and structures; Phanerozoic stratigraphy and syngenetic mineral deposits of Ontario. *Mailing Add:* 2010 Islington Toronto ON M9W 3W5 Can

PYE, EDWARD KENDALL, b Bexleyheath, Eng, Jan 16, 38; m 61; c 2. BIOCHEMISTRY. *Educ:* Univ Manchester, BSc, 61, PhD(biochem), 64. *Prof Exp:* Assoc, 67-68, asst prof, 68-73, ASSOC PROF BIOCHEM & BIOPHYS, SCH MED, UNIV PA, 73- *Concurrent Pos:* NIH fel biophys, Johnson Res Found, Univ Pa, 65-67, NIH res career develop award, 70-75. *Mem:* AAAS; Am Chem Soc. *Res:* Metabolic control, particularly carbohydrate metabolism; oscillating biochemical reactions; fuels from biomass; affinity chromatography; biological rhythms and timing mechanisms; enzyme technology; use of immobilized enzymes; fermentation. *Mailing Add:* Repap Tech Inc PO Box 766 Valley Forge PA 19482

PYE, LENWOOD D(AVID), b Little Falls, NY, May 16, 37; m 58; c 4. CERAMICS, CHEMISTRY. *Educ:* Alfred Univ, BS, 59, PhD(ceramics), 68. *Prof Exp:* Proj leader, Pittsburgh Plate Glass Industs, Inc, 59-60 & Bausch & Lomb Inc, 60-64; assoc prof, 68-80, PROF GLASS SCI, NY STATE COL CERAMICS, ALFRED UNIV, 80-, CO-DIR, INST GLASS SCI & ENG. *Concurrent Pos:* Vis prof, Univ Erlangen-Nurnberg, 75 & 80. *Mem:* Am Ceramic Soc; Nat Inst Ceramic Engrs; Sigma Xi; Am Soc Eng Educ. *Res:* Radiation damage in glasses; nuclear waste disposal; physical optics; crystallization phenomena. *Mailing Add:* NY State Col Ceramics Alfred Univ Alfred NY 14802

PYE, ORREA F, BIOCHEMISTRY. *Educ:* Columbia Univ, PhD(nutrit biochem), 43. *Prof Exp:* Prof nutrit, Columbia Univ, 72-88; RETIRED. *Mailing Add:* Heritage Hills of Westchester Unit 157-A Somers NY 10589

PYKE, RONALD, b Hamilton, Ont, Nov 24, 31; m 53; c 4. MATHEMATICAL STATISTICS, PROBABILITY. *Educ:* McMaster Univ, BA, 53; Univ Wash, Seattle, MS, 55, PhD(math), 56. *Prof Exp:* Res assoc, Stanford Univ, 56-57, vis asst prof math, 57-58; asst prof, Columbia Univ, 58-60; from asst prof to assoc prof, 60-66, PROF MATH, UNIV WASH, 66- *Concurrent Pos:* Prin investr, NSF, 61-; vis prof, Cambridge Univ, 64-65 & Imp Col, London, Eng, 70-71 & Technion, Israel, 88; ed, Ann of Probability, Inst Math Statist, 72-75; mem, Bd Math Sci, Nat Res Coun, 84-88. *Mem:* Am Math Soc; Math Asn Am; fel Am Statist Asn; fel Inst Math Statist (pres, 86-87); Int Statist Inst (vpres, 89-91); Can Statist Soc. *Res:* Stochastic processes; limit theorems and weak convergence; multiparameter and set-indexed processes; renewal theory; nonparametric statistics. *Mailing Add:* Dept Math GN-50 Univ Wash Seattle WA 98195

PYKE, THOMAS NICHOLAS, JR, b Washington, DC, July 16, 42; m 68; c 2. AERONAUTICAL & ASTRONAUTICAL ENGINEERING. *Educ:* Carnegie Inst Technol, BS, 64; Univ Pa, MSE, 65. *Prof Exp:* Electronic engr res & develop, Nat Bur Standards, 64-69, chief, Comput Networking Sect, 69-75, chief, Comput Systs Eng Div Standards-Res, 75-79, dir, Ctr Comput Systs Eng, 79-81, dir, Ctr Prog Sci & Technol, 81-86; ASST ADMINR, SATELLITE & INFO SERV, NAT OCEANIC & ATMOSPHERIC ADMIN, 86- *Concurrent Pos:* Bd gov, Inst Elec & Electronic Engrs Computer Soc, 71-73, 75-77 & vchair, Tech Comt Personal Comput, 82-86; bd dir, Am Fedn Processing Soc, 74-76; mem, Presidential Adv Comt Networking Struct & Function, 80. *Honors & Awards:* Eng Sci Award, Wash Acad Sci, 74; Award for Exemplary Achievement in Pub Admin, William A Jump Found, 75 & 76; Exec Excellence Award in Info Resources Mgt, Nat Environ Sci Data Policy & Prog Planning, 91. *Mem:* Sr mem Inst Elec & Electronics Engrs; Asn Comput Mach; AAAS; Sigma Xi. *Res:* Computer networking, computer system architecture and data communications, distributed systems; computer network protocols, performance measurement and network access techniques. *Mailing Add:* NOAA Nat Environ Satellite Data & Info Serv Washington DC 20233

PYKE, THOMAS RICHARD, b Center, Ind, Jan 8, 32; m 51; c 3. MICROBIOLOGY, BIOCHEMISTRY. *Educ:* Purdue Univ, BS, 54, MS, 58, PhD, 61. *Prof Exp:* Sr res microbiologist, Squibb Inst Med Res, 60-62; res assoc infectious dis, Upjohn Co, 62-72, res head, Anal Methods Res & Develop & Microbiol Testing, 72-75, res assoc, Fermentation Res & Develop, 75-81, res mgr, culture improv & fermentation prep, 81-; DIR, CTR PROG SCI & TECHNOL. *Mem:* AAAS; Am Soc Microbiol. *Res:* Physiology of microsclerotia of verticillium albo-atrum; antibiotics and pharmacologically active compounds produced by fungi and actinomycetes; diseases caused by Streptococcus and Neisseria; microbial genetics. *Mailing Add:* 7169 Hazelwood Richland MI 49083

PYLE, JAMES JOHNSTON, b Calgary, Alta, Apr 26, 14; nat US; m 35; c 3. PLASTICS CHEMISTRY. *Educ:* Univ BC, BA, 35, MA, 37; McGill Univ, PhD(org chem), 39. *Prof Exp:* Asst chem, Univ BC, 35-37 & McGill Univ, 37-39; develop chemist, Gen Elec Co, Mass, 39, res chemist, 39-40, group leader laminated plastics, 40-41, res group leader, 41-43, dir plastics lab, 43-49, mgr new prod develop lab, 49, mgr eng, laminated prod dept, Chem & Metall Div, 49-66, mgr overseas bus, 66-68; CONSULT PLASTICS TECHNOL, 68- *Concurrent Pos:* Res chemist, Can Fishing Co, BC, 36-37; civilian with AEC, 46-47. *Mem:* AAAS; Am Chem Soc. *Res:* Organic, lignin and cellulose biochemistry; flotation of ores; plastics; rubber; ion exchange; fibers. *Mailing Add:* 1677 Bayberry Lane Coshocton OH 43812-3120

PYLE, JAMES L, b Wilmington, Ohio, Aug 5, 38; m 60; c 5. ORGANIC CHEMISTRY, ENVIRONMENTAL CHEMISTRY. *Educ:* Univ Ohio, BS, 60; Brown Univ, PhD(chem), 67. *Prof Exp:* Res assoc chem, Univ Calif, Davis, 65-66; from asst prof to assoc prof chem & dir res, Miami Univ, 66-83; DIR OF RES, BALL STATE UNIV, 83- *Mem:* Am Chem Soc; AAAS; Soc Res Admin. *Res:* University research administration; synthesis of chlorinated aromatic compounds. *Mailing Add:* Off Res Ball State Univ 2000 University Ave Muncie IN 47306

PYLE, JOHN TILLMAN, b Atlanta, Ga, May 3, 35; m 71; c 1. ANALYTICAL CHEMISTRY. *Educ:* Ga State Col, BS, 59; Univ Ga, PhD(chem), 64. *Prof Exp:* Assoc prof chem, Mars Hill Col, 64-69; assoc prof, 69-75, PROF CHEM, COLUMBUS COL, 75- *Mem:* Am Chem Soc. *Res:* Spectrophotometric determinations of trace quantities of platinum group metals using various derivatives of dithio-oxamide as color producing reagents. *Mailing Add:* Dept Chem & Geol Columbus Col Columbus GA 31993-2399

PYLE, K ROGER, b Shanghai, China, Nov 5, 41; m 65; c 2. COSMIC RAY PARTICLE PHYSICS. *Educ:* Mass Inst Technol, SB, 63; Univ Chicago, MS, 65, PhD(physics), 73. *Prof Exp:* Physicist, 72-82, SR RES ASSOC, ENRICO FERMI INST, UNIV CHICAGO, 82- *Mem:* Am Geophys Union; Sigma Xi; Am Phys Soc. *Res:* Space physics, primarily interplanetary, focusing on cosmic ray and solar energetic charged particles, their origin and propagation; applications of computers to the space sciences. *Mailing Add:* 933 E 56th St Chicago IL 60637

PYLE, LEONARD DUANE, b Crawfordsville, Ind, July 13, 30; m 54; c 4. COMPUTER SCIENCE, MATHEMATICS. *Educ:* Rose Polytech Inst, BS, 52; Purdue Univ, MS, 54, PhD(math), 60. *Prof Exp:* Res asst, Statist & Comput Lab, Purdue Univ, Lafayette, 54-58, actg head comput div, 58-60, head, 60-61, from asst prof to assoc prof math & comput sci, 60-71, asst dir, Comput Sci Ctr, 61-65, asst head dept comput sci, 65-69; chmn dept, 71-74, PROF COMPUT SCI UNIV HOUSTON, 71- *Concurrent Pos:* Consult, Ind State Dept Data Processing, 61 & Stewart Warner Corp, 62; hon fel, Math Res Ctr, Univ Wis, 65; vis prof, Univ Tex, Austin, 69-70; vis scholar, Univ Tex, Austin, 79. *Mem:* Asn Comput Mach; Soc Indust & Appl Math. *Res:* Linear and nonlinear programming; acceleration of vector sequences; theory and applications involving the generalized inverse of a matrix; numerical analysis; applications of vector computers. *Mailing Add:* 11905 Knippwood Rd Houston TX 77024

PYLE, ROBERT LAWRENCE, b Wilmington, Del, Aug 27, 23; m 54; c 4. ORNITHOLOGY, METEOROLOGY. *Educ:* NY Univ, BS, 43; Univ Calif, Los Angeles, MA, 51; Univ Wash, PhD, 58. *Prof Exp:* Asst res geophysicist, Inst Geophys, Univ Calif, Los Angeles, 53-56; meteorologist Bur Commercial Fisheries, US Fish & Wildlife Serv, 58-60 & res & develop, Meteorol Satellite Lab, US Weather Bur, 60-64; tech asst to dir, Nat Environ Satellite Ctr, Environ Sci Serv Admin, 64-66; field dir, Pac Prog, Div Birds, Smithsonian Inst, 66-69; tech asst to dir, Washington, DC, 69-75, mgr satellite field serv sta, Honolulu, Nat Environ Satellite Serv, Nat Oceanic & Atmospheric Admin, US Dept Com, 75-84; CURATORIAL ASST BIRDS, BISHOP MUSEUM, HONOLULU, 84- *Concurrent Pos:* Regional ed, Am Birds Mag, 77- *Mem:* Am Meteorol Soc; Sigma Xi; Am Ornith Union. *Res:* Occurrence and status of birds in Hawaii. *Mailing Add:* 741 N Kalaheo Ave Kailua HI 96734

PYLE, ROBERT V, b Washington, DC, Oct 4, 23; m 50. PLASMA PHYSICS. *Educ:* Univ Calif, Los Angeles, AB, 44; Univ Calif, Berkeley, MA & PhD(physics), 51. *Prof Exp:* SR SCIENTIST, LAWRENCE BERKELEY LAB, UNIV CALIF, 51-, PROF-IN-RESIDENCE, BERKELEY NUCLEAR ENG DEPT, 65- *Mem:* Fel Am Phys Soc; Am Vacuum Soc; Sigma Xi. *Res:* Controlled thermonuclear energy physics. *Mailing Add:* 920 17th St Bellingham WA 98225

PYLE, ROBERT WENDELL, b Phoenixville, Pa, Mar 3, 08; m 33; c 2. ZOOLOGY, GENERAL BIOLOGY. *Educ:* Univ Pa, BS, 33, MA, 39; Harvard Univ, PhD(biol), 41. *Prof Exp:* Mem fac, Georgetown Spec Sch, Del, 34-37 & Caesar Rodney Spec Sch, 37-38; zool librn, Univ Pa, 38-39; teaching fel, Radcliffe Col, 41; fel & asst, Nat Defense Res Comt Proj, Oceanog Inst, Woods Hole, Mass, 41-42; instr biol, Rensselaer Polytech Inst, 42-44; assoc prof biol, Mary Washington Col, 44-50; from asst prof to prof biol, State Univ NY Col New Paltz, 50-59 & 67-73, dean of col, 59-65, emer prof biol, 73; RETIRED. *Concurrent Pos:* Vis prof educ, Teachers Col, Columbia Univ & specialist teacher educ, Univ Team, India, 65-67; biologist, Brewster Conserv Comn, Brewster, Mass, 79-85; field observer, Manomet Mass Bird Observ-Int Shore Bird Migration, 77-84; trustee, Cape Cod Mus Natural Hist, 76-82 & 85-88, treas, 85-87. *Mem:* AAAS; Am Soc Zoologists; Ecol Soc Am; Sigma Xi; NY Acad Sci. *Res:* Histology and cytology of invertebrates, primarily Arthropods; plant taxonomy; cytology; ecology; entomology. *Mailing Add:* 150 Susan Lane Brewster MA 02631-1728

PYLE, ROBERT WENDELL, JR, b Lewes, Del, Mar 11, 36; m 63; c 4. PHYSICS. *Educ:* Harvard Univ, AB, 57, AM, 58, PhD(appl physics), 63. *Prof Exp:* Res fel acoust, Harvard Univ, 63-65; consult archit acoust, 65-66, SR SCIENTIST, BOLT BERANEK & NEWMAN, INC, 66- *Mem:* AAAS; Acoust Soc Am; Audio Eng Soc; Inst Elec & Electronics Engrs. *Res:* Physical acoustics, including underwater sound, noise and vibration and ultrasonics; musical and architectural acoustics. *Mailing Add:* 11 Holworthy Pl Cambridge MA 02138

PYLE, THOMAS EDWARD, b New York, NY, Oct 31, 41; m 63; c 2. MARINE GEOLOGY. *Educ:* Columbia Col, BA, 63; Tex A&M Univ, MS, 66, PhD(oceanog), 72. *Prof Exp:* From asst prof to assoc prof marine sci, Univ S Fla, 69-76; prog dir marine geol & geophys, Off Naval Res, 76-81; dep dir, Nat Ocean Surv, 81-82, chief scientist, Nat Oceanic & Atmospheric Admin, 82-86; DIR, OCEAN DRILLING PROGS, JOINT OCEANOG INSTS INC, 86-, VPRES, 88- *Concurrent Pos:* Geologist, US Geol Surv, 70-72; grants, Off Naval Res, NSF, Dept of Interior & Fla Dept Natural Resources. *Mem:* Am Geophys Union; AAAS; Oceanog Soc. *Res:* Structural aspects of marine geology and geophysics; geodetic control of sea level measurement. *Mailing Add:* 8601 Nanlee Dr Springfield VA 22152

PYLER, RICHARD ERNST, b Chicago, Ill, May 26, 41; m 71; c 2. CEREAL CHEMISTRY, BREWING CHEMISTRY. *Educ:* Univ Chicago, BS, 63; Kans State Univ, MS, 65; Purdue Univ, PhD(biochem), 69. *Prof Exp:* Sr res chemist, Anheuser-Busch, Inc, Mo, 69-72; asst prof cereal chem, NDak State Univ, 72-73; ed, Baker's Digest & tech ed, Brewer's Digest, 73-77; assoc prof cereal chem, NDak State Univ, 77-84; MGR, BREWING CHEM, ADOLPH COORS CO, 84- *Mem:* Am Asn Cereal Chemists; Am Soc Brewing Chemists; Master Brewers Asn Am; Am Soc Bakery Engrs. *Res:* Synthesis of monosaccharide derivatives; biochemical constituents of cereal grains, especially polysaccharides; malting and brewing science. *Mailing Add:* Dept BC-600 Adolph Coors Brewing Co Golden CO 80401-1295

PYNADATH, THOMAS I, b Karukutty, India, Apr 24, 29; m 55; c 4. BIOCHEMISTRY, ORGANIC CHEMISTRY. *Educ:* Madras Univ, BS, 49, MS, 53; Georgetown Univ, PhD(biochem), 63. *Prof Exp:* Lectr chem, Madras Univ, 52-56; sr lectr, Kerala Univ, 56-59; res fel biol chem, Univ Calif, Los Angeles, 63-65; asst prof chem, Georgetown Univ, 65-66; asst prof, 66-70, assoc prof, 70- PROF CHEM, KENT STATE UNIV. *Mem:* AAAS; Am Chem Soc. *Res:* Metabolism of lipids; biosynthesis of glycerides and fatty acids; metabolism of nucleic acids in Neurospora crassa; lipid metabolism in obese and atherosclerotic animals. *Mailing Add:* Dept Chem Kent State Univ Main Campus Kent OH 44242

PYNES, GENE DALE, b Zwolle, La, Dec 25, 33; m 58; c 3. BIOCHEMISTRY, MICROBIOLOGY. *Educ:* Hendrix Col, BA, 56; Univ Ark, Little Rock, MS, 59, PhD(biochem), 67; Univ Ark Med Sci, Col Pharm, BS, 78. *Prof Exp:* Asst prof, 67-71, ASSOC PROF MED CHEM, COL PHARM, UNIV ARK MED SCI, LITTLE ROCK, 71- *Mem:* Sigma Xi; Am Asn Col Pharm. *Res:* Inorganic pyrophosphatase; kinetic properties of the enzyme from erythrocytes, pancreas and yeast; pharmacy. *Mailing Add:* 7008 Morgan Dr Little Rock AR 72209

PYNN, ROGER, b Maidstone, Eng, Feb 15, 45; m 71; c 1. NEUTRON SCATTERING STUDIES OF CONDENSED MATTER. *Educ:* Univ Cambridge, Pa, 66, MA, 69 & PhD(physics), 69. *Prof Exp:* Fel, AB Atomenergi, Sweden, 70-71, Inst Atomenergi, Norway, 71-73; assoc physicist, Brookhaven Nat Lab, 73-75; staff scientist, Inst Laue-Langevin, France, 75-87; DIR, LOS ALAMOS NEUTRON SCATTERING CTR, 87- *Concurrent Pos:* Guest scientist, Riso Nat Lab, Denmark, 78, Brookshaven Nat Lab, 80, Yamada Sci Found, Japan, 85; consult, Los Alamos Nat Lab, 82-86, Nat Bur Standards, 83; adj prof, Univ Calif Santa Barbara, 88- *Res:* Neutron scattering studies of condensed matter systems including phase transitions, low- dimensional magnets, aggregates, magnetic colloids, liquid crystals, magnetic superconductors and surfaces. *Mailing Add:* Los Alamos Neutron Scattering Ctr Los Alamos Nat Lab Los Alamos NM 87545

PYNNONEN, BRUCE W, b Ishpeming, Mich, Aug 10, 54; m 79; c 2. PSEUDO-MOVING BED CHROMATOGRAPHY, CHROMATOGRAPHIC ANALYSIS OF CARBOHYDRATES. *Educ:* Univ Mich, Ann Arbor, BS, 76, MS, 84. *Prof Exp:* Chemist, Anal Labs, Dow Chem Co, 76-78, sr res chemist, Styrene Polymers Res & Develop, 78-82, proj leader, Chem Res, 82- 86, DEPT LEADER, SEPARATIONS SYSTS, DOW CHEM CO, 86- *Mem:* Am Inst Chem Engrs; Am Soc Sugar Beet Technologists. *Res:* Optimizing the efficiency of production processes; improving industrial-scale chromatographic separations and applying these processes to carbohydrate purifications. *Mailing Add:* 113 Helen St Midland MI 48640

PYPER, GORDON R(ICHARDSON), b Cedar Falls, Iowa, Sept 1, 24; m 46; c 3. ENVIRONMENTAL ENGINEERING, WATER RESOURCES. *Educ:* Brown Univ, ScB, 48; Univ Mich, MSE, 49, PhD, 70. *Prof Exp:* Sanit engr, Ill State Dept Pub Health, 49-53; from instr to prof sanit & civil eng, Norwich Univ, 53-73, head dept, 65-73; comnr water resources, State of Vt Dept Water Resources, 73-77; head res & develop & sr environ engr, Dufresne-Henry Inc, 77-78, head, Water Resources Dept & Res & Develop sr engr, 78-80, managing dir, Res & Spec Proj, 80-88; RETIRED. *Concurrent Pos:* Comnr, NE Interstate Water Pollution Cent Comn, 73-77. *Mem:* Water Pollution Control Fedn; Am Soc Civil Engrs; Am Soc Eng Educ; Nat Soc Prof Engrs; Am Acad Environ Engrs. *Res:* Civil engineering; environmental engineering; Ability of fresh water mussels to remove solids from and improve the quality of aerated lagoon waste treatment effluent; removal of Giardia, cysts and other contaminants from water by slow sand filtration; compact slow sand filtration. *Mailing Add:* PO Box 41 Allen Pt South Hero UT 05486

PYPER, JAMES WILLIAM, b Wells, Nev, Sept 5, 34; m 57; c 11. PHYSICAL CHEMISTRY, MASS SPECTROMETRY. *Educ:* Brigham Young Univ, BS, 58, MS, 60; Cornell Univ, PhD(phys chem), 64. *Prof Exp:* Res chemist, Lawrence Livermore Nat Lab, Univ Calif, 63-90, group leader mass spectros, 73-75, group leader appl phys chem group, 79-80, sect leader, Anal Chem Sect, 80-84, assoc div leader opers, 87-89, qual assurance mgr, 89-90; RETIRED. *Concurrent Pos:* Dep sect leader, Anal Chem Sect, Lawrence Livermore Nat Labs, Univ Calif, 84-86. *Mem:* Am Soc Mass Spectrometry; Sigma Xi. *Res:* Isotopic exchange reactions; mass spectroscopy; physical chemistry of the hydrogen isotopes at low temperatures; moisture analysis. *Mailing Add:* 17500 W Von Sosten Rd Tracy CA 95376

PYRCIOCH, EUGENE JOSEPH, b Chicago, Ill, Aug 10, 20. CHEMICAL ENGINEERING. *Educ:* Northwestern Univ, BSChE, 48; Ill Inst Technol, MS, 58. *Prof Exp:* Chem engr, Ill Inst Technol, 48-53, supvr, 53-62, scientist, 48-83; RETIRED. *Mem:* Am Chem Soc; Am Inst Chem Engrs; Am Meteorol Soc; fel Am Inst Chem; Sigma Xi. *Res:* Process design and development of synthetic gas processes; pilot plant design and operation; fossil fuel gasification and processing; solids fluidization; high pressure technology; particulate solids feeding, grinding and drying. *Mailing Add:* 3700 N Lockwood Ave Chicago IL 60641

PYSH, JOSEPH JOHN, b Olyphant, Pa, Nov 14, 35; m 69. NEUROCYTOLOGY, ELECTRON MICROSCOPY. *Educ:* Chicago Col Osteopath Med, DO, 62, Wayne State Univ, BA, 63; Northwestern Univ, PhD(anat), 67. *Prof Exp:* From instr to asst prof anat, Med & Dent Schs, Northwestern Univ, Chicago, 66-86, actg chmn dept, 78-81, resident physician neurol, 83-86; ASSOC PROF NEUROL, MICH STATE UNIV, 86- *Concurrent Pos:* NSF grant reviewer, 74-; mem, NIH Neurol Study Sect, 76-77; manuscript referee, J Cell Biol, Am J Anat, Exp Neurol, Brain Res Bull & Sci; NIH res grants. *Mem:* AAAS; Am Asn Anat; Am Soc Cell Biol; Electron Micros Soc Am; Soc Neurosci; Am Acad Neurol. *Res:* Developmental neurobiology; morphology and physiology of synaptic transmission. *Mailing Add:* Dept Internal Med Mich State Univ B311 W Fee Hall East Lansing MI 48824

PYTKOWICZ, RICARDO MARCOS, physical chemistry, oceanography; deceased, see previous edition for last biography

PYTLEWSKI, LOUIS LAWRENCE, b Philadelphia, Pa, May 11, 32; m 57; c 3. INORGANIC CHEMISTRY. *Educ:* Univ Pa, BA, 54, MS, 56, PhD(inorg chem), 60. *Prof Exp:* Instr chem, St Joseph's Col, Pa, 58-59; res chemist, Res & Develop Lab, Armstrong Cork Co, 59-62; asst prof, 62-67, ASSOC PROF INORG CHEM, DREXEL UNIV, 67- *Concurrent Pos:* Consult, Astromech Div, Giannini Controls Corp, 65-68 & Edgewood Arsenal, US Army, 71-; prin scientist, Franklin Res Ctr, Philadelphia, 77-78. *Mem:* AAAS; Am Chem Soc. *Res:* Synthesis and characterization of metal salt organophosphorus and organonitrogen complexes and polymers; thermal and hydrolytic decomposition of metalammine complexes. *Mailing Add:* 9603 Leon St Philadelphia PA 19114-2818

PYTTE, AGNAR, b Norway, Dec 23, 32; nat US; m 55; c 3. THEORETICAL PHYSICS. *Educ:* Princeton Univ, AB, 53; Harvard Univ, PhD(physics), 58. *Prof Exp:* From instr to assoc prof, 57-67, chmn dept, 71-75, dean grad study & assoc dean fac sci, 75-78, PROF PHYSICS, DARTMOUTH COL, 67-, PROVOST, 82- *Concurrent Pos:* Proj Matterhorn, Princeton Univ, 59-60; NSF sci fac fel, Univ Brussels, 66-67; mem staff, Princeton Plasma Physics Lab, 78-79. *Mem:* Am Phys Soc; Europ Phys Soc. *Res:* Plasma physics. *Mailing Add:* Case Western Reserve Univ 2040 Adelbert Rd Cleveland OH 44106

PYTTE, ERLING, b Hvittingfoss, Norway, July 29, 37; m 63; c 3. SOLID STATE PHYSICS. *Educ:* Princeton Univ, AB, 59; Harvard Univ, MA, 61, PhD(physics), 64. *Prof Exp:* Res assoc physics, Univ Calif, Berkeley, 63-64; asst prof, Univ Oslo, 64-65; res physicist, IBM Res Lab, Zurich, Switz, 65-69, MGR THEORET PHYSICS, T J WATSON RES CTR, IBM CORP, 69- *Mem:* Fel Am Phys Soc. *Res:* Theoretical solid state physics. *Mailing Add:* T J Watson Res Ctr IBM Corp PO Box 218 Yorktown Heights NY 10598

PYUN, CHONG WHA, b Korea, May 11, 30; m 57; c 2. PHYSICAL CHEMISTRY, POLYMER CHEMISTRY. *Educ:* Seoul Nat Univ, BS, 55, MS, 58; Brown Univ, PhD(chem), 64. *Prof Exp:* Instr chem, Seoul Nat Univ, 58-60; res assoc, Inst Theoret Sci, Univ Ore, 63-65; res fel, Univ Minn, 65-66; fel, Mellon Inst, Carnegie-Mellon Univ, 66-70; from asst prof to assoc prof, 70-78, PROF CHEM, UNIV LOWELL, 78- *Concurrent Pos:* Lectr, Dept Chem, Carnegie-Mellon Univ, 67-70; vis prof, Dept Chem, Seoul Nat Univ, 77-78, Pohang Inst Sci & Technol, Pohang, Korea, 86-87. *Mem:* Am Chem Soc; Am Phys Soc; Korean Chem Soc; Sigma Xi; Korean Scientists & Engrs Asn Am (pres, 76-77). *Res:* Physical chemistry of high polymers and chemical kinetics; polymer chain dynamics; comonomer sequence and sterosequence distribution statistics; polymerization kinetics; gas-phase reactions and multistate kinetics. *Mailing Add:* Dept of Chem Univ of Lowell Lowell MA 01854